THE

Electrical

Engineering

HANDBOOK

Editor-in-Chief
RICHARD C. DORF
University of California, Davis

CRC PRESS
Boca Raton   Ann Arbor   London   Tokyo

Library of Congress Cataloging-in-Publication Data

The electrical engineering handbook / editor-in-chief, Richard C. Dorf.
    p.   cm.
Includes bibliographical references and index.
ISBN 0-8493-0185-8
    1. Electric engineering   I. Dorf, Richard C.
TK145.E354 1993
621.3—dc20   92-35356
                                                                CIP

Preface

Purpose

The purpose of *The Electrical Engineering Handbook* is to provide in a single volume a ready reference for the practicing engineer in industry, government, and academia. The book in its comprehensive format is divided into twelve sections which encompass the field of electrical engineering. The goal is to provide the most up-to-date information in the classical fields of circuits, signal processing, electronics, electromagnetic fields, energy devices, and electrical effects and devices, while covering the emerging fields of communications, digital devices, computer engineering, systems, and biomedical engineering. In addition, the final section provides a complete compendium of information regarding physical, chemical, and materials data, as well as widely inclusive information on mathematics.

Organization

The fundamentals of electrical engineering have evolved to include a wide range of knowledge, substantial empirical data, and a broad range of practice. The focus of the handbook is on the key concepts, models, and equations that enable the electrical engineer to analyze, design, and predict the behavior of complex electrical devices, circuits, instruments, and systems. While data and formulae are summarized, the main focus is the provision of the underlying theories and concepts and the appropriate application of these theories to the field of electrical engineering. Thus, the reader will find the key concepts defined, described, and illustrated in order to serve the needs of the engineer over many years. With equal emphasis on electronics, circuits, power systems, instruments, materials, effects and devices, systems, and control, the engineer should encounter a wide range of concepts and considerable depth of exploration of these concepts as they lead to application and design.

The level of conceptual development of each topic is challenging, but tutorial and relatively fundamental. Each article, of which there are more than 200, is written to enlighten the expert, refresh the knowledge of the mature engineer, and educate the novice.

The information is organized into twelve major sections. The first eleven sections encompass 109 chapters and the last section summarizes the applicable mathematics, symbols, and physical constants. Each section contains one or more historical vignettes that serve to enliven and illuminate the history of the subject of that section. Furthermore, each section is preceded by a photo of a device, circuit, or system that demonstrates an application illustrative of the material in the section.

Each article includes three important and useful categories: defining terms, references, and further information. *Defining terms* are key definitions and the first occurrence of each term defined is indicated in boldface in the text. The definitions of these terms are summarized as a list at the end of each chapter or article. The *references* provide a list of useful books and articles for follow-up reading. Finally, *further information* provides some general and useful sources of additional information on the topic.

Locating Your Topic

Numerous avenues of access to information contained in the handbook are provided. A complete table of contents is presented at the front of the book. In addition, an individual table of contents precedes each of the twelve sections. Finally, each chapter begins with its own table of contents. The reader should look over these tables of contents to become familiar with the structure, organization, and content of the book. For example, see Section II: Signal Processing, then Chapter 16: Multidimensional Signal Processing, and then Chapter 16.2: Video Signal Processing. This tree-and-branch table of contents enables the reader to move up the tree to locate information on the topic of interest.

Five indexes have been compiled to provide multiple means of accessing information. Three indexes are listed in alphabetical order: (1) subject index, (2) index of basic equations by title or name, and (3) index of contributing authors. The subject index can also be used to locate key definitions. The page on which the definition appears for each key (defining) term is clearly identified in the subject index. Two additional indexes are sequenced by order of appearance: (1) index of key tables of data or information and (2) index of key illustrations.

The Electrical Engineering Handbook is designed to provide answers to most inquiries and direct the inquirer to further sources and references. We hope that this handbook will be referred to often and that informational requirements will be satisfied effectively.

Acknowledgments

This handbook is testimony to the dedication of the Board of Advisors, the publishers, and my editorial associates. I particularly wish to acknowledge at CRC Press Joel Claypool, Publisher; Kristen Peterson, Developmental Editor; and Sandy Pearlman, Production Coordinator. Finally, I am indebted to the assistance of Aysha Griffin and Sara Hare, who served as editorial assistants.

Richard C. Dorf
Editor-in-Chief

Editor-in-Chief

Richard C. Dorf, professor of electrical and computer engineering at the University of California, Davis, teaches graduate and undergraduate courses in electrical engineering in the fields of circuits and control systems. He earned a Ph.D. in electrical engineering from the U.S. Naval Postgraduate School, an M.S. from the University of Colorado, and a B.S. from Clarkson University. Highly concerned with the discipline of electrical engineering and its wide value to social and economic needs, he has written and lectured internationally on the contributions and advances in electrical engineering.

Professor Dorf has extensive experience with education and industry and is professionally active in the fields of robotics, automation, electric circuits, and communications. He has served as a visiting professor at the University of Edinburgh, Scotland; the Massachusetts Institute of Technology; Stanford University; and the University of California, Berkeley.

A Fellow of The Institute of Electrical and Electronics Engineers, Dr. Dorf is widely known to the profession for his *Modern Control Systems,* 6th edition (Addison-Wesley, 1992) and *The International Encyclopedia of Robotics* (Wiley, 1988). Dr. Dorf is also the co-author of *Circuits, Devices and Systems* (with Ralph Smith), 5th edition (Wiley, 1992).

Advisory Board

Contributors

M. Abdelguerfi
University of New Orleans
New Orleans, Louisiana

Samuel O. Agbo
California Polytechnic State
   University
San Luis Obispo, California

Ian P. Allebach
Purdue University
West Lafayette, Indiana

Ahmed Amin
Texas Instruments Incorporated
Attleboro, Massachusetts

Kristinn Andersen
Vanderbilt University
Nashville, Tennessee

Nick Angelopoulos
Gould Shawmut
Toronto, Canada

Carl A. Argila
Software Engineering Consultant
Pico Rivera, California

A. Terry Bahill
The University of Arizona
Tucson, Arizona

Inder J. Bahl
ITT-Gallium Arsenide Technology
   Center
Roanoke, Virginia

Norman Balabanian
Syracuse University
Syracuse, New York

Glen Ballou
Ballou Associates
Guilford, Connecticut

Brian R. Bannister
University of Hull
Hull, United Kingdom

Joseph Bannister
The Aerospace Corporation
El Segundo, California

Avram Bar-Cohen
University of Minnesota
Minneapolis, Minnesota

Matthew F. Baretich
University Hospital
University of Colorado
Denver, Colorado

Robert Joel Barnett
Vanderbilt University
Nashville, Tennessee

R. Bartnikas
Institut de Recherche d'Hydro-
   Québec
Varennes, Canada

Stella N. Batalama
University of Virginia
Charlottesville, Virginia

Geoffrey Bate
Santa Clara University
Santa Clara, California

Behnam Bavarian
Printrak International Inc.
Anaheim, California

R. A. Becker
Integrated Optical Circuit
   Consultants
Cupertino, California

Melvin L. Belcher
Georgia Institute of Technology
Atlanta, Georgia

Edward J. Berbari
University of Oklahoma
Oklahoma City, Oklahoma

Ashoka K. S. Bhat
University of Victoria
Victoria, Canada

Imran A. Bhutta
Virginia Polytechnic Institute and
   State University
Blacksburg, Virginia

Theodore A. Bickart
Michigan State University
East Lansing, Michigan

Bill Bitler
Philips Components
Slatersville, Rhode Island

Theodore F. Bogart, Jr.
University of Southern Mississippi
Hattiesburg, Mississippi

Martin Bolton
SGS-Thomson Microelectronics
Grenoble, France

Bruce W. Bomar
University of Tennessee Space
   Institute
Tullahoma, Tennessee

Anjan Bose
Arizona State University
Tempe, Arizona

Bimal K. Bose
University of Tennessee
Knoxville, Tennessee

N. K. Bose
The Pennsylvania State University
University Park, Pennsylvania

Charles A. Bouman
Purdue University
West Lafayette, Indiana

Joseph Boykin
GTE Laboratories
Waltham, Massachusetts

John R. Brews
The University of Arizona
Tucson, Arizona

William L. Brogan
University of Nevada
Las Vegas, Nevada

Joseph D. Bronzino
Trinity College
Hartford, Connecticut

Marcia A. Bush
Xerox Palo Alto Research Center
Palo Alto, California

James A. Cadzow
Vanderbilt University
Nashville, Tennessee

Gordon L. Carpenter
California State University Long
Beach
Long Beach, California

Bill D. Carroll
The University of Texas at
Arlington
Arlington, Texas

G. Clifford Carter
NUWC Detachment
New London, Connecticut

Shu-Park Chan
Santa Clara University
Santa Clara, California

Rulph Chassaing
Roger Williams University
Bristol, Rhode Island

William Check
GTE Spacenet
McLean, Virginia

Kao Chen
Carlsons Consulting Engineers
San Diego, California

Mo-Shing Chen
University of Texas at Arlington
Arlington, Texas

Sue-Ling Chen
Wavefront Technologies
Santa Barbara, California

Wai-Kai Chen
University of Illinois at Chicago
Chicago, Illinois

Allen H. Cherin
AT&T Laboratories
Norcross, Georgia

John Choma, Jr.
University of Southern California
Los Angeles, California

Michael D. Ciletti
University of Colorado
Colorado Springs, Colorado

Gerald A. Clapp
Naval Command Control and
Ocean Surveillance Center
San Diego, California

Almon H. Clegg
CCi
Rockaway, New Jersey

J. R. Cogdell
The University of Texas at Austin
Austin, Texas

Richard C. Compton
Cornell University
Ithaca, New York

George E. Cook
Vanderbilt University
Nashville, Tennessee

J. Arlin Cooper
Sandia National Laboratories
Albuquerque, New Mexico

Thomas M. Cover
Stanford University
Stanford, California

Edward W. Czeck
Northeastern University
Boston, Massachusetts

John N. Daigle
The Mitre Corporation
McLean, Virginia

Thomas E. Darcie
AT&T Bell Laboratories
Holmdel, New Jersey

Yadin David
Texas Children's Hospital
Houston, Texas

Kevin A. Delin
Massachusetts Institute of
Technology Lincoln Laboratory
Lexington, Massachusetts

Edward J. Delp
Purdue University
West Lafayette, Indiana

Kenneth Demarest
University of Kansas
Lawrence, Kansas

Bülent I. Dervisoğlu
Hewlett Packard
Chelmsford, Massachusetts

Daniel F. DiFonzo
Planar Communications
Corporation
Rockville, Maryland

Dennis F. Doelitzsch
3-D Communications
Corporation
Marion, Illinois

Robert C. Durbeck
IBM Corporation (retired)
San Jose, California

Alexander C. Ehrlich
U.S. Naval Research Laboratory
Washington, D.C.

Mohamed E. El-Hawary
Technical University of Nova
Scotia
Halifax, Nova Scotia

Aicha Elshabini-Riad
Virginia Polytechnic Institute and
    State University
Blacksburg, Virginia

Yariv Ephraim
AT&T Bell Laboratories
Murray Hill, New Jersey

Delores M. Etter
University of Colorado
Boulder, Colorado

K. F. Etzold
IBM T. J. Watson Research Center
Yorktown Heights, New York

Gerald W. Farnell
McGill University
Montreal, Canada

William M. Feaster
Auburn University
Auburn, Alabama

Lyle D. Feisel
State University of New York,
    Binghamton
Binghamton, New York

James M. Feldman
Northeastern University
Boston, Massachusetts

Tse-yun Feng
The Pennsylvania State University
State College, Pennsylvania

J. Patrick Fitch
Lawrence Livermore National
    Laboratory
Livermore, California

Martin D. Fox
University of Connecticut
Storrs, Connecticut

James F. Frenzel
University of Idaho
Moscow, Idaho

Leon A. Frizzell
University of Illinois
Urbana, Illinois

Jesse W. Fussell
The Johns Hopkins University
Baltimore, Maryland

Donald Galler
FaAA Electrical Corporation
Westborough, Massachusetts

Susan A. R. Garrod
Purdue University
West Lafayette, Indiana

L. A. Geddes
Purdue University
Lafayette, Indiana

Boris Gelmont
University of Virginia
Charlottesville, Virginia

Mario Gerla
University of California, Los
    Angeles
Los Angeles, California

Jerry D. Gibson
Texas A&M University
College Station, Texas

Gennady Sh. Gildenblat
The Pennsylvania State University
University Park, Pennsylvania

Gerald L. Ginsberg
Component Data Associates, Inc.
Lafayette Hill, Pennsylvania

Jay C. Giri
ESCA Corporation
Bellevue, Washington

J. Duncan Glover
FaAA Electrical Corporation
Westborough, Massachusetts

James Richard Goodman
University of Wisconsin-Madison
Madison, Wisconsin

Peter J. Graham
Florida Atlantic University
Boca Raton, Florida

Charles A. Gross
Auburn University
Auburn, Alabama

R. B. Gungor
University of South Alabama
Mobile, Alabama

Chris G. Guy
University of Reading
Reading, United Kingdom

V. Carl Hamacher
Queen's University
Kingston, Canada

Royce D. Harbor
University of West Florida
Pensacola, Florida

Jeff Hecht
Laser Focus World
Auburndale, Massachusetts

Leland H. Hemming
McDonnell Douglas
    Technologies, Inc.
San Diego, California

H. Scott Hinton
McGill University
Montreal, Canada

Conrad H. Hoeppner
The Johns Hopkins University
Baltimore, Maryland

S. Ratnajeevan H. Hoole
Harvey Mudd College
Claremont, California

Tien C. Hsia
University of California, Davis
Davis, California

Manfred N. Huber
Siemens
Munich, Germany

Jerry L. Hudgins
University of South Carolina
Columbia, South Carolina

J. David Irwin
Auburn University
Auburn, Alabama

Raymond G. Jacquot
University of Wyoming
Laramie, Wyoming

Barry W. Johnson
University of Virginia
Charlottesville, Virginia

David E. Johnson
Birmingham-Southern College
Birmingham, Alabama

Capers Jones
Software Productivity
    Research, Inc.
Burlington, Massachusetts

Ivan P. Kaminow
AT&T Bell Laboratories
Holmdel, New Jersey

George G. Karady
Arizona State University
Tempe, Arizona

Randy H. Katz
University of California
Berkeley, California

Myron Kayton
Kayton Engineering Co.
Santa Monica, California

Dimitri Kazakos
University of Virginia
Charlottesville, Virginia

E. J. Kennedy
University of Tennessee
Knoxville, Tennessee

William H. Kersting
New Mexico State University
Las Cruces, New Mexico

William J. Kerwin
The University of Arizona
Tucson, Arizona

Nicholas J. Kolias
Cornell University
Ithaca, New York

Jin Au Kong
Massachusetts Institute of
Technology
Cambridge, Massachusetts

Kurt L. Kosbar
University of Missouri–Rolla
Rolla, Missouri

Allan D. Kraus
Naval Postgraduate School
Monterey, California

Mark H. Kryder
Carnegie Mellon University
Pittsburgh, Pennsylvania

Luis G. Kun
Cedars-Sinai Medical Center
Los Angeles, California

Benjamin C. Kuo
University of Illinois at Urbana-
Champaign
Urbana, Illinois

Dhammika Kurumbalapitiya
Harvey Mudd College
Claremont, California

Kuang-Chung Lai
University of Texas at Arlington
Arlington, Texas

Pradeep Lall
University of Maryland
College Park, Maryland

Ty A. Lasky
University of California, Davis
Davis, California

Gordon K. F. Lee
North Carolina State University
Raleigh, North Carolina

Peter A. Lee
Department of Trade and
Industry
London, England

William C. Y. Lee
PacTel Corporation
Walnut Creek, California

Cornelius T. Leondes
University of California, San
Diego
La Jolla, California

Ted G. Lewis
Oregon State University
Corvallis, Oregon

Jay Liebowitz
George Washington University
Washington, D.C.

Michael Lightner
University of Colorado
Boulder, Colorado

Jefferson F. Lindsey III
Southern Illinois University at
Carbondale
Carbondale, Illinois

Chen-Ching Liu
University of Washington
Seattle, Washington

Carl Grant Looney
University of Nevada
Reno, Nevada

Steven L. Maddy
RLM Research
Boulder, Colorado

Donald C. Malocha
University of Central Florida
Orlando, Florida

Masud Mansuripur
University of Arizona
Tucson, Arizona

Robert J. Marks II
University of Washington
Seattle, Washington

André F. Martin
Hughes Display Products
Lexington, Kentucky

Johannes J. Martin
University of New Orleans
New Orleans, Louisiana

Daniel A. Martinec
Aeronautical Radio, Inc.
Annapolis, Maryland

Robert E. Massara
University of Essex
Colchester, Essex, United
Kingdom

Kartikeya Mayaram
AT&T Bell Laboratories
Allentown, Pennsylvania

John E. McInroy
University of Wyoming
Laramie, Wyoming

Duane McRuer
Systems Technology, Inc.
Hawthorne, California

Sanjay K. Mehta
NUWE Detachment
New London, Connecticut

Miran Milkovic
Georgia Institute of Technology
Atlanta, Georgia

Edmund K. Miller
Los Alamos National Laboratory
Los Alamos, New Mexico

James E. Morris
State University of New York
Binghamton, New York

Gregory L. Moss
Purdue University
West Lafayette, Indiana

Wayne Needham
Intel Corporation
Chandler, Arizona

P. S. Neelakanta
Florida Atlantic University
Boca Raton, Florida

Josh T. Nessmith
Georgia Institute of Technology
Atlanta, Georgia

Paul Neudorfer
Seattle University
Seattle, Washington

Michael R. Neuman
Case Western Reserve University
Cleveland, Ohio

Nicholas G. Odrey
Lehigh University
Bethlehem, Pennsylvania

Vojin G. Oklobdzija
University of California, Davis
Davis, California

John V. Oldfield
Syracuse University
Syracuse, New York

Terry P. Orlando
Massachusetts Institute of
    Technology
Cambridge, Massachusetts

Mil Ovan
Motorola, Inc.
Arlington Heights, Illinois

Joseph C. Palais
Arizona State University
Tempe, Arizona

Keshab K. Parhi
University of Minnesota
Minneapolis, Minnesota

Harold G. Parks
The University of Arizona, Tucson
Tucson, Arizona

Clayton R. Paul
University of Kentucky
Lexington, Kentucky

Michael Pecht
University of Maryland
College Park, Maryland

Arun G. Phadke
Virginia Polytechnic Institute and
    State University
Blacksburg, Virginia

Charles L. Phillips
Auburn University
Auburn, Alabama

S. Unnikrishna Pillai
Polytechnic University
Brooklyn, New York

Charles Polk
University of Rhode Island
Kingston, Rhode Island

H. Vincent Poor
Princeton University
Princeton, New Jersey

Alexander D. Poularikas
University of Alabama in
    Huntsville
Huntsville, Alabama

Franco P. Preparata
Brown University
Providence, Rhode Island

W. David Pricer
IBM
Essex Junction, Vermont

Yuan Pu
University of California, Irvine
Irvine, California

Sarah A. Rajala
North Carolina State University
Raleigh, North Carolina

S. Rajaram
AT&T Bell Laboratories
Whippany, New Jersey

Kaushik Rajashekara
Delco Remy
Anderson, Indiana

R. Ramakumar
Oklahoma State University
Stillwater, Oklahoma

Abdul Hamid Rana
GTE Spacenet
McLean, Virginia

Banmali S. Rawat
University of Nevada, Reno
Reno, Nevada

Jacques Raymond
Ottawa University
Ottawa, Canada

J. Patrick Reilly
The Johns Hopkins University
Laurel, Maryland

Thomas G. Robertazzi
State University of New York at
    Stony Brook
Stony Brook, New York

Charles J. Robinson
University of Pittsburgh
Pittsburgh, Pennsylvania

Richard B. Robrock II
Bell Communications Research
Piscataway, New Jersey

Martin S. Roden
California State University
Los Angeles, California

Peter H. Rogers
Georgia Institute of Technology
Atlanta, Georgia

John M. Roman
Florida Atlantic University
Boca Raton, Florida

Evelyn P. Rozanski
Rochester Institute of Technology
Rochester, New York

Marcos Rubinstein
University of Florida
Gainesville, Florida

Matthew N. O. Sadiku
Temple University
Philadelphia, Pennsylvania

Andrew P. Sage
George Mason University
Fairfax, Virginia

Stanley Salek
Hammett & Edison, Inc.
Burlingame, California

Richard S. Sandige
University of Wyoming
Laramie, Wyoming

C. Sankaran
Machine Design Engineers, Inc.
Seattle, Washington

John L. Schmalzel
University of Texas, San Antonio
San Antonio, Texas

Juergen Schroeter
Acoustics Research Department
AT&T Bell Laboratories
Murray Hill, New Jersey

J. Leland Seely (retired)
Bonneville Microelectronics, Inc.
Salt Lake City, Utah

Micaela Serra
University of Victoria
Victoria, Canada

Leonard Shaw
Polytechnic University
Brooklyn, New York

Solomon Sherr
Westland Electronics
Old Chatham, New York

Theodore I. Shim
Polytechnic University
Brooklyn, New York

L. H. Sibul
The Pennsylvania State University
University Park, Pennsylvania

L. Montgomery Smith
University of Tennessee Space
    Institute
Tullahoma, Tennessee

Rosemary L. Smith
University of California, Davis
Davis, California

Sidney Soclof
California State University
Los Angeles, California

Gurindar S. Sohi
University of Wisconsin–Madison
Madison, Wisconsin

René Spée
Oregon State University
Corvallis, Oregon

Cary R. Spitzer
NASA Langley Research Center
Hampton, Virginia

K. Neil Stanton
ESCA Corporation
Bellevue, Washington

John Staudhammer
University of Florida
Gainesville, Florida

J. W. Steadman
University of Wyoming
Laramie, Wyoming

Michael B. Steer
North Carolina State University
Raleigh, North Carolina

F. W. Stephenson
Virginia Polytechnic Institute and
    State University
Blacksburg, Virginia

David Sworder
University of California, San
    Diego
San Diego, California

Ferenc Szidarovszky
The University of Arizona
Tucson, Arizona

Ronald J. Tallarida
Temple University
Philadelphia, Pennsylvania

Basant K. Tariyal
AT&T Laboratories
Research Triangle Park, North
    Carolina

Rao S. Thallam
Salt River Project
Phoenix, Arizona

Joy A. Thomas
IBM T. J. Watson Research Center
Yorktown Heights, New York

Richard F. Tinder
Washington State University
Pullman, Washington

Spyros Tragoudas
Southern Illinois University at
    Carbondale
Carbondale, Illinois

William H. Tranter
University of Missouri–Rolla
Rolla, Missouri

Robert J. Trew
North Carolina State University
Raleigh, North Carolina

C. W. Trowbridge
Vector Fields, Ltd.
Oxford, England

R. Lal Tummala
Michigan State University
East Lansing, Michigan

Martin A. Uman
University of Florida
Gainesville, Florida

Vichate Ungvichian
Florida Atlantic University
Boca Raton, Florida

Zvonko G. Vranesic
University of Toronto
Toronto, Canada

Khoi Tien Vu
Clemson University
Clemson, South Carolina

John V. Wait
The University of Arizona
Tucson, Arizona

Alan K. Wallace
Oregon State University
Corvallis, Oregon

Zhen Wan
University of California, Davis
Davis, California

Chih-Lin Wang
Electro-Optek Corporation
Torrance, California

Laurence S. Watkins
AT&T Bell Laboratories
Princeton, New Jersey

Joseph Watson
University of Wales, Swansea
Swansea, United Kingdom

Larry F. Weber
Plasmaco
Highland, New York

Roger W. Whatmore
GEC Marconi Materials
 Technology, Ltd.
Towcester, United Kingdom

Jerry Whitaker
Technical Writer
Beaverton, Oregon

Donald G. Whitehead
University of Hull
Hull, United Kingdom

B. M. Wilamowski
University of Wyoming
Laramie, Wyoming

Lynn D. Wilcox
Xerox Palo Alto Research Center
Palo Alto, California

James C. Wiltse
Georgia Institute of Technology
Atlanta, Georgia

Phillip J. Windley
University of Idaho
Moscow, Idaho

David Young
Rockwell International
Newport Beach, California

Yixin Yu
Tianjin University
Tianjin, People's Republic of
 China

Safwat G. Zaky
University of Toronto
Toronto, Canada

Mehdi R. Zargham
Southern Illinois University at
 Carbondale
Carbondale, Illinois

Rodger E. Ziemer
University of Colorado at
 Colorado Springs
Colorado Springs, Colorado

Contents

SECTION I   Circuits

Introduction   *Shu-Park Chan* 1

1   Passive Components
1.1   Resistors   *Michael Pecht and Pradeep Lall* 5
1.2   Capacitors and Inductors   *Glen Ballou* 15
1.3   Transformers   *C. Sankaran* 33
1.4   Electrical Fuses   *Nick Angelopoulos* 39

2   Voltage and Current Sources
2.1   Step, Impulse, Ramp, Sinusoidal, and DC Signals
       Richard C. Dorf and Zhen Wan 47
2.2   Ideal and Practical Sources   *Clayton R. Paul* 50
2.3   Controlled Sources   *J. R. Cogdell* 53

3   Linear Circuit Analysis
3.1   Voltage and Current Laws   *Michael D. Ciletti* 58
3.2   Node and Mesh Analysis   *J. David Irwin* 63
3.3   Network Theorems   *Allan D. Kraus* 69
3.4   Power and Energy   *Norman Balabanian and Theodore A. Bickart* 79
3.5   Three-Phase Circuits   *Norman Balabanian* 88
3.6   Graph Theory   *Shu-Park Chan* 92

4   Passive Signal Processing   *William J. Kerwin* 111

5   Nonlinear Circuits
5.1   Diodes and Rectifiers   *Jerry L. Hudgins* 126
5.2   Limiters   *Theodore F. Bogart, Jr.* 132
5.3   Distortion   *Kartikeya Mayaram* 139

6   Laplace Transform
6.1   Definitions and Properties   *Richard C. Dorf and Zhen Wan* 149
6.2   Applications   *David E. Johnson* 158

7   State Variables   *Wai-Kai Chen* 169

8   The *z*-Transform   *Richard C. Dorf and Zhen Wan* 178

9   T-Π Equivalent Networks   *Zhen Wan and Richard C. Dorf* 184

10   Transfer Functions of Filters   *Richard C. Dorf and Zhen Wan* 189

11   Frequency Response   *Paul Neudorfer* 198

12   Stability Analysis   *Ferenc Szidarovszky and A. Terry Bahill* 207

SECTION II   Signal Processing

Introduction   *Delores M. Etter* 225

13   Digital Signal Processing
13.1 Transforms and Fast Algorithms   *Alexander D. Poularikas* 229
13.2 Design and Implementation of Digital Filters
      Bruce W. Bomar and L. Montgomery Smith 238
13.3 Signal Restoration   *James A. Cadzow* 251

14   Speech Signal Processing
14.1 Coding, Transmission, and Storage   *Jerry D. Gibson* 279
14.2 Speech Enhancement and Noise Reduction   *Yariv Ephraim* 287
14.3 Analysis and Synthesis   *Jesse W. Fussell* 298
14.4 Speech Recognition   *Lynn D. Wilcox and Marcia A. Bush* 306

15   Spectral Estimation and Modeling
15.1 Spectral Analysis   *S. Unnikrishna Pillai and Theodore I. Shim* 315
15.2 Parameter Estimation   *Stella N. Batalama and Dimitri Kazakos* 321

16   Multidimensional Signal Processing
16.1 Digital Image Processing   *Edward J. Delp, Jan Allebach, and*
      Charles A. Bouman 329
16.2 Video Signal Processing   *Sarah A. Rajala* 345
16.3 Sensor Array Processing   *N. K. Bose and L. H. Sibul* 359

17   VLSI for Signal Processing
17.1 Special Architectures   *Keshab K. Parhi* 370
17.2 Signal Processing Chips and Applications   *Rulph Chassaing and*
      Bill Bitler 385

18   Acoustic Signal Processing
18.1 Digital Signal Processing in Audio and Electroacoustics
      Juergen Schroeter 395
18.2 Underwater Acoustical Signal Processing   *Sanjay K. Mehta and*
      G. Clifford Carter 406

19   Neural Networks   *Behnam Bavarian* 420

SECTION III Electronics

Introduction *John W. Steadman* 431

20 Semiconductors
20.1 Physical Properties *Gennady Sh. Gildenblat and Boris Gelmont* 435
20.2 Diodes *Miran Milkovic* 447
20.3 Electrical Equivalent Circuit Models and Device Simulators for
      Semiconductor Devices *Aicha Elshabini-Riad, F. W. Stephenson, and*
      Imran A. Bhutta 460

21 Semiconductor Manufacturing
21.1 Processes *Harold G. Parks* 475
21.2 Testing *Wayne Needham* 490
21.3 Electrical Characterization of Interconnections *S. Rajaram* 499

22 Transistors
22.1 Junction Field-Effect Transistors *Sidney Soclof* 530
22.2 Bipolar Transistors *J. Watson* 545
22.3 The Metal-Oxide Semiconductor Field-Effect Transistor (MOSFET)
      John R. Brews 567

23 Integrated Circuits
23.1 Layout, Placement, and Routing *Mehdi R. Zargham and*
      Spyros Tragoudas 581
23.2 Application-Specific Integrated Circuits (ASICs) *J. Leland Seely* 591

24 Surface Mount Technology *Gerald L. Ginsberg* 603

25 Operational Amplifiers
25.1 Ideal and Practical Models *E. J. Kennedy* 616
25.2 Applications *John V. Wait* 625

26 Amplifiers
26.1 Large Signal Analysis *Gordon L. Carpenter* 634
26.2 Small Signal Analysis *John Choma, Jr.* 639

27 Computer-Aided Circuit Simulation *Michael Lightner* 653

28 Active Filters
28.1 Synthesis of Low-Pass Forms *Robert E. Massara* 674
28.2 Realization *J. W. Steadman and B. M. Wilamowski* 683

29 Power Electronics
29.1 Power Semiconductor Devices *Kaushik Rajashekara* 694
29.2 Power Conversion *Kaushik Rajashekara* 702
29.3 Power Supplies *Ashoka K. S. Bhat* 711
29.4 Converter Control of Machines *Bimal K. Bose* 729

30 Optoelectronics
30.1 Lasers *Jeff Hecht* .... 73█
30.2 Sources and Detectors *Laurence S. Watkins* .... 74█
30.3 Circuits *R. A. Becker* .... 75█

31 D/A and A/D Converters *Susan A. R. Garrod* .... 77█

32 Thermal Management of Electronics *Avram Bar-Cohen* .... 78█

SECTION IV Electromagnetics

Introduction *Banmali S. Rawat* .... 79█

33 Electromagnetic Fields *Jin Au Kong* .... 80█

34 Magnetism and Magnetic Fields
34.1 Magnetism *Geoffrey Bate* .... 81█
34.2 Magnetic Recording *Mark H. Kryder* .... 82█

35 Wave Propagation
35.1 Space Propagation *Matthew N. O. Sadiku* .... 83█
35.2 Waveguides *Kenneth Demarest* .... 84█

36 Antennas
36.1 Wire *N. J. Kolias and R. C. Compton* .... 86█
36.2 Aperture *J. Patrick Fitch* .... 87█

37 Microwave Devices
37.1 Passive Microwave Devices *Michael B. Steer* .... 88█
37.2 Active Microwave Devices *Robert J. Trew* .... 89█

38 Compatibility
38.1 Grounding, Shielding, and Filtering *Leland H. Hemming* .... 90█
38.2 Spectrum, Specifications, and Measurement Techniques
      Vichate Ungvichian and John M. Roman .... 91█
38.3 Lightning *Martin Uman and Marcos Rubinstein* .... 93█

39 Radar
39.1 Pulse Radar *Melvin L. Belcher and Josh T. Nessmith* .... 94█
39.2 Continuous Wave Radar *James C. Wiltse* .... 96█

40 Lightwave
40.1 Lightwave Waveguides *Samuel O. Agbo* .... 97█
40.2 Optical Fibers and Cables *Allen H. Cherin and Basant K. Tariyal* .... 98█

41 Solid State Circuits *I. J. Bahl* .... 100█

42 Three-Dimensional Analysis   *C. W. Trowbridge* 1018

43 Computational Electromagnetics   *E. K. Miller* 1028

SECTION V   Electrical Effects and Devices

Introduction   *Lyle D. Feisel* 1051

44 Electroacoustic Devices   *Peter H. Rogers* 1055

45 Surface Acoustic Wave Filters   *Donald C. Malocha* 1062

46 Ultrasound   *Gerald W. Farnell* 1077

47 Ferroelectric and Piezoelectric Materials   *K. F. Etzold* 1087

48 Piezoresistivity   *Ahmed Amin* 1099

49 The Hall Effect   *Alexander C. Ehrlich* 1106

50 Superconductivity   *Kevin A. Delin and Terry P. Orlando* 1114

51 Pyroelectric Materials and Devices   *Roger W. Whatmore* 1126

52 Dielectrics and Insulators   *R. Bartnikas* 1132

53 Sensors   *Rosemary L. Smith* 1152

54 Magnetooptics   *David Young, Chih-Lin Wang, and Yuan Pu* 1162

55 Smart Materials   *P. S. Neelakanta* 1173

SECTION VI   Energy

Introduction   *William H. Kersting* 1191

56 Conventional Power Generation   *George G. Karady* 1193

57 Distributed Power Generation   *R. Ramakumar* 1207

58 Transmission
   58.1 Alternating Current Overhead: Line Parameters, Models, Standard
        Voltages, Insulators   *Mo-Shing Chen* 1217
   58.2 Alternating Current Underground: Line Parameters, Models, Standard
        Voltages, Cables   *Mo-Shing Chen and K. C. Lai* 1223
   58.3 High-Voltage Direct-Current Transmission   *Rao S. Thallam* 1227
   58.4 Compensation   *Mohamed E. El-Hawary* 1242

58.5 Fault Analysis in Power Systems   *Charles Gross* 1252
58.6 Protection   *Arun G. Phadke* 1268
58.7 Transient Operation   *R. B. Gungor* 1279
58.8 Planning   *J. Duncan Glover* 1287

59 Power Transformers   *Charles A. Gross and William M. Feaster* 1296

60 Energy Distribution   *George G. Karady* 1310

61 Electrical Machines
61.1 Generators   *Chen-Ching Liu, Khoi Tien Vu, and Yixin Yu* 1321
61.2 Motors   *Donald Galler* 1333

62 Energy Management   *K. Neil Stanton, Jay C. Giri, and Anjan Bose* 1344

SECTION VII   Communications

Introduction   *Leonard Shaw* 1355

63 Broadcasting
63.1 Modulation and Demodulation   *Richard C. Dorf and Zhen Wan* 1359
63.2 Radio   *Jefferson F. Lindsey III and Dennis F. Doelitzsch* 1367
63.3 Television Systems   *Jerry Whitaker* 1379
63.4 High-Definition Television   *Martin S. Roden* 1394
63.5 Digital Audio Broadcasting   *Stanley Salek and Almon H. Clegg* 1397

64 Digital Communication
64.1 Coding   *Richard C. Dorf and Zhen Wan* 1405
64.2 Equalization   *Richard C. Dorf and Zhen Wan* 1410

65 Optical Communication
65.1 Lightwave Technology for Video Transmission   *T. E. Darcie* 1417
65.2 Long Distance   *Joseph C. Palais* 1427
65.3 Photonic Networks   *Ivan P. Kaminow* 1434

66 Networks
66.1 B-ISDN   *Manfred N. Huber* 1441
66.2 Computer Communication Networks   *J. N. Daigle* 1447
66.3 Local-Area Networks   *Joseph Bannister and Mario Gerla* 1460
66.4 The Intelligent Network   *Richard B. Robrock II* 1468

67 Information Theory
67.1 Signal Detection   *H. Vincent Poor* 1478
67.2 Noise   *Carl G. Looney* 1488
67.3 Stochastic Processes   *Carl G. Looney* 1499
67.4 The Sampling Theorem   *R. J. Marks II* 1510
67.5 Data Compression   *Joy A. Thomas and Thomas M. Cover* 1517

68 Satellites and Aerospace   *Daniel F. DiFonzo* 1532

69 Personal and Office

69.1 Mobile Radio and Cellular Communications *William C. Y. Lee* 1546
69.2 Facsimile *Rodger E. Ziemer* 1554
69.3 Wireless Local-Area Networks for the 1990s *Mil Ovan* 1557

70 Phase-Locked Loop *Steven L. Maddy* 1567

71 Telemetry *Conrad H. Hoeppner* 1578

72 Computer-Aided Design and Analysis of Communication Systems *William H. Tranter and Kurt L. Kosbar* 1593

SECTION VIII   Digital Devices

Introduction *Richard S. Sandige* 1611

73 Logic Elements

73.1 IC Logic Family Operation and Characteristics *Gregory L. Moss* 1613
73.2 Logic Gates (IC) *Peter Graham* 1622
73.3 Bistable Devices *Richard S. Sandige* 1635
73.4 Optical Devices *H. S. Hinton* 1641

74 Memory Devices

74.1 Integrated Circuits (RAM, ROM) *W. David Pricer* 1651
74.2 Basic Disk System Architectures *Randy H. Katz* 1658
74.3 Magnetic Tape *Peter A. Lee* 1670
74.4 Magneto-Optical Disk Data Storage *M. Mansuripur* 1675

75 Logical Devices

75.1 Combinational Networks and Switching Algebra *Franco P. Preparata* 1695
75.2 Logic Circuits *Richard S. Sandige* 1711
75.3 Registers *B. R. Bannister and D. G. Whitehead* 1721
75.4 Programmable Arrays *Martin Bolton* 1735
75.5 Arithmetic Logic Units *Bill D. Carroll* 1741

76 Microprocessors

76.1 Practical Microprocessors *John Staudhammer and Sue-Ling Chen* 1748
76.2 Applications *Phillip J. Windley and James F. Frenzel* 1753

77 Displays

77.1 Light-Emitting Diodes *James E. Morris* 1763
77.2 Liquid-Crystal Displays *James E. Morris* 1772
77.3 The Cathode Ray Tube *André Martin* 1778
77.4 Plasma Displays *Larry F. Weber* 1786

78 Data Acquisition *Dhammika Kurumbalapitiya and S. Ratnajeevan H. Hoole* 1799

79 Testing
79.1 Digital IC Testing *Micaela Serra* 1808
79.2 Design for Test *Bulent I. Dervisoglu* 1816

SECTION IX Computer Engineering

Introduction *John V. Oldfield* 1839

80 Organization
80.1 Number Systems *Richard F. Tinder* 1843
80.2 Computer Arithmetic *Vojin G. Oklobdzija* 1858
80.3 Architecture *V. Carl Hamacher, Zvonko G. Vranesic, and
Safwat G. Zaky* 1865
80.4 Microprogramming *Jacques Raymond* 1870

81 Programming
81.1 Assembly Language *James M. Feldman and Edward W. Czeck* 1878
81.2 High-Level Languages *Ted G. Lewis* 1902
81.3 Data Types and Data Structures *Johannes J. Martin* 1915

82 Memory Systems *James R. Goodman and Gurindar S. Sohi* 1927

83 Input and Output
83.1 Input Devices *Solomon Sherr* 1938
83.2 Computer Output Printer Technologies *Robert C. Durbeck* 1958

84 Software Engineering
84.1 Tools and Techniques *Carl A. Argila* 1976
84.2 Testing, Debugging, and Verification *Capers Jones* 1985
84.3 Programming Methodology *Johannes J. Martin* 1996

85 Computer Graphics *Evelyn P. Rozanski* 2004

86 Computer Networks *Thomas G. Robertazzi* 2015

87 Fault Tolerance *Barry W. Johnson* 2020

88 Knowledge Engineering
88.1 Databases *M. Abdelguerfi* 2032
88.2 Rule-Based Expert Systems *Jay Liebowitz* 2048

89 Parallel Processors *Tse-yun Feng* 2052

90 Operating Systems *Joseph Boykin* 2061

91 Computer Security *J. Arlin Cooper* 2072

92 Computer Reliability *Chris G. Guy* 2087

SECTION X   Systems

Introduction   *Richard C. Dorf* 2097

93   Control Systems
93.1   Models   *William L. Brogan* 2099
93.2   Dynamic Response   *Gordon K. F. Lee* 2106
93.3   Frequency Response Methods: Bode Diagram Approach
        Andrew P. Sage 2113
93.4   Root Locus   *Benjamin C. Kuo* 2131
93.5   Compensation   *Charles L. Phillips and Royce D. Harbor* 2139
93.6   Digital Control Systems   *Raymond G. Jacquot and John E. McInroy* 2147

94   Robotics
94.1   Robot Configuration   *Ty A. Lasky and Tien C. Hsia* 2154
94.2   Dynamics and Control   *R. Lal Tummala* 2163
94.3   Applications   *Nicholas Odrey* 2175

95   Aerospace Systems
95.1   Avionics Systems   *Cary R. Spitzer, Daniel A. Martinec, and
        Cornelius T. Leondes* 2188
95.2   Communications Satellite Systems: Applications   *Abdul Hamid Rana
        and William Check* 2194

96   Command, Control, and Communications (C³)   *G. Clapp and
        D. Sworder* 2211

97   Industrial Systems
97.1   Welding and Bonding   *George E. Cook, Kristinn Andersen, and
        Robert Joel Barnett* 2223
97.2   Large Drives   *Alan K. Wallace and René Spée* 2237

98   Man-Machine Systems   *Duane McRuer* 2247

99   Vehicular Systems   *Richard C. Dorf* 2255

100   Industrial Illuminating Systems   *Kao Chen* 2257

101   Instruments   *John L. Schmalzel* 2277

102   Navigation Systems   *Myron Kayton* 2285

SECTION XI   Biomedical Systems

Introduction   *Joseph D. Bronzino* 2297

103   Bioelectricity
103.1   Neuroelectric Principles   *J. Patrick Reilly* 2301

103.2   Bioelectric Events   *L. A. Geddes* 2311

103.3   Application of Electric and Magnetic Fields in Bone and
Soft Tissue Repair   *C. Polk* 2329

104   **Biomedical Sensors**   *Michael R. Neuman* 2342

105   **Bioelectronics and Instruments**

105.1   Quantitative Analysis of the Electroencephalograms
Joseph D. Bronzino 2351

105.2   The Electrocardiograph   *Edward J. Berbari* 2362

106   **Medical Imaging**

106.1   Tomography   *M. D. Fox* 2374

106.2   Ultrasound   *Leon A. Frizzell* 2380

107   **Rehabilitation Engineering**   *Charles J. Robinson* 2387

108   **Biocomputing**

108.1   Clinical Information Systems   *Luis Kun* 2397

108.2   Hospital Information Systems   *Matthew F. Baretich* 2405

109   **Safety and Risk-Control Issues**   *Yadin David* 2408

SECTION XII   Mathematics, Symbols, and Physical Constants

Introduction   *Ronald J. Tallarida* 2415

Greek Alphabet 2419

International System of Units (SI) 2419

Conversion Constants and Multipliers 2422

Physical Constants 2424

Symbols and Terminology for Physical and Chemical Quantities 2425

Elementary Algebra and Geometry 2430

Determinants, Matrices, and Linear Systems of Equations 2436

Trigonometry 2442

Analytic Geometry 2446

Series 2454

Differential Calculus 2461

Integral Calculus 2466

Vector Analysis 2472

Special Functions 2475

Statistics 2486

Tables of Probability and Statistics 2490

Table of Derivatives 2496

Integrals 2497

The Fourier Transforms 2500

Numerical Methods 2507

Probability 2517

Positional Notation 2520

Fundamental Physical Constants 2524
Periodic Table of the Elements 2525
Classification of Electromagnetic Radiation 2526
Electrical Resistivity 2527
Dielectric Constants 2532
Properties of Semiconductors 2533
Properties of Magnetic Alloys 2536
Resistance of Wires 2543

Associations and Societies 2547

Indexes

Author Index 2559
Index of Key Tables 2562
Index of Key Figures 2564
Index of Key Equations 2570
Subject Index 2577

THE
Electrical Engineering
HANDBOOK

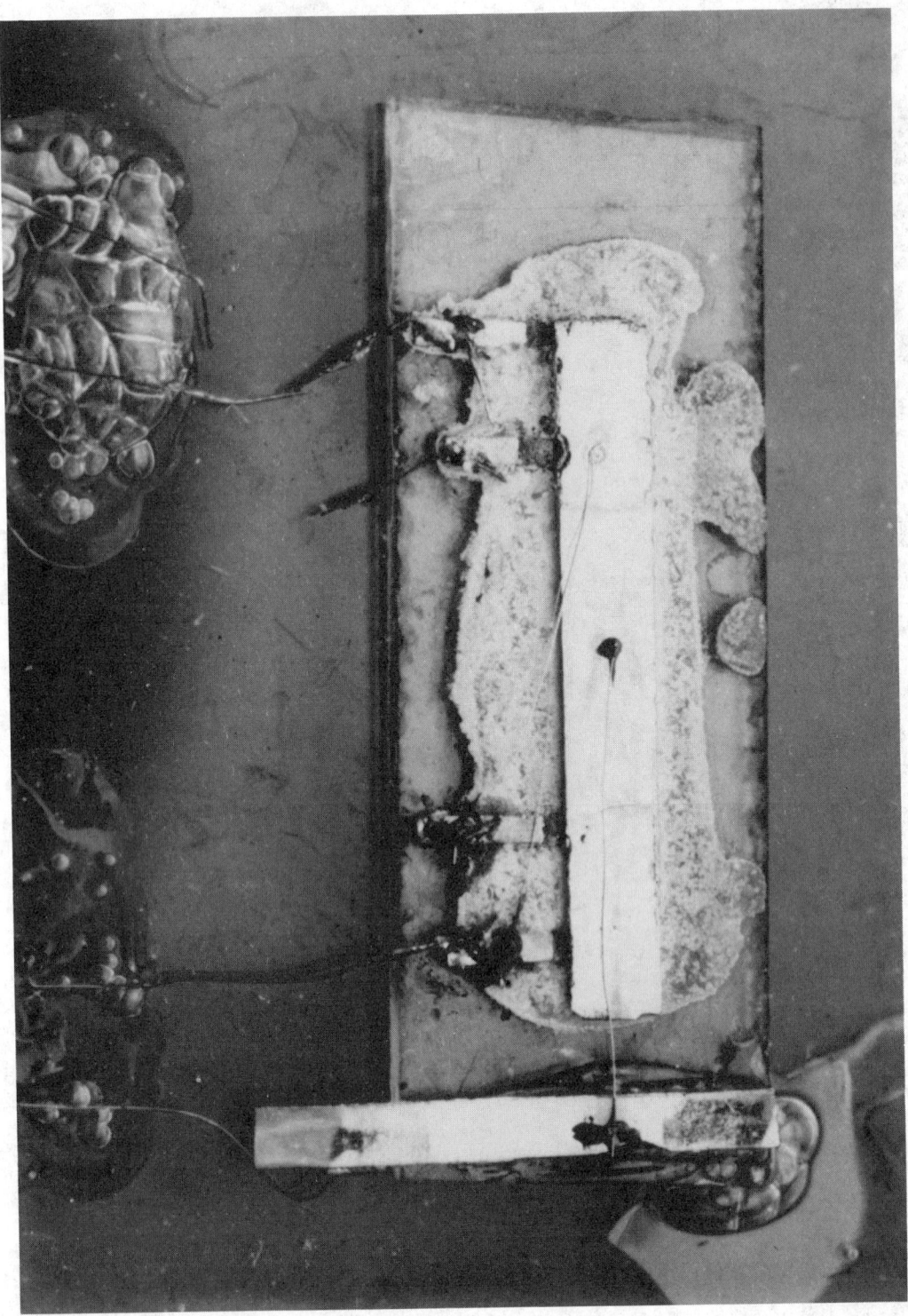

The world's first commercial integrated circuit was invented by Jack Kilby of Texas Instruments in 1958. The invention of the integrated circuit was considered so significant that Jack Kilby was later presented the highest award for technological achievement in the United States, the National Medal of Science. (Photo provided by William A. Gordon. Courtesy of Texas Instruments.)

I

Circuits

Shu-Park Chan
Santa Clara University

1 **Passive Components** *M. Pecht, P. Lall, G. Ballou, C. Sankaran, N. Angelopoulos* 5
Resistors • Capacitors and Inductors • Transformers • Electrical Fuses

2 **Voltage and Current Sources** *R. Dorf, Z. Wan, C. Paul, J. Cogdell* 47
Step, Impulse, Ramp, Sinusoidal, and DC Signals • Ideal and Practical Sources • Controlled
Sources

3 **Linear Circuit Analysis** *M. Ciletti, J. Irwin, A. Kraus, N. Balabanian, T. Bickart,
S. Chan* 58
Voltage and Current Laws • Node and Mesh Analysis • Network Theorems • Power and
Energy • Three-Phase Circuits • Graph Theory

4 **Passive Signal Processing** *W. Kerwin* 111
Low-Pass Filter Functions • Low-Pass Filters • Filter Design

5 **Nonlinear Circuits** *J. Hudgins, T. Bogart, Jr., K. Mayaram* 126
Diodes and Rectifiers • Limiters • Distortion

6 **Laplace Transform** *R. Dorf, Z. Wan, D. Johnson* 149
Definitions and Properties • Applications

7 **State Variables** *W. Chen* 169
State Equations in Normal Form • The Concept of State and State Variables and Normal
Tree • Systematic Procedure in Writing State Equations • State Equations for Networks De-
scribed by Scalar Differential Equations • Extension to Time-Varying and Nonlinear Networks

8 **The z-Transform** *R. Dorf, Z. Wan* 178
Properties of the z-Transform • Unilateral z-Transform • z-Transform Inversion • Sampled
Data

9 **T-Π Equivalent Networks** *Z. Wan, R. Dorf* 184
Three-Phase Connections • Wye ⇔ Delta Transformations

10 **Transfer Functions of Filters** *R. Dorf, Z. Wan* 189
Ideal Filters • The Ideal Linear-Phase Low-Pass Filter • Ideal Linear-Phase Bandpass Filters • Causal Filters • Butterworth Filters • Chebyshev Filters

11 **Frequency Response** *P. Neudorfer* 198
Linear Frequency Response Plotting • Bode Diagrams • A Comparison of Methods

12 **Stability Analysis** *F. Szidarovszky, A. Bahill* 207
Using the State of the System to Determine Stability • Lyapunov Stability Theory • Stability of
Time-Invariant Linear Systems • BIBO Stability • Physical Examples

THIS SECTION PROVIDES A BRIEF REVIEW of the definitions and fundamental concepts used in the study of linear circuits and systems. We can describe a *circuit* or *system,* in a broad sense, as a collection of objects called *elements* (*components, parts,* or *subsystems*) which form an entity governed by certain laws or constraints. Thus, a physical system is an entity made up of physical objects as its elements or components. A subsystem of a given system can also be considered as a system itself.

A mathematical model describes the behavior of a physical system or device in terms of a set of equations, together with a schematic diagram of the device containing the symbols of its elements, their connections, and numerical values. As an example, a physical electrical system can be represented graphically by a network which includes resistors, inductors, and capacitors, etc. as its components. Such an illustration, together with a set of linear differential equations, is referred to as a model system.

Electrical circuits may be classified into various categories. Four of the more familiar classifications are (a) linear and nonlinear circuits, (b) time-invariant and time-varying circuits, (c) passive and active circuits, and (d) lumped and distributed circuits. A *linear* circuit can be described by a set of linear (differential) equations; otherwise it is a nonlinear circuit. A *time-invariant* circuit or system implies that none of the components of the circuit have parameters that vary with time; otherwise it is a *time-variant* system. If the total energy delivered to a given circuit is nonnegative at any instant of time, the circuit is said to be *passive*; otherwise it is *active*. Finally, if the dimensions of the components of the circuit are small compared to the wavelength of the highest of the signal frequencies applied to the circuit, it is called a *lumped* circuit; otherwise it is referred to as a *distributed* circuit.

There are, of course, other ways of classifying circuits. For example, one might wish to classify circuits according to the number of accessible terminals or terminal pairs (ports). Thus, terms such as *n-terminal circuit* and *n-port* are commonly used in circuit theory. Another method of classification is based on circuit configurations (topology),[1] which gives rise to such terms as *ladders, lattices, bridged-T circuits*, etc.

As indicated earlier, although the words *circuit* and *system* are synonymous and will be used interchangeably throughout the text, the terms *circuit theory* and *system theory* sometimes denote different points of view in the study of circuits or systems. Roughly speaking, *circuit theory* is mainly concerned with interconnections of components (circuit topology) within a given system, whereas *system theory* attempts to attain generality by means of abstraction through a generalized (input-output state) model.

One of the goals of this section is to present a unified treatment on the study of linear circuits and systems. That is, while the study of linear circuits with regard to their topological properties is treated as an important phase of the entire development of the theory, a generality can be attained from such a study.

The subject of circuit theory can be divided into two main parts, namely, analysis and synthesis. In a broad sense, *analysis* may be defined as "the separating of any material or abstract entity [system] into its constituent elements;" on the other hand, *synthesis* is "the combining of the constituent elements of separate materials or abstract entities into a single or unified entity [system]."[2]

It is worth noting that in an analysis problem, the solution is always *unique* no matter how difficult it may be, whereas in a synthesis problem there might exist an infinite number of solutions or, sometimes, *none at all!*

It should also be noted that in some network theory texts the words *synthesis* and *design* might be used interchangeably throughout the entire discussion of the subject. However, the term *synthesis* is generally used to describe *analytical* procedures that can usually be carried out step by step,

---

[1]Circuit topology or graph theory deals with the way in which the circuit elements are interconnected. A detailed discussion on elementary applied graph theory is given in Chapter 3.6.

[2]The definitions of analysis and synthesis are quoted directly from *The Random House Dictionary of the English Language,* 2nd ed., Unabridged, New York: Random House, 1987.

whereas the term *design* includes practical (design) procedures (such as trial-and-error techniques which are based, to a great extent, on the experience of the designer) as well as analytical methods.

In analyzing the behavior of a given physical system, the first step is to establish a mathematical model. This model is usually in the form of a set of either differential or difference equations (or a combination of them), the solution of which accurately describes the motion of the physical systems. There is, of course, no exception to this in the field of electrical engineering. A physical electrical system such as an amplifier circuit, for example, is first represented by a circuit drawn on paper. The circuit is composed of resistors, capacitors, inductors, and voltage and/or current sources,[3] and each of these circuit elements is given a symbol together with a mathematical expression (i.e., the voltage-current or simply *v-i* relation) relating its terminal voltage and current at every instant of time. Once the network and the *v-i* relation for each element is specified, Kirchhoff's voltage and current laws can be applied, possibly together with the physical principles to be introduced in Chapter 3.1, to establish the mathematical model in the form of differential equations.

In Section I, focus is on analysis only (leaving coverage of synthesis and design to Section III, "Electronics"). Specifically, the passive circuit elements—resistors, capacitors, inductors, transformers, and fuses—as well as voltage and current sources (active elements) are discussed. This is followed by a brief discussion on the elements of linear circuit analysis. Next, some popularly used passive filters and nonlinear circuits are introduced. Then, Laplace transform, state variables, *z*-transform, and T and π configurations are covered. Finally, transfer functions, frequency response, and stability analysis are discussed.

Nomenclature

Symbol	Quantity	Unit	Symbol	Quantity	Unit
A	area	m^2	ω	angular frequency	rad/s
B	magnetic flux density	Tesla	P	power	W
C	capacitance	F	PF	power factor	
e	induced voltage	V	q	charge	C
ε	dielectric constant	F/m	Q	selectivity	
ε	ripple factor		R	resistance	Ω
f	frequency	Hz	$R(T)$	temperature coefficient	$\Omega/°C$
F	force	Newton		of resistance	
ϕ	magnetic flux	weber	ρ	resistivity	Ωm
I	current	A	s	Laplace operator	
J	Jacobian		τ	damping factor	
k	Boltzmann constant	1.38×10^{-23} J/K	θ	phase angle	degree
k	dielectric coefficient		v	velocity	m/s
K	coupling coefficient		V	voltage	V
L	inductance	H	W	energy	J
λ	eigenvalue		X	reactance	Ω
M	mutual inductance	H	Y	admittance	S
n	turns ratio		Z	impedance	Ω
n	filter order				

[3]Here, of course, active elements such as transistors are represented by their equivalent circuits as combinations of resistors and dependent sources.

1

Passive Components

Michael Pecht
University of Maryland

Pradeep Lall
University of Maryland

Glen Ballou
Ballou Associates

C. Sankaran
Machine Design Engineers, Inc.

Nick Angelopoulos
Gould Shawmut Company

1.1 Resistors 5
    Resistor Characteristics • Resistor Types
1.2 Capacitors and Inductors 15
    Capacitors • Types of Capacitors • Inductors
1.3 Transformers 33
    Types of Transformers • Principle of Transformation • Electro-
    magnetic Equation • Transformer Core • Transformer Losses •
    Transformer Connections • Transformer Impedance
1.4 Electrical Fuses 39
    Ratings • Fuse Performance • Selective Coordination • Standards •
    Products • Standard—Class H • HRC • Trends

1.1 Resistors

Michael Pecht and Pradeep Lall

The resistor is an electrical device whose primary function is to introduce resistance to the flow of electric current. The magnitude of opposition to the flow of current is called the resistance of the resistor. A larger resistance value indicates a greater opposition to current flow.

The resistance is measured in ohms. An ohm is the resistance that arises when a current of one ampere is passed through a resistor subjected to one volt across its terminals.

The various uses of resistors include setting biases, controlling gain, fixing time constants, matching and loading circuits, voltage division, and heat generation. The following sections discuss resistor characteristics and various resistor types.

Resistor Characteristics

Voltage and Current Characteristics of Resistors

The resistance of a resistor is directly proportional to the **resistivity** of the material and the length of the resistor and inversely proportional to the cross-sectional area perpendicular to the direction of current flow. The resistance R of a resistor is given by

$$R = \frac{\rho l}{A} \tag{1.1}$$

where ρ is the resistivity of the resistor material ($\Omega \cdot$ cm), l is the length of the resistor along direction of current flow (cm), and A is the cross-sectional area perpendicular to current flow (cm^2) (Fig. 1.1). Resistivity is an inherent property of materials. Good resistor materials typically have resistivities between 2×10^{-6} and 200×10^{-6} $\Omega \cdot$ cm.

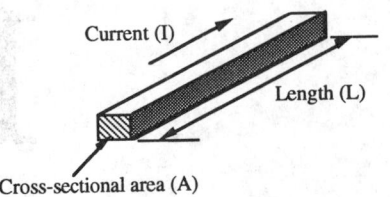

FIGURE 1.1 Resistance of a rectangular cross-section resistor with cross-sectional area A and length L.

The resistance can also be defined in terms of sheet resistivity. If the sheet resistivity is used, a standard sheet thickness is assumed and factored into resistivity. Typically, resistors are rectangular in shape; therefore the length l divided by the width w gives the number of squares within the resistor (Fig. 1.2). The number of squares multiplied by the resistivity is the resistance.

$$R_{sheet} = \rho_{sheet}\frac{l}{w} \qquad (1.2)$$

where ρ_{sheet} is the sheet resistivity (Ω/square), l is the length of resistor (cm), w is the width of the resistor (cm), and R_{sheet} is the sheet resistance (Ω).

The resistance of a resistor can be defined in terms of the **voltage drop** across the resistor and current through the resistor related by Ohm's law,

$$R = \frac{V}{I} \qquad (1.3)$$

where R is the resistance (Ω), V is the voltage across the resistor (V), and I is the current through the resistor (A). Whenever a current is passed through a resistor, a voltage is dropped across the ends of the resistor. Figure 1.3 depicts the symbol of the resistor with the Ohm's law relation.

All resistors dissipate power when a voltage is applied. The power dissipated by the resistor is represented by

$$P = \frac{V^2}{R} \qquad (1.4)$$

where P is the power dissipated (W), V is the voltage across the resistor (V), and R is the resistance (Ω). An ideal resistor dissipates electric energy without storing electric or magnetic energy.

Resistor Networks

Resistors may be joined to form networks. If resistors are joined in series, the effective resistance (R_T) is the sum of the individual resistances (Fig. 1.4).

$$R_T = \sum_{i=1}^{n} R_i \qquad (1.5)$$

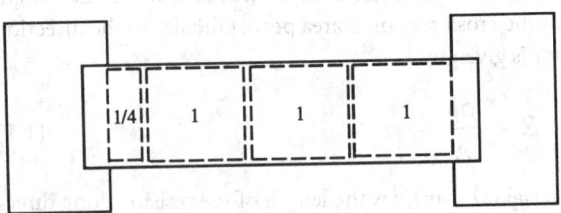

THE ABOVE RESISTOR IS 3.25 SQUARES
IF $\rho = 100\ \Omega/\square$, THEN R = 3.25 $\square$ X 100 $\Omega/\square$ = 325 Ω

FIGURE 1.2 Number of squares in a rectangular resistor.

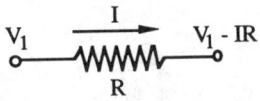

FIGURE 1.3 A resistor with resistance R having a current I flowing through it will have a voltage drop of IR across it.

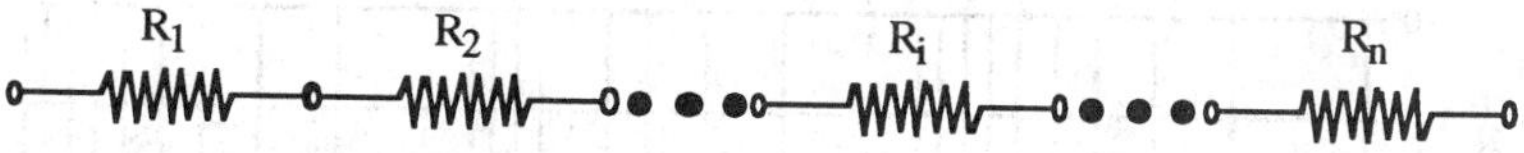

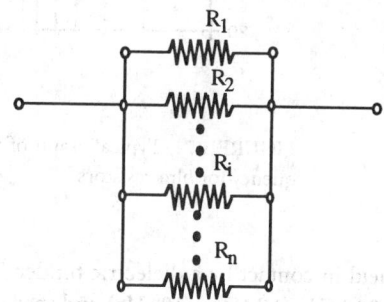

FIGURE 1.4   Resistors connected in series.

If resistors are joined in parallel, the effective resistance (R_T) is the reciprocal of the sum of the reciprocals of individual resistances (Fig. 1.5).

$$\frac{1}{R_T} = \sum_{i=1}^{n} \frac{1}{R_i}$$ (1.6)

Temperature Coefficient of Electrical Resistance

The resistance for most resistors changes with temperature. The temperature coefficient of electrical resistance is the change in electrical resistance of a resistor per unit change in temperature. The **temperature coefficient of resistance** is measured in $\Omega/°C$. The temperature coefficient of resistors may be either positive or negative. A positive temperature coefficient denotes a rise in resistance with a rise in temperature; a negative temperature coefficient of resistance denotes a decrease in resistance with a rise in temperature. Pure metals typically have a positive temperature coefficient of resistance, while some metal alloys such as constantin and manganin have a zero temperature coefficient of resistance. Carbon and graphite mixed with binders usually exhibit negative temperature coefficients, although certain choices of binders and process variations may yield positive temperature coefficients. The temperature coefficient of resistance is given by

FIGURE 1.5   Resistors connected in parallel.

$$R(T_2) = R(T_1)[1 + \alpha_{T_1}(T_2 - T_1)]$$ (1.7)

where α_{T_1} is the temperature coefficient of electrical resistance at reference temperature T_1, $R(T_2)$ is the resistance at temperature T_2 (Ω), and $R(T_1)$ is the resistance at temperature T_1 (Ω). The reference temperature is usually taken to be 20°C. Because the variation in resistance between any two temperatures is usually not linear as predicted by Eq. (1.7), common practice is to apply the equation between temperature increments and then to plot the resistance change versus temperature for a number of incremental temperatures.

High-Frequency Effects

Resistors show a change in their resistance value when subjected to ac voltages. The change in resistance with voltage frequency is known as the *Boella effect*. The effect occurs because all resistors have some inductance and capacitance along with the resistive component and thus can be approximated by an equivalent circuit shown in Fig. 1.6. Even though the definition of useful frequency range is application dependent, typically, the useful range of the resistor is the highest frequency at which the impedance differs from the resistance by more than the tolerance of the resistor.

The frequency effect on resistance varies with the resistor construction. Wire-wound resistors typically exhibit an increase in their impedance with frequency. In composition resistors the capacitances are formed by the many conducting particles which are

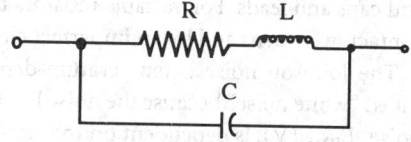

FIGURE 1.6   Equivalent circuit for a resistor.

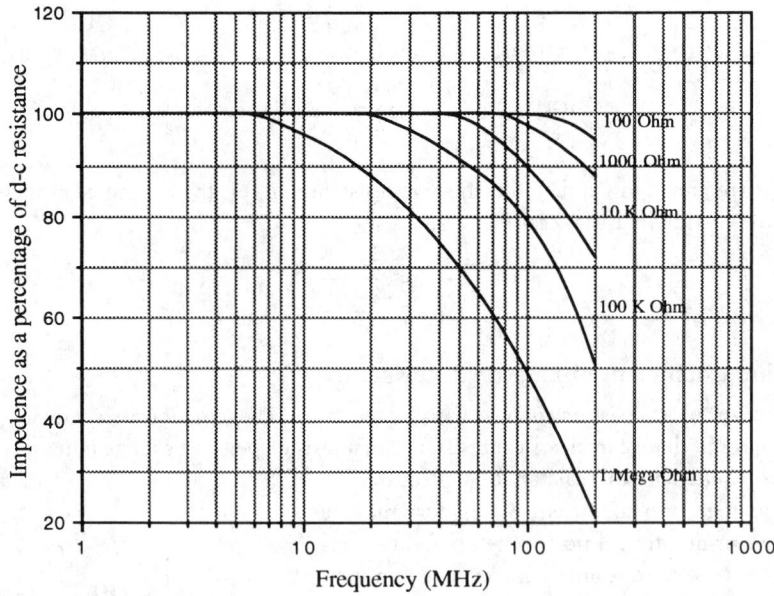

FIGURE 1.7   Typical graph of impedance as a percentage of dc resistance versus frequency for film resistors.

held in contact by a dielectric binder. The ac impedance for film resistors remains constant until 100 MHz (1 MHz = 10^6 Hz) and then decreases at higher frequencies (Fig. 1.7). For film resistors, the decrease in dc resistance at higher frequencies decreases with increase in resistance. Film resistors have the most stable high-frequency performance.

The smaller the diameter of the resistor the better is its frequency response. Most high-frequency resistors have a length to diameter ratio between 4:1 to 10:1. Dielectric losses are kept to a minimum by proper choice of base material.

Voltage Coefficient of Resistance

Resistance is not always independent of the applied voltage. The **voltage coefficient of resistance** is the change in resistance per unit change in voltage, expressed as a percentage of the resistance at 10% of rated voltage. The voltage coefficient is given by the relationship

$$\text{Voltage coefficient} = \frac{100(R_1 - R_2)}{R_2(V_1 - V_2)} \tag{1.8}$$

where R_1 is the resistance at the rated voltage V_1 and R_2 is the resistance at 10% of rated voltage V_2.

Noise

Resistors exhibit electrical noise in the form of small ac voltage fluctuations when dc voltage is applied. Noise in a resistor is a function of the applied voltage, physical dimensions, and materials. The total noise is a sum of Johnson noise, current flow noise, noise due to cracked bodies, and loose end caps and leads. For variable resistors the noise can also be caused by the jumping of a moving contact over turns and by an imperfect electrical path between the contact and resistance element.

The Johnson noise is temperature-dependent thermal noise (Fig. 1.8). Thermal noise is also called "white noise" because the noise level is the same at all frequencies. The magnitude of thermal noise, E_{RMS} (V), is dependent on the resistance value and the temperature of the resistance due to thermal agitation.

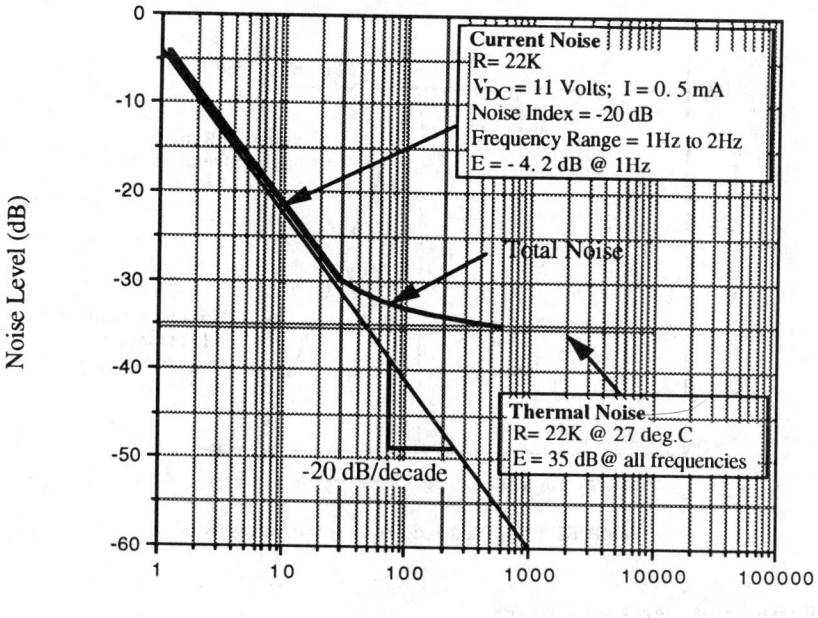

FIGURE 1.8 The total resistor noise is the sum of current noise and thermal noise. The current noise approaches the thermal noise at higher frequencies. (*Source:* Phillips Components, Discrete Products Division, *1990–91 Resistor/Capacitor Data Book,* 1991. With permission.)

$$E_{RMS} = \sqrt{4kRT\Delta f} \qquad (1.9)$$

where E_{RMS} is the root-mean-square value of the noise voltage (V), R is the resistance (Ω), K is the Boltzmann constant (1.38×10^{-23} J/K), T is the temperature (K), and Δf is the bandwidth (Hz) over which the noise energy is measured.

Figure 1.8 shows the variation in current noise versus voltage frequency. Current noise varies inversely with frequency and is a function of the current flowing through the resistor and the value of the resistor. The magnitude of current noise is directly proportional to the square root of current. The current noise magnitude is usually expressed by a noise index given as the ratio of the root-mean-square current noise voltage (E_{RMS}) over one decade bandwidth to the average voltage caused by a specified constant current passed through the resistor at a specified hot-spot temperature [Phillips, 1991].

$$\text{N.I.} = 20 \ \log_{10}\left(\frac{\text{Noise voltage}}{\text{dc voltage}}\right) \qquad (1.10)$$

$$E_{RMS} = V_{dc} \times 10^{\text{N.I.}/20} \sqrt{\log\left(\frac{f_2}{f_1}\right)} \qquad (1.11)$$

where N.I. is the noise index, V_{dc} is the dc voltage drop across the resistor, and f_1 and f_2 represent the frequency range over which the noise is being computed. Units of noise index are μV/V. At higher frequencies, the current noise becomes less dominant compared to Johnson noise.

Precision film resistors have extremely low noise. Composition resistors show some degree of noise due to internal electrical contacts between the conducting particles held together with the binder. Wire-wound resistors are essentially free of electrical noise unless resistor terminations are faulty.

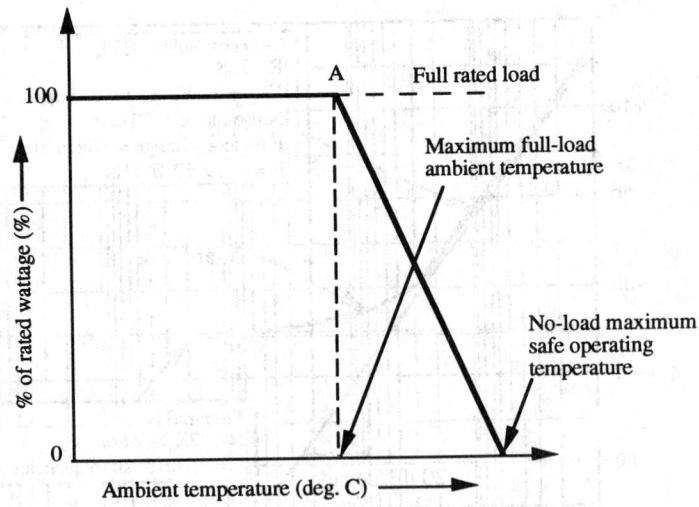

FIGURE 1.9 Typical derating curve for resistors.

Power Rating and Derating Curves

Resistors must be operated within specified temperature limits to avoid permanent damage to the materials. The temperature limit is defined in terms of the maximum power, called the *power rating*, and derating curve. The power rating of a resistor is the maximum power in watts which the resistor can dissipate. The maximum power rating is a function of resistor material, maximum voltage rating, resistor dimensions, and maximum allowable hot-spot temperature. The maximum hot-spot temperature is the temperature of the hottest part on the resistor when dissipating full-rated power at rated ambient temperature.

The maximum allowable power rating as a function of the ambient temperature is given by the derating curve. Figure 1.9 shows a typical power rating curve for a resistor. The derating curve is usually linearly drawn from the full-rated load temperature to the maximum allowable no-load temperature. A resistor may be operated at ambient temperatures above the maximum full-load ambient temperature if operating at lower than full-rated power capacity. The maximum allowable no-load temperature is also the maximum storage temperature for the resistor.

Voltage Rating of Resistors

The maximum voltage that may be applied to the resistor is called the **voltage rating** and is related to the power rating by

$$V = \sqrt{PR} \tag{1.12}$$

where V is the voltage rating (V), P is the power rating (W), and R is the resistance (Ω). For a given value of voltage and power rating, a critical value of resistance can be calculated. For values of resistance below the critical value, the maximum voltage is never reached; for values of resistance above the critical value, the power dissipated is lower than the rated power (Fig. 1.10).

Color Coding of Resistors

Resistors are generally identified by color coding or direct digital marking. The color code is given in Table 1.1. The color code is commonly used in composition resistors and film resistors. The color code essentially consists of four bands of different colors. The first band is the most significant figure, the second band is the second significant figure, the third band is the multiplier or the number of zeros that have to be added after the first two significant figures, and the fourth band is the tolerance on the resistance value. If the fourth band is not present, the resistor tolerance is the

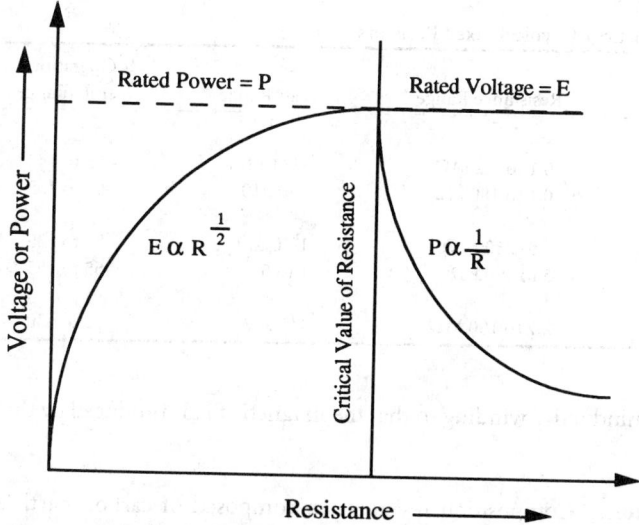

FIGURE 1.10   Relationship of applied voltage and power above and below the critical value of resistance.

Table 1.1   Color Code Table for Resistors

Color	First Band	Second Band	Third Band	Fourth Band Tolerance, %
Black	0	0	1	
Brown	1	1	10	
Red	2	2	100	
Orange	3	3	1,000	
Yellow	4	4	10,000	
Green	5	5	100,000	
Blue	6	6	1,000,000	
Violet	7	7	10,000,000	
Gray	8	8	100,000,000	
White	9	9	1,000,000,000	
Gold			0.1	5%
Silver			0.01	10%
No band				20%

Blanks in the table represent situations which do not exist in the color code.

standard 20% above and below the rated value. When the color code is used on fixed wire-wound resistors, the first band is applied in double width.

Resistor Types

Resistors can be broadly categorized as fixed, variable, and special-purpose. Each of these resistor types is discussed in detail with typical ranges of their characteristics.

Fixed Resistors

The fixed resistors are those whose value cannot be varied after manufacture. Fixed resistors are classified into composition resistors, wire-wound resistors, and metal-film resistors. Table 1.2 outlines the characteristics of some typical fixed resistors.

Wire-Wound Resistors.   Wire-wound resistors are made by winding wire of nickel-chromium alloy on a ceramic tube covering with a vitreous coating. The spiral winding has inductive and capacitive characteristics that make it unsuitable for operation above 50 kHz. The frequency limit

Table 1.2   Characteristics of Typical Fixed Resistors

Resistor Types	Resistance Range	Watt Range	Operating Temp. Range	α, ppm/°C
Wire-wound resistor				
Precision	0.1 to 1.2 MΩ	1/8 to 1/4	−55 to 145	10
Power	0.1 to 180 kΩ	1 to 210	−55 to 275	260
Metal-film resistor				
Precision	1 to 250 MΩ	1/20 to 1	−55 to 125	50–100
Power	5 to 100 kΩ	1 to 5	−55 to 155	20–100
Composition resistor				
General purpose	2.7 to 100 MΩ	1/8 to 2	−55 to 130	1500

can be raised by noninductive winding so that the magnetic fields produced by the two parts of the winding cancel.

Composition Resistors.   Composition resistors are composed of carbon particles mixed with a binder. This mixture is molded into a cylindrical shape and hardened by baking. Leads are attached axially to each end, and the assembly is encapsulated in a protective encapsulation coating. Color bands on the outer surface indicate the resistance value and tolerance. Composition resistors are economical and exhibit low noise levels for resistances above 1 MΩ. Composition resistors are usually rated for temperatures in the neighborhood of 70°C for power ranging from 1/8 to 2 W. Composition resistors have end-to-end shunted capacitance that may be noticed at frequencies in the neighborhood of 100 kHz, especially for resistance values above 0.3 MΩ.

Metal-Film Resistors.   Metal-film resistors are commonly made of nichrome, tin-oxide, or tantalum nitride, either hermetically sealed or using molded-phenolic cases. Metal-film resistors are not as stable as the wire-wound resistors. Depending on the application, fixed resistors are manufactured as precision resistors, semiprecision resistors, standard general-purpose resistors, or power resistors. Precision resistors have low voltage and power coefficients, excellent temperature and **time stabilities**, low noise, and very low reactance. These resistors are available in metal-film or wire constructions and are typically designed for circuits having very close resistance tolerances on values. Semiprecision resistors are smaller than precision resistors and are primarily used for current-limiting or voltage-dropping functions in circuit applications. Semiprecision resistors have long-term temperature stability. General-purpose resistors are used in circuits that do not require tight resistance tolerances or long-term stability. For general-purpose resistors, initial resistance variation may be in the neighborhood of 5% and the variation in resistance under full-rated power may approach 20%. Typically, general-purpose resistors have a high coefficient of resistance and high noise levels. Power resistors are used for power supplies, control circuits, and voltage dividers where operational stability of 5% is acceptable. Power resistors are available in wire-wound and film constructions. Film-type power resistors have the advantage of stability at high frequencies and have higher resistance values than wire-wound resistors for a given size.

Variable Resistors

Potentiometers.   The potentiometer is a special form of variable resistor with three terminals. Two terminals are connected to the opposite sides of the resistive element, and the third connects to a sliding contact that can be adjusted as a voltage divider.

Potentiometers are usually circular in form with the movable contact attached to a shaft that rotates. Potentiometers are manufactured as carbon composition, metallic film, and wire-wound resistors available in single-turn or multiturn units. The movable contact does not go all the way toward the end of the resistive element, and a small resistance called the *hop-off* resistance is present to prevent accidental burning of the resistive element.

Rheostat.   The rheostat is a current-setting device in which one terminal is connected to the resistive element and the second terminal is connected to a movable contact to place a selected section of the resistive element into the circuit. Typically, rheostats are wire-wound resistors used as speed controls for motors, ovens, and heater controls and in applications where adjustments on the voltage and current levels are required, such as voltage dividers and bleeder circuits.

Special-Purpose Resistors

Integrated Circuit Resistors.   Integrated circuit resistors are classified into two general categories: semiconductor resistors and deposited film resistors. Semiconductor resistors use the bulk resistivity of **doped** semiconductor regions to obtain the desired resistance value. Deposited film resistors are formed by depositing resistance films on an insulating substrate which are etched and patterned to form the desired resistive network. Depending on the thickness and dimensions of the deposited films, the resistors are classified into thick-film and thin-film resistors.

Semiconductor resistors can be divided into four types: diffused, bulk, pinched, and ion-implanted. Table 1.3 shows some typical resistor properties for semiconductor resistors. Diffused semiconductor resistors use resistivity of the diffused region in the semiconductor substrate to introduce a resistance in the circuit. Both *n*-type and *p*-type diffusions are used to form the diffused resistor.

A bulk resistor uses the bulk resistivity of the semiconductor to introduce a resistance into the circuit. Mathematically the sheet resistance of a bulk resistor is given by

$$R_{\text{sheet}} = \frac{\rho_e}{d} \tag{1.13}$$

where R_{sheet} is the sheet resistance in (Ω/square), ρ_e is the sheet resistivity (Ω/square), and d is the depth of the *n*-type **epitaxial layer**.

Pinched resistors are formed by reducing the effective cross-sectional area of diffused resistors. The reduced cross section of the diffused length results in extremely high sheet resistivities from ordinary diffused resistors.

Ion-implanted resistors are formed by implanting ions on the semiconductor surface by bombarding the silicon lattice with high-energy ions. The implanted ions lie in a very shallow layer along the surface (0.1 to 0.8 μm). For similar thicknesses ion-implanted resistors yield sheet resistivities 20 times greater than diffused resistors. Table 1.3 shows typical properties of diffused, bulk, pinched, and ion-implanted resistors. Typical sheet resistance values range from 80 to 250 Ω/square.

Table 1.3   Typical Characteristics of Integrated Circuit Resistors

Resistor Type	Sheet Resistivity (per square)	Temperature Coefficient (ppm/°C)
Semiconductor		
Diffused	0.8 to 260 Ω	1100 to 2000
Bulk	0.003 to 10 kΩ	2900 to 5000
Pinched	0.001 to 10 kΩ	3000 to 6000
Ion-implanted	0.5 to 20 kΩ	100 to 1300
Deposited resistors		
Thin-film		
Tantalum	0.01 to 1 kΩ	$\mp$100
SnO$_2$	0.08 to 4 kΩ	-1500 to 0
Ni-Cr	40 to 450 Ω	$\mp$100
Cermet (Cr-SiO)	0.03 to 2.5 kΩ	$\mp$150
Thick-film		
Ruthenium-silver	10 Ω to 10 MΩ	$\mp$200
Palladium-silver	0.01 to 100 kΩ	-500 to 150

Varistors.   Varistors are voltage-dependent resistors that show a high degree of nonlinearity between their resistance value and applied voltage. They are composed of a nonhomogeneous material that provides a rectifying action. Varistors are used for protection of electronic circuits, semiconductor components, collectors of motors, and relay contacts against overvoltage.

The relationship between the voltage and current of a varistor is given by

$$V = kI^\beta \tag{1.14}$$

where V is the voltage (V), I is the current (A), and k and β are constants that depend on the materials and manufacturing process. The electrical characteristics of a varistor are specified by its β and k values.

Varistors in Series.   The resultant k value of n varistors connected in series is nk. This can be derived by considering n varistors connected in series and a voltage nV applied across the ends. The current through each varistor remains the same as for V volts over one varistor. Mathematically, the voltage and current are expressed as

$$nV = k_1 I^\beta \tag{1.15}$$

Equating the expressions (1.14) and (1.15), the equivalent constant k_1 for the series combination of varistors is given as

$$k_1 = nk \tag{1.16}$$

Varistors in Parallel.   The equivalent k value for a parallel combination of varistors can be obtained by connecting n varistors in parallel and applying a voltage V across the terminals. The current through the varistors will still be n times the current through a single varistor with a voltage V across it. Mathematically the current and voltage are related as

$$V = k_2 (nI)^\beta \tag{1.17}$$

From Eqs. (1.14) and (1.17) the equivalent constant k_2 for the series combination of varistors is given as

$$k_2 = \frac{k}{n^\beta} \tag{1.18}$$

Thermistors.   Thermistors are resistors that change their resistance exponentially with changes in temperature. If the resistance decreases with increase in temperature, the resistor is called a negative temperature coefficient (NTC) resistor. If the resistance increases with temperature, the resistor is called a positive temperature coefficient (PTC) resistor.

NTC thermistors are ceramic semiconductors made by sintering mixtures of heavy metal oxides such as manganese, nickel, cobalt, copper, and iron. The resistance temperature relationship for NTC thermistors is

$$R_T = Ae^{B/T} \tag{1.19}$$

where T is temperature (K), R_T is the resistance (Ω), and A, B are constants whose values are determined by conducting experiments at two temperatures and solving the equations simultaneously.

PTC thermistors are prepared from $BaTiO_3$ or solid solutions of $PbTiO_3$ or $SrTiO_3$. The resistance temperature relationship for PTC thermistors is

$$R_T = A + Ce^{BT} \tag{1.20}$$

where T is temperature (K), R_T is the resistance (Ω), and A, B are constants determined by conducting experiments at two temperatures and solving the equations simultaneously. Positive thermistors have a PTC only between certain temperature ranges. Outside this range the temperature is either zero or negative. Typically, the absolute value of the temperature coefficient of resistance for PTC resistors is much higher than for NTC resistors.

Defining Terms

Doping: The intrinsic carrier concentration of semiconductors (e.g., Si) is too low to allow controlled charge transport. For this reason some impurities called dopants are purposely added to the semiconductor. The process of adding dopants is called doping. Dopants may belong to group IIIA (e.g., boron) or group VA (e.g., phosphorus) in the periodic table. If the elements belong to the group IIIA, the resulting semiconductor is called a *p*-type semiconductor. On the other hand, if the elements belong to the group VA, the resulting semiconductor is called an *n*-type semiconductor.

Epitaxial layer: Epitaxy refers to processes used to grow a thin crystalline layer on a crystalline substrate. In the epitaxial process the wafer acts as a seed crystal. The layer grown by this process is called an epitaxial layer.

Resistivity: The resistance of a conductor with unit length and unit cross-sectional area.

Temperature coefficient of resistance: The change in electrical resistance of a resistor per unit change in temperature.

Time stability: The degree to which the initial value of resistance is maintained to a stated degree of certainty under stated conditions of use over a stated period of time. Time stability is usually expressed as a percent or parts per million change in resistance per 1000 hours of continuous use.

Voltage coefficient of resistance: The change in resistance per unit change in voltage, expressed as a percentage of the resistance at 10% of rated voltage.

Voltage drop: The difference in potential between the two ends of the resistor measured in the direction of flow of current. The voltage drop is $V = IR$, where V is the voltage across the resistor, I is the current through the resistor, and R is the resistance.

Voltage rating: The maximum voltage that may be applied to the resistor.

References

Phillips Components, Discrete Products Division, *1990–91 Resistor/Capacitor Data Book,* 1991.

C.C. Wellard, *Resistance and Resistors,* New York: McGraw-Hill, 1960.

Further Information

IEEE Transactions on Electron Devices and *IEEE Electron Device Letters:* Published monthly by the Institute of Electrical and Electronics Engineers.

IEEE Components, Hybrids and Manufacturing Technology: Published quarterly by the Institute of Electrical and Electronics Engineers.

G.W.A. Dummer, *Materials for Conductive and Resistive Functions,* New York: Hayden Book Co., 1970.

H.F. Littlejohn and C.E. Burckel, *Handbook of Power Resistors,* Mount Vernon, N.Y.: Ward Leonard Electric Company, 1951.

I.R. Sinclair, *Passive Components: A User's Guide,* Oxford: Heinmann Newnes, 1990.

1.2 Capacitors and Inductors

Glen Ballou

Capacitors

If a potential difference is found between two points, an electric **field** exists that is the result of the separation of unlike charges. The strength of the field will depend on the amount the charges have been separated.

Capacitance is the concept of energy storage in an electric field and is restricted to the area, shape, and spacing of the **capacitor** plates and the property of the material separating them.

When electrical current flows into a capacitor, a force is established between two parallel plates separated by a **dielectric.** This energy is stored and remains even after the input is removed. By connecting a **conductor** (a resistor, hard wire, or even air) across the capacitor, the charged capacitor can regain electron balance, that is, discharge its stored energy.

The value of a parallel-plate capacitor can be found with the equation

$$C = \frac{x\epsilon[(N-1)A]}{d} \times 10^{-13} \qquad (1.21)$$

where C = capacitance, F; ϵ = dielectric constant of insulation; d = spacing between plates; N = number of plates; A = area of plates; and x = 0.0885 when A and d are in centimeters, and x = 0.225 when A and d are in inches.

The work necessary to transport a unit charge from one plate to the other is

$$e = kg \qquad (1.22)$$

where e = volts expressing energy per unit charge, g = coulombs of charge already transported, and k = proportionality factor between work necessary to carry a unit charge between the two plates and charge already transported. It is equal to $1/C$, where C is the capacitance, F.

The value of a capacitor can now be calculated from the equation

$$C = \frac{q}{e} \qquad (1.23)$$

where q = charge (C) and e is found with Eq. (1.22).

The energy stored in a capacitor is

$$W = \frac{CV^2}{2} \qquad (1.24)$$

where W = energy, J; C = capacitance, F; and V = applied voltage, V.

The **dielectric constant** of a material determines the electrostatic energy which may be stored in that material per unit volume for a given voltage. The value of the dielectric constant expresses the ratio of a capacitor in a vacuum to one using a given dielectric. The dielectric of air is 1, the reference unit employed for expressing the dielectric constant. As the dielectric constant is increased or decreased, the capacitance will increase or decrease, respectively. Table 1.4 lists the dielectric constants of various materials.

The dielectric constant of most materials is affected by both temperature and frequency, except for quartz, Styrofoam, and Teflon, whose dielectric constants remain essentially constant.

The equation for calculating the *force of attraction* between two plates is

$$F = \frac{AV^2}{k(1504S)^2} \qquad (1.25)$$

Table 1.4 Comparison of Capacitor Dielectric Constants

Dielectric	K (Dielectric Constant)
Air or vacuum	1.0
Paper	2.0–6.0
Plastic	2.1–6.0
Mineral oil	2.2–2.3
Silicone oil	2.7–2.8
Quartz	3.8–4.4
Glass	4.8–8.0
Porcelain	5.1–5.9
Mica	5.4–8.7
Aluminum oxide	8.4
Tantalum pentoxide	26
Ceramic	12–400,000

Source: G. Ballou, *Handbook for Sound Engineers, The New Audio Cyclopedia,* Carmel, Ind.: Macmillan Computer Publishing Company, 1991. With permission.

where F = attraction force, dyn; A = area of one plate, cm^2; V = potential energy difference, V; k = dielectric coefficient; and S = separation between plates, cm.

The Q for a capacitor when the resistance and capacitance is in series is

$$Q = \frac{1}{2\pi fRC} \tag{1.26}$$

where Q = ratio expressing the factor of merit; f = frequency, Hz; R = resistance, Ω; and C = capacitance, F.

When capacitors are connected in *series,* the total capacitance is

$$C_T = \frac{1}{1/C_1 + 1/C_2 + \cdots + 1/C_n} \tag{1.27}$$

and is always less than the value of the smallest capacitor.

When capacitors are connected in *parallel,* the total capacitance is

$$C_T = C_1 + C_2 + \cdots + C_n \tag{1.28}$$

and is always larger than the largest capacitor.

When a voltage is applied across a group of capacitors connected in series, the voltage drop across the combination is equal to the applied voltage. The drop across each individual capacitor is inversely proportional to its capacitance.

$$V_C = \frac{V_A C_X}{C_T} \tag{1.29}$$

where V_C = voltage across the individual capacitor in the series ($C_1, C_2, \ldots, C_n$), V; V_A = applied voltage, V; C_T = total capacitance of the series combination, F; and C_X = capacitance of individual capacitor under consideration, F.

In an ac circuit, the **capacitive reactance**, or the **impedance**, of the capacitor is

$$X_C = \frac{1}{2\pi fC} \tag{1.30}$$

where X_C = capacitive reactance, Ω; f = frequency, Hz; and C = capacitance, F. The current will lead the voltage by 90° in a circuit with a pure capacitor.

When a dc voltage is connected across a capacitor, a time t is required to charge the capacitor to the applied voltage. This is called a **time constant** and is calculated with the equation

$$t = RC \tag{1.31}$$

where t = time, s; R = resistance, Ω; and C = capacitance, F.

In a circuit consisting of pure resistance and capacitance, the *time constant t* is defined as the time required to charge the capacitor to 63.2% of the applied voltage.

During the next time constant, the capacitor charges to 63.2% of the remaining difference of full value, or to 86.5% of the full value. The charge on a capacitor can never actually reach 100% but is considered to be 100% after five time constants. When the voltage is removed, the capacitor discharges to 63.2% of the full value.

Capacitance is expressed in microfarads (μF, or 10^{-6} F) or picofarads (pF, or 10^{-12} F) with a stated accuracy or tolerance. Tolerance may also be stated as GMV (guaranteed minimum value), sometimes referred to as MRV (minimum rated value).

All capacitors have a *maximum working voltage* that must not be exceeded and is a combination of the dc value plus the peak ac value which may be applied during operation.

Quality Factor (Q)

Quality factor is the ratio of the capacitor's **reactance** to its resistance at a specified frequency and is found by the equation

$$Q = \frac{1}{2\pi fCR}$$

$$= \frac{1}{PF} \tag{1.32}$$

where Q = quality factor; f = frequency, Hz; C = value of capacitance, F; R = internal resistance, Ω; and PF = power factor

Power Factor (PF)

Power factor is the preferred measurement in describing capacitive losses in ac circuits. It is the fraction of input volt-amperes (or power) dissipated in the capacitor dielectric and is virtually independent of the capacitance, applied voltage, and frequency.

Equivalent Series Resistance (ESR)

Equivalent series resistance is expressed in ohms or milliohms (Ω, mΩ) and is derived from lead resistance, termination losses, and dissipation in the dielectric material.

Equivalent Series Inductance (ESL)

The *equivalent series inductance* can be useful or detrimental. It reduces high-frequency performance; however, it can be used in conjunction with the internal capacitance to form a resonant circuit.

Dissipation Factor (DF)

The **dissipation factor** in percentage is the ratio of the effective series resistance of a capacitor to its reactance at a specified frequency. It is the reciprocal of *quality factor* (Q) and an indication of power loss within the capacitor. It should be as low as possible.

Insulation Resistance

Insulation resistance is the resistance of the dielectric material and determines the time a capacitor, once charged, will hold its charge. A discharged capacitor has a low insulation resistance; however once charged to its rated value, it increases to megohms. The leakage in electrolytic capacitors should not exceed

$$I_L = 0.04C + 0.30 \tag{1.33}$$

where I_L = leakage current, μA, and C = capacitance, μF.

Dielectric Absorption (DA)

The *dielectric absorption* is a reluctance of the dielectric to give up stored electrons when the capacitor is discharged. This is often called "memory" because if a capacitor is discharged through a resistance and the resistance is removed, the electrons that remained in the dielectric will reconvene on the electrode, causing a voltage to appear across the capacitor. DA is tested by charging the capacitor for 5 min, discharging it for 5 s, then having an open circuit for 1 min after which the recovery voltage is read. The percentage of DA is defined as the ratio of recovery to charging voltage times 100.

Types of Capacitors

Capacitors are used to filter, couple, tune, block dc, pass ac, bypass, shift phase, compensate, feed through, isolate, store energy, suppress noise, and start motors. They must also be small, lightweight, reliable, and withstand adverse conditions.

Capacitors are grouped according to their dielectric material and mechanical configuration.

Ceramic Capacitors

Ceramic capacitors are used most often for bypass and coupling applications (Fig. 1.11). Ceramic capacitors can be produced with a variety of K values (dielectric constant). A high K value translates to small size and less stability. High-K capacitors with a dielectric constant >3000 are physically small and have values between 0.001 to several microfarads.

Good temperature stability requires capacitors to have a K value between 10 and 200. If high Q is also required, the capacitor will be physically larger. Ceramic capacitors with a zero temperature change are called **negative-positive-zero (NPO)** and come in a capacitance range of 1.0 pF to 0.033 μF.

An N750 temperature-compensated capacitor is used when accurate capacitance is required over a large temperature range. The 750 indicates a 750-ppm decrease in capacitance with a 1°C increase in temperature (750 ppm/°C). This equates to a 1.5% decrease in capacitance for a 20°C temperature increase. N750 capacitors come in values between 4.0 and 680 pF.

Film Capacitors

Film capacitors consist of alternate layers of metal foil and one or more layers of a flexible plastic insulating material (dielectric) in ribbon form rolled and encapsulated (see Fig. 1.12).

Mica Capacitors

Mica capacitors have small capacitance values and are usually used in high-frequency circuits. They are constructed as alternate layers of metal foil and mica insulation, which are stacked and encapsulated, or are silvered mica, where a silver electrode is screened on the mica insulators.

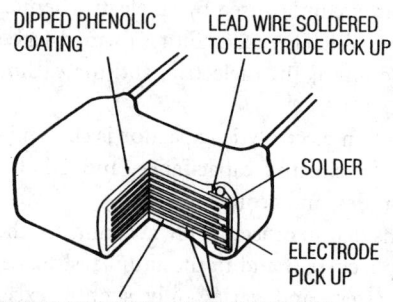

DIPPED PHENOLIC COATING

LEAD WIRE SOLDERED TO ELECTRODE PICK UP

SOLDER

ELECTRODE PICK UP

CERAMIC DIELECTRIC   METAL ELECTRODES

(ALTERNATELY DEPOSITED LAYERS OF CERAMIC DIELECTRIC MATERIAL AND METAL ELECTRODES FIRED INTO A RUGGED, SOLID BLOCK)

Voltage Ratings: 50 and 100 WVDC
Capacitance Range: 1.0 pF to 4.7 μF
Size Range: 0.150" x 0.150" 0.100" to 0.500" x 0.500" x 0.125"
Primary Applications: Used where capacitors with EIA
  Characteristics Z5U, X7R, and C0G must be selected
  to meet specific requirements.

FIGURE 1.11 Monolythic® multilayer ceramic capacitors. (Courtesy of Sprague Electric Company.)

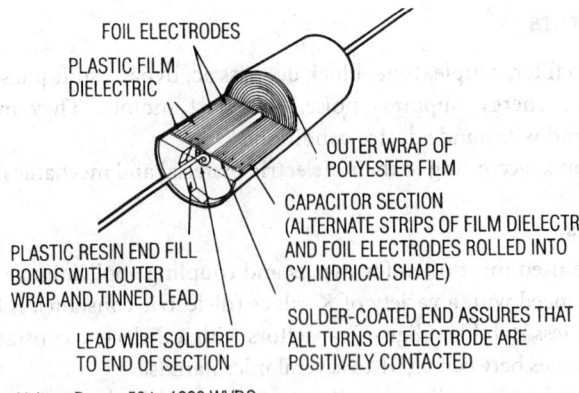

FOIL ELECTRODES
PLASTIC FILM DIELECTRIC
OUTER WRAP OF POLYESTER FILM
CAPACITOR SECTION
(ALTERNATE STRIPS OF FILM DIELECTRIC AND FOIL ELECTRODES ROLLED INTO CYLINDRICAL SHAPE)
PLASTIC RESIN END FILL BONDS WITH OUTER WRAP AND TINNED LEAD
SOLDER-COATED END ASSURES THAT ALL TURNS OF ELECTRODE ARE POSITIVELY CONTACTED
LEAD WIRE SOLDERED TO END OF SECTION

Voltage Range: 50 to 1000 WVDC
Capacitance Range: 0.001 to 1.0 µF
Size Range: 0.190" dia. x 0.450" long to 0.890" dia. x 2.125" long
Primary Applications: Radio, communications gear, and other commercial equipment where high IR, size, weight, and cost are important considerations.

FIGURE 1.12 Film-wrapped film capacitors. (Courtesy of Sprague Electric Company.)

Paper-Foil-Filled Capacitors

Paper-foil-filled capacitors are often used as motor capacitors and are rated at 60 Hz. They are made of alternate layers of aluminum and paper saturated with oil that are rolled together. The assembly is mounted in an oil-filled, hermetically sealed metal case.

Electrolytic Capacitors

Electrolytic capacitors provide high capacitance in a tolerable size; however, they do have drawbacks. Low temperatures reduce performance, while high temperatures dry them out. The **electrolytes** themselves can leak and corrode the equipment. Repeated surges above the rated working voltage, excessive ripple currents, and high operating temperature reduce performance and shorten capacitor life.

Electrolytic capacitors are manufactured by an electrochemical formation of an oxide film on a metal surface. The metal on which the oxide film is formed serves as the **anode** or positive terminal of the capacitor; the oxide film is the dielectric, and the **cathode** or negative terminal is either a conducting liquid or a gel.

The equivalent circuit of an electrolytic capacitor is shown in Fig. 1.13, where *A* and *B* are the capacitor terminals, *C* is the effective capacitance, and *L* is the self-inductance of the capacitor caused by terminals, electrodes, and geometry.

The shunt resistance (insulation resistance) R_s accounts for the dc leakage current. Heat is generated in the ESR from ripple current and in the shunt resistance by voltage. The ESR is due to the spacer-electrolyte-oxide system and varies only slightly except at low temperature, where it increases greatly.

The *impedance* of a capacitor (Fig. 1.14) is frequency-dependent. The initial downward slope is caused by the capacitive reactance X_C. The trough (lowest impedance) is almost totally resistive, and the upward slope is due to the capacitor's self-inductance X_L. An ESR plot would show an ESR decrease to about 5–10 kHz, remaining relatively constant thereafter.

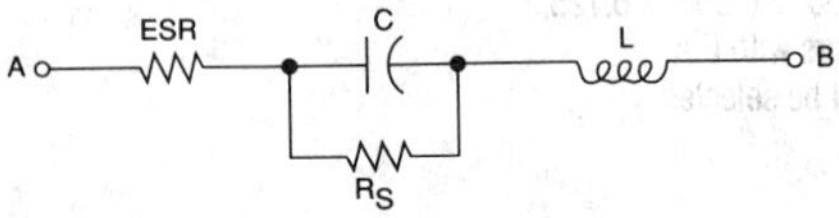

Leakage current is the direct current that passes through a capacitor when a correctly polarized dc voltage is applied to its terminals. It is proportional to temperature, becoming increasingly important at

FIGURE 1.13 Simplified equivalent circuit of an electrolytic capacitor.

elevated **ambient temperatures**. Leakage current decreases slowly after voltage is applied, reaching steady-state conditions in about 10 min.

If a capacitor is connected with reverse polarity, the oxide film is forward-biased, offering very little resistance to current flow. This causes overheating and self-destruction of the capacitor.

The total heat generated within a capacitor is the sum of the heat created by the $I_{leakage} \times V_{applied}$ and the I^2R losses in the ESR.

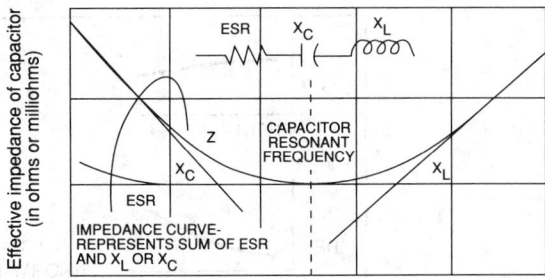

FIGURE 1.14   Impedance characteristics of a capacitor.

The ac **ripple current** rating is very important in filter applications because excessive current produces temperature rise, shortening capacitor life. The maximum permissible rms ripple current is limited by the internal temperature and the rate of heat dissipation from the capacitor. Lower ESR and longer enclosures increase the ripple current rating.

Capacitor life expectancy is doubled for each 10°C decrease in operating temperature, so a capacitor operating at room temperature will have a life expectancy 64 times that of the same capacitor operating at 85°C (185°F).

The *surge voltage* specification of a capacitor determines its ability to withstand high transient voltages that generally occur during the starting up period of equipment. Standard tests generally specify a short on and long off period for an interval of 24 h or more, and the allowable surge voltage levels are generally 10% above the rated voltage of the capacitor.

Figure 1.15 shows how temperature, frequency, time, and applied voltage affect electrolytic capacitors.

Aluminum Electrolytic Capacitors. Aluminum electrolytic capacitors use aluminum as the base material (Fig. 1.16). The surface is often etched to increase the surface area as much as 100 times that of unetched foil, resulting in higher capacitance in the same volume.

Aluminum electrolytic capacitors can withstand up to 1.5 V of reverse voltage without detriment. Higher reverse voltages, when applied over extended periods, lead to loss of capacitance. Excess reverse voltages applied for short periods cause some change in capacitance but not to capacitor failure.

Large-value capacitors are often used to filter dc power supplies. After a capacitor is charged, the rectifier stops conducting and the capacitor discharges into the load, as shown in Fig. 1.17, until the next cycle. Then the capacitor recharges again to the peak voltage. The Δe is equal to the total peak-to-peak ripple voltage and is a complex wave containing many harmonics of the fundamental ripple frequency, causing the noticeable heating of the capacitor.

Tantalum Capacitors. Tantalum electrolytics are the preferred type where high reliability and long service life are paramount considerations.

Tantalum capacitors have as much as three times better capacitance per volume efficiency than aluminum electrolytic capacitors, because tantalum pentoxide has a dielectric constant three times greater than that of aluminum oxide (see Table 1.4).

The capacitance of any capacitor is determined by the surface area of the two conducting plates, the distance between the plates, and the dielectric constant of the insulating material between the plates [see Eq. (1.21)].

In tantalum electrolytics, the distance between the plates is the thickness of the tantalum pentoxide film, and since the dielectric constant of the tantalum pentoxide is high, the capacitance of a tantalum capacitor is high.

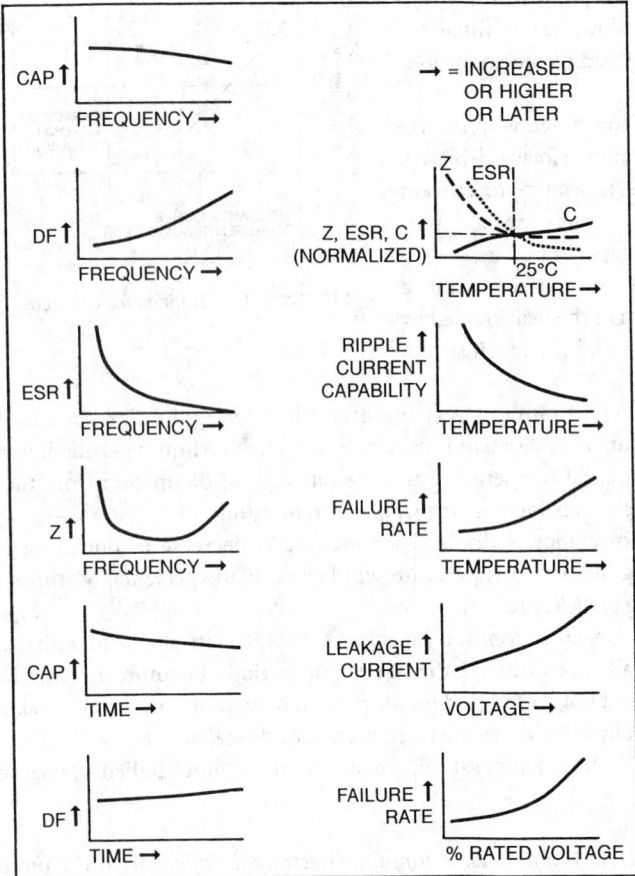

FIGURE 1.15 Variations in aluminum electrolytic characteristics caused by temperature, frequency, time, and applied voltage. (Courtesy of Sprague Electric Company.)

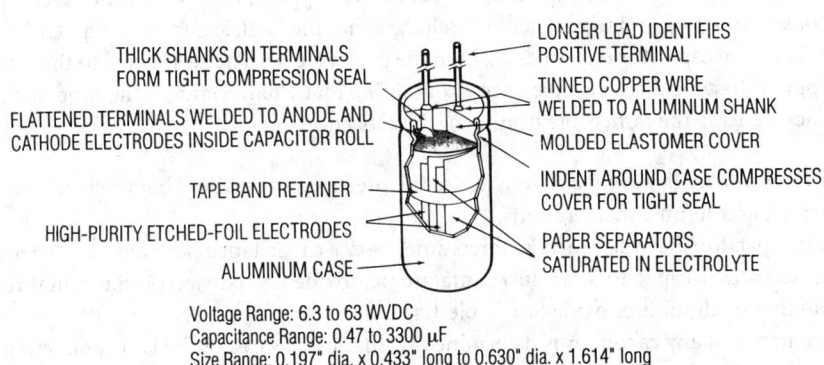

Voltage Range: 6.3 to 63 WVDC
Capacitance Range: 0.47 to 3300 µF
Size Range: 0.197" dia. x 0.433" long to 0.630" dia. x 1.614" long
Primary Applications: Coupling, decoupling, bypass, and filtering.
Vertical installation on high-density printed wiring boards in transistorized radios, portable TV sets, auto radios, tape recorders, etc.

FIGURE 1.16 Verti-lytic® miniature single-ended aluminum electrolytic capacitor. (Courtesy of Sprague Electric Company.)

Tantalum capacitors contain either liquid or solid electrolytes. The liquid electrolyte in wet-slug and foil capacitors, generally sulfuric acid, forms the cathode (negative) plate. In solid-electrolyte capacitors, a dry material, manganese dioxide, forms the cathode plate.

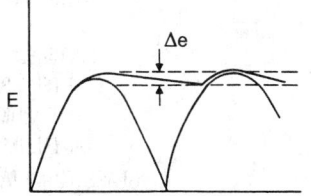

FIGURE 1.17 Full-wave capacitor charge and discharge.

Foil Tantalum Capacitors. *Foil tantalum capacitors* can be designed to voltage values up to 300 V dc. Of the three types of tantalum electrolytic capacitors, the foil design has the lowest capacitance per unit volume and is best suited for the higher voltages primarily found in older designs of equipment. It is expensive and used only where neither a solid-electrolyte (Fig. 1.18) nor a wet-slug (Fig. 1.19) tantalum capacitor can be employed.

Foil tantalum capacitors are generally designed for operation over the temperature range of –55 to +125°C (–67 to +257°F) and are found primarily in industrial and military electronics equipment.

Solid-electrolyte sintered-anode tantalum capacitors differ from the wet versions in their electrolyte, which is manganese dioxide.

Another variation of the solid-electrolyte tantalum capacitor encases the element in plastic resins, such as epoxy materials offering excellent reliability and high stability for consumer and commercial electronics with the added feature of low cost.

Still other designs of "solid tantalum" capacitors use plastic film or sleeving as the encasing material, and others use metal shells that are backfilled with an epoxy resin. Finally, there are small tubular and rectangular molded plastic encasements.

Wet-electrolyte sintered-anode tantalum capacitors, often called "wet-slug" tantalum capacitors, use a pellet of sintered tantalum powder to which a lead has been attached, as shown in Fig. 1.19. This anode has an enormous surface area for its size.

Wet-slug tantalum capacitors are manufactured in a voltage range to 125 V dc.

Use Considerations. Foil tantalum capacitors are used only where high-voltage constructions are required or where there is substantial reverse voltage applied to a capacitor during circuit operation.

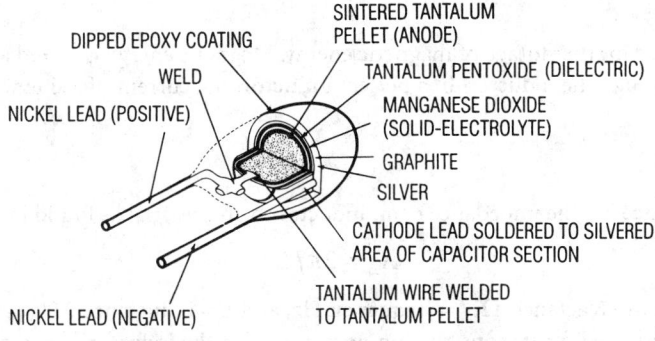

Voltage Range: 3 to 50 WVDC
Capacitance Range: 0.10 to 680 μF
Size Range: 0.175" dia. x 0.280" high to 0.400" dia. x 0.750" high
Primary Applications: For printed wiring boards applications where low cost, small size, high stability, low d-c leakage , and low dissipation factor are important.

FIGURE 1.18 Tantalex® solid electrolyte tantalum capacitor. (Courtesy of Sprague Electric Company.)

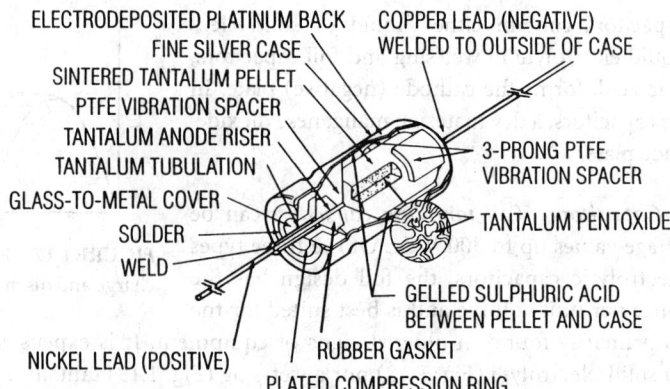

ELECTRODEPOSITED PLATINUM BACK
FINE SILVER CASE
SINTERED TANTALUM PELLET
PTFE VIBRATION SPACER
TANTALUM ANODE RISER
TANTALUM TUBULATION
GLASS-TO-METAL COVER
SOLDER
WELD
NICKEL LEAD (POSITIVE)

COPPER LEAD (NEGATIVE)
WELDED TO OUTSIDE OF CASE
3-PRONG PTFE
VIBRATION SPACER
TANTALUM PENTOXIDE
GELLED SULPHURIC ACID
BETWEEN PELLET AND CASE
RUBBER GASKET
PLATED COMPRESSION RING

Voltage Range: 6 to 125 WVDC
Capacitance Range: 1.7 to 1200 μF
Size Range: 0.188" dia. x 0.453" long to 0.375" dia. x 1.062" long
Primary Applications: Industrial and military equipment where
   reliability and premium performance with respect to low d-c
   leakage current, high inrush current capability, and high
   volumetric efficiency.

FIGURE 1.19 Hermetically sealed sintered-anode tantalum capacitor. (Courtesy of Sprague Electric Company.)

Wet sintered-anode capacitors, or "wet-slug" tantalum capacitors, are used where low dc leakage is required. The conventional "silver can" design will not tolerate reverse voltage. In military or aerospace applications where utmost reliability is desired, tantalum cases are used instead of silver cases. The tantalum-cased wet-slug units withstand up to 3 V reverse voltage and operate under higher ripple currents and at temperatures up to 200°C (392°F).

Solid-electrolyte designs are the least expensive for a given rating and are used where their very small size is important. They will typically withstand a reverse voltage up to 15% of the rated dc working voltage. They also have good low-temperature performance characteristics and freedom from corrosive electrolytes.

Inductors

Inductance is used for the storage of magnetic energy. Magnetic energy is stored as long as current keeps flowing through the inductor. In a perfect **inductor**, the current of a sine wave lags the voltage by 90°.

Impedance

Inductive reactance X_L, the impedance of an inductor to an ac signal, is found by the equation

$$X_L = 2\pi f L \qquad (1.34)$$

where X_L = inductive reactance, Ω; f = frequency, Hz; and L = inductance, H.

The type of wire used for its construction does not affect the inductance of a **coil**. Q of the coil will be governed by the resistance of the wire. Therefore coils wound with silver or gold wire have the highest Q for a given design.

To increase inductance, inductors are connected in series. The total inductance will always be greater than the largest inductor.

$$L_T = L_1 + L_2 + \cdots + L_n \qquad (1.35)$$

To reduce inductance, inductors are connected in parallel.

$$L_T = \frac{1}{1/L_1 + 1/L_2 + \cdots + 1/L_n} \tag{1.36}$$

The total inductance will always be less than the value of the lowest inductor.

Mutual Inductance

Mutual inductance is the property that exists between two conductors carrying current when their magnetic lines of force link together.

The mutual inductance of two coils with fields interacting can be determined by the equation

$$M = \frac{L_A - L_B}{4} \tag{1.37}$$

where M = mutual inductance of L_A and L_B, H; L_A = total inductance, H, of coils L_1 and L_2 with fields aiding; and L_B = total inductance, H, of coils L_1 and L_2 with fields opposing.

The *coupled inductance* can be determined by the following equations. In parallel with fields aiding,

$$L_T = \frac{1}{\dfrac{1}{L_1 + M} + \dfrac{1}{L_2 + M}} \tag{1.38}$$

In parallel with fields opposing,

$$L_T = \frac{1}{\dfrac{1}{L_1 - M} - \dfrac{1}{L_2 - M}} \tag{1.39}$$

In series with fields aiding,

$$L_T = L_1 + L_2 + 2M \tag{1.40}$$

In series with fields opposing,

$$L_T = L_1 + L_2 - 2M \tag{1.41}$$

where L_T = total inductance, H; L_1 and L_2 = inductances of the individual coils, H; and M = mutual inductance, H.

When two coils are inductively coupled to give transformer action, the coupling coefficient is determined by

$$K = \frac{M}{\sqrt{L_1 L_2}} \tag{1.42}$$

where K = coupling coefficient; M = mutual inductance, H; and L_1 and L_2 = inductances of the two coils, H.

An inductor in a circuit has a reactance equal to $j2\pi f L$ Ω. Mutual inductance in a circuit has a reactance equal to $j2\pi f M$ Ω. The operator j denotes that the reactance dissipates no energy; however, it does oppose current flow.

The energy stored in an inductor can be determined by the equation

$$W = \frac{LI^2}{2} \tag{1.43}$$

where W = energy, J (W $\cdot$ s); L = inductance, H; and I = current, A.

Coil Inductance

Inductance is related to the turns in a coil as follows:

1. The inductance is proportional to the square of the turns.
2. The inductance increases as the length of the **winding** is increased.
3. A shorted turn decreases the inductance, affects the frequency response, and increases the insertion loss.
4. The inductance increases as the permeability of the core material increases.
5. The inductance increases with an increase in the cross-sectional area of the core material.
6. Inductance is increased by inserting an iron core into the coil.
7. Introducing an air gap into a choke reduces the inductance.

A conductor moving at any angle to the lines of force cuts a number of lines of force proportional to the sine of the angles. Thus,

$$V = \beta L v \sin \theta \times 10^{-8} \tag{1.44}$$

where β = flux density; L = length of the conductor, cm; and v = velocity, cm/s, of conductor moving at an angle θ.

The maximum voltage induced in a conductor moving in a magnetic field is proportional to the number of magnetic lines of force cut by that conductor. When a conductor moves parallel to the lines of force, it cuts no lines of force; therefore, no current is generated in the conductor. A conductor that moves at right angles to the lines of force cuts the maximum number of lines per inch per second, therefore creating a maximum voltage. The right-hand rule determines direction of the induced electromotive force (emf). The emf is in the direction in which the axis of a right-hand screw, when turned with the velocity vector, moves through the smallest angle toward the flux density vector.

The **magnetomotive force** (mmf) in **ampere-turns** produced by a coil is found by multiplying the number of turns of wire in the coil by the current flowing through it.

$$\text{Ampere-turns} = T\left(\frac{V}{R}\right) \tag{1.45}$$

$$= TI$$

where T = number of turns; V = voltage, V; and R = resistance, Ω.

The inductance of a single layer, a spiral, and multilayer coils can be calculated by using either Wheeler's or Nagaoka's equations. The accuracy of the calculation will vary between 1 and 5%. The inductance of a single-layer coil can be calculated using Wheeler's equation:

$$L = \frac{B^2 N^2}{9B + 10A} \ \mu H \tag{1.46}$$

For the multilayer coil,

$$L = \frac{0.8B^2 N^2}{6B + 9A + 10C} \ \mu H \tag{1.47}$$

For the spiral coil,

$$L = \frac{B^2 N^2}{8B + 11C} \ \mu H \tag{1.48}$$

where B = radius of the winding, N = number of turns in the coil, A = length of the winding, and C = thickness of the winding.

Q

Q is the ratio of the inductive reactance to the internal resistance of the coil and is affected by frequency, inductance, dc resistance, inductive reactance, the type of winding, the core losses, the distributed capacity, and the permeability of the core material.

The Q for a coil where R and L are in series is

$$Q = \frac{2\pi fL}{R} \tag{1.49}$$

where f = frequency, Hz; L = inductance, H; and R = resistance, Ω.

The Q of the coil can be measured using the circuit of Fig. 1.20 for frequencies up to 1 MHz. The voltage across the inductance (L) at resonance equals $Q(V)$ (where V is the voltage developed by the oscillator); therefore, it is only necessary to measure the output voltage from the oscillator and the voltage across the inductance.

The oscillator voltage is driven across a low value of resistance, R, about 1/100 of the anticipated rf resistance of the LC combination, to assure that the measurement will not be in error by more than 1%. For most measurements, R will be about 0.10 Ω and should have a voltage of 0.1 V. Most oscillators cannot be operated into this low impedance, so a step-down matching transformer must be employed. Make C as large as convenient to minimize the ratio of the impedance looking from the voltmeter to the impedance of the test circuit. The LC circuit is then tuned to resonate and the resultant voltage measured. The value of Q may then be equated

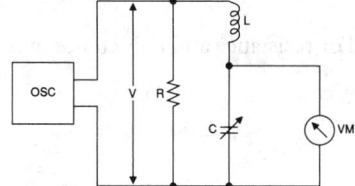

FIGURE 1.20 Circuit for measuring the Q of a coil.

$$Q = \frac{\text{resonant voltage across } C}{\text{voltage across } R} \tag{1.50}$$

The Q of any coil may be approximated by the equation

$$Q = \frac{2\pi fL}{R} \\ = \frac{X_L}{R} \tag{1.51}$$

where f = the frequency, Hz; L = the inductance, H; R = the dc resistance, Ω (as measured by an ohmmeter); and X_L = the inductive reactance of the coil.

Time Constant

When a dc voltage is applied to an RL circuit, a certain amount of time is required to change the circuit [see text with Eq. (1.31)]. The time constant can be determined with the equation

$$T = \frac{L}{R} \tag{1.52}$$

where R = resistance, Ω; L = inductance, H; and T = time, s.

The *right-hand rule* is used to determine the direction of a magnetic field around a conductor carrying a direct current. Grasp the conductor in the right hand with the thumb extending along the conductor pointing in the direction of the current. With the fingers partly closed, the finger tips will point in the direction of the magnetic field.

Maxwell's rule states, "If the direction of travel of a right-handed corkscrew represents the direction of the current in a straight conductor, the direction of rotation of the corkscrew will represent the direction of the magnetic lines of force."

Impedance

The total impedance created by resistors, capacitors, and inductors in circuits can be determined with the following equations.

For resistance and capacitance in series,

$$Z = \sqrt{R^2 + X_C^2} \tag{1.53}$$

$$\theta = \arctan \frac{X_C}{R} \tag{1.54}$$

For resistance and inductance in series,

$$Z = \sqrt{R^2 + X_L^2} \tag{1.55}$$

$$\theta = \arctan \frac{X_L}{R} \tag{1.56}$$

For inductance and capacitance in series,

$$Z = \begin{cases} X_L - X_C & \text{when } X_L > X_C \tag{1.57} \\ X_C - X_L & \text{when } X_C > X_L \tag{1.58} \end{cases}$$

For resistance, inductance, and capacitance in series,

$$Z = \sqrt{R^2 + (X_L - X_C)^2} \tag{1.59}$$

$$\theta = \arctan \frac{X_L - X_C}{R} \tag{1.60}$$

For capacitance and resistance in parallel,

$$Z = \frac{RX_C}{\sqrt{R^2 + X_C^2}} \tag{1.61}$$

For resistance and inductance in parallel,

$$Z = \frac{RX_L}{\sqrt{R^2 + X_L^2}} \tag{1.62}$$

For capacitance and inductance in parallel,

$$Z = \begin{cases} \dfrac{X_L X_C}{X_L - X_C} & \text{when } X_L > X_C & (1.63) \\[4mm] \dfrac{X_C X_L}{X_C - X_L} & \text{when } X_C > X_L & (1.64) \end{cases}$$

For inductance, capacitance, and resistance in parallel,

$$Z = \frac{R X_L X_C}{\sqrt{X_L^2 X_C^2 + R^2 (X_L - X_C)^2}} \qquad (1.65)$$

$$\theta = \arctan \frac{R(X_L - X_C)}{X_L X_C} \qquad (1.66)$$

For inductance and series resistance in parallel with resistance,

$$Z = R_2 \sqrt{\frac{R_1^2 + X_L^2}{(R_1 + R_2)^2 + X_L^2}} \qquad (1.67)$$

$$\theta = \arctan \frac{X_L R_2}{R_1^2 + X_L^2 + R_1 R_2} \qquad (1.68)$$

For inductance and series resistance in parallel with capacitance,

$$Z = X_C \sqrt{\frac{R^2 + X_L^2}{R^2 + (X_L - X_C)^2}} \qquad (1.69)$$

$$\theta = \arctan \frac{X_L (X_C - X_L) - R^2}{R X_C} \qquad (1.70)$$

For capacitance and series resistance in parallel with inductance and series resistance,

$$Z = \sqrt{\frac{(R_1^2 + X_L^2)(R_2^2 + X_C^2)}{(R_1 + R_2)^2 + (X_L - X_C)^2}} \qquad (1.71)$$

$$\theta = \arctan \frac{X_L(R_2^2 + X_C^2) - X_C(R_1^2 + X_L^2)}{R_1(R_2^2 + X_C^2) + R_2(R_1^2 + X_L^2)} \qquad (1.72)$$

where Z = impedance, Ω; R = resistance, Ω; L = inductance, H; X_L = inductive reactance, Ω; X_C = capacitive reactance, Ω; and θ = **phase** angle, degrees, by which current leads voltage in a capacitive circuit or lags voltage in an inductive circuit (0° indicates an in-phase condition).

Resonant Frequency

When an inductor and capacitor are connected in series or parallel, they form a resonant circuit. The **resonant frequency** can be determined from the equation

$$f = \frac{1}{2\pi\sqrt{LC}}$$

$$= \frac{1}{2\pi C X_C} \tag{1.73}$$

$$= \frac{X_L}{2\pi L}$$

where f = frequency, Hz; L = inductance, H; C = capacitance, F; and X_L, X_C = impedance, Ω.

The resonant frequency can also be determined through the use of a reactance chart developed by the Bell Telephone Laboratories (Fig. 1.21). This chart can be used for solving problems of inductance, capacitance, frequency, and impedance. If two of the values are known, the third and fourth values may be found with its use.

Defining Terms

Air capacitor: A fixed or variable capacitor in which air is the dielectric material between the capacitor's plates.

Ambient temperature: The temperature of the air or liquid surrounding any electrical part or device. Usually refers to the effect of such temperature in aiding or retarding removal of heat by radiation and convection from the part or device in question.

Ampere-turns: The magnetomotive force produced by a coil, derived by multiplying the number of turns of wire in a coil by the current (A) flowing through it.

Anode: The positive electrode of a capacitor.

Capacitive reactance: The opposition offered to the flow of an alternating or pulsating current by capacitance measured in ohms.

Capacitor: An electrical device capable of storing electrical energy and releasing it at some predetermined rate at some predetermined time. It consists essentially of two conducting surfaces (electrodes) separated by an insulating material or dielectric. A capacitor stores electrical energy, blocks the flow of direct current, and permits the flow of alternating current to a degree dependent essentially upon capacitance and frequency. The amount of energy stored, $E = 0.5\ CV^2$.

Cathode: The capacitor's negative electrode.

Coil: A number of turns of wire in the form of a spiral. The spiral may be wrapped around an iron core or an insulating form, or it may be self-supporting. A coil offers considerable opposition to ac current but very little to dc current.

Conductor: A bare or insulated wire or combination of wires not insulated from one another, suitable for carrying an electric current.

Dielectric: The insulating (nonconducting) medium between the two electrodes (plates) of a capacitor.

Dielectric constant: The ratio of the capacitance of a capacitor with a given dielectric to that of the same capacitor having a vacuum dielectric.

Disk capacitor: A small single-layer ceramic capacitor with a dielectric insulator consisting of conductively silvered opposing surfaces.

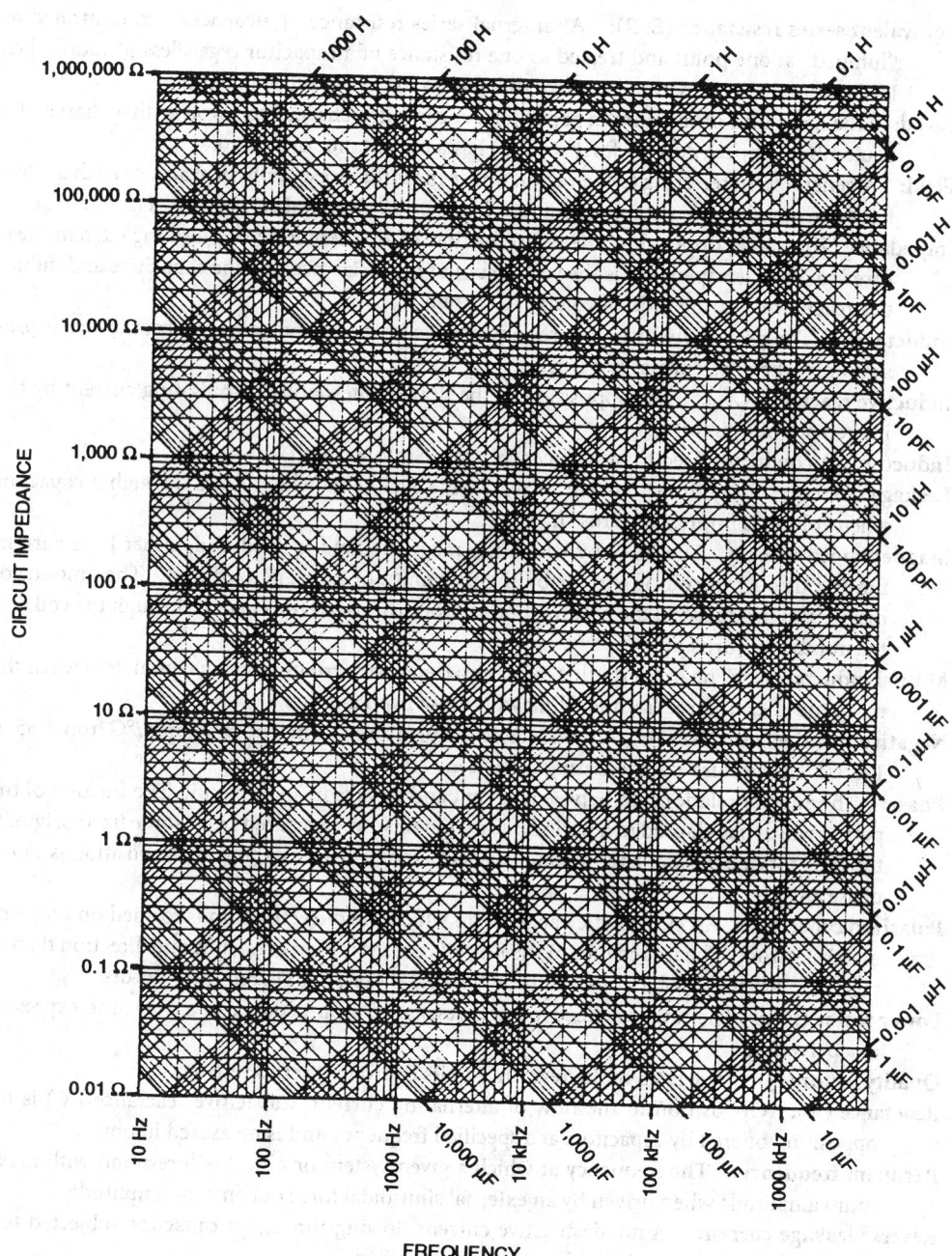

FIGURE 1.21   Reactance chart. (Courtesy AT&T Bell Laboratories.)

Dissipation factor (DF):   The ratio of the effective series resistance of a capacitor to its reactance at a specified frequency measured in percent.
Electrolyte: Current-conducting solution between two electrodes or plates of a capacitor, at least one of which is covered by a dielectric.
Electrolytic capacitor: A capacitor solution between two electrodes or plates of a capacitor, at least one of which is covered by a dielectric.

Equivalent series resistance (ESR): All internal series resistance of a capacitor concentrated or "lumped" at one point and treated as one resistance of a capacitor regardless of source, i.e., lead resistance, termination losses, or dissipation in the dielectric material.

Farad: The basic unit of measure in capacitors. A capacitor charged to 1 volt with a charge of 1 coulomb (1 ampere flowing for 1 second) has a capacitance of 1 farad.

Field: A general term referring to the region under the influence of a physical agency such as electricity, magnetism, or a combination produced by an electrical charged object.

Impedance (Z): Total opposition offered to the flow of an alternating or pulsating current measured in ohms. (Impedance is the vector sum of the resistance and the capacitive and inductive reactance, i.e., the ratio of voltage to current.)

Inductance: The property which opposes any change in the existing current. Inductance is present only when the current is changing.

Inductive reactance (X_L): The opposition to the flow of alternating or pulsating current by the inductance of a circuit.

Inductor: A conductor used to introduce inductance into a circuit.

Leakage current: Stray direct current of relatively small value which flows through a capacitor when voltage is impressed across it.

Magnetomotive force: The force by which the magnetic field is produced, either by a current flowing through a coil of wire or by the proximity of a magnetized body. The amount of magnetism produced in the first method is proportional to the current through the coil and the number of turns in it.

Mutual inductance: The property that exists between two current-carrying conductors when the magnetic lines of force from one link with those from another.

Negative-positive-zero (NPO): An ultrastable temperature coefficient (± 30 ppm/°C from −55 to 125°C) temperature-compensating capacitor.

Phase: The angular relationship between current and voltage in an ac circuit. The fraction of the period which has elapsed in a periodic function or wave measured from some fixed origin. If the time for one period is represented as 360° along a time axis, the phase position is called phase angle.

Polarized capacitor: An electrolytic capacitor in which the dielectric film is formed on only one metal electrode. The impedance to the flow of current is then greater in one direction than in the other. Reversed polarity can damage the part if excessive current flow occurs.

Power factor (PF): The ratio of effective series resistance to impedance of a capacitor, expressed as a percentage.

Quality factor (Q): The ratio of the reactance to its equivalent series resistance.

Reactance (X): Opposition to the flow of alternating current. Capacitive reactance (X_c) is the opposition offered by capacitors at a specified frequency and is measured in ohms.

Resonant frequency: The frequency at which a given system or object will respond with maximum amplitude when driven by an external sinusoidal force of constant amplitude.

Reverse leakage current: A nondestructive current flowing through a capacitor subjected to a voltage of polarity opposite to that normally specified.

Ripple current: The total amount of alternating and direct current that may be applied to an electrolytic capacitor under stated conditions.

Temperature coefficient (TC): A capacitor's change in capacitance per degree change in temperature. May be positive, negative, or zero and is usually expressed in parts per million per degree Celsius (ppm/°C) if the characteristics are linear. For nonlinear types, TC is expressed as a percentage of room temperature (25°C) capacitance.

Time constant: In a capacitor-resistor circuit, the number of seconds required for the capacitor to reach 63.2% of its full charge after a voltage is applied. The time constant of a capacitor with a capacitance (C) in farads in series with a resistance (R) in ohms is equal to $R \times C$ seconds.

Winding: A conductive path, usually wire, inductively coupled to a magnetic core or cell.

References

1. Exploring the capacitor, *Hewlett-Packard Bench Briefs*, September/October 1979. Sections reprinted with permission from *Bench Briefs*, a Hewlett-Packard service publication.
2. Capacitors, *1979 Electronic Buyer's Handbook*, vol. 1, November 1978. Copyright 1978 by CMP Publications, Inc. Reprinted with permission.
3. W. G. Jung and R. March, "Picking capacitors," *Audio*, March 1980.
4. "Electrolytic capacitors: Past, present and future," and "What is an electrolytic capacitor," *Electron. Des.*, May 28, 1981.
5. R. F. Graf, "Introduction To Aluminum Capacitors," Sprague Electric Company. Parts reprinted with permission.
6. "Introduction To Aluminum Capacitors," Sprague Electric Company. Parts reprinted with permission.
7. *Handbook of Electronics Tables and Formulas*, 6th ed., Indianapolis: Sams, 1986.

1.3   Transformers

C. Sankaran

The electrical transformer was invented by an American electrical engineer, William Stanley, in 1885 and was used in the first ac lighting installation at Great Barrington, Massachusetts. The first transformer was used to step up the power from 500 to 3000 V and transmitted for a distance of 1219 m (4000 ft). At the receiving end the voltage was stepped down to 500 V to power street and office lighting. By comparison, present transformers are designed to transmit hundreds of megawatts of power at voltages of 700 kV and beyond for distances of several hundred miles.

Transformation of power from one voltage level to another is a vital operation in any transmission, distribution, and utilization network. Normally, power is generated at a voltage that takes into consideration the cost of generators in relation to their operating voltage. Generated power is transmitted by overhead lines many miles and undergoes several voltage transformations before it is made available to the actual user. Figure 1.22 shows a typical power flow line diagram.

Types of Transformers

Transformers are broadly grouped into two main categories: dry-type and liquid-filled transformers. Dry-type transformers are cooled by natural or forced circulation of air or inert gas through or around the transformer enclosure. Dry-type transformers are further subdivided into ventilated,

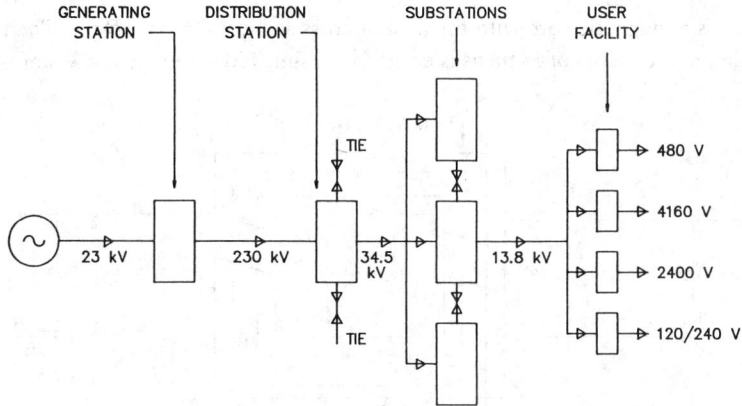

FIGURE 1.22   Power flow line diagram.

sealed, or encapsulated types depending upon the construction of the transformer. Dry transformers are extensively used in industrial power distribution for rating up to 5000 kVA and 34.5 kV.

Liquid-filled transformers are cooled by natural or forced circulation of a liquid coolant through the windings of the transformer. This liquid also serves as a **dielectric** to provide superior voltage-withstand characteristics. The most commonly used liquid in a transformer is a mineral oil known as transformer oil that has a continuous operating temperature rating of 105°C, a flash point of 150°C, and a fire point of 180°C. A good grade transformer oil has a **breakdown strength** of 86.6 kV/cm (220 kV/in.) that is far higher than the breakdown strength of air, which is 9.84 kV/cm (25 kV/in.) at atmospheric pressure.

Silicone fluid is used as an alternative to mineral oil. The breakdown strength of silicone liquid is over 118 kV/cm (300 kV/in.) and it has a flash point of 300°C and a fire point of 360°C. Silicone-fluid-filled transformers are classified as less flammable. The high dielectric strengths and superior thermal conductivities of liquid coolants make them ideally suited for large high-voltage power transformers that are used in modern power generation and distribution.

Principle of Transformation

The actual process of transfer of electrical power from a voltage of V_1 to a voltage of V_2 is explained with the aid of the simplified transformer representation shown in Fig. 1.23. Application of voltage across the primary winding of the transformer results in a **magnetic field** of ϕ_1 Wb in the magnetic core, which in turn induces a voltage of V_2 at the secondary terminals. V_1 and V_2 are related by the expression $V_1/V_2 = N_1/N_2$, where N_1 and N_2 are the number of turns in the primary and secondary windings, respectively. If a load current of I_2 A is drawn from the secondary terminals, the load current establishes a magnetic field of ϕ_2 Wb in the core and in the direction shown. Since the effect of load current is to reduce the amount of primary magnetic field, the reduction in ϕ_1 results in an increase in the primary current I_1 so that the net magnetic field is almost restored to the initial value and the slight reduction in the field is due to leakage **magnetic flux**. The currents in the two windings are related by the expression $I_1/I_2 = N_2/N_1$. Since $V_1/V_2 = N_1/N_2 = I_2/I_1$, we have the expression $V_1 \cdot I_1 = V_2 \cdot I_2$. Therefore, the voltamperes in the two windings are equal in theory. In reality, there is a slight loss of power during transformation that is due to the energy necessary to set up the magnetic field and to overcome the losses in the transformer core and windings. Transformers are static power conversion devices and are therefore highly efficient. Transformer efficiencies are about 95% for small units (15 kVA and less), and the efficiency can be higher than 99% for units rated above 5 MVA.

Electromagnetic Equation

Figure 1.24 shows a magnetic core with the area of cross section $A = W \cdot D$ m². The transformer primary winding that consists of N turns is excited by a sinusoidal voltage $v = V \sin(\omega t)$, where ω

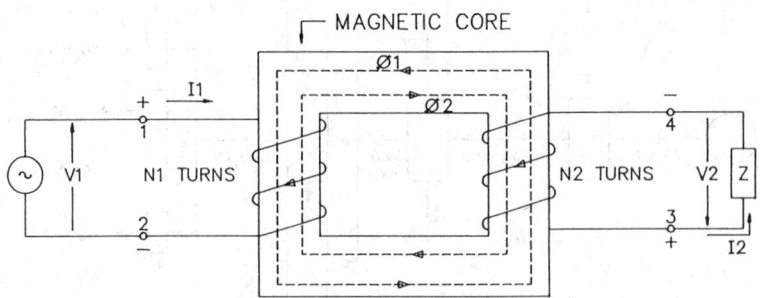

FIGURE 1.23   Electrical power transfer.

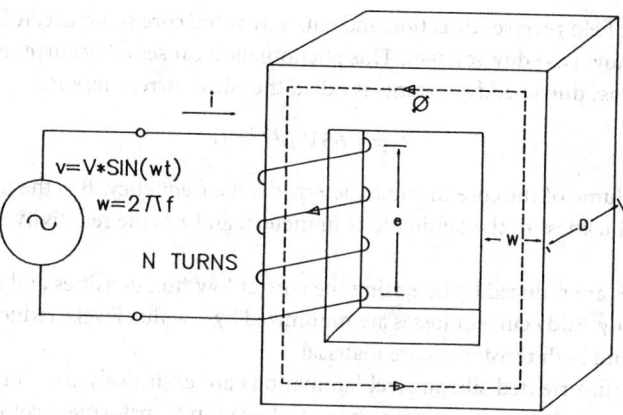

FIGURE 1.24 Electromagnetic relation.

is the angular frequency given by the expression $\omega = 2\pi f$ and f is the frequency of the applied voltage waveform. ϕ is magnetic field in the core due to the excitation current i:

$$\phi = \Phi \sin\left(\omega t - \frac{\pi}{2}\right) = -\Phi \cos(\omega t)$$

Induced voltage in the winding

$$e = -N\frac{d\phi}{dt} = N\frac{d[\Phi \cos(\omega t)]}{dt} = -N\omega\Phi \sin(\omega t)$$

Maximum value of the induced voltage

$$E = N\omega\Phi$$

The root-mean-square value

$$E_{rms} = \frac{E}{\sqrt{2}} = \frac{2\pi f N\Phi}{\sqrt{2}} = 4.44 f \, NBA$$

where flux Φ (webers) is replaced by the product of the flux density B (teslas) and the area of cross section of the core.

This fundamental design equation determines the size of the transformer for any given voltage and frequency. Power transformers are normally operated at flux density levels of 1.5 T.

Transformer Core

The transformer core is the medium that enables the transfer of power from the primary to the secondary to occur in a transformer. In order that the transformation of power may occur with the least amount of loss, the magnetic core is made up of laminations which have the highest permeability, permeability being a measure of the ease with which the magnetic field is set up in the core.

The magnetic field reverses direction every one half cycle of the applied voltage and energy is expended in the core to accomplish the cyclic reversals of the field. This loss component is known as the hysteresis loss P_h:

$$P_h = 150.7 V_e f B^{1.6} \quad W$$

where V_e is the volume of the core in cubic meters, f is the frequency, and B is the maximum flux density in teslas.

As the magnetic field reverses direction and cuts across the core structure, it induces a voltage in the laminations known as eddy voltages. This phenomenon causes eddy currents to circulate in the laminations. The loss due to eddy currents is called the eddy current loss P_e:

$$P_e = 1.65 V_e B^2 f^2 t^2 / r$$

where V_e is the volume of the core in cubic meters, f is the frequency, B is the maximum flux density in teslas, t is thickness of the laminations in meters, and r is the resistivity of the core material in ohm-meters.

Hysteresis losses are reduced by operating the core at low flux densities and using core material of high permeability. Eddy current losses are minimized by low flux levels, reduction in thickness of the laminations, and high resistivity core material.

Cold-rolled, grain-oriented silicon steel laminations are exclusively used in large power transformers to reduce core losses. A typical silicon steel used in transformers contains 95% iron, 3% silicon, 1% manganese, 0.2% phosphor, 0.06% carbon, 0.025% sulphur, and traces of other impurities.

Transformer Losses

The heat developed in a transformer is a function of the losses that occur during transformation. Therefore, the transformer losses must be minimized and the heat due to the losses must be efficiently conducted away from the core, the windings, and the cooling medium. The losses in a transformer are grouped into two categories: (1) no-load losses and (2) load losses. The no-load losses are the losses in the core due to excitation and are mostly composed of hysteresis and eddy current losses. The load losses are grouped into three categories: (1) winding I^2R losses, (2) winding eddy current losses, and (3) other stray losses. The winding I^2R losses are the result of the flow of load current through the resistance of the primary and secondary windings. The winding eddy current losses are caused by the magnetic field set up by the winding current, due to formation of eddy voltages in the conductors. The winding eddy losses are proportional to the square of the rms value of the current and to the square of the frequency of the current. When transformers are required to supply loads that are rich in **harmonic frequency** components, the eddy loss factor must be given extra consideration. The other stray loss component is the result of induced currents in the buswork, core clamps, and tank walls by the magnetic field set up by the load current.

Transformer Connections

A single-phase transformer has one input (primary) winding and one output (secondary) winding. A conventional three-phase transformer has three input and three output windings. The three windings can be connected in one of several different configurations to obtain three-phase connections that are distinct. Each form of connection has its own merits and demerits.

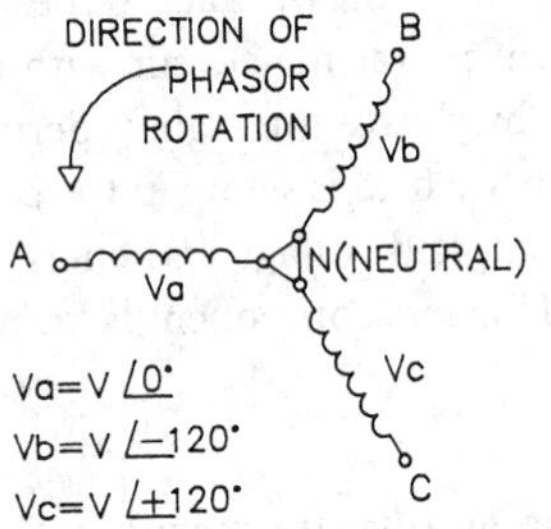

Y Connection (Fig. 1.25)

In the Y connection, one end of each of the three windings is connected together to form a Y, or a neutral point. This point is normally grounded, which limits the maximum potential to ground in the transformer to the line to neutral voltage of the power system. The grounded neutral also limits transient overvoltages in the transformer when subjected to lightning or switching surges. Availability of the neutral point allows the transformer to supply line to neutral single-phase loads in addition to normal three-phase loads. Each phase of the Y-connected

FIGURE 1.25 Y connection.

winding must be designed to carry the full line current, whereas the phase voltages are only 57.7% of the line voltages.

Delta Connection (Fig. 1.26)

In the delta connection, the finish point of each winding is connected to the start point of the adjacent winding to form a closed triangle, or delta. A delta winding in the transformer tends to balance out unbalanced loads that are present on the system. Each phase of the delta winding only carries 57.7% of the line current, whereas the phase voltages are equal to the line voltages.

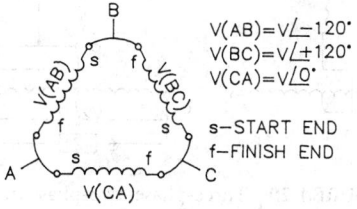

Large power transformers are designed so that the high-voltage side is connected in Y and the low-voltage side is connected in delta. Distribution transformers that are required to supply single-phase loads are designed in the opposite configuration so that the neutral point is available at the low-voltage end.

FIGURE 1.26   Delta connection.

Open-Delta Connection (Fig. 1.27)

An open-delta connection is used to deliver three-phase power if one phase of a three-phase bank of transformers fails in service. When the failed unit is removed from service, the remaining units can still supply three-phase power but at a reduced rating. An open-delta connection is also used as an economical means to deliver three-phase power using only two single-phase transformers. If P is the total three-phase kVA, then each transformer of the open-delta bank must have a rating of $P/\sqrt{3}$ kVA. The disadvantage of the open-delta connection is the unequal **regulation** of the three phases of the transformer.

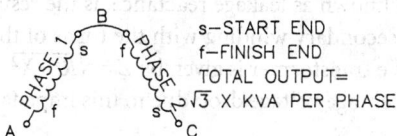

FIGURE 1.27   Open-delta connection.

T Connection (Fig. 1.28)

The T connection is used for three-phase power transformation when two separate single-phase transformers with special configurations are available. If a voltage transformation from V_1 to V_2 volts is required, one of the units (main transformer) must have a voltage ratio of V_1/V_2 with the midpoint of each winding brought out. The other unit must have a ratio of $0.866V_1/0.866V_2$ with the neutral point brought out, if needed.

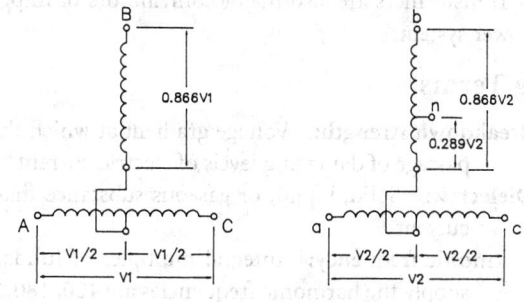

The Scott connection is a special type of T connection used to transform three-phase power to two-phase power for operation of electric furnaces and two-phase motors. It is shown in Fig. 1.29.

FIGURE 1.28   T connection.

Zigzag Connection (Fig. 1.30)

This connection is also called the interconnected star connection where the winding of each phase is divided into two halves and interconnected to form a zigzag configuration. The zigzag connection is mostly used to derive a neutral point for grounding purposes in three-phase, three-wire systems. The neutral point can be used to (1) supply single-phase loads, (2) provide a safety ground, and (3) sense and limit ground fault currents.

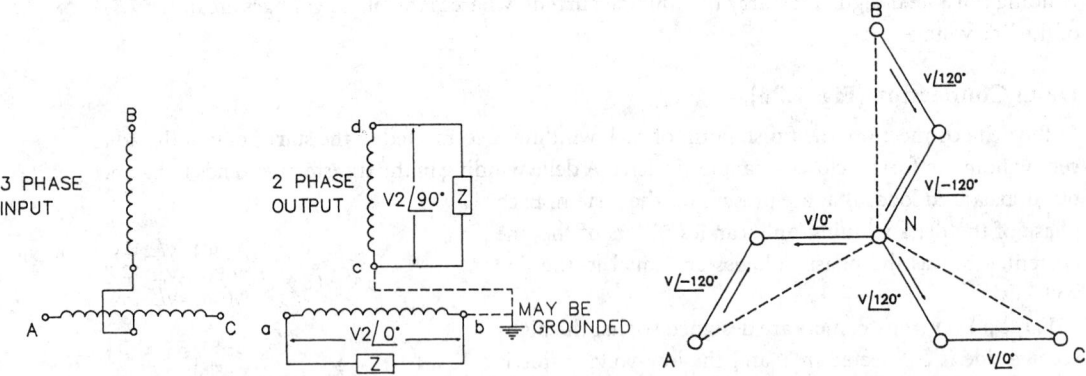

FIGURE 1.29 Three-phase–two-phase transformation.   **FIGURE 1.30** Zigzag connection.

Transformer Impedance

Impedance is an inherent property in a transformer that results in a voltage drop as power is transferred from the primary to the secondary side of the power system. The impedance of a transformer consists of two parts: resistance (R) and reactance (X). The resistance component is due to the resistance of the material of the winding and the percentage value of the voltage drop due to resistance becomes less as the rating of the transformer increases. The reactive component, which is also known as leakage reactance, is the result of incomplete linkage of the magnetic field set up by the secondary winding with the turns of the primary winding, and vice versa. The net impedance of the transformer is given by $Z = \sqrt{R^2 + X^2}$. The impedance value marked on the transformer is the percentage voltage drop due to this impedance under full-load operating conditions:

$$\% \text{ impedance } z = IZ\left(\frac{100}{V}\right)$$

where I is the full-load current of the transformer, Z is the impedance in ohms of the transformer, and V is the voltage rating of the transformer winding. It should be noted that the values of I and Z must be referred to the same side of the transformer as the voltage V.

Transformers are also major contributors of impedance to limit the fault currents in electrical power systems.

Defining Terms

Breakdown strength: Voltage gradient at which the molecules of medium break down to allow passage of damaging levels of electric current.

Dielectric: Solid, liquid, or gaseous substance that acts as an insulation to the flow of electric current.

Harmonic frequency: Integral multiples of fundamental frequency. For example, for a 60-Hz supply the harmonic frequencies are 120, 180, 240, 300, . . .

Magnetic field: Magnetic force field where lines of magnetism exist.

Magnetic flux: Term for lines of magnetism.

Regulation: The change in voltage from no-load to full-load expressed as a percentage of full-load voltage.

References and Further Information

Bean, Chackan, Moore and Wentz, *Transformers for the Electric Power Industry*, New York: McGraw-Hill, 1966.

General Electric, *Transformer Connections*, 1960.

A. Gray, *Electrical Machine Design*, New York: McGraw-Hill.

IEEE, *C57 Standards on Transformers*, New York: IEEE Press, 1992.

IEEE Transactions on Industry Applications.

R. R. Lawrence, *Principles of Alternating Current Machinery*, New York: McGraw-Hill, 1920.

Power Engineering Review.

C. Sankaran, *Introduction to Transformers*, New York: IEEE Press, 1992.

S. A. Stigant and A.C. Franklin, T*he J & P Transformer Book*, London: Newnes-Butterworths, 1973.

1.4 Electrical Fuses

Nick Angelopoulos

The fuse is a simple and reliable safety device. It is second to none in its ease of application and its ability to protect people and equipment.

The fuse is a current-sensitive device. It has a conductor with a reduced cross section (element) normally surrounded by an arc-quenching and heat-conducting material (filler). The entire unit is enclosed in a body fitted with end contacts. A basic fuse element design is illustrated in Fig. 1.32.

Ratings

Most fuses have three electrical ratings: ampere rating, voltage rating, and **interrupting rating**. The ampere rating indicates the current the fuse can carry without melting or exceeding specific temperature rise limits. The voltage rating, ac or dc, usually indicates the maximum system voltage that can be applied to the fuse. The interrupting rating (I.R.) defines the maximum short-circuit current that a fuse can safely interrupt. If a fault current higher than the interrupting rating causes the fuse to operate, the high internal pressure may cause the fuse to rupture. It is imperative, therefore, to install a fuse, or any other type of protective device, that has an interrupting rating not less than the available short-circuit current. A violent explosion may occur if the interrupting rating of any protective device is inadequate.

A fuse must perform two functions. The first, the "passive" function, is one that tends to be taken for granted. In fact, if the fuse performs the passive function well, we tend to forget that the fuse

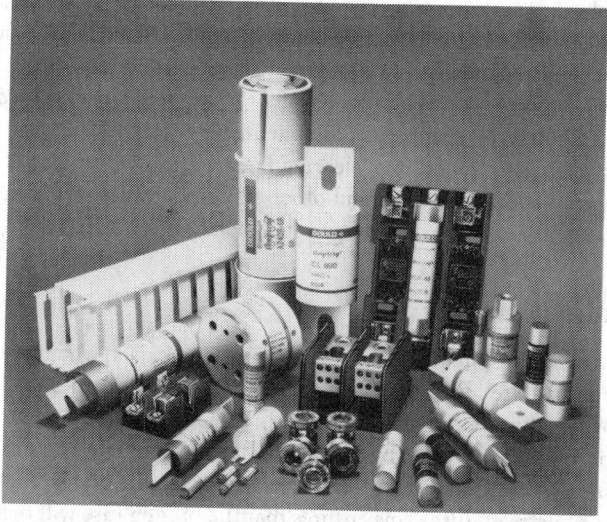

FIGURE 1.31 A variety of plug, cartridge, and blade type fuses.

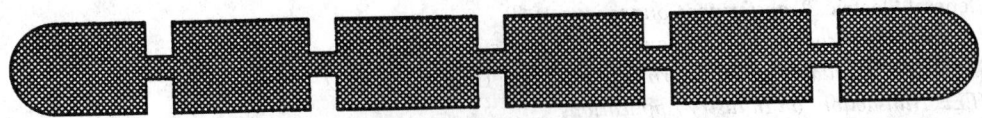

FIGURE 1.32   Basic fuse element.

exists at all. The passive function simply entails that the fuse can carry up to its normal load current without aging or overheating. Once the current level exceeds predetermined limits, the "active" function comes into play and the fuse operates. It is when the fuse is performing its active function that we become aware of its existence.

In most cases, the fuse will perform its active function in response to two types of circuit conditions. The first is an overload condition, for instance, when a hair dryer, teakettle, toaster, and radio are plugged into the same circuit. This overload condition will eventually cause the element to melt. The second condition is the overcurrent condition, commonly called the short circuit or the fault condition. This can produce a drastic, almost instantaneous, rise in current, causing the element to melt usually in less than a quarter of a cycle. Factors that can lead to a fault condition include rodents in the electrical system, loose connections, dirt and moisture, breakdown of insulation, foreign contaminants, and personal mistakes. Preventive maintenance and care can reduce these causes. Unfortunately, none of us are perfect and faults can occur in virtually every electrical system—we must protect against them.

Fuse Performance

Fuse performance characteristics under overload conditions are published in the form of *average melting time–current characteristic curves,* or simply *time-current curves.* Fuses are tested with a variety of currents, and the melting times are recorded. The result is a graph of time versus current coordinates that are plotted on log-log scale, as illustrated in Fig. 1.33.

Under short-circuit conditions the fuse operates and fully opens the circuit in less than 0.01 s. At 50 or 60 Hz, this represents operation within the first half cycle. The current waveform let-through by the fuse is the shaded, almost triangular, portion shown in Fig. 1.34(a). This depicts a fraction of the current that would have been let through into the circuit had a fuse not been installed.

Fuse short-circuit performance characteristics are published in the form of peak let-through (I_p) graphs and I^2t graphs. I_p (peak current) is simply the peak of the shaded triangular waveform, which increases as the fault current increases, as shown in Fig. 1.34(b). The electromagnetic forces, which can cause mechanical damage to equipment, are proportional to I_p^2.

I^2t represents heat energy measured in units of A^2 s (ampere squared seconds) and is documented on I^2t graphs. These I^2t graphs, as illustrated in Fig. 1.34(c), provide three values of I^2t: minimum melting I^2t, arcing I^2t, and total clearing I^2t. I^2t and I_p short-circuit performance characteristics can be used to coordinate fuses and other equipment. In particular, I^2t values are often used to selectively coordinate fuses in a distribution system.

Selective Coordination

In any power distribution system, selective coordination exists when the fuse immediately upstream from a fault operates, leaving all other fuses further upstream unaffected. This increases system reliability by isolating the faulted branch while maintaining power to all other branches. Selective coordination is easily assessed by comparing the I^2t characteristics for feeder and branch circuit fuses. The branch fuse should have a total clearing I^2t value that is less than the melting I^2t value of the feeder or upstream fuse. This ensures that the branch fuse will melt, arc, and clear the fault before the feeder fuse begins to melt.

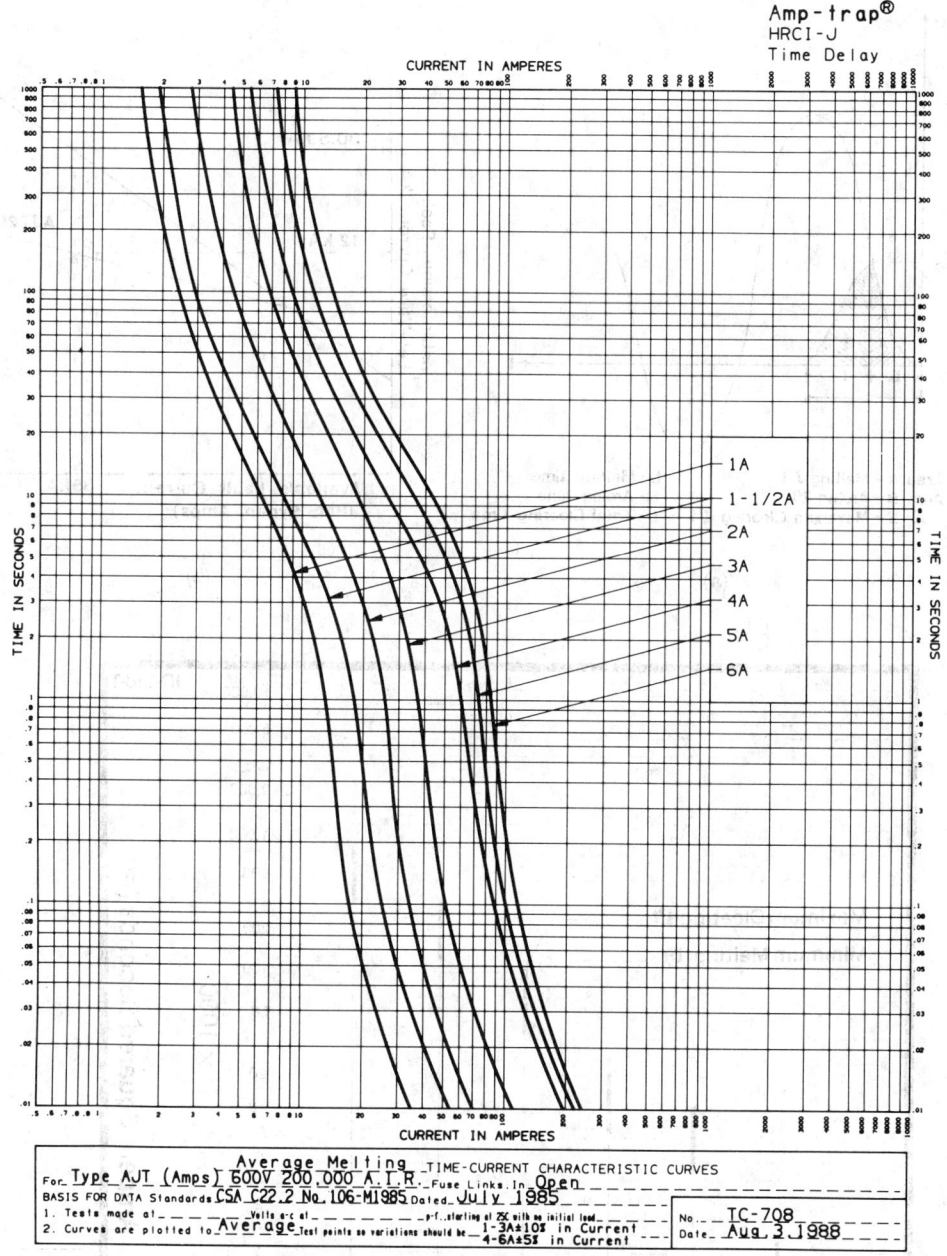

FIGURE 1.33 Time-current characteristic curves.

Standards

Overload and short-circuit characteristics are well documented by fuse manufacturers. These characteristics are standardized by product standards written in most cases by safety organizations such as CSA (Canadian Standards Association) and UL (Underwriters Laboratories). CSA standards and UL specify product designations, dimensions, performance characteristics, and temperature rise limits. These standards are used in conjunction with national code regulations such as CEC (Canadian Electrical Code) and NEC (National Electrical Code) that specify how the product is applied.

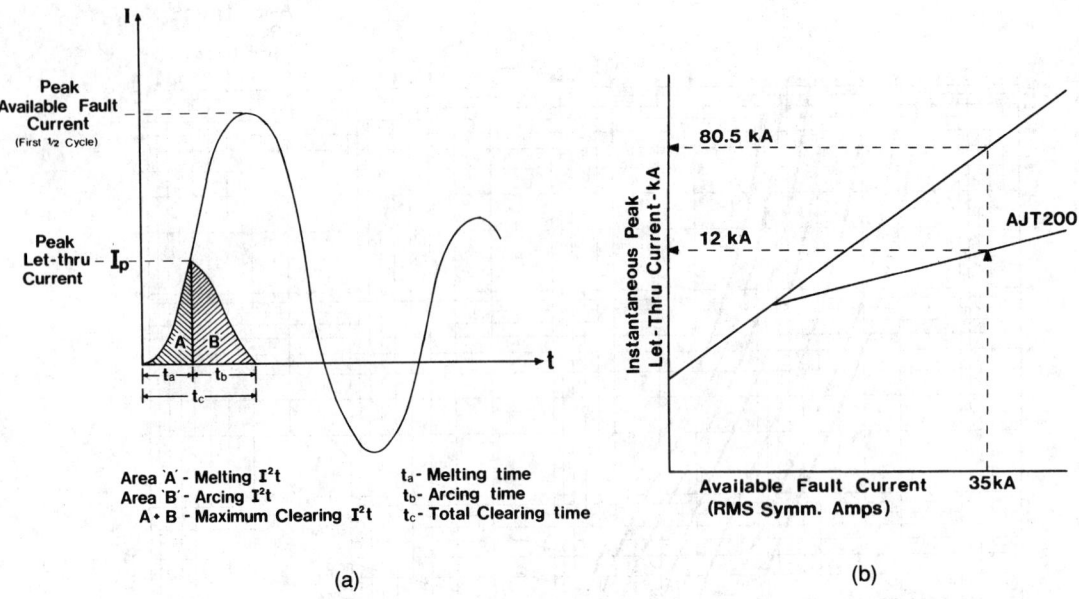

Area 'A' - Melting I^2t t_a - Melting time
Area 'B' - Arcing I^2t t_b - Arcing time
A + B - Maximum Clearing I^2t t_c - Total Clearing time

(a)   (b)

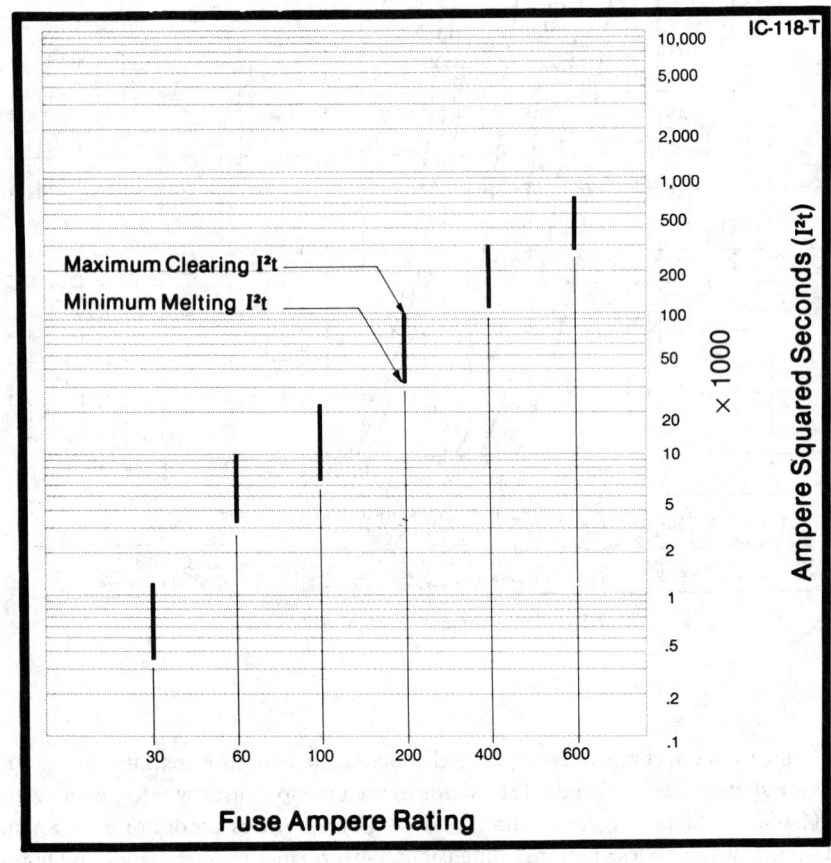

(c)

FIGURE 1.34   (a) Fuse short-circuit operation. (b) Variation of fuse peak let-through current I_p. (c) I^2t graph.

IEC (International Electrotechnical Commission—Geneva, Switzerland) was founded to harmonize international electrical standards to increase international trade in electrical products. Any country can become a member and participate in the standards-writing activities of IEC. Unlike CSA and UL, IEC is not a certifying body that certifies or approves products. IEC publishes consensus standards for national standards authorities such as CSA (Canada), UL (USA), BSI (UK) and DIN (Germany) to adopt as their own national standards.

Products

North American low-voltage distribution fuses can be classified under two types: Standard or Class H, as referred to in the United States, and **HRC (high rupturing capacity)** or current-limiting fuses, as referred to in Canada. It is the interrupting rating that essentially differentiates one type from the other.

Most Standard or Class H fuses have an interrupting rating of 10,000 A. They are not classified as HRC or current-limiting fuses, which usually have an interrupting rating of 200,000 A. Selection is often based on the calculated available short-circuit current.

In general, short-circuit currents in excess of 10,000 A do not exist in residential applications. In commercial and industrial installations, short-circuit currents in excess of 10,000 A are very common. Use of HRC fuses usually means that a fault current assessment is not required.

Standard—Class H

In North America, Standard or Class H fuses are available in 250- and 600-V ratings with ampere ratings up to 600 A. There are primarily three types: one-time, time-delay, and renewable. Rating for rating, they are all constructed to the same dimensions and are physically interchangeable in standard-type fusible switches and fuse blocks.

One-time fuses are not reusable once blown. They are used for general-purpose resistive loads such as lighting, feeders, and cables.

Time-delay fuses have a specified delay in their overload characteristics and are designed for motor circuits. When started, motors typically draw six times their full load current for approximately 3 to 4 seconds. This surge then decreases to a level within the motor full-load current rating. Time-delay fuse overload characteristics are designed to allow for motor starting conditions.

Renewable fuses are constructed with replaceable links or elements. This feature minimizes the cost of replacing fuses. However, the concept of replacing fuse elements in the field is not acceptable to most users today because of the potential risk of improper replacement.

HRC

HRC or current-limiting fuses have an interrupting rating of 200 kA and are recognized by a letter designation system common to North American fuses. In the United States they are known as Class J, Class L, Class R, etc., and in Canada they are known as HRCI-J, HRC-L, HRCI-R, and so forth. HRC fuses are available in ratings up to 600 V and 6000 A. The main differences among the various types are their dimensions and their short-circuit performance (I_p and I^2t) characteristics.

One type of HRC fuse found in Canada, but not in the United States, is the HRCII-C or Class C fuse. This fuse was developed originally in England and is constructed with bolt-on-type blade contacts. It is available in a voltage rating of 600 V with ampere ratings from 2 to 600 A. Some higher ampere ratings are also available but are not as common. HRCII-C fuses are primarily regarded as providing short-circuit protection only. Therefore, they should be used in conjunction with an overload device.

HRCI-R or Class R fuses were developed in the United States. Originally constructed to Standard or Class H fuse dimensions, they were classified as Class K and are available in the United States with two levels of short-circuit performance characteristics: Class K1 and Class K5. However, they are not recognized in Canadian Standards. Under fault conditions, Class K1 fuses limit the I_p and I^2t to lower levels than do Class K5 fuses. Since both Class K1 and K5 are constructed to Standard or Class H fuse dimensions, problems with interchangeability occur. As a result, a second generation of these K fuses was therefore introduced with a rejection feature incorporated in the end caps and blade contacts. This rejection feature, when used in conjunction with rejection-style fuse clips, prevents replacement of these fuses with Standard or Class H 10-kA I.R. fuses. These rejection style fuses are known as Class RK1 and Class RK5. They are available with time-delay or non-time-delay characteristics and with voltage ratings of 250 or 600 V and ampere ratings up to 600 A. In Canada, CSA has only one classification for these fuses, HRCI-R, which have the same maximum I_p and I^2t current-limiting levels as specified by UL for Class RK5 fuses.

HRCI-J or Class J fuses are a more recent development. In Canada, they have become the most popular HRC fuse specified for new installations. Both time-delay and non-time-delay characteristics are available in ratings of 600 V with ampere ratings up to 600 A. They are constructed with dimensions much smaller than HRCI-R or Class R fuses and have end caps or blade contacts which fit into 600-V Standard or Class H-type fuse clips.

However, the fuse clips must be mounted closer together to accommodate the shorter fuse length. Its shorter length, therefore, becomes an inherent rejection feature that does not allow insertion of Standard or HRCI-R fuses. The blade contacts are also drilled to allow bolt-on mounting if required. CSA and UL specify these fuses to have maximum short-circuit current-limiting I_p and I^2t limits lower than those specified for HRCI-R and HRCII-C fuses. HRCI-J fuses may be used for a wide variety of applications. The time-delay type is commonly used in motor circuits sized at approximately 125 to 150% of motor full-load current.

HRC-L or Class L fuses are unique in dimension but may be considered as an extension of the HRCI-J fuses for ampere ratings above 600 A. They are rated at 600 V with ampere ratings from 601 to 6000 A. They are physically larger and are constructed with bolt-on-type blade contacts. These fuses are generally used in low-voltage distribution systems where supply transformers are capable of delivering more than 600 A.

In addition to Standard and HRC fuses, there are many other types designed for specific applications. For example, there are medium- or high-voltage fuses to protect power distribution transformers and medium-voltage motors. There are fuses used to protect sensitive semiconductor devices such as diodes, SCRs, and triacs. These fuses are designed to be extremely fast under short-circuit conditions. There is also a wide variety of dedicated fuses designed for protection of specific equipment requirements such as electric welders, capacitors, and circuit breakers, to name a few.

Trends

Ultimately, it is the electrical equipment being protected that dictates the type of fuse needed for proper protection. This equipment is forever changing and tends to get smaller as new technology becomes available. Present trends indicate that fuses also must become smaller and faster under fault conditions, particularly as available short-circuit fault currents are tending to increase.

With free trade and the globalization of industry, a greater need for harmonizing product standards exists. The North American fuse industry is taking big steps toward harmonizing CSA and UL fuse standards, and at the same time is participating in the IEC standards process. Standardization will help the electrical industry to identify and select the best fuse for the job—anywhere in the world.

Defining Terms

HRC (high rupturing capacity): A term used to denote fuses having a high interrupting rating. Most low-voltage HRC-type fuses have an interrupting rating of 200 kA rms symmetrical.

I^2t **(ampere squared seconds):** A convenient way of indicating the heating effect or thermal energy which is produced during a fault condition before the circuit protective device has opened the circuit. As a protective device, the HRC or current-limiting fuse lets through far less damaging I^2t than other protective devices.

Interrupting rating (I.R.): The maximum value of short-circuit current that a fuse can safely interrupt.

References

R. K. Clidero and K. H. Sharpe, *Application of Electrical Construction*, Ontario, Canada: General Publishing Co. Ltd., 1982.

Gould Inc., *Shawmut Advisor*, Circuit Protection Division, Newburyport, Mass.

C. A. Gross, *Power Systems Analysis*, 2nd ed., New York: Wiley 1986.

E. Jacks, *High Rupturing Capacity Fuses*, New York: Wiley, 1975.

A. Wright and P. G. Newbery, *Electric Fuses*, London: Peter Peregrinus Ltd., 1984.

Further Information

For greater detail the "Shawmut Advisor" (Gould, Inc., 374 Merrimac Street, Newburyport MA 01950) or the "Fuse Technology Course Notes" (Gould Shawmut Company, 88 Horner Avenue, Toronto, Canada M8Z-5Y3) may be referred to for fuse performance and application.

RALPH V. L. HARTLEY

Over fifty years ago, the PROCEEDINGS OF THE IRE (Institute of Radio Engineers) included a paper by Ralph V. L. Hartley of Bell Telephone Laboratories. Hartley, who was to receive the IRE Medal of Honor in 1946 in recognition of his early work on vacuum-tube circuits and contributions to information theory, is perhaps most remembered for his invention of the oscillator bearing his name. His 1942 paper concerned an improved way to use the Fourier integral in the solution of transmission problems.

Ralph Vinton Lyon Hartley was born in Nevada in 1888 and graduated from the University of Utah in 1909. He was selected as a Rhodes Scholar, which enabled him to study at Oxford University, from which he received additional degrees in 1912 and 1913. He then returned to the United States to join the research laboratories of the Western Electric Company in New York City.

Hartley joined a group of engineers and scientists who were just beginning to develop vacuum-tube circuits for use in telephonic communication. In particular, he participated in the development of radio receivers used in long-distance radio telephone experiments in 1915. By using several hundred vacuum tubes in parallel, Hartley and his colleagues were able to transmit signals from Arlington, Virginia, to Paris, France, in October 1915. It was during this period that the Hartley oscillator was reduced to practice.

Information about the oscillator and circuit diagrams were included in the classic 1920 book *The Thermionic Vacuum Tube and Its Applications* by Hartley's former colleague Hendrik J. van der Bijl. By using an iron-core transformer, the oscillator could produce frequencies of less than one hertz. It came to be used commonly as the local oscillator in both AM and FM radio receivers.

In February 1925, Hartley published an

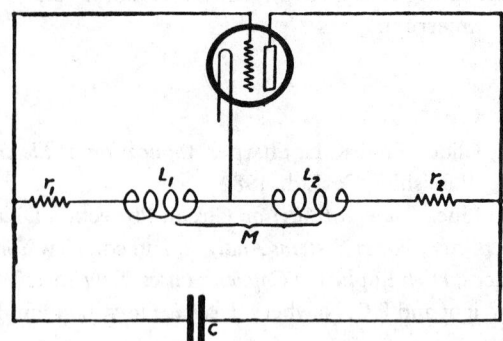

The Hartley oscillator, as illustrated in the classic book *The Thermionic Vacuum Tube and Its Applications* by H. J. van der Bijl (McGraw-Hill, 1920).

important paper on sidebands in radio. He explained the advantages of single-sideband transmission and carrier suppression. He noted that this technique made possible "a very great saving in the frequency range required per channel," pointing out that, although it had been developed for wire telephony, he saw no reason why it could not be equally beneficial in radio.

Hartley became a staff member of Bell Telephone Laboratories when it was established in 1925. He is credited with formulating the law "that the total amount of information that can be transmitted is proportional to frequency range transmitted and the time of transmission."

An illness forced his temporary retirement beginning in 1934, but Hartley returned to the Bell Labs in 1939 and during World War II worked on problems relating to servo control systems. In 1945 he wrote a memorandum on the transmission of guided beams of light by means of internal reflections in transparent rods. He again retired in 1950 and died in 1970.

Source: **Adapted from J. E. Brittain, *Proc. IEEE*, vol. 80, no. 3, p. 463, March 1992. © 1992 IEEE.**

$\Large 2$

Voltage and
Current Sources

Richard C. Dorf
University of California, Davis

Zhen Wan
University of California, Davis

Clayton R. Paul
*University of Kentucky,
Lexington*

J. R. Cogdell
University of Texas at Austin

2.1   Step, Impulse, Ramp, Sinusoidal, and DC Signals 47
Step Function • The Impulse • Ramp Function • Sinusoidal Function
• DC Signal

2.2   Ideal and Practical Sources 50
Ideal Sources • Practical Sources

2.3   Controlled Sources 53
What Are Controlled Sources? • What Is the Significance of
Controlled Sources? • How Does the Presence of Controlled Sources
Affect Circuit Analysis?

2.1   Step, Impulse, Ramp, Sinusoidal, and DC Signals

Richard C. Dorf and Zhen Wan

The important signals for circuits include the step, impulse, **ramp**, **sinusoid**, and dc signals. These
signals are widely used and are described here in the time domain. All of these signals have a
Laplace transform.

Step Function

The **unit-step** function $u(t)$ is defined mathematically by

$$u(t) = \begin{cases} 1, & t \geq 0 \\ 0, & t < 0 \end{cases}$$

Here *unit step* means that the amplitude of $u(t)$ is equal to 1 for $t \geq 0$. Note that we are following the
convention that $u(0) = 1$. From a strict mathematical standpoint, $u(t)$ is not defined at $t = 0$.
Nevertheless, we usually take $u(0) = 1$. If A is an arbitrary nonzero number, $Au(t)$ is the step function with amplitude A for $t \geq 0$. The unit step function is plotted in Fig. 2.1.

The Impulse

The **unit impulse** $\delta(t)$, also called the *delta function* or the *Dirac distribution*, is defined by

$$\delta(t) = 0, \qquad t \neq 0$$

$$\int_{-\varepsilon}^{\varepsilon} \delta(\lambda)\, d\lambda = 1, \qquad \text{for any real number } \varepsilon > 0$$

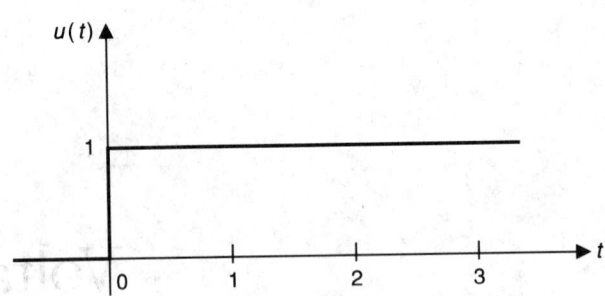

FIGURE 2.1   Unit-step function.

The first condition states that $\delta(t)$ is zero for all nonzero values of t, while the second condition states that the area under the impulse is 1, so $\delta(t)$ has unit area. It is important to point out that the value $\delta(0)$ of $\delta(t)$ at $t = 0$ is not defined; in particular, $\delta(0)$ is not equal to infinity. For any real number K, $K\delta(t)$ is the impulse with area K. It is defined by

$$K\delta(t) = 0, \qquad t \neq 0$$

$$\int_{-\varepsilon}^{\varepsilon} K\delta(\lambda)\, d\lambda = K, \qquad \text{for any real number } \varepsilon > 0$$

The graphical representation of $K\delta(t)$ is shown in Fig. 2.2. The notation K in the figure refers to the area of the impulse $K\delta(t)$.

The unit-step function $u(t)$ is equal to the integral of the unit impulse $\delta(t)$; more precisely, we have

$$u(t) = \int_{-\infty}^{t} \delta(\lambda)\, d\lambda, \qquad \text{all } t \text{ except } t = 0$$

Ramp Function

The *unit-ramp function* $r(t)$ is defined mathematically by

$$r(t) = \begin{cases} t, & t \geq 0 \\ 0, & t < 0 \end{cases}$$

Note that for $t \geq 0$, the slope of $r(t)$ is 1. Thus, $r(t)$ has *unit slope*, which is the reason $r(t)$ is called the unit-ramp function. If K is an arbitrary nonzero scalar (real number), the ramp function $Kr(t)$ has slope K for $t \geq 0$. The unit-ramp function is plotted in Fig. 2.3.

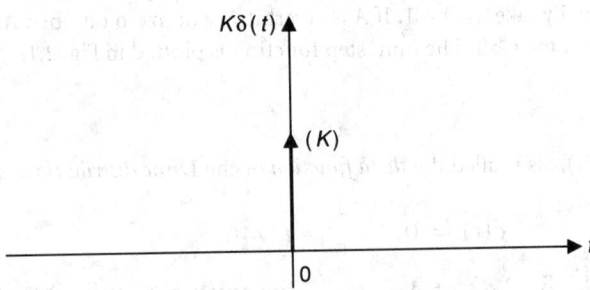

FIGURE 2.2   Graphical representation of the impulse $K\delta(t)$.

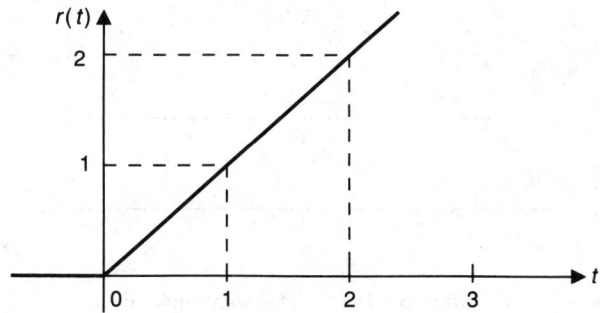

FIGURE 2.3 Unit-ramp function.

The unit-ramp function $r(t)$ is equal to the integral of the unit-step function $u(t)$; that is,

$$r(t) = \int_{-\infty}^{t} u(\lambda)\, d\lambda$$

Conversely, the first derivative of $r(t)$ with respect to t is equal to $u(t)$, except at $t = 0$, where the derivative of $r(t)$ is not defined.

Sinusoidal Function

The sinusoid is a continuous-time signal: $A \cos(\omega t + \theta)$.

Here A is the amplitude, ω is the frequency in radians per second (rad/s), and θ is the phase in radians. The frequency f in cycles per second, or hertz (Hz), is $f = \omega/2\pi$. The sinusoid is a periodic signal with period $2\pi/\omega$. The sinusoid is plotted in Fig. 2.4.

DC Signal

The direct current signal (dc signal) can be defined mathematically by

$$i(t) = K \qquad -\infty < t < +\infty$$

Here, K is any nonzero number. The dc signal remains a constant value of K for any $-\infty < t < \infty$. The dc signal is plotted in Fig. 2.5.

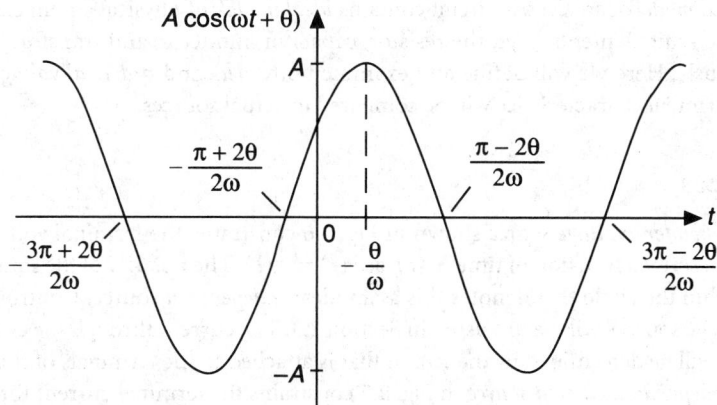

FIGURE 2.4 The sinusoid $A \cos(\omega t + \theta)$ with $-\pi/2 < \theta < 0$.

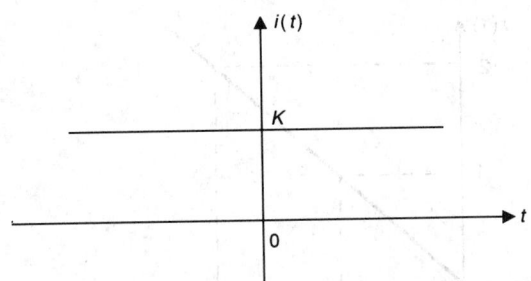

FIGURE 2.5   The dc signal with amplitude K.

Defining Terms

Ramp: A continually growing signal such that its value is zero for $t \leq 0$ and proportional to time t for $t > 0$.

Sinusoid: A periodic signal $x(t) = A\cos(\omega t + \theta)$ where $\omega = 2\pi f$ with frequency in hertz.

Unit impulse: A very short pulse such that its value is zero for $t \neq 0$ and the integral of the pulse is 1.

Unit step: Function of time that is zero for $t < t_0$ and unity for $t > t_0$. At $t = t_0$ the magnitude changes from zero to one. The unit step is dimensionless.

References

R. C. Dorf, *Introduction to Electric Circuits,* 2nd ed., New York: Wiley, 1992.

R. E. Ziemer, *Signals and Systems,* 2nd ed., New York: Macmillan, 1989.

Further Information

IEEE Transactions on Circuits and Systems

IEEE Transactions on Education

2.2 Ideal and Practical Sources

Clayton R. Paul

A *mathematical model* of an electric circuit contains *ideal models* of physical circuit elements. Some of these ideal circuit elements (e.g., the resistor, capacitor, inductor, and transformer) were discussed previously. Here we will define and examine both *ideal* and *practical voltage and current sources*. The terminal characteristics will be compared to actual sources.

Ideal Sources

The *ideal independent voltage source* shown in Fig. 2.6 constrains the terminal voltage across the element to a prescribed function of time, $v_S(t)$, as $v(t) = v_S(t)$. The polarity of the source is denoted by $\pm$ signs within the circle that denotes this as an ideal *independent* source. Controlled or *dependent* ideal voltage sources will be discussed in Section 2.3. The current through the element is as yet unknown but will be determined by the circuit that is attached to the terminals of this source.

The *ideal independent current source* in Fig. 2.7 constrains the terminal current through the element to a prescribed function of time, $i_S(t)$, as $i(t) = i_S(t)$. The polarity of the source is denoted by an arrow within the circle that also denotes this as an ideal *independent* source. The voltage across

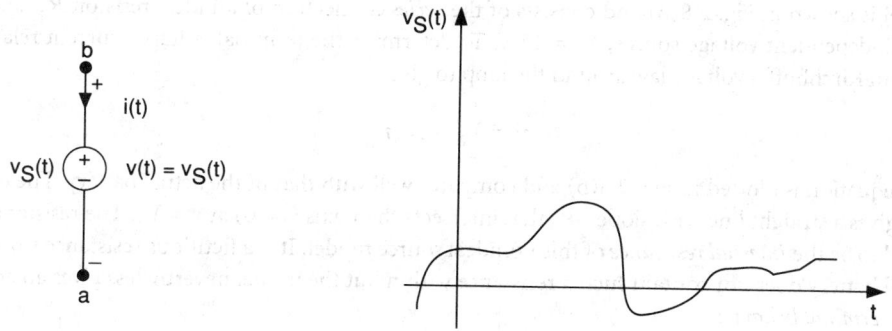

FIGURE 2.6   Ideal independent voltage source.

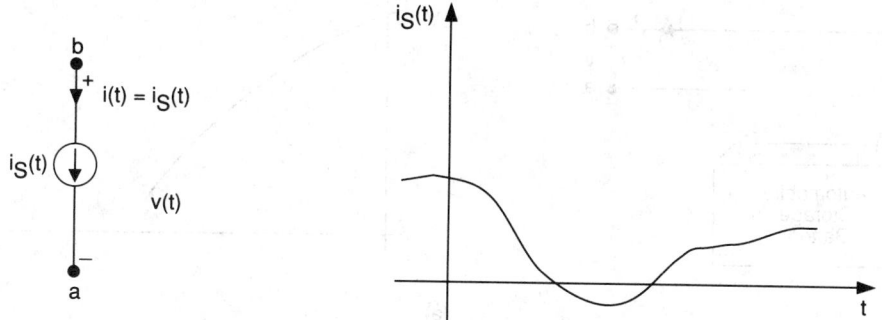

FIGURE 2.7   Ideal independent current source.

the element is as yet unknown but will be determined by the circuit that is attached to the terminals of this source.

Numerous functional forms are useful in describing the source variation with time. These were discussed in Section 2.1—the step, impulse, ramp, sinusoidal, and dc functions. For example, an ideal independent dc voltage source is described by $v_S(t) = V_S$, where V_S is a constant. An ideal independent sinusoidal current source is described by $i_S(t) = I_S \sin(\omega t + \phi)$ or $i_S(t) = I_S \cos(\omega t + \phi)$, where I_S is a constant, $\omega = 2\pi f$ with f the *frequency* in hertz and ϕ is a phase angle. These ideal sources will be used to model actual sources such as temperature transducers, phonograph cartridges, and electric power generators. Thus, the time form of the output cannot generally be described with a simple, basic function such as dc, sinusoidal, ramp, step, or impulse waveforms. We often, however, represent the more complicated waveforms as a linear combination of more basic functions.

Practical Sources

The preceding ideal independent sources constrain the terminal voltage or current to a *known* function of time *independent of the circuit that may be placed across its terminals.* Actual sources such as batteries have their terminal voltage (current) dependent upon the terminal current (voltage) caused by the circuit attached to the source terminals. A simple example of this is an automobile storage battery. The battery's terminal voltage is approximately 12 V when no load is connected across its terminals. When the battery is applied across the terminals of the starter by activating the ignition switch, a current of approximately 40 A is drawn from its terminals. During starting, its terminal voltage drops to about 9 V as illustrated in Fig. 2.8(a). How shall we construct a *circuit model* using the ideal elements discussed thus far to *simulate* this nonideal behavior? A satisfactory model is shown in Fig. 2.8(b) and consists of the *series* connection of an ideal resistor, R_S, and an ideal independent voltage source, $V_S = 12$ V. To determine the terminal voltage–current relation, we sum Kirchhoff's voltage law around the loop to give

$$v = V_S - R_S i \tag{2.1}$$

This equation is plotted in Fig. 2.8(b) and compares well with that of the actual battery. The equation gives a straight line with slope $-R_S$ that intersects the v axis ($i = 0$) at $v = V_S$. The resistance R_S is said to be the *internal resistance* of this nonideal source model. It is a fictitious resistance since the actual battery does not contain such a resistance within but the model nevertheless gives an equivalent *terminal behavior*.

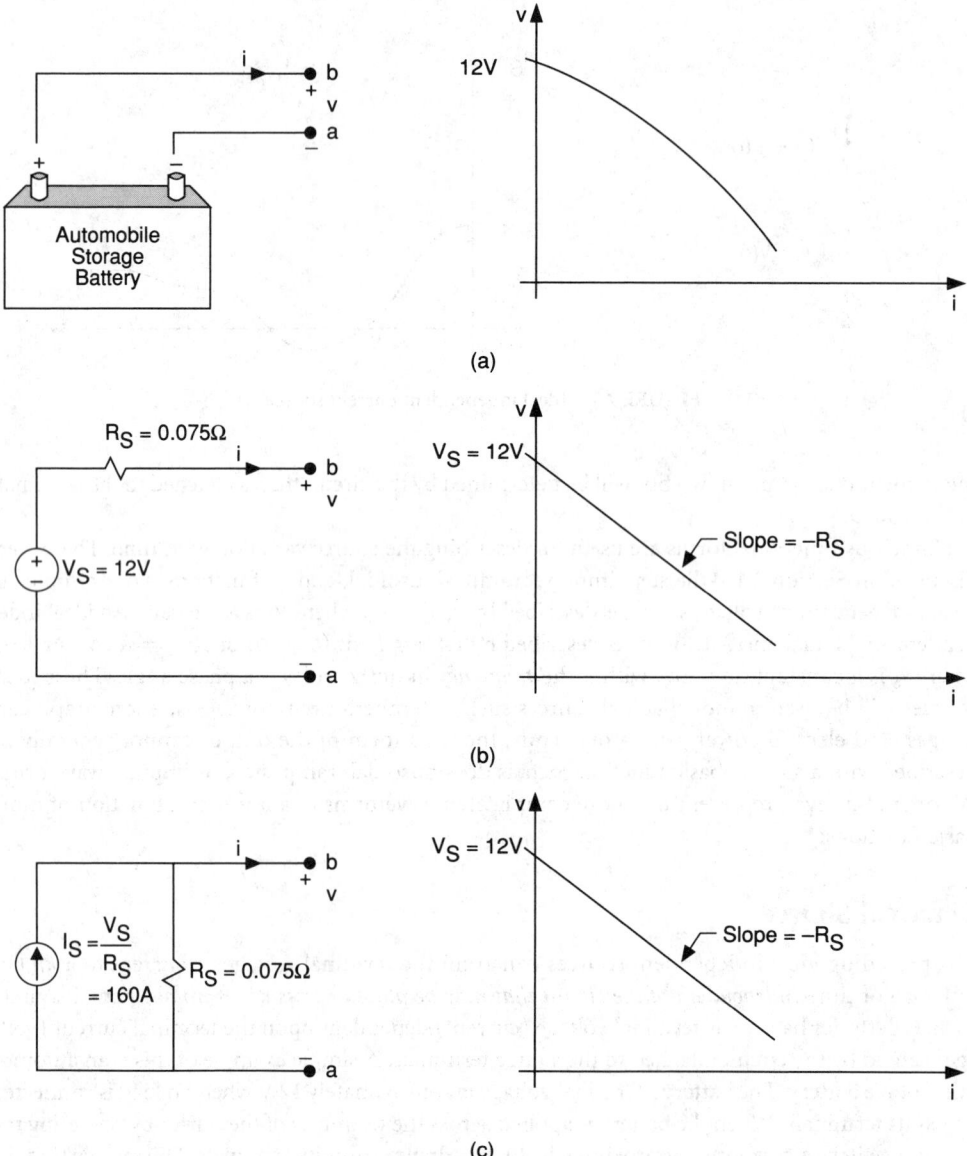

FIGURE 2.8   Practical sources. (a) Terminal *v-i* characteristic; (b) approximation by an ideal voltage source; (c) approximation by an ideal current source.

Although we have derived a satisfactory but approximate model of an actual source, other equivalent forms may be obtained. An alternative form is shown in Fig. 2.8(c) and consists of the *parallel* combination of an ideal independent current source, $I_S = V_S/R_S$, and the same resistance, R_S, used in the previous model. Although it may seem strange to model an automobile battery using a current source, the model is completely equivalent to the series voltage source–resistor model of Fig. 2.8(b) *at the output terminals a–b*. This is shown by writing Kirchhoff's current law at the upper node to give

$$i = I_S - \frac{1}{R_S}v \tag{2.2}$$

Rewriting this equation gives

$$v = R_S I_S - R_S i \tag{2.3}$$

Comparing Eq. (2.3) to Eq. (2.1) shows that

$$V_S = R_S I_S \tag{2.4}$$

Therefore, we can convert from one form (voltage source in series with a resistor) to another form (current source in parallel with a resistor) very simply.

An ideal voltage source is represented by the model of Fig. 2.8(b) with $R_S = 0$. An actual battery therefore provides a close approximation of an ideal voltage source since the source resistance R_S is usually quite small. An ideal current source is represented by the model of Fig. 2.8(c) with $R_S = \infty$. This is very closely represented by the bipolar junction transistor (BJT).

Defining Term

Ideal source: An ideal model of an actual source that assumes that the parameters of the source, such as its magnitude, are independent of other circuit variables.

Reference

C. R. Paul, *Analysis of Linear Circuits,* New York: McGraw-Hill, 1989.

2.3 Controlled Sources

J. R. Cogdell

When the analysis of electronic (nonreciprocal) circuits became important in circuit theory, **controlled sources** were added to the family of circuit elements. Table 2.1 shows the four types of controlled sources. In this section, we will address the questions: What are controlled sources? Why are controlled sources important? How do controlled sources affect methods of circuit analysis?

What Are Controlled Sources?

By *source* we mean a voltage or current source in the usual sense. By *controlled* we mean that the strength of such a source is controlled by some circuit variable(s) elsewhere in the circuit. Figure 2.9 illustrates a simple circuit containing an (independent) current source, i_s, two resistors, and a controlled voltage source, whose magnitude is controlled by the current i_1. Thus, i_1 determines two voltages in the circuit, the voltage across R_1 via Ohm's law and the controlled voltage source via some unspecified effect.

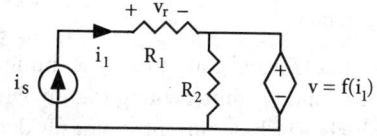

FIGURE 2.9 A simple circuit containing a controlled source.

Table 2.1 Names, Circuit Symbols, and Definitions for the Four Possible Types of Controlled Sources

Name	Circuit Symbol	Definition and Units
Current-controlled voltage source (CCVS)	i_1 $r_m i_1$ v_2	$v_2 = r_m i_1$ r_m = transresistance units, ohms
Current-controlled current source (CCCS)	i_1 i_2 βi_1	$i_2 = \beta i_1$ β, current gain, dimensionless
Voltage-controlled voltage source (VCVS)	v_1 μv_1 v_2	$v_2 = \mu v_1$ μ, voltage gain, dimensionless
Voltage-controlled current source (VCCS)	v_1 i_2 $g_m v_1$	$i_2 = g_m v_1$ g_m, transconductance units, Siemans (mhos)

A controlled source may be controlled by more than one circuit variable, but we will discuss those having a single controlling variable since multiple controlling variables require no new ideas. Similarly, we will deal only with resistive elements, since inductors and capacitors introduce no new concepts. The controlled voltage or current source may depend on the controlling variable in a linear or nonlinear manner. When the relationship is nonlinear, however, the equations are frequently linearized to examine the effects of small variations about some dc values. When we linearize, we will use the customary notation of small letters to represent general and time-variable voltages and currents and large letters to represent constants such as the dc value or the peak value of a sinusoid. On subscripts, large letters represent the total voltage or current and small letters represent the **small-signal** component. Thus, the equation $i_B = I_B + I_b \cos \omega t$ means that the total base current is the sum of a constant and a small-signal component, which is sinusoidal with an amplitude of I_b.

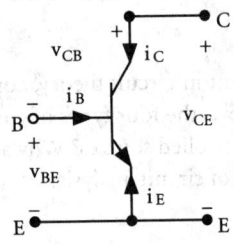

FIGURE 2.10 An *npn* BJT in the common emitter configuration.

To introduce the context and use of controlled sources we will consider a circuit model for the bipolar junction transistor (BJT). In Fig. 2.10 we show the standard symbol for an *npn* BJT with base (*B*), emitter (*E*), and collector (*C*) identified, and voltage and current variables defined. We have shown the common emitter configuration, with the emitter terminal shared to make input and output terminals. The base current, i_B, ideally depends upon the base-emitter voltage, v_{BE}, by the relationship

$$i_B = I_0 \left\{ \exp\left[\frac{v_{BE}}{V_T} \right] - 1 \right\} \tag{2.5}$$

where I_0 and V_T are constants. We note that the base current depends on the base-emitter voltage only, but in a nonlinear manner. We can represent this current by a voltage-controlled current source, but the more common representation would be that of a nonlinear conductance, $G_{BE}(v_{BE})$, where

$$G_{BE}(v_{BE}) = \frac{i_B}{v_{BE}}$$

Let us model the effects of small changes in the base current. If the changes are small, the nonlinear nature of the conductance can be ignored and the circuit model becomes a linear conductance (or resistor). Mathematically this conductance arises from a first-order expansion of the nonlinear function. Thus, if $v_{BE} = V_{BE} + v_{be}$, where v_{BE} is the total base-emitter voltage, V_{BE} is a (large) constant voltage and v_{be} is a (small) variation in the base-emitter voltage, then the first two terms in an expansion are

$$i_B = I_0\left\{\exp\left[\frac{V_{BE} + v_{be}}{V_T}\right] - 1\right\} \cong I_0\left\{\exp\left[\frac{V_{BE}}{V_T}\right] - 1\right\} + \frac{I_0}{V_T}\exp\left[\frac{V_{BE}}{V_T}\right]v_{be} \qquad (2.6)$$

We note that the base current is approximated by the sum of a constant term and a term that is first order in the small variation in base-emitter voltage, v_{be}. The multiplier of this small voltage is the linearized conductance, g_{be}. If we were interested only in small changes in currents and voltages, only this conductance would be required in the model. Thus, the input (base-emitter) circuit can be represented for the small-signal base variables, i_b and v_{be}, by either equivalent circuit in Fig. 2.11.

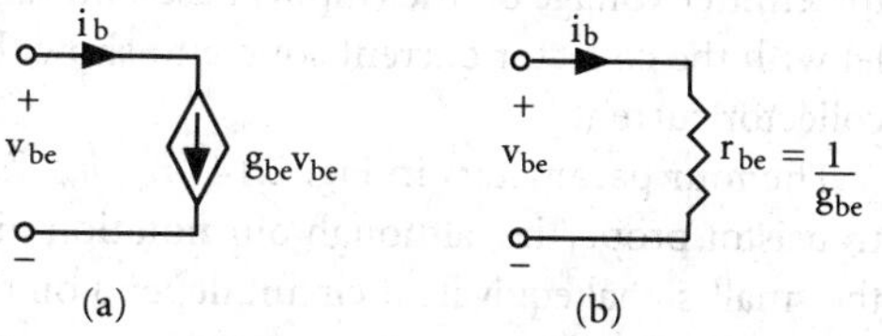

The voltage-controlled current source, $g_{be}v_{be}$, can be replaced by a simple resistor because the small-signal voltage and current flow in the same branch. The process of **linearization** is important to the modeling of the collector-emitter characteristic, to which we now turn.

FIGURE 2.11 Equivalent circuits for the base circuit: (a) uses a controlled source and (b) uses a resistor.

The collector current, I_c, can be represented by one of the Eber and Moll equations

$$i_C = \beta I_0\left\{\exp\left[\frac{v_{BE}}{V_T}\right] - 1\right\} - I_0'\left\{\exp\left[\frac{v_{BC}}{V_T}\right] - 1\right\} \qquad (2.7)$$

where β and I_0' are constants. If we restrict our model to the amplifying region of the transistor, the second term is negligible and we may express the collector current as

$$i_C = \beta I_0\left\{\exp\left[\frac{v_{BE}}{V_T}\right] - 1\right\} = \beta i_B \qquad (2.8)$$

Thus, for the ideal transistor, the collector-emitter circuit may be modeled by a current-controlled current source, which may be combined with the results expressed in Eq. (2.5) to give the model shown in Fig. 2.12.

Using the technique of small-signal analysis, we may derive either of the small-signal equivalent circuits shown in Fig. 2.13.

The small-signal characteristics of the *npn* transistor in its amplifying region is better represented by the equivalent circuit shown in Fig. 2.14. Note we

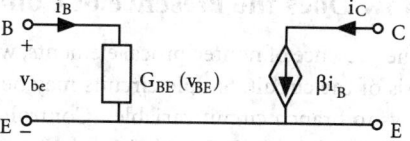

FIGURE 2.12 Equivalent circuit for BJT.

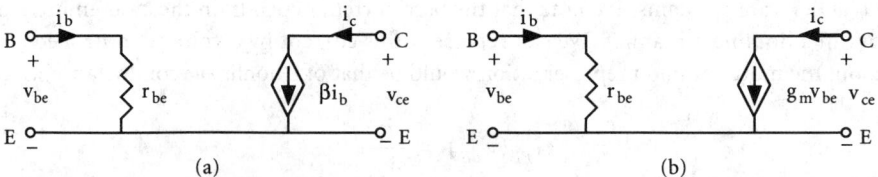

FIGURE 2.13   Two BJT small-signal equivalent circuits ($g_m = \beta/r_{be}$): (a) uses a CCCS and (b) uses a VCCS.

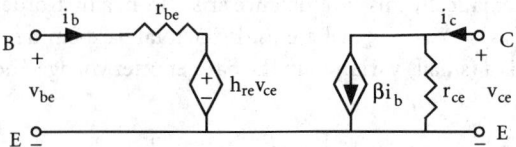

FIGURE 2.14   Full hybrid parameter model for small-signal BJT.

have introduced a voltage-controlled voltage source to model the influence of the (output) collector-emitter voltage on the (input) base-emitter voltage, and we have placed a resistor, r_{ce}, in parallel with the collector current source to show the influence of the collector-emitter voltage on the collector current.

The four parameters in Fig. 2.14 (r_{be}, h_{re}, β, and r_{ce}) are the hybrid parameters describing the transistor properties, although our notation differs from that commonly used. The parameters in the small-signal equivalent circuit depend on the operating point of the device, which is set by the time-average voltages and currents (V_{BE}, I_C, etc.) applied to the device. All of the parameters are readily measured for a given transistor and operating point, and manufacturers commonly specify ranges for the various parameters for a type of transistor.

What Is the Significance of Controlled Sources?

Commonplace wisdom in engineering education and practice is that information and techniques that are presented visually are more useful than abstract, mathematical forms. Equivalent circuits are universally used in describing electrical engineering systems and devices because circuits portray interactions in a universal, pictorial language. This is true generally, and it is doubly necessary when circuit variables interact through the mysterious coupling modeled by controlled sources. This is the primary significance of controlled sources: that they represent unusual couplings of circuit variables in the universal, visual language of circuits.

A second significance is illustrated by our equivalent circuit of the *npn* bipolar transistor, namely, the characterization of a class of similar devices. For example, the parameter β in Eq. (2.8) gives important information about a single transistor, and similarly for the range of β for a type of transistor. In this connection, controlled sources lead to a vocabulary for discussing some property of a class of systems or devices, in this case the current gain of an *npn* BJT.

How Does the Presence of Controlled Sources Affect Circuit Analysis?

The presence of nonreciprocal elements, which are modeled by controlled sources, affects the analysis of the circuit. Simple circuits may be analyzed through the direct application of Kirchhoff's laws to branch circuit variables. Controlled sources enter this process similar to the constitutive relations defining R, L, and C, i.e., in defining relationships between branch circuit variables. Thus, controlled sources add no complexity to this basic technique.

The presence of controlled sources negates the advantages of the method that uses series and parallel combinations of resistors for voltage and current dividers. The problem is that the couplings between circuit variables that are expressed by controlled sources make all the familiar formulas unreliable.

When superposition is used, the controlled sources are left on in all cases as independent sources are turned on and off, thus reflecting the kinship of controlled sources to the circuit elements. In principle, little complexity is added; in practice, the repeated solutions required by superposition entail much additional work when controlled sources are involved.

The classical methods of nodal and loop (mesh) analysis incorporate controlled sources without great difficulty. For purposes of determining the number of independent variables required, that is, in establishing the topology of the circuit, the controlled sources are treated as ordinary voltage or current sources. The equations are then written according to the usual procedures. Before the equations are solved, however, the controlling variables must be expressed in terms of the unknowns of the problem. For example, let us say we are performing a nodal analysis on a circuit containing a current-controlled current source. For purposes of counting independent nodes, the controlled current source is treated as an open circuit. After equations are written for the unknown node voltages, the current source will introduce into at least one equation its controlling current, which is not one of the nodal variables. The additional step required by the controlled source is that of expressing the controlling current in terms of the nodal variables.

The parameters introduced into the circuit equations by the controlled sources end up on the left side of the equations with the resistors rather than on the right side with the independent sources. Furthermore, the symmetries that normally exist among the coefficients are disturbed by the presence of controlled sources.

The methods of Thévenin and Norton equivalent circuits continue to be very powerful with controlled sources in the circuits, but some complications arise. The controlled sources must be left on for calculation of the Thévenin (open-circuit) voltage or Norton (short-circuit) current and also for the calculation of the output impedance of the circuit. This usually eliminates the method of combining elements in series or parallel to determine the output impedance of the circuit, and one must either determine the output impedance from the ratio of the Thévenin voltage to the Norton current or else excite the circuit with an external source and calculate the response.

Defining Terms

Controlled source (dependent source): A voltage or current source whose intensity is controlled by a circuit voltage or current elsewhere in the circuit.

Linearization: Approximating nonlinear relationships by linear relationships derived from the first-order terms in a power series expansion of the nonlinear relationships. Normally the linearized equations are useful for a limited range of the voltage and current variables.

Small-signal: Small-signal variables are those first-order variables used in a linearized circuit. A small-signal equivalent circuit is a linearized circuit picturing the relationships between the small-signal voltages and currents in a linearized circuit.

References

E. J. Angelo, Jr., *Electronic Circuits*, 2nd ed., New York: McGraw-Hill, 1964.

N. Balabanian and T. Bickart, *Linear Network Theory*, Chesterland, Ohio: Matrix Publishers, 1981.

L. O. Chua, *Introduction to Nonlinear Network Theory*, New York: McGraw-Hill, 1969.

B. Friedland, O. Wing, and R. Ash, *Principles of Linear Networks*, New York: McGraw-Hill, 1961.

L. P. Huelsman, *Basic Circuit Theory*, 3rd ed., Englewood Cliffs, N.J.: Prentice-Hall, 1981.

Linear Circuit Analysis

Michael D. Ciletti
University of Colorado

J. David Irwin
Auburn University

Allan D. Kraus
Naval Postgraduate School

Norman Balabanian
Syracuse University

Theodore A. Bickart
Michigan State University

Shu-Park Chan
Santa Clara University

3.1 Voltage and Current Laws 58
    Kirchhoff's Current Law • Kirchhoff's Current Law in the Complex
    Domain • Kirchhoff's Voltage Law • Kirchhoff's Voltage Law in the
    Complex Domain • Importance of KVL and KCL
3.2 Node and Mesh Analysis 63
    Node Analysis • Mesh Analysis • Summary
3.3 Network Theorems 69
    Linearity and Superposition • The Network Theorems of Thévenin
    and Norton • Tellegen's Theorem • The Reciprocity Theorem
3.4 Power and Energy 79
    Tellegen's Theorem • AC Steady-State Power • Maximum Power
    Transfer • Measuring AC Power and Energy
3.5 Three-Phase Circuits 88
3.6 Graph Theory 92
    The k-Tree Approach • The Flowgraph Approach • The k-Tree
    Approach Versus the Flowgraph Approach • Some Topological
    Applications in Network Analysis and Design

3.1 Voltage and Current Laws

Michael D. Ciletti

Analysis of linear circuits rests on two fundamental physical laws that describe how the voltages and currents in a circuit must behave. This behavior results from whatever voltage sources, current sources, and energy storage elements are connected to the circuit. A voltage source imposes a constraint on the evolution of the voltage between a pair of **nodes**; a current source imposes a constraint on the evolution of the current in a **branch** of the circuit. The energy storage elements (capacitors and inductors) impose initial conditions on currents and voltages in the circuit; they also establish a dynamic relationship between the voltage and the current at their terminals.

Regardless of how a linear circuit is stimulated, every node voltage and every **branch current**, at every instant of time, must be consistent with Kirchhoff's voltage and current laws. These two laws govern even the most complex linear circuits. (They also apply to a broad category of nonlinear circuits that are modeled by point models of voltage and current.)

A circuit can be considered to have a topological (or graph) view, consisting of a labeled set of nodes and a labeled set of edges. Each edge is associated with a pair of nodes. A node is drawn as a *dot* and represents a connection between two or more physical components; an edge is drawn as a *line* and represents a path, or branch, for current flow through a component (see Fig. 3.1).

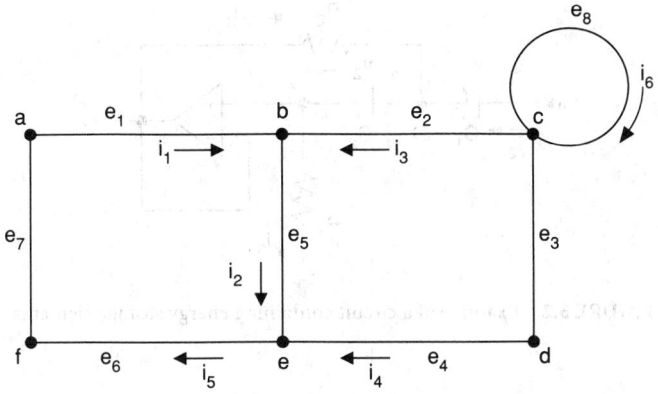

FIGURE 3.1 Graph representation of a linear circuit.

The edges, or branches, of the graph are assigned current labels, $i_1, i_2, \ldots, i_m$. Each current has a designated direction, usually denoted by an *arrow* symbol. If the arrow is drawn toward a node, the associated current is said to be *entering* the node; if the arrow is drawn away from the node, the current is said to be *leaving* the node. The current i_1 is entering node b in Fig. 3.1; the current i_5 is leaving node e.

Given a branch, the pair of nodes to which the branch is attached defines the convention for measuring voltages in the circuit. Given the ordered pair of nodes (a, b), a voltage measurement is formed as follows:

$$v_{ab} = v_a - v_b$$

where v_a and v_b are the absolute electrical potentials at the respective nodes, taken relative to some reference node. Typically, one node of the circuit is labeled as *ground*, or reference node; the remaining nodes are assigned voltage labels. The measured quantity, v_{ab}, is called the *voltage drop* from node a to node b. We note that

$$v_{ab} = - v_{ba}$$

and that

$$v_{ba} = v_b - v_a$$

is called the *voltage rise* from a to b. Each node voltage implicitly defines the voltage drop between the respective node and the ground node.

The pair of nodes to which an edge is attached may be written as (a,b) or (b,a). Given an ordered pair of nodes (a, b), a path *from a to b* is a directed sequence of edges in which the first edge in the sequence contains node label a, the last edge in the sequence contains node label b, and the node indices of any two adjacent members of the sequence have at least one node label in common. In Fig. 3.1, the edge sequence $\{e_1, e_2, e_4\}$ is not a path, because e_2 and e_4 do not share a common node label. The sequence $\{e_1, e_2\}$ is a path from node a to node c.

A path is said to be *closed* if the first node index of its first edge is identical to the second node index of its last edge. The following edge sequence forms a closed path in the graph given in Fig. 3.1: $\{e_1, e_2, e_3, e_4, e_6, e_7\}$. Note that the edge sequences $\{e_8\}$ and $\{e_1, e_1\}$ are closed paths.

Kirchhoff's Current Law

Kirchhoff's current law (KCL) imposes constraints on the currents in the branches that are attached to each node of a circuit. In simplest terms, KCL states that the sum of the currents that are entering a given node must equal the sum of the currents that are leaving the node. Thus, the set of currents in branches attached to a given node can be partitioned into two groups according to

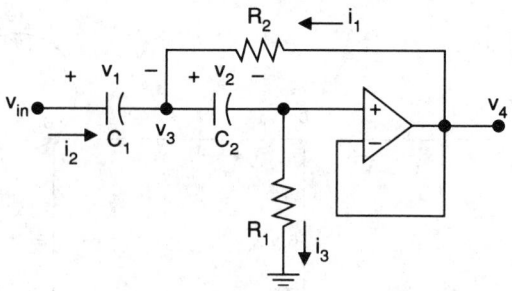

FIGURE 3.2   Example of a circuit containing energy storage elements.

their orientation. The groups must contain the same net current. Applying KCL at node *b* in Fig. 3.1 gives

$$i_1(t) + i_3(t) = i_2(t)$$

A connection of water pipes that has no leaks is a physical analogy of this situation. The net rate at which water is flowing into a joint of two or more pipes must equal the net rate at which water is flowing away from the joint. The joint itself has the property that it only connects the pipes and thereby imposes a structure on the flow of water, but it cannot store water. This is true regardless of when the flow is measured. Likewise, the nodes of a circuit are modeled as though they cannot store charge. (Physical circuits are sometimes modeled for the purpose of simulation as though they store charge, but these nodes implicitly have a capacitor that provides the physical mechanism for *storing* the charge. Thus, KCL is ultimately satisfied.)

KCL can be stated alternatively as: "the algebraic sum of the branch currents entering (or leaving) any node of a circuit at any instant of time must be zero." In this form, the label of any current whose orientation is away from the node is preceded by a minus sign. The currents *entering* node *b* in Fig. 3.1 must satisfy

$$i_1(t) - i_2(t) + i_3(t) = 0$$

In general, the currents entering or leaving each node *m* of a circuit must satisfy

$$\sum i_{km}(t) = 0$$

where $i_{km}(t)$ is understood to be the current in branch *k* attached to node *m*. The currents used in this expression are understood to be the currents that would be measured in the branches attached to the node, and their values include a magnitude and an algebraic sign. If the measurement convention is oriented for the case where currents are entering the node, then the actual current in a branch has a positive or negative sign, depending on whether the current is truly flowing toward the node in question.

Once KCL has been written for the nodes of a circuit, the equations can be rewritten by substituting into the equations the voltage-current relationships of the individual components. If a circuit is resistive, the resulting equations will be algebraic. If capacitors or inductors are included in the circuit, the substitution will produce a differential equation. For example, writing KCL at the node for v_3 in Fig. 3.2 produces

$$i_1 + i_2 - i_3 = 0$$

and

$$C_1 \frac{dv_1}{dt} + \frac{v_4 - v_3}{R_2} - C_2 \frac{dv_2}{dt} = 0$$

Writing KCL for v_2 and eliminating variables would lead to a solution describing the capacitor voltages. The capacitor voltages, together with the applied voltage source, determine the remaining voltages and currents in the circuit. Nodal analysis (see Section 3.2) treats the systematic modeling and analysis of a circuit under the influence of its sources and energy storage elements.

Kirchhoff's Current Law in the Complex Domain

Kirchhoff's current law is ordinarily stated in terms of the real (time-domain) currents flowing in a circuit, because it actually describes physical quantities, at least in a macroscopic, statistical sense. It also applies, however, to a variety of purely mathematical models that are commonly used to analyze circuits in the so-called complex domain.

For example, if a linear circuit is in the sinusoidal steady state, all of the currents and voltages in the circuit are sinusoidal. Thus, each voltage has the form

$$v(t) = A\sin(\omega t + \phi)$$

and each current has the form

$$i(t) = B\sin(\omega t + \theta)$$

where the positive coefficients A and B are called the magnitudes of the signals, and ϕ and θ are the phase angles of the signals. These mathematical models describe the physical behavior of electrical quantities, and instrumentation, such as an oscilloscope, can display the actual waveforms represented by the mathematical model. Although methods exist for manipulating the models of circuits to obtain the magnitude and phase coefficients that uniquely determine the waveform of each voltage and current, the manipulations are cumbersome and not easily extended to address other issues in circuit analysis.

Steinmetz [Smith and Dorf, 1992] found a way to exploit complex algebra to create an elegant framework for representing signals and analyzing circuits when they are in the steady state. In this approach, a model is developed in which each physical signal is replaced by a "complex" mathematical signal. This complex signal in polar, or exponential, form is represented as

$$v_c(t) = Ae^{(j\omega t + \phi)}$$

The algebra of complex exponential signals allows us to write this as

$$v_c(t) = Ae^{j\phi}e^{j\omega t}$$

and Euler's identity gives the equivalent rectangular form:

$$v_c(t) = A[\cos(\omega t + \phi) + j\sin(\omega t + \phi)]$$

So we see that a physical signal is either the real (cosine) or the imaginary (sine) component of an abstract, complex mathematical signal. The additional mathematics required for treatment of complex numbers allows us to associate a phasor, or complex amplitude, with a sinuosoidal signal. The time-invariant phasor associated with $v(t)$ is the quantity

$$\mathbf{v}_c = Ae^{j\phi}$$

Notice that the phasor $\mathbf{v}_c$ is an algebraic constant and that it incorporates the parameters A and ϕ of the corresponding time-domain sinusoidal signal.

Phasors can be thought of as being vectors in a two-dimensional plane. If the vector is allowed to rotate about the origin in the counterclockwise direction with frequency ω, the projection of its tip onto the horizontal (real) axis defines the time-domain signal corresponding to the real part of $v_c(t)$, i.e., $A\cos[\omega t + \phi]$, and its projection onto the vertical (imaginary) axis defines the time-domain signal corresponding to the imaginary part of $v_c(t)$, i.e., $A\sin[\omega t + \phi]$.

The composite signal $v_c(t)$ is a mathematical entity; it cannot be seen with an oscilloscope. Its value lies in the fact that when a circuit is in the steady state, its voltages and currents are uniquely determined by their corresponding phasors, and these in turn satisfy Kirchhoff's voltage and current laws! Thus, we are able to write

$$\sum_j I_{km} = 0$$

where I_{km} is the phasor of $i_{km}(t)$, the sinusoidal current in branch k attached to node m. An equation of this form can be written at each node of the circuit. For example, at node b in Fig. 3.1 KCL would have the form

$$I_1 - I_2 + I_3 = 0$$

Consequently, a set of linear, algebraic equations describe the phasors of the currents and voltages in a circuit in the sinusoidal steady state, i.e., the notion of time is suppressed (see Section 3.2). The solution of the set of equations yields the phasor of each voltage and current in the circuit, from which the actual time-domain expressions can be extracted.

It can also be shown that KCL can be extended to apply to the Fourier transforms and the Laplace transforms of the currents in a circuit. Thus, a single relationship between the currents at the nodes of a circuit applies to all of the known mathematical representations of the currents [Ciletti, 1988].

Kirchhoff's Voltage Law

Kirchhoff's voltage law (KVL) describes the behavior of the voltages in any closed, connected path formed by the branches of a circuit. It states that the sum of the voltage drops across the branches of any closed path in a circuit must be zero. Here, a **branch voltage** is the voltage drop measured across the branch.

The polarity of each voltage measurement in KVL is such that the absolute electrical potential at the node exited by the path is subtracted from the absolute electrical potential at the node entered by the path. In Fig. 3.1, the path enters node a and exits node b. Thus, the drop across the branch, from a to b, is denoted by v_{ab}. Alternatively, the voltage rise across the branch is denoted by v_{ba}, and $v_{ab} = -v_{ba}$.

Kirchhoff's voltage law, like Kirchhoff's current law, is true at any time. KVL can also be stated in terms of voltage rises instead of voltage drops.

KVL can be expressed mathematically as "the algebraic sum of the voltages drops around any closed path of a circuit at any instant of time is zero." This statement can also be cast as an equation:

$$\sum_j v_{km}(t) = 0$$

where $v_{km}(t)$ is the instantaneous voltage drop measured across branch k of path m. By convention, the voltage drop is taken in the direction of the edge sequence that forms the path.

The edge sequence $\{e_1, e_2, e_3, e_4, e_6, e_7\}$ forms a closed path in Fig. 3.1. The sum of the voltage drops taken around the path must satisfy KVL:

$$v_{ab}(t) + v_{bc}(t) + v_{cd}(t) + v_{de}(t) + v_{ef}(t) + v_{fa}(t) = 0$$

Since $v_{af}(t) = -v_{fa}(t)$, we can also write

$$v_{af}(t) = v_{ab}(t) + v_{bc}(t) + v_{cd} + v_{de}(t) + v_{ef}(t)$$

Had we chosen the path corresponding to the edge sequence $\{e_1, e_5, e_6, e_7\}$ for the path, we would have obtained

$$v_{af}(t) = v_{ab}(t) + v_{be}(t) + v_{ef}(t) + v_{fa}(t)$$

This demonstrates how KVL can be used to determine the voltage between a pair of nodes. It also reveals the fact that the voltage between a pair of nodes is independent of the path between the nodes on which the voltages are measured.

Kirchhoff's Voltage Law in the Complex Domain

Kirchhoff's voltage law also applies to the phasors of the voltages in a circuit in steady state and to the Fourier transforms and Laplace transforms of the voltages in a circuit.

Importance of KVL and KCL

Kirchhoff's current law is used extensively in nodal analysis because it is amenable to computer-based implementation and supports a systematic approach to circuit analysis. Nodal analysis leads to a set of algebraic equations in which the variables are the voltages at the nodes of the circuit. This formulation is popular in CAD programs because the variables correspond directly to physical quantities that can be measured easily.

Kirchhoff's voltage law can be used to completely analyze a circuit, but it is seldom used in large-scale circuit simulation programs. The basic reason is that the currents that correspond to a loop of a circuit do not necessarily correspond to the currents in the individual branches of the circuit. Nonetheless, KVL is frequently used to troubleshoot a circuit by measuring voltage drops across selected components.

Defining Terms

Branch: A symbol representing a path for current through a component in an electrical circuit.
Branch current: The current in a branch of a circuit.
Branch voltage: The voltage across a branch of a circuit.
Independent source: A voltage (current) source whose voltage (current) does not depend on any other voltage or current in the circuit.
Node: A symbol representing a physical connection between two electrical components in a circuit.
Node voltage: The voltage between a node and a reference node (usually ground).

References

M.D. Ciletti, *Introduction to Circuit Analysis and Design,* New York: Holt, Rinehart and Winston, 1988.
R.H. Smith and R.C. Dorf, *Circuits, Devices and Systems,* New York: Wiley, 1992.

Further Information

Kirchhoff's laws form the foundation of modern computer software for analyzing electrical circuits. The interested reader might consider the issue of determining the minimum number of algebraic equations that fully characterizes the circuit. Is it determined by KCL, KVL, or some mixture of the two?

3.2 Node and Mesh Analysis

J. David Irwin

In this section **Kirchhoff's current law (KCL)** and **Kirchhoff's voltage law (KVL)** will be used to determine currents and voltages throughout a network. For simplicity, we will first illustrate the

basic principles of both **node analysis** and **mesh analysis** using only **dc** circuits. Once the fundamental concepts have been explained and illustrated, we will demonstrate the generality of both
analysis techniques through an **ac** circuit example.

Node Analysis

In a node analysis, the node voltages are the variables in a circuit, and KCL is the vehicle used to
determine them. One node in the network is selected as a
reference node, and then all other node voltages are
defined with respect to that particular node. This reference node is typically referred to as *ground* using the symbol ($\perp$), indicating that it is at ground-zero potential.

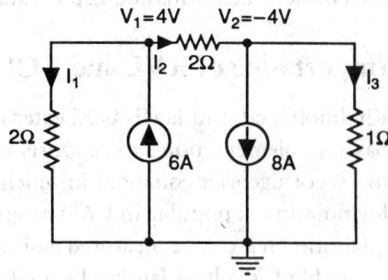

Consider the network shown in Fig. 3.3. The network
has three nodes, and the node at the bottom of the circuit
has been selected as the reference node. Therefore the two
remaining nodes, labeled V_1 and V_2, are measured with
respect to this reference node.

Suppose that the node voltages V_1 and V_2 have some-

FIGURE 3.3   A three-node network.

how been determined, i.e., $V_1 = 4$ V and $V_2 = -4$ V. Once these node voltages are known, **Ohm's law**
can be used to find all branch currents. For example,

$$I_1 = \frac{V_1 - 0}{2} = 2\,\text{A}$$

$$I_2 = \frac{V_1 - V_2}{2} = \frac{4 - (-4)}{2} = 4\,\text{A}$$

$$I_3 = \frac{V_2 - 0}{1} = \frac{-4}{1} = -4\,\text{A}$$

Note that KCL is satisfied at every node, i.e.,

$$I_1 - 6 + I_2 = 0$$

$$-I_2 + 8 + I_3 = 0$$

$$-I_1 + 6 - 8 - I_3 = 0$$

Therefore, as a general rule, if the node voltages are known, all branch currents in the network
can be immediately determined.

In order to determine the node voltages in a network, we apply KCL to every node in the network
except the reference node. Therefore, given an N-node
circuit, we employ $N-1$ linearly independent simultaneous equations to determine the $N-1$ unknown node
voltages. Graph theory, which is covered in Section 3.6,
can be used to prove that exactly $N-1$ linearly independent KCL equations are required to find the $N-1$
unknown node voltages in a network.

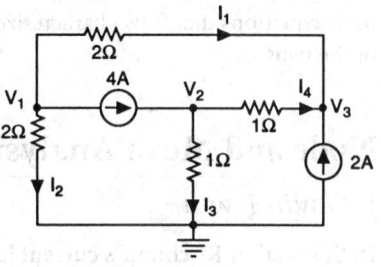

Let us now demonstrate the use of KCL in determining
the node voltages in a network. For the network shown in
Fig. 3.4, the bottom node is selected as the reference and

FIGURE 3.4   A four-node network.

the three remaining nodes, labeled V_1, V_2, and V_3, are measured with respect to that node. All unknown branch currents are also labeled. The KCL equations for the three nonreference nodes are

$$I_1 + 4 + I_2 = 0$$

$$-4 + I_3 + I_4 = 0$$

$$-I_1 - I_4 - 2 = 0$$

Using Ohm's law these equations can be expressed as

$$\frac{V_1 - V_3}{2} + 4 + \frac{V_1}{2} = 0$$

$$-4 + \frac{V_2}{1} + \frac{V_2 - V_3}{1} = 0$$

$$-\frac{(V_1 - V_3)}{2} - \frac{(V_2 - V_3)}{1} - 2 = 0$$

Solving these equations, using any convenient method, yields $V_1 = -8/3$ V, $V_2 = 10/3$ V, and $V_3 = 8/3$ V. Applying Ohm's law we find that the branch currents are $I_1 = -16/6$ A, $I_2 = -8/6$ A, $I_3 = 20/6$ A, and $I_4 = 4/6$ A. A quick check indicates that KCL is satisfied at every node.

The circuits examined thus far have contained only current sources and resistors. In order to expand our capabilities, we next examine a circuit containing voltage sources. The circuit shown in Fig. 3.5 has three nonreference nodes labeled V_1, V_2, and V_3. However, we do not have three unknown node voltages. Since known voltage sources exist between the reference node and nodes V_1 and V_3, these two node voltages are known, i.e., $V_1 = 12$ V and $V_3 = -4$ V. Therefore, we have only one unknown node voltage, V_2. The equations for this network are then

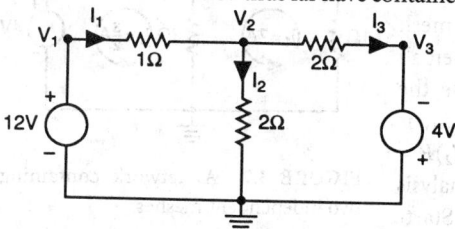

FIGURE 3.5 A four-node network containing voltage sources.

$$V_1 = 12$$

$$V_3 = -4$$

and

$$-I_1 + I_2 + I_3 = 0$$

The KCL equation for node V_2 written using Ohm's law is

$$-\frac{(12 - V_2)}{1} + \frac{V_2}{2} + \frac{V_2 - (-4)}{2} = 0$$

Solving this equation yields $V_2 = 5$ V, $I_1 = 7$ A, $I_2 = 5/2$ A, and $I_3 = 9/2$ A. Therefore, KCL is satisfied at every node.

Thus, the presence of a voltage source in the network actually simplifies a node analysis. In an attempt to generalize this idea, consider the network in Fig. 3.6. Note that in this case $V_1 = 12$ V and the difference between node voltages V_3 and V_2 is constrained to be 6 V. Hence, two of the three equations needed to solve for the node voltages in the network are

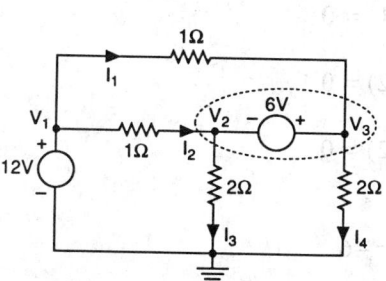

FIGURE 3.6 A four-node network used to illustrate a supernode.

$$V_1 = 12$$

$$V_3 - V_2 = 6$$

To obtain the third required equation, we form what is called a **supernode**, indicated by the dotted enclosure in the network. Just as KCL must be satisfied at any node in the network, it must be satisfied at the supernode as well. Therefore, summing all the currents leaving the supernode yields the equation

$$\frac{V_2 - V_1}{1} + \frac{V_2}{2} + \frac{V_3 - V_1}{1} + \frac{V_3}{2} = 0$$

The three equations yield the node voltages $V_1 = 12$ V, $V_2 = 5$ V, and $V_3 = 11$ V, and therefore $I_1 = 1$ A, $I_2 = 7$ A, $I_3 = 5/2$ A, and $I_4 = 11/2$ A.

Mesh Analysis

In a mesh analysis the mesh currents in the network are the variables and KVL is the mechanism used to determine them. Once all the mesh currents have been determined, Ohm's law will yield the voltages anywhere in the circuit. If the network contains N independent meshes, then graph theory can be used to prove that N independent linear simultaneous equations will be required to determine the N mesh currents.

The network shown in Fig. 3.7 has two independent meshes. They are labeled I_1 and I_2, as shown. If the mesh currents are known to be $I_1 = 7$ A and $I_2 = 5/2$ A, then all voltages in the network can be calculated. For example, the voltage V_1, i.e., the voltage across the 1-Ω resistor, is $V_1 = -I_1 R = -(7)(1) = -7$ V. Likewise $V_2 = (I_1 - I_2)R = (7 - 5/2)(2) = 9$ V. Furthermore, we can check our analysis by showing that KVL is satisfied around every mesh. Starting at the lower left-hand corner and applying KVL to the left-hand mesh we obtain

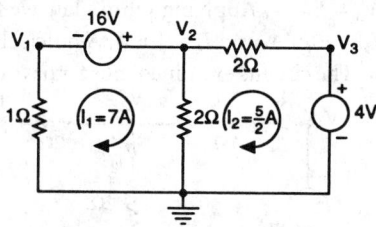

FIGURE 3.7   A network containing two independent meshes.

$$-(7)(1) + 16 - (7 - 5/2)(2) = 0$$

where we have assumed that increases in energy level are positive and decreases in energy level are negative.

Consider now the network in Fig. 3.8. Once again, if we assume that an increase in energy level is positive and a decrease in energy level is negative, the three KVL equations for the three meshes defined are

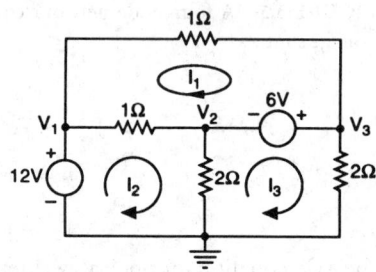

FIGURE 3.8   A three-mesh network.

$$-I_1(1) - 6 - (I_1 - I_2)(1) = 0$$

$$+12 - (I_2 - I_1)(1) - (I_2 - I_3)(2) = 0$$

$$-(I_3 - I_2)(2) + 6 - I_3(2) = 0$$

These equations can be written as

$$2I_1 - I_2 = -6$$

$$-I_1 + 3I_2 - 2I_3 = 12$$

$$-2I_2 + 4I_3 = 6$$

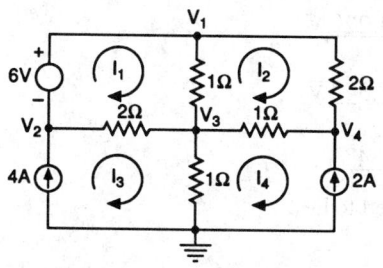

FIGURE 3.9 A four-mesh network containing current sources.

Solving these equations using any convenient method yields $I_1 = 1$ A, $I_2 = 8$ A, and $I_3 = 11/2$ A. Any voltage in the network can now be easily calculated, e.g., $V_2 = (I_2 - I_3)(2) = 5$ V and $V_3 = I_3(2) = 11$ V.

Just as in the node analysis discussion, we now expand our capabilities by considering circuits which contain current sources. In this case, we will show that for mesh analysis, the presence of current sources makes the solution easier.

The network in Fig. 3.9 has four meshes which are labeled I_1, I_2, I_3, and I_4. However, since two of these currents, i.e., I_3 and I_4, pass directly through a current source, two of the four linearly independent equations required to solve the network are

$$I_3 = 4$$

$$I_4 = -2$$

The two remaining KVL equations for the meshes defined by I_1 and I_2 are

$$+6 - (I_1 - I_2)(1) - (I_1 - I_3)(2) = 0$$

$$-(I_2 - I_1)(1) - I_2(2) - (I_2 - I_4)(1) = 0$$

Solving these equations for I_1 and I_2 yields $I_1 = 54/11$ A and $I_2 = 8/11$ A. A quick check will show that KCL is satisfied at every node. Furthermore, we can calculate any node voltage in the network. For example, $V_3 = (I_3 - I_4)(1) = 6$ V and $V_1 = V_3 + (I_1 - I_2)(1) = 112/11$ V.

Summary

Both node analysis and mesh analysis have been presented and discussed. Although the methods have been presented within the framework of dc circuits with only independent sources, the techniques are applicable to ac analysis and circuits containing dependent sources.

To illustrate the applicability of the two techniques to ac circuit analysis, consider the network in Fig. 3.10. All voltages and currents are phasors and the impedance of each passive element is known.

In the node analysis case, the voltage $\mathbf{V}_4$ is known and the voltage between $\mathbf{V}_2$ and $\mathbf{V}_3$ is constrained. Therefore, two of the four required equations are

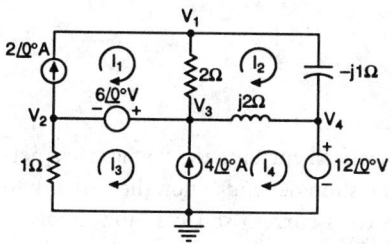

FIGURE 3.10 A network containing five nodes and four meshes.

$$\mathbf{V}_4 = 12 \underline{/0^\circ}$$

$$\mathbf{V}_2 + 6\underline{/0^\circ} = V_3$$

KCL for the node labeled $\mathbf{V}_1$ and the supernode containing the nodes labeled $\mathbf{V}_2$ and $\mathbf{V}_3$ is

$$\frac{\mathbf{V}_1 - \mathbf{V}_3}{2} + \frac{\mathbf{V}_1 - \mathbf{V}_4}{-j1} = 2 \underline{/0^\circ}$$

$$\frac{\mathbf{V}_2}{1} + 2 \underline{/0^\circ} + \frac{\mathbf{V}_3 - \mathbf{V}_1}{2} + \frac{\mathbf{V}_3 - \mathbf{V}_4}{j2} = 4 \underline{/0^\circ} = 0$$

Solving these equations yields the remaining unknown node voltages.

$$\mathbf{V}_1 = 11.9 - j0.88 = 11.93\underline{/-4.22°}\,\text{V}$$

$$\mathbf{V}_2 = 3.66 - j1.07 = 3.91\underline{/-16.34°}\,\text{V}$$

$$\mathbf{V}_3 = 9.66 - j1.07 = 9.72\underline{/-6.34°}\,\text{V}$$

In the mesh analysis case, the currents $\mathbf{I}_1$ and $\mathbf{I}_3$ are constrained to be

$$\mathbf{I}_1 = 2\underline{/0°}$$

$$\mathbf{I}_4 - \mathbf{I}_3 = -4\underline{/0°}$$

The two remaining KVL equations are obtained from the mesh defined by mesh current $\mathbf{I}_2$ and the loop which encompasses the meshes defined by mesh currents $\mathbf{I}_3$ and $\mathbf{I}_4$.

$$-2(I_2 - I_1) - (-j1)I_2 - j2(I_2 - I_4) = 0$$

$$-(1I_3 + 6\underline{/0°} - j2(I_4 - I_2) - 12\underline{/0°} = 0$$

Solving these equations yields the remaining unknown mesh currents

$$\mathbf{I}_2 = 0.88\underline{/-6.34°}\,\text{A}$$

$$\mathbf{I}_3 = 3.91\underline{/163.66°}\,\text{A}$$

$$\mathbf{I}_4 = 1.13\underline{/72.35°}\,\text{A}$$

As a quick check we can use these currents to compute the node voltages. For example, if we calculate

$$\mathbf{V}_2 = -1(\mathbf{I}_3)$$

and

$$\mathbf{V}_1 = -j1(\mathbf{I}_2) + 12\underline{/0°}$$

we obtain the answers computed earlier.

As a final point, because both node and mesh analysis will yield all currents and voltages in a network, which technique should be used? The answer to this question depends upon the network to be analyzed. If the network contains more voltage sources than current sources, node analysis might be the easier technique. If, however, the network contains more current sources than voltage sources, mesh analysis may be the easiest approach.

Defining Terms

ac: An abbreviation for alternating current.

dc: An abbreviation for direct current.

Kirchhoff's current law (KCL):   This law states that the algebraic sum of the currents either entering or leaving a node must be zero. Alternatively, the law states that the sum of the currents entering a node must be equal to the sum of the currents leaving that node.

Kirchhoff's voltage law (KVL):   This law states that the algebraic sum of the voltages around any loop is zero. A loop is any closed path through the circuit in which no node is encountered more than once.

Mesh analysis: A circuit analysis technique in which KVL is used to determine the mesh currents in a network. A mesh is a loop that does not contain any loops within it.

Node analysis: A circuit analysis technique in which KCL is used to determine the node voltages in a network.

Ohm's law: A fundamental law which states that the voltage across a resistance is directly proportional to the current flowing through it.

Reference node: One node in a network that is selected to be a common point, and all other node voltages are measured with respect to that point.

Supernode: A cluster of nodes, interconnected with voltage sources, such that the voltage between any two nodes in the group is known.

Reference

J. D. Irwin, *Basic Engineering Circuit Analysis,* 4th ed., New York: Macmillan, 1993.

3.3 Network Theorems

Allan D. Kraus

Linearity and Superposition

Linearity

Consider a system (which may consist of a single network element) represented by a block, as shown in Fig. 3.11, and observe that the system has an input designated by e (for excitation) and an output designated by r (for response). The system is considered to be **linear** if it satisfies the *homogeneity* and *superposition* conditions.

The homogeneity condition: If an arbitrary input to the system, e, causes a response, r, then if ce is the input, the output is cr where c is some arbitrary constant.

The superposition condition: If the input to the system, e_1, causes a response, r_1, and if an input to the system, e_2, causes a response, r_2, then a response, $r_1 + r_2$, will occur when the input is $e_1 + e_2$.

If neither the homogeneity condition nor the superposition condition is satisfied, the system is said to be *nonlinear.*

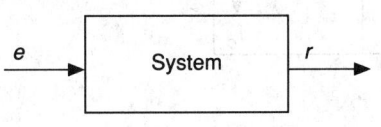

FIGURE 3.11 A simple system.

The Superposition Theorem

While both the homogeneity and superposition conditions are necessary for linearity, the superposition condition, in itself, provides the basis for the superposition theorem:

> If cause and effect are linearly related, the total effect due to several causes acting simultaneously is equal to the sum of the individual effects due to each of the causes acting one at a time.

Example 3.1. Consider the network driven by a current source at the left and a voltage source at the top, as shown in Fig. 3.12(a). The current phasor indicated by $\hat{I}$ is to be determined. According to the superposition theorem, the current $\hat{I}$ will be the sum of the two current components $\hat{I}_V$ due to the voltage source acting alone as shown in Fig. 3.12(b) and $\hat{I}_C$ due to the current source acting alone shown in Fig. 3.12(c).

$$\hat{I} = \hat{I}_V + \hat{I}_C$$

Figures 3.12(b) and (c) follow from the methods of removing the effects of independent voltage and current sources. Voltage sources are nulled in a network by replacing them with short circuits and current sources are nulled in a network by replacing them with open circuits.

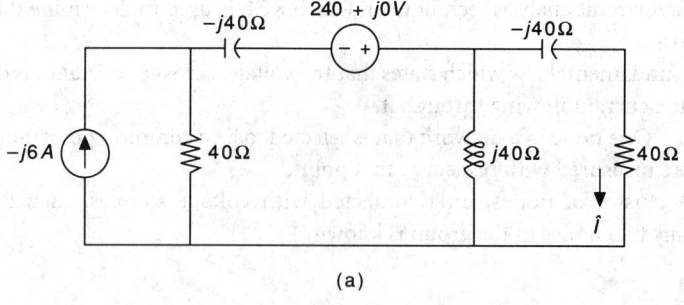

(a)

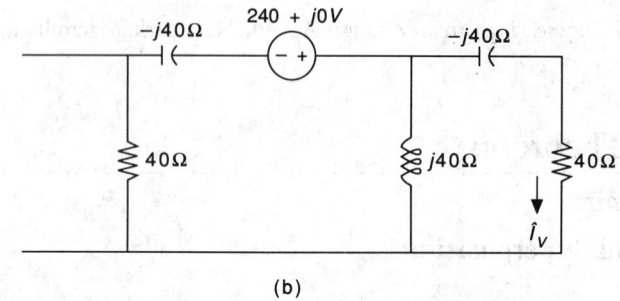

(b)

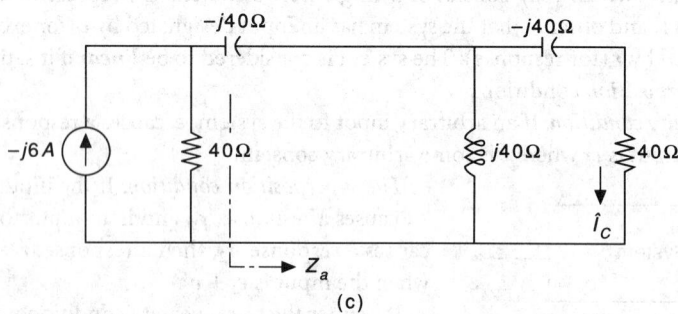

(c)

FIGURE 3.12   (a) A network to be solved by using superposition; (b) the network with the current source nulled; and (c) the network with the voltage source nulled.

The networks displayed in Figs. 3.12(b) and (c) are simple ladder networks in the phasor domain, and the strategy is to first determine the equivalent impedances presented to the voltage and current sources. In Fig. 3.12(b), the group of three impedances to the right of the voltage source are in series-parallel and possess an impedance of

$$Z_P = \frac{(40 - j40)(j40)}{40 + j40 - j40} = 40 + j40 \ \Omega$$

and the total impedance presented to the voltage source is

$$Z = Z_P + 40 - j40 = 40 + j40 + 40 - j40 = 80 \ \Omega$$

Then $\hat{I}_1$, the current leaving the voltage source, is

$$\hat{I}_1 = \frac{240 + j0}{80} = 3 + j0 \, \text{A}$$

and by a current division

$$\hat{I}_V = \left[\frac{j40}{40 - j40 + j40} \right](3 + j0) = j(3 + j0) = 0 + j3 \, \text{A}$$

In Fig. 3.12(c), the current source delivers current to the 40-Ω resistor and to an impedance consisting of the capacitor and Z_P. Call this impedance Z_a so that

$$Z_a = -j40 + Z_P = -j40 + 40 + j40 = 40 \, \Omega$$

Then, two current divisions give $\hat{I}_C$

$$\hat{I}_C = \left[\frac{40}{40 + 40} \right]\left[\frac{j40}{40 - j40 + j40} \right](0 - j6) = \frac{j}{2}(0 - j6) = 3 + j0 \, \text{A}$$

The current $\hat{I}$ in the circuit of Fig. 3.12(a) is

$$\hat{I} = \hat{I}_V + \hat{I}_C = 0 + j3 + (3 + j0) = 3 + j3 \, \text{A}$$

The Network Theorems of Thévenin and Norton

If interest is to be focused on the voltages and across the currents through a small portion of a network such as network B in Fig. 3.13(a), it is convenient to replace network A, which is complicated and of little interest, by a simple equivalent. The simple equivalent may contain a single, equivalent, voltage source in series with an equivalent impedance in series as displayed in Fig. 3.13(b). In this case, the equivalent is called a *Thévenin equivalent*. Alternatively, the simple equivalent may consist

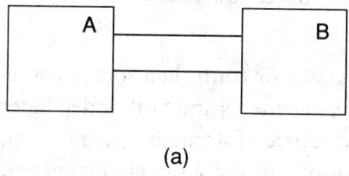

(a)

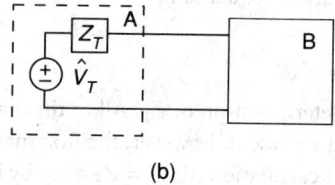

(b)

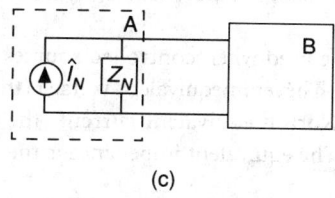

(c)

FIGURE 3.13 (a) Two one-port networks; (b) the Thévenin equivalent for network a; and (c) the Norton equivalent for network a.

of an equivalent current source in parallel with an equivalent impedance. This equivalent, shown in Fig. 3.13(c), is called a *Norton equivalent*. Observe that as long as Z_T (subscript T for Thévenin) is equal to Z_N (subscript N for Norton), the two equivalents may be obtained from one another by a simple source transformation.

Conditions of Application

The Thévenin and Norton network equivalents are only valid at the terminals of network A in Fig. 3.13(a) and they do not extend to its interior. In addition, there are certain restrictions on networks A and B. Network A may contain only linear elements but may contain both independent and dependent sources. Network B, on the other hand, is not restricted to linear elements; it may contain nonlinear or time-varying elements and may also contain both independent and dependent sources. Together, there can be no controlled source coupling or magnetic coupling between networks A and B.

The Thévenin Theorem

The statement of the **Thévenin theorem** is based on Fig. 3.13(b):

Insofar as a load which has no magnetic or controlled source coupling to a one-port is concerned, a network containing linear elements and both independent and controlled sources may be replaced by an ideal voltage source of strength, $\hat{V}_T$, and an equivalent impedance, Z_T, in series with the source. The value of $\hat{V}_T$ is the open-circuit voltage, $\hat{V}_{OC}$, appearing across the terminals of the network and Z_T is the driving point impedance at the terminals of the network, obtained with all independent sources set equal to zero.

The Norton Theorem

The **Norton theorem** involves a current source equivalent. The statement of the Norton theorem is based on Fig. 3.13(c):

Insofar as a load which has no magnetic or controlled source coupling to a one-port is concerned, the network containing linear elements and both independent and controlled sources may be replaced by an ideal current source of strength, $\hat{I}_N$, and an equivalent impedance, Z_N, in parallel with the source. The value of $\hat{I}_N$ is the short-circuit current, $\hat{I}_{SC}$, which results when the terminals of the network are shorted and Z_N is the driving point impedance at the terminals when all independent sources are set equal to zero.

The Equivalent Impedance, $Z_T = Z_N$

Three methods are available for the determination of Z_T. All of them are applicable at the analyst's discretion. When controlled sources are present, however, the first method cannot be used.

The first method involves the direct calculation of $Z_{eq} = Z_T = Z_N$ by looking into the terminals of the network after all independent sources have been nulled. Independent sources are nulled in a network by replacing all independent voltage sources with a short circuit and all independent current sources with an open circuit.

The second method, which may be used when controlled sources are present in the network, requires the computation of both the Thévenin equivalent voltage (the open-circuit voltage at the terminals of the network) and the Norton equivalent current (the current through the short-circuited terminals of the network). The equivalent impedance is the ratio of these two quantities

$$Z_T = Z_N = Z_{eq} = \frac{\hat{V}_T}{\hat{I}_N} = \frac{\hat{V}_{OC}}{\hat{I}_{SC}}$$

· The third method may also be used when controlled sources are present within the network. A test voltage may be placed across the terminals with a resulting current calculated or measured. Alternatively, a test current may be injected into the terminals with a resulting voltage determined. In either case, the equivalent resistance can be obtained from the value of the ratio of the test voltage $\hat{V}_o$ to the resulting current $\hat{I}_o$

$$Z_T = \frac{\hat{V}_o}{\hat{I}_o}$$

Example 3.2. The current through the capacitor with impedance $-j35\ \Omega$ in Fig. 3.14(a) may be found using Thévenin's theorem. The first step is to remove the $-j35$-Ω capacitor and consider it as the load. When this is done, the network in Fig. 3.14(b) results.

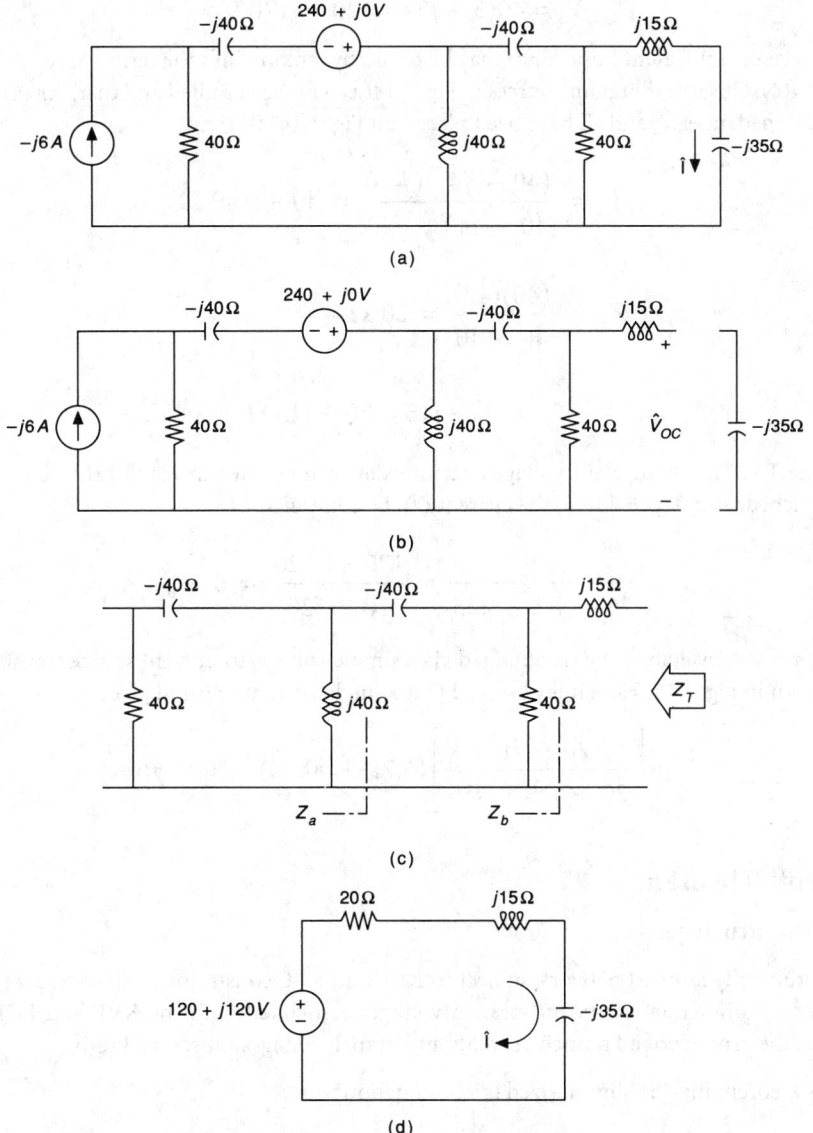

(a)

(b)

(c)

(d)

FIGURE 3.14 (a) A network in the phasor domain; (b) the network with the load removed; (c) the network for the computation of the Thévenin equivalent impedance; and (d) the Thévenin equivalent.

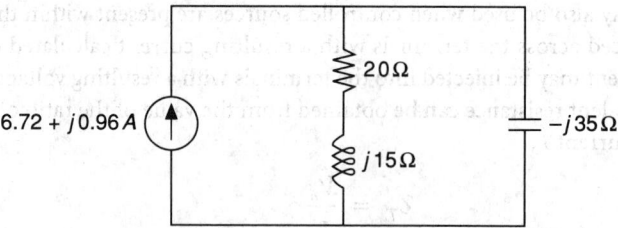

FIGURE 3.15   The Norton equivalent of Fig. 3.14(d).

The Thévenin equivalent voltage is the voltage across the 40-Ω resistor. The current through the 40-Ω resistor was found in Example 3.1 to be $\hat{I} = 3 + j3\ \Omega$. Thus,

$$\hat{V}_T = 40(3 + j3) = 120 + j120 \text{ V}$$

The Thévenin equivalent impedance may be found by looking into the terminals of the network in Fig. 3.14(c). Observe that both sources in Fig. 3.14(a) have been nulled and that, for ease of computation, impedances Z_a and Z_b have been placed on Fig. 3.14(c). Here,

$$Z_a = \frac{(40 - j40)(j40)}{40 + j40 - j40} = 40 + j40\ \Omega$$

$$Z_b = \frac{(40)(40)}{40 + 40} = 20\ \Omega$$

and

$$Z_T = Z_b + j15 = 20 + j15\ \Omega$$

Both the Thévenin equivalent voltage and impedance are shown in Fig. 3.14(d), and when the load is attached, as in Fig. 3.14(d), the current can be computed as

$$\hat{I} = \frac{\hat{V}_T}{20 + j15 - j35} = \frac{120 + j120}{20 - j20} = 0 + j6 \text{ A}$$

The Norton equivalent circuit is obtained via a simple voltage-to-current source transformation and is shown in Fig. 3.15. Here it is observed that a single current division gives

$$\hat{I} = \left[\frac{20 + j15}{20 + j15 - j35} \right](6.72 + j0.96) = 0 + j6 \text{ A}$$

Tellegen's Theorem

Tellegen's theorem states:

> In an arbitrarily lumped network subject to KVL and KCL constraints, with reference directions of the branch currents and branch voltages associated with the KVL and KCL constraints, the product of all branch currents and branch voltages must equal zero.

Tellegen's theorem may be summarized by the equation

$$\sum_{k=1}^{b} v_k j_k = 0$$

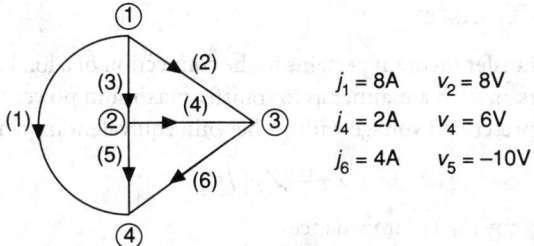

FIGURE 3.16 An oriented graph of a particular network with some known branch currents and branch voltages.

where the lower case letters v and j represent instantaneous values of the branch voltages and branch currents, respectively, and where b is the total number of branches. A matrix representation employing the branch current and branch voltage vectors also exists. Because $\mathbf{V}$ and $\mathbf{J}$ are column vectors

$$\mathbf{V} \cdot \mathbf{J} = \mathbf{V}^T \mathbf{J} = \mathbf{J}^T \mathbf{V}$$

The prerequisite concerning the KVL and KCL constraints in the statement of Tellegen's theorem is of crucial importance.

Example 3.3. Figure 3.16 displays an oriented graph of a particular network in which there are six branches labeled with numbers within parentheses and four nodes labeled by numbers within circles. Several known branch currents and branch voltages are indicated. Because the type of elements or their values is not germane to the construction of the graph, the other branch currents and branch voltages may be evaluated from repeated applications of KCL and KVL. KCL may be used first at the various nodes.

$$\text{node 3:} \quad j_2 = j_6 - j_4 = 4 - 2 = 2 \text{ A}$$

$$\text{node 1:} \quad j_3 = -j_1 - j_2 = -8 - 2 = -10 \text{ A}$$

$$\text{node 2:} \quad j_5 = j_3 - j_4 = -10 - 2 = -12 \text{ A}$$

Then KVL gives

$$v_3 = v_2 - v_4 = 8 - 6 = 2 \text{ V}$$

$$v_6 = v_5 - v_4 = -10 - 6 = -16 \text{ V}$$

$$v_1 = v_2 + v_6 = 8 - 16 = -8 \text{ V}$$

The transposes of the branch voltage and current vectors are

$$\mathbf{V}^T = [-8 \quad 8 \quad 2 \quad 6 \quad -10 \quad -16] \text{ V}$$

and

$$\mathbf{J}^T = [8 \quad 2 \quad -10 \quad 2 \quad -12 \quad 4] \text{ V}$$

The scalar product of $\mathbf{V}$ and $\mathbf{J}$ gives

$$-8(8) + 8(2) + 2(-10) + 6(2) + (-10)(-12) + (-16)(4) = -148 + 148 = 0$$

and Tellegen's theorem is confirmed.

Maximum Power Transfer

The maximum power transfer theorem pertains to the connection of a load to the Thévenin equivalent of a source network in such a manner as to transfer maximum power to the load. For a given network operating at a prescribed voltage with a Thévenin equivalent impedance

$$Z_T = |Z_T| \underline{/\theta_T}$$

the real power drawn by any load of impedance

$$Z_o = |Z_o| \underline{/\theta_o}$$

is a function of just two variables, $|Z_o|$ and θ_o. If the power is to be a maximum, there are three alternatives to the selection of $|Z_o|$ and θ_o:

(1) Both $|Z_o|$ and θ_o are at the designer's discretion and both are allowed to vary in any manner in order to achieve the desired result. In this case, the load should be selected to be the complex conjugate of the Thévenin equivalent impedance

$$Z_o = Z_T^*$$

(2) The angle, θ_o, is fixed but the magnitude, $|Z_o|$, is allowed to vary. For example, the analyst may select and fix $\theta_o = 0°$. This requires that the load be resistive (Z is entirely real). In this case, the value of the load resistance should be selected to be equal to the magnitude of the Thévenin equivalent impedance

$$R_o = |Z_T|$$

(3) The magnitude of the load impedance, $|Z_o|$, can be fixed, but the impedance angle, θ_o, is allowed to vary. In this case, the value of the load impedance angle should be

$$\theta_o = \arcsin\left[-\frac{2|Z_o||Z_T|\sin\theta_T}{|Z_o|^2 + |Z_T|^2}\right]$$

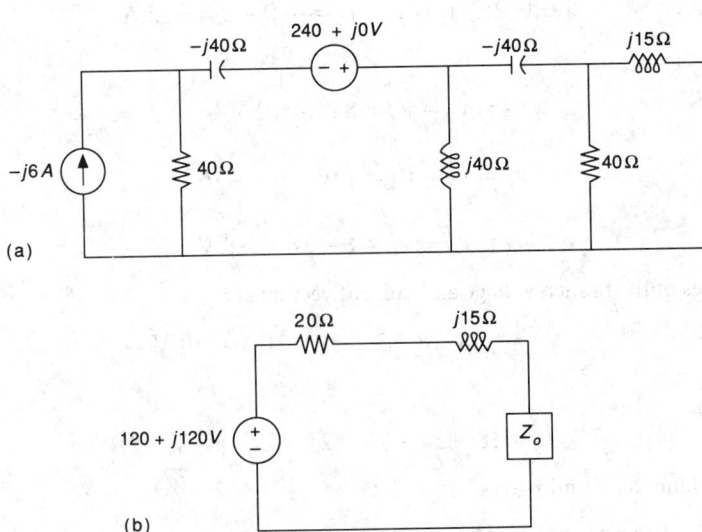

(a)

(b)

FIGURE 3.17   (a) A network for which the load, Z_o, is to be selected for maximum power transfer, and (b) the Thévenin equivalent of the network.

Example 3.4. Figure 3.17(a) is identical to Fig. 3.14(b) with the exception of a load, Z_o, substituted for the capacitive load. The Thévenin equivalent is shown in Fig. 3.17(b). The value of Z_o to transfer maximum power is to be found if its elements are unrestricted, if it is to be a single resistor, or if the magnitude of Z_o must be 20 Ω but its angle is adjustable.

For **maximum power transfer** to Z_o when the elements of Z_o are completely at the discretion of the network designer, Z_o must be the complex conjugate of Z_T

$$Z_o = Z_T^* = 20 - j15 \ \Omega$$

If Z_o is to be a single resistor, R_o, then the magnitude of $Z_o = R_o$ must be equal to the magnitude of Z_T. Here

$$Z_T = 20 + j15 = 25\underline{/36.87°}$$

so that

$$R_o = |Z_o| = 25 \ \Omega$$

If the magnitude of Z_o must be 20 Ω but the angle is adjustable, the required angle is calculated from

$$\theta_o = \arcsin\left[-\frac{2\,|Z_o|\,|Z_T|}{|Z_o|^2 + |Z_T|^2} \sin \theta_T\right]$$

$$= \arcsin\left[-\frac{2(20)(25)}{(20)^2 + (25)^2} \sin \underline{/36.87°}\right]$$

$$= \arcsin(-0.585) = -35.83°$$

This makes Z_o

$$Z_o = 20\underline{/-35.83°} = 16.22 - j11.71 \ \Omega$$

The Reciprocity Theorem

The **reciprocity theorem** is a useful general theorem that applies to all linear, passive, and bilateral networks. However, it applies only to cases where current and voltage are involved.

The ratio of a single excitation applied at one point to an observed response at another is invariant with respect to an interchange of the points of excitation and observation.

The reciprocity principle also applies if the excitation is a current and the observed response is a voltage. It will not apply, in general, for voltage–voltage and current–current situations, and, of course, it is not applicable to network models of nonlinear devices.

Example 3.5. It is easily shown that the positions of v_s and i in Fig. 3.18(a) may be interchanged as in Fig. 3.18(b) without changing the value of the current i.

In Fig. 3.18(a), the resistance presented to the voltage source is

$$R = 4 + \frac{3(6)}{3 + 6} = 4 + 2 = 6 \ \Omega$$

Then

$$i_a = \frac{v_s}{R} = \frac{36}{6} = 6 \ \text{A}$$

and by current division

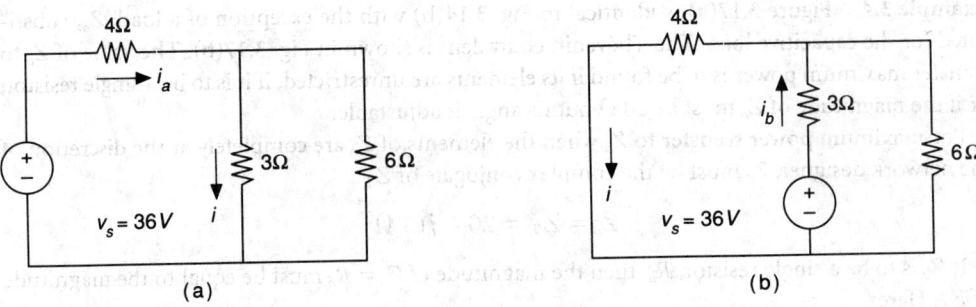

FIGURE 3.18   Two networks which can be used to illustrate the reciprocity principle.

$$i_a = \frac{6}{6+3} i_a = \left(\frac{2}{3}\right) 6 = 4 \text{ A}$$

In Fig. 3.18(b), the resistance presented to the voltage source is

$$R = 3 + \frac{6(4)}{6+4} = 3 + \frac{12}{5} = \frac{27}{5} \ \Omega$$

Then

$$i_b = \frac{v_s}{R} = \frac{36}{27/5} = \frac{180}{27} = \frac{20}{3} \text{ A}$$

and again, by current division

$$i = \frac{6}{4+6} i_b = \left(\frac{3}{5}\right) \frac{20}{3} = 4 \text{ A}$$

The network is reciprocal.

The Substitution and Compensation Theorems

The Substitution Theorem

Any branch in a network with branch voltage, v_k, and branch current, i_k, can be replaced by another branch provided it also has branch voltage, v_k, and branch current, i_k.

The Compensation Theorem

In a linear network, if the impedance of a branch carrying a current $\hat{I}$ is changed from Z to $Z + \Delta Z$, then the corresponding change of any voltage or current elsewhere in the network will be due to a compensating voltage source, $\Delta Z \hat{I}$, placed in series with $Z + \Delta Z$ with polarity such that the source, $\Delta Z \hat{I}$, is opposing the current $\hat{I}$.

Defining Terms

Linear network: A network in which the parameters of resistance, inductance, and capacitance are constant with respect to voltage or current or the rate of change of voltage or current and in which the voltage or current of sources is either independent of or proportional to other voltages or currents, or their derivatives.

Maximum power transfer theorem: In any electrical network which carries direct or alternating current, the maximum possible power transferred from one section to another occurs when the impedance of the section acting as the load is the complex conjugate of the impedance of the section that acts as the source. Here, both impedances are measured across the pair of terminals at which the power is transferred with the other part of the network disconnected.

Norton theorem: The voltage across an element that is connected to two terminals of a linear, bilateral network is equal to the short-circuit current between these terminals in the absence of the element, divided by the admittance of the network looking back from the terminals into the network, with all generators replaced by their internal admittances.

Principle of superposition: In a linear electrical network, the voltage or current in any element resulting from several sources acting together is the sum of the voltages or currents from each source acting alone.

Reciprocity theorem: In a network consisting of linear, passive impedances, the ratio of the voltage introduced into any branch to the current in any other branch is equal in magnitude and phase to the ratio that results if the positions of the voltage and current are interchanged.

Thévenin theorem: The current flowing in any impedance connected to two terminals of a linear, bilateral network containing generators is equal to the current flowing in the same impedance when it is connected to a voltage generator whose voltage is the voltage at the open-circuited terminals in question and whose series impedance is the impedance of the network looking back from the terminals into the network, with all generators replaced by their internal impedances.

References

J. D. Irwin, *Basic Engineering Circuit Analysis*, 3rd ed., New York: Macmillan, 1989.

A. D. Kraus, *Circuit Analysis*, St. Paul: West Publishing, 1991.

J. W. Nilsson, *Electric Circuits*, 4th ed., Reading, Mass.: Addison-Wesley, 1992.

Further Information

Three texts listed in the References have achieved widespread usage and contain more details on the material contained in this section.

3.4 Power and Energy

Norman Balabanian and Theodore A. Bickart

The concept of the voltage v between two points was introduced in Section 3.1 as the energy w expended per unit charge in moving the charge between the two points. Coupled with the definition of current i as the time rate of charge motion and that of **power** p as the time rate of change of **energy**, this leads to the following fundamental relationship between the power delivered to a two-terminal electrical component and the voltage and current of that component, with standard references (meaning that the voltage reference plus is at the tail of the current reference arrow) as shown in Fig. 3.19:

$$p = vi \tag{3.1}$$

Assuming that the voltage and current are in volts and amperes, respectively, the power is in *watts*. This relationship applies to any two-terminal component or network, whether linear or nonlinear.

The power delivered to the basic linear resistive, inductive, and capacitive elements is obtained by inserting the v-i relationships into this expression. Then, using the relationship between power and energy (power as the time derivative of energy and energy, therefore, as the integral of power), the energy stored in the capacitor and inductor is also obtained:

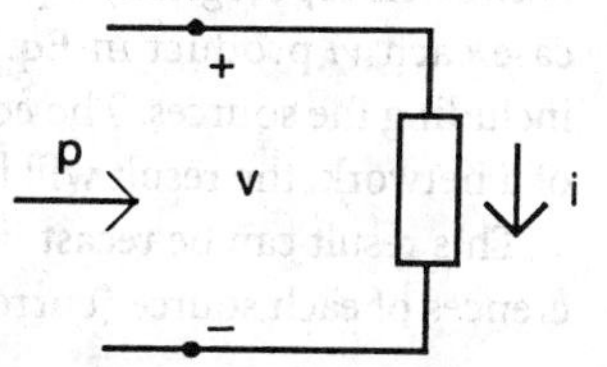

FIGURE 3.19 Power delivered to a circuit.

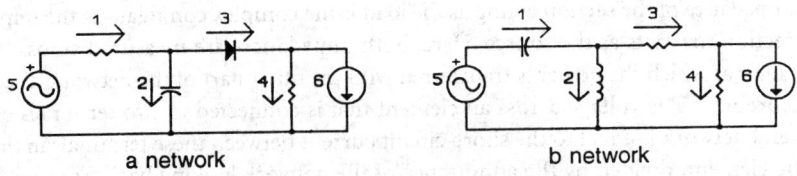

FIGURE 3.20 Topologically equivalent networks.

$$p_R = v_R i_R = Ri^2$$

$$p_C = v_C i_C = Cv_C \frac{dv_C}{dt} \qquad w_C(t) = \int_0^t Cv_C \frac{dv_C}{dt}\, dt = \frac{1}{2} Cv_C^2(t)$$

$$(3.2)$$

$$p_L = v_L i_L = Li_L \frac{di_L}{dt} \qquad w_L(t) = \int_0^t Li_L \frac{di_L}{dt}\, dt = \frac{1}{2} Li_L^2(t)$$

where the origin of time ($t = 0$) is chosen as the time when the capacitor voltage (respectively, the inductor current) is zero.

Tellegen's Theorem

A result that has far-reaching consequences in electrical engineering is **Tellegen's theorem**. It will be stated in terms of the networks shown in Fig. 3.20. These two are said to be topologically equivalent; that is, they are represented by the same graph but the components that constitute the branches of the graph are not necessarily the same in the two networks. They can even be nonlinear, as illustrated by the diode in one of the networks. Assuming all branches have standard references, including the source branches, Tellegen's theorem states that

$$\sum_{\text{all } j} v_{bj} i_{aj} = 0$$

$$\mathbf{v}_b' \mathbf{i}_a = 0$$

$$(3.3)$$

In the second line, the variables are vectors and the prime stands for the transpose. The a and b subscripts refer to the two networks.

This is an amazing result. It can be easily proved with the use of Kirchhoff's two laws.[1] The products of v and i are reminiscent of power as in Eq. (3.1). However, the product of the voltage of a branch in one network and the current of its topologically corresponding branch (which may not even be the same type of component) in another network does not constitute power in either branch. Furthermore, the variables in one network might be functions of time, while those of the other network might be steady-state phasors or Laplace transforms.

Nevertheless, some conclusions about power can be derived from Tellegen's theorem. Since a network is topologically equivalent to itself, the b network can be the same as the a network. In that case each vi product in Eq. (3.3) represents the power delivered to the corresponding branch, including the sources. The equation then says that if we add the power delivered to all the branches of a network, the result will be zero.

This result can be recast if the sources are separated from the other branches and one of the references of each source (current reference for each v-source and voltage reference for each i-source)

[1] See, for example, N. Balabanian and T. A. Bickart, *Linear Network Theory*, Matrix Publishers, Chesterland, Ohio, 1981, chap. 9.

is reversed. Then the *vi* product for each source, with new references, will enter Eq. (3.3) with a negative sign and will represent the power supplied by this source. When these terms are transposed to the right side of the equation, their signs are changed. The new equation will state in mathematical form that

> In any electrical network, the sum of the power supplied by the sources is equal to the sum of the power delivered to all the nonsource branches.

This is not very surprising since it is equivalent to the law of conservation of energy, a fundamental principle of science.

AC Steady-State Power

Let us now consider the **ac steady-state** case, where all voltages and currents are sinusoidal. Thus, in the two-terminal circuit of Fig. 3.19:

$$v(t) = \sqrt{2}\, |V| \cos(\omega t + \alpha) \leftrightarrow V = |V|\, e^{j\alpha}$$

$$i(t) = \sqrt{2}\, |I| \cos(\omega t + \beta) \leftrightarrow I = |I|\, e^{j\beta}$$

(3.4)

The capital *V* and *I* are phasors representing the voltage and current, and their magnitudes are the corresponding rms values. The power delivered to the network at any instant of time is given by:

$$p(t) = v(t)i(t) = 2\, |V|\,|I| \cos(\omega t + \alpha)\cos(\omega t + \beta)$$

$$= \Big[|V|\,|I| \cos(\alpha - \beta)\Big] + \Big[|V|\,|I| \cos(2\omega t + \alpha + \beta)\Big]$$

(3.5)

The last form is obtained by using trigonometric identities for the sum and difference of two angles. Whereas both the voltage and the current are sinusoidal, the instantaneous power contains a constant term (independent of time) in addition to a sinusoidal term. Furthermore, the frequency of the sinusoidal term is twice that of the voltage or current. Plots of *v*, *i*, and *p* are shown in Fig. 3.21 for specific values of α and β. The power is sometimes positive, sometimes negative. This means that power is sometimes delivered to the terminals and sometimes extracted from them.

The energy which is transmitted into the network over some interval of time is found by integrating the power over this interval. If the area under the positive part of the power curve were the same as the area under the negative part, the net energy transmitted over one cycle would be zero. For the values of α and β used in the figure, however, the positive area is greater, so there is a net

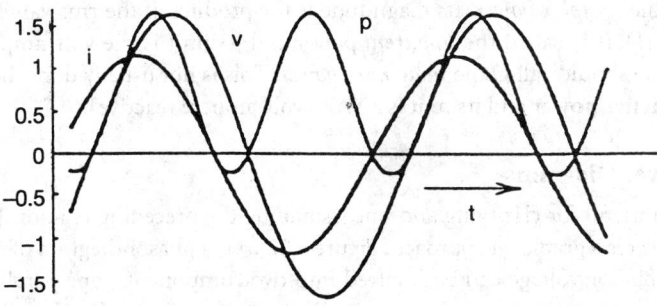

FIGURE 3.21 Instantaneous voltage, current, and power.

transmission of energy toward the network. The energy flows back from the network to the source over part of the cycle, but on the average, more energy flows toward the network than away from it.

In Terms of RMS Values and Phase Difference

Consider the question from another point of view. The preceding equation shows the power to consist of a constant term and a sinusoid. The average value of a sinusoid is zero, so this term will contribute nothing to the net energy transmitted. Only the constant term will contribute. This constant term is the average value of the power, as can be seen either from the preceding figure or by integrating the preceding equation over one cycle. Denoting the average power by P and letting $\theta = \alpha - \beta$, which is the angle of the network impedance, the average power becomes:

$$\begin{aligned}
P &= |V||I| \cos \theta \\
&= |V||I| \, \mathrm{Re}(e^{j\theta}) = \mathrm{Re}\big[|V||I| \, e^{j(\alpha - \beta)}\big] \\
&= \mathrm{Re} \,\big[(|V|e^{j\alpha}) \, (|I|^* e^{-j\beta})\big] \\
&= \mathrm{Re} \, (VI^*)
\end{aligned} \tag{3.6}$$

The third line is obtained by breaking up the exponential in the previous line by the law of exponents. The first factor between square brackets in this line is identified as the phasor voltage and the second factor as the conjugate of the phasor current. The last line then follows. It expresses the average power in terms of the voltage and current phasors and is sometimes more convenient to use.

Complex and Reactive Power

The average ac power is found to be the real part of a complex quantity VI^*, labeled S, that in rectangular form is

$$\begin{aligned}
S = VI^* &= |V||I|e^{j\theta} = |V||I| \cos \theta + j|V||I| \sin \theta \\
&= P + jQ
\end{aligned} \tag{3.7}$$

where

$$P = |V||I| \cos \theta \quad \text{(a)}$$
$$Q = |V||I| \sin \theta \quad \text{(b)} \tag{3.8}$$
$$|S| = |V||I| \quad\quad \text{(c)}$$

We already know P to be the average power. Since it is the real part of some complex quantity, it would be reasonable to call it the **real power**. The complex quantity S of which P is the real part is, therefore, called the *complex power*. Its magnitude is the product of the rms values of voltage and current: $|S| = |V||I|$. It is called the *apparent power* and its unit is the volt-ampere (VA). To be consistent, then, we should call Q the *imaginary power*. This is not usually done, however; instead, Q is called the **reactive power** and its unit is a VAR (volt-ampere reactive).

Phasor and Power Diagrams

An interpretation useful for clarifying and understanding the preceding relationships and for the calculation of power is a graphical approach. Figure 3.22(a) is a phasor diagram of V and I in a particular case. The phasor voltage can be resolved into two components, one parallel to the phasor current (or in phase with I) and another perpendicular to the current (or in quadrature with it). This is illustrated in Fig. 3.22(b). Hence, the average power P is the magnitude of phasor I multi-

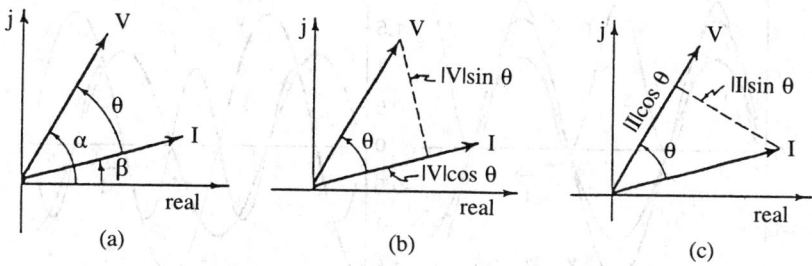

FIGURE 3.22   In-phase and quadrature components of V and I.

plied by the in-phase component of V; the reactive power Q is the magnitude of I multiplied by the quadrature component of V.

Alternatively, one can imagine resolving phasor I into two components, one in phase with V and one in quadrature with it, as illustrated in Fig. 3.22(c). Then P is the product of the magnitude of V with the in-phase component of I, and Q is the product of the magnitude of V with the quadrature component of I. Real power is produced only by the in-phase components of V and I. The quadrature components contribute only to the reactive power.

The in-phase or quadrature components of V and I do not depend on the specific values of the angles of each, but on their phase difference. One can imagine the two phasors in the preceding diagram to be rigidly held together and rotated around the origin by any angle. As long as the angle θ is held fixed, all of the discussion of this section will still apply. It is common to take the current phasor as the reference for angle; that is, to choose $\beta = 0$ so that phasor I lies along the real axis. Then $\theta = \alpha$.

Power Factor

For any given circuit it is useful to know what part of the total complex power is real (average) power and what part is reactive power. This is usually expressed in terms of the **power factor** F_p, defined as the ratio of real power to apparent power:

$$\text{Power factor} \doteq F_p = \frac{P}{|S|} = \frac{P}{|V||I|} \tag{3.9}$$

Not counting the right side, this is a general relationship, although we developed it here for sinusoidal excitations. With $P = |V||I| \cos \theta$, we find that the power factor is simply $\cos \theta$. Because of this, θ itself is called the power factor angle.

Since the cosine is an even function $[\cos(-\theta) = \cos \theta]$, specifying the power factor does not reveal the sign of θ. Remember that θ is the angle of the impedance. If θ is positive, this means that the current lags the voltage; we say that the power factor is a *lagging* power factor. On the other hand, if θ is negative, the current leads the voltage and we say this represents a *leading* power factor.

The power factor will reach its maximum value, unity, when the voltage and current are in phase. This will happen in a purely resistive circuit, of course. It will also happen in more general circuits for specific element values and a specific frequency.

We can now obtain a physical interpretation for the reactive power. When the power factor is unity, the voltage and current are in phase and $\sin \theta = 0$. Hence, the reactive power is zero. In this case, the instantaneous power is never negative. This case is illustrated by the current, voltage, and power waveforms in Fig. 3.23; the power curve never dips below the axis, and there is no exchange of energy between the source and the circuit. At the other extreme, when the power factor is zero, the voltage and current are 90° out of phase and $\sin \theta = 1$. Now the reactive power is a maximum and the average power is zero. In this case, the instantaneous power is positive over half a cycle (of

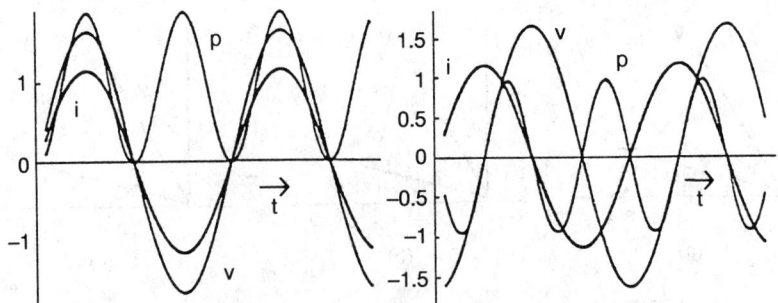

FIGURE 3.23 Power waveform for unity and zero power factors.

the voltage) and negative over the other half. All the energy delivered by the source over half a cycle is returned to the source by the circuit over the other half.

It is clear, then, that the reactive power is a measure of the exchange of energy between the source and the circuit without being used by the circuit. Although none of this exchanged energy is dissipated by or stored in the circuit, and it is returned unused to the source, nevertheless it is temporarily made available to the circuit by the source.[2]

Average Stored Energy

The average ac energy stored in an inductor or a capacitor can be established by using the expressions for the instantaneous stored energy for arbitrary time functions in Eq. (3.2), specifying the time function to be sinusoidal, and taking the average value of the result.

$$W_L = \frac{1}{2} L \left|I\right|^2 \qquad W_C = \frac{1}{2} C \left|V\right|^2 \qquad (3.10)$$

Application of Tellegen's Theorem to Complex Power

An example of two topologically equivalent networks was shown in Fig. 3.20. Let us now specify that two such networks are linear, all sources are same-frequency sinusoids, they are operating in the steady state, and all variables are phasors. Furthermore, suppose the two networks are the same, except that the sources of network *b* have phasors that are the complex conjugates of those of network *a*. Then, if **V** and **I** denote the vectors of branch voltages and currents of network *a*, Tellegen's theorem in Eq. (3.3) becomes:

$$\sum_{\text{all } j} V_j^* I_j = V^* I = 0 \qquad (3.11)$$

where V^* is the conjugate transpose of vector **V**.

This result states that the sum of the complex power delivered to all branches of a linear circuit operating in the ac steady state is zero. Alternatively stated, the total complex power delivered to a

---

[2] Power companies charge their industrial customers not only for the average power they use but for the reactive power they return. There is a reason for this. Suppose a given power system is to deliver a fixed amount of average power at a constant voltage amplitude. Since $P = \left|V\right|\left|I\right| \cos \theta$, the current will be inversely proportional to the power factor. If the reactive power is high, the power factor will be low and a high current will be required to deliver the given power. To carry a large current, the conductors carrying it to the customer must be correspondingly larger and better insulated, which means a larger capital investment in physical plant and facilities. It may be cost effective for customers to try to reduce the reactive power they require, even if they have to buy additional equipment to do so.

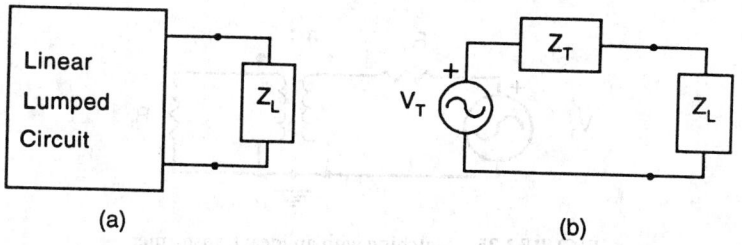

FIGURE 3.24   A linear circuit delivering power to a load in the steady state.

network by its sources equals the sum of the complex power delivered to its nonsource branches. Again, this result is not surprising. Since, if a complex quantity is zero, both the real and imaginary parts must be zero, the same result can be stated for the average power and for the reactive power.

Maximum Power Transfer

The diagram in Fig. 3.24 illustrates a two-terminal linear circuit at whose terminals an impedance Z_L is connected. The circuit is assumed to be operating in the ac steady state. The problem to be addressed is this: given the two-terminal circuit, how can the impedance connected to it be adjusted so that the maximum possible average power is transferred from the circuit to the impedance?

The first step is to replace the circuit by its Thévenin equivalent, as shown in Fig. 3.24(b). The current phasor in this circuit is $I = V_T/(Z_T + Z_L)$. The average power transferred by the circuit to the impedance is:

$$P = |I|^2 \, \text{Re}(Z_L) = \frac{|V_T|^2 \, \text{Re}(Z_L)}{|Z_T + Z_L|^2}$$

$$= \frac{|V_T|^2 \, R_L}{(R_T + R_L)^2 + (X_T + X_L)^2} \tag{3.12}$$

In this expression, only the load (that is, R_L and X_L) can be varied. The preceding equation, then, expresses a dependent variable (P) in terms of two independent ones (R_L and X_L).

What is required is to maximize P. For a function of more than one variable, this is done by setting the partial derivatives with respect to each of the independent variables equal to zero; that is, $\partial P/\partial R_L = 0$ and $\partial P/\partial X_L = 0$. Carrying out these differentiations leads to the result that maximum power will be transferred when the load impedance is the conjugate of the Thévenin impedance of the circuit: $Z_L = Z_T^*$. If the Thévenin impedance is purely resistive, then the load resistance must equal the Thévenin resistance.

In some cases, both the load impedance and the Thévenin impedance of the source may be fixed. In such a case, the matching for maximum power transfer can be achieved by using a transformer, as illustrated in Fig. 3.25, where the impedances are both resistive. The transformer is assumed to be ideal, with turns ratio n. Maximum power is transferred if $n^2 = R_T/R_L$.

Measuring AC Power and Energy

With ac steady-state average power given in the first line of Eq. (3.6), measuring the average power requires measuring the rms values of voltage and current, as well as the power factor. This is accomplished by the arrangement shown in Fig. 3.26, which includes a breakout of an electro-dynamometer-type wattmeter. The current in the high-resistance pivoted coil is proportional to

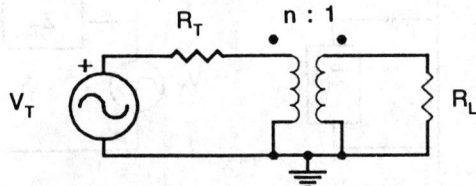

FIGURE 3.25   Matching with an ideal transformer.

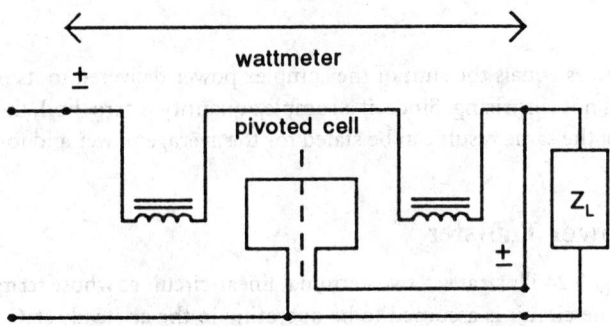

FIGURE 3.26   A wattmeter connected to a load.

the voltage across the load. The current to the load and the pivoted coil together through the ener-gizing coil of the electromagnet establishes a proportional magnetic field across the cylinder of rotation of the pivoted coil. The torque on the pivoted coil is proportional to the product of the magnetic field strength and the current in the pivoted coil. If the current in the pivoted coil is neg-ligible compared to that in the load, then the torque becomes essentially proportional to the prod-uct of the voltage across the load (equal to that across the pivoted coil) and the current in the load (essentially equal to that through the energizing coil of the electromagnet). The dynamics of the pivoted coil together with the restraining spring, at ac power frequencies, ensures that the angular displacement of the pivoted coil becomes proportional to the average of the torque or, equivalently, the average power.

One of the most ubiquitous of electrical instruments is the induction-type watthour meter, which measures the energy delivered to a load. Every customer of an electrical utility has one, for example. In this instance the pivoted coil is replaced by a rotating conducting (usually aluminum) disk as shown in Fig. 3.27. An induced eddy current in the disk replaces the pivoted coil current in

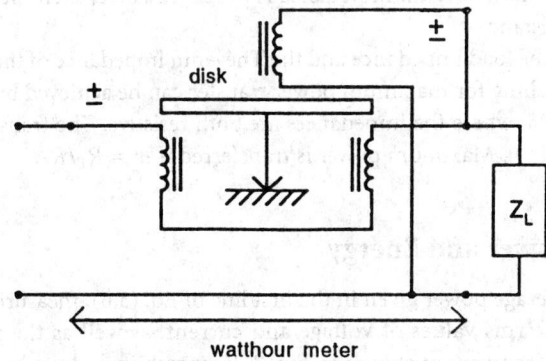

FIGURE 3.27   A watthour meter connected to a load.

interaction with the load-current-established magnetic field. After compensating for the less than ideal nature of the electrical elements making up the meter as just described, the result is that the disk rotates at a rate proportional to the average power to the load and the rotational count is proportional to the energy delivered to the load.

At frequencies above the ac power frequencies and, in some instances, at the ac power frequencies, electronic instruments are available to measure power and energy. They are not a cost-effective substitute for these meters in the monitoring of power and energy delivered to most of the millions upon millions of homes and small businesses.

Defining Terms

AC steady-state power: Consider an ac source connected at a pair of terminals to an otherwise isolated network. Let $\sqrt{2} \cdot |V|$ and $\sqrt{2} \cdot |I|$ denote the peak values, respectively, of the ac steady-state voltage and current at the terminals. Furthermore, let θ denote the phase angle by which the voltage leads the current. Then the average power delivered by the source to the network would be expressed as $P = |V| \cdot |I| \cos(\theta)$.

Power and energy: Consider an electrical source connected at a pair of terminals to an otherwise isolated network. Power, denoted by p, is the time rate of change in the energy delivered to the network by the source. This can be expressed as $p = vi$, where v, the voltage across the terminals, is the energy expended per unit charge in moving the charge between the pair of terminals and i, the current through the terminals, is the time rate of charge motion.

Power factor: Consider an ac source connected at a pair of terminals to an otherwise isolated network. The power factor, the ratio of the real power to the apparent power $|V| \cdot |I|$, is easily established to be $\cos(\theta)$, where θ is the power factor angle.

Reactive power: Consider an ac source connected at a pair of terminals to an otherwise isolated network. The reactive power is a measure of the energy exchanged between the source and the network without being dissipated in the network. The reactive power delivered would be expressed as $Q = |V| \cdot |I| \sin(\theta)$.

Real power: Consider an ac source connected at a pair of terminals to an otherwise isolated network. The real power, equal to the average power, is the power dissipated by the source in the network.

Tellegen's theorem: Two networks, here including all sources, are topologically equivalent if they are similar structurally, component by component. Tellegen's theorem states that the sum over all products of the product of the current of a component of one network, network a, and of the voltage of the corresponding component of the other network, network b, is zero. This would be expressed as $\sum_{\text{all } j} v_{bj} i_{aj} = 0$. From this general relationship it follows that in any electrical network, the sum of the power supplied by the sources is equal to the sum of the power delivered to all the nonsource components.

References

A. E. Fitzgerald, D. E. Higginbotham, and A. Grabel, *Basic Electrical Engineering*, 5th ed., New York: McGraw-Hill, 1981.

W. H. Hayt, Jr. and J. E. Kemmerly, *Engineering Circuit Analysis*, 4th ed., New York: McGraw-Hill, 1986.

J. D. Irwin, *Basic Engineering Circuit Analysis*, New York: Macmillan, 1987.

D. E. Johnson, J. L. Hilburn, and J. R. Johnson, *Basic Electric Circuit Analysis*, 2nd ed., Englewood Cliffs, N.J.: Prentice-Hall, 1984.

R. J. Smith, *Circuits, Devices, and Systems*, 2nd ed., New York: John Wiley, 1971.

T. N. Trick, *Introduction to Circuit Analysis*, New York: John Wiley, 1977.

3.5 Three-Phase Circuits

Norman Balabanian

Figure 3.28(a) represents the basic circuit for considering the flow of power from a single sinusoidal source to a load. The power can be thought to cross an imaginary boundary surface (represented by the dotted line in the figure) separating the source from the load. Suppose that:

$$v(t) = \sqrt{2}\,|V|\cos(\omega t + \alpha)$$

$$i(t) = \sqrt{2}\,|I|\cos(\omega t + \beta)$$

(3.13)

Then the power to the load at any instant of time is

$$p(t) = |V||I|\big[\cos(\alpha - \beta) + \cos(2\omega t + \alpha + \beta)\big]$$

(3.14)

The instantaneous power has a constant term and a sinusoidal term at twice the frequency. The quantity in brackets fluctuates between a minimum value of $\cos(\alpha - \beta) - 1$ and a maximum value of $\cos(\alpha - \beta) + 1$. This fluctuation of power delivered to the load has certain disadvantages in some situations where the transmission of power is the purpose of a system. An electric motor, for example, operates by receiving electric power and transmitting mechanical (rotational) power at its shaft. If the electric power is delivered to the motor in spurts, the motor is likely to vibrate. In order to run satisfactorily, a physically larger motor will be needed, with a larger shaft and flywheel, to provide inertia than would be the case if the delivered power were constant.

This problem is overcome in practice by the use of what is called a *three-phase* system. This section will provide a brief discussion of three-phase systems.

Consider the circuit in Fig 3.28(b). This arrangement is similar to a combination of three of the simple circuits in Fig 3.28(a) connected in such a way that each one shares the return connection from O to N. The three sources can be viewed collectively as a single source and the three loads—which are assumed to be identical—can be viewed collectively as a single load. Then, as before, the dotted line represents a surface separating the source from the load. Each of the individual sources and loads is referred to as one *phase* of the three-phase system.

The three sources are assumed to have the same frequency; they are said to be *synchronized*. It is also assumed that the three voltages have the same rms values and the phase difference between each pair of voltages is $\pm120°$ ($2\pi/3$ rad). Thus, they can be written:

$$v_a = \sqrt{2}\,|V|\cos(\omega t + \alpha_1) \leftrightarrow V_a = |V|\,e^{j0°}$$

$$v_b = \sqrt{2}\,|V|\cos(\omega t + \alpha_2) \leftrightarrow V_b = |V|\,e^{-j120°}$$

$$v_c = \sqrt{2}\,|V|\cos(\omega t + \alpha_3) \leftrightarrow V_c = |V|\,e^{j120°}$$

(3.15)

The **phasors** representing the sinusoids have also been shown. For convenience, the angle of v_a has been chosen as the reference for angles; v_b lags v_a by 120° and v_c leads v_a by 120°.

Because the loads are identical, the rms values of the three currents shown in the figure will also be the same and the phase difference between each pair of them will be $\pm120°$. Thus, the currents can be written:

$$i_1 = \sqrt{2}\,|I|\cos(\omega t + \beta_1) \leftrightarrow I_1 = |I|\,e^{j\beta_1}$$

$$i_2 = \sqrt{2}\,|I|\cos(\omega t + \beta_2) \leftrightarrow I_2 = |I|\,e^{j(\beta_1 - 120°)}$$

$$i_3 = \sqrt{2}\,|I|\cos(\omega t + \beta_3) \leftrightarrow I_3 = |I|\,e^{j(\beta_1 + 120°)}$$

(3.16)

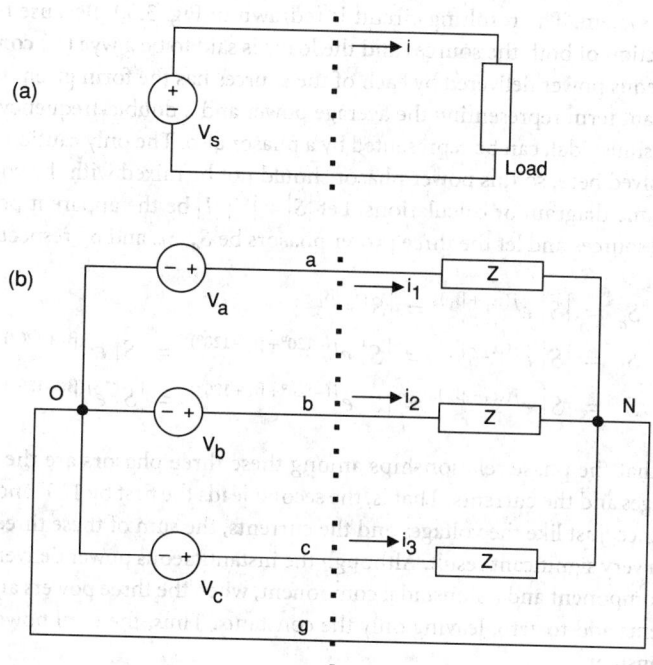

FIGURE 3.28   Flow of power from source to load.

Perhaps a better form for visualizing the voltages and currents is a graphical one. Phasor diagrams for the voltages separately and the currents separately are shown in Fig. 3.29. The value of angle β_1 will depend on the load. An interesting result is clear from these diagrams. First, V_2 and V_3 are each other's conjugates. So if we add them, the imaginary parts cancel and the sum will be real, as illustrated by the construction in the voltage diagram. Furthermore, the construction shows that this real part is negative and equal in size to V_1. Hence, the sum of the three voltages is zero. The same is true of the sum of the three currents, as can be established graphically by a similar construction.

By Kirchhoff's current law applied at node N in Fig. 3.28(b), we find that the current in the return line is the sum of the three currents in Eq. (3.16). However, since this sum was found to be zero, the return line carries no current. Hence it can be removed entirely without affecting the

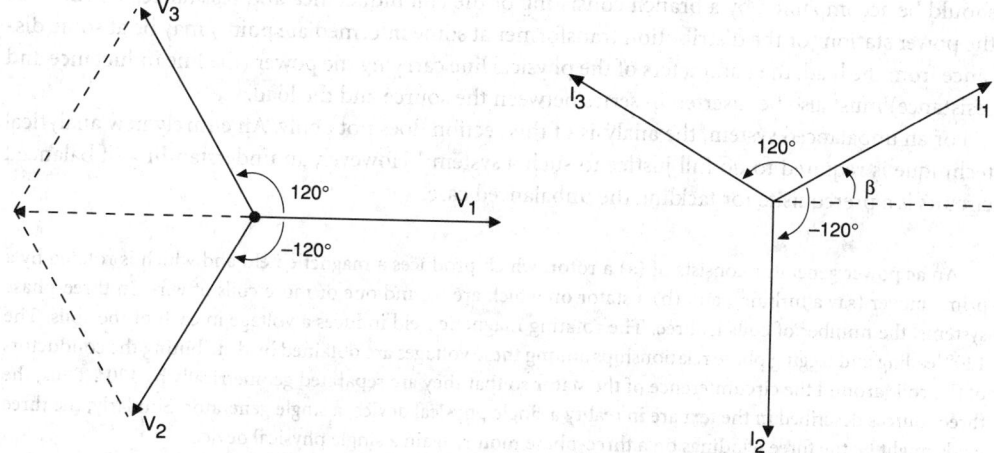

FIGURE 3.29   Voltage and current phasor diagrams.

operation of the system. The resulting circuit is redrawn in Fig. 3.30. Because of its geometrical form, this connection of both the sources and the loads is said to be a **wye (Y) connection**.

The instantaneous power delivered by each of the sources has the form given in Eq. (3.14), consisting of a constant term representing the average power and a double-frequency sinusoidal term. The latter, being sinusoidal, can be represented by a phasor also. The only caution is that a different frequency is involved here, so this power phasor should not be mixed with the voltage and current phasors in the same diagram or calculations. Let $|S| = |V| |I|$ be the apparent power delivered by each of the three sources and let the three power phasors be S_a, S_b, and S_c, respectively. Then:

$$
\begin{aligned}
S_a &= |S|\, e^{j(\alpha_1 + \beta_1)} = |S|\, e^{j\beta_1} \\
S_b &= |S|\, e^{j(\alpha_2 + \beta_2)} = |S|\, e^{j(-120° + \beta_1 - 120°)} = |S|\, e^{j(\beta_1 + 120°)} \\
S_c &= |S|\, e^{j(\alpha_3 + \beta_3)} = |S|\, e^{j(-120° + \beta_1 + 120°)} = |S|\, e^{j(\beta_1 - 120°)}
\end{aligned}
\tag{3.17}
$$

It is evident that the phase relationships among these three phasors are the same as the ones among the voltages and the currents. That is, the second leads the first by 120° and the third lags the first by 120°. Hence, just like the voltages and the currents, the sum of these three phasors will also be zero. This is a very significant result. Although the instantaneous power delivered by each source has a constant component and a sinusoidal component, when the three powers are added, the sinusoidal components add to zero, leaving only the constants. Thus, the total power delivered to the three loads is constant.

To determine the value of this constant power, use Eq. (3.14) as a model. The contribution of the kth source to the total (constant) power is $|S| \cos(\alpha_k - \beta_k)$. One can easily verify that $\alpha_k - \beta_k = \alpha_1 - \beta_1 = -\beta_1$. The first equality follows from the relationships among the α's from Eq. (3.15) and among the β's from Eq. (3.16). The choice of $\alpha_1 = 0$ leads to the last equality. Hence, the constant terms contributed to the power by each source are the same. If P is the total average power, then:

$$
P = P_a + P_b + P_c = 3P_a = 3\,|V||I|\cos(\alpha_1 - \beta_1)
\tag{3.18}
$$

Although the angle α_1 has been set equal to zero, for the sake of generality we have shown it explicitly in this equation.

What has just been described is a *balanced* three-phase three-wire power system. The three sources in practice are not three independent sources but consist of three different parts of the same generator. The same is true of the loads.[3] What has been described is ideal in a number of ways. First, the circuit can be *unbalanced*—for example, by the loads being somewhat unequal. Second, since the real devices whose ideal model is a voltage source are coils of wire, each source should be accompanied by a branch consisting of the coil inductance and resistance. Third, since the power station (or the distribution transformer at some intermediate point) may be at some distance from the load, the parameters of the physical line carrying the power (the line inductance and resistance) must also be inserted in series between the source and the load.

For an unbalanced system, the analysis of this section does not apply. An entirely new analytical technique is required to do full justice to such a system.[4] However, an understanding of balanced circuits is a prerequisite for tackling the unbalanced case.

---

[3] An ac power generator consists of (a) a rotor, which produces a magnetic field and which is rotated by a prime mover (say a turbine), and (b) a stator on which are wound one or more coils of wire. In three-phase systems, the number of coils is three. The rotating magnetic field induces a voltage in each of the coils. The 120° leading and lagging phase relationships among these voltages are obtained by distributing the conductors of the coils around the circumference of the stator so that they are separated geometrically by 120°. Thus, the three sources described in the text are in reality a single physical device, a single generator. Similarly, the three loads might be the three windings on a three-phase motor, again a single physical device.

[4] The technique for analyzing unbalanced circuits utilizes what are called *symmetrical components*.

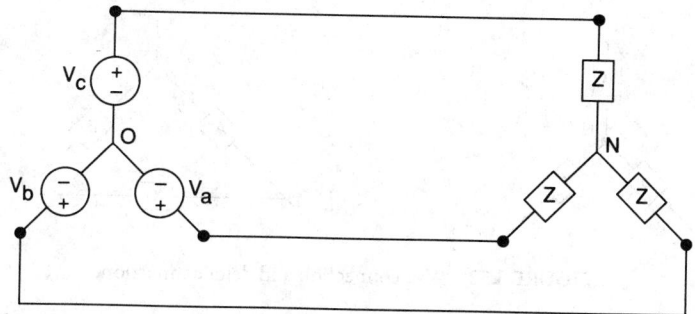

FIGURE 3.30 Wye-connected three-phase system.

The last two of the conditions that make the circuit less than ideal (line and source impedances) introduce algebraic complications, but nothing fundamental is changed in the preceding theory. If these two conditions are taken into account, the appropriate circuit takes the form shown in Fig. 3.31. Here the internal impedance of a source and the line impedance connecting that source to its load are both connected in series with the corresponding load. Thus, instead of the impedance in each phase being Z, it is $Z + Z_w + Z_l$, where w and l are subscripts standing for "winding" and "line," respectively. Hence, the rms value of each current is

$$|I| = \frac{|V|}{|Z + Z_w + Z_l|} \tag{3.19}$$

instead of $|V|/|Z|$. All other results we had arrived at remain unchanged, namely, that the sum of the phase currents is zero and that the sum of the phase powers is a constant. The detailed calculations simply become a little more complicated.

One other point, illustrated for the loads in Fig. 3.32, should be mentioned. Given wye-connected sources or loads, the wye and the **delta** can be made equivalent by proper selection of the arms of the delta. Thus, either the sources in Fig. 3.30 or the loads, or both, can be replaced by a delta equivalent; thus we can conceive of four different three-phase circuits: wye-wye, delta-wye, wye-delta, and delta-delta. Not only can we conceive of them, they are extensively used in practice.

It is not worthwhile to carry out detailed calculations for these four cases. Once the basic properties described here are understood, one should be able to make the calculations. Observe, how-

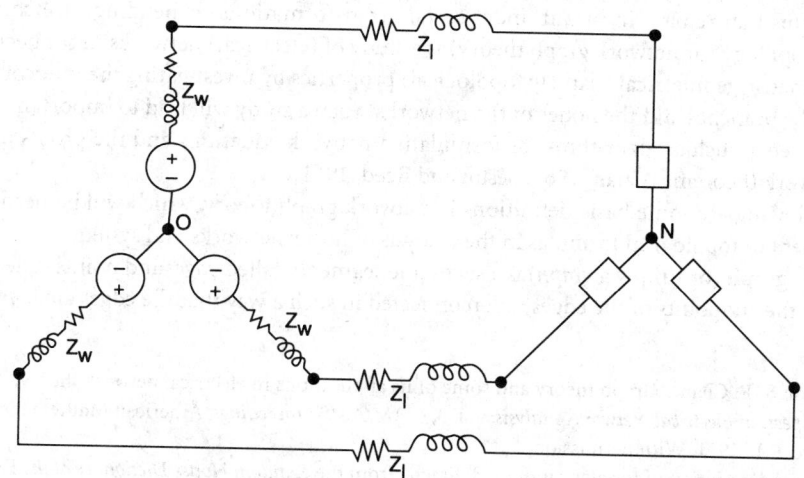

FIGURE 3.31 Three-phase circuit with nonzero winding and line impedances.

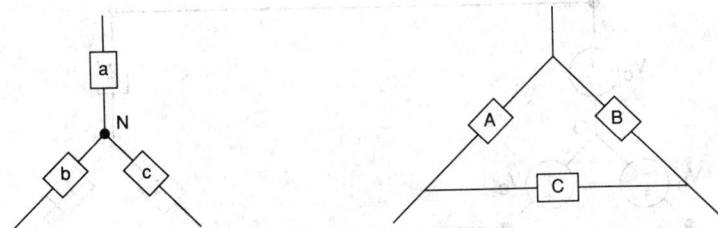

FIGURE 3.32   Wye connection and delta connection.

ever, that in the delta structure, there is no neutral connection, so the phase voltages cannot be measured. The only voltages that can be measured are the *line-to-line* or simply the *line* voltages. These are the differences of the phase voltages taken in pairs, as is evident from Fig. 3.31.

Defining Terms

Delta connection: The sources or loads in a three-phase system connected end-to-end, forming a closed path, like the Greek letter Δ.

Phasor: A complex number representing a sinusoid; its magnitude and angle are the rms value and phase of the sinusoid, respectively.

Wye connection: The three sources or loads in a three-phase system connected to have one common point, like the letter Y.

References

V. del Toro, *Electric Power Systems*, Englewood Cliffs, N.J.: Prentice-Hall, 1992.

B.R. Gungor, *Power Systems*, San Diego: Harcourt Brace Jovanovich, 1988.

P.Z. Peebles and T.A. Giuma, *Principles of Electrical Engineering*, New York: McGraw-Hill, 1991.

3.6 Graph Theory[5]

Shu-Park Chan

Topology is a branch of mathematics; it may be described as "the study of those properties of geometric forms that remain invariant under certain transformations, as bending, stretching, etc."[6] Network topology (or network graph theory) is a study of (electrical) networks in connection with their nonmetric geometrical (namely topological) properties by investigating the interconnections between the branches and the nodes of the networks. Such a study will lead to important results in network theory such as algorithms for formulating network equations and the proofs of various basic network theorems [Chan, 1969; Seshu and Reed, 1961].

The following are some basic definitions in network graph theory, which will be needed in the development of topological formulas in the analysis of linear networks and systems.

A **linear graph** (or simply a *graph*) is a set of line segments called *edges* and points called *vertices*, which are the endpoints of the edges, interconnected in such a way that the edges are connected to

---

[5]Based on S.-P. Chan, "Graph theory and some of its applications in electrical network theory," in *Mathematical Aspects of Electrical Network Analysis*, vol. 3, *SIAM/AMS Proceedings*, American Mathematical Society, Providence, R.I., 1971. With permission.

[6]This brief description of topology is quoted directly from the *Random House Dictionary of the English Language*, Random House, New York, 1967.

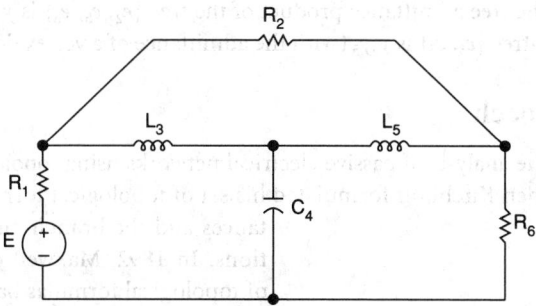

FIGURE 3.33 A passive network *N* with a voltage driver *E*.

(or *incident* with) the vertices. The *degree* of a vertex of a graph is the number of edges incident with that vertex.

A subset G_i of the edges of a given graph *G* is called a **subgraph** of *G*. If G_i does not contain all of the edges of *G*, it is a **proper subgraph** of *G*. A **path** is a subgraph having all vertices of degree 2 except for the two endpoints, which are of degree 1 and are called the *terminals* of the path. The set of all edges in a path constitutes a **path-set**. If the two terminals of a path coincide, the path is a closed path and is called a **circuit** (or **loop**). The set of all edges contained in a circuit is called a **circuit-set** (or **loop-set**).

A graph or subgraph is said to be **connected** if there is at least one path between *every* pair of its vertices. A **tree** of a connected graph *G* is a connected subgraph which contains all the vertices of *G* but no circuits. The edges contained in a tree are called the **branches of the tree**. A 2-tree of a connected graph *G* is a (proper) subgraph of *G* consisting of two unconnected circuitless subgraphs, each subgraph itself being connected, which together contain all the vertices of *G*. Similarly, a *k*-**tree** is a subgraph of *k* unconnected circuitless subgraphs, each subgraph being connected, which together include all the vertices of *G*. The *k*-**tree admittance product of a** *k*-tree is the product of the admittances of all the branches of the *k*-tree.

Example 3.5. The graph *G* shown in Fig. 3.34 is the graph of the network *N* of Fig. 3.33. The edges of *G* are e_1, e_2, e_4, e_5, and e_6; the vertices of *G* are V_1, V_2, V_3, and V_4. A path of *G* is the subgraph G_1 consisting of edges e_2, e_3, and e_6 with vertices V_2 and V_4 as terminals. Thus, the set $\{e_2, e_3, e_6\}$ is a path-set. With edge e_4 added to G_1, we form another subgraph G_2, which is a circuit since as far as G_2 is concerned all its vertices are of degree 2. Hence the set $\{e_2, e_3, e_4, e_6\}$ is a circuit-set. Obviously, *G* is a connected graph since there exists a path between every pair of vertices of *G*. A tree of *G* may be the subgraph consisting of edges e_1, e_4, and e_6. Two other trees of *G* are $\{e_2, e_5, e_6\}$ and $\{e_3, e_4, e_5\}$. A 2-tree of *G* is $\{e_2, e_4\}$; another one is $\{e_3, e_6\}$; and still another one is $\{e_3, e_5\}$. Note that both $\{e_2, e_4\}$ and $\{e_3, e_6\}$ are subgraphs which obviously satisfy the definition of a 2-tree in the sense that each contains two disjoint circuitless connected subgraphs, both of which include all the four vertices of *G*. Thus, $\{e_3, e_5\}$ does not seem to be a 2-tree. However, if we agree to consider $\{e_3, e_5\}$ as a subgraph which contains edges e_3 and e_5 plus the isolated vertex V_4, we see that $\{e_3, e_5\}$ will satisfy the definition of a 2-tree since it now has two circuitless connected subgraphs with e_3 and e_5 forming one of them and the vertex V_4 alone forming the other. Moreover, both subgraphs together indeed include all the four vertices of *G*. It is worth noting that a 2-tree is obtained from a tree by removing *any one* of the branches from the tree; in general, a *k*-tree is obtained from a $(k-1)$-tree by removing from it any one of

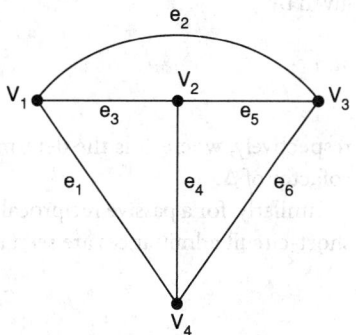

FIGURE 3.34 The graph *G* of the network *N* of Fig. 3.33.

its branches. Finally, the tree admittance product of the tree $\{e_2, e_5, e_6\}$ is $y_2 y_5 y_6$; the 2-tree admittance product of the 2-tree $\{e_3, e_5\}$ is $y_3 y_5$ (with the admittance of a vertex defined to be 1).

The k-Tree Approach

The development of the analysis of passive electrical networks using topological concepts may be dated back to 1847 when Kirchhoff formulated his set of topological formulas in terms of resistances and the branch-current system of equations. In 1892, Maxwell developed another set of topological formulas based on the k-tree concept, which are the duals of Kirchhoff's. These two sets of formulas were supported mainly by heuristic reasoning and no formal proofs were then available.

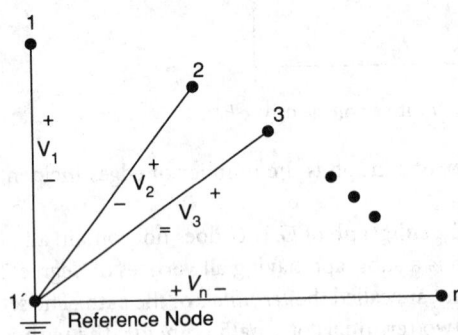

FIGURE 3.35 A network N with n independent nodes.

In the following we shall discuss only Maxwell's topological formulas for linear networks without mutual inductances.

Consider a network N with n independent nodes as shown in Fig. 3.35. The node $1'$ is taken as the reference (datum) node. The voltages V_1, $V_2, \ldots, V_n$ (which are functions of s) are the transforms of the node-pair voltages (or simply node voltages) $v_1, v_2, \ldots, v_n$ (which are functions of t) between the n nodes and the reference node $1'$ with the plus polarity marks at the n nodes. It can be shown [Aitken, 1956] that the matrix equation for the n independent nodes of N is given by

$$\begin{bmatrix} y_{11} & y_{12} & \cdots & y_{1n} \\ y_{21} & y_{22} & \cdots & y_{2n} \\ \vdots & \vdots & \vdots & \vdots \\ y_{n1} & y_{n2} & \cdots & y_{nn} \end{bmatrix} \begin{bmatrix} V_1 \\ V_2 \\ \vdots \\ V_n \end{bmatrix} = \begin{bmatrix} I_1 \\ I_2 \\ \vdots \\ I_n \end{bmatrix} \tag{3.20}$$

or, in abbreviated matrix notation,

$$Y_n V_n = I_n \tag{3.21}$$

where Y_n is the node admittance matrix, V_n the $n \times 1$ matrix of the node voltage transforms, and I_n the $n \times 1$ matrix of the transforms of the known current sources.

For a relaxed passive one-port (with zero initial conditions) shown in Fig. 3.36, the driving-point impedance function $Z_d(s)$ and its reciprocal, namely driving-point admittance function $Y_d(s)$, are given by

$$Z_d(s) = V_1 / I_1 = \Delta_{11} / \Delta$$

and

$$Y_d(s) = 1 / Z_d(s) = \Delta / \Delta_{11}$$

respectively, where Δ is the determinant of the node admittance matrix Y_n, and Δ_{11} is the $(1,1)$-cofactor of Δ.

Similarly, for a passive reciprocal RLC two-port (Fig. 3.37), the open-circuit impedances and the short-circuit admittances are seen to be

$$z_{11} = \Delta_{11} / \Delta \tag{3.22a}$$

$$z_{12} = z_{21} = (\Delta_{12} - \Delta_{12'}) / \Delta \tag{3.22b}$$

$$z_{22} = (\Delta_{22} + \Delta_{2'2'} - 2\Delta_{22'}) / \Delta \tag{3.22c}$$

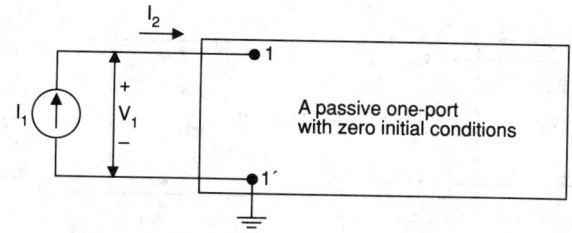

FIGURE 3.36 The network N driven by a single current source.

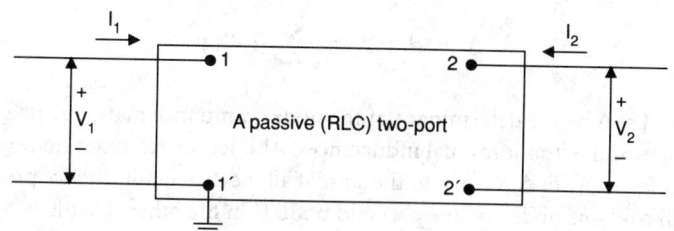

FIGURE 3.37 A passive two-port.

and

$$y_{11} = (\Delta_{22} + \Delta_{2'2'} - 2\Delta_{22'})/(\Delta_{1122} + \Delta_{112'2'} - 2\Delta_{1122'}) \tag{3.23a}$$

$$y_{12} = y_{21} = \Delta_{12'} - \Delta_{12}/(\Delta_{1122} + \Delta_{112'2'} - 2\Delta_{1122'}) \tag{3.23b}$$

$$y_{22} = \Delta_{11}/(\Delta_{1122} + \Delta_{112'2'} - 2\Delta_{1122'}) \tag{3.23c}$$

respectively, where Δ_{ij} is the (i,j)-cofactor of Δ, and Δ_{ijkm} is the cofactor of Δ by deleting rows i and k and columns j and m from Δ [Aitken, 1956].

Expressions in terms of network determinants and cofactors for other network transfer functions are given by (Fig. 3.38):

$$z_{12} = \frac{V_2}{I_1} = \frac{\Delta_{12} - \Delta_{12'}}{\Delta} \qquad \text{(transfer impedance function)} \tag{3.24a}$$

$$G_{12} = \frac{V_2}{V_1} = \frac{\Delta_{12} - \Delta_{12'}}{\Delta_{11}} \qquad \text{(voltage-ratio transfer function)} \tag{3.24b}$$

$$Y_{12} = Y_L G_{12} = Y_L\left(\frac{\Delta_{12} - \Delta_{12'}}{\Delta_{11}}\right) \qquad \text{(transfer admittance function)} \tag{3.24c}$$

$$\alpha_{12} = Y_1 Z_{12} = Y_L\left(\frac{\Delta_{12} - \Delta_{12'}}{\Delta}\right) \qquad \text{(current-ratio transfer function)} \tag{3.24d}$$

The topological formulas for the various network functions of a passive one-port or two-port are derived from the following theorems which are stated without proof [Chan, 1969].

Theorem 3.1.   Let N be a passive network without mutual inductances. The determinant Δ of the node admittance matrix Y_n is equal to the sum of all tree-admittances of N, where a tree-admittance product $T^{(i)}(y)$ is defined to be the product of the admittance of all the branches of the tree $T^{(i)}$. That is,

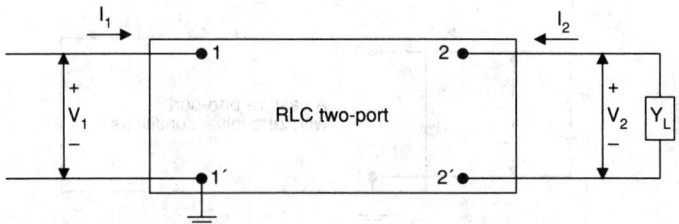

FIGURE 3.38 A loaded passive two-port.

$$\Delta = \det Y_n = \sum_i T^{(i)}(y) \tag{3.25}$$

Theorem 3.2. Let Δ be the determinant of the node admittance matrix Y_n of a passive network N with $n + 1$ nodes and without mutual inductances. Also let the reference node be denoted by $1'$. Then the (j,j)-cofactor Δ_{jj} of Δ is equal to the sum of all the 2-tree-admittance products $T_{2_{j,1'}}(y)$ of N, each of which contains node j in one part and node $1'$ in the other. That is

$$\Delta_{jj} = \sum_k T^{(k)}_{2_{j,1'}}(y) \tag{3.26}$$

where the summation is taken over all the 2-tree-admittance products of the form $T_{2_{j,1'}}(y)$.

Theorem 3.3. The (i,j)-cofactor Δ_{ij} of Δ of a relaxed passive network N with n independent nodes (with node $1'$ as the reference node) and without mutual inductances is given by

$$\Delta_{ij} = \sum_k T^{(k)}_{2_{ij,1'}}(y) \tag{3.27}$$

where the summation is taken over all the 2-tree-admittance products of the form $T_{2_{ij,1'}}(y)$ with each containing nodes i and j in one connected port and the reference node $1'$ in the other.

For example, the topological formulas for the driving-point function of a passive one-port can be readily obtained from Eqs. (3.25) and (3.26) in Theorems 3.1 and 3.2 as stated in the next theorem.

Theorem 3.4. With the same notation as in Theorems 3.1 and 3.2, the driving-point admittance $Y_d(s)$ and the driving-point impedance $Z_d(s)$ of a passive one-port containing no mutual inductances at terminals 1 and $1'$ are given by

$$Y_d(s) = \frac{\Delta}{\Delta_{11}} = \frac{\sum_i T^{(i)}(y)}{\sum_k T^{(k)}_{2_{1,1}}(y)} \quad \text{and} \quad Z_d(s) = \frac{\Delta_{11}}{\Delta} = \frac{\sum_k T^{(k)}_{2_{1,1}}(y)}{\sum_i T^{(i)}(y)} \tag{3.28}$$

respectively.

For convenience we define the following shorthand notation:

(a) $V(Y) \equiv \sum_i T^{(i)}(y) =$ sum of all tree-admittance products

(b) $W_{j,r}(y) \equiv \sum_k T_{2_{j,r}}(y) =$ sum of all 2-tree-admittance products with $\qquad$ (3.29)
node j and the reference node r contained in different parts

Thus Eq. (3.28) may be written as

$$Y_d(s) = V(Y)/W_{1,1'}(Y) \quad \text{and} \quad Z_d(s) = W_{1,1'}(Y)/V(Y) \tag{3.30}$$

In a two-port network N, there are four nodes to be specified, namely, 1 and $1'$ at the input port $(1,1')$ and nodes 2 and $2'$ at the output port $(2,2')$, as illustrated in Fig. 3.38. However, for a 2-tree of the type $T_{2_{ij,1'}}$, only three nodes have been used, thus leaving the fourth one unidentified.

With very little effort, it can be shown that, in general, the following relationship holds:

$$W_{ij,1'}(Y) = W_{ijk,1'}(Y) + W_{ij,k1'}(Y)$$

or simply

$$W_{ij,1'} = W_{ijk,1'} + W_{ij,k1'} \tag{3.31}$$

where i, j, k, and $1'$ are the four terminals of N with $1'$ denoting the datum (reference) node. The symbol $W_{ijk,1'}$ denotes the sum of all the 2-tree-admittance products, each containing nodes i, j, and k in one connected part and the reference node, $1'$, in the other.

We now state the next theorem.

Theorem 3.5. With the same hypothesis and notation as stated earlier in this section,

$$\Delta_{12} - \Delta_{12'} = W_{12,12'}(Y) - W_{12',1'2}(Y) \tag{3.32}$$

It is interesting to note that Eq. (3.32) is stated by Percival [1953] in the following descriptive fashion:

$$\Delta_{12} - \Delta_{12'} = W_{12,12'} - W_{12',1'2} = \begin{pmatrix} 1 \circ\!\!-\!\!\circ 2 \\ \\ 1'\circ\!\!-\!\!\circ 2' \end{pmatrix} - \begin{pmatrix} 1\circ \quad \circ 2 \\ \times \\ 1'\circ \quad \circ 2' \end{pmatrix}$$

which illustrates the two types of 2-trees involved in the formula. Hence, we state the topological formulas for z_{11}, z_{12}, and z_{22} in the following theorem.

Theorem 3.6. With the same hypothesis and notation as stated earlier in this section

$$z_{11} = W_{1,1'}(Y)/V(Y) \tag{3.33a}$$

$$z_{12} = z_{21} = \{W_{12,1'2'}(Y) - W_{12',1'2}(Y)\}/V(Y) \tag{3.33b}$$

$$z_{22} = W_{2,2'}(Y)/V(Y) \tag{3.33c}$$

We shall now develop the topological expressions for the short-circuit admittance functions. Let us denote by $U_{a,b,c}(Y)$ the sum of all 3-tree admittance products of the form $T_{3_{a,b,c}}(Y)$ with identical subscripts in both symbols to represent the same specified distribution of vertices. Then, following arguments similar to those of Theorem 3.5, we readily see that

$$\Delta_{1122} = \sum_i T_{3_{1,2,1'}}^{(i)}(y) \equiv U_{1,2,1'}(Y) \tag{3.34a}$$

$$\Delta_{112'2'} = \sum_j T_{3_{1,2',1'}}^{(j)}(y) \equiv U_{1,2',1'}(Y) \tag{3.34b}$$

$$\Delta_{1122'} = \sum_k T_{3_{1,22',1'}}^{(k)}(y) \equiv U_{1,22',1'}(U) \tag{3.34c}$$

where $1,1',2,2'$ are the four terminals of the two-port with $1'$ denoting the reference node (Fig. 3.39). However, we note that in Eqs. (3.34a) and (3.34b) only three of the four terminals have been specified. We can therefore further expand $U_{1,2,1'}$ and $U_{1,2',1'}$ to obtain the following:

$$\Delta_{1122} + \Delta_{112'2'} - 2\Delta_{1122'} = U_{12',2,1'} + U_{1,2,1'2'} + U_{12,2',1'} + U_{1,2',1'2} \tag{3.35}$$

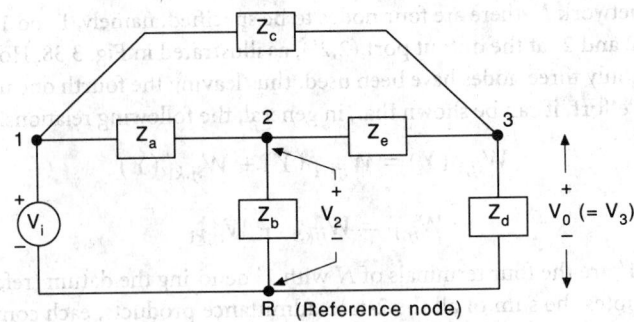

FIGURE 3.39   The network *N* of Example 3.7.

For convenience, we shall use the shorthand notation ΣU to denote the sum of the right of Eq. (3.35). Thus, we define

$$\sum U = U_{12',2,1'} + U_{1,2,1'2'} + U_{12,2',1'} + U_{1,2',1'2} \tag{3.36}$$

Hence, we obtain the topological formulas for the short-circuit admittances as stated in the following theorem.

Theorem 3.7.   The short-circuit admittance functions y_{11}, y_{12}, and y_{22} of a passive two-port network with no mutual inductances are given by

$$y_{11} = W_{2,2'} / \sum U \tag{3.37a}$$

$$y_{12} = y_{21} = (W_{12',1'2} - W_{12,1'2'}) / \sum U \tag{3.37b}$$

$$y_{22} = W_{1,1'} / \sum U \tag{3.37c}$$

where ΣU is defined in Eq. (3.36) above.

Finally, following similar developments, other network functions are stated in Theorem 3.8.

Theorem 3.8.   With the same notation as before,

$$Z_{12}(s) = \frac{W_{12,1'2'} - W_{12',1'2}}{V} \tag{3.38a}$$

$$G_{12}(s) = \frac{W_{12,1'2'} - W_{12',1'2}}{W_{1,1'}} \tag{3.38b}$$

$$Y_{12}(s) = Y_L \frac{W_{12,1'2'} - W_{12',1'2}}{W_{1,1'}} \tag{3.38c}$$

$$\alpha_{12}(s) = Y_L \frac{W_{12,1'2'} - W_{12',1'2}}{V} \tag{3.38d}$$

The Flowgraph Approach

Mathematically speaking, a linear electrical network or, more generally, a linear system can be described by a set of simultaneous linear equations. Solutions to these equations can be obtained either by the method of successive substitutions (elimination theory), by the method of determinants (Cramer's rule), or by any of the topological techniques such as Maxwell's k-tree approach discussed in the preceding subsection and the flowgraph techniques represented by the works of Mason [1953, 1956], and Coates [1959].

Although the methods using algebraic manipulations can be amended and executed by a computer, they do not reveal the physical situations existing in the system. The flowgraph techniques, on the other hand, show intuitively the causal relationships between the variables of the system of interest and hence enable the network analyst to have an excellent physical insight into the problem.

In the following, two of the more well-known flowgraph techniques are discussed, namely, the **signal-flowgraph** technique devised by Mason and the method based on the flowgraph of Coates and recently modified by Chan and Bapna [1967].

A *signal-flowgraph* G_m of a system S of n independent linear (algrebraic) equations in n unknowns

$$\sum_{j=1}^{n} a_{ij}x_j = b_i \qquad i = 1, 2, \ldots, n \qquad (3.39)$$

is a graph with junction points called *nodes* which are connected by directed line segments called *branches* with signals traveling along the branches only in the direction described by the arrows of the branches. A signal x_k traveling along a branch between x_k and x_j is multiplied by the gain of the branches g_{kj}, so that a signal of $g_{kj}x_k$ is delivered at node x_j. An *input node* (*source*) is a node which contains only outgoing branches; an *output node* (*sink*) is a node which has only incoming branches. A *path* is a continuous unidirectional succession of branches, all of which are traveling in the same direction; a *forward path* is a path from the input node to the output node along which all nodes are encountered exactly once; and a *feedback path* (*loop*) is a closed path which originates from and terminates at the same node, and along which all other nodes are encountered exactly once (the trivial case is a *self-loop* which contains exactly one node and one branch). A *path gain* is the product of all the branch-gains of the path; similarly, a loop gain is the product of all the branch gains of the branches in a loop.

The procedure for obtaining the Mason graph from a system of linear algebraic equations may be described in the following steps:

a. Arrange all the equations of the system in such a way that the jth dependent (output) variable x_j in the jth equation is expressed explicitly in terms of the other variables. Thus, if the system under study is given by Eq. (3.39), namely

$$a_{11}x_1 + a_{12}x_2 + \cdots + a_{1n}x_n = b_1$$
$$a_{21}x_1 + a_{22}x_2 + \cdots + a_{2n}x_n = b_2$$
$$\vdots \qquad \vdots \qquad \vdots \qquad \vdots \qquad \vdots \qquad (3.40)$$
$$a_{n1}x_1 + a_{n2}x_2 + \cdots + a_{nn}x_n = b_n$$

where $b_1, b_2, \ldots, b_n$ are inputs (sources) and $x_1, x_2, \ldots, x_n$ are outputs, the equations may be rewritten as

$$x_1 = \frac{1}{a_{11}} b_1 - \frac{a_{12}}{a_{11}} x_2 - \frac{a_{13}}{a_{11}} x_3 - \cdots - \frac{a_{1n}}{a_{11}} x_n$$

$$x_2 = \frac{1}{a_{22}} b_2 - \frac{a_{21}}{a_{22}} x_1 - \frac{a_{23}}{a_{22}} x_3 - \cdots - \frac{a_{2n}}{a_{22}} x_n$$

$$\vdots \qquad \vdots \qquad \vdots \qquad \vdots \qquad \vdots \qquad \vdots \qquad (3.41)$$

$$x_n = \frac{1}{a_{nn}} b_n - \frac{a_{n1}}{a_{nn}} x_1 - \frac{a_{n2}}{a_{nn}} x_2 - \cdots - \frac{a_{n-1,n-1}}{a_{nn}} x_{n-1}$$

b. The number of input nodes in the flowgraph is equal to the number of nonzero b's. That is, each of the source nodes corresponds to a nonzero b_j.

c. To each of the output nodes is associated one of the dependent variables $x_1, x_2, \ldots, x_n$.

d. The value of the variable represented by a node is equal to the sum of all the incoming signals.

e. The value of the variable represented by any node is transmitted onto all branches leaving the node.

It is a simple matter to write the equations from the flowgraph since every node, except the source nodes of the graph, represents an equation, and the equation associated with node k, for example, is obtained by equating to x_k the sum of all incoming branch gains multiplied by the values of the variables from which these branches originate.

Mason's general gain formula is now stated in the following theorem.

Theorem 3.9.   Let G be the overall graph gain and G_k be the gain of the kth forward path from the source to the sink. Then

$$G = \frac{1}{\Delta} \sum_k G_k \Delta_k \qquad\qquad (3.42)$$

where

$$\Delta = 1 \sum_m P_{m1} + \sum_m P_{m2} - \sum_m P_{m3} + \cdots + (-1)^j \sum_m P_{mj}$$

P_{m1} = loop gain (the product of all the branch gains around a loop)
P_{m2} = product of the loop gains of the mth set of two nontouching loops
P_{m3} = product of the loop gains of the mth set of three nontouching loops, and in general
P_{mj} = product of the loop gains of the mth set of j nontouching loops
Δ_k = the value of Δ for that subgraph of the graph obtained by removing the kth forward path
          along with those branches touching the path

Mason's signal-flowgraphs constitute a very useful graphical technique for the analysis of linear systems. This technique not only retains the intuitive character of the block diagrams but at the same time allows one to obtain the gain between an input node and an output node of a signal-flowgraph by inspection. However, the derivation of the gain formula [Eq. (3.42)] is by no means simple, and, more importantly, if more than one input is present in the system, the gain cannot be obtained directly; that is, the principle of superposition must be applied to determine the gain due to the presence of more than one input. Thus, by slight modification of the conventions involved in Mason's signal-flowgraph, Coates [1959] was able to introduce the so-called "flowgraphs" which are suitable for direct calculation of gain.

Recently, Chan and Bapna [1967] further modified Coates's flowgraphs and developed a simpler gain formula based on the modified graphs. The definitions and the gain formula based on the modified Coates graphs are presented in the following discussion.

The **flowgraph** G_I (called the *modified Coates graph*) of a system S of n independent linear equations in n unknowns

$$\sum_{j=1}^n a_{ij} x_j = b_i \qquad\qquad i = 1, 2, \ldots, n$$

is an oriented graph such that the variable x_j in S is represented by a *node* (also denoted by x_j) in G_I, and the coefficient a_{ij} of the variable x_j in S by a *branch* with a branch gain a_{ij} connected between nodes x_i and x_j in G_I and directed from x_j to x_i. Furthermore, a *source node* 1 is included in G_I such that for each constant b_k in S there is a node with gain b_k in G_l from node 1 to node s_k. Graph G_{l0} is the subgraph of G_l obtained by deleting the source node 1 and all the branches connected to it. Graph G_{lj} is the subgraph of G_l obtained by first removing all the outgoing branches from node x_j and then short-circuiting node 1 to node x_j. A *loop set* l is a subgraph of G_{l0} that contains all the nodes of G_{l0} with each node having exactly one incoming and one outgoing branch. The product p of the gains of all the branches in l is called a *loop-set product*. A 2-*loop-set* l_2 is a subgraph of G_{lj} containing all the nodes of G_{lj} with each node having exactly one incoming and one outgoing branch. The product p_2 of the gains of all the branches in l_2 is called a 2-*loop-set product*.

The modified Coates gain formula is now stated in the following theorem.

Theorem 3.10. In a system of n independent linear equations in n unknowns

$$a_{ij}x_j = b_i \qquad i = 1, 2, \ldots, n$$

the value of the variable x_j is given by

$$x_j = \sum_{(\text{all } p_2)} (-1)^{N_{l_2}} p_2 \Big/ \sum_{(\text{all } p_2)} (-1)^{N_l} p \qquad (3.43)$$

where N_{l_2} is the number of loops in a 2-loop-set l_2 and N_l is the number of loops in a loop set l.

Since both the Mason graph G_m and the modified Coates graph G_l are topological representations of a system of equations, it is logical that certain interrelationships exist between the two graphs so that one can be transformed into the other. Such interrelationships have been noted [Chan, 1969], and the transformations are briefly stated as follows:

A. *Transformation of G_m into G_l.* Graph G_m can be transformed into an equivalent Coates graph G_l (representing an equivalent system of equations) by the following steps:
   a. Subtract 1 from the gain of each existing self-loop.
   b. Add a self-loop with a gain of -1 to each branch devoid of self-loop.
   c. Multiply by $-b_k$ the gain of the branch at the kth source node b_k ($k = 1, 2, \ldots, r$, r being the number of source nodes) and then combine all the (r) nodes into one source node (now denoted by 1).

B. *Transformation of G_l into G_m.* Graph G_l can be transformed into G_m by the following steps:
   a. Add 1 to the gain of each existing self-loop.
   b. Add a self-loop with a gain of 1 to each node devoid of self-loop except the source node 1.
   c. Break the source node 1 into r source nodes (r being the number of branches connected to the source node 1 before breaking), and identify the r new source nodes by $b_1, b_2, \ldots, b_r$ with the gain of the corresponding r branches multipled by $-1/b_1, -1/b_2, \ldots, -1/b_r$, respectively, so that the new gains of these branches are all equal to 1, keeping the edge orientations unchanged.

The gain formulas of Mason and Coates are the classical ones in the theory of flowgraphs. From the systems viewpoint, the Mason technique provides an excellent physical insight as one can visualize the signal flow through the subgraphs (forward paths and feedback loops) of G_m. The graph reduction technique based on the Mason graph enables one to obtain the gain expression using a step-by-step approach and at the same time observe the cause-and-effect relationships in each step. However, since the Mason formula computes the ratio of a specified output over *one* particular input, the principle of superposition must be used in order to obtain the overall gain of the system if more than one input is present. The Coates formula, on the other hand, computes the output directly regardless of the number of inputs present in the system, but because of such a direct computation of a given output, the graph reduction rules of Mason cannot be applied to a Coates graph since the Coates graph is *not* based on the same cause-effect formulation of equations as Mason's.

The *k*-Tree Approach Versus the Flowgraph Approach

When a linear network is given, loop or node equations can be written from the network, and the analysis of the network can be accomplished by means of either Coates's or Mason's technique.

However, it has been shown [Chan, 1969] that if the Maxwell *k*-tree approach is employed in solving a linear network, the redundancy inherent either in the direct expansion of determinants or in the flowgraph techniques described above can be either completely eliminated for passive networks or greatly reduced for active networks. This point and others will be illustrated in the following example.

Example 3.7. Consider the network N as shown in Fig. 3.39. Let us determine the voltage gain, $G_{12} = V_0/V_1$, using (1) Mason's method, (2) Coates's method, and (3) the *k*-tree method.

The two node equations for the network are given by

$$\text{for node 2:} \quad (Y_a + Y_b + Y_e)V_2 + (-Y_e)V_0 = Y_aV_i$$
$$\text{for node 3:} \quad (-Y_e)V_2 + (Y_c + Y_d + Y_e)V_0 = Y_cV_i \tag{3.44}$$

where

$$Y_a = 1/Z_a, \; Y_b = 1/Z_b, \; Y_c = 1/Z_c, \; Y_d = 1/Z_d \text{ and } Y_e = 1/Z_e$$

(1) *Mason's approach.* Rewrite the system of two equations (3.44) as follows:

$$V_2 = \left(\frac{Y_a}{Y_a + Y_b + Y_e}\right)V_i + \left(\frac{Y_e}{Y_a + Y_b + Y_e}\right)V_0$$

$$V_0 = \left(\frac{Y_c}{Y_c + Y_d + Y_e}\right)V_i + \left(\frac{Y_e}{Y_c + Y_d + Y_e}\right)V_2 \tag{3.45}$$

or

$$V_2 = AV_i + BV_0 \quad V_0 = CV_i + DV_2 \tag{3.46}$$

where

$$A = \frac{Y_a}{Y_a + Y_b + Y_e} \qquad B = \frac{Y_e}{Y_a + Y_b + Y_e}$$

$$C = \frac{Y_c}{Y_c + Y_d + Y_e} \qquad D = \frac{Y_e}{Y_c + Y_d + Y_e}$$

The Mason graph of system (3.46) is shown in Fig. 3.40, and according to the Mason graph formula (3.42), we have

$$\Delta = 1 - BD$$

$$G_C = C \qquad \Delta_C = 1$$

$$G_{AD} = AD \qquad \Delta_{AD} = 1$$

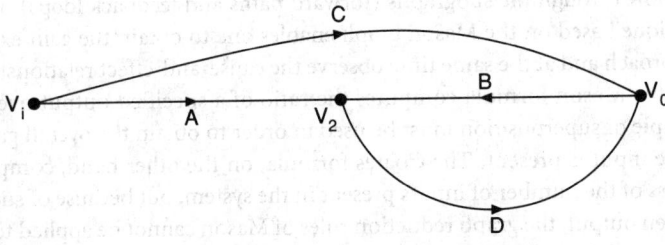

FIGURE 3.40 The Mason graph of N.

and hence

$$G_{12} = \frac{V_0}{V_1} = \frac{1}{\Delta} \sum_k G_k \Delta_k = \frac{1}{1 - BD} (C + AD)$$

$$= \frac{Y_c /(Y_c + Y_d + Y_e) + Y_a /(Y_a + Y_b + Y_e)(Y_c + Y_d + Y_e)}{1 - Y_e^2 /(Y_a + Y_b + Y_e)(Y_c + Y_d + Y_e)}$$

Upon cancelation and rearrangement of terms

$$G_{12} = \frac{Y_a Y_c + Y_a Y_e + Y_b Y_c + Y_c Y_e}{Y_a Y_c + Y_a Y_d + Y_a Y_e + Y_b Y_c + Y_b Y_d + Y_b Y_e + Y_c Y_e + Y_d Y_e} \qquad (3.47)$$

(2) *Coates's approach.* From (3.44) we obtain the Coates graphs G_l, G_{l0}, and G_{l3} as shown in Fig. 3.41(a), (b), and (c), respectively. The set of all loop-sets of G_{l0} is shown in Fig. 3.42, and the set of all 2-loop-sets of G_{l3} is shown in Fig. 3.43. Thus, by Eq. (3.43),

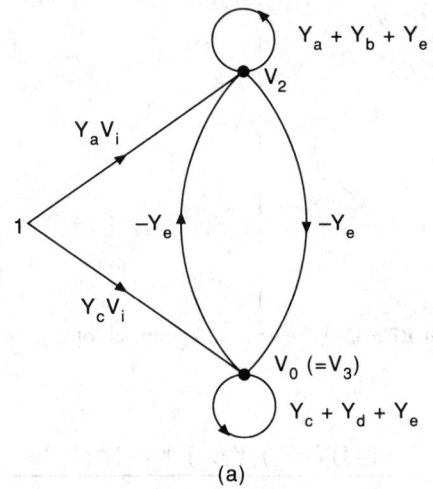

(a)

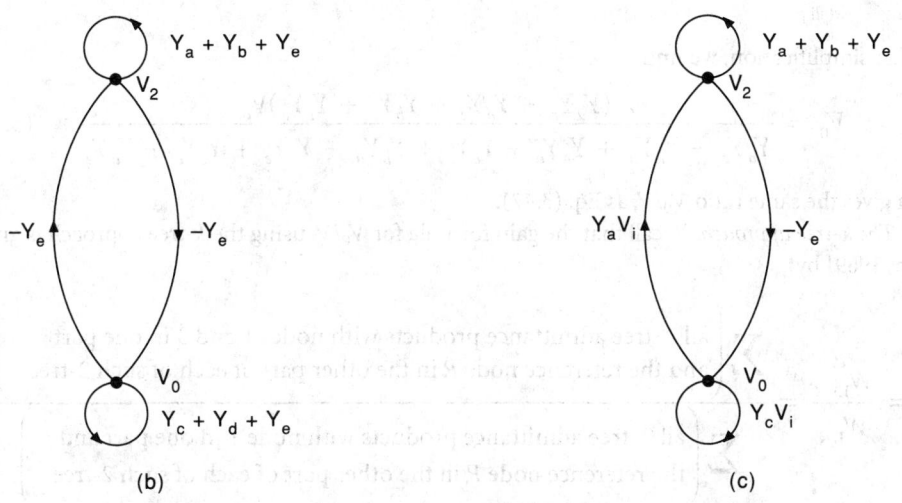

(b)   (c)

FIGURE 3.41   The Coates graphs: (a) G_l, (b) G_{l0}, and (c) G_{l3}.

FIGURE 3.42   The set of all loop-sets of G_{10}.

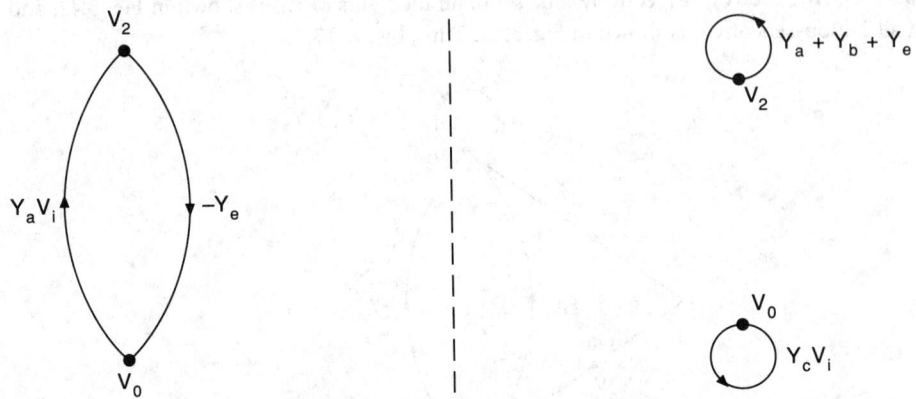

FIGURE 3.43   The set of all 2-loop-sets of G_{13}.

$$V_0 = \frac{\displaystyle\sum_{(\text{all } p_2)} (-1)^{N_{l_2}} p_2}{\displaystyle\sum_{(\text{all } p)} (-1)^{N_l} p} = \frac{(-1)^1 (-Y_e)(Y_a V_i) + (-1)^2 (Y_a + Y_b + Y_e)(Y_c V_i)}{(-1)^1 (-Y_e)(-Y_e) + (-1)^2 (Y_a + Y_b + Y_e)(Y_c + Y_d + Y_e)}$$

Or, after simplification, we find

$$V_0 = \frac{(Y_a Y_c + Y_a Y_e + Y_b Y_c + Y_c Y_e) V_i}{Y_a Y_c + Y_a Y_d + Y_a Y_e + Y_b Y_c + Y_b Y_d + Y_b Y_e + Y_c Y_e + Y_d Y_e} \tag{3.48}$$

which gives the same ratio V_0/V_i as Eq. (3.47).

(3) *The k-tree approach.* Recall that the gain formula for V_0/V_i using the *k*-tree approach is given [Chan, 1969] by

$$\frac{V_0}{V_i} = \frac{\Delta_{13}}{\Delta_{11}} = \frac{W_{13,R}}{W_{1,R}} = \frac{\displaystyle\sum \left(\begin{array}{l} \text{all 2-tree admittance products with nodes 1 and 3 in one part} \\ \text{and the reference node } R \text{ in the other part of each of such 2-tree} \end{array} \right)}{\displaystyle\sum \left(\begin{array}{l} \text{all 2-tree admittance products with node 1 in one part and} \\ \text{the reference node } R \text{ in the other part of each of such 2-tree} \end{array} \right)}$$

$$\tag{3.49}$$

where Δ_{13} and Δ_{11} are cofactors of the determinant Δ of the node admittance matrix of the network. Furthermore, it is noted that the 2-trees corresponding to Δ_{ii} may be obtained by finding all the trees of the modified graph G_i, which is obtained from the graph G of the network by short-circuiting node i (i being any node other than R) to the reference node R, and that the 2-trees corresponding to Δ_{ij} can be found by taking all those 2-trees each of which is a tree of both G_i and G_j [Chan, 1969]. Thus, for Δ_{11}, we first find G and G_1 (Fig. 3.44), and then find the set S_1 of all trees of G_1 (Fig. 3.45); then for Δ_{13}, we find G_3 (Fig. 3.46) and the set S_3 of all trees of G_3 (Fig. 3.47) and then from S_1 and S_3 we find all the terms common to both sets (which correspond to the set of all trees common to G_1 and G_3) as shown in Fig. 3.48. Finally we form the ratio of 2-tree admittance products according to Eq. (3.49). Thus from Figs. 3.45 and 3.48, we find

$$\frac{V_0}{V_i} = \frac{Y_a Y_c + Y_a Y_e + Y_b Y_c + Y_c Y_e}{Y_a Y_c + Y_a Y_d + Y_a Y_e + Y_b Y_c + Y_b Y_d + Y_b Y_e + Y_c Y_e + Y_d Y_e}$$

which is identical to the results obtained by the flowgraph techniques.

From the above discussions and Example 3.7 we see that the Mason approach is the best from the systems viewpoint, especially when a single source is involved. It gives an excellent physical insight to the system and reveals the cause-effect relationships at various stages when graph reduction technique is employed. While the Coates approach enables one to compute the output directly regardless of the number of inputs involved in the system, thus overcoming one of the difficulties associated with Mason's approach, it does not allow one to reduce the graph step-by-step toward the final solution as Mason's does. However, it is interesting to note that in the modified Coates technique the introduction of the loop-sets (analogous to trees) and the 2-loop-sets (analogous to 2-trees) brings together the two different concepts—the flowgraph approach and the k-tree approach.

From the networks point of view, the Maxwell k-tree approach not only enables one to express the solution in terms of the topology (namely the trees and 2-trees in Example 3.7) of the network but also avoids the cancelation problem inherent in all the flowgraph techniques since, as evident from Example 3.7, the trees and the 2-trees in the gain expression by the k-tree approach correspond (one-to-one) to the *uncanceled terms* in the final expressions of the gain by the flowgraph techniques. Finally, it should be obvious that the k-tree approach depends upon the knowledge of the graph of a given network. Thus, if in a network problem only the system of (loop or node) equations is given and the network is not known, or more generally, if a system is characterized by

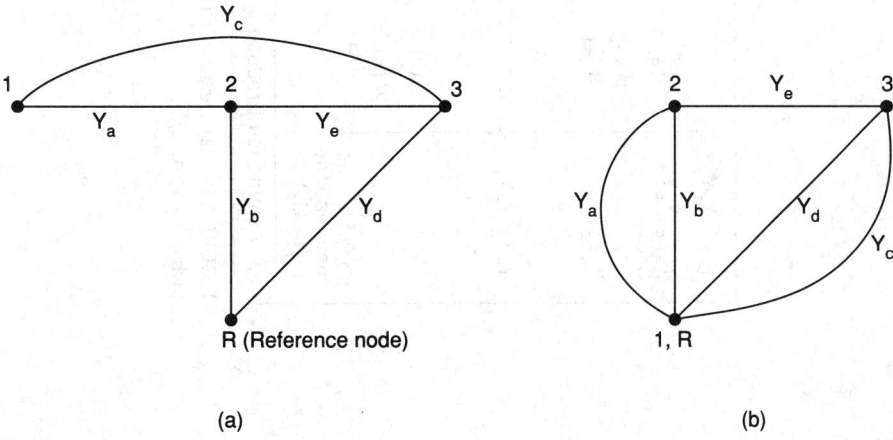

(a) (b)

FIGURE 3.44 (a) Graph G, and (b) the modified graph G_1 of G.

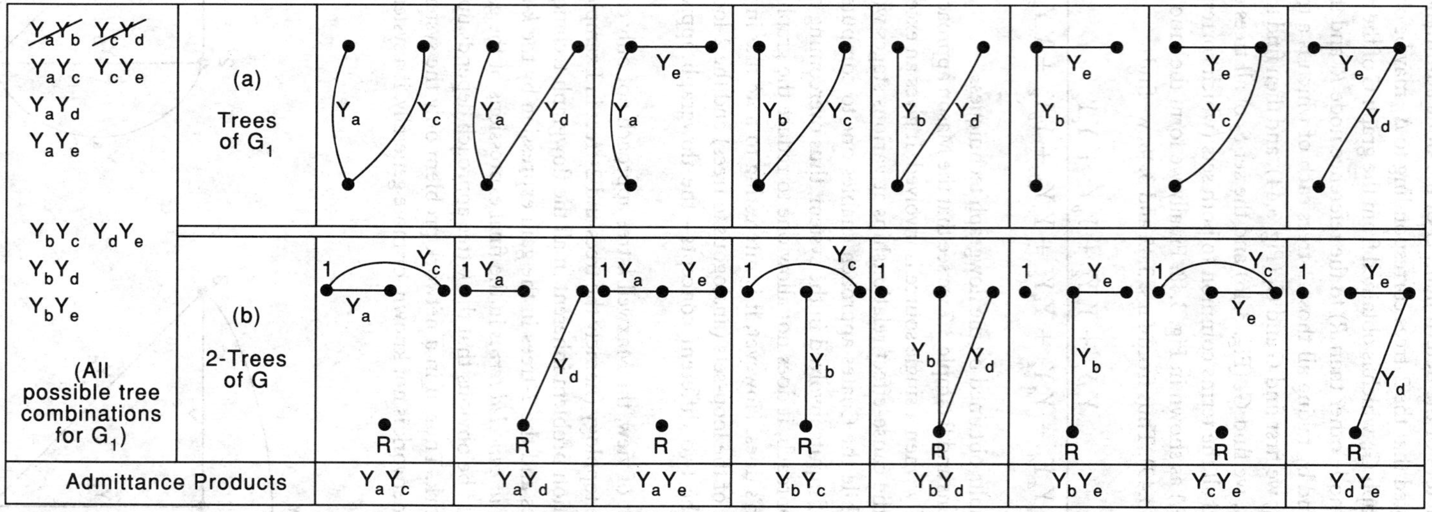

(a) Trees of G_1	$Y_a Y_c$	$Y_a Y_d$	Y_a	$Y_b Y_c$	$Y_b Y_d$	Y_e Y_b	Y_e Y_d
(b) 2-Trees of G	$Y_a Y_c$	$Y_a Y_d$	$Y_a Y_e$	$Y_b Y_c$	$Y_b Y_d$	$Y_b Y_e$	$Y_d Y_e$

Admittance Products	(All possible tree combinations for G_1)
	$Y_a Y_b$ ~~$Y_c Y_d$~~ $Y_c Y_e$ $Y_a Y_c$ $Y_a Y_d$ $Y_a Y_e$
	$Y_b Y_c$ $Y_d Y_e$ $Y_b Y_d$ $Y_b Y_e$

FIGURE 3.45 (a) The set of all trees of the modified graph G_1 which corresponds to (b) the set of all 2-trees of G (with nodes 1 and R in separate parts in each of such 2-tree).

106

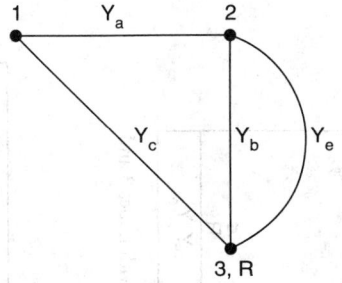

FIGURE 3.46 The modified graph G_3 of G.

a block diagram or a system of equations, the k-tree approach cannot be applied and one must resort to the flowgraph techniques between the two approaches.

Some Topological Applications in Network Analysis and Design

In practice a circuit designer often has to make approximations and analyze the same network structure many times with different sets of component values before the final network realization is obtained. Conventional analysis techniques which require the evaluation of high-order determinants are undesirable even on a digital computer because of the large amount of redundancy inherent in the determinant expansion process. The extra calculation in the evaluation (expansion of determinants) and simplification (cancellation of terms) is time consuming and costly and thereby contributes much to the undesirability of such methods.

The k-tree topological formulas presented in this section, on the other hand, eliminate completely the cancellation of terms. Also, they are particularly suited for digital computation when the size of the network is not exceedingly large. All of the terms involved in the formulas can be computed by means of a digital computer using a single "tree-finding" program [Chan, 1969]. Thus, the application of topological formulas in analyzing a network with the aid of a digital computer can mean a saving of a considerable amount of time and cost to the circuit designer, especially true when it is necessary to repeat the same analysis procedure a large number of times.

In a preliminary system design, the designer usually seeks one or more concepts which will meet the specifications, and in engineering practice each concept is generally subjected to some form of analysis. For linear systems, the signal flowgraph of Mason is widely used in this activity. The flowgraph analysis is popular because it depicts the relationships existing between system variables, and the graphical structure may be manipulated using Mason's formulas to obtain system transfer functions in symbolic or symbolic/numerical form.

Although the preliminary design problems are usually of limited size (several variables), hand derivation of transfer functions is nonetheless difficult and often prone to error arising from the omission of terms. The recent introduction of remote, time-shared computers into modern design areas offers a means to handle such problems swiftly and effectively.

An efficient algorithm suitable for digital computation of transfer functions from the signal flowgraph description of a linear system has been developed [Dunn and Chan, 1969] which provides a powerful analytical tool in the conceptual phases of linear system design.

In the past several decades, graph theory has been widely used in electrical engineering, computer science, social science, and in the solution of economic problems [Swamy and Thulasiraman, 1981; Chen, 1990]. Finally, the application of graph theory in conjunction with symbolic network analysis and computer-aided simulation of electronic circuits has been well recognized in recent years [Lin, 1991].

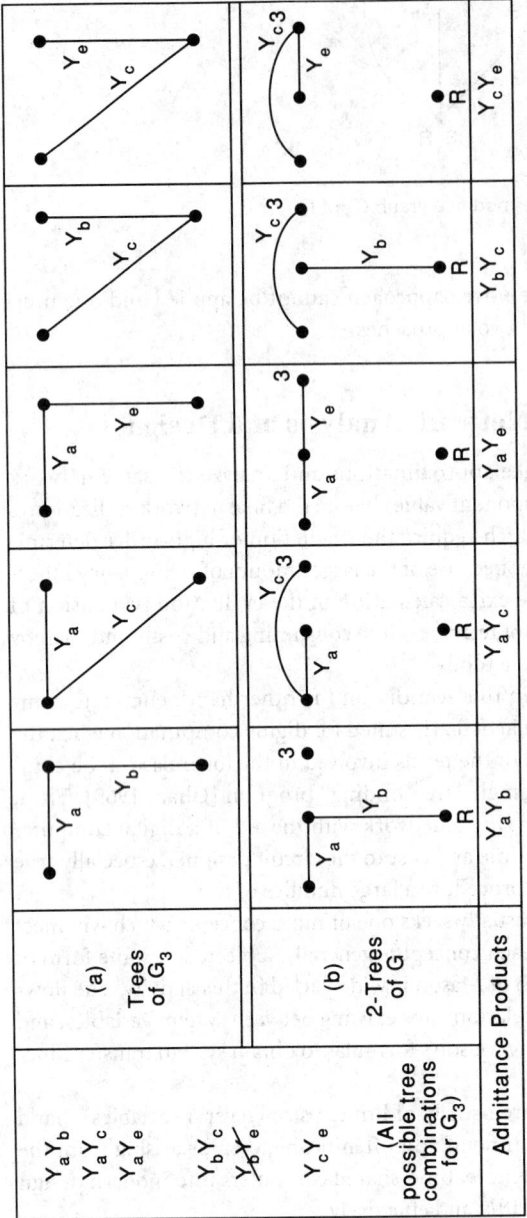

FIGURE 3.47 (a) The set of all trees of the modified graph G_3, which corresponds to (b) the set of all 2-trees of G (with nodes 3 and R in separate parts in each of such 2-tree).

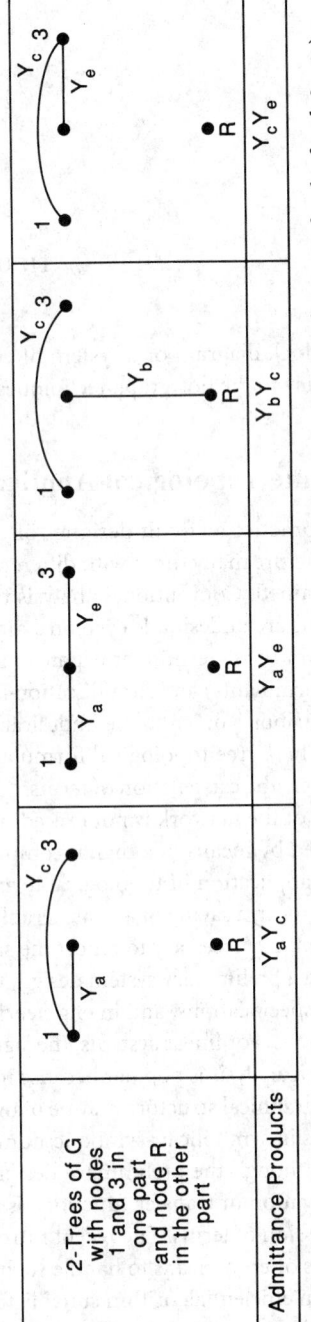

FIGURE 3.48 The set of all 2-trees of G (with nodes 1 and 3 in one part and the reference node R in the other part of each of such 2-tree).

Defining Terms

Branches of a tree: The edges contained in a tree.

Circuit (or loop): A closed path where all vertices are of degree 2, thus having *no* endpoints in the path.

Circuit-set (or loop-set): The set of all edges contained in a circuit (loop).

Connectedness: A graph or subgraph is said to be connected if there is at least one path between *every* pair of its vertices.

Flowgraph G_l (or modified Coates graph G_l): The flowgraph G_l (called *the modified Coates graph*) of a system S of n independent linear equations in n unknowns

$$\sum_{j=1}^{n} a_{ij}x_j = b_i \qquad i = 1, 2, \ldots, n$$

is an oriented graph such that the variable x_j in S is represented by a *node* (also denoted by x_j) in G_l, and the coefficient a_{ij} of the variable x_j in S by a *branch* with a branch gain a_{ij} connected between nodes x_i and x_j in G_l and directed from x_j to x_i. Furthermore, a *source node* 1 is included in G_l such that for each constant b_k in S there is a node with gain b_k in G_l from node 1 to node s_k. Graph G_{l0} is the subgraph of G_l obtained by deleting the source node 1 and all the branches connected to it. Graph G_{lj} is the subgraph of G_l obtained by first removing all the outgoing branches from node x_j and then short-circuiting node 1 to node x_j. A *loop set l* is a subgraph of G_{l0} that contains all the nodes of G_{l0} with each node having exactly one incoming and one outgoing branch. The product p of the gains of all the branches in l is called a *loop-set product*. A 2-*loop-set* l_2 is a subgraph of G_{lj} containing all the nodes of G_{lj} with each node having exactly one incoming and one outgoing branch. The product p_2 of the gains of all the branches in l_2 is called a 2-*loop-set product*.

k-tree admittance product of a k-tree: The product of the admittances of all the branches of the k-tree.

k-tree of a connected graph G: A proper subgraph of G consisting of k unconnected circuitless subgraphs, each subgraph itself being connected, which together contain all the vertices of G.

Linear graph: A set of line segments called edges and points called vertices, which are the endpoints of the edges, interconnected in such a way that the edges are connected to (or incident with) the vertices. The degree of a vertex of a graph is the number of edges incident with that vertex.

Path: A subgraph having all vertices of degree 2 except for the two endpoints which are of degree 1 and are called the terminals of the path, where the degree of a vertex is the number of edges connected to the vertex in the subgraph.

Path-set: The set of all edges in a path.

Proper subgraph: A subgraph which does not contain all of the edges of the given graph.

Signal-flowgraph G_m (or Mason's graph G_m): A signal-flowgraph G_m of a system S of n independent linear (algebraic) equations in n unknowns

$$\sum_{j=1}^{n} a_{ij}x_j = b_i \qquad i = 1, 2, \ldots, n$$

is a graph with junction points called *nodes* which are connected by directed line segments called *branches* with signals traveling along the branches only in the direction described by the arrows of the branches. A signal x_k traveling along a branch between x_k and x_j is multiplied by the gain of the branches g_{kj}, so that a signal $g_{kj}x_k$ is delivered at node x_j. An *input node* (*source*) is a node which contains only outgoing branches; an *output node* (*sink*) is a node which has only incoming branches. A *path* is a continuous unidirectional succession of branches, all of which are traveling in the same direction; a *forward path* is a path from the input node to the output node along which all nodes are encountered exactly once; and a *feedback path* (*loop*) is a closed path which originates from and terminates at the same node, and along with all other nodes are encountered exactly once (the trivial case is a *self-loop* which contains exactly one node and one branch). A *path* gain is the product of all the branch gains of the branches in a loop.

Subgraph: A subset of the edges of a given graph.

Tree: A connected subgraph of a given connected graph *G* which contains all the vertices of *G* but no circuits.

References

A.C. Aitken, *Determinants and Matrices*, 9th ed., New York: Interscience, 1956.

S.P. Chan, *Introductory Topological Analysis of Electrical Networks*, New York: Holt, Rinehart and Winston, 1969.

S.P. Chan and B.H. Bapna, "A modification of the Coates gain formula for the analysis of linear systems," *Inst. J. Control*, vol. 5, pp. 483–495, 1967.

S.P. Chan and S.G. Chan, "Modifications of topological formulas," *IEEE Trans. Circuit Theory*, vol. CT-15, pp. 84–86, 1968.

W.K. Chen, *Theory of Nets: Flows in Networks*, New York: Wiley Interscience, 1990.

C.L. Coates, "Flow-graph solutions of linear algebraic equations," *IRE Trans. Circuit Theory*, vol. CT-6, pp. 170–187, 1959.

W.R. Dunn, Jr., and S.P. Chan, "Flowgraph analysis of linear systems using remote timeshared computation, *J. Franklin Inst.*, vol. 288, pp. 337–349, 1969.

G. Kirchhoff, "Über die Auflösung der Gleichungen, auf welche man bei der Untersuchung der linearen Vertheilung galvanischer Ströme, geführt wird," *Ann. Physik Chemie*, vol. 72, pp. 497–508, 1847; English transl., *IRE Trans. Circuit Theory*, vol. CT-5, pp. 4–7, 1958.

P.M. Lin, *Symbolic Network Analysis*, New York: Elsevier, 1991.

S.J. Mason, "Feedback theory—Some properties of signal flow graphs," *Proc. IRE*, vol. 41, pp. 1144–1156, 1953.

S.J. Mason, "Feedback theory—Further properties of signal flow graphs," *Proc. IRE*, vol. 44, pp. 920–926, 1956.

J.C. Maxwell, *Electricity and Magnetism*, Oxford: Clarendon Press, 1892.

W.S. Percival, "Solution of passive electrical networks by means of mathematical trees," *Proc. IEE*, vol. 100, pp. 143–150, 1953.

S. Seshu and M.B. Reed, *Linear Graphs and Electrical Networks*, Reading, Mass.: Addison-Wesley, 1961.

M.N.S. Swamy and K. Thulasiraman, *Graphs, Networks, and Algorithms*, New York: Wiley, 1981.

Further Information

All defining terms used in this section can be found in S. P. Chan, *Introductory Topological Analysis of Electrical Networks*, Holt, Rinehart and Winston, New York, 1969. Also an excellent reference for the applications of graph theory in electrical engineering (i.e., network analysis and design) is S. Seshu and M. B. Reed, *Linear Graphs and Electrical Networks*, Addison-Wesley, Reading, Mass., 1961.

For applications of graph theory in computer science, see M. N. S. Swamy and K. Thulasiraman, *Graphs, Networks, and Algorithms*, Wiley, New York, 1981.

For flowgraph applications, see W. K. Chen, *Theory of Nets: Flows in Networks*, Wiley Interscience, New York, 1990.

For applications of graph theory in symbolic network analysis, see P. M. Lin, *Symbolic Network Analysis*, Elsevier, New York, 1991.

4

Passive Signal Processing

4.1 Introduction 111
    Laplace Transform • Transfer Functions
4.2 Low-Pass Filter Functions 113
    Thomson Functions • Chebyshev Functions • Inverse
    Chebyshev Function
4.3 Low-Pass Filters 117
    Introduction • Butterworth Filters • Thomson Filters •
    Chebyshev Filters • Inverse Chebyshev Filters
4.4 Filter Design 121
    Scaling Laws and a Design Example • Transformation Rules, Passive
    Circuits

William J. Kerwin
University of Arizona

4.1 Introduction

We will be concerned here with systems having an electrical input and output. Very powerful filtering techniques are available for this case, and frequently the electrical system is the lowest in cost.

We can specify either a phase characteristic or a magnitude characteristic, but only rarely can both be specified. The greater the filtering requirement—that is, the steeper the slope in the magnitude beyond the cut-off frequency—the greater will be the cost. If we need to eliminate specific finite frequencies, then we must use a more sophisticated and more costly filter having response zeros at finite frequencies. In addition, as we obtain a steeper slope beyond cut-off, we worsen the time domain response; that is, the overshoot to a step input is greater and the ringing continues for a longer time, leading to a very long settling time. It may therefore be necessary to temper the steady-state filtering requirements in order to achieve acceptable time domain response. We will start with some details of notation and some examples.

Laplace Transform

We will use the Laplace operator, $s = \sigma + j\omega$. Steady-state impedance is thus Ls and $1/Cs$, respectively, for an inductor (L) and a capacitor (C), and admittance is $1/Ls$ and Cs. In steady state $\sigma = 0$ and therefore $s = j\omega$.

Transfer Functions

We will consider only lumped, linear, constant, bilateral elements, and we will define the **transfer function** $T(s)$ as response over excitation.

Reprinted from *Instrumentation and Control: Fundamentals and Applications,* edited by Chester L. Nachtigal, pp. 487–497, copyright 1990, John Wiley and Sons, Inc. Reproduced by permission of John Wiley and Sons, Inc.

$$T(s) = \frac{\text{signal output}}{\text{signal input}} = \frac{N(s)}{D(s)}$$

The roots of the numerator polynomial $N(s)$ are the zeros of the system, and the roots of the denominator $D(s)$ are the poles of the system (the points of infinite response). If we substitute $s = j\omega$ into $T(s)$ and separate the result into real and imaginary parts (numerator and denominator) we obtain

$$T(j\omega) = \frac{A_1 + jB_1}{A_2 + jB_2} \tag{4.1}$$

Then the magnitude of the function, $|T(j\omega)|$, is

$$|T(j\omega)| = \left(\frac{A_1^2 + B_1^2}{A_2^2 + B_2^2}\right)^{\frac{1}{2}} \tag{4.2}$$

and the phase $\overline{T(j\omega)}$ is

$$\overline{T(j\omega)} = \tan^{-1}\frac{B_1}{A_1} - \tan^{-1}\frac{B_2}{A_2} \tag{4.3}$$

Analysis

Although mesh or nodal analysis can always be used, since we will consider only ladder networks we will use a method commonly called *linearity*, or *working your way through*. The method starts at the output and assumes either 1 volt or 1 ampere as appropriate and uses Ohm's law and Kirchhoff's current law only.

Example 4.1.   Determine the transfer function of the circuit of Fig. 4.1.

Let $v_o = 1$ V; then

$$i_5 = 1, \; i_4 = s, \; i_3 = 1 + s$$
$$v_1 = v_o + i_3(2s) = 1 + 2s(1 + s)$$
$$i_2 = sv_1 = 2s^3 + 2s^2 + s$$
$$i_1 = i_2 + i_3 = 2s^3 + 2s^2 + 2s + 1$$
$$v_i = v_1 + i_1 = 2s^3 + 4s^2 + 4s + 2$$
$$\frac{v_o}{v_i} = T(s) = \frac{1}{2s^3 + 4s^2 + 4s + 2}$$

Example 4.2   Determine the steady-state magnitude and phase of the transfer function derived in Example 4.1.

$$\frac{v_o}{v_i} = T(s) = \frac{1}{2s^3 + 4s^2 + 4s + 2} = \frac{1}{2}\left(\frac{1}{s^3 + 2s^2 + 2s + 1}\right)$$

Substituting $s = j\omega$ (steady state)

$$T(j\omega) = \frac{1}{2}\left(\frac{1}{1 - 2\omega^2 + j(2\omega - \omega^3)}\right)$$

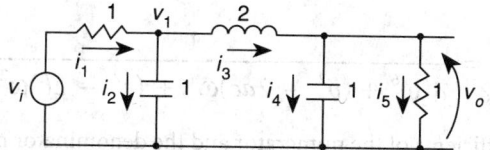

FIGURE 4.1   Doubly terminated third-order low-pass filter (in Ω, H, F).

and from Eq. (4.2)

$$|T(j\omega)| = \frac{1}{2}\left(\frac{1}{\sqrt{[(1-2\omega^2)^2+(2\omega-\omega^3)^2]}}\right) = \frac{1}{2}\left(\frac{1}{\sqrt{(\omega^6+1)}}\right)$$

and from Eq. (4.3)

$$\angle T(j\omega) = 0° - \tan^{-1}\frac{2\omega-\omega^3}{1-2\omega^2}$$

The values used for the circuit of Fig. 4.1 were normalized; that is, they are all near unity in ohms, henrys, and farads. These values simplify computation and, as we will see later, can easily be scaled to any desired set of actual element values. In addition, this circuit is low-pass because of the shunt capacitors and the series inductor. By low-pass we mean a circuit that passes the lower frequencies and attenuates higher frequencies. The cut-off frequency is the point at which the magnitude is 0.707 (–3 dB) of the dc level and is the dividing line between the **passband** and the **stopband**. In the above example we see that the magnitude of v_o/v_i at $\omega = 0$ (dc) is 0.50 and that at $\omega = 1$ rad/s we have

$$|T(j\omega)| = \frac{1}{2}\frac{1}{\sqrt{(\omega^6+1)}} = 0.3535 \qquad (4.4)$$

Thus, we see that the normalized element values used here give us a cut-off frequency of 1 rad/s.

4.2 Low-Pass Filter Functions*

The most common function in signal processing is the Butterworth. It is a function that has only poles (i.e., no finite zeros) and has the flattest magnitude possible in the passband. This function is also called **maximally flat magnitude (MFM)**. The derivation of this function is illustrated by taking a general all-pole function of third-order with a dc gain of 1 as follows:

$$T(s) = \frac{1}{as^3 + bs^2 + cs + 1} \qquad (4.5)$$

The squared magnitude is

$$|T(j\omega)|^2 = \frac{1}{(1-b\omega^2)^2+(c\omega-a\omega^3)^2} \qquad (4.6)$$

*Reprinted from *Handbook of Measurement Science*, edited by Peter Sydenham, copyright 1982, John Wiley and Sons Limited. Reproduced by permission of John Wiley and Sons Limited.

or

$$|T(j\omega)|^2 = \frac{1}{a^2\omega^6 + (b^2 - 2ac)\omega^4 + (c^2 - 2b)\omega^2 + 1} \tag{4.7}$$

MFM requires that the coefficients of the numerator and the denominator match term by term (or be in the same ratio) except for the highest power. Since the numerator and denominator dc values are equal (both unity) we have the matching condition for this case.

Therefore

$$c^2 - 2b = 0; \qquad b^2 - 2ac = 0 \tag{4.8}$$

We will also impose a normalized cut-off (−3 dB) at $\omega = 1$ rad/s; that is,

$$|T(j\omega)|_{\omega=1} = \frac{1}{\sqrt{(a^2 + 1)}} = 0.707 \tag{4.9}$$

Thus, we find $a = 1$, then $b = 2$, $c = 2$ are solutions to the flat magnitude conditions of Eq. 4.8 and our third-order Butterworth function is

$$T(s) = \frac{1}{s^3 + 2s^2 + 2s + 1} \tag{4.10}$$

Table 4.1 gives the Butterworth denominator polynomials up to $n = 5$.

In general, for all Butterworth functions the normalized magnitude is

$$|T(j\omega)| = \frac{1}{\sqrt{(\omega^{2n} + 1)}} \tag{4.11}$$

Note that this is down 3 dB at $\omega = 1$ rad/s for all n.

This may, of course, be multiplied by any constant less than one for circuits whose dc gain is deliberately set to be less than one. The circuit of Fig. 4.1 has a dc gain of 0.5.

Example 4.3. A low-pass Butterworth filter is required whose cut-off frequency (−3 dB) is 4 kHz and in which the response must be down 45 dB at 10 kHz. Normalizing to a cut-off frequency of 1 rad/s, the −45-dB frequency is

$$\frac{10 \text{ kHz}}{4 \text{ kHz}} = 2.5 \text{ rad/s}$$

thus

$$-45 = 20 \log \frac{1}{\sqrt{2.5^{2n} + 1}}$$

therefore $n = 5.65$. Since n must be an integer, a sixth-order filter is required for this specification.

There is an extremely important difference between the singly terminated (dc gain = 1) and the doubly terminated filters (dc gain = 0.5). As was shown by John Orchard, the sensitivity in the passband (ideally at maximum output) to all L, C components in an L, C filter with *equal* terminations is *zero*. This is true regardless of the circuit.

Table 4.1 Butterworth Denominator Polynomials

$$s + 1$$
$$s^2 + \sqrt{2}s + 1$$
$$s^3 + 2s^2 + 2s + 1$$
$$s^4 + 2.6131s^3 + 3.4142s^2 + 2.6131s + 1$$
$$s^5 + 3.2361s^4 + 5.2361s^3 + 5.2361s^2 + 3.2361s + 1$$

Source: Handbook of Measurement Science, edited by Peter Sydenham, copyright 1982, John Wiley and Sons Limited. Reproduced by permission of John Wiley and Sons Limited.

This, of course, means component tolerances and temperature coefficients are of much less importance in the doubly terminated (equal values) filter. For this type of Butterworth low-pass filter (normalized to equal 1-Ω terminations), Takahasi has shown that the normalized element values are exactly given by

$$L, C = 2 \sin\left(\frac{(2k-1)\pi}{2n}\right) \tag{4.12}$$

for any order n, where k is the L or C element from 1 to n.

Example 4.4. Design a normalized ($\omega_{-3dB} = 1$ rad/s) doubly terminated (i.e., source and load = 1 Ω) Butterworth low-pass filter of order 6; that is, $n = 6$. The form of the filter is shown in Fig. 4.2. The element values from Eq. (4.12) are

$$L_1 = 2 \sin\frac{(2-1)\pi}{12} = 0.5176 \text{ H}$$

$$C_2 = 2 \sin\frac{(4-1)\pi}{12} = 1.4141 \text{ F}$$

$$L_3 = 2 \sin\frac{(6-1)\pi}{12} = 1.9319 \text{ H}$$

The values repeat for C_4, L_5, C_6 so that

$$C_4 = L_3, L_5 = C_2, C_6 = L_1$$

Thomson Functions

The Thomson function is one in which the time delay of the network is made maximally flat. This implies a linear phase characteristic since the steady-state time delay is the negative of the derivative of the phase. This function has excellent time domain characteristics and is used wherever excellent step response is required. These functions have very little overshoot to a step input and have far superior settling times compared to the Butterworth functions. The slope near cut-off is more gradual than the Butterworth. Table 4.2 gives the Thomson denominator polynomials. The numerator is a constant equal to the dc gain of the circuit multiplied by the denominator constant. The cut-off frequencies are *not* all 1 rad/s. They are given in Table 4.2.

Chebyshev Functions

A second function defined in terms of magnitude, the Chebyshev, has an **equal ripple** character within the passband. The ripple is determined by ϵ.

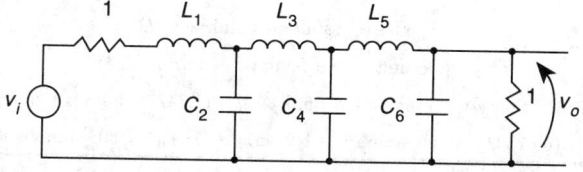

FIGURE 4.2   Doubly terminated sixth-order low-pass filter (in Ω, H, F).

Table 4.2 Thomson Denominator Polynomials

	ω_{-3dB} (rad/s)
$s + 1$	1.0000
$s^2 + 3s + 3$	1.3617
$s^3 + 6s^2 + 15s + 15$	1.7557
$s^4 + 10s^3 + 45s^2 + 105s + 105$	2.1139
$s^5 + 15s^4 + 105s^3 + 420s^2 + 945s + 945$	2.4274

Source: Handbook of Measurement Science, edited by Peter Sydenham, copyright 1982, John Wiley and Sons Limited. Reproduced by permission of John Wiley and Sons Limited.

$$\epsilon = \sqrt{(10^{A/10} - 1)} \qquad (4.13)$$

where A = decibels of ripple; then for a given order n, we define v.

$$v = \frac{1}{n} \sinh^{-1}\left(\frac{1}{\epsilon}\right) \qquad (4.14)$$

Table 4.3 gives denominator polynomials for the Chebyshev functions. In all cases, the cut-off frequency (defined as the end of the ripple) is 1 rad/s. The −3-dB frequency for the Chebyshev function is

$$\omega_{-3dB} = \cosh\left[\frac{\cosh^{-1}(1/\epsilon)}{n}\right] \qquad (4.15)$$

The magnitude in the *stopband* ($\omega > 1$ rad/s) for the normalized filter is

$$|T(j\omega)|^2 = \frac{1}{1 + \epsilon^2 \cosh^2(n \cosh^{-1} \omega)} \qquad (4.16)$$

for the singly terminated filter. For equal terminations the above magnitude is multiplied by one-half [1/4 in Eq. (4.16)].

Example 4.5. What order of singly terminated Chebyshev filter having 0.25-dB ripple (A) is required if the magnitude must be −60 dB at 15 kHz and the cut-off frequency (−0.25 dB) is to be 3 kHz? The normalized frequency for a magnitude of −60 dB is

$$\frac{15 \text{ kHz}}{3 \text{ kHz}} = 5 \text{ rad/s}$$

Table 4.3 Chebyshev Denominator Polynomials

$$s + \sinh v$$
$$s^2 + (\sqrt{2} \sinh v)s + \sinh^2 v + 1/2$$
$$(s + \sinh v)[s^2 + (\sinh v)s + \sinh^2 v + 3/4]$$
$$[s^2 + (0.75637 \sinh v)s + \sinh^2 v + 0.85355] \times [s^2 + (1.84776 \sinh v)s + \sinh^2 v + 0.14645]$$
$$(s + \sinh v)[s^2 + (0.61803 \sinh v)s + \sinh^2 v + 0.90451] \times [s^2 + (1.61803 \sinh v)s + \sinh^2 v + 0.34549]$$

Source: Handbook of Measurement Science, edited by Peter Sydenham, copyright 1982, John Wiley and Sons Limited. Reproduced by permission of John Wiley and Sons Limited.

Thus, for a ripple of $A = 0.25$ dB, we have from Eq. (4.13)

$$\epsilon = \sqrt{(10^{A/10} - 1)} = 0.2434$$

and solving Eq. (4.16) for n with $\omega = 5$ rad/s and $|T(j\omega)| = -60$ dB, we obtain $n = 3.93$. Therefore we must use $n = 4$ to meet these specifications.

Inverse Chebyshev Function

The inverse Chebyshev function has an MFM passband and an equal ripple stopband. This requires finite $j\omega$-axis zeros. We will consider only the two-zero function since it is adequate for most requirements. This function has two advantages over the Butterworth function: (1) a steeper cut-off slope for a given order, and (2) the zero can be placed at any specific frequency in the stopband where, for example, a particular signal must be highly attenuated.

4.3 Low-Pass Filters*

Introduction

Normalized element values are given here for both singly and doubly terminated filters. The source and load resistors are normalized to 1 Ω. Scaling rules will be given in Section 4.4 that will allow these values to be modified to any specified impedance value and to any cut-off frequency desired. In addition, we will cover the **transformation** of these low-pass filters to **high-pass** or **bandpass filters**.

Butterworth Filters

For $n = 2$, 3, 4, or 5, Fig. 4.3 gives the element values for the singly terminated filters and Fig. 4.4 gives the element values for the doubly terminated filters. All cut-off frequencies (–3 dB) are 1 rad/s.

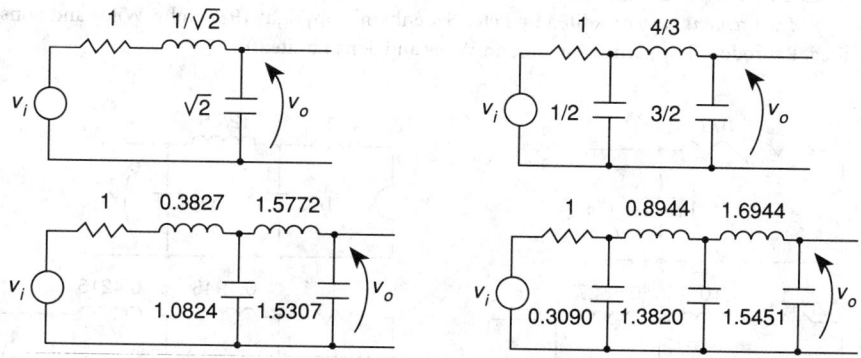

FIGURE 4.3 Singly terminated Butterworth filter element values (in Ω, H, F). (*Source: Handbook of Measurement Science*, edited by Peter Sydenham, copyright 1982, John Wiley and Sons Limited. Reproduced by permission of John Wiley and Sons Limited.)

Thomson Filters

Singly and doubly terminated Thomson filters of order $n = 2, 3, 4, 5$ are shown in Figs. 4.5 and 4.6. All time delays are 1 s. The cut-off frequencies are given in Table 4.2.

Chebyshev Filters

The amount of ripple can be specified as desired, so that only a selective sample can be given here. We will use 0.1 dB, 0.25 dB, and 0.5 dB. All cut-off frequencies (end of ripple for the Chebyshev function) are at 1 rad/s. Since the maximum power transfer condition precludes the existence of an equally terminated even-order filter, only odd orders are given for the doubly terminated case. Figure 4.7 gives the singly terminated Chebyshev filters for $n = 2, 3, 4$, and 5 and Fig. 4.8 gives the doubly terminated Chebyshev filters for $n = 3$ and $n = 5$.

Inverse Chebyshev Filters

For the two-zero, three-pole case shown in Fig. 4.9, the element values are given in Table 4.4 for various zero positions. In addition, the stopband peak return magnitude is also given. Note that R_L is either 1 Ω or ∞, so that both singly and doubly terminated filters are included.

The two-zero, four-pole filter is shown in Fig. 4.10 and the element values for the various zero positions are shown in Table 4.5.

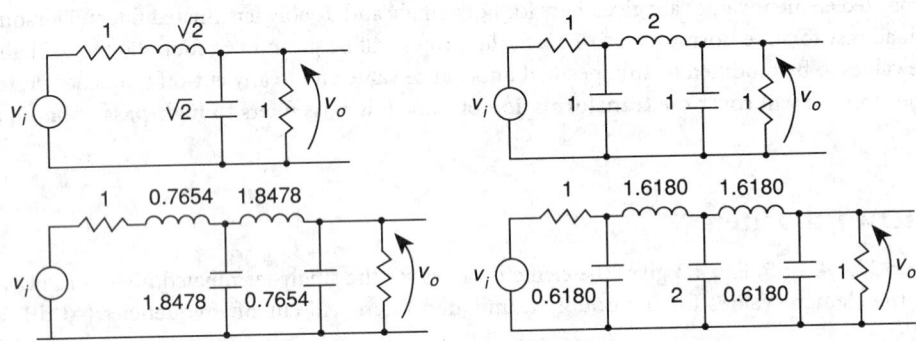

FIGURE 4.4 Doubly terminated Butterworth filter element values (in Ω, H, F). (*Source: Handbook of Measurement Science*, edited by Peter Sydenham, copyright 1982, John Wiley and Sons Limited. Reproduced by permission of John Wiley and Sons Limited.)

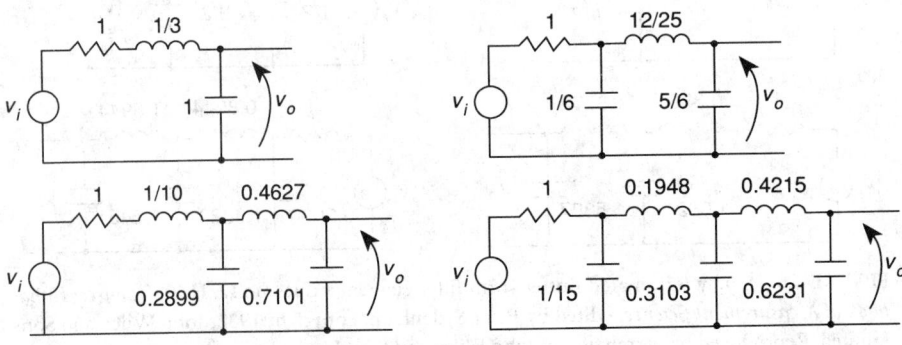

FIGURE 4.5 Singly terminated Thomson filter element values (in Ω, H, F). (*Source: Handbook of Measurement Science*, edited by Peter Sydenham, copyright 1982, John Wiley and Sons Limited. Reproduced by permission of John Wiley and Sons Limited.)

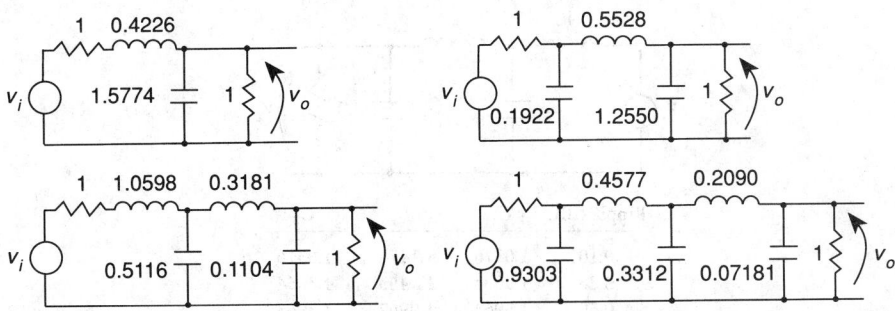

FIGURE 4.6 Doubly terminated Thomson filter element values (in Ω, H, F). (*Source: Handbook of Measurement Science,* edited by Peter Sydenham, copyright 1982, John Wiley and Sons Limited. Reproduced by permission of John Wiley and Sons Limited.)

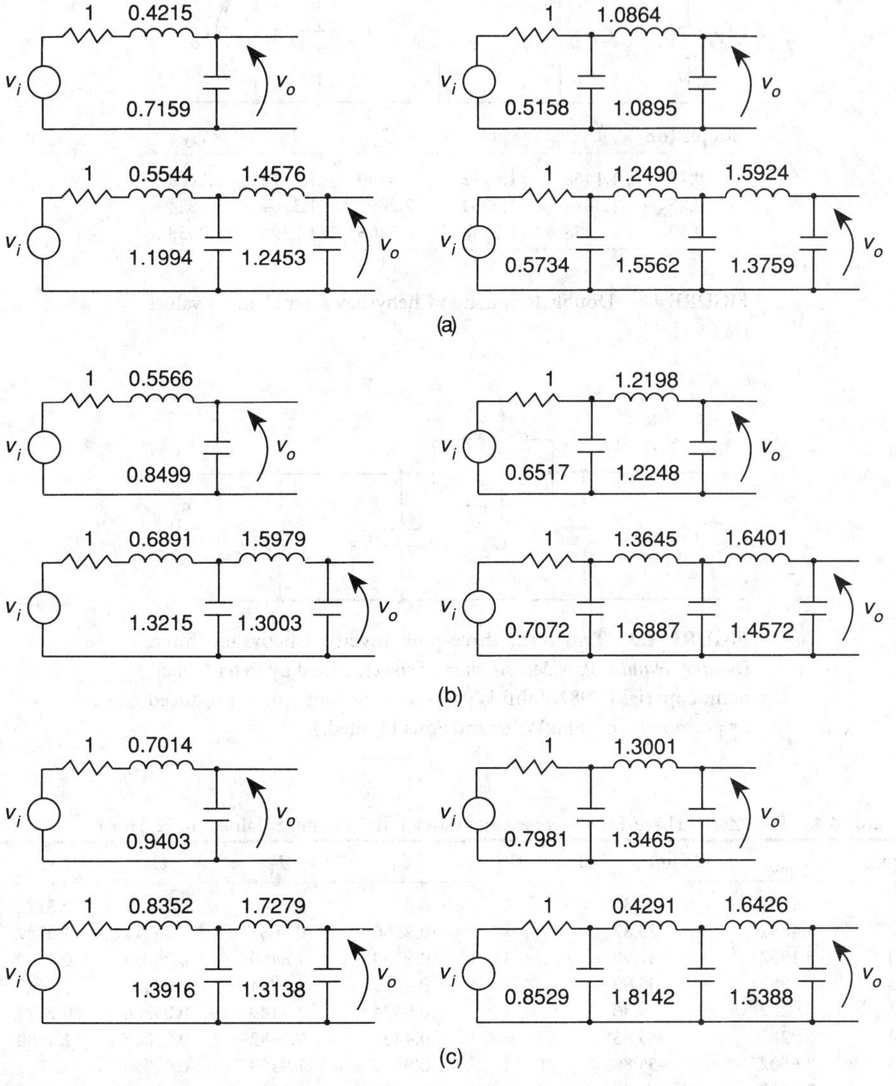

FIGURE 4.7 Singly terminated Chebyshev filter element values (in Ω, H, F): (a) 0.1-dB ripple; (b) 0.25-dB ripple; (c) 0.50-dB ripple. (*Source: Handbook of Measurement Science,* edited by Peter Sydenham, copyright 1982, John Wiley and Sons Limited. Reproduced by permission of John Wiley and Sons Limited.)

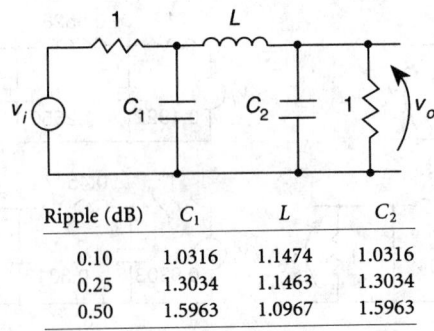

Ripple (dB)	C_1	L	C_2
0.10	1.0316	1.1474	1.0316
0.25	1.3034	1.1463	1.3034
0.50	1.5963	1.0967	1.5963

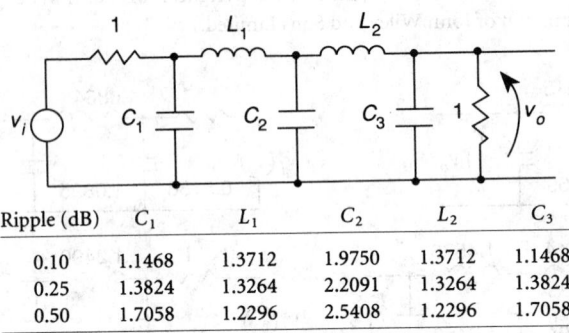

Ripple (dB)	C_1	L_1	C_2	L_2	C_3
0.10	1.1468	1.3712	1.9750	1.3712	1.1468
0.25	1.3824	1.3264	2.2091	1.3264	1.3824
0.50	1.7058	1.2296	2.5408	1.2296	1.7058

FIGURE 4.8 Doubly terminated Chebyshev filter element values (in Ω, H, F).

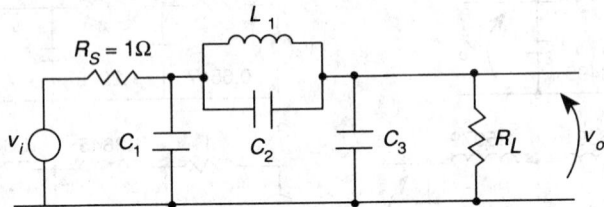

FIGURE 4.9 Two-zero, three-pole inverse Chebyshev filter. (*Source: Handbook of Measurement Science*, edited by Peter Sydenham, copyright 1982, John Wiley and Sons Limited. Reproduced by permission of John Wiley and Sons Limited.)

Table 4.4 Two-Zero, Three-Pole Inverse Chebyshev Filter Element Values (in Ω, H, F)

$\omega_{-\infty}$	ω_{max}	$|T(j\omega)|_{\omega_{max}}$ (dB)	R_L	C_1	L_1	C_2	C_3
2	3.4641	−29.87	1	0.8172	1.6344	0.1530	0.8172
2	3.4641	−23.87	∞	0.2556	0.9687	0.2581	1.3787
3	5.1962	−41.90	1	0.9230	1.8460	0.06019	0.9230
3	5.1962	−35.90	∞	0.4013	1.1794	0.09421	1.4447
4	6.9282	−49.86	1	0.9574	1.9149	0.03264	0.9574
4	6.9282	−43.86	∞	0.4461	1.2482	0.05007	1.4688
5	8.6602	−55.88	1	0.9730	1.9459	0.02056	0.9730
5	8.6602	−49.88	∞	0.4659	1.2793	0.03127	1.4800

Source: Handbook of Measurement Science, edited by Peter Sydenham, copyright 1982, John Wiley and Sons Limited. Reproduced by permission of John Wiley and Sons Limited.

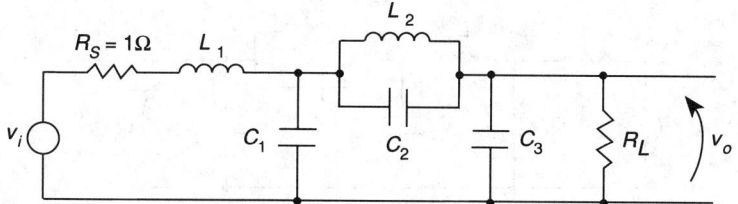

FIGURE 4.10 Two-zero, four-pole inverse Chebyshev filter. (*Source: Handbook of Measurement Science*, edited by Peter Sydenham, copyright 1982, John Wiley and Sons Limited. Reproduced by permission of John Wiley and Sons Limited.)

Table 4.5 Two-Zero, Four-Pole Inverse Chebyshev Filter Element Values (in Ω, H, F)

$\omega_{-\infty}$	ω_{max}	$|T(j\omega)|_{\omega_{max}}$(dB)	R_L	L_1	C_1	L_2	C_2	C_3
2	2.8284	−39.63	1	0.7350	1.6723	1.5189	0.16460	0.5816
2	2.8284	−33.63	∞	0.3666	0.8649	1.2338	0.20260	1.3890
3	4.2426	−55.19	1	0.7500	1.7703	1.7121	0.06490	0.6918
3	4.2426	−49.19	∞	0.3761	0.9928	1.4327	0.07756	1.4693
4	5.6568	−65.65	1	0.7570	1.8050	1.7726	0.03526	0.7247
4	5.6568	−59.65	∞	0.3791	1.0332	1.4973	0.04174	1.4965
5	7.0711	−73.60	1	0.7600	1.8205	1.8001	0.02222	0.7396
5	7.0711	−67.60	∞	0.3804	1.0512	1.5265	0.02620	1.5089

Source: Handbook of Measurement Science, edited by Peter Sydenham, copyright 1982, John Wiley and Sons Limited. Reproduced by permission of John Wiley and Sons Limited.

The two-zero, five-pole filter is shown in Fig. 4.11 and the element values for various zero positions are shown in Table 4.6.

Interpolation to find element values for zero frequencies between those given in Tables 4.4, 4.5, and 4.6 is quite adequate, particularly for the doubly terminated (equal values) case.

4.4 Filter Design

We now consider the steps necessary to convert normalized filters into actual filters by scaling both in frequency and in impedance. In addition, we will cover the transformation laws that convert low-pass filters to high-pass filters and low-pass to bandpass filters.

Scaling Laws and a Design Example

Since all data previously given are for normalized filters, it is necessary to use the scaling rules to design a low-pass filter for a specific signal processing application.

Rule 1. All impedances may be multiplied by any constant without affecting the transfer voltage ratio.

Rule 2. To modify the cut-off frequency, divide all inductors and capacitors by the ratio of the desired frequency to the normalized frequency.

Example 4.6. Design a low-pass filter of MFM type (Butterworth) to operate from a 600-Ω source into a 600-Ω load, with a cut-off frequency of 500 Hz. The filter must be at least 36 dB below the dc level at 2 kHz, that is, −42 dB (dc level is −6 dB).

Since 2 kHz is four times 500 Hz, it corresponds to $\omega = 4$ rad/s in the normalized filter. Thus at $\omega = 4$ rad/s we have

$$-42 \text{ dB} = 20 \log \frac{1}{2}\left[\frac{1}{\sqrt{4^{2n} + 1}}\right]$$

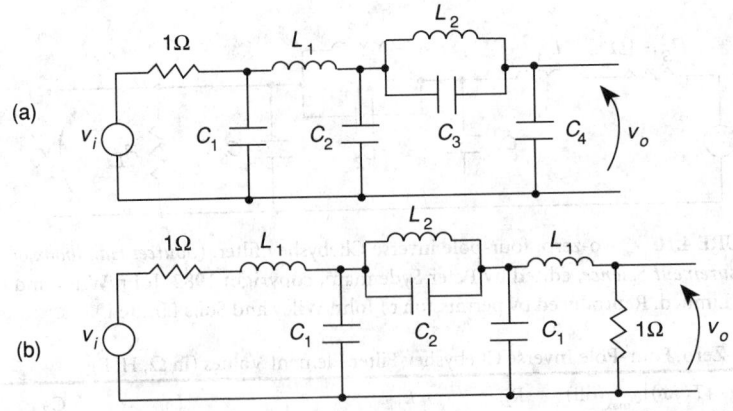

FIGURE 4.11 Two-zero, five-pole inverse Chebyshev filter: (a) singly terminated; (b) doubly terminated. (*Source:* (a) *Handbook of Measurement Science*, edited by Peter Sydenham, copyright 1982, John Wiley and Sons Limited. Reproduced by permission of John Wiley and Sons Limited.)

Table 4.6 Two-Zero, Five-Pole Inverse Chebyshev Filter Element Values (in Ω, H, F)

$\omega_{-\infty}$	ω_{max}	$|T(j\omega)|_{\omega_{max}}$(dB)	C_1	L_1	C_2	L_2	C_3	C_4
(a) Singly terminated, Fig. 4.11(a)								
2	2.5820	−42.22	0.2981	0.8649	1.18240	1.3678	0.18280	1.3996
3	3.8730	−61.30	0.3045	0.8824	1.29910	1.5568	0.07137	1.4828
4	5.1640	−74.26	0.3066	0.8878	1.33640	1.6184	0.03862	1.5105
5	6.4550	−84.16	0.3075	0.8902	1.35300	1.6461	0.02430	1.5231
(b) Doubly terminated, Fig. 4.11(b)								
2	2.5820	−48.24	1.4401	0.5962	0.14812	1.6878		
3	3.8730	−67.32	1.5432	0.6091	0.05947	1.8683		
4	5.1640	−80.28	1.5767	0.6131	0.03243	1.9272		
5	6.4550	−90.18	1.5918	0.6149	0.02047	1.9537		

therefore, $n = 2.99$, so $n = 3$ must be chosen. The 1/2 is present because this is a doubly terminated (equal values) filter so that the dc gain is 1/2.

Thus a third-order, doubly terminated Butterworth filter is required. From Fig. 4.4 we obtain the normalized network shown in Fig. 4.12(a).

The **impedance scaling** factor is $600/1 = 600$ and the **frequency scaling** factor is $2\pi500/1 = 2\pi500$: that is, the ratio of the desired radian cut-off frequency to the normalized cut-off frequency (1 rad/s). Note that the impedance scaling factor increases the size of the resistors and inductors, but reduces the size of the capacitors. The result is shown in Fig. 4.12(b).

Transformation Rules, Passive Circuits

All information given so far applies only to low-pass filters, yet we frequently need high-pass or bandpass filters in signal processing.

Low-Pass to High-Pass Transformation

To transform a low-pass filter to high-pass, we first scale it to a cut-off frequency of 1 rad/s if it is not already at 1 rad/s. This allows a simple frequency rotation about 1 rad/s of $s \to 1/s$. All L's become C's, all C's become L's, and all values reciprocate. The cut-off frequency does not change.

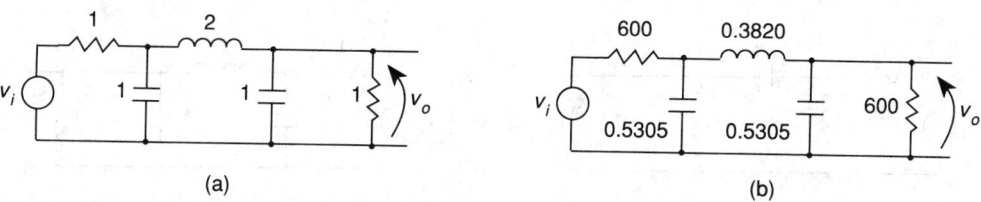

FIGURE 4.12   Third-order Butterworth low-pass filter: (a) normalized (in Ω, H, F); (b) scaled in (in Ω, H, F).

Example 4.7.   Design a third-order, high-pass Butterworth filter to operate from a 600-Ω source to a 600-Ω load with a cut-off frequency of 500 Hz.

Starting with the normalized third-order low-pass filter of Fig. 4.4 for which $\omega_{-3} = 1$ rad/s, we reciprocate all elements and all values to obtain the filter shown in Fig. 4.13(a) for which $\omega_{-3} = 1$ rad/s.

Now we apply the scaling rules to raise all impedances to 600 Ω and the radian cut-off frequency to 2π500 rad/s as shown in Fig. 4.13(b).

Low-Pass to Bandpass Transformation

To transform a low-pass filter to a bandpass filter we must first scale the low-pass filter so that the cut-off frequency is equal to the bandwidth of the normalized bandpass filter. The normalized center frequency of the bandpass filter is $\omega_0 = 1$ rad/s. Then we apply the transformation $s \rightarrow s + 1/s$. For an inductor

$$Z = Ls \text{ transforms to } Z = L\left(s + \frac{1}{s}\right)$$

For a capacitor

$$Y = Cs \text{ transforms to } Y = C\left(s + \frac{1}{s}\right)$$

The first step is then to determine the Q of the bandpass filter where

$$Q = \frac{f_0}{B} = \frac{\omega_0}{B_r}$$

(f_0 is the center frequency in Hz and B is the 3-dB bandwidth in Hz). Now we scale the low-pass filter to a cut-off frequency of $1/Q$ rad/s, then series tune every inductor, L, with a capacitor of value $1/L$ and parallel tune every capacitor, C, with an inductor of value $1/C$.

Example 4.8.   Design a bandpass filter centered at 100 kHz having a 3-dB bandwidth of 10 kHz starting with a third-order Butterworth low-pass filter. The source and load resistors are each to be 600 Ω.

The Q required is

$$Q = \frac{100 \text{ kHz}}{10 \text{ kHz}} = 10, \text{ or } \frac{1}{Q} = 0.1$$

Scaling the normalized third-order low-pass filter of Fig. 4.14(a) to $\omega_{-3dB} = 1/Q = 0.1$ rad/s, we obtain the filter of Fig. 4.14(b).

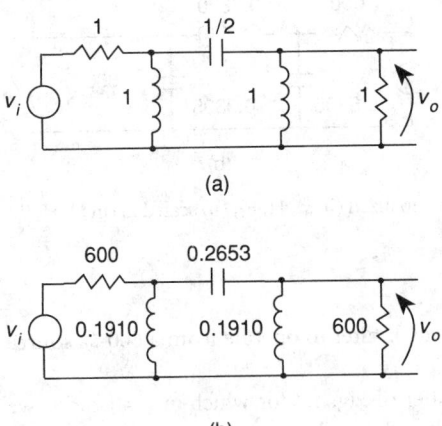

(a)

(b)

FIGURE 4.13 Third-order Butterworth high-pass filter: (a) normalized (in Ω, H, F); (b) scaled (in Ω, H, F).

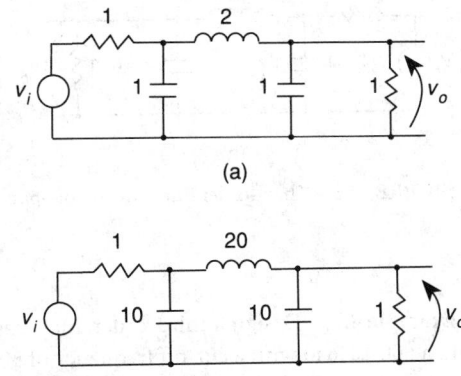

(a)

(b)

FIGURE 4.14 Third-order Butterworth low-pass filter: (a) normalized (in Ω, H, F); (b) scaled in (in Ω, H, F).

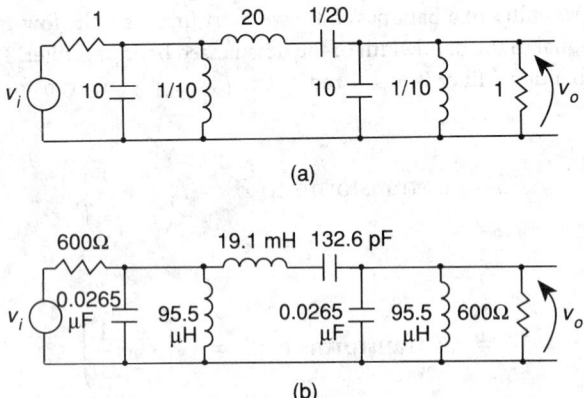

(a)

(b)

FIGURE 4.15 Sixth-order Butterworth bandpass filter ($Q = 10$): (a) normalized, $\omega_0 = 1$ rad/s (in Ω, H, F); (b) scaled.

Now converting to bandpass with $\omega_0 = 1$ rad/s, we obtain the normalized bandpass filter of Fig. 4.15(a). Next, scaling to an impedance of 600 Ω and to a center frequency of $f_0 = 100$ kHz ($\omega_0 = 2\pi 100$ k rad/s), we obtain the filter of Fig. 4.15(b).

Defining Terms

Bandpass filter: A filter whose passband extends from a finite lower cut-off frequency to a finite upper cut-off frequency.

Equal ripple: A frequency response function whose magnitude has equal maxima and equal minima in the passband.

Frequency scaling: The process of modifying a filter to change from a normalized set of element values to other usually more practical values by dividing all L, C elements by a constant equal to the ratio of the scaled (cut-off) frequency desired to the normalized cut-off frequency.

High-pass filter: A filter whose band extends from some finite cut-off frequency to infinity.

Impedance scaling: Modifying a filter circuit to change from a normalized set of element values to other usually more practical element values by multiplying all impedances by a constant equal to the ratio of the desired (scaled) impedance to the normalized impedance.

Low-pass filter: A filter whose passband extends from dc to some finite cut-off frequency.

Maximally flat delay (MFD) filter: A filter having a time delay that is as flat (constant) as possible versus frequency while maintaining a monotonic characteristic.

Maximally flat magnitude (MFM) filter: A filter having a magnitude that is as flat as possible versus frequency while maintaining a monotonic characteristic.

Octave: A frequency ratio of two.

Passband: A frequency region of signal transmission usually within 3 dB of the maximum transmission.

Sinusoidal steady-state response: Response of a circuit to a sine wave input as $t \to \infty$. The output then has two components, a magnitude and a phase, that are the magnitude and the phase of the transfer function itself for $s = j\omega$.

Stopband: The frequency response region in which the signal is attenuated, usually by more than 3 dB from the maximum transmission.

Transfer function: The Laplace transform of the response (output voltage) divided by the Laplace transform of the excitation (input voltage).

Transformation: The modification of a low-pass filter to convert it to an equivalent high-pass or bandpass filter.

References

A. Budak, *Passive and Active Network Analysis and Synthesis*, Boston: Houghton Mifflin, 1974.

C. Nachtigal, Ed., *Instrumentation and Control: Fundamentals and Applications*, New York: John Wiley, 1990.

H.-J. Orchard, "Inductorless filters," *Electron. Lett.*, vol. 2, pp. 224–225, 1966.

P. Sydenham, Ed., *Handbook of Measurement Science*, Chichester, U.K.: John Wiley, 1982.

W. E. Thomson, "Maximally flat delay networks," *IRE Transactions*, vol. CT-6, p. 235, 1959.

L. Weinberg, *Network Analysis and Synthesis*, New York: McGraw-Hill, 1962.

L. Weinberg and P. Slepian, "Takahasi's results on Tchebycheff and Butterworth ladder networks," *IRE Transactions, Professional Group on Circuit Theory*, vol. CT-7, no. 2, pp. 88–101, 1960.

5
Nonlinear Circuits

Jerry L. Hudgins
University of South Carolina

Theodore F. Bogart, Jr.
University of Southern Mississippi

Kartikeya Mayaram
AT&T Bell Laboratories

5.1 Diodes and Rectifiers 126
    Diodes • Rectifiers
5.2 Limiters 132
    Limiting Circuits • Precision Rectifying Circuits
5.3 Distortion 139
    Harmonic Distortion • Power-Series Method • Differential-Error
    Method • Three-Point Method • Five-Point Method • Intermodu-
    lation Distortion • Triple-Beat Distortion • Cross Modulation •
    Crossover Distortion • Failure-to-Follow Distortion • Frequency
    Distortion • Phase Distortion • Computer Simulation of Distortion
    Components

5.1 Diodes and Rectifiers

Jerry L. Hudgins

A **diode** generally refers to a two-terminal solid-state semiconductor device that presents a low impedance to current flow in one direction and a high impedance to current flow in the opposite direction. These properties allow the diode to be used as a one-way current valve in electronic circuits. *Rectifiers* are a class of circuits whose purpose is to convert ac waveforms (usually sinusoidal and with zero average value) into a waveform that has a significant non-zero average value (dc component). Simply stated, rectifiers are ac-to-dc energy converter circuits. Most rectifier circuits employ diodes as the principal elements in the energy conversion process; thus the almost inseparable notions of diodes and rectifiers. The general electrical characteristics of common diodes and some simple rectifier topologies incorporating diodes are discussed.

Diodes

Most diodes are made from a host crystal of silicon (Si) with appropriate impurity elements introduced to modify, in a controlled manner, the electrical characteristics of the device. These diodes are the typical ***pn*-junction** (or **bipolar**) devices used in electronic circuits. Another type is the **Schottky diode** (unipolar), produced by placing a metal layer directly onto the semiconductor [Schottky, 1938; Mott, 1938]. The metal-semiconductor interface serves the same function as the *pn*-junction in the common diode structure. Other semiconductor materials such as gallium-arsenide (GaAs) and silicon-carbide (SiC) are also in use for new and specialized applications of diodes. Detailed discussions of various diode structures, fabrication technology, and the physics of their operation can be found in later sections of this paper.

The electrical circuit symbol for a bipolar diode is shown in Fig. 5.1. The polarities associated with the forward voltage drop for forward current flow are also included. Current or voltage oppo-

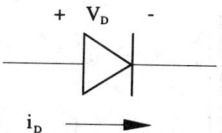

+ V_D -

i_D ———→

FIGURE 5.1 Circuit symbol for a bipolar diode indicating the polarity associated with the forward voltage and current directions.

site to the polarities indicated in Fig. 5.1 are considered to be negative values with respect to the diode conventions shown.

The characteristic curve shown in Fig. 5.2 is representative of the current-voltage dependencies of typical diodes. The diode conducts forward current with a small forward voltage drop across the device, simulating a closed switch. The relationship between the forward current and forward voltage is approximately given by the Shockley diode equation [Shockley, 1949]:

$$i_D = I_s \left[\exp\left(\frac{qV_D}{nkT} \right) - 1 \right] \tag{5.1}$$

where I_s is the leakage current through the diode, q is the electronic charge, n is a correction factor, k is Boltzmann's constant, and T is the temperature of the semiconductor. Around the knee of the curve in Fig. 5.2 is a positive voltage that is termed the turn-on or sometimes the threshold voltage for the diode. This value is an approximate voltage above which the diode is considered turned "on" and can be modeled to first degree as a closed switch with constant forward drop. Below the threshold voltage value the diode is considered weakly conducting and approximated as an open switch. The exponential relationship shown in Eq. (5.1) means that the diode forward current can change by orders of magnitude before there is a large change in diode voltage, thus providing the simple circuit model during conduction. The nonlinear relationship of Eq. (5.1) also provides a means of frequency mixing for applications in modulation circuits.

Reverse voltage applied to the diode causes a small leakage current (negative according to the sign convention) to flow that is typically orders of magnitude lower than current in the forward direction. The diode can withstand reverse voltages up to a limit determined by its physical construction and the semiconductor material used. Beyond this value the reverse voltage imparts enough energy to the charge carriers to cause large increases in current. The mechanisms by which this current increase occurs are impact ionization (avalanche) [McKay, 1954] and a tunneling phenomenon (Zener breakdown) [Moll, 1964]. Avalanche breakdown results in large power dissipation in the diode, is generally destructive, and should be avoided at all times. Both breakdown regions are superimposed in Fig. 5.2 for comparison of their effects on the shape of the diode characteristic curve. Avalanche breakdown occurs for reverse applied voltages in the range of volts to kilovolts depending on the exact design of the diode. Zener breakdown occurs at much lower volt-

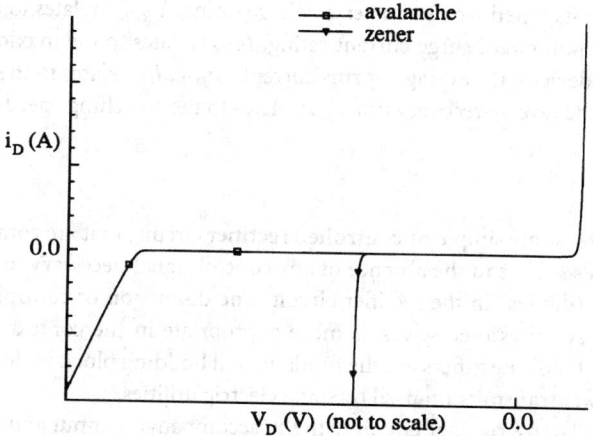

FIGURE 5.2 A typical diode dc characteristic curve showing the current dependence on voltage.

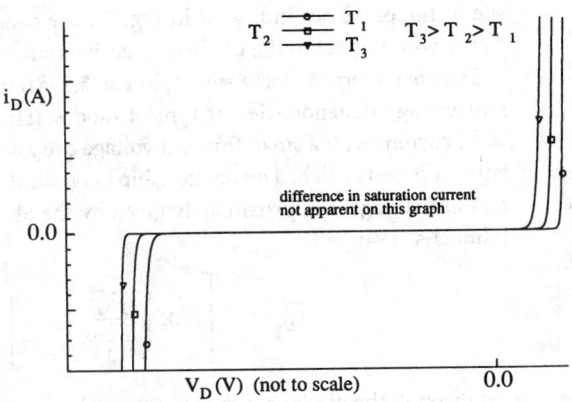

FIGURE 5.3 The effects of temperature variations on the forward voltage drop and the avalanche breakdown voltage in a bipolar diode.

ages than the avalanche mechanism. Diodes specifically designed to operate in the Zener breakdown mode are used extensively as voltage regulators in regulator integrated circuits and as discrete components in large regulated power supplies.

During forward conduction the power loss in the diode can become excessive for large current flow. Schottky diodes have an inherently lower turn-on voltage than *pn*-junction diodes and are therefore more desirable in applications where the energy losses in the diodes are significant (such as output rectifiers in switching power supplies). Other considerations such as recovery characteristics from forward conduction to reverse blocking may also make one diode type more desirable than another. Schottky diodes conduct current with one type of charge carrier and are therefore inherently faster to turn off than bipolar diodes. However, one of the limitations of Schottky diodes is their excessive forward voltage drop when designed to support reverse biases above about 200 V. Therefore, high-voltage diodes are the *pn*-junction type.

The effects due to an increase in the temperature in a bipolar diode are many. The forward voltage drop during conduction will decrease over a large current range, the reverse leakage current will increase, and the reverse avalanche breakdown voltage (V_{BD}) will increase as the device temperature climbs. A family of static characteristic curves highlighting these effects is shown in Fig. 5.3 where $T_3 > T_2 > T_1$. In addition, a major effect on the switching characteristic is the increase in the reverse recovery time during turn-off. Some of the key parameters to be aware of when choosing a diode are its repetitive peak inverse voltage rating, V_{RRM} (relates to the avalanche breakdown value), the peak forward surge current rating, I_{FSM} (relates to the maximum allowable transient heating in the device), the average or rms current rating, I_O (relates to the steady-state heating in the device), and the reverse recovery time, t_{rr} (relates to the switching speed of the device).

Rectifiers

This section discusses some simple **uncontrolled rectifier** circuits that are commonly encountered. The term *uncontrolled* refers to the absence of any control signal necessary to operate the primary switching elements (diodes) in the rectifier circuit. The discussion of controlled rectifier circuits, and the controlled switches themselves, is more appropriate in the context of power electronics applications [Hoft, 1986]. Rectifiers are the fundamental building block in dc power supplies of all types and in dc power transmission used by some electric utilities.

A single-phase full-wave rectifier circuit with the accompanying input and output voltage waveforms is shown in Fig. 5.4. This topology makes use of a center-tapped transformer with each diode conducting on opposite half-cycles of the input voltage. The forward drop across the diodes is

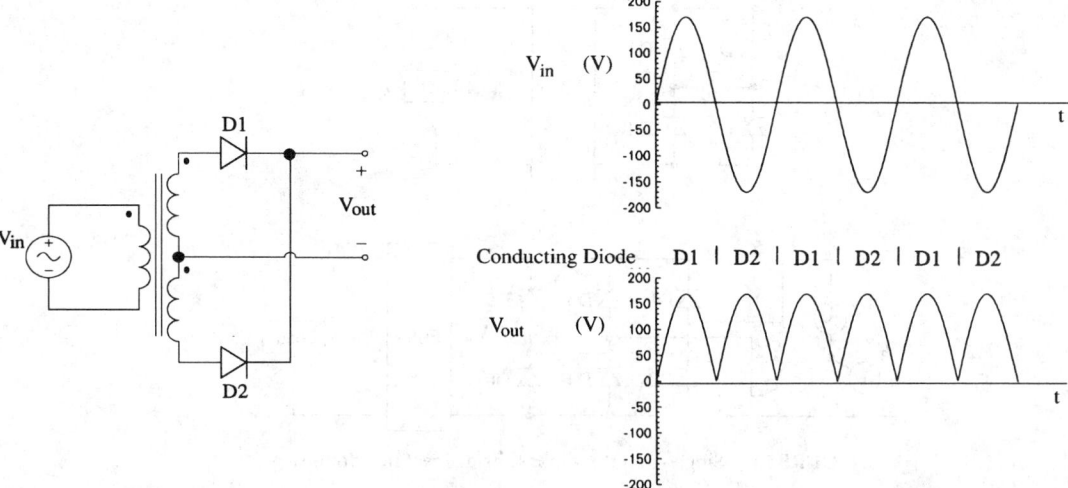

FIGURE 5.4 A single-phase full-wave rectifier circuit using a center-tapped transformer with the associated input and output waveforms.

ignored on the output graph, which is a valid approximation if the peak voltages of the input and output are large compared to 1 V. The circuit changes a sinusoidal waveform with no dc component (zero average value) to one with a dc component of $2V_{peak}/\pi$. The rms value of the output is $0.707V_{peak}$.

The dc value can be increased further by adding a low-pass filter in cascade with the output. The usual form of this filter is a shunt capacitor or an LC filter as shown in Fig. 5.5. The resonant frequency of the LC filter should be lower than the fundamental frequency of the rectifier output for effective performance. The ac portion of the output signal is reduced while the dc and rms values are increased by adding the filter. The remaining ac portion of the output is called the **ripple**. Though somewhat confusing, the transformer, diodes, and filter are often collectively called the rectifier circuit.

Another circuit topology commonly encountered is the bridge rectifier. Figure 5.6 illustrates single- and three-phase versions of the circuit. In the single-phase circuit diodes D1 and D4 conduct on the positive half-cycle of the input while D2 and D3 conduct on the negative half-cycle of the input. Alternate pairs of diodes conduct in the three-phase circuit depending on the relative amplitude of the source signals.

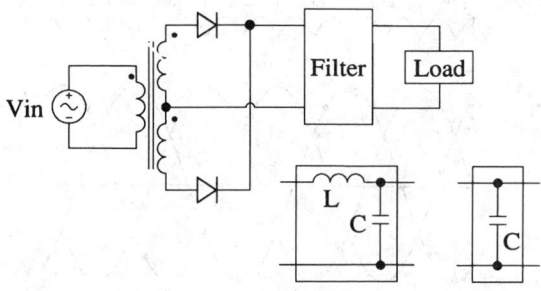

Two Types of Filters

FIGURE 5.5 A single-phase full-wave rectifier with the addition of an output filter.

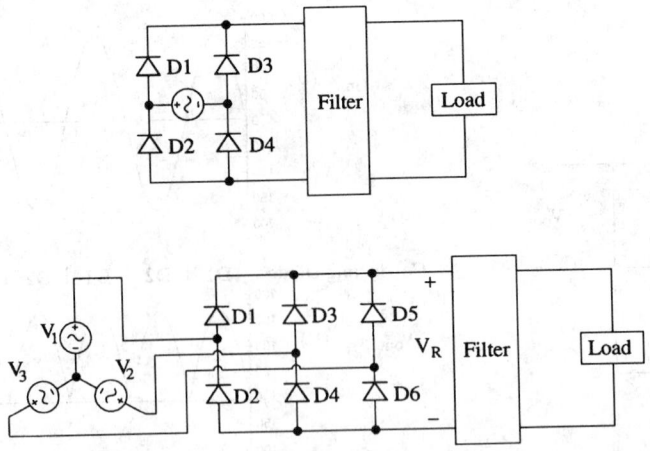

FIGURE 5.6   Single- and three-phase bridge rectifier circuits.

The three-phase inputs with the associated rectifier output voltage are shown in Fig. 5.7 as they would appear without the low-pass filter section. The three-phase bridge rectifier has a reduced ripple content of 4% as compared to a ripple content of 47% in the single-phase bridge rectifier [Milnes, 1980]. The corresponding diodes that conduct are also shown at the top of the figure. This output waveform assumes a purely resistive load connected as shown in Fig. 5.6. Most loads (motors, transformers, etc.) and many sources (power grid) include some inductance, and in fact may be dominated by inductive properties. This causes phase shifts between the input and output waveforms. The rectifier output may thus vary in shape and phase considerably from that shown in Fig. 5.7 [Kassakian *et al.*, 1991]. When other types of switches are used in these circuits the inductive elements can induce large voltages that may damage sensitive or expensive components. Diodes are used regularly in such circuits to shunt current and clamp induced voltages at low levels to protect expensive components such as electronic switches.

One variation of the typical rectifier is the Cockroft-Walton circuit used to obtain high voltages without the necessity of providing a high-voltage transformer. The circuit in Fig. 5.8 multiplies the peak secondary voltage by a factor of six. The steady-state voltage level at each filter capacitor node is shown in the figure. Adding additional stages increases the load voltage further. As in other rectifier circuits, the value of the capacitors will determine the amount of ripple in the output waveform

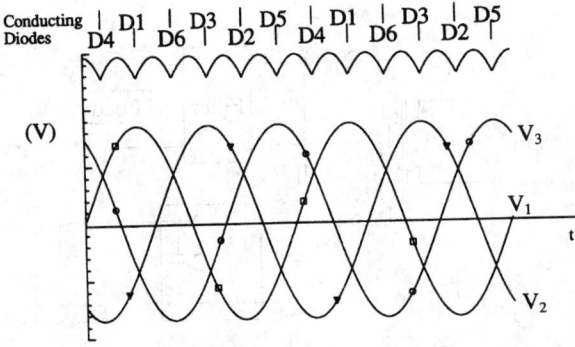

FIGURE 5.7   Three-phase rectifier output compared to the input signals. The input signals as well as the conducting diode labels are those referenced to Fig. 5.6.

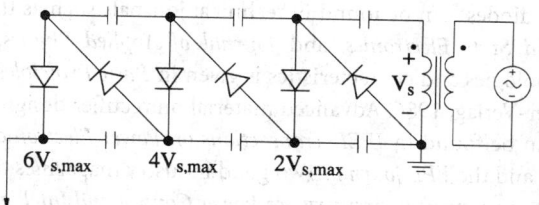

FIGURE 5.8   Cockroft-Walton circuit used for voltage multiplication.

for given load resistance values. In general, the capacitors in a lower voltage stage should be larger than in the next highest voltage stage.

Defining Terms

Bipolar device: Semiconductor electronic device that uses positive and negative charge carriers to conduct electric current.

Diode: Two-terminal solid-state semiconductor device that presents a low impedance to current flow in one direction and a high impedance to current flow in the opposite direction.

***pn*-junction:**   Metallurgical interface of two regions in a semiconductor where one region contains impurity elements that create equivalent positive charge carriers (*p*-type) and the other semiconductor region contains impurities that create negative charge carriers (*n*-type).

Ripple: The ac (time-varying) portion of the output signal from a rectifier circuit.

Schottky diode: A diode formed by placing a metal layer directly onto a unipolar semiconductor substrate.

Uncontrolled rectifier: A rectifier circuit employing switches that do not require control signals to operate them in their "on" or "off" states.

References

R.G. Hoft, *Semiconductor Power Electronics,* New York: Van Nostrand Reinhold, 1986.

J.G. Kassakian, M.F. Schlecht, and G.C. Verghese, *Principles of Power Electronics,* Reading, Mass.: Addison-Wesley, 1991.

K.G. McKay, "Avalanche breakdown in silicon," *Physical Review,* vol. 94, p. 877, 1954.

A.G. Milnes, *Semiconductor Devices and Integrated Electronics,* New York: Van Nostrand Reinhold, 1980.

J.L. Moll, *Physics of Semiconductors,* New York: McGraw-Hill, 1964.

N.F. Mott, "Note on the contact between a metal and an insulator or semiconductor," *Proc. Cambridge Philos. Soc.,* vol. 34, p. 568, 1938.

W. Schottky, "Halbleitertheorie der Sperrschicht," *Naturwissenschaften,* vol. 26, p. 843, 1938.

W. Shockley, "The theory of p-n junctions in semiconductors and p-n junction transistors," *Bell System Tech. J.,* vol. 28, p. 435, 1949.

Further Information

A good introduction to solid-state electronic devices with a minimum of mathematics and physics is *Solid State Electronic Devices,* 3rd edition, by B.G. Streetman, Prentice-Hall, 1989. A rigorous and more detailed discussion is provided in *Physics of Semiconductor Devices,* 2nd edition, by S.M. Sze, John Wiley & Sons, 1981. Both of these books discuss many specialized diode structures as well as other semiconductor devices. Advanced material on the most recent developments in semiconductor devices, including diodes, can be found in technical journals such as the *IEEE Transactions on Electron Devices, Solid State Electronics,* and *Journal of Applied Physics.* A good summary of advanced rectifier topologies and characteristics is given in *Basic Principles of Power Electronics* by K. Heumann, Springer-Verlag, 1986. Advanced material on rectifier designs as well as other power electronics circuits can be found in *IEEE Transactions on Power Electronics, IEEE Transactions on Industry Applications,* and the *EPE Journal.* Two good industry magazines that cover power devices such as diodes and power converter circuitry are *Power Control and Intelligent Motion* (PCIM) and *Power Technics.*

5.2 Limiters*

Theodore F. Bogart, Jr.

Limiters are named for their ability to limit voltage excursions at the output of a circuit whose input may undergo unrestricted variations. They are also called *clipping circuits* because waveforms having rounded peaks that exceed the limit(s) imposed by such circuits appear, after limiting, to have their peaks flattened, or "clipped" off. Limiters may be designed to clip positive voltages at a certain level, negative voltages at a different level, or to do both. The simplest types consist simply of diodes and dc voltage sources, while more elaborate designs incorporate operational amplifiers.

Limiting Circuits

Figure 5.9 shows how the transfer characteristics of limiting circuits reflect the fact that outputs are clipped at certain levels. In each of the examples shown, note that the characteristic becomes horizontal at the output level where clipping occurs. The horizontal line means that the output remains constant regardless of the input level in that region. Outside of the clipping region, the transfer characteristic is simply a line whose slope equals the gain of the device. This is the region of linear operation. In these examples, the devices are assumed to have unity gain, so the slope of each line in the linear region is 1.

Figure 5.10 illustrates a somewhat different kind of limiting action. Instead of the positive or negative peaks being clipped, the output follows the input when the signal is

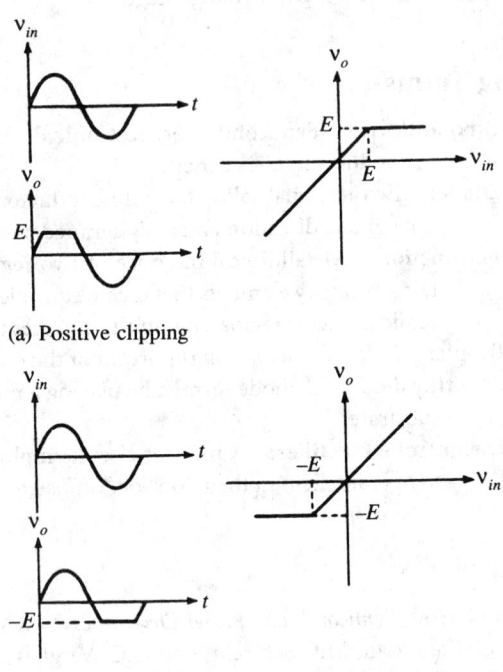

(a) Positive clipping

(b) Negative clipping

(c) Positive and negative clipping

FIGURE 5.9 Waveforms and transfer characteristics of limiting circuits. (*Source:* T.F. Bogart, Jr., *Electronic Devices and Circuits,* 2nd ed., Columbus, Ohio: Macmillan/Merrill, 1990, p. 676. With permission.)

*Excerpted from T.F. Bogart, Jr., *Electronic Devices and Circuits,* 2nd ed., Columbus, Ohio: Macmillan/ Merrill, 1990, pp. 675–683. With permission.

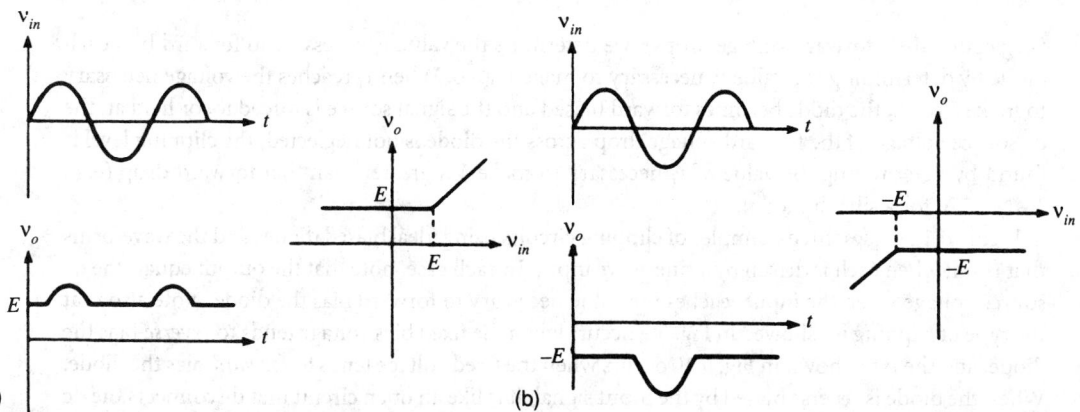

FIGURE 5.10 Another form of clipping. Compare with Fig. 5.9. (*Source:* T. F. Bogart, Jr., *Electronic Devices and Circuits,* 2nd ed., Columbus, Ohio: Macmillan/Merrill, 1990, p. 677. With permission.)

above or below a certain level. The transfer characteristics show that linear operation occurs only when certain signal levels are reached and that the output remains constant below those levels. This form of limiting can also be thought of as a special case of that shown in Fig. 5.9. Imagine, for example, that the clipping level in Fig. 5.9(b) is raised to a positive value; then the result is the same as Fig. 5.10(a).

Limiting can be accomplished using **biased diodes.** Such circuits rely on the fact that diodes have very low impedances when they are forward biased and are essentially open circuits when reverse biased. If a certain point in a circuit, such as the output of an amplifier, is connected through a very small impedance to a *constant* voltage, then the voltage at the circuit point cannot differ significantly from the constant voltage. We say in this case that the point is *clamped* to the fixed voltage. An ideal, forward-biased diode is like a closed switch, so if it is connected between a point in a circuit and a fixed voltage source, the diode very effectively holds the point to the fixed voltage. Diodes can be connected in operational amplifier circuits, as well as other circuits, in such a way that they become forward biased when a signal reaches a certain voltage. When the forward-biasing level is reached, the diode serves to hold the output to a fixed voltage and thereby establishes a clipping level.

A biased diode is simply a diode connected to a fixed voltage source. The value and polarity of the voltage source determine what value of total voltage across the combination is necessary to forward bias the diode. Figure 5.11 shows several examples. (In practice, a series resistor would be connected in each circuit to limit current flow when the diode is forward biased.) In each part of the figure, we can write Kirchhoff's voltage law around the loop to determine the value of input voltage v_i that is necessary to forward bias the diode. Assuming that the diodes are ideal

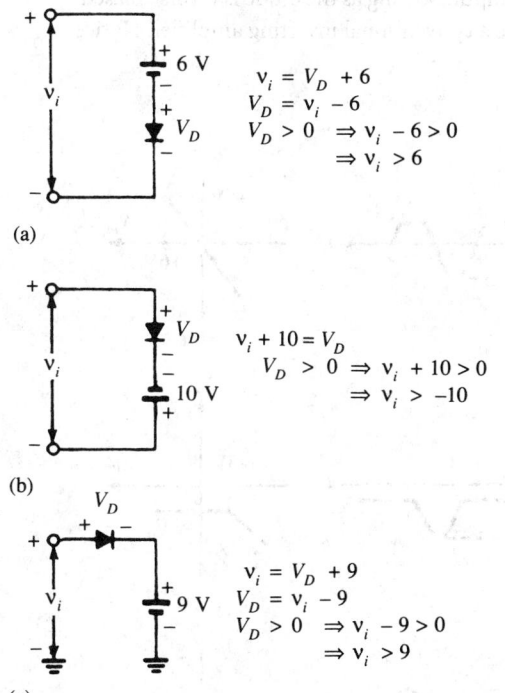

(a)

$$v_i = V_D + 6$$
$$V_D = v_i - 6$$
$$V_D > 0 \Rightarrow v_i - 6 > 0$$
$$\Rightarrow v_i > 6$$

(b)

$$v_i + 10 = V_D$$
$$V_D > 0 \Rightarrow v_i + 10 > 0$$
$$\Rightarrow v_i > -10$$

(c)

$$v_i = V_D + 9$$
$$V_D = v_i - 9$$
$$V_D > 0 \Rightarrow v_i - 9 > 0$$
$$\Rightarrow v_i > 9$$

FIGURE 5.11 Examples of biased diodes and the signal voltages v_i required to forward bias them. (Ideal diodes are assumed.) In each case, we solve for the value of v_i that is necessary to make $V_D > 0$. (*Source:* T. F. Bogart, Jr., *Electronic Devices and Circuits,* 2nd ed., Columbus, Ohio: Macmillan/Merrill, 1990, p. 678. With permission.)

(neglecting their forward voltage drops), we determine the value v_i necessary to forward bias each diode by determining the value v_i necessary to make $v_D > 0$. When v_i reaches the voltage necessary to make $V_D > 0$, the diode becomes forward biased and the signal source is forced to, or held at, the dc source voltage. If the forward voltage drop across the diode is not neglected, the clipping level is found by determining the value of v_i necessary to make V_D greater than that forward drop (e.g., $V_D > 0.7$ V for a silicon diode).

Figure 5.12 shows three examples of clipping circuits using ideal biased diodes and the waveforms that result when each is driven by a sine-wave input. In each case, note that the output equals the dc source voltage when the input reaches the value necessary to forward bias the diode. Note also that the type of clipping we showed in Fig. 5.9 occurs when the fixed bias voltage tends to *reverse* bias the diode, and the type shown in Fig. 5.10 occurs when the fixed voltage tends to *forward* bias the diode. When the diode is reverse biased by the input signal, it is like an open circuit that disconnects the dc source, and the output follows the input. These circuits are called *parallel* clippers because the biased diode is in parallel with the output. Although the circuits behave the same way whether or not one side of the dc voltage source is connected to the common (low) side of the input and output, the connections shown in Fig. 5.12(a) and (c) are preferred to that in (b), because the latter uses a floating source.

Figure 5.13 shows a biased diode connected in the feedback path of an inverting operational amplifier. The diode is in parallel with the feedback resistor and forms a parallel clipping circuit like that shown in Fig. 5.12. In an operational amplifier circuit, $v^- \approx v^+$, and since $v^+ = 0$ V in this circuit, v^- is approximately 0 V (virtual ground). Thus, the voltage across R_f is the same as the output voltage v_o. Therefore, when the output voltage reaches the bias voltage E, the output is held at E volts. Figure 5.13(b) illustrates this fact for a sinusoidal input. So long as the diode is reverse biased, it acts like an open circuit and the amplifier behaves like a conventional inverting amplifier. Notice

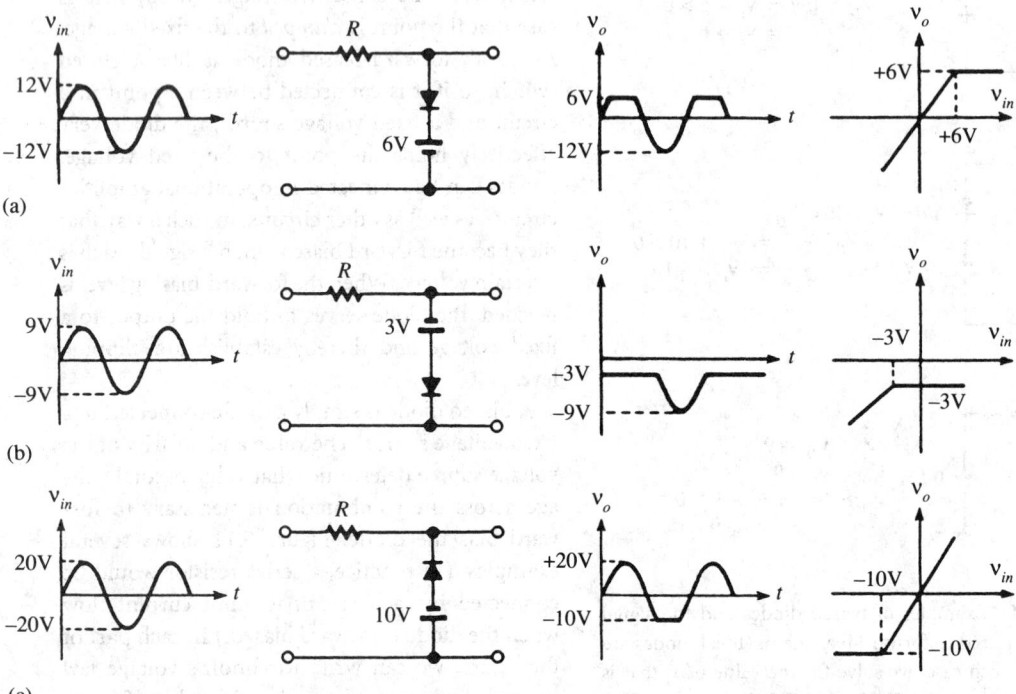

FIGURE 5.12 Examples of parallel clipping circuits. (*Source:* T.F. Bogart, Jr., *Electronic Devices and Circuits,* 2nd ed., Columbus, Ohio: Macmillan/Merrill, 1990, p. 678. With permission.)

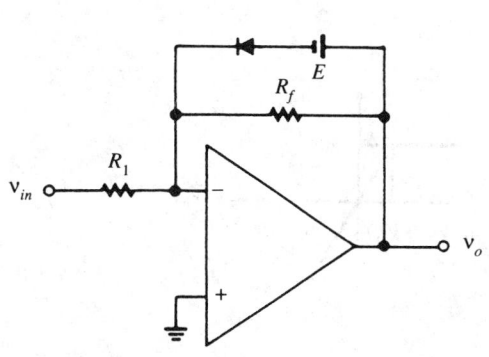

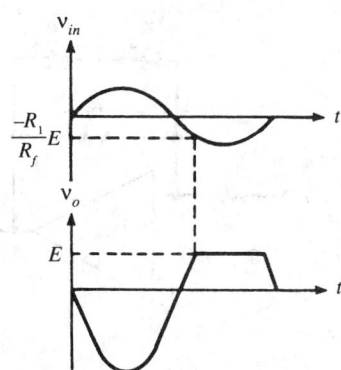

(a) The biased diode in the feedback path provides (parallel) clipping of the output at E volts.

(b) The output clamps at E volts when the input reaches $\dfrac{-R_1}{R_f}E$ volts.

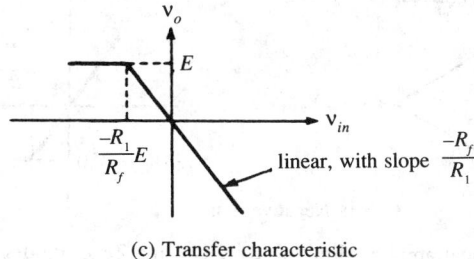

(c) Transfer characteristic

FIGURE 5.13 An operational amplifier limiting circuit. (*Source:* T.F. Bogart, Jr., *Electronic Devices and Circuits,* 2nd ed., Columbus, Ohio: Macmillan/Merrill, 1990, p. 679. With permission.)

that output clipping occurs at *input* voltage $-(R_1/R_f)E$, since the amplifier inverts and has closed-loop gain magnitude R_f/R_1. The resulting transfer characteristic is shown in Fig. 5.13(c).

In practice, the biased diode shown in the feedback of Fig. 5.13(a) is often replaced by a *Zener* diode in series with a conventional diode. This arrangement eliminates the need for a floating volt-age source. Zener diodes are in many respects functionally equivalent to biased diodes. Figure 5.14 shows two operational amplifier clipping circuits using Zener diodes. The Zener diode conducts like a conventional diode when it is forward biased, so it is necessary to connect a reversed diode in series with it to prevent shorting of R_f. When the reverse voltage across the Zener diode reaches V_Z, the diode breaks down and conducts heavily, while maintaining an essentially constant voltage, V_Z, across it. Under those conditions, the total voltage across R_f, i.e., v_o, equals V_Z plus the forward drop, V_D, across the conventional diode.

Figure 5.15 shows *double-ended* limiting circuits, in which both positive and negative peaks of the output waveform are clipped. Figure 5.15(a) shows the conventional parallel clipping circuit and (b) shows how double-ended limiting is accomplished in an operational amplifier circuit. In each circuit, note that no more than one diode is forward biased at any given time and that both diodes are reverse biased for $-E_1 < v_o < E_2$, the linear region.

Figure 5.16 shows a double-ended limiting circuit using back-to-back Zener diodes. Operation is similar to that shown in Fig. 5.14, but no conventional diode is required. Note that diode D_1 is con-ducting in a forward direction when D_2 conducts in its reverse breakdown (Zener) region, while D_2 is forward biased when D_1 is conducting in its reverse breakdown region. Neither diode conducts when $-(V_{Z2} + 0.7) < v_o < (V_{Z1} + 0.7)$, which is the region of linear amplifier operation.

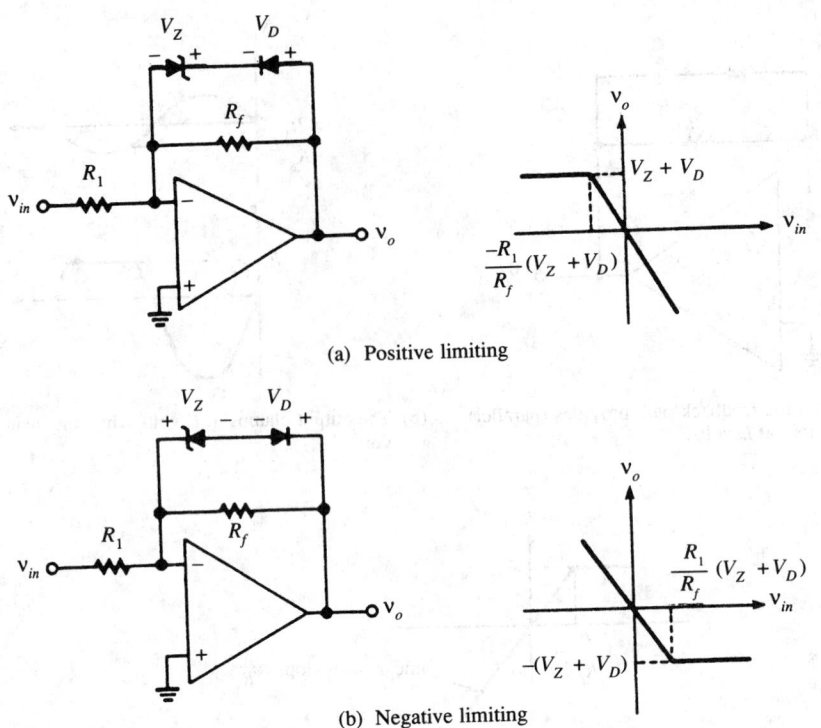

(a) Positive limiting

(b) Negative limiting

FIGURE 5.14 Operational amplifier limiting circuits using Zener diodes. (*Source:* T.F. Bogart, Jr., *Electronic Devices and Circuits,* 2nd ed., Columbus, Ohio: Macmillan/Merrill, 1990, p. 680. With permission.)

Precision Rectifying Circuits

A *rectifier* is a device that allows current to pass through it in one direction only. A diode can serve as a rectifier because it permits generous current flow in only one direction—the direction of forward bias. Rectification is the same as limiting at the 0-V level: all of the waveform below (or above) the zero-axis is eliminated. However, a diode rectifier has certain intervals of nonconduction and produces resulting "gaps" at the zero-crossing points of the output voltage, due to the fact that the input must overcome the diode drop (0.7 V for silicon) before conduction begins. In power-supply applications, where input voltages are quite large, these gaps are of no concern. However, in many other applications, especially in instrumentation, the 0.7-V drop can be a significant portion of the total input voltage swing and can seriously affect circuit performance. For example, most ac instruments rectify ac inputs so they can be measured by a device that responds to dc levels. It is obvious that small ac signals could not be measured if it were always necessary for them to reach 0.7 V before rectification could begin. For these applications, *precision* rectifiers are necessary.

Figure 5.17 shows one way to obtain precision rectification using an operational amplifier and a diode. The circuit is essentially a noninverting voltage follower (whose output follows, or duplicates, its input) when the diode is forward biased. When v_{in} is positive, the output of the amplifier, v_o, is positive, the diode is forward biased, and a low-resistance path is established between v_o and v^-, as necessary for a voltage follower. The load voltage, v_L, then follows the positive variations of $v_{in} = v^+$. Note that even a very small positive value of v_{in} will cause this result, because of the large differential gain of the amplifier. That is, the large gain and the action of the feedback cause the usual result that $v^+ \approx v^-$. Note also that the drop across the diode does not appear in v_L.

When the input goes negative, v_o becomes negative, and the diode is reverse biased. This effec-

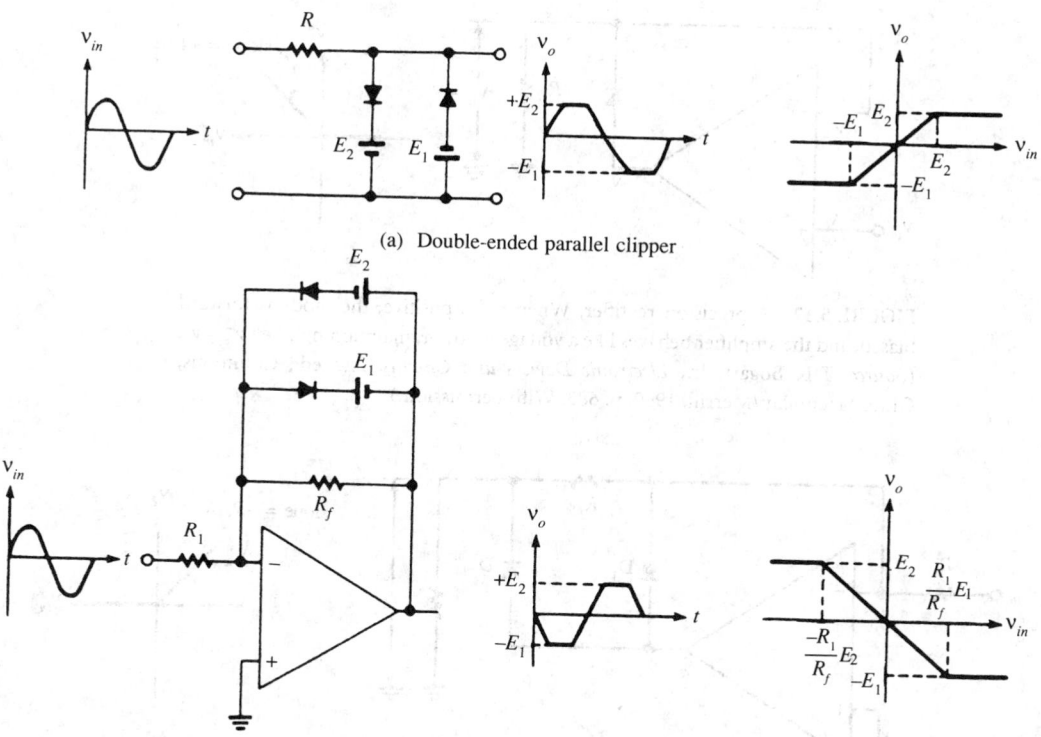

(a) Double-ended parallel clipper

(b) Operational amplifier with double-ended clipping

FIGURE 5.15 Double-ended clipping, or limiting. (*Source:* T.F. Bogart, Jr., *Electronic Devices and Circuits*, 2nd ed., Columbus, Ohio: Macmillan/Merrill, 1990, p. 681. With permission.)

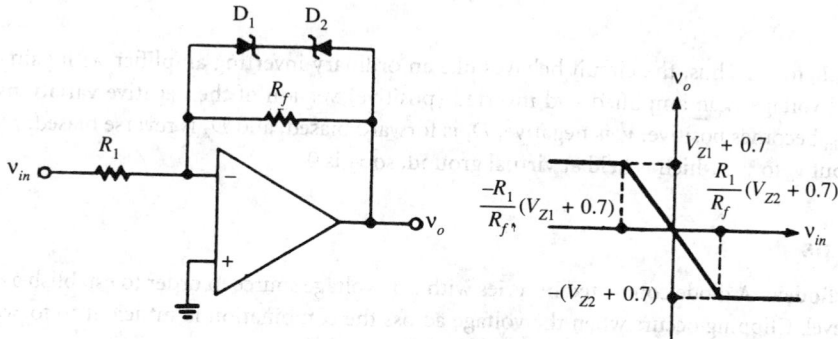

FIGURE 5.16 A double-ended limiting circuit using Zener diodes. (*Source:* T.F. Bogart, Jr., *Electronic Devices and Circuits*, 2nd ed., Columbus, Ohio: Macmillan/Merrill, 1990, p. 682. With permission.)

tively opens the feedback loop, so v_L no longer follows v_{in}. The amplifier itself, now operating open-loop, is quickly driven to its maximum negative output, thus holding the diode well into reverse bias.

Another precision rectifier circuit is shown in Fig. 5.18. In this circuit, the load voltage is an amplified and inverted version of the *negative* variations in the input signal, and is 0 when the input is positive. Also in contrast with the previous circuit, the amplifier in this rectifier is not driven to one of its output extremes. When v_{in} is negative, the amplifier output, v_o, is positive, so diode D_1 is reverse biased and diode D_2 is forward biased. D_1 is open and D_2 connects the amplifier output

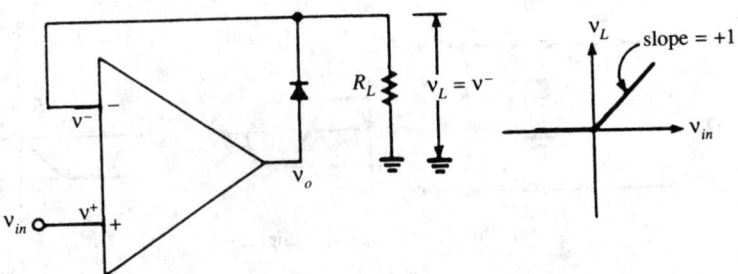

FIGURE 5.17 A precision rectifier. When v_{in} is positive, the diode is forward biased, and the amplifier behaves like a voltage follower, maintaining $v^+ \approx v^- = v_L$. (*Source:* T.F. Bogart, Jr., *Electronic Devices and Circuits*, 2nd ed., Columbus, Ohio: Macmillan/Merrill, 1990, p. 683. With permission.)

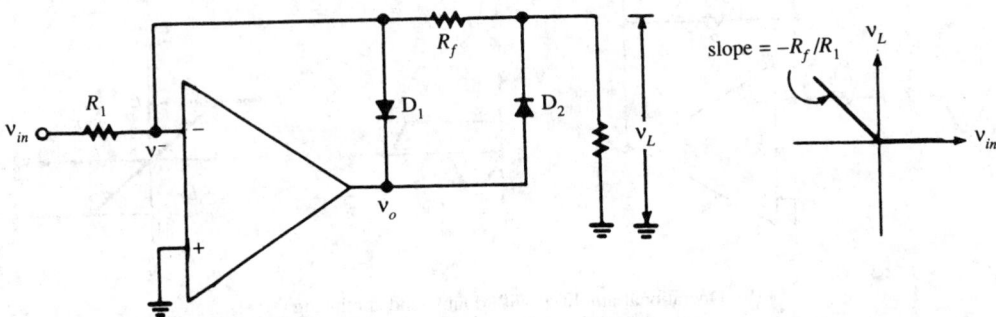

FIGURE 5.18 A precision rectifier circuit that amplifies and inverts the negative variations in the input voltage. (*Source:* T. F. Bogart, Jr., *Electronic Devices and Circuits*, 2nd ed., Columbus, Ohio: Macmillan/Merrill, 1990, p. 683. With permission.)

through R_f to v^-. Thus, the circuit behaves like an ordinary inverting amplifier with gain $-R_f/R_1$. The load voltage is an amplified and inverted (positive) version of the negative variations in v_{in}. When v_{in} becomes positive, v_o is negative, D_1 is forward biased, and D_2 is reverse biased. D_1 shorts the output v_o to v^-, which is held at virtual ground, so v_L is 0.

Defining Terms

Biased diode: A diode connected in series with a dc voltage source in order to establish a clipping level. Clipping occurs when the voltage across the combination is sufficient to forward bias the diode.

Limiter: A device or circuit that restricts voltage excursions to prescribed level(s). Also called a clipping circuit.

References

W. H. Baumgartner, *Pulse Fundamentals and Small-Scale Digital Circuits*, Reston, Va.: Reston Publishing, 1985.

T. F. Bogart, Jr., *Electronic Devices and Circuits*, 3rd ed., Columbus, Ohio: Macmillan/Merrill, 1993.

R.A. Gayakwad, *Op-Amps and Linear Integrated Circuit Technology*, Englewood Cliffs, N.J.: Prentice-Hall, 1983.

A.S. Sedra and K.C. Smith, *Microelectronic Circuits*, New York: CBS College Publishing, 1982.

H. Zanger, *Semiconductor Devices and Circuits*, New York: John Wiley & Sons, 1984.

5.3 Distortion

Kartikeya Mayaram

The diode was introduced in the previous sections as a nonlinear device that is used in rectifiers and limiters. These are applications that depend on the nonlinear nature of the diode. Typical electronic systems are composed not only of diodes but also of other nonlinear devices such as transistors (Chapter 22). In analog applications transistors are used to amplify weak signals (amplifiers) and to drive large loads (output stages). For such situations it is desirable that the output be an amplified true reproduction of the input signal; therefore, the transistors must operate as linear devices. However, the inherent nonlinearity of transistors results in an output which is a "distorted" version of the input.

The distortion due to a nonlinear device is illustrated in Fig. 5.19. For an input X the output is $Y = F(X)$ where F denotes the nonlinear transfer characteristics of the device; the dc operating point is given by X_0. Sinusoidal input signals of two different amplitudes are applied and the output responses corresponding to these inputs are also shown.

For an input signal of small amplitude the output faithfully follows the input, whereas for large-amplitude signals the output is distorted; a flattening occurs at the negative peak value. The distortion in amplitude results in the output having frequency components that are integer multiples of the input frequency, *harmonics*, and this type of distortion is referred to as **harmonic distortion.**

The distortion level places a restriction on the amplitude of the input signal that can be applied to an electronic system. Therefore, it is essential to characterize the distortion in a circuit. In this section different types of distortion are defined and techniques for distortion calculation are presented. These techniques are applicable to simple circuit configurations. For larger circuits a circuit simulation program is invaluable.

Harmonic Distortion

When a sinusoidal signal of a single frequency is applied at the input of a nonlinear device or circuit, the resulting output contains frequency components that are integer multiples of the input signal. These harmonics are generated by the nonlinearity of the circuit and the *harmonic distortion* is measured by comparing the magnitudes of the harmonics with the fundamental component (input frequency) of the output.

Consider the input signal to be of the form:

$$x(t) = X_1 \cos \omega_1 t \tag{5.2}$$

where $f_1 = \omega_1/2\pi$ is the frequency and X_1 is the amplitude of the input signal. Let the output of the nonlinear circuit be

$$y(t) = Y_0 + Y_1 \cos \omega_1 t + Y_2 \cos 2\omega_1 t + Y_3 \cos 3\omega_1 t + \dots \tag{5.3}$$

where Y_0 is the dc component of the output, Y_1 is the amplitude of the fundamental component, and Y_2, Y_3 are the amplitudes of the second and third harmonic components. The *second harmonic distortion factor* (HD_2), the *third harmonic distortion factor* (HD_3), and the *nth harmonic distortion factor* (HD_n) are defined as

$$HD_2 = \frac{|Y_2|}{|Y_1|} \tag{5.4}$$

$$HD_3 = \frac{|Y_3|}{|Y_1|} \tag{5.5}$$

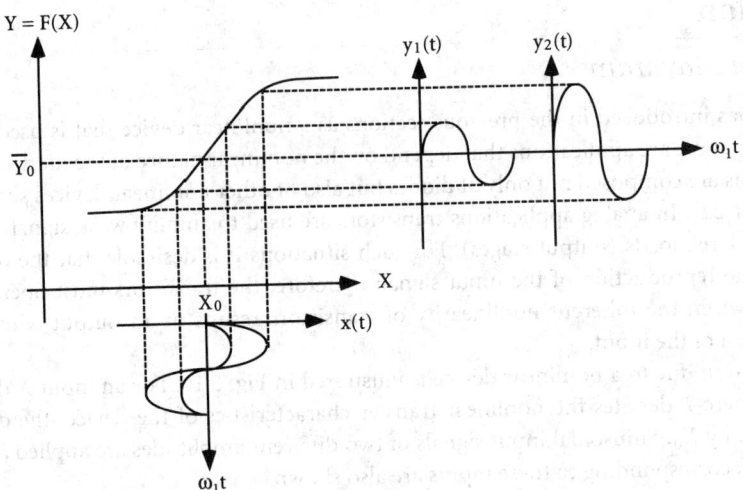

FIGURE 5.19 DC transfer characteristics of a nonlinear circuit and the input and output waveforms. For a large input amplitude the output is distorted.

$$\mathrm{HD}_n = \frac{|Y_n|}{|Y_1|} \tag{5.6}$$

The **total harmonic distortion** (THD) of a waveform is defined to be the ratio of the rms (root-mean-square) value of the harmonics to the amplitude of the fundamental component.

$$\mathrm{THD} = \frac{\sqrt{Y_2^2 + Y_3^2 + \cdots + Y_n^2}}{|Y_1|} \tag{5.7}$$

THD can be expressed in terms of the individual **harmonic distortion factors**

$$\mathrm{THD} = \sqrt{\mathrm{HD}_2^2 + \mathrm{HD}_3^2 + \cdots + \mathrm{HD}_n^2} \tag{5.8}$$

Various methods for computing the harmonic distortion factors are described next.

Power-Series Method

In this method a truncated power-series expansion of the dc transfer characteristics of a nonlinear circuit is used. Therefore, the method is suitable only when energy storage effects in the nonlinear circuit are negligible and the input signal is small. In general, the input and output signals comprise both dc and time-varying components. For distortion calculation we are interested in the time-varying or incremental components around a quiescent[1] operating point. For the transfer characteristic of Fig. 5.19, denote the quiescent operating conditions by X_0 and $\overline{Y}_0$ and the incremental variables by $x(t)$ and $y(t)$, at the input and output, respectively. The output can be expressed as a function of the input using a series expansion

$$\overline{Y}_0 + y = F(X_0 + x) = a_0 + a_1 x + a_2 x^2 + a_3 x^3 + \cdots \tag{5.9}$$

where $a_0 = \overline{Y}_0 = F(X_0)$ is the output at the dc operating point. The incremental output is

$$y = a_1 x + a_2 x^2 + a_3 x^3 + \cdots \tag{5.10}$$

[1] Defined as the operating condition when the input has no time-varying component.

Depending on the amplitude of the input signal, the series can be truncated at an appropriate term. Typically only the first few terms are used, which makes this technique applicable only to small input signals. For a pure sinusoidal input [Eq. (5.2)], the distortion in the output can be estimated by substituting for x in Eq. (5.10) and by use of trigonometric identities one can arrive at the form given by Eq. (5.3). For a series expansion that is truncated after the cubic term

$$Y_0 = \frac{a_2 X_1^2}{2}$$

$$Y_1 = a_1 X_1 + \frac{3a_3 X_1^3}{4} \cong a_1 X_1$$

$$Y_2 = \frac{a_2 X_1^2}{2} \tag{5.11}$$

$$Y_3 = \frac{a_3 X_1^3}{4}$$

Notice that a dc term Y_0 is present in the output (produced by the even-powered terms) which results in a shift of the operating point of the circuit due to distortion. In addition, depending on the sign of a_3 there can be an *expansion* or *compression* of the fundamental component. The harmonic distortion factors (assuming $Y_1 = a_1 X_1$) are

$$HD_2 = \frac{|Y_2|}{|Y_1|} = \frac{1}{2} \left| \frac{a_2}{a_1} X_1 \right|$$

$$HD_3 = \frac{|Y_3|}{|Y_1|} = \frac{1}{4} \left| \frac{a_3}{a_1} X_1^2 \right| \tag{5.12}$$

As an example, choose as the transfer function $Y = F(X) = \exp(X)$; then, $a_1 = 1$, $a_2 = 1/2$, $a_3 = 1/6$. For an input signal amplitude of 0.1, $HD_2 = 2.5\%$ and $HD_3 = 0.04\%$.

Differential-Error Method

This technique is also applicable to nonlinear circuits in which energy storage effects can be neglected. The method is valuable for circuits that have small distortion levels and relies on one's ability to calculate the small-signal gain of the nonlinear function at the quiescent operating point and at the maximum and minimum excursions of the input signal. Again the power-series expansion provides the basis for developing this technique. The small-signal gain[2] at the quiescent state ($x = 0$) is a_1. At the extreme values of the input signal X_1 (positive peak) and $-X_1$ (negative peak) let the small-signal gains be a^+ and a^-, respectively. By defining two new parameters, the differential errors, E^+ and E^-, as

$$E^+ = \frac{a^+ - a_1}{a_1} \qquad E^- = \frac{a^- - a_1}{a_1} \tag{5.13}$$

the distortion factors are given by

$$HD_2 = \frac{E^+ - E^-}{8}$$

$$HD_3 = \frac{E^+ + E^-}{24} \tag{5.14}$$

[2] Small-signal gain $= dy/dx = a_1 + 2a_2 x + 3a_3 x^2 + \cdots$

The advantage of this method is that the transfer characteristics of a nonlinear circuit can be directly used; an explicit power-series expansion is not required. Both the power-series and the differential-error techniques cannot be applied when only the output waveform is known. In such a situation the distortion factors are calculated from the output signal waveform by a simplified Fourier analysis as described in the next section.

Three-Point Method

The three-point method is a simplified analysis applicable to small levels of distortion and can only be used to calculate HD_2. The output is written directly as a Fourier cosine series as in Eq. (5.3) where only terms up to the second harmonic are retained. The dc component includes the quiescent state and the contribution due to distortion that results in a shift of the dc operating point. The output waveform values at $\omega_1 t = 0$ (F_0), $\omega_1 t = \pi/2$ $(F_{\pi/2})$, $\omega_1 t = \pi$ (F_π), as shown in Fig. 5.20, are used to calculate Y_0, Y_1, and Y_2.

$$Y_0 = \frac{F_0 + 2F_{\pi/2} + F_\pi}{4}$$

$$Y_1 = \frac{F_0 - F_\pi}{2}$$

$$Y_2 = \frac{F_0 - 2F_{\pi/2} + F_\pi}{4} \tag{5.15}$$

The second harmonic distortion is calculated from the definition. From Fig. 5.20, $F_0 = 5$, $F_{\pi/2} = 3.2$, $F_\pi = 1$, $Y_0 = 3.1$, $Y_1 = 2.0$, $Y_2 = -0.1$, and $HD_2 = 5.0\%$.

Five-Point Method

The five-point method is an extension of the above technique and allows calculation of third and fourth harmonic distortion factors. For distortion calculation the output is expressed as a Fourier cosine series with terms up to the fourth harmonic where the dc component includes the quiescent state and the shift due to distortion. The output waveform values at $\omega_1 t = 0$ (F_0), $\omega_1 t = \pi/3$ $(F_{\pi/3})$, $\omega_1 t = \pi/2$ $(F_{\pi/2})$, $\omega_1 t = 2\pi/3$ $(F_{2\pi/3})$, $\omega_1 t = \pi$ (F_π), as shown in Fig. 5.20, are used to calculate Y_0, Y_1, Y_2, Y_3, and Y_4.

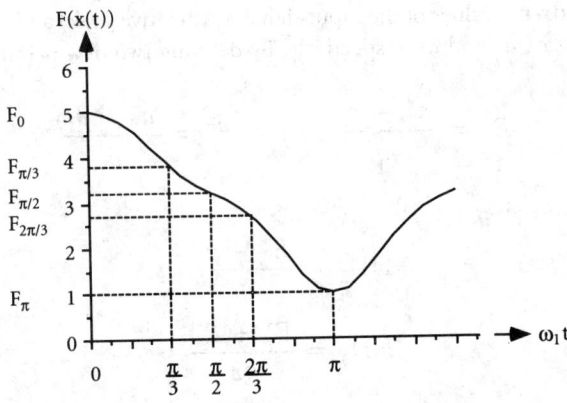

FIGURE 5.20   Output waveform from a nonlinear circuit.

$$Y_0 = \frac{F_0 + 2F_{\pi/3} + 2F_{2\pi/3} + F_\pi}{6}$$

$$Y_1 = \frac{F_0 + F_{\pi/3} - F_{2\pi/3} - F_\pi}{3}$$

$$Y_2 = \frac{F_0 - 2F_{\pi/2} + F_\pi}{4} \tag{5.16}$$

$$Y_3 = \frac{F_0 - 2F_{\pi/3} + 2F_{2\pi/3} - F_\pi}{6}$$

$$Y_4 = \frac{F_0 - 4F_{\pi/3} + 6F_{\pi/2} - 4F_{2\pi/3} + F_\pi}{12}$$

For $F_0 = 5$, $F_{\pi/3} = 3.8$, $F_{\pi/2} = 3.2$, $F_{2\pi/3} = 2.7$, $F_\pi = 1$, $Y_0 = 3.17$, $Y_1 = 1.7$, $Y_2 = -0.1$, $Y_3 = 0.3$, $Y_4 = -0.07$, and $HD_2 = 5.9\%$, $HD_3 = 17.6\%$. This particular method allows calculation of HD_3 and also gives a better estimate of HD_2. To obtain higher-order harmonics a detailed Fourier series analysis is required and for such applications a circuit simulator, such as SPICE, should be used.

Intermodulation Distortion

The previous sections have examined the effect of nonlinear device characteristics when a single-frequency sinusoidal signal is applied at the input. However, if there are two or more sinusoidal inputs, then the nonlinearity results in not only the fundamental and harmonics but also additional frequencies called the *beat frequencies* at the output. The distortion due to the components at the beat frequencies is called **intermodulation distortion.** To characterize this type of distortion consider the incremental output given by Eq. (5.10) and the input signal to be

$$x(t) = X_1 \cos \omega_1 t + X_2 \cos \omega_2 t \tag{5.17}$$

where $f_1 = \omega_1/2\pi$ and $f_2 = \omega_2/2\pi$ are the two input frequencies. The output frequency spectrum due to the quadratic term is shown in Table 5.1.

Table 5.1 Output Frequency Spectrum Due to the Quadratic Term

Frequency	0	$2f_1$	$2f_2$	$f_1 \pm f_2$
Amplitude	$\frac{a_2}{2}[X_1^2 + X_2^2]$	$\frac{a_2}{2}X_1^2$	$\frac{a_2}{2}X_2^2$	$a_2 X_1 X_2$

In addition to the dc term and the second harmonics of the two frequencies, there are additional terms at the sum and difference frequencies, $f_1 + f_2$, $f_1 - f_2$, which are the beat frequencies. The *second-order intermodulation distortion* (IM_2) is defined as the ratio of the amplitude at a beat frequency to the amplitude of the fundamental component.

$$IM_2 = \left| \frac{a_2 X_1 X_2}{a_1 X_1} \right| = \left| \frac{a_2 X_2}{a_1} \right| \tag{5.18}$$

where it has been assumed that the contribution to second-order intermodulation by higher-order terms is negligible. In defining IM_2 the input signals are assumed to be of equal amplitude and for this particular condition $IM_2 = 2\,HD_2$ [Eq. (5.12)].

The cubic term of the series expansion for the nonlinear circuit gives rise to components at frequencies $f_1 + 2f_2$, $2f_2 + f_1$, $2f_1 - f_2$, $2f_2 - f_1$, and these terms result in *third-order intermodulation distortion* (IM$_3$). The frequency spectrum obtained from the cubic term is shown in Table 5.2.

Table 5.2   Output Frequency Spectrum Due to the Cubic Term

Frequency	f_1	f_2	$2f_1 \pm f_2$	$2f_2 \pm f_1$	$3f_1$	$3f_2$
Amplitude	$\dfrac{3a_3}{4}[X_1^3 + X_1 X_2^2]$	$\dfrac{3a_3}{4}[X_2^3 + X_1^2 X_2]$	$\dfrac{3}{4} a_3 X_1^2 X_2$	$\dfrac{3}{4} a_3 X_1 X_2^2$	$\dfrac{1}{4} a_3 X_1^3$	$\dfrac{1}{4} a_3 X_2^3$

For definition purposes the two input signals are assumed to be of equal amplitude and IM$_3$ is given by (assuming negligible contribution to the fundamental by the cubic term)

$$\text{IM}_3 = \frac{3}{4} \left| \frac{a_3 X_1^3}{a_1 X_1} \right| = \frac{3}{4} \left| \frac{a_3 X_1^2}{a_1} \right| \tag{5.19}$$

Under these conditions IM$_3$ = 3 HD$_3$ [Eq. (5.12)]. When f_1 and f_2 are close to one another, then the third-order intermodulation components, $2f_1 - f_2$, $2f_2 - f_1$, are close to the fundamental and are difficult to filter out.

Triple-Beat Distortion

When three sinusoidal signals are applied at the input then the output consists of components at the triple-beat frequencies. The cubic term in the nonlinearity results in the triple-beat terms

$$\frac{3}{2} a_3 X_1 X_2 X_2 \cos[\omega_1 \pm \omega_2 \pm \omega_3]t \tag{5.20}$$

and the *triple-beat distortion factor* (TB) is defined for equal amplitude input signals.

$$\text{TB} = \frac{3}{2} \left| \frac{a_3 X_1^2}{a_1} \right| \tag{5.21}$$

From the above definition TB = 2 IM$_3$. If all of the frequencies are close to one another, the triple beats will be close to the fundamental and cannot be easily removed.

Cross Modulation

Another form of distortion that occurs in amplitude-modulated (AM) systems (Chapter 63) due to the circuit nonlinearity is **cross modulation**. The modulation from an unwanted AM signal is transferred to the signal of interest and results in distortion. Consider an AM signal

$$x(t) = X_1 \cos \omega_1 t + X_2[1 + m \cos \omega_m t] \cos \omega_2 t \tag{5.22}$$

where $m < 1$ is the modulation index. Due to the cubic term of the nonlinearity the modulation from the second signal is transferred to the first and the modulated component corresponding to the fundamental is

$$a_1 X_1 \left[1 + \frac{3a_3 X_2^2 m}{a_1} \cos \omega_m t \right] \cos \omega_1 t \tag{5.23}$$

The *cross-modulation factor* (CM) is defined as the ratio of the transferred modulation index to the original modulation.

$$CM = 3 \left| \frac{a_3 X_2^2}{a_1} \right| \tag{5.24}$$

The cross modulation is a factor of four larger than IM_3 and twelve times as large as HD_3.

Crossover Distortion

This type of distortion occurs in circuits that use devices operating in a "push-pull" manner. The devices are used in pairs and each device operates only for half a cycle of the input signal (Class AB operation). One advantage of such an arrangement is the cancellation of even harmonic terms resulting in smaller total harmonic distortion. However, if the circuit is not designed to achieve a smooth crossover or transition from one device to another, then there is a region of the transfer characteristics when the output is zero. The resulting distortion is called **crossover distortion.**

Failure-to-Follow Distortion

When a properly designed peak detector circuit is used for AM demodulation the output follows the envelope of the input signal whereby the original modulation signal is recovered. A simple peak detector is a diode in series with a low-pass RC filter. The critical component of such a circuit is a linear element, the filter capacitance C. If C is large, then the output fails to follow the envelope of the input signal, resulting in **failure-to-follow distortion.**

Frequency Distortion

Ideally an amplifier circuit should provide the same amplification for all input frequencies. However, due to the presence of energy storage elements the gain of the amplifier is frequency dependent. Consequently different frequency components have different amplifications resulting in **frequency distortion.** The distortion is specified by a frequency response curve in which the amplifier output is plotted as a function of frequency. An ideal amplifier has a flat frequency response over the frequency range of interest.

Phase Distortion

When the phase shift (θ) in the output signal of an amplifier is not proportional to the frequency, the output does not preserve the form of the input signal, resulting in **phase distortion.** If the phase shift is proportional to frequency, different frequency components have a constant delay time (θ/ω) and no distortion is observed. In TV applications phase distortion can result in a smeared picture.

Computer Simulation of Distortion Components

Distortion characterization is important for nonlinear circuits. However, the techniques presented for distortion calculation can only be used for simple circuit configurations and at best to determine the second and third harmonic distortion factors. In order to determine the distortion generation in actual circuits one must fabricate the circuit and then use a harmonic analyzer for sine curve inputs to determine the harmonics present in the output. An attractive alternative is the use of circuit simulation programs that allow one to investigate circuit performance before fabricating the circuit. In this section a brief overview of the techniques used in circuit simulators for distortion characterization is provided.

The simplest approach is to simulate the time-domain output for a circuit with a specified sinusoidal input signal and then perform a Fourier analysis of the output waveform. The simulation program SPICE2 provides a capability for computing the Fourier components of any waveform using a *.FOUR* command and specifying the voltage or current for which the analysis has to be performed. A simple diode circuit, the SPICE input file, and transient voltage waveforms for an input signal frequency of 1 MHz and amplitudes of 10 and 100 mV are shown in Fig. 5.21. The Fourier components of the resistor voltage are shown in Fig. 5.22; only the fundamental and first two significant harmonics are shown (SPICE provides information to the ninth harmonic).

In this particular example the input signal frequency is 1 MHz, and this is the frequency at which the Fourier analysis is requested. Since there are no energy storage elements in the circuit another

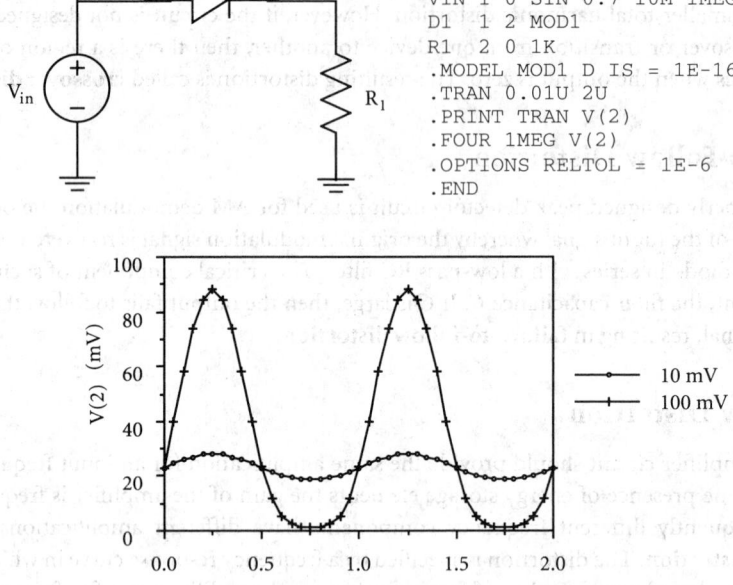

```
SIMPLE DIODE CIRCUIT
VIN 1 0 SIN 0.7 10M 1MEG
D1 1 2 MOD1
R1 2 0 1K
.MODEL MOD1 D IS = 1E-16
.TRAN 0.01U 2U
.PRINT TRAN V(2)
.FOUR 1MEG V(2)
.OPTIONS RELTOL = 1E-6
.END
```

FIGURE 5.21   Simple diode circuit, SPICE input file, and output voltage waveforms.

```
FOURIER COMPONENTS OF TRANSIENT RESPONSE V(2)                          VIN = 10mV
DC COMPONENT =      2.330D-02
  HARMONIC    FREQUENCY        FOURIER        NORMALIZED        PHASE      NORMALIZED
    NO          (HZ)          COMPONENT       COMPONENT         (DEG)      PHASE (DEG)
     1        1.000D + 06      4.695D-03       1.000000        -0.001        0.000
     2        2.000D + 06      1.242D-04       0.026462       -89.989      -89.988
     3        3.000D+06        1.705D-06       0.000363        -3.241       -3.239

TOTAL HARMONIC DISTORTION =      2.646409 PERCENT


FOURIER COMPONENTS OF TRANSIENT RESPONSE V(2)                          VIN = 100mV
DC COMPONENT =      3.445D-02
  HARMONIC    FREQUENCY        FOURIER        NORMALIZED        PHASE      NORMALIZED
    NO          (HZ)          COMPONENT       COMPONENT         (DEG)      PHASE (DEG)
     1        1.000D + 06      4.402D-02       1.000000        -0.011        0.000
     2        2.000D + 06      1.059-02        0.240634       -89.993      -89.983
     3        3.000D+06        1.658D-04       0.015127        -0.686       -0.675

TOTAL HARMONIC DISTORTION =     24.132679 PERCENT
```

FIGURE 5.22   Fourier components of the resistor voltage for input amplitudes of 10 and 100 mV, respectively.

frequency would have given identical results. To determine the Fourier components accurately a small value of the parameter RELTOL is used and a sufficient number of points for transient analysis are specified. From the output voltage waveforms and the Fourier analysis it is seen that the harmonic distortion increases significantly when the input voltage amplitude is increased from 10 mV to 100 mV.

The transient approach can be computationally expensive for circuits that reach their periodic steady state after a long simulation time. Results from the Fourier analysis are meaningful only in the periodic steady state, and although this approach works well for large levels of distortion it is inaccurate for small distortion levels.

For small distortion levels accurate distortion analysis can be performed by use of the Volterra series method. This technique is a generalization of the power-series method and is useful for analyzing harmonic and intermodulation distortion due to frequency-dependent nonlinearities. The SPICE3 program supports this analysis technique (in addition to the Fourier analysis of SPICE2) whereby the second and third harmonic and intermodulation components can be efficiently obtained by three small-signal analyses of the circuit.

An approach based on the *harmonic balance* technique available in the simulation program SPECTRE is applicable to both large and small levels of distortion. The program determines the periodic steady state of a circuit with a sinusoidal input. The unknowns are the magnitudes of the circuit variables at the fundamental frequency and at all the significant harmonics of the fundamental. The distortion levels can be simply calculated by taking the ratios of the magnitudes of the appropriate harmonics to the fundamental.

Defining Terms

Cross modulation: Occurs in amplitude-modulated systems when the modulation of one signal is transferred to another by the nonlinearity of the system.

Crossover distortion: Present in circuits that use devices operating in a push-pull arrangement such that one device conducts when the other is off. Crossover distortion results if the transition or crossover from one device to the other is not smooth.

Failure-to-follow distortion: Can occur during demodulation of an amplitude-modulated signal by a peak detector circuit. If the capacitance of the low-pass RC filter of the peak detector is large, then the output fails to follow the envelope of the input signal, resulting in failure-to-follow distortion.

Frequency distortion: Caused by the presence of energy storage elements in an amplifier circuit. Different frequency components have different amplifications, resulting in frequency distortion and the distortion is specified by a frequency response curve.

Harmonic distortion: Caused by the nonlinear transfer characteristics of a device or circuit. When a sinusoidal signal of a single frequency (the *fundamental* frequency) is applied at the input of a nonlinear circuit, the output contains frequency components that are integer multiples of the fundamental frequency (*harmonics*). The resulting distortion is called harmonic distortion.

Harmonic distortion factors: A measure of the harmonic content of the output. The nth *harmonic distortion factor* is the ratio of the amplitude of the nth harmonic to the amplitude of the fundamental component of the output.

Intermodulation distortion: Distortion caused by the mixing or beating of two or more sinusoidal inputs due to the nonlinearity of a device. The output contains terms at the sum and difference frequencies called the *beat frequencies*.

Phase distortion: Occurs when the phase shift in the output signal of an amplifier is not proportional to the frequency.

Total harmonic distortion: The ratio of the root-mean-square value of the harmonics to the amplitude of the fundamental component of a waveform.

References

K. K. Clarke and D. T. Hess, *Communication Circuits: Analysis and Design*, Reading, Mass.: Addison-Wesley, 1971.

P. R. Gray and R. G. Meyer, *Analysis and Design of Analog Integrated Circuits*, New York: John Wiley and Sons, 1992.

K. S. Kundert, *Spectre User's Guide: A Frequency Domain Simulator for Nonlinear Circuits*, EECS Industrial Liaison Program Office, University of California, Berkeley, 1987.

L. W. Nagel, "SPICE2: A Computer Program to Simulate Semiconductor Circuits," Memo No. ERL-M520, Electronics Research Laboratory, University of California, Berkeley, 1975.

D. O. Pederson and K. Mayaram, *Analog Integrated Circuits for Communication: Principles, Simulation and Design*, Boston: Kluwer Academic Publishers, 1991.

T. L. Quarles, *SPICE3C.1 User's Guide*, EECS Industrial Liaison Program Office, University of California, Berkeley, 1989.

J. S. Roychowdhury, "SPICE 3 Distortion Analysis," Memo No. UCB/ERL M89/48, Electronics Research Laboratory, University of California, Berkeley, 1989.

D. D. Weiner and J. F. Spina, *Sinusoidal Analysis and Modeling of Weakly Nonlinear Circuits*, New York: Van Nostrand Reinhold Company, 1980.

Further Information

Characterization and simulation of distortion in a wide variety of electronic circuits (with and without feedback) is presented in detail in Pederson and Mayaram [1991]. Also derivations for the simple analysis techniques are provided and verified using SPICE2 simulations. Algorithms for computer-aided analysis of distortion are available in Weiner and Spina [1980], Nagel [1975], Roychowdhury [1989], and Kundert [1987].

$$6$$

Laplace Transform

Richard C. Dorf
University of California, Davis

Zhen Wan
University of California, Davis

David E. Johnson
Birmingham-Southern College

6.1 Definitions and Properties 149
Laplace Transform Integral • Region of Absolute Convergence •
Properties of Laplace Transform • Time-Convolution Property •
Time-Correlation Property • Inverse Laplace Transform

6.2 Applications 158
Differentiation Theorems • Applications to Integrodifferential
Equations • Applications to Electric Circuits • The Transformed Cir-
cuit • Thévenin's and Norton's Theorems • Network Functions •
Step and Impulse Responses • Stability

6.1 Definitions and Properties

Richard C. Dorf and Zhen Wan

The **Laplace transform** is a useful analytical tool for converting time-domain signal descriptions into functions of a complex variable. This *complex domain* description of a signal provides new insight into the analysis of signals and systems. In addition, the Laplace transform method often simplifies the calculations involved in obtaining system response signals.

Laplace Transform Integral

The Laplace transform completely characterizes the exponential response of a time-invariant linear function. This transformation is formally generated through the process of multiplying the linear characteristic signal $x(t)$ by the signal e^{-st} and then integrating that product over the time interval $(-\infty, +\infty)$. This systematic procedure is more generally known as *taking the Laplace transform* of the signal $x(t)$.

Definition: The Laplace transform of the continuous-time signal $x(t)$ is

$$X(s) = \int_{-\infty}^{+\infty} x(t)e^{-st}dt$$

The variable s that appears in this integrand exponential is generally complex valued and is therefore often expressed in terms of its rectangular coordinates

$$s = \sigma + j\omega$$

where $\sigma = \mathrm{Re}(s)$ and $\omega = \mathrm{Im}(s)$ are referred to as the *real* and *imaginary* components of s, respectively.

The signal $x(t)$ and its associated Laplace transform $X(s)$ are said to form a *Laplace transform pair.* This reflects a form of equivalency between the two apparently different entities $x(t)$ and $X(s)$. We may symbolize this interrelationship in the following suggestive manner:

$$X(s) = \mathscr{L}[x(t)]$$

where the operator notation $\mathscr{L}$ means to multiply the signal $x(t)$ being operated upon by the complex exponential e^{-st} and then to integrate that product over the time interval $(-\infty, +\infty)$.

Region of Absolute Convergence

In evaluating the Laplace transform integral that corresponds to a given signal, it is generally found that this integral will exist (that is, the integral has finite magnitude) for only a restricted set of s values.

The definition of **region of absolute convergence** is as follows. The set of complex numbers s for which the magnitude of the Laplace transform integral is finite is said to constitute the region of absolute convergence for that integral transform. This region of convergence is always expressible as

$$\sigma_+ < \text{Re}(s) < \sigma_-$$

where σ_+ and σ_- denote real parameters that are related to the causal and anticausal components, respectively, of the signal whose Laplace transform is being sought.

Laplace Transform Pair Tables

It is convenient to display the Laplace transforms of standard signals in one table. Table 6.1 displays the time signal $x(t)$ and its corresponding Laplace transform and region of absolute convergence and is sufficient for our needs.

Example. To find the Laplace transform of the first-order causal exponential signal

$$x_1(t) = e^{-at}u(t)$$

where the constant a can in general be a complex number.

The Laplace transform of this general exponential signal is determined upon evaluating the associated Laplace transform integral

$$X_1(s) = \int_{-\infty}^{+\infty} e^{-at}u(t)e^{-st}dt = \int_{0}^{+\infty} e^{-(s+a)t}dt$$

$$= \frac{e^{-(s+a)t}}{-(s+a)}\bigg|_0^{+\infty} \tag{6.1}$$

In order for $X_1(s)$ to exist, it must follow that the real part of the exponential argument be positive, that is,

$$\text{Re}(s+a) = \text{Re}(s) + \text{Re}(a) > 0$$

If this were not the case, the evaluation of expression (6.1) at the upper limit $t = +\infty$ would either be unbounded if $\text{Re}(s) + \text{Re}(a) < 0$ or undefined when $\text{Re}(s) + \text{Re}(a) = 0$. On the other hand, the upper limit evaluation is zero when $\text{Re}(s) + \text{Re}(a) > 0$, as is already apparent. The lower limit evaluation at $t = 0$ is equal to $1/(s+a)$ for all choices of the variable s.

The Laplace transform of exponential signal $e^{-at}u(t)$ has therefore been found and is given by

$$\mathscr{L}[e^{-at}u(t)] = \frac{1}{s+a} \quad \text{for } \text{Re}(s) > -\text{Re}(a)$$

Table 6.1   Laplace Transform Pairs

Time Signal $x(t)$	Laplace Transform $X(s)$	Region of Absolute Convergence
1. $e^{-at}u(t)$	$\dfrac{1}{s+a}$	$\mathrm{Re}(s) > -\mathrm{Re}(a)$
2. $t^k e^{-at}u(-t)$	$\dfrac{k!}{(s+a)^{k+1}}$	$\mathrm{Re}(s) > -\mathrm{Re}(a)$
3. $-e^{-at}u(-t)$	$\dfrac{1}{(s+a)}$	$\mathrm{Re}(s) < -\mathrm{Re}(a)$
4. $(-t)^k e^{-at}u(-t)$	$\dfrac{k!}{(s+a)^{k+1}}$	$\mathrm{Re}(s) < -\mathrm{Re}(a)$
5. $u(t)$	$\dfrac{1}{s}$	$\mathrm{Re}(s) > 0$
6. $\delta(t)$	1	all s
7. $\dfrac{d^k \delta(t)}{dt^k}$	s^k	all s
8. $t^k u(t)$	$\dfrac{k!}{s^{k+1}}$	$\mathrm{Re}(s) > 0$
9. $\mathrm{sgn}\, t = \begin{cases} 1, t \geq 0 \\ -1, t < 0 \end{cases}$	$\dfrac{2}{s}$	$\mathrm{Re}(s) = 0$
10. $\sin \omega_0 t\, u(t)$	$\dfrac{\omega_0}{s^2 + \omega_0^2}$	$\mathrm{Re}(s) > 0$
11. $\cos \omega_0 t\, u(t)$	$\dfrac{s}{s^2 + \omega_0^2}$	$\mathrm{Re}(s) > 0$
12. $e^{-at} \sin \omega_0 t\, u(t)$	$\dfrac{\omega}{(s+a)^2 + \omega_0^2}$	$\mathrm{Re}(s) > -\mathrm{Re}(a)$
13. $e^{-at} \cos \omega_0 t\, u(t)$	$\dfrac{s+a}{(s+a)^2 + \omega_0^2}$	$\mathrm{Re}(s) > -\mathrm{Re}(a)$

Source: J. A. Cadzow and H. F. Van Landingham, *Signals, Systems, and Transforms,* Englewood Cliffs, N.J.: Prentice-Hall, 1985, p. 133. With permission.

Properties of Laplace Transform

Linearity

Let us obtain the Laplace transform of a signal, $x(t)$, that is composed of a linear combination of two other signals,

$$x(t) = \alpha_1 x_1(t) + \alpha_2 x_2(t)$$

where α_1 and α_2 are constants.

The linearity property indicates that

$$\mathscr{L}[\alpha_1 x_1(t) + \alpha_2 x_2(t)] = \alpha_1 X_1(s) + \alpha_2 X_2(s)$$

and the region of absolute convergence is *at least as large* as that given by the expression

$$\max(\sigma_+^1; \sigma_+^2) < \mathrm{Re}(s) < \min(\sigma_-^1; \sigma_-^2)$$

where the pairs $(\sigma_+^1; \sigma_+^2) < \mathrm{Re}(s) < \min(\sigma_-^1; \sigma_-^2)$ identify the regions of convergence for the Laplace transforms $X_1(s)$ and $X_2(s)$, respectively.

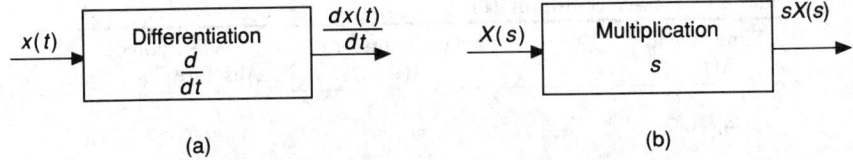

(a)

(b)

FIGURE 6.1 Equivalent operations in the (a) time-domain operation and (b) Laplace transform-domain operation. (*Source:* J. A. Cadzow and H. F. Van Landingham, *Signals, Systems, and Transforms,* Englewood Cliffs, N.J.: Prentice-Hall, 1985, p. 138. With permission.)

Time-Domain Differentiation

The operation of time-domain differentiation has then been found to correspond to a multiplication by s in the Laplace variable s domain.

The Laplace transform of differentiated signal $dx(t)/dt$ is

$$\mathscr{L}\left[\frac{dx(t)}{dt}\right] = sX(s)$$

Furthermore, it is clear that the region of absolute convergence of $dx(t)/dt$ is at least as large as that of $x(t)$. This property may be envisioned as shown in Fig. 6.1.

Time Shift

The signal $x(t - t_0)$ is said to be a version of the signal $x(t)$ right shifted (or delayed) by t_0 seconds. Right shifting (delaying) a signal by a t_0 second duration in the time domain is seen to correspond to a multiplication by e^{-st_0} in the Laplace transform domain. The desired Laplace transform relationship is

$$\mathscr{L}\left[x(t - t_0)\right] = e^{-st_0}X(s)$$

where $X(s)$ denotes the Laplace transform of the unshifted signal $x(t)$. As a general rule, any time a term of the form e^{-st_0} appears in $X(s)$, this implies some form of time shift in the time domain. This most important property is depicted in Fig. 6.2. It should be further noted that the regions of absolute convergence for the signals $x(t)$ and $x(t - t_0)$ are identical.

Time-Convolution Property

The convolution integral signal $y(t)$ can be expressed as

$$y(t) = \int_{-\infty}^{\infty} h(\tau)x(t - \tau)d\tau$$

where $x(t)$ denotes the input signal, the $h(t)$ characteristic signal identifying the operation process.

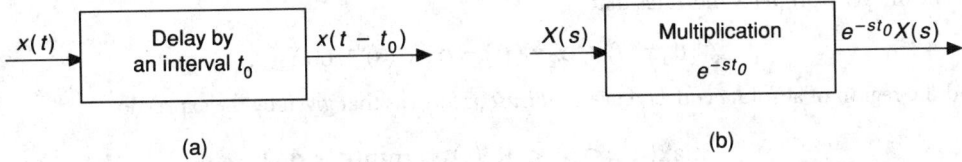

(a)

(b)

FIGURE 6.2 Equivalent operations in (a) the time domain and (b) the Laplace transform domain. (*Source:* J. A. Cadzow and H. F. Van Landingham, *Signals, Systems, and Transforms,* Englewood Cliffs, N.J.: Prentice-Hall, 1985, p. 140. With permission.)

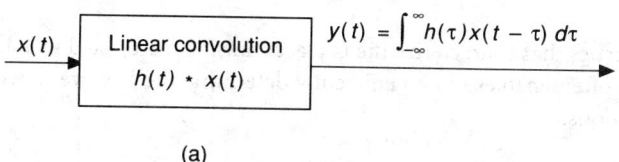

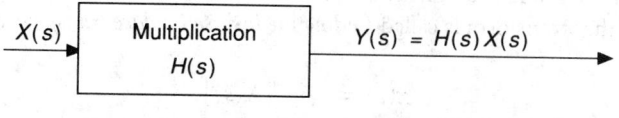

FIGURE 6.3 Representation of a time-invariant linear operator in (a) the time domain and (b) the *s*-domain. (*Source:* J. A. Cadzow and H. F. Van Landingham, *Signals, Systems, and Transforms,* Englewood Cliffs, N.J.: Prentice-Hall, 1985, p. 144. With permission.)

The Laplace transform of the response signal is simply given by

$$Y(s) = H(s)X(s)$$

where $H(s) = \mathcal{L}[h(t)]$ and $X(s) = \mathcal{L}[x(t)]$. Thus, the convolution of two time-domain signals is seen to correspond to the multiplication of their respective Laplace transforms in the *s*-domain. This property may be envisioned as shown in Fig. 6.3.

Time-Correlation Property

The operation of correlating two signals $x(t)$ and $y(t)$ is formally defined by the integral relationship

$$\phi_{xy}(\tau) = \int_{-\infty}^{\infty} x(t)y(t + \tau)dt$$

The Laplace transform property of the correlation function $\phi_{xy}(\tau)$ is

$$\Phi_{xy}(s) = X(-s)Y(s)$$

in which the region of absolute convergence is given by

$$\max(-\sigma_{x^-}, \sigma_{y^+}) < \mathrm{Re}(s) < \min(-\sigma_{x^+}, \sigma_{y^-})$$

Autocorrelation Function

The autocorrelation function of the signal $x(t)$ is formally defined by

$$\phi_{xx}(\tau) = \int_{-\infty}^{\infty} x(t)x(t + \tau)dt$$

The Laplace transform of the autocorrelation function is

$$\Phi_{xx}(s) = X(-s)X(s)$$

and the corresponding region of absolute convergence is

$$\max(-\sigma_{x^-}, \sigma_{y^+}) < \mathrm{Re}(s) < \min(-\sigma_{x^+}, \sigma_{y^-})$$

Other Properties

A number of properties that characterize the Laplace transform are listed in Table 6.2. Application of these properties often enables one to efficiently determine the Laplace transform of seemingly complex time functions.

Inverse Laplace Transform

Given a transform function $X(s)$ and its region of convergence, the procedure for finding the signal $x(t)$ that generated that transform is called *finding the inverse Laplace transform* and is symbolically denoted as

$$x(t) = \mathscr{L}^{-1}[X(s)]$$

Table 6.2   Laplace Transform Properties

Property	Signal $x(t)$ Time Domain	Laplace Transform $X(s)$ s Domain	Region of Convergence of $X(s)$ $\sigma_+ < \mathrm{Re}(s) < \sigma_-$
Linearity	$\alpha_1 x_1(t) + \alpha_2 x_2(t)$	$\alpha_1 X_1(s) + \alpha_2 X_2(s)$	At least the intersection of the regions of convergence of $X_1(s)$ and $X_2(s)$
Time differentiation	$\dfrac{dx(t)}{dt}$	$sX(s)$	At least $\sigma_+ < \mathrm{Re}(s)$
Time shift	$x(t - t_0)$	$e^{-st_0} X(s)$	$\sigma_+ < \mathrm{Re}(s) < \sigma_-$
Time convolution	$\displaystyle\int_{-\infty}^{\infty} h(\tau)x(t-\tau)d\tau$	$H(s)X(s)$	At least the intersection of the regions of convergence of $H(s)$ and $X(s)$
Time scaling	$x(at)$	$\dfrac{1}{\lvert a \rvert} X\!\left(\dfrac{s}{a}\right)$	$\sigma_+ < \mathrm{Re}\!\left(\dfrac{s}{a}\right) < \sigma_-$
Frequency shift	$e^{-at}x(t)$	$X(s + a)$	$\sigma_+ - \mathrm{Re}(a) < \mathrm{Re}(s) < \sigma_- - \mathrm{Re}(a)$
Multiplication (frequency convolution)	$x_1(t)x_2(t)$	$\dfrac{1}{2\pi j}\displaystyle\int_{c-j\infty}^{c+j\infty} X_1(u)X_2(s-u)du$	$\sigma_+^{(1)} + \sigma_+^{(2)} < \mathrm{Re}(s) < \sigma_-^{(1)} + \sigma_-^{(2)}$ $\sigma_+^{(1)} + \sigma_+^{(2)} < c < \sigma_-^{(1)} + \sigma_-^{(2)}$
Time integration	$\displaystyle\int_{-\infty}^{t} x(\tau)d\tau$	$\dfrac{1}{s} X(s)$ for $X(0) = 0$	At least $\sigma_+ < \mathrm{Re}(s) < \sigma_-$
Frequency differentiation	$(-t)^k x(t)$	$\dfrac{d^k X(s)}{ds^k}$	At least $\sigma_+ < \mathrm{Re}(s) < \sigma_-$
Time correlation	$\displaystyle\int_{-\infty}^{+\infty} x(t)y(t+z)dt$	$X(-s)Y(s)$	$\max(-\sigma_{x-}, \sigma_{y+}) < \mathrm{Re}(s) < \min(-\sigma_{x+}, \sigma_{y-})$
Autocorrelation function	$\displaystyle\int_{-\infty}^{+\infty} x(t)x(t+z)dt$	$X(-s)X(s)$	$\max(-\sigma_{x-}, \sigma_{x+}) < \mathrm{Re}(s) < \min(-\sigma_{x+}, \sigma_{x-})$

Source: J. A. Cadzow and H.F. Van Landingham, *Signals, Systems, and Transforms*, Englewood Cliffs, N.J.: Prentice-Hall, 1985. With permission.

The signal $x(t)$ can be recovered by means of the relationship

$$x(t) = \frac{1}{2\pi j} \int_{c-j\infty}^{c+j\infty} X(s)e^{st} ds$$

In this integral, the real number c is to be selected so that the complex number $c + j\omega$ lies entirely within the region of convergence of $X(s)$ for all values of the imaginary component ω. For the important class of rational Laplace transform functions, there exists an effective alternate procedure that does not necessitate directly evaluating this integral. This procedure is generally known as the *partial-fraction expansion method.*

Partial Fraction Expansion Method

As just indicated, the partial fraction expansion method provides a convenient technique for reacquiring the signal that generates a given rational Laplace transform. Recall that a transform function is said to be rational if it is expressible as a ratio of polynomial in s, that is,

$$X(s) = \frac{B(s)}{A(s)} = \frac{b_m s^m + b_{m-1} s^{m-1} + \cdots + b_1 s + b_0}{s^n + a_{n-1} s^{n-1} + \cdots + a_1 s + a_0}$$

The partial fraction expansion method is based on the appealing notion of equivalently expressing this rational transform as a sum of n elementary transforms whose corresponding inverse Laplace transforms (i.e., generating signals) are readily found in standard Laplace transform pair tables. This method entails the simple five-step process as outlined in Table 6.3. A description of each of these steps and their implementation is now given.

I. Proper Form for Rational Transform. This division process yields an expression in the proper form as given by

$$X(s) = \frac{B(s)}{A(s)}$$

$$= Q(s) + \frac{R(s)}{A(s)}$$

in which $Q(s)$ and $R(s)$ are the quotient and remainder polynomials, respectively, with the division made so that the degree of $R(s)$ is less than or equal to that of $A(s)$.

II. Factorization of Denominator Polynomial. The next step of the partial fraction expansion method entails the factorizing of the nth-order denominator polynomial $A(s)$ into a product of n first-order factors. This factorization is always possible and results in the equivalent representation of $A(s)$ as given by

$$A(s) = (s - p_1)(s - p_2) \cdots (s - p_n)$$

Table 6.3 Partial Fraction Expansion Method for Determining the Inverse Laplace Transform

I. Put rational transform into proper form whereby the degree of the numerator polynomial is less than or equal to that of the denominator polynomial.

II. Factor the denominator polynomial.

III. Perform a partial fraction expansion.

IV. Separate partial fraction expansion terms into causal and anticausal components using the associated region of absolute convergence for this purpose.

V. Using a Laplace transform pair table, obtain the inverse Laplace transform.

Source: J. A. Cadzow and H. F. Van Landingham, *Signals, Systems, and Transforms,* Englewood Cliffs, N.J.: Prentice-Hall, 1985, p. 153. With permission.

The terms $p_1, p_2, \ldots, p_n$ constituting this factorization are called the *roots of polynomial A(s)*, or the *poles of X(s)*.

III. Partial Fraction Expansion. With this factorization of the denominator polynomial accomplished, the rational Laplace transform $X(s)$ can be expressed as

$$X(s) = \frac{B(s)}{A(s)} = \frac{b_n s^n + b_{n-1} s^{n-1} + \cdots + b_0}{(s - p_1)(s - p_2) \cdots (s - p_n)} \tag{6.2}$$

We shall now *equivalently represent* this transform function as a linear combination of elementary transform functions.

Case 1: A(s) Has Distinct Roots.

$$X(s) = \alpha_0 + \frac{\alpha_1}{s - p_1} + \frac{\alpha_2}{s - p_2} + \cdots + \frac{\alpha_n}{s - p_n}$$

where the α_k are constants that identify the expansion and must be properly chosen for a valid representation.

$$\alpha_k = (s - p_k)X(s)\big|_{s=p_k} \quad \text{for } k = 1, 2, \ldots, n$$

and

$$\alpha_0 = b_n$$

The expression for parameter α_0 is obtained by letting s become unbounded (i.e., $s = +\infty$) in expansion (6.2).

Case 2: A(s) Has Multiple Roots.

$$X(s) = \frac{B(s)}{A(s)} = \frac{B(s)}{(s - p_1)^q A_1(s)}$$

The appropriate partial fraction expansion of this rational function is then given by

$$X(s) = \alpha_0 + \frac{\alpha_1}{(s - p_1)^1} + \cdots + \frac{\alpha_q}{(s - p_1)^q} + (n - q) \quad \substack{\text{other elementary terms} \\ \text{due to the roots of } A_1(s)}$$

The coefficient α_0 may be expediently evaluated by letting s approach infinity, whereby each term on the right side goes to zero except α_0. Thus,

$$\alpha_0 = \lim_{s \to +\infty} X(s) = 0$$

The α_q coefficient is given by the convenient expression

$$\alpha_q = (s - p_1)^q X(s)\big|_{s=p_1}$$

$$= \frac{B(p_1)}{A_1(p_1)} \tag{6.3}$$

The remaining coefficients $\alpha_1, \alpha_2, \ldots, \alpha_{q-1}$ associated with the multiple root p_1 may be evaluated by solving Eq. (6.3) by setting s to a specific value.

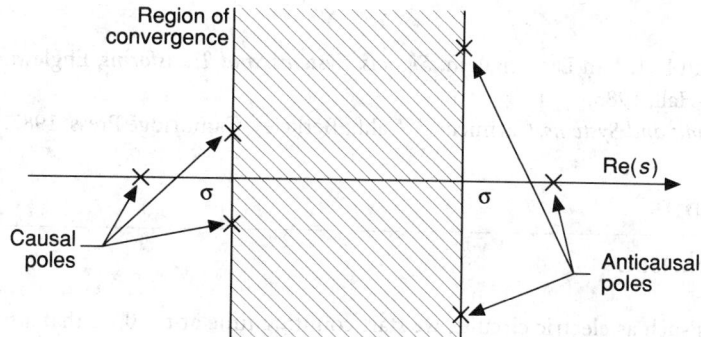

FIGURE 6.4 Location of causal and anticausal poles of a rational transform. (*Source:* J. A. Cadzow and H. F. Van Landingham, *Signals, Systems, and Transforms*, Englewood Cliffs, N.J.: Prentice-Hall, 1985, p. 161. With permission.)

IV. Causal and Anticausal Components. In a partial fraction expansion of a rational Laplace transform $X(s)$ whose region of absolute convergence is given by

$$\sigma_+ < \text{Re}(s) < \sigma_-$$

it is possible to decompose the expansion's elementary transform functions into causal and anticausal functions (and possibly impulse-generated terms). Any elementary function is interpreted as being (1) *causal* if the real component of its pole is less than or equal to σ_+ and (2) *anticausal* if the real component of its pole is greater than or equal to σ_-.

The poles of the rational transform that lie to the left (right) of the associated region of absolute convergence correspond to the causal (anticausal) component of that transform. Figure 6.4 shows the location of causal and anticausal poles of rational transform.

V. Table Look-Up of Inverse Laplace Transform. To complete the inverse Laplace transform procedure, one need simply refer to a standard Laplace transform function table to determine the time signals that generate each of the elementary transform functions. The required time signal is then equal to the same linear combination of the inverse Laplace transforms of these elementary transform functions.

Defining Terms

Laplace transform: A transformation of a function $f(t)$ from the time domain into the complex frequency domain yielding $F(s)$.

$$F(s) = \int_{-\infty}^{\infty} f(t)e^{-st}dt$$

where $s = \sigma + j\omega$.

Region of absolute convergence: The set of complex numbers s for which the magnitude of the Laplace transform integral is finite. The region can be expressed as

$$\sigma_+ < \text{Re}(s) < \sigma_-$$

where σ_+ and σ_- denote real parameters that are related to the causal and anticausal components, respectively, of the signal whose Laplace transform is being sought.

References

J. A. Cadzow and H. F. Van Landingham, *Signals, Systems, and Transforms*, Englewood Cliffs, N.J.: Prentice-Hall, 1985.

B. P. Lathi, *Signals and Systems*, Carmichael, Calif.: Berkeley-Cambridge Press, 1987.

6.2 Applications[*]

David E. Johnson

In applications such as electric circuits, we start counting time at $t = 0$, so that a typical function $f(t)$ has the property $f(t) = 0$, $t < 0$. Its transform is given therefore by

$$F(s) = \int_0^\infty f(t)e^{-st}\, dt$$

which is sometimes called the *one-sided Laplace transform*. Since $f(t)$ is like $x(t)u(t)$ we may still use Table 6.1 of the previous section to look up the transforms, but for simplicity we will omit the factor $u(t)$, which is understood to be present.

Differentiation Theorems

If we replace $f(t)$ in the one-sided transform by its derivative $f'(t)$ and integrate by parts, we have the transform of the derivative,

$$\mathscr{L}[f'(t)] = sF(s) - f(0) \tag{6.4}$$

We may formally replace f by f' to obtain

$$\mathscr{L}[f''(t)] = s\mathscr{L}[f'(t)] - f'(0)$$

or by (6.4),

$$\mathscr{L}[f''(t)] = s^2 F(s) - sf(0) - f'(0) \tag{6.5}$$

We may replace f by f' again in (6.5) to obtain $\mathscr{L}[f'''(t)]$, and so forth, obtaining the general result,

$$\mathscr{L}[f^{(n)}(t)] = s^n F(s) - s^{n-1} f(0) - s^{n-2} f'(0) - \cdots - f^{(n-1)}(0) \tag{6.6}$$

where $f^{(n)}$ is the nth derivative. The functions $f, f', \ldots, f^{(n-1)}$ are assumed to be continuous on $(0, \infty)$, and $f^{(n)}$ is continuous except possibly for a finite number of finite discontinuities.

As an example, let $f(t) = t^n$, for n a nonnegative integer. Then $f^{(n)}(t) = n!$ and $f(0) = f(0) = \cdots = f^{(n-1)}(0) = 0$. Therefore, we have

$$\mathscr{L}[n!] = s^n \mathscr{L}[t^n]$$

or

$$\mathscr{L}[t^n] = \frac{1}{s^n} \mathscr{L}[n!] = \frac{n!}{s^{n+1}}; \quad n = 0, 1, 2, \ldots \tag{6.7}$$

As another example, let us invert the transform

$$F(s) = \frac{8}{s^3(s + 2)}$$

[*]Based on D.E. Johnson, J.R. Johnson, and J.L. Hilburn, *Electric Circuit Analysis*, 2nd ed., Englewood Cliffs, N.J.: Prentice-Hall, 1992, chapters 19 and 20. With permission.

which has the partial fraction expansion

$$F(s) = \frac{A}{s^3} + \frac{B}{s^2} + \frac{C}{s} + \frac{D}{s+2}$$

where

$$A = s^3 F(s)\big|_{s=0} = 4$$

and

$$D = (s+2)F(s)\big|_{s=-2} = -1$$

To obtain B and C, we clear $F(s)$ of fractions, resulting in

$$8 = 4(s+2) + Bs(s+2) + Cs^2(s+2) - s^3$$

Equating coefficients of s^3 yields $C = 1$, and equating those of s^2 yields $B = -2$. The transform is therefore

$$F(s) = 2\frac{2!}{s^3} - 2\frac{1!}{s^2} + \frac{1}{s} - \frac{1}{s+2}$$

so that

$$f(t) = 2t^2 - 2t + 1 - e^{-2t}$$

Frequency-domain differentiation formulas may be obtained by differentiating the Laplace transform with respect to s. That is, if $F(s) = \mathcal{L}[f(t)]$,

$$\frac{dF(s)}{ds} = \frac{d}{ds}\int_0^\infty f(t)e^{-st}\,dt$$

Assuming that the operations of differentiation and integration may be interchanged, we have

$$\frac{dF(s)}{ds} = \int_0^\infty \frac{d}{ds}[f(t)e^{-st}]\,dt$$

$$= \int_0^\infty [-t f(t)]\,e^{-st}\,dt$$

From the last integral it follows by definition of the transform that

$$\mathcal{L}[t f(t)] = -\frac{dF(s)}{ds} \tag{6.8}$$

As an example, if $f(t) = \cos kt$, then $F(s) = s/(s^2 + k^2)$, and we have

$$\mathcal{L}[t \cos kt] = -\frac{d}{ds}\left(\frac{s}{s^2+k^2}\right) = \frac{s^2-k^2}{(s^2+k^2)^2}$$

We may repeatedly differentiate the transform to obtain the general case

$$\frac{d^n F(s)}{ds^n} = \int_0^\infty [(-t)^n f(t)]\,e^{-st}\,dt$$

Table 6.4   One-Sided Laplace Transform Properties

	$f(t)$	$F(s)$
1.	$cf(t)$	$cF(s)$
2.	$f_1(t) + f_2(t)$	$F_1(s) + F_2(s)$
3.	$\dfrac{df(t)}{dt}$	$sF(s) - f(0)$
4.	$\dfrac{d^n f(t)}{dt^n}$	$s^n F(s) - s^{n-1} f(0) - s^{n-2} f'(0)$ $- s^{n-1} f''(0) - \cdots - f^{n-1}(0)$
5.	$\displaystyle\int_0^t f(\tau)d\tau$	$\dfrac{F(s)}{s}$
6.	$e^{-at} f(t)$	$F(s + a)$
7.	$f(t - \tau)u(t - \tau)$	$e^{-s\tau} F(s)$
8.	$f * g = \displaystyle\int_0^t f(\tau)g(t - \tau)d\tau$	$F(s)G(s)$
9.	$f(ct), \ c > 0$	$\dfrac{1}{c} F\left(\dfrac{s}{c}\right)$
10.	$t^n f(t), \ n = 0, 1, 2, \ldots$	$(-1)^n F^{(n)}(s)$

from which we conclude that

$$\mathscr{L}[t^n f(t)] = (-1)^n \frac{d^n F(s)}{ds^n}; \quad n = 0, 1, 2, \ldots \tag{6.9}$$

Properties of the Laplace transform obtained in this and the previous section are listed in Table 6.4.

Applications to Integrodifferential Equations

If we transform both members of a linear differential equation with constant coefficients, the result will be an algebraic equation in the transform of the unknown variable. This follows from Eq. (6.6), which also shows that the initial conditions are automatically taken into account. The transformed equation may then be solved for the transform of the unknown and inverted to obtain the time-domain answer.

As an example, let us find the solution $x(t)$, for $t > 0$, of the system of equations

$$x'' + 4x' + 3x = e^{-2t}$$

$$x(0) = 1, \quad x'(0) = 2$$

Transforming, we have

$$s^2 X(s) - s - 2 + 4[sX(s) - 1] + 3X(s) = \frac{1}{s + 2}$$

from which

$$X(s) = \frac{s^2 + 8s + 13}{(s + 1)(s + 2)(s + 3)}$$

The partial fraction expansion is

$$X(s) = \frac{3}{s + 1} - \frac{1}{s + 2} - \frac{1}{s + 3}$$

from which

$$x(t) = 3e^{-t} - e^{-2t} - e^{-3t}$$

Certain integrodifferential equations may be transformed directly without first differentiating to remove the integrals. We need only transform the integrals by means of

$$\mathscr{L}\left[\int_0^t f(\tau)\, d\tau\right] = \frac{F(s)}{s}$$

As an example, the current $i(t)$ in Fig. 6.5, with no initial stored energy, satisfies the system of equations,

$$\frac{di}{dt} + 2i + 5\int_0^t i\, dt = u(t)$$

$$i(0) = 0$$

Transforming yields

$$sI(s) + 2I(s) + \frac{5}{s}I(s) = \frac{1}{s}$$

or

$$I(s) = \frac{1}{s^2 + 2s + 5} = \frac{1}{2}\left[\frac{2}{(s+1)^2 + 4}\right]$$

Therefore the current is

$$i(t) = 0.5e^{-t}\sin 2t \quad \text{A}$$

Applications to Electric Circuits

As the foregoing example shows, the Laplace transform method is an elegant procedure than can be used for solving electric circuits by transforming their describing integrodifferential equations into algebraic equations and applying the rules of algebra. If there is more than one loop or nodal equation, their transformed equations are solved simultaneously for the desired circuit current or voltage transforms, which are then inverted to obtain the time-domain answers. Superposition is not necessary because the various source functions appearing in the equations are simply transformed into algebraic quantities.

As an example, let us find v for $t > 0$ in the circuit of Fig. 6.6, given that $v(0) = 0$. We note that the circuit has two sources,

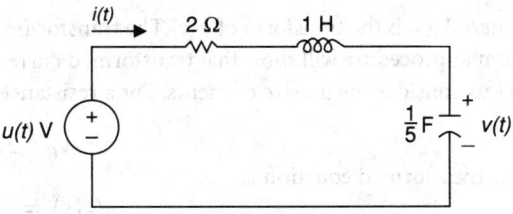

FIGURE 6.5 An *RLC* circuit.

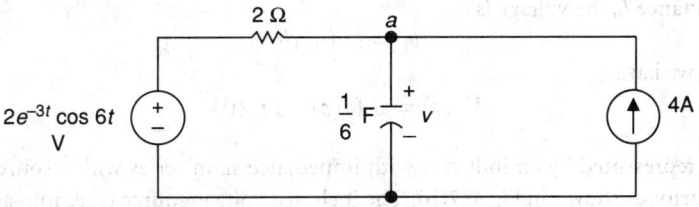

FIGURE 6.6 Circuit with two sources.

each with a different frequency, which poses no difficulty with the Laplace transform method. The nodal equation at node a is

$$\frac{v - 2e^{-3t} \cos 6t}{2} + \frac{1}{6} \frac{dv}{dt} = 4$$

Transforming yields

$$3V(s) - \frac{6(s + 3)}{(s + 3)^2 + 36} + sV(s) = \frac{24}{s}$$

from which we have

$$V(s) = \frac{24}{s(s + 3)} + \frac{6}{(s + 3)^2 + 36}$$

Expanding the first term on the right into partial fractions yields

$$V(s) = \frac{8}{s} - \frac{8}{s + 3} + \frac{6}{(s + 3)^2 + 6^2}$$

whose inverse transform is

$$v = 8 - 8e^{-3t} + e^{-3t} \sin 6t \quad \text{V}$$

The Transformed Circuit

Instead of writing the describing circuit equations, transforming the results, and solving for the transform of the circuit current or voltage, we may go directly to a **transformed circuit**, which is the original circuit with the currents, voltages, sources, and passive elements replaced by transformed equivalents. The current or voltage transforms are then found using ordinary circuit theory and the results inverted to the time-domain answers.

First, let us note that if we transform Kirchhoff's voltage law,

$$v_1(t) + v_2(t) + \cdots + v_n(t) = 0$$

we have

$$V_1(s) + V_2(s) + \cdots + V_n(s) = 0$$

where $V_i(s)$ is the transform of $v_i(t)$. The transformed voltages thus satisfy Kirchhoff's voltage law. A similar procedure will show that transformed currents satisfy Kirchhoff's current law, as well. Next, let us consider the passive elements. For a resistance R, with current i_R and voltage v_R, for which

$$v_R = Ri_R$$

the transformed equation is

$$V_R(s) = RI_R(s) \tag{6.10}$$

This result may be represented by the transformed resistor element of Fig. 6.7(a).

For an inductance L, the voltage is

$$v_L = L \, di_L/dt$$

Transforming, we have

$$V_L(s) = sLI_L(s) - Li_L(0) \tag{6.11}$$

which may be represented by an inductor with impedance sL in series with a source, $Li_L(0)$, with the proper polarity, as shown in Fig. 6.7(b). The included voltage source takes into account the initial condition $i_L(0)$.

In the case of a capacitance C we have

$$v_C = \frac{1}{C}\int_0^t i_C \; dt + v_C(0)$$

which transforms to

$$V_C(s) = \frac{1}{sC}I_C(s) + \frac{1}{s}v_C(0) \tag{6.12}$$

This is represented in Fig. 6.7(c) as a capacitor with impedance $1/sC$ in series with a source, $v_C(0)/s$, accounting for the initial condition.

We may solve Eqs. (6.10), (6.11), and (6.12) for the transformed currents and use the results to obtain alternate transformed elements useful for nodal analysis, as opposed to those of Fig. 6.7, which are ideal for loop analysis. The alternate elements are shown in Fig. 6.8.

Independent sources are simply labeled with their transforms in the transformed circuit. Dependent sources are transformed in the same way as passive elements. For example, a controlled voltage source defined by

$$v_1(t) = Kv_2(t)$$

transforms to

$$V_1(s) = KV_2(s)$$

which in the transformed circuit is the transformed source controlled by a transformed variable. Since Kirchhoff's laws hold and the rules for impedance hold, the transformed circuit may be analyzed exactly as we would an ordinary resistive circuit.

To illustrate, let us find $i(t)$ in Fig. 6.9(a), given that $i(0) = 4$ A and $v(0) = 8$ V. The transformed circuit is shown in Fig. 6.9(b), from which we have

$$I(s) = \frac{[2/(s+3)] + 4 - (8/s)}{3 + s + (2/s)}$$

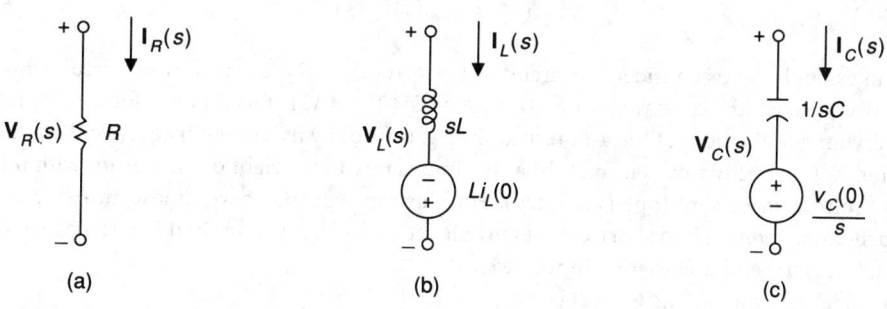

(a)   (b)   (c)

FIGURE 6.7 Transformed circuit elements.

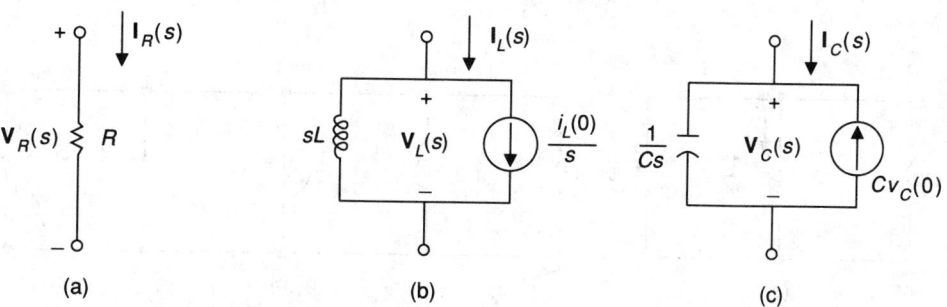

(a)   (b)   (c)

FIGURE 6.8 Transformed elements useful for nodal analysis.

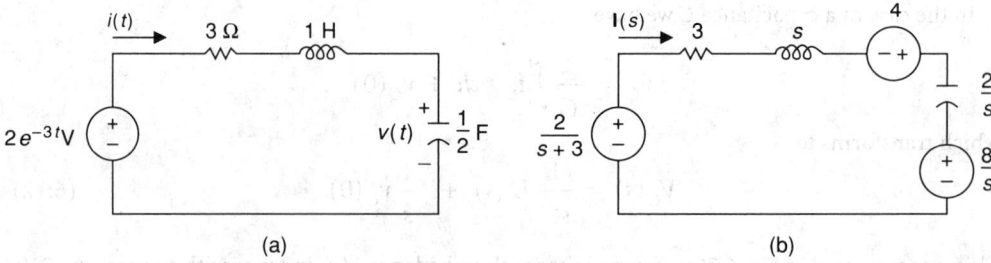

FIGURE 6.9   (a) A circuit and (b) its transformed counterpart.

This may be written

$$I(s) = -\frac{13}{s+1} + \frac{20}{s+2} - \frac{3}{s+3}$$

so that

$$i(t) = -13e^{-t} + 20e^{-2t} - 3e^{-3t}$$

Thévenin's and Norton's Theorems

Since the procedure using transformed circuits is identical to that using the phasor equivalent circuits in the ac steady-state case, we may obtain transformed Thévenin and Norton equivalent circuits exactly as in the phasor case. That is, the Thévenin impedance will be $Z_{th}(s)$ seen at the terminals of the transformed circuit with the sources made zero, and the open-circuit voltage and the short-circuit current will be $V_{oc}(s)$ and $I_{sc}(s)$, respectively, at the circuit terminals. The procedure is exactly like that for resistive circuits, except that in the transformed circuit the quantities involved are functions of s. Also, as in the resistor and phasor cases, the open-circuit voltage and short-circuit current are related by

$$V_{oc}(s) = Z_{th}(s)I_{sc}(s) \tag{6.13}$$

As an example, let us consider the circuit of Fig. 6.10(a) with the transformed circuit shown in Fig. 6.10(b). The initial conditions are $i(0) = 1$ A and $v(0) = 4$ V. Let us find $v(t)$ for $t > 0$ by replacing everything to the right of the 4-Ω resistor in Fig. 6.10(b) by its Thévenin equivalent circuit. We may find $Z_{th}(s)$ directly from Fig. 6.10(b) as the impedance to the right of the resistor with the two current sources made zero (open circuited). For illustrative purposes we choose, however, to find the open-circuit voltage and short-circuit current shown in Figs. 6.11(a) and (b), respectively, and use Eq. (6.13) to get the Thévenin impedance.

The nodal equation in Fig. 6.11(a) is

$$\frac{V_{oc}(s)}{3s} + \frac{1}{s} + \frac{s}{24}V_{oc}(s) = \frac{1}{6}$$

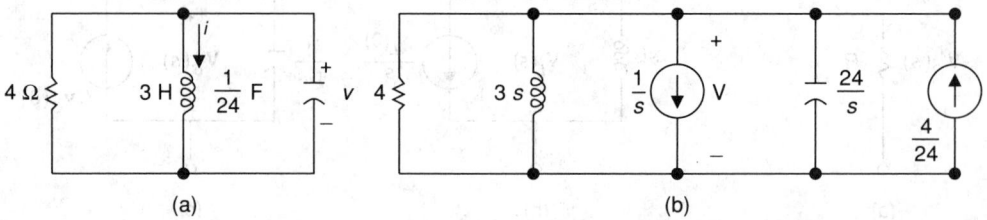

FIGURE 6.10   (a) An *RLC* parallel circuit and (b) its transformed circuit.

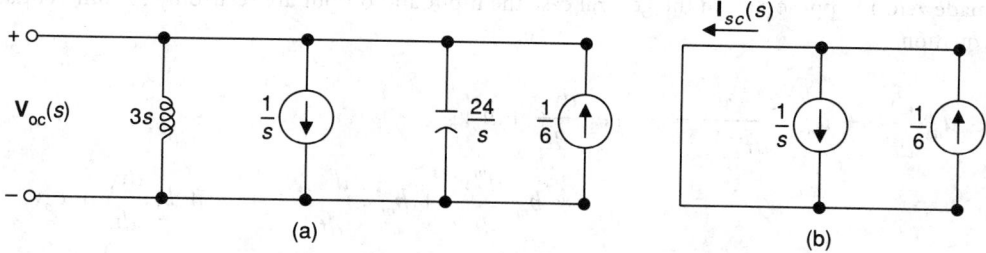

FIGURE 6.11 Circuit for obtaining (a) $V_{oc}(s)$ and (b) $I_{sc}(s)$.

from which we have

$$V_{oc}(s) = \frac{4(s-6)}{s^2 + 8}$$

From Fig. 6.11(b)

$$I_{sc}(s) = \frac{s-6}{6s}$$

The Thévenin impedance is therefore

$$Z_{th}(s) = \frac{V_{oc}(s)}{I_{sc}(s)} = \frac{\left[\dfrac{4(s-6)}{s^2+8}\right]}{\left[\dfrac{s-6}{6s}\right]} = \frac{24s}{s^2+8}$$

and the Thévenin equivalent circuit, with the 4 Ω connected, is shown in Fig. 6.12. From this circuit we find the transform

$$V(s) = \frac{4(s-6)}{(s+2)(s+4)} = \frac{-16}{s+2} + \frac{20}{s+4}$$

from which

$$v(t) = -16e^{-2t} + 20e^{-4t} \quad V$$

Network Functions

A **network function** or **transfer function** is the ratio $H(s)$ of the Laplace transform of the output function, say $v_o(t)$, to the Laplace transform of the input, say $v_i(t)$, assuming that there is only one input. (If there are multiple inputs, the transfer function is based on one of them with the others

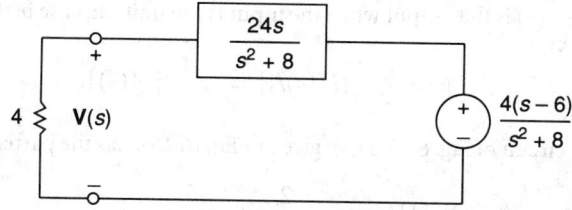

FIGURE 6.12 Thévenin equivalent circuit terminated in a resistor.

made zero.) Suppose that in the general case the input and output are related by the differential equation

$$a_n \frac{d^n v_o}{dt^n} + a_{n-1} \frac{d^{n-1} v_o}{dt^{n-1}} + \cdots + a_1 \frac{dv_o}{dt} + a_0 v_o$$

$$= b_m \frac{d^m v_i}{dt^m} + b_{m-1} \frac{d^{m-1} v_i}{dt^{m-1}} + \cdots + b_1 \frac{dv_i}{dt} + b_0 v_i$$

and that the initial conditions are all zero; that is,

$$v_o(0) = \frac{dv_o(0)}{dt} = \cdots = \frac{d^{n-1} v_o(0)}{dt^{n-1}} = v_i(0) = \frac{dv_i(0)}{dt} = \cdots = \frac{d^{m-1} v_i(0)}{dt^{m-1}} = 0$$

Then, transforming the differential equation results in

$$(a_n s^n + a_{n-1} s^{n-1} + \cdots + a_1 s + a_0) V_o(s)$$

$$= (b_m s^m + b_{m-1} s^{m-1} + \cdots + b_1 s + b_0) V_i(s)$$

from which the network function, or transfer function, is given by

$$H(s) = \frac{V_o(s)}{V_i(s)} = \frac{b_m s^m + b_{m-1} s^{m-1} + \cdots + b_1 s + b_0}{a_n s^n + a_{n-1} s^{n-1} + \cdots + a_1 s + a_0} \tag{6.14}$$

As an example, let us find the transfer function for the transformed circuit of Fig. 6.13, where the transfer function is $V_o(s)/V_i(s)$. By voltage division we have

$$H(s) = \frac{V_o(s)}{V_i(s)} = \frac{4}{s + 4 + (3/s)} = \frac{4s}{(s + 1)(s + 3)} \tag{6.15}$$

Step and Impulse Responses

In general, if $Y(s)$ and $X(s)$ are the transformed output and input, respectively, then the network function is $H(s) = Y(s)/X(s)$ and the output is

$$Y(s) = H(s)X(s) \tag{6.16}$$

The **step response** $r(t)$ is the output of a circuit when the input is the unit step function $u(t)$, with transform $1/s$. Therefore, the transform of the step response $R(s)$ is given by

$$R(s) = H(s)/s \tag{6.17}$$

The **impulse response** $h(t)$ is the output when the input is the unit impulse $\delta(t)$. Since $\mathscr{L}[\delta(t)] = 1$, we have from Eq. (6.16),

$$h(t) = \mathscr{L}^{-1}[H(s)/1] = \mathscr{L}^{-1}[H(s)] \tag{6.18}$$

As an example, for the circuit of Fig. 6.13, $H(s)$, given in Eq. (6.15), has the partial fraction expansion,

$$H(s) = \frac{-2}{s + 1} + \frac{6}{s + 3}$$

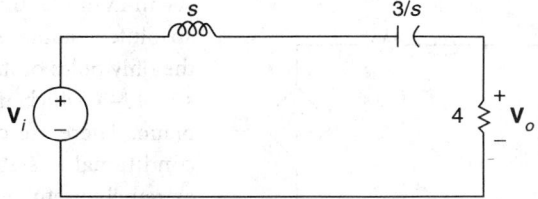

FIGURE 6.13 An *RLC* circuit.

so that

$$h(t) = -2e^{-t} + 6e^{-3t} \quad V$$

If we know the impulse response, we can find the transfer function,

$$H(s) = \mathcal{L}[h(t)]$$

from which we can find the response to *any* input. In the case of the step and impulse responses, it is understood that there are no other inputs except the step or the impulse. Otherwise, the transfer function would not be defined.

Stability

An important concern in circuit theory is whether the output signal remains bounded or increases indefinitely following the application of an input signal. An unbounded output could damage or even destroy the circuit, and thus it is important to know before applying the input if the circuit can accommodate the expected output. This question can be answered by determining the *stability* of the circuit.

A circuit is defined to have **bounded input–bounded output** (BIBO) stability if any bounded input results in a bounded output. The circuit in this case is said to be **absolutely stable** or *unconditionally stable*. BIBO stability can be determined by examining the *poles* of the network function (6.14).

If the denominator of $H(s)$ in Eq. (6.14) contains a factor $(s - p)^n$, then p is said to be a pole of $H(s)$ of *order n*. The output $V_o(s)$ would also contain this factor, and its partial fraction expansion would contain the term $K/(s - p)^n$. Thus, the inverse transform $v_o(t)$ is of the form

$$v_o(t) = A_n t^{n-1} e^{pt} + A_{n-1} t^{n-2} e^{pt} + \cdots + A_1 e^{pt} + v_1(t) \tag{6.19}$$

where $v_1(t)$ results from other poles of $V_o(s)$. If p is a real positive number or a complex number with a positive real part, $v_o(t)$ is unbounded because e^{pt} is a growing exponential. Therefore, for absolute stability there can be no pole of $V_o(s)$ that is positive or has a positive real part. This is equivalent to saying that $V_o(s)$ has no poles in the right half of the s-plane. Since $v_i(t)$ is bounded, $V_i(s)$ has no poles in the right half-plane. Therefore, since the only poles of $V_o(s)$ are those of $H(s)$ and $V_i(s)$, no pole of $H(s)$ for an absolutely stable circuit can be in the right-half of the s-plane.

From Eq. (6.19) we see that $v_i(t)$ is bounded, as far as pole p is concerned, if p is a *simple* pole (of order 1) and is purely imaginary. That is, $p = j\omega$, for which

$$e^{pt} = \cos \omega t + j \sin \omega t$$

which has a bounded magnitude. Unless $V_i(s)$ contributes an identical pole $j\omega$, $v_o(t)$ is bounded. Thus, $v_o(t)$ is bounded on the *condition* that any $j\omega$ pole of $H(s)$ is simple.

In summary, a network is *absolutely stable* if its network function $H(s)$ has only left half-plane poles. It is **conditionally stable** if $H(s)$ has only simple $j\omega$-axis poles and possibly left half-plane poles. It is *unstable* otherwise (right half-plane or multiple $j\omega$-axis poles).

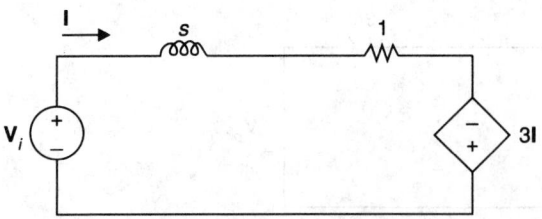

FIGURE 6.14   Unstable circuit.

As an example, the circuit of Fig. 6.13 is absolutely stable, since from Eq. (6.15) the only poles of its transfer function are $s = -1, -3$, which are both in the left half-plane. There are countless examples of conditionally stable circuits that are extremely useful, for example, a network consisting of a single capacitor with $C = 1$ F with input current $I(s)$ and output voltage $V(s)$. The transfer function is $H(s) = Z(s) = 1/Cs = 1/s$, which has the simple pole $s = 0$ on the $j\omega$-axis. Figure 6.14 illustrates a circuit which is unstable. The transfer function is

$$H(s) = I(s)/V_i(s) = 1/(s - 2)$$

which has the right half-plane pole $s = 2$.

Defining Terms

Absolute stability: When the network function $H(s)$ has only left half-plane poles.

Bounded input–bounded output stability: When any bounded input results in a bounded output.

Conditional stability: When the network function $H(s)$ has only simple $j\omega$-axis poles and possibly left half-plane poles.

Impulse response, $h(t)$:   The output when the input is the unit impulse $\delta(t)$.

Network or transfer function: The ratio $H(s)$ of the Laplace transform of the output function to the Laplace transform of the input function.

Step response, $r(t)$:   The output of a circuit when the input is the unit step function $u(t)$, with transform $1/s$.

Transformed circuit: An original circuit with the currents, voltages, sources, and passive elements replaced by transformed equivalents.

References

J. D. Irwin, *Basic Engineering Circuit Analysis,* 2nd ed., New York: Macmillan, 1987.

D. E. Johnson, J. R. Johnson, and J. L. Hilburn, *Electric Circuit Analysis,* 2nd ed., Englewood Cliffs, N.J.: Prentice-Hall, 1992.

J. W. Nilsson, *Electric Circuits,* 3rd ed., Reading, Mass.: Addison-Wesley, 1990.

7

State Variables

7.1 Introduction 169
7.2 State Equations in Normal Form 170
7.3 The Concept of State and State Variables and Normal Tree 171
7.4 Systematic Procedure in Writing State Equations 171
7.5 State Equations for Networks Described by Scalar
Differential Equations 174
7.6 Extension to Time-Varying and Nonlinear Networks 175

Wai-Kai Chen
University of Illinois, Chicago

7.1 Introduction

An electrical network is describable by a system of algebraic and differential equations known as the **primary system of equations** obtained by applying the Kirchhoff's current and voltage laws and the element *v-i* relations. In the case of linear networks, these equations can be transformed into a system of linear algebraic equations by means of the Laplace transformation, which is relatively simple to manipulate. The main drawback is that it contains a large number equations. To reduce this number, three **secondary systems of equations** are available: the **nodal system**, the **cutset system**, and the **loop system**. If a network has n nodes, b branches, and c components, there are $n - c$ linearly independent equations in nodal or **cutset** analysis and $b - n + c$ linearly independent equations in loop analysis. These equations can then be solved to yield the Laplace transformed solution. To obtain the final time-domain solution, we must take the inverse Laplace transformation. For most practical networks, the procedure is usually long and complicated and requires an excessive amount of computer time.

As an alternative we can formulate the network equations in the time domain as a system of first-order differential equations, which describe the dynamic behavior of the network. Some advantages of representing the network equations in this form are the following. First, such a system has been widely studied in mathematics, and its solution, both analytic and numerical, is readily known and available. Second, the representation can easily and naturally be extended to time-varying, nonlinear networks. In fact, computer-aided solution of time-varying, nonlinear network problems is almost always accomplished using the state-variable approach. Finally, the first-order differential equations can easily be programmed for a digital computer or simulated on an analog computer. Even if it were not for the above reasons, the approach provides an alternative view of the physical behavior of the network.

The term **state** is an abstract concept that may be represented in many ways. If we call the set of instantaneous values of all the branch currents and voltages as the *state* of the network, then the knowledge of the instantaneous values of all these variables determines this instantaneous state. Not all of these instantaneous values are required in order to determine the instantaneous state, however, because some can be calculated from the others. A set of data qualifies to be called the *state* of a system if it fulfills the following two requirements:

1. The state of any time, say, t_0, and the input to the system from t_0 on determine uniquely the state at any time $t > t_0$.
2. The state at time t and the inputs together with some of their derivatives at time t determine uniquely the value of any system variable at the time t.

The state may be regarded as a vector, the components of which are **state variables**. Network variables that are candidates for the state variables are the branch currents and voltages. Our problem is to choose state variables in order to formulate the **state equations**. Like the nodal, cutset, or loop system of equations, the state equations are formulated from the primary system of equations. For our purposes, we shall focus our attention on how to obtain state equations for linear systems.

7.2 State Equations in Normal Form

For a linear network containing k energy storage elements and h independent sources, our objective is to write a system of k first-order differential equations from the primary system of equations, as follows:

$$\dot{x}_i(t) = \sum_{j=1}^{k} a_{ij} x_j(t) + \sum_{j=1}^{h} b_{ij} u_j(t), \qquad (i = 1, 2, \ldots, k) \qquad (7.1)$$

In matrix notation, Eq. (7.1) becomes

$$
\begin{bmatrix} \dot{x}_1(t) \\ \dot{x}_2(t) \\ \cdot \\ \cdot \\ \cdot \\ \dot{x}_k(t) \end{bmatrix}
=
\begin{bmatrix}
a_{11} & a_{12} & \cdot & \cdot & a_{1k} \\
a_{21} & a_{22} & \cdot & \cdot & a_{2k} \\
\cdot & \cdot & \cdot & \cdot & \cdot \\
\cdot & \cdot & \cdot & \cdot & \cdot \\
\cdot & \cdot & \cdot & \cdot & \cdot \\
a_{k1} & a_{k2} & \cdot & \cdot & a_{kk}
\end{bmatrix}
\begin{bmatrix} x_1(t) \\ x_2(t) \\ \cdot \\ \cdot \\ \cdot \\ x_k(t) \end{bmatrix}
$$

$$(7.2)$$

$$
+
\begin{bmatrix}
b_{11} & b_{12} & \cdot & \cdot & b_{1h} \\
b_{21} & b_{22} & \cdot & \cdot & b_{2h} \\
\cdot & \cdot & \cdot & \cdot & \cdot \\
\cdot & \cdot & \cdot & \cdot & \cdot \\
b_{k1} & b_{k2} & \cdot & \cdot & b_{kh}
\end{bmatrix}
\begin{bmatrix} u_1(t) \\ u_2(t) \\ \cdot \\ \cdot \\ \cdot \\ u_h(t) \end{bmatrix}
$$

or, more compactly,

$$\dot{\mathbf{x}}(t) = \mathbf{A}\mathbf{x}(t) + \mathbf{B}\mathbf{u}(t) \qquad (7.3)$$

The real functions $x_1(t), x_2(t), \ldots, x_k(t)$ of the time t are called the state variables, and the k-vector $\mathbf{x}(t)$ formed by the state variables is known as the **state vector**. The h-vector $\mathbf{u}(t)$ formed by the h known forcing functions or excitations $u_j(t)$ is referred to as the **input vector**. The coefficient matrices $\mathbf{A}$ and $\mathbf{B}$, depending only upon the network parameters, are of orders $k \times k$ and $k \times h$, respectively. Equation (7.3) is usually called the **state equation in normal form**.

The state variables x_j may or may not be the desired output variables. We therefore must express the desired output variables in terms of the state variables and excitations. In general, if there are q

output variables $y_j(t)$ ($j = 1, 2, \ldots, q$) and h input excitations, the **output vector** $\mathbf{y}(t)$ formed by the q output variables $y_j(t)$ can be expressed in terms of the state vector $\mathbf{x}(t)$ and the input vector $\mathbf{u}(t)$ by the matrix equation

$$\mathbf{y}(t) = \mathbf{Cx}(t) + \mathbf{Du}(t) \tag{7.4}$$

where the known coefficient matrices $\mathbf{C}$ and $\mathbf{D}$, depending only on the network parameters, are of orders $q \times k$ and $q \times h$, respectively. Equation (7.4) is called the **output equation**. The state equation, Eq. (7.3), and the output equation, Eq. (7.4), together are known as the *state equations*.

7.3 The Concept of State and State Variables and Normal Tree

Our immediate problem is to choose the network variables as the state variables in order to formulate the state equations. If we call the set of instantaneous values of all the branch currents and voltages the *state* of the network, then the knowledge of the instantaneous values of all these variables determines this instantaneous state. Not all of these instantaneous values are required in order to determine the instantaneous state, however, because some can be calculated from the others. For example, the instantaneous voltage of a resistor can be obtained from its instantaneous current through Ohm's law. The question arises as to the minimum number of instantaneous values of branch voltages and currents that are sufficient to determine completely the instantaneous state of the network.

In a given network, a minimal set of its branch variables is said to be a **complete set of state variables** if their instantaneous values are sufficient to determine completely the instantaneous values of all the branch variables. For a linear time-invariant nondegenerate network, it is convenient to choose the capacitor voltages and inductor currents as the state variables. A **nondegenerate network** is one that contains neither a circuit composed only of capacitors and/or independent or dependent voltage sources nor a cutset composed only of inductors and/or independent or dependent current sources, where a cutset is a minimal subnetwork the removal of which cuts the original network into two connected pieces. Thus, not all the capacitor voltages and inductor currents of a degenerate network can be state variables. To help systematically select the state variables, we introduce the notion of normal tree.

A **tree** of a connected network is a connected subnetwork that contains all the nodes but does not contain any circuit. A **normal tree** of a connected network is a tree that contains all the independent voltage sources, the maximum number of capacitors, the minimum number of inductors, and none of the independent current sources. This definition excludes the possibility of having unconnected networks. In the case of unconnected networks, we can consider the normal trees of the individual components. We remark that the representation of the state of a network is generally not unique, but the state of a network itself is.

7.4 Systematic Procedure in Writing State Equations

In the following we present a systematic step-by-step procedure for writing the state equation for a network. They are a systematic way to eliminate the unwanted variables in the primary system of equations.

1. In a given network N, assign the voltage and current references of its branches.
2. In N select a normal tree T and choose as the state variables the capacitor voltages of T and the inductor currents of the **cotree** $\overline{T}$, the complement of T in N.
3. Assign each branch of T a voltage symbol, and assign each element of $\overline{T}$, called the **link**, a current symbol.
4. Using Kirchhoff's current law, express each tree-branch current as a sum of cotree-link currents, and indicate it in N if necessary.

5. Using Kirchhoff's voltage law, express each cotree-link voltage as a sum of tree-branch voltages, and indicate it in N if necessary.
6. Write the element v-i equations for the passive elements and separate these equations into two groups:
   a. Those element v-i equations for the tree-branch capacitors and the cotree-link inductors
   b. Those element v-i equations for all other passive elements
7. Eliminate the nonstate variables among the equations obtained in the preceding step. **Nonstate variables** are defined as those variables that are neither state variables nor known independent sources.
8. Rearrange the terms and write the resulting equations in normal form.

We illustrate the preceding steps by the following examples.

Example 1

We write the state equations for the network N of Fig. 7.1 by following the eight steps outlined above.

Step 1

The voltage and current references of the branches of the active network N are as indicated in Fig. 7.1.

Step 2

Select a normal tree T consisting of the branches R_1, C_3, and v_g. The subnetwork $C_3 i_5 v_g$ is another example of a normal tree.

Step 3

The tree branches R_1, C_3, and v_g are assigned the voltage symbols v_1, v_3, and v_g; and the cotree-links R_2, L_4, i_5, and i_g are assigned the current symbols i_2, i_4, i_3, and i_g, respectively. The controlled current source i_5 is given the current symbol i_3 because its current is controlled by the current of the branch C_3, which is i_3.

Step 4

Applying Kirchhoff's current law, the branch currents i_1, i_3, and i_7 can each be expressed as the sums of cotree-link currents:

$$i_1 = i_4 + i_g - i_3 \qquad (7.5a)$$

$$i_3 = i_2 - i_4 \qquad (7.5b)$$

$$i_7 = -i_2 \qquad (7.5c)$$

Step 5

Applying Kirchhoff's voltage law, the cotree-link voltages v_2, v_4, v_5, and v_6 can each be expressed as the sums of tree-branch voltages:

$$v_2 = v_g - v_3 \qquad (7.6a)$$

$$v_4 = v_3 - v_1 \qquad (7.6b)$$

$$v_5 = v_1 \qquad (7.6c)$$

$$v_6 = -v_1 \qquad (7.6d)$$

Step 6

The element v-i equations for the tree-branch capacitor and the cotree-link inductor are found to be

$$C_3 \dot{v}_3 = i_3 = i_2 - i_4 \qquad (7.7a)$$

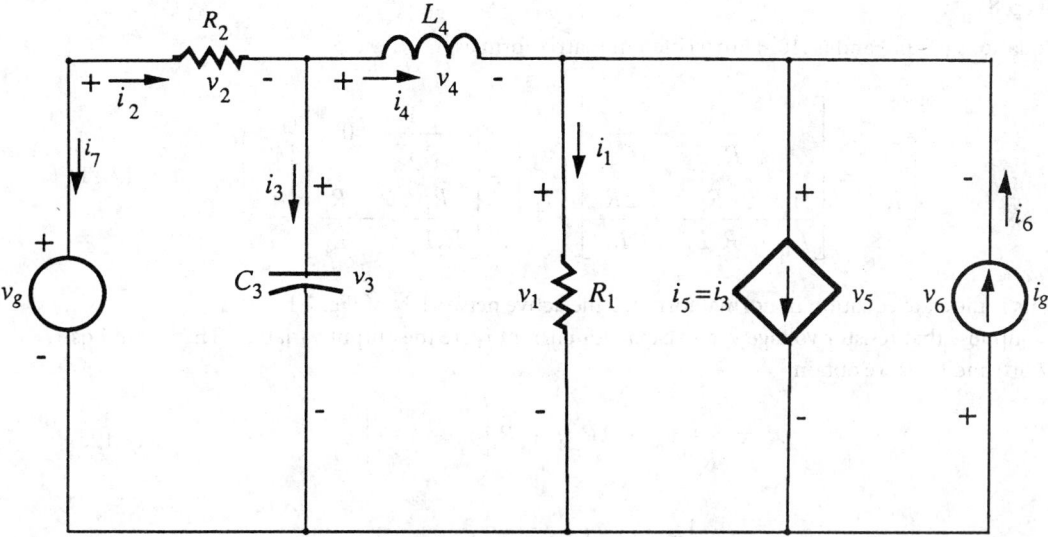

FIGURE 7.1 An active network used to illustrate the procedure for writing the state equations in normal form.

$$L_4\dot{i}_4 = v_4 = v_3 - v_1 \tag{7.7b}$$

Likewise, the element v-i equations for other passive elements are obtained as

$$v_1 = R_1 i_1 = R_1(i_4 + i_g - i_3) \tag{7.8a}$$

$$i_2 = \frac{v_2}{R_2} = \frac{v_g - v_3}{R_2} \tag{7.8b}$$

Step 7

The state variables are the capacitor voltage v_3 and inductor current i_4, and the known independent sources are i_g and v_g. To obtain the state equation, we must eliminate the nonstate variables v_1 and i_2 in Eq. (7.7). From Eqs. (7.5b) and (7.8) we express v_1 and i_2 in terms of the state variables and obtain

$$v_1 = R_1\left(2i_4 + i_g + \frac{v_3}{R_2} - \frac{v_g}{R_2}\right) \tag{7.9a}$$

$$i_2 = \frac{v_g - v_3}{R_2} \tag{7.9b}$$

Substituting these in Eq. (7.7) yields

$$C_3\dot{v}_3 = \frac{v_g - v_3}{R_2} - i_4 \tag{7.10a}$$

$$L_4\dot{i}_4 = \left(1 - \frac{R_1}{R_2}\right)v_3 - 2R_1 i_4 - R_1 i_g + \frac{R_1 v_g}{R_2} \tag{7.10b}$$

Step 8

Equations (7.10a) and (7.10b) are written in matrix form as

$$
\begin{bmatrix} \dot{v}_3 \\ \dot{i}_4 \end{bmatrix} = \begin{bmatrix} -\dfrac{1}{R_2 C_3} & -\dfrac{1}{C_3} \\ \dfrac{1}{L_4} - \dfrac{R_1}{R_2 L_4} & -\dfrac{2R_1}{L_4} \end{bmatrix} \begin{bmatrix} v_3 \\ i_4 \end{bmatrix} + \begin{bmatrix} \dfrac{1}{R_2 C_3} & 0 \\ \dfrac{R_1}{R_2 L_4} & -\dfrac{R_1}{L_4} \end{bmatrix} \begin{bmatrix} v_g \\ i_g \end{bmatrix} \tag{7.11}
$$

This is the state equation in normal form for the active network N of Fig. 7.1.

Suppose that resistor voltage v_1 and capacitor current i_3 are the output variables. Then from Eqs. (7.5b) and (7.9) we obtain

$$
v_1 = \frac{R_1}{R_2} v_3 + 2R_1 i_4 + R_1 \left(i_g - \frac{v_g}{R_2} \right) \tag{7.12a}
$$

$$
i_3 = -\frac{v_3}{R_2} - i_4 + \frac{v_g}{R_2} \tag{7.12b}
$$

In matrix form, the output equation of the network becomes

$$
\begin{bmatrix} v_1 \\ i_3 \end{bmatrix} = \begin{bmatrix} \dfrac{R_1}{R_2} & 2R_1 \\ -\dfrac{1}{R_2} & -1 \end{bmatrix} \begin{bmatrix} v_3 \\ i_4 \end{bmatrix} + \begin{bmatrix} -\dfrac{R_1}{R_2} & R_1 \\ \dfrac{1}{R_2} & 0 \end{bmatrix} \begin{bmatrix} v_g \\ i_g \end{bmatrix} \tag{7.13}
$$

Equations (7.11) and (7.13) together are the state equations of the active network of Fig. 7.1.

7.5 State Equations for Networks Described by Scalar Differential Equations

In many situations we are faced with networks that are described by scalar differential equations of order higher than one. Our purpose here is to show that these networks can also be represented by the state equations in normal.

Consider a network that can be described by the nth-order linear differential equation

$$
\frac{d^n y}{dt^n} + a_1 \frac{d^{n-1} y}{dt^{n-1}} + a_2 \frac{d^{n-2} y}{dt^{n-2}} + \cdots + a_{n-1} \frac{dy}{dt} + a_n y = bu \tag{7.14}
$$

Then its state equation can be obtained by defining

$$
\begin{aligned}
x_1 &= y \\
x_2 &= \dot{x}_1 \\
&\cdot \\
&\cdot \\
&\cdot \\
x_n &= \dot{x}_{n-1}
\end{aligned} \tag{7.15}
$$

showing that the nth-order linear differential Eq. (7.14) is equivalent to

$$\dot{x}_1 = x_2$$
$$\dot{x}_2 = x_3$$

$$\cdot$$
$$\cdot$$
$$\cdot \qquad (7.16)$$
$$\cdot$$

$$\dot{x}_{n-1} = x_n$$
$$\dot{x}_n = -a_n x_1 - a_{n-1} x_2 - \quad \cdots \quad - a_2 x_{n-1} - a_1 x_n + bu$$

or, in matrix form,

$$
\begin{bmatrix} \dot{x}_1 \\ \dot{x}_2 \\ \cdot \\ \cdot \\ \cdot \\ \dot{x}_{n-1} \\ \dot{x}_n \end{bmatrix}
=
\begin{bmatrix}
0 & 1 & 0 & \cdots & 0 \\
0 & 0 & 1 & \cdots & 0 \\
\cdot & \cdot & \cdot & \cdots & \cdot \\
\cdot & \cdot & \cdot & \cdots & \cdot \\
\cdot & \cdot & \cdot & \cdots & \cdot \\
0 & 0 & 0 & \cdots & 1 \\
-a_n & -a_{n-1} & -a_{n-2} & \cdots & -a_1
\end{bmatrix}
\begin{bmatrix} x_1 \\ x_2 \\ \cdot \\ \cdot \\ \cdot \\ x_{n-1} \\ x_n \end{bmatrix}
+
\begin{bmatrix} 0 \\ 0 \\ \cdot \\ \cdot \\ \cdot \\ 0 \\ b \end{bmatrix}
\begin{bmatrix} u \end{bmatrix}
\qquad (7.17)
$$

More compactly, Eq. (7.17) can be written as

$$\dot{\mathbf{x}}(t) = \mathbf{A}\mathbf{x}(t) + \mathbf{B}\mathbf{u}(t) \qquad (7.18)$$

The coefficient matrix $\mathbf{A}$ is called the **companion matrix** of Eq. (7.14), and Eq. (7.17) is the state-equation representation of the network describable by the linear differential equation (7.14).

7.6 Extension to Time-Varying and Nonlinear Networks

A great advantage in the state-variable approach to network analysis is that it can easily be extended to time-varying and nonlinear networks, which are often not readily amenable to the conventional methods of analysis. In these cases, it is more convenient to choose the capacitor charges and inductor flux as the the state variables instead of capacitor voltages and inductor currents.

In the case of a linear time-varying network, its state equations can be written the same as before except that now the coefficient matrices are time-dependent:

$$\dot{\mathbf{x}}(t) = \mathbf{A}(t)\mathbf{x}(t) + \mathbf{B}(t)\mathbf{u}(t) \qquad (7.19a)$$

$$\mathbf{y}(t) = \mathbf{C}(t)\mathbf{x}(t) + \mathbf{D}(t)\mathbf{u}(t) \qquad (7.19b)$$

Thus, with the state-variable approach, it is no more difficult to write the governing equations for a linear time-varying network than it is for a linear time-invariant network. Their solutions are, of course, a different matter.

For a nonlinear network, its state equation in normal form is describable by a coupled set of first-order differential equations:

$$\dot{\mathbf{x}} = \mathbf{f}(\mathbf{x}, \mathbf{u}, t) \qquad (7.20)$$

If the function **f** satisfies the familiar Lipshitz condition with respect to **x** in a given domain, then for every set of initial conditions $\mathbf{x}_0(t_0)$ and every input **u** there exists a unique solution $\mathbf{x}(t)$, the components of which are the state variables of the network.

Defining Terms

Companion matrix: The coefficient matrix in the state-equation representation of the network describable by a linear differential equation.

Complete set of state variables: A minimal set of network variables, the instantaneous values of which are sufficient to determine completely the instantaneous values of all the network variables.

Cotree: The complement of a tree in a network.

Cutset: A minimal subnetwork, the removal of which cuts the original network into two connected pieces.

Cutset system: A secondary system of equations using cutset voltages as variables.

Input vector: A vector formed by the input variables to a network.

Link: An element of a cotree.

Loop system: A secondary system of equations using loop currents as variables.

Nodal system: A secondary system of equations using nodal voltages as variables.

Nondegenerate network: A network that contains neither a circuit composed only of capacitors and/or independent or dependent voltage sources nor a cutset composed only of inductors and/or independent or dependent current sources.

Nonstate variables: Network variables that are not state variables.

Normal tree: A tree that contains all the independent voltage sources, the maximum number of capacitors, the minimum number of inductors, and none of the independent current sources.

Output equation: An equation expressing the output vector in terms of the state vector and the input vector.

Output vector: A vector formed by the output variables of a network.

Primary system of equations: A system of algebraic and differential equations obtained by applying the Kirchhoff's current and voltage laws and the element v-i relations.

Secondary system of equations: A system of algebraic and differential equations obtained from the primary system of equations by transformation of network variables.

State: A set of data, the values of which at any time t, together with the input to the system at the time, determine uniquely the value of any network variable at the time t.

State equation in normal form: A system of first-order differential equations that describes the dynamic behavior of a network and that is put into a standard form.

State equations: Equations formed by the state equation and the output equation.

State variables: Network variables used to describe the state.

State vector: A vector formed by the state variables.

Tree: A connected subnetwork that contains all the nodes of the original network but does not contain any circuit.

References

W. K. Chen, *Linear Networks and Systems: Algorithms and Computer-Aided Implementations,* Singapore: World Scientific Publishing, 1990.

W. K. Chen, *Active Network Analysis,* Singapore: World Scientific Publishing, 1991.

L. O. Chua and P.M. Lin, *Computer-Aided Analysis of Electronics Circuits: Algorithms & Computational Techniques,* Englewood Cliffs, N.J.: Prentice-Hall, 1975.

E. S. Kuh and R.A. Rohrer, "State-variables approach to network analysis," *Proc. IEEE,* vol. 53, pp. 672–686, July 1965.

Further Information

An expository paper on the application of the state-variables technique to network analysis was originally written by E. S. Kuh and R. A. Rohrer ("State-variables approach to network analysis," *Proc. IEEE,* vol. 53, pp. 672–686, July 1965). A computer-aided network analysis based on state-variables approach is extensively discussed in the book by Wai-Kai Chen, *Linear Networks and Systems: Algorithms and Computer-Aided Implementations* (World Scientific Publishing Co., Singapore, 1990). The use of state variables in the analysis of electronics circuits and nonlinear networks is treated in the book by L. O. Chua and P. M. Lin, *Computer-Aided Analysis of Electronics Circuits: Algorithms & Computational Techniques* (Prentice-Hall, Englewood Cliffs, N.J., 1975). The application of state-variables technique to active network analysis is contained in the book by Wai-Kai Chen, *Active Network Analysis* (World Scientific Publishing Co., Singapore, 1991).

8

The z-Transform

Richard C. Dorf
University of California, Davis

Zhen Wan
University of California, Davis

8.1 Introduction 178
8.2 Properties of the z-Transform 178
    Linearity • Translation • Convolution • Multiplication by a^n • Time Reversal
8.3 Unilateral z-Transform 180
    Time Advance • Initial Signal Value • Final Value
8.4 z-Transform Inversion 181
    Method 1 • Method 2 • Inverse Transform Formula (Method 2)
8.5 Sampled Data 182

8.1 Introduction

Discrete-time signals can be represented as sequences of numbers. Thus, if x is a discrete-time signal, its values can, in general, be indexed by n as follows:

$$x = \{\ldots, x(-2), x(-1), x(0), x(1), x(2), \ldots, x(n), \ldots\}$$

In order to work within a transform domain for discrete-time signals, we define the **z-transform** as follows. The z-transform of the sequence x in the previous equation is

$$Z\{x(n)\} = X(z) = \sum_{n=-\infty}^{\infty} x(n)z^{-n}$$

in which the variable z can be interpreted as being either a time-position marker or a complex-valued variable, and the script Z is the z-transform operator. If the former interpretation is employed, the number multiplying the marker z^{-n} is identified as being the nth element of the x sequence, i.e., $x(n)$. It will be generally beneficial to take z to be a complex-valued variable. The z-transforms of some useful sequences are listed in Table 8.1.

8.2 Properties of the z-Transform

Linearity

Both the direct and inverse z-transform obey the property of linearity. Thus, if $Z\{f(n)\}$ and $Z\{g(n)\}$ are denoted by $F(z)$ and $G(z)$, respectively, then

$$Z\{af(n) + bg(n)\} = aF(z) + bG(z)$$

where a and b are constant multipliers.

Table 8.1 Partial-Fraction Equivalents Listing Causal and Anticausal z-Transform Pairs

z-Domain: $F(z)$	Sequence Domain: $f(n)$
1a. $\dfrac{1}{z-a}$, for $\lvert z \rvert > \lvert a \rvert$	$a^{n-1}u(n-1) = \left\{0,\ 1,\ a,\ a^2,\ \dots\right\}$
1b. $\dfrac{1}{z-a}$, for $\lvert z \rvert < \lvert a \rvert$	$-a^{n-1}u(-n) = \left\{\dots,\ \dfrac{-1}{a^3},\ \dfrac{-1}{a^2},\ \dfrac{-1}{a}\right\}$
2a. $\dfrac{1}{(z-a)^2}$, for $\lvert z \rvert > \lvert a \rvert$	$(n-1)a^{n-2}u(n-1) = \left\{0,\ 1,\ 2a,\ 3a^2,\ \dots\right\}$
2b. $\dfrac{1}{(z-a)^2}$, for $\lvert z \rvert < \lvert a \rvert$	$-(n-1)a^{n-2}u(-n) = \left\{\dots,\ \dfrac{3}{a^4},\ \dfrac{2}{a^3},\ \dfrac{1}{a^2}\right\}$
3a. $\dfrac{1}{(z-a)^3}$, for $\lvert z \rvert > \lvert a \rvert$	$\dfrac{1}{2}(n-1)(n-2)a^{n-3}u(n-1) = \left\{0,\ 0,\ 1,\ 3a,\ 6a^2,\ \dots\right\}$
3b. $\dfrac{1}{(z-a)^3}$, for $\lvert z \rvert < \lvert a \rvert$	$\dfrac{-1}{2}(n-1)(n-2)a^{n-3}u(-n) = \left\{\dots,\ \dfrac{-6}{a^5},\ \dfrac{-3}{a^4},\ \dfrac{-1}{a^3}\right\}$
4a. $\dfrac{1}{(z-a)^m}$, for $\lvert z \rvert > \lvert a \rvert$	$\dfrac{1}{(m-1)!}\displaystyle\prod_{k=1}^{m-1}(n-k)a^{n-m}u(n-1)$
4b. $\dfrac{1}{(z-a)^m}$, for $\lvert z \rvert < \lvert a \rvert$	$\dfrac{-1}{(m-1)!}\displaystyle\prod_{k=1}^{m-1}(n-k)a^{n-m}u(-n)$
5a. z^{-m}, for $z \neq 0$, $m \geq 0$	$\delta(n-m) = \left\{\dots,\ 0,\ 0,\ \dots,\underset{\underset{m}{\uparrow}}{1},\ 0,\ \dots,\ 0,\ \dots\right\}$
5b. z^{+m}, for $\lvert z \rvert < \infty$, $m \geq 0$	$\delta(n+m) = \left\{\dots,\ 0,\ 0,\ \dots,\underset{\underset{-m}{\uparrow}}{1},\ \dots,\ 0,\ \dots,\ 0,\ \dots\right\}$

Source: J. A. Cadzow and H. F. Van Landingham, *Signals, Systems and Transforms,* Englewood Cliffs, N.J.: Prentice-Hall, 1985, p. 191. With permission.

Translation

An important property when transforming terms of a difference equation is the *z-transform* of a sequence shifted in time. For a constant shift, we have

$$\mathcal{Z}\{f(n+k)\} = z^k F(z)$$

for positive or negative integer k. The region of convergence of $z^k F(z)$ is the same as for $F(z)$ for positive k; only the point $z = 0$ need be eliminated from the convergence region of $F(z)$ for negative k.

Convolution

In the z-domain, the time-domain convolution operation becomes a simple product of the corresponding transforms, that is,

$$Z\{f(n) * g(n)\} = F(z)G(z)$$

Multiplication by a^n

This operation corresponds to a rescaling of the z-plane. For $a > 0$,

$$Z\{a^n f(n)\} = F\left(\frac{z}{a}\right) \quad \text{for } aR_1 < |z| < aR_2$$

where $F(z)$ is defined for $R_1 < |z| < R_2$.

Time Reversal

$$Z\{f(-n)\} = F(z^{-1}) \quad \text{for } R_2^{-1} < |z| < R_1^{-1}$$

where $F(z)$ is defined for $R_1 < |z| < R_2$.

8.3 Unilateral z-Transform

The unilateral z-transform is defined as

$$Z_+\{x(n)\} = X(z) = \sum_{n=0}^{\infty} x(n)z^{-n} \quad \text{for } |z| > R$$

where it is called single-sided since $n \geq 0$, just as if the sequence $x(n)$ was in fact single-sided. If there is no ambiguity in the sequel, the subscript plus is omitted and we use the expression *z-transform* to mean either the double- or the single-sided transform. It is usually clear from the context which is meant. By restricting signals to be single-sided, the following useful properties can be proved.

Time Advance

For a single-sided signal $f(n)$,

$$Z_+\{f(n + 1)\} = zF(z) - zf(0)$$

More generally,

$$Z_+\{f(n + k)\} = z^k F(z) - z^k f(0) - z^{k-1} f(1) - \cdots - z f(k - 1)$$

This result can be used to solve linear constant-coefficient difference equations. Occasionally, it is desirable to calculate the initial or final value of a single-sided sequence without a complete inversion. The following two properties present these results.

Initial Signal Value

If $f(n) = 0$ for $n < 0$,

$$f(0) = \lim_{z \Rightarrow \infty} F(z)$$

where $F(z) = Z\{f(n)\}$ for $|z| > R$.

Final Value

If $f(n) = 0$ for $n < 0$ and $Z\{f(n)\} = F(z)$ is a rational function with all its denominator roots (poles) strictly inside the unit circle except possibly for a first-order pole at $z = 1$,

$$f(\infty) = \lim_{n \Rightarrow \infty} f(n) = \lim_{z \Rightarrow \infty}(1 - z^{-1})F(z)$$

8.4 z-Transform Inversion

We operationally denote the inverse transform of $F(z)$ in the form

$$f(n) = Z^{-1}\{F(z)\}$$

There are three useful methods for inverting a transformed signal. They are:

1. Expansion into a series of terms in the variables z and z^{-1}
2. Complex integration by the method of residues
3. Partial-fraction expansion and table look-up

We discuss two of these methods in turn.

Method 1

For the expansion of $F(z)$ into a series, the theory of functions of a complex variable provides a practical basis for developing our inverse transform techniques. As we have seen, the general region of convergence for a transform function $F(z)$ is of the form $a < |z| < b$, i.e., an annulus centered at the origin of the z-plane. This first method is to obtain a series expression of the form

$$F(z) = \sum_{n=-\infty}^{\infty} c_n z^{-n}$$

which is valid in the annulus of convergence. This double-sided series is called a *Laurent series*. When $F(z)$ has been expanded as in the previous equation, that is, when the coefficients c_n, $n = 0$, $\pm 1, \pm 2, \ldots$ have been found, the corresponding sequence is specified by $f(n) = c_n$ by uniqueness of the transform.

Method 2

We evaluate the inverse transform of $F(z)$ by the method of residues. The method involves the calculation of residues of a function both inside and outside of a simple closed path that lies inside the region of convergence. A number of key concepts are necessary in order to describe the required procedure.

A complex-valued function $G(z)$ has a pole of order k at $z = z_0$ if it can be expressed as

$$G(z) = \frac{G_1(z_0)}{(z - z_0)^k}$$

where $G_1(z_0)$ is finite.

The residue of a complex function $G(z)$ at a pole of order k at $z = z_0$ is defined by

$$\text{Res}[G(z)]\Big|_{z=z_0} = \frac{1}{(k-1)!} \frac{d^{k-1}}{dz^{k-1}}[(z - z_0)^k G(z)]\Big|_{z=z_0}$$

Inverse Transform Formula (Method 2)

If $F(z)$ is convergent in the annulus $0 < a < |z| < b$ as shown in Fig. 8.1 and C is the closed path shown (the path C must lie entirely within the annulus of convergence), then

$$f(n) \begin{cases} \text{sum of residues of } F(z)z^{n-1} \text{ at poles of } F(z) \text{ inside } C, & m \geq 0 \\ -(\text{sum of residues of } F(z)z^{n-1} \text{ at poles of } F(z) \text{ outside } C), & m < 0 \end{cases}$$

where m is the least power of z in the numerator of $F(z)z^{n-1}$, e.g., m might equal $n - 1$. Figure 8.1 illustrates the previous equation.

8.5 Sampled Data

Data obtained for a signal only at discrete intervals (**sampling period**) is called **sampled data**. One advantage of working with sampled data is the ability to represent sequences as combinations of sampled time signals. Table 8.2 provides some key z-transform pairs. So that the table can serve a multiple purpose, there are three items per line: the first is an indicated sampled continuous-time signal, the second is the Laplace transform of the continuous-time signal, and the third is the z-transform of the uniformly sampled continous-time signal. To illustrate the interrelation of these entries, consider Fig. 8.2. For simplicity, only single-sided signals have been used in Table 8.2. Con-

Table 8.2 z-Transforms for Sampled Data

$f(t), t = nT,$ $n = 0, 1, 2, \ldots$	$F(s), \operatorname{Re}[s] > \sigma_0$	$F(z), |z| > \rho_0$
1. 1 (unit step)	$\dfrac{1}{s}$	$\dfrac{z}{z-1}$
2. t (unit ramp)	$\dfrac{1}{s^2}$	$\dfrac{Tz}{(z-1)^2}$
3. t^2	$\dfrac{2}{s^3}$	$\dfrac{T^2 z(z+1)}{(z-1)^3}$
4. e^{-at}	$\dfrac{1}{s+a}$	$\dfrac{z}{z-e^{-aT}}$
5. te^{-at}	$\dfrac{1}{(s+a)^2}$	$\dfrac{Tze^{-aT}}{(z-e^{-aT})^2}$
6. $\sin \omega t$	$\dfrac{\omega}{s^2+\omega^2}$	$\dfrac{z \sin \omega T}{z^2 - 2z \cos \omega T + 1}$
7. $\cos \omega t$	$\dfrac{s}{s^2+\omega^2}$	$\dfrac{z(z - \cos \omega T)}{z^2 - 2z \cos \omega T + 1}$
8. $e^{-at} \sin \omega t$	$\dfrac{\omega}{(s+a)^2 + \omega^2}$	$\dfrac{ze^{-aT} \sin \omega T}{z^2 - 2ze^{-aT} \cos \omega T + e^{-2aT}}$
9. $e^{-at} \cos \omega t$	$\dfrac{s+a}{(s+a)^2 + \omega^2}$	$\dfrac{z(z - e^{-aT} \cos \omega T)}{z^2 - 2ze^{-aT} \cos \omega T + e^{-2aT}}$

Source: J.A. Cadzow and H.F. Landingham, *Signals, Systems and Transforms,* Englewood Cliffs, N.J.: Prentice-Hall, 1985, p. 191. With permission.

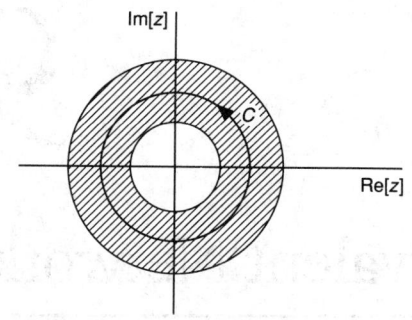

FIGURE 8.1 Typical convergence region for a transformed discrete-time signal (*Source:* J. A. Cadzow and H. F. Van Landingham, *Signals, Systems and Transforms*, Englewood Cliffs, N.J.: Prentice-Hall, 1985, p. 191. With permission.)

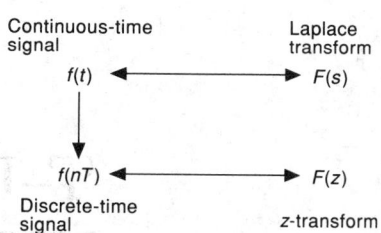

FIGURE 8.2 Signal and transform relationships for Table 8.2.

sequently, the convergence regions are understood in this context to be $\text{Re}[s] < \sigma_0$ and $|z| > \rho_0$ for the Laplace and z-transforms, respectively. The parameters σ_0 and ρ_0 depend on the actual transformed functions; in factor z, the inverse sequence would begin at $n = 0$. Thus, we use a modified partial-fraction expansion whose terms have this extra z-factor.

Defining Terms

Sampled data: Data obtained for a variable only at discrete intervals. Data is obtained every sampling period.

Sampling period: The period for which the sampled variable is held constant.

z-transform: A transform from the s-domain to the z-domain by $z = e^{st}$.

References

R. C. Dorf, *Modern Control Systems*, 6th ed. Reading, Mass.: Addison-Wesley, 1992.
R. E. Ziemer, *Signals and Systems*, 2nd ed., New York: MacMillan, 1989.

Further Information

IEEE Transactions on Education
IEEE Transactions on Automatic Control
Contact IEEE, Piscataway, N.J. 08855-1313

9

T–Π Equivalent Networks

Zhen Wan
University of California, Davis

Richard C. Dorf
University of California, Davis

9.1 Introduction 184
9.2 Three-Phase Connections 185
9.3 Wye ⇔ Delta Transformations 186

9.1 Introduction

Two very important two-ports are the **T and Π networks** shown in Fig. 9.1. Because we encounter these two geometrical forms often in two-port analyses, it is useful to determine the conditions under which these two networks are equivalent. In order to determine the equivalence relationship, we will examine Z-parameter equations for the T network and the Y-parameter equations for the Π network.

For the T network the equations are

$$\mathbf{V}_1 = (\mathbf{Z}_1 + \mathbf{Z}_3)\mathbf{I}_1 + \mathbf{Z}_3\mathbf{I}_2$$

$$\mathbf{V}_2 = \mathbf{Z}_3\mathbf{I}_1 + (\mathbf{Z}_2 + \mathbf{Z}_3)\mathbf{I}_2$$

and for the Π network the equations are

$$\mathbf{I}_1 = (\mathbf{Y}_a + \mathbf{Y}_b)\mathbf{V}_1 - \mathbf{Y}_b\mathbf{V}_2$$

$$\mathbf{I}_2 = -\mathbf{Y}_b\mathbf{V}_1 + (\mathbf{Y}_b + \mathbf{Y}_c)\mathbf{V}_2$$

Solving the equations for the T network in terms of $\mathbf{I}_1$ and $\mathbf{I}_2$, we obtain

$$\mathbf{I}_1 = \left(\frac{\mathbf{Z}_2 + \mathbf{Z}_3}{\mathbf{D}_1}\right)\mathbf{V}_1 - \frac{\mathbf{Z}_3\mathbf{V}_2}{\mathbf{D}_1}$$

$$\mathbf{I}_2 = -\frac{\mathbf{Z}_3\mathbf{V}_1}{\mathbf{D}_1} + \left(\frac{\mathbf{Z}_1 + \mathbf{Z}_3}{\mathbf{D}_1}\right)\mathbf{V}_2$$

where $\mathbf{D}_1 = \mathbf{Z}_1\mathbf{Z}_2 + \mathbf{Z}_2\mathbf{Z}_3 + \mathbf{Z}_1\mathbf{Z}_3$. Comparing these equations with those for the Π network, we find that

$$\mathbf{Y}_a = \frac{\mathbf{Z}_2}{\mathbf{D}_1}$$

$$\mathbf{Y}_b = \frac{\mathbf{Z}_3}{\mathbf{D}_1}$$

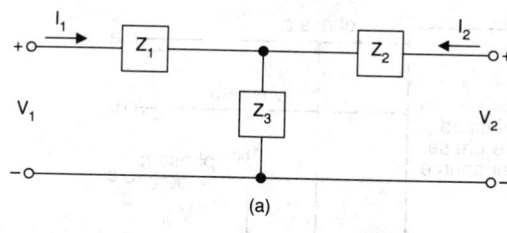

(a)

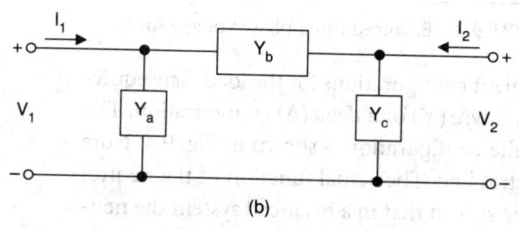

(b)

FIGURE 9.1   T and Π two-port networks.

$$Y_c = \frac{Z_1}{D_1}$$

or in terms of the impedances of the Π network

$$Z_a = \frac{D_1}{Z_2}$$

$$Z_b = \frac{D_1}{Z_3}$$

$$Z_c = \frac{D_1}{Z_1}$$

If we reverse this procedure and solve the equations for the Π network in terms of V_1 and V_2 and then compare the resultant equations with those for the T network, we find that

$$Z_1 = \frac{Y_c}{D_2}$$

$$Z_2 = \frac{Y_a}{D_2} \qquad (9.1)$$

$$Z_3 = \frac{Y_b}{D_2}$$

where $D_2 = Y_a Y_b + Y_b Y_c + Y_a Y_c$. Equation (9.1) can also be written in the form

$$Z_1 = \frac{Z_a Z_b}{Z_a + Z_b + Z_c}$$

$$Z_2 = \frac{Z_b Z_c}{Z_a + Z_b + Z_c}$$

$$Z_3 = \frac{Z_a Z_c}{Z_a + Z_b + Z_c}$$

The T is a wye-connected network and the Π is a delta-connected network, as we discuss in the next section.

9.2 Three-Phase Connections

By far the most important polyphase voltage source is the balanced three-phase source. This source, as illustrated by Fig. 9.2, has the following properties. The phase voltages, that is, the voltage from each line a, b, and c to the neutral n, are given by

$$V_{an} = V_p \angle 0°$$

$$V_{bn} = V_p \angle -120°$$

$$V_{cn} = V_p \angle +120° \qquad (9.2)$$

An important property of the balanced voltage set is that

$$\mathbf{V}_{an} + \mathbf{V}_{bn} + \mathbf{V}_{cn} = 0 \qquad (9.3)$$

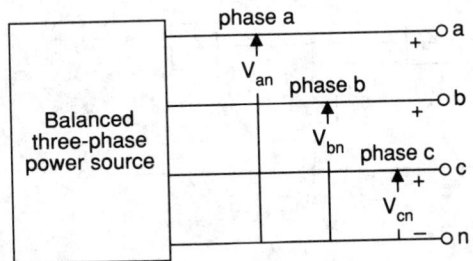

FIGURE 9.2   Balanced three-phase voltage source.

From the standpoint of the user who connects a load to the balanced three-phase voltage source, it is not important how the voltages are generated. It is important to note, however, that if the load currents generated by connecting a load to the power source shown in Fig. 9.2 are also *balanced*, there are two possible equivalent configurations for the load. The equivalent load can be considered as being connected in either a *wye* (*Y*) or a *delta* (Δ) configuration. The balanced wye configuration is shown in Fig. 9.3. The delta configuration is shown in Fig. 9.4. Note that in the case of the delta connection, there is no neutral line. The actual function of the neutral line in the wye connection will be examined and it will be shown that in a balanced system the neutral line carries no current and therefore may be omitted.

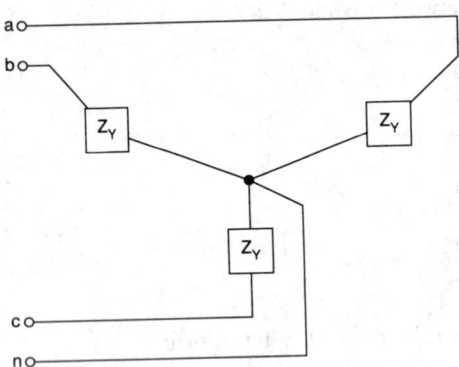

FIGURE 9.3   Wye (*Y*)-connected loads.

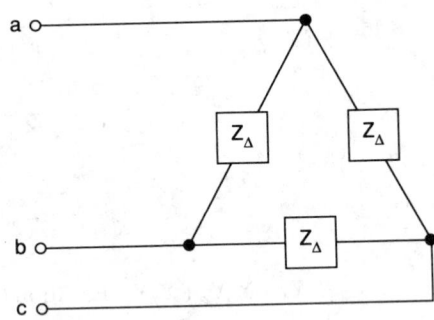

FIGURE 9.4   Delta (Δ)-connected loads.

9.3 Wye ⇔ Delta Transformations

For a balanced system, the equivalent load configuration may be either wye or delta. If both of these configurations are connected at only three terminals, it would be very advantageous if an equivalence could be established between them. It is, in fact, possible characteristics are the same. Consider, for example, the two networks shown in Fig. 9.5. For these two networks to be equivalent at each corresponding pair of terminals it is necessary that the input impedances at the corresponding terminals be equal, for example, if at terminals *a* and *b*, with *c* open-circuited, the impedance is the same for both configurations. Equating the impedances at each port yields

$$\mathbf{Z}_{ab} = \mathbf{Z}_a + \mathbf{Z}_b = \frac{\mathbf{Z}_1(\mathbf{Z}_2 + \mathbf{Z}_3)}{\mathbf{Z}_1 + \mathbf{Z}_2 + \mathbf{Z}_3}$$

$$\mathbf{Z}_{bc} = \mathbf{Z}_b + \mathbf{Z}_c = \frac{\mathbf{Z}_3(\mathbf{Z}_1 + \mathbf{Z}_2)}{\mathbf{Z}_1 + \mathbf{Z}_2 + \mathbf{Z}_3} \qquad (9.4)$$

$$\mathbf{Z}_{ca} = \mathbf{Z}_c + \mathbf{Z}_a = \frac{\mathbf{Z}_2(\mathbf{Z}_1 + \mathbf{Z}_3)}{\mathbf{Z}_1 + \mathbf{Z}_2 + \mathbf{Z}_3}$$

Solving this set of equations for $\mathbf{Z}_a$, $\mathbf{Z}_b$, and $\mathbf{Z}_c$ yields

Table 9.1 Current-Voltage Relationships for the Wye and
Delta Load Configurations

Parameter	Wye Configuration	Delta Configuration
Voltage	$V_{\text{line to line}} = \sqrt{3}V_Y$	$V_{\text{line to line}} = V_\Delta$
Current	$I_{\text{line}} = I_Y$	$I_{\text{line}} = \sqrt{3}I_\Delta$

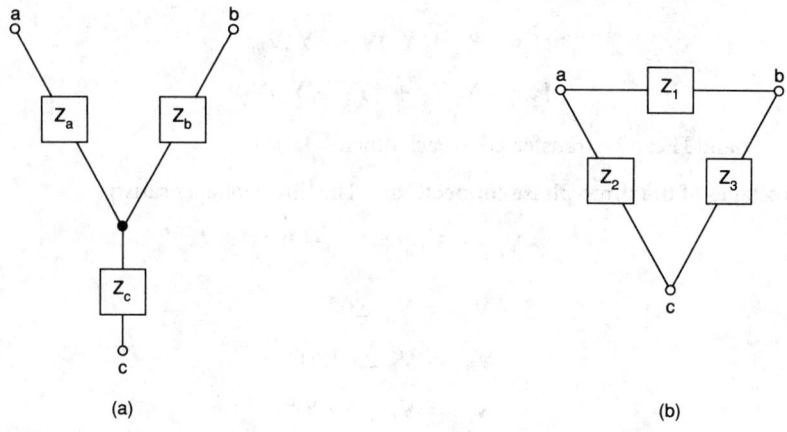

FIGURE 9.5 General wye- and delta-connected loads.

$$Z_a = \frac{Z_1 Z_2}{Z_1 + Z_2 + Z_3}$$

$$Z_b = \frac{Z_1 Z_3}{Z_1 + Z_2 + Z_3} \tag{9.5}$$

$$Z_c = \frac{Z_2 Z_3}{Z_1 + Z_2 + Z_3}$$

Similary, if we solve Eq. (9.4) for Z_1, Z_2, and Z_3, we obtain

$$Z_1 = \frac{Z_a Z_b + Z_b Z_c + Z_c Z_a}{Z_c}$$

$$Z_2 = \frac{Z_a Z_b + Z_b Z_c + Z_c Z_a}{Z_b} \tag{9.6}$$

$$Z_3 = \frac{Z_a Z_b + Z_b Z_c + Z_c Z_a}{Z_a}$$

Equations (9.5) and (9.6) are general relationships and apply to any set of impedances connected in a wye or delta configuration. For the balanced case where $Z_a = Z_b = Z_c$ and $Z_1 = Z_2 = Z_3$, the equations above reduce to

$$Z_Y = \frac{1}{3}Z_\Delta \tag{9.7}$$

and

$$Z_\Delta = 3Z_Y \tag{9.8}$$

Defining Terms

T network: The equations of the T network are

$$V_1 = (Z_1 + Z_3)I_1 + Z_3I_2$$

$$V_2 = Z_3I_1 + (Z_2 + Z_3)I_2$$

Π network: The equations of Π network are

$$I_1 = (Y_a + Y_b)V_1 - Y_bV_2$$

$$I_2 = -Y_bV_1 + (Y_b + Y_c)V_2$$

T and Π can be transferred to each other.

Balanced voltages of the three-phase connection: The three voltages satisfy

$$V_{an} + V_{bn} + V_{cn} = 0$$

where

$$V_{an} = V_p \angle 0°$$

$$V_{bn} = V_p \angle -120°$$

$$V_{cn} = V_p \angle +120°$$

References

J. D. Irwin, *Basic Engineering Circuit Analysis*, 3rd ed., New York: MacMillan, 1989.

R. C. Dorf, *Introduction to Electric Circuits*, 2nd ed., New York: John Wiley and Sons, 1993.

Further Information

IEEE Transactions on Power Systems

IEEE Transactions on Circuits and Systems, Part II: Analog and Digital Signal Processing

Transfer Functions
of Filters

Richard C. Dorf
University of California, Davis

Zhen Wan
University of California, Davis

10.1 Introduction 189
10.2 Ideal Filters 189
10.3 The Ideal Linear-Phase Low-Pass Filter 190
10.4 Ideal Linear-Phase Bandpass Filters 191
10.5 Causal Filters 191
10.6 Butterworth Filters 192
10.7 Chebyshev Filters 193

10.1 Introduction

Filters are widely used to pass signals at selected frequencies and reject signals at other frequencies. An *electrical filter* is a circuit that is designed to introduce gain or loss over a prescribed range of frequencies. In this section, we will describe ideal filters and then a selected set of practical filters.

10.2 Ideal Filters

An **ideal filter** is a system that completely rejects sinusoidal inputs of the form $x(t) = A \cos \omega t$, $-\infty < t < \infty$, for ω in certain frequency ranges and does not attenuate sinusoidal inputs whose frequencies are outside these ranges. There are four basic types of ideal filters: low-pass, high-pass, bandpass, and bandstop. The magnitude functions of these four types of filters are displayed in Fig. 10.1. Mathematical expressions for these magnitude functions are as follows:

$$\text{Ideal low-pass:} \quad |H(\omega)| = \begin{cases} 1, & -B \leq \omega \leq B \\ 0, & |\omega| > B \end{cases} \tag{10.1}$$

$$\text{Ideal high-pass:} \quad |H(\omega)| = \begin{cases} 0, & -B < \omega < B \\ 1, & |\omega| \geq B \end{cases} \tag{10.2}$$

$$\text{Ideal bandpass:} \quad |H(\omega)| = \begin{cases} 1, & B_1 \leq |\omega| \leq B_2 \\ 0, & \text{all other } \omega \end{cases} \tag{10.3}$$

$$\text{Ideal bandstop:} \quad |H(\omega)| = \begin{cases} 0, & B_1 \leq |\omega| \leq B_2 \\ 1, & \text{all other } \omega \end{cases} \tag{10.4}$$

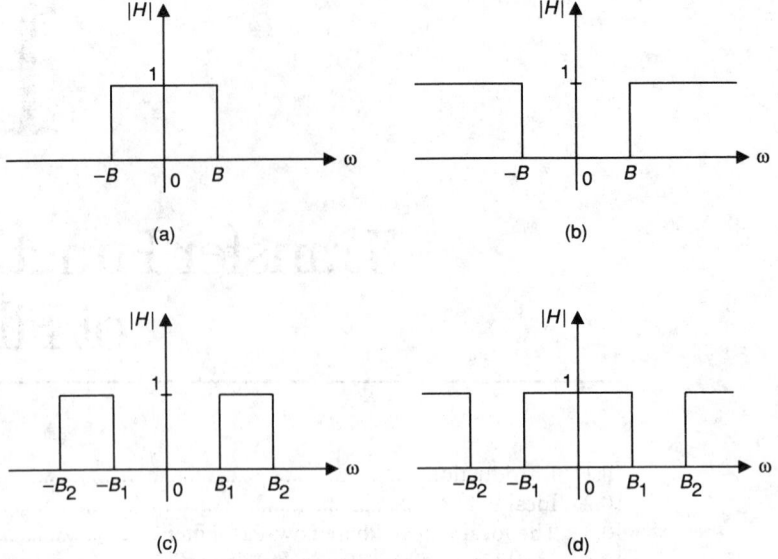

FIGURE 10.1   Magnitude functions of ideal filters: (a) low-pass; (b) high-pass; (c) bandpass; (d) bandstop.

The stopband of an ideal filter is defined to be the set of all frequencies ω for which the filter completely stops the sinusoidal input $x(t) = A \cos \omega t$, $-\infty < t < \infty$. The passband of the filter is the set of all frequencies ω for which the input $x(t)$ is passed without attenuation.

More complicated examples of ideal filters can be constructed by cascading ideal low-pass, high-pass, bandpass, and bandstop filters. For instance, by cascading bandpass filters with different values of B_1 and B_2, we can construct an ideal comb filter, whose magnitude function is illustrated in Fig. 10.2.

10.3 The Ideal Linear-Phase Low-Pass Filter

Consider the ideal low-pass filter with the frequency function

$$H(\omega) = \begin{cases} e^{-j\omega t_d}, & -B \leq \omega \leq B \\ 0, & \omega < -B, \quad \omega > B \end{cases} \tag{10.5}$$

where t_d is a positive real number. Equation (10.5) is the polar-form representation of $H(\omega)$. From Eq. (10.5) we have

$$|H(\omega)| = \begin{cases} 1, & -B \leq \omega \leq B \\ 0, & \omega < -B, \quad \omega > B \end{cases}$$

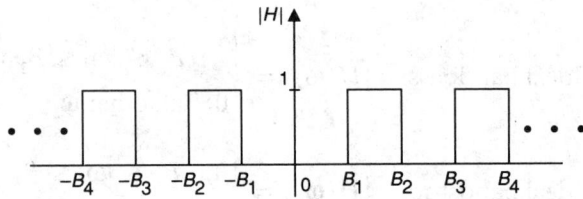

FIGURE 10.2   Magnitude function of an ideal comb filter.

and

$$\underline{/H(\omega)} = \begin{cases} -\omega t_d, & -B \leq \omega \leq B \\ 0, & \omega < -B, \quad \omega > B \end{cases}$$

The phase function $\underline{/H(\omega)}$ of the filter is plotted in Fig. 10.3. Note that over the frequency range 0 to B, the phase function of the system is linear with slope equal to $-t_d$.

The impulse response of the low-pass filter defined by Eq. (10.5) can be computed by taking the inverse Fourier transform of the frequency function $H(\omega)$. The impulse response of the ideal low-pass filter is

$$h(t) = \frac{B}{\pi} Sa[B(t - t_d)], \quad -\infty < t < \infty \tag{10.6}$$

where $Sa(x) = (\sin x)/x$. The impulse response $h(t)$ of the ideal low-pass filter is not zero for $t < 0$. Thus, the filter has a response before the impulse at $t = 0$ and is said to be noncausal. As a result, it is not possible to build an ideal low-pass filter.

10.4 Ideal Linear-Phase Bandpass Filters

One can extend the analysis to ideal linear-phase bandpass filters. The frequency function of an ideal linear-phase bandpass filter is given by

$$H(\omega) = \begin{cases} e^{-j\omega t_d}, & B_1 \leq |\omega| \leq B_2 \\ 0, & \text{all other } \omega \end{cases}$$

where t_d, B_1, and B_2 are positive real numbers. The magnitude function is plotted in Fig. 10.1(c) and the phase function is plotted in Fig. 10.4. The passband of the filter is from B_1 to B_2. The filter will pass the signal within the band with no distortion, although there will be a time delay of t_d seconds.

10.5 Causal Filters

As observed in the preceding section, ideal filters cannot be utilized in real-time filtering applications, since they are noncausal. In such applications, one must use **causal filters**, which are necessarily nonideal; that is, the transition from the passband to the stopband (and vice versa) is gradual. In particular, the magnitude functions of causal versions of low-pass, high-pass, bandpass, and bandstop filters have gradual transitions from the passband to the stopband. Examples of magnitude functions for these basic types of filters are shown in Fig. 10.5.

For a causal filter with frequency function $H(\omega)$, the passband is defined as the set of all frequencies ω for which

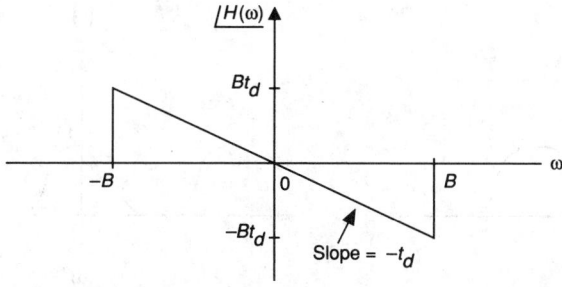

FIGURE 10.3 Phase function of ideal low-pass filter defined by Eq. (10.5).

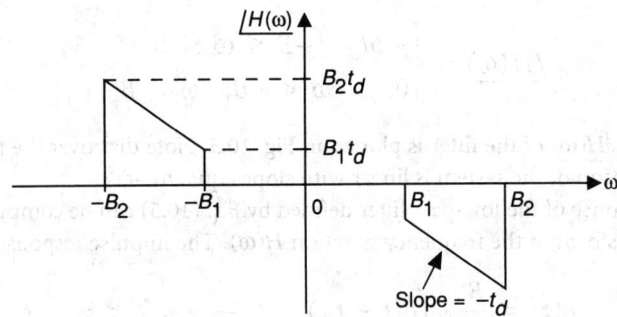

FIGURE 10.4 Phase function of ideal linear-phase bandpass filter.

$$|H(\omega)| \geq \frac{1}{\sqrt{2}} \, |H(\omega_p)| \qquad (10.7)$$

where ω_p is the value of ω for which $|H(\omega)|$ is maximum. Note that Eq. (10.7) is equivalent to the condition that $|H(\omega)|_{dB}$ is less than 3 dB down from the peak value $|H(\omega_p)|_{dB}$. For low-pass or band-pass filters, the width of the passband is called the **3-dB bandwidth**.

A stopband in a causal filter is a set of frequencies ω for which $|H(\omega)|_{dB}$ is down some desired amount (e.g., 40 or 50 dB) from the peak value $|H(\omega_p)|_{dB}$. The range of frequencies between a pass-band and a stopband is called a **transition region**. In causal filter design, a key objective is to have the transition regions be suitably small in extent.

10.6 Butterworth Filters

The transfer function of the two-pole Butterworth filter is

$$H(s) = \frac{\omega_n^2}{s^2 + \sqrt{2} \, \omega_n s + \omega_n^2}$$

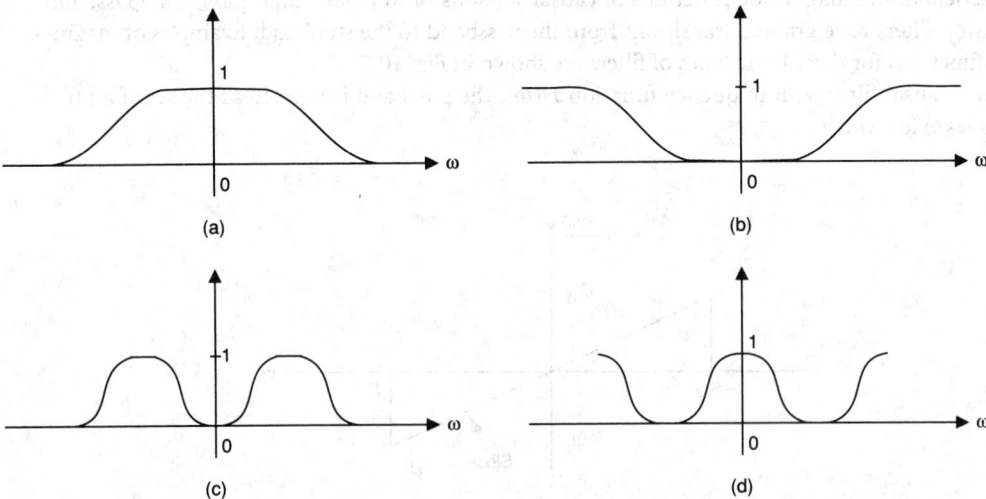

FIGURE 10.5 Causal filter magnitude functions: (a) low-pass; (b) high-pass; (c) bandpass; (d) bandstop.

Factoring the denominator of $H(s)$, we see that the poles are located at

$$s = -\frac{\omega_n}{\sqrt{2}} \pm j\frac{\omega_n}{\sqrt{2}}$$

Note that the magnitude of each of the poles is equal to ω_n.

Setting $s = j\omega$ in $H(s)$, we have that the magnitude function of the two-pole Butterworth filter is

$$|H(\omega)| = \frac{1}{\sqrt{1 + (\omega/\omega_n)^4}} \tag{10.8}$$

From Eq. (10.8) we see that the 3-dB bandwidth of the Butterworth filter is equal to ω_n. For the case $\omega_n = 2$ rad/s, the frequency response curves of the Butterworth filter are plotted in Fig. 10.6. Also displayed are the frequency response curves for the one-pole low-pass filter with transfer function $H(s) = 2/(s + 2)$, and the two-pole low-pass filter with $\zeta = 1$ and with 3-dB bandwidth equal to 2 rad/s. Note that the Butterworth filter has the sharpest cutoff of all three filters.

10.7 Chebyshev Filters

The magnitude function of the n-pole Butterworth filter has a monotone characteristic in both the passband and stopband of the filter. Here *monotone* means that the magnitude curve is gradually decreasing over the passband and stopband. In contrast to the Butterworth filter, the magnitude function of a type 1 Chebyshev filter has ripple in the passband and is monotone decreasing in the stopband (a type 2 Chebyshev filter has the opposite characteristic). By allowing ripple in the passband or stopband, we are able to achieve a sharper transition between the passband and stopband in comparison with the Butterworth filter.

The n-pole type 1 Chebyshev filter is given by the frequency function

$$|H(\omega)| = \frac{1}{\sqrt{1 + \epsilon^2 T_n^2(\omega/\omega_1)}} \tag{10.9}$$

where $T_n(\omega/\omega_1)$ is the nth-order Chebyshev polynomial. The Chebyshev polynomials can be generated from the recursion

$$T_n(x) = 2xT_{n-1}(x) - T_{n-2}(x)$$

where $T_0(x) = 1$ and $T_1(x) = x$. The polynomials for $n = 2, 3, 4, 5$ are

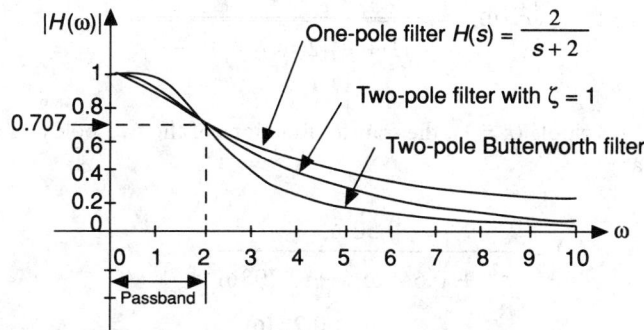

FIGURE 10.6　Magnitude curves of one- and two-pole low-pass filters.

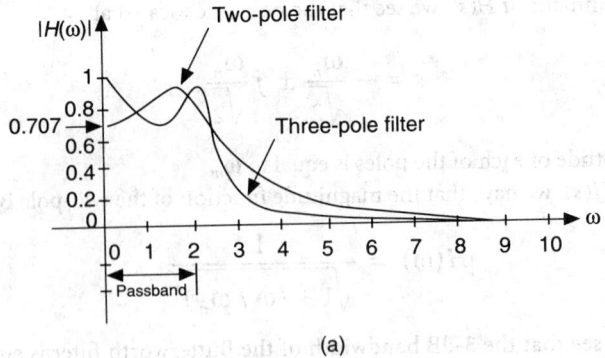

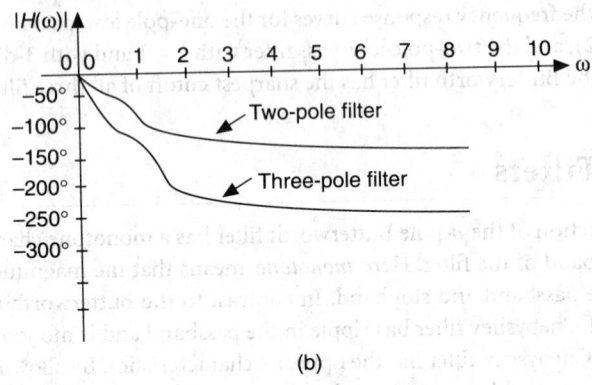

FIGURE 10.7   Frequency curves of two- and three-pole Chebyshev filters with $\omega_c = 2$ rad/s: (a) magnitude curves; (b) phase curves.

$$T_2(x) = 2x(x) - 1 = 2x^2 - 1$$
$$T_3(x) = 2x(2x^2 - 1) - x = 4x^3 - 3x \qquad\qquad (10.10)$$
$$T_4(x) = 2x(4x^3 - 3x) - (2x^2 - 1) = 8x^4 - 8x^2 + 1$$
$$T_5(x) = 2x(8x^4 - 8x^2 + 1) - (4x^3 - 3x) = 16x^5 - 20x^3 + 5x$$

Using Eq. (10.10), the two-pole type 1 Chebyshev filter has the following frequency function

$$\left|H(\omega)\right| = \frac{1}{\sqrt{1 + \epsilon^2 [2(\omega / \omega_1)^2 - 1]^2}}$$

For the case of a 3-dB ripple ($\epsilon = 1$), the transfer functions of the two-pole and three-pole type 1 Chebyshev filters are

$$H(s) = \frac{0.50\omega_c^2}{s^2 + 0.645\omega_c s + 0.708\omega_c^2}$$

$$H(s) = \frac{0.251\omega_c^3}{s^3 + 0.597\omega_c s^2 + 0.928\omega_c^2 s + 0.251\omega_c^3}$$

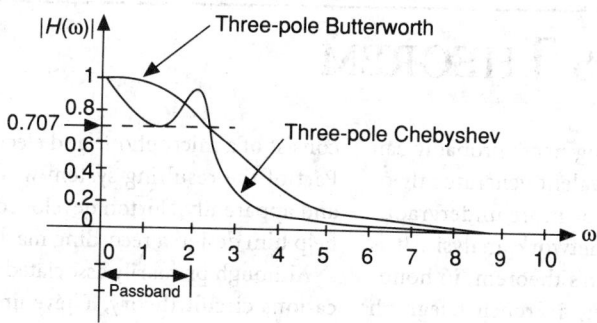

FIGURE 10.8 Magnitude curves of three-pole Butterworth and three-pole Chebyshev filters with 3-dB bandwidth equal to 2 rad/s.

where ω_c = 3-dB bandwidth. The frequency curves for these two filters are plotted in Fig. 10.7 for the case ω_c = 2 rad.

The magnitude response functions of the three-pole Butterworth filter and the three-pole type 1 Chebyshev filter are compared in Fig. 10.8 with the 3-dB bandwidth of both filters equal to 2 rad. Note that the transition from passband to stopband is sharper in the Chebyshev filter; however, the Chebyshev filter does have the 3-dB ripple over the passband.

Defining Terms

Causal filter: A filter of which the transition from the passband to the stopband is gradual, not ideal. This filter is realizable.

Ideal filter: An ideal filter is a system that completely rejects sinusoidal inputs of the form $x(t) = A \cos \omega t$, $-\infty < t < \infty$, for ω within a certain frequency range, and does not attenuate sinusoidal inputs whose frequencies are outside this range. There are four basic types of ideal filters: low-pass, high-pass, bandpass, and bandstop.

3-dB bandwidth: For a causal low-pass or bandpass filter with a frequency function $H(j\omega)$: the frequency at which $|H(\omega)|_{dB}$ is less than 3 dB down from the peak value $|H(\omega_p)|_{dB}$.

Transition region: The range of frequencies of a filter between a passband and a stopband.

References

R. C. Dorf, *Introduction to Electrical Circuits*, 2nd ed., New York: Wiley, 1993.

E. W. Kamen, *Introduction to Signals and Systems*, 2nd ed., New York: Macmillan, 1990.

G. R. Cooper and C. D. McGillem, *Modern Communications and Spread Spectrum*, New York: McGraw-Hill, 1986.

Further Information

IEEE Transactions on Circuits and Systems, Part I: Fundamental Theory and Applications.
IEEE Transactions on Circuits and Systems, Part II: Analog and Digital Signal Processing.
Available from IEEE.

THÉVENIN'S THEOREM

Most electrical engineers probably can recall the equivalent generator theorem from one or more undergraduate courses in circuit or network analysis. It is commonly called Thévenin's theorem, in honor of Léon Charles Thévenin, a French telegraph engineer and educator who proposed it in 1883. But in fact Hermann von Helmholtz proposed it first, in an 1853 paper.

Although originally introduced to facilitate the analysis of linear networks of resistances and voltage sources, the theorem subsequently was defined in terms of impedances and voltage sources. As a tool for circuit analysis, it is allied to the superposition theorem.

The evidence suggests that in the United States AT&T Co. engineers pioneered in using the theorem. And in 1926, one of them, Edward L. Norton, proposed a dual of Thévenin's equivalent circuit—an equivalent circuit using a current source and an equivalent resistance (also proposed by Hans Mayer in 1926). In other words, the Norton equivalent is the source transformation of the Thévenin equivalent.

Sidney Darlington, who worked for Norton at Bell Laboratories in 1929, told Paul J. Nahin of the University of New Hampshire, in Durham, in 1988 that he recalled Norton as a real electromechanical genius. While working at Bell, Norton had been assigned to a Victor Co. project on the first electric phonograph recording apparatus to consist of a microphone and electronic amplifier. Part of the resulting system was current driven and apparently Norton developed his theorem to help him design a recording machine.

Although primarily associated with communications circuit theory, Thévenin's and Norton's theorems have been used in other applications. For example, Thévenin's theorem has been used in teaching induction motors.

The historical context in which Thévenin's theorem was conceived and disseminated is an interesting case study in the intellectual and social history of engineering. Thévenin was born in Meaux, France, and graduated from the Ecole Polytechnique in 1876. He became one of the first students to enroll in the Ecole Supérieure de Télégraphie (EST), a school founded in 1878 to prepare engineers for a career in the Government-owned telegraph service.

The two-year program at the EST included two courses on electrical measurement taught by Jules Raynaud, who introduced his students to Gustav Kirchhoff's laws of circuit analysis. Thévenin completed the EST program and was affiliated with the corps of telegraph engineers until 1914. His duties included educational as well as administrative activities, and it was in connection with his teaching that he undertook an investigation of Kirchhoff's Laws as applied to electric networks. This study resulted in 1883 in Thévenin's formulation of the equivalent generator theorem, which was published in the French Academy of Sciences' *Comptes Rendus* in December 1883.

A 1943 book on electric circuits written by members of the electrical engineering department at the Massachusetts Institute of Technology in Cambridge included Thévenin's theorem, but also a footnote concerning Helmholtz's 1853 publication of the theorem. A translated version of a brief biographical sketch of Thévenin by Charles Suchet was published in the IEEE's *Electrical Engineering* in 1949. This article stimulated two letters to the editor, again pointing out Helmholtz's priority with regard to the equivalent generator theorem. In a letter published in the November 1988 issue of *IEEE Spectrum*, Harry E. Stockman again raised this issue along with the suggestion that future circuitry textbooks credit the theorem to Helmholtz.

But why was the Helmholtz publication on the theorem overlooked by American circuit theorists of the early 20th century? The explanation may be that his proposal of the theorem occurred not in the context of a telecommunication system but was embedded in an investigation of animal electricity. In contrast, the investigation that led Thévenin to rediscover the theorem occurred in the context of his interest in analysis of communication circuits and networks. This interest was shared by the AT&T engineers who encountered the theorem and named it for him.

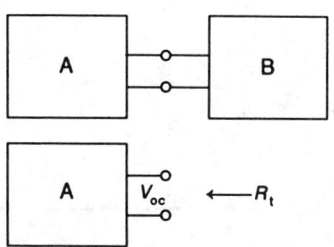

Thévenin's theorem: Divide a circuit into two parts, A and B, connected at a pair of terminals. Determine V_{oc} as the open-circuit voltage of A with B disconnected. Then the equivalent circuit of A is a source voltage V_{oc} in series with R_t where R_t is the resistance at the terminals when all the independent sources are deactivated.

Superposition theorem: In a linear circuit containing independent sources, the voltage across (or the current through) any element may be obtained by adding algebraically all the individual voltages (or currents) caused by each independent source acting alone, with all other independent voltage sources replaced by short circuits and all other independent current sources replaced by open circuits.

Norton's theorem: Divide a linear circuit into two parts, A and B. For circuit A, determine its short circuit currents at its terminals. The equivalent circuit is a current source I_{sc} in parallel with a resistance R_n, where R_n is the resistance calculated with all its independent sources deactivated.

These theorem definitions are from *Introduction to Electric Circuits*, by Richard C. Dorf, John Wiley & Sons, New York, 1989.

Source: Adapted from J. E. Brittain, *IEEE Spectrum*, p. 42, March 1990. © 1990 IEEE.

11

Frequency Response

Paul Neudorfer
Seattle University

11.1 Introduction 198
11.2 Linear Frequency Response Plotting 199
11.3 Bode Diagrams 200
11.4 A Comparison of Methods 206

11.1 Introduction

The Institute of Electrical and Electronics Engineers defines **frequency response** in stable, linear systems to be "the frequency-dependent relation in both gain and phase difference between steady-state sinusoidal inputs and the resultant steady-state sinusoidal outputs" [IEEE, 1988]. In certain specialized applications, the term *frequency response* may be used with more restrictive meanings. However, all such uses can be related back to the fundamental definition. The frequency response characteristics of a system can be directly found from its transfer function. A single-input/single-output linear time-invariant system is shown in Fig. 11.1.

For dynamic linear systems with no time delay, the transfer function $H(s)$ is in the form of a ratio of polynomials in the complex frequency s,

$$H(s) = K \frac{N(s)}{D(s)}$$

where K is a frequency-independent constant. For a system in the sinusoidal steady state, s is replaced by the sinusoidal frequency $j\omega$ ($j = \sqrt{-1}$) and the system function becomes

$$H(j\omega) = K \frac{N(j\omega)}{D(j\omega)} = |H(j\omega)| e^{j \arg H(j\omega)}$$

$H(j\omega)$ is a complex quantity. Its magnitude, $|H(j\omega)|$, and its argument or angle, $\arg H(j\omega)$, relate, respectively, the amplitudes and phase angles of sinusoidal steady-state input and output signals. Using the terminology of Fig. 11.1, if the input and output signals are, in general form,

$$x(t) = X \cos (\omega t + \Theta_x)$$
$$y(t) = Y \cos (\omega t + \Theta_y)$$

then the output's amplitude Y and phase angle Θ_y are related to those of the input by the two equations

$$Y = |H(j\omega)|X$$

$$\Theta_y = \arg H(j\omega) + \Theta_x$$

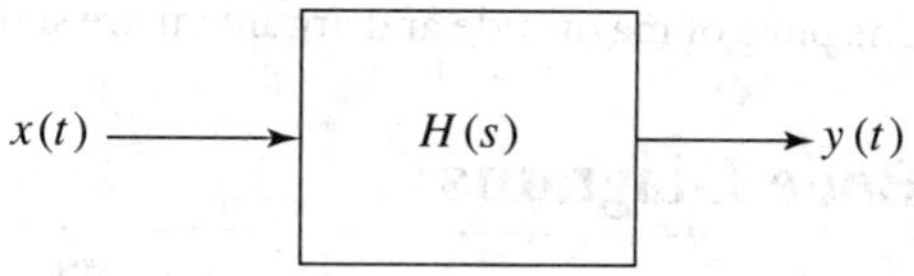

$x(t) \longrightarrow \boxed{H(s)} \longrightarrow y(t)$

The phrase *frequency response characteristics* usually implies a complete description of a system's sinusoidal steady-state behavior as a function of frequency. Because $H(j\omega)$ is complex and, therefore, two dimensional in nature, fre-

FIGURE 11.1 A single-input/single-output linear system.

quency response characteristics cannot be graphically displayed as a single curve plotted with respect to frequency. Instead, the magnitude and argument of $H(j\omega)$ can be separately plotted as functions of frequency. Often, only the magnitude curve is presented as a concise way of characterizing the system's behavior, but this must be viewed as an incomplete description. The most common form for such plots is the **Bode diagram** (developed by H. W. Bode of Bell Laboratories), which uses a logarithmic scale for frequency. Other forms of frequency response plots have also been developed. In the **Nyquist plot** (Harry Nyquist, also of Bell Labs), $H(j\omega)$ is displayed on the complex plane, $\text{Re}[H(j\omega)]$ being on the horizontal axis and $\text{Im}[H(j\omega)]$ being on the vertical. Frequency is a parameter of such curves. It is sometimes numerically identified at selected points of the curve and sometimes omitted. The **Nichols chart** (N. B. Nichols) graphs magnitude versus phase for the system function. Frequency again is a parameter of the resultant curve, sometimes shown and sometimes not.

Frequency response techniques are used in many areas of engineering. They are most obviously applicable to such topics as communications and filters, where the frequency response behaviors of systems are central to an understanding of their operations. It is, however, in the area of control systems where frequency response techniques are most fully developed as analytical and design tools. The Nichols chart, for instance, is used exclusively in the analysis and design of classical feedback control systems.

The remaining sections of this chapter describe several frequency response plotting methods. Applications of the methods can be found in other chapters throughout the *Handbook*.

11.2 Linear Frequency Response Plotting

Frequency response plots are prepared most directly by computing the magnitude and argument of $H(j\omega)$ and graphing each as a function of frequency (either f or ω), the frequency axis being scaled linearly. As an example, consider the transfer function

$$H(s) = \frac{160,000}{s^2 + 220s + 160,000}$$

Formally, the complex frequency variable s is replaced by the sinusoidal frequency $j\omega$ and the magnitude and argument found.

$$H(j\omega) = \frac{160,000}{(j\omega)^2 + 220(j\omega) + 160,000}$$

$$|H(j\omega)| = \frac{160,000}{\sqrt{(160,000 - \omega^2)^2 + (220\omega)^2}}$$

$$\arg H(j\omega) = -\tan^{-1}\frac{220\omega}{160,000 - \omega^2}$$

The plots of magnitude and argument are shown in Fig. 11.2.

11.3 Bode Diagrams

A Bode diagram consists of plots of the gain and phase of a transfer function, each with respect to logarithmically scaled frequency axes. In addition, the gain of the transfer function is scaled in **decibels** according to the definition

$$|H|_{dB} = H_{dB} = 20 \log_{10}|H(j\omega)|$$

This method of plotting frequency response information was popularized by H. W. Bode in the 1930s. There are two main advantages of the Bode approach. The first is that, with it, the gain and phase curves can be easily and accurately drawn. Second, once drawn, features of the curves can be identified both qualitatively and quantitatively with relative ease, even when those features occur over a wide dynamic range. Digital computers have rendered the first advantage obsolete. Ease of interpretation, however, remains a powerful advantage, and the Bode diagram is today the most common method chosen for the display of frequency response data.

A Bode diagram is drawn by applying a set of simple rules or procedures to a transfer function. The rules relate directly to the set of poles and zeros and/or time constants of the function. Before constructing a Bode diagram, the transfer function is normalized so that each pole or zero term (except those at $s = 0$) has a dc gain of one. For instance:

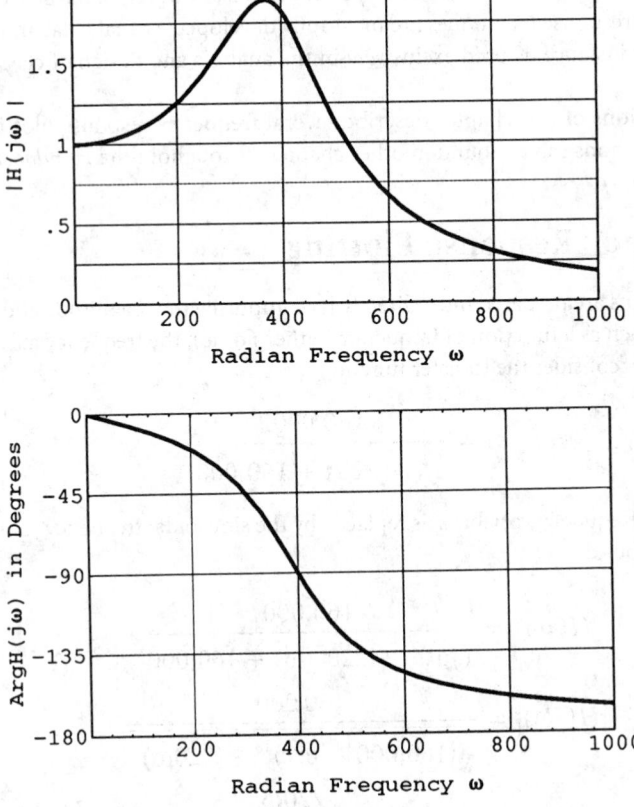

FIGURE 11.2   Linear frequency response curves of $H(j\omega)$.

$$H(s) = K\frac{s + \omega_z}{s(s + \omega_p)} = \frac{K\omega_z}{\omega_p}\frac{s/\omega_z + 1}{s(s/\omega_p + 1)} = K'\frac{s\tau_z + 1}{s(s\tau_p + 1)}$$

In the last form of the expression, $\tau_z = 1/\omega_z$ and $\tau_p = 1/\omega_p$. τ_p is a time constant of the system and $s = -\omega_p$ is the corresponding natural frequency. Because it is understood that Bode diagrams are limited to sinusoidal steady-state frequency response analysis, one can work directly from the transfer function $H(s)$ rather than resorting to the formalism of making the substitution $s = j\omega$.

Bode frequency response curves (gain and phase) for $H(s)$ are generated from the individual contributions of the four terms K', $s\tau_z + 1$, $1/s$, and $1/(s\tau_p + 1)$. As described in the following paragraph, the frequency response effects of these individual terms are easily drawn. To obtain the overall frequency response curves for the transfer function, the curves for the individual terms are added together.

The terms used as the basis for drawing Bode diagrams are found from factoring $N(s)$ and $D(s)$, the numerator and denominator polynomials of the transfer function. The factorization results in four standard forms. These are (1) a constant K; (2) a simple s term corresponding to either a zero (if in the numerator) or a pole (if in the denominator) at the origin; (3) a term such as $(s\tau + 1)$ corresponding to a real valued (nonzero) pole or zero; and (4) a quadratic term with a possible standard form of $[(s/\omega_n)^2 + (2\zeta/\omega_n)s + 1]$ corresponding to a pair of complex conjugate poles or zeros. The Bode magnitude and phase curves for these possibilities are displayed in Figs. 11.3–11.5. Note that both decibel magnitude and phase are plotted semilogarithmically. The frequency axis is logarithmically scaled so that every tenfold, or **decade**, change in frequency occurs over an equal distance. The magnitude axis is given in decibels. Customarily, this axis is marked in 20-dB increments. Positive decibel magnitudes correspond to amplifications between input and output that are greater than one (output amplitude larger than input). Negative decibel gains correspond to attenuation between input and output.

Figure 11.3 shows three separate magnitude functions. Curve 1 is trivial; the Bode magnitude of a constant K is simply the decibel-scaled constant $20 \log_{10} K$, shown for an arbitrary value of $K = 5$ ($20 \log_5 = 13.98$). Phase is not shown. However, a constant of $K > 0$ has a phase contribution of $0°$ for all frequencies. For $K < 0$, the contribution would be $\pm 180°$. Curve 2 shows the magnitude frequency response curve for a pole at the origin ($1/s$). It is a straight line with a slope of -20 dB/decade. The line passes through 0 dB at $\omega = 0$ rad/s. The phase contribution of a simple pole at the origin is a constant $-90°$, independent of frequency. The effect of a zero at the origin (s) is shown in Curve 3. It is again a straight line that passes through 0 dB at $\omega = 0$ rad/s; however, the slope is $+20$ dB/decade. The phase contribution of a simple zero at $s = 0$ is $+90°$, independent of frequency.

Note from Fig. 11.3 and the foregoing discussion that in Bode diagrams the effect of a pole term at a given location is simply the negative of that of a zero term at the same location. This is true both of magnitude and phase curves.

Figure 11.4 shows the magnitude and phase curves for a zero term of the form $(s/\omega_z + 1)$ and a pole term of the form $1/(s/\omega_p + 1)$. Exact plots of the magnitude and phase curves are shown as dashed lines. *Straight line approximations* to these curves are shown as solid lines. Note that the straight line approximations are so good that they obscure the exact curves at most frequencies. For this reason, some of the curves in this and later figures have been displaced slightly to enhance clarity. The greatest error between the exact and approximate magnitude curves is ± 3 dB. The approximation for phase is always within $7°$ of the exact curve and usually much closer. The approximations for magnitude consist of two straight lines. The points of intersection between these two lines ($\omega = \omega_z$ for the zero term and $\omega = \omega_p$ for the pole) are **breakpoints** of the curves. Breakpoints of Bode gain curves always correspond to locations of poles and zeros in the transfer function.

In Bode analysis complex conjugate poles or zeros are always treated as pairs in the correspond-

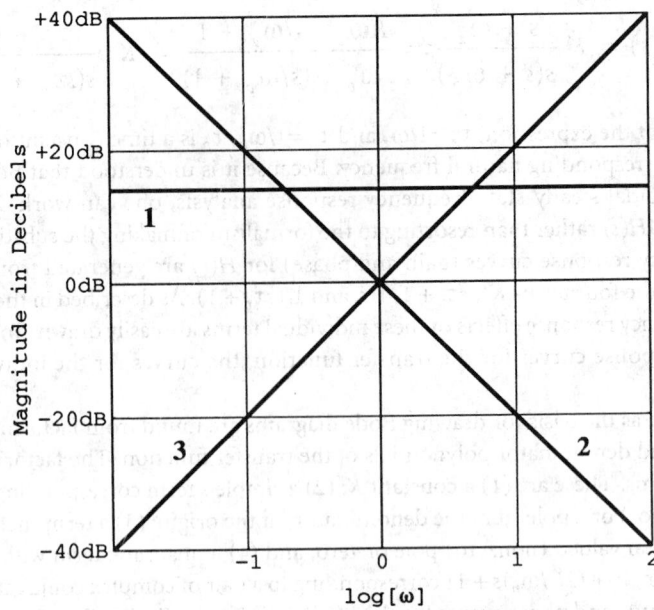

FIGURE 11.3 Bode magnitude functions for (1) $K = 5$, (2) $1/s$, and (3) s.

ing quadratic form $[(s/\omega_n)^2 + (2\zeta/\omega_n)s + 1]$.[1] For quadratic terms in stable, minimum phase systems, the **damping ratio** ζ (Greek letter zeta) is within the range $0 < \zeta < 1$. Quadratic terms cannot always be adequately represented by straight line approximations. This is especially true for lightly damped systems (small ζ). The traditional approach was to draw a preliminary representation of the contribution. This consists of a straight line of 0 dB from dc up to the breakpoint at ω_n followed by a straight line of slope ± 40 dB/decade beyond the breakpoint, depending on whether the plot refers to a pair of poles or a pair of zeros. Then, referring to a family of curves as shown in Fig. 11.5, the preliminary representation was improved based on the value of ζ. The phase contribution of the quadratic term was similarly constructed. Note that Fig. 11.5 presents frequency response contributions for a quadratic pair of poles. For zeros in the corresponding locations, both the magnitude and phase curves would be negated. Digital computer applications programs render this procedure unnecessary for purposes of constructing frequency response curves. Knowledge of the technique is still valuable, however, in the qualitative and quantitative interpretation of frequency response curves. Localized peaking in the gain curve is a reflection of the existence of **resonance** in a system. The height of such a peak (and the corresponding value of ζ) is a direct indication of the degree of resonance.

Bode diagrams are easily constructed because, with the exception of lightly damped quadratic terms, each contribution can be reasonably approximated with straight lines. Also, the overall frequency response curve is found by adding the individual contributions. Two examples follow.

Example 1

$$A(s) = \frac{10^4 s}{s^2 + 1100s + 10^5} = \frac{10^4 s}{(s + 100)(s + 1000)} = 10^{-1}\frac{s}{(s/100 + 1)(s/1000 + 1)}$$

[1]Several such standard forms are used. This is the one most commonly encountered in controls applications.

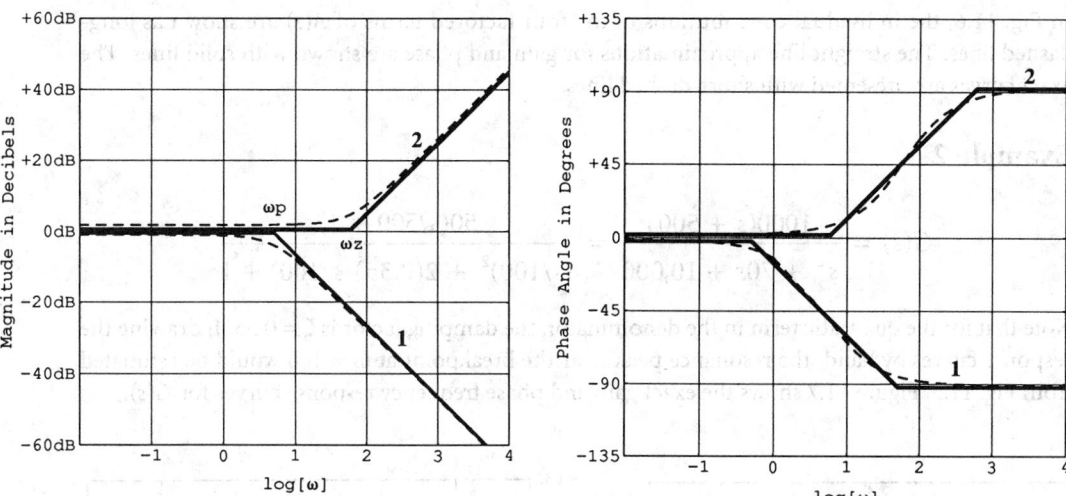

FIGURE 11.4 Bode curves for (1) a simple pole at $s = -\omega_p$ and (2) a simple zero at $s = -\omega_z$.

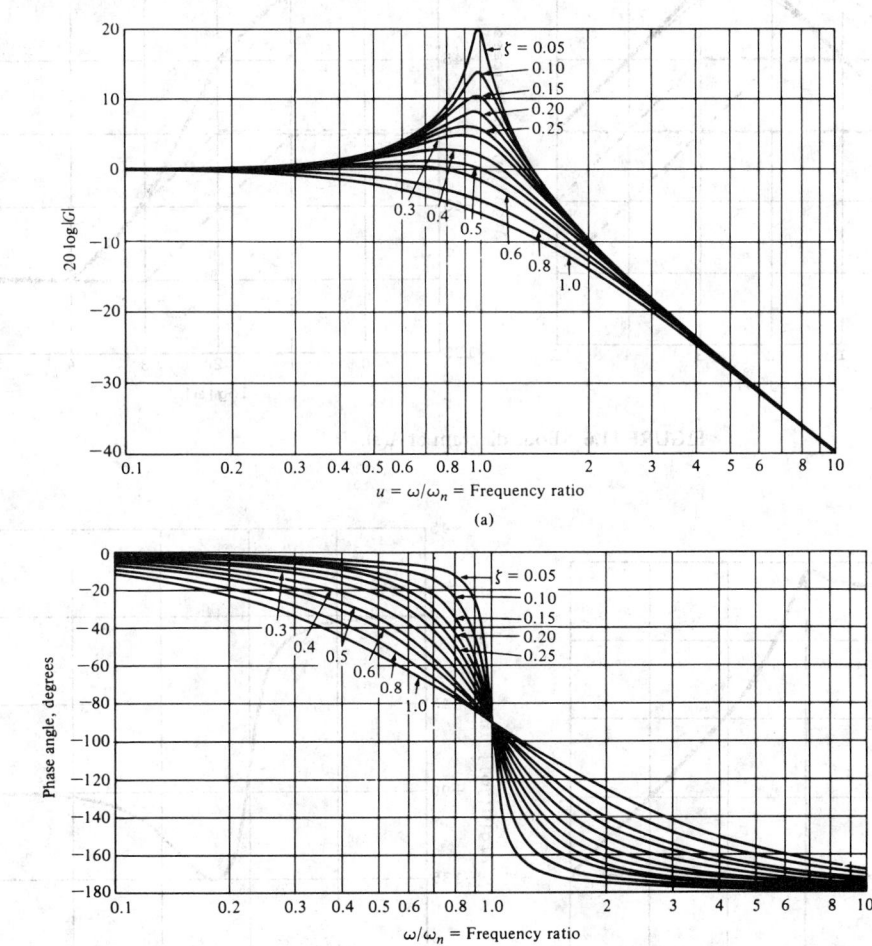

FIGURE 11.5 Bode diagram of $1/[(s/\omega_n)^2 + (2\zeta/\omega_n)s + 1]$. (*Source:* R. C. Dorf, *Modern Control Systems,* 4th ed., Reading, Mass.: Addison-Wesley, 1986, p. 258. With permission.)

In Fig. 11.6, the individual contributions of the four factored terms of $A(s)$ are shown as long-dashed lines. The straight line approximations for gain and phase are shown with solid lines. The exact curves are presented with short-dashed lines.

Example 2

$$G(s) = \frac{1000(s + 500)}{s^2 + 70s + 10,000} = \frac{50(s/500 + 1)}{(s/100)^2 + 2(0.35)(s/100) + 1}$$

Note that for the quadratic term in the denominator, the damping factor is $\zeta = 0.35$. If drawing the response curves by hand, the resonance peak near the breakpoint at $\omega = 100$ would be estimated from Fig. 11.5. Figure 11.7 shows the exact gain and phase frequency response curves for $G(s)$.

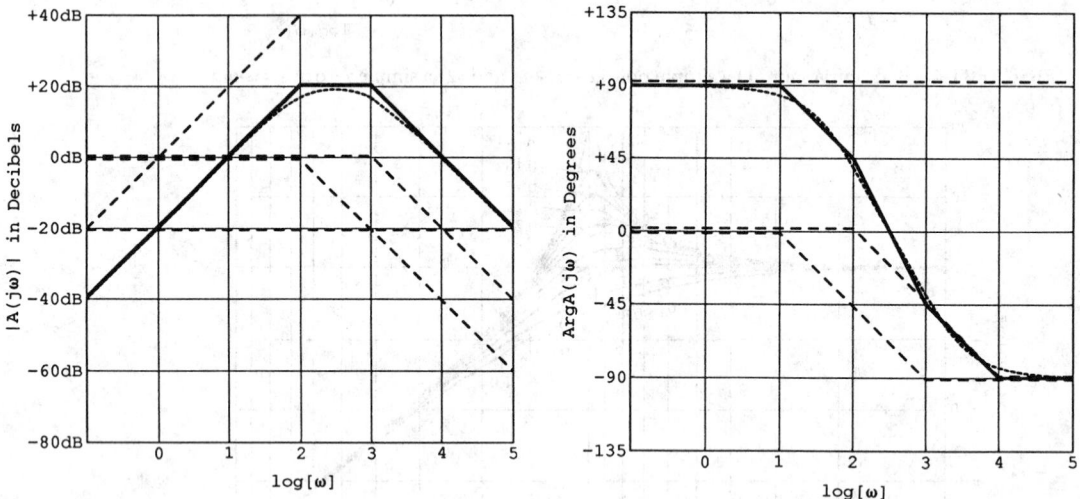

FIGURE 11.6   Bode diagram of $A(s)$.

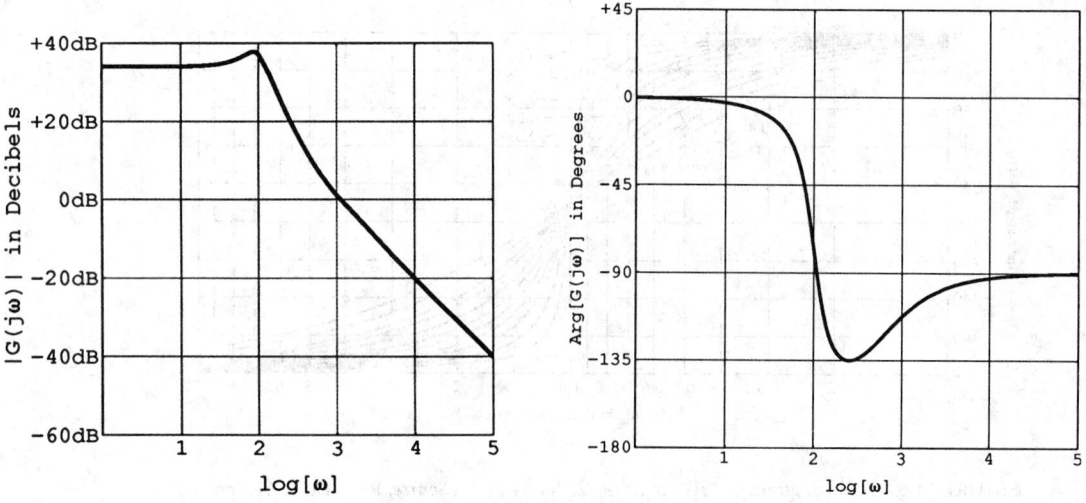

FIGURE 11.7   Bode diagram of $G(s)$.

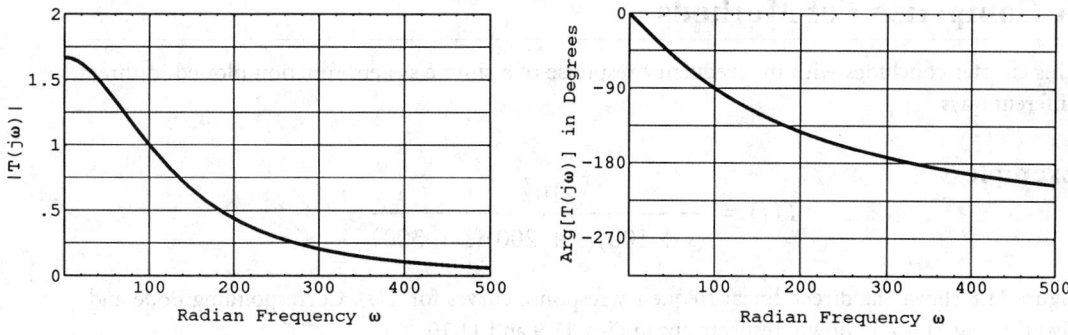

FIGURE 11.8 Linear frequency response plot of $T(s)$.

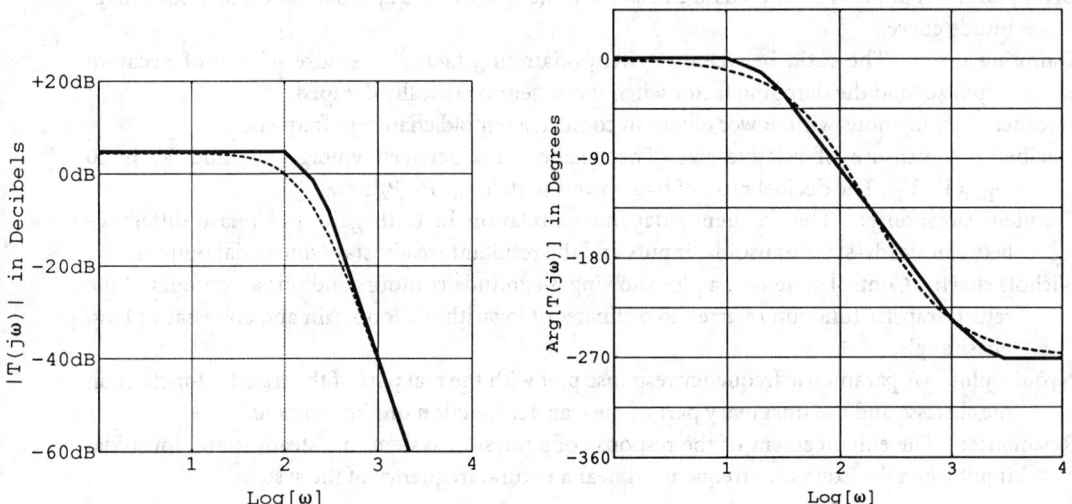

FIGURE 11.9 Bode diagram of $T(s)$.

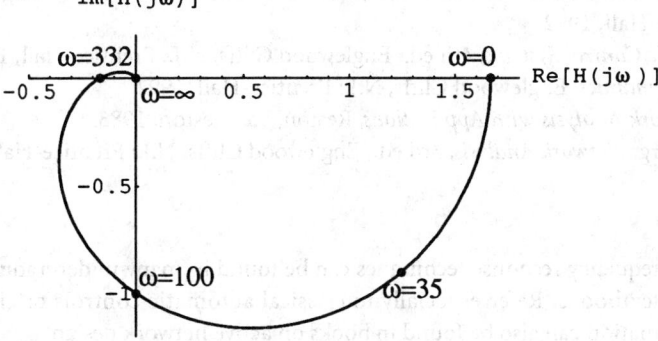

FIGURE 11.10 Nyquist plot of $T(s)$.

11.4 A Comparison of Methods

This chapter concludes with the frequency response of a simple system function plotted in three different ways.

Example 3

$$T(s) = \frac{10^7}{(s + 100)(s + 200)(s + 300)}$$

Figure 11.8 shows the direct, linear frequency response curves for $T(s)$. Corresponding Bode and Nyquist diagrams are shown, respectively, in Figs. 11.9 and 11.10.

Defining Terms

Bode diagram: A frequency response plot of 20 log gain and phase angle on a log-frequency base.

Breakpoint: A point of abrupt change in slope in the straight line approximation of a Bode magnitude curve.

Damping ratio: The ratio between a system's damping factor (measure of rate of decay of response) and the damping factor when the system is critically damped.

Decade: Synonymous with power of ten. In context, a tenfold change in frequency.

Decibel: A measure of relative size. The decibel gain between voltages V_1 and V_2 is 20 $\log_{10}(V_1/V_2)$. The decibel ratio of two powers is 10 $\log_{10}(P_1/P_2)$.

Frequency response: The frequency-dependent relation in both gain and phase difference between steady-state sinusoidal inputs and the resultant steady-state sinusoidal outputs.

Nichols chart: Control systems—a plot showing magnitude contours and phase contours of the return transfer function referred to ordinates of logarithmic loop gain and abscissae of loop phase angle.

Nyquist plot: A parametric frequency response plot with the real part of the transfer function on the abscissa and the imaginary part of the transfer function on the ordinate.

Resonance: The enhancement of the response of a physical system to a steady-state sinusoidal input when the excitation frequency is near a natural frequency of the system.

References

R.C. Dorf, *Modern Control Systems,* 4th ed., Reading, Mass.: Addison-Wesley, 1986.

IEEE Standard Dictionary of Electrical and Electronics Terms, 4th ed., The Institute of Electrical and Electronics Engineers, 1988.

D.E. Johnson, J.R. Johnson, and J.L. Hilburn, *Electric Circuit Analysis,* 2nd ed., Englewood Cliffs, N.J.: Prentice-Hall, 1992.

B.C. Kuo, *Automatic Control Systems,* 4th ed., Englewood Cliffs, N.J.: Prentice-Hall, 1982.

K. Ogata, *System Dynamics,* Englewood Cliffs, N.J.: Prentice-Hall, 1992.

W.D. Stanley, *Network Analysis with Applications,* Reston, Va.: Reston, 1985.

M.E. Van Valkenburg, *Network Analysis,* 3rd ed., Englewood Cliffs, N.J.: Prentice-Hall, 1974.

Further Information

Good coverage of frequency response techniques can be found in many undergraduate-level electrical engineering textbooks. Refer especially to classical automatic controls or circuit analysis books. Useful information can also be found in books on active network design.

Examples of the application of frequency response methods abound in journal articles ranging over such diverse topics as controls, acoustics, electronics, and communications.

12

Stability Analysis

Ferenc Szidarovszky
University of Arizona

A. Terry Bahill
University of Arizona

12.1 Introduction 207
12.2 Using the State of the System to Determine Stability 208
12.3 Lyapunov Stability Theory 209
12.4 Stability of Time-Invariant Linear Systems 210
      Stability Analysis with State-Space Notation • The Transfer
      Function Approach
12.5 BIBO Stability 216
12.6 Physical Examples 218

12.1 Introduction

In this chapter, which is based on Szidarovszky and Bahill [1992], we first discuss **stability** in general and then present four techniques for assessing the stability of a system: (1) Lyapunov functions, (2) finding the eigenvalues for state-space notation, (3) finding the location in the complex frequency plane of the poles of the closed-loop transfer function, and (4) proving bounded outputs for all bounded inputs. Proving stability with Lyapunov functions is very general: it works for nonlinear and time-varying systems. It is also good for doing proofs. Proving the stability of a system with Lyapunov functions is difficult, however, and failure to find a Lyapunov function that proves a system is stable does not prove that the system is unstable. The next techniques we present, finding the eigenvalues or the poles of the transfer function, are sometimes difficult, because they require factoring high-order polynomials. Many commercial software packages are now available for this task, however. We think most engineers would benefit by having one of these computer programs. Jamshidi *et al.* [1992] and advertisements in technical publications such as the *IEEE Control Systems Magazine* and *IEEE Spectrum* describe many appropriate software packages. The last technique we present, bounded-input, bounded-output stability, is also quite general.

Let us begin our discussion of stability and **instability** of systems informally. In an *unstable system* the state can have large variations, and small inputs or small changes in the initial state may produce large variations in the output. A common example of an unstable system is illustrated by someone pointing the microphone of a public address (PA) system at a speaker; a loud high-pitched tone results. Often instabilities are caused by too much gain, so to quiet the PA system, decrease the gain by pointing the microphone away from the speaker. Discrete systems can also be unstable. A friend of ours once provided an example. She was sitting in a chair reading and she got cold. So she went over and turned up the thermostat on the heater. The house warmed up. She got hot, so she got up and turned down the thermostat. The house cooled off. She got cold and turned up the thermostat. This process continued until someone finally suggested that she put on a sweater (reducing the gain of her heat loss system). She did, and was much more comfortable. We called this a discrete system, because she seemed to sample the environment and produce outputs at discrete intervals about 15 minutes apart.

12.2 Using the State of the System to Determine Stability

The stability of a system is defined with respect to a given equilibrium point in state space. If the initial state x_0 is selected at an equilibrium state $\bar{x}$ of the system, then the state will remain at $\bar{x}$ for all future time. When the initial state is selected close to an equilibrium state, the system might remain close to the equilibrium state or it might move away. In this section we introduce conditions that guarantee that whenever the system starts near an equilibrium state, it remains near it, perhaps even converging to the equilibrium state as time increases. For simplicity, only time-invariant systems are considered in this section. Time-variant systems are discussed in Section 12.5.

Continuous, time-invariant systems have the form

$$\dot{x}(t) = f(x(t)) \tag{12.1}$$

and discrete, time-invariant systems are modeled by the difference equation

$$x(t + 1) = f(x(t)) \tag{12.2}$$

Here we assume that $f: X \to \mathbf{R}^n$, where $X \subseteq \mathbf{R}^n$ is the state space. We also assume that function f is continuous; furthermore, for arbitrary initial state $x_0 \in X$, there is a unique solution of the corresponding initial value problem $x(t_0) = x_0$, and the entire trajectory $x(t)$ is in X. Assume furthermore that t_0 denotes the initial time period of the system.

It is also known that a vector $\bar{x} \in X$ is an equilibrium state of the continuous system, Eq. (12.1), if and only if $f(\bar{x}) = 0$, and it is an equilibrium state of the discrete system, Eq. (12.2), if and only if $\bar{x} = f(\bar{x})$. In this chapter the equilibrium of a system will always mean the equilibrium *state*, if it is not specified otherwise. In analyzing the dependence of the state trajectory $x(t)$ on the selection of the initial state x_0 nearby the equilibrium, the following stability types are considered.

Definition 12.1

1. An equilibrium state $\bar{x}$ is stable if there is an $\varepsilon_0 > 0$ with the following property: For all ε_1, $0 < \varepsilon_1 < \varepsilon_0$, there is an $\varepsilon > 0$ such that if $\| \bar{x} - x_0 \| < \varepsilon$, then $\| \bar{x} - x(t) \| < \varepsilon_1$, for all $t > t_0$.

2. An equilibrium state $\bar{x}$ is **asymptotically stable** if it is stable and there is an $\varepsilon > 0$ such that whenever $\| \bar{x} - x_0 \| < \varepsilon$, then $x(t) \to \bar{x}$ as $t \to \infty$.

3. An equilibrium state $\bar{x}$ is **globally asymptotically stable** if it is stable and with arbitrary initial state $x_0 \in X$, $x(t) \to \bar{x}$ as $t \to \infty$.

The first definition says an equilibrium state $\bar{x}$ is stable if the entire trajectory $x(t)$ is closer to the equilibrium state than any small ε_1, if the initial state x_0 is selected close enough to the equilibrium state. For asymptotic stability, in addition, $x(t)$ has to converge to the equilibrium state as $t \to \infty$. If an equilibrium state is globally asymptotically stable, then $x(t)$ converges to the equilibrium state regardless of how the initial state x_0 is selected.

These stability concepts are called **internal**, because they represent properties of the state of the system. They are illustrated in Fig. 12.1.

In the electrical engineering literature sometimes our stability definition is called marginal stability, and our asymptotic stability is called stability.

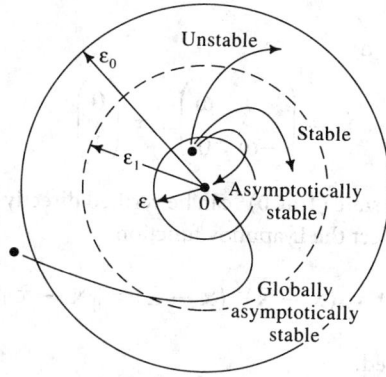

FIGURE 12.1 Stability concepts. (*Source:* F. Szidarovszky and A.T. Bahill, *Linear Systems Theory*, Boca Raton, Fla.: CRC Press, 1992, p. 168. With permission.)

12.3 Lyapunov Stability Theory

Assume that $\bar{x}$ is an equilibrium state of a continuous or discrete system, and let Ω denote a subset of the state space X such that $\bar{x} \in \Omega$.

Definition 12.2

A real-valued function V defined on Ω is called a Lyapunov function, if

1. V is continuous;
2. V has a unique global minimum at $\bar{x}$ with respect to all other points in Ω;
3. for any state trajectory $x(t)$ contained in Ω, $V(x(t))$ is nonincreasing in t.

The Lyapunov function can be interpreted as the generalization of the energy function in electrical systems. The first requirement simply means that the graph of V has no breaks. The second requirement means that the graph of V has its lowest point at the equilibrium, and the third requirement generalizes the well-known fact of electrical systems, that the energy in a free electrical system with resistance always decreases, unless the system is at rest.

Theorem 12.1

Assume that there exists a Lyapunov function V on the spherical region

$$\Omega = \{x \mid \| x - \bar{x} \| < \varepsilon_0 \} \tag{12.3}$$

where $\varepsilon_0 > 0$ is given; furthermore $\Omega \subseteq X$. Then the equilibrium state is stable.

Theorem 12.2

Assume that in addition to the conditions of Theorem 12.1, the Lyapunov function $V(x(t))$ is strictly decreasing in t, unless $x(t) = \bar{x}$. Then the equilibrium state is asymptotically stable.

Theorem 12.3

Assume that the Lyapunov function is defined on the entire state space X, $V(x(t))$ is strictly decreasing in t unless $x(t) = \bar{x}$; furthermore, $V(x)$ tends to infinity as any component of x gets arbitrarily large in magnitude. Then the equilibrium state is globally asymptotically stable.

Example 12.1

Consider the differential equation

$$\dot{\mathbf{x}} = \begin{pmatrix} 0 & \omega \\ -\omega & 0 \end{pmatrix} \mathbf{x} + \begin{pmatrix} 0 \\ 1 \end{pmatrix}$$

The stability of the equilibrium state $(1/\omega, 0)^T$ can be verified directly by using Theorem 12.1 without computing the solution. Select the Lyapunov function

$$V(\mathbf{x}) = (\mathbf{x} - \bar{\mathbf{x}})^T (\mathbf{x} - \bar{\mathbf{x}}) = \| \mathbf{x} - \bar{\mathbf{x}} \|_2^2$$

where the Euclidian norm is used.

This is continuous in $\mathbf{x}$; furthermore, it has its minimal (zero) value at $\mathbf{x} = \bar{\mathbf{x}}$. Therefore, to establish the stability of the equilibrium state we have to show only that $V(\mathbf{x}(t))$ is decreasing. Simple differentiation shows that

$$\frac{d}{dt} V(\mathbf{x}(t)) = 2(\mathbf{x} - \bar{\mathbf{x}})^T \cdot \dot{\mathbf{x}} = 2(\mathbf{x} - \bar{\mathbf{x}})^T (\mathbf{A}\mathbf{x} + \mathbf{b})$$

with

$$\mathbf{A} = \begin{pmatrix} 0 & \omega \\ -\omega & 0 \end{pmatrix} \quad \text{and} \quad \mathbf{b} = \begin{pmatrix} 0 \\ 1 \end{pmatrix}$$

That is, with $\mathbf{x} = (x_1, x_2)^T$,

$$\frac{d}{dt} V(\mathbf{x}(t)) = 2 \left(x_1 - \frac{1}{\omega}, \quad x_2 \right) \begin{pmatrix} \omega x_2 \\ -\omega x_1 + 1 \end{pmatrix}$$

$$= 2(\omega x_1 x_2 - x_2 - \omega x_1 x_2 + x_2) = 0$$

Therefore, function $V(\mathbf{x}(t))$ is a constant, which is a (not strictly) decreasing function. That is, all conditions of Theorem 12.1 are satisfied, which implies the stability of the equilibrium state.

Theorems 12.1, 12.2, and 12.3 guarantee, respectively, the stability, asymptotic stability, and global asymptotic stability of the equilibrium state, if a Lyapunov function is found. Failure to find such a Lyapunov function does not mean that the system is unstable or that the stability is not asymptotic or globally asymptotic. It only means that you were not clever enough to find a Lyapunov function that proved stability.

12.4 Stability of Time-Invariant Linear Systems

This section is divided into two subsections. In the first subsection the stability of linear time-invariant systems given in state-space notation is analyzed. In the second subsection, methods based on transfer functions are discussed.

Stability Analysis with State-Space Notation

Consider the time-invariant continuous linear system

$$\dot{\mathbf{x}} = \mathbf{A}\mathbf{x} + \mathbf{b} \tag{12.4}$$

and the time-invariant discrete linear system

$$\mathbf{x}(t + 1) = \mathbf{A}\mathbf{x}(t) + \mathbf{b} \tag{12.5}$$

Assume that $\bar{\mathbf{x}}$ is an equilibrium state, and let $\phi(t,t_0)$ denote the fundamental matrix.

Theorem 12.4

1. The equilibrium state $\bar{\mathbf{x}}$ is stable if and only if $\phi(t,t_0)$ is bounded for $t \geq t_0$.
2. The equilibrium state $\bar{\mathbf{x}}$ is asymptotically stable if and only if $\phi(t,t_0)$ is bounded and tends to zero as $t \to \infty$.

We use the symbol s to denote complex frequency, i.e., $s = \sigma + j\omega$. For specific values of s, such as eigenvalues and poles, we use the symbol λ.

Theorem 12.5

1. Assume that for all eigenvalues λ_i of $\mathbf{A}$, Re $\lambda_i \leq 0$ in the continuous case (or $|\lambda_i| \leq 1$ in the discrete case), and all eigenvalues with the property Re $\lambda_i = 0$ (or $|\lambda_i| = 1$) have single multiplicity; then the equilibrium state is stable.
2. The stability is asymptotic if and only if for all i, Re $\lambda_i < 0$ (or $|\lambda_i| < 1$).

Remark 1. Note that Part 1 gives only sufficient conditions for the stability of the equilibrium state. As the following example shows, these conditions are not necessary.

Example 12.2

Consider first the continuous system $\dot{\mathbf{x}} = \mathbf{O}\mathbf{x}$, where $\mathbf{O}$ is the zero matrix. Note that all constant functions $\mathbf{x}(t) \equiv \bar{\mathbf{x}}$ are solutions and also equilibrium states. Since

$$\phi(t, t_0) = e^{\mathbf{O}(t-t_0)} = \mathbf{I}$$

is bounded (being independent of t), all equilibrium states are stable, but $\mathbf{O}$ has only one eigenvalue $\lambda_1 = 0$ with zero real part and multiplicity n, where n is the order of the system.

Consider next the discrete systems $\mathbf{x}(t+1) = \mathbf{I}\mathbf{x}(t)$, when all constant functions $\mathbf{x}(t) \equiv \bar{\mathbf{x}}$ are also solutions and equilibrium states. Furthermore,

$$\phi(t, 0) = \mathbf{A}^{t-t_0} = \mathbf{I}^{t-t_0} = \mathbf{I}$$

which is obviously bounded. Therefore, all equilibrium states are stable, but the condition of Part 1 of the theorem is violated again.

Remark 2. The following extension of Theorem 12.5 can be proven. The equilibrium state is stable if and only if for all eigenvalues of $\mathbf{A}$, Re $\lambda_i \leq 0$ (or $|\lambda_i| \leq 1$), and if λ_i is a repeated eigenvalue of $\mathbf{A}$ such that Re $\lambda_i = 0$ (or $|\lambda_i| = 1$), then the size of each block containing λ_i in the Jordan canonical form of $\mathbf{A}$ is 1×1. If for at least one eigenvalue of $\mathbf{A}$, Re $\lambda_i > 0$ (or $|\lambda_i| > 1$), then the system is unstable.

Remark 3. The equilibrium states of inhomogeneous equations are stable or asymptotically stable if and only if the same holds for the equilibrium states of the corresponding homogeneous equations.

Example 12.3

Consider again the continuous system

$$\dot{\mathbf{x}} = \begin{pmatrix} 0 & \omega \\ -\omega & 0 \end{pmatrix} \mathbf{x} + \begin{pmatrix} 0 \\ 1 \end{pmatrix}$$

the stability of which was analyzed earlier in Example 12.1 by using the Lyapunov function method. The characteristic polynomial of the coefficient matrix is

$$\varphi(s) = \det \begin{pmatrix} -s & \omega \\ -\omega & -s \end{pmatrix} = s^2 + \omega^2$$

therefore, the eigenvalues are $\lambda_1 = j\omega$ and $\lambda_2 = -j\omega$. Both eigenvalues have single multiplicities, and Re $\lambda_1 = $ Re $\lambda_2 = 0$. Hence, the conditions of Part 1 are satisfied, and therefore the equilibrium state is stable. The conditions of Part 2 do not hold. Consequently, the system is not asymptotically stable.

If a time-invariant system is nonlinear, then the Lyapunov method is the most popular choice for stability analysis. If the system is linear, then the direct application of Theorem 12.5 is more attractive, since the eigenvalues of the coefficient matrix $\mathbf{A}$ can be obtained by standard methods. In addition, several conditions are known from the literature that guarantee the asymptotic stability of time-invariant discrete and continuous systems even without computing the eigenvalues. For examining asymptotic stability, linearization is an alternative approach to the Lyapunov method as is shown here. Consider the time-invariant continuous and discrete systems

$$\dot{\mathbf{x}}(t) = \mathbf{f}(\mathbf{x}(t))$$

and

$$\mathbf{x}(t+1) = \mathbf{f}(\mathbf{x}(t))$$

Let $\mathbf{J}(\mathbf{x})$ denote the Jacobian of $\mathbf{f}(\mathbf{x})$, and let $\bar{\mathbf{x}}$ be an equilibrium state of the system. It is known that the method of linearization around the equilibrium state results in the time-invariant linear systems

$$\dot{\mathbf{x}}_\delta(t) = \mathbf{J}(\bar{\mathbf{x}})\mathbf{x}_\delta(t)$$

and

$$\mathbf{x}_\delta(t+1) = \mathbf{J}(\bar{\mathbf{x}})\mathbf{x}_\delta(t)$$

where $\mathbf{x}_\delta(t) = \mathbf{x}(t) - \bar{\mathbf{x}}$. It is also known from the theory of ordinary differential equations that the asymptotic stability of the zero vector in the linearized system implies the asymptotic stability of the equilibrium state $\bar{\mathbf{x}}$ in the original nonlinear system.

For continuous systems the following result has a special importance.

Theorem 12.6

The equilibrium state of a continuous system [Eq. (12.4)] is asymptotically stable if and only if equation

$$\mathbf{A}^T\mathbf{Q} + \mathbf{Q}\mathbf{A} = -\mathbf{M} \tag{12.6}$$

has positive definite solution $\mathbf{Q}$ with some positive definite matrix $\mathbf{M}$.

We note that in practical applications the identity matrix is almost always selected for $\mathbf{M}$.

Theorem 12.7

Let $\varphi(\lambda) = \lambda^n + p_{n-1}\lambda^{n-1} + \cdots + p_1\lambda + p_0$ be the characteristic polynomial of matrix $\mathbf{A}$. Assume that all eigenvalues of matrix $\mathbf{A}$ have negative real parts. Then $p_i > 0$ $(i = 0, 1, \ldots, n-1)$.

Corollary. If any of the coefficients p_i is negative or zero, the equilibrium state of the system with coefficient matrix $\mathbf{A}$ cannot be asymptotically stable. This result can be used as an initial stability test. However, the conditions of the theorem do not imply that the eigenvalues of **A** have negative real parts.

Example 12.4

For matrix

$$\mathbf{A} = \begin{pmatrix} 0 & \omega \\ -\omega & 0 \end{pmatrix}$$

the characteristic polynominal is $\varphi(s) = s^2 + \omega^2$. Since the coefficient of s^1 is zero, the system of Example 12.3 is not asymptotically stable.

The Transfer Function Approach

The transfer function of the continuous system

$$\dot{\mathbf{x}} = \mathbf{A}\mathbf{x} + \mathbf{B}\mathbf{u}$$
$$\mathbf{y} = \mathbf{C}\mathbf{x} \tag{12.7}$$

and that of the discrete system

$$\mathbf{x}(t+1) = \mathbf{A}\mathbf{x}(t) + \mathbf{B}\mathbf{u}(t)$$
$$\mathbf{y}(t) = \mathbf{C}\mathbf{x}(t) \tag{12.8}$$

have the common form

$$\mathbf{TF}(s) = \mathbf{C}(s\mathbf{I} - \mathbf{A})^{-1}\mathbf{B}$$

If both the input and output are single, then

$$\mathbf{TF}(s) = \frac{\mathbf{Y}(s)}{\mathbf{U}(s)}$$

or in the familiar electrical engineering notation

$$\mathbf{TF}(s) = \frac{KG(s)}{1 + KG(s)H(s)} \tag{12.9}$$

where K is the gain term in the forward loop, $\mathbf{G}(s)$ represents the dynamics of the forward loop, or the plant, and $\mathbf{H}(s)$ models the dynamics in the feedback loop. We note that in the case of continuous systems s is the variable of the transfer function, and for discrete systems the variable is denoted by z.

After the Second World War systems and control theory flourished. The transfer function representation was the most popular representation for systems. To determine the stability of a system we merely had to factor the denominator of the transfer function (12.9) and see if any of the poles were in the left half of the complex frequency plane. However, with manual techniques, factoring polynomials of large order is difficult. So engineers, being naturally lazy people, developed several ways to determine the stability of a system without factoring the polynomials [Dorf, 1992]. First, we have the methods of Routh and Hurwitz, developed a century ago, that looked at the coefficients of the characteristic polynomial. These methods showed whether the system was stable or not, but they did not show how close the system was to being stable.

What we want to know is for what value of gain, K, and at what frequency, ω, will the denominator of the transfer function (12.9) become zero. Or, when will $KGH = -1$, meaning when will the magnitude of KGH equal 1 with a phase angle of -180 degrees? These parameters can be determined easily with a Bode diagram. Construct a Bode diagram for KGH of the system, look at the frequency where the phase angle equals -180 degrees, and look up at the magnitude plot. If it is smaller than 1.0, then the system is stable. If it is larger than 1.0, then the system is unstable. Bode diagram techniques are discussed in Chapter 93.3.

The quantity $KG(s)H(s)$ is called the open-loop transfer function of the system, because it is the effect that would be encountered by a signal in one loop around the system if the feedback loop were artificially opened [Bahill, 1981].

To gain some intuition, think of a closed-loop negative feedback system. Apply a small sinusoid at frequency ω to the input. Assume that the gain around the loop, KGH, is 1 or more, and that the phase lag is 180 degrees. The summing junction will flip over the fed back signal and add it to the original signal. The result is a signal that is bigger than what came in. This signal will circulate around this loop, getting bigger and bigger until the real system no longer matches the model. This is what we call instability.

The question of stability can also be answered with Nyquist diagrams. They are related to Bode diagrams, but they give more information. A simple way to construct a Nyquist diagram is to make a polar plot on the complex frequency plane of the Bode diagram. Simply stated, if this contour encircles the -1 point in the complex frequency plane, then the system is unstable. The two advantages of the Nyquist technique are (1) in addition to the information on Bode diagrams, there are about a dozen rules that can be used to help construct Nyquist diagrams, and (2) Nyquist diagrams handle bizarre systems better, as is shown in the following rigorous statement of the Nyquist stability criterion. The number of clockwise encirclements minus the number of counterclockwise encirclements of the point $s = -1 + j0$ by the Nyquist plot of $KG(s)H(s)$ is equal to the number of poles of $Y(s)/U(s)$ minus the number of poles of $KG(s)H(s)$ in the right half of the s-plane.

The root-locus technique was another popular technique for assessing stability. It furthermore allowed the engineer to see the effects of small changes in the gain, K, on the stability of the system. The root-locus diagram shows the location in the s-plane of the poles of the closed-loop transfer function, $Y(s)/U(s)$. All branches of the root-locus diagram start on poles of the open-loop transfer function, KGH, and end either on zeros of the open-loop transfer function, KGH, or at infinity. There are about a dozen rules to help draw these trajectories. The root-locus technique is discussed in Chapter 93.4.

We consider all these techniques to be old fashioned. They were developed to help answer the question of stability without factoring the characteristic polynomial. However, many computer programs are currently available that factor polynomials. We recommend that engineers merely buy one of these computer packages and find the roots of the closed-loop transfer function to assess the stability of a system.

The poles of a system are defined as all values of s such that $s\mathbf{I} - \mathbf{A}$ is singular. The poles of a closed-loop transfer function are exactly the same as the eigenvalues of the system: engineers prefer the term *poles* and the symbol s, and mathematicians prefer the term *eigenvalues* and the symbol λ. We will use s for complex frequency and λ for specific values of s.

Sometimes, some poles could be canceled in the rational function form of $\mathbf{TF}(s)$ so that they would not be explicitly shown. However, even if some poles could be canceled by zeros, we still have to consider all poles in the following criteria. The equilibrium state of the continuous system (12.7) with constant input is stable if all poles of $\mathbf{TF}(s)$ have nonpositive real parts and all poles with zero real parts are single. The equilibrium state is asymptotically stable if and only if all poles of $\mathbf{TF}(s)$ have negative real parts; that is, all poles are in the left half of the s-plane. Similarly, the equilibrium state of the discrete system (12.8) with constant input is stable if all poles of $\mathbf{TF}(z)$ have absolute values less than or equal to one and all poles with unit absolute values are single. The equilibrium state is asymptotically stable if and only if all poles of $\mathbf{TF}(z)$ have absolute values less than one; that is, the poles are all inside the unit circle of the z-plane.

Example 12.5

Consider again the system

$$\dot{\mathbf{x}} = \begin{pmatrix} 0 & \omega \\ -\omega & 0 \end{pmatrix} \mathbf{x} + \begin{pmatrix} 0 \\ 1 \end{pmatrix}$$

which was discussed earlier. Assume that the output equation has the form

$$y = (1, 1)\mathbf{x}$$

Then

$$\mathbf{TF}(s) = \frac{s + \omega}{s^2 + \omega^2}$$

The poles are $j\omega$ and $-j\omega$, which have zero real parts; that is, they are on the imaginary axis of the *s*-plane. Consequently, the equilibrium state is stable but not asymptotically stable. A system such as this would produce constant amplitude sinusoids at frequency ω. So it seems natural to assume that such systems would be used to build sinusoidal signal generators and to model oscillating systems. However, this is not the case, because (1) zero resistance circuits are hard to make; therefore, most function generators use other techniques to produce sinusoids; and (2) such systems are not good models for oscillating systems, because most real-world oscillating systems (i.e., biological systems) have energy dissipation elements in them.

More generally, real-world function generators are seldom made from closed-loop feedback control systems with 180 degrees of phase shift, because (1) it would be difficult to get a broad range of frequencies and several waveforms from such systems, (2) precise frequency selection would require expensive high-precision components, and (3) it would be difficult to maintain a constant frequency in such circuits in the face of changing temperatures and power supply variations. Likewise, closed-loop feedback control systems with 180 degrees of phase shift are not good models for oscillating biological systems, because most biological systems oscillate because of nonlinear network properties.

A special stability criterion for single-input, single-output time-invariant continuous systems will be introduced next. Consider the system

$$\dot{\mathbf{x}} = \mathbf{A}\mathbf{x} + \mathbf{b}u \quad \text{and} \quad y = \mathbf{c}^T\mathbf{x} \tag{12.10}$$

where $\mathbf{A}$ is an $n \times n$ constant matrix, and $\mathbf{b}$ and $\mathbf{c}$ are constant n-dimensional vectors. The transfer function of this system is

$$TF_1(s) = \mathbf{c}^T(s\mathbf{I} - \mathbf{A})^{-1}\mathbf{b}$$

which is obviously a rational function of s. Now let us add negative feedback around this system so that $u = ky$, where k is a constant. The resulting system can be described by the differential equation

$$\dot{\mathbf{x}} = \mathbf{A}\mathbf{x} + k\mathbf{b}\mathbf{c}^T\mathbf{x} = (\mathbf{A} + k\mathbf{b}\mathbf{c}^T)\mathbf{x} \tag{12.11}$$

The transfer function of this feedback system is

$$TF(s) = \frac{TF_1(s)}{1 - kTF_1(s)} \tag{12.12}$$

To help show the connection between the asymptotic stability of systems (12.10) and (12.11), we introduce the following definition.

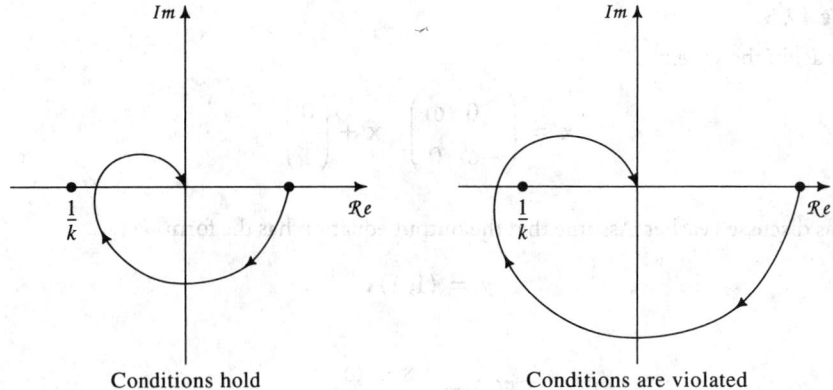

FIGURE 12.2   Illustration of Nyquist stability criteria. (*Source:* F. Szidarovszky and A. T. Bahill, *Linear Systems Theory,* Boca Raton, Fla.: CRC Press, 1992, p.184. With permission.)

Definition 12.3

Let $r(s)$ be a rational function of s. Then the locus of points

$$L(r) = \{a + jb \,|\, a = Re(r(jv)), \quad b = Im(r(jv)), \quad v \in R\}$$

is called the *response diagram of r*. Note that $L(r)$ is the image of the imaginary line $Re(s) = 0$ under the mapping r. We shall assume that $L(r)$ is bounded, which is the case if and only if the degree of the denominator is not less than that of the numerator and r has no poles on the line $Re(s) = 0$.

Theorem 12.8

The Nyquist stability criterion.   Assume that TF_1 has a bounded response diagram $L(TF_1)$. If TF_1 has v poles in the right half of the s-plane, where $Re(s) > 0$, then H has $\rho + v$ poles in the right half of the s-plane where $Re(s) > 0$ if the point $1/k + j \cdot 0$ is not on $L(TF_1)$, and $L(TF_1)$ encircles $1/k + j \cdot 0$ ρ times in the clockwise sense.

Corollary.   Assume that system (12.10) is asymptotically stable with constant input and that $L(TF_1)$ is bounded and traversed in the direction of increasing v and has the point $1/k + j \cdot 0$ on its left. Then the feedback system (12.11) is also asymptotically stable.

This result has many applications, since feedback systems have a crucial role in constructing stabilizers, observers, and filters for given systems. Figure 12.2 illustrates the conditions of the corollary. The application of this result is especially convenient, if system (12.10) is given and only appropriate values k of the feedback are to be determined. In such cases the locus $L(TF_1)$ has to be computed first, and then the region of all appropriate k values can be determined easily from the graph of $L(TF_1)$.

This analysis has dealt with the closed-loop transfer function, whereas the techniques of Bode, root-locus, etc. use the open-loop transfer function. This should cause little confusion as long as the distinction is kept in mind.

12.5 BIBO Stability

In the previous sections, internal stability of time-invariant systems was examined, i.e., the stability of the state was investigated. In this section the **external stability** of systems is discussed; this is usually called the **BIBO** (*bounded-input, bounded-output*) **stability.** Here we drop the simplifying assumption of the previous section that the system is time-invariant: we will examine time-variant systems.

Definition 12.4

A system is called BIBO stable if for zero initial conditions, a bounded input always evokes a bounded output.

For continuous systems a necessary and sufficient condition for BIBO stability can be formulated as follows.

Theorem 12.9

Let $\mathbf{T}(t, \tau) = (t_{ij}(t, \tau))$ be the weighting pattern, $\mathbf{C}(t)\phi(t, \tau)\mathbf{B}(\tau)$, of the system. Then the continuous time-variant linear system is BIBO stable if and only if the integral

$$\int_{t_0}^{t} \left| t_{ij}(t, \tau) \right| d\tau \tag{12.13}$$

is bounded for all $t > t_0$, i and j.

Corollary. Integrals (12.13) are all bounded if and only if

$$I(t) = \int_{t_0}^{t} \sum_i \sum_j \left| t_{ij}(t, \tau) \right| d\tau \tag{12.14}$$

is bounded for $t \geq t_0$. Therefore, it is sufficient to show the boundedness of only one integral in order to establish BIBO stability.

The discrete counterpart of this theorem can be given in the following way.

Theorem 12.10

Let $\mathbf{T}(t, \tau) = (t_{ij}(t, \tau))$ be the weighting pattern of the discrete linear system. Then it is BIBO stable if and only if the sum

$$I(t) = \sum_{\tau=t_0}^{t-1} \left| t_{ij}(t, \tau) \right| \tag{12.15}$$

is bounded for all $t > t_0$, i and j.

Corollary. The sums (12.15) are all bounded if and only if

$$\sum_{\tau=t_0}^{t-1} \sum_i \sum_j \left| t_{ij}(t, \tau) \right| \tag{12.16}$$

is bounded. Therefore it is sufficient to verify the boundedness of only one sum in order to establish BIBO stability.

Consider next the time-invariant case, when $\mathbf{A}(t) \equiv \mathbf{A}$, $\mathbf{B}(t) \equiv \mathbf{B}$ and $\mathbf{C}(t) \equiv \mathbf{C}$. From the foregoing theorems and the definition of $\mathbf{T}(t, \tau)$ we have immediately the following sufficient condition.

Theorem 12.11

Assume that for all eigenvalues λ_i of $\mathbf{A}$, Re $\lambda_i < 0$ (or $|\lambda_i| < 1$). Then the time-invariant linear continuous (or discrete) system is BIBO stable.

Finally, we note that BIBO stability is not implied by an observation that a certain bounded input generates bounded output. All bounded inputs must generate bounded outputs in order to guarantee BIBO stability.

Adaptive-control systems are time-varying systems. Therefore, it is usually difficult to prove that they are stable. Szidarovszky *et al.* [1990], however, show a technique for doing this. This new result gives a necessary and sufficient condition for the existence of an asymptotically stable model-following adaptive-control system, and in the case of the existence of such systems they present an algorithm for finding the appropriate feedback parameters.

12.6 Physical Examples

In this section we show some examples of stability analysis of physical systems.

1. Consider a simple *harmonic oscillator* constructed of a mass and an ideal spring. Its dynamic response is summarized with

$$\dot{\mathbf{x}} = \begin{pmatrix} 0 & \omega \\ -\omega & 0 \end{pmatrix} \mathbf{x} + \begin{pmatrix} 0 \\ 1 \end{pmatrix} u$$

In Example 12.3 we showed that this system is stable but not asymptotically stable. This means that if we leave it alone in its equilibrium state, it will remain stationary, but if we jerk on the mass it will oscillate forever. There is no damping term to remove the energy, so the energy will be transferred back and forth between potential energy in the spring and kinetic energy in the moving mass. A good approximation of such a harmonic oscillator is a pendulum clock. The more expensive it is (i.e., the smaller the damping), the less often we have to wind it (i.e., add energy).

2. A *linear second-order electrical system* composed of a series connection of an input voltage source, an inductor, a resistor, and a capacitor, with the output defined as the voltage across the capacitor, can be characterized by the second-order equation

$$\frac{V_{out}}{V_{in}} = \frac{1}{LCs^2 + RCs + 1}$$

For convenience, let us define

$$\omega_n = \sqrt{\frac{1}{LC}} \quad \text{and} \quad \zeta = \frac{R}{2}\sqrt{\frac{C}{L}}$$

and assume that $\zeta < 1$. With these parameters the transfer function becomes

$$\frac{V_{out}}{V_{in}} = \frac{\omega_n^2}{s^2 + 2\zeta\omega_n + \omega_n^2}$$

Is this system stable? The roots of the characteristic equation are

$$\lambda_{1,2} = -\zeta\omega_n \pm j\omega_n\sqrt{1 - \zeta^2}$$

If $\zeta > 0$, the poles are in the left half of the *s*-plane, and therefore the system is asymptotically stable. If $\zeta = 0$, as in the previous example, the poles are on the imaginary axis; therefore, the system is stable but not asymptotically stable. If $\zeta < 0$, the poles are in the right half of the *s*-plane and the system is unstable.

3. An *electrical system* is shown in Fig. 12.3. Simple calculation shows that by introducing the variables

$$x_1 = i_L, \quad x_2 = v_c, \quad \text{and} \quad u = v_s$$

the system can be described by the differential equations

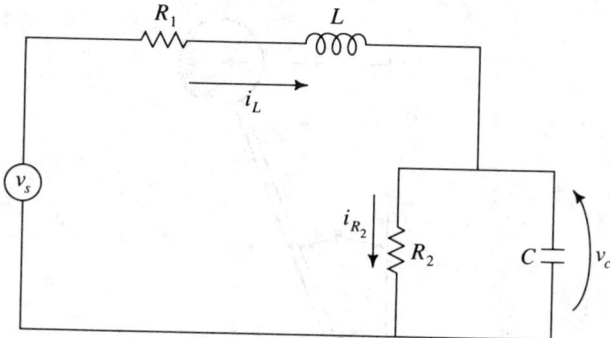

FIGURE 12.3 A simple electrical system. (*Source:* F. Szidarovszky and A. T. Bahill, *Linear Systems Theory,* Boca Raton, Fla.: CRC Press, 1992, p. 125. With permission.)

$$\dot{x}_1 = -\frac{R_1}{L}x_1 - \frac{1}{L}x_2 + \frac{1}{L}u$$

$$\dot{x}_2 = \frac{1}{C}x_1 - \frac{1}{CR_2}x_2$$

The characteristic equation has the form

$$\left(-s - \frac{R_1}{L}\right)\left(-s - \frac{1}{CR_2}\right) + \frac{1}{LC} = 0$$

which simplifies as

$$s^2 + s\left(\frac{R_1}{L} + \frac{1}{CR_2}\right) + \left(\frac{R_1}{LCR_2} + \frac{1}{LC}\right) = 0$$

Since R_1, R_2, L, and C are positive numbers, the coefficients of this equation are all positive. The constant term equals $\lambda_1\lambda_2$, and the coefficient of s^1 is $-(\lambda_1 + \lambda_2)$. Therefore

$$\lambda_1 + \lambda_2 < 0 \quad \text{and} \quad \lambda_1\lambda_2 > 0$$

If the eigenvalues are real, then these relations hold if and only if both eigenvalues are negative. If they were positive, then $\lambda_1 + \lambda_2 > 0$. If they had different signs, then $\lambda_1\lambda_2 < 0$. Furthermore, if at least one eigenvalue is zero, then $\lambda_1\lambda_2 = 0$. Assume next that the eigenvalues are complex:

$$\lambda_{1,2} = Re\ s \pm j\ Im\ s$$

Then

$$\lambda_1 + \lambda_2 = 2Re\ s$$

and

$$\lambda_1\lambda_2 = (Re\ s)^2 + (Im\ s)^2$$

Hence $\lambda_1 + \lambda_2 < 0$ implies that $Re\ s < 0$.

In summary, the system is asymptotically stable, since in both the real and complex cases the eigenvalues have negative values and negative real parts, respectively.

4. The classical *stick balancing* problem is shown in Fig. 12.4. Simple analysis shows that $y(t)$ satisfies the second-order equation

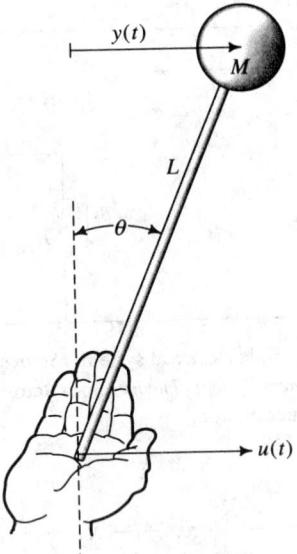

FIGURE 12.4 Stick balancing. (*Source:* F. Szidarovszky and A. T. Bahill, *Linear Systems Theory,* Boca Raton, Fla.: CRC Press, 1992, p. 127. With permission.)

$$\ddot{y} = \frac{g}{L}(y - u)$$

If one selects $L = 1$, then the characteristic equation has the form

$$s^2 - g = 0$$

So, the eigenvalues are

$$\lambda_{1,2} = \pm\sqrt{g}$$

One is in the right half of the *s*-plane and the other is in the left half of the *s*-plane, so the system is unstable. This instability is understandable, since without an input to control the system, if the stick is not upright with zero velocity, it will fall over.

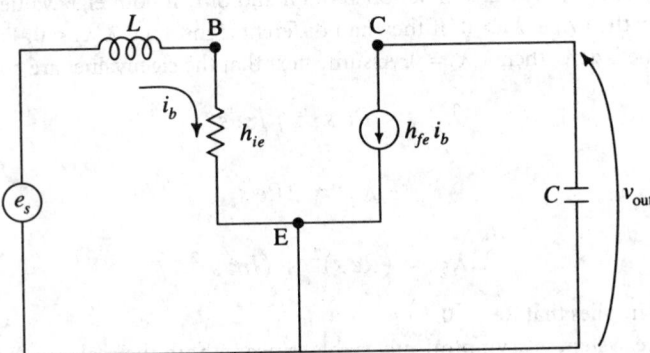

FIGURE 12.5 A model for a simple transistor circuit. (*Source:* F. Szidarovszky and A. T. Bahill, *Linear Systems Theory,* Boca Raton, Fla.: CRC Press 1992, p. 127. With permission.)

5. A sample *transistor circuit* can be modeled as shown in Fig. 12.5. The state variables are related to the input and output of the circuit: the base current, i_b, is x_1 and the output voltage, v_{out}, is x_2. Therefore,

$$\dot{\mathbf{x}} = \begin{pmatrix} -\dfrac{h_{ie}}{L} & 0 \\ \dfrac{h_{fe}}{C} & 0 \end{pmatrix} \mathbf{x} + \begin{pmatrix} \dfrac{1}{L} \\ 0 \end{pmatrix} e_s \text{ and } \mathbf{c}^T = (0, 1)$$

The **A** matrix looks strange with a column of all zeros, and indeed the circuit does exhibit odd behavior. For example, as we will show, there is no equilibrium state for a unit step input of e_s. This is reasonable, however, because the model is for mid-frequencies, and a unit step does not qualify. In response to a unit step the output voltage will increase linearly until the model is no longer valid. If e_s is considered to be the input, then the system is

$$\dot{\mathbf{x}} = \begin{pmatrix} -\dfrac{h_{ie}}{L} & 0 \\ \dfrac{h_{fe}}{C} & 0 \end{pmatrix} \mathbf{x} + \begin{pmatrix} \dfrac{1}{L} \\ 0 \end{pmatrix} u$$

If $u(t) \equiv 1$, then at the equilibrium state:

$$\begin{pmatrix} -\dfrac{h_{ie}}{L} & 0 \\ \dfrac{h_{fe}}{C} & 0 \end{pmatrix} \begin{pmatrix} \overline{x}_1 \\ \overline{x}_2 \end{pmatrix} + \begin{pmatrix} \dfrac{1}{L} \\ 0 \end{pmatrix} = \begin{pmatrix} 0 \\ 0 \end{pmatrix}$$

That is,

$$-\dfrac{h_{ie}}{L} \overline{x}_1 + \dfrac{1}{L} = 0$$

$$\dfrac{h_{fe}}{C} \overline{x}_1 = 0$$

Since $h_{fe}/C \neq 0$, the second equation implies that $\overline{x}_1 = 0$, and by substituting this value into the first equation we get the obvious contradiction $1/L = 0$. Hence, with nonzero constant input *no* equilibrium state exists.

Let us now investigate the stability of this system. First let $\widetilde{\mathbf{x}}(t)$ denote a fixed trajectory of this system, and let $\mathbf{x}(t)$ be an arbitrary solution. Then the difference $\delta\mathbf{x}(t) = \mathbf{x}(t) - \widetilde{\mathbf{x}}(t)$ satisfies the homogeneous equation

$$\delta\dot{\mathbf{x}} = \begin{pmatrix} -\dfrac{h_{ie}}{L} & 0 \\ \dfrac{h_{fe}}{C} & 0 \end{pmatrix} \delta\mathbf{x}$$

This system has an equilibrium $\delta\mathbf{x}(t) = 0$. Next, the stability of this equilibrium is examined by solving for the poles of the closed-loop transfer function. The characteristic equation is

$$
\det \begin{pmatrix} -\dfrac{h_{ie}}{L} - s & 0 \\ \dfrac{h_{fe}}{C} & -s \end{pmatrix} = 0
$$

which can be simplified as

$$
s^2 + s\,\frac{h_{ie}}{L} + 0 = 0
$$

The roots are

$$
\lambda_1 = 0 \quad \text{and} \quad \lambda_2 = -\frac{h_{ie}}{L}
$$

Therefore, the system is stable but not asymptotically stable. This stability means that for small changes in the initial state the entire trajectory $x(t)$ remains close to $\tilde{x}(t)$.

Defining Terms

Asymptotic stability: An equilibrium state $\bar{x}$ of a system is asymptotically stable if, in addition to being stable, there is an $\varepsilon > 0$ such that whenever $\|\bar{x} - x_0\| < \varepsilon$, then $x(t) \to \bar{x}$ as $t \to \infty$. A system is asymptotically stable if all the poles of the closed-loop transfer function are in the left half of the s-plane (inside the unit circle of the z-plane for discrete systems). This is sometimes defined as *stability*.

BIBO stability: A system is BIBO stable if for zero initial conditions a bounded input always evokes a bounded output.

External stability: Stability concepts related to the input-output behavior of the system.

Global asymptotic stability: An equilibrium state $\bar{x}$ of a system is globally asymptotically stable if it is stable and with arbitrary initial state $x_0 \in X$, $x(t) \to \bar{x}$ as $t \to \infty$.

Instability: An equilibrium state of a system is unstable if it is not stable. A system is unstable if at least one pole of the closed-loop transfer function is in the right half of the s-plane (outside the unit circle of the z-plane for discrete systems).

Internal stability: Stability concepts related to the state of the system.

Stability: An equilibrium state $\bar{x}$ of a system is stable if there is an $\varepsilon_0 > 0$ with the following property: for all ε_1, $0 < \varepsilon_1 < \varepsilon_0$, there is an $\varepsilon > 0$ such that if $\|\bar{x} - x_0\| < \varepsilon$, then $\|\bar{x} - x(t)\| < \varepsilon_1$ for all $t > t_0$. A system is stable if the poles of its closed-loop transfer function are (1) in the left half of the complex frequency plane, called the s-plane (inside the unit circle of the z-plane for discrete systems), or (2) on the imaginary axis, and all of the poles on the imaginary axis are single (on the unit circle and all such poles are single for discrete systems). Stability for a system with repeated poles on the $j\omega$ axis (the unit circle) is complicated and is examined in the discussion after Theorem 12.5. In the electrical engineering literature, this definition of stability is sometimes called *marginal stability* and sometimes *stability in the sense of Lyapunov*.

References

A. T. Bahill, *Bioengineering: Biomedical, Medical and Clinical Engineering*, Englewood Cliffs, N.J.: Prentice-Hall, 1981, pp. 214–215, 250–252.

R. C. Dorf, *Modern Control Systems*, 6th ed., Reading, Mass.: Addison-Wesley, 1992.

M. Jamshidi, M. Tarokh, and B. Shafai, *Computer-Aided Analysis and Design of Linear Control Systems*, Englewood Cliffs, N.J.: Prentice-Hall, 1992.

F. Szidarovszky and A. T. Bahill, *Linear Systems Theory*, Boca Raton, Fla.: CRC Press, 1992.

F. Szidarovszky, A. T. Bahill, and S. Molnar, "On stable adaptive control systems," *Pure Math. and Appl.*, vol. 1, ser. B, no. 2–3, pp. 115–121, 1990.

Further Information

For further information consult the textbooks *Modern Control Systems* by Dorf [1992] or *Linear Systems Theory* by Szidarovszky and Bahill [1992].

Thomas A. Edison in his laboratory. He is shown with his Edison lamps, discovered in 1879. (Courtesy of Edison National Historical Site.)

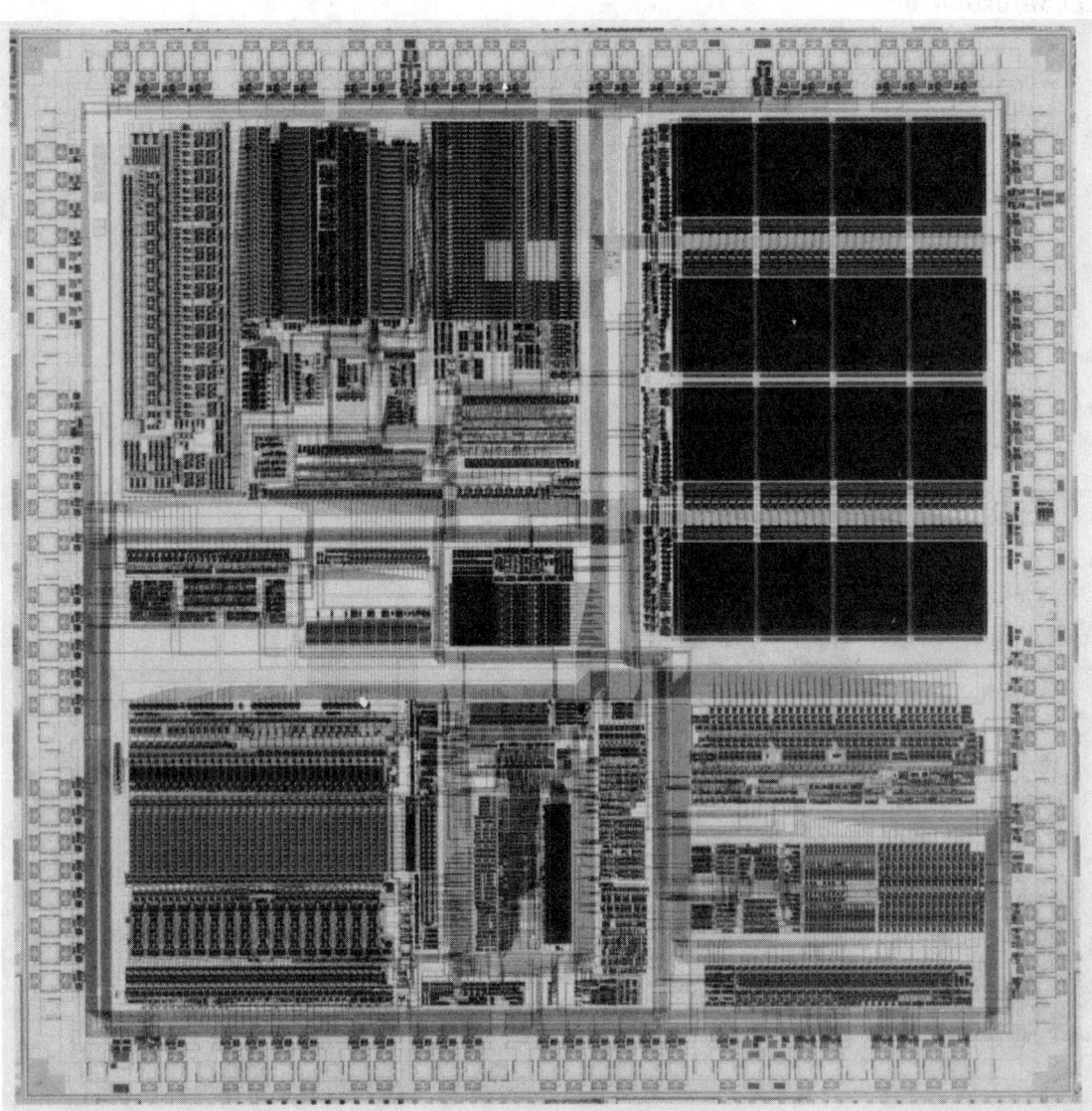

The AT&T DSP3210 Digital Signal Processor is specifically designed to bring real-time multimedia applications to low-cost personal computers. The DSP3210 operates as an asymmetric parallel processor with the PC's host CPU while sharing system memory. In addition to its high-performance floating-point signal processing unit, the DSP3210 includes a RISC CPU, 8 kbytes of RAM, and a bus interface compatible with various 32-bit microprocessor busses. A wide variety of personal computers are currently being introduced that include the DSP3210 as their multimedia processor to enable voice recognition, voice synthesis, modem, audio and image compression applications. (Photo courtesy of AT&T.)

Signal Processing

Delores M. Etter
University of Colorado, Boulder

13 Digital Signal Processing *A. Poularikas, B. Bomar, L. Smith, J. Cadzow* 229
Transforms and Fast Algorithms • Design and Implementation of Digital Filters • Signal Restoration

14 Speech Signal Processing *J. Gibson, Y. Ephraim, J. Fussell, L. Wilcox, M. Bush* 279
Coding, Transmission, and Storage • Speech Enhancement and Noise Reduction • Analysis and Synthesis • Speech Recognition

15 Spectral Estimation and Modeling *S. Pillai, T. Shim, S. Batalama, D. Kazakos* 315
Spectral Analysis • Parameter Estimation

16 Multidimensional Signal Processing *E. Delp, J. Allebach, C. Bouman, S. Rajala, N. Bose, L. Sibul* 329
Digital Image Processing • Video Signal Processing • Sensor Array Processing

17 VLSI for Signal Processing *K. Parhi, R. Chassaing, B. Bitler* 370
Special Architectures • Signal Processing Chips and Applications

18 Acoustic Signal Processing *J. Schroeter, S. Mehta, G. Carter* 395
Digital Signal Processing in Audio and Electroacoustics • Underwater Acoustical Signal Processing

19 Neural Networks *B. Bavarian* 420
Perceptrons • Hopfield Network • Topology-Preserving Network • Adaptive Resonance Theory

S IGNAL PROCESSING WAS DEFINED at a meeting in 1991 of the National Science Foundation's MIPS (Microelectronics and Information Processing Systems) Advisory Committee as "the extraction of information-bearing attributes from measured data, and any subsequent transformation of those attributes for the purposes of detection, estimation, classification, or waveform synthesis." If we expand this concise definition, we observe that the signals we typically use in signal processing are functions of time, such as temperature measurements, velocity measurements, voltages, blood pressures, earth motion, and speech signals. Most of these signals are initially continuous signals (also called analog signals) which are measured by sensors that convert energy to electricity. Some of the common types of sensors used for collecting data are microphones, which measure acoustic or sound data; seismometers, which measure earth motion; photocells, which measure light intensity; thermistors, which measure temperature; and oscilloscopes, which measure voltage. When we work with the continuous electrical signals collected by sensors, we often convert the continuous signal to a digital signal (a sequence of values) with a piece of hardware called an analog-to-digital (A/D) converter. Once we have collected the digital signal, we are ready to use the computer to apply digital signal processing (DSP) techniques to it. These DSP techniques can be designed to perform a number of operations such as:

- Removing noise that is distorting the signal, such as static on a communication line.
- Extracting information from the signal, such as the average value and the power in a signal.
- Separating components of the signal, such as the separation of a band of frequencies that represent the television signal for a specific channel.
- Encoding the information in a more efficient way for transmission, such as the encoding of speech signals into digital signals for transmitting across telephone lines.
- Detecting information in a signal, such as the detection of a surface ship in a sonar signal.

These are just a few of the types of operations that can be performed by signal processing techniques. For some applications, an analog or continuous output signal is needed, and thus a digital-to-analog (D/A) converter is used to convert the modified digital signal to a continuous signal. Another device called a transducer can be used to convert the continuous electrical signal to another form; for example, a speaker converts a continuous electrical signal to an acoustical signal.

In this section the variety and diversity of signal processing is presented from a theoretical point of view, from an implementation point of view, and from an applications point of view. The theoretical point of view includes the development of mathematical models and the development of software algorithms and computer simulations to evaluate and analyze the models both with simulated data and with real data. High-level software tools are important in both the development of new theoretical results and in establishing the validity of the results when applied to real data. The applications determine the way in which the theory is implemented; a key element in the implementation of a signal processing technique relates to whether the technique is applied in real-time (or close to real-time) or whether the processing can be handled off-line. Real-time implementation can use VLSI (very large scale integration) techniques, with commercial DSP chips, or it can involve custom design of chips, MCMs (multichip modules), or ASICs (application-specific integrated circuits). The selection of topics in this section covers the three points of view (theoretical, application, implementation) but should not be assumed to include a complete summary of these topics.

Nomenclature

Symbol	Quantity	Unit	Symbol	Quantity	Unit
AG	array gain	dB	$\phi(K)$	sampled degree phase spectrum	
$A(k)$	sampled amplitude spectrum				
C	compression rate		$G(e^{j\omega})$	spectral gain function	
DFT	discrete Fourier transform		H	entropy	
δ_p	passband ripple		$H(e^{j\omega})$	transfer function of discrete time system	
δ_s	stopband attenuation				
$\delta(t)$	dirac or impulse function		$h(n)$	impulse response	
$\Delta\omega$	transition bandwidth	Hz	η	learning rate parameter	
$E(e^{j\omega})$	Fourier transform of error sequence		$I_n(x)$	modified Bessel function of order n	
f	analog frequency	Hz	L	length of continuous function	s
$f(n)$	sequence		$\mu_x(t)$	ensemble average	
$f(t)$	continuous signal		N	number of sample values	
FFT	fast Fourier transform		ω	digital frequency	rad/s
ϕ	azimuthal angle		Ω	angular frequency	rad

Symbol	Quantity	Unit	Symbol	Quantity	Unit
p_n	probability		T	sampling time interval	s
$P(k)$	sampled power spectrum		$w(n)$	window function	
q	quantization step size		$x(n)$	input sequence	
Q	number of quantization levels		$y(n)$	output sequence	
σ	elevation angle				

13

Digital Signal Processing

Alexander D. Poularikas
University of Alabama in Huntsville

Bruce W. Bomar
University of Tennessee Space Institute

L. Montgomery Smith
University of Tennessee Space Institute

James A. Cadzow
Vanderbilt University

13.1 Transforms and Fast Algorithms 229
Properties of the DFT • Relation between DFT and Fourier Transform • Power, Amplitude, and Phase Spectra • Observations • Data Windowing • Fast Fourier Transform • Computation of the Inverse DFT

13.2 Design and Implementation of Digital Filters 238
Finite Impulse Response Filter Design • Infinite Impulse Response Filter Design • Finite Impulse Response Filter Implementation • Infinite Impulse Response Filter Implementation

13.3 Signal Restoration 251
Introduction • Attribute Sets: Closed Subspaces • Attribute Sets: Closed Convex Sets • Closed Projection Operators • Algebraic Properties of Matrices • Structural Properties of Matrices • Nonnegative Sequence Approximation • Exponential Signals and the Data Matrix • Recursive Modeling of Data

13.1 Transforms and Fast Algorithms

Alexander D. Poularikas

One of the methods, and one that is used extensively, calls for replacing continuous Fourier transforms by an equivalent *discrete Fourier transform* (DFT) and then evaluating the DFT using the discrete data. However, evaluating a DFT with 512 samples (a small number in most cases) requires more than 1.5×10^6 mathematical operations. It was the development of the *fast Fourier transform* (**FFT**), a computational technique that reduces the number of mathematical operations in the evaluation of the DFT to $N \log_2 (N)$ (approximately 2.5×10^4 operations for the 512-point case mentioned above), that makes DFT an extremely useful tool in most all fields of science and engineering.

A data sequence is available only with a finite time window from $n = 0$ to $n = N - 1$. The transform is discretized for N values by taking samples at the frequencies $2\pi/NT$, where T is the time interval between sample points. Hence, we define the DFT of a sequence of N samples for $0 \leq k \leq N - 1$ by the relation

$$F(k\Omega) \doteq \mathcal{F}_d\{f(nT)\} = T \sum_{n=0}^{N-1} f(nT)e^{-j2\pi nkT/NT}$$

$$= T \sum_{n=0}^{N-1} f(nT)e^{-j\Omega Tnk} \quad n = 0, 1, \dots, N - 1 \tag{13.1}$$

where N = number of sample values, T = sampling time interval, $(N-1)T$ = signal length, $f(nT) =$

229

sampled form of $f(t)$ at points nT, $\Omega = (2\pi/T)1/N = \omega_s/N =$ frequency sampling interval, $e^{-i\Omega T} =$ Nth principal root of unity, and $j = \sqrt{-1}$. The inverse DFT is given by

$$f(nT) \doteq \mathcal{F}_d^{-1}\{F(k\Omega)\} = \frac{1}{NT} \sum_{k=0}^{N-1} F(k\Omega)e^{j2\pi nkT/NT}$$

$$= \frac{1}{NT} \sum_{k=0}^{N-1} F(k\Omega)e^{i\Omega Tnk}$$

(13.2)

The sequence $f(nT)$ can be viewed as representing N consecutive samples $f(n)$ of the continuous signal, while the sequence $F(k\Omega)$ can be considered as representing N consecutive samples $F(k)$ in the frequency domain. Therefore, Eqs. (13.1) and (13.2) take the compact form

$$F(k) \doteq \mathcal{F}_d\{f(n)\} = \sum_{n=0}^{N-1} f(n)e^{-j2\pi nk/N}$$

$$= \sum_{n=0}^{N-1} f(n)W_N^{nk} \quad k = 0, \ldots, N-1$$

(13.3)

$$f(n) \doteq \mathcal{F}_d^{-1}\{F(k)\} = \frac{1}{N} \sum_{k=0}^{N-1} F(k)e^{j2\pi nk/N}$$

$$= \sum_{k=0}^{N-1} F(k)W_N^{-nk} \quad k = 0, \ldots, N-1$$

(13.4)

where

$$W_N = e^{-j2\pi/N} \qquad j = \sqrt{-1}$$

An important property of the DFT is that $f(n)$ and $F(k)$ are uniquely related by the transform pair (13.3) and (13.4).

We observe that the functions W^{kn} are N-periodic; that is,

$$W_N^{kn} = W_N^{k(n+N)} \qquad k, n = 1, \pm 1, \pm 2, \ldots$$

(13.5)

As a consequence, the sequences $f(n)$ and $F(k)$ as defined by (13.3) and (13.4) are also N-periodic.

It is generally convenient to adopt the convention

$$\{f(n)\} \leftrightarrow \{F(k)\}$$

(13.6)

to represent the transform pair (13.3) and (13.4).

Properties of the DFT

A detailed discussion of the properties of DFT can be found in the cited references at the end of this section. In what follows we consider a few of these properties that are of value for the development of the FFT.

1. *Linearity:*

$$\{af(n) + by(n)\} \leftrightarrow \{aF(k)\} \leftrightarrow \{bY(k)\}$$

(13.7)

2. *Complex conjugate:* If $N/2$ is an integer and $\{f(n)\} \leftrightarrow \{F(k)\}$, then

$$F\left(\frac{N}{2} + l\right) = F^*\left(\frac{N}{2} - l\right) \quad l = 0, 1, \ldots, \frac{N}{2} \tag{13.8}$$

where $F^*(k)$ denotes the complex conjugate of $F(k)$. The preceding identity shows the folding property of the DFT.

3. *Reversal:*

$$\{f(-n)\} \leftrightarrow \{F(-k)\} \tag{13.9}$$

4. *Time shifting:*

$$\{f(n + l)\} \leftrightarrow \{W^{-lk} F(k)\} \tag{13.10}$$

5. *Convolution of real sequences:* If

$$y(n) = \frac{1}{N} \sum_{l=0}^{N-1} f(l)h(n - l) \quad n = 0, 1, \ldots, N - 1 \tag{13.11}$$

then

$$\{y(n)\} \leftrightarrow \{F(k)\, H(k)\} \tag{13.12}$$

6. *Correlation of real sequences:* If

$$y(n) = \frac{1}{N} \sum_{l=0}^{N-1} f(l)h(n + l) \quad n = 0, 1, \ldots, N - 1 \tag{13.13}$$

then

$$\{y(n)\} \leftrightarrow \{F(r)\, H^*(k)\} \tag{13.14}$$

7. *Symmetry:*

$$\left\{\frac{1}{N} F(n)\right\} \leftrightarrow \{f(-k)\} \tag{13.15}$$

8. Parseval's theorem:

$$\sum_{n=0}^{N-1} f^2(n) = \frac{1}{N} \sum_{k=0}^{N-1} \left|F(k)\right|^2 \tag{13.16}$$

where $|F(k)| = F(k)\, F^*(k)$.

Example 1

Verify Parseval's theorem for the sequence $\{f(n)\} = \{1, 2, -1, 3\}$.

Solution. With the help of (13.3) we obtain

$$F(k)\Big|_{k=0} = F(0) = \sum_{n=0}^{3} f(n)e^{-j(2\pi/4)kn}\Big|_{k=0}$$

$$= (1e^{-j(\pi/2)0\cdot0} + 2e^{-j(\pi/2)0\cdot1} - e^{-j(\pi/2)0\cdot2} + 3e^{-j(\pi/2)0\cdot3})$$

$$= 5$$

Similarly, we find

$$F(1) = 2 + j \quad F(2) = -5 \quad F(3) = 2 - j$$

Introducing these values in (13.16) we obtain

$$1^2 + 2^2 + (-1)^2 + 3^2 = 1/4[5^2 + (2 + j)(2 - j) + 5^2 + (2 - j)(2 + j)] \quad \text{or} \quad 15 = 60/4$$

which is an identity, as it should have been.

Relation between DFT and Fourier Transform

The sampled form of a continuous function $f(t)$ can be represented by N equally spaced sampled values $f(n)$ such that

$$f(n) = f(nT) \quad n = 0, 1, \ldots, N-1 \tag{13.17}$$

where T is the sampling interval. The length of the continuous function is $L = NT$, where $f(N) = f(0)$.

We denote the sampled version of $f(t)$ by $f_s(t)$, which may be represented by a sequence of impulses. Mathematically it is represented by the expression

$$f_s(t) = \sum_{n=0}^{N-1} [Tf(n)]\delta(t - nT) \tag{13.18}$$

where $\delta(t)$ is the Dirac or impulse function.

Taking the Fourier transform of $f_s(t)$ in (13.18) we obtain

$$F_s(\omega) = T\int_{\infty}^{\infty} \sum_{n=0}^{N-1} f(n)\delta(t - nT)e^{-j\omega t}\, dt$$

$$= T\sum_{n=0}^{N-1} f(n)\int_{\infty}^{\infty} \delta(t - nT)e^{-j\omega t}\, dt \tag{13.19}$$

$$= T\sum_{n=0}^{N-1} f(n)e^{-j\omega nT}$$

Equation (13.19) yields $F_s(\omega)$ for all values of ω. However, if we are only interested in the values of $F_s(\omega)$ at a set of discrete equidistant points, then (13.19) is expressed in the form [see also (13.1)]

$$F_s(k\Omega) = T\sum_{n=0}^{N-1} f(n)e^{-jkn\Omega T} \quad k = 0, \pm 1, \pm 2, \ldots, \pm N/2 \tag{13.20}$$

where $\Omega = 2\pi/L = 2\pi/NT$. Therefore, comparing (13.3) and (13.20) we observe that we can find $F(\omega)$ from $F_s(\omega)$ using the relation

$$F(k) = F_s(\omega)\big|_{\omega=k\Omega} \tag{13.21}$$

Power, Amplitude, and Phase Spectra

If $f(t)$ represents voltage or current waveform supplying a load of 1 Ω, the left-hand side of Parseval's theorem (13.16) represents the power dissipated in the 1-Ω resistor. Therefore, the right-hand side represents the power contributed by each harmonic of the spectrum. Thus the DFT **power spectrum** is defined as

$$P(k) = F(k)F^*(k) = |F(k)|^2 \quad k = 0, 1, \ldots, N-1 \tag{13.22}$$

For real $f(n)$ there are only $(N/2 + 1)$ independent DFT spectral points as the complex conjugate property shows (13.8). Hence we write

$$P(k) = |F(k)|^2 \quad k = 0, 1, \ldots, N/2 \tag{13.23}$$

The *amplitude spectrum* is readily found from that of a power spectrum, and it is defined as

$$A(k) = |F(k)| \quad k = 0, 1, \ldots, N-1 \tag{13.24}$$

The power and amplitude spectra are invariant with respect to shifts of the data sequence $\{f(n)\}$.
The **phase spectrum** of a sequence $\{f(n)\}$ is defined as

$$\phi_f(k) = \tan^{-1} \frac{\text{Im}\{F(k)\}}{\text{Re}\{F(k)\}} \qquad k = 0, 1, \ldots, N - 1 \tag{13.25}$$

As in the case of the power spectrum, only $(N/2 + 1)$ of the DFT phase spectral points are independent for real $\{f(n)\}$. For a real sequence $\{f(n)\}$ the power spectrum is an *even function* about the point $k = N/2$ and the phase spectrum is an *odd function* about the point $k = N/2$.

Observations

1. The frequency spacing $\Delta\omega$ between coefficients is

$$\Delta\omega = \Omega = \frac{2\pi}{NT} = \frac{\omega_s}{N} \quad \text{or} \quad \Delta f = \frac{1}{NT} = \frac{f_s}{N} = \frac{1}{T_0} \tag{13.26}$$

2. The reciprocal of the record length defines the frequency resolution.
3. If the number of samples N is fixed and the sampling time is increased, the record length and the precision of frequency resolution is increased. When the sampling time is decreased, the opposite is true.
4. If the record length is fixed and the sampling time is decreased (N increases), the resolution stays the same and the computed accuracy of $F(n\Omega)$ increases.
5. If the record length is fixed and the sampling time is increased (N decreases), the resolution stays the same and the computed accuracy of $F(n\Omega)$ decreases.

Data Windowing

To produce more accurate frequency spectra it is recommended that the data are weighted by a **window** function. Hence, the new data set will be of the form $\{f(n) \, w(n)\}$. The following are the most commonly used windows:

1. Triangle (Fejer, Bartlet) window:

$$w(n) = \begin{cases} \dfrac{n}{N/2} & n = 0, 1, \ldots, \dfrac{N}{2} \\[2ex] w(N - n) & n = \dfrac{N}{2}, \ldots, N - 1 \end{cases} \tag{13.27}$$

2. $\text{Cos}^\alpha(x)$ windows:

$$w(n) = \sin^2\left(\frac{n}{N}\pi\right)$$

$$= 0.5\left[1 - \cos\left(\frac{2n}{N}\pi\right)\right] \qquad n = 0, 1 \ldots, N - 1 \qquad \alpha = 2 \tag{13.28}$$

This window is also called the raised cosine or Hamming window.

3. Hamming window:

$$w(n) = 0.54 - 0.46 \, \cos\left(\frac{2\pi}{N}n\right) \qquad n = 0, 1, \ldots, N - 1 \tag{13.29}$$

4. Blackman window:

$$w(n) = \sum_{m=0}^{K} (-1)^m a_m \cos\left(2\pi m \frac{n}{N}\right) \qquad n = 0, 1, \ldots, N - 1 \qquad K \leq \frac{N}{2} \tag{13.30}$$

for $K = 2$, $a_0 = 0.42$, $a_1 = 0.50$, and $a_2 = 0.08$.

5. Blackman-Harris window. Harris used a gradient search technique to find three- and four-term expansion of (13.30) that either minimized the maximum sidelobe level for fixed mainlobe width, or traded mainlobe width versus minimum sidelobe level (see Table 13.1)

6. Centered Gaussian window:

$$w(n) = \exp\left[-\frac{1}{2}\alpha\left(\frac{n}{N/2}\right)^2\right] \qquad 0 \leq |n| \leq \frac{N}{2} \qquad \alpha = 2, 3, \ldots \tag{13.31}$$

As α increases, the mainlobe of the frequency spectrum becomes broader and the sidelobe peaks become lower.

7. Centered Kaiser-Bessel window:

$$w(n) = \frac{I_0\left[\pi a\sqrt{1.0 - \left(\frac{n}{N/2}\right)^2}\right]}{I_0(\pi\alpha)} \qquad 0 \leq |n| \leq \frac{N}{2} \tag{13.32}$$

where

$$I_0(x) = \text{zero-order modified Bessel function}$$

$$= \sum_{k=0}^{\infty}\left(\frac{(x/2)^k}{k!}\right)^2 \tag{13.33}$$

$$k! = 1 \times 2 \times 3 \times \cdots \times k$$

$$\alpha = 2, 2.5, 3 \qquad \text{(typical values)}$$

Fast Fourier Transform

One of the approaches to speed the computation of the DFT of a sequence is the *decimation-in-time* method. This approach is one of breaking the N-point transform into two $(N/2)$-point transforms, breaking each $(N/2)$-point transform into two $(N/4)$-point transforms, and continuing the above process until we obtain the two-point transform. We start with the DFT expression and factor it into two DFTs of length $N/2$:

Table 13.1

No. of Terms in (13.30)	Maximum Sidelobe, dB	Parameter Values			
		a_0	a_1	a_2	a_3
3	−70.83	0.42323	0.49755	0.07922	
3	−62.05	0.44959	0.49364	0.05677	
4	−92	0.35875	0.48829	0.14128	0.01168
4	−74.39	0.40217	0.49703	0.09892	0.00188

$$F(k) = \sum_{n=0}^{N-2} f(n)W_N^{kn} \qquad n \text{ even}$$

$$+ \sum_{n=1}^{N-1} f(n)W_N^{kn} \qquad n \text{ odd} \tag{13.34}$$

Letting $n = 2m$ in the first sum and $n = 2m + 1$ in the second, (13.34) becomes

$$F(k) = \sum_{m=0}^{(N/2)-1} f(2m)W_N^{2mk} + \sum_{m=0}^{(N/2)-1} f(2m + 1)W_N^{(2m+1)k} \tag{13.35}$$

However, because of the identities

$$W_N^{2mk} = (W_N^2)^{mk} = e^{-j(2\pi/N)2mk} = e^{-j(4\pi mk/N)} = W_{N/2}^{mk} \tag{13.36}$$

and the substitution $f(2m) = f_1(m)$ and $f(2m + 1) = f_2(m)$, $m = 0, 1, \ldots, N/2 - 1$, takes the form

$$F(k) = \sum_{m=0}^{(N/2)-1} f_1(m)W_{N/2}^{mk} \qquad \frac{N}{2} - \text{point DFT of even-indexed sequence}$$

$$+ W_N^k \sum_{m=0}^{(N/2)-1} f_2(m)W_{N/2}^{mk} \qquad \frac{N}{2} - \text{point DFT of odd-indexed sequence} \tag{13.37}$$

$$k = 0, \ldots, N/2 - 1$$

We can also write (13.37) in the form

$$F(k) = F_1(k) + W_N^k F_2(k) \qquad k = 0, 1, \ldots, N/2 - 1$$

$$F\left(k + \frac{N}{2}\right) = F_1(k) + W_N^{k+N/2}F_2(k) \tag{13.38}$$

$$= F_1(k) - W_N^k F_2(k) \qquad k = 0, 1, \ldots, N/2 - 1$$

where $W_N^{k+N/2} = -W_N^k$ and $W_{N/2}^{m(k+N/2)} = W_{N/2}^{mk}$. Since the DFT is periodic, $F_1(k) = F_1(k+N/2)$ and $F_2(k) = F_2(k + N/2)$.

We next apply the same procedure to each $N/2$ samples, where $f_{11}(m) = f_1(2m)$ and $f_{21}(m) = f_2(2m + 1)$, $m = 0, 1, \ldots, (N/4) - 1$. Hence,

$$F_1(k) = \sum_{m=0}^{(N/4)-1} f_{11}(m)W_{N/4}^{mk} + W_N^{2k} \sum_{m=0}^{(N/4)-1} f_{21}(m)W_{N/4}^{mk} \tag{13.39}$$

$$k = 0, 1, \ldots, N/4 - 1$$

or

$$F_1(k) = F_{11}(k) + W_N^{2k}F_{21}(k)$$

$$F_1\left(k + \frac{N}{4}\right) = F_{11}(k) - W_N^{2k}F_{21}(k) \qquad k = 0, 1, \ldots, N/4 - 1 \tag{13.40}$$

Therefore, each one of the sequences f_1 and f_2 has been split into two DFTs of length $N/4$.

Example 2

To find the FFT of the sequence {2, 3, 4, 5} we first bit reverse the position of the elements from their priority {00, 01, 10, 11} to {00, 10, 01, 11} position. The new sequence is {2, 4, 3, 5} (see also Fig. 13.1). Using (13.37) and (13.38) we obtain

$$F_1(0) = \sum_{m=0}^{1} f_1(m)W_2^{m0} \quad = f_1(0)W_2^0 + f_1(1)W_2^0 \quad = f(0) \cdot 1 + f(2) \cdot 1$$

$$F_1(1) = \sum_{m=0}^{1} f_1(m)W_2^{m\cdot1} \quad = f_1(0)W_2^{0\cdot1} + f_1(1)W_2^1 \quad = f(0) + f(2)(-j)$$

$$F_2(0) = W_4^0 \sum_{m=0}^{1} f_2(m)W_2^{m\cdot0} \quad = f_2(0)W_2^0 + f_2(1)W_2^0 \quad = f(1) + f(3)$$

$$F_2(1) = W_4^1 \sum_{m=0}^{1} f_2(m)W_2^{m\cdot1} \quad = W_4^1\Big[f(1)W_2^0 + f(3)W_2^1\Big] \quad = W_4^1 f(1) - W_4^1 f(3)$$

From (13.38) the output is

$$F(0) = F_1(0) + W_4^0 F_2(0)$$

$$F(1) = F_1(1) + W_4^1 F_2(1)$$

$$F(2) = F_1(0) - W_4^0 F_2(0)$$

$$F(3) = F_1(1) - W_4^1 F_2(1)$$

Computation of the Inverse DFT

To find the inverse FFT using an FFT algorithm, we use the relation

$$f(n) = \frac{[FFT(F^*(k))]^*}{N} \tag{13.41}$$

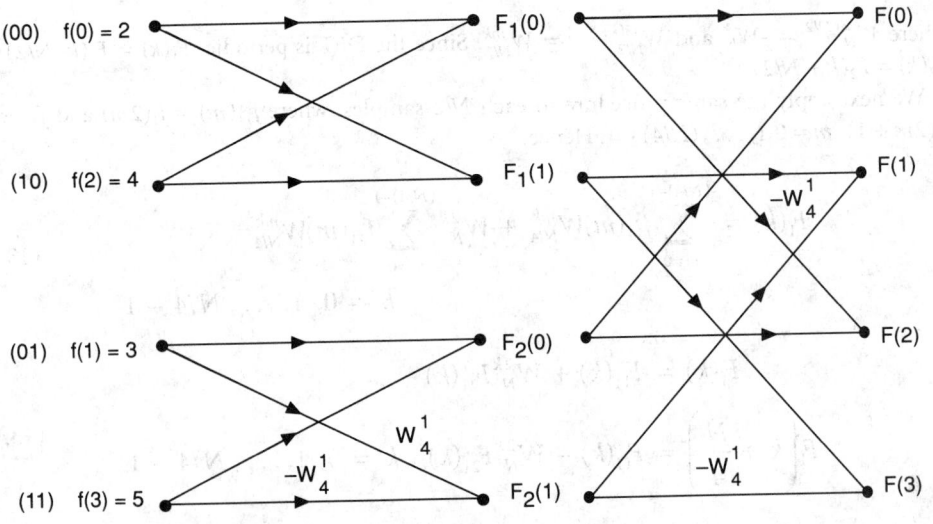

FIGURE 13.1   Illustration of Example 2.

Table 13.2 FFT Subroutine

```
SUBROUTINE FOUR1 (DATA, NN, ISIGN)
        Replaces DATA by its discrete Fourier transform, if SIGN is input as 1; or replaces DATA by NN
        times its inverse discrete Fourier transform, if ISIGN is input as −1. DATA is a complex array
        of length NN or, equivalently, a real array of length 2*NN. NN must be an integer power of 2.
REAL*8 WR, WI, WPR, WPI, WTEMP, THETA          Double precision for the trigonometric recurrences.
DIMENSION DATA (2*NN)
N=2*NN
J=1
DO 11 I=1, N, 2
        IF (J.GT.I) THEN                        This is the bit-reversal section of the routine.
                TEMPR=DATA(J)                   Exchange the two complex numbers.
                TEMPI=DATA(J+1)
                DATA(J)=DATA(I)
                DATA(J+1)=DATA(I+1)
                DATA(I)=TEMPR
                DATA(I+1)=TEMPI
        ENDIF
        M=N/2
1       IF ((M.GE.2).AND. (J.GT.M)) THEN
                J=J-M
                M=M/2
        GO TO 1
        ENDIF
        J=J+M
11   CONTINUE
     MMAX=2                                     Here begins the Danielson-Lanczos section of the routine.
2        IF (N.GT.MMAX) THEN                    Outer loop executed log₂ NN times.
             ISTEP=2*MMAX
             THETA=6.28318530717959D0/(ISIGN*MMAX)   Initialize for the trigonometric recurrence.
             WPR=-2.D0*DSIN(0.5D0*THETA)**2
             WPI=DSIN(THETA)
             WR=1.D0
             WI=0.D0
             DO 13 M=1,MMAX,2
                 DO 12 I=M,N,ISTEP             Here are the two nested inner loops.
                 J=I+MMAX                       This is the Danielson-Lanczos formula:
                 TEMPR=SNGL(WR)*DATA(J)-SNGL(WI)*DATA(J+1)
                 TEMPI=SNGL(WR)*DATA(J+1)+SNGL(WI)*DATA(J)
                 DATA(J)=DATA(I)-TEMPR
                 DATA(J+1)=DATA(I+1)-TEMPI
                 DATA(I)=DATA(I)+TEMPR
                 DATA(I+1)=DATA(I+1)+TEMPI
12           CONTINUE
             WTEMP=WR
             WR=WR*WPR-WI*WPI+WR                Trigonometric recurrence.
             WI=WI*WPR+WTEMP*WPI+WI
13       CONTINUE
         MMAX=STEP
         GO TO 2
     ENDIF
     RETURN
     END
```

Source: W. H. Press, B. P. Flannery, S. A. Teukolsky, and W. T. Vetterling, *Numerical Recipes,* New York: Cambridge University Press, 1986. With permission.

For other transforms and their fast algorithms the reader should consult the references given at the end of this section.

Table 13.2 gives the FFT subroutine for fast implementation of the DFT of a finite sequence.

Defining Terms

FFT: A computational technique that reduces the number of mathematical operations in the evaluation of the discrete Fourier transform (DFT) to $N \log_2 N$.

Phase spectrum: All phases associated with the spectrum harmonics.

Power spectrum: A power contributed by each harmonic of the spectrum.

Window: Any appropriate function that multiplies the data with the intent to minimize the distortions of the Fourier spectra.

References

A. Ahmed and K. R. Rao, *Orthogonal Transforms for Digital Signal Processing*, New York: Springer-Verlag, 1975.

E. R. Blahut, *Fast Algorithms for Digital Signal Processing*, Reading, Mass.: Addison-Wesley, 1987.

E. O. Bringham, *The Fast Fourier Transform*, Englewood Cliffs, N.J.: Prentice-Hall, 1974.

F. D. Elliot, *Fast Transforms, Algorithms, Analysis, Applications*, New York: Academic Press, 1982.

H. J. Nussbaumer, *Fast Fourier Transform and Convolution Algorithms*, New York: Springer-Verlag, 1982.

A. D. Poularikas and S. Seely, *Signals and System*. Boston: PWS-Kent, 1992.

Further Information

A historical overview of the fast Fourier transform can be found in J. W. Cooley, P. A. W. Lewis, and P. D. Welch, "Historical notes on the fast Fourier transform," *IEEE Trans. Audio Electroacoust.*, vol. AV-15, pp. 76–79, June 1967.

Fast algorithms appear frequently in the monthly magazine *Signal Processing*, published by The Institute of Electrical and Electronics Engineers.

13.2 Design and Implementation of Digital Filters

Bruce W. Bomar and L. Montgomery Smith

A *digital filter* is a linear, shift-invariant system for computing a **discrete output sequence** from a **discrete input sequence**. The input/output relationship is defined by the *convolution sum*

$$y(n) = \sum_{m=-\infty}^{\infty} h(m)x(n-m)$$

where $x(n)$ is the input sequence, $y(n)$ is the output sequence, and $h(n)$ is the *impulse response* of the filter. The filter is often conveniently described in terms of its frequency characteristics that are given by the *transfer function* $H(e^{j\omega})$. The impulse response and transfer function are a Fourier transform pair:

$$H(e^{j\omega}) = \sum_{n=-\infty}^{\infty} h(n)e^{-j\omega n} \qquad -\pi \leq \omega \leq \pi$$

$$h(n) = \frac{1}{2\pi} \int_{-\pi}^{\pi} H(e^{j\omega})e^{j\omega n} \, d\omega \qquad -\infty \leq n \leq \infty$$

Closely related to the Fourier transform of $h(n)$ is the *z*-transform defined by

$$H(z) = \sum_{n=-\infty}^{\infty} h(n)z^{-n}$$

The Fourier transform is then the *z*-transform evaluated on the unit circle in the *z*-plane ($z = e^{j\omega}$). An important property of the *z*-transform is that $z^{-1} H(z)$ corresponds to $h(n-1)$, so z^{-1} represents a one-sample delay, termed a *unit delay*.

In this section, attention will be restricted to *frequency-selective* filters. These filters are intended to pass frequency components of the input sequence in a given band of the spectrum while blocking the rest. Typical frequency-selective filter types are *low-pass, high-pass, bandpass,* and *band-reject*. Other special-purpose filters exist, but their design is an advanced topic that will not be addressed here. In addition, special attention is given to *causal* filters, that is, those for which the impulse response is identically zero for negative n and thus can be realized in real time. Digital filters are further separated into two classes depending on whether the impulse response contains a finite or infinite number of nonzero terms.

Finite Impulse Response Filter Design

The objective of **finite impulse response (FIR) filter** design is to determine $N + 1$ coefficients

$$h(0), h(1), \dots, h(N)$$

so that the transfer function $H(e^{j\omega})$ approximates a desired frequency characteristic $H_d(e^{j\omega})$. All other impulse response coefficients are zero. An important property of FIR filters for practical applications is that they can be designed to be *linear phase*; that is, the transfer function has the form

$$H(e^{j\omega}) = A(e^{j\omega})e^{-j\omega N/2}$$

where the amplitude $A(e^{j\omega})$ is a real function of frequency. The desired transfer function can be similarly written

$$H_d(e^{j\omega}) = A_d(e^{j\omega})e^{-j\omega N/2}$$

where $A_d(e^{j\omega})$ describes the amplitude of the desired frequency-selective characteristics. For example, the amplitude frequency characteristics of an ideal low-pass filter are given by

$$A_d(e^{j\omega}) = \begin{cases} 1 & \text{for } |\omega| \leq \omega_c \\ 0 & \text{otherwise} \end{cases}$$

where ω_c is the *cutoff frequency* of the filter.

A linear phase characteristic ensures that a filter has a constant group delay independent of frequency. Thus, all frequency components in the signal are delayed by the same amount, and the only signal distortion introduced is that imposed by the filter's frequency-selective characteristics. Since a FIR filter can only approximate a desired frequency-selective characteristic, some measures of the accuracy of approximation are needed to describe the quality of the design. These are the *passband ripple δ_p*, the *stopband attenuation δ_s*, and the *transition bandwidth $\Delta\omega$*. These quantities are illustrated in Fig. 13.2 for a prototype low-pass filter. The passband ripple gives the maximum deviation from the desired amplitude (typically unity) in the region where the input signal spectral components are desired to be passed unattenuated. The stopband attenuation gives the maximum deviation from zero in the region where the input signal spectral components are desired to be blocked. The transition bandwidth gives the width of the spectral region in which the frequency characteristics of the transfer function change from the passband to the stopband values. Often, the passband ripple and stopband attenuation are specified in decibels, in which case their values are related to the quantities δ_p and δ_s by

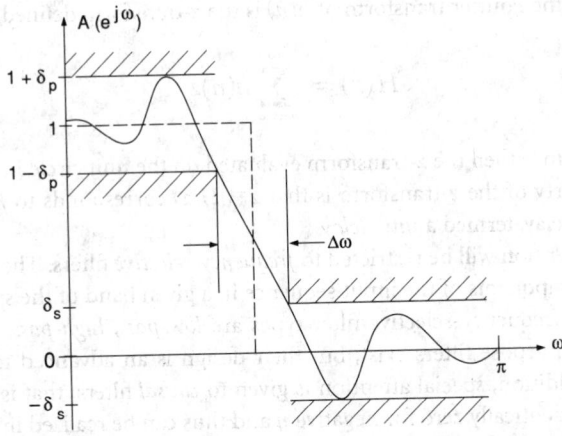

FIGURE 13.2 Amplitude frequency characteristics of a FIR low-pass filter showing definitions of passband ripple δ_p, stopband attenuation δ_s, and transition bandwidth $\Delta\omega$.

$$\text{Passband ripple (dB)} = P = -20 \log_{10} (1 - \delta_p)$$

$$\text{Stopband attenuation (dB)} = S = -20 \log_{10} \delta_s$$

FIR Filter Design by Windowing

The windowing design method is a computationally efficient technique for producing nonoptimal filters. Filters designed in this manner have equal passband ripple and stopband attenuation:

$$\delta_p = \delta_s = \delta$$

The method begins by finding the impulse response of the desired filter from

$$h_d(n) = \frac{1}{2\pi} \int_{-\pi}^{\pi} A_d(e^{j\omega}) e^{j\omega (n - N/2)} d\omega$$

For ideal low-pass, high-pass, bandpass, and band-reject frequency-selective filters, the integral can be solved in closed form. The impulse response of the filter is then found by multiplying this ideal impulse response with a window $w(n)$ that is identically zero for $n < 0$ and for $n > N$:

$$h(n) = h_d(n)w(n) \quad n = 0, 1, \ldots, N$$

Some commonly used windows are defined as follows:

1. Rectangular (truncation)

$$w(n) = \begin{cases} 1 & \text{for } 0 \le n \le N \\ 0 & \text{otherwise} \end{cases}$$

2. Hamming

$$w(n) = \begin{cases} 0.54 - 0.46 \cos \dfrac{2\pi n}{N} & \text{for } 0 \le n \le N \\ 0 & \text{otherwise} \end{cases}$$

3. Kaiser

$$w(n) = \begin{cases} \dfrac{I_0\left(\beta\sqrt{1 - [(2n - N)/N]^2}\right)}{I_0(\beta)} & \text{for } 0 \leq n \leq N \\[3mm] 0 & \text{otherwise} \end{cases}$$

In general, windows that slowly taper the impulse response to zero result in lower passband ripple and a wider transition bandwidth. Other windows (e.g., Hamming, Blackman) are also sometimes used but not as often as those shown above.

Of particular note is the Kaiser window where $I_0(.)$ is the 0th-order modified Bessel function of the first kind and β is a shape parameter. The proper choice of N and β allows the designer to meet given passband ripple/stopband attenuation and transition bandwidth specifications. Specifically, using S, the stopband attenuation in dB, the filter order must satisfy

$$N = \frac{S - 8}{2.285\Delta\omega}$$

Then, the required value of the shape parameter is given by

$$\beta = \begin{cases} 0 & \text{for } S < 21 \\ 0.5842(S - 21)^{0.4} + 0.07886(S - 21) & \text{for } 21 \leq S \leq 50 \\ 0.1102(S - 8.7) & \text{for } S > 50 \end{cases}$$

As an example of this design technique, consider a low-pass filter with a cutoff frequency of $\omega_c = 0.4\pi$. The ideal impulse response for this filter is given by

$$h_d(n) = \frac{\sin[0.4\pi(n - N/2)]}{\pi(n - N/2)}$$

Choosing $N = 8$ and a Kaiser window with a shape parameter of $\beta = 0.5$ yields the following impulse response coefficients:

$$h(0) = h(8) = -0.07568267$$
$$h(1) = h(7) = -0.06236596$$
$$h(2) = h(6) = 0.09354892$$
$$h(3) = h(5) = 0.30273070$$
$$h(4) = 0.40000000$$

Design of Optimal FIR Filters

The accepted standard criterion for the design of optimal FIR filters is to minimize the maximum value of the error function

$$E(e^{j\omega}) = W_d(e^{j\omega})|A_d(e^{j\omega}) - A(e^{j\omega})|$$

over the full range of $-\pi \leq \omega \leq \pi$. $W_d(e^{j\omega})$ is a desired weighting function used to emphasize specifications in a given frequency band. The ratio of the deviation in any two bands is inversely proportional to the ratio of their respective weighting.

A consequence of this optimization criterion is that the frequency characteristics of optimal filters are *equiripple:* although the maximum deviation from the desired characteristic is minimized, it is reached several times in each band. Thus, the passband and stopband deviations oscillate about the desired values with equal amplitude in each band. Such approximations are frequently referred to as *minimax* or *Chebyshev* approximations. In contrast, the maximum deviations occur near the band edges for filters designed by windowing.

Equiripple FIR filters are usually designed using the *Parks-McClellan* computer program [Parks and Burrus, 1987], which uses the *Remez exchange algorithm* to determine iteratively the *extremal frequencies* at which the maximum deviations in the error function occur. A listing of this program along with a detailed description of its use is available in several references including Parks and Burrus [1987] and DSP Committee [1979]. The program is executed by specifying as inputs the desired band edges, gain for each band (usually 0 or 1), band weighting, and FIR length. If the resulting filter has too much ripple in some bands, those bands can be weighted more heavily and the filter redesigned. Details on this design procedure are discussed in Rabiner [1973], along with approximate design relationships which aid in selecting the filter length needed to meet a given set of specifications.

Although we have focused attention on the design of frequency-selective filters, other types of FIR filters exist. For example, the Parks-McClellan program will also design linear-phase FIR filters for differentiating broadband signals and for approximating the Hilbert transform of such signals. A simple modification to this program permits arbitrary magnitude responses to be approximated with linear-phase filters. Other design techniques are available that permit the design of FIR filters which approximate an arbitrary complex response [Parks and Burrus, 1987; Chen and Parks, 1987], and, in cases where a nonlinear phase response is acceptable, design techniques are available that give a shorter impulse response length than would be required by a linear-phase design [Goldberg *et al.*, 1981].

As an example of an equiripple filter design, an 8th-order low-pass filter with a passband $0 \leq \omega \leq 0.3\pi$, a stopband $0.5\pi \leq \omega \leq \pi$, and equal weighting for each band was designed. The impulse response coefficients generated by the Parks-McClellan program were as follows:

$$h(0) = h(8) = -0.06367859$$
$$h(1) = h(7) = -0.06912276$$
$$h(2) = h(6) = 0.10104360$$
$$h(3) = h(5) = 0.28574990$$
$$h(4) = 0.41073000$$

These values can be compared to those for the similarly specified filter designed in the previous subsection using the windowing method.

Infinite Impulse Response Filter Design

An **infinite impulse response (IIR) digital filter** requires less computation to implement than a FIR digital filter with a corresponding frequency response. However, IIR filters cannot generally achieve a perfect linear-phase response and are more susceptible to **finite wordlength effects.**

Techniques for the design of IIR analog filters are well established. For this reason, the most important class of IIR digital filter design techniques is based on forcing a digital filter to behave like a reference analog filter. This can be done in several different ways. For example, if the analog filter impulse response is $h_a(t)$ and the digital filter impulse response is $h(n)$, then it is possible to make $h(n) = h_a(nT)$, where T is the sample spacing of the digital filter. Such designs are referred to as *impulse-invariant* [Parks and Burrus, 1987]. Likewise, if $g_a(t)$ is the unit step response of the analog filter and $g(n)$ is the unit step response of the digital filter, it is possible to make $g(n) = g_a(nT)$, which gives a *step-invariant* design [Parks and Burrus, 1987].

The step-invariant and impulse-invariant techniques perform a time domain matching of the analog and digital filters but can produce aliasing in the frequency domain. For frequency-selective filters it is better to attempt matching frequency responses. This task is complicated by the fact that the analog filter response is defined for an infinite range of frequencies ($\Omega = 0$ to ∞), while the digital filter response is defined for a finite range of frequencies ($\omega = 0$ to π). Therefore, a method for mapping the infinite range of analog frequencies Ω into the finite range from $\omega = 0$ to π, termed the *bilinear transform,* is employed.

Bilinear Transform Design of IIR Filters

Let $H_a(s)$ be the Laplace transform transfer function of an analog filter with frequency response $H_a(j\Omega)$. The bilinear transform method obtains the digital filter transfer function $H(z)$ from $H_a(s)$ using the substitution

$$s = \frac{2(1 - z^{-1})}{T(1 + z^{-1})}$$

That is,

$$H(z) = H_a(s)\Big|s = \frac{2}{T} \frac{1 - z^{-1}}{1 + z^{-1}}$$

This maps analog frequency Ω to digital frequency ω according to

$$\omega = 2 \tan^{-1} \frac{\Omega T}{2}$$

thereby warping the frequency response $H_a(j\Omega)$ and forcing it to lie between 0 and π for $H(e^{j\omega})$. Therefore, to obtain a digital filter with a cutoff frequency of ω_c it is necessary to design an analog filter with cutoff frequency

$$\Omega_c = \frac{2}{T} \tan \frac{\omega_c}{2}$$

This process is referred to as *prewarping* the analog filter frequency response to compensate for the warping of the bilinear transform. Applying the bilinear transform substitution to this analog filter will then give a digital filter that has the desired cutoff frequency.

Analog filters and hence IIR digital filters are typically specified in a slightly different fashion than FIR filters. Figure 13.3 illustrates how analog and IIR digital filters are usually specified. Notice by comparison to Fig. 13.2 that the passband ripple in this case never goes above unity, whereas in the FIR case the passband ripple is specified about unity.

Four basic types of analog filters are generally used to design digital filters: (1) Butterworth filters that are maximally flat in the passband and decrease monotonically outside the passband, (2) Chebyshev filters that are equiripple in the passband and decrease monotonically outside the passband, (3) inverse Chebyshev filters that are flat in the passband and equiripple in the stopband, and (4) elliptic filters that are equiripple in both the passband and stopband. Techniques for designing these analog filters are covered elsewhere [see, for example, Van Valkenberg, 1982] and will not be considered here.

To illustrate the design of an IIR digital filter using the bilinear transform, consider the design of a second-order Chebyshev low-pass filter with 0.5 dB of passband ripple and a cutoff frequency of $\omega_c = 0.4\,\pi$. The sample rate of the digital filter is to be 5 Hz, giving $T = 0.2$ s. To design this filter we first design an analog Chebyshev low-pass filter with a cutoff frequency of

$$\Omega_c = \frac{2}{0.2} \tan 0.2\pi = 7.2654 \text{ rad/s}$$

This filter has a transfer function

$$H(s) = \frac{0.9441}{1 + 0.1249s + 0.01249s^2}$$

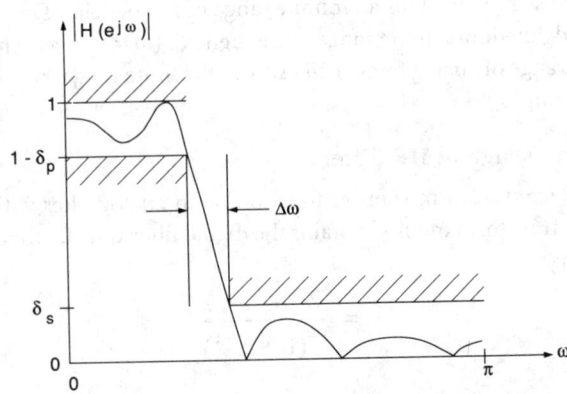

FIGURE 13.3 Frequency characteristics of an IIR digital low-pass filter showing definitions of passband ripple δ_p, stopband attenuation δ_s, and transition bandwith $\Delta\omega$.

Substituting

$$s = \frac{2}{0.2}\, \frac{z-1}{z+1}$$

gives

$$H(z) = \frac{0.2665(z+1)^2}{z^2 - 0.1406z + 0.2695}$$

Computer programs are available that accept specifications on a digital filter and carry out all steps required to design the filter, including prewarping frequencies, designing the analog filter, and performing the bilinear transform. Two such programs are given in the references [Parks and Burrus, 1987; Antoniou, 1979].

Design of Other IIR Filters

For frequency-selective filters, the bilinear transformation of an elliptic analog filter provides an optimal equiripple design. However, if a design other than standard low-pass, high-pass, bandpass, or bandstop is needed or if it is desired to approximate an arbitrary magnitude or group delay characteristic, some other design technique is needed. Unlike the FIR case, there is no standard IIR design program for obtaining optimal approximations to an arbitrary response.

Four techniques that have been used for designing optimal equiripple IIR digital filters are [Parks and Burrus, 1987] (1) minimizing the L_p norm of the weighted difference between the desired and actual responses, (2) linear programming, (3) iteratively using the Remez exchange algorithm on the numerator and denominator of the transfer function, and (4) the differential correction algorithm. A computer program for implementing the first method is available in DSP Committee [1979].

Finite Impulse Response Filter Implementation

For FIR filters, the convolution sum represents a computable process, and so filters can be implemented by directly programming the arithmetic operations. Nevertheless, some options are available that may be preferable for a given processor architecture, and means for reducing computational loads exist. This section outlines some of these methods and presents schemes for FIR filter realization.

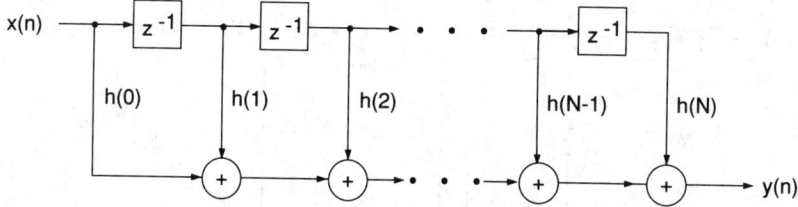

FIGURE 13.4   A direct-form implementation of a FIR filter.

Direct Convolution Methods

The most obvious method for the implementation of FIR filters is to directly evaluate the sum of products in the convolution sum:

$$y(n) = h(0)x(n) + h(1)x(n-1) + \cdots + h(N)x(n-N)$$

The block diagram for this is shown in Fig. 13.4. This method involves storing the present and previous N values of the input, multiplying each sample by the corresponding impulse response coefficient, and summing the products to compute the output. This method is referred to as a *tapped delay line* structure.

A modification to this approach is suggested by writing the convolution as

$$y(n) = h(0)x(n) + \sum_{m=1}^{N} h(m)x(n-m)$$

In this approach, the output is computed by adding the product of $h(0)$ with the present input sample to a previously computed sum of products and updating a set of N sums of products with the present input sample value. The signal flow graph for this method is shown in Fig. 13.5.

FIR filters designed to have linear phase are usually obtained by enforcing the symmetry constraint

$$h(n) = h(N-n)$$

For these filters, the convolution sum can be written

$$y(n) = \begin{cases} \displaystyle\sum_{m=0}^{N/2-1} h(m)[x(n-m) + x(n+m-N)] + h\left(\frac{N}{2}\right)x\left(n-\frac{N}{2}\right) & N \text{ even} \\[4mm] \displaystyle\sum_{m=0}^{(N-1)/2} h(m)[x(n-m) + x(n+m-N)] & N \text{ odd} \end{cases}$$

Implementation of the filter according to these formulas reduces the number of multiplications by approximately a factor of 2 over direct-form methods. The block diagrams for these filter structures are shown in Figs. 13.6 and 13.7.

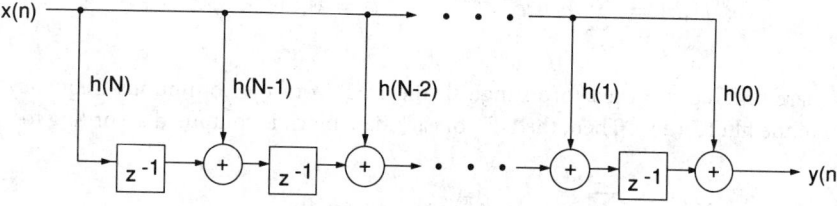

FIGURE 13.5   Another direct-form implementation of a FIR filter.

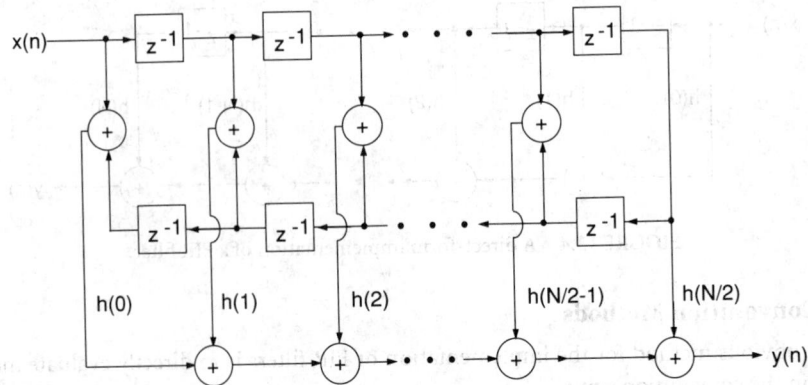

FIGURE 13.6   Implementation of a linear-phase FIR filter for even N.

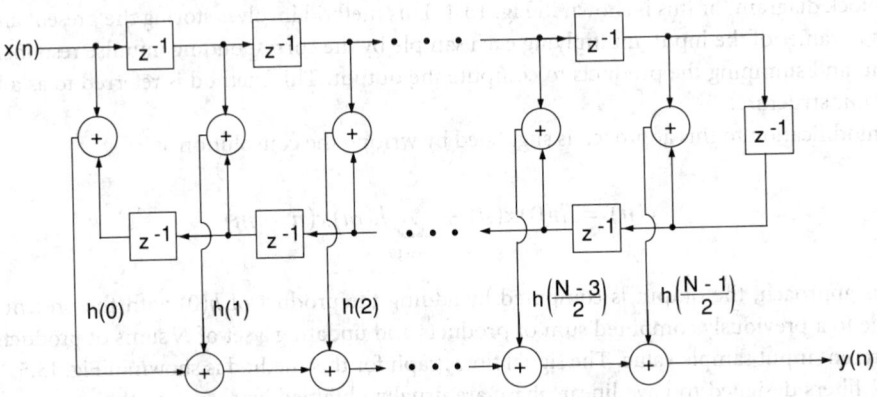

FIGURE 13.7   Implementation of a linear-phase FIR filter for odd N.

Implementation of FIR Filters Using the Discrete Fourier Transform

A method for implementing FIR filters that can have computational advantages over direct-form convolution involves processing the input data in blocks using the discrete Fourier transform (DFT) via the *overlap-save* method. The computational advantage arises primarily from use of the fast Fourier transform (FFT) algorithm (discussed in Section 13.1) to compute the DFTs of the individual data blocks. In this method, the input data sequence $\{x(n); -\infty < n < \infty\}$ is divided into L-point blocks

$$x_i(n) \quad 0 \leq n \leq L - 1 \quad -\infty < i < \infty$$

where $L > N + 1$, the length of the FIR filter. The L-point DFT of the impulse response is precomputed from

$$H[k] = \sum_{n=0}^{L-1} h(n)e^{-j2\pi kn/L} \quad k = 0, 1, \ldots, L - 1$$

where square brackets are used to distinguish the DFT from the continuous-frequency transfer function of the filter $H(e^{j\omega})$. Then, the DFT of each data block is computed according to

$$X_i[k] = \sum_{n=0}^{L-1} x_i(n)e^{-j2\pi kn/L} \quad k = 0, 1, \ldots, L - 1$$

These two complex sequences are multiplied together term by term to form the DFT of the output data block:

$$Y_i[k] = H[k]X_i[k] \quad k = 0, 1, \ldots, L-1$$

and the output data block is computed by the inverse DFT:

$$y_i(n) = \frac{1}{L} \sum_{k=0}^{L-1} Y_i[k]e^{j2\pi kn/L} \quad n = 0, 1, \ldots, L-1$$

However, the output data block computed in this manner is the *circular convolution* of the impulse response of the filter and the input data block given by

$$y_i(n) = \sum_{m=0}^{N} h(m)x_i((n-m) \text{ modulo } L)$$

Thus, only the output samples from $n = N$ to $n = L-1$ are the same as those that would result from the convolution of the impulse response with the infinite-length data sequence $x(n)$. The first N data points are corrupted and must therefore be discarded. So that the output data sequence does not have N-point "gaps" in it, it is therefore necessary to *overlap* the data in adjacent input data blocks. In carrying out the processing, samples from block to block are *saved* so that the last N points of the ith data block $x_i(n)$ are the same as the first N points of the following data block $x_{i+1}(n)$. Each processed L-point data block thus produces $L - N$ output samples.

Another technique of block processing of data using DFTs is the *overlap-add* method in which $L - N$-point blocks of input data are zero-padded to L points, the resulting output blocks are overlapped by N points, and corresponding samples added together. This method requires more computation than the overlap-save method and is somewhat more difficult to program. Therefore, its usage is not as widespread as the overlap-save method.

Infinite Impulse Response Filter Implementation

Direct-Form Realizations

For an IIR filter the convolution sum does not represent a computable process. Therefore, it is necessary to examine the general transfer function, which is given by

$$H(z) = \frac{Y(z)}{X(z)} = \frac{\gamma_0 + \gamma_1 z^{-1} + \gamma_2 z^{-2} + \cdots + \gamma_M z^{-M}}{1 + \beta_1 z^{-1} + \beta_2 z^{-2} + \cdots + \beta_N z^{-N}}$$

where $Y(z)$ is the z-transform of the filter output $y(n)$ and $X(z)$ is the z-transform of the filter input $x(n)$. The unit-delay characteristic of z^{-1} then gives the following *difference equation* for implementing the filter:

$$y(n) = \gamma_0 x(n) + \gamma_1 x(n-1) + \cdots + \gamma_M x(n-M) - \beta_1 y(n-1) - \cdots - \beta_N y(n-N)$$

When calculating $y(0)$, the values of $y(-1)$, $y(-2)$, $\ldots$, $y(-N)$ represent initial conditions on the filter. If the filter is started in an initially relaxed state, then these initial conditions are zero.

Figure 13.8 gives a block diagram realizing the filter's difference equation. This structure is referred to as the *direct-form I* realization. Notice that this block diagram can be separated into two parts, giving two cascaded networks, one of which realizes the filter zeros and the other the filter poles. The order of these networks can be reversed without changing the transfer function. This results in a structure where the two strings of delays are storing the same values, so a single string of delays of length *max(M, N)* is sufficient, as shown in Fig. 13.9. The realization of Fig. 13.9 requires the minimum number of z^{-1} delay operations and is referred to as the *direct-form II* realization.

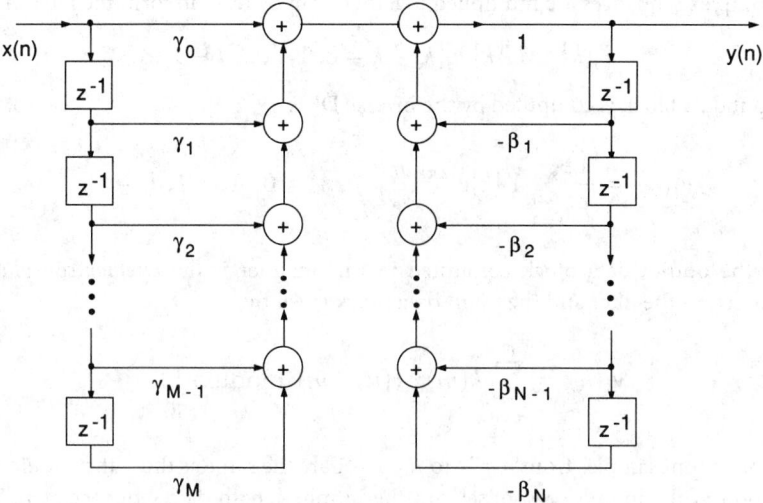

FIGURE 13.8   Direct-form I realization.

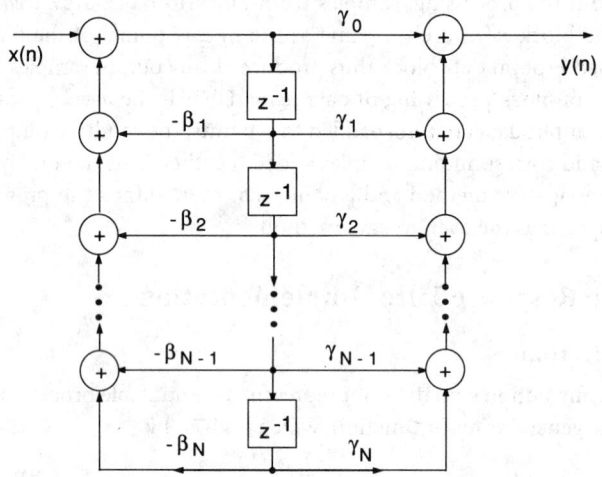

FIGURE 13.9   Direct-form II realization.

Cascade and Parallel Realizations

The transfer function of an IIR filter can always be factored into the product of second-order transfer functions as

$$H(z) = C\prod_{k=1}^{K} \frac{1 + a_{1k}z^{-1} + a_{2k}z^{-2}}{1 + b_{1k}z^{-1} + b_{2k}z^{-2}} = C\prod_{k=1}^{K} H_k(z)$$

where we have assumed $M = N$ in the original transfer function and where K is the largest integer contained in $(N + 1)/2$. If N is odd, the values of a_{2k} and b_{2k} in one term are zero. The realization corresponding to this transfer function factorization is shown in Fig. 13.10. Each second-order $H_k(z)$ term in this realization is referred to as a *biquad*. The digital filter design programs in Parks and Burrus [1987] and Antoniou [1979] give the filter transfer function in factored form.

If the transfer function of an IIR filter is written as a partial-fraction expansion and first-order sections with complex-conjugate poles are combined, $H(z)$ can be expressed in the form

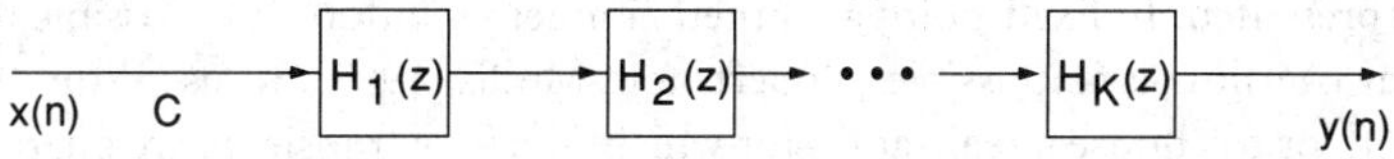

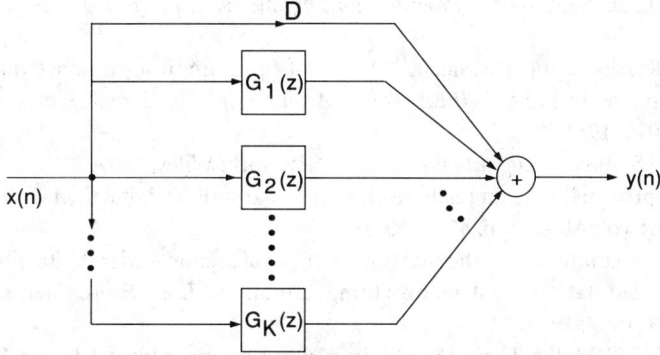

$$x(n) \quad C \quad \boxed{H_1(z)} \longrightarrow \boxed{H_2(z)} \longrightarrow \cdots \longrightarrow \boxed{H_K(z)} \longrightarrow \atop y(n)$$

FIGURE 13.10   Cascade realization of an IIR filter.

$$H(z) = D + \sum_{k=1}^{K} \frac{\alpha_{0k} + \alpha_{1k}z^{-1}}{1 + b_{1k}z^{-1} + b_{2k}z^{-2}} = D + \sum_{k=1}^{K} G_k(z)$$

This results in the parallel realization of Fig. 13.11.

Finite Wordlength Effects in IIR Filters

Since practical digital filters must be implemented with limited-precision arithmetic, four types of finite wordlength effects result: (1) roundoff noise, (2) coefficient quantization error, (3) overflow oscillations, and (4) limit cycles. *Round-off noise* is that error in the filter output which results from rounding (or truncating) calculations within the filter. This error appears as low-level noise at the filter output. *Coefficient quantization error* refers to the deviation of a practical filter's frequency response from the ideal due to the filter's coefficients being represented with finite precision. The term *overflow oscillation*, sometimes also referred to as *adder overflow limit cycle*, refers to a high-level oscillation that can exist in an otherwise stable filter due to the nonlinearity associated with the overflow of internal filter calculations. A *limit cycle*, sometimes referred to as a *multiplier round-off limit cycle*, is a low-level oscillation that can exist in an otherwise stable filter as a result of the nonlinearity associated with rounding (or truncating) internal filter calculations. Overflow oscillations and limit cycles require recursion to exist and do not occur in nonrecursive FIR filters.

The direct-form I and direct-form II IIR filter realizations generally have very poor performance in terms of all finite wordlength effects. Therefore, alternative realizations are usually employed. The most common alternatives are the cascade and parallel realizations where the direct-form II realization is used for each second-order section. By simply factoring or expanding the original transfer function, round-off noise and coefficient quantization error are significantly reduced. A further improvement is possible by implementing the cascade or parallel sections using *state-space* realizations [Roberts and Mullis, 1987]. The price paid for using state-space realizations is an increase in the computation required to implement each section. Another realization that has been used to reduce round-off noise and coefficient quantization error is the *lattice realization* [Parks and Burrus, 1987] which is usually formed directly from the unfactored and unexpanded transfer function.

Overflow oscillations can be prevented in several different ways. One technique is to employ floating-point arithmetic that renders overflow virtually impossible due to the large dynamic range

FIGURE 13.11   Parallel realization of an IIR filter.

which can be represented. In fixed-point arithmetic implementations it is possible to *scale* the calculations so that overflow is impossible [Roberts and Mullis, 1987], to use saturation arithmetic [Ritzerfeld, 1989], or to choose a realization for which overflow transients are guaranteed to decay to zero [Roberts and Mullis, 1987].

Limit cycles can exist in both fixed-point and floating-point digital filter implementations. Many techniques have been proposed for testing a realization for limit cycles and for bounding their amplitude when they do exist. In fixed-point realizations it is possible to prevent limit cycles by choosing a state-space realization for which any internal transient is guaranteed to decay to zero and then using magnitude truncation of internal calculations in place of rounding [Diniz and Antoniou, 1986].

Defining Terms

Discrete sequence: A set of values constituting a signal whose values are known only at distinct sampled points. Also called a digital signal.

Filter design: The process of determining the coefficients of a difference equation to meet a given frequency or time response characteristic.

Filter implementation: The numerical method or algorithm by which the output sequence is computed from the input sequence.

Finite impulse response (FIR) filter: A filter whose output in response to a unit impulse function is identically zero after a given bounded number of samples. A FIR filter is defined by a linear constant-coefficient difference equation in which the output depends only on the present and previous sample values of the input.

Finite wordlength effects: Perturbations of a digital filter output due to the use of finite precision arithmetic in implementing the filter calculations. Also called quantization effects.

Infinite impulse response (IIR) filter: A filter whose output in response to a unit impulse function remains nonzero for indefinitely many samples. An IIR filter is defined by a linear constant-coefficient difference equation in which the output depends on the present and previous samples of the input and on previously computed samples of the output.

References

A. Antoniou, *Digital Filters: Analysis and Design*, New York: McGraw-Hill, 1979.

X. Chen and T. W. Parks, "Design of FIR filters in the complex domain," *IEEE Trans. Acoust., Speech, Signal Process.*, vol. ASSP-35, pp. 144–153, 1987.

P. S. R. Diniz and A. Antoniou, "More economical state-space digital filter structures which are free of constant-input limit cycles," *IEEE Trans. Acoust., Speech, Signal Process.*, vol. ASSP-34, pp. 807–815, 1986.

DSP Committee, IEEE ASSP (eds.), *Programs for Digital Signal Processing*, New York: IEEE Press, 1979.

E. Goldberg, R. Kurshan, and D. Malah, "Design of finite impulse response digital filters with nonlinear phase response," *IEEE Trans. Acoust., Speech, Signal Process.*, vol. ASSP-29, pp. 1003–1010, 1981.

T. W. Parks and C. S. Burrus, *Digital Filter Design*, New York: Wiley, 1987.

L. R. Rabiner, "Approximate design relationships for low-pass FIR digital filters," *IEEE Trans. Audio Electroacoust.* vol. AU-21, pp. 456–460, 1973.

J. H. F. Ritzerfeld, "A condition for the overflow stability of second-order digital filters that is satisfied by all scaled state-space structures using saturation," *IEEE Trans. Circuits Syst.*, vol. CAS-36, pp. 1049–1057, 1989.

R. A. Roberts and C. T. Mullis, *Digital Signal Processing*, Reading, Mass.: Addison-Wesley, 1987.

M. E. Van Valkenberg, *Analog Filter Design*, New York: Holt, Rinehart, Winston, 1982.

Further Information

The monthly magazine *IEEE Transactions on Circuits and Systems II* routinely publishes articles on the design and implementation of digital filters. Finite wordlength effects are discussed in articles published in the April 1988 issue (pp. 365–374) and in the February 1992 issue (pp. 90–98).

Another journal containing articles on digital filters is the *IEEE Transactions on Signal Processing*. Overflow oscillations and limit cycles are discussed in the August 1978 issue (pp. 334–338).

The bimonthly journal *IEEE Transactions on Instrumentation and Measurement* also contains related information. The use of digital filters for integration and differentiation is discussed in the December 1990 issue (pp. 923–927).

13.3 Signal Restoration

James A. Cadzow

The concept of **signal restoration** has been applied with success to a number of fundamental problems found in interdisciplinary applications. In the typical signal restoration problem, there is given empirical data as symbolically designated by $\mathbf{x}$ that corresponds to corrupted measurements made on an underlying signal being monitored. By using attributes (properties) known to be or hypothesized as being possessed by the monitored signal, it is possible to devise signal restoration algorithms that effectively remove the corruption to give useful approximations of the underlying signal being monitored. Many of the more widely used restoration algorithms take the form of a sequence of *successive projections* as specified by

$$\mathbf{x}_n = P_1 P_2 \ldots P_m \mathbf{x}_{n-1}$$

In this algorithm, P_k designates the projection operator corresponding to the set of signals possessing attribute k while $\mathbf{x}_n$ denotes the enhanced signal at the nth iteration in which the initial point is $\mathbf{x}_o = \mathbf{x}$. In this section, we will give a historical perspective on the evolution of the **method of successive projections** as well as provide implementations of some of the most important projection operators.

Members of the class of signal restoration algorithms that are describable by a successive projection operator are primarily distinguished by the restrictions placed on the **attribute sets**. The earliest versions of these algorithms were highly restrictive and required that the attribute sets be closed subspaces. Unfortunately, this requirement severely limited the types of restoration problems that could be addressed. In recognition of this limitation, subsequent methods of successive projections algorithms eased the requirements on the attribute sets to that of being **closed convex sets** and eventually to the projection operators being closed mappings. This progression of less restrictive requirements significantly expands the types of signal processing problems that are amenable to signal restoration by successive projections.

Introduction

In a typical signal processing application, one is given empirically gathered data that arises from measurements made on a signal(s) characterizing a phenomenon under study. These measurements are invariably corrupted due to imperfections arising from the measurement instrumentation and environmental influences. To recover a reasonable facsimile of the original signal being monitored, it is generally necessary to process the measurement data in a manner that takes into account all information known about the monitored signal, the instrument dynamics, and the nature of the corruption. Although it is generally impossible to obtain a perfect recovery, remarkable approximations can be made in several important applications by employing the concept of *signal restoration* (or *signal recovery*). In signal restoration, *a priori* knowledge concerning the underlying signal's intrinsic nature may be effectively used to strip away the corruption in the measurement data. The philosophy behind this approach is to modify the measurement data to the smallest extent possible so that the modified data possesses the prescribed properties known (or hypothesized) as being possessed by the underlying signal(s). This modification then serves as a cleansing process whereby the measurement corruption is effectively removed.

Metric Space Formulation

To avail ourselves of the extensive analysis tools found in algebra, it is useful to formulate the basic signal restoration problem in a general metric space setting. This has the desirable effect of enabling us to treat a wide variety of applications in a single setting. With this in mind, the measurement signals are taken to lie in a *metric space* which is composed of a set of elements designated by X and a metric $d(\mathbf{x}, \mathbf{y})$ that measures the distance between any two elements $\mathbf{x}, \mathbf{y} \in X$. The elements of the set X are taken to possess a common form such as being composed of all $n \times 1$ real-valued vectors or all complex-valued continuous functions defined on a given interval. Moreover, the distance metric identifying this space must satisfy the axioms associated with a distance measure.[1] We will interchangeably refer to the elements of the metric space as vectors or signals. Depending on the nature of a particular application, the signals can take on such disparate forms as being composed of all real- (or complex-) valued $n \times 1$ vectors, $m \times n$ matrices, infinite-length sequences, continuous-time functions, and so forth.

Example 1. In digital signal processing, the two most commonly employed metric spaces correspond to the set of all real-valued $n \times 1$ vectors as designated by R^n and the set of all real-valued $m \times n$ matrices as designated by $R^{m \times n}$. The elements of a vector contained in R^n typically correspond to samples made of a one-dimensional signal. On the other hand, the elements of a matrix contained in $R^{m \times n}$ might correspond to the brightness levels of the pixels in a rectangular image or the entries of a data matrix formed from samples of a one-dimensional signal. It often happens in engineering applications that the data under analysis is complex valued. To treat such cases, we will have need to consider the set of all complex-valued $n \times 1$ tuples as designated by C^n and the set of all complex-valued $m \times n$ matrices as denoted by $C^{m \times n}$.

To complete the metric space description of spaces R^n and C^n, it is necessary to introduce a distance metric. The most commonly employed distance measure for either space is the Euclidean-induced metric

$$d(\mathbf{x}, \mathbf{y}) = \sqrt{\sum_{k=1}^{n} |y(k) - x(k)|^2} \tag{13.42}$$

In a similar fashion, the *Frobenius norm* distance metric on the spaces $R^{m \times n}$ and $C^{m \times n}$ is commonly used where

$$d(\mathbf{X}, \mathbf{Y}) = \sqrt{\sum_{i=1}^{m} \sum_{j=1}^{n} |Y(i,j) - X(i,j)|^2} \tag{13.43}$$

It is a simple matter to show that either of these measures satisfies the axioms of a distance metric.

Attribute Sets

As indicated previously, the concept of signal restoration is directed towards applications in which a measurement process introduces inevitable distortion in a signal that is being monitored. The

---

[1] The mapping $X \times X \to R$ designated by $d(\mathbf{x}, \mathbf{y})$ is said to be a distance metric if it satisfies the four axioms: (1) $d(\mathbf{x}, \mathbf{y})$ is nonnegative real-valued, (2) $d(\mathbf{x}, \mathbf{y}) = 0$ if and only if $\mathbf{x} = \mathbf{y}$, (3) $d(\mathbf{x}, \mathbf{y}) = d(\mathbf{y}, \mathbf{x})$, and (4) $d(\mathbf{x}, \mathbf{y}) \leq d(\mathbf{x}, \mathbf{z}) + d(\mathbf{z}, \mathbf{y})$.

fundamental philosophy underlying signal restoration is based on the hypothesis that the signal under observation is known or presumed to lie in a restricted *attribute set* $C \subset X$. This attribute set is composed of all signals in X that possess a prescribed set of attributes (or properties). The measurement process, however, results in a measurement signal that is perturbed outside this set. The set C employed in a signal restoration problem is often decomposable as the intersection of a finite number of basic attribute sets that are each describable by elementary attributes (or properties). We will designate these basic attribute sets as

$$C_k = \{\mathbf{x} \in X : \mathbf{x} \text{ has attribute } \mathcal{A}_k\} \qquad \text{for } 1 \le k \le m \tag{13.44}$$

We will hereafter refer to the intersection of these basic attribute subsets as the composite attribute set since its constituent elements satisfy all the required attributes believed to be possessed by the signal being monitored.

$$C = C_1 \cap C_2 \cap \ldots \cap C_m \tag{13.45}$$

The usefulness of signal restoration is critically dependent on one's ability to identify all essential basic sets C_k describing the underlying information signal.

Once the composite attribute set C has been identified, the fundamental signal restoration problem entails finding a signal contained in this set that lies closest to the measurement signal $\mathbf{x}$ in the underlying distance metric sense. This gives rise to the following optimization problem:

$$\min_{\mathbf{y} \in C} d(\mathbf{x}, \mathbf{y}) \tag{13.46}$$

In finding that signal contained in C which lies closest to $\mathbf{x}$, we are in effect seeking the smallest perturbation of the measured signal $\mathbf{x}$ which results in a modified signal that possesses the hypothesized properties of the signal being monitored. Implicit in this approach is the assumption that the measurement process introduces the smallest distortion possible compatible with the measured data. Although this assumption is generally violated in most applications, it is reasonably accurate so that the solution to signal restoration problem (13.46) typically provides for a useful reconstruction of the signal being monitored.

Example 2. To illustrate the concept of signal properties, let us consider the autocorrelation sequence associated with the wide-sense stationary time series $\{x(n)\}$. This two-sided sequence is formally defined by

$$r_{xx}(n) = E\{x(n+m)\,\bar{x}(m)\} \qquad \text{for } n = 0, \pm 1, \pm 2 \ldots \tag{13.47}$$

where E designates the *expected value operator*. It is well known that this autocorrelation sequence satisfies the two attributes of being conjugate symmetric and having a nonnegative Fourier transform. The set of signals associated with these two attributes are then formally given by

$$C_{cs} = \{\text{set of sequences } \{x(n)\} \text{ such that } x(n) = \bar{x}(-n)\}$$

$$C_{nnd} = \{\text{set of sequences } \{x(n)\} \text{ such that } X(\omega) = \sum_{n=-\infty}^{\infty} x(n)e^{-j\omega n} \ge 0\}$$

It can be shown that the set of conjugate-symmetric sequences C_{cs} is a closed vector space. Moreover, the set of nonnegative definite sequences C_{nnd} is a closed convex cone subset of C_{cs}.

Normed Vector Space Formulation

The primary objective of this paper is to present several algorithms which have been proposed for solving optimization problem (13.46) using an iterative approach. These algorithms take on a similar form and are distinguished by the algebraic restrictions placed on the underlying attribute subsets. To provide for a satisfactory mathematical characterization of the restoration problem and its solution, it is essential that we provide an algebraic structure to the underlying metric space. In particular, the ability to add vectors and multiply vectors by scalars provides a useful means for interpreting the measurement signal and more importantly considerably increases the arsenal of analysis tools at our disposal. Fortunately, little loss in generality is incurred by introducing these two algebraic operations since in most signal processing applications of interest there exists an intuitively obvious means for their implementation. For example, if the metric space is taken to be R^n (or C^n), then the sum of any two vectors $\mathbf{x}, \mathbf{y} \in R^n$ (or C^n) has as its kth component $x(k) + y(k)$, while the scalar product $a\mathbf{x}$ has as its kth component $ax(k)$ for $1 \leq k \leq n$.

One of the most important benefits of posing the signal restoration problem in a vector space setting is that of providing a widely invoked model for the measured signal (vector) as specified by

$$\mathbf{x} = \mathbf{s} + \mathbf{w} \tag{13.48}$$

where $\mathbf{s}$ designates the signal being monitored and $\mathbf{w}$ represents measurement error. From our previous discussion, it has been hypothesized that the monitored signal lies in each of the attribute sets so that $\mathbf{s} \in C_k$ for $1 \leq k \leq m$. To effectively recover $\mathbf{s}$ from $\mathbf{x}$ using signal restoration, it is tacitly assumed that the measurement error vector $\mathbf{w}$ has features which are distinctly different from those specified by these m attribute subsets.

The metric needed to measure the distance between any two elements in vector space X often follows in a natural fashion from the basic structure of the vector space. In particular, many of the vector spaces encountered in typical applications have an underlying vector norm measure.[2] A natural choice of a distance metric on a normed vector space is specified by

$$d(\mathbf{x}, \mathbf{y}) = \| \mathbf{x} - \mathbf{y} \| \tag{13.49}$$

This distance metric is said to be induced by the norm defined on the underlying normed vector space X. A solution to the signal restoration problem in a normed vector space setting therefore requires solving the minimization problem

$$\min_{\mathbf{y} \in C} \| \mathbf{x} - \mathbf{y} \| \tag{13.50}$$

Example 3. As indicated previously, the vector spaces R^n and C^n play prominent roles in signal processing applications. The class of l_p norms as defined by

$$\| \mathbf{x} \|_p = \left[\sum_{k=1}^{n} | x(k) |^p \right]^{1/p} \tag{13.51}$$

are commonly used where the number p is restricted to lie in the interval $1 \leq p < \infty$. Three of the most widely employed choices of the norm index are $p = 1$, 2, and ∞. In a similar manner, the l_p-induced norm for any matrix $A \in R^{m \times n}$ (or $C^{m \times n}$) is specified by

$$\| A \|_p = \max_{\mathbf{x}*\mathbf{x}=1} \| A\mathbf{x} \|_p \tag{13.52}$$

where the number p is again restricted to lie in the interval $1 \leq p < \infty$.

---

[2]A mapping between X and R as designated by $\| \mathbf{x} \|$ is said to be a *norm* if it satisfies the axioms (1) $\| \mathbf{x} \| \geq 0$ for all $\mathbf{x} \in X$ and $\| \mathbf{x} \| = 0$ if and only if $\mathbf{x} = \mathbf{0}$; (2) $\| \mathbf{x} + \mathbf{y} \| \leq \| \mathbf{x} \| + \| \mathbf{y} \|$ for all $\mathbf{x}, \mathbf{y} \in X$; (3) $\| \alpha \mathbf{x} \| = | \alpha | \, \| \mathbf{x} \|$ for all scalars α and all $\mathbf{x} \in X$.

Hilbert Space Setting

Much of the original research in signal restoration assumed that the underlying set of signals X is a complete inner product (i.e., a Hilbert) space. An inner product space is a vector space upon which there is defined an inner product $<\mathbf{x},\mathbf{y}>$ which maps any two vectors $\mathbf{x},\mathbf{y} \in X$ into a scalar such that the axioms of an inner product axiom are satisfied.[3] This inner product induces a natural norm distance metric as specified by

$$\|\mathbf{x}\| = \sqrt{\langle \mathbf{x}, \mathbf{x} \rangle} \tag{13.53}$$

An inner product space is said to be complete if every Cauchy sequence contained in that space converges to an element of that space.[4] Our interest in Hilbert spaces arises in recognition of the fact that many important signal processing problems can be naturally formulated in such a setting.

A variety of algorithms have been proposed for iteratively solving minimization problem (13.50) when the signals are taken to lie in a Hilbert space. These algorithms differentiate themselves by the assumptions made on the characteristics of the attribute sets $C_1, C_2, \ldots, C_m$ in which the unknown signal being monitored is assumed to be contained. Of particular interest is the situation in which these sets are constrained to be made up exclusively of (1) closed subspaces or translated closed subspaces (i.e., linear varieties) and (2) closed convex sets. Some of the more important theoretical results characterizing these cases are examined in the next two subsections.

Attribute Sets: Closed Subspaces

Much of the original research in signal restoration was concerned with the highly restrictive case in which each of the individual attribute sets C_k is a closed subspace. The composite attribute set C as formed from their set intersection (13.45) must therefore also be a closed subspace. For this special case there exists a useful analytical characterization of the solution to the signal restoration problem (13.50). The notion of vector orthogonality is central to this discussion. In particular, the two vectors $\mathbf{x}, \mathbf{y} \in X$ are said to be orthogonal if their inner product is zero, that is,

$$<\mathbf{x},\mathbf{y}> = 0 \tag{13.54}$$

Furthermore, if $\mathbf{x}$ and $\mathbf{y}$ are orthogonal, it follows that the squared inner product-induced norm (13.53) of their vector difference satisfies

$$\|\mathbf{x} - \mathbf{y}\|^2 = \|\mathbf{x}\|^2 + \|\mathbf{y}\|^2 \tag{13.55}$$

which is commonly known as the general *Pythagorean theorem*. This theorem is readily proven by direct substitution and using the orthogonality of the two vectors. With these preliminaries completed, the celebrated projection theorem is now given [Luenberger, 1969].

Theorem 1. Let C be a closed subspace of Hilbert space X. Corresponding to any vector $\mathbf{x} \in X$, there is a unique vector $\mathbf{x}^o \in C$ such that $\|\mathbf{x} - \mathbf{x}^o\| \leq \|\mathbf{x} - \mathbf{y}\|$ for all $\mathbf{y} \in C$. Furthermore, a necessary and sufficient condition that $\mathbf{x}^o \in C$ be the unique minimization vector is that $\mathbf{x} - \mathbf{x}^o$ is orthogonal to every vector in C.

When the attribute sets C_k are each closed spaces, it follows that their intersection gives rise to a composite attribute set C which is also a closed subspace. The above theorem indicates that the solution to the signal restoration problem (13.50) is unique. It will be useful to interpret this solution from an algebraic viewpoint. Specifically, the concept of orthogonal projection operator plays

---

[3] The axioms of an inner product are (1) $<\mathbf{x},\mathbf{y}> = <\overline{\mathbf{y},\mathbf{x}}>$, (2) $<\mathbf{x} + \mathbf{y}, \mathbf{z}> = <\mathbf{x}, \mathbf{z}> + <\mathbf{y}, \mathbf{z}>$, (3) $<\alpha\mathbf{x},\mathbf{y}> = \alpha<\mathbf{x},\mathbf{y}>$, and (4) $<\mathbf{x},\mathbf{x}> \geq 0$ and $<\mathbf{x},\mathbf{x}> = 0$ if and only if $\mathbf{x} = \mathbf{0}$.

[4] The sequence $\{\mathbf{x}_n\}$ contained in X is said to be Cauchy if for every $\epsilon > 0$ there is an integer $N(\epsilon)$ such that $d(\mathbf{x}_n, \mathbf{x}_m) < \epsilon$ for all m and $n > N(\epsilon)$.

a central role in characterizing the basic nature of the vector $\mathbf{x}^o$ described in Theorem 1. The association of the given vector $\mathbf{x}$ with its unique approximating vector $\mathbf{x}^o \in C$ is notationally specified by

$$\mathbf{x}^o = P_C \mathbf{x} \tag{13.56}$$

It is straightforwardly shown that this one-to-one association P_C possesses the three properties of being:

1. Linear so that $P_C(\alpha \mathbf{x} + \beta \mathbf{y}) = \alpha P_C \mathbf{x} + \beta P_C \mathbf{y}$
2. Idempotent so that $P_C^2 = P_C$
3. Self-adjacent so that $P_C^* = P_C$

A mapping that possesses these three properties is commonly referred to as an *orthogonal projection operator*. The mapping P_C designates the *orthogonal projection* of vector space X onto the closed subspace C. The term orthogonal arises from the observation that every vector in subspace C is orthogonal to the associated error vector $\mathbf{x} - P_C\mathbf{x}$. The concept of orthogonal projection operators is of fundamental importance in optimization theory and has many important practical and theoretical implications.

There exists a convenient means for obtaining a solution to the signal restoration problem (13.50) when the composite attribute set C is a finite-dimensional subspace. The following theorem summarizes the main points of this solution procedure.

Theorem 2. Let the nonempty closed subspace C of Hilbert space X be composed of all vectors which are linear combinations of the vectors $\mathbf{x}_1, \mathbf{x}_2, \ldots, \mathbf{x}_q$. Corresponding to any vector $\mathbf{x} \in X$, the unique vector that minimizes $\| \mathbf{x} - \mathbf{y} \|$ for all $\mathbf{y} \in C$ is of the form

$$\mathbf{x}^o = \sum_{k=1}^{q} a_k \mathbf{x}_k \tag{13.57}$$

The a_k coefficients in this linear combination are the components of any solution to the consistent system of linear equations

$$G\mathbf{a} = \mathbf{b} \tag{13.58}$$

where G is the $q \times q$ Gram matrix whose (i,j)th component is specified by $<\mathbf{x}_j,\mathbf{x}_i>$ and $\mathbf{b}$ is the $q \times 1$ vector whose ith component is given by $<\mathbf{x},\mathbf{x}_i>$.

If the vectors $\mathbf{x}_1, \mathbf{x}_2, \ldots, \mathbf{x}_q$, that span subspace C are linearly independent, then the Gram matrix G is invertible and the linear system of equations (13.58) has a unique solution for the coefficient vector $\mathbf{a}$. On the other hand, this linear system of equations will have an infinity of coefficient vector solutions when the $\{\mathbf{x}_k\}$ vectors are linearly dependent. Nevertheless, each of these coefficient vector solutions lead to the same unique optimum solution (13.57).

Example 4. One of the most commonly studied signal restoration and signal recovery problems is concerned with the task of estimating the signal component in a noise-contaminated measurement as described by the relationship

$$\mathbf{x} = H\mathbf{a} + \mathbf{w} \tag{13.59}$$

In this expression the measurement signal $\mathbf{x}$ is assumed to lie in ϵC^m, H is a known matrix contained in $C^{m \times n}$ of full rank n, and $\mathbf{w} \in C^m$ is an unobserved noise vector. It is now desired to find a vector of form $\mathbf{y} = H\mathbf{a}$ that best approximates the measurement $\mathbf{x}$ in the sense of minimizing the quadratic criterion

$$f(\mathbf{a}) = [\mathbf{x} - H\mathbf{a}]^* W[\mathbf{x} - H\mathbf{a}] \tag{13.60}$$

In this criterion W is a positive-definite Hermitian matrix, and the asterisk symbol designates the operation of complex transposition. It is to be noted that if $\mathbf{w}$ is a Gaussian random vector with covariance matrix W^{-1}, then the minimization of functional (13.60) corresponds to the maximum likelihood estimation of the signal component $H\mathbf{a}$.

Examination of this problem formulation reveals that it is of the same form treated in Theorem 1 in which the Hilbert space is C^n. In view of criterion (13.60) that is to be minimized, the inner product identifying this Hilbert space is taken to be $\langle \mathbf{x}, \mathbf{y} \rangle \equiv \mathbf{x}^* W \mathbf{y}$. It is a simple matter to show that this measure satisfies the three axioms of an inner product. The closed subspace C corresponds to the set of all vectors that are expressible as $H\mathbf{a}$ (i.e., the range space of matrix H). Since matrix H is taken to have full column rank n, it follows that C has dimension n. If in a given application the elements of the coefficient vector $\mathbf{a}$ can take on complex values, it is readily shown that the orthogonal projection matrix P_C is specified by

$$P_C = H[H^* W H]^{-1} H^* W \tag{13.61}$$

Furthermore, the unique vector $\mathbf{a}^o \in C^n$ which minimizes functional (13.60) is obtained by using the projection relationship $H\mathbf{a}^o = P_C \mathbf{x}$ to yield

$$\mathbf{a}^o = [H^* W H]^{-1} H^* W \mathbf{x} \tag{13.62}$$

On the other hand, if the coefficient vector $\mathbf{a}$ is restricted to lie in R^n, the required orthogonal projection operation takes the form

$$P_C \mathbf{x} = H[\mathrm{Re}\{H^* W H\}]^{-1} \mathrm{Re}\{H^* W \mathbf{x}\} \tag{13.63}$$

where Re designates the "real part of" operator. Moreover, the unique real-valued coefficient vector minimizing functional (13.60) is specified by

$$\mathbf{a}^o = [\mathrm{Re}\{H^* W H\}]^{-1} \mathrm{Re}\{H^* W \mathbf{x}\} \tag{13.64}$$

A solution to the general signal restoration problem for the case in which the individual attribute sets are closed subspaces formally entails determining the orthogonal projection operator P_C on the closed composite attribute subspace (13.45). Unfortunately, an analytical expression for P_C is typically intractable even when the orthogonal projection operators P_k defined on each of the individual attribute subspaces C_k are readily constructed. The natural question arises as to whether it is possible to use these individual orthogonal projection operators to generate P_C. In recognition of this need, J. von Neumann developed a method for iteratively constructing the required projection operator [von Neumann, 1950] for the case of two ($m = 2$) closed subspaces (p. 55, theorem 13.7). This result was later extended by Halperin [1962] for the multiple attribute closed subspace case as described in the following theorem.

Theorem 3. Let P_k denote the orthogonal projection operators onto the closed subspaces C_k for $k = 1, 2, \ldots, m$ of Hilbert space X. Moreover, let C designate the nonempty closed subspace formed by the intersection of these closed subspaces so that $C = C_1 \cap C_2 \cap \ldots \cap C_m$. If P_C designates the orthogonal projection matrix onto closed subspace C and $T = P_1 P_2 \ldots P_m$, it then follows that T^n converges strongly to P_C, that is,

$$\lim_{n \to \infty} \| T^n - P_C \| = 0 \tag{13.65}$$

This theorem indicates that repeated applications of operator T converge to the required orthogonal projection operator P_C. The practicality of this result arises from the observation that it is often possible to synthesize the individual orthogonal projection operators P_k but not the composite orthogonal projection operator P_C. This capability was illustrated in Theorem 2 for the case in which the subspaces were finite dimensional. To solve the signal restoration problem (13.50) when the attribute sets are each closed subspaces, we then simply use the following iterative scheme

$$\mathbf{x}_n = T\mathbf{x}_{n-1} \quad \text{for } n = 1, 2, 3, \ldots \tag{13.66}$$

in which the initial point is taken to be the measurement signal so that $\mathbf{x}_o = \mathbf{x}$. The sequence of signals thereby generated is guaranteed to converge to the unique solution of the signal restoration problem.

Linear Variety Property Sets

Theorem 3 is readily generalized to the case in which the individual attribute sets are closed linear varieties. A set contained in vector space X is said to be a linear variety if it is a translation of a subspace contained in X. More specifically, if C is a subspace of X and $\mathbf{u}$ is a fixed vector contained in X, then the associated linear variety is specified by

$$V = \mathbf{u} + C \tag{13.67}$$

$$= \{\mathbf{x} \in X : \mathbf{x} = \mathbf{u} + \mathbf{y} \text{ for all } \mathbf{y} \in C\}$$

It is to be noted that the vector $\mathbf{u}$ used in this linear variety formulation is not unique. In fact, any vector contained in V could have been used in place of $\mathbf{u}$. When subspace C is closed, there exists a unique vector contained in V of minimum norm that is of particular interest in many applications. It is formally specified by

$$\mathbf{u}^o = P_C \mathbf{u} \tag{13.68}$$

where P_C designates the orthogonal projection operator onto the closed subspace C. Vector $\mathbf{u}^o$ represents that unique vector contained in the closed linear variety V which lies closest to the origin in the inner product-induced norm sense. With these thoughts in mind, the following lemma is readily proven.

Lemma 1. Let $V_k = \mathbf{u}_k + C_k$ be closed linear varieties associated with the closed subspaces C_k and vectors $\mathbf{u}_k$ contained in Hilbert space X for $k = 1, 2, \ldots, m$. Moreover, let V designate the nonempty closed linear variety formed by the intersection of these closed linear varieties so that $V = V_1 \cap V_2 \cap \ldots \cap V_m$. Corresponding to any vector $\mathbf{x} \in X$, the vector contained in V that lies closest to $\mathbf{x}$ in the inner product-induced sense is the limit of the sequence generated according to

$$\mathbf{x}_n = T_1 T_2 \ldots T_m \mathbf{x}_{n-1} \qquad \text{for } n = 1, 2, 3, \ldots \tag{13.69}$$

where $\mathbf{x}_o = \mathbf{x}$. The operators appearing in this expression are formally defined by

$$T_k \mathbf{y} = P_{C_k} \mathbf{y} + \{I - P_{C_k}\} \mathbf{u}_k \qquad \text{for } k = 1, 2, \ldots, m \tag{13.70}$$

Attribute Sets: Closed Convex Sets

Although the case in which the signal attribute sets are closed subspaces is of theoretical interest, it is typically found to be too restrictive for most practical applications. As we will now see, however, it is possible to extend these concepts to the more general case of closed convex attribute sets. The set C is said to be convex if for any two vectors $\mathbf{x}, \mathbf{y} \in C$ their convex sum as defined by $\lambda \mathbf{x} + (1 - \lambda)\mathbf{y}$ is also contained in C for all $0 \leq \lambda \leq 1$. The ability to broaden the class of attribute sets to include closed convex sets significantly expands the class of problems that can be treated by signal restoration. The following well-known functional analysis theorem provides the framework for this development.

Theorem 4. Let C be a nonempty closed convex set of Hilbert space X. Corresponding to any vector $\mathbf{x} \in X$, there is a unique vector $\mathbf{x}^o \in C$ such that $\| \mathbf{x} - \mathbf{x}^o \| \leq \| \mathbf{x} - \mathbf{y} \|$ for all $\mathbf{y} \in C$. Furthermore, a necessary and sufficient condition that $\mathbf{x}^o$ be the unique minimizing vector is that $<\mathbf{x} - \mathbf{x}^o, \mathbf{y} - \mathbf{x}^o>$ ≤ 0 for all $\mathbf{y} \in C$.

From this theorem it is seen that there exists a one-to-one correspondence between a general vector $\mathbf{x} \in X$ and its closest approximation in the closed convex set C. This mapping is here operationally designated by

$$\mathbf{x}^o = P_C \mathbf{x} \qquad (13.71)$$

and we will refer to P_C as the projection operator onto the closed convex set C. It is important to note that if the closed convex set C is not a subspace or a linear variety, then this associated projection operator is no longer linear. Moreover, if $\mathbf{x} \in C$, then $P_C \mathbf{x} = \mathbf{x}$, so that every vector of C is a *fixed point* of operator P_C.[5] Thus, the fixed points of P_C and the closed convex set C are equivalent. These concepts were used by Bregman [1965] and others to prove a type of convergence of the successive projection algorithm. The following theorem summarizes their results.

Theorem 5. Let P_k designate the projection operators onto the closed convex sets C_k for $k = 1, 2, \ldots, m$ of Hilbert space X and $C = C_1 \cap C_2 \cap \ldots \cap C_m$ be their nonempty closed convex set intersection. Furthermore, for every $\mathbf{x} \in X$ consider the *sequence of successive projections* as generated according to

$$\mathbf{x}_n = P_1 P_2 \ldots P_m \mathbf{x}_{n-1} \text{ for } n = 1, 2, 3, \ldots \qquad (13.72)$$

in which $\mathbf{x}^o = \mathbf{x}$. It then follows that this sequence converges:

1. Weakly to a point in the set intersection C[6]
2. Strongly to a point in the set intersection C if at least one of the sets C_k is bounded and compact[7]

The weak-point convergence theorem (1) was first proven by Bregman [1965], while the strong-point convergence theorem (2) was developed by Stiles [1965]. It is important to appreciate what this theorem does and does not say. Specifically, although it ensures that sequence (13.72) converges to a vector contained in the set intersection C, this convergent point need not minimize the original signal restoration criterion (13.50). This is the price paid when considering the more general case of closed convex attribute sets. Nonetheless, it is found that the vector to which sequence (13.72) converges often provides a satisfactory approximation.

To improve the convergence of algorithm (13.72), Gubin *et al.* introduced an overrelaxation modification that extends the projections beyond the boundary of the attribute sets [1967]. This overrelaxation approach was also adopted by Youla [1978] who proposed the algorithm

$$\mathbf{x}_n = T_1 T_2 \ldots T_m \mathbf{x}_{n-1} \qquad \text{for } n = 1, 2, 3, \ldots \qquad (13.73)$$

in which operators $T_k = [I + \lambda_k(P_k - I)]$ are employed. They proved that the sequence so generated converges weakly to a point of C for any choice of the relaxation constants λ_k in the open interval $0 < \lambda_k < 2$. Moreover, if at least one of the closed convex sets C_k is contained in a finite-dimensional subspace, then the convergence is strong.

In summary, the successive projection algorithm provides a useful signal-processing tool for the case in which the individual attribute sets are closed convex sets. Its primary deficiency is that although the signal sequence (13.73) so generated converges to an element of C, it need not converge to the closest approximation of the data signal $\mathbf{x}$ contained in C. Thus, the successive projection algorithm generally fails to provide a solution to the signal restoration problem (13.50). In recognition of this shortcoming, Dykstra [1983] developed an algorithm which does provide an algorithmic solution. This algorithm was further studied by Dykstra and Boyle [1987], Han [1988], and Gaffke and Mathar [1989]. The formulation as given by Gaffke and Mathar is now summarized.

Theorem 6. Let P_k designate the projection operators onto the closed convex sets C_k for $k = 1, 2, \ldots, m$ of Hilbert space X in which the closed convex set intersection $C = C_1 \cap C_2 \cap \ldots \cap C_m$ is

---

[5] The vector $\mathbf{x}$ is said to be a fixed point of operator T if $T\mathbf{x} = \mathbf{x}$.

[6] The sequence $\{\mathbf{y}_n\}$ is said to converge weakly to $\mathbf{y}$ if $<\mathbf{y}_n - \mathbf{y}, \mathbf{z}>$ converges to zero for every $\mathbf{z} \in X$ as n becomes unbounded.

[7] The sequence $\{\mathbf{y}_n\}$ is said to converge strongly to $\mathbf{y}$ if $\| y_n - \mathbf{y} \|$ approaches zero as n becomes unbounded.

nonempty. Let the two sets of m vector sequences $\mathbf{x}_1^{(k)}, \mathbf{x}_2^{(k)}, \ldots, \mathbf{x}_m^{(k)}$ and $\mathbf{y}_1^{(k)}, \mathbf{y}_2^{(k)}, \ldots, \mathbf{y}_m^{(k)}$ for $k = 1, 2, 3, \ldots$ be generated according to

$$\mathbf{x}_i^{(k)} = P_i\left(\mathbf{x}_{i-1}^{(k)} - \mathbf{y}_i^{(k-1)}\right) \qquad \text{for } i = 1, 2, \ldots, m$$

$$\mathbf{x}_0^{(k+1)} = \mathbf{x}_m^{(k)} \tag{13.74}$$

$$\mathbf{y}_i^{(k)} = \mathbf{x}_i^{(k)} - \left(\mathbf{x}_{i-1}^{(k)} - d_i^{(k-1)}\right) \qquad \text{for } i = 1, 2, \ldots, m$$

in which the initial conditions are taken to be $\mathbf{y}_1^{(o)} = \mathbf{y}_2^{(o)} = \cdots = \mathbf{y}_m^{(o)} = \mathbf{0}$ and $\mathbf{x}_m^{(o)} = \mathbf{x}$. It then follows that the sequence $\{\mathbf{x}_m^k\}$ converges to the unique point in C that lies closest to $\mathbf{x}$ in the sense of minimizing functional (13.50).

This algorithm provides a useful means for iteratively finding a solution to the signal restoration problem for the case of closed convex sets.

Closed Projection Operators

A solution to the signal restoration problem (13.50) is generally intractable unless restrictive assumptions are made on the constituent attribute sets. In the previous two subsections, the method of successive projections and its variations were presented for iteratively finding a solution when the underlying attribute sets are closed subspaces, closed linear varieties, or closed convex sets. Unfortunately, some of the more important attribute sets encountered in signal processing do not fall into any of these categories. This is illustrated by the case in which the Hilbert space is taken to be $C^{m \times n}$ and one of the attribute sets corresponds to all $m \times n$ matrices which have rank q where $q < \min(m,n)$. It is readily shown that this set is neither a subspace nor a linear variety, nor is it convex. Thus, use of the extremely important rank q attribute set cannot be justified for any of the algorithms considered up to this point. This is a serious shortcoming when it is realized that this attribute set is used so extensively in many contemporary signal processing applications.

To provide a viable method for approximating a solution to the signal restoration problem for nonconvex attribute sets, we shall now broaden the approach taken. The signals are again assumed to lie in a metric space X with distance metric $d(\mathbf{x},\mathbf{y})$. Furthermore, it is assumed that it is possible to solve each of the individual *projection operator* problems as defined by

$$P_k(\mathbf{x}) = \{\mathbf{y} : d(x,y) = \min_{\mathbf{z} \in C_k} d(\mathbf{x},\mathbf{z})\} \qquad \text{for } 1 \leq k \leq m \tag{13.75}$$

for an arbitrary signal $\mathbf{x} \in X$. These sets consist of all elements in C_k that lie closest to $\mathbf{x}$ in the distance-metric sense for $1 \leq k \leq m$. It is to be noted that the projection operators $P_k(\mathbf{x})$ are generally nonlinear and not one to one as was the case in the previous two subsections. When determining the individual projection operators (13.75), the fundamental issues of the existence and uniqueness of solution need to be addressed. It is tacitly assumed that the signal attributes and metrics under consideration are such that at least one solution exists and the solution(s) may be obtained in a reasonably simple fashion. Fortunately, the generation of solutions imposes no serious restrictions for many commonly invoked attribute sets. Moreover, many relevant signal attributes are characterized by the fact that more than one solution to optimization problems (13.75) exists.

The projection operators (13.75) are unusual in that they need not be of the traditional point-to-point variety as was the case when the attribute set is a closed subspace, closed linear variety or closed convex set. For general C_k sets, P_k is a point-to-set mapping. The concept of a *closed mapping* is of importance when extending the notion of signal restoration to nonconvex sets. A closed mapping is a generalization of the notion of continuity as applied to standard point-to-point mappings [Zangwill, 1969], that is:

Definition 1. The point-to-set **projection mapping** P is said to be closed at $\mathbf{x} \in X$ if the assumptions

1. $\mathbf{x}_k \to \mathbf{x}$ with $\mathbf{x}_k \in X$
2. $\mathbf{y}_k \to \mathbf{y}$ with $\mathbf{y}_k \in P(\mathbf{x}_k)$

imply that $\mathbf{y} \in P(\mathbf{x})$. The point-to-set projection mapping P is said to be closed on the set X_1 if it is closed at each point in X_1.

Signal Restoration Algorithm

We are now in a position to describe the signal restoration algorithm for the case of general attribute sets. This shall be formally done in the format of the following theorem [Cadzow, 1988].

Theorem 7. Let P_k be the projection operators associated with the attribute sets C_k contained in metric space X for $k = 1, 2, \ldots, m$. For any signal $\mathbf{x} \in X$, let the sequence $\{\mathbf{x}_k\}$ be generated according to

$$\mathbf{x}_k \in P_1 P_2 \ldots P_m(\mathbf{x}_{k-1}) \qquad \text{for } k \geq 1 \tag{13.76}$$

in which the initial signal is specified by $\mathbf{x}_o = \mathbf{x}$. A subsequence of this sequence always exists which converges to an element of the set intersection $C = C_1 \cap C_2 \cap \ldots \cap C_m$ provided that: (1) the $d(\mathbf{x}_k, \mathbf{x}_r) < d(\mathbf{x}_{k-1}, \mathbf{x}_r)$, where $\mathbf{x}_r \in X$ designates a reference signal which is often the origin of X, and (2) the set of signals $\mathbf{y} \in X$ that satisfy the inequality $d(\mathbf{y}, \mathbf{x}_r) \leq d(\mathbf{x}, \mathbf{x}_r)$ defines a closed and bounded set.

A casual examination of signal restoration algorithm (13.76) indicates that it is of the same form as the sequence of projections algorithms described in the previous two subsections. It distinguishes itself from those algorithms in that the attribute sets C_k need not be closed subspaces, closed linear varieties, or closed convex sets. The proposed algorithm also distinguishes itself from several other signal restoration algorithms in that the metric d need not be inner product induced. These can be important considerations in specific applications. As an example, it has been conjectured by several authors that the l_1 norm provides for a more effective error measure when the data set has outliers (i.e., unrepresentative data). Signal restoration algorithm (13.76) can be directly applied to such problems since we have not restricted the metric. It must be observed, however, that the nature of the individual projection operators P_k is typically most easily characterized when the metric employed is inner product induced.

It is useful to represent the multiple mapping signal restoration algorithm by the *composite mapping* as defined by

$$P = P_1 P_2 \ldots P_m \tag{13.77}$$

The process of generating the solution set $P(\mathbf{x}_{k-1})$ from $\mathbf{x}_{k-1}$ is to be interpreted in the following sequential manner. First, the set $P_m(\mathbf{x}_{k-1})$ is found. This set consists of all signals possessing the mth attribute that lie closest to $\mathbf{x}_{k-1}$ in the given signal metric. Next, the set $P_{m-1}(P_m(\mathbf{x}_{k-1}))$ is formed and consists of all signals possessing the $(m-1)$th attribute that lie closest to each of the signals in set $P_m(\mathbf{x}_{k-1})$. It is to be noted that although each of the signals in $P_{m-1}(P_m(\mathbf{x}_{k-1}))$ possess the $(m-1)$th attribute, they need not possess the mth attribute. This process is continued in this fashion until the set $P(\mathbf{x}_{k-1})$ is generated. Finally, we arbitrarily select one signal from $P(\mathbf{x}_{k-1})$ to be equal to $\mathbf{x}_k$. When the individual projection mappings P_k are each point-to-point mappings, the signal $\mathbf{x}_k$ generated in this fashion will be unique.

Signal restoration algorithm (13.76) has been applied to many fundamental signal processing problems. It has produced effective results that often exceed those achieved by more traditional methods. The ultimate utility of signal restoration is dependent on the user's innovativeness in generating signal attributes that distinguish the underlying signal from the corruption in the data. In many applications, matrix descriptions of the data under analysis arise in a natural manner. With this in mind, we will now explore some salient matrix properties and how they can be used in typical signal restoration applications.

Algebraic Properties of Matrices

Many of the more important and interesting applications of signal restoration are related to the vector space $C^{m \times n}$. Matrices contained in $C^{m \times n}$ may occur in a natural manner as exemplified by digital images where the nonnegative elements of the matrix correspond to the brightness levels of associated pixels. The underlying signal restoration problem in such cases is commonly referred to as *image reconstruction* or *image restoration*. In other examples, however, the matrix under consideration may be a by-product of an associated data analysis solution routine. For example, in approximating a finite-length time series as a sum of weighted exponentials, one often forms an associated *data matrix* from the time series elements. Whatever the case, the matrix under analysis is typically corrupted in some manner, and it is desired to remove this corruption in order to recover the underlying information-bearing matrix. In using signal restoration for this purpose, it is necessary to employ attributes associated with the information-bearing signal. These attributes are normally of an algebraic or a structural description. In this subsection we will examine two of the more widely invoked algebraic attributes, and some commonly employed structural attributes are examined in the next subsection.

Singular Value Decomposition

The **singular value decomposition** (SVD) of a real- or complex-valued matrix plays an increasingly important role in contemporary signal processing applications. For a generally complex-valued matrix $A \in C^{m \times n}$, its associated SVD representation takes the form of the following sum of r weighted outer products,

$$A = \sum_{k=1}^{r} \sigma_k \mathbf{u}_k \mathbf{v}_k^* \tag{13.78}$$

where r designates the rank of matrix A. In this SVD representation, the σ_k are positive *singular values* that are ordered in the monotonic fashion $\sigma_{k+1} \leq \sigma_k$, while the $\mathbf{u}_k$ are the $m \times 1$ pairwise orthogonal *left singular vectors* and the $\mathbf{v}_k$ are the $n \times 1$ pairwise orthogonal *right singular vectors*. Upon examination of SVD representation (13.78) it is seen that the mn components of matrix A are equivalently replaced by the $r(m + n + 1)$ elements corresponding to the SVD singular values and singular vectors. For low-rank matrices [i.e., $r(1 + m + n) < mn$], the SVD provides for a more efficient representation of a matrix. This observation has been effectively used for the data compression of digital images. Furthermore, the concept of a low rank data matrix plays a vital role in the modeling of empirical data as a sum of weighted exponential signals. With these thoughts in mind, the important concept of low rank approximation of matrices is now examined.

Reduced-Rank Approximation

In various applications, it is frequently desired to find a matrix of rank q that best approximates a given matrix $A \in C^{m \times n}$, where $q < \text{rank} \{A\}$. If the Frobenius metric (13.43) is used to measure the distance between two matrices, it is well known that the best rank-q matrix approximation of A is obtained by simply dropping all but the q largest singular-valued weighted outer products in SVD representation (13.78), that is,

$$A^{(q)} = \sum_{k=1}^{q} \sigma_k \mathbf{u}_k \mathbf{v}_k^* \tag{13.79}$$

where it is tacitly assumed that $q \leq r$. From this expression, it follows that the mapping from A into $A^{(q)}$ is one-to-one when $\sigma_q > \sigma_{q+1}$ but is one-to-many (infinity) if $\sigma_q = \sigma_{q+1}$. The special case in which $\sigma_q = \sigma_{q+1}$ therefore results in a point-to-set mapping and gives rise to subtle issues which are addressed in the following theorem [Mittelmann and Cadzow, 1987].

Theorem 8. Let the $m \times n$ matrix $A \in C^{m \times n}$ have the SVD decomposition (13.78) in which $r = \text{rank}(A)$. The best rank-q Frobenius norm approximation of A as given by expression (13.79) is unique if and only if $\sigma_q > \sigma_{q+1}$. The projection operator $P^{(q)}$ from A into $A^{(q)}$ as designated by

$$A^{(q)} = P^{(q)} A \qquad (13.80)$$

is nonlinear and closed. Furthermore, this mapping is continuous for $\sigma_q \neq \sigma_{q+1}$ and is not continuous when $\sigma_q = \sigma_{q+1}$.

When applying the reduced-rank approximation of a matrix as specified by relationship (13.79), it is desirable that the gap between the so-called *signal-level* singular values and *noise-level* singular values be large (i.e., $\sigma_q - \sigma_{q+1} \gg 0$). If this is true, then the issues of nonuniqueness and continuity of mapping do not arise. Unfortunately, in many challenging applications, this gap is often small, and one must carefully examine the consequences of this fact on the underlying problem being addressed. For example, in modeling empirical data as a sum of exponentials, this gap is typically small, which in turn leads to potential undesirable algorithmic sensitivities. We will examine the exponential modeling problem in a later subsection.

Positive-Semidefinite Matrices

Positive-semidefinite matrices frequently arise in applications related to random and deterministic time series. For example, if $\mathbf{x} \in C^n$ is a vector whose components are random variables, it is well known that the associated $n \times n$ *correlation matrix* as defined by $R_{xx} = E\{\mathbf{x}\mathbf{x}^*\}$ is positive-semidefinite Hermitian, where E designates the expected value operator. Similarly, *orthogonal projection* matrices are often used to describe vector subspaces that identify signals present in empirical data. An orthogonal projection matrix is a positive-semidefinite matrix which has the additional requirements of being idempotent (i.e., $A^2 = A$) and Hermitian (i.e., $A^* = A$). It is well known that the eigenvalues associated with an orthogonal projection matrix are exclusively equal to zero and one. With these examples serving as motivation, we shall now examine some of the salient algebraic characteristics of positive-semidefinite matrices.

The $n \times n$ matrix $A \in C^{n \times n}$ is said to be positive semidefinite if the associated quadratic form inequality as specified by

$$\mathbf{x}^* A \mathbf{x} \geq 0 \qquad \text{for all } \mathbf{x} \in C^n \qquad (13.81)$$

is satisfied for all vectors $\mathbf{x} \in C^n$. Furthermore, if the only vector that causes this quadratic form to be zero is the zero vector, then A is said to be *positive definite*. Since this quadratic form is real valued, we can further infer that any positive-semidefinite matrix must be Hermitian so that $A^* = A$. Moreover, using elementary reasoning it directly follows that the set of positive-definite matrices contained in C^n is a closed convex cone.

In many practical applications, there is given empirical time series data to be analyzed. This analysis is often predicated on one's having knowledge of the time series-associated correlation matrices or orthogonal projection matrices. Since such knowledge is generally unknown, these matrices must be estimated from the empirical data under analysis. These estimates, however, are almost always in error. For example, the estimate $\hat{R}_{xx}$ of a correlation matrix R_{xx} is often Hermitian but not positive semidefinite. To mitigate the effects of these errors, an intuitively appealing procedure would be to find a matrix lying close to $\hat{R}_{xx}$ that possesses the two prerequisite properties of being (1) Hermitian and (2) positive semidefinite. The concept of signal restoration can be used for this purpose if it is possible to develop a closed-form expression for the operator that maps a general Hermitian matrix into the closest positive-semidefinite Hermitian matrix in the Frobenius matrix norm sense. As is now shown, a simple expression for this operator is available using the SVD of the Hermitian matrix being approximated.

With these thoughts in mind, we will now consider the generic problem of finding a positive-semidefinite matrix that lies closest to a given Hermitian matrix $A \in C^{n \times n}$. In those applications where matrix A is not Hermitian, then this matrix is first replaced by its Hermitian component as defined by $(A + A^*)/2$ and then the closest positive-semidefinite matrix to this Hermitian component is found.[8] The problem at hand is readily solved by first making an eigenanalysis of the Hermitian matrix A, that is,

$$A\mathbf{x}_k = \lambda_k \mathbf{x}_k \qquad \text{for } 1 \le k \le n \tag{13.82}$$

Since A is Hermitian, its eigenvalues must all be real and there exists a full set of n associated eigenvectors that can always be chosen orthonormal [i.e., $\mathbf{x}_k^* \mathbf{x}_m = \delta(k - m)$]. With this characterization, the following theorem is readily proven.

Theorem 9. Let A be a Hermitian matrix contained in $C^{n \times n}$ whose eigenanalysis is specified by relationship (13.82). Furthermore, let the eigenvalues be ordered in the monotonically nonincreasing fashion $\lambda_k \ge \lambda_{k+1}$ in which the first p eigenvalues are positive and the last $n - p$ are nonpositive. It then follows that the SVD of Hermitian matrix A can be decomposed as

$$A = \sum_{k=1}^{p} \lambda_k \mathbf{x}_k \mathbf{x}_k^* + \sum_{k=p+1}^{n} \lambda_k \mathbf{x}_k \mathbf{x}_k^* \tag{13.83}$$

$$= A^+ + A^-$$

The Hermitian matrix A^+ is positive semidefinite of rank p, while the Hermitian matrix A^- is negative semidefinite. Furthermore, the unique positive-semidefinite matrix that lies closest to A in the Frobenius and Euclidean norm sense is given by the truncated SVD mapping

$$A^+ = P^+(A) = \sum_{k=1}^{p} \lambda_k \mathbf{x}_k \mathbf{x}_k^* \tag{13.84}$$

The projection operator P^+ is closed and continuous. Furthermore, an idempotent Hermitian (i.e., an orthogonal projection) matrix which lies closest to A in the Frobenius and Euclidean norm sense is specified by

$$A^{op} = P^{op}(A) = \sum_{k:\lambda_k \ge 0.5} \mathbf{x}_k \mathbf{x}_k^* \tag{13.85}$$

This closest idempotent Hermitian matrix is unique provided that none of the eigenvalues of A are equal to 0.5. Moreover, projection operator P^{op} is closed for any distribution of eigenvalues.

Examination of this theorem indicates that the left and right singular vectors of a Hermitian matrix corresponding to its positive eigenvalues are each equal to the associated eigenvector while those corresponding to its negative eigenvalues are equal to the associated eigenvector and its negative image. Furthermore, any Hermitian matrix may be uniquely decomposed into the sum of a positive- and negative-semidefinite Hermitian matrix as specified by (13.83).

This theorem's proof is a direct consequence of the fact that the Frobenius and Euclidean norm of the matrix A and Q^*AQ are equal for any unitary matrix Q. Upon setting Q equal to the $n \times n$ matrix whose columns are equal to the n orthonormal eigenvectors of matrix A, it follows that the Frobenius (Euclidean) norm of the matrices $A - B$ and $Q^*[A - B]Q$ are equal. From this equality the optimality of positive-definite matrix (13.84) immediately follows since Q^*AQ is equal to the diagonal matrix with the eigenvalues of A as its diagonal components. The closest positive-semi-

---

[8] Any matrix A can be represented as the sum of a Hermitian matrix and a skew Hermitian matrix using the decomposition $A = (A + A^*)/2 + (A - A^*)/2$.

Table 13.3 Structured Matrices

Matrix Class	Matrix Elements
Hermitian	$A(i,j) = \bar{A}(j,i)$
Toeplitz	$A(i+1, j+1) = A(i,j)$
Hankel	$A(i+1, j) = A(i, j+1)$
Circulant	$A(i+1, j) = A(i, j-1)$ with $A(i+1, 1) = A(i,n)$
Vandermonde	$A(i,j) = A(2,j)^{i-1}$

definite matrix (13.84) is obtained by simply truncating the SVD to the positive singular value outer products. Similarly, the closest orthogonal projection matrix is obtained by replacing each singular value by 1 if the singular value is greater than or equal to 0.5 and by 0 otherwise.

Structural Properties of Matrices

In various applications, a matrix under consideration is known to have its elements functionally dependent on a relatively small set of parameters. A brief listing of some of the more commonly used matrix classes so characterized is given in Table 13.3. In each case, there exists a relatively simple relationship for the elements of the matrix. For example, an $m \times n$ Toeplitz matrix is completely specified by the $m + n - 1$ elements of its first row and first column. We now formalize this concept.

Definition 2. Let $a_{ij}(\theta_1, \theta_2, \ldots, \theta_p)$ for $1 \le i \le m$ and $1 \le j \le n$ designate a set of mn functions that are dependent on a set of real-valued parameters $\theta_1, \theta_2, \ldots, \theta_p$ in which $p < mn$. Furthermore, consider the matrix $A \in C^{m \times n}$ whose elements are given by

$$A(i,j) = a_{ij}(\theta_1, \theta_2, \ldots, \theta_p) \qquad \text{for } 1 \le i \le m \text{ and } 1 \le j \le n+1 \qquad (13.86)$$

for a specific choice of the parameters $\theta_1, \theta_2, \ldots, \theta_p$. These p parameters shall be compactly represented by the parameter vector $\theta \in R^p$. The set of all matrices that can be represented in this fashion is designated by $\mathcal{M}$ and is said to have a structure induced by the functions $a_{ij}(\theta)$ and to have p degrees of freedom. If the functions $a_{ij}(\theta)$ are linearly dependent on the parameters, the matrix class $\mathcal{M}$ is said to have a linear structure.

In what is to follow, we will be concerned with the task of optimally approximating a given matrix $B \in C^{m \times n}$ by a matrix with a specific linear structure. For purposes of description, the specific class of matrices to be considered is denoted by $\mathcal{L}$ and its constituent matrices are functionally dependent on the real-valued parameter vector $\theta \in R^p$. The matrix approximation problem is formally expressed as

$$\min_{\theta \in R^p} \| B - A(\theta) \|_F \qquad (13.87)$$

where $A(\theta)$ designates a general matrix contained in $\mathcal{L}$ that is linearly dependent on the parameter vector θ.

It is possible to utilize standard analysis techniques to obtain a closed-form solution to matrix approximation problem (13.87). To begin this analysis, it is useful to represent matrix B by the $mn \times 1$ vector $\mathbf{b}_c$ whose elements are formed by concatenating the column vectors of B. This concatenation mapping is one to one and onto and is therefore invertible. Namely, given B there exists a unique concatenated vector image $\mathbf{b}_c$, and vice versa. It further follows that the Euclidean norm of $\mathbf{b}_c$ and the Frobenius norm of B are equal, that is,

$$\| b_c \|_2 = \| B \|_F \qquad (13.88)$$

Using this norm equivalency, it follows that the original matrix approximation problem (13.87) can be equivalently expressed as

$$\min_{\boldsymbol{\theta} \in R^p} \left\| B - A(\boldsymbol{\theta}) \right\|_F = \min_{\boldsymbol{\theta} \in R^p} \left\| \mathbf{b}_c - \mathbf{a}_c(\boldsymbol{\theta}) \right\|_2 \tag{13.89}$$

where $\mathbf{b}_c$ and $\mathbf{a}_c(\boldsymbol{\theta})$ designate the concatenated vector representations for matrices B and $A(\boldsymbol{\theta})$, respectively. Since each element of matrix $A(\boldsymbol{\theta}) \in \mathscr{L}$ is linearly dependent on the parameter vector $\boldsymbol{\theta}$, it follows that there exists a unique $mn \times p$ matrix L such that

$$\mathbf{a}_c(\boldsymbol{\theta}) = L\boldsymbol{\theta} \tag{13.90}$$

is the concatenated representation for matrix $A(\boldsymbol{\theta}) \in \mathscr{L}$. Thus, the original matrix approximation problem can be equivalently expressed as

$$\min_{\boldsymbol{\theta} \in R^p} \left\| \mathbf{b}_c - L\boldsymbol{\theta} \right\|_2 \tag{13.91}$$

This problem, however, is seen to be quadratic in the parameter vector $\boldsymbol{\theta}$, and an optimum parameter vector is obtained by solving the associated consistent system of normal equations

$$L^*L\boldsymbol{\theta}^o = L^*\mathbf{b}_c \tag{13.92}$$

In many cases, the matrix product L^*L is invertible, thereby rendering a unique solution to these equations. Whatever the case, the associated vector representation for the optimum matrix contained in $\mathscr{L}$ is given by $\mathbf{a}_c(\boldsymbol{\theta}^o) = L\boldsymbol{\theta}^o$. Finally, the corresponding optimum approximating matrix $A(\boldsymbol{\theta}^o)$ is simply obtained by reversing the column vector concatenation mapping that generates $\mathbf{a}_c(\boldsymbol{\theta}^o)$.

Example 5. To illustrate the above procedure, let us consider the specific case of the class of real 3×2 Toeplitz matrices. A general parametric representation for a matrix in this class and its associated concatenated vector equivalent is given by

$$A(\boldsymbol{\theta}) = \begin{bmatrix} \theta_1 & \theta_2 \\ \theta_3 & \theta_1 \\ \theta_4 & \theta_3 \end{bmatrix} \leftrightarrow \mathbf{a}_c(\boldsymbol{\theta}) = \begin{bmatrix} \theta_1 \\ \theta_3 \\ \theta_4 \\ \theta_2 \\ \theta_1 \\ \theta_3 \end{bmatrix} \tag{13.93}$$

It then follows that the matrix mapping the parameter vector $\boldsymbol{\theta}$ into $\mathbf{a}_c(\boldsymbol{\theta})$ is given by

$$\mathbf{a}_c(\boldsymbol{\theta}) = L\boldsymbol{\theta} = \begin{bmatrix} 1 & 0 & 0 & 0 \\ 0 & 0 & 1 & 0 \\ 0 & 0 & 0 & 1 \\ 0 & 1 & 0 & 0 \\ 1 & 0 & 0 & 0 \\ 0 & 0 & 1 & 0 \end{bmatrix} \begin{bmatrix} \theta_1 \\ \theta_2 \\ \theta_3 \\ \theta_4 \end{bmatrix} \tag{13.94}$$

It is seen that the structure matrix mapping L has a full column rank of four. Finally, the unique solution to relationship (13.92) for the optimum parameter vector used for representing the closest Euclidean norm approximation is specified by

$$\boldsymbol{\theta}^o = [L^*L]^{-1}L^*\mathbf{b}_c = \begin{bmatrix} \frac{1}{2}[B(1,1) + B(2,2)] \\ B(1,2) \\ \frac{1}{2}[B(2,1) + B(3,3)] \\ B(3,2) \end{bmatrix} \tag{13.95}$$

It is clear that the entities $A(\boldsymbol{\theta})$ and $\mathbf{a}_c(\boldsymbol{\theta})$ are equivalent. Moreover, the class of real 3×2 Toeplitz matrices is seen to have four degrees of freedom (i.e., the real parameters θ_1, θ_2, θ_3, and θ_4).

It is readily established using the above arguments that the best Toeplitz approximation to a matrix B is obtained by first determining the mean value of each of its diagonals and then using these mean values as entries of the best approximating Toeplitz diagonals. Let us formalize this important result.

Theorem 10. Let C_T and C_H designate the set of all Toeplitz and Hankel matrices contained in the space $C^{m \times n}$, respectively. It follows that C_T and C_H are each complex $(m + n - 1)$-dimensional subspaces of $C^{n \times n}$. Furthermore, the Toeplitz matrix A_T which best approximates $A \in C^{m \times n}$ in the Frobenius norm sense has the constant element along its kth diagonal specified by

$$\alpha_k = \text{mean}[\mathbf{d}_k] \qquad \text{for } -n + 1 \le k \le m - 1 \tag{13.96}$$

In this expression, mean $[\mathbf{d}_k]$ designates the arithmetic mean of vector $\mathbf{d}_k$ whose components correspond to the elements of matrix A along its kth diagonal. In particular, vector $\mathbf{d}_o$ has as its components the elements of the main diagonal [i.e, elements $A(1,1)$, $A(2,2)$, etc.], vector $\mathbf{d}_1$ has as its components the elements of the diagonal immediately below the main diagonal, vector $\mathbf{d}_{-1}$ has as its components the elements of the diagonal immediately above the main diagonal, and so forth. The projection operator P_T that maps A into A_T as governed by relationship (13.96) is designated by

$$A_T = P_T A \tag{13.97}$$

and is linear and one to one.

Similarly, the Hankel matrix A_H that lies closest to A in the Frobenius norm sense has the constant element along its kth antidiagonal specified by

$$\beta_k = \text{mean}[\mathbf{a}_k] \qquad \text{for } -n + 1 \le k \le m - 1 \tag{13.98}$$

The components of $\mathbf{a}_k$ correspond to the elements of matrix A along its kth antidiagonal in which vector $\mathbf{a}_o$ corresponds to the main antidiagonal [i.e., elements $A(1,n)$, $A(2,n-1)$, etc.], vector $\mathbf{a}_1$ to the antidiagonal immediately below the main antidiagonal, and so forth. The projection operator P_H mapping A into A_H as governed by relationship (13.98) is designated by

$$A_H = P_H A \tag{13.99}$$

and is linear and one to one.

It is interesting to note that relationships (13.97) and (13.99) which identify the closest approximating Toeplitz and Hankel matrices are very much dependent on the Frobenius measure of matrix size. If a different metric had been incorporated, then different expressions for the best approximating Toeplitz and Hankel matrix approximations would have arisen. For example, in applications in which data outliers are anticipated, it is often beneficial to use the l_1-induced norm. In this case, the elements α_k and β_k are replaced by the median value of the kth diagonal and antidiagonal of matrix A, respectively. Similarly, if the l_∞-induced norm were used, the elements α_k and β_k would be replaced by the midpoint of the largest and smallest elements of the kth diagonal and the kth antidiagonal of matrix A, respectively.

In pure sinusoidal modeling applications, the concept of forward and backward prediction is often used. The data matrix arising from a forward-backward modeling will then have a block Toeplitz-Hankel structure. This in turn gives rise to the signal restoration task of finding a matrix of block Toeplitz-Hankel structure that most closely approximates a given matrix. The results of Theorem 10 can be trivially extended to treat this case.

Nonnegative Sequence Approximation

Two related fundamental problems which arise in various signal processing applications are that of finding (1) a nonnegative-definite sequence which lies closest to a given sequence, or (2) a nonnegative-definite matrix which lies closest to a given matrix. For example, in many commonly employed spectral estimation algorithms, estimates of a time series autocorrelation sequence are either explicitly or implicitly computed from a finite-length sample of the time series. It is well known that the autocorrelation sequence associated with a wide-sense stationary time series has a nonnegative-definite Fourier transform. However, the process of forming the autocorrelation lag estimates from empirical data often results in lag estimates whose Fourier transform can be negative. With this application (and others) in mind, we will now briefly explore some basic theory related to nonnegative-definite sequences and then employ the signal restoration algorithm to solve the second problem posed above.

To begin our development, the sequence $\{x(n)\}$ is said to be *nonnegative definite* if its Fourier transform is real nonnegative, that is,

$$X(\omega) = \sum_{n=-\infty}^{\infty} x(n)e^{-j\omega n} \geq 0 \quad \text{for } \omega \in [0, 2\pi] \tag{13.100}$$

As one might suspect, nonnegative-definite time series possess a number of salient properties which distinguish themselves from more general time series. The following theorem provides an insight into some of the more important properties.

Theorem 11. Let C_{nnd} designate the set of all finite- and infinite-length nonnegative-definite time series. It follows that C_{nnd} is a closed convex cone whose vertex is located at the origin. Moreover, every time series contained in this cone is conjugate symmetric so that $x(-n) = \bar{x}(n)$ for all integers n. Furthermore, the data matrix of order k formed from any nonnegative-definite time series $\{x(n)\}$ $\in C_{nnd}$ and having the Hermitian-Toeplitz structure given by

$$X_k = \begin{bmatrix} x(0) & \bar{x}(1) & \cdots & \bar{x}(k) \\ x(1) & x(0) & \cdots & \bar{x}(k-1) \\ \vdots & \vdots & \vdots & \vdots \\ x(k) & x(k-1) & \cdots & x(0) \end{bmatrix} \tag{13.101}$$

is positive semidefinite for all nonnegative integer values of the order parameter k.

If $\{x(n)\}$ is a nonnegative sequence of length $2q + 1$, then the zeros of its z-transform always occur in conjugate reciprocal pairs, that is,

$$X(z) = \sum_{n=-q}^{q} x(n)z^{-n} = \alpha \prod_{k=1}^{q} [1 - z_k z][1 - \bar{z}_k z^{-1}] \tag{13.102}$$

where α is a positive scalar. In addition, there will exist scalars $b_o, b_1, \ldots, b_q$ such that this $(2q + 1)$-length nonnegative sequence can be represented by

$$x(n) = \sum_{k=0}^{q-n} b_k b_{k+n} \qquad \text{for } 1 \le n \le q \qquad (13.103)$$

A more detailed treatment of this topic is found in Cadzow and Sun [1986]. This theorem must be carefully interpreted in order not to inappropriately infer properties possessed by nonnegative sequences. For instance, it is not true that the positive semidefiniteness of a finite number of data matrices X_k implies that the underlying time series is nonnegative. As an example, the length-three sequence with elements $x(0) = 3$, $x(1) = x(-1) = 2$ is not positive although the data matrix X_1 is positive definite. In a similar fashion, this theorem does not indicate that the symmetric truncation of a nonnegative sequence is itself nonnegative.

Nonnegative-Definite Toeplitz-Hermitian Matrix Approximation

It is possible to employ the concept of signal restoration in a straightforward fashion to iteratively solve the two problems considered at the beginning of this section. We will demonstrate a solution procedure by treating the problem in which it is desired to find that unique nonnegative-definite Toeplitz-Hermitian matrix which lies closest to a given matrix $X \in C^{n \times n}$. To employ the concept of signal restoration to this problem, we now identify the two attribute sets in which the required approximating matrix must lie, namely,

$$C_+ = \{Y \in C^{n \times n} \text{ which are nonnegative}\}$$
$$C_{TH} = \{Y \in C^{n \times n} \text{ for which } Y \text{ is Toeplitz and Hermitian}\} \qquad (13.104)$$

Relationship (13.84) provides the mapping corresponding to the attribute set C_+. Implementation of the mapping associated with attribute set C_{TH} is a straightforward modification of Toeplitz mapping (13.97). In particular, the mean value of the elements of the two diagonals equispaced k units above and below the main diagonal is employed to determine the constant element used in the Toeplitz-Hermitian approximation where $0 \le k \le n-1$.

Since the attribute sets C_+ and C_{TH} are both closed convex sets, the sequence of successive projections algorithm as specified by

$$X_k = P_+ P_{TH} X_{k-1} \qquad \text{for } k = 1, 2, \ldots \qquad (13.105)$$

can be employed. The initial matrix for this algorithm is set equal to the matrix being approximated (i.e., $X_o = X$). This algorithm produces a matrix sequence that is guaranteed to converge to a positive-semidefinite matrix with the required Toeplitz-Hermitian structure. Unfortunately, this solution need not be the closest matrix contained in $C_+ \cap C_{TH}$ that lies closest to X in the Frobenius norm sense. To obtain the optimum approximating matrix contained in $C_+ \cap C_{TH}$, we can alternatively employ the algorithm described by Relationship (13.74).

Example 6. We will now illustrate the point that the sequence of vectors generated by the successive projections algorithm need not always converge to the closest vector in a closed convex set. This will be accomplished by considering the task of finding a positive-semidefinite Toeplitz matrix that lies closest to given matrix. The given matrix X is here taken to be

$$X = \begin{bmatrix} 2 & 4 \\ 2 & 4 \end{bmatrix}$$

Although this matrix is positive semidefinite, it is not Toeplitz, thereby necessitating the use of a signal restoration algorithm for finding a suitably close positive-semidefinite Toeplitz-Hermitian matrix approximation. The sequence of successive projections algorithm (13.105) has guaranteed convergence to a positive-semidefinite Toeplitz matrix approximation. To be assured of finding the positive-semidefinite Toeplitz matrix that lies closest to X in the Frobenius norm sense, however, it is generally necessary to employ the Dykstra algorithm (13.74). Using the given matrix X as the initial condition, the positive-semidefinite Toeplitz matrices to which these two algorithms converge are found to be

$$X_{ssp} = \begin{bmatrix} 3.0811 & 2.9230 \\ 2.9230 & 3.0811 \end{bmatrix} \quad \text{and} \quad X_{dyk} = \begin{bmatrix} 3 & 3 \\ 3 & 3 \end{bmatrix}$$

Clearly, the positive-semidefinite Toeplitz matrix approximation X_{dyk} lies slightly closer to X in the Frobenius norm sense than does X_{ssp}. Convergence was deemed to have occurred when the normed matrix error $\|X_n - X\|/\|X\|$ became less than 10^{-9}. The successive projection algorithms and Dykstra's algorithm took two and three iterations, respectively, to reach this normed error level.

Exponential Signals and the Data Matrix

In various applications, the basic objective is to approximate a finite sample of a time series by a linear combination of real- and/or complex-valued exponential signals. The set of data to be modeled is taken to be

$$x(1), x(2), \ldots, x(N) \tag{13.106}$$

where N designates the length of the data. It is well known that this data set can be exactly modeled by an exponential signal of order p or less if and only if there exists a nontrivial set of coefficients $a_o, a_1, \ldots, a_p$ such that the following homogeneous relationship is satisfied:

$$a_o x_n + a_1 x_{n-1} + \cdots + a_p x_{n-p} = 0 \quad \text{for } p + 1 \leq n \leq N \tag{13.107}$$

Upon examination of these relationships, it is clear that nontrivial solutions will always exist when the number of equations is fewer than the number of unknowns (i.e., $N - p < p$). Most data modeling applications, however, are concerned with the distinctly overdetermined case in which $N - p >> p$.

From the above comments, it is apparent that a characterization of the exponential data modeling problem can be obtained by analyzing the linear homogeneous relationships (13.107). It will be convenient to compactly represent these ideal relationships in the vector format

$$Xa = 0 \tag{13.108}$$

where a is the $(p + 1) \times 1$ coefficient vector with elements a_k and X is the corresponding $(N - p) \times (p + 1)$ *data matrix* as specified by

$$X = \begin{bmatrix} x(p + 1) & x(p) & \cdots & x(1) \\ x(p + 2) & x(p + 1) & \cdots & x(2) \\ \vdots & \vdots & \vdots & \vdots \\ x(N) & x(N - 1) & \cdots & x(N - p) \end{bmatrix} \tag{13.109}$$

This data matrix is seen to have a *Toeplitz* structure since the elements along each of its diagonals are equal. Furthermore, if relationship (13.108) is to have a nontrivial solution, it is clear that the rank of data matrix X must be equal to or less than p. These salient attributes play a critical role in various exponential modeling algorithms, and they are now formalized.

Theorem 12. The data set $\{x_1, x_2, \ldots, x_N\}$ is exactly representable as a qth-order exponential signal if and only if the associated $(N - p) \times (p + 1)$ Toeplitz-structured data matrix (13.109) has exactly q nonzero singular values provided that $q \leq p$ and $N - p > p$.

The exponential modeling characterization spelled out in this theorem is only applicable to data that is exactly represented by an exponential model. In most practical applications, however, it is found that the data being analyzed can only be approximately represented by an exponential model of reasonably small order. For such situations, it is conceptually possible to employ the concept of signal restoration to slightly perturb the given data set so that the perturbed data set is exactly represented by a qth-order exponential model. To achieve this objective, we need to introduce signal attributes that facilitate this goal. From Theorem 12, it is apparent that the ideal data matrix should be contained in the two attribute sets

$$C^{(q)} = \{Y \in C^{(N-p) \times (p+1)} \quad \text{which have rank } q\}$$

$$C_T = \{Y \in C^{(N-p) \times (p+1)} \quad \text{which have Toeplitz structure}\}$$

The attribute set C_T is a closed subspace and therefore possesses a prerequisite property needed for signal restoration. On the other hand, attribute set $C^{(q)}$ is not convex, which seemingly precludes us from using the sequence of successive projections algorithm for signal restoration. Theorem 8, however, indicates that the associated rank-q operator $P^{(q)}$ is closed. We may therefore employ the sequence of successive projections algorithm to effect the desired signal restoration. This algorithm takes the form

$$X_k = P_T P^{(q)}(X_{k-1}) \qquad \text{for } k \geq 1 \tag{13.110}$$

where the projection operators P_T and $P^{(q)}$ are described in Theorems 10 and 8, respectively. The initial data matrix used in this iterative scheme is set equal to the given data matrix (13.109), that is, $X_o = X$.

To implement algorithm (13.110), we first generate the rank-q approximation of the data matrix X. The corresponding matrix $P^{(q)}(X)$ is generally found to be non-Toeplitz in structure. To recover the prerequisite Toeplitz structure, we next apply projection operator P_T to matrix $P^{(q)}(X)$ to complete the first iteration of the signal restoration algorithm. It is generally found that this new Toeplitz-structured data matrix $X_1 = P_T P^{(q)}(X)$ has full rank. It is closer to a rank-q matrix, however, than was the original data matrix X. The first iteration has therefore led to a data matrix whose elements comprise a data sequence that is more compatible with a qth-order exponential model. Often, this first iteration is sufficient in many modeling applications.

To obtain a data sequence that is exactly representable by a qth-order exponential model, we continue this iterative process in an obvious manner. In particular, one sequentially computes the data matrices $X_{k+1} = P_T P^{(q)}(X_k)$ for $k = 0, 1, 2, \ldots$ until the rank of data matrix X_{k+1} is deemed sufficiently close to q. Since the projection operator $P^{(q)}$ and P_T are each closed, we are assured that this iterative process will eventually converge to a Toeplitz-structured data matrix of rank q. It has been empirically determined that the algorithmic process converges in a rapid fashion and typically takes from three to ten iterations for small-dimensioned matrices. Furthermore, the resulting enhanced data matrix has data elements that generally provide a better representation of the underlying signal components than did the original data. The restoration process has therefore effectively stripped away noise that contaminates the original data. We will now examine a special case of data restoration that has important practical applications.

Sinusoidal Signal Identification

In a surprisingly large number of important signal processing applications, the primary objective is that of identifying sinusoidal components in noise-corrupted data. For example, multiple plane waves incident on an equispaced linear array produce complex sinusoidal steering vectors. To identify sinusoidal signals in data, a widely employed procedure is to first form the data matrix whose upper and lower halves correspond to the forward and backward prediction equations associated with the data, respectively. If the data under analysis is specified by $x(1), x(2), \ldots, x(N)$, the associated forward-backward data matrix then takes the form

$$X_{fb} = \begin{bmatrix} X \\ \cdots \\ J_{N-p}\overline{X}J_{p+1} \end{bmatrix} \qquad (13.111)$$

In this expression, the forward data matrix X is given by (13.109) while J_n designates the *order reversal matrix* whose elements are all zero except for ones which appear along its main antidiagonal [i.e., $J_n(i,j) = \delta(i,n+1-j)$]. The matrix $J_{N-p}\overline{X}J_{p+1}$ appearing in the lower half of the data matrix (13.111) corresponds to the backward prediction equations. The matrices X and $J_{N-p}\overline{X}J_{p+1}$ are seen to have a Toeplitz and Hankel structure, respectively. The combined forward-backward data matrix X_{fb} is therefore said to have a block Toeplitz-Hankel structure.

If the data is noise-free and composed of q complex sinusoids, then the block Toeplitz-Hankel data matrix X_{fb} has rank q. Various procedures for identifying the q (with $q < p$) complex sinusoidal signal components when noise is present have been proposed. Two related SVD-based methods that appeared at the same time have proven effective for this purpose and are now briefly described. In each method, the forward-backward data matrix (13.111) is first decomposed as

$$X_{fb} = [\mathbf{x}_1 \ X_r] \qquad (13.112)$$

where $\mathbf{x}_1$ denotes the first column of X_{fb} and X_r its remaining p columns. In the method developed by the author [Cadzow, 1982], the rank-q approximation of the total forward-backward data matrix X is first determined using the truncated SVD (i.e., $X_{fb}^{(q)}$). Finally, the related coefficient vector is then specified by

$$\mathbf{a}_c^o = -[X_r^{(q)}]^\dagger \mathbf{x}_1^{(q)} \qquad (13.113)$$

where $\dagger$ designates the pseudo matrix inverse operator while $\mathbf{x}_1^{(q)}$ and $X_r^{(q)}$ are the first and remaining p columns, respectively, of the rank-q approximation matrix $X^{(q)}$. In a very similar fashion, the Tufts-Kumaresan method is obtained by first determining the rank-q approximation of submatrix X_r which is here denoted by $X_{rkt}^{(q)}$ [Tufts and Kumaresan, 1982]. The corresponding coefficient vector is then given by

$$\mathbf{a}_{kt}^o = -[X_{rkt}^{(q)}]^\dagger \mathbf{x}_1 \qquad (13.114)$$

It is to be noted that although these two coefficient vectors are similar, the latter approach excludes the first column of X in the rank-q approximation. As such, it does not achieve the full benefits of the SVD decomposition and therefore typically yields marginally poorer performance, as the example to follow illustrates. In both methods, the component sinusoids may be graphically identified by peaks that appear in the *detection functional*

$$d(f) = \frac{1}{\left| \sum_{n=0}^p a_n^o e^{j2\pi fn} \right|} \qquad (13.115)$$

Rank-Reduced Data Matrix Enhancement

Although these algorithms are effective in identifying sinusoidal components, the application of signal restoration can improve their performance. In particular, one simply applies the signal restoration algorithm (13.110) with mapping P_T replaced by P_{TH}. The restoration algorithm for determining that rank-q data matrix with the block Toeplitz-Hankel structure (13.111) that approximates X is then given by

$$X_k = P_{TH}P^{(q)}(X_{k-1}) \qquad \text{for } k \geq 1 \qquad (13.116)$$

The mapping $P_{TH}(X)$ determines the block Toeplitz-Hankel matrix that lies closest to matrix X in the Frobenius norm. Implementation of $P_{TH}(X)$ is realized in a fashion similar to $P_T(X)$. The mod-

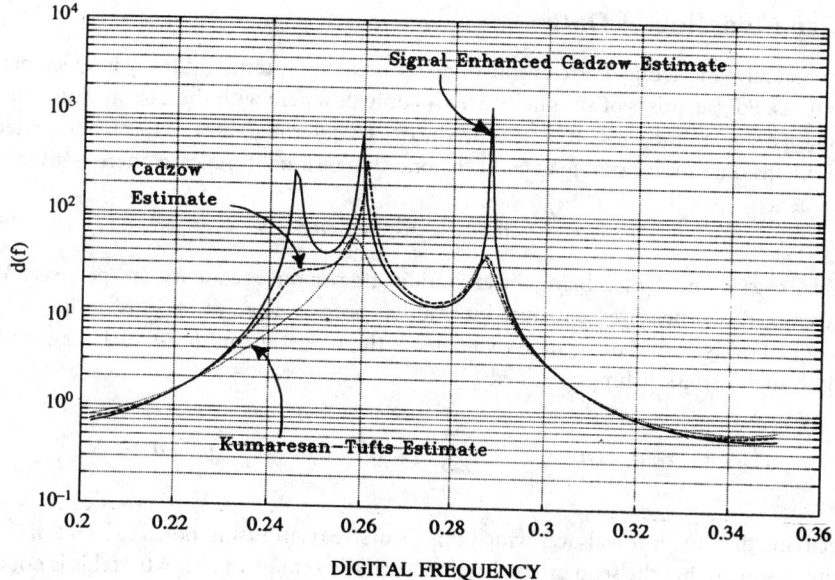

FIGURE 13.12   Sinusoid estimates.

ified matrix achieved through iteration (13.116) is then used in expression (13.113) or (13.114) to provide an enhanced coefficient vector estimate.

Example 7.   To illustrate the effectiveness of signal restoration preprocessing let us consider the following data set

$$x(n) = e^{j2\pi(0.25)n} + e^{j2\pi(0.26)n} + e^{j2\pi(0.29)n} + w(n) \qquad 1 \le n \le 24 \qquad (13.117)$$

where $w(n)$ is Gaussian white noise whose real and imaginary components have standard deviation 0.05. When estimation procedures (13.113) and (13.114) are applied to the original data with $p = 17$ (the choice advocated by Tufts and Kumaresan [1982]) and $m = 3$, the spectral estimates shown in Fig. 13.12 arise. Each estimate produces two clear peaks with the Cadzow estimate also providing a hint of a third sinusoidal peak. When signal restoration is applied to the original block Toeplitz-Hankel data matrix (13.111), the enhanced Cadzow estimate also shown in Fig. 13.12 arises. This enhanced estimate clearly identifies the three sinusoids and their corresponding frequency estimates in an accurate fashion. The advantages accrued through signal restoration are made evident by this example.

Subsequence Restoration

It is possible to obtain further performance improvements in modeling data as a linear combination of exponentials by employing the concept of *data decimation*. As an example, any data sequence may be decomposed into two subsequences that are composed of its even and odd samples, respectively. If the data under analysis is exactly modeled as a linear combination of exponential signals, it is a simple matter to establish that the associated even and odd decimated subsequences are similarly characterized. The component exponentials in the even and odd decimated subsequences are found to equal the square of those in the original data. This decomposition procedure can be continued in an obvious fashion to generate an additional three subsequences, each of which is composed of every third sample of the original data, and so forth. This data decimation procedure has been combined with the signal restoration technique presented in this subsection to effect improved estimation performance. The interested reader will find this approach described in Cadzow and Wilkes [1991].

Recursive Modeling of Data

The linear recursive modeling of excitation-response data is of interdisciplinary interest in a variety of applications. For purposes of simplicity, we will only deal here with the case in which the data is dependent on a single time variable. The procedure to be described, however, is readily extended to the multidimensional time variable case. In the one-dimensional time case, there is given the pair of data sequences

$$(x(n), y(n)) \qquad \text{for } 0 \le n \le N \tag{13.118}$$

We will refer to $x(n)$ and $y(n)$ as being the excitation and response sequences, respectively. Without loss of generality, the measurement time interval has been selected to be $[0, N]$. The pair of data sequences (13.118) is said to be recursively related if there exist a_k and b_k coefficients such that the following recursive relationship is satisfied:

$$y(n) + \sum_{k=1}^{p} a_k y(n - k) = \sum_{k=0}^{q} b_k x(n - k) \qquad \text{for } 0 \le n \le N \tag{13.119}$$

In specifying the time interval over which this recursive relationship holds to be $0 \le n \le N$, it has been tacitly assumed that the sequence pairs are identically zero prior to $n = 0$. If this is not the case, then the time interval over which relationship (13.119) holds must be changed to $\max(p,q) \le n \le N$. In the analysis to follow, it is assumed that the appropriate time interval is $0 \le n \le N$. Modification of this analysis for the time interval $\max(p,q) \le n \le N$ is straightforward and not given.

It will be convenient to represent recursive relationships (13.119) in a matrix format so as to draw upon algebraic attributes that characterize an associated data matrix. This matrix representation takes the form

$$
\begin{bmatrix}
y(0) & 0 & \cdots & 0 \\
y(1) & y(0) & \cdots & 0 \\
\vdots & \vdots & \vdots & \vdots \\
y(N) & y(N-1) & \cdots & y(N-p)
\end{bmatrix}
\begin{bmatrix}
1 \\
a_1 \\
\vdots \\
a_p
\end{bmatrix}
=
\begin{bmatrix}
x(0) & 0 & \cdots & 0 \\
x(1) & x(0) & \cdots & 0 \\
\vdots & \vdots & \vdots & \vdots \\
x(N) & x(N-1) & \cdots & x(N-q)
\end{bmatrix}
\begin{bmatrix}
b_o \\
b_1 \\
\vdots \\
b_q
\end{bmatrix}
$$

or, equivalently,

$$Y_p \mathbf{a}_p = X_q \mathbf{b}_q \tag{13.120}$$

In this latter representation, Y_p and X_q are referred to as the $(N+1) \times (p+1)$ *response matrix* and the $(N+1) \times (q+1)$ *excitation matrix*, respectively. Similarly, $\mathbf{a}_p$ and $\mathbf{b}_q$ are the recursive coefficient vectors identifying the recursive operator. With this preliminary development, the basic attributes characterizing recursively related data are now formally spelled out [see Cadzow and Solomon, 1988].

Theorem 13. Let the excitation-response data $(x(n), y(n))$ for $0 \le n \le N$ be related through a reduced-order recursive relationship $(\bar{p}, \bar{q})$ in which $\bar{p} \le p$ and $\bar{q} \le q$. It then follows that the extended-order recursive relationship (13.120) will always have a solution. Moreover, if the excitation and response matrices are full rank so that

$$\text{rank}[X_q] = q + 1 \qquad \text{and} \qquad \text{rank}[Y_p] = p + 1 \tag{13.121}$$

then all solutions are expressible as

$$
\begin{bmatrix}
\mathbf{a}_p \\
\cdots \\
\mathbf{b}_q
\end{bmatrix}
= \sum_{k=1}^{s} \alpha_k \mathbf{v}_k \tag{13.122}
$$

with the α_k parameters selected to ensure that the first component of $\mathbf{a}_p$ is one as required. The upper limit in this sum is given by $s = 1 + \min(p - \tilde{p}, q - \tilde{q})$ while the vectors $\mathbf{v}_k$ correspond to the s eigenvectors associated with zero eigenvalue of multiplicity s of matrix $D_{p,q}^* D_{p,q}$ where $D_{p,q}$ is the $(N+1) \times (p+q+2)$ composite data matrix

$$D_{p,q} = [Y_p \, \vdots \, -X_q] \tag{13.123}$$

Furthermore, the transfer function associated with any solution to the system of equations (13.122) reduces to (after pole-zero cancelation) the underlying reduced-order transfer function of order $(\tilde{p}, \tilde{q})$.

When using the algebraic characteristics of the composite matrix to form a rational model of empirical data pairs, there is much to be gained by using an over-ordered model. By taking an over-ordered approach, the recursive model parameters are made less sensitive to quirks in the empirical data. A more detailed explanation of this concept is found in Cadzow and Solomon [1988].

Signal-Enhanced Data Modeling

From the above development, it follows that when the observed data $\{(x(n), y(n))\}$ are perfectly represented by a recursive relationship of order $(\tilde{p}, \tilde{q})$, the composite data matrix will satisfy the two attributes

- $D_{p,q}$ is a block Toeplitz matrix
- $D_{p,q}$ has nullity $s = 1 + \min(p - \tilde{p}, q - \tilde{p})$

In most practical applications, the given data observations are not perfectly represented by a low-order recursive relationship. This is typically manifested in the composite data matrix being full rank. To use the concept of signal restoration to achieve a suitably good approximate recursive model, we could suitably modify the given excitation-response data so that the modified data has an associated composite data matrix which satisfies the above two attributes. The signal restoration algorithm associated with this objective is given by

$$D_k = P_T P^{(p+q+2-s)}(D_{k-1}) \qquad \text{for } k \geq 1 \tag{13.124}$$

where the initial composite matrix $D_o = [Y_p \, \vdots \, -X_q]$ has the given original excitation-response data as entries. We have dropped the subscript p,q in the composite data matrix to simplify the notation.

The signal restoration theorem ensures that the composite data matrix sequence (13.124) will contain a subsequence that converges to a composite data matrix which satisfies the prerequisite block Toeplitz–nullity s attributes. The recursive coefficient vectors as specified by (13.122) when applied to the convergent composite data matrix will typically give a satisfactory model of the data. It should be noted that in some applications it is known that either the excitation or the response data is accurate and should not be perturbed when applying the operator $P^{(p+q+2-s)}$. This is readily accomplished by inserting the original block after the rank reduction projection mapping $P^{(p+q+2-s)}$ has been applied to D_{k-1}. This is illustrated in the following numerical example.

Example 8. Let us apply the above signal restoration procedure to model recursively an unknown system when the input signal $x(n)$ and a noisy observation of the output signal $y(n)$ are available. The previously described signal-enhanced data modeling technique will be used with $p = q$ for simplicity. Clearly, if the excitation-response data were noiseless and the unknown system could be modeled by an autoregressive moving average (ARMA) (p,p) system, $D_{p,p}$ would not have full rank. The presence of noise in the response data will cause $D_{p,p}$ to have full rank, and the signal restoration algorithm will be applied to produce a block Toeplitz matrix having nullity $s = 1$. Since the input data is known exactly, its block will be inserted after each low rank approximation step. From (13.122) above it is clear that the resulting solution for the $\mathbf{a}_p$ and $\mathbf{b}_p$ coefficients will consist of the eigenvector associated with the zero eigenvalue of $D_{p,p}^* D_{p,p}$.

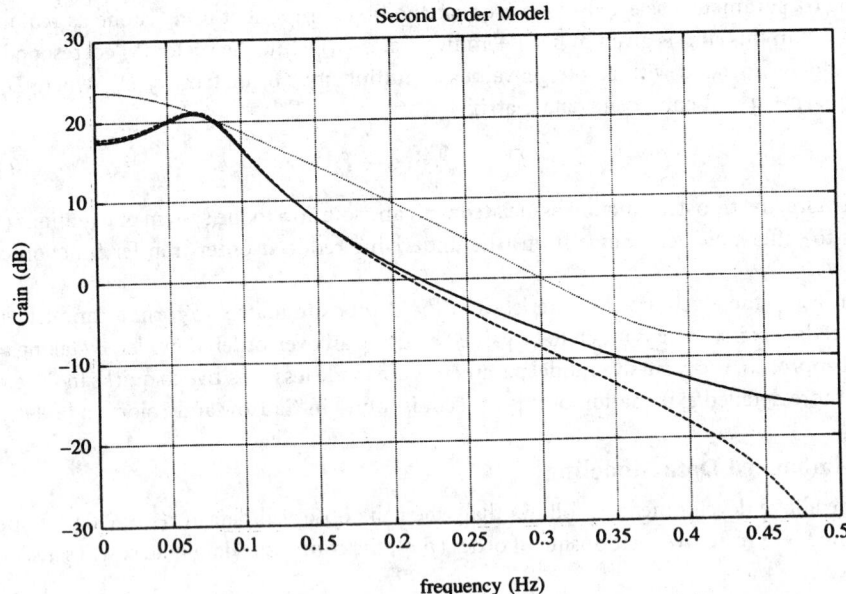

FIGURE 13.13   Second-order model results. — : true; - - - : signal enhanced; ···· : no enhancement.

The system to be identified has the following ARMA relationship:

$$y_a(n) - 1.5y_a(n-1) + 0.7y_a(n-2) = x(n-1) + 0.5x(n-2) \qquad (13.125)$$

and the observed output is $y(n) = y_a(n) + w(n)$, where $w(n)$ is the measurement noise at the output. In this example, the input signal is zero-mean unit variance white Gaussian noise. The signal-to-noise ratio at the output is 12 dB, and 300 samples of the input and output signals are used. The results for $p = 2$ are shown in Fig. 13.13; the true frequency response is given by the solid line, the dotted line is the solution that would result if no signal restoration were performed, and the dashed line depicts the solution after signal restoration (25 iterations).

Conclusion

The signal restoration algorithm has been shown to provide a useful means for solving a variety of important signal processing problems. In addition to the problems described in this chapter, it has been successfully applied to the missing data problem, deconvolution, and high-dimensional filter synthesis. Very useful results can be achieved by innovatively introducing signal attributes that characterize the underlying information signals.

Defining Terms

Attribute set: A set of vectors (signals) lying in a metric space that possess prescribed properties.

Closed convex sets: A set of vectors C such that if $\mathbf{x}, \mathbf{y} \in C$ then $\lambda \mathbf{x} + (1 - \lambda)\mathbf{y} \in C$ for all $0 \leq \lambda \leq 1$.

Method of successive projections: An iterative procedure for modifying a signal so that the modified signal has properties which match an ideal objective.

Projection mapping: A mathematical procedure for determining a vector (signal) lying in a prescribed set that lies closest to a given vector.

Signal restoration: The restoring of data that has been corrupted by instrumentation dynamics and noise.

Singular value decomposition: A procedure for representing a matrix as a sum of positive weighted orthogonal outer products.

Structured matrix set: A set of common dimensioned matrices that have a prescribed algebraic structure (e.g., Toeplitz, Hankel, Hermitian).

References

L. M. Bregman, "The method of successive projection for finding the common point of convex sets," *Soviet Mathematics-Doklady*, vol. 6, pp. 688–692, 1965.

J. A. Cadzow, "Signal enhancement: A composite property mapping algorithm," *IEEE Trans. Acoustics, Speech and Signal Processing*, vol. ASSP-36, no. 1, pp. 49–62, January 1988.

J. A. Cadzow, "Spectral estimation: An overdetermined rational model equation approach," *Proc. IEEE, Special Issue on Spectral Analysis*, pp. 907–939, September 1982.

J. A. Cadzow and O. M. Solomon, "Algebraic approach to system identification," *IEEE Trans. Acoustics, Speech and Signal Processing*, vol. ASSP-34, no. 3, pp. 462–469, June 1988.

J. A. Cadzow and Y. Sun, "Sequences with positive semidefinite Fourier transforms," *IEEE Trans. Acoustics, Speech and Signal Processing*, vol. ASSP-34, no. 6, pp. 1502–1510, December 1986.

J. A. Cadzow and D. M. Wilkes, "Enhanced rational signal modeling," *Signal Processing*, vol. 25, no. 2, pp. 171–188, November 1991.

P. L. Combettes and H. J. Trussel, "Method of successive projections for finding a common point of sets in metric spaces," *JOTA*, vol. 67, no. 3, pp. 487–507, December 1990.

R. L. Dykstra, "An algorithm for restricted least squares regression," *Journal Amer. Stat. Assoc.*, vol. 78, pp. 837–842, 1983.

R. L. Dykstra and J. P. Boyle, "An algorithm for least squares projection onto the intersection of translated, convex cones," *Journal Statistical Plann. Inference*, vol. 15, pp. 391–399, 1987.

N. Gaffke and R. Mathar, "A cyclic projection algorithm via duality," *Metrika*, vol. 36, pp. 29–54, 1989.

L. G. Gubin, B. T. Polyak, and E. V. Raik, "The method of projections for finding the common point of sets," *USSR Computational Mathematics and Mathematical Physics*, vol. 7, pp. 1–24, 1967.

I. Halperin, "The product of projection operators," *Acta Scientiarum Mathematicarum*, vol. 23, pp. 96–99, 1962.

S.-P. Han, "A successive projection method," *Mathematical Programming*, vol. 40, pp. 1–14, 1988.

D. G. Luenberger, *Optimization by Vector Space Methods*, New York: John Wiley, 1969.

H. D. Mittelmann and J. A. Cadzow, "Continuity of closest rank-p approximations to matrices," *IEEE Transactions on Acoustics, Speech, and Signal Processing*, vol. ASSP-35, no. 8, pp. 1211–1212, August 1987.

N. Ottavy, "Strong convergence of projection-like methods in Hilbert spaces," *J. Optimization Theory and Applications*, vol. 56, pp. 433–461, 1988.

W. J. Stiles, "Closest point maps and their product," *Nieuw Archief voor Wiskunde*, vol. 13, pp. 212–225, 1965.

H. J. Trussel and M. R. Civanlar, "The feasible solution in signal restoration," *IEEE Trans. Acoustics, Speech and Signal Processing*, vol. ASSP-32, pp. 201–212, 1984.

D. W. Tufts and R. Kumaresan, "Estimation of frequencies of multiple sinusoids: Making linear prediction perform like maximum likelihood," *Proc. IEEE, Special Issue on Spectral Analysis*, pp. 975–989, September 1982.

J. von Neumann, *Functional Operators*, vol. 2 (Annals of Mathematics Studies, no. 22), Princeton, N.J., 1950. Reprinted from mimeographed lecture notes first distributed in 1933.

D. C. Youla, "Generalized image restoration by the method of alternating orthogonal projections," *IEEE Trans. Circuits and Systems*, vol. CAS-25, pp. 694–702, Sept. 1978.

W. I. Zangwill, *Nonlinear Programming: A Unified Approach*, Englewood Cliffs, N.J.: Prentice-Hall, 1969.

Further Information

The monthly *IEEE Transactions on Acoustics, Speech, and Signal Processing* frequently publishes articles on the theory and application of signal restoration and recovery. Signal restoration concepts were discussed in articles published in the January 1988 issue (pp. 49–62), in the March 1989 issue (pp. 393–401), and in the May 1990 issue (pp. 778–786).

The *IEEE Transactions on Circuits and Systems* also publishes signal restoration application papers. Examples are to be found in the September 1975 issue (pp. 735–742) and the September 1978 issue (pp. 694–702).

Image restoration articles appear in the *IEEE Transactions on Medical Imaging* as illustrated by the articles that appeared in the October 1992 issue (pp. 81–94) and the January 1984 issue (pp. 91–98).

14

Speech Signal Processing

Jerry D. Gibson
Texas A&M University

Yariv Ephraim
AT&T Bell Laboratories
George Mason University

Jesse W. Fussell
The Johns Hopkins University

Lynn D. Wilcox
Xerox Palo Alto Research Center

Marcia A. Bush
Xerox Palo Alto Research Center

14.1 Coding, Transmission, and Storage 279
   Speech Quality and Intelligibility • Predictive Coders • Frequency Domain Coders • Speech Coding Standards • Performance Comparisons • Summary and Conclusions
14.2 Speech Enhancement and Noise Reduction 287
   Models and Performance Measures • Signal Estimation • Source Coding • Signal Classification • Comments
14.3 Analysis and Synthesis 298
   Analysis of Excitation • Fourier Analysis • Linear Predictive Analysis • Homomorphic (Cepstral) Analysis • Speech Synthesis
14.4 Speech Recognition 306
   Speech Recognition System Architecture • Signal Pre-Processing • Dynamic Time Warping • Hidden Markov Models • State-of-the-Art Recognition Systems

14.1 Coding, Transmission, and Storage

Jerry D. Gibson

Interest in speech coding is motivated by a wide range of applications, including commercial telephony, digital cellular mobile radio, military communications, voice mail, speech storage, and future personal communications networks. The goal of speech coding is to represent speech in digital form with as few bits as possible while maintaining the intelligibility and quality required for the particular application. At higher bit rates, such as 64 and 32 kilobits/second (kbits/s), achieving good quality and intelligibility is not too difficult, but as the desired bit rate is lowered to 16 kbits/s and below, the problem becomes increasingly challenging. Furthermore, there are almost always other constraints. Certainly, complexity is an omnipresent issue, but depending upon the application, other parameters become important.

For example, for the 32-kbits/s speech coding **standard,** the CCITT (International Telephone and Telegraph Consultative Committee), not only required highly intelligible, high-quality speech, but the coder also had to have low delay, withstand independent bit error rates up to 10^{-2}, have acceptable performance degradation for several synchronous or asynchronous tandem connections, and pass some voiceband modem signals. Other applications may have different criteria. Digital cellular mobile radio in the United States has no low delay or voiceband modem signal requirements, but the data rate needed is 8 kbits/s and the transmission medium (usually called the channel) can be very noisy and have relatively long fades. These considerations can have a great effect on the speech coder chosen for a particular application.

As the speech coder data rates drop to 16 kbits/s and below, perceptual criteria that take into account human auditory response begin to play a prominent role. For time domain coders, which are currently **analysis-by-synthesis** predictive coders, the perceptual effects are incorporated by using a frequency weighted error criterion. The frequency domain coders, which tend to be subband coders or transform-based coders, try to include perceptual effects by allocating the available bit rate nonuniformly to the various frequency bands or components. The message is clear—modeling of the human auditory response can substantially improve speech coder performance at 4–16 kbits/s.

In this section we describe the basic components of time domain predictive coders and frequency domain coders and survey several of the important speech coding standards. We begin with a discussion of the various measures used to indicate speech quality and intelligibility.

Speech Quality and Intelligibility

To compare the performance of two speech coders, it is necessary to have some indicator of the intelligibility and quality of the speech produced by each coder. The term *intelligibility* usually refers to whether the output speech is easily understandable, while the term *quality* ideally is an indicator of how natural the speech sounds. It is possible for a coder to produce highly intelligible speech that is low quality in that the speech may sound very machine-like and the speaker is not identifiable. On the other hand, it is unlikely that unintelligible speech would be called high quality, but there are situations in which perceptually pleasing speech does not have high intelligibility. We briefly discuss here the most common measures of intelligibility and quality used in formal tests of speech coders.

The mean opinion score (MOS) is a performance measure often used by a number of workers, including AT&T [Jayant and Noll, 1984]. To establish a MOS for a coder, listeners are asked to classify the speech coder output as to whether it is excellent (5), good (4), fair (3), poor (2), or bad (1). Alternatively, the listeners may be asked to classify the coded speech according to the perceptible distortion present by associating the output with one of the impairment categories: imperceptible (5), barely perceptible but not annoying (4), perceptible and annoying (3), annoying but not objectionable (2), or very annoying and objectionable (1). The numbers in parentheses are used to assign a numerical value to the subjective evaluations, and the numerical ratings of all listeners are averaged to produce a MOS for the coder. Many times, the standard deviation of the numerical ratings is also computed to aid one in assessing the utility of the MOS thus obtained. A MOS between 4.0 and 4.5 usually indicates high quality. For example, 8-bit $\mu = 255$ log-PCM (pulse code modulation) was recently judged to have a MOS of about 4.5 with a standard deviation of near 0.6.

It is important to compute the variance of MOS values since large variances indicate unreliability of the test. A large variance can occur because of listeners not knowing what such categories as *good* and *bad* mean, and it is sometimes useful to present examples of good and bad speech to the listeners before the test to calibrate the 5-point scale [Papamichalis, 1987]. One study revealed that MOS scores taken under similar circuit conditions in different countries with native languages and listeners did not easily translate between locations; that is, the MOS needed to be adjusted to get a reliable indicator of quality [Goodman and Nash, 1982]. The MOS values for a variety of speech coders and noise conditions are given by Daumer [1982].

The modulated noise reference unit (MNRU)/opinion equivalent Q is being used more often recently because of its appearance in a CCITT recommendation for waveform coder evaluation [CCITT, 1988; Kitawaki and Nagabuchi, 1988]. In a paired comparison or opinion test, the coded speech is compared to a reference signal with a fixed, but adjustable, level of speech-correlated noise generated using the MNRU system described in CCITT, 1988. Reference signals with different signal to speech-correlated noise ratios, which is the Q in decibels, are obtained by adjusting the relative gains of the attenuator/amplifiers in two different paths, one path for the direct speech and one path for the noise-modulated speech. That Q value for which the subjective match of the MNRU output with the coded speech is best is the quantitative performance indicator. There are narrowband and wideband MNRU systems, and the narrowband Q is denoted by Q_N. This is a fairly accurate method for speech quality assessment for waveform coders at 16 kbits/s and above, since the MNRU distortion tends to mimic waveform coder noise. For coders with other types of distortion, such as the analysis-by-synthesis predictive coders at 8 kbits/s and lower rates, it is not useful.

The diagnostic rhyme test (DRT) was devised by Voiers [1977a] to test the intelligibility of coders known to produce speech of lower quality. Rhyme tests are so named because the listener must determine which consonant was spoken when presented with a pair of rhyming words; that is, the listener is asked to distinguish between word pairs such as meat-beat, pool-tool, saw-thaw, and caught-taught. Each pair of words differs in only one of six phonemic attributes: voicing, nasality, sustention, sibilation, graveness, and compactness. Specifically, the listener is presented with one spoken word from a pair and asked to decide which word was spoken. The final DRT score is the percent responses computed according to

$$P = \frac{R - W}{T} \times 100 \tag{14.1}$$

where R is the number correctly chosen, W is the number of incorrect choices, and T is the total of word pairs tested. Usually, $75 \leq DRT \leq 95$, with a *good* being about 90 [Papamichalis, 1987].

The diagnostic acceptability measure (DAM) developed by Dynastat [Voiers, 1977b] is an attempt to make the measurement of speech quality more systematic. For the DAM, it is critical that the listener crews be highly trained and repeatedly calibrated in order to get meaningful results. The listeners are each presented with sentences taken from the Harvard 1965 list of phonetically balanced sentences, such as "Cats and dogs each hate the other" and "The pipe began to rust while new," that have been processed by the speech coder of interest. The listener is asked to assign a number between 0 and 100 to characteristics in three classifications—signal qualities, background qualities, and total effect. The ratings of each characteristic are weighted and used in a multiple nonlinear regression. Finally, adjustments are made to compensate for listener performance. A typical DAM score is 45–55%, with 50% corresponding to a *good* system [Papamichalis, 1987].

Predictive Coders

Predictive coders attempt to model the speech signal at any time instant as a predicted value plus an error term. The most familiar predictive coder is the differential pulse code modulation (DPCM) system shown in Fig. 14.1. In DPCM the predicted value at time instant k, $\hat{s}(k \mid k-1)$, is subtracted from the input signal at time k, $s(k)$, to produce the prediction error signal $e(k)$. The prediction error is then quantized and the quantized prediction error, $e_q(k)$, is coded (represented as a binary number) for transmission to the receiver. Simultaneously with the coding, $e_q(k)$ is summed with

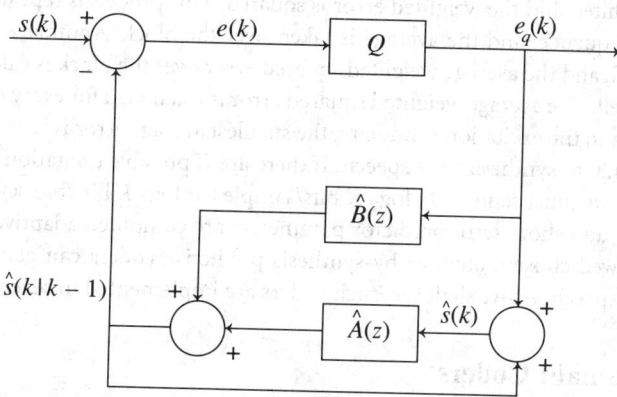

FIGURE 14.1(a) Differential encoder transmitter with a pole-zero predictor.

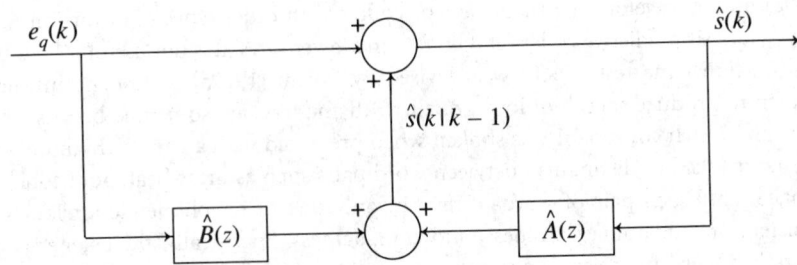

FIGURE 14.1(b) Receiver for a differential encoder transmitter with a pole-zero predictor.

$\hat{s}(k \mid k - 1)$ to yield a reconstructed version of the input sample, $\hat{s}(k)$. Assuming no channel errors, an identical reconstruction is accomplished at the receiver. At both the transmitter and receiver, the predicted value at time instant $k + 1$ is derived using reconstructed values up through time k, and the procedure is repeated.

The first DPCM systems had $\hat{B}(z) = 0$ and $\hat{A}(z) = \sum_{i=1}^{N} a_i z^{-i}$, where z^{-1} represents unit delay, so that the predicted value was

$$\hat{s}(k \mid k - 1) = \sum_{i=1}^{N} a_i \hat{s}(k - i) \tag{14.2}$$

a weighted linear combination of past reconstructed values. Later work showed that letting $\hat{B}(z) = \sum_{j=1}^{M} b_j z^{-j}$, so that the predicted value is

$$\hat{s}(k \mid k - 1) = \sum_{i=1}^{N} a_i \hat{s}(k - i) + \sum_{j=1}^{M} b_j e_q(k - j) \tag{14.3}$$

improves the reconstructed speech. To produce high-quality, highly intelligible speech, it is necessary that the quantizer step size and the predictor coefficients be adaptive. Indeed, such an approach is the basis for the CCITT standard at 32 kbits/s, G.721 [Jayant and Noll, 1984].

To improve the speech quality further or to lower the bit rate some, a long-term (or pitch) predictor can be added, but to produce good quality speech at 16 kbits/s and below, it has been necessary to consider the class of analysis-by-synthesis coders. The encoding step in analysis-by-synthesis coders is illustrated in Fig. 14.2, where a candidate excitation sequence of some block length is applied to a long-term predictor cascaded with a short-term predictor. The reconstructed speech sample at time n is subtracted from the corresponding input speech sample, passed through a perceptual weighting filter, and the weighted error is squared. This process is repeated for each sample in the excitation sequence, and the average is taken over the block. Another candidate excitation sequence is applied, and the average weighted squared error over the block is calculated for the second sequence as well. The average weighted squared error is calculated for every excitation sequence in a restricted set, and the excitation producing the smallest average error is sent to the receiver, represented by Fig. 14.3, to synthesize the speech. If there are K possible excitation sequences of block length L in the set, we must send $(1/L) \log_2 K$ bits/sample (or $\log_2 K$ bits/block) to the receiver.

When the long- and short-term predictor parameters are computed adaptively and the perceptual weighting is well chosen, analysis-by-synthesis predictive coders can generate good-quality, highly intelligible speech. As we shall see, such coders are implemented in several recent standards.

Frequency Domain Coders

The two most prominent types of frequency domain coders today are transform coders and subband coders. Transform coders take a discrete transform of a frame of input speech samples, allo-

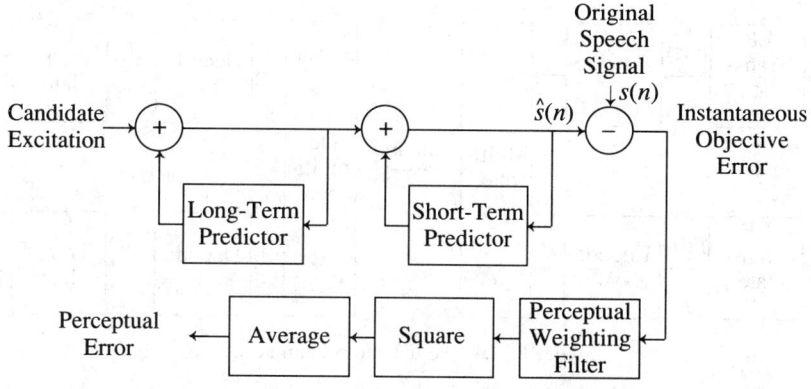

FIGURE 14.2 Analysis-by-synthesis predictive coder.

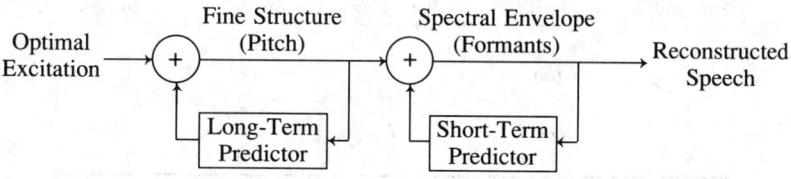

FIGURE 14.3 Receiver for Fig. 14.2.

cate bits to the various transform coefficients, quantize/code these coefficients, and transmit these quantized coefficients to the receiver. At the receiver, the frame of speech is reconstructed by taking the inverse discrete transform. Note that if the number of bits available over the frame is small (that is, the transmitted bit rate is low), some coefficients will be allocated 0 bits (discarded) and some coefficients will be coarsely quantized. This is what produces distortion in the output. However, the bit allocation procedure allows perceptually important frequency components to be quantized/coded more accurately than others. The most commonly used transform for these applications is the discrete cosine transform [Jayant and Noll, 1984].

Subband coders, represented in Fig. 14.4, digitally filter the speech into nonoverlapping (as nearly as possible) bands, each band is lowpass translated and decimated (effectively sampled at a lower rate) and each band is coded separately using PCM, DPCM, or some other method. At the receiver, the bands are decoded, upsampled, translated to their original frequency range, and summed to reconstruct the speech. By allocating a different number of bits per sample to the subbands, the perceptually more important frequency bands can be coded with greater accuracy. The design and implementation of subband coders and the speech quality produced have been greatly improved by the development of digital filters called quadrature mirror filters (QMFs). These filters allow subband overlap at the encoder, which causes aliasing, but the reconstruction filters at the receiver can be chosen to eliminate the aliasing.

The frequency bands, sampling rates, and bit allocations for a subband coder used in AT&T voice store and forward products are listed in Table 14.1. Note that the sampling rates of the subbands are only dependent on the subband bandwidth, not the full speech band. Also, attention is called to the nonuniform allocation of bits to the subbands.

Speech Coding Standards

Since the early 1980s, there has been a concerted effort toward establishing speech coding standards for a variety of applications, both nationally and internationally. Several of the most recent stan-

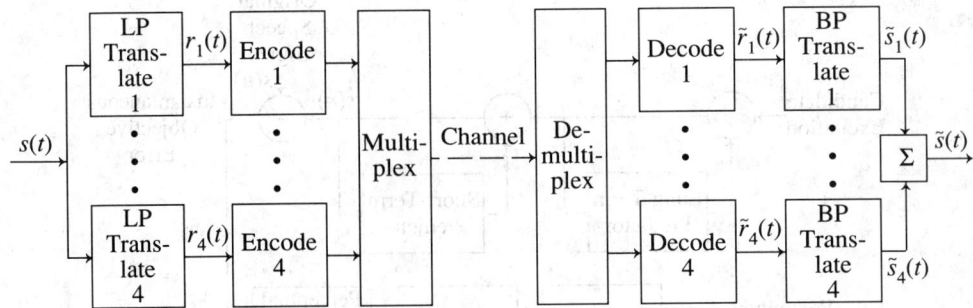

FIGURE 14.4   Four-band subband coder.

Table 14.1   Example Subband Coder Characteristics

Subband	Frequency Range (kHz)	Sampling Rate (kHz)	Bit Allocation 16 kbits/s	24 kbits/s
1	0–0.5	1	4	5
2	0.5–1.0	1	4	5
3	1.0–2.0	2	2	4
4	2.0–3.0	2	2	3
5	3.0–4.0	2	0	0

Source: J. G. Josenhans *et al.*, "Speech Processing Application Standards," *AT&T Technical Journal*, pp. 23–33, Sept./Oct. 1986. With permission.

dards are listed in Table 14.2. All of these standards, except G.711, are based upon some form of **predictive coding**. We briefly survey these standards and highlight some of their features.

The CCITT G.721 standard is an adaptive differential pulse code modulation (ADPCM) system that operates at 32 kbits/s (8000 samples/s × 4 bits/sample). The predictor has two poles and six zeros that are adaptively recomputed after every sample. The quantizer has 15 levels (not exactly 4 bits/sample), and the step size is adapted after every sample. Both the predictor parameters and the quantizer step size are calculated using only the coded quantizer output, which means that there is no need to transmit this information separately and that the coder has low delay. The G.721 ADPCM system produces highly intelligible, high-quality speech, has low delay, maintains good performance for up to four asynchronous tandems, passes V.26 2400-bits/s and V.27 4800-bits/s voiceband data, and is robust to noisy channels [Benvenuto *et al.*, 1990].

For network applications, it would be helpful if one or two of the least significant bits in a coded speech sample could be dropped, as network conditions indicate, without the speech coder transmitter being informed. Speech coders with this property are called *embedded coders*, and the G.721 ADPCM system is not an embedded coder. However, ADPCM can be modified to obtain an embedded structure, and embedded ADPCM at various rates is the G.727 standard.

Table 14.2   Speech Coding Standards

Standard	Rate (kbits/s)	Coder
CCITT G.711	64	log PCM
CCITT G.721	32	ADPCM
CCITT G.723	24, 40	ADPCM
CCITT G.727	16, 24, 32, 40	Embedded ADPCM
CCITT G.728	16	LD-CELP
GSM	13	RPE-LTP
CTIA	8	VSELP
Fed. Std. 1016	4.8	CELP
Fed. Std. 1015	2.4	LPC

The emerging CCITT standard at 16 kbits/s, designated G.728 in Table 14.2, uses *code-excited linear prediction* (CELP), which is an analysis-by-synthesis predictive coder. Low delay of less than 2.5 ms is achieved by using excitation sequences only 5 samples long and adapting the predictor parameters based only upon the transmitted excitation, which is encoded to 10-bit accuracy (using a vector quantizer). There is no long-term predictor, and the short-term predictor has 50 coefficients. The coder is not required to pass voiceband data but it is expected to operate over channels with an independent bit error rate as high as 10^{-2}.

The Group Speciale Mobile (GSM) standard is for the European digital mobile radio system. The coder is based on an analysis-by-synthesis technique called regular pulse excitation (RPE), and it has both a short-term and a long-term predictor (hence, LTP). For this coder the predictor parameters are calculated based upon the input speech over a 20-ms frame and are transmitted to the receiver along with the excitation. This coder is not low delay since it has a delay of at least 20 ms. Because the digital mobile radio channel can be very noisy and has deep fades, error control coding is added on in addition to the 13 kbits/s and the speech coder must be very robust to channel errors.

The Cellular Telecommunications Industry Association (CTIA) standard at 8 kbits/s is for digital cellular mobile radio in the United States. The speech coder is called *vector sum excited linear prediction* (VSELP), and it is an analysis-by-synthesis predictive coder. It has both long- and short-term predictors, and the coder parameters are calculated on frames of input data and transmitted to the receiver. The coder is not low delay. Selected bits are error protected and the coder must be robust to channel errors because the cellular radio channel can be poor at times. The innovation in VSELP with respect to CELP is that there are two sets of excitation sequences, called codebooks, rather than one. This reduces the complexity of the encoding step.

U.S. Federal Standard 1016 is a 4.8-kbits/s CELP system. It has both long- and short-term predictors, and their parameters are computed from input speech. The coder has a relatively large delay of 100 ms. This coder produces highly intelligible, good-quality speech in a variety of environments and is robust to independent bit errors.

U.S. Federal Standard 1015 uses an older technique called *linear predictive coding* (LPC). It is not low delay, and it produces intelligible but not necessarily good-quality speech.

Performance Comparisons

As noted earlier, no single performance measure has gained wide acceptance as an indicator of the quality and intelligibility of speech produced by a coder. Further, there is no substitute for subjective listening tests under the actual environmental conditions expected in a particular application. As a rough guide to the performance of some of the coders discussed here, we present the DRT, DAM, and MOS values in Table 14.3, which is adapted from Jayant [1990].

Table 14.3 Speech Coder Performance Comparison

Coder	DRT	DAM	MOS
64 kbits/s log PCM	95	73	4.3
32 kbits/s ADPCM	94	68	4.1
16 kbits/s LD-CELP	94[a]	70[a]	4.0
8 kbits/s CELP	93+[b]	68[a]	3.7
4.8 kbits/s CELP	93−	67−	3.0[a]
2.4 kbits/s LPC	90	54	2.5[a]

[a]Estimated value.

[b]Plus and minus signs denote lower and upper bounds, respectively.

Source: N. S. Jayant, "High-Quality Coding of Telephone Speech and Wideband Audio," *IEEE Commun. Magazine*, vol. 28, pp. 10–20. © 1990 IEEE. With permission.

From the table, it is evident that at 8 kbits/s and above, performance is quite good and that the 4.8-kbits/s CELP has substantially better performance than LPC. The CTIA 8-kbits/s VSELP is not included in the table, but it is usually stated to have a MOS of 3.5 to 4.0.

Summary and Conclusions

The availability of digital signal processing chips and the ever-widening interest in digital communications have led to an increasing demand for speech coders. The worldwide desire to establish standards in a host of applications is a primary driving force for speech coder research and development. The speech coders that are available today for operation at 16 kbits/s and below are conceptually quite exotic compared with products available less than 10 years ago.

Standards to be established in the near term include the European (GSM) and U.S. (CTIA) half-rate speech coders for digital cellular mobile radio. For the longer term, the specification of standards for forthcoming mobile personal communications networks will be a primary focus in the next 5 to 10 years.

In the preface to their book, Jayant and Noll [1984] state that "our understanding of speech and image coding has now reached a very mature point. . . ." As of 1992, this statement rings truer than ever. The field is a dynamic one, however, and the wide range of commerical applications demands continual progress.

Defining Terms

Analysis-by-synthesis: Constructing several versions of a waveform and choosing the best match.
Predictive coding: Coding of waveforms based upon a (usually) linear prediction model.
Standard: An encoding technique adopted by an industry to be used in a particular application.

References

N. Benvenuto *et al.*, "The 32-kb/s ADPCM Coding Standard," *AT&T Tech. J.*, vol. 65, pp. 12–22, Sept./Oct.1990.

CCITT, "Recommendations of the *P* series," *Red Book*, vol. V, Malaga-Torremolinos, 1984, amended Melbourne, 1988, pp. 198–203, 1988.

W. R. Daumer, "Subjective evaluation of several efficient speech coders," *IEEE Trans. Commun.*, vol. COM-30, pp. 655–662, April 1982.

D. J. Goodman and R. D. Nash, "Subjective quality of the same speech transmission conditions in seven different countries," *IEEE Trans. Commun.*, vol. COM-30, pp. 642–654, April 1982.

N. S. Jayant, "High-quality coding of telephone speech and wideband audio," *IEEE Commun. Mag.*, vol. 28, pp. 10–20, Jan. 1990.

N. S. Jayant and P. Noll, *Digital Coding of Waveforms*, Englewood Cliffs, N.J.: Prentice-Hall, 1984.

J. G. Josenhans *et al.*, "Speech processing application standards," *AT&T Tech. J.*, pp. 23–33, Sept./Oct. 1986.

N. Kitawaki and H. Nagabuchi, "Quality assessment of speech coding and speech synthesis systems," *IEEE Commun. Mag.*, pp. 36–44, Oct. 1988.

P. E. Papamichalis, *Practical Approaches to Speech Coding*, Englewood Cliffs, N.J.: Prentice-Hall, 1987.

W. D. Voiers, "Diagnostic evaluation of speech intelligibility," in *Speech Intelligibility and Recognition*, M. E. Hawley, Ed., Stroudsburg, Penn.: Dowden, Hutchinson, and Ross, 1977a.

W. D. Voiers, "Diagnostic acceptibility measure for speech communication systems," *Proceedings, IEEE ICASSP*, pp. 204–207, 1977b.

Further Information

For further information on the state of the art in speech coding, see *Advances in Speech Coding*, B. S. Atal, V. Cuperman, and A. Gersho, Eds., Kluwer, 1991. For more details on standards, consult "CCITT standards on digital speech processing," *IEEE J. Selected Areas in Communications*, vol. 6, pp. 227–234, Feb. 1988.

14.2 Speech Enhancement and Noise Reduction

Yariv Ephraim

Voice communication systems are susceptible to interfering **signals** normally referred to as **noise.** The interfering signals may have harmful effects on the performance of any speech communication system. These effects depend on the specific system being used, on the nature of the noise and the way it interacts with the clean signal, and on the relative intensity of the noise compared to that of the signal. The latter is usually measured by the **signal-to-noise ratio** (SNR), which is the ratio of the power of the signal to the power of the noise.

The speech communication system may simply be a recording which was performed in a noisy environment, a standard digital or analog communication system, or a speech recognition system for human-machine communication. The noise may be present at the input of the communication system, in the channel, or at the receiving end. The noise may be correlated or uncorrelated with the signal. It may accompany the clean signal in an additive, multiplicative, or any other more general manner. Examples of noise sources include competitive speech; background sounds like music, a fan, machines, door slamming, wind, and traffic; room reverberation; and white Gaussian channel noise.

The ultimate goal of **speech enhancement** is to minimize the effects of the noise on the performance of speech communication systems. The performance measure is system dependent. For systems which comprise recordings of noisy speech, or standard analog communication systems, the goal of speech enhancement is to improve perceptual aspects of the noisy signal. For example, improving the **quality** and **intelligibility** of the noisy signal are common goals. Quality is a subjective measure which reflects on the pleasantness of the speech or on the amount of effort needed to understand the speech material. Intelligibility, on the other hand, is an objective measure which signifies the amount of speech material correctly understood. For standard digital communication systems, the goal of speech enhancement is to improve perceptual aspects of the encoded speech signal. For human-machine speech communication systems, the goal of speech enhancement is to reduce the error rate in recognizing the noisy speech signals.

To demonstrate the above ideas, consider a "hands-free" cellular radio telephone communication system. In this system, the transmitted signal is composed of the original speech and the background noise in the car. The background noise is generated by an engine, fan, traffic, wind, etc. The transmitted signal is also affected by the radio channel noise. As a result, noisy speech with low quality and intelligibility is delivered by such systems. The background noise may have additional devastating effects on the performance of this system. Specifically, if the system encodes the signal prior to its transmission, then the performance of the speech coder may significantly deteriorate in the presence of the noise. The reason is that speech coders rely on some statistical model for the clean signal, and this model becomes invalid when the signal is noisy. For a similar reason, if the cellular radio system is equipped with a speech recognizer for automatic dialing, then the error rate of such recognizer will be elevated in the presence of the background noise. The goals of speech enhancement in this example are to improve perceptual aspects of the transmitted noisy speech signals as well as to reduce the speech recognizer error rate.

Other important applications of speech enhancement include improving the performance of:

1. Pay phones located in noisy environments (e.g., airports)
2. Air-ground communication systems in which the cockpit noise corrupts the pilot's speech
3. Teleconferencing systems where noise sources in one location may be broadcasted to all other locations
4. Long distance communication over noisy radio channels

The problem of speech enhancement has been a challenge for many researchers for almost three decades. Different solutions with various degrees of success have been proposed over the years. An excellent introduction to the problem, and review of the systems developed up until 1979, can be found in the landmark paper by Lim and Oppenheim [1979]. A panel of the National Academy of Sciences discussed in 1988 the problem and various ways to evaluate speech enhancement systems. The panel's findings were summarized in Makhoul *et al.* [1989]. Modern statistical approaches for speech enhancement were recently reviewed in Boll [1992] and Ephraim [1992].

In this section the principles and performance of the major speech enhancement approaches are reviewed, and the advantages and disadvantages of each approach are discussed. The signal is assumed to be corrupted by additive statistically independent noise. Only a single noisy version of the clean signal is assumed available for enhancement. Furthermore, it is assumed that the clean signal cannot be preprocessed to increase its robustness prior to being affected by the noise. Speech enhancement systems which can either preprocess the clean speech signal or which have access to multiple versions of the noisy signal obtained from a number of microphones are discussed in Lim [1983].

This presentation is organized as follows. In the second section the speech enhancement problem is formulated and commonly used models and performance measures are presented. In the next section signal estimation for improving perceptual aspects of the noisy signal is discussed. In the fourth section source coding techniques for noisy signals are summarized, and the last section deals with recognition of noisy speech signals. Due to the limited number of references (10) allowed in this publication, tutorial papers are mainly referenced. Appropriate credit will be given by pointing to the tutorial papers which reference the original papers.

Models and Performance Measures

The goals of speech enhancement as stated in the first section are to improve perceptual aspects of the noisy signal whether the signal is transmitted through analog or digital channels and to reduce the error rate in recognizing noisy speech signals. Improving perceptual aspects of the noisy signal can be accomplished by estimating the clean signal from the noisy signal using perceptually meaningful estimation performance measures. If the signal has to be encoded for transmission over digital channels, then source coding techniques can be applied to the given noisy signal. In this case, a perceptually meaningful fidelity measure between the clean signal and the encoded noisy signal must be used. Reducing error rate in speech communication systems can be accomplished by applying optimal signal classification approaches to the given noisy signals. Thus the speech enhancement problem is essentially a set of signal estimation, source coding, and signal classification problems.

The probabilistic approach for solving these problems requires explicit knowledge of the performance measure as well as the probability laws of the clean signal and noise process. Such knowledge, however, is not explicitly available. Hence, mathematically tractable performance measures and statistical models which are believed to be meaningful are used. In this section we briefly review the most commonly used statistical models and performance measures.

The most fundamental model for speech signals is the Gaussian **autoregressive (AR) model**. This model assumes that each 20- to 40-msec segment of the signal is generated from an excitation signal which is applied to a linear time-invariant all-pole filter. The excitation signal comprises a mixture of white Gaussian noise and a periodic sequence of impulses. The period of that sequence is determined by the pitch period of the speech signal. This model is described in Fig. 14.5. Generally, the

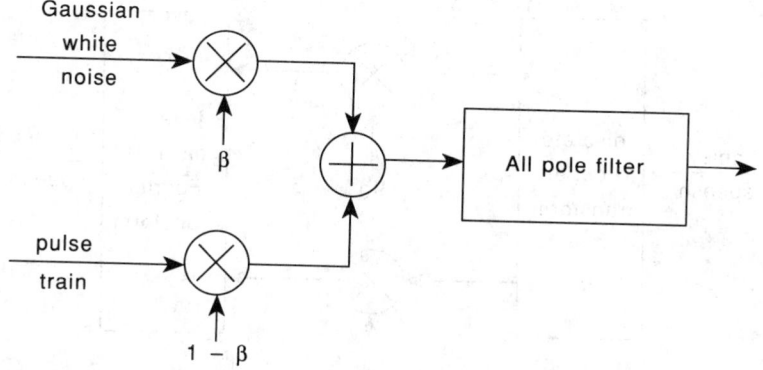

FIGURE 14.5 Gaussian autoregressive speech model.

excitation signal represents the flow of air through the vocal cords and the all-pole filter represents the vocal tract. The model for a given sample function of speech signals, which is composed of several consecutive 20- to 40-msec segments of that signal, is obtained from the sequence of AR models for the individual segments. Thus, a linear time-varying AR model is assumed for each sample function of the speech signal. This model, however, is slowly varying in accordance with the slow temporal variation of the articulatory system. It was found that a set of approximately 2048 prototype AR models can reliably represent all segments of speech signals. The AR models are useful in representing the short time spectrum of the signal, since the spectrum of the excitation signal is white. Thus, the set of AR models represents a set of 2048 spectral prototypes for the speech signal.

The time-varying AR model for speech signals lacks the "memory" which assigns preference to one AR model to follow another AR model. This memory could be incorporated, for example, by assuming that the individual AR models are chosen in a Markovian manner. That is, given an AR model for the current segment of speech, certain AR models for the following segment of speech will be more likely than others. This results in the so-called *composite source model* (CSM) for the speech signal.

A block diagram of a CSM is shown in Fig. 14.6. In general, this model is composed of a set of M vector subsources which are controlled by a switch. The position of the switch at each time instant is chosen randomly, and the output of one subsource is provided. The position of the switch defines the state of the source at each time instant. CSMs for speech signals assume that the subsources are Gaussian AR sources, and the switch is controlled by a Markov chain. Furthermore, the subsources are usually assumed statistically independent and the vectors generated from each sub-

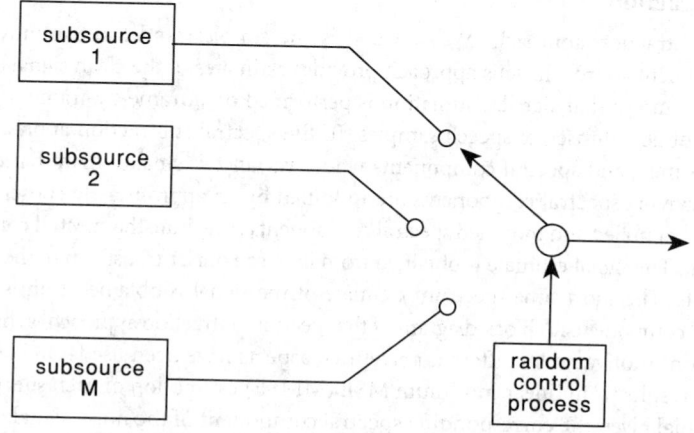

FIGURE 14.6 Composite source model.

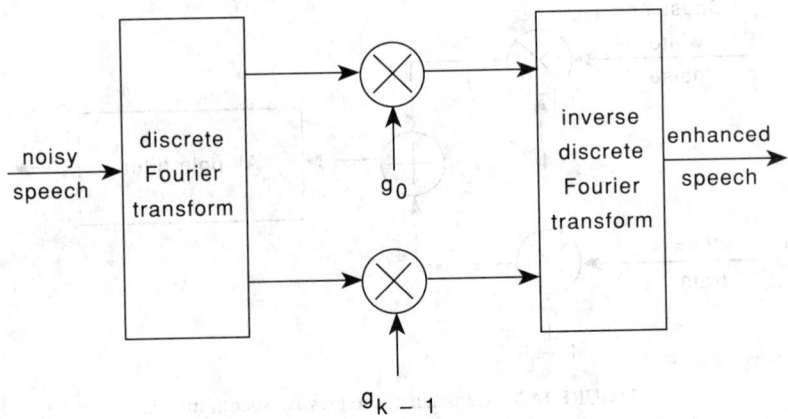

FIGURE 14.7   Spectral subtraction signal estimator.

source are also assumed statistically independent. The resulting model is known as a **hidden Markov model** (HMM) [Rabiner, 1989] since the output of the model does not contain the states of the Markovian switch.

The performance measure for speech enhancement is task dependent. For signal estimation and coding, this measure is given in terms of a distortion measure between the clean signal and the estimated or the encoded signals, respectively. For signal classification applications the performance measure is normally the probability of misclassification. Commonly used distortion measures are the mean-squared error (MSE) and the Itakura-Saito distortion measures. The Itakura-Saito distortion measure is a measure between two power spectral densities, of which one is usually that of the clean signal and the other of a model for that signal [Gersho and Gray, 1991]. This distortion measure is normally used in designing speech coding systems and it is believed to be perceptually meaningful. Both measures are mathematically tractable and lead to intuitive estimation and coding schemes. Systems designed using these two measures need not be optimal only in the MSE and the Itakura-Saito sense, but they may as well be optimal in other more meaningful senses (see a discussion in Ephraim [1992]).

Signal Estimation

In this section we review the major approaches for speech signal estimation given noisy signals.

Spectral Subtraction

The spectral subtraction approach [Weiss, 1974] is the simplest and most intuitive and popular speech enhancement approach. This approach provides estimates of the clean signal as well as of the short time spectrum of that signal. Estimation is performed on a frame-by-frame basis, where each frame consists of 20–40 msec of speech samples. In the spectral subtraction approach the signal is Fourier transformed, and spectral components whose variance is smaller than that of the noise are nulled. The surviving spectral components are modified by an appropriately chosen gain function. The resulting set of nulled and modified spectral components constitute the spectral components of the enhanced signal. The signal estimate is obtained from inverse Fourier transform of the enhanced spectral components. The short time spectrum estimate of the signal is obtained from squaring the enhanced spectral components. A block diagram of the spectral subtraction approach is shown in Fig. 14.7.

Gain functions motivated by different perceptual aspects have been used. One of the most popular functions results from linear minimum MSE (MMSE) estimation of each spectral component of the clean signal given the corresponding spectral component of the noisy signal. In this case, the value of the gain function for a given spectral component constitutes the ratio of the variances of the clean and noisy spectral components. The variance of the clean spectral component is obtained by subtracting an assumed known variance of the noise spectral component from the variance of the noisy spectral component. The resulting variance is guaranteed to be positive by the nulling process mentioned above. The variances of the spectral components of the noise process are normally estimated from silence portions of the noisy signal.

A family of spectral gain functions proposed in Lim and Oppenheim [1979] is given by

$$g_n = \left(\frac{|Z_n|^a - bE\left\{|V_n|^a\right\}}{|Z_n|^a} \right)^c \qquad n = 1, \ldots, N \qquad (14.4)$$

where Z_n and V_n denote the nth spectral components of the noisy signal and the noise process, respectively, and $a > 0$, $b \geq 0$, $c > 0$. The MMSE gain function is obtained when $a = 2$, $b = 1$, and $c = 1$. Another commonly used gain function in the spectral subtraction approach is obtained from using $a = 2$, $b = 1$, and $c = 1/2$. This gain function results from estimating the spectral magnitude of the signal and combining the resulting estimate with the phase of the noisy signal. This choice of gain function is motivated by the relative importance of the spectral magnitude of the signal compared to its phase. Since both cannot be simultaneously optimally estimated [Ephraim, 1992], only the spectral magnitude is optimally estimated, and combined with an estimate of the complex exponential of the phase which does not affect the spectral magnitude estimate. The resulting estimate of the phase can be shown to be the phase of the noisy signal within the HMM statistical framework. Normally, the spectral subtraction approach is used with $b = 2$, which corresponds to an artificially elevated noise level.

The spectral subtraction approach has been very popular since it is relatively easy to implement; it makes minimal assumptions about the signal and noise; and when carefully implemented, it results in reasonably clear enhanced signals. A major drawback of the spectral subtraction enhancement approach, however, is that the residual noise has annoying tonal characteristics referred to as "musical noise." This noise consists of narrowband signals with time-varying frequencies and amplitudes. Another major drawback of the spectral subtraction approach is that its optimality in any given sense has never been proven. Thus, no systematic methodology for improving the performance of this approach has been developed, and all attempts to achieve this goal have been based on purely heuristic arguments. As a result, a family of spectral subtraction speech enhancement approaches have been developed and experimentally optimized.

Empirical Averages

This approach attempts to estimate the clean signal from the noisy signal in the MMSE sense. The conditional mean estimator is implemented using the conditional sample average of the clean signal given the noisy signal. The sample average is obtained from appropriate training sequences of the clean and noisy signals. This is equivalent to using the sample distribution or the histogram estimate of the probability density function (pdf) of the clean signal given the noisy signal. The sample average approach is applicable for estimating the signal as well as functionals of that signal, e.g., the spectrum, the logarithm of the spectrum, and the spectral magnitude.

Let $\{Y_t, t = 0, \ldots, T\}$ be a training data from the clean signal, where Y_t is a K-dimensional vector in the Euclidean space R^K. Let $\{Z_t, t = 0, \ldots, T\}$ be a training data from the noisy signal, where $Z_t, \in R^K$. The sequence $\{Z_t\}$ can be obtained by adding a noise training sequence $\{V_t, t = 0, \ldots, T\}$ to the sequence of clean signals $\{Y_t\}$. Let $z \in R^K$ be a vector of the noisy signal from which the vector y of the clean signal is estimated. Let $\mathbf{Y}(z) = \{Y_t: Z_t = z, t = 0, \ldots, T\}$ be the set of all clean vectors from the training data of the clean signal which could have resulted in the given noisy observation z. The cardinality of this set is denoted by $|\mathbf{Y}(z)|$. Then, the sample average estimate of the conditional mean of the clean signal y given the noisy signal z is given by

$$\hat{y} = E\{y \mid z\}$$

$$= \frac{\int y p(y,z) dy}{\int p(y,z) dy} \tag{14.5}$$

$$= \frac{1}{|\mathbf{Y}(z)|} \sum_{\{Y_t \in \mathbf{Y}(z)\}} Y_t$$

Obviously, this approach is only applicable for signals with finite alphabet since otherwise the set $\mathbf{Y}(z)$ is empty with probability one. For signals with continuous pdf's, the approach can be applied only if those signals are appropriately quantized.

The sample average approach was first applied for enhancing speech signals by Porter and Boll in 1984 [Boll, 1992]. They, however, considered a simpler situation in which the noise true pdf was assumed known. In this case, enhanced signals with residual noise characterized as being a blend of wideband noise and musical noise were obtained. The balance between the two types of residual noise depended on the functional of the clean signal which was estimated.

The advantages of the sample average approach are that it is conceptually simple and it does not require *a priori* assumptions about the form of the pdf's of the signal and noise. Hence, it is a non-parametric estimation approach. This approach, however, has three major disadvantages. First, the estimator does not utilize any speech specific information such as the periodicity of the signal and the signal's AR model. Second, the training sequences from the signal and noise must be available at the speech enhancement unit. Furthermore, these training sequences must be applied for each newly observed vector of the noisy signal. Since the training sequences are normally very long, the speech enhancement unit must have extensive memory and computational resources. These problems are addressed in the model-based approach described next.

Model-Based Approach

The model-based approach [Ephraim, 1992] is a Bayesian approach for estimating the clean signal or any functional of that signal from the observed noisy signal. This approach assumes CSMs for the clean signal and noise process. The models are estimated from training sequences of those processes using the maximum likelihood (ML) estimation approach. Under ideal conditions the ML model estimate is consistent and asymptotically efficient. The ML model estimation is performed using the expectation-maximization (EM) or the Baum iterative algorithm [Rabiner, 1989; Ephraim, 1992]. Given the CSMs for the signal and noise, the clean signal is estimated by minimizing the expected value of the chosen distortion measure. The model-based approach uses significantly more statistical knowledge about the signal and noise compared to either the spectral subtraction or the sample average approaches.

The MMSE signal estimator is obtained from the conditional mean of the clean signal given the noisy signal. If $y_t \in R^K$ denotes the vector of the speech signal at time t, and z_0^t denotes the sequence of K-dimensional vectors of noisy signals $\{z_0, \dots, z_t\}$ from time $\tau = 0$ to $\tau = t$, then the MMSE estimator of y_t is given by

$$\hat{y}_t = E\{y_t \mid z_0^t\}$$

$$= \sum_{\bar{x}_t} P(\bar{x}_t \mid z_0^t) E\{y_t \mid z_t, \bar{x}_t\} \tag{14.6}$$

where $\bar{x}_t$ denotes the composite state of the noisy signal at time t. This state is given by $\bar{x}_t \triangleq (x_t, \tilde{x}_t)$, where x_t is the Markov state of the clean signal at time t and $\tilde{x}_t$ denotes the Markov state of the noise process at the same time instant t. The MMSE estimator, Eq. (14.6), comprises a weighted sum of conditional mean estimators for the composite states of the noisy signal, where the weights are the probabilities of those states given the noisy observed signal. A block diagram of this estimator is shown in Fig. 14.8.

The probability $P(\bar{x}_t|z_0^t)$ can be efficiently calculated using the forward recursion associated with HMMs. For CSMs with Gaussian subsources, the conditional mean $E\{y_t|z_t, \bar{x}_t\}$ is a linear function of the noisy vector z_t, given by

$$E(y_t|z_t, \bar{x}_t) = S_{x_t}(S_{x_t} + S_{\tilde{x}_t})^{-1}z_t \triangleq H_{\bar{x}_t}z_t \qquad (14.7)$$

where S_{x_t} and $S_{\tilde{x}_t}$ denote the covariance matrices of the Gaussian subsources associated with the Markov states x_t and $\bar{x}_t$, respectively. Since, however, $P(\bar{x}_t|z_0^t)$ is a nonlinear function of the noisy signal z_0^t, the MMSE signal estimator $\hat{y}_t$ is a nonlinear function of the noisy signal z_0^t.

The MMSE estimator, Eq. (14.6), is intuitively appealing. It uses a predesigned set of filters $\{H_{\bar{x}_t}\}$ obtained from training data of speech and noise. Each filter is optimal for a pair of subsources of the CSMs for the clean signal and the noise process. Since each subsource represents a subset of signals from the corresponding source, each filter is optimal for a pair of signal subsets from the speech and noise. The set of predesigned filters covers all possible pairs of speech and noise signal subsets. Hence, for each noisy vector of speech there must exist an optimal filter in the set of predesigned filters. Since, however, a vector of the noisy signal could possibly be generated from any pair of subsources of the clean signal and noise, the most appropriate filter for a given noisy vector is not known. Consequently, in estimating the signal vector at each time instant, all filters are tried and their outputs are weighted by the probabilities of the filters to be correct for the given noisy signal. Other strategies for utilizing the predesigned set of filters are possible. For example, at each time instant only the most likely filter can be applied to the noisy signal. This approach is more intuitive than that of the MMSE estimation. It was first proposed in Drucker [1968] for a five-state model which comprises subsources for fricatives, stops, vowels, glides, and nasals. This approach was shown by Ephraim and Merhav [Ephraim, 1992] to be optimal only in an asymptotic MMSE sense.

The model-based MMSE approach provides reasonably good enhanced speech quality with significantly less structured residual noise than the spectral subtraction approach. This performance was achieved for white Gaussian input noise at 10 dB input SNR using 512-2048 filters. An improvement of 5–6 dB in SNR was achieved by this approach. The model-based approach, however, is more elaborate than the spectral subtraction approach, since it involves two steps of training and estimation, and training must be performed on sufficiently long data. The MMSE estimation approach is usually superior to the asymptotic MMSE enhancement approach. The reason is

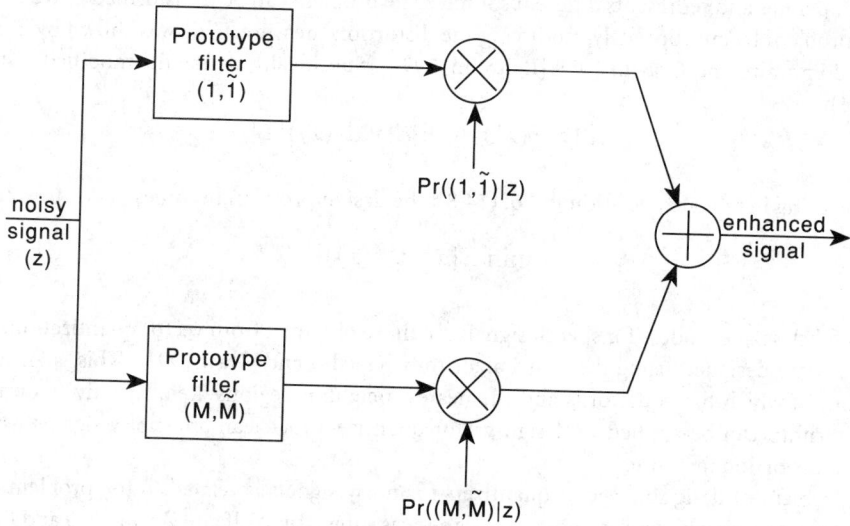

FIGURE 14.8 HMM-based MMSE signal estimator.

that the MMSE approach applies a "soft decision" rather than a "hard decision" in choosing the most appropriate filter for a given vector of the noisy signal.

A two-state version of the MMSE estimator was first applied to speech enhancement by McAulay and Malpass in 1980 [Ephraim, 1992]. The two states corresponded to speech presence and speech absence (silence) in the noisy observations. The estimator for the signal given that it is present in the noisy observations was implemented by the spectral subtraction approach. The estimator for the signal in the "silence state" is obviously equal to zero. This approach significantly improved the performance of the spectral subtraction approach.

Source Coding

An **encoder** for the clean signal maps vectors of that signal onto a finite set of representative signal vectors referred to as codewords. The mapping is performed by assigning each signal vector to its nearest neighbor codeword. The index of the chosen codeword is transmitted to the receiver in a signal communication system, and the signal is reconstructed using a copy of the chosen codeword. The codewords are designed to minimize the average distortion resulting from the nearest neighbor mapping. The codewords may simply represent waveform vectors of the signal. In another important application of low bit-rate speech coding, the codewords represent a set of parameter vectors of the AR model for the speech signal. Such coding systems synthesize the signal using the speech model in Fig. 14.5. The synthesis is performed using the encoded vector of AR coefficients as well as the parameters of the excitation signal. Reasonably good speech quality can be obtained using this coding approach at rates as low as 2400–4800 bits/sample [Gersho and Gray, 1991].

When only noisy signals are available for coding, the encoder operates on the noisy signal while representing the clean signal. In this case, the encoder is designed by minimizing the average distortion between the clean signal and the encoded signal. Specifically, let y denote the vector of clean signal to be encoded. Let z denote the corresponding given vector of the noisy signal. Let q denote the encoder. Let d denote a distortion measure. Then, the optimal encoder is designed by

$$\min_{q} E\{d(y, q(z))\} \tag{14.8}$$

When the clean signal is available for encoding the design problem is similarly defined, and it is obtained from Eq. (14.8) using $z = y$. The design problem in Eq. 14.8 is not standard since the encoder operates and represents different sources. The problem can be transformed into a standard coding problem by appropriately modifying the distortion measure. This was shown by Berger in 1971 and Ephraim and Gray in 1988 [Ephraim, 1992]. Specifically, define the modified distortion measure by

$$d'(z, q(z)) \triangleq E\{d(y, q(z))|z\} \tag{14.9}$$

Then, by using iterated expectation in Eq. (14.8), the design problem becomes

$$\min_{q} E\{d'(z, q(z))\} \tag{14.10}$$

A useful class of encoders for speech signals are those obtained from vector quantization. Vector quantizers are designed using the Lloyd algorithm [Gersho and Gray, 1991]. This is an iterative algorithm in which the codewords and the nearest neighbor regions are alternatively optimized. This algorithm can be applied to design vector quantizers for clean and noisy signals using the modified distortion measure.

The problem of designing vector quantizers for noisy signals is related to the problem of estimating the clean signals from the noisy signals, as was shown by Wolf and Ziv in 1970 and Ephraim and Gray in 1988 [Ephraim, 1992]. Specifically, optimal waveform vector quantizers in the MMSE sense can be designed by first estimating the clean signal and then quantizing the estimated signal. Both estimation and quantization are performed in the MMSE sense. Similarly, optimal quantization of the vector of parameters of the AR model for the speech signal in the Itakura-Saito sense can be performed in two steps of estimation and quantization. Specifically, the autocorrelation function of the clean signal, which approximately constitutes the sufficient statistics of that signal for estimating the AR model, is first estimated in the MMSE sense. Then, optimal vector quantization in the Itakura-Saito sense is applied to the estimated autocorrelation.

The estimation-quantization approach has been most popular in designing encoders for speech signals given noisy signals. Since such design requires explicit knowledge of the statistics of the clean signal and the noise process, but this knowledge is not available as argued in the second section, a variety of suboptimal encoders were proposed. Most of the research in this area focused on designing encoders for the AR model of the signal due to the importance of such encoders in low bit-rate speech coding. The proposed encoders mainly differ in the estimators they used and the functionals of the speech signal these estimators have been applied to. Important examples of functionals which have commonly been estimated include the signal waveform, autocorrelation, and the spectral magnitude. The primarily set of estimators used for this application were obtained from the spectral subtraction approach and its derivatives. A version of the sample average estimator was also developed and applied to AR modeling by Juang and Rabiner in 1987 [Ephraim, 1992]. Recently, the HMM-based estimator of the autocorrelation function of the clean signal was used in AR model vector quantization [Ephraim, 1992].

Designing of AR model-based encoders from noisy signals has been a very successful application of speech enhancement. In this case both the quality and intelligibility of the encoded signal can be improved compared to the case where the encoder is designed for the clean signal and the input noise is simply ignored. The reason is that the input noise has devastating effects of the performance of AR model-based speech coders, and any "reasonable" estimation approach can significantly improve the performance of those coders in noisy environments.

Signal Classification

In recognition of clean speech signals a sample function of the signal is associated with one of the words in the vocabulary. The association or decision rule is designed to minimize the probability of classification error. When only noisy speech signals are available for recognition a very similar problem results. Specifically, a sample function of the noisy signal is now associated with one of the words in the vocabulary in a way which minimizes the probability of classification error. The only difference between the two problems is that the sample functions of the clean and noisy signals from a given word have different statistics. The problem in both cases is that of partitioning the sample space of the given acoustic signals from all words in the vocabulary into L partition cells, where L is the number of words in the vocabulary.

Let $\{W_i, i = 1, \ldots, L\}$ denote a set of words in a given vocabulary. Let z denote the acoustic noisy signal from some word in the vocabulary. Let $\Omega \triangleq \{\omega_1, \ldots, \omega_L\}$ be a partition of the sample space of the noisy signals. The probability of error associated with this partition is given by

$$P_e(\Omega) = \sum_{i=1}^{L} P(W_i) \int_{z \notin \omega_i} p(z \mid W_i) dz \qquad (14.11)$$

where $P(W_i)$ is the *a priori* probability of occurrence of the ith word, and $p(z \mid W_i)$ is the pdf of the noisy signal from the ith word. The minimization of $P_e(\Omega)$ is achieved by the well-known maximum *a posteriori* (MAP) decision rule. Specifically, z is associated with the word W_i for which $p(z \mid W_i)P(W_i) > p(z \mid W_j)P(W_j)$ for all $j = 1, \ldots, L$ and $j \neq i$. Ties are arbitrarily broken. In the absence of noise, the noisy signal z becomes a clean signal y, and the optimal recognizer is obtained by using the same decision rule with $z = y$. Hence, the only difference between recognition of clean signals and recognition of noisy signals is that in the first case the pdf's $\{p(y\,|\,W_i)\}$ are used in the decision rule, while in the second case the pdf's $\{p(z\,|\,W_i)\}$ are used in the same decision rule.

Note that optimal recognition of noisy signals requires explicit knowledge of the statistics of the clean signal and noise. Neither the clean signal nor any function of that signal needs to be estimated. Since, however, the statistics of the signal and noise are not explicitly available as argued in the second section, parametric models are usually assumed for these pdf's and their parameters are estimated from appropriate training data. Normally, HMMs with mixture of Gaussian pdf's at each state are attributed to both the clean signal and noise process. It can be shown (similarly to the case of classification of clean signals dealt with by Merhav and Ephraim in 1991 [Ephraim, 1992]) that if the pdf's of the signal and noise are precisely HMMs and the training sequences are significantly longer than the test data, then the MAP decision rule which uses estimates of the pdf's of the signal and noise is asymptotically optimal.

A key issue in applying hidden Markov modeling for recognition of speech signals is the matching of the energy contour of the signal to the energy contour of the model for that signal. Energy matching is required for two main reasons. First, speech signals are not strictly stationary and hence their energy contours cannot be reliably estimated from training data. Second, recording conditions during training and testing vary. An approach for gain adaptation was developed [Ephraim, 1992]. In this approach, HMMs for gain-normalized clean signals are designed and used together with gain contour estimates obtained from the noisy signals. The gain adaptation approach is implemented using the EM algorithm. This approach provides robust speech recognition at input SNRs greater than or equal to 10 dB.

The relation between signal classification and estimation was established in Kailath [1969] for *continuous* time signals contaminated by additive statistically independent Gaussian white noise. It was shown that minimum probability of error classification can be achieved by applying the MAP decision rule to the *causal* MMSE estimator of the clean signal. This interesting theoretical result provides the intuitive basis for a popular approach for recognition of noisy speech signals. In this approach, the clean signal or some feature vector of the signal is first estimated and then recognition is applied. In the statistical framework of hidden Markov modeling, however, the direct recognition approach presented earlier is significantly simpler since both the clean signal and the noisy signal are HMMs [Ephraim, 1992]. Hence, the complexity of recognizing the estimated signal is the same as that of recognizing the noisy signal directly.

Other commonly used approaches for recognition of noisy speech signals were developed for systems which are based on pattern recognition. When clean signals are available for recognition, these systems match the input signal to the nearest neighbor acoustic templet which represents some word in the vocabulary. The templets mainly comprise spectral prototypes of the clean signals. The matching is performed using a distance measure between the clean input signal and the templet. When only noisy signals are available for recognition, several modifications of the pattern matching approach were proposed. Specifically, adapting the templets of the clean signal to reflect the presence of the noise was proposed by Roe in 1987 [Ephraim, 1992]; choosing templets for the noisy signal which are more robust than those obtained from adaptation of the templets for the clean signal was often proposed; and using distance measures which are robust to noise, such as the projection measure proposed by Mansour and Juang in 1989 [Ephraim, 1992]. These approaches along with the prefiltering approach in the sampled signal case are fairly intuitive and are relatively easy to implement. It is difficult, however, to establish their optimality in any well-defined sense. Another interesting approach based on robust statistics was developed by Merhav and Lee [Ephraim, 1992]. This approach was shown asymptotically optimal in the minimum probability of error sense within the hidden Markov modeling framework.

The speech recognition problem in noisy environments has also been a successful application of speech enhancement. Significant reduction in the error rate due to the noise presence was achieved by the various approaches mentioned above.

Comments

Three major aspects of speech enhancement were reviewed. These comprise improving the perception of speech signals in noisy environments and increasing the robustness of speech coders and recognition systems in noisy environments. The inherent difficulties associated with these problems were discussed, and the main solutions along with their strengths and weaknesses were presented. This section is an introductory presentation to the speech enhancement problem. A comprehensive treatment of the subject can be found in Lim [1979], Makhoul *et al.* [1989], Boll [1992], and Ephraim [1992]. Significant progress in understanding the problem and in developing new speech enhancement systems was made during the 1980s with the introduction of statistical model-based approaches. The speech enhancement problem, however, is far from being solved, and major progress is still needed. In particular, no speech enhancement system which is capable of simultaneously improving both the quality and intelligibility of the noisy signal is currently known. Progress in this direction can be made if more reliable statistical models for the speech signal and noise process as well as meaningful distortion measures can be found.

Defining Terms

Autoregressive model: Statistical model for resonant signals.

Classifier: Maps signal utterances into a finite set of word units, e.g., syllables.

Encoder: Maps signal vectors into a finite set of codewords. A vector quantizer is a particular type of encoder.

Hidden Markov model: Statistical model comprised of several subsources controlled by Markovian process.

Intelligibility: Objective quantitative measure of speech perception.

Noise: Any interfering signal adversely affecting the communication of the clean signal.

Quality: Subjective descriptive measure of speech perception.

Signal: Clean speech sample to be communicated with human or machine.

Signal-to-noise ratio: Ratio of the signal power to the noise power measured in decibels.

Speech enhancement: Improvement of perceptual aspects of speech signals.

References

S. F. Boll, "Speech enhancement in the 1980's: noise suppression with pattern matching," in *Advances in Speech Signal Processing,* S. Furui and M. M. Sonhdi, Eds., New York: Marcel Dekker, 1992.

H. Drucker, "Speech processing in a high ambient noise environment," *IEEE Trans. Audio Electroacoust.,* vol. 16, 1968.

Y. Ephraim, "Statistical model based speech enhancement systems," *Proc. IEEE,* vol. 80, 1992.

A. Gersho and R. M. Gray, *Vector Quantization and Signal Compression,* Boston: Kluwer Academic Publishers, 1991.

T. Kailath, "A general likelihood-ratio formula for random signals in Gaussian noise," *IEEE Trans. on Inform Theory,* vol. 15, 1969.

J. S. Lim, Ed., *Speech Enhancement,* Englewood Cliffs, N.J.: Prentice-Hall, 1983.

J. S. Lim and A. V. Oppenheim, "Enhancement and bandwidth compression of noisy speech," *Proc. IEEE,* vol. 67, 1979.

J. Makhoul, T. H. Crystal, D. M. Green, D. Hogan, R. J. McAulay, D. B. Pisoni, R. D. Sorkin, and T. G. Stockham, *Removal of Noise From Noise-Degraded Speech Signals,* Washington, D.C.: National Academy Press, 1989.

L. R. Rabiner, "A tutorial on hidden Markov models and selected applications in speech recognition," *Proc. IEEE,* vol. 77, 1989.

M. R. Weiss, E. Aschkenasy, and T. W. Parsons, "Processing speech signals to attenuate interference," in *IEEE Symp. on Speech Recognition,* Pittsburgh, 1974.

Further Information

A comprehensive treatment of the speech enhancement problem can be found in the four tutorial papers and book listed below.

J. S. Lim and A. V. Oppenheim, "Enhancement and bandwidth compression of noisy speech," *Proc. IEEE*, vol. 67, 1979.

J. Makhoul, T. H. Crystal, D. M. Green, D. Hogan, R. J. McAulay, D. B. Pisoni, R. D. Sorkin, and T. G. Stockham, *Removal of Noise From Noise-Degraded Speech Signals*, Washington, D.C.: National Academy Press, 1989.

S. F. Boll, "Speech enhancement in the 1980's: noise suppression with pattern matching," in *Advances in Speech Signal Processing*, S. Furui and M. M. Sonhdi, Eds., New York: Marcel Dekker, 1992.

Y. Ephraim, "Statistical model based speech enhancement systems," *Proc. IEEE*, vol. 80, 1992.

J. S. Lim, Ed., *Speech Enhancement*, Englewood Cliffs, N.J.: Prentice-Hall, 1983.

14.3 Analysis and Synthesis

Jesse W. Fussell

After an acoustic speech signal is converted to an electrical signal by a microphone, it may be desirable to analyze the electrical signal to estimate some time-varying parameters which provide information about a model of the speech production mechanism. **Speech analysis** is the process of estimating such parameters. Similarly, given some parametric model of speech production and a sequence of parameters for that model, **speech synthesis** is the process of creating an electrical signal which approximates speech. While analysis and synthesis techniques may be done either on the continuous signal or on a sampled version of the signal, most modern analysis and synthesis methods are based on digital signal processing.

A typical speech production model is shown in Fig. 14.9. In this model the output of the excitation function is scaled by the gain parameter and then filtered to produce speech. All of these functions are time-varying. For many models, the parameters are varied at a periodic rate, typically 50 to 100 times per second. Most speech information is contained in the portion of the signal below about 4 kHz.

The excitation is usually modeled as either a mixture or a choice of random noise and periodic waveform. For human speech, voiced excitation occurs when the vocal folds in the larynx vibrate; unvoiced excitation occurs at constrictions in the vocal tract which create turbulent air flow [Flanagan, 1965]. The relative mix of these two types of excitation is termed "**voicing.**" In addition, the periodic excitation is characterized by a fundamental frequency, termed **pitch** or F0. The excitation is scaled by a factor designed to produce the proper amplitude or level of the speech signal. The scaled excitation function is then filtered to produce the proper spectral characteristics. While the filter may be nonlinear, it is usually modeled as a linear function.

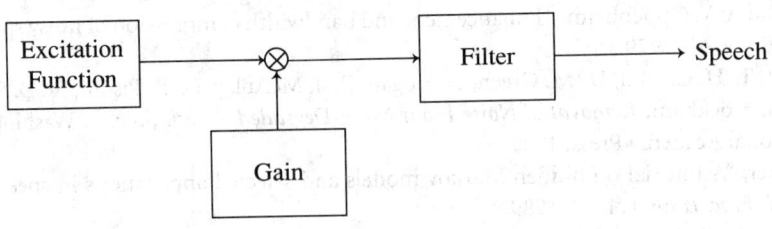

FIGURE 14.9   A general speech production model.

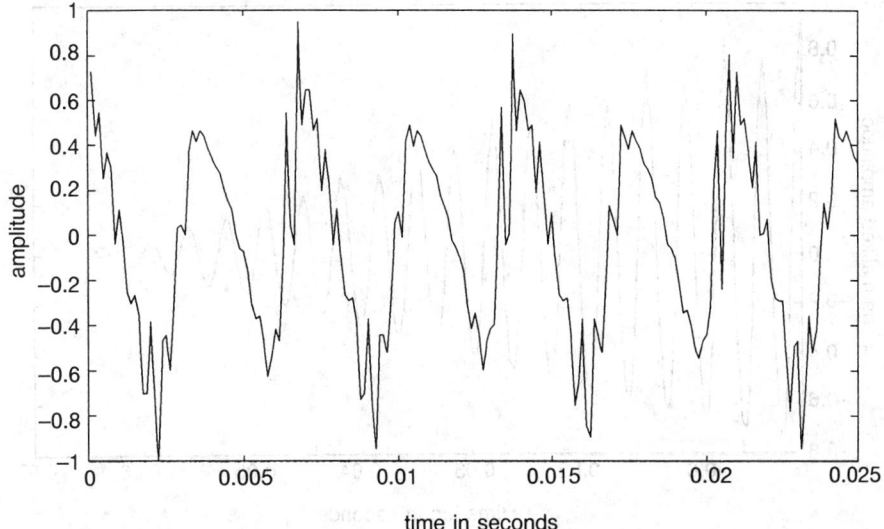

FIGURE 14.10   Waveform of a spoken phoneme /i/ as in beet.

Analysis of Excitation

In a simplified form, the excitation function may be considered to be purely periodic, for voiced speech, or purely random, for unvoiced. These two states correspond to voiced phonetic classes such as vowels and nasals and unvoiced sounds such as unvoiced fricatives. This binary voicing model is an oversimplification for sounds such as voiced fricatives, which consist of a mixture of periodic and random components. Figure 14.10 is an example of a time waveform of a spoken /i/ phoneme, which is well modeled by only periodic excitation.

Both time domain and frequency domain analysis techniques have been used to estimate the degree of voicing for a short segment or frame of speech. One time domain feature, termed the zero crossing rate, is the number of times the signal changes sign in a short interval. As shown in Fig. 14.10, the zero crossing rate for voiced sounds is relatively low. Since unvoiced speech typically has a larger proportion of high-frequency energy than voiced speech, the ratio of high-frequency to low-frequency energy is a frequency domain technique that provides information on voicing.

Another measure used to estimate the degree of voicing is the autocorrelation function, which is defined for a sampled speech segment, S, as

$$\text{ACF}(\tau) = \frac{1}{N} \sum_{n=0}^{N-1} s(n)s(n - \tau) \tag{14.12}$$

where $s(n)$ is the value of the nth sample within the segment of length N. Since the autocorrelation function of a periodic function is itself periodic, voicing can be estimated from the degree of periodicity of the autocorrelation function. Figure 14.11 is a graph of the nonnegative terms of the autocorrelation function for a 64-ms frame of the waveform of Fig. 14.10. Except for the decrease in amplitude with increasing lag, which results from the rectangular window function which delimits the segment, the autocorrelation function is seen to be quite periodic for this voiced utterance.

If an analysis of the voicing of the speech signal indicates a voiced or periodic component is present, another step in the analysis process may be to estimate the frequency (or period) of the voiced component. There are a number of ways in which this may be done. One is to measure the time lapse between peaks in the time domain signal. For example in Fig. 14.10 the major peaks are separated by about 0.0071 s, for a fundamental frequency of about 141 Hz. Note, it would be quite

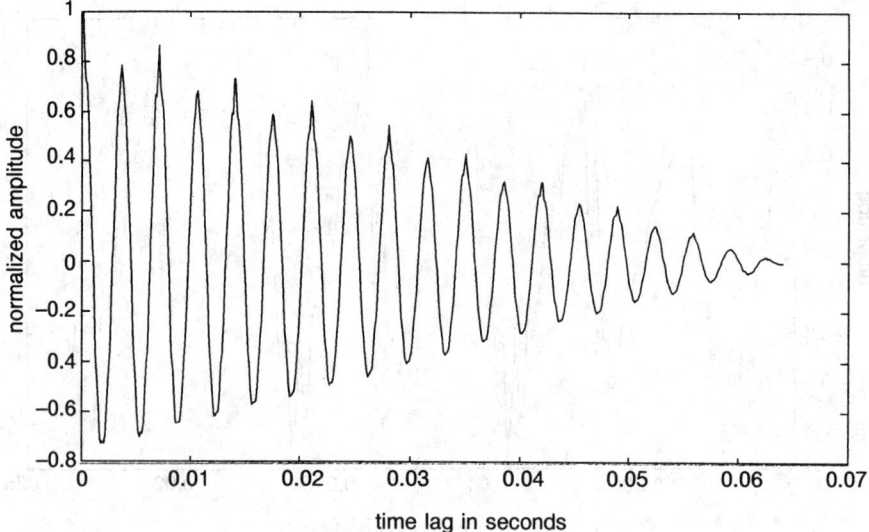

FIGURE 14.11 Autocorrelation function of one frame of /i/.

possible to err in the estimate of fundamental frequency by mistaking the smaller peaks that occur between the major peaks for the major peaks. These smaller peaks are produced by resonance in the vocal tract which, in this example, happen to be at about twice the excitation frequency. This type of error would result in an estimate of pitch approximately twice the correct frequency.

The distance between major peaks of the autocorrelation function is a closely related feature that is frequently used to estimate the pitch period. In Fig. 14.11, the distance between the major peaks in the autocorrelation function is about 0.0071 s. Estimates of pitch from the autocorrelation function are also susceptible to mistaking the first vocal track resonance for the glottal excitation frequency.

The absolute magnitude difference function (AMDF), defined as,

$$\text{AMDF}(\tau) = \frac{1}{N} \sum_{n=0}^{N-1} \left| s(n) - s(n-\tau) \right| \tag{14.13}$$

is another function which is often used in estimating the pitch of voiced speech. An example of the AMDF is shown in Fig. 14.12 for the same 64-ms frame of the /i/ phoneme. However, the minima of the AMDF is used as an indicator of the pitch period. The AMDF has been shown to be a good pitch period indicator [Ross *et al.*, 1974] and does not require multiplications.

Fourier Analysis

One of the more common processes for estimating the spectrum of a segment of speech is the Fourier transform [Oppenheim and Schafer, 1975]. The Fourier transform of a sequence is mathematically defined as

$$S(e^{j\omega}) = \sum_{n=-\infty}^{\infty} s(n)e^{-j\omega n} \tag{14.14}$$

where $s(n)$ represents the terms of the sequence. The short-time Fourier transform of a sequence is a time-dependent function, defined as

$$S_m(e^{j\omega}) = \sum_{n=-\infty}^{\infty} w(m-n) \, s(n)e^{-j\omega n} \tag{14.15}$$

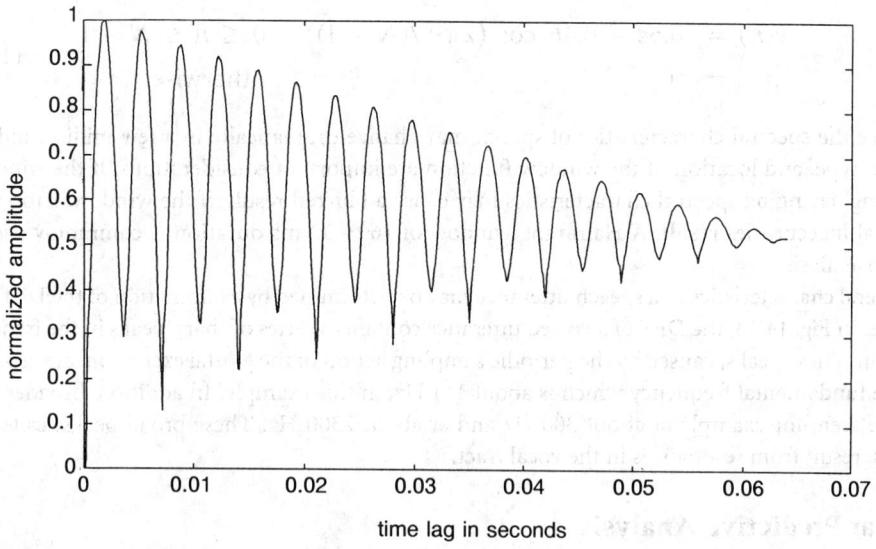

FIGURE 14.12   Absolute magnitude difference function of one frame of /i/.

where the window function $w(n)$ is usually zero except for some finite range, and the variable m is used to select the section of the sequence for analysis. The discrete Fourier transform (DFT) is obtained by uniformly sampling the short-time Fourier transform in the frequency dimension. Thus an N-point DFT is computed using Eq. (14.16),

$$S(k) = \sum_{n=0}^{N-1} s(n)e^{-j2\pi nk/N} \qquad (14.16)$$

where the set of N samples, $s(n)$, may have first been multiplied by a window function. An example of the magnitude of a 512-point DFT of the waveform of the /i/ from Fig. 14.10 is shown in Fig. 14.13. Note for this figure, the 512 points in the sequence have been multiplied by a Hamming window defined by

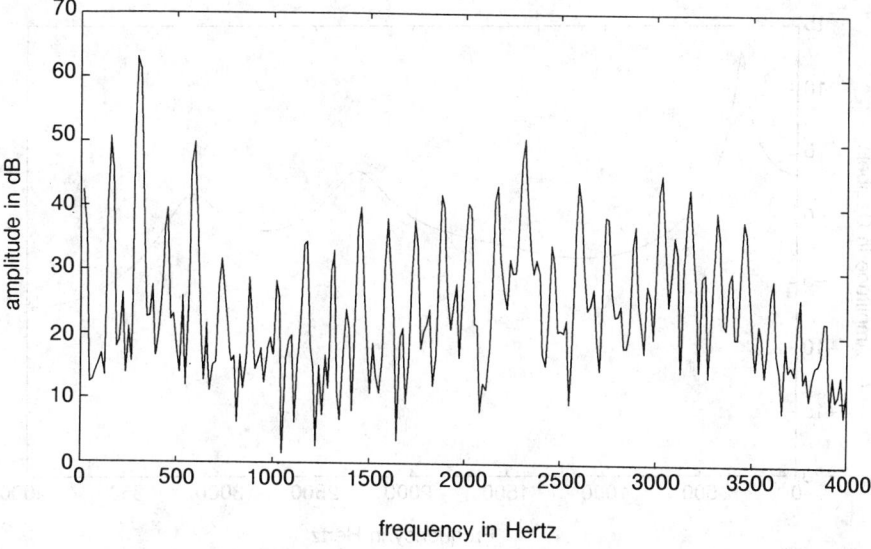

FIGURE 14.13   Magnitude of 512-point FFT of Hamming windowed /i/.

$$w(n) = 0.54 - 0.46 \cos(2\pi n/(N-1)) \quad 0 \le n \le N - 1$$
$$= 0 \qquad\qquad\qquad\qquad\qquad\qquad \text{otherwise}$$

(14.17)

Since the spectral characteristics of speech may change dramatically in a few milliseconds, the length, type, and location of the window function are important considerations. If the window is too long, changing spectral characteristics may cause a blurred result; if the window is too short, spectral inaccuracies result. A Hamming window of 16 to 32 ms duration is commonly used for speech analysis.

Several characteristics of a speech utterance may be determined by examination of the DFT magnitude. In Fig. 14.13, the DFT of a voiced utterance contains a series of sharp peaks in the frequency domain. These peaks, caused by the periodic sampling action of the glottal excitation, are separated by the fundamental frequency which is about 141 Hz, in this example. In addition, broader peaks can be seen, for example at about 300 Hz and at about 2300 Hz. These broad peaks, called formants, result from resonances in the vocal tract.

Linear Predictive Analysis

Given a sampled (discrete-time) signal $s(n)$, a powerful and general parametric model for time series analysis is

$$s(n) = -\sum_{k=1}^{p} a(k)s(n-k) + G\sum_{l=0}^{q} b(l)\, u(n-l)$$

(14.18)

where $s(n)$ is the output and $u(n)$ is the input (perhaps unknown). The model parameters are $a(k)$ for $k = 1, p$, $b(l)$ for $l = 1, q$, and G. $b(0)$ is assumed to be unity. This model, described as an autoregressive moving average (ARMA) or pole-zero model, forms the foundation for the analysis method termed linear prediction. An autoregressive (AR) or all-pole model, for which all of the "b" coefficients except $b(0)$ are zero, is frequently used for speech analysis [Markel and Gray, 1976].

In the standard AR formulation of linear prediction, the model parameters are selected to minimize the mean-squared error between the model and the speech data. In one of the variants of linear prediction, the autocorrelation method, the minimization is carried out for a windowed segment of data. In the autocorrelation method, minimizing the mean-square error of the time

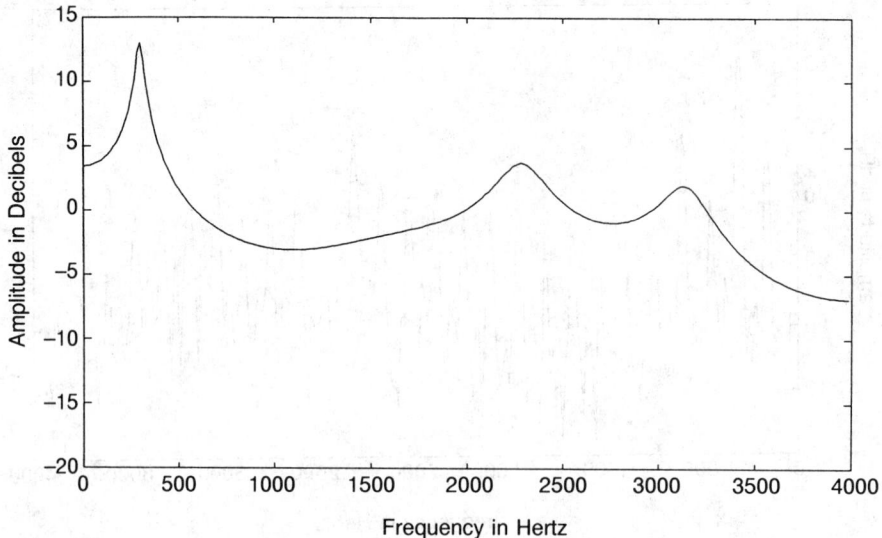

FIGURE 14.14   Eighth-order linear predictive analysis of an "i".

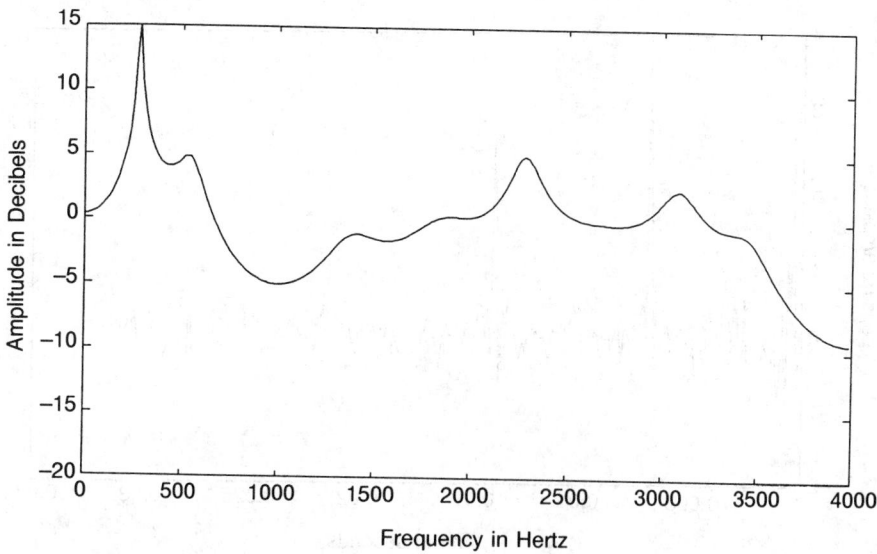

FIGURE 14.15 Sixteenth-order linear predictive analysis of an "i".

domain samples is equivalent to minimizing the integrated ratio of the signal spectrum to the spectrum of the all-pole model. Thus, linear predictive analysis is a good method for spectral analysis whenever the signal is produced by an all-pole system. Most speech sounds fit this model well.

One key consideration for linear predictive analysis is the order of the model, p. For speech, if the order is too small, the formant structure is not well represented. If the order is too large, pitch pulses as well as formants begin to be represented. Tenth- or twelfth-order analysis is typical for speech. Figures 14.14 and 14.15 provide examples of the spectrum produced by eighth-order and sixteenth-order linear predictive analysis of the /i/ waveform of Fig. 14.10. Figure 14.14 shows there to be three formants at frequencies of about 300, 2300, and 3200 Hz, which are typical for an /i/.

Homomorphic (Cepstral) Analysis

For the speech model of Fig. 14.9, the excitation and filter impulse response are convolved to produce the speech. One of the problems of speech analysis is to separate or deconvolve the speech into these two components. One such technique is called homomorphic filtering [Oppenheim and Schafer, 1968]. The characteristic system for a system for homomorphic deconvolution converts a convolution operation to an addition operation. The output of such a characteristic system is called the complex **cepstrum.** The complex cepstrum is defined as the inverse Fourier transform of the complex logarithm of the Fourier transform of the input. If the input sequence is minimum phase (i.e., the z-transform of the input sequence has no poles or zeros outside the unit circle), the sequence can be represented by the real portion of the transforms. Thus, the real cepstrum can be computed by calculating the inverse Fourier transform of the log-spectrum of the input.

Figure 14.16 shows an example of the cepstrum for the voiced /i/ utterance from Fig. 14.10. The cepstrum of such a voiced utterance is characterized by relatively large values in the first one or two milliseconds as well as by pulses of decaying amplitudes at multiples of the pitch period. The first two of these pulses can clearly be seen in Fig. 14.16 at time lags of 7.1 and 14.2 ms. The location and amplitudes of these pulses may be used to estimate pitch and voicing [Rabiner and Schafer, 1978].

In addition to pitch and voicing estimation, a smooth log magnitude function may be obtained by windowing or "liftering" the cepstrum to eliminate the terms which contain the pitch information. Figure 14.17 is one such smoothed spectrum. It was obtained from the DFT of the cepstrum of Fig 14.16 after first setting all terms of the cepstrum to zero except for the first 16.

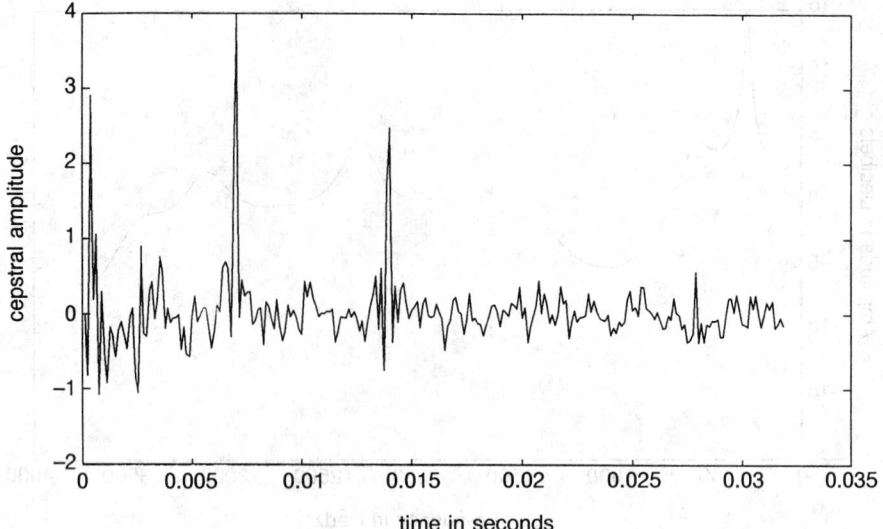

FIGURE 14.16   Real cepstrum of /i/.

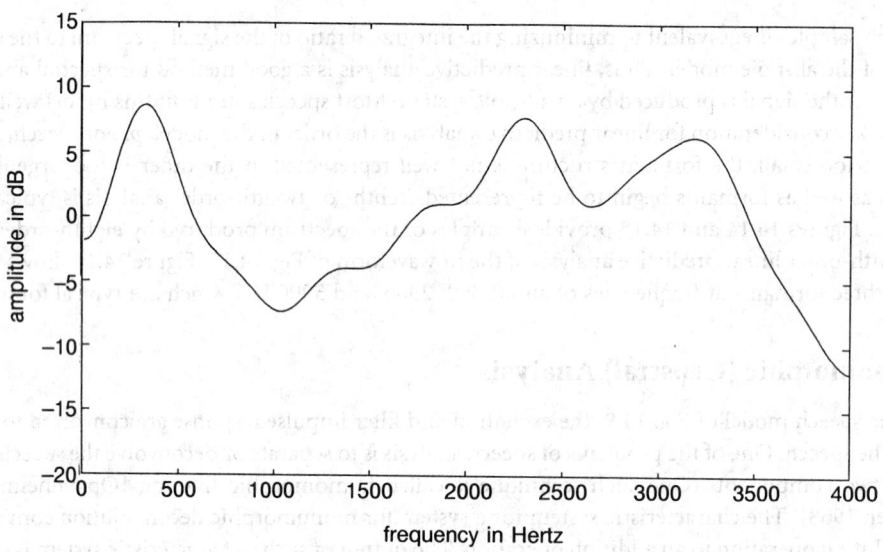

FIGURE 14.17   Smoothed spectrum of /i/ from 16 points of cepstrum.

Speech Synthesis

Speech synthesis is the creation of speech-like waveforms from textual words or symbols. In general, the speech synthesis process may be divided into three levels of processing [Klatt, 1982]. The first level transforms the text into a series of acoustic phonetic symbols, the second transforms those symbols to smoothed synthesis parameters, and the third level generates the speech waveform from the parameters. While speech synthesizers have been designed for a variety of languages and the processes described here apply to several languages, the examples given are for English text-to-speech.

In the first level of processing, abbreviations such as "Dr." (which could mean "doctor" or "drive"), numbers ("1492" could be a year or a quantity), special symbols such as "$", upper case acronyms (e.g., NASA), and nonspoken symbols such as "'" (apostrophe) are converted to a standard form. Next prefixes and perhaps suffixes are removed from the body of words prior to search-ing for the root word in a lexicon, which defines the phonetic representation for the word. The lex-icon includes words which do not obey the normal rules of pronunciation, such as "of". If the word is not contained in the lexicon, it is processed by an algorithm which contains a large set of rules of pronunciation.

In the second level, the sequences of words consisting of phrases or sentences are analyzed for grammar and syntax. This analysis provides information to another set of rules which determine the stress, duration, and pitch to be added to the phonemic representation. This level of process-ing may also alter the phonemic representation of individual words to account for coarticulation effects. Finally, the sequences of parameters which specify the pronunciation are smoothed in an attempt to mimic the smooth movements of the human articulators (lips, jaw, velum, and tongue).

The last processing level converts the smoothed parameters into a time waveform. Many vari-eties of waveform synthesizers have been used, including formant, linear predictive, and filter-bank versions. These waveform synthesizers generally correspond to the synthesizers used in speech cod-ing systems which are described in the first section of this chapter.

Defining Terms

Cepstrum: Inverse Fourier transform of the logarithm of the Fourier power spectrum of a signal. The complex cepstrum is the inverse Fourier transform of the complex logarithm of the Fourier tranform of the complex logarithm of the Fourier transform of the signal.

Pitch: Frequency of glottal vibration of a voiced utterance.

Spectrum or power density spectrum: Amplitude of a signal as a function of frequency, fre-quently defined as the Fourier transform of the autocovariance of the signal.

Speech analysis: Process of extracting time-varying parameters from the speech signal which represent a model for speech production.

Speech synthesis: Production of a speech signal from a model for speech production and a set of time-varying parameters of that model.

Voicing: Classification of a speech segment as being voiced (i.e., produced by glottal excitation), unvoiced (i.e., produced by turbulent air flow at a constriction) or some mix of those two.

References

J. Allen, "Synthesis of speech from unrestricted text," *Proc. IEEE,* vol. 64, no. 4, pp. 433–442, 1976.

J. L. Flanagan, *Speech Analysis, Synthesis and Perception,* Berlin: Springer-Verlag, 1965.

D. H. Klatt, "The Klattalk Text-to-Speech System" IEEE Int. Conf. on Acoustics, Speech and Signal Proc., pp. 1589–1592, Paris, 1982.

J. D. Markel and A. H. Gray, Jr., *Linear Prediction of Speech,* Berlin: Springer-Verlag, 1976.

A. V. Oppenheim and R. W. Schafer, "Homomorphic analysis of speech," *IEEE Trans. Audio Electroacoust.,* pp. 221–226, 1968.

A. V. Oppenheim and R. W. Schafer, *Digital Signal Processing,* Englewood Cliffs, N.J.: Prentice-Hall, 1975.

D. O'Shaughnessy, *Speech Communication,* Reading, Mass.: Addison-Wesley, 1987.

L. R. Rabiner and R. W. Schafer, *Digital Processing of Speech Signals,* Englewood Cliffs, N.J.: Pren-tice-Hall, 1978.

M. J. Ross, H . L. Shaffer, A. Cohen, R. Freudberg, and H. J. Manley, "Average magnitude difference function pitch extractor," *IEEE Trans. Acoustics, Speech and Signal Proc.,* vol. ASSP-22, pp. 353–362, 1974.

R. W. Schafer and J. D. Markel, *Speech Analysis,* New York: IEEE Press, 1979.

Further Information

The monthly magazine *IEEE Transactions on Signal Processing*, formerly *IEEE Transactions on Acoustics, Speech and Signal Processing*, frequently contains articles on speech analysis and synthesis. In addition, the annual conference of the IEEE Signal Processing Society, the International Conference on Acoustics, Speech, and Signal Processing, is a rich source of papers on the subject.

14.4 Speech Recognition

Lynn D. Wilcox and Marcia A. Bush

Speech recognition is the process of translating an acoustic signal into a linguistic message. In certain applications, the desired form of the message is a verbatim transcription of a sequence of spoken words. For example, in using speech recognition technology to automate dictation or data entry to a computer, transcription accuracy is of prime importance. In other cases, such as when speech recognition is used as an interface to a database query system or to index by keyword into audio recordings, word-for-word transcription is less critical. Rather, the message must contain only enough information to reliably communicate the speaker's goal. The use of speech recognition technology to facilitate a dialog between person and computer is often referred to as "spoken language processing."

Speech recognition by machine has proven an extremely difficult task. One complicating factor is that, unlike written text, no clear spacing exists between spoken words; speakers typically utter full phrases or sentences without pause. Furthermore, acoustic variability in the speech signal typically precludes an unambiguous mapping to a sequence of words or subword units, such as phones.[1] One major source of variability in speech is coarticulation, or the tendency for the acoustic characteristics of a given speech sound or phone to differ depending upon the phonetic context in which it is produced. Other sources of acoustic variability include differences in vocal-tract size, dialect, speaking rate, speaking style, and communication channel.

Speech recognition systems can be constrained along a number of dimensions in order to make the recognition problem more tractable. Training the parameters of a recognizer to the speech of the user is one way of reducing variability and, thus, increasing recognition accuracy. Recognizers are categorized as speaker-dependent or speaker-independent, depending upon whether or not full training is required by each new user. Speaker-adaptive systems adjust automatically to the voice of a new talker, either on the basis of a relatively small amount of training data or on a continuing basis while the system is in use.

Recognizers can also be categorized by the speaking styles, vocabularies, and language models they accommodate. **Isolated word recognizers** require speakers to insert brief pauses between individual words. **Continuous speech recognizers** operate on fluent speech, but typically employ strict language models, or grammars, to limit the number of allowable word sequences. **Wordspotters** also accept fluent speech as input. However, rather than providing full transcription, wordspotters selectively locate relevant words or phrases in an utterance. Wordspotting is useful both in information-retrieval tasks based on keyword indexing and as an alternative to isolated word recogniton in voice command applications.

Speech Recognition System Architecture

Figure 14.18 shows a block diagram of a speech recognition system. Speech is typically input to the system using an analog transducer, such as a microphone, and converted to digital form. **Signal pre-processing** consists of computing a sequence of acoustic feature vectors by processing the

---

[1]Phones correspond roughly to pronunciations of consonants and vowels.

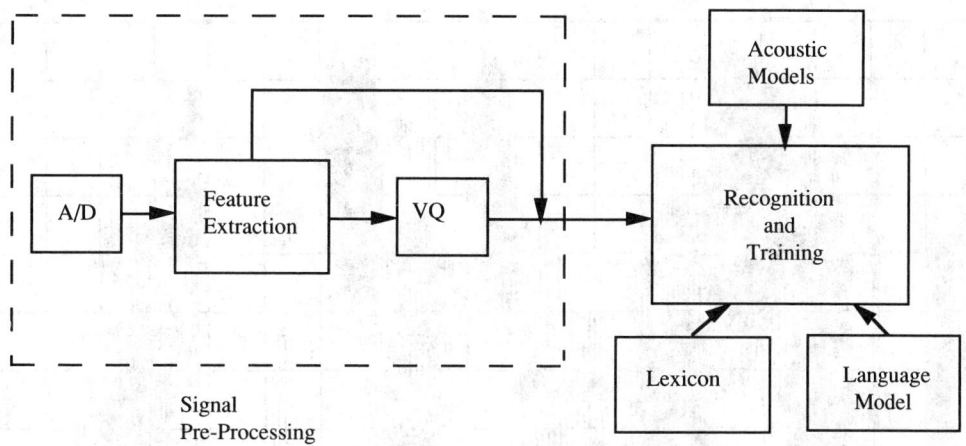

FIGURE 14.18   Architecture for a speech recognition system.

speech samples in successive time intervals. In some systems, a clustering technique known as vector quantization is used to convert these continuous-valued features to a sequence of discrete codewords drawn from a codebook of acoustic prototypes. Recognition of an unknown utterance involves transforming the sequence of feature vectors, or codewords, into an appropriate message. The recognition process is typically constrained by a set of acoustic models which correspond to the basic units of speech employed in the recognizer, a lexicon which defines the vocabulary of the recognizer in terms of these basic units, and a language model which specifies allowable sequences of vocabulary items. The acoustic models, and in some cases the language model and lexicon, are learned from a set of representative training data. These components are discussed in greater detail in the remainder of this chapter, as are the two recognition paradigms most frequently employed in speech recognition: **dynamic time warping** and **hidden Markov models.**

Signal Pre-Processing

An amplitude waveform and speech spectrogram of the sentence "Two plus seven is less than ten" is shown in Fig. 14.19. The spectrogram represents the time evolution (horizontal axis) of the frequency spectrum (vertical axis) of the speech signal, with darkness corresponding to high energy. In this example, the speech has been digitized at a sampling rate of 16 kHz, or roughly twice the highest frequency of relevant energy in a high-quality speech signal. In general, the appropriate sampling rate is a function of the communication channel. In telecommunications, for example, a bandwidth of 4 kHz, and, thus, a Nyquist sampling rate of 8 kHz, is standard.

The speech spectrum can be viewed as the product of a source spectrum and the transfer function of a linear, time-varying filter which represents the changing configuration of the vocal tract. The transfer function determines the shape, or envelope, of the spectrum, which carries phonetic information in speech. When excited by a voicing source, the formants, or natural resonant frequencies of the vocal tract, appear as black bands running horizontally through regions of the speech spectrogram. These regions represent voiced segments of speech and correspond primarily to vowels. Regions characterized by broadband high-frequency energy, and by extremely low energy, result from noise excitation and vocal-tract closures, respectively, and are associated with the articulation of consonantal sounds.

Feature extraction for speech recognition involves computing sequences of numeric measurements, or feature vectors, which typically approximate the envelope of the speech spectrum. Spectral features can be extracted directly from the discrete Fourier transform (DFT) or computed using linear predictive coding (LPC) techniques. Cepstral analysis can also be used to deconvolve the spectral

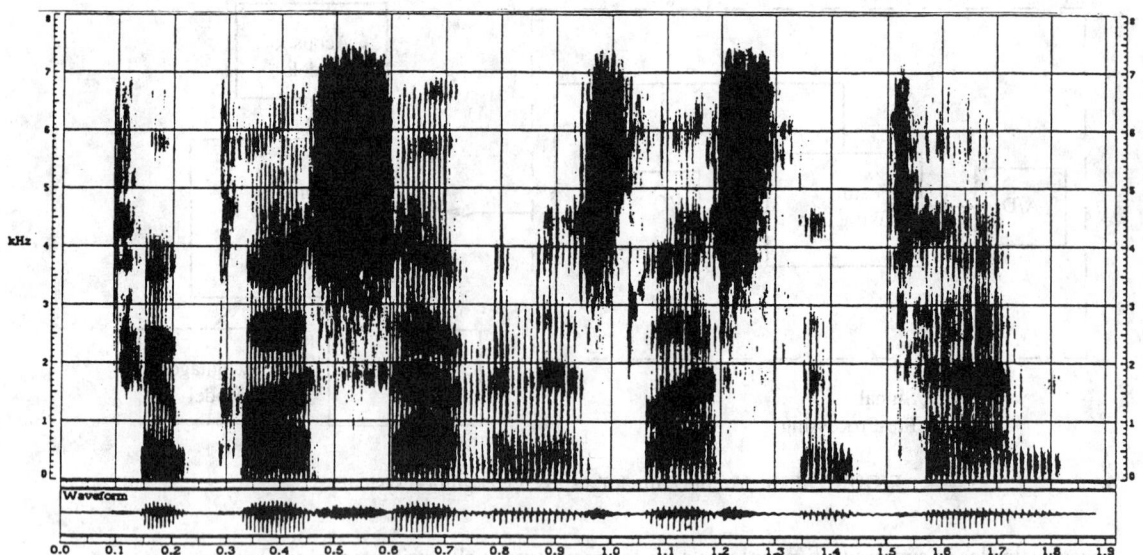

FIGURE 14.19 Speech spectrogram of the utterance "Two plus seven is less than ten." (*Source:* V.W. Zue, "The use of speech knowledge in automatic speech recognition," *Proc. IEEE*, vol. 73, no. 11, pp. 1602–1615, © 1985 IEEE. With permission.)

envelope and the periodic voicing source. Each feature vector is computed from a frame of speech data defined by windowing N samples of the signal. While a better spectral estimate can be obtained using more samples, the interval must be short enough so that the windowed signal is roughly stationary. For speech data, N is chosen such that the length of the interval covered by the window is approximately 25 to 30 msec. The feature vectors are typically computed at a frame rate of 10 to 20 msec by shifting the window forward in time. Tapered windowing functions, such as the Hamming window, are used to reduce dependence of the spectral estimate on the exact temporal position of the window. Spectral features are often augmented with a measure of the short time energy of the signal, as well as with measures of energy and spectral change over time [Lee, 1988].

For recognition systems which use discrete features, vector quantization can be used to quantize continuous-valued feature vectors into a set or codebook of K discrete symbols, or codewords [Gray, 1984]. The K codewords are characterized by prototypes $y^1 \ldots y^K$. A feature vector x is quantized to the kth codeword if the distance from x to y^k, or $d(x,y^k)$, is less than the distance from x to any other codeword. The distance $d(x,y)$ depends on the type of features being quantized. For features derived from the short-time spectrum and cepstrum, this distance is typically Euclidean or weighted Euclidean. For LPC-based features, the Itakura metric, which is based on spectral distortion, is typically used [Furui, 1989].

Dynamic Time Warping

Dynamic time warping (DTW) is a technique for nonlinear time alignment of pairs of spoken utterances. DTW-based speech recognition, often referred to as "template matching," involves aligning feature vectors extracted from an unknown utterance with those from a set of exemplars or templates obtained from training data. Nonlinear feature alignment is necessitated by nonlinear time-scale warping associated with variations in speaking rate.

Figure 14.20 illustrates the time correspondence between two utterances, A and B, represented as feature-vector sequences of unequal length. The time warping function consists of a sequence of points $F = c_1, \ldots, c_K$ in the plane spanned by A and B, where $c_k = (i_k, j_k)$. The local distance between the feature vectors a_i and b_j on the warping path at point $c = (i,j)$ is given as

$$d(c) = d(a_i, b_j) \tag{14.19}$$

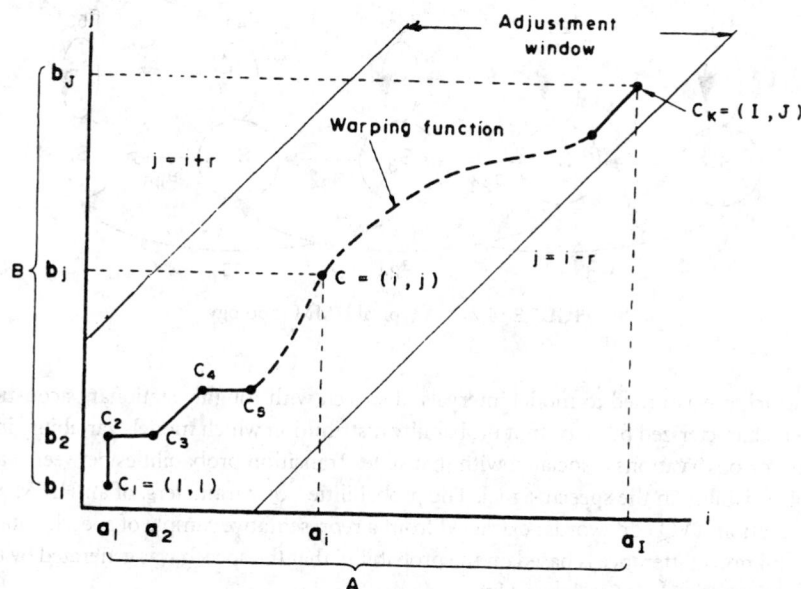

FIGURE 14.20 Dynamic time warping of utterances A and B. (*Source:* S. Furui, *Digital Speech Processing, Synthesis and Recognition*, New York: Marcel Dekker, 1989. With permission.)

The distance between A and B aligned with warping function F is a weighted sum of the local distances along the path,

$$D(F) = \frac{1}{N} \sum_{k=1}^{K} d(c_k) w_k \tag{14.20}$$

where w_k is a nonnegative weighting function and N is the sum of the weights. Path constraints and weighting functions are chosen to control whether or not the distance $D(F)$ is symmetric and the allowable degree of warping in each direction. Dynamic programming is used to efficiently determine the optimal time alignment between two feature-vector sequences [Sakoe and Chiba, 1978].

In DTW-based recognition, one or more templates are generated for each word in the recognition vocabulary. For speaker-dependent recognition tasks, templates are typically created by aligning and averaging the feature vectors corresponding to several repetitions of a word. For speaker-independent tasks, clustering techniques can be used to generate templates which better model pronunciation variability across talkers. In isolated word recognition, the distance $D(F)$ is computed between the feature-vector sequence for the unknown word and the templates corresponding to each vocabulary item. The unknown is recognized as that word for which $D(F)$ is a minimum. DTW can be extended to connected word recognition by aligning the input utterance to all possible concatenations of reference templates. Efficient algorithms for computing such alignments have been developed [Furui, 1989]; however, in general, DTW has proved most applicable to isolated word recognition tasks.

Hidden Markov Models[2]

Hidden Markov modeling is a probabilistic pattern matching technique which is more robust than DTW at modeling acoustic variability in speech and more readily extensible to continuous speech recognition. As shown in Fig. 14.21, hidden Markov models (HMMs) represent speech as a sequence

---

[2] Although the discussion here is limited to HMMs with discrete observations, output distributions such as Gaussians can be defined for continuous-valued features.

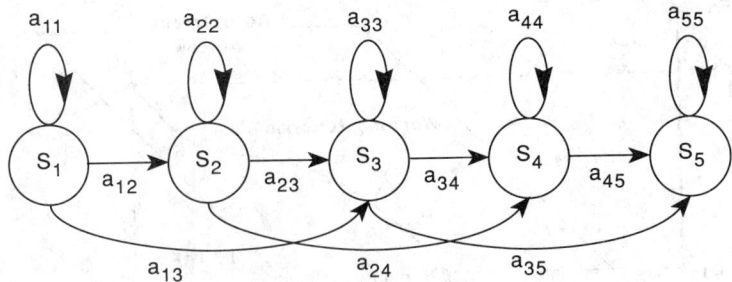

FIGURE 14.21   A typical HMM topology.

of states, which are assumed to model intervals of speech with roughly stationary acoustic features. Each state is characterized by an output probability distribution which models variability in the spectral features or observations associated with that state. Transition probabilities between states model durational variability in the speech signal. The probabilities, or parameters, of an HMM are trained using observations (VQ codewords) extracted from a representative sample of speech data. Recognition of an unknown utterance is based on the probability that the speech was generated by the HMM.

More precisely, an HMM is defined by:

1. A set of N states $\{S_1 \ldots S_N\}$, where q_t is the state at time t.
2. A set of K observation symbols $\{v_1 \ldots v_K\}$, where O_t is the observation at time t.
3. A state transition probability matrix $A = \{a_{ij}\}$, where the probability of transitioning from state S_i at time t to state S_j at time $t + 1$ is $a_{ij} = P(q_{t+1} = S_j | q_t = S_i)$.
4. A set of output probability distributions B, where for each state j, $b_j(k) = P(O_t = v_k | q_t = S_j)$.
5. An initial state distribution $\pi = \{\pi_i\}$, where $\pi_i = P(q_1 = S_i)$.

At each time t a transition to a new state is made, and an observation is generated. State transitions have the Markov property, in that the probability of transitioning to a state at time t depends only on the state at time $t - 1$. The observations are conditionally independent given the state, and the transition probabilites are not dependent on time. The model is called hidden because the identity of the state at time t is unknown; only the output of the state is observed. It is common to specify an HMM by its parameters $\lambda = (A, B, \pi)$.

The basic acoustic unit modeled by the HMM can be either a word or a subword unit. For small recognition vocabularies, the lexicon typically consists of whole-word models similar to the model shown in Fig. 14.21. The number of states in such a model can either be fixed or be made to depend on word duration. For larger vocabularies, words are more often defined in the lexicon as concatenations of phone or triphone models. Triphones are phone models with left and right context specified [Lee, 1988]; they are used to model acoustic variability which results from the coarticulation of adjacent speech sounds.

In isolated word recognition tasks, an HMM is created for each word in the recognition vocabulary. In continuous speech recognition, on the other hand, a single HMM network is generated by expressing allowable word strings or sentences as concatenations of word models, as shown in Fig. 14.22. In wordspotting, the HMM network consists of a parallel connection of keyword models and a background model which represents the speech within which the keywords are embedded. Background models, in turn, typically consist of parallel connections of subword acoustic units such as phones [Wilcox and Bush, 1992].

The language model or grammar of a recognition system defines the sequences of vocabulary items which are allowed. For simple tasks, deterministic finite-state grammars can be used to define all allowable word sequences. Typically, however, recognizers make use of stochastic grammars based on n-gram statistics [Jelinek, 1985]. A bigram language model, for example, specifies the probability of a vocabulary item given the item which precedes it.

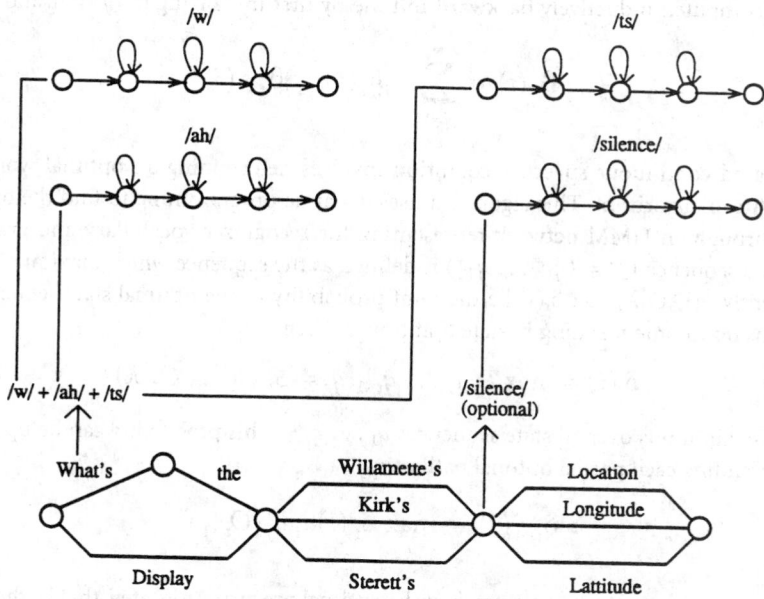

FIGURE 14.22 Language model, lexicon, and HMM phone models for a continuous speech recognition system. (*Source:* K.F. Lee, "Large-Vocabulary Speaker-Independent Continuous Speech Recognition: The SPHINX System," Ph.D. Dissertation, Computer Science Dept., Carnegie Mellon, April 1988. With permission.)

Isolated word recognition using HMMs involves computing, for each word in the recognition vocabulary, the probability $P(O|\lambda)$ of the observation sequence $O = O_1 \ldots O_T$. The unknown utterance is recognized as the word which maximizes this probability. The probability $P(O|\lambda)$ is the sum over all possible state sequences $Q = q_1 \ldots q_T$ of the probability of O and Q given λ, or

$$p(O|\lambda) = \sum_Q P(O,Q|\lambda) = \sum_Q P(O|Q,\lambda)P(Q|\lambda) = \sum_{q_1 \cdots q_T} \pi_{q_1} b_{q_1}(O_1) a_{q_1 q_2} b_{q_2}(O_2) \cdots \quad (14.21)$$

Direct computation of this sum is computationally infeasible for even a moderate number of states and observations. However, an iterative algorithm known as the **forward-backward** procedure [Rabiner, 1989] makes this computation possible. Defining the forward variable α as

$$\alpha_t(i) = P(O_1 \ldots O_t, q_t = S_i | \lambda) \quad (14.22)$$

and initializing $\alpha_1(i) = \pi_i b_i(O_1)$, subsequent $\alpha_t(i)$ are computed inductively as

$$\alpha_{t+1}(j) = \sum_{i=1}^{N} \alpha_t(i) a_{ij} b_j(O_{t+1}) \quad (14.23)$$

By definition, the desired probability of the observation sequence given the model λ is

$$P(O|\lambda) = \sum_{i=1}^{N} \alpha_T(i) \quad (14.24)$$

Similarly, the backward variable β can be defined

$$\beta_t(i) = P(O_{t+1} \ldots O_T | q_t = S_i, \lambda) \quad (14.25)$$

The βs are computed inductively backward in time by first initializing $\beta_T(j) = 1$ and computing

$$\beta_t(i) = \sum_{j=1}^{N} a_{ij} b_j(O_{t+1}) \beta_{t+1}(j) \tag{14.26}$$

HMM-based continuous speech recognition involves determining an optimal word sequence using the **Viterbi** algorithm. This algorithm uses dynamic programming to find the optimal state sequence through an HMM network representing the recognizer vocabulary and grammar. The optimal state sequence $Q^* = (q_1^* \ldots q_T^*)$ is defined as the sequence which maximizes $P(Q|O,\lambda)$, or equivalently $P(Q,O|\lambda)$. Let $\delta_t(i)$ be the joint probability of the optimal state sequence and the observations up to time t, ending in state S_i at time t. Then

$$\delta_t(i) = \max P(q_1 \ldots q_{t-1}, q_t = S_i, O_1 \ldots O_t | \lambda) \tag{14.27}$$

where the maximum is over all state sequences $q_1 \ldots q_{t-1}$. This probability can be updated recursively by extending each partial optimal path using

$$\delta_{t+1}(j) = \max_i \delta_t(i) a_{ij} b_j(O_{t+1}) \tag{14.28}$$

At each time t, it is necessary to keep track of the optimal precursor to state j, that is, the state which maximized the above probability. Then, at the end of the utterance, the optimal state sequence can be retrieved by backtracking through the precursor list.

Training HMM-based recognizers involves estimating the parameters for the word or phone models used in the system. As with DTW, several repetitions of each word in the recognition vocabulary are used to train HMM-based isolated word recognizers. For continuous speech recognition, word or phone exemplars are typically extracted from word strings or sentences [Lee, 1988]. Parameters for the models are chosen based on a maximum likelihood criterion; that is, the parameters λ maximize the likelihood of the training data O, $P(O|\lambda)$. This maximization is performed using the **Baum-Welch** algorithm [Rabiner, 1989], a re-estimation technique based on first aligning the training data O with the current models, and then updating the parameters of the models based on this alignment.

Let $\xi_t(i,j)$ be the probability of being in state S_i at time t and state S_j at time $t + 1$ and observing the observation sequence O. Using the forward and backward variables $\alpha_t(i)$ and $\beta_t(j)$, $\xi_t(i,j)$ can be written as

$$\xi_t(i,j) = P(q_t = S_i, q_{t+1} = S_j | O, \lambda) = \frac{\alpha_t(i) a_{ij} \beta_{t+1}(j) b_j(O_{t+1})}{\sum_{ij=1}^{N} \alpha_t(i) a_{ij} \beta_{t+1}(j) b_j(O_{t+1})} \tag{14.29}$$

An estimate of a_{ij} is given by the expected number of transitions from state S_i to state S_j divided by the expected number of transitions from state S_i. Define $\gamma_t(i)$ as the probability of being in state S_i at time t, given the observation sequence O

$$\gamma_t(i) = P(q_t = S_i | O, \lambda) = \sum_{j=1}^{N} \xi_t(i,j) \tag{14.30}$$

Summing $\gamma_t(i)$ over t yields a quantity which can be interpreted as the expected number of transitions from state S_i. Summing $\xi_t(i,j)$ over t gives the expected number of transitions from state i to state j. An estimate of a_{ij} can then be computed as the ratio of these two sums. Similarly, an estimate of $b_j(k)$ is obtained as the expected number of times being in state j and observing symbol v_k divided by the expected number of times in state j.

$$\hat{a}_{ij} = \frac{\displaystyle\sum_{t=1}^{T-1} \xi_t(i, j)}{\displaystyle\sum_{t=1}^{T} \gamma_t(i)} \qquad \hat{b}_j(k) = \frac{\displaystyle\sum_{t:O_t=y_k} \gamma_t(j)}{\displaystyle\sum_{t=1}^{T} \gamma_t(j)} \tag{14.31}$$

State-of-the-Art Recognition Systems

Dictation-oriented recognizers which accommodate isolated word vocabularies of many thousands of words in speaker-adaptive manner are currently available commercially. So too are speaker-independent, continuous speech recognizers for small vocabularies, such as the digits; similar products for larger (1000-word) vocabularies with constrained grammars are imminent. Speech recognition research is aimed, in part, at the development of more robust pattern classification techniques, including some based on neural networks [Lippmann, 1989] and on the development of systems which accommodate more natural spoken language dialogs between human and machine.

Defining Terms

Baum-Welch: A re-estimation technique for computing optimal values for HMM state transition and output probabilities.

Continuous speech recognition: Recognition of fluently spoken utterances.

Dynamic time warping (DTW): A recognition technique based on nonlinear time alignment of unknown utterances with reference templates.

Forward-backward: An efficient algorithm for computing the probability of an observation sequence from an HMM.

Hidden Markov model (HMM): A stochastic model which uses state transition and output probabilities to generate observation sequences.

Isolated word recognition: Recognition of words or short phrases preceded and followed by silence.

Signal pre-processing: Conversion of an analog speech signal into a sequence of numeric feature vectors or observations.

Viterbi: An algorithm for finding the optimal state sequence through an HMM given a particular observation sequence.

Wordspotting: Detection or location of keywords in the context of fluent speech.

References

S. Furui, *Digital Speech Processing, Synthesis, and Recognition,* New York: Marcel Dekker, 1989.

R. M. Gray, "Vector quantization," *IEEE ASSP Magazine,* vol. 1, no. 2, pp. 4–29, April 1984.

F. Jelinek, "The development of an experimental discrete dictation recognizer," *Proc. IEEE,* vol. 73, no. 11, pp. 1616–1624, Nov. 1985.

K. F. Lee, "Large-Vocabulary Speaker-Independent Continuous Speech Recognition: The SPHINX System," Ph.D. Dissertation, Computer Science Department, Carnegie Mellon University, April 1988.

R. P. Lippmann, "Review of neural networks for speech recognition," *Neural Computation,* vol. 1, pp. 1–38, 1989.

L. R. Rabiner, "A tutorial on hidden Markov models and selected applications in speech recognition," *Proc. IEEE,* vol. 77, no. 2, pp. 257–285, Feb. 1989.

H. Sakoe and S. Chiba, "Dynamic programming algorithm optimization for spoken word recognition," *IEEE Transactions on Acoustics, Speech and Signal Processing,* vol. 26, no. 1, pp. 43–49, Feb. 1978.

L. D. Wilcox and M. A. Bush, "Training and search algorithms for an interactive wordspotting system," in Proceedings, International Conference on Acoustics, Speech and Signal Processing, San Francisco, March 1992, pp. II-97–II-100.

V. W. Zue, "The use of speech knowledge in automatic speech recognition," *Proc. IEEE*, vol. 73, no. 11, pp. 1602–1615, Nov. 1985.

Further Information

Papers on speech recognition are regularly published in the *IEEE Speech and Audio Transactions* (formerly part of the *IEEE Transactions on Acoustics, Speech and Signal Processing*) and in the journal *Computer Speech and Language*. Speech recognition research and technical exhibits are presented at the annual IEEE International Conference on Acoustics, Speech and Signal Processing (ICASSP), the biannual European Conference on Speech Communication and Technology (Eurospeech), and the biannual International Conference on Spoken Language Processing (ICSLP), all of which publish proceedings. Commercial applications of speech recognition technology are featured at annual American Voice Input-Output Society (AVIOS) and Speech Systems Worldwide meetings. A variety of standardized databases for speech recognition system development are available from the National Institute of Standards and Technology in Gaithersburg, MD.

15

Spectral Estimation and Modeling

S. Unnikrishna Pillai
Polytechnic University

Theodore I. Shim
Polytechnic University

Stella N. Batalama
University of Virginia

Dimitri Kazakos
University of Virginia

15.1 Spectral Analysis 315
Historical Perspective • Modern Spectral Analysis
15.2 Parameter Estimation 321
Bayesian Estimation Scheme • Mean-Square Estimation •
Minimax Estimation Scheme • Maximum Likelihood Estimation
Scheme • Other Parameter Estimation Schemes

15.1 Spectral Analysis

S. Unnikrishna Pillai and Theodore I. Shim

Historical Perspective

Modern spectral analysis dates back at least to Sir Isaac Newton [Newton, 1671], whose prism experiments with sunlight led him to discover that each color represented a particular wavelength of light and that the sunlight contained all wavelengths. Newton used the word *spectrum*, a variant of the Latin word *specter*, to describe the band of visible light colors.

In the early eighteenth century, Bernoulli discovered that the solution to the wave equation describing a vibrating string can be expressed as an infinite sum containing weighted sine and cosine terms. Later, the French engineer Joseph Fourier in his *Analytical Theory of Heat* [Fourier, 1822] extended Bernoulli's wave equation results to arbitrary periodic functions that might contain a finite number of jump discontinuities. Thus, for some $T_0 > 0$, if $f(t) = f(t + T_0)$ for all t, then $f(t)$ represents a periodic signal with period T_0 and in the case of real signals, it has the Fourier series representation

$$f(t) = A_0 + 2 \sum_{k=1}^{\infty} (A_k \cos k\omega_0 t + B_k \sin k\omega_0 t)$$

where $\omega_0 = 2\pi/T_0$, and

$$A_k = \frac{1}{T_0} \int_0^{T_0} f(t) \cos k\omega_0 t \, dt, \qquad B_k = \frac{1}{T_0} \int_0^{T_0} f(t) \sin k\omega_0 t \, dt$$

with A_0 representing the dc term ($k = 0$). Moreover, the infinite sum on the right-hand side of the above expression converges to $[f(t_{-0}) + f(t_{+0})]/2$. The total power P of the periodic function satisfies the relation

$$P = \frac{1}{T_0} \int_0^{T_0} |f(t)|^2 \, dt = A_0^2 + 2\sum_{k=1}^{\infty} (A_k^2 + B_k^2)$$

implying that the total power is distributed only among the dc term, the fundamental frequency $\omega_0 = 2\pi/T_0$ and its harmonics $k\omega_0$, $k \geq 1$, with $2(A_k^2 + B_k^2)$ representing the power contained at the harmonic $k\omega_0$. For every periodic signal with finite power, since $A_k \to 0$, $B_k \to 0$, eventually the overharmonics become of decreasing importance.

The British physicist Schuster [Schuster, 1898] used this observation to suggest that the partial power $P_k = 2(A_k^2 + B_k^2)$ at frequency $k\omega_0$, $k = 0 \to \infty$, be displayed as the spectrum. Schuster termed this method the *periodogram*, and information over a multiple of periods was used to compute the Fourier coefficients and/or to smooth the periodogram, since depending on the starting time, the periodogram may contain irregular and spurious peaks. A notable exception to periodogram was the linear regression analysis introduced by the British statistician Yule [Yule, 1927] to obtain a more accurate description of the periodicities in noisy data. Because the sampled periodic process $x(k) = \cos k\omega_0 T$ containing a single harmonic component satisfies the recursive relation

$$x(k) = ax(k-1) - x(k-2)$$

where $a = 2 \cos \omega_0 T$ represents the harmonic component, its noisy version $x(k) + n(k)$ satisfies the recursion

$$x(k) = ax(k-1) - x(k-2) + n(k)$$

Yule interpreted this time series model as a recursive harmonic process driven by a noise process and used this form to determine the periodicity in the sequence of sunspot numbers. Yule further generalized the above recursion to

$$x(k) = ax(k-1) + bx(k-2) + n(k)$$

where a and b are arbitrary, to describe a truly autoregressive process and since for the right choice of a, b the least-square solution to the above autoregressive equation is a damped sinusoid, this generalization forms the basis for the modern day parametric methods.

Modern Spectral Analysis

Norbert Wiener's classic work on Generalized Harmonic Analysis [Wiener, 1930] gave random processes a firm statistical foundation, and with the notion of ensemble average several key concepts were then introduced. The formalization of modern day probability theory by Kolmogorov and his school also played an indispensable part in this development. Thus, if $x(t)$ represents a continuous-time stochastic (random) process, then for every fixed t, it behaves like a **random variable** with some **probability density function** $f_x(x, t)$. The ensemble average or **expected value** of the process is given by

$$\mu_x(t) = E[x(t)] = \int_{-\infty}^{\infty} x f_x(x, t) \, dx$$

and the statistical correlation between two time instants t_1 and t_2 of the random process is described through its **autocorrelation function**

$$R_{xx}(t_1, t_2) = E[x(t_1)x^*(t_2)] = \int_{-\infty}^{\infty} \int_{-\infty}^{-\infty} x_1 x_2^* f_{x_1 x_2}(x_1, x_2, t_1, t_2) dx_1 dx_2 = R_{xx}^*(t_2, t_1)$$

where $f_{x_1 x_2}(x_1, x_2, t_1, t_2)$ represents the joint probability density function of the random variable $x_1 = x(t_1)$ and $x_2 = x(t_2)$ and $*$ denotes the complex conjugate transpose in general. Processes with autocorrelation functions that depend only upon the difference of the time intervals t_1 and t_2 are known as wide sense stationary processes. Thus, if $x(t)$ is wide sense stationary, then

$$E[x(t + \tau)x^*(t)] = R_{xx}(\tau) = R_{xx}^*(-\tau)$$

To obtain the distribution of power versus frequency in the case of a **stochastic process**, one can make use of the Fourier transform based on a finite segment of the data. Letting

$$P_T(\omega) = \frac{1}{2T} \left| \int_{-T}^{T} x(t)e^{-j\omega t} \, dt \right|^2$$

represent the power contained in a typical realization over the interval $(-T, T)$, its ensemble average value as $T \to \infty$ represents the true power contained at frequency ω. Thus, for wide sense stationary processes

$$S(\omega) = \lim_{T \to \infty} E[P_T(\omega)] = \lim_{T \to \infty} \int_{-T}^{T} \int_{-T}^{T} R_{xx}(t_1 - t_2) e^{-j\omega(t_1 - t_2)} dt_1 dt_2$$

$$= \lim_{T \to \infty} \int_{-2T}^{2T} R_{xx}(\tau) \left(1 - \frac{|\tau|}{2T} \right) e^{-j\omega \tau} d\tau = \int_{-\infty}^{\infty} R_{xx}(\tau) e^{-j\omega \tau} d\tau \geq 0$$

$$(15.1)$$

Moreover, the inverse relation gives

$$R_{xx}(\tau) = \frac{1}{2\pi} \int_{-\infty}^{\infty} S(\omega)e^{j\omega \tau} d\omega \qquad (15.2)$$

and hence

$$R_{xx}(0) = E\left[|x(t)|^2\right] = P = \frac{1}{2\pi} \int_{-\infty}^{\infty} S(\omega) \, d\omega$$

Thus $S(\omega)$ represents the power spectral density and from Eqs. (15.1) and (15.2), the power spectral density and the autocorrelation function form a Fourier transform pair, the well-known Wiener–Khinchin theorem.

If $x(kT)$ represents a discrete-time wide sense stationary stochastic process, then

$$r_k = E\{x((n + k)T)x^*(nT)\} = r_{-k}^*$$

and the power spectral density is given by

$$S(\omega) = \sum_{k=-\infty}^{\infty} r_k e^{-jk\omega T}$$

or in terms of the normalized variable $\theta = \omega T$,

$$S(\theta) = \sum_{k=-\infty}^{\infty} r_k e^{-jk\theta} = S(\theta + 2\pi k) \geq 0 \qquad (15.3)$$

and the inverse relation gives the autocorrelations to be

$$r_k = \frac{1}{2\pi} \int_{-\pi}^{\pi} S(\theta)e^{jk\theta} d\theta = r_{-k}^*$$

Thus, the power spectral density of a discrete-time process is periodic. Such a process can be obtained by sampling a continuous-time process at $t = kT$, $|k| = 0 \to \infty$, and if the original continuous-time process is band-limited with a two-sided bandwidth equal to $2\omega_b = 2\pi/T$, then the set of discrete samples so obtained is equivalent to the original process in a mean-square sense.

As Schur [Schur, 1917] has observed, for discrete-time stationary processes the nonnegativity of the **power spectrum** is equivalent to the nonnegative definiteness of the Hermitian Toeplitz matrices, i.e.,

$$S(\theta) \geq 0 \Leftrightarrow \mathbf{T}_k = \begin{pmatrix} r_0 & r_1 & \cdots & r_k \\ r_1^* & r_0 & \cdots & r_{k-1} \\ \vdots & \vdots & \ddots & \vdots \\ r_k^* & r_{k-1} & \cdots & r_0 \end{pmatrix} = \mathbf{T}_k^* \geq 0, \qquad k = 0 \to \infty \qquad (15.4)$$

If $x(nT)$ is the output of a discrete-time linear time-invariant causal system driven by $w(nT)$, then we have the following representation:

$$w(nT) \to \boxed{H(z) = \sum_{k=0}^{\infty} h(kT)z^k} \to x(nT) = \sum_{k=0}^{\infty} h(kT)w\big((n-k)T\big) \qquad (15.5)$$

In the case of a stationary input, the output is also stationary, and its power spectral density is given by

$$S_x(\theta) = |H(e^{j\theta})|^2 S_w(\theta) \qquad (15.6)$$

where $S_w(\theta)$ represents the power spectral density of the input process. If the input is a white noise process, then $S_w(\theta) = \sigma^2$ and

$$S_x(\theta) = \sigma^2 |H(e^{j\theta})|^2$$

Clearly if $H(z)$ is rational, so is $S_x(\theta)$. Conversely, given a power spectral density

$$S_x(\theta) = \sum_{k=-\infty}^{\infty} r_k e^{jk\theta} \geq 0 \qquad (15.7)$$

that satisfies the integrability condition

$$\int_{-\pi}^{\pi} S_x(\theta)\, d\theta < \infty \qquad (15.8)$$

and the physical realizability (Paley–Wiener) criterion

$$\int_{-\pi}^{\pi} \ln S_x(\theta)\, d\theta > -\infty \qquad (15.9)$$

there exists a unique function $H(z)$ that is analytic together with its inverse in $|z| < 1$ (minimum phase factor) such that

$$H(z) = \sum_{k=0}^{\infty} b_k z^k, \qquad |z| < 1 \qquad (15.10)$$

and

$$S_x(\theta) = \lim_{r \to 1-0} |H(re^{j\theta})|^2 = |H(e^{j\theta})|^2, \ a.e.$$

$H(z)$ is known as the Wiener factor associated with $S_x(\theta)$ and as Eq. (15.5) shows, when driven by white noise, it generates a stochastic process $x(nT)$ from past samples and its power spectral density matches with the given $S_x(\theta)$.

In this context, given a finite set of autocorrelations $r_0, r_1, \ldots, r_n$, the spectral extension problem is to obtain the class of all extensions that match the given data, i.e., such an extension $K(\theta)$ must automatically satisfy

$$K(\theta) \geq 0$$

and

$$\frac{1}{2\pi} \int_{-\pi}^{\pi} K(\theta) e^{jk\theta} d\theta = r_k, \quad k = 0 \to n$$

in addition to satisfying Eqs. (15.8) and (15.9).

The solution to this problem is closely related to the trigonometric moment problem, and it has a long and continuing history through the works of Schur [1917]; Nevanlinna, Akheizer and Krein [Akheizer and Krein, 1962]; Geronimus [1954]; and Shohat and Tamarkin [1970], to name a few. If the given autocorrelations are such that the matrix T_n in Eq. (15.4) is singular, then there exists an $m \leq n$ such that T_{m-1} is positive definite ($T_{m-1} > 0$) and T_m is singular [det $T_m = 0$, det $(\cdot)$ representing the determinant of $(\cdot)$]. In that case, there exists a unique vector $X = (x_0, x_1, \cdots, x_m)^T$ such that $T_m X = 0$ and further, the autocorrelations have a unique extension given by

$$c_k = \sum_{i=1}^{m} P_i e^{jk\theta_i}, \quad |k| = 0 \to \infty \tag{15.11}$$

where $e^{j\theta_i}$, $i = 1 \to m$ are the m zeros of the polynomial $x_0 + x_1 z + \cdots + x_m z^m$ and $P_i > 0$. This gives

$$T_{m-1} = A \begin{pmatrix} P_1 & 0 & \cdots & 0 \\ 0 & P_2 & \cdots & 0 \\ \vdots & \vdots & \ddots & \vdots \\ 0 & 0 & \cdots & P_m \end{pmatrix} A^* \tag{15.12}$$

where A is an $m \times m$ Vandermonde matrix given by

$$A = \begin{pmatrix} 1 & 1 & \cdots & 1 \\ \lambda_1 & \lambda_2 & \cdots & \lambda_m \\ \lambda_1^2 & \lambda_2^2 & \cdots & \lambda_m^2 \\ \vdots & \vdots & \cdots & \vdots \\ \lambda_1^{m-1} & \lambda_2^{m-1} & \cdots & \lambda_m^{m-1} \end{pmatrix}, \quad \lambda_i = e^{j\theta_i}, \quad i = 1 \to m$$

and Eq. (15.12) can be used to determine $P_k > 0$, $k = 1 \to m$. The power spectrum associated with Eq. (15.11) is given by

$$S(\theta) = \sum_{k=1}^{m} P_k \delta(\theta - \theta_k)$$

and it represents a discrete spectrum that corresponds to pure uncorrelated sinusoids with signal powers $P_1, P_2, \ldots, P_m$.

If the given autocorrelations satisfy $\mathbf{T}_n > 0$, from Eq. (15.4), every unknown r_k, $k \geq n + 1$, must be selected so as to satisfy $\mathbf{T}_k > 0$, and this gives

$$|r_{k+1} - \zeta_k|^2 \leq R_k^2 \qquad (15.13)$$

where $\zeta_k = \mathbf{f}_k^T \mathbf{T}_k^{-1} \mathbf{b}_k$, $\mathbf{f}_k = (r_1, r_2, \cdots, r_k)^T$, $\mathbf{b}_k = (r_k, r_{k-1}, \cdots, r_1)$ and $R_k = \det \mathbf{T}_k / \det \mathbf{T}_{k-1}$.

From Eq. (15.13), the unknowns could be anywhere inside a sequence of circles with center ζ_k and radius R_k, and as a result, there are an infinite number of solutions to this problem. Schur and Nevanlinna have given an analytic characterization to these solutions in terms of bounded function extensions. A bounded function $\rho(z)$ is analytic in $|z| < 1$ and satisfies the inequality $|\rho(z)| \leq 1$ everywhere in $|z| < 1$.

In a network theory context, Youla [1980] has also given a closed form parametrization to this class of solutions. In that case, given $r_0, r_1, \cdots, r_n$, the minimum phase transfer functions satisfying Eqs. (15.8) and (15.9) are given by

$$H_\rho(z) = \frac{\Gamma(z)}{P_n(z) - z\rho(z)\tilde{P}_n(z)} \qquad (15.14)$$

where $\rho(z)$ is an *arbitrary* bounded function that satisfies the inequality (Paley-Wiener criterion)

$$\int_{-\pi}^{\pi} \ln\left[1 - |\rho(e^{j\theta})|^2\right] d\theta > -\infty$$

and $\Gamma(z)$ is the minimum phase factor obtained from the factorization

$$1 - |\rho(e^{j\theta})|^2 = |\Gamma(e^{j\theta})|^2$$

Further, $P_n(z)$ represents the Levinson polynomial generated from $r_0 \to r_n$ through the recursion

$$\sqrt{1 - |s_n|^2}\, P_n(z) = P_{n-1}(z) - zs_n\tilde{P}_{n-1}(z)$$

that starts with $P_0(z) = 1/\sqrt{r_0}$, where

$$s_n = \left\{ P_{n-1}(z) \sum_{k=1}^{n} r_k z^k \right\}_n P_{n-1}(0) \qquad (15.15)$$

represents the reflection coefficient at stage n. Here, $\{\ \}_n$ denotes the coefficient of z^n in the expansion $\{\ \}$, and $\tilde{P}_n(z) \triangleq z^n P_n^*(1/z^*)$ represents the polynomial reciprocal to $P_n(z)$. Notice that the given information $r_0 \to r_n$ enters $P_n(z)$ through Eq. (15.5). The power spectral density

$$K(\theta) = |H_\rho(e^{j\theta})|^2$$

associated with Eq. (15.14) satisfies all the interpolation properties described before. In Eq. (15.14), the solution $\rho(z) \equiv 0$ gives $H(z) = 1/P_n(z)$, a pure AR(n) system that coincides with Burg's maximum entropy extension. Clearly, if $H_\rho(z)$ is rational, then $\rho(z)$ must be rational and, more interestingly, every rational system must follow from Eq. (15.14) for a specific rational bounded function $\rho(z)$. Of course, the choice of $\rho(z)$ brings in extra freedom, and this can be profitably used for system identification as well as rational and stable approximation of nonrational systems [Pillai and Shim, 1993].

Defining Terms

Autocorrelation function: The expected value of the product of two random variables generated from a random process for two time instants; it represents their interdependence.

Expected value (or mean) of a random variable: Ensemble average value of a random variable that is given by integrating the random variable after scaling by its probability density function (weighted average) over the entire range.

Power spectrum: A nonnegative function that describes the distribution of power versus frequency. For wide sense stationary processes, the power spectrum and the autocorrelation function form a Fourier transform pair.

Probability density function: The probability of the random variable taking values between two real numbers x_1 and x_2 is given by the area under the nonnegative probability density function between those two points.

Random variable: A continuous or discrete valued variable that maps the set of all outcomes of an experiment into the real line (or complex plane). Because the outcomes of an experiment are inherently random, the final value of the variable cannot be predetermined.

Stochastic process: A real valued function of time t, which for every fixed t behaves like a random variable.

References

N. I. Akheizer and M. Krein, *Some Questions in the Theory of Moments,* American Math. Soc. Monogr., 2, 1962.

J. B. J. Fourier, *Theorie Analytique de la Chaleur (Analytical Theory of Heat),* Paris, 1822.

L. Y. Geronimus, *Polynomials Orthogonal on a Circle and Their Applications,* American Math. Soc., Translation, 104, 1954.

I. Newton, *Philos. Trans.,* vol. IV, p. 3076, 1671.

S. U. Pillai and T. I. Shim, *Spectrum Estimation and System Identification,* New York: Springer-Verlag, 1993.

I. Schur, "Uber Potenzreihen, die im Innern des Einheitzkreises Beschrankt Sind," *Journal fur Reine und Angewandte Mathematik,* vol. 147, pp. 205–232, 1917.

J. A. Shohat and J.D. Tamarkin, *The Problem of Moments,* American Math. Soc., Math. Surveys, 1, 1970.

N. Wiener "Generalized harmonic analysis," *Acta Math.,* vol. 55, pp. 117–258, 1930.

D. C. Youla, "The FEE: A New Tunable High-Resolution Spectral Estimator: Part I," Technical note, no. 3, Dept. of Electrical Engineering, Polytechnic Institute of New York, Brooklyn, New York; also RADC Report, RADC-TR-81-397, AD A114996, 1982, 1980.

G. U. Yule, "On a method of investigating periodicities in disturbed series, with special reference to Wolfer's sunspot numbers," *Philos. Trans. R. Soc. London, Ser. A,* vol. 226, pp. 267–298, 1927.

15.2 Parameter Estimation

Stella N. Batalama and Dimitri Kazakos

Parameter estimation is the operation of assigning a value in a continuum of alternatives to an unknown parameter based on a set of observations involving some function of the parameter. **Estimate** is the value assigned to the parameter and **estimator** is the function of the observations that yields the estimate.

The basic elements in the parameter estimation are a vector parameter θ^m, a vector space $\mathscr{E}^m$ where θ^m takes its values, a stochastic process $X(t)$ parameterized by θ^m and a performance criterion or cost function. The estimate $\hat{\theta}^m(x^n)$ based on the observation vector $x^n = [x_1, x_2, \ldots, x_n]$ is a solution of some optimization problem according to the performance criterion. In the following, the function $f(x^n | \theta^m)$ will denote the conditional joint probability density function of the random variables $\mathbf{x}_1, \ldots, \mathbf{x}_n$.

There are several parameter estimation schemes. If the process $X(t)$ is parametrically known, i.e., if its conditional joint probability density functions are known for each fixed value θ^m of the vector parameter θ^m, then the corresponding parameter estimation scheme is called **parametric.** If the statistics of the process $X(t)$ are nonparametrically described, i.e., given $\theta^m \in \mathscr{E}^m$ any joint probability density function of the process is a member of some nonparametric class of probability density functions, then the **nonparametric** or **robust estimation** schemes arise.

Let Γ^n denote the n-dimensional observation space. Then an estimator $\hat{\theta}(x^n)$ of a vector parameter θ^m is a function from the observation space, Γ^n, to the parameter space, $\mathscr{E}^m$. Since this is a function of random variables, it is itself a random variable (or random vector).

There are certain stochastic properties of estimators that quantify somehow their quality. In this sense an estimator is said to be **unbiased** if its expected value is the true parameter value, i.e., if

$$E_\theta\{\hat{\theta}^m(\mathbf{x}^n)\} = \theta^m$$

where the subscript θ on the expectation symbol denotes that the expectation is taken according to the probability density function $f(x^n|\theta^m)$. In the case where the observation space is the $\mathfrak{R}^n$ and the parameter is a scalar, it is

$$E_\theta\{\hat{\theta}(\mathbf{x}^n)\} = \int_{R^n} \hat{\theta}(x^n)f(x^n|\theta^m)dx^n$$

The **bias** of the estimate is the Euclidean norm $\|\theta^m - E_\theta\{\theta^m(\mathbf{x}^n)\}\|^{1/2}$. Thus, the bias measures the distance between the expected value of the estimate and the true value of the parameter. Clearly, the estimator is unbiased when the bias is zero.

Usually it is of interest to know the conditional variance of an unbiased estimate. The bias of the estimate $\hat{\theta}^m(x^n)$ and the conditional variance

$$E_\theta\{\|\hat{\theta}^m(\mathbf{x}^n) - E_\theta\{\hat{\theta}^m(\mathbf{x}^n)\}\|^2|\theta^m\}$$

generally represent a trade-off. Indeed, an unbiased estimate may induce relatively large variance. On the other hand, the introduction of some low-level bias may then result in a significant reduction of the induced variance. In general, the bias versus variance trade-off should be studied carefully for the correct evaluation of any given parameter estimate. A parameter estimate is called **efficient** if the conditional variance equals a lower bound known as the Rao-Cramèr bound.

It will be useful to present briefly this bound.

The Rao-Cramèr bound gives a theoretical minimum for the variance of any estimate. More specifically, let $\hat{\theta}(x^n)$ be the estimate of a scalar parameter θ given the observation vector x^n. Let $f(x^n|\theta)$ be given, twice continuously differentiable with respect to θ, and satisfy also some other mild regularity conditions. Then,

$$E_\theta\left\{\left[\hat{\theta}(\mathbf{x}^n) - \theta\right]^2\right\} \geq E_\theta\left\{\left[\frac{\partial}{\partial\theta}\log f(x^n|\theta)\right]^2\right\}^{-1}$$

Sometimes we need to consider the case where the sample size n increases to infinity. In such a case, an estimator is said to be **consistent** if

$$\hat{\theta}^m(\mathbf{x}^n) \rightarrow \theta^m \text{ as } n \rightarrow \infty$$

Since the estimate $\hat{\theta}^m(\mathbf{x}^n)$ is a random variable, we have to specify in what sense the above holds. Thus, if the above limit holds w.p. 1, we say that $\hat{\theta}^m(x^n)$ is *strongly consistent* or *consistent w.p. 1.* In a similar way we can define a *weakly consistent* estimator.

As far as the asymptotic distribution of $\theta(x^n)$ as $n \to \infty$ is concerned, it turns out that the central limit theorem can often be applied to $\hat{\theta}(x^n)$ to infer that $\sqrt{n}[\hat{\theta}(x^n) - \theta]$ is asymptotically normal with zero mean as $n \to \infty$.

In order to examine certain parameter estimation schemes we need first to present the definition of some related functions. **Penalty or cost function** $c[\theta^m, \hat{\theta}^m(x^n)]$ is a scalar, nonnegative function whose values vary as θ^m varies in the parameter space $\mathscr{E}^m$ and as the sequence x^n takes different values in the observation space, Γ^n. *Conditional expected penalty* $c(\theta^m, \hat{\theta}^m)$ induced by the parameter estimate and the penalty function is a function defined as follows:

$$c(\theta^m, \hat{\theta}^m) = E_\theta\{c[\theta^m, \hat{\theta}^m(\mathbf{x}^n)]\}$$

If an *a priori* density function $p(\theta^m)$ is available, then the expected penalty $c(\hat{\theta}^m, p)$ can be evaluated.

The various existing parameter estimation schemes evolve as the solutions of optimization problems whose objective function is either the conditional expected penalty or the conditional density function $f(x^n|\theta^m)$.

Bayesian Estimation Scheme

In the **Bayesian estimation** scheme the available assets are:

1. A parametrically known stochastic process parameterized by θ^m, in other words, a given conditional joint density function $f(x^n|\theta^m)$ defined on the observation space Γ^n, where θ^m is a well-defined parameter vector.
2. A realization x^n from the underlying active process, where the implied assumption is that the process remains unchanged throughout the whole observation period.
3. A density function $p(\theta^m)$ defined on the parameter space $\mathscr{E}^m$.
4. For each data sequence x^n, parameter vector θ^m and parameter estimate $\hat{\theta}^m(x^n)$, a penalty scalar function $c[\theta^m, \hat{\theta}^m(x^n)]$ is given.
5. A performance criterion which is the minimization of the expected penalty $c(\theta^m, p)$.

The estimate $\hat{\theta}_0^m$ that minimizes the expected penalty is called *optimal Bayesian estimate at p*. Under some mild conditions the optimal Bayesian estimate $\hat{\theta}_0^m(x^n)$ is the conditional expectation $E\{\theta^m|x^n\}$.

If the penalty function has the form $c[\theta^m, \hat{\theta}^m] = 1 - \delta(\|\theta^m - \hat{\theta}^m\|)$, where $\delta()$ is the delta function, then the optimal Bayesian estimate is called *maximum a posteriori estimate* since it happens to maximize the *a posteriori* density $p(\theta^m|x^n)$.

Another special case of penalty function is the function $\|\theta^m - \hat{\theta}^m\|$. In this case the Bayesian estimate is called *minimum mean-square estimate* and equals the conditional expectation $E\{\theta^m|x^n\}$.

In the following we present some more details about **mean-square estimation** since it is one of the most popular schemes.

Mean-Square Estimation

For the simplicity of our discussion we consider the case of estimating a single continuous type random variable θ with density $p(\theta)$ instead of estimating a random vector. We also reduce the dimensionality of the observation space to one. In this framework the penalty function will be the square of the estimation error $(\theta - \hat{\theta})^2$ and the performance or optimality criterion will be the minimization of the mean (expected) square value of the estimation error.

We will first consider the case of estimating a random variable θ by a constant $\hat{\theta}$. This means that we wish to find a constant $\hat{\theta}$ such that the mean-square (MS) error

$$e = E\{(\theta - \hat{\theta})^2\} = \int\limits_{\infty}^{\infty} (\theta - \hat{\theta})^2 p(\theta) d\theta$$

is minimum. Since e depends on $\hat{\theta}$, it is minimum if

$$\frac{de}{d\theta} = \int\limits_{-\infty}^{\infty} 2(\theta - \hat{\theta}) p(\theta) d\theta = 0$$

i.e., if

$$\hat{\theta} = E\{\theta\} = \int\limits_{-\infty}^{\infty} \theta p(\theta) d\theta$$

The case where θ is to be estimated by a function $\hat{\theta}(\mathbf{x})$ of the random variable (observation) $\mathbf{x}$, and not by a constant, is examined next. In this case the MS error takes the form:

$$e = E\{[\theta - \hat{\theta}(\mathbf{x})]^2\} = \int\limits_{-\infty}^{\infty} \int\limits_{-\infty}^{\infty} [\theta - \hat{\theta}(x)]^2 \, p(\theta, x) dx dy$$

where $p(\theta, x)$ is the joint density of the random variables θ and $\mathbf{x}$. In this case we need to find that function $\hat{\theta}(\mathbf{x})$ which minimizes the MS error. It can be proved that the function that minimizes the MS error is

$$\hat{\theta}(x) = E\{\theta \mid x\} = \int\limits_{-\infty}^{\infty} \theta p(\theta \mid x) d\theta$$

The function $\hat{\theta}(\mathbf{x})$ is called **nonlinear MS estimate.**

As we have seen, when the penalty function is the quadratic function $(\theta - \hat{\theta})^2$, then the optimal Bayesian estimate is the conditional expectation $E\{\theta \mid x\}$. If $\mathbf{x}$ and θ are jointly Gaussian, then the above conditional expectation is a linear function of x. But when the above statistics are not Gaussian, then the optimal Bayesian estimate is generally a nonlinear function of x. Thus, to resolve this problem we introduce suboptimal Bayesian schemes for this quadratic penalty function. In particular we consider only the class of linear parameter estimates and we try to find that estimate which minimizes the expected quadratic penalty. This estimate is called linear MS estimate and it is used in many applications because of the simplicity of the solution.

The linear estimation problem is the estimation of a random variable θ in terms of a linear function $A\mathbf{x} + B$ of $\mathbf{x}$, i.e., $\hat{\theta}(x) = A\mathbf{x} + B$. In this case we need to find the constants A and B in order to minimize the MS error

$$e = E\{[\theta - (A\mathbf{x} + B)]^2\}$$

A fundamental principle in the MS estimation is the **orthogonality principle.** This principle states that the optimum linear MS estimate $A\mathbf{x} + B$ of θ is such that the estimation error $\theta - (A\mathbf{x} + B)$ is orthogonal to the data $\mathbf{x}$, i.e.,

$$E\{[\theta - (A\mathbf{x} + B)]\mathbf{x}\} = 0$$

Using the above principle, it can be proved that e is minimum if

where
$$A = \frac{r\sigma_\theta}{\sigma_x} \text{ and } B = \eta_\theta - A\eta_x$$

$$\eta_x = E\{\mathbf{x}\}, \qquad \eta_\theta = E\{\boldsymbol{\theta}\}$$

$$\sigma_x^2 = E\{(\mathbf{x} - \eta_x)^2\}, \quad \sigma_\theta^2 = E\{(\boldsymbol{\theta} - \eta_\theta)^2\}$$

$$r = \frac{E\{(\mathbf{x} - \eta_x)(\boldsymbol{\theta} - \eta_\theta)\}}{\sigma_x \sigma_\theta}$$

i.e., η_x, η_θ are the means of $\mathbf{x}$ and $\boldsymbol{\theta}$; σ_x^2, σ_θ^2 are the corresponding variances; σ_x, σ_θ is the standard deviation of $\mathbf{x}$ and $\boldsymbol{\theta}$; and r is the correlation coefficient of $\mathbf{x}$ and $\boldsymbol{\theta}$. Thus the MS error takes the form $e = \sigma_\theta^2 (1 - r^2)$.

The estimate

$$\hat{\theta}(\mathbf{x}) = Ax + B$$

is called the **nonhomogeneous linear estimate** of $\boldsymbol{\theta}$ in terms of $\mathbf{x}$. If $\boldsymbol{\theta}$ is estimated by a function $\hat{\theta}(\mathbf{x}) = \alpha\mathbf{x}$, the estimate is called **homogeneous.** It can be also shown by the orthogonality principle that for the homogeneous estimate

$$\alpha = \frac{E\{\mathbf{x}\boldsymbol{\theta}\}}{E\{\mathbf{x}^2\}}$$

Using the orthogonality principle it can be shown that if the random variables $\boldsymbol{\theta}$ and $\mathbf{x}$ are Gaussian zero mean, then the optimum nonlinear estimate of $\boldsymbol{\theta}$ equals the linear estimate. In other words if $\hat{\theta}(\mathbf{x}) = E\{\boldsymbol{\theta}|x\}$ is the optimum nonlinear estimate of $\boldsymbol{\theta}$ and $\hat{\theta} = \alpha x$ is the optimum linear estimate, then $\hat{\theta}(\mathbf{x}) = E\{\hat{\theta}|x\} = \hat{\theta} = \alpha x$.

This is true since the random variables $\boldsymbol{\theta}$ and $\mathbf{x}$ have zero mean, $E\{\boldsymbol{\theta}\} = E\{\mathbf{x}\} = 0$, and thus the linear estimate $\hat{\theta}$ has zero mean too, $E\{\hat{\theta}\} = 0$. This implies that the linear estimation error $\varepsilon = \boldsymbol{\theta} - \hat{\theta}$ also has zero mean, $E\{\varepsilon\} = E\{\boldsymbol{\theta} - \hat{\theta}\} = 0$.

On the other hand, the orthogonality principle can be applied, which implies that the linear estimation error ε is orthogonal to the data, $E\{\varepsilon x\} = 0$. Since ε is Gaussian, it is independent of $\mathbf{x}$ and thus $E\{\varepsilon|x\} = E\{\varepsilon\} = 0$, which is equivalent with the following:

$$E\{\boldsymbol{\theta} - \hat{\theta}|x\} = 0 \Rightarrow E\{\boldsymbol{\theta}|x\} - E\{\hat{\theta}|x\} = 0$$

$$\Rightarrow E\{\boldsymbol{\theta}|x\} = E\{\hat{\theta}|x\} \Rightarrow \hat{\theta}(\mathbf{x}) = \alpha x \Rightarrow \hat{\theta}(\mathbf{x}) = \hat{\theta}$$

i.e., the nonlinear and the linear estimates coincide.

In addition, since the linear estimation error $\varepsilon = \boldsymbol{\theta} - \hat{\theta}$ is independent of the data x, so is the square error, i.e.,

$$E\{(\boldsymbol{\theta} - \hat{\theta})^2|x\} = E\{(\boldsymbol{\theta} - \hat{\theta}^2\} = V$$

Thus, the conditional mean of $\boldsymbol{\theta}$ assuming the data x equals its MS estimate and the conditional variance the MS error. That simplifies the evaluation of conditional densities when Gaussian random variables are involved because since $f(\theta|x)$ is Gaussian, it has the form

$$f(\theta \,|\, X) = \frac{1}{\sqrt{2\pi V}} \exp\left\{\frac{-[\theta - \alpha x]^2}{2V}\right\}$$

Minimax Estimation Scheme

In the **minimax estimation** scheme the available assets are:

1. A parametrically known stochastic process parameterized by θ^m
2. A realization x^n from the underlying active process
3. A scalar penalty function $c[\theta^m, \hat{\theta}^m(x^n)]$ for each data sequence x^n, parameter vector θ^m, and parameter estimate $\hat{\theta}^m(x^n)$

The minimax schemes are solutions of saddle-point game formalizations, with payoff function the expected penalty $c(\hat{\theta}^m, p)$ and with variables the parameter estimate $\hat{\theta}^m$ and the *a priori* parameter density function p. If a minimax estimate $\hat{\theta}^m_0$ exists, it is an optimal Bayesian estimate, at some least favorable *a priori* distribution p_0.

Maximum Likelihood Estimation Scheme

Maximum likelihood estimation was first introduced by Fisher. It is a very powerful estimation procedure that yields many of the well-known estimation methods as special cases.

The essential difference between Bayesian and maximum likelihood parameter estimation is that in Bayesian the parameter θ^m is considered to be random with a given density function, while in the maximum likelihood framework it is unknown but fixed.

Consider a random process $X(t)$ parameterized by θ^m, where θ^m is an unknown fixed parameter vector of finite dimensionality m (e.g., $\theta^m \in \mathfrak{R}^m$). More specifically the conditional joint probability density function $f(x_1, ..., x_n | \theta^m)$ is well known for every θ^m, where $x^n = [x_1, ..., x_n]$ is a realization (or observation vector or sample vector) of the process $X(t)$.

The problem we are faced with is to find an estimate of the parameter vector θ^m based on the realization of $X(t)$. (Note that the dimensionality of the parameter vector θ^m in the joint probability density function is assumed to be fixed.)

The intuition behind the maximum likelihood method is that we choose those parameters $[\theta_1, ..., \theta_m]$ from which the actually observed sample vector is most likely to have come. This means that the estimator of θ^m is selected so that the observed sample vector becomes as "likely as possible."

In this sense we call the conditional joint probability density function $f(x^n | \theta^m)$ as likelihood function $\ell(\theta^m)$. The likelihood function $\ell(\theta^m)$ is a deterministic function of the parameter vector θ^m once the observed variables $x_1, ..., x_n$ are inserted. This means that θ^m is variable and the sample vector x^n is fixed, while the conditional joint probability density function is considered as a function of the observation vector x^n with θ^m fixed. The maximum likelihood estimator of θ^m is that value of the parameter vector for which the likelihood function is maximized.

In many cases it is more convenient to work with another function called log-likelihood function, $L(\theta^m)$, rather than the likelihood function. The log-likelihood function is the natural logarithm of $\ell(\theta^m)$. Since the logarithm is a monotone function, it follows that whenever $L(\theta)$ achieves its maximum value, $\ell(\theta^m)$ is maximized too, for the same value of the parameter vector θ^m. Thus the log-likelihood function is maximized for that value of the vector parameter θ^m for which the first partial derivatives with respect to θ_i, $i = 1, ..., m$ equal zero, i.e.,

$$\hat{\theta}_{ML} : \frac{\partial L(\theta^m)}{\partial \theta_i} = 0$$

where $\hat{\theta}_{ML}$ denotes the maximum likelihood estimate of the vector parameter θ^m.

It can be shown that when the process $X(t)$ is memoryless and stationary (i.e., when $x_1, ..., x_n$ are independent, identically distributed) then the ML estimators are consistent, asymptotically efficient, and asymptotically Gaussian.

Example: Let $\mathbf{x}_i$, $i = 1, ..., n$, be Gaussian independent random variables with mean θ and variance σ_i^2: $\mathbf{x}_i \in N(\theta, \sigma_i^2)$. The mean θ is to be estimated and the Rao-Cramèr bound is to be evaluated.

Since θ is unknown but fixed, we will use the maximum likelihood estimation scheme. The random variable $\mathbf{x}_i$ has the probability density function

$$\frac{1}{\sqrt{2\pi}\sigma_i} \exp\left\{-\frac{(x_i - \theta)^2}{2\sigma_i^2}\right\}$$

Since $\mathbf{x}_i$, $i = 1, \ldots, n$, are independent, the joint density function is

$$f(x_i, \ldots, x_n | \theta) = \prod_{i=1}^{n} \frac{1}{\sqrt{2\pi}\sigma_i} \exp\left\{-\frac{(x_i - \theta)^2}{2\sigma_i^2}\right\}$$

which is exactly the likelihood function for this estimation problem. The log-likelihood function is

$$\log f(x_1, \ldots, x_n | \theta) = -\frac{n}{2} \log(2\pi) - \sum_{i=1}^{n} \log \sigma_i - \frac{1}{2} \sum_{i=1}^{n} \frac{(x_i - \theta)^2}{\sigma_i^2}$$

We can maximize the log-likelihood function with respect to θ and find the maximizing value to be equal to

$$\hat{\theta}_{ML}(x^n) = \frac{1}{\sum_{i=1}^{n} \frac{1}{\sigma_i^2}} \sum_{i=1}^{n} \frac{x_i}{\sigma_i^2}$$

Note that for equal variances the maximum likelihood estimate coincides with the commonly used sample mean.

The Rao-Cramèr bound can be found as follows:

$$E_\theta\left\{\left[\frac{d}{d\theta} \log f(x^n | \theta)\right]^2\right\}^{-1} = -E_\theta\left\{\frac{d^2}{d\theta^2} \log f(x^n | \theta)\right\} = \sum_{i=1}^{n} \frac{1}{\sigma_i^2}$$

In conclusion, we see that for Gaussian data the sample mean estimate is efficient because it coincides with the maximum likelihood estimate.

When the data are contaminated with a fraction of data coming from an unknown probability density function, the so called *outliers*, the sample mean performs poorly even when the fraction of outliers is small. This observation gave birth to the branch of statistics called robust statistics.

Other Parameter Estimation Schemes

The Bayesian, minimax, and maximum likelihood estimation schemes described above make up the class of parametric parameter estimation procedures. The common characteristic of those procedures is the availability of some parametrically known stochastic process that generates the observation sequence x^n. When for every given parameter value θ^m the stochastic process that generates x^n is nonparametrically described, the nonparametric or robust estimation schemes arise. The latter schemes may evolve as the solutions of certain saddle-point games, whose payoff function originates from the parametric maximum likelihood formalizations. It is then assumed that, in addition to the nonparametrically described data-generating process, the only assets available are a realization x^n from the underlying active process and the parameter space $\mathscr{E}^m$. The **qualitative robustness** in parameter estimation corresponds to local performance stability for small deviations from a nominal, parametrically known, data-generating process.

Defining Terms

Bayesian estimation: An estimation scheme in which the parameter to be estimated is modeled as a random variable with known probability density function.

Bias: The norm of the difference between the true value of the estimate and its mean value.

Consistent estimator: An estimator whose value converges to the true parameter value as the sample size tends to infinity. If the convergence holds w.p. 1, then the estimator is called *strongly consistent* or *consistent w.p. 1*.

Efficient estimator: An estimator whose variance achieves the Rao-Cramèr bound.

Estimate: Our best guess of the parameter of interest based on a set of observations.

Estimator: A mapping from the data space to the parameter space that yields the estimate.

Homogeneous linear estimator: An estimator which is a homogeneous linear function of the data.

Maximum likelihood estimate: An estimate that maximizes the probability density function of the data conditioned on the parameter.

Mean-square estimation: An estimation scheme in which the cost function is the mean-square error.

Minimax estimate: The optimum estimate for the least favorable prior distribution.

Nonhomogeneous linear estimator: An estimator which is a nonhomogeneous linear function of the data.

Nonlinear MS estimate: The optimum estimate under the mean-square performance criterion.

Nonparametric estimation: An estimation scheme in which no parametric description of the statistical model is available.

Orthogonality principle: The fundamental principle for MS estimates. It states that the estimation error is orthogonal to the data.

Parameter estimation: The procedure by which we combine all available data to obtain our best guess about a parameter of interest.

Parametric estimation: An estimation scheme in which the statistical description of the data is given according to a parametric family of statistical models.

Penalty or cost function: A nonnegative scalar function which represents the cost incurred by an inaccurate value of the estimate.

Qualitative robustness: A geometric formulation of robust estimation.

Robust estimation: An estimation scheme in which we optimize performance for the least favorable statistical environment among a specified statistical class.

Unbiased estimator: An estimator whose mean value is equal to the true parameter value.

References

S. Haykin, *Adaptive Filter Theory,* Englewood Cliffs, N.J.: Prentice-Hall, 1991.

D. Kazakos and P. Papantoni-Kazakos, *Detection and Estimation,* New York: Computer Science Press, 1990.

L. Ljung and T. Söderström, *Theory and Practice of Recursive Identification,* Cambridge, Mass.: The MIT Press, 1983.

A. Papoulis, *Probability, Random Variables, and Stochastic Processes,* New York: McGraw-Hill, 1984.

Further Information

IEEE Transactions on Information Theory is a bimonthly journal that publishes papers on theoretical aspects of estimation theory. More specifically, the papers are concerned with the transmission, processing, and utilization of information.

IEEE Transactions on Signal Processing is a monthly journal which presents applications of estimation theory to image processing, speech recognition, speech processing, and acoustical signal processing.

IEEE Transactions on Communications is a monthly journal presenting applications of estimation theory to data communication problems, synchronization of communication systems, channel equalization, and image processing.

16

Multidimensional Signal Processing

Edward J. Delp
Purdue University

Jan Allebach
Purdue University

Charles A. Bouman
Purdue University

Sarah A. Rajala
North Carolina State University

N. K. Bose
Pennsylvania State University

L. H. Sibul
Pennsylvania State University

16.1 Digital Image Processing 329
Image Capture • Point Operations • Image Enhancement • Digital Image Compression • Reconstruction • Edge Detection • Analysis and Computer Vision

16.2 Video Signal Processing 345
Sampling • Quantization • Vector Quantization • Video Compression • Information-Preserving Coders • Predictive Coding • Motion-Compensated Predictive Coding • Transform Coding • Subband Coding • HDTV • Motion Estimation Techniques • Token Matching Methods • Image Quality and Visual Perception • Visual Perception

16.3 Sensor Array Processing 359
Spatial Arrays, Beamformers, and FIR Filters • Discrete Arrays for Beamforming • Discrete Arrays and Polynomials • Velocity Filtering

16.1 Digital Image Processing

Edward J. Delp, Jan Allebach, and Charles A. Bouman

What is a **digital image**? What is digital image processing? Why does the use of computers to process pictures seem to be everywhere? The space program, robots, and even people with personal computers are using digital image processing techniques. In this section we shall describe what a digital image is, how one obtains digital images, what the problems with digital images are (they are not trouble-free), and finally how these images are used by computers. A discussion of processing the images is presented later in the section. At the end of this section is a bibliography of selected references on digital image processing.

The use of computers to process pictures is about 30 years old. While some work was done more than 50 years ago, the year 1960 is usually the accepted date when serious work was started in such areas as optical character recognition, **image coding**, and the space program. NASA's Ranger moon mission was one of the first programs to return digital images from space. The Jet Propulsion Laboratory (JPL) established one of the early general-purpose image processing facilities using second-generation computer technology.

The early attempts at digital image processing were hampered because of the relatively slow computers used, i.e., the IBM 7094, the fact that computer time itself was expensive, and that image digitizers had to be built by the research centers. It was not until the late 1960s that image processing hardware was generally available (although expensive). Today it is possible to put together a small laboratory system for less than $60,000; a system based on a popular home computer can be assembled for about $5,000. As the cost of computer hardware decreases, more uses of digital image processing will appear in all facets of life. Some people have predicted that by the turn of the century at least 50% of the images we handle in our private and professional lives will have been processed on a computer.

Image Capture

A digital image is nothing more than a matrix of numbers. The question is how does this matrix represent a real image that one sees on a computer screen?

Like all imaging processes, whether they are analog or digital, one first starts with a sensor (or transducer) that converts the original imaging energy into an electrical signal. These sensors, for instance, could be the photomultiplier tubes used in an x-ray system that converts the x-ray energy into a *known* electrical voltage. The transducer system used in ultrasound imaging is an example where sound pressure is converted to electrical energy; a simple TV camera is perhaps the most ubiquitous example. An important fact to note is that the process of conversion from one energy form to an electrical signal is not necessarily a *linear* process. In other words, a proportional charge in the input energy to the sensor will not always cause the same proportional charge in the output electrical signal. In many cases calibration data are obtained in the laboratory so that the relationship between the input energy and output electrical signal is known. These data are necessary because some transducer performance characteristics change with age and other usage factors.

The sensor is not the only thing needed to form an image in an imaging system. The sensor must have some spatial extent before an image is formed. By spatial extent we mean that the sensor must not be a simple point source examining only one location of energy output. To explain this further, let us examine two types of imaging sensors used in imaging: a CCD video camera and the ultrasound transducer used in many medical imaging applications.

The CCD camera consists of an *array* of light sensors known as charge-coupled devices. The image is formed by examining the output of each sensor in a preset order for a finite time. The electronics of the system then forms an electrical signal which produces an image that is shown on a cathode-ray tube (CRT) display. The image is formed because there is an array of sensors, each one examining only one spatial location of the region to be sensed.

The process of **sampling** the output of the sensor array in a particular order is known as *scanning*. Scanning is the typical method used to convert a two-dimensional energy signal or image to a one-dimensional electrical signal that can be handled by the computer. (An image can be thought of as an energy field with spatial extent.) Another form of scanning is used in ultrasonic imaging. In this application there is *only* one sensor instead of an array of sensors. The ultrasound transducer is moved or steered (either mechanically or electrically) to various spatial locations on the patient's chest or stomach. As the sensor is moved to each location, the output electrical signal of the sensor is sampled and the electronics of the system then form a television-like signal which is displayed. Nearly all the transducers used in imaging form an image by either using an array of sensors or a single sensor that is moved to each spatial location.

One immediately observes that both of the approaches discussed above are equivalent in that the energy is sensed at various spatial locations of the object to be imaged. This energy is then converted to an electrical signal by the transducer. The image formation processes just described are classical analog image formation, with the distance between the sensor locations limiting the spatial resolution in the system. In the array sensors, resolution is determined by how close the sensors are located in the array. In the single-sensor approach, the spatial resolution is limited by how far the sensor is moved. In an actual system spatial resolution is also determined by the performance characteristics of the sensor. Here we are assuming for our purposes *perfect* sensors.

In digital image formation one is concerned about two processes: *spatial sampling* and **quantization**. Sampling is quite similar to scanning in analog image formation. The second process is known as *quantization* or *analog-to-digital conversion,* whereby at each spatial location a *number* is assigned to the amount of energy the transducer observes at that location. This number is usually proportional to the electrical signal at the output of the transducer. The overall process of sampling and quantization is known as *digitization*. Sometimes the digitization process is just referred to as analog-to-digital conversion, or A/D conversion; however, the reader should remember that digitization also includes spatial sampling.

The digital image formulation process is summarized in Fig. 16.1. The spatial sampling process can be considered as overlaying a grid on the object, with the sensor examining the energy output from each grid box and converting it to an electrical signal. The quantization process then assigns a number to the electrical signal; the result, which is a *matrix* of numbers, is the digital representation of the image. Each spatial location in the image (or grid) to which a number is assigned is known as a *picture element* or **pixel** (or pel). The size of the sampling grid is usually given by the number of pixels on each side of the grid, e.g., 256×256, 512×512, 488×380.

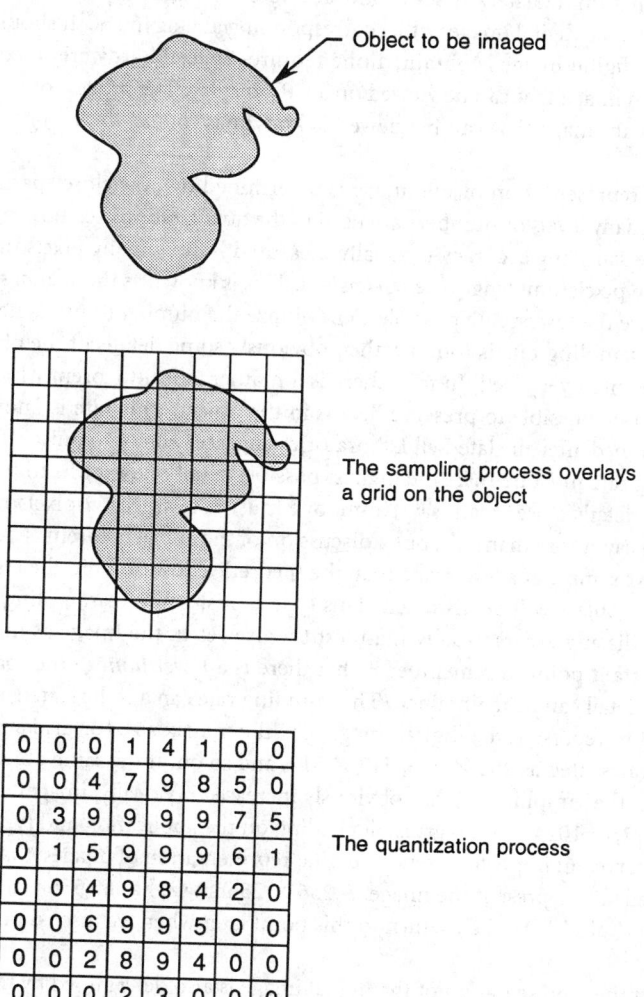

Object to be imaged

The sampling process overlays a grid on the object

The quantization process

FIGURE 16.1 Digital image formation: sampling and quantization.

The quantization process is necessary because all information to be processed using computers must be represented by numbers. The quantization process can be thought of as one where the input energy to the transducer is represented by a finite number of energy values. If the energy at a particular pixel location does not take on one of the finite energy values, it is assigned to the closest value. For instance, suppose that we assume *a priori* that only energy values of 10, 20, 50, and 110 will be represented (the units are of no concern in this example). Suppose at one pixel an energy of 23.5 was observed by the transducer. The A/D converter would then assign this pixel the energy value of 20 (the closest one). Notice that the quantization process makes mistakes; this error in assignment is known as *quantization* error or *quantization noise.*

In our example, each pixel is represented by one of four possible values. For ease of representation of the data, it would be simpler to assign to each pixel the index value 0, 1, 2, 3, instead of 10, 20, 50, 110. In fact, this is typically done by the quantization process. One needs a simple table to know that a pixel assigned the value 2 corresponds to an energy of 50. Also, the number of possible energy levels is typically some integer power of two to also aid in representation. This power is known as the number of *bits* needed to represent the energy of each pixel. In our example each pixel is represented by two bits.

One question that immediately arises is how accurate the digital representation of the image is when one compares the digital image with a corresponding analog image. It should first be pointed out that after the digital image is obtained one requires special hardware to convert the matrix of pixels back to an image that can be viewed on a CRT display. The process of converting the digital image back to an image that can be viewed is known as *digital-to-analog conversion,* or *D/A conversion.*

The quality of representation of the image is determined by how close spatially the pixels are located and how many levels or numbers are used in the quantization, i.e., how coarse or fine is the quantization. The sampling accuracy is usually measured in how many pixels there are in a given area and is cited in pixels/unit length, i.e., pixels/cm. This is known as the *spatial sampling rate.* One would desire to use the lowest rate possible to minimize the number of pixels needed to represent the object. If the sampling rate is too low, then obviously some details of the object to be imaged will not be represented very well. In fact, there is a mathematical theorem which determines the lowest sampling rate possible to preserve details in the object. This rate is known as the *Nyquist* sampling rate (named after the late Bell Laboratories engineer Harry Nyquist). The theorem states that the sampling rate must be *twice* the highest possible detail one expects to image in the object. If the object has details closer than, say 1 mm, one must take at least 2 pixels/mm. (The Nyquist theorem actually says more than this, but a discussion of the entire theorem is beyond the scope of this section.) If we sample at a lower rate than the theoretical lowest limit, the resulting digital representation of the object will be distorted. This type of distortion or sampling error is known as *aliasing* errors. Aliasing errors usually manifest themselves in the image as moiré patterns (Fig. 16.2). The important point to remember is that there is a *lower limit* to the spatial sampling rate such that object detail can be maintained. The sampling rate can also be stated as the total number of pixels needed to represent the digital image, i.e., the matrix size (or grid size). One often sees these sampling rates cited as 256×256, 512×512, and so on. If the same object is imaged with a large matrix size, the sampling rate has obviously increased. Typically, images are sampled on 256×256, 512×512, or 1024×1024 grids, depending on the application and type of modality. One immediately observes an important issue in digital representation of images: that of the large number of pixels needed to represent the image. A 256×256 image has 65,536 pixels and a 512×512 image has 262,144 pixels! We shall return to this point later when we discuss processing or storage of these images.

The quality of the representation of the digital image is also determined by the number of levels or shades of gray that are used in the quantization. If one has more levels, then fewer mistakes will be made in assigning values at the output of the transducer. Figure 16.3 demonstrates how the number of gray levels affects the digital representation of an artery. When a small number of levels

FIGURE 16.2 This image shows the effects of aliasing due to sampling the image at too low a rate. The image should be straight lines converging at a point. Because of under-sampling, it appears as if there are patterns in the lines at various angles. These are known as moiré patterns.

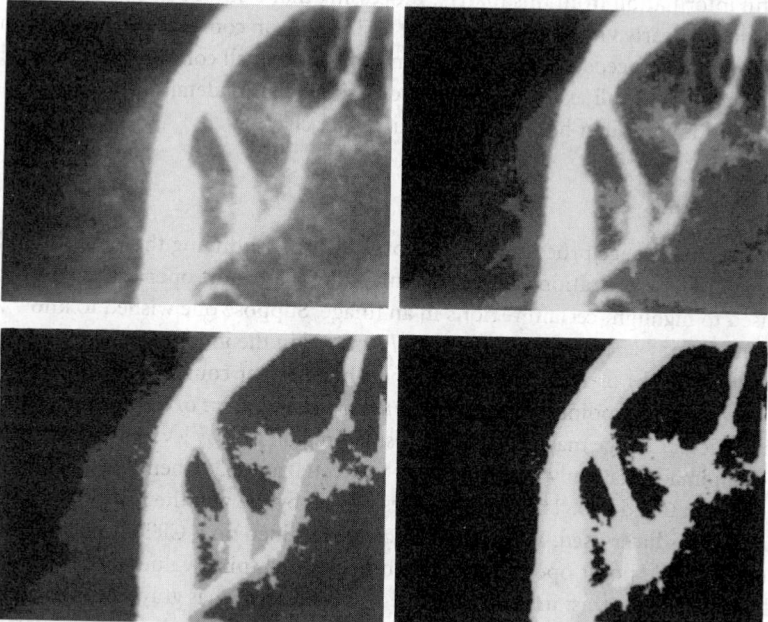

FIGURE 16.3 This image demonstrates the effects of quantization error. The upper left image is a coronary artery image with 8 bits (256 levels or shades of gray) per pixel. The upper right image has 4 bits/pixel (16 levels). The lower left image has 3 bits/pixel (8 levels). The lower right image has 2 bits/pixel (4 levels). Note the false contouring in the images as the number of possible levels in the pixel representation is reduced. This false contouring is the quantization error, and as the number of levels increases the quantization error decreases because fewer mistakes are being made in the representation.

are used, the quantization is coarse and the quantization error is large. The quantization error usually manifests itself in the digital image by the appearance of *false contouring* in the picture. One usually needs at least 6 bits or 64 gray levels to represent an image adequately. Higher-quality imaging systems use 8 bits (256 levels) or even as many as 10 bits (1024 levels) per pixel. In most applications, the human observer cannot distinguish quantization error when there are more than 256 levels. (Many times the number of gray levels is cited in bytes. One byte is 8 bits, i.e., high-quality monochrome digital imaging systems use one byte per pixel.)

One of the problems briefly mentioned previously is the large number of pixels needed to represent an image, which translates into a large amount of digital data needed for the representation. A 512×512 image with 8 bits/pixel (1 byte/pixel) of gray level representation requires 2,097,152 bits of computer data to describe it. A typical computer file that contains 1000 words usually requires only about 56,000 bits to describe it. The 512×512 image is 37 times larger! (A picture is truly worth more than 1000 words.) This data requirement is one of the major problems with digital imaging, given that the storage of digital images in a computer file system is expensive. Perhaps another example will demonstrate this problem. Many computers and word processing systems have the capability of transmitting information over telephone lines to other systems at data rates of 2400 bits per second. At this speed it would require nearly 15 minutes to transmit a 512×512 image! Moving objects are imaged digitally by taking *digital snapshots* of them, i.e., digital video. True digital imaging would acquire about 30 images/s to capture all the important motion in a scene. At 30 images/s, with each image sampled at 512×512 and with 8 bits/pixel, the system must handle 62,914,560 bits/s. Only very expensive acquisition systems are capable of handling these large data rates.

The greatest advantage of digital images is that they can be processed on a computer. Any type of operation that one can do on a computer can be done to a digital image. Recall that a digital image is just a (huge) matrix of numbers. Digital image processing is the process of using a computer to extract useful information from this matrix. Processing that cannot be done optically or with analog systems (such as early video systems) can be easily done on computers. The disadvantage is that a large amount of data needs to be processed and on some small computer systems this can take a long time (hours). We shall examine image processing in more detail in the next subsection and discuss some of the computer hardware issues in a later chapter.

Point Operations

Perhaps the simplest image processing operation is that of modifying the values of individual pixels in an image. These operations are commonly known as **point operations**. A point operation might be used to highlight certain regions in an image. Suppose one wished to know where all the pixels in a certain gray level region were *spatially* located in the image. One would modify all those pixel values to 0 (black) or 255 (white) such that the observer could see where they were located.

Another example of a point operation is *contrast enhancement* or *contrast stretching*. The pixel values in a particular image may occupy only a small region of gray level distribution. For instance, the pixels in an image may only take on values between 0 and 63, when they could nominally take on values between 0 and 255. This is sometimes caused by the way the image was digitized and/or by the type of transducer used. When this image is examined on a CRT display the contrast looks washed out. A simple point operation that multiplies each pixel value in the image by four will increase the apparent contrast in the image; the new image now has gray values between 0 and 252. This operation is shown in Fig. 16.4. Possibly the most widely used point operation in medical imaging is *pseudo-coloring*. In this point operation all the pixels in the image with a particular gray value are assigned a *color*. Various schemes have been proposed for appropriate pseudo-color tables that assign the gray values to colors. It should be mentioned that point operations are often cascaded, i.e., an image undergoes contrast enhancement and then pseudo-coloring.

The operations described above can be thought of as operations (or *algorithms*) that modify the range of the gray levels of the pixels. An important feature that describes a great deal about an

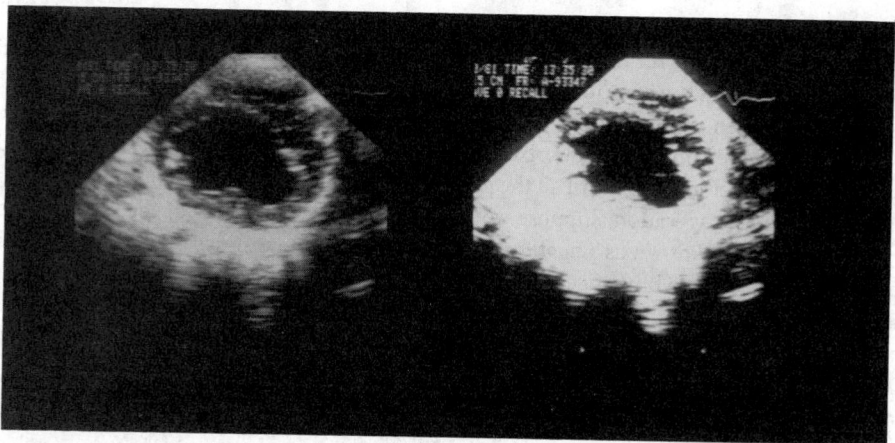

FIGURE 16.4 Contrast stretching. The image on the right has gray values between 0 and 63, causing the contrast to look washed out. The image on the right has been contrast enhanced by multiplying the gray levels by four.

image is the *histogram* of the pixel values. A histogram is a table that lists how many pixels in an image take on a particular gray value. These data are often plotted as a function of the gray value. Point operations are also known as *histogram modification* or *histogram stretching*. The contrast enhancement operation shown in Fig. 16.4 modifies the histogram of the resultant image by stretching the gray values from a range of 0–63 to a range of 0–252. Some point operations are such that the resulting histogram of the processed image has a particular shape. A popular form of histogram modification is known as *histogram equalization,* whereby the pixels are modified such that the histogram of the processed image is almost flat, i.e., all the pixel values occur equally.

It is impossible to list all possible types of point operations; however, the important thing to remember is that these operations process one pixel at a time by modifying the pixel based *only* on its gray level value and *not* where it is distributed spatially (i.e., location in the pixel matrix). These operations are performed to enhance the image, make it easier to see certain structures or regions in the image, or to force a particular shape to the histogram of the image. They are also used as initial operations in a more complicated image processing algorithm.

Image Enhancement

Image enhancement is the use of image processing algorithms to remove certain types of distortion in an image. The image is enhanced by removing noise, making the edge structures in the image stand out, or any other operation that makes the image *look* better.[1] Point operations discussed above are generally considered to be enhancement operations. Enhancement also includes operations that use groups of pixels and the spatial location of the pixels in the image.

The most widely used algorithms for enhancement are based on pixel functions that are known as **window operations.** A window operation performed on an image is nothing more than the process of examining the pixels in a certain region of the image, called the window region, and com-

---

[1]Image enhancement is often confused with *image restoration*. Image enhancement is the ad hoc application of various processing algorithms to enhance the appearance of the image. Image restoration is the application of algorithms that use knowledge of the degradation process to enhance or restore the image, i.e., deconvolution algorithms used to remove the effect of the aperture point spread function in blurred images. A discussion of image restoration is beyond the scope of this section.

puting some type of mathematical function derived from the pixels in the window. In most cases the windows are square or rectangle, although other shapes have been used. After the operation is performed, the result of the computation is placed in the center pixel of the window where a 3×3 pixel window has been extracted from the image. The values of the pixels in the window, labeled a_1, $a_2, \ldots, a_9$, are used to compute a new pixel value which replaces the value of a_5, and the window is moved to a new center location until all the pixels in the original image have been processed. As an example of a window operation, suppose we computed the average value of the pixels in the window. This operation is known as *smoothing* and will tend to reduce noise in the image, but unfortunately it will also tend to blur edge structures in the image.

Another window operation often used is the computation of a linear weighted sum of the pixel values. Let a_5' be the new pixel value that will replace a_5 in the original image. We then form

$$a_5' = \sum_{i=1}^{9} \alpha_i a_i \qquad (16.1)$$

where the α_i's are any real numbers. For the simple smoothing operation described above we set $\alpha_i = 1/9$ for all i. By changing the values of the α_i weights, one can perform different types of enhancement operations to an image. Any window operation that can be described by Eq. 16.1 is known as a *linear window operation* or *convolution* operator. If some of the α_i coefficients take on negative values, one can enhance the appearance of edge structures in the image.

It is possible to compute a nonlinear function of the pixels in the window. One of the more powerful nonlinear window operations is that of *median filtering*. In this operation all the pixels in the window are listed in descending magnitude and the middle, or *median*, pixel is obtained. The median pixel then is used to replace a_5. The median filter is used to remove noise from an image and at the same time preserve the edge structure in the image. More recently there has been a great deal of interest in *morphological operators*. These are also nonlinear window operations that can be used to extract or enhance shape information in an image.

In the preceding discussion, all of the window operations were described on 3×3 windows. The current research in window operations is directed at using large window sizes, i.e., 9×9, 13×13, or 21×21. The philosophy in this work is that small window sizes only use local information and what one really needs to use is information that is more global in nature.

Digital Image Compression

Image compression refers to the task of reducing the amount of data required to store or transmit a digital image. As discussed earlier, in its natural form, a digital image comprises an array of numbers. Each such number is the sampled value of the image at a pixel (picture element) location. These numbers are represented with finite precision using a fixed number of bits. Until recently, the dominant image size was 512×512 pixels with 8 bits or 1 byte per pixel. The total storage size for such an image is $512^2 \approx 0.25 \times 10^6$ bytes or 0.25 Mbytes. When digital image processing first emerged in the 1960s, this was considered to be a formidable amount of data, and so interest in developing ways to reduce this storage requirement arose immediately. Since that time, image compression has continued to be an active area of research. The recent emergence of standards for image coding algorithms and the commercial availability of very large scale integration (VLSI) chips that implement image coding algorithms is indicative of the present maturity of the field, although research activity continues apace.

With declining memory costs and increasing transmission bandwidths, 0.25 Mbytes is no longer considered to be the large amount of data that it once was. This might suggest that the need for image compression is not as great as previously. Unfortunately (or fortunately, depending on one's point of view), this is not the case because our appetite for image data has also grown enormously over the years. The old 512×512 pixels $\times$ 1 byte per pixel "standard" was a consequence of the spatial and gray scale resolution of sensors and displays that were commonly available until recently. At this time, displays with more than $10^3 \times 10^3$ pixels and 24 bits/pixel to allow full color representation (8 bits each for red, green, and blue) are becoming commonplace. Thus, our 0.25-Mbyte standard image size has grown to 3 Mbytes. This is just the tip of the iceberg, however. For example, in desktop printing applications, a 4-color (cyan, magenta, yellow, and black) image of an 8.5×11 in.^{2} page sampled at 600 dots per in. requires 134 Mbytes. In remote sensing applications, a typical hyperspectral image contains terrain irradiance measurements in each of 200 10-nm-wide spectral bands at 25-m intervals on the ground. Each measurement is recorded with 12-bit precision. Such data are acquired from aircraft or satellite and are used in agriculture, forestry, and other fields concerned with management of natural resources. Storage of these data from just a 10×10 km^2 area requires 4800 Mbytes.

Figure 16.5 shows the essential components of an image compression system. At the system input, the image is encoded into its compressed form by the image coder. The compressed image may then be subjected to further digital processing, such as error control coding, encryption, or multiplexing with other data sources, before being used to modulate the analog signal that is actually transmitted through the channel or stored in a storage medium. At the system output, the image is processed step by step to undo each of the operations that was performed on it at the system input. At the final step, the image is decoded into its original uncompressed form by the image decoder. Because of the role of the image encoder and decoder in an image compression system, **image coding** is often used as a synonym for image compression. If the reconstructed image is identical to the original image, the compression is said to be **lossless**. Otherwise, it is **lossy**.

Image compression algorithms depend for their success on two separate factors: redundancy and irrelevancy. *Redundancy* refers to the fact that each pixel in an image does not take on all possible values with equal probability, and the value that it does take on is not independent of that of the other pixels in the image. If this were not true, the image would appear as a white noise pattern such as that seen when a television receiver is tuned to an unused channel. From an information-theoretic point of view, such an image contains the maximum amount of information. From the point of view of a human or machine interpreter, however, it contains no information at all. *Irrelevancy* refers to the fact that not all the information in the image is required for its intended application. First, under typical viewing conditions, it is possible to remove some of the information in an image without producing a change that is perceptible to a human observer. This is because of the

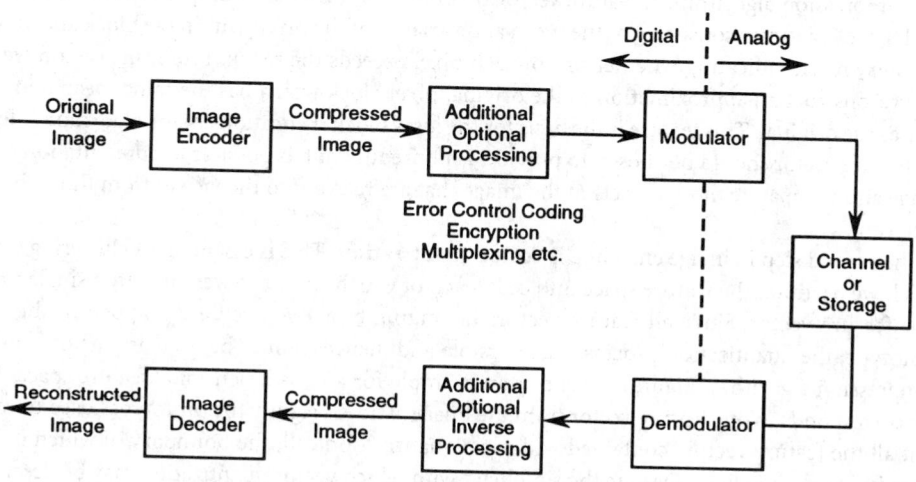

FIGURE 16.5   Overview of an image compression system.

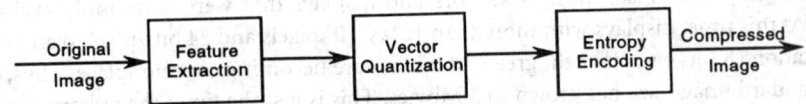

FIGURE 16.6 Key elements of an image encoder.

limited ability of the human viewer to detect small changes in luminance over a large area or larger changes in luminance over a very small area, especially in the presence of detail that may mask these changes. Second, even though some degradation in image quality may be observed as a result of image compression, the degradation may not be objectionable for a particular application, such as teleconferencing. Third, the degradation introduced by image compression may not interfere with the ability of a human or machine to extract the information from the image that is important for a particular application. Lossless compression algorithms can only exploit redundancy, whereas lossy methods may exploit both redundancy and irrelevancy.

A myriad of approaches have been proposed for image compression. To bring some semblance of order to the field, it is helpful to identify those key elements that provide a reasonably accurate description of most encoding algorithms. These are shown in Fig. 16.6. The first step is *feature extraction*. Here the image is partitioned into $N \times N$ blocks of pixels. Within each block, a feature vector is computed which is used to represent all the pixels within that block. If the feature vector provides a complete description of the block, i.e., the block of pixel values can be determined exactly from the feature vector, then the feature is suitable for use in a lossless compression algorithm. Otherwise, the algorithm will be lossy. For the simplest feature vector, we let the block size $N = 1$ and take the pixel values to be the features. Another important example for $N = 1$ is to let the feature be the error in the prediction of the pixel value based on the values of neighboring pixels which have already been encoded and, hence, whose values would be known as the decoder. This feature forms the basis for *predictive encoding*, of which differential pulse-code modulation (DPCM) is a special case. For larger size blocks, the most important example is to compute a two-dimensional (2-D) Fourier-like transform of the block of pixels and to use the N^2 transform coefficients as the feature vector. The widely used Joint Photographic Experts Group (JPEG) standard image coder is based on the discrete cosine transform (DCT) with a block size of $N = 8$. In all of the foregoing examples, the block of pixel values can be reconstructed exactly from the feature vector. In the last example, the inverse DCT is used. Hence, all these features may form the basis for a lossless compression algorithm. A feature vector that does not provide a complete description of the pixel block is a vector consisting of the mean and variance of the pixels within the block and an $N \times N$ binary mask indicating whether or not each pixel exceeds the mean. From this vector, we can only reconstruct an approximation to the original pixel block which has the same mean and variance as the original. This feature is the basis for the lossy block truncation coding algorithm. Ideally, the feature vector should be chosen to provide as nonredundant as possible a representation of the image and to separate those aspects of the image that are relevant to the viewer from those that are irrelevant.

The second step in image encoding is **vector quantization**. This is essentially a clustering step in which we partition the feature space into cells, each of which will be represented by a single prototype feature vector. Since all feature vectors belonging to a given cell are mapped to the same prototype, the quantization process is irreversible and, hence, cannot be used as part of a lossless compression algorithm. Figure 16.7 shows an example for a two-dimensional feature space. Each dot corresponds to one feature vector from the image. The X's signify the prototypes used to represent all the feature vectors contained within its quantization cell, the boundary of which is indicated by the dashed lines. Despite the simplicity with which vector quantization may be described, the implementation of a vector quantizer is a computationally complex task unless some structure

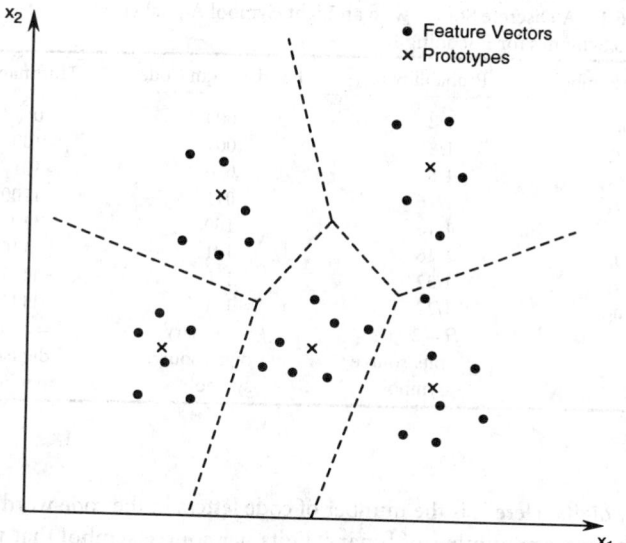

FIGURE 16.7 Vector quantization of a 2-D feature space.

is imposed on it. The clustering is based on minimizing the distortion between the original and quantized feature vectors, averaged over the entire image. The distortion measure can be chosen to account for the relative sensitivity of the human viewer to different kinds of degradation. In one dimension, the vector quantizer reduces to the Lloyd-Max scalar quantizer.

The final step in image encoding is *entropy coding*. Here we convert the stream of prototype feature vectors to a binary stream of 0's and 1's. Ideally, we would like to perform this conversion in a manner that yields the minimum average number of binary digits per prototype feature vector.

In 1948, Claude Shannon proved that it is possible to code a discrete memoryless source using on the average as few binary digits per source symbol as the *source entropy* defined as

$$H = -\sum_n p_n \log_2 p_n$$

Here p_n denotes the probability or relative frequency of occurrence of the nth symbol in the source alphabet, and $\log_2(x) = \ln(x)/\ln(2)$ is the base 2 logarithm of x. The units of H are bits/source symbol. The proof of Shannon's theorem is based on grouping the source symbols into large blocks and assigning binary code words of varying length to each block of source symbols. More probable blocks of source symbols are assigned shorter code words, whereas less probable blocks are assigned longer code words. As the block length approaches infinity, the bit rate tends to H. Huffman determined the *optimum* variable-length coding scheme for a discrete memoryless source using blocks of any finite length.

Table 16.1 provides an example illustrating the concept of source coding. The source alphabet contains eight symbols with the probabilities indicated. For convenience, these symbols have been labeled in order of decreasing probability. In the context of image encoding, the source alphabet would simply consist of the prototype feature vectors generated by the vector quantizer. The entropy of this source is 2.31 bits/source symbol. If we were to use a fixed-length code for this source, we would need to use three binary digits for each source symbol as shown in Table 16.1. On the other hand, the code words for the Huffman code contain from 1 to 4 code letters (binary digits). In this case, the average code word length

$$\bar{l} = \sum_n p_n l_n$$

Table 16.1   A Discrete Source with an Eight-Symbol Alphabet
and Two Schemes for Encoding It

Source Symbol	Probability p_n	Fixed-Length Code	Huffman Code
a_1	1/2	000	0
a_2	1/8	001	100
a_3	1/8	010	101
a_4	1/16	011	1100
a_5	1/16	100	1101
a_6	1/16	101	1110
a_7	1/32	110	11110
a_8	1/32	111	11111
	$H = 2.31$ bits/source symbol	$\bar{l} = 3$ binary digits/source symbol	$\bar{l} = 2.31$ binary digits/source symbol

is $\bar{l} = 2.31$ binary digits. Here l_n is the number of code letters in the code word for the source symbol a_n. This is the average number of binary digits per source symbol that would be needed to encode the source, and it is equal to the entropy. Thus, for this particular source, the Huffman code achieves the lower bound. It can be shown that in general the rate for the Huffman code will always be within 1 binary digit of the source entropy. By grouping source symbols into blocks of length L and assigning code words to each block, this maximum distance can be decreased to $1/L$ binary digits. Note the subtle distinction here between *bits*, which are units of information, a property of the source alone, and *binary digits*, which are units of code word length, and hence only a property of the code used to represent the source. Also note that the Huffman code satisfies the *prefix condition*, i.e., no code word is the prefix of another longer code word. This means that a stream of 0's and 1's may be uniquely decoded into the corresponding sequence of source symbols without the need for markers to delineate boundaries between code words.

The Huffman code is determined from the *binary tree* shown in Fig. 16.8. This tree is constructed recursively by combining the two least probable symbols in the alphabet into one composite sym-

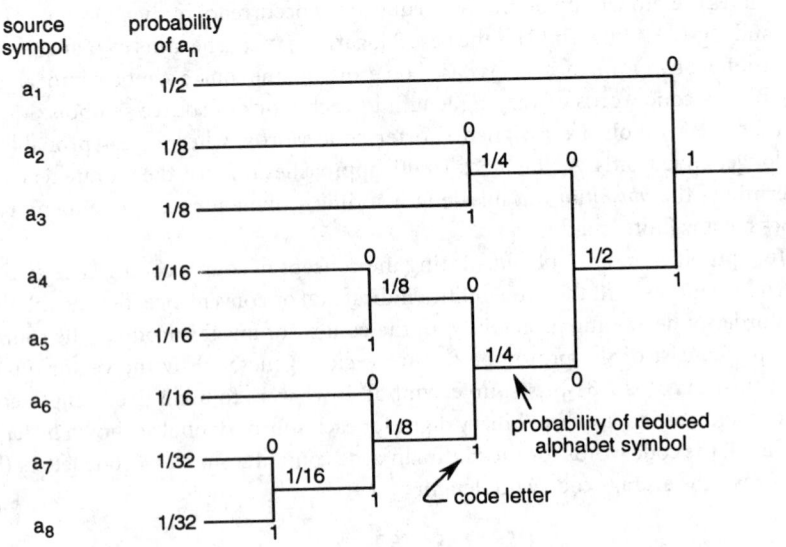

FIGURE 16.8   Binary tree used to generate the Huffman code for the source shown in Table 16.1.

bol whose probability of occurrence is the sum of the probabilities of the two symbols that it represents. The code words for these two symbols are the same as that of the composite symbol with a 0 or a 1 appended at the end to distinguish between them. This procedure is repeated until the reduced alphabet contains only a single code word. Then the code word for a particular source symbol is determined by traversing the tree from its root to the leaf node for that source symbol.

Reconstruction

The objective of **image reconstruction** is to compute an unknown image from many complex measurements of the image. Usually, each measurement depends on many pixels in the image which may be spatially distant from one another.

A typical reconstruction problem is **tomography**, in which each measurement is obtained by integrating the pixel values along a ray through the image. Figure 16.9 illustrates the measurement of these ray integrals in the **projection** process. For each angle θ a set of ray integrals is computed by varying the position t at which the ray passes through the image. The points along a ray are given by all the solutions (x,y) to the equation

$$t = x \cos \theta + y \sin \theta$$

We may therefore compute the ideal projection integrals by the following expression known as the Radon transform

$$p(\theta, t) = \int_{-\infty}^{\infty} \int_{-\infty}^{\infty} f(x, y)\delta(t - x \cos \theta - y \sin \theta)dxdy \tag{16.2}$$

where $\delta(t - x \cos \theta - y \sin \theta)$ is an impulse function that is nonzero along the projection ray.

In practice, these projection integrals may be measured using a variety of physical techniques. In transmission tomography, λ_T photons are emitted into an object under test. A detector then counts the number of photons, $\lambda(\theta,t)$, which pass through the object without being absorbed. Collimators are used to ensure the detected energy passes straight through the object along the desired path. Since the attenuation of energy as it passes through the object is exponentially related to the integral of the object's density, the projection integral may be computed from the formula

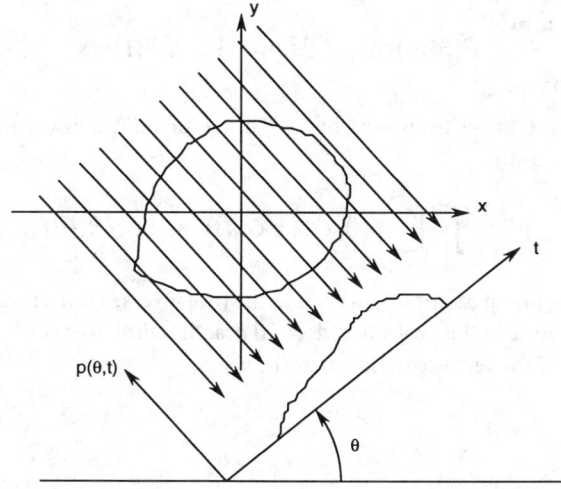

FIGURE 16.9   Projection data for angle θ, resulting in the one-dimensional function $p(\theta,t)$.

$$p(\theta, t) = -\log\left(\frac{\lambda(\theta, t)}{\lambda_T}\right)$$

In emission tomography, one wishes to measure the rate of photon emission at each pixel. In this case, various methods may be used to collect and count all the photons emitted along a ray passing through the object.

Once the projections $p(\theta, t)$ have been measured, the objective is to compute the unknown cross section $f(x, y)$. The image and projections may be related by first computing the Fourier transform of the 2-D image

$$F(\omega_x, \omega_y) = \int_{-\infty}^{\infty} \int_{-\infty}^{\infty} f(x, y) e^{j(\omega_x x + \omega_y y)} dx dy$$

and the 1-D projection for each angle

$$P(\theta, \omega) = \int_{-\infty}^{\infty} p(\theta, t) e^{j\omega t} dt$$

These two transforms are then related by the Fourier slice theorem.

$$F(\omega \cos \theta, \omega \sin \theta) = P(\theta, \omega)$$

In words, $P(\theta, \omega)$ corresponds to the value of the 2-D Fourier transform $F(\omega_x, \omega_y)$ along a 1-D line at an angle of θ passing through the origin.

The Fourier slice theorem may be used to develop two methods for inverting the Radon transform and thereby computing the image f. The first method, known as filtered back projection, computes the inverse Fourier transform in polar coordinates using the transformed projection data.

$$f(x, y) = \frac{1}{2\pi} \int_0^\pi \int_{-\infty}^{\infty} P(\theta, \omega) |\omega| e^{j\omega(x \cos \theta + y \sin \theta)} d\omega d\theta$$

Notice that the $|\omega|$ term accounts for the integration in polar coordinates.

A second inversion method results from performing all computations in the space domain rather than first transforming to the frequency domain ω. This can be done by expressing the inner integral of filtered back projection as a convolution in the space domain.

$$\frac{1}{2\pi} \int_{-\infty}^{\infty} P(\theta, \omega) |\omega| e^{j\omega s} d\omega = \int_{-\infty}^{\infty} p(\theta, t) h(s - t) dt$$

Here $h(t)$ is the inverse Fourier transform of $|\omega|$. This results in the inversion formula known as convolution back projection

$$f(x, y) = \int_0^\pi \int_{-\infty}^{\infty} p(\theta, t) h(x \cos \theta + y \sin \theta - t) dt d\theta$$

In practice, h must be a low-pass approximation to the true inverse Fourier transform of $|\omega|$. This is necessary to suppress noise in the projection data. In practice, the choice of h is the most important element in the design of the reconstruction algorithm.

Edge Detection

The ability to find gray level edge structures in images is an important image processing operation. We shall define an **edge** to be regions in the image where there is a large change in gray level over a relatively small spatial region. The process of finding edge locations in digital images is known as edge detection. Most edge detection operators, also known as edge operators, use a window operator to first enhance the edges in the image, followed by thresholding the enhanced image.

There has been a great deal of research performed in the area of edge detection. Some of the research issues include robust threshold selection, window size selection, noise response, edge linking, and the detection of edges in moving objects. While it is beyond the scope of this section to discuss these issues in detail, it is obvious that such things as threshold selection will greatly affect the performance of the edge detection algorithm. If the threshold is set too high, then many edge points will be missed; if set too low, then many "false" edge points will be obtained because of the inherent noise in the image. The investigation of the "optimal" choice of the threshold is an important research area. Selection of the particular window operation to enhance the edges of an image, as an initial step in edge detection, has recently been based on using models of the performance of the human visual system in detecting edges.

Analysis and Computer Vision

The process of extracting useful measurements from an image or sequence of images is known as *image analysis* or *computer vision*. Before analysis can be performed one must first determine pertinent features or attributes of the object in the scene and extract information about these features. The selection of which features in the image to measure must be chosen *a priori*, based on empirical results. Most features used consist of shape properties, shape change properties, shading, texture, motion, depth, and color. After the features are extracted, one must then use the feature measurements to determine scene characteristics such as object identification. In the past, simple pattern recognition algorithms, i.e., nearest-neighbor classification, have been used to compare the feature measurements of an image to a set of feature measurements that correspond to a known object. A decision is then made as to whether or not the features of the image match those of the known type.

Recently, there has been work in the application of *artificial intelligence* techniques to image analysis. These approaches are very much different from classical statistical pattern recognition in that the feature measurements are used in a different manner as part of a larger system that attempts to model the scene and then determine what is in it based on the model.

Defining Terms

Digital image: An array of numbers representing the spatial distribution of energy in a scene which is obtained by a process of sampling and quantization.

Edge: A localized region of rapid change in gray level in the image.

Entropy: A measure of the minimum amount of information required on the average to store or transmit each quantized feature vector.

Image compression or coding: The process of reducing the number of binary digits or bits required to represent the image.

Image enhancement: An image processing operation that is intended to improve the visual quality of the image or to emphasize certain features.

Image feature: An attribute of a block of image pixels.

Image reconstruction: The process of obtaining an image from nonimage data that characterizes that image.

Lossless vs. lossy compression: If the reconstructed or decoded image is identical to the original, the compression scheme is lossless. Otherwise, it is lossy.

Pixel: A single sample or picture element in the digital image which is located at specific spatial coordinates.

Point operation: `An image processing operation in which individual pixels are mapped to new values irrespective of the values of any neighboring pixels.

Projection: A set of parallel line integrals across the image oriented at a particular angle.

Quantization: The process of converting from a continuous-amplitude image to an image that takes on only a finite number of different amplitude values.

Sampling: The process of converting from a continuous-parameter image to a discrete-parameter image by discretizing the spatial coordinate.

Tomography: The process of reconstructing an image from projection data.

Vector quantization: The process of replacing an exact vector of features by a prototype vector that is used to represent all feature vectors contained within a cluster.

Window operation: An image processing operation in which the new value assigned to a given pixel depends on all the pixels within a window centered at that pixel location.

References

H. C. Andrews and B. R. Hunt, *Digital Image Restoration*, Englewood Cliffs, N.J.: Prentice-Hall, 1977.

D. H. Ballard and C. M. Brown, *Computer Vision*, Englewood Cliffs, N.J.: Prentice-Hall, 1982.

H. Barrow and J. Tenenbaum, "Computational vision," *Proc. IEEE*, vol. 69, pp. 572–595, May 1981.

A. Gersho and R. M. Gray, *Vector Quantization and Signal Compression*, Norwell, Mass.: Kluwer Academic Publishers, 1991.

R. C. Gonzalez and P. Wintz, *Digital Image Processing*, Reading, Mass.: Addison-Wesley, 1991.

G. T. Herman, *Image Reconstruction from Projections*, New York: Springer-Verlag, 1979.

T. S. Huang, *Image Sequence Analysis*, New York: Springer-Verlag, 1981.

A. K. Jain, *Fundamentals of Digital Image Processing*, Englewood Cliffs, N.J.: Prentice-Hall, 1989.

A. Kak and M. Slaney, *Principles of Computerized Tomographic Imaging*, New York: IEEE Press, 1988.

A. Macovski, *Medical Imaging Systems*, Englewood Cliffs, N.J.: Prentice-Hall, 1983.

M. D. McFarlane, "Digital pictures fifty years ago," *Proc. IEEE*, pp. 768–770, July 1972.

W. K. Pratt, *Digital Image Processing*, New York: Wiley, 1991.

A. Rosenfeld and A. Kak, *Digital Picture Processing*, vols. 1 and 2, San Diego: Academic Press, 1982.

J. Serra, *Image Analysis and Mathematical Morphology*, vols. 1 and 2, San Diego: Academic Press, 1982 and 1988.

Further Information

A number of textbooks are available that cover the broad area of image processing and several that focus on more specialized topics within this field. The texts by Gonzalez and Wintz [1991], Jain [1989], Pratt [1991], and Rosenfeld and Kak (Vol. 1) [1982] are quite broad in their scope. Gonzalez and Wintz's treatment is written at a somewhat lower level than that of the other texts. For a more detailed treatment of computed tomography and other medical imaging modalities, the reader may consult the texts by Herman [1979], Macovski [1983], and Kak and Slaney [1988]. To explore the field of computer vision, the reader is advised to consult the text by Ballard and Brown [1982]. Current research and applications of image processing are reported in a number of journals. Of particular note are the *IEEE Transactions on Image Processing*; the *IEEE Transactions on Pattern Analysis and Machine Intelligence*; the *IEEE Transactions on Geoscience and Remote Sensing*; the *IEEE Transactions on Medical Imaging*; the *Journal of the Optical Society of*

America, A: *Optical Engineering;* the *Journal of Electronic Imaging;* and *Computer Vision, Graphics, and Image Processing.*

16.2 Video Signal Processing

Sarah A. Rajala

Video signal processing is the area of specialization concerned with the processing of time sequences of image data, i.e., video. Because of the significant advances in computing power and increases in available transmission bandwidth, there has been a proliferation of potential applications in the area of video signal processing. Applications such as high-definition television, digital video, multimedia, video phone, interactive video, medical imaging, and information processing are the driving forces in the field today. As diverse as the applications may seem, it is possible to specify a set of fundamental principles and methods that can be used to develop the applications.

Considerable understanding of a video signal processing system can be gained by representing the system with the block diagram given in Fig. 16.10. Light from a real-world scene is captured by a **scanning system** and causes an image frame $f(x,y,t_0)$ to be formed on a focal plane. A video signal is a sequence of image frames that are created when a scanning system captures a new image frame at periodic intervals in time. In general, each frame of the video sequence is a function of two spatial variables x and y and one temporal variable t. An integral part of the scanning system is the process of converting the original analog signal into an appropriate digital representation. The conversion process includes the operations of sampling and quantization. **Sampling** is the process of converting a continuous-time/space signal into a discrete-time/space signal. **Quantization** is the process of converting a continuous-valued signal into a discrete-valued signal.

Once the video signal has been sampled and quantized, it can be processed digitally. Processing can be performed on special-purpose hardware or general-purpose computers. The type of processing performed depends on the particular application. For example, if the objective is to generate high-definition television, the processing would typically include compression and motion estimation. In fact, in most of the applications listed above these are the fundamental operations. **Compression** is the process of compactly representing the information contained in an image or video signal. **Motion estimation** is the process of estimating the displacement of the moving objects in a video sequence. The displacement information can then be used to interpolate missing frame data or to improve the performance of compression algorithms.

After the processing is complete, a video signal is ready for transmission over some channel or storage on some medium. If the signal is transmitted, the type of channel will vary depending on the application. For example, today analog television signals are transmitted one of three ways: via satellite, terrestrially, or by cable. All three channels have limited transmission bandwidths and can adversely affect the signals because of the imperfect frequency responses of the channels. Alternatively, with a digital channel, the primary limitation will be the bandwidth.

The final stage of the block diagram shown in Fig. 16.10 is the display. Of critical importance at this stage is the human observer. Understanding how humans respond to visual stimuli, i.e., the psychophysics of vision, will not only allow for better evaluation of the processed video signals but will also permit the design of better systems.

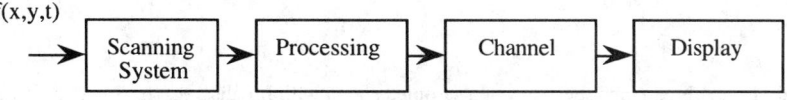

FIGURE 16.10   Video signal processing system block diagram.

Sampling

If a continuous-time video signal satisfies certain conditions, it can be exactly represented by and be reconstructed from its sample values. The conditions which must be satisfied are specified in the *sampling theorem*. The sampling theorem can be stated as follows:

Sampling Theorem:

Let $f(x,y,t)$ be a bandlimited signal with $F(\omega_x,\omega_y,\omega_t) = 0$ for $|\omega_x| > \omega_{xM}$, $|\omega_y| > \omega_{yM}$, and $|\omega_t| > \omega_{tM}$. Then $f(x,y,t)$ is uniquely determined by its samples $f(jX_S,kY_S,lT_S) = f(j,k,l)$, where $j,k,l = 0, \pm1, \pm2, \ldots$ if

$$\omega_{sx} > 2\omega_{xM}, \quad \omega_{sy} > 2\omega_{yM}, \quad \text{and} \quad \omega_{st} > 2\omega_{tM}$$

and

$$\omega_{sx} = 2\pi/X_S, \quad \omega_{sy} = 2\pi/Y_S, \quad \text{and} \quad \omega_{st} = 2\pi/T_S$$

X_S is the sampling period along the x direction, $\omega_x = 2\pi/X_S$ is the spatial sampling frequency along the x direction, Y_S is the sampling period along the y direction, $\omega_y = 2\pi/Y_S$ is the spatial sampling frequency along the y direction, T_S is the sampling period along the temporal direction, and $\omega_t = 2\pi/T_S$ is the temporal sampling frequency.

Given these samples, $f(x,y,t)$ can be reconstructed by generating a periodic impulse train in which successive impulses have amplitudes that are successive sample values. This impulse train is then processed through an ideal low-pass filter with appropriate gain and cut-off frequencies. The resulting output signal will be exactly equal to $f(x,y,t)$. (*Source:* Oppenheim *et al.*, 1983, p. 519.)

If the sampling theorem is not satisfied, **aliasing** will occur. Aliasing occurs when the signal is undersampled and therefore no longer recoverable by low-pass filtering. Figure 16.11(a) shows the frequency spectrum of a sampled bandlimited signal with no aliasing. Figure 16.11(b) shows the frequency response of the same signal with aliasing. The aliasing occurs at the points where there is overlap in the diamond-shaped regions. For video signals aliasing in the temporal direction will give rise to flicker on the display. For television systems, the standard temporal sampling rate is

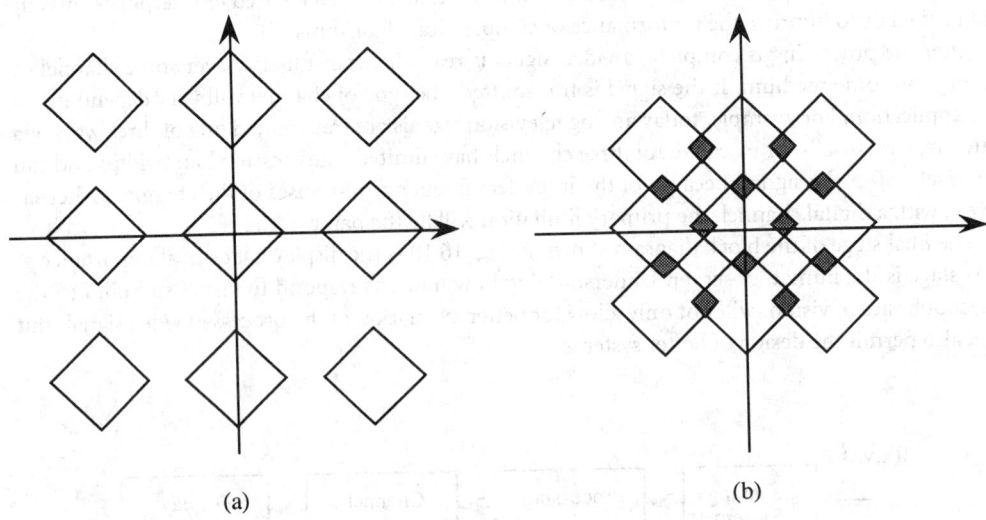

(a)   (b)

FIGURE 16.11   (a) Frequency spectrum of a sampled signal with no aliasing. (b) Frequency spectrum of a sampled signal with aliasing.

30 frames per second in the United States and Japan and 25 frames per second in Europe. However, these rates would be insufficient without the use of interlacing.

If the sampling rate (spatial and/or temporal) of a system is fixed, a standard approach for minimizing the effects of aliasing for signals that do not satisfy the sampling theorem is to use a presampling filter. Presampling filters are low-pass filters whose cut-off frequencies are chosen to be less than ω_M, ω_M, ω_M. Although the signal will still not be able to be reconstructed exactly, the degradations are less annoying. Another problem in a real system is the need for an ideal low-pass filter to reconstruct an analog signal. An ideal filter is not physically realizable, so in practice an approximation must be made. Several very simple filter structures are common in video systems: sample and hold, bilinear, and raised cosine.

Quantization

Quantization is the process of converting the continuous-valued amplitude of the video signal into a discrete-valued representation, i.e., a finite set of numbers. The output of the quantizer is characterized by quantities that are limited to a finite number of values. The process is a many-to-one mapping, and thus there is a loss of information. The quantized signal can be modeled as

$$f_q(j,k,l) = f(j,k,l) - e(j,k,l)$$

where $f_q(j,k,l)$ is the quantized video signal and $e(j,k,l)$ is the quantization noise. If too few bits per sample are used, the quantization noise will produce visible false contours in the image data.

The quantizer is a mapping operation which generally takes the form of a staircase function (see Fig. 16.12). A rule for quantization can be defined as follows: Let $\{d_k, k = 1, 2, \dots, N+1\}$ be the set of decision levels with d_1 the minimum amplitude value and d_N the maximum amplitude value of $f(j,k,l)$. If $f(j,k,l)$ is contained in the interval (d_k, d_{k+1}), then it is mapped to the kth reconstruction level r. Methods for designing quantizers can be broken into two categories: uniform and nonuniform. The input-output function for a typical uniform quantizer is shown in Fig. 16.12. The mean square value of the quantizing noise can be easily calculated if it is assumed that the amplitude probability distribution is constant within each quantization step. The quantization step size for a uniform quantizer is

$$q = \frac{d_{N+1} - d_1}{N}$$

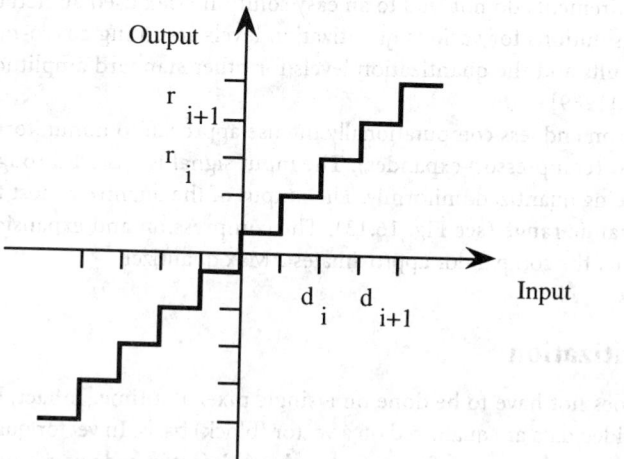

FIGURE 16.12   Characteristics of a uniform quantizer.

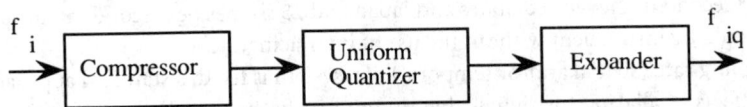

FIGURE 16.13   Nonuniform quantization using a compandor.

and all errors between $q/2$ and $-q/2$ are equally likely. The mean square quantization error is given by:

$$\left\langle e^2(j,k,l) \right\rangle = \int_{-q/2}^{q/2} \frac{f^2}{q}\, df = \frac{q^2}{12}$$

If one takes into account the exact amplitude probability distribution, an optimal quantizer can be designed. Here the objective is to choose a set of decision levels and reconstruction levels that will yield the minimum quantization error. If f has a probability density function $p_f(f)$, the mean square quantization error is

$$\left\langle e^2(j,k,l) \right\rangle = \sum_{i=1}^{N} \int_{di}^{di+1} (f - r_i)^2\, p_f(f) df$$

where N is the number of quantization levels. To minimize, the mean square quantization error is differentiated with respect to d_i and r_i. This results in the Max quantizer:

$$d_i = \frac{r_i + r_{i-1}}{2}$$

and

$$r_i = \frac{\int_{di}^{di+1} f p_f(f) df}{\int_{di}^{di+1} p_f(f) df}$$

Thus, the quantization levels need to be midway between the reconstruction levels, and the reconstruction levels are at the centroid of that portion of $p_f(f)$ between d_i and f_{d+1}. Unfortunately these requirements do not lead to an easy solution. Max used an iterative numerical technique to obtain solutions for various quantization levels assuming a zero-mean Gaussian input signal. These results and the quantization levels for other standard amplitude distributions can be found in Jain [1989].

A more common and less computationally intense approach to nonuniform quantization is to use a compandor (compressor–expander). The input signal is passed through a nonlinear compressor before being quantized uniformly. The output of the quantizer must then be expanded to the original dynamic range (see Fig. 16.13). The compression and expansion functions can be determined so that the compandor approximates a Max quantizer.

Vector Quantization

Quantization does not have to be done on a single pixel at a time. In fact, better results can be achieved if the video data are quantized on a vector (block) basis. In vector quantization, the image data are first processed into a set of vectors. A code book (set of code words or templates) that best matches the data to be quantized is then generated. Each input vector is then quantized to the closest code word. Compression is achieved by transmitting only the indices for the code words. At the receiver, the images are reconstructed using a table look-up procedure. Two areas of ongoing research are finding better methods for designing the code books and developing better search and update techniques for matching the input vectors to the code words.

Video Compression

Digital representations of video signals typically require a very large number of bits. If the video signal is to be transmitted and/or stored, compression is often required. Applications include conventional and high-definition television, video phone, video conferencing, multi-media, remote-sensed imaging, and magnetic resonance imaging. The objective of compression (source encoding) is to find a representation that maximizes picture quality while minimizing the data per picture element (pixel). A wealth of compression algorithms have been developed during the past 30 years for both image and video compression. However, the ultimate choice of an appropriate algorithm is application dependent. The following summary will provide some guidance in that selection process.

Compression algorithms can be divided into two major categories: information-preserving, or lossless, and lossy techniques. Information-preserving techniques introduce no errors in the encoding/decoding process; thus, the original signal can be reconstructed exactly. Unfortunately, the achievable compression rate, i.e., the reduction in bit rate, is quite small, typically on the order of 3:1. On the other hand, lossy techniques introduce errors in the coding/decoding process; thus, the received signal cannot be reconstructed exactly. The advantage of the lossy techniques is the ability to achieve much higher compression ratios. The limiting factor on the compression ratio is the required quality of the video signal in a specific application.

One approach to compression is to reduce the spatial and/or temporal sampling rate and the number of quantization levels. Unfortunately, if the sampling is too low and the quantization too coarse, aliasing, contouring, and flickering will occur. These distortions are often much greater than the distortions introduced by more sophisticated techniques at the same compression rate. Compression systems can generally be modeled by the block diagram shown in Fig. 16.14. The first stage of the compression system is the mapper. This is an operation in which the input pixels are mapped into a representation that can be more effectively encoded. This stage is generally reversible. The second stage is the quantizer and performs the same type of operation as described earlier. This stage is not reversible. The final stage attempts to remove any remaining statistical redundancy. This stage is reversible and is typically achieved with one of the information-preserving coders.

Information-Preserving Coders

The data rate required for an original digital video signal may not represent its average information rate. If the original signal is represented by M possible independent symbols with probabilities p_i, $i = 0, 1, \ldots, M-1$, then the information rate as given by the first-order entropy of the signal H is

$$H = -\sum_{i=1}^{M-1} p_i \log_2 p_i \text{ bits per sample}$$

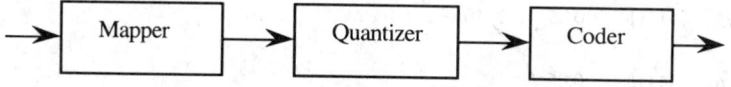

FIGURE 16.14 Three-stage model of an encoder.

According to Shannon's coding theorem [see Jain, 1989], it is possible to perform lossless coding of a source with entropy H bits per symbol using $H + \varepsilon$ bits per symbol. ε is a small positive quantity. The maximum obtainable compression rate C is then given by:

$$C = \frac{\text{average bit rate of the original data}}{\text{average bit rate of the encoded data}}$$

Huffman Coding

One of the most efficient information-preserving (entropy) coding methods is Huffman coding. Construction of a Huffman code involves arranging the symbol probabilities in decreasing order and considering them as leaf nodes of a tree. The tree is constructed by merging the two nodes with the smallest probability to form a new node. The probability of the new node is the sum of the two merged nodes. This process is continued until only two nodes remain. At this point, 1 and 0 are arbitrarily assigned to the two remaining nodes. The process now moves down the tree, decomposing probabilities and assigning 1's and 0's to each new pair. The process continues until all symbols have been assigned a code word (string of 1's and 0's). An example is given in Fig. 16.15. Many other types of information-preserving compression schemes exist (see, for example, Gonzalez and Wintz [1987]), including arithmetic coding, Lempel-Ziv algorithm, shift coding, and run-length coding.

Predictive Coding

Traditionally one of the most popular methods for reducing the bit rate has been predictive coding. In this class, differential pulse-code modulation (DPCM) has been used extensively. A block diagram for a basic DPCM system is shown in Fig. 16.16. In such a system the difference between the current pixel and a predicted version of that pixel gets quantized, coded, and transmitted to the receiver. This difference is referred to as the prediction error and is given by

$$e_i = f_i - \hat{f}_i$$

The prediction is based on previously transmitted and decoded spatial and/or temporal information and can be linear or nonlinear, fixed or adaptive. The difference signal e_i is then passed through a quantizer. The signal at the output of the quantizer is the quantized prediction error e_{i_q}, which is entropy encoded transmission. The first step at the receiver is to decode the quantized prediction

			Step 1		Step 2		Step 3		Step 4
w1	0	0.45	0	0.45	0	0.45	0	0.45	1
w2	10	0.25	10	0.25	10	0.25	11	0.30	0
w3	110	0.12	110	0.12	111	0.18	10	0.25	
w4	1110	0.08	1111	0.10	110	0.12			
w5	11111	0.05	1110	0.08					
w6	11110	0.05							

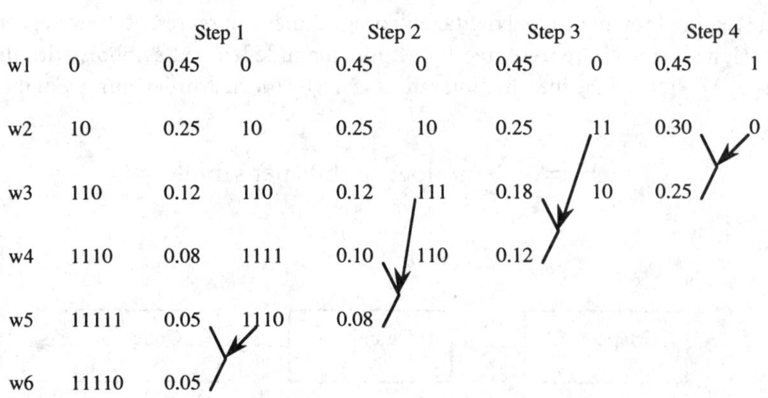

FIGURE 16.15   An example of constructing a Huffman code.

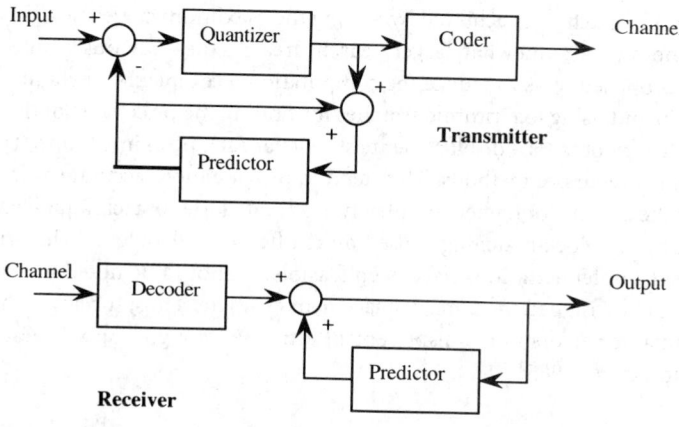

FIGURE 16.16 Block diagram of a basic DPCM system.

error. After decoding, d_{i_q} is added to the predicted value of the current pixel $\hat{f}_i$ to yield the reconstructed pixel value. Note that as long as a quantizer is included in the system, the output signal will not exactly equal the input signal.

The predictors can include pixels from the present frame as well as those from previous frames (see Fig. 16.17). If the motion and the spatial detail are not too high, frame (or field) prediction works well. If the motion is high and/or the spatial detail is high, intrafield prediction generally works better. A primary reason is that there is less correlation between frames and fields when the motion is high.

For more information on predictive coding, see Musmann *et al.* [1985] or Jain [1989].

Motion-Compensated Predictive Coding

Significant improvements in image quality, at a fixed compression rate, can be obtained when adaptive prediction algorithms take into account the frame-to-frame displacement of moving objects in the sequence. Alternatively, one could increase the compression rate for a fixed level of image quality. The amount of increase in performance will depend on one's ability to estimate the motion in the scene. Techniques for estimating the motion are described in a later subsection.

Motion-compensated prediction algorithms can be divided into two categories. One category estimates the motion on a block-by-block basis and the other estimates the motion one pixel at a time. For the block-based methods an estimate of the displacement is obtained for each block in

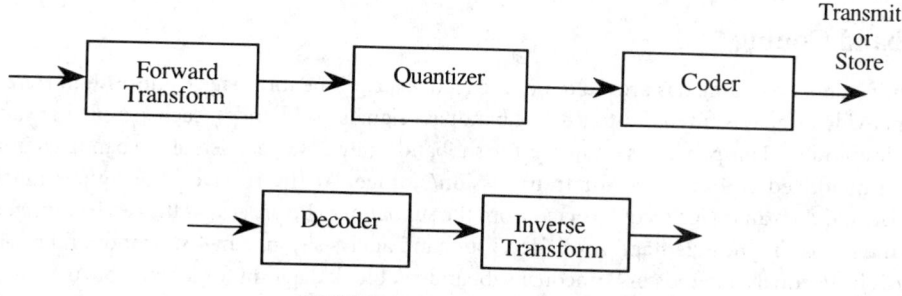

FIGURE 16.17 Transform coding system.

the image. The block matching is achieved by finding the maximum correlation between a block in the current frame and a somewhat larger search area in the previous frame. A number of researchers have proposed ways to reduce the computational complexity, including using a simple matching criterion and using logarithmic searches for finding the peak value of the correlation.

The second category obtains a displacement estimate at each pixel in a frame. These techniques are referred to as pel recursive methods. They tend to provide more accurate estimates of the displacement but at the expense of higher complexity. Both categories of techniques have been applied to video data; however, block matching is used more often in real systems. The primary reason is that more efficient implementations have been feasible. It should be noted, however, that every pixel in a block will be assigned the same displacement estimate. Thus, the larger the block size the greater the potential for errors in the displacement estimate for a given pixel. More details can be found in Musmann *et al.* [1985].

Transform Coding

In transform coding, the video signal $f(x,y,t)$ is subjected to an invertible transform, then quantized and encoded (see Fig. 16.17). The purpose of the transformation is to convert statistically dependent picture elements into a set of statistically independent coefficients. In practice, one of the separable fast transforms in the class of unitary transforms is used, e.g., cosine, Fourier, or Hadamard. In general, the transform coding algorithms can be implemented in 2-D or 3-D. However, because of the real-time constraints of many video signal processing applications, it is typically more efficient to combine a 2-D transform with a predictive algorithm in the temporal direction, e.g., motion compensation.

For 2-D transform coding the image data are first subdivided into blocks. Typical block sizes are 8×8 or 16×16. The transform independently maps each image block into a block of transform coefficients; thus, the processing of each block can be done in parallel. At this stage the data have been mapped into a new representation, but no compression has occurred. In fact, with the Fourier transform there is an expansion in the amount of data. This occurs because the transform generates coefficients that are complex-valued. To achieve compression the transform coefficients must be quantized and then coded to remove any remaining redundancy.

Two important issues in transform coding are the choice of transformation and the allocation of bits in the quantizer. The most commonly used transform is the discrete cosine transform (DCT). In fact, many of the proposed image and video standards utilize the DCT. The reasons for choosing a DCT include: its performance is superior to the other fast transforms and is very close to the optimal Karhunen-Loeve transform, it produces real-valued transform coefficients, and it has good symmetry properties, thus reducing the blocking artifacts inherent in block-based algorithms. One way to reduce these artifacts is by using a transform whose basis functions are even, i.e., the DCT, and another is to use overlapping blocks. For bit allocation, one can determine the variance of the transform coefficients and then assign the bits so the distortion is minimized. An example of a typical bit allocation map is shown in Fig. 16.18.

Subband Coding

Recently, subband coding has proved to be an effective technique for image compression. Here, the original video signal is filtered into a set of bandpass signals (subbands), each sampled at successively lower rates. This process is known as the subband analysis stage. Each of the bandpass images is then quantized and encoded for transmission/storage. At the receiver, the signals must be decoded and then an image reconstructed from the subbands. The process at the receiver is referred to as the subband synthesis stage. A one-level subband analysis results in 4 subbands and a 2-level analysis in 16 equal subbands or 7 unequal subbands. A block diagram for a separable two-dimensional subband analysis system is shown in Fig. 16.19.

```
8 7 6 5 3 3 4 4 4 1 1 1 1 1 0 0
7 6 5 4 3 3 2 2 1 1 1 1 1 0 0 0
6 5 4 3 3 2 2 2 1 1 1 1 1 0 0 0
5 4 3 3 3 2 2 2 1 1 1 1 1 0 0 0
3 3 3 3 2 2 2 1 1 1 1 1 0 0 0 0
3 3 2 2 2 2 2 1 1 1 1 1 0 0 0 0
2 2 2 2 2 2 1 1 1 1 1 0 0 0 0 0
2 2 2 2 1 1 1 1 1 1 1 0 0 0 0 0
2 1 1 1 1 1 1 1 1 1 0 0 0 0 0 0
1 1 1 1 1 1 1 1 1 0 0 0 0 0 0 0
1 1 1 1 1 1 1 1 0 0 0 0 0 0 0 0
1 1 1 1 1 0 0 0 0 0 0 0 0 0 0 0
1 1 1 1 0 0 0 0 0 0 0 0 0 0 0 0
1 0 0 0 0 0 0 0 0 0 0 0 0 0 0 0
0 0 0 0 0 0 0 0 0 0 0 0 0 0 0 0
0 0 0 0 0 0 0 0 0 0 0 0 0 0 0 0
```

FIGURE 16.18 A typical bit allocation for 16×16 block coding of an image using the DCT. (*Source:* A. K. Jain, *Fundamentals of Digital Image Processing*, Englewood Cliffs, N.J.: Prentice-Hall, 1989, p. 506. With permission.)

HDTV

High-definition television (HDTV) has received much attention in the past few years. With the recent push for all digital implementations of HDTV, the need for video signal processing techniques has become more obvious. In order for the digital HDTV signal to fit in the transmission bandwidth, there is a need for a compression ratio of approximately 10:1, with little or no degradation introduced. The goal of HDTV is to produce high-quality video signals by enhancing the detail, improving the aspect ratio and the viewing distance. The detail is enhanced by increasing the video bandwidth. The proposed aspect ratio of 16/9 will allow for a wide-screen format which is more consistent with the formats used in the motion-picture industry. The eye's ability to resolve fine detail is limited. To achieve full resolution of the detail, the HDTV image should be viewed at a distance of approximately three times the picture height. To accommodate typical home viewing environments, larger displays are needed.

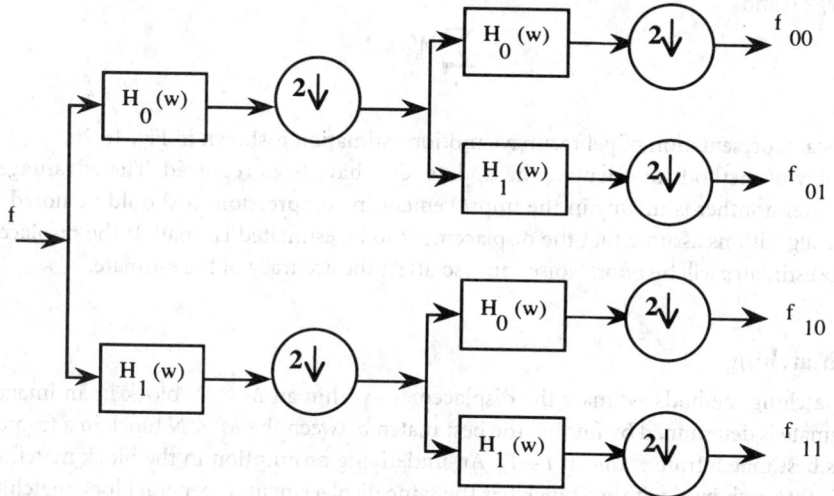

FIGURE 16.19 A two-dimensional subband analysis system for generating four equal subbands.

Motion Estimation Techniques

Frame-to-frame changes in luminance are generated when objects move in video sequences. The luminance changes can be used to estimate the displacement of the moving objects if an appropriate model of the motion is specified. A variety of motion models have been developed for dynamic scene analysis in machine vision and for video communications applications. In fact, motion estimates were first used as a control mechanism for the efficient coding of a sequence of images in an effort to reduce the temporal redundancy. Motion estimation algorithms can be classified in two broad categories: gradient or differential-based methods and token matching or correspondence methods. The gradient methods can be further divided into pel recursive, block matching, and optical flow methods.

Pel Recursive Methods

Netravali and Robbins [1979] developed the first pel recursive method for television signal compression. The algorithm begins with an initial estimate of the displacement, then iterates recursively to update the estimate. The iterations can be performed at a single pixel or at successive pixels along a scan line. The true displacement $\mathbf{D}$ at each pixel is estimated by

$$\hat{\mathbf{D}}^i = \hat{\mathbf{D}}^{i-1} + \mathbf{U}^i$$

where $\hat{\mathbf{D}}^i$ is the displacement estimate at the ith iteration and $\mathbf{U}^i$ is the update term. $\mathbf{U}^i$ is an estimate of $\mathbf{D} - \hat{\mathbf{D}}^{i-1}$. They then used the displaced frame difference (DFD):

$$DFD(x, y, \hat{\mathbf{D}}^{i-1}) = I(x, y, t) - I(x - \hat{\mathbf{D}}^{i-1}, t - T_S)$$

to obtain a relationship for the update term $\mathbf{U}^i$. In the previous equation, T_S is the temporal sample spacing. If the displacement estimate is updated from sample to sample using a steepest-descent algorithm to minimize the weighted sum of the squared displaced frame differences over a neighborhood, then $\hat{\mathbf{D}}^i$ becomes

$$\hat{\mathbf{D}}^i = \hat{\mathbf{D}}^{i-1} - \frac{\varepsilon}{2} \nabla \hat{\mathbf{D}}^i \left[\sum_j W_j [DFD(x_{k-j}, \hat{\mathbf{D}}^{i-1})]^2 \right]$$

where $W_j \geq 0$ and

$$\sum_j W_j = 1$$

A graphical representation of pel recursive motion estimation is shown in Fig. 16.20.

A variety of methods to calculate the update term have been reported. The advantage of one method over another is mainly in the improvement in compression. It should be noted that pel recursive algorithms assume that the displacement to be estimated is small. If the displacement is large, the estimates will be poor. Noise can also affect the accuracy of the estimate.

Block Matching

Block matching methods estimate the displacement within an $M \times N$ block in an image frame. The estimate is determined by finding the best match between the $M \times N$ block in a frame at time t and its best match from frame at $t - T_S$. An underlying assumption in the block matching techniques is that each pixel within a block has the same displacement. A general block matching algorithm is given as follows:

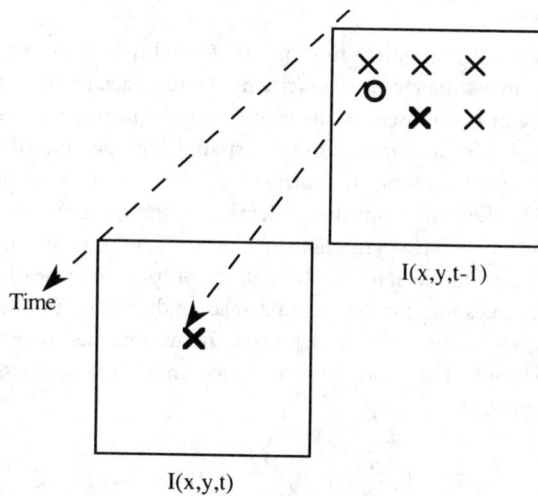

FIGURE 16.20 A graphical illustration of pel recursive motion estimation. The distance between the x and o pixels in the frame at $t-1$ is $\hat{\mathbf{D}}^{i-1}$.

1. Segment the image frame at time t into a fixed number of blocks of size $M \times N$.
2. Specify the size of the search area in the frame at time $t-1$. This depends on the maximum expected displacement. If D_{max} is the maximum displacement in either the horizontal or vertical direction, then the size of the search area, SA, is

$$SA = (M + 2D_{max}) \times (N + 2D_{max})$$

Figure 16.21 illustrates the search area in the frame at time $t-1$ for an $M \times N$ block at time t.
3. Using an appropriately defined matching criterion, e.g., mean-squared error or sum of absolute difference, find the best match for the $M \times N$ block.
4. Proceed to the next block in frame t and repeat step 3 until displacement estimates have been determined for all blocks in the image.

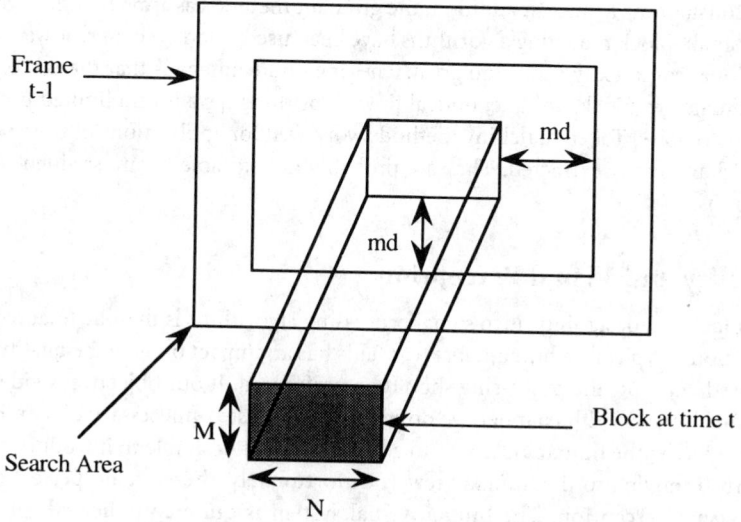

FIGURE 16.21 An illustration of block matching.

Optical Flow Methods

The optical flow is defined as the apparent motion of the brightness patterns from one frame to the next. The optical flow is an estimate of the velocity field and hence requires two equations to solve for it. Typically a constraint is imposed on the motion model to provide the necessary equations. Optical flow can give useful information about the spatial arrangement of the objects in a scene, as well as the rate of change of those objects. Horn [1986] also defines a motion field, which is a two-dimensional velocity field resulting from the projection of the three-dimensional velocity field of an object in the scene onto the image plane. The motion field and the optical flow are not the same.

In general, the optical flow has been found difficult to compute because of the algorithm sensitivity to noise. Also, the estimates may not be accurate at scene discontinuities. However, because of its importance in assigning a velocity vector at each pixel, there continues to be research in the field.

The optical flow equation is based on the assumption that the brightness of a pixel at location (x,y) is constant over time; thus,

$$I_x \frac{dx}{dt} + I_y \frac{dy}{dt} + I_t = 0$$

where dx/dt and dy/dt are the components of the optical flow. Several different constraints have been used with the optical flow equation to solve for dx/dt and dy/dt. A common constraint to impose is that the velocity field is smooth.

Token Matching Methods

Token matching methods are often referred to as discrete methods since the goal is to estimate the motion only at distinct image features (tokens). The result is a sparse velocity field. The algorithms attempt to match the set of discrete features in the frame at time $t-1$ with a set that best resembles them in the frame at time t. Most of the algorithms in this group assume that the estimation will be achieved in a two-step process. In the first step, the features are identified. The features could be points, corners, centers of mass, lines, or edges. This step typically requires segmentation and/or feature extraction. The second step determines the various velocity parameters. The velocity parameters include a translation component, a rotation component, and the rotation axis. The token matching algorithms fail if there are no distinct features to use.

All of the methods described in this subsection assume that the intensity at a given pixel location is reasonably constant over time. In addition, the gradient methods assume that the size of the displacements is small. Block matching algorithms have been used extensively in real systems, because the computational complexity is not too great. The one disadvantage is that there is only one displacement estimate per block. To date, optical flow algorithms have found limited use because of their sensitivity to noise. Token matching methods work well for applications in which the features are well defined and easily extracted. They are probably not suitable for most video communications applications.

Image Quality and Visual Perception

An important factor in designing video signal processing algorithms is that the final receiver of the video information is typically a human observer. This has an impact on how the quality of the final signal is assessed and how the processing should be performed. If our objective is video transmission over a limited bandwidth channel, we do not want to waste unnecessary bits on information that cannot be seen by the human observer. In addition, it is undesirable to introduce artifacts that are particularly annoying to the human viewer. Unfortunately, there are no perfect quantitative measures of visual perception. The human visual system is quite complicated. In spite of the advances that have been made, no complete model of human perception exists. Therefore, we often have to rely on subjective testing to evaluate picture quality. Although no comprehensive model of human vision exists, certain functions can be characterized and then used in designing improved solutions. For more information, see Netravali and Haskell [1988].

Subjective Quality Ratings

There are two primary categories of subjective testing: *category-judgment* (*rating-scale*) methods and *comparison* methods. Category-judgment methods ask the subjects to view a sequence of pictures and assign each picture (video sequence) to one of several categories. Categories may be based on overall quality or on visibility of impairment (see Table 16.2).

Comparison methods require the subjects to compare a distorted test picture with a reference picture. Distortion is added to the test picture until both pictures appear of the same quality to the subject. Viewing conditions can have a great impact on the results of such tests. Care must be taken in the experimental design to avoid biases in the results.

Visual Perception

In this subsection, a review of the major aspects of human psychophysics that have an impact in video signal processing is given. The phenomena of interest include light adaptation, visual thresholding and contrast sensitivity, masking, and temporal phenomena.

Light Adaptation

The human visual system (HVS) has two major classes of photoreceptors, the rods and the cones. Because these two types of receptors adapt to light differently, two different adaptation time constants exist. Furthermore, these receptors respond at different rates going from dark to light than from light to dark. It should also be noted that although the HVS has an ability to adapt to an enormous range of light intensity levels, on the order of 10^{10} in millilamberts, it does so adaptively. The simultaneous range is on the order of 10^3.

Visual Thresholding and Contrast Sensitivity

Determining how sensitive an observer is to small changes in luminance is important in the design of video systems. One's sensitivity will determine how visible noise will be and how accurately the luminance must be represented. The contrast sensitivity is determined by measuring the just-noticeable difference (JND) as a function of the brightness. The JND is the amount of additional brightness needed to distinguish a patch from the background. It is a visibility threshold. What is significant is that the JND is dependent on the background and surrounding luminances, the size of the background and surrounding areas, and the size of the patch, with the primary dependence being on the luminance of the background.

Table 16.2 Quality and Impairment Ratings

5 Excellent	5 Imperceptible	3 Much better
4 Good	4 Perceptible but not annoying	2 Better
3 Fair	3 Slightly annoying	1 Slightly better
2 Poor	2 Annoying	0 Same
1 Bad	1 Very annoying	−1 Slightly worse
		−2 Worse
		−3 Much worse

Source: A. N. Netravali and B. G. Haskell, *Digital Pictures: Representation and Compression*, New York: Plenum Press, 1988, p. 247. With permission.

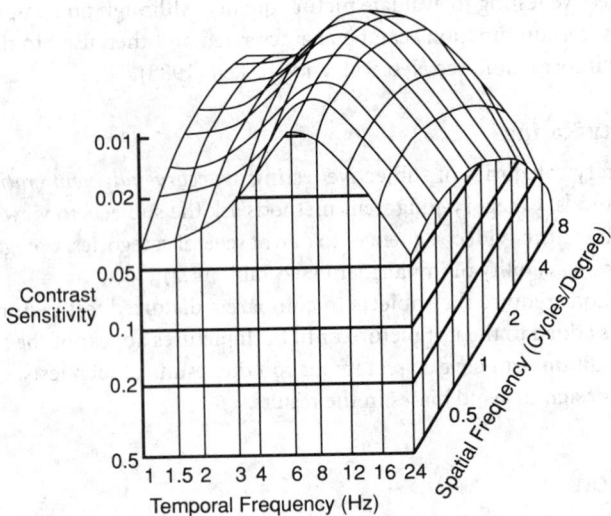

FIGURE 16.22 A perspective view of the spatio-temporal threshold surface. (*Source:* A. N. Netravali and B. G. Haskell, *Digital Pictures: Representation and Compression*, New York: Plenum Press, 1988, p. 273. With permission.)

Masking

The response to visual stimuli is greatly affected by what other visual stimuli are in the immediate neighborhood (spatially and temporally). An example is the reduced sensitivity of the HVS to noise in areas of high spatial activity. Another example is the masking of details in a new scene by what was present in the previous scene. In both cases, the masking phenomenon can be used to improve the quality of image compression systems.

Temporal Effects

One relevant temporal phenomenon is the flicker fusion frequency. This is a temporal threshold which determines the point at which the HVS fuses the motion in a sequence of frames. Unfortunately this frequency varies as a function of the average luminance. The HVS is more sensitive to flicker at high luminances than at low luminances. The spatial-temporal frequency response of the HVS is important in determining the sensitivity to small-amplitude stimuli. In both the temporal and spatial directions, the HVS responds as a bandpass filter (see Fig. 16.22). Also significant is the fact that the spatial and temporal properties are not independent of one another, especially at low frequencies.

For more details on image quality and visual perception see Schreiber [1991] and Netravali and Haskell [1988].

Defining Terms

Aliasing: Distortion introduced in a digital signal when it is undersampled.
Compression: Process of compactly representing the information contained in a signal.
Motion estimation: Process of estimating the displacement of moving objects in a scene.
Quantization: Process of converting a continuous-valued signal into a discrete-valued signal.
Sampling: Process of converting a continuous-time/space signal into a discrete-time/space signal.
Scanning system: System used to capture a new image at periodic intervals in time and to convert the image into a digital representation.

References

R. C. Gonzalez and P. Wintz, *Digital Image Processing*, Reading, Mass.: Addison-Wesley, 1987.

R. A. Haddad and T. W. Parsons, *Digital Signal Processing: Theory, Applications, and Hardware*, New York: Computer Science Press, 1991.

B. P. Horn, *Robot Vision*, Cambridge, Mass.: The MIT Press, 1986.

A. K. Jain, *Fundamentals of Digital Image Processing*, Englewood Cliffs, N.J.: Prentice-Hall, 1989.

N. Jayant, "Signal compression: Technology targets and research directions," *IEEE Journal on Selected Areas in Communications*, vol. 10, no. 5, pp. 796–818, 1992.

H. G. Musmann, P. Pirsch, and H.-J. Grallert, "Advances in picture coding," *Proc. IEEE*, vol. 73, no. 4, pp. 523–548, 1985.

A. N. Netravali and B. G. Haskell, *Digital Pictures: Representation and Compression*, New York: Plenum Press, 1988.

A. N. Netravali and J. D. Robbins, "Motion-compensated television coding: Part I," *Bell Syst. Tech. J.*, vol. 58, no. 3, pp. 631–670, 1979.

A. V. Oppenheim, A. S. Willsky, and I. T. Young, *Signals and Systems*, Englewood Cliffs, N.J.: Prentice-Hall, 1983.

W. F. Schreiber, *Fundamentals of Electronic Imaging Systems*, Berlin: Springer-Verlag, 1991.

Further Information

Other recommended sources of information include *IEEE Transactions on Circuits and Systems for Video Technology*, *IEEE Transactions on Image Processing*, and the *Proceedings of the IEEE*, April 1985, vol. 73.

16.3 Sensor Array Processing

N. K. Bose and L. H. Sibul

Multidimensional signal processing tools apply to aperture and sensor array processing. Planar sensor arrays can be considered to be sampled apertures. Three-dimensional or volumetric arrays can be viewed as multidimensional spatial filters. Therefore, the topics of sensor array processing, aperture processing, and multidimensional signal processing can be studied under a unified format. The basic function of the receiving array is transduction of propagating waves in the medium into electrical signals. Propagating waves are fundamental in radar, communication, optics, sonar, and geophysics. In electromagnetic applications, basic transducers are antennas and arrays of antennas. A large body of literature that exists on antennas and antenna arrays can be exploited in the areas of aperture and sensor array processing. Much of the antenna literature deals with transmitting antennas and their radiation patterns. Because of the reciprocity of transmitting and receiving transducers, key results that have been developed for transmitters can be used for analysis of receiver aperture and/or array processing. Transmitting transducers radiate energy in desired directions, whereas receiving apertures/arrays act as spatial filters that emphasize signals from a desired look direction while discriminating against interferences from other directions. The spatial filter **wavenumber** response is called the receiver beam pattern. Transmitting apertures are characterized by their radiation patterns.

Conventional beamforming deals with the design of fixed beam patterns for given specifications. Optimum beamforming is the design of beam patterns to meet a specified optimization criterion. It can be compared to optimum filtering, detection, and estimation. Adaptive **beamformers** sense their operating environment (for example, noise covariance matrix) and adjust beamformer parameters so that their performance is optimized [Monzingo and Miller, 1980]. Adaptive beamformers can be compared with adaptive filters.

Multidimensional signal processing techniques have found wide application in seismology—where a group of identical seismometers, called seismic arrays, are used for event location, studies of the earth's sedimentation structure, and separation of coherent signals from noise, which some-times may also propagate coherently across the array but with different horizontal velocities—by employing **velocity filtering** [Claerbout, 1976]. Velocity filtering is performed by multidimensional filters and allows also for the enhancement of signals which may occupy the same wavenumber range as noise or undesired signals do. In a broader context, beamforming can be used to separate signals received by sensor arrays based on frequency, wavenumber, and velocity (speed as well as direction) of propagation. Both the transfer and unit impulse-response functions of a velocity filter are two-dimensional functions in the case of one-dimensional arrays. The transfer function involves frequency and wavenumber (due to spatial sampling by equally spaced sensors) as independent variables, whereas the unit impulse response depends upon time and location within the array. Two-dimensional filtering is not limited to velocity filtering by means of seismic array. Two-dimensional spatial filters are frequently used, for example, in the interpretation of gravity and magnetic maps to differentiate between regional and local features. Input data for these filters may be observations in the survey of an area conducted over a planar grid over the earth's surface. Two-dimensional wavenumber digital filtering principles are useful for this purpose. Velocity filtering by means of two-dimensional arrays may be accomplished by properly shaping a three-dimensional response function $H(k_1,k_2,\omega)$. Velocity filtering by three-dimensional arrays may be accomplished through a four-dimensional function $H(k_1,k_2,k_3,\omega)$ as explained in the following subsection.

Spatial Arrays, Beamformers, and FIR Filters

A propagating plane wave, $s(\mathbf{x},t)$, is, in general, a function of the three-dimensional space variables and the time variable $(x_1,x_2,x_3) \triangleq \mathbf{x}$ and the time variable t. The 4-D Fourier transform of the stationary signal $s(\mathbf{x},t)$ is

$$S(\mathbf{k},\omega) = \int_{-\infty}^{\infty}\int_{-\infty}^{\infty}\int_{-\infty}^{\infty}\int_{-\infty}^{\infty} s(\mathbf{x},t)e^{-j(\omega t - \sum_{i=1}^{3}k_i x_i)}dx_1 dx_2 dx_3 dt \qquad (16.3)$$

which is referred to as the wavenumber–frequency spectrum of $s(\mathbf{x},t)$, and $(k_1,k_2,k_3) \triangleq \mathbf{k}$ denotes the wavenumber variables in radians per unit distance and ω is the frequency variable in radians per second. If c denotes the velocity of propagation of the plane wave, the following constraint must be satisfied

$$k_1^2 + k_2^2 + k_3^2 = \frac{\omega^2}{c^2}$$

If the 4-D Fourier transform of the unit impulse response $h(\mathbf{x},t)$ of a 4-D linear shift-invariant (LSI) filter is denoted by $H(k,\omega)$, then the response $y(\mathbf{x},t)$ of the filter to $s(\mathbf{x},t)$ is the 4-D linear convolution of $h(\mathbf{x},t)$ and $s(\mathbf{x},t)$, which is, uniquely, characterized by its 4-D Fourier transform

$$Y(\mathbf{k},\omega) = H(\mathbf{k},\omega)S(\mathbf{k},\omega) \qquad (16.4)$$

The inverse 4-D Fourier transform, which forms a 4-D Fourier transform pair with Eq. (16.3), is

$$s(\mathbf{x},t) = \frac{1}{(2\pi)^4}\int_{-\infty}^{\infty}\int_{-\infty}^{\infty}\int_{-\infty}^{\infty}\int_{-\infty}^{\infty} S(\mathbf{k},\omega)e^{j(\omega t - \sum_{i=1}^{3}k_i x_i)}dk_1 dk_2 dk_3 d\omega \qquad (16.5)$$

It is noted that $S(\mathbf{k},\omega)$ in Eq. (16.3) is product separable, i.e., expressible in the form

$$S(\mathbf{k},\omega) = S_1(k_1)S_2(k_2)S_3(k_3)S_4(\omega) \qquad (16.6)$$

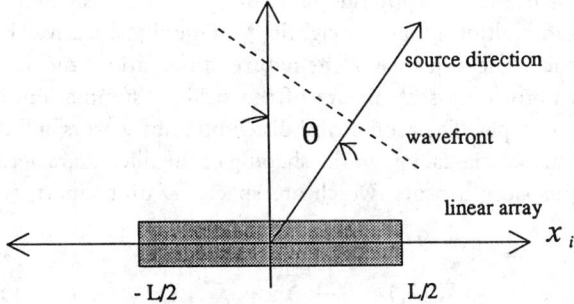

FIGURE 16.23 Uniformly weighted linear array.

where each function on the right-hand side is a univariate function of the respective independent variable, if and only if $s(\mathbf{x},t)$ in Eq. (16.3) is also product separable. In beamforming, $S_i(k_i)$ in Eq. (16.6) would be the far-field beam pattern of a linear array along the x_i-axis. For example, the normalized beam pattern of a uniformly weighted (shaded) linear array of length L is

$$
S(k,\theta) = \frac{\sin\left(\dfrac{kL \sin\theta}{2}\right)}{\left(\dfrac{kL}{2}\sin\theta\right)}
\tag{16.7}
$$

where $\lambda = (2\pi/k)$ is the wavelength of the propagating plane wave and θ is the angle of arrival at array site as shown in Fig. 16.23. Note that θ is explicitly admitted as a variable in $S(k,\theta)$ to allow for the possibility that for a fixed wavenumber, the beam pattern could be plotted as a function of the angle of arrival. In that case, when θ is zero, the wave impinges the array broadside and the normalized beam pattern evaluates to unity.

The counterpart, in aperture and sensor array processing, of the use of window functions in spectral analysis for reduction of sidelobes is the use of aperture shading. In aperture shading, one simply multiplies a uniformly weighted aperture by the shading function. The resulting beam pattern is, then, simply the convolution of the beam pattern of the uniformly shaded volumetric array and the beam pattern of the shading function. Fourier transform relationship between the stationary signal $s(\mathbf{x},t)$ and the wavenumber frequency spectrum $S(\mathbf{k},\omega)$ allows one to exploit high-resolution spectral analysis techniques for the high-resolution estimation of the direction of arrival [Pillai, 1989]. The superscript $*$, t, and H denote, respectively, complex conjugate, transpose, and conjugate transpose.

Discrete Arrays for Beamforming

An array of sensors could be distributed at distinct points in space in various ways. Line arrays, planar arrays, and volumetric arrays could be either uniformly spaced or nonuniformly spaced, including the possibility of placing sensors randomly according to some probability distribution function. Uniform spacing along each coordinate axis permits one to exploit the well-developed multidimensional signal processing techniques concerned with filter design, DFT computation via FFT, and high-resolution spectral analysis of sampled signals [Dudgeon, 1977]. Nonuniform spacing sometimes might be useful for reducing the number of sensors, which otherwise might be constrained to satisfy a maximum spacing between uniformly placed sensors to avoid **grating lobes** due to aliasing, as explained later. A discrete array, uniformly spaced, is convenient for the synthesis of a digital filter or beamformer by the performing of digital signal processing operations (namely delay, sum, and multiplication or weighting) on the signal received by a collection of sensors distributed in space. The sequence of the nature of operations dictates the types of beamformer. Common beamforming systems are of the straight summation, delay-and-sum, and weighted delay-and-sum types. The geometrical distribution of sensors and the weights w_i associated with each sensor are crucial factors in the shaping of the filter characteristics. In the case of a linear array of N equispaced sensors, which are spaced D units apart, starting at the origin $x_1 = 0$, the function

$$W(k_1) = \frac{1}{N} \sum_{n=0}^{N-1} w_n e^{-jk_1 nD} \tag{16.8}$$

becomes the **array pattern**, which may be viewed as the frequency response function for a finite impulse response (FIR) filter, characterized by the unit impulse response sequence $\{w_n\}$. In the case when $w_n = 1$, Eq. (16.8) simplifies to

$$W(k_1) = \frac{1}{N} \frac{\sin\left(\dfrac{k_1 ND}{2}\right)}{\sin\left(\dfrac{k_1 D}{2}\right)} \exp\left\{-j\frac{(N-1)k_1 D}{2}\right\} \tag{16.9}$$

If the N sensors are symmetrically placed on both sides of the origin, including one at the origin, and the sensor weights are $w_n = 1$, then the linear array pattern becomes

$$W(k_1) = \frac{1}{N} \frac{\sin\left(\dfrac{k_1 ND}{2}\right)}{\sin\left(\dfrac{k_1 D}{2}\right)}$$

For planar arrays, direct generalizations of the preceding linear array results can be obtained. To wit, if the sensors with unity weights are located at coordinates (kD, lD), where $k = 0, \pm 1, \pm 2, \ldots,$ $\pm[(N-1)/2]$, and $l = 0, \pm 1, \pm 2, \ldots, \pm[(M-1)/2]$, for odd integer values of N and M, then the array pattern function becomes

$$W(k_1, k_2) = \frac{1}{NM} \sum_{k=-\left(\frac{N-1}{2}\right)}^{\left(\frac{N-1}{2}\right)} \sum_{l=-\left(\frac{M-1}{2}\right)}^{\left(\frac{M-1}{2}\right)} \exp\left\{-j(k_1 kD + k_2 lD\right\}$$

$$\tag{16.10}$$

$$= \frac{1}{NM} \frac{\sin\left(\dfrac{k_1 ND}{2}\right)}{\sin\left(\dfrac{k_1 D}{2}\right)} \frac{\sin\left(\dfrac{k_2 MD}{2}\right)}{\sin\left(\dfrac{k_2 D}{2}\right)}$$

Routine generalizations to 3-D spatial arrays are also possible. The array pattern functions for other geometrical distributions may also be routinely generated. For example, if unit weight sensors are located at the six vertices and the center of a regular hexagon, each of whose sides is D units long, then the array pattern function can be shown to be

$$W(k_1, k_2) = \frac{1}{7}\left[1 + 2\cos k_1 D + 4\cos\frac{k_1 D}{2}\cos\frac{\sqrt{3}k_2 D}{2}\right] \tag{16.11}$$

The array pattern function reveals how selective a particular beamforming system is. In the case of a typical array function shown in Eq. (16.9), the beamwidth, which is the width of the main lobe of the array pattern, is inversely proportional to the array aperture. Because of the periodicity of the array pattern function, the main lobe is repeated at intervals of $2\pi/D$. These repetitive lobes are called grating lobes, whose existence may be interpreted in terms of spatial frequency aliasing resulting from a sampling interval D due to the N receiving sensors located at discrete points in space. If the spacing D between sensors satisfies

$$D \leq \frac{\lambda}{2} \tag{16.12}$$

where λ is the smallest wavelength component in the signal received by the array of sensors, then the grating lobes have no effect on the received signal. A plane wave of unit amplitude which is incident upon the array at bearing angle θ degrees, as shown in Fig. 16.23, produces outputs at the sensors given by the vector

$$\mathbf{s}(\theta) \triangleq \mathbf{s}_\theta = [\exp(j0)\ \exp(jk_1 D\sin\theta)\ \ldots\ \exp(jk_1(N-1)D\sin\theta)]^t \tag{16.13}$$

where $k_1 = 2\pi/\lambda$ is the wavenumber. In array processing, the array output y_θ may be viewed as the inner product of an array weight vector $\mathbf{w}$ and the steering vector $\mathbf{s}_\theta$. Thus, the beamformer response along a direction characterized by the angle θ is, treating $\mathbf{w}$ as complex,

$$y_\theta = \left\langle \mathbf{w}(\theta), \mathbf{s}_\theta \right\rangle = \sum_{k=0}^{N-1} w_k^* \exp(jk_1 kD\sin\theta) \tag{16.14}$$

The beamforming system is said to be robust if it performs satisfactorily despite certain perturbations [Ahmed and Evans, 1982]. It is possible for each component $s_{k\theta}$ of $\mathbf{s}_\theta$ to belong to an interval $[s_{k\theta} - \phi_{k\theta}, s_{k\theta} + \phi_{k\theta}]$, and a robust beamformer will require the existence of at least one weight vector $\mathbf{w}$ which will guarantee the output y_θ to belong to an output envelope for each $\mathbf{s}_\theta$ in the input envelope. The robust beamforming problem can be translated into an optimization problem, which may be tackled by minimizing the value of the array output power

$$P(\theta) = \mathbf{w}^H(\theta)R\mathbf{w}(\theta) \tag{16.15}$$

when the response to a unit amplitude plane wave incident at the steering direction θ is constrained to be unity, i.e., $\mathbf{w}^H(\theta)\mathbf{s}(\theta) = 1$, and R is the additive noise-corrupted signal autocorrelation matrix. The solution is called the minimum variance beamformer and is given by

$$\mathbf{w}_{MV}(\theta) = \frac{R^{-1}\mathbf{s}(\theta)}{\mathbf{s}^H(\theta)R^{-1}\mathbf{s}(\theta)} \tag{16.16}$$

and the corresponding power output is

$$P_{MV}(\theta) = \frac{1}{\mathbf{s}^H(\theta)R^{-1}\mathbf{s}(\theta)} \qquad (16.17)$$

The minimum variance power as a function of θ can be used as a form of the data-adaptive estimate of the directional power spectrum. However, in this mode of solution, the coefficient vector is unconstrained except at the steering direction. Consequently, a signal tends to be regarded as an unwanted interference and is, therefore, suppressed in the beamformed output unless it is almost exactly aligned with the steering direction. Therefore, it is desirable to broaden the signal acceptance angle while at the same time preserving the optimum beamformer's ability to reject noise and interference outside this region of angles. One way of achieving this is by the application of the principle of superdirectivity.

Discrete Arrays and Polynomials

It is common practice to relate discrete arrays to polynomials for array synthesis purposes [Steinberg, 1976]. For volumetric equispaced arrays (it is only necessary that the spacing be uniform along each coordinate axis so that the spatial sampling periods D_i and D_j along, respectively, the ith and jth coordinate axes could be different for $i \neq j$), the weight associated with sensors located at coordinate $(i_1 D_1, i_2 D_2, i_3 D_3)$ is denoted by $w(i_1, i_2, i_3)$. The function in the complex variables $(z_1, z_2,$ and $z_3)$ that is associated with the sequence $\{w(i_1, i_2, i_3)\}$ is the generating function for the sequence and is denoted by

$$W(z_1, z_2, z_3) = \sum_{i_1} \sum_{i_2} \sum_{i_3} w(i_1, i_2, i_3) z_1^{i_1} z_2^{i_2} z_3^{i_3} \qquad (16.18)$$

In the electrical engineering and geophysics literature, the generating function $W(z_1, z_2, z_3)$ is sometimes called the z-transform of the sequence $\{w(i_1, i_2, i_3)\}$. When there are a finite number of sensors, a realistic assumption for any physical discrete array, $W(z_1, z_2, z_3)$ becomes a trivariate polynomial. In the special case when $w(i_1, i_2, i_3)$ is product separable, the polynomial $W(z_1, z_2, z_3)$ is also product separable. Particularly, this separability property holds when the shading is uniform, i.e., $w(i_1, i_2, i_3) = 1$. When the support of the uniform shading function is defined by $i_1 = 0, 1, \ldots, N_1 - 1$, $i_2 = 0, 1, \ldots, N_2 - 1$, and $i_3 = 0, 1, \ldots, N_3 - 1$, the associated polynomial becomes

$$W(z_1, z_2, z_3) = \sum_{i_1=0}^{N_1-1} \sum_{i_2=0}^{N_2-1} \sum_{i_3=0}^{N_3-1} z_1^{i_1} z_2^{i_2} z_3^{i_3} = \prod_{i=1}^{3} \frac{z_i^{N_i} - 1}{z_i - 1} \qquad (16.19)$$

In this case, all results developed for the synthesis of linear arrays become directly applicable to the synthesis of volumetric arrays. For a linear uniform discrete array composed of N sensors with intersensor spacing D_1 starting at the origin and receiving a signal at a known fixed wavenumber k_1 at a receiving angle θ, the far-field beam pattern

$$S(k_1, \theta) \triangleq S(\theta) = \sum_{r=0}^{N-1} e^{jk_1 r D_1 \sin\theta}$$

may be associated with a polynomial $\sum_{r=0}^{N-1} z_1^r$, by setting $z_1 = e^{jk_1 D_1 \sin\theta}$. This polynomial has all its zeros on the unit circle in the z_1-plane. If the array just considered is not uniform but has a weighting factor w_r, for $r = 0, 1, \ldots, N_1 - 1$, the space factor,

$$Q(\theta) \triangleq \sum_{r=0}^{N_1-1} w_r e^{jk_1 D_1 r \sin\theta}$$

may again be associated with a polynomial $\sum_{r=0}^{N_1-1} w_r z_1^r$. By the pattern multiplication theorem, it is possible to get the polynomial associated with the total beam pattern of an array with weighted sensors by multiplying the polynomials associated with the array element pattern and the polynomial associated with the space factor $Q(\theta)$. The array factor $|Q(\theta)|^2$ may also be associated with the polynomial spectral factor

$$|Q(\theta)|^2 \leftrightarrow \sum_{r=0}^{N_1-1} w_r z_1^r \sum_{r=0}^{N_1-1} w_r^*(z_1^*)^r \qquad (16.20)$$

where the weighting (shading) factor is allowed to be complex. Uniformly distributed apertures and uniformly spaced volumetric arrays which admit product separable sensor weightings can be treated by using the well-developed theory of linear discrete arrays and their associated polynomial. When the product separability property does not hold, scopes exist for applying results from multidimensional systems theory [Bose, 1982] concerning multivariate polynomials to the synthesis problem of volumetric arrays.

Velocity Filtering

Combination of individual sensor outputs in a more sophisticated way than the delay-and-sum technique leads to the design of multichannel velocity filters for linear and planar as well as spatial arrays. Consider, first, a linear (1-D) array of sensors, which will be used to implement velocity discrimination. The pass and rejection zones are defined by straight lines in the (k_1, ω)-plane, where

$$k_1 = \frac{\omega}{V} = \frac{\omega}{(v/\sin \theta)}$$

is the wavenumber, ω the angular frequency in radians/second, V the apparent velocity on the earth's surface along the array line, v the velocity of wave propagation, and θ the horizontal arrival direction. The transfer function

$$H(\omega, k_1) = \begin{cases} 1, & -\dfrac{|\omega|}{V} \leq k_1 \leq \dfrac{|\omega|}{V} \\ 0, & \text{otherwise} \end{cases}$$

of a "pie-slice" or "fan" velocity filter [Bose, 1985] rejects totally wavenumbers outside the range $-|\omega|/V \leq k_1 \leq |\omega|/V$ and passes completely wavenumbers defined within that range. Thus, the transfer function defines a high-pass filter which passes signals with apparent velocities of magnitude greater than V at a fixed frequency ω. If the equispaced sensors are D units apart, the spatial sampling results in a periodic wavenumber response with period $k_1 = 1/(2D)$. Therefore, for a specified apparent velocity V, the resolvable wavenumber and frequency bands are, respectively, $-1/(2D) \leq k_1 \leq 1/(2D)$ and $-V/(2D) \leq \omega \leq V/(2D)$ where $\omega/(2D)$ represents the folding frequency in radians/second.

Linear arrays are subject to the limitation that the source is required to be located on the extended line of sensors so that plane wavefronts approaching the array site at a particular velocity excite the individual sensors, assumed equispaced, at arrival times which are also equispaced. In seismology, the equispaced interval between successive sensor arrival times is called a move-out or step-out and equals $(D \sin \theta)/v = D/V$. However, when the sensor-to-source azimuth varies, two or more independent signal move-outs may be present. Planar (2-D) arrays are then required to discriminate between velocities as well as azimuth. Spatial (3-D) arrays provide additional scope to the enhancement of discriminating capabilities when sensor/source locations are arbitrary. In such cases, an array origin is chosen and the mth sensor location is denoted by a vector $(x_{1m}x_{2m}x_{3m})^t$ and the frequency wavenumber response of an array of sensors is given by

$$H(\omega, k_1, k_2, k_3) = \frac{1}{N} \sum_{m=1}^{N} H_m(\omega) \exp\left[\sum_{i=1}^{3} -j2\pi k_i x_{im} \right]$$

where $H_m(\omega)$ denotes the frequency response of a filter associated with the mth recording device (sensor). The sum of all N filters provides flat frequency response so that waveforms arriving from the estimated directions of arrival at estimated velocities are passed undistorted and other waveforms are suppressed. In the planar specialization, the 2-D array of sensors leads to the theory of 3-D filtering involving a transfer function in the frequency wavenumber variables f, k_1, and k_2. The basic design equations for the optimum, in the least-mean-square error sense, frequency wavenumber filters have been developed [Burg, 1964]. This procedure of Burg can be routinely generalized to the 4-D filtering problem mentioned above.

Acknowledgment

N. K. Bose and L. H. Sibul acknowledge the support provided by the Office of Naval Research under, respectively, Contract N00014-92-J-1755 and the Fundamental Research Initiatives Program.

Defining Terms

Array pattern: Fourier transform of the receiver weighting function taking into account the positions of the receivers.

Beamformers: Systems commonly used for detecting and isolating signals that are propagating in a particular direction.

Grating lobes: Repeated main lobes in the array pattern interpretable in terms of spatial frequency aliasing.

Velocity filtering: Means for discriminating signals from noise or other undesired signals because of their different apparent velocities.

Wavenumber: 2π (spatial frequency in cycles per unit distance).

References

K. M. Ahmed and R. J. Evans, "Robust signal and array processing," *IEE Proceedings, F: Communications, Radar, and Signal Processing*, vol. 129, no. 4, pp. 297–302, 1982.

N. K. Bose, *Applied Multidimensional Systems Theory*, New York: Van Nostrand Reinhold, 1982.

N. K. Bose, *Digital Filters*, New York: Elsevier Science North-Holland, 1985. Reprint ed., Malabar, Fla.: Krieger Publishing, 1993.

J. P. Burg, "Three-dimensional filtering with an array of seismometers," *Geophysics*, vol. 23, no. 5, pp. 693–713, 1964.

J. F. Claerbout, *Fundamentals of Geophysical Data Processing*, New York: McGraw-Hill, 1976.

D. E. Dudgeon, "Fundamentals of digital array processing," *Proc. IEEE*, vol. 65, pp. 898–904, 1977.

R. A. Monzingo and T. W. Miller, *Introduction to Adaptive Arrays*, New York: Wiley, 1980.

S. M. Pillai, *Array Signal Processing*, New York: Springer-Verlag, 1989.

B. D. Steinberg, *Principles of Aperture and Array System Design*, New York: Wiley, 1976.

Further Information

Adaptive Signal Processing, edited by Leon H. Sibul, includes papers on adaptive arrays, adaptive algorithms and their properties, as well as other applications of adaptive signal processing techniques (IEEE Press, New York, 1987).

Adaptive Antennas: Concepts and Applications, by R. T. Compton, Jr., emphasizes adaptive antennas for electromagnetic wave propagation applications (Prentice-Hall, Englewood-Cliffs, N.J., 1988).

BEHIND THE LAPLACE TRANSFORM

The Laplace transform is perhaps *the* mathematical signature of the electrical engineer, having a long history of application to problems of electrical engineering. It changes some of the most important differential equations of physics into algebraic equations, which are generally far easier to solve.

Despite its Gaelic name, the transform originates with the Swiss mathematician, Leonhard Euler (1707–1783), who in 1744 wrote integrals that look much like the modern version. These were adapted by the Italian-French mathematical physicist, Joseph Louis Lagrange (1736–1813) to the needs of probability theory, and his work in turn influenced the Frenchman Pierre Simon Laplace (1749–1827).

By 1785, Laplace was writing the almost modern Laplace transform equation (Equation 1). Today it is still used, as the Mellin transform, to solve certain differential equations with variable coefficients. But an electrical engineer would write it differently (Equation 2).

Then in 1807, Laplace's fellow countryman, Joseph Fourier, published the first monograph describing the heat diffusion equation. Intrigued, Laplace tried his hand at solving it, obtaining results that in turn inspired Fourier's discovery of his own transform.

The connection between the Fourier transform and the Laplace transform is intimate, but they are not equivalent. While the Fourier transform is useful in finding the steady-state output of a linear circuit in response to a periodic output, the Laplace transform can provide both the steady-state *and* transient responses for periodic *and* aperiodic inputs.

For the modern EE, the lure of the Laplace transform is its ability to map the complicated operation of convolution into multiplication. This integral has for decades driven electrical engineering undergraduates to contemplate theology either for salvation or as an alternative career. The

$$y(s) = \int_0^\infty t^s \, y(t) \, dt \qquad (1)$$

$$Y(s) = \int_0^\infty e^{-st} y(t) \, dt \qquad (2)$$

$$y(t) = \int_0^\infty h(u) \, x(t-u) \, du \qquad (3)$$

$$Y(s) = H(s) \, X(s) \qquad (4)$$

equation states that if $x(t)$ is the input to a linear system with impulse response $h(t)$, then the output, $y(t)$, is $x(t)$ convolved with $h(t)$ (Equation 3). But if $Y(s)$ and $X(s)$ are the associated Laplace transforms, this imposing integral is completely tamed (Equation 4).

In one of those astonishing coincidences that could mean mathematics is not just a game of made-up rules, the convolution integral also plays a central role in the theory of random variables: if $x(t)$ and $h(t)$ are the probability density functions (pdfs) of two independent random variables, X and H, then the pdf, $y(t)$, of $Y = X + H$ is found by convolving the pdfs of X and H.

Over a century had to elapse, though, before these sophisticated applications of the Laplace transform could evolve from the form in which Laplace left it in 1827. For EEs, the next big advance was made by the eccentric Englishman Oliver Heaviside (1850–1925). He reduced differential equations directly to algebra by representing time differentiation as an operator. He used p, as in $px = dx/dt$ and $p^2x = d^2x/dt^2$, while $1/p(x) =$ the integral of $x\, dt$, and then manipulated these equations using any algebraic trick he could think of, including his famous Heaviside expansion theorem, which is essentially the partial fraction expansion of modern Laplace theory.

But Heaviside was notoriously unconcerned with rigor. For instance, he never blinked an eye when fractional operators, such as $p^{1/2}$, arose.

Until the end of the 1930s, the more advanced EEs continued to use Heaviside's *ad hoc* technique. But lesser analysts were all too often swallowed up in the dangers of methods a master like Heaviside could intuitively sense and sidestep.

Then, in 1937, the German mathematician Gustav Doetsch published his book *Theorie und Anwendung der Laplace-Transformation*. That same year, L. A. Pipes published the first explicit application of the method to electrical engineering problems in *Philosophical Magazine*, in a paper titled "Laplacian Transform Circuit Analysis." The technique quickly spread, and in the United States, *Transients in Linear Systems: Studied by the Laplace Transformation* — the still classic text, known to generations of EEs as "Gardner and Barnes" — was published in 1942.

More recently still, the development of computer programs that simulate even highly nonlinear circuits has threatened the practical importance of the Laplace transform. Software packages have made the Laplace transform more important for the theoretician than for the practical designer and analyst. Will the Laplace transform become one of these concepts learned in engineering school but never used on the job?

Source: P. J. Nahin, *IEEE Spectrum*, p. 60, March 1991. © 1991 IEEE.

17
VLSI for Signal Processing

17.1 Special Architectures 370

Pipelining • Parallel Processing • Retiming • Unfolding • Folding Transformation • Look-Ahead Technique • Associativity Transformation • Distributivity • Arithmetic Processor Architectures • Computer-Aided Design • Future VLSI DSP Systems

17.2 Signal Processing Chips and Applications 385

DSP Processors • Fixed-Point TMS320C25-Based Development System • Implementation of a Finite Impulse Response Filter with the TMS320C25 • Floating-Point TMS320C30-Based Development System • EVM Tools • Implementation of a Finite Impulse Response Filter with the TMS320C30 • FIR and IIR Implementation Using C and Assembly Code • Real-Time Applications • Conclusions and Future Directions

Keshab K. Parhi
University of Minnesota

Rulph Chassaing
Roger Williams University

Bill Bitler
Philips Components

17.1 Special Architectures

Keshab K. Parhi

Digital signal processing (DSP) is used in numerous applications. These applications include telephony, mobile radio, satellite communications, speech processing, video and image processing, biomedical applications, radar, and sonar. All these applications require different sample rates. Real-time implementations of DSP systems require design of hardware that can match the application sample rate to the hardware processing rate (which is related to the clock rate and the implementation style). Thus, real-time does not always mean high speed. Real-time architectures are capable of processing samples as they are received from the signal source, as opposed to storing them in buffers for later processing as done in batch processing. Furthermore, real-time architectures operate on an infinite time series (since the number of the samples of the signal source is so large that it can be considered infinite). While speech and sonar applications require lower sample rates, radar and video image processing applications require much higher sample rates. The sample rate information alone cannot be used to choose the architecture. The algorithm complexity is also an important consideration. For example, a very complex and computationally intensive algorithm for a low-sample-rate application and a computationally simple algorithm for a high-sample-rate application may require similar hardware speed and complexity. These ranges of algorithms and applications motivate us to study a wide variety of architecture styles.

Using very large scale integration (VLSI) technology, DSP algorithms can be prototyped in many ways. These options include (1) single or multiprocessor programmable digital signal processors, (2) use of core programmable digital signal processor with customized interface logic, (3) semicustom gate-array implementations, and (4) full-custom dedicated hardware implementation. The DSP algorithms are implemented in the programmable processors by translating the algorithm to the processor assembly code. This can require an extensive amount of time. On the other hand, high-level compilers for DSP can be used to generate the assembly code. Although this is currently feasible, the code generated by the compiler is not as efficient as hand-optimized code. Design of DSP compilers for generation of efficient code is still an active research topic. In the case of dedicated designs, the challenge lies in a thorough understanding of the DSP algorithms and theory of architectures. For example, just minimizing the number of multipliers in an algorithm as done in the old days may not lead to a better dedicated design. The area saved by the number of multipliers may be offset by the increase in control, routing, and placement costs.

Off-the-shelf programmable digital signal processors can lead to faster prototyping. These prototyped systems can prove very effective in fast simulation of computation-intensive algorithms (such as those encountered in speech recognition, video compression, and seismic signal processing) or in benchmarking and standardization. After standards are determined, it is more useful to implement the algorithms using dedicated circuits.

Design of dedicated circuits is not a simple task. Dedicated circuits provide limited or no programming flexibility. They require less silicon area and consume less power. However, the low production volume, high design cost, and long turnaround time are some of the difficulties associated with the design of dedicated systems. Another difficulty is the availability of appropriate computer-aided design (CAD) tools for DSP systems. As time progresses, however, the architectural design techniques will be better understood and can be incorporated into CAD tools, thus making the design of dedicated circuits easier. Hierarchical CAD tools can integrate the design at various levels in an automatic and efficient manner. Implementation of standards for signal and image processing using dedicated circuits will lead to higher volume production. As time progresses, dedicated designs will be more acceptable to customers of DSP.

Successful design of dedicated circuits requires careful algorithm and architecture considerations. For example, for a filtering application, different equivalent realizations may possess different levels of concurrency. Thus, some of these realizations may be suitable for a particular application while other realizations may not be able to meet the sample rate requirements of the application. The lower-level architecture may be implemented in word-serial or **word-parallel** manner. The arithmetic functional units may be implemented in **bit-serial** or **digit-serial** or bit-parallel manner. The synthesized architecture may be implemented with a dedicated data path or shared data path. The architecture may be systolic or nonsystolic.

Algorithm transformations play an important role in the design of dedicated architectures [Parhi, 1989]. This is because the transformed algorithms can be made to operate with better performance (where the performance may be measured in terms of speed, area, or power). Examples of these transformations include pipelining, parallel processing, retiming, unfolding, folding, look-ahead, associativity, and distributivity. These transformations and other architectural concepts are described in detail in subsequent sections.

Pipelining

Pipelining can increase the amount of concurrency (or the number of activities performed simultaneously) in an algorithm. Pipelining is accomplished by placing latches at appropriate intermediate points in a data flow graph that describes the algorithm. Each latch also refers to a storage unit or buffer or register. The latches can be placed at *feed-forward cutsets* of the data flow graph. In synchronous hardware implementations, pipelining can increase the clock rate of the system (and therefore the sample rate). The drawbacks associated with pipelining are the increase in system latency and the increase in the number of registers. To illustrate the speed increase using pipelining, consider the second-order three-tap finite impulse response (FIR) filter shown in Fig. 17.1(a). The signal $x(n)$ in this system can be sampled at a rate limited by the throughput of one multiplication and two additions. For simplicity, if we assume the multiplication time to be two times the addition time (T_{add}), the effective sample or clock rate of this system is $1/4 T_{add}$. By placing latches as shown in Fig. 17.1(b) at the cutset shown in the dashed line, the sample rate can be improved to the rate of

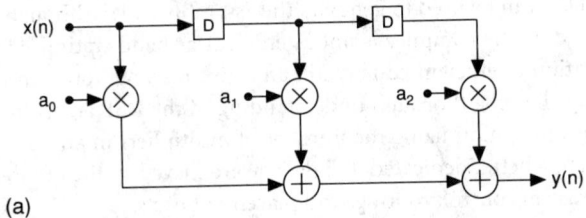

(a)

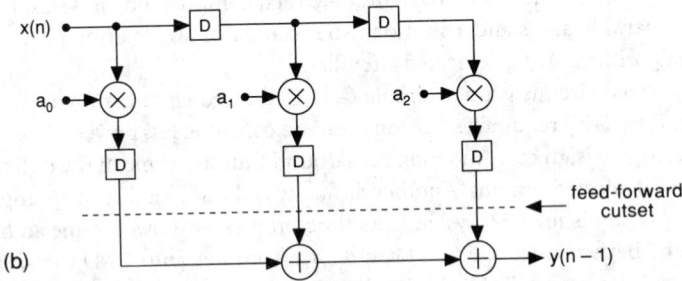

(b)

FIGURE 17.1 (a) A three-tap second-order nonrecursive digital filter; (b) the equivalent pipelined digital filter obtained by placing storage units at the intersection of the signal wires and the feed-forward cutset. If the multiplication and addition operations require 2 and 1 unit of time, respectively, then the maximum achievable sampling rates for the original and the pipelined architectures are 1/4 and 1/2 units, respectively.

one multiplication or two additions. While pipelining can be easily applied to all algorithms with no feedback loops by the appropriate placement of latches, it cannot easily be applied to algorithms with feedback loops. This is because the cutsets in feedback algorithms contain feed-forward and feedback data flow and cannot be considered as feed-forward cutsets.

Pipelining can also be used to improve the performance in software programmable multiprocessor systems. Most software programmable DSP processors are programmed using assembly code. The assembly code is generated by high-level compilers that perform scheduling. Schedulers typically use the acyclic precedence graph to construct schedules. The removal of all edges in the signal (or data) flow graph containing delay elements converts the signal flow graph to an acyclic precedence graph. By placing latches to pipeline a data flow graph, we can alter the acyclic precedence graph. In particular, the critical path of the acyclic precedence graph can be reduced. The new precedence graph can be used to construct schedules with lower iteration periods (although this may often require an increase in the number of processors).

Pipelining of algorithms can increase the sample rate of the system. Sometimes, for a constant sample rate, pipelining can also reduce the power consumed by the system. This is because the data paths in the pipelined system can be charged or discharged with lower supply voltage. Since the capacitance remains almost constant, the power can be reduced. Achieving low power can be important in many battery-powered applications [Chandrakasan *et al.*, 1992].

Parallel Processing

Parallel processing is related to pipelining but requires replication of hardware units. Pipelining exploits concurrency by breaking a large task into multiple smaller tasks and by separating these smaller tasks by storage units. On the other hand, parallelism exploits concurrency by performing multiple larger tasks simultaneously in separate hardware units.

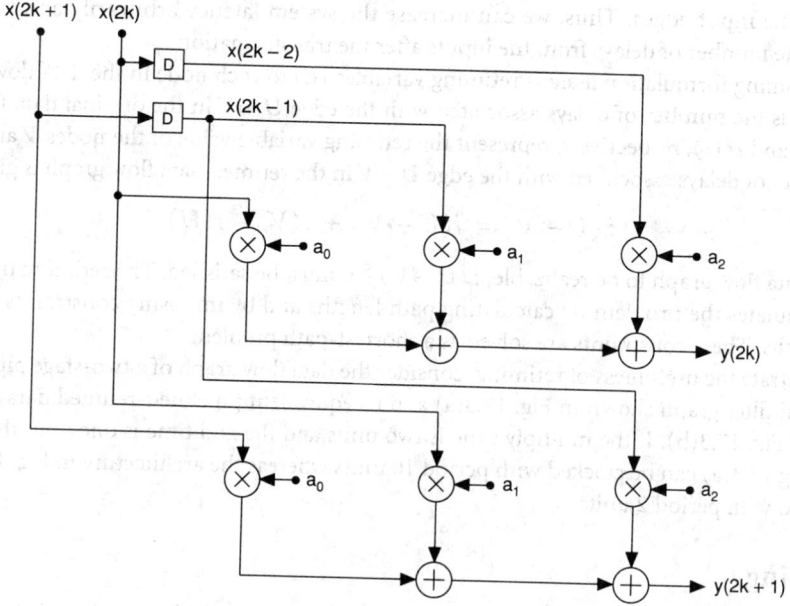

FIGURE 17.2 Twofold parallel realization of the three-tap filter of Fig. 17.1(a).

To illustrate the speed increase due to parallelism, consider the parallel implementation of the second-order three-tap FIR filter of Fig. 17.1(a) shown in Fig. 17.2. In the architecture of Fig. 17.2, two input samples are processed and two output samples are generated in each clock cycle period of four addition times. Because each clock cycle processes two samples, however, the effective sample rate is $1/2T_{add}$ which is the same as that of Fig. 17.1(b). The parallel architecture leads to the speed increase with significant hardware overhead. The entire data flow graph needs to be replicated with an increase in the amount of parallelism. Thus, it is more desirable to use pipelining as opposed to parallelism. However, parallelism may be useful if pipelining alone cannot meet the speed demand of the application or if the technology constraints (such as limitations on the clock rate by the I/O technology) limit the use of pipelining. In obvious ways, pipelining and parallelism can be combined also. Parallelism, like pipelining, can also lead to power reduction but with significant overhead in hardware requirements. Achieving pipelining and parallelism can be difficult for systems with feedback loops. Concurrency may be created in these systems by using the look-ahead transformation.

Retiming

Retiming is similar to pipelining but yet different in some ways [Leiserson *et al.*, 1983]. Retiming is the process of moving the delays around in the data flow graph. Removal of one delay from all input edges of a node and insertion of one delay to each outgoing edge of the same node is the simplest example of retiming. Unlike pipelining, retiming does not increase the latency of the system. However, retiming alters the number of delay elements in the system. Retiming can reduce the critical path of the data flow graph. As a result, it can lead to clock period reduction in hardware implementations or critical path of the acyclic precedence graph or the iteration period in programmable software system implementations.

The single host formulation of the retiming transformation preserves the latency of the algorithm. The retiming formulation with no constraints on latency (i.e., with separate input and output hosts) can also achieve *pipelining with no retiming* or *pipelining with retiming*. Pipelining with retiming is the most desirable transformation in DSP architecture design. Pipelining with retiming can be interpreted to be identical to retiming of the original algorithm with a large number of delays at the input edges. Thus, we can increase the system latency arbitrarily and remove the appropriate number of delays from the inputs after the transformation.

The retiming formulation assigns retiming variables $r(.)$ to each node in the data flow graph. If $i(U \rightarrow V)$ is the number of delays associated with the edge $U \rightarrow V$ in the original data flow graph and $r(V)$ and $r(U)$, respectively, represent the retiming variable value of the nodes V and U, then the number of delays associated with the edge $U \rightarrow V$ in the retimed data flow graph is given by

$$i_r(U \rightarrow V) = i(U \rightarrow V) + r(V) - r(U)$$

For the data flow graph to be realizable, $i_r(U \rightarrow V) \geq 0$ must be satisfied. The retiming transformation formulates the problem by calculating path lengths and by imposing constraints on certain path lengths. These constraints are solved as a shortest-path problem.

To illustrate the usefulness of retiming, consider the data flow graph of a two-stage pipelined lattice digital filter graph shown in Fig. 17.3(a) and its equivalent pipelined-retimed data flow graph shown in Fig. 17.3(b). If the multiply time is two units and the add time is one unit, the architecture in Fig 17.3(a) can be clocked with period 10 units whereas the architecture in Fig. 17.3(b) can be clocked with period 2 units.

Unfolding

The **unfolding** transformation is similar to loop unrolling. In J-unfolding, each node is replaced by J nodes and each edge is replaced by J edges. The J-unfolded data flow graph executes J iterations of the original algorithm [Parhi, 1991]

The unfolding transformation can unravel the hidden concurrency in a data flow program. The achievable iteration period for a J-unfolded data flow graph is $1/J$ times the critical path length of the unfolded data flow graph. By exploiting interiteration concurrency, unfolding can lead to a lower iteration period in the context of a software programmable multiprocessor implementation.

The unfolding transformation can also be applied in the context of hardware design. If we apply an unfolding transformation on a (word-serial) nonrecursive algorithm, the resulting data flow graph represents a word-parallel (or simply parallel) algorithm that processes multiple samples or words in parallel every clock cycle. If we apply 2-unfolding to the 3-tap FIR filter in Fig. 17.1(a), we can obtain the data flow graph of Fig. 17.2.

Because the unfolding algorithm is based on graph theoretic approach, it can also be applied at the bit level. Thus, unfolding of a bit-serial data flow program by a factor of J leads to a digit-serial program with digit size J. The *digit size* represents the number of bits processed per clock cycle. The digit-serial architecture is clocked at the same rate as the bit-serial (assuming that the clock rate is limited by the communication I/O bound much before reaching the computation bound of the bit-serial program). Because the digit-serial program processes J bits per clock cycle the effective bit rate of the digit-serial architecture is J times higher. A simple example of this unfolding is illustrated in Fig. 17.4, where the bit-serial adder in Fig. 17.4(a) is unfolded by a factor of 2 to obtain the digit-serial adder in Fig. 17.4(b) for digit size 2 for a word length of 4. In obvious ways, the unfolding transformation can be applied to both word level and bit level simultaneously to generate word-parallel digit-serial architectures. Such architectures process multiple words per clock cycle and process a digit of each word (not the entire word).

Folding Transformation

The **folding** transformation is the reverse of the unfolding transformation. While the unfolding transformation is simpler, the folding transformation is more difficult [Parhi *et al.*, 1992].

The folding transformation can be applied to fold a bit-parallel architecture to a digit-serial or bit-serial one or to fold a digit-serial architecture to a bit-serial one. It can also be applied to fold an algorithm data flow graph to a hardware data flow for a specified folding set. The folding set indicates the

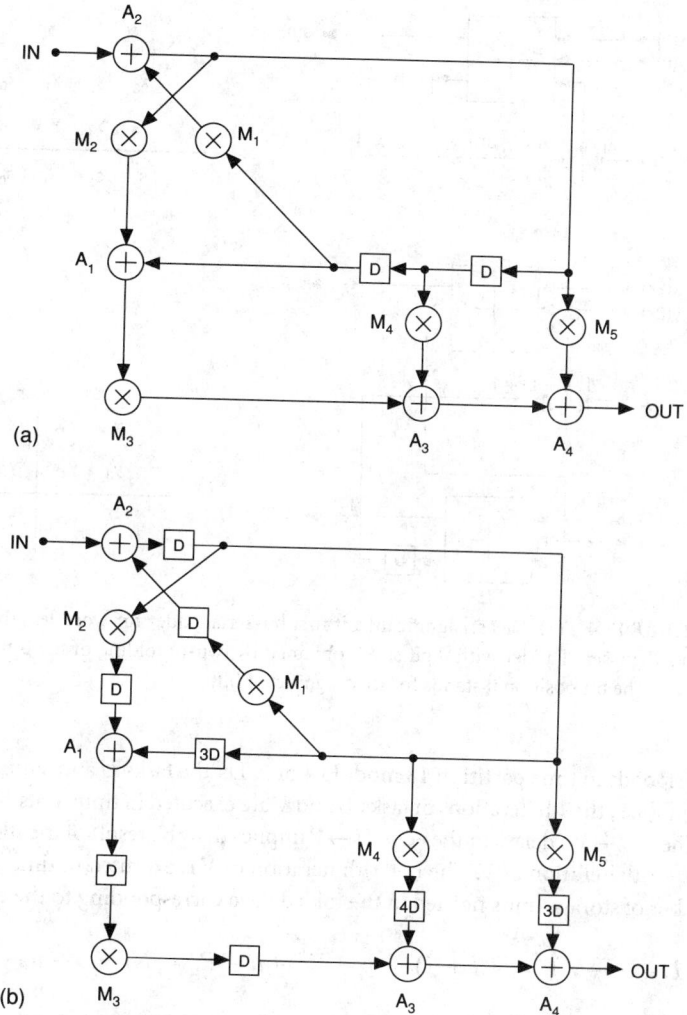

FIGURE 17.3 (a) A two-stage pipelinable time-invariant lattice digital filter. If multiplication and addition operations require 2 and 1 time units, respectively, then this data flow graph can achieve a sampling period of 10 time units (which corresponds to the critical path $M_1 \rightarrow A_2 \rightarrow M_2 \rightarrow A_1 \rightarrow M_3 \rightarrow A_3 \rightarrow A_4$). (b) The pipelined/retimed lattice digital filter can achieve a sampling period of 2 time units.

processor in which and the time partition at which a task is executed. A specified folding set may be infeasible, and this needs to be detected first. The folding transformation performs a preprocessing step to detect feasibility and in the feasible case transforms the algorithm data flow graph to an equivalent pipelined/retimed data flow graph that can be folded. For the special case of regular data flow graphs and for linear space–time mappings, the folding tranformation reduces to **systolic** array design.

In the folded architecture, each edge in the algorithm data flow graph is mapped to a communicating edge in the hardware architecture data flow graph. Consider an edge $U \rightarrow V$ in the algorithm data flow graph with associated number of delays $i(U \rightarrow V)$. Let the tasks U and V be mapped to the hardware units H_U and H_V, respectively. Assume that N time partitions are available, i.e., the iteration period is N. A modulo operation determines the time partition. For example, the time unit 18

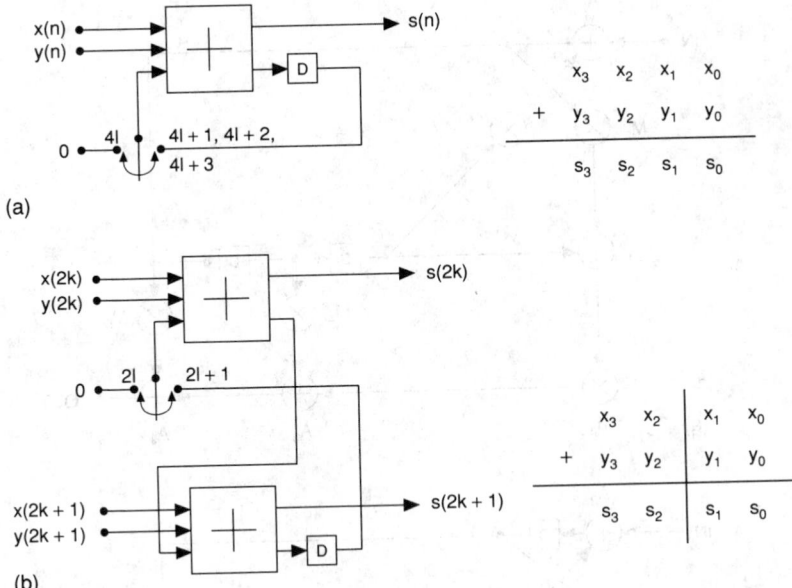

FIGURE 17.4 (a) A least-significant-bit first bit-serial adder for word length of 4;
(b) a digit-serial adder with digit size 2 obtained by two-unfolding of the bit-serial
adder. The bit position 0 stands for least significant bit.

for $N = 4$ corresponds to time partition 18 modulo 4 or 2. Let the tasks U and V be executed in time partitions u and v, i.e., the lth iterations of tasks U and V are executed in time units $Nl + u$ and $Nl + v$, respectively. The $i(U \rightarrow V)$ delays in the edge $U \rightarrow V$ implies that the result of the lth iteration of U is used for the $(l + i)$th iteration of V. The $(l + i)$th iteration of V is executed in time unit $N(l + i) + v$. Thus the number of storage units needed in the folded edge corresponding to the edge $U \rightarrow V$ is

$$D_F(U \rightarrow V) = N(l + i) + v - Nl - u - P_u = Ni + v - u - P_u$$

where P_u is the level of pipelining of the hardware operator H_U. The $D_F(U \rightarrow V)$ delays should be connected to the edge between H_U and H_V, and this signal should be switched to the input of H_V at time partition v. If the $D_F(U \rightarrow V)$'s as calculated here were always nonnegative for all edges $U \rightarrow V$, then the problem would be solved. However, some $D_F()$'s would be negative. The algorithm data flow graph needs to be pipelined and retimed such that all the $D_F()$'s are nonnegative. This can be formulated by simple inequalities using the retiming variables. The retiming formulation can be solved as a path problem, and the retiming variables can be determined if a solution exists. The algorithm data flow graph can be retimed for folding and the calculation of the $D_F()$'s can be repeated. The folded hardware architecture data flow graph can now be completed. The folding technique is illustrated in Fig. 17.5. The algorithm data flow graph of a two-stage pipelined lattice recursive digital filter of Fig. 17.3(a) is folded for the folding set shown in Fig. 17.5. Fig. 17.5(a) shows the pipelined/retimed data flow graph (preprocessed for folding) and Fig. 17.5(b) shows the hardware architecture data flow graph obtained after folding.

As indicated before, a special case of folding can address systolic array design for regular data flow graphs and for linear mappings. The systolic architectures make use of extensive pipelining and local communication and operate in a synchronous manner [Kung, 1988]. The systolic processors can also be made to operate in an asynchronous manner, and such systems are often referred to as wavefront processors. Systolic architectures have been designed for a variety of applications including convolution, matrix solvers, matrix decomposition, and filtering.

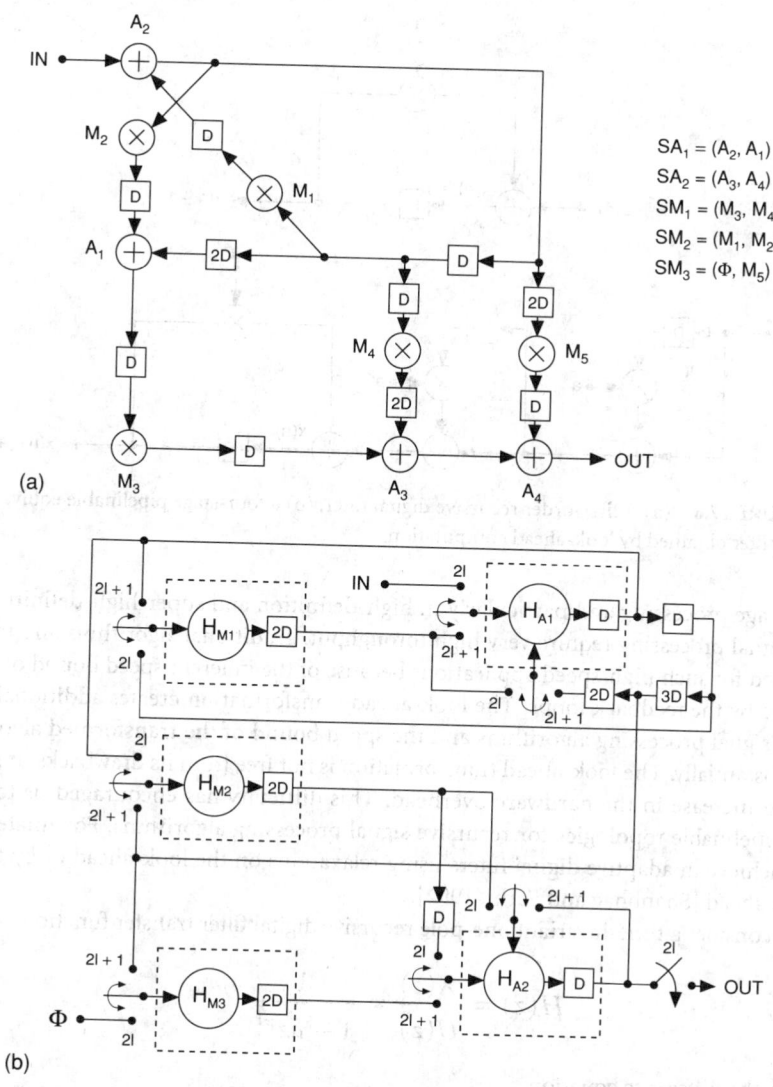

FIGURE 17.5 (a) A pipelined/retimed data flow graph obtained from Fig. 17.3(a) by preprocessing for folding; (b) the folded hardware architecture data flow graph. In our folding notation, the tasks are ordered within a set and the ordering represents the time partition in which the task is executed. For example, $SA_1 = (A_2, A_1)$ implies that A_2 and A_1 are, respectively, executed in even and odd time partitions in the same processor. The notation Φ represents a null operation.

Look-Ahead Technique

The **look-ahead** technique is a very powerful technique for pipelining of recursive signal processing algorithms [Parhi and Messerschmitt, 1989]. This technique can transform a sequential recursive algorithm to an equivalent concurrent one, which can then be realized using pipelining or parallel processing or both. This technique has been successfully applied to pipeline many signal processing algorithms, including recursive digital filters (in direct form and lattice form), adaptive lattice digital filters, two-dimensional recursive digital filters, Viterbi decoders, Huffman decoders, and finite state machines. This research demonstrated that the recursive signal processing algorithms can be operated at high speed. This is an important result since modern signal processing applications in

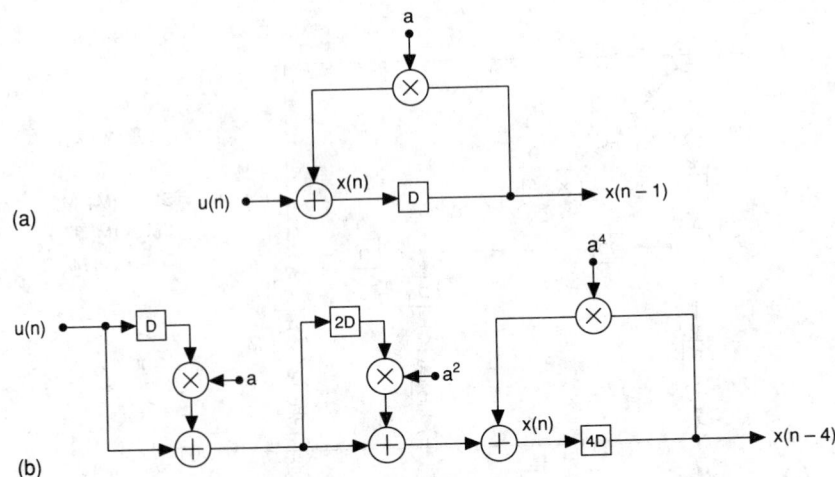

(a)

(b)

FIGURE 17.6   (a) A first-order recursive digital filter; (b) a four-stage pipelinable equivalent filter obtained by look-ahead computation.

radar and image processing and particularly in high-definition and super-high-definition television video signal processing require very high throughput. Traditional algorithms and topologies cannot be used for such high-speed applications because of the inherent speed bound of the algorithm created by the feedback loops. The look-ahead transformation creates additional concurrency in the signal processing algorithms and the speed bound of the transformed algorithms is increased substantially. The look-ahead transformation is not free from its drawbacks. It is accompanied by an increase in the hardware overhead. This difficulty has encouraged us to develop inherently pipelinable topologies for recursive signal processing algorithms. Fortunately, this is possible to achieve in adaptive digital filters using relaxations on the look-ahead or by the use of relaxed look-ahead [Shanbhag and Parhi, 1992].

To begin, consider a time-invariant one-pole recursive digital filter transfer function

$$H(z) = \frac{X(z)}{U(z)} = \frac{1}{1 - az^{-1}}$$

described by the difference equation

$$x(n) = ax(n-1) + u(n)$$

and shown in Fig. 17.6(a). The maximum achievable speed in this system is limited by the operating speed of one multiply–add operation. To increase the speed of this system by a factor of 2, we can express $x(n)$ in terms of $x(n-2)$ by substitution of one recursion within the other:

$$x(n) = a[ax(n-2) + u(n-1)] + u(n) = a^2 x(n-2) + au(n-1) + u(n)$$

The transfer function of the emulated second-order system is given by

$$H(z) = \frac{1 + az^{-1}}{1 - a^2 z^{-2}}$$

and is obtained by using a pole-zero cancellation at $-a$. In the modified system, $x(n)$ is computed using $x(n-2)$ as opposed to $x(n-1)$; thus we *look ahead*. The modified system has two delays in the multiply–add feedback loop, and these two delays can be distributed to pipeline the multiply–add operation by two stages. Of course, the additional multiply–add operation that represents one zero would also need to be pipelined by two stages to keep up with the sample rate of the system. To increase the speed by four times, we can rewrite the transfer function as:

$$H(z) = \frac{(1 + az^{-1})(1 + a^2 z^{-2})}{(1 - a^4 z^{-4})}$$

This system is shown in Fig. 17.6(b). Arbitrary speed increase is possible. However, for power-of-two speed increase the hardware complexity grows logarithmically with speed-up factor. The same technique can be applied to any higher-order system. For example, a second-order recursive filter with transfer function

$$H(z) = \frac{1}{1 - 2r \cos \theta z^{-1} + r^2 z^{-2}}$$

can be modified to

$$H(z) = \frac{1 + 2r \cos \theta z^{-1} + r^2 z^{-2}}{1 - 2r^2 \cos 2\theta z^{-2} + r^4 z^{-4}}$$

for a twofold increase in speed. In this example, the output $y(n)$ is computed using $y(n-2)$ and $y(n-4)$; thus, it is referred to as *scattered look-ahead*.

While look-ahead can transform any recursive digital filter transfer function to pipelined form, it leads to a hardware overhead proportional to $N \log_2 M$, where N is the filter order and M is the speed-up factor. Instead of starting with a sequential digital filter transfer function obtained by traditional design approaches and transforming it for pipelining, it is more desirable to use a constrained filter design program that can satisfy the filter spectrum and the pipelining constraint. The pipelining constraint is satisfied by expressing the denominator of the transfer function in scattered look-ahead form. Such filter design programs have now been developed in both time domain and frequency domain. The advantage of the constrained filter design approach is that we can obtain pipelined digital filters with marginal or zero hardware overhead compared with sequential digital filters. The pipelined transfer functions can also be mapped to pipelined lattice digital filters. The reader might note that the data flow graph of Fig. 17.3(a) was obtained by this approach.

The look-ahead pipelining can also be applied for the design of transversal and adaptive lattice digital filters. Although look-ahead transformation can be used to modify the adaptive filter recursions to create concurrency, this requires large hardware overhead. The adaptive filters are based on weight update operations, and the weights are adapted based on the current error. Finally, the error becomes close to zero and the filter coefficients have been adapted. Thus, making relaxations on the error can reduce the hardware overhead substantially without degradation of the convergence behavior of the adaptive filter. Three types of relaxations of look-ahead are possible. These are referred to as *sum relaxation, product relaxation*, and *delay relaxation*. To illustrate these three relaxations, consider the weight update recursion

$$w(n + 1) = a(n)w(n) + f(n)$$

where the term $a(n)$ is typically 1 for transversal least mean square (LMS) adaptive filters and of the form $(1 - \varepsilon(n))$ for lattice LMS adaptive digital filters, and $f(n) = \mu e(n)u(n)$ where μ is a constant, $e(n)$ is the error, and $u(n)$ is the input. The use of look-ahead transforms the above recursion to

$$w(n + M) = \prod_{i=0}^{M-1} a(n + M - i - 1) \; w(n)$$

$$+ \left[1a(n + M - 1) \prod_{i=0}^{1} a(n + M - i - 1) \ldots \prod_{i=0}^{M-2} a(n + M - i - 1) \right] \begin{bmatrix} f(n + M - 1) \\ f(n + M - 2) \\ \cdot \\ \cdot \\ \cdot \\ f(n) \end{bmatrix}$$

In sum relaxation, we only retain the single term dependent on the current input for the last term of the look-ahead recursion. The relaxed recursion after sum relaxation is given by

$$w(n + M) = \prod_{i=0}^{M-1} a(n + M - i - 1) \; w(n) + f(n + M - 1)$$

In lattice digital filters, the coefficient $a(n)$ is close to 1 for all n, since it can be expressed as $(1 - \varepsilon(n))$ and $\varepsilon(n)$ is close to zero for all n and is positive. The product relaxation on the above equation leads to

$$w(n + M) = (1 - M\varepsilon(n + M - 1)) \; w(n) + f(n + M - 1)$$

The delay relaxation assumes the signal to be slowly varying or to be constant over D samples and replaces the look-ahead by

$$w(n + M) = (1 - M\varepsilon(n + M - 1)) \; w(n) + f(n + M - D - 1)$$

These three types of relaxations make it possible to implement pipelined transversal and lattice adaptive digital filters with marginal increase in hardware overhead. Relaxations on the weight update operations change the convergence behavior of the adaptive filter, and we are forced to examine carefully the convergence behavior of the relaxed look-ahead adaptive digital filters. It has been shown that the relaxed look-ahead adaptive digital filters do not suffer from degradation in adaptation behavior. Futhermore, when coding, the use of pipelined adaptive filters could lead to a dramatic increase in pixel rate with no degradation in signal-to-noise ratio of the coded image and no increase in hardware overhead [Shanbhag and Parhi, 1992].

The concurrency created by look-ahead and relaxed look-ahead transformations can also be exploited in the form of parallel processing. Furthermore, for a constant speed, concurrent architectures (especially the pipelined architectures) can also lead to low power consumption.

Associativity Transformation

The addition operations in many signal processing algorithms can be interchanged since the add operations satisfy associativity. Thus, it is possible to move the add operations outside the critical loops to increase the maximum achievable speed of the system. As an example of the associative transformation, consider the realization of a second-order recursion $x(n) = 5/8x(n-1) - 3/4(n-2) + u(n)$. Two possible realizations are shown in Fig. 17.7(a). The realization on the left contains one multiplication and two add operations in the critical inner loop, whereas the realization on the right contains one multiplication and one add operation in the critical inner loop. The realization on the left can be transformed to the realization on the right using the associativity transformation. Figure 17.7(b) shows a bit-serial implementation of this second-order recursion for the realization on the right for a word length of 8. This bit-serial system can be operated in a functionally correct manner for any word length greater than or equal to 5 since the inner loop computation latency is

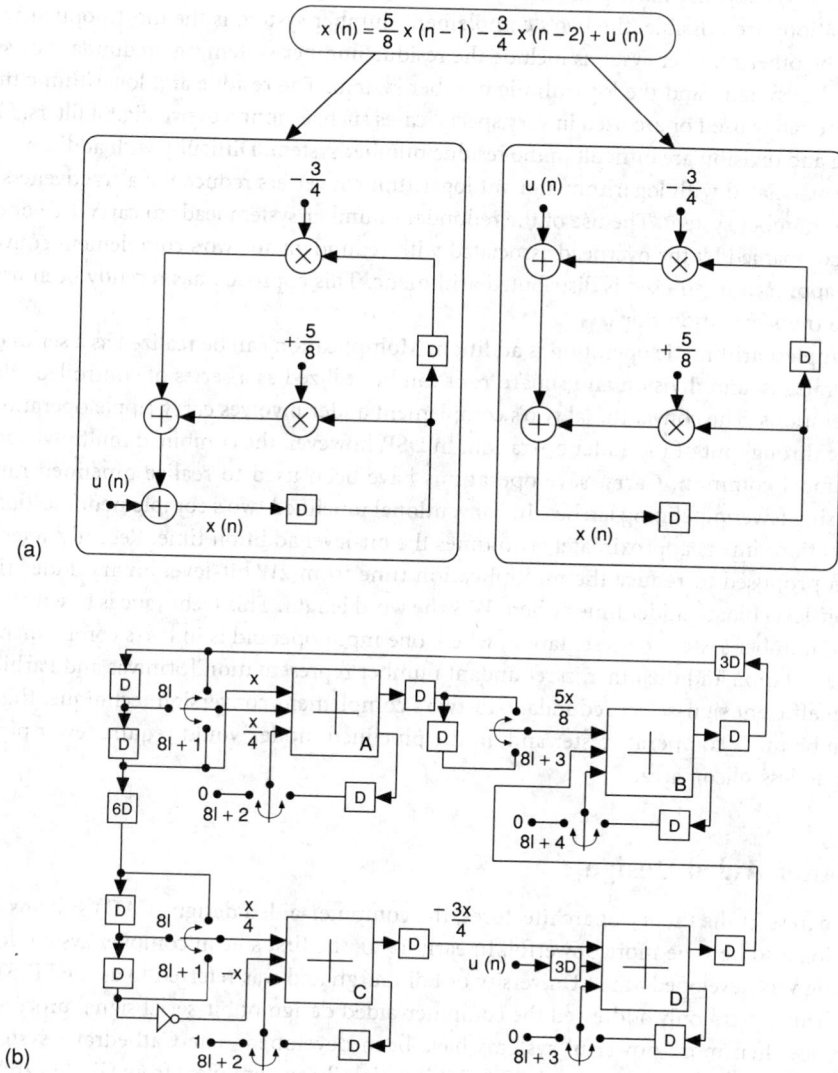

FIGURE 17.7 (a) Two associative realizations of a second-order recursion; (b) an efficient bit-serial realization of the recursion for a word length of 8.

5 cycles. On the other hand, if associativity were not exploited, then the minimum realizable word length would be 6. Thus, associativity can improve the achievable speed of the system.

Distributivity

Another local transformation that is often useful is distributivity. In this transformation, a computation $(A \times B) + (A \times C)$ may be reorganized as $A \times (B + C)$. Thus, the number of hardware units can be reduced from two multipliers and one adder to one multiplier and one adder.

Arithmetic Processor Architectures

In addition to algorithms and architecture designs, it is also important to address implementation styles and arithmetic processor architectures.

Most DSP systems use fixed-point hardware arithmetic operators. While many number system representations are possible, the two's complement number system is the most popular number system. The other number systems include the residue number system, the redundant or signed-digit number system, and the logarithmic number system. The residue and logarithmic number systems are rarely used or are used in very special cases such as nonrecursive digital filters. Shifting or scaling and division are difficult in the residue number system. Difficulty with addition and the overhead associated with logarithm and antilogarithm converters reduce the attractiveness of the logarithm number system. The use of the redundant number system leads to carry-free operation but is accompanied by the overhead associated with redundant-to-two's complement conversion. Another approach often used is distributed arithmetic. This approach has recently been used in a few video transformation chips.

The simplest arithmetic operation is addition. Multiplication can be realized as a series of add-shift operations, and division and square-root can be realized as a series of controlled add–sub-tract operations. The conventional two's complement adder involves carry ripple operation. This limits the throughput of the adder operation. In DSP, however, the combined multiply–add oper-ation is most common. Carry–save operations have been used to realize pipelined multiply-adders using fewer pipelining latches. In conventional pipelined two's complement multiplier, the multiplication time is approximately two times the bit-level addition time. Recently, a technique has been proposed to reduce the multiplication time from $2W$ bit-level binary adder times to $1.25W$ bit-level binary adder times where W is the word length. This technique is based on the use of hybrid number system representation, where one input operand is in two's complement num-ber representation and the other in redundant number representation [Srinivas and Parhi, 1992]. Using an efficient sign-select redundant-to-two's complement conversion technique, this multi-plier can be made to operate faster and, in the pipelined mode, would require fewer pipelining latches and less silicon area.

Computer-Aided Design

With progress in the theory of architectures, the computer-aided design (CAD) systems for DSP application also become more powerful. In early 1980, the first silicon compiler system for signal processing was developed at the University of Edinburgh and was referred to as the FIRST design system. This system only addressed the computer-aided design of bit-serial signal processing sys-tems. Since then more powerful systems have been developed. The Cathedral I system from Katholieke Universiteit Leuven and the BSSC (bit-serial silicon compiler) from GE Research Center in Schenectady, New York, also addressed synthesis of bit-serial circuits. The Cathedral system has now gone through many revisions, and the new versions can systhesize parallel multiprocessor data paths and can perform more powerful scheduling and allocation. The Lager design tool at the Uni-versity of California at Berkeley was developed to synthesize the DSP algorithms using parametriz-able macro building blocks (such as ALU, RAM, ROM). This system has also gone through many revisions. The Hyper system also developed at the University of California at Berkeley and the MARS design system developed at the University of Minnesota at Minneapolis perform higher level transformations and perform scheduling and allocation. These CAD tools are crucial to rapid prototyping of high-performance DSP integrated circuits.

Future VLSI DSP Systems

Future VLSI systems will make use of a combination of many types of architectures such as dedi-cated and programmable. These systems can be designed successfully with proper understanding of the algorithms, applications, theory of architectures, and with the use of advanced CAD systems.

Defining Terms

Bit serial: Processing of one bit per clock cycle. If word length is W, then one sample or word is processed in W clock cycles. In contrast, all W bits of a word are processed in the same clock cycle in a bit-parallel system.

Digit serial: Processing of more than one but not all bits in one clock cycle. If the digit size is W_1 and the word length is W, then the word is processed in W/W_1 clock cycles. If $W_1 = 1$, then the system is referred to as a bit-serial and if $W_1 = W$, then the system is referred to as a bit-parallel system. In general, the digit size W_1 need not be a divisor of the word length W, since the least and most significant bits of consecutive words can be overlapped and processed in the same clock cycle.

Folding: The technique of mapping many tasks to a single processor.

Look-ahead: The technique of computing a state $x(n)$ usng previous state $x(n - M)$ without requiring the intermediate states $x(n - 1)$ through $x(n - M + 1)$. This is referred to as a M-step look-ahead. In the case of higher-order computations, there are two forms of look-ahead: clustered look-ahead and scattered look-ahead. In clustered look-ahead, $x(n)$ is computed using the clustered states $x(n - M - N + 1)$ through $x(n - M)$ for an Nth order computation. In scattered look-ahead, $x(n)$ is computed using the scattered states $x(n - iM)$ where i varies from 1 to N.

Parallel processing: Processing of multiple tasks independently by different processors. This also increases the throughput.

Pipelining: A technique to increase throughput. A long task is divided into components, and each component is distributed to one processor. A new task can begin even though the former tasks have not been completed. In the pipelined operation, different components of different tasks are executed at the same time by different processors. Pipelining leads to an increase in the system latency, i.e., the time elapsed between the starting of a task and the completion of the task.

Retiming: The technique of moving the delays around the system. Retiming does not alter the latency of the system.

Systolic: Flow of data in a rhythmic fashion from a memory through many processors, returning to the memory just as blood flows from and to the heart.

Unfolding: The technique of transforming a program that describes one iteration of an algorithm to another equivalent program that describes multiple iterations of the same algorithm.

Word parallel: Processing of multiple words in the same clock cycle.

References

A.P. Chandrakasan, S. Sheng, and R.W. Brodersen, "Low-power CMOS digital design," *IEEE J. Solid State Circuits*, vol. 27(4), pp. 473–484, April 1992.

S.Y. Kung, *VLSI Array Processors*, Englewood Cliffs, N.J.: Prentice-Hall, 1988.

E.A. Lee and D.G. Messerschmitt, "Pipeline interleaved programmable DSP's," *IEEE Trans. Acoustics, Speech, Signal Processing*, vol. 35(9), pp. 1320–1345, September 1987.

C.E. Leiserson, F. Rose, and J. Saxe, "Optimizing synchronous circuitry by retiming," *Proc. 3rd Caltech Conf. VLSI*, Pasadena, Calif., pp. 87–116, March 1983.

K.K. Parhi, "Algorithm transformation techniques for concurrent processors," *Proc. IEEE*, vol. 77(12), pp. 1879–1895, December 1989.

K.K. Parhi, "Systematic approach for design of digit-serial processing architectures," *IEEE Trans. Circuits Systems*, vol. 38(4), pp. 358–375, April 1991.

K.K. Parhi and D.G. Messerschmitt, "Pipeline interleaving and parallelism in recursive digital filters," *IEEE Trans. Acoustics, Speech, Signal Processing*, vol. 37(7), pp. 1099–1135, July 1989.

K.K. Parhi, C.Y. Wang, and A.P. Brown, "Synthesis of control circuits in folded pipelined DSP architectures," *IEEE J. Solid State Circuits*, vol. 27(1), pp. 29–43, January 1992.

N.R. Shanbhag, and K.K. Parhi, "A pipelined adaptive lattice filter architecture," *Proc. 1992 IEEE Int. Symp. Circuits and Systems*, San Diego, May 1992.

H.R. Srinivas and K.K. Parhi, "High-speed VLSI arithmetic processor architectures using hybrid number representation," *J. VLSI Signal Processing*, vol. 4(2/3), pp. 177–198, 1992.

Further Information

A detailed video tutorial on "Implementation and Synthesis of VLSI Signal Processing Systems" presented by K.K. Parhi and J.M. Rabaey in March 1992 can be purchased from the customer service department of IEEE, 445 Hoes Lane, P.O. Box 1331, Piscataway, NJ 08855-1331.

Special architectures for video communications can be found in the book *VLSI Implementations for Image Communications*, published as the fourth volume of the series *Advances in Image Communications* (edited by Peter Pirsch) by the Elsevier Science Publishing Co. in 1993. The informative article "Research on VLSI for Digital Video Systems in Japan," published by K.K. Parhi in the fourth volume of the *1991 Office of Naval Research Asian Office Scientific Information Bulletin* (pages 93–98), provides examples of video codec designs using special architectures. For video programmable digital signal processor approaches, see I. Tamitani, H. Harasaki, and T. Nishitani, "A Real-Time HDTV Signal Processor: HD-VSP," published in *IEEE Transactions on Circuits and Systems for Video Technology*, March 1991, and T. Fujii, T. Sawabe, N. Ohta, and S. Ono, "Implementation of Super High-Definition Image Processing on HiPIPE," published in *1991 IEEE International Symposium on Circuits and Systems*, held in June 1991 in Singapore (pages 348–351).

The special issue of the *Proceedings of the IEEE* published in September 1987 provides numerous articles on hardware and software for signal processing.

The *IEEE Design and Test of Computers* published three special issues related to computer-aided design of special architectures; these issues were published in October 1990 (addressing high-level synthesis), December 1990 (addressing silicon compilations), and June 1991 (addressing rapid prototyping).

An early tutorial article entitled "Why Systolic Architectures?" by H.T. Kung, published in the January 1982 issue of *IEEE Computer* magazine, is good reading.

Descriptions of various CAD systems can be found in the following references. The description of the FIRST system can be found in the article "A Silicon Compiler for VLSI Signal Processing," by P. Denyer *et al.* in the *Proceedings of the ESSCIRC* conference held in Brussels in September 1982 (pages 215–218). The Cathedral system has been described in R. Jain *et al.*, "Custom Design of a VLSI PCM-FDM Transmultiplexor from System Specifications to Circuit Layout Using a Computer Aided Design System," published in *IEEE Journal of Solid State Circuits* in February 1986 (pages 73–85). The Lager system has been described in "An Integrated Automatic Layout Generation System for DSP Circuits," by J. Rabaey, S. Pope, and R. Brodersen, published in the July 1985 issue of the *IEEE Transactions on Computer Aided Design* (pages 285–296). The description of the MARS Design System can be found in C.-Y. Wang and K.K. Parhi, "High-Level DSP Synthesis Using MARS System," published in *Proceedings of the 1992 IEEE International Symposium on Circuits and Systems* in San Diego, May 1992. A tutorial article on high-level synthesis can be found in "The High-Level Synthesis of Digital Systems," by M.C. McFarland, A. Parker, and R. Composano, published in the February 1990 issue of the *Proceedings of the IEEE* (pages 310–318).

Articles on pipelined multipliers can be found in T.G. Noll *et al.*, "A Pipelined 330 MHZ Multiplier," *IEEE Journal of Solid State Circuits*, June 1986 (pages 411–416) and in M. Hatamian and G. Cash, "A 70-MHz 8-Bit × 8-Bit-Parallel Pipelined Multiplier in 2.5 μm CMOS," *IEEE Journal of Solid State Circuits*, 1986.

Technical articles on special architectures and chips for signal and image processing appear at different places, including proceedings of conferences such as IEEE Workshop on VLSI Signal Processing, IEEE International Conference on Acoustics, Speech, and Signal Processing, IEEE International Symposium on Circuits and Systems, IEEE International Solid State Circuits Conference, IEEE Customs Integrated Circuits Conference, IEEE International Conference on Computer Design, ACM/IEEE Design Automation Conference, ACM/IEEE International Conference on Computer Aided Design, International Conference on Application Specific Array Processors, and journals such as *IEEE Transactions on Signal Processing, IEEE Transactions on Image Processing, IEEE Transactions on Circuits and Systems: Part II: Analog and Digital Signal Processing, IEEE Transactions on Computers, IEEE Journal of Solid State Circuits, IEEE Signal Processing Magazine, IEEE Design and Test Magazine,* and *Journal of VLSI Signal Processing.*

17.2 Signal Processing Chips and Applications

Rulph Chassaing and Bill Bitler

Recent advances in very large scale integration (VLSI) have contributed to the current **digital signal processors.** These processors are just special-purpose fast microprocessors characterized by architectures and instructions suitable for real-time digital signal processing (DSP) applications. The commercial DSP processor, barely a decade old, has emerged because of the ever-increasing number of signal processing applications. DSP processors are now being utilized in a number of applications from communications and controls to speech and image processing. They have found their way into talking toys and music synthesizers. A number of texts [such as Chassaing and Horning, 1990] and articles [such as Ahmed and Kline, 1991] have been written, discussing the applications that use DSP processors and the recent advances in DSP systems.

DSP Processors

Digital signal processors are currently available from a number of companies, including Texas Instruments, Inc. (Texas), Motorola, Inc. (Arizona), Analog Devices, Inc. (Massachusetts), AT&T (New Jersey), and NEC (California). These processors are categorized as either **fixed-point** or **floating-point processors.** Several companies are now supporting both types of processors. **Special-purpose digital signal processors,** designed for a specific signal processing application such as for fast Fourier transform (FFT), have also emerged.

One of the first-generation digital signal processors is the (N-MOS technology) TMS32010, introduced by Texas Instruments (TI) in 1982. This first-generation fixed-point processor is based on the Harvard architecture, with a fast on-chip hardware multiplier/accumulator, and with data and instructions in separate memory spaces, allowing for concurrent accesses. This type of **pipelining feature** enables the processor to execute one instruction while fetching at the same time the next instruction. Other features include 144 (16-bit) words of on-chip data RAM and a 16-bit by 16-bit multiply operation in one instruction cycle time of 200 ns. Since many instructions can be executed in one single cycle, the TMS32010 is capable of executing 5 million instructions per second (MIPS). Major drawbacks of this first-generation processor are its limited **on-chip memory** size and much slower execution time for accessing external memory. Improved versions of this first-generation processor are now available in C-MOS technology, with a faster instruction cycle time of 160 ns.

The second-generation fixed-point processor TMS32020, introduced in 1985 by TI, was quickly followed by an improved C-MOS version TMS320C25 [Chassaing and Horning, 1990] in 1986. Features of the TMS320C25 include 544 (16-bit) words of on-chip data RAM, separate program and data memory spaces (each 64 K words), and an instruction cycle time of 100 ns, enabling the TMS320C25 to execute 10 MIPS.

The third-generation TMS320C30 (by TI) supports fixed- as well as floating-point operations [Chassaing, 1992]. Features of this processor include 32-bit by 32-bit floating-point multiply operations in one instruction cycle time of 60 ns. Since a number of instructions, such as load and store, multiply and add, can be performed in parallel (in one cycle time), the TMS320C30 can execute a pair of parallel instructions in 30 ns, allowing for 33.3 MIPS. The Harvard-based architecture of the fixed-point processors was abandoned for one allowing four levels of pipelining with three subsequent instructions being consequently fetched, decoded, and read while the current instruction is being executed. The TMS320C30 has 2 K words of on-chip memory and a total of 16 million words of addressable memory spaces for program, data, and input/output. Specialized instructions are available to make common DSP algorithms such as filtering and spectral analysis execute fast and efficiently. The architecture of the TMS320C30 was designed to take advantage of higher-level languages such as C and ADA.

We will discuss both the fixed-point TMS320C25 and the floating-point TMS320C30 digital signal processors, including the development tools available for each of these processors and DSP applications.

Fixed-Point TMS320C25-Based Development System

TMS320C25-based development systems are now available from a number of companies such as Hyperception Inc., Texas, and Atlanta Signal Processors, Inc., Georgia. The Software Development System (SWDS), available from TI includes a board containing the TMS320C25, which plugs into a slot on an IBM compatible PC. Within the SWDS environment, a program can be developed, assembled, and run. Debugging aids supported by the SWDS include single-stepping, setting of breakpoints, and display/modification of registers.

A typical workstation consists of:

1. An IBM compatible PC. A coprocessor is necessary to run many of the commercially available DSP packages (such as from Hyperception or Atlanta Signal Processors), which include a number of utilities and filter design techniques.
2. The SWDS package, which includes an assembler, a linker, a debug monitor, and a C compiler.
3. Input/output alternatives such as TI's analog interface board (AIB) or analog interface chip (AIC).

The AIB includes a 12-bit analog-to-digital converter (ADC) and a 12-bit digital-to-analog converter (DAC). A maximum sampling rate of 40 kHz can be obtained. With (input) antialiasing and (output) reconstruction filters mounted on a header on the AIB, different input/output (I/O) filter bandwidths can be achieved. Instructions such as **IN** and **OUT** can be used for input/output accesses. The AIC, which provides an inexpensive I/O alternative, includes 14-bit ADC and DAC, antialiasing/reconstruction filters, all on a single C-MOS chip. Two inputs and one output are available on the AIC. (A TMS320C25/AIC interface diagram and communication routines can be found in Chassaing and Horning, 1990.) The TLC32046 AIC is the newest member of the TLC32040 family of voiceband analog interface circuits, with a maximum sampling rate of 25 kHz.

Implementation of a Finite Impulse Response Filter with the TMS320C25

The convolution equation

$$y(n) = \sum_{k=0}^{N-1} h(k)x(n-k)$$

$$= h(0)x(n) + h(1)x(n-1) + \ldots + h(N-2)x(n-(N-2))$$

$$+ h(N-1)x(n-(N-1))$$

(17.1)

Table 17.1 TMS320C25 Memory Organization for Convolution

Coefficients	Time n	Input Samples Time $n+1$	Time $n+2$
PC $\rightarrow$ $h(N-1)$	$x(n)$	$x(n+1)$	$x(n+2)$
$h(N-2)$	$x(n-1)$	$x(n)$	$x(n+1)$
.	.	.	.
.	.	.	.
.	.	.	.
$h(2)$	$x(n-(N-3))$	$x(n-(N-4))$	$x(n-(N-5))$
$h(1)$	$x(n-(N-2))$	$x(n-(N-3))$	$x(n-(N-4))$
$h(0)$	AR1 $\rightarrow$ $x(n-(N-1))$	$x(n-(N-2))$	$x(n-(N-3))$

represents a finite impulse response (FIR) filter with length N. The memory organization for the coefficients $h(k)$ and the input samples $x(n-k)$ is shown in Table 17.1. The coefficients are placed within a specified internal program memory space and the input samples within a specified data memory space. The program counter (PC) initially points at the memory location that contains the last coefficient $h(N-1)$, for example at memory address FF00h (in hex). One of the (8) auxiliary registers points at the memory address of the last or least recent input sample. The most recent sample is represented by $x(n)$. The following program segment implements (17.1):

```
LARP      AR1
RPTK      N-1
MACD      FF00h,*-
APAC
```

The first instruction selects auxiliary register AR1, which will be used for indirect addressing. The second instruction RPTK causes the subsequent MACD instruction to execute N times (repeated $N-1$ times). The MACD instruction has the following functions:

1. Multiplies the coefficient value $h(N-1)$ by the input sample value $x(n-(N-1))$.
2. Accumulates any previous product stored in a special register (TR).
3. Copies the data memory sample value into the location of the next-higher memory. This "data move" is to model the input sample delays associated with the next unit of time $n+1$.

The last instruction APAC accumulates the last multiply operation $h(0)x(n)$.

At time $n+1$, the convolution Eq. (17.1) becomes

$$y(n+1) = h(0)x(n+1) + h(1)x(n) + \ldots$$
$$+ h(N-2)x(n-(N-3)) + h(N-1)x(n-(N-2)) \quad (17.2)$$

The previous program segment can be placed within a loop, with the PC and the auxiliary register AR1 reinitialized (see the memory organization of the samples $x(k)$ associated with time $n+1$ in Table 17.1). Note that the last multiply operation is $h(0)x(.)$, where $x(.)$ represents the newest sample. This process can be continuously repeated for time $n+2$, $n+3$, and so on.

The characteristics of a frequency selective FIR filter are specified by a set of coefficients that can be readily obtained using commercially available filter design packages. These coefficients can be placed within a generic FIR program. Within 5–10 minutes, an FIR filter can be implemented in real time. This includes finding the coefficients; assembling, linking and downloading the FIR program into the SWDS; and observing the desired frequency response displayed on a spectrum analyzer. A different FIR filter can be quickly obtained since the only necessary change in the generic program is to substitute a new set of coefficients.

The approach for modeling the sample delays involves moving the data. A different scheme is used with the floating-point TMS320C30 processor with a circular mode of addressing.

Floating-Point TMS320C30-Based Development System

TMS320C30-based DSP development systems are also currently available from a number of companies. The following are available from Texas Instruments:

1. An evaluation module (EVM). The EVM is a powerful, yet relatively inexpensive 8-bit card that plugs into a slot on an IBM AT compatible. It includes the third-generation TMS320C30, 16 K of user RAM, and an AIC for I/O. A serial port connector available on the EVM can be used to interface the TMS320C30 to other input/output devices (the TMS320C30 has two serial ports). An additional AIC can be interfaced to the TMS320C30 through this serial port connector. A very powerful, yet inexpensive, analog evaluation fixture, available from Burr-Brown (Arizona), can also be readily interfaced to the serial port on the EVM. This complete two-channel analog evaluation fixture includes an 18-bit DSP102 ADC, an 18-bit DSP202 DAC, antialiasing and reconstruction filters. The ADC has a maximum sampling rate of 200 kHz.

2. An XDS1000 emulator—powerful but quite expensive. A module can be readily built as a target system to interface to the XDS1000 [Chassaing, 1992]. This module contains the TMS320C30, 16 K of static RAM. Two connectors are included on this module, for interfacing to either an AIC module or to a second-generation analog interface board (AIB). The AIC was discussed in conjunction with the TMS320C25. The AIB includes Burr-Brown's 16-bit ADC and DAC with a maximum sampling rate of 58 kHz. An AIC is also included on this newer AIB version.

EVM Tools

The EVM package includes an assembler, a linker, a simulator, a C compiler, and a C source debugger. The second-generation TMS320C25 fixed-point processor is supported by C with some degrees of success. The architecture and instruction set of the third-generation TMS320C30 processor facilitate the development of high-level language compilers. An optimizer option is available with the C compiler for the TMS320C30. A C-code program can be readily compiled, assembled, linked, and downloaded into either a simulator or the EVM for real-time processing. A run-time support library of C functions, included with the EVM package, can be used during linking. During simulation, the input data can be retrieved from a file and the output data written into a file. Input and output port addresses can be appropriately specified. Within a real-time processing environment with the EVM, the C source debugger can be used. One can single-step through a C-code program while observing the equivalent step(s) through the assembly code. Both the C code and the corresponding assembly code can be viewed through the EVM windows. One can also monitor at the same time the contents of registers, memory locations, and so on.

Implementation of a Finite Impulse Response Filter with the TMS320C30

Consider again the convolution equation, Eq. (17.1), which represents an FIR filter. Table 17.2 shows the TMS320C30 memory organization used for the coefficients and the input samples. Initially, all the input samples can be set to zero. The newest sample $x(n)$, at time n, can be retrieved from an ADC using the following instructions:

```
FLOAT      *AR3,R3
STF        R3, *AR1++%
```

These two instructions cause an input value $x(n)$, retrieved from an input port address specified by auxiliary register AR3, to be loaded into a register R3 (one of eight 40-bit-wide extended precision registers), then stored in a memory location pointed by AR1 (AR1 would be first initialized to

Table 17.2 TMS320C30 Memory Organization for Convolution

Coefficients	Time n	Time $n+1$	Time $n+2$
AR0 $\rightarrow$ $h(N-1)$	AR1 $\rightarrow$ $x(n-(N-1))$	$x(n+1)$	$x(n+1)$
$h(N-2)$	$x(n-(N-2))$	AR1 $\rightarrow$ $x(n-(N-2))$	$x(n+2)$
$h(N-3)$	$x(n-(N-3))$	$x(n-(N-3))$	AR1 $\rightarrow$ $x(n-(N-3))$
.	.	.	.
.	.	.	.
.	.	.	.
$h(1)$	$x(n-1)$	$x(n-1)$	$x(n-1)$
$h(0)$	$x(n)$	$x(n)$	$x(n)$

point at the "bottom" or higher-memory address of the table for the input samples). AR1 is then postincremented in a circular fashion, designated with the modulo operator %, to point at the oldest sample $x(n-(N-1))$, as shown in Table 17.2. The size of the circular buffer must first be specified. The following program segment implements (17.1):

```
         RPTS     LENGTH-1
         MPYF     *AR0++%,*AR1++%,R0
    ||   ADDF     R0,R2,R2
         ADDF     R0,R2
```

The repeat "single" instruction RPTS causes the next (multiply) floating-point instruction MPYF to be executed LENGTH times (repeated LENGTH-1), where LENGTH is the length of the FIR filter. Furthermore, since the first ADDF addition instruction is in parallel (designated by ||) with the MPYF instruction, it is also executed LENGTH times. From Table 17.2, AR0, one of the eight available auxiliary registers, initially points at the memory address (a table address) which contains the coefficient $h(N-1)$, and a second auxiliary register AR1 now points to the address of the oldest input sample $x(n-(N-1))$. The second indirect addressing mode instruction multiplies the content in memory (address pointed by AR0) $h(N-1)$ by the content in memory (address pointed by AR1) $x(n-N-1))$, with the result stored in R0. Concurrently (in parallel), the content of R0 is added to the content of R2, with the result stored in R2. Initially R0 and R2 are set to zero; hence, the resulting value in R2 is *not* the product of the first multiply operation. After the first multiply operation, both AR0 and AR1 are incremented, and $h(N-2)$ is multiplied by $x(n-(N-2))$. Concurrently, the result of the first multiply operation (stored in R0) is accumulated into R2. The second addition instruction, executed only once, accumulates the last product $h(0)x(n)$ (similar to the APAC instruction associated with the fixed-point TMS320C25). The overall result yields an output value $y(n)$ at time n. After the last multiply operation, both AR0 and AR1 are postincremented to point at the "top" or lower-memory address of each circular buffer. The process can then be repeated for time $n+1$ in order to obtain a second output value $y(n+1)$. Note that the newest sample $x(n+1)$ would be retrieved from an ADC using the FLOAT and STF instructions, then placed at the top memory location of the buffer (table) containing the samples, overwriting the initial value $x(n-(N-1))$. AR1 is then incremented to point at the address containing

Table 17.3 Execution Time and Program Size of FIR Filter

FIR (45 samples)	Execution Time (msec)	Size (words)
C with modulo	4.16	122
C without modulo	0.338	116
C-called assembly	0.1666	74
Assembly	0.1652	27

Table 17.4 Execution Time and Program Size of 6th-Order IIR Filter

IIR (345 samples)	Execution Time (msec)	Size (words)
C	1.575	109
Assembly	1.18	29

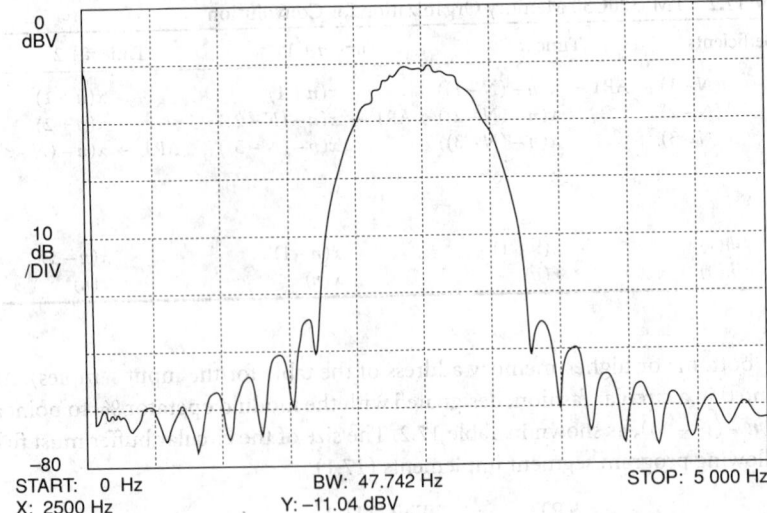

FIGURE 17.8   Frequency response of 41-coefficient FIR filter.

$x(n - (N - 2))$, and the previous four instructions can be repeated. The last multiply operation involves $h(0)$ and $x(.)$, where $x(.)$ is the newest sample $x(n + 1)$, at time $n + 1$. The foregoing procedure would be repeated to produce an output $y(n + 2)$, $y(n + 3)$, and so on. Each output value would be converted to a fixed-point equivalent value before being sent to a DAC. The frequency response of an FIR filter with 41 coefficients and a center frequency of 2.5 kHz, obtained from a signal analyzer, is displayed in Fig. 17.8.

FIR and IIR Implementation Using C and Assembly Code

A real-time implementation of a 45-coefficient bandpass FIR filter and a sixth-order IIR filter with 345 samples, using C code and TMS320C30 code, is discussed in Chassaing and Bitler [1991]. Tables 17.3 and 17.4 show a comparison of execution times of those two filters. The C language FIR filter, implemented without the modulo operator %, and compiled with a C compiler V4.1, executed two times slower[1] than an equivalent assembly language filter (which has a similar execution time as one implemented with a filter routine in assembly, called by a C program). The C language IIR filter ran 1.3 times slower than the corresponding assembly language IIR filter. These slower execution times may be acceptable for many applications. Where execution speed is crucial, a time-critical function may be written in assembly and called from a C program. In applications where speed is not absolutely crucial, C provides a better environment because of its portability and maintainability.

Real-Time Applications

A number of applications are discussed in Chassaing and Horning (1990) using TMS320C25 code and in Chassaing (1992) using TMS320C30 and C code. These applications include multirate and adaptive filtering, modulation techniques, and graphic and parametric equalizers. Two applications are briefly discussed here: a ten-band multirate filter and a video line rate analysis.

   1. The functional block diagram of the multirate filter is shown in Fig. 17.9. The multirate design provides a significant reduction in processing time and data storage, compared to an equivalent single-rate design. With multirate filtering, we can use a decimation operation in

---

[1]1.5 times slower using the new C compiler V4.4.

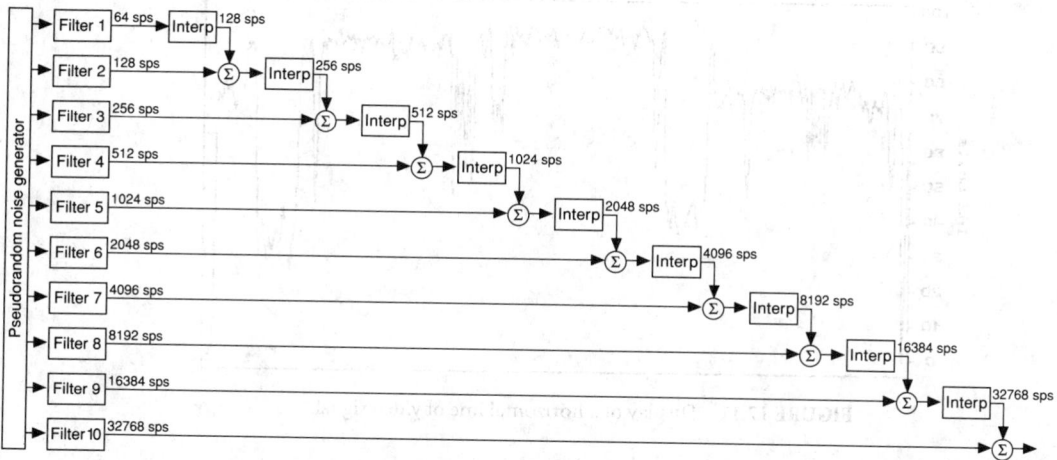

FIGURE 17.9 Multirate filter functional block diagram.

order to obtain a sample rate reduction or an interpolation operation (as shown in Fig. 17.9) in order to obtain a sample rate increase [Crochiere and Rabiner, 1983]. A pseudorandom noise generator implemented in software provides the input noise to the ten octave band filters. Each octave band filter consists of three 1/3-octave filters (each with 41 coefficients), which can be individually controlled. A controlled noise source can be obtained with this design. Since each 1/3-octave band filter can be turned *on* or *off*, the noise spectrum can be shaped accordingly. The interpolation filter is a low-pass FIR filter with a 2:1 data-rate increase, yielding two sample outputs for each input sample. The sample rate of the highest octave-band filter is set at 32,768 samples per second, with each successively lower band processing at half the rate of the next-higher band. The multirate filter (a nine-band version) was implemented with the TMS320C25 [Chassaing *et al.*, 1990]. Figure 17.10 shows the three 1/3-octave band filters of band 10 implemented with the EVM in conjunction with the two-channel analog fixture (made by Burr-Brown). The center frequency of the middle 1/3-octave band 10 filter is at approximately 8 kHz since the coefficients were designed for a cen-

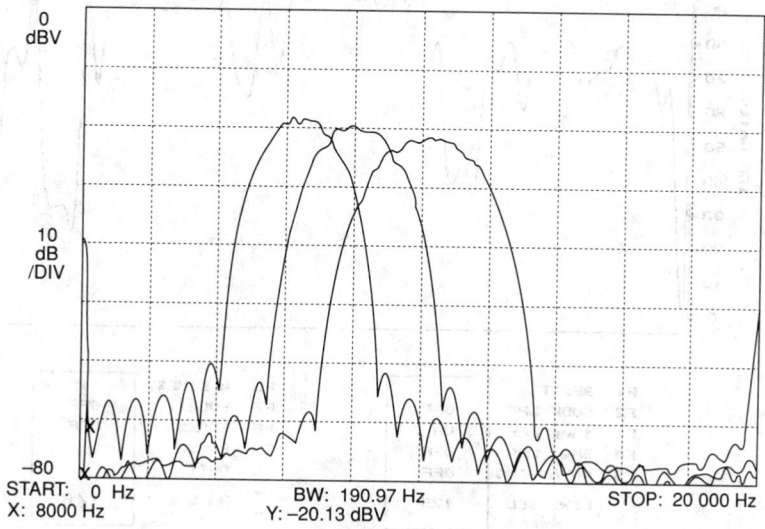

FIGURE 17.10 Frequency responses of the 1/3-octave band ten filters.

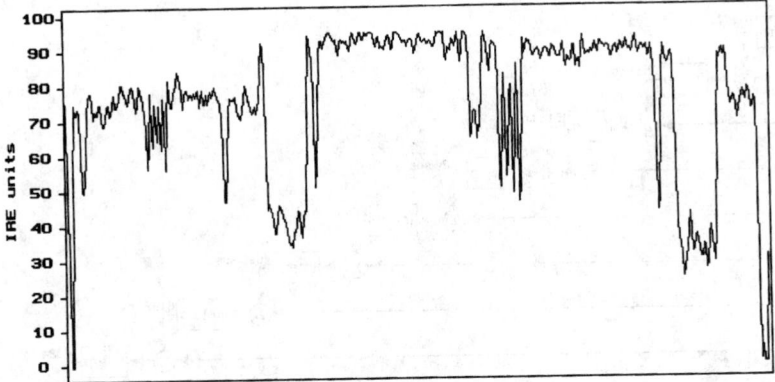

FIGURE 17.11   Display of a horizontal line of video signal.

ter frequency of 1/4 the sampling rate (the middle 1/3-octave band 9 filter would be centered at 4 kHz, the middle 1/3-octave band 8 filter at 2 kHz, and so on). Note that the center frequency of the middle 1/3-octave band 1 filter would be at 2 Hz if the highest sampling rate is set at 4 kHz. Observe from Fig. 17.10 that the crossover frequencies occur at the 3-dB points. Since the main processing time of the multirate filter (implemented in assembly code) was measured to be 8.8 ms, the maximum sampling rate was limited to 58 ksps.

2. A video line rate analysis implemented entirely in C code is discussed in Chassaing and Bitler [1992]. A module was built to sample a video line of information. This module included a 9.8-MHz clock, a high sampling rate 8-bit ADC and appropriate support circuitry (comparator, FIFO buffer, etc.). Interactive features allowed for the selection of one (out of 256) horizontal lines of information and the execution of algorithms for digital filtering, averaging, and edge enhancement, with the resulting effects displayed on the PC screen. Figure 17.11 shows the display of a horizontal line (line #125) of information obtained from a test chart with a charge coupled device (CCD) camera. The function key **F3** selects the 1-MHz low-pass filter resulting in the display shown in Fig. 17.12. The 3-MHz filter (with **F4**) would

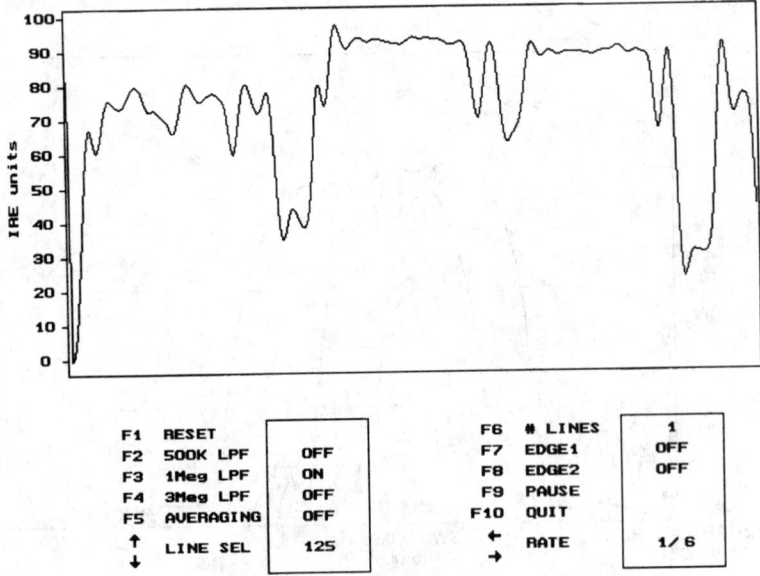

F1	RESET		F6	# LINES	1
F2	500K LPF	OFF	F7	EDGE1	OFF
F3	1Meg LPF	ON	F8	EDGE2	OFF
F4	3Meg LPF	OFF	F9	PAUSE	
F5	AVERAGING	OFF	F10	QUIT	
↑ ↓	LINE SEL	125	← →	RATE	1/6

FIGURE 17.12   Video line signal with 1-MHz filtering.

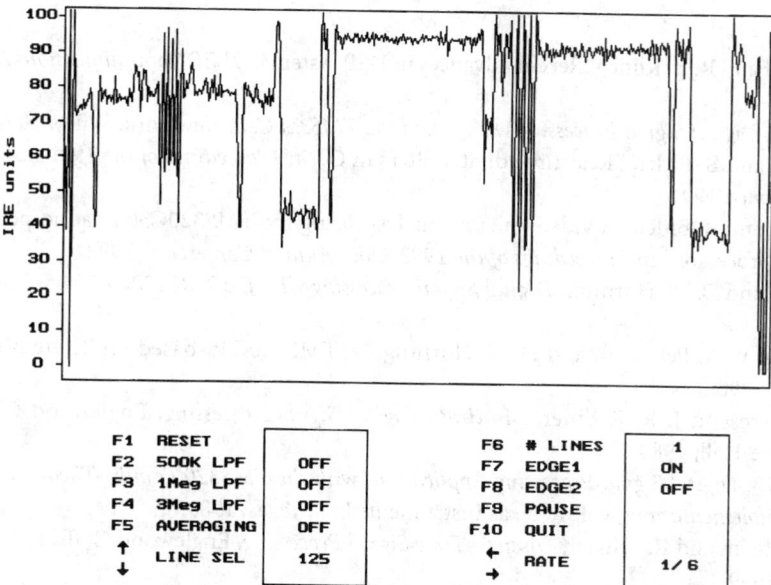

FIGURE 17.13   Video line signal with edge enhancement.

pass more of the higher-frequency components of the signal but with less noise reduction. **F5** implements the noise averaging algorithm. The effect of the edge enhancement algorithm (with **F7**) is displayed in Fig. 17.13.

Conclusions and Future Directions

DSP processors have beed used extensively in a number of applications, even in non-DSP applications such as graphics. Recently, TI introduced the fourth-generation floating-point TMS320C40, code compatible with the TMS320C30, with features such as an instruction cycle time of 40 ns and six serial ports. The recently announced fifth-generation fixed-point TMS320C50, code compatible with the first two generations of fixed-point processors, features an instruction cycle time of 35 ns and 10 K words (16-bit) of on-chip data and program memory. Currently, both the fixed-point and floating-point processors are being supported.

efining Terms

C compiler: Program that translates C code into assembly code.

Digital signal processor: Special-purpose microprocessor with an architecture suitable for fast execution of signal processing algorithms.

Fixed-point processor: A processor capable of operating on scaled integer and fractional data values.

Floating-point processor: Processor capable of operating on integers as well as on fractional data values without scaling.

On-chip memory: Internal memory available on the digital signal processor.

Pipelining feature: Feature that permits parallel operations of fetching, decoding, reading, and executing.

Special-purpose digital signal processor: Digital signal processor with special feature for handling a specific signal processing application, such as FFT.

References

H. M. Ahmed and R. B. Kline, "Recent advances in DSP systems," *IEEE Communications Magazine*, 1991.

R. Chassaing, *Digital Signal Processing with C and the TMS320C30*, New York: Wiley, 1992.

R. Chassaing and B. Bitler, "Real-time digital filters in C," in *Proceedings of the 1991 ASEE Annual Conference*, 1991.

R. Chassaing and B. Bitler, "A video line rate analysis using the TMS320C30 floating-point digital signal processor," in *Proceedings of the 1992 ASEE Annual Conference*, 1992.

R. Chassaing and D. W. Horning, *Digital Signal Processing with the TMS320C25*, New York: Wiley, 1990.

R. Chassaing, W. A. Peterson, and D. W. Horning, "A TMS320C25-based multirate filter," *IEEE Micro*, 1990.

R. E. Crochiere and L. R. Rabiner, *Multirate Digital Signal Processing*, Englewood Cliffs, N.J.: Prentice-Hall, 1983.

K. S. Lin (ed.), *Digital Signal Processing Applications with the TMS320 Family. Theory, Algorithms, and Implementations*, vol. 1, Texas Instruments Inc., Texas, 1989.

A. V. Oppenheim and R. Schafer, *Discrete-Time Signal Processing*, Englewood Cliffs, N.J.: Prentice-Hall, 1989.

P. Papamichalis (ed.), *Digital Signal Processing Applications with the TMS320 Family. Theory, Algorithms, and Implementations*, vol. 3, Texas Instruments, Inc., Texas, 1990.

Further Information

Rulph Chassaing teaches a three-day hands-on workshop on digital signal processing using C and the TMS320C30, offered at Roger Williams University in Bristol, RI, 02809. Workshops on the TMS320 family of digital signal processors are offered by Texas Instruments, Inc. at various locations.

Special issues on digital signal processing were published by the *IEEE Micro* in December 1988 and October 1990.

18

Acoustic Signal Processing

Juergen Schroeter
Acoustics Research Dept.,
AT&T Bell Laboratories

Sanjay K. Mehta
NUWC Detachment

G. Clifford Carter
NUWC Detachment

18.1 Digital Signal Processing in Audio and Electroacoustics 395
Steerable Microphone Arrays • Digital Hearing Aids • Spatial Processing • Audio Coding • Echo Cancellation • Active Noise and Sound Control

18.2 Underwater Acoustical Signal Processing 406
What Is Underwater Acoustical Signal Processing? • Technical Overview • Underwater Propagation • Processing Functions • Advanced Signal Processing • Application

18.1 Digital Signal Processing in Audio and Electroacoustics

Juergen Schroeter

In this section we will focus on advances in algorithms and technologies in digital signal processing (DSP) that have already had or, most likely, will soon have, a major impact on **audio** and **electroacoustics** (A&E). Because A&E embraces a wide range of topics, it is impossible for us to go here into any depth in any one of them. Instead, this section will try to give a compressed overview of the topics the author judges to be most important.

In the following, we will look into steerable microphone arrays, digital hearing aids, spatial processing, audio coding, echo cancellation, and active noise and sound control. We will *not* cover basic techniques in digital recording [Pohlmann, 1989] and computer music [Moore, 1990].

Steerable Microphone Arrays

Steerable microphone arrays have controllable directional characteristics. One important application is in teleconferencing. Here, sound pickup can be highly degraded by reverberation and room noise. One solution to this problem is to utilize highly directional microphones. Instead of pointing such a microphone manually to a desired talker, steerable microphone arrays can be used for reliable automatic tracking of speakers as they move around in a noisy room or auditorium, if combined with a suitable speech detection algorithm.

Figure 18.1 depicts the simplest kind of steerable array using N microphones that are uniformly spaced with distance d along the linear x-axis. It can be shown that the response of this system to a plane wave impinging at an angle θ is:

$$H(j\omega) = \sum_{n=0}^{N-1} a_n e^{-j(\omega/c)nd\cos\theta} \tag{18.1}$$

395

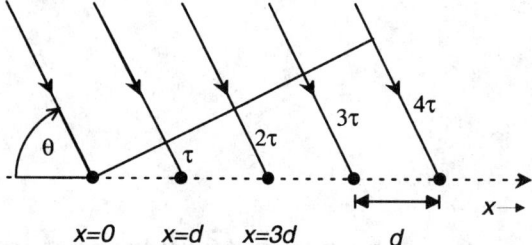

FIGURE 18.1 A linear array of N microphones (here, $N = 5$; $\tau = d/c \cos\theta$).

Here, $j = \sqrt{-1}$, ω is the radian frequency, and c is the speed of sound. Equation (18.1) is a spatial filter with coefficients a_n and the delay operator $z^{-1} = \exp(-jd\omega/c \cos\theta)$. Therefore, we can apply finite impulse response (FIR) filter theory. For example, we could taper the weights a_n to suppress sidelobes of the array. We also have to guard against spatial aliasing, that is, grating lobes that make the directional characteristic of the array ambiguous. The array is steered to an angle θ_0 by introducing appropriate delays into the N microphone lines. In Eq. (18.1), we can incorporate these delays by letting a_n

$$a_n = e^{-j\omega\tau_0} \; e^{+j(\omega/c)nd \cos\theta_0} \tag{18.2}$$

Here τ_0 is an overall delay equal to or larger than $Nd/c \cos\theta_0$ that ensures causality, while the second term in Eq. (18.2) cancels the corresponding term in Eq. (18.1) at $\theta = \theta_0$. Due to the axial symmetry of the one-dimensional (linear, 1-D) array, the directivity of the array is a figure of revolution around the x-axis. Therefore, in case we want the array to point to a *single* direction in space, we need a 2-D array.

Since most of the energy of typical room noise and the highest level of reverberation in a room is at low frequencies, one would like to use arrays that have their highest directivity (i.e., narrowest beamwidth) at low frequencies. Unfortunately, this need collides with the physics of arrays: the smaller the array relative to the wavelength, the wider the beam. (Again, the corresponding notion in filter theory is that systems with shorter impulse responses have wider bandwidth.) One solution to this problem is to superimpose different-size arrays and filter each output by the appropriate bandpass filter, similar to a crossover network used in two- or three-way loudspeaker designs. Such a superposition of three five-element arrays is shown in Fig. 18.2. Note that we only need nine microphones in this example, instead of $5 \times 3 = 15$.

Another interesting application is the use of an array to mitigate discrete noise sources in a room. For this, we need to attach an FIR filter to each of the microphone signal outputs. For any given frequency, one can show that N microphones can produce $N - 1$ nulls in the directional characteristic of the array. Similarly, attaching an M-point FIR filter to each of the microphones, we can get these zeros at $M - 1$ frequencies. The weights for these filters have to be adapted, usually under the constraint that the transfer function (frequency characteristic) of the array for the *desired* source is

FIGURE 18.2 Three superimposed linear arrays depicted by large, midsize, and small circles. The largest array covers the low frequencies, the midsize array covers the midrange frequencies, and the smallest covers the high frequencies.

optimally flat. In practical tests, systems of this kind work nicely in (almost) anechoic environments. Their performance degrades, however, with increasing reverberation.

More information on microphone arrays can be found in Flanagan *et al.* [1991]; in particular, they describe how to make arrays adapt to changing talker positions in a room by constantly scanning the room with a moving search beam and by switching the main beam accordingly. Current research issues are, among others, 3-D arrays and how to take advantage of low-order wall reflections.

Digital Hearing Aids

Commonly used hearing aids attempt to compensate for sensorineural (cochlear) hearing loss by delivering an amplifed acoustic signal to the external ear canal. As will be pointed out below, the most important problem is how to find the best aid for a given patient.

Historically, technology has been the limiting factor in hearing aids. Early on, carbon hearing aids provided a limited gain and a narrow, peaky frequency response. Nowadays, hearing aids have a broader bandwidth and a flatter frequency response. Consequently, more people can benefit from the improved technology. With the advent of digital technology, the promise is that even more people would be able to do so. Unfortunately, as will be pointed out below, we have not fulfilled this promise yet.

We distinguish between *analog, digitally controlled analog,* and *digital* hearing aids. Analog hearing aids contain only (low-power) pre-amp, filter(s), (optional) automatic gain control (AGC) or compressor, power amp, and output limiter. Digitally controlled aids have certain additional components: one kind adds a digital controller to monitor and adjust the analog components of the aid. Another kind contains switched-capacitor circuits that represent sampled signals in analog form, in effect allowing simple discrete-time processing (e.g., filtering). Aids with switched-capacitor circuits have a lower power consumption compared to digital aids. Digital aids—none are yet commercially available—contain A/D and D/A converters and at least one programmable digital signal processing (DSP) chip, allowing for the use of sophisticated DSP algorithms, (small) microphone arrays, speech enhancement in noise, etc. Experts disagree, however, as to the usefulness of these techniques. To date, the most successful approach seems to be to ensure that all parts of the signal get amplified so that they are clearly audible but not too loud and to "let the brain sort out signal and noise."

Hearing aids pose a tremendous challenge for the DSP engineer, as well as for the audiologist and acoustician. Due to the continuing progress in chip technology, the physical size of a digital aid should no longer be a serious problem in the near future; however, power consumption will still be a problem for quite some time. Besides the obvious necessity of avoiding howling (acoustic feedback), for example, by employing sophisticated models of the electroacoustic transducers, acoustic leaks, and ear canal to control the aid accordingly, there is a much more fundamental problem: since DSP allows complex schemes of splitting, filtering, compressing, and (re-) combining the signal, hearing aid performance is no longer limited by bottlenecks in technology. It is still limited, however, by the lack of basic knowledge about how to map an arbitrary input signal (i.e., speech from a desired speaker) onto the reduced capabilities of the auditory system of the targeted wearer of the aid. Hence, the *selection and fitting of an appropriate aid* becomes the most important issue. This serious problem is illustrated in Fig. 18.3.

It is important to note that for speech presented at a constant level, a linear (no compression) hearing aid can be tuned to do as well as a hearing aid with compression. However, if parameters like signal and background noise levels change dynamically, compression aids, in particular those with two bands or more, should have an advantage.

While a patient usually has no problem telling whether setting A or B is "clearer," adjusting more than just 2–3 (usually interdependent) parameters is very time consuming. For a multiparameter aid, an efficient fitting procedure that maximizes a certain objective is needed. Possible objectives

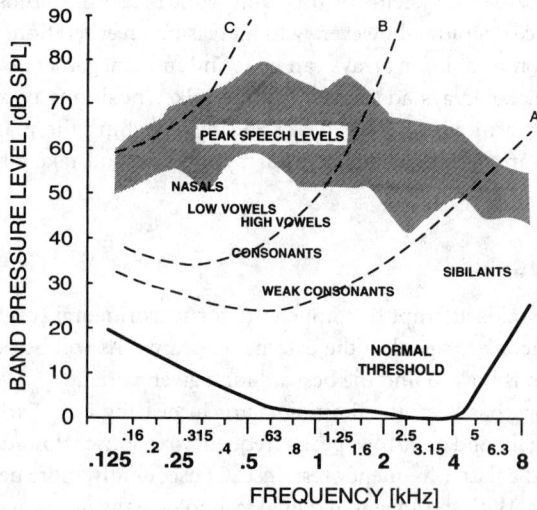

FIGURE 18.3   Peak third-octave band levels of normal to loud speech (hatched) and typical levels/dominant frequencies of speech sounds (identifiers). Both can be compared to the third-octave threshold of normal-hearing people (solid line), thresholds for a mildly hearing-impaired person (A), for a severely hearing-impaired person (B), and for a profoundly hearing-impaired person (C). For example, for person (A), sibilants and some weak consonants in a normal conversation cannot be perceived. (*Source:* H. Levitt, "Speech discrimination ability in the hearing impaired: spectrum considerations," in *The Vanderbilt Hearing-Aid Report: State of the Art-Research Needs*, G. A. Studebaker and F. H. Bess (Eds.), Monographs in Contemporary Audiology, Upper Darby, Pa., 1982, p. 34. With permission.)

are, for example, **intelligibility maximization** or **loudness restoration**. The latter objective is assumed in the following.

It is known that an impaired ear has a reduced dynamic range. Therefore, the procedure for fitting a patient with a hearing aid could estimate the so-called loudness-growth function (LGF) that relates the sound pressure level of a specific (band-limited) sound to its loudness. An efficient way of measuring the LGF is described by Allen *et al.* [1990]. Once the LGF of an impaired ear is known, a multiband hearing aid can implement the necessary compression for each band [Villchur, 1973]. Note, however, that this assumes that interactions between the bands can be neglected (problem of summation of partial loudnesses). This might not be valid for aids with a large number of bands. Other open questions include the choice of widths and filter shape of the bands, and optimization of dynamic aspects of the compression (e.g., time constants). For aids with just two bands, the crossover frequency is a crucial parameter that is difficult to optimize.

Spatial Processing

In spatial processing, audio signals are modified to give them new spatial attributes, such as, for example, the perception of having been recorded in a specific concert hall. The auditory system—using only the two ears as inputs—is capable of perceiving the direction and distance of a sound source with a high degree of accuracy, by exploiting **binaural** and **monaural** spectral cues. Wave propagation in the ear canal is essentially one-dimensional. Hence, the 3-D spatial information is coded by sound diffraction into spectral information before the sound enters the ear canal. The sound diffraction is caused by the head/torso (on the order of 20-dB and 600-µs **interaural** level difference and delay, respectively) and at the two pinnae (auriculae); see, for example, Shaw [1980]. Binaural techniques like the one discussed below can be used for evaluating room and concert-hall acoustics (optionally in reduced-scale model rooms using a miniature dummy head), for noise assessment (e.g., in cars), and for "Kunstkopfstereophonie" (dummy-head stereophony). In addition, there are techniques for *loudspeaker reproduction* (like "Q-Sound") that try to extend the range in horizontal angle of traditional stereo speakers by using interaural cross cancellation. Largely an open question is how to reproduce spatial information for large audiences, for example, in movie theaters.

Figure 18.4 illustrates the technique for filtering a single-channel source using measured head-related transfer functions, in effect, creating a virtual sound source in a given direction of the listener's auditory space (assuming plane waves, i.e., infinite source distance). On the left in this figure, the measurement of head-related transfer functions is shown. Focusing on the *left* ear for a moment (subscript l), we need to estimate the so-called free-field transfer function (subscript ff) for given angles of incidence in the horizontal plane (azimuth φ) and vertical plane (elevation δ):

$$H_{\mathrm{ff},l}(j\omega, \varphi, \delta) = P_{\mathrm{probe},l}(j\omega, \varphi, \delta)/P_{\mathrm{ref}}(j\omega) \tag{18.3}$$

where $P_{\mathrm{probe},l}$ is the Fourier transform of the sound pressure measured in the subject's left ear, and P_{ref} is the Fourier transform of the pressure measured at a suitable reference point in the free field without the subject being present (e.g., at the midpoint between the two ears). (Note that P_{ref} is independent of the direction of sound incidence since we assume an anechoic environment.) The middle of Fig. 18.4 depicts the convolution of any "dry" (e.g., mono, low reverberation) source with the stored $H_{\mathrm{ff},l}(j\omega, \varphi, \delta)$s and corresponding $H_{\mathrm{ff},r}(j\omega, \varphi, \delta)$s. On the right side in the figure, the resulting binaural signals are reproduced via equalized headphones. The equalization ensures that a sound source with a flat spectrum (e.g., white noise) does not suffer any perceivable coloration for any direction (φ, δ).

Implemented in a real-time "binaural mixing console," the above scheme can be used to create "virtual" sound sources. When combined with an appropriate scheme for interpolating head-related transfer functions, moving sound sources can be mimicked realistically. Furthermore, it is possible to superimpose early reflections of a hypothetical recording room, each filterered by the appropriate head-related transfer function. Such inclusion of a room in the simulation makes the

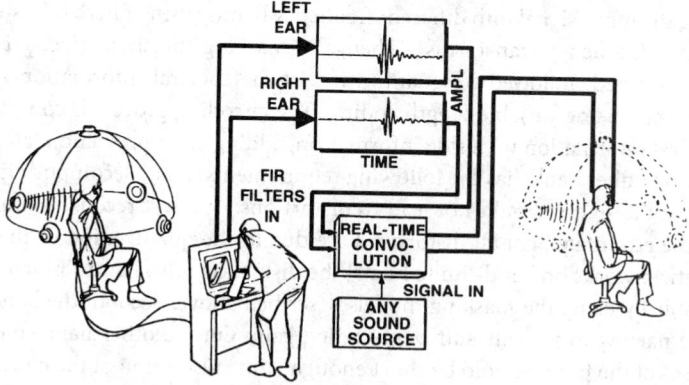

FIGURE 18.4 Measuring and using transfer functions of the external ear for binaural mixing (FIR = finite impulse response). (*Source:* E. M. Wenzel, Localization in virtual acoustic displays, *Presence,* vol. 1, p. 91, 1992. With permission.)

spatial reproduction more robust against individual differences between "recording" and "listening" ears, in particular, if the listener's head movements are fed back to the binaural mixing console. (Head movements are useful for disambiguating spatial cues.) Finally, such a system can be used to create "virtual acoustic displays," for example, for pilots and astronauts [Wenzel, 1992]. Other research issues are, for example, the required accuracy of the head-related transfer functions, intersubject variability, and psychoacoustic aspects of room simulations.

Audio Coding

Audio coding is concerned with compressing (reducing the bit rate) of audio signals. The uncompressed digital audio of compact disks (CDs) is recorded at a rate of 705.6 kbit/s for each of the two channels of a stereo signal (i.e., 16 bit/sample, 44.1-kHz sampling rate; 1411.2 kbit/s total). This is too high a bit rate for digital audio broadcasting (DAB) or for transmission via end-to-end digital telephone connections (integrated services digital network, ISDN). Current audio coding algorithms provide at least "better than FM" quality at a combined rate of 128 kbit/s for the two stereo channels (2 ISDN B channels!), "transparent coding" at rates of 96 to 128 kbit/s per mono channel, and "studio quality" at rates between 128 and 196 kbit/s per mono channel. (While a large number of people will be able to detect distortions in the first class of coders, even so-called "golden ears" should not be able to detect any differences between original and coded versions of known "critical" test signals; the highest quality category adds a safety margin for editing, filtering, and/or recoding.)

To compress audio signals by a factor as large as eleven while maintaining a quality exceeding that of a local FM radio station requires sophisticated algorithms for reducing the **irrelevance and redundancy** in a given signal. A large portion (but usually less than 50%) of the bit-rate reduction in an audio coder is due to the first of the two mechanisms. Eliminating irrelevant portions of an input signal is done with the help of psychoacoustic models. It is obvious that a coder can eliminate portions of the input signal that—when played back—will be below the threshold of hearing. More complicated is the case when we have multiple signal components that tend to cover each other, that is, when weaker components cannot be heard due to the presence of stronger components. This effect is called *masking*. To let a coder take advantage of masking effects, we need to use good masking models. Masking can be modeled in the time domain where we distinguish so-called simultaneous masking (masker and maskee occur at the *same* time), forward masking (masker occurs *before* maskee), and backward masking (masker occurs *after* maskee). Simultaneous masking usually is modeled in the frequency domain. This latter case is illustrated in Fig. 18.5.

Audio coders that employ common frequency-domain models of masking start out by splitting and subsampling the input signal into different frequency bands (using filterbanks such as subband filterbanks or time-frequency transforms). Then, the masking threshold (i.e., *predicted* masked threshold) is determined, followed by quantization of the spectral information and (optional) noiseless compression using variable-length coding. The encoding process is completed by multiplexing the spectral information with side information, adding error protection, etc.

The first stage, the filter bank, has the following requirements. First, decomposing and then simply reconstructing the signal should not lead to distortions ("perfect reconstruction filterbank"). This results in the advantage that all distortions are due to the quantization of the spectral data. Since each quantizer works on band-limited data, the distortion (also band-limited due to refiltering) is controllable by using the masking models described above. Second, the bandwidths of the filters should be narrow to provide sufficient coding gain. On the other hand, the length of the impulse responses of the filters should be short enough (time resolution of the coder!) to avoid so-called pre-echoes, that is, backward spreading of distortion components that result from sudden onsets (e.g., castanets). These two contradictory requirements, obviously, have to be worked out by a compromise. "**Critical band**" filters have the shortest impulse responses needed for coding of transient signals. On the other hand, the optimum frequency resolution (i.e., the one resulting in

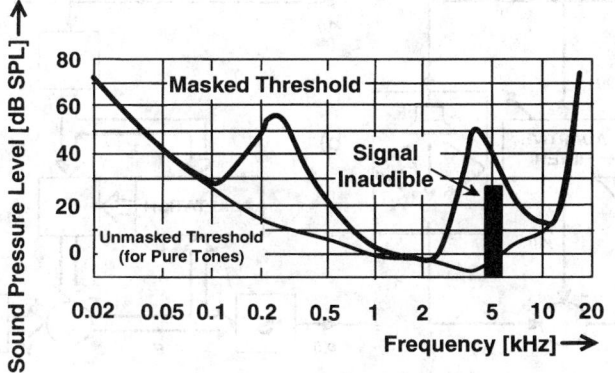

FIGURE 18.5 Masked threshold in the frequency domain for a hypothetical input signal. In the vicinity of high-level spectral components, signal components below the current masked threshold cannot be heard.

the highest coding gain) for a typical signal can be achieved by using, for example, a 2048-point modified discrete cosine transform (MDCT).

In the second stage, the (time-varying) masking threshold as determined by the psychoacoustic model usually controls an iterative analysis-by-synthesis quantization and coding loop. It can incorporate rules for masking of tones by noise and of noise by tones, though little is known in the psychoacoustic literature for more general signals. Quantizer step sizes can be set and bits can be allocated according to the known spectral estimate, by block companding with transmission of the scale factors as side information or iteratively in a variable-length coding loop (Huffman coding). In the latter case, one can low-pass filter the signal if the total required bit rate is too high.

The decoder has to invert the processing steps of the encoder, that is, do the error correction, perform Huffman decoding, and reconstruct the filter signals or the inverse-transformed time-domain signal. Since the decoder is significantly less complex than the encoder, it is usually implemented on a single DSP chip, while the encoder uses several DSP chips.

Current research topics encompass tonality measures and time-frequency representations of signals. More information can be found in Johnston and Brandenburg [1991].

Echo Cancellation

Echo cancellers were first deployed in the U.S. telephone network in 1979. Today, they are virtually ubiquitous in long-distance telephone circuits where they cancel so-called line echoes (i.e., *electrical* echoes) resulting from nonperfect hybrids (the devices that couple local two-wire to long-distance four-wire circuits). In satellite circuits, echoes bouncing back from the far end of a telephone connection with a round-trip delay of about 600 ms are very annoying and disruptive. *Acoustic* echo cancellation—where the echo path is characterized by the transfer function $H(z)$ between a loudspeaker and a microphone in a room (e.g., in a speakerphone)—is crucial for teleconferencing where two or more parties are connected via full-duplex links. Here, echo cancellation can also alleviate acoustic feedback ("howling").

The principle of acoustic echo cancellation is depicted in Fig. 18.6(a). The echo path $H(z)$ is cancelled by modeling $H(z)$ by an adaptive filter and subtracting the filter's output $\hat{y}(t)$ from the microphone signal $y(t)$. The adaptability of the filter is necessary since $H(z)$ changes appreciably with movement of people or objects in the room and because periodic measurements of the room would be impractical. *Acoustic* echo cancellation is more challenging than cancelling line echoes for several reasons. First, room impulse responses $h(t)$ are longer than 200 ms compared to less than

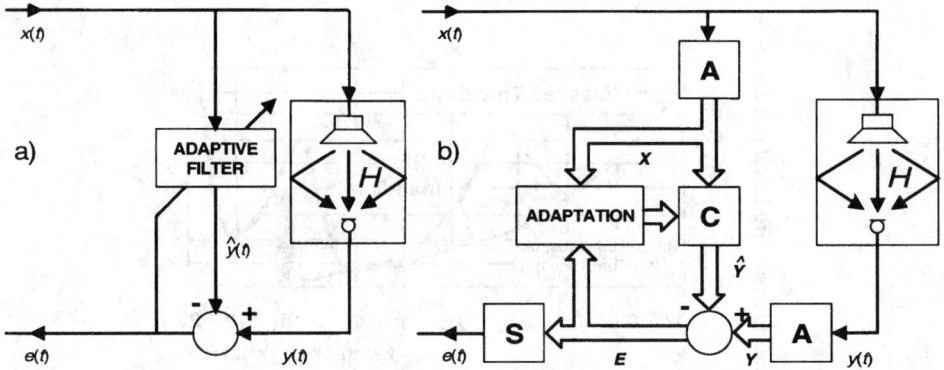

FIGURE 18.6 (a) Principle of using an echo canceller in teleconferencing. (b) Realization of the echo canceller in subbands. (After M. M. Sondhi and W. Kellermann, "Adaptive echo cancellation for speech signals," in *Advances in Speech Signal Processing*, S. Furui and M. M. Sondhi, Eds., New York: Marcel Dekker, 1991. By courtesy of Marcel Dekker, Inc.)

20 ms for line echo cancellers. Second, the echo path of a room $h(t)$ is likely to change constantly (note that even small changes in temperature can cause significant changes of h). Third, teleconferencing eventually will demand larger audio bandwidths (e.g., 7 kHz) compared to standard telephone connections (about 3.2 kHz). Finally, we note that echo cancellation in a stereo setup (*two* microphones and *two* loudspeakers at each end) is an even harder problem on which very little work has been done so far.

It is obvious that the initially unknown echo path $H(z)$ has to be "learned" by the canceller. It is also clear that for adaptation to work there needs to be a nonzero input signal $x(t)$ that excites all the eigenmodes of the system (resonances, or "peaks" of the system magnitude response $|H(j\omega)|$). Another important problem is how to handle double-talk (speakers at both ends are talking simultaneously). In such a case, the canceller could easily get confused by the speech from the near end that acts as an uncorrelated noise in the adaptation. Finally, the convergence rate, that is, how fast the canceller adapts to a change in the echo path, is an important measure to compare different algorithms.

Adaptive filter theory suggests several algorithms for use in echo cancellation. The most popular one is the so-called *least-mean square* (LMS) algorithm that models the echo path by an FIR filter with an impulse response $\hat{h}(t)$. Using vector notation **h** for the true echo path impulse response, $\hat{\mathbf{h}}$ for its estimate, and **x** for the excitation time signal, an estimate of the echo is obtained by $\hat{y}(t) = \hat{\mathbf{h}}'\mathbf{x}$, where the prime denotes vector transpose. A reasonable objective for a canceller is to minimize the instantaneous squared error $e^2(t)$, where $e(t) = y(t) - \hat{y}(t)$. The time derivative of $\hat{\mathbf{h}}$ can be set to

$$\frac{d\hat{\mathbf{h}}}{dt} = -\mu \nabla_{\hat{\mathbf{h}}} e^2(t) = -2\mu e(t) \nabla_{\hat{\mathbf{h}}} e(t) = 2\mu e(t)\mathbf{x} \qquad (18.4)$$

resulting in the simple update equation $\hat{h}_{k+1} = \hat{h}_k + \alpha e_k x_k$, where α (or μ) control the rate of change. In practice, whenever the far-end signal $x(t)$ is low in power, it is a good idea to freeze the canceller by setting $\alpha = 0$. Sophisticated logic is needed to detect double talk. When it occurs, then also set $\alpha = 0$. It can be shown that the spread of the eigenvalues of the autocorrelation matrix of $x(t)$ determines the convergence rate, where the slowest-converging eigenmode corresponds to the smallest eigenvalue. Since the eigenvalues themselves scale with the power of the predominant spectral components in $x(t)$, setting $\alpha = 2\mu/(\mathbf{x}'\mathbf{x})$ will make the convergence rate independent of the far-end power. This is the *normalized* LMS method. Even then, however, all eigenmodes will converge at the same rate only if $x(t)$ is white noise. Therefore, pre-whitening the far-end signal will help in speeding up convergence.

The LMS method is an iterative approach to echo cancellation. An example of a noniterative, block-oriented approach is the *least-squares* (LS) algorithm. Solving a system of equations to get $\hat{\mathbf{h}}$, however, is computationally more costly. This cost can be reduced considerably by running the LS method on a sample-by-sample basis and by taking advantage of the fact that the new signal vectors are the old vectors with the oldest sample dropped and one new sample added. This is the *recursive least-squares* (RLS) algorithm. It also has the advantage of normalizing $\mathbf{x}$ by multiplying it with the inverse of its autocorrelation matrix. This, in effect, equalizes the adaptation rate of all eigenmodes.

Another interesting approach is outlined in Fig. 18.6(b). As in subband coding (discussed earlier), splitting the signals x and y into subbands with analysis filterbanks $\mathbf{A}$, doing the cancellation in bands, and resynthesizing the outgoing ("error") signal e through a synthesis filterbank $\mathbf{S}$ also reduces the eigenvalue spread of each bandpass signal compared to the eigenvalue spread of the fullband signal. This is true for the eigenvalues that correspond to the "center" (i.e., unattenuated) portions of each band. It turns out, however, that the slowly converging "transition-band" eigenmodes get attenuated significantly by the synthesis filter $\mathbf{S}$. The main advantage of the subband approach is the reduction in computational complexity due to the down-sampling of the filterbank signals. The drawback of the subband approach, however, is the introduction of the combined delay of $\mathbf{A}$ and $\mathbf{S}$. Eliminating the analysis filterbank on $y(t)$ and moving the synthesis filterbank into the adaptation branch $\hat{\mathbf{Y}}$ will remove this delay with the result that the canceller will not be able to model the earliest portions of the echo-path impulse response $h(t)$. To alleviate this problem, we could add in parallel a fullband echo canceller with a short filter. Further information and an extensive bibliography can be found in Haensler [1992].

Active Noise and Sound Control

Active noise control (ANC) is a way to reduce the sound pressure level of a given noise source through electroacoustic means. ANC and echo cancellation are somewhat related. While even *acoustic* echo cancellation is actually done on electrical signals, ANC could be labeled "wave cancellation," since it involves using one or more secondary acoustic or vibrational sources. Another important difference is the fact that in ANC one usually would like to cancel a given noise in a whole *region* in space, while echo cancellation commonly involves only one microphone picking up the echo signal at a single *point* in space. Finally, the transfer function of the transducer used to generate a cancellation ("secondary source") signal needs to be considered in ANC.

Active sound control (ASC) can be viewed as an offspring of ANC. In ASC, instead of trying to cancel a given sound field, one tries to control specific spatial and temporal characteristics of the sound field. One application is in adaptive sound reproduction systems. Here, ASC aims at solving the large-audience spatial reproduction problem mentioned in the spatial processing section of this chapter.

Two important principles of ANC are depicted in Fig. 18.7. In the upper half [Fig. 18.7(a) and (b)], a feedback loop is formed between the controller $G(s)$ and the transfer function $C(s)$ of the secondary source, and the acoustic path to the error microphone. Control theory suggests that $E/Y = 1/[1 + C(s)G(s)]$, where $E(s)$ and $Y(s)$ are Laplace transforms of $e(t)$ and $y(t)$, respectively. Obviously, if we could make C a real constant and $G \to \infty$, we would get a "zone of quiet" around the error microphone. Unfortunately, in practice, $C(s)$ will introduce at least a delay, thus causing stability problems for too large a magnitude $|G|$ at high enough frequencies. The system can be kept stable, for example, by including a low-pass filter in G and by positioning the secondary source in close vicinity to the error microphone. A highly successful application of the feedback control in ANC is in *active* hearing protective devices (HPDs) and high-quality headsets and "motional-feedback" loudspeakers. *Passive* HPDs offer little or no noise attenuation at low frequencies due to inherent physical limitations. Since the volume enclosed by earmuffs is rather small, HPDs can

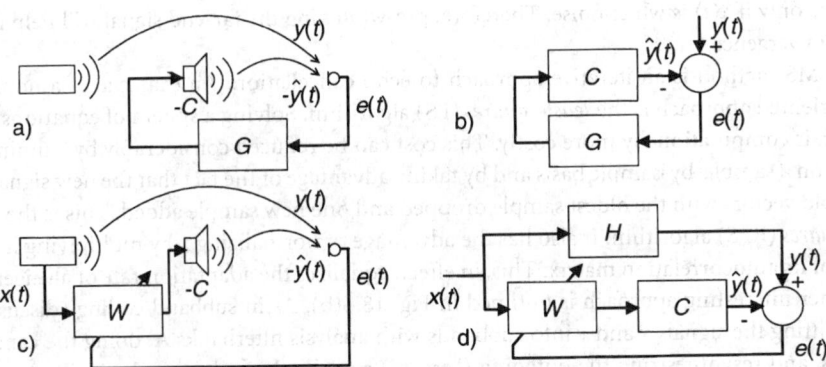

FIGURE 18.7 Two principles of active noise control. Feedback control system (a) and (b); feedforward control system (c) and (d). Physical block diagrams (a) and (c), and equivalent electrical forms (b) and (d). (After P. A. Nelson and S. J. Elliott, *Active Control of Sound*, London: Academic Press, 1992. With permission.)

benefit from the increase in low-frequency attenuation brought about by feedback-control ANC. Finally, note that the same circuit can be used for high-quality reproduction of a communications signal $s(t)$ fed into a headset by subtracting $s(t)$ electrically from $e(t)$. The resulting transfer function is $E/S = C(s)G(s)/[1 + C(s)G(s)]$ assuming $Y(s) = 0$. Thus, a high loop gain $|G(s)|$ will ensure both, a high noise attenuation at low frequencies and a faithful bass reproduction of the communications signal.

The principle of the feedforward control method in ANC is outlined in the lower half of Fig. 18.6(c) and (d). The obvious difference to the feedback control method is that the separate reference signal $x(t)$ is used. Here, cancellation is achieved for the filter transfer function $W = H(s)/C(s)$ which is most often implemented by an adaptive filter. The fact that $x(t)$ reaches the ANC system earlier than $e(t)$ allows for a causal filter, needed in broadband systems. However, a potential problem with this method is the possibility of feedback of the secondary source signal $\hat{y}(t)$ into the path of the reference signal $x(t)$. This is obviously the case when $x(t)$ is picked up by a microphone in a duct just upstream of the secondary source C. An elegant solution for ANC in a duct without explicit feedback cancellation is to use a recursive filter W.

Single error signal/single secondary source systems cannot achieve *global* cancellation or sound control in a room. An intuitive argument for this fact is that one needs *at least* as many secondary sources and error microphones as there are orthogonal wave modes in the room. Since the number of wave modes in a room below a given frequency is approximately proportional to the third power of this frequency, it is clear that ANC (and ASC) is practical only at low frequencies. In practice, using small (point-source) transducers, it turns out that one should use more error microphones than secondary sources. Examples of such multidimensional ANC systems are employed for cancelling the lowest few harmonics of the engine noise in an airplane cabin and in a passenger car. In both of these cases, the adaptive filter matrix is controlled by a multiple-error version of the LMS algorithm. Further information can be found in Nelson and Elliott [1992].

Summary and Acknowledgment

In this section, we have touched upon several topics in audio and electroacoustics. The reader may be reminded that the author's choice of these topics was biased by his background in communication acoustics (and by his lack of knowledge in music). Furthermore, ongoing efforts in integrating different communication modalities into systems for teleconferencing [see, e.g., Flanagan *et al.*, 1990] had a profound effect in focusing this contribution. Experts in topics covered in this contribution, like Jont Allen, David Berkley, Gary Elko, Joe Hall, Jim Johnston, Mead Killion, Harry Levitt, Dennis Morgan, and—last, but not least—Mohan Sondhi, are gratefully acknowledged for their patience and help.

Defining Terms

Audio: Science of processing signals that are within the frequency range of hearing, that is, roughly between 20 Hz and 20 kHz. Also name for this kind of signal.

Critical bands: Broadly used to refer to psychoacoustic phenomena of limited frequency resolution in the cochlea. More specifically, the concept of critical bands evolved in experiments on the audibility of a tone in noise of varying bandwidth, centered around the frequency of the tone. Increasing the noise bandwidth beyond a certain critical value has little effect on the audibility of the tone.

Electroacoustics: Science of interfacing between acoustical waves and corresponding electrical signals. This includes the engineering of transducers (e.g., loudspeakers and microphones), but also parts of the psychology of hearing, following the notion that it is not necessary to present to the ear signal components that cannot be perceived.

Intelligibility maximization and loudness restoration: Two different objectives in fitting hearing aids. Maximizing intelligibility involves conducting laborious intelligibility tests. Loudness restoration involves measuring the mapping between a given sound level and its perceived loudness. Here, we assume that recreating the loudness a normal hearing person would perceive is close to maximizing the intelligibility of speech.

Irrelevance and redundancy: In audio coding, irrelevant portions of an audio signal can be removed without perceptual effect. Once removed, however, they cannot be regenerated in the decoder. Contrary to this, redundant portions of a signal that have been removed in the encoder can be regenerated in the decoder. The "lacking" irrelevant parts of an original signal constitute the major cause for a (misleadingly) low signal-to-noise ratio (SNR) of the decoded signal while its subjective quality can still be high.

Monaural/interaural/binaural: *Monaural* attributes of ear input signals (e.g., timbre, loudness) require, in principle, only one ear to be detected. *Interaural* attributes of ear input signals (e.g., localization in the horizontal plane) depend on differences between, or ratios of measures of, the two ear input signals (e.g., delay and level differences). Psychoacoustic effects (e.g., cocktail-party effect) that depend on the fact that we have two ears are termed *binaural.*

References

J. B. Allen, J. L. Hall, and P. S. Jeng, "Loudness growth in 1/2-octave bands (LGOB) — A procedure for the assessment of loudness," *J. Acoust. Soc. Am.,* vol. 88, no. 2, pp. 745–753, 1990.

J. L. Flanagan, D. A. Berkley, and K. L. Shipley, "Integrated information modalities for human/machine communication: HuMaNet, an experimental system for conferencing," *J. of Visual Communication and Image Representation,* vol. 1, no. 2, pp. 113–126, 1990.

J. L. Flanagan, D. A. Berkley, G. W. Elko, J. E. West, and M. M. Sondhi, "Autodirective microphone systems," *Acustica,* vol. 73, pp. 58–71, 1991.

E. Haensler, "The hands-free telephone problem—An annotated bibliography," *Signal Processing,* vol. 27, pp. 259–271, 1992.

J. D. Johnston and K. Brandenburg, "Wideband coding—Perceptual considerations for speech and music," in *Advances in Speech Signal Processing,* S. Furui and M. M. Sondhi, Eds., New York: Marcel Dekker, 1991.

F. R. Moore, *Elements of Computer Music,* Englewood Cliffs, N.J.: Prentice-Hall, 1990.

P. A. Nelson and S. J. Elliott, *Active Control of Sound,* London: Academic Press, 1992.

K. C. Pohlmann, *Principles of Digital Audio*, 2nd ed., Carmel, Ind.: SAMS/Macmillan Computer Publishing, 1989.

E. A. G. Shaw, "The acoustics of the external ear," in *Acoustical Factors Affecting Hearing Aid Performance*, G. A. Studebaker and I. Hochberg, Eds., Baltimore, Md.: University Park Press, 1980.

E. Villchur, "Signal processing to improve speech intelligibility in perceptive deafness," *J. Acoust. Soc. Am.*, vol. 53, no. 6, pp. 1646–1657, 1973.

E. M. Wenzel, "Localization in virtual acoustic displays," *Presence*, vol. 1, pp. 80–107, 1992.

Further Information

A highly informative article that is complementary to this contribution is the one by P. J. Bloom, "High-quality digital audio in the entertainment industry: An overview of achievements and challenges," *IEEE-ASSP Magazine*, Oct. 1985. An excellent introduction to the fundamentals of audio, including music synthesis and digital recording, is contained in the 1992 book *Music Speech Audio*, by W. J. Strong and G. R. Plitnik, available from Soundprint, 2250 North 800 East, Provo, UT 84604 (ISBN 0-9611938-2-4). *Oversampling Delta-Sigma Data Converters* is a 1992 collection of papers edited by J. C. Candy and G. C. Temes. It is available from IEEE Press (IEEE order number PC0274-1). Specific issues of the *Journal of Rehabilitation Research and Development* (ISSN 007-506X), published by the Veterans Administration, are a good source of information on hearing aids, in particular the Fall 1987 issue. *Spatial Hearing* is the title of a 1982 book by J. Blauert, available from MIT Press (ISBN 0-262-02190-0). Anyone interested in *Psychoacoustics* should look into the 1990 book of this title by E. Zwicker and H. Fastl, available from Springer-Verlag (ISBN 0-387-52600-5).

The Institute of Electrical and Electronics Engineers (IEEE) *Transactions on Speech and Audio Processing* is keeping up-to-date on algorithms in audio. Every two to three years, a workshop on applications of signal processing to audio and electroacoustics covers the latest advances in areas introduced in this article. IEEE can be reached at 445 Hoes Lane, Piscataway, NJ 08855-1331, ph. (908) 981-0060. *The Journal of the Audio Engineering Society* (AES) is another useful source of information on audio. The AES can be reached at 60 East 42nd St., Suite 2520, New York, NY 10165-0075, ph. (212) 661-8528. *The Journal of the Acoustical Society of America* (ASA) contains information on physical, psychological, and physiological acoustics, as well as on acoustic signal processing, among other things. ASA's "Auditory Demonstrations" CD contains examples of signals demonstrating hearing-related phenomena ranging from "critical bands" over "pitch" to "binaural beats." ASA can be reached at 500 Sunnyside Blvd., Woodbury, NY 11797-2999, ph. (516) 576-2360.

18.2 Underwater Acoustical Signal Processing

Sanjay K. Mehta and G. Clifford Carter

What Is Underwater Acoustical Signal Processing?

The use of acoustical signals that have propagated through water to detect, classify, and localize underwater objects is referred to as underwater acoustical signal processing.

Why Exploit Sound for Underwater Applications?

It has been found that acoustic energy propagates better under water than other types of energy. For example, both light and radio waves (used for satellite or above-ground communications) are attenuated to a far greater degree under water than are sound waves. For this reason sound waves have generally been used to extract information about underwater objects. A typical underwater acoustical signal processing scenario is shown in Fig. 18.8.

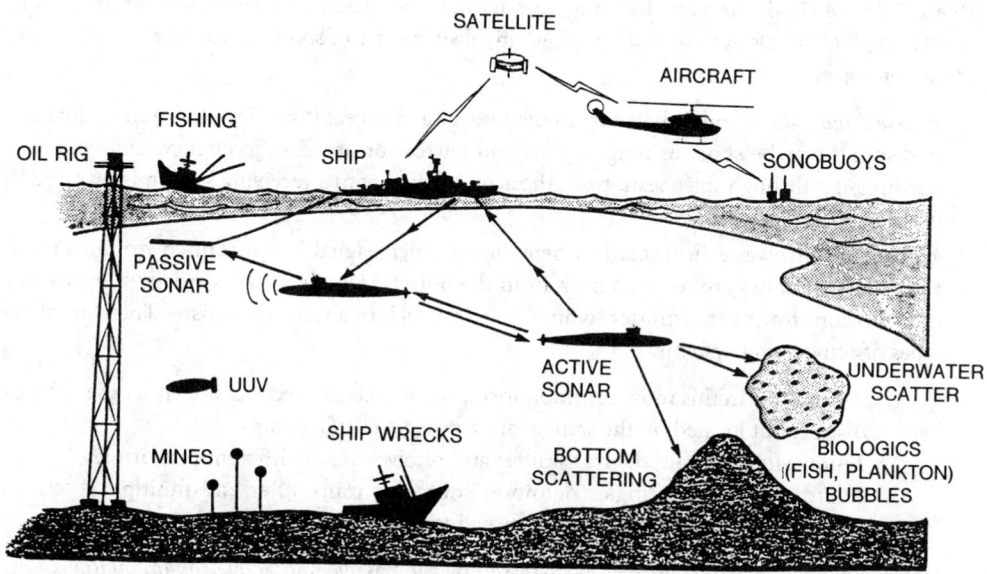

FIGURE 18.8 Active and passive underwater acoustical signal processing.

Technical Overview

In underwater acoustics, a number of units are used: distances of nautical miles (1852 m), yards (0.9144 m) and kiloyards; speeds of knots (nautical mile/h); depths of fathoms (6 ft or 1.8288 m); and bearing of degrees (0.1745 rad). However, in the past two decades there has been a conscious effort to be totally metric, i.e., to use MKS or Standard International units.

Underwater acoustic signals to be processed for detection, classification, and localization can be characterized from a statistical point of view. When time averages of each waveform are the same as the ensemble average of waveforms, the signals are ergodic. When the statistics do not change with time, the signals are said to be stationary. The spatial equivalent to stationary is homogeneous. For many introductory problems, only stationary signals and homogeneous noise are considered; more complex problems involve nonstationary, inhomogeneous environments.

Acoustic waveforms of interest have a probability density function (PDF); for example, the PDF may be Gaussian or in the case of clicking, sharp noise spikes, or crackling ice noise, the PDF may be non-Gaussian. In addition to being characterized by a PDF, signals can be characterized in the frequency domain by their power spectral density functions, which are Fourier transforms of the autocorrelation functions. White signals, which are uncorrelated from sample to sample, have a delta function autocorrelation or a flat (constant) power spectral density. Ocean signals in general are much more colorful and not limited to being stationary.

Passive sonar signals are primarily modeled as random signals. Their first-order PDFs are typically Gaussian; one exception is a stable sinusoidal signal that is non-Gaussian and has a power spectral density function that is a Dirac delta function in the frequency domain. However, in the ocean environment, an arbitrarily narrow frequency width is never observed, and signals have some finite narrow bandwidth. Indeed, the full spectrum of most underwater signals is quite "colorful."

Received active sonar signals can be viewed as consisting of the results of a deterministic component (known transmit waveform) convolved with the medium and reflector transfer functions and a random (noise) component. Moreover, the **Doppler** imparted (frequency shift) to the reflected signal makes the total system effect nonlinear, thereby complicating analysis and processing of these signals.

SONAR

SONAR, "SOund NAvigation and Ranging," the acronym adopted in the 1940s, similar to the popular RADAR, "RAdio Detection And Ranging," involves the use of sound to explore the ocean and underwater objects

- *Passive sonar* uses sound radiated from the underwater object itself. The duration of the radiated sound may be short or long in time and narrow or broad in frequency. Only one-way transmission through the ocean, from the acoustic source to a receiving sensor, is involved in this case.
- *Active sonar* involves echo-ranging where an acoustical signal is transmitted from a source, and reflected echoes are received back from the object. Here one is concerned with two-way transmissions from a transmitter to an object and back to a receiving sensor. There are three types of active sonar systems:

  1. *Monostatic:* In this most common form, the source and receiver are either identical or distinct but located on the same platform (e.g., a surface ship).
  2. *Bistatic:* In this form, the transmitter and receiver are on different platforms.
  3. *Multistatic:* Here, a single (or more) source or transmitter and multiple receivers, which can be located on different receiving platforms or ships, are used.

The performance of sonar systems can be assessed by the passive and active *sonar equations*. The major parameters in the sonar equation, measured in **decibels (dB)**, are as follows:

L_S = source level

L_N = noise level

N_{DI} = directivity index

N_{TS} = echo level or target strength

N_{RD} = recognition differential

Here, L_S is the target-radiated signal strength (for passive) or transmitted signal strength (for active), and L_N is the total background noise level. N_{DI}, or *DI*, is the directivity index, which is a measure of the capability of a receiving array to discriminate against unwanted noise. N_{TS} is the received echo level or target strength. Underwater objects with large values of N_{TS} are more easily detectable with active sonar than are those with small values of N_{TS}. In general, N_{TS} varies as a function of object size, aspect angle, (i.e., the direction at which impinging acoustic energy reaches the underwater object), and reflection angle (the direction at which the impinging acoustic energy is reflected off the underwater object). N_{RD} is the recognition differential of the processing system.

The **figure of merit** (FOM), a basic performance measure involving parameters of the sonar system, ocean, and target, is computed for active and passive sonar systems (in dBs) as follows:

For passive sonar,

$$\text{FOM}_P = L_S - (L_N - N_{DI}) - N_{RD} \tag{18.5}$$

For active sonar,

$$\text{FOM}_A = (L_S + N_{TS}) - (L_N - N_{DI}) - N_{RD} \tag{18.6}$$

Sonar systems, for a given set of parameters of the sonar equations, are designed so that the FOM exceeds the acoustic propagation loss. The amount above the FOM is called the *signal excess*. When two sonar systems are compared, the one with the largest signal excess is said to hold the *acoustic advantage*. However, it should be noted that the set of parameters in the preceding FOM equations is not unique. Depending on the design or parameter measurability conditions, different parameters can be combined or expanded in terms of quantities such as frequency dependency of the sonar system in particular ocean conditions, speed and bearing of the receiving or transmitting

Table 18.1 Expressions for Sound Speed in Meters per Second

Expression	Limits
$c = 1492.9 + 3(T-10) - 6 \times 10^{-3}(T-10)^2$	$-2 \le T \le 24.5°$
$\quad 4 \times 10^{-2}(T-18)^2 + 1.2(S-35)$	$30 \le S \le 42$
$\quad -10^{-2}(T-18)(S-35) + D/61$	$0 \le D \le 1.000$
$c = 1449.2 + 4.6T - 5.5 \times 10^{-2}T^2$	$0 \le T \le 35°$
$\quad + 2.9 \times 10^{-4}T^3 + (1.34 - 10^{-2}T)(S-35)$	$0 \le S \le 45$
$\quad + 1.6 \times 10^{-2}D$	$0 \le D \le 1,000$
$c = 1448.96 + 4.591T - 5.304 \times 10^{-2}T^2$	$0 \le T \le 30°$
$\quad + 2.374 \times 10^{-4}T^3 + 1.340(S-35)$	$30 \le S \le 40$
$\quad + 1.630 \times 10^{-2}D + 1.675 \times 10^{-7}D^2$	$0 \le D \le 8,000$
$\quad - 1.025 \times 10^{-2}T(S-35) - 7.139 \times 10^{-13}TD^3$	

D = depth, in meters. S = salinity, in parts per thousand. T = temperature, in degrees Celsius.

Source: R. J. Urick, *Principles of Underwater Sound,* New York: McGraw-Hill, 1983, p. 113. With permission.

platforms, reverberation loss, and so forth. Furthermore, due to multipaths, differences in sonar system equipment and operation, and the constantly changing nature of the ocean medium, the FOM parameters fluctuate with time. Thus, the FOM is not an absolute measure of performance but rather an expected value of performance over time in a stochastic sense [for details, see Urick, 1983].

Underwater Propagation

Speed/Velocity of Sound

Sound speed, c, in the ocean, in general lies between 1450–1540 m/s and varies as a function of several physical parameters, such as temperature, salinity, and pressure (depth). Variations in sound speed can significantly affect the propagation (range or quality) of sound in the ocean. Table 18.1 gives approximate expressions for sound speed as a function of these physical parameters.

Sound Velocity Profiles

Sound rays that are normal (perpendicular) to the acoustic wavefront can be traced from the source to the receiver by a process called ray tracing.[1] In general, the acoustic ray paths are not straight, but bend in a manner analogous to optical rays focused by a lens. In underwater acoustics, the ray paths are determined by the **sound velocity profile** (SVP) or *sound speed profile* (SSP), that is, the speed of sound in water as a function of water depth. The sound speed not only varies with depth but also varies in different regions of the ocean and with time as well. In deep water, the SVP fluctuates the most in the upper ocean due to variations of temperature and weather. Just below the sea surface is the *surface layer* where the sound speed is greatly affected by temperature and wind action. Below this layer lies the *seasonal thermocline* where the temperature and speed decrease with depth, and the variations are seasonal. In the next layer, the *main thermocline*, the temperature and speed decrease with depth and surface conditions or seasons have little effect. Finally, there is the *deep isothermal layer* where the temperature is nearly constant at 39°F, and the sound velocity

[1]Ray tracing models are used for high-frequency signals and in deep water. Generally, if the depth-to-wavelength ratio is 100 or more, ray tracing models are accurate. Below that, corrections must be made to the ray trace models. In shallow water or low frequencies, i.e., when depth-to-wavelength is about 30 or less, "mode theory" models are used.

increases almost linearly with depth. A typical deep water sound velocity profile as a function of depth is shown in Fig. 18.9.

If the sound speed is a minimum at a certain depth below the surface, then this depth is called axis of the underwater sound channel.[2] The sound velocity increases both above and below this axis. When the sound wave travels through a medium with a sound speed gradient, the direction of travel of sound wave is bent towards the area of lower sound speed.

Although the definition of shallow water can be signal dependent, in terms of depth-to-wavelength ratio, water depth of less than 1000 meters is generally referred to as shallow water. In shallow water the SVP is irregular and difficult to predict because of large surface temperature and salinity variations, wind effects, and multiple reflections of sound from the ocean bottom.

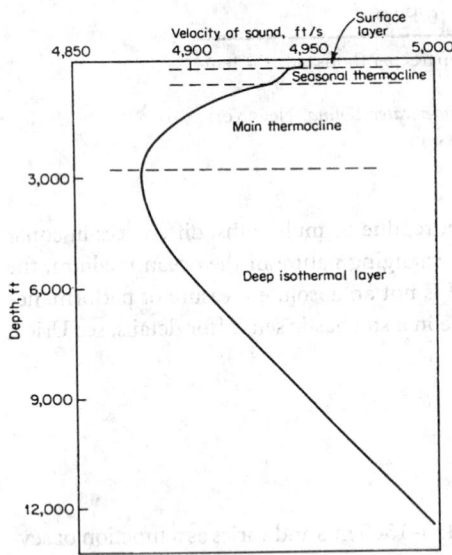

FIGURE 18.9 A typical sound velocity profile (SVP). (*Source: R. J. Urick, Principles of Underwater Sound,* New York, McGraw-Hill, 1983, p. 118. With permission.)

Propagation Modes

In general, there are three dominant propagation paths that depend on the distance or range between the acoustic source and the receiver (Fig. 18.10).

- *Direct Path:* Sound energy travels in (nominal) straight line path between the source and receiver, usually present at short ranges.
- *Bottom Bounce Path:* Sound energy is reflected from the ocean bottom (present at intermediate ranges).
- *Convergence Zone (CZ) Path:* Sound energy converges at longer ranges where multiple acoustic ray paths add or recombine coherently to reinforce the presence of acoustic energy from the radiating/reflecting source.

Figure 18.11 shows the propagation loss as a function of range for different frequencies of the signal. Note the recombination of energy at the convergence zones.

Multipaths

The ocean contains multiple acoustic paths that split the acoustic energy. When the receiving system can resolve these multiple paths (or multipaths), then they should be recombined by optimal signal processing to fully exploit the available acoustic energy for detection [Chan, 1989]. It is also theoretically possible to exploit the geometrical properties of multipaths present in the bottom bounce path by investigation of the apparent aperture created by the different path arrivals to localize the energy source. In the case of first-order bottom bounce transmission, i.e., only one bottom interaction, there are four paths (from source to receiver):

1. A bottom bounce ray path (B).
2. A surface interaction followed by a bottom interaction (SB).
3. A bottom bounce followed by a surface interaction (BS).
4. A path that first hits the surface, then the bottom, and finally the surface (SBS).

Typical first-order bottom bounce ocean propagation paths are depicted in Fig. 18.12.

---

[2]Often called the SOFAR (Sound Fixing and Ranging) channel.

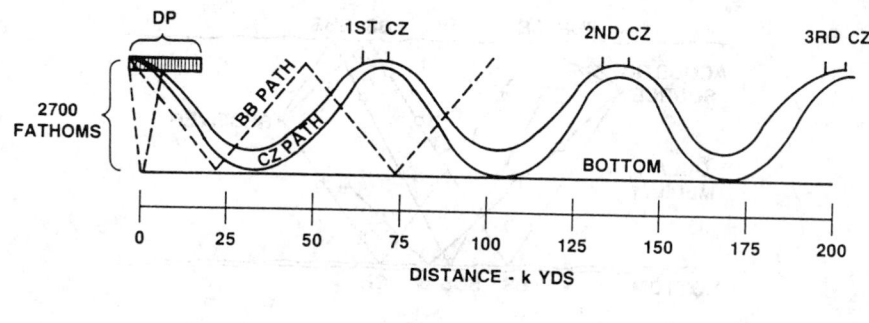

LEGEND

DP - DIRECT SOUND PATH
BB - BOTTOM BOUNCE SOUND PATH
CZ - CONVERGENCE ZONE SOUND PATH
AREA ASSUMED - MID NORTH ATLANTIC OCEAN

FIGURE 18.10 Typical sound paths between source and receiver.[3] (*Source: A.W. Cox, Sonar and Underwater Sound*, Lexington, Mass., Lexington Books, D.C. Health and Company, 1974, p. 25. With permission.)

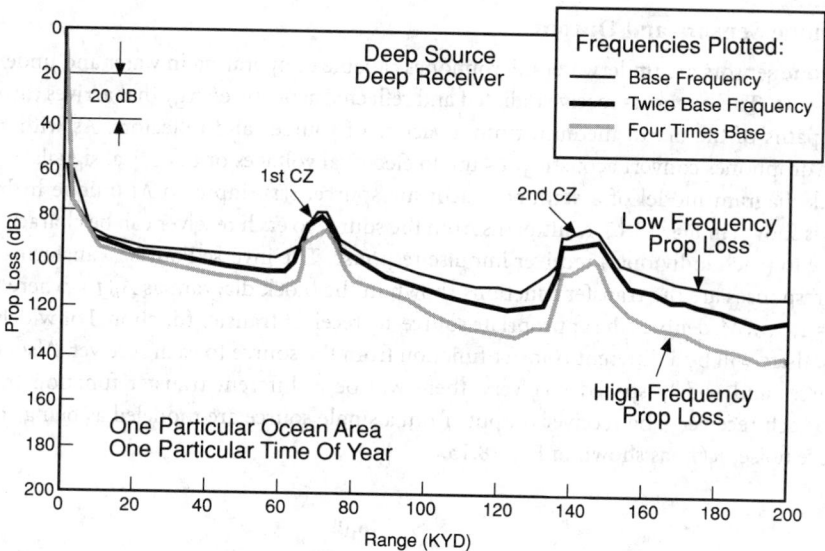

FIGURE 18.11 Propagation loss as a function of range.

Performance Limitations

In a typical reception of a signal wavefront, noise and interference can degrade the performance of a sonar system and limit the system's ability to detect signals in the underwater environment. The noise or interference could be sounds from a school of fish, shipping (surface or subsurface) noise, active transmission interference (e.g., jammers), or interference when multiple receivers or sonar systems are in operation simultaneously. Also, the ambient noise may have unusual vertical or horizontal directivity and in some environments, such as the Arctic, the noise due to ice motion may produce unfamiliar interference. Unwanted backscatters, similar to the headlights of a car driving in fog, can cause a signal-induced noise that degrades processing gain without proper processing.

[3]Fathom: unit of length or depth generally used for underwater measurements; 1 fathom = 6 ft.

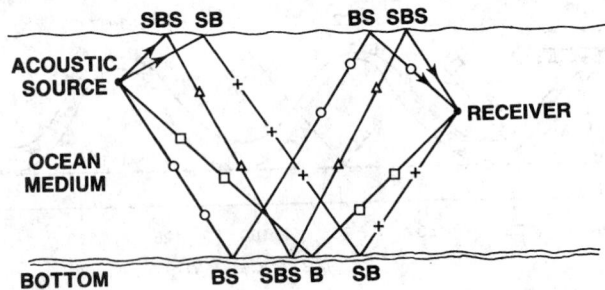

FIGURE 18.12 Multipaths for a first-order bottom bounce propagation model.

Some other performance-limiting factors are the loss of signal level and acoustic coherence due to boundary interaction as a function of grazing angle; the radiated pattern (signal level) of the object and its spatial coherence; the presence of surface, bottom, and volume **reverberation** (in active sonar); signal spreading owing to the modulating effect of surface motion; biologic noise as a function of time (both time of day and time of year); and statistics of the noise in the medium. (Does the noise arrive in the same or at different ray path angles as the signal?)

Hydrophone Sensors and Output

Hydrophone sensors are underwater microphones capable of operating in water and under hydrostatic pressure. These sensors receive radiated and reflected acoustic energy that arrives through the multiple paths of the ocean medium from a variety of sources and reflectors. As with a microphone, hydrophones convert acoustic pressure to electrical voltages or to optical signals.

A block diagram model of a stationary acoustic source, $s(t)$, input to M unique hydrophone receivers is shown in Fig. 18.13. Multipaths from the source to each receiver can be characterized by the source to (each individual) receiver impulse response. The inverse Fourier transforms of these impulse responses are the transfer functions shown in the block diagram as $A_j(f)$, where the subscript, $j = 1, \ldots, M$, denotes the appropriate source-to-receiver transfer function. For widely spaced receivers, there will be a different transfer function from the source to each receiver. Also, for multiple sources and widely spaced receivers, there will be a different transfer function from each source to each receiver. The receiver outputs from a single source are modeled as being corrupted by additive noise, $n_j(t)$, as shown in Fig. 18.13.

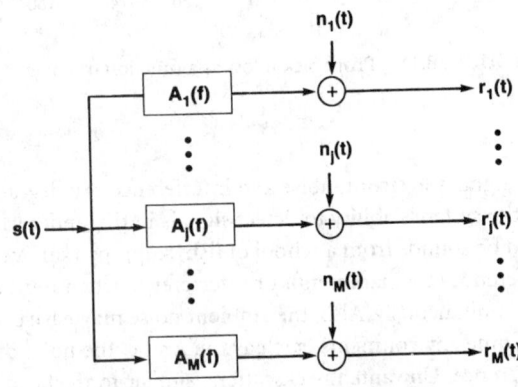

FIGURE 18.13 Hydrophone receiver model: source signal $s(t)$ through medium filter $A_j(t)$, corrupted by additive noise received at one of M hydrophones.

Processing Functions

Beamforming

Beamforming is a process in which outputs from the hydrophone sensors of an array are coherently combined by delaying and summing the outputs to provide enhanced detection and estimation. In underwater applications, one is trying to detect a directional (single direction) signal in the presence of normalized background noise that is ideally isotropic (nondirectional). By arranging the hydrophone (array) sensors in different physical geometries and electronically steering them in a particular direction, one can increase the signal-to-noise ratio (**SNR**) in a given direction by rejecting or canceling the noise in other directions. There are many different kinds of arrays (e.g., equally spaced line, continuous line, circular, cylindrical, spherical, or random sonobuoy arrays). The beam pattern specifies the response of these arrays to the variation in direction. In the simplest case, the increase in SNR due to the beamformer, called the *array gain* (in dB), is given by

$$AG = 10 \log \frac{SNR_{array\,(output)}}{SNR_{single\,sensor\,(input)}} \tag{18.7}$$

Detection

Detection of signals in the presence of noise, using classical Bayes or Neyman-Pearson decision criteria, is based on hypothesis testing. In the simplest binary hypothesis case, the detection problem is posed as two hypotheses:

- H_0: Signal is not present (referred to as the null hypothesis).
- H_1: Signal is present.

For a received wavefront, H_0 relates to the noise-only case and H_1 to the signal-plus-noise case. Complex hypotheses (M-hypotheses) can also be formed if detecting a signal among a variety of sources is required.

Probability is a measure, between zero and unity, of how likely an event is to occur. For a received wavefront the likelihood ratio, Λ, is the ratio of P_{H_1} (probability that hypothesis H_1 is true) to P_{H_0} (probability that hypothesis H_0 is true). A decision (detection) is made by comparing the likelihood, or logarithm of the likelihood ratio called the log-likelihood ratio, to a predetermined threshold η. That is, if $\Lambda = P_{H_1}/P_{H_0} > \eta$, a decision is made that the signal is present.

Probability of detection, P_D, measures the likelihood of detecting an event or object when the event does occur. *Probability of false alarm*, P_{fa}, is a measure of the likelihood of saying something happened when the event did NOT occur. **Receiver operating characteristics (ROC)** curves plot P_D versus P_{fa} for a particular (sonar signal) processing system. A single plot of P_D versus P_{fa} for one system must fix the SNR and processing time. The threshold η is varied to sweep out the ROC curve. The curve is often plotted on either log-log scale or "probability" scale. In comparing a variety of processing systems one would like to select the system (or develop a new one) that maximizes the P_D for every given P_{fa}. Processing systems must operate on their ROC curves, but most processing systems allow the operator to select where on the ROC curve the system is operated by adjusting a threshold; low thresholds ensure a high probability of detection at the expense of high false alarm rate. A sketch of two monotonically increasing ROC curves is given in Fig. 18.14. By proper adjustment of the decision threshold, one can trade off detection performance for false alarm performance. Since the points (0,0) and (1,1) are on all ROC curves, one can always guarantee 100% probability of detection with an arbitrarily low threshold (albeit at the expense of 100% probability of false alarm) or 0% probability of false alarm with an arbitrarily high threshold (albeit at the expense of 0% probability of detection). The (log) *likelihood detector* is a detector that achieves the maximum probability of detection for fixed probability of false alarm; it is shown in Fig. 18.15 for

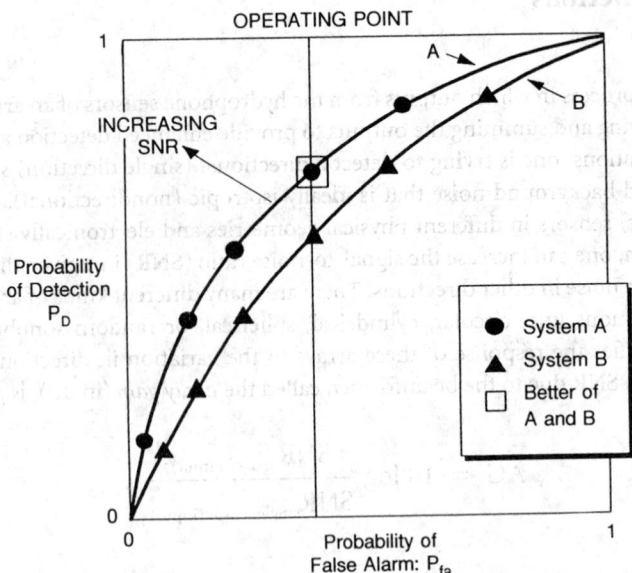

FIGURE 18.14   Typical ROC curves. Note points (0,0) and (1,1) are on all ROC curves; upper curve represents higher P_D for fixed P_{fa} and hence better performance by having higher SNR or processing time.

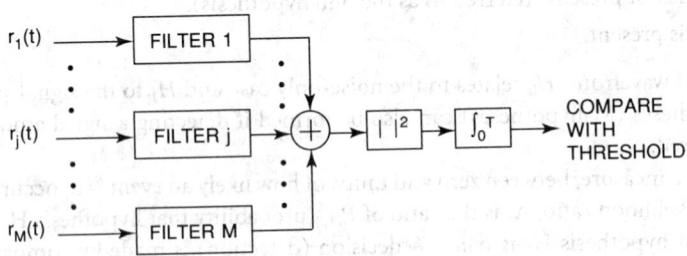

FIGURE 18.15   Log likelihood detector structure for uncorrelated Gaussian noise in the received signal $r_j(t)$, $j = 1, \ldots, M$.

detecting Gaussian signals reflected or radiated from the stationary objects modeled in Fig. 18.13. For spiky non-Gaussian noise, clipping prior to filtering improves detection performance.

In active sonar, the filters are matched to the known transmitted waveforms. If the object (acoustic reflector) has motion, it will induce Doppler on the reflected signal, and the receiver will be complicated by the addition of a bank of Doppler compensators. Returns from a moving object are shifted in frequency by $\Delta f = (2v/c)f$, where v is the relative velocity (range rate) between the source and object, c is the speed of sound in water, and f is the operating frequency of the source transmitter.

In passive sonar, at low SNR, the optimal filters in Fig. 18.15 (so-called Eckart filters) are functions of $G_{ss}^{1/2}(f)/G_{nn}(f)$, where f is frequency in hertz, $G_{ss}(f)$ is the signal power spectrum, and $G_{nn}(f)$ is the noise power spectrum.

Estimation/Localization

The second function of underwater acoustic signal processing estimates the parameters that localize the position of the detected object. The source position is estimated in range, bearing, and depth, typically from the underlying parameter of **time delay** associated with the acoustic wavefront. The statistical uncertainty of the positional estimates is important. Knowledge of the first order probability density function or its first- and second-order moments, the mean (expected value) and the variance, are vital to understanding the expected performance of the processing system. In the passive case, the ability to estimate range is extremely limited by the geometry of the measurements; indeed, the variance of passive range estimates can be extremely large, especially when the true range to the acoustic source is long when compared with the aperture length of the receiving array. Figure 18.16 depicts direct path passive ranging uncertainty from a collinear array with sensors clustered so as to minimize the bearing and uncertainty region. Beyond the direct path, multipath signals can be processed to estimate source depth covertly. Range estimation accuracy is not difficult with the active sonar, but active sonar is not covert, which for some applications can be important.

Classification

The third function of sonar signal processing is classification. This function determines the type of object that has radiated or reflected acoustic energy. For example, was the sonar signal return from a school of fish or a reflection from the ocean bottom? The action one takes is highly dependent upon this important function. The amount of radiated or reflected signal power relative to the background noise (that is, SNR) necessary to achieve good classification may be higher than for detection. Also, the type of signal processing required for classification may be different than the type of processing for detection. Processing methods that are developed on the basis of detection might not have the requisite SNR to adequately perform the classification function. Classifiers are, in general, divided into feature (or clue) extractors followed by a classifier decision box. A key to successful classification is feature extraction. Performance of classifiers is plotted as in ROC detection curves as probability of deciding on class A, given A was actually present, or $P(A/A)$, versus the probability of deciding on class B, given that A was present, i.e., $P(B/A)$, for two different classes of objects, A and B. Of course, for the same class of objects, one could also plot $P(B/B)$ versus $P(A/B)$.

Motion Analysis or Tracking

The fourth function of underwater acoustic signal processing is to perform contact (or target) motion analysis (TMA), that is, to estimate parameters of bearing and speed. Generally, nonlinear filtering methods, including Kalman-Bucy filters, are applied; typically these methods rely on a

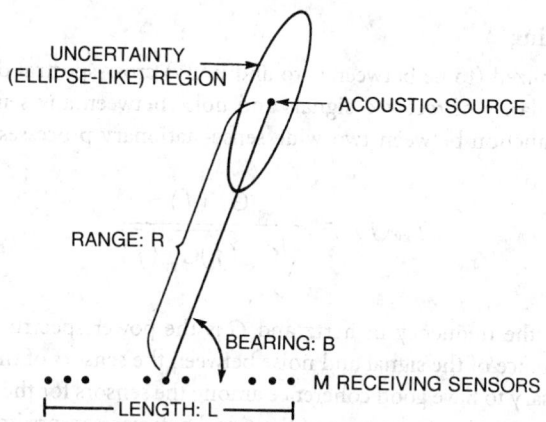

FIGURE 18.16 Array geometry used to estimate source position. (*Source:* G.C. Carter, "Coherence and time delay estimation," *Proceedings IEEE*, vol. 75, no. 2, p. 251, © 1987 IEEE. With permission.)

state space model for the motion of the contact. For example, the underlying model of motion could assume a straight-line course and constant speed of the contact of interest. When the acoustic source of interest behaves like the model, then results consistent with the basic theory can be expected. It is also possible to incorporate motion compensation into the signal processing detection function. For example, in the active sonar case, proper waveform selection and processing can reduce the degradation of detector performance caused by uncompensated Doppler. Moreover, joint detection and estimation can provide clues to the TMA and classification processes. For example, if the processor simultaneously estimates depth in the process of performing detection, then a submerged object would not be classified as a surface object. Also, joint detection and estimation using Doppler for detection can directly improve contact motion estimates.

Normalization

Another important signal processing function for the detection of weak signals in the presence of unknown and (temporal and spatial) varying noise is normalization. The statistics of noise or reverberation for oceans typically varies in time, frequency, and/or bearing from measurement to measurement and location to location. To detect a weak signal in a broadband, nonstationary, and inhomogeneous background, it is usually desirable to make the noise background statistics as uniform as possible for the variations in time, frequency, and/or bearing. The noise background estimates are first obtained from a window of resolution cells (which usually surrounds the test data cell). These estimates are then used to normalize the test cell, thus reducing the effects of the background noise on detection. Window length and distance from the test cell are two of the parameters that can be adjusted to obtain accurate estimates of the different types of stationary or nonstationary noise.

Advanced Signal Processing

Adaptive Beamforming

Beamforming was discussed in an earlier section. The cancellation of noise through beamforming can also be done adaptively, which can improve the array gain further. Some of the various adaptive beamforming techniques are [Knight *et al.*, 1981], Dicanne, sidelobe cancellers, maximum entropy array processing, and maximum-likelihood (ML) array processing.

Coherence Processing

Coherence is a normalized (to lie between zero and unity) cross-spectral density function that is a measure of the similarity of received signals and noise between any sensors of the array. The complex coherence function between two wide-sense-stationary processes x and y is defined by

$$\gamma_{xy}(f) = \frac{G_{xy}(f)}{\sqrt{G_{xx}(f)G_{yy}(f)}} \tag{18.8}$$

where, as before, f is the frequency in hertz and G is the power spectrum function. Array gain depends on the coherence of the signal and noise between the sensors of the array. To increase the array gain, it is necessary to have good coherence among the sensors for the signal, but poor coherence (incoherent) for the noise. Coherence of the signal between sensors improves with decreasing separation between the sensors, frequency of the received waveform, total bandwidth, and integration time. Loss of coherence of the signal could be due to ocean motion, object motion, multipaths, reverberation, or scattering. The coherence function has many uses, including measurement of SNR or array gain, system identification, and determination of time delays [Carter, 1987].

Acoustic Data Fusion

Acoustic data fusion is a technique that combines information from multiple receivers or receiving platforms about a common object or channel. Instead of each receiver making a decision, relevant information from the different receivers is sent to a common control unit where the acoustic data is combined and processed (hence the name *data fusion*). After fusion, a decision can be relayed or "fed" back to each of the receivers. If data transmission is a concern, due to time constraints, cost, or security, other techniques can be used in which each receiver makes a decision and transmits only the decision. The control unit makes a global decision based on the decisions of all the receivers and relays this global decision back to the receivers. This is called "distributed detection." The receivers can then be asked to re-evaluate their individual decisions based on the new global decision. This process could continue until all the receivers are in agreement or could be terminated whenever an acceptable level of consensus is attained.

An advantage of data fusion is that the receivers can be located at different ranges (e.g., on two different ships), in different mediums (shallow or deep water, or even at the surface), and at different bearings from the object, thus giving comprehensive information about the object or the underwater acoustic channel.

Application

Since World War II, in addition to military applications, there has been an expansion in commercial and industrial underwater acoustics applications. Table 18.2 lists the military and nonmilitary functions of sonar along with some of the current applications.

Table 18.2 Underwater Acoustics Applications

Function	Description
Military	
Detection	Deciding if a target is present or not.
Classification	Deciding if a detected target does or does not belong to a specific class.
Localization	Measuring at least one of the instantaneous positions and velocity components of a target (either relative or absolute), such as range, bearing, range rate, or bearing rate.
Navigation	Determining, controlling, and/or steering a course through a medium (includes avoidance of obstacles and the boundaries of the medium).
Communications	Instead of a wire link, transmitting and receiving acoustic power and information.
Control	Using a sound-activated release mechanism.
Position marking	Transmitting a sound signal continuously (beacons) or transmitting only when suitably interrogated (transponders).
Depth sounding	Sending short pulses downward and timing the bottom return.
Acoustic-speedometers	Using pairs of transducers pointing obliquely downwards to obtain speed over the bottom from the Doppler shift of the bottom return.

Commercial Applications:

Industrial	Oceanological
Fish finders/fish herding	Subbottom geological mapping
Oil and mineral explorations	Ocean topography
River flow meter	Bathyvelocimeter
Acoustic holography	Emergency telephone
Viscosimeter	Seismic simulation and measurement
Acoustic ship docking system	Biological signal and noise measurement
Ultrasonic grinding/drilling	Sonar calibration

Source: R. J. Urick, *Principles of Underwater Sound,* New York, McGraw-Hill, 1983, p. 8. and A.W. Cox, *Sonar and Underwater Sound,* Lexington, Mass.: Lexington Books, D.C. Health and Company, 1974, p. 2. With permission.

Defining Terms

Decibels (dB): Logarithmic scale of representing the ratio of two quantities given as $10 \log_{10}(P_1/P_0)$ for power level ratios and $20 \log_{10}(V_1/V_0)$ for comparing acoustic pressure or voltage ratios. A standard reference pressure or intensity level in SI units is equal to 1 micropascal (1 pascal = 1 newton per square meter = 10 dyne per square centimeter).

Doppler shift: Shift in frequency of transmitted waveform due to the relative motion between the source and object.

Figure of merit/sonar equation: Performance evaluation measure for the various target and equipment parameters of a sonar system. It is a subset of the broader sonar performance given by the *sonar equations*, which includes reverberation effects.

Hydrophone: Receiving sensors that convert sound energy into electrical or optical energy (analogous to underwater microphones).

Receiver operating characteristics (ROC) curves: Plots of the probability of detection (likelihood of detecting the object when the object is present) versus the probability of false alarm (likelihood of detecting the object when the object is not present) for a particular processing system.

Reverberation/clutter: Inhomogeneities, such as dust, sea organisms, schools of fish, sea mounds on the bottom of the sea, form mass density discontinuities in the ocean medium. When an acoustic wave strikes these inhomogeneities, some of the acoustic energy is reflected and reradiated. The sum total of all such reradiations is called reverberation. Reverberation is present only in active sonar, and in the case where the object echoes are completely masked by reverberation, the sonar system is said to be "reverberation limited."

SNR: The signal-to-noise (power) ratios, usually measured in decibels (dB).

SONAR: Acronym for "SOund NAvigation and Ranging," adopted in the 1940s, involves the use of sound to explore the ocean and underwater objects.

Sound velocity profile (SVP): Description of the speed of sound in water as a function of water depth.

Time delay: The time (delay) difference in seconds from when an acoustic wavefront impinges on one hydrophone or receiver until it strikes another.

References

L. Brekhovskikh and Yu. Lysanov, *Fundamentals of Ocean Acoustics*, New York.: Springer-Verlag, 1982.

W. S. Burdic, *Underwater Acoustic System Analysis*, Englewood Cliffs, N.J.: Prentice-Hall, 1984.

G. C. Carter, "Coherence and time delay estimation," *Proceedings of the IEEE*, vol. 75, no. 2, pp. 236–255, Feb. 1987.

Y. T. Chan, Ed., *Digital Signal Processing for Sonar Underwater Acoustic Signal Processing*, NATO ASI Series, Series E: Applied Sciences, vol. 161, Kluwer Academic Publishers, 1989.

A. W. Cox, *Sonar and Underwater Sound*, Lexington, Mass.: Lexington Books, D.C. Health and Company, 1974.

W. C. Knight, R. G. Pridham, and S. M. Kay, "Digital signal processing for sonar," *Proceedings of the IEEE*, vol. 69, no. 11, pp. 1451–1506, Nov. 1981.

R. O. Nielsen, *Sonar Signal Processing*, Boston: Artech House, 1991.

A. V. Oppenheim, Ed., *Applications of Digital Signal Processing*, Englewood Cliffs, N.J.: Prentice-Hall, 1980.

R. J. Urick, *Principles of Underwater Sound*, New York.: McGraw-Hill, 1983.

H. L. Van Trees, *Detection, Estimation, and Modulation Theory*, New York: John Wiley & Sons, 1968.

L. J. Ziomek, *Underwater Acoustics, A Linear Systems Theory Approach*, New York: Academic Press, 1985.

Further Information

Journal of Acoustical Society of America (JASA), *IEEE Transactions on Signal Processing* (formerly the *IEEE Transactions on Acoustics, Speech and Signal Processing*), and *IEEE Journal of Oceanic Engineering* are professional journals providing current information on underwater acoustical signal processing.

The annual meetings of the International Conference on Acoustics, Speech and Signal Processing, sponsored by the IEEE, and the biannual meetings of the Acoustical Society of America are a good source for current trends and technologies.

A detailed tutorial on *Digital Signal Processing for Sonar* by W.C. Knight *et al.* is an informative and detailed tutorial on the subject [Knight *et al.*, 1981].

Hewlett-Packard's first product, the model 200A audio oscillator (preproduction version). William Hewlett and David Packard built an audio oscillator in 1938, from which the famous firm grew. (Courtesy of Hewlett-Packard Company.)

19

Neural Networks

19.1 Introduction 420
19.2 Perceptrons 421
19.3 Hopfield Network 423
19.4 Topology-Preserving Network 424
19.5 Adaptive Resonance Theory 426

Behnam Bavarian
Printrak International

19.1 Introduction

A *neural network* is a parallel, distributed information processing structure consisting of **processing elements** (which can possess a local memory and can carry out localized information processing operations) interconnected via unidirectional signal channels called connections. Each processing element has many inputs and a single output that fans out into as many collateral connections as desired. Figure 19.1 shows a processing element which is also referred to as a neuron in the literature. The processing element output signal can be of any mathematical type desired, such as

$$\dot{x}_i = g_i\left(\sum_{j=1}^{n} w_{ij}v_j\right) \tag{19.1}$$

where x_i is the internal excitation state of the neuron, w_{ij} is the **connection weight** of the jth input to the ith neuron, v_j is the output of the neuron defined by $v_j = f(x_j)$ where f is a nonlinear function which maps the internal excitation signal to the output, and g_i defines the general functional behavior of the neuron. The input/output behavior of the processing element depends on the weight parameters w_{ij}. These are also referred to as synaptic efficacies in the literature. The process of learning a given task by a neural network is the weight adaptation, given in general by

$$\dot{w}_{ij} = h(w_{ij}, x_i, v_j) \tag{19.2}$$

where h is a general nonlinear function.

The information processing that goes on within each processing element can be defined arbitrarily with the restriction that it must be completely local; that is, it must depend only on the current values of the input signals arriving at the processing element's local memory. This is clear from the processing element Eqs. (19.1) and (19.2). From a system point of view, a neural network is a large-dimensional nonlinear dynamic system which is defined by a set of first-order nonlinear differential or difference equations. These equations are represented in the form of connected elementary processing units as in a graph. To summarize the above discussion an artificial neural network is defined by

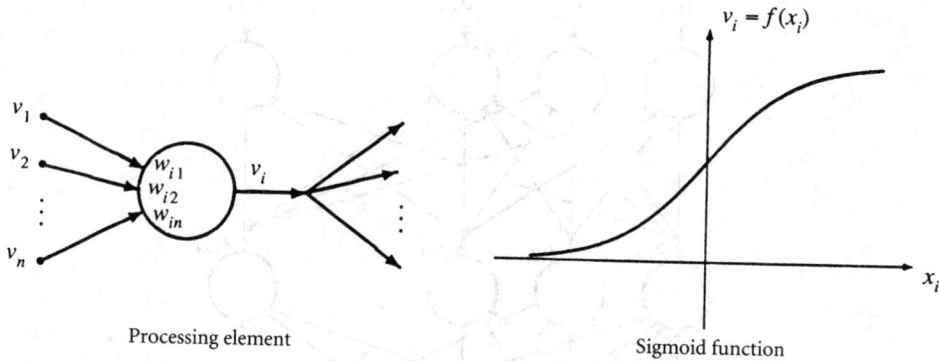

FIGURE 19.1   Example of a processing element and the typical sigmoid nonlinearity used at the output of the neuron.

- The processing element or the neuron characteristics, Eq. (19.1) (i.e., the equations defining what the neuron will do)
- **Learning rule,** Eq. (19.2) (i.e., how the connection weights change upon stimulus)
- **Neural network topology** (i.e., how the neurons are connected)

The state of the neurons is referred to as *short-term memory* (STM), since they represent the current state of the neural network. Upon an initial stimulus, the STMs converge to recall and to carry the learned action. The approximate time constant of Eq. (19.1) determines how fast the recall is done. In the case of biological neural networks, this reaction time varies from 20 to 200 ms. The state of the connection weights (or synaptic efficacies), Eq. (19.2), is referred to as *long-term memory* (LTM). The associations between objects in response to their environment are encoded in the space defined by the weight vectors. The approximate time constant of the weight adaptation equation varies from minutes to days through extensive training in the environment.

In this chapter four neural network paradigms are described.

19.2 Perceptrons

The perceptron neural networks are the original networks developed by Rosenblatt based on the neuron models proposed in the neurodynamics studies. The topology of the network is **feedforward multilayer**. A three-layer perceptron is shown in Fig. 19.2. This topology represents an evolution of the previous two-layer network introduced by Rosenblatt. The network shown has N processing elements in the input layer, P processing elements in the output layer, and a variable number of processing elements in the middle or hidden layer.

The characteristic equation for all processing elements (PEs) is the same. For the ith processing element in the mth layer, the output $o_{m,i}$ is defined by

$$o_{m,i}(t+1) = f\left(\sum_k o_{m-1,k}(t)\, W_{m,k,i}(t)\right) \tag{19.3}$$

where t is discrete time index, m is 2 for hidden-layer PEs and 3 for output-layer PEs, $W_{m,k,i}$ is the weight of interconnection between PEs in the previous layer to this layer, and $f(\cdot)$ is the nonlinear activation function. This function is typically selected as a monotonically increasing bounded function like a sigmoid function, which is shown in Fig. 19.1.

The learning law for the perceptron is a simple error feedback. The network learns the associations between input and output patterns by being exposed to many "lessons." The weights are

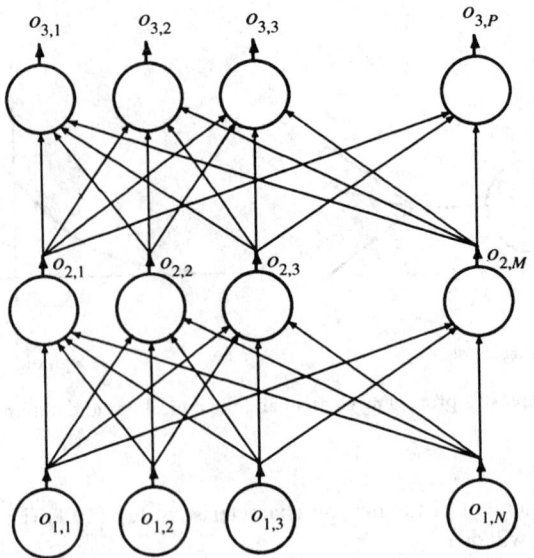

FIGURE 19.2 N input/P output three-layer perceptron with M neurons in the middle layer.

adjusted until the desired target output is produced. This weight adaptation is referred to as error backpropagation learning law. At the output layer, the error associated with the jth PE is

$$e_{3,j} = d_j - o_{3,j} \tag{19.4}$$

where d_j is the desired output of the jth output PE, and $o_{3,j}$ is its actual output. The error for the jth neuron in the hidden layer is defined as

$$e_{2,j} = \sum_k e_{3,k} W_{3,k,j} \tag{19.5}$$

The weight adaptation law is then given by

$$W_{m,i,j}(t+1) = W_{m,i,j}(t) + \Delta W_{m,i,j} \tag{19.6}$$

where

$$\Delta W_{m,i,j} = \eta e_{m,j} o_{m,i} \dot{f}\left(\sum_k o_{m-1,k}(t) W_{m-1,k,j}(t) \right) \tag{19.7}$$

where $\dot{f}(\cdot)$ is the derivative with respect to the function argument, and η is a parameter of the learning rate. It is shown that for a sigmoid function the adaptation law is an implementation of the gradient descent in the output error. Hence the convergence of the learning is established.

The three-layer perceptron can only recall static pattern association. One can add P PEs to the input layer with the values which are the previous values of the output PEs, hence providing feedback around the network. This model is then capable of realizing movable partition of the subspaces which define the learned classes.

The perceptron network provides a very simple approach for universal function approximation and is the most widely used neural network. Any unknown process can be modeled easily by the perceptron as long as there are many input/output training pairs available. It has been applied to many problems, such as stock market prediction and general pattern recognition. One of the most

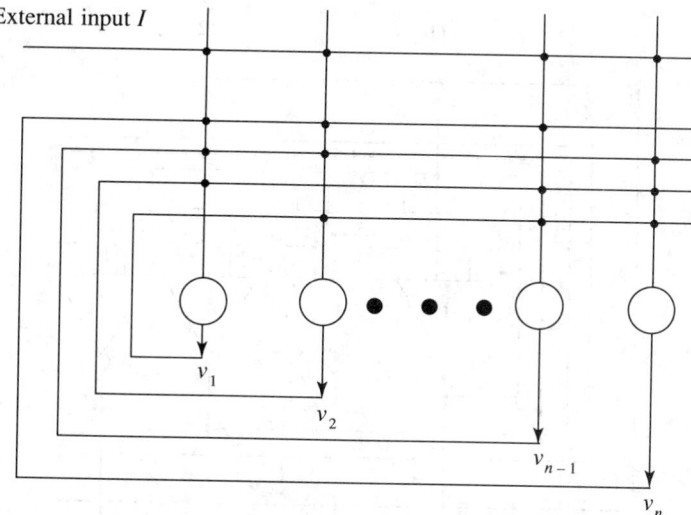

External input I

v_1

v_2

v_{n-1}

v_n

FIGURE 19.3 The Hopfield crossbar neural network.

popular applications is a network which learns to read text. In general the learning algorithm has problems similar to the gradient descent approach, such as slow convergence and being trapped in local minima. There are many robust techniques that one can use to improve the performance, such as using extended Kalman filter weight adaptation which greatly improves the speed of convergence and is inherently noise tolerant.

19.3 Hopfield Network

The original Hopfield model of the biological neuron was a logical decision element described by a two-state (binary) variable. The output nonlinearity is a threshold function which determines active output (state 1), if the input weighted sum value is greater than the threshold, or inactive output (state 0) otherwise. The neural networks suggested by Hopfield introduced the feedback connectivity which distinguished them from perceptron-like feedforward networks. Figure 19.3 shows the basic structure of the Hopfield crossbar neural network. The solid circles on the interconnection crossings represent weight of connection. In general, for a specific problem, a "computational energy," E, is defined, and the rate of change of the activation level of the neurons is set equal to the negative of the partial derivative of the energy with respect to the neuron variable (differential gradient optimization method). Then, by assigning the resulting equations to the processing elements, the optimization problem is solved in a parallel distributed fashion. Motivated by graded response of neurons, Hopfield has extended his binary crossbar network to have neurons with continuous variable output and a neuron characteristic defined by a first-order differential equation given by

$$C_i \frac{dx_i}{dt} = -\frac{x_i}{R_i} + \sum_{j=1}^{n} w_{ij} v_j + I_i \tag{19.8}$$

where x_i is the internal activation potential of the neuron, $R_i C_i$ are resistance/capacitance values defining the time constant of the active RC circuit, w_{ij} is the connection weight (conductance) of the output of neuron j as input to neuron i, I_i is the external input, and v_j defines the output of the neuron which is related to the internal activation potential through a function $f(\cdot)$,

$$v_j = f(x_j)$$

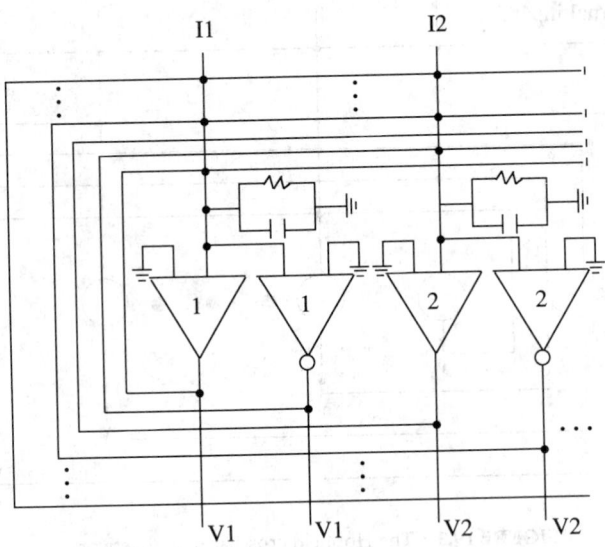

FIGURE 19.4   Architecture of Hopfield neural circuit.

and, in general, is a monotonically increasing bounded function, like the sigmoid function shown in Fig. 19.1.

The Hopfield continuous crossbar network can be realized by first-order active circuits. Figure 19.4 shows such an implementation. The outputs of the operational amplifiers are both positive and negative, to realize the general quadratic energy function which is used to prove the global convergence of the network,

$$E = -\frac{1}{2}\sum_{i=1}^{n}\sum_{j=1}^{n}w_{ij}v_iv_j - \sum_{i=1}^{n}I_iv_i + \sum_{i=1}^{n}\left(\frac{1}{R_i}\right)\int_0^{v_i}f^{-1}(v)dv \qquad (19.9)$$

where $f^{-1}(v)$ is the inverse of the sigmoid nonlinearity. The conditions for convergence depend on the symmetry of the weight connection matrix and the monotonically increasing, bounded sigmoid nonlinearity. Again, upon definition of an energy or cost function for a specific problem, the neuron differential equations or the weight connection matrix are computed by taking the gradient of Eq. (19.9).

The state of neurons evolves in time, moving in the direction of decreasing energy surface to converge to a local minimum. So, for a given problem, an energy function is constructed such that the network state configurations represent possible solutions at the local minima. For a very steep sigmoid function, the third term in Eq. (19.9) drops out, which makes it easy to construct the energy or the cost function. One of the first applications of the Hopfield network was the analog-to-digtal conversion circuit. It has also been applied to solve combinatorially difficult optimization problems such as the traveling salesman problem and, in general, scheduling and decision support systems.

19.4 Topology-Preserving Network

The topology-preserving neural network is a model of the learning and self-organization in the brain where the outside world, as experienced by a sensory device, is mapped into an ordered internal representation. The maps are characterized by the fact that excitations at nearby positions on the cortical plane are caused by similar sensory signals. It was originally proposed by Kohonen [1984]. The weight adaptation is an unsupervised algorithm for generating a mapping of an input

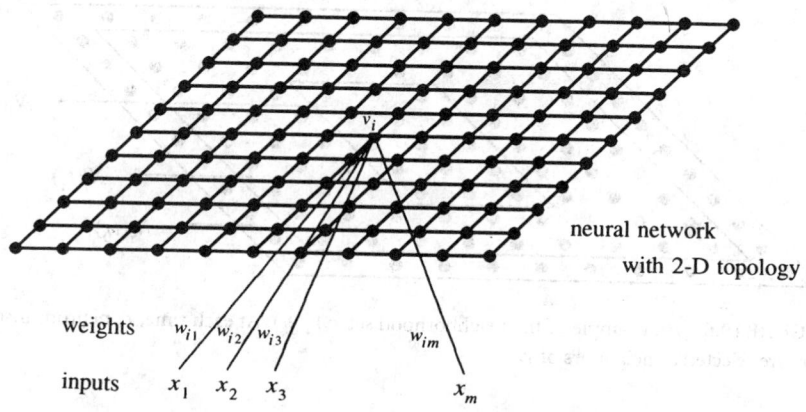

FIGURE 19.5 Two-dimensional self-organizing neural network.

signal vector $\mathbf{x}$ from a high-dimensional space $\Re^n$ onto a one- or two-dimensional topological space of a neural network. Figure 19.5 shows a one-layer topology-preserving neural network, arranged in a two-dimensional topology. Each neuron has eight topological neighbors in this network. Alternatively, one can design a one-dimensional network in which each neuron has only two neighbors. The selection of the topology of the network depends on the application problem.

The fundamental property of this network is its ability to learn the underlying statistical distribution of a pattern mapping. This is done by the self-organization of the weight space of the neurons after a period of learning iterations. Let w_{ij} represent the connection weight of the jth input to the ith neuron and let $\mathbf{w}$ be the matrix of weights. So each row of $\mathbf{w}$, say $\mathbf{w}_i = [w_{i1}, W_{i2}, \ldots, w_{in}]$, is the weight vector for neuron v_i. Let the n inputs be denoted by $\mathbf{x} = [x_1, x_2, \ldots, x_n]$. Here both $\mathbf{w}_i$ and $\mathbf{x}$ are time variables updated at every time t_k, $k = 1, 2, \ldots$, i.e., $\mathbf{w}_i = \mathbf{w}_i(t_k)$ and $\mathbf{x} = \mathbf{x}(t_k)$.

Two important parameters should be defined for this network. The first is the definition of a neighborhood function, $\mathcal{N}_i(t_k)$. This is a set of neurons considered to be in the neighborhood of neuron i at time t_k. Figure 19.6 shows an example of $\mathcal{N}_i(t_k)$ for the network with a two-dimensional structure. Typically the initial values of $\mathcal{N}_i(t_k)$ are large and then slowly decrease over time.

The next parameter is the learning gain function, which appears in the weight adaptation equation and is denoted by $\alpha(t_k)$ below. Initially, when the weight space is randomly oriented, the value of the learning gain function is set close to unity and decreases gradually as learning proceeds.

The network weight spaces, $\mathbf{w}_i$'s, are initially set to small random vectors. Next a new input vector, $\mathbf{x}(t_k)$, is presented to the network. The closest neuron in the sense of l_2 norm, i.e.,

$$\min_i \| \mathbf{x}(t_k) - \mathbf{w}_i(t_k) \|$$

is selected and the weight vectors of the v_i neuron and its neighbors are updated according to

$$\mathbf{w}_i(t_{k+1}) = \mathbf{w}_i(t_k) + \alpha(t_k)(\mathbf{x}(t_k) - \mathbf{w}_i(t_k)) \quad \text{for } i \in \mathcal{N}_i \tag{19.10}$$

otherwise

$$\mathbf{w}_i(t_{k+1}) = \mathbf{w}_i(t_k) \tag{19.11}$$

This weight updating is repeated for many training sample points from the input vector space. As learning proceeds, the network self-organizes itself to map the underlying distribution. Figure 19.7 shows the simulation of a two-dimensional network for learning a two-dimensional uniform distribution. Figure 19.7 shows four snapshots of the weight space of a two-dimensional (32×32) network learning a uniform square distribution. Initially the weight vectors are bunched together and, as training goes on, the shape of the distribution is evolved.

The topology-preserving neural networks are used in adaptive clustering problems, vector quan-

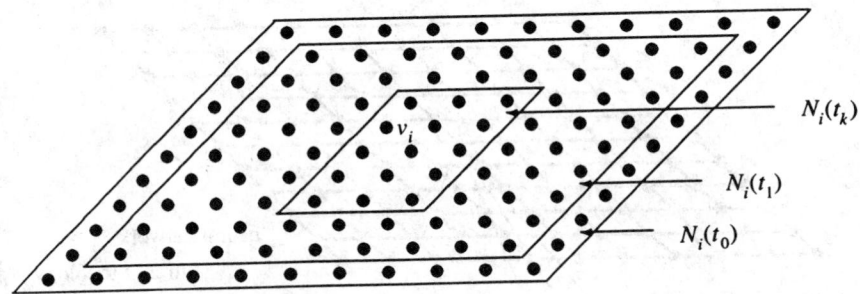

FIGURE 19.6   An example of the neighborhood set $\mathcal{N}_i(t_k)$. At each time, t_k neurons inside a box are selected as neighbors of v_i.

tization problems with application in image compression, and environment-learning problems for autonomous mobile robots.

19.5 Adaptive Resonance Theory

Adaptive resonance theory (ART) was first proposed by Grossberg in 1976. The original motivation for the theory was to solve the instability problem of competitive networks having only feedforward connections. The theory explains why feedback connections are required to overcome this instability problem, which more generally is called the "stability-plasticity dilemma" of adaptive systems. This psychophysiological theory also explains how the system makes expectations, focuses attention, and learns temporal/spatial or sequence patterns.

Although some applications of the ART were worked out after the theory was proposed, those were theoretical works without any computer simulations. Recently, Carpenter and Grossberg proposed a neural network called "ART1" in which the theory was embedded. Basically, the ART1 is an implementation of the previous works by Grossberg. However, some new mechanics and learning rules were added to the original work. Furthermore, Carpenter and Grossberg proved mathematically that the ART1 self-organizes and self-stabilizes its learning of cognitive codes in response to many arbitrarily chosen binary input patterns. For continuous input patterns including binary patterns, a network called "ART2" was proposed and its self-organization and self-stability were analyzed by Carpenter and Grossberg.

A typical model of the ART2, which is similar to the ART1 network, is shown in Fig. 19.8. There are two layers, F_1 and F_2, and the orienting subsystem to reset the F_2. The first layer F_1 is further composed of three layers so that each analog signal, the bottom-up input I_i and the top-down signal

$$V_i = z_{ij}g(x_{2j}) \tag{19.12}$$

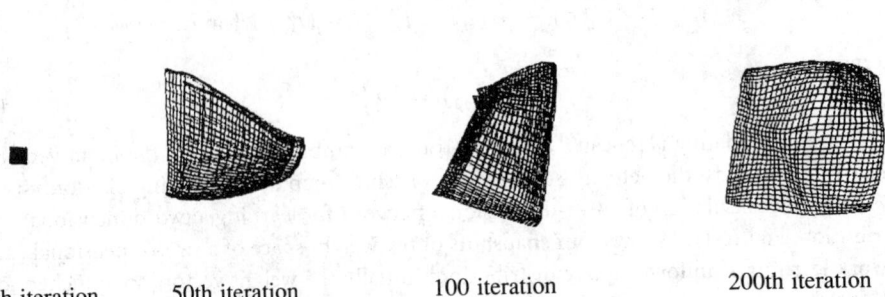

0th iteration   50th iteration   100 iteration   200th iteration

FIGURE 19.7   Two-dimensional self-organizing network learning square shape distribution.

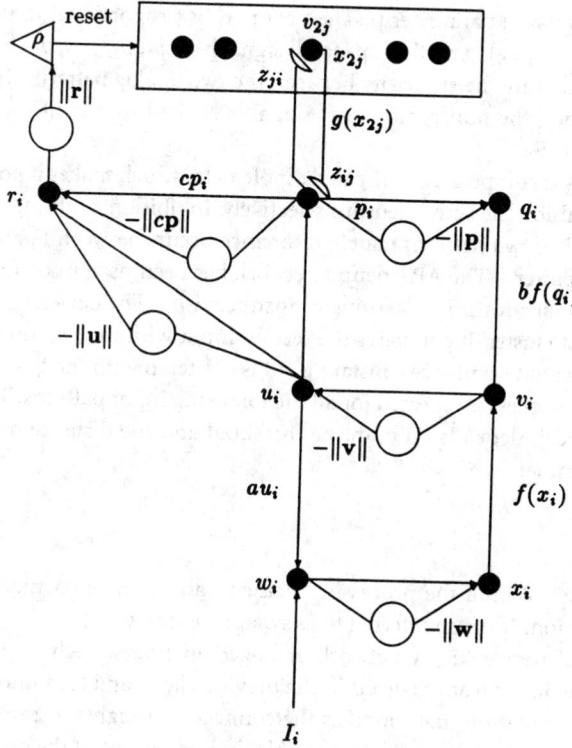

FIGURE 19.8 Structure of ART2. The circles indicate the nonspecific signals.

can be normalized separately. In Fig. 19.8, an arrow with an empty circle represents a nonspecific signal, and an arrow alone represents a specific signal. For example, x_i receives both the specific signal w_i and the nonspecific signal $-\|\mathbf{w}\|$ where $\|\mathbf{w}\|$ is the norm of the vector $\mathbf{w} = [w_1, w_2, \ldots, w_N]^T$. It can be shown that the normalization $\mathbf{x}_i \simeq \mathbf{w}_i / \|\mathbf{w}\|$ is implemented by this structure. Physiologically, the naturally realizable norm of $\mathbf{w}$ is

$$\|\mathbf{w}\| = \sum_{k=1}^{N} w_i \tag{19.13}$$

which is just the sum of all signals w_i; however, the l^2 norm is employed by the ART2, i.e.,

$$\|\mathbf{w}\| = \sqrt{\sum_{k=1}^{N} w_k^2} \tag{19.14}$$

Both the normalized signals x_i and q_i go through a nonlinear function $f(\cdot)$ separately and are added to each other to form v_i. The v_i is again normalized to be u_i which returns to both the input and the top-down signals separately. Simply speaking, the closed loop $(w_i \rightarrow x_i \rightarrow v_i \rightarrow u_i \rightarrow w_i)$ with the nonlinear function $f(x_i)$ forms a positive feedback loop to suppress the noise and to enhance the contrast in the signals. Also, the pass $(u_i \rightarrow p_i)$ makes the bridge from input I_i to the LTMs through noise suppression and contrast enhancement mechanisms.

The comparison of the normalized input signal with the top-down signal is implemented by the orienting subsystem in Fig. 19.8. Each node i in F_1 has a localized subsystem by which r_i is made, and the global orienting subsystem measures the norm $\|\mathbf{r}\|$ of the vector $\mathbf{r} = [r_1, r_2, \ldots, r_N]^T$ to compare with the vigilance parameter ρ. The norm $\|\mathbf{r}\|$ is proportional to the cosine of the angle between $\mathbf{u} = [u_1, u_2, \ldots, u_N]^T$ and the top-down signal $\mathbf{p} = [p_1, p_2, \ldots, p_N]^T$; therefore, the orienting subsystem can measure the similarity between the two analog patterns. If two patterns are different from each other, the nonspecific reset signal is elicited to inhibit the active neuron which may be misselected in F_2.

The upper layer F_2 is composed of the gated dipole field, which makes it possible for the nonspecific reset signal to inhibit the active neurons selectively. Inhibition continues until the presentation of a new input signal. If two patterns match, resonance occurs between F_1 and F_2 so that the LTMs can learn the input pattern. The ART neural network has been used in pattern clustering and pattern classification applications. The learning is **unsupervised**. The network selects the first input as a member of the first cluster. It compares the second input with the first one; if the distance to the first is less than threshold (vigilance constant ρ), it is clustered with the first; otherwise a new cluster is generated. The process is repeated for all the following input patterns. The number of clusters grows with time, and it depends on both the threshold and the distance metric used to compare inputs to cluster members.

Defining Terms

Connection weight: Within the processing element, an adaptive coefficient associated with an input connection. It is also referred to as synaptic efficacy.

Feedforward neural network: A network arranged in layers, each layer with many neurons which only get input from neurons in the previous layer and feed into the next layer.

Learning rule: An equation that modifies the connection weights in response to input, current state of the processing element, and possible desired output of the processing element.

Neural network topology: The way the processing elements are arranged and connected, e.g., layers connected with feedforward connections.

Processing element: An artificial neuron in a neural network, consisting of a small amount of local memory and processing power.

Recurrent neural network: A network in which the neurons arranged in layers have feedforward, feedback, and possible lateral connections.

Supervised learning: Teaching the neural network by training samples with the input and desired output.

Unsupervised learning: Teaching the neural network by training samples with the input only. The neural network will perform a general statistical clustering of the information in the input space.

References

B. Bavarian, "Introduction to neural networks for intelligent control," *IEEE Control System Magazine*, vol. 8. no. 2, 1988.

S. Grossberg, Ed., *Neural Networks and Natural Intelligence,* Cambridge, Mass.: MIT Press, 1988.

D.O. Hebb, *The Organization of Behavior,* New York: Wiley, 1949.

R. Hecht Nielsen, *Neurocomputing,* Reading, Mass.: Addison-Wesley, 1990.

J.J. Hopfield, "Neural networks and physical systems with emergent collective computational abilities," in *Proceedings of the National Academy of Sciences,* pp. 2554–2558, 1979.

T. Kohonen, *Self-Organization and Associative Memory,* New York: Springer-Verlag, 1984.

B. Kosko, *Neural Networks for Signal Processing,* Englewood Cliffs, N.J.: Prentice-Hall, 1992.

W.T. Miller, III, R.S. Sutton, and P.J. Werbos, *Neural Networks for Control,* Cambridge, Mass.: MIT Press, 1990.

J. McClelland, D.G. Rumelhart, and PDP Research Group, *Parallel Distributed Processing,* Cambridge, Mass.: MIT Press, 1986.

Further Information

The *IEEE Transactions on Neural Networks* publishes papers on the theory, design, and application of neural networks, ranging from software to hardware. Other magazines from IEEE societies, such as Control Society, System, Man, and Cybernetics Society, and Robotics and Automation Society, also publish papers on the applications of neural networks.

Another source is the official journal of the International Neural Network Society (INNS), *Neural Networks,* which is published bimonthly by Pergamon Press. This magazine publishes articles concerned with the modeling of brain and behavioral processes and the application of these models to computer and related technologies.

Both IEEE and INNS sponsor international conferences every year which produce multivolume proceedings with the latest research and technology information on neural networks.

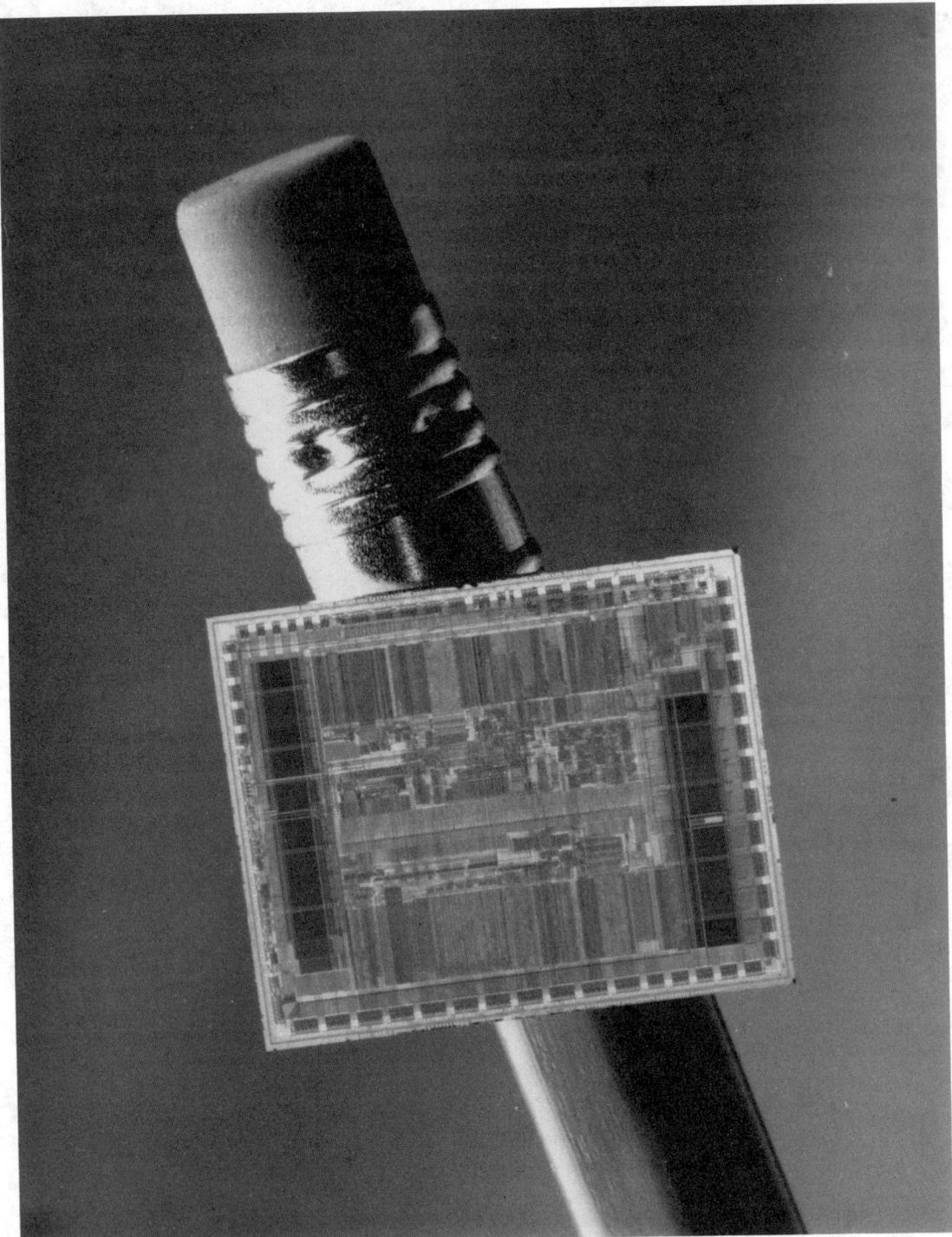

Digital's 21064 chip is the leading 64-bit processor in the industry. The tiny package is deceiving: the chip can operate at speeds up to 200 MHz. Evidencing open business practices, Digital is making this powerful chip available for licensing by other vendors. Cray Research, Inc., has stepped forward as the first company to release its plans to incorporate the 64-bit processor. Cray intends to build a massively parallel processing (MPP) system based on the 21064 chip. (Photo courtesy of Digital Equipment Corporation. Photo: Monty Abbott.)

III

Electronics

John W. Steadman
University of Wyoming

20 **Semiconductors**   G. Gildenblat, B. Gelmont, M. Milkovic, A. Elshabini-Riad, F. Stephenson, I. Bhutta 435
Physical Properties • Diodes • Electrical Equivalent Circuit Models and Device Simulators for Semiconductor Devices

21 **Semiconductor Manufacturing**   H. Parks, W. Needham, S. Rajaram 475
Processes • Testing • Electrical Characterization of Interconnections

22 **Transistors**   S. Soclof, J. Watson, J. Brews 530
Junction Field-Effect Transistors • Bipolar Transistors • The Metal-Oxide Semiconductor Field-Effect Transistor (MOSFET)

23 **Integrated Circuits**   M. Zargham, S. Tragoudas, J. Seely 581
Layout, Placement, and Routing • Application-Specific Integrated Circuits (ASICs)

24 **Surface Mount Technology**   G. Ginsberg 603
Packaged-Component Subassemblies • Technology Overview • Electronic Circuit Components • SMT Assembly Processing

25 **Operational Amplifiers**   E. Kennedy, J. Wait 616
Ideal and Practical Models • Applications

26 **Amplifiers**   G. Carpenter, J. Choma, Jr. 634
Large Signal Analysis • Small Signal Analysis

27 **Computer-Aided Circuit Simulation**   M. Lightner 653
Hierarchy of Modeling and Simulation • Modeling and Basic Circuit Equations • Equation Formulation • Solving Linear Equations • Solving Nonlinear Equations • Newton-Raphson Applied to Circuit Simulation • Solving Differential Equations • Techniques for Large Circuits • Specialized Techniques for MOS Circuits

28 **Active Filters**   R. Massara, J. Steadman, B. Wilamowski 674
Synthesis of Low-Pass Forms • Realization

29 **Power Electronics**   K. Rajashekara, A. Bhat, B. Bose 694
Power Semiconductor Devices • Power Conversion • Power Supplies • Converter Control of Machines

30 **Optoelectronics**   J. Hecht, L. Watkins, R. Becker 738
Lasers • Sources and Detectors • Circuits

31 **D/A and A/D Converters**   S. Garrod 771
Integrated Circuits

32 **Thermal Management of Electronics**   A. Bar-Cohen 784
Heat Transfer Fundamentals • Chip Module Thermal Resistance

431

T HE TRULY INCREDIBLE CHANGES in the technology associated with electronics over the past three decades have certainly been the driving force for most of the growth in the field of electrical engineering. Recall that 30 years ago the transistor was a novel device and that the majority of electronic systems still used vacuum tubes. Then look at the section headings in the following chapters and appreciate the range of ways that electronics has impacted electrical engineering. Amplifiers, integrated circuits, filters, power electronics, and optoelectronics are examples of how electronics transformed the practice of electrical engineering in such diverse fields as power generation and distribution, communications, signal processing, and computers.

The various contributors to this section have done an outstanding job of providing concise and practical coverage of this immense field. By necessity, the content ranges from rather theoretical considerations, such as physical principles of semiconductors, to quite practical issues such as printed circuit board technology and circuits for active filter realizations. There are areas of overlap with other chapters in the *Handbook*, such as those covering electrical effects and devices, biomedical electronics, digital devices, and computers. The contributors to this section, however, have maintained a focus on providing practical and useful information directly related to electronics as needed by a practicing electrical engineer.

The author(s) of each chapter was given the task of providing broad coverage of the field while being restricted to only a few pages of text. As a result, the information content is quite high and tends to treat the main principles or most useful topics in each area without giving the details or extensions of the subject. This practice, followed throughout the *Handbook*, is what makes it a valuable new work in electrical engineering. In most cases the information here will be complete enough. When this is not the case, the references will point the way to whatever added information is necessary.

Nomenclature

Symbol	Quantity	Unit	Symbol	Quantity	Unit
A	area	m^2	h_{re}	small-signal current gain	
A_i	current gain		η	quantum efficiency	
A_v	terminal voltage gain		i_b	incremental base current	A
α_i	ionization coefficient		I	illuminance	lumen/cm
B	bandwidth	Hz	I_B	direct base current	A
C	velocity of light in vacuum	2.998×10^8 m/s	I_D	diode forward current	A
			I_E	direct emitter current	A
C	specific heat	W/kg K	I_s	reverse saturation current	A
C_c	coupling capacitor		J	current density	A/m^2
C_E	emitter bypass capacitor		k	Boltzmann constant	1.38×10^{-23} J/K
C_j	junction capacitance	F	k	wavenumber	rad/m
E	energy	J	k	wave vector	
ϵ_o	permittivity constant	8.85×10^{-12} F/m	k	attenuation	
f	focal length	m	k	thermal conductivity	W/m K
F	luminous flux	lumen	λ	carrier mean free path	m
F	radiational factor		λ	wavelength	m
ϕ	pn-junction contact potential	V	μ	magnetic permeability	H/m
			μ	viscosity	kg/ms
g_m	transconductance	S	μ_n	electron mobility	
h	Planck's constant	6.626×10^{-34} J·s	n	electron density	electrons/cm^3
h	heat transfer coefficient		n	refractive index	
h_{FE}	common-emitter direct current gain		ν	light frequency	Hz
			p	hole density	holes/cm^3

Symbol	Quantity	Unit
Pr	Prandtl number	
ψ_{bk}	Bloch wave function	
q	electronic charge	1.6×10^{-19} C
q	heat flow	W
R_B	base resistor	
Re	Reynolds number	
R_g	generator internal resistance	Ω
R_G	total resistance	Ω
σ	conductivity	S
σ	Stefan-Boltzmann constant	5.67×10^{-8} W/m^2 K^4

Symbol	Quantity	Unit
T	absolute temperature	K
τ	momentum relaxation time	s
θ	volumetric flow rate	m^3/s
v	electron velocity	m/s
V_{BE}	direct base-emitter voltage	V
V_{CC}	direct voltage supply	V
V_T	thermal voltage	mV
V_Z	Zener voltage	V
W	power	W
Z_o	characteristic impedance	Ω

20

Semiconductors

Gennady Sh.
Gildenblat
*The Pennsylvania State
University*

Boris Gelmont
University of Virginia

Miran Milkovic
Georgia Institute of Technology

Aicha Elshabini-Riad
*Virginia Polytechnic Institute
and State University*

F. W. Stephenson
*Virginia Polytechnic Institute
and State University*

Imran A. Bhutta
*Virginia Polytechnic Institute
and State University*

20.1   Physical Properties 435
Energy Bands • Electrons and Holes • Transport Properties •
Hall Effect • Electrical Breakdown • Optical Properties and Recombination Processes • Nanostructure Engineering • Disordered Semiconductors

20.2   Diodes 447
pn-Junction Diode • *pn*-Junction with Applied Voltage • For-
ward-Biased Diode • I_D-V_D Characteristic • DC and Large-Signal
Model • High Forward Current Effects • Large-Signal Piecewise
Linear Model • Small-Signal Incremental Model • Large-Signal
Switching Behavior of a *pn*-Diode • Diode Reverse Breakdown •
Zener and Avalanche Diodes • Varactor Diodes • Tunnel Diodes •
Photodiodes and Solar Cells • Schottky Barrier Diode

20.3   Electrical Equivalent Circuit Models and Device Simulators
for Semiconductor Devices 460
Overview of Equivalent Circuit Models • Overview of Semiconductor Device Simulators

20.1 Physical Properties

Gennady Sh. Gildenblat and Boris Gelmont

Electronic applications of semiconductors are based on our ability to vary their properties on a very small scale. In conventional semiconductor devices, one can easily alter charge carrier concentrations, fields, and current densities over distances of 0.1–10 μm. Even smaller characteristic lengths of 10–100 nm are feasible in materials with an engineered band structure. This section reviews the essential physics underlying modern semiconductor technology.

Energy Bands

In crystalline semiconductors atoms are arranged in periodic arrays known as crystalline lattices. The lattice structure of silicon is shown in Fig. 20.1. Germanium and diamond have the same structure but with different interatomic distances. As a consequence of this periodic arrangement, the allowed energy levels of electrons are grouped into **energy bands,** as shown in Fig. 20.2. The probability that an electron will occupy an allowed quantum state with energy E is

$$f = [1 + \exp(E - F)/k_B T]^{-1} \qquad (20.1)$$

435

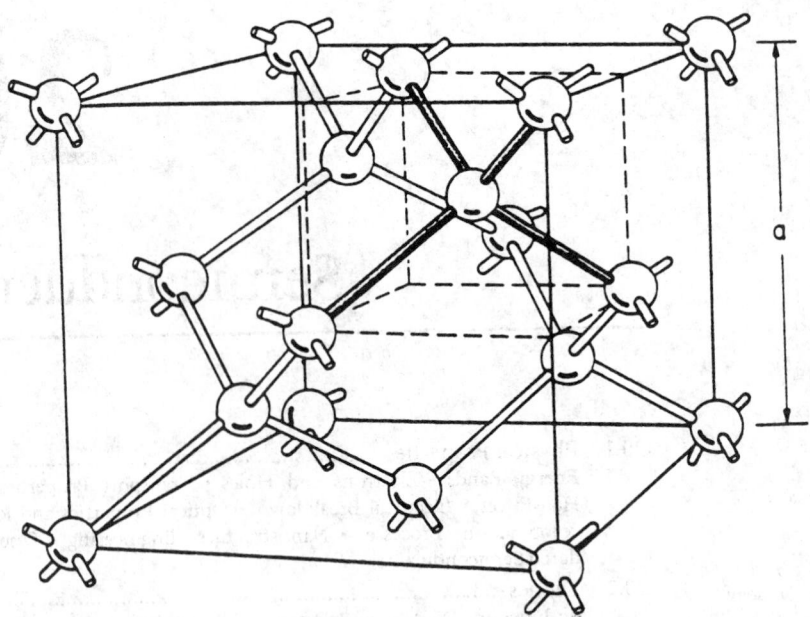

FIGURE 20.1   Crystalline lattice of silicon, $a = 5.43$ Å at 300°C. (*Source:* S.M. Sze, *Semiconductor Devices: Physics and Technology,* New York: John Wiley & Sons, 1985, p. 5. With permission.)

Here $k_B = 1/11,606$ eV/K denotes the Boltzmann constant, T is the absolute temperature, and F is a parameter known as the Fermi level. If the energy $E > F + 3k_BT$, then $f(E) < 0.05$ and these states are mostly empty. Similarly, the states with $E < F - 3k_BT$ are mostly occupied by electrons. In a typical metal [Fig. 20.2(a)], the energy level $E = F$ is allowed, and only one energy band is partially filled. (In metals like aluminum, the partially filled band in Fig. 20.2(a) may actually represent a combination of several overlapping bands.) The remaining energy bands are either completely filled or totally empty. Obviously, the empty energy bands do not contribute to the charge transfer. It is a fundamental result of solid-state physics that energy bands that are completely filled also do not contribute. What happens is that in the filled bands the average velocity of electrons is equal to zero. In semiconductors (and insulators) the Fermi level falls within a forbidden **energy gap** so that two of the energy bands are partially filled by electrons and may give rise to electron current. The upper partially filled band is called the **conduction band** while the lower is known as the **valence band**. The number of electrons in the conduction band of a semiconductor is relatively small and can be easily changed by adding impurities. In metals, the number of free carriers is large and is not sensitive to doping.

A more detailed description of energy bands in a crystalline semiconductor is based on the Bloch theorem, which states that an electron wave function has the form (Bloch wave)

$$\mathbf{\Psi}_{b\mathbf{k}} = u_{b\mathbf{k}}(\mathbf{r}) \exp(i\mathbf{k}\mathbf{r}) \tag{20.2}$$

where $\mathbf{r}$ is the radius vector, the modulating function $u_{b\mathbf{k}}(\mathbf{r})$ has the periodicity of the lattice, and the quantum state is characterized by wave vector $\mathbf{k}$ and the band number b. Physically, Eq. (20.2) means that an electron wave propagates through a periodic lattice without attenuation. For each energy band one can consider the dispersion law $E = E_b(\mathbf{k})$. Since (see Fig. 20.2) in the conduction band only the states with energies close to the bottom, E_c, are occupied, it suffices to consider the $E(\mathbf{k})$ dependence near E_c. The simplified band diagrams of Si and GaAs are shown in Fig. 20.3.

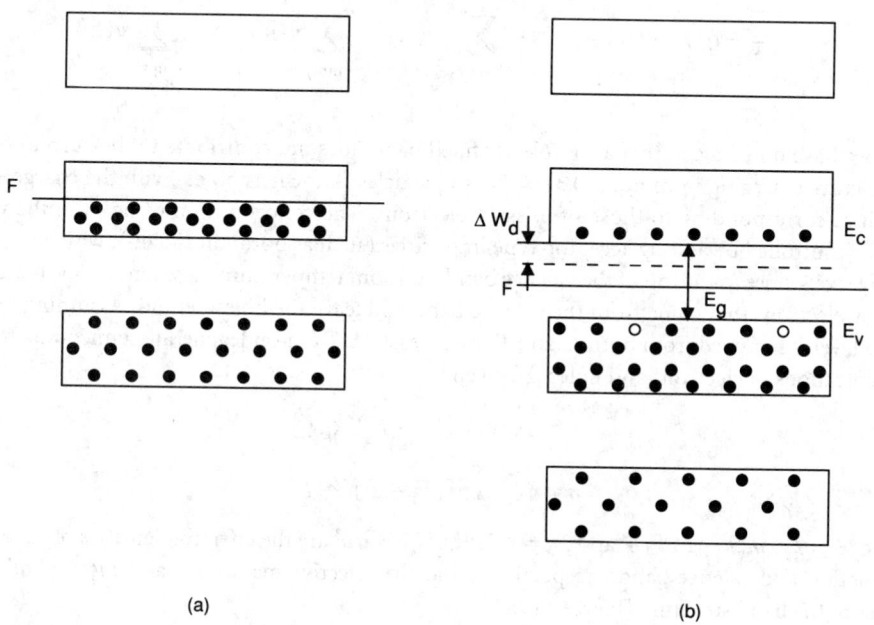

FIGURE 20.2 Band diagrams of metal (a) and semiconductor (b); ●, electron; ○, missing electron (hole).

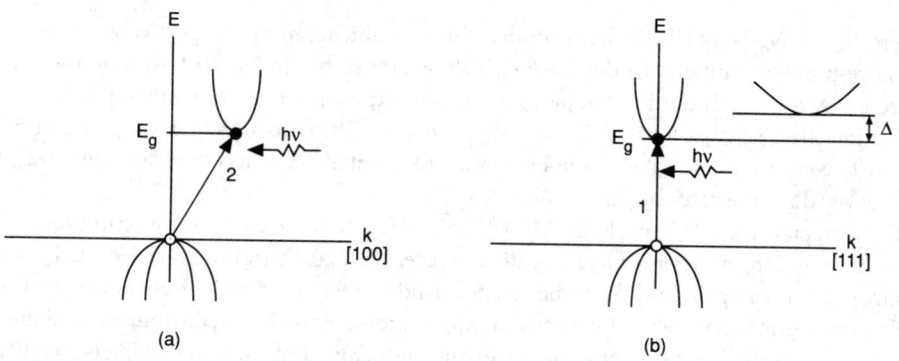

FIGURE 20.3 Simplified $E(\mathbf{k})$ dependence for Si (a) and GaAs (b). At room temperature E_g(Si) = 1.12 eV, E_g(GaAs) = 1.43 eV, and Δ = 0.31 eV; (1) and (2) indicate direct and indirect band-to-band transitions.

Electrons and Holes

The concentration of electrons in the valence band can be controlled by introducing impurity atoms. For example, the substitutional doping of Si with As results in a local energy level with an energy about $\Delta W_d \approx 45$ meV below the conduction band edge, E_c [Fig. 20.2(b)]. At room temperature this impurity center is readily ionized, and (in the absence of other impurities) the concentration of electrons is close to the concentration of As atoms. Impurities of this type are known as **donors**.

While considering the contribution $\mathbf{j}_p$ of the predominantly filled valence band to the current density, it is convenient to concentrate on the few missing electrons. This is achieved as follows:

$$\mathbf{j}_p = -q \sum_{\substack{\text{filled} \\ \text{states}}} \mathbf{v}(\mathbf{k}) = -q \left[\sum_{\text{all states}} \mathbf{v}(\mathbf{k}) - \sum_{\substack{\text{empty} \\ \text{states}}} \mathbf{v}(\mathbf{k}) \right] = q \sum_{\substack{\text{empty} \\ \text{states}}} \mathbf{v}(\mathbf{k}) \qquad (20.3)$$

Here we have noted again that a completely filled band does not contribute to the current density. The picture emerging from Eq. (20.3) is that of particles (known as **holes**) with the charge $+q$ and velocities corresponding to those of missing electrons. The concentration of holes in the valence band is controlled by adding acceptor-type impurities (such as boron in silicon), which form local energy levels close to the top of the valence band. At room temperature these energy levels are occupied by electrons that come from the valence band and leave the holes behind. Assuming that the Fermi level is removed from both E_c and E_v by at least $3k_BT$ (a nondegenerate semiconductor), the concentrations of electrons and holes are given by

$$n = N_c \exp[(F - E_c)/k_BT] \qquad (20.4)$$

and

$$p = N_v \exp[(E_v - F)/k_BT] \qquad (20.5)$$

where $N_c = 2\,(2m_n^*\pi k_BT)^{3/2}/h^3$ and $N_v = 2(2m_p^*\pi k_BT)^{3/2}/h^3$ are the effective densities of states in the conduction and valence bands, respectively, and the effective masses m_n^* and m_p^* depend on the details of the band structure [Pierret, 1987].

In a nondegenerate semiconductor, $np = N_cN_v \exp(-E_g/k_BT) \triangleq n_i^2$ is independent of the doping level. The neutrality condition can be used to show that in an n-type $(n > p)$ semiconductor at or below room temperature

$$n(n + N_a)(N_d - N_a - n)^{-1} = (N_c/2)\exp(-\Delta W_d/k_BT) \qquad (20.6)$$

where N_d and N_a denote the concentrations of donors and **acceptors**, respectively.

Corresponding temperature dependence is shown for silicon in Fig. 20.4. Around room temperature $n = N_d - N_a$, while at low temperatures n is an exponential function of temperature with the activation energy $\Delta W_d/2$ for $n > N_a$ and ΔW_d for $n < N_a$. The reduction of n compared with the net impurity concentration $N_d - N_a$ is known as a freeze-out effect. This effect does not take place in the heavily doped semiconductors.

For temperatures $T > T_i = (E_g/2k_B)/\ln[\sqrt{N_cN_v}/(N_d - N_a)]$ the electron concentration $n \approx n_i \gg N_d - N_a$ is no longer dependent on the doping level (Fig. 20.4). In this so-called intrinsic regime electrons are coming directly from the valence band. A loss of technological control over n and p makes this regime unattractive for electronic applications. Since $T_i \propto E_g$ the transition to the intrinsic region can be delayed by using widegap semiconductors. Both silicon carbide (several types of SiC with different lattice structures are available with $E_g = 2.2$–2.86 eV) and diamond ($E_g = 5.5$ eV) have been used to fabricate diodes and transistors operating in the 300–700°C temperature range.

Transport Properties

In a semiconductor the motion of an electron is affected by frequent collisions with **phonons** (quanta of lattice vibrations), impurities, and crystal imperfections. In weak uniform electric fields, $\mathscr{E}$, the carrier drift velocity, $\mathbf{v}_d$, is determined by the balance of the electric and collision forces:

$$m_n^*\mathbf{v}_d/\tau = -q\mathscr{E} \qquad (20.7)$$

where τ is the momentum relaxation time. Consequently $\mathbf{v}_d = -\mu_n\,\mathscr{E}$, where $\mu_n = q\tau/m_n^*$ is the electron mobility. For an n-type semiconductor with uniform electron density, n, the current density $\mathbf{j}_n = -qn\mathbf{v}_d$ and we obtain Ohm's law $\mathbf{j}_n = \sigma\mathscr{E}$ with the conductivity $\sigma = qn\mu_n$. The momentum relaxation time can be approximately expressed as

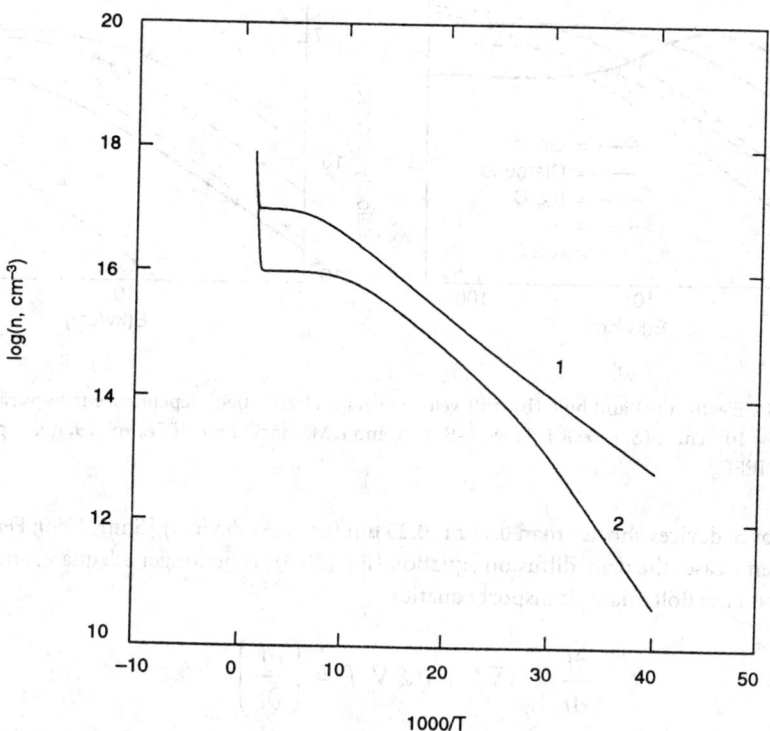

FIGURE 20.4 The inverse temperature dependence of electron concentration in Si; 1: $N_d = 10^{17}$ cm^{-3}, $N_a = 0$; 2: $N_d = 10^{16}$ cm^{-3}, $N_a = 10^{14}$ cm^{-3}.

$$1/\tau = 1/\tau_{ii} + 1/\tau_{ni} + 1/\tau_{ac} + 1/\tau_{npo} + 1/\tau_{po} + 1/\tau_{pe} + \cdots \qquad (20.8)$$

where τ_{ii}, τ_{ni}, τ_{ac}, τ_{npo}, τ_{po}, τ_{pe} are the relaxation times due to ionized impurity, neutral impurity, acoustic phonon, nonpolar optical, polar optical, and piezoelectric scattering, respectively.

In the presence of concentration gradients, electron current density is given by the drift-diffusion equation

$$\mathbf{j}_n = qn\mu_n \mathscr{E} + qD_n \nabla n \qquad (20.9)$$

where the diffusion coefficient is related to mobility by the Einstein relation $D_n = (k_B T/q)\mu_n$.

Similar equations can be written for holes and the total current density is $\mathbf{j} = \mathbf{j}_n + \mathbf{j}_p$. The right-hand side of Eq. (20.9) may contain additional terms corresponding to temperature gradient and compositional nonuniformity of the material [Wolfe et al., 1989].

In sufficiently strong electric fields the drift velocity is no longer proportional to the electric field. Typical velocity–field dependencies for several semiconductors are shown in Fig. 20.5. In GaAs $v_d(\mathscr{E})$ dependence is not monotonic, which results in negative differential conductivity. Physically, this effect is related to the transfer of electrons from the conduction band to a secondary valley (see Fig. 20.3).

The limiting value v_s of the drift velocity in a strong electric field is known as the saturation velocity and is usually within the 10^7–$3\cdot10^7$ cm/s range. As semiconductor device dimensions are scaled down to the submicrometer range, v_s becomes an important parameter that determines the upper limits of device performance. The curves shown in Fig. 20.5 were obtained for uniform semiconductors under steady-state conditions. Strictly speaking, this is not the case with actual semiconductor devices, where velocity can "overshoot" the value shown in Fig. 20.5. This effect is

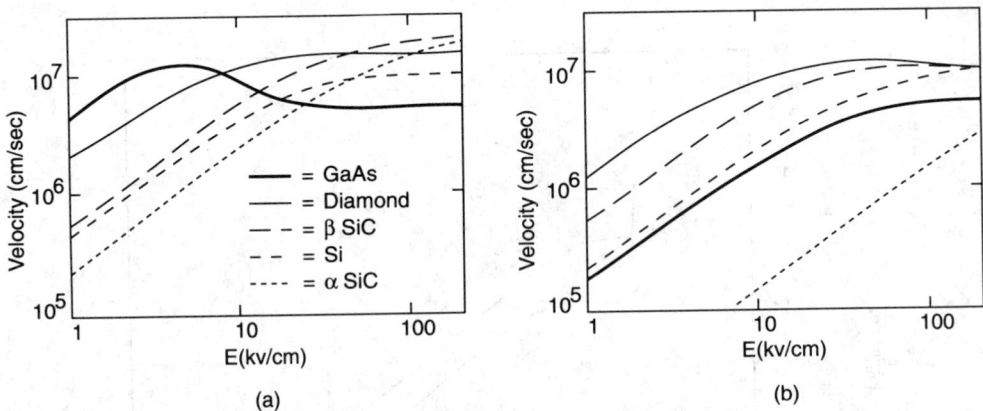

FIGURE 20.5 Electron (a) and hole (b) drift velocity versus electric field dependence for several semiconductors at $N_d = 10^{17}$ cm^{-3}. (*Source:* R.J. Trew, J.-B. Yan, and L.M. Mack, *Proc. IEEE*, vol. 79, no. 5, p. 602, May 1991. © 1991 IEEE.)

important for Si devices shorter than $0.1\,\mu$m ($0.25\,\mu$m for GaAs devices) [Shur, 1990; Ferry, 1991]. In such extreme cases the drift-diffusion equation [Eq. (20.9)] is no longer adequate, and the analysis is based on the Boltzmann transport equation

$$\frac{\partial f}{\partial t} + \mathbf{v}\nabla f + q\mathscr{E}\,\nabla_p f = \left(\frac{\partial f}{\partial t}\right)_{\text{coll}} \tag{20.10}$$

Here f denotes the distribution function (number of electrons per unit volume of the phase space), $\mathbf{v}$ is electron velocity, $\mathbf{p}$ is momentum, and $(\partial f/\partial t)_{\text{coll}}$ is the "collision integral" describing the change of f caused by collision processes described earlier. For the purpose of semiconductor modeling, Eq. (20.10) can be solved directly using various numerical techniques, including the method of moments (hydrodynamic modeling) or Monte Carlo approach. The drift-diffusion equation [Eq. (20.9)] follows from Eq. (20.10) as a special case. For even shorter devices quantum effects become important and device modeling may involve quantum transport theory [Ferry, 1991].

Hall Effect

In a uniform magnetic field electrons move along circular orbits in a plane normal to the magnetic field $\mathbf{B}$ with the angular (cyclotron) frequency $\omega_c = qB/m_n^*$. For a uniform semiconductor the current density satisfies the equation

$$\mathbf{j} = \sigma(\mathscr{E} + R_H[\mathbf{jB}]) \tag{20.11}$$

In the usual weak-field limit $\omega_c\tau \ll 1$ the Hall coefficient $R_H = -r/nq$ and the Hall factor r depend on the dominating scattering mode. It varies between $3\pi/8 \approx 1.18$ (acoustic phonon scattering) and $315\pi/518 \approx 1.93$ (ionized impurity scattering).

The Hall coefficient can be measured as $R_H = V_y d/I_x B$ using the test structure shown in Fig. 20.6. In this expression V_y is the Hall voltage corresponding to $I_y = 0$ and d denotes the film thickness.

Combining the results of the Hall and conductivity measurements one can extract the carrier concentration type (the signs of V_y are opposite for n-type and p-type semiconductors) and Hall mobility $\mu_H = r\mu$:

$$\mu_H = -R_H\sigma, \quad n = -r/qR_H \tag{20.12}$$

Measurements of this type are routinely used to extract concentration and mobility in doped semiconductors. The weak-field Hall effect is also used for the purpose of magnetic field measurements.

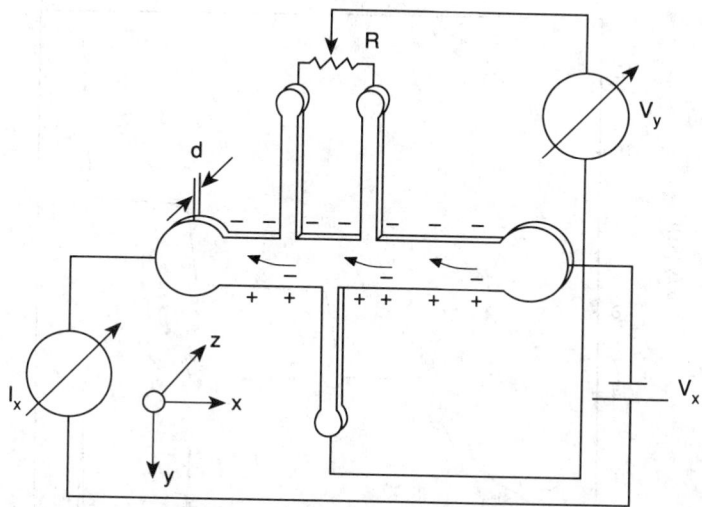

FIGURE 20.6 Experimental setup for Hall effect measurements in a long two-dimensional sample. The Hall angle is determined by a setting of the rheostat that renders $j_y = 0$. Magnetic field $B = B_z$. (*Source:* K.W. Böer, *Surveys of Semiconductor Physics*, New York: Van Nostrand Reinhold, 1990, p. 760. With permission.)

In strong magnetic fields $\omega_c \tau \gg 1$ and on the average an electron completes several circular orbits without a collision. Instead of the conventional $E_b(\mathbf{k})$ dependence, the allowed electron energy levels in the magnetic field are given by ($s = 0, 1, 2, \ldots$)

$$E_s = \hbar\omega_c\,(s + 1/2) + \hbar^2 k_z^2 / 2m_n^* \tag{20.13}$$

The first term in Eq. (20.13) describes the so-called Landau levels, while the second corresponds to the kinetic energy of motion along the magnetic field $B = B_z$. In a pseudo-two-dimensional system like the channel of a field-effect transistor the second term in Eq. (20.13) does not appear, since the motion of electrons takes place in the plane perpendicular to the magnetic field.[1] In such a structure the electron density of states (number of allowed quantum states per unit energy interval) is peaked at the Landau level. Since $\omega_c \propto B$, the positions of these peaks relative to the Fermi level are controlled by the magnetic field.

The most striking consequence of this phenomenon is the quantum Hall effect, which manifests itself as a stepwise change of the Hall resistance $\rho_{xy} = V_y/I_x$ as a function of magnetic field (see Fig. 20.7). At low temperature (required to establish the condition $\tau \ll \omega_c^{-1}$) it can be shown [von Klitzing, 1986] that

$$\rho_{xy} = h/sq^2 \tag{20.14}$$

where s is the number of the highest occupied Landau level. Accordingly, when the increased magnetic field pushes the sth Landau level above the Fermi level, ρ_{xy} changes from h/sq^2 to $h/(s-1)q^2$. This stepwise change of ρ_{xy} is clearly seen in Fig. 20.7. Localized states produced by crystal defects determine the shape of the $\rho_{xy}(B)$ dependence between the plateaus given by Eq. (20.14). They are also responsible for the disappearance of $\rho_{xx} = V_x/I_x$ between the transition points (see Fig. 20.7). The quantized Hall resistance ρ_{xy} is expressed in terms of fundamental constants and can be used as a resistance standard that permits one to measure an electrical resistance with better accuracy

[1] To simplify the matter we do not discuss surface subbands, which is justified as long as only the lowest of them is occupied.

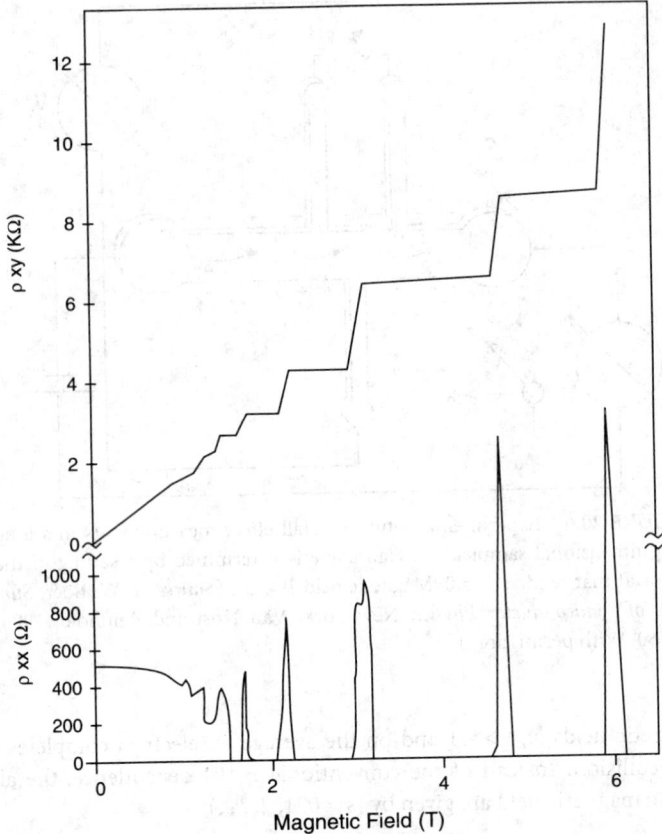

FIGURE 20.7 Experimental curves for the Hall resistance $\rho_{xy} = \mathscr{E}_y/j_x$ and the resistivity $\rho_{xx} = \mathscr{E}_x/j_x$ of a heterostructure as a function of the magnetic field at a fixed carrier density. (*Source:* K. von Klitzing, *Rev. Modern Phys.*, vol. 58, no. 3, p. 525, 1986. With permission.)

than any wire resistor standard. In an ultraquantum magnetic field, i.e., when only the lowest Landau level is occupied, plateaus of the Hall resistance are also observed at fractional s (the fractional quantum Hall effect). These plateaus are related to the Coulomb interaction of electrons.

Electrical Breakdown

In sufficiently strong electric fields a measurable fraction of electrons (or holes) acquires the energy required to break the valence bond. Such an event (called impact ionization) results in the creation of an electron–hole pair by the energetic electron. Both the primary and secondary electrons as well as the hole are accelerated by the electric field and may participate in further acts of impact ionization. Usually, the impact ionization is balanced by recombination processes. If the applied voltage is high enough, however, the process of electron multiplication leads to avalanche breakdown. The threshold energy E_{th} (the minimum electron energy required to produce an electron–hole pair) is determined by energy and momentum conservation laws. The latter usually results in $E_{th} > E_g$, as shown in Table 20.1.

The field dependence of the impact ionization is usually described in terms of the impact ionization coefficient α_i, defined as the average number of electron–hole pairs created by a charge carrier per unit distance traveled. A simple analytical expression for α_i [Okuto and Crowell, 1972] can be written as

Table 20.1   Impact Ionization Threshold Energy (eV)

Semiconductor	Si	Ge	GaAs	GaP	InSb
Energy gap, E_g	1.1	0.7	1.4	2.3	0.2
E_{th}, electron-initiated	1.18	0.76	1.7	2.6	0.2
E_{th}, hole-initiated	1.71	0.88	1.4	2.3	0.2

$$\alpha_i = (\lambda/x) \exp\left(a - \sqrt{a^2 + x^2} \right) \tag{20.15}$$

where $x = q \, \mathscr{E} \lambda / E_{th}$, $a = 0.217 \, (E_{th}/E_{opt})^{1.14}$, λ is the carrier mean free path, and E_{opt} is the optical phonon energy ($E_{opt} = 0.063$ eV for Si at 300°C).

An alternative breakdown mechanism is tunneling breakdown, which occurs in highly doped semiconductors when electrons may tunnel from occupied states in the valence band into the empty states of the conduction band.

Optical Properties and Recombination Processes

If the energy of an incident **photon** $\hbar\omega > E_g$, the energy conservation law permits a direct band-to-band transition, as indicated in Fig. 20.2(b). Because the photon's momentum is negligible compared to that of an electron or hole, the electron's momentum $\hbar\mathbf{k}$ does not change in a direct transition. Consequently, direct transitions are possible only in direct-gap semiconductors where the conduction band minimum and the valence band maximum occur at the same $\mathbf{k}$. The same is true for the reverse transition, where the electron is transferred from the conduction to the valence band and a photon is emitted. Direct-gap semiconductors (e.g., GaAs) are widely used in optoelectronics.

In indirect-band materials [e.g., Si, see Fig. 20.3(a)], a band-to-band transition requires a change of momentum that cannot be accomplished by absorption or emission of a photon. Indirect band-to-band transitions require the emission or absorption of a phonon and are much less probable than direct transitions.

For $\hbar\omega < E_g$ [i.e., for $\lambda > \lambda_c = 1.24$ μm/E_g (eV) – cutoff wavelength] band-to-band transitions do not occur, but light can be absorbed by a variety of the so-called subgap processes. These processes include the absorption by free carriers, formation of excitons (bound electron–hole pairs whose formation requires less energy than the creation of a free electron and a free hole), transitions involving localized states (e.g., from an acceptor state to the conduction band), and phonon absorption. Both band-to-band and subgap processes may be responsible for the increase of the free charge carriers concentration. The resulting reduction of the resistivity of illuminated semiconductors is called *photoconductivity* and is used in photodetectors.

In a strong magnetic field ($\omega_c\tau \gg 1$) the absorption of the microwave radiation is peaked at $\omega = \omega_c$. At this frequency the photon energy is equal to the distance between two Landau levels, i.e., $\hbar\omega = E_{S+1} - E_S$ with reference to Eq. (20.13). This effect, known as cyclotron resonance, is used to measure the effective masses of charge carriers in semiconductors [in a simplest case of isotropic $E(\mathbf{k})$ dependence, $m_n^* = qB/\omega_c$].

In indirect-gap materials like silicon, the generation and annihilation (or recombination) of electron–hole pairs is often a two-step process. First, an electron (or a hole) is trapped in a localized state (called a recombination center) with the energy near the center of the energy gap. In a second step, the electron (or hole) is transferred to the valence (conduction) band. The net rate of recombination per unit volume per unit time is given by the Shockley–Read–Hall theory as

$$R = \frac{np - n_i^2}{\tau_n(p + p_1) + \tau_p(n + n_1)} \tag{20.16}$$

where τ_n, τ_p, p_1, and n_1 are parameters depending on the concentration and the physical nature of recombination centers and temperature. Note that the sign of R indicates the tendency of a semiconductor toward equilibrium (where $np = n_i^2$, and $R = 0$). For example, in the depleted region $np < n_i^2$ and $R < 0$, so that charge carriers are generated.

Shockley–Read–Hall recombination is the dominating recombination mechanism in moderately doped silicon. Other recombination mechanisms (e.g., Auger) become important in heavily doped semiconductors [Wolfe *et al.*, 1989; Shur, 1990; Ferry, 1991].

The recombination processes are of prime importance for semiconductor device theory, where they are usually modeled using the continuity equation

$$\frac{\partial n}{\partial t} = div \frac{j_n}{q} - R \tag{20.17}$$

Nanostructure Engineering

Epitaxial growth techniques, especially molecular beam epitaxy and metal-organic chemical vapor deposition, allow one to achieve a monolayer control in the chemical composition process. Both single thin layers and superlattices can be obtained by such methods. The electronic properties of these structures are of interest for potential device applications. In a single quantum well, electrons are bound in the confining well potential. For example, in a rectangular quantum well of width b with infinite walls, the allowed energy levels are

$$E_s(\mathbf{k}) = \pi^2 s^2 \hbar^2/(2m_n^* b^2) + \hbar^2 k^2/(2m_n^*), \quad s = 1, 2, 3, \ldots \tag{20.18}$$

where $\mathbf{k}$ is the electron wave vector parallel to the plane of the semiconductor layer. The charge carriers in quantum wells exhibit confined particle behavior. Since $E_s \propto b^{-2}$, well structures can be grown with distance between energy levels equal to a desired photon energy. Furthermore, the photoluminescence intensity is enhanced because of carrier confinement. These properties are advantageous in fabrication of lasers and photodetectors.

If a quantum well is placed between two thin barriers, the tunneling probability is greatly enhanced when the energy level in the quantum well coincides with the Fermi energy (resonant tunneling). The distance between this "resonant" energy level and the Fermi level is controlled by the applied voltage. Consequently, the current is peaked at the voltage corresponding to the resonant tunneling condition. The resulting negative differential resistance effect has been used to fabricate microwave generators operating at both room and cryogenic temperatures.

Two kinds of superlattices are possible: compositional and doping. Compositional superlattices are made of alternating layers of semiconductors with different energy gaps. Doping superlattices consist of alternating n- and p-type layers of the same semiconductor. The potential is modulated by electric fields arising from the charged dopants. Compositional superlattices can be grown as lattice matched or as strained layers. The latter are used for modification of the band structure, which depends on the lattice constant to produce desirable properties.

In superlattices energy levels of individual quantum wells are split into minibands as a result of electron tunneling through the wide-bandgap layers. This occurs if the electron mean free path is larger than the superlattice period. In such structures the electron motion perpendicular to the layer is quantized. In a one-dimensional tight binding approximation the miniband can be described as

$$E(k) = E_o[1 - \cos(ka)] \tag{20.19}$$

where a is the superlattice period and E_o is the half-width of the energy band. The electron group velocity

$$v = \hbar^{-1}\partial E(k)/\partial k = (E_o a/\hbar)\sin(ka) \qquad (20.20)$$

is a decreasing function of k (and hence of energy) for $k > \pi/2a$. The higher energy states with $k > \pi/2a$ may become occupied if the electrons are heated by the external field. As a result, a negative differential resistance can be achieved at high electric fields. The weak-field mobility in a superlattice may exceed that of the bulk material because of the separation of dopants if only barriers are doped. In such modulated structures, the increased spatial separation between electrons and holes is also responsible for a strong increase in the recombination lifetimes.

Disordered Semiconductors

Both amorphous and heavily doped semiconductors are finding increasing applications in semiconductor technology. The electronic processes in these materials have specific features arising from the lack of long-range order.

Amorphous semiconductors do not have a crystalline lattice, and their properties are determined by the arrangement of the nearest neighboring atoms. Even so, experimental data show that the forbidden energy band concept can be applied to characterize their electrical properties. However, the disordered nature of these materials results in a large number of localized quantum states with energies within the energy gap. The localized states in the upper and lower half of the gap behave like acceptors and donors, respectively. As an example, consider the density of states in hydrogenated amorphous silicon (a-Si) shown in Fig. 20.8. The distribution of the localized states is not symmetrical with respect to the middle of the energy gap. In particular, the undoped hydrogenated amorphous silicon is an n-type semiconductor.

Usually amorphous semiconductors are not sensitive to the presence of impurity atoms, which saturate all their chemical bonds in the flexible network of the host atoms. (Compare this with a situation in crystalline silicon where an arsenic impurity can form only four chemical bonds with the host lattice, leaving the fifth responsible for the formation of the donor state.) Consequently, the doping of amorphous semiconductors is difficult to accomplish. However, in hydrogenated a-Si (which can be prepared by the glow discharge decomposition of silane), the density of the localized states is considerably reduced and the conductivity of this material can be controlled by doping. As in crystalline semiconductors, the charge carrier concentration in hydrogenated a-Si can also be affected by light and strong field effects. The a-Si is used in applications that require deposition of thin-film semiconductors over large areas [xerography, solar cells, thin-film transistors (TFT) for liquid-crystal displays]. The a-Si device performance degrades with time under electric stress (TFTs) or under illumination (Staebler–Wronski effect) because of the creation of new localized states.

An impurity band in crystalline semiconductors is another example of a disordered system. Indeed, the impurity atoms are randomly distributed within the host lattice. For lightly doped semiconductors at room temperature, the random potential associated with charged impurities can usually be ignored. As the doping level increases, however, a single energy level of a donor or an acceptor is transformed into an energy band with a width determined by impurity concentrations. Unless the degree of compensation is unusually high, this results in the reduction of the activation energy compared to lightly doped semiconductors. The activation energy is further reduced because of the overlap of the wave functions associated with the individual donor or acceptor states.

For sufficiently heavy doping, i.e., for $N_d > N_{dc} = (0.2/a_B)^3$, the ionization energy is reduced to zero, and the transition to metal-type conductivity (the Anderson–Mott transition) takes place. In this expression the effective electron Bohr radius $a_B = \hbar/\sqrt{2m_n^* E_i}$, where E_i is the ionization energy of the donor state. For silicon, $N_{dc} \approx 3.8 \cdot 10^{18}$ cm^{-3}. This effect explains the absence of freeze-out in heavily doped semiconductors.

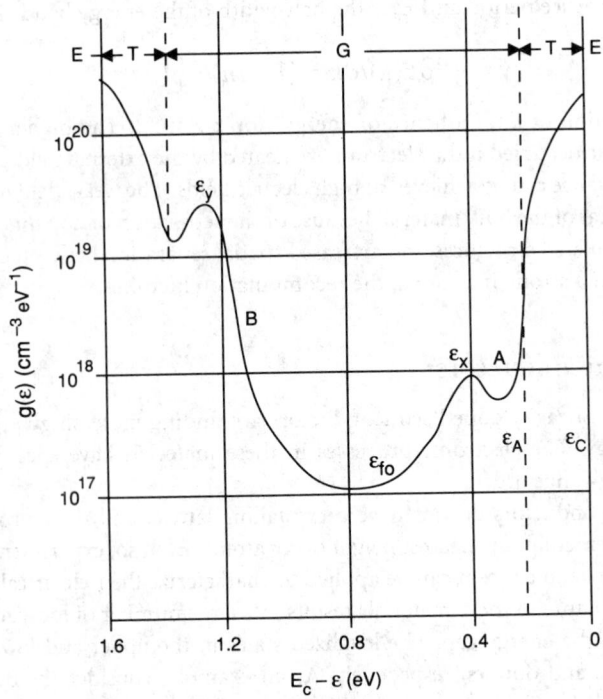

FIGURE 20.8 Experimentally determined density of states for a-Si. *A* and *B* are acceptor-like and donor-like states, respectively. The arrow marks the position of the Fermi level ε_{fo} in undoped hydrogenated a-Si. The energy spectrum is divided into extended states *E*, band-tail states *T*, and gap states *G*. (*Source:* M.H. Brodsky, Ed., *Amorphous Semiconductors*, 2nd ed., Berlin: Springer-Verlag, 1985. With permission.)

Defining Terms

Conduction/valence band: The upper/lower of the two partially filled bands in a semiconductor.

Donors/acceptors: Impurities that can be used to increase the concentration of electrons/holes in a semiconductor.

Energy band: Continuous interval of energy levels that are allowed in the periodic potential field of the crystalline lattice.

Energy gap: The width of the energy interval between the top of the valence band and the bottom of the conduction band.

Hole: Fictitious positive charge representing the motion of electrons in the valence band of a semiconductor; the number of holes equals the number of unoccupied quantum states in the valence band.

Phonon: Quantum of lattice vibration.

Photon: Quantum of electromagnetic radiation.

References

D. K. Ferry, *Semiconductors*, New York: Macmillan, 1991.

Y. Okuto and C. R. Crowell, *Phys. Rev.*, vol. B6, p. 3076, 1972.

R. F. Pierret, *Advanced Semiconductor Fundamentals*, Reading, Mass.: Addison-Wesley, 1987.

M. Shur, *Physics of Semiconductor Devices*, Englewood Cliffs, N.J.: Prentice Hall, 1990.

K. von Klitzing, *Rev. Modern Phys.*, vol. 58, p. 519, 1986.

C.M. Wolfe, N. Holonyak, and G.E. Stilman, *Physical Properties of Semiconductors*, Englewood Cliffs, N.J.: Prentice Hall, 1989.

Further Information

Engineering aspects of semiconductor physics are often discussed in the *IEEE Transactions on Electron Devices, Journal of Applied Physics*, and *Solid-State Electronics*.

20.2 Diodes

Miran Milkovic

Diodes are the most widely used devices in low- and high-speed electronic circuits and in rectifiers and power supplies. Other applications are in voltage regulators, detectors, and demodulators. Rectifier diodes are capable of conducting several hundred amperes in the forward direction and less than 1 μA in the reverse direction. Zener diodes are ordinary diodes operated in the Zener or avalanche region and are used as voltage regulators. Varactor diodes are ordinary diodes used in reverse biasing as voltage-dependent capacitors. Tunnel diodes and quantum well devices have a negative differential resistance and are capable of operating in the upper gigahertz region. Photodiodes are ordinary diodes operated in the reverse direction. They are sensitive to light and are used as light sensors. Solar cells are diodes which convert light energy into electrical energy. Schottky diodes, also known as metal-semiconductor diodes, are extremely fast because they are **majority carrier** devices.

pn-Junction Diode

Modern planar semiconductor *pn*-junction diodes are fabricated by **diffusion** or implantation of impurities into a doped semiconductor. An *n*-type semiconductor has a relatively large density of free electrons to conduct electric current, and the *p*-type semiconductor has a relatively large concentration of "free" holes to conduct electric current. The *pn*-junction is formed during the fabrication process. There is a large concentration of holes in the *p*-semiconductor and a large concentration of electrons in the *n*-semiconductor. Because of their large concentration gradients, holes and electrons start to diffuse across the junction. As holes move across the junction, negative immobile charges (**acceptors**) are uncovered on the *p* side, and positive immobile charges (**donors**) are uncovered on the *n* side. When a sufficient number of the immobile charges on both sides of the junction are uncovered, a potential energy barrier V_0 is created by the uncovered acceptors and donors. This **barrier voltage** prevents further diffusion of holes and electrons across the junction. The charge distribution of acceptors and donors establishes an opposing electric field, E, which at equilibrium prevents a further diffusion of carriers. This equilibrium can be regarded as the flow of two equal and opposite currents across the junction, such that the net current across the junction is equal to zero. Thus, one component represents the diffusion of carriers across the junction and the other component represents the **drift** of carriers across the junction due to the electric field E in the junction. The barrier voltage V_0 is, according to the **Boltzmann relation**,

$$V_0 = V_T \ln[p_p/p_n] \qquad (20.21)$$

In this equation, p_p is the concentration of holes in the *p*-material and p_n is the concentration of electrons in the *n*-material. V_T is the thermal voltage. $V_T = 26$ mV at room temperature (300 K). With

$$p_p \approx N_A \quad \text{and} \quad p_n \approx N_D$$

the barrier voltage becomes

$$V_0 = V_T \ln[N_A N_D / n^2] \qquad (20.22)$$

Here N_A is the concentration of immobile acceptors on the p side of the junction and N_D is the concentration of immobile donors on the n side of the junction. A depletion layer of immobile acceptors and donors causes an electric field E across the junction. For silicon, V_0 is typically $V_0 = 0.7$ V for an abrupt junction with $N_A = 10^{17}$ at/cm^3 and $N_D = 10^{15}$ at/cm^3. The depletion layer width is typically about 4 μm, and the electric field E is about $E = 60$ kV/cm. Note the magnitude of the electric field across the junction.

pn-Junction with Applied Voltage

If the externally applied voltage V_D to the diode is opposite to the barrier voltage V_0, then the Boltzmann relation in Eq. (20.21) is altered to

$$p_p = p_n(x = 0) \exp(V_0 - V_D)/V_T \qquad (20.23)$$

This implies that the effective barrier voltage V_0 is reduced and the diffusion of carriers is increased. According to Eq. (20.21) the concentration of diffusing holes into the n material is at $x = 0$,

$$p_n(x = 0) = p_n \exp V_D/V_T \qquad (20.24)$$

and accordingly the concentration of electrons

$$n_n(x = 0) = n_n \exp V_D/V_T \qquad (20.25)$$

Most modern planar diodes are unsymmetrical. Figure 20.9 shows a *pn*-diode with the n region W_n much shorter than the diffusion length L_{pn} of holes. This results in a linear **concentration gradient** of injected diffusing holes in the n region given by

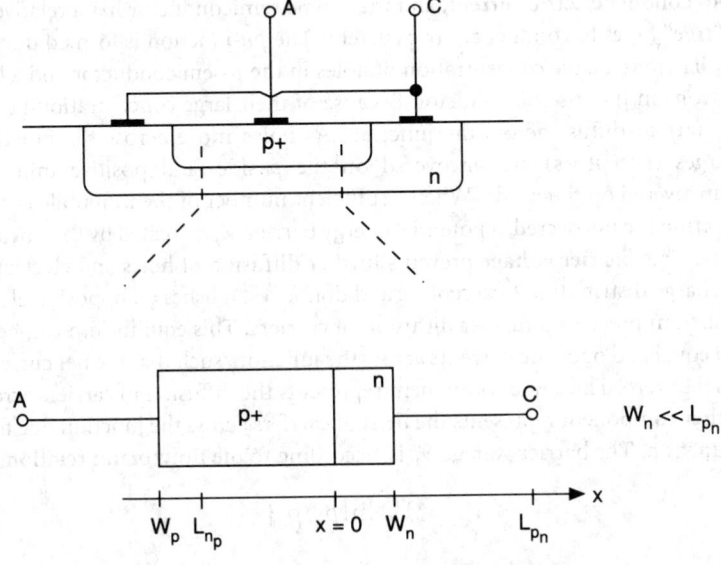

FIGURE 20.9 Planar diodes are fabricated in planar technology. Most modern diodes are unsymmetrical; thus $W_n \ll L_{pn}$. The p-type region is more highly doped than the n region.

$$dp/dx = -(p_n \exp V_D/V_T - p_n)/W_n \tag{20.26}$$

The diffusion gradient is negative since the concentration of holes decreases with distance due to the hole–electron recombinations. By using this equation with the equation for the hole diffusion current

$$I_p = -qA_j D_p \, dp/dx \tag{20.27}$$

A_j is the junction area, D_p is the **diffusion constant** for holes, and q is the elementary charge. Since $dp = p_n \exp V_0/V_T$ and $dx = W_n$, we get

$$I_p = (qA_j D_p p_n/W_n) \, (\exp V_D/V_T - 1) \tag{20.28}$$

In a p-semiconductor we assume that $L_{np} \ll W_p$; then

$$dn/dx = n_p \exp(V_D/V_T - 1) \tag{20.29}$$

By substituting this into the electron diffusion equation,

$$I_n = qA_j D_n \, dn/dx \tag{20.30}$$

or

$$I_n = (qA_j D_n n_p)/L_{np}(\exp V_D/V_T - 1) \tag{20.31}$$

Thus, the total junction diffusion current is

$$I_D = I_p + I_n = \{qA_j D_p p_n /W_n + qA_j D_n n_p/L_{np}\} \, (\exp V_D/V_T - 1) \tag{20.32}$$

Thus, the recombination of the injected carriers establishes a diffusion gradient which in turn yields a flow of current proportional to the slope. For $|-V_D| \gg V_T$, i.e., $V_D = -0.1$ V,

$$I_S = -(qA_i D_p p_n /W_n + qAD_n \, n_p/L_{np}) \tag{20.33}$$

Here I_S is the **reverse saturation current**. In practical junctions, the p region is usually much more heavily doped than the n region; thus $n_p \ll p_n$. Also, since $W_n \ll L_{np}$ in Eq. (20.33),

$$I_S = -qA_j D_p p_n/W_n = -qA_j D_p \, n_i^2/W_n N_D \tag{20.34}$$

The reverse saturation current in short diodes is mainly determined by the diffusion constant D_p and the width W_n of the n region, by intrinsic concentration n_i, by the doping concentration N_D in the n region, and by the diode area A_j. (In reality, I_S is also slightly dependent on the reverse voltage [Phillips, 1962].)

If V_D is made positive, the exponential term in Eq. (20.32) rapidly becomes larger than one; thus

$$I_D = I_S \exp V_D/V_T \tag{20.35}$$

where I_D is the diode forward current and I_S is the reverse saturation current.

Another mechanism predominates the reverse current I_S in silicon. Due to the recombination centers in the depletion region, generation-recombination hole–electron current I_G is generated in the depletion region [Phillips, 1962].

$$I_G = KqA_j eX_d \tag{20.36a}$$

Here e is the generation rate unit volume, A_j is the junction area, q is the elementary charge, X_d is the depletion layer thickness, and K is a dimensional constant. I_G is proportional to the thickness X_d of the depletion layer and to the junction area A_j. Since X_d increases with the square root of the reverse voltage, I_G increases accordingly, yielding a slight slope in the reverse I-V characteristic. The forward I-V characteristic of the practical diode is only slightly affected (slope $m = 2$) at very small forward currents ($I_D = 1$ nA to 1 μA). In practical diodes $n \approx 1$ at small to medium currents

($I_D = 1$ μA to 10 mA). At large currents ($I_D > 10$ mA), $m = 1$ to 2 due to the high current effects [Phillips, 1962] and due to the series bulk resistance of the diode.

The reverse current I_R in silicon is voltage dependent. The predominant effect is the voltage dependence of the generation-recombination current I_G and to a smaller extent the voltage dependence of I_S.

The total reverse current of the diode is thus equal to

$$I_R = I_G + I_S \tag{20.36b}$$

Forward-Biased Diode

For most practical applications

$$I_D = I_S \exp V_D/mV_T \tag{20.37}$$

where I_S is the reverse saturation current (about 10^{-14} A for a small-signal diode); $V_T = kT/q$ is the temperature voltage equal to 26 mV at room temperature; k = Boltzmann's constant, $1.38 \cdot 10^{-23}$ J/K; T is the absolute temperature in kelvin; q is the elementary charge $1.602 \cdot 10^{-19}$ C; m is the **ideality factor**, $m = 1$ for medium currents, $m = 2$ for very small and very large currents; I_S is part of the total reverse current I_R of the diode $I_R = I_S + I_G$; and I_S is the reverse saturation current and I_G is the generation-recombination current, also called diode leakage current because I_G is not a part of the carrier diffusion process in the diode. I_D is exponentially related to V_D in Fig. 20.10.

Temperature Dependence of V_D

Equation (20.37) solved for V_D yields

$$V_D = mV_T \ln(I_D/I_S) \tag{20.38}$$

at constant current I_D, V_D is temperature dependent because V_T and I_S are temperature dependent. Assume $m = 1$. The reverse saturation current I_S from Eq. (20.34) is

$$I_S = qA_j n_i^2 D_p/W_n N_D = B_1 n_i^2 D_p = B_2 n_i^2 \mu_p$$

where $D_p = V_T \mu_p$. With $\mu_p = B_3 T^{-n}$ and for n_i^2 [Gray and Meyer, 1984]

$$n_i^2 = B_4 T^{\gamma} \exp(-V_{G0}/V_T) \tag{20.39}$$

where $\gamma = 4 - n$, and V_{G0} is the extrapolated **bandgap energy** [Gray and Meyer, 1984]. With Eq. (20.39) into Eq. (20.38), the derivative dV_D/dT for $I_D = $ const yields

$$dV_D/dT = (V_D - V_{G0})/T - \gamma k/q \tag{20.40}$$

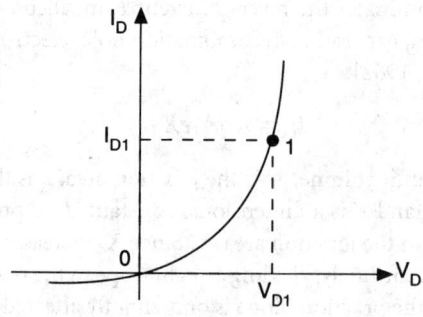

FIGURE 20.10 I_D versus V_D of a diode.

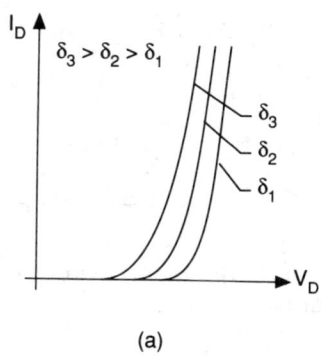

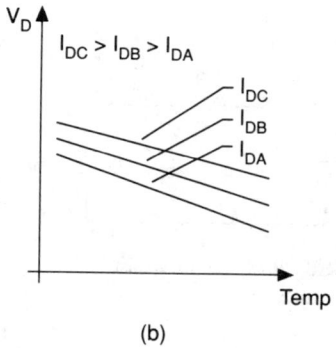

(a) (b)

FIGURE 20.11 (a) I_D versus V_D of a diode at three different temperatures $\delta_3 > \delta_2 > \delta_1$. (b) $V_D = f(\text{Temp})$, $I_{DC} > I_{DB} > I_{DA}$.

At room temperature ($T = 300$ K), and $V_D = 0.65$ V, $V_{G0} = 1.2$ V, $\gamma = 3$, $V_T = 26$ mV, and $k/q = 86$ μV/degree, one gets $dV_D/dT \approx -2.1$ mV/degree. The **temperature coefficient** T_C of V_D is thus

$$T_C = dV_D/V_D \, dT = 1/T - V_{G0}/V_D T - \gamma k/q V_D \tag{20.41}$$

For the above case $T_C \approx -0.32\%$/degree. In practical applications it is more convenient to use the expression

$$V_D(\delta_2) = V_D(\delta_1) - T_C(\delta_2 - \delta_1) \tag{20.42}$$

where δ_1 and δ_2 are temperatures in degrees Celsius. For $T_C = -0.32\%$/degree and $V_D = 0.65$ V at $\delta_1 = 27°$C, $V_D = 0.618$ V at $\delta_2 = 37°$C. Both dV_D/dT and T_C are I_D dependent. At higher I_D, both dV_D/dT and T_C are smaller than a lower I_D, as shown in Fig. 20.11.

I_D-V_D Characteristic

From the I_D-V_D characteristic of the diode one can find for $m = 1$

$$I_{D1} = I_S \exp(V_{D1}/V_T) \quad \text{and} \quad I_{D2} = I_S \exp(V_{D2}/V_T) \tag{20.43}$$

Thus, the ratio of currents is

$$I_{D2}/I_{D1} = \exp(V_{D_2} - V_{D_1})/V_T \tag{20.44}$$

or the difference voltage

$$V_{D2} - V_{D1} = V_T \ln(I_{D2}/I_{D1}) \tag{20.45}$$

in terms of base 10 logarithm

$$V_{D2} - V_{D1} = V_T \, 2.3 \log(I_{D2}/I_{D1}) \tag{20.46}$$

For $\log(I_{D2}/I_{D1}) = 10$ (one decade), $V_{D2} - V_{D1} = {\sim}60$ mV, or $V_{D2} - V_{D1} = 17.4$ mV for $\log(I_{D2}/I_{D1}) = 2$. In a practical example, for $m = 1$, $V_D = 0.67$ V at $I_D = 100$ μA. At $I_D = 200$ μA, $V_D = 0.67$ V $+ 17.4$ mV $= 0.687$ V. The diode cut-in voltage V_{D0} is defined as the voltage V_D at a very small current I_D typically at about 1 nA. For silicon diodes this voltage is typically $V_{D0} = 0.6$ V.

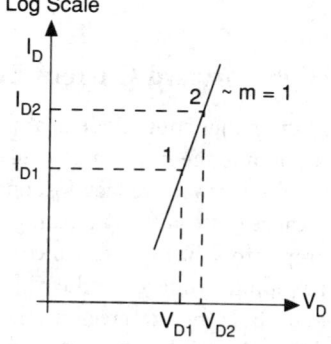

FIGURE 20.12 I_D versus V_D of a diode on a semi-logarithmic plot.

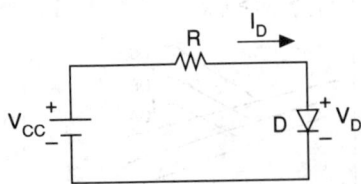

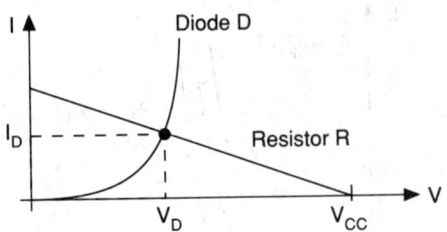

FIGURE 20.13   Diode-resistor biasing circuit.

FIGURE 20.14   Graphical analysis of a diode-
resistor circuit.

DC and Large-Signal Model

The diode equation in Eq. (20.37) is widely utilized in diode circuit design. I_S and m are usually not specified. They can be found from the data book or they can be determined experimentally. From two measurements of I_D and V_D, for example, $I_D = 0.2$ mA at $V_D = 0.670$ V and $I_D = 10$ mA at $V_D = 0.772$ V, one can find $m = 1.012$ and $I_S = 1.78 \cdot 10^{-15}$ A for the particular diode. A practical application of the large-signal diode model is shown in Fig. 20.13. Here, the current I_D through the series resistor R and a diode D is to be found,

$$I_D = (V_{CC} - V_D)/R \qquad (20.47)$$

The equation is implicit and cannot be solved for I_D since V_D is a function of I_D. Here, V_D and I_D are determined by using iteration. By assuming $V_D = V_{D0} = 0.6$ V (cut-in voltage), the first iteration yields

$$I_D(1) = (5 \text{ V} - 0.6 \text{ V})/1 \text{ k}\Omega = 4.4 \text{ mA}$$

Next, the first iteration voltage $V_D(1)$ is calculated (by using m and I_S above and $I_{D1} = 4.4$ mA), thus

$$V_D(1) = mV_T[\ln I_D(1)/I_S] = 1.012 \times 26 \text{ mV } \ln(4.4 \text{ mA}/1.78 \cdot 10^{-15} \text{ A})$$

$$= 0.751 \text{ V}$$

From the second iteration $I_D(2) = [V_{CC} - V_D(1)]/R = 4.25$ mA and thus $V_D(2) = 0.75$ V. The third iteration yields $I_D(3) = 4.25$ mA, and $V_D(3) = 0.75$ V. These are the actual values of I_D and V_D for the above example, since the second and the third iterations are equal.

Graphical analysis (in Fig. 20.14) is another way to analyze the circuit in Fig. 20.13. Here the load line R is drawn with the diode I-V characteristic, where $V_{CC} = V_D + I_D R$. This type of analysis is illustrative but not well suited for a numerical analysis.

High Forward Current Effects

In the *pn*-junction diode analysis it was assumed that the density of injected carriers from the *p* region into the *n* region is small compared to the density of majority carriers in that region. Thus, all of the forward voltage V_D appears across the junction. Therefore, the injected carriers move only because of the diffusion. At high forward currents this is not the case anymore. When the voltage drop across the bulk resistance becomes comparable with the voltage across the junction, the effective applied voltage is reduced [Phillips, 1962]. Due to the electric field created by the voltage drop in the bulk (neutral) regions, the current is not only a diffusion current anymore. The drift current due to the voltage drop across the bulk region opposes the diffusion current. The net effect is that, first, the current becomes proportional to twice the diffusion constant, second, the high-level current becomes independent of resistivity, and, third, the magnitude of the exponent is reduced by a

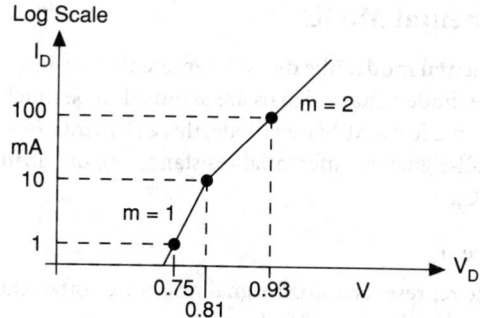

FIGURE 20.15 I_D versus V_D of a diode at low and high forward currents.

factor of two in Eq. (20.37). The effect of high forward current on the I-V characteristic is shown in Fig. 20.15. In all practical designs, $m \approx 2$ at $I_D \geq 20$ mA in small-signal silicon diodes.

Large-Signal Piecewise Linear Model

Piecewise linear model of a diode is a very useful tool for quick circuit design containing diodes. Here, the diode is represented by asymptotes and not by the exponential I-V curve. The simplest piecewise linear model is shown in Fig. 20.16(a). Here D_i is an ideal diode with $V_D = 0$ at $I_D \geq 0$, in series with V_{D0}, where V_{D0} is the diode threshold voltage, $V_{D0} = 0.60$ V for silicon. The current in the diode will start to flow at $V_D \geq 0.60$ V.

An improved model is shown in Fig. 20.16(b), where V_{D0} is the diode voltage at a very small current I_{D0}, r_D is the extrapolated diode resistance, and I_{D1} is the diode current in operating point 1. Thus, the diode voltage is

$$V_{D1} = V_{D0} + I_{D1}\, r_D \tag{20.48}$$

where V_{D1} is the diode voltage at I_{D1}. V_{D0} for silicon is about 0.60 V. r_D is estimated from the fact that V_D in a real diode is changing per decade of current by $m\, 2.3\, V_T$. V_D changes about 60 mV per decade of current for $m = 1$. Thus for 0.1 to 10 mA current change, V_D changes about 120 mV, which corresponds to an $r_D \approx 120$ mV/10 mA = 12 Ω.

The foregoing method is a gross approximation; however, it is quite practical for first-hand calculations. To compare this with the above iterative approach let us assume $m = 1$, $V_{D0} = 0.60$ V, $r_D = 12$ Ω, $V_{CC} = 5$ V, $R = 1$ kΩ. The current $I_{D1} = [V_{CC} - V_{D0}]/(R + r_D) = 4.34$ mA compared with $I_{D1} = 4.25$ mA in the iterative approach.

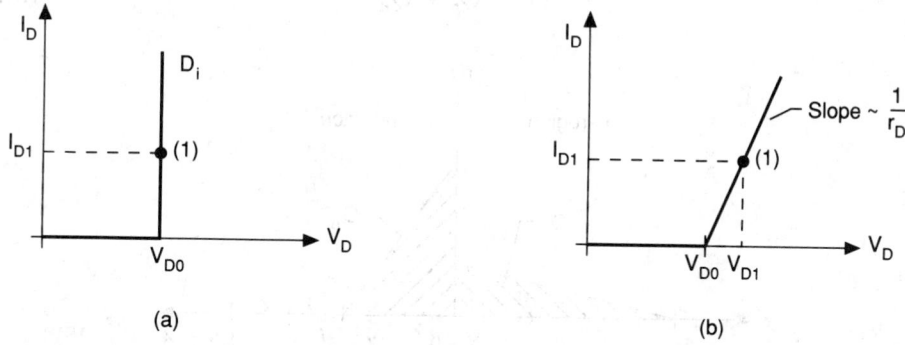

(a) (b)

FIGURE 20.16 (a) Simplified piecewise linear model of a diode; (b) improved piecewise linear model of a diode.

Small-Signal Incremental Model

In the small-signal **incremental model**, the diode is represented by linear elements. In small-signal (incremental) analysis, the diode voltage signals are assumed to be much smaller than the dc voltage V_D across the diode. In the forward-biased diode, three elements are of practical interest: **incremental resistance** (or small-signal or differential resistance) r_d, the **diffusion capacitance** C_d, and the **junction capacitance** C_j.

Incremental Resistance, r_d

For small signals the diode represents a small-signal resistance (often called incremental or differential resistance) r_d in the operating point (I_D, V_D) where

$$r_d = dV_D/dI_D = mV_T/I_S \exp(V_D/mV_T) = mV_T/I_D \qquad (20.49)$$

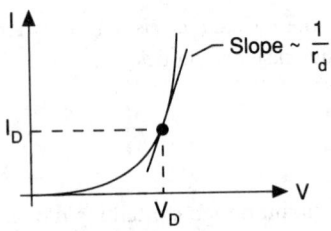

FIGURE 20.17 Small-signal incremental resistance r_d of a diode.

In Fig. 20.17, r_d is shown as the tangent in the dc operating point (V_{D1}, I_{D1}). Note that r_d is independent of the geometry of the device and inversely proportional to the diode dc current. Thus for $I_D = 1$ mA, $m = 1$ and $V_T = 26$ mV, the incremental resistance is $r_d = 26 \, \Omega$.

Diffusion Capacitance, C_d

C_d is associated with the injection of holes and electrons in the forward-biased diode. In steady state, holes and electrons are injected across the junction. Hole and electron currents flow due to the diffusion gradients on both sides of the junction in Fig. 20.18. In a short diode, holes are traveling a distance $W_n << L_{p_n}$. For injected holes

$$I_p = dq_p/dt = dq_p \, v/dx \qquad (20.50)$$

or

$$I_p \int_0^{W_p} dx = v \int_0^{Q_p} dq_p$$

and the hole charge Q_p

$$Q_p = I_p W_p/v = I_p \tau_p \qquad (20.51)$$

$\tau_p = W_p/V$ is the transit time holes travel the distance W_p. Similarly, for electron charge Q_n

$$Q_n = I_n L_{n_p}/v = I_n \tau_n \qquad (20.52)$$

The total diffusion charge Q_d is

$$Q_d = Q_p + Q_n \qquad (20.53)$$

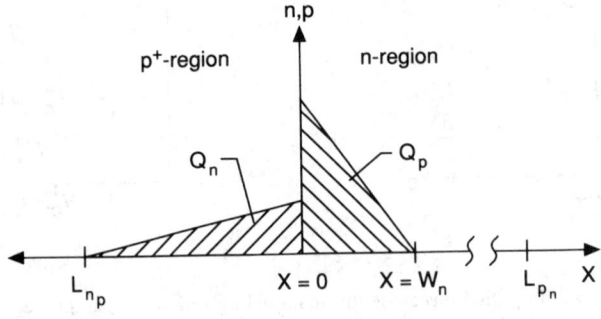

FIGURE 20.18 Minority carrier charge injection in a diode.

The total transit time is

$$\tau_F = \tau_p + \tau_n \tag{20.54}$$

and with $I_p + I_n = I_D = I_S \exp V_D/mV_T$ and Eqs. (20.51), (20.52), and (20.54)

$$Q_d = \tau_F I_S \exp V_D/mV_T = \tau_F I_D \tag{20.55}$$

the diffusion capacitance is

$$C_d = C_p + C_n = dQ_d/dV_D = Q_d/mV_T \tag{20.56}$$

and from Eqs. (20.55) and (20.56)

$$C_d = I_D \tau_F/mV_T \tag{20.57}$$

C_d is thus directly proportional to I_D and to the carrier transit time τ_F. For an unsymmetrical diode with $W_n \ll L_{p_n}$ and $N_A \gg N_D$ [Gray and Meyer, 1984]

$$\tau_F \approx W_n^2/2D_p \tag{20.58}$$

For $W_n = 6\ \mu$ and $D_p = 14\ cm^2/s$, $\tau_F \approx 13\ ns$, $I_D = 1\ mA$, $V_T = 26\ mV$, and $m = 1$, the diffusion capacitance is $C_d = 500\ pF$.

Depletion Capacitance, C_j

The depletion region is always present in a *pn*-diode. Because of the immobile ions in the depletion region, the junction acts as a voltage-dependent plate capacitor C_j [Gray and Meyer, 1984]

$$C_j = C_{j0}/\sqrt{V_0 - V_D} \tag{20.59}$$

V_D is the diode voltage (positive value for forward biasing, negative value for reverse biasing), and C_{j0} is the zero bias depletion capacitance; A_j is the junction diode area:

$$C_{j0} = KA_j \tag{20.60}$$

K is a proportionality constant dependent on diode doping, and A_j is the diode area. C_j is voltage dependent. As V_D increases, C_j increases in a forward-biased diode in Fig. 20.19. For $V_0 = 0.7\ V$ and $V_D = -10\ V$ and $C_{j0} = 3\ pF$, the diode depletion capacitance is $C_j = 0.75\ pF$. In Fig. 20.20 the small-signal model of the diode is shown. The total small-signal time constant τ_d is thus (by neglecting the bulk series diode resistance R_{BB})

$$\tau_d = r_d(C_d + C_j) = r_d C_d + r_d C_j = \tau_F + r_d C_j \tag{20.61}$$

τ_d is thus current dependent. At small I_D the $r_d C_j$ product is predominant. For high-speed operation $r_d C_j$ must be kept much smaller than τ_F. This is achieved by a large operating current I_D. The

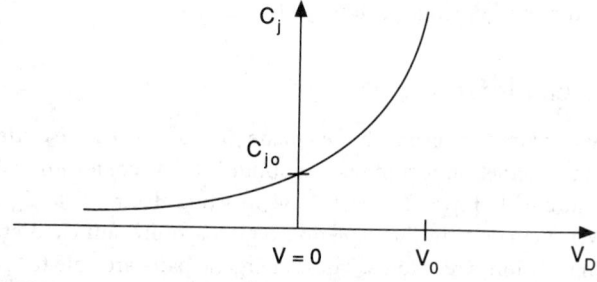

FIGURE 20.19 Depletion capacitance C_j of a diode versus diode voltage V_R.

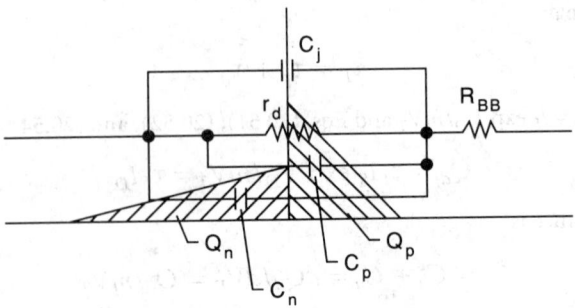

FIGURE 20.20 Simplified small-signal model of a diode.

diode behaves to a first approximation as a frequency-dependent element. In the reverse operation, the diode behaves as a high ohmic resistor $R_p \approx V_R/I_G$ in parallel with the capacitor C_j. In forward small-signal operation, the diode behaves as a resistor r_d in parallel with the capacitors C_j and C_d (R_p is neglected). Thus, the diode is in a first approximation, a low-pass network.

Large-Signal Switching Behavior of a *pn*-Diode

When a forward-biased diode is switched from the forward into the reverse direction, the stored charge Q_d of **minority carriers** must first be removed. The charge of minority carriers in the forward-biased unsymmetrical diode is approximately

$$Q_d = I_D \tau_F = I_D W_n^2 / 2D_p \tag{20.62}$$

where $W_n \ll L_{p_n}$ is assumed. τ_F is minimized by making W_n very small. Very low-lifetime τ_F is required for high-speed diodes. **Carrier lifetime** τ_F is reduced by adding a large concentration of recombination centers into the junction. This is common practice in the fabrication of high-speed computer diodes [Phillips, 1962]. The charge Q_d is stored mainly in the n region in the form of a concentration gradient of holes in Fig. 20.21(a). The diode is turned off by moving the switch from position (a) into position (b) [Fig. 20.21(a)]. The removal of carriers is done in three time intervals. During the time interval t_1, also called the recovery phase, a constant reverse current $|I_R| = V_R/R$ flows in the diode. During the time interval $t_2 - t_1$ the charge in the diode is reduced by about 1/2 of the original charge. During the third interval $t_3 - t_2$, the residual charge is removed.

If during the interval t_1, $|I_R| \gg I_D$, then Q_d is reduced only by flow of reverse diffusion current; no holes arrive at the metal contact [Gugenbuehl *et al.*, 1962], and

$$t_1 \approx \tau_F (I_D/|I_R|)^2 \tag{20.63}$$

During time interval $t_2 - t_1$, when $|I_R| = I_D$, in Fig. 20.21(b),

$$t_2 - t_1 \approx \tau_F I_D / |I_R| \tag{20.64}$$

The residual charge is removed during the time $t_3 - t_2 \approx 0.5 \tau_F$.

Diode Reverse Breakdown

Avalanche breakdown occurs in a reverse-biased plane junction when the critical electric field E_{crt} at the junction within the depletion region reaches about $3 \cdot 10^5$ V/cm for junction doping densities of about 10^{15} to 10^{16} at/cm³ [Gray and Meyer, 1984]. At this electric field E_{crt}, the minority carriers traveling (as reverse current) in the depletion region acquire sufficient energy to create new hole–electron pairs in collision with atoms. These energetic pairs are able to create new pairs, etc. This process is called the avalanche process and leads to a sudden increase of the reverse current I_R

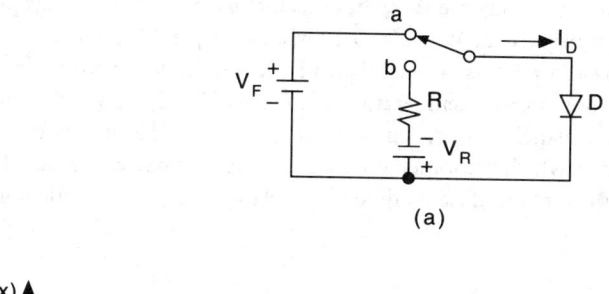

(a)

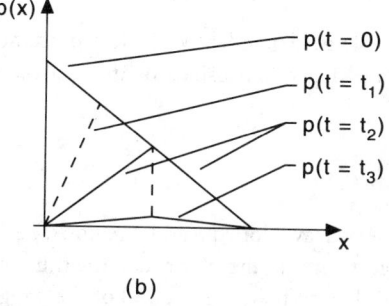

(b)

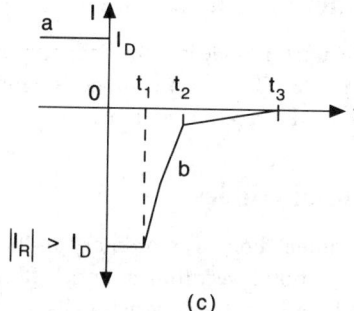

(c)

FIGURE 20.21 (a) Diode is switched from forward into reverse direction; (b) concentration of holes in the *n* region; (c) diode turns off in three time intervals.

in a diode. The current is then limited only by the external circuitry. The avalanche current is not destructive as long as the local junction temperature does not create local hot spots, i.e., melting of material at the junction. Figure 20.22 shows a typical *I-V* characteristic for a junction diode in the avalanche breakdown. The effect of breakdown is seen by the large increase of the reverse current I_R when V_R reaches $-BV$. Here BV is the actual breakdown voltage. It was found that $I_{RA} = M I_R$, where I_{RA} is the avalanche reverse current at BV, M is the multiplication factor, and I_R is the reverse current not in the breakdown region. M is defined as

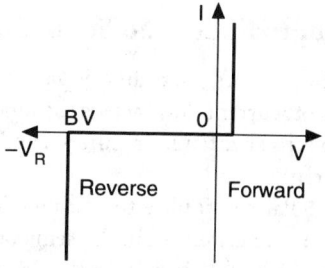

FIGURE 20.22 Reverse breakdown voltage of a diode at $-V_R = BV$.

$$M = 1/[1 - V_R/BV]^n \qquad (20.65)$$

where $n = 3$ to 6. As $V_R = BV$, $M \to \infty$ and $I_{RA} \to \infty$. The above BV is valid for a strictly plane junction without any curvature. However, in a real planar diode as shown in Fig. 20.9, the *p*-diffusion has a curvature with a finite radius x_j. If the diode is doped unsymmetrically, thus $\sigma_p \gg \sigma_n$, then the depletion area penetrates mostly into the *n* region. Because of the finite radius, the breakdown occurs at the radius x_j, rather than in a plane junction [Grove, 1967]. The breakdown voltage is significantly reduced due to the curvature. In very shallow planar diodes, the avalanche breakdown voltage BV can be much smaller than 10 V.

Zener and Avalanche Diodes

Zener diodes (ZD) and avalanche diodes are *pn*-diodes specially built to operate in reverse breakdown. They operate in the reverse direction; however, their operating mechanism is different. In a Zener diode the hole–electron pairs are generated by the electric field by direct transition of carriers from valence band into the conductance band. In an avalanche diode, the hole–electron pairs are generated by impact ionization due to high-energy holes and electrons.

Avalanche and Zener diodes are extensively used as voltage regulators and as overvoltage protection devices. T_C of Zener diodes is negative at $V_Z \leq 3.5$ to 4.5 V and is equal to zero at about $V_Z \approx$ 5 V. T_C of a Zener diode operating above 5 V is in general positive. Above 10 V the *pn*-diodes operate as avalanche diodes with a strong positive temperature coefficient. The T_C of a Zener diode is more predictable than that of the avalanche diode. Temperature-compensated Zener diodes utilize the positive T_C of a 7-V Zener diode, which is compensated with a series-connected forward-biased diode with a negative T_C. The disadvantage of Zener diodes is a relatively large electronic noise.

Varactor Diodes

The varactor diode is an ordinary *pn*-diode that uses the voltage-dependent variable capacitance of the diode. The varactor diode is widely used as a voltage-dependent capacitor in electronically tuned radio receivers and in TV.

Tunnel Diodes

The tunnel diode is an ordinary *pn*-junction diode with very heavy doped *n* and *p* regions. Because the junction is very thin, a tunnel effect takes place. An electron can tunnel through the thin depletion layer from the conduction band of the *n* region directly into the valence band of the *p* region. Tunnel diodes create a negative differential resistance in the forward direction, due to the tunnel effect. Tunnel diodes are used as mixers, oscillators, amplifiers, and detectors. They operate at very high frequencies in the gigahertz bands.

Photodiodes and Solar Cells

Photodiodes are ordinary *pn*-diodes that generate hole–electron pairs when exposed to light. A photocurrent flows across the junction, if the diode is reverse biased. Silicon *pn*-junctions are used to sense light at near-infrared and visible spectra around 0.9 μm. Other materials are used for different spectra.

Solar cells utilize the *pn*-junction to convert light energy into electrical energy. Hole–electron pairs are generated in the semiconductor material by light photons. The carriers are separated by the high electric field in the depletion region across the *pn*-junction. The electric field forces the holes into the *p* region and the electrons into the *n* region. This displacement of mobile charges creates a voltage difference between the two semiconductor regions. Electric power is generated in an external load connected between the terminals to the *p* and *n* regions. The conversion efficiency is relatively low, around 10 to 12%. With the use of new materials, an efficiency of about 30% has been reported. Efficiency up to 45% was achieved by using monochromatic light.

Schottky Barrier Diode

The Schottky barrier diode is a metal-semiconductor diode. Majority carriers carry the electric current. No minority carrier injection takes place. When the diode is forward biased, carriers are injected into the metal, where they reside as majority carriers at an energy level that is higher than the Fermi level in metals. The *I-V* characteristic is similar to conventional diodes. The barrier voltage is small, about 0.2 V for silicon. Since no minority carrier charge exists, the Schottky barrier diodes are very fast. They are used in high-speed electronic circuitry.

Defining Terms

Acceptor: Ionized, negative-charged immobile dopant atom (ion) in a *p*-type semiconductor after the release of a hole.

Avalanche breakdown: In the reverse-biased diode, hole–electron pairs are generated in the depletion region by ionization, thus by the lattice collision with energetic electrons and holes.

Bandgap energy: Energy difference between the conduction band and the valence band in a semiconductor.

Barrier voltage: A voltage which develops across the junction due to uncovered immobile ions on both sides of the junction. Ions are uncovered due to the diffusion of carriers across the junction.

Boltzmann relation: Relates the density of particles in one region to that in an adjacent region, with the potential energy between both regions.

Carrier lifetime: Time an injected minority carrier travels before its recombination with a majority carrier.

Concentration gradient: Difference in carrier concentration.

Diffusion: Movement of free carriers in a semiconductor caused by the difference in carrier densities (concentration gradient).

Diffusion capacitance: Change in charge of injected carriers corresponding to change in forward bias voltage in a diode.

Diffusion constant: Product of the thermal voltage and the mobility in a semiconductor.

Donor: Ionized, positive-charged immobile dopant atom (ion) in an n-type semiconductor after the release of an electron.

Drift: Movement of free carriers in a semiconductor due to the electric field.

Ideality factor: The factor determining the deviation from the ideal diode characteristic $m = 1$. At small and large currents $m \approx 2$.

Incremental model: Small-signal differential (incremental) semiconductor diode equivalent RC circuit of a diode, biased in a dc operating point.

Incremental resistance: Small-signal differential (incremental) resistance of a diode, biased in a dc operating point.

Junction capacitance: Change in charge of immobile ions in the depletion region of a diode corresponding to a change in reverse bias voltage on a diode.

Majority carriers: Holes are in majority in a p-type semiconductor; electrons are in majority in an n-type semiconductor.

Minority carriers: Electrons in a p-type semiconductor are in minority; holes are in majority. Similarly, holes are in minority in an n-type semiconductor and electrons are in majority.

Reverse breakdown: At the reverse breakdown voltage the diode can conduct a large current in the reverse direction.

Reverse generation-recombination current: Part of the reverse current in a diode caused by the generation of hole–electron pairs in the depletion region. This current is voltage dependent because the depletion region width is voltage dependent.

Reverse saturation current: Part of the reverse current in a diode which is caused by diffusion of minority carriers from the neutral regions to the depletion region. This current is almost independent of the reverse voltage.

Temperature coefficient: Relative variation $\Delta X/X$ of a value X over a temperature range, divided by the difference in temperature ΔT.

Zener breakdown: In the reverse-biased diode, hole–electron pairs are generated by a large electric field in the depletion region.

References

P. R. Gray and R. G. Meyer, *Analysis and Design of Analog Integrated Circuits*, New York: John Wiley & Sons, 1984.

A. S. Grove, *Physics and Technology of Semiconductor Devices*, New York: John Wiley & Sons, 1967.

W. Gugenbuehl, M.J.O. Strutt, and W. Wunderlin, *Semiconductor Elements*, Basel: Birkhauser Verlag, 1962.

A.B. Phillips, *Transistor Engineering*, New York: McGraw-Hill, 1962.

Further Information

A good introduction to diodes is found in P. E. Gray and C. L. Searle, *Electronic Principles*, New York: Wiley, 1969, and in S. M. Sze, *Semiconductor Devices, Physics and Technology*, New York: Wiley, 1985. Other sources include S. Soclof, *Applications of Analog Integrated Circuits*, Englewood Cliffs, N.J.: Prentice-Hall, 1985 and E. J. Angelo, Jr., *Electronics: BJT's, FET's and Microcircuits*, New York: McGraw-Hill, 1969.

20.3 Electrical Equivalent Circuit Models and Device Simulators for Semiconductor Devices

Aicha Elshabini-Riad, F. W. Stephenson, and Imran A. Bhutta

In the past 15 years, the electronics industry has seen a tremendous surge in the development of new semiconductor materials, novel devices, and circuits. For the designer to bring these circuits or devices to the market in a timely fashion, he or she must have design tools capable of predicting the device behavior in a variety of circuit configurations and environmental conditions. Equivalent circuit models and semiconductor device simulators represent such design tools.

Overview of Equivalent Circuit Models

Circuit analysis is an important tool in circuit design. It saves considerable time, at the circuit design stage, by providing the designer with a tool for predicting the circuit behavior without actually processing the circuit.

An electronic circuit usually contains active devices, in addition to passive components. While the current and voltage behavior of passive devices is defined by simple relationships, the equivalent relationships in active devices are quite complicated in nature. Therefore, in order to analyze an active circuit, the devices are replaced by equivalent circuit models that give the same output characteristics as the active device itself. These models are made up of passive elements, voltage sources, and current sources. Equivalent circuit models provide the designer with reasonably accurate values for frequencies below 1 GHz for bipolar junction transistors (BJTs), and their use is quite popular in circuit analysis software. Some field-effect transistor (FET) models are accurate up to 10 GHz. As the analysis frequency increases, however, so does the model complexity. Since the equivalent circuit models are based on some fundamental equations describing the device behavior, they can also be used to predict the characteristics of the device itself.

When performing circuit analysis, two important factors that must be taken into account are the speed and accuracy of computation. Sometimes the computation speed can be considerably improved by simplifying the equivalent circuit model, without significant loss in computation accuracy. For this reason, there are a number of equivalent circuit models, depending on the device application and related conditions. Equivalent circuit models have been developed for diodes, BJTs, and FETs. In this overview, the equivalent circuit models for BJT and FET devices are presented.

Most of the equivalent circuits for BJTs are based on the Ebers–Moll model [1954] or the Gummel–Poon model [1970]. The original Ebers–Moll model was a large signal, nonlinear dc model for BJTs. Since then, a number of improvements have been incorporated to make the model more accurate for various applications. In addition, an accurate model has been introduced by Gummel and Poon.

There are three main types of equivalent circuit models, depending on the device signal strength. On this basis, the models can be classified as follows:

1. Large-signal equivalent circuit model
2. Small-signal equivalent circuit model
3. DC equivalent circuit model

Use of the large-signal or small-signal model depends on the magnitude of the driving source. In applications where the driving currents or the driving voltages have large amplitudes, large-signal models are used. In circuits where the signal does not deviate much from the dc biasing point, small-signal models are more suitable. For dc conditions and very-low-frequency applications, dc equivalent circuit models are used. For dc and very-low-frequency analysis, the circuit element values can be assumed to be lumped, whereas in high-frequency analysis, incremental element values give much more precise results.

Large-Signal Equivalent Circuit Model

Depending on the frequency of operation, large-signal equivalent circuit models can be further classified as (1) high-frequency large-signal equivalent circuit model and (2) low-frequency large-signal equivalent circuit model.

High-Frequency Large-Signal Equivalent Circuit Model of a BJT.   In this context, high-frequency denotes frequencies above 10 kHz. In the equivalent circuit model, the transistor is assumed to be composed of two back-to-back diodes. Two current-dependent current sources are added to model the current flowing through the reverse-biased base-collector junction and the forward-biased base-emitter junction. Two junction capacitances, C_{jE} and C_{jC}, model the fixed charges in the emitter-base space charge region and base-collector space charge region, respectively. Two diffusion capacitances, C_{DE} and C_{DC}, model the corresponding charge associated with mobile carriers, while the base resistance, r_b, represents the voltage drop in the base region. All the above circuit elements are very strong functions of operating frequency, signal strength, and bias voltage.

The high-frequency large-signal equivalent circuit model of an *npn* BJT is shown in Fig. 20.23, where the capacitances C_{jE}, C_{jC}, C_{DE}, C_{DC} are defined as follows:

$$C_{jE}(V_{B'E'}) = \frac{C_{jEO}}{\left(1 - \dfrac{V_{B'E'}}{\phi_E}\right)^{m_E}} \tag{20.66}$$

$$C_{jC}(V_{B'C'}) = \frac{C_{jCO}}{\left(1 - \dfrac{V_{B'C'}}{\phi_C}\right)^{m_C}} \tag{20.67}$$

$$C_{DE} = \frac{\tau_F I_{CC}}{V_{B'E'}} \tag{20.68}$$

and

$$C_{DC} = \frac{\tau_R I_{EC}}{V_{B'E'}} \tag{20.69}$$

In these equations, $V_{B'E'}$ is the internal base-emitter voltage, C_{jEO} is the base-emitter junction capacitance at $V_{B'E'} = 0$, ϕ_E is the base-emitter barrier potential, and m_E is the base-emitter capaci-

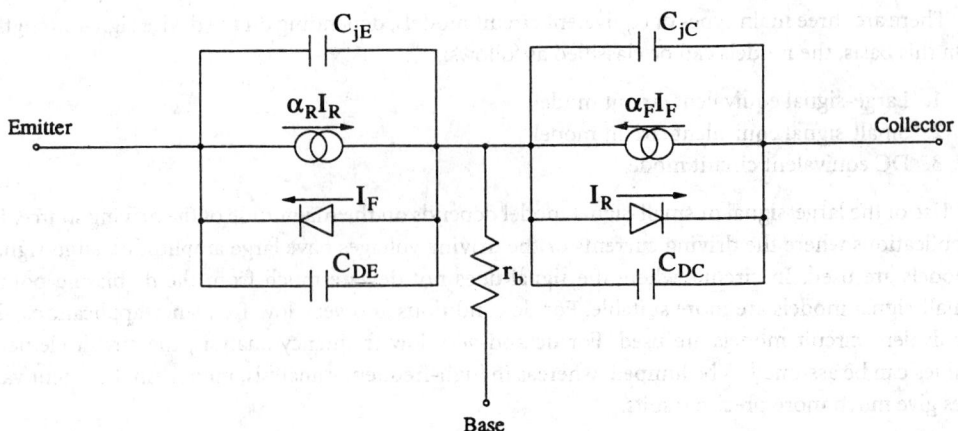

FIGURE 20.23   High-frequency large-signal equivalent circuit model of an *npn* BJT.

tance gradient factor. Similarly, $V_{B'C'}$ is the internal base-collector voltage, C_{jCO} is the base-collector junction capacitance at $V_{B'C'} = 0$, ϕ_C is the base-collector barrier potential, and m_C is the base-collector capacitance gradient factor. I_{CC} and I_{EC} denote the collector and emitter reference currents, respectively, while τ_F is the total forward transit time, and τ_R is the total reverse transit time. α_R and α_F are the large-signal reverse and forward current gains of a common base transistor, respectively.

This circuit can be made linear by replacing the forward-biased base-emitter diode with a low-value resistor, r_π, while the reverse-biased base-collector diode is replaced with a high-value resistor, r_μ. The junction and diffusion capacitors are lumped together to form C_π and C_μ, while the two current sources are lumped into one source ($g_{mF}V_F - g_{mR}V_R$), where g_{mF} and g_{mR} are the transistor forward and reverse transconductances, respectively. V_F and V_R are the voltages across the forward- and reverse-biased diodes, represented by r_π and r_μ, respectively. r_π is typically about 3 kΩ, while r_μ is more than a few megohms, and C_π is about 120 pF. The linear circuit representation is illustrated in Fig. 20.24.

The Gummel–Poon representation is very similar to the high-frequency large-signal linear circuit model of Fig. 20.24. However, the terms describing the elements are different and a little more involved.

High-Frequency Large-Signal Equivalent Circuit Model of a FET.   In the high-frequency large-signal equivalent circuit model of a FET, the fixed charge stored between the gate and the source

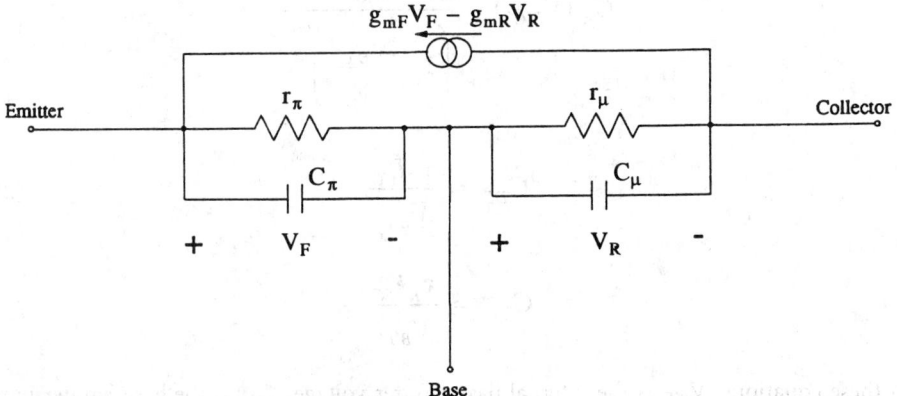

FIGURE 20.24   High-frequency large-signal equivalent circuit model (linear) of an *npn* BJT.

and between the gate and the drain is modeled by the gate-to-source and the gate-to-drain capacitances, C_{GS} and C_{GD}, respectively. The **mobile charges** between the drain and the source are modeled by the drain-to-source capacitance, C_{DS}. The voltage drop through the active channel is modeled by the drain-to-source resistance, R_{DS}. The current through the channel is modeled by a voltage-controlled current source. For large signals, the gate is sometimes driven into the forward region, and thus the conductance through the gate is modeled by the gate conductance, G_g. The conductance from the gate to the drain and from the gate to the source is modeled by the gate-to-drain and gate-to-source resistances, R_{GD} and R_{GS}, respectively. A variable resistor, R_i, is added to model the gate charging time such that the time constant given by R_iC_{GS} holds the following relationship

$$R_iC_{GS} = \text{constant} \tag{20.70}$$

For MOSFETs, typical element values are: C_{GS} and C_{GD} are in the range of 1–10 pF, C_{DS} is in the range of 0.1–1 pF, R_{DS} is in the range of 1–50 kΩ, R_{GD} is more than 10^{14} Ω, R_{GS} is more than 10^{10} Ω, and g_m is in the range of 0.1–20 mA/V.

Figure 20.25 illustrates the high-frequency large-signal equivalent model of a FET.

Low-Frequency Large-Signal Equivalent Circuit Model of a BJT. In this case, low frequency denotes frequencies below 10 kHz. The low-frequency large-signal equivalent circuit model of a BJT is based on its dc characteristics. Whereas at high frequencies one has to take incremental values to obtain accurate analysis, at low frequencies the average of these incremental values yields the same level of accuracy in the analysis. Therefore, in low-frequency analysis, the circuit elements of the high-frequency model are replaced by their average values. The low-frequency large-signal equivalent circuit model is shown in Fig. 20.26.

Low-Frequency Large-Signal Equivalent Circuit Model of a FET. Because of their high reactance values, the gate-to-source, gate-to-drain, and drain-to-source capacitances can be assumed to be open circuits, at low frequencies. Therefore, the low-frequency large-signal model is similar to the high-frequency large-signal model, except that it has no capacitances. The resulting circuit describing low-frequency operation is shown in Fig. 20.27.

Small-Signal Equivalent Circuit Model

In a small-signal equivalent circuit model, the signal variations around the dc-bias operating point are very small. Just as for the large-signal model, there are two types of small-signal models,

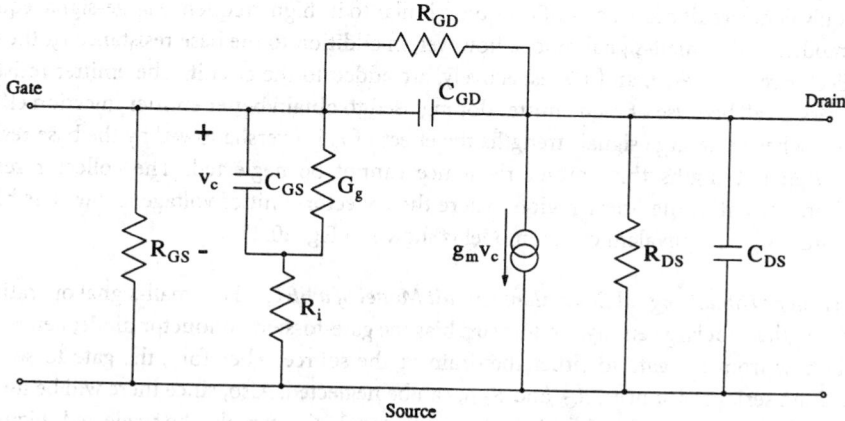

FIGURE 20.25 High-frequency large-signal equivalent circuit model of a FET.

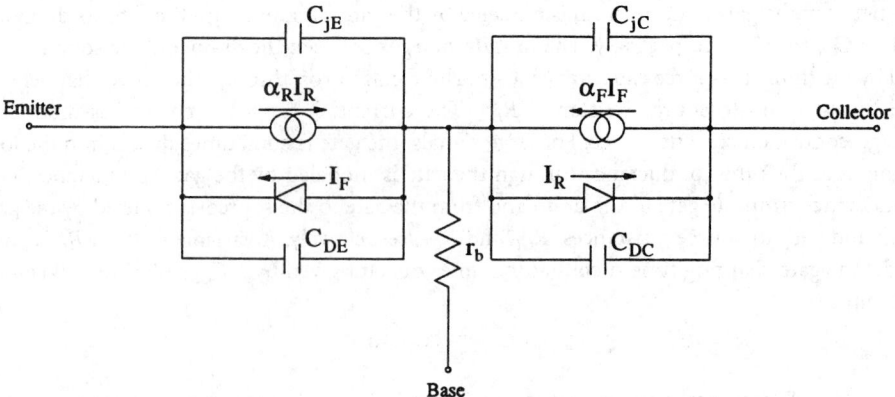

FIGURE 20.26   Low-frequency large-signal equivalent circuit model of an *npn* BJT.

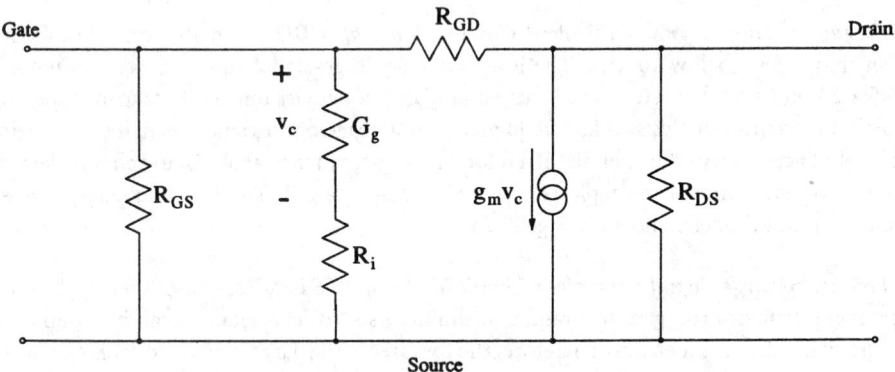

FIGURE 20.27   Low-frequency large-signal equivalent circuit model of a FET.

depending upon the operating frequency: (1) the high-frequency small-signal equivalent circuit model and (2) the low-frequency small-signal equivalent circuit model.

High-Frequency Small-Signal Equivalent Circuit Model of a BJT.   The high-frequency small-signal equivalent circuit model of a BJT is quite similar to its high-frequency large-signal equivalent circuit model. In the small-signal model, however, in addition to the base resistance r_b, the emitter and collector resistances, r_e and r_c, respectively, are added to the circuit. The emitter resistance is usually very small because of high emitter doping used to obtain better emitter injection efficiency. Therefore, whereas at large signal strengths the effect of r_e is overshadowed by the base resistance, at small signal strengths this emitter resistance cannot be neglected. The collector resistance becomes important in the linear region, where the collector-emitter voltage is low. The high-frequency small-signal equivalent circuit model is shown in Fig. 20.28.

High-Frequency Small-Signal Equivalent Circuit Model of a FET.   For small-signal operations, the signal strength is not large enough to forward bias the gate-to-semiconductor diode; hence, no current will flow from the gate to either the drain or the source. Therefore, the gate-to-source and gate-to-drain series resistances, R_{GS} and R_{GD}, can be neglected. Also, since there will be no current flow from the gate to the channel, the gate conductance, G_g, can also be neglected. Figure 20.29 illustrates the high-frequency small-signal equivalent circuit model of a FET.

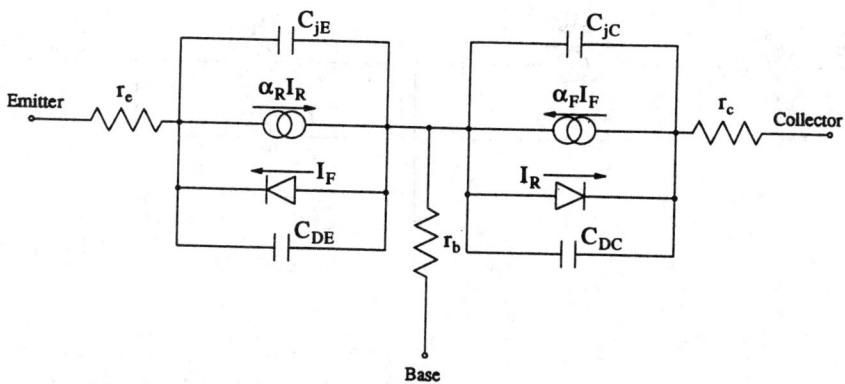

FIGURE 20.28 High-frequency small-signal equivalent circuit model of an *npn* BJT.

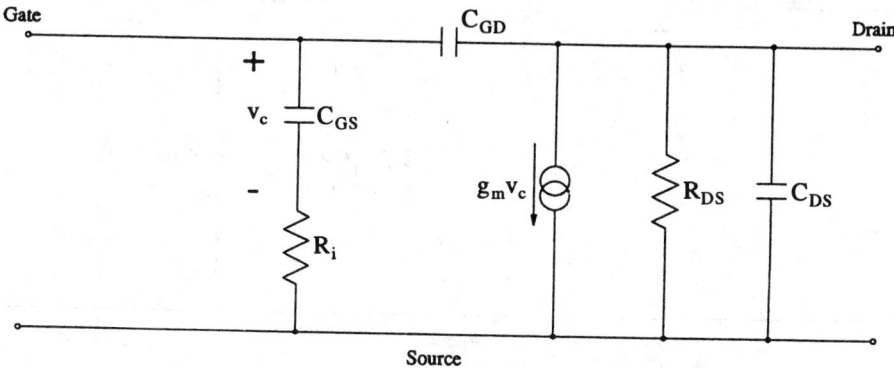

FIGURE 20.29 High-frequency small-signal equivalent circuit model of a FET.

Low-Frequency Small-Signal Equivalent Circuit Model of a BJT. As in the low-frequency large-signal model, the junction capacitances, C_{jC} and C_{jE}, and the diffusion capacitances, C_{DE} and C_{DC}, can be neglected. Furthermore, the base resistance, r_b, can also be neglected, because the voltage drop across the base is not significant and the variations in the base width caused by changes in the collector-base voltage are also very small. The low-frequency small-signal equivalent circuit model is shown in Fig. 20.30.

Low-Frequency Small-Signal Equivalent Circuit Model of a FET. Because the reactances associated with all the capacitances are very high, one can neglect the capacitances for low-frequency analysis. The gate conductance as well as the gate-to-source and gate-to-drain resistances can also be neglected in small-signal operation. The resulting low-frequency equivalent circuit model of a FET is shown in Fig. 20.31.

DC Equivalent Circuit Model

DC Equivalent Circuit Model of a BJT. The dc equivalent circuit model of a BJT is based on the original Ebers–Moll model. Such models are used when the transistor is operated at dc or in applications where the operating frequency is below 1 kHz.

There are two versions of the dc equivalent circuit model—the *injection version* and the *transport version*. The difference between the two versions lies in the choice of the reference current. In the

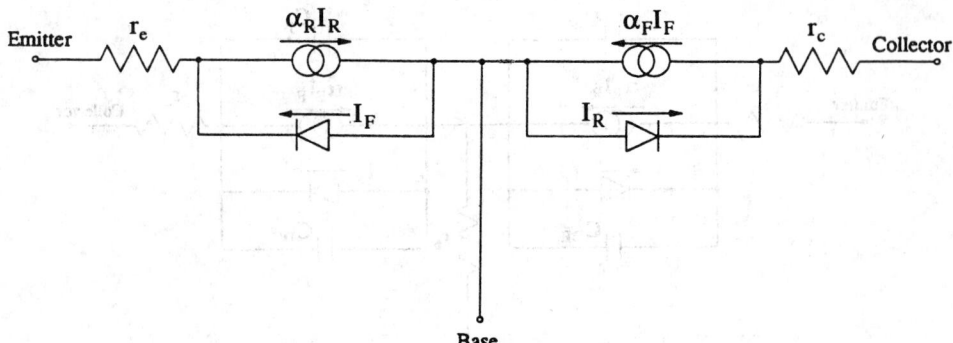

FIGURE 20.30   Low-frequency small-signal equivalent circuit model of an *npn* BJT.

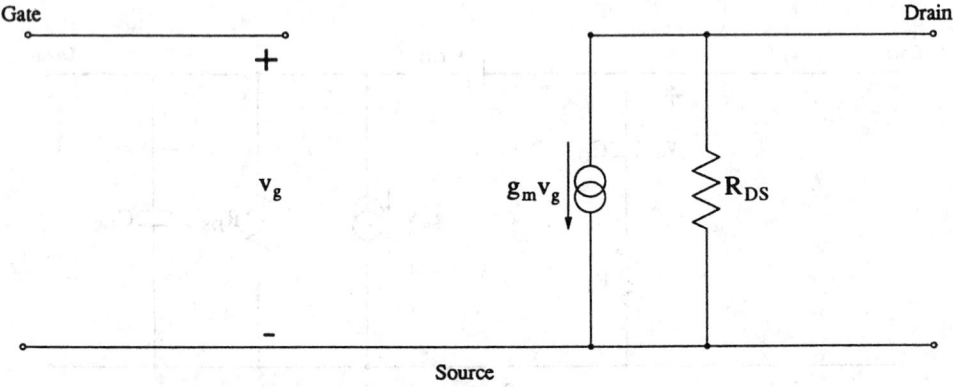

FIGURE 20.31   Low-frequency small-signal equivalent circuit model of a FET.

injection version, the reference currents are I_F and I_R, the forward- and reverse-biased diode currents, respectively. In the *transport version*, the reference currents are the collector transport current, I_{CC}, and the emitter transport current, I_{CE}. These currents are of the form:

$$I_F = I_{ES}\left[\exp\left(\frac{qV_{BE}}{kT}\right) - 1\right] \tag{20.71}$$

$$I_R = I_{CS}\left[\exp\left(\frac{qV_{BC}}{kT}\right) - 1\right] \tag{20.72}$$

$$I_{CC} = I_S\left[\exp\left(\frac{qV_{BE}}{kT}\right) - 1\right] \tag{20.73}$$

and

$$I_{EC} = I_S\left[\exp\left(\frac{qV_{BC}}{kT}\right) - 1\right] \tag{20.74}$$

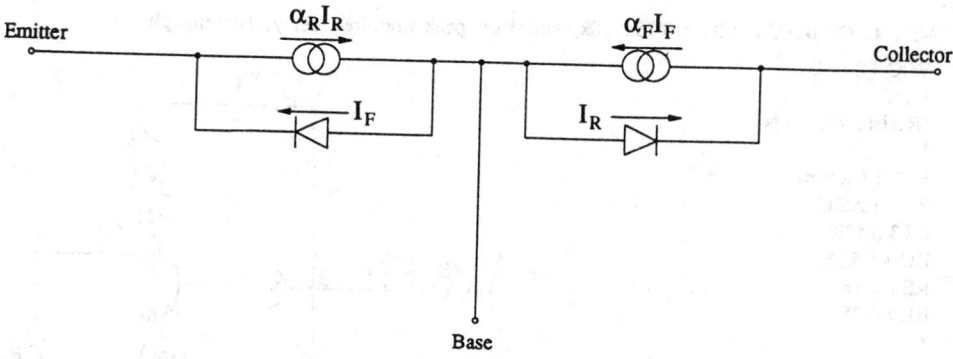

FIGURE 20.32 DC equivalent circuit model (injection version) of an *npn* BJT.

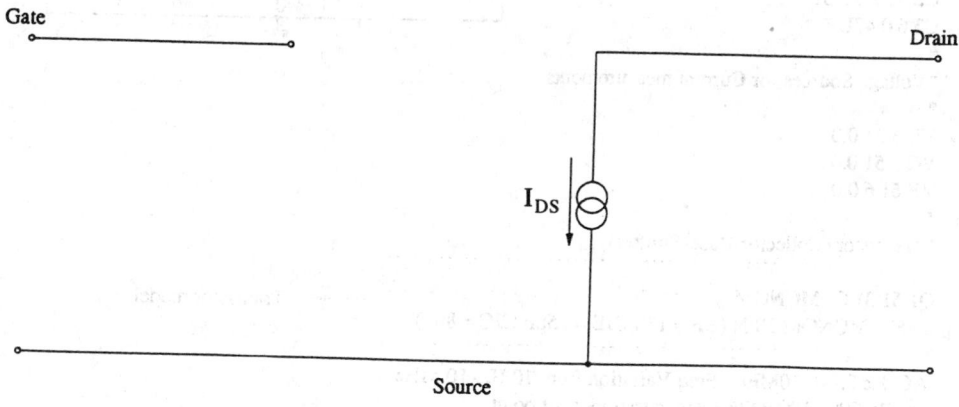

FIGURE 20.33 DC equivalent circuit model of a FET.

In these equations, I_{ES} and I_{CS} are the base-emitter saturation current and the base-collector saturation current, respectively. I_S denotes the saturation current.

In most computer simulations, the *transport version* is usually preferred because of the following conditions:

1. I_{CC} and I_{EC} are ideal over many decades.
2. I_S can specify both reference currents at any given voltage.

The dc equivalent circuit model of a BJT is shown in Fig. 20.32.

DC Equivalent Circuit Model of a FET. In the dc equivalent circuit model of a FET, the gate is considered isolated because the gate-semiconductor interface is formed as a reverse-biased diode and therefore is open circuited. All capacitances are also assumed to represent open circuits. R_{GS}, R_{GD}, and R_{DS} are neglected because there is no conductance through the gate and, because this is a dc analysis, there are no charging effects associated with the gate. The dc equivalent circuit of a FET is illustrated in Fig. 20.33.

Commercially Available Packages

A number of circuit analysis software packages are commercially available, one of the most widely used being SPICE. In this package, the BJT models are a combination of the Gummel–Poon and the modified Ebers–Moll models. Figure 20.34 shows a common emitter transistor circuit and a

VS 1 0 DC 0.0 AC 75e-3 sin(0 75m 10k); sin(offset, peak amp, freq, delay, damping, phase)

VCC 4 0 12.0
*

*Resistor elements
*

Rs 1 2 1 Kohm
R1 4 3 225K
R2 3 0 47K
RC 4 5 5.1K
RE 6 0 1K
RL 7 0 2K
*

*Capacitor elements
*

C1 2 3 3.3UFd
C2 51 7 3.3UF
C3 6 0 47UF
*

*Voltage Sources for Current measurements
*

VB 3 31 0.0
VC 5 51 0.0
VE 61 6 0.0
*

*Transistor (Collector-Base-Emitter)
*

Q1 51 31 61 MQNOM
.model MQNOM NPN (BF = 130 CJE = 25pF CJC = 8pF)
*

.AC dec 25 10 10MEG; Freq Variation from 10 Hz- 10 MHz
.tran 1u 200u 100u; Step size, duration, start point
.Probe
.End

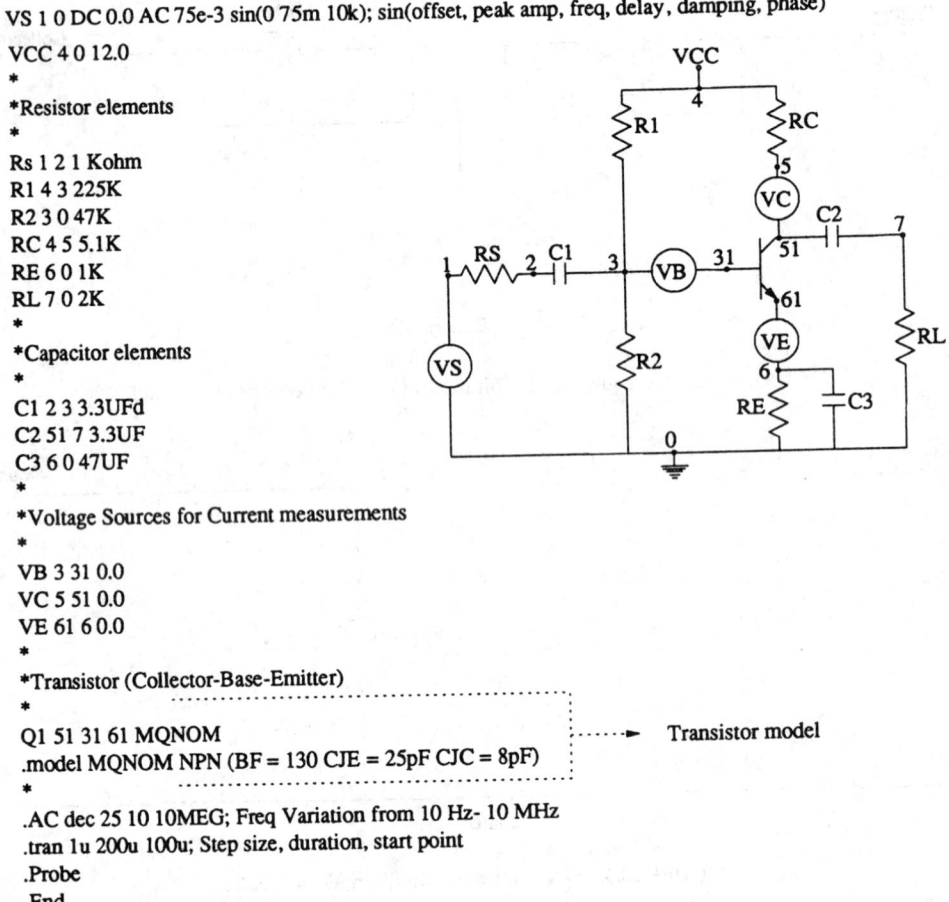

FIGURE 20.34   Common emitter transistor circuit and SPICE circuit file.

SPICE input file containing the transistor model. Some other available packages are SLIC, SINC, and SITCAP.

Equivalent circuit models are basically used to replace the semiconductor device in an electronic circuit. These models are developed from an understanding of the device's current and voltage behavior for novel devices where internal device operation is not well understood. For such situations the designer has another tool available, the semiconductor device simulator.

Overview of Semiconductor Device Simulators

Device simulators are based on the physics of semiconductor devices. The input to the simulator takes the form of information about the device under consideration such as material type, device, dimensions, doping concentrations, and operating conditions. Based on this information, the device simulator computes the electric field inside the device and thus predicts carrier concentrations in the different regions of the device. Device simulators can also predict transient behavior, including quantities such as current–voltage characteristics and frequency bandwidth. The three basic approaches to device simulation are (1) the classical approach, (2) the semiclassical approach, and (3) the quantum mechanical approach.

Device Simulators Based on the Classical Approach

The *classical approach* is based on the solution of Poisson's equation and the current continuity equations. The current consists of the drift and the diffusion current components.

Assumptions. The equations for the classical approach can be obtained by making the following approximations to the Boltzmann transport equation:

1. Carrier temperature is the same throughout the device and is assumed to be equal to the lattice temperature.
2. Quasi steady-state conditions exist.
3. Carrier **mean free path** must be smaller than the distance over which the **quasi-Fermi level** is changing by kT/q.
4. The impurity concentration is constant or varies very slowly along the mean free path of the carrier.
5. The energy band is parabolic.
6. The influence of the boundary conditions is negligible.

For general purposes, even with these assumptions and limitations, the models based on the classical approach give fairly accurate results. The model assumes that the driving force for the carriers is the quasi-Fermi potential gradient, which is also dependent upon the electric field value. Therefore, in some simulators, the quasi-Fermi level distributions are computed and the carrier distribution is estimated from this information.

Equations to Be Solved. With the assumption of a quasi-steady-state condition, the operating wavelength is much larger than the device dimensions. Hence, Maxwell's equations can be reduced to the more familiar Poisson's equation:

$$\nabla^2 \psi = -\frac{\rho}{\epsilon} \tag{20.75}$$

and, for a nonhomogeneous medium,

$$\nabla \cdot \epsilon \, (\nabla \psi) = -\rho \tag{20.76}$$

where ψ denotes the potential of the region under simulation, ϵ denotes the permittivity, and ρ denotes the charge enclosed by this region.

Also from Maxwell's equations, we can determine the current continuity equations for a homogeneous medium as:

$$\nabla \cdot J_n - q \left(\frac{\partial n}{\partial t} \right) = +qU \tag{20.77}$$

where

$$J_n = q\mu_n E + qD_n \nabla \cdot n \tag{20.78}$$

and

$$\nabla \cdot J_p + q \left(\frac{\partial p}{\partial t} \right) = -qU \tag{20.79}$$

where

$$J_p = q\mu_p pE - qD_p \nabla \cdot p \tag{20.80}$$

For nonhomogeneous media, the electric field term in the current expressions is modified to account for the nonuniform **density of states** and the bandgap variation [Lundstrom and Schuelke, 1983].

In the classical approach, the objective is to calculate the potential and the carrier distribution inside the device. Poisson's equation is solved to yield the potential distribution inside the device from which the electric field can be approximated. The electric field distribution is then used in the current continuity equations to obtain the carrier distribution and the current densities. The diffusion coefficients and carrier mobilities are usually field as well as spatially dependent.

The generation-recombination term U is usually specified by the Shockley–Read–Hall relationship [Yoshi *et al.*, 1982]:

$$Rn = \frac{p\,n - n_{ie}^2}{\tau_p(n + n_t) + \tau_n(p + p_t)} \tag{20.81}$$

where p and n are the hole and electron concentrations, respectively, n_{ie} is the effective intrinsic carrier density, τ_p and τ_n are the hole and electron lifetimes, and p_t and n_t are the hole and electron trap densities, respectively.

The electron and hole mobilities are usually specified by the Scharfetter–Gummel empirical formula, as

$$\mu = \mu_0\left[1 + \frac{N}{(N/a) + b} + \frac{(E/c)^2}{(E/c) + d} + (E/e)^2\right]^{-1/2} \tag{20.82}$$

where N is the total ionized impurity concentration, E is the electric field, and a, b, c, d, and e are defined constants [Scharfetter and Gummel, 1969] that have different values for electrons and holes.

Boundary Conditions. Boundary conditions have a large effect on the final solution, and their specific choice is a very important issue. For ohmic contacts, infinite recombination velocities and space charge neutrality conditions are assumed. Therefore, for a *p*-type material, the ohmic boundary conditions take the form

$$\psi = V_{\text{appl}} + \frac{kT}{q}\ln\left(\frac{n_{ie}}{p}\right) \tag{20.83}$$

$$p = \left[\left(\frac{N_D^+ - N_A^-}{2}\right)^2 + n_{ie}^2\right]^{1/2} - \left(\frac{N_D^+ - N_A^-}{2}\right) \tag{20.84}$$

and

$$n = \frac{n_{ie}^2}{p} \tag{20.85}$$

where V_{appl} is the applied voltage, k is Boltzmann's constant, and N_D^+ and N_A^- are the donor and acceptor ionized impurity concentrations, respectively.

For **Schottky contacts**, the boundary conditions take the form

$$\psi = V_{\text{appl}} + \frac{E_G}{2} - \phi_B \tag{20.86}$$

and

$$n = n_{ie} \exp\left(\frac{(E_G/2) - \phi_B}{kT/q}\right) \tag{20.87}$$

where E_G is the semiconductor bandgap and ϕ_B is the barrier potential. For other boundaries with no current flow across them, the boundary conditions are of the form

$$\frac{\partial \psi}{\partial n} = \frac{\partial \varphi_n}{\partial n} = \frac{\partial \varphi_p}{\partial p} = 0 \tag{20.88}$$

where φ_n and φ_p are the electron and hole quasi-Fermi levels.

For field-effect devices, the potential under the gate may be obtained either by setting the gradient of the potential near the semiconductor-oxide interface equal to the gradient of potential inside the oxide [Kasai *et al.*, 1982], or by solving Laplace's equation in the oxide layer, or by assuming a Dirichlet boundary condition at the oxide-gate interface and determining the potential at the semiconductor-oxide interface as:

$$\epsilon_{Si} \left.\frac{\partial \psi}{\partial y}\right|_{Si} = \epsilon_{Ox} \frac{\psi_G - \psi_S^*(x,z)}{T(z)} \tag{20.89}$$

where ϵ_{Si} and ϵ_{Ox} are the permittivities of silicon and the oxide, respectively, ψ_G is the potential at the top of the gate, $\psi_S^*(x,z)$ is the potential of the gate near the interface, and $T(z)$ is the thickness of the gate metal.

Solution Methods. Two of the most popular methods of solving the above equations are finite difference method (FDM) and finite element method (FEM).

In FDM, the region under simulation is divided into rectangular or triangular areas for two-dimensional cases or into cubic or tetrahedron volumes in three-dimensional cases. Each corner or vertex is considered as a node. The differential equations are modified using finite difference approximations, and a set of equations is constructed in matrix form. The finite difference equations are solved iteratively at only these nodes. The most commonly used solvers are Gauss–Seidel/Jacobi (G-S/J) techniques or Newton's technique (NT) [Banks *et al.*, 1983]. FDM has the disadvantage of requiring more nodes than the FEM for the same structure. A new variation of FDM, namely the finite boxes scheme [Franz *et al.*, 1983], however, overcomes this problem by enabling local area refinement. The advantage of FDM is that its computational memory requirement is less than that required for FEM because of the band structure of the matrix.

In FEM, the region under simulation is divided into triangular and quadrilateral regions in two dimensions or into tetrahedra in three dimensions. The regions are placed to have the maximum number of vertices in areas where there is expected to be a large variation of composition or a large variation in the solution. The equations in FEM are modified by multiplying them with some shape function and integrating over the simulated region. In triangular meshes, the shape function is dependent on the area of the triangle and the spatial location of the node. The value of the spatial function is between 0 and 1. The solution at one node is the sum of all the solutions, resulting from the nearby nodes, multiplied by their respective shape functions. The number of nodes required to simulate a region is less than that in FDM; however, the memory requirement is greater.

Device Simulators Based on the Semiclassical Approach

The *semiclassical* approach is based upon the Boltzmann transport equation (BTE) [Engl, 1986] which can be written as

$$\frac{df}{dt} = \frac{\partial f}{\partial t} + v \cdot \nabla_r \pm \frac{q}{(h/2\pi)} E \cdot \nabla_k f = \left(\frac{\partial f}{\partial t}\right)_{coll} \qquad (20.90)$$

where f represents the carrier distribution in the volume under consideration at any time t, v is the group velocity, E is the electric field, and q and h are the electronic charge and Planck's constant, respectively.

BTE is a simplified form of the Liouville–Von Neumann equation for the density matrix. In this approach, the free flight between two consecutive collisions of the carrier is considered to be under the influence of the electric field, whereas different scattering mechanisms determine how and when the carrier will undergo a collision.

Assumptions. The assumptions for the semiclassical model can be summarized as follows:

1. Carrier-to-carrier interactions are considered to be very weak.
2. Particles cannot gain energy from the electric field during collision.
3. Scattering probability is independent of the electric field.
4. Magnetic field effects are neglected.
5. No electron-to-electron interaction occurs in the collision term.
6. Electric field varies very slowly, i.e., electric field is considered constant for a wave packet describing the particle's motion.
7. The electron and hole gas is not degenerate.
8. Band theory and effective-mass theorems apply to the semiconductor.

Equations to Be Solved. As a starting point, Poisson's equation is solved to obtain the electric field inside the device. Using the Monte Carlo technique (MCT), the BTE is solved to obtain the carrier distribution function, f. In the MCT, the path of one or more carriers, under the influence of external forces, is followed, and from this information the carrier distribution function is determined. BTE can also be solved by the momentum and energy balance equations.

The carrier distribution function gives the carrier concentrations in the different regions of the device and can also be used to obtain the electron and hole currents, using the following expressions:

$$J_n = -q \int_k v f(r,k,t) d^3k \qquad (20.91)$$

and

$$J_p = +q \int_k v f(r,k,t) d^3k \qquad (20.92)$$

Device Simulators Based on the Quantum Mechanical Approach

The *quantum mechanical approach* is based on the solution of the Schrodinger wave equation (SWE), which, in its time-independent form, can be represented as

$$\frac{(h/2\pi)^2}{2m} \nabla^2 \varphi_n + (E_n + qV)\ \varphi_n = 0 \qquad (20.93)$$

where φ_n is the wave function corresponding to the subband n whose minimum energy is E_n, V is the potential of the region, m is the particle mass, and h and q are Planck's constant and the electronic charge, respectively.

Equations to Be Solved. In this approach, the potential distribution inside the device is calculated using Poisson's equation. This potential distribution is then used in the SWE to yield the electron wave vector, which in turn is used to calculate the carrier distribution, using the following expression:

$$n = \sum_n N_n \left| \varphi_n \right|^2 \tag{20.94}$$

where n is the electron concentration and N_n is the concentration of the subband n.

This carrier concentration is again used in Poisson's equation, and new values of φ_n, E_n, and n are calculated. This process is repeated until a self-consistent solution is obtained. The final wave vector is invoked to determine the scattering matrix, after which MCT is used to yield the carrier distribution and current densities.

Commercially Available Device Simulation Packages

The classical approach is the most commonly used procedure since it is the easiest to implement and, in most cases, the fastest technique. Simulators based on the classical approach are available in two-dimensional forms like FEDAS, HESPER, PISCES-II, PISCES-2B, MINIMOS, and BAMBI or three-dimensional forms like TRANAL, SIERRA, FIELDAY, DAVINCI, and CADDETH.

Large-dimension devices, where the carriers travel far from the boundaries, can be simulated based on a one-dimensional approach. Most currently used devices, however, do not fit into this category, and therefore one has to resort to either two- or three-dimensional simulators.

FEDAS (Field Effect Device Analysis System) is a two-dimensional device simulator that simulates MOSFETs, JFETs, and MESFETs by considering only those carriers that form the channel. The Poisson equation is solved everywhere except in the oxide region. Instead of carrying the potential calculation within the oxide region, the potential at the semiconductor-oxide interface is calculated by assuming a mixed boundary condition. FEDAS uses FDM to solve the set of linear equations. A three-dimensional variation of FEDAS is available for the simulation of small geometry MOSFETs.

HESPER (HEterostructure device Simulation Program to Estimate the performance Rigorously) is a two-dimensional device simulator that can be used to simulate heterostructure photodiodes, HBTs, and HEMTs. The simulation starts with the solution of Poisson's equation in which the electron and hole concentrations are described as functions of the composition (composition dependent). The recombination rate is given by the Shockley–Read–Hall relationship. Lifetimes of both types of carriers are assumed to be equal in this model.

PISCES-2B is a two-dimensional device simulator for simulation of diodes, BJTs, MOSFETs, JFETs, and MESFETs. Besides steady-state analysis, transient and ac small-signal analysis can also be performed.

Conclusion

The decision to use an equivalent circuit model or a device simulator depends upon the designer and the required accuracy of prediction. To save computational time, one should use as simple a model as accuracy will allow. At this time, however, the trend is toward developing quantum mechanical models that are more accurate, and with faster computers available, the computational time for these simulators has been considerably reduced.

Defining Terms

Density of states: The total number of charged carrier states per unit volume.
Fermi levels: The energy level at which there is a 50% probability of finding a charged carrier.
Mean free path: The distance traveled by the charged carrier between two collisions.
Mobile charge: The charge due to the free electrons and holes.
Quasi-Fermi levels: Energy levels that specify the carrier concentration inside a semiconductor under nonequilibrium conditions.

Schottky contact: A metal-to-semiconductor contact where, in order to align the Fermi levels on both sides of the junction, the energy band forms a barrier in the majority carrier path.

References

R. E. Banks, D. J. Rose, and W. Fitchner, "Numerical methods for semiconductor device simulation," *IEEE Trans. Electron Devices*, vol. ED-30, no. 9, pp. 1031–1041, 1983.

J. J. Ebers and J. L. Moll, "Large signal behavior of junction transistors," *Proc. IRE*, vol. 42, pp. 1761–1772, Dec. 1954.

W. L. Engl, *Process and Device Modeling*, Amsterdam: North-Holland, 1986.

A. F. Franz, G. A. Franz, S. Selberherr, C. Ringhofer, and P. Markowich, "Finite boxes—A generalization of the finite-difference method suitable for semiconductor device simulation," *IEEE Trans. Electron Devices*, vol. ED-30, no. 9, pp. 1070–1082, 1983.

H. K. Gummel and H. C. Poon, "An integral charge control model of bipolar transistors," *Bell Syst. Tech. J.*, vol. 49, pp. 827–852, May-June 1970.

R. Kasai, K. Yokoyama, A. Yoshii, and T. Sudo, "Threshold-voltage analysis of short- and narrow-channel MOSFETs by three-dimensional computer simulation," *IEEE Trans. Electron Devices*, vol. ED-21, no. 5, pp. 870–876, 1982.

M. S. Lundstrom and R. J. Schuelke, "Numerical analysis of heterostructure semiconductor devices," *IEEE Trans. Electron Devices*, vol. ED-30, no. 9, pp. 1151–1159, 1983.

D. L. Scharfetter and H. K. Gummel, "Large-signal analysis of a silicon read diode oscillator," *IEEE Trans. Electron Devices*, vol. ED-16, no. 1, pp. 64–77, 1969.

A. Yoshii, H. Kitazawa, M. Tomzawa, S. Horiguchi, and T. Sudo, "A three dimensional analysis of semiconductor devices," *IEEE Trans. Electron Devices*, vol. ED-29, no. 2, pp. 184–189, 1982.

Further Information

Further information about semiconductor device simulation and equivalent circuit modeling, as well as about the different software packages available, can be found in the following articles and books:

C. M. Snowden, *Semiconductor Device Modeling*, London: Peter Peregrinus Ltd., 1988.

C. M. Snowden, *Introduction to Semiconductor Device Modeling*, Teaneck, N.J.: World Scientific, 1986.

W. L. Engl, *Process and Device Modeling*, Amsterdam: North-Holland, 1986.

J.-H. Chern, J. T. Maeda, L. A. Arledge, Jr., and P. Yang, "SIERRA: A 3-D device simulator for reliability modeling," *IEEE Trans. Computer-Aided Design*, vol. CAD-8, no. 5, pp. 516–527, 1989.

T. Toyabe, H. Masuda, Y. Aoki, H. Shukuri, and T. Hagiwara, "Three-dimensional device simulator CADDETH with highly convergent matrix solution algorithms," *IEEE Trans. Electron Devices*, vol. ED-32, no. 10, pp. 2038–2044, 1985.

PISCES-2B and DAVINCI are softwares developed by TMA Inc., Palo Alto, California 94301.

21

Semiconductor Manufacturing

Harold G. Parks
The University of Arizona, Tucson

Wayne Needham
Intel Corporation

S. Rajaram
AT&T Bell Laboratories

21.1 Processes 475
      Thermal Oxidation · Diffusion · Ion Implantation · Deposition
      · Lithography and Pattern Transfer
21.2 Testing 490
      Built-In Self-Test · Scan · Direct Access Testing · Joint Test
      Action Group · Pattern Generation for Functional Test Using
      Unit Delay · Pattern Generation for Timing · Temperature,
      Voltage, and Processing Effects · Fault Grading · Test Program
      Flow
21.3 Electrical Characterization of Interconnections 499
      Interconnection Metrics · Interconnection Electrical Parameters

21.1 Processes

Harold G. Parks

Integrated circuit (IC) fabrication consists of a sequence of processing steps referred to as unit step processes that result in the devices contained on today's microchips. These unit step processes provide the methodology for introducing and transporting dopants to change the conductivity of the semiconductor substrate, growing thermal oxides for inter- and intra-level isolation, depositing insulating and conducting films, and patterning and etching the various layers in the formation of the IC. Many of these unit steps have essentially remained the same since discrete component processing, whereas many have originated and grown with the integrated circuit evolution from small-scale integration (SSI) with less than 50 components per chip through very large scale integration (VLSI) with up to one million devices per chip. As the ultra large scale integration (ULSI) era, with more than a million devices per chip, proceeds to billion-device chips shortly after the turn of the century, new processes and further modification of the current unit step processes will be required. In this section the unit step processes for silicon IC processing as they exist today with an eye toward the future are presented. How they are combined to form the actual IC process will be discussed in a later section. Due to space limitations only silicon processes are discussed. This author does not feel this is a major limitation, as many of the steps are used in processing other types of semiconductors, and perhaps more than 98% of all ICs today and in the near future are and will be silicon. Furthermore, only the highlights of the unit steps can be presented in this space, with ample references provided for a more thorough presentation. Specifically, the referenced processing textbooks provide detailed discussion of all processes.

Thermal Oxidation

Silicon dioxide (SiO_2) layers are important in integrated circuit technology for surface passivation, as a diffusion barrier and as a surface dielectric. The fact that silicon readily forms a high-quality, dense, natural oxide is the major reason it is the dominant integrated circuit technology today. If a silicon wafer is exposed to air, it will grow a thin (≈ 45 Å) oxide in a relatively short time. To achieve the thicknesses of SiO_2 used in integrated circuit technology (100 Å to 2 µm) alternative steps must be taken. **Thermal oxidation** is an extension of the natural oxide growth at an elevated temperature (800 to 1200°C). The temperature is usually selected out of compromise, i.e., it must be high enough to grow the oxide in a reasonable time and it must be as low as practical to minimize crystal damage and unwanted **diffusion** of dopants already in the wafer.

The Oxidation Process

Thermal oxidation is usually accomplished by placing wafers in a slotted quartz carrier which is inserted into a quartz furnace tube. The tube is surrounded by a resistance heater and has provisions for controlled flow of an inert gas such as nitrogen and the oxidant. A vented cap is placed over the input end of the tube. The gas flows in the back end of the tube, over the wafers, and is exhausted through the vented cap. The wafer zone has a flat temperature profile to within 1/2°C and can handle up to 50 parallel stacked wafers. Modern furnaces are computer controlled and programmable. Wafers are usually loaded, in an inert environment, ramped to temperature, and switched to the oxidant for a programmed time. When the oxidation is complete the gas is switched back to the inert gas and the temperature is ramped down to the unload temperature. All these complications in the process are to minimize thermal stress damage to the wafers and the procedures can vary considerably. Detailed discussions of the equipment and procedures can be found in references [Sze, 1983].

The two most common oxidizing environments are dry and wet. As the name implies, dry oxides are grown in dry O_2 gas following the reaction:

$$Si + O_2 \rightarrow SiO_2 \tag{21.1}$$

Wet oxides were originally grown by bubbling the dry oxygen gas through water at 95°C. Most "wet" oxides today are accomplished by the pyrogenic reaction of H_2 and O_2 gas to form steam, and are referred to as steam oxidations. In either case the reaction is essentially the same at the wafer:

$$Si + H_2O \rightarrow SiO_2 + 2H_2 \tag{21.2}$$

The oxidation process can be modeled as shown in Fig. 21.1 The position X_0 represents the Si/SiO_2 interface which is a moving boundary. The volume density of oxidizing species in the bulk gas, N_G, is depleted at the oxide surface, N_S, due to an amount, N_0, being incorporated in the oxide layer. The oxidizing species then diffuses across the growing oxide layer where it reacts with the silicon at the moving interface to form SiO_2. F_G represents the flux of oxidant transported by diffusion from the bulk gas to the oxide surface. The oxidizing species that enters the SiO_2 diffuses across the growing SiO_2 layer with a flux, F_{ox}. A reaction takes place at the Si/SiO_2 interface that consumes some or all of the oxidizing species, as represented by the flux, F_I.

In steady state these three flux terms are equal and can be used to solve for the concentrations N_I

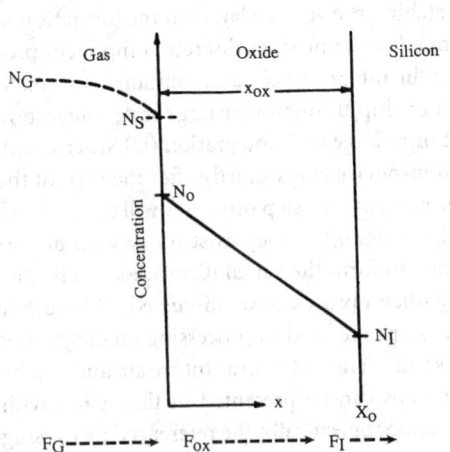

FIGURE 21.1 Model of the oxidation process.

and N_0 in terms of the reaction rate and diffusion coefficient of the oxidizing species. This in turn specifies the flux terms which can be used in the solution of the differential equation:

$$\frac{dx}{dt} = \frac{F}{N_{ox}} \tag{21.3}$$

for the oxide growth, x. In this equation N_{ox} is the number of oxidant molecules per unit volume of oxide. An excellent derivation of the growth equation is given in Grove [1967]. Here we give the result which can be represented by:

$$x_{ox} = \frac{A}{2}\left[\sqrt{1 + \frac{4B}{A^2}(t + \tau)} - 1\right] \tag{21.4}$$

where x_{ox} is the oxide thickness, B is the parabolic rate constant, B/A is the linear rate constant, t is the oxidation time, and τ represents the initial oxide thickness.

Referring to Eq. (21.4) we see there are two regimes of oxide growth. For thin oxides or short times, i.e., the initial phase of the oxidation process, the equation reduces to:

$$x_{ox} = \frac{B}{A}(t + \tau) \tag{21.5}$$

and the growth is a linear function of time, limited by the surface reaction at the Si/SiO$_2$ interface.

For thicker oxides and longer times the reaction is limited by the diffusion of the oxidizing species across the growing oxide layer, and the limiting form of Eq. (21.4) is

$$x_{ox} = \sqrt{Bt} \tag{21.6}$$

Oxidation Rate Dependencies

Typical oxidation curves showing oxide thickness as a function of time with temperature as a parameter for wet and dry oxidation of <100> silicon are shown in Fig. 21.2. This type of curve is qualitatively similar for all oxidations. The oxidation rates are strongly temperature dependent as both the linear and parabolic rate constants show an Arrhenius relationship with temperature. The linear rate is dominated by the temperature dependence of the interfacial growth reaction and the parabolic rate is dominated by the temperature dependence of the diffusion coefficient of the oxidizing species in SiO$_2$.

Wet oxides grow faster than dry oxides. Both the linear and parabolic rate constants are proportional to equilibrium concentration of the oxidant in the oxide. The solubility of H$_2$O in SiO$_2$ is greater than that of O$_2$ and hence the oxidation rate is enhanced for wet oxides.

Oxidation rate depends on substrate orientation [Ghandhi, 1968]. This effect is related to the surface atom density of the substrate, i.e., the higher the density, the faster the oxidation rate. Oxidation rate also depends on pressure. The linear and parabolic rates are dependent on the equilibrium concentration of the oxidizing species in the SiO$_2$ which is directly proportional to the partial pressure of the oxidant in the ambient.

Oxide growth rate shows a doping dependence for heavily doped substrates ($>10^{20}$ cm^{-3}). Boron increases the parabolic rate constant and phosphorus enhances the linear rate constant [Wolf and Tauber, 1986].

Oxide Characteristics

Dry oxides grow more slowly than wet oxides, resulting in higher density, higher breakdown field strengths, and more controlled growth, making them ideal for metal-oxide semiconductor (MOS) gate dielectrics.

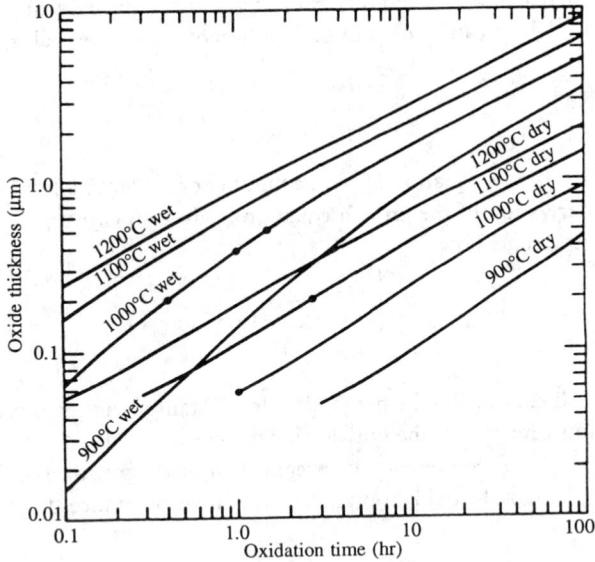

FIGURE 21.2 Thermal silicon dioxide growth on <100> silicon for wet and dry oxides. (*Source:* R. C. Jaeger, *Introduction to Microelectronic Fabrication*, vol. 5 in the Modular Series on Solid State Devices, G. W. Neudek and R. F. Pierret, Eds., Reading, Mass.: Addison-Wesley, 1988, p. 35. With permission.)

Wet oxidation is used for forming thick oxides for field isolation and masking implants and diffusions. The slight degradation in oxide density is more than compensated for by the thickness in these applications.

<100> substrates have fewer dangling bonds at the surface, which results in lower fixed oxide charge and interface traps and therefore higher quality MOS devices.

Conventional dopants (B, P, As, and Sb) diffuse slowly in both wet and dry oxides and hence these oxides provide a good barrier for masking diffusions in integrated circuit fabrication.

High-pressure steam oxidations provide a means for growing relatively thick oxides in reasonable times at low temperatures to avoid dopant diffusion. Conversely, low-pressure oxidations show promise for forming controlled ultra thin gate growth for ULSI technologies.

Chlorine added to gate oxides [Sze, 1988] has been shown to reduce mobile ions, reduce oxide defects, increase breakdown voltage, reduce fixed oxide charge and interface traps, reduce oxygen-induced stacking faults, and increase substrate minority carrier lifetime. Chlorine is introduced into dry oxidations in less than 5% concentrations as anhydrous HCl gas or by trichloroethylene (TCE) or trichloroethane (TCA).

Dopant Segregation and Redistribution

Since silicon is consumed during the oxidation process, the dopant in the substrate will redistribute due to segregation [Wolf and Tauber, 1986]. The boundary condition across the Si/SiO$_2$ interface is that the chemical potential of the dopant is the same on both sides. This results in the definition of a segregation coefficient, m, as the ratio of the equilibrium concentration of dopant in Si to the equilibrium concentration of dopant in SiO$_2$. Depending on the value of m (i.e., less than or greater than 1) and the diffusion properties of the dopant in SiO$_2$, various redistributions are possible. For example, $m \approx 0.3$ for boron and it is a slow diffuser in SiO$_2$, so it tends to deplete from the Si surface and accumulate in the oxide at the Si/SiO$_2$ interface. Phosphorus, on the other hand, has $m \approx 10$, is also a slow diffuser in SiO$_2$, and tends to pile up in the Si at the Si/SiO$_2$ interface. Antimony and arsenic behave similarly to phosphorus.

Diffusion

Diffusion was the traditional way dopants were introduced into silicon wafers to create junctions and control the resistivity of layers. **Ion implantation** has now superseded diffusion for this purpose. The principles and concepts of diffusion theory, however, remain important since they describe the movement and transport of dopants and impurities during the high-temperature processing steps of integrated circuit manufacture.

Diffusion Mechanism

Consider a silicon wafer with a high concentration of an impurity on its surface. At any temperature there are a certain number of vacancies in the Si lattice. If the wafer is subjected to an elevated temperature, the number of vacancies in the silicon will increase and the impurity will enter the wafer moving from the high surface concentration to redistribute in the bulk. The redistribution mechanism is diffusion, and depending on the impurity type it will either be substitutional or interstitial [Ghandhi, 1982].

For substitutional diffusion the impurity atom substitutes for a silicon atom at a vacancy lattice site and then progresses into the wafer by hopping from lattice site to lattice site via the vacancies. Clearly, the hopping can be in a random direction; however, since the impurity is present initially in high concentration on the surface only, there is a net flow of the impurity from the surface into the bulk.

In the case of interstitial diffusion the impurity diffuses by squeezing between the lattice atoms and taking residence in the interstitial space between lattice sites. Since this mechanism does not require the presence of a vacancy, it proceeds much faster than substitutional diffusion.

Conventional dopants such as B, P, As, and Sb diffuse by the substitutional method. This is beneficial in that the diffusion process is much slower and can therefore be controlled more easily in the manufacturing process. Many of the undesired impurities such as Fe, Cu, and other heavy metals diffuse by the interstitial mechanism and therefore the process is extremely fast. This again is beneficial in that at the temperatures used, and in the duration of fabrication processes, the unwanted metals can diffuse completely through the Si wafer. Gettering creates trapping sites on the back surface of the wafer for these impurities that would otherwise remain in the silicon and cause adverse device effects.

Regardless of the diffusion mechanism, it can be formalized mathematically in the same way by introducing a diffusion coefficient, D (cm^2/sec), that accounts for the diffusion rate. The diffusion constants follow an Arrhenius behavior according to the equation:

$$D = D_0 \exp\left[-\frac{E_A}{kT}\right] \tag{21.7}$$

where D_0 is the prefactor, E_A the activation energy, k Boltzmann's constant, and T the absolute temperature. Conventional silicon dopants (substitutional diffusers) have diffusion coefficients on the order of 10^{-14} to 10^{-12} at 1100°C, whereas heavy metal interstitial diffusers (Fe, Au, and Cu) have diffusion coefficients of 10^{-6} to 10^{-5} at this temperature.

The diffusion process can be described using Fick's Laws. Fick's first law says that the flux of impurity, F, crossing any plane is related to the impurity distribution, $N(x,t)$ per cm^3, by:

$$F = D\frac{\partial N}{\partial x} \tag{21.8}$$

in the one-dimensional case. Fick's second law states that the time rate of change of the particle density in turn is related to the divergence of the particle flux:

$$\frac{\partial N}{\partial t} = \frac{\partial F}{\partial x} \tag{21.9}$$

Combining these two equations gives:

$$\frac{\partial N}{\partial t} = \frac{\partial}{\partial x}\left(D \frac{\partial N}{\partial x}\right) = D \frac{\partial^2 N}{\partial x^2} \tag{21.10}$$

in the case of a constant diffusion coefficient as is often assumed. This partial differential equation can be solved by separation of variables or by Laplace transform techniques for specified boundary conditions.

For a constant source diffusion the impurity concentration at the surface of the wafer is held constant throughout the diffusion process. Solution of Eq. (21.10) under these boundary conditions, assuming a semi-infinite wafer, results in a complementary error function diffusion profile:

$$N_{(x,t)} = N_0 \, \text{erfc}\left(\frac{x}{2\sqrt{Dt}}\right) \tag{21.11}$$

Here, N_0 is the impurity concentration at the surface of the wafer, x the distance into the wafer, and t the diffusion time. As time progresses the impurity profile penetrates deeper into the wafer while maintaining a constant surface concentration. The total number of impurity atoms/cm^2 in the wafer is the dose, Q, and continually increases with time:

$$Q = \int_0^\infty N_{(x,t)}dx = 2N_0\sqrt{\frac{Dt}{\pi}} \tag{21.12}$$

For a limited source diffusion an impulse of impurity of dose Q is assumed to be deposited on the wafer surface. Solution of Eq. (21.10) under these boundary conditions, assuming a semi-infinite wafer with no loss of impurity, results in a Gaussian diffusion profile:

$$N_{(x,t)} = \frac{Q}{\sqrt{\pi Dt}} \exp\left[-\left(\frac{x}{2\sqrt{Dt}}\right)^2\right] \tag{21.13}$$

In this case, as time progresses the impurity penetrates more deeply into the wafer and the surface concentration falls so as to maintain a constant dose in the wafer.

Practical Diffusions

Most real diffusions follow a two-step procedure, where the dopant is applied to the wafer with a short constant source diffusion, then driven in with a limited source diffusion. The reason for this is that in order to control the dose, a constant source diffusion must be done at the solid solubility limit of the impurity in the Si, which is on the order of 10^{20} for most dopants. If only a constant source diffusion were done, this would result in only very high surface concentrations. Therefore, to achieve lower concentrations, a short constant source diffusion to get a controlled dose of impurities in a near surface layer is done first. This diffusion is known as the *predeposition* or *predep* step. Then the source is removed and the dose is diffused into the wafer, simulating a limited source diffusion in the subsequent *drive-in* step.

If the Dt product for the drive-in step is much greater than the Dt product for the predep, the resulting profile is very close to Gaussian. In this case the dose can be calculated by Eq. (21.12) for the predep time and diffusion coefficient. This dose is then used in the limited source Eq. (21.13) to describe the final profile based on the time and diffusion coefficient for the drive-in. If these Dt criteria are not met, then an integral solution exists for the evaluation of the resulting profiles [Ghandhi, 1968].

Further Profile Considerations

A wafer typically goes through many temperature cycles during fabrication, which can alter the impurity profile. The effects of many thermal cycles that take place at different times and temperatures are accounted for by calculating a total Dt product for the diffusion that is equal to the sum of the individual process Dt products:

$$(Dt)_{tot} = \sum_i D_i t_i \tag{21.14}$$

Here D_i and t_i are the diffusion coefficient and time that pertain to the ith process step.

Many diffusions are used to form junctions by diffusing an impurity opposite in type to the substrate. At the metallurgical junction, x_j, the impurity diffusion profile has the same concentration as the substrate. For a junction with a surface concentration N_0 and substrate doping N_B the metallurgical junction for a Gaussian profile is

$$x_j = 2\sqrt{Dt \, \ln\left(\frac{N_0}{N_B}\right)} \tag{21.15}$$

and for a complementary error function profile is

$$x_j = 2\sqrt{Dt} \, \text{erfc}^{-1}\left(\frac{N_B}{N_0}\right) \tag{21.16}$$

So far we have considered just vertical diffusion. In practical IC fabrication, usually only small regions are affected by the diffusion by using an oxide mask and making a cut in it where specific diffusion is to occur. Hence, we also have to be concerned with lateral diffusion of the dopant so as not to affect adjacent devices. Two-dimensional numerical solutions exist for solving this problem [Jaeger, 1988]; however, a useful rule of thumb is that the lateral junction, y_j, is $0.8x_j$.

Another parameter of interest is the sheet resistance of the diffused layer. This has been numerically evaluated for various profiles and presented as general-purpose graphs known as Irvin's curves. For a given profile type, such as n-type Gaussian, Irvin's curves plot surface dopant concentration versus the product of sheet resistance and junction depth with substrate doping as a parameter. Thus, given a calculated diffusion profile one could estimate the sheet resistivity for the diffused layer. Alternatively, given the measured junction depth and sheet resistance, one could estimate the surface concentration for a given profile and substrate doping. Most processing books [e.g., Jaeger, 1988] contain Irvin's curves.

Ion Implantation

Diffusion places severe limits on device design, such as hard to control low-dose diffusions, no tailored profiles, and appreciable lateral diffusion at mask edges. Ion implantation overcomes all of these drawbacks and is an alternative approach to diffusion used in the majority of production doping applications today. Although many different elements can be implanted, IC manufacture is primarily interested in B, P, As, and Sb.

Ion Implant Technology

A schematic drawing of an ion implanter is shown in Fig. 21.3. The ion source operates at relatively high voltage (≈ 20–25 kV) and for conventional dopants is usually a gaseous type which extracts the ions from a plasma. The ions are mass separated with a 90 degree analyzer magnet that directs the selected species through a resolving aperture focused and accelerated to the desired implant energy. At the other end of the implanter is the target chamber where the wafer is placed in the beam

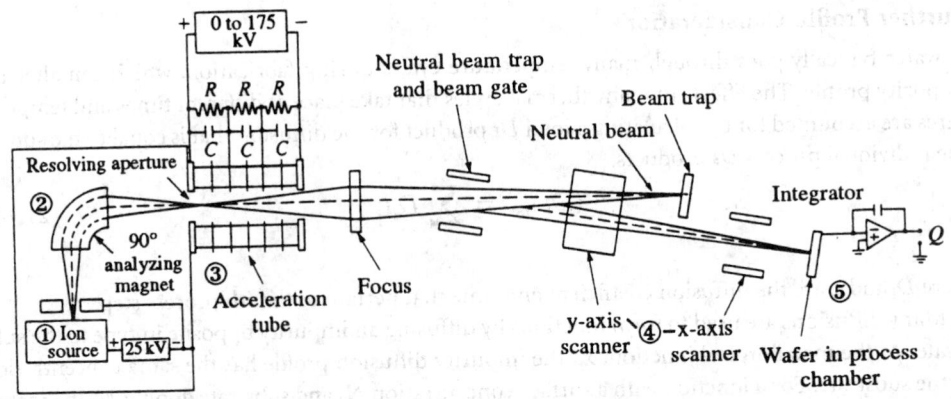

FIGURE 21.3 Schematic drawing of an ion implanter. (*Source:* R. C. Jaeger, *Introduction to Microelectronic Fabrication*, vol. 5 in the Modular Series on Solid State Devices, G. W. Neudek and R. F. Pierret, Eds., Reading, Mass: Addison-Wesley, 1988, p. 90. With permission.)

path. The beam line following the final accelerator and the target chamber are held at or near ground potential for safety reasons. After final acceleration the beam is bent slightly off axis to trap neutrals and is asynchronously scanned in the X and Y directions over the wafer to maintain dose uniformity. This is often accompanied by rotation and sometimes translation of the target wafer also.

The implant parameters of interest are the ion species, implant energy, and dose. The ion species can consist of singly ionized elements, doubly ionized elements, or ionized molecules. The molecular species are of interest in forming shallow junctions with light ions, i.e., B, using BF_2^+. The beam energy is

$$E = nqV \tag{21.17}$$

where n represents the ionization state (1 for singly and 2 for doubly ionized species), q the electronic charge, and V the total acceleration potential (source + acceleration tube) seen by the beam. The dose, Q, from the implanter is

$$Q = \int_0^{t_I} \frac{I}{nqA}\, dt \tag{21.18}$$

where I is the beam current in amperes, A the wafer area in cm^2, t_I the implant time in sec, and n the ionization state.

Ion Implant Profiles

Ions impinge on the surface of the wafer at a certain energy and give up that energy in a series of electronic and nuclear interactions with the target atoms before coming to rest. As a result the ions do not travel in a straight line but follow a zigzag path resulting in a statistical distribution of final placement. To first order the ion distribution can be described with a Gaussian distribution:

$$N_{(x)} = N_p \exp\left[-\frac{(x - R_p)^2}{2(\Delta R_p)^2}\right] \tag{21.19}$$

R_p is the projected range which is the average depth of an implanted ion. The peak concentration, N_p, occurs at R_p and the ions are distributed about the peak with a standard deviation ΔR_p known as the straggle. Curves for projected range and straggle taken from Lindhard, Scharff, and Schiott (LSS) theory [Gibbons *et al.*, 1975] are shown in Figs. 21.4 and 21.5, respectively, for the conventional dopants.

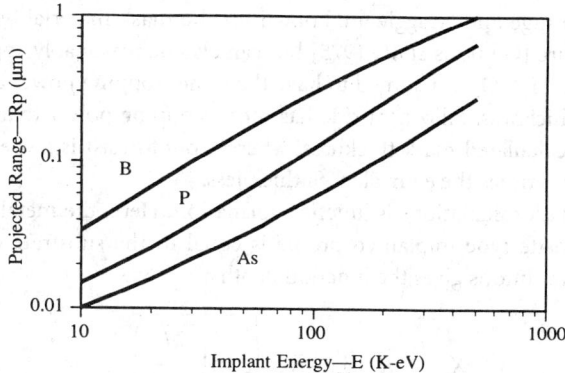

FIGURE 21.4 Projected range for B, P, and As based on LSS calculations.

The area under the implanted distribution represents the dose as given by:

$$Q = \int_0^\infty N_{(x)} dx = \sqrt{2\pi} N_p \Delta R_p \quad (21.20)$$

which can be related to the implant conditions by Eq. (21.18). Implant doses can range from 10^{10} to 10^{18} per cm^2 and can be controlled within a few percent.

The mathematical representation of the implant profile just presented really pertains to an amorphous substrate. Silicon wafers are crystalline and therefore present the opportunity for the ions to travel much deeper into the substrate by a process

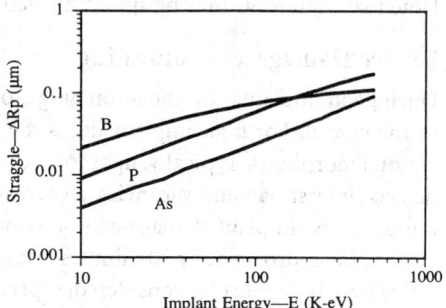

FIGURE 21.5 Implant straggle for B, P, and As based on LSS calculations.

known as channeling. The regular arrangement of atoms in the crystalline lattice leaves large amounts of open space that appear as channels into the bulk when viewed from the major orientation directions, i.e, <110>, <100>, and <111>. Practical implants are usually done through a thin oxide with the wafers tilted off normal by a small angle (typically 7 degrees) and rotated by 30 degrees to make the surface atoms appear more random. Implants with these conditions agree well with the projected range curves of Fig. 21.4, indicating the wafers do appear amorphous.

Actual implant profiles deviate from the simple Gaussian profiles described in the previous paragraphs. Light ions tend to backscatter from target atoms and fill in the distribution on the surface side of the peak. Heavy atoms tend to forward scatter from the target atoms and fill in the profile on the substrate side of the peak. This behavior has been modeled with distributions such as the Pearson Type-IV distribution [Jaeger, 1988]. However, for implant energies below 200 keV and first-order calculations, the Gaussian model will more than suffice.

Masking and Junction Formation

Usually it is desired to implant species only in selected areas of the wafer to alter or create device properties, and hence the implant must be masked. This is done by putting a thick layer of silicon dioxide, silicon nitride, or photoresist on the wafer and patterning and opening the layer where the implant is desired. To prevent significant alteration of the substrate doping in the mask regions the implant concentration at the Si/mask interface, X_0, must be less than 1/10 of the substrate doping, N_B. Under these conditions Eq. (21.19) can be solved for the required mask thickness as:

$$X_0 = R_p + \Delta R_p \sqrt{2 \ln\left(\frac{10 N_p}{N_B}\right)} \quad (21.21)$$

This implies that the range and straggle are known for the mask material being used. These are available in the literature [Gibbons *et al.*, 1975] but can also be reasonably approximated by making the calculations for Si. SiO$_2$ is assumed to have the same stopping power as Si and thus would have the same mask thickness. Silicon nitride has more stopping power than SiO$_2$ and therefore requires only 85% of calculated mask thickness, whereas photoresist is less effective for stopping the ions and requires 1.8 times the equivalent Si thickness.

Analogous to the mask calculations is junction formation. Here, the metallurgical junction, x_j, occurs when the opposite type implanted profile is equal to the substrate doping, N_B. Solving Eq. (21.19) for these conditions gives the junction depth as:

$$X_j = R_p \pm \Delta R_p \sqrt{2 \ln\left(\frac{N_p}{N_B}\right)} \tag{21.22}$$

Note that both roots may be applicable depending on the depth of the implant.

Lattice Damage and Annealing

During ion implantation the impinging atoms can displace Si atoms in the lattice, causing damage to the crystal. For high implant doses the damage can be severe enough to make the implanted region amorphous. Typically, light ions, or light doses of heavy ions, will cause primary crystalline defects (interstitials and vacancies), whereas medium to heavy doses of heavy ions will cause amorphous layers. Implant damage can be removed by annealing the wafers in an inert gas at 800 to 1000°C for approximately 30 minutes. Annealing cycles at high temperatures can cause appreciable diffusion which must be considered, especially in the newer technologies where shallow junctions are required. Rapid thermal annealing (RTA) can be successfully applied to prevent undesirable diffusion in these cases [Wolf and Tauber, 1986].

Deposition

During IC fabrication, thin films of dielectrics (SiO$_2$, Si$_3$N$_4$, etc.), polysilicon, and metal conductors are deposited on the wafer surface to form devices and circuits. The techniques used for forming these thin films are **physical vapor deposition** (PVD), **chemical vapor deposition** (CVD), and epitaxy, which is just a special case of CVD.

Physical Vapor Deposition

Vacuum evaporation and sputtering are the two methods of physical vapor deposition used in the fabrication of integrated circuits. Both of these processes are carried out in a vacuum to prevent contamination of the substrate and to provide a reasonable mean free path for the material being deposited. In the early days of integrated circuits aluminum was used exclusively for IC metallization and evaporation was used for its deposition. As IC technology matured, the need for metal alloys, alternative metals, and various insulating thin films stimulated the development and acceptance of sputter deposition as the PVD method of choice.

When the temperature is raised high enough to melt a solid some of the atoms have enough internal energy to break the surface and escape into the surrounding atmosphere. These evaporated atoms strike the wafer and condense into a thin film. Typical film thickness used in the IC industry are in the few thousand angstroms to 1 μm range. Heat is provided by resistance heating, electron beam heating, or by rf inductive heating. Discussions of these techniques and their historical significance can be found in most processing books [e.g., Sze, 1983; Wolf and Tauber, 1986]. The most commonly used system employs a focussed electron beam scanned over a crucible of metal as illustrated in Fig. 21.6(a). Contamination levels can be quite low because only electrons come in contact with the melted metal. A high intensity electron beam, typically 15 keV, bent through a 270° angle to shield the wafers from filament contamination, provides the heating for evaporation of the

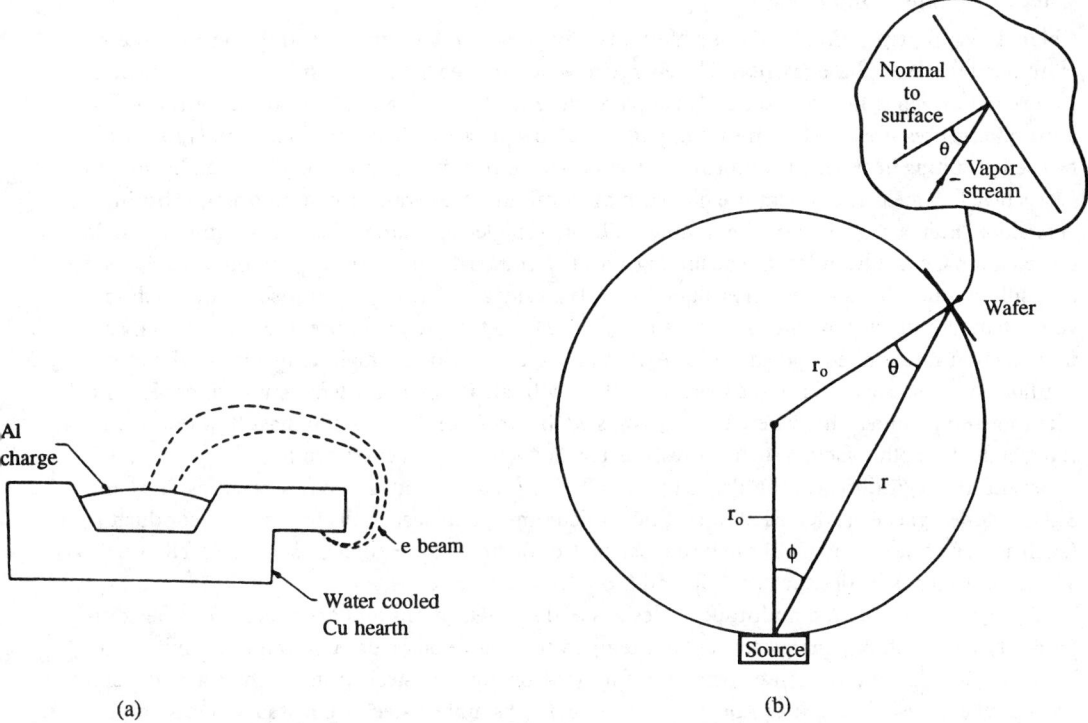

FIGURE 21.6 (a) Electron beam evaporation source. (b) Geometry for a planetary holder. (*Source:* R. C. Jaeger, *Introduction to Microelectronic Fabrication*, vol. 5 in the Modular Series on Solid State Devices, G. W. Neudek and R. F. Pierret, Eds., Reading, Mass: Addison-Wesley, 1988, pp. 110, 112. With permission.)

metal. The wafers are mounted above the source on a planetary substrate holder that rotates around the source during the deposition to insure uniform step coverage. For the planetary substrate holder, shown schematically in Fig. 21.6(b), the growth rate is independent of substrate position. The relatively large size of the crucible provides an ample supply of source material for the depositions. The deposition rate is controlled by changing the current and energy of the electron beam. Dual beams with dual targets can be used to coevaporate composite films. Device degradation due to X-ray radiation, generated by the electron beam system, is of great concern in MOS processing. Because of this, sputtering has replaced e-beam evaporation in many process lines.

Sputtering is accomplished by bombarding a target surface with energetic ions that dislodge target atoms from the surface by direct momentum transfer. Under proper conditions the sputtered atoms are transported to the wafer surface where they deposit to form a thin film. Sputter deposition takes place in a chamber that is evacuated and then backfilled with an inert gas at roughly 10 mtorr pressure. A glow discharge between two electrodes, one of which is the target, within the gas creates a plasma that provides the source of ions for the sputter process. Metals can be sputtered in a simple dc parallel plate reactor with the most negative electrode being the target and the most positive electrode (usually ground) being the substrate holder. If the dc voltage is replaced by an rf voltage, insulators as well as conductors can be sputter deposited. Magnetron systems incorporate a magnetic field that enhances the efficiency of the discharge by trapping secondary electrons in the plasma and lowers substrate heating by reducing the energy of electrons reaching the wafer. Circular magnetrons (or S-guns) virtually eliminate substrate heating by electron bombardment due to a ring-shaped cathode/anode combination with the substrate a nonparticipating system electrode [Sze, 1983]. Besides insulators and conductors, sputtering can be used to deposit alloy films with the same composition as an alloy target.

Chemical Vapor Deposition

Chemical vapor deposition (CVD) is a method of forming thin films on a substrate in which energy is supplied for a gas phase reaction. The energy may be supplied by heat, plasma excitation, or optical excitation. Since the reaction can take place close to the substrate, CVD can be performed at atmospheric pressure (i.e., low mean free path) or at low pressures. Relatively high temperatures can be used, resulting in excellent conformal step coverage, or relatively low temperatures can be used to passivate a low melting temperature film such as aluminum. Substrates can be amorphous or single crystalline. Epitaxial growth of Si is simply CVD on a single crystalline substrate resulting in single crystal layer. Generally, unless the qualifying adjective epitaxial is used, the depositions are assumed to result in amorphous or polycrystalline films. Typically all CVD depositions show a growth rate versus temperature dependence as illustrated in Fig. 21.7. At low temperatures the deposition is surface reaction rate limited and shows an Arrhenius type behavior. At high temperatures the rate is dominated by mass transfer of the reactant and growth rate is essentially temperature independent. The transition temperature where the reaction switches from reaction rate to mass flow dominated is dependent on other factors such as pressure, reactant species, and energy source.

Several types of films can be deposited by CVD. Insulators and dielectrics (such as SiO_2 and Si_3N_4), doped glasses (PSG, BPSG), and nonstoichiometric dielectrics as well as semiconductors (such as Si, Ge, GaAs, and GaP) can be deposited. Conductors of pure metals, such as Al, Ni, Au, Pt, Ti, etc., as well as silicides such as WSi_2 and $MoSi_2$ can also be deposited.

CVD systems come in a multitude of designs and configurations and are selected for the compatibility with method, energy source, and temperature range being used. Reaction chambers can be quartz, similar to diffusion or oxidation furnaces, or stainless steel. Wafer holders also depend on the type of reaction and can be graphite, quartz, or stainless steel. Cold wall systems employ direct heating of the substrate or wafer holder by induction or radiation. As such, the reaction takes place right at the wafer surface and is usually cleaner because the film does not build up on the chamber walls. In a hot wall deposition system the reaction takes place in the gas stream and the reaction product is deposited on every surface in the system, including the walls. Such unwanted depositions build up and flake off of these surfaces in time. Without proper cleaning and maintenance procedures they become a source of contamination.

Atmospheric deposition systems were first used in the deposition of epitaxial Si in bipolar processes (i.e., buried layer formation). Early systems were horizontal rf induction heated systems using graphite substrate holders. Atmospheric pressure systems rely on flow dynamics in the chamber to produce uniform films. Since the reactant species is being depleted from the gas stream, it is imperative that the system design (flow pattern and rate) ensure that all wafers receive the same amount of deposit. Because of this, most atmospheric depositions are carried out using a horizontal or near horizontal wafer holder.

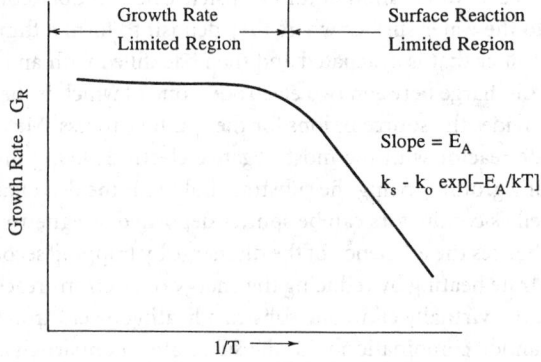

FIGURE 21.7 Typical chemical vapor deposition growth rate versus reciprocal temperature characteristics.

Low-pressure CVD depositions are done at pressures in the 0.5 to 1 torr range and most often in a horizontal hot wall reactor. Wafers are held vertically and the system is operated in the reaction rate limited regime. A multi-zone furnace (typically three) is used to allow the temperature to be increased along the wafer holder to compensate for gas depletion by the deposition along the flow path. LPCVD is typically used to deposit SiO_2 ($\approx925°C$), Si_3N_4 ($\approx850°C$), and polycrystalline Si ($\approx630°C$). Due to the higher temperatures excellent conformal step coverage can be obtained.

Plasma-enhanced CVD uses a plasma to supply the required reaction energy. Typically, rf energy is used to produce a glow discharge such that the gas constituents are in a highly reactive state. Because of the ability of the plasma to impart high energy to the reaction at low temperature, these depositions maintain many of the excellent features of LPCVD, such as step coverage, with low temperature attributes, such as reduced wafer warpage, less impurity diffusion, and less film stress. Plasma-enhanced deposition results in nonstoichiometric highly hydrogenated films such as SiO-H, SiN-H, and amorphous Si-H.

Lithography and Pattern Transfer

In the production of integrated circuits various thin films are fabricated in the Si wafer or grown or deposited on the surface. Each of these layers has a functionality as either an active part of a device, a barrier or mask, or an inter- or intra-layer isolation. To perform its intended function, each of these layers has to be located in specific regions on the wafer surface. This is accomplished by either patterning and etching a thin film to serve as a mask to form the desired function by a blanket process or by forming the desired layer and then patterning and etching it to provide the device function in local areas.

Wafer Patterning

Lithography is a transfer process where the pattern on a mask is replicated in a radiation-sensitive layer on the wafer surface. Typically, this has been accomplished with UV light as the radiation source and photoresist, or resist in conventional terminology, which is a UV-sensitive polymer, as the mask layer. The wafers with the layer to be patterned are cleaned and prebaked (400–800°C for 20–30 min) to drive off moisture and promote resist adhesion. Many processes also add an adhesion promoter, such as hexamethydisilazane (HMDS), at this point, which functions by removing unwanted surface radicals that prevent adhesion. Then, a few drops of resist are deposited on a wafer which is spinning at a slow rate to produce a uniform coating and the spin speed is increased to enhance drying. The wafer with resist is softbaked at 80–90°C for 10–30 min to drive off the remaining solvents. The wafers are then put in an exposure system and the mask pattern to be transferred is aligned to any existing wafer patterns. Present-day exposure systems are step and repeat cameras where the mask is a 5–10× enlargement of a single chip pattern. Earlier systems used a 1× mask for the whole wafer and were of the contact, proximity, or 1× projection variety [Anner, 1990]. The resist is exposed through the mask to UV radiation that changes its structure, depending on whether the resist type is positive or negative. Negative resist becomes polymerized (i.e., cross linked) in areas exposed to the radiation, whereas in a positive resist the polymer bonds are scissioned upon exposure. The resist is not affected in regions where the mask is opaque in either case. After full wafer exposure the resist is developed such that the unpolymerized regions are selectively dissolved in an appropriate solvent. The polymerized portion of the resist remains intact on the wafer surface, replicating the opaque features of the mask in a positive resist and just the opposite for a negative resist.

As the minimum feature size approaches the wavelength of light used in optical exposure systems (≈4000 Å), resolution is lost due to diffraction limitations. Alternatives are being developed to overcome these limitations [Okazaki, 1991], the simplest of which is using shorter wavelength, i.e., deep UV, radiation to reduce the diffraction limit. A further extension of this technique introduces optical phase shifting in the photomask itself to extend the diffraction limit even further. These techniques are eventually limited, somewhere in the range of 2000 Å, and alternative techniques such as electron beam or X-ray radiation will have to replace the UV radiation source. These techniques are quite analogous to UV systems but use resist specifically tailored for sensitivity in their appropriate wavelength range.

Pattern Transfer

After the resist pattern is formed it is then transferred to the surface layer of the wafer. Sometimes this is an invisible transfer, such as ion implantation, but more often than not it is a physical transfer of the pattern by etching the surface layer, using the resist as a mask. This results in the desired structure or produces a more etch-resistant mask for further pattern transfer operations.

Historically, the most common etch processes used wet chemicals. During **wet etching**, wafers with resist (or a resist transferred mask) are immersed in a temperature-controlled etchant for a fixed period of time. The etch rate is dependent on the strength of the etchant, temperature, and material being etched. Such chemical etches are isotopic, which means the vertical and lateral etch rates are the same. Thus, the thicker the layer being etched, the more undercutting of the mask pattern. Most wet etches are stopped with an underlying etchstop layer that is impervious to the etchant used to remove the top layer. In order to ensure the layer is totally removed, an over etch is allowed, which exacerbates the undercutting.

Modern high-density, small-feature processes require anisotropic etch processes; this requirement has driven the development of **dry etching** techniques. Plasma etching is a dry etch technique that uses an rf plasma to generate chemically active etchants that form volatile etch species with the substrate. Typically chlorine or fluorine compounds, most notably being CCl_4 and CF_4, have been tailored for etching polysilicon, SiO_2, Si_3N_4, and metals. The etch rate can be significantly enhanced by adding 5–10% O_2 to the etch gas. This, however, increases erosion of resist masks.

Early plasma etch systems were barrel reactors which used a perforated metal cylinder to confine the plasma in a region exterior to the wafers. In such reactors the etch species are electrically neutral so the reaction is entirely chemical and just as isotropic as wet chemical etch. A similar result occurs for a planar parallel plate rf reactor where the wafer is placed on the grounded electrode. However, some anisotropic etching is achieved in this configuration because ions can reach the wafer. This is further enhanced in reactive ion etching (RIE) where the wafer is placed on the rf electrode in a planar parallel plate reactor. Here the ions experience a considerable acceleration to the wafer by the dc potential developed between plasma and cathode that results in anisotropic etching. The etch processes require significant characterization and development through optimization of pressure, gas flow rate, gas mixture, and power to produce the desired etch rate, anisotropy, selectivity, uniformity and resist erosion. Nevertheless, this has become the etch process of choice in modern ULSI processes. A less popular version uses a beam of reactive species for the etch process and is called reactive ion beam etching (RIBE).

Finally, it should be noted that near total anisotropic dry etching can be achieved with ion etching. This is done with an inactive species (e.g., Ar ions) either in a beam or with a parallel plate sputtering system with the wafer on the rf electrode. This results in an etch process that is entirely physical through momentum transfer. The etch rate is primarily controlled by the sputtering efficiency for various materials and thus does not differ significantly from material to material. Hence, such processes suffer from poor selectivity. Resist or mask erosion is also a problem with these techniques. Because of these limitations, virtually the only application of pure ion etching in VLSI/ULSI is for sputter cleaning of wafers before deposition.

Defining Terms

Chemical vapor deposition: A process in which insulating or conducting films are deposited on a substrate by use of reactant gases and an energy source to produce a gas-phase chemical reaction. The energy source may be thermal, optical, or plasma in nature.

Deposition: An operation in which a film is placed on a wafer surface, usually without a chemical reaction with the underlying layer.

Diffusion: A high-temperature process in which impurities on or in a wafer are redistributed within the silicon. If the impurities are desired dopants, this technique is often used to form specific device structures. If the impurities are undesired contaminants, diffusion often results in undesired device degradation.

Dry etching: Processes that use gas-phase reactants, inert or active ionic species, or a mixture of these to remove unprotected layers of a substrate by chemical processes, physical processes, or a mixture of these, respectively.

Ion implantation: A high-energy process, usually greater than 10 keV, that injects ionized species into a semiconductor substrate. Often done for introducing dopants for device fabrication into silicon with boron, phosphorus, or arsenic ions.

Lithography: A patterning process in which a mask pattern is transferred by a radiation source to a radiation-sensitive coating that covers the substrate.

Physical vapor deposition: A process in which a conductive or insulating film is deposited on a wafer surface without the assistance of a chemical reaction. Examples are vacuum evaporation and sputtering.

Thermal oxidation of silicon: A high-temperature chemical reaction, typically greater than 800°C, in which the silicon of the wafer surface reacts with oxygen or water vapor to form silicon dioxide.

Wet etching: A process that uses liquid chemical reactions with unprotected regions of a wafer to remove specific layers of the substrate.

References

G. E. Anner, *Planar Processing Primer*, New York: Van Nostrand Reinhold, 1990.

S. K. Ghandhi, *The Theory and Practice of Microelectronics*, New York: John Wiley & Sons, 1968.

S. K. Ghandhi, *VLSI Fabrication Principles—Silicon and Gallium Arsenide*, New York: John Wiley & Sons, 1982.

J. F. Gibbons, W. S. Johnson, and S. M. Mylroie, Projected range statistics, in *Semiconductors and Related Materials*, vol. 2, 2nd ed., Dowden, Hutchinson, and Ross, Eds., New York: Academic Press, 1975.

A. S. Grove, *Physics and Technology of Semiconductor Devices*, New York: John Wiley & Sons, 1967.

R. C. Jaeger, *Introduction to Microelectronic Fabrication*, vol. 5 in the Modular Series on Solid State Devices, G. W. Neudek and R. F. Pierret, Eds., Reading, Mass.: Addison-Wesley, 1988.

S. Okazaki, "Lithographic technology for future ULSI's," *Solid State Technology*, vol. 34, no. 11, p. 77, November 1991.

S. M. Sze, Ed., *VLSI Technology*, 1st ed., New York: McGraw-Hill, 1983.

S. M. Sze, Ed., *VLSI Technology*, 2nd ed., New York: McGraw-Hill, 1988.

S. Wolf and R. N. Tauber, *Silicon Processing for the VLSI ERA*, vol. 1, *Process Technology*, Sunset Beach, Calif.: Lattice Press, 1986.

Further Information

The references given in this section have been chosen to provide more detail than is possible to provide in the limited space allocation for this section. Specifically, the referenced processing textbooks provide detailed discussion of all unit step processes presented. Further details and more recent process developments can be found in several magazines/journals, such as *Semiconductor International*, *Solid State Technology*, *IEEE Transactions on Semiconductor Manufacturing*, *IEEE Transactions on Electron Devices*, *IEEE Electron Device Letters*, *Journal of Applied Physics*, and the *Journal of the Electrochemical Society*.

21.2 Testing*

Wayne Needham

The function of test of a semiconductor device is twofold. First is design debug, to understand the failing section of the device, identify areas for changes and verify correct modes of operation. The second major area is to simply separate good devices from bad devices in a production test environment. Data collection and information analysis are equally important in both types of test, but for different reasons. The first case is obvious, debug requires understanding of the part. In the second case, data collected from a test program may be used for **yield** enhancement. This is done by finding marginal areas of the device or in the fabrication process and then making improvements to raise yields and therefore lower cost.

In this section, we will look at methods of testing which can be used for data collection, analysis, and debug of a new device. No discussion of test would be useful if the test strategy was not thought out clearly before design implementation. Design for test (access for control and observation) is a requirement for successful debug and testing of any semiconductor device.

The basis for all testing of complex integrated circuits is a comparison of known good patterns to the response of a **DUT** (device under test). The simulation of the devices is done with input stimuli, and those same input stimuli (**vectors**) are presented on the DUT. Comparisons are made cycle by cycle with an option to ignore certain pins, times, or patterns. If the device response and the expected response are not in agreement, the devices are usually considered defective.

This section will cover common techniques for testing. Details of generation of test programs, simulation of devices, and tester restrictions are not covered here.

Built-In Self-Test

Self-testing (**built-in self-test** or BIST) is essentially the implementation of logic built into the circuitry to do testing without the use of the **tester** for pattern generation and comparison purposes. A tester is still needed to categorize failures and to separate good from bad units. In this case the test system supplies clocks to the device and determines pass/fail from the outputs of the device. The sequential elements are run with a known data pattern, and a signature is generated. The signature can be a simple go or no-go signal presented on one pin of the part, or the signal may be a polynomial generated during testing. This polynomial has some significance to actual states of the part. Figure 21.8 shows a typical self-testing technique implemented on a large block of random logic, inside a device. The number of unique inputs to the logic block should be limited to 20 bits so the total vectors are less than one million. This keeps test time to less than one second in most cases.

Self-test capability can be implemented on virtually any size block. It is important to understand the tradeoffs between the extra logic added to implement self-testing and the inherent logic of the block to be tested. For instance, adding self-testing capability to a RAM requires adding counters and read, write, and multiplexor circuitry to allow access to the part. The access is needed not only by the self-test circuitry, but by the circuitry that would normally access the RAM.

When implementing self-testing on blocks such as RAMs and ROMs, it is worthwhile to note the typical failure mode of semiconductor devices. Single-bit defects can easily be detected using self-testing techniques. Single-point defects in the manufacturing process can show up as a single transistor failure in a RAM or ROM, or they may be somewhat more complex. If a single-point defect happens to be in the decoder section or in a row or column within the RAM, a full section of the device may be nonfunctional.

The problem with this failure mode is that RAMs and ROMs are typically laid out in a square or

* Portions reproduced from W. M. Needham, *Designer's Guide to Testable ASIC Devices*, New York: Van Nostrand Reinhold, 1991. With permission.

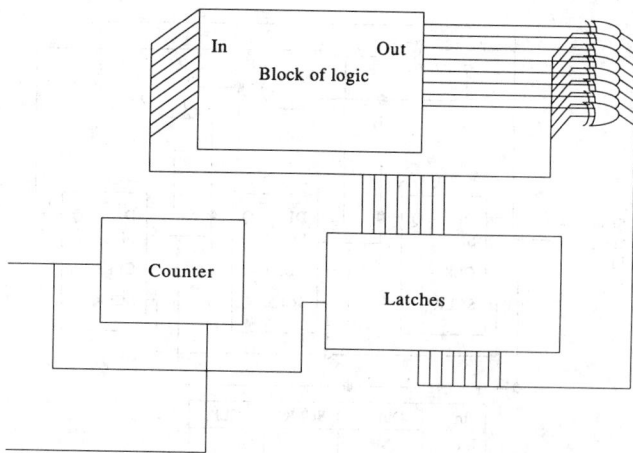

FIGURE 21.8 Example of self-test in circuit. (*Source:* W. M. Needham, *Designer's Guide to Testable ASIC Devices*, New York: Van Nostrand Reinhold, 1991, p. 88. With permission.)

rectangular array. They are usually decoded in powers of 2, such as a 32 ∗ 64 or a 256 ∗ 512 array. If the self-testing circuitry is an 8-bit-wide counter or linear feedback shift register, there may be problems. There are 256 possible combinations within the states of the counter, and this may be a multiple of the row or columns. Notice that 256 possible rows or multiples of 256 rows in the array and 256 states in the counter make for a potential error-masking combination. If the right type of failure modes occur and a full row is nonfunctional, it may be masked. The implementation of the counter or shift register must be done with full-column or row failure modes in mind, or it can easily mask failures. This masking of the failure gives a false result that the device is passing the test.

Scan

Scan is a test technique that ties together some or all the latches in the part to form a path through the part to shift data. Figure 21.9 shows the implementation of a scan latch in a circuit. Note that

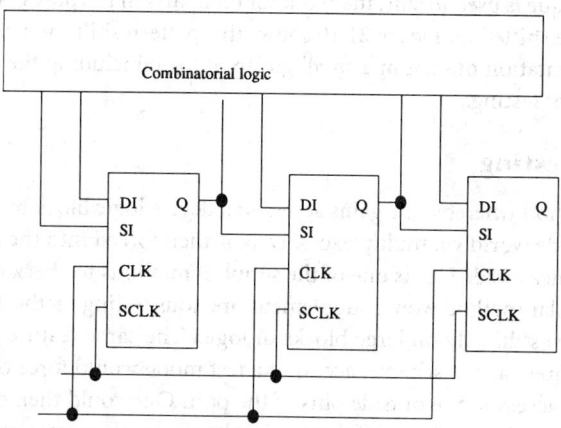

FIGURE 21.9 Scan test implementation. (*Source:* W. M. Needham, *Designer's Guide to Testable ASIC Devices*, New York: Van Nostrand Reinhold, 1991, p. 94. With permission.)

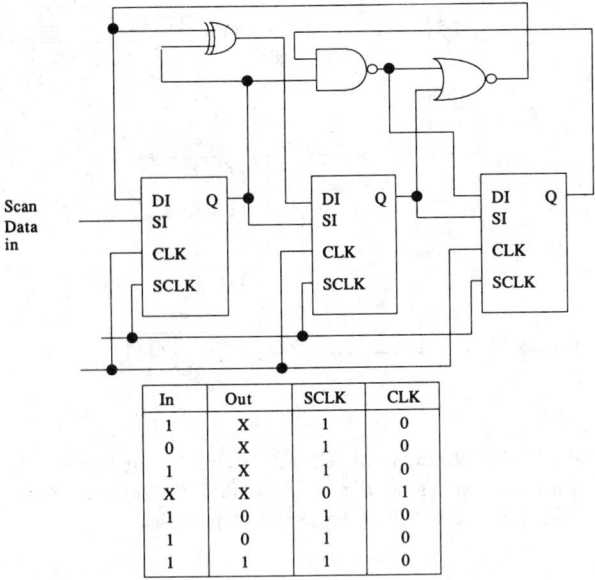

In	Out	SCLK	CLK
1	X	1	0
0	X	1	0
1	X	1	0
X	X	0	1
1	0	1	0
1	0	1	0
1	1	1	0

FIGURE 21.10 Scan test example and patterns. (*Source:* W. M. Needham, *Designer's Guide to Testable ASIC Devices*, New York: Van Nostrand Reinhold, 1991, p. 95. With permission.)

the latch has a dual mode of operation. In the normal mode the latches act like normal flip-flops for data storage. In the scan mode the latches act like a shift register connecting one element to another. This is a basic implementation of a shift register or scan latch in a block of logic. Data can be shifted in via the shift pin or can be clocked in from the data pin. Clock line is the normal system clock and SCLK is the shift clock for scan operations.

Data for testing is shifted in on the serial data in pins of the device. Patterns for the exercise of the combinatorial logic could be generated by truth table or by random generation of patterns. These patterns are then clocked a single time to store the results of the combinatorial logic. The latches now contain the results of the combinatorial logic operations. Testing of the logic becomes quite easy, as the depth into the part is of no significance to the designer. Once the patterns are latched, the same serial technique is used to shift them out for comparison purposes. At the time of outward shift, new patterns are shifted in. Figure 21.10 shows this pattern shift for a circuit using scan. This is the actual implementation of scan in a small group of logic including the truth table associated with it and one state of testing.

Direct Access Testing

Direct access is a method whereby one gains access to a device logic block by bringing signals from the block to the outside world via multiplexors. Data is then forced into the block directly and the outputs are directly measured. This is one of the simplest methods to check devices for logic functionality. This particular method would supplement previous testing methods by allowing the user to impose data patterns directly on large blocks of logic. The same feature holds true for output observation. In the direct access scheme, access of a test mode would force certain logic blocks via multiplexors to have access to the outside pins of the part. One could then drive the data patterns to the input pins, compare output pins of the part, and measure the access, status, and logical functionality of the block directly.

Figure 21.11 shows implementation of direct access test techniques on a block of logic. During normal operation, the logic block B has inputs driven by the logic block A, and outputs are con-

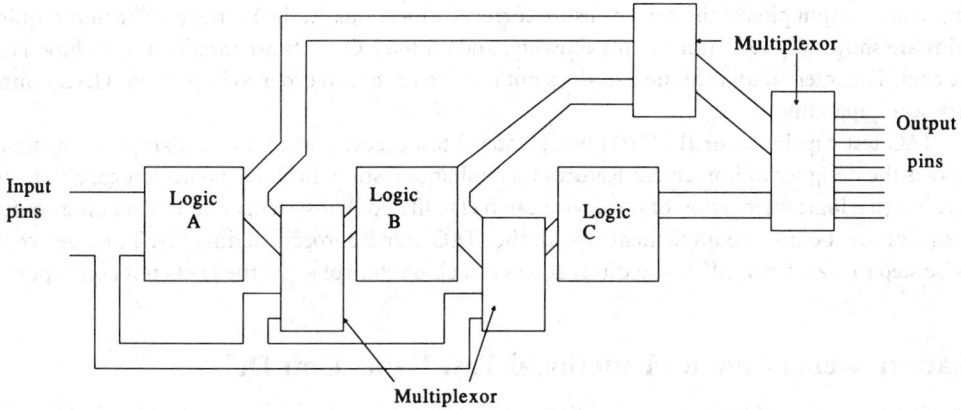

FIGURE 21.11 Direct access implementation. (*Source:* W. M. Needham, *Designer's Guide to Testable ASIC Devices*, New York: Van Nostrand Reinhold, 1991, p. 104. With permission.)

nected to the logic block C. In the test mode, the two multiplexors are switched so that the input and output pins of the device can control and observe the logic block B directly.

Joint Test Action Group (JTAG)

When implemented in a device, **JTAG** (Joint Test Action Group), or IEEE 1149.1, allows rapid and accurate measurement of the direct connections from one device to another on a PC board. This specification defines a test access port for internal and external IC testing. Figure 21.12 shows the JTAG technique in a small device, thus allowing accurate measurement and detection of solder connections and bridging on a PC board. This technique allows the shifting of data through the

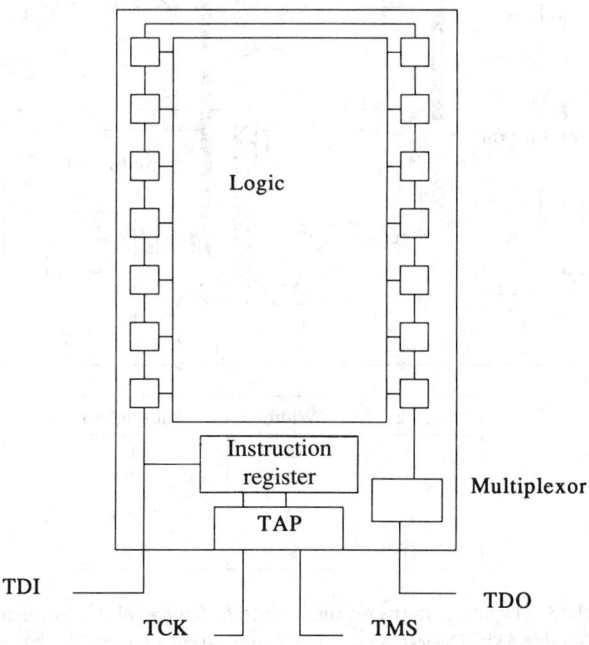

FIGURE 21.12 JTAG circuit implementation.

input and output pins of the part to ensure correct connections on the PC board. The four required pins are shown at the bottom of the drawing, and all the I/O ports are modified to include JTAG latches. The internal logic of the part does not need to change in order to implement JTAG boundary scan capability.

JTAG test capability, or IEEE P1149.1 standard test access port and boundary scan capability, allows the designer to implement features that enhance testing. Both PC board test capability and the internal logic verification of the device can be facilitated. For systems where remaining components on the boards are implemented with the JTAG scan approach, adding JTAG to a device is a wise step to take. Scan, BIST, and direct access can all be controlled by the JTAG test access port.

Pattern Generation for Functional Test Using Unit Delay

During simulation of the function portion of the device, patterns are captured and stored for exercise of the system. Figure 21.13 shows typical patterns for each block and the width of the data for test. The patterns check the functionality of the logic and ensure that the logic implemented performs the desired function in the device. The patterns can be generated by several methods if they were not generated by self-test. First is exercising of the system by use of code such as assemblers, macros, and high-level inputs that ensure the operation. Usually this is done by comparison of a high-level model to the actual logic. A second method of pattern generation is random number generation; in this case random numbers of ones and zeros are impressed on the logic and the results compared to a high-level model. Finally, coding of ones and zeros for logic checking is an alternative, but this is typically prohibitive in today's technology where devices may contain millions of logic gates.

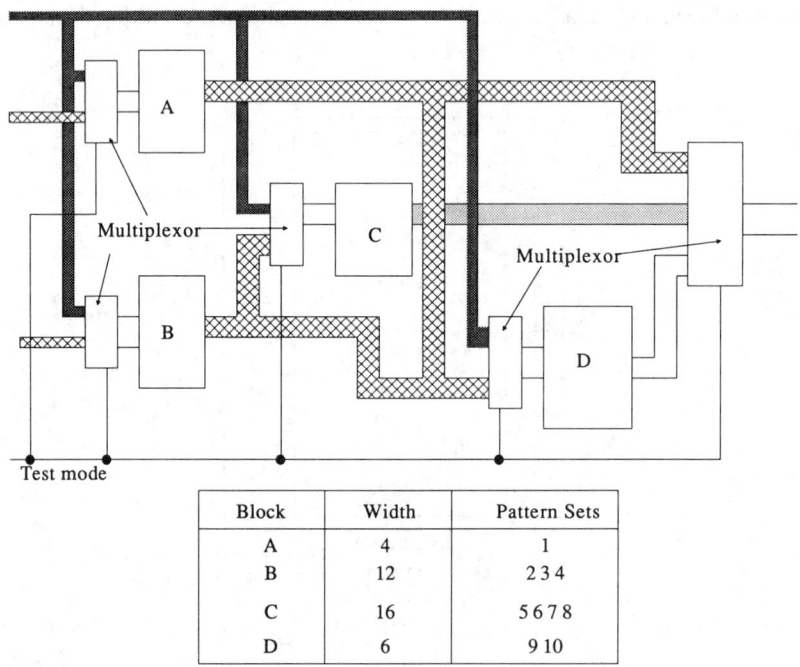

Block	Width	Pattern Sets
A	4	1
B	12	2 3 4
C	16	5 6 7 8
D	6	9 10

FIGURE 21.13   Typical patterns for function test. (*Source:* W. M. Needham, *Designer's Guide to Testable ASIC Devices,* New York: Van Nostrand Reinhold, 1991, p. 129. With permission.)

Table 21.1   Typical Devices Simulation and Test Files

Time (ns)	1	2	3	4	5	6	7	8	9	10	11	12	13	14	15	16	17	18	19	20
31	1	0	1	1	1	0	1	1	1	1	1	1	1	1	1	0	1	1	1	1
52	1	0	1	1	1	0	1	1	0	1	1	1	1	1	0	1	1	0	1	0
97	1	0	1	1	1	0	1	1	1	1	1	1	1	1	0	1	1	1	0	0
101	1	0	1	1	1	0	1	1	1	1	1	1	1	1	0	1	1	1	0	0
156	1	0	1	1	1	0	1	1	1	1	1	1	1	1	0	1	1	1	0	0
207	1	0	1	1	1	1	0	1	1	0	1	1	0	0	0	0	0	0	0	0
229	1	0	1	1	1	0	1	1	0	1	1	1	1	1	0	0	0	0	1	1

Vector	Time	Data																				
1	0	1	0	1	1	1	0	1	1	1	1	1	1	1	1	1	0	1	1	1	1	
2	50	1	0	1	1	1	0	1	1	0	1	1	1	1	1	0	1	1	0	1	1	N=NRZ
3	100	1	0	1	1	1	0	1	1	1	1	1	1	1	1	0	1	1	1	0	0	R=RZ
4	150	1	0	1	1	1	0	1	1	1	1	1	1	1	1	0	1	1	1	0	0	1=R1
5	200	1	0	1	1	1	1	0	1	1	0	1	1	0	0	0	0	0	0	0	0	
Format		N	N	N	N	N	R	1	1	1	1	N	N	1	1	R	N	N	1	N	N	

Pattern Generation for Timing

After the functional patterns are complete, specific tests for timing may be generated. Timing tests are test procedures to verify the correction operation to the timing specification of the device. Typical timing tests include outputs relative to the clocks, propagation delays, set-up and hold times, access times, minimum and maximum speed of operation, rise and fall time, and others. These tests are captured in simulation and used in the test system to verify performance of the device. Table 21.1 shows the relationship between time-based simulation and cycle-based test files.

Temperature, Voltage, and Processing Effects

Figure 21.14 shows the impact on speed of process variations, and as a result of voltage or temperature variations. It is important to simulate with the total variation over the entire process, temperature, and voltage range to ensure testability. Remember that if the system design of the logic was not done with some guardbanding, there may be no margin. If the library used for the design did not include some amount of margin, there may be a need to add guardbanding by optimizing logic for speed or choosing faster gates. There must be a delta placed between the testing parameters

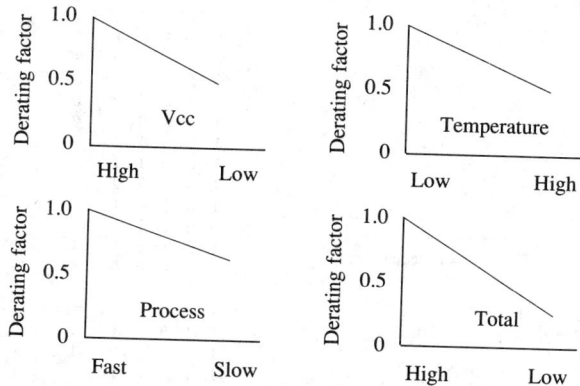

FIGURE 21.14   Effects of voltage, process, and temperature on test. (*Source:* W. M. Needham, *Designer's Guide to Testable ASIC Devices*, New York: Van Nostrand Reinhold, 1991, p. 172. With permission.)

used for initial production testing and final quality assurance (QA) test of the devices. This will ensure that they are electrically stable and manufacturable.

It is very important to review the critical paths, timing parameters, and the early simulations versus the timing parameters in the final simulation. The effect of parasitic capacitance and resistance will show up now. This is the time to look back to see whether the performance of the device will meet all the system specifications.

If the design meets all the device specifications and all the system specifications, now is the time to sign off and accept this simulation. If the simulations do not match the system requirements, this is the time to fix them once again.

Fault Grading

Fault grading is a measure of test coverage of a given set of stimuli and response for a given circuit. This figure of merit is used to ensure that the device is fully testable and measurable. Once the test patterns are developed for the device, artificial faults are put in via simulation to ensure that the part is testable and observable.

Figure 21.15 shows a single stuck-at-one faults (S@1) injected into logic circuitry with data patterns for checking the functionality. Notice in the truth table that faults on one gate show up on an output pin of the part or test observation point. Gaining high fault coverage is a very desirable method for ensuring that the device will function correctly in the system. Semiconductor failure modes include not only stuck-at-zeros and -ones but several other types.

It is also important to mention that although fault grading using single stuck-at-zero and single stuck-at-one faults covers many aspects of semiconductor failures, it does not cover all of them.

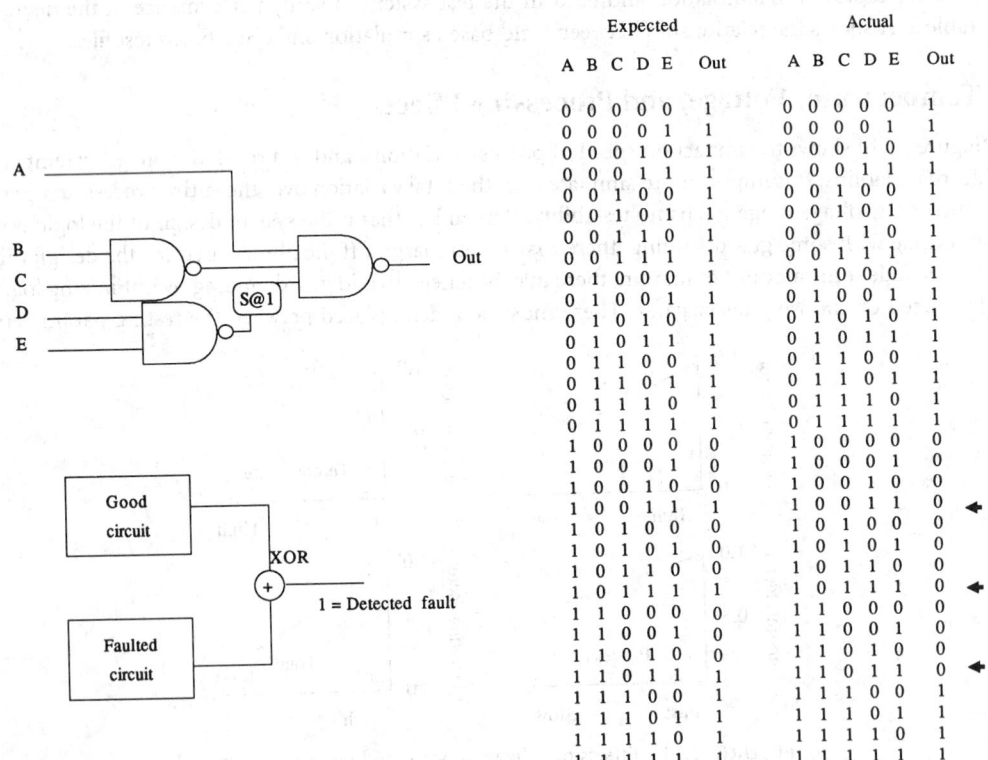

Expected						Actual						
A	B	C	D	E	Out	A	B	C	D	E	Out	
0	0	0	0	0	1	0	0	0	0	0	1	
0	0	0	0	1	1	0	0	0	0	1	1	
0	0	0	1	0	1	0	0	0	1	0	1	
0	0	0	1	1	1	0	0	0	1	1	1	
0	0	1	0	0	1	0	0	1	0	0	1	
0	0	1	0	1	1	0	0	1	0	1	1	
0	0	1	1	0	1	0	0	1	1	0	1	
0	0	1	1	1	1	0	0	1	1	1	1	
0	1	0	0	0	1	0	1	0	0	0	1	
0	1	0	0	1	1	0	1	0	0	1	1	
0	1	0	1	0	1	0	1	0	1	0	1	
0	1	0	1	1	1	0	1	0	1	1	1	
0	1	1	0	0	1	0	1	1	0	0	1	
0	1	1	0	1	1	0	1	1	0	1	1	
0	1	1	1	0	1	0	1	1	1	0	1	
0	1	1	1	1	1	0	1	1	1	1	1	
1	0	0	0	0	0	1	0	0	0	0	0	
1	0	0	0	1	0	1	0	0	0	1	0	
1	0	0	1	0	0	1	0	0	1	0	0	
1	0	0	1	1	1	1	0	0	1	1	0	←
1	0	1	0	0	0	1	0	1	0	0	0	
1	0	1	0	1	0	1	0	1	0	1	0	
1	0	1	1	0	0	1	0	1	1	0	0	
1	0	1	1	1	1	1	0	1	1	1	0	←
1	1	0	0	0	0	1	1	0	0	0	0	
1	1	0	0	1	0	1	1	0	0	1	0	
1	1	0	1	0	0	1	1	0	1	0	0	
1	1	0	1	1	1	1	1	0	1	1	0	←
1	1	1	0	0	1	1	1	1	0	0	1	
1	1	1	0	1	1	1	1	1	0	1	1	
1	1	1	1	0	1	1	1	1	1	0	1	
1	1	1	1	1	1	1	1	1	1	1	1	

FIGURE 21.15 Stuck-at-faults test sequences. (*Source:* W. M. Needham, *Designer's Guide to Testable ASIC Devices,* New York: Van Nostrand Reinhold, 1991, p. 152. With permission.)

For example, bridging faults are one of the more common types of failures within semiconductor devices; they may be caught with fault-graded patterns if the patterns are done correctly.

Test Program Flow

Data patterns that were generated during simulation are then converted into a functional **test program**. Table 21.2 shows a typical test flow used by a semiconductor vendor for the exercising of the device. It describes the basic test done along with the portions of the specification of the device tested in each step. The flow is from loose testing of the device to detailed checking of specifications; this basic flow is both efficient and helpful in debugging failures encountered during test. Table 21.3 shows the approximate test time for a 100-pin device as executed on a verification system. Notice that a large amount of tests can be done in a very short time. The test times are approximate and will vary based on the type of system used by the type of test executed, length of patterns, and pin count.

The cost of testing is very much related to test time. In production, package handlers, wafer probe equipment, people, and perhaps a computer network are all needed to run production tests. When comparing test time on a small system to a large system, total cost must also be compared. In this case, total cost is the cost of the test equipment along with the needed support equipment. Table 21.4 shows the same test time for the same device on a large production test system. Again the test time is approximate and will vary depending on options, program flow, and test details. The biggest reason for the difference in test time from a verification system to a production system is due to pattern reload time.

Regardless of the techniques that were used for the generation of the data patterns, it will be necessary to include those patterns into a test program. This is similar to that shown in Table 21.5. This is the same flow with the addition of information from the net list and vector simulations. Vectors are used for functional tests and some portions of I/O testing. Net list description and selection of input pads and power connections are used in the parametric tests. Although the flow may vary from one manufacturer to another, most of them perform essentially the same kinds of tests.

Even if the device used special testing techniques such as JTAG, BIST, scan, or direct access, the flow is always the same. It is basically always done with a setup sequence or a preconditioning of the part, and a formal measurement of the data on the device under test. The testing may be on a cycle-by-cycle basis or as a pass or fail conclusion at the end of a long routine. There may be a burst of data to set up the part and a burst of data to measure it. Finally, information is interpreted by the program to categorize or bin the device as a pass or fail rating.

Table 21.2 Typical Test Flow

Test	Function
Shorts	Checks if the adjacent pins are shorted
Open	Checks to see if the p-n junction exits (pad protection device)
Basic function test	Checks functionality of the part, uses most vectors, loose timing and nominal voltages
dc spec test	Checks the inputs and outputs compared to spec for dc levels
ac spec and margin test	Checks the vectors with timing set to spec; checks at minimum and maximum voltage

Source: W. M. Needham, *Designer's Guide to Testable ASIC Devices*, New York: Van Nostrand Reinhold, 1991, p. 181. With permission.

Table 21.3 Test Execution Time on a Verification Tester

Test	Number of Test	Type of Test	Time to Execute
Opens	100	Parametric	1 s
Shorts	100	Parametric	1 s
Basic function test	40,000	Vector pattern	0.04 s*
ac spec test	300–500	Parametric-functional	3–5 s
ac spec and margin test	100,000	Vector pattern	0.1 s*

*Execution time only; vector reload time may be between 10 and 500 s.

Source: W. M. Needham, *Designer's Guide to Testable ASIC Devices*, New York: Van Nostrand Reinhold, 1991, p. 182. With permission.

Table 21.4 Test Execution Time on a Production Tester

Test	Number of Test	Type of Test	Time to Execute
Opens/shorts	2	Parametric	0.05 s
Basic function			
test	40,000	Vector pattern	0.04 s*
ac spec test	300–500	Parametric-functional	0.5 s
ac spec and	100,000	Vector pattern	0.75 s*
margin test			

*Execution time only; vector reload time may be between 10 and 500 s.

Source: W. M. Needham, *Designer's Guide to Testable ASIC Devices,* New York: Van Nostrand Reinhold, 1991, p. 183. With permission.

Table 21.5 Test Data Sources

Test	Source of Data
Opens	Process description and net list for I/O pads
Shorts	Process description and net list for I/O pads
Basic function test	Simulations done for test
	Internal library for cores selected by the user
dc spec test	Net list for types of pads
	Automatic place and route data for placement
	Simulation data for vector setup sequences
ac spec and	Simulations done for test and performance
margin test	Internal library for cores selected by the user

Source: W. M. Needham, *Designer's Guide to Testable ASIC Devices,* New York: Van Nostrand Reinhold, 1991, p. 184. With permission.

Defining Terms

Built-in self-test: A design process where logic is added to the part to perform testing. Advantages include easy testing of buried logic. The biggest negative is patterns are set in hardware and the design must be changed to improve test coverage.

DUT: Device under test

Fault coverage: A metric of test pattern coverage. Calculated by measuring faults tested, divided by total possible faults in the circuits.

JTAG: A specification from the IEEE 1149.1 that defines a test access port.

Scan: A method of connection of latches within an integrated circuit that allows (a) patterns to be shifted in, (b) combinatorial logic check to be exercised, and (c) patterns to be shifted out and compared to known simulation results.

T_0: A timing reference point used in simulation and on the tester. This point T_0 is where all timing is referenced.

Test program: A software routine consisting of patterns, flow information, voltage and timing control, and decision processes to make certain the DUT is correct.

Tester: A piece of equipment used to verify the device, often called Automate Test Equipment (ATE).

Vector: A series of ones and zeros that describe the input and output states of the device.

Yield: A metric of good devices after test divided by total tested.

References

J. M. Acken, *Deriving Accurate Fault Models,* Palo Alto, Calif.: Stanford University, Computer Systems Laboratory, 1989.

V. D. Agrawal and S. C. Seth, "Fault coverage requirements in production testing of LSI circuits," *IEEE Journal of Solid-State Circuits*, vol. SC-17, no. 1, pp. 57–61, February 1982.

E. J. McCluskey, *Logic Design Principles: With Emphasis on Testable Semicustom Circuits*, Englewood Cliffs, N.J.: Prentice-Hall, 1986.

W. M. Needham, *Designer's Guide to Testable ASIC Devices*, New York: Van Nostrand Reinhold, 1991.

F. Tsui, *LSI/VLSI Testability Design*, New York: McGraw-Hill, 1987.

Further Information

IEEE Design and Test Computers is a bi-monthly publication focusing on test of digital circuitry.

Proceedings of the IEEE Test Conference—This conference occurs each year in the late summer. The conference is the major focal point for test equipment vendors, test technology, and a forum for advanced test papers.

Design Automation Conference—This conference is held annually in the late spring. The focus is the design process, but many of the papers and vendors at the conference have programs for design for test.

Vendor material is available from design and test equipment vendors.

21.3 Electrical Characterization of Interconnections

S. Rajaram

Semiconductor technology provides electronic system designers with high-speed integrated circuit (IC) devices that operate at switching speeds approaching 1 ns and lower. For the past 25 years, the performance and cost of digital electronic systems have improved continuously because of the technological advances in manufacturing physically smaller devices. With the reduction in the feature sizes of the gates and cells that make up these devices, there has also been a tremendous increase in the density of IC integration. These physically smaller devices with their intrinsic faster switching speeds and lower power per gate have given rise to the very large scale integration (VLSI) technology. We therefore have a situation today in which the IC technology includes both high density and high speed. The unprecedented gains in the increasing scales of integration in IC technology are shown in Fig. 21.16.

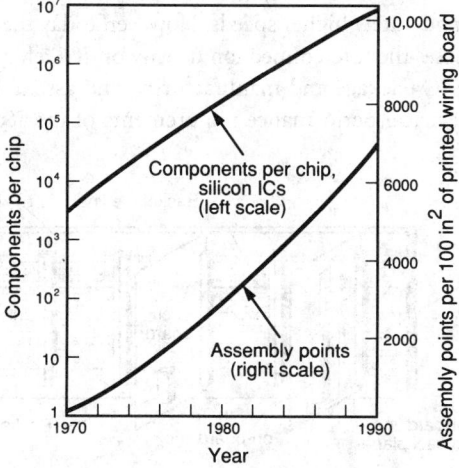

FIGURE 21.16 The history of silicon integrated circuit scale of integration and its effect on interconnection density.

For the leading-edge technology, the maximum number of components (transistors) per chip increases by a factor of 100 per decade. It is tempting to believe that these increases may be extended indefinitely, leading to ever smaller devices with increasingly larger electronic functionality. However, the limiting factor for such spectacular gains in IC technology is the electrical performance capability of its associated **interconnections and packaging (I&P)**. Thus, interconnections and packaging is the bottleneck to IC performance. The signal originates from an IC device referred to as the driver and is received by another IC device, called the receiver. The path between the driver and the receiver is the I&P medium. The various elements of I&P are shown in Fig. 21.17.

The elements that make up I&P are the IC package (including its wirebonds inside the package and the leads or pins external to the package), printed circuit board (PCB) with conductor (copper) traces, electrical PCB connector, backplane or motherboard, and external cables. The driver and receiver may be on the same PCB or on different PCBs within the same equipment or in different equipments. I&P should be designed so as to cause minimal signal degradation from the driver to the receiver. Thus, there are several interfaces within I&P, and signal degradation occurs at every interface. The elements of I&P between the driver and the receiver resemble the mechanical links in a chain.

The key role of I&P is to enable electronic system designers to fully exploit the advancements taking place in device and manufacturing technologies related to ICs. I&P technology is dictated by three main external forces. These are silicon IC technology, automated systems for assembly and testing, and the fundamental architecture of electronic systems [Hoover *et al.*, 1987]. We have already seen the 100-fold increase in the component density at the IC level. Most of the electronic functions are built into these dense ICs. However, the system functionality is contained in several ICs that must "talk" to each other. The medium connecting various ICs in a system is generically referred to as I&P. Therefore, in order to attain the performance capability of the IC, it is necessary to have an I&P medium to match its electrical performance. The electrical parameters of the I&P must be tailored to meet IC performance. This becomes the crux of any high-speed electronic system design. To optimize IC performance, therefore, it is also necessary to minimize the size of all interconnections. However, to optimize manufacturing costs I&P must be large enough to attain reliable and high-yield assembly operations, test access, repair, and maintenance. Thus, these requirements imply automated assembly and test equipment for high-quality manufacture of electronic systems with optimal cost. Finally, I&P is also dictated by system architecture requirements.

Historically, telecommunications systems have had system requirements dictating high density of interconnection and input/output (I/O) signals at moderately low speeds and low interconnection density or low I/O at moderately higher speeds. However, today the trend is towards high I/O and high speed which increases the interconnection density on I&P elements. These have caused a fundamental change in the IC package and manufacturing and assembly technologies associated with electronic systems. Electrical performance requirements of the I&P have forced the change

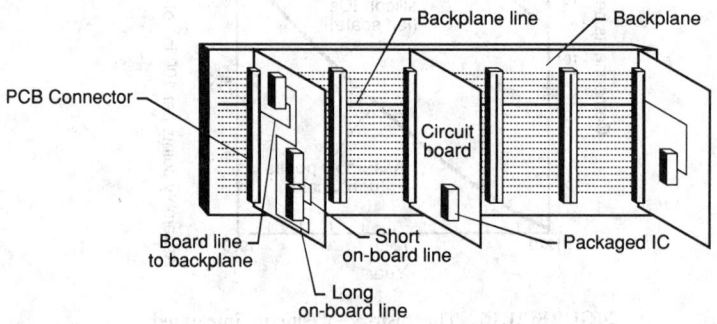

FIGURE 21.17 Chip-to-chip interconnections on same board and between boards through backplane.

from dual inline package (DIP) ICs to surface mount (SM). The SM ICs have much smaller package outlines and leads than the corresponding DIPs. While the DIPs are soldered into holes drilled through the PCB, SM components are soldered onto pads on the surface of the PCB. While the pitch on the leads of a DIP is 100 mils (0.100 inch), the pitch on SM IC package leads is typically 25 to 50 mils. I&P technology has therefore evolved rapidly as the driving forces behind it have advanced at a rapid pace. It is fairly accurate to state that the ultimate quality and reliability of many complex electronic systems are determined primarily by the reliability of the I&P of the system. The stringent electrical and density requirements impose severe constraints and challenges on mechanical design and manufacturing capability. Most failure mechanisms today are related to interconnections in the system.

Interconnection Metrics

In Fig. 21.16, we observed the tremendous increase in the IC scale of integration. We have also made the statement that the increased IC component density has led to increased interconnection density. It is important to understand the relationship between the scale of component integration inside the IC package and the interconnection density requirements outside the IC package. Figure 21.16 also shows the evolution of the number of pins interconnected (assembly points) on a PCB within an area of 100 square inches, and the trend is obvious. This data is derived from CAD packages used within Bell Labs for PCB layout. IC circuits are usually described in terms of number of gates (1 gate ≈ 4 transistors). The relationship between the number of gates of an IC (G) to the number of signal I/O pins in the package (P) is given empirically by Rent's rule [Hoover *et al.*, 1987]. For random logic, Rent's rule states that $P = KG^\alpha$. The value of K ranges from 3 to 6 and α typically ranges from 0.4 to 0.55. For a functionally complete chip, $K \approx 7$, and $\alpha \approx 0.2$ [Moresco, 1990]. The empirical Rent's rule has been found to apply remarkably well over a large range of circuit sizes as shown in Fig. 21.18.

This simple empirical relationship is important in understanding fundamental trade-offs between interconnection costs, reliability, and increased scales of integration. Let's assume that a certain electrical function requires the use of 100,000 gates. Let's partition the circuitry into ICs with 100, 1000, and 10,000 gates per chip (device). Let us apply Rent's rule by taking $K = 4$, and $\alpha = 0.5$, and examine

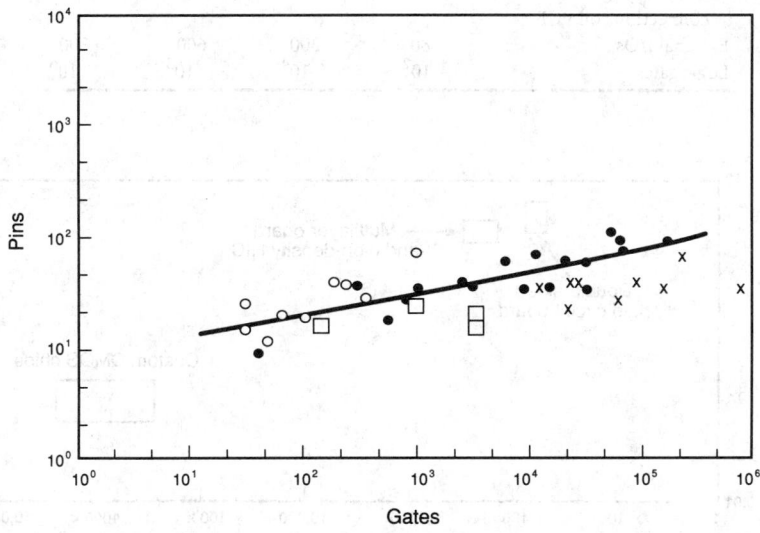

FIGURE 21.18 Rent's rule for LSI and VLSI chips. (●) Silicon microprocessors and random logic, (X) silicon memories, (O) GaAs functional chips, (□) GaAs memory.

Table 21.6 Distribution of Interconnections for 100,000-Gate Random Logic Circuit

	Gates per IC		
Interconnections	100	1,000	10,000
On silicon	160,000	187,400	196,000
On PCB	40,000	12,600	4,000

the I/O pins, and therefore the interconnection requirements for the three circuit partition choices, as shown in Table 21.6 [Hoover *et al.*, 1987].

The most important observation to be made from Table 21.6 is that increasing the IC scale of integration by a factor of 100 moves 90 percent of the external interconnections into silicon. The density of interconnection at the silicon increases, but the overall cost of interconnections for the system decreases since the relative costs per interconnection in silicon are 1 to 2 orders of magnitude cheaper than those on PCBs and hybrid integrated circuits (HICs). This is clearly shown in Fig. 21.19.

This is the driving force for integrating as much functionality in silicon, with reduced external interconnections. However, the problem is compounded by the fact that many such dense ICs are now placed on a board, and these have to be interconnected on a dense multilayer board. This has led to the requirement for fine-line PCB technology with increasing number of layers. The dense traces on the board lead to high I/O (pin count) connectors requiring as many as 400 I/Os on an 8-inch board edge. Thus, there is an increasing density of interconnection at every level, both internal and external to silicon, as shown in Table 21.7.

Another important aspect of increasing scale of integration at the IC level is the improved overall reliability of the system. Practical experience indicates that the FIT (Failure unIT) count of an IC is approximately constant for a given IC technology and manufacturer and is almost independent of the number of gates per chip. A FIT is defined as one failure in 10^9 hours. Therefore, system FIT count due to devices should decrease inversely in proportion to scale of integration. Once again let us consider the example of partitioning 100,000 gates into devices with 100, 1000 and 10,000 gates per IC. Thus, from 100 gates per IC, we have increased the scale of integration by a factor of 10 and

Table 21.7 Interconnection Limits for Telecommunications Systems

Packaging Technologies	1970	1978	1986	1990s
PCB area, square inch	50	100	200	200
Number of layers in PCB	2	6	8	12–14
Line width, mils	25	7	6	4
Number of device terminal connections on PCB	100	2,000	9,000	16,000
External I/Os	80	300	600	> 800
Logic gates	10^2	10^4	10^5	10^6

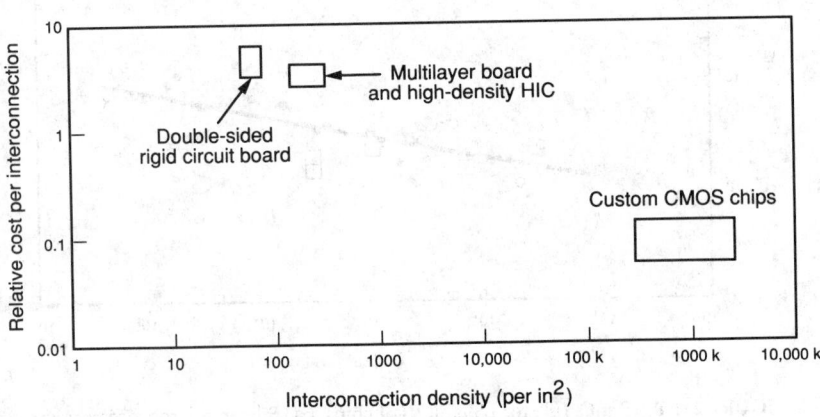

FIGURE 21.19   Interconnections cost trade-off.

Table 21.8 Effect of Increasing Scale of Integration on System Reliability

	Gates per IC		
	100	1,000	10,000
Number of ICs	1,000	100	10
IC FIT count, total	50,000	5,000	500
Number of external interconnections	40,000	12,600	4,000
Interconnection FIT count, total	4,000	750	200
System FIT count	54,000	5,750	700

100 as we go to 1000 and 10,000 gates per chip, respectively. The FIT count for an IC is taken as 50. The total system FIT count is the sum total of the IC and interconnections FIT count. These are shown in Table 21.8 where the FIT counts for interconnections are those appropriate for the PCB types for the levels of integration based on experience at AT&T [Hoover *et al.*, 1987].

The important message is that there is a tremendous improvement in system reliability (FIT count) with increasing scales of integration in silicon. It is important to remember that the example shown in Table 21.8 is an ideal situation. In any practical situation, the circuit is partitioned into many devices with different levels of integration, with a few VLSI, but 50 or more small scale integration devices. Typically, the total system FIT count including devices and interconnections may be in the range of 3000 to 5000. The remarkable attribute of IC technology is that as the scale of integration increases, the total system cost decreases, while the overall system reliability improves considerably. It is this remarkable attribute that has led to a continuous decrease in the cost of electronics products with increased functionality with time. However, the miniaturization and the consequent increased scales of integration and interconnections density at the packaging level together with increasing switching speeds have resulted in a whole range of interesting electrical problems. Although these problems appear to be new, solutions to many of them are available from extensive work that has been well documented in the microwave area. In the sections that follow we will discuss the significance of the important electrical parameters and techniques to evaluate them.

Interconnection Electrical Parameters

The interconnection medium is the basic path for transmitting the pulse from the driver to the receiver. As the speed of the circuits goes beyond 10 MHz, it is necessary to use high-frequency techniques developed by RF engineers. The fundamental electrical parameters of the circuit such as inductance L and capacitance C, behave as lumped elements at low frequencies and as distributed parameters at high frequencies where transmission line techniques must be employed. The transition from lumped element to transmission line behavior depends upon the risetime of the pulse T_r and on the total delay in the pulse transmission through the interconnection T_d. In the lumped element mode, the inductance and capacitance appear to the pulse to be concentrated at a point. On the other hand, in the transmission line mode, inductance and capacitance appear to be uniformly distributed throughout the interconnection, and as far as the pulse is concerned, the medium is infinite in length, and all the characteristics of wave propagation must be taken into consideration. Interestingly, similar techniques may be used for electrical characterization of all of the interconnection elements such as wirebonds, package leads, PCBs, connectors, and cables. The basic pulse transmission parameters of interest are propagation delay (T_d), characteristic impedance (Z_0), reflection coefficient (Γ), **crosstalk** (X), and **risetime degradation** (T_{dr}).

Propagation Delay (T_d)

For a pulse being transmitted through a medium of length l and wave velocity v, delay $T_d = l/v$. The speed of pulse transmission depends upon the dielectric constant of the material ε_r and is given by $v = c_0/\sqrt{\varepsilon_r}$, where c_0 is the speed of light in air $= 3 \times 10^8$ m/s. Thus, **propagation delay** is proportional to length and also the square root of the dielectric constant. In order to reduce delay and reduce machine cycle time and increase speed, it is necessary to reduce the dielectric constant of the material. There is a tremendous interest in developing low dielectric constant materials for this reason.

Characteristic Impedance (Z_0)

In general, a transmission line is any structure that propagates an electromagnetic wave from one point to another. However, in the world of microwave, RF, or high-speed digital designs, the use of the term *transmission line* is far more restrictive. In order for a structure to be a transmission line, the electrical length of the (transmission) line must be much larger than the wavelength at the frequency of interest. For an interconnection medium which behaves as a transmission line as shown in Fig. 21.20, the **characteristic impedance** is defined as

$$Z_0 = \sqrt{\frac{R + j\omega L}{G + j\omega C}} \quad (\Omega) \tag{21.23}$$

where R = series resistance (Ω/m), L = series inductance (H/m), G = shunt conductance ($\Omega^{-1}\,\mathrm{m}^{-1}$), C = shunt capacitance (F/m), and $\omega = 2\pi f$ = radian frequency. In general, Z_0 is complex, and the distributed per unit length quantities R, L, G, and C must be determined from the material and structural characteristics of the transmission line. For most applications $R \ll L$, and $G \ll C$, and

$$Z_0 = \sqrt{\frac{L}{C}} \tag{21.24}$$

The speed of wave propagation through the transmission line is given by $v = 1/\sqrt{LC}$. Therefore we can also define $Z_0 = 1/vC = vL$.

A digital pulse propagating in a circuit consists of voltages and currents of different frequencies (ac components). A digital pulse is therefore an electromagnetic wave of many frequencies propagating down the transmission line. In any circuit with transients (ac components), the current and voltage are not in phase because of inductance and capacitance. In a pure inductor, voltage leads current by 90 degrees, and in a pure capacitor, voltage lags behind current by 90 degrees. The total energy (W_T) of the electromagnetic wave is made up of the magnetic energy (W_m) and the electric energy (W_e). Magnetic energy is stored in inductance and is given by $W_m = 1/2LI^2$, where I is the current. Similarly, electric energy is stored in its capacitance and is given by $1/2CV^2$. Therefore, the total energy is

$$W_T = W_m + W_e = \frac{1}{2}LI^2 + \frac{1}{2}CV^2 \tag{21.25}$$

In an alternating field, the total energy is continually being swapped between the magnetic and electrical elements, one at the expense of the other. Because of the phase relationships discussed earlier, magnetic energy is a maximum when electric energy is 0, and vice versa. Since it is the same stored energy which appears alternately as magnetic and electrical energy, we can write

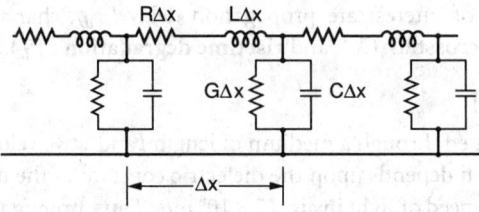

FIGURE 21.20 Representation of a short section of transmission line.

$$W_T = \frac{1}{2} LI_{max}^2 = \frac{1}{2} CV_{max}^2 \qquad (21.26)$$

Therefore, $Z_0 = V_{max}/I_{max} = \sqrt{L/C}$.

Therefore, characteristic impedance Z_0 gives the relationship between the maximum voltage and maximum current in a transmission line and has the units of impedance (Ω). Thus, it is important to note that the current required to drive a transmission line is determined by its characteristic impedance. Z_0 is really the impedance (resistance) to energy transfer associated with electromagnetic wave propagation. In fact, characteristic impedance is not unique to electromagnetic waves alone. Characteristic impedance Z_0 is an important parameter associated with propagation of waves in a medium. Some examples are given below:

- Electromagnetic waves: $Z_0 = \sqrt{L/C} = vL = 1/vC$
- Transverse vibrating string: $Z_0 = \lambda_0 v = \sqrt{\lambda_0 T_0}$, where λ_0 = mass/unit length and T_0 = force
- Longitudinal waves in a rod: $Z_0 = \rho_0 v = \sqrt{\rho_0 E}$, where ρ_0 = mass/unit volume and E = Young's modulus
- Plane acoustic waves: $Z_0 = \rho_0 v = \sqrt{B\rho_0}$, where ρ_0 = density and B = bulk modulus

Note that Z_0 is a unique characteristic of material properties and geometry alone.

The performance of an interconnect that behaves as a transmission line is measured in terms of the efficiency of energy (information) transfer from input to output, with minimal loss and dispersion effects. At high speeds, the interconnect must maintain a uniform Z_0 along the length of the signal. Any mismatch in characteristic impedance across interconnect interfaces will cause reflection of the signal at the interface, which can cause errors in digital circuits. Reflections are part of the losses in interconnect and lead to loss of information. These reflections are sent back to the signal source. Therefore multiple reflections from interfaces can distort the input signal. The magnitude of reflection is defined by the reflection coefficient Γ, given by

$$\Gamma = \frac{Z_L - Z_0}{Z_L + Z_0} \qquad (21.27)$$

where Z_L is the load impedance. Note that when the load and impedances are matched, $Z_0 = Z_L$, reflection coefficient $\Gamma = 0$, and the wave (energy, information) is transmitted without loss. For an open circuit, $Z_L = \infty$, and $K = 1$. Thus, the entire pulse is reflected back to the source. For a short circuit, $Z_L = 0$, and $K = -1$. Here, the pulse is reflected back to the source with the same amplitude, but with the sign reversed. For practical purposes, Z_0 of PCBs may be considered to be independent of frequency up to nearly 1 GHz. In other words, PCBs may be considered lossless transmission lines for impedance calculations. Beyond 1 GHz, skin effect in conductors, and dielectric and dispersion losses, cause signal degradation. Z_0 is a parameter associated with PCBs and cables but not with connectors at lower speeds, because of the definition of transmission line. The length of an electrical path through a connector is approximately 1 inch, which is very small compared to the wavelength at lower frequencies. However, as speeds increase, the connector also has a Z_0 associated with it. Otherwise, a connector acts as a lumped L and C in a transmission path with its associated losses such as reflection and degradation. Z_0 values for rectangular and circular transmission lines with and without ground planes applicable for cables are given in Everitt [1970]. We shall now look at evaluation of Z_0 for typical PCB structures.

Z_0 of PCB Structures. The two commonly used PCB structures in electronic circuits are the microstrip and stripline designs, shown in Figs. 21.21 and 21.22. In a pure microstrip, there is only one ground plane below the conductor. The space below the conductor is filled with a dielectric

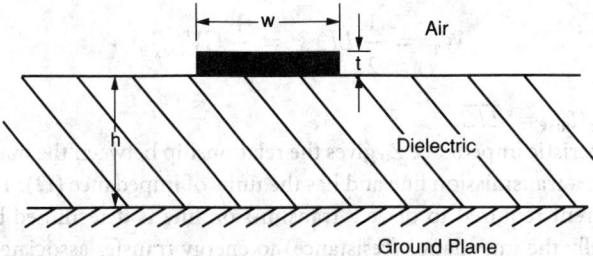

FIGURE 21.21   Classical microstrip.

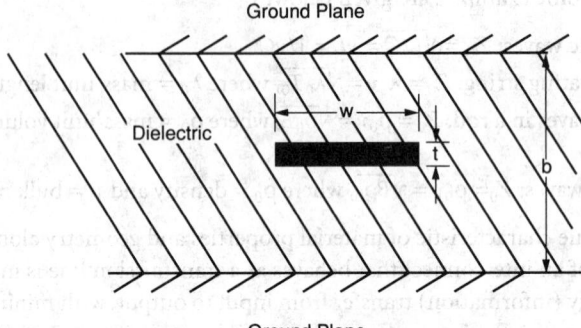

FIGURE 21.22   Classical stripline.

material, and above the conductor it is air. In most applications, however, the conductor traces are protected with a solder mask coating of 2–3 mils thickness above it, as in Fig. 21.23. This solder mask has the effect of reducing Z_0 of the microstrip. In a pure microstrip, the electromagnetic wave is transmitted partially in the dielectric and partially in air. Therefore the effective dielectric constant is a weighted average between that of air and the dielectric material. In a stripline design, the conductor is placed symmetrically between two ground planes and the space filled with dielectric material. Thus, in a stripline, the wave is completely transmitted in the dielectric. A variation of the stripline design is the asymmetric stripline where the conductor is closer to one ground plane than the other, as in Fig. 21.24.

For a microstrip transmission line structure, the characteristic impedance Z_0 is given by [Kaupp, 1967]

$$Z_0 = \frac{60}{\sqrt{0.475\,\varepsilon_r + 0.67}} \ln\left[\frac{4h}{0.67(0.8w + t)}\right]$$

$$= \frac{87}{\sqrt{\varepsilon_r + 1.41}} \ln\left[\frac{5.98h}{(0.8w + t)}\right] \ \Omega \tag{21.28}$$

The effective dielectric constant $\varepsilon_{reff} = \sqrt{0.475\varepsilon_r + 0.67}$. Experimental measurements with fiberglass epoxy boards have shown that Eq. (21.28) predicts Z_0 accurately for most practical applications.

For stripline configuration, characteristic impedance is given by [Howe, 1974]

For $w/b \leq 0.35$, $$Z_0 = \frac{60}{\sqrt{\varepsilon_r}} \ln\left[\frac{4b}{\pi d}\right] \ \Omega \tag{21.29a}$$

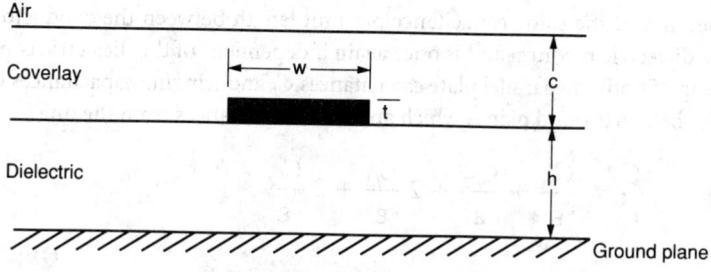

FIGURE 21.23   Covered microstrip.

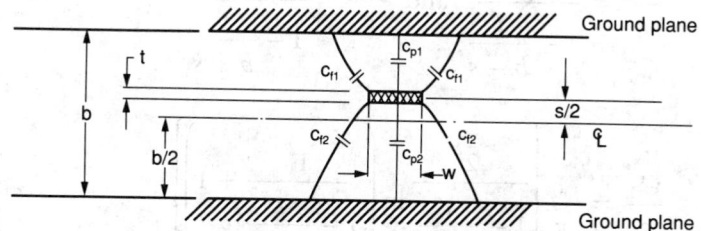

FIGURE 21.24   Asymmetric stripline.

$$d = \frac{w}{2}\left[1 + \frac{t}{\pi w}\left(1 + \ln\frac{4\pi w}{t} + 0.51\pi\left(\frac{t}{w}\right)^2\right)\right] \tag{21.29b}$$

For $w/b \geq 0.35$,
$$Z_0 = \frac{94.15}{\sqrt{\varepsilon_r}}\frac{1}{\dfrac{w}{b\left(1-\dfrac{t}{b}\right)} + \dfrac{C_f'}{\varepsilon}}\ \Omega \tag{21.29c}$$

$$\frac{C_f'}{\varepsilon} = \frac{1}{\pi}\left\{\frac{2}{1-\dfrac{t}{b}}\ln\left(\frac{1}{1-\dfrac{t}{b}}+1\right)\right\}$$

$$-\frac{1}{\pi}\left\{\left(\frac{1}{1-\dfrac{t}{b}}-1\right)\ln\left(\frac{1}{\left(1-\dfrac{t}{b}\right)^2}-1\right)\right\} \tag{21.29d}$$

where C_f'/ε is the ratio of the static fringing capacitance per unit length between conductors to the permittivity (in the same units) of the dielectric material. This ratio is independent of the dielectric constant. $\varepsilon = \varepsilon_r\varepsilon_0$, where ε_0 is the permittivity of air 8.854*1.0e-12 F/m. For the asymmetric stripline, where the conductor is closer to one ground plane than the other,

$$Z_0 = \frac{376.7}{\sqrt{\varepsilon_r}\dfrac{C}{\varepsilon}} \tag{21.30a}$$

Here C/ε is the ratio of the static capacitance per unit length between the conductors to the permittivity of the dielectric medium and is once again independent of the dielectric constant. Capacitance is made up of both the parallel plate capacitances C_p and fringing capacitances C_f. Subscripts 1 and 2 refer to the two ground planes which are different distances from the line.

$$\frac{C}{\varepsilon} = \frac{C_{p1}}{\varepsilon} + \frac{C_{p2}}{\varepsilon} + 2\frac{C_{f1}}{\varepsilon} + 2\frac{C_{f2}}{\varepsilon} \tag{21.30b}$$

$$\frac{C_{p1}}{\varepsilon} = \frac{\dfrac{2w}{b-s}}{1-\dfrac{t}{b-s}} \quad \text{and} \quad \frac{C_{p2}}{\varepsilon} = \frac{\dfrac{2w}{b+s}}{1-\dfrac{t}{b+s}} \tag{21.30c}$$

$$\frac{C_{f1}}{\varepsilon} = \frac{1}{\pi}\left\{ \frac{2}{1-\dfrac{t}{b-s}} \ln\left(\frac{1}{1-\dfrac{t}{b-s}} + 1 \right) \right\}$$

$$\tag{21.30d}$$

$$- \frac{1}{\pi}\left\{ \left(\frac{1}{1-\dfrac{t}{b-s}} - 1 \right) \ln\left(\frac{1}{\left(1-\dfrac{t}{b-s}\right)^2} - 1 \right) \right\}$$

$$\frac{C_{f2}}{\varepsilon} = \frac{1}{\pi}\left\{ \frac{2}{1-\dfrac{t}{b+s}} \ln\left(\frac{1}{1-\dfrac{t}{b+s}} + 1 \right) \right\}$$

$$\tag{21.30e}$$

$$- \frac{1}{\pi}\left\{ \left(\frac{1}{1-\dfrac{t}{b+s}} - 1 \right) \ln\left(\frac{1}{\left(1-\dfrac{t}{b+s}\right)^2} - 1 \right) \right\}$$

The formulas given above are for Z_0 of a single (isolated) line. As we shall see later, the presence of adjacent lines alters the value of Z_0. The appropriate line widths and the dielectric thicknesses required for Z_0 in the range of 50 to 75 Ω in glass epoxy FR4 boards with $\varepsilon_r = 4.2$ are shown in Tables 21.9 and 21.10 for microstrip and stripline. Also shown are the maximum crosstalk noise X_{max} between two conductors in percent, for a given spacing s between conductors in the same signal layer as in Figs. 21.27 and 21.38b.

An important observation from Tables 21.9 and 21.10 is that the thickness of the dielectric for a microstrip is considerably less than that of the stripline for the same Z_0. Thus microstrip structures lend themselves to much thinner PCBs, and this is an advantage from a manufacturing point of view, since thicker boards are more expensive to make. In practical PCBs, the traces are protected by a coat of solder mask (cover coat). This cover coat is usually a dry film or is screen printed and has an $\varepsilon_r \approx 4$.

Table 21.9 Microstrip in Glass Epoxy, $\varepsilon_r = 4.2$, $t = 1.2$ mil

w (mil)	s (mil)	h (mil)	w/h	s/h	Z_0 (Ω)	X_{max} (%)
8	8	5	1.6	1.6	50.3	4.1
8	8	10	0.8	0.8	75.8	9.5
6	6	7.5	0.8	0.8	73.9	9.7

Table 21.10 Stripline in Glass Epoxy, $\varepsilon_r = 4.2$, $t = 1.2$ mil

w (mil)	s (mil)	b (mil)	w/b	s/b	Z_0 (Ω)	X_{max} (%)
8	8	22	0.363	0.363	50.5	6.8
8	8	40	0.2	0.2	67.7	14.6
6	6	18.0	0.33	0.33	51.2	7.8
6	6	35	0.17	0.17	70.6	16.9

Thus, we never have a classical microstrip with air at the top, but a covered microstrip. The effect of the cover layer is to increase the effective dielectric constant. Consequently, the capacitance of the line increases, and Z_0 decreases. The reduction in Z_0 with cover layer thickness is shown in Fig. 21.25.

In practical situations, the thickness of cover layer is ≈3–4 mils, which can lead to a reduction in Z_0 of ≈10%. The covered microstrip can be considered to be a three dielectric layer problem, and the appropriate line parameters may be evaluated by techniques given in Das and Prasad [1984]. For most practical designs today, the line width is typically 6 or 8 mils. There are some high-speed designs where 4-mil lines have been used, where the technology is migrating due to the demand for high circuit and interconnection density. There are some innovative design techniques which enable the user to attain high interconnection (circuit) density without increasing the board thickness by a large amount. We noted earlier that the microstrip structure lends itself to thinner dielectrics. However, the stripline provides a pure transverse electromagnetic (EM) wave and the protection of two ground planes. We can combine the advantages of both of these structures by using the asymmetric stripline.

To design a stripline with $Z_0 = 50$ Ω in FR4 material ($\varepsilon_r ≈ 4.2$) for a line width of 8 mils, we require a dielectric thickness of 22 mils. Now consider a classical stripline with $w = 6$ mils and $b = 18$ mils. In FR4, $Z_0 ≈ 51$ Ω. If we now start moving one of the ground planes away from the line, Z_0 increases up to a certain distance, beyond which the line is not influenced by the presence of this

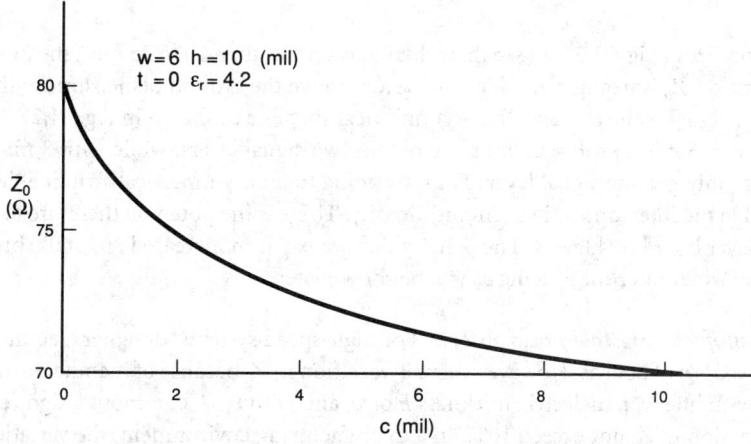

FIGURE 21.25 Z_0 reduction due to coverlay.

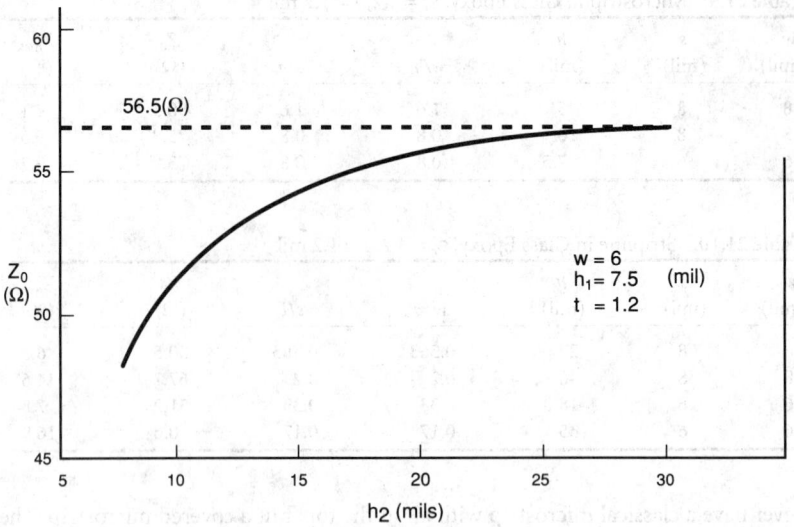

FIGURE 21.26 Increase in Z_0 by moving second groundplate.

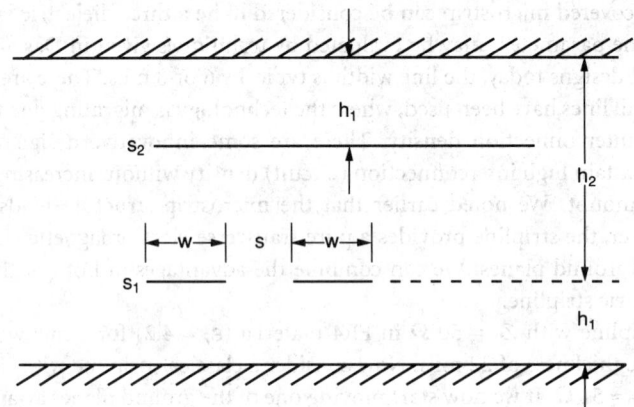

FIGURE 21.27 Two signal layers S_1 and S_2 in asymmetric stripline. S_1 and S_2 are orthogonal lines.

ground plane. From Fig. 21.26 we see that when b_2 is greater than 20 to 25 mils, the influence of this ground plane on Z_0 is negligible. We can therefore move the ground plane sufficiently far so as to have another signal plane to obtain an asymmetrical stripline as shown in Fig. 21.27.

Now for $w = 6$ mils and $b = 22$ mils, we obtain two signal layers, while with 8 mils and $b = 22$ mils, we can only get one signal layer. Thus, by going to an asymmetrical stripline design, we can almost double the interconnection (circuit) density. There is the potential therefore to reduce an 8-layer multilayer board to 4 layers. The penalty that we pay is in increased crosstalk, but that can be addressed by wider interline spacing as will be shown later.

Effect of Manufacturing Tolerances on Z_0.   For high-speed systems, designers require a tight control on impedance of boards. However, there is variation in Z_0 because of manufacturing tolerances associated with line w, t, dielectric thickness h or b, and with ε_r. A common design requirement is that the variation in Z_0 not exceed 10%. In a manufacturing environment, the variation in the relevant parameters is random in nature. We may therefore express the variation in Z_0 as

$$\Delta Z_0 \approx \sqrt{\left(\frac{\partial Z_0}{\partial w}\right)^2 \Delta w^2 + \left(\frac{\partial Z_0}{\partial h}\right)^2 \Delta h^2 + \left(\frac{\partial Z_0}{\partial t}\right)^2 \Delta t^2 + \left(\frac{\partial Z_0}{\partial \varepsilon_r}\right)^2 \Delta \varepsilon_r^2} \qquad (21.31)$$

The partial derivatives may be evaluated from the appropriate expressions given in Eqs. (21.28), (21.29) and (21.30). They are shown in Table 21.11 for $Z_0 = 50$ and $75\ \Omega$ for classical microstrip and stripline.

However, the worst-case tolerances or the limits of the tolerance are defined by a Taylor series as

$$\Delta Z_0 \approx \left(\frac{\partial Z_0}{\partial w}\right)\Delta w + \left(\frac{\partial Z_0}{\partial h}\right)\Delta h + \left(\frac{\partial Z_0}{\partial t}\right)\Delta t + \left(\frac{\partial Z_0}{\partial \varepsilon_r}\right)\Delta \varepsilon_r \qquad (21.32)$$

These tolerance limits are shown in Figs. 21.28 and 21.29 for a microstrip of $Z_0 = 75\ \Omega$ and for a stripline of $50\ \Omega$. The line width is 6 mils ($\approx 150\ \mu m$) and the values chosen for Z_0 are of practical interest. The limits of variation in Z_0 are shown for a change in dielectric constant $\Delta \varepsilon_r = 0.1$, which is very realistic for variation in material properties. Also, dielectric constant for FR4 is not a con-

Table 21.11 Variation in Z_0 Due To Parameter Changes

Design	Z_0 (Ω)	w (mil)	$\partial Z_0/\partial h$ or $\partial Z_0/\partial b$ (Ω/mil)	$\partial Z_0/\partial w$ (Ω/mil)	$\partial Z_0/\partial \varepsilon_r$ (Ω)
Microstrip	50	8	7.5	−3.8	−4.6
Microstrip	50	6	9.5	−4.8	−4.6
Microstrip	75	8	3.8	−3.8	−6.9
Microstrip	75	6	4.8	−4.8	−6.9
Stripline	50	8	1.3	−2.7	−6.2
Stripline	50	6	1.6	−3.4	−6.2
Stripline	50	4	2.1	−4.6	−6.2

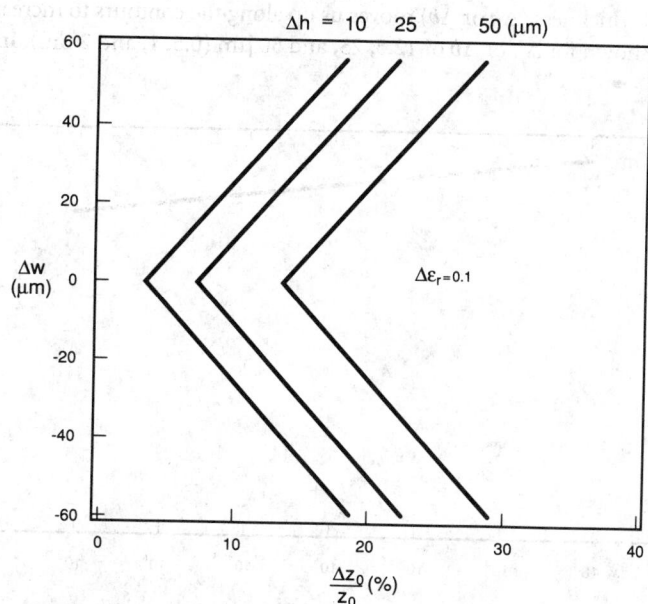

FIGURE 21.28 Tolerance limits for microstrip line. $Z_0 = 75\ \Omega$, $w = 6$ mil, $t = 1.4$ mil, $\varepsilon_r = 4.2$ (1 mil = 25 μm).

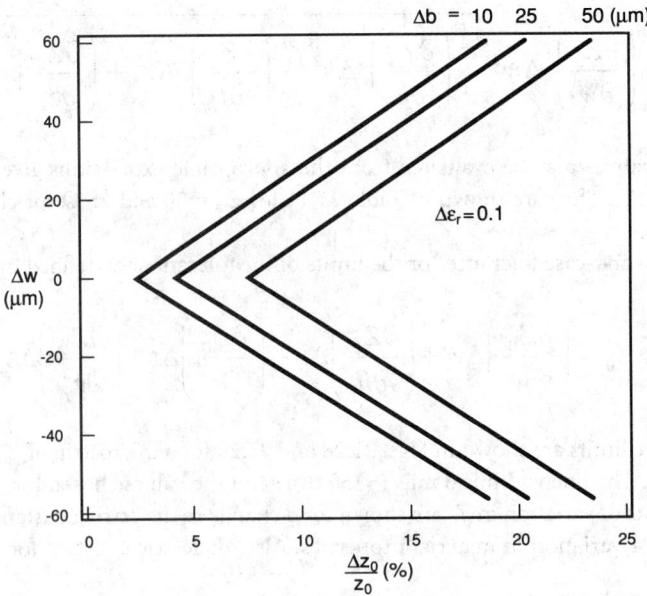

FIGURE 21.29   Tolerance limits for stripline. $Z_0 = 50\ \Omega$, $w = 6$ mil, $t = 1.4$ mil, $\varepsilon_r = 4.2$.

stant, but varies with frequency as shown in Fig. 21.30, taken from measurements made by S. Mumby at Bell Laboratories.

We observe that ε_r decreases with frequency, and that in the range of 100 Hz to 1 GHz, ε_r decreases from 4.8 to 4, a change of $\approx 20\%$. From the equations for Z_0, we observe that it scales approximately as $1/\sqrt{\varepsilon_r}$. Thus, a change of 20% in ε_r will result in $\approx 14\%$ change in Z_0 as we go from 100 Hz to 1 GHz. This is just to point out that ε_r is not a constant but does vary slightly with frequency. In Figs. 21.29 and 21.30, it is clear that in addition to a change of 0.1 in ε_r, any variation in w (Δw) and dielectric thickness (Δh or Δb) moves us up along the contours to increase $\Delta Z_0 / Z_0$. The tolerance limits are shown for Δh or Δb of 12.5, 25, and 50 μm (0.5, 1, and 2 mil). In order to limit

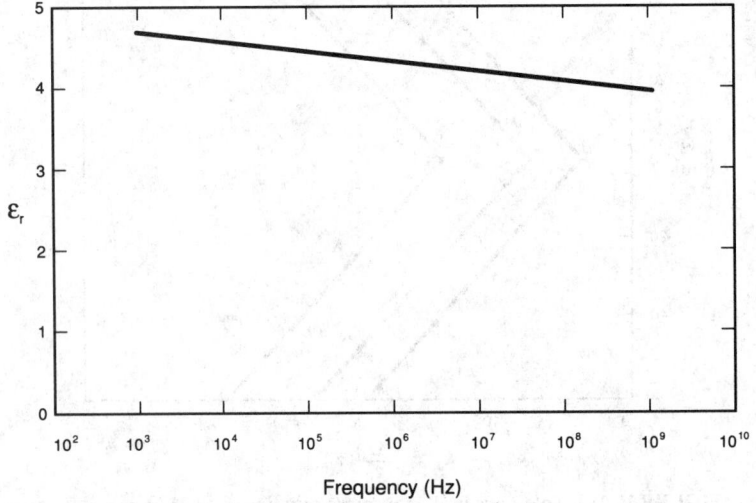

FIGURE 21.30   Variation of ε_r with frequency for FR4.

the Z_0 tolerance to 10%, we need tight control over the manufacturing process. Designers should be aware of the manufacturing tolerances of PCB manufacturing to determine the ability to meet their controlled impedance requirements. These aspects are discussed in Ritz [1988].

Crosstalk (*X*)

Crosstalk may be defined as noise that occurs on idle lines due to interactions with stray EM fields that originate from active (pulsed) lines. This interaction is shown pictorially in Figs. 21.31 and 21.32.

There are two components to crosstalk. They are inductive crosstalk due to mutual inductance, L_m, and capacitive crosstalk due to mutual capacitance, C_m. Inductive crosstalk is proportional to $L_m \, dI/dt$, and capacitive crosstalk is proportional to $C_m \, dV/dt$. In any conductor in which we have a transient current propagating perpendicular to the plane of the paper, circular magnetic field lines are generated in space as shown in Fig. 21.31. Any conductor that interacts with this magnetic field has crosstalk noise imposed on it due to mutual inductive coupling L_m between the pulsed and the idle lines. Similarly, a transient voltage pulse in the conductor generates radial electric fields emanating from the conductor as shown in Fig. 21.32. An idle line that interacts with this electric field has crosstalk noise imposed on it due to mutual capacitance coupling between the pulsed and idle lines. Thus, to predict crosstalk, we must be able to evaluate the mutual coupling coefficients L_m and C_m.

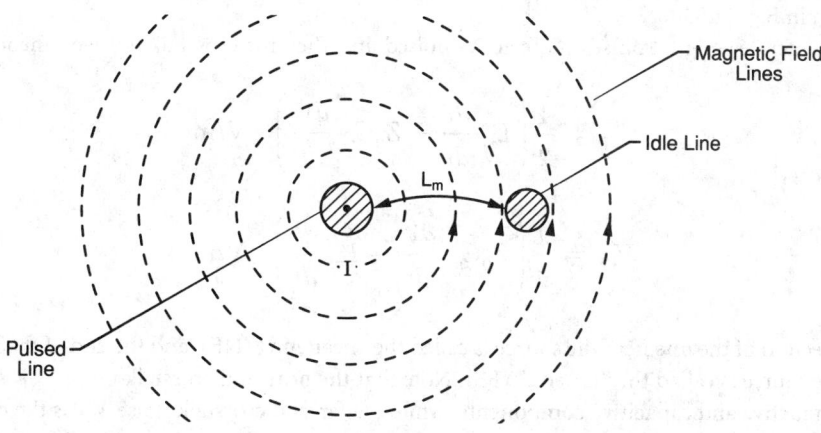

FIGURE 21.31   Schematic of inductive crosstalk coupling.

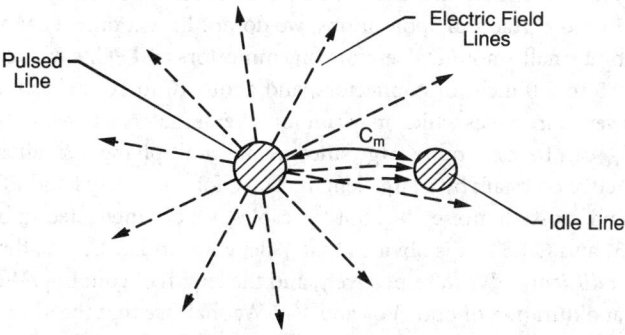

FIGURE 21.32   Schematic of capacitive crosstalk coupling.

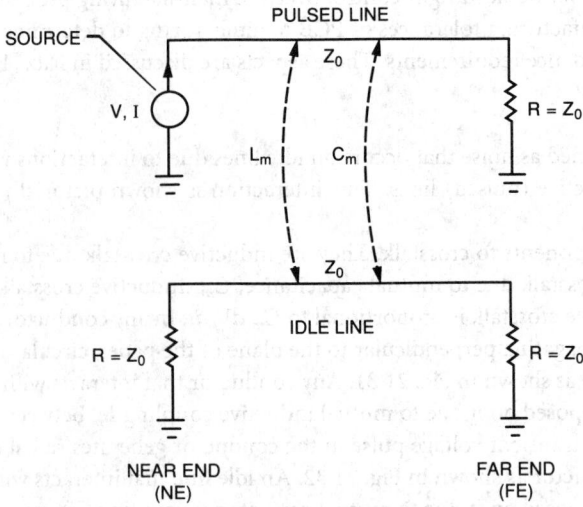

FIGURE 21.33   Crosstalk model showing near end and far end.

Consider a pulsed line and an idle line which are terminated in their characteristic impedances as shown in Fig. 21.33.

If we generate a digital pulse on the active pulsed line, then the crosstalk noise on the idle line is given by

$$V_{\text{NE}} = \frac{1}{2}\left(L_m \frac{dI}{dt} + Z_0 C_m \frac{dV}{dt}\right) \quad \text{V/m} \tag{21.33}$$

$$V_{\text{FE}} = \frac{1}{2}\left(Z_0 C_m \frac{dV}{dt} - L_m \frac{dI}{dt}\right) \quad \text{V/m} \tag{21.34}$$

The section of the line near the source is called the "near end" (NE), and the end of the line away from the source is called the "far end" (FE). Note that the near-end crosstalk noise V_{NE} is the sum of the inductive and capacitive components, while the far-end crosstalk noise V_{FE} is the difference between the capacitive and inductive components. The crosstalk noise given by Eqs. (21.33) and (21.34) is per unit length of the line. Therefore, to evaluate the total crosstalk, we must multiply V_{NE} and V_{FE} by the length of coupling. This is generally true for PCBs and cables where L_m and C_m are expressed in nH/m and pF/m. For a pure transverse EM (TEM) wave propagation $Z_0 = \sqrt{L_m/C_m}$. Therefore, the inductive and capacitive crosstalk noise components are equal, and $V_{\text{FE}} = 0$. However, in most practical applications, we do not have a pure TEM wave, and the two components differ by a small amount. Therefore, in connectors and PCBs where the length of coupling ranges from 0.5 to 1.0 inch for connectors, and about 10 to 20 inches for PCBs and backplanes, $V_{\text{FE}} \approx 0$. However, in cables which may run for several meters between equipment, the total far-end crosstalk V_{FE} can become quite large since we are multiplying a small difference between capacitive and inductive crosstalk by a large length. Thus, for connectors and PCBs we are mainly interested in near-end crosstalk noise V_{NE}, but for cables, we are interested in both V_{NE} and V_{FE}.

From Eqs. (21.33) and (21.34) it is obvious that V depends on L_m, C_m, Z_0, the rate of change of current and voltage, dI/dt and dV/dt, respectively, and the length of coupling. What is of interest is to know the shape and duration of both V_{NE} and V_{FE}. We shall see that the answer to this question depends upon the relative magnitudes of the propagation delay in the line T_d (which is proportional to length) and the risetime of the pulse T_r in digital systems, or the frequency of the pulse for

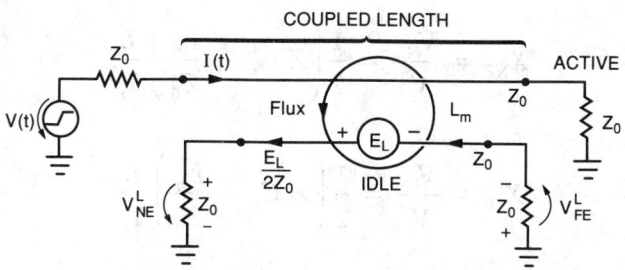

FIGURE 21.34   Inductive noise on idle line.

analog systems. The sign of the inductive and capacitive noise components may be easily under-stood by looking at the simple mechanical analogy of the spring-mass system. The inductive noise given by $L_m \, dI/dt$ has the effect of inserting a voltage source V_L in the idle line as shown in Fig. 21.34.

However, inductance has the effect of opposing the source current pulse. The source creates a clockwise loop of current (action). Inductance behaves like the mechanical spring which produces a reaction $-kx$ to the displacement x (action). So, as a reaction to the clockwise loop of current in the source line (action), mutual inductance induces a counterclockwise loop of current in the idle line (reaction) to oppose the source pulse. This therefore determines the polarity of the voltage source on the idle line as shown in Fig. 21.34. It is this voltage source that generates the counter-clockwise loop of current. Therefore, we observe from Fig. 21.34 that since the current flows from the near end to ground, it must be positive with respect to ground. Similarly, since the direction of current is from ground to far end, it must be negative with respect to ground. Thus, we may observe from Eqs. (22.33) and (22.34) that the inductive coupling term is positive at the near end and negative at the far end. The current produced by mutual inductance is $E_L/2Z_0$. Therefore, the inductive coupling noise at the near end and far end is given by $V_{NE}^L = -V_{FE}^L = E_L/2$, where $E_L = L_m \, dI/dt$. For the source line, we may write $dI/dt \approx V_s/Z_0 T_r$. Here, V_s is the magnitude of the source pulse, and T_r is the risetime of the pulse.

Capacitive coupling has the effect of inserting a mutual capacitor C_m between the source and idle lines as shown in Fig. 21.35. The current that passes through C_m divides evenly in the idle line, with half going to ground through the near end and the other half going to ground through the far end. Thus, the capacitive crosstalk voltages at the near end and the far end are given by $V_{NE}^C = V_{FE}^C = I_C Z_0/2$. The coupled capacitive current $I_C = C_m \, dV/dt \approx C_m V_s/T_r$. Therefore, mutual capacitance may be considered to be like mass (inertia) in the mechanical spring-mass analogy. Mass (inertia) tends to keep going in the direction of displacement. Similarly, mutual capacitance induces the same noise in both the near and far end as can be seen in Eqs. (21.33) and (21.34), and the sign of the noise is the same as dV/dt (inertia). The total crosstalk noise (X) is due to inductive (X_L) and capacitive (X_C) noise and is usually expressed as a fraction of the input source pulse V_s as $X = V/V_s$. Therefore, for a coupling length of L, we may write crosstalk noise as

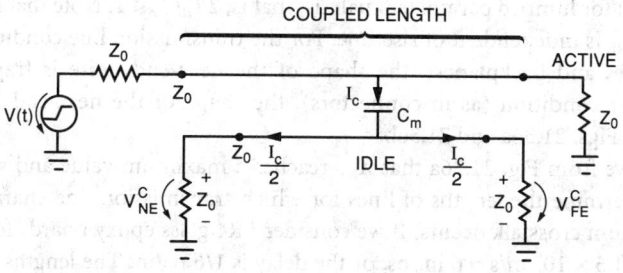

FIGURE 21.35   Capacitive noise on idle line.

$$X_{NE} = \frac{V_{NE}}{V_s} = \frac{1}{2}\left(Z_0 C_m + \frac{L_m}{Z_0}\right)\frac{L}{T_r} \tag{21.35}$$

$$X_{FE} = \frac{V_{FE}}{V_s} = \frac{1}{2}\left(Z_0 C_m - \frac{L_m}{Z_0}\right)\frac{L}{T_r} \tag{21.36}$$

But we know that length $L = vT_d$. Therefore

$$X_{NE} = \frac{V_{NE}}{V_s} = \frac{v}{4}\left(Z_0 C_m + \frac{L_m}{Z_0}\right)\frac{2T_d}{T_r} = K_{NE}\left(\frac{2T_d}{T_r}\right) \tag{21.37}$$

$$X_{FE} = \frac{V_{FE}}{V_s} = \frac{1}{2}\left(Z_0 C_m - \frac{L_m}{Z_0}\right)\frac{L}{T_r} = K_{FE}\left(\frac{L}{T_r}\right) \tag{21.38}$$

$$K_{NE} = \frac{v}{4}\left(Z_0 C_m + \frac{L_m}{Z_0}\right) \tag{21.39}$$

$$K_{FE} = \frac{1}{2}\left(Z_0 C_m - \frac{L_m}{Z_0}\right) \tag{21.40}$$

The expressions derived here are for noise in digital systems. It can be shown analytically and experimentally [Rainal, 1979] that near-end noise, X_{NE} increases with length of coupling, reaches a maximum value and saturates there. Any increase in coupling length beyond a critical length will not increase the value of X_{NE}. This critical coupling length for maximum crosstalk X_{max} depends upon the delay and signal risetime and is given by

$$X_{NE} = K_{NE} = X_{max} \quad \text{for } \frac{2T_d}{T_r} \geq 1 \quad \text{(maximum crosstalk)} \tag{21.41}$$

$$X_{NE} = K_{NE}\left(\frac{2T_d}{T_r}\right) \quad \text{for } \frac{2T_d}{T_r} < 1 \tag{21.42}$$

Equation (21.41) is the condition for transmission line behavior, that is, $2T_d/T_r \geq 1$. Equation (21.42) is the limit for lumped parameter analysis, that is, $2T_d/T_r < 1$. Note that in the transmission line condition, X_{NE} is independent of risetime. For the transmission line condition (long coupling length as in cables and backplanes), the shape of the near-end noise is trapezoidal, while for lumped parameter condition (as in connectors), the shape of the near-end noise is triangular. These are clear in Figs. 21.36a and 21.36b.

Also, we observe from Fig. 21.36a that X_{NE} reaches a maximum value and saturates there. It is instructive to determine the lengths of lines for which transmission line characteristics are valid and when maximum crosstalk occurs. If we consider FR4 glass epoxy boards for which $\varepsilon_r \approx 4$, the wave speed v is $\approx 1.5 \times 10^8$ m/s ≈ 6 in./ns, or the delay is 1/6 ns/in. The lengths for which $2T_d \approx T_r$ are shown in Table 21.12 for FR4 boards.

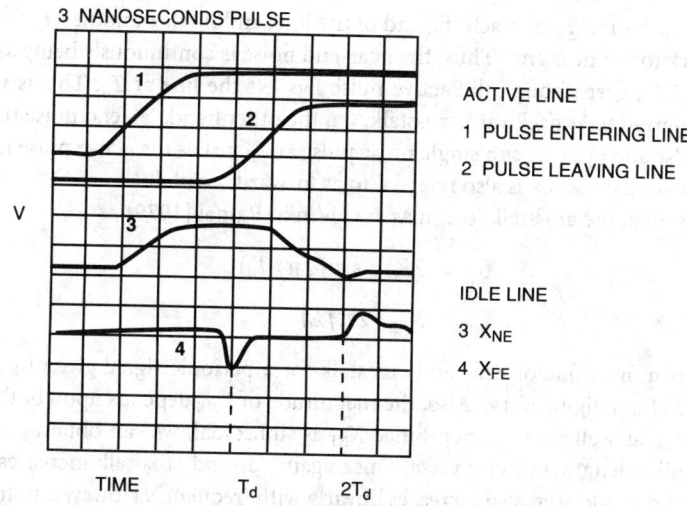

FIGURE 21.36a Typical cable crosstalk noise X_{NE} and X_{FE}.

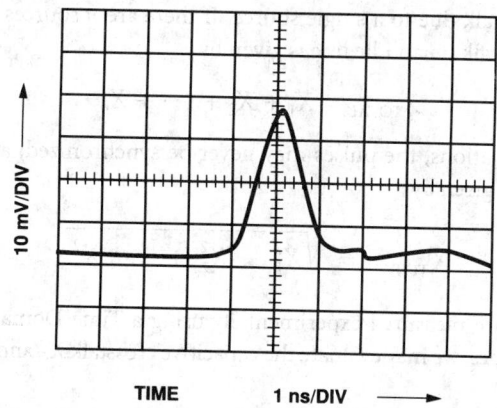

FIGURE 21.36b Typical connector near-end crosstalk X_{NE}.

Table 21.12 Lengths for Maximum Crosstalk in FR4 Boards

T_r (ns)	$T_d = T_r/2$ (ns)	L (in.)
0.2	0.1	0.5
0.5	0.25	1.5
1.0	0.5	3
2.0	1.0	6
5.0	2.5	15
10	5	30

Note that for a risetime of 1ns, it only takes 3 inches of coupling for maximum crosstalk to occur. Thus, while the near-end noise X_{NE} has a maximum limit, far-end noise X_{FE} given by Eq. (21.38) increases linearly with length. This is the reason that X_{FE} is important for cables running over long distances. This is obvious from Fig. 21.36a, where near-end and far-end noise are almost of the same magnitude. Another very important observation to be made from Fig. 21.36a is that while the active pulse (source) leaves the source line at time T_d, the near-end noise has a width of $2T_d$. Thus, the crosstalk noise stays around for an extra time delay T_d, even though the active pulse causing the noise has already left the source line. But, observe that far-end noise occurs as a single pulse at time T_d. The reason for this phenomenon is that near-end coupling occurs the moment the pulse enters the active line. As the source pulse travels on the active line, the coupled noise on the idle line is sent back continuously to the near-end from every point of coupling between the two lines. Therefore, there is a delay in the noise as it travels from a point of coupling back to the near-end. As the coupled noise travels back to the near-end, the last portion of the noise that coupled at the end of the active line (far end) has to travel back all the way to the near-end. It took the active pulse a time T_d to reach the end of the line, and it takes another T_d for the coupled noise to travel back to the near end. Thus, the near-end noise is continuously being sent back for a duration of time $2T_d$, even though the active pulse has left the line at T_d. This is why near-end crosstalk is also referred to as backward crosstalk. On the other hand, far-end noise travels forward with the active pulse and appears as a single noise pulse at T_d just as the active pulse leaves the line. For this reason, far-end crosstalk is also referred to as forward crosstalk.

For a periodic signal, the crosstalk formulas are given in Rainal [1979] as

$$X_{NE} = 2K_{NE} \sin(2\pi f T_d) \tag{21.43}$$

$$X_{FE} = K_{FE} (2\pi f L) \tag{21.44}$$

Note that the maximum value of near-end crosstalk for a periodic signal given by Eq. (21.43) is twice that obtained for a digital pulse. Also, the magnitude of X_{NE} depends upon both the length of coupling (delay T_d), as well as frequency. Since X_{NE} is sinusoidal, we can obtain greater crosstalk with smaller coupling length and vice versa. Once again, far-end crosstalk increases linearly with coupling length, and in addition also increases linearly with frequency. However, note that for both analog and digital signals, crosstalk noise X_{NE} and X_{FE} depend upon the mutual coupling coefficients L_m and C_m. Thus, the key to reducing crosstalk is to reduce both L_m and C_m. The formulas given so far are for crosstalk due to a single source. If there are n sources pulsing simultaneously, then the worst-case crosstalk on an idle line is given by

$$X_{TOTAL} = X_1 + X_2 + \cdots + X_n \tag{21.45a}$$

For most practical applications, the pulses may never be synchronized, and a more realistic estimate of crosstalk may be given by

$$X_{TOTAL} = \sqrt{X_1^2 + X_2^2 + \cdots + X_n^2} \tag{21.45b}$$

Normally, X_{NE} and X_{FE} are measured experimentally using a Time Domain Reflectometer (TDR). From these measured values, we may evaluate the capacitive crosstalk X_C and inductive crosstalk X_L as

$$X_C = \frac{1}{2}\left(X_{NE} + X_{FE}\right) \tag{21.46a}$$

$$X_L = \frac{1}{2}\left(X_{NE} - X_{FE}\right) \tag{21.46b}$$

From these, we can evaluate the mutual coefficients L_m and C_m which enable us to scale the results for crosstalk noise for different values of T_r, T_d, L and T_r, or f. L_m and C_m may also be evaluated analytically for many practical situations, and these will now be considered.

Expressions for L_m and C_m. In order to analytically evaluate L_m and C_m, we use the coupled lines techniques developed by microwave engineers. For this, we consider the two coupled lines to be in the odd and even mode configurations as shown in Figs. 21.37a and 21.37b for a microstrip design.

In the odd mode configuration, the potential on one line is the negative potential of the other line, while in the even mode, both lines have the same potential imposed upon them. Let C_a be the capacitance of an isolated line to ground. Then, the odd and even mode capacitances, inductances, and characteristic impedances are

$$C_{odd} = C_a + 2C_m, \qquad L_{odd} = L_a - L_m, \qquad Z_{0odd} = \sqrt{\frac{L_{odd}}{C_{odd}}} \tag{21.47a}$$

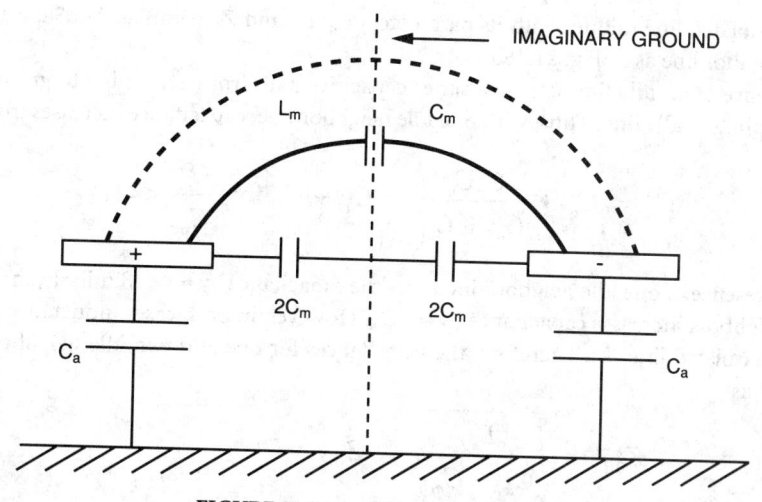

FIGURE 21.37a Odd mode microstrip.

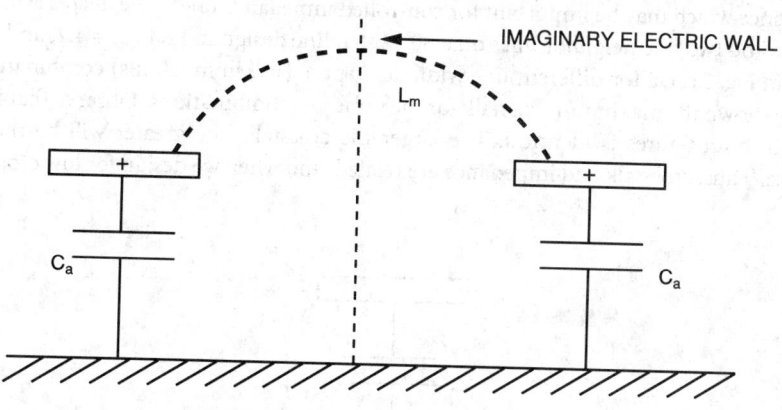

FIGURE 21.37b Even mode microstrip.

$$C_{even} = C_a, \qquad L_{even} = L_a + L_m, \qquad Z_{0even} = \sqrt{\frac{L_{even}}{C_{even}}} \qquad (21.47b)$$

The characteristic impedance for a balanced or differential line Z_{0b} is given by

$$Z_{0b} = 2Z_{0odd} \qquad (21.47c)$$

Expressions for C_{odd} and C_{even}, and Z_{0odd} and Z_{0even} are given in many references (such as Davis, 1990). From these, we can then evaluate L_{odd} and L_{even}, and we may express the mutual coefficients as

$$C_m = \frac{C_{odd} - C_{even}}{2}, \qquad L_m = \frac{L_{even} - L_{odd}}{2} \qquad (21.47d)$$

Mathematically, the odd and even mode impedances are the minimum and maximum values of impedances for coupled lines. Observe that this change in impedance is due to the mutual coefficients. This has important implications in high-speed and high-density designs where many lines are closely spaced. The very presence of an idle line in proximity alters Z_0 of the line. Consider a completely isolated line (alone) with its parameters C_a, L_a, and Z_a as in Fig. 21.38a. Next consider one idle neighbor line as in Fig. 21.38b.

The presence of an idle line sets up a series capacitive path from active line to ground through mutual coupling to idle line. Thus, with one idle neighbor line, capacitance increases from C_a to C_{1i} given by

$$C_{1i} = C_a + \frac{C_a C_m}{C_a + C_m} \approx C_a + C_m \text{ for } \frac{C_m}{C_a} \ll 1 \qquad (21.48)$$

Thus, the presence of one idle neighbor increases the capacitance by $\approx C_m$. Similarly, the presence of two idle neighbors increases capacitance by $\approx 2C_m$. However, in both cases inductance remains the same as the isolated line, L_a. Therefore, the impedances for one and two idle neighbors are given respectively as

$$Z_{0i1} \approx \sqrt{\frac{L_a}{C_a + C_m}}, \quad Z_{0i2} \approx \sqrt{\frac{L_a}{C_a + 2C_m}} \qquad (21.49)$$

Therefore, since the presence of idle lines increases capacitance, there is a reduction in characteristic impedance which may be important for controlled impedance lines. The decrease in impedance due to one and two idle neighbor lines on a 50-Ω stripline design in FR4 ($\varepsilon_r = 4.2$, and $t = 1.2$ mil) is shown in Fig. 21.39a for different line width and space (w/s in mils/mils) combinations. In Fig. 21.39b are shown the maximum crosstalk for the same w/s combinations. Observe that the shape of the lines in both figures is identical. The larger the crosstalk, the greater will be the change in impedance. Thus, crosstalk and impedance are related, and when we design for low crosstalk we are

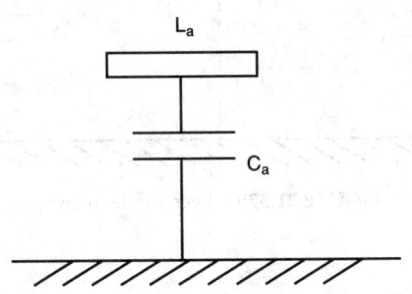

FIGURE 21.38a Isolated line.

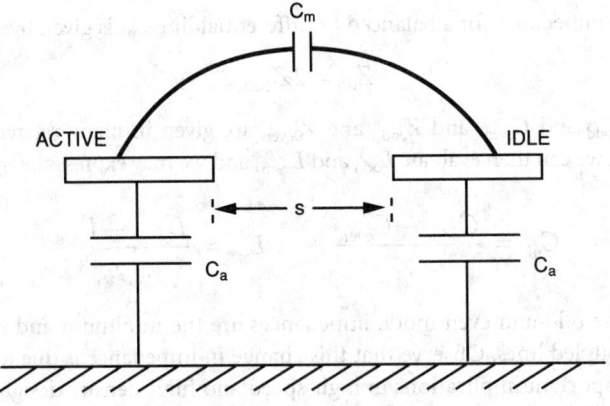

FIGURE 21.38b Line with one idle neighbor.

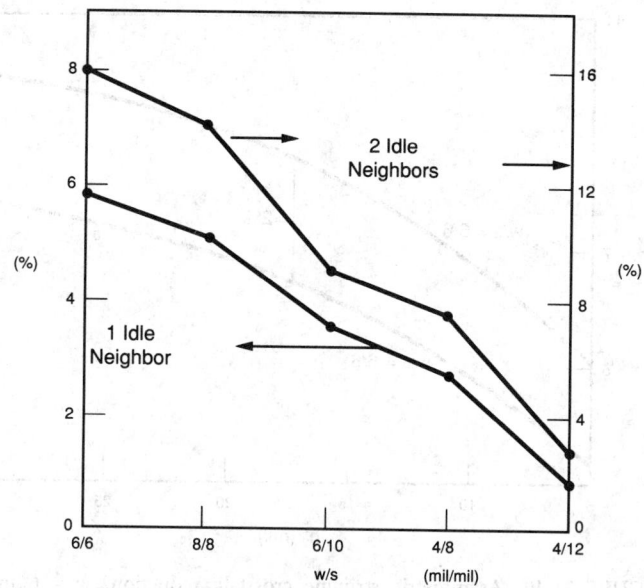

FIGURE 21.39a Reduction in Z_0 due to presence of adjacent lines. Stripline nominal $Z_0 = 50\ \Omega$, $\varepsilon_r = 4.2$, $t = 1.2$ mil.

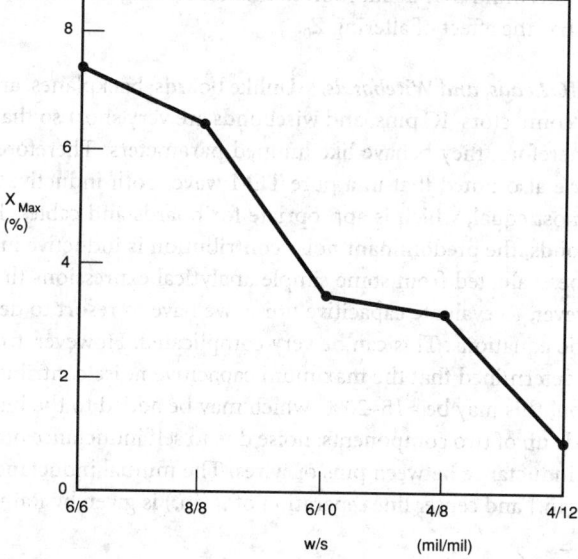

FIGURE 21.39b Maximum crosstalk for various w/s combinations. Stripline nominal $Z_0 = 50\ \Omega$, $\varepsilon_r = 4.2$, $t = 1.2$ mil.

automatically designing for good impedance control. Also, from Fig. 21.39b we observe that to reduce crosstalk we must increase the space s between lines. As we go from a 6/6 design to a 6/10 design, maximum crosstalk reduces from 7.5% to 3.5%, a reduction of 4%. This has relevance to Fig. 21.27 where we showed that the asymmetric stripline design lends itself to higher interconnection density. We also noted that the penalty to be paid for moving one ground plane farther away is increased crosstalk. However, by increasing s, we can reduce crosstalk and reduce this penalty. As shown in Fig. 21.40, as we change from a w/s design of 6/6 to 6/8, we get an almost uniform reduc-

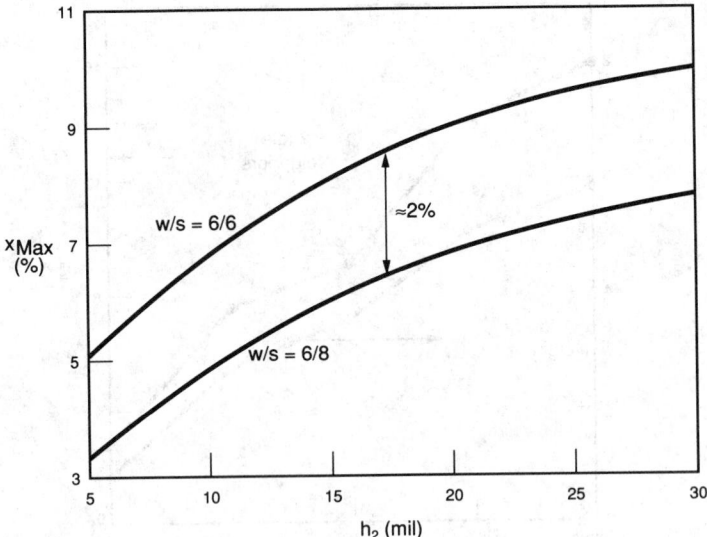

FIGURE 21.40   Asymmetric stripline crosstalk reduction. $w = 6$ mil, $t = 1.2$ mil, $h_1 = 7.5$ mil.

tion of 2% in maximum crosstalk. Thus, we can adjust the spacing to achieve crosstalk immunity. We should always keep in mind that in addition to manufacturing tolerances, presence of adjacent lines in dense boards has the effect of altering Z_0.

Noise in Connectors, IC Leads, and Wirebonds.   Unlike boards, backplanes, and cables, the lengths of the signal paths in connectors, IC pins, and wirebonds are very short so that delay is small compared to risetime. Therefore, they behave like lumped parameters. Therefore, near-end crosstalk noise is important. We also noted that in a pure TEM wave, both inductive and capacitive noise contributions are almost equal, which is appropriate for boards and cables. However, in connectors, pins, and wirebonds, the predominant noise contribution is inductive in nature. Fortunately, inductive noise can be evaluated from some simple analytical expressions that have been verified experimentally. However, to evaluate capacitive noise, we have to resort to detailed field solutions of the electromagnetic equations. This can be very complicated. However, from measurements in connectors, we have determined that the maximum capacitive noise contribution is $\approx 30\%$ of X_{NE}. In a practical situation, this may be ≈ 15–20%, which may be added to the inductive noise. Inductive noise itself is made up of two components: noise due to self inductance of the pins or wires and noise due to mutual inductance between pins or wires. The mutual inductance, L_m (nH), between two wires of length l (in.) and center line separation of d (in.) is given by Rainal [1984].

$$L_m \approx 5l\left[\ln\left(\frac{l}{d} + \sqrt{1 + (l/d)^2}\right) - \sqrt{1 + (d/l)^2} + d/l\right] \qquad \text{nH} \qquad (21.50)$$

Similarly, the self inductance L_s (nH) of a pin of radius ρ (in.) and length l (in.) is given by

$$L_s = L_m\big|_{d=\rho} + \frac{5l}{4} \qquad \text{nH} \qquad (21.51)$$

Thus, total self inductance is the sum of the contributions from both external and internal magnetic fields. For a rectangular conductor of perimeter p (in.), self inductance is given by

$$L_s \approx 5l \left[\ln\left(\frac{4l}{p}\right) + \frac{1}{2} \right] \quad \text{nH} \tag{21.52}$$

Thus, inductive noise due to self inductance can be significant even in the presence of mutual inductance. It is this phenomenon that gives rise to ground noise in connector pins and wire-bonds in IC packages and may be as high as 100 mV or more [Rainal, 1984]. This becomes an important issue in the allocation of ground return leads or wires during the signal layout of connectors and ICs. The layout of a connector or IC leads (wirebonds) consists of arrays or clusters of conductors. Of these, a certain number are allocated to signals and the remaining to ground. In connectors, we are mainly interested in X_{NE} on an idle signal pin, when one or more of the adjacent signal lines are pulsed. If an array of n conductors are active, then the total noise on the idle pin due to mutual coupling is given by

$$V_m = \sum_{i=1}^{i=n} L_{mi}(dI/dt)_i \tag{21.53}$$

Noise on the ground pins is the sum of self and mutual inductances and is given by

$$V_n = -\sum_{i=1}^{i=n} (L_s - L_{mi})(dI/dt)_i \tag{21.54}$$

The return currents in the ground pins or leads will vary according to the inductive field around it. Since the ground has shifted, the near-end inductive crosstalk on the idle pin is given by

$$V_t = \frac{V_m - V_n}{2} = V_{NE}^L \tag{21.55}$$

The model for noise evaluation in connectors and wirebonds is shown in Fig. 21.41.

Note that ground noise becomes an important part of inductive noise. This phenomenon, also called "ground bounce," can be a significant problem when 16 or 32 signal bits are switched simultaneously in a chip. For high-speed logic designs, ground noise must be kept to a minimum. This requires as many ground return leads as possible. In addition, ground and signal leads must be

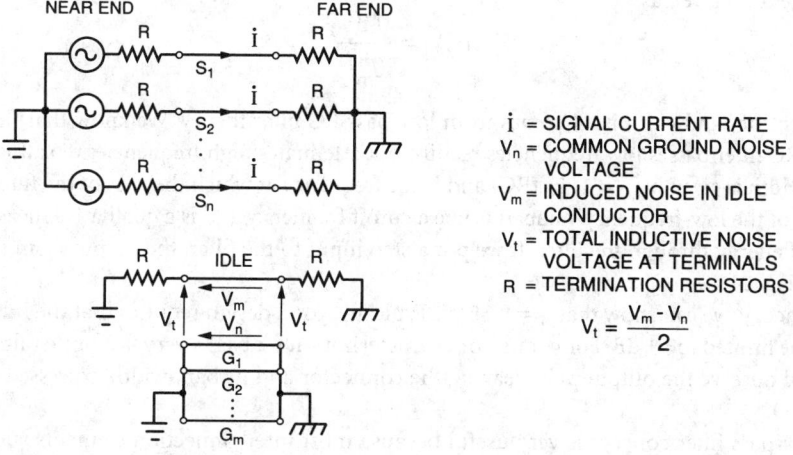

FIGURE 21.41 Inductive noise model for an array of signal (S_i), ground (G_i), and idle conductors.

closely alternated in a checkerboard pattern. Keeping all the signals clustered in one region and all the grounds clustered in another region will cause significant ground noise problems [Rainal, 1984].

Risetime Degradation

When a pulse is passed through an interconnection element, there is an increase in the risetime of the pulse, and this slowing down of the wave is referred to as risetime degradation. If the input risetime is T_{ri} and the output risetime is T_{ro}, then risetime degradation T_{rd} is defined as

$$T_{rd} = \sqrt{T_{ro}^2 - T_{ri}^2} \qquad (21.56)$$

Thus, if we have a step input, the output risetime is T_{rd}. If there are n interconnection elements, each with risetime degradation T_{rd1}, T_{rd2}, etc., then the risetime at output of n interconnection elements is

$$T_{ro}^2 = T_{ri}^2 + T_{rd1}^2 + T_{rd2}^2 + \cdots + T_{rdn}^2 \qquad (21.57)$$

This relationship is important because risetime degradation is not just the sum, but the square root of the sum of squares. Thus, if T_{rd1} is 0.2, and T_{rd2} is 0.5, it is almost 6 times (0.25/0.04) more important than T_{rd1}. Therefore, attention should be paid to those elements that are most important. In general, risetime degradation occurs due to resistive, inductive, and capacitive effects of circuits. In PCBs with long traces as transmission lines, T_{rd} is due to a combination of dc (IR loss) and skin effect. At very low frequencies, the current fills the entire cross section of the conductor. As we go to higher frequencies, the current is concentrated in a very thin layer (skin) around the perimeter of the conductor. This thin layer is called skin depth, $\delta = 1/\sqrt{\pi f \sigma \mu}$, where σ is the conductivity, μ the permeability, and f the frequency. For copper, $\delta = 2.09$ μm at 1 GHz. Skin effect adds a series resistance to the line, and for a rectangular conductor, the first-order approximation is

$$R_s = \frac{1}{[\sigma \times \text{perimeter} \times \delta]} \qquad (21.58)$$

Skin effect slows down the signal and also lowers the magnitude of the pulse. For a copper conductor with $w = 4$ mils and $t = 2$ mils, for a risetime of 0.5 ns, skin effect becomes critical when the length of the trace approaches 40 in. [Chang, 1988]. Therefore, for most practical board designs today, the lossless line is a good approximation.

Risetime degradation is important because it is related to the **bandwidth** (BW) of the interconnection medium given as

$$BW = \frac{0.35}{T_{rd}} \qquad (21.59)$$

The concept of bandwidth itself comes from low-pass RC filter theory. We know that the classical low-pass RC filter passes low frequencies readily, but attenuates high frequencies. For this RC filter, the bandwidth is given by $f_2 = 1/2(\pi RC)$ and is the frequency at which the gain of the filter falls to 3 dB (70%) of the low-frequency value. It is like a cutoff frequency and is a qualitative measure of the transfer of energy through the filter. If we put a step input to the filter, the output from the filter is shown in Fig. 21.42.

From theory, we can show that $f_2 = 0.35/t_r$. Typically, good design requires that the interconnection loss be limited to ≈ 1 dB. For connector characterization we pass a very fast pulse (almost a step input) and observe the output pulse leaving the connector and its bandwidth expressed according to Eq. 21.59.

The low-pass filter concept is very useful because most interconnection elements such as connectors, short PCB traces, and IC package pins behave as low-pass filters. Since their lengths are relatively small compared to pulse risetime, they act as lumped parameters and discontinuities along

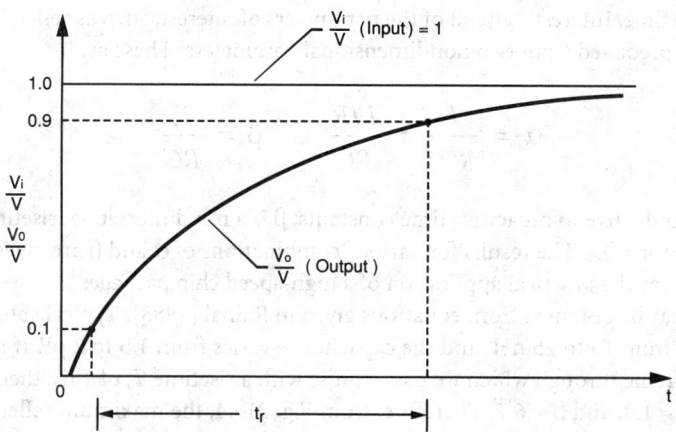

FIGURE 21.42 Low-pass RC circuit response to step input.

the transmission path. The degradation of the signal is mainly due to capacitance and inductance of the pins and leads. The appropriate model for such an interconnection discontinuity at a PCB IC package interface is shown in Fig. 21.43 [Rainal, 1988].

This generic model is valid for any interface including connector and wirebonds or short PCB traces which act as lumped elements. The interface is dominated by both parasitic capacitance and inductance of the signal and ground leads. As the signal enters the interface, part of the energy is absorbed, part of it is reflected due to the discontinuity, and the remaining is transmitted. This combination of events causes both an attenuation of the magnitude of the pulse, as well as degra-

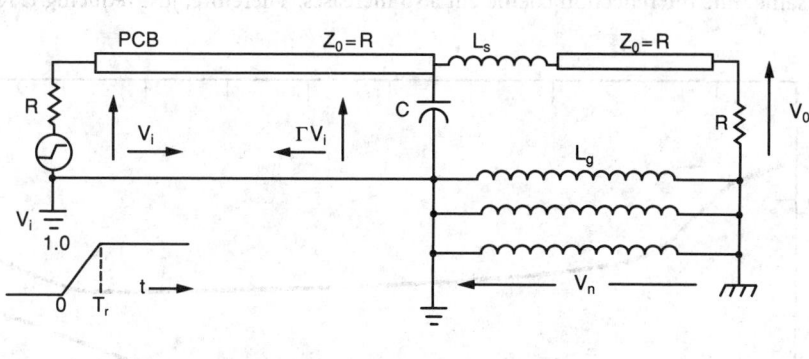

V_i = INPUT RAMP WAVEFORM
V_0 = WAVEFORM AT CHIP
ΓV_i = REFLECTED WAVEFORM
V_n = INDUCTIVE NOISE
R = TERMINATION RESISTORS
C = TOTAL PARASITIC CAPACITANCE AT INTERFACE
L_s = SELF INDUCTANCE OF SIGNAL LEAD AT INTERFACE
L_g = EFFECTIVE INDUCTANCE OF GROUND RETURNS
L = $L_s + L_g$ = TOTAL PARASITIC LOOP INDUCTANCE
Z_0 = CHARACTERISTIC IMPEDANCE OF PRINTED TRANSMISSION LINE
T_r = INPUT RISETIME

FIGURE 21.43 Electrical model of the transmission path from a printed wiring board to a matched high-speed chip. (*Source:* A. J. Rainal, "Performance limits of electrical interconnections to a high-speed chip," *IEEE Trans. CHMT*, vol. 11, no. 3, pp. 260–266, Sept. 1988. © 1988 IEEE.)

dation of the risetime. Interestingly, all of the parameters of interest, such as reflection, magnitude, and T_{rd}, may be predicted from two nondimensional parameters. These are

$$\alpha = \frac{L}{R^2 C} = \frac{L/R}{RC}, \qquad \beta = \frac{T_r}{RC} \tag{21.60}$$

α is the ratio of inductive to capacitive time constants, β is a nondimensional risetime, and R is the terminating resistor $= Z_0$. The results for various combinations of α and β are given in Figs. 21.44, 21.45, and 21.46 for the practical application of a high-speed chip package.

Other cases may be obtained from equations given in Rainal [1988]. Typical values of L for PCB connectors vary from 15 to 25 nH, and the capacitance varies from 1.5 to 2 pF. If the connector is placed on a 75-Ω line through which we pass a pulse with a risetime T_r of 1 ns, then for $L = 20$ nH, and $C = 2$ pF, $\alpha \approx 1.8$, and $\beta \approx 6.7$. Therefore, from Fig. 21.44, the maximum reflection coefficient from the connector is 0.1 (10%). Also, from Fig. 21.45, for a 1-V input pulse, the output from the connector will only be 0.75 V (75%). Thus, this connector configuration may not be suitable for high-speed designs as we would like the reflection coefficient to be < 5% and the output voltage at least 80%. In addition, there is also the degradation of the risetime. Note that this degradation of signal is just at the connector alone. In any interconnection medium, there may be many interfaces at which signal distortion also occurs. The goal is to minimize distortion at every interconnecting interface. Observe that the minimum reflection occurs for $\alpha = 1$, when $L/R = RC$, or $R = \sqrt{L/C} = Z_0$, when the inductive and capacitive time constants are equal. This may be considered as a "matched" discontinuity. For $\alpha > 1$, the interface may be considered inductive in nature, while it is capacitive for $\alpha < 1$. Thus, the reflection will be positive in one case and negative in the other. The $\alpha - \beta$ theory given above also implies some important considerations. From Fig. 21.45, we observe that β must be large to have a high value for the output voltage. β may be made large by reducing C, but when we reduce C, we increase α, and we observe that both the output voltage falls as α increases, and at the same time the reflection coefficient also increases. Therefore, just reducing C results in

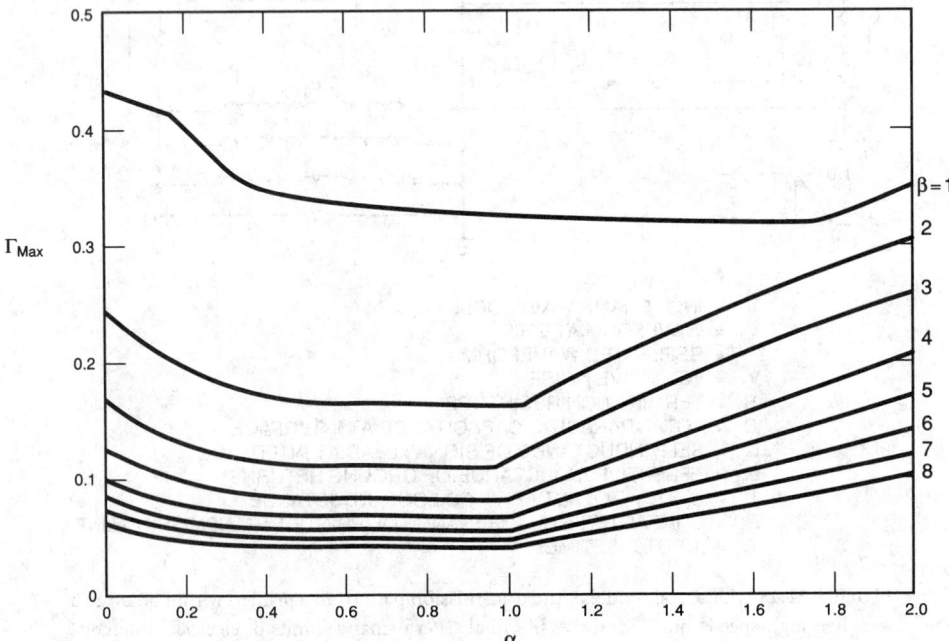

FIGURE 21.44 Maximum reflection coefficient at interface.

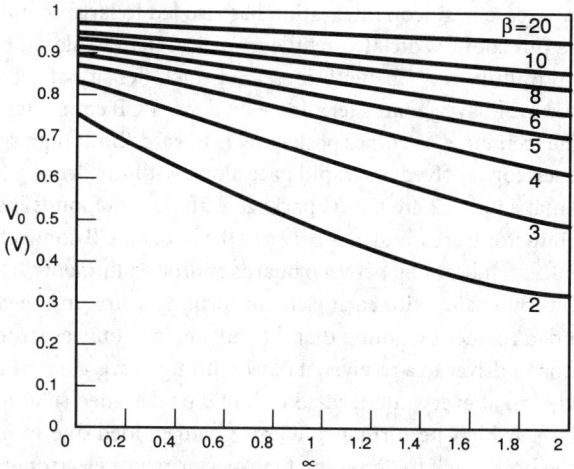

FIGURE 21.45 For a 1-V ramp input, V_0 denotes the output response at the interface for $t = \beta RC$. (*Source:* A. J. Rainal, "Performance limits of electrical interconnections to a high-speed chip," *IEEE Trans. CHMT,* vol. 11, no. 3, pp. 260–266, Sept. 1988. © 1988 IEEE.)

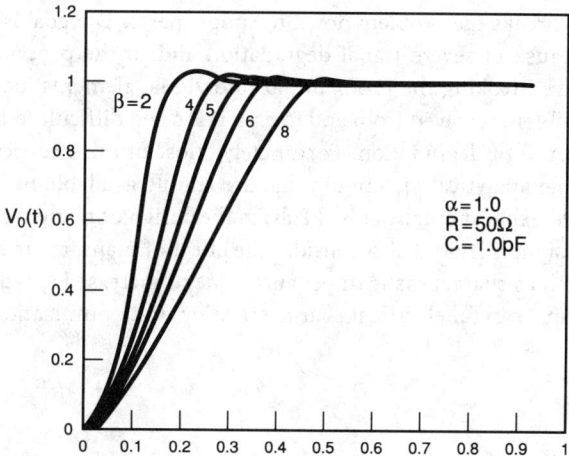

FIGURE 21.46 Output waveform at interface for a 1-V ramp input. (*Source:* A. J. Rainal, "Performance limits of electrical interconnections to a high-speed chip," *IEEE Trans. CHMT,* vol. 11, no. 3, pp. 260–266, Sept. 1988. © 1988 IEEE.)

signal degradation. It is not possible to maintain signal integrity by tuning either L or C alone. We need to specify joint bounds for L and C to minimize signal degradation. These are discussed in Rainal [1988]. This simple electrical model is a very practical tool for evaluating interconnection elements for signal integrity.

Conclusions

This article is an overview of electrical characterization of interconnections and packaging (I&P) of electronic systems. It is shown that I&P technology is mainly driven by advances in silicon IC technology, such as VLSI where there has been a hundred-fold increase in component density per decade. Such a rapid increase in silicon integration has also led to larger connections outside of the IC package, where the pin count is dictated by the empirical Rent's rule. Integration of such dense ICs in turn has led to multilayer PCBs with increased track density and layer count. The signal traces from these dense multilayer boards terminate on dense PCB connectors with 300 to 400 I/Os on an 8-in. height. The trend in electronics packaging is towards both high speed and high density. I&P technology has therefore evolved at a rapid pace along with the driving forces behind it.

The elements that make up I&P are the IC package with the wirebonds inside and the pins outside, PCB with the conductor traces, and vias between the layers, PCB connector, motherboard with pins and conductor traces, and cabling between boards and/or equipments. The elements of I&P are analogous to a mechanical chain, with each element forming a link in the chain. The mechanical analogy may be extended further by noting that the I&P chain is only as strong as the weakest link. As the signal travels from a driver to a receiver, it passes through several interfaces between elements. There is signal degradation at every interface. I&P should be designed so as to minimize the degradation at each interface. A high-performance IC may be degraded due to improper interconnections. Today, I&P is the bottleneck to designing high-performance electronic systems. Reliability of the system depends critically upon the integrity of the interconnection scheme.

The important electrical parameters characterizing I&P are delay, characteristic impedance, crosstalk, and risetime degradation. The article outlines the significance of each of these parameters and techniques to evaluate them. The electrical transmission path from driver to receiver is a very complicated one. It is extremely difficult to model the problem electrically from end to end. The best approach is to electrically characterize each interface (element) individually for signal integrity. This breaks the problem down to simple parts. The goal is to address those elements that are the cause of severe signal degradation and in the process optimize the entire interconnection chain. Breaking the problem into individual elements identifies the important links. If the entire problem is solved from end to end, it is often difficult to isolate those elements that are the major cause of degradation. Fortunately, most of the electrical parameters can be evaluated using simple analytical techniques that are readily available in literature. These have been used and verified experimentally at Bell Labs in the design of many systems. In the design of high-speed PCBs, designers should also consider the effect of manufacturing tolerances on electrical parameters, such as characteristic impedance. Due to increased speed and density requirements of many systems, careful electrical characterization and optimization of I&P is the key to system performance.

Defining Terms

Bandwidth (BW):   Related to risetime degradation and the frequency at which the energy transfer through I&P falls to −3 dB below the low-frequency value. It is only a qualitative measure to compare different options, and the concept comes from classical RC low-pass filter theory.

Characteristic impedance (Z_0):   The impedance (resistance) to energy transfer associated with wave propagation in a line that is much larger than the wavelength. It gives the relationship between the maximum voltage and maximum current in a line.

Crosstalk (X):   Noise that appears on a line due to interactions with stray electromagnetic fields that originate from adjacent pulsed lines.

Interconnections and packaging (I&P):   Elements of the electrical signal transmission path from the driver chip to the receiver chip. Various elements that make up I&P are chip wirebonds and package pins, circuit boards, connectors, motherboards, and cables.

Propagation delay (T_d):   Time required by a signal to travel from source to receiver.

Risetime degradation (T_{rd}):   A measure of the slowing down of the pulse as it passes through an I&P element. It includes both the increase in risetime of the pulse, as well as loss in amplitude.

References

C.S. Chang, "Electrical design of signal lines for multilayer printed circuit boards," *IBM J. Res. Develop.*, vol. 32, no. 5, pp. 647–657, Sept. 1988.

B.N. Das and K.V.S.V.R. Prasad, "A generalized formulation of electromagnetically coupled striplines," *IEEE Trans. Microwave Theory and Techniques*, vol. MTT-32, no. 11, pp. 1427–1433, Nov. 1984.

W.A. Davis, *Microwave Semiconductor Circuit Design*, New York: Van Nostrand Reinhold, 1990.

W.L. Everitt (Ed.), *Physical Design Of Electronic Systems, Volume 1, Design Technology*, Englewood Cliffs, N.J.: Prentice-Hall, 1970, pp. 362–363.

C.W. Hoover, W.L. Harrod, and M.I. Cohen, "The technology of interconnection," *AT&T Technical Journal*, vol. 66, issue 4, pp. 2–12, July/August 1987.

H. Howe, *Stripline Circuit Design*, Burlington, Mass.: Microwave Associates, 1974.

H.R. Kaupp, "Characteristics of microstrip transmission lines," *IEEE Trans. Electronic Computers*, vol. EC-16, no. 2, pp. 185–193, April 1967.

L.L. Moresco, "Electronic system packaging: The search for manufacturing the optimum in a sea of constraints," *IEEE Trans. CHMT*, vol. 13, no. 3, pp. 494–508, Sept. 1990.

A.J. Rainal, "Transmission properties of various styles of printed wiring boards," *Bell System Tech. Journal*, vol. 58, no. 5, pp. 995–1025, May-June 1979.

A.J. Rainal, "Computing inductive noise of chip packages," *AT&T Bell Laboratories Tech. Journal*, vol. 63, no. 1, pp. 177–195, 1984.

A.J. Rainal, "Performance limits of electrical interconnections to a high-speed chip," *IEEE Trans. CHMT*, vol. 11, no. 3, pp. 260–266, Sept. 1988.

K. Ritz, "Manufacturing tolerances for high-speed PCBs," *Circuit World*, vol. 15, no. 1, pp. 54–56, 1988.

Further Information

The issue of the *AT&T Technical Journal* in Hoover *et al.* [1987] focuses on the technology of electronic interconnections. It is an excellent source of information on important areas of interconnections and packaging. It begins with an overview and includes areas such as systems integration and architecture, materials and media, computer-aided design (CAD) tools, reliability evaluation, and standardized systems packaging techniques. The book *Microelectronics Packaging Handbook*, edited by Rao R. Tummala and Eugene J. Rymaszewski and published by Van Nostrand Reinhold, is a standard reference for packaging engineers, covering all major areas in detail. Most of the work on electrical characterization was done by researchers in the microwave area. These studies have been extensively published in the *IEEE Transactions on Microwave Theory and Techniques*, a publication of the IEEE Microwave Theory and Techniques Society. Parameters for electrical package design and evaluation have also been published in the *IBM Journal of Research and Development*. A source for the general areas of components, connector technologies, and manufacturing aspects related to packaging is the *IEEE Transactions on Components, Hybrids, and Manufacturing Technology*.

22

Transistors

22.1 Junction Field-Effect Transistors 530
JFET Biasing • Transfer Characteristics • JFET Output Resistance
• Source Follower • Frequency and Time-Domain Response •
Voltage-Variable Resistor

22.2 Bipolar Transistors 545
Biasing the Bipolar Transistor • Small-Signal Operation • A Small-
Signal Equivalent Circuit • Low-Frequency Performance • The
Emitter-Follower or Common-Collector (CC) Circuit • The Com-
mon-Emitter Bypass Capacitor C_E • High-Frequency Response
• Complete Response • Design Comments • Integrated Circuits
• The Degenerate Common-Emitter Stage • The Difference Ampli-
fier • The Current Mirror • The Difference Stage with Current
Mirror Biasing • The Current Mirror as a Load

22.3 The Metal-Oxide Semiconductor Field-Effect Transistor
(MOSFET) 567
Current-Voltage Characteristics • Important Device Parameters •
Limitations upon Miniaturization

Sidney Soclof
California State University,
Los Angeles

J. Watson
University of Wales, Swansea

John R. Brews
The University of Arizona

22.1 Junction Field-Effect Transistors

Sidney Soclof

A junction field-effect transistor, or JFET, is a type of transistor in which the current flow through
the device between the drain and source electrodes is controlled by the voltage applied to the gate
electrode. A simple physical model of the JFET is shown in Fig. 22.1. In this JFET an *n*-type conducting channel exists between drain and source. The gate is a p^+ region that surrounds the *n*-type
channel. The gate-to-channel *pn* junction is normally kept reverse-biased. As the reverse bias voltage between gate and channel increases, the depletion region width increases, as shown in Fig. 22.2.
The depletion region extends mostly into the *n*-type channel because of the heavy doping on the p^+
side. The depletion region is depleted of mobile charge carriers and thus cannot contribute to the
conduction of current between drain and source. Thus as the gate voltage increases, the cross-
sectional areas of the *n*-type channel available for current flow decreases. This reduces the current
flow between drain and source. As the gate voltage increases, the channel gets further constricted,
and the current flow gets smaller. Finally when the depletion regions meet in the middle of the
channel, as shown in Fig. 22.3, the channel is pinched off in its entirety between source and drain.
At this point the current flow between drain and source is reduced to essentially zero. This voltage
is called the **pinch-off voltage**, V_P. The pinch-off voltage is also represented by V_{GS} (off) as being
the gate-to-source voltage that turns the drain-to-source current I_{DS} off. We have been considering
here an *n*-channel JFET. The complementary device is the *p*-channel JFET that has an n^+ gate

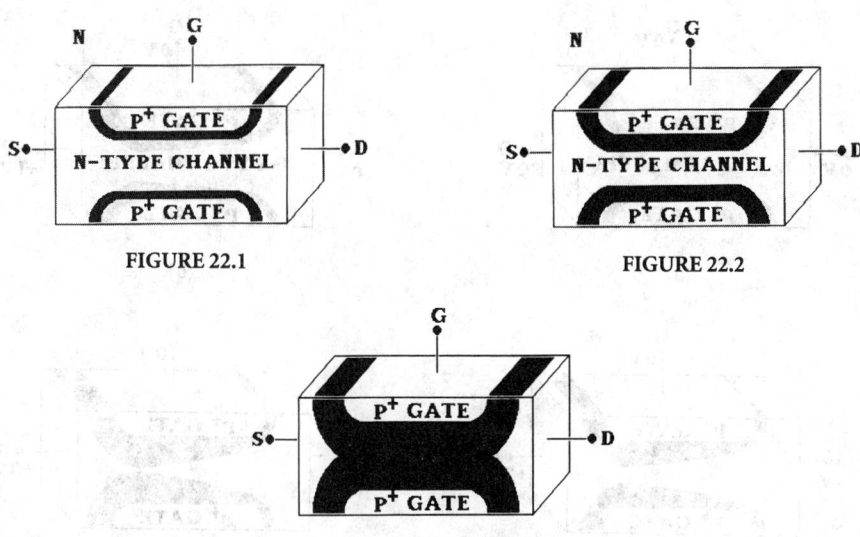

FIGURE 22.1   FIGURE 22.2

FIGURE 22.3

region surrounding a *p*-type channel. The operation of a *p*-channel JFET is the same as for an *n*-channel device, except the algebraic signs of all dc voltages and currents are reversed.

We have been considering the case for V_{DS} small compared to the pinch-off voltage such that the channel is essentially uniform from drain to source, as shown in Fig. 22.4(a). Now let's see what happens as V_{DS} increases. As an example let's assume an *n*-channel JFET with a pinch-off voltage of $V_P = -4$ V. We will see what happens for the case of $V_{GS} = 0$ as V_{DS} increases. In Fig. 22.4(a) the situation is shown for the case of $V_{DS} = 0$ in which the JFET is fully "on" and there is a uniform channel from source to drain. This is at point *A* on the I_{DS} vs. V_{DS} curve of Fig. 22.5. The drain-to-source conductance is at its maximum value of g_{ds} (on), and the drain-to-source resistance is correspondingly at its minimum value of r_{ds} (on). Now let's consider the case of $V_{DS} = +1$ V, as shown in Fig. 22.4(b). The gate-to-channel bias voltage at the source end is still $V_{GS} = 0$. The gate-to-channel bias voltage at the drain end is $V_{GD} = V_{GS} - V_{DS} = -1$ V, so the depletion region will be wider at the drain end of the channel than at the source end. The channel will thus be narrower at the drain end than at the source end, and this will result in a decrease in the channel conductance g_{ds} and, correspondingly, an increase in the channel resistance r_{ds}. So the slope of the I_{DS} vs. V_{DS} curve that corresponds to the channel conductance will be smaller at $V_{DS} = 1$ V than it was at $V_{DS} = 0$, as shown at point *B* on the I_{DS} vs. V_{DS} curve of Fig. 22.5.

In Fig. 22.4(c) the situation for $V_{DS} = +2$ V is shown. The gate-to-channel bias voltage at the source end is still $V_{GS} = 0$, but the gate-to-channel bias voltage at the drain end is now $V_{GD} = V_{GS} - V_{DS} = -2$ V, so the depletion region will now be substantially wider at the drain end of the channel than at the source end. This leads to a further constriction of the channel at the drain end, and this will again result in a decrease in the channel conductance g_{ds} and, correspondingly, an increase in the channel resistance r_{ds}. So the slope of the I_{DS} vs. V_{DS} curve will be smaller at $V_{DS} = 2$ V than it was at $V_{DS} = 1$ V, as shown at point *C* on the I_{DS} vs. V_{DS} curve of Fig. 22.5.

In Fig. 22.4(d) the situation for $V_{DS} = +3$ V is shown, and this corresponds to point *D* on the I_{DS} vs. V_{DS} curve of Fig. 22.5.

When $V_{DS} = +4$ V, the gate-to-channel bias voltage will be $V_{GD} = V_{GS} - V_{DS} = 0 - 4$ V $= -4$ V $= V_P$. As a result the channel is now pinched off at the drain end but is still wide open at the source end since $V_{GS} = 0$, as shown in Fig. 22.4(e). It is very important to note that the channel is pinched off just for a very short distance at the drain end so that the drain-to-source current I_{DS} can still continue to flow. This is not at all the same situation as for the case of $V_{GS} = V_P$, where the channel is pinched off in its entirety, all the way from source to drain. When this happens, it is like having a

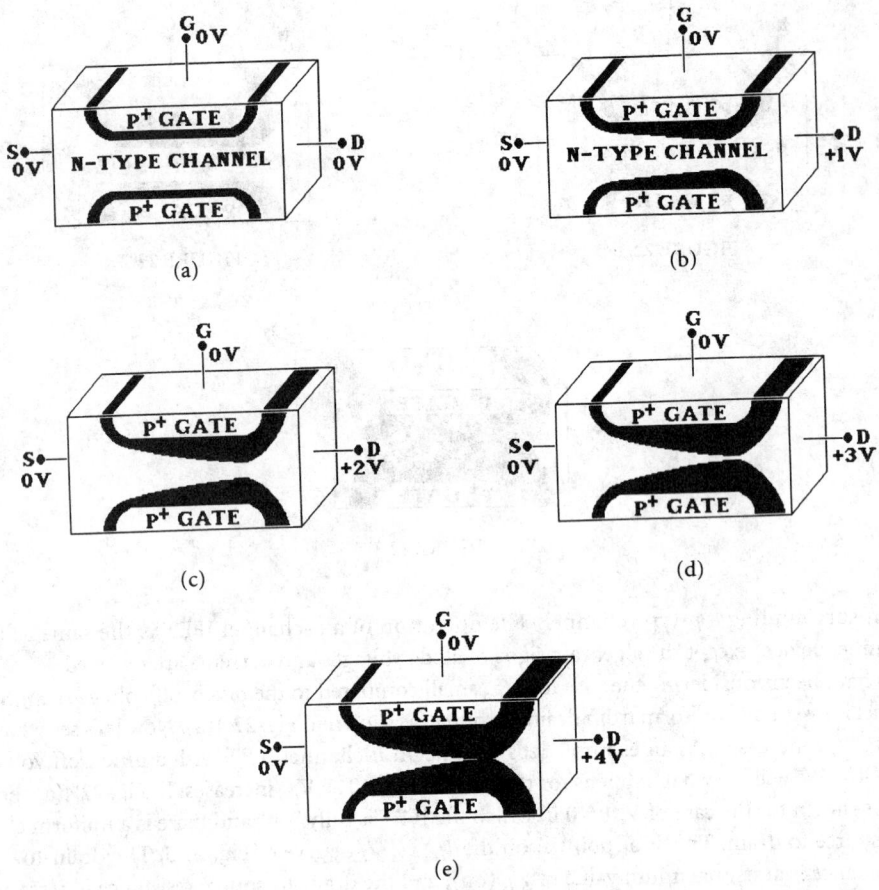

FIGURE 22.4

big block of insulator the entire distance between source and drain, and I_{DS} is reduced to essentially zero. The situation for $V_{DS} = +4$ V $= -V_P$ is shown at point E on the I_{DS} vs. V_{DS} curve of Fig. 22.5.

For $V_{DS} > +4$ V, the current essentially saturates and doesn't increase much with further increases in V_{DS}. As V_{DS} increases above $+4$ V, the pinched-off region at the drain end of the channel gets wider, which increases r_{ds}. This increase in r_{ds} essentially counterbalances the increase in V_{DS} such that I_{DS} does not increase much. This region of the I_{DS} vs. V_{DS} curve in which the channel is pinched off at the drain end is called the **active region** and is also known as the *saturated region*. It is called the active region because when the JFET is to be used as an amplifier, it should be biased and operated in this region. The saturated value of drain current up in the active region for the case of $V_{GS} = 0$ is called **the drain saturation current**, I_{DSS} (the third subscript S refers to I_{DS} under the condition of the gate *shorted* to the source). Since there is not really a true saturation of current in the active region, I_{DSS} is usually specified at some value of V_{DS}. For most JFETs, the values of I_{DSS} fall in the range of 1 to 30 mA.

The region below the active region where $V_{DS} < +4$ V $= -V_P$ has several names. It is called the **nonsaturated region**, the **triode region**, and the **ohmic region**. The term *triode region* apparently originates from the similarity of the shape of the curves to that of the vacuum tube triode. The term *ohmic region* is due to the variation of

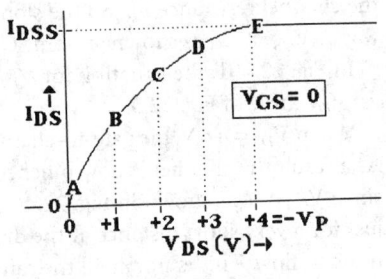

FIGURE 22.5

I_{DS} with V_{DS} as in Ohm's law, although this variation is nonlinear except for the region of V_{DS} that is small compared to the pinch-off voltage where I_{DS} will have an approximately linear variation with V_{DS}.

The upper limit of the active region is marked by the onset of the breakdown of the gate-to-channel *pn* junction. This will occur at the drain end at a voltage designated as BV_{DG}, or BV_{DS}, since $V_{GS} = 0$. This breakdown voltage is generally in the 30- to 150-V range for most JFETs.

So far we have looked at the I_{DS} vs. V_{DS} curve only for the case of $V_{GS} = 0$. In Fig. 22.6 a family of curves of I_{DS} vs. V_{DS} for various constant values of V_{GS} is presented. This is called the *drain characteristics*, also known as the *output characteristics*, since the output side of the JFET is usually the drain side.

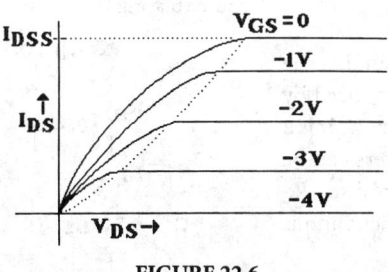

FIGURE 22.6

In the active region where I_{DS} is relatively independent of V_{DS}, a simple approximate equation relating I_{DS} to V_{GS} is the square-law *transfer equation* as given by $I_{DS} = I_{DSS}[1 - (V_{GS}/V_P)]^2$. When $V_{GS} = 0$, $I_{DS} = I_{DSS}$ as expected, and as $V_{GS} \rightarrow V_P$, $I_{DS} \rightarrow 0$. The lower boundary of the active region is controlled by the condition that the channel be pinched off at the drain end. To meet this condition the basic requirement is that the gate-to-channel bias voltage at the drain end of the channel, V_{GD}, be greater than the pinch-off voltage V_P. For the example under consideration with $V_P = -4$ V, this means that $V_{GD} = V_{GS} - V_{DS}$ must be more negative than –4 V. Therefore, $V_{DS} - V_{GS} \geq +4$ V. Thus, for $V_{GS} = 0$, the active region will begin at $V_{DS} = +4$ V. When $V_{GS} = -1$ V, the active region will begin at $V_{DS} = +3$ V, for now $V_{GD} = -4$ V. When $V_{GS} = -2$ V, the active region begins at $V_{DS} = +2$ V, and when $V_{GS} = -3$ V, the active region begins at $V_{DS} = +1$ V. The dotted line in Fig. 22.6 marks the boundary between the nonsaturated and active regions.

The upper boundary of the active region is marked by the onset of the avalanche breakdown of the gate-to-channel *pn* junction. When $V_{GS} = 0$, this occurs at $V_{DS} = BV_{DS} = BV_{DG}$. Since $V_{DG} = V_{DS} - V_{GS}$ and breakdown occurs when $V_{DG} = BV_{DG}$, as V_{GS} increases the breakdown voltage decreases, as given by $BV_{DG} = BV_{DS} - V_{GS}$. Thus $BV_{DS} = BV_{DG} + V_{GS}$. For example, if the gate-to-channel breakdown voltage is 50 V, the V_{DS} breakdown voltage will start off at 50 V when $V_{GS} = 0$ but decrease to 46 V when $V_{GS} = -4$ V.

In the nonsaturated region I_{DS} is a function of both V_{GS} and V_{DS}, and in the lower portion of the nonsaturated region where V_{DS} is small compared to V_P, I_{DS} becomes an approximately linear function of V_{DS}. This linear portion of the nonsaturated is called the *voltage-variable resistance* (VVR) region, for in this region the JFET acts like a linear resistance element between source and drain. The resistance is variable in that it is controlled by the gate voltage. This region and VVR application will be discussed in a later section. The JFET can also be operated in this region as a switch, and this will also be discussed in a later section.

JFET Biasing

Voltage Source Biasing

Now we will consider the biasing of JFETs for operation in the active region. The simplest biasing method is shown in Fig. 22.7, in which a voltage source V_{GG} is used to provide the quiescent gate-to-source bias voltage V_{GSQ}. In the active region the transfer equation for the JFET has been given as $I_{DS} = I_{DSS}[1 - (V_{GS}/V_P)]^2$, so for a quiescent drain current of I_{DSQ} the corresponding gate voltage will be given by $V_{GSQ} = V_P(1 - \sqrt{I_{DSQ}/I_{DSS}})$. For a Q point in the middle of the active region, we have that $I_{DSQ} = I_{DSS}/2$, so $V_{GSQ} = V_P(1 - \sqrt{1/2}) = 0.293V_P$.

The voltage source method of biasing has several major drawbacks. Since V_P will have the opposite polarity of the drain supply voltage V_{DD}, the gate bias voltage will require a second power supply. For the case of an *n*-channel JFET, V_{DD} will come from a positive supply voltage and V_{GG} must come from a separate negative power supply voltage or battery. A second, and perhaps more serious, problem is the "open-loop" nature of this biasing method. The JFET parameters of I_{DDS} and V_P will exhibit very substantial unit-to-unit variations, often by as much as a 2:1 factor. There is also a significant temperature dependence of I_{DDS} and V_P. These variations will lead to major shifts in the position of the Q point and the resulting distortion of the signal. A much better biasing method is shown in Fig. 22.8.

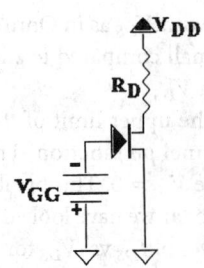

FIGURE 22.7 Voltage source biasing.

Self-Biasing

The biasing circuit of Fig. 22.8 is called a *self-biasing* circuit in that the gate-to-source voltage is derived from the voltage drop produced by the flow of drain current through the source biasing resistor R_S. It is a closed-loop system in that variations in the JFET parameters can be partially compensated for by the biasing circuit. The gate resistor R_G is used to provide a dc return path for the gate leakage current and is generally up in the megohm range.

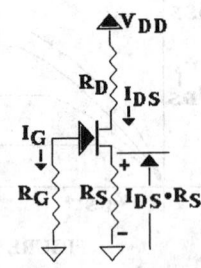

FIGURE 22.8 Self-biasing.

The voltage drop across R_S is given by $V_S = I_{DS} \cdot R_S$. The voltage drop across the gate resistor R_G is $V_G = I_G \cdot R_G$. Since I_G is usually in the low nanoampere or even picoampere range, as long as R_G is not extremely large the voltage drop across R_G can be neglected, so $V_G \cong 0$. Thus, we have that $V_{GS} = V_G - V_S \cong -V_S = -I_{DS} \cdot R_S$. For example, if $I_{DSS} = 10$ mA and $V_P = -4$ V, and for a Q point in the middle of the active region with $I_{DSQ} = I_{DSS}/2 = 5$ mA, we have that $V_{GSQ} = 0.293 V_P = -1.17$ V. Therefore the required value for the source biasing resistor is given by $R_S = -V_{GS}/I_{DSQ} = 1.17$ V/5 mA = 234 Ω. This produces a more stable quiescent point than voltage source biasing, and no separate negative power supply is required.

The closed-loop nature of this biasing circuit can be seen by noting that if changes in the JFET parameters were to cause I_{DS} to increase, the voltage drop across R_S would also increase. This will produce an increase in V_{GS} (in the negative direction for an *n*-channel JFET), which will act to reduce the increase in I_{DS}. Thus the net increase in I_{DS} will be less due to the feedback voltage drop produced by the flow of I_{DS} through R_S. The same basic action would, of course, occur for changes in the JFET parameters that would cause I_{DS} to decrease.

Bias Stability

Now let's examine the stability of the Q point. We will start again with the basic transfer equation as given by $I_{DS} = I_{DSS}[1 - (V_{GS}/V_P)]^2$. From this equation the change in the drain current, ΔI_{DS}, due to changes in I_{DSS}, V_{GS}, and V_P can be written as

$$\Delta I_{DS} = g_m \Delta V_{GS} - g_m \frac{V_{GS}}{V_P} \Delta V_P + \frac{I_{DS}}{I_{DSS}} \Delta I_{DSS}$$

Since $V_{GS} = -I_{DS} \cdot R_S$, $\Delta V_{GS} = -R_S \cdot \Delta I_{DS}$, we obtain that

$$\Delta I_{DS} = -g_m R_S \Delta I_{DS} - g_m \frac{V_{GS}}{V_P} \Delta V_P + \frac{I_{DS}}{I_{DSS}} \Delta I_{DSS}$$

Collecting terms in ΔI_{DS} on the left side gives

$$\Delta I_{DS}(1 + g_m R_S) = -g_m \frac{V_{GS}}{V_P} \Delta V_P + \frac{I_{DS}}{I_{DSS}} \Delta I_{DSS}$$

Now solving this for ΔI_{DS} yields

$$\Delta I_{DS} = \frac{-g_m(V_{GS}/V_P)\Delta V_P + \dfrac{I_{DS}}{I_{DSS}}\Delta I_{DSS}}{1 + g_m R_S}$$

From this we see that the shift in the quiescent drain current, ΔI_{DS}, is reduced by the presence of R_S by a factor of $1 + g_m R_S$.

If $I_{DS} = I_{DSS}/2$, then

$$g_m = \frac{2\sqrt{I_{DS} \cdot I_{DSS}}}{-V_P} = \frac{2\sqrt{I_{DS} \cdot 2I_{DS}}}{-V_P} = \frac{2\sqrt{2}\, I_{DS}}{-V_P}$$

Since $V_{GS} = 0.293 V_P$, the source biasing resistor will be $R_S = -V_{GS}/I_{DS} = -0.293 V_P/I_{DS}$. Thus

$$g_m R_S = \frac{2\sqrt{2}\, I_{DS}}{-V_P} \times \frac{-0.293 V_P}{I_{DS}} = 2\sqrt{2} \times 0.293 = 0.83$$

so $1 + g_m R_S = 1.83$. Thus the sensitivity of I_{DS} due to changes in V_P and I_{DSS} is reduced by a factor of 1.83.

The equation for ΔI_{DS} can now be written in the following form for the fractional change in I_{DS}:

$$\frac{\Delta I_{DS}}{I_{DS}} = \frac{-0.83(\Delta V_P/V_P) + 1.41(\Delta I_{DSS}/I_{DSS})}{1.83}$$

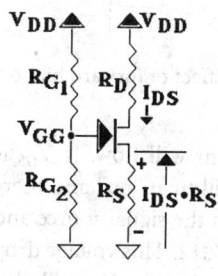

FIGURE 22.9

so $\Delta I_{DS}/I_{DS} = -0.45\,(\Delta V_P/V_P) + 0.77\,(\Delta I_{DSS}/I_{DSS})$, and thus a 10% change in V_P will result in approximately a 4.5% change in I_{DS}, and a 10% change in I_{DSS} will result in an 8% change in I_{DS}. Thus, although the situation is improved with the self-biasing circuit using R_S, there will still be a substantial variation in the quiescent current with changes in the JFET parameters.

A further improvement in bias stability can be obtained by the use of the biasing methods of Figs. 22.9 and 22.10. In Fig. 22.9 a gate bias voltage V_{GG} is obtained from the V_{DD} supply voltage by means of the R_{G1}–R_{G2} voltage divider. The gate-to-source voltage is now $V_{GS} = V_G - V_S = V_{GG} - I_{DS}R_S$. So now for R_S we have $R_S = (V_{GG} - V_{GS})/I_{DS}$. Since V_{GS} is of opposite polarity to V_{GG}, this will result in a larger value for R_S than before. This in turn will result in a larger value for the $g_m R_S$ product and hence improved bias stability. If we continue with the preceding examples and now let $V_{GG} = V_{DD}/2 = +10$ V, we have that $R_S = (V_{GG} - V_{GS})/I_{DS} = [+10\text{V} - (-1.17\text{V})]/5$ mA = 2.234 kΩ, as compared to $R_S = 234\ \Omega$ that was obtained before. For g_m we have $g_m = 2\sqrt{I_{DS} \cdot I_{DSS}}/(-V_P) = 3.54$ mS, so $g_m R_S = 3.54$ mS $\cdot$ 2.234 k$\Omega = 7.90$. Since $1 + g_m R_S = 8.90$, we now have an improvement by a factor of 8.9 over the open-loop voltage source biasing and by a factor of 4.9 over the self-biasing method without the V_{GG} biasing of the gate.

Another biasing method that can lead to similar results is the method shown in Fig. 22.10. In this method the bottom end of the source biasing resistor goes to a negative supply voltage V_{SS} instead of to ground. The gate-to-source bias voltage is now given by $V_{GS} = V_G - V_S = 0 - (I_{DS} \cdot R_S + V_{SS})$ so that for R_S we now have $R_S = (-V_{GS} - V_{SS})/I_{DS}$. If $V_{SS} = -10$ V, and as before $I_{DS} = 5$ mA and $V_{GS} = -1.17$ V, we have $R_S = 11.7$ V/5 mA = 2.34 kΩ, and thus $g_m R_S = 7.9$ as in the preceding exam-

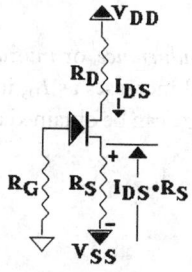

FIGURE 22.10

ple. So this method does indeed lead to results similar to that for the R_S and V_{GG} combination biasing. With either of these two methods the change in I_{DS} due to a 10% change in V_P will be only 0.9%, and the change in I_{DS} due to a 10% change in I_{DSS} will be only 1.6%.

The biasing circuits under consideration here can be applied directly to the common-source (CS) amplifier configuration, and can also be used for the common-drain (CD), or source-follower, and common-gate (CG) JFET configurations.

Transfer Characteristics

Transfer Equation

Now we will consider the *transfer characteristics* of the JFET, which is a graph of the output current I_{DS} vs. the input voltage V_{GS} in the active region. In Fig. 22.11 a transfer characteristic curve for a JFET with $V_P = -4$ V and $I_{DSS} = +10$ mA is given. This is approximately a square-law relationship as given by $I_{DS} = I_{DSS}[1 - (V_{GS}/V_P)]^2$. This equation is not valid for V_{GS} beyond V_P (i.e., $V_{GS} < V_P$), for in this region the channel is pinched off and $I_{DS} \cong 0$.

At $V_{GS} = 0$, $I_{DS} = I_{DSS}$. This equation and the corresponding transfer curve can actually be extended up to the point where $V_{GS} \cong +0.5$ V. In the region where $0 < V_{GS} < +0.5$ V, the gate-to-channel pn junction is *forward-biased* and the depletion region width is reduced below the width under zero bias conditions. This reduction in the depletion region width leads to a corresponding expansion of the conducting channel and thus an increase in I_{DS} above I_{DSS}. As long as the gate-to-channel forward bias voltage is less than about

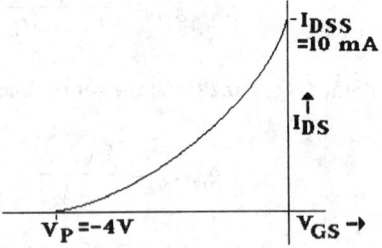

FIGURE 22.11   Transfer characteristic.

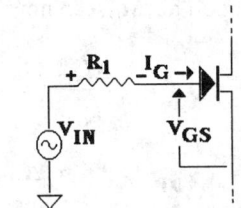

FIGURE 22.12   Effect of forward bias on V_{GS}.

0.5 V, the pn junction will be essentially "off" and very little gate current will flow. If V_{GS} is increased much above +0.5 V, however, the gate-to-channel pn junction will turn "on" and there will be a substantial flow of gate voltage I_G. This gate current will load down the signal source and produce a voltage drop across the signal source resistance, as shown in Fig. 22.12. This voltage drop can cause V_{GS} to be much smaller than the signal source voltage V_{in}. As V_{in} increases, V_{GS} will ultimately level off at a forward bias voltage of about +0.7 V, and the signal source will lose control over V_{GS}, and hence over I_{DS}. This can result in severe distortion of the input signal in the form of clipping, and thus this situation should be avoided. Thus, although it is possible to increase I_{DS} above I_{DSS} by allowing the gate-to-channel junction to become forward-biased by a small amount (≤ 0.5 V), the possible benefits are generally far outweighed by the risk of signal distortion. Therefore, JFETs are almost always operated with the gate-to-channel pn junction reverse-biased.

Transfer Conductance

The slope of the transfer curve, dI_{DS}/dV_{GS}, is the *dynamic forward transfer conductance*, or *mutual transfer conductance*, g_m. We see that g_m starts off at zero when $V_{GS} = V_P$ and increases as I_{DS} increases, reaching a maximum when $I_{DS} = I_{DSS}$. Since $I_{DS} = I_{DSS}[1 - (V_{GS}/V_P)]^2$, g_m can be obtained as

$$g_m = \frac{dI_{DS}}{dV_{GS}} = 2I_{DSS}\frac{\left(1 - \dfrac{V_{GS}}{V_P}\right)}{-V_P}$$

Since

$$1 - \left(\frac{V_{GS}}{V_P}\right) = \sqrt{\frac{I_{DS}}{I_{DSS}}}$$

we have that

$$g_m = 2I_{DSS}\frac{\sqrt{I_{DS}/I_{DSS}}}{-V_P} = 2\frac{\sqrt{I_{DS} \cdot I_{DSS}}}{-V_P}$$

The maximum value of g_m is obtained when $V_{GS} = 0$ ($I_{DS} = I_{DSS}$) and is given by $g_m(V_{GS} = 0) = g_{m0} = 2I_{DS}/(-V_P)$.

Small-Signal AC Voltage Gain

Let's consider the CS amplifier circuit of Fig. 22.13. The input ac signal is applied between gate and source, and the output ac voltage is taken between drain and source. Thus the source electrode of

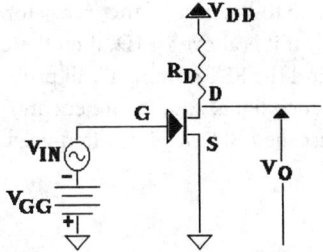

FIGURE 22.13 Common-source amplifier.

this triode device is common to input and output, hence the designation of this JFET configuration as a CS amplifier.

A good choice of the dc operating point or quiescent point (Q point) for an amplifier is in the middle of the active region at $I_{DS} = I_{DSS}/2$. This allows for the maximum symmetrical drain current swing, from the quiescent level of $I_{DSQ} = I_{DSS}/2$, down to a minimum of $I_{DS} \cong 0$, and up to a maximum of $I_{DS} = I_{DSS}$. This choice for the Q point is also a good one from the standpoint of allowing for an adequate safety margin for the location of the actual Q point due to the inevitable variations in device and component characteristics and values. This safety

margin should keep the Q point well away from the extreme limits of the active region, and thus ensure operation of the JFET in the active region under most conditions. If $I_{DSS} = +10$ mA, then a good choice for the Q point would thus be around +5.0 mA. If $V_P = -4$ V, then

$$g_m = \frac{2\sqrt{I_{DS} \cdot I_{DSS}}}{-V_P} = \frac{2\sqrt{5 \text{ mA} \cdot 10 \text{ mA}}}{4 \text{ V}} = 3.54 \text{ mA/V} = 3.54 \text{ mS}$$

If a small ac signal voltage v_{GS} is superimposed on the dc gate bias voltage V_{GS}, only a small segment of the transfer characteristic adjacent to the Q point will be traversed, as shown in Fig. 22.14. This small segment will be close to a straight line, and as a result the ac drain current i_{ds} will have a

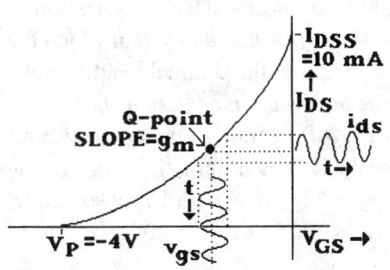

FIGURE 22.14 Transfer characteristic.

waveform close to that of the ac voltage applied to the gate. The ratio of i_{ds} to v_{GS} will be the slope of the transfer curve as given by $i_{ds}/v_{GS} \cong dI_{DS}/dV_{GS} = g_m$. Thus $i_{ds} \cong g_m v_{GS}$. If the net load driven by the drain of the JFET is the drain load resistor R_D, as shown in Fig. 22.13, then the ac drain current i_{ds} will produce an ac drain voltage of $v_{ds} = -i_{ds} \cdot R_D$. Since $i_{ds} = g_m v_{GS}$, this becomes $v_{ds} = -g_m v_{GS} \cdot R_D$. The ac small-signal voltage gain from gate to drain thus becomes $A_V = v_O/v_{in} = v_{ds}/v_{GS} = -g_m \cdot R_D$. The negative sign indicates signal inversion as is the case for a CS amplifier.

If the dc drain supply voltage is $V_{DD} = +20$ V, a quiescent drain-to-source voltage of $V_{DSQ} = V_{DD}/2 = +10$ V will result in the JFET being biased in the middle of the active region. Since $I_{DSQ} = +5$ mA in the example under consideration, the voltage drop across the drain load resistor R_D is 10 V. Thus $R_D = 10$ V/5 mA = 2 kΩ. The ac small-signal voltage gain A_V thus becomes $A_V = -g_m \cdot R_D = -3.54$ mS $\cdot$ 2 k$\Omega = -7.07$. Note that the voltage gain is relatively modest as compared to the much larger voltage gains that can be obtained in a bipolar-junction transistor (BJT) common-emitter amplifier. This is due to the lower transfer conductance of both JFETs and MOSFETs (metal-oxide semiconductor field-effect transistors) as compared to BJTs. For a BJT the transfer conductance is given by $g_m = I_C/V_T$, where I_C is the quiescent collector current and $V_T = kT/q \cong 25$ mV is the thermal voltage. At $I_C = 5$ mA, $g_m = 5$ mA/25 mV = 200 mS, as compared to only 3.5 mS for the JFET in this example. With a net load of 2 kΩ, the BJT voltage gain will be –400 as compared to the JFET voltage gain of only 7.1. Thus FETs do have the disadvantage of a much lower transfer conductance, and therefore voltage gain, than BJTs operating under similar quiescent current levels, but they do have the major advantage of a much higher input impedance and a much lower input current. In the case of a JFET the input signal is applied to the *reverse-biased* gate-to-channel *pn* junction and thus sees a very high impedance. In the case of a common-emitter BJT amplifier, the input signal is applied to the *forward-biased* base-emitter junction, and the input impedance is given approximately by $r_{in} = r_{BE} \cong 1.5 \cdot \beta \cdot V_T/I_C$. If $I_C = 5$ mA and $\beta = 200$, for example, then $r_{in} \cong 1500$ Ω. This moderate input resistance value of 1.5 kΩ is certainly no problem if the signal source resistance is less than around 100 Ω. However, if the source resistance is above 1 kΩ, then there will be a substantial signal loss in the coupling of the signal from the signal source to the base of the transistor. If the source resistance is in the range of above 100 kΩ, and certainly if it is above 1 MΩ, then there will be severe signal attenuation due to the BJT input impedance, and the FET amplifier will probably offer a greater overall voltage gain. Indeed, when high-impedance signal sources are encountered, a multistage amplifier with a FET input stage followed by cascaded BJT stages is often used.

JFET Output Resistance

Dynamic Drain-to-Source Conductance

For the JFET in the active region the drain current I_{DS} is a strong function of the gate-to-source voltage V_{GS} but is relatively independent of the drain-to-source voltage V_{DS}. The transfer equation has previously been stated as $I_{DS} = I_{DSS} [1 - (V_{GS}/V_P)]^2$.

The drain current will, however, increase slowly with increasing V_{DS}. To take this dependence of I_{DS} on V_{DS} into account, the transfer equation can be modified to give

$$I_{DS} = I_{DSS}\left(1 - \frac{V_{GS}}{V_P}\right)^2\left(1 + \frac{V_{DS}}{V_A}\right)$$

where V_A is a constant called the *Early voltage* and is a parameter of the transistor with units of volts. The early voltage V_A is generally in the range of 30 to 300 V for most JFETs. The variation of the drain current with drain voltage is the result of the *channel length modulation effect* in which the channel length decreases as the drain voltage increases. This decrease in the channel length results in an increase in the drain current. In BJTs a similar effect is the *base width modulation effect*.

The *dynamic drain-to-source conductance* is defined as $g_{ds} = dI_{DS}/dV_{DS}$ and can be obtained from the modified transfer equation $I_{DS} = I_{DSS}[1 - (V_{GS}/V_P)]^2 [1 + V_{DS}/V_A]$ as simply $g_{ds} = I_{DS}/V_A$. The reciprocal of g_{ds} is *dynamic drain-to-source resistance* r_{ds}, so $r_{ds} = 1/g_{ds} = V_A/I_{DS}$. If, for example, $V_A = 100$ V, we have that $r_{ds} = 100$ V/I_{DS}. At $I_{DS} = 1$ mA, $r_{ds} = 100$ V/1 mA = 100 kΩ, and at $I_{DS} = 10$ mA, $r_{ds} = 10$ kΩ.

Equivalent Circuit Model of CS Amplifier Stage

A small-signal equivalent circuit model of a CS FET amplifier stage is shown in Fig. 22.15. The ac small-signal voltage gain is given by $A_V = -g_m \cdot R_{net}$, where $R_{net} = [r_{ds}\|R_D\|R_L]$ is the net load driven by the drain for the FET and includes the dynamic drain-to-source resistance r_{ds}. Since r_{ds} is generally much larger than $[R_D\|R_L]$, it will usually be the case that $R_{net} \cong [R_D\|R_L]$, and r_{ds} can be

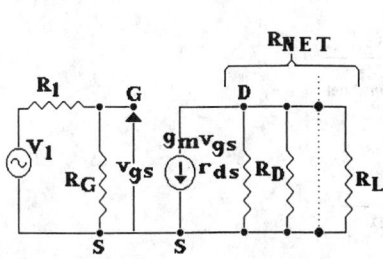

FIGURE 22.15 Effect of r_{ds} on R_{net}.

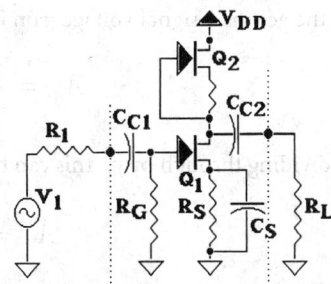

FIGURE 22.16 Active load circuit.

neglected. There are, however, some cases in which r_{ds} must be taken into account. This is especially true for the case in which an active load is used, as shown in Fig. 22.16. For this case $R_{net} = [r_{ds1} \| r_{ds2} \| R_L]$, and r_{ds} can be a limiting factor in determining the voltage gain.

Consider an example for the active load circuit of Fig. 22.16 for the case of identical JFETs with the same quiescent current. Assume that $R_L \gg r_{ds}$ so that $R_{net} \cong [r_{ds1} \| r_{ds2}] = V_A/(2I_{DSQ})$. Let $I_{DSQ} = I_{DSS}/2$, so $g_m = -2\sqrt{I_{DSS} \cdot I_{DSQ}}/(-V_P) = 2\sqrt{2}\, I_{DSQ}/(-V_P)$. The voltage gain is

$$A_V = -g_m \cdot R_{net} = \frac{2\sqrt{2}I_{DSQ}}{V_P} \times \frac{V_A}{2I_{DSQ}} = \sqrt{2}\,\frac{V_A}{V_P}$$

If $V_A = 100$ V and $V_P = -2$ V, we obtain $A_V = -70$, so we see that with active loads relatively large voltage gains can be obtained with FETs.

Another circuit in which the dynamic drain-to-source resistance r_{ds} is important is the constant-current source or current regulator diode. In this case the current regulation is directly proportional to the dynamic drain-to-source resistance.

Source Follower

Source-Follower Voltage Gain

We will now consider the CD JFET configuration, which is also known as the source follower. A

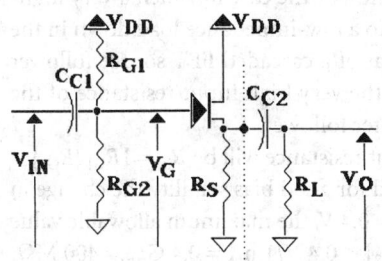

FIGURE 22.17 Source follower.

basic CD circuit is shown in Fig. 22.17. The input signal is supplied to the gate of the JFET. The output is taken from the source of the JFET, and the drain is connected directly to the V_{DD} supply voltage, which is ac ground.

For the JFET in the active region we have that $i_{ds} = g_m v_{GS}$. For this CD circuit we also have that $v_{GS} = v_G - v_S$ and $v_S = i_{ds}R_{net}$, where $R_{net} = [R_S \| R_L]$ is the net load resistance driven by the transistor. Since $v_{GS} = i_{ds}/g_m$, we have that $i_{ds}/g_m = v_G - i_{ds}R_{net}$. Collecting terms in i_{ds} on the left side yields $i_{ds}[(1/g_m) + R_{net}] = v_G$, so

$$i_{ds} = \frac{v_G}{(1/g_m) + R_{net}} = \frac{g_m v_G}{1 + g_m R_{net}}$$

The output voltage is

$$v_O = v_S = i_{ds}R_{net} = \frac{g_m R_{net} v_G}{1 + g_m R_{net}}$$

and thus the ac small-signal voltage gain is

$$A_V = \frac{v_O}{v_G} = \frac{g_m R_{\text{net}}}{1 + g_m R_{\text{net}}}$$

Upon dividing through by g_m this can be rewritten as

$$A_V = \frac{R_{\text{net}}}{(1/g_m) + R_{\text{net}}}$$

From this we see that the voltage gain will be positive, and thus the source follower is a non-inverting amplifier. We also note that A_V will always be less than unity, although for the usual case of $R_{\text{net}} \gg 1/g_m$, the voltage gain will be close to unity.

The source follower can be represented as an amplifier with an open-circuit (i.e., no load) voltage transfer ratio of unity and an output resistance of $r_O = 1/g_m$. The equation for A_V can be expressed as $A_V = R_{\text{net}}/(R_{\text{net}} + r_O)$, which is the voltage division ratio of the $r_O = R_{\text{net}}$ circuit.

Source-Follower Examples

Let's consider an example of a JFET with $I_{DSS} = 10$ mA and $V_P = -4$ V. Let $V_{DD} = +20$ V and $I_{DSQ} = I_{DSS}/2 = 5$ mA. For $I_{DS} = I_{DSS}/2$ the value of V_{GS} is -1.17 V. To bias the JFET in the middle of the active region, we will let $V_{GQ} = V_{DD}/2 = +10$ V, so $V_{SQ} = V_{GQ} - V_{GS} = +10$ V $- (-1.17$ V$) = +11.17$ V. Thus $R_S = V_{SQ}/I_{DSQ} = 11.17$ V/5 mA $= 2.23$ kΩ.

The transfer conductance at $I_{DS} = 5$ mA is 3.54 mS so that $r_O = 1/g_m = 283$ Ω. Since $g_m R_S = 7.9$, good bias stability will be obtained. If $R_L \gg R_S$, then $A_V \cong R_S/(r_O + R_S) = 2.23$ k$\Omega/(283$ Ω + 2.23 kΩ) = 0.887. If $R_L = 1$ kΩ, then $R_{\text{net}} = 690$ Ω, and A_V drops to 0.709, and if $R_L = 300$ Ω, $R_{\text{net}} = 264$ Ω and A_V is down to 0.483. A BJT emitter-follower circuit has the same equations for the voltage gain as the FET source follower. For the BJT case, $r_O = 1/g_m = V_T/I_C$, where $V_T =$ thermal voltage $= kT/q \cong 25$ mV and I_C is the quiescent collector current. For $I_C = 5$ mA, we get $r_O \cong 25$ mV/5 mA $= 5$ Ω as compared to $r_O = 283$ Ω for the JFET case at the same quiescent current level. So the emitter follower does have a major advantage over the source follower since it has a much lower output resistance r_O and can thus drive very small load resistances with a voltage gain close to unity. For example, with $R_L = 100$ Ω, we get $A_V \cong 0.26$ for the source follower as compared to $A_V \cong 0.95$ for the emitter follower.

The FET source follower does, however, offer substantial advantages over the emitter follower of a much higher input resistance and a much lower input current. For the case in which a very high-impedance source, up in the megohm range, is to be coupled to a low-impedance load down in the range of 100 Ω or less, a good combination to consider is that of a cascaded FET source follower followed by a BJT emitter follower. This combination offers the very high input resistance of the source follower and the very low output resistance of the emitter follower.

For the source-follower circuit under consideration the input resistance will be $R_{\text{in}} = [R_{G1} \| R_{G2}] = 10$ MΩ. If the JFET gate current is specified as 1 nA (max), and for good bias stability the change in gate voltage due to the gate current should not exceed $|V_P|/10 = 0.4$ V, the maximum allowable value for $[R_{G1} \| R_{G2}]$ is given by $I_G \cdot [R_{G1} \| R_{G2}] < 0.4$ V. Thus $[R_{G1} \| R_{G2}] < 0.4$ V/1 nA $= 0.4$ G$\Omega = 400$ MΩ. Therefore R_{G1} and R_{G2} can each be allowed to be as large as 800 MΩ, and very large values for R_{in} can thus be obtained. At higher frequencies the input capacitance C_{in} must be considered, and C_{in} will ultimately limit the input impedance of the circuit. Since the input capacitance of the FET will be comparable to that of the BJT, the advantage of the FET source follower over the BJT emitter follower from the standpoint of input impedance will be obtained only at relatively low frequencies.

Source-Follower Frequency Response

The input capacitance of the source follower is given by $C_{\text{in}} = C_{GD} + (1 - A_V)C_{GS}$. Since A_V is close to unity, C_{in} will be approximately given by $C_{\text{in}} \cong C_{GD}$. The source-follower input capacitance can, however, be reduced below C_{GD} by a bootstrapping circuit in which the drain voltage is made to follow the gate voltage. Let's consider a representative example in which $C_{GD} = 5$ pF, and let the signal-source output resistance be $R_1 = 100$ kΩ. The input circuit is in the form of a simple *RC* low-pass network. The *RC* time constant is

$$\tau = [R\|R_{G1}\|R_{G2}] \cdot C_{in} \cong R_1 \cdot C_{in} \cong R_1 \cdot C_{GD}$$

Thus $\tau \cong 100$ kΩ $\cdot$ 5 pF $= 500$ ns $= 0.5$ μs. The corresponding 3-dB or half-power frequency is $f_H = 1/(2\,\pi\tau) = 318$ kHz. If $R_1 = 1$ MΩ, the 3-dB frequency will be down to about 30 kHz. Thus we see indeed the limitation on the frequency response that is due to the input capacitance.

Frequency and Time-Domain Response

Small-Signal CS Model for High-Frequency Response

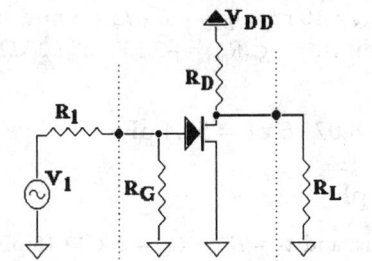

FIGURE 22.18   Common-source amplifier.

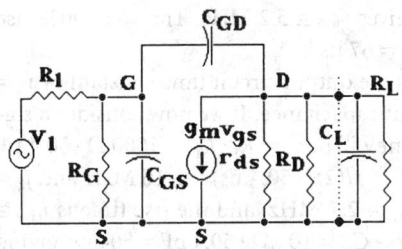

FIGURE 22.19   AC small-signal model.

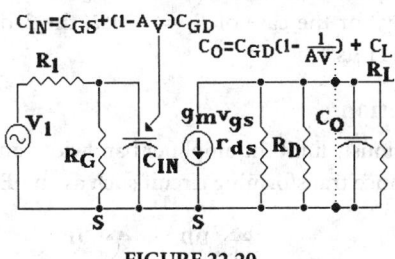

FIGURE 22.20

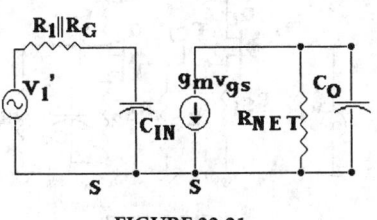

FIGURE 22.21

We will now consider the frequency- and time-domain response of the JFET CS amplifier. In Fig. 22.18 an ac representation of a CS amplifier is shown, the dc biasing not being shown. In Fig. 22.19 the JFET small-signal ac equivalent circuit model is shown including the junction capacitances C_{GS} and C_{GD}. The gate-to-drain capacitance C_{GD} is a feedback capacitance in that it is connected between output (drain) and input (gate). Using Miller's theorem for shunt feedback this feedback capacitance can be transformed into an equivalent input capacitance $C_{GD'} = (1 - A_V)C_{GD}$ and an equivalent output capacitance $C_{GD''} = (1 - 1/A_V)C_{GD}$, as shown in Fig. 22.20. The net input capacitance is now $C_{in} = C_{GS} + (1 - A_V)C_{GD}$ and the net output capacitance is $C_O = (1 - 1/A_V)C_{GD} + C_L$. Since the voltage gain A_V is given by $A_V = -g_mR_{net}$, where R_{net} represents the net load resistance, the equations for C_{in} and C_O can be written approximately as $C_{in} = C_{GS} + (1 + g_mR_{net})C_{GD}$ and $C_O = [1 + 1/(g_mR_{net})]C_{GD} + C_L$. Since usually $A_V = g_mR_{net} >> 1$, C_O can be written as $C_O \cong C_{GD} + C_L$. Note that the voltage gain given by $A_V = -g_mR_{net}$ is not valid in the higher frequency, where A_V will decrease with increasing frequency. Therefore the expressions for C_{in} and C_O will not be exact but will still be a useful approximation for the determination of the frequency- and time-domain responses. We also note that the contribution of C_{GD} to the input capacitance is increased by the Miller effect factor of $1 + g_mR_{net}$.

The circuit in Fig. 22.21 is in the form of two cascaded RC low-pass networks. The *RC* time constant on the input side is $\tau_1 = [R_1\|R_G] \cdot C_{in} \cong R_1 \cdot C_{in}$, where R_1 is the signal-source resistance. The *RC* time constant on the output side is given by $\tau_2 = R_{net} \cdot C_O$. The corresponding breakpoint frequencies are

$$f_1 = \frac{1}{2\pi\tau_1} = \frac{1}{2\pi R_1 \cdot C_{\text{in}}}$$

and

$$f_2 = \frac{1}{2\pi\tau_2} = \frac{1}{2\pi R_{\text{net}} \cdot C_O}$$

The 3-dB or half-power frequency of this amplifier stage will be a function of f_1 and f_2. If these two breakpoint frequencies are separated by at least a decade (i.e., 10:1 ratio), the 3-dB frequency will be approximately equal to the lower of the two breakpoint frequencies. If the breakpoint frequencies are not well separated, then the 3-dB frequency can be obtained from the following approximate relationship: $(1/f_{3\text{dB}})^2 \cong (1/f_1)^2 + (1/f_2)^2$. The time-domain response as expressed in terms of the 10 to 90% rise time is related to the frequency-domain response by the approximate relationship that $t_{\text{rise}} \cong 0.35/f_{3\text{dB}}$.

We will now consider a representative example. We will let $C_{GS} = 10$ pF and $C_{GD} = 5$ pF. We will assume that the net load driven by the drain of the transistors is $R_{\text{net}} = 2$ kΩ and $C_L = 10$ pF. The signal-source resistance $R_1 = 100$ Ω. The JFET will have $I_{DSS} = 10$ mA, $I_{DSQ} = I_{DSS}/2 = 5$ mA, and $V_P = -4$ V, so $g_m = 3.535$ mS. Thus the midfrequency gain is $A_V = -g_m R_{\text{net}} = -3.535$ mS $\cdot$ 2 k$\Omega = -7.07$. Therefore we have that

$$C_{\text{in}} \cong C_{GS} + (1 + g_m R_{\text{net}}) C_{GD} = 10 \text{ pF} + 8.07 \cdot 5 \text{ pF} = 50.4 \text{ pF}$$

and

$$C_O \cong C_{GD} + C_L = 15 \text{ pF}$$

Thus $\tau_1 = R_1 \cdot C_{\text{in}} = 100$ $\Omega \cdot 50.4$ pF $= 5040$ ps $= 5.04$ ns, and $\tau_2 = R_{\text{net}} \cdot C_O = 2$ k$\Omega \cdot 15$ pF $= 30$ ns. The corresponding breakpoint frequencies are $f_1 = 1/(2\pi \cdot 5.04 \text{ ns}) = 31.6$ MHz and $f_2 = 1/(2\pi \cdot 30 \text{ ns}) = 5.3$ MHz. The 3-dB frequency of the amplifier can be obtained from $(1/f_{3\text{dB}})^2 \cong (1/f_1)^2 + (1/f_2)^2 = (1/31.6 \text{ MHz})^2 + (1/5.3 \text{ MHz})^2$, which gives $f_{3\text{dB}} \cong 5.2$ MHz. The 10 to 90% rise time can be obtained from $t_{\text{rise}} \cong 0.35/f_{3\text{dB}} = 0.35/5.2$ MHz $= 67$ ns.

In the preceding example the dominant time constant is the output circuit time constant of $\tau_2 = 30$ ns due to the combination of load resistance and output capacitance. If we now consider a signal-source resistance of 1 kΩ, the input circuit time constant will be $\tau_1 = R_1 \cdot C_{\text{in}} = 1000$ $\Omega \cdot 50.4$ pF $= 50.4$ ns. The corresponding breakpoint frequencies are $f_1 = 1/(2\pi \cdot 50.4 \text{ ns}) = 3.16$ MHz and $f_2 = 1/(2\pi \cdot 30 \text{ ns}) = 5.3$ MHz. The 3-dB frequency is now $f_{3\text{dB}} \cong 2.7$ MHz, and the rise time is $t_{\text{rise}} \cong 129$ ns. If R_1 is further increased to 10 kΩ, we obtain $\tau_1 = R_1 \cdot C_{\text{in}} = 10$ k$\Omega \cdot 50.4$ pF $= 504$ ns, giving breakpoint frequencies of $f_1 = 1/(2\pi \cdot 504 \text{ ns}) = 316$ kHz and $f_2 = 1/(2\pi \cdot 30 \text{ ns}) = 5.3$ MHz. Now τ_1 is clearly the dominant time constant, the 3-dB frequency is now down to $f_{3\text{dB}} \cong f_1 = 316$ kHz, and the rise time is up to $t_{\text{rise}} \cong 1.1$ μs. Finally, for the case of $R_1 = 1$ MΩ, the 3-dB frequency will be only 3.16 kHz and the rise time will be 111 μs.

Use of Source Follower for Impedance Transformation

We see that large values of signal-source resistance can seriously limit the amplifier bandwidth and increase the rise time. In these cases, the use of an impedance transforming circuit such as an FET source follower or a BJT emitter follower can be very useful. Let's consider the use of a source follower as shown in Fig. 22.22. We will assume that both FETs are identical to the one in the preceding examples and are biased at $I_{DSQ} = 5$ mA. The source follower Q_1 will have an input capacitance of $C_{\text{in}} = C_{GD} + (1 - A_{V1})C_{GS} \cong C_{GD} = 5$ pF, since A_V will be very close to unity for a source follower that is driving a CS amplifier. The source-follower output resistance will be $r_O = 1/g_m = 1/3.535$ mS $= 283$ Ω. Let's again consider the case of $R_1 = 1$ MΩ. The

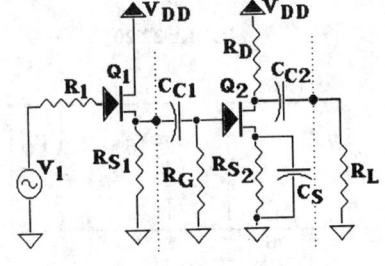

FIGURE 22.22

time constant due to the combination of R_1 and the input capacitance of the source follower is τ_{SF} = 1 M$\Omega \cdot$ 5 pf = 5 μs. The time constant due to the combination of the source-follower output resistance r_O and the input capacitance of the CS stage is $\tau_1 = r_O \cdot C_{in}$ = 283 $\Omega \cdot$ 50.4 pF = 14 ns, and the time constant of the output circuit is τ_2 = 30 ns, as before. The breakpoint frequencies are f_{SF} = 31.8 kHz, f_1 = 11 MHz, and f_2 = 5.3 MHz. The 3-dB frequency of the system is now $f_{3dB} \cong f_{SF}$ = 31.8 kHz, and the rise time is $t_{rise} \cong 11$ μs. The use of the source follower thus results in an improvement by a factor of 10:1 over the preceding circuit.

Voltage-Variable Resistor

Operation of a JFET as a Voltage-Variable Resistor

We will now consider the operation of a JFET as a voltage-variable resistor (VVR). A JFET can be used as a VVR in which the drain-to-source resistance r_{ds} of the JFET can be varied by variation of V_{GS}. For values of $V_{DS} << V_P$ the I_{DS} vs. V_{DS} characteristics are approximately linear, so the JFET looks like a resistor, the resistance value of which can be varied by the gate voltage as shown in Fig. 22.23

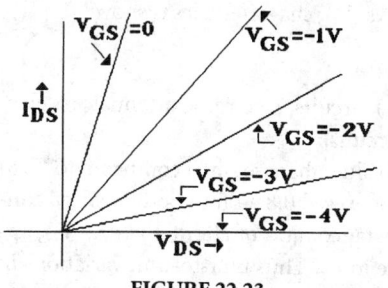

FIGURE 22.23

The channel conductance in the region where $V_{DS} << V_P$ is given by $g_{ds} = A\sigma/L = WH\sigma/L$, where the channel height H is given by $H = H_0 - 2W_D$. In this equation W_D is the depletion region width and H_0 is the value of H as $W_D \to 0$. The depletion region width is given by $W_D = K\sqrt{V_J} = K\sqrt{V_{GS} + \phi}$, where K is a constant, V_J is the junction voltage, and ϕ is the pn-junction contact potential (typically around 0.8 to 1.0 V). As V_{GS} increases, W_D increases and the channel height H decreases as given by $H = H_0 - 2K\sqrt{V_{GS} + \phi}$. When $V_{GS} = V_P$, the channel is completely pinched off, so $H = 0$ and thus $2K\sqrt{V_P + \phi} = H_0$. Therefore $2K = H_0/\sqrt{V_P + \phi}$, and thus

$$H = H_0 - H_0 \frac{\sqrt{V_{GS} + \phi}}{\sqrt{V_P + \phi}} = H_0 \left(1 - \frac{\sqrt{V_{GS} + \phi}}{\sqrt{V_P + \phi}}\right)$$

For g_{ds} we now have

$$g_{ds} = \frac{\sigma WH}{L} = \sigma \frac{WH_0}{L} \left(1 - \frac{\sqrt{V_{GS} + \phi}}{\sqrt{V_P + \phi}}\right)$$

When $V_{GS} = 0$, the channel is fully open or "on," and

$$g_{ds} = g_{ds}(\text{on}) = \sigma \frac{WH_0}{L} \left(1 - \frac{\sqrt{\phi}}{\sqrt{V_P + \phi}}\right)$$

The drain-to-source conductance can now be expressed as

$$g_{ds} = g_{ds}(\text{on}) \frac{1 - \left(\sqrt{V_{GS} + \phi}/\sqrt{V_P + \phi}\right)}{1 - \left(\sqrt{\phi}/\sqrt{V_P + \phi}\right)}$$

The reciprocal quantity is the drain-to-source resistance r_{ds} as given by $r_{ds} = 1/g_{ds}$, and $r_{ds}(\text{on}) =$ $1/g_{ds}(\text{on})$, so

$$r_{ds} = r_{ds}(\text{on}) \; \frac{1 - \left(\sqrt{\phi}/\sqrt{V_P + \phi}\right)}{1 - \left(\sqrt{V_{GS} + \phi}/\sqrt{V_P + \phi}\right)}$$

As $V_{GS} \to 0$, $r_{ds} \to r_{ds}(\text{on})$, and as $V_{GS} \to V_P$, $r_{ds} \to \infty$. This latter condition corresponds to the channel being pinched off in its entirety all the way from source to drain. This is like having a big block of insulator (i.e., the depletion region) between source and drain. When $V_{GS} = 0$, r_{ds} is reduced to its minimum value of $r_{ds}(\text{on})$, which for most JFETs is in the 20- to 400-Ω range. At the other extreme, when $V_{GS} > V_P$, the drain-to-source current I_{DS} is reduced to a very small value, generally down into the low nanoampere or even picoampere range. The corresponding value of r_{ds} is not really infinite but is very large, generally well up into the gigaohm (1000 MΩ) range. Thus by variation of V_{GS}, the drain-to-source resistance can be varied over a very wide range. As long as the gate-to-channel junction is reverse-biased, the gate current will be very small, generally down in the low nanoampere or even picoampere range, so the gate as a control electrode draws very little current. Since V_P is generally in the 2- to 5-V range for most JFETs, the V_{DS} values required to operate the JFET in the VVR range are generally <0.1 V. In Fig. 22.23 the VVR region of the JFET I_{DS} vs. V_{DS} characteristics is shown.

VVR Applications

Applications of VVRs include automatic gain control (AGC) circuits, electronic attenuators, electronically variable filters, and oscillator amplitude control circuits.

When using a JFET as a VVR, it is necessary to limit V_{DS} to values that are small compared to V_P to maintain good linearity. In addition V_{GS} should preferably not exceed 0.8 V_P for good linearity, control, and stability. This limitation corresponds to an r_{ds} resistance ratio of about 10:1. As V_{GS} approaches V_P, a small change in V_P can produce a large change in r_{ds}. Thus unit-to-unit variations in V_P as well as changes in V_P with temperature can result in large changes in r_{ds} as V_{GS} approaches V_P.

The drain-to-source resistance r_{ds} will have a temperature coefficient (TC) due to two causes: (1) the variation of the channel resistivity with temperature and (2) the temperature variation of V_P. The TC of the channel resistivity is positive, whereas the TC of V_P is negative due to the negative TC of the contact potential ϕ. The positive TC of the channel resistivity will contribute to a positive TC of r_{ds}. The negative TC of V_P will contribute to a negative TC of r_{ds}. At small values of V_{GS}, the dominant contribution to the TC is the positive TC of the channel resistivity, so r_{ds} will have a positive TC. As V_{GS} gets larger, the negative TC contribution of V_P becomes increasingly important, and there will be a value of V_{GS} at which the net TC of r_{ds} is zero, and above this value of V_{GS} the TC will be negative. The TC of $r_{ds}(\text{on})$ is typically +0.3%/°C for n-channel JFETs and +0.7%/°C for p-channel JFETs. For example, for a typical JFET with an $r_{ds}(\text{on}) = 500$ Ω at 25°C and $V_P = 2.6$ V, the zero TC point will occur at $V_{GS} = 2.0$ V. Any JFET can be used as a VVR, although there are JFETs that are specifically made for this application.

A simple example of a VVR application is the electronic gain control circuit of Fig. 22.24. The voltage gain is given by $A_V = 1 + (R_F/r_{ds})$. If, for example, $R_F = 19$ kΩ and $r_{ds}(\text{on}) = 1$ kΩ, then the maximum gain will be $A_{V\max} = 1 + [R_F/r_{ds}(\text{on})] = 20$. As V_{GS} approaches V_P, r_{ds} will increase and become very large such that $r_{ds} \gg R_F$, so that A_V will decrease to a minimum value of close to unity. Thus the gain can be varied over a 20:1 ratio. Note that $V_{DS} \cong V_{\text{in}}$, so to minimize distortion the input signal amplitude should be small compared to V_P.

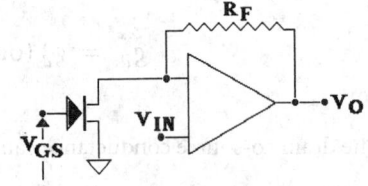

FIGURE 22.24 Electronic gain control.

Defining Terms

Active region: The region of JFET operation in which the channel is pinched off at the drain end but still open at the source end such that the drain-to-source current I_{DS} approximately saturates. The condition for this is that $|V_{GS}| < |V_P|$ and $|V_{DS}| > |V_P|$. The active region is also known as the saturated region.

Drain saturation current, I_{DSS}:   The drain-to-source current flow through the JFET under the conditions that $V_{GS} = 0$ and $|V_{DS}| > |V_P|$ such that the JFET is operating in the active or saturated region.

Ohmic, nonsaturated, or triode region: The three terms all refer to the region of JFET operation in which a conducting channel exists all the way between source and drain. In this region the drain current varies with both V_{GS} and V_{DS}.

Pinch-off voltage, V_P: The voltage that when applied across the gate-to-channel *pn* junction will cause the conducting channel between drain and source to become pinched off. This is also represented as $V_{GS}(\text{off})$.

References

R. Mauro, *Engineering Electronics*, Englewood Cliffs, N.J.: Prentice-Hall, 1989, pp. 199–260.

J. Millman and A. Grabel, *Microelectronics*, 2nd ed., New York: McGraw-Hill, 1987, pp. 133–167, 425–429.

F. H. Mitchell, Jr. and F. H. Mitchell, Sr., *Introduction to Electronics Design*, 2nd ed., Englewood Cliffs, N.J.: Prentice-Hall, 1992, pp. 275–328.

C. J. Savant, M. S. Roden, and G. L. Carpenter, *Electronic Design*, 2nd ed., Menlo Park, Calif.: Benjamin-Cummings, 1991, pp. 171–208.

A. S. Sedra and K. C. Smith, *Microelectronic Circuits*, 3rd ed., Philadelphia: Saunders, 1991, pp. 322–361.

22.2 Bipolar Transistors

J. Watson

Modern amplifiers abound in the form of *integrated circuits* (ICs), which contain transistors, diodes, and other structures diffused into single-crystal silicon *dice*. As an introduction to these ICs, it is convenient to examine first the characteristics of a single-transistor amplifier circuit.

There are two basic forms of transistor, the *bipolar* family and the *field-effect* family, and both appear in ICs. They differ in their modes of operation but may be incorporated into circuits in quite similar ways. To understand elementary circuits, there is no need to become too familiar with the physics of transistors, but some basic facts about their electrical properties must be known.

Consider the bipolar transistor, of which there are two types, *npn* and *pnp*. Electrically, they differ only in terms of current direction and voltage polarity. Figure 22.25(a) illustrates the idealized structure of an *npn* transistor, and diagram (b) implies that it corresponds to a pair of diodes with three leads. This representation does *not* convey sufficient information about the actual operation of the transistor, but it does make the point that the flow of conventional current (positive to negative) is easy from the *base* to the *emitter*, since it passes through a *forward-biased diode*, but difficult from the *collector* to the *base*, because flow is prevented by a *reverse-biased diode*.

Figure 22.25(c) gives the standard symbol for the *npn* transistor, and diagram (d) defines the direction of current flow and the voltage polarities observed when the device is in operation. Finally, diagram (e) shows that for the *pnp* transistor, all these directions are reversed and the polarities are inverted.

For a transistor, there is a main current flow between the collector and the emitter, and a very much smaller current flow between the base and the emitter. So, the following relations may be written:

$$I_E = I_C + I_B \tag{22.1}$$

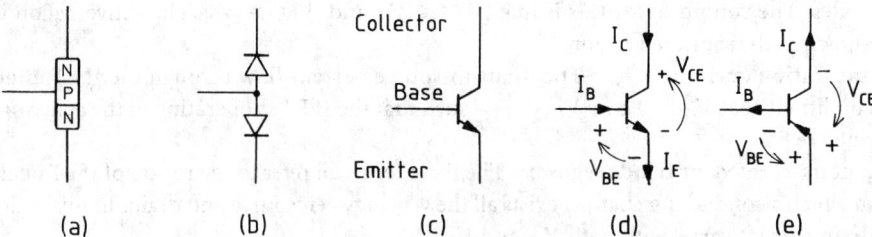

FIGURE 22.25   The bipolar transistor. (a) to (d) *npn* transistor; (e) *pnp* transistor.

(Note that the arrow on the transistor symbol defines the emitter and the direction of current flow—*out* for the *npn* device, and *in* for the *pnp*.) Also

$$I_C/I_B = h_{FE} \tag{22.2}$$

Here, h_{FE} is called the *dc common-emitter current gain*, and because $I_C \gg I_B$, then h_{FE} is large, typically 50 to 300. The implication of this may be seen immediately: if the small current I_B can be used to control the large current I_C, then the transistor may obviously be used as a current amplifier. [This is why Fig. 22.25(b) is inadequate—it completely neglects this all-important current-gain property of the transistor.] Furthermore, if a load resistance is connected into the collector circuit, it will become a voltage amplifier, too.

Unfortunately, h_{FE} is an ill-defined quantity and varies not only from transistor to transistor but also changes with temperature. The relationship between the base-emitter voltage V_{BE} and the collector current is much better defined and follows an exponential law closely over at least eight decades. This relationship is shown in both linear and logarithmic form in Fig. 22.26. Because the output current I_C is dependent upon the input voltage V_{BE}, the plot must be a transfer conductance or *transconductance* characteristic. The relevant law is

$$I_C = I_{ES}(e^{(q/kT)V_{BE}} - 1) \tag{22.3}$$

Here, I_{ES} is an extremely small leakage current internal to the transistor, q is the electronic charge, k is Boltzmann's constant, and T is the absolute temperature in kelvins. Usually, kT/q is called V_T and is about 26 mV at a room temperature of 25°C. This implies that for any value of V_{BE} over about 100 mV, then $\exp(V_{BE}/V_T) \gg 1$, and for all normal operating conditions, Eq. (22.3) reduces to

$$I_C = I_{ES}e^{V_{BE}/V_T} \quad \text{for } V_{BE} > 100 \text{ mV} \tag{22.4}$$

The term "normal operating conditions" is easily interpreted from Fig. 22.26(a), which shows that when V_{BE} has reached about 0.6 to 0.7 V, any small fluctuations in its value cause major fluctuations in I_C. This situation is illustrated by the dashed lines enclosing ΔV_{BE} and ΔI_C, and it implies that to use the transistor as an amplifier, working values of V_{BE} and I_C must be established, after which signals may be regarded as fluctuations around these values.

Under these *quiescent*, *operating*, or *working* conditions,

$$I_C = I_Q \quad \text{and} \quad V_{CE} = V_Q$$

and methods of defining these quiescent or operating conditions are called *biasing*.

Biasing the Bipolar Transistor

A fairly obvious way to bias the transistor is to first establish a constant voltage V_B using a potential divider $R1$ and $R2$ as shown in the **biasing circuit** of Fig. 22.27. Here,

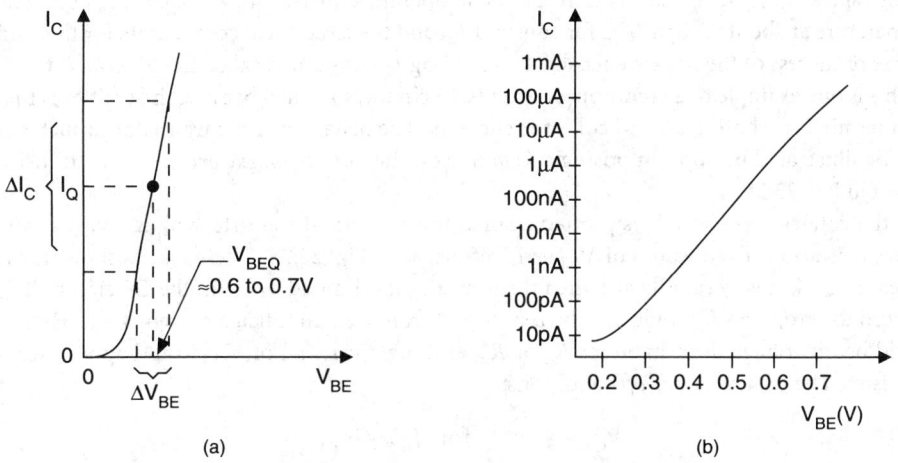

FIGURE 22.26   The transconductance curve for a transistor on (a) linear and (b) logarithmic axes.

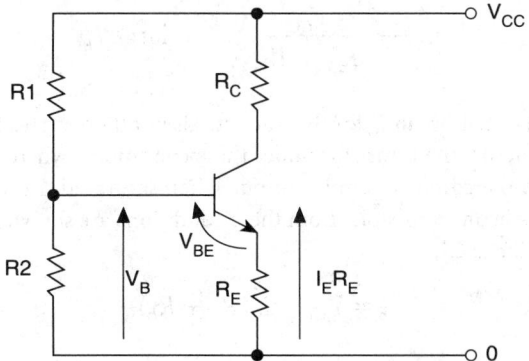

FIGURE 22.27   A transistor biasing circuit.

$$V_B \simeq \frac{V_{CC}R2}{R1 + R2}$$

if I_B is very small compared with the current through $R2$, which is usual. If it is not, this fact must be taken into account.

This voltage will be much greater than V_{BE} if a realistic power supply is used along with realistic values of $R1$ and $R2$. Hence, when the transistor is connected into the circuit, an emitter resistor must also be included so that

$$V_{BE} = V_B - I_E R_E \tag{22.5}$$

Now consider what happens when the power supply is connected. As V_B appears, a current I_B flows into the base and produces a much larger current $I_C = h_{FE}I_B$ in the collector. These currents add in the emitter to give

$$I_E = I_B + h_{FE}I_B = (1 + h_{FE})I_B \simeq h_{FE}I_B \tag{22.6}$$

Clearly, I_E will build up until a fixed or quiescent value of base-emitter voltage V_{BEQ} appears. Should I_E try to build up further, V_{BE} will fall according to Eq. (22.5) and, hence, so will I_E. Conversely, should I_E not build up enough, V_{BE} will increase until it does so.

This is actually a case of current-derived negative feedback, and it successfully holds the collector current near the quiescent value I_Q. Furthermore, it does so in spite of different transistors with different values of h_{FE} being used and in spite of temperature variations. Actually, V_{BE} itself falls with temperature at about −2.2 mV/°C for constant I_C, and the circuit will compensate for this, too. The degree of success of the negative feedback in holding I_Q constant is called the *bias stability*.

This is one example of a **common-emitter** (CE) circuit, so-called because the emitter is the common terminal for both base and collector currents. The behavior of the transistor in such a circuit may be illustrated by superimposing a *load line* on the *output characteristics* of the transistor, as shown in Fig. 22.28.

If the collector current I_C is plotted against the collector-to-emitter voltage V_{CE}, a family of curves for various fixed values of V_{BE} or I_B results, as in Fig. 22.28. These curves show that as V_{CE} increases, I_C rises very rapidly and then turns over as it is limited by I_B. In the CE circuit, if I_B were reduced to zero, then I_C would also be zero (apart from a small leakage current I_{CE0}). Hence there would be no voltage drop in either R_C or R_E, and practically all of V_{CC} would appear across the transistor. That is, under *cut-off* conditions,

$$V_{CE} \rightarrow V_{CC} \quad \text{for} \quad I_B = 0 \tag{22.7}$$

Conversely, if I_B were large, I_C would be very large, almost all of V_{CC} would be dropped across $R_C + R_E$ and

$$I_C \rightarrow \frac{V_{CC}}{R_C + R_E} \quad \text{for large } I_B \tag{22.8}$$

Actually, because the initial rise in I_C for the transistor is not quite vertical, there is always a small *saturation voltage* V_{CES} across the transistor under these conditions, where V_{CES} means the voltage across the transistor in the common-emitter mode when saturated. In this saturated condition $V_{CES} \simeq 0.3$ V for small silicon transistors. Both these conditions are shown in Fig. 22.28.

From the circuit of Fig. 22.27,

$$V_{CE} = V_{CC} - I_C(R_C + R_E) \tag{22.9a}$$

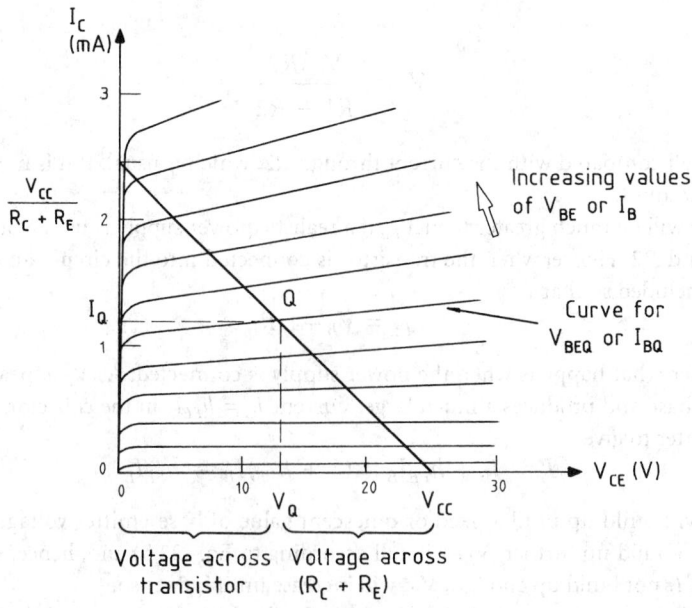

FIGURE 22.28   The load-line diagram.

which may be rewritten as

$$I_C = -V_{CE}/(R_C + R_E) + V_{CC}/(R_C + R_E) \qquad (22.9b)$$

This is the straight-line equation to the *dc load-line* (compare $y = mx + c$), showing that its slope is $-1/(R_C + R_E)$ and that it crosses the I_C axis at $V_{CC}/(R_C + R_E)$ as expected. The actual position of a point is determined by where this load line crosses the output characteristic in use, that is, by what value of V_{BE} or I_B is chosen. For example, the quiescent point for the transistor is where the load line crosses the output curve defined by $V_{BE} = V_{BEQ}$ (or $I_B = I_{BQ}$) to give $V_{CE} = V_Q$ and $I_C = I_Q$.

Note that because the transistor is *nonohmic* (that is, it does not obey Ohm's law), the voltage across it may only be determined by using the (ohmic) voltage drop across the resistors R_C and R_E according to Eq. (22.9). At the quiescent point this is

$$V_Q = V_{CC} - I_Q(R_C + R_E)$$

A design example will illustrate typical values involved with a small-transistor CE stage.

Example 1

A transistor is to be biased at a collector current of 1 mA when a 12-V power supply is applied. Using the circuit of Fig. 22.27, determine the values of R1, R2, and R_E if 3.4 V is to be dropped across R_E and if the current through R2 is to be 10 I_{BQ}. Assume that for the transistor used, $V_{BEQ} = 0.6$ V and $h_{FE} = 100$.

Solution. In this circuit $I_Q = 1$ mA $\simeq I_E$ (because $I_B \ll I_C$). Hence

$$R_E = \frac{V_{R_E}}{I_Q} = \frac{3.4}{1} = 3.4 \text{ k}\Omega$$

Also, $V_B = V_{R_E} + V_{BE} = 3.4 + 0.6 = 4$ V. This gives

$$R2 = \frac{V_B}{10\,I_{BQ}}$$

where $I_{BQ} = I_Q/h_{FE} = 1/100 = 0.01$ mA, so

$$R2 = \frac{4}{10 \times 0.01} = 40 \text{ k}\Omega$$

Now $V_{R1} = V_{CC} - V_B = 12 - 4 = 8$ V, and the current through R1 is $10\,I_{BQ} + I_{BQ} = 11\,I_{BQ}$, so

$$R1 = \frac{V_{R1}}{I_{R1}} = \frac{8}{11 \times 0.01} = 72.7 \text{ k}\Omega$$

In the above design example, the base current I_{BQ} has been included in the current passing through R1. Had this not been done, R1 would have worked out at 80 kΩ. Usually, this difference is not very important because *discrete* (or individual) resistors are available only in a series of nominal values, and each of these is subject to a *tolerance*, including 10, 5, 2, and 1%.

In the present case, the following (5%) values could reasonably be chosen:

$$R_E = 3.3 \text{ k}\Omega \qquad R1 = 75 \text{ k}\Omega \qquad R2 = 39 \text{ k}\Omega$$

All this means that I_Q cannot be predetermined very accurately, but the circuit nevertheless settles down to a value close to the chosen one, and, most importantly, stays there almost irrespective of the transistor used and the ambient temperature encountered.

Having biased the transistor into an operating condition, it is possible to consider *small-signal operation*.

Small-Signal Operation

In the biasing circuit of Fig. 22.27, the collector resistor R_C had no discernible function, because it is simply the load resistor across which the signal output voltage is developed. However, it was included because it also drops a voltage due to the bias current flowing through it. This means that its value must not be so large that it robs the transistor of adequate operating voltage; that is, it must not be responsible for moving the operating point too far to the left in Fig. 22.28.

If the chosen bias current and voltage are I_Q and V_Q, then small signals are actually only fluctuations in these bias (or average) values that can be separated from them using coupling capacitors.

To inject an input signal to the base, causing V_{BE} and I_B to fluctuate by v_{be} and i_b, a signal source must be connected between the base and the common or zero line (also usually called ground or earth whether it is actually connected to ground or not!). However, most signal sources present a resistive path through themselves, which would shunt $R2$ and so change, or even destroy, the bias conditions. Hence, a coupling capacitor C_c must be included, as shown in Fig. 22.29, in series with a signal source represented by a Thévenin equivalent.

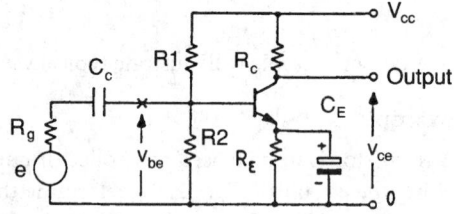

FIGURE 22.29 A complete common-emitter stage.

The emitter resistor R_E was included for biasing reasons (although there are other bias circuits that omit it), but for signal amplification purposes it must be shunted by a high-value capacitor C_E so that the signal current can flow down to ground without producing a signal voltage drop leading to negative feedback (as did the bias current). The value of C_E must be much greater than is apparent at first sight, and this point will be developed later; for the present, it will be assumed that it is large enough to constitute a short circuit at all the signal frequencies of interest. So, for ac signals R_E is short-circuited and only R_C acts as a load. This implies that a *signal* or *ac load line* comes into operation with a slope of $-1/R_C$, as shown in Fig. 22.30.

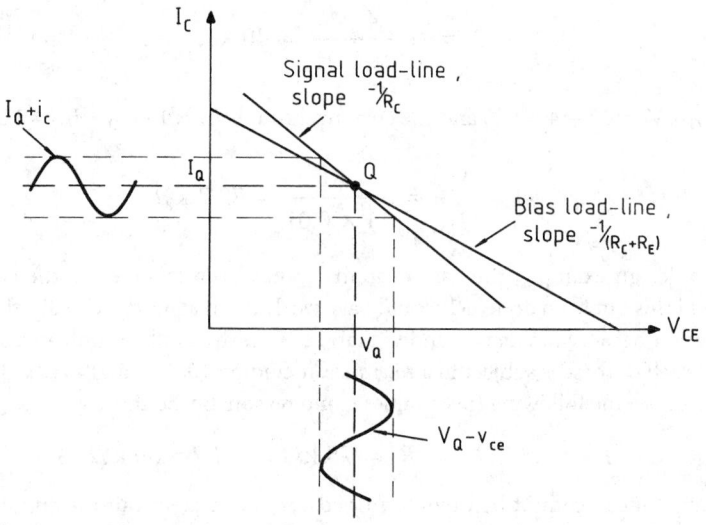

FIGURE 22.30 The signal or ac load line.

The ways in which the small-signal quantities fluctuate may now be examined. If v_{be} goes positive, this actually means that V_{BE} increases a little. This in turn implies that I_C increases by an amount i_c, so the voltage drop in R_C increases by v_{ce}. Keeping in mind that the top of R_C is held at a constant voltage, this means that the voltage at the bottom of R_C must fall by v_{ce}. This very important point shows that because v_{ce} falls as v_{be} rises, there is 180° phase shift through the stage. That is, the CE stage is an *inverting voltage amplifier*. However, because i_c increases into the collector as i_b increases into the base, it is also a *noninverting current amplifier*.

Now consider the amount by which v_{ce} changes with v_{be}, which is the *terminal voltage gain* of the stage. In Fig. 22.26, the slope of the transconductance curve at any point defines by how much I_C changes with a fluctuation in V_{BE}. That is, it gives the ratio i_c/v_{be} at any operating point Q. Equation (22.4) is

$$I_C = I_{ES} e^{V_{BE}/V_T}$$

so that

$$\frac{dI_C}{dV_{BE}} = \frac{1}{V_T} I_{ES} e^{V_{BE}/V_T}$$

or

$$\frac{i_c}{v_{be}} = \frac{I_C}{V_T} = g_m = \text{the transconductance} \tag{22.10}$$

Now the signal output voltage is

$$v_{ce} \simeq -i_c R_C$$

(Here, the approximation sign is because the collector-emitter path within the transistor does present a large resistance r_{ce} through which a very small part of i_c flows.)

The terminal voltage gain is therefore

$$A_v = \frac{v_{ce}}{v_{be}} \simeq \frac{-i_c R_C}{v_{be}} = -g_m R_C \tag{22.11}$$

where the negative sign implies signal inversion.

In practice, $V_T \simeq 26$ mV at room temperature, as has been mentioned, and this leads to a very simple numerical approximation. From Eq. (22.10) and using $I_C = I_Q$,

$$g_m = \frac{I_Q}{V_T} \simeq \frac{I_Q}{0.026} \simeq 39 I_Q \quad \text{mA/V}$$

if I_Q is in mA and at room temperature. This shows that irrespective of the transistor used, the transconductance may be approximated knowing only the quiescent collector current.

The magnitude and phase relationships between v_{ce} and i_c can easily be seen by including them on the signal load-line diagram as shown in Fig. 22.30, where the output characteristics of the transistor have been omitted for clarity. Sinusoidal output signals have been inserted, and either may be obtained from the other by following the signal load-line locus.

Now consider the small-signal current gain. Because the value of h_{FE} is not quite linear on the I_C/I_B graph, its slope too must be used for small-signal work. However, the departure from linearity is not great over normal working conditions, and the small-signal value h_{fe} is usually quite close to that of h_{FE}. Hence,

$$A_i = \frac{i_c}{i_b} \simeq h_{fe} \tag{22.12}$$

The small-signal or incremental input resistance to the base itself (to the right of point X in Fig. 22.29) may now be found:

$$R_{\text{in}} = \frac{v_{be}}{i_b} = \frac{v_{be}}{i_c} \frac{i_c}{i_b} \simeq \frac{h_{fe}}{g_m} \qquad (22.13)$$

Three of the four main (midfrequency) parameters for the CE stage have now been derived, all from a rather primitive understanding of the transistor itself. The fourth, R_{out}, is the dynamic, incremental, or small-signal resistance of the transistor from collector to emitter, which is the slope of the output characteristic at the working point r_{ce}. Being associated with a reverse-biased (CB) junction, this is high—typically about 0.5 MΩ—so that the transistor acts as a current source feeding a comparatively low load resistance R_C. Summarizing, at mid frequency,

$$A_i \simeq h_{fe} \qquad A_v \simeq -g_m R_C \qquad R_{\text{in}} \simeq \frac{h_{fe}}{g_m} \qquad R_{\text{out}} \simeq r_{ce}$$

Example 2

Using the biasing values for $R1$, $R2$, and R_E already obtained in Example 1, calculate the value of R_C to give a terminal voltage gain of −150. Then determine the input resistance R_{in} if h_{fe} for the transistor is 10% higher than h_{FE}.

Solution. Because $I_Q = 1$ mA, $g_m \simeq 39 \times 1 = 39$ mA/V. Hence $A_v \simeq -g_m R_C$ or $-150 \simeq -39 R_C$, giving

$$R_C = 150/39 \simeq 3.9 \text{ k}\Omega$$

(*Note:* This value *must* be checked to determine that it is reasonable insofar as biasing is concerned. In this case, it will drop $I_Q R_C = 1 \times 3.9 = 3.9$ V. Because $V_{R_E} = 3.4$ V, this leaves $12 - 3.9 - 3.4 = 4.7$ V across the transistor, which is reasonable.)

Finally,

$$R_{\text{in}} \simeq \frac{h_{fe}}{g_m} = \frac{110}{39} = 2.8 \text{ k}\Omega$$

A Small-Signal Equivalent Circuit

The conclusions reached above regarding the performance of the bipolar transistor are sufficient for the development of a basic equivalent circuit, or model, relevant *only* to small-signal operation. Taking the operating CE amplifier, this may be done by first "looking into" the base, shown as b in Fig. 22.31. Between this point and the actual active part of the base region b', it is reasonable to suppose that the intervening (inactive) semiconductor material will present a small resistance $r_{bb'}$. This is called the *base spreading resistance,* and it is also shown in Fig. 22.31.

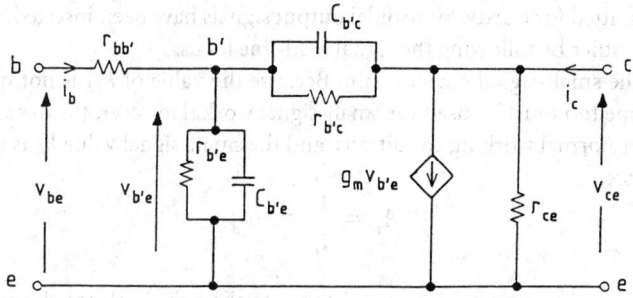

FIGURE 22.31 The hybrid-π small-signal transistor equivalent circuit or model.

From b' to the emitter e, there will be a dynamic or incremental resistance given by

$$r_{b'e} = \frac{v_{b'e}}{i_b} = \frac{v_{b'e}}{i_c}\frac{i_c}{i_b} = \frac{h_{fe}}{g_m} \qquad (22.14)$$

so that the full resistance from the base to the emitter must be

$$R_{in} = r_{bb'} + r_{b'e} = r_{bb'} + \frac{h_{fe}}{g_m} \simeq \frac{h_{fe}}{g_m} \qquad (22.15)$$

because $r_{bb'}$ is only about 10 to 100 Ω, which is small compared with $r_{b'e}$, this being several kilohms (as shown by the last example). It will now be understood why Eq. (22.13) gave $R_{in} \simeq h_{fe}/g_m$.

The reverse-biased junction that exists from b' to the collector ensures that the associated dynamic resistance $r_{b'c}$ will be very large indeed, which is fortunate, otherwise signal feedback from the output to the input would modify the gain characteristics of the amplifier. Typically, $r_{b'c}$ will be some tens of megohms.

However, because of transistor action, the dynamic resistance from collector to emitter, r_{ce}, will be smaller than $r_{b'c}$ and will typically be below a megohm. This "transistor action" may be represented by a current source from collector to emitter that is dependent upon either i_b or $v_{b'e}$. That is, it will be either $h_{fe}i_b$ or $g_m v_{b'e}$. The latter leads to the well-known hybrid-π model, and it is this which is shown in Fig. 22.31.

Where junctions or interfaces of any sort exist, there will always be distributed capacitances associated with them, and to make these easy to handle analytically, they may be "lumped" into single capacitances. In the present context, two lumped capacitances have been incorporated into the hybrid-π model, $C_{b'e}$ from base to emitter and $C_{b'c}$ from base to collector, respectively. These now complete the model, and it will be appreciated that they make it possible to analyze high-frequency performance. Typically, $C_{b'e}$ will be a few picofarads and will always be rather larger than $C_{b'c}$.

Figure 22.31 is the hybrid-π small-signal, dynamic, or incremental model for a bipolar transistor, and when external components are added and simplifications made, it makes possible the determination of the performance of an amplifier using that transistor not only at midfrequencies but at high and low frequencies, too.

Low-Frequency Performance

In Fig. 22.32 both a source and a load have been added to the hybrid-π equivalent circuit to model the complete CE stage of Fig. 22.29. Here, both $C_{b'e}$ and $C_{b'c}$ have been omitted because they are too small to affect the low-frequency performance, as has $r_{b'c}$ because it is large and so neither loads the source significantly compared to $r_{bb'} + r_{b'e}$ nor applies much feedback.

The signal source has been represented by a Thévenin equivalent that applies a signal via a coupling capacitor C_c. Note that this signal source has been returned to the emitter, which implies that the emitter resistor bypass capacitor C_E has been treated as a short circuit at all signal frequencies for the purposes of this analysis.

Because the top of biasing resistor $R1$ (Fig. 22.29) is taken to ground via the power supply insofar as the signal is concerned, it appears in parallel with $R2$, and the emitter is also grounded to the signal via C_E. That is, a composite biasing resistance to ground R_B appears:

$$R_B = \frac{R1 \cdot R2}{R1 + R2}$$

Finally, the collector load is taken to ground via the power supply and hence to the emitter via C_E.

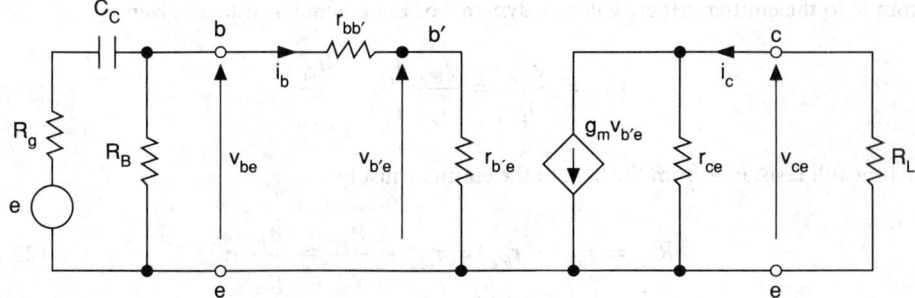

FIGURE 22.32   The loaded hybrid-π model for low frequencies.

Figure 22.32 shows that v_{be} is amplified independently of frequency, so the terminal voltage gain A_v may easily be determined:

$$A_v = \frac{v_{ce}}{v_{be}} = \frac{-i_c R_L}{v_{be}}$$

Now

$$v_{be} = \frac{v_{b'e}(r_{bb'} + r_{b'e})}{r_{b'e}} \simeq v_{b'e} \quad \text{and} \quad i_c = \frac{g_m v_{b'e} r_{ce}}{r_{ce} + R_L} \simeq g_m v_{b'e}$$

because $r_{b'e} \gg r_{bb'}$ and $r_{ce} \gg R_L$. So,

$$A_v \simeq -g_m R_L$$

which is as expected.

The model shows that v_{be} is amplified independently of frequency because there are no capacitances to its right, so an analysis of low-frequency response devolves down to determining v_{be} in terms of e. Here, part of e will appear across the capacitive reactance X_{C_c}, and the remainder is v_{be}. So, to make the concept of reactance valid, a sinusoidal signal E must be postulated, giving a sinusoidal value for $v_{be} = V_{be}$.

At midfrequencies, where the reactance of C_c is small, the signal input voltage is

$$V_{be}(f_m) = \frac{E \cdot R_{BP}}{R_g + R_{BP}} \tag{22.16}$$

where $R_{BP} = R_B R_{in}/(R_B + R_{in})$ and $R_{in} = r_{bb'} + r_{b'e}$ as before.

At low frequencies, where the reactance of C_c is significant,

$$V_{be}(f_{low}) = \frac{E \cdot R_{BP}}{\sqrt{(R_g + R_{BP})^2 + X_{C_c}^2}} \tag{22.17}$$

Dividing (22.16) by (22.17) gives

$$\frac{V_{be}(f_m)}{V_{be}(f_{low})} = \frac{\sqrt{(R_g + R_{BP})^2 + X_{C_c}^2}}{R_g + R_{BP}}$$

There will be a frequency f_L at which $|X_{C_c}| = R_g + R_{BP}$ given by

$$\frac{1}{2\pi f_L C_c} = R_G + R_{BP} \quad \text{or} \quad f_L = \frac{1}{2\pi C_c(R_g + R_{BP})} \tag{22.18}$$

At this frequency, $V_{BE}(f_m)/V_{BE}(f_L) = \sqrt{2}$ or $V_{be}(f_L)$ is 3 dB lower than $V_{be}(f_m)$.

Example 3

Using the circuit components of the previous examples along with a signal source having an internal resistance of $R_G = 5$ kΩ, find the value of a coupling capacitor that will define a low-frequency −3-dB point at 42 Hz.

Solution. Using Eq. (22.18),

$$C_c = \frac{1}{2\pi(R_g + R_{BP})f_L}$$

where $R_{BP} = R1\|R2\|R_{\text{in}} = 75\|39\|2.8 = 2.5$ kΩ. That is,

$$C = \frac{10^6}{2\pi(5000 + 2500)(42)} \simeq 0.5 \ \mu\text{F}$$

Since a single RC time constant is involved, the voltage gain of the CE stage will appear to fall at 6 dB/octave as the frequency is reduced because more and more of the signal is dropped across C_c. However, even if C_E is very large, it too will contribute to a fall in gain as it allows more and more of the output signal to be dropped across the $R_C\|X_{C_E}$ combination, this being applied also to the input loop, resulting in negative feedback. So, at very low frequencies, the gain roll-off will tend to 12 dB/octave. The question therefore arises of how large C_E should be, and this can be conveniently answered by considering a second basic form of transistor connection as follows.

The Emitter-Follower or Common-Collector (CC) Circuit

Suppose that R_C is short-circuited in the circuit of Fig. 22.29. This will not affect the biasing because the collector voltage may take any value (the output characteristic is nearly horizontal, as seen in Fig. 22.28). However, the small-signal output voltage ceases to exist because there is now no load resistor across which it can be developed, though the output current i_c will continue to flow as before.

If now C_E is removed, i_c flows entirely through R_E and develops a voltage which can be observed at the emitter $i_e R_E$ ($\simeq i_c R_E$). Consider the magnitude of this voltage. Figure 22.26(a) shows that for a normally operating transistor, the signal component of the base-emitter voltage ΔV_{BE} (or v_{be}) is very small indeed, whereas the constant component needed for biasing is normally about 0.6 to 0.7 V. That is, $v_{be} \ll V_{BE}$. This implies that the emitter voltage must always *follow* the base voltage but at a dc level about 0.6 to 0.7 V below it. So, if an output signal is taken from the emitter, it is almost the same as the input signal at the base. In other words, *the voltage gain of an **emitter follower** is almost unity.*

If this is the case, what is the use of the emitter follower? The answer is that because the signal *current gain* is unchanged at $i_e/i_b = (h_{fe} + 1) \simeq h_{fe}$, then the power gain must also be about h_{fe}. This means in turn that the output resistance must be the resistance "looking into" the transistor from the emitter, *divided* by h_{fe}. If the parallel combination of R_g and the bias resistors is R_G, then

$$R_{\text{out}(CC)} = \frac{R_G + r_{bb'} + r_{b'e}}{h_{fe}} \tag{22.19}$$

where $R_G = R_g\|R1\|R2$ (or $R_g\|R_B$).

If a voltage generator with zero internal resistance ($R_g = 0$) were applied to the input, then this would become

$$R_{\text{out}(CC)} = \frac{r_{bb'} + r_{b'e}}{h_{fe}}$$

and if $r_{b'e} \gg r_{bb'}$ (which is usual), then

$$R_{out(CC)} \simeq \frac{r_{b'e}}{h_{fe}} = \frac{1}{g_m} \qquad (22.20)$$

Consider the numerical implications of this: if $I_C = 1$ mA, then $g_m \simeq 39$ mA/V (at room temperature), so $1/g_m \simeq 26\ \Omega$, which is a very low output resistance indeed. In fact, though it appears in parallel with R_E, it is unlikely that R_E will make any significant contribution because it is usually hundreds or thousands of ohms.

Example 4

Using the same bias resistors as for the CE examples, find the output resistance at the emitter of a CC stage.

Solution.   The parallel resistances to the left of the base are

$$R_G = R_g \| R1 \| R2 = 5 \| 75 \| 39 \approx 4.2\ \text{k}\Omega$$

Using Eq. (22.19),

$$R_{out} \simeq \frac{R_G + r_{b'e}}{h_{fe}} = \frac{R_G}{h_{fe}} + \frac{1}{g_m} \qquad \text{(neglecting } r_{bb'}\text{)}$$

where $g_m \approx 39I_C$, $I_Q = 1$ mA, and $h_{fe} = 110$, so

$$R_{out(CC)} \simeq \frac{4200}{110} + \frac{1000}{39} \approx 63.8\ \Omega$$

From values like this, it is clear that the output of an emitter follower can be thought of as a good practical dependent voltage source of very low internal resistance.

The converse is also true: the input at the base presents a high resistance. This is simply because whereas much the same signal voltage appears at the base as at the emitter, the base signal current i_b is smaller than the emitter signal current i_e by a factor of $(h_{fe} + 1) \simeq h_{fe}$. Hence, the apparent resistance at the base must be at least $h_{fe}R_E$. To this must be added $r_{bb'} + r_{b'e}$ so that

$$R_{in(CC)} \simeq r_{bb'} + r_{b'e} + h_{fe}R_E \qquad (22.21a)$$

Now h_{fe} is rarely less than about 100, so $h_{fe}R_E$ is usually predominant and

$$R_{in(CC)} \simeq h_{fe}R_E \qquad (22.21b)$$

The emitter-follower circuit is therefore a *buffer stage* because it can accept a signal at a high resistance level without significant attenuation and reproduce it at a low resistance level and with *no phase shift* (except at high frequencies).

In this configuration, the unbypassed emitter resistor R_E is obviously in series with the input circuit as well as the output circuit. Hence, it is actually a feedback resistor and so may be given the alternative symbol R_F, as in Fig. 22.33. Because all the output signal voltage is fed back in series with the input, this represents 100% voltage-derived series negative feedback.

The hybrid-π model for the bipolar transistor may now be inserted into the emitter-follower circuit of Fig. 22.33, resulting in Fig. 22.34, from which the four midfrequency parameters may be obtained. As an example of the procedures involved, consider the derivation of the voltage gain expression.

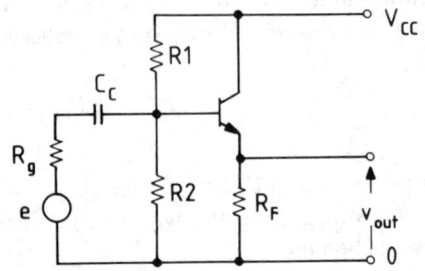

FIGURE 22.33   The emitter follower (or CC stage).

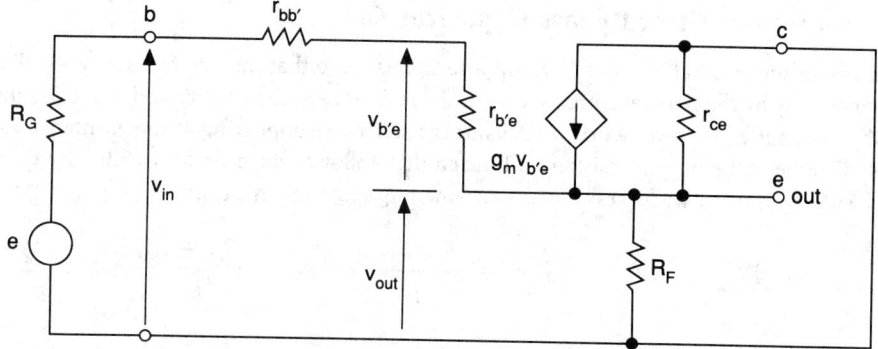

FIGURE 22.34 An emitter-follower equivalent circuit for low frequencies.

Summing signal currents at the emitter,

$$v_{out}\left(\frac{1}{R_F} + \frac{1}{r_{ce}}\right) = v_{b'e}\left(\frac{1}{r_{b'e}} + g_m\right)$$

Now $1/r_{ce} \ll 1/R_F$ and so may be neglected, and $v_{b'e} = v_{in} - v_{out}$, so

$$v_{out}\left(\frac{1}{R_F}\right) = (v_{in} - v_{out})\left(\frac{1}{r_{b'e}} + g_m\right)$$

or

$$v_{out}\left(\frac{1}{R_F} + \frac{1}{r_{b'e}} + g_m\right) = v_{in}\left(\frac{1}{r_{b'e}} + g_m\right)$$

giving

$$A_{v(CC)} = \frac{v_{out}}{v_{in}} = \frac{1/r_{b'e} + g_m}{1/r_{b'e} + g_m + 1/R_F} = \frac{1 + g_m r_{b'e}}{1 + g_m r_{b'e} + r_{b'e}/R_F}$$

$$\simeq \frac{g_m r_{b'e}}{g_m r_{b'e} + r_{b'e}/R_F} = \frac{g_m R_F}{g_m R_F + 1} \qquad (22.22)$$

which is a little less than unity as expected.

Similar derivations based on the equivalent circuit of Fig. 22.34 result in the other three basic midband operating parameters for the emitter follower, and all may be listed:

$$A_{i(CC)} \simeq h_{fe} \qquad A_{v(CC)} \to +1$$

$$R_{in(CC)} \simeq r_{bb'} + r_{b'e} + h_{fe}R_F \simeq h_{fe}R_F$$

and

$$R_{out(CC)} \simeq \frac{R_G + r_{bb'} + r_{b'e}}{h_{fe}} \,\|\, R_F \simeq \frac{R_G + r_{bb'} + r_{b'e}}{h_{fe}}$$

$$\simeq \frac{1}{g_m} \qquad \text{if } R_g \to 0 \text{ and } r_{bb'} \ll r_{b'e}$$

The Common-Emitter Bypass Capacitor C_E

In a CE circuit such as that of Fig. 22.29, suppose C_c is large so that the low-frequency –3-dB point f_L is defined only by the parallel combination of the resistance at the emitter and C_E. It will now be seen why the emitter-follower work is relevant: the resistance appearing at the emitter of the CE stage is the same as the output resistance of the emitter-follower stage, and this will now appear in parallel with R_E. If this parallel resistance is renamed $R_{emitter}$, then, neglecting $r_{bb'}$,

$$R_{emitter} = R_{out(CC)} \,\|\, R_E \simeq \frac{R_G + r_{b'e}}{h_{fe}} \,\|\, R_E \simeq \frac{R_G + r_{b'e}}{h_{fe}}$$

and if C_E were to define f_L, then

$$f_L = \frac{1}{2\pi R_{emitter} C_E} \tag{22.23a}$$

For design purposes, C_E can be extracted for any given value of f_L:

$$C_E = \frac{1}{2\pi R_{emitter} f_L} \tag{22.23b}$$

Example 5

In Example 4, let C_c be large so that only C_E defines f_L at 42 Hz, and find the value of C_E.

Solution. In the emitter-follower example, where $R_g = 5$ kΩ, $R_{out(CC)}$ was found to be 63.8 Ω, and this is the same as $R_{emitter}$ in the present case. Therefore,

$$C_E = \frac{10^6}{2\pi 63.8 \times 42} \simeq 60 \text{ }\mu\text{F}$$

This is the value of C_E that would define f_L if C_c were large. However, if C_E is to act as a short circuit at this frequency, so allowing C_c to define f_L, then its value would have to be one or two orders of magnitude greater, that is, 600 to 6000 μF.

Summarizing, three possibilities exist:

1. If C_E is very large, C_c defines f_L and a 6-dB/octave roll-off results.
2. If C_c is large, C_E defines f_L and again a 6-dB/octave roll-off results.
3. If both C_c and C_E act together, a 12-dB/octave roll-off results.

In point of fact, at frequencies much less than f_L, both conditions (1) and (2) eventually produce 12-dB/octave roll-offs as the alternate "large" capacitors come into play at very low frequencies, but since the amplifier will not still have a useful gain at such frequencies, this is of little importance.

High-Frequency Response

Unlike the low-frequency response situation, the high-frequency response is governed by the small distributed capacitances inside the transistor structure, and these have been lumped together in the hybrid-π model of Fig. 22.31 as $C_{b'e}$ and $C_{b'c}$. At high frequencies, $r_{b'c}$ may be neglected in comparison with the reactance of $C_{b'c}$, so the model may be simplified as in Fig. 22.35(a). From this it will be seen that $C_{b'c}$ is a capacitance which appears from the output to the input so that it may be converted by the Miller Effect into a capacitance at the input of value:

$$C_{b'c}(1 - A_v) = C_{b'c}(1 + g_m R_C)$$

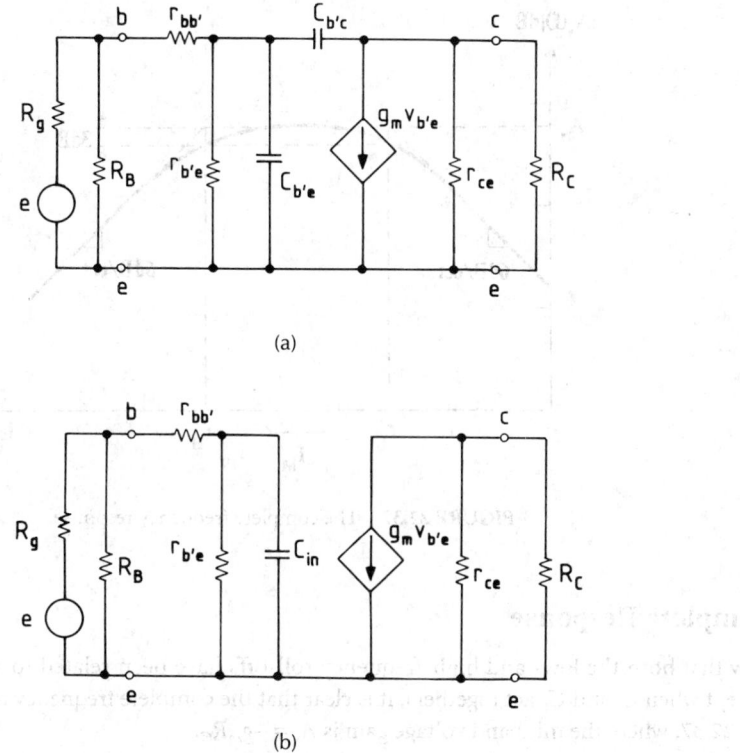

(a)

(b)

FIGURE 22.35   (a) The high-frequency hybrid-π model and (b) its simplification.

This will now add to $C_{b'e}$ to give C_{in}:

$$C_{in} = C_{b'e} + C_{b'c}(1 + g_m R_C) \tag{22.24}$$

This simplification is shown in Fig. 22.35(b), where C_{in} is seen to be shunted by the input parts of the model. These input parts may be reduced by sequential use of Thévenin–Norton transformations to result in Fig. 22.36, which is a simple parallel RC circuit driven by a current source. The actual value of this current source is immaterial—what matters is that the input signal to be amplified, $v_{b'e}$, will be progressively reduced as the frequency rises and the reactance of C_{in} falls.

Using a sinusoidal source, $V_{b'e}$ will be 3 dB down when $R = |X_{C_{in}}|$, which gives

$$R = \frac{1}{2\pi f_H C_{in}} \quad \text{or} \quad f_H = \frac{1}{2\pi R C_{in}} \tag{22.25}$$

where $R = (R_G + r_{bb'}) \| r_{b'e}$ from the circuit reduction.

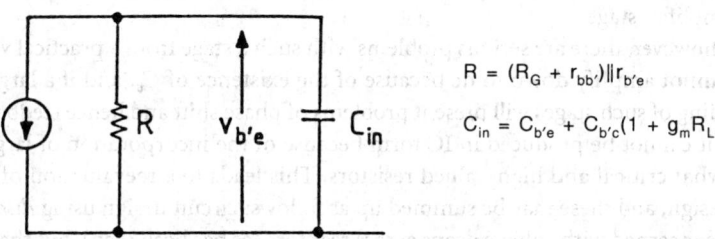

$$R = (R_G + r_{bb'}) \| r_{b'e}$$

$$C_{in} = C_{b'e} + C_{b'c}(1 + g_m R_L)$$

FIGURE 22.36   Simplification of the input part of the high-frequency hybrid-π model.

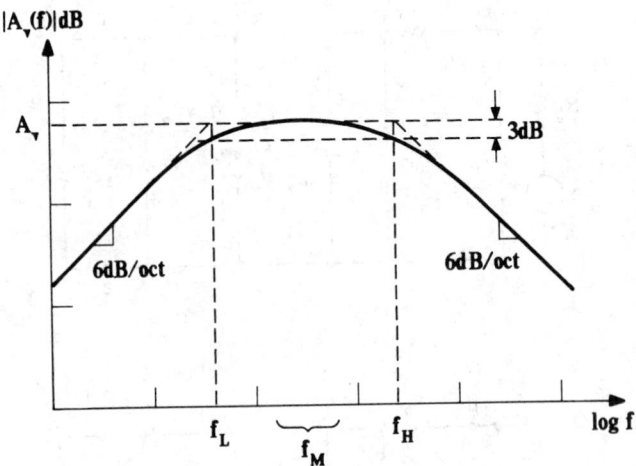

FIGURE 22.37   The complete frequency response.

Complete Response

Now that both the low- and high-frequency roll-offs have been related to single time constants (except when C_c and C_E act together), it is clear that the complete frequency response will look like Fig. 22.37, where the midband voltage gain is $A_v = -g_m R_C$.

Design Comments

The design of a simple single-transistor amplifier stage has now been covered in terms of both biasing and small-signal performance. These two concepts have been kept separate, but it will have been noticed that they are bridged by the transconductance, because $g_m = (q/kT)I_Q$ ($\simeq 39 I_Q$ at room temperature). That is, when I_Q has been determined, then the small-signal performance follows from expressions involving g_m.

In fact, once the quiescent voltage across the load resistor of a CE stage has been determined, the voltage gain follows from this irrespective of the values of I_Q and R_C.

If the quiescent voltage at the collector is V_{out}, then in dc biasing terms,

$$V_{R_C} = I_Q R_C = (V_{CC} - V_{out})$$

and in small-signal terms,

$$A_v = -g_m R_C \cong -39 I_Q R_C \qquad \text{(at 25°C)}$$
$$= -39(V_{CC} - V_{out})$$

Thus, g_m really does act as a bridge between the bias and the small-signal conditions for the bipolar transistor amplifier stage.

Unfortunately, however, there are serious problems with such a stage from a practical viewpoint. For example, it cannot amplify down to dc because of the existence of C_c, and if a larger gain is needed, the cascading of such stages will present problems of phase shift and hence feedback stability. Furthermore, it cannot be produced in IC form because of the incorporation of large capacitances and somewhat critical and high-valued resistors. This leads to a reevaluation of the basic tenets of circuit design, and these may be summed up as follows: circuit design using *discrete* components is largely concerned with voltage drops across resistors (as has been seen), but the design of ICs depends extensively on *currents* and *current sources and sinks*.

Integrated Circuits

Monolithic ICs are fabricated on single chips of silicon or *dice* (the singular being *die*). This means that the active and passive structures on the chips are manufactured all at the same time, so it is easy to ensure that a large number of such structures are identical, or bear some fixed ratio to one another, but it is more difficult to establish precise values for such sets of structures. For example, a set of transistors may all exhibit almost the same values of h_{FE}, but the actual numerical value of h_{FE} may be subject to wider tolerances. Similarly, many pairs of resistors may bear a ratio $n{:}1$ to each other, but the actual values of these resistors are more difficult to define. So, in IC design, it is very desirable to exploit the close similarity of devices (or close ratios) rather than depend upon their having predictable absolute values. This approach has led to two ubiquitous circuit configurations, both of which depend upon device similarity: the **long-tailed pair** or **difference amplifier** (often called the *differential amplifier*), and the **current mirror**. This section will treat both, and the former is best introduced by considering the **degenerate common-emitter** stage.

The Degenerate Common-Emitter Stage

Consider two CE stages which are identical in every respect but which have no emitter resistor bypass capacitors, as shown in Fig. 22.38. Also, notice that in these diagrams, two power supply rails have been used, a positive one at V_{CC}^+ and a negative one at V_{CC}^-. The reason for this latter, negative rail is that the bases may be operated via signal sources referred to a common line or ground. (If, for example, V_{CC}^+ and V_{CC}^- are obtained from batteries as shown, then the common line is simply the junction of the two batteries, as is also shown.) The absence of capacitors now means that amplification down to dc is possible.

It is now very easy to find the quiescent collector currents I_Q, because from a dc bias point of view the bases are connected to ground via resistances R_g, which will be taken as having low values so that they drop negligibly small voltages. Hence,

$$I_Q = \frac{|V_{CC}^-| - V_{BE}}{R_F} \qquad (22.26)$$

[For example, if industry-standard supplies of ±15 V are used, $V_{BE} = 0.6$ V, and for $R_F = 15$ kΩ, then $I_Q = (15 - 0.6)/15 = 0.96 \approx 1$ mA.]

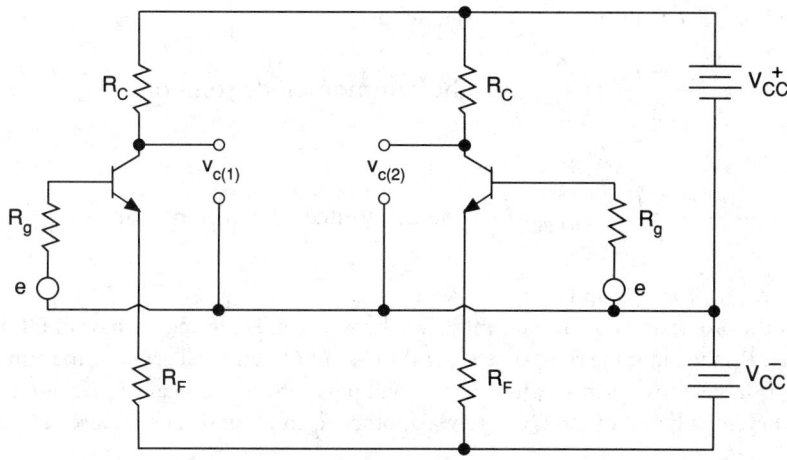

FIGURE 22.38 Two degenerate CE stages.

Now suppose that identical signals e are applied. At each collector, this will result in an output signal voltage v_c, where $v_c = -i_c R_C$. Also, at each emitter, the output signal voltage will be $v_e = i_e R_F \simeq i_c R_F$. That is,

$$\frac{v_c}{v_e} \simeq \frac{-i_c R_c}{i_c R_F} = -\frac{R_C}{R_F}$$

If the voltage gain from base to collector of a degenerate CE stage is $A_{v(dCE)}$ and the voltage gain from base to emitter is simply the emitter-follower gain $A_{v(CC)}$, then

$$v_c = A_{v(dCE)}e \qquad \text{and} \qquad v_e = A_{v(CC)}e$$

giving

$$\frac{A_{v(dCE)}}{A_{v(CC)}} \simeq -\frac{R_C}{R_F}$$

Now $A_{v(CC)}$ is known from Eq. (22.22) so that

$$A_{v(dCE)} \simeq -A_{v(CC)} \frac{R_C}{R_F} = -\frac{g_m R_C}{1 + g_m R_F} \simeq -\frac{R_C}{R_F} \tag{22.27}$$

Note that the input resistance to each base is as for the emitter-follower stage:

$$R_{in(dCE)} = r_{bb'} + r_{b'e} + h_{fe}R_F \simeq h_{fe}R_F$$

Now consider what happens if the emitters are connected together as in Fig. 22.39, where the two resistors R_F have now become R_X, where $R_X = \frac{1}{2}R_F$.

The two quiescent emitter currents now combine to give $I_X = 2I_E \simeq 2I_C$, and otherwise the circuit currents and voltages remain undisturbed. So, if the two input signals are identical, then the two output signals will also be identical. This circuit is now called a *difference amplifier*, and the reason will become obvious as soon as the two input signals differ.

The Difference Amplifier

In Fig. 22.39, if $e_1 = e_2$, these are called common-mode input signals, $e_{in(CM)}$, and they will be amplified by $-R_C/R_F$ as for the degenerate CE stage. However, if $e_1 \neq e_2$, then $e_1 - e_2 = e_{in}$, the difference input signal. The following definitions now apply:

$$\frac{e_1 + e_2}{2} = e_{in(CM)} \qquad \text{the common mode component}$$

and

$$\frac{\pm(e_1 - e_2)}{2} = e_{in(diff)} \qquad \text{the difference component, or } \tfrac{1}{2}e_{in}$$

Hence, $e_1 = e_{in(CM)} + e_{in(diff)}$ and $e_2 = e_{in(CM)} - e_{in(diff)}$.

Consider the progress of a signal current driven by $e_1 - e_2$ and entering the base of Q1. It will first pass through R_g, then into the resistance R_{in} at the base of Q1, and will arrive at the emitter of Q2. Here, if R_X is large, most of this signal current will pass into the resistance presented by the Q2 emitter and eventually out of the Q2 base via another R_g to ground. The total series resistance is therefore

$$R_g + R_{in} = R_g + r_{bb'} + r_{b'e} + h_{fe}R_{emitter(2)}$$

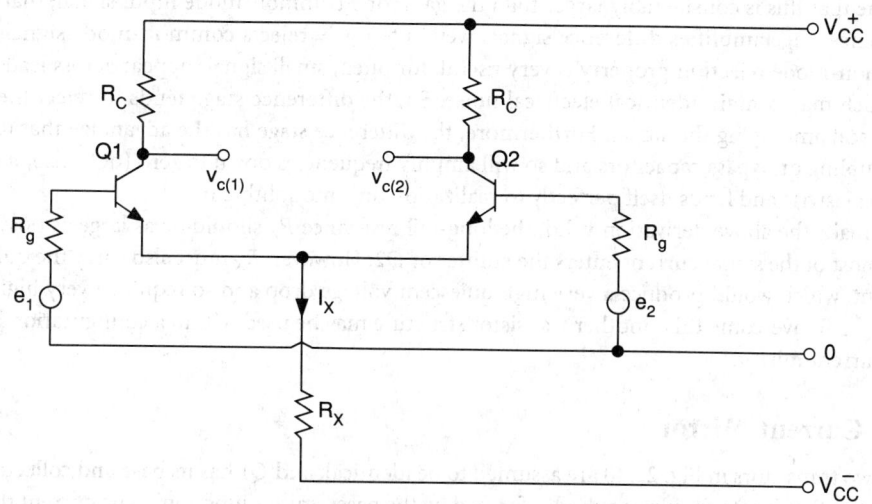

FIGURE 22.39 The difference amplifier.

But

$$R_{emitter(2)} = \frac{R_g + r_{bb'} + r_{b'e}}{h_{fe}}$$

so

$$R_g + R_{in} = 2(R_g + r_{bb'} + r_{b'e})$$

which is the resistance between the two signal sources. Hence,

$$i_{b(1)} = -i_{b(2)} = \frac{e_1 - e_2}{2(R_g + r_{bb'} + r_{b'e})}$$

giving

$$v_{c(1)} = -v_{c(2)} = \frac{h_{fe}R_C(e_1 - e_2)}{2(R_g + r_{bb'} + r_{b'e})}$$

so that the overall difference voltage gain to each collector is

$$A_{ov} = \frac{v_c}{e_1 - e_2} = \frac{\pm h_{fe}R_C}{2(R_g + r_{bb'} + r_{b'e})} \qquad (22.28a)$$

If the voltage gain with the input signal measured between the actual bases is needed, R_g may be removed to give

$$A_v = \frac{\pm h_{fe}R_C}{2(r_{bb'} + r_{b'e})} \qquad (22.28b)$$

Finally, if the output signal is measured between the collectors (which will be twice that at each collector because they are in antiphase), the difference-in–to–difference-out voltage gain will be

$$A_{v(diff)} = \frac{h_{fe}R_C}{r_{bb'} + r_{b'e}} \simeq \frac{h_{fe}R_C}{r_{b'e}} = g_m R_C \qquad (22.28c)$$

which is the same as for a single CE stage.

Note that this is considerably larger than the gain for a common-mode input signal; that is, the difference stage amplifies difference signals well but largely rejects common-mode signals. This common-mode rejection property is very useful, for often, small signals appear across leads, both of which may contain identical electrical noise. So, the difference stage tends to reject the noise while still amplifying the signal. Furthermore, the difference stage has the advantage that it needs no coupling or bypass capacitors and so will amplify frequencies down to zero (dc). Also, it is very stable biaswise and lends itself perfectly to realization on a monolithic IC.

To make the above derivation valid, the long-tail resistance R_X should be as large as possible so that most of the signal current enters the emitter of Q2. However, R_X must also carry the quiescent current, which would produce a very high quiescent voltage drop and so require a very high value of V_{CC-}. To overcome this, another transistor structure may be used within a configuration known as a current mirror.

The Current Mirror

The two transistors in Fig. 22.40 are assumed to be identical, and Q1 has its base and collector connected so that it acts simply as a diode (formed by the base-emitter junction). The current through it is therefore

$$I = \frac{V_{CC} - V_{BE}}{R} \tag{22.29}$$

The voltage drop V_{BE} so produced is applied to Q2 as shown so that it is forced to carry the same collector current I; that is, it mirrors the current in Q1.

The transistor Q2 is now a device that carries a dc $I_{C(2)} = I$ but presents a large incremental resistance r_{ce} at its collector. This is exactly what is required by the difference amplifier pair, so it may be used in place of R_X.

The Difference Stage with Current Mirror Biasing

Figure 22.41 shows a complete difference stage complete with a current mirror substituting for the long-tail resistor R_X, where the emitter quiescent currents combine to give I_X:

$$I_X = \frac{V_{CC^+} + \left| V_{CC^-} \right| - V_{BE(3)}}{R} = I \tag{22.30}$$

This quiescent or bias current is very stable, because the change in $V_{BE(3)}$ due to temperature variations is exactly matched by that required by Q4 to produce the same current. The difference gain will be as discussed above, but the common-mode gain will be extremely low because of the high incremental resistance r_{ce} presented by the long-tail transistor.

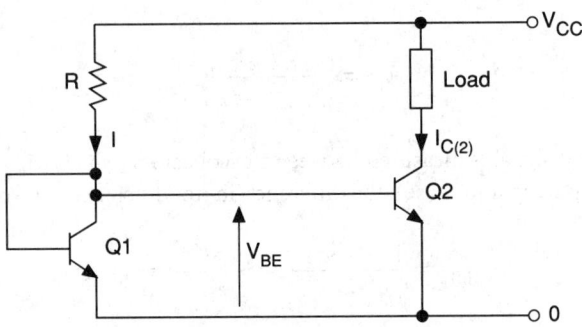

FIGURE 22.40   The current mirror.

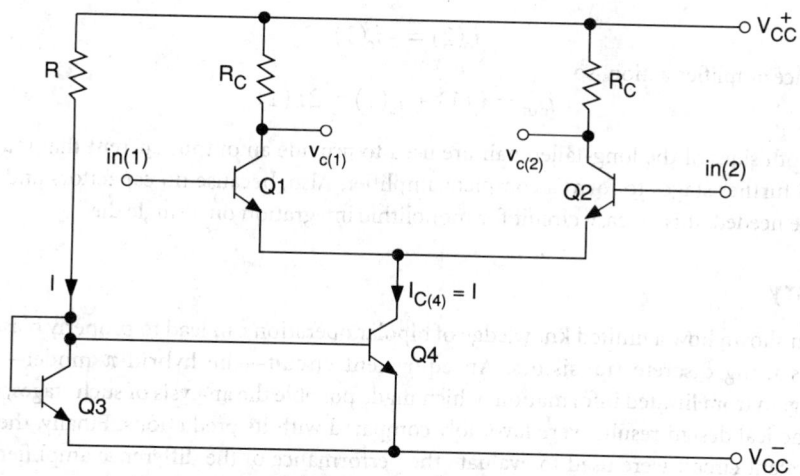

FIGURE 22.41 Current mirror biasing.

The Current Mirror as a Load

A second current mirror may be used as a load for the difference amplifier, as shown in Fig. 22.42. This must utilize *pnp* transistor structures so that the Q6 collector loads the Q2 collector with a large incremental resistance $r_{ce}(6)$, making for an extremely high voltage gain. Furthermore, Q5 and Q6 combine the signal output currents of both Q1 and Q2 to perform a double-ended–to–single-ended conversion as follows. Taking signal currents,

$$i_{out} = i_c(6) - i_c(2)$$

But

$$i_c(6) = i_c(5)$$

by current mirror action, and

$$i_c(5) = i_c(1)$$

so

$$i_c(6) = i_c(1)$$

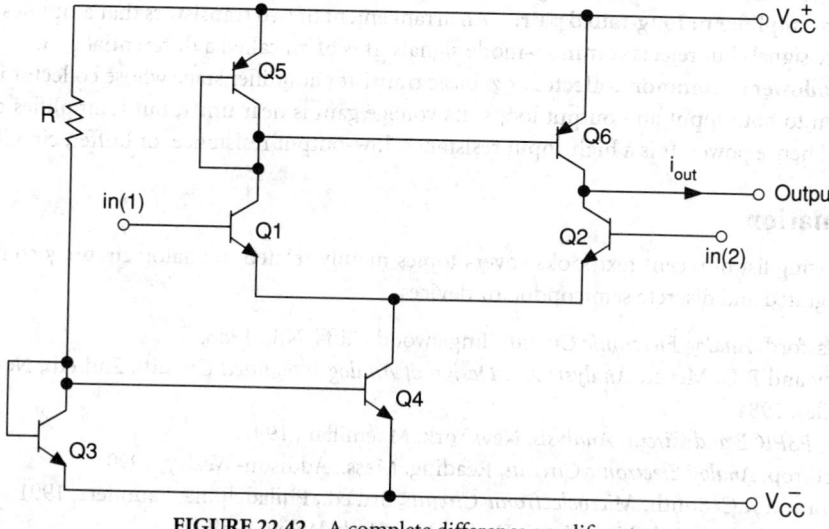

FIGURE 22.42 A complete difference amplifier stage.

Also,
$$i_c(2) = -i_c(1)$$
by difference amplifier action, so
$$i_{\text{out}} = i_c(1) + i_c(1) = 2i_c(1) \tag{22.31}$$

Thus, both sides of the long-tailed pair are used to provide an output current that may then be applied to further stages to form a complete amplifier. Also, because no capacitors and only one resistor are needed, it is an easy circuit for monolithic integration on a single die.

Summary

It has been shown how a limited knowledge of bipolar operation can lead to properly biased amplifier stages using discrete transistors. An equivalent circuit—the hybrid-π model—was then derived, again from limited information, which made possible the analysis of such stages, and some purely practical design results were favorably compared with its predictions. Finally, the tenets of this equivalent circuit were used to evaluate the performance of the difference amplifier and current mirror circuits, which are the cornerstones of modern electronic circuit design in a very wide variety of its manifestations. These circuits are, in fact, the classic transconductance and translinear elements that are ubiquitous in modern IC signal conditioning and function networks.

It should be recognized that there are many models other than the one introduced here, from the simple but very common *h*-parameter version to complex and comprehensive versions developed for computer-aided design (CAD) methods. However, the present elementary approach has been from a design rather than an analytical direction, for it is obvious that powerful modern computer-oriented methods such as the SPICE variants become useful only when a basic circuit configuration has been established, and at the time of writing, this is still the province of the human designer.

Defining Terms

Biasing circuit: A circuit that holds a transistor in an operating condition ready to receive signals.

Common emitter: A basic transistor amplifier stage whose emitter is common to both input and output loops. It amplifies voltage, current, and hence power.

Current mirror: An arrangement of two (or more) transistors such that a defined current passing into one is mirrored in another at a high resistance level.

Degenerate common emitter: A combination of the common-emitter and emitter-follower stages with a very well-defined gain.

Difference amplifier or long-tailed pair: An arrangement of two transistors that amplifies difference signals but rejects common-mode signals. It is often called a differential pair.

Emitter follower or common collector: A basic transistor amplifier stage whose collector is common to both input and output loops. Its voltage gain is near unity, but it amplifies current and hence power. It is a high-input resistance, low-output resistance, or buffer, circuit.

Further Information

The following list of recent textbooks covers topics mainly related to analog circuitry containing both integrated and discrete semiconductor devices.

G. M. Glasford, *Analog Electronic Circuits,* Englewood Cliffs, N.J.: 1986.

P. R. Gray and R.G. Meyer, *Analysis and Design of Analog Integrated Circuits,* 2nd ed., New York: Wiley, 1984.

J. Keown, *PSPICE and Circuit Analysis,* New York: Macmillan, 1991.

R. B. Northrop, *Analog Electronic Circuits,* Reading, Mass.: Addison-Wesley, 1990.

A. S. Sedra and K.C. Smith, *Microelectronic Circuits,* 3rd ed., Philadelphia: Saunders, 1991.

J. Watson, *Analog and Switching Circuit Design,* New York: Wiley, 1989.

22.3 The Metal-Oxide Semiconductor Field-Effect Transistor (MOSFET)

John R. Brews

The MOSFET is a transistor that uses a control electrode, the **gate**, to capacitively modulate the conductance of a surface **channel** joining two end contacts, the **source** and the **drain**. The gate is separated from the semiconductor **body** underlying the gate by a thin *gate insulator*, usually silicon dioxide. The surface channel is formed at the interface between the semiconductor body and the gate insulator (see Fig. 22.43).

The MOSFET can be understood by contrast with other field-effect devices, like the *JFET*, or junction field-effect transistor, and the *MESFET*, or metal semiconductor field-effect transistor [Hollis and Murphy, 1990]. These other transistors modulate the conductance of a *majority-carrier* path between two *ohmic* contacts by capacitive control of its cross section. (Majority carriers are those in greatest abundance in a field-free semiconductor, electrons in *n*-type material and holes in *p*-type material.) This modulation of the cross section can take place at any point along the length of the channel, so the gate electrode can be positioned anywhere and need not extend the entire length of the channel.

Analogous to these field-effect devices is the *buried-channel, depletion-mode,* or *normally on*

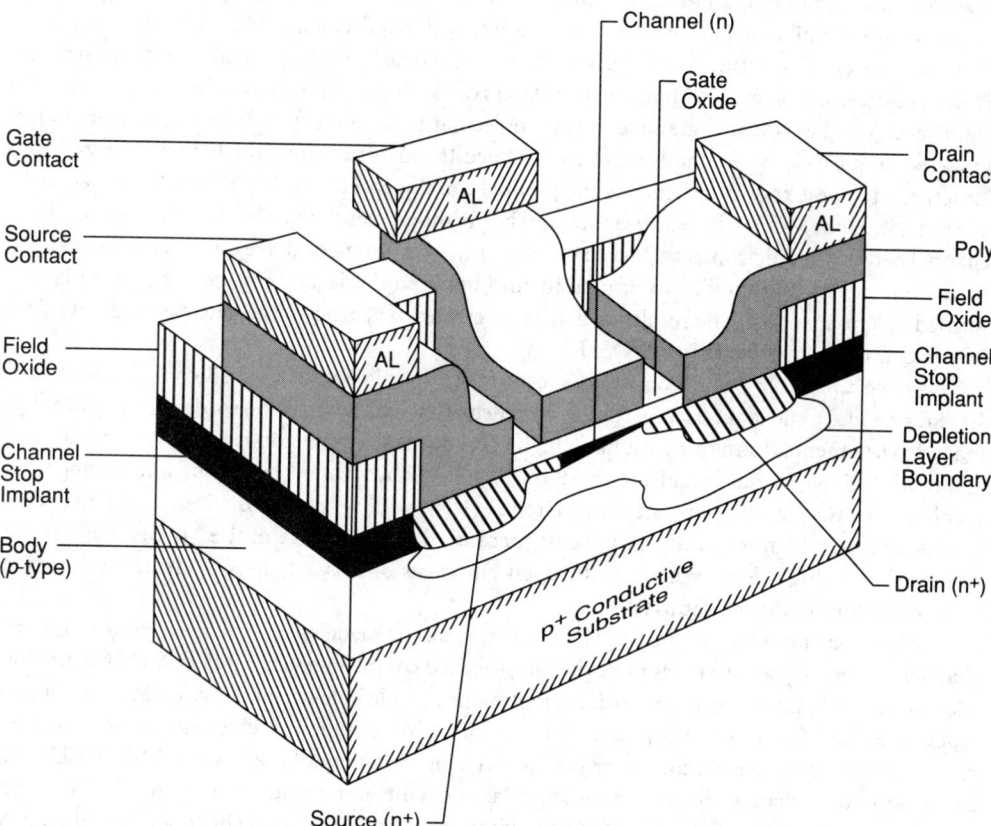

FIGURE 22.43 A high-performance *n*-channel MOSFET. The device is isolated from its neighbors by a surrounding thick *field oxide* under which is a heavily doped *channel stop* implant intended to suppress accidental channel formation that could couple the device to its neighbors. The drain contacts are placed over the field oxide to reduce the capacitance to the body, a parasitic that slows response times. These structural details are described later. (*Source:* After Brews, 1990.)

MOSFET, which contains a surface layer of the same doping type as the source and drain (opposite type to the semiconductor body of the device). As a result, it has a built-in or normally on channel from source to drain with a conductance that is reduced when the gate depletes the majority carriers.

In contrast, the true MOSFET is an *enhancement-mode* or *normally off* device. The device is normally off because the body forms *pn* junctions with both the source and the drain, so no majority-carrier current can flow between them. Instead, *minority-carrier* current can flow, provided minority carriers are available. As discussed later, for gate biases that are sufficiently attractive, above **threshold**, minority carriers are drawn into a surface channel, forming a conducting path from source to drain. The gate and channel then form two sides of a capacitor separated by the gate insulator. As additional attractive charges are placed on the gate side, the channel side of the capacitor draws a balancing charge of minority carriers from the source and the drain. The more charges on the gate, the more populated the channel, and the larger the conductance. Because the gate *creates* the channel, to ensure electrical continuity the gate must extend over the entire length of the separation between source and drain.

The MOSFET channel is created by attraction to the gate and relies upon the insulating layer between the channel and the gate to prevent leakage of minority carriers to the gate. As a result, MOSFETs can be made only in material systems that provide very good gate insulators, and the best system known is the silicon–silicon dioxide combination. This requirement for a good gate insulator is not so important for JFETs and MESFETs, where the role of the gate is to *push away* majority carriers rather than to attract minority carriers. Thus, in GaAs systems where good insulators are incompatible with other device or fabricational requirements, MESFETs are used.

A more recent development in GaAs systems is the heterostructure field-effect transistor, or HFET [Pearton and Shaw, 1990], made up of layers of varying compositions of Al, Ga, and As or In, Ga, P, and As. These devices are made using molecular beam epitaxy or by organometallic vapor phase epitaxy, expensive methods still being refined for manufacture. HFETs include a variety of structures, the best known of which is the modulation doped FET, or MODFET. HFETs are field-effect devices, not MOSFETs, because the gate simply modulates the carrier density in a preexistent channel between ohmic contacts. The channel is formed spontaneously, regardless of the quality of the gate insulator as a condition of equilibrium between the layers, just as a depletion layer is formed in a *pn* junction. The resulting channel is created very near to the gate electrode, resulting in gate control as effective as in a MOSFET.

The silicon-based MOSFET has been successful primarily because the silicon–silicon dioxide system provides a stable interface with low trap densities, and because the oxide is impermeable to many environmental contaminants, has a high breakdown strength, and is easy to grow uniformly and reproducibly [Nicollian and Brews, 1982]. These attributes allow easy fabrication using lithographic processes, resulting in integrated circuits (ICs), with very small devices, very large device counts, and very high reliability at low cost. Because the importance of the MOSFET lies in this relationship to high-density manufacture, an emphasis of this article is to describe the issues involved in continuing miniaturization.

An additional advantage of the MOSFET is that it can be made using either electrons or holes as channel carrier. Using both types of devices in so-called complementary MOS (CMOS) technology allows circuits that draw no *dc* power if current paths include at least one series connection of both types of device because, in steady state, only one or the other type conducts, not both at once. Of course, in exercising the circuit, power is drawn during switching of the devices. This flexibility in choosing *n*- or *p*-channel devices has enabled large circuits to be made that use low power levels. Hence, complex systems can be manufactured without expensive packaging or cooling requirements.

Current-Voltage Characteristics

The derivation of the current-voltage characteristics of the MOSFET can be found in several sources [Annaratone, 1986; Brews, 1981; Pierret, 1990]. Here a qualitative discussion is provided.

Strong-Inversion Characteristics

In Fig. 22.44 the source-drain current I_D is plotted versus drain-to-source voltage V_D (the *I-V* curves for the MOSFET). At low V_D the current increases approximately linearly with increased V_D, behaving like a simple resistor with a resistance that is controlled by the gate voltage V_G: as the gate voltage is made more attractive for channel carriers, the channel becomes stronger, more carriers are contained in the channel, and its resistance R_{ch} drops. Hence, at larger V_G the current is larger.

At large V_D the curves flatten out, and the current is less sensitive to drain bias. The MOSFET is said to be in *saturation*. There are different reasons for this behavior, depending upon the field along the channel caused by the drain voltage. If the source-drain separation is short, near or below a micrometer, the usual drain voltage is sufficient to create fields along the channel of more than a few $\times 10^4$ V/cm. In this case the carrier energy is sufficient for carriers to lose energy by causing vibrations of the silicon atoms composing the crystal (optical phonon emission). Consequently, the carrier velocity does not increase much with increased field, saturating at a value $\upsilon_{sat} \approx 10^7$ cm/s in silicon MOSFETs. Because the carriers do not move faster with increased V_D, the current also saturates.

For longer devices the current-voltage curves saturate for a different reason. Consider the potential along the insulator-channel interface, the surface potential. Whatever the surface potential is at the source end of the channel, it varies from the source end to a value larger at the drain end by V_D because the drain potential is V_D higher than the source. The gate, on the other hand, is at the same potential everywhere. Thus, the difference in potential between the gate and the source is larger

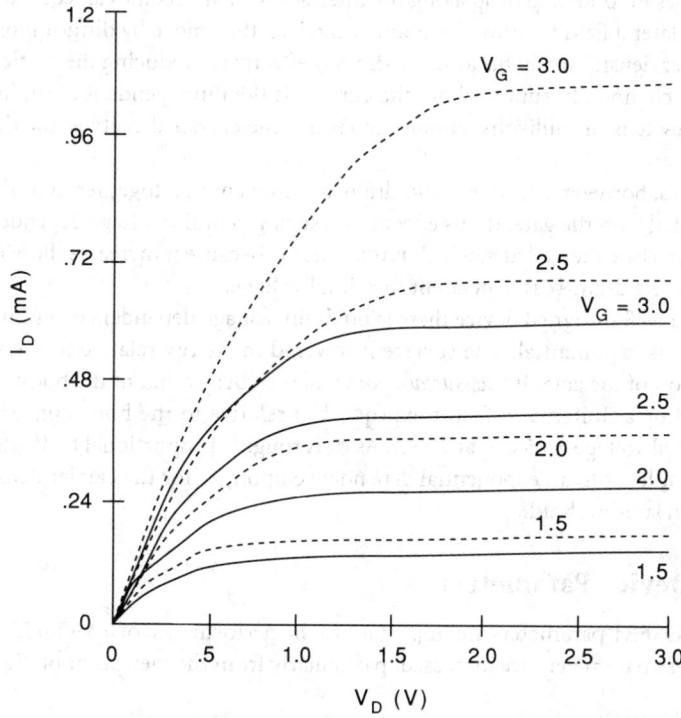

FIGURE 22.44 Drain current I_D versus drain voltage V_D for various choices of gate bias V_G. The dashed-line curves are for a long-channel device for which the current in saturation increases quadratically with gate bias. The solid-line curves are for a *short-channel* device that is approaching *velocity saturation* and thus exhibits a more linear increase in saturation current with gate bias, as discussed in the text.

than that between the gate and the drain. Correspondingly, the oxide field at the source is larger than that at the drain, and as a result less charge can be supported at the drain. This reduction in attractive power of the gate reduces the number of carriers in the channel at the drain end, increasing channel resistance. In short, we have $I_D \approx V_D / R_{ch}$, but the channel resistance $R_{ch} = R_{ch}(V_D)$ is increasing with V_D. As a result, the current-voltage curves do not continue along the initial straight line, but bend over and saturate.

Another difference between the current-voltage curves for short devices and those for long devices is the dependence on gate voltage. For long devices, the current level in saturation, $I_{D,sat}$, increases quadratically with gate bias. The reason is that the number of carriers in the channel is proportional to $V_G - V_{TH}$ (where V_{TH} is the *threshold voltage*), as is discussed later, the channel resistance $R_{ch} \propto 1/(V_G - V_{TH})$, and the drain bias in saturation is approximately V_G. Thus, $I_{D,sat} = V_D/R_{ch} \propto (V_G - V_{TH})^2$, and we have quadratic dependence. When the carrier velocity is saturated, however, the dependence of the current on drain bias is suppressed because the speed of the carriers is fixed at v_{sat}, and $I_{D,sat} \propto v_{sat}/R_{ch} \propto (V_G - V_{TH}) v_{sat}$, a linear gate-voltage dependence. As a result, the current available from a short device is not as large as would be expected if we assumed it behaved like a long device.

Subthreshold Characteristics

Quite different current-voltage behavior is seen in **subthreshold**, that is, for gate biases so low that the channel is in *weak inversion*. In this case the number of carriers in the channel is so small that their charge does not affect the potential, and channel carriers simply must adapt to the potential set up by the electrodes and the dopant ions. Likewise, in subthreshold any flow of current is so small that it causes no potential drop along the interface, which becomes an equipotential.

As there is no lateral field to move the channel carriers, they move by diffusion only, driven by a gradient in carrier density set up because the drain is effective in reducing the carrier density at the drain end of the channel. In subthreshold the current is then independent of drain bias once this bias exceeds a few tens of millivolts, enough to reduce the carrier density at the drain end of the channel to near zero.

In short devices, however, the source and drain are close enough together to begin to share control of the potential with the gate. If this effect is too strong, a drain-voltage dependence of the subthreshold characteristic then occurs, which is undesirable because it increases the MOSFET off current, and can cause a drain-bias dependent threshold voltage.

Although for a well-designed device there is no drain-voltage dependence in subthreshold, gate-bias dependence is exponential. The surface is lowered in energy relative to the semiconductor body by the action of the gate. If this *surface potential* is ϕ_S below that of the body, the carrier density is enhanced by a Boltzmann factor $\exp(q\phi_S/kT)$ relative to the body concentration, where $kT/q =$ the thermal voltage ≈ 25 mV at 290 K. As ϕ_S is roughly proportional to V_G, this exponential dependence on ϕ_S leads to an exponential dependence upon V_G for the carrier density and, hence, for the current in subthreshold.

Important Device Parameters

A number of MOSFET parameters are important to the performance of a MOSFET. In this subsection some of these parameters are discussed, particularly from the viewpoint of digital ICs.

Threshold Voltage

The threshold voltage is vaguely defined as the gate voltage V_{TH} at which the channel begins to form. At this voltage devices begin to switch from "off" to "on," and circuits depend on a voltage swing that straddles this value. Thus, threshold voltage helps in deciding the necessary supply voltage for circuit operation, and also it helps in determining the leakage or "off" current that flows when the device is in the off state.

Threshold voltage is controlled by oxide thickness d and by body doping. To control the body doping, ion implantation is used so that the dopant-ion density is not simply a uniform extension of the bulk, background level N_B ions/unit volume but has superposed upon it an implanted-ion density. To estimate the threshold voltage, we need a picture of what happens in the semiconductor under the gate as the gate voltage is changed from its off level toward threshold.

If we imagine changing the gate bias from its off condition toward threshold, at first the result is to repel majority carriers, forming a surface *depletion layer* (refer to Fig. 22.43). In the depletion layer there are almost no carriers present, but there are dopant ions. In n-type material these dopant ions are positive donor impurities that cannot move under fields because they are locked in the silicon lattice, where they have been deliberately introduced to replace silicon atoms. In p-type material these dopant ions are negative acceptors. Thus, each charge added to the gate electrode to bring the gate voltage closer to threshold causes an increase in the depletion-layer width sufficient to balance the gate charge by an equal but opposite charge of dopant ions in the silicon depletion layer.

This expansion of the depletion layer continues to balance the addition of gate charge until threshold is reached. Then this charge response changes: above threshold any additional gate charge is balanced by an increasingly strong inversion layer or channel. The border between a depletion-layer and an inversion-layer response, threshold, should occur when

$$\frac{dqN_{inv}}{d\phi_S} = \frac{dQ_D}{d\phi_S} \tag{22.32}$$

where $d\phi_S$ is the small change in surface potential that corresponds to our incremental change in gate charge, qN_{inv} is the inversion-layer charge/unit area, and Q_D is the depletion-layer charge/unit area. According to Eq. (22.32), the two types of response are equal at threshold, so one is larger than the other on either side of this condition. To be more quantitative, the rate of increase in qN_{inv} is exponential; that is, its rate of change is proportional to qN_{inv}, so as qN_{inv} increases, so does the left side of Eq. (22.32). On the other hand, Q_D has a square-root dependence on ϕ_S, which means its rate of change becomes smaller as Q_D increases. Thus, as surface potential is increased, the left side of Eq. (22.32) increases proportional to qN_{inv} until, at threshold, Eq. (22.32) is satisfied. Then, beyond threshold, the exponential increase in qN_{inv} with ϕ_S swamps Q_D, making change in qN_{inv} the dominant response. Likewise, below threshold, the exponential decrease in qN_{inv} with decreasing ϕ_S makes qN_{inv} negligible and change in Q_D becomes the dominant response. The abruptness of this change in behavior is the reason for the term *threshold* to describe MOSFET switching.

To use Eq. (22.32) to find a formula for threshold voltage, we need expressions for N_{inv} and Q_D. Assuming the interface is held at a lower energy than the bulk due to the charge on the gate, the minority-carrier density at the interface is larger than in the bulk semiconductor, even below threshold. Below threshold and even up to the threshold of Eq. (22.32), the number of charges in the channel/unit area N_{inv} is given for n-channel devices approximately by [Brews, 1981]:

$$N_{inv} \approx d_{INV} \frac{n_i^2}{N_B} e^{q(\phi_S - V_S)/kT} \tag{22.33}$$

where the various symbols are defined as follows: n_i = intrinsic carrier density/unit volume $\approx 10^{10}$/cm^3 in silicon at 290 K, and V_S = body reverse bias, if any. The first factor, d_{INV}, is an effective depth of minority carriers from the interface given by

$$d_{INV} = \frac{\varepsilon_s kT/q}{Q_D} \tag{22.34}$$

where Q_D = depletion-layer charge/unit area due to charged dopant ions in the region where there are no carriers and ε_s is the dielectric permittivity of the semiconductor.

Equation (22.33) expresses the net minority-carrier density/unit area as the product of the bulk minority-carrier density/unit volume, n_i^2/N_B, with the depth of the minority-carrier distribution d_{INV} multiplied in turn by the customary Boltzmann factor $\exp[q(\phi_S - V_S)/kT]$ expressing the enhancement of the interface density over the bulk due to lower energy at the interface. The depth d_{INV} is related to the carrier distribution near the interface using the approximation (valid in *weak inversion*) that the minority-carrier density decays exponentially with distance from the oxide-silicon surface. In this approximation, d_{INV} is the *centroid* of the minority-carrier density. For example, for a uniform bulk doping of 10^{16} dopant ions /cm^3 at 290 K, using Eq. (22.33) and the surface potential at threshold from Eq. (22.38) below ($\phi_{TH} = 0.69$ V), there are $Q_D/q = 3 \times 10^{11}$ charges/cm^2 in the depletion layer at threshold. This Q_D corresponds to a $d_{INV} = 5.4$ nm and a carrier density at threshold of $N_{inv} = 5.4 \times 10^9$ charges/cm^2.

The next step in using the definition of threshold, Eq. (22.32), is to introduce the depletion-layer charge/unit area Q_D. For the ion-implanted case, Q_D is made up of two terms [Brews, 1981]:

$$Q_D = qN_B L_B \left\{ 2[q\phi_{TH}/(kT) - m_1 - 1] \right\}^{1/2} + qD_I \qquad (22.35)$$

where the first term is Q_B, the depletion-layer charge from bulk dopant atoms in the depletion layer with a width that has been reduced by the first moment of the implant, namely, m_1 given in terms of the centroid of the implant x_C by

$$m_1 = \frac{D_I x_C}{N_B L_B^2} \qquad (22.36)$$

The second term is the additional charge due to the implanted-ion density within the depletion layer, D_I unit area. The Debye length L_B is defined as

$$L_B^2 \equiv \frac{kT}{q} \frac{\varepsilon_s}{qN_B} \qquad (22.37)$$

where ε_s is the dielectric permittivity of the semiconductor. The Debye length is a measure of how deeply a variation of surface potential penetrates into the body when $D_I = 0$ and the depletion layer is of zero width.

Approximating qN_{inv} by Eq. (22.33) and Q_D by Eq. (22.35), Eq. (22.32) determines the surface potential at threshold, ϕ_{TH}, to be

$$\phi_{TH} = 2\frac{kT}{q} \ln \frac{N_B}{n_i} + \frac{kT}{q} \ln\left(1 + \frac{qD_I}{Q_B}\right) \qquad (22.38)$$

where the new symbols are defined as follows: Q_B = depletion-layer charge/unit area due to bulk body dopant N_B in the depletion layer, and qD_I = depletion-layer charge/unit area due to implanted ions in the depletion layer between the inversion-layer edge and the depletion-layer edge. Because even a small increase in ϕ_S above ϕ_{TH} causes a large increase in qN_{inv}, which can balance a rather large change in gate charge or gate voltage, ϕ_S does not increase much as $V_G - V_{TH}$ increases. Nonetheless, in strong inversion $N_{inv} \approx 10^{12}$ charges/cm^2, so in strong inversion ϕ_S will be about $10\ kT/q$ larger than ϕ_{TH}.

Equation (22.38) indicates for uniform doping (no implant, $D_I = 0$) that threshold occurs approximately for $\phi_S = \phi_{TH} = 2(kT/q)\ln(N_B/n_i) \equiv 2\phi_B$, but for the nonuniformly doped case a larger surface potential is needed, assuming the case of a normal implant where D_I is positive, increasing the dopant density. The implant increases the required surface potential because the field at the surface is larger, narrowing the inversion layer, and reducing the channel strength for $\phi_S = 2\phi_B$. Hence, a somewhat larger surface potential is needed to increase qN_{inv} to the point that Eq. (22.32) is satisfied.

Equation (22.38) would not apply if a significant fraction of the implant were confined to lie within the inversion layer itself. However, no realistic implant can be confined within a distance comparable to an inversion-layer thickness (a few tens of nanometers), so Eq. (22.38) covers practical cases.

With the surface potential ϕ_{TH} known, the potential on the gate at threshold Φ_{TH} can be found if we know the oxide field F_{ox} by simply adding the potential drop across the semiconductor to that across the oxide. That is, $\Phi_{TH} = \phi_{TH} + F_{ox}\,d$, with d = oxide thickness and F_{ox} given by Gauss's law as

$$\varepsilon_{ox}F_{ox} = Q_D \tag{22.39}$$

There are two more complications in finding the threshold voltage. First, the *gate voltage* V_{TH} usually differs from the gate potential Φ_{TH} at threshold because of a work-function difference between the body and the gate material. This difference causes a spontaneous charge exchange between the two materials as soon as the MOSFET is placed in a circuit allowing charge transfer to occur. Thus, even before any *voltage* is applied to the device, a *potential* difference exists between the gate and the body due to spontaneous charge transfer. The second complication affecting threshold voltage is the existence of charges in the insulator and at the insulator-semiconductor interface. These nonideal contributions to the overall charge balance are due to traps and fixed charges incorporated during the device processing.

Ordinarily interface-trap charge is negligible ($<10^{10}/cm^2$ in silicon MOSFETs), and the other nonideal effects upon threshold voltage are accounted for by introducing the *flatband voltage* V_{FB}, which corrects the gate bias for these contributions. Then, using Eq. (22.39) with $F_{ox} = (V_{TH} - V_{FB} - \phi_{TH})/d$ we find

$$V_{TH} = V_{FB} + \phi_{TH} + Q_D\,\frac{d}{\varepsilon_{ox}} \tag{22.40}$$

which determines V_{TH} even for the nonuniformly doped case, using Eq. (22.38) for ϕ_{TH} and Q_D at threshold from Eq. (22.35). If interface-trap charge/unit area is not negligible, then terms in the interface-trap charge/unit area Q_{IT} must be added to Q_D in Eq. (22.40).

From Eqs. (22.35) and (22.38), the threshold voltage depends upon the implanted dopant-ion profile only through two parameters, the net charge introduced by the implant in the region between the inversion layer and the depletion-layer edge qD_I, and the centroid of this portion of the implanted charge x_C. As a result, a variety of implants can result in the same threshold, ranging from the extreme of a δ-function spike implant of dose D_I/unit area located at the centroid x_C, to a box-type rectangular distribution with the same dose and centroid, namely, a rectangular distribution of width $x_W = 2x_C$ and volume density D_I/x_W. (Of course, x_W must be no larger than the depletion-layer width at threshold for this equivalence to hold true, and x_C must not lie within the inversion layer.) This weak dependence on the details of the profile leaves flexibility to satisfy other requirements, such as control of off current.

As already said, for gate biases $V_G > V_{TH}$ any gate charge above the threshold value is balanced mainly by inversion-layer charge. Thus, the additional oxide field, given by $(V_G - V_{TH})/d$, is related by Gauss's law to the inversion-layer carrier density approximately by

$$\varepsilon_{ox}\,\frac{V_G - V_{TH}}{d} \approx qN_{inv} \tag{22.41}$$

which shows that channel strength above threshold is proportional to $V_G - V_{TH}$, an approximation often used in this article. Thus, the switch in balancing gate charge from the depletion layer to the inversion layer causes N_{inv} to switch from an exponential gate-voltage dependence in subthreshold to a linear dependence above threshold.

For circuit analysis Eq. (22.41) is a convenient *definition* of V_{TH} because it fits current-voltage curves. If this definition is chosen instead of the charge-balance definition of Eq. (22.32), then Eqs. (22.32) and (22.38) result in an *approximation* to ϕ_{TH}.

Driving Ability and $I_{D,\text{sat}}$

The driving ability of the MOSFET is proportional to the current it can provide at a given gate bias. One might anticipate that the larger this current, the faster the circuit. Here this current is used to find some response times governing MOSFET circuits.

MOSFET current is dependent upon the carrier density in the channel, or upon $V_G - V_{TH}$, see Eq. (22.41). For a long-channel device, driving ability depends also on channel length. The shorter the channel length L, the greater the driving ability, because the channel resistance is directly proportional to the channel length. Supposing that the MOSFET is primarily in saturation during the driving of its load, driving ability is proportional to current in saturation, or to

$$I_{D,\text{sat}} = \frac{\varepsilon_{ox} Z \mu}{2dL} (V_G - V_{TH})^2 \tag{22.42}$$

where the factor of two results from the saturating behavior of the *I-V* curves at large drain biases and Z is the width of the channel normal to the direction of current flow. Evidently, for long devices driving ability is quadratic in $V_G - V_{TH}$, and inversely proportional to d.

The result of Eq. (22.42) holds for long devices. For short-channel devices, as explained for Fig. 22.44, the larger fields exerted by the drain electrode cause *velocity saturation* and, as a result, $I_{D,\text{sat}}$ is given roughly by [Einspruch and Gildenblat, 1989]

$$I_{D,\text{sat}} \approx \frac{\varepsilon_{ox} Z \upsilon_{sat}}{d} \frac{(V_G - V_{TH})^2}{V_G - V_{TH} + F_{sat}L} \tag{22.43}$$

where υ_{sat} is the carrier saturation velocity, about 10^7 cm/s for silicon at 290 K, F_{sat} is the field at which velocity saturation sets in, about 5×10^4 V/cm for electrons and not well established as $\geq 10^5$ V/cm for holes in silicon MOSFETs. For Eq. (22.43) to agree with Eq. (22.42) at long L, we need $\mu \approx 2\upsilon_{sat}/F_{sat} \approx 400$ cm^2/(V·s) for electrons in silicon MOSFETs, which is only roughly correct. Nonetheless, we can see that for devices in the submicron channel length regime, $I_{D,\text{sat}}$ tends to become independent of channel length L and becomes more linear with $V_G - V_{TH}$ and less quadratic (see Fig. 22.44). Equation (22.43) shows that velocity saturation is significant when $V_G/L \geq F_{sat}$ for example, when $L \leq 0.7$ μm if $V_G = 3.3$ V.

To relate $I_{D,\text{sat}}$ to a gate response time, τ_G, consider one MOSFET driving an identical MOSFET as load capacitance. Then the current from Eq. (22.43) charges this capacitance to a voltage V_G in a gate response time τ_G given by [Shoji, 1988]

$$\begin{aligned}
\tau_G &= \frac{C_G V_G}{I_{D,\text{sat}}} \\
&= \frac{L}{\upsilon_{sat}} \left(1 + \frac{C_{par}}{C_{ox}} \right) \frac{V_G(V_G - V_{TH} + F_{sat}L)}{(V_G - V_{TH})^2}
\end{aligned} \tag{22.44}$$

where C_G is the MOSFET gate capacitance $C_G = C_{ox} + C_{par}$, with $C_{ox} = \varepsilon_{ox} ZL/d$ the MOSFET oxide capacitance, and C_{par} the parasitic component of the gate capacitance [Chen, 1990]. The parasitic capacitance C_{par} is due mainly to overlap of the gate electrode over the source and drain and partly to fringing-field and channel-edge capacitances. For short-channel lengths, C_{par} is a significant part of C_G, and keeping C_{par} under control as L reduced is an objective of gate-drain alignment technology. Typically, $V_{TH} \approx V_G/4$, so

$$\tau_G \approx \left(\frac{L}{\upsilon_{sat}} \right) \left(1 + \frac{C_{par}}{C_{ox}} \right) \left(1.3 + 1.8 \frac{F_{sat}L}{V_G} \right) \tag{22.45}$$

Thus, on an intrinsic level, the gate response time is closely related to the transit time of an electron from source to drain, which is L/υ_{sat} in velocity saturation. At shorter L, a linear reduction in delay with L is predicted, while for longer devices the improvement can be quadratic in L, depending upon how V_G is scaled as L is reduced.

The gate response time is not the only delay in device switching, because the drain-body *pn* junction also must charge or discharge for the MOSFET to change state [Shoji, 1988]. Hence, we must also consider a drain response time τ_D. Following Eq. (22.44), we suppose that the drain capacitance C_D is charged by the supply voltage through a MOSFET in saturation so that

$$\tau_D = \frac{C_D V_G}{I_{D,sat}} = \frac{C_D}{C_G}\,\tau_G \tag{22.46}$$

Equation (22.46) suggests that τ_D will show a similar improvement to τ_G as L is reduced, provided that C_D/C_G does not increase as L is reduced. However, $C_{ox} \propto L/d$, and the major component of C_{par}, namely, the overlap capacitance contribution, leads to $C_{par} \propto L_{ovlp}/d$ where L_{ovlp} is roughly three times the length of overlap of the gate over the source or drain [Chen, 1990]. Then $C_G \propto (L + L_{ovlp})/d$ and, to keep the C_D/C_G ratio from increasing as L is reduced, either C_D or oxide-thickness d must be reduced along with L.

Clever design can reduce C_D. For example, various *raised-drain* designs reduce the drain-to-body capacitance by separating much of the drain area from the body using a thick oxide layer. The contribution to drain capacitance stemming from the sidewall depletion-layer width next to the channel region is more difficult to handle, because the sidewall depletion layer is deliberately reduced during miniaturization to avoid *short-channel* effects, that is, drain influence upon the channel in competition with gate control. As a result this sidewall contribution to the drain capacitance tends to increase with miniaturization unless junction depth can be shrunk.

Equations (22.45) and (22.46) predict reduction of response times by reduction in channel length L. Decreasing oxide thickness leads to no improvement in τ_G, but Eq. (22.46) shows a possibility of improvement in τ_D. The *ring oscillator*, a closed loop of an odd number of inverters, is a test circuit whose performance depends primarily on τ_G and τ_D. Gate delay/stage for ring oscillators is found to be near 12 ps/stage at 0.1 μm channel length, and 60 ps/stage at 0.5 μm.

For circuits, interconnection capacitances and fan-out (multiple MOSFET loads) will increase response times beyond the device response time, even when parasitics are taken into account. Thus, we are led to consider interconnection delay, τ_{INT}. Although a lumped model suggests, as with Eq. (22.46), that $\tau_{INT} \approx (C_{INT}/C_G)\,\tau_G$, the length of interconnections requires a *distributed* model. Interconnection delay is then

$$\tau_{INT} = \frac{R_{INT} C_{INT}}{2} + R_{INT} C_G + \left(1 + \frac{C_{INT}}{C_G}\right)\tau_G \tag{22.47}$$

where the new symbols are R_{INT} = interconnection resistance, C_{INT} = interconnection capacitance, and we have assumed that the interconnection joins a MOSFET driver in saturation to a MOSFET load C_G. For small R_{INT}, τ_{INT} is dominated by the last term, which resembles Eqs. (22.44) and (22.46). However, unlike the ratio C_D/C_G in Eq. (22.46), it is difficult to reduce or even maintain the ratio C_{INT}/C_G in Eq. (22.47) as L is reduced. Remember, $C_G \propto Z(L + L_{ovlp})/d$. Reduction of L therefore tends to increase C_{INT}/C_G, especially because interconnect cross sections cannot be reduced without impractical increases in R_{INT}. What is worse, along with reduction in L, chip sizes usually increase, making line lengths longer, increasing R_{INT} even at constant cross section. As a result, interconnection delay becomes a major problem as L is reduced. The obvious way to keep C_{INT}/C_G under control is to increase the device width Z so that $C_G \propto Z(L + L_{ovlp})/d$ remains constant as L is reduced. A better way is to cascade drivers of increasing Z [Chen, 1990; Shoji, 1988]. Either solution requires extra area, however, reducing the packing density that is a major objective in decreasing L in the first place. An alternative is to reduce the oxide thickness d, a major technology objective today.

Transconductance

Another important device parameter is the small-signal transconductance g_m [Sedra and Smith, 1991; Haznedar, 1991], which determines the amount of output current swing at the drain that results from a given input voltage variation at the gate, that is, the small-signal gain:

$$g_m = \frac{\partial I_D}{\partial V_G}\bigg|_{V_D=\text{const}} \tag{22.48}$$

Using the chain rule of differentiation, the transconductance in saturation can be related to the small-signal *transition* or *unity-gain frequency*, which determines at how high a frequency ω the small-signal current gain $|\iota_{\text{out}}/\iota_{\text{in}}| = g_m/(\omega C_G)$ drops to unity. Using the chain rule,

$$g_m = \frac{\partial I_{D,\text{sat}}}{\partial Q_G}\frac{\partial Q_G}{\partial V_G} = \omega_T C_G \tag{22.49}$$

where C_G is the oxide capacitance of the device, $C_G = \partial Q_G/\partial V_G|_{V_D}$ with $Q_G =$ the charge on the gate electrode. The frequency ω_T is a measure of the small-signal, high-frequency speed of the device, neglecting parasitic resistances. Using Eq. (22.43) in Eq. (22.49) we find that the transition frequency also is related to the transit time L/υ_{sat} of Eq. (22.45), so that both the digital and small-signal circuit speeds are related to this parameter.

Output Resistance and Drain Conductance

For small-signal circuits the output resistance r_o of the MOSFET [Sedra and Smith, 1991] is important in limiting the gain of amplifiers. This resistance is related to the small-signal drain conductance g_D in saturation by

$$r_o = \frac{1}{g_D} = \frac{\partial V_D}{\partial I_{D,\text{sat}}}\bigg|_{V_G=\text{const}} \tag{22.50}$$

If the MOSFET is used alone as a simple amplifier with a load line set by a resistor R_L, the gain becomes

$$\left|\frac{\upsilon_o}{\upsilon_{\text{in}}}\right| = g_m \frac{R_L r_o}{R_L + r_o} \leq g_m R_L \tag{22.51}$$

showing how gain is reduced if r_o is reduced to a value approaching R_L.

As devices are miniaturized, r_o is decreased, g_D increased, due to several factors. At moderate drain biases, the main factor is channel-length modulation, the reduction of the channel length with increasing drain voltage that results when the depletion region around the drain expands toward the source, causing L to become drain-bias dependent. At larger drain biases, a second factor is drain control of the inversion-layer charge density, which can compete with gate control in short devices. This is the same mechanism discussed later in the context of subthreshold behavior. At rather high drain bias, carrier multiplication further lowers r_o.

In a digital inverter, a lower r_o widens the voltage swing needed to cause a transition in output voltage. This widening increases power loss due to current spiking during the transition, and reduces noise margins [Annaratone, 1986]. It is not, however, a first-order concern in device miniaturization for digital applications. Because small-signal circuits are more sensitive to r_o than digital circuits, MOSFETs designed for small-signal applications cannot be made as small as those for digital applications.

Limitations upon Miniaturization

A major factor in the success of the MOSFET has been its compatibility with processing useful down to very small dimensions. Today channel lengths (source-to-drain spacings) of 0.5 μm are manufacturable, and further reduction to 0.1 μm has been achieved for limited numbers of devices in test circuits such as ring oscillators. In this section some of the limits that must be considered in miniaturization are outlined [Brews, 1990].

Subthreshold Control

When a MOSFET is in the "off" condition, that is, when the MOSFET is in *subthreshold*, the off current drawn with the drain at supply voltage must not be too large in order to avoid power consumption and discharge of ostensibly isolated nodes [Shoji, 1988]. In small devices, however, the source and drain are closely spaced, so there exists a danger of direct interaction of the drain with the source, rather than an interaction mediated by the gate and channel. In an extreme case, the drain may draw current directly from the source, even though the gate is "off" (*punchthrough*). A less extreme but also undesirable case occurs when the drain and gate jointly control the carrier density in the channel (*drain-induced barrier lowering*, or drain control of threshold voltage). In such a case, the on–off behavior of the MOSFET is not controlled by the gate alone, and switching can occur over a range of gate voltages dependent on the drain voltage. Reliable circuit design under these circumstances is very complicated, and testing for design errors is prohibitive. Hence, in designing MOSFETs, a drain-bias independent subthreshold behavior is necessary.

A measure of the range of influence of the source and drain is the depletion-layer width of the associated *pn* junctions. The depletion layer of such a junction is the region in which all carriers have been depleted, or pushed away, due to the potential drop across the junction. This potential drop includes the applied bias across the junction and a spontaneous *built-in* potential drop induced by spontaneous charge exchange when *p* and *n* regions are brought into contact. The depletion-layer width W of an abrupt junction is related to potential drop V and dopant-ion concentration/unit volume N by

$$W = \left(\frac{2\varepsilon_s V}{qN}\right)^{1/2}$$

(22.52)

To avoid subthreshold problems, a commonly used rule of thumb is to make sure that the channel length is longer than a minimum length L_{min} related to the junction depth r_j, the oxide thickness d, and the depletion-layer widths W_S and W_D of the source and drain by [Brews, 1990]

$$L_{min} = A[r_j d(W_S + W_D)^2]^{1/3}$$

(22.53)

where the empirical constant $A = 0.88 \ nm^{-1/3}$ if r_j, W_S, and W_D are in micrometers and d is in nanometers.

Equation (22.53) shows that smaller devices require shallower junctions (smaller r_j), thinner oxides (smaller d), or smaller depletion-layer widths (smaller voltage levels or heavier doping). These requirements introduce side effects that are difficult to control. For example, if the oxide is made thinner while voltages are not reduced proportionately, then oxide fields increase, requiring better oxides. If junction depths are reduced, better control of processing is required, and the junction resistance is increased due to smaller cross sections. To control this resistance, various *self-aligned contact* schemes have been developed to bring the source and drain contacts closer to the gate [Brews, 1990; Einspruch and Gildenblat, 1989], reducing the resistance of these connections. If depletion-layer widths are reduced by increasing the dopant-ion density the *driving ability* of the MOSFET suffers because the threshold voltage increases. That is, Q_D increases in Eq. (22.40), reducing $V_G - V_{TH}$. Thus, particularly for devices that are not velocity-saturated, that is, devices where $V_G/L \lesssim F_{sat}$, increasing V_{TH} results in slower circuits.

As secondary consequences of increasing dopant-ion density, channel conductance is further reduced due to the combined effects of increased scattering of electrons from the dopant atoms and increased oxide fields that pin carriers in the inversion layer closer to the insulator–semiconductor interface, increasing scattering at the interface. These effects also reduce driving ability, although for shorter devices they are important only in the linear region (that is, below saturation), assuming that mobility μ is more strongly affected than saturation velocity υ_{sat}.

Hot-Electron Effects

Another limit upon how small a MOSFET can be made is a direct result of the larger fields in small devices. Let us digress to consider why proportionately larger voltages, and thus larger fields, are used in smaller devices. First, according to Eq. (22.45), τ_G is shortened if voltages are increased, at least so long as $V_G/L \lesssim F_{sat} \sim 5 \times 10^4$ V/cm. If τ_G is shortened this way, then so are τ_D and τ_{INT}, Eqs. (22.46) and (22.47). Thus, faster response is gained by increasing voltages into the velocity saturation region. Second, the fabricational control of smaller devices has not improved proportionately as L has shrunk, so there is a larger percentage variation in device parameters with smaller devices. Thus, disproportionately larger voltages are needed to ensure that all devices operate in the circuit, to overcome this increased fabricational "noise." Thus, to increase speed and to cope with fabricational variations, fields go up in smaller devices.

As a result of these larger fields along the channel direction, a small fraction of the channel carriers have enough energy to enter the insulating layer near the drain. In silicon-based p-channel MOSFETs, energetic holes can become trapped in the oxide, leading to a positive oxide charge near the drain that reduces the strength of the channel, degrading device behavior. In n-channel MOSFETs, energetic electrons entering the oxide create interface traps and oxide wear-out, eventually leading to gate-to-drain shorts [Pimbley *et al.*, 1989].

To cope with these problems "drain-engineering" has been tried, the most common solution being the *lightly doped drain* [Chen, 1990; Einspruch and Gildenblat, 1989; Pimbley *et al.*, 1989]. In this design, a lightly doped extension of the drain is inserted between the channel and the drain proper. To keep the field moderate and reduce any peaks in the field, the lightly doped drain extension is designed to spread the drain-to-channel voltage drop as evenly as possible. The aim is to smooth out the field at a value close to F_{sat} so that energetic carriers are kept to a minimum. The expense of this solution is an increase in drain resistance and a decreased gain. To increase packing density, this lightly doped drain extension can be stacked vertically alongside the gate, rather than laterally under the gate, to control the overall device area.

Thin Oxides

According to Eq. (22.53), thinner oxides allow shorter devices and therefore higher packing densities for devices. In addition, driving ability is increased, shortening response times for capacitive loads, and output resistance and transconductance are increased. There are some basic limitations upon how thin the oxide can be made. For instance, there is a maximum oxide field that the insulator can withstand. It is thought that the intrinsic breakdown voltage of SiO_2 is of the order of 10^7 V/cm, a field that can support $\approx 2 \times 10^{13}$ charges/cm^2, a large enough value to make this field limitation secondary. Unfortunately, as they are presently manufactured, the intrinsic breakdown of MOSFET oxides is much less likely to limit fields than defect-related leakage or breakdown, and control of these defects has limited reduction of oxide thicknesses in manufacture to about 10 nm to date.

If defect-related problems could be avoided, the thinnest useful oxide would probably be about 3 nm, limited by direct tunneling of channel carriers to the gate. This tunneling limit is not well established, and also is subject to oxide-defect enhancement due to tunneling through intermediate defect levels. Thus, the manufacture of thin oxides is a very active area of exploration.

Dopant-Ion Control

As devices are made smaller, the precise positioning of dopant inside the device is critical. At high temperatures during processing, dopant ions can move. For example, source and drain dopants can enter the channel region, causing position dependence of threshold voltage. Similar problems occur in isolation structures that separate one device from another [Pimbley *et al.*, 1989; Einspruch and Gildenblat, 1989].

To control these thermal effects, process sequences are carefully designed to limit high-temperature steps. This design effort is shortened and improved by the use of computer modeling of the processes. Dopant-ion movement is complex, however, and its theory is made more difficult by the growing trend to use *rapid thermal processing* that involves short-time heat treatments. As a result, dopant response is not steady state, but transient. Computer models of transient response are primitive, forcing further advance in small-device design to be more empirical.

Other Limitations

Besides limitations directly related to the MOSFET, there are some broader difficulties in using MOSFETs of smaller dimension in chips involving even greater numbers of devices. Already mentioned is the increased delay due to interconnections that are lengthening due to increasing chip area and increasing complexity of connection. The capacitive loading of MOSFETs that must drive signals down these lines can slow circuit response, requiring extra circuitry to compensate. Another limitation is the need to isolate devices from each other [Brews, 1990; Chen 1990; Einspruch and Gildenblat, 1989; Pimbley *et al.*, 1989], so their actions remain uncoupled by parasitics. As isolation structures are reduced in size to increase device densities, new parasitics are discovered. A developing solution to this problem is the manufacture of circuits on insulating substrates, silicon-on-insulator technology [Colinge, 1991]. To succeed, this approach must deal with new problems, such as the electrical quality of the underlying silicon–insulator interface and the defect densities in the silicon layer on top of this insulator.

Defining Terms

Channel: The conducting region in a MOSFET between source and drain. In an *enhancement-mode* (or normally off) MOSFET, the channel is an inversion layer formed by attraction of minority carriers toward the gate. These carriers form a thin conducting layer that is prevented from reaching the gate by a thin *gate-oxide* isulating layer when the gate bias exceeds *threshold*. In a *buried-channel*, or *depletion-mode* (or normally on) MOSFET, the channel is present even at zero gate bias, and the gate serves to increase the channel resistance when its bias is nonzero. Thus, this device is based on majority-carrier modulation, like a MESFET.

Gate: The control electrode of a MOSFET. The voltage on the gate capacitively modulates the resistance of the connecting channel between the source and drain.

Source, drain: The two output contacts of a MOSFET, usually formed as *pn* junctions with the *substrate* or *body* of the device.

Strong inversion: The range of gate biases corresponding to the "on" condition of the MOSFET. At a fixed gate bias in this region, for low drain-to-source biases the MOSFET behaves as a simple gate-controlled resistor. At larger drain biases, the channel resistance can increase with drain bias, even to the point that the current *saturates*, or becomes independent of drain bias.

Substrate or body: The portion of the MOSFET that lies between the *source* and *drain* and under the *gate*. The gate is separated from the body by a thin *gate insulator*, usually silicon dioxide. The gate modulates the conductivity of the body, providing a gate-controlled resistance between the source and drain. The body is sometimes *dc*-biased to adjust overall circuit operation. In some circuits the body voltage can swing up and down as a result of input signals, leading to "body-effect" or "back-gate bias" effects that must be controlled for reliable circuit response.

Subthreshold: The range of gate biases corresponding to the "off" condition of the MOSFET. In this regime the MOSFET is not perfectly "off" but conducts a leakage current that must be controlled to avoid circuit errors and power consumption.

Threshold: The gate bias of a MOSFET that marks the boundary between "on" and "off" conditions.

References

The following references are not to the original sources of the ideas discussed in this article, but have been chosen to be generally useful to the reader.

M. Annaratone, *Digital CMOS Circuit Design,* Boston: Kluwer Academic, 1986.

J. R. Brews, "Physics of the MOS transistor" in *Applied Solid State Science, Supplement 2A,* D. Kahng, Ed., New York: Academic, 1981.

J. R. Brews, "The submicron MOSFET" in *High-Speed Semiconductor Devices,* S. M. Sze, Ed., New York: Wiley, 1990, pp. 139–210.

J. Y. Chen, *CMOS Devices and Technology for VLSI,* Englewood Cliffs, N.J.: Prentice-Hall, 1990.

J.-P. Colinge, *Silicon-on-Insulator Technology: Materials to VLSI,* Boston: Kluwer Academic, 1991.

H. Haznedar, *Digital Microelectronics,* Redwood City, Calif.: Benjamin-Cummings, 1991.

M. A. Hollis and R. A. Murphy, "Homogeneous field-effect transistors," in *High-Speed Semiconductor Devices,* S. M. Sze, Ed., New York: Wiley, 1990, pp. 211–282.

N. G. Einspruch and G. Sh. Gildenblat, Eds., *VLSI Microstructure Science,* vol. 18, *Advanced MOS Device Physics,* New York: Academic, 1989.

E. H. Nicollian and J. R. Brews, *MOS Physics and Technology,* New York: Wiley, 1982, chap. 1.

S. J. Pearton and N. J. Shaw, "Heterostructure field-effect transistors" in *High-Speed Semiconductor Devices,* S. M. Sze, Ed., New York: Wiley, 1990, pp. 283–334.

R. F. Pierret, *Modular Series on Solid State Devices, Field Effect Devices,* 2nd ed., vol. 4, Reading, Mass.: Addison-Wesley, 1990.

J. M. Pimbley, M. Ghezzo, H. G. Parks, and D. M. Brown, *VLSI Electronics Microstructure Science, Advanced CMOS Process Technology,* vol. 19, N. G. Einspruch, Ed., New York: Academic, 1989.

S. S. Sedra and K. C. Smith, *Microelectronic Circuits,* 3rd ed., Philadelphia: Saunders, 1991.

M. Shoji, *CMOS Digital Circuit Technology,* Englewood Cliffs, N.J.: Prentice-Hall, 1988.

Further Information

The references given in this section have been chosen to provide more detail than is possible to provide in the limited space of this article. In particular, Annaratone [1986] and Shoji [1988] provide much more detail about device and circuit behavior. Chen [1990] and Pimbley *et al.* [1989] provide many technological details of processing and its device impact. Haznedar [1991] and Sedra and Smith [1991] provide much information about circuits. Brews [1981] and Pierret [1990] provide good discussions of the derivation of the device current-voltage curves and device behavior in all bias regions.

23

Integrated Circuits

Mehdi R. Zargham
Southern Illinois University

Spyros Tragoudas
Southern Illinois University

J. Leland Seely
Bonneville Microelectronics, Inc.

23.1 Layout, Placement, and Routing 581
   What Is Layout? • Placement Techniques • Routing Techniques
23.2 Application-Specific Integrated Circuits (ASICs) 591
   ASIC Design Methods Available • When and What Type of an
   ASIC Should Be Used • ASIC Design Tools • The PPL Technology

23.1 Layout, Placement, and Routing

Mehdi R. Zargham and Spyros Tragoudas

Very large scale integrated (VLSI) electronics presents a challenge, not only to those involved in the development of fabrication technology, but also to computer scientists, computer engineers, and electrical engineers. The ways in which digital systems are structured, the procedures used to design them, the trade-offs between hardware and software, and the design of computational algorithms will all be greatly affected by the coming changes in integrated electronics.

A VLSI chip can today contain millions of transistors and is expected to contain more than 100 million transistors in the year 2000. One of the main factors contributing to this increase is the effort that has been invested in the development of computer-aided design (CAD) systems for VLSI design. The VLSI CAD systems are able to simplify the design process by hiding the low-level circuit theory and device physics details from the designer, and allowing him or her to concentrate on the functionality of the design and on ways of optimizing it.

A VLSI CAD system supports descriptions of hardware at many levels of abstraction, such as system, subsystem, register, gate, circuit, and layout levels. It allows designers to design a hardware device at an abstract level and progressively work down to the layout level. A layout is a complete geometric representation (a set of rectangles) from which the latest fabrication technologies directly produce reliable, working chips. A VLSI CAD system also supports verification, synthesis, and testing of the design. Using a CAD system, the designer can make sure that all of the parts work before actually implementing the design.

A variety of VLSI CAD systems are commercially available that perform all or some of the levels of abstraction of design. Most of these systems support a *layout editor* for designing a circuit **layout**. A layout-editor is software that provides commands for drawing lines and boxes, copying objects, moving objects, erasing unwanted objects, and so on. The output of such an editor is a design file that describes the layout. Usually, the design file is represented in a standard format, called Caltech Intermediate Form (CIF), which is accepted by the fabrication industry.

What Is Layout?

For a specific circuit, a layout specifies the position and dimension of the different layers of materials as they would be laid on the silicon wafer. However, the layout description is only a symbolic

581

representation, which simplifies the description of the actual fabrication process. For example, the layout representation does not explicitly indicate the thickness of the layers, thickness of oxide coating, amount of ionization in the transistors channels, etc., but these factors are implicitly understood in the fabrication process. Some of the main layers used in any layout description are *n*-diffusion, *p*-diffusion, poly, metal-1, and metal-2. Each of these layers is represented by a polygon of a particular color or pattern. As an example, Fig. 23.1 presents a specific pattern for each layer that will be used through the rest of this section.

As is shown in Fig. 23.2, an *n*-diffusion layer crossing a poly layer implies an nMOS transistor, and a *p*-diffusion crossing poly implies a pMOS transistor.

Note that the widths of diffusion and poly are represented with a scalable parameter called *lambda*. These measurements, referred to as *design rules*, are introduced to prevent errors on the chip, such as preventing thin lines from opening (disconnecting) and short circuiting.

Implementing the design rules based on lambda makes the design process independent of the fabrication process. This allows the design to be rescaled as the fabrication process improves.

Metal layers are used as wires for connections between the components. This is because metal has the lowest propagation delay compared to the other layers. However, sometimes a poly layer is also used for short wires in order to reduce the complexity of the wire routing. Any wire can cross another wire without getting electrically affected as long as they are in different layers. Two different layers can be electrically connected together using *contacts*. The fabrication process of the contacts depends on types of the layers that are to be connected. Therefore, a layout editor supports different types of contacts by using different patterns.

From the circuit layout, the actual chip is fabricated. Based on the layers in the layout, various layers of materials, one on top of the others, are laid down on a silicon wafer. Typically, the pro-

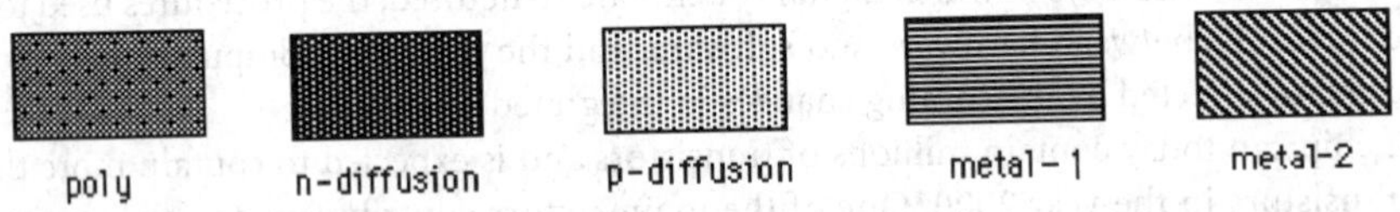

| poly | n-diffusion | p-diffusion | metal-1 | metal-2 |

FIGURE 23.1   Different layers.

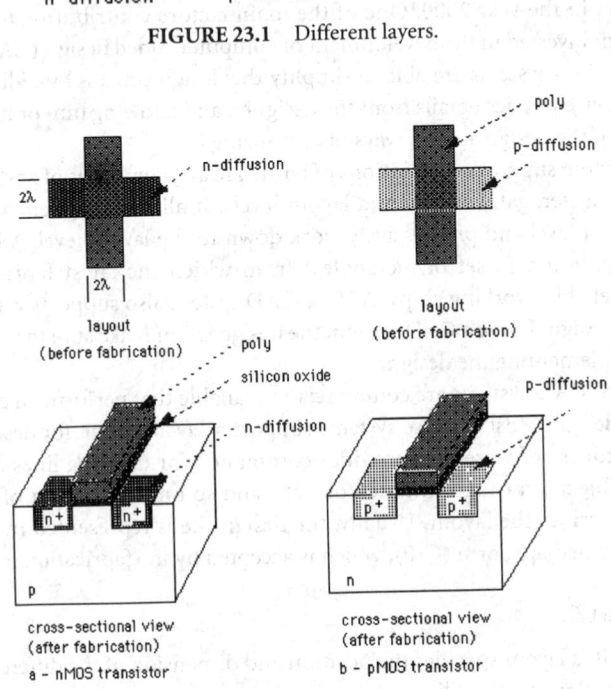

FIGURE 23.2   Layout and fabrication of MOS transistors.

cessing of laying down each of these materials involves several steps, such as masking, oxide coating, lithography and etching [Mead and Conway, 1980]. For example, as shown in Fig. 23.3(a), for fabricating an nMOS transistor, first two masks, one for poly and one for *n*-diffusion, are obtained from the circuit layout. Next, the *n*-diffusion mask is used to create a layer of silicon oxide on the wafer [see Fig. 23.3(b)]. The wafer will be covered with a thin layer of oxide in places where the transistors are supposed to be placed as opposed to a thick layer in other places. The poly mask is used to place a layer of polysilicon on top of the oxide layer to define the gate terminals of the transistor [see Fig. 23.3(c)]. Finally, the *n*-diffusion regions are made to form the source and drain terminals of the transistor [see Fig. 23.3(d)].

To better illustrate the concept of layout design, the design of an inverter in the CMOS technology is shown in Fig. 23.4. An inverter produces an output voltage that is the logical inverse of its input. Considering the circuit diagram of Fig. 23.4(a), when the input is 1, the lower nMOS is on, but the upper pMOS is off. Thus, the output becomes 0 by becoming connected to the ground through the nMOS. On the other hand, if the input is 0, the pMOS is on and the nMOS is off, so the output must find a charge-up path through the pMOS to the supply and therefore becomes 1. Figure 23.4(b) represents a layout for such an inverter. As can be seen from this figure, the problem of a layout design is essentially reduced to drawing and painting a set of polygons. Layout editors provide commands for drawing such polygons. The commands are usually entered at the keyboard or with a mouse and, in some menu-driven packages, can be selected as options from a pull-down menu.

In addition to the drawing commands, often a layout system provides tools for minimizing the overall area of the layout (i.e., size of the chip). Today a VLSI chip consists of a lot of individual cells, with each one laid out separately. A cell can be an inverter, a NAND gate, a multiplier, a memory unit, etc. The designer can make the layout of a cell and then store it in a file called the *cell library*. Later, each time the designer wants to design a circuit that requires the stored cell, he or she simply copies the layout from the cell library. A layout may consist of many cells. Most of the layout systems provide routines, called **placement** and **routing** routines, for placing the cells and then

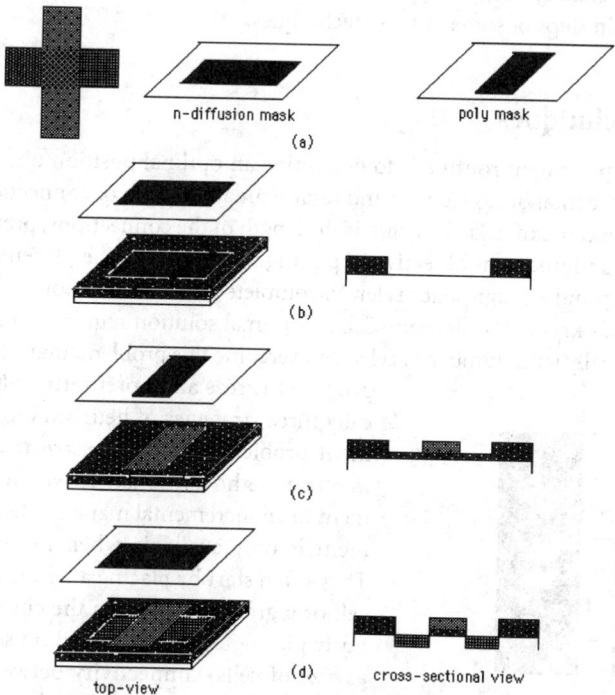

n-diffusion mask   poly mask

(a)

(b)

(c)

(d)

top-view   cross-sectional view

FIGURE 23.3 Fabrication steps for an nMOS transistor.

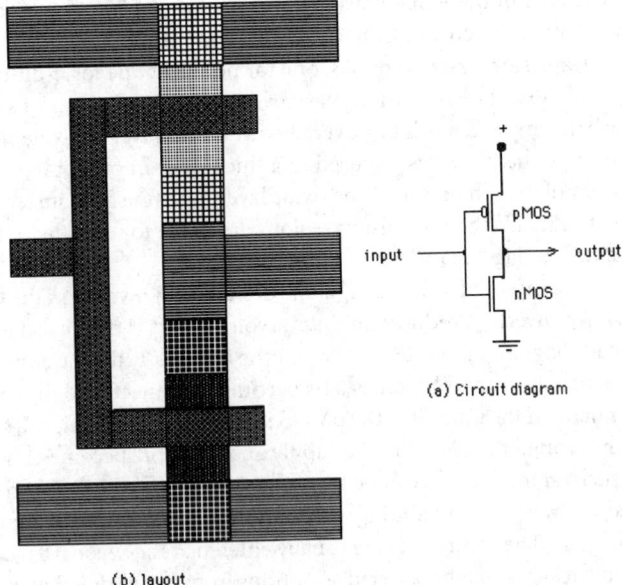

(a) Circuit diagram

(b) layout

FIGURE 23.4   An inverter.

interconnecting them with wires in such a way that minimizes the layout area. As an example, Fig. 23.5 presents the placement of three cells. The area between the cells is used for routing. The entire routing surface is divided into a set of rectangular routing areas called channels. The sides of each channel consist of a set of terminals. A wire that connects the terminals with the same ID is called a net. The router finds a location for the wire segments of each net within the channel. The following sections classify various types of placement and routing techniques and provide an overview of the main steps of some of these techniques.

Placement Techniques

The objective of a placement routine is to determine an optimal position on the chip for a set of cells in a way that the total occupied area and total estimated length of connections are minimized. Given that the main cause of delay in a chip is the length of the connections, providing shorter connections becomes an important objective in placing a set of cells. The placement should be such that no cells overlap and enough space is left to complete all the connections.

All exact methods known for determining an optimal solution require a computing effort that increases exponentially with number of cells. To overcome this problem, many heuristics have been proposed [Preas and Lorenzetti, 1988]. There are basically three strategies of heuristics for solving the placement problem, namely, *constructive*, *partitioning*, and *iterative* methods. Constructive methods create placement in an incremental manner where a complete placement is only available when the method terminates. They often start by placing a *seed* (a seed can be a single cell or a group of cells) on the chip and then continuously placing other cells based on some heuristics such as size of cells, connectivity between the cells, design condition for connection lengths, or size of chip. This process continues until all the cells are placed on the

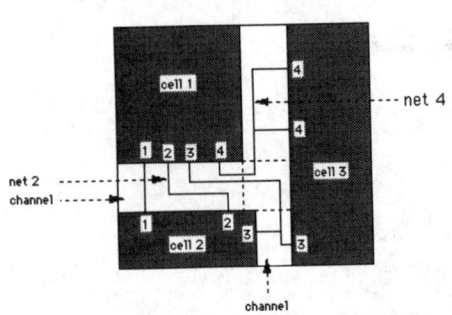

FIGURE 23.5   Placement and routing.

chip. Partitioning methods divide the cells into two or more partitions so that the number of connections that cross the partition boundaries is minimized. The process of dividing is continued until the number of cells per partition becomes less than a certain small number. Iterative methods seek to improve an initial placement by repeatedly modifying it. Improvement might be made by transforming one cell to a new position or switching positions of two or more cells. After a change is made to the current placement configuration based on some cost function, a decision is made to see whether to accept the new configuration. This process continues until an optimal (in most cases a near optimal) solution is obtained. Often the constructive methods are used to create initial placement on which an iterative method subsequently improves.

Constructive Method

In most of the constructive methods, at each step an unplaced cell is selected and then located in the proper area. There are different strategies for selecting a cell from the collection of unplaced cells [Wimer and Koren, 1988]. One strategy is to select the cell that is most strongly connected to already placed cells. For each unplaced cell, we find the total of its connections to all of the already placed cells. Then we select the unplaced cell that has the maximum number of connections. As an example consider the cells in Fig. 23.6. Assume that cells c_1 and c_2 are already placed on the chip. In Fig. 23.7 we see that cell c_5 has been selected as the next cell to be placed. This is because cell c_5 has the largest number of connections (i.e., three) to cells c_1 and c_2.

The foregoing strategy does not consider area as a factor and thus results in fragmentation of the available free area; this may make it difficult to place some of the large unplaced cells later. This problem can be overcome, however, by considering the product of the number of connections and the area of the cell as a criteria for the selection. Figure 23.8 presents an example of such a strategy. Cell c_3 is selected as the next choice since the product of its area and its connections to c_1 and c_2 combine to associate with the maximum value.

Partitioning Method

The approaches for the partitioning method can be classified as quadratic and sliced bisection. In both approaches the layout is divided into two subareas, A and B, each having a size within a pre-defined range. Each cell is assigned to one of these subareas. This assignment is such that the number of interconnections between the two subareas is minimal. For example, Fig. 23.9 presents successive steps for the quadratic and sliced-bisection methods. As shown in Fig. 23.9(a), in the first

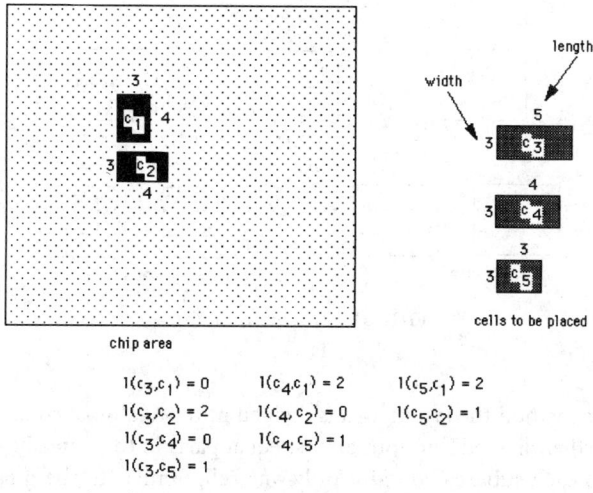

$$l(c_3,c_1) = 0 \qquad l(c_4,c_1) = 2 \qquad l(c_5,c_1) = 2$$
$$l(c_3,c_2) = 2 \qquad l(c_4,c_2) = 0 \qquad l(c_5,c_2) = 1$$
$$l(c_3,c_4) = 0 \qquad l(c_4,c_5) = 1$$
$$l(c_3,c_5) = 1$$

$l(c_i, c_j)$ represents the number of connections between cell c_i and c_j

FIGURE 23.6 Initial configuration.

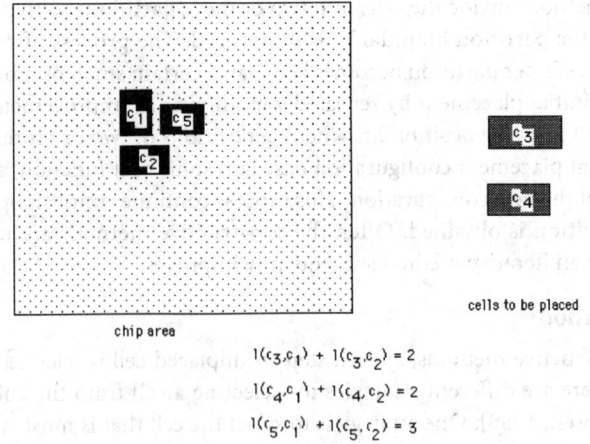

FIGURE 23.7   Selection based on the number of connections.

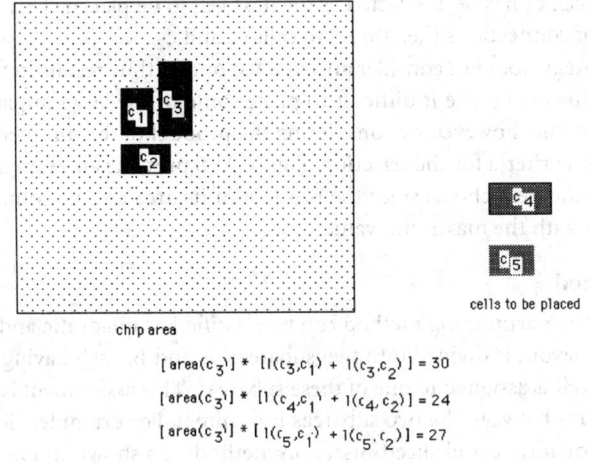

FIGURE 23.8   Selection based on the number of connections and area.

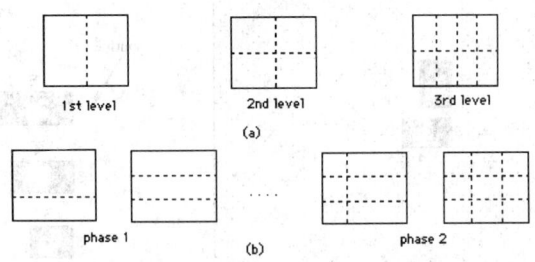

FIGURE 23.9   Partitioning.

step of the quadratic method the layout area is divided into two almost equal parts; in the second step the layout is further divided into four almost equal parts in the opposite direction. This process continues until each subarea contains only one cell. Similar to the quadratic method, the sliced-bisection also divides the layout area into several subareas.

The sliced-bisection method has two phases. In the first phase, the layout area is iteratively divided into a certain number of almost equal subareas in the same direction. In this way, we end up with a set of slices [see Fig. 23.9(b)]. Similarly, the second phase divides the area into a certain number of subareas; however, the slicing is done in the opposite direction.

Several heuristics have been proposed for each of the preceding partitioning methods. Here, for example, we emphasize the work of Fiduccia–Mattheyses [Fiduccia and Mattheyses, 1982], which uses the quadratic method. For simplicity, their algorithm is only explained for one step of this method. Initially the set of cells is randomly divided into two sets, A and B. Each set represents a subarea of the layout and has size equal to the area it represents. A cell is selected from one of these sets to be moved to the other set. The selection of the cell depends on three criteria. The cell should be free, i.e., the cell must have a minimum gain among the gains of all other free cells. A cell c has gain g if the number of interconnections between the cells of the two sets decreases by g units when c is moved from its current set to the other. Finally, the selected set's move should not violate a pre-defined balancing criterion that guarantees that the sizes of the two sets are almost equal. After moving the selected cell from its current set to the complementary set, it is no longer free. A new partition, which corresponds to the new instance of the two sets A and B, is created. The cost of a partition is defined as the number of interconnections between cells in the two sets of the partition. The Fiduccia–Mattheyses algorithm keeps track of the best partition encountered so far, i.e., the partition with the minimum cost. The algorithm will move the selected cell to the complementary set even if its gain is negative. In this case, the new partition is worse than the previous one, but the move can eventually lead to better partitions.

This process of generating new partitions and keeping track of the best encountered partition is repeated until no free cells are left. At this point of the process, which is called a pass, the algorithm returns the best partition and terminates. To obtain better partitions, the algorithm can be modified such that more passes occur. This can easily be done by selecting the best partition of a pass as the initial partition of the next pass. In this partition, however, all cells are free. The modified algorithm terminates whenever a new pass returns a partition that is no better than the partition returned by the previous pass. This way, the number of passes generated will never be more than the number of the interconnections in the circuit and the algorithm always terminates.

The balancing criterion in the Fiduccia–Mattheyses algorithm is easily maintained when the cells have uniform areas. The only way a pass can be implemented to satisfy the criterion is to start with a random initial partition in which the two sets differ by one cell, each time select the cell of maximum gain from the larger sized set, move the cell and generate a new partition, and repeat until no free cells are left.

In the example of Fig. 23.10, the areas of the cells are nonuniform. However, the assigned cell areas ensure that the previously described operation occurs so that the balancing criterion is satisfied. (The cell areas are omitted in this figure, but they correspond to the ones given in Fig. 23.6.) Figure 23.10 illustrates a pass of the Fiduccia–Mattheyses algorithm. During this pass, six different partitions are generated, i.e., the initial partition and five additional ones. Note that according to the description of the pass, the number of additional partitions equals the number of cells in the circuit.

Each partition consists of the cells of the circuit (colored according to the set to which they belong and labeled with an integer), the gain value associated with each free cell in the set from which the selected cell will be moved (this value can be a negative number), and the nets (labeled with letters). In the figure a cell that is no longer free is distinguished by an empty circle placed inside the rectangle that represents that cell.

The initial partition has cost 5 since nets a, b, h, g, f connect the cells in the two sets. Then the algorithm selects cell 1, which has the maximum gain. The new partition has cost 3 (nets e, g, f), and cell 1 is no longer free. The final partition has no free cells. The best partition in this pass has cost 3.

Iterative Method

Many iterative techniques have been proposed. Here, we emphasize one of these techniques called simulated annealing. Simulated annealing, as proposed by Kirkpatrick *et al.* [1983], makes the con-

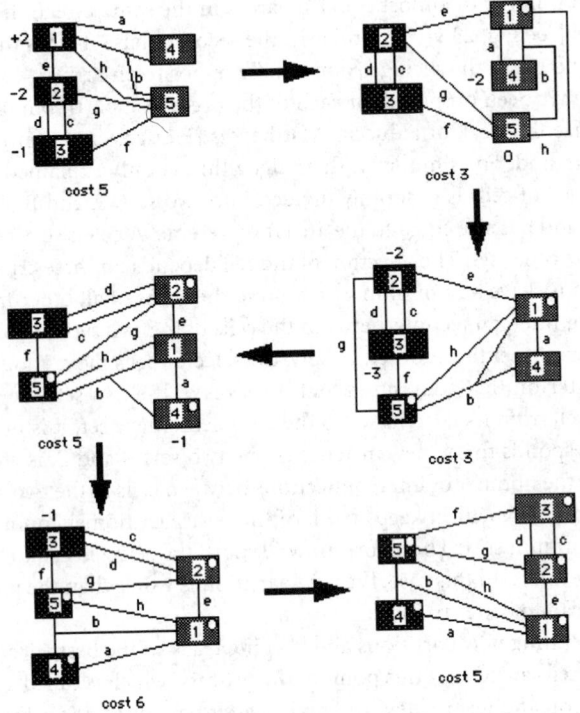

FIGURE 23.10   Illustration of a pass.

nection between statistical mechanics and combinatorial optimization problems. The main advantage with simulated annealing is its hill-climbing ability, which allows it to back out of inferior local solutions and find better solutions.

Sechen has applied simulated annealing to the placement problem and has obtained good solutions [Sechen, 1990]. The method basically involves the following steps:

BEGIN
1. Find an initial configuration by placing the cells randomly. Set the initial temperature, T, and the maximum number of iterations.
2. Calculate the cost of the initial configuration.
   A general form of the cost function may be: $Cost = c_1 * Area\ of\ layout + c_2 * Total\ interconnection\ length$ where c_1 and c_2 are tuning factors.
3. While (stopping criterion is not satisfied)
   {587
   a. For (maximum number of iteration)
      {
      1. Transform the old configuration into a new configuration.
         This transformation can be in the form of exchange of positions of two randomly selected cells or change of position of a randomly selected cell.
      2. Calculate the cost of the new configuration.
      3. If (new cost < old cost) accept the iteration, else check if the new iteration could be accepted with the probability: e (| new cost − old cost | / T).
         There are also other options for the probability function.
      }
   b. Update the temperature.
   }

END

The parameter *T* is called *temperature*; it is initially set to a very large value, so that the probability of accepting "uphill" moves is very close to 1, that it is slowly decreasing toward zero, according to a rule called the cooling schedule. Usually, the new reduced temperature is calculated as follows:

$$\text{New temperature} = (\text{cooling rate}) \times (\text{Old temperature})$$

Using a faster cooling rate can result in getting stuck at local minima; however, a cooling rate that is too slow can pass over the possible global minima. In general, the cooling rate is taken from approximately 0.80 to 0.95.

Usually, the stopping criterion for the while-loop is implemented by recording the cost function's value at the end of each temperature stage of the annealing process. The stopping criterion is satisfied when the cost function's value has not changed for a number of consecutive stages.

Though simulated annealing is not the ultimate solution to placement problems, it gives very good results compared to the other popular techniques. The long execution time of this algorithm is its major disadvantage. Although a great deal of research has been done in improving this technique, substantial improvements have not been achieved.

Routing Techniques

Given a collection of cells placed on a chip, the routing problem is to connect the terminals (or ports) of these cells for a specific design requirement. Routing is often done in two steps: first by a global router and then by a detailed router. The global router considers all the nets on the chip in order to determine for each net the set of channels that it should pass through. For every channel that a net passes through, the net's id is placed on the sides of the channel as channel terminals. Within each channel, the detailed router connects all the terminals with the same id by a set of wire segments. (A wire segment is a piece of material described by a layer, two end-points, and a width.)

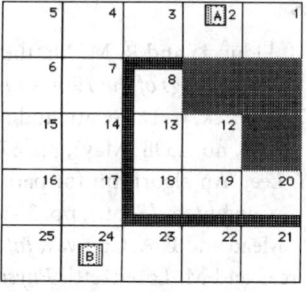

FIGURE 23.11 Initial configuration.

There are numerous algorithms to solve the routing problem. For example, one of the most widely known of these algorithms is called the maze router. The maze router can be used as a global router and/or detailed router. It finds the shortest rectilinear path by propagating a wavefront from a source point toward a destination point [Lee, 1969]. Considering the routing surface as a rectangular array of cells, the algorithm starts by marking the source cell as visited. In successive steps, it visits all the unvisited neighbors of visited cells. This continues until the destination cell is visited. For example, consider the cell configuration given in Fig. 23.11.

We would like to find a minimal-crossing path from source cell A (cell 2) to destination cell B (cell 24). (The minimal-crossing path is defined as a path that crosses over the fewest number of existing paths.) The algorithm begins by assigning the start cell 2 to a list, denoted as L, i.e., L consists of the single entry {2}. For each entry in L, its immediate neighbors (which are not blocked) will be added to an auxiliary list L'. (The auxiliary cell list L' is provided for momentary storage of names of cells.) Therefore, list L' contains entries {1,3}. To these cells a chain coordinate and a weight are assigned, as denoted in Fig. 23.12. For example, in cell 3, we have the pair $(0, \rightarrow)$, meaning that the chain coordinate is toward right and the cell weight is 0. The weight for a cell represents the number of wires that should be crossed in order to reach that cell from the source cell. The cells with minimum weight in list L' are appended to list L. Thus, cells 1 and 3 are appended to list L. Moreover, cell 2 is erased from list L. Appending the immediate neighbors of the

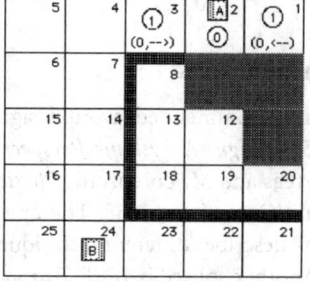

FIGURE 23.12 First step.

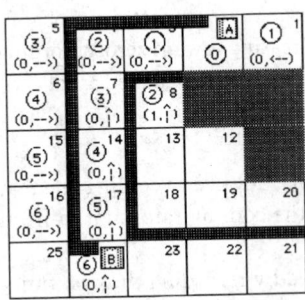

FIGURE 23.13 Final step.

cells in L to L', we find that list L' now contains entries 4 and 8. Note that cell 8 has a weight of 1; this is because a wire must be crossed in order to reach to this cell. Again the cells with minimum weight in list L' are appended to list L, and cell 3 and cell 1 are erased from L. Now L contains entry {4}. The above procedure is repeated until it reaches to the final cell B. Then a solution is found by tracing the chain coordinated from cell B to cell A as shown in Fig. 23.13.

The importance of Lee's algorithm is that it always finds the shortest path between two points. Since it routes one net at a time, however, there is a possibility of having some nets unrouted at the end of the routing process. The other weak points of this technique are the requirements of a large memory space and long execution time. For this reason, the maze router is often used as a side router for the routing of critical nets and/or routing of leftover unrouted nets.

Defining Terms

Layout: Specifies the position and dimension of the different layers of materials as they would be layed on the silicon wafer.

Placement: A placement routine determines an optimal position on the chip for a set of cells in a way that the total occupied area and the total estimated length of connections are minimized.

Routing: Given a collection of cells placed on a chip, the routing routine connects the terminals of these cells for a specific design requirement.

References

C. M. Fiduccia and R. M. Mattheyses, "A linear-time heuristic for improving network partitions," *Proceedings of the 19th Annual Design Automation Conference*, (July), pp. 175–181, 1982.

S. Kirkpatrick, C. D. Gelatt, and M. P. Vecchi, "Optimization by simulated annealing," *Science*, vol. 220, no. 4598 (May), pp. 671–680, 1983.

C. Y. Lee, "An algorithm for path connections and its application," *IRE Transactions on Electronic Computers*, (Sept.), pp. 346–365, 1969.

C. A. Mead and L. A. Conway, *Introduction to VLSI Systems*, Reading, Mass.: Addison-Wesley, 1980.

B. Preas and M. Lorenzetti, *Physical Design Automation of VLSI Systems*, Menlo Park, Calif.: Benjamin/Cummings, 1988.

C. Sechen, "Chip-planning, placement and global routing of macro-cell integrated circuits using simulated annealing," *International Journal of Computer Aided VLSI Design 2*, pp. 127–158, 1990.

S. Wimer and I. Koren, "Analysis of strategies for constructive general block placement," *IEEE Transactions on Computer-Aided Design*, vol. 7, no. 3 (March), pp. 371–377, 1988.

Further Information

Other recommended layout design publications include Weste and Eshraghian, *Principles of CMOS VLSI Design: A Systems Perspective*, Reading, Mass.: Addison-Wesley, 1988, and the book by B. Preas and M. Lorenzetti, *Physical Design Automation of VLSI Systems*, Menlo Park, Calif.: Benjamin/Cummings, 1988. The first book describes the design and analysis of a layout. The second book describes different techniques for development of CAD systems.

Another source is *IEEE Transactions on Computer-Aided Design of Integrated Circuits and Systems*, which is published monthly by the Institute of Electrical and Electronics Engineers.

23.2 Application-Specific Integrated Circuits (ASICs)

J. Leland Seely

This section will discuss the various **ASIC** design methods which are available, give criteria for determining when an ASIC be should be used, and discuss a variety of ASIC design tools that are commercially available. Obviously, not every method or design tool can be covered in this section.

ASIC Design Methods Available

Application-specific integrated circuit (ASIC) is a designation applied to a variety of devices fabricated in a variety of semiconductor technologies. ASIC, as the name implies, means the device is designed to perform a specific function, as opposed to a microprocessor which can be programmed to perform a variety of functions. The major advantages of the ASIC are typically lower unit cost and higher performance. This is the result of eliminating the circuitry from the **chip** needed to make it programmable and incorporating just those logic functions needed for the application. This makes a smaller, less costly chip. The higher performance comes from directly implementing the logic instead of using an instruction set requiring multiple clock cycles to execute.

ASICs can be divided into two categories, i.e., standard and custom. Standard ASICs are generally commodity items which have been designed by a semiconductor manufacturer and which are sold in large volume. The semiconductor manufacturer amortizes his development costs over a large number of devices, thus making such devices available at a relatively low cost. The price of such devices, however, will be determined by the market conditions as seen by the manufacturer. Custom ASICs are generally designed by or for a specific customer and are not available to the public. This allows the developer of a custom ASIC to uniquely exploit the advantages of his device. The disadvantage of custom ASICs is typically a high nonrecurring engineering (NRE) cost which may or may not be amortized over large volume.

ASICs are normally fabricated in some form of **CMOS** technology using **custom, standard cell**, physical placement of logic (**PPL**), gate array, or field programmable gate array (**FPGA**) design methods. The gate array and FPGA design methods are often referred to as semi-custom devices because they contain a fixed set of prediffused gate structures which, in the case of gate arrays, are programmed by defining the **interconnection pattern** using customized **masks** and, in the case of FPGAs, by defining the interconnection pattern electrically. Some publications do not recognize gate arrays and FPGAs as ASICs because they typically contain unused gates. They will be considered as ASICs in this document.

When and What Type of an ASIC Should Be Used

The obvious time to use an ASIC is when design objectives of performance, space, or power cannot be met with a collection of standard parts. More often than not, however, the decision to use an ASIC is based on economics, i.e., is the overall product cost reduced if an ASIC is used instead of standard products. This is such an important aspect of the decision-making process that it will be considered here in some detail. As already stated, cost is broken into two parts, i.e., NRE and unit price. The NRE is amortized over the number of parts purchased and is added to the unit price. The NRE itself consists of two parts, i.e., the engineering charge and the tooling charge. The engineering charge includes the cost of doing the logic design, the physical (geometrical) **layout**, the simulations, and the checking of the layout. The tooling charge is for the set of masks used to manufacture the **wafers** and the cost of the prototype wafer fabrication run. The cost of the mask set as of this writing is $2,000 per layer with 11 or 12 layers typical for a set. The prototype run is typically $1,000 per wafer for a 12-wafer run. Thus the total tooling cost runs approximately $35,000 and is roughly independent of which type of ASIC is used. However, the way the tooling cost is allocated to an individual customer is critically dependent on the type of ASIC. We will now discuss the NRE associated with the various types of ASICs.

FPGA NRE

It is widely advertised that FPGAs have zero NRE because the user programs the device himself. The same tooling is used for every customer and there is no "prototype run." The cost of the original tooling and the wafer fabrication is included in the unit price. However, the cost of the software and hardware needed to do the programming must be taken into account. At the time of this writing, the cost is typically around $10,000. To this cost, the cost of the engineer's time to do the actual programming must be added. The programming basically consists of converting the circuit into a netlist of the appropriate format for the specific device which is to be used. This format is dependent upon the type of basic cell or cells used internally in the device. The format is not standardized amongst FPGA vendors. The amount of time needed to do this programming obviously depends on the complexity of the required circuit and the familiarity of the designer with the design tools. It also can be dramatically affected by the required performance of the device and the degree of utilization of the usable gates contained in the device. The reasons for this have to do with timing requirements and the routing algorithm

Gate Array NRE

Gate arrays are produced from wafers which typically use seven masks which are identical regardless of what ASIC is being produced and four masks which personalize the array to perform a specific function. Thus the user pays directly for only the four masks. In addition, because the personalization involves only four masking steps, prototype runs for various ASIC customers are combined for most of the processing steps with the end result that only two wafers may be charged to a given customer's prototype run. Semiconductor vendors quote NREs for gate arrays over a huge range, the lowest being around $7,000, typical being $15,000 and some exceeding $50,000 for very complex devices. To this cost, one must again add the cost of engineering time required to get the netlist into the format required by the vendor. This is typically done using a **schematic capture** program. The schematic capture program must itself be programmed to use only devices contained in the **cell library** of the vendor. Again, these libraries are not standardized so care must be exercised in the purchase of a schematic entry program to also purchase an appropriate cell library. Cost of such schematic entry programs ranges from $1,000 to $15,000. Cost of libraries may be only a few hundred dollars for basic gates up to hundreds of thousands of dollars for libraries containing complete microprocessors. The time needed to create the netlist is that needed to draw the schematic using the symbols provided by the library. The symbols must be given an instance number so as to uniquely define the symbol (which is likely used more than once in the schematic). Inputs and outputs must be wired and the wires must be given names.

The schematic entry is the responsibility of the designer. The netlist obtained from the schematic entry program is given to the semiconductor vendor who does the physical interconnection of the gate array using an auto-place-and-route program. The physical layout of the gate array affects its performance, so the semiconductor vendor will run an "extract" program which will extract the electrical parameters of the proposed circuit which are needed to run a timing simulation. If the results of the timing simulation are acceptable to the designer, the designer spends no more time on the design. If, however, the timing results are not acceptable, the designer must modify his schematic until they are. This makes the final NRE somewhat unpredictable.

Custom NRE

Historically, in the case of custom devices, the entire cost of tooling has been charged to the customer because every mask level is unique to the given ASIC. The entire cost of the prototype run was also charged to the customer. This meant that the minimum tooling charge was $35,000 for prototypes. This is still the minimum cost for production tooling but new services from certain semiconductor companies have resulted in dramatically lower prototype costs. This service, called "shared silicon," combines many different ASIC designs into the same mask set and divides the tooling and prototype costs amongst the users. One service, called MOSIS, provides four packaged, untested devices, approximately 2.3 mm × 2.3 mm, fabricated in 2-micron CMOS technology for only $500.00. Another service, called FORESIGHT, provides 12 unpackaged, untested devices of the same size but fabricated in 1.2-micron CMOS technology for $1,500. These services make what was historically the highest prototype tooling cost into the lowest tooling cost available.

The engineering portion of the NRE still remains the highest of any of the ASIC types. This is because in a full custom device, every transistor and every interconnection is individually designed and fitted into the device in a manner designed to take up the absolute minimum chip area. Detailed **design rules** governing the placement of each geometrical shape must be understood and adhered to by the designer. Over 40 man-years of effort can go into a very complex design. Less complex designs usually cost anywhere from $50,000 to $150,000. Such time and effort is readily justified if millions of identical devices are to be produced, as is the case for a popular microprocessor. For the majority of applications, full custom design has not been the answer.

Standard Cell NRE

In order to reduce the engineering time associated with a custom layout, standard cell design methods were developed. In this design method, a set of logic devices such as gates of various types with various numbers of inputs, flip-flops, adders, etc. is generated. An effort is made to include all the functions necessary to build an ASIC. This typically results in about 250 devices in a cell library. The devices are physically laid out using design rules from the targeted wafer fabricator. The standard cells are usually designed to be of constant height and variable width with interconnection points located along the bottom and possibly the top of the cell. This is done to facilitate the use of an auto-place-and-route program. Designs using standard cells are usually entered by means of a schematic capture program to generate a netlist which drives the auto-place-and-route program. The end result is a dramatic reduction in layout engineering time. Unfortunately, there is a corresponding dramatic increase in chip size, which, as will be shown later, means a dramatic increase in unit cost. Given a netlist from an appropriate schematic capture tool, the engineering portion of the NRE can be as little as $10,000 to $15,000. As with the gate array, the engineering time spent by the customer in doing the schematic capture and the cost of the schematic capture software and hardware must be taken into consideration. Tooling charges are the same as for custom.

PPL NRE

The PPL design method is a compromise between standard cell and full custom methods. It uses more primitive cells than do typical standard cell libraries, but in so doing it allows the designer to construct any desired logic function very quickly from fewer cells and with **transistor densities** close to that of full custom designs. The engineering is done entirely by the design engineer so the entire engineering portion of the NRE is the cost of the in-house engineer. The amount of time he spends doing the design is typically less than he would spend doing just schematic capture. The cost of the software programs used for PPL design is about $15,000 at the time of this writing. The tooling portion of the NRE is the same as for custom, i.e., $500 to $1,500 for small circuits using shared silicon services.

Unit Cost

Wafer processing is a batch process which basically results in a fixed cost per wafer regardless of the number of good chips contained on that wafer. There are semiconductor manufacturers willing to sell wafers at a fixed cost, depending on the volume of wafers committed for purchase. The price as of this writing is $800 to $1000 per 6-in. diameter wafer. The cost per good chip is determined by this price and the number of good chips per wafer. The trick then is to estimate the number of good chips per wafer.

The ratio of the number of good chips compared to the total number of chips on a wafer is called yield and is designated by y. Murphy of Bell Labs was the first to give a good theoretical treatment of yield. He arrived at Eq. (23.1)

$$y = [(1 - \exp(-D_0 A))/D_0 A]^2 \tag{23.1}$$

where D_0 is the number of defects per unit area and A is the active area of the circuit. Figure 23.14 shows a plot of this equation for various values of D_0. This equation was verified by physically counting the number of defects and the number of good chips on a wafer. In the early days of semiconductor manufacturing, agreement between theory and actual was good. Later, as clean room technology, mask-making technology, and manufacturing practices improved, agreement between physically counted defect density and defect density implied by curve fitting became increasingly poor. Finally, defect density became a hypothetical number obtained by curve fitting the simple exponential equation shown as Eq. (23.2).

$$y = \exp(-D_0 A) \tag{23.2}$$

The value of D_0 typically runs between 8 and 25 defects per square inch. The actual value can be obtained from the manufacturer. Care should be exercised to make sure that Eq. (23.2) is being used as the reference equation and that defect density is expressed in the same units as A.

The cost of the unpackaged, untested die, C_d, can now be easily calculated as follows:

$$C_d = C_w/(yN) \tag{23.3}$$

where C_w is the fixed wafer cost, y is yield, and N is the total number of die on the wafer.

The good die must be separated from the bad die on the wafer. This is done at a testing operation called **wafer sort**. Wafer sort involves a test fixture called a probe station capable of making temporary electrical contact to the pads of a particular die, applying electrical signals to the input pads and comparing the signals from output pads to the expected values. Such test fixtures typically cost from $500,000 to $1,000,000 so an attempt is made to minimize the time needed to test each individual die in order to reduce wafer sort cost. If the test is made too short, bad die will be allowed to pass and will incur packaging costs just to be thrown away at final test. Typical wafer sort test times

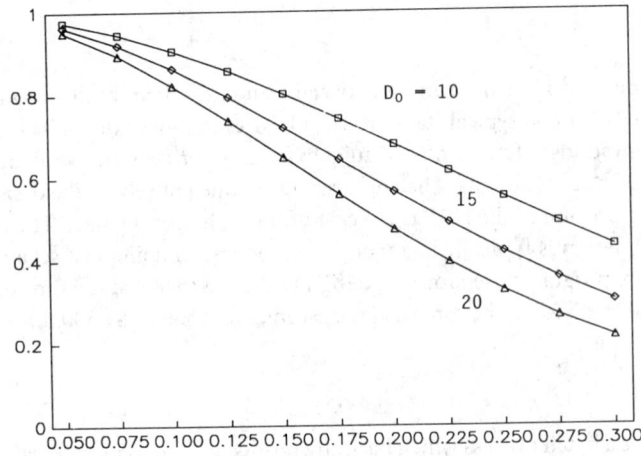

FIGURE 23.14 Yield versus chip dimension in inches.

are about 0.5 seconds per die and typical wafer sort equipment is charged at $100 per hour. Cost to sort a complete wafer, C_s, is then

$$C_s = C_t t_s N \tag{23.4}$$

where C_t is the cost per unit time for the wafer sort station and t_s is the sort time per die.

The die that pass wafer sort must be packaged. The packaging operation is called assembly. Assembly results in the loss of some of the good die. The number of packaged devices out of assembly divided by the number of die put into assembly is called the assembly yield, y_a. The packaging cost is usually expressed in cents per lead and refers to standard, high-volume plastic packages. The price is typically 1 to 2 cents per lead. Equation (23.5) expresses the total assembly cost, C_a.

$$C_a = lC_l yN \tag{23.5}$$

where l is the number of leads and C_l is the cost per lead.

The last major step is **final test**. This is done using automatic test equipment (ATE). ATE for digital circuits costs around $1 million to $2 million. For analog circuits, the cost can be up to $5 million. The high cost of this equipment means that test times must be made as short as possible consistent with insuring that only an acceptably low percentage of bad devices pass the test. Typical final test times are around 3 seconds. Independent test houses will quote an hourly rate for use of a given tester, usually between $75.00 and $200.00 per hour. Equation (23.6) gives the total final test cost, C_t, for testing all of the devices from one wafer which come out of assembly.

$$C_t = C_h t_t y y_a N \tag{23.6}$$

where C_h is the cost per hour for testing and t_t is the time for testing. Again, not all devices that go into final test come out as good. The final test yield is designated y_t.

The total cost of the wafer, wafer sort, assembly, and test is then given by Eq. (23.7).

$$C = C_w + C_s + C_a + C_t \tag{23.7}$$

The total number of good assembled and tested devices is

$$n = yy_a y_t N$$

Finally, the cost per good finished device is simply

$$\text{total cost per device} = C/n \tag{23.8}$$

Figure 23.15 shows a plot of total cost versus chip dimension and Fig. 23.16 shows a plot of total cost per unit area versus chip dimension assuming the following parameters.

$$C_w = \$1,000 \text{ for a 6-in. diameter wafer}$$
$$D_0 = 15 \text{ defects per square inch}$$
$$C_t = \$100 \text{ per hour}$$
$$t_s = 0.5 \text{ s}$$
$$l = 16 \text{ leads}$$
$$C_l = 1 \text{ cent per lead}$$
$$y_a = 95\%$$
$$C_h = \$150 \text{ per hour}$$
$$t_t = 3 \text{ s}$$
$$y_t = 90\%$$

Table 23.1 summarizes typical prototype and production costs for a 7000-gate device using the

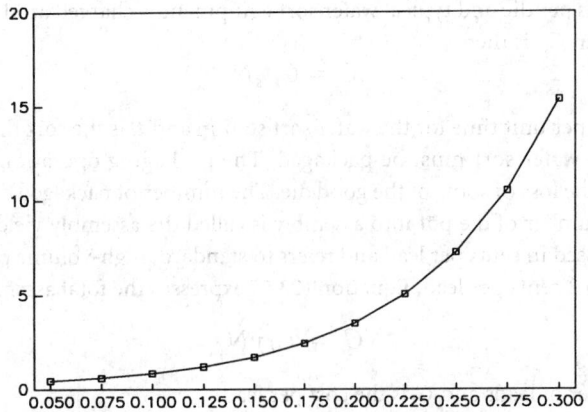

FIGURE 23.15 Cost of finished device in dollars versus chip dimension in inches.

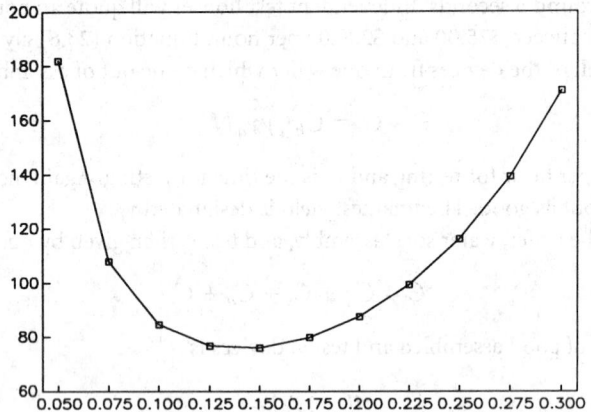

FIGURE 23.16 Cost in dollars per square inch of area of device versus chip dimension in inches.

Table 23.1 Typical Costs for Prototypes and Production Unit

Design Method	Software Cost	In-House Man-Weeks	NRE	Prototype Costs	Production Tooling	Unit Cost
FPGA	$ 10,000	9	$ 0	$80.00	$ 0	$60.00
Gate array	6,000	7	10,000	Included in NRE	20,000	15.00
Custom	6,000	7	100,000	3,000	35,000	5.00
	100,000	52	0	3,000	35,000	3.50
Standard cell	6,000	7	15,000	6,800	35,000	10.00
PPL	15,000	7	0	3,000	35,000	5.00

various ASIC options. Keep in mind that price and cost may differ dramatically if the customer does not own the tooling!

Development Time

Another factor in determining which type of ASIC should be used is the amount of time available for the design and creation of prototype and production units. There are often strong market-driven pressures to create a prototype in a very short time. FPGAs are the device of choice in this case but care must be taken that the path to production is not overlooked. To convert an FPGA to a more cost-effective ASIC can be difficult, expensive, and time consuming if such things as timing simulations and **test vectors** are omitted during the initial FPGA development.

ASIC Design Tools

Schematic Entry

Figure 23.17 shows a typical schematic entry design system which uses five different programs. In addition to the programs, a library of cells usable by the targeted silicon foundry and technical data relating to the cell performance must be programmed into the schematic capture tool. The design system is then used in the following way.

The block diagram of the desired ASIC has been created by the system designer and has been implemented by placing logical elements from the library supported by the target foundry onto the screen of a workstation, uniquely defining the elements by giving them an instance number and then drawing in the connections and naming the wires. The operator then runs the schematic entry program which produces a netlist. Note that in producing the netlist, all of the positional information contained in the schematic drawing, i.e., what element goes next to what and where the critical signal paths are, is lost. Since the netlist was produced from a schematic, there is no actual information as to stray capacitance, inductive coupling, transmission line length, etc. These parameters must be estimated and added to the netlist in a format usable by a timing simulator.

In addition to the netlist from the schematic entry program, test vectors must be generated to stimulate the netlist. Again, in order to bridge the interface between the schematic entry program and the **simulator** program, the test vectors must be in the format required by the simulator that is to be used. The simulator program itself must be programmed to handle all of the logic elements contained in the foundry library.

When the simulation is completed, the results must be examined. This requires another program to take the output of the simulator and display it in the format desired.

The next interface to bridge is that between the schematic entry program and the auto-place-and-route program. Again, the router needs to be programmed with the appropriate standard cell library and the output from the schematic entry tool must be in a format acceptable to the router. Obviously, this can only be done if the silicon foundry has already been selected.

The router outputs the geometrical data defining the structures and locations of the structures used in the fabrication of the silicon. These structures define data paths and capacitive loading which may be critical to the performance of the circuit. Since this is left to the caprice of the router,

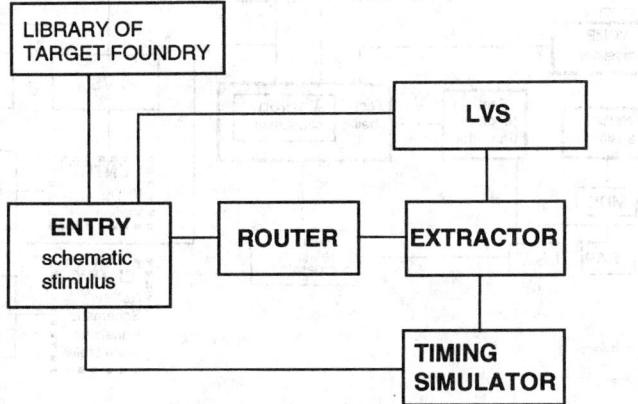

FIGURE 23.17   Schematic entry design system.

the performance may not meet specification. It is therefore necessary to run an extraction program on the output of the router. The extraction program traces the layout and constructs a netlist of the actual layout and can be programmed to calculate the actual node loading. This constitutes yet another interface to be bridged by programming the extractor with the appropriate data relating to the electrical properties of the various structural layers and making sure that the output format of the router is usable by the input to the extractor and that the output of the extractor is usable by the input to the timing simulator. The netlist containing the actual loading obtained by extraction from the layout is called back-annotated data. The timing simulation obtained from stimulating the back-annotated netlist is expected to represent the actual circuit performance. If this performance does not meet specification, then some sort of human interaction with the router is required. In the worst case, the router has no such capability. In the best case, it is another program that the engineer must master.

The next interface occurs between the netlist provided by the extractor—the real netlist—and that provided by the schematic entry—the intended netlist. These should be compared to see that the real netlist is the same as the intended netlist. This is done by a layout-versus-schematic (LVS) program. Inputs to the LVS program coming from the extractor and schematic entry programs must be compatible.

Each of the five programs must be programmed with information from the library of the target foundry. The design system shown here is actually simplified because no "output" programs such as a layout editor program for viewing the geometries is included and no schematic generator programs are included.

Hardware Description Language

Many design systems are based on the use of a Hardware Description Language such as VHDL or Verilog and may use a **silicon compiler** and **synthesizer**. Such a design system is shown in Fig. 23.18. This system was put together by an actual design group and consisted of a collection of commercially available **CAE** tools strapped together with internally developed translators and interfaces.

Assuming the system has been described in VHDL and partitioned (if needed) into one or more ASICs, the resulting VHDL description must be converted to a detailed specification. The way the

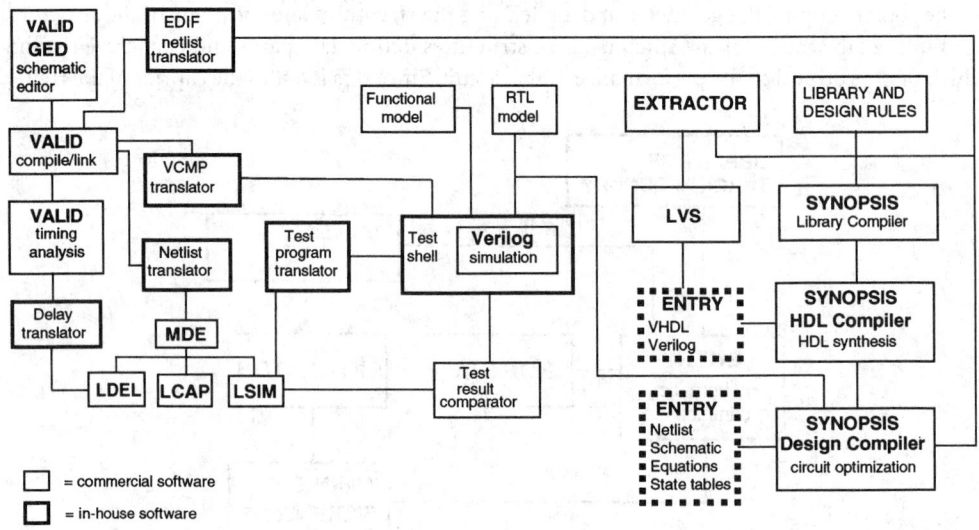

FIGURE 23.18 Hardware description language design system.

specification is written can be dependent upon the design system which will be used to implement the design. In our example, a synthesizer program is to be used to synthesize the ASIC from the VHDL description. In this case, the VHDL description must be expanded to include specified timing. VHDL is a C-like language which not every system engineer speaks fluently. VHDL is certainly not intuitive, except perhaps to a programmer. To determine if the specification is correct, the VHDL description must be simulated. This means that test vectors described in the required stimulus format for a VHDL simulator must be written and someone in the organization must be able to run a VHDL simulator. Once the specification is determined to be correct, then a synthesizer program is needed. Programming a synthesizer is a nontrivial task. It needs to be set up with a library of cells and a set of design rules appropriate for the silicon foundry targeted for wafer fabrication.

Just as was the case for schematic entry, the timing simulator, extractor, and LVS programs are required and the interfaces must be compatible. Also, most of the 13 commercial programs incorporated in this system must themselves be programmed with the library and technical information from the target foundry. In the actual operation of this design system, it was found necessary to employ a full-time librarian to maintain the Synopsys, Verilog, and Valid libraries because of the difficulties of such a diverse environment. UNIX source control utilities to provide version control, locking, dependency analysis, system builds, and checkout were also required. The effect of using a Hardware Description Language is to increase the number of programs and interfaces. At the present time, the complexity and cost of this type of design system makes it suitable for well-established and well-staffed professional design departments, not for entry level designers.

The situation becomes even more complex if the designer wishes to combine both digital and analog circuitry on the same ASIC chip or if performance dictates that the designer control the position of his critical data paths or analog components or if he wishes to make use of generators for the automatic synthesis of random access memory or read-only memory, etc. The use of different tools makes the design entry more complex. Today's higher level tools tend to be very specialized, doing only a small portion of the overall design, and they tend to be less understandable to others on the design team.

The PPL Technology

PPL, meaning physical placement of logic, is an entirely new concept in integrated circuit design. PPL was developed for use by the "**occasional designer**," i.e., an engineer who is not a full-time ASIC designer. PPL departs significantly from traditional IC design methods which grew from incremental improvements in the original manual design methods. With PPL, the systems engineer works from the block diagram. In a single step, he designs the circuit logic and specifies its complete physical layout and interconnect.

The key to the system is the design entry. It is at a high enough level to handle large designs. It is intuitive so that it is easy to learn. The user of the system does not have to learn the detailed operations of the many different programs involved. The many interfaces between the programs involved are transparent to the user, i.e., the translator programs are in place. The unique information associated with the physical layout is available during design entry, which allows immediate timing simulation of the actual layout, not an estimated layout.

The design entry is done using **structured cells**, not standard cells. These cells must be carefully designed to meet conflicting requirements. If they are too high level, they become too specific in function and require too large a library. If the cells are too low level, they can be used to construct any desired logic function but too much effort is required to construct the logic function. The correct place in the hierarchy must be chosen to allow enough flexibility in the cells to construct any required logic function with the minimum effort.

A cell need not be a complete logic function by itself if it is useful in constructing a group of logic functions. The cell representation is reduced to a single alpha-numeric character and automatically generated "space markers" which show the relative size of the cell. These are the two fundamental properties needed to completely specify its function and physical properties. These cells are referred to as physical logic elements.

The cells connect automatically on all four sides. To do this, a predefined **grid** of interconnect wires is used. Wires running in the X direction are at one level of metallization and wires running in the Y direction are on a second level so that the wires do not connect to each other. The grid looks like the wire mesh in a screen door. This is the departure point from the conventional method. The interconnect is defined before the logic elements instead of after. This places the major emphasis on the interconnect instead of active devices. The logical function performed by the physical logic elements is determined by placing them next to each other, either above or below. Thus the connections and the logic function performed by the cells are determined by the adjacent cells. The physical area occupied by the logic element is also determined at the time the cells are placed.

The important principles of the physical logic design are:

1. The wiring including signal paths gets first priority over transistor placement. Both circuit density and performance are decided by how well the wiring is done.
2. The software takes care of the cell environment by providing power buses, wells, and automatic connections, thus freeing the designer from much of the routine work and letting him concentrate on the engineering.
3. The placed screen characters define logic functions, physical location of the logic functions, schematic icons of the logic functions, physical layout of the logic functions, all electrical parameters of the interconnected layout—stray capacitance, line lengths, etc.
4. Logic design always includes the physical placement. The netlist is always extracted from the physical layout and always contains updated back-annotated data so that timing simulation is not estimated but is actual.

Table 23.2   Commercially Available Design Programs

Program	Price	Vendor	Program	Price	Vendor
PPL	$15,000	Bonneville Microelectronics Inc.	**Analog Simulation**		
			Hspice	$13,000	MetaSoftware
			Pspice	950	Microsim
Schematic Capture			Mspice	19,800	Mentor
Microwave Musician	15,000	Cadence	spice s/w	790	Intusoft
Workview Sun	10,000	Viewlogic	Analog Systems Lab	45,000	Valid Logic
Workview PC	6,000	Viewlogic	**Analog/Digital Simulation**		
Orcad SDT	1,000	Orcad	Delux A/D Sun	14,900	Microsim
Schema III	495	Omation	Delux A/D PC	4,950	Microsim
Capfast	995	Phase Three Logic	Smash	3,950	Dolphin Integration
Schematic Capture + Sim	75,000	LSI Logic			
Tango-Schematic Series II	495	Tango	**Autoroute**		
Design Compiler	35,000	Synopsys	Tancell	50,000	Cadence
			Parade	300,000	Mentor
Digital Simulation			Easy Route	695	AMS
Hilo	23,000	GenRad	Schema Route	749	Omation
Susie 6	3,995	Aldec	**Extraction**		
VST	995	Orcad	Dracula	75,000	Cadence
Silos II/386	2,500	Simucad	Explorer Checkmate	98,500	Mentor
Gatesim	1,295	Tanner Research	Layout		
Msim	23,800	Mentor	Edge	100,000	Cadence
Library	20,000	Mentor	Ledit Sun	4,000	Tanner Research
Viewsim/SD	19,000	Viewlogic	Ledit PC	995	Tanner Research
			HDL Compiler	22,500	Synopsys
			GDT Cell Compiler	90,000	Mentor
			VHDL System Sim	24,000	Synopsys

5. Schematic generation is automatic by simply replacing the screen character with its schematic icon and letting the software automatically remove unused wires.
6. The logic function of a cell may be altered by an adjacent cell. For example, placing a "U" cell under an "F" cell changes the "F" cell from a flip-flop to one stage of an "Up counter".

PPL solves the problems associated with needing to program the ASIC design tools with specific standard libraries from the target foundry because the tools are already programmed with the PPL cells—the foundry library is not used. However, the library of PPL physical logic elements must be laid out to the rules of the target foundry. This turns out to be a manageable problem because of the small number of such cells in the library—especially compared to the size of standard cell libraries—and because of the availability of a software program which can take the **GDS2** data which is the geometrical description of the cells and migrate the data to that which describes the same set of cells drawn to a different set of design rules.

Examples of commercially available programs are given in Table 23.2.

Defining Terms

ASIC: Application-specific integrated circuit. Device designed to perform a specific function as opposed to a device which can be programmed to perform a variety of functions.

Assembly: The operation of putting a chip into a package.

CAE: Computer-aided engineering. Software tools for use by engineers.

Cell library: A collection of about 250 logic elements which have been designed, laid out to a specific set of design rules for a specific silicon foundry, and simulated for performance.

Chip: The rectangular piece of silicon containing a single integrated circuit. Also called a die.

CMOS: Complementary metal-oxide silicon. Refers to a popular technology used to fabricate chips.

Custom: A design method wherein every transistor and every interconnection is individually designed and fitted onto the chip in a way that requires the absolute minimum chip area.

Design rules: Detailed statements of the geometrical relationships of the various layers comprising the structures of the transistors and interconnections of the chip such as metal-to-metal spacings, minimum width, metal overlap of vias, etc. These rules are dictated by the technology, equipment, and capabilities of the silicon foundry.

Final test: Electrical test performed after assembly to separate "good" devices from "bad."

FPGA: Field programmable gate array. An assembled chip containing an array of logic elements which can be interconnected by the user in the user's facility by means of some special hardware and software.

GDS2:   A common format for the geometrical data which completely describes the chip. The masks used in the actual wafer fabrication are made from this data. Another common format is CIF.

Grid: Refers to the regular array of vertical and horizontal wires used for interconnecting the chip.

Interconnection pattern: The segments of the metal grid and associated vias actually used in the connection of logic elements for a specific function.

Layout: The creation of the geometrical description of the device structure, typically eleven layers superimposed on each other to give relative position as well as shape.

Masks: Typically a set of eleven photographic plates, each containing multiple images of one layer of the device structure. The images on the plates are transferred to the wafers whereon subsequent wafer processing steps are used to create the actual device structures on the wafer.

Occasional designer: An engineer who needs ASIC, and knows how to do the logical design of an ASIC, but who is not a full-time ASIC designer.

PPL: Physical placement of logic. A design entry method between full custom and standard cell. It begins at the block diagram level where the detailed block specification and the corresponding layout are done simultaneously. PPL is targeted at the occasional designer.

Schematic capture: A design entry method wherein the designer draws the schematic of the desired ASIC using a library of standard cells. The program outputs a netlist of the schematic.

Silicon compiler: A set of software programs intended to start with design equations and output the corresponding GDS2 data. Silicon compilers are currently used to translate a standard cell design from one set of design rules to another or to create a new set of standard cells.

Simulator: A program used to predict the behavior of a circuit. Simulators may be transistor level, gate level, behavioral level, analog, digital, unit delay, timing, or various combinations.

Standard cell: An element of a standard cell library designed using rules from the targeted wafer fabricator. Standard cells are usually designed to be of constant height and variable width with interconnection points located along the bottom and possibly the top of the cell. This is done to facilitate use of an auto-place-and-route program.

Structured cell: An element of the PPL cell library designed using rules from the targeted wafer fabricator. Structured cells are integral multiples of a unit cell with interconnection points on all four sides of the cell. Structured cells normally interconnect simply by being placed next to another structured cell. Unwanted connections are broken as opposed to desired connections being made.

Synthesizer: A software program which creates GDS2 data from a hardware description language specification such as VHDL or Verilog.

Test vectors: The input voltages used to test an ASIC expressed as a series of 1s and 0s for each of the inputs to the circuit along with the corresponding output voltages expected expressed in the same way.

Transistor density: Number of transistors per unit area. Useful for comparing relative unit cost of ASICs designed with different design methods. Higher density means lower cost.

Wafer: The substrate upon which the ASIC is fabricated. CMOS ASICs use silicon wafers, typically circular with a 6-in. diameter. Other technologies use GaAs or sapphire.

Wafer sort: A preliminary electrical test of each die while still on the wafer to eliminate most of the bad die before they are assembled.

Further Information

Because of the diversity of ASIC design methods and tools, it is suggested that parties contact directly a vendor of the specific type of tool of interest. The vendors are listed in Table 23.2. For locations of the vendors parties may contact author Seely or Brian Taylor of Bonneville Microelectronics, Inc. at (801) 467-4698.

24

Surface Mount Technology

24.1 Introduction 603
24.2 Packaged-Component Subassemblies 603
24.3 Technology Overview 604
24.4 Electronic Circuit Components 606
     Integrated Circuit Packages • Chip Carriers • Flat Packs • Small-Outline Integrated Circuits • Discrete Leadless Components
24.5 SMT Assembly Processing 611
     Substrate Preparation • Component Placement • Soldering • Cleaning • Repairing/Reworking

Gerald L. Ginsberg
Component Data Associates

24.1 Introduction

Until recently, packaged-component subassembly technologies were based on the use of printed wiring boards and through-hole mounted **components**. However, during the last several years there has been considerable activity in developing new **integrated circuit (IC)** device packages and improving the density and performance of enhanced printed wiring boards, i.e., **packaging and interconnecting (P&I) structures**, that mount and interconnect them. This intense activity basically has been promoted by:

- The relatively static state of interconnection technology in the 1970s
- The emergence of new package types for the complex large scale integration (LSI) and very large scale integration (VLSI) devices
- The basic packaging advantages of using **"surface mount"** attachments

24.2 Packaged-Component Subassemblies

The selection of an appropriate packaged-component assembly technique should initially include the requirements of the end-product equipment and subassembly from the viewpoint of form, fit, and function with respect to cost effectiveness, performance, and marketability issues. After characterizing the assembly in this manner, it is then necessary to select an implementation technique based on specific electrical and mechanical functions. Determining factors will include packaging density, assembly profile height, development time, development cost, circuit element factors, manufacturing costs, thermal considerations, reliability, etc. and specific related implementation details, as listed in Table 24.1. The steps of the assembly process differ according to the type of

This chapter is reprinted from G. L. Ginsberg, *Electronic Equipment Packaging Technology*, New York: Van Nostrand Reinhold, 1991. With permission.

Table 24.1 Integrated Circuit Packaging Technology Comparison

Characteristics	Through-Hole	Leaded Surface Mount	Leadless Surface Mount	Bare Die
Packaging density	Low	Moderate	Good	High
Standardization	Very good	Good	Good	Limited
Thermal performance	Moderate	Good	Very good	Fair
Substrate choices	Very good	Very good	Good	Limited
Fab investment	Low	Low	Moderate	High
Assembly investment	Moderate	Low	Moderate	High
Support investment	Low	Moderate	Moderate	High
External assembly services	Very high	High	Moderate	Limited
Maintenance skills	Low	Moderate	Moderate	High
Change risk	Very low	Low	Moderate	High
Assembly test	Very easy	Easy	Moderate	Complex
Documentation	Easy	Easy	Moderate	Complex
Logistics support	Field change	Field change	Field change	Factory only
Inspect (circuit)	100% test	100% test	100% test	Lot accept
Component burn-in	Easy	Easy	Easy	Impractical
Pretest	Very easy	Very easy	Easy	Impractical
Change and repair	Easy	Easy	Easy	Difficult
Component availability	Excellent	Very good	Good	Limited
Multiple sourcing	Excellent	Very good	Good	Limited
Footprint commonality	Excellent	Very good	Good	Poor
Profile	High	High	Moderate	Low

Source: IPC, 1987a; G. L. Ginsberg, *Electronic Equipment Packaging Technology*, New York: Van Nostrand Reinhold, 1991, p. 107. With permission.

product being assembled, i.e., through-hole, surface mount, or **mixed technology**. They also vary according to manufacturer expertise, experience, and preference. Table 24.2 compares some of the possible process flow sequences for various types of products.

24.3 Technology Overview

The development of through-hole packaged-component assembly technology [IPC, 1988] has essentially reached its limits as far as improvements in end-product cost, packaging density, circuit performance, and reliability are concerned. To realize further benefits in these areas the trend is toward the increased usage of surface mount technology (SMT) or a mixed SMT/through-hole technology. As the name implies, SMT terminates packaged circuit components in a planar manner on the surface of the printed wiring board, as illustrated in Fig. 24.1. The increased usage of SMT has a profound impact on packaging technology due to its use of relatively smaller component packages and by not requiring **component mounting** holes in the interconnecting substrates. This translates into packaging advantages such as:

- Reduced subassembly and ultimately equipment, size, and volume through the use of increased component placement densities, finer-pitch component terminals, and the ability to have components mounted on both sides of the assembly.
- Reduced component mounting costs by eliminating the need for packaged-component lead forming. This also facilitates the increase in component placement rates.
- Improved high-speed/high-frequency circuit performance with shorter interconnection wiring lengths and lower inductances, capacitances, and resistances.
- Improved shock and vibration environmental stability through the use of smaller (lower mass) component packages.
- The availability of some high-speed IC devices only in surface mount configurations.

Table 24.2 Integrated Circuit Assembly Process Flow Comparison

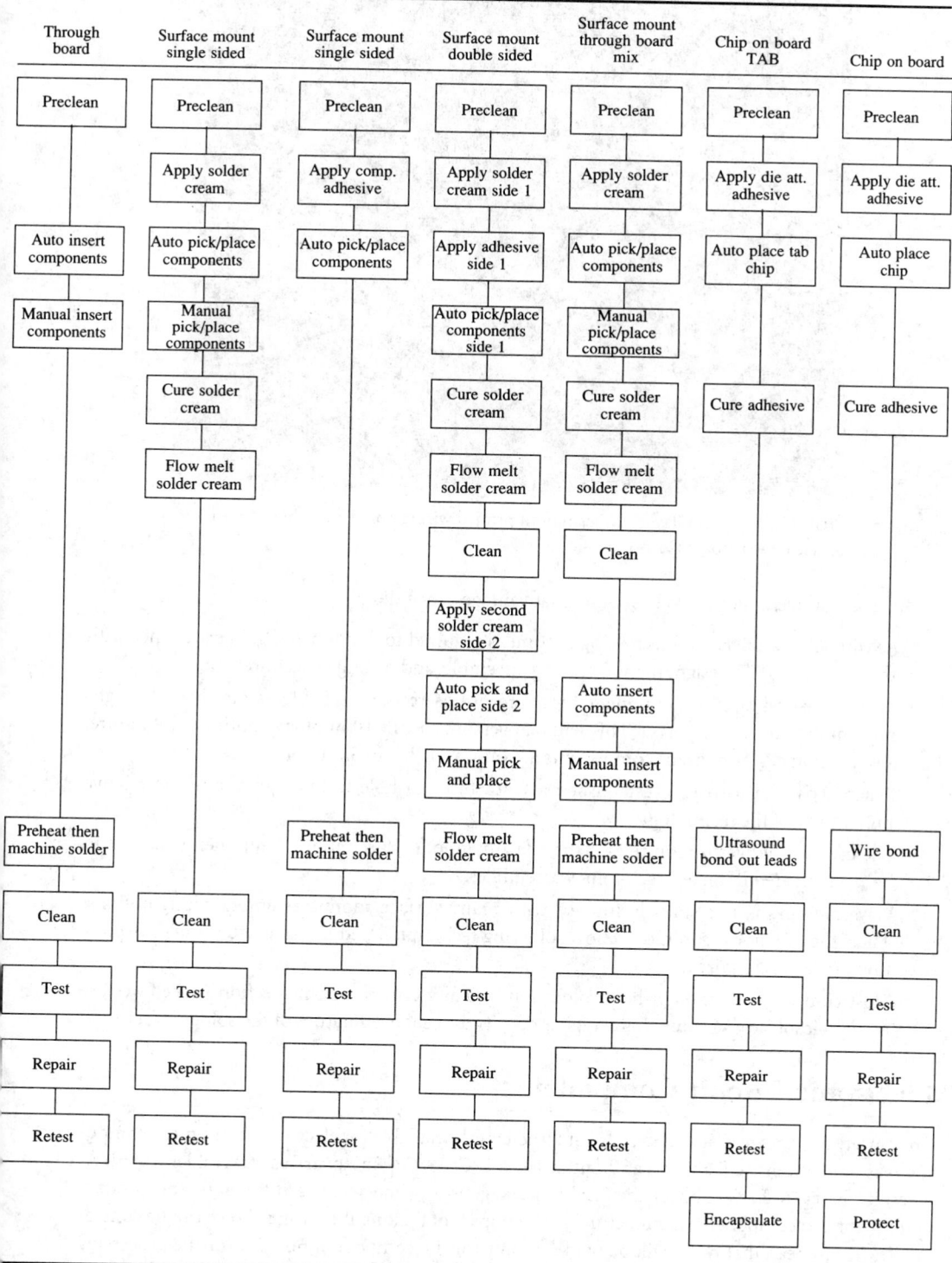

Through board	Surface mount single sided	Surface mount single sided	Surface mount double sided	Surface mount through board mix	Chip on board TAB	Chip on board
Preclean	Preclean	Preclean	Preclean	Preclean	Preclean	Preclean
	Apply solder cream	Apply comp. adhesive	Apply solder cream side 1	Apply solder cream	Apply die att. adhesive	Apply die att. adhesive
Auto insert components	Auto pick/place components	Auto pick/place components	Apply adhesive side 1	Auto pick/place components	Auto place tab chip	Auto place chip
Manual insert components	Manual pick/place components		Auto pick/place components side 1	Manual pick/place components		
	Cure solder cream		Cure solder cream	Cure solder cream	Cure adhesive	Cure adhesive
	Flow melt solder cream		Flow melt solder cream	Flow melt solder cream		
			Clean	Clean		
			Apply second solder cream side 2			
			Auto pick and place side 2	Auto insert components		
			Manual pick and place	Manual insert components		
Preheat then machine solder		Preheat then machine solder	Flow melt solder cream	Preheat then machine solder	Ultrasound bond out leads	Wire bond
Clean	Clean	Clean	Clean	Clean	Clean	Clean
Test	Test	Test	Test	Test	Test	Test
Repair	Repair	Repair	Repair	Repair	Repair	Repair
Retest	Retest	Retest	Retest	Retest	Retest	Retest
					Encapsulate	Protect

Source: IPC, 1987b; G. L. Ginsberg, *Electronic Equipment Packaging Technology*, New York: Van Nostrand Reinhold, 1991, p. 108. With permission.

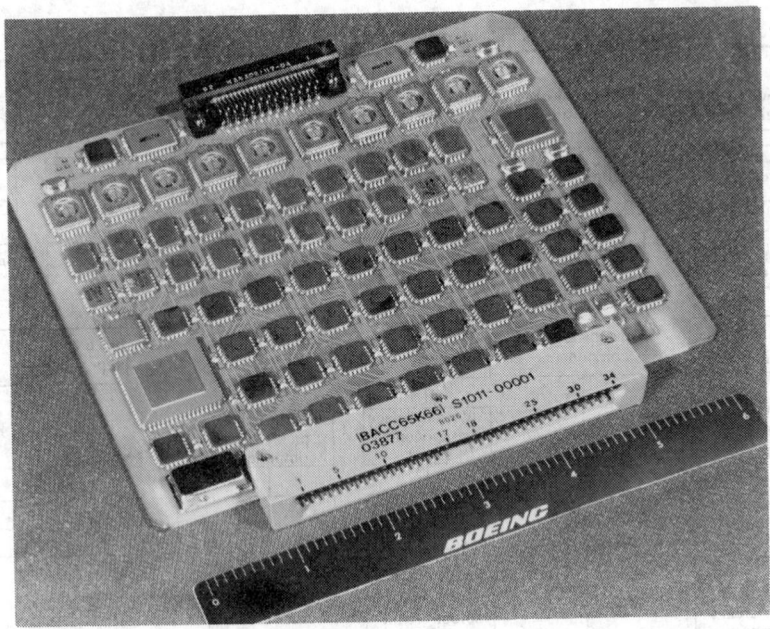

FIGURE 24.1 Multilayer surface mount printed wiring board assembly. (Courtesy of Boeing Aerospace Corp.)

However this technology also has some disadvantages, including:

- As with all new packaging technologies, time is required to develop and gain experience with the necessary SMT design, manufacturing, assembly, and testing procedures.
- Some packaged circuit component types are not as readily available as are their through-hole–mounted counterparts. (This will be overcome as the technology continues to mature. Also, as noted, some circuit components are only available in SMT configurations.)
- Reliability is not as readily established as with through-hole technology due to the relative immaturity of the technology.
- The use of adhesives is required for the attachment of surface mount components that are to be wavesoldered in mixed-technology assemblies.
- Wavesoldering is not suitable for use with many surface mount assembly configurations. Thus, the use of less common reflow soldering techniques, and the associated solder deposition process, is required.
- Most conventional through-hole solder joint quality criteria and inspection procedures are not applicable to SMT due to the higher density and changed nature of the solder joints.

24.4 Electronic Circuit Components

The variety of package types, materials, and lead counts available for electronic equipment components is quite extensive [Johnson and Luthi, 1989; IPC, 1987a; Rosengarth and Winkler, 1986]. As a result, the careful selection of the "right" package is a primary concern for both end-product equipment and component manufacturers. The impact of making the wrong choice can be considerable. It can mean that more will be paid not only for the component but also for the processing needed to assemble it. Thus, all components should be qualified for the assembly process to be used. The physical dimensions of the component should be compatible with the assembly-handling equipment. Also, the part must not be excessively degraded physically or electrically by the assembly and end-product environments to which it will be exposed.

Electronic component packages serve to protect the devices within them from the environment, provide communication links with other components, remove heat, and provide a means for handling and testing. With large, high-density ICs, these functions are not only important, but they are a challenge to meet. At the same time, there is an emphasis on maintaining or decreasing the procurement/assembled costs of the more complex packages. Also, the material composition, finish, and configuration of both the component package body and its terminations must be considered in the choice of the assembly method. This can be achieved by increases in functional density and performance by the increased use of enhanced memory, data processing, and application-specific integrated circuits (ASICs).

There are several interrelated electronic component package design factors, some of which strongly impact the others. In the package design stage, one such factor may be emphasized over another for a given application. Examples of this might be emphasizing package electrical performance over cost for a high-speed device or emphasizing cost over performance for a high-quantity application. Therefore, the characteristics of each package must be clearly understood in order to make the optimum selection. In addition to the component package cost itself, it is important to consider the sum of the direct and indirect cost increments associated with component procurement (multiple sourcing), handling, assembly, testing, repair/rework, inventory, and reliability criteria. In other words, the electronic component packaging cost factor must be reviewed, not only in terms of related design trade-offs but also in terms of the effect on overall end-product cost.

Integrated Circuit Packages

For many years the **dual inline package (DIP)** predominated for use with IC components. However, the steady increase in IC package [Johnson and Luthi, 1989; Waltersdorf, 1986] input/output (I/O) terminal count requirements and the emergence of SMT resulted in the use of a wide variety of IC component packages. The integrated circuit packages come in the form of **leadless chip carriers** and **leaded chip carriers**, **flat packs**, and small-outline ICs for surface-mounting applications and pin-**grid** arrays for through-hole mounting, as shown in Table 24.3 [Richards, 1989]. Besides taking up far less space on printed circuit boards than does the comparable DIP, the new packages can accommodate larger IC dies and significantly more I/O terminals (and reduced terminal pitch) with improved electrical performance. However, due to the variety of packages available and the wide range of their costs, each type of IC package comes with its own advantages and limitations. Thus, the selection process becomes a trade-off that takes into account many factors, including:

- Through-hole versus surface mounting
- I/O terminal capacity (and pitch)
- Ceramic (hermetic) versus plastic dielectric
- Leaded versus leadless terminations
- Electrical performance
- Environmental stability
- Thermal management features
- Procurement and assembly costs

Chip Carriers

Chip carriers can be generally described as being low-profile, rectangular (usually square), surface mount IC packages with I/O terminals on all four sides. The I/O terminals consist of either metallized features on "leadless" versions or discrete leads formed around or attached to the side of the

Table 24.3 Characteristics of Common Integrated Circuit Packages

Package Type	Range of Physical Dimensions	Electrical Characteristics[1]	Thermal Characteristics (°C/W)	Usable Gates[3]	Relative Cost (per pin)
Through-hole DIP	Number of pins: 16 to 64 Pin pitch: 100 mils Body length: 75 to 2.3 in. Body width: 0.300 to 0.700 in.	R: Medium L: High C: Low	Ceramic/plastic θJA:[2] 70–40/ 120–80	Up to 17,000 gates	1
Surface mount SOIC	Number of pins: 16 to 28 Pin pitch: 10 mils Body length: 50 to 70 mils Body width: 0.300 to 0.400 in.	R: Medium L: Medium C: Low	Ceramic/plastic θJA:[2] 110–80/ 130–105	Up to 6,500 gates	6-Ceramic 2.5-Plastic
Surface mount OFPT	Number of pins: 48 to 260 Pin pitch: 10 mils Body width: 0.65 to 1.7 in.	R: Medium L: Medium C: Low	Plastic θJA:[2] 95–60	Up to 17,000 gates	6
Surface mount CLCC	Number of pins: 28 to 84 Pin pitch: 40 to 50 mils Body width: 0.45 to 0.97 in.	R: Medium L: Medium C: Medium	Ceramic θJA:[2] 70–45	Up to 25,000 gates	30
Surface mount PLCC	Number of pins: 28 to 84 Pin pitch: 50 mils Body width: 0.49 to 1.19 in.	R: Medium L: Medium C: Low	Plastic θJA:[2] 65–50	Up to 17,000 gates	2
Through-hole PGA	Number of pins: 64 to 299 Pin pitch: 100 mils, 70 mils Body width: 1.033 to 1.7 in.	Ceramic/plastic R: Low/low L: Low/low C: High/low	Ceramic/plastic θJA:[2] 40–19/ 46–38	Up to 75,000 gates	60-Ceramic 12-Plastic

[1]R = resistance, L = inductance, C = capacitance.
[2]Assuming static airflow.
[3]Assuming 1.5-μm CMOS technology.
Source: Fujitsu Microelectronics; G. L. Ginsberg, *Electronic Equipment Packaging Technology*, New York: Van Nostrand Reinhold, 1991, p. 46. With permission.

package on the "leaded" versions. The leadless chip carrier usually has a ceramic body, while the leaded types are usually plastic.

The Joint Electronic Device Engineering Council (JEDEC) established configurations for the original chip carriers that allowed for multiple design approaches, manufacturing techniques, and attachment means in order to allow for the choice of package that is best tailored for the application. JEDEC standardized both leadless and leaded chip carriers in two basic styles, one with 1.27-mm (0.050-in.) terminal centers and another with 1.0-mm (0.040-in.) spacing, and provided interchangeability within similar terminal spacing outlines.

The original "50-mil" center family contained six square-package variations with up to 156 I/O terminals. Four leadless versions, types A (see Fig. 24.2, top left), B, C, and D, mount in different orientations depending on the type, mounting substrate, and preferred thermal orientation. (A rectangular type E was subsequently added for use primarily with memory devices with up to 32 I/O terminals.) These packages are ceramic, with hermetically sealed metal or ceramic lids. There are two leaded versions. Type A is a premolded or postmolded J-lead plastic package (Fig. 24.2, top right). The other version, type B, is the leadless type A package with clip-type leads attached to it to facilitate soldering to the interconnecting substrate.

Flat Packs

Flat packs are among the oldest types of IC packages for surface mount applications. The typical dual-row flat pack has up to 50 flat ribbon leads that come straight out from its body on 1.27-mm

FIGURE 24.2 Integrated circuit packages: dual inline packages (bottom), leadless type A chip carrier (top left), cavity-down pin-grid array (top center), J-lead plastic chip carrier (top right). (Courtesy of Siemens Corp.)

(0.050-in.) centers. Thus, the leads have to be formed downward, usually in a "gullwing" fashion, to facilitate surface mount assembly. A newer version is the plastic quad flat pack (PQFP). The JEDEC-approved PQFP (Fig. 24.3) is a high-density package that has up to 244 gullwing leads that are preformed by the IC manufacturer on 0.63-mm (0.025-in.) centers. The PQFP also features molded "bumpers" in its corners to help protect the leads from damage during handling and assembly.

Other four-row, plastic flat packs with higher I/O counts and smaller terminal pitch are also available. One version uses a relatively low-cost, easy-to-test, straight-lead, tape-automated bond-

FIGURE 24.3 Plastic quad gullwing flat pack. (Courtesy of Intel Corp.)

ing (TAB) construction. As supplied by the IC manufacturer, the tested/molded die is ready for excising from its own carrier frame and lead forming by the printed board assembler. Such devices are projected to be used with ICs with from 40 to 350 or more leads.

Small-Outline Integrated Circuits

With all of the recognition being given to surface mount chip carriers and flat packs, small-outline integrated circuits (SOICs) are quite often used for low lead-count applications, i.e., devices with up to 28 leads. In appearance, SOICs resemble miniature versions of molded plastic DIPs with either gullwing or J-lead terminals. The primary advantage of the SOIC is its small size and its suitability for surface mounting. As compared to the DIP, as shown in Table 24.4, the package is about one-third the size and uses about one-quarter of the substrate mounting area than does the equivalent I/O DIP; its body thickness, height, and weight are also proportionately less. Small-outline transistors (SOTs) are also available that compare in a similar manner to conventionally packaged transistors (see Fig. 24.4, bottom). Consequently, small-outline components are quite often used in applications where space and weight are at a premium.

Discrete Leadless Components

Electronic circuit components vary both in type and shape [IPC, 1988; IPC, 1987a; Bos *et al.*, 1989]. In general, these components are usually selected for electrical, thermal, and/or mechanical characteristics that are determined by the requirements of the end product. Often the selection of such components is also dependent on the invoked specifications, availability, and/or cost. The material composition, finish, and configuration of both the body and the component's I/O terminals must be considered in the choice of the circuit assembly methods. Thus, many discrete electronic components, such as resistors and capacitors, are available in several configurations in order to facilitate their uses in both through-hole and surface mount applications.

Discrete leadless "chip" components suitable for surface mounting are basically miniature axial-lead components with the leads replaced by metallized terminals. They are available in both flat-rectangular and cylindrical shapes (see Fig. 24.4, top) for packaging resistors, capacitors, diodes, and transistors. Chip resistors are available in either a solderable wraparound or wire bonding termination configuration. The wraparound end-metallization resistor is terminated "face-up" so that post-attachment inspection and/or trimming is practical. More importantly, the resistive element itself is away from the solder joint and **land pattern**. The wire-bonding type of leadless resis-

Table 24.4   Small-Outline Integrated Circuit and Dual Inline Package Comparison

Item	8-Pin		16-Pin		28-Pin	
	SO	DIP	SO	DIP	SO	DIP
Body size, L × W, mm²	20	70	40	140	140	500
		(3.5)		(3.5)		(3.5)
Board area, mm²	31	80	62	175	192	590
		(2.6)		(2.8)		(3.0)
Body thickness, mm	1.45	3.1	1.45	3.6	2.45	3.9
Height above board, mm	1.75	4.2	1.75	5.1	2.65	5.1
Weight, mg	60	600	130	1200	700	4300
		(10)		(9)		(6)

Ratios are shown in parentheses.

Source: IPC, 1987a; G. L. Ginsberg, *Electronic Equipment Packaging Technology*, New York: Van Nostrand Reinhold, 1991, p. 51. With permission.

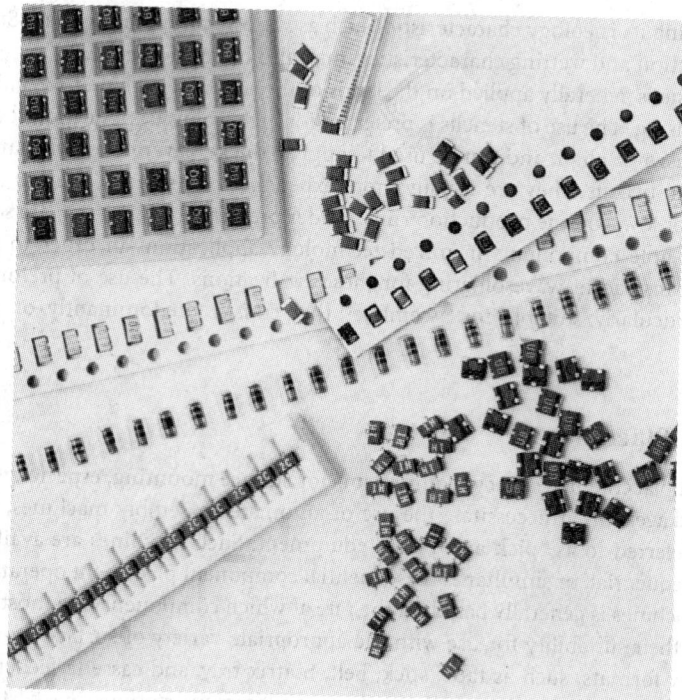

FIGURE 24.4   Discrete electronic component packages: small-outline transistor (bottom), leadless chip components (top). (Courtesy of Rohm Corp.)

tor has all of the termination metallization on the top of the chip. A chip-attach adhesive is usually used to hold the component in place as part of the assembly operation. The most popular leadless chip capacitors have ceramic bodies, although silicon, porcelain, tantalum, and glass capacitors are also available. However, the ceramic types are used more extensively because of their ruggedness, ease of handling, wide range of values, high volumetric efficiency, and relatively low cost.

24.5 SMT Assembly Processing

As shown in Table 24.2, there are several different ways to process surface mount assemblies [IPC, 1987b]. Basically, the differences pertain to the use of SMT components on one (**single-sided**) or both (**double-sided**) sides of the assembly, whether there is a mixed SMT/through-hole configuration, and whether or not special components must be mounted and/or soldered manually.

Substrate Preparation

Unlike through-hole technology, surface mounting requires additional printed wiring board substrate preparation prior to component mounting. When mixed-technology wavesoldering is used, this includes the deposition of an adhesive in the location where the wavesoldered surface components are to be placed. One of the most significant departures from through-hole technology is the need for a solder-deposition process for all single-sided and double-sided SMT assemblies, on the side opposite the wave of mixed-technology assemblies. Thus, the use of a solder paste (cream) is an important part of surface mount substrate preparation prior to reflow soldering. The solder paste acts partially as an adhesive before reflow, and its surface tension helps to align skewed parts during soldering. It contains the flux, solvent, suspending agent, and solder that is traditionally supplied by the wavesoldering machine. Therefore, the selection of a particular solder paste involves optimizing its rheology characteristics such as viscosity, flow, and spread. Susceptibility to solder ball formation and wetting characteristics must also be considered.

The solder paste is generally applied on the lands of the substrate by either screening, stenciling, or syringe dispensing. The use of stencils is preferred for high-volume applications because they are more durable, easier to align, and can be used to apply a thicker layer of solder than the screening process. However, because they are usually more expensive than screens, the use of stencils may not be suitable for low-volume production runs. Solder preforms (doughnuts) are sometimes used for the through-hole components in mixed-technology applications in order to be able to use reflow soldering instead of wavesoldering for these applications. The use of preforms and reflow soldering is particularly suitable for assemblies that consist predominantly of surface mount components.

Component Placement

The component placement accuracy requirements for surface mounting, especially with fine-pitch component packages, often necessitate the use of automated assembly machines, as in Fig. 24.5, that are often referred to as "pick-and-place" equipment. Such machines are available for inline, simultaneous, sequential, or simultaneous/sequential component placement operations. The selection of these machines is generally based on the rate at which components are most cost-effectively assembled and their suitability for use with the appropriate variety of surface mount component package delivery formats, such as tape, stick, belt, matrix tray, and cassette feeders. In line pick-and-place machines employ a series of component placement stations. Each station places its respective component as the substrate moves through it down the line. Inline component placement times can vary from 1.8 to 4.5 seconds per assembly. Simultaneous placement equipment

FIGURE 24.5 Surface mount "pick-and-place" assembly machine. (Courtesy of Emhart/Dynapert Corp.)

mounts an entire array of components on the substrate at one time. Typical simultaneous component placement times vary from 7 to 10 seconds per assembly. Sequential SMT component mounting units typically utilize a software-controlled x-y–axis moving table system. The components are individually placed on the substrate in this type of assembly. Typical sequential component placement times vary from 0.3 to 1.8 seconds per component. Sequential/simultaneous pick-and-place machines also feature a software-controlled x-y–axis moving table system. However, the components are individually placed on the substrate in succession from multiple-component feeding heads. Simultaneous/sequential component placement times are typically about 0.2 seconds per component.

Soldering

The selection of a soldering process depends upon the type of components being assembled and, as previously mentioned, whether or not a mixed-technology or all-surface-mount assembly is being manufactured. Depending on the component mix, the candidate technologies are wavesoldering and the variety of reflow soldering techniques that are based on the use of either vapor-phase (condensation) energy, infrared (IR) energy, lasers, hot-belt conduction, or hot gases. Each process has a corresponding set of manufacturing parameters for which it is best suited. Another consideration is for the use of a low-volume batch processing unit or a high-volume inline machine as shown in Fig. 24.6.

Cleaning

The cleaning of surface mount assemblies is harder to perform than it is with through-hole technology because of the higher-density nature of the assembly, especially with respect to removing flux that has been entrapped under the component packages. This has become further complicated by the concerns about worldwide ozone depletion that is eliminating the use of the traditional chlorofluorocarbon cleaning solvents. Flux entrapment may cause potential reliability problems if the assembly is not properly cleaned. However, new solder paste formulations are being developed that might result in the use of "no-clean" or "no-flux" soldering technologies in order to avoid these problems.

FIGURE 24.6   Inline vapor-phase soldering machine. (Courtesy of Centech Corp.)

Repairing/Reworking

The repair or rework of surface mount assemblies is generally easier than with all through-hole assemblies due to the absence (or minimization with mixed technology) of component mounting holes. However, again, because of the high-density nature of the assembly, special tools are used for this purpose that carefully direct the reflow soldering energy so as to minimize the amount of heat applied to the assembly during this operation. The manual component removal/replacement tools include the use of fork-like soldering iron tips and resistance-heated tweezers. Various types of hot-air devices are also available.

Defining Terms

Blind via: A via connected to either the primary side or secondary side and one or more internal layers of a multilayer packaging and interconnecting structure.

Buried via: A via connected to neither the primary side nor the secondary side of a multilayer packaging and interconnecting structure, i.e., it connects only internal layers.

Castellations: Recessed metallized features on the edges of a chip carrier which interconnect conducting surfaces or planes within or on the chip carrier.

Chip carrier: A low-profile rectangular component package, usually square, whose semiconductor chip cavity or mounting area is a large fraction of the package size and whose external connections are usually on all four sides of the package.

Coefficient of thermal expansion mismatch (CTE): The difference between the coefficients of thermal expansion of two components, i.e., the difference in linear thermal expansion per unit change in temperature. (This term is not to be confused with *thermal expansion mismatch.*)

Component: A separable part of a printed board assembly which performs a circuit function (e.g., a resistor, capacitor, transistor, etc.).

Component mounting site: A location on a packaging and interconnecting structure, consisting of a land pattern and conductor fan-out to additional lands for testing or vias, used for mounting a single component.

Constraining core: An internal supporting plane in a packaging and interconnecting structure, used to alter the structure's coefficient of thermal expansion.

Double-sided assembly: A packaging and interconnecting structure with components mounted on both the primary and secondary sides.

Dual inline package (DIP): A component which terminates in two straight, parallel rows of pins or lead wires.

Flat pack: A component with two straight rows of leads (normally on 0.050-in. centers) which are parallel to the component body.

Grid: An orthogonal network of two sets of parallel equidistant lines used for locating points on a printed board.

Integrated circuit (IC): An assembly of miniature electronic components simultaneously produced in batch processing, on or within a single substrate, which performs an electronic circuit function.

Land pattern: A combination of lands intended for the mounting, interconnection, and testing of a particular component.

Leaded chip carrier: A chip carrier whose external connections consist of leads around and down the sides of the package.

Leadless chip carrier: A chip carrier whose external connections consist of metallized terminations.

Mixed mounting technology: A component mounting technology that uses both through-hole and surface-mounting technologies on the same packaging and interconnecting structure.

Packaging and interconnecting (P&I) structure: The generic term for a completely processed combination of substrates, metal planes or constraining cores, and interconnection wiring, used for mounting components.

Primary side: That side of the packaging and interconnecting structure closest to layer number one. (Also called the component side in through-hole component mounting technology.)

Secondary side: That side of the packaging and interconnecting structure farthest from layer number one. (Also called the solder side in through-hole component mounting technology.)

Single inline package (SIP): A component which terminates in one straight row of pins and lead wires.

Single-sided assembly: A packaging and interconnecting structure with components mounted only on the primary side.

Supporting plane: A planar structure that is an external support for a packaging and interconnecting structure, used to alter the structure's coefficient of thermal expansion.

Surface mounting: The electrical connection of components to the surface of a conductive pattern without component lead holes.

Thermal expansion mismatch: The absolute difference in thermal expansion of two components.

Through via: A via that connects the primary side and secondary side of a packaging and interconnecting structure.

Via: A plated through-hole used as a through connection, into which no component lead or other reinforcing material is inserted.

References

L. Bos, W. Winkelmann, and B. Robbins, "An overview of SMT resistor packaging," *Surface Mount Technology,* pp. 69–71, October 1989.

IPC-CM-770, Revision C, "Printed Board Component Mounting," Institute for Interconnecting and Packaging Electronic Circuits, January 1987a.

IPC-CM-780, "Component Packaging and Interconnecting with Emphasis on Surface Mounting," Institute for Interconnecting and Packaging Electronic Circuits, July 1988.

IPC-SM-782, "Surface Mount Land Patterns (Configurations and Design Rules)," Institute for Interconnecting and Packaging Electronic Circuits, March 1987b.

B. Johnson and R.C. Luthi, "A review of integrated circuit packaging options," *Microelectronic Manufacturing and Testing,* pp. 112–113, March 1989.

R. Richards, "Trends in semiconductor packaging," *Electronic Products,* pp. 59–62, November 1989.

K.W. Rosengarth, Jr. and E.R. Winkler, "Surface mounting fine-pitch chip carriers," *Electronic Packaging & Production,* pp. 121–123, January 1986.

H.R. Waltersdorf, "Choosing packages wisely pays off in I/O, speed, space," *Electronic Design,* vol. 32, no. 6, pp. 107–111, June 19, 1986.

25

Operational Amplifiers

E. J. Kennedy
University of Tennessee

John V. Wait
University of Arizona

25.1 Ideal and Practical Models 616
 The Ideal Op Amp • Practical Op Amps • SPICE Computer Models

25.2 Applications 625
 Noninverting Circuits

25.1 Ideal and Practical Models

E. J. Kennedy

The concept of the **operational amplifier** (usually referred to as an *op amp*) originated at the beginning of the Second World War with the use of vacuum tubes in dc amplifier designs developed by the George A. Philbrick Co. [some of the early history of operational amplifiers is found in Williams, 1991]. The op amp was the basic building block for early electronic servomechanisms, for synthesizers, and in particular for analog computers used to solve differential equations. With the advent of the first monolithic integrated-circuit (IC) op amp in 1965 (the µA709, designed by the late Bob Widlar, then with Fairchild Semiconductor), the availability of op amps was no longer a factor, while within a few years the cost of these devices (which had been as high as $200 each) rapidly plummeted to close to that of individual discrete transistors.

Although the digital computer has now largely supplanted the analog computer in mathematically intensive applications, the use of inexpensive operational amplifiers in instrumentation applications, in pulse shaping, in filtering, and in signal processing applications in general has continued to grow. There are currently many commercial manufacturers whose main products are high-quality op amps. This competitiveness has ensured a marketplace featuring a wide range of relatively inexpensive devices suitable for use by electronic engineers, physicists, chemists, biologists, and almost any discipline that requires obtaining quantitative analog data from instrumented experiments.

Most operational amplifier circuits can be analyzed, at least for first-order calculations, by considering the op amp to be an "ideal" device. For more quantitative information, however, and particularly when frequency response and dc offsets are important, one must refer to a more "practical" model that includes the internal limitations of the device. If the op amp is characterized by a really complete model, the resulting circuit may be quite complex, leading to rather laborious calculations. Fortunately, however, computer analysis using the program **SPICE** significantly reduces the problem to one of a simple input specification to the computer. Today, nearly all the op amp manufacturers provide SPICE models for their line of devices, with excellent correlation obtained between the computer simulation and the actual measured results.

The Ideal Op Amp

An **ideal operational amplifier** is a dc-coupled amplifier having two inputs and normally one output (although in a few infrequent cases there may be a differential output). The inputs are designated as noninverting (designated + or NI) and inverting (designated – or Inv.). The amplified signal is the *differential* signal, v_ε, between the two inputs, so that the output voltage as indicated in Fig. 25.1 is

$$v_{\text{out}} = A_{OL}(v_B - v_A) \tag{25.1}$$

The general characteristics of an ideal op amp can be summarized as follows:

1. The open-loop gain A_{OL} is infinite. Or, since the output signal v_{out} is finite, then the differential input signal v_ε must approach zero.
2. The input resistance R_{IN} is infinite, while the output resistance R_O is zero.
3. The amplifier has zero current at the input (i_A and i_B in Fig. 25.1 are zero), but the op amp can either sink or source an infinite current at the output.
4. The op amp is not sensitive to a common signal on both inputs (i.e., $v_A = v_B$); thus, the output voltage change due to a common input signal will be zero. This common signal is referred to as a common-mode signal, and manufacturers specify this effect by an op amp's *common-mode rejection ratio* (CMRR), which relates the ratio of the open-loop gain (A_{OL}) of the op amp to the common-mode gain (A_{CM}). Hence, for an ideal op amp CMRR = ∞.
5. A somewhat analogous specification to the CMRR is the *power-supply rejection ratio* (PSRR), which relates the ratio of a power supply voltage change to an equivalent input voltage change produced by the change in the power supply. Because an ideal op amp can operate with any power supply, without restriction, then for the ideal device PSRR = ∞.
6. The gain of the op amp is not a function of frequency. This implies an infinite bandwidth.

Although the foregoing requirements for an ideal op amp appear to be impossible to achieve practically, modern devices can quite closely approximate many of these conditions. An op amp with a field-effect transistor (FET) on the input would certainly not have zero input current and infinite input resistance, but a current of <10 pA and an $R_{IN} = 10^{12} \Omega$ is obtainable and is a reasonable approximation to the ideal conditions. Further, although a CMRR and PSRR of infinity are not possible, there are several commercial op amps available with values of 140 dB (i.e., a ratio of 10^7). Open-loop gains of several precision op amps now have reached values of $>10^7$, although certainly not infinity. The two most difficult ideal conditions to approach are the ability to handle large output currents and the requirement of a gain independence with frequency.

Using the ideal model conditions it is quite simple to evaluate the two basic op amp circuit configurations, (1) the inverting amplifier and (2) the noninverting amplifier, as designated in Fig. 25.2.

For the ideal inverting amplifier, since the open-loop gain is infinite and since the output voltage v_o is finite, then the input differential voltage (often referred to as the *error signal*) v_ε must approach zero, or the input current is

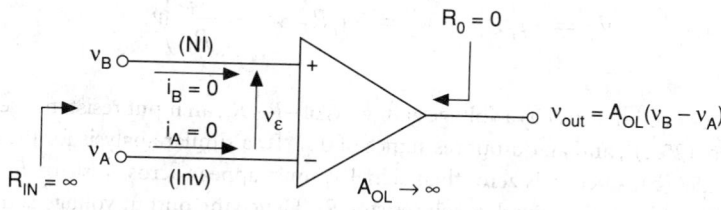

FIGURE 25.1 Configuration for an ideal op amp.

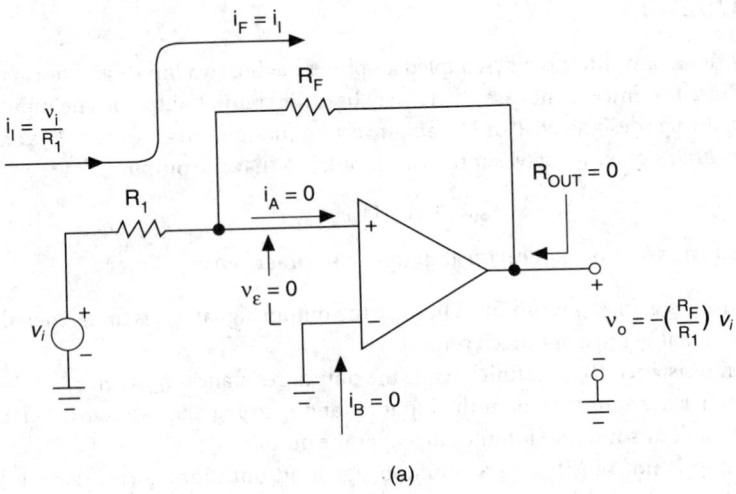

(a)

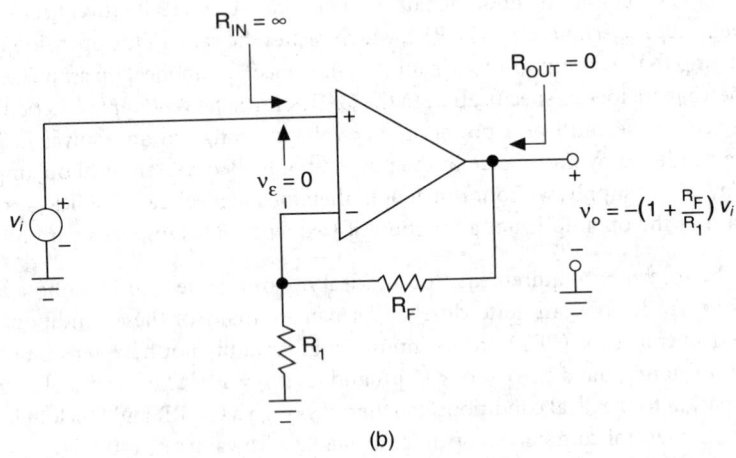

(b)

FIGURE 25.2 Illustration of (a) the inverting amplifier and (b) the noninverting amplifier. (*Source:* E.J. Kennedy, *Operational Amplifier Circuits, Theory and Applications,* New York: Holt, Rinehart and Winston, 1988, pp. 4, 6. With permission.)

$$i_I = \frac{v_I - v_\varepsilon}{R_1} = \frac{v_I - 0}{R_1} \tag{25.2}$$

The feedback current i_F must equal i_I, and the output voltage must then be due to the voltage drop across R_F, or

$$v_o = -i_F R_F + v_\varepsilon = -i_I R_F = -\left(\frac{R_F}{R_1}\right) v_I \tag{25.3}$$

The inverting connection thus has a voltage gain v_o/v_I of $-R_F/R_1$, an input resistance seen by v_I of R_1 ohms [from Eq. (25.2)], and an output resistance of 0 Ω. By a similar analysis for the noninverting circuit of Fig. 25.2(b), since v_ε is zero, then signal v_I must appear across resistor R_1, producing a current of v_I/R_1, which must flow through resistor R_F. Hence the output voltage is the sum of the voltage drops across R_F and R_1, or

$$v_o = R_F \left(\frac{v_I}{R_1} \right) + v_I = \left(1 + \frac{R_F}{R_1} \right) v_I \qquad (25.4)$$

As opposed to the inverting connection, the input resistance seen by the source v_I is now equal to an infinite resistance, since R_{IN} for the ideal op amp is infinite.

Practical Op Amps

A nonideal op amp is characterized not only by finite open-loop gain, input and output resistance, finite currents, and frequency bandwidths, but also by various nonidealities due to the construction of the op amp circuit or external connections. A complete model for a practical op amp is illustrated in Fig. 25.3. The nonideal effects of the PSRR and CMRR are represented by the input series voltage sources of $\Delta V_{\text{supply}}/\text{PSRR}$ and V_{CM}/CMRR, where ΔV_{supply} would be any total change of the two power supply voltages, V_{dc}^+ and V_{dc}^-, from their nominal values, while V_{CM} is the common

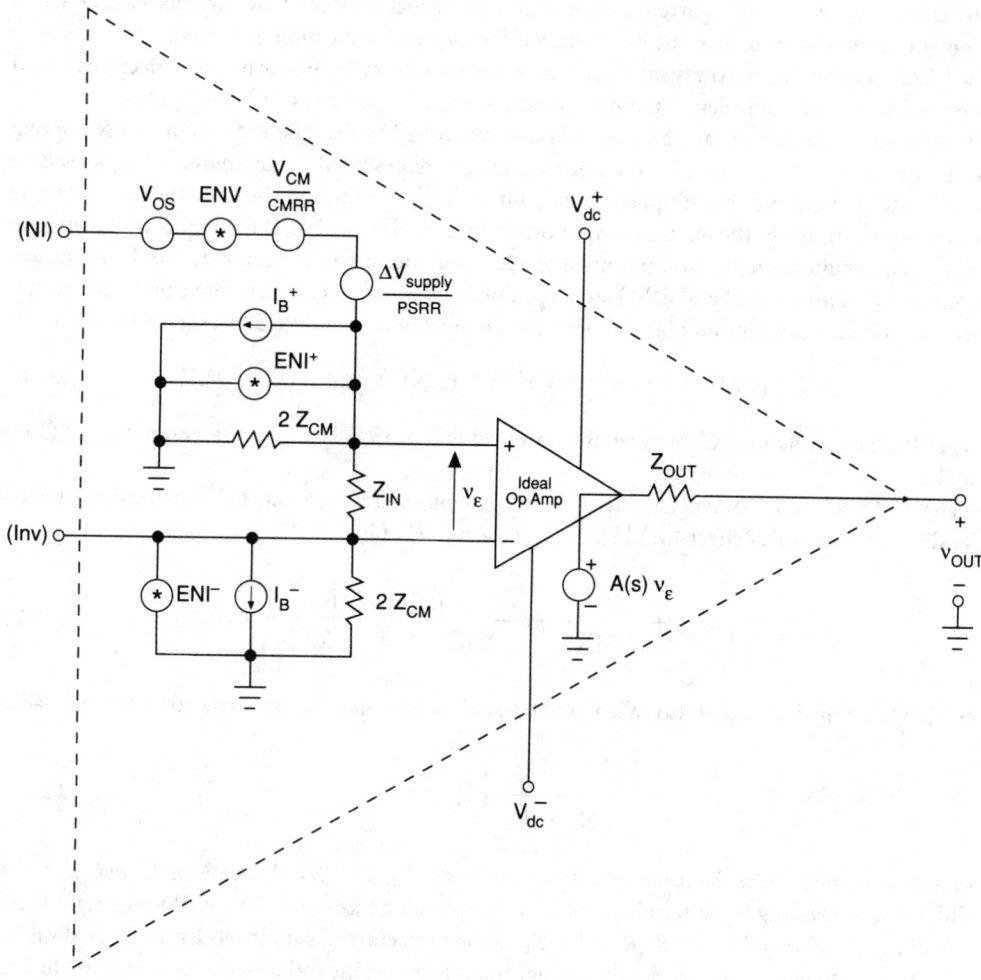

FIGURE 25.3 A model for a practical op amp illustrating nonideal effects. (*Source:* E. J. Kennedy, *Operational Amplifier Circuits, Theory and Applications,* New York: Holt, Rinehart and Winston, 1988, pp. 53, 126. With permission.)

voltage present at the input of the op amp. The open-loop gain of the op amp is no longer infinite but is modeled by a network of the output impedance Z_{out} (which may be merely a resistor but could also be a series R-L network) in series with a source $A(s)$, which includes all the open-loop poles and zeroes of the op amp as

$$A(s) = \frac{A_{OL}\left(1 + \dfrac{s}{\omega_{Z_1}}\right)(1 + \cdots)}{\left(1 + \dfrac{s}{\omega_{p1}}\right)\left(1 + \dfrac{s}{\omega_{p2}}\right)(1 + \cdots)} \tag{25.5}$$

where A_{OL} is the finite dc open-loop gain, while poles are at frequencies $\omega_{p1}, \omega_{p2}, \ldots$ and zeroes are at ω_{Z_1}, etc. The differential input resistance is Z_{IN}, which is typically a resistance R_{IN} in parallel with a capacitor C_{IN}. Similarly, the common-mode input impedance Z_{CM} is established by placing an impedance $2Z_{CM}$ in parallel with each input terminal. Normally, Z_{CM} is best represented by a parallel resistance and capacitance of $2R_{CM}$ (which is $>> R_{IN}$) and $C_{CM}/2$. The dc bias currents at the input are represented by I_B^+ and I_B^- current sources that would equal the input base currents if a differential bipolar transistor were used as the input stage of the op amp, or the input gate currents if FETs were used. The fact that the two transistors of the input stage of the op amp may not be perfectly balanced is represented by an equivalent input *offset voltage* source, V_{OS}, in series with the input.

The smallest signal that can be amplified is always limited by the inherent random noise internal to the op amp itself. In Fig. 25.3 the noise effects are represented by an *equivalent input voltage source* (ENV), which when multiplied by the gain of the op amp would equal the total output noise present if the inputs to the op amp were shorted. In a similar fashion, if the inputs to the op amp were open circuited, the total output noise would equal the sum of the noise due to the *equivalent input current sources* (ENI$^+$ and ENI$^-$), each multiplied by their respective current gain to the output. Because noise is a random variable, this summation must be accomplished in a squared fashion, i.e.,

$$E_o^2(\text{rms volt}^2/\text{Hz}) = (\text{ENV})^2 A_v^2 + (\text{ENI}^+)^2 A_{I1}^2 + (\text{ENI}^-)^2 A_{I2}^2 \tag{25.6}$$

Typically, the correlation (C) between the ENV and ENI sources is low, so the assumption of $C \approx 0$ can be made.

For the basic circuits of Fig. 25.2(a) or (b), if the signal source v_I is shorted then the output voltage due to the nonideal effects would be (using the model of Fig. 25.3)

$$v_o = \left(V_{OS} + \frac{V_{CM}}{\text{CMRR}} + \frac{\Delta V_{\text{supply}}}{\text{PSRR}}\right)\left(1 + \frac{R_F}{R_1}\right) + I_B^- R_F \tag{25.7}$$

provided that the loop gain (also called loop transmission in many texts) is related by the inequality

$$\left(\frac{R_1}{R_1 + R_F}\right) A(s) >> 1 \tag{25.8}$$

Inherent in Eq. (25.8) is the usual condition that $R_1 << Z_{IN}$ and Z_{CM}. If a resistor R_2 were in series with the noninverting input terminal, then a corresponding term must be added to the right hand side of Eq. (25.7) of value $-I_B^+ R_2(R_1 + R_F)/R_1$. On manufacturers' data sheets the individual values of I_B^+ and I_B^- are not stated; instead the average input bias current and offset current are specified as

$$I_B = \frac{I_B^+ + I_B^-}{2}; \qquad I_{\text{offset}} = |I_B^+ - I_B^-| \tag{25.9}$$

The output noise effects can be obtained using the model of Fig. 25.3 along with the circuits of Fig. 25.2 as

$$E_{out}^2 \text{(rms volts}^2/\text{Hz)} = E_1^2 \left(\frac{R_F}{R_1}\right)^2 + E_F^2 + (\text{ENV}^2 + E_2^2) \times$$

$$\left(1 + \frac{R_F}{R_1}\right)^2 + (\text{ENI}^-)^2 R_F^2 + (\text{ENI}^+)^2 R_2^2 \left(1 + \frac{R_F}{R_1}\right)^2 \quad (25.10)$$

where it is assumed that a resistor R_2 is also in series with the noninverting input of either Fig. 25.2(a) or (b). The thermal noise (often called Johnson or Nyquist noise) due to the resistors R_1, R_2, and R_F is given by (in rms volt^2/Hz)

$$E_1^2 = 4kTR_1$$

$$E_2^2 = 4kTR_2 \quad (25.11)$$

$$E_F^2 = 4kTR_F$$

where k is Boltzmann's constant and T is absolute temperature (Kelvin). To obtain the total output noise, one must multiply the E_{out}^2 expression of Eq. (25.10) by the noise bandwidth of the circuit, which typically is equal to $\pi/2$ times the –3-dB signal bandwidth, for a single-pole response system [Kennedy, 1988].

SPICE Computer Models

The use of op amps can be considerably simplified by computer-aided analysis using the program SPICE. SPICE originated with the University of California, Berkeley, in 1975 [Nagel, 1975], although more recent user-friendly commercial versions are now available such as HSPICE, HPSPICE, IS-SPICE, PSPICE, and ZSPICE, to mention a few of those most widely used. A simple macromodel for a near-ideal op amp could be simply stated with the SPICE subcircuit file (* indicates a comment that is not processed by the file)

```
.SUBCKT IDEALOA 1 2 3
*A near-ideal op amp: (1) is noninv, (2) is inv, and (3) is output.
RIN 1 2 1E12
E1 (3, 0) (1, 2) 1E8
.ENDS IDEALOA                                    (25.12)
```

The circuit model for IDEALOA would appear as in Fig. 25.4(a). A more complete model, but not including nonideal offset effects, could be constructed for the 741 op amp as the subcircuit file OA741, shown in Fig. 25.4(b).

```
.SUBCKT OA741 1 2 6
*A linear model for the 741 op amp: (1) is noninv, (2) is inv, and
*(6) is output. RIN = 2MEG, AOL = 200,000, ROUT = 75 ohm,
*Dominant open - loop pole at 5 Hz, gain - bandwidth product
*is 1 MHz.
RIN 1 2 2MEG
E1 (3, 0) (1, 2) 2E5
R1 3 4 100K
C1 4 0 0.318UF ; R1 × C1 = 5HZPOLE
E2 (5, 0) (4, 0) 1.0
ROUT 5 6 75
.ENDS OA741                                      (25.13)
```

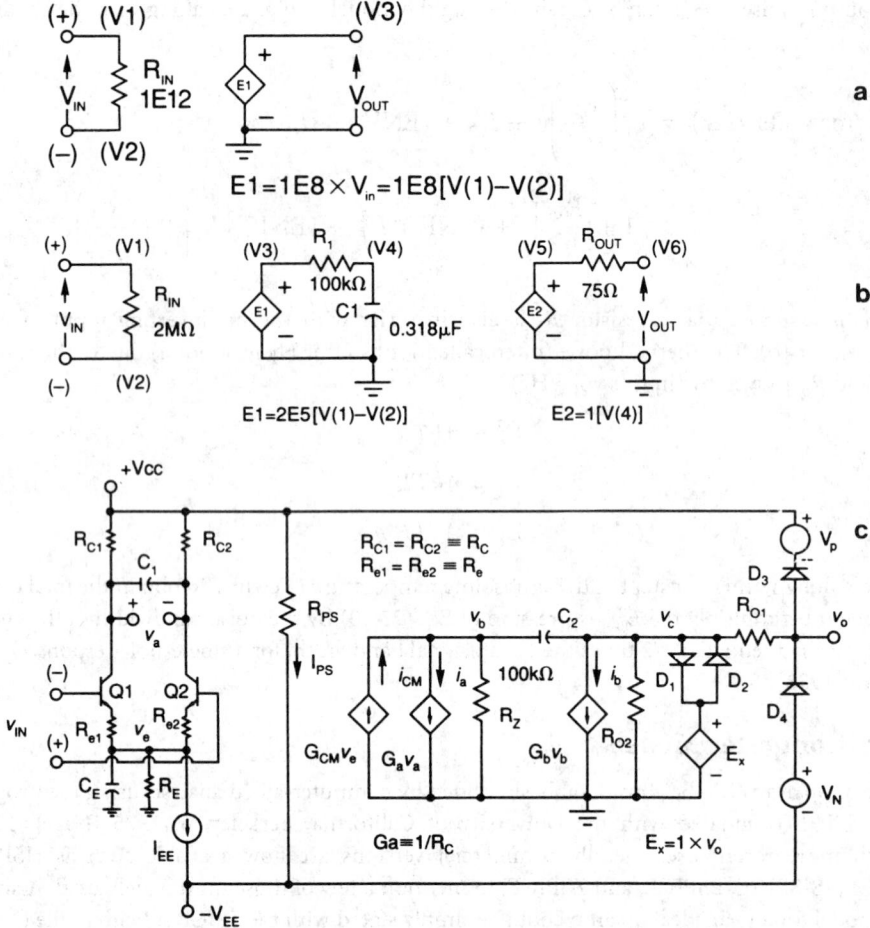

FIGURE 25.4 Some simple SPICE macromodels. (a) A near ideal op amp. (b) A linear model for a 741 op amp. (c) The *Boyle* macromodel. (*Source:* Reprinted, with permission, from J. Williams (Ed.), *Analog Circuit Design*, Stoneham, Mass.: Butterworth-Heinemann, 1991, p. 304.)

The most widely used op amp macromodel that includes dc offset effects is the **Boyle model** [Boyle *et al.*, 1974]. Most op amp manufacturers use this model, usually with additions to add more poles (and perhaps zeroes). The various resistor and capacitor values, as well as transistor, and current and voltage generator, values are intimately related to the specifications of the op amp, as shown earlier in the nonideal model of Fig. 25.3. The appropriate equations are too involved to list here; instead, the interested reader is referred to the article by Boyle in the listed references. The Boyle model does not accurately model noise effects, nor does it fully model PSRR and CMRR effects.

A more circuits-oriented approach to modeling op amps can be obtained if the input transistors are removed and a model formed by using passive components along with both fixed and dependent voltage and current sources. Such a model is shown in Fig. 25.5. This model not only includes all the basic nonideal effects of the op amp, allowing for multiple poles and zeroes, but can also accurately include ENV and ENI noise effects. The circuits-approach macromodel can also be easily adapted to current-feedback op amp designs, whose input impedance at the noninverting input is much greater than that at the inverting input [see Williams, 1991]. The interested reader is referred to the text edited by J. Williams, listed in the references, as well as the recent SPICE modeling book by Connelly and Choi [1992].

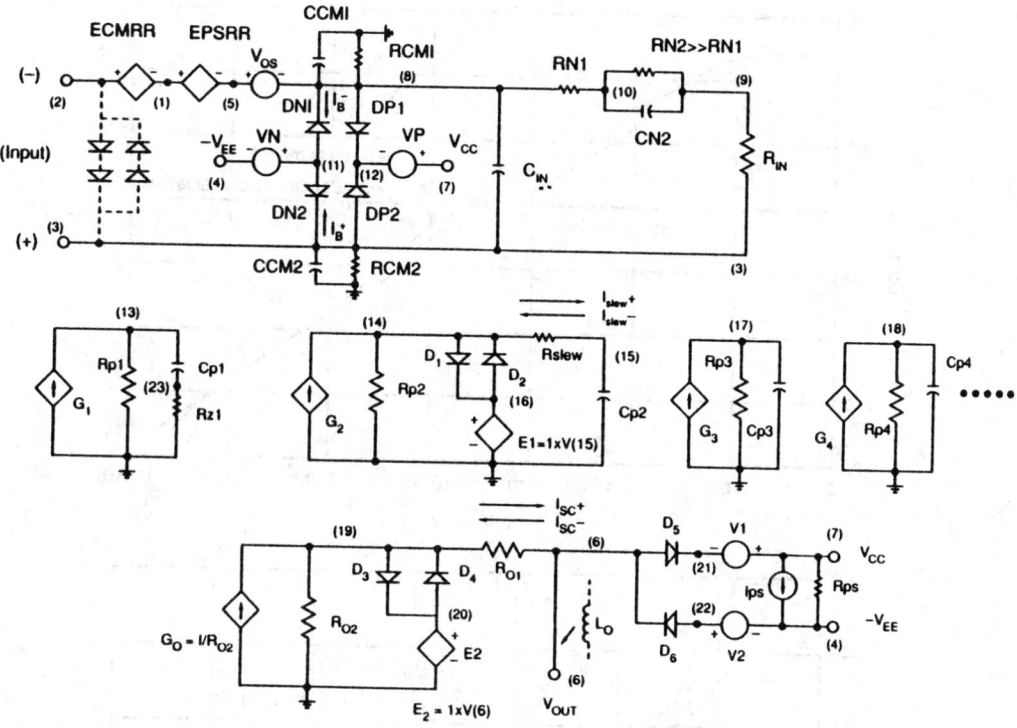

FIGURE 25.5 A SPICE circuits-approach macromodel. (*Source:* Reprinted, with permission, from J. Williams (Ed.), *Analog Circuit Design*, Stoneham, Mass.: Butterworth-Heinemann, 1991, p. 314.)

A comparison of the SPICE macromodels with actual manufacturer's data for the case of an LM318 op amp is demonstrated in Fig. 25.6, for the open-loop gain versus frequency specification.

Defining Terms

Boyle macromodel: A SPICE computer model for an op amp. Developed by G.R. Boyle in 1974.

Equivalent noise current (ENI): A noise current source that is effectively in parallel with either the noninverting input terminal (ENI$^+$) or the inverting input terminal (ENI$^-$) and represents the total noise contributed by the op amp if either input terminal is open circuited.

Equivalent noise voltage (ENV): A noise voltage source that is effectively in series with either the inverting or noninverting input terminal of the op amp and represents the total noise contributed by the op amp if the inputs were shorted.

Ideal operational amplifier: An op amp having infinite gain from input to output, with infinite input resistance and zero output resistance and insensitive to the frequency of the signal. An ideal op amp is useful in first-order analysis of circuits.

Operational amplifier (op amp): A dc amplifier having both an inverting and noninverting input and normally one output, with a very large gain from input to output.

SPICE: A computer simulation program developed by the University of California, Berkeley, in 1975. Versions are available from several companies. The program is particularly advantageous for electronic circuit analysis, since dc, ac, transient, noise, and statistical analysis is possible.

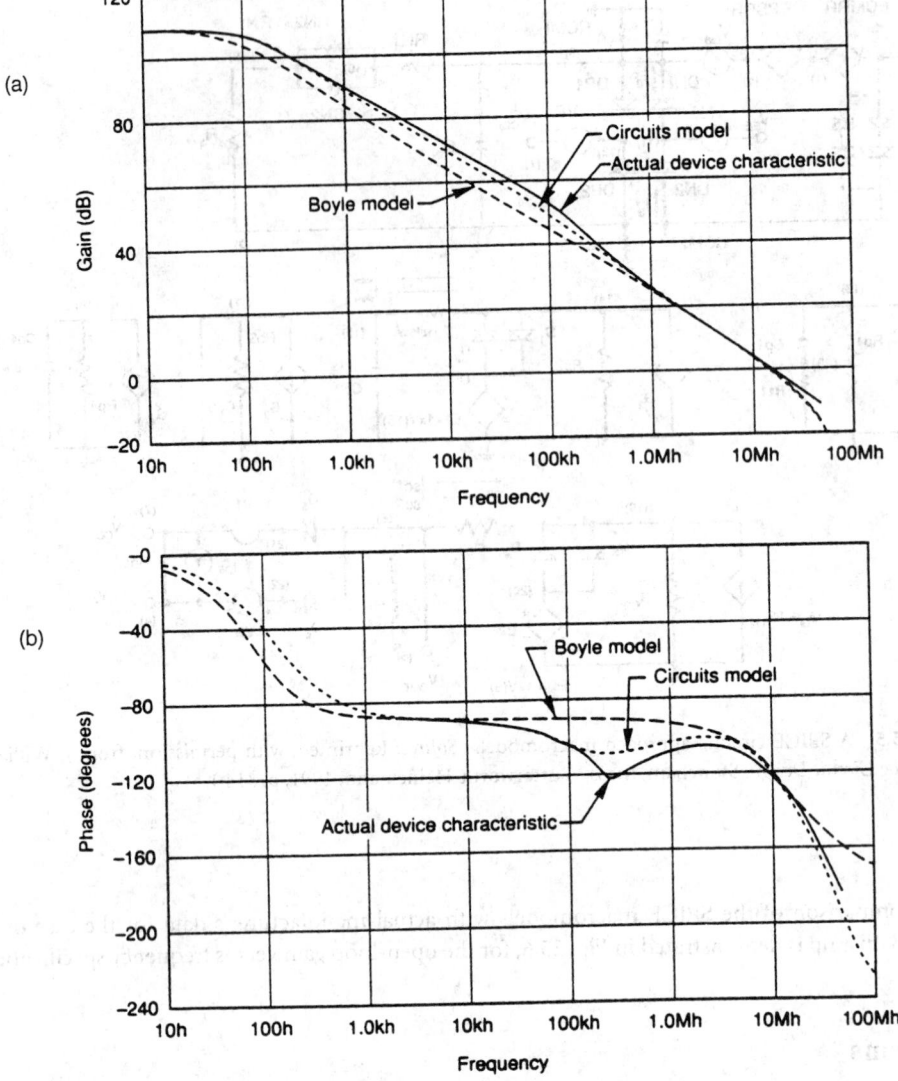

FIGURE 25.6 Comparison between manufacturer's data and the SPICE macromodels. (*Source:* Reprinted, with permission, from J. Williams (Ed.), *Analog Circuit Design*, Stoneham, Mass.: Butterworth-Heinemann, 1991, p. 319.)

References

G. R. Boyle *et al.,* "Macromodeling of integrated circuit operational amplifiers," *IEEE J. S. S. Circuits,* pp. 353–363, 1974.

J. A. Connelly and P. Choi, *Macromodeling with SPICE,* Englewood Cliffs, N.J.: Prentice-Hall, 1992.

E. J. Kennedy, *Operational Amplifier Circuits, Theory and Applications,* New York: Holt, Rinehart and Winston, 1988.

L. W. Nagel, *SPICE 2: A Computer Program to Simulate Semiconductor Circuits,* ERL-M520, University of California, Berkeley, 1975.

J. Williams (ed.), *Analog Circuit Design,* Boston: Butterworth-Heinemann, 1991.

25.2 Applications

John V. Wait

In microminiature form (epoxy or metal packages or as part of a VLSI mask layout) the **operational amplifier** (op amp) is usually fabricated in integrated circuit (IC) form. The general environment is shown in Fig. 25.7. A pair of + and – regulated power supplies (or batteries) may supply all of the op amp in a system, typically with ±10–±15 V. The ground and power supply buses are usually assumed, and an individual op-amp symbol is shown in Fig. 25.8. Such amplifiers feature:

1. A high voltage gain, down to and including dc, and a dc open loop gain of perhaps 10^5 (100 dB) or more
2. An inverting (–) and noninverting (+) symbol
3. Minimized dc offsets, a high input impedance, and a low output impedance
4. An output stage able to deliver or absorb currents over a dynamic range approaching the power supply voltages

It is important *never* to use the op amp without feedback between the output and inverting terminals at all frequencies. A simple inverting amplifier is shown in Fig. 25.9. Here the voltage gain is

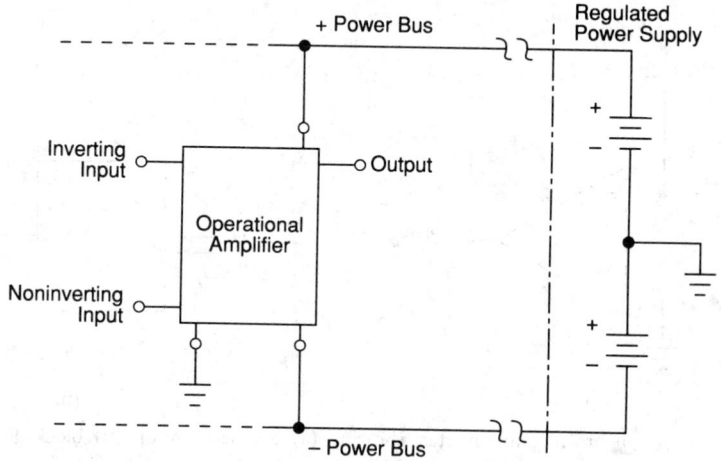

FIGURE 25.7 Typical operational amplifier environment. (*Source:* J.V. Wait, L.P. Huelsman, and G.A. Korn, *Introduction to Operational Amplifier Theory and Applications,* 2nd ed., New York: McGraw-Hill, 1992. With permission.)

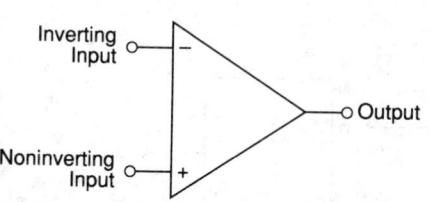

FIGURE 25.8 Conventional operational amplifier symbol. Only active signal lines are shown, and all signals are referenced to ground. (*Source:* J.V. Wait, L.P. Huelsman, and G.A. Korn, *Introduction to Operational Amplifier Theory and Applications,* 2nd ed., New York: McGraw-Hill, 1992. With permission.)

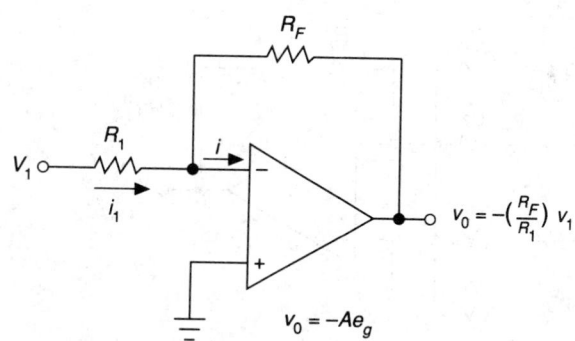

FIGURE 25.9 Simple resistive inverter-amplifier. (*Source:* J.V. Wait, L.P. Huelsman, and G.A. Korn, *Introduction to Operational Amplifier Theory and Applications,* 2nd ed., New York: McGraw-Hill, 1992. With permission.)

$$V_{out}/V_{in} = -K = -R_F/R_1$$

The circuit gain is determined essentially by the external resistances, within the bandwidth and output-driving capabilities of the op amp (more later). If $R_F = R_1 = R$, we have the simple *unity gain inverter* of Fig. 25.10.

Figure 25.11 shows a more flexible *summer-inverter* circuit with

$$v_0 = -(K_1 v_1 + K_2 v_2 + \cdots + K_n v_n)$$

where $K_i = R_F/R_i$.

The summer-inverter is generally useful for precisely combining or mixing signals, e.g., summing and inverting. The signal levels must be appropriately limited but may generally be *bipolar* (+/−).

The resistance values should be in a proper range since (a) too low resistance values draw excessive current from the signal source, and (b) too high resistance values make the circuit performance too sensitive to stray capacitances and dc offset effects.

Typical values are from 1 MΩ and 10 kΩ. The circuit of Fig. 25.12 shows a circuit to implement

$$v_0 = -4v_1 - 2v_2$$

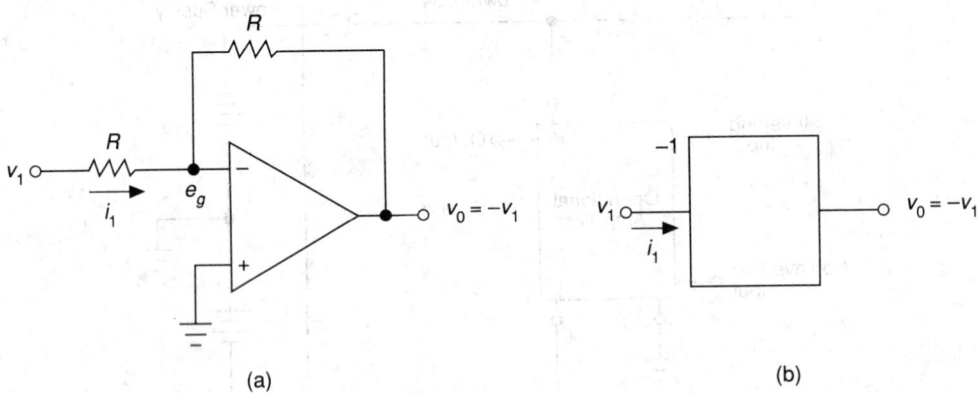

(a) (b)

FIGURE 25.10 A simple unity gain inverter, showing (a) detailed circuit; (b) block-diagram symbol. (*Source:* J.V. Wait, L.P. Huelsman, and G.A. Korn, *Introduction to Operational Amplifier Theory and Applications,* 2nd ed., New York: McGraw-Hill, 1992. With permission.)

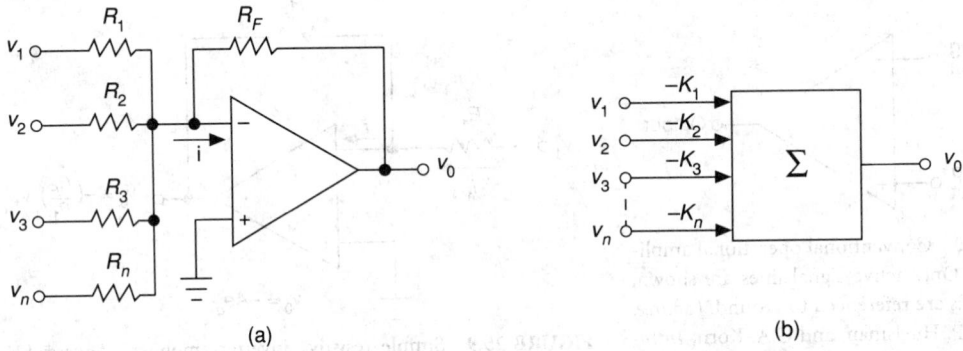

(a) (b)

FIGURE 25.11 The summer-inverter circuit, showing (a) complete circuit; (b) block-diagram symbol. (*Source:* J.V. Wait, L.P. Huelsman, and G.A. Korn, *Introduction to Operational Amplifier Theory and Applications,* 2nd ed., New York: McGraw-Hill, 1992. With permission.)

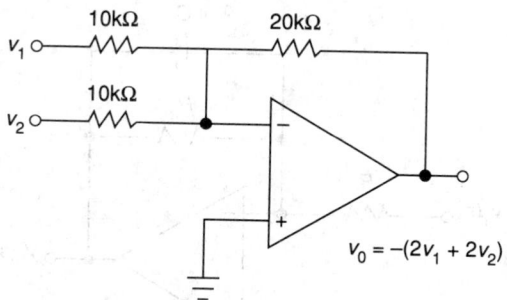

FIGURE 25.12 Simple summer-inverter.

Noninverting Circuits

Figure 25.13(a) shows the useful *noninverting* amplifier circuit. It has a voltage gain

$$V_0/V_1 = (R_2 + R_1)/R_1$$

$$= 1 + (R_2/R_1)$$

Figure 25.13(b) shows the important unity gain follower circuit, which has a *very high input impedance,* which lightly loads the signal source but which can provide a reasonable amount of output current milliamps.

It is fairly easy to show that the *inverting first-order low-pass filter* of Fig. 25.14 has a dc gain or $-R_2/R_1$ and a -3-dB frequency $= 1/(2\pi R_2 C)$.

Figure 25.15 shows a two-amplifier differentiator and high-pass filter circuit with a resistive input impedance and a low-frequency cutoff determined by R_1 and C.

Op amps provide good differential amplifier circuits. Figure 25.16 is a single amplifier circuit with a differential gain

$$A_d = R_0/R_1$$

Good resistance matching is required to have good common-mode rejection of unwanted common-mode signals (static, 60-Hz hum, etc.). The one-amplifier circuit of Fig. 25.16 has a differential input impedance of $2R_1$. R_1 may be chosen to provide a good load for a microphone, phono-pickup, etc.

The improved three-amplifier instrumentation amplifier circuit of Fig. 25.17, which several manufacturers provide in a single module, provides

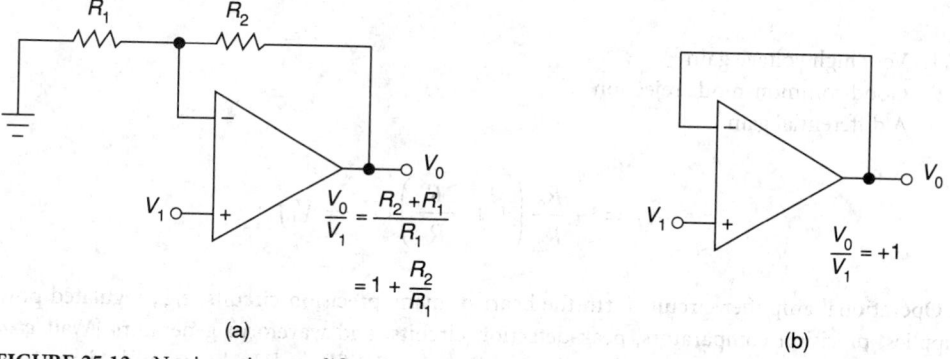

FIGURE 25.13 Noninverting amplifier circuit with resistive elements. (a) General circuit; (b) simple unity gain follower. (*Source:* J.V. Wait, L.P. Huelsman, and G.A. Korn, *Introduction to Operational Amplifier Theory and Applications,* 2nd ed., New York: McGraw-Hill, 1992. With permission.)

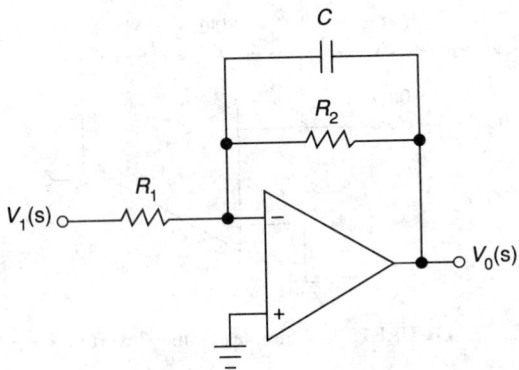

FIGURE 25.14 First-order low-pass filter circuit. (*Source:* J.V. Wait, L.P. Huelsman, and G.A. Korn, *Introduction to Operational Amplifier Theory and Applications,* 2nd ed., New York: McGraw-Hill, 1992. With permission.)

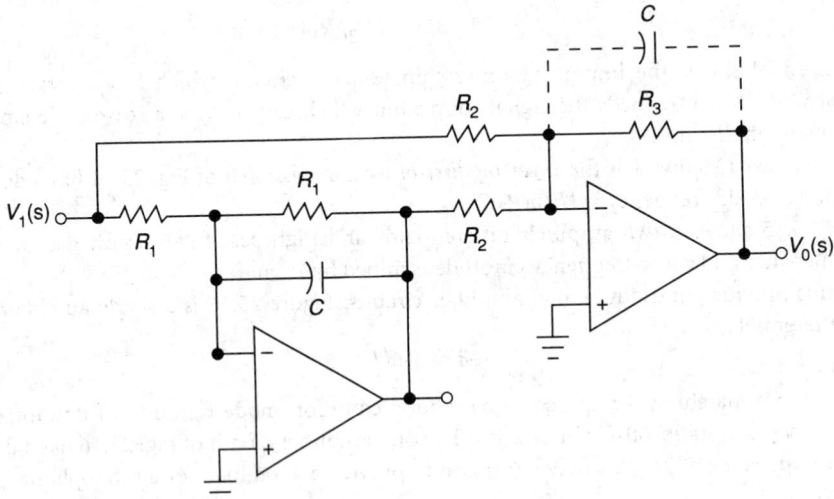

FIGURE 25.15 A two-amplifier high-pass circuit. (*Source:* J.V. Wait, L.P. Huelsman, and G.A. Korn, *Introduction to Operational Amplifier Theory and Applications,* 2nd ed., New York: McGraw-Hill, 1992. With permission.)

1. Very high voltage gain
2. Good common-mode rejection
3. A differential gain

$$A_d = -\frac{R_0}{R_2}\left(1 + \frac{2R_1}{R}\right)\left(V_2 - V_1\right)$$

Operational amplifier circuits form the heart of many precision circuits, e.g., regulated power supplies, precision comparators, peak-detection circuits, and waveform generators [Wait *et al.,* 1992]. Another important area of application is **active RC filters** [Huelsman and Allen, 1980]. Microminiature electronic circuits seldom use inductors. Through the use of op amps, resistors, and capacitors, one can implement precise filter circuits (low-pass, high-pass, and bandpass).

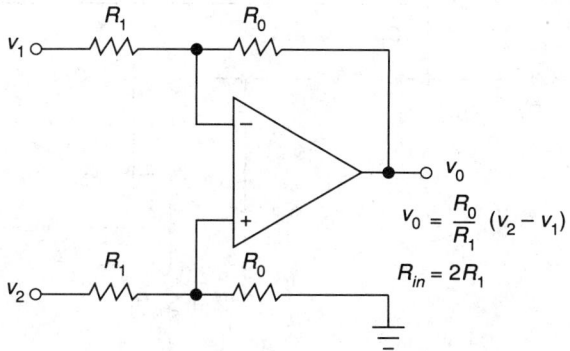

FIGURE 25.16 Single-output differential-input amplifier circuit. (*Source:* J.V. Wait, L.P. Huelsman, and G.A. Korn, *Introduction to Operational Amplifier Theory and Applications,* 2nd ed., New York: McGraw-Hill, 1992. With permission.)

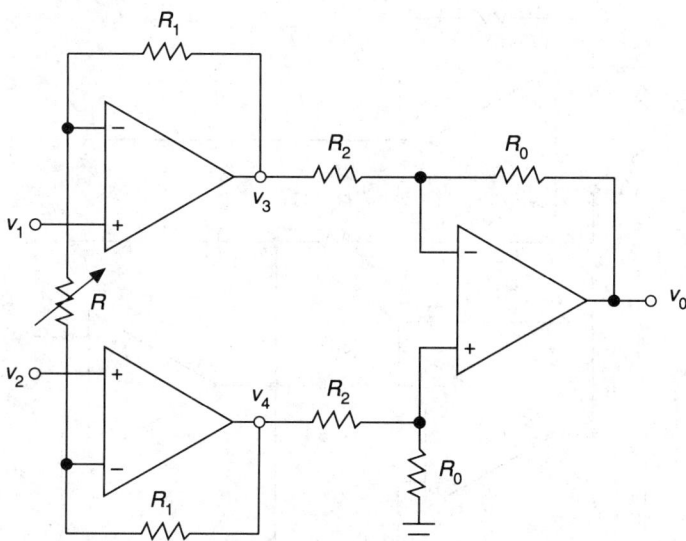

FIGURE 25.17 A three-amplifier differential-input instrumentation amplifier featuring high input impedance and easily adjustable gain. (*Source:* J.V. Wait, L.P. Huelsman, and G.A. Korn, *Introduction to Operational Amplifier Theory and Applications,* 2nd ed., New York: McGraw-Hill, 1992. With permission.)

Figures 25.18 and 25.19 show second-order low-pass and bandpass filter circuits that feature relatively low sensitivity of filter performance to component values. Details are provided in Wait *et al.* [1992] and Huelsman and Allen [1980].

Of course, the op amp does not have infinite bandwidth and gain. An important op-amp parameter is the unity-gain frequency, f_u. For example, it is fairly easy to show the actual bandwidth of a constant gain amplifier of nominal gain G is approximately

$$f_{-3\text{ dB}} = f_u/G$$

Thus, an op amp with $f_u = 1$ MHz will provide an amplifier gain of 20 up to about 50 kHz.

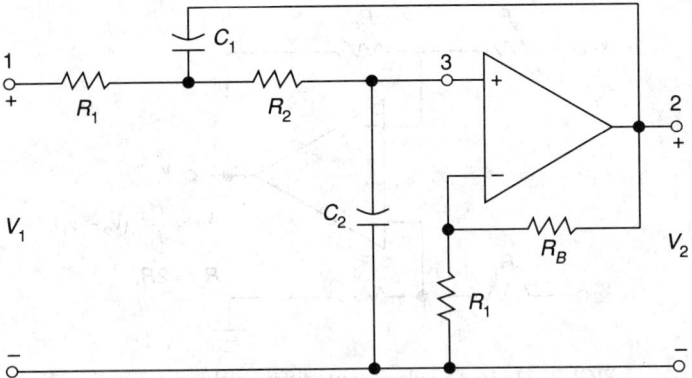

FIGURE 25.18 Sallen and Key low-pass filter. (*Source:* J.V. Wait, L.P. Huelsman, and G.A. Korn, *Introduction to Operational Amplifier Theory and Applications,* 2nd ed., New York: McGraw-Hill, 1992. With permission.)

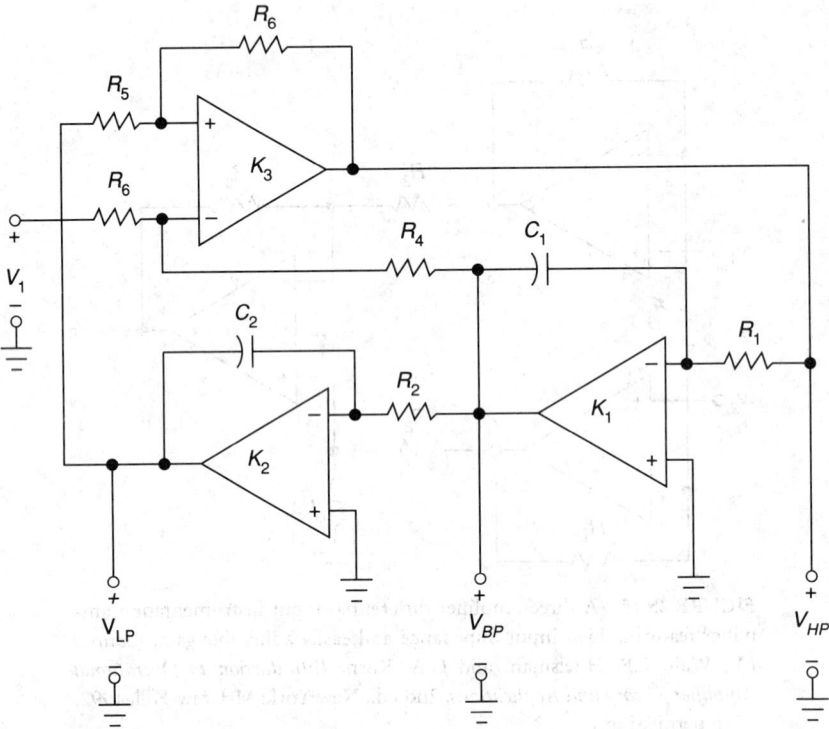

FIGURE 25.19 State-variable filter. (*Source:* J.V. Wait, L.P. Huelsman, and G.A. Korn, *Introduction to Operational Amplifier Theory and Applications,* 2nd ed., New York: McGraw-Hill, 1992. With permission.)

When a circuit designer needs to accurately explore the performance of an op-amp circuit design, modern circuit simulation programs (SPICE, PSPICE, and MICRO-CAP) permit a thorough study of circuit design, as related to op-amp performance parameters. We have not here treated nonlinear op-amp performance limitations such as *slew rate, full-power bandwidth,* and *rated output.* Surely, the op-amp circuit designer must be careful not to exceed the output rating of the op amp, as related to *maximum output voltage* and *current* and *output rate-of-change.*

Nevertheless, op-amp circuits provide the circuit designer with a handy and straightforward way to complete electronic system designs with the use of only a few basic circuit components plus, of course, the operational amplifier.

Defining Terms

Active RC filter: An electronic circuit made up of resistors, capacitors, and operational amplifiers that provide well-controlled linear frequency-dependent functions, e.g., low-, high-, and bandpass filters.

Analog-to-digital converter (ADC):   An electronic circuit that receives a magnitude-scaled analog voltage and generates a binary-coded number proportional to the analog input, which is delivered to an interface subsystem to a digital computer.

Digital-to-analog converter (DAC):   An electronic circuit that receives an *n*-bit digital word from an interface circuit and generates an analog voltage proportional to it.

Electronic switch: An electronic circuit that controls analog signals with digital (binary) signals.

Interface: A collection of electronic modules that provide data transfer between analog and digital systems.

Operational amplifier: A small (usually integrated circuit) electronic module with a bipolar (+/–) output terminal and a pair of differential input terminals. It is provided with power and external components, e.g., resistors, capacitors, and semiconductors, to make amplifiers, filters, and wave-shaping circuits with well-controlled performance characteristics, relatively immune to environmental effects.

References

Electronic Design, Hasbrouk Heights, N.J.: Hayden Publishing Co.; a biweekly journal for electronics engineers. (In particular, see the articles in the Technology section.)

Electronics, New York: McGraw-Hill; a biweekly journal for electronic engineers. (In particular, see the circuit design features.)

J. G. Graeme, *Applications of Operational Amplifiers*, New York: McGraw-Hill, 1973.

L. P. Huelsman and P. E. Allen, *Introduction to the Theory and Design of Active Filters*. New York: McGraw-Hill, 1980.

J. Till, "Flexible Op-Amp Model Improves SPICE," *Electronic Design*, June 22, 1989.

G. E. Tobey, J. G. Graeme, and L. P. Huelsman, *Operational Amplifiers*, New York: McGraw-Hill, 1971.

J. V. Wait, L. P. Huelsman, and G. A. Korn, *Introduction to Operational Amplifier Theory and Applications*, 2nd ed., New York: McGraw-Hill, 1992.

Further Information

For further information see J. V. Wait, L. P. Huelsman, and G. A. Korn, *Introduction to Operational Amplifier Theory and Applications*, 2nd ed., New York: McGraw-Hill, 1992, a general textbook on the design of operational amplifier circuits, including the SPICE model of operational amplifiers; and L.P. Huelsman and P.E. Allen, *Introduction to the Theory and Design of Active Filters*, New York: McGraw-Hill, 1980, a general textbook of design considerations and configurations of active RC filters.

KILBY AND THE IC

How he documented the famous experiments

The records that Jack Kilby, developer of the monolithic integrated circuit, kept of his progress 30 years ago appear in one of Texas Instruments Inc.'s official laboratory notebooks. It was probably Kilby's first TI notebook, as he signed for it on June 12, 1958, only a few weeks after joining the company.

TI gave him no guidance as to what information, in what format, should be entered on its pages. Instead, Kilby recalls, the employer "just shoved a notebook at you and suggested you use it" and turn it in to the company library when you left.

With no rules to follow, Kilby's record keeping was often haphazard. Sometimes he put off recording his efforts so long that nothing got written up. Later, 10 years went by with nary a trace of an entry.

But 1958 was different. Kilby knew he was on to something important. He also knew there were others in the race to integrate an entire circuit on a single semiconductor substrate, for the possibility had been mooted as far back as 1952, at the Electronic Components Conference held in Washington, D.C. To cover himself, Kilby told *Spectrum,* "I was pretty religious about writing every time I thought I had something significant."

Kilby worked steadily through the summer of 1958. First, to demonstrate the feasibility of a semiconductor circuit, he designed and built a circuit with all its components cut from a piece of semiconductor material. Then he laid out his first IC, on a phase-shift oscillator.

On Sept. 12, having completed three of his phase-shift oscillators, Kilby successfully tested one of them in front of a group that included Mark Shepherd, then vice president and general manager of TI's semiconductor components division, and Willis Adcock, then manager of the development department of that division.

Recording that test on p. 20 of his notebook, Kilby wrote:

"A wafer of germanium has been prepared as shown to form a phase shift oscillator. The bulk resistance of the germanium was used for resistors, and a p-n junction for a capacitor. The p type Ge wafer was diffused by conventional techniques, and an aluminum emitter dot was evaporated and alloyed. Gold was evaporated and alloyed to provide connections to the transistor base and to the capacitor area. Plateaus were formed by etching for the transistor and capacitor. Tabs were attached to make contact with the Germanium wafer as shown. The wafer was mounted on a glass slide with Saureisen cement, and gold was bonded thermally to make the nec-

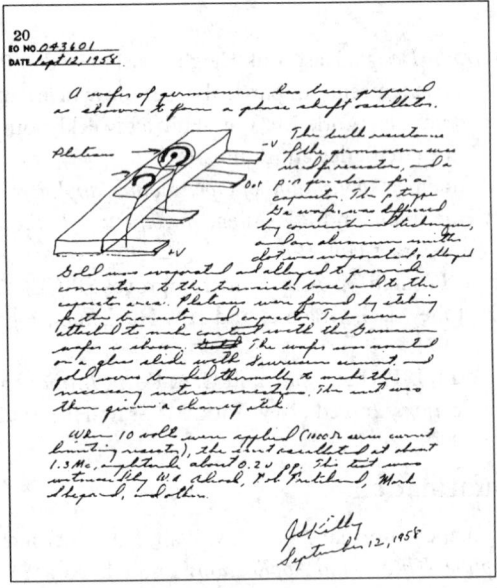

On Sept. 12, 1958, Jack Kilby documented the successful demonstration of the first working semiconductor integrated circuit, a phase-shift oscillator. This showed that the IC concept was applicable to linear circuits.

essary interconnections. The unit was then given a clean-up etch. When 10 volts were applied (1000 ohm series current limiting resistor) the unit oscillated at about 1.3 Mc, amplitude about 0.2v pp. This test was witnessed by W. A. Adcock, Bob Pritchard, Mark Shepard, and others."

J.S. Kilby, September 12, 1958.

That test proved the IC concept worked for linear circuits. More significantly for future developments, a Sept. 19 experiment with a flip-flop recorded on the next page of the notebook showed the concept also applied to digital circuits.

While a second signature is appended to some other entries in Kilby's notebooks, on the page for Sept. 12 Kilby's stands alone. "There was a suggestion from TI that you should have a second signature," Kilby said, "but I ignored it in this case."

TI's general patent counsel, Melvin Sharp, prefers that engineers obtain a witness's signature. Sharp, who worked on the defense of Kilby's patent, said that if there is no second signature, the lawyers must find corroborating witnesses to testify, which here, he says, were easy to find.

Kilby's 1958 notebook was used in preparation of the patent application—filed in February 1959—and again in the mid-1960s to defend parts of that patent against at least three challengers, including Fairchild Semiconductor Corp.'s Robert Noyce, Sharp recalls. Kilby's notebook pages were "more complete than we are used to, which became very important," Sharp said. "You look to see that all the essential elements that go into making the invention had been thought through and set forth."

According to Washington, D.C., patent attorney Robert J. Frank, "neatness, orderly presentation, and omission of unsuccessful results are *not* usually helpful" in defending a patent. What is important, Frank said, is that notebook entries be made on a daily basis and "record *all* failures, foolish statements, and wasteful diversions as well as the invention."

Kilby has no recollection of ever going back to this notebook for his own reference. Either, he recalls, "I remember what is in there or I can't find it when I need it."

TI still hands out laboratory notebooks—some 40,000 are currently signed out to employees and some 30,000 are stored in the corporate technical information library. But, said Sharp, notebooks have been evolving. Unlike Kilby's detailed notes and careful drawings, today's engineering notebooks contain only brief handwritten entries and bulge with supporting documentation, such as photographs and printouts of test results.

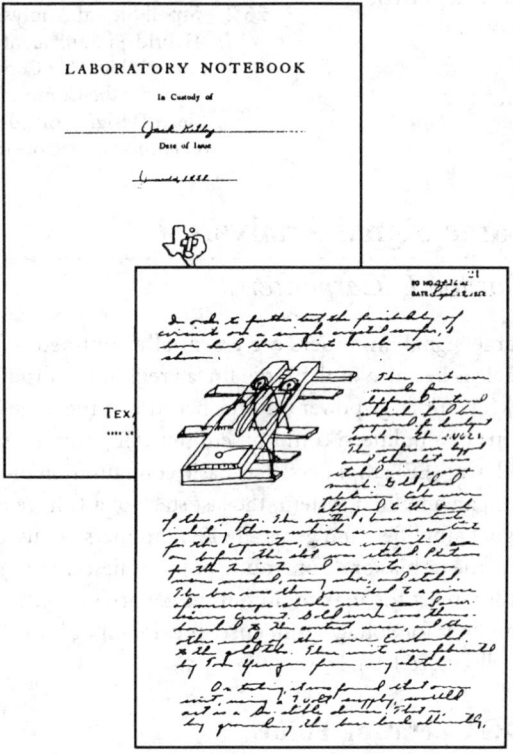

Kilby was issued this notebook from Texas Instruments, Inc. on June 12, 1958, about a month after starting his job. His entry dated Sept. 19 of that year described a digital integrated circuit, a flip-flop.

Source: Adapted from T.S. Perry, *IEEE Spectrum,* pp. 40–41, December 1988. © 1988 IEEE.

26

Amplifiers

Gordon L. Carpenter
California State University,
Long Beach

John Choma, Jr.
University of Southern
California

26.1 Large Signal Analysis 634
  DC Operating Point • Graphical Approach • Power Amplifiers
26.2 Small Signal Analysis 639
  Hybrid-Pi Equivalent Circuit • Hybrid-Pi Equivalent Circuit of a
  Monolithic BJT • Common Emitter Amplifier • Design Considera-
  tions for the Common Emitter Amplifier • Common Base Ampli-
  fier • Design Considerations for the Common Base Amplifier •
  Common Collector Amplifier

26.1 Large Signal Analysis

Gordon L. Carpenter

Large signal amplifiers are essentially confined to using bipolar transistors as their solid state devices because of the large linear region of amplification required. The exception is the use of VMOS for large power outputs because of the large linear region. There are three basic configurations of amplifiers: common emitter (CE) amplifiers, common base (CB) amplifiers, and emitter follower (EF) amplifiers. The basic configuration of each is shown in Fig. 26.1.

In an amplifier system, the last stage of a voltage amplifier string has to be considered as a large signal amplifier, and generally EF amplifiers are used as large signal amplifiers. This then requires that the dc bias or dc operating point (quiescent point) be located near the center of the load line in order to get the maximum output voltage swing. Small signal analysis can be used to evaluate the amplifier for voltage gain, current gain, input impedance, and output impedance, all of which will be discussed later.

DC Operating Point

Each transistor when connected in an amplifier configuration has a set of characteristic curves based on different values of base current and collector to emitter voltages. An example of this is shown in Fig. 26.2.

When amplifiers are coupled together with capacitors, the configuration is as shown in Fig. 26.3. The load resistor is really the input impedance of the next stage. To be able to evaluate this amplifier, a dc equivalent circuit needs to be developed as shown in Fig. 26.4. This will result in the following dc bias equation:

$$I_{CQ} = \frac{V_{BB} - V_{BE}}{R_B/\text{beta} + R_E}$$

where beta (h_{fe}) is the current gain of the transistor and V_{BE} is the conducting voltage across the base-emitter junction. This equation is the same for all amplifier configurations.

634

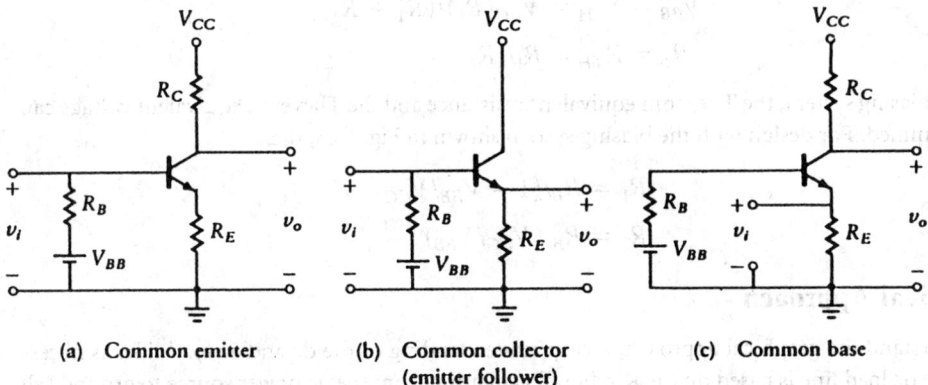

FIGURE 26.1 Amplifier circuits. (*Source:* C. J. Savant, M. Roden, and G. Carpenter, *Electronic Design, Circuits and Systems,* 2nd ed., Redwood City, Calif.: Benjamin-Cummings, 1991, p. 80. With permission.)

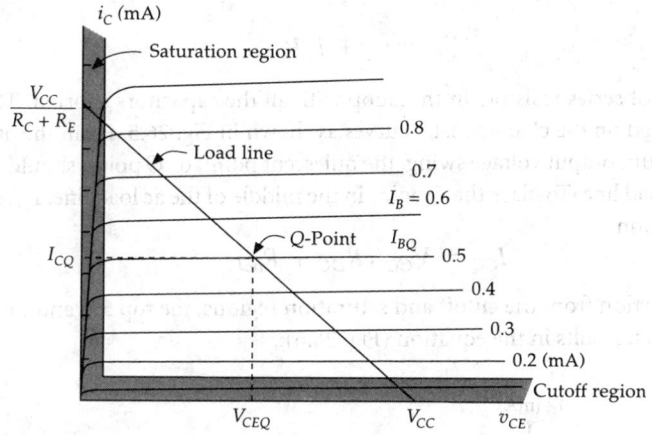

FIGURE 26.2 Transistor characteristic curves. (*Source:* C. J. Savant, M. Roden, and G. Carpenter, *Electronic Design, Circuits and Systems,* 2nd ed., Redwood City, Calif.: Benjamin-Cummings, 1991, p. 82. With permission.)

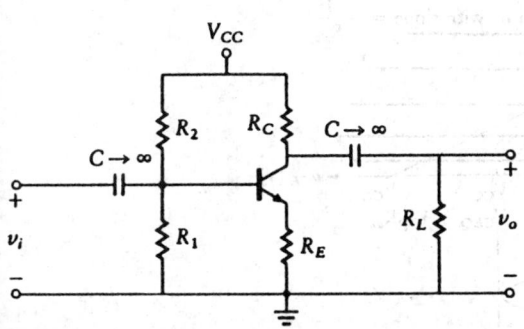

FIGURE 26.3 Amplifier circuit. (*Source:* C. J. Savant, M. Roden, and G. Carpenter, *Electronic Design, Circuits and Systems,* 2nd ed., Redwood City, Calif.: Benjamin-Cummings, 1991, p. 92. With permission.)

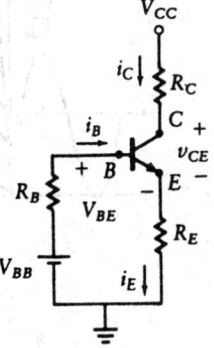

FIGURE 26.4 Amplifier equivalent circuit. (*Source:* C. J. Savant, M. Roden, and G. Carpenter, *Electronic Design, Circuits and Systems,* 2nd ed., Redwood City, Calif.: Benjamin-Cummings, 1991, p. 82. With permission.)

$$V_{BB} = V_{TH} = V_{CC}(R_1)/(R_1 + R_2)$$

$$R_B = R_{TH} = R_1 // R_2$$

For this biasing system, the Thévenin equivalent resistance and the Thévenin equivalent voltage can be determined. For design with the biasing system shown in Fig. 26.3, then:

$$R_1 = R_B/(1 - V_{BB}/V_{CC})$$

$$R_2 = R_B (V_{CC}/V_{BB})$$

Graphical Approach

To understand the graphical approach, a clear understanding of the dc and ac load lines is necessary. The dc load line is based on the Kirchhoff's equation from the dc power source to ground (all capacitors open)

$$V_{CC} = v_{CE} + i_C R_{DC}$$

where R_{DC} is the sum of the resistors in this loop.

The ac load line is the loop, assuming the transistor is the ac source and the source voltage is zero, then

$$V'_{CC} = v_{ce} + i_C R_{ac}$$

where R_{ac} is the sum of series resistors in that loop with all the capacitors shorted. The load lines then can be constructed on the characteristic curves as shown in Fig. 26.5. From this it can be seen that to get the maximum output voltage swing, the quiescent point, or Q point, should be located in the middle of the ac load line. To place the Q point in the middle of the ac load line, I_{CQ} can be determined from the equation

$$I_{CQ} = V_{CC}/(R_{DC} + R_{ac})$$

To avoid getting distortion from the cutoff and saturation regions, the top 5% and the bottom 5% are discarded. This then results in the equation (Fig. 26.6):

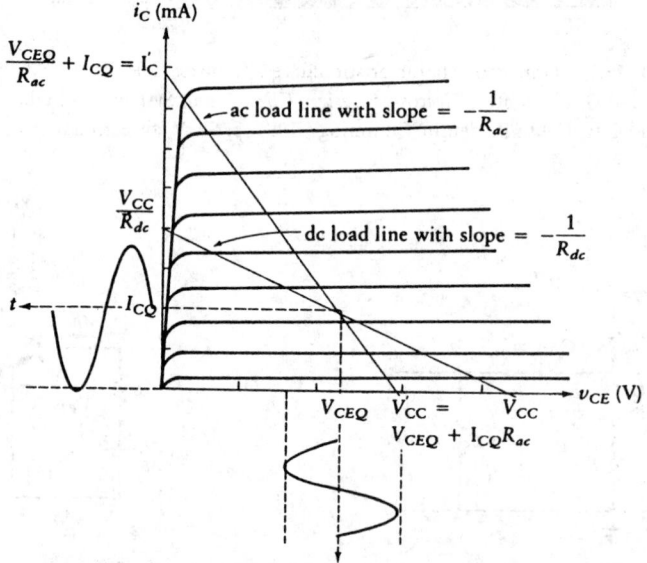

FIGURE 26.5 Load lines. (*Source:* C. J. Savant, M. Roden, and G. Carpenter, *Electronic Design, Circuits and Systems*, 2nd ed., Redwood City, Calif.: Benjamin-Cummings, 1991, p. 94. With permission.)

$$V_o \text{ (peak to peak)} = 2 \ (0.9) \ I_{CQ} \ (R_C//R_L)$$

If, however, the Q point is not in the middle of the ac load line, the output voltage swing will be reduced. Below the middle of the ac load line [Fig. 26.7(a)]:

$$V_o \text{ (peak to peak)} = 2 \ (I_{CQ} - 0.05 \ I_{CMax}) \ R_C//R_L$$

Above the middle of the ac load line [Fig. 26.7(b)]:

$$V_o \text{ (peak to peak)} = 2 \ (0.95 \ I_{CMax} - I_{CQ}) \ R_C//R_L$$

These values allow the highest allowable input signal to be used to avoid any distortion by dividing the voltage gain of the amplifier into the maximum output voltage swing. The preceding equations are the same for the CB configuration. For the EF configurations, the R_C is changed to R_E in the equations.

Power Amplifiers

Emitter follower amplifiers are considered power amplifiers because they are essentially current gain amplifiers. Using the standard linear EF amplifier for a maximum output voltage swing provides less than 25% efficiency (ratio of power in to power out). The dc current carrying the ac signal is where the loss of efficiency occurs. To avoid this power loss, the Q point is placed at I_{CQ} equal to zero, thus

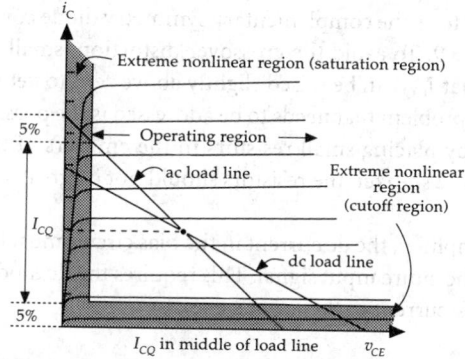

FIGURE 26.6 Q point in middle of load line. (*Source:* C. J. Savant, M. Roden, and G. Carpenter, *Electronic Design, Circuits and Systems,* 2nd ed., Redwood City, Calif.: Benjamin-Cummings, 1991, p. 135. With permission.)

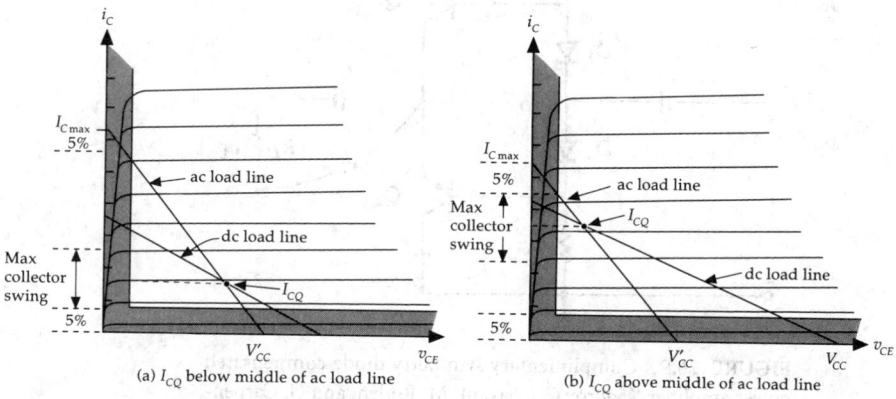

FIGURE 26.7 Reduced output voltage swing. (*Source:* C. J. Savant, M. Roden, and G. Carpenter, *Electronic Design, Circuits and Systems,* 2nd ed., Redwood City, Calif.: Benjamin-Cummings, 1991, p. 136. With permission.)

using the majority of the power for the output signal. This allows the efficiency to increase up toward 70%. Full signal amplification requires one transistor to amplify the positive portion of the input signal and another transistor to amplify the negative portion of the input signal. In the past, this was referred to as push-pull operation. A better system is to use an NPN transistor for the positive part of the input signal and a PNP transistor for the negative part. This type of operation is referred to as Class B complementary symmetry operation (Fig. 26.8).

In Fig. 26.8, the dc voltage drop across R_1 provides the voltage to bias the transistor at cutoff. Because these are power transistors, the temperature will change based on the amount of power the transistor is absorbing. This means the base-emitter junction voltage will have to change to keep $I_{CQ} = 0$. To compensate for this change in temperature, the R_1 resistors are replaced with diodes or transistors connected as diodes with the same turn-on characteristics as the power transistors. This type

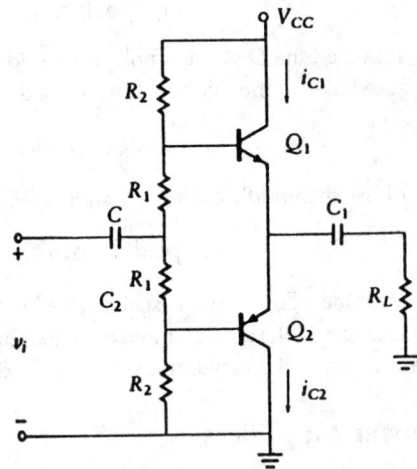

FIGURE 26.8 Complementary symmetry power amplifier. (*Source:* C. J. Savant, M. Roden, and G. Carpenter, *Electronic Design, Circuits and Systems,* 2nd ed., Redwood City, Calif.: Benjamin-Cummings, 1991, p. 248. With permission.)

of configuration is referred to as the complementary symmetry diode compensated (CSDC) amplifier and is shown in Fig. 26.9. To avoid the crossover distortion, small resistors can be placed in series with the diodes so that I_{CQ} can be raised slightly above zero to get increased amplification in the cutoff region. Another problem that needs to be addressed is the possibility of thermal runaway. This can be easily solved by placing small resistors in the emitters of the power transistors. For example, if the load is an 8-Ω speaker, the resistors should not be greater than 0.47 Ω to avoid output signal loss.

To design this type of amplifier, the dc current in the bias circuit must be large enough so that the diodes remain on during the entire input signal. This requires the dc diode current to be equal to or larger than the zero to peak current of the input signal, or

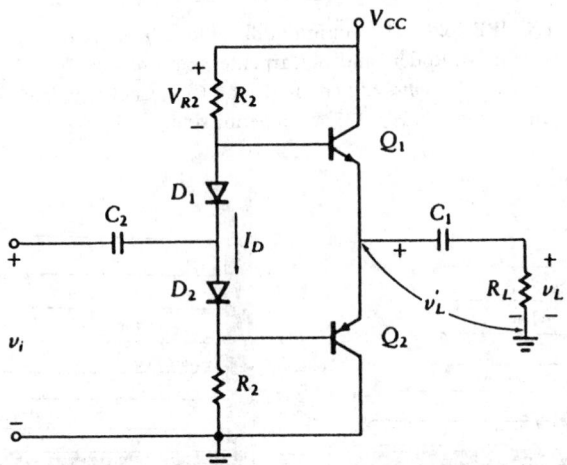

FIGURE 26.9 Complimentary symmetry diode compensated power amplifier. (*Source:* C. J. Savant, M. Roden, and G. Carpenter, *Electronic Design, Circuits and Systems,* 2nd ed., Redwood City, Calif.: Benjamin-Cummings, 1991, p. 251. With permission.)

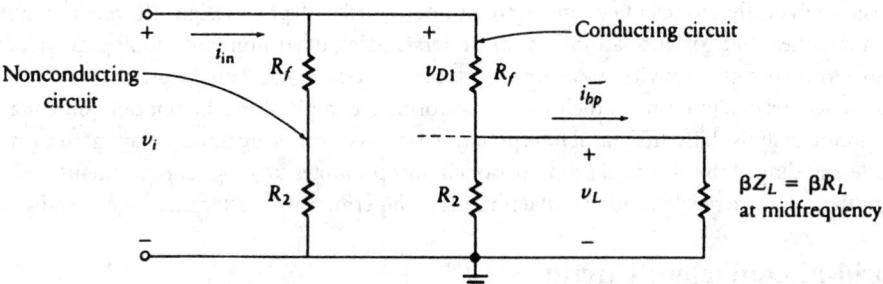

FIGURE 26.10 AC equivalent circuit of the CSDC amplifier. (*Source:* C. J. Savant, M. Roden, and G. Carpenter, *Electronic Design, Circuits and Systems,* 2nd ed., Redwood City, Calif.: Benjamin-Cummings, 1991, p. 255. With permission.)

$$I_D \geq I_{ac} \text{ (0 to peak)}$$

$$(V_{CC}/2 - V_{BE})/R_2 = I_B \text{(0 to peak)} + V_L \text{(0 to peak)}/R_2$$

When designing to a specific power, both I_B and V_L can be determined. This allows the selection of the value of R_2 and the equivalent circuit shown in Fig. 26.10 can be developed. Using this equivalent circuit, both the input resistance and the current gain can be shown. R_f is the forward resistance of the diodes.

$$R_{in} = (R_f + R_2)//[R_f + (R_2//\text{Beta} R_L)]$$

$$P_o = I_{C\max} R_L/2$$

The power rating of the transistors to be used in this circuit should be greater than

$$P_{rating} = V_{CC}^2/(4\text{Pi}^2 R_L)$$

$$C_1 = 1/(2\text{Pi} f_{low} R_L)$$

$$C_2 = 10/[2\text{Pi} f_{low}(R_{in} + R_i)]$$

where R_i is the output impedance of the previous stage and f_{low} is the desired low frequency cutoff of the amplifier.

References

P. R. Gray and R. G. Meyer, *Analysis and Design of Analog Integrated Circuits,* New York: Wiley, 1984.

J. Millman and A. Grabel, *Microelectronics,* New York: McGraw-Hill, 1987.

P. O. Neudorfer and M. Hassul, *Introduction to Circuit Analysis,* Needham Heights, Mass.: Allyn and Bacon, 1990.

C. J. Savant, M. Roden, and G. Carpenter, *Electronic Design, Circuits and Systems,* 2nd ed., Redwood City, Calif.: Benjamin-Cummings, 1991.

D. L. Schilling and C. Belove, *Electronic Circuits,* New York: McGraw-Hill, 1989.

26.2 Small Signal Analysis

John Choma, Jr.

This section introduces the reader to the analytical methodologies that underlie the design of small signal, analog bipolar junction transistor (BJT) amplifiers. Analog circuit and system design entails complementing basic circuit analysis skills with the art of architecting a circuit topology that produces acceptable input-to-output (I/O) electrical characteristics. Because design is not the inverse of analysis, analytically proficient engineers are not necessarily adept at design. However, circuit and system analyses that conduce an insightful understanding of meaningful topological structures arguably foster design creativity. Accordingly, this section focuses more on the problems of interpreting analytical results in terms of their circuit performance implications than it does on enhancing basic circuit analysis skills. Insightful interpretation breeds engineering understanding. In turn, such an understanding of the electrical properties of circuits promotes topological refinements and innovations that produce reliable and manufacturable, high performance electronic circuits and systems.

Hybrid-Pi Equivalent Circuit

In order for a BJT to function properly in linear amplifier applications, it must operate in the forward active region of its volt–ampere characteristic curves. Two conditions ensure BJT operation in the forward domain. First, the applied emitter-base terminal voltage must forward bias the intrinsic emitter-base junction diode at all times. Second, the instantaneous voltage established across the base-collector terminals of the transistor must preclude a forward biased intrinsic base-collector diode. The simultaneous satisfaction of these two conditions requires appropriate biasing subcircuits, and it imposes restrictions on the amplitudes of applied input signals [Clarke and Hess, 1978].

The most commonly used BJT equivalent circuit for investigating the dynamical responses to small input signals is the **hybrid-pi model** offered in Fig. 26.11 [Sedra and Smith, 1987]. In this model, R_b, R_c, and R_e, respectively, represent the *internal base, collector,* and *emitter resistances* of the considered BJT. Although these series resistances vary somewhat with quiescent operating point [de Graaf, 1969], they can be viewed as constants in first-order manual analyses.

The *emitter-base junction diffusion resistance,* R_π, is the small signal resistance of the emitter-base junction diode. It represents the inverse of the slope of the common emitter static input characteristic curves. Analytically, R_π is given by

$$R_\pi = \frac{h_{FE}N_F V_T}{I_{CQ}} \tag{26.1}$$

where h_{FE} is the *static common emitter current gain* of the BJT, N_F is the *emitter-base junction injection coefficient,* V_T is the *Boltzmann voltage corresponding to an absolute junction operating temperature of T,* and I_{CQ} is the *quiescent collector current.*

The expression for the resistance, R_o, which accounts for *conductivity modulation* in the neutral base, is

$$R_o = \frac{V'_{CEQ} + V_{AF}}{I_{CQ}\left(1 - \dfrac{I_{CQ}}{I_{KF}}\right)} \tag{26.2}$$

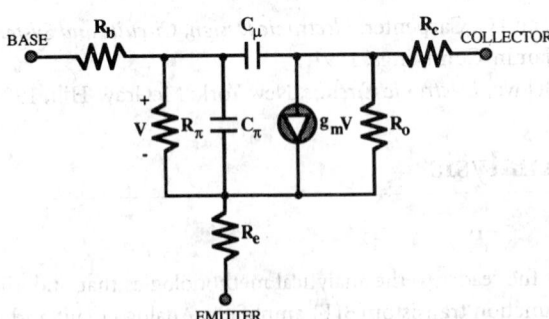

FIGURE 26.11   The small signal equivalent circuit (hybrid-pi model) of a bipolar junction transistor.

where V_{AF} is the *forward Early voltage*, V'_{CEQ} is the quiescent voltage developed across the internal collector-emitter terminals, and I_{KF} symbolizes the *forward knee current*. The knee current is a measure of the onset of *high injection effects* [Gummel and Poon, 1970] in the base. In particular, a collector current numerically equal to I_{KF} implies that the forward biasing of the emitter-base junction promotes a net minority carrier charge injected into the base from the emitter that is equal to the background majority charge in the neutral base. The Early voltage is an inverse measure of the slope of the common emitter output characteristic curves.

The final low frequency parameter of the hybrid-pi model is the *forward transconductance, g_m*. This parameter, which is a measure of the forward small signal gain available at a quiescent operating point, is given by

$$g_m = \frac{I_{CQ}}{N_F V_T} \left(\frac{1 - \dfrac{I_{CQ}}{I_{KF}}}{1 + \dfrac{V_{CEQ'}}{V_{AF}}} \right) \tag{26.3}$$

Two capacitances, C_π and C_μ, are incorporated in the small signal model to provide a first-order approximation of steady-state transistor behavior at high signal frequencies. The capacitance, C_π, is the *net capacitance of the emitter-base junction diode* and is given by

$$C_\pi = \frac{C_{JE}}{\left(1 - \dfrac{V_E}{V_{JE} - 2V_T} \right)^{M_{JE}}} + \tau_f g_m \tag{26.4}$$

where the first term on the right-hand side represents the *depletion component* and the second term is the *diffusion component* of C_π. In Eq. (26.4), τ_f is the *average forward transit time of minority carriers in the field-neutral base*, C_{JE} is the *zero bias value of emitter-base junction depletion capacitance*, V_{JE} is the *built-in potential of the junction*, V_E is the *forward biasing voltage developed across the intrinsic emitter-base junction*, and M_{JE} is the *grading coefficient of the junction*. The capacitance, C_μ, has only a depletion component, owing to the reverse (or at most zero) bias impressed across the internal base-collector junction. Accordingly, its analytical form is analogous to the first term on the right-hand side of Eq. (26.4). Specifically,

$$C_\mu = \frac{C_{JC}}{\left(1 - \dfrac{V_C}{V_{JC} - 2V_T} \right)^{M_{JC}}} \tag{26.5}$$

where the physical interpretation of C_{JC}, V_{JC}, and M_{JC} is analogous to C_{JE}, V_{JE}, and M_{JE}, respectively.

A commonly invoked figure of merit for assessing the high speed, small signal performance attributes of a BJT is the *common emitter*, **short circuit gain-bandwidth product, ω_T**, which is given by

$$\omega_T = \frac{g_m}{C_\pi + C_\mu} \tag{26.6}$$

The significance of Eq. (26.6) is best appreciated by studying the simple circuit diagram of Fig. 26.12(a), which depicts the grounded emitter configuration of a BJT biased for linear operation at a quiescent base current of I_{BQ} and a quiescent collector-emitter voltage of V_{CEQ}. Note that the battery supplying V_{CEQ} grounds the collector for small signal conditions. The small signal

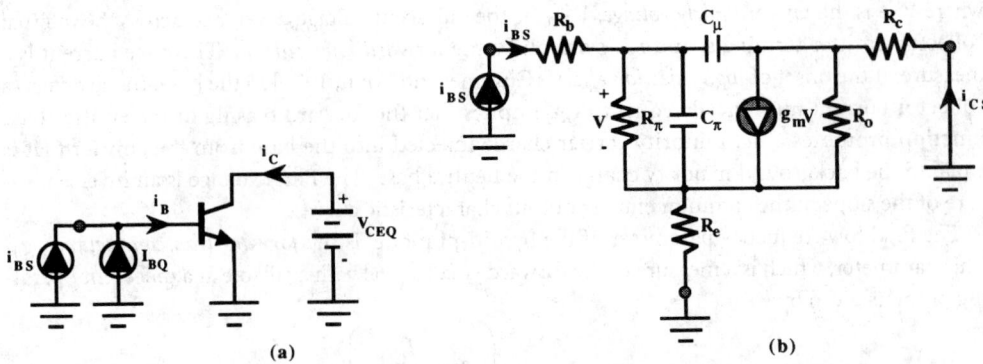

FIGURE 26.12 (a) Schematic diagram pertinent to the evaluation of the short circuit, common emitter, small signal current gain. (b) High frequency small signal model of the circuit in part (a).

model of the circuit at hand is resultantly seen to be the topology offered in Fig. 26.12(b), where i_{BS} and i_{CS}, respectively, denote the signal components of the net instantaneous base current, i_B, and the net instantaneous collector current, i_C.

For negligibly small internal collector (R_c) and emitter (R_e) resistances, it can be shown that the *small signal, short circuit, high frequency common emitter current gain,* $\beta_{ac}(j\omega)$, is expressible as

$$\beta_{ac}(j\omega) \triangleq \frac{i_{CS}}{i_{BS}} = \frac{\beta_{ac}\left(1 - \dfrac{j\omega C_\mu}{g_m}\right)}{1 + \dfrac{j\omega}{\omega_\beta}} \tag{26.7}$$

where β_{ac}, the low frequency value of $\beta_{ac}(j\omega)$, or simply the *low frequency beta*, is

$$\beta_{ac} = \beta_{ac}(0) = g_m R_\pi \tag{26.8}$$

and

$$\omega_\beta = \frac{1}{R_\pi(C_\pi + C_\mu)} \tag{26.9}$$

symbolizes the so-called *beta cutoff frequency* of the BJT. Because the frequency, g_m/C_μ, is typically much larger than ω_β, ω_β is the approximate **3-dB bandwidth** of $\beta_{ac}(j\omega)$; that is,

$$|\beta_{ac}(j\omega_\beta)| \cong \frac{\beta_{ac}}{\sqrt{2}} \tag{26.10}$$

It follows that the corresponding *gain-bandwidth product,* ω_T, is the product of β_{ac} and ω_β, which, recalling Eq. (26.8), leads directly to the expression in Eq. (26.6). Moreover, in the neighborhood of ω_T,

$$\beta_{ac}(j\omega) \cong \frac{\beta_{ac}\omega_\beta}{j\omega} = \frac{\omega_T}{j\omega} \tag{26.11}$$

which suggests that ω_T is the approximate frequency at which the magnitude of the small signal, short circuit, common emitter current gain degrades to unity.

Hybrid-Pi Equivalent Circuit of a Monolithic BJT

The conventional hyprid-pi model in Fig. 26.11 generally fails to provide sufficiently accurate predictions of the high frequency response of monolithic diffused or implanted BJTs. One reason for this modeling inaccuracy is that the hybrid-pi equivalent circuit does not reflect the fact that monolithic transistors are often fabricated on lightly doped, noninsulating substrates that establish a distributed, large area, *pn* junction with the collector region. Since the substrate-collector *pn* junction is back biased in linear applications of a BJT, negligible static and low frequency signal currents flow from the collector to the substrate. At high frequencies, however, the depletion capacitance associated with the reverse biased substrate-collector junction can cause significant susceptive loading of the collector port. In Fig. 26.13, the lumped capacitance, C_{bb}, whose mathematical definition is similar to that of C_μ in Eq. (26.5), provides a first-order account of this collector loading. Observe that this substrate capacitance appears in series with a substrate resistance, R_{bb}, which reflects the light doping nature of the substrate material. For monolithic transistors fabricated on insulating or semi-insulating substrates, R_{bb} is a very large resistance, thereby rendering C_{bb} unimportant with respect to the problem of predicting steady-state transistor responses at high signal frequencies.

A problem that is even more significant than parasitic substrate dynamics stems from the fact that the hybrid-pi equivalent circuit in Fig. 26.11 is premised on a uniform transistor structure whose emitter-base and base-collector junction areas are identical. In a monolithic device, however, the effective base-collector junction area is much larger than that of the emitter-base junction because the base region is diffused or implanted into the collector [Glaser and Subak-Sharpe, 1977]. The effect of such a geometry is twofold. First, the actual value of C_μ is larger than the value predicated on the physical considerations that surround a simplified uniform structure BJT. Second, C_μ is not a single lumped capacitance that is incident with only the intrinsic base-collector junction. Rather, the effective value of C_μ is distributed between the intrinsic collector and the entire base-collector junction interface. A first-order account of this capacitance distribution entails partitioning C_μ in Fig. 26.11 into two capacitances, say $C_{\mu1}$ and $C_{\mu2}$, as indicated in Fig. 26.13. In general, $C_{\mu2}$ is 3 to 5 times larger than $C_{\mu1}$. Whereas $C_{\mu1}$ is proportional to the emitter-base junction area, $C_{\mu2}$ is proportional to the net base-collector junction area, less the area of the emitter-base junction.

Just as $C_{\mu1}$ and $C_{\mu2}$ superimpose to yield the original C_μ in the simplified high frequency model of a BJT, the effective base resistances, R_{b1} and R_{b2}, sum to yield the original base resistance, R_b. The resistance, R_{b1}, is the *contact resistance* associated with the base lead and the inactive BJT base region. It is inversely proportional to the surface area of the base contact. On the other hand, R_{b2}, which is referred to as the *active base resistance*, is nominally an inverse function of emitter finger length. Because of submicron base widths and the relatively light average doping concentrations of active base regions, R_{b2} is significantly larger than R_{b1}.

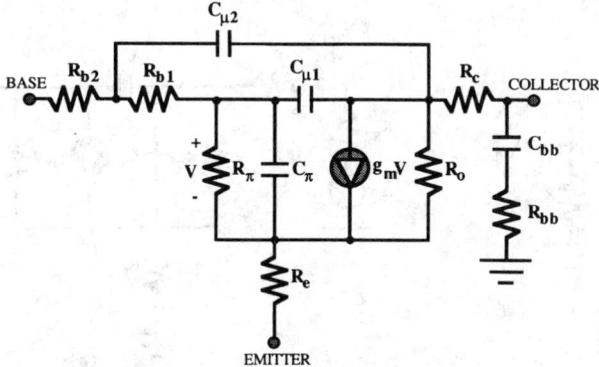

FIGURE 26.13 The hybrid-pi equivalent circuit of a monolithic bipolar junction transistor.

Common Emitter Amplifier

The most commonly used canonic cell of linear BJT amplifiers is the *common emitter amplifier*, whose basic circuit schematic diagram is depicted in Fig. 26.14(a). In this diagram, R_{ST} is the Thévenin resistance of the applied signal source, V_{ST}, and R_{LT} is the effective, or Thévenin, load resistance driven by the amplifier. The signal source has zero average, or dc, value. Although requisite biasing is not shown in the figure, it is tacitly assumed that the transistor is biased for linear operation. Hence, the diagram at hand is actually the **ac schematic diagram**; that is, it delineates only the signal paths of the circuit. Note that in the common emitter orientation, the input signal is applied to the base of the transistor, while the resultant small signal voltage response, V_{OS}, is extracted at the transistor collector.

The hybrid-pi model of Fig. 26.11 forms the basis for the small signal equivalent circuit of the common emitter cell, which is given in Fig. 26.14(b). In this configuration, the capacitance, C_o, represents an effective output port capacitance that accounts for both substrate loading of the collector port (if the BJT is a monolithic device) and the net effective shunt capacitance associated with the load.

At low signal frequencies, the capacitors, C_π, C_μ, and C_o in the model of Fig. 26.14(b), can be replaced by open circuits. A straightforward circuit analysis of the resultantly simplified equivalent circuit produces analytical expressions for the low frequency values of the small signal *voltage gain*, $A_{vCE} = V_{OS}/V_{ST}$; the *driving point input impedance*, Z_{inCE}; and the *driving point output impedance*, Z_{outCE}. Because the Early resistance, R_o, is invariably much larger than the resistance sum $(R_c + R_e + R_{LT})$, the low frequency voltage gain of the common emitter cell is expressible as

$$A_{vCE}(0) \cong -\left[\frac{\beta_{ac}R_{LT}}{R_{ST} + R_b + R_\pi + (\beta_{ac} + 1)R_e}\right] \quad (26.12)$$

For large R_o, conventional circuit analyses also produce a low frequency driving point input resistance of

$$R_{inCE} = Z_{inCE}(0) \cong R_b + R_\pi + (\beta_{ac} + 1)R_e \quad (26.13)$$

and a low frequency driving point output resistance of

$$R_{outCE} = Z_{outCE}(0) \cong \left(\frac{\beta_{ac}R_e}{R_e + R_b + R_\pi + R_{ST}} + 1\right)R_o \quad (26.14)$$

At high signal frequencies, the capacitors in the small signal equivalent circuit of Fig. 26.14(b) produce a third-order voltage gain frequency response whose analytical formulation is algebraically

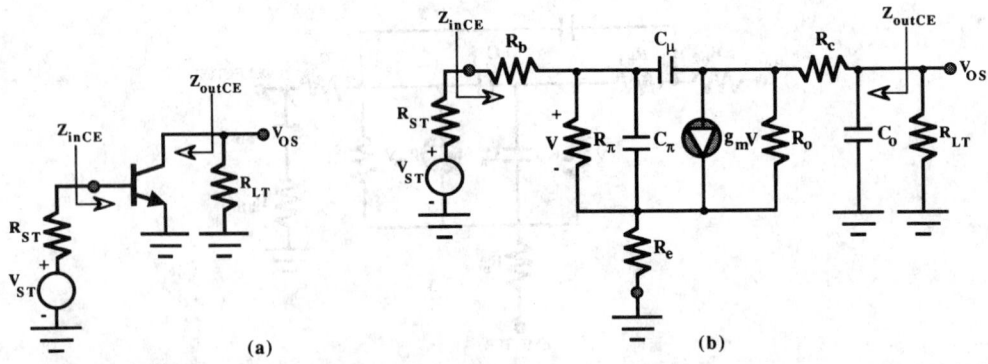

FIGURE 26.14 (a) AC schematic diagram of a common emitter amplifier. (b) Modified small signal, high frequency equivalent circuit of common emitter amplifier.

cumbersome [Singhal and Vlach, 1977; Haley, 1988]. However, because the poles produced by these capacitors are real, lie in the left half complex frequency plane, and generally have widely separated frequency values, the dominant pole approximation provides an adequate estimate of high frequency common emitter amplifier response in the usable passband of the amplifier. Accordingly, the high frequency voltage gain, say $A_{vCE}(s)$, of the common emitter amplifier can be approximated as

$$A_{vCE}(s) \cong A_{vCE}(0)\left[\frac{1 + sT_{zCE}}{1 + sT_{pCE}}\right]$$
(26.15)

In this expression, T_{pCE} is of the form,

$$T_{pCE} = R_{C\pi}C_\pi + R_{C\mu}C_\mu + R_{Co}C_o$$
(26.16)

where $R_{C\pi}$, $R_{C\mu}$, and R_{Co}, respectively, represent the Thévenin resistances seen by the capacitors, C_π, C_μ, and C_o, under the conditions that (1) all capacitors are supplanted by open circuits and (2) the independent signal generator, V_{ST}, is reduced to zero. Analogously, T_{zCE} is of the form

$$T_{zCE} = R_{C\pi o}C_\pi + R_{C\mu o}C_\mu + R_{Coo}C_o$$
(26.17)

where $R_{C\pi o}$, $R_{C\mu o}$, and R_{Coo}, respectively, represent the Thévenin resistances seen by the capacitors, C_π, C_μ, and C_o, under the conditions that (1) all capacitors are supplanted by open circuits and (2) the output voltage response, V_{OS}, is constrained to zero while maintaining nonzero input signal source voltage. It can be shown that when R_o is very large and R_c is negligibly small,

$$R_{C\pi} = \frac{R_\pi \| (R_{ST} + R_b + R_e)}{1 + \dfrac{\beta_{ac}R_e}{R_{ST} + R_b + R_\pi + R_e}}$$
(26.18)

$$R_{C\mu} = (R_{LT} + R_c) + \{(R_{ST} + R_b)\|[R_\pi$$
$$+(\beta_{ac} + 1)R_e]\}\left[1 + \frac{\beta_{ac}(R_{LT} + R_c)}{R_\pi + (\beta_{ac} + 1)R_e}\right]$$
(26.19)

and

$$R_{Co} = R_{LT}$$
(26.20)

Additionally, $R_{C\pi o} = R_{Coo} = 0$, and

$$R_{C\mu o} = -\frac{R_\pi + (\beta_{ac} + 1)R_e}{\beta_{ac}}$$
(26.21)

Once T_{pCE} and T_{zCE} are determined, the 3-dB voltage gain bandwidth, B_{CE}, of the common emitter amplifier can be estimated in accordance with

$$B_{CE} \cong \frac{1}{T_{pCE}\sqrt{1 - 2\left(\dfrac{T_{zCE}}{T_{pCE}}\right)^2}}$$
(26.22)

The high frequency behavior of both the driving point input and output impedances, $Z_{inCE}(s)$ and $Z_{outCE}(s)$, respectively, can be approximated by mathematical functions whose forms are analogous to the gain expression in Eq. (26.15). In particular,

$$Z_{inCE}(s) \cong R_{inCE}\left[\frac{1 + sT_{zCE1}}{1 + sT_{pCE1}}\right] \tag{26.23}$$

and

$$Z_{outCE}(s) \cong R_{outCE}\left[\frac{1 + sT_{zCE1}}{1 + sT_{pCE2}}\right] \tag{26.24}$$

where R_{inCE} and R_{outCE} are defined by Eqs. (26.13) and (26.14). The dominant time constants, T_{pCE1}, T_{zCE1}, T_{pCE2}, and T_{zCE2}, derive directly from Eqs. (26.16) and (26.17) in accordance with [Choma and Witherspoon, 1990]

$$T_{pCE1} = \lim_{R_{ST} \to \infty} [T_{pCE}] \tag{26.25}$$

$$T_{zCE1} = \lim_{R_{ST} \to 0} [T_{pCE}] \tag{26.26}$$

$$T_{pCE2} = \lim_{R_{LT} \to \infty} [T_{pCE}] \tag{26.27}$$

and

$$T_{zCE2} = \lim_{R_{LT} \to 0} [T_{pCE}] \tag{26.28}$$

For reasonable values of transistor model parameters and terminating resistances, $T_{pCE1} > T_{zCE1}$, and $T_{pCE2} > T_{zCE2}$. It follows that both the input and output ports of a common emitter canonic cell are capacitive at high signal frequencies.

Design Considerations for the Common Emitter Amplifier

Equation (26.12) underscores a serious shortcoming of the canonical common emitter configuration. In particular, since the internal emitter resistance of a BJT is small, the low frequency voltage gain is sensitive to the processing uncertainties that accompany the numerical value of the small signal beta. The problem can be rectified at the price of a diminished voltage gain magnitude by inserting an *emitter degeneration resistance*, R_{EE}, in series with the emitter lead, as shown in Fig. 26.15(a). Since R_{EE} appears in series with the internal emitter resistance, R_e, as suggested in Fig. 26.15(b), the impact of emitter degeneration can be assessed analytically by replacing R_e in Eqs. (26.12) through (26.28) by the resistance sum $(R_e + R_{EE})$. For sufficiently large R_{EE}, such that

$$R_e + R_{EE} \cong R_{EE} >> \frac{R_{ST} + R_b + R_\pi}{\beta_{ac} + 1} \tag{26.29}$$

the low frequency voltage gain becomes

$$A_{vCE}(0) \cong - \frac{\alpha_{ac}R_{LT}}{R_{EE}} \tag{26.30}$$

where α_{ac}, which symbolizes the *small signal, short circuit, common base current gain*, or simply the *ac alpha*, of the transistor is given by

$$\alpha_{ac} = \frac{\beta_{ac}}{\beta_{ac} + 1} \tag{26.31}$$

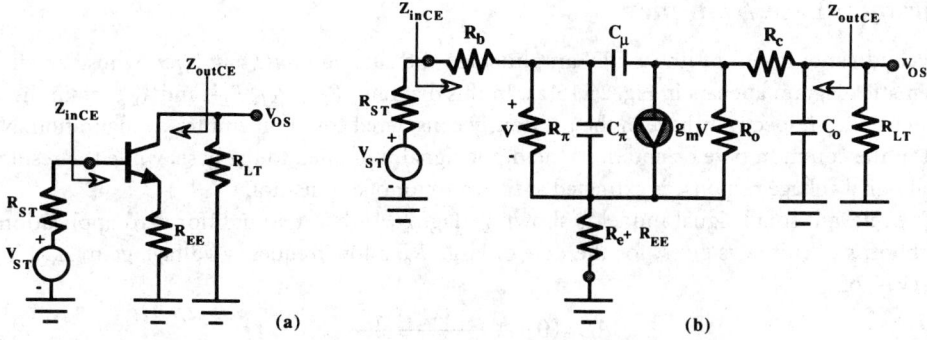

FIGURE 26.15 (a) AC schematic diagram of a common emitter amplifier using an emitter degeneration resistance. (b) Small signal, high frequency equivalent circuit of amplifier in part (a).

Despite numerical uncertainties in β_{ac}, minimum values of β_{ac} are much larger than one, thereby rendering the voltage gain in Eq. (26.30) almost completely independent of small signal BJT parameters.

A second effect of emitter degeneration is an increase in both the low frequency driving point input and output resistances. This contention is confirmed by Eq. (26.13), which shows that if R_o remains much larger than $(R_c + R_e + R_{EE} + R_{LT})$, a resistance in the amount of $(\beta_{ac} + 1)R_{EE}$ is added to the input resistance established when the emitter of a common emitter amplifier is returned to signal ground. Likewise, Eq. (26.14) verifies that emitter degeneration increases the low frequency driving point output resistance. In fact, a very large value of R_{EE} produces an output resistance that approaches a limiting value of $(\beta_{ac} + 1)R_o$. It follows that a common emitter amplifier that exploits emitter degeneration behaves as a voltage-to-current converter at low signal frequencies. In particular, its high input resistance does not incur an appreciable load on signal voltage sources that are characterized by even moderately large Thévenin resistances, while its large output resistance comprises an almost ideal current source at its output port.

A third effect of emitter degeneration is a decrease in the effective pole time constant, T_{pCE}, as well as an increase in the effective zero time constant, T_{zCE}, which can be confirmed by reinvestigating Eqs. (26.18) through (26.21) for the case of R_e replaced by the resistance sum $(R_e + R_{EE})$. The use of an emitter degeneration resistance therefore promotes an increased 3-dB circuit bandwidth. Unfortunately, it also yields a diminished circuit gain-bandwidth product; that is, a given emitter degeneration resistance causes a degradation in the low frequency gain magnitude that is larger than the corresponding bandwidth increase promoted by this resistance. This deterioration of circuit gain-bandwidth product is a property of all *negative feedback circuits* [Choma, 1984].

For reasonable values of the emitter degeneration resistance, R_{EE}, the Thévenin time constant, $R_{C\mu}C_\mu$, is likely to be the dominant contribution to the effective first-order time constant, T_{pCE}, attributed to the poles of a common emitter amplifier. Hence, C_μ is the likely device capacitance that dominantly imposes an upper limit to the achievable 3-dB bandwidth of a common emitter cell. The reason for this substantial bandwidth sensitivity to C_μ is the so-called Miller multiplication factor, say M, which appears as the last bracketed term on the right-hand side of Eq. (26.19), namely,

$$M = 1 + \frac{\beta_{ac}(R_{LT} + R_c)}{R_\pi + (\beta_{ac} + 1)R_e} \tag{26.32}$$

The Miller factor, M, which effectively multiplies C_μ in the expression for $R_{C\mu}C_\mu$, increases sharply with the load resistance, R_{LT}, and hence with the gain magnitude of the common emitter amplifier. Note that in the limit of a large emitter degeneration resistance (which adds directly to R_e), Eq. (26.30) reduces Eq. (26.32) to the factor

$$M \cong 1 + \left| A_{vCE}(0) \right| \tag{26.33}$$

Common Base Amplifier

A second canonic cell of linear BJT amplifiers is the *common base amplifier*, whose ac circuit schematic diagram appears in Fig. 26.16(a). In this diagram, R_{ST}, V_{ST}, R_{LT}, and V_{OS} retain the significance they respectively have in the previously considered common emitter configuration. Note that in the common base orientation, the input signal is applied to the base, while the resultant small signal voltage response is extracted at the collector of a transistor.

The relevant small signal model is shown in Fig. 26.16(b). A straightforward application of Kirchhoff's circuit laws gives, for the case of large R_o, a low frequency voltage gain, $A_{vCB}(0) = V_{OS}/V_{ST}$, of

$$A_{vCB}(0) \cong \frac{\alpha_{ac} R_{LT}}{R_{ST} + R_{inCB}} \qquad (26.34)$$

where R_{inCB} is the low frequency value of the common base driving point input impedance,

$$R_{inCB} = Z_{inCB}(0) \cong R_e + \frac{R_b + R_\pi}{\beta_{ac} + 1} \qquad (26.35)$$

Moreover, it can be shown that the low frequency driving point output resistance is

$$R_{outCB} = Z_{outCB}(0) \cong \left[\frac{\beta_{ac}(R_e + R_{ST})}{R_e + R_b + R_\pi + R_{ST}} + 1 \right] R_o \qquad (26.36)$$

The preceding three equations underscore several operating characteristics that distinguish the common base amplifier from its common emitter counterpart. For example, Eq. (26.35) suggests a low frequency input resistance that is significantly smaller than that of a common emitter unit. To underscore this contention, consider the case of two identical transistors, one used in a common emitter amplifier and the other used in a common base configuration, that are biased at identical quiescent operating points. Under this circumstance, Eqs. (26.35) and (26.13) combine to deliver

$$R_{inCB} \cong \frac{R_{inCE}}{\beta_{ac} + 1} \qquad (26.37)$$

which shows that the common base input resistance is a factor of $(\beta_{ac} + 1)$ times smaller than the input resistance of the common emitter cell. The resistance reflection factor, $(\beta_{ac} + 1)$, in Eq. (26.37) represents the ratio of small signal emitter current to small signal base current. Accordingly,

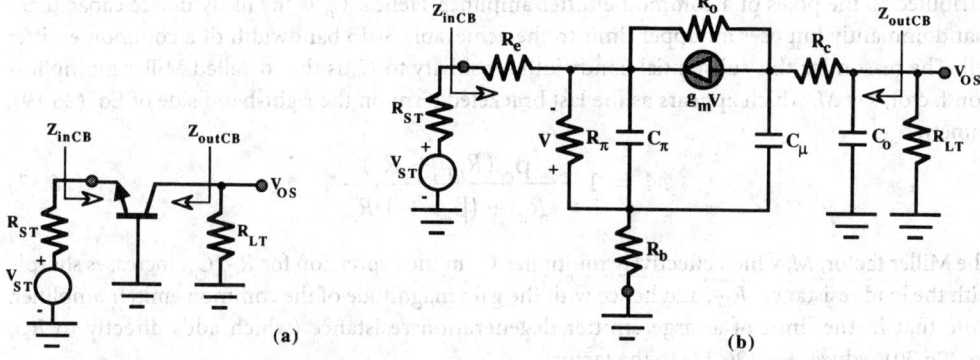

FIGURE 26.16   (a) AC schematic diagram of a common base amplifier. (b) Small signal, high frequency equivalent circuit of amplifier in part (a).

Eq. (26.37) is self-evident when it is noted that the input resistance of a common base stage is referred to an input emitter current, whereas the input resistance of its common emitter counterpart is referred to an input base current.

A second difference between the common emitter and common base amplifiers is that the voltage gain of the latter displays no phase inversion between source and response voltages. Moreover, for the same load and source terminations and for identical transistors biased identically, the voltage gain of the common base cell is likely to be much smaller than that of the common emitter unit. This contention is verified by substituting Eq. (26.37) into Eq. (26.34) and using Eqs. (26.31), (26.13), and (26.12) to write

$$A_{vCB}(0) \cong \frac{|A_{vCE}(0)|}{1 + \dfrac{\beta_{ac} R_{ST}}{R_{ST} + R_{inCE}}} \qquad (26.38)$$

At high signal frequencies, the voltage gain, driving point input impedance, and driving point output impedance can be approximated by functions whose analytical forms mirror those of Eqs. (26.15), (26.23), and (26.24). Let T_{pCB} and T_{zCB} designate the time constants of the effective dominant pole and the effective dominant zero, respectively, of the common base cell. An analysis of the structure of Fig. 26.16(b) resultantly produces, with R_o and R_c ignored,

$$T_{pCB} = R_{C\pi} C_\pi + R_{C\mu} C_\mu + R_{Co} C_o \qquad (26.39)$$

where

$$R_{C\pi} = \frac{R_\pi \,\|\, (R_{ST} + R_b + R_e)}{1 + \dfrac{\beta_{ac}(R_{ST} + R_e)}{R_{ST} + R_b + R_\pi + R_e}} \qquad (26.40)$$

$$R_{C\mu} = R_b \,\|\, [R_\pi + (\beta_{ac} + 1)(R_{ST} + R_e)] \\ + R_{LT}\left[1 + \frac{\beta_{ac} R_b}{R_b + R_\pi + (\beta_{ac} + 1)(R_{ST} + R_e)}\right] \qquad (26.41)$$

and R_{Co} remains given by Eq. (26.20). Moreover,

$$T_{zCB} = \frac{R_b C_\mu}{\alpha_{ac}} \qquad (26.42)$$

Design Considerations for the Common Base Amplifier

An adaptation of Eqs. (26.25) through (26.28) to the common base stage confirms that the driving point input impedance is capacitive at high signal frequencies. On the other hand, $g_m R_b > 1$ renders a common base driving point input impedance that is inductive at high frequencies. This impedance property can be gainfully exploited to realize monolithic shunt peaked amplifiers in which the requisite circuit inductance is synthesized as the driving point input impedance of a common base stage (or the driving point output impedance of a common collector cell) [Grebene, 1984].

The common base stage is often used to broadband the common emitter amplifier by forming the *common emitter–common base cascode,* whose ac schematic diagram is given in Fig. 26.17. The broadbanding afforded by the cascode structure stems from the fact that the effective low frequency load resistance, say R_{Le}, seen by the common emitter transistor, QE, is the small driving point input resistance of the common base amplifier, QB. This effective load resistance, as witnessed by C_μ of the common emitter transistor, is much smaller than the actual load resistance that terminates the output port of the amplifier, thereby decreasing the Miller multiplication of the C_μ in QE. If the time constant savings afforded by decreased Miller multiplication is larger than the sum of the additional time constants presented to the circuit by the common base transistor, an enhancement of common emitter bandwidth occurs. Note that such bandwidth enhancement is realized without compromising the

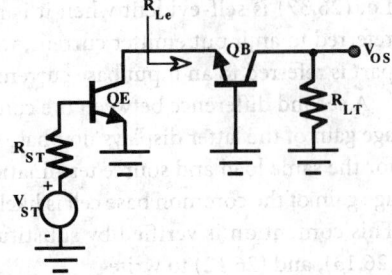

FIGURE 26.17   AC schematic diagram of a common emitter–common base cascode amplifier.

common emitter gain-bandwidth product, since the voltage gain of the common emitter–common base unit is almost identical to that of the common emitter amplifier alone.

Common Collector Amplifier

The final canonic cell of linear BJT amplifiers is the *common collector amplifier*. The ac schematic diagram of this stage, which is often referred to as an *emitter follower*, is given in Fig. 26.18(a). In emitter followers, the input signal is applied to the base, and the resultant small signal output voltage is extracted at the transistor emitter.

The small signal equivalent circuit corresponding to the amplifier in Fig. 26.18(a) is shown in Fig. 26.18(b). A straightforward circuit analysis gives, for the case of large R_o, a low frequency voltage gain, $A_{vCC}(0) = V_{OS}/V_{ST}$, of

$$A_{vCC}(0) \cong \frac{R_{LT}}{R_{LT} + R_{outCC}} \tag{26.43}$$

where R_{outCC} is the low frequency value of the driving point output impedance,

$$R_{outCC} = Z_{outCC}(0) \cong R_e + \frac{R_b + R_\pi + R_{ST}}{\beta_{ac} + 1} \tag{26.44}$$

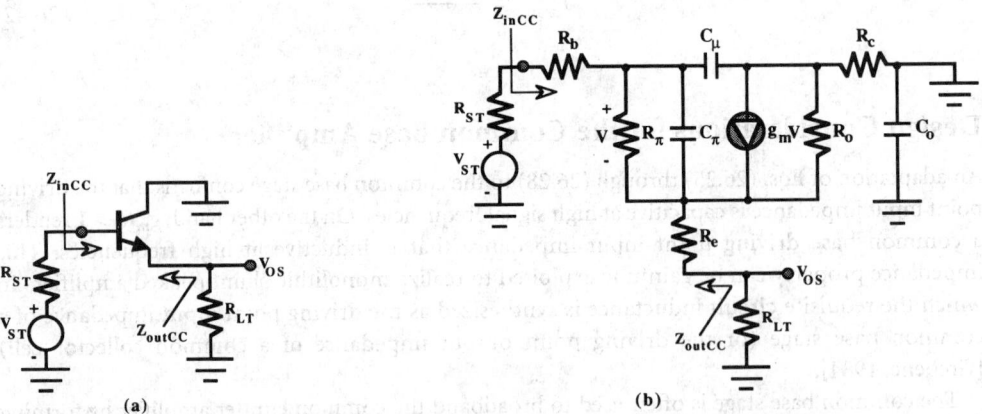

FIGURE 26.18   (a) AC schematic diagram of a common collector (emitter follower) amplifier. (b) Small signal, high frequency equivalent circuit of amplifier in part (a).

The low frequency driving point output resistance is

$$R_{\text{in}CC} = Z_{\text{in}CC}(0) \cong R_b + R_\pi + (\beta_{ac} + 1)(R_e + R_{LT}) \tag{26.45}$$

The facts that the voltage gain is less than one and is without phase inversion, the output resistance is small, and the input resistance is large make the emitter follower an excellent candidate for impedance buffering applications.

As in the cases of the common emitter and the common base amplifiers, the high frequency voltage gain, driving point input resistance, and driving point output resistance can be approximated by functions having analytical forms that are similar to those of Eqs. (26.15), (26.23), and (26.24). Let T_{pCC} and T_{zCC} designate the time constants of the effective dominant pole and the effective dominant zero, respectively, of the emitter follower. Since the output port capacitance, C_o, appears across a short circuit, T_{pCC} is expressible as

$$T_{pCC} = R_{C\pi}C_\pi + R_{C\mu}C_\mu \tag{26.46}$$

With R_o ignored,

$$R_{C\pi} = \cfrac{R_\pi \,\|\, (R_{ST} + R_b + R_{LT} + R_e)}{1 + \cfrac{\beta_{ac}(R_{LT} + R_e)}{R_{ST} + R_b + R_\pi + R_{LT} + R_e}} \tag{26.47}$$

and

$$R_{C\mu} = (R_{ST} + R_b) \,\|\, [R_\pi + (\beta_{ac} + 1)(R_{LT} + R_e) \\ + \left[1 + \cfrac{\beta_{ac}(R_{ST} + R_b)}{R_{ST} + R_b + R_\pi + (\beta_{ac} + 1)(R_{LT} + R_e)}\right] R_c \tag{26.48}$$

The time constant of the effective dominant zero is

$$T_{zCC} = \frac{R_\pi C_\pi}{\beta_{ac} + 1} \tag{26.49}$$

Although the emitter follower possesses excellent wideband response characteristics, it should be noted in Eq. (26.48) that the internal collector resistance, R_c, incurs some Miller multiplication of the base-collector junction capacitance, C_μ. For this reason, monolithic common collector amplifiers work best in broadband impedance buffering applications when they exploit transistors that have collector sinker diffusions and buried collector layers, which collectively serve to minimize the parasitic internal collector resistance.

Defining Terms

ac schematic diagram: A circuit schematic diagram, divorced of biasing subcircuits, that depicts only the dynamic signal flow paths of an electronic circuit.

Driving point impedance: The effective impedance presented at a port of a circuit under the condition that all other circuit ports are terminated in the resistances actually used in the design realization.

Hybrid-pi model: A two-pole linear circuit used to model the small signal responses of bipolar circuits and circuits fabricated in other device technologies.

Miller effect: The deterioration of the effective input impedance caused by the presence of feedback from the output port to the input port of a phase-inverting voltage amplifier.

Short circuit gain-bandwidth product: A measure of the frequency response capability of an electronic circuit. When applied to bipolar circuits, it is nominally the signal frequency at which the magnitude of the current gain degrades to one.

Three-decibel bandwidth: A measure of the frequency response capability of low-pass and bandpass electronic circuits. It is the range of signal frequencies over which the maximum gain of the circuit is constant to within a factor of the square root of two.

References

J. Choma, "A generalized bandwidth estimation theory for feedback amplifiers," *IEEE Transactions on Circuits and Systems*, vol. CAS-31, Oct. 1984.

J. Choma and S. Witherspoon, "Computationally efficient estimation of frequency response and driving point impedances in wide-band analog amplifiers," *IEEE Transactions on Circuits and Systems*, vol. CAS-37, June 1990.

K. K. Clarke and D. T. Hess, *Communication Circuits: Analysis and Design*, Reading, Mass.: Addison-Wesley, 1978.

H. C. de Graaf, "Two New Methods for Determining the Collector Series Resistance in Bipolar Transistors With Lightly Doped Collectors," Phillips Research Report, 24, 1969.

A. B. Glaser and G. E. Subak-Sharpe, *Integrated Circuit Engineering: Design, Fabrication, and Applications*, Reading, Mass.: Addison-Wesley, 1977.

A. B. Grebene, *Bipolar and MOS Analog Integrated Circuit Design*, New York: Wiley Interscience, 1984.

H. K. Gummel and H. C. Poon, "An integral charge-control model of bipolar transistors," *Bell System Technical Journal*, 49, May–June 1970.

S. B. Haley, "The general eigenproblem: pole-zero computation," *Proc. IEEE*, 76, Feb. 1988.

A. S. Sedra and K. C. Smith, *Microelectronic Circuits*, 2nd ed., New York: Holt, Rinehart and Winston, 1987.

K. Singhal and J. Vlach, "Symbolic analysis of analog and digital circuits," *IEEE Transactions on Circuits and Systems*, vol. CAS-24, Nov. 1977.

Further Information

The *IEEE Journal of Solid-State Circuits* publishes state-of-the-art articles on all aspects of integrated electronic circuit design. The December issue of this journal focuses on analog electronics.

The *IEEE Transactions on Circuits and Systems* also publishes circuit design articles. Unlike the *IEEE Journal of Solid-State Circuits*, this journal addresses passive and active, discrete component circuits, as well as integrated circuits and systems, and it features theoretic research that underpins circuit design strategies.

The *Journal of Analog Integrated Circuits and Signal Processing* publishes design-oriented papers with emphasis on design methodologies and design results.

Nobel Prize winners John Bardeen, William Shockley, and Walter H. Brattain (left to right), shown at Bell Telephone Laboratories in 1948 with apparatus used in the first investigations that led to the invention of the transistor. The trio received the 1956 Nobel Prize in physics for their invention of the transistor, which was announced by Bell Laboratories in 1948. (Courtesy of Bell Telephone Laboratories.)

27

Computer-Aided Circuit Simulation

27.1	Introduction	653
27.2	Hierarchy of Modeling and Simulation	654
27.3	Modeling and Basic Circuit Equations	654
27.4	Equation Formulation	655
	The Tableau Approach • The Modified Nodal Approach	
27.5	Solving Linear Equations	657
	Direct Methods for Full Systems • Solving Sparse Linear Systems	
27.6	Solving Nonlinear Equations	661
27.7	Newton-Raphson Applied to Circuit Simulation	663
27.8	Solving Differential Equations	666
	Truncation Error for Linear Multistep Methods • Stability of Linear Multistep Methods • Stiff Systems • Time Step Control for Multistep Methods	
27.9	Techniques for Large Circuits	670
27.10	Specialized Techniques for MOS Circuits	671

Michael Lightner
University of Colorado

27.1 Introduction

The simulation of circuits at the electrical level has been the primary **CAD** activity of many designers. Electrical level simulation is used in both digital and analog system design. Its main use is in determining the timing behavior of circuits and in the study of the behavior of circuits under various environmental and parameter variations. Electrical level simulation is also one of the most costly CAD activities involved in the design of circuits. Because of the importance and expense of electrical simulation it has been widely studied and a huge literature exists of exact and approximate techniques. In this chapter we will briefly outline the ideas and algorithms of exact electrical level simulation and in the next chapter we will study approximate methods for simulating the same circuits. Specifically, we will only consider the transient solution of nonlinear circuits, as this is the most important area of electrical simulation.

Before we begin the discussion of the details it is useful to see the big picture of how a program such as **SPICE** operates when solving for the transient response of a circuit. First the program must read in the description of the circuit and form the equations that will be solved. Actually, what SPICE does is read in the data associated with the circuit, the elements and values, the transistor models, and the element interconnections. In addition, controls such as the time range over which to solve the equations, the accuracy requirements, initial conditions, etc. are obtained from the user.

Since we assume that the program will be solving nonlinear differential equations there is a set procedure (Table 27.1). These steps are common to solving for the transient response of any circuit.

Table 27.1   Standard Steps in Solving Circuit Equations

Find the initial operating point (dc solution).
Choose a first time step (use the default).
  Discretize the derivatives (C's and L's) forming approximate models.
    Linearize the nonlinear elements forming approximate models.
      Form the *linear* equations associated with the approximate circuit.
      Solve the linear equations.
    Iterate until have solution or reduce the time step and try again.
  Check for accuracy.
If accurate continue forward in time else reduce time step and try again.

It is these steps that are built into the SPICE program. In fact, any program that produces exact solutions will use a procedure basically similar to this one. It is these steps which we will study in this chapter. Notice that the equations that are written and solved are linear equations. Since this forms the heart of the program we will begin with equation formulation and linear equation solution and follow that with the approximations used to deal with nonlinear and differential elements.

27.2 Hierarchy of Modeling and Simulation

Modern integrated circuits are exceptionally large and complex. It is not possible to perform circuit simulation on the entire chip; indeed, besides taking much too much time and memory the data generated would overload any engineer's ability to assimilate information. In order to deal with the complexity of modern designs a hierarchy of models and associated simulators is used.

At the highest level, a behavioral model of the entire circuit is written, often in a standard programming language, and the large-scale input/output behavior of the system is studied. This level of modeling is done for both analog and digital circuits. Depending on the type of circuit, i.e., analog or digital, other more complex levels of modeling and simulation are used in the design process. At the lowest level typically dealt with by designers we find interconnections of individual transistors.

The transistor level is what concerns us in this chapter. We will assume that the higher levels of modeling and simulation have already been done. In addition, we will assume that the integrated circuit process which will be used in building the chip is well characterized and that models for the individual transistors and the interconnect are available. Thus the task that we face is the simulation of a collection of transistors in order to study the details of the voltage and current waveforms and their timing.

Typically, for a digital design, transistor level simulation has two major uses: first, the characterization of cell libraries for use in higher levels of modeling and simulation and, second, the study of critical portions of the circuit to determine key timing-related performance parameters. For these two applications the number of transistors being simulated at one time varies from 10 to 1000, with a typical number being about 500. It is this range of circuit size and these applications that we will be concerned with in this chapter.

27.3 Modeling and Basic Circuit Equations

It is clear that no matter how accurate and efficient a simulation program is, the most important factor is the models that are used to represent the circuit. These include models for transistors, interconnect, passive component and process and temperature dependencies. We will concentrate on the solution of network equations given the basic models. For detailed discussion of modeling the reader is referred to Chapters 1, 3, 5, and 8 in this handbook.

27.4 Equation Formulation

There are two common methods that are used to write circuit equations for simulators, the tableau approach [Hachtel *et al.*, 1971] and the modified nodal method [Ho *et al.*, 1975]. Both methods will be discussed briefly in this section.

The Tableau Approach

The simplest way to write the equations associated with a circuit is to use the tableau approach. In this technique the Kirchhoff voltage and current laws (**KVL** and **KCL**) and the individual branch/element equations are written in one large matrix. Although there are several ways of writing the KVL and KCL equations we will concentrate on one.

To begin we will assume that we are using the **associated reference direction** to align current and voltage for each element. This reference scheme is shown in Fig. 27.1.

We will further assume that each element has an orientation of + and − associated with it. In SPICE the first node is the + node and the second the −, for a two-terminal element. Finally, we assume that each node in the network has a unique identifier (we will use numbers) and that the ground node has been identified and has reference number 0. For convenience, let

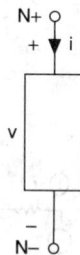

FIGURE 27.1 Associated reference directions.

n be the number of nodes including the ground node and b be the number of branches (assuming two terminal branches). With these assumptions we can now form the node-branch incidence matrix A. This matrix, A, is $(n-1) \times b$ with entries:

$$A_{ij} = 1 \text{ if the } + \text{ of element } j \text{ is connected to node } i$$
$$= -1 \text{ if the } - \text{ of element } j \text{ is connected to node } i$$
$$= 0 \text{ otherwise}$$

Now if we let i_b be the vector of branch currents, then KCL can be expressed as:

$$A\, i_b = 0$$

Further, if v_b is the vector of branch voltages and v_n is the vector of node voltages, not including the reference node, then KVL can be expressed:

$$v_b = A^T v_n$$

Finally, we need to express the **branch relationships**. For complete generality in dealing with two-terminal elements, including independent and dependent sources, we need a form of the branch equations such as:

$$R\, i_b + G\, v_b = s$$

Where R is a matrix of resistances, G is a matrix of conductances, and s is a vector of source values. For example, if the kth element was a resistor, R_2, then $R_{kk} = R_2$, $G_{kk} = -1$, and $s_k = 0$. Further, if the nth component was a voltage source of value V_n, then $R_{nn} = 0$, $G_{nn} = 1$, and $s_n = V_n$. If we arrange these equations in one large matrix, we have the **tableau formulation** which has the form:

$$\begin{bmatrix} A & 0 & 0 \\ 0 & -I & A^T \\ R & G & 0 \end{bmatrix} \begin{bmatrix} i_b \\ v_b \\ v_n \end{bmatrix} = \begin{bmatrix} 0 \\ 0 \\ s \end{bmatrix}$$

A simple example is given in Fig. 27.2.

For Fig. 27.2 the entries to the tableau are:

$$A = \begin{bmatrix} 1 & 1 & 0 & 0 \\ 0 & -1 & 1 & 1 \end{bmatrix}$$

$$R = \begin{bmatrix} 0 & & & \\ & R_1 & & \\ & & -1 & \\ & & & R_2 \end{bmatrix} \quad G = \begin{bmatrix} 1 & & & \\ & -1 & & \\ & & G_1 & \\ & & & -1 \end{bmatrix} \quad s = \begin{bmatrix} V \\ 0 \\ 0 \\ 0 \end{bmatrix}$$

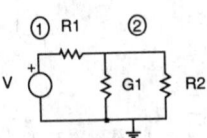

FIGURE 27.2 Example circuit for tableau equations.

This method of formulating equations is quite easy to implement. The one drawback is that the set of equations tends to be quite large. For example, if a small chip had 1000 nodes, not including ground, and there were 2500 branches, and an average of 2.5 branches per node, then the tableau equations would be a system of equations 6000 by 6000. In addition, most of the entries would be zero; thus sparse matrix techniques would be required to solve the equations. However, note that any element can be modeled using this approach and there are no limitations as there are with the standard nodal and loop analysis methods.

The Modified Nodal Approach

Another method of writing circuit equations which is less demanding of space and almost as flexible is called the **modified nodal approach.** In this approach, the typical nodal equations are written. However, for elements that are a function of current, e.g., inductors and certain controlled sources, extra variables and equations are added. This approach allows a broad range of elements to be modeled and yet keeps the size of the matrix relatively small. In order to describe, in a simple way, the modified nodal method we will consider networks with two terminal elements only (this is not a restriction of the method, but simply a convenience). We will label the elements as either being current controlled or voltage controlled. For example, a resistor, inductor, and independent voltage source will be considered current controlled, while conductors, capacitors, and independent current sources will be considered voltage controlled. Specifically, if the branch relation is written

$$i = f(v)$$

the element is considered voltage controlled. If the branch relation is written

$$v = g(i)$$

the element is considered current controlled. Note that the standard nodal equation approach only deals with voltage controlled elements. Consider S_I to be the set of current controlled elements and S_V to be the set of voltage controlled elements. Now we label each current controlled element, $l \in S_I$, with a new unknown, I_l, the current through the element and we write the KCL at node k in the following way:

$$\sum_{j \in S_V, \, j \text{ connected to node } k} f_j(v) \quad + \quad \sum_{j \in S_I, \, j \text{ connected to node } k} I_j \quad =$$

$$- \sum_{j \in S_V, \text{ current source currents connected to } k} I_j$$

The summations are all algebraic sums taking the reference directions of the current into account with current flow leaving the node considered positive. This is, of course, the normal way to write a nodal equation except that the unknown currents through current controlled elements have been added as unknowns. This generates a set of equations with more unknowns than equations. In order to provide the necessary additional equations we add the branch relations for each element in S_I to the previous set of equations. There are methods for easily generating these sets of equations. For our purposes a simple example will illustrate the important principles.

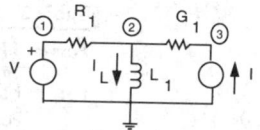

FIGURE 27.3 Example circuit for modified nodal equations.

For the circuit in Fig. 27.3 the modified nodal equations are given by

$$
\begin{bmatrix}
0 & 0 & 0 & 1 & 1 & 0 \\
0 & G_1 & -G_1 & 0 & -1 & 1 \\
0 & -G_1 & G_1 & 0 & 0 & 0 \\
1 & 0 & 0 & 0 & 0 & 0 \\
1 & -1 & 0 & 0 & -R_1 & 0 \\
0 & 1 & 0 & 0 & 0 & -L_1 \dfrac{d}{dt}
\end{bmatrix}
\begin{bmatrix}
V_1 \\ V_2 \\ V_3 \\ I_V \\ I_R \\ I_L
\end{bmatrix}
=
\begin{bmatrix}
0 \\ 0 \\ I \\ V \\ 0 \\ 0
\end{bmatrix}
$$

27.5 Solving Linear Equations

The general form of the equations we generated in the previous section is

$$Ax = b$$

This is a set of linear algebraic equations and for our applications the system is completely specified with a unique solution. The goal of our discussion in this section is to explore some of the methods for solving this type of system of equations.

In general, for a system of linear equations we distinguish between direct and iterative methods of solution. Direct methods, the ones we shall examine in this section, are concerned with the calculation of the exact solution of the equation in a single computational step. In contrast, the iterative techniques, sometimes called relaxation techniques, use a sequence of steps to solve the equations and have to specify the conditions under which the computational sequence will converge to the correct answer (or, indeed, converge at all).

We will not be concerned to any great extent with accuracy, however, the reader should know that there is a tremendous amount of literature on all facets of these methods.

Direct Methods for Full Systems

We will now consider the classical **Gaussian elimination** scheme for solving

$$Ax = b$$

We will assume that the matrix is relatively full and that whenever we choose a matrix location to use as a divisor that the entry in that location will be nonzero. Methods to deal with relaxing these assumptions are discussed later and in detail in the references.

For an *n* by *n* system of equations the standard Gaussian elimination procedure is given in Table 27.2.

Table 27.2 Gaussian Elimination

Forward Elimination	Backward Substitution
For $i = 1$ to $n - 1$	For $i = n$ to 1
For $j = i + 1$ to n	
For $k = i$ to n	$x_i = b_i - \displaystyle\sum_{j=i+1}^{n} a_{ij} x_j$
	where
$a_{jk} = a_{jk} - \dfrac{a_{ji}}{a_{ii}} a_{ik}$	$\displaystyle\sum_{k=n+1}^{n} = 0$
End for	End For
$b_j = b_j - \dfrac{b_i}{a_{ii}} a_{ji}$	
End For	
End For	

A simple example of this procedure is given below. Forward elimination produces the sequence of matrices given below.

$$\begin{bmatrix} 2 & 2 & 1 \\ 3 & 1 & 0 \\ 0 & 1 & 2 \end{bmatrix} \begin{bmatrix} x_1 \\ x_2 \\ x_3 \end{bmatrix} = \begin{bmatrix} 1 \\ 2 \\ 1 \end{bmatrix}$$

$$\begin{bmatrix} 2 & 2 & 1 \\ 0 & -2 & -3/2 \\ 0 & 1 & 2 \end{bmatrix} \begin{bmatrix} x_1 \\ x_2 \\ x_3 \end{bmatrix} = \begin{bmatrix} 1 \\ 1/2 \\ 1 \end{bmatrix}$$

$$\begin{bmatrix} 2 & 2 & 1 \\ 0 & -2 & -3/2 \\ 0 & 0 & 5/4 \end{bmatrix} \begin{bmatrix} x_1 \\ x_2 \\ x_3 \end{bmatrix} = \begin{bmatrix} 1 \\ 1/2 \\ 5/4 \end{bmatrix}$$

Backward substitution yields:

$$x_3 = 1$$
$$x_2 = (\tfrac{1}{2} + \tfrac{3}{2} \cdot 1)/2 = -1$$
$$x_1 = (1 - 1 - 2(-1)/2 = 1$$

This is the simple procedure which you may have seen before. The element which is used as the divisor in the forward elimination is called the *pivot element*. The procedure obviously depends on the pivot elements being nonzero. Furthermore, there is unnecessary work that is done. In particular, there is no need to zero out the column underneath the pivot element. Modifications based upon these observations are common. We note that the complexity of Gaussian elimination is $O(n^3)$ and that this complexity comes from the three nested loops in the forward elimination portion of the algorithm.

In solving linear systems it is often the case that the same equation will be solved with several right-hand sides. In this case, there is a procedure that can result in considerable savings. This procedure is known as *LU factorization* and also incorporates the savings from not zeroing out the column below the pivot element.

In LU factorization, the A matrix is decomposed into the product of a lower triangular matrix, L, and an upper triangular matrix, U. Thus becomes

$$Ax = b$$

$$LUx = b$$

With this decomposition we can now perform a forward substitution and a backward substitution to solve the system of equations. Specifically, we solve

$$Ly = b$$

using a forward substitution and

$$Ux = y$$

using a backward substitution. Both of these systems are triangular and are easy to solve requiring $O(n^2)$ operations each. Furthermore, if the right-hand-side vector is changed it is a simple task to perform the substitutions to find the new answer.

Thus the question becomes: How do we decompose the A matrix into LU? Although we will not analyze this carefully it will turn out that the steps required for the decomposition are the same as those for the forward elimination portion of Gaussian elimination except that we will not zero out the elements below the diagonal. In fact, the procedure given in Table 27.3 will write the LU matrices on top of the A matrix. Also, because the substitution process is straightforward we will only give the procedure for generating the LU decomposition.

For illustration purposes we present a matrix A and its LU factors. If

Table 27.3 LU Decomposition

For $i = 1$ to n
    For $j = i$ to n
    $$u_{ij} = a_{ij} - \sum_{p=1}^{i-1} l_{ip} u_{pj}$$
    $$l_{ji} = a_{ji} - \sum_{p=1}^{j-1} l_{jp} u_{pi} / u_{ii}$$
    End For
End For
where
$$\sum_{p=1}^{0} = 0$$

$$A = \begin{bmatrix} 3 & 5 & 1 \\ 1 & 4 & 3 \\ 2 & 2 & 3 \end{bmatrix}$$

then

$$L = \begin{bmatrix} 3 & 0 & 0 \\ 1/3 & 1 & 0 \\ 2/3 & -4/7 & 1 \end{bmatrix}, \quad U = \begin{bmatrix} 3 & 0 & 1 \\ 0 & 7/3 & 5/3 \\ 0 & 0 & 69/21 \end{bmatrix}$$

Pivoting for Accuracy

One subject that we have ignored in our discussion of Gaussian elimination and LU factorization is what to do when the element on the diagonal is zero or very small. In that case we cannot simply divide by the diagonal element without incurring significant numerical error. The choice of which element to use as the divisor in each row is known as *pivot selection* and the element chosen is known as a *pivot element*.

Significant research has gone into the subject of pivot selection, which we will not discuss in this chapter. However, we do wish to point out that the discussion of **pivoting** in most references will leave a suggestion that after the pivot element is selected there will be a row and column permutation of the matrix to bring that element to the diagonal. In point of fact this is not necessary. In an efficient implementation there would be no data movement, merely a keeping track of the proper order in which to perform the operations. This extra bookkeeping is much cheaper than repeated data motion.

Finally, note that moving columns causes the order of the variables to change and moving rows causes the right-hand side to move. This information should be part of the bookkeeping involved in the solution process.

Solving Sparse Linear Systems

As pointed out earlier, most of the entries in the matrices associated with circuit equations are zero. That is, the matrices are very **sparse**. There are two issues of importance in dealing with sparse matrices: the storage scheme and the pivot strategy.

If proper care is taken in solving sparse systems, the computational cost can, in practice, be reduced from $O(n^3)$ to $O(n^{1.3})$. This is a significant savings and well worth the programming investment when we are solving the systems repeatedly as is the case in circuit simulation.

Storage Scheme for Sparse Matrices

If the matrix is very sparse, it is wasteful of space to store using a n^2 array. In fact, for circuit equations there tends to be a small constant number of nonzeros per row independent of the size of the system. Typically, this is between 3 and 10. Thus in a medium-size circuit having 1000 nodes there will be about 1% nonzeros.

Many storage schemes have been proposed for sparse matrices. The general trade-off is space versus ease of operation. Remember that we still have to search the rows and columns of the matrix during the elimination process. A bidirectional doubly linked list is a very easy data structure to traverse and modify and is a common one for use in storing sparse matrices. Note that this data structure has a relatively high degree of overhead and thus requires a very sparse system to make viable.

Figure 27.4 illustrates the bidirectional doubly linked list structure used to store a sparse matrix.

It is quite possible in the course of the Gaussian elimination process for an entry in the matrix that was zero to become nonzero. This is known as **fill-in.** The example shown below is an extreme case. For an array with the zero/nonzero structure given by

$$
A = \begin{bmatrix} x & x & x & x \\ x & x & & \\ x & & x & \\ x & & & x \end{bmatrix}
$$

after the first step of Gaussian elimination the zero/nonzero structure becomes

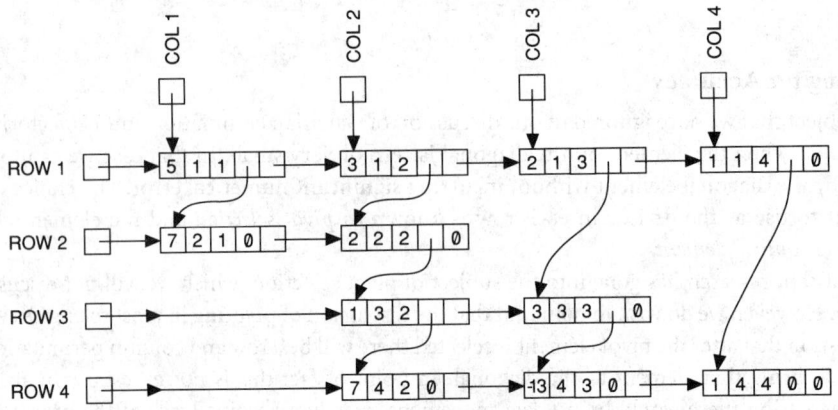

FIGURE 27.4   Bidirectional doubly linked list pivoting for sparsity.

$$\begin{bmatrix} x & x & x & x \\ x & x & x & x \\ x & x & x & x \\ x & x & x & x \end{bmatrix}$$

All of the zeros have become nonzero. In this case there will be $O(n^3)$ operations. This was not necessary. If the matrix had been pivoted to the following form:

$$\begin{bmatrix} x & & & x \\ & x & & x \\ & & x & x \\ x & x & x & x \end{bmatrix}$$

there would have been no fill-in during the elimination process. Clearly there is a requirement that the solution order be chosen to preserve the sparse structure of the equations to the extent possible.

Significant research has been done to find the best pivoting strategy to preserve sparseness. However, of all the methods available, one of the earliest is still robust and used widely. The *Markowitz criteria* is a simple method used to determine the pivot order.

In describing the Markowitz criteria, remember that we are always working on a portion of the matrix, the whole matrix at the first step and reduced portions thereafter. Markowitz looks at each nonzero, a_{ij} in the reduced matrix and counts the number of nonzeros in the associated row, r_i, and column, c_j, and forms the quantity, $(r_i - 1)(c_j - 1)$ (known as the Markowitz count). Then the pivot element is chosen as that nonzero with the smallest Markowitz count (ties are broken arbitrarily).

To illustrate this process we take the arrow head matrix shown above and attach the Markowitz count as superscripts to each of the nonzeros.

$$A = \begin{bmatrix} x^9 & x^3 & x^3 & x^3 \\ x^3 & x^1 & & \\ x^3 & & x^1 & \\ x^3 & & & x^1 \end{bmatrix}$$

Clearly, the (1,1) position would be the last chosen or, in other words, is the worst possible choice. Note that the count has to be recalculated at every step of the elimination process, which emphasizes the need for easy traversal of the data structure.

27.6 Solving Nonlinear Equations

Obviously we must deal with nonlinear elements when simulating integrated circuits. These elements arise from the nonlinearities in transistor device models. For the sake of simplicity we will first consider two terminal nonlinearities.

For a simple nonlinear equation

$$f(x) = 0$$

the usual method of solution is some variant on the **Newton-Raphson** (NR) procedure.

In order to understand the NR procedure we must realize that the entire procedure is based upon a linear approximation to the function at the current point, x_c. Thus, if we expand the function in a Taylor series about the current point we have

$$f(x_+) = f(x_c) + \left.\frac{\partial f}{\partial x}\right|_{x=x_c} (x_+ - x_c)$$

Of course, we are ignoring the error involved in the approximation. Now, if we further assume that the x_+ is the solution, i.e.,

$$f(x_+) = 0$$

then we have

$$x_+ = x_c - f(x_c)/(\partial f/\partial x)|_{x=x_c}$$

which is the basic NR step.

The process of repeated calculation of linear approximations and their solution is illustrated in Fig. 27.5.

The difficulties associated with this scheme are numerous. For example, if the partial derivative (the normal derivative in the single variable case) is zero, then the step fails. Further, we have no guarantee of convergence for the general case. In addition, with the applications to electrical circuits with exponential nonlinearities, numerical accuracy can be a tremendous problem. Significant research has be done to deal with all of these cases, and more. Yet the nonlinear solution problem, especially the initial dc solution, continues to be the least robust portion of the electrical level simulation package.

The question of convergence of the NR algorithm is answered by the following theorem.

Table 27.4   Newton-Raphson

Error = ∞
Accuracy = ε (user supplied)
x_c = current estimate of the solution, initially provided by user
While (Error ≥ Accuracy)
  calculate an update of the estimated solution

$$x_+ = x_c - f(x_c)/(\partial f/\partial x)|_{x=x_c}$$

  calculate the new error

$$Error = |x_+ - x_c|$$

  update the estimate of the solution

$$x_c = x_+$$

End While

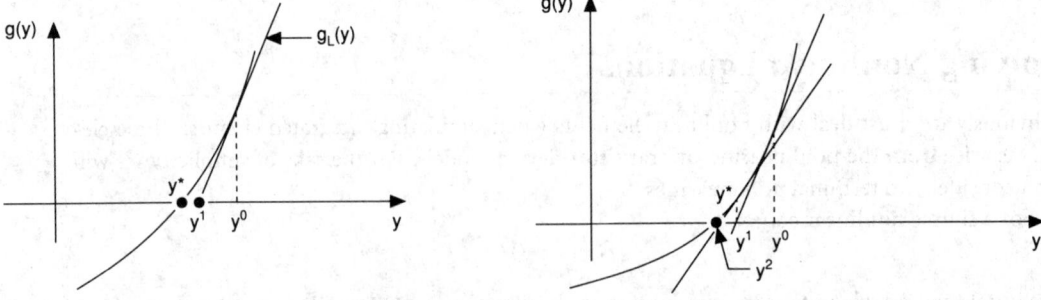

FIGURE 27.5   Newton-Raphson example.

Theorem: If f is twice continuously differentiable, $[\partial f(x)/\partial x]|_{x\,=\,x^*} \neq 0$, i.e., the derivative is nonzero at the solution, x^*, and the initial guess x_0 is close enough to the solution, then the NR method always converges to the solution. Furthermore, letting

$$\varepsilon_k = |x_k - x^*|$$

be the error at the kth iteration, then

$$\varepsilon_{k+1} \leq C\varepsilon_k^2$$

where C is a positive constant, i.e., the NR methods have quadratic convergence.

Notice that there is no mechanism for determining how close is close enough in the previous theorem. This type of convergence is known as *local convergence;* if the iteration converges from *any* staring point, the algorithm is said to possess *global convergence*. In addition, the quadratic convergence property is very important because it shows that if the error is ever less than one that it will be driven to zero very quickly. In fact, once the error is below one, the number of significant decimal places effectively doubles at each iteration.

27.7 Newton-Raphson Applied to Circuit Simulation

Given the general picture of the NR scheme, how is this applied to circuits? The first difficulty is that nonlinear circuit equations are not just a function of a single variable. Further, there is a set of equations and not just a single one. This problem is easily handled by extending the one-dimensional NR scheme to multidimensions using vector calculus. The update formula now becomes

$$x_+ = x_c + \left[\frac{\partial f}{\partial x}\right]^{-1}_{x=x_c} f(x_c)$$

where $\partial f/\partial x$ is known as the *Jacobian* of the set of circuit equations and is the matrix of partial derivatives.

$$\frac{\partial f}{\partial x} = \begin{bmatrix} \dfrac{\partial f_1}{\partial x_1} & \cdots & \dfrac{\partial f_1}{\partial x_n} \\ \vdots & \ddots & \vdots \\ \dfrac{\partial f_n}{\partial x_1} & \cdots & \dfrac{\partial f_n}{\partial x_n} \end{bmatrix}$$

Rewriting this set of equations we have

$$\frac{\partial f}{\partial x}\bigg|_{x=x_c} (x_+ - x_c) = f(x_c)$$

This is precisely in the form

$$Ax = b$$

which we have studied in the previous section. A similar convergence condition applies and the reader is referred to the literature. Most importantly, we note that local convergence and a quadratic rate of convergence apply in the multidimensional case as well as the single-dimensional case.

The final difficulty we face is how to generate the equations in the first place and then how to take the partial derivative of this large set of equations. What can be shown without too much difficulty,

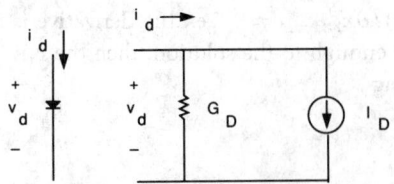

FIGURE 27.6 Diode linearized model.

The linearized form of this becomes

although we will skip the demonstration, is that the individual elements can be linearized and the equations written for this linear network and we still have the same set of equations required for NR.

To illustrate this we will consider the network in Fig. 27.6 with two diodes. We will assume a simple form for the diode equation

$$i = f(v)$$

$$i_+ = i_c + \left.\frac{\partial f}{\partial v}\right|_{v=v_c} (v_+ - v_c)$$

If we rewrite this grouping the known and unknown quantities we have

$$i_+ = \left[i_c - \left.\frac{\partial f}{\partial v}\right|_{v=v_c} v_c \right] + \left.\frac{\partial f}{\partial v}\right|_{v=v_c} v_+$$

This can be interpreted as an electrical model of the linearized diode that has a current source of value

$$I_D = \left[i_c - \left.\frac{\partial f}{\partial v}\right|_{v=v_c} v_c \right]$$

in parallel with a conductance of value

$$G_D = \left.\frac{\partial f}{\partial v}\right|_{v=v_c}$$

This model is illustrated in Figure 27.7.

With this model the original circuit can be transformed into a linearized model as shown in Fig. 27.8.

Now we can use the normal equation formulation techniques to generate the set of linear circuit equations that represent the Jacobian of the nonlinear network equations. This is the process used in most circuit simulation programs.

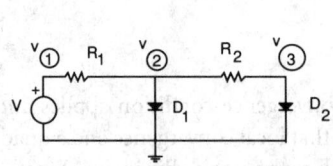

FIGURE 27.7 Nonlinear network used
for NR example.

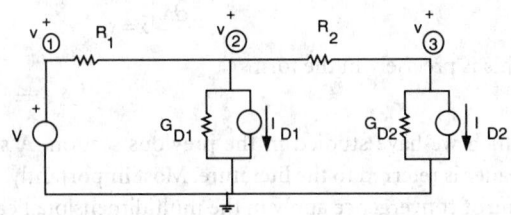

FIGURE 27.8 Linearized network.

Accuracy Control in Circuit Simulation with NR

Two major areas of concern in using NR for solving circuit equations are

- Numerical overflow during the iterations
- Nonconvergence of the iterations

The first problem is primarily due to the nature of the nonlinearities in the circuit itself. This is especially a problem with the exponential nonlinearity associated with pn junction models. For example, in the circuit shown in Fig. 27.9 if the initial guess is v_0, then the result of the first iteration is v_1 and to continue the iteration we have to evaluate the diode current for a voltage of 10 V; this requires evaluating e^{400} (these iterations are shown in Fig. 27.10). Clearly overflow can be a problem.

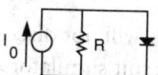

FIGURE 27.9 Circuit example with numerical overflow problems.

Two classes of methods have been proposed to alleviate this problem. Both limit the step in voltage that is actually taken and are known as *limiting methods*. The first group of methods which we will not discuss in detail require some arbitrary limitation of the NR step. Although some of these can work well, they are quite heuristic.

The second class of limiting methods which are used more often in solving circuit equations involve the changing of the variable of iteration. We have been writing the diode equation as

$$i = f(v)$$

but we could have just as well written

$$v = g(i)$$

As we know, the function g is a logarithmic function and so is much less prone to numerical difficulties than the exponential function f. The decision on which form to represent the function can be made during the iterations and can change from step to step. The difference between these two steps is illustrated in Fig. 27.10.

The second class of difficulties, nonconvergence, can have several causes:

- The equations may not have a solution due to error in their formulation.
- The first derivatives of the elements' branch relationships lead to a singular Jacobian.
- The initial guess is not sufficiently close to the solution.
- The branch relationships may not have continuous derivatives.

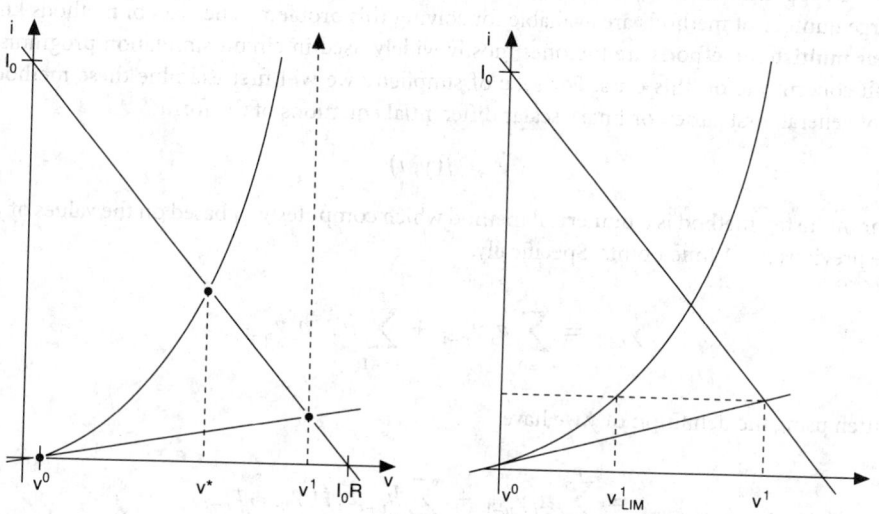

FIGURE 27.10 Alternating V and I as unknowns in NR.

Difficulties with model equations can only be solved with a refinement of the equations themselves and so this is out of our range of discussion.

There are two broad classes of methods to attempt to force convergence of the NR equations:

- Optimization-based methods
- Continuation methods

We will not discuss the first class of methods. The second class of methods is best characterized in circuit simulators as *source stepping methods*. Source stepping methods are generally used to help find the initial dc solution of the circuit. First we note that if all the dc sources are set to zero, then we know the solution to the network—zero. So if we increase the source values slightly from zero to their final value, we have a very good guess at the solution—zero. This process is continued until the sources reach their final value.

There is much known on how to improve the numerical behavior of NR in circuit simulation. Unfortunately, it is mostly known to practitioners and is not in the public domain. This is primarily because good nonlinear solution techniques (especially for the dc solution) are valuable and are not lightly shared from one company to the next.

27.8 Solving Differential Equations

The last task we face in constructing and understanding electrical level simulation is how to deal with elements with memory such as capacitors and inductors. For these problems we could write the tableau equations in the following general form

$$F(\dot{x}, x, T) = 0 \quad x(0) = X_0 \text{ for } 0 \le t \le T$$

This is known as an initial value problem and has no closed form solution in general.

In attempting to solve this problem we will find approximate solutions at a finite set of points over the time interval of interest

$$t_0 = 0, \ t_N = T, \ t_{n+1} = t_n + h_{n+1}, \ n = 0, 1, \dots, N$$

The h_{n+1} are called *time steps* and their values are called **step sizes.** At each point, t_n, we are going to compute an approximation, x_n, to the exact solution, $x(t_n)$.

A large number of methods are available for solving this problem. The class of methods known as **linear multistep methods** are the ones mostly widely used in circuit simulation programs and we shall concentrate on this class. For sake of simplicity we will first examine these methods in terms of general, first-order, ordinary scalar differential equations of the form

$$\dot{y} = f(y, t)$$

A linear multistep method is a numerical method which computes y_{n+1} based on the values of y and $\dot{y}$ at the previous $p + 1$ time points. Specifically,

$$y_{n+1} = \sum_{i=0}^{p} a_i y_{n-i} + \sum_{i=-1}^{p} h_{n-1} b_i \dot{y}_{n-1}$$

or written using the definition of $\dot{y}$ we have

$$y_{n+1} = \sum_{i=0}^{p} a_i y_{n-1} + \sum_{i=-1}^{p} h_{n-i} b_i f(y_{n-i}, t_{n-i})$$

A simple example of a multistep method is the forward Euler method (FE) which is given by

$$y_{n+1} = y_n + h_{n+1}\dot{y}_n$$

where $p = 0$, $a_0 = 1$, $b_0 = 1$ and all other coefficients are zero. This method can also be considered as a Taylor series expansion about t_n truncated at its first term.

Other well-known multistep methods with $p = 0$, also called single-step methods, are

$$\text{Backward Euler (BE)} \quad y_{n+1} = y_n + h_{n+1}\dot{y}_{n+1}$$

and

$$\text{Trapezoidal (TR)} \quad y_{n+1} = y_n = \left(\frac{h_{n+1}}{2}\right)(\dot{y}_{n+1} + \dot{y}_n)$$

Each of these three methods has a simple interpretation. First, note that the solution we are calculating is really the integral of $\dot{y}$. So that if we plot $f(y, t)$ and then look at estimating the area under the curve between t_n and t_{n+1} we generate a simple graphical interpretation of the methods given in Fig. 27.11.

Truncation Error for Linear Multistep Methods

A major concern in solving systems of differential equations is the accuracy of the computed solution. In particular, what is the difference between the computed solution at t_n, y_n, and the exact solution, $y(t_n)$? This particular type of error is known as the **truncation error.**

There are two types of truncation error, local and global. Local truncation error (LTE) of a

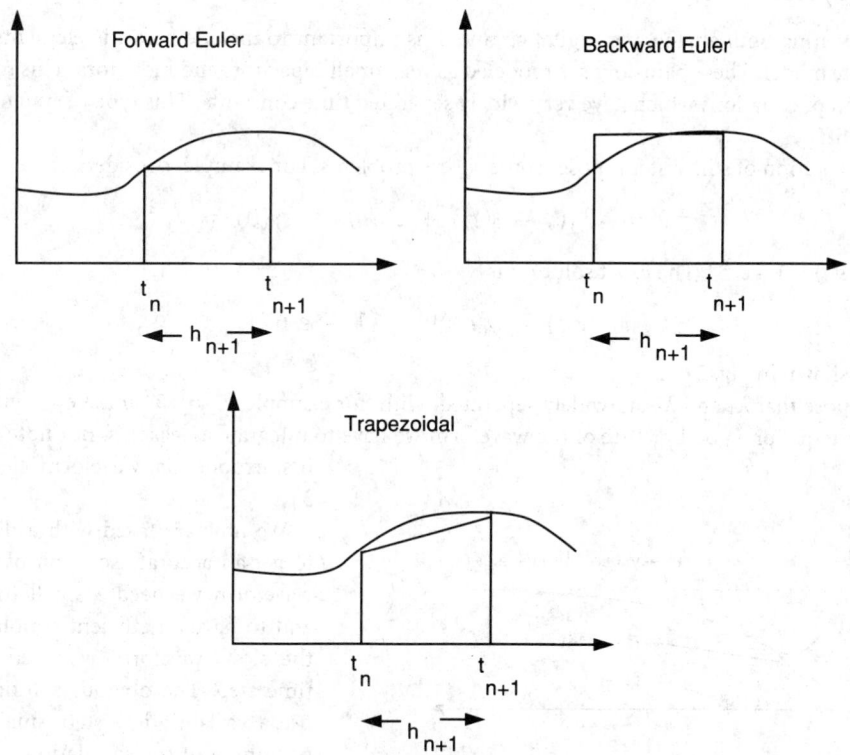

FIGURE 27.11   Geometrical interpretation of integration methods.

method at t_{n+1} is the difference between the computed value y_{n+1} and the exact value, $y(t_{n+1})$ assuming that *no previous error has been made*, that is, that

$$y(t_i) = y_i, \qquad i = 0, 1, 2, \ldots, n$$

Stability of Linear Multistep Methods

In order to study the global behavior of the multistep methods we must examine how any errors made accumulate. This allows us to gain an understanding of the global truncation error. This type of study addresses the **stability** of the method.

To illustrate that there might be a problem consider the differential equation

$$\dot{y} = -y, \qquad y(0) = 1$$

If we apply FE to this equation with step size $h = 1$, we generate the sequence

$$0, \ 0, \ 0, \ldots$$

Although not a very accurate answer for the beginning, at least it does not blow-up and, indeed, the difference between this sequence and the correct answer goes to zero as time increases. However, if we apply FE with a step size of $h = 3$, we generate the sequence

$$-2, \ 4, \ -8, \ 16, \ -32, \ldots$$

which shows a growing oscillation. So whatever error is being made in the calculation accumulates in a disastrous manner.

Applying BE to the same equation with $h = 3$ yields a well-behaved sequence.

Stiff Systems

In simulating both analog and digital circuits it is important to include parasitic elements for an accurate model. These parasitics are modeled as very small capacitors and inductors. This results in a system of equations which have very widely separated time constants. This type of system is said to be **stiff.**

The solution of stiff systems poses some severe problems. For example, consider

$$\dot{y} = -\lambda_1(y - s(t)) + ds/dt, \qquad y(0) = y_0$$

where $s(t) = 1 - e^{-\lambda_2 t}$. The exact solution is

$$y(t) = y_0 e^{-\lambda_1 t} + (1 - e^{-\lambda_2 t})$$

and is shown in Fig. 27.12.

Suppose that λ_1 and λ_2 are widely separated, with, for example, $\lambda_1 = 10^6$ and $\lambda_2 = 1$ yielding a stiff system. For a good picture of the waveform we have to integrate at least 5 s. But note that the first exponential waveform dies out in 5 µs.

We now are faced with a dilemma; to get an accurate solution of the fast waveform we need a small time step and to have an efficient simulation of the slow waveform we need a large time step. The obvious solution is to use a variable time step, small at the beginning of the simulation and large at the end.

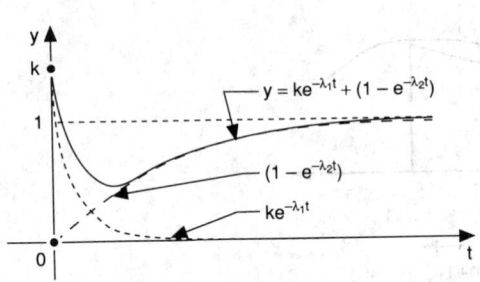

FIGURE 27.12   Exact solution to stiff system.

However, varying the time step changes the analysis of the regions of stability of the linear multistep methods (recall that we assumed a constant step size in our analysis). If we apply the FE method to the previous example with an initial step size of 10^{-6} and after five steps we change to a step size of 10^{-4}, the behavior shown in Fig. 27.13 is obtained.

It turns out that the region of absolute stability for FE can be found by looking at the fastest time constant. In Fig. 27.13 y_{n+1} computed by FE is based upon the first derivative of the solution *passing through* y_n. If y_n is not very close to the exact solution, then FE picks up a spurious fast transient which causes instability if h is not chosen small enough.

Thus we see that

- The analysis of stiff systems requires variable time step.
- Not all linear multistep methods can be efficiently used to integrate stiff equations.

Clearly, to be useful for stiff systems the integration methods have to have a region of absolute stability which allows for a large step size for large time constants without being constrained by the small time constants. Thus A-stable methods are fine (remember that FE is not A-stable).

Time Step Control for Multistep Methods

Now that we know that the time step must be changed for efficiency and we are aware of the importance of different orders of integration schemes, how can we control these in a circuit simulator?

If we let E_{n+1} be a bound on the absolute value of the LTE at t_{n+1}, we know for a multistep method of order k that

$$\left| \text{LTE}_{n+1} \right| = \left| \left[C_{k+1} h_{n+1}^{k+1} / (k+1)! \right] y^{(k+1)}(t_{n+1}) \right|$$

Thus we must have

$$h_{n+1} \leq \left| \frac{(k+1)! \, E_{n+1}}{C_{k+1} y^{(k+1)}(t_{n+1})} \right|^{\frac{1}{k+1}}$$

in order to meet the error requirement. Since the $(k+1)^{st}$ derivative of y is not available we will have to form some approximation of it.

The scheme used in SPICE is the method of *divided differences* (DD). The DD are defined recursively as

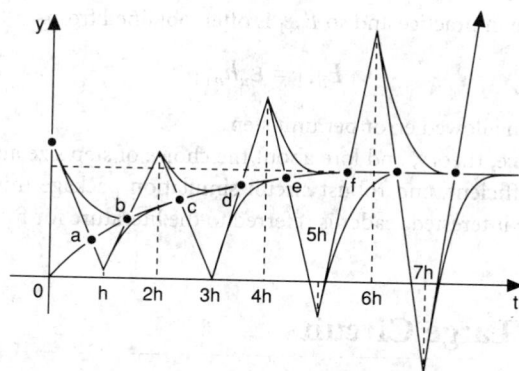

FIGURE 27.13 Numerical instability in FE applied to stiff system.

$$DD_1(t_{n+1}) = \frac{y_{n+1} - y_n}{h_{n+1}}$$

$$DD_2(t_{n+1}) = \frac{DD_1(t_{n+1}) - DD_1(t_n)}{h_{n+1} - h_n}$$

$$\vdots$$

$$DD_{k+1}(t_{n+1}) = \frac{DD_k(t_{n+1}) - DD_k(t_n)}{\displaystyle\sum_{i=-1}^{k-1} h_{n-1}}$$

It is relatively easy to show that

$$y^{(k+1)}(t_{n+1}) \cong (k+1)! DD_{k+1}(t_{n+1})$$

Thus we have an algorithm for time step control.

Given a step h_{n+1}, y_{n+1} is computed according to the method chosen for integration. Then, y_{n+1} and h_{n+1} are used to compute $DD_{k+1}(t_{n+1})$ and h_{n+1} is checked using

$$h_{n+1} \leq \left[\frac{E_{n+1}}{C_{k+1} DD_{k+1}(t_{n+1})} \right]^{\frac{1}{k+1}}$$

If h_{n+1} satisfies the test, it is accepted: otherwise it is rejected and a new h_{n+1} given by the right-hand side is used.

After h_{n+1} is accepted h_{n+2} must be chosen. A common strategy is to use the right-hand side of the error checking equation as the new step size.

An important factor in the selection of the step size is the choice of E_{n+1}. In general, a circuit designer has a rough idea of the precision which is required in the simulation. This is generally related to the GTE and not the LTE. Thus a mechanism to transfer the GTE estimate to the LTE is required.

It is often assumed that the LTE accumulates linearly thus giving the following estimate

$$|GTE_{n+1}| \leq \sum_{i=1}^{n+1} |LTE_i|$$

This is quite conservative in practice and so E_{n+1} is often obtained from

$$E_{n+1} = \varepsilon_u h_{n+1}$$

where ε_u is the maximum allowed error per unit step.

There is a large practice, theory, and lore about the choice of step size and integration method. To construct a useful, efficient, and robust circuit simulation package this information must be tapped and utilized. The interested reader is referred to the literature for further information.

27.9 Techniques for Large Circuits

The techniques that we have been discussing are used in most standard circuit simulators, e.g., Spice, PSpice, HSpice. Although very common, these techniques have limitations when solving very large circuits. We have indicated that the most common size circuits to simulate with a Spice-type simulator would have less than 1000 transistors and more often less than 500. When very large circuits must be simulated special techniques must be used.

There are three types of specialized techniques that have been used: techniques for partitioning large circuits into smaller ones and combining the solutions, techniques which exploit special characteristics of certain classes of circuits (see the next section), and techniques which utilize parallel and vector high-speed computers.

All of these techniques are beyond the scope of this chapter. However, the reader is referred to the references for further information.

27.10 Specialized Techniques for MOS Circuits

MOS circuits have special characteristics which allow simpler and faster simulation techniques to be developed. In particular, the nearly infinite input impedance of MOS transistors effectively isolates the input of the circuit from the output. This isolation means that the equations describing MOS networks, especially digital networks, have a highly unidirectional character and are not tightly coupled. These characteristics have been exploited in a number of simulators using simpler methods for solving linear and nonlinear equations such as relaxation techniques.

The simplified methods have been taken to an extreme in switch level simulators where the MOS transistor is modeled as a switch with either an ideal or real conductance. Switch level simulators have been very popular in replacing logic simulators for complex custom MOS designs. However, the models of time and the results of timing simulation using switch simulators have not been competitive with circuit simulators.

To pursue specialized MOS circuit simulation techniques see Saleh and Newton [1990]. Further information on switch level simulation can be found in Bryant [1987] and Rao *et al.* [1989].

Defining Terms

Associated reference directions: A method of assigning the current and voltage directions to an electrical element so that a positive current-voltage product always means that the element is absorbing power from the network and a negative product always means that the element is delivering power to the network. This method of assigning directions is used in most circuit simulation programs.

Branch relation: The relationship between voltage and current for electrical components. Common branch relations are Ohm's law and the lumped equations for capacitors and inductors. More complex branch relationships would be transistor models.

CAD: Computer-aided-design for the electronics industry is concerned with producing new algorithms/programs which aid the designer in the complex tasks associated with designing and building an integrated circuit. There are many subfields of electrical CAD: simulation, synthesis, physical design, testing, packaging, and semiconductor process support.

Circuit simulation: Constructing a mathematical model of an electrical circuit and then solving that model to find the behavior of the model. Since the model usually cannot be solved in closed form, the equations are evaluated for a specific set of input conditions to produce a sequence of voltages and currents which approximates the true solution.

Fill-in: When solving a set of sparse linear equations using Gaussian elimination it is possible for a zero location to become nonzero. This new nonzero is termed a fill-in.

Gaussian elimination: The standard direct method for solving a set of linear equations. It is termed direct because it does not involve iterative solutions. Variations of this scheme are used in most circuit simulators.

KVL, KCL: Kirchhoff voltage and current laws provide the basic physical constraints on voltage and current in an electrical circuit.

Linear multistep method: This is a class of techniques for solving ordinary differential equations which is widely used in circuit simulators.

Modified nodal formulation: A modification of the classical nodal formulation which allows any network to be described. The modification consists of adding extra equations and unknowns when an element not normally modeled in classical nodal analysis is encountered.

Newton-Raphson: A numerical method for finding the solution to a set of simultaneous nonlinear equations. Variations of this method are commonly used in circuit simulation programs.

Pivoting: When applying Gaussian elimination to solve a set of simultaneous linear equations the natural solution order is sometimes varied. The process of varying the natural solution order is termed pivoting. Pivoting is used to avoid fill-in and to maintain the accuracy of a solution.

Sparse equations: When a set of linear simultaneous equations has very few nonzeros in any row the system is said to be sparse. Normally for a system to be considered sparse less than 10% of the possible entries should be nonzero. For large integrated circuits less than 1% of the possible entries are nonzero.

SPICE: The most widely used circuit simulation program. Originally developed at the University of California, Berkeley, this program is available from several commercial sources. SPICE is an acronym for Simulation Program with Integrated Circuit Emphasis.

Stability: In the context of circuit simulation the main stability concern is in the solution of differential equations. Specifically, if small errors made by the integration method—such as a linear multistep method—are amplified as the solution progresses, the integration method is exhibiting unstable behavior. Controlling the step size is one of the primary mechanisms for maintaining stability.

Step size: When solving for the transient behavior of an electrical circuit the associated differential equations are solved at specific points in time. The difference between two adjacent solution time points is known as the step size.

Stiff system: When an electrical circuit has widely separated time constants the circuit is said to be stiff. The system of equations associated with the circuit is known as a stiff system and special numerical methods must be used to maintain stability and accuracy when simulating a stiff system.

Tableau formulation: A method for formulating the equations governing the behavior of electrical networks. The tableau method simply groups the KVL, KCL, and branch relationships into one huge set of equations.

Truncation error: When numerically solving the differential equations associated with electrical circuits, approximation techniques are used. The errors associated with the use of these methods are termed truncation error. Controlling the local and global truncation errors is an important part of a circuit simulator's task. Limits on these errors are often given by the user of the program.

References

W. Banzhaf, *Computer-Aided Circuit Analysis Using PSpice*, Englewood Cliffs, N.J.: Prentice-Hall, 1992.

R. E. Bryant, "A survey of switch-level algorithms," *IEEE Design and Test of Computers*, vol. 4, no. 4, pp. 26–40, 1987.

L. O. Chua and P.-M. Lin, *Computer-Aided Analysis of Electronic Circuits: Algorithms and Computational Techniques*, Englewood Cliffs, N.J.: Prentice-Hall, 1975.

S. W. Director, *Circuit Theory: A Computational Approach*, New York: John Wiley & Sons, 1975.

G. F. Forsythe, M. A. Malcom, and C. B. Moler, *Computer Methods for Mathematical Computations*, Englewood Cliffs, N.J.: Prentice-Hall, 1977.

G. D. Hachtel, R. K. Brayton, and F. G. Gustavson, "The sparse tableau approach to network analysis and design," *IEEE Trans. Circuit Theory*, CT-18, vol. 1, pp. 101–113, 1971.

C. H. Ho, A. E. Ruehli, and P. A. Brennan, "The modified nodal approach to network analysis," *IEEE Trans. Circuits and Systems*, CAS-22, vol. 6, pp. 504–509, 1975.

A. Jennings, *Matrix Computation for Engineers and Scientists*, New York: John Wiley & Sons, 1977.

J. D. Lambert, *Computational Methods in Ordinary Differential Equations*, New York: John Wiley & Sons, 1973.

J. M. Ortega and W. C. Rheinboldt, *Iterative Solutions of Nonlinear Equations in Several Variables*, New York: Academic Press, 1970.

V. B. Rao, D. V. Overhauser, T. N. Trick, and I. B. Hajj, *Switch-Level Timing Simulation of MOS VLSI Circuits*, Dordrecht, Netherlands: Kluwer Academic Publishers, 1989.

R. A. Saleh and A. R. Newton, *Mixed-Mode Simulation*, Dordrecht, Netherlands: Kluwer Academic Publishers, 1990.

A. F. Schwarz, *Computer-Aided Design of Microelectronic Circuits and Systems* (two volumes), New York: Academic Press, 1987.

R. Spence and J. P. Burgess, *Circuit Analysis by Computer—From Algorithms to Package*, Englewood Cliffs, N.J.: Prentice-Hall, 1986.

J. Vlach and K. Singhal, *Computer Methods for Circuit Analysis and Design*, New York: Van Nostrand Reinhold, 1983.

Further Information

Simulation of circuits at all levels of abstraction is an active research area. The interested reader is encouraged to examine the *IEEE Transactions on Computer-Aided Design*, the *IEEE Transactions on Circuits and Systems*, and the *IEEE Transactions on Computers*. In addition to journal articles there are a number of conferences held each year at which the latest results on circuit simulation are presented. The proceedings of the following conferences have a wealth of information on circuit simulation developments: *Proceedings of the Design Automation Conference (DAC)*, *Proceedings of the International Conference on Computer-Aided Design (ICCAD)*, *Proceeding of the International Conference on Computer Design (ICCD)*, *Proceedings of the European Design Automation Conference (EDAC)*, and the *Proceedings of Euro-DAC*. Finally, there are a large number of numerical analysis, computer science, and applied mathematics conferences and journals in which results related to circuit simulation are published.

Active Filters

Robert E. Massara
University of Essex

J. W. Steadman
University of Wyoming

B. M. Wilamowski
University of Wyoming

28.1 Synthesis of Low-Pass Forms 674
   Passive and Active Filters • Active Filter Classification and
   Sensitivity • Cascaded Second-Order Sections • Passive Ladder
   Simulation • Active Filters for ICs
28.2 Realization 683
   Transformation from Low-Pass to Other Filter Types • Circuit
   Realizations

28.1 Synthesis of Low-Pass Forms

Robert E. Massara

Passive and Active Filters

There are formal definitions of activity and passivity in electronics, but it is sufficient to observe that passive filters are built from passive components; resistors, capacitors, and inductors are the commonly encountered building blocks although distributed RC components, quartz crystals, and surface acoustic wave devices are used in filters working in the high-megahertz regions. **Active filters** also use resistors and capacitors, but the inductors are replaced by active devices capable of producing power gain. These devices can range from single transistors to integrated circuit (IC) -controlled sources such as the operational amplifier (op amp), and more exotic devices, such as the operational transconductance amplifier (OTA), the generalized impedance converter (GIC), and the frequency-dependent negative resistor (FDNR).

The theory of filter synthesis, whether active or passive, involves the determination of a suitable circuit topology and the computation of the circuit component values within the topology, such that a required network response is obtained. This response is most commonly a voltage transfer function (VTF) specified in the frequency domain. Circuit analysis will allow the performance of a filter to be evaluated, and this can be done by obtaining the VTF, $H(s)$, which is, in general, a rational function of s, the complex frequency variable. The *poles* of a VTF correspond to the roots of its denominator polynomial. It was established early in the history of filter theory that a network capable of yielding complex-conjugate transfer function (TF) pole-pairs is required to achieve high selectivity. A highly selective network is one that gives a rapid transition between passband and stopband regions of the frequency response. Figure 28.1(a) gives an example of a passive low-pass LCR ladder network capable of producing a VTF with the necessary pole pattern.

The network of Fig. 28.1(a) yields a VTF of the form

$$H(s) = \frac{V_{\text{out}}(s)}{V_{\text{in}}(s)} = \frac{1}{a_5 s^5 + a_4 s^4 + a_3 s^3 + a_2 s^2 + a_1 s + a_0} \tag{28.1}$$

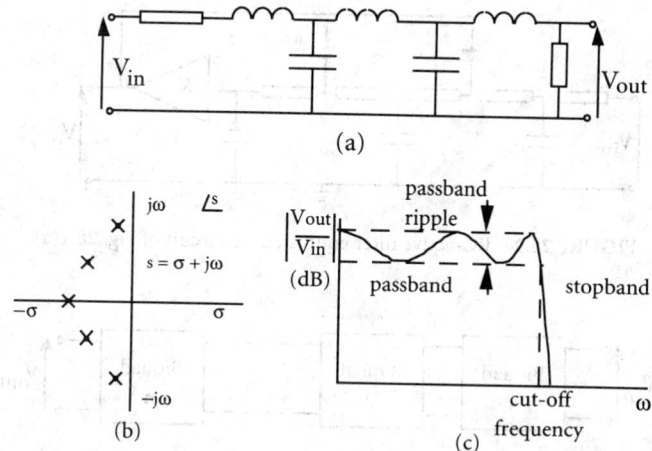

FIGURE 28.1   (a) Passive LCR filter; (b) typical pole plot; (c) typical frequency response.

Figure 28.1(b) shows a typical pole plot for the fifth-order VTF produced by this circuit. Figure 28.1(c) gives a sample sinusoidal steady-state frequency response plot. The frequency response is found by setting $s = j\omega$ in Eq. (28.1) and taking $|H(j\omega)|$. The LCR low-pass ladder structure of Fig. 28.1(a) can be altered to higher or lower order simply by adding or subtracting reactances, preserving the series-inductor/shunt-capacitor pattern. In general terms, the higher the filter order, the greater the selectivity.

This simple circuit structure is associated with a well-established design theory and might appear the perfect solution to the filter synthesis problem. Unfortunately, the problems introduced by the use of the inductor as a circuit component proved a serious difficulty from the outset. Inductors are intrinsically nonideal components, and the lower the frequency range of operation, the greater these problems become. Problems include significant series resistance associated with the physical structure of the inductor as a coil of wire, its ability to couple by electromagnetic induction into fields emanating from external components and sources and from other inductors within the filter, its physical size, and potential mechanical instability. Added to these problems is the fact that the inductor tends not to be an off-the-shelf component but has instead to be fabricated to the required value as a bespoke device. These serious practical difficulties created an early pressure to develop alternative approaches to electrical filtering. After the emergence of the electronic amplifier based on thermionic valves, it was discovered that networks involving resistors, capacitors, and amplifiers—*RC-active filters*—were capable of producing TFs exactly equivalent to those of LCR ladders. Figure 28.2 shows a single-amplifier multiloop ladder structure that can produce a fifth-order response identical to that of the circuit of Fig. 28.1(a).

The early active filters, based as they were on valve amplifiers, did not constitute any significant advance over their passive counterparts. It required the advent of solid-state active devices to make the RC-active filter a viable alternative. Over the subsequent three decades, active filter theory has developed to an advanced state, and this development continues as new IC technologies create opportunities for novel network structures and applications.

Active Filter Classification and Sensitivity

There are two major approaches to the synthesis of RC-active filters. In the first approach, a TF specification is factored into a product of second-order terms. Each of these terms is realized by a separate RC-active subnetwork designed to allow for non-interactive interconnection. The subnet-

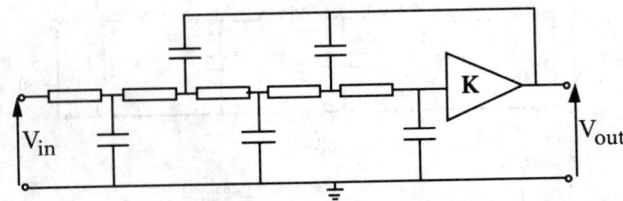

FIGURE 28.2   RC-active filter equivalent to circuit of Fig. 28.1(a).

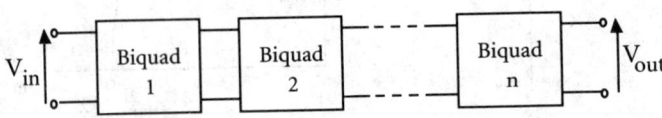

FIGURE 28.3   Biquad cascade realizing high-order filter.

works are then connected in cascade to realize the required overall TF, as shown in Fig. 28.3. A first-order section is also required to realize odd-order TF specifications. These second-order sections may, depending on the exact form of the overall TF specification, be required to realize numerator terms of up to second order. An RC-active network capable of realizing a biquadratic TF (that is, one whose numerator *and* denominator polynomials are second-order) is called a **biquad**.

This scheme has the advantage of design ease since simple equations can be derived relating the components of each section to the coefficients of each factor in the VTF. Also, each biquad can be independently adjusted relatively easily to give the correct performance. Because of these important practical merits, a large number of alternative biquad structures have been proposed, and the newcomer may easily find the choice overwhelming.

The second approach to active filter synthesis involves the use of RC-active circuits to simulate passive LCR ladders. This has two important advantages. First, the design process can be very straightforward: the wealth of design data published for passive ladder filters (see Further Information) can be used directly so that the sometimes difficult process of component value synthesis from specification is eliminated. Second, the LCR ladder offers optimal **sensitivity** properties [Orchard, 1966], and RC-active filters designed by ladder simulation share the same low sensitivity features. Chapter 4 of Bowron and Stephenson [1979] gives an excellent introduction to the formal treatment of circuit sensitivity.

Sensitivity plays a vital role in the characterization of RC-active filters. It provides a measure of the extent to which a change in the value of any given component affects the response of the filter. High sensitivity in an RC-active filter should also alert the designer to the possibility of oscillation. A nominally stable design will be unstable in practical realization if sensitivities are such that component value errors cause one or more pairs of poles to migrate into the RHP. Because any practical filter will be built with components that are not exactly nominal in value, sensitivity information provides a practical and useful indication of how different filter structures will react and provides a basis for comparison.

Cascaded Second-Order Sections

This section will introduce the cascade approach to active filter design. As noted earlier, there are a great many second-order RC-active sections to choose from, and the present treatment aims only to convey some of the main ideas involved in this strategy. The references provided at the end of this section point the reader to several comprehensive treatments of the subject.

Sallen and Key Section

This is an early and simple example of a second-order section building block [Sallen and Key, 1955]. It remains a commonly used filter despite its age, and it will serve to illustrate some key stages in the design of all such RC-active sections. The circuit is shown in Fig. 28.4. A straightforward analysis of this circuit yields a VTF

$$H(s) = \frac{K \dfrac{1}{C_1 C_2 R_1 R_2}}{s^2 + s\left[\dfrac{1}{C_2 R_2} + \dfrac{1}{C_2 R_1} + \dfrac{1-K}{C_1 R_1}\right] + \dfrac{1}{C_1 C_2 R_1 R_2}} \qquad (28.2)$$

This is an all-pole low-pass form since the numerator involves only a constant term.

Specifications for an all-pole second-order section may arise in coefficient form, where the required s-domain VTF is given as

$$H(s) = \frac{k}{s^2 + a_1 s + a_0} \qquad (28.3)$$

or in Q-ω_0 standard second-order form

$$H(s) = \frac{k}{s^2 + \dfrac{\omega_0}{Q} s + \omega_0} \qquad (28.4)$$

Figure 28.5 shows the relationship between these VTF forms.

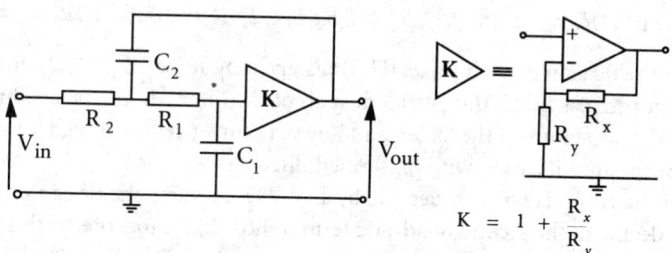

FIGURE 28.4 Sallen and Key second-order filter section.

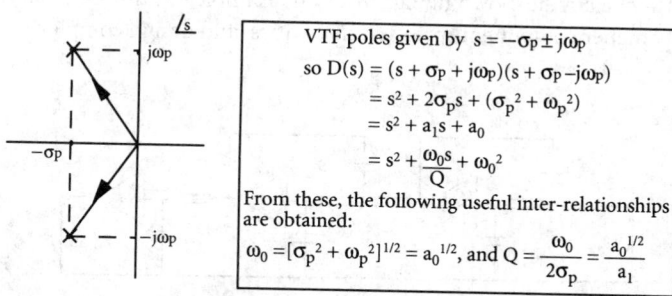

VTF poles given by $s = -\sigma_P \pm j\omega_P$

so $D(s) = (s + \sigma_P + j\omega_P)(s + \sigma_P - j\omega_P)$

$\qquad = s^2 + 2\sigma_P s + (\sigma_P^2 + \omega_P^2)$

$\qquad = s^2 + a_1 s + a_0$

$\qquad = s^2 + \dfrac{\omega_0 s}{Q} + \omega_0^2$

From these, the following useful inter-relationships are obtained:

$\omega_0 = [\sigma_P^2 + \omega_P^2]^{1/2} = a_0^{1/2}$, and $Q = \dfrac{\omega_0}{2\sigma_P} = \dfrac{a_0^{1/2}}{a_1}$

FIGURE 28.5 VTF pole relationships.

As a design example, the VTF for an all-pole fifth-order Chebyshev filter with 0.5-dB passband ripple [see Fig. 28.1(c)] has the factored-form denominator

$$D(s) = (s + 0.36232)(s^2 + 0.22393s + 1.0358)(s^2 + 0.58625s + 0.47677) \quad (28.5)$$

Taking the first of the quadratic factors in Eq. (28.5) and comparing like coefficients from Eq. (28.2) gives the following design equations:

$$\frac{1}{C_1 C_2 R_1 R_2} = 1.0358; \qquad \frac{1}{C_2 R_2} + \frac{1}{C_2 R_1} + \frac{1 - K}{C_1 R_1} = 0.22393 \quad (28.6)$$

Clearly, the designer has some degrees of freedom here since there are two equations in five unknowns. Choosing to set both (normalized) capacitor values to unity, and fixing the dc stage gain $K = 5$, gives

$$C_1 = C_2 = 1\text{F}; \quad R_1 = 1.8134\ \Omega; \quad R_2 = 1.3705\ \Omega; \quad R_x = 4\ \Omega; \quad R_y = 1\ \Omega$$

Note that Eq. (28.5) is a normalized specification giving a filter cut-off frequency of 1 rad s^{-1}. These normalized component values can now be denormalized to give a required cut-off frequency and practical component values. Suppose that the filter is, in fact, required to give a cut-off frequency $f_c = 1$ kHz. The necessary shift is produced by multiplying all the capacitors (leaving the resistors fixed) by the factor ω_N/ω_D where ω_N is the normalized cut-off frequency (1 rad s^{-1} here) and ω_D is the required denormalized cut-off frequency ($2\pi \times 1000$ rad s^{-1}). Applying this results in denormalized capacitor values of 159.2 µF. A useful rule of thumb [Waters, 1991] advises that capacitor values should be on the order of magnitude of $(10/f_c)$ µF, which suggests that the capacitors should be further scaled to around 10 nF. This can be achieved without altering of the filter's f_c, by means of the impedance scaling property of electrical circuits. Providing all circuit impedances are scaled by the same amount, current and voltage TFs are preserved. In an RC-active circuit, this requires that all resistances are multiplied by some factor while all capacitances are divided by it (since capacitive impedance is proportional to $1/C$). Applying this process yields final values as follows:

$$C_1, C_2 = 10\text{ nF}; \quad R_1 = 28.86\text{ k}\Omega; \quad R_2 = 21.81\text{ k}\Omega; \quad R_x = 63.66\text{ k}\Omega; \quad R_y = 15.92\text{ k}\Omega$$

Note also that the dc gain of each stage, $|H(0)|$, is given by K [see Eq. (28.2) and Fig. 28.4] and, when several stages are cascaded, the overall dc gain of the filter will be the product of these individual stage gains. This feature of the Sallen and Key structure gives the designer the ability to combine easy-to-manage amplification with prescribed filtering.

Realization of the complete fifth-order Chebyshev VTF requires the design of another second-order section to deal with the second quadratic term in Eq. (28.5), together with a simple circuit to realize the first-order term arising because this is an odd-order VTF. Figure 28.6 shows the form of the overall cascade. Note that the op amps at the output of each stage provide the necessary interstage isolation. It is finally worth noting that an extended single-amplifier form of the Sallen and Key network exists—the circuit shown in Fig. 28.2 is an example of this—but that the saving in op amps is paid for by higher component spreads, sensitivities, and design complexity.

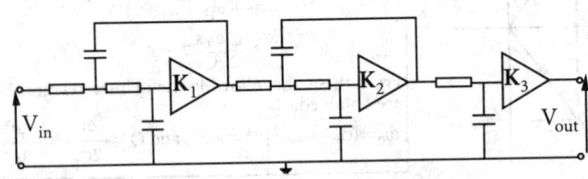

FIGURE 28.6 Form of fifth-order Sallen and Key cascade.

State-Variable Biquad

The simple Sallen and Key filter provides only an all-pole TF; many commonly encountered filter specifications are of this form—the Butterworth and Chebyshev approximations are notable examples—so this is not a serious limitation. In general, however, it will be necessary to produce sections capable of realizing a second-order denominator together with a numerator polynomial of up to second-order:

$$H(s) = \frac{b_2 s^2 + b_1 s + b_0}{s^2 + a_1 s + a_0} \tag{28.7}$$

The other major filter approximation in common use—the elliptic (or Cauer) function filter—involves quadratic numerator terms in which the b_1 coefficient in Eq. (28.7) is missing. The resulting numerator polynomial, of the form $b_2 s^2 + b_0$, gives rise to s-plane zeros on the $j\omega$ axis corresponding to points in the stopband of the sinusoidal frequency response where the filter's transmission goes to zero. These notches or *transmission zeros* account for the elliptic's very rapid transition from passband to stopband and, hence, its optimal selectivity.

A filter structure capable of producing a VTF of the form of Eq. (28.7) was introduced as a state-variable realization by its originators [Kerwin *et al.*, 1967]. The structure comprises two op amp integrators and an op amp summer connected in a loop and was based on the integrator-summer analog computer used in control/analog systems analysis, where system state is characterized by some set of so-called state variables. It is also often referred to as a ring-of-three structure. Many subsequent refinements of this design have appeared (Schaumann *et al.*, [1990] gives a useful treatment of some of these developments) and the state-variable biquad has achieved considerable popularity as the basis of many commercial universal packaged active filter building blocks. By selecting appropriate chip/package output terminals, and with the use of external trimming components, a very wide range of filter responses can be obtained.

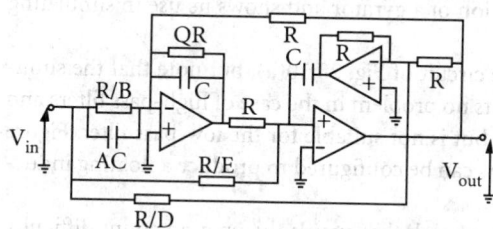

FIGURE 28.7 Circuit schematic for state-variable biquad.

Figure 28.7 shows a circuit developed from this basic state-variable network and described in Schaumann *et al.* [1990]. The circuit yields a VTF

$$H(s) = \frac{V_{out}(s)}{V_{in}(s)} = -\frac{A s^2 + \omega_0 (B - D)s + E\omega_0^2}{s^2 + \frac{\omega_0}{Q}s + \omega_0^2}, \text{ with } \omega_0 \triangleq 1/RC \tag{28.8}$$

By an appropriate choice of the circuit component values, a desired VTF of the form of Eq. (28.8) can be realized.

Consider, for example, a specification requirement for a second-order elliptic filter cutting off at 10 kHz. Assume that a suitable normalized specification for the VTF is

$$H(s) = -\frac{0.15677(s^2 + 7.464)}{s^2 + 0.9989s + 1.1701} \tag{28.9}$$

From Eq. (28.8) and Eq. (28.9), and referring to Fig. 28.7, normalized values for the components are computed as follows. As the s term in the numerator is to be zero, set $B = D = 0$ (which obtains if resistors R/B and R/D are simply removed from the circuit). Setting $C = 1$ F gives the following results:

$$AC = 0.15677F; \quad R = 1/C\omega_0 = 0.92446 \ \Omega; \quad QR = 1.08290 \ \Omega; \quad R/E = 0.92446 \ \Omega$$

Removing the normalization and setting $C = (10/10 \ k) \ \mu F = 1$ nF requires capacitors to be multiplied by 10^{-9} and resistors to be multiplied by 15.9155×10^3. Final denormalized component values for the 10-kHz filter are thus:

$$C = 1 \ nF; \quad AC = 0.15677 \ nF; \quad R = R/E = 14.713 \ k\Omega; \quad QR = 17.235 \ k\Omega$$

Passive Ladder Simulation

As for the biquad approach, numerous different ladder-based design methods have been proposed. Two representative schemes will be considered here: inductance simulation and ladder transformation.

Inductance Simulation

In the inductance simulation approach, use is made of impedance converter/inverter networks. Figure 28.8 gives a classification of the various generic forms of device. The NIC enjoyed prominence in the early days of active filters but was found to be prone to instability. Two classes of device that have proved more useful in the longer term are the GIC and the gyrator.

Figure 28.9 introduces the symbolic representation of a gyrator and shows its use in simulating an inductor.

The gyrator can conveniently be realized by the circuit of Fig. 28.10(a), but note that the simulated inductor is grounded at one end. This presents no problem in the case of high-pass filters and other forms requiring a grounded shunt inductor but is not suitable for the low-pass filter. Figure 28.10(b) shows how a pair of back-to-back gyrators can be configured to produce a floating inductance, but this involves four op amps per inductor.

The next section will introduce an alternative approach that avoids the op amp count difficulty associated with simulating the floating inductors directly.

Ladder Transformation

The other main approach to the RC-active simulation of passive ladders involves the transformation of a prototype ladder into a form suitable for active realization. A most effective method of this class

converter: $Z_{in} = K(s) \cdot Z_L$

inverter: $Z_{in} = \dfrac{K(s)}{Z_L}$

converter classes:

$K(s)$ { real, positive: Positive Impedance Converter (PIC)
real, negative: Negative Impedance Converter (NIC)
complex: Generalized Impedance Converter (GIC)

inverter classes:

$K(s)$ { real, positive: Positive Impedance Inverter (PII or Gyrator)
real, negative: Negative Impedance Inverter (NII)

FIGURE 28.8   Generic impedance converter/inverter networks.

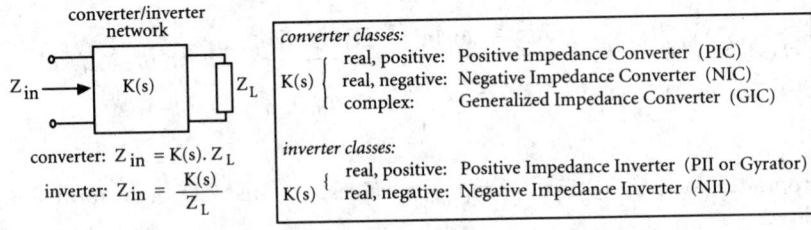

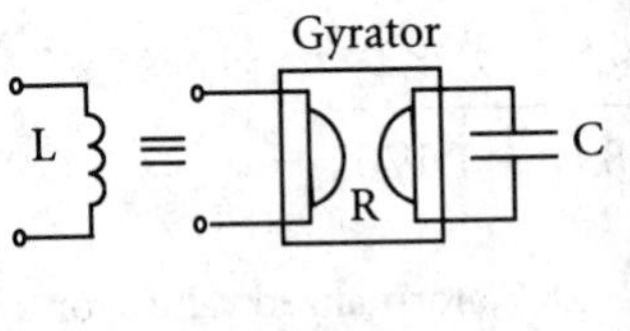

$$Z_{in} = \frac{K(s)}{Z_L} = \frac{R^2}{(1/sC)}$$
$$= sCR^2$$

- indicating an inductor of effective value CR^2 henries. R is known as the gyration resistance of the gyrator

FIGURE 28.9   Gyrator simulation of an inductor.

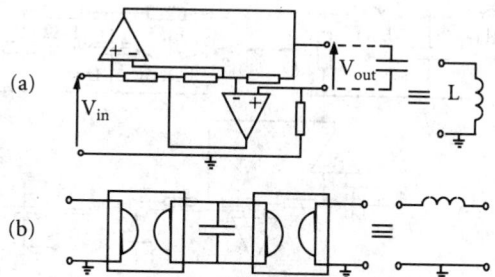

FIGURE 28.10 (a) Practical gyrator and (b) simulation of floating inductor. (*Source:* A. Antoniou, *Proc. IEE*, vol. 116, pp. 1838–1850, 1969. With permission.)

is based on the use of the Bruton transformation [Bruton, 1969], which involves the complex impedance scaling of a prototype passive LCR ladder network. All prototype circuit impedances $Z(s)$ are transformed to $Z_T(s)$ with

$$Z_T(s) = \frac{K}{s} \cdot Z(s) \tag{28.10}$$

where K is a constant chosen by the designer and which provides the capacity to scale component values in the final filter. Since impedance transformations do not affect voltage and current transfer ratios, the VTF remains unaltered by this change. The Bruton transformation is applied directly to the elements in the prototype network, and it follows from Eq. (28.10) that a resistance R transforms into a capacitance $C = K/R$, while an inductance L transforms into a resistance $R = KL$. The elimination of inductors in favor of resistors is the key purpose of the Bruton transform method. Applying the Bruton transform to a prototype circuit capacitance C gives

$$Z_T(s) = \frac{K}{s} \cdot \frac{1}{sC} = \frac{K}{s^2 C} = \frac{1}{s^2 D} \tag{28.11}$$

where $D = C/K$ is the parameter value of a new component produced by the transformation, which is usually referred to as a frequency-dependent negative resistance (FDNR). This name results from the fact that the sinusoidal steady-state impedance $Z_T(j\omega) = -(1/\omega^2 D)$ is frequency-dependent, negative, and real, hence, resistive. In practice, the FDNR elements are realized by RC-active subnetworks using op amps, normally two per FDNR. Figure 28.11(a) and (b) shows the sequence of circuit changes involved in transforming from a third-order LCR prototype ladder to an FDNR circuit. Figure 28.11(c) gives an RC-active realization for the FDNR based on the use of a GIC, introduced in the previous subsection.

Active Filters for ICs

It was noted earlier that the advent of the IC op amp made the RC-active filter a practical reality. A typical state-of-the-art 1960–70s active filter would involve a printed circuit board-mounted circuit comprising discrete passive components together with IC op amps. Also appearing at this time were hybrid implementations, which involve special-purpose discrete components and op amp ICs interconnected on a ceramic or glass substrate. It was recognized, however, that there were considerable benefits to be had from producing an all-IC active filter.

Production of a quality on-chip capacitor involves substantial chip area, so the scaling techniques referred to earlier must be used to keep capacitance values down to the low picofarad range. The consequence of this is that, unfortunately, the circuit resistance values become proportionately

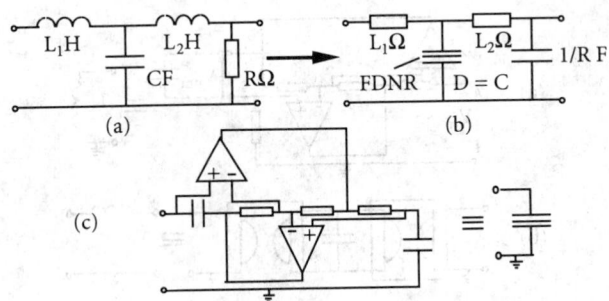

FIGURE 28.11   FDNR active filter.

large so that, again, there is a chip-area problem. The solution to this dilemma emerged in the late 1970s/early 1980s with the advent of the switched-capacitor (SC) active filter. This device, a development of the active-RC filter that is specifically intended for use in IC form, replaces prototype circuit resistors with arrangements of switches and capacitors that can be shown to simulate resistances, under certain circumstances. The great merit of the scheme is that the values of the capacitors involved in this process of resistor simulation are inversely proportional to the values of the prototype resistors; thus, the final IC structure involves principal and switched capacitors that are small in magnitude and hence ideal for IC realization. A good account of SC filters is given, for example, in Schaumann *et al.* [1990]. Commonly encountered techniques for SC filter design are based on the two major design styles (biquads and ladder simulation) that have been introduced in this section.

Many commercial IC active filters are based on SC techniques, and it is also becoming usual to find custom and semicustom IC design systems that include active filter modules as components within a macrocell library that the system-level design can simply invoke where analog filtering is required within an all-analog or mixed-signal analog/digital system.

Defining Terms

Active filter: An electronic filter whose design includes one or more active devices.

Biquad: An active filter whose transfer function comprises a ratio of second-order numerator and denominator polynomials in the frequency variable.

Electronic filter: An electronic circuit designed to transmit some range of signal frequencies while rejecting others. Phase and time-domain specifications may also occur.

Sensitivity: A measure of the extent to which a given circuit performance measure is affected by a given component within the circuit.

References

A. Antoniou, "Realization of gyrators using operational amplifiers and their use in RC-active network synthesis," *Proc. IEE*, vol. 116, pp. 1838–1850, 1969.

P. Bowron and F. W. Stephenson, *Active Filters for Communications and Instrumentation*, New York: McGraw-Hill, 1979.

L. T. Bruton, "Network transfer functions using the concept of frequency dependent negative resistance," *IEEE Trans.*, vol. CT-18, pp. 406–408, 1969.

W. J. Kerwin, L. P. Huelsman, and R. W. Newcomb, "State-variable synthesis for insensitive integrated circuit transfer functions," *IEEE J.*, vol. SC-2, pp. 87–92, 1967.

H. J. Orchard, "Inductorless filters," *Electron. Letters*, vol. 2, pp. 224–225, 1966.

P. R. Sallen and E. L. Key, "A practical method of designing RC active filters," *IRE Trans.*, vol. CT-2, pp. 74–85, 1955.

R. Schaumann, M.S. Ghausi, and K.R. Laker, *Design of Analog Filters*, Englewood Cliffs, N.J: Prentice-Hall, 1990.

A. Waters, *Active Filter Design*, New York: Macmillan, 1991.

Further Information

Tabulations of representative standard filter specification functions appear in the sources in the References by Schaumann *et al.* [1990] and Bowron and Stephenson [1979], but more extensive tabulations, including prototype passive filter component values, are given in A. I. Zverev, *Handbook of Filter Synthesis* (New York: John Wiley, 1967). More generally, the Schaumann text provides an admirable, up-to-date coverage of filter design with an extensive list of references.

The field of active filter design remains active, and new developments appear in *IEEE Transactions on Circuits and Systems* and *IEE Proceedings Part G (Circuits and Systems)*. The IEE publication *Electronic Letters* provides for short contributions. A number of international conferences (whose proceedings can be borrowed through technical libraries) feature active filter and related sessions, notably the *IEEE International Symposium on Circuits and Systems* (ISCAS) and the *European Conference on Circuit Theory and Design* (ECCTD).

28.2 Realization

J. W. Steadman and B. M. Wilamowski

After the appropriate low-pass form of a given **filter** has been synthesized, the designer must address the realization of the filter using **operational amplifiers**. If the required filter is not low-pass but high-pass, bandpass, or bandstop, transformation of the prototype function is also required [Budak, 1974; Van Valkenburg, 1982]. While a detailed treatment of the various transformations is beyond the scope of this work, most of the filter designs encountered in practice can be accomplished using the techniques given here.

When the desired filter function has been determined, the corresponding electronic circuit must be designed. Many different circuits can be used to realize any given transfer function. For purposes of this handbook, we present several of the most popular types of realizations. Much more detailed information on various circuit realizations and the advantages of each may be found in the literature, in particular Van Valkenburg [1982], Huelseman and Allen [1980], and Chen [1986]. Generally the design trade-offs in making the choice of circuit to be used for the realization involve considerations of the number of elements required, the sensitivity of the circuit to changes in component values, and the ease of tuning the circuit to given specifications. Accordingly, limited information is included about these characteristics of the example circuits in this section.

Each of the circuits described here is commonly used in the realization of **active filters**. When implemented as shown and used in the appropriate gain and bandwidth specifications of the amplifiers, they will provide excellent performance. Computer-aided filter design programs are available which simplify the process of obtaining proper element values and simulation of the resulting circuits [Krobe *et al.*, 1989; Wilamowski *et al.*, 1992].

Transformation from Low-Pass to Other Filter Types

To obtain a high-pass, bandpass, or bandstop filter function from a low-pass prototype, one of two general methods can be used. In one of these, the circuit is realized and then individual circuit elements are replaced by other elements or subcircuits. This method is more useful in **passive filter** designs and is not discussed further here. In the other approach, the transfer function of the low-pass prototype is transformed into the required form for the desired filter. Then a circuit is chosen to realize the new filter function. We give a brief description of the transformation in this section, then give examples of circuit realizations in the following sections.

Low-Pass to High-Pass Transformation

Suppose the desired filter is, for example, a high-pass Butterworth. Begin with the low-pass Butterworth transfer function of the desired order and then *transform* each pole of the original function using the formula

$$\frac{1}{S - S_j} \rightarrow \frac{Hs}{s - s_j} \tag{28.12}$$

which results in one complex pole and one zero at the origin for each pole in the original function. Similarly, each zero of the original function is transformed using the formula

$$S - S_j \rightarrow \frac{s - s_j}{Hs} \tag{28.13}$$

which results in one zero on the imaginary axis and one pole at the origin. In both equations, the scaling factors used are

$$H = \frac{1}{S_j} \quad \text{and} \quad s_j = \frac{\omega_0}{S_j} \tag{28.14}$$

where ω_0 is the desired cut-off frequency in radians per second.

Low-Pass to Bandpass Transformation

Begin with the low-pass prototype function in factored, or *pole-zero*, form. Then each pole is transformed using the formula

$$\frac{1}{S - S_j} \rightarrow \frac{Hs}{(s - s_1)(s - s_2)} \tag{28.15}$$

resulting in one zero at the origin and two conjugate poles. Each zero is transformed using the formula

$$S - S_j \rightarrow \frac{(s - s_1)(s - s_2)}{Hs} \tag{28.16}$$

resulting in one pole at origin and two conjugate zeros. In Eqs. (28.15) and (28.16)

$$H = -B; \quad s_{1,2} = \omega_c\left(\alpha \pm \sqrt{\alpha^2 - 1}\right); \quad \text{and} \quad \alpha = \frac{BS_j}{2\omega_c} \tag{28.17}$$

where ω_c is the center frequency and B is the bandwidth of the bandpass function.

Low-Pass to Bandstop Transformation

Begin with the low-pass prototype function in factored, or pole-zero, form. Then each pole is transformed using the formula

$$\frac{1}{S - S_j} \rightarrow \frac{H(s - s_1)(s - s_2)}{(s - s_3)(s - s_4)} \tag{28.18}$$

transforming each pole into two zeros on the imaginary axis and into two conjugate poles. Similarly, each zero is transformed into two poles on the imaginary axis and into two conjugate zeros using the formula

$$S - S_j \rightarrow \frac{(s - s_3)(s - s_4)}{H(s - s_1)(s - s_2)} \qquad (28.19)$$

where

$$H = \frac{1}{S_j}; \; s_{1,2} = \pm j\omega_c; \; s_{3,4} = \omega_c\left(\beta \pm \sqrt{\beta^2 - 1}\right); \; \text{and } \beta = \frac{B}{2\omega_c S_j} \qquad (28.20)$$

Once the desired transfer function has been obtained through obtaining the appropriate low-pass prototype and transformation, if necessary, to the associated high-pass, bandpass or bandstop function, all that remains is to obtain a circuit and the element values to realize the transfer function.

Circuit Realizations

Various electronic circuits can be found to implement any given transfer function. Cascade filters and ladder filters are two of the basic approaches for obtaining a practical circuit. Cascade realizations are much easier to find and to tune, but ladder filters are less sensitive to element variations. In cascade realizations, the transfer function is simply factored into first- and second-order parts. Circuits are built for the individual parts and then cascaded to produce the overall filter. For simple to moderately complex filter designs, this is the most common method, and the remainder of this section is devoted to several examples of the circuits used to obtain the first- and second-order filters. For very high-order transfer functions, ladder filters should be considered, and further information can be obtained by consulting the literature.

In order to simplify the circuit synthesis procedure, very often ω_0 is assumed to be equal to one and then after a circuit is found, the values of all capacitances in the circuit are divided by ω_0. In general, the following magnitude and frequency transformations are allowed:

$$R_{new} = K_M R_{old} \text{ and } C_{new} = \frac{1}{K_F K_M} C_{old} \qquad (28.21)$$

where K_M and K_F are magnitude and frequency scaling factors, respectively.

Cascade filter designs require the transfer function to be expressed as a product of first- and second-order terms. For each of these terms a practical circuit can be implemented. Examples of these circuits are presented in Figs. 28.12–28.22. In general the following first- and second-order terms can be distinguished:

(a) First-order low-pass:

$$T(s) = \frac{H\omega_0}{s + \omega_0}$$

Assumption : $r_1 = 1$

$$c_1 = \frac{1}{\omega_0} \qquad r_2 = |H|\,\omega_0$$

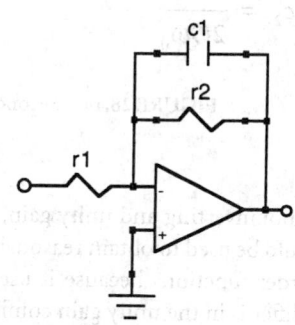

FIGURE 28.12 First-order low-pass filter.

This filter is inverting, i.e., H must be negative, and the scaling factors shown in Eq. (28.21) should be used to obtain reasonable values for the components.

(b) First-order high-pass:

$$T(s) = \frac{Hs}{s + \omega_0}$$

Assumption : $r_1 = 1$

$$c_1 = \frac{1}{\omega_0} \qquad r_2 = |H|$$

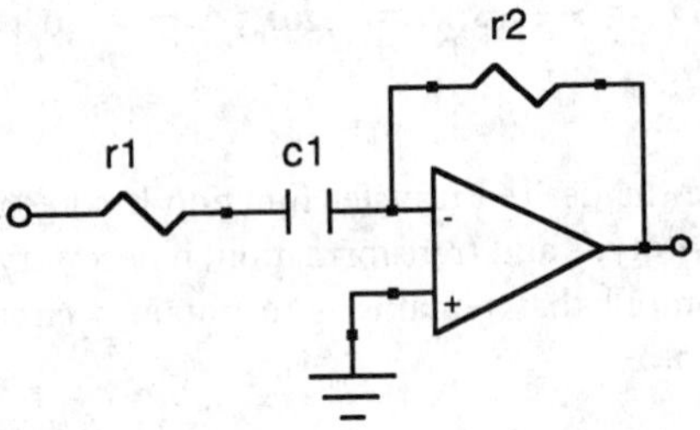

FIGURE 28.13   First-order high-pass filter.

This filter is inverting, i.e., H must be negative, and the scaling factors shown in Eq. (28.21) should be used to obtain reasonable values for the components.

While several passive realizations of first-order filters are possible (low-pass, high-pass, and lead-lag), the active circuits shown here are inexpensive and avoid any loading of the other filter sections when the individual circuits are cascaded. Consequently, these circuits are preferred unless there is some reason to avoid the use of the additional operational amplifier. Note that a second-order filter can be realized using one operational amplifer as shown in the following paragraphs, so it is common practice to choose even-order transfer functions, thus avoiding the use of any first-order filters.

(c) There are several second-order low-pass circuits:

$$T(s) = \frac{H\omega_0^2}{s^2 + \dfrac{\omega_0}{Q} s + \omega_0^2}$$

Assumption : $r_1 = r_2 = 1$

$$c_1 = \frac{2Q}{\omega_0} \qquad c_2 = \frac{1}{2Q\omega_0}$$

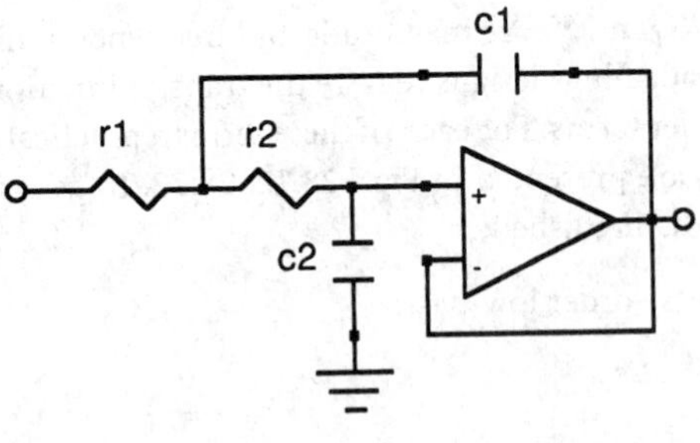

FIGURE 28.14   Second-order low-pass Sallen-Key filter.

This filter is noninverting and unity gain, i.e., H must be one, and the scaling factors shown in Eq. (28.21) should be used to obtain reasonable element values. This is a very popular filter for realizing second-order functions because it uses a minimum number of components and since the operation amplifier is in the unity gain configuration it has very good bandwidth.

Another useful configuration for second-order low-pass filters uses the operational amplifier in its inverting "infinite gain" mode as shown in Fig. 28.15.

$$T(s) = \cfrac{H\omega_0^2}{s^2 + \cfrac{\omega_0}{Q}s + \omega_0^2}$$

Assumption : $r_1 = r_2 = r_3 = 1$

$$c_1 = \frac{3Q}{\omega_0} \qquad c_2 = \frac{1}{3Q\omega_0}$$

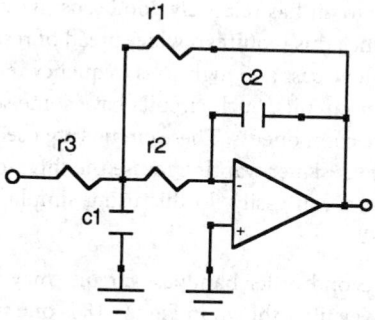

FIGURE 28.15 Second-order low-pass filter using the inverting circuit.

This circuit has the advantage of relatively low sensitivity of ω_0 and Q to variations in component values. In this configuration the operational amplifier's gain-bandwidth product may become a limitation for high-Q and high-frequency applications [Budak, 1974]. There are several other circuit configurations for low-pass filters. The references given at the end of the section will guide the designer to alternatives and the advantages of each.

(d) Second-order high-pass filters may be designed using circuits very much like those shown for the low-pass realizations. For example, the Sallen-Key low-pass filter is shown in Fig. 28.16.

$$T(s) = \cfrac{Hs^2}{s^2 + \cfrac{\omega_0}{Q}s + \omega_0^2}$$

Assumption : $r_3 = 1$

$$c_1 = c_2 = 1$$

$$r_1 = r_2 = \frac{1}{\omega_0} \qquad r_4 = 2 - \frac{1}{Q}$$

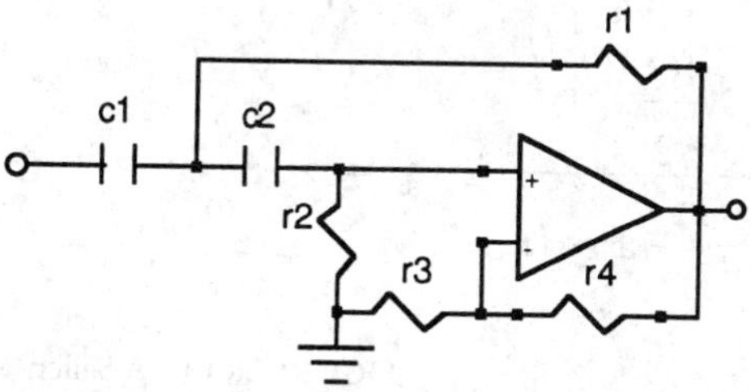

FIGURE 28.16 A second-order high-pass Sallen-Key filter.

As in the case of the low-pass Sallen-Key filter, this circuit is noninverting and requires very little gain from the operational amplifier. For low to moderate values of Q, the **sensitivity functions** are reasonable and the circuit performs well.

The inverting *infinite gain* high-pass circuit is shown in Fig. 28.17 and is similar to the corresponding low-pass circuit.

$$T(s) = \cfrac{Hs^2}{s^2 + \cfrac{\omega_0}{Q}s + \omega_0^2}$$

Assumption : $r_1 = 1$

$$r_2 = 9Q^2 \qquad c_1 = c_2 = c_3 = \frac{1}{3Q^2}$$

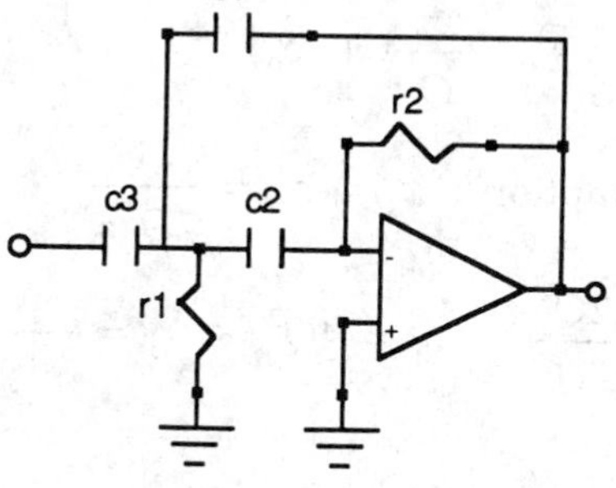

FIGURE 28.17 An inverting second-order high-pass circuit.

This circuit has relatively good sensitivity figures. The principal limitation occurs with high-Q filters since this requires a wide spread of resistor values.

Both low-pass and high-pass frequency response circuits can be achieved using three operational amplifier circuits. Such circuits have some sensitivity function and tuning advantages but require far more components. These circuits are used in the sections describing bandpass and bandstop filters. The designer wanting to use the three-operational-amplifier realization for low-pass or high-pass filters can easily do this using simple modifications of the circuits shown in the following sections.

(e) Second-order bandpass circuits may be realized using only one operational amplifier. The Sallen-Key filter shown in Fig. 28.18 is one such circuit.

$$T(s) = \frac{H \dfrac{\omega_0}{Q} s}{s^2 + \dfrac{\omega_0}{Q} s + \omega_0^2}$$

Assumption : $c_1 = c_2 = 1; \quad r_5 = 1$

$$r_2 = r_3 = \frac{\sqrt{2}}{\omega_0} \qquad r_1 = \frac{\dfrac{4Q}{\sqrt{2}} - 1}{H}$$

$$r_4 = \frac{\dfrac{4Q}{\sqrt{2}} - 1}{\dfrac{4Q}{\sqrt{2}} - 1 - H} \qquad r_6 = 3 - \frac{\sqrt{2}}{\omega_0}$$

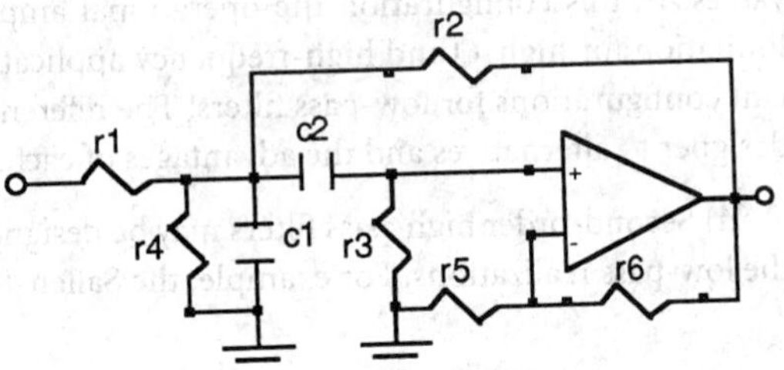

FIGURE 28.18 A Sallen-Key bandpass filter.

This is a noninverting amplifier which works well for low- to moderate-Q filters and is easily tuned [Budak, 1974]. For high-Q filters the sensitivity of Q to element values becomes high, and alternative circuits are recommended. One of these is the bandpass version of the inverting amplifier filter as shown in Fig. 28.19.

$$T(s) = \frac{H \dfrac{\omega_0}{Q} s}{s^2 + \dfrac{\omega_0}{Q} s + \omega_0^2}$$

Assumption : $c_1 = c_2 = \dfrac{1}{2Q\omega_0}$

$$r_1 = \frac{2Q^2}{H} \qquad r_2 = 4Q^2 \qquad r_3 = \frac{1}{1 - \dfrac{H}{2Q^2}}$$

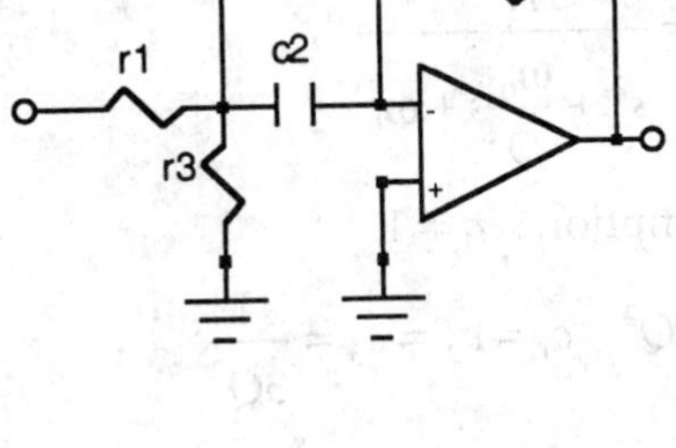

FIGURE 28.19 The inverting amplifier bandpass filter.

This circuit has few components and relatively small sensitivity of ω_0 and Q to variations in element values. For high-Q circuits, the range of resistor values is quite large as r_1 and r_2 are much larger than r_3.

When ease of tuning and small sensitivities are more important than the circuit complexity, the three-operational-amplifier circuit of Fig. 28.20 may be used to implement the bandpass transfer function.

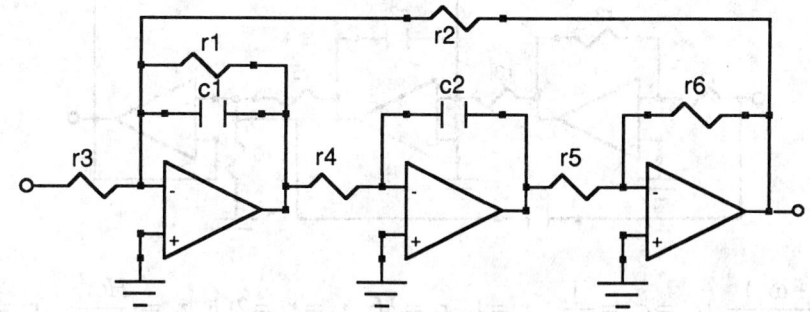

$$T(s) = \cfrac{H\dfrac{\omega_0}{Q}s}{s^2 + \dfrac{\omega_0}{Q}s + \omega_0^2} \qquad c_1 = c_2 = \frac{1}{\omega_0} \qquad r_1 = Q \qquad r_2 = r_4 = r_5 = r_6 = 1 \qquad r_3 = \frac{Q}{|H|}$$

FIGURE 28.20   The three-operational-amplifier bandpass filter.

The filter as shown in Fig. 28.20 is inverting. For a noninverting realization, simply take the output from the middle amplifier rather than the right one. This same configuration can be used for a three-operational-amplifier low-pass filter by putting the input into the summing junction of the middle amplifier and taking the output from the left operational amplifier. Note that Q may be changed in this circuit by varying r_1 and that this will not alter ω_0. Similarly, ω_0 can be adjusted by varying c_1 or c_2 and this will not change Q. If only variable resistors are to be used, the filter can be tuned by setting ω_0 using any of the resistors other than r_1 and then setting Q using r_1.

(f) Second-order bandstop filters are very useful in rejecting unwanted signals such as line noise or carrier frequencies in instrumentation applications. Such filters are implemented with methods very similar to the bandpass filters just discussed. In most cases, the frequency of the zeros is to be the same as the frequency of the poles. For this application, the circuit shown in Fig. 28.21 can be used.

$$T(s) = \frac{H(s^2 + \omega_z^2)}{s^2 + \dfrac{\omega_0}{Q}s + \omega_0^2}$$

Assumption : $c_1 = c_2 = 1$

$$r_1 = \frac{1}{2Q\omega_0} \qquad r_3 = \frac{1}{Q\omega_0} \qquad r_2 = r_4 = \frac{2Q}{\omega_0}$$

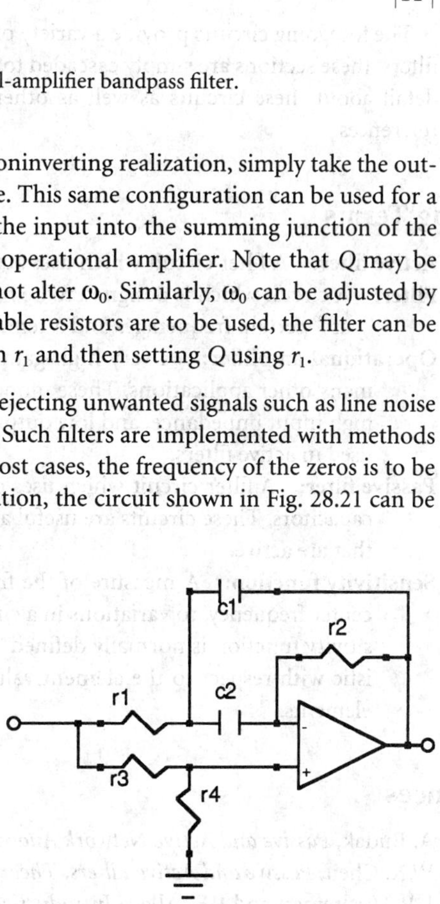

FIGURE 28.21   A single operational-amplifier bandstop filter.

The primary advantage of this circuit is that it requires a minimum number of components. For applications where no tuning is required and the Q is low, this circuit works very well. When the bandstop filter must be tuned, the three-operational-amplifier circuit is preferable.

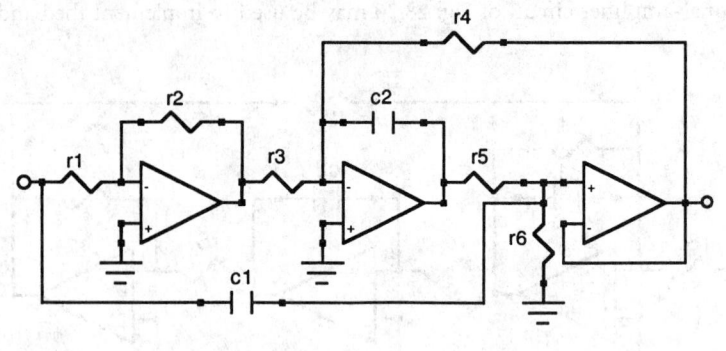

$$T(s) = \frac{H(s^2 + \omega_z^2)}{s^2 + \dfrac{\omega_0}{Q}s + \omega_0^2} \quad c_1 = c_2 = \frac{1}{\omega_0} \quad r_1 = 1 \quad r_2 = H \quad r_5 = r_6 = 2Q \quad r_3 = \frac{H\omega_0^2}{2Q\omega_z^2} \quad r_4 = \frac{1}{2Q}$$

FIGURE 28.22   A three-operational-amplifier bandstop filter.

The foregoing circuits provide a variety of useful first- and second-order filters. For higher-order filters, these sections are simply cascaded to realize the overall transfer function desired. Additional detail about these circuits as well as other circuits used for active filters may be found in the references.

Defining Terms

Active filter: A filter circuit which uses active components, usually operational amplifiers.

Filter: A circuit which is designed to be frequency selective. That is, the circuit will emphasize or "pass" certain frequencies and attenuate or "stop" others.

Operational amplifier: A very high-gain differential amplifier used in active filter circuits and many other applications. These monolithic integrated circuits typically have such high gain, high input impedance, and low output impedance that they can be considered "ideal" when used in active filters.

Passive filter: A filter circuit which uses only passive components, i.e., resistors, inductors, and capacitors. These circuits are useful at higher frequencies and as prototypes for ladder filters that are active.

Sensitivity function: A measure of the fractional change in some circuit characteristic, such as center frequency, to variations in a circuit parameter, such as the value of a resistor. The sensitivity function is normally defined as the partial derivative of the desired circuit characteristic with respect to the element value and is usually evaluated at the nominal value of all elements.

References

A. Budak, *Passive and Active Network Analysis and Synthesis*, Boston: Houghton Mifflin, 1974.

W.K. Chen, *Passive and Active Filters, Theory and Implementations*, New York: Wiley, 1986.

L.P. Huelseman and P.E. Allen, *Introduction to the Theory and Design of Active Filters*, New York: McGraw-Hill, 1980.

M.R. Krobe, J. Ramirez-Angulo, and E. Sanchez-Sinencio, "FIESTA—A filter educational synthesis teaching aid," *IEEE Trans. on Education*, vol. 12, no. 3, pp. 280–286, August 1989.

M.E. Van Valkenburg, *Analog Filter Design*, New York: Holt, Rinehart and Winston, 1982.

B.M. Wilamowski, S.F. Legowski, and J.W. Steadman, "Personal computer support for teaching analog filter analysis and design," *IEEE Trans. on Education*, vol. 35, no. 4, November 1992.

Further Information

The monthly journal *IEEE Transactions on Circuits and Systems* is one of the best sources of information on new active filter functions and associated circuits.

The British journal *Electronics Letters* also often publishes articles about active circuits.

The *IEEE Transactions on Education* has carried articles on innovative approaches to active filter synthesis as well as computer programs for assisting in the design of active filters.

Wanlass's CMOS Circuit

Electronics engineers of a certain age remember the 1960s as the days of freewheeling experimentation when—at some companies, at least—bright young Ph.D.'s were given latitude to see what they could create without interference from corporate managers.

One bright young Ph.D. of that time was Frank Wanlass, who received the 1991 IEEE Solid-State Circuits Award for his invention of complementary-MOS (CMOS) logic circuitry.

Wanlass's interest in MOS technology dates to 1962. Upon reading about the Radio Corporation of America's work with thin-film cadmium sulfide field-effect transistors (FETs), he became intrigued by the simple structure of the devices, which he thought would make it easy to design fairly complex integrated circuits. But the devices were unstable.

Wanlass believed that making the FETs in silicon would solve the problem. Fairchild Semiconductor Research and Development had made some very stable and reliable bipolar transistors in silicon using its planar process, so why not use the same material to make stable MOSFETs?

So he went to work for a Fairchild Semiconductor subsidiary in Palo Alto, Calif., in August 1962. A few months later, he made his first p-channel silicon MOSFETs, which, like all early MOSFETs, were a great disappointment. At 10–20 volts, their threshold voltages were very high and very unstable.

Wanlass speculated that the aluminum gate

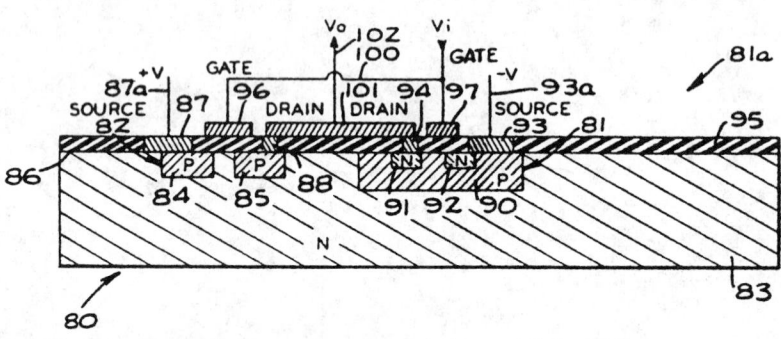

Dec. 5, 1967   F. M. WANLASS   3,356,858

LOW STAND-BY POWER COMPLEMENTARY FIELD EFFECT CIRCUITRY

Filed June 18, 1963   5 Sheets—Sheet 5

Wanlass's patent portrayed an integrated CMOS inverter.

electrode was diffusing into the gate oxide. While investigating the possibility of using a more inert metal, he found to his surprise that aluminum electrodes deposited by an electron beam evaporation machine yielded quite stable devices. The problem, he began to suspect, was not aluminum diffusion after all, but contamination.

The usual way to deposit aluminum in those days was to evaporate it by placing an aluminum wire in contact with a heated tungsten filament. Wanlass reasoned that the process was introducing positive ions into the aluminum and thence the gate oxide.

Further investigation inculpated sodium contamination from both the tungsten and the aluminum. Electron beam evaporation solved the problem because the electron-beam apparatus had a shutter mechanism that protected the silicon wafers from a carbon crucible of molten aluminum until the aluminum was at evaporation temperature. The sodium, having a much lower boiling point, boiled away before the shutter opened.

Wanlass next had the idea for CMOS. "It occurred to me," he said, "that a complementary circuit of NMOS and PMOS devices, if it could be made, would use very little power. In standby, it would draw practically nothing—just the leakage current."

His boss, Gordon Moore, now the chairman of Intel Corp., gave him a free hand to pursue his idea. At first, Wanlass tried to build a CMOS circuit monolithically, but that was so difficult he decided to prove the concept with discrete p-channel and n-channel MOSFETs instead.

The CMOS concept requires that both of its transistors be enhancement-mode devices. But whereas PMOS transistors were inherently of that type, the n-channel MOSFET was not. It would be years before MOS surface physics was well enough understood to permit the fabrication of such devices. Consequently, Wanlass made depletion-mode n-channel MOSFETs and back-biased their bodies negative with respect to their sources to turn them into enhancement-mode units.

The concept worked. The first demonstration circuit, a two-transistor inverter, consumed just a few nanowatts of standby power. CMOS shrank the standby power consumption by six orders of magnitude!

The speed was impressive enough, too. Propagation delay times were on the order of 100 nanoseconds—about half the speed of bipolar, but almost an order of magnitude faster than PMOS.

On June 18, 1963, Frank Wanlass applied for patent protection for his CMOS concept and was granted U.S. patent no. 3 356 858, the rights to which became part of Fairchild's patent portfolio. That patent described the overall concept and three specific circuits—an inverter, a NOR gate, and a NAND gate—from which any digital function can be built. In addition to the discrete implementations that were actually built, the patent includes the representation of an integrated CMOS inverter shown here.

Neither Wanlass nor Fairchild Semiconductor grew rich from the invention. Still, the integrated CMOS inverter shown here, although never built, is the progenitor of all CMOS ICs today.

Source: Adapted from M.J. Riezenman, *IEEE Spectrum*, p. 44, May 1991. © 1991 IEEE.

29

Power Electronics

29.1 Power Semiconductor Devices 694
Thyristor and Triac • Gate Turn-Off Thyristor (GTO) • Reverse-
Conducting Thyristor (RCT) and Asymmetrical Silicon-
Controlled Rectifier (ASCR) • Power Transistor • Power MOSFET
• Insulated-Gate Bipolar Transistor (IGBT) • MOS Controlled
Thyristor (MCT)

29.2 Power Conversion 702
AC-DC Converters • Cycloconverters • DC-to-AC Converters
• DC-DC Converters

29.3 Power Supplies 711
DC Power Supplies • AC Power Supplies • Special Power Supplies

29.4 Converter Control of Machines 729
Converter Control of DC Machines • Converter Control of AC
Machines

Kaushik Rajashekara
Delco Remy

Ashoka K. S. Bhat
University of Victoria

Bimal K. Bose
University of Tennessee

29.1 Power Semiconductor Devices

Kaushik Rajashekara

The modern age of power electronics began with the introduction of thyristors in the late 1950s. Now there are several types of power devices available for high-power and high-frequency applications. The most notable power devices are gate turn-off thyristors, power Darlington transistors, power MOSFETs, and insulated-gate bipolar transistors (IGBTs). Power semiconductor devices are the most important functional elements in all power conversion applications. The power devices are mainly used as switches to convert power from one form to another. They are used in motor control systems, uninterrupted power supplies, high-voltage dc transmission, power supplies, induction heating, and in many other power conversion applications. A review of the basic characteristics of these power devices is presented in this section.

Thyristor and Triac

The thyristor, also called a silicon-controlled rectifier (SCR), is basically a four-layer three-junction *pnpn* device. It has three terminals: anode, cathode, and gate. The device is turned on by applying a short pulse across the gate and cathode. Once the device turns on, the gate loses its control to turn off the device. The turn-off is achieved by applying a **reverse voltage** across the anode and cathode. The thyristor symbol and its volt-ampere characteristics are shown in Fig. 29.1. There are basically two classifications of thyristors: converter grade and inverter grade. The difference between a converter-grade and an inverter-grade thyristor is the low turn-off time (on the order of a few

694

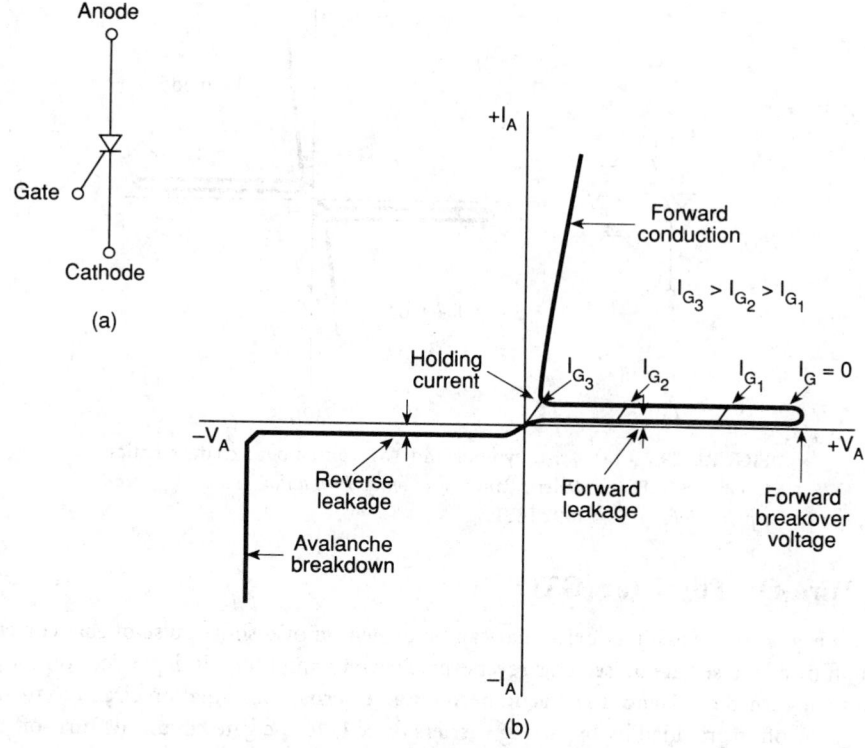

FIGURE 29.1 (a) Thyristor symbol and (b) volt-ampere characteristics. (*Source:* B.K. Bose, *Modern Power Electronics: Evaluation, Technology, and Applications,* p. 5. © 1992 IEEE.)

microseconds) for the latter. The converter-grade thyristors are slow type and are used in natural commutation (or phase-controlled) applications. Inverter-grade thyristors are used in forced commutation applications such as dc-dc choppers and dc-ac inverters. The inverter-grade thyristors are turned off by forcing the current to zero using an external commutation circuit. This requires additional commutating components, thus resulting in additional losses in the inverter.

Thyristors are highly rugged devices in terms of transient currents, ***di/dt,*** and ***dv/dt*** capability. The **forward voltage** drop in thyristors is about 1.5 to 2 V, and even at higher currents of the order of 500 A, it seldom exceeds 3 V. While the forward voltage determines the on-state power loss of the device at any given current, the switching power loss becomes a dominating factor affecting the device junction temperature at high operating frequencies. Because of this, the maximum switching frequencies possible using thyristors are limited in comparison with other power devices considered in this section.

Thyristors have I^2t withstand capability and can be protected by fuses. The nonrepetitive surge current capability for thyristors is about 10 times their rated root mean square (rms) current. They must be protected by snubber networks for *dv/dt* and *di/dt* effects. If the specified *dv/dt* is exceeded, thyristors may start conducting without applying a gate pulse. In dc-to-ac conversion applications it is necessary to use an antiparallel diode of similar rating across each main thyristor. Thyristors are available up to 6000 V, 3500 A.

A triac is functionally a pair of converter-grade thyristors connected in antiparallel. The triac symbol and volt-ampere characteristics are shown in Fig. 29.2. Because of the integration, the triac has poor reapplied *dv/dt,* poor gate current sensitivity at turn-on, and longer turn-off time. Triacs are mainly used in phase control applications such as in ac regulators for lighting and fan control and in solid-state ac relays.

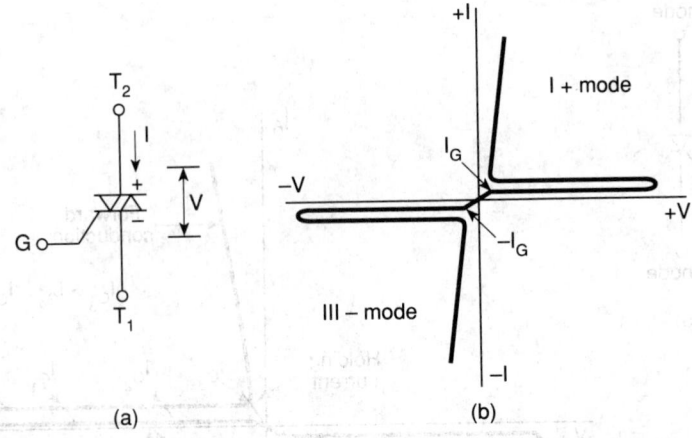

FIGURE 29.2 (a) Triac symbol and (b) volt-ampere characteristics.
(*Source:* B.K. Bose, *Modern Power Electronics: Evaluation, Technology, and Applications,* p. 5. © 1992 IEEE.)

Gate Turn-Off Thyristor (GTO)

The GTO is a power switching device that can be turned on by a short pulse of gate current and turned off by a reverse gate pulse. This reverse gate current amplitude is dependent on the anode current to be turned off. Hence there is no need for an external commutation circuit to turn it off. Because turn-off is provided by bypassing carriers directly to the gate circuit, its turn-off time is short, thus giving it more capability for high-frequency operation than thyristors. The GTO symbol and turn-off characteristics are shown in Fig. 29.3.

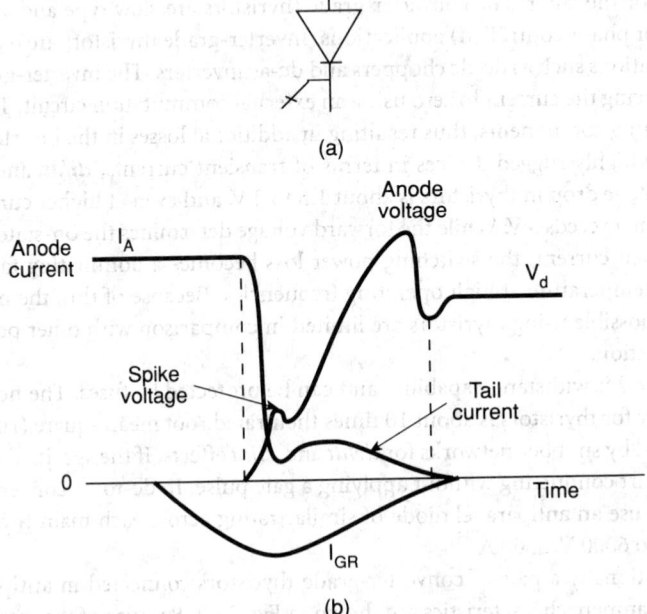

FIGURE 29.3 (a) GTO symbol and (b) turn-off characteristics.
(*Source:* B.K. Bose, *Modern Power Electronics: Evaluation, Technology, and Applications,* p. 5. © 1992 IEEE.)

GTOs have the I^2t withstand capability and hence can be protected by semiconductor fuses. For reliable operation of GTOs, the critical aspects are proper design of the gate turn-off circuit and the snubber circuit. A GTO has a poor turn-off current gain of the order of 4 to 5. For example, a 2000-A peak current GTO may require as high as 500 A of reverse gate current. Also, a GTO has the tendency to latch at temperatures above 125°C. GTOs are available up to about 4500 V, 2500 A.

Reverse-Conducting Thyristor (RCT) and Asymmetrical Silicon-Controlled Rectifier (ASCR)

Normally in inverter applications, a diode in antiparallel is connected to the thyristor for commutation/freewheeling purposes. In RCTs, the diode is integrated with a fast switching thyristor in a single silicon chip. Thus, the number of power devices could be reduced. This integration brings forth a substantial improvement of the static and dynamic characteristics as well as its overall circuit performance.

The RCTs are designed mainly for specific applications such as traction drives. The antiparallel diode limits the reverse voltage across the thyristor to 1 to 2 V. Also, because of the reverse recovery behavior of the diodes, the thyristor may see very high reapplied *dv/dt* when the diode recovers from its reverse voltage. This necessitates use of large RC snubber networks to suppress voltage transients. As the range of application of thyristors and diodes extends into higher frequencies, their reverse recovery charge becomes increasingly important. High reverse recovery charge results in high power dissipation during switching.

The ASCR has a similar forward blocking capability as an inverter-grade thyristor, but it has a limited reverse blocking (about 20–30 V) capability. It has an on-state voltage drop of about 25% less than an inverter-grade thyristor of a similar rating. The ASCR features a fast turn-off time; thus it can work at a higher frequency than an SCR. Since the turn-off time is down by a factor of nearly 2, the size of the commutating components can be halved. Because of this, the switching losses will also be low.

Gate-assisted turn-off techniques are used to even further reduce the turn-off time of an ASCR. The application of a negative voltage to the gate during turn-off helps to evacuate stored charge in the device and aids the recovery mechanisms. This will in effect reduce the turn-off time by a factor of up to 2 over the conventional device.

Power Transistor

Power transistors are used in applications ranging from a few to several hundred kilowatts and switching frequencies up to about 10 kHz. Power transistors used in power conversion applications are generally *npn* type. The power transistor is turned on by supplying sufficient base current, and this base drive has to be maintained throughout its conduction period. It is turned off by removing the base drive and making the base voltage slightly negative (within $-V_{BE(max)}$). The saturation voltage of the device is normally 0.5 to 2.5 V and increases as the current increases. Hence the on-state losses increase more than proportionately with current. The transistor off-state losses are much lower than the on-state losses because the leakage current of the device is of the order of a few milliamperes. Because of relatively larger switching times, the switching loss significantly increases with switching frequency. Power transistors can block only forward voltages. The reverse peak voltage rating of these devices is as low as 5 to 10 V.

Power transistors do not have I^2t withstand capability. In other words, they can absorb only very little energy before breakdown. Therefore, they cannot be protected by semiconductor fuses, and thus an electronic protection method has to be used.

To eliminate high base current requirements, Darlington configurations are commonly used. They are available in monolithic or in isolated packages. The basic Darlington configuration is shown schematically in Fig. 29.4. The Darlington configuration presents a specific advantage in that it can considerably increase the current switched by the transistor for a given base drive. The $V_{CE(sat)}$ for the Darlington is generally more than that of a single transistor of similar rating with corresponding increase in on-state power loss. During switching, the reverse-biased collector junction may show hot spot breakdown effects that are specified by reverse-bias safe operating area (RBSOA) and forward bias safe operating area (FBSOA). Modern devices with highly interdigited emitter base geometry force more uniform current distribution and therefore considerably improve second breakdown effects. Normally, a well-designed switching aid network constrains the device operation well within the SOAs.

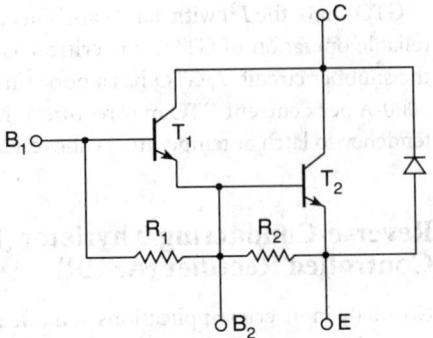

FIGURE 29.4 A two-stage Darlington transistor with bypass diode. (*Source*: B.K. Bose, *Modern Power Electronics: Evaluation, Technology, and Applications*, p. 6. © 1992 IEEE.)

Power MOSFET

Power MOSFETs are marketed by different manufacturers with differences in internal geometry and with different names such as MegaMOS, HEXFET, SIPMOS, and TMOS. They have unique features that make them potentially attractive for switching applications. They are essentially voltage-driven rather than current-driven devices, unlike bipolar transistors.

The gate of a MOSFET is isolated electrically from the source by a layer of silicon oxide. The gate draws only a minute leakage current of the order of nanoamperes. Hence the gate drive circuit is simple and power loss in the gate control circuit is practically negligible. Although in steady state the gate draws virtually no current, this is not so under transient conditions. The gate-to-source and gate-to-drain capacitances have to be charged and discharged appropriately to obtain the desired switching speed, and the drive circuit must have a sufficiently low output impedance to supply the required charging and discharging currents. The circuit symbol of a power MOSFET is shown in Fig. 29.5.

Power MOSFETs are majority carrier devices, and there is no minority carrier storage time. Hence they have exceptionally fast rise and fall times. They are essentially resistive devices when turned on, while bipolar transistors present a more or less constant $V_{CE(sat)}$ over the normal operating range. Power dissipation in MOSFETs is $I_d^2 R_{DS(on)}$, and in bipolars it is $I_C V_{CE(sat)}$. At low currents, therefore, a power MOSFET may have a lower conduction loss than a comparable bipolar device, but at higher currents, the conduction loss will exceed that of bipolars. Also, the $R_{DS(on)}$ increases with temperature.

An important feature of a power MOSFET is the absence of a secondary breakdown effect, which is present in a bipolar transistor, and as a result, it has an extremely rugged switching performance. In MOSFETs, $R_{DS(on)}$ increases with temperature, and thus the current is automatically diverted away from the hot spot. The drain body junction appears as an antiparallel diode between source and drain. Thus power MOSFETs will not support

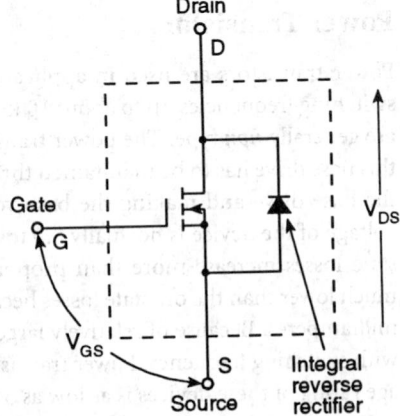

FIGURE 29.5 Power MOSFET circuit symbol. (*Source*: B.K. Bose, *Modern Power Electronics: Evaluation, Technology, and Applications*, p. 7. © 1992 IEEE.)

voltage in the reverse direction. Although this inverse diode is relatively fast, it is slow by comparison with the MOSFET. Recent devices have the diode recovery time as low as 100 ns. Since MOSFETs cannot be protected by fuses, an electronic protection technique has to be used.

Rugged MOSFETs

With the advancement in MOS technology, ruggedized MOSFETs are replacing conventional MOSFETs. Theoretically, the second breakdown mechanism is absent in power MOSFETs. In the real world of vertical conductive power MOSFETs, however, second breakdown exists due to the presence of a parasitic *npn* transistor, thus making chip designers develop the ruggedness concept to thwart such failures. The need to ruggedize power MOSFETs is related to device reliability. If a MOSFET is operated within its specification range at all times, its chances for failing catastrophically are minimal. However, if its absolute maximum ratings are exceeded, failure probability increases dramatically. Under actual operating conditions, a MOSFET may be subjected to transients either externally from the power bus supplying the circuit or from the circuit itself due, for example, to inductive kicks going beyond the absolute maximum ratings. Such conditions are likely in almost every application and in most cases are beyond a designer's control. Rugged devices are made to be more tolerant for overvoltage transients. Ruggedness is the ability of a MOSFET to operate in an environment of dynamic electrical stresses, without activating any of the parasitic bipolar junction transistors. These dynamic electrical stresses can be subdivided into three distinct types: (1) avalanche energy injection, (2) diode recovery dv/dt, and (3) static dv/dt.

The difference between a ruggedized MOSFET and a conventional device is that the ruggedized version is rated to withstand a specific amount of unclamped avalanche energy when operated at voltages above its maximum drain-to-source breakdown voltage (BV_{DSS}). In effect, the manufacturer guarantees that MOSFET will not fail catastrophically up to a specified amount of avalanche energy. Thus, ruggedizing enhances power MOSFET reliability. The rugged device can withstand higher levels of diode recovery dv/dt and static dv/dt.

Insulated-Gate Bipolar Transistor (IGBT)

The IGBT has the high input impedance and high-speed characteristics of a MOSFET with the conductivity characteristic (low saturation voltage) of a bipolar transistor. The equivalent circuit of an IGBT and its circuit symbol are shown in Fig. 29.6. The IGBT is turned on by applying a positive voltage between the gate and emitter, and as in the MOSFET, it is turned off by making the gate signal zero or slightly negative. The IGBT has a much lower on-resistance than a MOSFET. It is more like a thyristor and MOSFET. For a given IGBT, there is a critical value of drain current that will cause a large enough voltage drop to activate the thyristor. Hence, the device manufacturer specifies the peak allowable drain current that can flow without latch-up occurring. There is also a corresponding gate source voltage that permits this current to flow that should not be exceeded. Once the IGBT is in latch-up, the gate no longer has any control of the drain current. The only way to turn off the IGBT in this situation is by forced commutation of the current, exactly the same as for a thyristor. If latch-up is not terminated quickly, the IGBT will be destroyed by the excessive power dissipation. Under dynamic conditions, when the IGBT is switching from on to off, it may latch up at drain current values less than the current values described above (static latch-up current value). The improvements in design have increased the latching current to workable values.

Like the power MOSFET, the IGBT does not exhibit the secondary breakdown phenomenon common to bipolar transistors. However, care should be taken not to exceed the maximum power dissipation and specified maximum junction temperature of the device under all conditions for guaranteed reliable operation.

The on-state voltage of the IGBT is heavily dependent on the gate voltage. To obtain a low on-state voltage, a sufficiently high gate voltage must be applied. The on-state voltage also increases with temperature. Compared to a MOSFET structure, the IGBT is generally smaller for the same

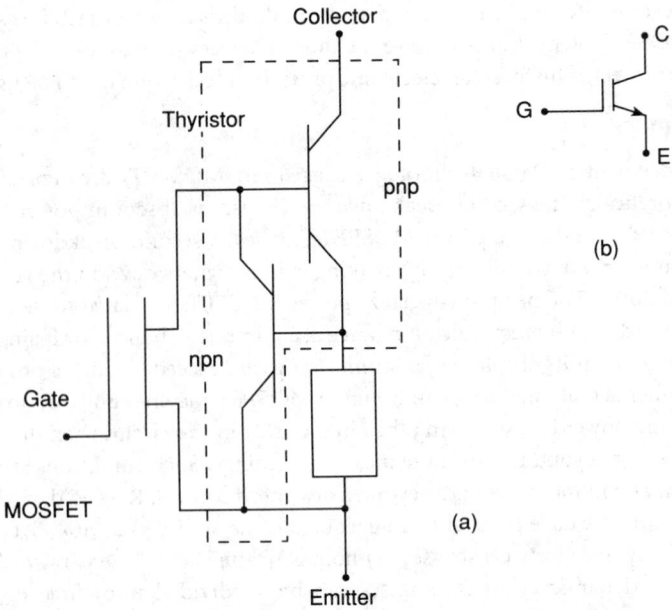

FIGURE 29.6 (a) IGBT equivalent circuit and (b) circuit symbol. (*Source:* B.K. Bose, *Modern Power Electronics: Evaluation, Technology, and Applications,* p. 7. © 1992 IEEE.)

current rating. At voltages above 400 V, an IGBT can be one-third the size of a MOSFET. The bipolar action in the IGBT slows down the speed of the device so that it exhibits a much lower frequency than the MOSFET. Typical switching frequency can be as high as 50 kHz for a standard device; higher frequencies are achievable but at the expense of higher losses.

The IGBTs cannot be as easily paralleled as MOSFETs. The factors that inhibit current sharing of parallel-connected IGBTs are (1) on-state current unbalance, caused by $V_{CE(sat)}$ distribution and main circuit wiring resistance distribution, and (2) current unbalance at turn-on and turn-off, caused by switching time difference of the parallel-connected devices and circuit wiring inductance distribution. If IGBTs having different turn-on times are paralleled, the current is hogged by the device having the shorter turn-on time. If IGBTs having different turn-off times are paralleled, current is hogged by the device having the longer turn-off time. The time differences at turn-off must be controlled carefully because turn-off times are larger than those of turn-on.

MOS-Controlled Thyristor (MCT)

The MCT is basically a thyristor with built-in MOSFETs to turn on and turn off. It is a high-power, high-frequency, low-conduction-drop and a rugged device, which is more likely to be used in future medium- and high-power applications. The MCT was announced by General Electric R&D Center on November 30, 1988. Harris is in the process of producing these devices and has supplied samples of 600-V, 75-A MCTs for evaluation. The MCT equivalent circuit and circuit symbol are shown in Fig. 29.7. The MCT has thyristor type three junctions and *pnpn* layers between the anode and cathode. MCTs are turned on by a negative voltage pulse at the gate with respect to the anode and turned off by a positive voltage pulse. Further research work is going on to make these devices turn on and off by applying a positive pulse to the gate to turn on and a negative pulse to turn off the device.

MCTs can operate at higher junction temperatures than BJTs, IGBTs, and MOSFETs. They have been tested up to 300°C at the expense of high leakage current and high turn-off current gain. MCTs have relatively low switching times and storage time. The MCT is capable of high current

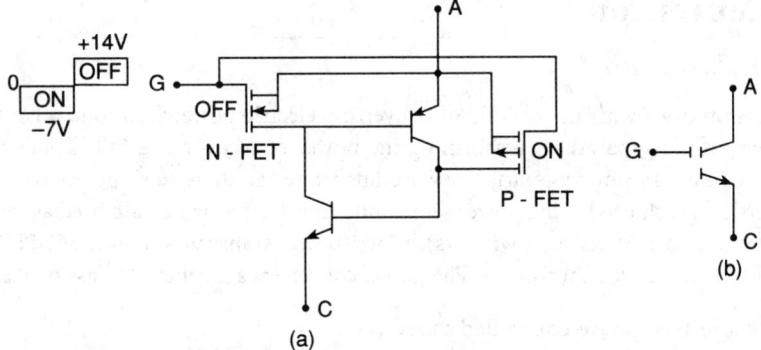

FIGURE 29.7 (a) MCT equivalent circuit and (b) circuit symbol. (*Source:* B.K. Bose, *Modern Power Electronics: Evaluation, Technology, and Applications,* p. 9. © 1992 IEEE.)

densities and blocking voltages in both directions. Since the power gain of an MCT is extremely high, it could be driven directly from logic gates. The MCT has high di/dt (of the order of 2500 A/µs) and high dv/dt (of the order of 20,000 V/µs).

The MCT, because of its superior characteristics, shows a tremendous possibility for applications such as motor drives, uninterrupted power supplies, static VAR compensators, and high-power active power line conditioners.

Defining Terms

di/dt **limit:**   Maximum allowed rate of change of current through a device. If this limit is exceeded, the device may not be guaranteed to work reliably.

*dv/dt***:**   Rate of change of voltage withstand capability without spurious turn-on of the device.

Forward voltage: The voltage across the device when the anode is positive with respect to the cathode.

I^2t**:**   Represents available thermal energy resulting from current flow.

Reverse voltage: The voltage across the device when the anode is negative with respect to the cathode.

References

B.K. Bose, *Modern Power Electronics: Evaluation, Technology, and Applications,* New York: IEEE Press, 1992.

N. Mohan and T. Undeland, *Power Electronics: Converters, Applications, and Design,* New York: John Wiley, 1989.

J. Wojslawowicz, "Ruggedized transistors emerging as power MOSFET standard-bearers," *Power Technics Magazine,* pp. 29–32, January 1988.

Further Information

B.M. Bird and K.G. King, *An Introduction to Power Electronics,* New York: Wiley-Interscience, 1984.

R. Sittig and P. Roggwiller, *Semiconductor Devices for Power Conditioning,* New York: Plenum, 1982.

V.A.K. Temple, "Advances in MOS controlled thyristor technology and capability," *Power Conversion,* pp. 544–554, Oct. 1989.

B.W. Williams, *Power Electronics, Devices, Drivers and Applications,* New York: John Wiley, 1987.

29.2 Power Conversion

Kaushik Rajashekara

Power conversion deals with the process of converting electric power from one form to another. The power electronic apparatuses performing the power conversion are called *power converters*. Because they contain no moving parts, they are often referred to as *static* power converters. The power conversion is achieved using power semiconductor devices, which are used as switches. The power devices used are **SCRs** (or thyristors), triacs, power transistors, power MOSFETs, **IGBTs**, and MCTs (MOS-controlled thyristors). The power converters are generally classified as:

1. ac-dc converters (**phase-controlled converters**)
2. direct ac-ac converters (cycloconverters)
3. dc-ac converters (inverters)
4. dc-dc converters (choppers, buck and boost converters)

AC-DC Converters

The basic function of a phase-controlled converter is to convert an alternating voltage of variable amplitude and frequency to a variable dc voltage. The power devices used for this application are generally SCRs. The average value of the output voltage is controlled by varying the conduction time of the SCRs. The turn-on of the SCR is achieved by providing a gate pulse when it is forward-biased. The turn-off is achieved by the **commutation** of current from one device to another at the instant the incoming ac voltage has a higher instantaneous potential than that of the outgoing wave. Thus there is a natural tendency for current to be commutated from the outgoing to the incoming SCR, without the aid of any external commutation circuitry. This commutation process is often referred to as *natural commutation*.

A single-phase half-wave converter is shown in Fig. 29.8. When the SCR is turned on at an angle α, full supply voltage (neglecting the SCR drop) is applied to the load. For a purely resistive load, during the positive half cycle, the output voltage waveform follows the input ac voltage waveform. During the negative half cycle, the SCR is turned off. In the case of inductive load, the energy stored in the inductance causes the current to flow in the load circuit even after the reversal of the supply voltage, as shown in Fig. 29.8(b). If there is no freewheeling diode D_F, the load current is discontinuous. A freewheeling diode is connected across the load to turn off the SCR as soon as the input voltage polarity reverses, as shown in Fig. 28.9(c). When the SCR is off, the load current will freewheel through the diode. The power flows from the input to the load only when the SCR is conducting. If there is no freewheeling diode, during the negative portion of the supply voltage, SCR returns the energy stored in the load inductance to the supply. The freewheeling diode improves the input power factor.

The controlled full-wave dc output may be obtained by using either a center tap transformer (Fig. 29.9) or by bridge configuration (Fig. 29.10). The bridge configuration is often used when a transformer is undesirable and the magnitude of the supply voltage properly meets the load voltage requirements. The average output voltage of a single-phase full-wave converter is given by

$$v_{d\alpha} = 2\,\frac{E_m}{\pi}\cos\alpha$$

where E_m is the peak value of the input voltage and α is the firing angle. The output voltage of a single-phase bridge circuit is the same as that shown in Fig. 29.9. Various configurations of the single-phase bridge circuit can be obtained if, instead of four SCRs, two diodes and two SCRs are used, with or without freewheeling diodes.

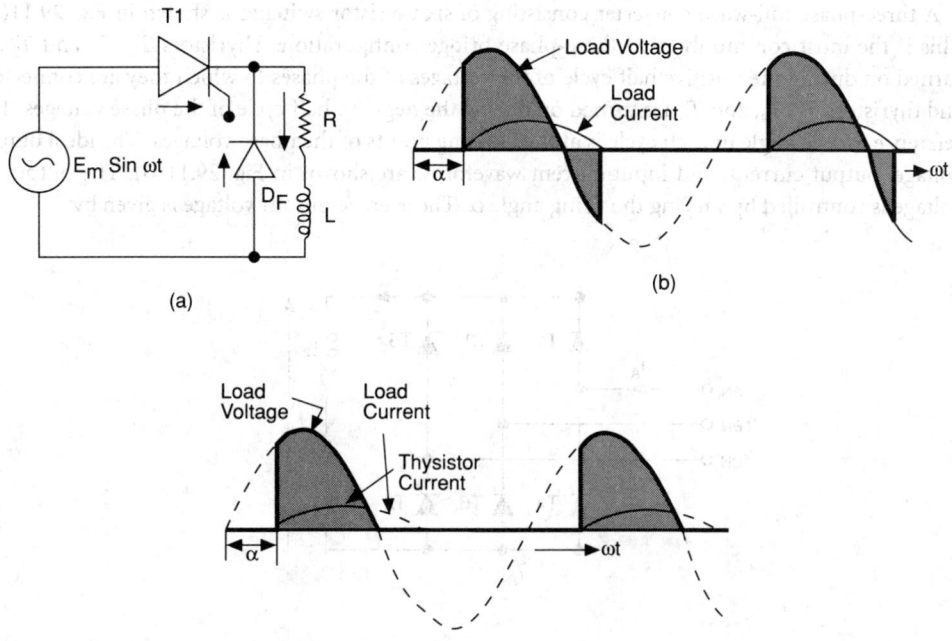

FIGURE 29.8 Single-phase half-wave converter with freewheeling diode. (a) Circuit diagram; (b) waveform for inductive load with no freewheeling diode; (c) waveform with freewheeling diode.

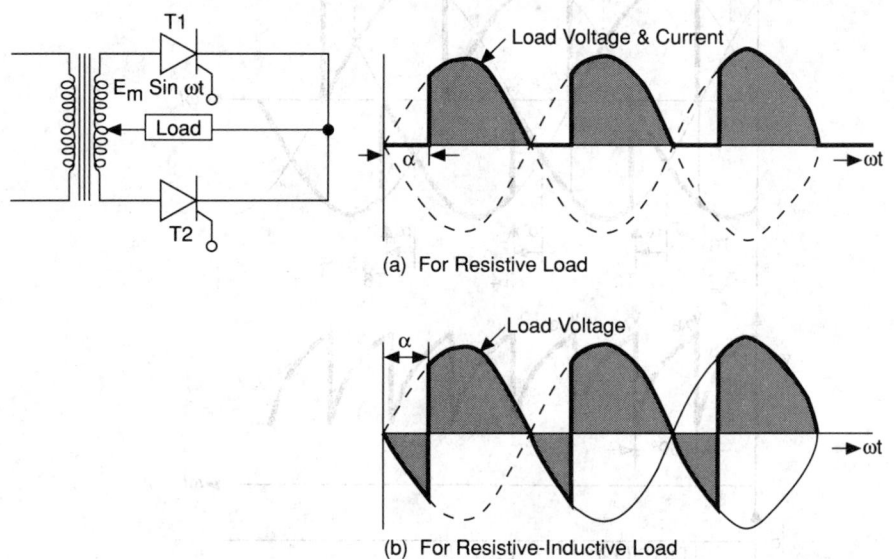

FIGURE 29.9 Single-phase full-wave converter with transformer.

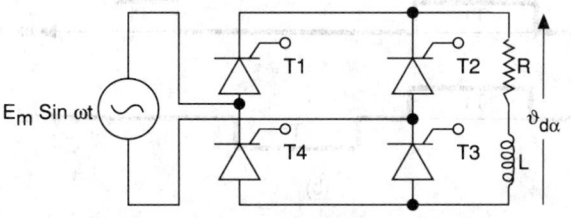

FIGURE 29.10 Single-phase bridge converter.

A three-phase full-wave converter consisting of six thyristor switches is shown in Fig. 29.11(a). This is the most commonly used three-phase bridge configuration. Thyristors T_1, T_3, and T_5 are turned on during the positive half cycle of the voltages of the phases to which they are connected, and thyristors T_2, T_4, and T_6 are turned on during the negative half cycle of the phase voltages. The reference for the angle in each cycle is at the crossing points of the phase voltages. The ideal output voltage, output current, and input current waveforms are shown in Fig. 29.11(b). The output dc voltage is controlled by varying the firing angle α. The average output voltage is given by

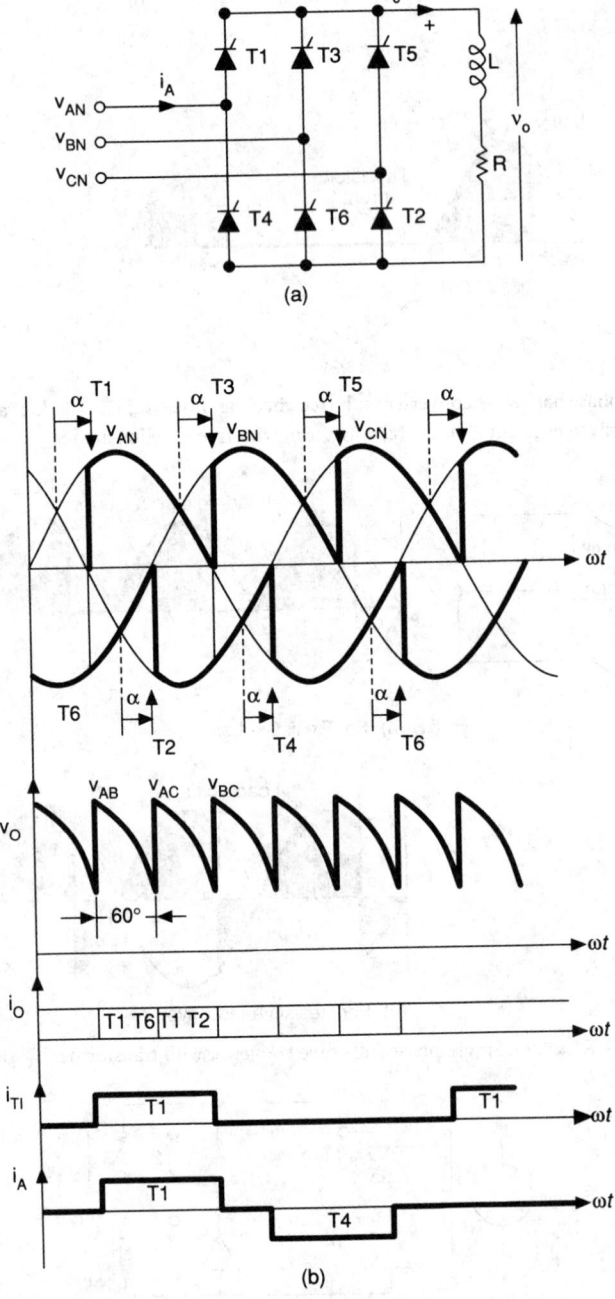

FIGURE 29.11 (a) Three-phase thyristor full bridge configuration; (b) output voltage and current waveforms.

$$v_o = \frac{3\sqrt{3}}{\pi} E_m \cos \alpha$$

where E_m is the peak value of the phase voltage. At $\alpha = 90°$, the output voltage is zero. For $0 < \alpha < 90°$, v_o is positive and power flows from ac supply to the load. For $90° < \alpha < 180°$, v_o is negative and the converter operates in the inversion mode. If the load is a dc motor, the power can be transferred from the motor to the ac supply, a process known as *regeneration*.

In Fig. 29.11(a), the top or bottom thyristors could be replaced by diodes. The resulting topology is called a *thyristor semiconverter*. With this configuration, the input power factor is improved, but the regeneration is not possible.

Cycloconverters

Cycloconverters are direct ac-to-ac frequency changers. The term *direct conversion* means that the energy does not appear in any form other than the ac input or ac output. The output frequency is lower than the input frequency and is generally an integral multiple of the input frequency. A cycloconverter permits energy to be fed back into the utility network without any additional measures. Also, the phase sequence of the output voltage can be easily reversed by the control system. Cycloconverters have found applications in aircraft systems and industrial drives. These cycloconverters are suitable for synchronous and induction motor control. The operation of the cycloconverter is illustrated in Section 29.4 of this chapter.

DC-to-AC Converters

The dc-to-ac converters are generally called *inverters*. The ac supply is first converted to dc, which is then converted to a variable-voltage and variable-frequency power supply. This generally consists of a three-phase bridge connected to the ac power source, a dc link with a filter, and the three-phase inverter bridge connected to the load. In the case of battery-operated systems, there is no intermediate dc link. Inverters can be classified as voltage source inverters (VSIs) and current source inverters (CSIs). A voltage source inverter is fed by a stiff dc voltage, whereas a current source inverter is fed by a stiff current source. A voltage source can be converted to a current source by connecting a series inductance and then varying the voltage to obtain the desired current. A VSI can also be operated in current-controlled mode, and similarly a CSI can also be operated in the voltage-control mode. The inverters are used in variable frequency ac motor drives, uninterrupted power supplies, induction heating, static VAR compensators, etc.

Voltage Source Inverter

A three-phase voltage source inverter configuration is shown in Fig. 29.12(a). The VSIs are controlled either in square-wave mode or in pulsewidth-modulated (PWM) mode. In square-wave mode, the frequency of the output voltage is controlled within the inverter, the devices being used to switch the output circuit between the plus and minus bus. Each device conducts for 180 degrees, and each of the outputs is displaced 120 degrees to generate a six-step waveform, as shown in Fig. 29.12(b). The amplitude of the output voltage is controlled by varying the dc link voltage. This is done by varying the firing angle of the thyristors of the three-phase bridge converter at the input. The square-wave-type VSI is not suitable if the dc source is a battery. The six-step output voltage is rich in harmonics and thus needs heavy filtering.

In PWM inverters, the output voltage and frequency are controlled within the inverter by varying the width of the output pulses. Hence at the front end, instead of a phase-controlled thyristor converter, a diode bridge rectifier can be used. A very popular method of controlling the voltage and frequency is by sinusoidal pulsewidth modulation. In this method, a high-frequency triangle carrier wave is compared with a three-phase sinusoidal waveform, as shown in Fig. 29.13. The

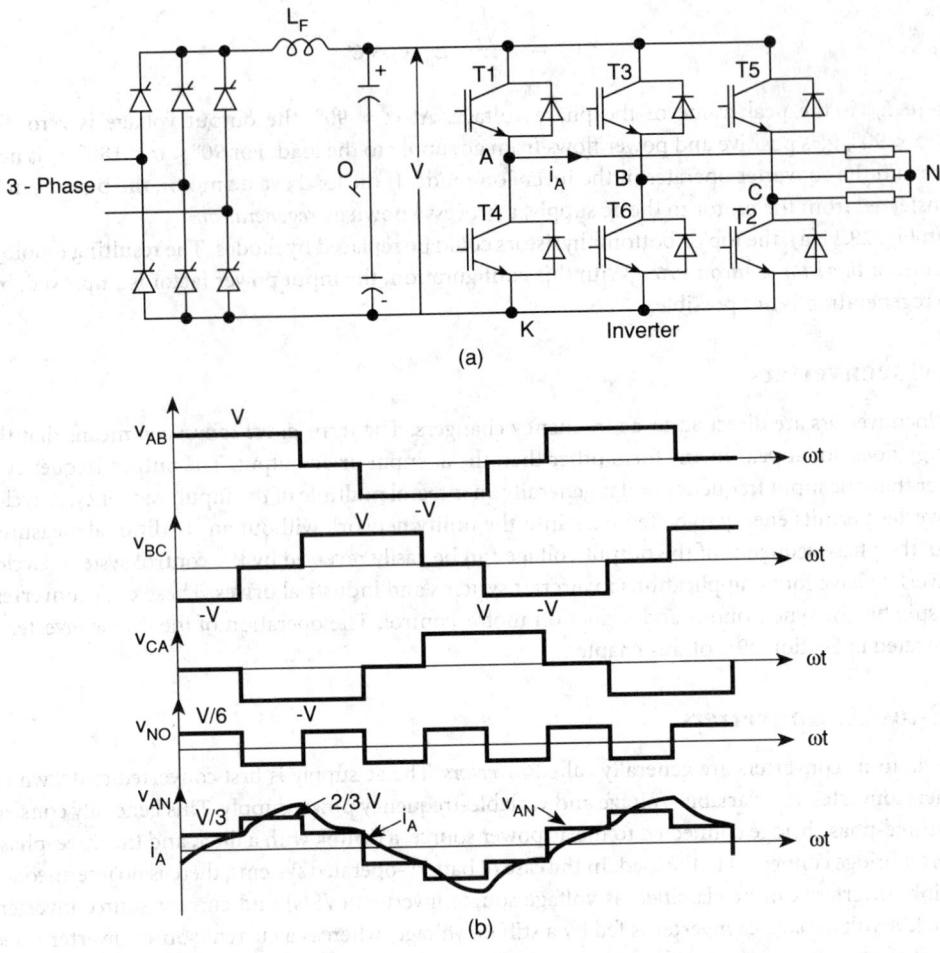

FIGURE 29.12   (a) Three-phase converter and voltage source inverter configuration; (b) three-phase
square-wave inverter waveforms.

power devices in each phase are switched on at the intersection of sine and triangle waves. The
amplitude and frequency of the output voltage are varied, respectively, by varying the amplitude
and frequency of the reference sine waves. The ratio of the amplitude of the sine wave to the
amplitude of the carrier wave is called the *modulation index.*

The harmonic components in a PWM wave are easily filtered because they are shifted to a
higher-frequency region. It is desirable to have a high ratio of carrier frequency to fundamental frequency to reduce the harmonics of lower-frequency components. There are several other PWM
techniques mentioned in the literature. The most notable ones are selected harmonic elimination,
delta modulation, hysteresis controller, and space vector PWM technique.

In inverters, if SCRs are used as power switching devices, an external forced commutation circuit
has to be used to turn off the devices. Now, with the availability of IGBTs up to 750-A, 600-V ratings, they are being used in applications up to 150- to 200-kW motor drives. Above this power rating, GTOs are generally used. Power Darlington transistors, which are available up to 800 A,
1200 V, could also be used for inverter applications.

Current Source Inverter

Contrary to the voltage source inverter where the voltage of the dc link is imposed on the motor
windings, in the current source inverter the current is imposed into the motor. Here the amplitude

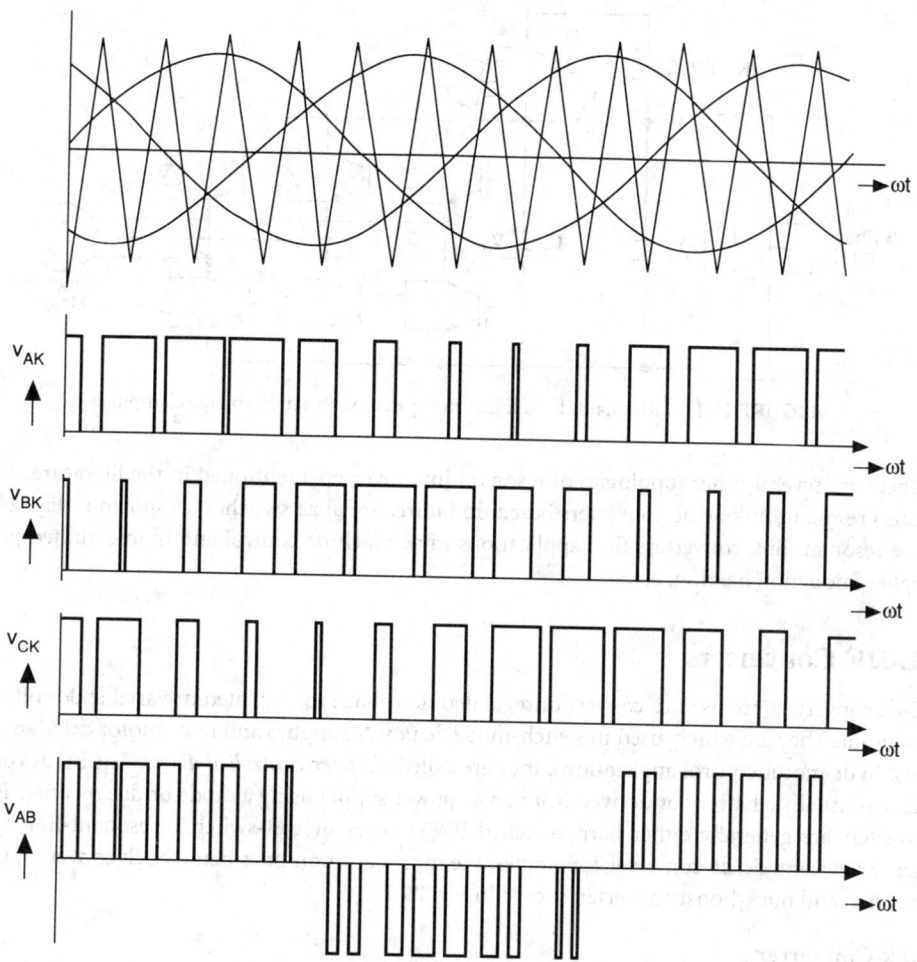

FIGURE 29.13 Three-phase sinusoidal PWM inverter waveforms.

and phase angle of the motor voltage depend on the load conditions of the motor. The current source inverter is described in detail in Section 29.4.

Resonant-Link Inverters

The use of resonant switching techniques can be applied to inverter topologies to reduce the switching losses in the power devices. They also permit high switching frequency operation to reduce the size of the magnetic components in the inverter unit. In the resonant dc-link inverter shown in Fig. 29.14, a resonant circuit is added at the inverter input to convert a fixed dc voltage to a pulsating dc voltage. This resonant circuit enables the devices to be turned on and turned off during the zero voltage interval. Zero voltage or zero current switching is often termed *soft switching*. Under soft switching, the switching losses in the power devices are almost eliminated. The electromagnetic interference (EMI) problem is less severe because resonant voltage pulses have lower *dv/dt* compared to those of hard-switched PWM inverters. Also, the machine insulation is less stretched because of lower *dv/dt* resonant voltage pulses. In Fig. 29.14, all the inverter devices are turned on simultaneously to initiate a resonant cycle. The commutation from one device to another is initiated at the zero dc-link voltage. The inverter output voltage is formed by the integral numbers of quasi-sinusoidal pulses. The circuit consisting of devices Q, D, and the capacitor C acts as an active clamp to limit the dc voltage to about 1.4 times the diode rectifier voltage V_s.

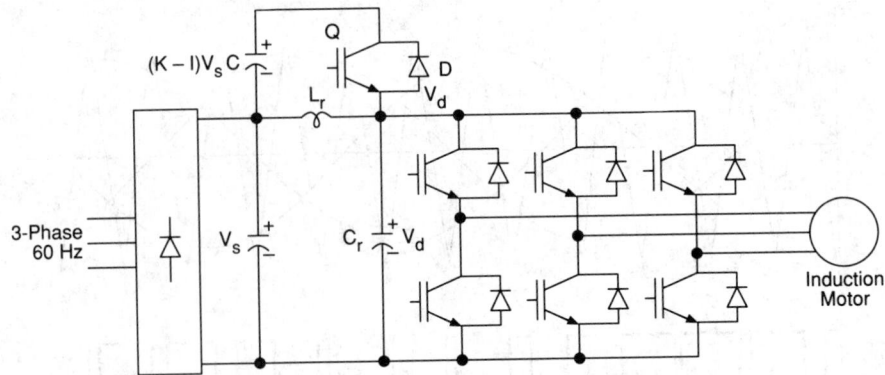

FIGURE 29.14 Resonant dc-link inverter system with active voltage clamping.

There are several other topologies of resonant link inverters mentioned in the literature. There are also resonant link ac-ac converters based on bidirectional ac switches, as shown in Fig. 29.15. These resonant link converters find applications in ac machine control and uninterrupted power supplies, induction heating, etc.

DC-DC Converters

DC-dc converters are used to convert unregulated dc voltage to regulated or variable dc voltage at the output. They are widely used in switch-mode dc power supplies and in dc motor drive applications. In dc motor control applications, they are called *chopper-controlled drives*. The input voltage source is usually a battery or derived from an ac power supply using a diode bridge rectifier. These converters are generally either hard-switched PWM types or soft-switched resonant-link types. There are several dc-dc converter topologies, the most common ones being buck converter, boost converter, and buck-boost converter, shown in Fig. 29.16.

Buck Converter

A buck converter is also called a *step-down* converter. Its principle of operation is illustrated by referring to Fig. 29.16(a). The IGBT acts as a high-frequency switch. The IGBT is repetitively closed for a time t_{on} and opened for a time t_{off}. During t_{on}, the supply terminals are connected to the load,

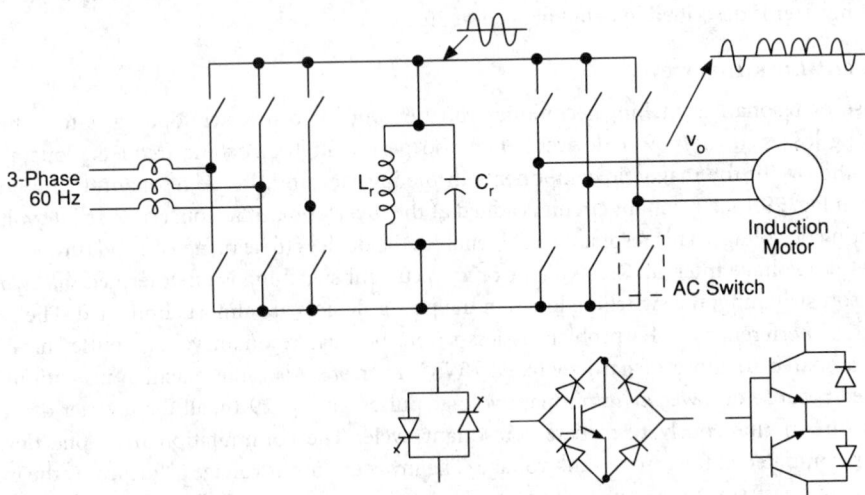

FIGURE 29.15 Resonant ac-link converter system showing configuration of ac switches.

and power flows from supply to the load. During t_{off}, load current flows through the freewheeling diode D_1, and the load voltage is ideally zero. The average output voltage is given by

$$V_{out} = DV_{in}$$

where D is the **duty cycle** of the switch and is given by $D = t_{on}/T$, where T is the time for one period. $1/T$ is the switching frequency of the power device IGBT.

Boost Converter

A boost converter is also called a *step-up* converter. Its principle of operation is illustrated by referring to Fig. 29.16(b). This converter is used to produce higher voltage at the load than the supply voltage. When the power switch is on, the inductor is connected to the dc source and the energy from the supply is stored in it. When the device is off, the inductor current is forced to flow through the diode and the load. The induced voltage across the inductor is negative. The inductor adds to the source voltage to force the inductor current into the load. The output voltage is given by

$$V_{out} = \frac{V_{in}}{1 - D}$$

Thus for variation of D in the range $0 < D < 1$, the load voltage V_{out} will vary in the range $V_{in} < V_{out} < \infty$.

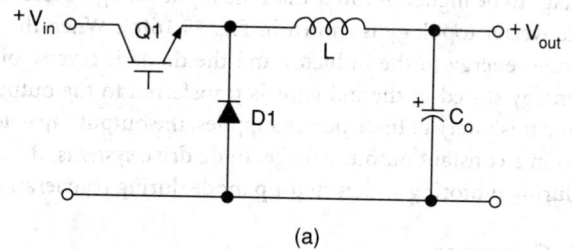

(a)

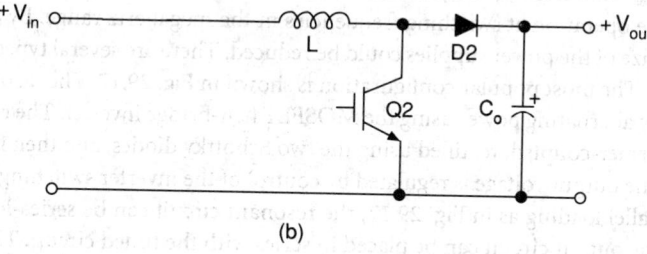

(b)

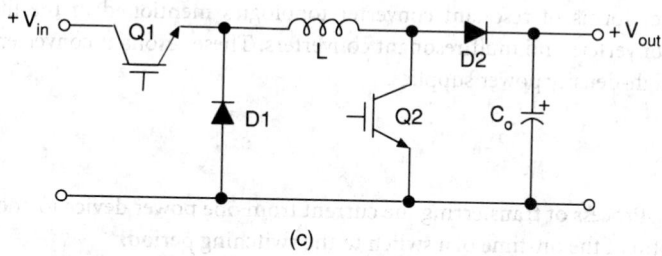

(c)

FIGURE 29.16 DC-DC converter configurations: (a) buck converter; (b) boost converter; (c) buck-boost converter.

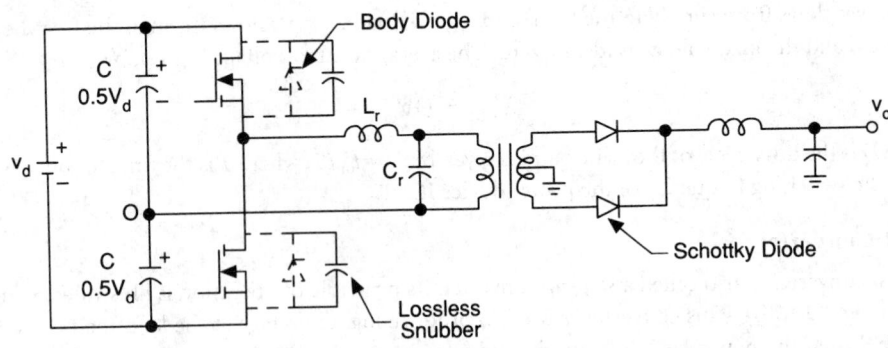

FIGURE 29.17 Resonant-link dc-dc converter.

Buck-Boost Converter

A buck-boost converter can be obtained by the cascade connection of the buck and the boost converter. The steady-state output voltage V_{out} is given by

$$V_{out} = V_{in} \frac{D}{1 - D}$$

This allows the output voltage to be higher or lower than the input voltage, based on the duty cycle D. A typical buck-boost converter topology is shown in Fig. 29.16(c). When the power device is turned on, the input provides energy to the inductor and the diode is reverse biased. When the device is turned off, the energy stored in the inductor is transferred to the output. No energy is supplied by the input during this interval. In dc power supplies, the output capacitor is assumed to be very large, which results in a constant output voltage. In dc drive systems, the chopper is operated in step-down mode during motoring and in step-up mode during regeneration operation.

Resonant-Link DC-DC Converters

The use of resonant converter topologies would help to reduce the switching losses in dc-dc converters and enable the operation at switching frequencies in the megahertz range. By operating at high frequencies, the size of the power supplies could be reduced. There are several types of resonant converter topologies. The most popular configuration is shown in Fig. 29.17. The dc power is converted to high-frequency alternating power using the MOSFET half-bridge inverter. The resonant capacitor voltage is transformer-coupled, rectified using the two Schottky diodes, and then filtered to get output dc voltage. The output voltage is regulated by control of the inverter switching frequency.

Instead of parallel loading as in Fig. 29.17, the resonant circuit can be series-loaded; that is, the transformer in the output circuit can be placed in series with the tuned circuit. The series resonant circuit provides the short-circuit limiting feature.

There are other forms of resonant converter topologies mentioned in the literature such as quasi-resonant converters and multiresonant converters. These resonant converter topologies find applications in high-density power supplies.

Defining Terms

Commutation: Process of transferring the current from one power device to another.
Duty cycle: Ratio of the on-time of a switch to the switching period.
Full-wave control: Both the positive and negative half cycle of the waveforms are controlled.
IGBT: Insulated-gate bipolar transistor.

Phase-controlled converter: Converter in which the power devices are turned off at the natural crossing of zero voltage in ac to dc conversion applications.

SCR: Silicon-controlled rectifier.

References

B.K. Bose, *Modern Power Electronics,* New York: IEEE Press, 1992.

Motorola, *Linear/Switchmode Voltage Regulator Handbook,* 1989.

K.S. Rajashekara, H. Le-Huy, *et al.,* "Resonant DC Link Inverter-Fed AC Machines Control," IEEE Power Electronics Specialists Conference, 1987, pp. 491–496.

P.C. Sen, *Thyristor DC Drives,* New York: John Wiley, 1981.

G. Venkataramanan and D. Divan, "Pulse Width Modulation with Resonant DC Link Converters," IEEE IAS Annual Meeting, 1990, pp. 984–990.

Further Information

B.K. Bose, *Power Electronics & AC Drives,* Englewood Cliffs, N.J.: Prentice-Hall, 1986.

R. Hoft, *Semiconductor Power Electronics,* New York: Van Nostrand Reinhold, 1986.

B.R. Pelly, *Thyristor Phase Controlled Converters and Cycloconverters,* New York: Wiley-Interscience, 1971.

A.I. Pressman, *Switching and Linear Power Supply, Power Converter Design,* Carmel, Ind.: Hayden Book Company, 1977.

M.H. Rashid, *Power Electronics, Circuits, Devices and Applications,* Englewood Cliffs, N.J.: Prentice-Hall, 1988.

29.3 Power Supplies

Ashoka K. S. Bhat

Power supplies are used in many industrial and aerospace applications and also in consumer products. Some of the requirements of power supplies are small size, light weight, low cost, and high power conversion efficiency. In addition to these, some power supplies require the following: electrical isolation between the source and load, low harmonic distortion for the input and output waveforms, and high power factor (PF) if the source is ac voltage. Some special power supplies require controlled direction of power flow.

Basically two types of power supplies are required: dc power supplies and ac power supplies. The output of dc power supplies is **regulated** or controllable dc, whereas the output for ac power supplies is ac. The input to these power supplies can be ac or dc.

DC Power Supplies

If an ac source is used, then ac-to-dc **converters** explained in Section 29.2 can be used. In these converters, electrical isolation can only be provided by bulky line frequency transformers. The ac source can be rectified with a diode rectifier to get an uncontrolled dc, and then a dc-to-dc converter can be used to get a controlled dc output. Electrical isolation between the input source and the output load can be provided in the dc-to-dc converter using a high-frequency (HF) transformer. Such HF transformers have small size, light weight, and low cost compared to bulky line frequency transformers. Whether the input source is dc (e.g., battery) or ac, dc-to-dc converters form an important part of dc power supplies, and they are explained in this subsection.

DC power supplies can be broadly classified as linear and switching power supplies.

A *linear power supply* is the oldest and simplest type of power supply. The output voltage is regulated by dropping the extra input voltage across a series transistor (therefore, also referred to as a series regulator). They have very small output ripple, theoretically zero noise, large hold-up time (typically 1–2 ms), and fast response. Linear power supplies have the following disadvantages: very low efficiency, electrical isolation can only be on 60-Hz ac side, larger volume and weight, and, in general, only a single output possible. However, they are still used in very small regulated power supplies and in some special applications (e.g., magnet power supplies). Three terminal linear regulator integrated circuits (ICs) are readily available (e.g., μA7815 has +15-V, 1-A output), are easy to use, and have built-in load short-circuit protection.

Switching power supplies use power semiconductor switches in the *on* and *off* switching states resulting in high efficiency, small size, and light weight. With the availability of fast switching devices, HF magnetics and capacitors, and high-speed control ICs, switching power supplies have become very popular. They can be further classified as **pulsewidth-modulated (PWM) converters** and **resonant converters**, and they are explained below.

Pulsewidth-Modulated Converters

These converters employ square-wave pulsewidth modulation to achieve voltage regulation. The average output voltage is varied by varying the duty cycle of the power semiconductor switch. The voltage waveform across the switch and at the output are square wave in nature [refer to Fig. 29.12(b)] and they generally result in higher switching losses when the switching frequency is increased. Also, the switching stresses are high with the generation of large electromagnetic interference (EMI), which is difficult to filter. However, these converters are easy to control, well understood, and have wide load control range.

The methods of control of PWM converters are discussed next.

The Methods of Control.   The PWM converters operate with a fixed-frequency, variable duty cycle. Depending on the duty cycle, they can operate in either continuous current mode (CCM) or discontinuous current mode (DCM). If the current through the output inductor never reaches zero (refer to Fig. 29.12), then the converter operates in CCM; otherwise DCM occurs.

The three possible control methods [Severns and Bloom, 1988; Hnatek, 1981; Unitrode Corporation, 1984; Motorola, 1989; Philips Semiconductors, 1991] are briefly explained below.

1. *Direct duty cycle control* is the simplest control method. A fixed-frequency ramp is compared with the control voltage [Fig. 29.18(a)] to obtain a variable duty cycle base drive signal for the transistor. This is the simplest method of control. Disadvantages of this method are (a) provides no voltage feedforward to anticipate the effects of input voltage changes, slow response to sudden input changes, poor audio susceptibility, poor open-loop line regulation, requiring higher loop gain to achieve specifications; (b) poor dynamic response.

2. *Voltage feedforward control.* In this case the ramp amplitude varies in direct proportion to the input voltage [Fig. 29.18(b)]. The open-loop regulation is very good, and the problems in 1(a) above are corrected.

3. *Current mode control.* In this method, a second inner control loop compares the peak inductor current with the control voltage which provides improved open-loop line regulation [Fig. 29.18(c)]. All the problems of the direct duty cycle control method 1 above are corrected with this method. An additional advantage of this method is that the two-pole second-order filter is reduced to a single-pole (the filter capacitor) first-order filter, resulting in simpler compensation networks.

The above control methods can be used in all the PWM converter configurations explained below. PWM converters can be classified as single-ended and double-ended converters. These converters may or may not have a high-frequency transformer for isolation.

Nonisolated Single-Ended PWM Converters.   The basic nonisolated single-ended converters are (a) buck (step-down), (b) boost (step-up), (c) buck-boost (step-up or step-down, also referred to as

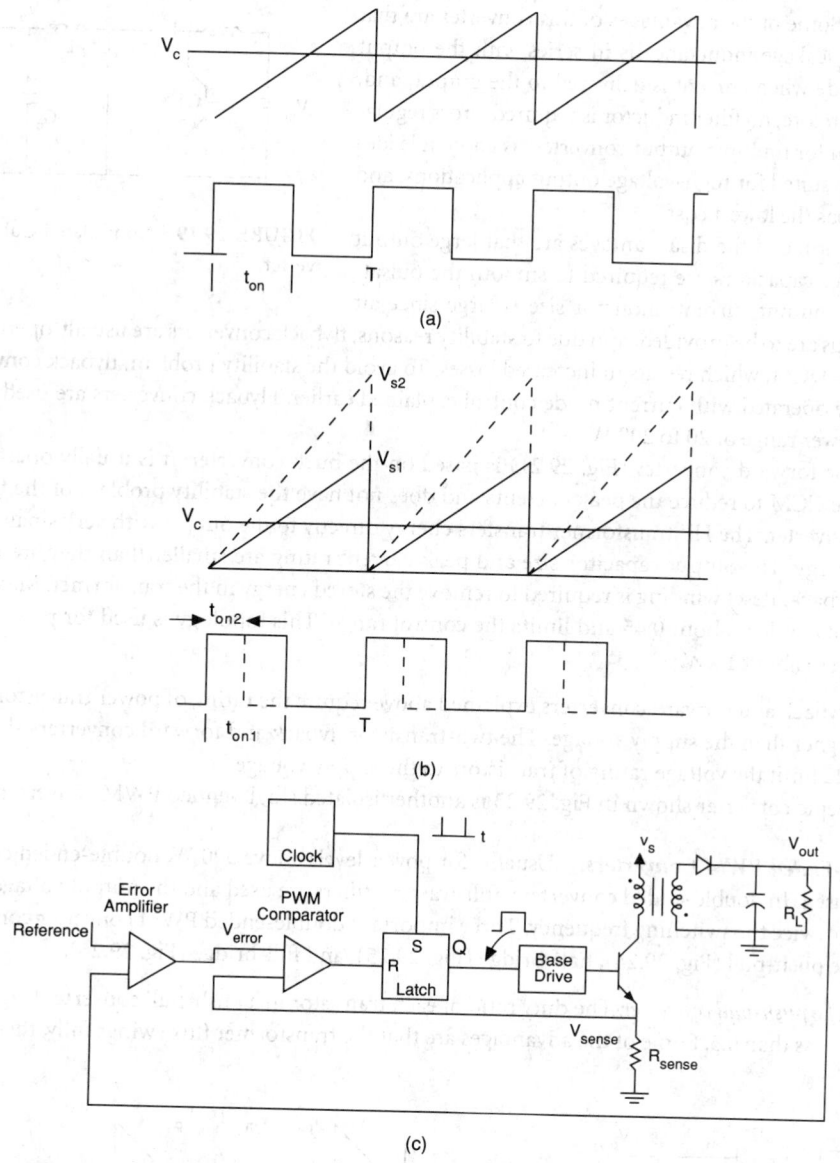

FIGURE 29.18 PWM converter control methods: (a) direct duty cycle control; (b) voltage feedforward control; (c) current mode control (illustrated for flyback converter).

flyback), and (d) Ćuk converters (Fig. 29.19). The first three of these converters have been discussed in Section 29.2. The Ćuk converter provides the advantage of nonpulsating input-output current ripple requiring smaller size external filters. Output voltage expression is the same as the buck-boost converter (refer to Section 29.2) and can be less than or greater than the input voltage. There are many variations of the above basic nonisolated converters, and most of them use a high-frequency transformer for ohmic isolation between the input and the output. Some of them are discussed below.

Isolated Single-Ended Topologies

1. The flyback converter (Fig. 29.20) is an **isolated** version of the buck-boost converter. In this converter (Fig. 29.20), when the transistor is on, energy is stored in the coupled inductor (not a transformer), and this energy is transferred to the load when the switch is off.

Some of the advantages of this converter are that the leakage inductance is in series with the output diode when current is delivered to the output, and, therefore, no filter inductor is required; cross regulation for multiple output converters is good; it is ideally suited for high-voltage output applications; and it has the lowest cost.

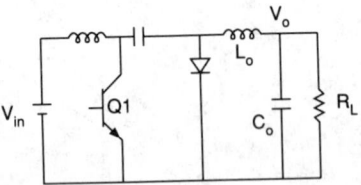

FIGURE 29.19 Nonisolated Ćuk converter.

Some of the disadvantages are that large output filter capacitors are required to smooth the pulsating output current; inductor size is large since air gaps are to be provided; and due to stability reasons, flyback converters are usually operated in the DCM, which results in increased losses. To avoid the stability problem, flyback converters are operated with current mode control explained earlier. Flyback converters are used in the power range of 20 to 200 W.

2. The forward converter (Fig. 29.21) is based on the buck converter. It is usually operated in the CCM to reduce the peak currents and does not have the stability problem of the flyback converter. The HF transformer transfers energy directly to the output with very small stored energy. The output capacitor size and peak current rating are smaller than they are for the flyback. Reset winding is required to remove the stored energy in the transformer. Maximum duty cycle is about 0.45 and limits the control range. This topology is used for power levels up to about 1 kW.

The flyback and forward converters explained above require the rating of power transistors to be much higher than the supply voltage. The two-transistor flyback and forward converters shown in Fig. 29.22 limit the voltage rating of transistors to the supply voltage.

The Sepic converter shown in Fig. 29.23 is another isolated single-ended PWM converter.

Double-Ended PWM Converters. Usually, for power levels above 300 W, double-ended converters are used. In double-ended converters, full-wave rectifiers are used and the output voltage ripple will have twice the switching frequency. Three important double-ended PWM converter configurations are push-pull (Fig. 29.24), half-bridge (Fig. 29.25), and full-bridge (Fig. 29.26).

1. *The push-pull converter.* The duty ratio of each transistor in a push-pull converter (Fig. 29.24) is less than 0.5. Some of the advantages are that the transformer flux swings fully, thereby the

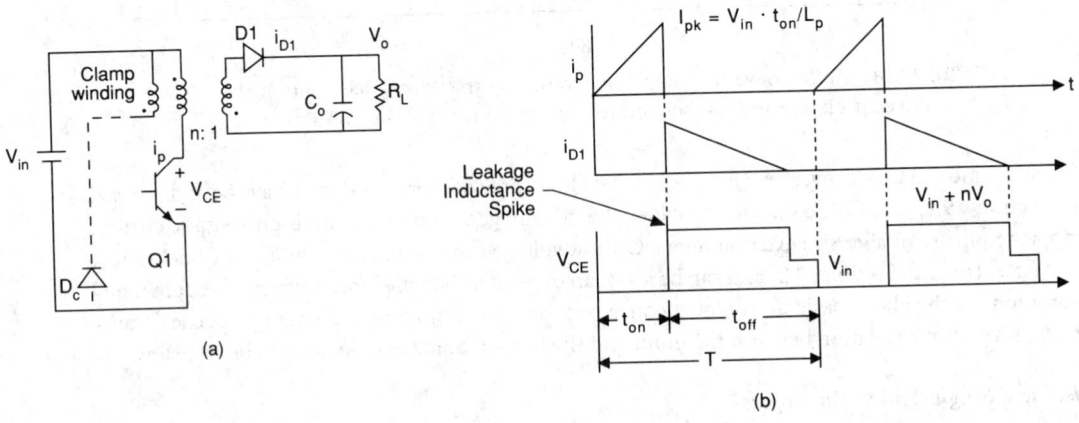

FIGURE 29.20 (a) Flyback converter. The clamp winding shown is optional and is used to clamp the transistor voltage stress to $V_{in} + nV_o$. (b) Flyback converter waveforms without the clamp winding. The leakage inductance spikes vanish with the clamp winding.

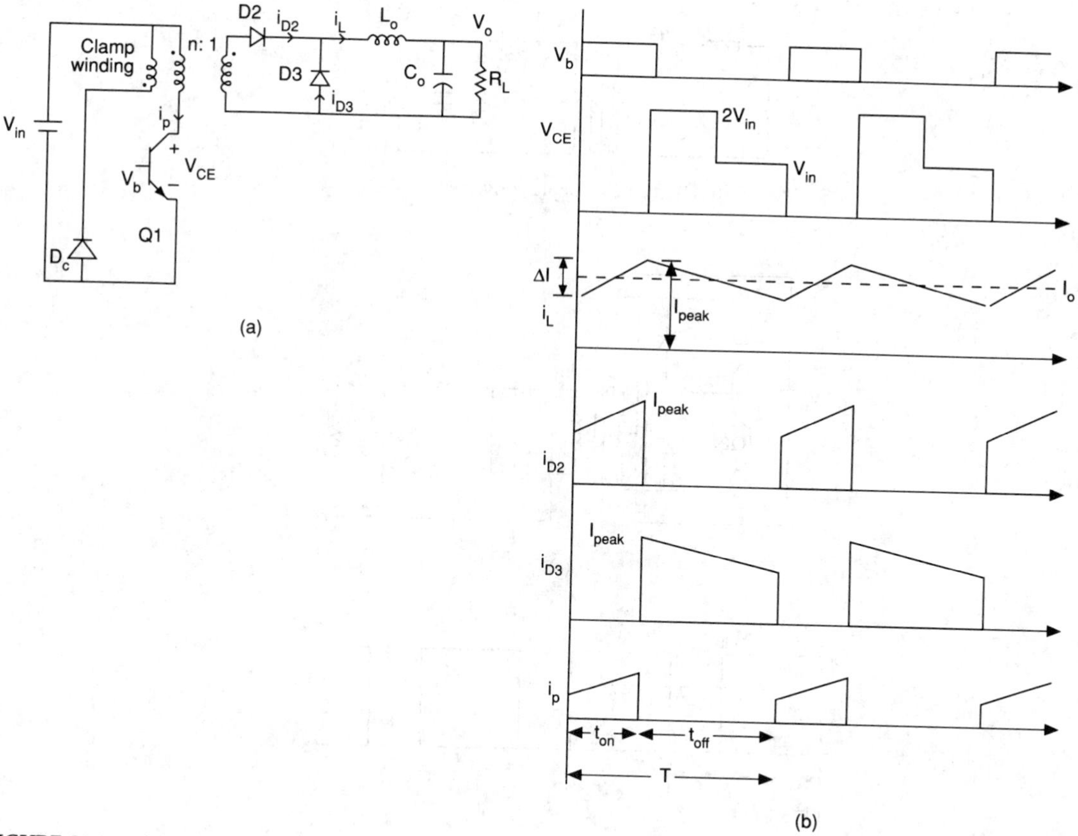

FIGURE 29.21 (a) Forward converter. The clamp winding shown is required for operation. (b) Forward converter waveforms.

size of the transformer is much smaller (typically half the size) than single-ended converters, and output ripple is twice the switching frequency of transistors, allowing smaller filters.

Some of the disadvantages of this configuration are that transistors must block twice the supply voltage, flux symmetry imbalance can cause transformer saturation and special control circuitry is required to avoid this problem, and use of center-tap transformer requires extra copper resulting in higher volt-ampere (VA) rating.

Current mode control (for the primary current) can be used to overcome the flux imbalance. This configuration is used in 100- to 500-W output range.

2. *The half-bridge.* In the half-bridge configuration (Fig. 29.25) center-tapped dc source is created by two smoothing capacitors (C_{in}), and this configuration utilizes the transformer core efficiently. The voltage across each transistor is equal to the supply voltage (half of push-pull) and, therefore, is suitable for high-voltage inputs. One salient feature of this configuration

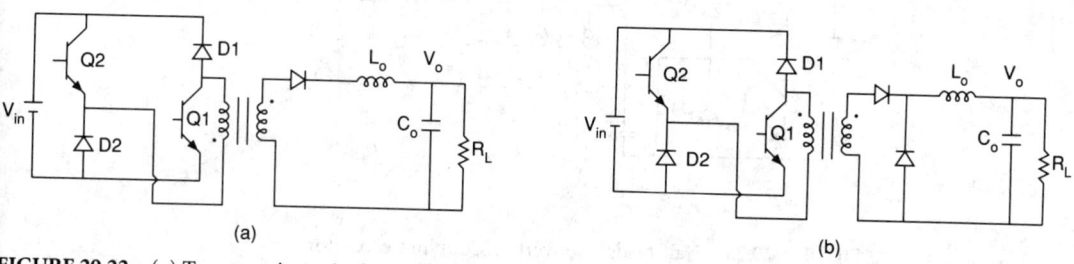

FIGURE 29.22 (a) Two-transistor single-ended flyback converter. (b) Two-transistor single-ended forward converter.

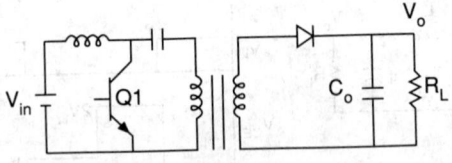

FIGURE 29.23 Sepic converter.

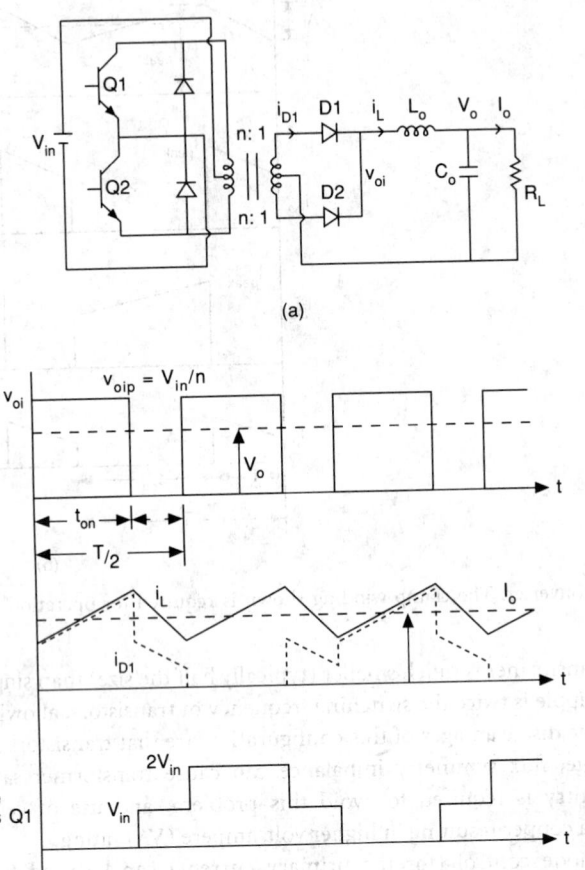

(a)

(b)

FIGURE 29.24 (a) Push-pull converter and (b) its operating waveforms.

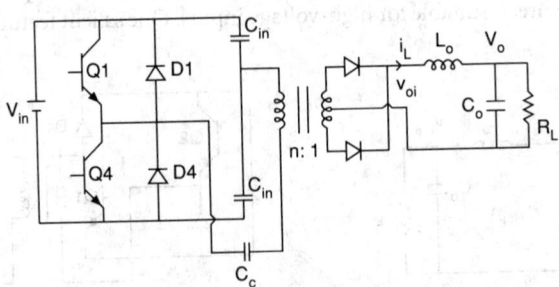

FIGURE 29.25 Half-bridge converter. Coupling capacitor C_c is used to avoid transformer saturation.

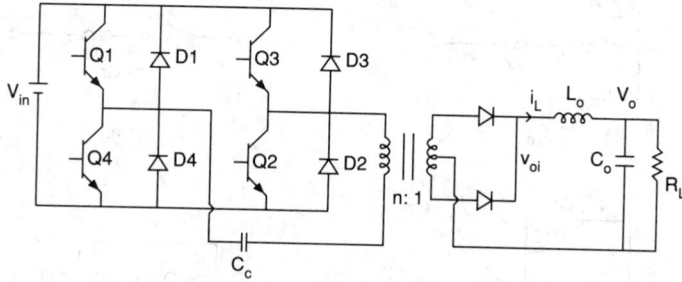

FIGURE 29.26   Full-bridge converter.

is that the input filter capacitors can be used to change between 110/220-V mains as selectable inputs to the supply.

The disadvantage of this configuration is the requirement for large-size input filter capacitors. The half-bridge configuration is used for power levels of the order of 500 to 1000 W.

3. The full-bridge configuration (Fig. 29.26) requires only one smoothing capacitor, and for the same transistor type as that of half-bridge, output power can be doubled. It is usually used for power levels above 1 kW, and the design is more costly due to increased number of components (uses four transistors compared to two in push-pull and half-bridge converters).

One of the salient features of a full-bridge converter is that by using proper control technique it can be operated in zero-voltage switching (ZVS) mode. This type of operation results in negligible switching losses. However, at reduced load currents, the ZVS property is lost. Recently, there has been a lot of effort to overcome this problem.

Resonant Power Supplies

Similar to the PWM converters, there are two types of resonant converters: single-ended and double-ended. Resonant converter configurations are obtained from the PWM converters explained earlier by adding LC (inductor-capacitor) resonating elements to obtain sinusoidally varying voltage and/or current waveforms. This approach reduces the switching losses and the switch stresses, enabling the converter to operate at high switching frequencies, resulting in reduced size, weight, and cost. Some other advantages of resonant converters are that leakage inductances of HF transformers and the junction capacitances of semiconductors can be used profitably in the resonant circuit, and reduced EMI. The major disadvantage of resonant converters is increased peak current (or voltage) stress.

Single-Ended Resonant Converters.   They are referred to as quasi-resonant converters (QRCs) since the voltage (or current) waveforms are quasi-sinusoidal in nature. The QRCs can operate with

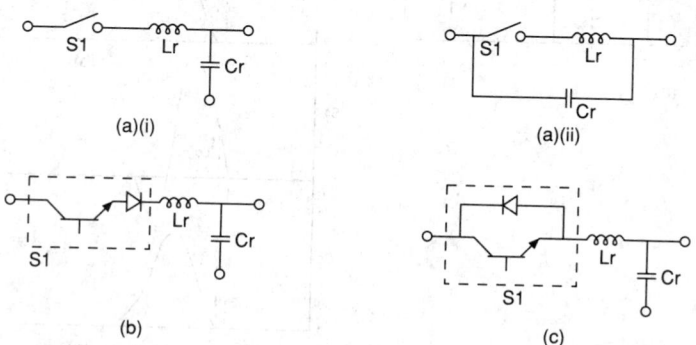

FIGURE 29.27   (a) Zero-current resonant switch: (i) L-type and (ii) M-type. (b) Half-wave configuration using L-type ZC resonant switch. (c) Full-wave configuration using L-type ZC resonant switch.

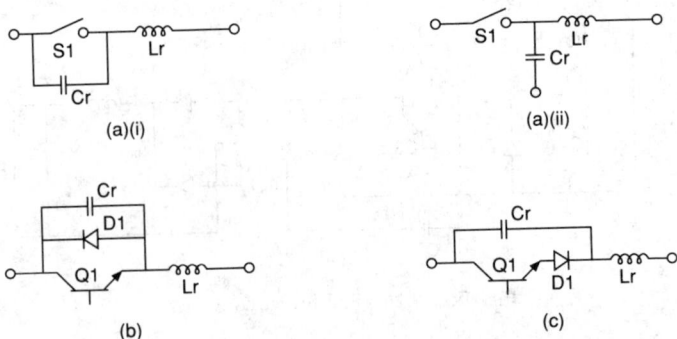

FIGURE 29.28   (a) Zero-voltage resonant switches. (b) Half-wave configuration using ZV resonant switch shown in Fig. (a)(i). (c) Full-wave configuration using ZV resonant switch shown in Fig. (a)(i).

zero-current switching (ZCS) or ZVS or both. All the QRC configurations can be generated by replacing the conventional switches by the resonant switches shown in Figs. 29.27 and 29.28. A number of configurations are realizable. Basic principles of ZCS and ZVS are explained briefly below.

1. *Zero-current switching QRCs* [Sum, 1988; Liu *et al.*, 1985].   Figure 29.29(a) shows an example of a ZCS QR buck converter implemented using a ZC resonant switch. Depending on whether the resonant switch is half-wave type or full-wave type, the resonating current will be only half-wave sinusoidal [Fig. 29.29(b)] or a full sine-wave [Fig. 29.29(c)]. The device currents are shaped sinusoidally, and, therefore, the switching losses are almost negligible with low turn-on and turn-off stresses. ZCS QRCs can operate at frequencies of the order of 2 MHz. The major problems with this type of converter are high peak currents through the switch and capacitive turn-on losses.

2. *Zero-voltage switching QRCs* [Sum, 1988; Liu and Lee, 1986].   ZVS QRCs are duals of ZCS QRCs. The auxiliary LC elements are used to shape the switching device's voltage waveform

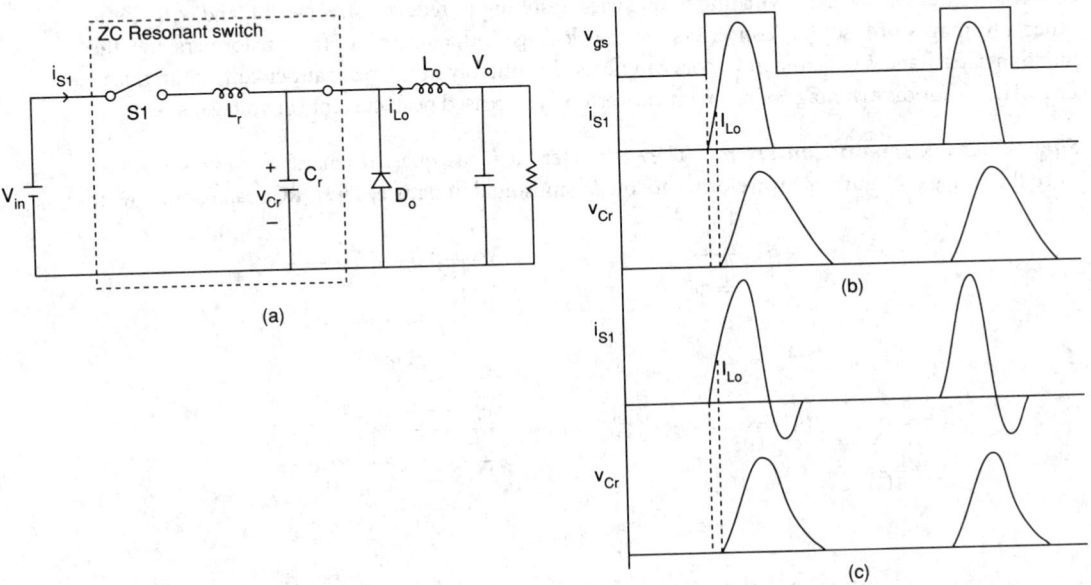

FIGURE 29.29   (a) Implementation of ZCS QR buck converter using L-type resonant switch. (b) Operating waveforms for half-wave mode. (c) Operating waveforms for full-wave mode.

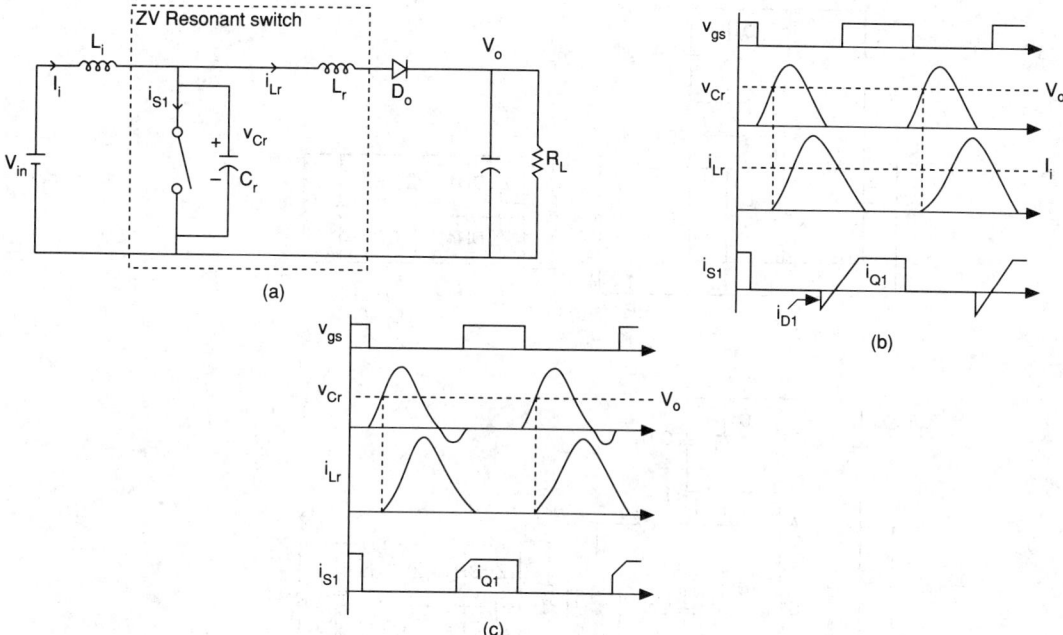

FIGURE 29.30 (a) Implementation of ZVS QR buck converter using resonant switch shown in Fig. 29.28(a)(i). (b) Operating waveforms for half-wave mode. (c) Operating waveforms for full-wave mode.

at off time in order to create a zero-voltage condition for the device to turn on. Fig. 29.30(a) shows an example of ZVS QR boost converter implemented using a ZV resonant switch. The circuit can operate in the half-wave mode [Fig. 29.30(b)] or in the full-wave mode [Fig. 29.30(c)] depending on whether a half-wave or full-wave ZV resonant switch is used, and the name comes from the capacitor voltage waveform. The full-wave mode ZVS circuit suffers from capacitive turn-on losses. The ZVS QRCs suffer from increased voltage stress on the switch. However, they can be operated at much higher frequencies compared to ZCS QRCs.

Double-Ended Resonant Converters. These converters [Sum, 1988; Bhat, 1991; Steigerwald, 1988; Bhat, 1992] use full-wave rectifiers at the output, and they are generally referred to as resonant converters. A number of resonant converter configurations are realizable by using different resonant tank circuits, and the three most popular configurations, namely, the series resonant converter (SRC), the parallel resonant converter (PRC), and the series-parallel resonant converter (SPRC) (also called LCC-type PRC), are shown in Fig. 29.31.

Series resonant converters [Fig. 29.31(a)] have high efficiency from full load to part load. Transformer saturation is avoided due to the series blocking resonating capacitor. The major problems with the SRC are that it requires a very wide change in switching frequency to regulate the load voltage and the output filter capacitor must carry high ripple current (a major problem especially in low output voltage, high output current applications).

Parallel resonant converters [Fig. 29.31(b)] are suitable for low output voltage, high output current applications due to the use of filter inductance at the output with low ripple current requirements for the filter capacitor. The major disadvantage of the PRC is that the device currents do not decrease with the load current, resulting in reduced efficiency at reduced load currents.

The SPRC [Fig. 29.31(c)] takes the desirable features of SRC and PRC.

Load voltage regulation in resonant converters for input supply variations and load changes is achieved by either varying the switching frequency or using fixed-frequency (variable pulsewidth) control.

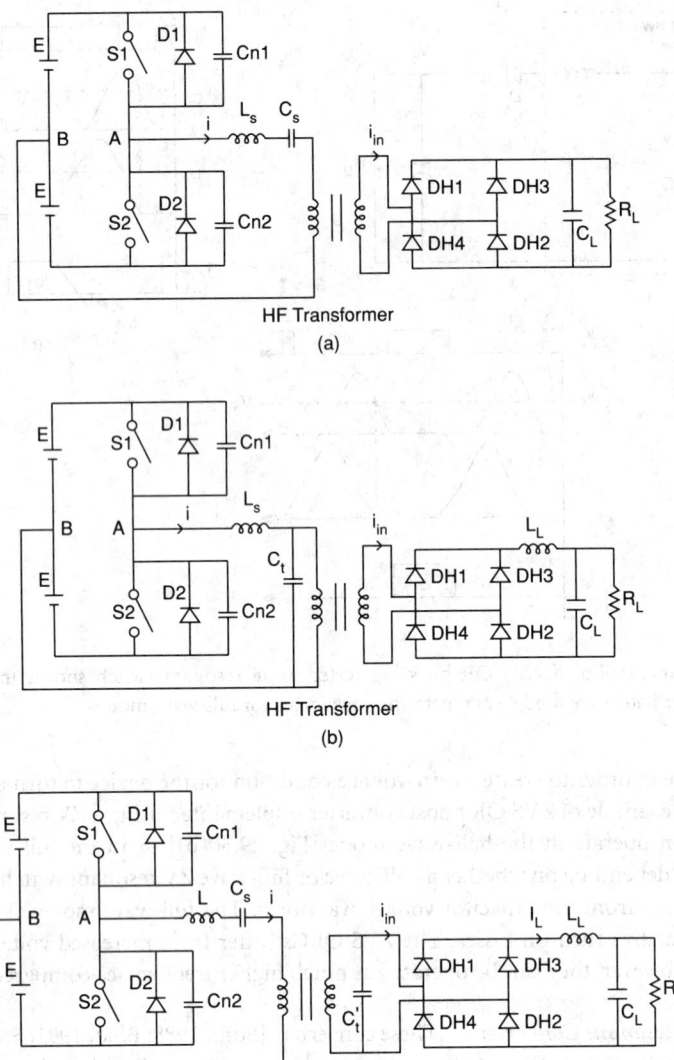

FIGURE 29.31 High-frequency resonant converter (half-bridge version) configurations suitable for operation above resonance. *Cn*1 and *Cn*2 are the snubber capacitors. (*Note:* For operation below resonance, *di/dt* limiting inductors and RC snubbers are required. For operation above resonance, only capacitive snubbers are required as shown.) (a) Series resonant converter. Leakage inductances of the HF transformer can be part of resonant inductance. (b) Parallel resonant converter. (c) Series-parallel (or LCC-type) resonant converter with capacitor C_t placed on the secondary side of the HF transformer.

1. *Variable-frequency operation.* Depending on whether the switching frequency is below or above the natural resonance frequency (ω_r), the converter can operate in different operating modes as explained below.

    a. Below-resonance (leading PF) mode. When the switching frequency is below the natural resonance frequency, the converter operates in a below-resonance mode (Fig. 29.32). The equivalent impedance across *AB* presents a leading PF so that natural turn-off of the switches is assured and any type of fast turn-off switch (including asymmetric SCRs) can be used. Depending on the instant of turn-on of switches S_1 and S_2, the converter can enter into two modes of operation, namely, continuous and discontinuous current modes. The steady-state operation in continuous current mode (CCM) [Fig. 29.32(a)] is explained briefly as follows.

Assume that diode D_2 was conducting and switch S_1 is turned on. The current carried by D_2 will be transferred to S_1 almost instantaneously (except for a small time of recovery of D_2 during which input supply is shorted through D_2 and S_1, and the current is limited by the *di/dt* limiting inductors). The current *i* then oscillates sinusoidally and goes to zero in the natural way. The current tries to reverse, and the path for this current is provided by the diode D_1. Conduction of D_1 feeds the reactive energy in the load and the tank circuit back to the supply. The on-state of D_1 also provides a reverse voltage across S_1, allowing it to turn off. After providing a time equal to or greater than the turn-off time of S_1, switch S_2 can be turned on to initiate the second half cycle. The process is similar to the first half cycle, with the voltage across v_{AB} being of opposite polarity, and the functions of D_1, S_1 will be assumed by D_2, S_2. With this type of operation, the converter works in the continuous current mode as the switches are turned on before the currents in the diodes reach zero. If the switching on of S_1 and S_2 is delayed such that the currents through the previously conducting diodes reach zero, then there are zero current intervals and the **inverter** operates in the DCM [Fig. 29.32(b)].

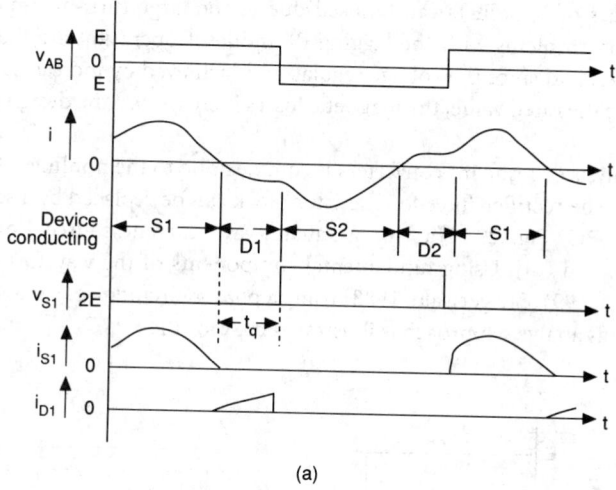

(a)

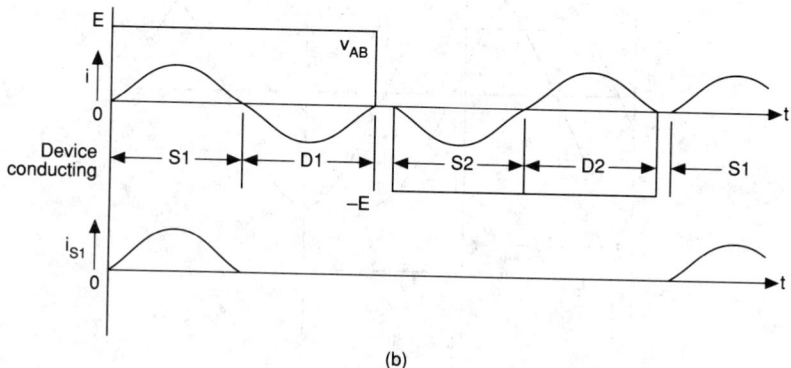

(b)

FIGURE 29.32 Typical waveforms at different points of a resonant converter operating below resonance (a) in continuous current mode and (b) in discontinuous current mode.

Load voltage regulation is achieved by decreasing the switching frequency below the rated value. Since the inverter output current i leads the inverter output voltage v_{AB}, this type of operation is also called a leading PF mode of operation. If transistors are used as the switching devices, then for operation in DCM, the pulsewidth can be kept constant while decreasing the switching frequency to avoid CCM operation. DCM operation has the advantages of negligible switching losses due to ZCS, lower di/dt and dv/dt stresses, and simple control circuitry. However, DCM operation results in higher switch peak currents.

From the waveforms shown in Fig. 29.32, the following problems can be identified for operation in the below-resonance mode: requirement of di/dt inductors to limit the large turn-on switch currents and a need for lossy RC snubbers and fast recovery diodes. Since the switching frequency is decreased to control the load power, the HF transformer and magnetics must be designed for the lowest switching frequency, resulting in increased size of the converter.

b. Above-resonance (lagging PF) mode. If switches capable of gate or base turn-off (e.g., MOSFETs, bipolar transistors) are used, then the converter can operate in the above-resonance mode (lagging PF mode). Figure 29.33 shows some typical operating waveforms for such type of operation, and it can be noticed that the current i lags the voltage v_{AB}. Since the switch takes current from its own diode across it at zero-current point, there is no need for di/dt limiting inductance, and a simple capacitive snubber can be used. In addition, the internal diodes of MOSFETs can be used due to the large turn-off time available for the diodes. Major problems with the lagging PF mode of operation are that there are switch turn-off losses, and since the voltage regulation is achieved by increasing the switching frequency above the rated value, the magnetic losses increase and the design of a control circuit is difficult.

Exact analysis of resonant converters is complex due to the nonlinear loading on the resonant tanks. The rectifier-filter-load resistor block can be replaced by a square-wave voltage source [for SRC, Fig. 29.31(a)] or a square-wave current source [for PRC and SPRC, Fig. 29.31(b) and (c)]. Using fundamental components of the waveforms, an approximate analysis [Bhat, 1991; Steigerwald, 1988] using a phasor circuit gives a reasonably good design approach. This analysis approach is illustrated next for the SPRC.

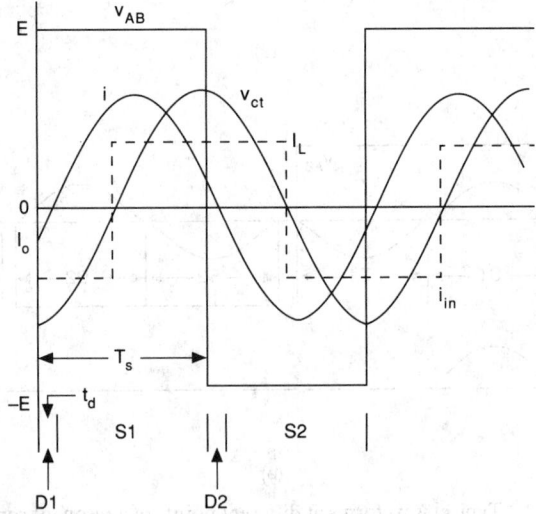

FIGURE 29.33 Typical operating waveforms at different points of an SPRC operating above resonance.

2. *Approximate analysis of SPRC.* Figure 29.34 shows the equivalent circuit at the output of the inverter and the phasor circuit used for the analysis. All the equations are normalized using the base quantities

$$\text{Base voltage } V_B = E_{min}$$

$$\text{Base impedance } Z_B = R'_L = n^2 R_L$$

$$\text{Base current } I_B = V_B / I_B$$

The converter gain [normalized output voltage in per unit (p.u.) referred to the primary-side] can be derived as [Bhat, 1991; Steigerwald, 1988]

$$M = \frac{1}{\left\{ \left(\frac{\pi^2}{8} \right)^2 \left[1 + \left(\frac{C_t}{C_s} \right) \left(1 - y_s^2 \right) \right]^2 + Q_s^2 \left[y_s - \left(\frac{1}{y_s} \right) \right]^2 \right\}^{1/2}} \text{ p.u.} \tag{29.1}$$

where

$$Q_s = \frac{(L_s / C_s)^{1/2}}{R'_L} \tag{29.2}$$

$$y_s = \frac{f_s}{f_r} \tag{29.3}$$

and

$$f_s = \text{switching frequency}$$

$$f_r = \text{series resonance frequency}$$

$$= \frac{\omega_r}{2\pi} = \frac{1}{2\pi (L_s C_s)^{1/2}} \tag{29.4}$$

The equivalent impedance looking into the terminals *AB* is given by

$$Z_{eq} = \frac{B_1 + jB_2}{B_3} \text{ p.u.} \tag{29.5}$$

where

$$B_1 = \left(\frac{8}{\pi^2} \right) \left(\frac{C_s}{C_t} \right)^2 \left(\frac{Q_s}{y_s} \right)^2 \tag{29.6}$$

$$B_2 = Q_s \left(y_s - \frac{1}{y_s} \right) \left[1 + \left(\frac{8}{\pi^2} \right)^2 \left(\frac{C_s}{C_t} \right)^2 \left(\frac{Q_s}{y_s} \right)^2 \right] - \left(\frac{C_s}{C_t} \right) \left(\frac{Q_s}{y_s} \right) \tag{29.7}$$

$$B_3 = 1 + \left(\frac{8}{\pi^2} \right)^2 \left(\frac{C_s}{C_t} \right)^2 \left(\frac{Q_s}{y_s} \right)^2 \tag{29.8}$$

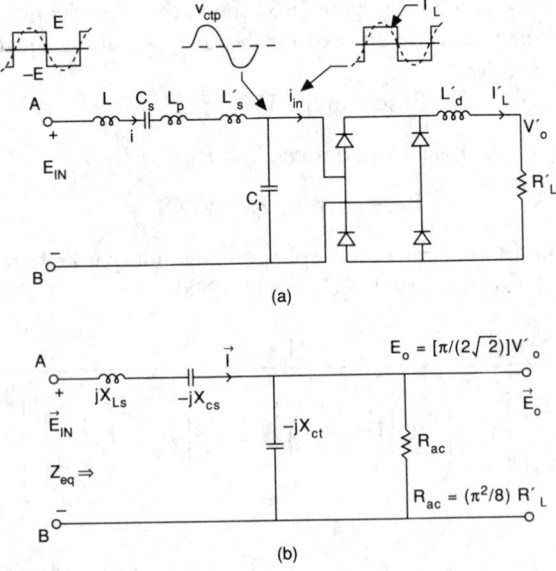

FIGURE 29.34 (a) Equivalent circuit for a SPRC at the output of the inverter terminals (across AB) of Fig. 29.31(c), L_p and L'_s are the leakage inductance of the primary and primary referred leakage inductance of the secondary, respectively. (b) Phasor circuit model used for the analysis of the SPRC converter.

The peak inverter output (resonant inductor) current can be calculated using

$$I_p = \frac{4}{\pi |Z_{eq}|} \quad \text{p.u.} \tag{29.9}$$

The same current flows through the switching devices.

The value of initial current I_0 is given by

$$I_0 = I_p \sin(-\phi) \quad \text{p.u.} \tag{29.10}$$

where $\phi = \tan^{-1}(B_2/B_1)$ rad. B_1 and B_2 are given by Eqs. (29.6) and (29.7), respectively.

If I_0 is negative, then forced commutation is necessary and the converter is operating in the lagging PF mode. The peak voltage across the capacitor C'_t (on the secondary side) is

$$v_{ctp} = \frac{\pi}{2} V_o \quad \text{V} \tag{29.11}$$

The peak voltage across C_s and the peak current through C'_t are given by

$$V_{csp} = \frac{Q_s}{y_s} I_p \quad \text{p.u.} \tag{29.12}$$

$$I_{ctp} = \frac{V_{ctp}}{X_{cptu} R_L} \quad \text{A} \tag{29.13}$$

$$X_{ctpu} = \left(\frac{C_s}{C_t}\right)\left(\frac{Q_s}{y_s}\right) \quad \text{p.u.} \tag{29.14}$$

The plot of converter gain versus the switching frequency ratio y_s, obtained using (29.1), is shown for $C_s/C_t = 1$ in Fig. 29.35, for the lagging PF mode of operation. If the ratio C_s/C_t increases, then the converter takes the characteristics of SRC and the load voltage regulation requires a very wide range in the frequency change. Lower values of C_s/C_t take the characteristics of a PRC. Therefore, a compromised value of $C_s/C_t = 1$ is chosen.

It is possible to realize higher-order resonant converters with improved characteristics and many of them are presented in Bhat [1991].

3. *Fixed-frequency operation.* To overcome some of the problems associated with the variable frequency control of resonant converters, they are operated with fixed frequency [Sum, 1988; Bhat, 1992]. A number of configurations and control methods for fixed-frequency operation are available in the literature (Bhat [1992] gives a list of papers). One of the most popular methods of control is the phase-shift control (also called clamped-mode or PWM operation) method. Figure 29.36 illustrates the clamped-mode fixed-frequency operation of the SPRC. The load power control is achieved by changing the phase-shift angle ϕ between the gating signals to vary the pulsewidth of v_{AB}.

4. *Design example.* Design a 500-W output SPRC (half-bridge version) with secondary-side resonance (operation in lagging PF mode and variable-frequency control) with the following specifications:

$$\text{Minimum input supply voltage} = 2E_{\min} = 230 \text{ V}$$

$$\text{Load voltage, } V_o = 48 \text{ V}$$

$$\text{Switching frequency, } f_s = 100 \text{ kHz}$$

$$\text{Maximum load current} = 10.42 \text{ A}$$

As explained in item 2, $C_s/C_t = 1$ is chosen. Using the constraints (1) minimum kVA rating of tank circuit per kW output power, (2) minimum inverter output peak current, and (3) enough turn-off time for the switches, it can be shown that [Bhat, 1991] $Q_s = 4$ and $y_s = 1.1$ satisfy the design constraints. From Fig. 29.35, $M = 0.8$ p.u.

Average load voltage referred to the primary side of the HF transformer = 0.8×115 V = 92 V. Therefore, the transformer turns ratio required ≈ 1.84.

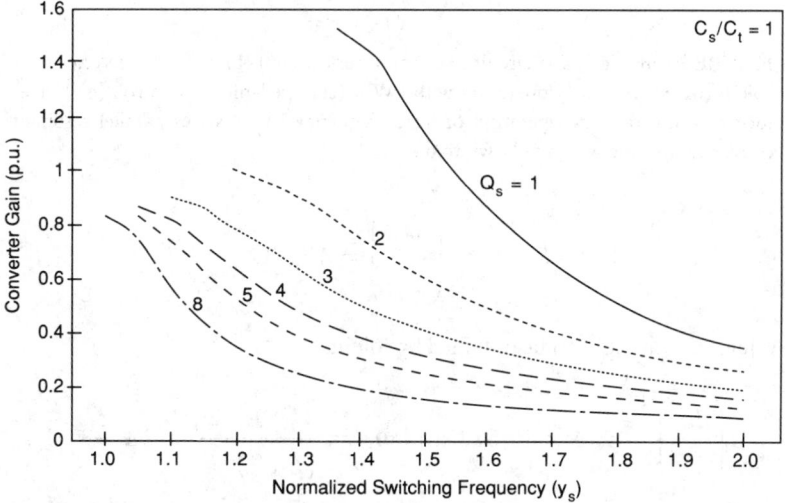

FIGURE 29.35 The converter gain M (p.u.) (normalized output voltage) versus normalized switching frequency y_s of SPRC operating above resonance for $C_s/C_t = 1$.

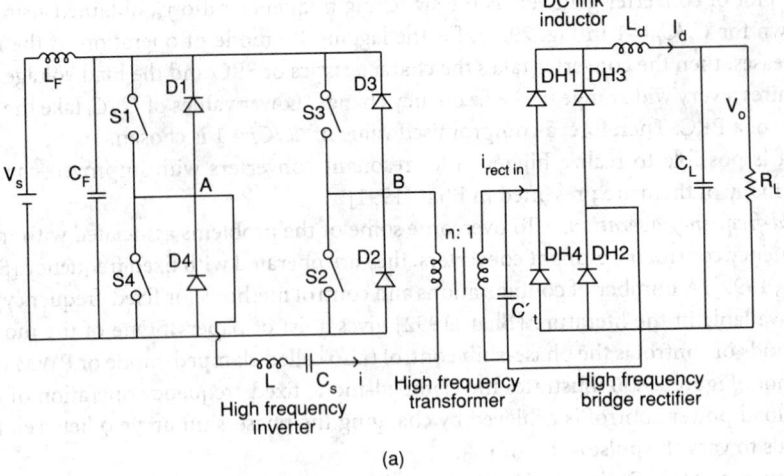

(a)

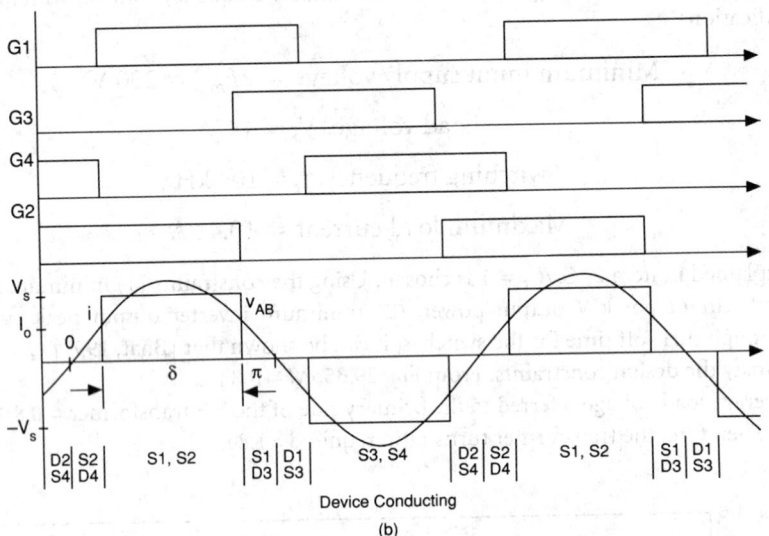

(b)

FIGURE 29.36 (a) Basic circuit diagram of series-parallel resonant converter suitable for fixed-frequency operation with PWM (clamped-mode) control. (b) Waveforms to illustrate the operation of fixed-frequency PWM series-parallel resonant converter working with a pulsewidth δ.

$$R_L' = n^2 \left(\frac{V_o^2}{P_o} \right) = 15.6 \ \Omega$$

The values of L_s and C_s can be obtained by solving

$$\left(\frac{L_s}{C_s} \right)^{1/2} = 4 \times 15.6 \ \Omega \quad \text{and} \quad \omega_r = \frac{1}{(L_s C_s)^{1/2}} = 2\pi \frac{f_s}{y_s}$$

Solving the above equations gives $L_s = 109 \ \mu\text{H}$ and $C_s = 0.0281 \ \mu\text{F}$. Leakage inductance $(L_p + L_s')$ of the HF transformer can be used as part of L_s. Typical value for a 100-kHz

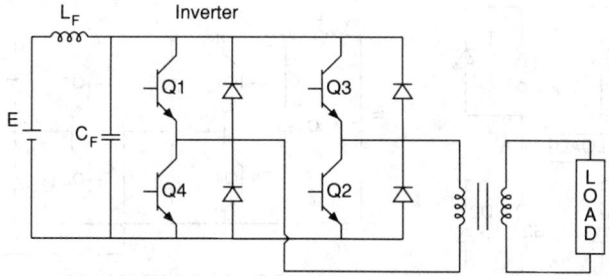

FIGURE 29.37 An inverter circuit to obtain variable-voltage, variable-frequency ac source. Using sinusoidal pulsewidth modulation control scheme, sine-wave ac output voltage can be obtained.

practical transformer (using Tokin Mn-Zn 2500B2 Ferrite, E-I type core) for this application is about 5 μH. Therefore, the external resonant inductance required is $L = 104$ μH.

Since $C_s/C_t = 1$ is chosen, $C_t = 0.0281$ μF. The actual value of C_t used on the secondary side of the HF transformer = $(1.84)^2 \times 0.0281 = 0.09514$ μF. The resonating capacitors must be HF type (e.g., polypropylene) and must be capable of withstanding the voltage and current ratings obtained above (enough safety margin must be provided).

Using Eqs. (29.9) and (29.11) to (29.13):

$$\text{Peak current through switches} = 7.6 \text{ A}$$

$$\text{Peak voltage across } C_s, \ V_{csp} = 430 \text{ V}$$

$$\text{Peak voltage (on secondary side) across } C_t', \ V_{ctp} = 76 \text{ V}$$

$$\text{Peak current through capacitor } C_t' \text{ (on secondary side)}, \ I_{ctp} = 4.54 \text{ A}$$

A simple control circuit can be built using PWM IC SG3525 and TSC429 MOSFET driver ICs.

With the development of digital ICs operating on low-voltage (of the order of 3 V) supplies, use of MOSFETs as *synchronous rectifiers* with very low voltage drop (~0.2 V) has become essential [Motorola, 1989] to increase the efficiency of the power supply.

AC Power Supplies

Some applications of ac power supplies are ac motor drives, **uninterruptible power supply (UPS)** used as a standby ac source for critical loads (e.g., in hospitals, computers), and dc source-to-utility interface (either to meet peak power demands or to augment energy by connecting unconventional energy sources like photovoltaic arrays to the utility line). In ac induction motor drives, the ac power main is rectified and filtered to obtain a smooth dc source, and then an inverter (single-phase version is shown in Fig. 29.37) is used to obtain a variable-frequency, variable-voltage ac source. The sinusoidal pulsewidth modulation technique described in Section 29.2 can be used to obtain a sinusoidal output voltage. Some other methods used to get sinusoidal voltage output are [Rashid, 1988] a number of phase-shifted inverter outputs summed in an output transformer to get a stepped waveform that approximates a sine wave and the use of a bang-bang controller in Fig. 29.37. All these methods use line-frequency (60 Hz) transformers for voltage translation and isolation purposes. To reduce the size, weight, and cost of such systems, one can use dc-to-dc converters (discussed earlier) as an intermediate stage. Figure 29.38 shows such a system in block schematic form. One can use an HF inverter circuit (discussed earlier) followed by a **cycloconverter** stage. The major problem with

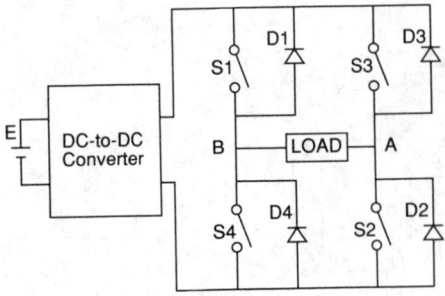

FIGURE 29.38 AC power supplies using HF switching (PWM or resonant) dc-to-dc converter as an input stage. HF transformer isolated dc-to-dc converters can be used to reduce the size and weight of the power supply. Sinusoidal voltage output can be obtained using the modulation in the output inverter stage or in the dc-to-dc converter.

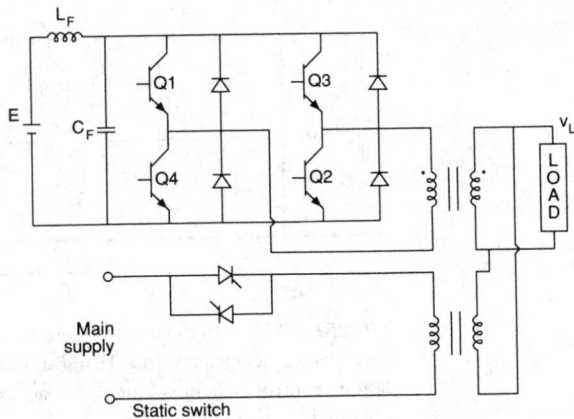

FIGURE 29.39 A typical arrangement of UPS system. The load gets power through the static switch when the ac main supply is present. The inverter supplies power when the main supply fails.

these schemes is the reduction in efficiency due to the extra power stage. Figure 29.39 shows a typical UPS scheme. The battery shown has to be charged by a separate rectifier circuit.

AC-to-ac conversion can also be achieved using cycloconverters [e.g., Rashid, 1988].

Special Power Supplies

Using the inverters and cycloconverters, it is possible to realize bidirectional ac and dc power supplies. In these power supplies [Rashid, 1988], power can flow in both directions, i.e., from input to output or from output to input. It is also possible to control the ac-to-dc converters to obtain sinusoidal line current with unity PF and low harmonic distortion at the ac source.

Defining Terms

Converter: A circuit that converts ac to dc, or vice versa.

Cycloconverter: A power electronic circuit that converts ac input to ac output (generally) of lower frequency than the input source without using any intermediate dc state.

Inverter: A power electronic circuit that converts dc input to ac output.

Isolated: A power electronic circuit that has ohmic isolation between the input source and the load circuit.

Pulsewidth-modulated (PWM) converters: A power electronic converter that employs square-wave switching waveforms with variation of pulsewidth for controlling the load voltage.

Regulated output: Output load voltage is kept at the required value for changes in either the load or the input supply voltage.

Resonant converters: A power electronic converter that employs "LC resonant circuits" to obtain sinusoidal switching waveforms.

Uninterruptible power supply (UPS): A stand-by dc-to-ac inverter used mostly to provide an emergency power to loads at mains frequency (50/60 Hz) in the event of a mains failure.

References

A.K.S. Bhat, "A unified approach for the steady-state analysis of resonant converters," *IEEE Trans. Industrial Electronics*, vol. 38, no. 4, pp. 251–259, Aug. 1991.

A.K.S. Bhat, "Fixed frequency PWM series-parallel resonant converter," *IEEE Trans. Industry Applications*, vol. 28, no. 5, pp. 1002–1009, 1992.

E.R. Hnatek, *Design of Solid-State Power Supplies*, 2nd ed., New York: Van Nostrand Reinhold, 1981.

K.H. Liu and F.C. Lee, "Zero-Voltage Switching Technique In DC/DC Converters," IEEE Power Electronics Specialists Conference Record, 1986, pp. 58–70.

K.H. Liu, R. Oruganti, and F.C. Lee, "Resonant Switches—Topologies and Characteristics," IEEE Power Electronics Specialists Conference Record, 1985, pp. 106–116.

Motorola, *Linear/Switchmode Voltage Regulator Handbook*, 1989.

Philips Semiconductors, *Power Semiconductor Applications*, 1991.

M.H. Rashid, *Power Electronics: Circuits, Devices, and Applications*, Englewood Cliffs, N.J.: Prentice-Hall, 1988.

R. Severns and G. Bloom, *Modern Switching DC-to-DC Converters*, New York: Van Nostrand Reinhold, 1988.

R.L. Steigerwald, "A comparison of half-bridge resonant converter topologies," *IEEE Trans. Power Electron.*, vol. PE-3, no. 2, pp. 174–182, April 1988.

K.K. Sum, *Recent Developments in Resonant Power Conversion*, Calif.: Intertech Communications, 1988.

Unitrode Switching Regulated Power Supply Design Seminar Manual, Lexington, Mass.: Unitrode Corporation, 1984.

Further Information

The following monthly magazines and conference records publish papers on the analysis, design, and experimental aspects of power supply configurations and their applications:

IEEE Transactions on Power Electronics, IEEE Transactions on Industrial Electronics, IEEE Transactions on Industry Applications, and *IEEE Transactions on Aerospace and Electronic Systems.*

IEEE Power Electronics Specialists Conference Records, IEEE Applied Power Electronics Conference Records, IEEE Industry Applications Conference Records, and *IEEE International Telecommunications Energy Conference Records.*

29.4 Converter Control of Machines

Bimal K. Bose

Converter-controlled electrical machine drives are very important in modern industrial applications. Some examples in the high-power range are metal rolling mills, cement mills, and gas line compressors. In the medium-power range are textile mills, paper mills, and subway car propulsion. Machine tools and computer peripherals are examples of converter-controlled electrical machine drive applications in the low-power range. The converter normally provides a variable-voltage dc power source for a dc motor drive and a variable-frequency, variable-voltage ac power source for an ac motor drive. The drive system efficiency is high because the converter operates in switching mode using power semiconductor devices. The primary control variable of the machine may be torque, speed, or position, or the converter can operate as a solid-state starter of the machine. The recent evolution of high-frequency power semiconductor devices and high-density and economical microelectronic chips, coupled with converter and control technology developments, is providing a tremendous boost in the applications of drives.

Converter Control of DC Machines

The speed of a dc motor can be controlled by controlling the dc voltage across its armature terminals. A phase-controlled thyristor converter can provide this dc voltage source. For a low-power drive, a single-phase bridge converter can be used, whereas for a high-power drive, a three-phase bridge cir-

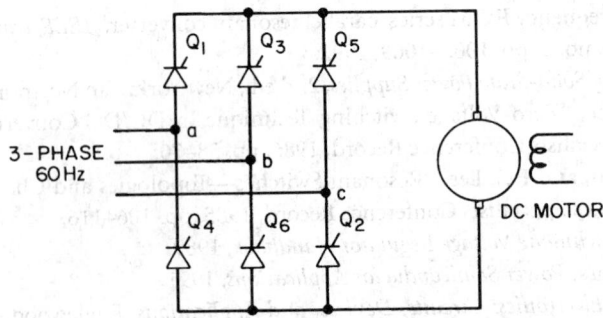

FIGURE 29.40   Three-phase thyristor bridge converter control of a dc machine.

cuit is preferred. The machine can be a permanent magnet or wound field type. The wound field type permits variation and reversal of field and is normally preferred in large power machines.

Phase-Controlled Converter DC Drive

Figure 29.40 shows a dc drive using a three-phase thyristor bridge converter. The converter rectifies line ac voltage to variable dc output voltage by controlling the firing angle of the thyristors. With rated field excitation, as the armature voltage is increased, the machine will develop speed in the forward direction until the rated, or base, speed is developed at full voltage when the firing angle is zero. The motor speed can be increased further by weakening the field excitation. Below the base speed, the machine is said to operate in constant torque region, whereas the field weakening mode is defined as the constant power region. At any operating speed, the field can be reversed and the converter firing angle can be controlled beyond 90 degrees for **regenerative braking** mode operation of the drive. In this mode, the motor acts as a generator (with negative induced voltage) and the converter acts as an inverter so that the mechanical energy stored in the inertia is converted to electrical energy and pumped back to the source. Such **two-quadrant** operation gives improved efficiency if the drive accelerates and decelerates frequently. The speed of the machine can be controlled with precision by a feedback loop where the command speed is compared with the machine speed measured by a tachometer. The speed loop error generally generates the armature current command through a compensator. The current is then feedback controlled with the firing angle control in the inner loop. Since torque is proportional to armature current (with fixed field), a current loop provides direct torque control, and the drive can accelerate or decelerate with the rated torque. A second bridge converter can be connected in antiparallel so that the dual converter can control the machine speed in all the four quadrants (motoring and regeneration in forward and reverse speeds).

Pulsewidth Modulation Converter DC Machine Drive

Four-quadrant speed control of a dc drive is also possible using an H-bridge pulsewidth modulation (PWM) converter as shown in Fig. 29.41. Such drives (using a permanent magnet dc motor) are popular in low-power applications, such as robotic and instrumentation drives. The dc source can be a battery or may be obtained from ac supply through a diode rectifier and filter. With PWM operation, the drive response is very fast and the armature current ripple is small, giving less harmonic heating and torque pulsation. Four-quadrant operation can be summarized as follows:

Quadrant 1: Forward motoring (buck or step-down converter mode)
   Q_1—on
   Q_3, Q_4—off
   Q_2—chopping
   Current freewheeling through D_3 and Q_1
Quadrant 2: Forward regeneration (boost or step-up converter mode)

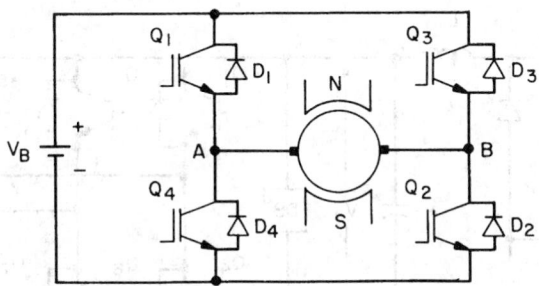

FIGURE 29.41 Four-quadrant dc motor drive using an H-bridge converter.

Q_1, Q_2, Q_3—off
Q_4—chopping
Current freewheeling through D_1 and D_2
Quadrant 3: Reverse motoring (buck converter mode)
Q_3—on
Q_1, Q_2—off
Q_4—chopping
Current freewheeling through D_1 and Q_3
Quadrant 4: Reverse regeneration (boost converter mode)
Q_1, Q_3, Q_4—off
Q_2—chopping
Current freewheeling through D_3 and D_4

Often a drive may need only a one- or two-quadrant mode of operation. In such a case, the converter topology can be simple. For example, in one-quadrant drive, only Q_2 chopping and D_3 freewheeling devices are required, and the terminal A is connected to the supply positive. Similarly, a two-quadrant drive will need only one leg of the bridge, where the upper device can be controlled for motoring mode and the lower device can be controlled for regeneration mode.

Converter Control of AC Machines

Although application of dc drives is quite common, disadvantages are that the machines are bulky and expensive, and the commutators and brushes require frequent maintenance. In fact, commutator sparking prevents machine application in an unclean environment, at high speed, and at high elevation. AC machines, particularly the cage-type induction motor, are favorable when compared with all the features of dc machines. Although converter system, control, and signal processing of ac drives is definitely complex, the evolution of ac drive technology in the past two decades has permitted more economical and higher performance ac drives. Consequently, ac drives are finding expanding applications, pushing dc drives towards obsolescence.

Voltage-Fed Inverter Induction Motor Drive

A simple and popular converter system for speed control of an induction motor is shown in Fig. 29.42. The front-end diode rectifier converts 60 Hz ac to dc, which is then filtered to remove the ripple. The dc voltage is then converted to variable-frequency, variable-voltage output for the machine through a PWM bridge inverter. Among a number of PWM techniques, the sinusoidal PWM is common, and it is illustrated in Fig. 29.43 for one phase only. The stator sinusoidal reference phase voltage signal is compared with a high-frequency carrier wave, and the comparator logic output controls switching of the upper and lower transistors in a phase leg. The phase voltage wave shown refers to the fictitious center tap of the filter capacitor. With the PWM technique, the fundamental voltage and frequency can be easily varied. The stator voltage wave contains high-frequency

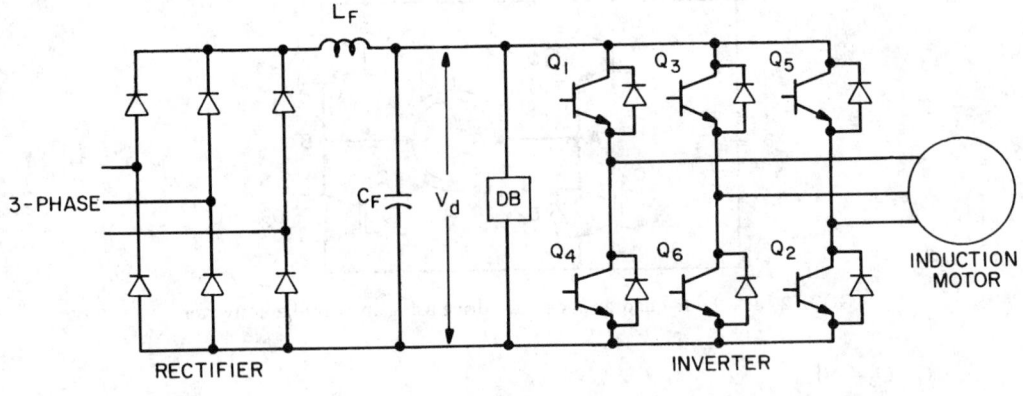

FIGURE 29.42   Diode rectifier PWM inverter control of an induction motor.

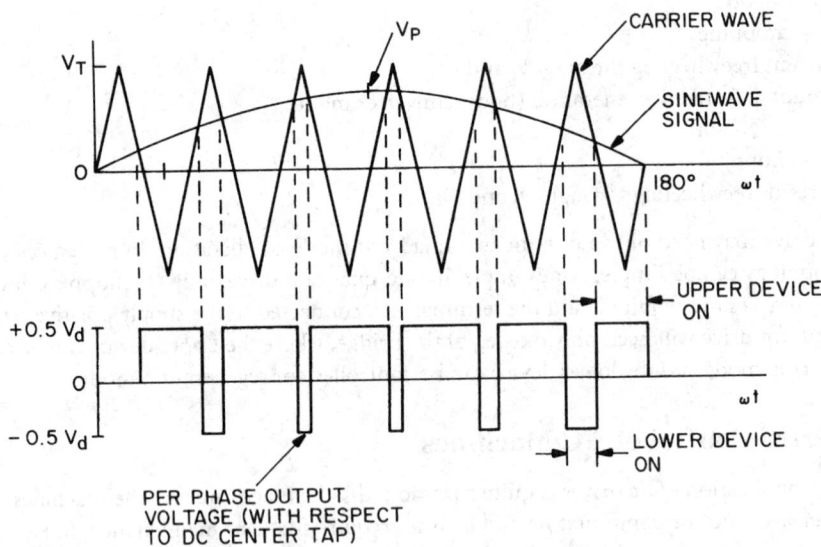

FIGURE 29.43   Sinusoidal pulse width modulation principle.

ripple, which is easily filtered by the machine leakage inductance. The voltage-to-frequency ratio is kept constant to provide constant airgap flux in the machine. The machine voltage-frequency relation, and the corresponding torque, stator current, and slip, are shown in Fig. 29.44. Up to the base or rated frequency ω_b, the machine can develop constant torque. Then, the field flux weakens as the frequency is increased at constant voltage. The speed of the machine can be controlled in a simple open-loop manner by controlling the frequency and maintaining the proportionality between the voltage and frequency. During acceleration, machine-developed torque should be limited so that the inverter current rating is not exceeded. By controlling the frequency, the operation can be extended in the field weakening region. If the supply frequency is controlled to be lower than the machine speed (equivalent frequency), the motor will act as a generator and the inverter will act as a rectifier, and energy from the motor will be pumped back to the dc link. The **dynamic brake** shown is nothing but a buck converter with resistive load that dissipates excess power to maintain the dc bus voltage constant. When the motor speed is reduced to zero, the phase sequence of the inverter can be reversed for speed reversal. Therefore, the machine speed can be easily controlled in all four quadrants.

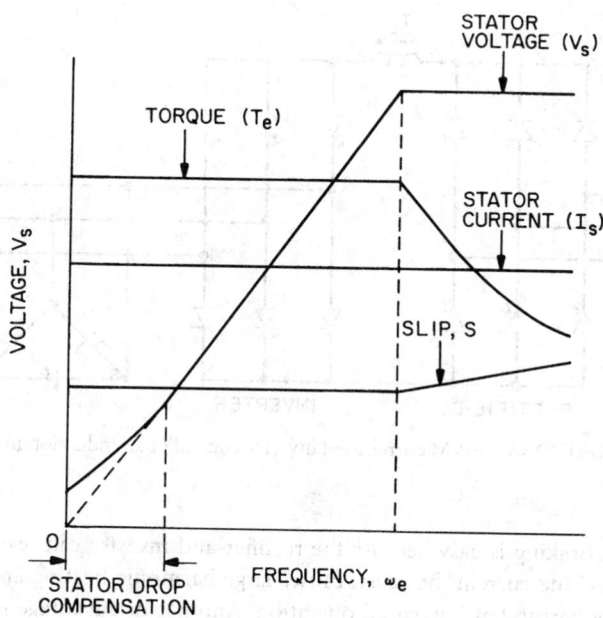

FIGURE 29.44 Voltage-frequency relation of an induction motor.

Current-Fed Inverter Induction Motor Drive

The speed of a machine can be controlled by a current-fed inverter as shown in Fig. 29.45. The front-end thyristor rectifier generates a variable dc current source in the dc link inductor. The dc current is then converted to six-step machine current wave through the inverter. The basic mode of operation of the inverter is the same as that of the rectifier, except that it is **force-commutated,** that is, the capacitors and series diodes help commutation of the thyristors. One advantage of the drive

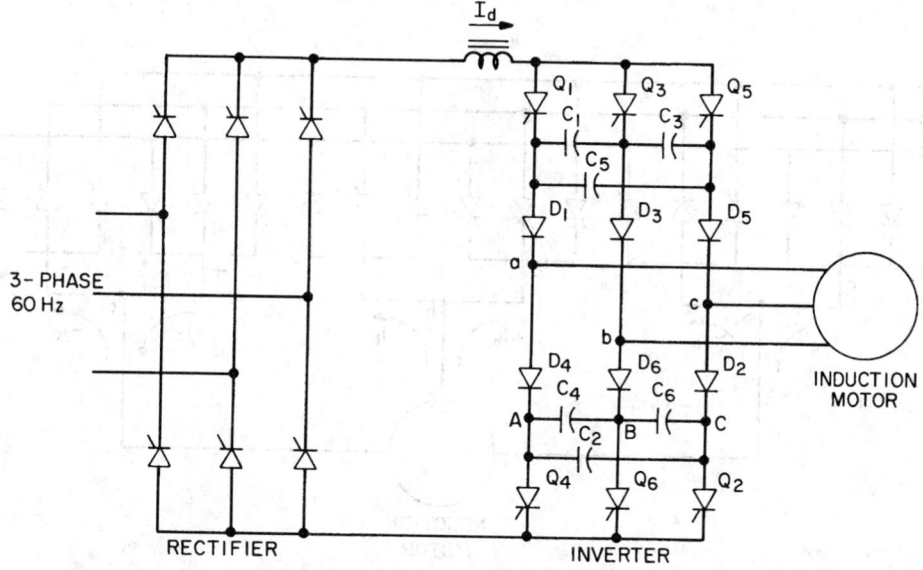

FIGURE 29.45 Force-commutated current-fed inverter control of an induction motor.

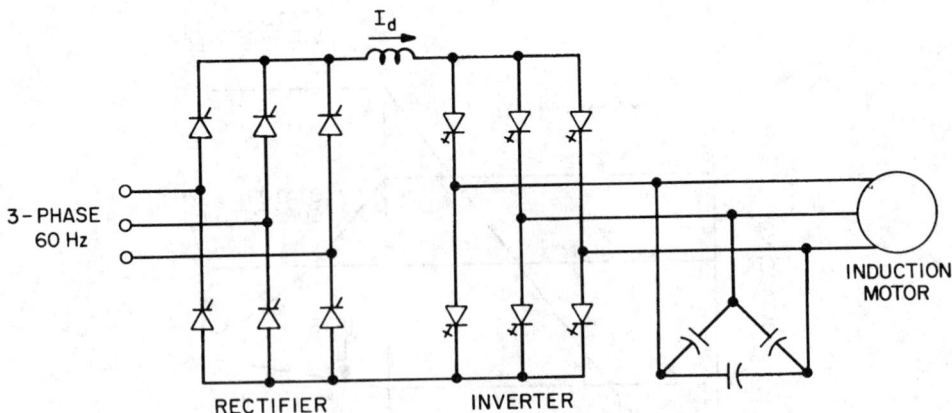

FIGURE 29.46   PWM current-fed inverter control of an induction motor.

is that regenerative braking is easy because the rectifier and inverter can reverse their operation modes. Six-step machine current, however, causes large harmonic heating and torque pulsation, which may be quite harmful at low-speed operation. Another disadvantage is that the converter system cannot be controlled in open loop like a voltage-fed inverter.

Current-Fed PWM Inverter Induction Motor Drive

The force-commutated thyristor inverter in Fig. 29.45 can be replaced by a **self-commutating** gate turn-off (GTO) thyristor PWM inverter as shown in Fig. 29.46. The output capacitor bank shown has two functions: (1) it permits PWM switching of the GTO by diverting the load inductive current, and (2) it acts as a low-pass filter causing sinusoidal machine current. The second function improves machine efficiency and attenuates the irritating magnetic noise. Note that the fundamental machine current is controlled by the front-end rectifier, and the fixed PWM pattern is for controlling the harmonics only. The GTO is to be the reverse-blocking type. Such drives are popular in the multimegawatt power range. For lower power, an **insulated gate bipolar transistor (IGBT)** or transistor can be used with a series diode.

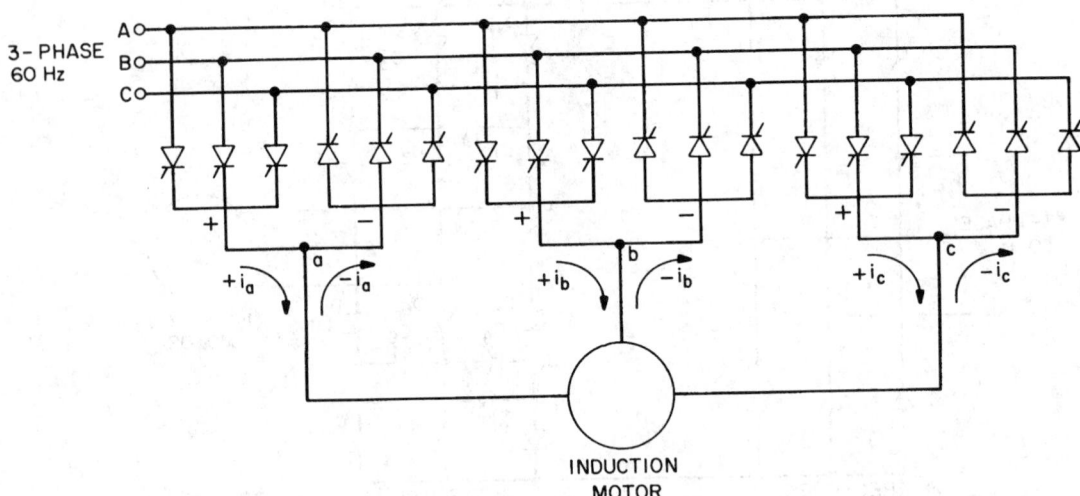

FIGURE 29.47   Cycloconverter control of an induction motor.

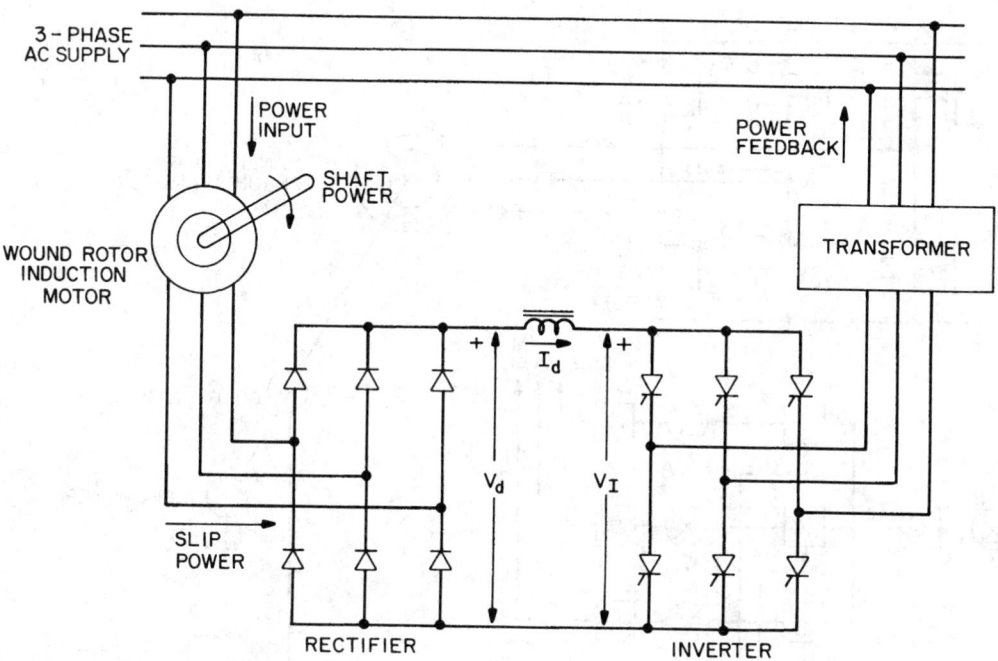

FIGURE 29.48 Slip power recovery control of a wound rotor induction motor.

Cycloconverter Induction Motor Drive

A phase-controlled cycloconverter can be used for speed control of an ac machine. Figure 29.47 shows a drive using a three-pulse half-wave or 18-thyristor cycloconverter. Each output phase group consists of positive and negative converter components which permit bidirectional current flow. The firing angle of each converter is sinusoidally modulated to generate the variable-frequency, variable-voltage output required for ac machine drive. Speed reversal and regenerative mode operation are easy. The cycloconverter can be operated in blocking or circulating current mode. In blocking mode, the positive or negative converter is enabled, depending on the polarity of the load current. In circulating current mode, the converter components are always enabled to permit circulating current through them. The circulating current reactor between the positive and negative converter prevents short circuits due to ripple voltage. The circulating current mode gives simple control and a higher range of output frequency with lower harmonic distortion.

Slip Power Recovery Drive of Induction Motor

In a cage-type induction motor, the rotor current at slip frequency reacting with the airgap flux develops the torque. The corresponding slip power is dissipated in the rotor resistance. In a wound rotor induction motor, the slip power can be controlled to control the torque and speed of a machine. Figure 29.48 shows a popular slip power-controlled drive, known as a static Kramer drive. The slip power is rectified to dc with a diode rectifier and is then pumped back to an ac line through a thyristor phase-controlled inverter. The method permits speed control in the subsynchronous speed range. It can be shown that the developed machine torque is proportional to the dc link current I_d and the voltage V_d varies directly with speed deviation from the synchronous speed. The current I_d is controlled by the firing angle of the inverter. Since V_d and V_I voltages balance at steady state, at synchronous speed the voltage V_d is zero and the firing angle is 90 degrees. The firing angle increases as the speed falls, and at 50% synchronous speed the firing angle is near 180 degrees. This is practically the lowest speed in static Kramer drive. The transformer steps down the

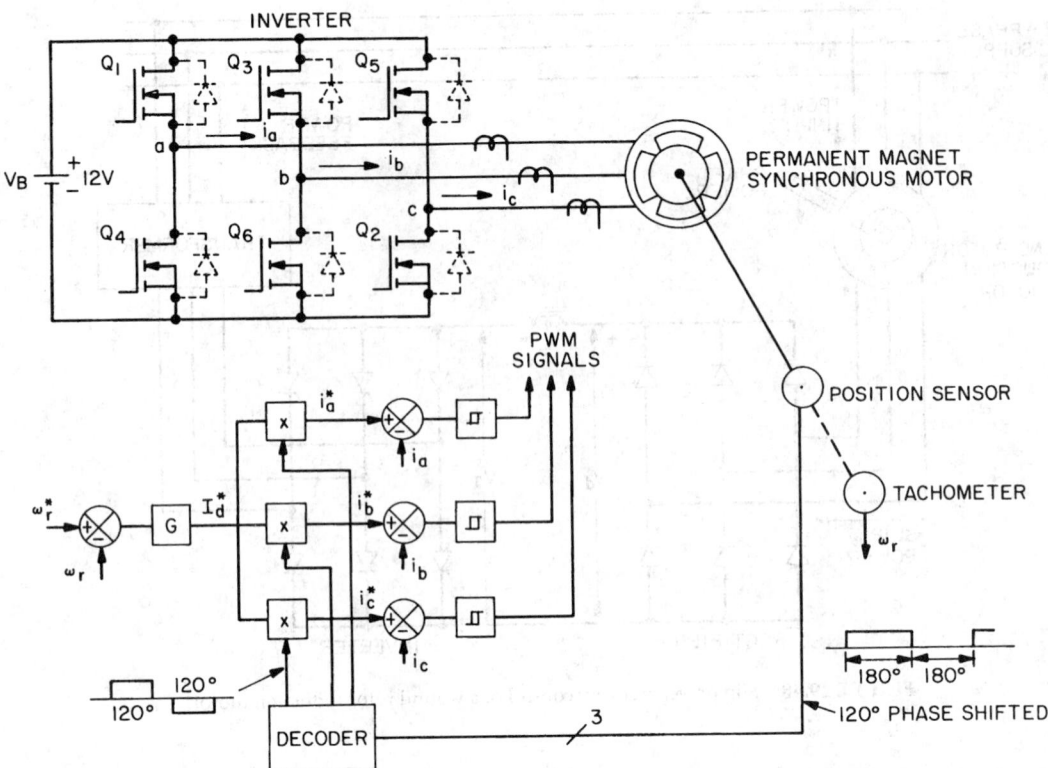

FIGURE 29.49 Permanent magnet synchronous motor control with PWM inverter.

inverter input voltage to get a 180-degree firing angle at lowest speed. The advantage of this drive is that the converter rating is low compared with the machine rating. Disadvantages are that the line power factor is low and the machine is expensive. For limited speed range applications, this drive has been popular.

Wound Field Synchronous Motor Drive

The speed of a wound field synchronous machine can be controlled by a current-fed converter scheme as shown in Fig. 29.45, except that the forced-commutation elements can be removed. The machine is operated at leading power factor by overexcitation so that the inverter can be load commutated. Because of the simplicity of converter topology and control, such a drive is popular in the multimegawatt range.

Permanent Magnet Synchronous Motor Drive

Permanent magnet (PM) machine drives are quite popular in the low-power range. A PM machine can have sinusoidal or concentrated winding, giving the corresponding sinusoidal or trapezoidal induced stator voltage wave. Figure 29.49 shows the speed control system using a trapezoidal machine, and Fig. 29.50 explains the wave forms. The power MOSFET inverter supplies variable-frequency, variable-magnitude six-step current wave to the stator. The inverter is self-controlled, that is, the firing pulses are generated by the machine position sensor through a decoder. It can be shown that such a drive has the features of dc drive and is normally defined as *brushless dc drive*. The speed control loop generates the dc current command, which is then controlled by the **hysteresis-band** method to construct the six-step phase current waves in correct phase relation with the induced voltage waves as shown in Fig. 29.50. The drive can easily operate in four-quadrant mode.

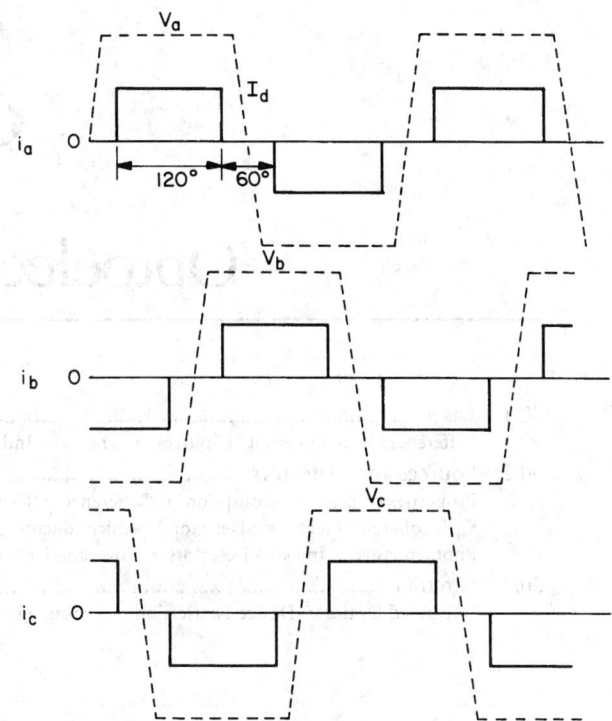

FIGURE 29.50 Phase voltage and current waves in brushless dc drive.

Defining Terms

Dynamic brake: The braking operation of a machine by extracting electrical energy and then dissipating it in a resistor.

Forced-commutation: Switching off a power semiconductor device by external circuit transient.

Four-quadrant: A drive that can operate as a motor as well as a generator in both directions.

Hysteresis-band: A method of controlling current where the instantaneous current can vary within a band.

Insulated gate bipolar transistor (IGBT): A device that combines the features of a power transistor and MOSFET.

Regenerative braking: The braking operation of a machine by converting its mechanical energy into electrical form and then pumping it back to the source.

Self-commutation: Switching off a power semiconductor device by its gate or base drive.

Two-quadrant: A drive that can operate as a motor as well as a generator in one direction.

References

B.K. Bose, *Power Electronics and AC Drives*, Englewood Cliffs, N.J.: Prentice-Hall, 1986.

B.K. Bose, "Adjustable speed AC drives—A technology status review," *Proc. IEEE*, vol. 70, pp. 116–135, Feb. 1982.

B.K. Bose, *Modern Power Electronics*, New York: IEEE Press, 1992.

J.M.D. Murphy and F.G. Turnbull, *Power Electronic Control of AC Motors*, New York: Pergamon Press, 1988.

P.C. Sen, *Thyristor DC Drives*, New York: John Wiley, 1981.

30

Optoelectronics

Jeff Hecht
Laser Focus World

Laurence S. Watkins
AT&T Bell Laboratories

R. A. Becker
*Integrated Optical Circuit
Consultants*

30.1 Lasers 738
  Differences from Other Light Sources • The Laser Industry
30.2 Sources and Detectors 742
  Properties of Light • Absorption • Coherence • Geometric Optics
  • Incoherent Light • Detectors, Semiconductor • Detectors,
  Photoemissive • Imaging Detectors • Noise and Detectivity
30.3 Circuits 759
  Integrated Optics • Device Fabrication • Packaging • Applications

30.1 Lasers*

Jeff Hecht

The word *laser* is an acronym for "light amplification by the stimulated emission of radiation," a phrase that covers most, though not all, of the key physical processes inside a laser. Unfortunately, that concise definition may not be very enlightening to the nonspecialist who wants to *use* a laser and cares less about its internal physics than its external characteristics. From a practical standpoint, a laser can be considered as a source of a narrow beam of **monochromatic**, coherent light in the visible, infrared, or ultraviolet parts of the spectrum. The power in a continuous beam can range from a fraction of a milliwatt to around 25 kilowatts (kW) in commercial lasers, and up to more than a megawatt in special military lasers. Pulsed lasers can deliver much higher peak powers during a pulse, although the power averaged over intervals while the laser is off and on is comparable to that of continuous lasers.

The range of laser devices is broad. The **laser medium**, or material emitting the laser beam, can be a gas, liquid, glass, crystalline solid, or semiconductor crystal and can range in size from a grain of salt to filling the inside of a moderate-sized building. Not every laser produces a narrow beam of monochromatic, coherent light. Semiconductor diode lasers, for example, produce beams that spread out over an angle of 20 to 40 degrees, hardly a pencil-thin beam. Liquid dye lasers emit at a broad or narrow range of wavelengths, depending on the optics used with them. Other types emit at a number of spectral lines, producing light that is neither truly monochromatic nor coherent. Table 30.1 summarizes important commercial lasers.

Practically speaking, lasers contain three key elements. One is the laser medium itself, which generates the laser light. A second in the power supply, which delivers energy to the laser medium in the form needed to excite it to emit light. The third is the optical cavity or **resonator**, which concentrates the light to stimulate the emission of laser radiation. All three elements can take various forms, and although they are not always immediately evident in all types of lasers, their functions are essential. Figure 30.1 shows these elements in a ruby and a helium-neon laser.

*Modified from J. Hecht, *The Laser Guidebook,* 2nd ed., New York: McGraw-Hill, 1991. With permission.

Table 30.1 Important Commercial Lasers

Wavelength (μm)	Type	Output Type and Power
0.152	Molecular fluorine (F_2)	Pulsed, avg. to a few watts
0.192	ArF excimer	Pulsed, avg. to tens of watts
0.2–0.35	Doubled dye	Pulsed
0.235–0.3	Tripled Ti-sapphire	Pulsed
0.248	KrF excimer	Pulsed, avg. to over 100 W
0.266	Quadrupled Nd	Pulsed, watts
0.275–0.306	Argon-ion	CW, 1-W range
0.308	XeCl excimer	Pulsed, to tens of watts
0.32–1.0	Pulsed dye	Pulsed, to tens of watts
0.325	He-Cd	CW, to tens of milliwatts
0.33–0.38	Neon	CW, 1-W range
0.337	Nitrogen	Pulsed, under 1 W avg.
0.35–0.47	Doubled Ti-sapphire	Pulsed
0.351	XeF excimer	Pulsed, to tens of watts
0.355	Tripled Nd	Pulsed, to tens of watts
0.36–0.4	Doubled alexandrite	Pulsed, watts
0.37–1.0	CW dye	CW, to a few watts
0.442	He-Cd	CW, to over 0.1 W
0.45–0.52	Ar-ion	CW, to tens of watts
0.51	Copper vapor	Pulsed, tens of watts
0.523	Doubled Nd-YLF	Pulsed, watts
0.532	Doubled Nd-YAG	Pulsed to 50 W, or CW to watts
0.5435	He-Ne	CW, 1-mW range
0.578	Copper vapor	Pulsed, tens of watts
0.594	He-Ne	CW, to several milliwatts
0.612	He-Ne	CW, to several milliwatts
0.628	Gold vapor	Pulsed
0.6328	He-Ne	CW, to about 50 mW
0.635–0.66	InGaAlP diode	CW, milliwatts
0.647	Krypton ion	CW, to several watts
0.67	GaInP diode	CW, to 10 mW
0.68–1.13	Ti-sapphire	CW, watts
0.694	Ruby	Pulsed, to a few watts
0.72–0.8	Alexandrite	Pulsed, to tens of watts (CW in lab)
0.75–0.9	GaAlAs diode	CW, to many watts in arrays
0.98	InGaAs diode	CW, to 50 mW
1.047 or 1.053	Nd-YLF	CW or pulsed, to tens of watts
1.061	Nd-glass	Pulsed, to 100 W
1.064	Nd-YAG	CW or pulsed, to kilowatts
1.15	He-Ne	CW, milliwatts
1.2–1.6	InGaAsP diode	CW, to 100 mW
1.313	Nd-YLF	CW or pulsed, to 0.1 W
1.32	Nd-YAG	Pulsed or CW, to a few watts
1.4–1.6	Color center	CW, under 1 W
1.523	He-Ne	CW, milliwatts
1.54	Erbium-glass (bulk)	Pulsed, to 1 W
1.54	Erbium-fiber (amplifier)	CW, milliwatts
1.75–2.5	Cobalt-MgF_2	Pulsed, 1-W range
2.3–3.3	Color center	CW, under 1 W
2.6–3.0	HF chemical	CW or pulsed, to hundreds of watts
3.3–29	Lead-salt diode	CW, milliwatt range
3.39	He-Ne	CW, to tens of milliwatts
3.6–4.0	DF chemical	CW or pulsed, to hundreds of watts
5–6	Carbon monoxide	CW, to tens of watts
9–11	Carbon dioxide	CW or pulsed, to tens of kilowatts
40–100	Far-infrared gas	CW, generally under 1 W

Source: J. Hecht, *The Laser Guidebook*, 2nd ed., New York: McGraw-Hill, 1991. With permission.

Several general characteristics are common to most lasers that new users may not expect. Like most other light sources, lasers are inefficient in converting input energy into light. Efficiencies range from less than 0.001 to more than 30%, but few types are much above 1% efficient. These low efficiencies can lead to special cooling requirements and duty-cycle limitations, particularly for high-power lasers. In some cases, special equipment may be needed to produce the right conditions for laser operation, such as cryogenic temperatures for lead salt semiconductor lasers. Operating characteristics of individual lasers depend strongly on structural components such as cavity optics, and in many cases a wide range is possible. Packaging can also have a strong impact on laser characteristics and the use of lasers for certain applications. Thus, there are wide ranges of possible characteristics, although single devices will have much more limited ranges of operation.

Differences from Other Light Sources

The basic differences between lasers and other light sources are the characteristics often used to describe a laser: the output beam is narrow, the light is monochromatic, and the emission is coherent. Each of these features is important for certain applications and deserves more explanation.

Most gas or **solid-state lasers** emit beams with divergence angle of about a milliradian, meaning that they spread to about 1 m in diameter after traveling a kilometer. (Semiconductor lasers have much larger beam divergence, but suitable optics can reshape the beam to make it much narrower.) The actual beam divergence depends on the type of laser and the optics used with it. The fact that laser light is contained in a beam serves to concentrate the output power onto a small area. Thus, a modest laser power can produce a high intensity inside the small area of the laser beam; the intensity of light in a 1-mW helium-neon laser beam is comparable to that of sunlight on a clear day, for

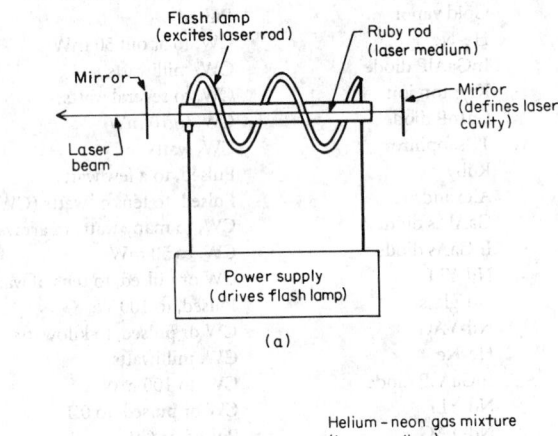

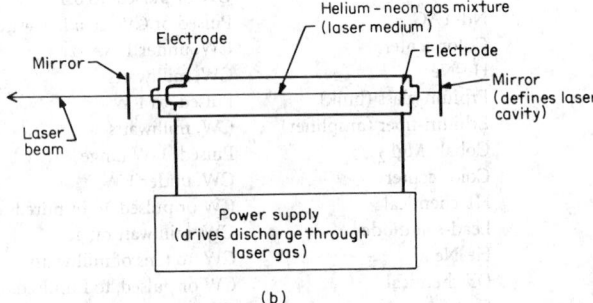

FIGURE 30.1 Simplified views of two common lasers, (a) ruby and (b) helium-neon, showing the basic components that make a laser. (*Source:* J. Hecht, *The Laser Guidebook,* 2nd ed., New York: McGraw-Hill, 1991. With permission.)

example. The beams from high-power lasers, delivering tens of watts or more of continuous power or higher peak powers in pulses, can be concentrated to high enough intensities that they can weld, drill, or cut many materials.

The laser beam's concentrated light delivers energy only where it is focused. For example, a tightly focused laser beam can write a spot on a light-sensitive material without exposing the adjacent area, allowing high-resolution printing. Similarly, the beam from a surgical laser can be focused onto a tiny spot for microsurgery, without heating or damaging surrounding tissue. Lenses can focus the parallel rays in a laser beam to a much smaller spot than they can the diverging rays from a point source, a factor that helps compensate for the limited light-production efficiency of lasers.

Most lasers deliver a beam that contains only a narrow range of wavelengths, and thus the beam can be considered monochromatic for all practical purposes. Conventional light sources, in contrast, emit light over much of the visible and infrared spectrum. For most applications, the range of wavelengths emitted by lasers is narrow enough to make life easier for designers by avoiding the need for achromatic optics and simplifying the task of understanding the interactions between laser beam and target. For some applications in spectroscopy and communications, however, that range of wavelengths is not narrow enough, and special line-narrowing options may be required.

One of the beam's unique properties is its **coherence**, the property that the light waves it contains are in phase with one another. Strictly speaking, all light sources have a finite coherence length, or distance over which the light they produce is in phase. However, for conventional light sources that distance is essentially zero. For many common lasers, it is a fraction of a meter or more, allowing their use for applications requiring coherent light. The most important of these applications is probably holography, although coherence is useful in some types of spectroscopy, and there is growing interest in communications using coherent light.

Some types of lasers have two other advantages over other light sources: higher power and longer lifetime. For some high-power semiconductor lasers, lifetime must be traded off against higher power, but for most others the life vs. power trade-off is minimal. The combination of high power and strong directionality makes certain lasers the logical choice to deliver high light intensities to small areas. For some applications, lasers offer longer lifetimes than do other light sources of comparable brightness and cost. In addition, despite their low efficiency, some lasers may be more efficient in converting energy to light than other light sources.

The Laser Industry

Commercial Lasers

There is a big difference between the world of laser research and the world of the commercial laser industry. Unfortunately, many text and reference books fail to differentiate between types of lasers that can be built in the laboratory and those that are readily available commercially. That distinction is a crucial one for laser users.

Laser emission has been obtained from hundreds of materials at many thousands of emission lines in laboratories around the world. Extensive tabulations of these laser lines are available [Weber, 1982], and even today researchers are adding more lines to the list. However, most of these laser lines are of purely academic interest. Many are weak lines close to much stronger lines that dominate the emission in practical lasers. Most of the lasers that have been demonstrated in the laboratory have proved to be cumbersome to operate, low in power, inefficient, and/or simply less practical to use than other types.

Only a couple of dozen types of lasers have proved to be commercially viable on any significant scale; these are summarized in Table 30.1. Some of these types, notably the ruby and helium-neon lasers, have been around since the beginning of the laser era. Others, such as vibronic solid-state, are promising newcomers. The family of commercial lasers is expanding slowly, as new types such as titanium-sapphire come on the market, but with the economics of production a factor to be considered, the number of commercially viable lasers will always be limited.

There are many possible reasons why certain lasers do not find their way onto the market. Some require exotic operating conditions or laser media, such as high temperatures or highly reactive metal vapors. Some emit only feeble powers. Others have only limited applications, particularly lasers emitting low powers in the far-infrared or in parts of the infrared where the atmosphere is opaque. Some simply cannot compete with materials already on the market.

Defining Terms

Coherence: The condition of light waves that stay in the same phase relative to each other; they must have the same wavelength.

Continuous wave (CW): A laser that emits a steady beam rather than pulses.

Laser medium: The material in a laser that emits light; it may be a gas, solid, or liquid.

Monochromatic: Of a single wavelength or frequency.

Resonator: Mirrors that reflect light back and forth through a laser medium, usually on opposite ends of a rod, tube, or semiconductor wafer. One mirror lets some light escape to form the laser beam.

Solid-state laser: A laser in which light is emitted by atoms in a glass or crystalline matrix. Laser specialists do not consider semiconductor lasers to be solid-state types.

References

J. Hecht, *The Laser Guidebook,* 2nd ed., New York: McGraw-Hill, 1991; this section is excerpted from the introduction.

M. J. Weber (ed.), *CRC Handbook of Laser Science and Technology* (2 vols.), Boca Raton, Fla.: CRC Press, 1982.

M. J. Weber (ed.), *CRC Handbook of Laser Science and Technology, Supplement 1,* Boca Raton, Fla.: CRC Press, 1989; other supplements are in preparation.

Further Information

Several excellent introductory college texts are available that concentrate on laser principles. These include: Anthony E. Siegman, *Lasers,* University Science Books, Mill Valley, Calif., 1986, and Orzio Svelto, *Principles of Lasers,* 3rd ed., Plenum, New York, 1989.

Three trade magazines serve the laser field; each publishes an annual directory issue. For further information contact: *Laser Focus World,* PennWell Publishing, PO Box 989, Westford, Mass. 01886; *Lasers & Optronics,* PO Box 650, Morris Plains, N.J. 07950-0650; or *Photonics Spectra,* Laurin Publishing Co., Berkshire Common, PO Box 1146, Pittsfield, Mass. 01202. Write the publishers for information.

30.2 Sources and Detectors

Laurence S. Watkins

Properties of Light

The strict definition of light is electromagnetic radiation to which the eye is sensitive. Optical devices, however, can operate over a larger range of the electromagnetic spectrum, and so the term usually refers to devices which can operate in some part of the spectrum from the near ultraviolet

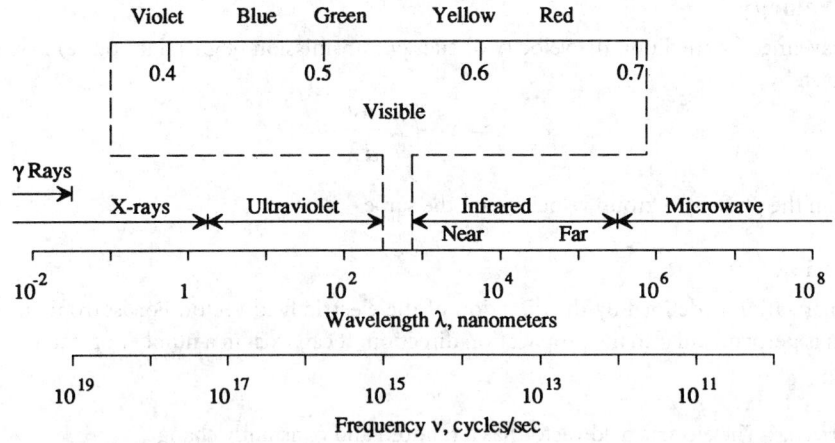

FIGURE 30.2 Electromagnetic spectrum showing visible and optical wavelengths.

(UV) through the visible range to the near infrared. Figure 30.2 shows the whole spectrum and delineates these ranges.

Optical radiation is electromagnetic radiation and so obeys and can be completely described by Maxwell's equations. We will not discuss this analysis here but just review the important properties of light.

Phase Velocity

In isotropic media light propagates as transverse electromagnetic (TEM) waves. The electric and magnetic field vectors are perpendicular to the propagation direction and orthogonal to each other. The velocity of light propagation in a medium (the velocity of planes of constant phase, i.e., wave-fronts) is given by

$$v = \frac{c}{\sqrt{\varepsilon\mu}} \tag{30.1}$$

where c is the velocity of light in a vacuum ($c = 299,796$ km/s). The denominator in Eq. (30.1) is a term in optics called the refractive index of the medium

$$n = \sqrt{\varepsilon\mu} \tag{30.2}$$

where ε is the dielectric constant (permittivity) and μ is the magnetic permeability. The wavelength of light, λ, which is the distance between phase fronts is

$$\lambda = \frac{\lambda_0}{n} = \frac{v}{\upsilon} \tag{30.3}$$

where λ_0 is the wavelength in vacuum and υ is the light frequency. The refractive index varies with wavelength, and this is referred to as the dispersive property of a medium.

Another parameter used to describe light frequency is wave number. This is given by

$$\sigma = \frac{1}{\lambda} \tag{30.4}$$

and is usually expressed in cm^{-1}, giving the number of waves in a 1-cm path.

Group Velocity

When traveling in a medium, the velocity of energy transmission (e.g., a light pulse) is less than c and is given by

$$u = v - \lambda \frac{dv}{d\lambda} \tag{30.5}$$

In vacuum the phase and group velocities are the same.

Polarization

Light polarization is defined by the direction of the electric field vector. For isotropic media this direction is perpendicular to the propagation direction. It can exist in a number of states, described as follows.

Unpolarized. The electric field vector has a random and constantly changing direction, and when there are multiple frequencies the vector directions are different for each frequency.

Linear. The electric field vector is confined to one direction.

Elliptical. The electric field vector rotates, either left hand or right hand, at the light frequency. The magnitude of the vector (intensity of the light) traces out an ellipse.

Circular. Circular is the special case of the above where the electric field vector traces out a circle.

Absorption

Light in traveling through media can be absorbed. This can be represented in two ways. The light flux propagating through a medium can be written as

$$I = I_0 e^{-\alpha x} \tag{30.6}$$

where x is the distance through the medium with incident light flux I_0. α is the absorption coefficient, usually stated in cm^{-1}. An alternative way of describing absorption is to use the imaginary term in the media refractive index. The complex refractive index is

$$\bar{n} = n(1 + ik) \tag{30.7}$$

where k is the attenuation index. α and k are related as

$$\alpha = \frac{4\pi}{\lambda_0} nk \tag{30.8}$$

Coherence

Light can be partially or fully coherent or incoherent, depending on the source and subsequent filtering operations. Common sources of light are incoherent because they consist of many independent radiators. An example of this is the fluorescent lamp in which each excited atom radiates light independently. There is no fixed phase relationship between the waves from these atoms. In a laser the light is generated in a resonant cavity using a light amplifier and the resulting coherent light has well-defined phase fronts and frequency characteristics.

Spatial and Temporal Coherence. Spatial coherence describes the phase front properties of light. A beam from a single-mode laser which has one well-defined phase front is fully spatially coherent. A collection of light waves from a number of light emitters is incoherent because the resulting phase front has a randomly indefinable form. Temporal coherence describes the frequency properties of light. A single-frequency laser output is fully temporally coherent. White light, which contains many frequency components, is incoherent, and a narrow band of frequencies is partially coherent.

Laser Beam Focusing

The radial intensity profile of a collimated single-mode TEM$_{00}$ (Gaussian) beam from a laser is given by

$$I(r) = I_0 \exp\left[2\left(\frac{-r^2}{w_0^2}\right)\right] \tag{30.9}$$

where w_0 is the beam radius ($1/e^2$ intensity). This beam will diverge as it propagates out from the laser, and the half angle of the divergence is given by

$$\theta_{1/2} = \frac{\lambda}{\pi w_0} \tag{30.10}$$

When this beam is focused by a lens the resulting light spot radius is given by

$$w_f = \frac{\lambda l}{\pi w_d} \tag{30.11}$$

where l is the distance from the lens to the position of the focused spot and w_d is the beam radius entering the lens. It should be noted that $l \cong f$, the lens focal length, for a collimated beam entering the lens. However, l will be a greater distance than f if the beam is diverging when entering the lens.

Geometric Optics

The wavelength of light can be approximated to zero for many situations. This permits light to be described in terms of light rays which travel in the direction of the wave normal. This branch of optics is referred to geometric optics.

Properties of Light Rays

Refraction. When light travels from one medium into another it changes propagation velocity, Eq. (30.1). This results in refraction (bending) of the light as shown in Fig. 30.3.

The change in propagation direction of the light ray is given by Snell's law:

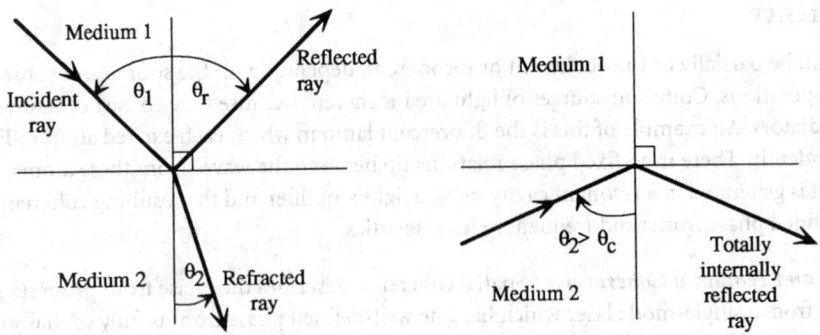

FIGURE 30.3 (a) Diagram of a light ray in medium 1 incident at angle θ_1 on the surface to medium 2. The ray is refracted at angle θ_2. (b) Diagram of the situation when the ray in medium 2 is incident at an angle greater than the critical angle θ_c and totally internally reflected.

$$n_1 \sin \theta_1 = n_2 \sin \theta_2 \qquad (30.12)$$

where n_1 and n_2 are the refractive indices of media 1 and 2, respectively.

Critical Angle.   When a light ray traveling in a medium is incident on a surface of a less dense medium, there is an incidence angle θ_2, where $\sin \theta_1 = 1$. This is the critical angle; for light incident at angles greater than θ_2 the light is totally internally reflected as shown in Fig. 30.3(b). The critical angle is given by $\theta_c = \sin^{-1}(n_1/n_2)$.

Image Formation with a Lens

Many applications require a lens to focus light or to form an image onto a detector. A well-corrected lens usually consists of a number of lens elements in a mount, and this can be treated as a black box system. The characteristics of this lens are known as the cardinal points. Figure 30.4 shows how a lens is used to form an image from an illuminated object.

The equation which relates the object, image, and lens system is

$$\frac{1}{f} = \frac{1}{s_1} + \frac{1}{s_2} \qquad (30.13)$$

The image magnification is given by $M = s_2/s_1$. When the object is very far away s_1 is infinite and the image is formed at the back focal plane.

Incoherent Light

When two or more incoherent light beams are combined, the resulting light flux is the sum of their energies. For coherent light this is not necessarily true and the resulting light intensity depends on the phase relationships between the electric fields of the two beams, as well as the degree of coherence.

Brightness and Illumination

The flux density of a light beam emitted from a **point source** decreases with the square of distance from it. Light sources are typically extended sources (being larger than point sources). The illumination of a surface from light emitted from an extended source can be calculated using Fig. 30.5.

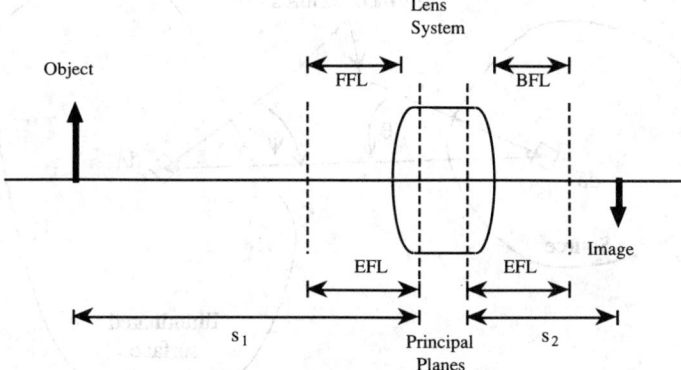

FIGURE 30.4 Schematic of an optical system forming an image of an object. Light rays from the object are captured by the lens which focuses them to form the image. EFL, effective focal length, f, of the lens; FFL and BFL, distances from the focal points to the outer lens surface. Principal planes are the positions to which the focal points, object distance, and image distance are measured; in a simple lens they are coincident.

The flux incident on a surface element dA from a source element dS is given by

$$dE = \frac{B \, dA \cos \theta \, dS \cos \psi}{r^2} \tag{30.14}$$

The constant B is called the luminance or photometric brightness of the source. Its units are candles per square meter (1 stilb = π lamberts) and dE is the luminous flux in lumens. The total illumination E of the surface element is calculated by integrating over the source. The illuminance or flux density on the surface is thus

$$I = \frac{E}{dA} \quad (\text{lumens}/\text{cm}^2) \tag{30.15}$$

Two methods are commonly used for quantifying light energy, namely, the radiometric unit of watts and the photometric unit of candelas. The candela is an energy unit which is derived from light emission from a blackbody source. The two can be related using the relative visibility curve $V(\lambda)$, which describes the eye's sensitivity to the visible light spectrum, it being maximum near a wavelength of 550 nm. The constant which relates lumens to watts at this wavelength is 685 lm/W. The luminous flux emitted by a source can therefore be written as

$$F = 685 \int V(\lambda) \, P(\lambda) \, d\lambda \quad (\text{lumens}) \tag{30.16}$$

where V is the spectral response of the eye and P is the source radiant intensity in watts.

The source radiance is normally stated as luminance in candle per square centimeter (1 lumen per steradian per square centimeter) or radiance in watts per square centimeter per steradian per nanometer. The lumen is defined as the luminous flux emitted into a solid angle of a steradian by a point source of intensity 1/60th that of a 1-cm^2 blackbody source held at 2042 K temperature (molten platinum).

Thermal Sources

Objects emit and absorb radiation, and as their temperature is increased the amount of radiation emitted increases. In addition, the spectral distribution changes, with proportionally more radia-

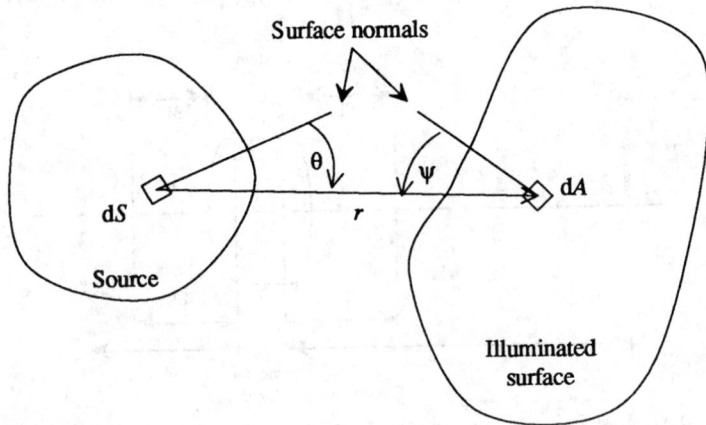

FIGURE 30.5 Surface being illuminated by an extended source. Illumination of surface element dA is calculated by summing the effects of elements dS.

tion emitted at shorter wavelengths. A blackbody is defined as a surface which absorbs all radiation incident upon it, and Kirchhoff's law of radiation is given by

$$\frac{W}{a} = \text{constant} = WB \qquad (30.17)$$

stating that the ratio of emitted to absorbed radiation is a constant a at a given temperature.

The energy or wavelength distribution for a blackbody is given by Planck's law

$$W = \frac{c_1}{\lambda^5}\left[\exp\left(\frac{c_2}{\lambda T}\right) - 1\right]^{-1} \quad (\text{watts/cm}^2 \text{ area per } \mu m \text{ wavelength})$$

$$c_1 = 3.7413 \times 10^4 \qquad (30.18)$$
$$c_2 = 1.4380 \times 10^4$$

T is in degrees Kelvin, λ is in micrometers, and W is the power emitted into a hemisphere direction. Blackbody radiation is incoherent, with atoms or molecules emitting radiation independently. Figure 30.6 is a plot of the blackbody radiation spectrum for a series of temperatures.

Very few materials are true blackbodies; carbon lampblack is one. For this reason a surface emissivity is used which describes the ratio of actual radiation emitted to that from a perfect blackbody. Table 30.2 is a listing of emissivities for some common materials.

Tungsten Filament Lamp

In the standard incandescent lamp a tungsten filament is heated to greater than 2000°C, and it is protected from oxidation and vaporization by an inert gas. In a quartz halogen lamp the envelope is quartz, which allows the filament to run at a higher temperature. This increases the light output and gives a whiter wavelength spectrum with proportionally more visible radiation to infrared.

Table 30.2 Emissivities of Some Common Materials

Material	Temperature (°C)	Emissivity
Tungsten	2000	0.28
Nickel-chromium (80-20)	600	0.87
Lampblack	20–400	0.96
Polished silver	200	0.02
Glass	1000	0.72
Platinum	600	0.1
Graphite	3600	0.8
Aluminum (oxidized)	600	0.16
Carbon filament	1400	0.53

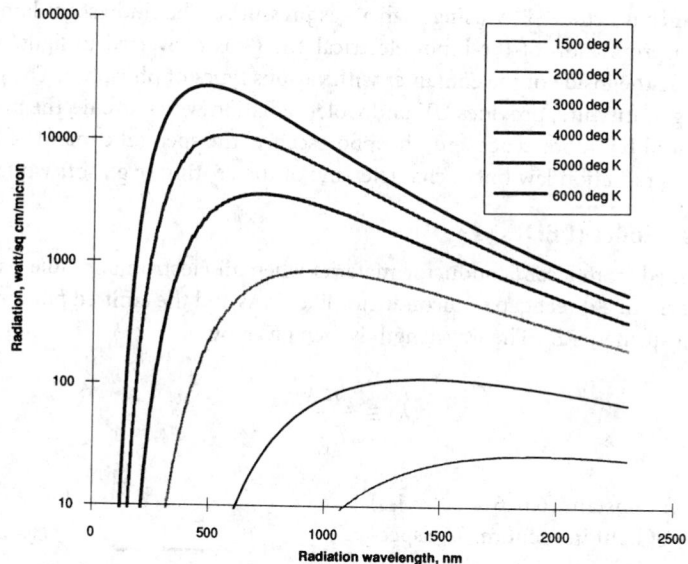

FIGURE 30.6 Plot of blackbody radiation for a series of temperatures. Radiation is in watts into a hemisphere direction from a 1-cm^2 of surface in a 1-μm wavelength band.

Standard Light Source—Equivalent Black Blackbody

Because the emissivity of incandescent materials is less than 1, an equivalent source is needed for measurement and calibration purposes. This is formed by using an enclosed space which has a small opening in it. Provided the opening is much smaller than the enclosed area, the radiation from the opening will be nearly equal to that from a blackbody at the same temperature, as long as the interior surface emissivity is > 0.5. Blackbody radiation from such a source at the melting point of platinum is defined as 1/60 cd/cm^2.

Arc Lamp

A gas can be heated to temperatures of 6000 K or more by generating an electric arc between two electrodes. The actual resulting temperature is dependent on the current flowing through the arc, the gas pressure and its composition, and other factors. This does provide a light source which is close to the temperature of the sun. Using an inert xenon gas results in essentially a white light spectrum. The use of a gas such as mercury gives more light in the UV as well as a number of strong peak light intensities at certain wavelengths. This is due to excitation and fluorescence of the mercury atoms.

Fluorescent Lamp

A fluorescent source is a container (transparent envelope) in which a gas is excited by either a dc discharge or an RF excitation. The excitation causes the electrons of the gas to move to higher energy orbits, raising the atoms to a higher excited state. When the atoms relax to lower states they give off energy, and some of this energy can be light. The wavelength of the light is characteristically related to the energy levels of the excited states of the gas involved. Typically a number of different wavelengths are associated with a particular gas.

Low-pressure lamps have relatively low luminance but provide light with narrow linewidths and stable spectral wavelengths. If only one wavelength is required, then optical filters can be used to isolate it by blocking the unwanted wavelengths.

Higher luminance is achieved by using higher gas pressures. The fluorescent lamp is very efficient since a high proportion of the input electrical energy is converted to light. White light is achieved by coating the inside of the container with various types of phosphor. The gas, for example a mercury–argon mixture, provides UV and violet radiation which excites the phosphor. Since the light is produced by fluorescence and phosphorescence, the spectral content of the light does not follow Planck's radiation law but is characteristic of the coating (e.g., soft white, cool white).

Light-Emitting Diodes (LED)

Light can be emitted from a semiconductor material when an electron and hole pair recombine. This is most efficient in a direct gap semiconductor like GaAs and the emitted photons have energy close to the bandgap energy E_g. The wavelength is then given by

$$\lambda \cong \frac{hc}{E_g} \tag{30.19}$$

where h is Planck's constant (6.626×10^{-34} J-s) and c the velocity of light in vacuum. The spectral width of the emission is quite broad, a few hundred nanometers, and is a function of the density of states, transition probabilities, and temperature.

For **light emission** to occur, the conduction band must be populated with many electrons. This is achieved by forward biasing a *pn* junction to inject electrons and holes into the junction region as shown in Fig. 30.7.

Figure 30.8(a) shows the cross section of a surface emitting LED with an integral lens fabricated into the surface. The light from the LED is incoherent and emitted in all directions. The lens and the bottom reflecting surface increase the amount of light transmitted out of the front of the device. The output from the LED is

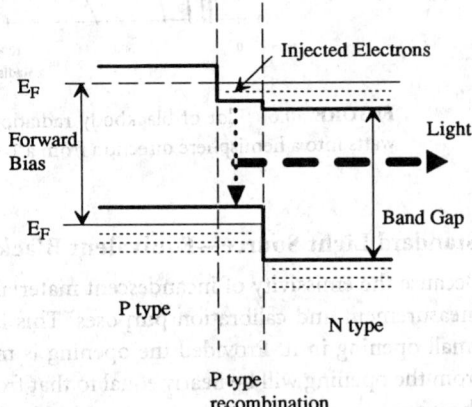

FIGURE 30.7 Band structure of a double heterostructure LED. Forward bias injects holes and electrons into the junction region where they recombine and emit light.

approximately linear with current but does decrease with increasing junction temperature.

Figure 30.8(b) shows an edge emitting LED. Here the light is generated in a waveguide region

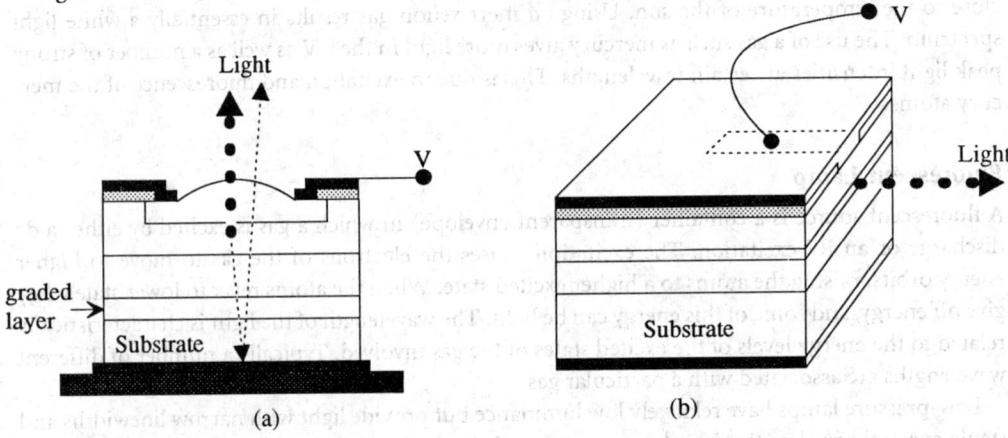

FIGURE 30.8 Cross-sectional diagrams of (a) surface emitting LED and (b) edge emitting LED. The light output from the edge emitter is more directional because of confinement by the junction guide region.

Table 30.3 Common Light-Emitting
Diode Compounds and Wavelengths

Compound	Wavelength (nm)	Color
GaP	565	Green
GaAsP	590	Yellow
GaAsP	632	Orange
GaAsP	649	Red
GaAlAs	850	Near IR
GaAs	940	Near IR
InGaAs	1060	Near IR
InGaAsP	1300	Near IR
InGaAsP	1550	Near IR

which confines the light, giving a more directional output beam.

Various wavelengths are available and are obtained by using different bandgap semiconductors. This is done by choosing different binary, ternary, and quaternary compositions. Table 30.3 is a listing of the more common ones.

The output power is usually specified in milliwatts per milliamp current obtained in a given measurement situation, e.g., into a fiber or with a 0.5 numerical aperture large area detector. Other parameters are peak wavelength, wavelength band (usually full width half max), and temperature characteristics.

Detectors, Semiconductor

When light interacts electronically with a medium, by changing the energy of electrons or creating carriers, for example, it interacts in a quantized manner. The light energy can be quantized according to Planck's theory

$$E = h\upsilon \tag{30.20}$$

where υ is the light frequency and h is Planck's constant. The energy of each photon is very small; however, it does increase with shorter wavelengths.

Photoconductors

Semiconductors can act as photoconductors, where incident light increases the carrier density, thus increasing the conductivity. There are two basic types, intrinsic and extrinsic. Figure 30.9 shows a simple energy diagram containing conduction and valence bands. Also indicated are the levels which occur with the introduction of donor and acceptor impurities.

Intrinsic photoconduction effect is when a photon with energy $h\upsilon$, which is greater than the bandgap energy, excites an electron from the valence band into the conduction band, creating a hole–electron pair. This increases the conductivity of the material. The spectral response of this type of detector is governed by the bandgap of the semiconductor.

In an extrinsic photoconductor (see Fig. 30.9), the photon excites an electron from the valence band into the acceptor level corresponding to the hole of the acceptor atom. The resulting energy

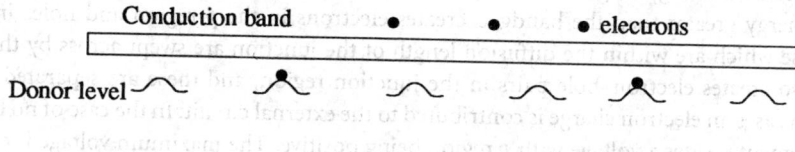

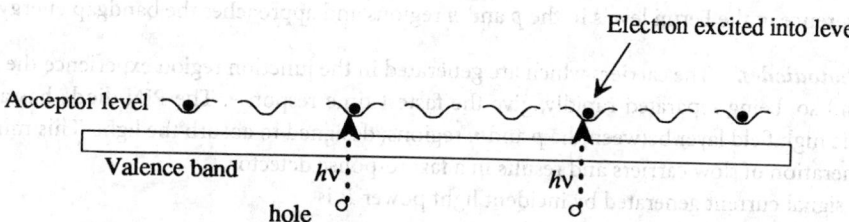

FIGURE 30.9 A simplified energy diagram for a photoconductive semiconductor, showing extrinsic effect of electrons into the acceptor level.

$h\upsilon$ is much smaller than the bandgap and is the reason why these detectors have applications for long wavelength infrared sensors. Table 30.4 is a list of commercial photoconductors and their peak wavelength sensitivities.

Table 30.4 Photoconductor Materials and Their Peak Wavelength Sensitivity

Photoconductor	Peak Wavelength (μm)
PbS	3
PbSe	5
HgCd	4
HgCaTe	10
HgCdTe	11
Si:Ga (4.2 K)	11
Si:As (4.2 K)	20
Si:Sb (4.2 K)	28

The doping material in the semiconductor determines the acceptor energy level, and so both the host material and the dopant are named. Since the energy level is quite small it can be populated by a considerable amount by thermal excitation. Thus, for useful detection sensitivity the devices are normally operated at liquid nitrogen and sometimes liquid helium temperatures. The current response, i, of a photoconductor can be written as

$$i = \frac{P\eta\tau_0 ev}{h\upsilon d} \tag{30.21}$$

where P is the optical power at frequency υ; h is Planck's constant; v is drift velocity $= \mu E$, where μ is mobility and E is electric field; η is quantum efficiency (at frequency υ); τ_0 is lifetime of carriers; and e is charge on electron.

Charge Amplification.   For semiconductor photoconductors like CdS there can be traps. These are holes, which under the influence of a bias field will be captured for a period of time. This allows electrons to move to the anode instead of recombining with a hole, resulting in a longer period for the conduction increase. This provides a photoconductive gain which is equal to the mean time the hole is trapped divided by the electron transit time in the photoconductor. Gains of 10^4 are typical.

The charge amplification can be written as

$$\frac{\tau_0}{\tau_d} \tag{30.22}$$

where $\tau_d = d/v$, the drift time for a carrier to go across the semiconductor. The response time of this type of sensor is consequently slow, ~10 ms, and the output is quite nonlinear.

Junction Photodiodes

In a simple junction photodiode a *pn* junction is fabricated in a semiconductor material. Figure 30.10 shows the energy diagram of such a device with a reverse voltage bias applied. Incident light with energy greater than the bandgap creates electrons in the *p* region and holes in the *n* region. Those which are within the diffusion length of the junction are swept across by the field. The light also creates electron–hole pairs in the junction region, and these are separated by the field. In both cases an electron charge is contributed to the external circuit. In the case of no bias the carrier movement creates a voltage with *p* region being positive. The maximum voltage is equal to the difference in the Fermi levels in the *p* and *n* regions and approaches the bandgap energy E_g.

PIN Photodiodes.   The carriers which are generated in the junction region experience the highest field and so, being separated rapidly, give the fastest time response. The PIN diode has an extra intrinsic high field layer between the *p* and *n* regions, designed to absorb the light. This minimizes the generation of slow carriers and results in a fast response detector.

The signal current generated by incident light power P is

$$i = \frac{Pe\eta}{h\upsilon} + \text{dark current} \tag{30.23}$$

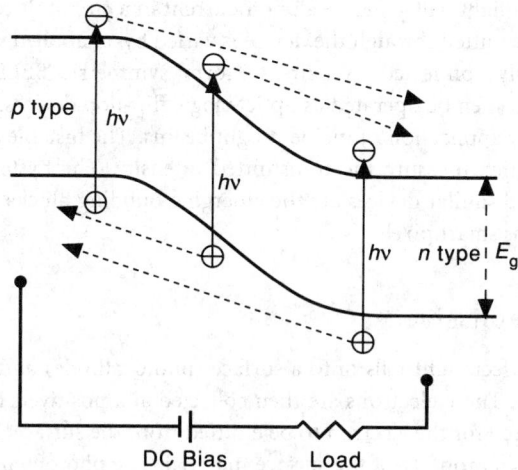

FIGURE 30.10 Energy diagram of a *pn* junction photo-diode showing the three ways electron–hole pairs are created by absorbing photons and the contribution to current flow in the circuit.

The output current is linear with incident power plus a constant dark current due to thermal generation of carriers; η is the quantum efficiency.

Avalanche Photodiodes

When the reverse bias of a photodiode is increased to near the breakdown voltage, carriers in the depletion region can be accelerated to the point where they will excite electrons from the valence band into the conduction band, creating more carriers. This current multiplication is called avalanche gain, and typical gains of 50 are available. Avalanche diodes are specially designed to have uniform junction regions to handle the high applied fields.

Solar Cells

Solar cells are large-area *pn* junction photodiodes, usually in silicon, which are optimized to convert light to electrical power. They are normally operated in the photovoltaic mode without a reverse voltage bias being applied.

Linear Position Sensors

Large-area photodiodes can be made into single axis and two axis position sensors. The single axis device is a long strip detector, and the two axis is normally square. In the single axis device the common terminal is in the middle and there are two signal terminals, one at each end. When a light beam is directed onto the detector, the relative output current from each signal terminal depends on how close the beam is to the terminal. The sum of the output currents from both terminals is proportional to the light intensity.

Phototransistors

For bipolar devices the light generates carriers which inject current into the base of the transistor. This modulates the collector base current, providing a higher output signal. For a field effect device the light generates carriers which create a gate voltage. PhotoFETs can have very high sensitivities.

SEEDs

A self-electro-optic effect device (SEED) is a multiple quantum well semiconductor optical pin device and forms the combination of a photodiode and a modulator. It can operate as a photodetector where incident light will generate a photocurrent in a circuit. It can also act as a modulator where the light transmitted through the device is varied by an applied voltage.

Devices are normally connected in pairs to form symmetric SEEDs as demonstrated in Fig. 30.11(a). These can then be operated as optical logic flip-flop devices. They can be set in one of two bistable states by application of incident light beams. The bistable state can be read out by similar light beams which measure the transmitted intensity. The hysteresis curve is shown in Fig. 30.11(b). These and similar devices are the emerging building blocks for optical logic and are sometimes referred to as smart pixels.

Detectors, Photoemissive

In the photoemissive effect, light falls onto a surface (photocathode) and the light energy causes electrons to be emitted. These electrons are then collected at a positively biased anode. There is a threshold energy required for the electron to be emitted from the surface. This energy is called the work function, f, and is a property of the surface material. The photon energy $h\upsilon$ must be greater than f, and this determines the longest wavelength sensitivity of the photocathode.

Vacuum Photodiodes

A vacuum photodiode comprises a negatively biased photocathode and a positive anode in a vacuum envelope. Light falling on the cathode causes electrons to be emitted, and these electrons are collected at the anode. Not all photons cause photoelectrons to be emitted, and quantum efficiencies, η, typically run 0.5–20%. These devices are not very sensitive; however, they have very good linearity of current to incident light power, P. They are also high-speed devices, with rise time being limited by the transit time fluctuations of electrons arriving at the anode. The photocurrent is given by

$$i = \frac{Pe\eta}{h\upsilon} + \text{dark current} \qquad (30.24)$$

This kind of detector exhibits excellent short-term stability. The emissive surface can fatigue with exposure to light but will recover if the illumination is not excessive. Because of these properties, these devices have been used for accurate light measurement, although in many cases semiconductor devices are now supplanting them.

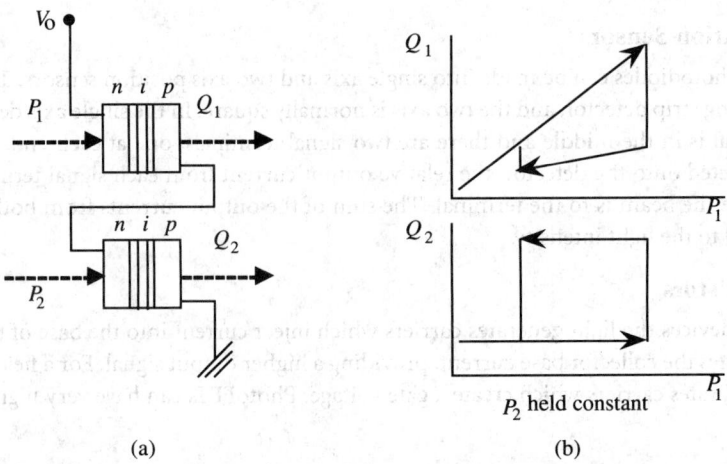

(a)   (b)

FIGURE 30.11 (a) S-SEED with voltage bias applied; (b) bistable outputs Q as a result of varying the input light power P_1 holding input power P_2 constant.

Gas-Filled Tubes

The light sensitivity of vacuum phototubes can be increased by adding 0.1 mm pressure of argon. The photoelectrons under the influence of the anode voltage accelerate and ionize the gas, creating more electrons. Gains of 5–10 can be realized. These devices are both low frequency, in the 10-kHz range, and nonlinear and are suitable only for simple light sensors. Semiconductor devices again are displacing these devices for most applications.

Photomultiplier Tubes

Photomultiplier tubes are the most sensitive light sensors, especially for visible radiation. Figure 30.12 is a schematic showing the electrical circuit used to bias it and form the output voltage signal. Light is incident on the photocathode, and the resulting photoelectrons are accelerated to a series of dynodes to generate secondary electrons and through this **electron multiplication** amplify the signal. Gains of 10^8 can be achieved with only minor degradation of the linearity and speed of vacuum photodiodes. The spectral response is governed by the emission properties of the photocathode.

There are various types of photomultipliers with different physical arrangements to optimize for a specific application. The high voltage supply ranges from 700 to 3000 V, and the electron multiplication gain is normally adjusted by varying the supply voltage. The linearity of a photomultiplier is very good, typically 3% over 3 decades of light level. Saturation is normally encountered at high anode currents caused by space charge effects at the last dynode where most of the current is generated. The decoupling capacitors, C_1, on the last few dynodes are used for high-frequency response and to prevent saturation from the dynode resistors.

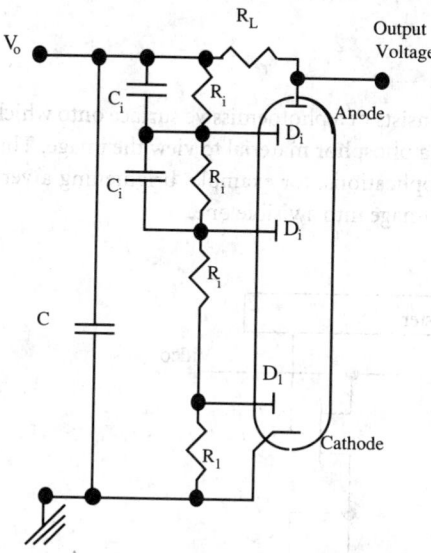

FIGURE 30.12 The basic layout of a photomultiplier tube showing the dynodes and the electrical circuit to bias them.

Photon Counting

For the detection of very low light levels and for measuring the statistical properties of light, photon counting can be done using photomultipliers. A pulse of up to 10^8 electrons can be generated for each photoelectron emitted from the cathode, and so the arrival of individual photons can be detected. There is a considerable field of study into the statistical properties of light fields as measured by photon counting statistics.

Imaging Detectors

A natural extension to single photodetectors is to arrange them in arrays, both linear single dimension and two dimensions. Imaging detectors are made from both semiconductors and vacuum phototubes.

Semiconductor Detector Arrays

Detector arrays have been made using either photodiodes or photoconductors. The applications are for visible and infrared imaging devices. For small-sized arrays each detector is individually connected to an electrical lead on the package. This becomes impossible for large arrays, however, and these contain additional electronic switching circuits to provide sequential access to each diode.

Figure 30.13 show an example of a **charge-coupled device** (CCD) linear photodiode array. The device consists of a linear array of *pn* junction photodiodes. Each diode has capacitance associated with it, and when light falls on the detector the resulting photocurrent charges this capacitance. The charge is thus the time integral of the light intensity falling on the diode. The CCD periodically and sequentially switches the charge to the video line, resulting in a series of pulses. These pulses can be converted to a voltage signal which represents the light pattern incident on the array.

The location of the diodes is accurately defined by the lithographic fabrication process and, being solid state, is also a rugged detector. These devices are thus very suitable for linear or two-dimensional optical image measurement. The devices can be quite sensitive and can have variable sensitivity by adjusting the CCD scan speed since the diode integrates the current until accessed by the CCD switch. The spectral sensitivity is that of the semiconductor photodiode, and the majority of devices now available are silicon. Smaller arrays are becoming more available in many types of semiconductors, however.

Image-Intensifier Tubes

An image-intensifier tube is a vacuum device which consists of a photoemissive surface onto which a light image is projected, an electron accelerator, and a phosphor material to view the image. This device, shown in Fig. 30.14, can have a number of applications, for example, brightening a very weak image for night vision or converting an infrared image into a visible one.

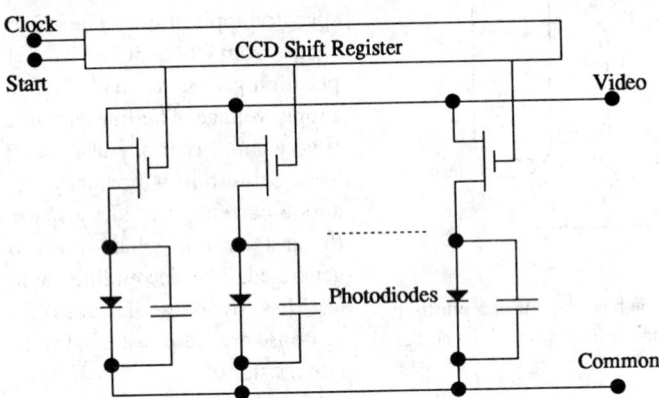

FIGURE 30.13 Schematic diagram of a linear CCD diode array sensor. CCD shift register sequentially clocks out charge from each photodiode to the video line.

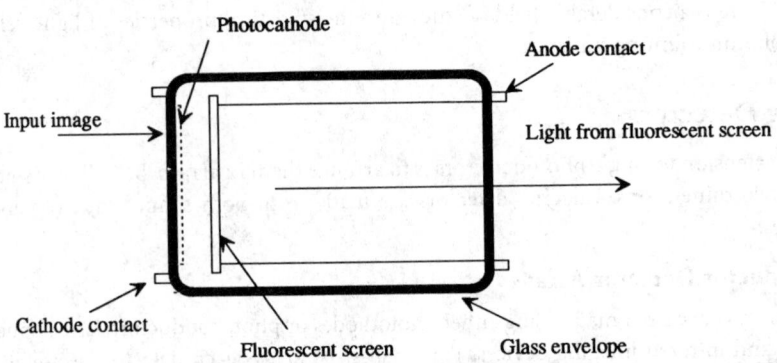

FIGURE 30.14 Diagram of a simple image-intensifier tube. More complex ones use improved electron optics.

Light falling on the cathode causes electrons to be emitted in proportion to the light intensity. These electrons are accelerated and focused by the applied electric field onto the fluorescent screen to form a visible image. Luminance gains of 50–100 times can be achieved, and a sequence of devices can be used to magnify the gain even more.

Image Orthicon Tube (TV Camera)

There are two basic types of **television** (TV) camera tubes, the orthicon and the vidicon. The orthicon uses the photoemissive effect. A light image is focused onto the photocathode, and the electrons emitted are attracted toward a positively based target (see Fig. 30.15). The target is a wire mesh, and the electrons pass through it to be collected on a glass electron target screen. This also causes secondary electrons to be emitted, and they also collect on the screen. This results in a positive charge image which replicates the light image on the photocathode.

A low-velocity electron beam is raster scanned across the target to neutralize the charge. The surplus electrons return to the electron multiplier and generate a current for the signal output. The output current is thus inversely proportional to the light level at the scanning position of the beam. The orthicon tube is very sensitive because there is both charge accumulation between scans and gain from the electron multiplier.

Vidicon Camera Tube

A simple TV camera tube is the vidicon. This is the type used in camcorders and for many video applications where a rugged, simple, and inexpensive camera is required. Figure 30.16 is a schematic of a vidicon tube; the optical image is formed on the surface of a large-area photoconductor, causing corresponding variations in the conductivity. This causes the rear surface to charge toward the bias voltage V_b in relation to the conductivity image. The scanning electron beam periodically recharges the rear side to 0 V, resulting in a recharging current flow in the out-

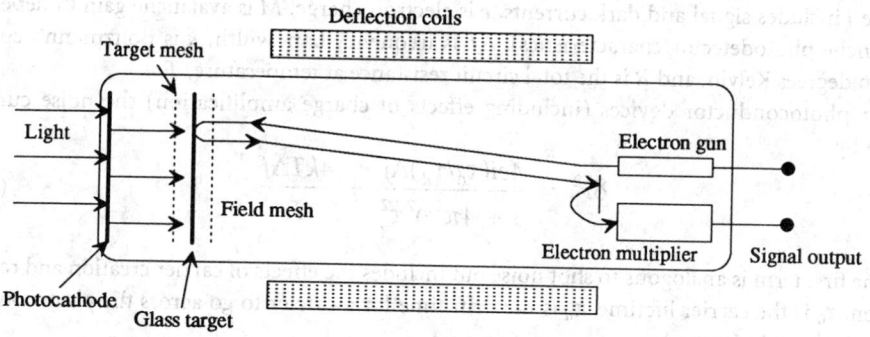

FIGURE 30.15 Schematic diagram of an image orthicon TV camera tube.

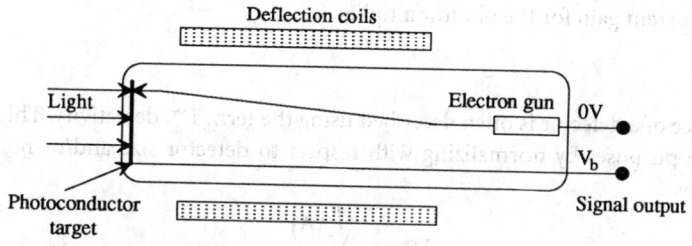

FIGURE 30.16 Schematic of a vidicon TV camera tube.

put. The output signal is a current signal proportional to the light incident at the position of the scanning electron beam.

The primary disadvantages of the vidicon are its longer response time and smaller dynamic range. The recent availability of longer wavelength photoconducting films has resulted in new infrared cameras becoming available.

A recent advance in these types of image sensor is to replace the photoconductor with a dense array of very small semiconductor photodiodes. Photocurrent in the diode charges a capacitor connected to it. The raster scanned electron beam discharges this capacitor in the same way.

Image Dissector Tube

The image dissector tube is a photosensitive device which uses an electron deflection lens to image the electron from the cathode onto a pinhole in front of an electron multiplier. The image can be deflected around in front of the pinhole in a random access manner. The primary application of this kind of device is for tracking purposes.

Noise and Detectivity

Noise

There are two primary sources of noise in photodetectors: Johnson noise due to thermal effects in the resistive components of the device and its circuits, and shot noise or its equivalent, which is due to the quantized nature of electro-optic interactions.

In semiconductor devices noise is usually given in terms of noise current,

$$\delta \bar{i}^2 = 2eiM^{2+x}\Delta f + \frac{4kT\Delta f}{R} \qquad (30.25)$$

where i includes signal and dark currents, e is electron charge, M is avalanche gain (x depends on avalanche photodetector characteristics), Δf is frequency bandwidth, k is Boltzmann's constant, T is in degrees Kelvin, and R is the total circuit resistance at temperature, T.

For photoconductor devices (including effects of charge amplification) the noise current is given by

$$\delta \bar{i}^2 = \frac{4ei(\tau_0/\tau_d)\Delta f}{1 + 4\pi^2 v^2 \tau_0^2} + \frac{4kT\Delta f}{R} \qquad (30.26)$$

The first term is analogous to shot noise but includes the effects of carrier creation and recombination. τ_0 is the carrier lifetime, τ_d is the drift time for a carrier to go across the photoconductor, and v is the light frequency.

The noise for photoemissive devices is usually written as a noise voltage and is given by

$$\delta_v^2 = 2ei\, G^2 \Delta f R^2 + 4kT\Delta f R \qquad (30.27)$$

where G is the current gain for the photomultiplier.

Detectivity

The performance of a detector is often described using the term D^*, detectivity. This term is useful for comparison purposes by normalizing with respect to detector size and/or noise bandwidth. This is written as

$$D^* = \frac{\sqrt{A\Delta f}}{\text{NEP}} \qquad (30.28)$$

where NEP is the noise equivalent power (for signal-to-noise ratio equal to 1) and A is detector area. The term $D^*(\lambda)$ is used for quoting the result using a single-wavelength light source and $D^*(T)$ is used for the unfiltered blackbody radiation source.

Defining Terms

Charge-coupled device (CCD): A series of electronic logic cells in a device in which a signal is represented and stored as an electronic charge on a capacitor. The signal is moved from one cell (memory position or register) to an adjacent cell by electronically switching the charge between the capacitors.

Electron multiplication: The phenomenon where a high-energy electron strikes a surface and causes additional electrons to be emitted from the surface. Energy from the incident electron transfers to the other electrons to cause this. The result is electron gain which is proportional to the incident electron energy.

Extended source: A light source with finite size where the source size and shape can be determined from the emitted light characteristics. The light is spatially incoherent.

Light detection: The conversion of light energy into an electrical signal, either current or voltage.

Light emission: The creation or emission of light from a surface or device.

Point source: A light source which is so small that its size and shape cannot be determined from the characteristics of the light emanating from it. The light emitted has a spherical wave front and is spatially coherent.

Television (TV): The process of detecting an image and converting it to a serial electronic representation. A detector raster scans the image, producing a voltage proportional to the light intensity. The time axis represents the distance along the raster scan. Several hundred horizontal scans make up the image starting at the top. The raster scan is repeated to provide a continuing sequence of images.

References

B. Crosignani, P. DiPorto, and M. Bartolotti, *Statistical Properties of Scattered Light*, New York: Academic Press, 1975.

A.L. Lentine *et al.*, "A 2 kbit array of symmetric self-electrooptic effect devices," *IEEE Photonics Technol. Lett.*, vol. 2, no. 1, 1990.

Reticon Corp., subsidiary of EG&G, Inc., Application notes #101.

Further Information

W. J. Smith, *Modern Optical Engineering*, New York: McGraw Hill, 1966.

M. J. Howes and D.V. Morgan, *Gallium Arsenide Materials, Devices and Circuits*, New York: John Wiley, 1985.

M. K. Baroski, *Fundamentals of Optical Fiber Communications*, New York: Academic Press, 1981.

C. Y. Wyatt, *Electro-Optic System Design for Information Processes*, New York: McGraw-Hill, 1991.

S. Ungar, *Fibre Optics—Theory and Applications*, New York: John Wiley, 1990.

30.3 Circuits

R. A. Becker

In 1969, Stewart Miller of AT&T Bell Laboratories published his landmark article on **integrated optics**. This article laid the foundation for what has now developed into optoelectronic circuits. In it he described the concepts of planar **optical guided-wave devices** formed as thin films on various substrates using fabrication techniques similar to those used in the semiconductor integrated circuit (IC) industry. The attributes of these new circuits included small size, weight, power consumption, and mechanical robustness because all components were integrated on a single substrate. The field of optoelectronic circuits began as a hybrid implementation where optical sources (laser diodes) and detectors have historically been fabricated on separate semiconductor substrates, and waveguide devices, such as modulators and switches, have been fabricated on electro-optic single-crystal oxides such as **lithium niobate** ($LiNbO_3$). Often, the two dissimilar substrates have been connected using single-mode polarization preserving optical fiber. Now, although the hybrid concept is finding commercial applications, most active research is performed on monolithic implementations, where all devices are fabricated on a common semiconductor substrate. Our discussion will deal exclusively with the more mature hybrid implementation of optoelectronic circuits based on $LiNbO_3$.

Because sources and detectors have been covered in previous sections, in this section, the devices that are utilized in between, i.e., modulators and switches, will be discussed.

Integrated Optics

Integrated optics can be defined as the monolithic integration of one or more optical guided-wave structures on a common substrate. These structures can be passive, such as a fixed optical power splitter, or active, such as an optical switch. Active devices are realized by placing metal electrodes in close proximity to the optical waveguides. Applying a voltage to the electrodes changes the velocity of the light within the waveguide. Depending on the waveguide geometry and the electrode placement, a wide variety of technologically useful devices and circuits can be realized.

The technological significance of integrated optics stems from its natural compatibility with two other rapidly expanding technologies: fiber optics and semiconductor laser diodes. These technologies have moved in the past 10 years from laboratory curiosities to large-scale commercial ventures. Integrated optic devices typically use laser diode optical sources, diode-pumped yttrium, aluminum, garnet (YAG) lasers, and transmit the modified optical output on a single-mode optical fiber. Integrated optic devices are typically very high speed, compact, and require only moderate control voltages compared to their bulk-optical counterparts.

In integrated optic devices, the optical channel waveguides are formed on a thin, planar, optically polished substrate using photolithographic techniques similar to those used in the semiconductor IC industry. Waveguide routing is accomplished by the mask used in the photolithographic process, similar to the way electrically conductive paths are defined in semiconductor ICs. The photolithographic nature of device fabrication offers the potential of readily scaling the technology to large volumes, as is done in the semiconductor IC industry. For example, the typical device is 0.75 in. × 0.078 in. in size. Dividing the substrate size by the typical device size and assuming a 50% area usage indicates that one can achieve 50 devices per 3-in. wafer.

The typical substrate material in integrated optics is not silicon, but the synthetic, widely available crystal, lithium niobate ($LiNbO_3$), which has been commercially produced in volume for more than 20 years. This material is transparent to optical wavelengths between 400 and 4500 nm, has a hardness similar to glass, and is nontoxic. In addition, devices have also been fabricated in a number of other materials, including glass and semiconductor materials such as GaAs and InP. Glass integrated optic technology is relatively mature; however, it is severely limited in application since only passive devices, or low-frequency thermally operated active devices, can be implemented. The area of semiconductor-based integrated optics has attracted much attention worldwide, because it offers the potential of integrating electronic circuitry, optical sources and detectors, and optical waveguides on a single substrate. While being quite promising, the technology is still 5 years away from commercialization. Technological problems in semiconductor-based integrated optics include low electro-optic coefficient, higher optical waveguide attenuation, and an incompatibility of the processing steps needed to fabricate the various types of devices on a single substrate. However, considerable attention is being paid to these problems, and improvements are continually occurring.

In constrast, LiNbO₃-based devices have been commercially available since 1985 and have been incorporated in a large number of experimental systems. The basic LiNbO₃ waveguide fabrication technique was developed in 1974 and has been continually refined and improved during subsequent years. The material itself finds wide application in a number of electrical and optical devices because of its excellent optical, electrical, acoustic, and electro- and acousto-optic properties. For example, almost all color television sets manufactured today incorporate a surface-acoustic-wave (SAW) electrical filter based on LiNbO₃.

In LiNbO₃-based integrated optics, optical waveguides are formed in one of two ways. The first uses photolithographically patterned lines of titanium (Ti), several hundred angstroms thick, on the substrate surface. The titanium is then diffused into the substrate surface at a temperature of about 1000°C for several hours. This process locally raises the refractive index in the regions where titanium has been diffused, forming high-refractive index stripes that will confine and guide light. Because the diffusion is done at exceedingly high temperatures, the waveguide stability is excellent. The waveguide mechanism used is similar to that used in fiber optics, where the higher-index, doped cores guide the light. The exact titanium stripe width, the titanium thickness, and diffusion process are critical parameters in implementing a low-loss single-mode waveguide. Different fabrication recipes are required to optimize the waveguides for operation at the three standard diode laser wavelengths: 800 nm, 1300 nm, and 1500 nm. The second approach uses a technique known as proton exchange. In this approach, a mask is used to define regions of the substrate where hydrogen will be exchanged for lithium, resulting in an increase in the refractive index. This reaction takes place at lower temperatures (200–250°C) but has been found to produce stable waveguides if an anneal at 350–400°C is performed. Waveguides formed using the proton exchange method support only one polarized mode of propagation, whereas those formed using Ti indiffusion support two. Proton exchange waveguides are also capable of handling much higher optical power densities, especially at the shorter wavelengths, than are those formed by Ti indiffusion. More fabrication detail will be provided later.

Light modulation is realized via the electro-optic effect, i.e., inducing a small change in the waveguide refractive index by applying an electric field within the waveguide. On an atomic scale the applied electric field causes slight changes in the basic crystal unit cell dimensions, which changes the crystal's refractive index. The magnitude of this change depends on the orientation of the applied electric field and the optical polarization. As a result, only certain crystallographic orientations are useful for device fabrication and devices are typically polarization dependent. The electro-optic coefficients of LiNbO₃ are among the highest (30.8 pm/V) of any inorganic material, making the material very attractive for integrated optic applications.

Combining the concepts of optical waveguides and electro-optic modulation with the geometric freedom of photolithographic techniques leads to an extremely diverse array of passive and active devices.

Passive components do not require any electric fields and are used for power splitting and combining functions. Two types of passive power division structures have been fabricated: Y-junctions and directional couplers. A single waveguide can be split into two by fabricating a shallow-angle Y-junction as shown in Fig. 30.17.

An optical signal entering from the single-waveguide side of the junction is split into two optical signals with the same relative phase but one-half the original intensity. Conversely, light incident on the two-waveguide side of the junction will be combined into the single waveguide with a phase and intensity dependent on the original inputs. Directional couplers consist of two or more waveguides fabricated in close proximity to each other so that the optical fields overlap as shown in Fig. 30.18. As a result, optical power is transferred between the waveguides. The extent of the power transfer is dependent on the waveguide characteristics, the waveguide spacing, and the interaction length.

A different type of passive component is an optical polarizer, which can be made using several different techniques. One such method is the metal-clad, dielectric-buffered waveguide shown in Fig. 30.19. In this passive device, the TM polarization state is coupled into the absorbing metal and

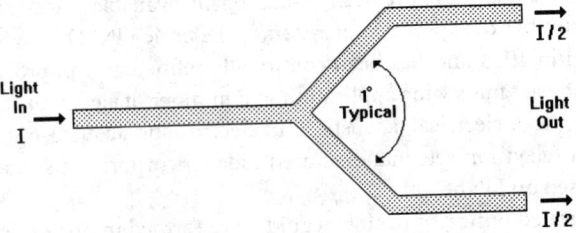

FIGURE 30.17   Passive Y-splitter.

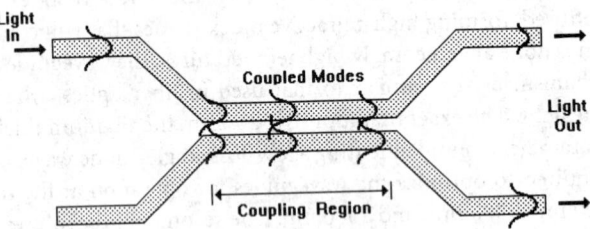

FIGURE 30.18   Directional coupler power splitter.

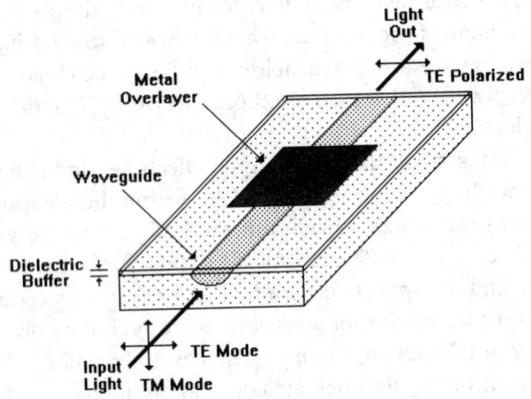

FIGURE 30.19   Thin-film optical polarizer.

is thus attenuated, while the TE polarization is virtually unaffected. Measurements of a 2-mm-long polarizer of this type have demonstrated TM attenuations exceeding 50 dB (100,000:1). Polarizers can also be fabricated in others ways. One interesting technique involves the diffusion of hydrogen ions into the LiNbO$_3$. This results in a waveguide which, as discussed earlier, will only support the TE-polarized mode and, thus, is a natural polarizer.

Active components are realized by placing electrodes in close proximity to the waveguide structures. Depending on the substrate crystallographic orientation, the waveguide geometry, and the electrode geometry, a wide variety of components can be demonstrated. The simplest active device is the **phase modulator**, which is a single waveguide with electrodes on either side as shown in Fig. 30.20. Applying a voltage across the electrodes induces an electric field across the waveguide, which changes its refractive index via the electro-optic effect. For 800-nm wavelength operation, a typical phase modulator would be 6 mm long and would induce a π-phase shift for an applied voltage of 4 V. The transfer function (light out versus voltage in) can be expressed as

$$I_0(V) = I_i \exp(j\omega t + \pi V/V_\pi) \tag{30.29}$$

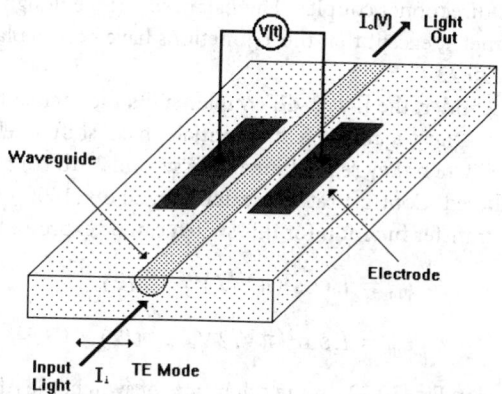

FIGURE 30.20   Electro-optic integrated optic phase modulator.

where V_π is the voltage required to cause a 180-degree phase shift. Note that there is no change in the intensity of the light. Coherent techniques are used to measure the amount of phase change.

Optical **intensity modulators** can be fabricated by combining two passive Y-junctions with a phase modulator situated between them. The result, which is shown in Fig. 30.21, is a guided-wave implementation of the classic Mach–Zehnder interferometer. In this device the incoming light is split into two equal components by the first Y-junction. An electrically controlled differential phase shift is then introduced by the phase modulator, and the two optical signals are recombined in the second Y-junction. If the two signals are exactly in phase, then they recombine to excite the lowest-order mode of the output waveguide and the intensity modulator is turned fully on. If instead there exists a π-phase shift between the two signals, then they recombine to form the second mode, which is radiated into the substrate and the modulator is turned fully off. Contrast ratios greater than 25 dB (300:1) are routinely achieved in commercial devices. The transfer function for the Mach–Zehnder modulator can be expressed as

$$I_0(V) = I_i \cos^2(\pi V/2 V_\pi + \phi) \qquad (30.30)$$

where V_π is the voltage required to turn the modulator from on to off, and ϕ is any static phase imbalance between the interferometer arms. This transfer function is shown graphically in Fig. 30.21. Note that the modulator shown in Fig. 30.21 has push-pull electrodes. This means that when a voltage is applied, the refractive index is changed in opposite directions in the two arms, yielding a twice-as-efficient modulation.

Optical switches can be realized using a number of different waveguide, electrode, and substrate orientations. Two different designs are used in commercially available optical switches: the bal-

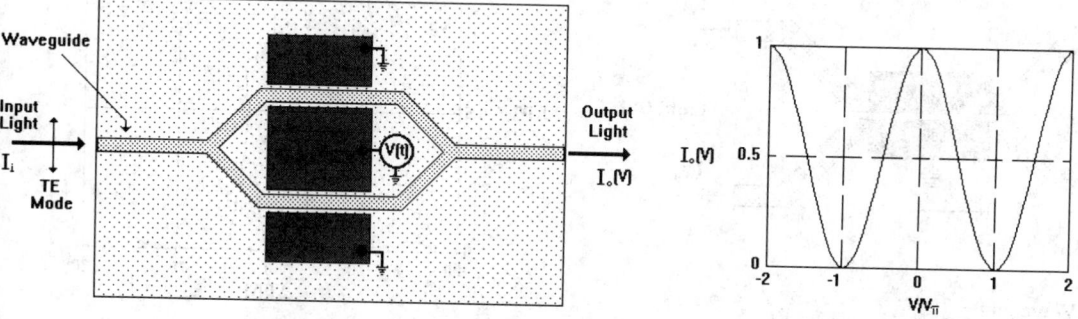

FIGURE 30.21   Mach–Zehnder intensity modulator and transfer function.

anced-bridge and the $\Delta\beta$ directional coupler. The balanced-bridge design is similar to that of the Mach–Zehnder interferometer, except that the Y-junctions have been replaced by 3-dB directional couplers as shown in Fig. 30.22.

Similar to the Mach–Zehnder, the first 3-dB coupler splits the incident signal into two signals, ideally of equal intensity. Once again, if a differential phase shift is electro-optically induced between these signals, then when they recombine in the second 3-dB coupler, the ratio of power in the two outputs will be altered. Contrast ratios greater than 20 dB (100:1) are routinely achieved in commercial devices. The transfer function for this switch can be expressed as

$$I_{0a} = I_i \cos^2(\pi V/2V_\pi + \pi/2) \tag{30.31}$$

$$I_{0b} = I_i \sin^2(\pi V/2V_\pi + \pi/2) \tag{30.32}$$

and is graphically depicted in Fig. 30.22. In the other type of switch, the $\Delta\beta$ directional coupler, the electrodes are placed directly over the directional coupler as shown in Fig. 30.23. The applied electric field alters the power transfer between the two adjacent waveguides. Research versions of this switch have demonstrated contrast ratios greater than 40 dB (10,000:1); however, commercial versions typically achieve 20 dB, which is competitive with that achieved with the balanced-bridge switch. The transfer function for the $\Delta\beta$ directional coupler switch can be expressed as

$$I_{0a} = \sin^2 \kappa L^* \mathrm{sqrt}(1 + (\Delta\beta/2\kappa)^2)/(1 + (\Delta\beta/2\kappa)^2) \tag{30.33}$$

$$I_{0b} = 1 - I_{0a} \tag{30.34}$$

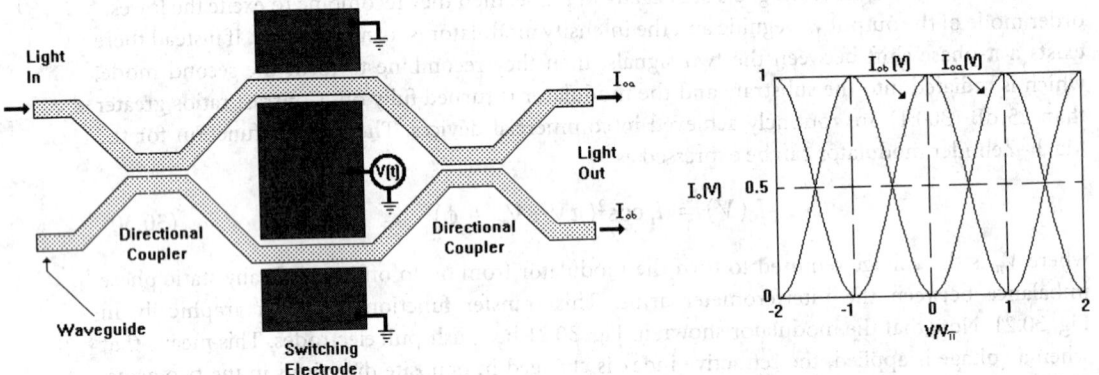

FIGURE 30.22 Balanced-bridge modulator/switch and transfer function.

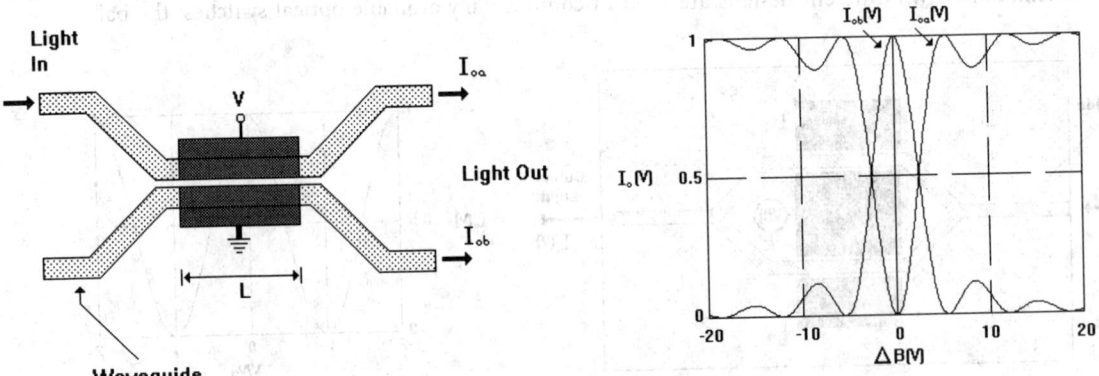

FIGURE 30.23 Directional coupler switch and transfer function.

where κ is the coupling constant and Δβ is the voltage-induced change in the propagation constant. This transfer function is depicted in Fig. 30.23.

Another type of active component that has recently become available commercially is the **polarization controller.** This component allows the incoming optical polarization to be continuously adjustable. The device functions as an electrically variable optical waveplate, where both the birefringence and axes orientation can be controlled. The controller is realized by using a three-electrode configuration as shown in Fig. 30.24 on a substrate orientation where the TE and TM optical polarizations have almost equal velocities. Typical performance values are TE/TM conversion of greater than 99% with less than 50 V.

One of the great strengths of integrated optic technology is the possibility of integrating different types or multiple copies of the same type of device on a single substrate. While this concept is routinely used in the semiconductor IC industry, its application in the optical domain is novel. The scale of integration in integrated optics is quite modest by semiconductor standards. To date the most complex component demonstrated is an 8 × 8 optical switch matrix that uses 64 identical 2 × 2 optical switches. The most device diversity on a given substrate is found in fiber gyro applications. Here, components incorporating six phase modulators, two electrically tunable directional couplers, and two passive directional couplers have been demonstrated.

Device Fabrication

The fabrication of an integrated optic device uses the same techniques as used in the semiconductor IC industry. Device designs are first entered into a computer-aided design (CAD) system for accurate feature placement and dimensional control. This design is then output as a digitized tape that will control a pattern generation system for fabrication of the chrome masks that are used in device fabrication. A variety of equipment such as step-and-repeat and E-beam systems has been developed for the semiconductor IC industry for the generation of chrome masks. These same systems are used today for generation of masks for integrated optic devices.

The waveguides can be fabricated by using either the Ti indiffusion method or the proton exchange method. The first step in fabricating a waveguide device using Ti indiffusion is the patterning in titanium. The bare LiNbO₃ surface is first cleaned and then coated with photoresist. Next, the coated substrate is exposed using the waveguide-layer chrome mask. The photoresist is then developed. The areas that have been exposed are removed in the development cycle. The patterned substrates are then coated with titanium in a vacuum evaporator. The titanium covers the exposed regions of the substrate as well as the surface of the remaining photoresist. The substrate is next soaked in a photoresist solvent. This causes all the residual photoresist (with titanium on top)

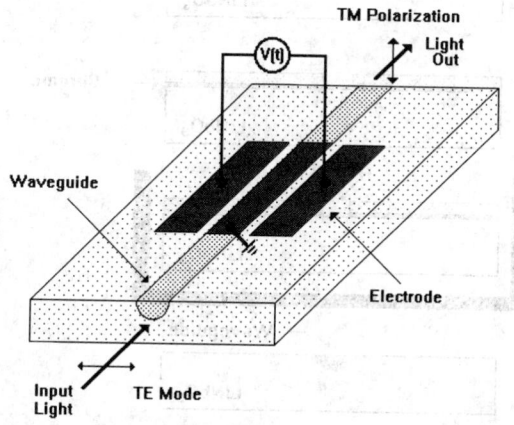

FIGURE 30.24 Guided-wave polarization controller.

to be removed, leaving only the titanium that coated the bare regions of the substrate. This process is known as *lift-off*. Finally, the substrate, which is now patterned with titanium, is placed in a diffusion system. At temperatures above 1000°C the titanium diffuses into the substrate, slightly raising the refractive index of these regions. This process typically takes less than 10 hours. This sequence of steps is depicted in Fig. 30.25. The proton exchange method is depicted in Fig. 30.26.

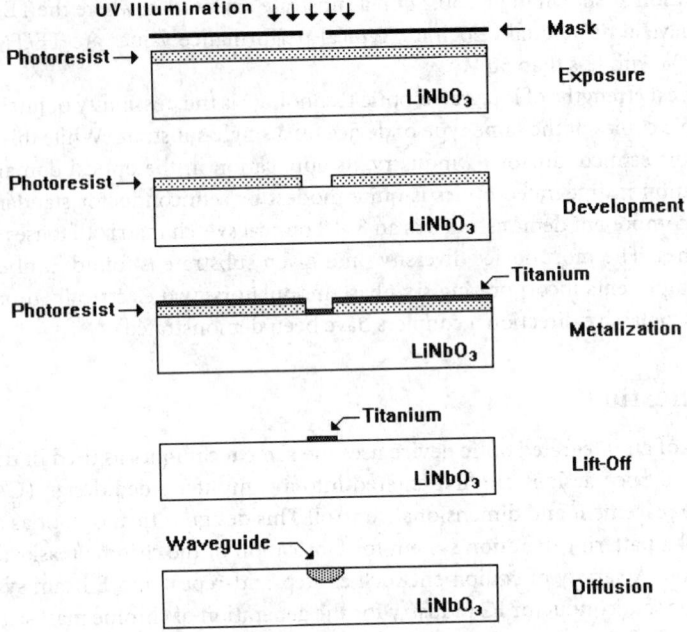

FIGURE 30.25   Ti-indiffused LiNbO$_3$ waveguide fabrication.

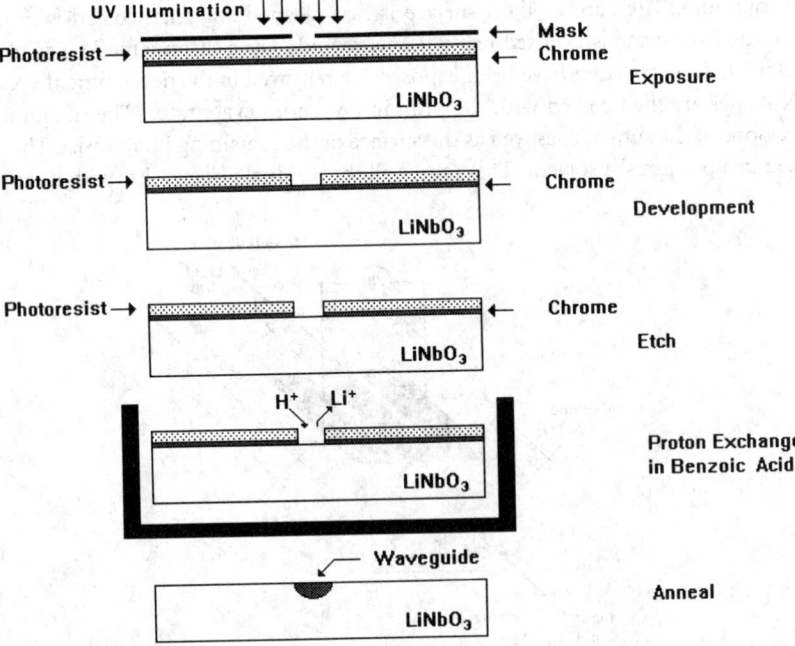

FIGURE 30.26   Proton exchange LiNbO$_3$ waveguide fabrication.

Here a chrome masking layer is first deposited on the LiNbO$_3$ substrate. It is patterned using photoresist and etching. Next, the substrate is submerged in hot benzoic acid. Finally, the chrome mask is removed and the substrate is annealed. The regions that have been exposed to the benzoic acid will have an increased refractive index and will guide light.

If the devices being fabricated are to be active (i.e., voltage controlled), then an electrode fabrication step is also required. This sequence of steps parallels the waveguide fabrication sequence. The only differences are that an electrode mask is used and the vacuum-deposited metal used is chrome/gold or chrome/aluminum. This sequence of steps is shown in Fig. 30.27.

In order to get the light in and out of the waveguide, the endfaces have to be lapped and polished flat with chip-free knife edges. This is currently accomplished using standard lapping and polishing techniques. After this step, the substrate can be diced into as many devices as were included on the substrate. Finally, the diced parts need to be electrically and optically packaged.

Packaging

To get the light in and out of an integrated optic waveguide requires a tiny optical window to be polished onto the waveguide's end. Currently, the entire endface of the substrate is polished to a sharp, nearly perfect corner, making the whole endface into an optical window. An optical fiber can then be aligned to the waveguide end and attached. Typically, centration of the fiber axis to the waveguide axis must be better than 0.2 μm. Some devices require multiple inputs and outputs. In this case the fibers are prealigned in silicon V-grooves. These V-grooves are fabricated by anisotropic etching of a photolithographically defined pattern on the silicon. The center-to-center spacing of the fiber V-groove array can be made to closely match that of the multiple waveguide inputs and outputs.

Integrated optic devices built on LiNbO$_3$ are inherently single-mode devices. This means that the light is confined in a cross-sectional area of approximately 30 μm^2. The optical mode has a near-field pattern that is 5 to 10 μm across and 3 to 6 μm deep, depending on the wavelength. These mode spot sizes set limits on how light can be coupled in and out. There are a number of methods

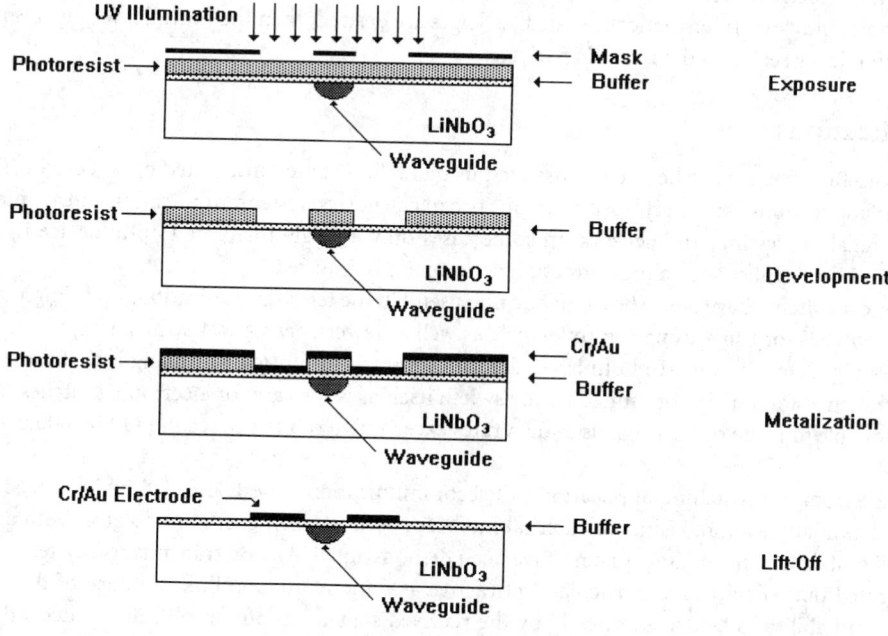

FIGURE 30.27   Electrode fabrication via lift-off.

that can be used to couple the light into LiNbO$_3$ waveguides. These include prism coupling, grating coupling, end-fire coupling with lenses, and end-fire coupling with single-mode optical fibers. In general, most of these techniques are only useful for laboratory purposes. The most practical real-world technique is end-fire coupling with an optical fiber. In this case the optical fiber is aligned to the waveguide end. This is an excellent practical method since integrated optic devices are most often used in fiber optic systems. Therefore, the coupling problem is one of aligning and fixing a single-mode fiber to the single-mode LiNbO$_3$ waveguide. The size of the single-mode radiation pattern and its angular divergence set the alignment tolerances. A low-loss connection between a fiber and a LiNbO$_3$ waveguide requires <1/20 of a mode spot diameter (0.25–0.5 μm) in transverse offset, angular tilt of <2 degrees, and a longitudinal offset of <1 mode spot diameter (5–10 μm). These are very stringent alignment requirements, especially if they have to be maintained over a wide temperature range.

Another aspect of the problem is that many integrated optic devices require a single, well-defined linearly polarized input. Ordinary single-mode fiber is not suitable in this case. The solution is to use polarization preserving fiber. This fiber is made such that it will maintain a single linearly polarized input over long distances. The use of polarization preserving fiber, however, adds another requirement to the coupling problem. This requirement is that the fiber must be rotationally aligned about its cylindrical axis so that the linearly polarized light coincides with the desired rotational axis of the LiNbO$_3$ waveguide. The rotational precision needed is <0.5 degrees.

Many LiNbO$_3$ devices, such as fiber gyro components, require multiple input and/or output optical connections. Thus, the packaging must be able to accommodate multiple inputs/outputs and maintain strict alignment for all connections.

The method of end-fire coupling optical fibers to the LiNbO$_3$ waveguide is commonly called *pig-tailing*. This is the only practical packaging method now used for integrated optic devices that operate in a real system and outside the laboratory. The reasons for this are quite logical. The end user installs the device in his system by connecting to the fiber pigtails. The connection can be made with single-mode connectors or by splicing. Flexibility is one of the big advantages of using fiber pigtails.

The typical LiNbO$_3$ device is packaged in a metallic case with optical fiber pigtails connected at both ends. Electrical connections are provided by RF connectors or pins, which are common in the electronics industry. If hermetically sealed packages are desired, then the optical fiber pigtails must be hermetically sealed to the metallic package.

Applications

Many useful systems have been demonstrated using LiNbO$_3$-based integrated optic devices. These system applications can be grouped into four broad categories: telecommunications, instrumentation, signal processing, and sensors. In some cases only a single integrated optic device is used, while in other applications a multifunction component is required.

Optical switches have been shown to be quite useful in the telecommunications area. High-speed 2 × 2 switches for time-domain mux/demux as well as lower-speed 4 × 4 switch arrays have been successfully demonstrated. In both cases a major advantage of optical switching is that the switch data transmission rate is not limited by the switch itself as is the case for electronic switches. Thus, it is possible to route optical signals at data rates exceeding terabits/per second (1,000,000,000,000 bits/s).

Aside from the switching application in telecommunications, there also is the high-speed laser modulation application. Using an external LiNbO$_3$-based optic intensity modulator, both analog and digital data transmission systems have been demonstrated. Analog transmission systems using integrated optic devices are particularly attractive as remote antenna links because of their high speed and ability to be driven directly by the received signal without amplification. Recently, the use of high-power diode-pumped YAG lasers operating at 1300 nm and external intensity modulators based on LiNbO$_3$ have found wide application in the cable TV industry. In addition, the ability of the Mach–Zehnder intensity modulator to control intensity with a controlled wavelength change (i.e., chirp) has allowed its use in long-haul telecommunications systems.

Another demonstrated application of integrated optics in the telecommunications area is in coherent communication systems. These systems require both phase modulators and polarization controllers. Current optical fiber transmission systems rely on intensity modulated data transmission schemes. Coherent communication systems are attractive because of the promise of higher bit rates, wavelength division multiplexing capability, and greater noise immunity. In coherent communication systems the information is coded by varying either the phase or frequency of the optical carrier with a phase modulator. At the receiver, a polarization controller is used to ensure a good signal-to-noise ratio in the heterodyne detection system.

One promising application of integrated optic devices in instrumentation is a high-speed, polarization-independent optical switch for use in optical-time-domain reflectometers (OTDR). OTDRs are used to locate breaks or poor splices in fiber optic networks. The instruments work by sending out an optical pulse and measuring the backscattered radiation returning to the instrument as a function of time. The next generation of OTDRs will possibly employ an optical switch, which will be used to rapidly switch the optical fiber under test from the pulsed light source to the OTDR receiver. Such an instrument could detect faults closer to the OTDR than currently possible, which is important in the short-haul systems now being installed. This feature is necessary for local area network (LAN) installations.

Several types of sensors using integrated optic devices have been demonstrated. Two of the most promising are electric/magnetic field or voltage sensors and rotation sensors (fiber optic gyro). Electric field sensors typically consist of either a Mach–Zehnder intensity modulator or an optical switch that is biased midway between the on and off states. For small modulation depths about this midpoint the induced optical modulation is linear with respect to applied voltage. Linear dynamic ranges in excess of 80 dB have been accomplished. This is larger than that obtained using any other known technology.

Perhaps the most promising near-term application of integrated optic devices in the field of rotation sensing is as a key component in optical fiber gyroscopes. A typical fiber optic gyro component is shown in Fig. 30.28. The device consists of a polarizer, a Y-junction, and two phase modulators, all integrated on a single substrate. In fiber gyro systems, the integrated optic component replaces individual, fiber-based components that perform the same function. The integrated optic component offers a greatly improved performance, at a significant reduction in cost, compared to

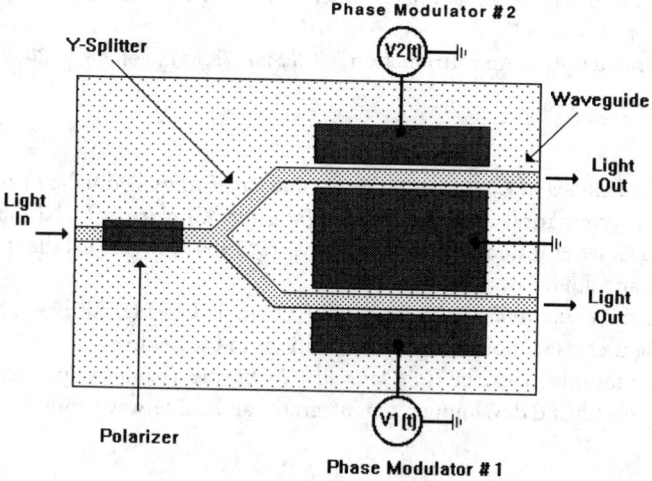

FIGURE 30.28   Fiber-optic gyro chip.

the fiber-based components. Most fiber optic gyro development teams have done away with the fiber components and have adopted LiNbO$_3$-based components as the technology of choice.

Defining Terms

Integrated optics: The monolithic integration of one or more optical guided-wave devices on a common substrate.

Intensity modulator: A modulator that alters only the intensity of the incident light.

Lithium niobate (LiNbO$_3$): A single-crystal oxide that displays electro-optic, acousto-optic, piezoelectric, and pyroelectric properties that is often the substrate of choice for surface acoustic wave devices and integrate optical devices.

Optical guided-wave device: An optical device that transmits or modifies light while it is confined in a thin-film optical waveguide.

Phase modulator: A modulator that alters only the phase of the incident light.

Polarization controller: A device that alters only the polarization state of the incident light.

References

R. Alferness, "Waveguide electrooptic modulators," *IEEE Trans. Microwave. Theory Tech.*, vol. MTT-30, p. 1121, 1982.

R. A. Becker, "Commercially available integrated optics products and services," *SPIE*, vol. 993, p. 246, 1988.

R. Childs and V. O'Byrne, "Predistortion Linearization of Directly Modulated DFB Lasers and External Modulators for AM Video Transmission," *OFC'90 Tech. Dig.*, Paper WHG, 1990, p. 79.

C. Cox, G. Betts, and L. Johnson, "An analytic and experimental comparison of direct and external modulation in analog fiber-optic links," *IEEE Trans. Microwave Theory Tech.*, vol. 38, p. 501, 1990.

T. Findakly and M. Bramson, "High-performance integrated-optical chip for a broad range of fiber-optic gyro applications," *Opt. Lett.*, vol. 15, p. 673, 1990.

P. Granestrand, B. Stoltz, L. Thylen, K. Bergvall, W. Doldisen, H. Heinrich, and D. Hoffmann, "Strictly nonblocking 8 × 8 integrated optical switch matrix," *Electron. Lett.*, vol. 22, p. 816, 1986.

M. Howerton, C. Bulmer, and W. Burns, "Effect of intrinsic phase mismatch on linear modulator performance of the 1 × 2 directional coupler and Mach–Zehnder interferometer," *J. Lightw. Tech.*, vol. 8, p. 1177, 1990.

S. E. Miller, "Integrated optics: An introduction," *Bell Syst. Tech. J.*, vol. 48, p. 2059, 1969.

Further Information

Integrated Optical Circuits and Components, Design and Applications, edited by Lynn D. Hutcheson (Marcel Dekker, Inc., New York, 1987) and *Optical Integrated Circuits*, by H. Nishihara, M. Haruna, and T. Suhara (McGraw-Hill Book Company, New York, 1989) provide excellent overviews of the field of integrated and guided-wave optics.

Integrated Optics: Devices and Applications, edited by J. T. Boyd (IEEE Press, New York, 1991) provides an excellent cross section of recent publications in the field.

In addition, the monthly magazine *IEEE Journal of Lightwave Technology* provides many publications on current research and development on integrated and guide-wave optic devices and systems.

31

D/A and A/D Converters

31.1 Integrated Circuits 771
D/A and A/D Converter Performance Criteria • D/A Conversion
Processes • D/A Converter ICs • A/D Conversion Processes •
A/D Converter ICs • Grounding and Bypassing on D/A and A/D
ICs • Selection Criteria for D/A and A/D Converter ICs

Susan A. R. Garrod
Purdue University

31.1 Integrated Circuits *

Digital-to-analog (D/A) conversion is the process of converting digital codes into a continuous range of analog signal levels. *Analog-to-digital (A/D) conversion* is the process of converting a continuous range of analog signal levels into digital codes. Such conversion processes are necessary to interface real-world systems, which typically monitor continuously varying analog signals, with digital systems that process, store, interpret, and manipulate the analog values.

D/A and A/D conversion circuits are available as integrated circuits (ICs) from many manufacturers. A huge array of ICs exists, consisting of not only the D/A or A/D conversion circuits, but also closely related circuits such as sample-and-hold amplifiers, analog multiplexers, voltage-to-frequency and frequency-to-voltage converters, voltage references, calibrators, operation amplifiers, isolation amplifiers, instrumentation amplifiers, active filters, dc-to-dc converters, analog interfaces to digital signal processing systems, and data acquisition subsystems. Data books from the IC manufacturers contain an enormous amount of information about these devices and their applications to assist the design engineer.

The ICs discussed in this chapter will be strictly the D/A and A/D conversion circuits. Table 31.1 lists a small sample of the variety of the D/A and A/D converters currently available. The ICs usually perform either D/A or A/D conversion. There are serial interface ICs, however, typically for high-performance audio and digital signal processing applications, that perform both A/D and D/A processes.

D/A and A/D Converter Performance Criteria

The major factors that determine the quality of performance of D/A and A/D converters are *resolution, sampling rate, speed,* and *linearity.*

The *resolution* of a D/A circuit is the smallest change in the output analog signal. In an A/D system, the resolution is the smallest change in voltage that can be detected by the system and that can

---

*Excerpts from *Digital Logic: Analysis, Application and Design,* by Susan Garrod and Robert Borns, copyright © 1991 by Saunders College Publishing, reprinted by permission of the publisher. Material regarding the advanced technical details of the ICs has been taken from data books that are listed in the reference section at the end of this chapter.

Table 31.1   D/A and A/D Integrated Circuits

D/A Converter ICs

IC	Resolution (bits)	Multiplying vs. Fixed Reference	Settling Time (μs)	Input Data Format
Analog Devices AD558	8	Fixed reference	3	Parallel
Analog Devices AD7524	8	Multiplying	0.400	Parallel
Analog Devices AD390	Quad, 12	Fixed reference	8	Parallel
Analog Devices AD1856	16	Fixed reference	1.5	Serial
Burr-Brown DAC729	18	Fixed reference	8	Parallel
DATEL DAC-08B	8	Multiplying	0.085	Parallel
National DAC0800	8	Multiplying	0.1	Parallel
TI AD7533 or TLC7533	10	Multiplying	0.15	Parallel

A/D Converter ICs

IC	Resolution (bits)	Signal Inputs	Conversion Speed (μs)	Output Data Format
Analog Devices AD572	12	1	25	Serial & parallel
Burr-Brown ADC804	12	1	17	Serial
Burr-Brown ADC700	16	1	17	Serial & parallel
DATEL ADC-208	8	1	35	Parallel
DATEL ADC-830	8	1	100	Parallel
National ADC1005B	10	1	50	Parallel
TI, National ADC0808	8	8	100	Parallel
TI, National ADC0834	8	4	84	Serial
TI TLC0820	8	1	1	Parallel
TI TLC1540	10	11	31	Serial

A/D and D/A Interface ICs

IC	Resolution (bits)	On-Board Filters	Sampling Rate (kHz)	Data Format
TI TLC32040	14	Yes	19.2 (programmable)	Serial
TI 2914 PCM codec & filter	8	Yes	8	Serial

produce a change in the digital code. The resolution determines the total number of digital codes, or *quantization levels*, that will be recognized or produced by the circuit.

The *resolution* of a D/A or A/D IC is usually specified in terms of the bits in the digital code or in terms of the least significant bit (LSB) of the system. An n-bit code allows for 2^n quantization levels, or $2^n - 1$ steps between quantization levels, as shown in Fig. 31.1. As the number of bits increases, the step size between quantization levels decreases, therefore increasing the accuracy of the system when a conversion is made between an analog and digital signal. The system resolution can be specified also as the voltage step size between quantization levels. For A/D circuits, the resolution is the smallest input voltage that is detected by the system.

The *speed* of a D/A or A/D converter is determined by the time it takes to perform the conversion process. For D/A converters, the speed is specified as the *settling time*. For A/D converters, the speed is specified as the *conversion time*. The settling time for D/A converters will vary with supply voltage and transition in the digital code; thus, it is specified in the data sheet with the appropriate conditions stated.

A/D converters have a maximum *sampling rate* that limits the speed at which they can perform continuous conversions. The sampling rate is the number of times per second that the analog signal can be sampled and converted into a digital code. For proper A/D conversion, the minimum sampling rate must be at least two times the highest frequency of the analog signal being sampled to satisfy the Nyquist sampling criterion. The conversion speed and other timing factors must be taken into consideration to determine the maximum sampling rate of an A/D converter. **Nyquist A/D converters** use a sampling rate that is slightly more than twice the highest frequency in the analog signal. **Oversampling A/D converters** use sampling rates of N times this rate, where N typically ranges from 2 to 64.

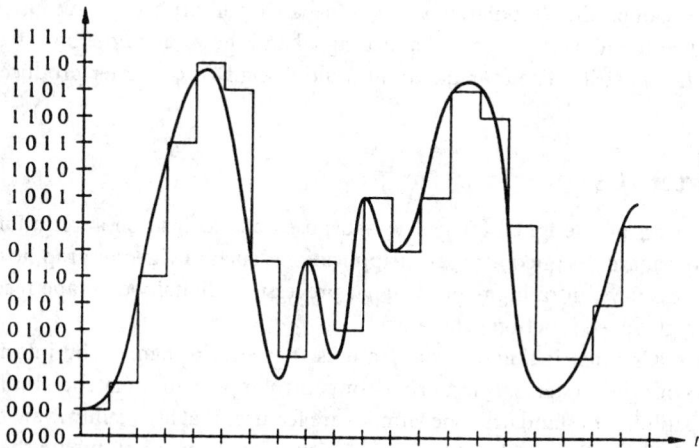

FIGURE 31.1 A/D conversion—4-bit resolution. (*Source: Digital Logic: Analysis, Application, and Design* by Susan Garrod and Robert Borns, copyright © 1991 by Saunders College Publishing, p. 893. Reprinted by permission of the publisher.)

Both D/A and A/D converters require a voltage reference in order to achieve absolute measurement accuracy. Some conversion ICs have internal voltage references, while others accept external voltage references. For high-performance systems, an external precision reference is needed to ensure long-term stability, load regulation, and control over temperature fluctuations. External precision voltage reference ICs can be found in manufacturers' data books.

Measurement accuracy is specified by the converter's *linearity*. *Integral linearity* is a measure of linearity over the entire conversion range. It is often defined as the deviation from a straight line drawn between the endpoints and through zero (or the offset value) of the conversion range. Integral linearity is also referred to as *relative accuracy*. The *offset* value is the reference level required to establish the zero or midpoint of the conversion range. *Differential linearity* is the linearity between code transitions. Differential linearity is a measure of the *monotonicity* of the converter. A converter is said to be monotonic if increasing input values result in increasing output values.

The accuracy and linearity values of a converter are specified in the data sheet in units of the LSB of the code. The linearity can vary with temperature, so the values are often specified at +25°C as well as over the entire temperature range of the device.

D/A Conversion Processes

Digital codes are typically converted to analog voltages by assigning a voltage weight to each bit in the digital code and then summing the voltage weights of the entire code. A general D/A converter consists of a network of precision resistors, input switches, and level shifters to activate the switches to convert a digital code to an analog current or voltage. D/A ICs that produce an analog current output usually have a faster settling time and better linearity than those that produce a voltage output. When the output current is available, the designer can convert this to a voltage through the selection of an appropriate output amplifier to achieve the necessary response speed for the given application.

D/A converters commonly have a fixed or variable reference level. The reference level determines the switching threshold of the precision switches that form a controlled impedance network, which in turn controls the value of the output signal. **Fixed reference D/A converters** produce an output signal that is proportional to the digital input. **Multiplying D/A converters** produce an output signal that is proportional to the product of a varying reference level times a digital code.

D/A converters can produce bipolar, positive, or negative polarity signals. A four-quadrant multiplying D/A converter allows both the reference signal and the value of the binary code to have a positive or negative polarity. The four-quadrant multiplying D/A converter produces bipolar output signals.

D/A Converter ICs

Most D/A converters are designed for general-purpose data acquisition applications. Some D/A converters, however, are designed for special applications, such as video or graphic outputs, high-definition video displays, ultra high-speed signal processing, digital video tape recording, digital attenuators, or high-speed function generators.

D/A converter ICs often include special features that enable them to be interfaced easily to microprocessors or other systems. Microprocessor control inputs, input latches, buffers, input registers, and compatibility to standard logic families are features that are readily available in D/A ICs. In addition, the ICs usually have laser-trimmed precision resistors to eliminate the need for user trimming to achieve full-scale performance.

A small sample of specific D/A converter ICs will be discussed in this section to illustrate their operation. Numerous D/A converter ICs exist for D/A applications that should be investigated by examining the data books published by the IC manufacturers.

AD558 D/A Converter by Analog Devices

The AD558 is a fixed reference D/A converter by Analog Devices that produces an output voltage proportional to the digital input code. The output voltage is calibrated over two voltage ranges: 0 to 2.56 V and 0 to 10.00 V. The functional block diagram for the AD558 is shown in Fig. 31.2.

The AD558 is an 8-bit D/A converter. The functions shown in the block diagram in Fig. 31.2 enable the AD558 to be interfaced directly to microprocessors or other circuitry, because of its input latches, control signals, internal precision voltage reference, and output amplifier. The input latches for the digital code can be controlled for microprocessor interfacing to accept a new digital code or to latch the given digital code, or they can be disabled for direct D/A conversion interfacing.

The AD558 is designed to operate with a single power supply of +4.5 to +16.5 V. The output voltage range is unipolar, from 0 to +2.56 V or from 0 to +10 V, depending on the power supply and output amplifier configuration.

The output settling time varies with the output voltage range, the output load, and whether the

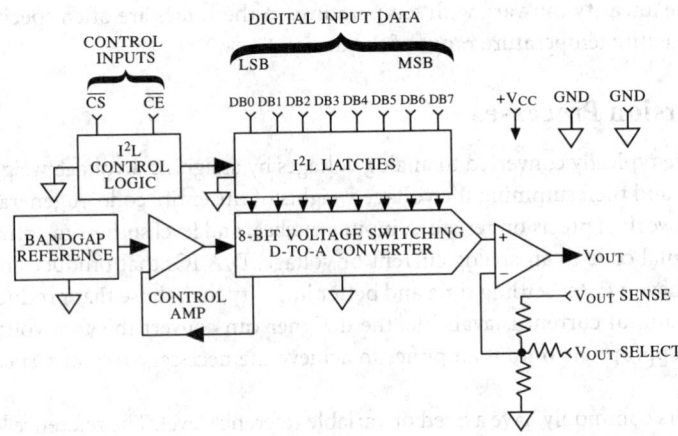

FIGURE 31.2 D/A converter: AD558 functional block diagram. (*Source:* AD558 D/A Converter Data Sheet, *Analog Devices Data Conversion Products Databook*, Analog Devices, Inc., Norwood, Mass., 1989/90, p. 2–50. With permission.)

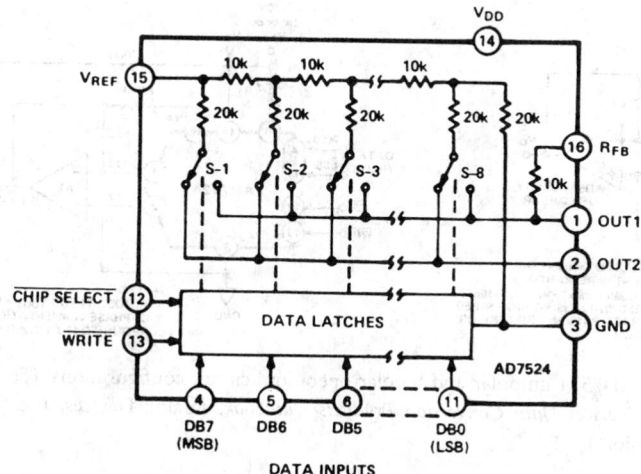

FIGURE 31.3 D/A converter: AD7524 functional block diagram. (*Source:* AD7524 D/A Converter Data Sheet, *Analog Devices Data Conversion Products Databook,* Analog Devices, Inc., Norwood, Mass., 1989/90, p. 2–235. With permission.)

output is positive-going or negative-going. For a positive-going full-scale to ±LSB output, the maximum settling time is 1.5 μs for the 2.56-V output range and 3 μs for the 10-V output range.

Relative accuracy varies from ±0.25 to ±0.75 LSB, depending on the temperature range and laser trimming of the device. Full-scale accuracy is ±1.5 LSB at +25°C and ±2.5 LSB over the entire temperature range. The AD558 is available in packages rated for performance over a temperature range of 0 to +70°C or the range −55 to +125°C.

AD7524 D/A Converter by Analog Devices

The AD7524 varying reference D/A converter IC by Analog Devices produces an output current that is a product of the 8-bit digital input code and an input reference voltage. The functional block diagram for the AD7524 is shown in Fig. 31.3. The AD7524 is capable of 2- or 4-quadrant multiplication when the reference voltage is varied.

As shown in Fig. 31.3, the AD7524 consists of an R/2R resistor ladder, data latches, and microprocessor interface logic. The AD7524 is often used with a voltage reference and an output amplifier to produce an output voltage. The manufacturer recommends this circuit for general-purpose microprocessor-controlled applications such as function generators, attenuators, and precision automatic gain control circuits.

The AD7524 is designed to operate with a power supply of ±4.5 to ±18 V. The output can be unipolar or bipolar, depending on the output amplifier configuration, as shown in Fig. 31.4. The output settling time varies with the input reference voltage, digital code transition, and output load. The maximum settling time, with a V_{DD} of +5 V, is 400 ns for 0.5 LSB accuracy over the full scale.

A/D Conversion Processes

Analog signals can be converted to digital codes by many methods, including integration, **succesive approximation,** parallel (flash) conversion, **delta modulation, pulse code modulation,** and **sigma-delta conversion.** Two of the most common A/D conversion processes are successive approximation A/D conversion and parallel or flash A/D conversion. Very high-resolution digital audio or video systems require specialized A/D techniques that often incorporate one of these general techniques as well as specialized A/D conversion processes. Examples of specialized A/D conversion techniques are

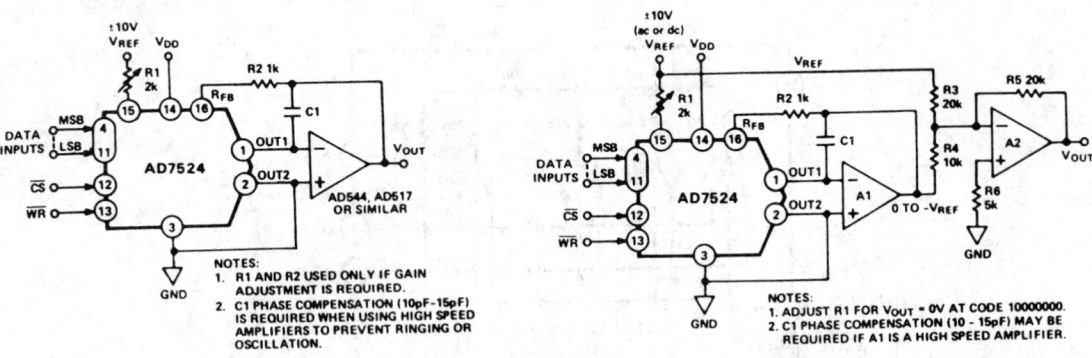

FIGURE 31.4 D/A converter: AD7524 unipolar and bipolar operation, circuit configurations. (*Source:* AD7524 D/A Converter Data Sheet, *Analog Devices Data Conversion Products Databook,* Analog Devices, Inc., Norwood, Mass., 1989/90, p. 2–239. With permission.)

pulse code modulation (PCM), and sigma-delta conversion. PCM is a common voice encoding scheme used not only by the audio industry in digital audio recordings but also by the telecommunications industry for voice encoding and multiplexing. Sigma-delta conversion is an oversampling A/D conversion where signals are sampled at very high frequencies. It has very high resolution and low distortion and is being used in the digital audio recording industry.

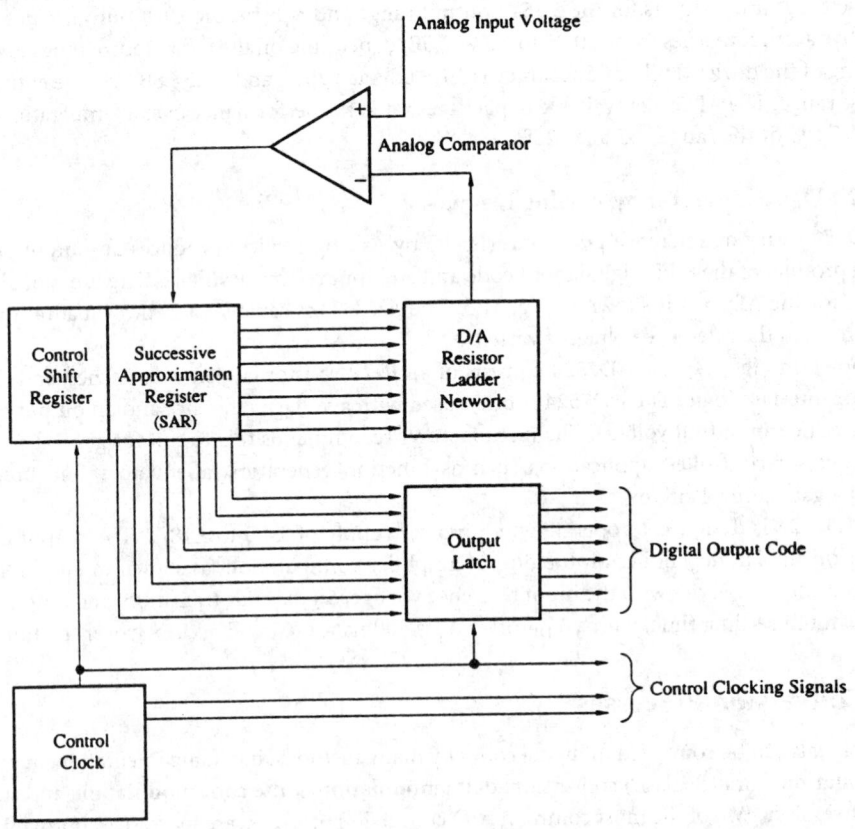

FIGURE 31.5 Successive approximation A/D converter block diagram. (*Source: Digital Logic: Analysis, Application, and Design* by Susan Garrod and Robert Borns, copyright © 1991 by Saunders College Publishing, p. 919. Reprinted by permission of the publisher.)

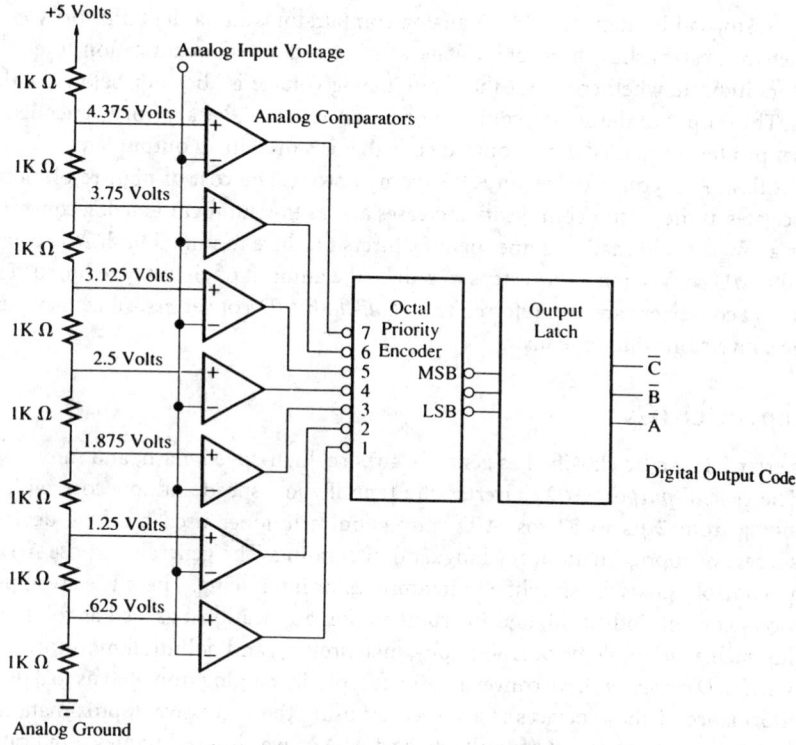

FIGURE 31.6 Flash A/D converter block diagram. (*Source: Digital Logic: Analysis, Application, and Design* by Susan Garrod and Robert Borns, copyright © 1991 by Saunders College Publishing, p. 928. Reprinted by permission of the publisher.)

Successive approximation A/D conversion is a technique that is commonly used in medium- to high-speed data acquisition applications. It is one of the fastest A/D conversion techniques that requires a minimum amount of circuitry. The conversion times for successive approximation A/D conversion typically range from 10 to 300 μs for 8-bit systems.

The successive approximation A/D converter can approximate the analog signal to form an n-bit digital code in n steps. The successive approximation register (SAR) individually compares an analog input voltage to the midpoint of one of n ranges to determine the value of one bit. This process is repeated a total of n times, using n ranges, to determine the n bits in the code. The comparison is accomplished as follows: The SAR determines if the analog input is above or below the midpoint and sets the bit of the digital code accordingly. The SAR assigns the bits beginning with the most significant bit. The bit is set to a 1 if the analog input is greater than the midpoint voltage, or it is set to a 0 if it is less than the midpoint voltage. The SAR then moves to the next bit and sets it to a 1 or a 0 based on the results of comparing the analog input with the midpoint of the next allowed range. Because the SAR must perform one approximation for each bit in the digital code, an n-bit code requires n approximations.

A successive approximation A/D converter consists of four functional blocks, as shown in Fig. 31.5: the SAR, the analog comparator, a D/A converter, and a clock.

Parallel or flash A/D conversion is used in high-speed applications such as video signal processing, medical imaging, and radar detection systems. A flash A/D converter simultaneously compares the input analog voltage to $2^n - 1$ threshold voltages to produce an n-bit digital code representing the analog voltage. Typical flash A/D converters with 8-bit resolution operate at 20 to 100 MHz.

The functional blocks of a flash A/D converter are shown in Fig. 31.6. The circuitry consists of a precision resistor ladder network, $2^n - 1$ analog comparators, and a digital priority encoder. The resistor network establishes threshold voltages for each allowed quantization level. The analog comparators indicate whether or not the input analog voltage is above or below the threshold at each level. The output of the analog comparators is input to the digital priority encoder. The priority encoder produces the final digital output code that is stored in an output latch.

An 8-bit flash A/D converter requires 255 comparators. The cost of high-resolution A/D converters escalates as the circuit complexity increases and as the number of analog converters rises by $2^n - 1$. As a low-cost alternative, some manufacturers produce modified flash A/D converters that perform the A/D conversion in two steps to reduce the amount of circuitry required. These modified flash A/D converters are also referred to as *half-flash* A/D converters, since they perform only half of the conversion simultaneously.

A/D Converter ICs

A/D converter ICs can be classified as general-purpose, high-speed, flash, and sampling A/D converters. The *general-purpose A/D converters* are typically low speed and low cost, with conversion times ranging from 2 µs to 33 ms. A/D conversion techniques used by these devices typically include successive approximation, tracking, and integrating. The general-purpose A/D converters often have control signals for simplified microprocessor interfacing. These ICs are appropriate for many process control, industrial, and instrumentation applications, as well as for environmental monitoring such as seismology, oceanography, meteorology, and pollution monitoring.

High-speed A/D converters have conversion times typically ranging from 400 ns to 3 µs. The higher speed performance of these devices is achieved by using the successive approximation technique, modified flash techniques, and statistically derived A/D conversion techniques. Applications appropriate for these A/D ICs include fast Fourier transform (FFT) analysis, radar digitization, medical instrumentation, and multiplexed data acquisition. Some ICs have been manufactured with an extremely high degree of linearity, to be appropriate for specialized applications in digital spectrum analysis, vibration analysis, geological research, sonar digitizing, and medical imaging.

Flash A/D converters have conversion times ranging typically from 10 to 50 ns. Flash A/D conversion techniques enable these ICs to be used in many specialized high-speed data acquisition applications such as TV video digitizing (encoding), radar analysis, transient analysis, high-speed digital oscilloscopes, medical ultrasound imaging, high-energy physics, and robotic vision applications.

Sampling A/D converters have a sample-and-hold amplifier circuit built into the IC. This eliminates the need for an external sample-and-hold circuit. The throughput of these A/D converter ICs ranges typically from 35 kHz to 100 MHz. The speed of the system is dependent on the A/D technique used by the sampling A/D converter.

A/D converter ICs produce digital codes in a serial or parallel format, and some ICs offer the designer both formats. The digital outputs are compatible with standard logic families to facilitate interfacing to other digital systems. In addition, some A/D converter ICs have a built-in analog multiplexer and therefore can accept more than one analog input signal.

Pulse code modulation (PCM) ICs are high-precision A/D converters. The PCM IC is often referred to as a PCM *codec* with both encoder and decoder functions. The encoder portion of the codec performs the A/D conversion, and the decoder portion of the codec performs the D/A conversion. The digital code is usually formatted as a serial data stream for ease of interfacing to digital transmission and multiplexing systems.

PCM is a technique where an analog signal is sampled, quantized, and then encoded as a digital word. The PCM IC can include successive approximation techniques or other techniques to accomplish the PCM encoding. In addition, the PCM codec may employ nonlinear data compression techniques, such as companding, if it is necessary to minimize the number of bits in the output digital code. Companding is a logarithmic technique used to compress a code to fewer bits before transmission. The inverse logarithmic function is then used to expand the code to its origi-

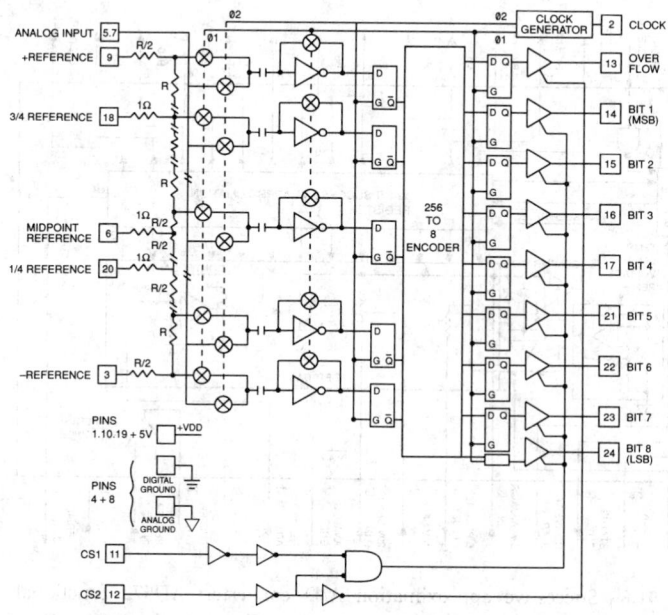

FIGURE 31.7 Flash A/D converter: ADC-208 simplified block diagram. (*Source:* ADC-208 A/D Converter Data Sheet, *DATEL Data Conversion Catalog*, DATEL, Inc., Mansfield, Mass., 1988, p. 1–11. With permission)

nal number of bits before coverting it to the analog signal. Companding is typically used in telecommunications transmission systems to minimize data transmission rates without degrading the resolution of low-amplitude signals. Linear PCM conversion is used in high-fidelity audio systems to preserve the integrity of the audio signal throughout the entire analog range.

Digital signal processing (DSP) techniques provide another type of A/D conversion ICs. Specialized A/D conversion such as *adaptive differential pulse code modulation* (ADPCM), *sigma-delta modulation, speech subband encoding, adaptive predictive speech encoding,* and *speech recognition* can be accomplished through the use of DSP systems. Some DSP systems require analog front ends that employ traditional PCM codec ICs or DSP interface ICs. These ICs can interface to a digital signal processor for advanced A/D applications. Some manufacturers have incorporated DSP techniques on board the single-chip A/D IC, as in the case of the DSP56ACD16 sigma-delta modulation IC by Motorola.

A small sample of specific A/D converter ICs will be discussed in this section to illustrate their operation. As was the case with D/A converters, numerous A/D converter ICs exist that should be investigated for the specific A/D application in question.

ADC-208 Flash A/D Converter by DATEL

The ADC-208 is an 8-bit flash A/D converter. The simplified block diagram for the ADC-208 is shown in Fig. 31.7. The ADC-208 consists of 256 comparators that perform the parallel or flash conversion of the analog input signal. The output data are available in three-state output latches with an overflow detection indicator.

The ADC-208 has a switched capacitor input stage, and therefore its input impedance is dependent on its sampling frequency. It is recommended by the manufacturer that an external input buffer be used because of this varying input impedance.

The ADC-208 is designed to operate with a single power supply of +5 V. The allowed analog input voltage range is 0 to +5 V. The sampling rate of the ADC-208 is 20 mega samples per second. The full

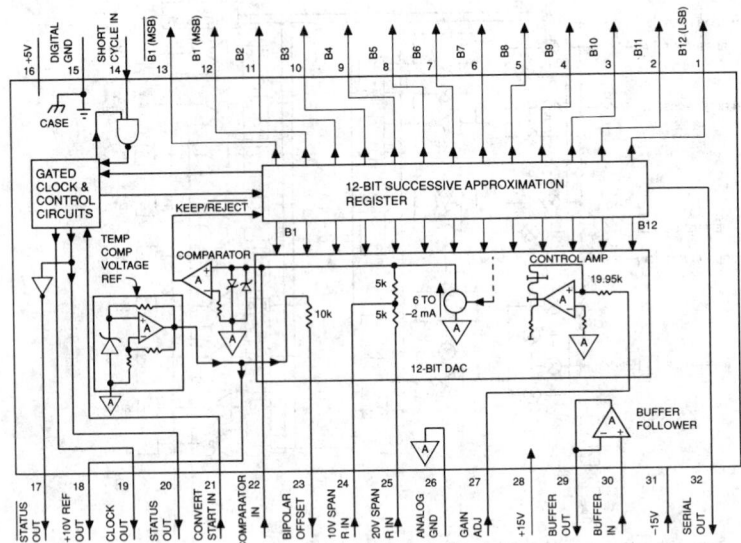

FIGURE 31.8 Successive approximation A/D converter: AD572 functional block diagram. (*Source:* AD572 A/D Converter Data Sheet, *Analog Devices Data Conversion Products Databook,* Analog Devices, Inc., Norwood, Mass., 1989/90, p. 3–21. With permission.)

power bandwidth is 10 MHz, adhering to the Nyquist sampling criteria. The integral linearity is ±0.5 LSB between endpoints at ±25°C and ±1 LSB over the temperature range. Differential linearity is ±0.75 LSB between code transitions at ±25°C and ±1 LSB over the temperature range.

The digital output is a straight binary 8-bit code. The code range is adjustable over a voltage range defined by the positive and negative reference voltage inputs. Two ADC-208s can be cascaded to achieve 9-bit resolution over the defined voltage range.

AD571 Successive Approximation A/D Converter by Analog Devices

The AD572 is a 12-bit A/D converter that uses the successive approximation method for conversion. The functional block diagram of the AD572 is shown in Fig. 31.8. The output data are available in both a serial and parallel format.

The AD572 12-bit successive approximation A/D converter has an internal clock, a +10-V voltage reference, and an input buffer amplifier. Analog scaling resistors enable the AD572 to be operated for analog input ranges of ±2.5 V, ±5.0 V, ±10 V, 0 to +5 V, or 0 to +10 V. The internal +10-V precision voltage reference can be used for external applications. The AD572 requires a ±15-V and +5-V power supply.

The AD572 requires a conversion time of 25 μs. The internal clock has a maximum output frequency of 500 kHz. Thirteen clock cycles are required for each conversion.

The maximum linearity error is 0.012% of full scale. The differential linearity error is ±0.5 LSB. The unipolar offset error is ±0.05% of full scale, and the bipolar offset error is ±0.1% of full scale. A sample-and-hold amplifier is recommended for use between the analog source and the input to the AD572 if the input voltage can change by more than 0.5 LSB during a conversion.

Grounding and Bypassing on D/A and A/D ICs

D/A and A/D converter ICs require correct grounding and capacitive bypassing in order to operate according to performance specifications. The digital signals can severely impair analog signals. To combat the electromagnetic interference induced by the digital signals, the analog and digital grounds should be kept separate and should have only one common point on the circuit board. If possible, this common point should be the connection to the power supply.

Bypass capacitors are required at the power connections to the IC, the reference signal inputs, and the analog inputs to minimize noise that is induced by the digital signals. Each manufacturer specifies the recommended bypass capacitor locations and values in the data sheet. The 1-μF tantalum capacitors are commonly recommended, with additional high-frequency power supply decoupling sometimes being recommended through the use of ceramic disc shunt capacitors. The manufacturers' recommendations should be followed to ensure proper performance.

Selection Criteria for D/A and A/D Converter ICs

Hundreds of D/A and A/D converter ICs are available, with prices ranging from a few dollars to several hundred dollars each. The selection of the appropriate type of converter is based on the application requirements of the system, the performance requirements, and cost. The following issues should be considered in order to select the appropriate converter.

1. What are the input and output requirements of the system? Specify all signal current and voltage ranges, logic levels, input and output impedances, digital codes, data rates, and data formats.
2. What level of accuracy is required? Determine the resolution needed throughout the analog voltage range, the dynamic response, the degree of linearity, and the number of bits encoding.
3. What speed is required? Determine the maximum analog input frequency for sampling in an A/D system, the number of bits for encoding each analog signal, and the rate of change of input digital codes in a D/A system.
4. What is the operating environment of the system? Obtain information on the temperature range and power supply to select a converter that is accurate over the operating range.

Final selection of D/A and A/D converter ICs should be made by consulting manufacturers to obtain their technical specifications of the devices. Major manufacturers of D/A and A/D converters include Analog Devices, Burr-Brown, DATEL, Maxim, National, Phillips Components, Precision Monolithics, Signetics, Sony, Texas Instruments, Ultra Analog, and Yamaha. Information on contacting these manufacturers and others can be found in an *IC Master Catalog*.

Defining Terms

Delta modulation: An A/D conversion process where the digital code out represents the change, or slope, of the analog input signal, rather than the absolute value of the analog input signal. A 1 indicates a rising slope of the input signal. A 0 indicates a falling slope of the input signal. The sampling rate is dependent on the derivative of the signal, since a rapidly changing signal would require a rapid sampling rate for acceptable performance.

Fixed reference D/A converter: The analog output is proportional to a fixed (nonvarying) reference signal.

Flash A/D: The fastest A/D conversion process available to date, also referred to as parallel A/D conversion. The analog signal is simultaneously evaluated by $2^n - 1$ comparators to produce an n-bit digital code in one step. Because of the large number of comparators required, the circuitry for flash A/D converters can be very expensive. This technique is commonly used in digital video systems.

Integrating A/D: The analog input signal is integrated over time to produce a digital signal that represents the area under the curve, or the integral.

Multiplying D/A: A D/A conversion process where the output signal is the product of a digital code multiplied times an analog input reference signal. This allows the analog reference signal to be scaled by a digital code.

Nyquist A/D converters: A/D converters that sample analog signals that have a maximum frequency that is less than the Nyquist frequency. The Nyquist frequency is defined as one-half of the sampling frequency. If a signal has frequencies above the Nyquist frequency, a distortion called *aliasing* occurs. To prevent aliasing, an *antialiasing filter* with a flat passband and very sharp roll-off is required.

Oversampling converters: A/D converters that sample frequencies at a rate much higher than the Nyquist frequency. Typical oversampling rates are 32 and 64 times the sampling rate that would be required with the Nyquist converters.

Pulse code modulation (PCM):   An A/D conversion process requiring three steps: the analog signal is sampled, quantized, and encoded into a fixed length digital code. This technique is used in many digital voice and audio systems. The reverse process reconstructs an analog signal from the PCM code.

Sigma-delta A/D conversion: An *oversampling* A/D conversion process where the analog signal is sampled at rates much higher (typically 64 times) than the sampling rates that would be required with a Nyquist converter. Sigma-delta modulators integrate the analog signal before performing the delta modulation. The integral of the analog signal is encoded rather than the change in the analog signal, as is the case for traditional delta modulation. A digital sample rate reduction filter (also called a digital decimation filter) is used to provide an output sampling rate at twice the Nyquist frequency of the signal. The overall result of oversampling and digital sample rate reduction is greater resolution and less distortion compared to a Nyquist converter process.

Successive approximation: An A/D conversion process that systematically evaluates the analog signal in *n* steps to produce an *n*-bit digital code. The analog signal is successively compared to determine the digital code, beginning with the determination of the most significant bit of the code.

References

Analog Devices Data Conversion Products Data Book, Analog Devices, Inc., Norwood, Mass., 1989.

Burr-Brown Integrated Circuits Data Book, Burr-Brown, Tucson, Arizona, 1989.

DATEL Data Conversion Catalog, DATEL, Inc., Mansfield, Mass., 1988.

S. Garrod and R. Borns, *Digital Logic: Analysis, Application and Design*, Philadelphia: Saunders College Publishing, 1991, Chapter 16.

J.M. Jacob, *Industrial Control Electronics*, Englewood Cliffs, N.J.: Prentice-Hall, 1989, Chapter 6.

Motorola Telecommunications Data Book, Motorola, Inc., 1989.

National Semiconductor Data Acquisition Linear Devices Data Book, National Semiconductor Corp., Santa Clara, Calif., 1989.

S. Park, *Principles of Sigma-Delta Modulation for Analog-to-Digital Converters*, Motorola, Inc., Phoenix, Arizona, 1990.

Texas Instruments Digital Signal Processing Applications with the TMS320 Family, Texas Instruments, Dallas, Texas, 1986.

Texas Instruments Linear Circuits Data Acquisition and Conversion Data Book, Texas Instruments, Dallas, Texas, 1989.

Further Information

Analog Devices, Inc. has edited or published several technical handbooks to assist design engineers with their data acquisition system requirements. These references should be consulted for extensive technical information and depth. The publications include *Analog-Digital Conversion Handbook*, by the engineering staff of Analog Devices, published by Prentice-Hall, Englewood Cliffs, N.J.,

1986; *Nonlinear Circuits Handbook, Transducer Interfacing Handbook,* and *Synchro and Resolver Conversion,* all published by Analog Devices Inc., Norwood, Mass.

Engineering trade journals and design publications often have articles describing recent A/D and D/A circuits and their applications. These publications include *EDN Magazine, EE Times,* and *IEEE Spectrum.*

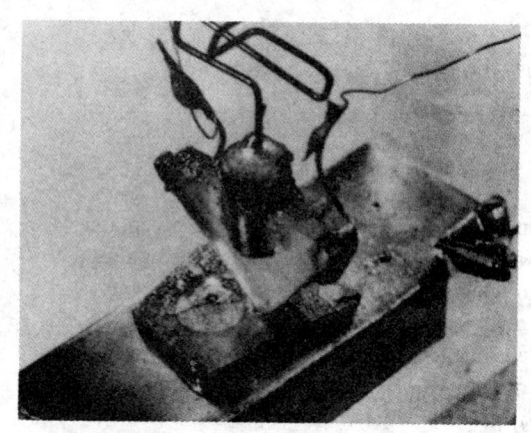

The first transistors assembled by their inventors at Bell Laboratories (in 1947) were primitive by today's standards. Yet they revolutionized the electronics industry and changed our way of life. The first transistor, a "point-contact" type, amplified electrical signals by passing them through a solid semiconductor material, basically the same operation as performed by present "junction" transistors. The three terminal wires can be seen on the top of the transistor. The actual record of the first transistor operation was December 23, 1947. (Courtesy of Bell Telephone Laboratories.)

32

Thermal Management of Electronics

32.1 Introduction 784
    Motivation • Requirements
32.2 Heat Transfer Fundamentals 787
32.3 Chip Module Thermal Resistance 789
    Definition • Internal Resistance • External Resistance • Total Resistance • Multichip Modules

Avram Bar-Cohen
University of Minnesota

32.1 Introduction

Motivation

In the thermal control of microelectronic components, it is necessary to provide an acceptable *microclimate* for a diversity of devices and packages, which vary widely in size, power dissipation, and sensitivity to temperature. Although the **thermal management** of all electronic components is motivated by a common set of concerns, this diversity often leads to the design and development of distinct thermal control systems for different types of electronic equipment. Moreover, due to substantial variations in the performance, cost, and environmental specifications across product categories, the thermal control of similar components may require widely differing thermal management strategies.

The prevention of **catastrophic thermal failure,** defined as an immediate, thermally induced, total loss of electronic function, must be viewed as the primary and foremost aim of electronics thermal control. Catastrophic failure may result from a significant deterioration in the performance of the component/system or from a loss of structural integrity at one of the relevant packaging levels. In early microelectronic systems, catastrophic failure was primarily *functional* and thought to result from changes in the bias voltage, *thermal runaway* produced by regenerative heating, and dopant migration, all occurring at elevated transistor junction temperatures. While these failure modes may still occur during the device development process, improved silicon simulation tools and thermally compensated integrated circuits have largely quieted these concerns and substantially broadened the operating temperature range of today's silicon-based logic and memory devices. Similar concerns do still exist in the use of CMOS devices for high-performance systems. Because of the dependence of CMOS circuit speed on temperature, it may be necessary to limit the maximum chip temperature to achieve a desired cycle time and/or to maintain timing margins in the system.

More generally, however, thermal design in the 1990s is aimed at preventing thermally induced physical failures, through reduction of the temperature rise above ambient and minimization of temperature variations within the packaging structure(s). The use of many low-temperature materials and the structural complexity of chip packages and printed circuit boards has increased the risk of catastrophic failures associated with the vaporization of organic materials, the melting of solders, and thermal-stress fracture of leads, joints, and seals, as well as the fatigue-induced fracture or creep-induced deformation of encapsulants and laminates. To prevent catastrophic thermal failure, the designer must know the maximum allowable temperatures, acceptable internal temperature differences, and the power consumption/dissipation of the various components. This information can be used to select the appropriate fluid, heat transfer mode, and inlet temperature for the coolant and to thus establish the thermal control strategy early in the design process.

After the selection of an appropriate thermal control strategy, attention can be turned to meeting the desired system-level reliability and the target failure rates of each component and subassembly. Individual solid-state electronic devices are inherently reliable and can typically be expected to operate, at room temperature, for some 100,000 years, i.e., with a base failure rate of 1 FIT (failures in 10^9 h). However, since the number of devices in a typical logic component is rapidly approaching 1 million and since an electronic system may consist of many tens to several hundreds of such components, achieving a system Mean Time Between Failures of several thousand hours in military equipment and 40,000–60,000 hours in commercial computers is a most formidable task.

Many of the failure mechanisms, which are activated by prolonged operation of electronic components, are related to the local temperature and/or temperature gradients, as well as the thermal history of the package [Pecht *et al.*, 1992]. Device-related functional failures often exhibit a strong relationship between failure rate and operating temperature. This dependence, illustrated in Fig. 32.1, is exponential in nature and commonly represented in the form of an Arrhenius relation, with unique, empirically determined coefficients for each component type. In the normal operating range of microelectronic components, a 10–20°C increase in chip temperature is thought to double the component failure rate, and even a 1°C decrease may lower the failure rate associated with such mechanisms by 2–4% [Morrison *et al.*, 1982].

Unfortunately, it is not generally possible to characterize thermally induced structural failures in

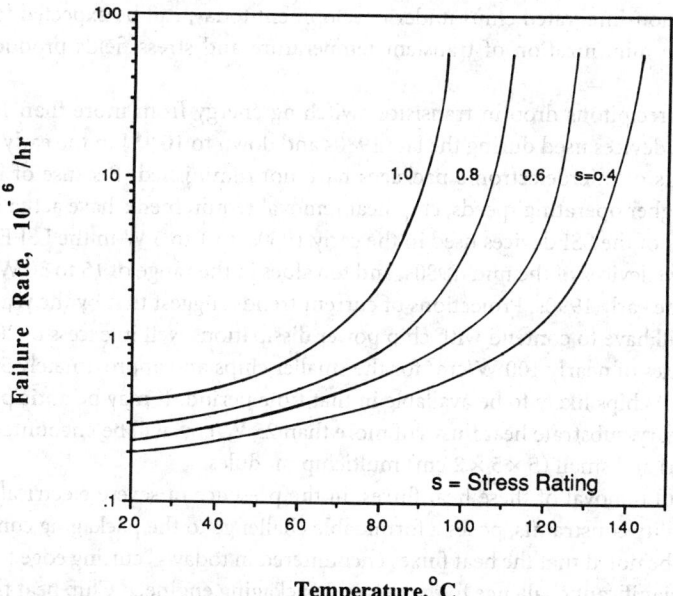

FIGURE 32.1 Exponential dependence of failure rate on component temperature.

the form of an Arrhenius relation. The mechanical stresses, which develop as a result of differential thermal expansion among the materials constituting an electronic package, may well increase as the temperature of the component is elevated. Thermal stress failures are, by their nature, dependent on the details of the local temperature fields, as well as the assembly, attachment, and local operating history of the component. Furthermore, thermal stress generation in packaging materials and structures is exacerbated by power transients, as well as by the periodically varying environmental temperatures experienced by most electronic systems, during both qualification tests and actual operation. However, stress variations in the elastic domain or in the range below the fatigue limit may have little effect on the component failure rate. Consequently, the minimization or elimination of thermally induced failures often requires careful attention to both the temperature and stress fields in the electronic components and the empirical validation of proposed thermostructural design criteria.

Requirements

Consideration of the Arrhenius relationship has resulted in peak allowable temperatures of 110–120°C for most military equipment [Morrison *et al.*, 1982] and has led designers of commercial equipment to specify average chip operating temperatures in the 65–85°C range [Bar-Cohen, 1987, 1988]. Theoretical predictions of dramatic reductions in component failure rates have been used to justify the use of refrigerated avionics [Morrison, 1982] and cryogenic electronics [Jaeger, 1986; Vacca *et al.*, 1987].

The stabilization of component temperature and minimization of the temperature differences between adjacent devices, components, and various packaging levels have long been known to reduce failure rates in electronic systems [Hilbert and Kube, 1969]. In layered structures, such as chip packages and printed circuit boards, and in the joints of surface mounted components, temperature nonuniformities, in all but the most clever designs, accentuate the differences in the thermal expansion coefficients among the various materials and can frequently result in thermal stresses that threaten the integrity of these components and joints [Englemier, 1984; Suhir, 1988]. The growing integration on a single chip of functionally distinct and thermally diverse devices, as in the power-integrated chips of the late 1980s and in the microsensor, RF, and in optical communication-integrated chips under development today, can be expected to focus renewed attention on the minimization of transient temperature and stress fields produced by localized heat sources.

Despite the precipitous drop in transistor switching energy from more than 10^{-9} J in 1960 to nearly 10^{-13} J in devices used during the late 1980s and down to 10^{-14} J in the early 1990s, the cooling requirements of microelectronic packages have not diminished. Because of increased device densities and higher operating speeds, chip heat removal requirements have actually risen from 0.1 to 0.3 W, typical of the SSI devices used in the early 1960s, to 1 to 5 W in the LSI ECL components and VLSI CMOS devices of the mid-1980s, and to values in the range of 15 to 30 W for commercial equipment in the early 1990s. Projections of current trends suggest that by the year 2000, the thermal designer will have to contend with chip power dissipations well in excess of 200 W, producing surface heat fluxes of nearly 100 W/cm^2 for the smaller chips and approximately 50 W/cm^2 for the larger (2×2 cm) chips likely to be available in that time period. It may be anticipated that, by the turn of the century, substrate heat fluxes of more than 25 W/cm^2 will be encountered in both large ($30 \times 30 \times 5$ cm) and small ($5 \times 5 \times 2$ cm) multichip modules.

The successful removal of these heat fluxes, in the presence of severe electrical, manufacturing cost, and reliability constraints, poses a formidable challenge to the packaging community. Nevertheless, it must be noted that the heat fluxes encountered in today's "cutting edge technology" chips already pose a significant challenge to the thermal packaging engineer. Chip heat fluxes in the mid-1980s typically ranged from 5 W/cm^2 to nearly 30 W/cm^2, for both single-chip packages and multichip modules [Bar-Cohen, 1987]. Recently released commercial computers often include chips dissipating 15 to 30 W/cm² [e.g., Kaneko *et al.*, 1990; Pei *et al.*, 1990], and laboratory prototypes have extended the chip heat flux range to nearly 65 W/cm². These heat fluxes are comparable, at the upper end, to the thermal loading experienced by reentry vehicles and even, at the lower end, to heat fluxes imposed on rocket motor cases. The anticipated peak heat flux in the year 2000 of approximately 100 W/cm² is in the range of thermal loadings associated with nuclear blasts.

32.2 Heat Transfer Fundamentals

To determine the temperature differences encountered in the flow of heat within electronic systems, it is necessary to recognize several different heat transfer mechanisms and their governing relations. In a typical system, heat removal from the active regions of the chip(s) requires the use of several mechanisms, some operating in series and others in parallel, to transport the generated heat to the coolant.

Thermal transport through solids is governed by the Fourier equation, which, in one-dimensional form, is expressible as

$$q = kAdT/dx \text{ [W]} \tag{32.1}$$

where q is the heat flow, k is the thermal conductivity of the medium, A is the cross-sectional area for heat flow, and dT/dx the temperature gradient.

The temperature difference resulting from the conduction of heat is thus related to the thermal conductivity of the material, the cross-sectional area, and the path length, l, or

$$(T_1 - T_2)_{\text{cond}} = q(1/kA) \text{ [K]} \tag{32.2}$$

The form of this equation suggests that, by analogy to electrical current flow in a conductor, it is possible to define a conduction thermal resistance as [Kraus, 1958]

$$R_{\text{cond}} = (T_1 - T_2)/q = 1/kA \text{ [K/W]} \tag{32.3}$$

Thermal transport from a surface to a fluid in motion is called **convective heat transfer** and can be related to the **heat transfer coefficient**, h, the surface-to-fluid temperature difference, and the "wetted" area, in the form

$$q = hA(T_{\text{surf}} - T_{\text{fluid}}) \text{ [W]} \tag{32.4}$$

The differences among convection to a fast-moving fluid, a slowly flowing fluid, and a stagnant fluid, as well as variations in the convective heat transfer rate among various fluids, are reflected in the value of h. Some theoretical and many empirical correlations are available for determining this convective heat transfer coefficient (e.g., Kraus and Bar-Cohen, 1983). Using Eq. (32.4), it is possible to define the convective **thermal resistance,** as

$$R_{\text{conv}} = (hA)^{-1} \text{ [K/W]} \tag{32.5}$$

Unlike conduction and convection, **radiative heat transfer** between two surfaces or between a surface and its surroundings is not linearly dependent on the temperature difference and is expressed instead as

$$q = \sigma AF(T_1^4 - T_2^4) \text{ [W]} \tag{32.6}$$

where F includes the effect of surface properties and geometry and σ is the Stefan-Boltzmann constant, which equals 5.67×10^{-8} W/m²K⁴.

For modest temperature differences, this equation can be linearized to the form

$$q_r = h_r A(T_1 - T_2) \text{ [W]} \tag{32.7}$$

where h_r is the effective "radiation" heat transfer coefficient and is approximately equal to $4\sigma\ F\ (T_1 T_2)^{1.5}$. It is of interest to note that for temperature differences on the order of 10 K, the radiative heat transfer coefficient, h_r, for a radiationally active (or "black") surface in an absorbing environment, is approximately equal to the heat transfer coefficient in natural convection of air. Noting the form of Eq. (32.7), the radiational thermal resistance, analogous to the convective resistance, is seen to equal $(h_r A)^{-1}$.

Ebullient thermal transport displays a complex dependence on the temperature difference between the heated surface and the saturation temperature (boiling point) of the liquid. In nucleate boiling, the primary region of interest, the **ebullient heat transfer** rate can be approximated by a relation of the form

$$q_b = C'_{sf} A(T_{surf} - T_{sat})^3\ [W] \tag{32.8}$$

where C'_{sf} is a function of the surface/fluid combination and T_{sat} is the boiling point of the liquid. For comparison purposes, it is possible to define a boiling heat transfer coefficient, h_b, equal to $C'_{sf}(T_1 - T_{sat})^2$, which, however, will vary strongly with surface temperature.

In the thermal design of electronic equipment, frequent use is made of finned or *extended* surfaces. While such finning can substantially increase the surface area in contact with the coolant, conduction in the **thermal fin** reduces the average temperature of the exposed surface relative to the fin base. In the analysis of such finned surfaces, it is thus common to define a **fin efficiency**, η, equal to the ratio of the average temperature rise of the fin (above the coolant) to the temperature rise of the fin base. Using this approach, heat transfer by a fin or fin structure can be expressed in the form

$$q_f = hA\eta(T_o - T_f)\ [W] \tag{32.9}$$

where T_o is the temperature of the fin base. The thermal resistance of a finned surface is given by $(\eta hA)^{-1}$.

The transfer of heat to a flowing gas or liquid not undergoing a phase change results in an increase in the coolant temperature, according to

$$q = \dot{m}c_p(T_{out} - T_{in}) = \rho \mathcal{Q} c_p(T_{out} - T_{in})\ [W] \tag{32.10}$$

Based on this relation, it is possible to define an effective flow resistance, R_f, as

$$R_f = 1/\dot{m}c_p\ [K/W] \tag{32.11}$$

In a first-order thermal model, it is generally appropriate to relate the heat source temperature to the average (rather than the outlet) coolant temperature. In such calculations the flow resistance should be taken to equal one-half of the value given by Eq. (32.11).

The expression of the governing heat transfer relations in the form of thermal resistances greatly simplifies the first-order thermal analysis of electronic systems. Following the established rules for resistance networks, thermal resistances that occur sequentially along a thermal path can be simply summed to establish the overall thermal resistance for that path. Similarly, the reciprocal of the effective overall resistance of several parallel heat transfer paths can be found by summing the reciprocals of the individual resistances. In refining the thermal design of an electronic system, prime attention should, then, be devoted to reducing the largest resistances along a specified thermal path and/or providing parallel paths for heat removal from a critical area.

While the thermal resistances associated with various paths and thermal transport mechanisms constitute the building blocks in performing a detailed thermal analysis, they have also found widespread application as figures-of-merit in evaluating and comparing the thermal efficacy of various packaging techniques and thermal management strategies. The determination of the relevant thermal resistances is, thus, the key task in the thermal design of an electronic system.

32.3 Chip Module Thermal Resistance

Definition

The thermal performance of chip packaging techniques is commonly compared on the basis of the overall (junction-to-coolant) thermal resistance, R_T. This packaging figure-of-merit is generally defined in a purely empirical fashion to equal

$$R_T = (T_j - T_f)/q_c \ [\text{K/W}] \tag{32.12}$$

where T_j and T_f are the junction and coolant (fluid) temperatures, respectively, and q_c is the chip heat dissipation.

Unfortunately, however, most measurement techniques are incapable of detecting the actual junction temperature, i.e., the temperature of the small volume at the interface of p-type and n-type semiconductors, and, hence, this term generally refers to the average temperature or a representative temperature on the chip. Because many of the failure mechanisms of integrated circuits are accelerated by an increase in the average chip temperature, low thermal resistances are to be preferred in nearly all categories of electronic packaging.

Single-chip packages can be characterized by their internal, or so-called junction-to-case, resistance. The convective heat removal techniques applied to the external surfaces of the package, including the effect of finned heat sinks and other thermal enhancements, can be compared on the basis of the external thermal resistance. The complexity of heat flow and coolant flow paths in a multichip module generally requires that the thermal capability of these packaging configurations be examined on the basis of overall, or chip-to-coolant, thermal resistance.

Examination of various packaging techniques reveals that the junction-to-coolant thermal resistance is, in fact, composed of an internal, largely conductive, resistance and an external, primarily convective, resistance. As shown in Fig. 32.2, the internal resistance, R_{jc}, is encountered in the flow of dissipated heat from the active chip surface, through the materials used to support and bond the chip, and on to the case of the integrated circuit package. The flow of heat from the case directly to the coolant, or indirectly through a fin structure and then to the coolant, must overcome the external resistance, R_{ex}.

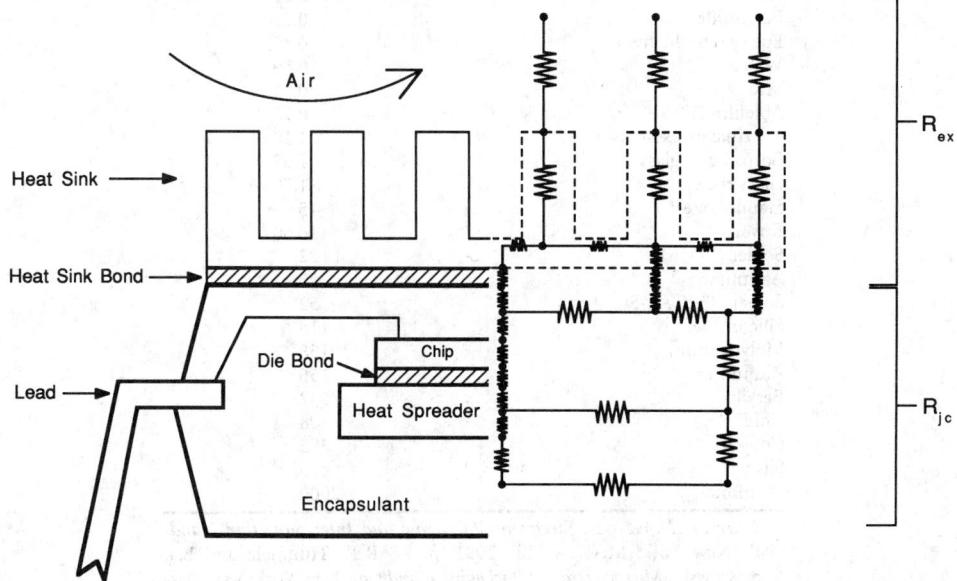

FIGURE 32.2   Thermal resistances in a single-chip package.

Internal Resistance

As previously noted, conductive thermal transport is governed by the Fourier equation (Eq. 32.1). For composite, rectilinear structures, as encountered in many chip modules, the Fourier equation (with temperature and time invariant properties), takes the form

$$q = (T_i - T_e) \Big/ \sum_p (\Delta x / kA) \ \ [\mathrm{W}] \tag{32.13}$$

where T_i and T_e are the temperatures internal and external to the composite structure, respectively, Δx is the thickness of each material in the direction of heat flow, and the summation sign pertains to p distinct layers of material. The thermal conductivities of typical packaging materials are tabulated in Table 32.1.

Assuming that power is dissipated uniformly across the chip surface and that heat flow is largely one-dimensional, Eq. (32.13) can be used to provide a first-order approximation for the internal chip module resistance, as

$$R_{jc} = (T_j - T_c)/q_c = \sum_p (\Delta x / kA) \ \ [\mathrm{K/W}] \tag{32.14}$$

where the summed terms represent the thermal resistances of the individual layers of silicon, solder, copper, alumina, etc. It is to be noted that the contact resistances that occur at the interfaces

Table 32.1   Thermal Conductivities of
Typical Packaging Materials at Room Temperature

Materials	Thermal Conductivity (W/m K)
Air	0.024
Mylar	0.19
Silicone rubber	0.19
Solder mask	0.21
Epoxy (dielectric)	0.23
Ablefilm 550 dielectric	0.24
Nylon	0.24
Polytetrafluorethylene	0.24
RTV	0.31
Polyimide	0.33
Epoxy (conductive)	0.35
Water	0.59
Mica	0.71
Ablefilm 550 K	0.78
Thermal greases/pastes	1.10
Borosilicate glass	1.67
Glass epoxy	1.70
Stainless steel	15
Kovar	16.60
Solder (Pb-In)	22
Alumina	25
Solder 80-20 Au-Sn	52
Silicon	118
Molybdenum	138
Aluminum	156
Beryllia	242
Gold	298
Copper	395
Silver	419
Diamond	2000

Source: C.A. Harper, *Electronic Packaging and Interconnection Handbook,* New York: McGraw-Hill, 1991, p. 27. R.R. Tummala and E.J. Rymaszewski, *Microelectronics Packaging Handbook,* New York: Van Nostrand Reinhold, 1989, p. 174.

between pairs of materials can be added, as appropriate, to this summation. As the thickness of each layer decreases and/or the thermal conductivity and cross-sectional area increase, the resistance of the individual layers decreases.

In chip packages that provide for lateral spreading of the heat generated in the chips, the increasing cross-sectional area for heat flow at successive layers reduces the internal thermal resistance. Unfortunately, however, there is an additional resistance associated with the lateral flow of the heat, which must be taken into account in determining the chip-to-case temperature difference.

Following Yovanovich and Antonetti [1988], the spreading resistance for a small heat source on a thick substrate (typically 3–5 times thicker than the square root of the heat source area) can be expressed as

$$R_c = (0.475 - 0.62\varepsilon + 0.13\varepsilon^3)/k \, (A_c)^{0.5} \, [\text{K/W}] \qquad (32.15)$$

where ε is the square root of the ratio of the heat source area to the substrate area, k the thermal conductivity of the substrate, and A_c the area of the heat source. For relatively thin layers on thicker substrates, such as encountered in the use of thin lead frames or heat spreaders interposed between the chip and substrate, Eq. (32.15) cannot provide an acceptable prediction of R_c. Instead, use can be made of the numerical results plotted in Fig. 32.3 to obtain the requisite value of the spreading resistance.

The internal thermal resistance of a chip package can be expected to vary from approximately 80 K/W for a plastic package with no heat spreader, to 15 to 20 K/W for a plastic package with heat spreader, and to 5 to 10 K/W for a ceramic package or a specially designed plastic chip package. Carefully designed chip packages can attain even lower values of R_{jc}, but the conductive thermal resistances at the interfaces, between materials, and especially along the chip surfaces where the heat fluxes are highest are often quite significant and impose a lower bound on the internal package resistance. Typical specific thermal resistances, expressed as (K/W)/cm² and based on values in Nakayama [1988] for lightly loaded interfaces, are shown in Fig. 32.4.

External Resistance

To determine the resistance to thermal transport from a surface to a fluid in motion, i.e., the convective resistance, it is necessary to quantify the heat transfer coefficient, h. For a particular geome-

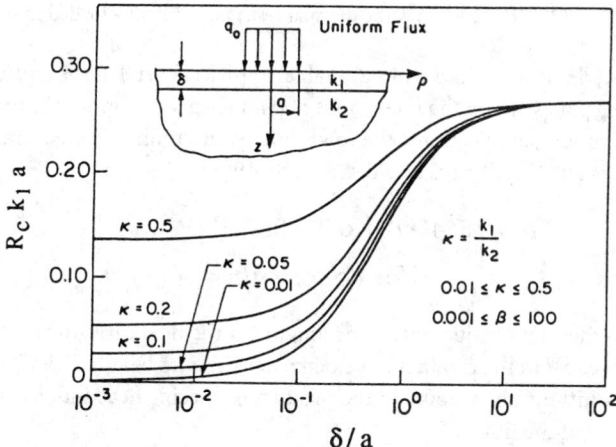

FIGURE 32.3 The thermal resistance for a circular heat source on a two-layer substrate. (*Source:* M.M. Yovanovich and V.W. Antonetti, "Application of Thermal Contact Resistance Theory to Electronic Packages," in *Advances in Thermal Modeling of Electronic Components and Systems,* vol. 1, A. Bar-Cohen and A.D. Kraus, Eds., New York: Hemisphere, 1988, pp. 79–128. With permission.)

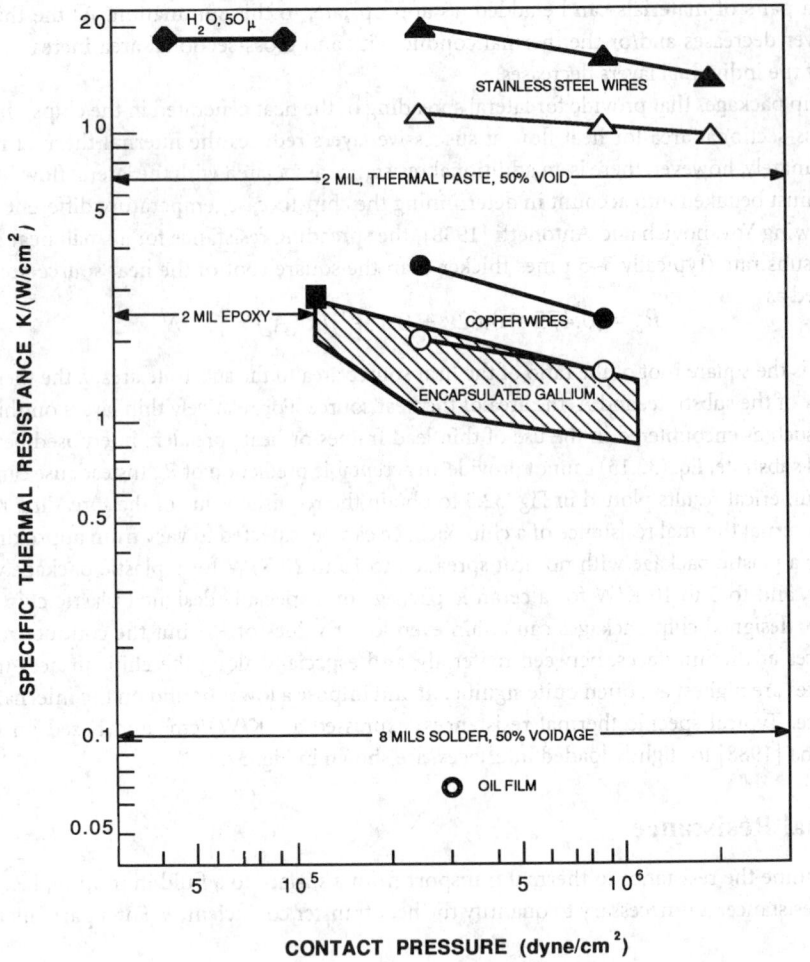

FIGURE 32.4 Specific thermal resistance at lightly loaded interfaces.

try and flow regime, *h* may be found from available empirical correlations and/or theoretical relations. For flow along plates and in the inlet zones of parallel-plate channels, as may well be encountered in electronic cooling applications, the low velocity, or laminar flow, average convective heat transfer coefficient is given by [Kraus and Bar-Cohen, 1983]

$$h = 0.664(k/l)(\mathrm{Re})^{0.5}(\mathrm{Pr})^{0.333} \ [\mathrm{W/m^2K}] \qquad (32.16)$$

$$\text{for } \mathrm{Re} < 2 \times 10^5$$

where *k* is the fluid thermal conductivity, *l* the length (in the flow direction) of the surface, Re the **Reynolds number** (equal to the product of velocity, density, and length divided by the fluid viscosity), and Pr the **Prandtl number** (equal to the product of specific heat and viscosity divided by the thermal conductivity of the fluid).

Inserting the various parameters associated with the Re and Pr in Eq. (32.16), the laminar heat transfer coefficient is found to be directly proportional to the square root of fluid velocity and inversely proportional to the square root of the length. Furthermore, increases in the thermal conductivity of the fluid and in the Pr, as are encountered in replacing air with a liquid coolant, can be expected to result in higher heat transfer coefficients. In studies of low-velocity convective air cooling of simulated integrated circuit packages, *h* has been found to depend somewhat more strongly on Re than suggested in Eq. (32.16), and to display an Re exponent of 0.54–0.72 [Buller and Kilburn, 1981; Sparrow *et al.*, 1982; Wirtz and Dykshoorn, 1984].

In higher velocity turbulent flow, the dependence of the convective heat transfer coefficient on the Re increases and is typically given by [Kraus and Bar-Cohen, 1983]:

$$h = 0.036(k/l) \, (Re)^{0.8} \, (Pr)^{0.333}, \, W/m^2K \tag{32.17}$$
$$\text{for } Re > 3 \times 10^5$$

In this flow regime, the convective heat transfer coefficient is, thus, found to vary directly with the velocity to the 0.8 power and inversely with the characteristic dimension to the 0.2 power. The dependence on fluid conductivity and Pr remains unchanged.

Applying Eq. (32.16) or (32.17) to the transfer of heat from the case of a chip module to the coolant, the external resistance, $R_{ex} = 1/hA$, is found to be inversely proportional to the wetted area and to the coolant velocity to the 0.5 to 0.8 power and directly proportional to the length in the flow direction to the 0.5 to 0.2 power. It may, thus, be observed that the external resistance can be strongly influenced by the fluid velocity and package dimensions and that these factors must be addressed in any meaningful evaluation of the external thermal resistances offered by various packaging technologies.

Values of the external resistance, for a variety of coolants and heat transfer mechanisms, are shown in Fig. 32.5 for a typical component wetted area of 10 cm² and a velocity range of 2 m/s to 8 m/s. They are seen to vary from a nominal 100 K/W for natural convection in air, to 33 K/W for forced convection in air, to 1 K/W in fluorocarbon liquid forced convection, and to less than 0.5 K/W for boiling in fluorocarbon liquids. Clearly, larger chip packages will experience proportionately lower external resistances than the tabulated values, and conduction of heat through the leads and package base into the printed circuit board or substrate will serve to further reduce the effective thermal resistance.

When the direct cooling of the package surface is inadequate to maintain the desired chip temperature, it is common to attach finned heat sinks, or compact heat exchangers, to the chip package. These heat sinks can considerably increase the wetted area but may act to reduce the convective heat transfer coefficient and most definitely introduce additional conductive resistances in the adhesive used to bond the heat sink to the package and in the body of the heat sink. Typical air-cooled heat sinks can reduce the external resistance to approximately 15 K/W in natural convection and as low as 5 K/W for moderate forced convection velocities.

When a heat sink or compact heat exchanger is attached to the package, the external resistance can be modified to account for the bond-layer conduction and fin efficiency as

$$R_{ex} = (T_c - T_f)/q_c = (\Delta x/kA)_b + (1/\eta hA) \tag{32.18}$$

In an optimally designed fin structure, η can be expected to fall in the range of 0.5 to 0.7 [Kraus and Bar-Cohen, 1983]. Relatively thick fins in a low velocity flow of gas are likely to yield fin efficiencies approaching unity. This same unity value would be appropriate, as well, for an unfinned surface and, thus, serve to generalize the use of Eq. (32.18) to all package configurations.

Total Resistance

Based on the accuracy of the assumptions used in the preceding development, the overall chip module resistance, relating the chip temperature to the inlet temperature of the coolant, can be found by summing the internal, external, and flow resistances to yield:

$$R_T = R_{jc} + R_{ex} + R_f = \sum_p R_c + (\Delta x/kA) + (\eta hA)^{-1} + 1/2 \, \dot{m}c_p \tag{32.19}$$

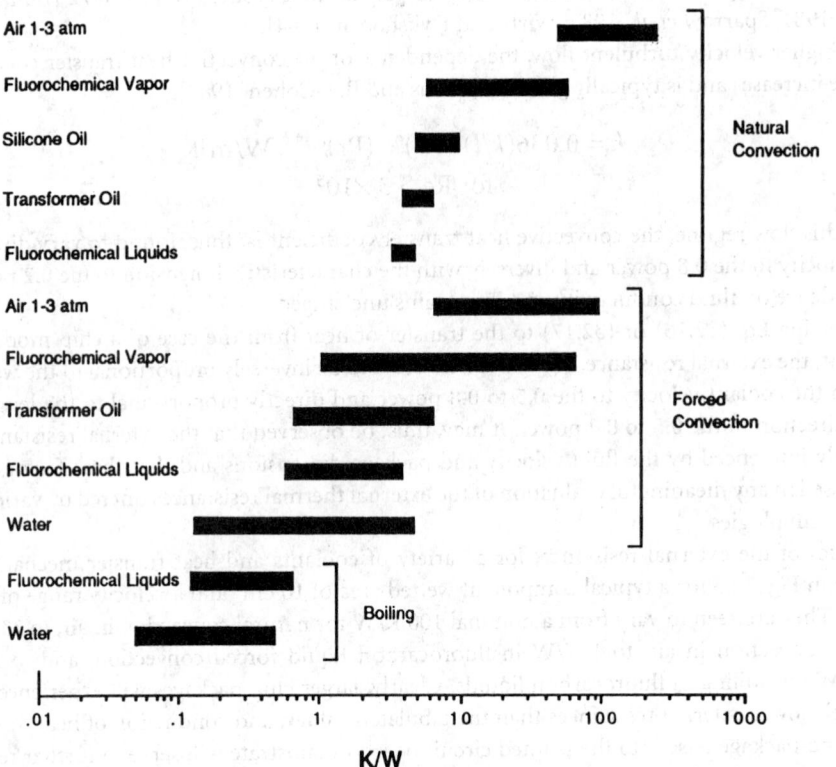

NOTE: FOR WETTED AREA = 10 cm^2

FIGURE 32.5 External thermal resistances for various fluids and cooling modes.

In evaluating the thermal resistance by this relation, care must be taken to determine the effective cross-sectional area for heat flow at each layer in the module. For single-chip modules, the requisite areas can be readily obtained, although care must be taken to consider possible voidage in solder and adhesive layers. As previously noted in the development of the relations for external and internal resistances, Eq. (32.19) shows R_T to be a strong function of the convective heat transfer coefficient, the flowing heat capacity of the coolant, and geometric parameters (thickness and cross-sectional area of each layer). Thus, the introduction of a superior coolant, use of thermal enhancement techniques that increase the local heat transfer coefficient, and selection of a heat transfer mode with inherently high heat transfer coefficients (e.g., boiling) will all be reflected in appropriately lower external and total thermal resistances. Similarly, improvements in the thermal conductivity of and reduction in the thickness of the relatively low conductivity bonding materials (e.g., soft solder, epoxy, silicone) would act to reduce the internal and total thermal resistances.

Frequently, however, even more dramatic reductions in the total resistance can be achieved simply by increasing the cross-sectional area for heat flow, within the chip module (e.g., chip, substrate, heat spreader) as well as along the wetted exterior surface. The implementation of such a scale-up generally results in a larger module footprint and/or lower volumetric packaging density, both of which are highly undesirable, and yet is rewarded with a lower thermal resistance. In evaluating packaging approaches, it must, therefore, be understood that the thermal resistance is a somewhat flawed figure-of-merit and that a better reflection of the efficacy of a thermal management technique be obtained by normalizing R_T with respect to the packaging density, using the number of chips on a substrate or number of chips/packages on a printed circuit board.

Multichip Modules

The thermostructural complexity of the multichip modules in current use hampers effective thermal characterization and introduces significant uncertainty in any attempt to compare the thermal performance of these packaging configurations. Variations in heat generation patterns across the active chips (e.g., devices versus drivers), as well as nonuniformities in heat dissipation among the chips assembled in a single module, further complicate this task. To establish a common, though approximate, basis for comparison of multichip modules, it is possible to neglect these variations and consider that the heat generated by each chip flows through a unit cell of the module structure to the external coolant [Bar-Cohen, 1987, 1988]. For a given structure, increasing the area of the unit cell allows heat to spread from the chip to a larger cross section, reducing the heat flux at some of the thermally critical interfaces and at the convectively cooled surfaces. Consequently, the thermal performance of a multichip module can be best represented by the area-specific thermal resistance, i.e., the temperature difference between the chip and the coolant divided by the substrate heat flux. This figure-of-merit is equivalent to the inverse of the overall heat transfer coefficient, U, commonly used in the compact heat exchanger literature. Examination of Fig. 32.6 reveals that, despite significant variation in design and fabrication, all the late-1980s water-cooled modules and one air-cooled module provide a specific thermal resistance of approximately 20°C for every watt per square centimeter at the substrate. In cutting-edge multichip modules in use in the early 1990s, their value has been reduced to 5–10 K/(W/cm^2).

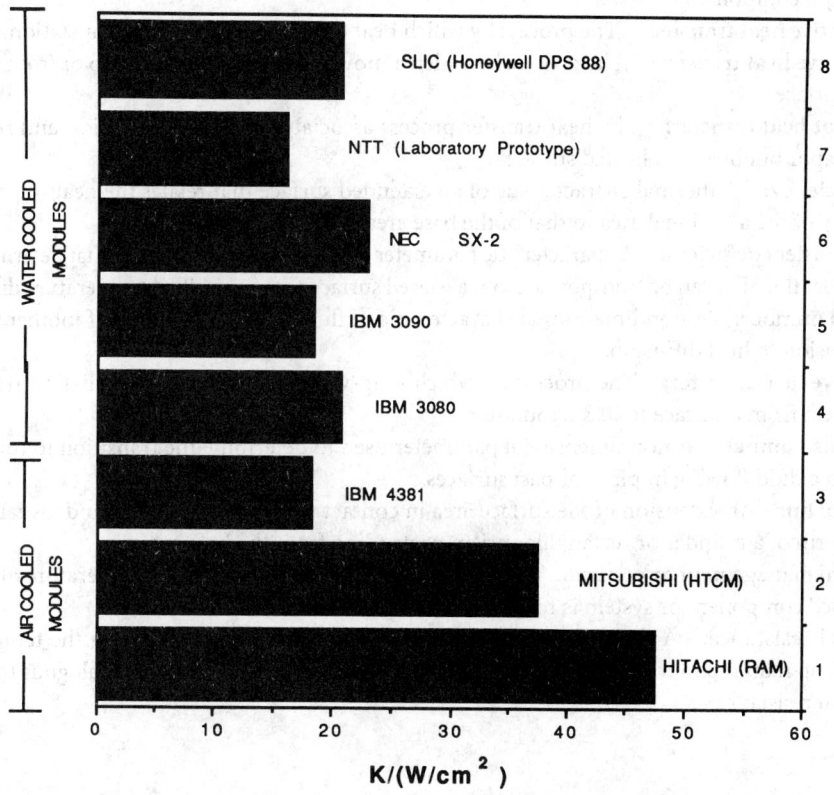

FIGURE 32.6 Specific thermal resistance of commercial multichip modules.

Nomenclature

A	area, m^2	$\mathcal{Q}$	volumetric flow rate, m^3/s
c_p	specific heat, W/kgK	ε	ratio of heater to substrate size
C'_{sf}	boiling surface parameter, W/m^{2}K^3	μ	viscosity, kg/ms
F	radiational factor	**Subscripts**	
h	heat transfer coefficient, W/m^{2}K	b	bond layer
k	thermal conductivity, W/mK	b	boiling
l	path length, m	cond	conduction
m	mass flow, kg/s	conv	convection
Pr	Prandtl Number, $\cong c_p\mu/k$	f, fluid	fluid
q	heat flow, W	i	internal
R	thermal resistance, K/W	in	inlet
Re	Reynolds Number, $\equiv \rho vl/\mu$	j	junction
T	temperature, K	jc	junction to case
v	velocity, m/s	o	base, external
x	length, m	out	outlet
σ	Stefan-Boltzmann constant,	r	radiation
	5.67×10^{-8} W/m^{2}K^4	sat	thermodynamic saturation
η	fin efficiency	surf	surface
ρ	density, kg/m^3	T	total

Defining Terms

Catastrophic thermal failure: An immediate, thermally induced total loss of electronic function by a component or system.

Conductive heat transfer: The process by which heat diffuses through a solid or stationary fluid.

Convective heat transfer: The process by which a moving fluid transfers heat to or from a wetted surface.

Ebullient heat transfer: The heat transfer process associated with the formation and release of vapor bubbles on a heated surface.

Fin efficiency: A thermal characteristic of an extended surface that relates the heat transfer ability of the additional area to that of the base area.

Heat transfer coefficient: A characteristic parameter of convective heat transfer that determines the heat flux that can be transported from a wetted surface with a specified temperature difference.

Prandtl number: A nondimensional characteristic of fluids, relating the rate of momentum diffusion to heat diffusion.

Radiative heat transfer: The process by which long-wave electromagnetic radiation transports heat from a surface to its surroundings.

Reynolds number: A nondimensional parameter used to determine the transition to turbulence in a fluid flowing in pipes or past surfaces.

Thermal fin: An extension of the surface area in contact with a heat transfer fluid, usually in the form of a cylinder or rectangular prism protruding from the base surface.

Thermal management or control: The process or processes by which the temperature of a specified component or system is maintained at the desired level.

Thermal resistance: A thermal characteristic of a heat flow path, establishing the temperature drop required to transport heat across the specified segment or surface; analogous to electrical resistance.

References

A. Bar-Cohen, "Thermal management of air- and liquid-cooled multichip modules," *IEEE CHMT Transactions,* vol. CHMT-10, no. 2, pp. 159–175, 1987.

A. Bar-Cohen, "Addendum and correction to thermal management of air- and liquid-cooled multichip modules," *IEEE CHMT Transactions,* vol. CHMT-11, no. 3, pp. 333–334, 1988.

M.L. Buller and R.F. Kilburn, "Evaluation of surface heat transfer coefficients for electronic module packages," *Heat Transfer in Electronic Equipment,* vol. HTD-20, ASME, New York, 1981.

W. Englemier, "Functional cycles and surface mounting attachment reliability," in *Thermal Management Concepts in Microelectronic Packaging,* R.T. Howard, *et al.,* Eds., ISHM Technical Monograph Series, 6984-003, ISHM, New York, 1984, pp. 83–109.

C.A. Harper, Ed., *Electronic Packaging and Interconnection Handbook,* New York: McGraw-Hill, 1991, p. 2.7.

W.F. Hilbert and F.H. Kube, "Effects on electronic equipment reliability of temperature cycling in equipment," Final Report, Grumman Aircraft Engineering Co., Report No. EC-69-400, Bethpage, N.Y., 1969.

R.C. Jaeger, "Development of low temperature CMOS for high performance computer systems," *IEEE International Conference on Computer Design: VLSI in Computers,* 1986, pp. 128–130.

A. Kaneko, K. Seyama, and M. Suzuki, "LSI packaging and cooling technologies for Fujitsu VP-2000 Series," *Fujitsu,* vol. 41, no. 1, pp. 12–19, 1990.

A.D. Kraus, "The use of steady state electrical network analysis in solving heat flow problems," 2nd National Heat Transfer Conference, Chicago, Ill., 1958.

A.D. Kraus and A. Bar-Cohen, *Thermal Analysis and Control of Electronic Equipment,* New York: Hemisphere Publishing Corporation, 1983.

G.N. Morrison, J.M. Kallis, L.A. Strattan, I.R. Jones, and A.L. Lena, "RADC thermal guide for reliability engineers," Report Number RADC-TR-82-172, Rome Air Development Center, Air Force Systems Command, Griffis Air Force Base, New York, 1982.

R.A. Morrison, "Improved avionics reliability through phase change conductive cooling," *Proceedings, IEEE National Telesystems Conference,* pp. B5.6.1–B5.6.5, 1982.

W. Nakayama, "Thermal management of electronic equipment: A review of technology and research topics," in *Advances in Thermal Modeling of Electronic Components and Systems,* vol. 1, A. Bar-Cohen and A.D. Kraus, Eds., New York: Hemisphere Publishing Corporation, 1988, pp. 1–78.

M. Pecht, P. Lall, and E. Hakim, "The influence of temperature on integrated circuit failure mechanisms," *Advances in Thermal Modeling of Electronic Components and Systems,* vol. 3, A. Bar-Cohen and A.D. Kraus, Eds., New York: ASME Press, 1992.

J. Pei, S. Heng, R. Charlantini, and P. Gildea, "Cooling components used in the Vax 9000 family of computers," *Proceedings, 1990 International Electronic Packaging Society Conference,* 1990, pp. 587–601.

E.M. Sparrow, J.E. Niethhammer, and A. Chaboki, "Heat transfer and pressure drop characteristics of arrays of rectangular modules encountered in electronic equipment," *Int. J. Heat & Mass Transfer,* vol. 25, no. 7, pp. 961–973, 1982.

E. Suhir, "Thermal stress in electronic components," in *Advances in Thermal Modeling of Electronic Components and Systems,* vol. 1, A. Bar-Cohen and A.D. Kraus, Eds., New York: Hemisphere Publishing Corporation, 1988, pp. 337–412.

R.R. Tummala and E.J. Rymaszewski, *Microelectronics Packaging Handbook,* New York: Van Nostrand Reinhold, 1989, p. 174.

A. Vacca, D. Resnick, D. Frankel, R. Bach, J. Kreilich, and D. Carlson, "A cryogenically cooled CMOS VLSI supercomputer," *VLSI Systems Design,* pp. 80–88, 1987.

R.A. Wirtz and P. Dykshoorn, "Heat transfer from arrays of flat packs in channel flow," *Proceedings, 4th International Electronic Packaging Society Conference,* New York, 1984, pp. 318–326.

M.M. Yovanovich and V.W. Antonetti, "Application of thermal contact resistance theory to electronic packages," in *Advances in Thermal Modeling of Electronic Components and Systems,* vol. 1, A. Bar-Cohen and A.D. Kraus, Eds., New York: Hemisphere Publishing Corporation, 1988, pp. 79–128.

AT&T glass strands. The loops of a hair-thin glass fiber illuminated by laser light represent the transmission medium for lightwave systems. Typically 12 fibers are embedded between 2 strips of plastic in a flat ribbon and as many as 12 ribbons are stacked in a cable that can carry more than 40,000 voice channels. (Photo courtesy of AT&T Archives.)

IV

Electromagnetics

Banmali S. Rawat
University of Nevada, Reno

33 **Electromagnetic Fields** *J. Kong* 803
Maxwell Equations • Constitutive Relations • Wave Equations and Wave Solutions

34 **Magnetism and Magnetic Fields** *G. Bate, M. Kryder* 811
Magnetism • Magnetic Recording

35 **Wave Propagation** *M. Sadiku, K. Demarest* 837
Space Propagation • Waveguides

36 **Antennas** *N. Kolias, R. Compton, J. Fitch* 861
Wire • Aperture

37 **Microwave Devices** *M. Steer, R. Trew* 882
Passive Microwave Devices • Active Microwave Devices

38 **Compatibility** *L. Hemming, V. Ungvichian, J. Roman, M. Uman, M. Rubinstein* 903
Grounding, Shielding, and Filtering • Spectrum, Specifications, and Measurement Techniques • Lightning

39 **Radar** *M. Belcher, J. Nessmith, J. Wiltse* 949
Pulse Radar • Continuous Wave Radar

40 **Lightwave** *S. Agbo, A. Cherin, B. Tariyal* 975
Lightwave Waveguides • Optical Fibers and Cables

41 **Solid State Circuits** *I. Bahl* 1004
Amplifiers • Oscillators • Multipliers • Mixers • Control Circuits

42 **Three-Dimensional Analysis** *C. Trowbridge* 1018
The Field Equations • Numerical Methods • The Modern Design Environment

43 **Computational Electromagnetics** *E. Miller* 1028
Background Discussion • Analytical Issues in Developing a Computer Model • Numerical Issues
in Developing a Computer Model • Some Practical Considerations • Ways of Decreasing Computer Time • Validation, Error Checking, and Error Analysis

E LECTRIC AND MAGNETIC FORCES are among the five original forces in the universe. These forces are important as we are affected by them almost every instant. Electromagnetics is the combined effect of electric and magnetic fields. Today's scientific development to a great extent is based on the electromagnetic fields, their propagation, and varying effects under different boundary conditions. Very few subjects are understood as thoroughly as electromagnetics and have such wide applications. Electricity, telephones, radio, television, datalinks, medical electronics, radar, remote sensing, etc.—all have considerable impact on human life. Now that impact is being carried out further with optical fiber technology, which is also based on the concept of electromagnetic wave propagation. All of human society has been revolutionized by electromagnetics, but still our understanding is not complete. As H. G. Wells once wrote, and this is still true, "The past is but a beginning of a beginning, and all that is and has been is but the twilight of the dawn."

This section focuses on the basic electromagnetic field concepts, wave propagation, devices, circuits, and other applications. The electric fields which are produced by stationary or moving charges are described in Chapter 33. Maxwell's equations and their solutions under different boundary conditions help in determining the electric field components and resulting effects. The next chapter describes the magnetic fields and magnetic effects due to moving charges or current. These magnetic fields are also governed by Maxwell's equations and their solutions are obtained for different boundary conditions. Particular magnetic materials with an assemblage of ferromagnetic particles in a nonferromagnetic matrix are useful as audio or video tapes. This subject is investigated in Chapter 34 to provide insight into the recording mechanism of the music we hear all the time. The time-varying electromagnetic field propagation in space or in transmission lines provides the concept of radio communication as discussed in Chapter 35. Another article in the chapter analyzes the transmission of energy through waveguides and microstriplines. Microstriplines have become the basic building blocks for microwave integrated circuits (MICS). For the propagation of electromagnetic fields in space, properly matched antennas between generator and space are required, as described in Chapter 36. Wire and aperture antennas are also described.

The high-frequency or microwave-frequency electromagnetic field concepts are helpful in studying the microwave devices as discussed in Chapter 37. The electromagnetic compatibility (EMC) study in the following chapter is important for proper functioning of microwave devices and circuits. The important application of electromagnetic radiation in the form of radar, discussed in Chapter 39, is useful not only for defense but in remote sensing and weather forecasting also. The next chapter explains the propagation of light through waveguides and optical fibers/cables. Optical fiber technology is an emerging technology and is affecting every facet of human life. Microwave circuits are the practical realization of electromagnetic field concepts and are discussed in Chapter 40. With the arrival of sophisticated software packages and high-speed computers, now it is possible and worthwhile to do 3-D analysis and computer modeling of electromagnetic fields in the circuits or devices, as discussed in the last two chapters of this section. This is helpful in the accurate design of microwave components and circuits. All the topics mentioned in this introduction are discussed in detail in their respective chapters.

Nomenclature

Symbol	Quantity	Unit	Symbol	Quantity	Unit
A_e	actual effective aperture of antenna	m^2	D	divergent factor	
			D	directivity of antenna	dB
A_{em}	maximum effective aperture of antenna	m^2	δ	penetrating depth	m
			E	electric field intensity	V/m
α	attenuation constant	neper/m	ϵ	permittivity	F/m
b	Doppler filter bandwidth	Hz	ϵ_0	8.854×10^{-12}	F/m
B	magnetic flux density	Wb/m^2	f_D	Doppler frequency	Hz
β	phase constant	rad/m	F	receiver noise figure	dB
c	velocity of light in vacuum	2.998×10^8 m/s	g_m	transconductance	S
			G	gain of antenna	dB
D	electric displacement	C/m^2	γ	propagation constant	m^{-1}

Symbol	Quantity	Unit	Symbol	Quantity	Unit
Γ	Fresnel reflection coefficient		μ_0	$4\pi \times 10^{-7}$	H/m
H	magnetic field intensity	A/m	**P**	Poynting vector	W/m^2
η	intrinsic impedance	Ω	P_T	average power	W
J	electric current density	A/m^2	Ψ	grazing angle	degree
J	electric charge density	C/m^2	q	electronic charge	1.6×10^{-19} C
k	wavenumber		R	detection range of target	m
k	radiation efficiency factor		ρ_s	roughness coefficient	
L	antenna loss	dB	$S(\theta)$	shadowing function	
λ	wavelength	m	U	unilateral power gain	dB
μ	permeability	H/m			

33

Electromagnetic Fields

Jin Au Kong
Massachusetts Institute of Technology

33.1 Maxwell Equations 803
33.2 Constitutive Relations 804
Anisotropic and Bianisotropic Media • Biisotropic Media • Constitutive Matrices
33.3 Wave Equations and Wave Solutions 807
Wave Solution • Wave Vector $\bar{k}$ • Wavenumbers k

33.1 Maxwell Equations

The fundamental equations of electromagnetic theory were established by James Clerk Maxwell in 1873. In three-dimensional vector notation, the Maxwell equations are

$$\nabla \times \bar{E}(\bar{r},t) + \frac{\partial}{\partial t} \bar{B}(\bar{r},t) = 0 \tag{33.1}$$

$$\nabla \times \bar{H}(\bar{r},t) - \frac{\partial}{\partial t} \bar{D}(\bar{r},t) = \bar{J}(\bar{r},t) \tag{33.2}$$

$$\nabla \cdot \bar{B}(\bar{r},t) = 0 \tag{33.3}$$

$$\nabla \cdot \bar{D}(\bar{r},t) = \rho(\bar{r},t) \tag{33.4}$$

where $\bar{E}, \bar{B}, \bar{H}, \bar{D}, \bar{J}$, and ρ are real functions of position and time.

$$\bar{E}(\bar{r},t) = \text{electric field strength} \quad \text{(volts/m)}$$

$$\bar{B}(\bar{r},t) = \text{magnetic flux density} \quad \text{(webers/m}^2\text{)}$$

$$\bar{H}(\bar{r},t) = \text{magnetic field strength} \quad \text{(amperes/m)}$$

$$\bar{D}(\bar{r},t) = \text{electric displacement} \quad \text{(coulombs/m}^2\text{)}$$

$$\bar{J}(\bar{r},t) = \text{electric current density} \quad \text{(amperes/m}^2\text{)}$$

$$\rho(\bar{r},t) = \text{electric charge density} \quad \text{(coulombs/m}^3\text{)}$$

This chapter is an abridged version of Chapter 1 in *Electromagnetic Wave Theory* (J. A. Kong), New York: Wiley-Interscience, 1990.

Equation (33.1) is Faraday's induction law. Equation (33.2) is the generalized Ampere's circuit law. Equations (33.3) and (33.4) are Gauss' laws for **magnetic and electric fields.** Taking the divergence of (33.2) and introducing (33.4), we find that

$$\nabla \cdot \bar{J}(\bar{r},t) + \frac{\partial}{\partial t}\, \rho(\bar{r},t) = 0 \qquad (33.5)$$

This is the conservation law for electric charge and current densities. Regarding (33.5) as a fundamental equation, we can use it to derive (33.4) by taking the divergence of (33.2). Equation (33.3) can also be derived by taking the divergence of (33.1) which gives $\partial(\nabla \cdot \bar{B}(\bar{r},t))/\partial t = 0$ or that $\nabla \cdot \bar{B}(\bar{r},t)$ is a constant independent of time. Such a constant, if not zero, then implies the existence of magnetic monopoles similar to free electric charges. Since magnetic monopoles have not been found to exist, this constant must be zero and we arrive at (33.3).

33.2 Constitutive Relations

The Maxwell equations are fundamental laws governing the behavior of electromagnetic fields in free space and in media. We have so far made no reference to the various material properties that provide connections to other disciplines of physics, such as plasma physics, continuum mechanics, solid-state physics, fluid dynamics, statistical physics, thermodynamics, biophysics, etc., all of which interact in one way or another with electromagnetic fields. We did not even mention the Lorentz force law, which constitutes a direct link to mechanics. It is time to state how we are going to account for this vast "outside" world. From the electromagnetic wave point of view, we shall be interested in how electromagnetic fields behave in the presence of media, whether the wave is diffracted, refracted, or scattered. Whatever happens to a medium, whether it is moved or deformed, is of secondary interest. Thus we shall characterize material media by the so-called constitutive relations that can be classified according to the various properties of the media.

The necessity of using constitutive relations to supplement the Maxwell equations is clear from the following mathematical observations. In most problems we shall assume that sources of electromagnetic fields are given. Thus $\bar{J}$ and ρ are known and they satisfy the conservation law (33.5). Let us examine the Maxwell equations and see if there are enough equations for the number of unknown quantities. There are a total of 12 scalar unknowns for the four field vectors $\bar{E}$, $\bar{H}$, $\bar{B}$, and $\bar{D}$. As we have learned, Eqs. (33.3) and (33.4) are not independent equations; they can be derived from Eqs. (33.1), (33.2), and (33.5). The independent equations are Eqs. (33.1) and (33.2), which constitute six scalar equations. Thus we need six more scalar equations. These are the constitutive relations.

The constitutive relations for an isotropic medium can be written simply as

$$\bar{D} = \epsilon E \quad \text{where } \epsilon = \text{permittivity} \qquad (33.6b)$$

$$\bar{B} = \mu\bar{H} \quad \text{where } \mu = \text{permeability} \qquad (33.6b)$$

By isotropy we mean that the field vector $\bar{E}$ is parallel to $\bar{D}$ and the field vector $\bar{H}$ is parallel to $\bar{B}$. In free space void of any matter, $\mu = \mu_o$ and $\epsilon = \epsilon_o$,

$$\mu_o = 4\pi \times 10^{-7} \qquad \text{henry/meter}$$

$$\epsilon_o \approx 8.85 \times 10^{-12} \quad \text{farad/meter}$$

Inside a material medium, the permittivity ϵ is determined by the electrical properties of the medium and the permeability μ by the magnetic properties of the medium.

A dielectric material can be described by a free-space part and a part that is due to the material alone. The material part can be characterized by a polarization vector $\bar{P}$ such that $\bar{D} = \epsilon_o\bar{E} + \bar{P}$. The polarization $\bar{P}$ symbolizes the electric dipole moment per unit volume of the dielectric material. In the presence of an external electric field, the polarization vector may be caused by induced dipole moments, alignment of the permanent dipole moments of the medium, or migration of ionic charges.

A magnetic material can also be described by a free-space part and a part characterized by a magnetization vector $\overline{M}$ such that $\overline{B} = \mu_o\overline{H} + \mu_o\overline{M}$. A medium is diamagnetic if $\mu \leq \mu_o$ and paramagnetic if $\mu \geq \mu_o$. Diamagnetism is caused by induced magnetic moments that tend to oppose the externally applied magnetic field. Paramagnetism is due to alignment of magnetic moments. When placed in an inhomogeneous magnetic field, a diamagnetic material tends to move toward regions of weaker magnetic field and a paramagnetic material toward regions of stronger magnetic field. Ferromagnetism and antiferromagnetism are highly nonlinear effects. Ferromagnetic substances are characterized by spontaneous magnetization below the Curie temperature. The medium also depends on the history of applied fields, and in many instances the magnetization curve forms a hysteresis loop. In an antiferromagnetic material, the spins form sublattices that become spontaneously magnetized in an antiparallel arrangement below the Néel temperature.

Anisotropic and Bianisotropic Media

The constitutive relations for anisotropic media are usually written as

$$\overline{D} = \overline{\overline{\epsilon}} \cdot \overline{E} \quad \text{where } \overline{\overline{\epsilon}} = \text{permittivity tensor} \tag{33.7a}$$

$$\overline{B} = \overline{\overline{\mu}} \cdot \overline{H} \quad \text{where } \overline{\overline{\mu}} = \text{permeability tensor} \tag{33.7b}$$

The field vector $\overline{E}$ is no longer parallel to $\overline{D}$, and the field vector $\overline{H}$ is no longer parallel to $\overline{B}$. A medium is *electrically anisotropic* if it is described by the permittivity tensor $\overline{\overline{\epsilon}}$ and a scalar permeability μ, and *magnetically anisotropic* if it is described by the permeability tensor $\overline{\overline{\mu}}$ and a scalar permittivity ϵ. Note that a medium can be both electrically and magnetically anisotropic as described by both $\overline{\overline{\epsilon}}$ and $\overline{\overline{\mu}}$ in Eq. (33.7).

Crystals are described in general by symmetric permittivity tensors. There always exists a coordinate transformation that transforms a symmetric matrix into a diagonal matrix. In this coordinate system, called the *principal system*,

$$\overline{\overline{\epsilon}} = \begin{bmatrix} \epsilon_x & 0 & 0 \\ 0 & \epsilon_y & 0 \\ 0 & 0 & \epsilon_z \end{bmatrix} \tag{33.8}$$

The three coordinate axes are referred to as the principal axes of the crystal. For cubic crystals, $\epsilon_x = \epsilon_y = \epsilon_z$ and they are isotropic. In tetragonal, hexagonal, and rhombohedral crystals, two of the three parameters are equal. Such crystals are *uniaxial*. Here there is a two-dimensional degeneracy; the principal axis that exhibits this anisotropy is called the optic axis. For a uniaxial crystal with

$$\overline{\overline{\epsilon}} = \begin{bmatrix} \epsilon & 0 & 0 \\ 0 & \epsilon & 0 \\ 0 & 0 & \epsilon_z \end{bmatrix} \tag{33.9}$$

the z axis is the optic axis. The crystal is *positive uniaxial* if $\epsilon_z > \epsilon$; it is *negative uniaxial* if $\epsilon_z < \epsilon$. In orthorhombic, monoclinic, and triclinic crystals, all three crystallographic axes are unequal. We have $\epsilon_x \neq \epsilon_y \neq \epsilon_z$, and the medium is *biaxial*.

For isotropic or anisotropic media, the constitutive relations relate the two electric field vectors and the two magnetic field vectors by either a scalar or a tensor. Such media become polarized when placed in an electric field and become magnetized when placed in a magnetic field. A bianisotropic medium provides the cross coupling between the electric and magnetic fields. The constitutive relations for a bianisotropic medium can be written as

$$\bar{D} = \bar{\bar{\epsilon}} \cdot \bar{E} + \bar{\bar{\xi}} \cdot \bar{H} \tag{33.10a}$$

$$\bar{B} = \bar{\bar{\zeta}} \cdot \bar{E} + \bar{\bar{\mu}} \cdot \bar{H} \tag{33.10b}$$

When placed in an electric or a magnetic field, a bianisotropic medium becomes both polarized and magnetized.

Magnetoelectric materials, theoretically predicted by Dzyaloshinskii and by Landau and Lifshitz, were observed experimentally in 1960 by Astrov in antiferromagnetic chromium oxide. The constitutive relations that Dzyaloshinskii proposed for chromium oxide have the following form:

$$\bar{D} = \begin{bmatrix} \epsilon & 0 & 0 \\ 0 & \epsilon & 0 \\ 0 & 0 & \epsilon_z \end{bmatrix} \cdot \bar{E} + \begin{bmatrix} \xi & 0 & 0 \\ 0 & \xi & 0 \\ 0 & 0 & \xi_z \end{bmatrix} \cdot \bar{H} \tag{33.11a}$$

$$\bar{B} = \begin{bmatrix} \xi & 0 & 0 \\ 0 & \xi & 0 \\ 0 & 0 & \xi_z \end{bmatrix} \cdot \bar{E} + \begin{bmatrix} \mu & 0 & 0 \\ 0 & \mu & 0 \\ 0 & 0 & \mu_z \end{bmatrix} \cdot \bar{H} \tag{33.11b}$$

It was then shown by Indenbom and by Birss that 58 magnetic crystal classes can exhibit the magnetoelectric effect. Rado proved that the effect is not restricted to antiferromagnetics; ferromagnetic gallium iron oxide is also magnetoelectric.

Biisotropic Media

In 1948, the gyrator was introduced by Tellegen as a new element, in addition to the resistor, the capacitor, the inductor, and the ideal transformer, for describing a network. To realize his new network element, Tellegen conceived of a medium possessing constitutive relations of the form

$$\bar{D} = \epsilon \bar{E} + \xi \bar{H} \tag{33.12a}$$

$$\bar{B} = \xi \bar{E} + \mu \bar{H} \tag{33.12b}$$

where $\xi^2/\mu\epsilon$ is nearly equal to 1. Tellegen considered that the model of the medium had elements possessing permanent electric and magnetic dipoles parallel or antiparallel to each other, so that an applied electric field that aligns the electric dipoles simultaneously aligns the magnetic dipoles, and a magnetic field that aligns the magnetic dipoles simultaneously aligns the electric dipoles. Tellegen also wrote general constitutive relations Eq. (33.10) and examined the symmetry properties by energy conservation.

Chiral media, which include many classes of sugar solutions, amino acids, DNA, and natural substances, have the following constitutive relations

$$\bar{D} = \epsilon \bar{E} - \chi \frac{\partial \bar{H}}{\partial t} \tag{33.13a}$$

$$\bar{B} = \mu \bar{H} + \chi \frac{\partial \bar{E}}{\partial t} \tag{33.13b}$$

where χ is the chiral parameter. Media characterized by the constitutive relations, Eqs. (33.12) and (33.13), are biisotropic media.

Media in motion were the first bianisotropic media to receive attention in electromagnetic theory. In 1888, Roentgen discovered that a moving dielectric becomes magnetized when it is placed in an electric field. In 1905, Wilson showed that a moving dielectric in a uniform magnetic field becomes electrically polarized. Almost any medium becomes bianisotropic when it is in motion.

The bianisotropic description of material has fundamental importance from the point of view of relativity. The principle of relativity postulates that all physical laws of nature must be characterized by mathematical equations that are form-invariant from one observer to the other. For electromagnetic theory, the Maxwell equations are form-invariant with respect to all observers, although the numerical values of the field quantities may vary from one observer to another. The constitutive relations are form-invariant when they are written in bianisotropic form.

Constitutive Matrices

Constitutive relations in the most general form can be written as

$$c\bar{D} = \bar{\bar{P}} \cdot \bar{E} + \bar{\bar{L}} \cdot c\bar{B} \tag{33.14a}$$

$$\bar{H} = \bar{\bar{M}} \cdot \bar{E} + \bar{\bar{Q}} \cdot c\bar{B} \tag{33.14b}$$

where $c = 3 \times 10^8$ m/s is the velocity of light in vacuum, and $\bar{\bar{P}}, \bar{\bar{Q}}, \bar{\bar{L}}$, and $\bar{\bar{M}}$ are all 3×3 matrices. Their elements are called *constitutive parameters*. In the definition of the constitutive relations, the constitutive matrices $\bar{\bar{L}}$ and $\bar{\bar{M}}$ relate electric and magnetic fields. When $\bar{\bar{L}}$ and $\bar{\bar{M}}$ are not identically zero, the medium is *bianisotropic*. When there is no coupling between electric and magnetic fields, $\bar{\bar{L}} = \bar{\bar{M}} = 0$ and the medium is *anisotropic*. For an anisotropic medium, if $\bar{\bar{P}} = c\epsilon \bar{\bar{I}}$ and $\bar{\bar{Q}} = (1/c\mu)\bar{\bar{I}}$ with $\bar{I}$ denoting the 3×3 unit matrix, the medium is *isotropic*. The reason that we write constitutive relations in the present form is based on relativistic considerations. First, the fields $\bar{E}$ and $c\bar{B}$ form a single tensor in four-dimensional space, and so do $c\bar{D}$ and $\bar{H}$. Second, constitutive relations written in the form Eq. (33.14) are Lorentz-covariant.

Equation (33.14) can be rewritten in the form

$$\begin{bmatrix} c\bar{D} \\ \bar{H} \end{bmatrix} = \bar{\bar{C}} \cdot \begin{bmatrix} \bar{E} \\ c\bar{B} \end{bmatrix} \tag{33.15a}$$

and $\bar{\bar{C}}$ is a 6×6 constitutive matrix:

$$\bar{\bar{C}} = \begin{bmatrix} \bar{\bar{P}} & \bar{\bar{L}} \\ \bar{\bar{M}} & \bar{\bar{Q}} \end{bmatrix} \tag{33.15b}$$

which has the dimension of admittance.

The constitutive matrix $\bar{\bar{C}}$ may be functions of space–time coordinates, thermodynamical and continuum-mechanical variables, or electromagnetic field strengths. According to the functional dependence of $\bar{\bar{C}}$, we can classify the various media as (1) inhomogeneous if $\bar{\bar{C}}$ is a function of space coordinates, (2) nonstationary if $\bar{\bar{C}}$ is a function of time, (3) time-dispersive if $\bar{\bar{C}}$ is a function of time derivatives, (4) spatial-dispersive if $\bar{\bar{C}}$ is a function of spatial derivatives, (5) nonlinear if $\bar{\bar{C}}$ is a function of the electromagnetic field, and so forth. In the general case $\bar{\bar{C}}$ may be a function of integral-differential operators and coupled to fundamental equations of other physical disciplines.

33.3 Wave Equations and Wave Solutions

The Maxwell equations in differential form are valid at all times for every point in space. First we shall investigate solutions to the Maxwell equations in regions void of source, namely, in regions where $\bar{J} = \rho = 0$. This, of course, does not mean that there is no source anywhere in all space. Sources must exist outside the regions of interest in order to produce fields in these regions. From the source-free Maxwell equations, a wave equation for the electric field $\bar{E}$ can be easily derived for isotropic permittivity ϵ and permeability μ

$$\nabla^2 \bar{E} - \mu\epsilon \frac{\partial^2}{\partial t^2} \bar{E} = 0 \tag{33.16}$$

The Laplacian operator ∇^2 in a rectangular coordinate system is

$$\nabla^2 = \frac{\partial^2}{\partial x^2} + \frac{\partial^2}{\partial y^2} + \frac{\partial^2}{\partial z^2}$$

The wave Eq. (33.16) is a second-order partial differential equation of space and time coordinates x, y, z, and t.

Wave Solution

The simplest solution to Eq. (33.16) for the electric field $\bar{E}$ is

$$\bar{E} = \hat{x}E_0 \cos(kz - \omega t) = \hat{x}E_x(z, t) \tag{33.17}$$

Substituting Eqs. (33.17) in (33.16) we find that the following equation, called the dispersion relation, which relates ω and k, must be satisfied:

$$k^2 = \omega^2 \mu\epsilon \tag{33.18}$$

There are two points of view useful in the study of a space–time varying quantity such as $E_x(z, t)$. The first is to examine the time variation at fixed points in space. The second is to examine spatial variation at fixed times, a process that amounts to taking a series of pictures.

We first fix our attention to one particular point in space, say $z = 0$. We then have the electric vector $E_x(z, t) = E_0 \cos \omega t$. Plotted as a function of time, we find that the waveform repeats itself in time as $\omega t = 2m\pi$ for any integer m. The period is defined as the time T for which $\omega T = 2\pi$. The frequency f is defined as $f = 1/T$ which gives

$$f = \frac{\omega}{2\pi}$$

Since $\omega = 2\pi f$, ω is the angular frequency of the wave.

To examine wave behavior from the other point of view, we let $\omega t = 0$ and plot $E_x(z, t)$. The waveform repeats itself in space when $kz = 2m\pi$ for integer values of m. The wavelength λ is defined as the distance for which $k\lambda = 2\pi$. Thus $\lambda = 2\pi/k$, or

$$k = \frac{2\pi}{\lambda}$$

We call k the wavenumber which is equal to the number of wavelengths in a distance of 2π and has the dimension inverse length.

Wave Vector $\bar{k}$

The solution for the electric field $\bar{E}$ in Eq. (33.17) represents an electromagnetic wave propagating in the $\hat{z}$-direction. For a wave propagating in a general direction, we define a wave vector

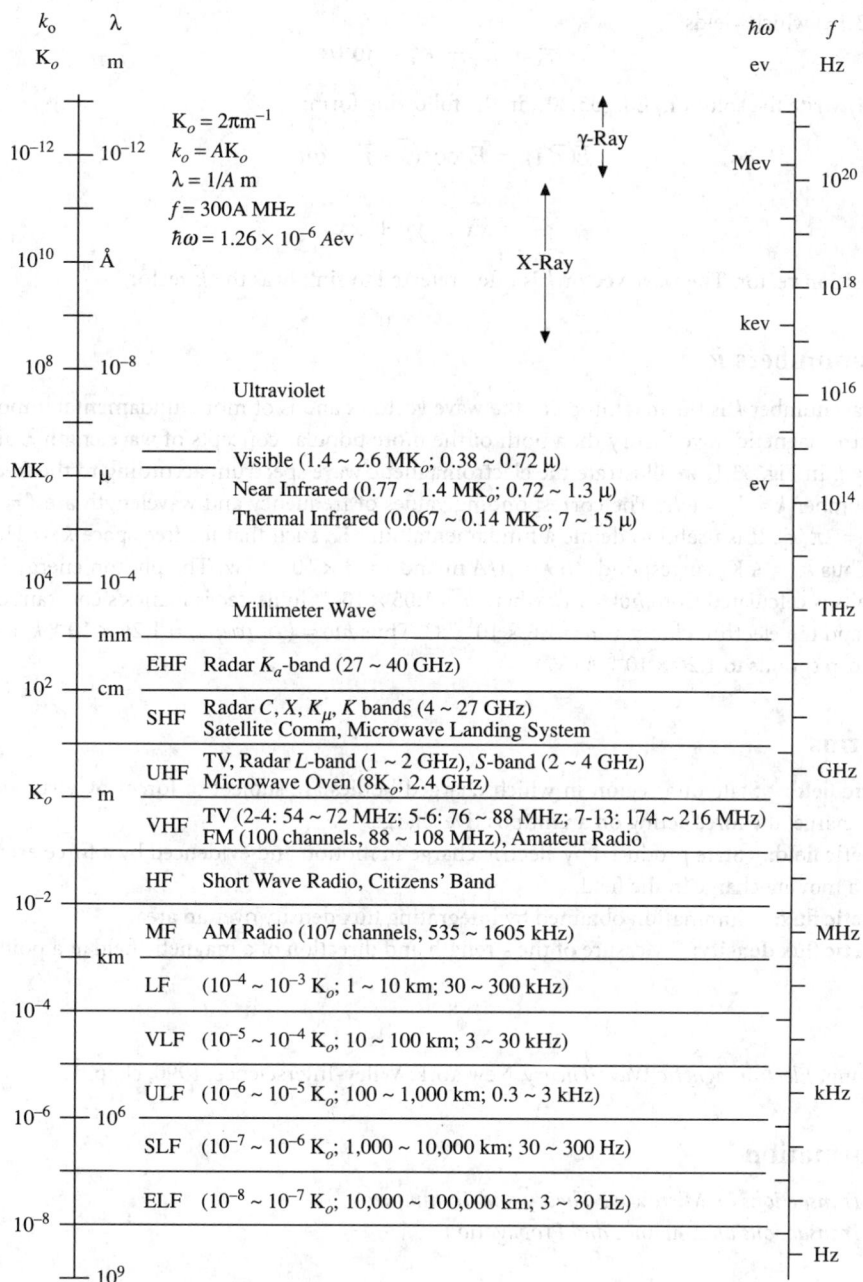

k_o	λ				$\hbar\omega$	f
K_o	m				ev	Hz

$K_o = 2\pi m^{-1}$
$k_o = AK_o$
$\lambda = 1/A$ m
$f = 300A$ MHz
$\hbar\omega = 1.26 \times 10^{-6} A$ ev

γ-Ray

X-Ray

Ultraviolet

Visible (1.4 ~ 2.6 MK$_o$; 0.38 ~ 0.72 μ)
Near Infrared (0.77 ~ 1.4 MK$_o$; 0.72 ~ 1.3 μ)
Thermal Infrared (0.067 ~ 0.14 MK$_o$; 7 ~ 15 μ)

Millimeter Wave

EHF   Radar K_a-band (27 ~ 40 GHz)

SHF   Radar C, X, K_μ, K bands (4 ~ 27 GHz)
      Satellite Comm, Microwave Landing System

UHF   TV, Radar L-band (1 ~ 2 GHz), S-band (2 ~ 4 GHz)
      Microwave Oven (8K$_o$; 2.4 GHz)

VHF   TV (2-4: 54 ~ 72 MHz; 5-6: 76 ~ 88 MHz; 7-13: 174 ~ 216 MHz)
      FM (100 channels, 88 ~ 108 MHz), Amateur Radio

HF   Short Wave Radio, Citizens' Band

MF   AM Radio (107 channels, 535 ~ 1605 kHz)

LF   (10^{-4} ~ 10^{-3} K$_o$; 1 ~ 10 km; 30 ~ 300 kHz)

VLF   (10^{-5} ~ 10^{-4} K$_o$; 10 ~ 100 km; 3 ~ 30 kHz)

ULF   (10^{-6} ~ 10^{-5} K$_o$; 100 ~ 1,000 km; 0.3 ~ 3 kHz)

SLF   (10^{-7} ~ 10^{-6} K$_o$; 1,000 ~ 10,000 km; 30 ~ 300 Hz)

ELF   (10^{-8} ~ 10^{-7} K$_o$; 10,000 ~ 100,000 km; 3 ~ 30 Hz)

FIGURE 33.1   Electromagnetic wave spectrum.

$$\overline{k} = \hat{x}k_x + \hat{y}k_y + \hat{z}k_z \tag{33.19}$$

It is easily verified that the electric field

$$\overline{E}(\overline{r}, t) = \overline{E}_0 \cos(k_x x + k_y y + k_z z - \omega t) \tag{33.20}$$

is a solution to Eq. (33.16), where $\overline{E}_0$ is a constant vector.

The dispersion relation corresponding to Eq. (33.18) is obtained by substituting Eq. (33.20) in Eq. (33.16) which yields

$$k_x^2 + k_y^2 + k_z^2 = \omega^2 \mu \epsilon$$

We may write the solution, Eq. (33.20), in the following form:

$$\overline{E}(\overline{r}, t) = \overline{E}_0 \cos(\overline{k} \cdot \overline{r} - \omega t)$$

where

$$\overline{r} = \hat{x}x + \hat{y}y + \hat{x}x$$

is a position vector. The wave vector $\overline{k}$ is often referred to simply as the $\overline{k}$ vector.

Wavenumbers k

The wavenumber k is the magnitude of the wave vector $\overline{k}$ and is of more fundamental importance in electromagnetic wave theory than both of the more popular concepts of wavelength λ and frequency f. In Fig. 33.1, we illustrate the electromagnetic wave spectrum according to the free space wavenumber $k = k_o = \omega/c$. The corresponding values of frequency and wavelength are $f = ck_o/2\pi$ and $\lambda = 2\pi/k_o$. It is useful to define a fundamental unit K_o such that for free space $k_o = 1K_o = 2\pi$ m^{-1}. Thus $k_o = A\,K_o$ corresponds to $\lambda = 1/A$ m and $f = 3 \times 10^8\,A$ Hz. The photon energy in electronvolts is calculated from $\hbar\omega = \hbar ck$ where $\hbar = 1.05 \times 10^{-34}$ Joule-sec is Planck's constant divided by 2π and the electron charge is $q = 1.6 \times 10^{-19}$ C. Thus $\hbar\omega = (2\pi\hbar c/q)k_o \approx 1.26 \times 10^{-6}\,k_o$ and $k_o = A\,K_o$ corresponds to $1.26 \times 10^{-6}\,A$ eV.

Defining Terms

Electric field: State of a region in which charged bodies are subject to forces by virtue of their charge, the force acting on a unit positive charge.

Magnetic field: State produced by electric charge in motion and evidenced by a force exerted on a moving charge in the field.

Magnetic flux: Summation obtained by integrating flux density over an area.

Magnetic flux density: Measure of the strength and direction of a magnetic field at a point.

Reference

·   J.A. Kong, *Electromagnetic Wave Theory*, New York: Wiley-Interscience, 1990, chap. 1.

Further Information

IEEE Transactions on Microwave Theory and Techniques

IEEE Transactions on Antennas and Propagation

34

Magnetism and Magnetic Fields

34.1 Magnetism 811
      Static Magnetic Fields • Time-Dependent Electric and Magnetic
      Fields • Magnetic Flux Density • Relative Permeabilities • Forces on
      a Moving Charge • Time-Varying Magnetic Fields • Maxwell's
      Equations • Dia- and Paramagnetism • Ferromagnetism and
      Ferrimagnetism • Intrinsic Magnetic Properties • Extrinsic
      Magnetic Properties • Amorphous Magnetic Materials
34.2 Magnetic Recording 826
      Fundamentals of Magnetic Recording • The Recording Process •
      The Readback Process • Magnetic Recording Media • Magnetic
      Recording Heads • Conclusions

Geoffrey Bate
Santa Clara University

Mark H. Kryder
Carnegie Mellon University

34.1 Magnetism

Geoffrey Bate

Static Magnetic Fields

To understand the phenomenon of magnetism we must also consider electricity and vice versa. A stationary electric charge produces, at a point a fixed distance from the charge, a static (i.e., time-invariant) electric field. A moving electric charge, i.e., a current, produces at the same point a time-dependent electric field and a magnetic field, d**H**, whose magnitude is constant if the electric current, I, represented by the moving electric charge, is constant.

Fields from Constant Currents

Figure 34.1 shows that the direction of the magnetic field is perpendicular both to the current I and to the line, **R**, from the element d**L** of the current to a point, P, where the magnetic field, d**H**, is being calculated or measured.

$$d\mathbf{H} = I\,d\mathbf{L} \times \mathbf{R}/4\pi R^3 \quad \text{A/m when } I \text{ is in amps and } d\mathbf{L} \text{ and } R \text{ are in meters}$$

If the thumb of the right hand points in the direction of the current, then the fingers of the hand curl in the direction of the magnetic field. Thus, the stream lines of H, i.e., the lines representing at any point the direction of the H field, will be an infinite set of circles having the current as center. The magnitude of the field $H_\phi = I/2\pi R$ A/m. The line integral of H about any closed path around

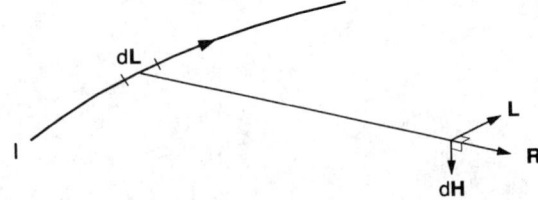

FIGURE 34.1 A current *I* flowing through a small segment dL of a wire produces at a distance **R** a magnetic field whose direction dH is perpendicular both to **R** and dL.

the current is $\oint \mathbf{H} \cdot \mathbf{dL} = I$. This relationship (known as Ampère's circuital law) allows one to find formulas for the magnetic field strength for a variety of symmetrical coil geometries, e.g.,

1. At a radius, ρ, between the conductors of a coaxial cable

$$H_\phi = I/2\pi\rho \quad \text{A/m}$$

2. Between two infinite current sheets in which the current, **K**, flows in opposite directions

$$\mathbf{H} = \mathbf{K} \times \mathbf{a_n}$$

where $\mathbf{a_n}$ is the unit vector normal to the current sheets

3. Inside an infinitely long, straight solenoid of diameter *d*, having *N* turns closely wound

$$\mathbf{H} = NI/d \quad \text{A/m}$$

4. Well inside a toroid of radius ρ, having *N* closely wound turns

$$\mathbf{H} = NI/2\pi\rho \cdot \mathbf{a_\phi} \quad \text{A/m}$$

Applying Stokes' theorem to Ampère's circuital law we find the point form of the latter.

$$\nabla \times \mathbf{H} = \mathbf{J}$$

where **J** is the current density in amps per square meter.

Time-Dependent Electric and Magnetic Fields

A constant current *I* produces a constant magnetic field **H** which, in turn, polarizes the medium containing **H**. While we cannot obtain isolated magnetic poles, it is possible to separate the "poles" by a small distance to create a magnetic dipole (i.e., to *polarize* the medium), and the dipole moment (the product of the pole strength and the separation of the poles) per unit volume is defined as the *magnetization M*. The units are emu/cc in the cgs system and amps per meter in the SI system of units. Because it is usually easier to determine the mass of a sample than to determine its volume, we also have a magnetization per unit mass, σ, whose units are emu/g or Am²/kg. The conversion factors between cgs and SI units in magnetism are shown in Table 34.1.

The effects of the static and time-varying currents may be summarized as follows:

$$\text{Static} \qquad [\mathbf{I}]_o \to [\mathbf{H}]_o \to [\mathbf{M}]_o$$
$$\text{motion}$$
$$\text{Time-varying} \qquad [\mathbf{I}]_t \to [\mathbf{H}]_t \to [\mathbf{M}]_t$$

where the suffixes "*o*" and "*t*" signify *static* and *time-dependent*, respectively.

Table 34.1 Units in Magnetism

Quality	Symbol	cgs Units	×	Factor	=	SI units
		$B = H + 4\pi M$				$B = \mu_o(H + M)$
Magnetic flux density	B	gauss (G)	×	10^{-4}	=	tesla (T), Wb/m^2
Magnetic flux	Φ	maxwell (Mx)	×	10^{-8}	=	webers (Wb)
		G · cm^2				
Magnetic potential difference (magnetomotive force)	U	gilbert (Gb)	×	$10/4\pi$	=	ampere (A)
Magnetic field strength	H	oersted (Oe)	×	$10^3/4\pi$	=	A/m
Magnetization (per volume)	M	emu/cc	×	10^3	=	A · m
Magnetization (per mass)	σ	emu/g	×	1	=	A · m^2/kg
Magnetic moment	m	emu	×	10^{-3}	=	A · m^2
Susceptibility (volume)	χ	dimensionless	×	4π	=	dimensionless
Susceptibility (mass)	κ	dimensionless	×	4π	=	dimensionless
Permeability (vacuum)	μ_o	dimensionless	×	$4\pi.10^{-7}$	=	Wb/A · m
Permeability (material)	μ	dimensionless	×	$4\pi.10^{-7}$	=	Wb/A · m
Bohr magneton	μ_B	$= 0.927 \times 10^{-20}$ erg/Oe	×	10^{-3}	=	Am^2
Demagnetizing factor	N	dimensionless	×	$1/4\pi$	=	dimensionless

Magnetic Flux Density

In the case of electric fields there is in addition to **E** an electric flux density field **D**, the lines of which begin on positive charges and end on negative charges. *D* is measured in coulombs per square meter and is associated with the electric field **E** (V/m) by the relation $\mathbf{D} = \epsilon_r\epsilon_o\mathbf{E}$ where ϵ_o is the *permittivity* of free space ($\epsilon_o = 8.854 \times 10^{-12}$ F/m) and ϵ_r is the (dimensionless) dielectric constant.

For magnetic fields there is a magnetic flux density **B** (Wb/m^2) $= \mu_r\mu_o\mathbf{H}$, where μ_o is the *permeability* of free space ($\mu_o = 4\pi \times 10^{-7}$ H/m) and μ_r is the (dimensionless) permeability. In contrast to the lines of the **D** field, lines of **B** are closed, having no beginning or ending. This is not surprising when we remember that while isolated positive and negative charges exist, no magnetic monopole has yet been discovered.

Relative Permeabilities

The range of the relative permeabilities covers about six orders of magnitude (Table 34.2) whereas the range of dielectric constants is only three orders of magnitude.

Forces on a Moving Charge

A charged particle, *q*, traveling with a velocity *v* and subjected to a magnetic field experiences a force

$$\mathbf{F} = q\mathbf{v} \times \mathbf{B}$$

This equation reveals how the Hall effect can be used to determine whether the majority current carriers in a sample of a semiconductor are (negatively charged) electrons flowing, say, in the negative direction or (positively charged) holes flowing in the positive direction. The (transverse) force (Fig. 34.2) will be in the same direction in either case, but the *sign* of the charge transported to the voltage probe will be positive for holes and negative for electrons.

In general, when both electric and magnetic fields are present, the force experienced by the carriers is given by

$$\mathbf{F} = q\,(\mathbf{E} + \mathbf{v} \times \mathbf{B})$$

Table 34.2 Relative Permeability, μ_r, of Some Diamagnetic, Paramagnetic, and Ferromagnetic Materials

Material	μ_r	M_s, A/m²
Diamagnetics		
Bismuth	0.999833	
Mercury	0.999968	
Silver	0.9999736	
Lead	0.9999831	
Copper	0.9999906	
Water	0.9999912	
Paraffin wax	0.99999942	
Paramagnetics		
Oxygen (s.t.p.)	1.000002	
Air	1.00000037	
Aluminum	1.000021	
Tungsten	1.00008	
Platinum	1.0003	
Manganese	1.001	
Ferromagnetics		
Purified iron: 99.96% Fe	280,000	2.158
Motor-grade iron: 99.6% Fe	5,000	2.12
Permalloy: 78.5% Ni, 21.5% Fe	70,000	2.00
Supermalloy: 79% Ni, 15% Fe,		
5% Mo, 0.5% Mn	1,000,000	0.79
Permendur: 49% Fe, 49% Ca, 2% V	5,000	2.36
Ferrimagnetics		
Manganese–zinc ferrite	750	0.34
	1,200	0.36
Nickel–zinc ferrite	650	0.29

Source: F. Brailsford, *Physical Principles of Magnetism*, London: Van Nostrand, 1966. With permission.

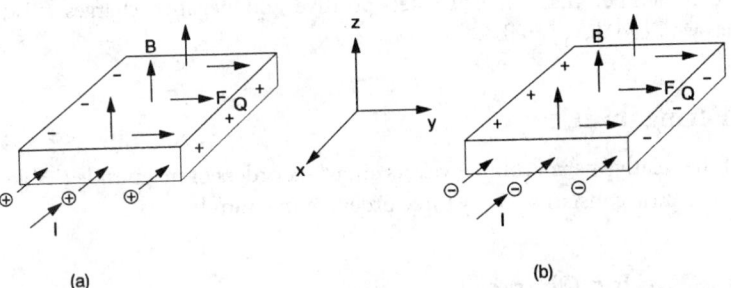

(a)   (b)

FIGURE 34.2 Hall effect. A magnetic field **B** applied to a block of semiconducting material through which a current I is flowing exerts a force $\mathbf{F} = \mathbf{V} \times \mathbf{B}$ on the current carriers (electrons or holes) and produces an electric charge on the right face of the block. The charge is positive if the carriers are holes and negative if the carriers are electrons.

The Hall effect is the basis of widely used and sensitive instruments for measuring the intensity of magnetic fields over a range of 10^{-5} to 2×10^6 A/m.

Time-Varying Magnetic Fields

In 1831, 11 years after Oersted demonstrated that a current produced a magnetic field which could deflect a compass needle, Faraday succeeded in showing the converse effect—that a magnetic field could produce a current. The reason for the delay between the two discoveries was that it is only when a magnetic field is changing that an emf is produced.

$$\text{emf} = -\frac{d\Phi}{dt} \text{ V}$$

where $\Phi = BS =$ flux density (in gauss) × area S. The time-changing flux, $d\Phi/dt$, can happen as a result of

1. A changing magnetic field within a stationary circuit
2. A circuit moving through a steady magnetic field
3. A combination of 1 and 2

The electrical circuit may have N turns and then

$$\text{emf} = -N\frac{d\Phi}{dt}$$

We can write emf = $\mathbf{E} \cdot d\mathbf{L}$ and in the presence of changing magnetic fields or a moving electrical circuit $\mathbf{E} \cdot d\mathbf{L}$ is no longer required to be equal to 0 as it was for stationary fields and circuits.

Maxwell's Equations

Because the flux Φ can be written $\int \mathbf{B} \cdot d\mathbf{s}$ we have emf = $\mathbf{E} \cdot d\mathbf{L} = -d/dt\, \mathbf{B} \cdot d\mathbf{s}$, and by using Stokes' theorem

$$(\nabla \times \mathbf{E}) \cdot d\mathbf{s} = -d\mathbf{B}/dt\, d\mathbf{s}$$

or

$$\nabla \times \mathbf{E} = -d\mathbf{B}/dt$$

That is, a spatially changing *electric field* produces a time-changing *magnetic field*. This is one of Maxwell's equations linking electric and magnetic fields.

By a similar argument it can be shown that

$$\nabla \times \mathbf{H} = \mathbf{J} + d\mathbf{D}/dt$$

This is another of Maxwell's equations and shows a spatially changing *magnetic field* produces a time-changing *electric field*. The latter $d\mathbf{D}/dt$ can be treated as an electric current which flows through a dielectric, e.g., in a capacitor, when an alternating potential is applied across the plates. This current is called the *displacement current* to distinguish it from the conduction current which flows in conductors. The *conduction current* involves the movement of electrons from one electrode to the other through the conductor (usually a metal). The *displacement current* involves no translation of electrons or holes but rather an alternating polarization throughout the dielectric material which is between the plates of the capacitor.

From the last two equations we see a key conclusion of Maxwell: that in electromagnetic fields a time-varying magnetic field produces a spatially varying electric field and a time-varying electric field produces a spatially varying magnetic field.

Maxwell's equations in point form, then, are

$$\nabla \times \mathbf{E} = d\mathbf{B}/dt$$

$$\nabla \times \mathbf{H} = \mathbf{J} + d\mathbf{D}/dt$$

$$\nabla \cdot \mathbf{D} = \rho_v$$

$$\nabla \cdot \mathbf{B} = 0$$

These equations are supported by the following auxiliary equations:

$\mathbf{D} = \epsilon\mathbf{E}$ (displacement = permittivity $\times$ electric field intensity)

$\mathbf{B} = \mu\mathbf{H}$ (flux density = permeability $\times$ magnetic field intensity)

$\mathbf{J} = \sigma\mathbf{E}$ (current density = conductivity $\times$ electric field strength)

$\mathbf{J} = \rho_v\mathbf{V}$ (current density = volume charge density $\times$ carrier velocity)

$\mathbf{D} = \epsilon_o\mathbf{E} + \mathbf{P}$ (displacement as function of electric field and polarization)

$\mathbf{B} = \mu_o(\mathbf{H} + \mathbf{M})$ (magnetic flux density as function of magnetic field strength and magnetization)

$\mathbf{P} = \chi_e\epsilon_o\mathbf{E}$ (polarization = electric susceptibility $\times$ permittivity of free space $\times$ electrical field strength)

$\mathbf{M} = \chi_m\mu_o\mathbf{H}$ (magnetization = magnetic susceptibility $\times$ permeability of free space $\times$ magnetic field strength)

The last two equations relate, respectively, the electric polarization $\mathbf{P}$ to the displacement $\mathbf{D} = \epsilon_o\mathbf{E}$ and the magnetic moment $\mathbf{M}$ to the flux density $\mathbf{B} = \mu_o\mathbf{H}$. They apply only to "linear" materials, i.e., those for which $\mathbf{P}$ is linearly related to $\mathbf{E}$ and $\mathbf{M}$ to $\mathbf{H}$. For magnetic materials we can say that non-linear materials are usually of greater practical interest.

Dia- and Paramagnetism

The phenomenon of magnetism arises ultimately from moving electrical charges (electrons). The movement may be orbital around the nucleus or the other degree of freedom possessed by electrons which, by analogy with the motion of the planets, is referred to as *spin*. In technologically important materials, i.e., ferromagnetics and ferrimagnetics, spin is more important than orbital motion. Each arrow in Fig. 34.3 represents the *total spin* of an atom.

An atom may have a permanent magnetic moment, in which case it is referred to as belonging to a paramagnetic material, or the atom may be magnetized only when in the presence of a magnetic field, in which case it is called *diamagnetic*. Diamagnetics are magnetized in the *opposite direction* to that of the applied magnetic field, i.e., they display *negative* susceptibility (a measure of the induced magnetization per unit of applied magnetic field). Paramagnetics are magnetized in the *same* direction as the applied magnetic field, i.e., they have *positive* susceptibility. *All* atoms are diamagnetic by virtue of their having electrons. *Some* atoms are also paramagnetic as well, but in this case they are called *paramagnetics* since paramagnetism is roughly a hundred times stronger than diamagnetism and overwhelms it. Faraday discovered that paramagnetics are attracted by a magnetic field and move toward the region of maximum field, whereas diamagnetics are repelled and move toward a field minimum.

The total magnetization of both paramagnetic and diamagnetic materials is zero in the absence of an applied field, i.e., they have zero **remanence**. Atomic paramagnetism is a necessary condition but not a sufficient condition for ferro- or ferrimagnetism, i.e., for materials having useful magnetic properties.

Ferromagnetism and Ferrimagnetism

To develop technologically useful materials, we need an additional force that ensures that the spins of the outermost (or almost outermost) electrons are mutually parallel. Slater showed that in iron, cobalt, and nickel this could happen if the distance apart of the atoms (D) was more than 1.5 times the diameter of the 3*d* electron shell (d). (These are the electrons, *near* the outside of atoms of iron,

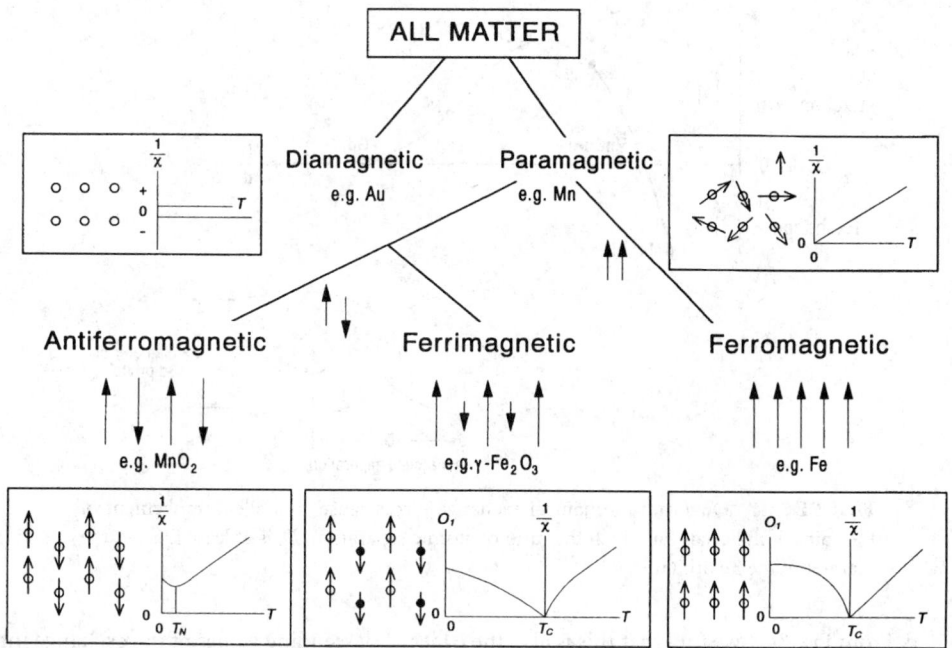

FIGURE 34.3 All matter consists of diamagnetic material (atoms having no permanent magnetic dipole moment) or paramagnetic material (atoms having magnetic dipole moment). Paramagnetic materials may be further divided into ferromagnetics, ferrimagnetics, and antiferromagnetics.

cobalt, and nickel, that are responsible for the strong paramagnetic moment of the atoms. Paramagnetism of the atoms is an essential prerequisite for ferro- or ferrimagnetism in a material.)

Slater's result suggested that, of these metals, iron, cobalt, nickel, and gadolinium should be ferromagnetic at room temperature, while chromium and manganese should not be ferromagnetic. This is in accordance with experiment. Gadolinium, one of the rare earth elements, is only weakly ferromagnetic in a cool room. Chromium and manganese in the elemental form narrowly miss being ferromagnetic. However, when manganese is alloyed with copper and aluminum ($Cu_{61}Mn_{24}Al_{15}$) to form what is known as a Heusler alloy [Crangle, 1962], it becomes ferromagnetic. The radius of the $3d$ electrons has not been changed by alloying, but the atomic spacing has been increased by a factor of $1.53/1.47$. This small change is sufficient to make the difference between positive exchange, parallel spins, and ferromagnetism and negative exchange, antiparallel spins, and antiferromagnetism.

For all ferromagnetic materials there exists a temperature (the **Curie temperature**) above which the thermal disordering forces are stronger than the exchange forces that cause the atomic spins to be parallel. From Table 34.3 we see that in order of descending Curie temperature we have Co, Fe,

Table 34.3 The Occurrence of Ferromagnetism

	Cr	Mn	Fe	Co	Ni	Gd
Atomic number	24	25	26	27	28	64
Atomic spacing/diameter	1.30	1.47	1.63	1.82	1.97	1.57
Ferromagnetic moment/mass (Am^2/kg)						
At 293 K	—	—	217.75	161	54.39	0
At 0 K	—	—	221.89	162.5	57.50	250
Curie point, Θ_c K	—	—	1,043	1,400	631	289
Néel temp., Θ_n K	475	100	—	—	—	—

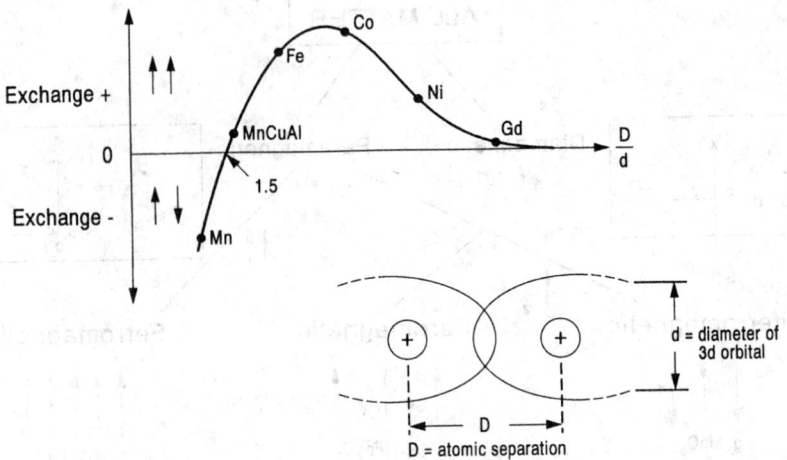

FIGURE 34.4 Quantum mechanical exchange forces cause a parallel arrangement of the spins of materials for which the ratio of atomic separation, D, is at least $1.5 \times d$, the diamter of the $3d$ orbital.

Ni, Gd. From Fig. 34.4 we find that this is also the order of descending values of the exchange integral, suggesting that high positive values of the exchange integral are indicative of high Curie temperatures rather than high magnetic intensity in ferromagnetic materials.

Negative values of exchange result in an antiparallel arrangement of the spins of adjacent atoms and in antiferromagnetic materials (Fig. 34.3). Until 5 years ago, it was true to say that antiferromagnetism had no practical application. Thin films on antiferromagnetic materials are now used to provide the bias field which is used to linearize the response of some magnetoresistive reading heads in magnetic disk drives. Ferrimagnetism, also illustrated in Fig. 34.3, is much more widely used. It can be produced as soft, i.e., low **coercivity**, ferrites for use in magnetic recording and reading heads or in the core of transformers operating at frequencies up to tens of megahertz. High-coercivity, single-domain particles (which are discussed later) are used in very large quantities to make magnetic recording tapes and flexible disks γ-Fe_2O_3 and cobalt-impregnated iron oxides and to make barium ferrite, the most widely used material for permanent magnets.

Intrinsic Magnetic Properties

Intrinsic magnetic properties are those properties that depend on the type of atoms and their composition and crystal structure, but not on the previous history of a particular sample. Examples of intrinsic magnetic properties are the *saturation* magnetization, Curie temperature, magnetocrystallic anisotropy, and magnetostriction.

Extrinsic magnetic properties depend on type, composition, and structure, but they also depend on the previous history of the sample, e.g., heat treatment. Examples of extrinsic magnetic properties include the technologically important properties of *remanent* magnetization, coercivity, and permeability. These properties can be substantially altered by heat treatment, quenching, cold-working the sample, or otherwise changing the size of the magnetic particle.

A ferromagnetic or ferrimagnetic material, on being heated, suffers a reduction of its magnetization (per unit mass, i.e., σ, and per unit volume, M). The slope of the curve of M_s vs. T increases with increasing temperature as shown in Fig. 34.5. This figure represents the conflict between the ordering tendency of the exchange interaction and the disordering effect of increasing temperature. At the Curie temperature, the order no longer exists and we have a paramagnetic material. The change from ferromagnetic or ferrimagnetic materials to paramagnetic is completely

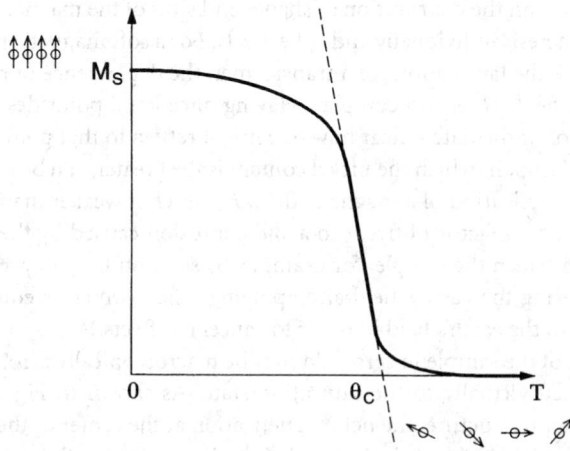

FIGURE 34.5 Ferro- and ferrimagnetic materials lose their spontaneous magnetic moment at temperatures above the Curie temperature, Θ_c.

reversible on reducing the temperature to its initial value. Curie temperatures are always lower than melting points.

A single crystal of iron has the body-centered structure at room temperature. If the magnetization as a function of applied magnetic field is measured, the shape of the curve is found to depend on the direction of the field. This phenomenon is *magnetocrystalline anisotropy*. Iron has body-centered structure at room temperature, and the "easy" directions of magnetization are those directions parallel to the cube edges [100], [010], and [001] or, collectively, <100>. The hard direction of magnetization for iron is the body diagonal [111]. At higher temperatures, the anisotropy becomes smaller and disappears above 300°C.

Nickel crystals (face-centered cubic) have an easy direction of [111] and a hard direction of [100]. Cobalt has the hexagonal close-packed (HCP) structure and the hexagonal axis is the easy direction at room temperature.

Magnetocrystalline anisotropy plays a very important part in determining the coercivity of ferro- or ferrimagnetic materials, i.e., the field value at which the direction of magnetization is reversed.

Many magnetic materials change dimensions on becoming magnetized: the phenomenon is known as *magnetostriction* and can be positive, i.e., length increases, or negative. Magnetostriction plays an important role in determining the preferred direction of magnetization of *soft*, i.e., low H_c, films such as those of alloys of nickel and iron, known as *Permalloy*.

The origin of both magnetocrystalline anisotropy and magnetostriction is *spin-orbit* coupling. The magnitude of the magnetization of the film is controlled by the electron spin as usual, but the preferred direction of that magnetization with respect to the crystal lattice is determined by the electron orbits which are large enough to interact with the atomic structure of the film.

Extrinsic Magnetic Properties

Extrinsic magnetic properties are those properties that depend not only on the shape and size of the sample, but also on the shape and size of the magnetic constituents of the sample. For example, if we measure the hysteresis loop like the one shown in Fig. 34.6 on a disk-shaped sample punched from a magnetic recording tape, the result will depend not only on the diameter and thickness of the disk coating but also on the distribution of shapes and sizes of the magnetic particles within the disk. They display hysteresis individually and collectively. For a soft magnetic material, i.e., one that might be used to make the laminations of a transformer, the dependence of magnetization, M, on the applied magnetic field, H, is also complex. Having once left a point described by the coordinates (H_1, M_1), it is not immediately clear how one might return to that point.

Alloys of nickel and iron, in which the nickel content is the greater, can be capable of a reversal of magnetization by the application of a magnetic field, H, which is weaker than the earth's magnetic field (0.5 Oe, 40 A/m) by a factor of five. (To avoid confusion caused by the geomagnetic field it would be necessary to screen the sample, for example, by surrounding it by a shield of equally soft material or by measuring the earth's field and applying a field which is equal in magnitude but opposite in direction to the earth's field in order to cancel its effects.)

The magnetization of the sample in zero field may be macroscopically zero, but locally the material may be magnetized virtually to the saturation state. As shown in Fig. 34.6, which shows a greatly simplified domain structure, the net magnetization at the center of the loop is zero because the magnetization of the four "domains" cancels in pairs. A domain is a region (not necessarily square or even of a regular shape, although the shape often is regular in Ni–Fe thin films or sheets) over which the magnetization is constant in magnitude and direction. Thus, the sample in Fig. 34.6 consists of four domains, initially separated from each other by "domain walls." If a magnetic field is applied in the direction of $+H$, that domain will grow whose direction of magnetization is closest to the field direction and the domains will shrink if their magnetization is opposed to the field. For small applied fields, the movement of the walls is reversible, i.e., on reducing the applied field to zero, the original domain configuration will be obtained. Beyond a certain field the movement of the walls is irreversible, and eventually near the knee of the magnetization curve all the domain walls have been swept away by the applied field. The sample is not yet in the saturated state since the direction of M is not quite the same as the direction of the applied field. However, a small increase in the strength of the applied field finally achieves the saturated state by rotating the magnetization of the whole sample into the field direction.

On removing the applied field, the sample does not retrace the magnetization curve, and when the applied field is zero, we can see that a considerable amount of magnetization remains, M_r. Appropriately, this is referred to as the remanent state, and M_r is the *remanent magnetization*. By

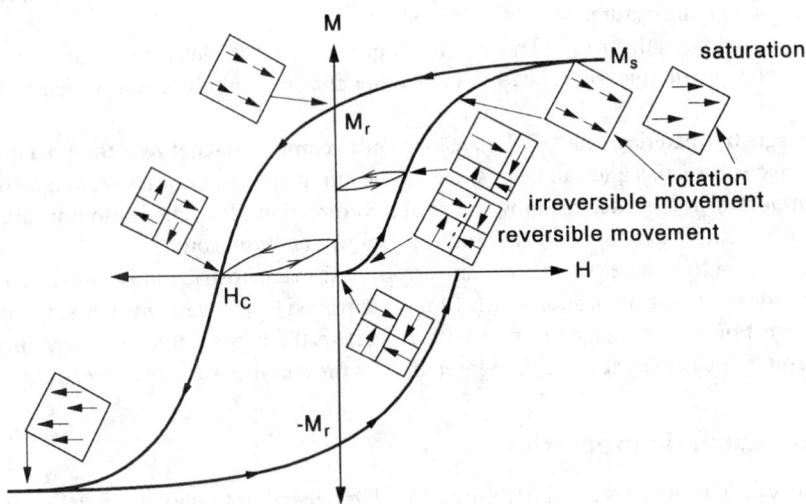

FIGURE 34.6 In soft magnetic materials domains form such that the total magnetization is zero. By applying small magnetic fields, domain walls move and the magnetization changes.

reversing the original direction of the applied field, domains reappear and the magnetization is eventually reduced to zero at the *coercive field*, H_c. It should be noticed that, at H_c, although the net magnetization is clearly zero, the individual domains may be magnetized in directions that are different from those at the starting point. Figure 34.6 shows an incomplete hysteresis loop. If the field H were increased beyond $+H_c$ the loop would be completed.

The differences between ideally magnetically soft materials (used in transformers and magnetic read/write heads) and magnetically hard materials (used in permanent magnets and in recording tapes and disks) are as follows:

Magnetically soft materials: H_c 0; M_r 0; M_s high value

Magnetically hard materials: H_c high value; M_r/M_s ("squareness"); M_s high value

Examples are given in Table 34.4.

It is noticeable that the differences between hard and soft magnetic materials are confined to the extrinsic properties, M_r, H_c, and permeability, μ. The latter is related to M and H as follows:

$$\mathbf{B} \text{ (G)} = \mathbf{H} \text{ (Oe)} + 4\pi\mathbf{M} \text{ (emu/cc)}$$

$$\mu = \mathbf{B/H} = 1 + 4\pi\kappa \text{ (cgs units)}$$

or

$$\mathbf{B} \text{ (Wb/m}^2\text{)} = \mu_o \text{ (}\mathbf{H} \text{ A/m} + \mathbf{M} \text{ A/m)}$$

$$= \mu_o\mu_r\mathbf{H} = \mu\mathbf{H} \text{ (SI units)}$$

Domain walls form in order to minimize the magnetic energy of the sample. The magnetic energy is $\mu H^2/8\pi$ cgs units or $1/2\ \mu H^2$ J/m^3 (SI units) and clearly depends on $\mathbf{H}$, the magnetic field emanating from the sample. In the *initial* domain configuration shown in Fig. 34.6, there is no net magnetization of the sample and thus no substantial $\mathbf{H}$ exists outside the sample and the magnetostatic energy is zero. Thus, the establishment of domains *reduces* the energy associated with H but it *increases* the energy needed to establish domain walls within the sample. A compromise is reached in which domain walls are formed until the establishment of one more wall would *increase*, rather than decrease, the total magnetic energy of the sample.

The wall energy depends on the area of the wall, i.e., L^2, while the energy associated with the external magnetic field depends on L^3, the volume of the sample. Clearly, as the size of single

Table 34.4 "Hard" and "Soft" Magnetic Materials

	High M_s	Low H_c	Low M_r	High μ
Soft				
Fe	1700 emu/cc	1 Oe	< 500	20,000
80 Ni 20 Fe	660	0.1	< 300	50,000
Mn Zn ferrite	400	0.02	< 200	5,000
$Co_{70}Fe_5Si_{15}B_{10}$	530	0.1	< 250	10,000

	High M_s	High H_c	High M_r	T_c
Hard				
Particles				
γ-Fe_2O_3	400	250–450	200–300	115–126
CrO_2	400	450–600	300	120
Fe	870–1100	1,100–1,500	435–550	768
$BaO.6Fe_2O_3$	238–370	800–3,000	143–260	320
Alloys				
$SmCo_5$	875	40,000	690	720
Sm_2Co_{17}	1,000	17,000	875	920
$Fe_{14}BNd_2$	1,020	12,000	980	310

particles becomes small, terms in L^2 are more important than terms in L^3, and so for small magnetic particles, the formation of domain walls may not be energetically feasible and a *single-domain particle* results. These are found in the particles of iron oxide, cobalt-modified iron oxide, chromium dioxide, iron, or barium ferrite, which are used to make magnetic recording tapes, and in barium ferrite, samarium cobalt, and neodymium iron boron, which are used to make powerful permanent magnets. In the latter cases, the very high coercivities are caused by domain walls being pinned at grain boundaries between the main phase grains and finely precipitated secondary phases. This is an example of *nucleation-controlled coercivity*.

The amount of available energy that can be stored in a permanent magnet is the area of the largest rectangle that can be drawn in the second quadrant of the *B* vs. *H* hysteresis loop. The *energy product* has grown remarkably by a factor of about 50 since 1900 [Strnat, 1986]. We see from the graph in Fig. 34.7 of *intrinsic* coercive force, i.e., the coercive force obtained from the graph of *M* vs. *H* (in contrast to the smaller coercive force obtained by plotting *B* vs. *H*), that increases in H_c (rather than increases in M_r) have been responsible for almost all the improvement in the energy product.

The key attributes of technologically important magnetic materials are

1. Large, spontaneous atomic magnetic moments
2. Large, positive exchange integrals
3. Magnetic anisotropy and heterogeneity which are small for soft magnetic materials and large for hard magnetic materials.

In single-domain materials, the magnetic particles are so small that reversal of the magnetization can only occur by rotation of the magnetization vector. This rotation can be resisted by combinations of three anisotropies: crystalline anisotropy, shape anisotropy, and magnetoelastic anisotropy (which depends on the magnetostrictive properties of the material).

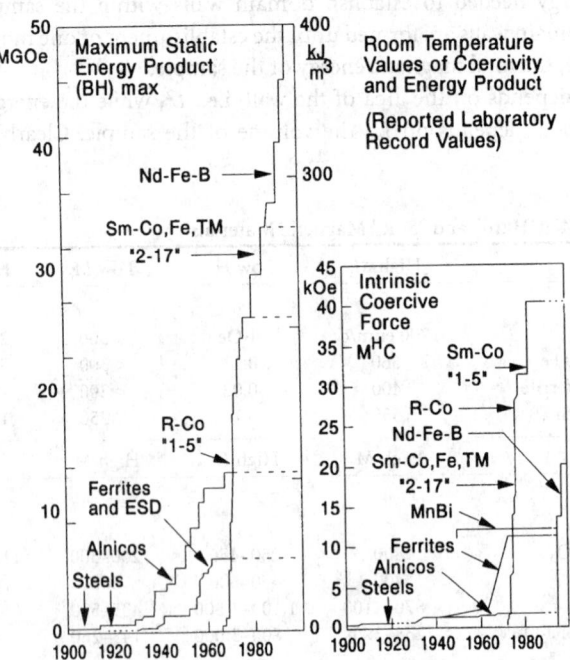

FIGURE 34.7 The development of magnetic materials for permanent magnets showing the increase in energy product and in intrinsic coercivity as a function of time.

Crystalline Anisotropy

Crystalline anisotropy arises from the existence of easy and hard directions of magnetization within the crystal structure of the material. For example, in iron the {100} directions are easy directions while the {111} directions are hard. In nickel crystals the reverse is true. In cobalt the hexagonal {00.1} directions are easy and the {10.0} directions in the basal plane are hard.

Hard and easy directions in crystalline materials come about as a result of spin-orbit coupling. The spin, as usual, determines the occurrence of ferro- or ferrimagnetization, while the orbital motion of the electrons (3d in the case of Fe, Co, and Ni) responds to the structure of the crystal lattice.

The maximum value of coercivity resulting from crystalline anisotropy is given by $H_c = 2K_1/M_s$, where K_1 is the first magnetocrystalline anisotropy constant and M_s is the saturation magnetization.

Shape Anisotropy

A spherical particle has no shape anisotropy, i.e., all directions are equally easy (or hard). For particles (having low crystalline anisotropy) of any other shape, the longest dimension is the easy direction and the shortest dimension is the hardest direction of magnetization. Thus, a needle-shaped (acicular) particle will tend to be magnetized along the long dimension, whereas a particle in the form of a disk will have the axis of the disk as its hard direction, while any direction in the plane of the disk will be equally easy (assuming that shape is the dominant anisotropy).

For an acicular particle, the maximum value of the particle's switching field is [Stoner and Wohlfarth, 1948]

$$H_c = (N_b - N_a) M_s$$

where N_b is the demagnetizing factor in the shorter dimension and N_a is the factor for the long axis of the particle. When the ratio $b/a \to \infty$, then

$$\left. \begin{array}{c} N_a \to 0 \\ N_b \to 2\pi \end{array} \right\} N_a + N_b + N_c \equiv 4\pi$$

and H_c for iron > 10,000 Oe (7.95×10^5 A/m), higher than has been achieved in the laboratory for single-domain iron particles. Particles of iron having $H_c \leq 2000$ Oe are widely used in high-quality audio and video tapes.

The reason for the discrepancy is that the simplest single-domain model makes the assumption that the spins on all the atoms in a particle rotate in the same direction and at the same time, i.e., are coherent. This seems to be improbable since switching may begin at different places in the single-domain particle at the same time. Jacobs and Bean [1955] proposed an incoherent mode, *fanning*, in which different segments on a longitudinal chain of atoms rotate in opposite directions. Shtrikman and Treves [1959] introduced another incoherent mode, *buckling*. These incoherent modes of magnetization reversal within single-domain particles not only predicted values of coercivity closer to the observed values, but also they could explain why the observed coercivity values for single-domain particles increased with decreasing particle size [Bate, 1980].

Shape anisotropy also plays an important role in determining the magnetization direction in thin magnetic films. It, of course, favors magnetization in the film plane.

Magnetoelastic Anisotropy

Spin-orbit coupling is also responsible for magnetostriction (the increase or decrease of the dimensions of a body on becoming magnetized or demagnetized). The magnetostriction coefficient $\lambda_s =$ fractional change of a dimension of the body. It can be positive or negative, and it varies with changes in the direction and magnitude of the applied stress (or internal stress) and of the applied magnetic field. It is highly sensitive to composition, to structure, and to the previous history of the sample. The maximum coercivity is given by the formula $H_c = 3\lambda_s T / M_s$ where λ_s is the saturation magnetostriction coefficient, T is the tension, and M_s the saturation magnetization. Magnetostriction has been put to practical use in the generation of sonar waves for the detection of schools of fish or submarines.

For samples made of single-domain particles, the maximum coercivity for three ferromagnetic metals and one ferrimagnetic oxide (widely used in magnetic recording) is calculated using the preceding formula. Table 34.5 shows maximum coercivity (Oe) for single-domain particles (coherent rotation).

The assumption is made that all the spins rotate so that they remain parallel at all times. This is known as *coherent rotation*. In the case of γ-Fe$_2$O$_3$, an incoherent mode of reversal probably occurs since the maximum observed coercivity is only 350 Oe. Several incoherent modes have been proposed, e.g., chain-of-spheres fanning [Jacobs and Bean, 1955], curling [Shtrikman and Treves, 1959]. Their characteristics and differences are discussed by Bate [1980]. The coercivity of particles of γ-Fe$_2$O$_3$ is increased (in order to make recording tapes of extended frequency response) by precipitating cobalt hydroxide on the surface of the particles. After gentle warming, the cobalt is incorporated on the surface of the particles and increases the coercivity to 650 Oe (51 · 73 kA/m).

Figure 34.8 illustrates two additional extrinsic magnetic properties of importance. They are the remanence coercivity, H_r, and the switching field distribution (SFD). These are of particular importance in magnetic particles used in magnetic tapes (audio, video, or data) or magnetic disks. The coercivity, H_c, of a magnetic material is the value of the magnetic field (the major loop) at which $M = 0$. However, if the applied field is allowed to go to zero, a small magnetization remains. It is necessary to increase the applied field from $H_c > H_r$ (Fig. 34.8) to achieve $M_r = 0$. H_r is the *remanence coercivity* and is more relevant than is H_c in discussing the writing process in magnetic

Table 34.5 Maximum Coercivity (Oe) for Single-Domain Particles (Coherent Rotation)

	Iron	Cobalt	Nickel	γ-Fe$_2$O$_3$
Crystalline	250	3,000	70	230
Strain	300	300	2,000	<10
Shape (10:1)	5,300	4,400	1,550	2,450

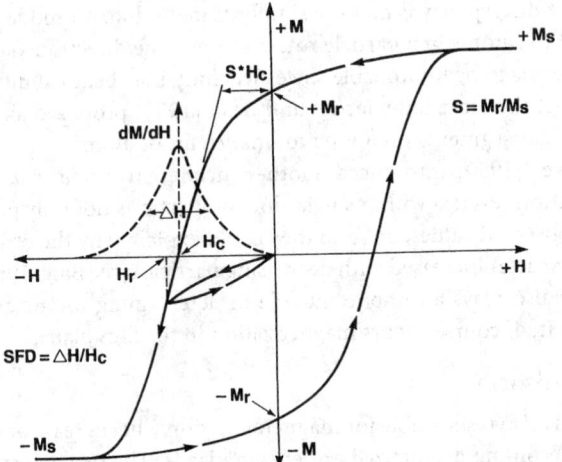

FIGURE 34.8 The switching field distribution (SFD) for a magnetic recording medium can be obtained in two ways: (1) SFD = $\Delta H / H_c$ or (2) SFD = $1 - S^*$.

recording since it corresponds to the center of the remanent magnetization transition on the recording medium.

Particles, for example, in a magnetic recording do not all reverse their magnetization at the same field; there is a distribution of switching fields, which can be found by differentiating M with respect to H around the point H_c. The result is shown as the broken curve on Fig. 34.8, where the SFD $= \Delta H / H_c$, ΔH being the width at half the maximum of the curve. Typically, SFD $= 0.2$–0.3 for a high-quality particulate medium and smaller than this for thin-film recording media.

Figure 34.8 also shows the construction used to find the parameter $S*$. The quantity $1 - S*$ is found to be very close to $\Delta H / H_c$, and it is quicker to evaluate. Either of these parameters can be used to determine the distribution of switching fields, small values of which are required to achieve high recording densities.

Amorphous Magnetic Materials

Before 1960, all the known ferro- and ferrimagnetic materials were crystalline. Because the occurrence of these magnetic states is known to depend on short-range interaction between atoms, there is no reason why amorphous materials (which have *only* short-range order) should not have useful magnetic properties. This was found to be true in 1960, when thin, amorphous ribbons of $Au_{81}Si_{19}$ were made by rapidly cooling the molten alloy through the melting point and the lower glass transition temperature. Because there is always at least one crystalline phase more stable than the amorphous state, the problem is to invent a production method that yields the amorphous phase rather than the crystalline one.

Most methods involve cooling the molten mixture so rapidly that there is insufficient time for crystals to form. Cooling rates of 10^5–10^6 degrees per second are needed and can be achieved in several ways:

1. Pouring the molten mixture from a silica crucible onto the edge of a rapidly rotating copper wheel. This yields a ribbon of the amorphous alloy, typically 1 mm wide and 25 μm thick.
2. Depositing a thin film from a metal vapor or a solution of metal ions.
3. Irradiating a thin sample of the metal with high-energy particles.

Once an amorphous alloy is formed, it will remain indefinitely in the *glassy* state at room temperature. The problem is that only *thin* films are obtained, and large areas are required to make, for example, the core of a transformer.

There are three main groups of amorphous films:

1. Metal-metalloid alloys, e.g., $Au_{81}Si_{19}$
2. Late transition–early transition metal alloys, e.g., $Ni_{60}Nb_{40}$
3. Simple metal alloys, e.g., $Cu_{65}Al_{35}$

When normal metals freeze, crystallization begins at a fixed temperature, the *liquidus*, T_e. In amorphous alloys *configurational freezing* occurs at a lower temperature, the *glass temperature*, T_g, which is not as well defined as T_e. There is an abrupt increase in the time required for the rearrangement of the atoms, from 10^{-12} s for liquids to 10^5 s (a day) for glasses. Not surprisingly, this increase in atomic rearrangement time is associated with an abrupt increase in viscosity, from 10^{-2} poise for liquids, e.g., water or mercury, to 10^{15} poise for glasses.

The principal difference between magnetic glasses and ferromagnetic alloys is that the glasses are completely isotropic (all directions of magnetization are very easy directions), and consequently, considering only the magnetic properties, soft amorphous alloys are almost ideally suited for use in the core of power transformers or magnetic recording heads, where almost zero remanence and coercivity are desired at frequencies up to megahertz. The limit on their performance seems to be the magnetic anisotropy which arises from strains generated during the manufacturing process.

When an amorphous material is required to store energy (as in a permanent magnet) or information (as in magnetic bubbles or thermomagneto-optic films), it must have magnetic anisotropy.

This is generally produced by applying a magnetic field at high temperatures to the amorphous material. The field and temperature must be high enough to allow a local rearrangement of atoms to take place in order to create the desired degree of magnetocrystalline anisotropy.

Amorphous materials will apparently play increasingly important roles as magnetic materials. To accelerate their use, we need to have answers to the questions "What governs the formation of amorphous materials?" and "What is the origin of their anisotropy and magnetostriction?".

Defining Terms

Coercivity, H_c (Oe, A/m):   The property of a magnetized body enabling it to resist reversal of its magnetization.

Compensation temperature, T_c (°C, K):   The temperature at which the magnetization of a material comprising ferromagnetic atoms (e.g., Fe, Co, Ni) and rare earth atoms (e.g., Gd, Tb) becomes zero because the magnetization of the sublattice of ferromagnetic atoms is canceled by the opposing magnetization of the rare earth sublattice.

Curie temperature, Θ_c (°C, K):   The temperature at which the spontaneous magnetization of a ferromagnetic or ferrimagnetic body becomes zero.

Remanence, M_r (emu/cc, A/m):   The property of a magnetized body enabling it to retain its magnetization.

References

G. Bate, in *Recording Materials in Ferromagnetic Materials*, vol. 2, Amsterdam: North-Holland, 1980, pp. 381–507.

G. Bate, *J. Magnetism and Magnetic Materials*, vol. 100, pp. 413–424, 1991.

F. Brailsford, *Physical Principles of Magnetism*, London: Van Nostrand, 1966.

J. Crangle, "Ferromagnetism and antiferromagnetism in non-ferrous metals and alloys," *Metallurgical Reviews*, pp. 133–174, 1962.

I. S. Jacobs and C. P. Bean, *Phys. Rev.*, vol. 100, p. 1060, 1955.

K. Moorjani and J. M. D. Coey, *Magnetic Glasses: Methods and Phenomena, Their Application in Science and Technology*, vol. 6, Amsterdam: Elsevier, 1984.

S. Shtrikman and D. Treves, *J. Phys. Radium*, vol. 20, p. 286, 1959.

J.C. Slater, *Phys. Rev.*, vol. 36, p. 57, 1930.

E. C. Stoner and E. P. Wohlfarth, *Phil. Trans. Roy. Soc.*, vol. A240, p. 599, 1948.

K. J. Strnat, *Proceedings of Symposium on Soft and Hard Magnetic Materials with Applications*, vol. 8617-005, Metals Park, Ohio: American Society of Metals, 1986.

Further Information

A substantial fraction of the papers published in English on the technologically important aspects of magnetism appear in the *IEEE Transactions on Magnetics* or in the *Journal of Magnetism and Magnetic Materials*.

The two major annual conferences are Intermag (proceedings published in the *IEEE Transactions on Magnetics*) and the Magnetism and Magnetic Materials Conference, MMM (proceedings published in the American Physical Society's *Journal of Applied Physics*).

34.2 Magnetic Recording

Mark H. Kryder

Magnetic recording is used in a wide variety of applications and formats, ranging from relatively low-density, low-cost floppy disk drives and audio recorders to high-density videocassette recorders, digital audio tape recorders, computer tape drives, rigid disk drives, and instrumentation recorders. The storage density of this technology has been advancing at a very rapid pace. With a storage density exceeding 100 Mbits/in.^{2}, magnetic recording media today can store the equivalent of about 6000 pages of text on one square inch. This is more than 50,000 times the storage density on the RAMAC, which was introduced in 1957 by IBM as the first disk drive for storage of digital information. The original Seagate 5.25-inch magnetic disk drive, introduced in 1980, stored just 5 Mbytes. Today, instead of storing megabytes, 5.25-inch drives store gigabytes, and drives as small as 1.3 inches store 40 Mbytes.

This astounding rate of progress shows no sign of slowing. Fundamental limits to magnetic recording density are still several orders of magnitude away, and recent product announcements and laboratory demonstrations indicate the industry is accelerating the rate of progress rather than approaching practical limits. In the past few years, both IBM [Tsang *et al.*, 1990] and Hitachi [Futamoto *et al.*, 1991] have demonstrated the feasibility of storing information at densities beyond 1 Gbit/in.^{2} on a disk. Similar advances can also be expected in audio and video recording.

Fundamentals of Magnetic Recording

Although magnetic recording is practiced in a wide variety of formats and serves a wide variety of applications, the fundamental principles by which it operates are similar in all cases. The fundamental magnetic recording configuration is illustrated in Fig. 34.9. The recording head consists of a toroidally shaped core of soft magnetic material with a few turns of conductor around it. The magnetic medium below the head could be either tape or disk, and the substrate could be either flexible (for tape and floppy disks) or rigid (for rigid disks). To record on the medium, current is applied to the coil around the core of the head, causing the high-permeability magnetic core to magnetize. Because of the gap in the recording head, magnetic flux emanates from the head and penetrates the medium. If the field produced by the head is sufficient to overcome the **coercive force** of the medium, the medium will be magnetized by the head field. Thus, a representation of the current waveform applied to the head is stored in the magnetization pattern in the medium.

Readout of previously recorded information is typically accomplished by using the head to sense the magnetic stray fields produced by the recorded patterns in the medium. The recorded patterns in the medium cause magnetic stray fields to emanate from the medium and to flow through the core of the head. Thus, if the medium is moved with respect to the head, the flux passing through the coil around the head will change in a manner which is representative of the recorded magnetization

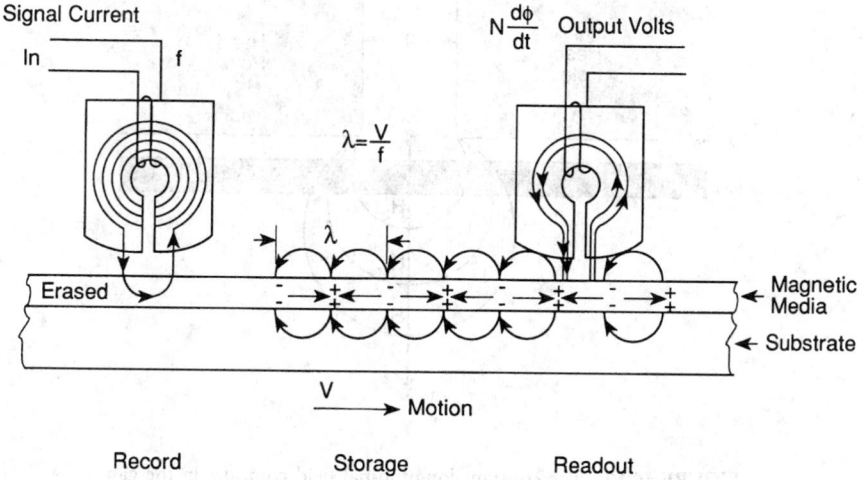

FIGURE 34.9 The fundamental magnetic recording configuration.

pattern in the medium. By Faraday's law of induction, a voltage representative of the recorded information is thus induced in the coil.

The Recording Process

During recording the head is used to produce large magnetic fields which magnetize the medium. It was shown by Karlqvist [1954] that, in the case where the track width and length of the poles along the gap are both large compared to the gap length, the fields produced by a recording head could be described by

$$H_x = \frac{NI}{2\pi g} \left[\tan^{-1}\left(\frac{x + g/2}{y} \right) - \tan^{-1}\left(\frac{x - g/2}{y} \right) \right] \tag{34.1a}$$

$$H_y = \frac{NI}{2\pi g} \ln \left[\frac{(x + g/2)^2 + y^2}{(x - g/2)^2 + y^2} \right] \tag{34.1b}$$

where H_x and H_y are the longitudinal and perpendicular components of field, as indicated by the coordinates in Fig. 34.10, N is the number of turns on the head, I is the current driving the head, and g is the gap width of the head. In this approximation, the contours of equal longitudinal field are described by circles which intersect the gap corners as shown in Fig. 34.10.

In digital or saturation recording, the recording head is driven with sufficiently large currents that a portion of the recording medium is driven into saturation. However, because of the gradient in the head fields, other portions of the medium see fields less than those required for saturation. This is illustrated in Fig. 34.10 where the contours for three different longitudinal fields are drawn. In this figure H_{cr} is the **remanence coercivity** or the field required to produce zero **remanent magnetization** in the medium after it was saturated in the opposite direction, and H_1 and H_2 are fields which would produce negative and positive remanent magnetization, respectively. Note that the head field gradient is the sharpest near the pole tips of the head. This means that smaller head-to-medium spacing and thinner medium both lead to narrower transitions being recorded.

Modeling the recording process involves convolving the head field contours with the very non-

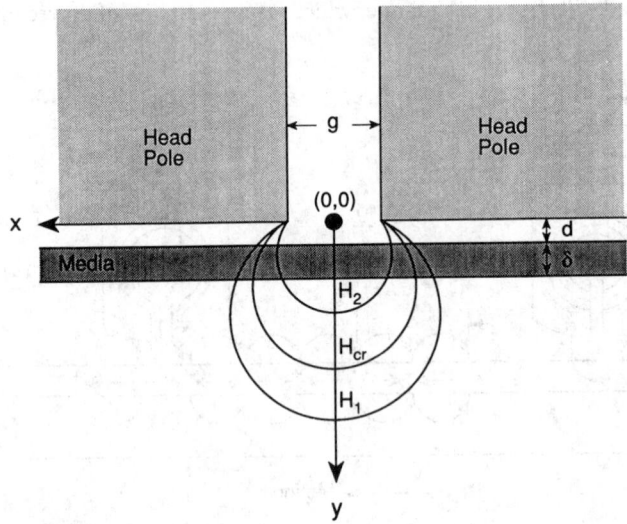

FIGURE 34.10 The constant longitudinal field contours in the gap region of a recording head.

linear and hysteretic magnetic properties of the recording medium. A typical M-H hysteresis loop for a longitudinal magnetic recording medium is shown in Fig. 34.11. Whether the medium has positive or negative magnetization depends upon not only the magnetic field applied but the past history of the magnetization. If the medium was previously saturated at $-M_s$, then when the magnetic field H is reduced to zero, the remanent magnetization will be $-M_r$; however, if it was previously saturated at $+M_s$, then the remanent magnetization would be $+M_r$. Similarly, if the medium was initially saturated to $-M_s$, then magnetized by a field $+H_1$, and finally allowed to go to a remanent state, the magnetization would go to value M_1. This hysteretic behavior is the basis for the use of the medium for long-term storage of information but makes the recording process highly nonlinear.

An additional complicating factor in determining the actual recorded pattern is the **demagnetizing field** of the medium itself. As shown in Fig. 34.9, transitions in the recorded magnetization direction produce effective magnetostatic charge given by

$$\rho_M = -\nabla \cdot \vec{M} \tag{34.2}$$

which in turn results in demagnetizing fields. The demagnetizing fields outside the medium are what is sensed by the head during readback, but demagnetizing fields also exist inside the medium and act to alter the total field seen by the medium during the recording process from that of the head field alone.

Taking into account the head field gradients, the nonlinear M-H loop characteristics of the medium and the demagnetizing fields, Williams and Comstock [1971] developed a model for the recording process. This model predicts the width of a recorded transition, in a material with a square hysteresis loop, to be

$$a = \sqrt{\frac{M_r \delta d}{\pi H_c}} \tag{34.3}$$

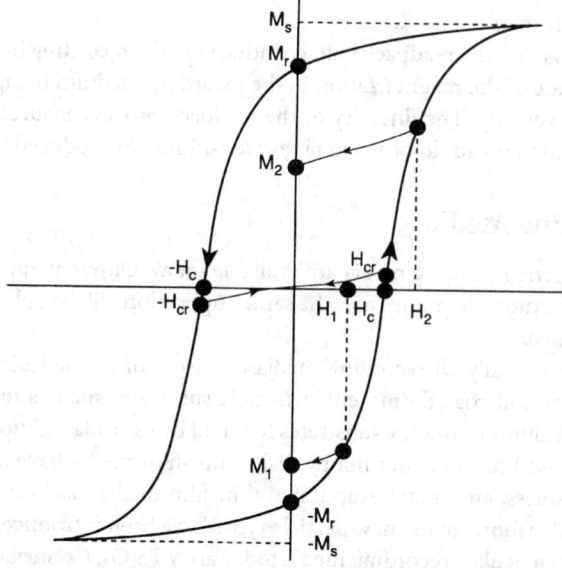

FIGURE 34.11 A remanent M-H hysteresis loop for a longitudinal recording medium.

where δ is the medium thickness, d is the head-medium spacing, M_r is the remanent magnetization of the medium, and H_c is the coercivity of the medium. That the transition widens with the product $M_r \cdot \delta$ is a result of the fact that the demagnetizing fields increase linearly with this quantity. Similarly, the transition narrows as H_c is increased, because with high coercivity, the medium can resist the transition broadening due to the demagnetizing fields. The increase in transition width with d is due to the fact that a poorer head field gradient is obtained with larger head-medium spacing.

The nonlinearities of the recording process can be largely removed by a technique referred to as ac bias recording. This is frequently used in analog recording in audio and video recorders. In this technique, a high-frequency ac bias is added to the signal to be recorded. This ac bias signal is ramped from a value much larger than the coercivity of the medium to zero. This removes the hysteretic behavior of the medium and causes it to assume a magnetization state which represents the minimum energy state determined by the amount of field produced by the signal to be recorded.

The Readback Process

As opposed to the recording process, the readback process can usually be modeled as a linear process. This is because the changes in magnetization which occur in either the head or the medium during readback are typically small.

The most common way to model the readback process is to use the principle of reciprocity, which states that the flux produced by the head through a cross section of an element of the medium, normalized by the number of ampere turns of current driving the head, is equal to the flux produced in the head by the element of medium, normalized by the equivalent current required to produce the magnetization of that element. For a magnetic recording head, which produces the longitudinal field $H_x(x,y)$ when driven by NI ampere turns of current, this principle leads to the following expression for the voltage induced in the head by a recording medium with magnetization $M(x,y)$ and moving with velocity v relative to the head:

$$e = \frac{\mu_o W v}{I} \int_d^{d+\delta} \int_{-\infty}^{\infty} \frac{\partial M_x(x - \bar{x}, y)}{\partial \bar{x}} H_x(x, y) dx dy \qquad (34.4)$$

where W is the track width of the head.

This expression shows that the readback voltage induced in the recording head is linearly dependent upon the magnitude of the magnetization in the recording medium being sensed and the relative head-to-medium velocity. The linearity of the readback process ensures that analog recordings such as those recorded on audio or video tapes are faithfully reproduced.

Magnetic Recording Media

A wide variety of magnetic recording media are available today. Different applications require different media, but furthermore, in many cases the same application will be able to utilize a variety of different competitive media.

Just a decade ago, essentially all recording media consisted of fine acicular magnetic particles embedded in a polymer and coated onto either flexible substrates such as mylar for floppy disks and tapes or onto rigid aluminum-alloy substrates for rigid disks. Today, although such particulate media are still widely used for tape and floppy disks, thin-film media have almost entirely taken over the rigid disk business, and metal-evaporated thin-film media has been introduced into the tape marketplace. Furthermore, many new particle types have been introduced.

The most common particulate recording media today are γ-Fe_2O_3, Co surface-modified γ-Fe_2O_3, CrO_2, and metal particle media. All of these particles are acicular in shape with aspect ratios on the order of 5 or 10 to 1. The particles are sufficiently small that it is energetically most favorable for them to remain in a single-domain saturated state. Because of demagnetizing effects caused by the acicular shape, the magnetization prefers to align along the long axis of the particle.

As was noted in the discussion of Eq. (34.3), to achieve higher recording densities requires media with higher coercivity. The coercivity of a particle is determined by the field required to cause the magnetization to switch by 180°. If the magnetization remained in a single-domain state during the switching process, then the coercivity should be given by

$$H_c = (N_a - N_b)M_s \qquad (34.5)$$

where N_a and N_b are the demagnetizing factors in the directions transverse and parallel to the particle axis, respectively. In practice the coercivity is measured to be less than this. This has been explained as being a result of the fact the magnetization does not remain uniform during the switching process, but switches inhomogeneously [White, 1984]. In addition to the effects which the shape anisotropy of the particles has on the coercivity, crystalline anisotropy can also be used to control coercivity.

The coercivity of the medium which is made from the particles is determined by the distribution of coercivities of the particles from which it is made, their orientation in the medium relative to the fields from the head, and their interactions among each other. The coercivities of a variety of particulate recording media are summarized in Table 34.6.

Although coercivity is indeed an important parameter for magnetic recording media, it is by no means the only one. Particle size affects the medium noise because, at any time, the head is sensing a fixed volume of the medium. Because the particles are quantized and there are statistical variations in their switching behavior, the medium power signal-to-noise ratio varies linearly with the number of particles contained in that volume. To reduce particulate medium noise, it is therefore generally desirable to use small particles.

There is a limit, however, to how small particles may be made and still remain stable. When the thermal energy kT is comparable in magnitude to the energy required to switch a particle, $M \cdot H_c$, the particle becomes unstable and may switch because of thermal excitation. This phenomenon is known as **superparamagnetism** and can lead to decay of recorded magnetization patterns over time.

The remanent magnetization of a medium is important because it directly affects the signal level during readback as shown by Eq. (34.4). The remanent magnetizations of several particulate media are listed in Table 34.6. Obtaining high remanent magnetization in particulate media requires the use of particles with high **saturation magnetization** and a high-volume packing fraction of particles in the polymer binder. Obtaining a high-volume packing fraction of particles in the binder, however, can lead to nonuniform distributions of particles and agglomerates of many particles, which switch together, also causing noise during readback.

Generally, then, to obtain good high-density particulate recording media it is desired to have adequate coercivity (to achieve the required recording density), small particles (for low noise), with a very narrow switching field distribution (to obtain a narrow transition), to have them oriented along the direction of recording (to obtain a large remanence), and to have them uniformly dispersed (to obtain low modulation noise), with high packing density (to obtain large signals).

Thin-film recording media generally have excellent magnetic properties for high-density recording. Because they are nearly 100% dense (voids at the grain boundaries reduce the density somewhat), they can be made to have the highest possible magnetization. Because of their high magnetization, they can be made extremely thin and still provide adequate signal during readback. This helps narrow the recorded transition since the head field gradient is sharper for thinner media, as was discussed in reference to Fig. 34.10.

Thin-film media can also be made extremely smooth. To achieve the smallest possible head-to-medium spacing and therefore the sharpest head field gradient and the least spacing loss, smooth media are required.

The coercivity of thin-film media can also be made very high. In volume production today are media with coercivities of 100 kA/m; however, media with coercivities to 250 kA/m have been

Table 34.6 Magnetic Material, Saturation Remanence $M_r(\infty)$, Coercivity H_c, Switching-Field Distribution Δh_r, and Number of Particles per Unit Volume, N, of Various Particulate Magnetic Recording Media

Application	Material	$M_r(\infty)$, kA/m (emu/cm^3)	H_c, kA/m (4π Oe)	Δh_r	N, $10^3/\mu m^3$
Reel-to-reel audio tape	γ-Fe$_2$O$_3$	100–120	23–28	0.30–0.35	0.3
Audio tape IEC I	γ-Fe$_2$O$_3$	120–140	27–32	0.25–0.35	0.6
Audio tape IEC II	CrO$_2$	120–140	38–42	0.25–0.35	1.4
	γ-Fe$_2$O$_3$ + Co	120–140	45–52	0.25–0.35	0.6
Audio tape IEC IV	Fe	230–260	80–95	0.30–0.37	3
Professional video tape	γ-Fe$_2$O$_3$	75	24	0.4	0.1
	CrO$_2$	110	42	0.3	1.5
	γ-Fe$_2$O$_3$ + Co	90	52	0.35	1
Home video tape	CrO$_2$	110	45–50	0.35	2
	γ-Fe$_2$O$_3$ + Co	105	52–57	0.35	1
	Fe	220	110–120	0.38	4
Instrumentation tape	γ-Fe$_2$O$_3$	90	27	0.35	0.6
	γ-Fe$_2$O$_3$ + Co	105	56	0.50	0.8
Computer tape	γ-Fe$_2$O$_3$	87	23	0.30	0.16
	CrO$_2$	120	40	0.29	1.4
Flexible disk	γ-Fe$_2$O$_3$	56	27	0.34	0.3
	γ-Fe$_2$O$_3$ + Co	60	50	0.34	0.5
Computer disk	γ-Fe$_2$O$_3$	56	26–30	0.30	0.3
	γ-Fe$_2$O$_3$ + Co	60	44–55	0.30	0.5

Source: E. Köster and T.C. Arnoldussen, "Recording media," in *Magnetic Recording*, C.D. Mee and E.D. Daniel, Eds., New York: McGraw-Hill, 1987. With permission.

made and appear promising [Velu and Lambeth, 1992]. Such high coercivities are adequate to achieve more than an order of magnitude higher recording density than today.

Numerical models indicate that noise in thin-film media increases when the grains in polycrystalline films are strongly exchange coupled [Zhu and Bertram, 1988]. Exchange coupled films tend to exhibit zigzag transitions, which produce considerable jitter in the transition position relative to the location where the record current in the head goes through zero. A variety of experimental studies have indicated that the introduction of nonmagnetic elements which segregate to the grain boundaries and careful control of the sputtering conditions to achieve a porous microstructure at the grain boundaries reduce such transition jitter [Chen and Yamashita, 1988].

Magnetic Recording Heads

Early recording heads consisted of toroids of magnetically soft ferrites, such as NiZn-ferrite and MnZn-ferrite, with a few turns of wire around them. For high-density recording applications, however, ferrite can no longer be used, because the saturation magnetization of ferrite is limited to about 400 kA/m. Saturation of the pole tips of a ferrite head begins to occur when the deep gap field in the head approaches one-half the saturation magnetization of the ferrite. Because the fields seen by a medium are one-half to one-quarter the deep gap field, media with coercivities above about 80 kA/m cannot be reliably written with a ferrite head. High-density thin-film disk media, metal particle media, and metal evaporated media, therefore, cannot be written with a ferrite head.

Magnetically soft alloys of metals such as Permalloy (NiFe) and Sendust (FeAlSi) have saturation magnetizations on the order of 800 kA/m, about twice that of ferrites, but because they are metallic may suffer from eddy current losses when operated at high frequencies. To overcome the limitations imposed by eddy currents, they are used in layers thinner than a skin depth at their operating frequency. To prevent saturation of the ferrite heads, the high magnetization metals are applied to the pole faces of the ferrite, making a so-called metal-in-gap or MiG recording head, as shown in Fig. 34.12. Since the corners of the pole faces are the first parts of a ferrite head to satu-

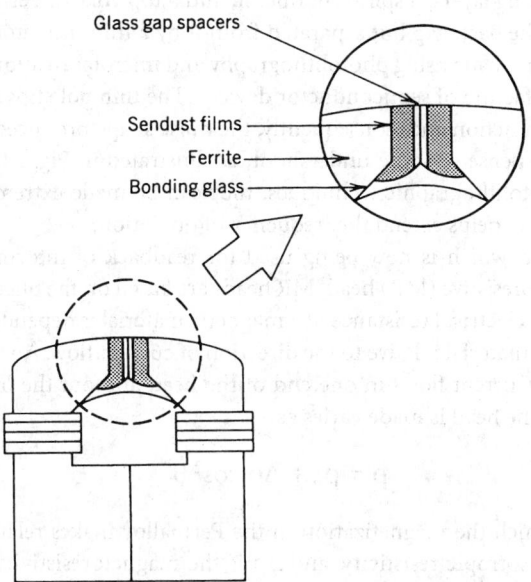

FIGURE 34.12 A diagram of a metal-in-gap or MiG recording head. (*Source:* A.S. Hoagland and J.E. Monson, *Digital Magnetic Recording*, 2nd ed., New York: Wiley-Interscience, 1991, p. 127. With permission.)

rate, the high magnetization metals enable these MiG heads to be operated to nearly twice the field to which a ferrite head can be operated. Because the layer of metal is thin, it can furthermore be less than a skin depth, and eddy current losses do not limit performance at high frequencies.

Yet another solution to the saturation problem of ferrite heads is to use thin-film heads. Thin-film heads are made of Permalloy and are therefore metallic, but the films are made sufficiently thin that they are thinner than the skin depth and, consequently, the heads operate well at high frequencies. A diagram of a thin-film head is shown in Fig. 34.13. It consists of a bottom yoke of

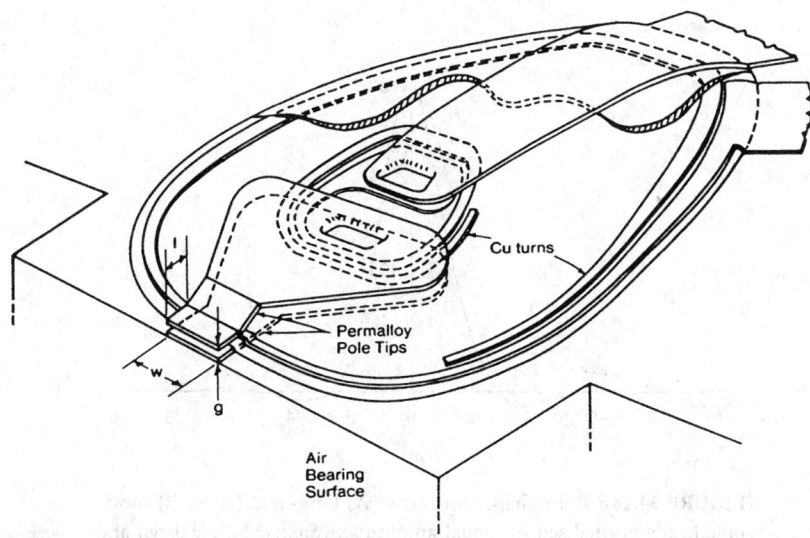

FIGURE 34.13 A thin-film head. (*Source:* R.M. White, Ed., *Introduction to Magnetic Recording*, New York: IEEE Press, p. 28. ©1985 IEEE.)

Permalloy, some insulating layers, a spiral conductor, and a top yoke of Permalloy, which is joined to the bottom yoke at the back gap but separated from it by a thin insulator at the recording gap. These thin-film heads are made using photolithography and microfabrication techniques similar to those used in the manufacture of semiconductor devices. The thin pole tips of these heads actually sharpen the head field function and, consequently, the pulse shape produced by an isolated transition, although at the expense of some undershoot, as illustrated in Fig. 34.14. Because thin-film heads are made by photolithographic techniques, they can be made extremely small and to have low inductance. This, too, helps extend the frequency of operation.

A relatively new head which is now being used for readback of information in high-density recording is the magnetoresistive (MR) head. MR heads are based on the phenomenon of **magneto-resistance**, in which the electrical resistance of a magnetic material is dependent upon the direction of magnetization in the material relative to the direction of current flow. An unshielded MR head is depicted in Fig. 34.15. Current flows in one end of the head and out the other. The resistivity of Permalloy from which the head is made varies as

$$\rho = \rho_o + \Delta\rho \cos^2 \theta \tag{34.6}$$

where θ is the angle which the magnetization in the Permalloy makes relative to the direction of current flow, ρ_o is the isotropic resistivity, and $\Delta\rho$ is the magnetoresistivity. When the recording medium with a changing magnetization pattern moves under the MR head, the stray fields from the medium cause a change in the direction of magnetization and, consequently, a change in resistance in the head. With a constant current source driving the head, the head will therefore exhibit a change in voltage across its terminals.

Magnetoresistive heads are typically more sensitive than inductive heads and therefore produce larger signal amplitudes during readback. The increased sensitivity and the fact that the read head is independent of the write head can be used to make a write/read head combination in which the write head writes a wider track than the read head senses. Thus, adjacent track interference is reduced during the readback process.

Another advantage of the MR head is that it senses magnetic flux ϕ, not the time rate of change of flux $d(\phi)/dt$ as an inductive head does. Consequently, whereas the inductive head output voltage is dependent upon the head-to-medium velocity as was shown by Eq. (34.4), the output voltage of an MR head is independent of velocity.

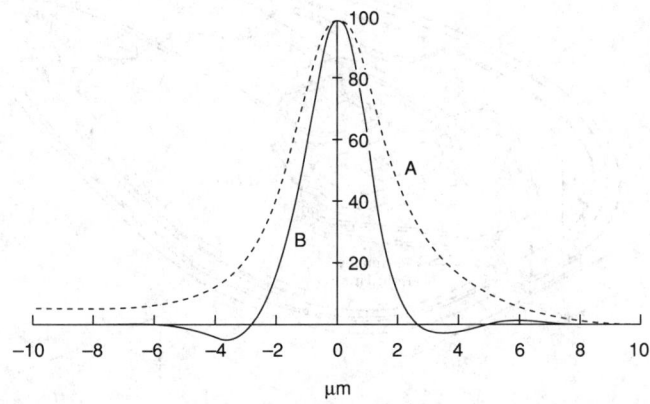

FIGURE 34.14 Pulse shapes for (curve A) long- and (curve B) short-pole heads normalized for equal amplitude. (*Source:* E.P. Valstyn and L.F. Shew, "Performance of single-turn film heads," *IEEE Trans. Magnet.*, vol. MAG-9, no. 3, p. 317. ©1973 IEEE.)

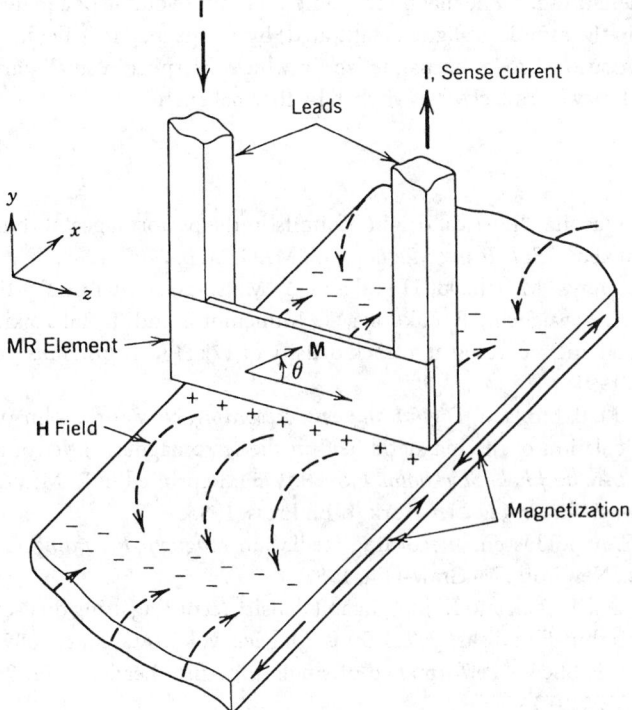

FIGURE 34.15 An unshielded magnetoresistive element. (*Source:* A.S. Hoagland and J.E. Monson, *Digital Magnetic Recording,* 2nd ed., New York: Wiley-Interscience, 1991, p. 131. With permission.)

Conclusions

Magnetic recording today is used in a wide variety of formats for a large number of applications. Formats range from tape, which has the highest volumetric packing density and lowest cost per bit stored, to rigid disks, which provide fast access to a large volume of data. Applications include computer data storage, audio and video recording, and collecting data from scientific instruments.

The technology has increased storage density by more than a factor of 50,000 over the past 35 years since it was first used in a disk format for computer data storage; however, fundamental limits, set by superparamagnetism, are estimated yet to be a factor of 50,000 from where we are today. Furthermore, recent product announcements and developments in research labs suggest that the rate of progress is likely to accelerate. Storage densities of over 1 Gbit/in.^{2} are likely by the end of this decade and densities of 10 Gbit/in.^{2} appear likely in the early twenty-first century.

Defining Terms

Coercive force or coercivity: The magnetic field required to reduce the mean magnetization of a sample to zero after it was saturated in the opposite direction.

Demagnetizing field: The magnetic field produced by divergences in the magnetization of a magnetic sample.

Magnetoresistance: The resistance change produced in a magnetic sample when its magnetization is changed.

Remanent coercivity: The magnetic field required to produce zero remanent magnetization in a material after the material was saturated in the opposite direction.

Remanence magnetization: The magnetic moment per unit volume of a material in zero field.

Saturation magnetization: The magnetic moment per unit volume of a material when the magnetization in the sample is aligned (saturated) by a large magnetic field.

Superparamagnetism: A form of magnetism in which the spins in small particles are exchange coupled but may be collectively switched by thermal energy.

References

T. Chen and T. Yamashita, "Physical origin of limits in the performance of thin-film longitudinal recording media," *IEEE Trans. Magnet.*, vol. MAG-24, p. 2700, 1988.

M. Futamoto, F. Kugiya, M. Suzuki, H. Takano, Y. Matsuda, N. Inaba, Y. Miyamura, K. Aragi, T. Nakao, H. Sawaguchi, F. Fukuoka, T. Munemoto, and T. Takagaki, "Investigation of 2 Gb/in² magnetic recording at a track density of 17kTPI," *IEEE Trans. Magnet.*, vol. MAG-27, p. 5280, 1991.

A. S. Hoagland and J. E. Monson, *Digital Magnetic Recording*, New York: John Wiley & Sons, 1991.

O. Karlqvist, "Calculation of the magnetic field in the ferromagnetic layer of a magnetic drum," *Trans. Roy. Inst. Technol., Stockholm*, No. 86, 1954. Reprinted in R. M. White, Ed., *Introduction to Magnetic Recording*, New York: IEEE Press, 1985.

E. Köster and T. C. Arnoldussen, "Recording media," in *Magnetic Recording*, C. D. Mee and E. D. Daniel, Eds., New York: McGraw-Hill, 1987.

M-M. Tsang, T. Chen, T. Yogi, and K. Ju, "Gigabit density recording using dual-element MR/inductive heads on thin-film disks," *IEEE Trans. Magnet.*, vol. MAG-26, p. 1689, 1990.

E. P. Valstyn and L. F. Shew, "Performance of single-turn film heads," *IEEE Trans. Magnet.*, vol. MAG-9, p. 317, 1973.

E. Velu and D. Lambeth, "High Density Recording on SmCo/Cr Thin Film Media," Paper KA-01, Intermag Conference, St. Louis, April 1992; to be published in *IEEE Trans. Magnet.*, vol. MAG-28, 1992.

R. M. White, *Introduction to Magnetic Recording*, New York: IEEE Press, 1984, p. 14.

M. L. Williams and R. L. Comstock, "An analytical model of the write process in digital magnetic recording," *AIP Conf. Proc.*, part 1, no. 5, pp. 738–742, 1971.

J-G. Zhu and H. N. Bertram, "Recording and transition noise simulations in thin film media," *IEEE Trans. Magnet.*, vol. MAG-24, p. 2706, 1988.

Further Information

There are several books which provide additional information on magnetic and magneto-optic recording. They include the following:

R. M. White, *Introduction to Magnetic Recording*, New York: IEEE Press, 1984.

C. D. Mee and E. D. Daniel, *Magnetic Recording*, New York: McGraw-Hill, 1987.

A. S. Hoagland and J. E. Monson, *Digital Magnetic Recording*, New York: John Wiley and Sons, 1991.

35

Wave Propagation

Matthew N. O. Sadiku
Temple University

Kenneth Demarest
University of Kansas

35.1 Space Propagation 837
   Propagation in Simple Media • Propagation in the Atmosphere
35.2 Waveguides 849
   Waveguide Modes • Rectangular Waveguides • Circular Wave-
   guides • Commercially Available Waveguides • Waveguide Losses •
   Mode Launching

35.1 Space Propagation

Matthew N. O. Sadiku

This section summarizes the basic principles of electromagnetic (EM) **wave propagation** in space. The principles essentially state how the characteristics of the earth and the atmosphere affect the propagation of EM waves. Understanding such principles is of practical interest to communication system engineers. Engineers cannot competently apply formulas or models for communication system design without an adequate knowledge of the propagation issue.

Propagation of an EM wave may be regarded as a means of transferring energy or information from one point (a transmitter) to another (a receiver). EM wave propagation is achieved through guided structures such as transmission lines and waveguides or through space. Wave propagation through waveguides and microstrip lines will be treated in Section 35.2. In this section, our major focus is on EM wave propagation in space and the power resident in the wave.

For a clear understanding of the phenomenon of EM wave propagation, it is expedient to break the discussion of propagation effects into categories represented by four broad frequency intervals [Collin, 1985]:

- Very low frequencies (VLF), 3–30 kHz
- Low-frequency (LF) band, 30–300 kHz
- High-frequency (HF) band, 3–30 MHz
- Above 50 MHz

In the first range, wave propagates as in a waveguide, using the earth's surface and the ionosphere as boundaries. Attenuation is comparatively low, and hence VLF propagation is useful for long-distance worldwide telegraphy and submarine communication. In the second frequency range, the availability of increased bandwidth makes standard AM broadcasting possible. Propagation in this band is by means of surface wave due to the presence of the ground. This third range is useful for long-range broadcasting services via sky wave reflection and refraction by the ionosphere. Basic problems in this band include fluctuations in the ionosphere and a limited usable frequency range. Frequencies above 50 MHz allow for line-of-sight space wave propagation, FM radio and TV channels, radar and navigation systems, and so on. In this band, due consideration must be given to reflection from the ground, refraction by the troposphere, scattering by atmospheric hydrometeors, and **multipath** effects of buildings, hills, trees, etc.

EM wave propagation can be described by two complementary models. The physicist attempts a theoretical model based on universal laws, which extends the field of application more widely than currently known. The engineer prefers an empirical model based on measurements, which can be used immediately. This section presents complementary standpoints by discussing theoretical factors affecting wave propagation and the semiempirical rules allowing handy engineering calculations. First, we consider wave propagation in idealistic simple media, with no obstacles. We later consider the more realistic case of wave propagation around the earth, as influenced by its curvature and by atmospheric conditions.

Propagation in Simple Media

The conventional propagation models, on which the basic calculation of radio links is based, result directly from Maxwell's equations:

$$\nabla \cdot \mathbf{D} = \rho_v \tag{35.1}$$

$$\nabla \cdot \mathbf{B} = 0 \tag{35.2}$$

$$\nabla \times \mathbf{E} = -\frac{\partial \mathbf{B}}{\partial t} \tag{35.3}$$

$$\nabla \times \mathbf{H} = \mathbf{J} + \frac{\partial \mathbf{D}}{\partial t} \tag{35.4}$$

In these equations, $\mathbf{E}$ is electric field strength in volts per meter, $\mathbf{H}$ is magnetic field strength in amperes per meter, $\mathbf{D}$ is electric flux density in coulombs per square meter, $\mathbf{B}$ is magnetic flux density in webers per square meter, $\mathbf{J}$ is conduction current density in amperes per square meter, and ρ_v is electric charge density in coulombs per cubic meter. These equations go hand in hand with the constitutive equations for the medium:

$$\mathbf{D} = \epsilon \mathbf{E} \tag{35.5}$$

$$\mathbf{B} = \mu \mathbf{H} \tag{35.6}$$

$$\mathbf{J} = \sigma \mathbf{E} \tag{35.7}$$

where $\epsilon = \epsilon_o \epsilon_r$, $\mu = \mu_o \mu_r$, and σ are the permittivity, the permeability, and the conductivity of the medium, respectively.

Consider the general case of a lossy medium which is charge-free ($\rho_v = 0$). Assuming time-harmonic fields and suppressing the time factor $e^{j\omega t}$, Eqs. (35.1) to (35.7) can be manipulated to yield Helmholtz's wave equations

$$\nabla^2 E - \gamma^2 E = 0 \tag{35.8}$$

$$\nabla^2 H - \gamma^2 H = 0 \tag{35.9}$$

where $\gamma = \alpha + j\beta$ is the **propagation constant**, α is the *attenuation constant* in nepers per meter or decibels per meter, and β is the *phase constant* in radians per meter. Constants α and β are given by

$$\alpha = \omega \sqrt{\frac{\mu \epsilon}{2} \left[\sqrt{1 + \left(\frac{\sigma}{\omega \epsilon}\right)^2} - 1 \right]} \tag{35.10}$$

$$\beta = \omega \sqrt{\frac{\mu\epsilon}{2}\left[\sqrt{1 + \left(\frac{\sigma}{\omega\epsilon}\right)^2} + 1\right]} \tag{35.11}$$

where $\omega = 2\pi f$ is the frequency of the wave. The wavelength λ and wave velocity u are given in terms of β as

$$\lambda = \frac{2\pi}{\beta} \tag{35.12}$$

$$u = \frac{\omega}{\beta} = f\lambda \tag{35.13}$$

Without loss of generality, if we assume that wave propagates in the z-direction and the wave is polarized in the x-direction, solving the wave equations (35.8) and (35.9) results in

$$\mathbf{E}(z,t) = E_o e^{-\alpha z} \cos(\omega t - \beta z)\mathbf{a}_x \tag{35.14}$$

$$\mathbf{H}(z,t) = \frac{E_o}{|\eta|} e^{-\alpha z} \cos(\omega t - \beta z - \theta_\eta)\mathbf{a}_y \tag{35.15}$$

where $\eta = |\eta| \angle\theta_\eta$ is the *intrinsic impedance* of the medium and is given by

$$|\eta| = \frac{\sqrt{\mu/\epsilon}}{\sqrt[4]{1 + \left(\frac{\sigma}{\omega\epsilon}\right)^2}}, \qquad \tan 2\theta_\eta = \frac{\sigma}{\omega\epsilon}, \qquad 0 \le \theta_\eta \le 45° \tag{35.16}$$

Equations (35.14) and (35.15) show that as the EM wave travels in the medium, its amplitude is attenuated according to $e^{-\alpha z}$, as illustrated in Fig. 35.1. The distance δ through which the wave amplitude is reduced by a factor of e^{-1} (about 37%) is called the *skin depth* or *penetration depth* of the medium, i.e.,

$$\delta = \frac{1}{\alpha} \tag{35.17}$$

The power density of the EM wave is obtained from the Poynting vector

$$\mathbf{P} = \mathbf{E} \times \mathbf{H} \tag{35.18}$$

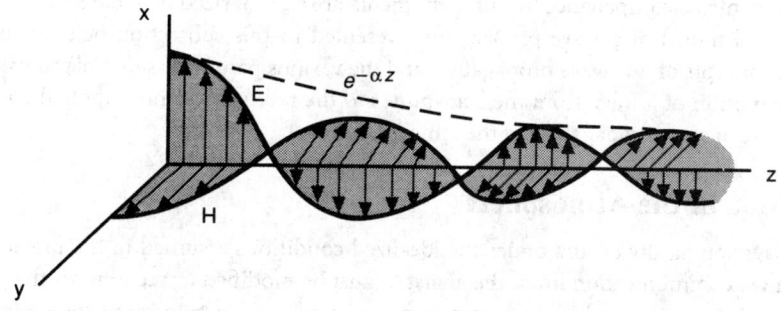

FIGURE 35.1 The magnetic and electric field components of a plane wave in a lossy medium.

with the time-average value of

$$P_{ave} = \frac{1}{2} \text{Re}(E \times H^*)$$

$$= \frac{E_o^2}{2|\eta|} e^{-2\alpha z} \cos \theta_\eta a_z$$

(35.19)

It should be noted from Eqs. (35.14) and (35.15) that **E** and **H** are everywhere perpendicular to each other and also to the direction of propagation. Thus, the wave described by Eqs. (35.14) and (35.15) is said to be *plane-polarized*, implying that the electric field is always parallel to the same plane (the *xz*-plane in this case) and is perpendicular to the direction of propagation. Also, as mentioned earlier, the wave decays as it travels in the *z*-direction because of loss. This loss is expressed in the *complex relative permittivity* of the medium

$$\epsilon_c = \epsilon_r' - j\epsilon_r'' = \epsilon_r\left(1 - j\frac{\sigma}{\omega\epsilon}\right)$$

(35.20)

and measured by the *loss tangent*, defined by

$$\tan \delta = \frac{\epsilon_r''}{\epsilon_r'} = \frac{\sigma}{\omega\epsilon}$$

(35.21)

The imaginary part $\epsilon_r'' = \sigma/\omega\epsilon_o$ corresponds to the losses in the medium. The refractive index of the medium *n* is given by

$$n = \sqrt{\epsilon_c}$$

(35.22)

Having considered the general case of wave propagation through a lossy medium, we now consider wave propagation in other types of media. A medium is said to be a good conductor if the loss tangent is large ($\sigma \gg \omega\epsilon$) or a lossless or good dielectric if the loss tangent is very small ($\sigma \ll \omega\epsilon$). Thus, the characteristics of wave propagation through other types of media can be obtained as special cases of wave propagation in a lossy medium as follows:

1. Good conductors: $\sigma \gg \omega\epsilon$, $\epsilon = \epsilon_o$, $\mu = \mu_o\mu_r$
2. Good dielectric: $\sigma \ll \omega\epsilon$, $\epsilon = \epsilon_o\epsilon_r$, $\mu = \mu_o\mu_r$
3. Free space: $\sigma = 0$, $\epsilon = \epsilon_o$, $\mu = \mu_o$

where $\epsilon_o = 8.854 \times 10^{-12}$ F/m is the free-space permittivity, and $\mu_o = 4\pi \times 10^{-7}$ H/m is the free-space permeability.

The conditions for each medium type are merely substituted in Eqs. (35.10) to (35.21) to obtain the wave properties for that medium. The formulas for calculating attenuation constant, phase constant, and intrinsic impedance for different media are summarized in Table 35.1.

The classical model of a wave propagation presented in this subsection helps us understand some basic concepts of EM wave propagation and the various parameters that play a part in determining the motion of a wave from the transmitter to the receiver. We now apply the ideas to the particular case of wave propagation in the atmosphere.

Propagation in the Atmosphere

Wave propagation hardly occurs under the idealized conditions assumed in the previous subsection. For most communication links, the analysis must be modified to account for the presence of the earth, the ionosphere, and atmospheric precipitates such as fog, raindrops, snow, and hail. This will be done in this subsection.

Table 35.1 Attenuation Constant, Phase Constant, and Intrinsic Impedance for Different Media

	Lossy Medium	Good Conductor $\sigma/\omega\epsilon \gg 1$	Good Dielectric $\sigma/\omega\epsilon \ll 1$	Free Space
Attenuation constant α	$\omega\sqrt{\dfrac{\mu\epsilon}{2}\left[\sqrt{1+\left(\dfrac{\sigma}{\omega\epsilon}\right)^2}-1\right]}$	$\sqrt{\dfrac{\omega\mu\sigma}{2}}$	$\simeq 0$	0
Phase constant β	$\omega\sqrt{\dfrac{\mu\epsilon}{2}\left[\sqrt{1+\left(\dfrac{\sigma}{\omega\epsilon}\right)^2}+1\right]}$	$\sqrt{\dfrac{\omega\mu\sigma}{2}}$	$\omega\sqrt{\mu\epsilon}$	$\omega\sqrt{\mu_o\epsilon_o}$
Intrinsic impedance η	$\sqrt{\dfrac{j\omega\mu}{\sigma+j\omega\epsilon}}$	$\sqrt{\dfrac{\omega\mu}{2\sigma}}(1+j)$	$\sqrt{\dfrac{\mu}{\epsilon}}$	377

The major regions of the earth's atmosphere that are of importance in radio wave propagation are the troposphere and the ionosphere. At radar frequencies (approximately 100 MHz to 300 GHz), the troposphere is by far the most important. It is the lower atmosphere consisting of a nonionized region extending from the earth's surface up to about 15 km. The ionosphere is the earth's upper atmosphere in the altitude region from 50 km to one earth radius (6370 km). Sufficient ionization exists in this region to influence wave propagation.

Wave propagation over the surface of the earth may assume one of the following three principal modes:

- Surface wave propagation along the surface of the earth
- Space wave propagation through the lower atmosphere
- Sky wave propagation by reflection from the upper atmosphere

These modes are portrayed in Fig. 35.2. The sky wave is directed toward the ionosphere, which

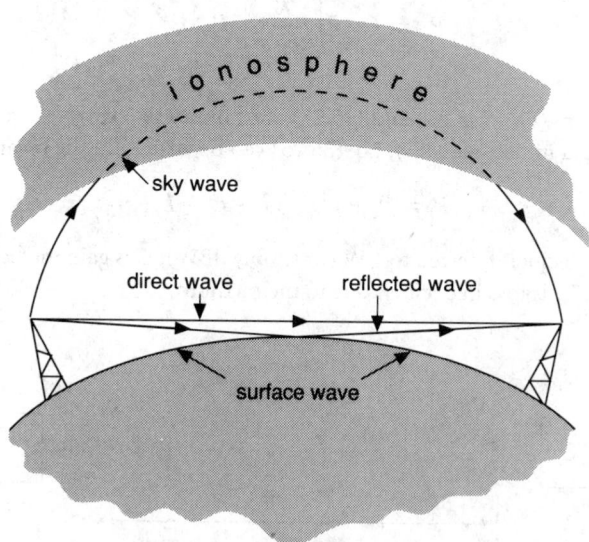

FIGURE 35.2 Modes of wave propagation.

bends the propagation path back toward the earth under certain conditions in a limited frequency range (0–50 MHz approximately). The surface wave is directed along the surface over which the wave is propagated. The space wave consists of the direct wave and the reflected wave. The direct wave travels from the transmitter to the receiver in nearly a straight path, while the reflected wave is due to ground reflection. The space wave obeys the optical laws in that direct and reflected wave components contribute to the total wave. Although the sky and surface waves are important in many applications, we will only consider space waves in this section.

Figure 35.3 depicts the electromagnetic energy transmission between two antennas in space. As the wave radiates from the transmitting antenna and propagates in space, its power density decreases, as expressed ideally in Eq. (35.19). Assuming that the antennas are in a lossless medium or free space, the power received by the receiving antenna is given by the *Friis transmission equation* [Liu and Fang, 1988]:

$$P_r = G_r G_t \left(\frac{\lambda}{4\pi r}\right)^2 P_t \tag{35.23}$$

where the subscripts t and r, respectively, refer to transmitting and receiving antennas. In Eq. (35.23), P is the power in watts, G is the antenna gain (dimensionless), r is the distance between the antennas in meters, and λ is the wavelength in meters. The Friis equation relates the power received by one antenna to the power transmitted by the other provided that the two antennas are separated by $r > 2d^2/\lambda$, where d is the largest dimension of either antenna. Thus, the Friis equation applies only when the two antennas are in the far-field of each other. In case the propagation path is not in free space, a correction factor F is included to account for the effect of the medium. This factor, known as the **propagation factor**, is simply the ratio of the electric field intensity E_m in the medium to the electric field intensity E_o in free space, i.e.,

$$F = \frac{E_m}{E_o} \tag{35.24}$$

The magnitude of F is always less than unity since E_m is always less than E_o. Thus, for a lossy medium, Eq. (35.23) becomes

$$P_r = G_r G_t \left(\frac{\lambda}{4\pi r}\right)^2 P_t \, |F|^2 \tag{35.25}$$

For practical reasons, Eqs. (35.23) and (35.25) are commonly expressed in logarithmic form. If all terms are expressed in decibels (dB), Eq. (35.25) can be written in the logarithmic form as

$$P_r = P_t + G_r + G_t - L_o - L_m \tag{35.26}$$

where P is power in decibels referred to 1 W (or simply dBW), G is gain in decibels, L_o is free-space loss in decibels, and L_m is loss in decibels due to the medium.

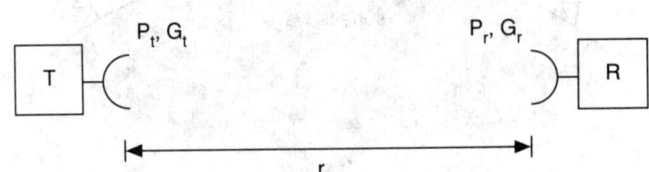

FIGURE 35.3   Transmitting and receiving antennas in free space.

The free-space loss is obtained from standard nomograph or directly from

$$L_o = 20 \log \left(\frac{4\pi r}{\lambda} \right) \qquad (35.27)$$

while the loss due to the medium is given by

$$L_m = -20 \log |F| \qquad (35.28)$$

Our major concern in the rest of the section is to determine L_o and L_m for two important cases of space propagation that differ considerably from the free-space conditions.

Effect of the Earth

The phenomenon of multipath propagation causes significant departures from free-space conditions. The term *multipath* denotes the possibility of EM wave propagation along various paths from the transmitter to the receiver. In multipath propagation of an EM wave over the earth's surface, two such paths exist: a direct path and a path via reflection and diffractions from the interface between the atmosphere and the earth. A simplified geometry of the multipath situation is shown in Fig. 35.4. The reflected and diffracted component is commonly separated into two parts, one specular (or coherent) and the other diffuse (or incoherent), that can be separately analyzed. The specular component is well defined in terms of its amplitude, phase, and incident direction. Its main characteristic is its conformance to Snell's law for reflection, which requires that the angles of incidence and reflection be equal and coplanar. It is a plane wave and, as such, is uniquely specified by its direction. The diffuse component, however, arises out of the random nature of the scattering surface and, as such, is nondeterministic. It is not a plane wave and does not obey Snell's law for reflection. It does not come from a given direction but from a continuum.

The loss factor F that accounts for the departures from free-space conditions is given by

$$F = 1 + \Gamma \rho_s D\, S(\theta) e^{-j\Delta} \qquad (35.29)$$

where Γ is the Fresnel reflection coefficient, ρ_s is the roughness coefficient, D is the divergence factor, $S(\theta)$ is the shadowing function, and Δ is the phase angle corresponding to the path difference. We now account for each of these terms.

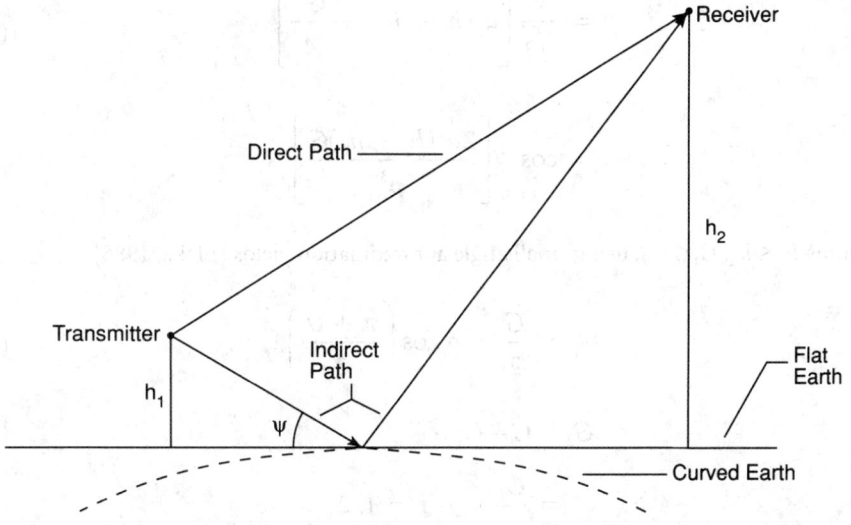

FIGURE 35.4   Multipath geometry.

The Fresnel reflection coefficient Γ accounts for the electrical properties of the earth's surface. Because the earth is a lossy medium, the value of the reflection coefficient depends on the complex relative permittivity ϵ_c of the surface, the grazing angle ψ, and the wave polarization. It is given by

$$\Gamma = \frac{\sin \psi - z}{\sin \psi + z} \tag{35.30}$$

where

$$z = \sqrt{\epsilon_c - \cos^2 \psi} \qquad \text{for horizontal polarization} \tag{35.31}$$

$$z = \frac{\sqrt{\epsilon_c - \cos^2 \psi}}{\epsilon_c} \qquad \text{for vertical polarization} \tag{35.32}$$

$$\epsilon_c = \epsilon_r - j \frac{\sigma}{\omega \epsilon_o} = \epsilon_r - j 60 \sigma \lambda \tag{35.33}$$

ϵ_r and σ are the dielectric constant and conductivity of the surface, ω and λ are the frequency and wavelength of the incident wave, and ψ is the grazing angle. It is apparent that $0 < |\Gamma| < 1$.

To account for the spreading (or divergence) of the reflected rays because of the earth's curvature, we introduce the divergence factor D. The curvature has a tendency to spread out the reflected energy more than a corresponding flat surface. The divergence factor is defined as the ratio of the reflected field from curved surface to the reflected field from flat surface [Kerr, 1951]. Using the geometry of Fig. 35.5, D is given by

$$D \simeq \left(1 + \frac{2 G_1 G_2}{a_e G \sin \psi}\right)^{-1/2} \tag{35.34}$$

where $G = G_1 + G_2$ is the total ground range and $a_e = 6370$ km is the effective earth radius. Given the transmitter height h_1, the receiver height h_2, and the total ground range G, we can determine G_1, G_2, and ψ. If we define

$$p = \frac{2}{\sqrt{3}} \left[a_e(h_1 + h_2) + \frac{G^2}{4} \right]^{1/2} \tag{35.35}$$

$$\alpha = \cos^{-1} \left[\frac{2 a_e (h_1 - h_2) G}{p^3} \right] \tag{35.36}$$

and assume $h_1 \leq h_2$, $G_1 \leq G_2$, using small angle approximation yields [Blake, 1986]

$$G_1 = \frac{G}{2} + p \cos \left(\frac{\pi + \alpha}{3} \right) \tag{35.37}$$

$$G_2 = G - G_1 \tag{35.38}$$

$$\phi_i = \frac{G_i}{a_e}, \qquad i = 1, 2 \tag{35.39}$$

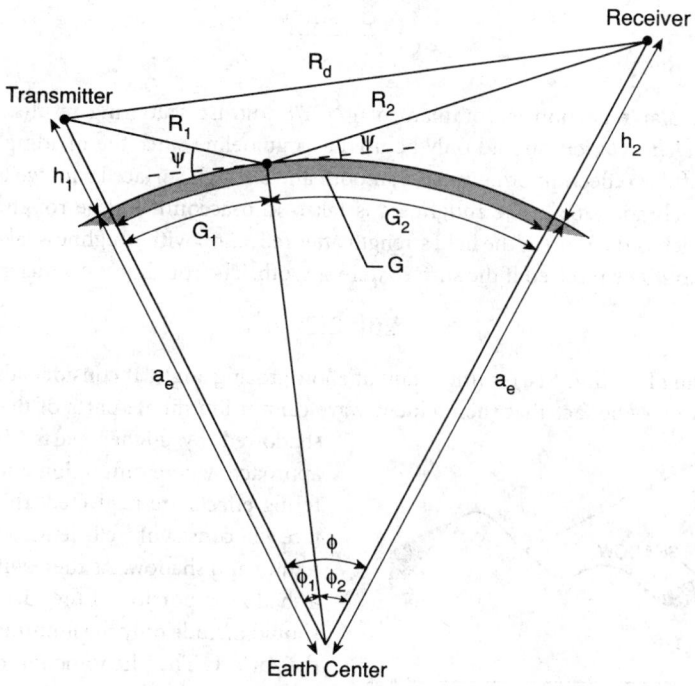

FIGURE 35.5 Geometry of spherical earth reflection.

$$R_i = [h_i^2 + 4a_e(a_e + h_i)\sin^2(\phi_i/2)]^{1/2} \qquad i = 1, 2 \tag{35.40}$$

The grazing angle is given by

$$\psi = \sin^{-1}\left[\frac{2a_e h_1 + h_1^2 - R_1^2}{2a_e R_1}\right] \tag{35.41}$$

or

$$\psi = \sin^{-1}\left[\frac{2a_e h_1 + h_1^2 + R_1^2}{2(a_e + h_1)R_1}\right] - \phi_1 \tag{35.42}$$

Although D varies from 0 to 1, in practice D is a significant factor at low grazing angle ψ (less than 0.1%).

The phase angle corresponding to the path difference between direct and reflected waves is given by

$$\Delta = \frac{2\pi}{\lambda}(R_1 + R_2 - R_d) \tag{35.43}$$

The roughness coefficient ρ_s takes care of the fact that the earth's surface is not sufficiently smooth to produce specular (mirrorlike) reflection except at a very low grazing angle. The earth's surface has a height distribution that is random in nature. The randomness arises out of the hills, structures, vegetation, and ocean waves. It is found that the distribution of the heights of the earth's surface is usually the Gaussian or normal distribution of probability theory. If σ_h is the standard deviation of the normal distribution of heights, we define the roughness parameters

$$g = \frac{\sigma_h \sin \psi}{\lambda} \tag{35.44}$$

If $g < 1/8$, specular reflection is dominant; if $g > 1/8$, diffuse scattering results. This criterion, known as *Rayleigh criterion*, should only be used as a guideline since the dividing line between a specular and diffuse reflection or between a smooth and a rough surface is not well defined [Beckman and Spizzichino, 1963]. The roughness is taken into account by the roughness coefficient $(0 < \rho_s < 1)$, which is the ratio of the field strength after reflection with roughness taken into account to that which would be received if the surface were smooth. The roughness coefficient is given by

$$\rho_s = \exp[-2(2\pi g)^2] \tag{35.45}$$

The shadowing function $S(\theta)$ is important at a low grazing angle. It considers the effect of geometric shadowing—the fact that the incident wave cannot illuminate parts of the earth's surface shadowed by higher parts. In a geometric approach, where diffraction and multiple scattering effects are neglected, the reflecting surface will consist of well-defined zones of illumination and shadow. As there will be no field on a shadowed portion of the surface, the analysis should include only the illuminated portions of the surface. The phenomenon of shadowing of a stationary surface was first investigated by Beckman in 1965 and subsequently refined by Smith [1967] and others. A pictorial representation of rough surfaces illuminated at angle of incidence $\theta\ (= 90° - \psi)$ is shown in Fig. 35.6. It is evident from the figure that the shadowing function $S(\theta)$ is equal to unity when $\theta = 0$ and zero when $\theta = \pi/2$. According to Smith [1967],

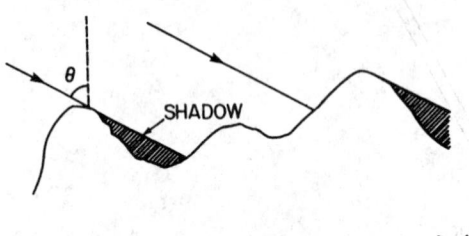

FIGURE 35.6 Rough surface illuminated at an angle of incidence θ.

$$S(\theta) \simeq \frac{\left[1 - \dfrac{1}{2}\, \mathrm{erfc}(a)\right]}{1 + 2B} \tag{35.46}$$

where $\mathrm{erfc}(x)$ is the complementary error function,

$$\mathrm{erfc}(x) = 1 - \mathrm{erf}(x) = \frac{2}{\sqrt{\pi}} \int_x^\infty e^{-t^2}\, dt \tag{35.47}$$

and

$$B = \frac{1}{4a}\left[\frac{1}{\sqrt{\pi}}\, e^{a^2} - a\, \mathrm{erfc}(a)\right] \tag{35.48}$$

$$a = \frac{\cot \theta}{2s} \tag{35.49}$$

$$s = \frac{\sigma_h}{\sigma_l} = \text{rms surface slope} \tag{35.50}$$

In Eq. (35.50) σ_h is the rms roughness height and σ_l is the correlation length. Alternative models for $S(\theta)$ are available in the literature. Using Eqs. (35.30) to (35.50), the loss factor in Eq. (35.29) can be calculated. Thus

$$L_o = 20 \log\left(\frac{4\pi R_d}{\lambda}\right) \tag{35.51}$$

$$L_m = -20 \log(1 + \Gamma \rho_s D S(\theta) e^{-j\Delta}) \tag{35.52}$$

Effect of Atmospheric Hydrometeors

The effect of atmospheric hydrometeors on satellite–earth propagation is of major concern at microwave frequencies. The problem of scattering of electromagnetic waves by atmospheric hydrometeors has attracted much interest since the late 1940s. The main hydrometeors that exist for long duration and have the greatest interaction with microwaves are rain and snow. At frequencies above 10 GHz, rain has been recognized as the most fundamental obstacle on the earth–space path. Rain has been known to cause attenuation, phase difference, and depolarization of radio waves. For analog signals, the effect of rain is more significant above 10 GHz, while for digital signals, rain effects can be significant down to 3 GHz. Attenuation of microwaves because of precipitation becomes severe owing to increased scattering and beam energy absorption by raindrops, thus impairing terrestrial as well as earth–satellite communication links. Cross-polarization distortion due to rain has also engaged the attention of researchers. This is of particular interest when frequency reuse employing signals with orthogonal polarizations is used for doubling the capacity of a communication system. A thorough review on the interaction of microwaves with hydrometeors has been given by Oguchi [1983].

The loss due to a rain-filled medium is given by

$$L_m = \gamma(R)\, \ell_e(R)\, p(R) \tag{35.53}$$

where γ is attenuation per unit length at rain rate R, ℓ is the equivalent path length at rain rate R, and $p(R)$ is the probability in percentage of rainfall rate R.

Attenuation is a function of the cumulative rain-rate distribution, drop-size distribution, refractive index of water, temperature, and other variables. A rigorous calculation of $\gamma(R)$ incorporating raindrop-size distribution, velocity of raindrops, and refractive index of water can be found in Sadiku [1992]. For practical engineering purposes, what is needed is a simple formula relating attenuation to rain parameters. Such is found in the aR^b empirical relationship, which has been used to calculate rain attenuation directly [Collin, 1985], i.e.,

$$\gamma(R) = aR^b \quad \text{dB/km} \tag{35.54}$$

where R is the rain rate and a and b are constants. At 0°C, the values of a and b are related to frequency f in gigahertz as follows:

$$a = G_a f^{E_a} \tag{35.55}$$

where $G_a = 6.39 \times 10^{-5}$, $E_a = 2.03$, for $f < 2.9$ GHz; $G_a = 4.21 \times 10^{-5}$, $E_a = 2.42$, for 2.9 GHz $\leq f \leq 54$ GHz; $G_a = 4.09 \times 10^{-2}$, $E_a = 0.699$, for 54 GHz $\leq f < 100$ GHz; $G_a = 3.38$, $E_a = -0.151$, for 180 GHz $< f$; and

$$b = G_b f^{E_b} \tag{35.56}$$

where $G_b = 0.851$, $E_b = 0.158$, for $f < 8.5$ GHz; $G_b = 1.41$, $E_b = -0.0779$, for 8.5 GHz $\leq f < 25$ GHz; $G_b = 2.63$, $E_b = -0.272$, for 25 GHz $\leq f < 164$ GHz; $G_b = 0.616$, $E_b = 0.0126$, for 164 GHz $\leq f$.

The effective length $\ell_e(R)$ through the medium is needed since rain intensity is not uniform over the path. Its actual value depends on the particular area of interest and therefore has a number of representations [Liu and Fang, 1988]. Based on data collected in western Europe and eastern North America, the effective path length has been approximated as [Hyde, 1984]

$$\ell_e(R) = [0.00741 R^{0.766} + (0.232 - 0.00018R) \sin \theta]^{-1} \qquad (35.57)$$

where θ is the elevation angle.

The cumulative probability in percentage of rainfall rate R is given by [Hyde, 1984]

$$p(R) = \frac{M}{87.66} [0.03\beta e^{-0.03R} + 0.2(1 - \beta)(e^{-0.258R} + 1.86 e^{-1.63R})] \qquad (35.58)$$

where M is the mean annual rainfall accumulation in millimeters and β is the Rice–Holmberg thunderstorm ratio.

The effect of other hydrometeors such as water vapor, fog, hail, snow, and ice is governed by similar fundamental principles as the effect of rain [Collin, 1985]. In most cases, however, their effects are at least an order of magnitude less than the effect of rain.

Other Effects

Besides hydrometeors, the atmosphere has the composition given in Table 35.2. While attenuation of EM waves by hydrometeors may result from both absorption and scattering, gases act only as absorbers. Although some of these gases do not absorb microwaves, some possess permanent electric and/or magnetic dipole moment and play some part in microwave absorption. For example, nitrogen molecules do not possess permanent electric or magnetic dipole moment and therefore play no part in microwave absorption. Oxygen has a small magnetic moment, which enables it to display weak absorption lines in the centimeter and millimeter wave regions. Water vapor is a molecular gas with a permanent electric dipole moment. It is more responsive to excitation by an EM field than is oxygen.

Table 35.2 Composition of Dry Atmosphere from Sea Level to about 90 km

Constituent	Percent by Volume	Percent by Weight
Nitrogen	78.088	75.527
Oxygen	20.949	23.143
Argon	0.93	1.282
Carbon dioxide	0.03	0.0456
Neon	1.8×10^{-3}	1.25×10^{-3}
Helium	5.24×10^{-4}	7.24×10^{-5}
Methane	1.4×10^{-4}	7.75×10^{-5}
Krypton	1.14×10^{-4}	3.30×10^{-4}
Nitrous oxide	5×10^{-5}	7.60×10^{-5}
Xenon	8.6×10^{-6}	3.90×10^{-5}
Hydrogen	5×10^{-5}	3.48×10^{-6}

Source: D.C. Livingston, *The Physics of Microwave Propagation*, Englewood Cliffs, N.J.: Prentice-Hall, 1970, p. 11. With permission.

Defining Terms

Multipath: Propagation of electromagnetic waves along various paths from the transmitter to the receiver.

Propagation constant: The negative of the partial logarithmic derivative, with respect to the distance in the direction of the wave normal, of the phasor quantity describing a traveling wave in a homogeneous medium.

Propagation factor: The ratio of the electric field intensity in a medium to its value if the propagation took place in free space.

Wave propagation: The transfer of energy by electromagnetic radiation.

References

P. Beckman and A. Spizzichino, *The Scattering of Electromagnetic Waves from Random Surfaces*, New York: Macmillan, 1963.

L.V. Blake, *Radar Range-Performance Analysis*, Norwood, Mass.: Artech House, 1986, pp. 253–271.

R. E. Collin, *Antennas and Radiowave Propagation*, New York: McGraw-Hill, 1985, pp. 339–456.

G. Hyde, "Microwave propagation," in *Antenna Engineering Handbook*, 2nd ed., R.C. Johnson and H. Jasik, Eds., New York: McGraw-Hill, 1984, pp. 45.1–45.17.

D. E. Kerr, *Propagation of Short Radio Waves*, New York: McGraw-Hill (republished by Peter Peregrinus, London, 1987), 1951, pp. 396–444.

C. H. Liu and D. J. Fang, "Propagation," in *Antenna Handbook: Theory, Applications, and Design*, Y.T. Lo and S.W. Lee, Eds., New York: Van Nostrand Reinhold, 1988, pp. 29.1–29.56.

T. Oguchi, "Electromagnetic wave propagation and scattering in rain and other hydrometeors," *Proc. IEEE*, vol. 71, pp. 1029–1078, 1983.

M. N. O. Sadiku, *Numerical Techniques in Electromagnetics*, Boca Raton, Fla.: CRC Press, 1992, pp. 96–116.

B. G. Smith, "Geometrical shadowing of a random rough surface," *IEEE Trans. Ant. Prog.*, vol. 15, pp. 668–671, 1967.

Further Information

There are several sources of information dealing with the theory and practice of wave propagation in space. Some of these are in the reference section. Journals such as *Radio Science, IEE Proceedings Part H*, and *IEEE Transactions on Antennas and Propagation* are devoted to EM wave propagation. *Radio Science* is available from the American Geophysical Union, 2000 Florida Avenue NW, Washington DC 20009; *IEE Proceedings Part H* from IEE Publishing Department, Michael Faraday House, 6 Hills Way, Stevenage, Herts, SG1 2AY, U.K.; and *IEEE Transactions on Antennas and Propagation* from IEEE, 445 Hoes Lane, P.O. Box 1331, Piscataway, NJ 08855-1331.

Other mechanisms that can affect EM wave propagation in space, not discussed in this section, include clouds, dust, and the ionosphere. The effect of the ionosphere is discussed in detail in standard texts.

35.2 Waveguides

Kenneth Demarest

Waveguide Modes

Any structure that guides electromagnetic waves can be considered a **waveguide**. Most often, however, this term refers to closed metal cylinders that maintain the same cross-sectional dimensions over long distances. Such a structure is shown in Fig. 35.7, which consists of a metal cylinder filled with a dielectric. When filled with low-loss dielectrics (such as air), waveguides typically exhibit lower losses than transmission lines, which makes them useful for transporting RF energy over relatively long distances. They are most often used for frequencies ranging from 1 to 150 GHz.

Every type of waveguide has an infinite number of distinct electromagnetic field configurations

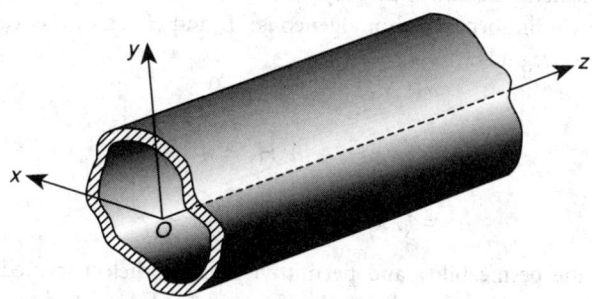

FIGURE 35.7 A uniform waveguide with arbitrary cross section.

that can exist inside it. Each of these configurations is called a **waveguide mode**. The characteristics of these modes depend upon the cross-sectional dimensions of the conducting cylinder, the type of dielectric material inside the waveguide, and the frequency of operation.

Waveguide modes are typically classed according to the nature of the electric and magnetic field components E_z and H_z. These components are called the longitudinal components of the fields. Several types of modes are possible in waveguides:

TE modes: Transverse-electric modes, sometimes called H modes. These modes have $E_z = 0$ at all points within the waveguide, which means that the electric field vector is always perpendicular (i.e., transverse) to the waveguide axis. These modes are always possible in waveguides with uniform dielectrics.

TM modes: Transverse-magnetic modes, sometimes called E modes. These modes have $H_z = 0$ at all points within the waveguide, which means that the magnetic field vector is perpendicular to the waveguide axis. Like TE modes, they are always possible in waveguides with uniform dielectrics.

EH modes: EH modes are hybrid modes in which neither E_z nor H_z are zero, but the characteristics of the transverse fields are controlled more by E_z than H_z. These modes are often possible in waveguides with inhomogeneous dielectrics.

HE modes: HE modes are hybrid modes in which neither E_z nor H_z are zero, but the characteristics of the transverse fields are controlled more by H_z than E_z. Like EH modes, these modes are often possible in waveguides with inhomogeneous dielectrics.

TEM modes Transverse-electromagnetic modes, often called transmission line modes. These modes can exist only when a second conductor exists within the waveguide, such as a center conductor on a coaxial cable. Because these modes cannot exist in single, closed conductor structures, they are not waveguide modes.

Waveguide modes are most easily determined by first computing the longitudinal field components, E_z and H_z, that can be supported by the waveguide. From these, the transverse components (such as E_x and E_y) can easily be found simply by taking spatial derivatives of the longitudinal fields [Collin, 1992].

When the waveguide properties are constant along the z axis, E_z and H_z vary in the longitudinal direction as E_z, $H_z \propto \exp(\omega t - \gamma z)$, where $\omega = 2\pi f$ is the radian frequency of operation and γ is a complex number of the form

$$\gamma = \alpha + j\beta \tag{35.59}$$

The parameters γ, α, and β are called the propagation, attenuation, and phase constants, respectively, and $j = \sqrt{-1}$. When there are no metal or dielectric losses, γ is always either purely real or imaginary. When γ is real, E_z and H_z have constant phase and decay exponentially with increasing z. When γ is imaginary, E_z and H_z vary in phase with increasing z but do not decay in amplitude. When this occurs, the fields are said to be propagating.

When the dielectric is uniform (i.e., homogeneous), E_z and H_z satisfy the scalar wave equation at all points within the waveguide:

$$\nabla_t^2 E_z + h^2 E_z = 0 \tag{35.60}$$

and

$$\nabla_t^2 H_z + h^2 H_z = 0 \tag{35.61}$$

where

$$h^2 = (2\pi f)^2 \mu\epsilon + \gamma^2 = k^2 + \gamma^2 \tag{35.62}$$

Here, μ and ϵ are the permeability and permittivity of the dielectric media, respectively, and $k = 2\pi f \sqrt{\mu\epsilon}$ is the wavenumber of the dielectric. The operator ∇_t^2 is called the transverse Laplacian operator. In Cartesian coordinates,

$$\nabla_t^2 = \frac{\partial^2}{\partial x^2} + \frac{\partial^2}{\partial y^2}$$

Most of the properties of the allowed modes in real waveguides can usually be found by assuming that the walls are perfectly conducting. Under this condition, $E_z = 0$ and $\partial H_z/\partial p = 0$ at the waveguide walls, where p is the direction perpendicular to the waveguide wall. When these conditions are imposed upon the general solutions of Eqs. (35.60) and (35.61), it is found that only certain values of h are allowed. These values are called the *modal eigenvalues* and are determined by the cross-sectional shape of the waveguide. Using Eq. (35.62), the propagation constant γ for each mode varies with frequency according to

$$\gamma = \alpha + j\beta = h\sqrt{1 - \left(\frac{f}{f_c}\right)^2} \tag{35.63}$$

where

$$f_c = \frac{h}{2\pi\sqrt{\mu\epsilon}} \tag{35.64}$$

The modal parameter f_c has units hertz and is called the **cut-off frequency** of the mode it is associated with. According to Eq. (35.63), when $f > f_c$, the propagation constant γ is imaginary and thus the mode is propagating. On the other hand, when $f < f_c$, γ is real, which means that the fields decay exponentially with increasing values of z. Modes operated at frequencies below their cut-off frequency are not able to propagate energy over long distances and are called evanescent modes.

The dominant mode of a waveguide is the one with the lowest cut-off frequency. Although higher-order modes are often useful for a variety of specialized uses of waveguides, signal distortion is usually minimized when a waveguide is operated in the frequency range where only the dominant mode exists. This range of frequencies is called the *dominant range* of the waveguide.

The distance over which the fields of propagating modes repeat themselves is called the **guide wavelength** λ_g. From Eq. (35.63), it can be shown that λ_g always varies with frequency according to

$$\lambda_g = \frac{\lambda_o}{\sqrt{1 - \left(\frac{f_c}{f}\right)^2}} \tag{35.65}$$

where $\lambda_o = 1/(f\sqrt{\mu\epsilon})$ is the wavelength of a plane wave of the same frequency in an infinite sample of the waveguide dielectric. For $f \gg f_c$, $\lambda_g \approx \lambda_o$. Also, $\lambda_g \to \infty$ as $f \to f_c$, which is one reason why it is usually undesirable to operate a waveguide mode near modal cut-off frequencies.

Although waveguide modes are not plane waves, the ratio of their transverse electric and magnetic field magnitudes is constant throughout the cross section of the waveguide, just as for plane waves. This ratio is called the modal **wave impedance** and has the following values for TE and TM modes:

$$Z_{TE} = \frac{E_T}{H_T} = \frac{j\omega\mu}{\gamma} \tag{35.66}$$

and

$$Z_{TM} = \frac{E_T}{H_T} = \frac{\gamma}{j\omega\epsilon} \tag{35.67}$$

where E_T and H_T are the magnitudes of the transverse electric and magnetic fields, respectively. In the limit as $f \to \infty$, both Z_{TE} and Z_{TM} approach $\sqrt{\mu/\epsilon}$, which is the intrinsic impedance of the dielectric medium. On the other hand, as $f \to f_c$, $Z_{TE} \to \infty$ and $Z_{TM} \to 0$, which means that the transverse electric fields are dominant in TE modes near cut-off and vice versa for TM modes.

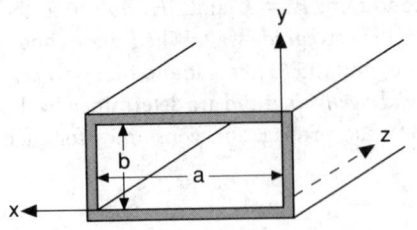

FIGURE 35.8 A rectangular waveguide.

Rectangular Waveguides

A rectangular waveguide is shown in Fig. 35.8. The conducting walls are formed such that the inner surfaces form a rectangular cross section, with dimensions a and b along the x and y coordinate axes, respectively.

If the walls are perfectly conducting and the dielectric material is lossless, the field components for the TE_{mn} modes are given by

$$E_x = H_0 \frac{j\omega\mu}{h_{mn}^2}\left(\frac{n\pi}{b}\right)\cos\left(\frac{m\pi}{a}x\right)\sin\left(\frac{n\pi}{b}y\right)\exp(j\omega t - \gamma_{mn}z) \tag{35.68}$$

$$E_y = -H_0 \frac{j\omega\mu}{h_{mn}^2}\left(\frac{m\pi}{a}\right)\sin\left(\frac{m\pi}{a}x\right)\cos\left(\frac{n\pi}{b}y\right)\exp(j\omega t - \gamma_{mn}z) \tag{35.69}$$

$$E_z = 0$$

$$H_x = H_0 \frac{\gamma_{mn}}{h_{mn}^2}\left(\frac{m\pi}{a}\right)\sin\left(\frac{m\pi}{a}x\right)\cos\left(\frac{n\pi}{b}y\right)\exp(j\omega t - \gamma_{mn}z) \tag{35.70}$$

$$H_y = H_0 \frac{\gamma_{mn}}{h_{mn}^2}\left(\frac{n\pi}{b}\right)\cos\left(\frac{m\pi}{a}x\right)\sin\left(\frac{n\pi}{b}y\right)\exp(j\omega t - \gamma_{mn}z) \tag{35.71}$$

$$H_z = H_0 \cos\left(\frac{m\pi}{a}x\right)\cos\left(\frac{n\pi}{b}y\right)\exp(j\omega t - \gamma_{mn}z) \tag{35.72}$$

where

$$h_{mn} = \sqrt{\left(\frac{m\pi}{a}\right)^2 + \left(\frac{n\pi}{b}\right)^2} = 2\pi f_{c_{mn}}\sqrt{\mu\epsilon} \tag{35.73}$$

For the TM_{mn} modes, m and n can be any positive integer value, including zero, as long as both are not zero.

The field components for the TM_{mn} modes are

$$E_x = -E_0 \frac{\gamma_{mn}}{h_{mn}^2}\left(\frac{m\pi}{a}\right)\cos\left(\frac{m\pi}{a}x\right)\sin\left(\frac{n\pi}{b}y\right)\exp(j\omega t - \gamma_{mn}z) \tag{35.74}$$

$$E_y = -E_0 \frac{\gamma_{mn}}{h_{mn}^2}\left(\frac{n\pi}{b}\right)\sin\left(\frac{m\pi}{a}x\right)\cos\left(\frac{n\pi}{b}y\right)\exp(j\omega t - \gamma_{mn}z) \tag{35.75}$$

$$E_z = E_0 \sin\left(\frac{m\pi}{a} x\right) \sin\left(\frac{n\pi}{b} y\right) \exp(j\omega t - \gamma_{mn} z) \tag{35.76}$$

$$H_x = E_0 \frac{j\omega\epsilon}{h_{mn}^2}\left(\frac{n\pi}{b}\right) \sin\left(\frac{m\pi}{a} x\right) \cos\left(\frac{n\pi}{b} y\right) \exp(j\omega t - \gamma_{mn} z) \tag{35.77}$$

$$H_y = -E_0 \frac{j\omega\epsilon}{h_{mn}^2}\left(\frac{m\pi}{a}\right) \cos\left(\frac{m\pi}{a} x\right) \sin\left(\frac{n\pi}{b} y\right) \exp(j\omega t - \gamma_{mn} z) \tag{35.78}$$

$$H_z = 0 \tag{35.79}$$

where the values of h_{mn} and $f_{c_{mn}}$ are given by Eq. (35.73). For the TM_{mn} modes, m and n can be any positive integer value except zero.

The dominant mode in a rectangular waveguide is the TE_{10} mode, which has a cut-off frequency

$$f_{c_{10}} = \frac{1}{2a\sqrt{\mu\epsilon}} = \frac{c}{2a} \tag{35.80}$$

where c is the speed of light in the dielectric media. The modal field patterns for this mode are shown in Fig. 35.9.

Table 35.3 shows the cut-off frequencies of the lowest-order rectangular waveguide modes (as referenced to the cut-off frequency of the dominant mode) when $a/b = 2.1$.

The modal field patterns for several lower-order modes are shown in Fig. 35.10.

Table 35.3 Cut-off Frequencies of the Lowest-Order Rectangular Waveguide Modes (Referenced to the Cut-off Frequency of the Dominant Mode) for a Rectangular Waveguide with $a/b = 2.1$

$f_c/f_{c_{10}}$	Modes
1.0	TE_{10}
2.0	TE_{20}
2.1	TE_{01}
2.326	TE_{11}, TM_{11}
2.9	TE_{21}, TM_{21}
3.0	TE_{30}
3.662	TE_{31}, TM_{31}
4.0	TE_{40}

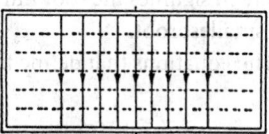

FIGURE 35.9 Field configurations for the TE_{10} (dominant) mode of a rectangular waveguide. Solid lines, E; dashed lines, H. (*Source:* Adapted from N. Marcuvitz, *Waveguide Handbook*, 2nd ed., London: Peter Peregrinus Ltd., and New York: McGraw-Hill, 1986, p. 63. With permission.)

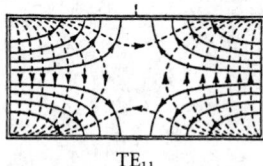

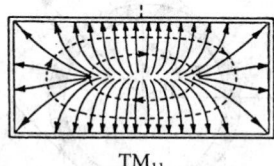

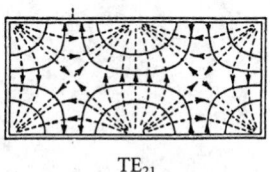

TE$_{11}$　　　　TM$_{11}$　　　　TE$_{21}$

FIGURE 35.10 Field configurations for the TE_{11}, TM_{11}, and the TE_{21} modes. Solid lines, E; dashed lines, H. (*Source:* Adapted from N. Marcuvitz, *Waveguide Handbook*, 2nd. ed., London: Peter Peregrinus Ltd., and New York: McGraw-Hill, 1986, p. 59. With permission.)

Circular Waveguides

A circular waveguide with inner radius a is shown in Fig. 35.11. Here the axis of the waveguide is aligned with the z axis of a circular-cylindrical coordinate system, where ρ and ϕ are the radial and azimuthal coordinates, respectively. If the walls are perfectly conducting and the dielectric material is lossless, the equations for the TE_{nm} modes are

$$E_\rho = H_0 \frac{j\omega\mu n}{h_{nm}^2 \rho} J_n(h_{nm}\rho) \sin(n\phi) \exp(j\omega t - \gamma_{nm} z) \tag{35.81}$$

$$E_\phi = H_0 \frac{j\omega\mu}{h_{nm}} J_n'(h_{nm}\rho) \cos(n\phi) \exp(j\omega t - \gamma_{nm} z) \tag{35.82}$$

$$E_z = 0 \tag{35.83}$$

$$H_\rho = -H_0 \frac{\gamma_{nm}}{h_{nm}} J_n'(h_{nm}\rho) \cos(n\phi) \exp(j\omega t - \gamma_{nm} z) \tag{35.84}$$

$$H_\phi = H_0 \frac{\gamma_{nm}}{h_{nm}^2 \rho} J_n(h_{nm}\rho) \sin(n\phi) \exp(j\omega t - \gamma_{nm} z) \tag{35.85}$$

$$H_z = H_0 J_n(h_{nm}\rho) \cos(n\phi) \exp(j\omega t - \gamma_{nm} z) \tag{35.86}$$

where n is any positive valued integer, including zero, and $J_n(x)$ and $J_n'(x)$ are the regular Bessel functions of order n and its first derivative, respectively. The allowed values of the modal eigenvalues h_{nm} satisfy

$$J_n'(h_{nm} a) = 0 \tag{35.87}$$

where m signifies the root number of Eq. (35.87). By convention, $1 < m < \infty$, where $m = 1$ indicates the smallest root.

The equations that define the TM_{nm} modes in circular waveguides are

$$E_\rho = -E_0 \frac{\gamma_{nm}}{h_{nm}} J_n'(h_{nm}\rho) \cos(n\phi) \exp(j\omega t - \gamma_{nm} z) \tag{35.88}$$

$$E_\phi = E_0 \frac{\gamma_{nm}}{h_{nm}^2 \rho} J_n(h_{nm}\rho) \sin(n\phi) \exp(j\omega t - \gamma_{nm} z) \tag{35.89}$$

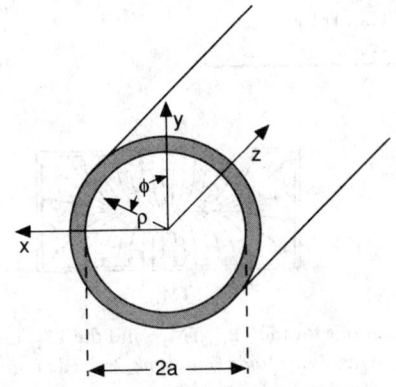

FIGURE 35.11 A circular waveguide.

$$E_z = E_0 J_n(h_{nm}\rho) \cos(n\phi) \exp(j\omega t - \gamma_{nm}z) \tag{35.90}$$

$$H_\rho = -E_0 \frac{j\omega \epsilon n}{h_{nm}^2 \rho} J_n(h_{nm}\rho) \sin(n\phi) \exp(j\omega t - \gamma_{nm}z) \tag{35.91}$$

$$H_\phi = -E_0 \frac{j\omega \epsilon}{h_{nm}} J_n'(h_{nm}\rho) \cos(n\phi) \exp(j\omega t - \gamma_{nm}z) \tag{35.92}$$

$$H_z = 0 \tag{35.93}$$

where n is any positive valued integer, including zero. For the TM_{nm} modes, the values of the modal eigenvalues are solutions of

$$J_n(h_{nm}a) = 0 \tag{35.94}$$

where m signifies the root number of Eq. (35.94). As in the case of the TE modes, $1 < m < \infty$.

The dominant mode in a circular waveguide is the TE_{11} mode, which has a cut-off frequency given by

$$f_{c_{11}} = \frac{0.293}{a\sqrt{\mu\epsilon}} \tag{35.95}$$

The configuration of the electric and magnetic fields of this mode is shown in Fig. 35.12.

Table 35.4 shows the cut-off frequencies of the lowest-order modes for circular waveguides, referenced to the cut-off frequency of the dominant mode.

The modal field patterns for several lower-order modes are shown in Fig. 35.13.

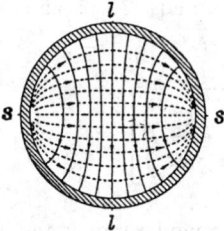

FIGURE 35.12 Field configuration for the TE_{11} (dominant) mode of a circular waveguide. Solid lines, E; dashed lines, H. (*Source:* Adapted from N. Marcuvitz, *Waveguide Handbook*, 2nd ed., London: Peter Peregrinus Ltd., and New York: McGraw-Hill, 1986, p. 68. With permission.)

Table 35.4 Cut-off Frequencies of the Lowest-Order Circular Waveguide Modes, Referenced to the Cut-off Frequency of the Dominant Mode

$f_c/f_{c_{11}}$	Modes
1.0	TE_{11}
1.307	TM_{01}
1.66	TE_{21}
2.083	TE_{01}, TM_{11}
2.283	TE_{31}
2.791	TM_{21}
2.89	TE_{41}
3.0	TE_{12}

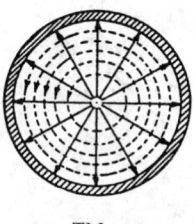

TM_{01}

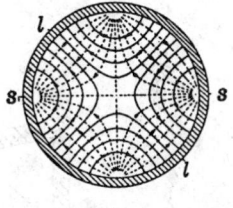

TE_{21}

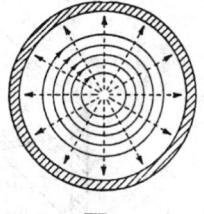

TE_{01}

FIGURE 35.13 Field configurations for the TM_{01}, TE_{21}, and TE_{01} circular waveguide modes. Solid lines, E; dashed lines, H. (*Source:* Adapted from N. Marcuvitz, *Waveguide Handbook*, 2nd ed., London: Peter Peregrinus Ltd., and New York: McGraw-Hill, 1986, p. 71. With permission.)

Commercially Available Waveguides

The dimensions of standard rectangular waveguides are given in Table 35.5.

In addition to rectangular and circular waveguides, several other waveguide types are commonly used in microwave applications. Among these are ridge waveguides and elliptical waveguides. The modes of elliptical waveguides can be expressed in terms of Mathieu functions [Kretzschmar, 1970] and are similar to those of circular waveguides but are less perturbed by minor twists and bends of the waveguide. This property makes them attractive for coupling to antennas.

Single-ridge and double-ridge waveguides are shown in Fig. 35.14. The modes of these waveguides bear similarities to those of rectangular guides, but can only be derived numerically [Montgomery, 1971]. Ridge waveguides are useful because their dominant ranges exceed those of rectangular waveguides. However, this range increase is obtained at the expense of higher losses.

Waveguides are also available in a number of constructions, including rigid, semirigid, and flexible. In applications where it is not necessary for the waveguide to bend, rigid construction is always the best since it exhibits the lowest loss. In general, the more flexible the waveguide construction, the higher the loss.

Waveguide Losses

There are two mechanisms that cause losses in waveguides: dielectric losses and metal losses. In both cases, these losses cause the amplitudes of the propagating modes to decay as $\exp(-\alpha z)$, where α is the attenuation constant, measured in units of nepers/meter. Typically, the attenuation constant is considered as the sum of two components: $\alpha = \alpha_{die} + \alpha_{met}$, where α_{die} and α_{met} are the dielectric and metal attenuation constants, respectively.

The attenuation constant α_{die} can be found directly from Eq. (35.63) simply by generalizing the dielectric wavenumber k to include the effect of the dielectric conductivity σ. For a lossy dielectric, the wavenumber is given by $k^2 = \omega^2 \mu \epsilon [1 + (\sigma/j\omega\epsilon)]$. Thus, from Eqs. (35.62) and (35.63) the attenuation constant α_{die} due to dielectric losses is given by

$$\alpha_{die} = \mathrm{real}\left[\sqrt{h^2 - \omega^2\mu\epsilon\left(1 + \frac{\sigma}{j\omega\epsilon}\right)}\right] \tag{35.96}$$

where the allowed values of h are given by Eq. (35.73) for rectangular modes and Eqs. (35.87) and (35.94) for circular modes.

The metal loss constant α_{met} is usually obtained by assuming that the wall conductivity is high enough to have only a negligible effect on the transverse properties of the modal field patterns. Using this assumption, the power loss in the walls per unit distance along the waveguide can then be calculated to obtain α_{met} [Marcuvitz, 1986]. Figure 35.15 shows the metal attenuation constants

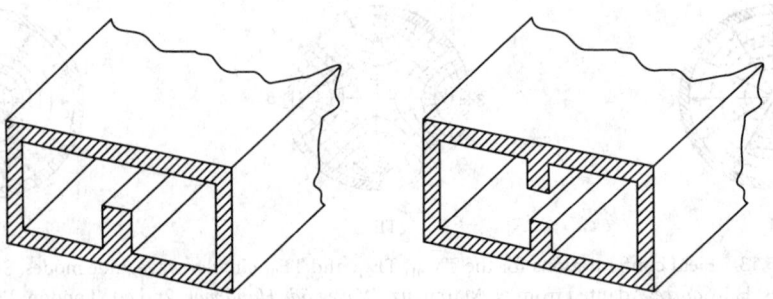

FIGURE 35.14   Single- and double-ridged waveguides.

Table 35.5 Standard Rectangular Waveguides

EIA[a] Designation WR[b]( )	Physical Dimensions				Cut-off Frequency for Air-filled Waveguide, GHz	Recommended Frequency Range for TE$_{10}$ Mode, GHZ
	Inside, cm (in.)		Outside, cm (in.)			
	Width	Height	Width	Height		
2300	58.420 (23.000)	29.210 (11.500)	59.055 (23.250)	29.845 (11.750)	0.257	0.32–0.49
2100	53.340 (21.000)	26.670 (10.500)	53.973 (21.250)	27.305 (10.750)	0.281	0.35–0.53
1800	45.720 (18.000)	22.860 (9.000)	46.350 (18.250)	23.495 (9.250)	0.328	0.41–0.62
1500	38.100 (15.000)	19.050 (7.500)	38.735 (15.250)	19.685 (7.750)	0.394	0.49–0.75
1150	29.210 (11.500)	14.605 (5.750)	29.845 (11.750)	15.240 (6.000)	0.514	0.64–0.98
975	24.765 (9.750)	12.383 (4.875)	25.400 (10.000)	13.018 (5.125)	0.606	0.76–1.15
770	19.550 (7.700)	9.779 (3.850)	20.244 (7.970)	10.414 (4.100)	0.767	0.96–1.46
650	16.510 (6.500)	8.255 (3.250)	16.916 (6.660)	8.661 (3.410)	0.909	1.14–1.73
510	12.954 (5.100)	6.477 (2.500)	13.360 (5.260)	6.883 (2.710)	1.158	1.45–2.20
430	10.922 (4.300)	5.461 (2.150)	11.328 (4.460)	5.867 (2.310)	1.373	1.72–2.61
340	8.636 (3.400)	4.318 (1.700)	9.042 (3.560)	4.724 (1.860)	1.737	2.17–3.30
284	7.214 (2.840)	3.404 (1.340)	7.620 (3.000)	3.810 (1.500)	2.079	2.60–3.95
229	5.817 (2.290)	2.908 (1.145)	6.142 (2.418)	3.233 (1.273)	2.579	3.22–4.90
187	4.755 (1.872)	2.215 (0.872)	5.080 (2.000)	2.540 (1.000)	3.155	3.94–5.99
159	4.039 (1.590)	2.019 (0.795)	4.364 (1.718)	2.344 (0.923)	3.714	4.64–7.05
137	3.485 (1.372)	1.580 (0.622)	3.810 (1.500)	1.905 (0.750)	4.304	5.38–8.17
112	2.850 (1.122)	1.262 (0.497)	3.175 (1.250)	1.588 (0.625)	5.263	6.57–9.99
90	2.286 (0.900)	1.016 (0.400)	2.540 (1.000)	1.270 (0.500)	6.562	8.20–12.50
75	1.905 (0.750)	0.953 (0.375)	2.159 (0.850)	1.207 (0.475)	7.874	9.84–15.00
62	1.580 (0.622)	0.790 (0.311)	1.783 (0.702)	0.993 (0.391)	9.494	11.90–18.00
51	1.295 (0.510)	0.648 (0.255)	1.499 (0.590)	0.851 (0.335)	11.583	14.50–22.00
42	1.067 (0.420)	0.432 (0.170)	1.270 (0.500)	0.635 (0.250)	14.058	17.60–26.70
34	0.864 (0.340)	0.432 (0.170)	1.067 (0.420)	0.635 (0.250)	17.361	21.70–33.00
28	0.711 (0.280)	0.356 (0.140)	0.914 (0.360)	0.559 (0.220)	21.097	26.40–40.00
22	0.569 (0.224)	0.284 (0.112)	0.772 (0.304)	0.488 (0.192)	26.362	32.90–50.10
19	0.478 (0.188)	0.239 (0.094)	0.681 (0.268)	0.442 (0.174)	31.381	39.20–59.60
15	0.376 (0.148)	0.188 (0.074)	0.579 (0.228)	0.391 (0.154)	39.894	49.80–75.80
12	0.310 (0.122)	0.155 (0.061)	0.513 (0.202)	0.358 (0.141)	48.387	60.50–91.90
10	0.254 (0.100)	0.127 (0.050)	0.457 (0.180)	0.330 (0.130)	59.055	73.80–112.00
8	0.203 (0.080)	0.102 (0.040)	0.406 (0.160)	0.305 (0.120)	73.892	92.20–140.00
7	0.165 (0.065)	0.084 (0.033)	0.343 (0.135)	0.262 (0.103)	90.909	114.00–173.00
5	0.130 (0.051)	0.066 (0.026)	0.257 (0.101)	0.193 (0.076)	115.385	145.00–220.00
4	0.109 (0.043)	0.056 (0.022)	0.211 (0.083)	0.157 (0.062)	137.615	172.00–261.00
3	0.086 (0.034)	0.043 (0.017)	0.163 (0.064)	0.119 (0.047)	174.419	217.00–333.00

[a]Electronic Industry Association.
[b]Rectangular waveguide.
Source: S.Y. Liao, *Microwave Devices and Circuits*, 3rd ed., Englewood Cliffs, N.J.: Prentice-Hall, 1990, p. 118. With permission.

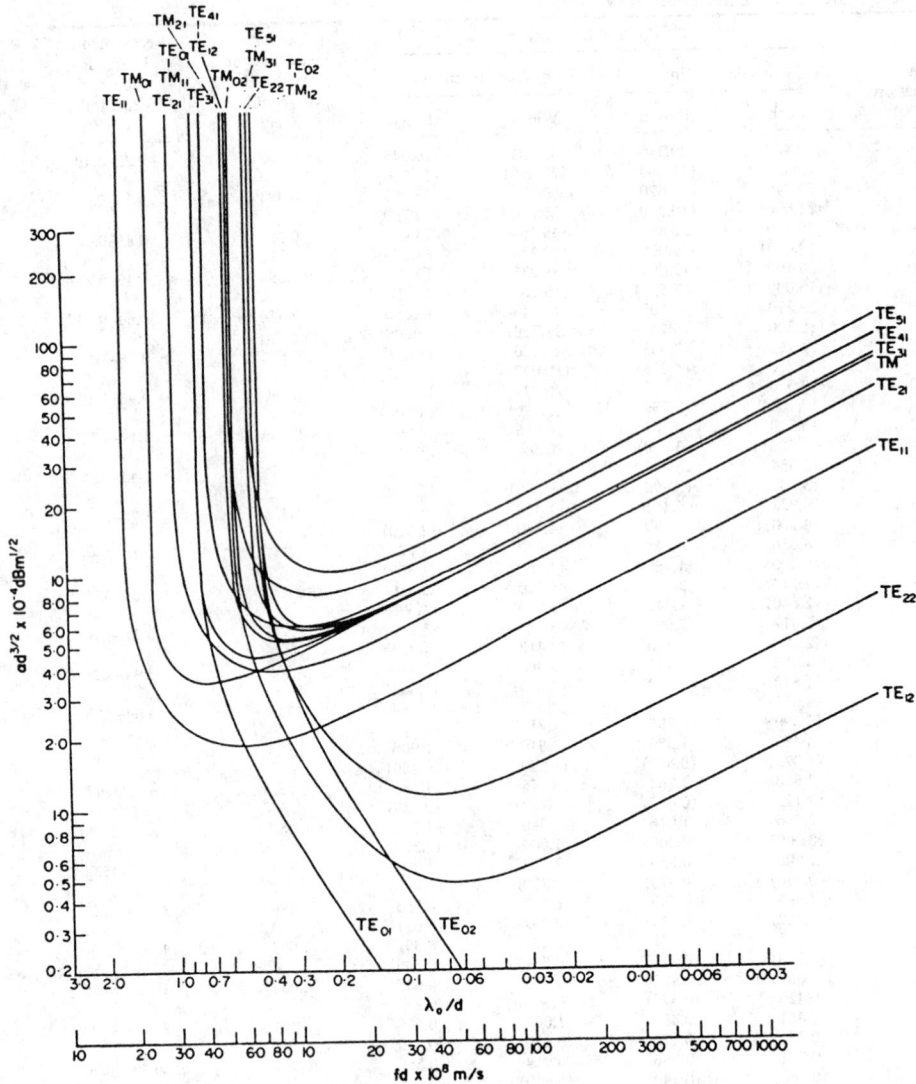

FIGURE 35.15 Values of metallic attenuation constant α for the first few waveguide modes in a circular waveguide of diameter d, plotted against normalized wavelength. (*Source:* A. J. Baden Fuller, *Microwaves,* 2nd ed., New York: Pergamon Press, 1979, p. 138. With permission.)

for several circular waveguide modes, each normalized to the resistivity R_s of the walls, where $R_s = \sqrt{(\pi f \mu / \sigma)}$ and where μ and σ are the permeability and conductivity of the metal walls, respectively. As can be seen from this figure, the TE_{0m} modes exhibit particularly low loss at frequencies significantly above their cut-off frequencies, making them useful for transporting microwave energy over large distances.

Mode Launching

When coupling electromagnetic energy into a waveguide, it is important to ensure that the desired modes are excited and that reflections back to the source are minimized. Similar concerns must be considered when coupling energy from a waveguide to a transmission line or circuit element. This is achieved by using launching (or coupling) structures that allow strong coupling between the desired modes on both structures.

Figure 35.16 shows a mode launching structure for coaxial cable to rectangular waveguide transitions.

This structure provides good coupling between the TEM (transmission line) mode on a coaxial cable and the TE_{10} mode in the waveguide because the antenna probe excites a strong transverse electric field in the center of the waveguide, directed between the broad walls. The distance between the probe and the short circuit back wall is chosen to be approximately $\lambda/4$, which allows the TE_{10} mode launched in this direction to reflect off the short circuit and arrive in phase with the mode launched towards the right.

Launching structures can also be devised to launch higher-order modes. Mode launchers that couple the transmission line mode on a coaxial cable to the TM_{11} and TM_{21} waveguide mode are shown in Fig. 35.17.

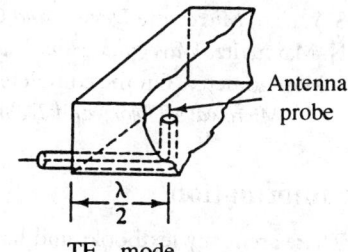

FIGURE 35.16 Coaxial to rectangular waveguide transition that couples the transmission line mode to the dominant waveguide mode (*Source:* S.Y. Liao, *Microwave Devices and Circuits,* 3rd ed., Englewood Cliffs, N.J.: Prentice-Hall, 1990, p. 117. With permission.)

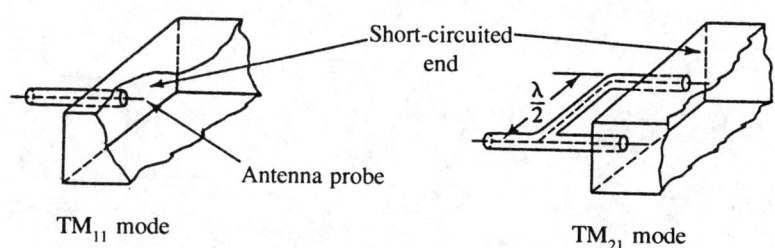

FIGURE 35.17 Coaxial to rectangular waveguide transitions that couple the transmission line mode to the TM_{11} and TM_{21} waveguide modes. (*Source:* S.Y. Liao, *Microwave Devices and Circuits,* 3rd ed., Englewood Cliffs, N.J.: Prentice-Hall, 1990, p. 117. With permission.)

Defining Terms

Cut-off frequency: The minimum frequency at which a waveguide mode will propagate energy with little or no attenuation.

Guide wavelength: The distance over which the fields of propagating modes repeat themselves in a waveguide.

Waveguide: A closed metal cylinder, filled with a dielectric, used to transport electromagnetic energy over short or long distances.

Waveguide modes: Unique electromagnetic field configurations supported by a waveguide that have distinct electrical characteristics.

Wave impedance: The ratio of the transverse electric and magnetic fields inside a waveguide.

References

A. J. Baden Fuller, *Microwaves,* 2nd ed., New York: Pergamon Press, 1979.

R. E. Collin, *Foundations for Microwave Engineering,* 2nd ed., New York: McGraw-Hill, 1992.

J. Kretzschmar, "Wave propagation in hollow conducting elliptical waveguides," *IEEE Transactions on Microwave Theory and Techniques,* vol. MTT-18, no. 9, pp. 547–554, Sept. 1970.

S. Y. Liao, *Microwave Devices and Circuits*, 3rd ed., Englewood Cliffs, N.J.: Prentice-Hall, 1990.

N. Marcuvitz, *Waveguide Handbook*, 2nd ed., London: Peter Peregrinus Ltd., 1986.

J. Montgomery, "On the complete eigenvalue solution of ridged waveguide," *IEEE Transactions on Microwave Theory and Techniques*, vol. MTT-19, no. 6, pp. 457–555, June 1971.

Further Information

There are many textbooks and handbooks that cover the subject of waveguides in great detail. In addition to the references cited above, others include

L. Lewin, *Theory of Waveguides*, New York: John Wiley, 1975.

Reference Data for Radio Engineers, Howard W. Sams Co., 1975.

R. E. Collin, *Field Theory of Guided Waves*, 2nd ed., Piscataway, N.J.: IEEE Press, 1991.

F. Gardiol, *Introduction to Microwaves*, Dedham, Mass.: Artech House, 1984.

S. Ramo, J. Whinnery, and T. Van Duzer, *Fields and Waves in Communication Electronics*, New York: John Wiley, 1965.

Antennas

N. J. Kolias
Cornell University

R. C. Compton
Cornell University

J. Patrick Fitch
Lawrence Livermore Laboratory

36.1 Wire 861
Short Dipole • Directivity • Magnetic Dipole • Input Impedance •
Arbitrary Wire Antennas • Resonant Half-Wavelength Antenna •
End Loading • Arrays of Wire Antennas • Analysis of General Arrays
• Arrays of Identical Elements • Equally Spaced Linear Arrays • Pla-
nar (2-D) Arrays • Yagi–Uda Arrays • Log-Periodic Dipole Arrays

36.2 Aperture 870
The Oscillator or Discrete Radiator • Synthetic Apertures • Geo-
metric Designs • Continuous Current Distributions (Fourier
Transform) • Antenna Parameters

36.1 Wire

N. J. Kolias and R. C. Compton

Antennas have been widely used in communication systems since the early 1900s. Over this span of time scientists and engineers have developed a vast number of different antennas. The radiative properties of each of these antennas are described by an antenna pattern. This is a plot, as a function of direction, of the power P_r per unit solid angle Ω radiated by the antenna. The antenna pattern, also called the **radiation pattern,** is usually plotted in spherical coordinates θ and φ. Often two orthogonal cross sections are plotted, one where the E-field lies in the plane of the slice (called the E-plane) and one where the H-field lies in the plane of the slice (called the H-plane).

Short Dipole

Antenna patterns for a short dipole are plotted in Fig. 36.1. In these plots the radial distance from the origin to the curve is proportional to the radiated power. Antenna plots are usually either on linear scales or decibel scales (10 log power).

The antenna pattern for a short dipole may be determined by first calculating the vector potential **A** [Collin, 1985; Balanis, 1982; Harrington, 1961; Lorrain and Corson, 1970]. Using Collin's notation, the vector potential in spherical coordinates is given by

$$\mathbf{A} = \mu_0 I \, dl \, \frac{e^{-jk_0 r}}{4\pi r} (\mathbf{a}_r \cos \theta - \mathbf{a}_\theta \sin \theta) \tag{36.1}$$

where $k_0 = 2\pi/\lambda_0$, and I is the current, assumed uniform, in the short dipole of length dl ($dl \ll \lambda_0$). Here the assumed time dependence $e^{j\omega t}$ has not been explicitly shown. The electric and magnetic fields may then be determined using

$$\mathbf{E} = -j\omega\mathbf{A} + \frac{\nabla\nabla \cdot \mathbf{A}}{j\omega\mu_0\varepsilon_0} \qquad \mathbf{H} = \frac{1}{\mu_0} \nabla \times \mathbf{A} \tag{36.2}$$

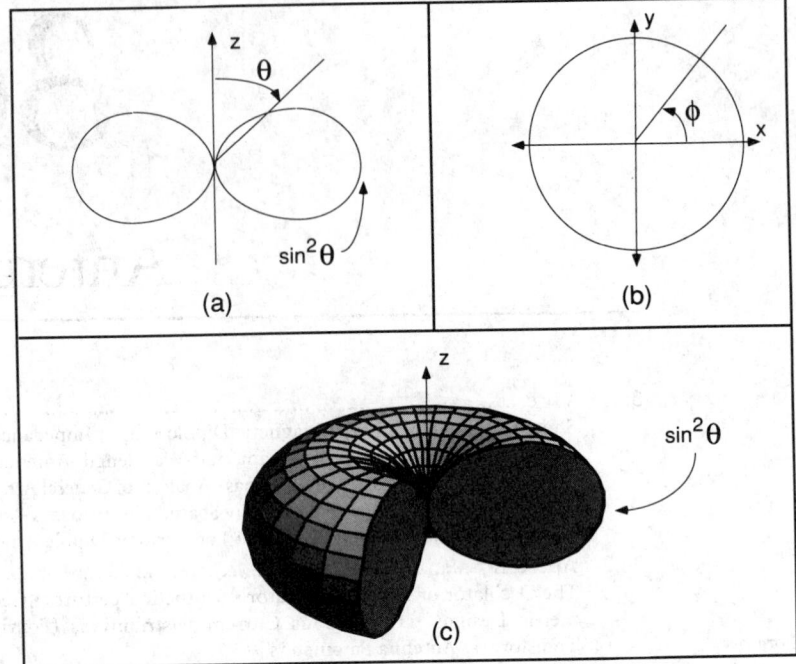

FIGURE 36.1   Radiation pattern for a short dipole of length dl $(dl << \lambda_0)$. These are plots of power density on linear scales. (a) *E*-plane; (b) *H*-plane; (c) three-dimensional view with cutout.

The radiated fields are obtained by calculating these fields in the so-called *far-field region* where $r >> \lambda$. Doing this for the short dipole yields

$$\mathbf{E} = jZ_0 Idl k_0 \sin\theta \frac{e^{-jk_0 r}}{4\pi r} \mathbf{a}_\theta$$

$$\mathbf{H} = jIdl k_0 \sin\theta \frac{e^{-jk_0 r}}{4\pi r} \mathbf{a}_\varphi \tag{36.3}$$

where $Z_0 = \sqrt{\mu_0/\varepsilon_0}$. The average radiated power per unit solid angle Ω can then be found to be

$$\frac{\Delta P_r(\theta,\varphi)}{\Delta \Omega} = \frac{1}{2} r^2 \Re\{\mathbf{E} \times \mathbf{H}^* \cdot \mathbf{a}_r\} = |I|^2 Z_0 (dl)^2 k_0^2 \frac{\sin^2\theta}{32\pi^2} \tag{36.4}$$

Directivity

The **directivity** $D(\theta,\varphi)$ and **gain** $G(\theta,\varphi)$ of an antenna are defined as

$$D(\theta,\varphi) = \frac{\text{Radiated power per solid angle}}{\text{Total radiated power}/4\pi} = \frac{\Delta P_r(\theta,\varphi)/\Delta\Omega}{P_r/4\pi}$$

$$G(\theta,\varphi) = \frac{\text{Radiated power per solid angle}}{\text{Total input power}/4\pi} = \frac{\Delta P_r(\theta,\varphi)/\Delta\Omega}{P_{in}/4\pi} \tag{36.5}$$

Antenna efficiency, η, is given by

$$\eta \equiv \frac{P_r}{P_{in}} = \frac{G(\theta,\varphi)}{D(\theta,\varphi)} \tag{36.6}$$

For many antennas $\eta \approx 1$ and so the words *gain* and *directivity* can be used interchangeably. For the short dipole

$$D(\theta, \varphi) = \frac{3}{2} \sin^2 \theta \tag{36.7}$$

The maximum directivity of the short dipole is 3/2. This single number is often abbreviated as the antenna directivity. By comparison, for an imaginary isotropic antenna which radiates equally in all directions, $D(\theta,\varphi) = 1$. The product of the maximum directivity with the total radiated power is called the *effective isotropic radiated power* (EIRP). It is the total radiated power that would be required for an isotropic radiator to produce the same signal as the original antenna in the direction of maximum directivity.

Magnetic Dipole

A small loop of current produces a *magnetic dipole*. The far fields for the magnetic dipole are dual to those of the electric dipole. They have the same angular dependence as the fields of the electric dipole, but the polarization orientations of **E** and **H** are interchanged.

$$\mathbf{H} = -Mk_0^2 \sin \theta \, \frac{e^{-jk_0 r}}{4\pi r} \, \mathbf{a}_\theta$$
$$\mathbf{E} = MZ_0 k_0^2 \sin \theta \, \frac{e^{-jk_0 r}}{4\pi r} \, \mathbf{a}_\varphi \tag{36.8}$$

where $M = \pi r_0^2 I$ for a loop with radius r_0 and uniform current I.

Input Impedance

At a given frequency the impedance at the feedpoint of an antenna can be represented as $Z_a = R_a + jX_a$. The real part corresponds to radiated fields plus losses, and the imaginary part arises from stored evanescent fields. The radiation resistance is obtained from $|I|^2 R_a/2 = P_r$. For electrically small electric and magnetic dipoles with uniform currents

$$R_a = 80\pi^2 \left(\frac{dl}{\lambda_0} \right)^2 \quad \text{electric dipole}$$
$$R_a = 320\pi^6 \left(\frac{r_0}{\lambda_0} \right)^4 \quad \text{magnetic dipole} \tag{36.9}$$

The reactive component of Z_a can be determined by integrating the evanescent fields. The reflection coefficient, Γ, of the antenna is just

$$\Gamma = \frac{Z_a - Z_0}{Z_a + Z_0} \tag{36.10}$$

where Z_0 is the characteristic impedance of the system used to measure the reflection coefficient.

Arbitrary Wire Antennas

An arbitrary wire antenna can be considered as a sum of small current dipole elements. The vector potential for each of these elements can be determined in the same way as for the short dipole. The total vector potential is then the sum over all these infinitesimal contributions and the resulting E in the far field can be found to be

$$\mathbf{E}(r) = jk_0Z_0\frac{e^{-jk_0r}}{4\pi r}\int_c [(\mathbf{a}_r \cdot \mathbf{a})\mathbf{a}_r - \mathbf{a}]I(l')e^{jk_0\mathbf{a}_r \cdot \mathbf{r}'}dl' \qquad (36.11)$$

where the integral is over the contour C of the wire, $\mathbf{a}$ is a unit vector tangential to the wire, and $\mathbf{r}'$ is the radial vector to the infinitesimal current element.

Resonant Half-Wavelength Antenna

The resonant half-wavelength antenna (commonly called the half-wave dipole) is used widely in antenna systems. Factors contributing to its popularity are its well-understood radiation pattern, its simple construction, its high efficiency, and its capability for easy impedance matching.

The electric and magnetic fields for the half-wave dipole can be calculated by substituting its current distribution, $I = I_0\cos(k_0z)$, into Eq. (36.11) to obtain

$$\mathbf{E} = jZ_0I_0\frac{\cos\left(\dfrac{\pi}{2}\cos\theta\right)}{\sin\theta}\frac{e^{-jk_0r}}{2\pi r}\mathbf{a}_\theta$$

$$\mathbf{H} = jI_0\frac{\cos\left(\dfrac{\pi}{2}\cos\theta\right)}{\sin\theta}\frac{e^{-jk_0r}}{2\pi r}\mathbf{a}_\varphi \qquad (36.12)$$

From the electric and magnetic fields, the radiation resistance can be determined to be

$$R_a = \frac{2P_r}{|I_0|^2} \approx 73\,\Omega \qquad (36.13)$$

This radiation resistance is considerably higher than the radiation resistance of a short dipole. For example, if we have a dipole of length 0.01λ, its radiation resistance will be approximately $0.08\,\Omega$ (from Eq. 36.9). This resistance is probably comparable to the ohmic resistance of the dipole, thereby resulting in a low efficiency. The half-wave dipole, having a much higher radiation resistance, will have much higher efficiency. The higher resistance of the half-wave dipole also makes impedance matching easier.

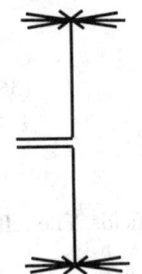

FIGURE 36.2 Using end loading to increase the effective electrical length of an electric dipole.

End Loading

At many frequencies of interest, for example, the broadcast band, a half-wavelength becomes unreasonably long. Figure 36.2 shows a way of increasing the effective length of the dipole without making it longer. Here, additional wires have been added to the ends of the dipoles. These wires increase the end capacitance of the dipole, thereby increasing the effective electrical length.

Arrays of Wire Antennas

Often it is advantageous to have several antennas operating together in an **array.** Arrays of antennas can be made to produce highly directional radiation patterns. Also, small antennas can be used in an array to obtain the level of performance of a large antenna at a fraction of the area.

The radiation pattern of an array depends on the number and type of antennas used, the spacing in the array, and the relative phase and magnitude of the excitation currents. The ability to control the phase of the exciting currents in each element of the array allows one to electronically scan the main radiated beam. An array that varies the phases of the exciting currents to scan the radiation pattern through space is called an electronically scanned **phased array.** Phased arrays are used extensively in radar applications.

Analysis of General Arrays

To obtain analytical expressions for the radiation fields due to an array one must first look at the fields produced by a single array element. For an isolated radiating element positioned as in Fig. 36.3, the electric field at a far-field point P is given by

$$\mathbf{E}_i = a_i \mathbf{K}_i(\theta, \varphi) e^{j[k_0(\mathbf{R}_i \cdot \mathbf{i}_p) - \alpha_i]} \tag{36.14}$$

where $\mathbf{K}_i(\theta, \varphi)$ is the electric field pattern of the individual element, $a_i e^{-j\alpha_i}$ is the excitation of the individual element, $\mathbf{R}_i$ is the position vector from the phase reference point to the element, $\mathbf{i}_p$ is a unit vector pointing toward the far-field point P, and k_0 is the free space wave vector.

Now, for an array of N of these arbitrary radiating elements the total E-field at position P is given by the vector sum

$$\mathbf{E}_{\text{tot}} = \sum_{i=0}^{N-1} \mathbf{E}_i = \sum_{i=0}^{N-1} a_i \mathbf{K}_i(\theta, \varphi) e^{j[k_0(\mathbf{R}_i \cdot \mathbf{i}_p) - \alpha_i]} \tag{36.15}$$

This equation may be used to calculate the total field for an array of antennas where the mutual coupling between the array elements can be neglected. For most practical antennas, however, there is mutual coupling, and the individual patterns will change when the element is placed in the array. Thus, Eq. (36.15) should be used with care.

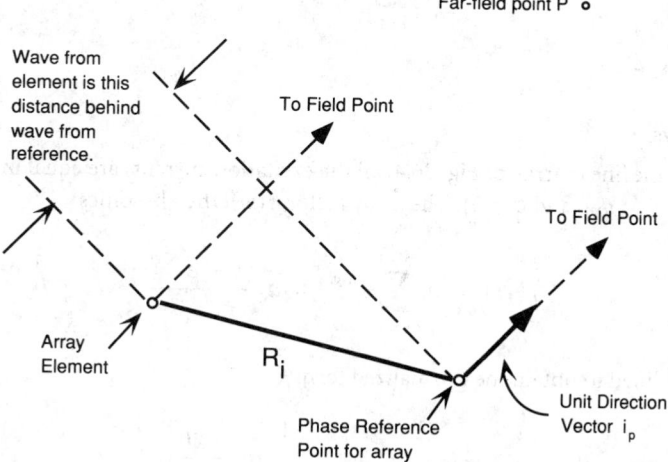

FIGURE 36.3 Diagram for determining the far field due to radiation from a single array element. (*Source: Reference Data for Radio Engineers,* Indianapolis: Howard W. Sams & Co., 1975, chap. 27–22. With permission.)

Arrays of Identical Elements

If all the radiating elements of an array are identical, then $K_i(\theta, \varphi)$ will be the same for each element and Eq. (36.15) can be rewritten as

$$\mathbf{E}_{tot} = \mathbf{K}(\theta, \varphi) \sum_{i=0}^{N-1} a_i e^{j[k_0(\mathbf{R}_i \cdot \mathbf{i}_p) - \alpha_i]} \tag{36.16}$$

This can also be written as

$$\mathbf{E}_{tot} = \mathbf{K}(\theta, \varphi) \, f(\theta, \varphi) \qquad \text{where} \quad f(\theta, \varphi) = \sum_{i=0}^{N-1} a_i e^{j[k_0(\mathbf{R}_i \cdot \mathbf{i}_p) - \alpha_i]} \tag{36.17}$$

The function $f(\theta, \varphi)$ is normally called the array factor or the array polynomial. Thus, one can find $\mathbf{E}_{tot}$ by just multiplying the individual element's electric field pattern, $\mathbf{K}(\theta, \varphi)$, by the array factor, $f(\theta, \varphi)$. This process is often referred to as pattern multiplication.

The average radiated power per unit solid angle is proportional to the square of $\mathbf{E}_{tot}$. Thus, for an array of identical elements

$$\frac{\Delta P_r(\theta, \varphi)}{\Delta \Omega} \sim |\mathbf{K}(\theta, \varphi)|^2 \, |f(\theta, \varphi)|^2 \tag{36.18}$$

Equally Spaced Linear Arrays

An important special case occurs when the array elements are identical and are arranged on a straight line with equal element spacing, d, as shown in Fig. 36.4. If a linear phase progression, α, is assumed for the excitation currents of the elements, then the total field at position P in Fig. 36.4 will be

$$\mathbf{E}_{tot} = \mathbf{K}(\theta, \varphi) \sum_{n=0}^{N-1} a_n e^{jn(k_0 d \cos \theta - \alpha)}$$

$$= \mathbf{K}(\theta, \varphi) \sum_{n=0}^{N-1} a_n e^{jn\psi} = \mathbf{K}(\theta, \varphi) f(\psi) \tag{36.19}$$

where $\psi = k_0 d \cos \theta - \alpha$.

Broadside Arrays

Suppose that, in the linear array of Fig. 36.4, all the excitation currents are equal in magnitude and phase ($a_0 = a_1 = \cdots = a_{N-1}$ and $\alpha = 0$). The array factor, $f(\psi)$, then becomes

$$f(\psi) = a_0 \sum_{n=0}^{N-1} e^{jn\psi} = a_0 \frac{1 - e^{jN\psi}}{1 - e^{j\psi}} \tag{36.20}$$

This can be simplified to obtain the normalized form

$$f'(\psi) = \left| \frac{f(\psi)}{a_0 N} \right| = \left| \frac{\sin \dfrac{N\psi}{2}}{N \sin \dfrac{\psi}{2}} \right| \tag{36.21}$$

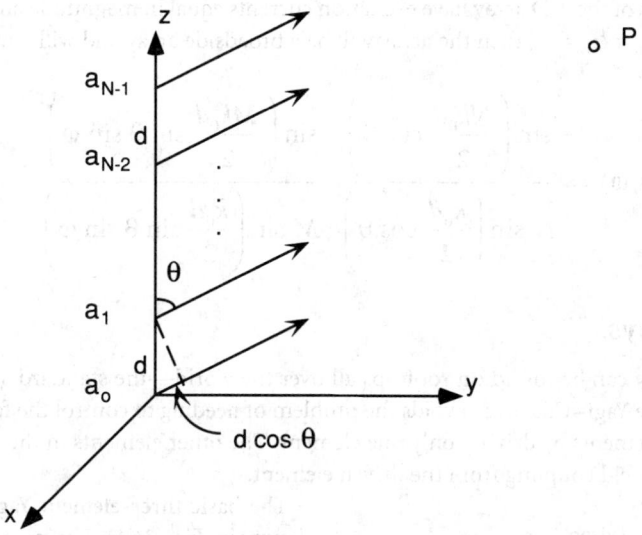

FIGURE 36.4 A linear array of equally spaced elements.

Note that $f'(\psi)$ is maximum when $\psi = 0$. For our case, with $\alpha = 0$, we have $\psi = k_0 d \cos \theta$. Thus $f'(\psi)$ will be maximized when $\theta = \pi/2$. This direction is perpendicular to the axis of the array (see Fig. 36.4), and so the resulting array is called a broadside array.

Phased Arrays

By adjusting the phase of the elements of the array it is possible to vary the direction of the maximum of the array's radiation pattern. For arrays where all the excitation currents are equal in magnitude but not necessarily phase, the array factor is a maximum when $\psi = 0$. From the definition of ψ, one can see that at the pattern maximum

$$k_0 d \cos \theta = \alpha$$

Thus, the direction of the array factor maximum is given by

$$\theta = \cos^{-1}\left(\frac{\alpha}{k_0 d}\right)$$

Note that if one is able to control the phase delay, α, the direction of the maximum can be scanned without physically moving the antenna.

Planar (2-D) Arrays

Suppose there are M linear arrays, all identical to the one pictured in Fig. 36.4, lying in the yz-plane with element spacing d in both the y and the z direction. Using the origin as the phase reference point, the array factor can be determined to be

$$f(\theta, \varphi) = \sum_{n=0}^{N-1} \sum_{m=0}^{M-1} a_{mn} e^{[jn(k_0 \, d \cos \theta - \alpha_z) + jm(k_0 \, d \sin \theta \sin \varphi - \alpha_y)]} \tag{36.22}$$

where α_y and α_z are the phase differences between the adjacent elements in the y and z directions, respectively. The formula can be derived by considering the 2-D array to be a 1-D array of subarrays, where each subarray has an antenna pattern given by Eq. (36.19).

If all the elements of the 2-D array have excitation currents equal in magnitude and phase (all the a_{mn} are equal and $\alpha_z = \alpha_y = 0$), then the array will be a broadside array and will have a normalized array factor given by

$$f'(\theta, \varphi) = \frac{\sin\left(\dfrac{Nk_0 d}{2} \cos\theta\right)}{N \sin\left(\dfrac{k_0 d}{2} \cos\theta\right)} \; \frac{\sin\left(\dfrac{Mk_0 d}{2} \sin\theta \sin\varphi\right)}{M \sin\left(\dfrac{k_0 d}{2} \sin\theta \sin\varphi\right)} \qquad (36.23)$$

Yagi–Uda Arrays

The Yagi–Uda array can be found on rooftops all over the world—the standard TV antenna is a Yagi–Uda array. The Yagi–Uda array avoids the problem of needing to control the feeding currents to all of the array elements by driving only one element. The other elements in the Yagi–Uda array are excited by near-field coupling from the driven element.

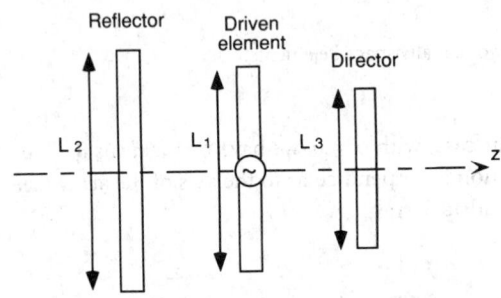

The basic three-element Yagi–Uda array is shown in Fig. 36.5. The array consists of a driven antenna of length l_1, a reflector element of length l_2, and a director element of length l_3. Typically, the director element is shorter than the driven element by 5% or more, while the reflector element is longer than the driven element by 5% or more [Stutzman and Thiele, 1981]. The radiation pattern for the array in Fig. 36.5 will have a maximum in the $+z$ direction.

One can increase the gain of the Yagi–Uda array by adding additional director elements. Adding additional reflector elements, however, has little effect because the field behind the first reflector element is small.

FIGURE 36.5 Three-element Yagi–Uda antenna. (*Source:* Shintaro Uda and Yasuto Mushiake, *Yagi–Uda Antenna,* Sendai, Japan: Sasaki Printing and Publishing Company, 1954, p. 100. With permission.)

Yagi–Uda arrays typically have directivities between 10 and 100, depending on the number of directors [Ramo *et al.*, 1984]. TV antennas usually have several directors.

Log-Periodic Dipole Arrays

Another variation of wire antenna arrays is the log-periodic dipole array. The log-periodic is popular in applications that require a broadband, frequency-independent antenna. An antenna will be independent of frequency if its dimensions, when measured in wavelengths, remain constant for all frequencies. If, however, an antenna is designed so that its characteristic dimensions are periodic with the logarithm of the frequency, and if the characteristic dimensions do not vary too much over a period of time, then the antenna will be essentially frequency independent. This is the basis for the log-periodic dipole array, shown in Fig. 36.6.

In Fig 36.6, the ratio of successive element positions equals the ratio of successive dipole lengths. This ratio is often called the scaling factor of the log-periodic array and is denoted by

$$\tau = \frac{z_{n+1}}{z_n} = \frac{L_{n+1}}{L_n} \qquad (36.24)$$

Also note that there is a mechanical phase reversal between successive elements in the array caused by the crossing over of the interconnecting feed lines. This phase reversal is necessary to obtain the proper phasing between adjacent array elements.

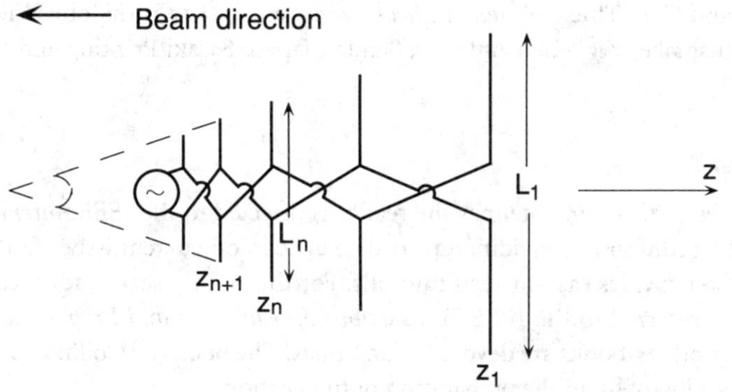

FIGURE 36.6 The log-periodic dipole array. (*Source:* D.G. Isbell, "Log periodic dipole arrays," *IRE Transactions on Antennas and Propagation,* vol. AP-8, p. 262, 1960. With permission.)

To get an idea of the operating range of the log-periodic antenna, note that for a given frequency within the operating range of the antenna, there will be one dipole in the array that is half-wave resonant or is nearly so. This half-wave resonant dipole and its immediate neighbors are called the active region of the log-periodic array. As the operating frequency changes, the active region shifts to a different part of the log-periodic. Hence, the frequency range for the log-periodic array is roughly given by the frequencies at which the longest and shortest dipoles in the array are half-wave resonant (wavelengths such that $2L_N < \lambda < 2L_1$) [Stutzman and Thiele, 1981].

Defining Terms

Antenna gain: The ratio of the actual radiated power per solid angle to the radiated power per solid angle that would result if the total input power were radiated isotropically.

Array: Several antennas arranged together in space and interconnected to produce a desired radiation pattern.

Directivity: The ratio of the actual radiated power per solid angle to the radiated power per solid angle that would result if the radiated power was radiated isotropically. Oftentimes the word *directivity* is used to refer to the maximum directivity.

Phased array: An array in which the phases of the exciting currents are varied to scan the radiation pattern through space.

Radiation pattern: A plot as a function of direction of the power per unit solid angle radiated in a given polarization by an antenna. The terms *radiation pattern* and *antenna pattern* can be used interchangeably.

References

C. A. Balanis, *Antenna Theory Analysis and Design,* New York: Harper and Row, 1982.

R. Carrel, "The design of log-periodic dipole antennas," *IRE International Convention Record* (part 1), 1961, pp. 61–75.

R. E. Collin, *Antennas and Radiowave Propagation,* New York: McGraw-Hill, 1985.

R. F. Harrington, *Time Harmonic Electromagnetic Fields,* New York: McGraw-Hill, 1961.

D. E. Isbell, "Log periodic dipole arrays," *IRE Transactions on Antennas and Propagation,* vol. AP-8, pp. 260–267, 1960.

P. Lorrain and D. R. Corson, *Electromagnetic Fields and Waves,* San Francisco: W. H. Freeman, 1970.

S. Ramo, J. R. Whinnery, and T. Van Duzer, *Fields and Waves in Communication Electronics,* New York: John Wiley & Sons, 1984.

W. L. Stutzman and G. A. Thiele, *Antenna Theory and Design*, New York: John Wiley & Sons, 1981.

S. Uda and Y. Mushiake, *Yagi–Uda Antenna*, Sendai, Japan: Sasaki Printing and Publishing Company, 1954.

Further Information

For general-interest articles on antennas the reader is directed to the *IEEE Antennas and Propagation Magazine*. In addition to providing up-to-date articles on current issues in the antenna field, this magazine also provides easy-to-read tutorials. For the latest research advances in the antenna field the reader is referred to the *IEEE Transactions on Antennas and Propagation*. In addition, a number of very good textbooks are devoted to antennas. The books by Collin and by Stutzman and Thiele were especially useful in the preparation of this section.

36.2 Aperture

J. Patrick Fitch

The main purpose of an **antenna** is to control a wave front at the boundary between two media: a source (or receiver) and the medium of propagation. The source can be a fiber, cable, waveguide, or other transmission line. The medium of propagation may be air, vacuum, water, concrete, metal, or tissue, depending on the application. Antenna aperture design is used in acoustic, optic, and electromagnetic systems for imaging, communications, radar, and spectroscopy applications.

There are many classes of antennas: wire, horn, slot, notch, reflector, lens, and **array**, to name a few (see Fig. 36.7). Within each class is a variety of subclasses. For instance, the horn antenna can be pyramidal or conical. The horn can also have flaring in only one direction (sectoral horn), asymmetric components, shaped edges, or a compound design of sectoral and pyramidal combined. For all antennas, the relevant design and analysis will depend on antenna aperture size and shape, the center wavelength λ, and the distance from the aperture to a point of interest (the range, R). This section covers discrete **oscillators**, arrays of oscillators, synthetic apertures, geometric design, Fourier analysis, and parameters of some typical antennas. The emphasis is on microwave-type designs.

The Oscillator or Discrete Radiator

The basic building block for antenna analysis is a linear conductor. Movement of electrons (current) in the conductor induces an electromagnetic field. When the electron motion is oscillatory—e.g., a dipole with periodic electron motion, the induced electric field, E, is proportional to $\cos(\omega t - kx + \phi)$, where ω is radian frequency of oscillation, t is time, k is wave number, x is distance from the oscillator, and ϕ is the phase associated with this oscillator (relative to the time and spatial coordinate origins). When the analysis is restricted to a fixed position x, the electric field can be expressed as

$$E(t) = A \cos(\omega t + \phi) \tag{36.25}$$

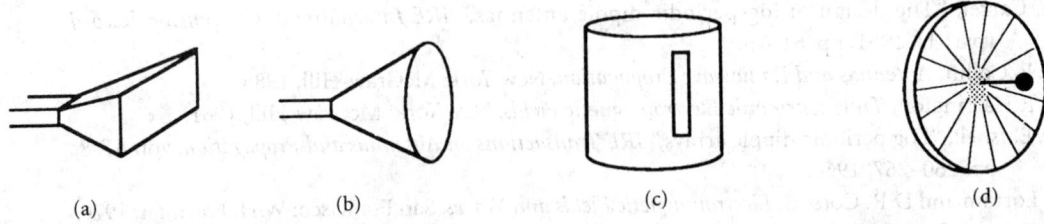

(a)   (b)   (c)   (d)

FIGURE 36.7 Examples of several types of antennas: (a) pyramidal horn, (b) conical horn, (c) axial slot on a cylinder, and (d) parabolic reflector.

where the phase term ϕ now includes the kx term, and all of the constants of proportionality are included in the amplitude A. Basically, the assumption is that oscillating currents produce oscillating fields. The description of a receiving antenna is analogous: an oscillating field induces a periodic current in the conductor.

The field from a pair of oscillators separated in phase by δ radians is

$$E_\delta(t) = A_1 \cos(\omega t + \phi) + A_2 \cos(\omega t + \phi + \delta) \tag{36.26}$$

Using phasor notation, $\tilde{E}_\delta$, the cosines are converted to complex exponentials and the radial frequency term, ωt, is suppressed,

$$\tilde{E}_\delta(t) = A_1 e^{i\phi} + A_2 e^{i(\phi+\delta)} \tag{36.27}$$

The amplitude of the sinusoidal modulation $E_\delta(t)$ can be calculated as $|\tilde{E}_\delta|$. The intensity is

$$I = |\tilde{E}_\delta|^2 = |A_1|^2 + |A_2|^2 + 2A_1 A_2 \cos(\delta) \tag{36.28}$$

When the oscillators are of the same amplitude, $A = A_1 = A_2$, then

$$\begin{aligned} E_\delta(t) &= A \cos(\omega t + \phi) + A \cos(\omega t + \phi + \delta) \\ &= 2A \cos\left(\frac{\delta}{2}\right) \cos\left(\omega t + \phi + \frac{\delta}{2}\right) \end{aligned} \tag{36.29}$$

For a series of n equal amplitude oscillators with equal phase spacing

$$E_{n\delta}(t) = \sum_{j=0}^{n-1} A \cos(\omega t + \phi + j\delta) \tag{36.30}$$

By using phasor arithmetic the intensity is given as

$$I_{n\delta}(t) = |\tilde{E}_{n\delta}|^2 = \left| A e^{i\phi} \sum_{j=0}^{n-1} e^{ij\delta} \right|^2 = A^2 \left| \frac{1 - e^{in\delta}}{1 - e^{i\delta}} \right|^2 = I_0 \frac{1 - \cos(n\delta)}{1 - \cos(\delta)}$$

$$= I_0 \frac{\sin^2(n\delta/2)}{\sin^2(\delta/2)} \tag{36.31}$$

where $I_0 = n^{-2}$ to normalize the intensity pattern at $\delta = 0$.

For an incoming plane wave which is tilted at an angle θ from the normal, the relative phase difference between two oscillators is $kd \sin\theta$, where d is the distance between oscillators and k is the wave number $2\pi/\lambda$ (see Fig. 36.8). For three evenly spaced oscillators, the phase difference between the end oscillators is $2kd \sin\theta$. In general, the end-to-end phase difference for n evenly spaced oscillators is $(n-1)kd \sin\theta$. This formulation is identical to the phase representation in Eq. (36.30) with $\delta = kd \sin\theta$. Therefore, the intensity as a function of incidence angle θ for an evenly spaced array of n elements is

$$I_{nL}(\theta) = I_0 \frac{\sin^2\left(\dfrac{1}{2} knd \sin\theta\right)}{\sin^2\left(\dfrac{1}{2} kd \sin\theta\right)} = I_0 \frac{\sin^2\left(\dfrac{1}{2} kL \sin\theta\right)}{\sin^2\left(\dfrac{1}{2n} kL \sin\theta\right)} = I_0 \frac{\sin^2\left(\dfrac{\pi L}{\lambda} \sin\theta\right)}{\sin^2\left(\dfrac{\pi L}{n\lambda} \sin\theta\right)} \tag{36.32}$$

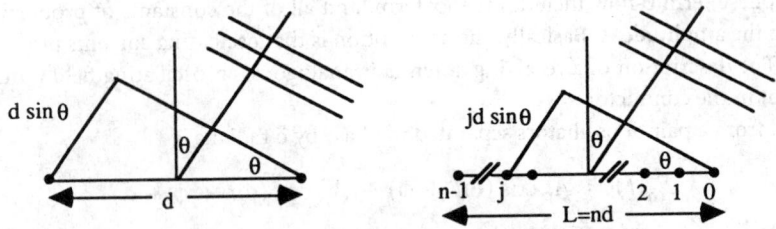

FIGURE 36.8   A two-element and an *n*-element array with equal spacing between elements. The propagation length difference between elements is $d \sin \theta$, which corresponds to a phase difference of $kd \sin \theta$, where k is the wave number $2\pi/\lambda$. The length L corresponds to a continuous aperture of length nd with the sample positions beginning $d/2$ from the ends.

where $L = nd$ corresponds to the physical dimension (length) of the aperture of oscillators. The zeros of this function occur at $kL \sin \theta = 2m\pi$, for any nonzero integer m. Equivalently, the zeros occur when $\sin \theta = m\lambda/L$. When the element spacing d is less than a wavelength, the number of zeros for $0 < \theta < \pi/2$ is given by the largest integer M such that $M \leq L/\lambda$. Therefore, the ratio of wavelength to largest dimension, λ/L, determines both the location (in θ space) and the number of zeros in the intensity pattern when $d \leq \lambda$. The number of oscillators controls the amplitude of the side lobes.

For $n = 1$, the intensity is constant—i.e., independent of angle. For $\lambda > L$, both the numerator and denominator of Eq. (36.32) have no zeros and as the length of an array shortens (relative to a wavelength), the intensity pattern converges to a constant ($n = 1$ case). As shown in Fig. 36.9, a separation of $\lambda/4$ has an intensity rolloff less than 1 dB over $\pi/2$ radians (a $\lambda/2$ separation rolls off 3 dB). This implies that placing antenna elements closer than $\lambda/4$ does not significantly change the intensity pattern. Many microwave antennas exploit this and use a mesh or parallel wire (for polarization sensitivity) design rather than covering the entire aperture with conductor. This reduces both weight and sensitivity to wind loading. Note that the analysis has not accounted for phase variations from position errors in the element placement where the required accuracy is typically better than $\lambda/10$.

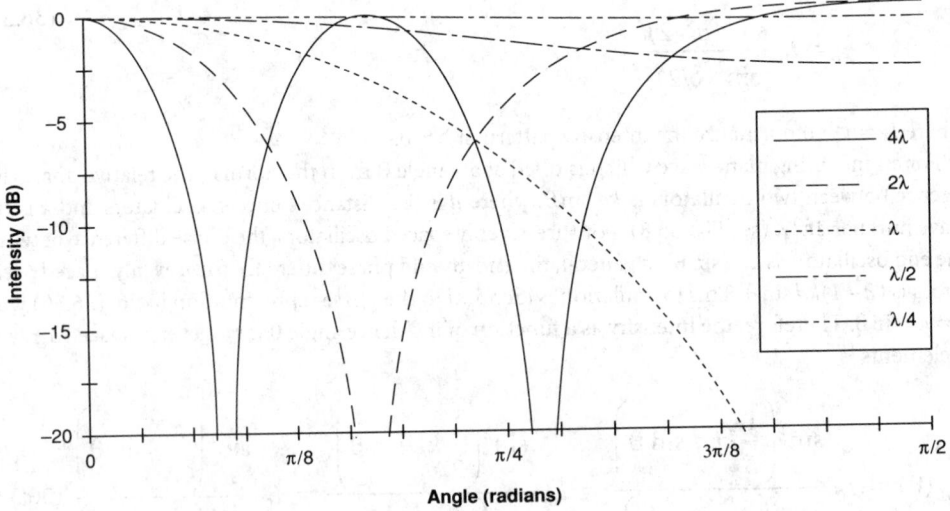

FIGURE 36.9   Normalized intensity pattern in decibels ($10 \log(I)$) for a two-element antenna with spacing 4λ, 2λ, λ, $\lambda/2$, and $\lambda/4$ between the elements.

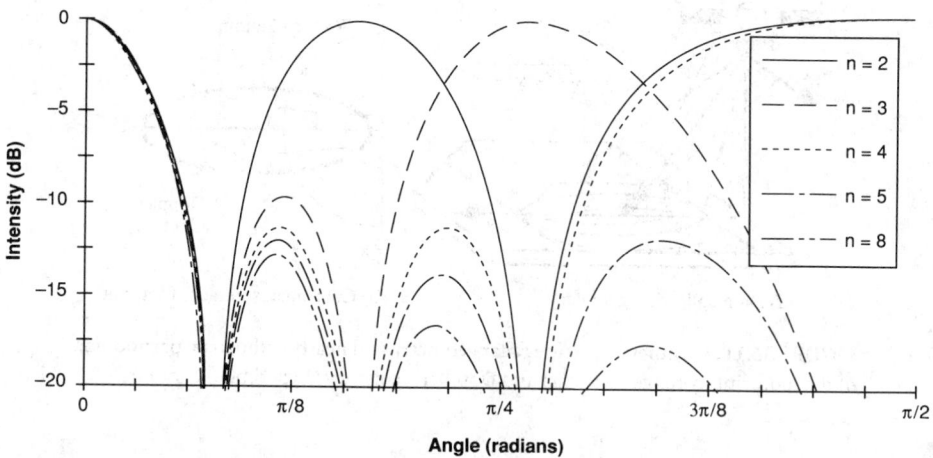

FIGURE 36.10 Normalized intensity pattern in decibels $(10 \log(I))$ for a length 4λ array with 2, 3, 4, 5, and 8 elements.

For $L >> \lambda$, $\sin\theta \approx \theta$, which implies that the first zero is at $\theta = \lambda/L$. The location of the first zero is known as the Rayleigh resolution criteria. That is, two plane waves separated by at least λ/L radians can be discriminated. For imaging applications, this corresponds roughly to the smallest detectable feature size. As shown in Fig. 36.10, the first zero occurs at approximately $\lambda/L = 0.25$ radians (the Rayleigh resolution). Note that there is no side lobe suppression until $d \leq \lambda$, when the location of the zeros becomes fixed. Having more than eight array elements (separation of less than a quarter wavelength) only moderately reduces the height of the maximum side lobe.

Synthetic Apertures

In applications such as air- and space-based radar, size and weight constraints prohibit the use of very large antennas. For instance, if the L-band (23.5-cm wavelength) radar imaging system on the Seasat satellite (800-km altitude, launched in 1978) had a minimum resolution specification of 23.5 m, then, using the Rayleigh resolution criteria, the aperture would need to be 8 km long. In order to attain the desired resolution, an aperture is "synthesized" from data collected with a physically small (10 m) antenna traversing an 8-km flight path. Basically, by using a stable oscillator on the spacecraft, both amplitude and phase are recorded, which allows postprocessing algorithms to combine the individual echoes in a manner analogous to an antenna array. From an antenna perspective, an individual scattering element produces a different round trip propagation path based on the position of the physical antenna—a synthetic antenna array. Using the geometry described in Fig. 36.11, the phase is

$$\phi(x) = \frac{2\pi}{\lambda} 2R(x) = \frac{2\pi}{\lambda} 2\sqrt{x^2 + y^2 + z^2} \qquad (36.33)$$

It is convenient to assume a straight-line flight path along the x-axis, a planar earth (x, y plane), and a constant velocity, v, with range and cross-range components $v_r(x)$ and $v_c(x)$, respectively. In many radar applications the broad side distance to the center of the footprint, R, is much larger than the size of the footprint. This allows the distance $R(x)$ to be expanded about R resulting in

$$\phi(t) = \frac{2\pi}{\lambda} 2R(vt) = 2\pi\left\{\frac{2R}{\lambda} + \frac{2v_r}{\lambda} t + \frac{v_c^2}{\lambda R} t^2\right\} \qquad (36.34)$$

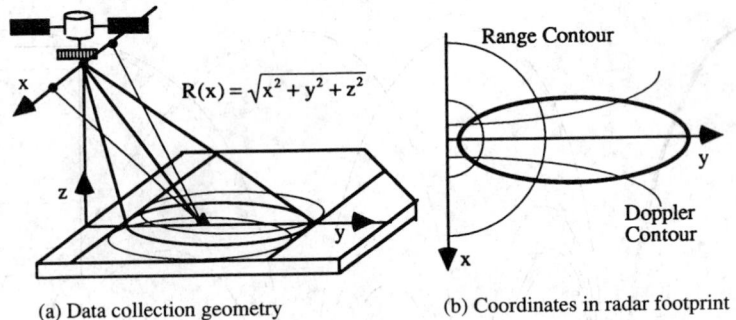

(a) Data collection geometry   (b) Coordinates in radar footprint

FIGURE 36.11 Synthetic aperture radar geometry and nearly orthogonal partitioning of the footprint by range (circular) and Doppler frequency (hyperbolic) contours.

The first term in Eq. (36.34) is a constant phase offset corresponding to the center of beam range bin and can be ignored from a resolution viewpoint. The second term, $2v_r/\lambda$, is the Doppler frequency shift due to the relative (radial) velocity between antenna and scattering element. The third term represents a quadratic correction of the linear flight path to approximate the constant range sphere from a scattering element. It is worth noting that synthetic aperture systems do not require the assumptions used here, but accurate position and motion compensation is required.

For an antenna with cross range dimension D and a scattering element at range R, the largest synthetic aperture that can be formed is of dimension $\lambda R/D$ (the width of the footprint). Because this data collection scenario is for round trip propagation, the phase shift at each collecting location is twice the shift at the edges of a single physical antenna. Therefore at a range R, the synthetic aperture resolution is

$$\frac{\lambda R}{D_{SA}} = \frac{\lambda R}{2\lambda R/D} = \frac{D}{2} \tag{36.35}$$

The standard radar interpretation for synthetic apertures is that information coded in the Doppler frequency shift can be decoded to produce high-resolution images. It is worth noting that the synthetic aperture can be formed even with no motion (zero Doppler shift). For the no-motion case the antenna array interpretation is appropriate. This approach has been used for acoustic signal processing in nondestructive evaluation systems as well as wave migration codes for seismic signal processing. When there is motion, the Doppler term in the expansion of the range dominates the phase shift and therefore becomes the useful metric for predicting resolution.

Geometric Designs

The phase difference in a linear array was caused by the spatial separation and allowed the discrimination of plane waves arriving at different angles. Desired phase patterns can be determined by using analytic geometry to position the elements. For example, if coherent superposition across a wave front is desired, the wave front can be directed (reflected, refracted, or diffracted) to the receiver in phase. For a planar wave front, this corresponds to a constant path length from any point on the reference plane to the receiver. Using the geometry in Fig. 36.12, the sum of the two lengths $(x, R + h)$ to (x, y) and (x, y) to $(0, h)$ must be a constant independent of x—which is $R + 2h$ for this geometry. This constraint on the length is

$$R + h - y + \sqrt{x^2 + (h - y)^2} = R + 2h \quad \text{or} \quad x^2 = 4hy \tag{36.36}$$

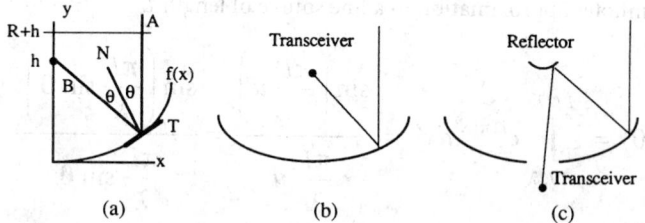

FIGURE 36.12   Parabolic reflector systems: (a) geometry for determining the function with a constant path length and specular reflection, (b) single-bounce parabolic reflector, (c) two-bounce reflector with a parabolic primary and hyperbolic secondary (Cassegrain).

This is the equation for a parabola. Losses would be minimized if the wave front were specularly reflected to the transceiver. Specular reflection occurs when the angles between the normal vector **N** [or equivalently the tangent vector $\mathbf{T} = (x, f'(x))] = (1, x/2h)$ and the vectors $\mathbf{A} = (0, -1)$ and $\mathbf{B} = (-x, h - y)$ are equal. This is the same as equality of the inner products of the normalized vectors, which is shown by

$$\hat{\mathbf{T}} \cdot \hat{\mathbf{A}} = \frac{(2h, x)}{\sqrt{x^2 + 4h^2}} \cdot (0, -1) = \frac{-x}{\sqrt{x^2 + 4h^2}} \tag{36.37}$$

$$\hat{\mathbf{T}} \cdot \hat{\mathbf{B}} = \frac{(2h, x)}{\sqrt{x^2 + 4h^2}} \cdot \frac{(-x, h - y)}{\sqrt{x^2 + (h - y)^2}} = \frac{-x(x^2 + 4h^2)}{(x^2 + 4h^2)^{3/2}} = \frac{-x}{\sqrt{x^2 + 4h^2}} \tag{36.38}$$

The constant path length and high gain make the parabolic antenna popular at many wavelengths including microwave and visible. More than one reflecting surface is allowed in the design. The surfaces are typically conical sections and may be designed to reduce a particular distortion or to provide better functionality. Compound designs often allow the active elements to be more accessible and eliminate long transmission lines. A two-bounce reflector with a parabolic primary and a hyperbolic secondary is known as a Cassegrain system. In all reflector systems it is important to account for the blockage ("shadow" of the feed, secondary reflector, and support structures) as well as the spillover (radiation propagating past the intended reflecting surface).

Continuous Current Distributions (Fourier Transform)

Ideally, antennas would be designed using solutions to Maxwell's equations. Unfortunately, in most cases exact analytic and numerical solutions to Maxwell's equations are difficult to obtain. Under certain conditions, approximations can be introduced that allow solution to the wave equations. Approximating spherical wave fronts as quadratics has been shown for the synthetic aperture application and is valid when the propagation distance is greater than $(\pi L^2/4\lambda)^{1/3}$, where L is the aperture size. In general, this is known as the **Fresnel** or **near-field** approximation. When the propagation distance is at least $2L^2/\lambda$, the angular radiation pattern can be approximated as independent of distance from the aperture. This pattern is known as the normalized **far-field** or **Fraunhofer** distribution, $E(\theta)$, and is related to the normalized current distributed across an antenna aperture, $i(x)$, by a Fourier transform:

$$E(u) = \int i(x')e^{i2\pi ux'} dx' \tag{36.39}$$

where $u = \sin\theta$ and $x' = x/\lambda$.

Applying the Fraunhofer approximation to a line source of length L

$$E_L(u = \sin \theta) = \int_{-L/2\lambda}^{L/2\lambda} e^{i2\pi ux'} dx' = \frac{\sin\left(\dfrac{\pi L}{\lambda} u\right)}{\dfrac{\pi L}{\lambda} u} = \frac{\sin\left(\dfrac{\pi L}{\lambda} \sin \theta\right)}{\dfrac{\pi L}{\lambda} \sin \theta} \qquad (36.40)$$

which is Eq. (36.32) when $n >> L/\lambda$. As with discrete arrays, the ratio L/λ is the important design parameter: $\sin \theta = \lambda/L$ is the first zero (no zeros for $\lambda > L$) and the number of zeros is the largest integer M such that $M \leq L/\lambda$.

In two dimensions, a rectangular aperture with uniform current distribution produces

$$E_R(u_1, u_2) = \frac{\sin\left(\dfrac{\pi}{\lambda} u_1 L_1\right)}{\dfrac{\pi}{\lambda} u_1 L_1} \frac{\sin\left(\dfrac{\pi}{\lambda} u_2 L_2\right)}{\dfrac{\pi}{\lambda} u_2 L_2} \quad \text{and} \quad I_R(u_1, u_2) = |E_L(u_1)|^2 |E_L(u_2)|^2 \qquad (36.41)$$

The field and intensity given in Eq. (36.41) are normalized. In practice, the field is proportional to the aperture area and inversely proportional to the wavelength and propagation distance.

The normalized far-field intensity distribution for a uniform current on a circular aperture is a circularly symmetric function given by

$$I_C(u) = \left[\frac{2J_1\left(\dfrac{\pi}{\lambda} uL\right)}{\dfrac{\pi}{\lambda} uL}\right]^2 \qquad (36.42)$$

where J_1 is the Bessel function of the first kind, order one. This far-field intensity is called the Airy pattern. As with the rectangular aperture, the far-field intensity is proportional to the square of the area and inversely proportional to the square of the wavelength and the propagation distance. The first zero (Rayleigh resolution criteria) of the Airy pattern occurs for $uL/\lambda = 1.22$ or $\sin \theta = 1.22\lambda/L$. As with linear and rectangular apertures, the resolution scales with λ/L.

Figure 36.13 shows a slice through the normalized far-field intensity of both a rectangular aperture and a circular aperture. The linearity of the Fourier transform allows apertures to be represented as the superposition of subapertures. The primary reflector, the obscurations from the support structures, and the secondary reflector of a Cassegrain-type antenna can be modeled. Numerical evaluation of the Fourier transform permits straightforward calculation of the intensity patterns, even for nonuniform current distributions.

Antenna Parameters

Direct solutions to Maxwell's equations or solutions dependent on approximations provide the analytic tools for designing antennas. Ultimately, the analysis must be confirmed with experiment. Increasingly sensitive radar and other antenna applications have resulted in much more attention to edge effects (from the primary aperture, secondary, and/or support structures). The geometric theory of diffraction as well as direct Maxwell solvers are making important contributions.

With the diversity of possible antenna designs, a collection of design rules of thumb are useful. The **directivity** and **gain** for a few popular antenna designs are given in Table 36.1. Directivity is the

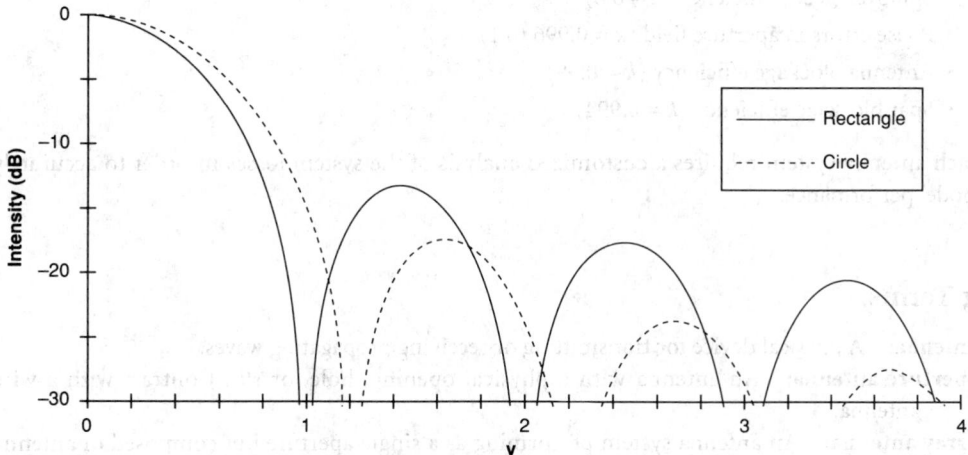

FIGURE 36.13 Normalized intensity pattern in decibels ($10 \log[I(v = uL/\lambda)]$) for a rectangular and a circular antenna aperture with uniform current distributions.

Table 36.1 Directivity and Gain of Some Higher Frequency Antennas

Antenna Type	Directivity[a]	Gain[a]
Uniform rectangular aperture	$\dfrac{4\pi}{\lambda^2} L_x L_y$	$\dfrac{4\pi}{\lambda^2} L_x L_y$
Large square aperture	$12.6 \left(\dfrac{L}{\lambda}\right)^2$	$7.7 \left(\dfrac{L}{\lambda}\right)^2$
Large circular aperture (parabolic reflector)	$9.87 \left(\dfrac{D}{\lambda}\right)^2$	$7 \left(\dfrac{D}{\lambda}\right)^2$
Pyramidal horn	$\left(\dfrac{4\pi}{\lambda^2}\right) L_x L_y$	$0.5 \left(\dfrac{4\pi}{\lambda^2}\right) L_x L_y$

[a]Directivity and gain are relative to a half-wave dipole.

ratio of the maximum to average radiation intensity. The gain is defined as the ratio of the maximum radiation intensity from the subject antenna to the maximum radiation intensity from a reference antenna with the same power input. The directivity, D, and gain, G, of an antenna can be expressed as

$$ D = \left(\frac{4\pi}{\lambda^2}\right) A_{em} \quad \text{and} \quad G = \left(\frac{4\pi}{\lambda^2}\right) A_e \qquad (36.43) $$

where A_{em} is the maximum effective aperture and A_e is the actual effective aperture of the antenna. Because of losses in the system, $A_e = kA_{em}$, where k is the radiation efficiency factor. The gain equals the directivity when there are no losses ($k = 1$), but is less than the directivity if there are any losses in the antenna ($k < 1$), that is, $G = kD$.

As an example, consider the parabolic reflector antenna where efficiency degradation includes

- Ohmic losses are small ($k = 1$)
- Aperture taper efficiency ($k = 0.975$)

- Spillover (feed) efficiency ($k = 0.8$)
- Phase errors in aperture field ($k = 0.996$ to 1)
- Antenna blockage efficiency ($k = 0.99$)
- Spar blockage efficiency ($k = 0.994$)

Each antenna system requires a customized analysis of the system losses in order to accurately model performance.

Defining Terms

Antenna: A physical device for transmitting or receiving propagating waves.

Aperture antenna: An antenna with a physical opening, hole, or slit. Contrast with a wire antenna.

Array antenna: An antenna system performing as a single aperture but composed of antenna subsystems.

Directivity: The ratio of the maximum to average radiation intensity.

Fraunhofer or far field: The propagation region where the normalized angular radiation pattern is independent of distance from the source. This typically occurs when the distance from the source is at least $2L^2/\lambda$, where L is the largest dimension of the antenna.

Fresnel or near field: The propagation region where the normalized radiation pattern can be calculated using quadratic approximations to the spherical Huygens' wavelet surfaces. The pattern can depend on distance from the source and is usually valid for distances greater than $(\pi/4\lambda)^{1/3} L^{2/3}$, where L is the largest dimension of the antenna.

Gain: The ratio of the maximum radiation intensity from the subject antenna to the maximum radiation intensity from a reference antenna with the same power input. Typical references are a lossless isotropic source and a lossless half-wave dipole.

Oscillator: A physical device that uses the periodic motion within the material to create propagating waves. In electromagnetics, an oscillator can be a conductor with a periodic current distribution.

Reactive near field: The region close to an antenna where the reactive components of the electromagnetic fields from charges on the antenna structure are very large compared to the radiating fields. Considered negligible at distances greater than a wavelength from the source (decay as the square or cube of distance). Reactive field is important at antenna edges and for electrically small antennas.

References

R. Feynman, R. B. Leighton, and M. L. Sands, *The Feynman Lectures on Physics*, Reading, Mass.: Addison-Wesley, 1989.

J. P. Fitch, *Synthetic Aperture Radar*, New York: Springer-Verlag, 1988.

J. W. Goodman, *Introduction to Fourier Optics*, New York: McGraw-Hill, 1968.

H. Jasik, *Antenna Engineering Handbook*, New York: McGraw-Hill, 1961.

R. W. P. King and G. S. Smith, *Antennas in Matter*, Cambridge: MIT Press, 1981.

J. D. Krause, *Antennas*, New York: McGraw-Hill, 1950.

Y. T. Lo and S. W. Lee, *Antenna Handbook*, New York: Van Nostrand Reinhold, 1988.

A. W. Rudge, K. Milne, A. D. Olver, and P. Knight, *The Handbook of Antenna Design*, London: Peter Peregrinus, 1982.

M. Skolnik, *Radar Handbook*, New York: McGraw-Hill, 1990.

B. D. Steinberg, *Principles of Aperture & Array System Design*, New York: John Wiley & Sons, 1976.

Further Information

The monthly *IEEE Transactions on Antennas and Propagation* as well as the proceedings of the annual *IEEE Antennas and Propagation International Symposium* provide information about recent developments in this field. Other publications of interest include the *IEEE Transactions on Microwave Theory and Techniques* and the *IEEE Transactions on Aerospace and Electronic Systems*.

Readers may also be interested in the "IEEE Standard Test Procedures for Antennas," The Institute for Electrical and Electronics Engineers, Inc., ANSI IEEE Std. 149-1979, 1979.

MAXWELL'S GRAND UNIFICATION

In the second half of the 19th century, theory was starting to grapple with electromagnetic phenomena. The battlefield soon drew two Scotsmen, both well armed with mathematical skills. One, William Thomson, was to fall by the wayside after some limited successes, though he later became Lord Kelvin. But the other, James Clerk Maxwell, succeeded.

In a series of letters to Thomson in the middle 1850s, Maxwell outlined his ideas on where to pick up the path to a theory of a "whole mass of confusion," as he called it. He felt the key lay in the *electrotonic state*, Michael Faraday's name for inductive effects. Maxwell's first step toward an electromagnetic theory, therefore, was a paper in 1856 on this vague, ill-formed concept. He wrote this paper, "On Faraday's Lines of Force," when he was just 24. He borrowed Thomson's 1847 idea of calculating a vector from another vector by means of the curl vector operation (the word *curl* being Maxwell's contribution), a move important to this day. Next followed "On Physical Lines of Force," published during 1861–62, which further clarified the electrotonic state in terms of a mechanical model and introduced yet more mathematical machinery, in particular, the integral theorems of Cambridge mathematician George Stokes, a friend of both Maxwell and Thomson. Finally, in 1865, came "The Dynamical Theory of the Electromagnetic Field."

In this third paper, with the mechanical model now gone, Maxwell presents his theory in essentially its final form. What had started in Faraday's wonderfully imaginative mind as the electrotonic state had become Maxwell's *electromagnetic momentum*. Today it is called the *vector potential*, a term first used by Maxwell in 1871. The curl of the vector potential is the magnetic field vector. Despite this, the third paper (like the second) would look well-nigh unintelligible to a modern

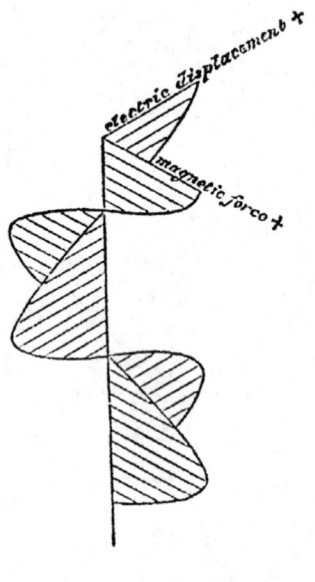

electrical engineer used to vectors because it presents the theory as 20 equations, in a hodgepodge mix of component and quaternionic notation.

PROPAGATING FIELD. But the physics is all there, and the conclusion is astounding: electromagnetic effects travel through space at the speed of light. Indeed, light itself is a propagating transverse electromagnetic field [see illustration]. By showing the science of light and optics as merely a branch of electromagnetism, Maxwell had achieved the second great unification in physics. (The first was Newton's unification of terrestrial and celestial mechanics.) Here, too, Maxwell stated that the energy of electromagnetic phenomena resides not just in electrified bodies, but also in the space surrounding such bodies.

Maxwell's famous equations, which summarize the electrical knowledge of his day, state that electric lines of force are created either by electric charge or by time-varying magnetic fields, while magnetic lines of force are created either by electric currents or by time-varying electric fields. This last part was uniquely Maxwell, as it represents his famous *displacement current*. That a time-varying electric field could produce a magnetic field, just like a conduction current in a wire, was an audacious statement because there was no evidence for it.

Today, electrical engineering and physics professors "derive" the displacement current by showing that, without it, the rest of the equations are inconsistent with the conservation of electric charge.

Any inconsistency with charge conservation is all but undetectable in a closed circuit—and only closed circuits had been studied as, after all, what sense could an open electrical circuit make? But with the displacement current, an open circuit does make sense, and the displacement current is what gives life to radio, television, and radar waves, light, and X-rays, all of which are propagating electromagnetic energy.

At the time of his death of cancer at 48, in 1879, Maxwell's theory of electricity and magnetism was one of several. Its correctness was established only in 1887, when the German Heinrich Hertz discovered electromagnetic radiation at microwave frequencies, as predicted by Maxwell. Others, though, had not had to wait for Hertz; these true believers were members of a small group that has become known as the Maxwellians. They included the Englishmen John Poynting and Oliver Heaviside who, in 1883, simultaneously discovered how Maxwell's theory predicts that a propagating electromagnetic field transports energy through space.

Since the pioneering work of the Maxwellians, Maxwell's equations have been studied for over a century, and have proved one of the most successful theories in the history of science.

Source: Adapted from P.J. Nahin, *IEEE Spectrum,* p. 45, March 1992. © 1992 IEEE.

37

Microwave Devices

37.1 Passive Microwave Devices 882
Characterization of Passive Elements • Transmission Line Sections
• Discontinuities • Impedance Transformers • Terminations
• Attenuators • Microwave Resonators • Tuning Elements • Hybrid
Circuits and Directional Couplers • Filters • Ferrite Components
• Passive Semiconductor Devices
37.2 Active Microwave Devices 891
Semiconductor Material Properties • Two-Terminal Active
Microwave Devices • Three-Terminal Active Microwave Devices

Michael B. Steer
North Carolina State University

Robert J. Trew
North Carolina State University

37.1 Passive Microwave Devices

Michael B. Steer

Wavelengths in air at microwave and millimeter-wave frequencies range from 1 m at 300 MHz to 1 mm at 300 GHz and are comparable to the physical dimensions of fabricated electrical components. For this reason circuit components commonly used at lower frequencies, such as resistors, capacitors, and inductors, are not readily available. The relationship between the wavelength and physical dimensions enables new classes of distributed components to be constructed that have no analogy at lower frequencies. Components are realized by disturbing the field structure on a transmission line, resulting in energy storage and thus reactive effects. Electric (E) field disturbances have a capacitive effect and the magnetic (H) field disturbances appear inductive. Microwave components are fabricated in waveguide, coaxial lines, and strip lines. The majority of circuits are constructed using strip lines as the cost is relatively low and they are highly reproducible due to the photolithographic techniques used. Fabrication of waveguide components requires precision machining but they can tolerate higher power levels and are more easily realized at millimeter-wave frequencies (30–300 GHz) than either coaxial or microstrip components.

Characterization of Passive Elements

Passive microwave elements are defined in terms of their reflection and transmission properties for an incident wave of electric field or voltage. In Fig. 37.1(a) a traveling voltage wave with phasor $\mathbf{V}_1^+$ is incident at port 1 of a two-port passive element. A voltage $\mathbf{V}_1^-$ is reflected and $\mathbf{V}_2^-$ is transmitted. In the absence of an incident voltage wave at port 2 (the voltage wave $\mathbf{V}_2^-$ is totally absorbed by Z_0), at port 1 the element has a voltage reflection coefficient

$$\Gamma_1 = \mathbf{V}_1^- / \mathbf{V}_1^+ \tag{37.1}$$

and transmission coefficient

$$T = \mathbf{V}_2^- / \mathbf{V}_1^+ \tag{37.2}$$

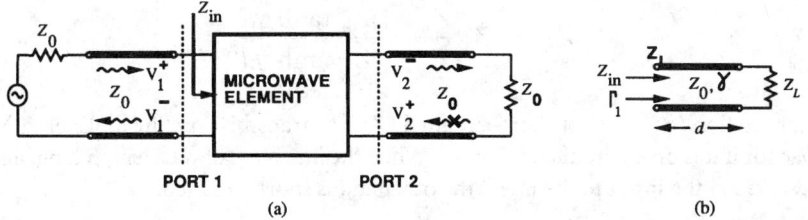

FIGURE 37.1 Incident, reflected, and transmitted traveling voltage waves at (a) a passive microwave element and (b) a transmission line.

More convenient measures of reflection and transmission performance are the **return loss** and **insertion loss** as they are relative measures of power in transmitted and reflected signals. In decibels

$$\text{Return loss} = -20 \log \Gamma_1 \ (\text{dB}) \quad \text{Insertion loss} = -20 \log \text{T} \ (\text{dB}) \tag{37.3}$$

The input impedance at port 1, Z_{in}, is related to Γ_1 by

$$Z_{in} = Z_0 \frac{1 + \Gamma_1}{1 - \Gamma_1} \tag{37.4}$$

The reflection characteristics are also described by the **voltage standing wave ratio** (VSWR), a quantity which is more easily measured. The VSWR is the ratio of the maximum voltage amplitude on a transmission line ($|V_1^+| + |V_1^-|$) to the minimum voltage amplitude ($|V_1^+| - |V_1^-|$). Thus

$$\text{VSWR} = \frac{1 + |\Gamma_1|}{1 - |\Gamma_1|} \tag{37.5}$$

These quantities will change if the loading conditions are changed. For this reason scattering parameters are used which are defined as the reflection and transmission coefficients with a load referred to as a **reference impedance.** Simple formulas relate the S parameters to other network parameters [Vendelin *et al.*, 1990, pp. 16–17]. Thus

$$S_{11} = \Gamma_1 \qquad S_{21} = \text{T} \tag{37.6}$$

S_{11} and S_{12} are similarly defined when a voltage wave is incident at port 2. For a multiport

$$S_{pq} = \mathbf{V}_p^-/\mathbf{V}_q^+ \tag{37.7}$$

with all of the ports terminated in the reference impedance. S parameters are the most convenient network parameters to use with distributed circuits as a change in line length results in a phase change. Also, they are the only network parameters that can be measured directly at microwave and millimeter-wave frequencies. Most passive devices, with the notable exception of ferrite devices, are reciprocal and so $S_{pq} = S_{qp}$. A loss-less passive device also satisfies the unitary condition: $\Sigma_p |S_{pq}|^2 = 1$, which is a statement of power conservation indicating that all power is either reflected or transmitted. A passive element is fully defined by its S parameters together with its reference impedance, here Z_0. In general the reference impedance at each port can be different.

Circuits are designed to minimize the reflected energy and maximize transmission at least over the frequency range of operation. Thus the return loss is high, and the VSWR $\simeq 1$ for well-designed circuits. Individual elements may have high reflection and the interaction of elements is used in design.

A terminated transmission line such as that in Fig. 37.1(b) has an input impedance

$$Z_{in} = Z_0 \frac{Z_L + jZ_0 \tanh \gamma d}{Z_0 + jZ_L \tanh \gamma d} \tag{37.8}$$

Thus a short section ($\gamma d \ll 1$) of short-circuited ($Z_L = 0$) transmission line looks like an inductor, and a capacitor if it is open circuited ($Z_L = \infty$). When the line is a half wavelength long, an open circuit is presented at the input to the line if the other end is short circuited.

Transmission Line Sections

The simplest microwave circuit element is a uniform section of transmission line which can be used to introduce a time delay or a frequency-dependent phase shift. Other line segments for interconnections include bends, corners, twists, and transitions between lines of different dimensions (see Fig. 37.2). The dimensions and shapes are designed to minimize reflections and so maximize return loss and minimize insertion loss.

Discontinuities

The waveguide discontinuities shown in Fig. 37.3(a)–(f) illustrate most clearly the use of E and H field disturbances to realize capacitive and inductive components. An E-plane discontinuity [Fig. 37.3(a)] can be modeled approximately by a frequency-dependent capacitor. H-plane discontinuities [Figs. 37.3(b) and (c)] resemble inductors as does the circular iris of Fig. 37.3(d). The resonant waveguide iris of Fig. 37.3(e) disturbs both the E and H fields and can be modeled by a parallel LC resonant circuit near the frequency of resonance. Posts in waveguide are used both as reactive elements [Fig. 37.3(f)] and to mount active devices [Fig. 37.3(g)]. The equivalent circuits of microstrip discontinuities [Figs. 37.3(k)–(o)] are again modeled by capacitive elements if the E field is interrupted and by inductive elements if the H field (or current) is interrupted. The stub shown in Fig. 37.3(j) presents a short circuit to the through transmission line when the length of the stub is $\lambda_g/4$. When the stubs are electrically short ($\ll \lambda_g/4$) they introduce shunt capacitances in the through transmission line.

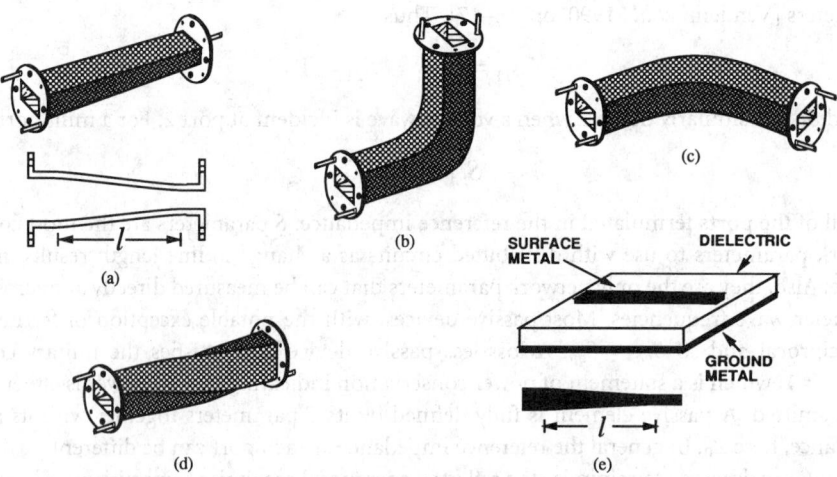

FIGURE 37.2 Sections of transmission lines used for interconnecting components: (a) waveguide tapered section, (b) waveguide E-plane bend, (c) waveguide H-plane bend, (d) waveguide twist, and (e) microstrip taper.

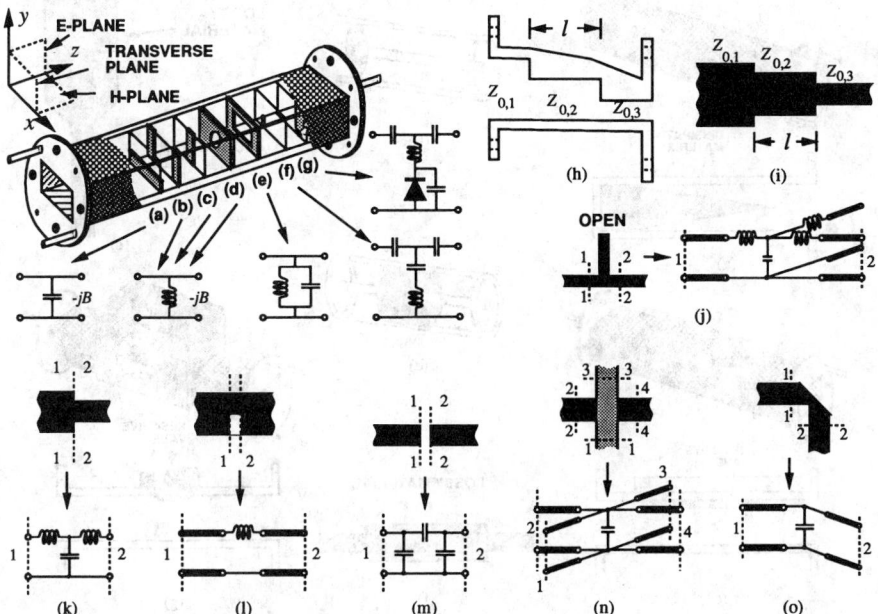

FIGURE 37.3 Discontinuities. Waveguide discontinuities: (a) capacitive E-plane discontinuity, (b) inductive H-plane discontinuity, (c) symmetrical inductive H-plane discontinuity, (d) inductive post discontinuity, (e) resonant window discontinuity, (f) capacitive post discontinuity, (g) diode post mount, and (h) quarter-wave impedance transformer. Microstrip discontinuities: (i) quarter-wave impedance transformer, (j) open microstrip stub, (k) step, (l) notch, (m) gap, (n) crossover, and (o) bend.

Impedance Transformers

Impedance transformers are used to interface two sections of line with different **characteristic impedances**. The smoothest transition and the one with the broadest bandwidth is a tapered line as shown in Fig. 37.2(a) and (e). This element tends to be very long and so step terminations called quarter-wave impedance transformers [see Fig. 37.3(h) and (i)] are sometimes used although their bandwidth is relatively small centered on the frequency at which $l = \lambda_g/4$. Ideally, $Z_{0,2} = \sqrt{Z_{0,1}Z_{0,3}}$.

Terminations

In a termination, power is absorbed by a length of lossy material at the end of a shorted piece of transmission line [Fig. 37.4 (a) and (c)]. This type of termination is called a matched load as power is absorbed and reflections are very small irrespective of the characteristic impedance of the transmission line. This is generally preferred as the characteristic impedance of transmission lines varies with frequency, particularly so for waveguides. When the characteristic impedance of a line does not vary much with frequency, as is the case with a coaxial line, a simpler smaller termination can be realized by placing a resistor to ground [Fig. 37.4(b)].

Attenuators

Attenuators reduce the level of a signal traveling along a transmission line. The basic construction is to make the line lossy but with a characteristic impedance approximating that of the connecting lines so as to reduce reflections. The line is made lossy by introducing a resistive vane in the case of a waveguide [Fig. 37.4(d)], replacing part of the outer conductor of a coaxial line by resistive material [Fig. 37.4(e)], or covering the line by resistive material in the case of a microstrip line [Fig. 37.4(f)]. If the amount of lossy material introduced into the transmission line is controlled, a variable attenuator is achieved, e.g., Fig. 37.4(d).

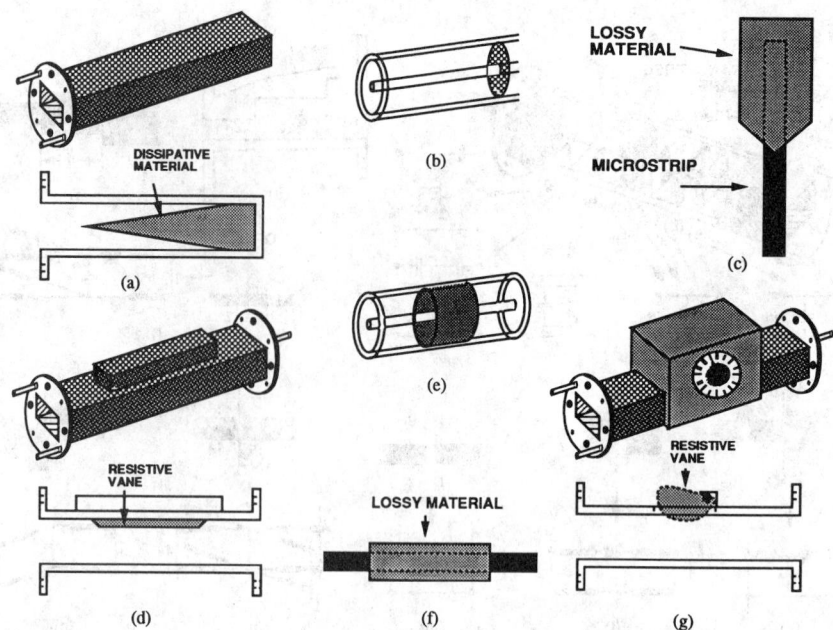

FIGURE 37.4 Terminations and attenuators: (a) waveguide matched load, (b) coaxial line resistive termination, (c) microstrip matched load, (d) waveguide fixed attenuator, (e) coaxial fixed attenuator, (f) microstrip attenuator, and (g) waveguide variable attenuator.

Microwave Resonators

In a lumped element resonant circuit, stored energy is transferred between an inductor which stores magnetic energy and a capacitor which stores electric energy, and back again every period. Microwave resonators function the same way, exchanging energy stored in electric and magnetic forms but with the energy stored spatially. Resonators are described in terms of their quality factor

$$Q = 2\pi f_0 \left(\frac{\text{Maximum energy stored in the resonator at } f_0}{\text{Power lost in the cavity}} \right) \qquad (37.9)$$

where f_0 is the resonant frequency. The Q is reduced and thus the resonator bandwidth is increased by the power lost due to coupling to the external circuit so that the loaded Q

$$Q_L = 2\pi f_0 \left(\frac{\text{Maximum energy stored in the resonator at } f_0}{\text{Power lost in the cavity and to the external circuit}} \right)$$

$$= \frac{1}{1/Q + 1/Q_{\text{ext}}} \qquad (37.10)$$

where Q_{ext} is called the external Q. Q_L accounts for the power extracted from the resonant circuit and is typically large. For the simple response shown in Fig. 37.5(a) the half power (3 dB) bandwidth is f_0/Q_L.

Near resonance the response of a microwave resonator is very similar to the resonance response of a parallel or series R, L, C resonant circuit [Fig. 37.5(f) and (g)]. These equivalent circuits can be used over a narrow frequency range.

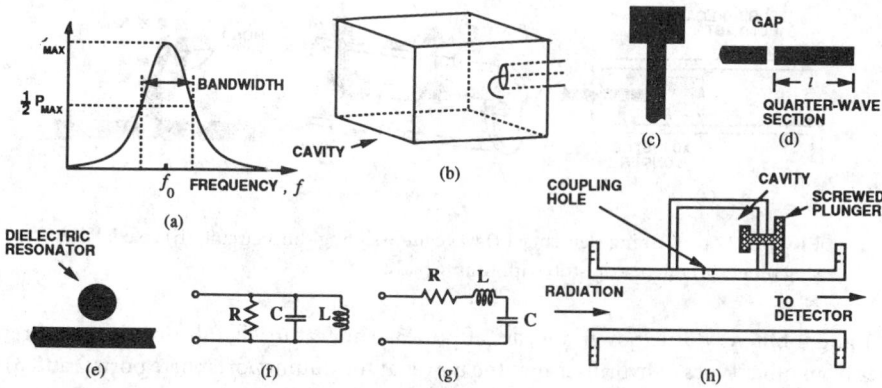

FIGURE 37.5 Microwave resonators: (a) resonator response, (b) rectangular cavity resonator, (c) microstrip patch resonator, (d) microstrip gap-coupled reflection resonator, (e) transmission dielectric transmission resonator in microstrip, (f) parallel equivalent circuits, (g) series equivalent circuits, and (h) waveguide wavemeter.

Several types of resonators are shown in Fig. 37.5. Figure 37.5(b) is a rectangular cavity resonator coupled to an external coaxial line by a small coupling loop. Figure 37.5(c) is a microstrip patch reflection resonator. This resonator has large coupling to the external circuit. The coupling can be reduced and photolithographically controlled by introducing a gap as shown in Fig. 37.5(d) for a microstrip gap-coupled transmission line reflection resonator. The Q of a resonator can be dramatically increased by using a high dielectric constant material as shown in Fig. 37.5(e) for a dielectric transmission resonator in microstrip. One simple application of a cavity resonator is the waveguide wavemeter [Fig. 37.5(h)]. Here the resonant frequency of a rectangular cavity is varied by changing the physical dimensions of the cavity with a null of the detector indicating that the frequency corresponds to the resonant cavity frequency.

Tuning Elements

In rectangular waveguide the basic adjustable tuning element is the sliding short shown in Fig. 37.6(a). Varying the position of the short will change resonance frequencies of cavities. It can be combined with hybrid tees to achieve a variety of tuning functions. The post in Fig. 37.3(f) can be replaced by a screw to obtain a screw tuner which is commonly used in waveguide filters. Sliding short circuits can be used in coaxial lines and in conjunction with branching elements to obtain stub tuners. Coaxial slug tuners are also used to provide adjustable matching at the input and output of active circuits. The slug is movable and changes the characteristic impedance of the transmission line. It is more difficult to achieve variable tuning in passive microstrip circuits. One solution is to provide a number of pads as shown in Fig. 37.6(c) which, in this case, can be bonded to the stub to obtain an adjustable stub length. Variable amounts of phase shift can be inserted by using a variable length of line called a line stretcher, or by a line with a variable propagation constant. One type of waveguide variable phase shifter is similar to the variable attenuator of Fig. 37.4(d) with the resistive material replaced by a low-loss dielectric.

Hybrid Circuits and Directional Couplers

Hybrid circuits are multiport components which preferentially route a signal incident at one port to the other ports. This property is called directivity. One type of hybrid is called a directional coupler, the schematic of which is shown in Fig. 37.7(a). Here the signal incident at port 1 is coupled to ports 2 and 3 while very little is coupled to port 4. Similarly, a signal incident at port 2 is coupled to

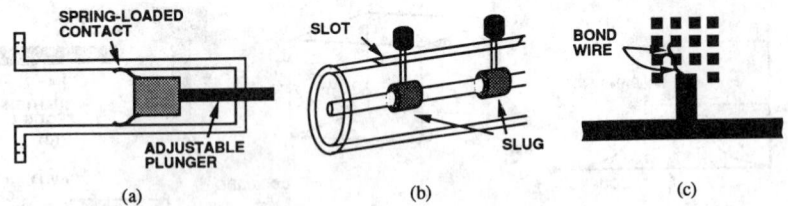

FIGURE 37.6 Tuning elements: (a) waveguide sliding short circuit, (b) coaxial line slug tuner, (c) microstrip stub with tuning pads.

ports 1 and 4 but very little power appears at port 3. The feature that distinguishes a directional coupler from other types of hybrids is that the power at the output ports (here ports 2 and 3) is different. The performance of a directional coupler is specified by three parameters:

$$\text{Coupling factor} = P_1/P_3$$
$$\text{Directivity} = P_3/P_4 \tag{37.11}$$
$$\text{Isolation} = P_1/P_4$$

Microstrip and waveguide realizations of directional couplers are shown in Figs. 37.7(b) and (c) where the microstrip coupler couples in the backward direction and the waveguide coupler couples in the forward direction. The powers at the output ports of the hybrids shown in Figs. 37.8 and 37.9 are equal and so the hybrids serve to split a signal into half as well as having directional sensitivity.

Filters

Filters are combinations of microwave passive elements designed to have a specified frequency response. Typically, a topology of a filter is chosen based on established lumped element filter design theory. Then computer-aided design techniques are used to optimize the response of the circuit to the desired response.

Ferrite Components

Ferrite components are nonreciprocal in that the insertion loss for a wave traveling from port A to port B is not the same as that from port B to port A.

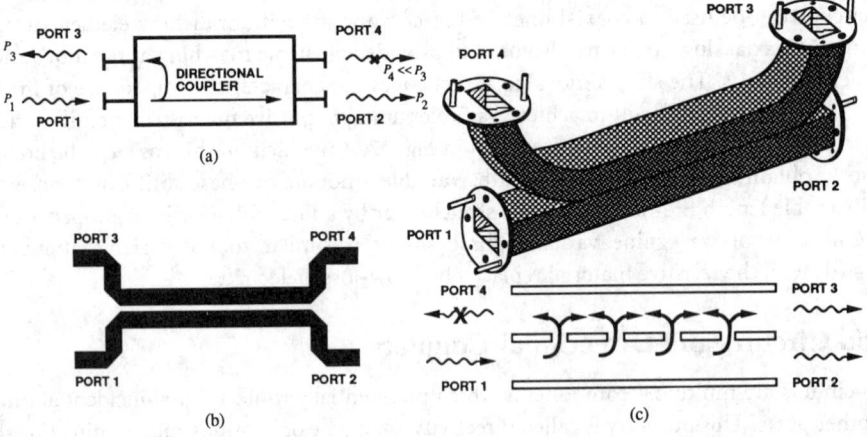

FIGURE 37.7 Directional couplers: (a) schematic, (b) backward-coupling microstrip directional coupler, (c) forward-coupling waveguide directional coupler.

Circulators and Isolators

The most important type of ferrite component is a circulator [Fig. 37.10(a) and (b)]. The essential element of a circulator is a piece of ferrite which when magnetized becomes nonreciprocal, preferring progression of electromagnetic fields in one circular direction. An ideal circulator has the scattering matrix

$$[S] = \begin{bmatrix} 0 & 0 & S_{13} \\ S_{21} & 0 & 0 \\ 0 & S_{32} & 0 \end{bmatrix} \qquad (37.12)$$

In addition to the insertion and return losses, the performance of a circulator is described by its isolation which is its insertion loss in the undesired direction. An isolator is just a three-port circulator with one of the ports terminated in a matched load as shown in the microstrip realization of Fig. 37.10(c). It is used in a transmission line to pass power in one direction but not in the reverse direction. It is commonly used to protect the output of equipment from high reflected signals. The heart of isolators and circulators is the nonreciprocal element. Electronic versions have been developed for MMICs. A four-port version is called a duplexer and is used in radar systems and to separate the received and transmitted signals in a transceiver.

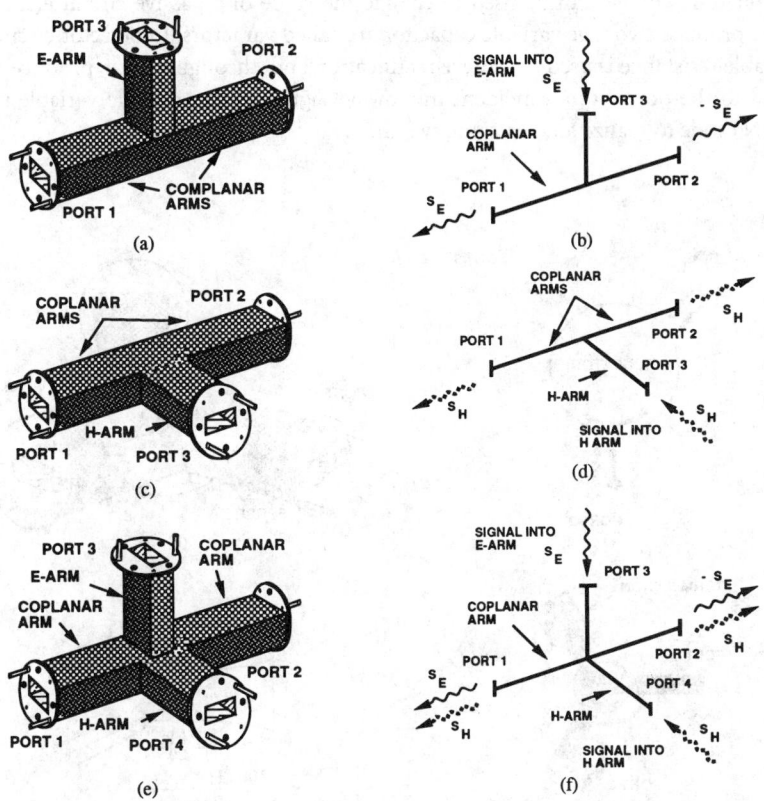

FIGURE 37.8   Waveguide hybrids: (a) E-plane tee and (b) its signal flow; (c) H-plane tee and (d) its signal flow; and (e) magic tee and (f) its signal flow. The negative sign indicates $180^{\to 0}$ phase reversal.

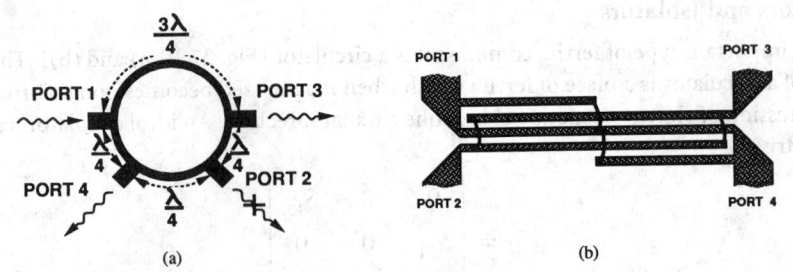

FIGURE 37.9   Microstrip hybrids: (a) rat race hybrid and (b) Lange coupler.

YIG Tuned Resonator

A magnetized YIG (yttrium iron garnet) sphere, shown in Fig. 37.10(d), provides coupling between two lines over a very narrow bandwidth. The center frequency of this bandpass filter can be adjusted by varying the magnetizing field.

Passive Semiconductor Devices

A semiconductor diode can be modeled by a voltage-dependent resistor and capacitor in shunt. Thus an applied dc voltage can be used to change the value of a passive circuit element. Diodes optimized to produce a voltage variable capacitor are called varactors. In detector circuits a diode's voltage variable resistance is used to achieve rectification and, through design, produce a dc voltage proportional to the power of an incident microwave signal. A controllable variable resistance is used in a PIN diode to realize an electronic switch.

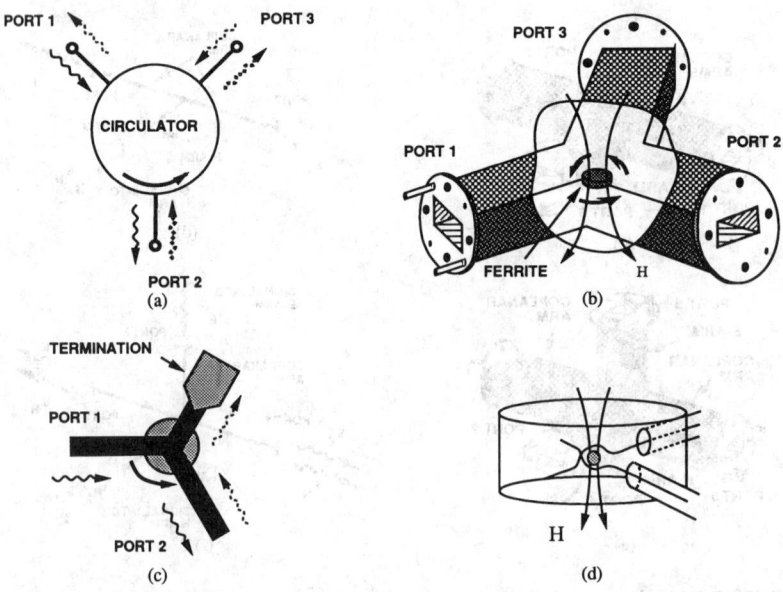

FIGURE 37.10   Ferrite components: (a) schematic of a circulator, (b) a waveguide circulator, (c) a microstrip isolator, and (d) a YIG tuned bandpass filter.

Defining Terms

Characteristic impedance: Ratio of the voltage and current on a transmission line when there are no reflections.

Insertion loss: Power lost when a signal passes through a device.

Reference impedance: Impedance to which scattering parameters are referenced.

Return loss: Power lost upon reflection from a device.

Voltage standing wave ratio (VSWR):   Ratio of the maximum voltage amplitude on a line to the minimum voltage amplitude.

Reference

G.D. Vendelin, A.M. Pavio, and U.L. Rohde, *Microwave Circuit Design Using Linear and Nonlinear Techniques*, New York: Wiley, 1990.

Further Information

The following books provide good overviews of passive microwave components: *Microwave Engineering Passive Circuits* by P.A. Rizzi, Prentice-Hall, Englewood Cliffs, N.J., 1988; *Microwave Devices and Circuits* by S.Y. Liao, 3rd ed., Prentice-Hall, Englewood Cliffs, N.J., 1990; *Microwave Theory, Components and Devices* by J.A. Seeger, Prentice-Hall, Englewood Cliffs, N.J., 1986; *Microwave Technology* by E. Pehl, Artech House, Dedham, Mass., 1985; *Microwave Engineering and Systems Applications* by E.A. Wolff and R. Kaul, Wiley, New York, 1988; and *Microwave Engineering* by T.K. Ishii, 2nd ed., Harcourt Brace Jovanovich, Orlando, Fla., 1989. *Microwave Circuit Design Using Linear and Nonlinear Techniques* by G.D. Vendelin, A.M. Pavio, and U.L. Rohde, Wiley, New York, 1990, provides a comprehensive treatment of computer-aided design techniques for both passive and active microwave circuits.

The monthly journals *IEEE Transactions on Microwave Theory Techniques, IEEE Microwave and Guided Wave Letters*, and *IEEE Transactions on Antennas and Propagation* publish articles on modeling and design of microwave passive circuit components. Articles in the first two journals are more circuit and component oriented while the third focuses on field theoretic analysis. These are published by The Institute of Electrical and Electronics Engineers, Inc. For subscription or ordering contact: IEEE Service Center, 445 Hoes Lane, P.O. Box 1331, Piscataway, New Jersey 08855-1331.

Articles can also be found in the biweekly magazine *Electronics Letters* and the bimonthly magazine *IEE Proceedings Part H—Microwave, Optics and Antennas*. Both are published by the Institute of Electrical Engineers and subscription inquiries should be sent to IEE Publication Sales, P.O. Box 96, Stenage, Herts. SG1 2SD, United Kingdom. Telephone number (0438) 313311.

The *International Journal of Microwave and Millimeter-Wave Computer-Aided Engineering* is a quarterly journal devoted to the computer-aided design aspects of microwave circuits and has articles on component modeling and computer-aided design techniques. It has a large number of review-type articles. For subscription information contact John Wiley & Sons, Inc., Periodicals Division, P.O. Box 7247-8491, Philadelphia, Pennsylvania 19170-8491.

37.2 Active Microwave Devices

Robert J. Trew

Active devices that can supply gain at microwave frequencies can be fabricated from a variety of semiconductor materials. The availability of such devices permits a wide variety of system components to be designed and fabricated. Systems are generally constructed from components such as filters, amplifiers, oscillators, mixers, phase shifters, switches, etc. Active devices are primarily required for the oscillator and amplifier components. For these functions, devices that can supply current, voltage, or power gain at the frequency of interest are embedded in circuits that are designed to provide the device with the proper environment to create the desired response. The operation of the component is dictated, therefore, by both the capabilities of the active device and its embedding circuit.

It is common to fabricate microwave integrated circuits using both hybrid and monolithic techniques. In the hybrid approach, discrete active devices are mounted in RF circuits that can be fabricated from waveguides or transmission lines fabricated using coaxial, microstrip, stripline, coplanar waveguide, or other such media. Monolithic circuits are fabricated with both the active device and the RF circuit fabricated in the same semiconductor chip. Interconnection lines and the embedding RF circuit are generally fabricated using microstrip or coplanar waveguide transmission lines.

Active microwave devices can be fabricated as **two-terminal devices** (diodes) or **three-terminal devices** (transistors). Generally, three-terminal devices are preferred for most applications since the third terminal provides a convenient means to control the RF performance of the device. The third terminal allows for inherent isolation between the input and output RF circuit. Amplifiers and oscillators can easily be designed by providing circuits with proper stabilization or feedback characteristics. Amplifiers and oscillators can also be designed using two-terminal devices (diodes), but input/output isolation is more difficult to achieve since only one RF port is available. In this case it is generally necessary to use RF isolators or circulators.

The most commonly used two-terminal active devices consist of Gunn, tunnel, and IMPATT diodes. These devices can be designed to provide useful gain from low gigahertz frequencies to high millimeter-wave frequencies. Three-terminal devices consist of bipolar (BJT), heterojunction bipolar (HBT), and field-effect transistors (MESFETs and HEMTs). These devices can also be operated from UHF to millimeter-wave frequencies.

Semiconductor Material Properties

Active device operation is strongly dependent upon the charge transport characteristics of the semiconductor materials from which the device is fabricated. Semiconductor materials can be grown in single crystals with very high purity. The electrical conductivity of the crystal can be precisely controlled by introduction of minute quantities of dopant impurities. When these impurities are electrically activated, they permit precise values of current flow through the crystals to be controlled by potentials applied to contacts, placed upon the crystals. By clever positioning of the metal contacts, various types of semiconductor devices are fabricated. In this section we will briefly discuss the important material characteristics.

Semiconductor material parameters of interest for device fabrication consist of those involved in charge transport through the crystal, as well as thermal and mechanical properties of the semiconductor. The charge transport properties describe the ease with which free charge can flow through the material. For example, the velocity-electric field characteristics for several commonly used semiconductors are shown in Fig. 37.11. At low values of electric field, the charge transport is ohmic and the charge velocity is directly proportional to the magnitude of the electric field. The proportionality constant is called the mobility and has units of cm^2/V-s. Above a critical value for the electric field, the charge velocity saturates and either becomes constant (e.g., Si) or decreases with increasing field (e.g., GaAs). Both of these behaviors have implications for device fabrication, especially for devices intended for high-frequency operation. Generally, a high velocity is desired since current is directly proportional to velocity. Also, a low value for the saturation electric field is desirable since this implies a high-charge mobility. High mobility implies low resistivity and, therefore, low values for parasitic and access resistances for semiconductor devices.

The decreasing electron velocity with electric field characteristic for compound semiconductors such as GaAs and InP makes possible active two-terminal devices called transferred electron devices (TEDs) or Gunn diodes. The negative slope of the velocity versus electric field characteristic implies a decreasing current with increasing voltage. That is, the device has a negative resistance.

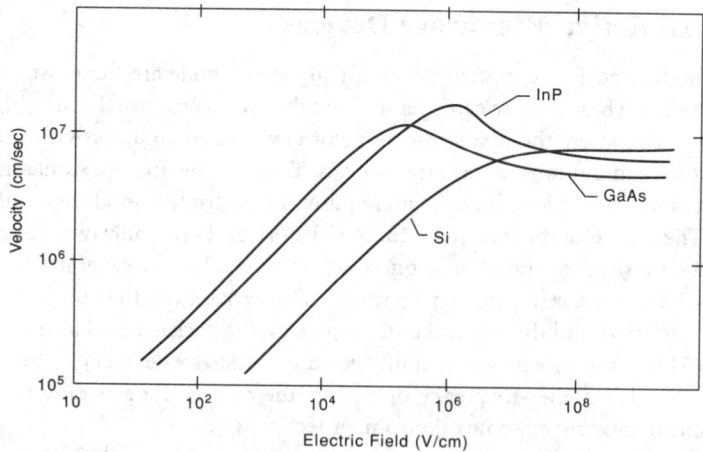

FIGURE 37.11 Electron velocity versus electric field for several semiconductors. This figure shows the electron velocity in several common semiconductors as a funcation of electric field strength. At low electric field the electron velocity is ohmic, as indicated by the linear characteristic. At higher electric field strength the electron velocity saturates and becomes nonlinear. Compound semiconductors such as GaAs and InP have highly nonlinear behavior at large electric fields.

When a properly sized piece of these materials is biased and placed in a resonant cavity, the device will be unstable up to very high frequencies. By proper selection of embedding impedances oscillators or amplifiers can be constructed.

Other semiconductor materials parameters of interest include thermal, dielectric constant, energy bandgap, electric breakdown characteristics, and minority carrier lifetime. The thermal conductivity of the material is important because it describes how easily heat can be extracted from the device. The thermal conductivity has units of W/cm-K. Generally, high thermal conductivity is desirable. Compound semiconductors, such as GaAs and InP, have relatively poor thermal conductivity compared to elemental semiconductors such as Si. Materials such as SiC have excellent thermal conductivity and have uses in high-power electronic devices. The dielectric constant is important since it affects the size of the semiconductor device. The larger the dielectric constant, the smaller the device. Electric breakdown characteristics are important since breakdown limits the magnitudes of the dc and RF voltages that can be applied to the device. This is turn limits the RF power that can be handled by the device. The electric breakdown for the material is generally described by the critical value of electric field that produces avalanche ionization. Minority carrier lifetime is important for bipolar devices, such as pn junction diodes, rectifiers, and bipolar junction transistors (BJTs). A low value for minority carrier lifetime is desirable for devices such as diode temperature sensors and switches where low reverse bias leakage current is desirable. A long minority carrier lifetime is desirable for devices such as bipolar transistors. For materials such as Si and SiC, the minority carrier lifetime can be varied by controlled impurity doping. A comparison of some of the important material parameters for several common semiconductors is presented in Table 37.1.

Table 37.1 Material Parameters for Several Semiconductors

Semiconductor	E_g (eV)	ε_r	κ (W/cm-K) @300 K	E_c (V/cm)	$\tau_{minority}$ (s)
Si	1.12	11.9	1.5	3×10^5	2.5×10^{-3}
GaAs	1.42	12.5	0.54	4×10^5	$\sim 10^{-8}$
InP	1.34	12.4	0.67	4.5×10^5	$\sim 10^{-8}$
α-SiC	2.86	10.0	4	$(1-5) \times 10^6$	$\sim(1-10) \times 10^{-9}$
β-SiC	2.2	9.7	4	$(1-5) \times 10^6$	$\sim(1-10) \times 10^{-9}$

Two-Terminal Active Microwave Devices

The IMPATT diode, transferred electron device, and tunnel diode are the most commonly used two-terminal devices. These devices can operate from the low microwave through high millimeter-wave frequencies. They were the first semiconductor devices that could provide useful RF power levels at microwave and millimeter-wave frequencies. The three devices are similar in that they are fabricated from blocks of semiconductors and require two electrodes (anode and cathode) for supplying dc bias. The same electrodes are used for the RF port, and since only two electrodes are available, the devices must be operated as a **one-port network**. This is generally accomplished by mounting the semiconductor in a pin-type package. The package can then be positioned in an RF circuit or resonant cavity and the top and bottom pins on the package used as the dc and RF electrical contacts. This arrangement works quite well and packaged devices can be operated up to about 90–100 GHz. For higher-frequency operation, the devices are generally mounted directly into circuits using microstrip or some other similar technology.

All three devices operate as negative immittance components. That is, their active characteristics can be described as either a negative resistance or a negative conductance. Which description to use is determined by the physical operating principles of the particular device.

Tunnel Diodes

Tunnel diodes [Sze, 1981] generate active characteristics by a mechanism involving the physical tunneling of electrons between energy bands in highly doped semiconductors. For example, if a pn junction diode is heavily doped, the conduction and valence bands will be located in close proximity and **charge carriers** can tunnel through the electrostatic barrier separating the p-type and n-type regions, rather than be thermionically emitted over the barrier as generally occurs in this type of diode. When the diode is biased (either forward or reverse bias) current immediately flows and junction conduction is basically ohmic. In the forward bias direction, conduction occurs until the applied bias forces the conduction and valence bands to separate. The tunnel current then decreases and normal junction conduction occurs. In the forward bias region where the tunnel current is decreasing with increasing bias voltage, a negative immittance characteristic is generated. The immittance is called "N-type" because the I-V characteristic "looks like" the letter N. This type of active element is short-circuit stable and is described by a negative conductance in shunt with a capacitance. Tunnel diodes are limited in operation frequency by the time it takes for charge carriers to tunnel through the junction. Since this time is very short (on the order of 10^{-12} s) operation frequency can be very high, approaching 1000 GHz. Tunnel diodes have been operated at hundreds of gigahertz, limited by practical packaging and parasitic impedance considerations. The RF power available from a tunnel diode is limited (hundreds of milliwatts level) since the maximum RF voltage swing that can be applied across the junction is limited by the forward turn-on characteristics of the device (typically 0.6–0.9 V). Increased RF power can only be obtained by increasing device area to increase RF current, but device area is limited by operation frequency according to an inverse scaling law. Tunnel diodes have moderate dc-to-RF conversion efficiency (<10%), very low **noise figures**, and are useful in low-noise systems applications, such as microwave and millimeter-wave receivers.

Transferred Electron Devices

Transferred electron devices (i.e., Gunn diodes) [Bosch and Engelmann, 1975] also have N-type active characteristics and can be modeled as a negative conductance in parallel with a capacitance. Device operation, however, is based upon a fundamentally different principle. The negative conductance derives from the complex conduction band structure of certain compound semiconductors, such as GaAs and InP. In these direct bandgap materials the central (or Γ) conduction band is in close energy-momentum proximity to secondary, higher-order conduction bands (i.e., the X and L valleys). The electron effective mass is determined by the shape of the conduction bands, and the effective mass is "light" in the Γ valley but "heavy" in the higher-order X and L valleys. When the crystal is biased, current flow is initially due to electrons in the light effective mass Γ valley and conduction is ohmic. However, as the bias field is increased, an increasing proportion of the free electrons are transferred into the X and L valleys where the electrons have heavier effective mass. The increased effective mass slows down the electrons, with a corresponding decrease in conduction current through the crystal. The net result is that the crystal displays a region of applied bias voltages where current decreases with increasing voltage. That is, a negative conductance is generated. The device is unstable and, when placed in an RF circuit or resonant cavity, oscillators or amplifiers can be fabricated. The device is not actually a diode since no pn or Schottky junction is used. The phenomenon is a characteristic of the bulk material and the special structure of the conduction bands in certain compound semiconductors. Most semiconductors do not have the conduction band structure necessary for the transferred electron effect. The term *Gunn diode* is actually a misnomer since the device is not a diode. TEDs are widely used in oscillators from the microwave through high millimeter-wave frequency bands. They have good RF output power capability (milliwatts to watts level), moderate efficiency (<20%), and excellent noise and bandwidth capability. Octave band tunable oscillators are easily fabricated using devices such as YIG (yttrium iron garnet) resonators or varactors as the tuning element. Most commercially available solid-state sources for 60- to 100-GHz operation generally use InP TEDs.

IMPATT Diodes

IMPATT (impact avalanche transit time) diodes [Bhartia and Bahl, 1984] are fabricated from pn or Schottky junctions. A typical pn junction device is shown in Fig. 37.12. For optimum RF performance the diode is separated, by use of specially designed layers of controlled impurity doping, into avalanche and drift regions. In operation the diode is reverse biased into avalanche breakdown. Due to the very sensitive I-V characteristic, it is best to bias the diode using a constant current source in which the magnitude of the current is limited. When the diode is placed in a microwave resonant circuit, RF voltage fluctuations in the bias circuit grow and are forced into a narrow frequency range by the impedance characteristics of the resonant circuit. Due to the avalanche process the RF current across the avalanche region lags the RF voltage by 90 degrees. This inductive delay is not sufficient, by itself, to produce active characteristics. However, when the 90 degrees phase shift is added to that arising from an additional inductive delay caused by the transit time of the carriers drifting through the remainder of the diode external to the avalanche region, a phase shift between the RF voltage and current greater than 90 degrees is obtained. A Fourier analysis of the resulting waveforms reveals a device impedance with a negative real part. That is, the device is active and can be used to generate or amplify RF signals. The device impedance has an "S-type" active characteristic and the device equivalent circuit consists of a negative resistance in series with an inductor. The device has significant pn junction capacitance that must be considered, and a complete equivalent circuit would include the device capacitance in parallel with the series negative resistance-inductance elements. For optimum performance the drift region is designed so that the electric field

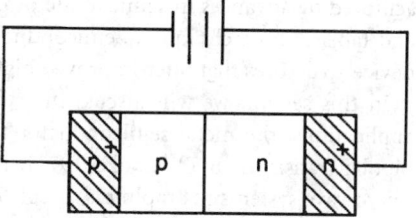

FIGURE 37.12 Diagram showing the structure for a typical pn junction IMPATT diode. This particular diode is called a double-drift device because avalanche breakdown occurs at the pn junction, which is located in the middle of the device. When operated in breakdown, electrons would travel through the n-type region towards the positive terminal of the bias source and holes would travel through the p-type region towards the negative terminal of the source. The diode, therefore, operates as two diodes connected in a back-to-back configuration. The frequency capability of the device is directly proportional to the width of the n and p regions.

throughout the RF cycle is sufficiently high to produce velocity saturation for the charge carriers. In order to achieve this, it is common to design complex structures consisting of alternating layers of highly doped and lightly doped semiconductor regions. These structures are called "high-low," "low-high-low," or "Read" diodes, after the man who first proposed their use. They can also be fabricated in a back-to-back arrangement to form double-drift structures. These devices are particularly attractive for millimeter-wave applications. IMPATT diodes can be fabricated from most semiconductors, but are generally fabricated from Si or GaAs. The devices are capable of good RF output power (mW to W) and good dc-to-RF conversion efficiency (~10–20%). They operate well into the millimeter-wave region and have been operated as high as 340 GHz. They have moderate bandwidth capability, but have relatively poor noise performance due to the impact ionization process.

Although the two-terminal active devices are used in many electronic systems, the one-port characteristic can introduce significant complexity into circuit design. Isolators and circulators are generally required, and these components are often large and bulky. They are often fabricated from magnetic materials, which can introduce thermal sensitivities. For these reasons three-terminal devices have replaced two-terminal devices in many practical applications. Generally, if two-terminal and three-terminal devices with comparable capability are available, the three-terminal device offers a more attractive design solution and will be selected. Two-terminal devices are generally only used when a comparable three-terminal device is not available. For this reason IMPATT and TED devices are used in millimeter-wave applications, where they retain an advantage in providing good RF power. Tunnel diodes are not often used, except in a few special applications where their low-noise and wide-bandwidth performance can be used to advantage.

Three-Terminal Active Microwave Devices

The high-frequency performance of three-terminal semiconductor devices has improved dramatically during the past two decades. Twenty years ago transistors that could provide useful **gain** at frequencies above 10 GHz were a laboratory curiosity. Today, such devices are readily available, and state-of-the-art transistors operate well above 100 GHz. This dramatic improvement has been achieved by advances in semiconductor growth technology, coupled with improved device design and fabrication techniques. Semiconductor materials technology continues to improve and new device structures that offer improved high-frequency performance are continually being reported.

In this section we will discuss the two most commonly employed transistors for microwave applications, the metal-semiconductor field-effect transistor (MESFET) [Liechti, 1976] and the bipolar transistor (BJT) [Cooke, 1971]. These two transistors are commonly employed in practical microwave systems as amplifiers, oscillators, and gain blocks. The transistors have replaced many two-terminal devices due to their improved performance and ease of use. Transistors are readily integrated into both hybrid and monolithic integrated circuit environments (MICs). This, in turn, has resulted in significantly reduced size, weight, and dc power consumption, as well as increased reliability and mean time to failure for systems that use these components. Transistors are easily biased and the **two-port network** configuration leads naturally to inherent separation between input and output networks.

Field-Effect Transistors

A cross-sectional view of a microwave MESFET is shown in Fig. 37.13. The device is conceptionally very simple. The MESFET has two ohmic contacts (the source and drain) separated by some distance, usually in the range of 3 to 10 μm. A rectifying Schottky contact (the gate) is located between the two ohmic contacts. Typically, the gate length is on the order of 0.1 to 2 μm for modern microwave devices. The width of the device scales with frequency and typically ranges from about 1 to 10 μm for power microwave devices to 50 μm for millimeter-wave devices. All three contacts are located on the surface of a thin conducting layer (the channel) which is located on top of a high-

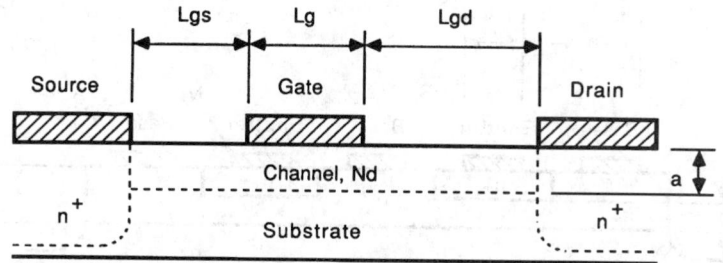

FIGURE 37.13   Cross-sectional view of a microwave MESFET. The cross-hatched areas indicate metal electrodes placed upon the semiconductor to provide for electrical connections. The areas indicated as n$^+$ are highly doped, highly conducting regions to reduce ohmic access resistances. The channel contains the region of current flow and the substrate is highly resistive and nonconducting so that the current flow is confined to the channel.

resistivity, nonconductive substrate to form the device. The channel region is typically very thin (on the order of 0.1–0.3 μm) and is fabricated by epitaxial growth or ion implantation. In operation, the drain contact is biased at a specified potential (positive drain potential for an n-channel device) and the source is grounded. The flow of current through the conducting channel is controlled by negative dc and superimposed RF potentials applied to the gate, which modulate the channel current and provide RF gain. The current flow is composed of only one type of charge carrier (generally electrons) and the device is termed *unipolar*. The MESFET can be fabricated from a variety of semiconductors, but is generally fabricated from GaAs. MESFETs fabricated from Si do not work at high frequencies as well as those fabricated from GaAs due to lower electron mobility in Si (e.g., μ$_n$ ~6000 cm^2/V-s for GaAs and 1450 cm^2/V-s for Si). The lower electron mobility in Si produces high source resistance, which seriously degrades the high-frequency gain possible from the device.

MESFETs can be optimized for small-signal, low-noise operation or for large-signal, RF power applications. Generally, low-noise operation requires short gate lengths, relatively narrow gate widths, and highly doped channels. Power devices generally have longer gate lengths, much wider gate widths, and lower doped channels. Low-noise devices can be fabricated that operate with good gain (~10 dB) and low noise figure (<3 dB) to above 100 GHz. Power devices can provide RF power levels on the order of watts (W) up to over 20 GHz.

The current gain of the MESFET is indicated by the f_T of the device, sometimes called the gain-bandwidth product. This parameter is defined as the frequency at which the short-circuited current gain is reduced to unity and can be expressed as

$$f_T = \frac{g_m}{2\pi C_{gs}}$$

(37.13)

where g_m is the device transconductance (a measure of gain capability) and C_{gs} is the gate source capacitance. High f_T is desirable and this is achieved with highly doped channels and low capacitance gates. The RF power gain is also of interest and this performance can be indicated by the unilateral power gain defined as

$$U = \frac{1}{4}\left(\frac{f_T}{f}\right)^2 \frac{R_{ds}}{R_g}$$

(37.14)

where U is the unilateral power gain, f is the operating frequency, R_{ds} is the drain-source resistance, and R_g is the gate resistance. As this expression indicates, large power gain requires a high f_T, and a large R_{ds}/R_g ratio. The highest frequency at which the device could be expected to produce power

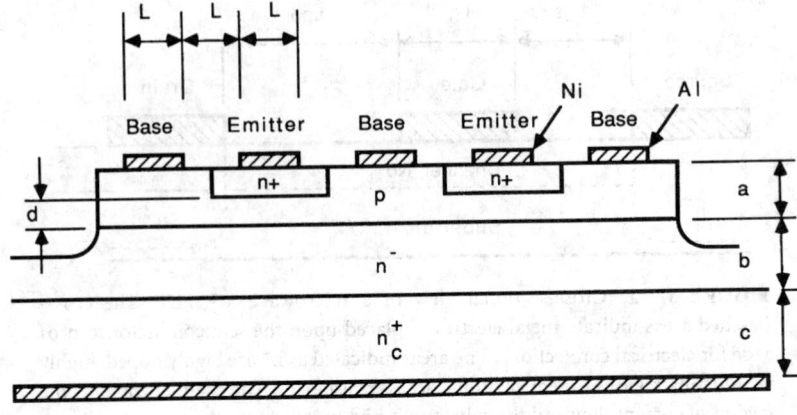

FIGURE 37.14 Cross-sectional view of a microwave bipolar transistor (BJT). The cross-hatched areas indicate metal electrodes. The electrode pattern on the surface is interdigitated with the base electrodes connected together at one end and the emitter electrodes connected together at the other end. Due to the interdigitated structure, there will always be one more base electrode than the number of emitter electrodes. The n^+, p, n^-, and n_c^+ designations indicate the impurity doping type and relative concentration level. This device has the collector electrode on the bottom of the device.

gain can be defined from the frequency, f, at which U goes to zero. This frequency is called the maximum frequency of oscillation, or f_{max}, and is defined as

$$f_{max} = \frac{f_T}{2} \sqrt{\frac{R_{ds}}{R_g}} \tag{37.15}$$

A different form of field-effect transistor can be fabricated by inserting a highly doped, wider-bandgap semiconductor between the conducting channel and the gate electrode [Drummond *et al.*, 1986]. The conducting channel is then fabricated from undoped semiconductor. The discontinuity in energy bandgaps between the two semiconductors, if properly designed, results in free charge transfer from the highly doped, wide-bandgap semiconductor into the undoped, lower-bandgap channel semiconductor. The charge accumulates at the interface and creates a two-dimensional electron gas (2DEG). The sheet charge is essentially two-dimensional and allows current to flow between the source and drain electrodes. The amount of charge in the 2DEG can be controlled by the potential applied to the gate electrode. In this manner the current flow through the device can be modulated by the gate and gain results. Since the charge flows at the interface between the two materials, but is confined in the undoped channel semiconductor, very little impurity scattering occurs and extremely high charge carrier mobility results. The device, therefore, has very high transconductance and is capable of very high frequency operation and very low noise figure operation. This type of device is called a high electron mobility transistor (HEMT). HEMTs can be fabricated from material systems such as AlGaAs/GaAs or AlInAs/GaInAs/InP. The latter material system produces devices that have f_T's above 300 GHz and have produced noise figures of about 1 dB at 100 GHz.

Bipolar Transistors

A cross-sectional view of a bipolar transistor is shown in Fig. 37.14. The bipolar transistor consists of back-to-back pn junctions arranged in a sandwich structure. The three regions are designated the emitter, base, and collector. This type of device differs from the field-effect transistors in that both electrons and holes are involved in the current transport process (thus the designation *bipolar*). Two structures are possible: pnp or npn, depending upon the conductivity type common to both pn junctions. Generally, for microwave applications the npn structure is used since device operation is controlled by electron flow. In general, electron transport is faster than that for holes, and npn transistors are capable of superior high-frequency performance compared to comparable pnp transistors. In operation, the base-emitter pn junction is forward biased and the collector-base pn junction is reverse biased. When an RF signal is applied to the base-emitter junction the junction allows a current to be injected into the base region. The current in the base region consists of minority charge carriers (i.e., carriers with the opposite polarity compared to the base material— electrons for an npn transistor). These charge carriers then diffuse across the base region to the base-collector junction, where they are swept across the junction by the large reverse bias electric field. The reverse bias electric field in the base-collector region is generally made sufficiently large that the carriers travel at their saturation velocity. The transit time of the charge carriers across this region is small, except for millimeter-wave transistors where the base-collector region transit time can be a significant fraction of the total time required for a charge carrier to travel from the emitter through the collector. The operation of the transistor is primarily controlled by the ability of the minority charge carriers to diffuse across the base region. For this reason microwave transistors are designed with narrow base regions in order to minimize the time required for the carriers to travel through this region. The base region transit time is generally the limiting factor in determining the high-frequency capability of the transistor. The gain of the transistor is also significantly affected by minority carrier behavior in the base region. The density of minority carriers is significantly smaller than the density of majority carriers (majority carrier density is approximately equal to the impurity doping density) for typical operating conditions and the probability that the minority charge will recombine with a majority carrier is high. If recombination occurs, the minority charge cannot reach the base-collector junction but appears as base current. This, in turn, reduces the current gain capability of the transistor. Narrow base regions reduce the semiconductor volume where recombination can occur and, therefore, result in increased **gain.** Modern microwave transistors typically have base regions on the order of 0.1–0.25 μm.

The frequency response of a bipolar transistor can be determined by an analysis of the total time it takes for a charge carrier to travel from the emitter through the collector. The total time can be expressed as

$$\tau_{ec} = \tau_e + \tau_b + \tau_c + \tau_c' \qquad (37.16)$$

where τ_{ec} is the total emitter-collector transit time, τ_e is the base-emitter junction capacitance charging time, τ_b is the base region transit time, τ_c is the base-collector junction capacitance charging time, and τ_c' is the base-collector region transit time. The total emitter-base time is related to the gain-bandwidth capability of the transistor according to the relation

$$f_T = \frac{1}{2\pi\tau_{ec}} \qquad (37.17)$$

Since the bipolar transistor has three terminals, it can be operated in various configurations, depending upon the electrode selected as the common terminal. The two most commonly employed are the common emitter (CE) and the common base (CB) configurations, although the common collector (CC) configuration can also be used. Small-signal amplifiers generally use the CE configuration and power amplifiers often use the CB configuration.

The current gain for a bipolar transistor is shown in Fig. 37.15. The current gains of the transistor operated in the CE and CB configurations are called β and α, respectively. As indicated in the figure, the CE current gain β is much larger than the CB current gain α, which is limited to values less than unity. For modern microwave transistors $\alpha_o \sim 0.98$–0.99 and $\beta_o \sim 50$–60.

A measure of the RF power gain for the transistor is indicated by the unilateral power gain, which can be expressed as

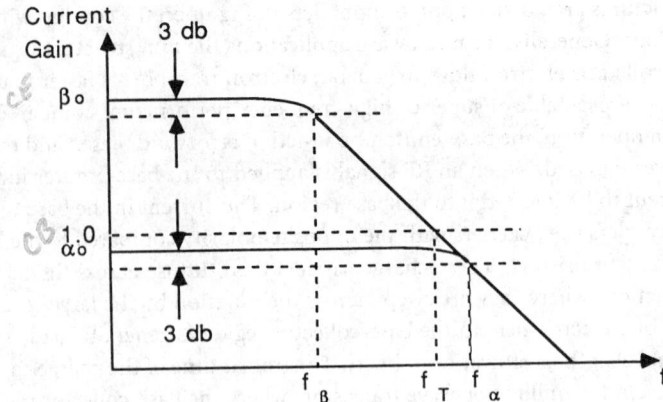

FIGURE 37.15 Current gains versus frequency for bipolar transistors. The common-emitter and common-base current gains are designated as β and α, respectively. The subscript "o" indicates the dc value. The gains decrease with frequency above a certain value. The frequencies where the gains are reduced by 3 dB from their dc values are indicated as the CE and CB cutoff frequencies, f_β and f_α, respectively. The frequency at which the CE current gain is reduced to unity is defined as the gain-bandwidth product f_T. Note that the CB current gain is restricted to values less than unity and that the CE current gain has values that significantly exceed unity.

$$U \cong \frac{\alpha_o}{16\pi^2 r_b C_c f^2 \left(\tau_{ec} + \dfrac{r_e C_c}{a_o} \right)} \qquad (37.18)$$

where U is the power gain, α_o is the dc CB current gain, r_b is the base resistance, C_c is the collector capacitance, τ_{ec} is the total emitter-to-collector transit-time, and r_e is the emitter resistance. The frequency at which U is reduced to unity (f_{max}) is the maximum frequency at which the device will have active characteristics. This frequency is

$$f_{max} = \left[\frac{f_T}{8\pi r_b C_c} \right]^{1/2} \qquad (37.19)$$

In order to maximize the high-frequency performance of a transistor, it is necessary to design the device so that it has high current gain (f_T), low base resistance (r_b), and low collector capacitance (C_c).

Bipolar transistors operating to about 20 GHz are generally fabricated from Si. These devices are easily fabricated and low cost. They are useful in moderate gain and low to high RF power applications. The have relatively high noise figure that varies from about 1 dB at 1 GHz to about 4–5 dB at 10 GHz.

An improved high-frequency bipolar transistor can be fabricated using heterostructures of compound semiconductors, such as AlGaAs/GaAs [Kroemer, 1982]. These devices have their emitters fabricated from a wide-bandgap semiconductor (such as AlGaAs) and the remainder of the device fabricated from the lower-bandgap semiconductor (GaAs). The wide-bandgap emitter results in improved charge injection efficiency across the base-emitter junction into the base region and much improved RF performance. While the operation of standard Si bipolar transistors is limited to frequencies less than about 40 GHz, the heterojunction bipolar transistors (HBTs) can operate in excess of 100 GHz. They are useful in both low-noise and high RF power applications. The heterostructure concept has recently been applied in Si-based devices using heterostructures using SiGe/Si compounds. These devices show consider promise for high-frequency applications and the transistors have demonstrated RF performance comparable to that obtained from the AlGaAs/GaAs HBTs.

Comparison of Bipolar Transistor and MESFET Noise Figures

In low-noise applications, GaAs MESFETs are generally preferred to Si bipolar transistors. The MESFET demonstrates a lower noise figure than the bipolar transistor throughout the microwave frequency range, and the advantage increases with frequency. This advantage is demonstrated by a comparison of the expressions for the minimum noise figure for the two devices. The bipolar transistor has a minimum noise figure that can be expressed as

$$F_{min} \cong 1 + bf^2 \left[1 + \sqrt{1 + \frac{2}{bf^2}} \right] \qquad (37.20)$$

where F_{min} is the noise figure and

$$b = \frac{40 I_c r_b}{f_T^2} \qquad (37.21)$$

where I_c is the collector current and the other terms are as previously defined. The minimum noise figure for the MESFET is

$$F_{min} \cong 1 + mf \qquad (37.22)$$

where

$$m = \frac{2.5}{f_T} \sqrt{g_m(R_g + R_s)} \qquad (37.23)$$

where g_m is the MESFET **transconductance**, R_g is the gate resistance, and R_s is the source resistance.

Comparing these expressions shows that the minimum noise figure increases with frequency quadratically for bipolar transistors and linearly for MESFETs. Therefore, as operating frequency increases, the MESFET demonstrates increasingly superior noise figure performance as compared to Si bipolar transistors.

Conclusions

Various active solid-state devices that are useful at microwave and millimeter-wave frequencies have been discussed. Both two-terminal and three-terminal devices were included. The most commonly used two-terminal devices are tunnel diodes, transferred-electron devices, and IMPATT diodes. Three-terminal devices consist of various forms of field-effect transistors and bipolar transistors. Recent advances employ heterostructures using combinations of different semiconductors to produce devices with improved RF performance, especially for high-frequency applications. Both two-terminal and three-terminal devices can provide useful gain at frequencies in excess of 100 GHz. Further improvements are likely as fabrication technology continues to improve.

Defining Terms

Active device: A device that can convert energy from a dc bias source to a signal at an RF frequency. Active devices are required in oscillators and amplifiers.

Charge carriers: Units of electrical charge that when moving produce current flow. In a semiconductor two types of charge carriers exist: electrons and holes. Electrons carry unit negative charge and have an effective mass that is determined by the shape of the conduction band in energy-momentum space. The effective mass of an electron in a semiconductor is generally significantly less than an electron in free space. Holes have unit positive charge. Holes have an effective mass that is determined by the shape of the valence band in energy-momentum space. The effective mass of a hole is generally significantly larger than that for an electron. For this reason electrons generally move much faster than holes when an electric field is applied to the semiconductor.

Gain: A measure of the ability of a network to increase the energy level of a signal. Gain is generally measured in decibels. For voltage or current gain: G (dB) $= 20 \log(S_{out}/S_{in})$, where S is the RF voltage or current out of and into the network. For power gain G (dB) $= 10 \log(P_{out}/P_{in})$. If the network has net loss, the gain will be negative.

Noise figure: A measure of the noise added by a network to an RF signal passing through it. Noise figure can be defined in terms of signal-to-noise ratios at the input and output ports of the network. Noise figure is generally measured in decibels and can be defined as F (dB) $= 10 \log[(S/N)_{in}/(S/N)_{out}]$.

One-port network: An electrical network that has only one RF port. This port must be used as both the input and output to the network. Two-terminal devices result in one-port networks.

Three-terminal device: An electronic device that has three contacts, such as a transistor.

Transconductance: A measure of the gain capability of a transistor. It is defined as the change in output current as a function of a change in input voltage.

Two-port network: An electrical network that has separate RF ports for the input and output. Three-terminal devices can be configured into two-port networks.

Two-terminal device: An electronic device, such as a diode, that has two contacts. The contacts are usually termed the cathode and anode.

References

P.B. Bhartia and I.J. Bahl, *Millimeter Wave Engineering and Applications*, New York: Wiley-Interscience, 1984.

B.G. Bosch and R.W. Engelmann, *Gunn-Effect Electronics*, New York: Halsted Press, 1975.

H.F. Cooke, "Microwave transistors: Theory and design," *Proc. IEEE*, vol. 59, pp. 1163–1181, Aug. 1971.

T.J. Drummond, W.T. Masselink, and H. Morkoc, "Modulation-doped GaAs/AlGaAs heterojunction field-effect transistors: MODFET's," *Proc. IEEE*, vol. 74, pp. 773–822, June 1986.

H. Kroemer, "Heterostructure bipolar transistors and integrated circuits," *Proc. IEEE*, vol. 70, pp. 13–25, Jan. 1982.

C.A. Liechti, "Microwave field-effect transistors—1976," *IEEE Trans. Microwave Theory and Tech.*, vol. MTT-24, pp. 128–149, June 1976.

S.M. Sze, *Physics of Semiconductor Devices*, 2nd ed., New York: Wiley-Interscience, 1981.

Further Information

Additional details on the various devices discussed in this chapter can be found in the following books:

I. Bahl and P. Bhartia, *Microwave Solid State Circuit Design*, New York: Wiley-Interscience, 1988.

M. Shur, *Physics of Semiconductor Devices*, Englewood Cliffs, N.J.: Prentice-Hall, 1990.

S.M. Sze, *High-Speed Semiconductor Devices*, New York: Wiley-Interscience, 1990.

S. Tiwari, *Compound Semiconductor Device Physics*, San Diego: Academic Press, 1992.

S. Wang, *Fundamentals of Semiconductor Theory and Device Physics*, Englewood Cliffs, N.J.: Prentice-Hall, 1989.

38

Leland H. Hemming
McDonnell Douglas Technologies, Inc.

Compatibility

Vichate Ungvichian
Florida Atlantic University

John M. Roman
Florida Atlantic University

Martin A. Uman
University of Florida, Gainesville

Marcos Rubinstein
University of Florida, Gainesville

38.1 Grounding, Shielding, and Filtering 903
    Grounding • Shielding • Filtering
38.2 Spectrum, Specifications, and Measurement Techniques 919
    Electromagnetic Spectrum • Specifications • Measurement Procedures
38.3 Lightning 935
    Terminology and Physics • Lightning Occurrence Statistics • Electric and Magnetic Fields • Modeling of the Return Stroke • Lightning-Overhead Wire Interactions

38.1 Grounding, Shielding, and Filtering

Leland H. Hemming

Electromagnetic interference (EMI) is defined to exist when undesirable voltages or currents are present to influence adversely the performance of an electronic circuit or system. Interference can be within the system (intrasystem), or it can be between systems (intersystem). The system is the equipment or circuit over which one exercises design or management control.

The cause of an EMI problem is an unplanned coupling between a source and a receptor by means of a transmission path. Transmission paths may be conducted or radiated. Conducted interference occurs by means of metallic paths. Radiated interference occurs by means of near- and far-field coupling. These different paths are illustrated in Fig. 38.1.

The control of EMI is best achieved by applying good interference control principles during the design process. These involve the selection of signal levels, impedance levels, frequencies, and circuit configurations that minimize conducted and radiated interference. In addition, signal levels should be selected to be as low as possible, while being consistent with the required signal-to-noise ratio. Impedance levels should be chosen to minimize undesirable capacitive and inductive coupling.

The frequency spectral content should be designed for the specific needs of the circuit, minimizing interference by constraining signals to desired paths, eliminating undesired paths, and separating signals from interference. Interference control is also achieved by physically separating leads carrying currents from different sources.

For optimum control, the three major methods of EMI suppression—grounding, shielding, and filtering—should be incorporated early in the design process. The control of EMI is first achieved by proper grounding, then by good shielding design, and finally by filtering.

Grounding is the process of electrically establishing a low impedance path between two or more points in a system. An ideal ground plane is a zero potential, zero impedance body that can be used as reference for all signals in the system. Associated with grounding is *bonding*, which is the establishment of a low impedance path between two metal surfaces.

Shielding is the process of confining radiated energy to the bounds of a specific volume or preventing radiated energy from reaching a specific volume. *Filtering* is the process of eliminating conducted interference by controlling the spectral content of the conducted path. Filtering is the last step in the EMI design process.

Grounding

Grounding Principles

The three fundamental grounding techniques—floating, single-point, and multiple-point—are illustrated in Fig. 38.2.

Floating grounds are used to isolate circuits or equipment from a common ground plane. Static charges are a hazard with this type of ground. Dangerous voltages may develop or a noise-producing discharge might occur. Generally, bleeder resistors are used to control the static problem. Floating grounds are useful only at low frequencies where capacitive coupling paths are negligible.

The single-point ground is a single physical point in a circuit. By connecting all grounds to a common point, no interference will be produced in the equipment because the configuration does not result in potential differences across the equipment. At high frequencies care must be taken to prevent capacitive coupling, which will result in interference.

A multipoint ground system exists when each ground connection is made directly to the ground plane at the closest available point on it, thus minimizing ground lead lengths. A large conductive body is chosen for the ground. Care must be taken to avoid ground loops.

Circuit grounding design is dependent on the function of each type of circuit. In unbalanced systems, care must be taken to reduce the potential of common mode noise. Differential devices are commonly used to suppress this form of noise. The use of high circuit impedances should be minimized. Where it cannot be avoided, all interconnecting leads should be shielded, with the shield well grounded. Power supply grounding must be done properly to minimize load inducted noise on a power supply bus. When electromechanical relays are used in a system, it is best that they be provided with their own power supplies.

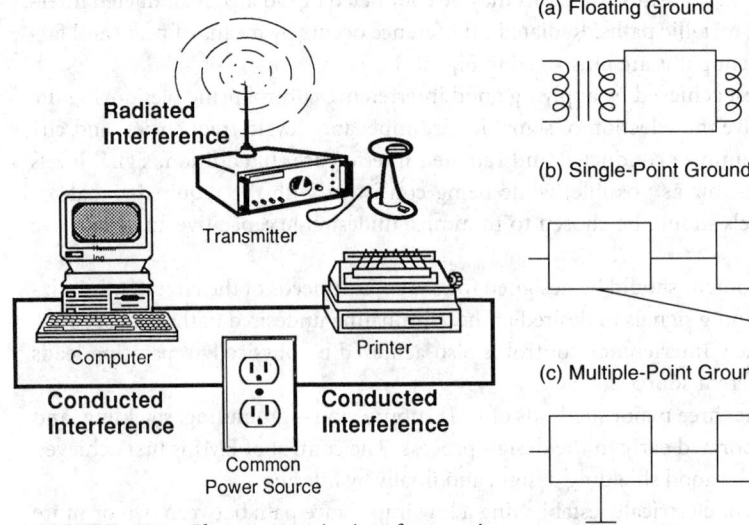

FIGURE 38.1 Electromagnetic interference is caused by uncontrolled conductive paths and radiated near/far fields.

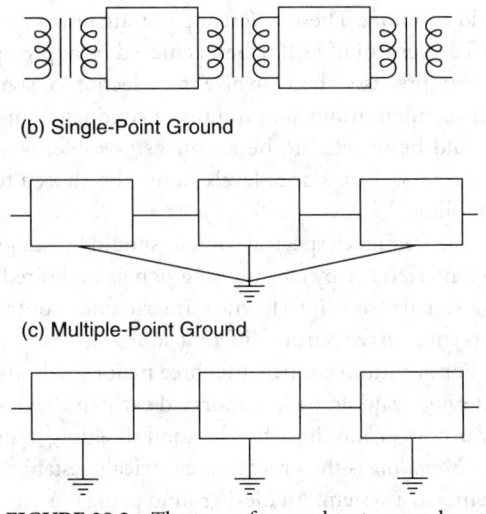

FIGURE 38.2 The type of ground system used must be selected carefully.

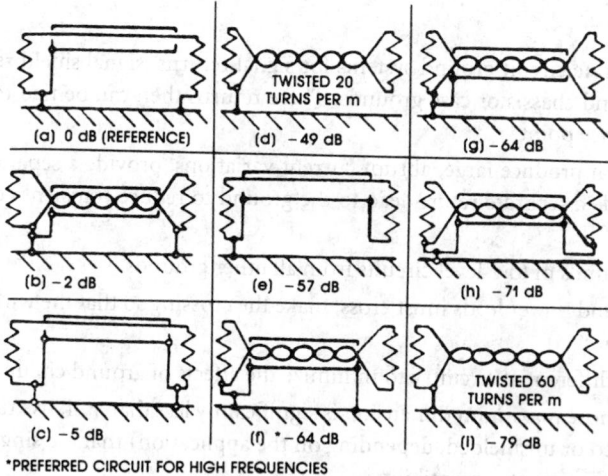

FIGURE 38.3 Shield termination is critical on transmission lines.

Cable shield grounding must be designed based upon the frequency range, impedance levels, (whether balanced or unbalanced) and operating voltage and/or current. Figure 38.3 illustrates a variety of termination methods and their relative susceptibility to magnetic interference. Cross talk between cables is a major problem and must be carefully considered during the design process.

Building facility grounds must be provided for electrical faults, signal, and lightning. The fault protection (green wire) subsystem is for the protection of personnel and equipment from the hazards of electrical power faults and static charge buildup. The lightning protection system consists of air terminals (lightning rods), heavy duty down-conductors, and ground rods. The **signal reference subsystem** provides a ground for signal circuits to control static charges and noise and to establish a common reference between signals and loads. Earth grounds may consist of vertical rods, horizontal grids or radials, plates, or incidental electrodes such as utility pipes or buried tanks. The latter must be constructed and tested to meet the design requirements of the facility.

Grounding Design Guidelines

The following design guidelines represent good practice but should be applied subject to the detailed design objectives of the system.

Fundamental Concepts

- Use single-point grounding for circuit dimensions less than 0.03 λ (wavelength) and multi-point grounding for dimensions greater than 0.15 λ.
- The type of grounding for circuit dimensions between 0.03 and 0.15 λ depends on the physical arrangement of the ground leads as well as the conducted emission and conducted susceptibility limits of the circuits to be grounded. Hybrid grounds may be needed for circuits that must handle a broad portion of the frequency spectrum.
- Apply floating ground isolation techniques (i.e., transformers) if ground loop problems occur.
- Keep all ground leads as short as possible.
- Design ground reference planes so that they have high electrical conductivity and can be maintained easily to retain good conductivity.

Safety Considerations

- Connect test equipment grounds directly to the grounds of the equipment being tested.
- Make certain the ground connections can handle fault currents that might flow unexpectedly.

Circuit Grounding

- Maintain separate circuit ground systems for signal returns, signal shield returns, power system returns, and chassis or case grounds. These returns then can be tied together at a single ground reference point.
- For circuits that produce large, abrupt current variations, provide a separate grounding system, or provide a separate return lead to the ground to reduce transient coupling into other circuits.
- Isolate the grounds of low-level circuits from all other grounds.
- Where signal and power leads must cross, make the crossing so that the wires are perpendicular to each other.
- Use balanced differential circuitry to minimize the effects of ground circuit interference.
- For circuits whose maximum dimension is significantly less than $\lambda/4$, use tightly twisted wires (either shielded or unshielded, depending on the application) that are single-point grounded to minimize equipment susceptibility.

Cable Grounding

- Avoid pigtails when terminating cable shields.
- When coaxial cable is needed for signal transmission, use the shield as the signal return and ground at the generator end for low-frequency circuits. Use multipoint grounding of the shield for high-frequency circuits.
- Provide multiple shields for low-level transmission lines. Single-point grounding of each shield is recommended.

Shielding

The control of near- and far-field coupling (radiation) is accomplished using shielding techniques. The first step in the design of a shield is to determine what undesired field level may exist at a point with no shielding and what the tolerable field level is. The difference between the two then is the needed **shielding effectiveness**. A methodology for accomplishing the required shielding effectiveness is illustrated in Fig. 38.4.

This section discusses the shielding effectiveness of various solid and nonsolid materials and their application to various shielding situations. **Penetrations** and their design are discussed so that the required shielding effectiveness is maintained. Finally, common shielding effectiveness testing methods are reviewed.

Enclosure Theory

The attenuation provided by a shield results from three loss mechanisms as illustrated in Fig. 38.5.

1. Incident energy is reflected (R) by the surface of the shield because of the impedance discontinuity of the air–metal boundary. This mechanism does not require a particular material thickness but simply an impedance discontinuity.
2. Energy that does cross the boundary (not reflected) is attenuated (A) in passing through the shield.
3. The energy that reaches the opposite face of the shield encounters another air–metal boundary and thus some of it is reflected (B) back into the shield. This term is only significant when $A < 15$ dB and is generally neglected because the barrier thickness is generally great enough to exceed the 15-dB loss rule of thumb.

Thus:
$$S = R + A + B \text{ dB} \tag{38.1}$$

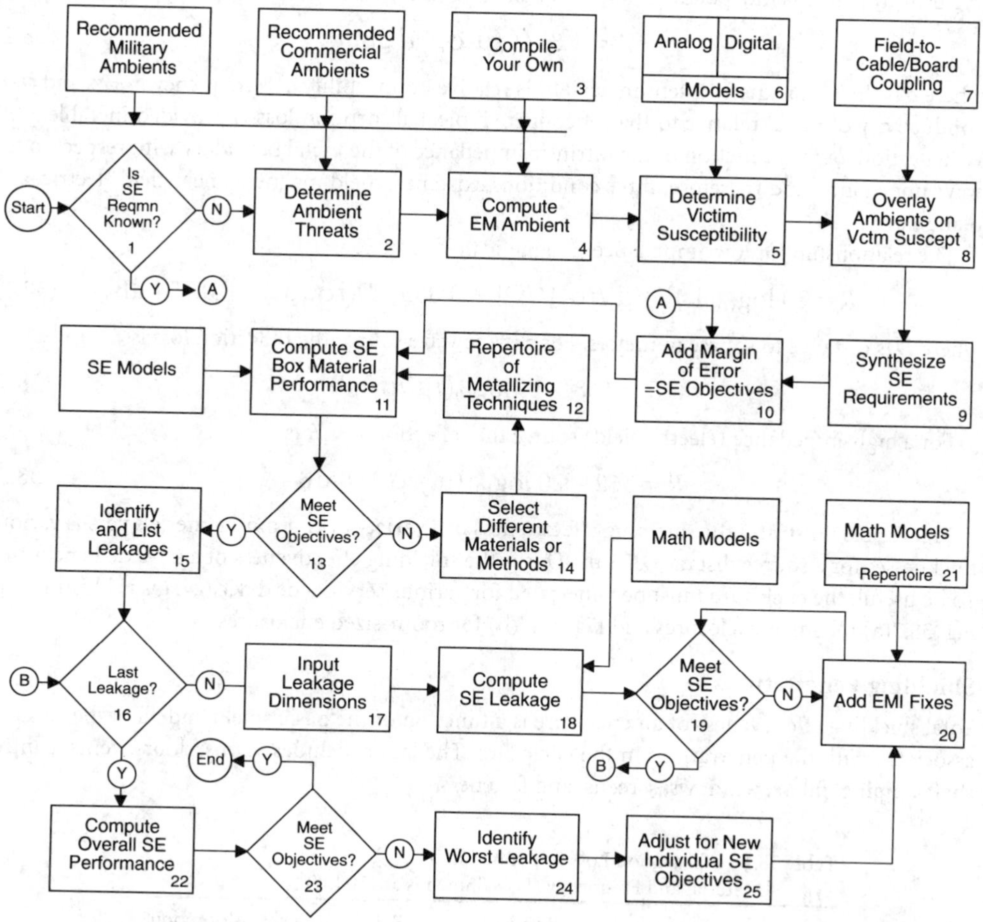

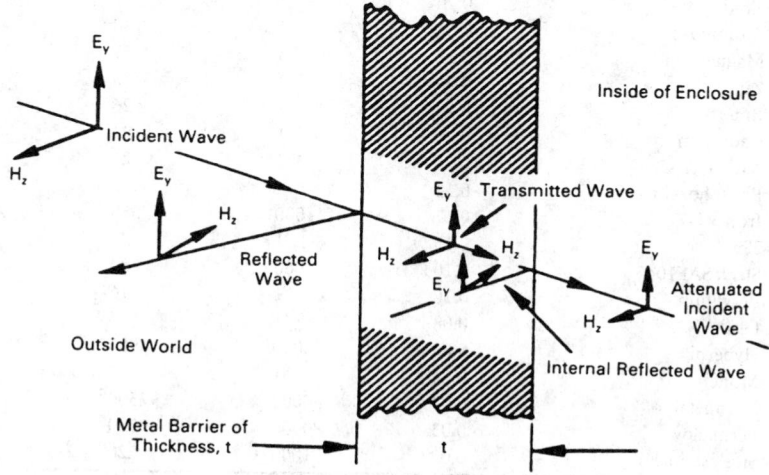

FIGURE 38.4 Shielding effectiveness requires careful analysis to achieve satisfactory results. (*Source:* D.R.J. White, *Shielding Design, Methodology and Procedures,* Interference Control Technologies, Springfield, Va., 1986. With permission.)

FIGURE 38.5 Shielding effectiveness is the result of three loss mechanisms.

Absorption loss is independent of the type of wave (electric/magnetic) and is given by

$$A = 1.314(f\mu_r\sigma_r)^{1/2}d \text{ dB} \tag{38.2}$$

where d is shield thickness in centimeters, μ_r is relative permeability, f is frequency in Hz, and σ_r is conductivity of metal relative to that of copper. Typical absorption loss is provided in Table 38.1.

Reflection loss is a function of the intrinsic impedance of the metal boundary with respect to the wave impedance, and therefore, three conditions exist: near-field magnetic, near-field electric, and plane wave.

The relationship for low-impedance (magnetic field) source is

$$R = 20 \log_{10}[1.173(\mu_r/f\sigma_r)^{1/2}/D] + 0.0535 \, D(f\sigma_r/\mu_r)^{1/2} + 0.354 \text{ dB} \tag{38.3}$$

where D is distance to source in meters. For a plane wave source the reflection loss is

$$R = 168 - 10 \log_{10}(f\mu_r/\sigma_r) \text{ dB} \tag{38.4}$$

For a high-impedance (electric field) source the reflection loss R is

$$R = 362 - 20 \log_{10}[(\mu_r f/\sigma_r)D] \text{ dB} \tag{38.5}$$

Figure 38.6 illustrates the shielding effectiveness of a variety of common materials versus various thicknesses for a source distance of 1 m. This is the shielding effectiveness of a six-sided enclosure. To be useful, the enclosure must be penetrated for various services or devices. This is illustrated in Fig. 38.7(a) for small enclosures and Fig. 38.7(b) for room-sized enclosures.

Shielding Penetrations

Total shielding effectiveness of an enclosure is a function of the basic shield and all of the leakages associated with the penetrations in the enclosure. The latter includes seams, doors, vents, control shafts, piping, filters, windows, screens, and fasteners.

Table 38.1 Absorption Loss Is a Function of Type of Material and Frequency (Loss Shown is at 150 kHz)

Metal	Relative Conductivity	Relative Permeability	Absorption Loss A, dB/mm
Silver	1.05	1	52
Copper—annealed	1.00	1	51
Copper—hard drawn	0.97	1	50
Gold	0.70	1	42
Aluminum	0.61	1	40
Magnesium	0.38	1	31
Zinc	0.29	1	28
Brass	0.26	1	26
Cadmium	0.23	1	24
Nickel	0.20	1	23
Phosphor–bronze	0.18	1	22
Iron	0.17	1000	650
Tin	0.15	1	20
Steel, SAE1045	0.10	1000	500
Beryllium	0.10	1	16
Lead	0.08	1	14
Hypernik	0.06	80000	3500[a]
Monel	0.04	1	10
Mu-metal	0.03	80000	2500[a]
Permalloy	0.03	80000	2500[a]
Steel, stainless	0.02	1000	220[a]

[a]Assuming that material is not saturated.

Source: MIL-HB-419A.

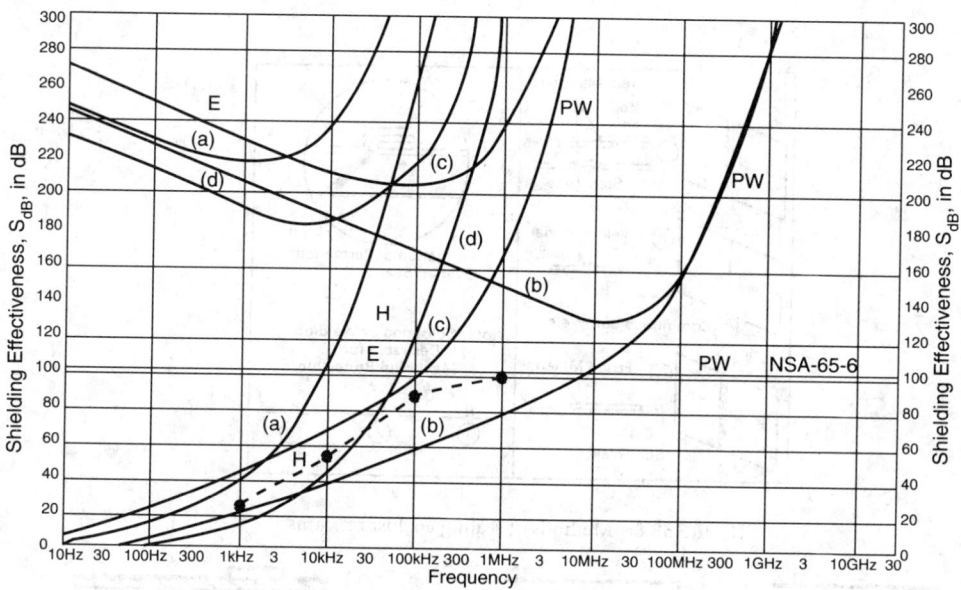

FIGURE 38.6 The shielding effectiveness of common sheet metals, 1 m separation. (a) 26-gage steel; (b) 3-oz. copper foil; (c) 0.030-in. aluminum sheet; (d) 0.003-in. Permalloy.

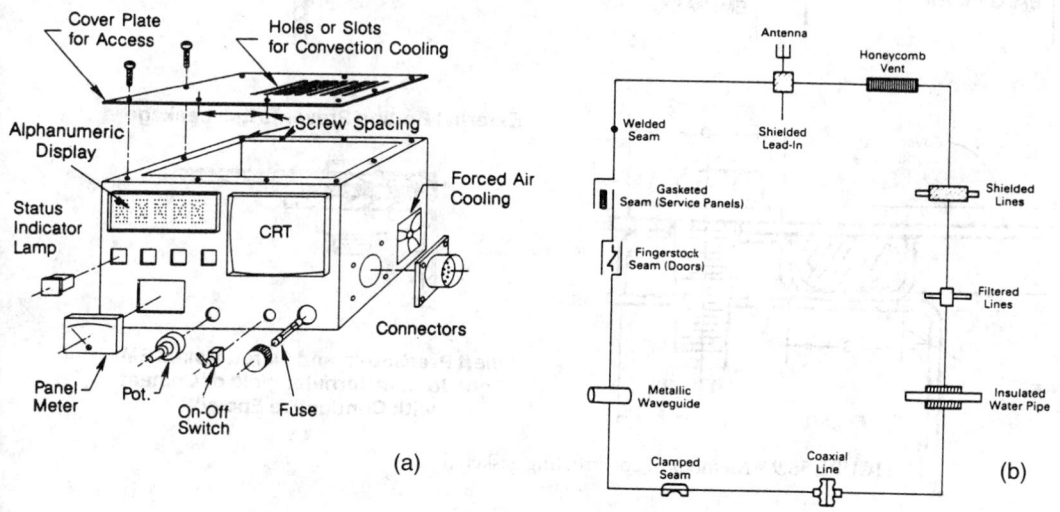

FIGURE 38.7 Penetrations in small (a) and large (b) enclosures.

The design of the seams is a function of the type of enclosure and the level and nature of the shielding effectiveness required. For small instruments, computers, and similar equipment, the typical shielding required is on the order of 60 dB for electric and plane wave shielding. EMI gaskets are commonly used to seal the openings in sheet metal construction. In some high-performance applications the shielding is achieved using very tight-fitting machined housings. Examples are IF strips and large dynamic range log amplifier circuits. Various methods of sealing joints are illustrated in Fig. 38.8. EMI gasketing methods are shown in Fig. 38.9. For large room-sized enclosures, the performance requirements typically range from 60 to 120 dB. Conductive EMI shielding tape is used in the 60-dB realm, clamped seams for 80–100 dB, and continuous welded seams for 120-dB performance. These are illustrated in Fig. 38.10.

A good electromagnetic shielded door design must meet a variety of physical and electrical requirements. Figure 38.11 illustrates a number of ways this is accomplished.

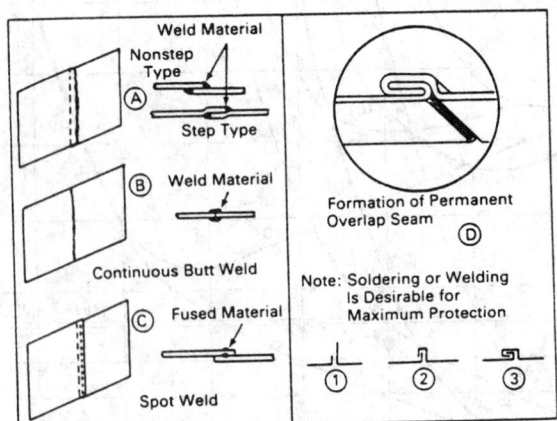

FIGURE 38.8 Methods of sealing enclosure seams.

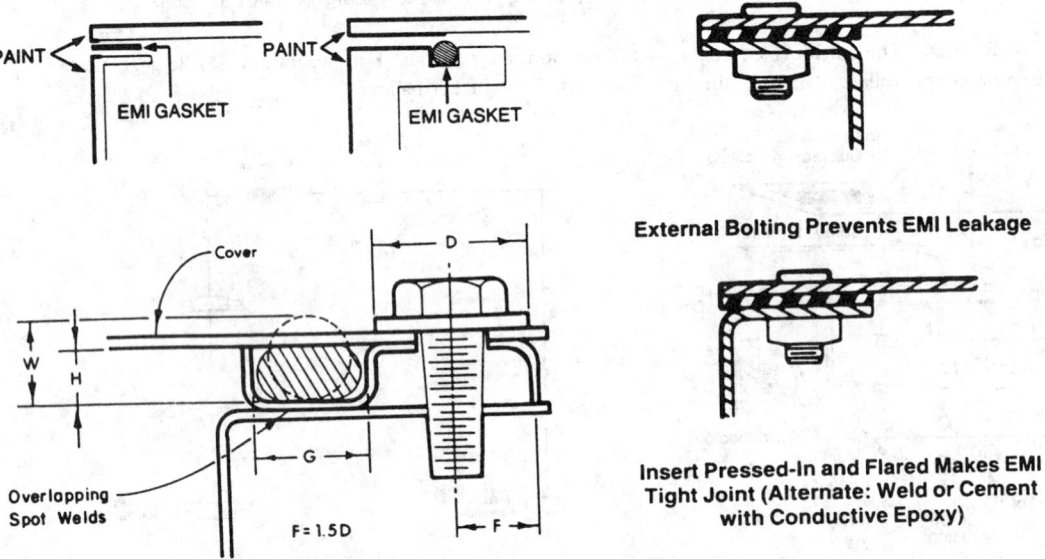

FIGURE 38.9 Methods of constructing gasketed joints.

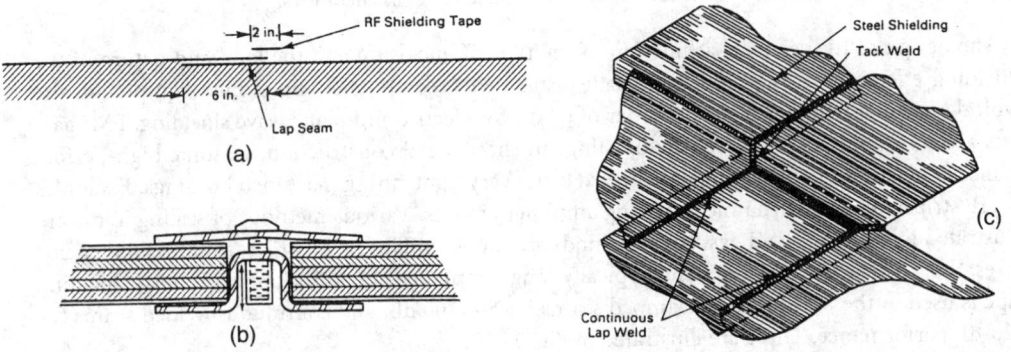

FIGURE 38.10 Most common seams in large enclosures. (a) Foil and shielding tape; (b) clamped; (c) welded.

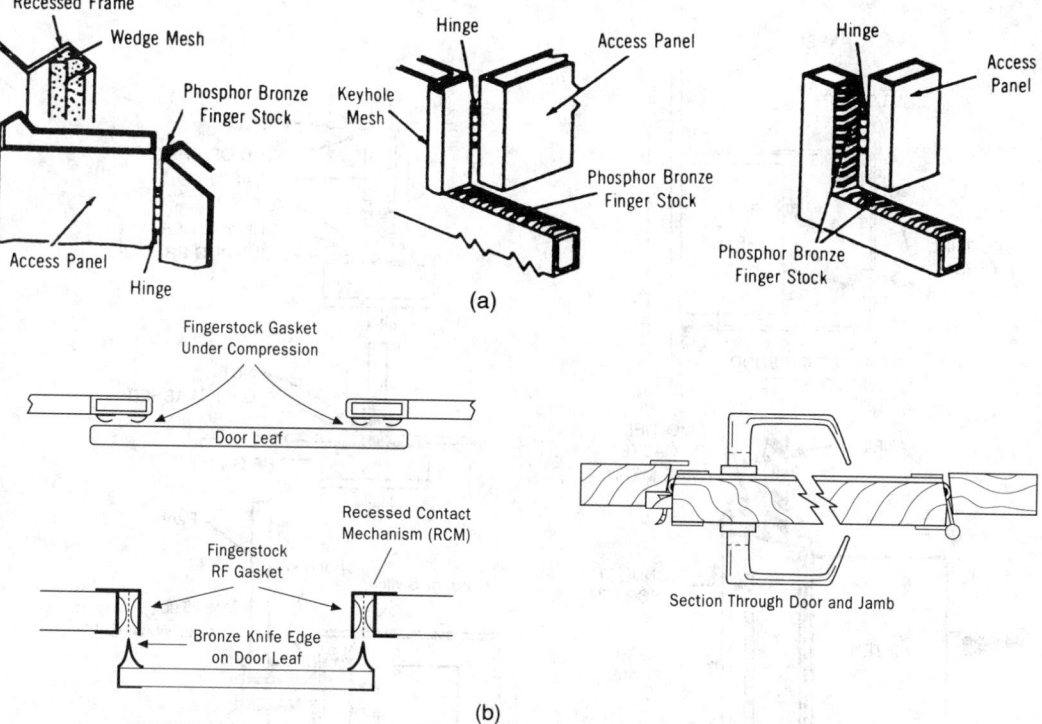

FIGURE 38.11 Methods of sealing seams in RF enclosure small (a) and large (b) doors.

For electronic equipment, a variety of penetrations must be made to make the shielded volume functional. These include control shafts, windows, lights, filters, and displays. How to maintain the shielding integrity for a variety of these devices is illustrated in Fig. 38.12.

Shield Testing

The most common specification used for shield evaluation is the procedure given in MIL-STD-285. This consists of establishing a reference level without the shield and then enclosing the receiver within the shield and determining the difference. The ratio is the shielding effectiveness. This applies regardless of materials used in the construction of the shield. Care must be taken in evaluating the results since the measured value is a function of a variety of factors, not all of which are definable.

Summary of Good Shielding Practice

Shielding Effectiveness

- Good conductors, such as copper and aluminum, should be used for electric field shields to obtain high reflection loss. A shielding material thick enough to support itself usually provides good electric shielding at all frequencies.
- Magnetic materials, such as iron and special high-permeability alloys, should be used for magnetic field shields to obtain high absorbtion loss.
- In the plane wave region, the sealing of all apertures is critical to good shielding practice.

Multiple Shields

- Multiple shields are quite useful where high degrees of shielding effectiveness are required.

Shield Seams

- All openings or discontinuities should be addressed in the design process to ensure achievement of the required shielding effectiveness. Shield material should be selected not only

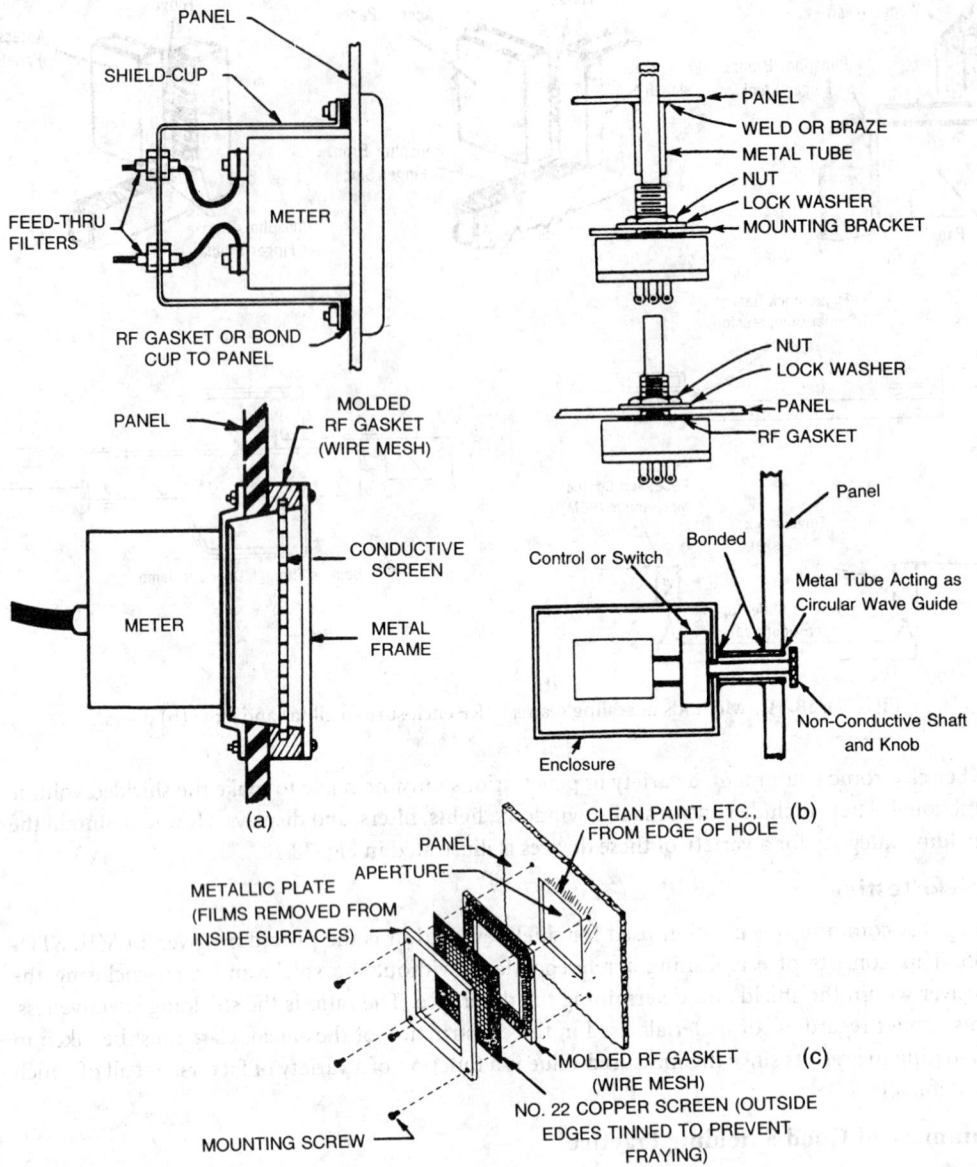

FIGURE 38.12 Maintaining shielding effectiveness around control and display devices. (a) Acceptable methods of shielding panel-mounted meters; (b) acceptable use of circular waveguide in a permanent aperture for control shaft; (c) method of mounting screen over a large aperture.

from a shielding requirement, but also from electrochemical corrosion and strength considerations.

- Whenever system design permits, use continuously overlapping welded seams. Obtain intimate contact between mating surfaces over as much of the seam as possible.
- Surfaces to be mated must be clean and free from nonconducting finishes, unless the bonding process positively and effectively cuts through the finish. When electromagnetic compatibility (**EMC**) and finish specifications conflict, the finishing requirements must be modified.

Case Construction

- Case material should have good shielding properties.
- Seams should be welded or overlapped.
- Panels and cover plates should be attached using conductive gasket material with closely spaced fasteners.
- Mating surfaces should be cleaned just before assembly to ensure good electrical contact and to minimize corrosion.
- A variety of special devices are available for sealing around doors, vents, and windows.
- Internal interference generating circuits must be isolated both electrically and physically. Electrical isolation is achieved by circuit design; physical isolation may be achieved by proper shielding.
- For components external to the case, use EMI boots on toggle switches, EMI rotary shaft seals on rotary shafts, and screening and shielding on meters and other indicator faces.

Cable Shields

- Cabling that penetrates a case should be shielded and the shield should be terminated in a peripheral bond at the point of entry. This peripheral bond should be made to the connector or adaptor shell.

Filtering

An electrical filter is a combination of lumped or distributed circuit elements arranged so that it has a frequency characteristic that passes some frequencies and blocks others.

Filters provide an effective means for the reduction and suppression of electromagnetic interference as they control the spectral content of signal paths. The application of filtering requires careful consideration of an extensive list of factors including insertion loss, impedance, power handling capability, signal distortion, tunability, cost, weight, size, and rejection of undesired signals. Often they are used as stopgap measures, but if suppression techniques are used early in the design process, then the complexity and cost of interference fixes can be minimized. There are many textbooks on filtering, which should be used for specific applications.

The types of filters are classified according to the band of frequencies to be transmitted or attenuated. The basic types illustrated in Fig. 38.13 include low-pass, high-pass, bandpass, and bandstop (reject).

Filters can be composed of lumped, distributed, or dissipative elements; the type used is mainly a function of frequency.

Filtering Guidance

- It is best to filter at the interference source.
- Suppress all spurious signals.
- Design nonsusceptible circuits.
- Ensure that all filter elements interface properly with other EMC elements, i.e., proper mounting of a filter in a shielded enclosure.

Filter Design

Filters using lumped and distributive elements generally are reflective, in that the various component combinations are designed for high series impedance and low shunt impedance in the stopband while providing low series impedance and high shunt impedance in the passband.

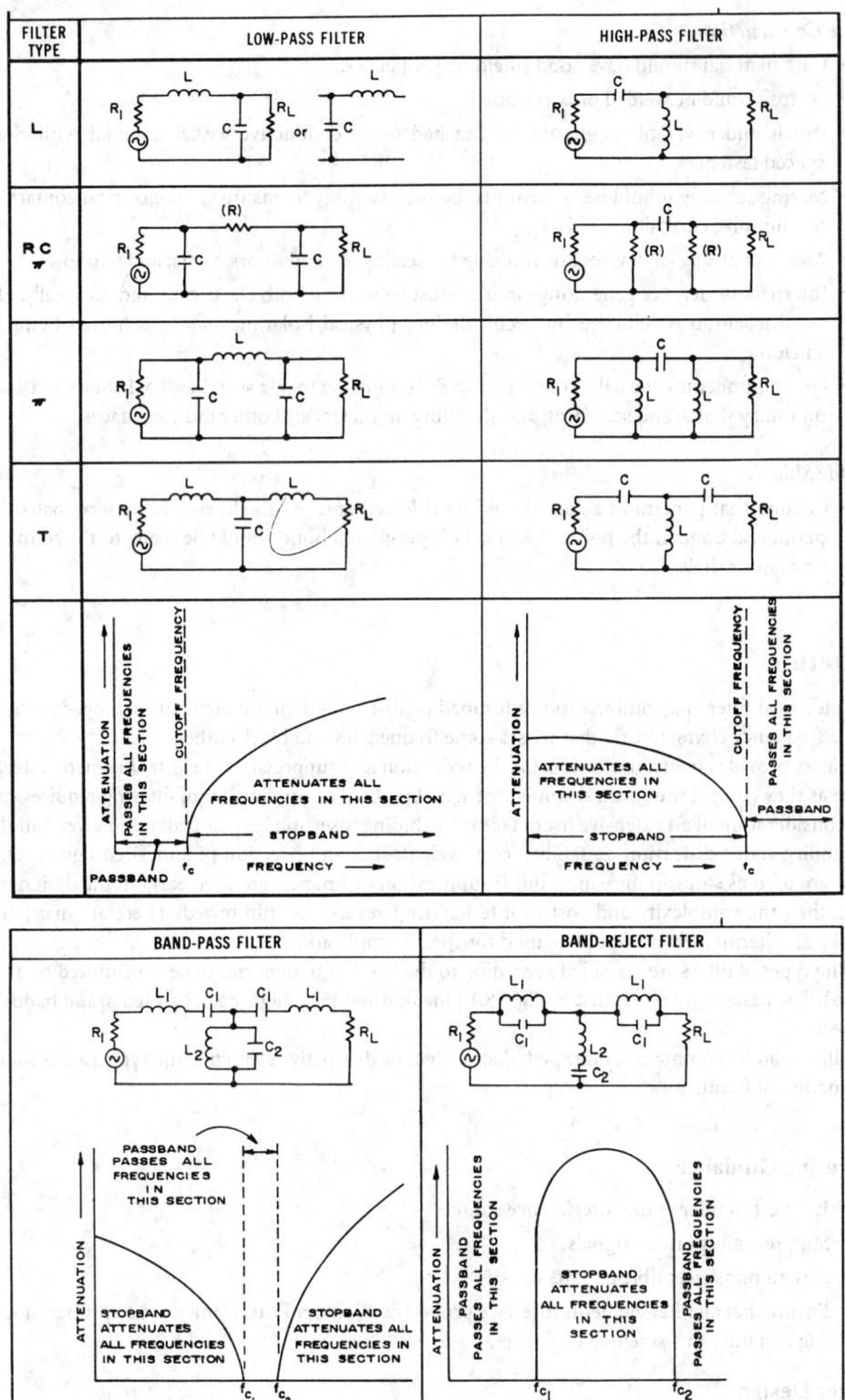

FIGURE 38.13 Filters provide a variety of frequency characteristics.

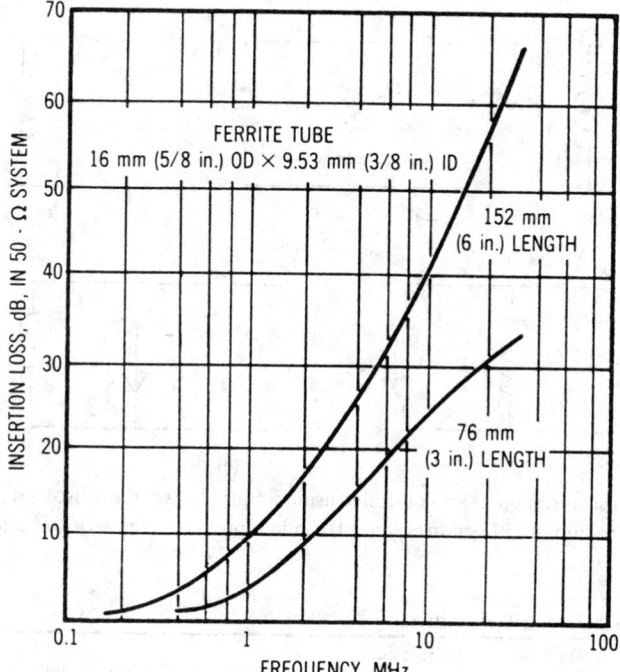

FIGURE 38.14 Ferrite provides a flexible means of achieving a low-pass filter with good high-frequency loss characteristics.

The impedance mismatches associated with the use of reflective filters can result in an increase of interference. In such cases, the use of dissipative elements is found to be useful. A broad range of ferrite components are available in the form of beads, tubes, connector shells, and pins. A very effective method of low-pass filtering is to form the ferrite into a coaxial geometry, the properties of which are proportional to the length of the ferrite, as shown in Fig. 38.14.

Application of filtering takes many forms. A common problem is transient suppression as illustrated in Fig. 38.15. All sources of transient interference should be treated at the source.

Power line filtering is recommended to eliminate conducted interference from reaching the powerline and adjacent equipment. Active filtering is very useful in that it can be built in as part of the circuit design and can be effective in passing only the design signals. A variety of noise blankers, cancelers, and limiter circuits are available for active cancellation of interference.

Special Filter Types

A variety of special-purpose filters are used in the design of electronic equipment. Transmitters require a variety of filters to achieve a noise-free output. A selection of these filters is listed in Table 38.2.

Receive preselectors play a useful role in interference rejection. Both distributed (cavity) and lumped element components are used. The various forms are given in Table 38.3.

IF filters control the selectivity of a receiving system and use a variety of mechanical and electrical filtering components. Table 38.4 summarizes some common applications.

Testing

The general requirements for electromagnetic filters are detailed in MIL-F-15733, MIL-F-18327, and MIL-F-25880. Insertion loss is measured in accordance with MIL-STD-220.

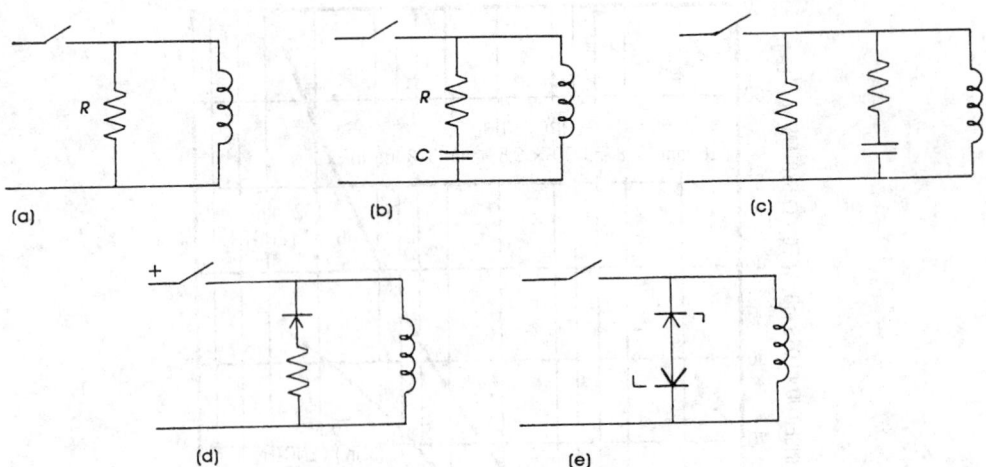

FIGURE 38.15 Transient responses are controlled using simple filters at the source. (a) Resistance damping; (b) capacitance suppression; (c) RC suppression; (d) diode suppression; (e) back-to-back diode suppression.

Table 38.2 A Selection of Filters Used in Transmitters

Type	Classification	Applicable Frequency Range (Hz)	Specific Design Frequency (Hz)	Power Capabilities (W)	Passband Loss (dB)	Stopband Attenuation (dB)	Tuning	Spurious
			Nonmicrowave					
Resonant LC circuits	Either band pass or reject	Up to ~100 MHz	—	500 kW	1	30	Mech.	Few
SRI HF filter	Bandpass	Up to ~100 MHz	8–15 MHz	20 kW	0.3	60	Mech. 8–15 MHz	Few
SRI VHF filter	Bandpass	Up to ~500 MHz	300 MHz	5–10 kW	0.3	80	Mech.	Few
			Microwave-Reflective					
Coupled resonators	Bandpass	Above 1 GHz	—	20% of waveguide	0.1	40	Mech.	Many
SRI three-cavity resonator	Bandpass	Above 1 GHz	1.25 to 1.35 GHz	1–9 MW	0.3	30–60	Mech.	Many
Serrated* ride waveguide	Bandpass	Above 1 GHz	—	Less than waveguide	0.1	40	Set	Few
Waffle iron	Bandpass	Above 1 GHz	1.25 to 1.35 GHz	1–4 MW	0.1	>40	Set	Few
Strip* line	Band reject	200 to 1000 MHz	250 MHz	50 MW	0.1	>50	Set	Few
			Microwave-Dissipation					
General leaky wall	Bandpass	Above 1 GHz	—	80% of waveguide	0.25	40	Set	Few
Straight* wall	Low pass	Above 1 GHz	3 GHz cutoff	80% of waveguide	0.5	10–30	Set	Few
Offset* wall	Low pass	Above 1 GHz	3 GHz cutoff	80% of waveguide	0.5	20–40	Set	Few
Serpentine wall*	Low pass	Above 1 GHz	3 GHz cutoff	80% of waveguide	2	5–55	Set	Many
Circular leaky wall	Low pass	Above 1 GHz	4.68 MHz cutoff	80% of waveguide	0.5	25–45	Set	Few
Circular* side waveguide	Low pass	Above 1 GHz	3 GHz cutoff	10–20 kW	0.5	>18	Set	Few
Coaxial* leaky wave	Low pass	1–3 GHz	1.7 GHz cutoff	140–280 kW	0.3	15–40	Set	Few
General directional coupler	Low pass	Above 1 GHz	—	80% of waveguide	0.3	30	Set	Few
Transfer* coupler	Low pass	Above 1 GHz	3 GHz cutoff	80% of waveguide	0.2	6	Set	Few
Ferrite filters	Band reject	Above 1 GHz	—	Low	0.1	>30	Set or elect.	Few
Leaky wall wiregrid*	Mode filter	Above 1 GHz	—	80% of waveguide	TE_{10} mode 0.1	TE_{20}, TE_{30} 30	Set	—
Periodic filter	Bandpass	Above 1 GHz	7 GHz	500 kW peak	1.4	>50	Set	None

* Experimental models only—not commercially available.

Source: AFSC DH1-4.

Table 38.3 Receive Preselectors Are Important in Interference Control

Techniques	Classification	Frequency Range (Hz)	Fractional Bandwidth (%)	Passband Insertion Loss (dB)	Stopband Attenuation (dB)	Tuning	Spurious Transmission Characteristics
LC resonant circuits	Bandpass or reject	Up to ~200 MHz	2	1 to 2	>30	Tunable over wide range	Medium
Spiral inductance tuner	Bandpass	200 to 400 MHz	1 to 2	~2	>30	Tunable over wide range	Medium
Helical resonator	Bandpass	200 to 400 MHz	0.4	1.5	>50	5 to 1 range	Medium
Butterfly resonator	Bandpass	100 to 1000 MHz	0.5	~2	>40	5 to 1 range	Poor
Butterfly resonator	Band reject	100 to 1000 MHz	0.2	1 to 2	26 to 44	2 to 1 range	Poor
Bridged tee filter	Band reject	Up to 1000 MHz	0.01	~3	60 to 100	Difficult and critical	Good
Cavity resonator	Bandpass	200 to 3000 MHz	0.2	1 to 2	>50	—	Medium
Coaxial cavity resonator	Bandpass	100 to 3000 MHz	0.2	~1.5	~60	—	Good
Ferrite resonator	Bandpass	2 to 7 GHz	0.2	~2	~40	Elec.	Medium
Ferrite resonator	Band reject	2 to 7 GHz	0.5	~0.5	10 to 15	Elec.	Medium
Backward wave amplifier	Bandpass	2 to 4 GHz	0.5	−20	20	Elec.	Poor
Modified hybrid	Band reject	500 to 4000 MHz	—	~2	13 to 40	Less than an octave	Medium
Waveguide cavity resonator	Bandpass	1 to 30 GHz	0.2	~0.5	~60	—	Good
Waveguide cavity resonator	Band reject	1 to 30 GHz	~0.1	~0.3	~60	—	Good
Diode up converter	Bandpass	Broad	—	−12	40	Elec.	Excellent

Source: AFSC DH1-4.

Table 38.4 Common IF Filters Used in Receiver Applications

Technique	Range of Available Frequencies (Hz)	Q	Spurious Resonances	Out of Band Rejection	Other Comments
Mechanical filters	60 to 500 kHz	10^4 to 10^6	Few	High	No adjustment other than initial matching
Crystal filters (bridge type)	20 kHz to 150 MHz	10^4 to 10^7	Many	High	None
Q multiplier	Very wide (at least 80 kHz to 100 MHz)	10^4	Few	High	May be unstable if large multiplication is required
Delay line controlled amplifier	Up to 300 MHz	Variable (up to 10^4)	Few	Variable	Provides gain in addition to selectivity
Negative feedback amplifier	Up to 300 MHz	High	Few	High	None
Automatic variable selectivity	Up to 300 MHz	Variable	Few	Varies	IF bandwidth can be varied for optimum
Lumped constant	Up to ~200 MHz	Medium	Few	Varies	None

Source: AFSC DH1-4.

Defining Terms

Earth electrode system: A network of electrically interconnected rods, plates, mats, or grids, installed for the purpose of establishing a low-resistance contact with earth. The design objective for resistance to earth of this subsystem should not exceed 10 Ω.

Electromagnetic compatibility (EMC): The capability of equipment or systems to be operated in their intended operational environment at designed levels of efficiency without causing or receiving degradation owing to unintentional electromagnetic interference. Electromagnetic compatibility is the result of an engineering planning process applied during the life cycle of the equipment. The process involves careful considerations of frequency allocation, design, procurement, production, site selection, installation, operation, and maintenance.

Electromagnetic pulse (EMP): A large impulsive-type electromagnetic wave generated by nuclear or chemical explosions.

Field strength: A general term that means the magnitude of the electric field vector (in volts per meter) or the magnitude of the magnetic field vector (in ampere-turns per meter). As used in the field of EMC/EMI, the term *field strength* shall be applied only to measurements made in the far field and shall be abbreviated as FS. For measurements made in the near field, the term *electric field strength* (EFS) or *magnetic field strength* (MFS) shall be used, according to whether the resultant electric or magnetic field, respectively, is measured.

Penetration: The passage through a partition or wall of an equipment or enclosure by a wire, cable, pipe, or other conductive object.

Radio frequency interference (RFI): Synonymous with *electromagnetic interference.*

Shielding effectiveness: A measure of the reduction or attenuation in the electromagnetic field strength at a point in space caused by the insertion of a shield between the source and that point.

Signal reference subsystem: This subsystem provides the reference points for all signal grounding to control static charges, noise, and interference. It may consist of any one or a combination of the lower frequency network, higher frequency network, or hybrid signal reference network.

TEMPEST: A code word (not an acronym) which encompasses the government/industrial program for controlling the emissions from systems processing classified data. Individual equipment may be *TEMPESTed* or commercial equipment may be placed in shielded enclosures.

References

AFSC Design Handbook, DH1-4, Electromagnetic Compatibility, 4th ed., U.S. Air Force, Wright-Patterson Air Force Base, Ohio, January 1991.

R. F. Ficchi, Ed., *Practical Design for Electromagnetic Compatibility*, Hayden, 1971.

E. R. Freeman, *Electromagnetic Compatibility Design Guide for Avionics and Related Ground Support Equipment*, Norwood, Mass.: Artech House, 1982.

L. H. Hemming, *Architectural Electromagnetic Shielding Handbook*, New York: IEEE Press, 1991.

B. Keiser, *Principles of Electromagnetic Compatibility*, 3rd ed., Norwood, Mass.: Artech House, 1987.

Y. J. Lubkin, *Filter Systems and Design: Electrical, Microwave, and Digital*, Reading, Mass.: Addison-Wesley, 1970.

MIL-HDBK-419A, Grounding, Bonding, and Shielding of Electronic Equipment and Facilities, U.S. Department of Defense, Washington, D.C., 1990.

R. Morrison, *Grounding and Shielding Techniques in Instrumentation*, New York: John Wiley, 1986.

R. Morrison and W. H. Lewis, *Grounding and Shielding Techniques in Facilities*, New York: John Wiley, 1990.

T. Rikitake, *Magnetic and Electromagnetic Shielding*, Amsterdam: D. Reidel, 1987.

N. O. N. Violetto, *Electromagnetic Compatibility Handbook*, New York: Van Nostrand Reinhold, 1987.

D. R. J. White, *Shielding Design, Methodology and Procedures*, Springfield, Va.: Interference Control Technologies, 1986.

D. R. J. White, *A Handbook on Electromagnetic Shielding Materials and Performance*, Springfield Va.: Interference Control Technologies, 1975.

Further Information

The annual publication *Interference Technology Engineers' Master (Item)*, published by R&B Enterprises, West Conshohocken, Pennsylvania, covers all aspects of EMI including an extensive product directory.

The periodical *IEEE Transactions on Electromagnetic Compatibility*, which is published by The Institute of Electrical and Electronics Engineers, Inc., provides theory and practice in the EMI field.

The periodical *EMC Test & Design*, published bimonthly by the Cardiff Publishing Company, is a good source for practical EMI design information.

The periodical *EMC Technology*, published bimonthly by Interference Control Technologies, Inc., is an excellent source for practical EMI information.

The periodical *Compliance Engineering*, published quarterly by Compliance Engineering, Inc., is a good source for information on EMC regulations and rules.

38.2 Spectrum, Specifications, and Measurement Techniques

Vichate Ungvichian and John M. Roman

Electromagnetic radiation is a form of energy at a particular frequency that can propagate through a medium. This intentionally or unintentionally generated electromagnetic energy is considered as **electromagnetic interference** (EMI) if it degrades the performance of electronic systems. The purposeful generation of electromagnetic energy for communications can be defined as intentionally generated EMI; unintentionally generated EMI can be created, for example, by the electrical signals in a computer and may be radiated into space by way of the interconnecting cables and/or by openings in the device enclosures.

All electrical devices create some form of electromagnetic energy that may potentially interfere with the operation of other electrical devices outside the system (inter-system) or within the system (intra-system). Due to the increasing man-made EMI generated around the globe, allowable limits as well as measurement techniques on RF noise/interference have been set at national and international levels. The Federal Communications Commission and the Military are the two governing bodies in the United States setting standards on EMI, whereas the International Electrotechnical Commission is the ruling body in Europe. These ruling bodies are concerned with only a fraction of the total electromagnetic spectrum.

Electromagnetic Spectrum

The frequency spectrum of electromagnetic energy can span from dc to gamma ray (10^{21} Hz) and beyond. Figure 38.16 shows the typical frequency spectrum chart over a fraction of hertz to 6×10^{22} Hz.

The spectrum for use in electromagnetic compatibility (EMC) purposes covers only from a few hertz (extreme low frequency, ELF) to 40 GHz (microwave bands). ELF has been in use mostly in the area of biological research and ELF communications. On the other side of the spectrum, the

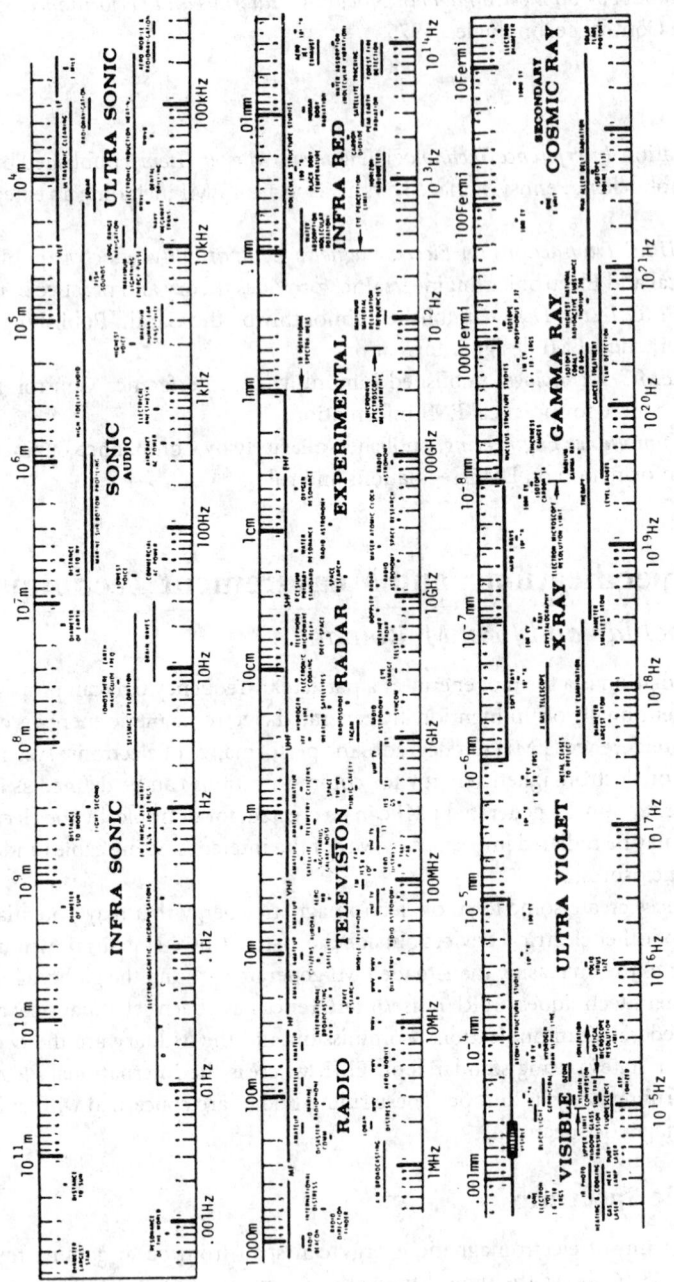

FIGURE 38.16 The frequency spectrum chart. (Contributed by Luther Monell, North America Rockwell Corp.)

electronic devices must function in a hostile environment, in military applications, over the giga-hertz frequency range.

Specifications

In the United States, the Federal Communications Commission (FCC) and the military (MIL) are the two regulating bodies governing the EMC standards for commercial and military-based electronic devices and systems. In Europe, each country has its own EMC governing body as well as its own standards. The Verband Deutscher Elektrotechniker (VDE), British Standards Institute (BSI), European Telecommunications Standards Institute (ETSI), and International Special Committee on Radio Interference (CISPR) standards are a very few examples of the acceptable standards in European countries such as Germany and Great Britain. However, in the near future, Europe may have common EMI/EMC standards and regulatory authority.

The United States of America

Federal Communications Commission. The FCC sets limits on the amount of electromagnetic radiation allowed to be emitted from commercial electronic equipment. Any electronic device capable of emitting radio frequency energy by **radiation** or **conduction** is defined by the Commission as a radio frequency device and is subject to comply with the standards. The relevant limits and some general measurement techniques, along with equipment authorization procedures, are given in the Code of Federal Regulations (CFR), Title 47 (Telecommunications).

Listed in the CFR, Title 47 are five different types of equipment authorization procedures, namely, type acceptance, type approval, notification, certification, and verification. Restrictions are placed on the marketing and sale of radio frequency devices until the appropriate equipment authorization criteria are met.

Devices and systems that require allocation of the frequency spectrum fall under either type acceptance or type approval equipment authorization procedures. These radio frequency devices usually radiate high powers as in radio or television broadcast transmitters. Radio frequency devices not within the allocated part of the RF spectrum would require either certification or verification equipment authorization. Some receivers, such as pagers, require notification equipment authorization. The requirements for these types of radio frequency devices are listed in Part 15 of the CFR, Title 47.

Part 15 contains three categories of equipment. Incidental radiators (such as dc motors, mechanical light switches, etc.) are not subject to FCC Part 15 emission control requirements. Unintentional radiators are radio frequency devices that intentionally generate radio frequency energy for use within the device, but which are not intended to emit RF energy by radiation or induction. Intentional radiators are radio frequency devices that intentionally radiate radio frequency energy by radiation or by induction.

Unintentional Radiators. There are two different classifications of digital devices listed under unintentional radiators. Class A digital devices are defined as devices that are intended for use in the commercial, industrial, or business environment. Class B digital devices are defined as devices that are intended to be used in a residential environment. Table 38.5 lists the different types of unintentional radiators and their corresponding equipment authorization procedures.

Unintentional Radiator Exempted Devices. There are some unintentional radiators that are listed as exempt devices. These devices are exempted from the technical requirements but are subject to the general requirements of Part 15, Sections 15.5 and 15.29, which state that if the devices cause harmful

Table 38.5 Unintentional Radiator Equipment Authorizations

Type of Device	Equipment Authorization Required[a]
TV broadcast receiver	Verification
FM broadcast receiver	Do
CB receiver	Certification
Superregenerative receiver	Do
Scanning receiver	Do
All other receivers subject to Part 15	Notification
TV interface device	Certification
Cable system terminal device	Notification
Stand-alone cable input selector switch	Verification
Class B personal computers and peripherals	Certification
Other Class B digital devices and peripherals	Verification
Class A digital devices and peripherals	Do
External switching power supplies	Do
All other devices	Do

[a]See additional provisions in CFR Part 15.101 and Part 15.103.

interference their operation must be ceased until such time that the interference is corrected, and that the device must be made available for inspection upon request by the Commission. It is also recommended (although not required) that these devices meet the technical specifications in Part 15. The exempted devices are as follows:

a. A digital device utilized exclusively in any transportation vehicle, including motor vehicles and aircraft.
b. A digital device used exclusively as an electronic control or power system utilized by a public utility or in an industrial plant. The term public utility includes equipment only to the extent that it is in a dedicated building or in a large room owned or leased by the utility and does not extend to equipment installed in a subscriber facility.
c. A digital device used exclusively as industrial, commercial, or medical test equipment.
d. A digital device utilized exclusively in an appliance, e.g., microwave oven, dishwasher, clothes dryer, air conditioner, etc.
e. Specialized medical digital devices (generally used at the direction of or under the supervision of a licensed health care practitioner) whether used in a patient's home or a health care facility. Nonspecialized medical devices, i.e., devices marketed through retail channels for use by the general public, are not exempted. This exemption also does not apply to digital devices used for recordkeeping or any purpose not directly connected with medical treatment.
f. Digital devices having a power consumption not exceeding 6 nW.
g. Joystick controllers, or similar devices such as a mouse, used with digital devices but which contain only nondigital circuitry or a simple circuit to convert the signal to the format required are viewed as passive add-on devices and are not directly subject to the technical standards or the equipment authorization requirements.
h. Digital devices in which the highest frequency generated and the highest frequency used are less than 1.705 MHz and which do not operate from the ac power lines or contain provisions to operate while connected to the ac power lines.
i. It should be noted that equipment containing more than one device is not exempt from the technical standards in Part 15 unless all of the devices meet the criteria for exemption.

Unintentional Radiator Conducted Emission Limits. The limits on the interference conducted back into the ac power distribution system are given in Fig. 38.17.

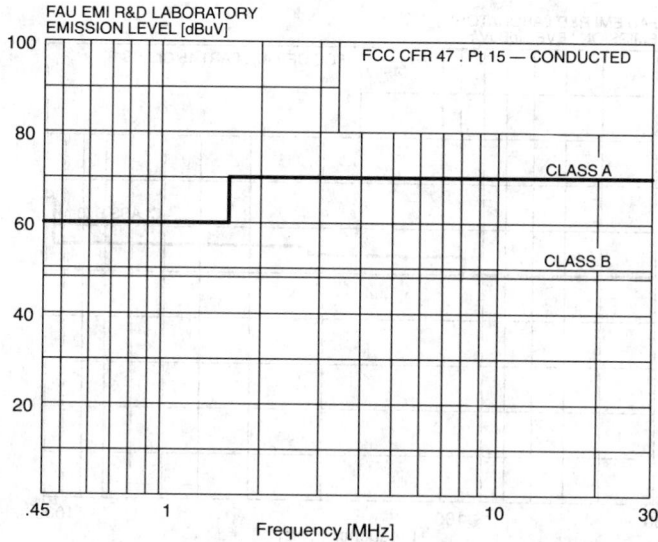

FIGURE 38.17 Class A and Class B conducted emission limits.

Unintentional Radiator Radiated Emission Limits. Unintentional radiator radiated emission limits for Classes A and B devices are given in Figs. 38.18(a) and (b). It should be noted that a 10-m distance is required for Class A limit whereas 3 m is needed for Class B limit.

Intentional Radiators. There are different limits on devices which intentionally radiate radio frequency energy. The authorization procedure required by the Commission is the same as a certification. In addition to the radiated and conducted emission limits, there are restricted bands of operation in which the devices may not intentionally radiate (spurious emissions are permitted in these bands); these bands are given in Table 38.6.

There is also a requirement that the antenna on the device be attached such that it may not be replaced by a different antenna.

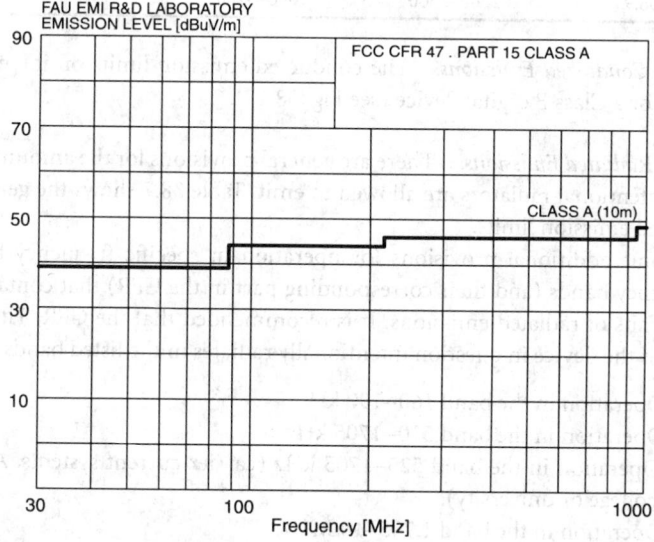

FIGURE 38.18(a) Unintentional (10 m) Class A radiated emission limit.

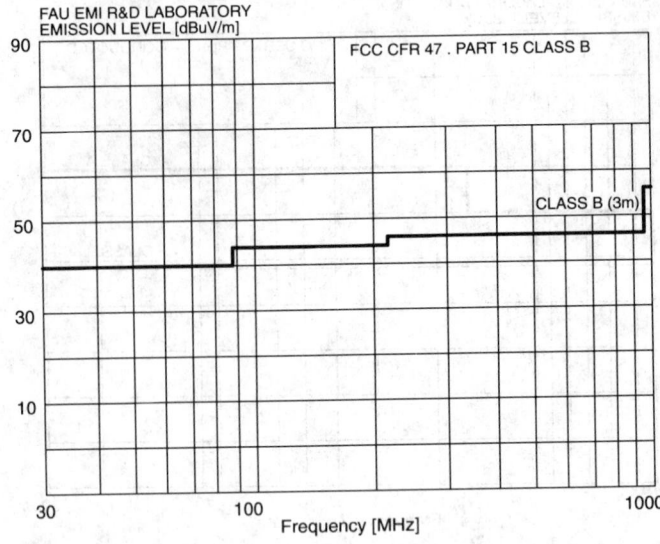

FIGURE 38.18(b)   Unintentional (3 m) Class B radiated emission limits.

Table 38.6   Restricted Bands of Operation

MHz	MHz	MHz	GHz
0.090–0.110	162.0125–167.17	2310–2390	9.3–9.5
0.49–0.51	167.72–173.2	2483.5–2500	10.6–12.7
2.1735–2.1905	240–285	2655–2900	13.25–13.4
8.362–8.366	322–335.4	3260–3267	14.47–14.5
13.36–13.41	399.9–410	3332–3339	15.35–16.2
25.5–25.67	608–614	3345.8–3358	17.7–21.4
37.5–38.25	960–1240	3600–4400	22.01–23.12
73–75.4	1300–1427	4500–5250	23.6–24.0
108–121.94	1435–1626.5	5350–5460	31.2–31.8
123–138	1660–1710	7250–7750	36.43–36.5
149.9–150.05	1718.8–1722.2	8025–8500	Above 38.6
156.7–156.9	2200–2300	9000–9200	

Intentional Radiator Conducted Emissions.   The conducted emission limits on intentional radiators are the same as for a Class B digital device (see Fig. 38.17).

Intentional Radiator Radiated Emissions.   There are general provisions for the amount of radio frequency energy the intentional radiators are allowed to emit. Table 38.7 shows the general requirements for the radiated emission limits.

There are also some additional provisions for operation in specific frequency bands. Listed below are the frequency bands (and their corresponding part in the CFR) that contain additional provisions on the limits of radiated emissions. It is recommended that the CFR, Title 47 be consulted for the limits if the device in question intentionally radiates in the listed bands.

1. Part 15.217: Operation in the band 160–190 kHz.
2. Part 15.219: Operation in the band 510–1705 kHz.
3. Part 15.221: Operation in the band 525–1705 kHz (carrier current systems, AM broadcast stations on a college or university).
4. Part 15.223: Operation in the band 1.705–10 MHz.
5. Part 15.225: Operation in the band 13.553–13.567 MHz.

Table 38.7 General Requirement Radiated Emission Limits

Frequency (MHz)	Field Strength (µV/m)	Measurement Distance (m)
0.009–0.490	2400/F (kHz)	300
0.490–1.705	24000/F (kHz)	30
1.705–30.0	30	30
30–88	100*	3
88–216	150*	3
216–960	200*	3
Above 960	500	3

*Except as provided in paragraph (g), fundamental emissions from intentional radiators operating under this Section shall not be located in the frequency bands 54–72 MHz, 76–88 MHz, 174–216 MHz, or 470–806 MHz. However, operation within these frequency bands is permitted under other sections of this part, e.g., §§ 15.231 and 15.241.

Note: All of the limits specified above are measured using a CISPR Quasi-Peak adapter except for the frequency bands 9–90 kHz, 110–490 kHz, and above 1000 MHz; these are specified for an average measurement.

6. Part 15.227: Operation in the band 26.96–27.28 MHz.
7. Part 15.229: Operation in the band 40.66–40.70 MHz.
8. Part 15.231: Periodic operation in the band 40.55–40.70 MHz and above 70 MHz (alarm systems, door openers, remote switches, etc.).
9. Part 15.233: Operations within the bands 46.60–46.98 MHz and 49.66–50.0 MHz (cordless telephones).
10. Part 15.235: Operation in the band 49.82–49.90 MHz.
11. Part 15.237: Operation in the band 72.0–73.0 MHz and 75.4–76.0 MHz.
12. Part 15.239: Operation in the band 88–108 MHz.
13. Part 15.241: Operation in the band 174–216 MHz (biomedical telemetry devices only).
14. Part 15.243: Operation in the band 890–940 MHz (devices that use radio frequency energy to measure the characteristics of materials only).
15. Part 15.245: Operation in the bands 902–928 MHz, 2435–2465 MHz, 5785–5815 MHz, 10500–10550 MHz, and 24075–24175 MHz (field disturbance sensors only, excluding perimeter protection systems).
16. Part 15.247: Operation in the bands 902–928 MHz, 2400–2483.5 MHz, and 5725–5850 MHz (certain frequency hopping and direct sequence spread spectrum intentional radiators).
17. Part 15.249: Operation within the bands 902–928 MHz, 2400–2483.5 MHz, 5725–5850 MHZ, and 24.0–24.25 GHz.
18. Part 15.251: Operation within the bands 2.9–3.26 GHz, 3.267–3.332 GHz, 3.339–3.3458 GHz, and 3.358–3.6 GHz.

Military Standards. The Military in the United States presents the standards on electromagnetic interference and susceptibility in the MIL-STD-461C document. Electromagnetic interference is defined as the radiated and conducted electromagnetic energy emitted from the device. Electromagnetic susceptibility is defined as the amount of radiated or conducted energy that the device can withstand without degrading its performance.

The standards are broken down into segments defined by the two-letter prefix code in the requirement name. These are conducted emissions (CE), conducted susceptibility (CS), radiated emissions (RE), radiated susceptibility (RS), and special limits encompassing a collection of the above (UM). Each segment is broken down further by adding a two-number suffix code to specify the frequency band, type of equipment, and application of test (see Table 38.8).

There are different equipment and subsystem classes defined for the various environments into which they are to be installed. Table 38.9 gives the descriptions of the different classes and environments. Some of the classes are refined further into categories or groupings. The refined categories for Class A1 and Class A2 devices are given given in Tables 38.10 and 38.11.

Table 38.8   MIL-STD-461C Emission and Susceptibility Requirements

Requirement	Description
CE01	Conducted emissions, power and interconnecting leads, low frequency (up to 15 kHz)
CE03	Conducted emissions, power and interconnecting leads, 0.015 to 50 MHz
CE06	Conducted emissions, antenna terminals, 10 kHz to 26 GHz
CE07	Conducted emissions, power leads, spikes, time domain
CS01	Conducted susceptibility, power leads, 30 Hz to 50 kHz
CS02	Conducted susceptibility, power and interconnecting control leads, 0.05 to 400 MHz
CS03	Intermodulation, 15 kHz to 10 GHz
CS04	Rejection of undesired signals, 30 Hz to 20 GHz
CS05	Cross-modulation, 30 Hz to 20 GHz
CS06	Conducted susceptibility, spikes, power leads
CS07	Conducted susceptibility, squelch circuits
CS09	Conducted susceptibility, structure (common mode) current, 60 Hz to 100 kHz
CS010	Conducted susceptibility, damped sinusoidal transients, pins and terminals, 10 kHz to 100 MHz
CS011	Conducted susceptibility, damped sinusoidal transients, cables, 10 kHz to 100 MHz
RE01	Radiated emissions, magnetic field, 0.03 to 50 kHz
RE02	Radiated emissions, electric field, 14 kHz to 10 GHz
RE03	Radiated emissions, spurious and harmonics, radiated technique
RS01	Radiated susceptibility, magnetic field, 0.03 to 50 kHz
RS02	Radiated susceptibility, magnetic and electric fields, spikes and power frequencies
RS03	Radiated susceptibility, electric field, 14 kHz to 40 GHz
RS05	Radiated susceptibility, electromagnetic pulse field transient
UM03	Radiated emissions and susceptibility, tactical and special-purpose vehicles and engine-driven equipment
UM04	Conducted emissions and radiated emissions and susceptibility, engine generators and associated components, UPS and MEP equipment
UM05	Conducted and radiated emissions, commercial electrical and electromechanical equipment

Table 38.9   Equipment and Subsystem Classes Versus Applicable Part for Emission and Susceptibility Requirements

Class	Description
A	Equipment and subsystems which must operate compatibly when installed in critical areas, such as the following platforms and installations:
A1	Aircraft (including associated ground support equipment)
A2	Spacecraft and launch vehicles (including associated ground support equipment)
A3	Ground facilities (fixed and mobile, including tracked and wheeled vehicles)
A4	Surface ships
A5	Submarines
B	Equipment and subsystems which support the Class A equipment and subsystems, but which will not be physically located in critical ground areas. Examples are electronic shop maintenance and test equipment used in noncritical areas, theodolites, navaids, and similar equipment used in isolated areas.
C	Miscellaneous, general-purpose equipment and subsystems not usually associated with a specific platform or installation. Specific items in this class are:
C1	Tactical and special-purpose vehicles and engine-driven equipment
C2	Engine generators and associated components, uninterruptible power sets (UPS) and mobile electric power (MEP) equipment supplying power to or used in critical areas
C3	Commercial electrical and electromechanical equipment

Table 38.10 Categories of Class A1 Equipment and Subsystems (for Air Force and Navy Use)

Category	Description
A1a	Air launched missiles
A1b	Equipment installed on aircraft (internal or external to airframe)
A1c	Aerospace ground equipment required for the checkout and launch of the aircraft, including electronic test and support equipment
A1d	Trainers and simulators
A1e	Portable medical equipment used for aeromedical airlift
A1f	Aerospace ground equipment used away from the flightline, such as engine test stands and hydraulic test fixtures
A1g	Jet engine accessories
A1h	Class A3 equipment procured for Air Force use

Table 38.11 Categories of Class A2 Equipment and Subsystems

Category	Description
A2a	Equipment installed on spacecraft or launch vehicle
A2b	Aerospace ground equipment required for the checkout and launch, including electronic test and support equipment
A2c	Trainers and simulators
A2d	Class A3 equipment procured for Air Force use

MIL-STD-461C gives the complete list of the applicable emission and susceptibility requirements per category of class of equipment and subsystem or individual requirement, as well as the corresponding limits associated with each requirement. In this handbook, only the radiated emission (RE02) limits and the conducted emission (CE03) limits will be described.

Given in Figs. 38.19 and 38.20(a) through (d) are the limits encompassing the RE02 and CE03 emission requirements. Table 38.12 describes the different RE02 limit curve numbers versus applicable class/category and which military branch they fall under. As an example, a Class A1a device intended for use by either the Air Force or Navy must meet the RE02 **broadband** and narrowband

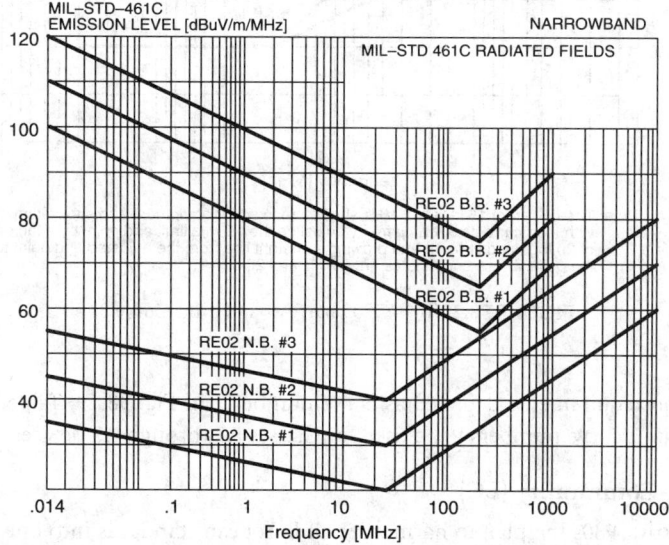

FIGURE 38.19 RE02 narrowband and broadband radiated emission limits.

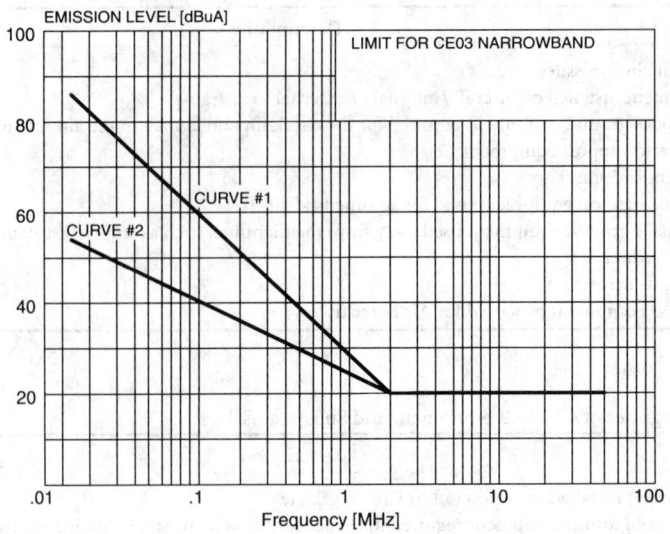

FIGURE 38.20(a)   CE03 narrowband limit Curves 1 and 2.

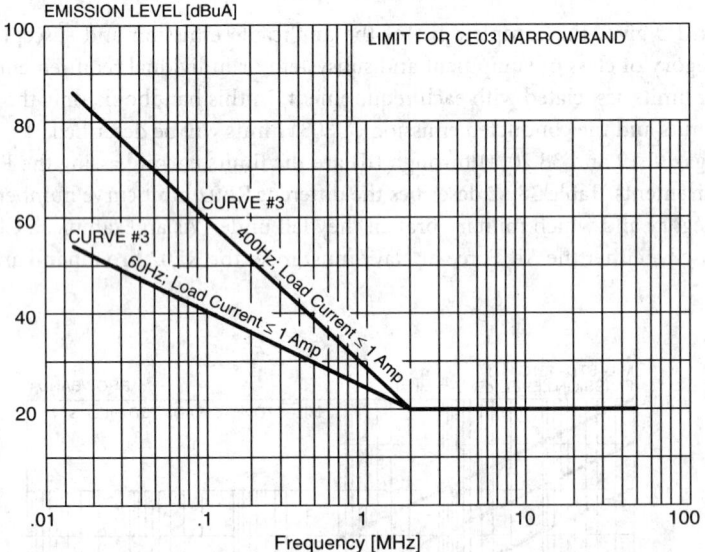

Note: The limit from 15 kHz to 2 MHz shall be relaxed for equipment and subsystems with load currents > 1 ampere by adjusting the 15 kHz limit end point by a factor of 20 log (load current) and drawing a straight line from the adjusted end point to the point whose coordinates are 2 MHz and 20 dBµA.

FIGURE 38.20(b)   CE03 narrowband limit Curve 3.

radiated emission requirements depicted as curve number 2 in Fig. 38.19. Table 38.13 shows the different CE03 limit curve numbers to be used for Classes A1 through A3 devices.

The European Community (EC)

During the early 1980s the plan to harmonize all European standards into one was introduced. There was a long deliberation in the process and, therefore, few acceptable directives have been

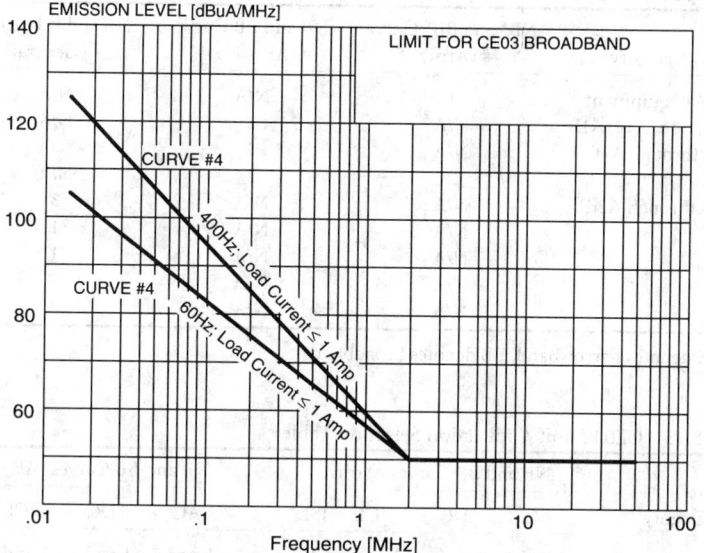

Note: The limit from 15 kHz to 2 MHz shall be relaxed for equipment and subsystems with load currents > 1 ampere by adjusting the 15 kHz limit end point by a factor of 20 log (load current) and drawing a straight line from the adjusted end point to the point whose coordinates are 2 MHz and 50 dBμA/MHz.

FIGURE 38.20(c)   CE03 broadband limit Curve 4.

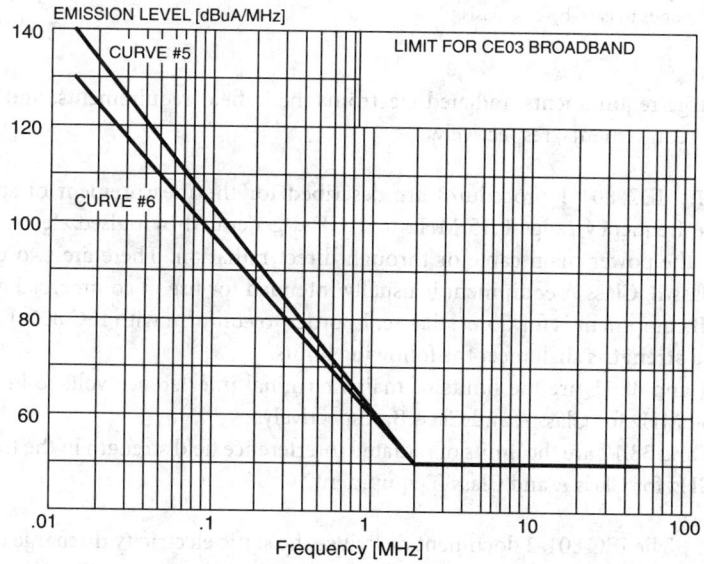

FIGURE 38.20(d)   CE03 broadband limit Curves 5 and 6.

established. Not until ten years later did the International Electrotechnical Commission (IEC) activate the plan and come up with the IEC 801 document.

IEC 801 Standards.   There are four parts described in the IEC 801 publication. Part 1 is also known as CISPR Publication 22 which involves limits and methods of measurement of radio interference characteristics of information technology equipment (ITE). Parts 2, 3, and 4 deal with elec-

Table 38.12   RE02 Limit Application Selection Table

Class/Category	NB* and BB* Curve # (Army)	NB and BB Curve # (AF + Navy)	NB and BB Curve # (General)
All A1 equipment	1	N/A	N/A
A1a, A1b, A1g, A1h	N/A	2	N/A
A1c through A1f	N/A	3	N/A
A2a	2	2	N/A
A2b through A2d	N/A	N/A	3
A3	N/A	N/A	1
A4	N/A	N/A	1
A5	N/A	N/A	1
B	N/A	N/A	3

*NB denotes narrowband, BB denotes broadband.

Table 38.13   CE03 Limit Application Selection Table

Class/Category	NB and BB Curves (Army)				NB and BB Curves (AF + Navy)			
	AC	DC	ICL	ISL	AC	DC	ICL	ISL
A1	1 and 6	1 and 6	1 and 6	*	1 and 5	1 and 5	1 and 5	*
A2	1 and 6	1 and 6	*	*	1 and 5	1 and 5	*	*
A3	1 and 6	1 and 6	1 and 6	*	3 and 4	1 and 6	*	*

AC corresponds to ac leads.
DC corresponds to dc leads.
ICL corresponds to interconnecting control leads.
ISL corresponds to interconnecting signal leads.
* corresponds to case-by-case basis.

trostatic discharge requirements, radiated electromagnetic field requirements, and electrical fast transient/burst requirements, respectively.

IEC 801 Part 1. IEC 801-1 procedures are described for the measurement of spurious signal strength, in the frequency range 0.15 MHz to 1 GHz, generated by pulsed electrical waveforms either through the power main cable or through direct radiation. There are two classes of ITE: Class A and Class B. Class A equipment is usually intended for use in commercial establishments whereas Class B equipment is for domestic use. In order to conform with IEC 801-1, the measured voltages or **field strengths** shall meet the following limits.

Tables 38.14 and 38.15 are the limits of mains terminal interference voltage in the frequency range 0.15 to 30 MHz for Class A and Class B, respectively.

Tables 38.16 and 38.17 are the limits of radiated interference field strength in the frequency range 30 MHz to 1 GHz for Class A and Class B equipment.

IEC 801 Part 2. The IEC 801-2 document describes the static electricity discharge (ESD) requirements. At the time of publication, the current limits tabulated herein are utilized; however, the limits are subject for a revision.

Since contact discharge method and air discharge method are acceptable for ESD tests, two limits have been specified and tabulated in Table 38.18(a) and (b).

IEC 801 Part 3. The IEC 801-3 document involves the susceptibility requirements for equipment under test (EUT). The primary concern is the degradation of EUT under the influence of the hand-held transceiver or other sources of radiation in the frequency range 27 to 500 MHz. Table 38.19 gives the tabulated limits of the susceptibility requirements.

Table 38.14 Limits of Mains Terminal Interference Voltage in the Frequency Range 0.15 to 30 MHz for Class A Equipment

Frequency Range (MHz)	Limits [dB (μV)]	
	Quasi-peak	Average
0.15 to 0.50	79	66
0.50 to 30	73	60

Table 38.15 Limits of Mains Terminal Interference Voltage in the Frequency Range 0.15 to 30 MHz for Class B Equipment

Frequency Range (MHz)	Limits [dB (μV)]	
	Quasi-peak	Average
0.15 to 0.50	66 to 56	56 to 46
0.50 to 5	56	46
5 to 30	60	50

The lower limit shall apply at the transition frequencies.

Note: In Table 30.15 the limit decreases linearly with the logarithm of the frequency in the range 0.15 to 0.50 MHz.

Table 38.16 Limits of Radiated Interference Field Strength in the Frequency Range 30 MHz to 1 GHz at a Test Distance of 30 m for Class A Equipment

Frequency Range (MHz)	Quasi-peak Limits [dB (μV/m)]
30 to 230	30
230 to 1000	37

Table 38.17 Limits of Radiated Interference Field Strength in the Frequency Range 30 MHz to 1 GHz at a Test Distance of 10 m for Class B Equipment

Frequency Range (MHz)	Quasi-peak Limits [dB (μV/m)]
30 to 230	30
230 to 1000	37

The lower limit shall apply at the transition frequencies.

Table 38.18(a) Contact Discharge Severity Levels

Level	Test Voltage Contact Discharge (kV)
1	2
2	4
3	6
4	8
x^1	Special

Table 38.18(b) Air Discharge Severity Levels

Level	Test Voltage Air Discharge (kV)
1	2
2	4
3	8
4	15
x^1	Special

[1]*x* is an open level. The level is subject to negotiations and has to be specified in the dedicated equipment specification. If higher voltages than those shown are specified, special test equipment may be needed.

Table 38.19 Severity Levels for Susceptibility Requirements

Level	Test Field Strength, (V/m)
1	1
2	3
3	10
x^1	Special

[1]*x* is an open class.

There are four classes of severity levels. Class 1 is for low-level electromagnetic radiation environments, such as levels typical of local radio/television stations located at more than 1 km and levels typical of low-power transceivers. Class 2 is for moderate electromagnetic radiation environments, such as portable transceivers that are close to the EUT but not closer than 1 m. Class 3 is for severe electromagnetic radiation environments, such as levels typical of high-power transceivers in close proximity to the EUT. Class 4 is an open class for situations involving very severe electromagnetic radiation environments. The level is subject to negotiation.

IEC 801 Part 4. The IEC 801-4 document involves electrical fast transient/burst requirements. The purpose is to evaluate the performance of EUT when switching transients with high repetition frequency couple into power main supply and external communication lines. Table 38.20 gives the test severity levels recommended at the time of publication.

Table 38.20   Severity Levels for the Fast Transient/Burst Requirements

| | Open Circuit Output Test Voltage ±10% (kV) | |
	On Power Supply	On I/O (Input/Output) Signal, Data and Control Lines
Level	On Power Supply	On I/O (Input/Output) Signal, Data and Control Lines
1	0.5	0.25
2	1	0.5
3	2	1
4	4	2
x^1	Special	Special

[1]x is an open level. The level is subject to negotiation between the user and the manufacturer or is specified by the manufacturer.

Measurement Procedures

To determine the emission or susceptibility levels, measurement procedures were established. There are many procedures existing around the world. Each country may adopt its own procedures. Due to vast existing procedures, only the FCC procedure will be described in detail. However, a list of the other procedures is also outlined.

FCC

The procedures used by the Commission to determine compliance are as follows:

1. FCC/OET MP-l: FCC Measurements for Determining Compliance of Radio Control and Security Alarm Devices and Associated Receivers.
2. FCC/OET MP-2: Measurement of UHF Noise Figures of TV Receivers.
3. FCC/OET MP-3: FCC Methods of Measurement of Output Signal Level, Output Terminal Conducted Spurious Emissions, Transfer Switch Characteristics, and Radio Noise Emissions from TV Interface Devices.
4. FCC/OET MP-4: FCC Procedure for Measuring RF Emissions from Computing Devices.
5. FCC/OET MP-9: FCC Procedure for Measuring Cable Television Switch Isolation.

In addition to the documents listed above, the FCC has outlined some broad measurement characteristics in Part 15, Sections 15.31 through 15.35 in the CFR, Title 47. The FCC is presently entertaining the possibility of adopting the American National Standards Institute (ANSI) Specification C63.4-1992 for methods and procedures of radio emissions noise measurements from electronic equipment. This document specifically defines some measurement setups not previously defined in the FCC MP-4 document, such as the specific measurement setup for floor standing devices. Shown below are some of the key points from the FCC MP-4 measurement procedure.

Listed in the FCC MP-4 measurement procedure document are the configuration setups for Class B computing devices and peripheral equipment. Some of the pertinent highlights are as follows:

- The equipment must be set up in a system configuration which includes the computer controller, a monitor, keyboard, serial device, parallel device, and any other device which may typically be connected to the system.
- The computer controller must be configured with the controller cards needed for typical operation (serial card, parallel card, video card, disk controller, memory), along with any other specialized cards defined in the typical setup to be marketed.
- A program to display, print, store, and/or send capital H characters to all of the pertinent devices (inclusive of the drives, CRT, printer, and any other data receiving devices) must be run for the duration of the evaluation.
- Typical cables and power cords are required for the test; the interface cables are bundled serpentine fashion in 30- to 40-cm bundle lengths, to an overall length of 1 meter.

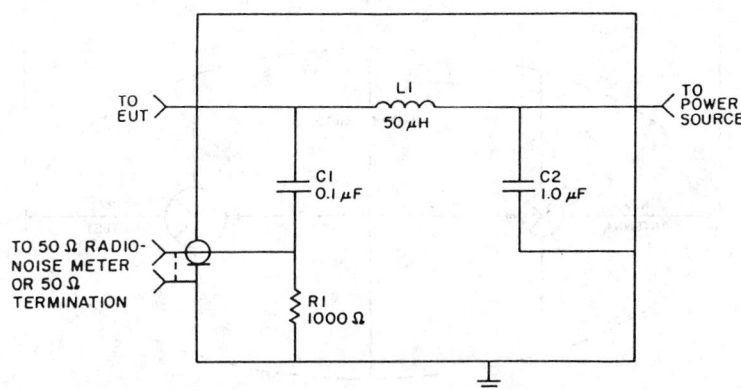

FIGURE 38.21   LISN circuit diagram.

Conducted Emission Testing.   Measurements are recommended to be performed inside of an RF shielded room in order to eliminate interference from ambient electromagnetic fields. The system units are placed on a nonconducting table 0.8 meters high and 2 meters from the rear wall of the RF shielded enclosure. The measurements are performed with a **line impedance stabilization network (LISN).** This type of network is specifically designed to present a known impedance to the device under test, filter the noise present on the power line, and to match impedances with the measurement receiver. Figure 38.21 shows the typical FCC LISN circuit, applicable for monitoring the noise present on either the phase or neutral line. Data is collected across phase and neutral to ground over the 450-kHz to 30-MHz frequency range and is compared with the aforementioned limits.

Radiated Emission Testing.   The radiated emissions are measured at an FCC listed site (either semi-anechoic or open field), which requires a metal ground plane over the floor (typically hardware cloth). The site must satisfy a certain minimum size criteria, depending on the prescribed measurement distance. The accepted criteria is based on the Fresnel ellipse, which is presented in Fig. 38.22. The procedure for listing a site with the FCC includes the submittal of the site attenuation measurements, the site description, and a list of measurement equipment. The qualified site should meet the ±4-dB variation from theoretical criteria.

The EUT is placed on a nonconducting table 1 m above the ground plane floor. The receiving antenna is placed at the prescribed measurement distance (R) from the system (3 m for Class B and 10 m for Class A) and is scanned from 1 to 4 m in height. It is also required that the EUT be rotated 360 degrees. The maximum emission data (per azimuth, elevation, and antenna orientation) is collected over the appropriate frequency range and reported.

Abbreviations

ANSI:	American National Standards Institute
BSI:	The British Standards Institute
CFR:	Code of Federal Regulations
CISPR:	International Special Committee on Radio Interference
EC:	European Community
EMC:	Electromagnetic compatibility
EMI:	Electromagnetic interference
ESD:	Static electricity discharge
EUT:	Equipment under test

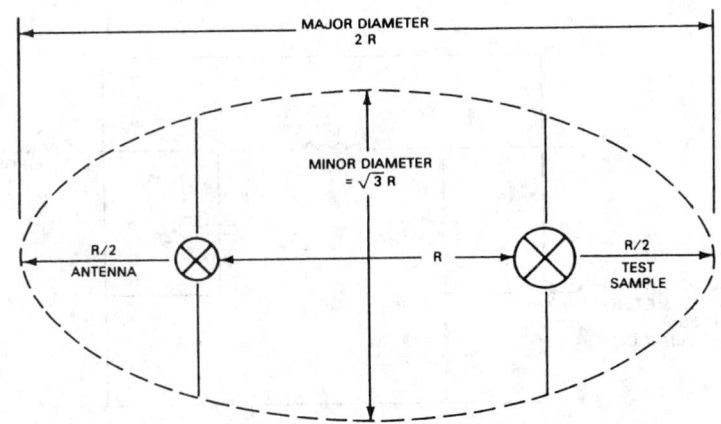

FIGURE 38.22　Minimum obstruction-free area for open field test site.

FCC:　　The Federal Communications Commission
IEC:　　International Electrotechnical Commission
IEEE:　　The Institute of Electrical and Electronics Engineers
ITE:　　Information technology equipment
LISN:　　Line impedance stabilization network
MIL:　　The U.S. Military

Defining Terms

Broadband emission:　An emission having a spectral distribution sufficiently broad in comparison to the response of a measuring receiver.

Conducted emission:　An RF current propagated through an electrical conductor.

Electromagnetic interference:　An unwanted electromagnetic signal which may degrade the performance of an electronic device.

Far field:　The region where the ratio of the electric to magnetic field is approximately equal to 377 Ω.

Field strength:　An amount of electric or magnetic field measured in far field and expressed in volts/meter or amps/meter.

Line impedance stabilization network (LISN):　A network designed to present a defined impedance at high frequency to a device under test, to filter any existing noise on the power mains, and to provide a 50-Ω impedance to the noise receiver.

Radiated emission:　An electromagnetic field propagated through space.

References

Air Force Systems Command Electromagnetic Compatibility Handbook, 3rd ed., January 5, 1975.

CEI International Standard, IEC 801-1 to 801-4, 2nd ed., 1991.

Code of Federal Regulations, Title 47, Telecommunications, Part 15, October, 1991.

Electromagnetic Interference and Compatibility Handbook, vol. 1, Donald White Consultants, Inc.

Military Standard 461C, Electromagnetic Emission and Susceptibility Requirements for the Control of Electromagnetic Interference, 4 August 1986.

C.R. Paul, *Introduction to Electromagnetic Compatibility*, New York: John Wiley & Sons, 1992.

D.R.J. White, *Electromagnetic Interference and Compatibility*, vol. 1, 1971.

Further Information

The aforementioned measurement procedures used by the FCC are available from the Government Printing Office, Washington, D.C., 20402. The ANSI C63.4 Specification is available from the IEEE, 345 East 47th Street, New York, NY, 10017-2394.

The procedures utilized for the measurements performed to military EMC specifications are given in the following documents which are available from the DOD Government Printing Office:

MIL-STD-285   Attenuation Measurements for Enclosure, EM Shielding, for Electronic Test Purposes, Method of

MIL-STD-462   Electromagnetic Emission and Susceptibility, Test Methods for

MIL-STD-463   Definitions and System of Units, Electromagnetic Interference and Electromagnetic Compatibility

MIL-STD-1377   Effectiveness of Cable, Connector and Weapon Enclosure Shielding and Filters in Precluding Hazards of Electromagnetic Radiation to Ordnance, Measurement of

The EC procedures and standards listed below are available from the Bureau Central de la Commission Electrotechnique Internationale 3, rue de Varembe, Geneve, Suisse.

IEC 801-1   Limits and Methods of Measurement of Radio Interference Characteristics of ITE

IEC 801-2   Electrostatic Discharge Requirements

IEC 801-3   Radiated Electromagnetic Field Requirements

IEC 801-4   Electrical Fast Transient/Burst Requirements

38.3 Lightning

Martin A. Uman and Marcos Rubinstein

An understanding of lightning and of the electric and magnetic fields produced by lightning is critical to an understanding of lightning-induced effects on electronic and electric power systems. This section begins with an overview of the terminology and physics of lightning. Then, statistics on lightning occurrence are given. Next the characteristics of the electric and the magnetic fields resulting from lightning charges and currents are examined, and the models used to describe that relationship are discussed. The section ends with a discussion on the coupling of the electric and magnetic fields from lightning to overhead wires.

Terminology and Physics

Lightning is a transient, high-current electric spark whose length is measured in kilometers. Lightning discharges can occur within a cloud, between clouds, from cloud to air, and from cloud to ground. All discharges except the latter are known as cloud discharges. The usual cloud-to-ground lightning is initiated in the cloud, lasts about half a second, and lowers to ground some 20 to 30 Coulombs of negative cloud charge. A less frequent type of cloud-to-ground discharge, accounting for less than 10% of all cloud-to-ground lightning, also begins in the cloud but lowers positive cloud charge. An even less frequent type of cloud-to-ground lightning is initiated in an upward direction from tall man-made structures such as TV towers or tall geographical features such as mountaintops. A complete lightning discharge of any type is called a **flash**. The usual negative cloud-to-ground lightning flash starts in the cloud when a so-called **preliminary breakdown**, a particular type of electric discharge in the cloud, occurs. This process is followed by a discharge, termed the **stepped leader**, that propagates towards the ground in a series of luminous steps tens of meters in length. In progressing toward the ground, the negatively charged stepped leader branches in a downward direction. When one or more leader branches approach within a hundred meters or so of the ground, after 10 to 20 ms of stepped leader travel at an average speed of 10^5 to 10^6 m/s, the electric field at the ground (or at objects on the ground) increases above the critical breakdown field of the surrounding air and one or more upward-going discharges is initiated, starting the **attachment process.** After traveling a few tens of meters, one of the upward-going discharges, which is essentially at ground potential, contacts the tip of one branch of the stepped leader, which is at a high negative potential, probably some tens of megavolts. From that point, ground potential propagates upward, discharging to ground some or all of the negative charge previously deposited along the channel by the stepped leader. This upward propagating potential discontinuity is called the **return stroke.** Its front is a region of high electric field that causes increased ionization, current, temperature, and pressure as it travels the 5-km or more length of the leader channel. That trip is made in about 100 μs at an initial speed of the order of one third to one half the speed of light, the speed decreasing with height. The current at ground associated with the negative first return stroke has a peak of typically 35 kA achieved in a few microseconds, has a maximum current derivative of about 10^{11} A/s and falls to half of peak value in some tens of microseconds. The cessation of the first return stroke current may or may not end the flash. If more cloud charge is made available to the first stroke channel by in-cloud discharges, another leader-return stroke sequence may ensue, typically after tens of milliseconds. Preceding and initiating a subsequent return stroke is a continuous leader lowering negative charge, called a **dart leader.** The dart leader typically propagates down the residual channel of the previous stroke, generally ignoring the first stroke branches, although in about 50% of cloud-to-ground flashes there is at least one dart leader which transforms to a stepped leader on the downward trip, creating a new path to ground. There are typically three or four leader-return stroke sequences per negative cloud-to-ground flash, but ten or more is not uncommon.

Of the many different processes that occur during the various phases of a negative cloud-to-ground lightning (e.g., the in-cloud K processes, in-cloud J processes, and cloud to ground M components that occur between strokes and after the final stroke and are not discussed here), the electric and magnetic fields associated with the return stroke described above generally are the largest and hence the most significant in inducing unwanted voltages in electronic and electric power systems. This is the case because the currents in all other lightning processes are generally smaller than return stroke currents and the ground strike point of the return stroke can be much closer to objects on the ground than are in-cloud discharges. Cloud discharges exhibit currents similar to those of the in-cloud processes occurring in ground discharges and hence produce similar relatively small fields at or near ground level.

Positive flashes to ground, those initiated in the cloud and lowering positive charge to earth, generally contain only one return stroke, which is preceded by a "pulsating" rather than the stepped leader characteristically preceding negative first strokes and is generally followed by a period of continuous current flow. Positive flashes contain a greater percentage of very large return stroke currents, in the 100- to 300-kA range, than do negative flashes. Positive flashes may represent half of all flashes to ground in winter storms which produce few total flashes and typically represent 1 to 20% of the overall flashes in summer storms, that percentage increasing with increasing latitude.

Lightning Occurrence Statistics

Lightning flash density is defined as the number of lightning flashes per unit time per unit area and is usually measured in units of lightning flashes, either cloud or cloud-to-ground or both, per square kilometer per year. The two most common techniques for directly measuring flash density are (1) the use of so-called flash counters, relatively crude devices which trigger on electric fields above a value of the order of 1 kV/m in a frequency band centered in the hundreds of hertz to kilohertz range, of which two models are extensively used, the CIGRE 10-kHz and the CIGRE 500-Hz and (2) the use of networks of wideband magnetic direction finders, such networks now covering the U.S., Japan, Korea, Taiwan, most of Europe, and parts of many other countries. The average flash density varies considerably with geographical location, generally increasing with decreasing latitude. Typical ground flash densities are 1 to 5 $km^{-2}\,yr^{-1}$, with the world's highest being 30 to 50 km^{-2} yr^{-1}. Significant variations in flash density are observed with changes in local meteorological conditions within distances of the order of 10 km, for example, perpendicular to and inland from the Florida coastline. A ground flash density map of the U.S. for 1989, obtained from the U.S. National Lightning Detection Network of 114 wideband magnetic direction finders, is given by Orville [1991]. Flash densities in the U.S. are maximum in Florida with 10 to 15 km^{-2} yr^{-1} and minimum along portions of the Pacific coast which has essentially no lightning.

An extensively measured parameter used to describe lightning activity worldwide is the thunderday or isokeraunic level, T_D, the number of days per year that thunder is heard at a given location. This parameter has been recorded by weather station observers worldwide for many decades, whereas the accurate direct measurement of flash density has been possible only recently. Commonly used relations to convert thunderday level to ground flash density N_g are of the form

$$N_g = a T_D^b \quad \text{km}^{-2} \text{ yr}^{-1} \tag{38.6}$$

where the value of a is near, and usually less than, 0.1 and the value of b is near, and usually greater than, 1. It should be noted that Eq. (38.6) is relatively inaccurate in that the data to which it is a fit is highly variable. The literature contains more than ten different values of a and b determined by different investigators.

Finally, from both worldwide thunderday and earth-orbiting satellite measurements, it has been estimated that there are about 100 total flashes, cloud and cloud-to-ground, per second over the whole earth. This number corresponds to an average global total flash density of 6 km^{-2} yr^{-1}.

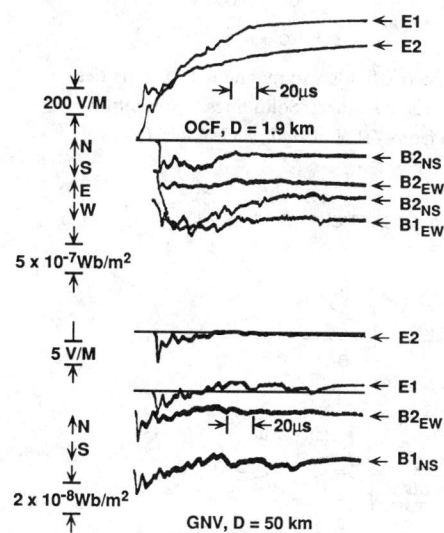

FIGURE 38.23 Simultaneously measured return stroke vertical electric field (E) and two horizontal magnetic flux densities (B_{EW} and B_{NS}) as observed about 2 and 50 km from a two-stroke flash, the first stroke being designated "1", the second "2". (*Source:* Adapted from Y.T. Lin *et al.*, *J. Geophys. Res.*, vol. 84, pp. 6307–6314, 1979. With permission.)

Electric and Magnetic Fields

For the usual negative return stroke, measurements of the vertical component of the electric field and the two horizontal components of the magnetic field at ground level using wideband systems with upper frequency 3-dB points in the 1- to 20-MHz range are well documented in the literature. Measured vertical electric field and horizontal magnetic field waveshapes are shown in Fig. 38.23. Sketches of typical electric and magnetic fields are given in Fig. 38.24 for lightning in the 1- to 5-km range and in Fig. 38.25 for lightning at 10, 15, 50, and 200 km. Measured vertical and measured horizontal electric fields near ground are shown in Fig. 38.26. The mean value of the initial peak vertical electric field, normalized to 100 km by assuming an inverse distance dependence, is about 7 V/m for negative first strokes and about 4 V/m for negative subsequent strokes.

The return stroke vertical electric field rise to peak is comprised of two distinguishable parts, evident in Fig 38.23: a slow front immediately followed by a fast transition to peak. For first strokes the slow front has a duration of a few microseconds and rises to typically half the peak amplitude, while for subsequent strokes the same slow front lasts less than 1 μs and rises only to typically 20% of the peak. The mean 10–90% fast transition time is about 200 ns regardless of stroke order for strokes observed over saltwater where there is minimal distortion of the waveform

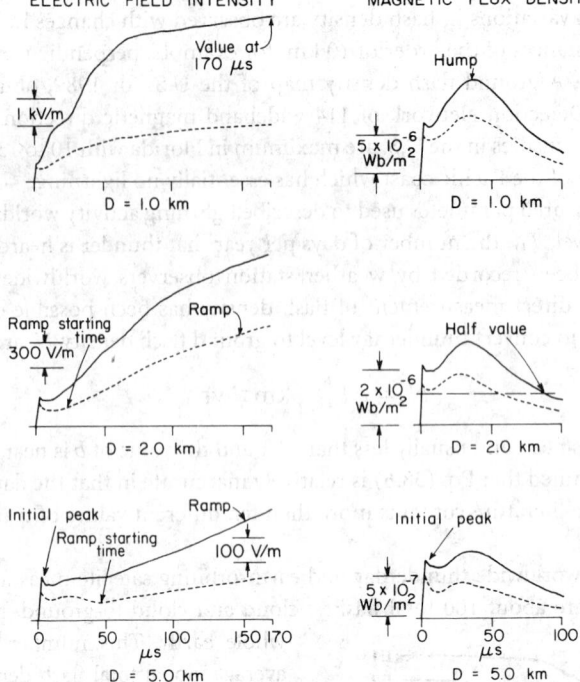

FIGURE 38.24 Drawings of typical return stroke electric fields and magnetic flux densities at 1, 2, and 5 km with definition of pertinent characteristic features. Solid lines represent first strokes, dotted subsequent strokes. (*Source:* Adapted from Y.T. Lin *et al.*, *J. Geophys. Res.*, vol. 84, pp. 6307–6314, 1979. With permission.)

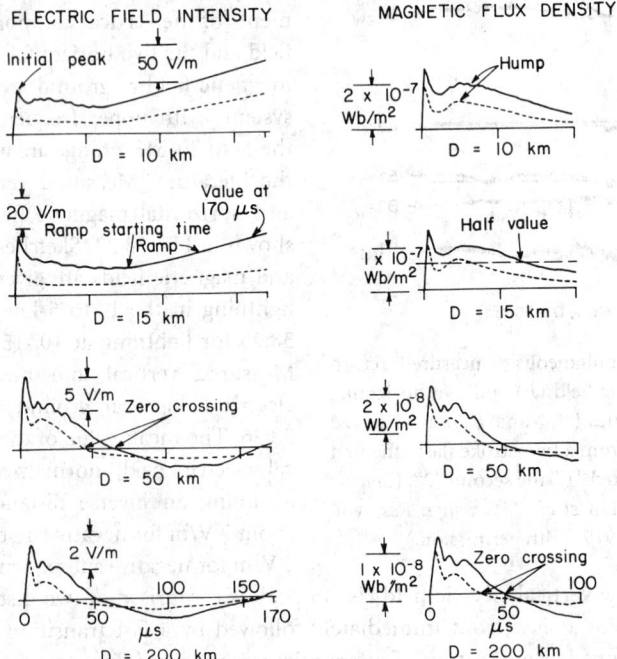

FIGURE 38.25 Drawings of typical return stroke electric fields and magnetic flux densities at 10, 15, 50, and 200 km; a continuation of Fig. 38.24. (*Source:* Adapted from Y.T. Lin *et al.*, *J. Geophys. Res.*, vol. 84, pp. 6307–6314, 1979. With permission.)

due to propagation. The waveforms in Fig. 38.23 have suffered distortion in propagating over land.

After the initial field peak, the waveshapes of the vertical electric field and the horizontal magnetic field for close lightning exhibit a valley followed by a hump in the case of the magnetic field

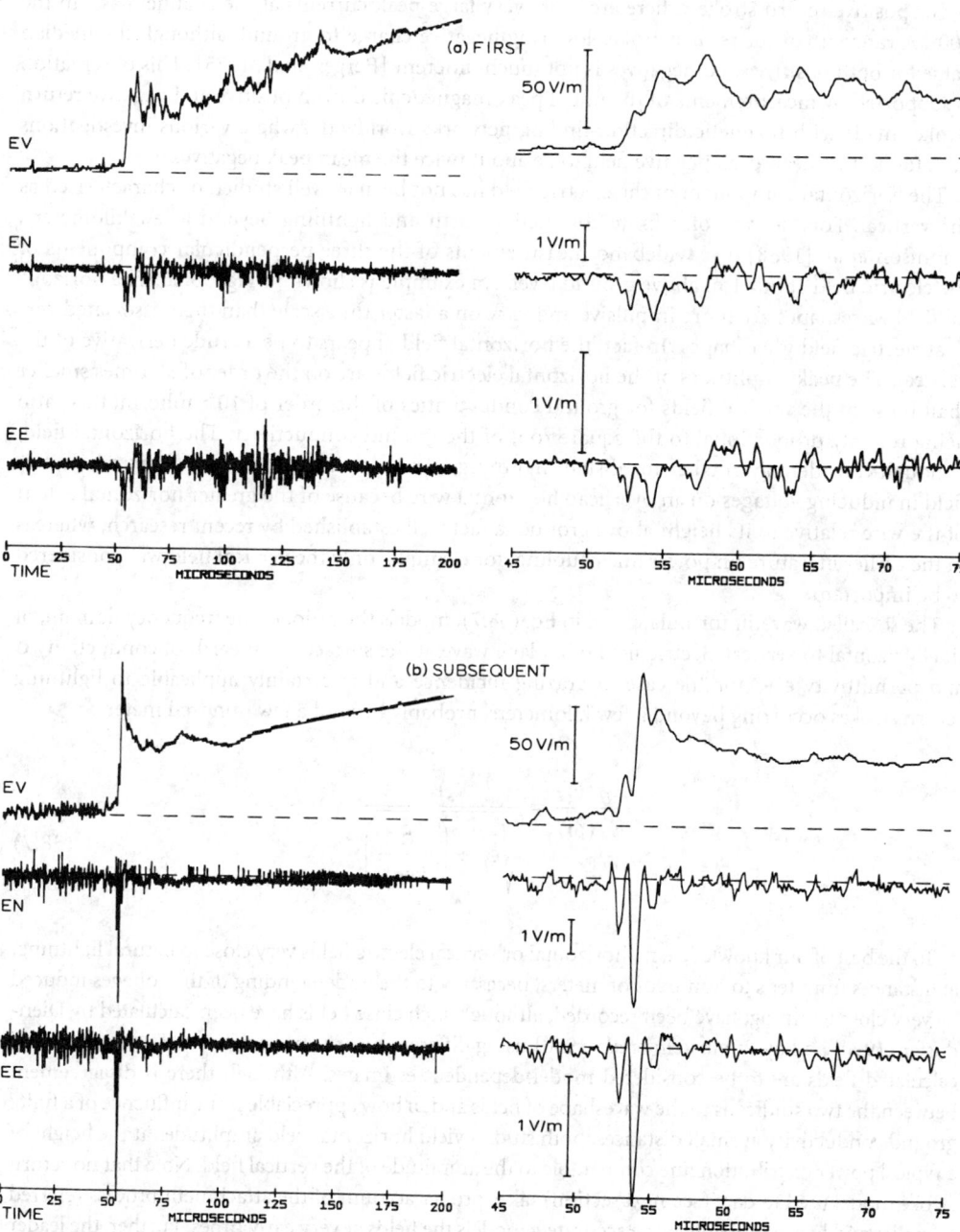

FIGURE 38.26 Measured horizontal electric field components (EN and EE) and vertical electric field (EV) one meter above ground for a first stroke (a) and a subsequent stroke (b) at a distance of 7 km presented on two time scales. (*Source:* Adapted from E.M. Thomson *et al., J. Geophys. Res.,* vol. 93, pp. 2429–2441, 1988. With permission.)

and by a ramp in the case of the electric field, as is evident from Figs. 38.23 through 38.26. Relative to the amplitude of the initial peak, the hump and the ramp decrease with increasing distance of the return stroke. For distances of 25 km or greater, the ramp in the electric field is no longer significant, and for distances of 50 km or more and for times of the order of 100 μs, the waveshapes of the electric and magnetic fields are nearly identical, exhibiting a zero crossing and polarity reversal at some tens of microseconds.

For positive return strokes, there are more very large peak currents at the channel base, in the 100-kA range, than for return strokes lowering negative charge to ground, although the median value for both positives and negatives is not much different [Berger *et al.*, 1975]. This observation is supported by measurements of the initial peak magnetic field from positive and negative return strokes made with magnetic direction-finding networks worldwide, where various investigations have found the mean peak positive field to be about twice the mean peak negative.

The horizontal component of the electric field has not been as well studied or characterized as the vertical. For the case of a finite-conducting earth and lightning beyond a few kilometers, Thomson *et al.* [1988] give wideband measurements of the three perpendicular components of the electric field about 1 m above ground level. An example is shown in Fig. 38.26. The horizontal field waveshapes are more impulsive and vary on a faster time scale than their associated vertical electric field waveshapes. In fact, the horizontal field appears to be a crude derivative of the vertical. The peak amplitudes of the horizontal electric fields are on the order of 30 times smaller than those of the vertical fields for ground conductivities of the order of 10^{-2} mho/m, this ratio being roughly proportional to the square root of the ground conductivity. The horizontal field, although considerably smaller for distant lightning, can be as important as the vertical electric field in inducing voltages on an overhead horizontal wire because of the greater horizontal extent of the wire relative to its height above ground, a fact well established by recent research, whereas in the earlier literature on power line coupling, for example, only the vertical field was considered to be important.

The so-called wavetilt formula, given in Eq. (38.7), models the ratio, in the frequency domain, of the horizontal to vertical electric field of a plane wave at the surface of an earth of conductivity σ and permittivity $\epsilon_r \epsilon_0$ for the case of grazing incidence and is certainly applicable to lightning return strokes occurring beyond a few kilometers, probably beyond a few hundred meters.

$$\frac{E_H(\omega)}{E_V(\omega)} = \frac{1}{\sqrt{\epsilon_r + \left(\dfrac{\sigma}{j\omega\epsilon_0}\right)}} \tag{38.7}$$

To the best of our knowledge, no horizontal or vertical electric fields very close to natural lightning, at distances from tens to hundreds of meters, necessary to the understanding of the voltages induced by very close lightning, have been recorded, although such close fields have been calculated by Diendorfer [1990] and by Rubinstein *et al.* [1990] using different return stroke models. These two sets of calculated fields are to be considered model-dependent estimates. Although there is disagreement between the two studies as to the waveshape of fields and in how appreciable is the influence of a finite ground conductivity at small distances, both studies yield horizontal field amplitudes at the height of a typical power distribution line comparable to the amplitude of the vertical field. Note that no return stroke model used to date (see next section) takes proper account of the attachment process referred to earlier and hence probably none accurately models the fields at very early times. Further, the leader fields preceding the return stroke field change are not taken into account in the existing models, although such fields at very close range are clearly important since it is the leader charge near ground that the return stroke discharges to ground, and hence the leader and return stroke electrostatic field changes should be of equivalent magnitude very close to the ground strike point.

Modeling of the Return Stroke

General

A number of return stroke current models are found in the literature from which, if the current at the channel base is specified (e.g., from measurement) along with the model parameters, the channel current can be calculated as a function of height and time: the Bruce–Golde (BG) model, the transmission line (TL) model, the modified transmission line (MTL) model, the traveling current source (TCS) model, the Lin–Uman–Standler (LUS) model, the Diendorfer–Uman (DU) model, and the modified DU model. Two assumptions are common to all of these models: that the lightning channel is perfectly straight and vertical and that the ground is a perfect conductor. Once the channel currents are determined as a function of height and time, the remote electric and magnetic fields can be calculated from Eqs. (38.8) through (38.14)

$$\bar{E} = \bar{E}_{ele} + \bar{E}_{ind} + \bar{E}_{rad} \tag{38.8}$$

$$\bar{E}_{ele} = \frac{1}{4\pi\epsilon_0} \int_{-h}^{h} \left\{ \frac{2\cos\theta' \hat{a}_R + \sin\theta' \hat{a}_{\theta'}}{R^3} \int_0^t i\left(|z'|, \tau - \frac{R}{c}\right) dt \right\} dz \tag{38.9}$$

$$\bar{E}_{ind} = \frac{1}{4\pi\epsilon_0} \int_{-h}^{h} \frac{2\cos\theta' \hat{a}_R + \sin\theta' \hat{a}_{\theta'}}{cR^2} i\left(|z'|, t - \frac{R}{c}\right) dz' \tag{38.10}$$

$$\bar{E}_{rad} = \frac{1}{4\pi\epsilon_0} \int_{-h}^{h} \frac{1}{c^2 R} \frac{\partial i\left(|z'|, t - \frac{R}{c}\right)}{\partial t} \hat{a}_{\theta'} dz' \tag{38.11}$$

$$\bar{B} = \bar{B}_{ind} + \bar{B}_{rad} \tag{38.12}$$

$$\bar{B}_{ind} = \frac{\mu_0}{4\pi} \int_{-h}^{h} \frac{\sin\theta'}{R^2} i\left(|z'|, t - \frac{R}{c}\right) \hat{a}_{\phi'} dz' \tag{38.13}$$

$$\bar{B}_{rad} = \frac{\mu_0}{4\pi} \int_{-h}^{h} \frac{\sin\theta'}{cR} \frac{\partial i\left(|z'|, t - \frac{R}{c}\right)}{\partial t} \hat{a}_{\phi'} dz' \tag{38.14}$$

where $i(z',t)$ is the current along the channel obtained from one of the return stroke current models mentioned above, and the geometry by which the above equations are to be interpreted is shown in Fig. 38.27. Note that the spatial integral includes the image current below the perfectly conducting ground plane so as to take account of reflections from the earth's surface. The three terms on the right-hand side of Eq. (38.8) [expanded in Eqs. (38.9) through (38.11)] are called, from left to right, the electrostatic, induction, and radiation terms. Similarly, the two terms on the right-hand side of Eq. (38.12) [expanded in Eqs. (38.13) and (38.14)] are termed the induction and radiation terms.

For large distances to the lightning channel, the radiation part of the electric and magnetic fields is dominant due to its $1/R$ dependence (as compared to the $1/R^2$ and $1/R^3$ dependencies of the induction and electrostatic terms, respectively). By a similar argument, for close distances, the dominant terms will be the electrostatic term in the case of the electric field and the induction term for the magnetic field. It can be readily shown from Eqs. (38.8) through (38.14) and the preceding discussion that for any individual lightning return stroke model, the waveshapes of the vertical electric field and the horizontal magnetic field are almost identical for great distances, and this fact is also evident in the experimental data (see Fig. 38.25). Moreover, it can be shown that for great distances, the ratio of the electric field intensity E to the magnetic flux density B is the speed of light c.

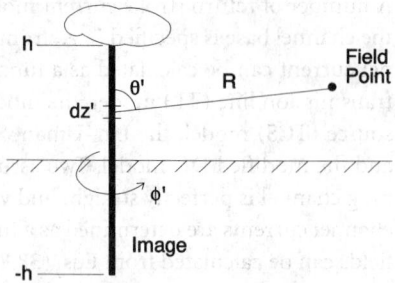

FIGURE 38.27 The geometry for the calculation of the fields using Eqs. (38.8) through (38.14).

A brief examination of return stroke current models follows. We discuss here only the transmission line (TL) model and the model of Diendorfer and Uman [1990] along with its modified version. More details of all models are found in Nucci *et al.* [1990], Diendorfer and Uman [1990], and Thottappillil *et al.* [1991], including fields calculated from the various models.

The Transmission Line Model

In the transmission line model it is assumed that the current waveform at the ground travels undistorted up the lightning channel at a constant speed v. Mathematically, this current is represented by

$$i(z',t) = i(0, t - z'/v) \qquad z' < vt$$
$$i(z',t) = 0 \qquad z' > vt \tag{38.15}$$

No charge is removed by the transmission line return stroke current along the channel since the charge entering the bottom of any section of the channel leaves the top when the current reaches it. All the charge is therefore transferred from the bottom to the top of the channel, an unrealistic situation given our knowledge of lightning physics.

Willett *et al.* [1989] have presented return stroke current, field, and speed data from artificially initiated (by firing small rockets trailing grounded wires) lightning in an attempt to validate the TL model. Using these data, Rakov *et al.* [1992] have shown, at least for subsequent strokes in artifically initiated lightning, that return stroke peak current can be derived from return stroke peak field by the expression $I = 1.5 - 0.037DE$ where the peak current I is in kA and is negative, the distance D is in km, and the peak electric field E is in V/m and is positive. Several investigators have published lightning peak current statistics derived from the magnetic radiation fields recorded by networks of magnetic direction finders by making use of the transmission line model. These studies are discussed by Rakov *et al.* [1992].

The Diendorfer–Uman (DU) Model and a Modification of It

The DU model [Diendorfer and Uman, 1990], is a physically reasonable model that can predict the salient features of the measured lightning electric and magnetic fields. Given the return stroke current at ground level, the channel current above ground is assumed to discharge the leader by way of two independent processes: (1) the discharge of the highly ionized core of the leader channel, termed the breakdown discharge process, with a time constant of 1 µs or less, and (2) the discharge of the corona envelope with a larger time constant. In both cases, the discharge at a height z' starts when the return stroke front, assumed to travel up at a constant speed v, arrives at z'. The liberated currents are assumed to flow to the ground at the speed of light.

For a current at ground $i(0,t)$, Diendorfer and Uman [1990] show that the current as a function of height and time is

$$i(z',t) = i(0,t_m) - i(0,z'/v*) \exp(-t_e/\tau) \tag{38.16}$$

where $t_m = (t + z'/v)$, $t_e = (t - z'/v)$, $v* = v(1 + v/c)$ and τ is the discharge time constant.

The Diendorfer–Uman model described above assumes that the return stroke propagates up the lightning channel at a constant speed and that the current from activated sections of the channel travels to ground at the speed of light. An analytical generalization of the DU model which allows for the return stroke speed and the downward current speed to be arbitrary functions of height has been presented by Thottappillil *et al.*[1991].

Lightning-Overhead Wire Interactions

General

Lightning interactions with overhead wires such as power distribution lines are a major source of electromagnetic compatibility problems, resulting in inferior power quality, power outages, and damaged electronics. Only a small fraction of all the cloud-to-ground lightning flashes directly strike overhead lines, making induced overvoltages a significant source of power disturbances. This section begins with a discussion of the appropriate transmission line equations. Then, examples of measured lightning-induced voltages on overhead lines as well as calculated voltages are presented.

Transmission Line Equations

The transmission line equations for a nonuniform electromagnetic field impinging on a system of horizontal wires have been derived in the time domain by Agrawal *et al.* [1980], who adapted the theory to the case of wires above an imperfectly conducting ground. The main advantage of a time domain model over an equivalent frequency domain model is its applicability to cases of time varying and nonlinear loads and its ability to account for multiple reflections on a line with two or more discontinuities. On the other hand, with a frequency domain model it is intrinsically easier to handle frequency-dependent parameters such as the ground impedance.

The derivation of the time domain coupling equations is conceptually simple: Maxwell's equations are first integrated over closed cylindrical surfaces and along closed rectangular paths. The resulting integral equations, which are in terms of electric and magnetic fields, are then recast in terms of voltages and currents. One version of the transmission line equations follows:

$$\frac{\partial V^s(x,t)}{\partial t} + Z_g * I(x,t) + L\frac{\partial I(x,t)}{\partial t} = E_x^i(x, z = h, t) \tag{38.17}$$

$$\frac{\partial I(x,t)}{\partial x} + C\frac{\partial V^s(x,t)}{\partial t} = 0 \tag{38.18}$$

$$V^t = V^i + V^s = -\int_0^h E_z^i(x,z,t)\, dz + V^s \tag{38.19}$$

where the superscript s identifies the "scattered" quantities, the superscript i identifies the "incident" quantities, the superscript t identifies the total, measurable quantities, and the asterisk is the convolution operator.

In these equations, the only source along the horizontal portion of the line is the horizontal component of the incident electric field. At the line terminations, the boundary condition and the termination current, I, are used to determine the end voltage. At those vertically oriented terminations, the vertical electric fields drive currents through the terminations into the line. The total voltage, $V^t(x,t)$, at the line terminations must equal $I_T{}^*Z_T$ at all times, where Z_T is the termination impedance. Equations (38.17) through (38.19) can be represented by the circuit model in Fig. 38.28.

Two basic assumptions are used to arrive at Eqs. (38.17) through (38.19): (1) The response of the power line (scattered voltages and currents) to the impinging EM wave (incident field) is quasi-TEM (i.e., the scattered fields can be approximated as transverse electromagnetic). This allows us to define a "static" voltage along the line and to relate the line current and the scattered magnetic flux by an inductance, as well as the line scattered voltage and charge by a capacitance. (2) The transverse dimensions of the line system are small compared to the minimum wavelength, λ_{min}, of the excitation wave, and the height of the line is much larger than the diameter of the wire.

Measured and Calculated Lightning-Induced Voltages on Overhead Wires

Several experiments have been carried out to test the theory of Agrawal *et al.* [1980] just presented [e.g., Georgiadis *et al.*, 1992]. The basic strategy is the same in each experiment: to measure the lightning electric and magnetic fields in the vicinity of an instrumented overhead line while simultaneously measuring the voltages induced on the line, the measured fields then being used as inputs to a computer program written to solve Eqs. (38.17) through (38.19) and the computer-calculated voltage waveforms being compared with the measured voltage waveforms. The following discussion illustrates the types of voltage waveforms induced on overhead wires by lightning beyond a few kilometers and the degree of agreement that has been obtained in the coupling-model calculations. Examples of voltages induced on a 450-m overhead line about 10 m above the ground are shown in Fig. 38.29. Each line end was either terminated in its characteristic impedance or open-circuited (four different cases), and voltages were measured simultaneously at each end. Figures 38.30 and 38.31 contain specific examples of measured and calculated voltage waveforms at each line end as well as the measured vertical electric field and calculated horizontal electric field via Eq. (38.7). It is clear from Figs. 38.29–38.31 that the induced voltage polarities and waveshapes are strongly dependent on the angle to the lightning and on the line end terminations. It is apparent also from Georgiadis *et al.* [1992] that while measured and calculated voltage waveshapes are in good agreement, the measured voltage amplitudes are, on average, a factor of three smaller than calculated voltages. This amplitude discrepancy remains unexplained but is probably due to the fact that the fields reaching the power line were shielded by trees along the line whereas the fields measured were in an open area and hence were unshielded.

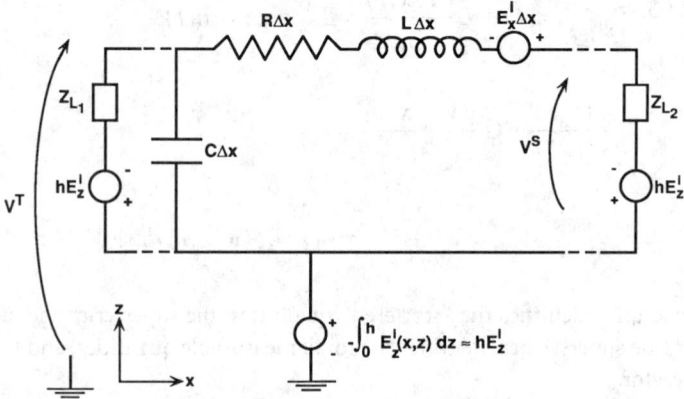

FIGURE 38.28 Equivalent circuit model obtained from Eqs. (38.17) through (38.19).

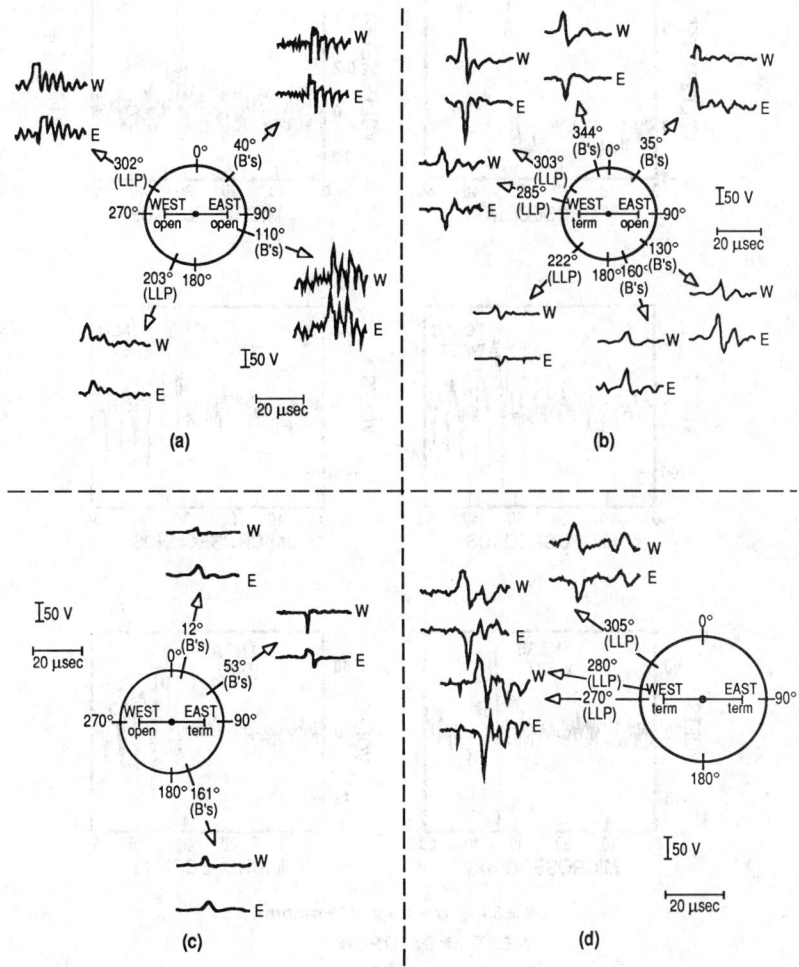

FIGURE 38.29 Examples of simultaneously measured lightning-induced voltages at the east end (E) and west end (W) of a 450-m line. Each line end is either open or terminated in its characteristic impedance, as noted. Directions to the lightning are determined from the ratio of the horizontal magnetic flux densities (*B*'s) or from a commercial lightning location system (LLP). (*Source:* Adapted from N. Georgiadis *et al.*, "Lightning-induced voltages at both ends of a 450-meter distribution line," *IEEE Trans. EMC,* vol. 34, pp. 451–460, 1992. ©1992 IEEE. With permission.)

Defining Terms

Attachment process: A process that occurs when one or more stepped leader branches approach within a hundred meters or so of the ground and the electric field at the ground increases above the critical breakdown field of the surrounding air. At that time one or more upward-going discharges is initiated. After traveling a few tens of meters, one of the upward discharges, which is essentially at ground potential, contracts the tip of one branch of the stepped leader, which is at a high potential, completing the leader path to ground.

Dart leader: A continuously moving leader lowering charge preceding a return stroke subsequent to the first. A dart leader typically propagates down the residual channel of the previous stroke.

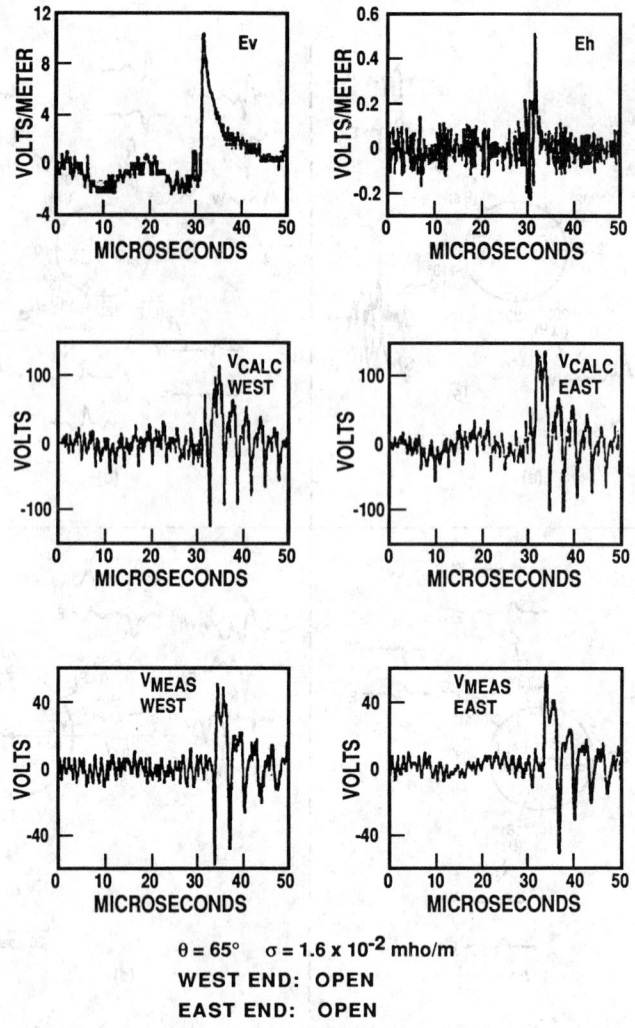

FIGURE 38.30 Measured and calculated voltages at the east and west ends of a 450-m line for both line ends open. Although the direction to the lightning as determined from the ratio of the two magnetic flux density (*B*) components was 40°, as shown in Fig. 38.29(a), the best calculated fit to the data was found for 65° as shown, the angular error apparently being caused by variation in the magnitudes of the magnetic flux density components due to nearby conductors as determined from comparing azimuths computed from the *B*'s and from a commercial lightning location system (LLP). (*Source:* Adapted from N. Georgiadis *et al.*, "Lightning-induced voltages at both ends of a 450-meter distribution line," *IEEE Trans. EMC*, vol. 34, pp. 451–460, 1992. ©1992 IEEE. With permission.)

Flash: A complete lightning discharge of any type.

Preliminary breakdown: An electrical discharge in the cloud that initiates a cloud-to-ground flash.

Return stroke: The upward propagating high-current, bright, potential discontinuity following the leader that discharges to the ground some or all of the charge previously deposited along the channel by the leader.

Stepped leader: A discharge following the preliminary breakdown that propagates from cloud towards the ground in a series of intermittant luminous steps with an average speed of 10^5 to

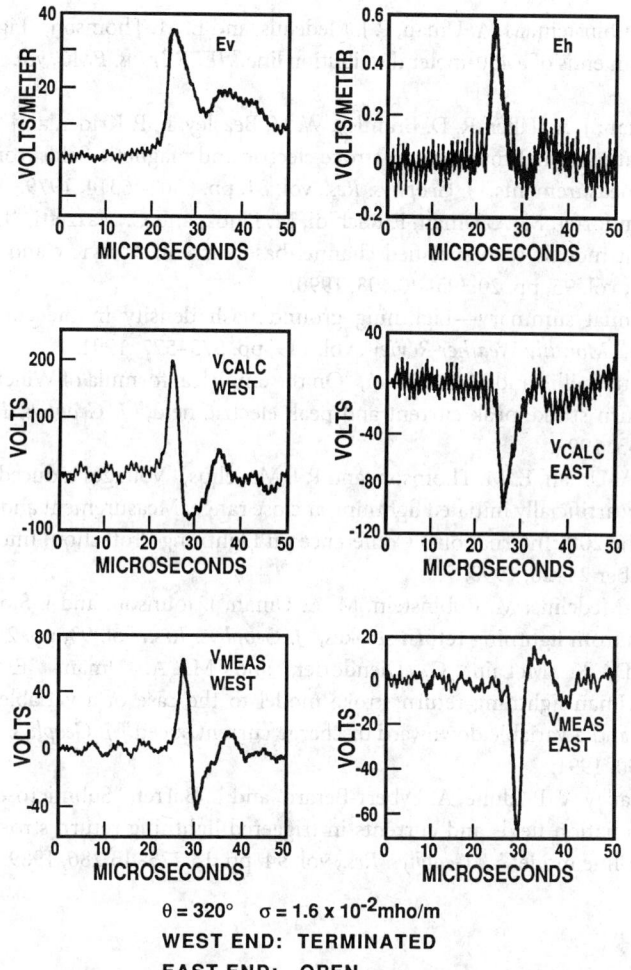

$\theta = 320°$ $\sigma = 1.6 \times 10^{-2}$ mho/m

WEST END: TERMINATED

EAST END: OPEN

FIGURE 38.31 Measured and calculated voltages at the east and west ends of a 450-m line with the west end terminated and the east end open. The azimuth was determined from LLP data and is shown in Fig. 38.29(b). (*Source:* Adapted from N. Georgiadis *et al.*, "Lightning-induced voltages at both ends of a 450-meter distribution line," *IEEE Trans. EMC*, vol. 34, pp. 451–460, 1992. ©1992 IEEE. With permission.)

10^6 m/s. Negatively charged leaders clearly step, while positively charged leaders are more pulsating than stepped.

References

A. K. Agrawal, H. J. Price, and S. H. Gurbaxani, "Transient response of multiconductor transmission lines excited by a non-uniform electromagnetic field," *IEEE Trans. EMC*, vol. EMC-22, pp. 119–129, 1980.

K. Berger, R. B. Anderson, and H. Kroninger, "Parameters of lightning flashes," *Electra*, vol. 80, pp. 23–37, 1975.

G. Diendorfer, "Induced voltage on an overhead line due to nearby lightning," *IEEE Trans. Electromag. Comp.*, vol. 32, pp. 292–299, 1990.

G. Diendorfer and M. A. Uman, "An improved return stroke model with specified channel-base current," *J. Geophys. Res.*, vol. 95, pp. 13,621–13,644, 1990.

N. Georgiadis, M. Rubinstein, M. A. Uman, P. J. Medelius, and E. M. Thomson, "Lightning-induced voltages at both ends of a 450-meter distribution line," *IEEE Trans. EMC*, vol. 34, pp. 451–460, 1992.

Y. T. Lin, M. A. Uman, J. A. Tiller, R. D. Brantley, W. H. Beasley, E. P. Krider, and C. D. Weidman, "Characterization of lightning return stroke electric and magnetic fields from simultaneous two-station measurements," *J. Geophys. Res.*, vol. 84, pp. 6307–6314, 1979.

C. A. Nucci, G. Diendorfer, M. A. Uman, F. Rachidi, M. Ianoz, and C. Mazzetti, "Lightning return stroke current models with specified channel-base current: A review and comparison," *J. Geophys. Res.*, vol. 95, pp. 20,395–20,408, 1990.

R. E. Orville, "Annual summary—Lightning ground flash density in the contiguous United States—1989," *Monthly Weather Review*, vol. 119, pp. 573–577, 1991.

V. A. Rakov, R. Thottappillil, and M. A. Uman, "On the empirical formula of Willett, *et al.*, relating lightning return stroke peak current and peak electric field," *J. Geophys. Res.*, vol. 97, pp. 11,527–11,533, 1992.

M. Rubinstein, M. A. Uman, E. M. Thomson, and P. J. Medelius, "Voltages induced on a test distribution line by artificially initiated lightning at close range: Measurement and theory," in Proceedings of the 20th International Conference on Lightning Protection, Interlaken, Switzerland, September 24–28, 1990.

E. M. Thomson, P. Medelius, M. Rubinstein, M. A. Uman, J. Johnson, and J. Stone, "Horizontal electric fields from lightning return strokes," *J. Geophys. Res.*, vol. 93, pp. 2429–2441, 1988.

R. Thottappillil, D. K. McLain, G. Diendorfer, and M. A. Uman, "Extension of the Diendorfer–Uman lightning return stroke model to the case of a variable upward return stroke speed and a variable downward discharge current speed," *J. Geophys. Res.*, vol. 96, pp. 17,143–17,150, 1991.

J. E. Willett, J. C. Bailey, V. P. Idone, A. Eybert-Berard, and L. Barret, "Submicrosecond intercomparison of radiation fields and currents in triggered lightning return strokes based on the transmission-line model," *J. Geophys. Res.*, vol. 94, pp. 13,275–13,286, 1989.

Further Information

For more details on the material presented here, see *The Lightning Discharge* (Academic Press, San Diego, 1987) by M. A. Uman and the review article "Natural and Artificially Initiated Lightning" (*Science*, vol. 246, 457–464, 1989) by M. A. Uman and E. P. Krider. For the most recent information on return stroke properties and references to previous work, see "Some Properties of Negative Cloud to Ground Lightning vs. Stroke Order" (*J. Geophys. Res.*, vol. 95, 5447–5453, 1990), by V. A. Rakov and M. A. Uman, and "Lightning Subsequent Stroke Electric Field Peak Greater than the First Stroke Peak and Multiple Ground Terminations" (*J. Geophys. Res.*, vol. 97, 7503–7509, 1992), by R. Thottappillil, V. A. Rakov, M. A. Uman, W. H. Beasley, M. J. Master, and D. V. Shelukhin.

For more information on lightning properties derived from networks of wideband magnetic direction finders, see "Cloud to Ground Lightning Flash Characteristics from June 1984 through May 1985" (*J. Geophys. Res.*, vol. 92, 5640–5644, 1992), by R. E. Orville, R. A. Weisman, R. B. Pyle, R. W. Henderson, and R. E. Orville, Jr., and "Calibration of a Magnetic Direction Finding Network Using Measured Triggered Lightning Return Stroke Peak Currents" (*J. Geophys. Res.*, vol. 96, 17,135–17,142, 1991), by R. E. Orville.

Radar

Melvin L. Belcher
Georgia Tech Research Institute

Josh T. Nessmith
Georgia Tech Research Institute

James C. Wiltse
Georgia Tech Research Institute

39.1 Pulse Radar 949
      Overview of Pulsed Radars • Critical Subsystem Design and
      Technology • Radar Performance Prediction • Radar Waveforms •
      Detection and Search • Estimation and Tracking
39.2 Continuous Wave Radar 964
      CW Doppler Radar • FM/CW Radar • Interrupted Frequency-
      Modulated CW (IFM/CW) • Applications • Summary Comments

39.1 Pulse Radar

Melvin L. Belcher and Josh T. Nessmith

Overview of Pulsed Radars

Basic Concept of Pulse Radar Operation

The basic operation of a pulse radar is depicted in Fig. 39.1. The radar transmits a pulse of RF energy and then receives returns (reflections) from desired and undesired targets. Desired targets may include space, airborne, and sea- and/or surface-based vehicles. They can also include the earth's surface and the atmosphere, depending on the application. Undesired targets are termed *clutter*. Clutter sources include the ground, natural and man-made objects, sea, atmospheric phenomena, and birds. Short-range/low-altitude radar operation is often constrained by clutter since the multitude of undesired returns masks returns from targets of interest such as aircraft.

The range, azimuth angle, elevation angle, and range rate can be directly measured from a return to estimate target position and velocity. Signature data can be extracted by measuring the amplitude, phase, and polarization of the return.

Pulse radar affords a great deal of design and operational flexibility. Pulse duration and pulse rate can be tailored to specific applications to provide optimal performance. Modern computer-controlled multiple-function radars exploit this capability by choosing the best waveform from a repertoire for a given operational mode and interference environment automatically.

Radar Applications

The breadth of pulse radar applications is summarized in Table 39.1. Radar applications can be grouped into search, track, and signature measurement applications. Search radars are used for tracking but have relatively large range and angle errors. The search functions favor broad beamwidths and low bandwidths in order to efficiently search over a large spatial volume. As indicated in Table 39.1, search is preferably performed in the lower frequency bands. The antenna pattern is narrow in azimuth and has a cosecant pattern in elevation to provide acceptable coverage from the horizon to the zenith.

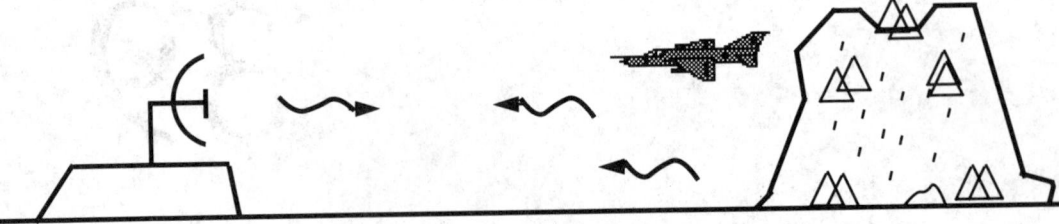

$$\text{Target Range} = \frac{\text{Two-Way-Time-Delay} \cdot \text{Speed-of-Light}}{2}$$

FIGURE 39.1 Pulse radar.

Table 39.1 Radar Bands

Band	Frequency Range	Principal Applications
HF	3–30 MHz	Over-the-horizon radar
VHF	30–300 MHz	Long-range search
UHF	300–1000 MHz	Long-range surveillance
L	1000–2000 MHz	Long-range surveillance
S	2000–4000 MHz	Surveillance
		Long-range weather characterization
		Terminal air traffic control
C	4000–8000 MHz	Fire control
		Instrumentation tracking
X	8–12 GHz	Fire control
		Air-to-air missile seeker
		Marine radar
		Airborne weather characterization
Ku	12–18 GHz	Short-range fire control
		Remote sensing
Ka	27–40 GHz	Remote sensing
		Weapon guidance
V	40–75 GHz	Remote sensing
		Weapon guidance
W	75–110 GHz	Remote sensing
		Weapon guidance

Tracking radars are typically characterized by a narrow beamwidth and moderate bandwidth in order to provide accurate range and angle measurements on a given target. The antenna pattern is a pencil beam with approximately the same dimensions in azimuth and elevation. Track is usually conducted at the higher frequency bands in order to minimize the beamwidth for a given antenna aperture. After each return from a target is received, the range and angle are measured and input into a track filter. Track filtering smooths the data to refine the estimate of target position and velocity. It also predicts the target's flight path to provide range gating and antenna pointing control to the radar system.

Signature measurement applications include remote sensing of the environment as well as the measurement of target characteristics. In some applications, synthetic aperture radar (SAR) imaging is conducted from aircraft or satellites to characterize land usage over broad areas. Moving targets that present changing aspect to the radar can be imaged from airborne or ground-based radars via inverse synthetic aperture radar (ISAR) techniques. As defined in the subsection "Resolution and Accuracy," cross-range resolution improves with increasing antenna extent. SAR/ISAR effectively substitutes an extended observation interval over which coherent returns are collected from different target aspect angles for a large antenna structure that would not be physically realizable in many instances.

In general, characterization performance improves with increasing frequency because of the associated improvement in range, range rate, and cross-range resolution. However, phenomenological characterization to support environmental remote sensing may require data collected across a broad swath of frequencies.

A multiple-function **phased array** radar generally integrates these functions to some degree. Its design is usually driven by the track function. Its operational frequency is generally a compromise between the lower frequency of the search radar and the higher frequency desired for the tracking radar. The degree of signature measurement implemented to support such functions as noncooperative target identification depends on the resolution capability of the radar as well as the operational user requirements. Multiple-function radar design represents a compromise among these different requirements. However, implementation constraints, multiple-target handling requirements, and reaction time requirements often dictate the use of phased array radar systems integrating search, track, and characterization functions.

Critical Subsystem Design and Technology

The major subsystems making up a pulse radar system are depicted in Fig. 39.2. The associated interaction between function and technology is summarized in this subsection.

Antenna

The radar antenna function is to first provide spatial directivity to the transmitted EM wave and then to intercept the scattering of that wave from a target. Most radar antennas may be categorized as mechanically scanning or electronically scanning. Mechanically scanned reflector antennas are used in applications where rapid beam scanning is not required. Electronic scanning antennas include phased arrays and frequency scanned antennas. Phased array beams can be steered to any point in their field-of-view, typically within 10 to 100 µs, depending on the latency of the beam steering subsystem and the switching time of the phase shifters. Phased arrays are desirable in multiple function radars since they can interleave search operations with multiple target tracks.

There is a Fourier transform relationship between the antenna illumination function and the far-field antenna pattern. Hence, tapering the illumination to concentrate power near the center of the antenna suppresses sidelobes while reducing the effective antenna aperture area. The phase and amplitude control of the antenna illumination determines the achievable sidelobe suppression and angle measurement accuracy.

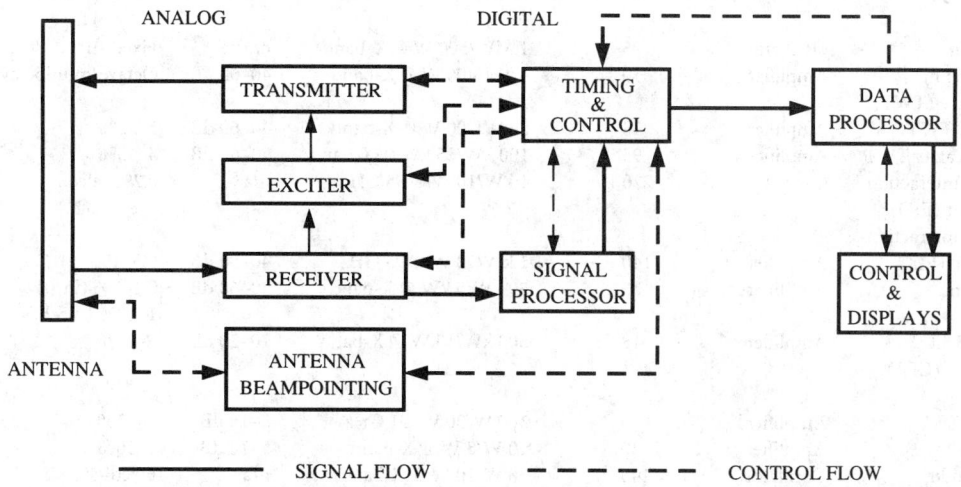

FIGURE 39.2   Radar system architecture.

Perturbations in the illumination due to the mechanical and electrical sources distort the illumination function and constrain performance in these areas. Mechanical illumination error sources include antenna shape deformation due to sag and thermal effects as well as manufacturing defects. Electrical illumination error is of particular concern in phased arrays where sources include beam steering computational error and phase shifter quantization. Control of both the mechanical and electrical perturbation errors is the key to both low sidelobes and highly accurate angle measurements. Control denotes that either tolerances are closely held and maintained or that there must be some means for monitoring and correction. Phased arrays are attractive for low sidelobe applications since they can provide element-level phase and amplitude control.

Transmitter

The transmitter function is to amplify waveforms to a power level sufficient for target detection and estimation. There is a general trend away from tube-based transmitters toward solid-state transmitters. In particular, solid-state transmit/receive modules appear attractive for constructing phased array radar systems. In this case, each radiating element is driven by a module that contains a solid-state transmitter, phase shifter, low-noise amplifier, and associated control components. Active arrays built from such modules appear to offer significant reliability advantages over radar systems driven from a single transmitter. However, microwave tube technology continues to offer substantial advantages in power output over solid-state technology. Transmitter technologies are summarized in Table 39.2.

Receiver and Exciter

This subsystem contains the precision timing and frequency reference source or sources used to derive the master oscillator and local oscillator reference frequencies. These reference frequencies are used to downconvert received signals in a multiple-stage superheterodyne architecture to accommodate signal amplification and interference rejection. The receiver front end is typically protected from overload during transmission through the combination of a circulator and a transmit/receive switch.

The exciter generates the waveforms for subsequent transmission. As in signal processing, the

Table 39.2 Radar Transmitter Technology

Technology	Mode of Operation	Maximum Frequency (GHz)	Demonstrated Peak/ Average Power (kW)	Typical Gain	Typical Bandwidth
Thermionic					
Magnetron	Oscillator	95	1 MW/500 W @ X-band	n/a	Fixed–10%
Helix traveling wave tube (TWT)	Amplifier	50	4 kW/400 W @ X-band	40–60 dB	Octave/multioctave
Ring-loop TWT	Amplifier	18	8 kW/200 W @ X-band	40–60 dB	5–15%
Coupled-cavity TWT	Amplifier	95	100 kW/25 kW @ X-band	40–60 dB	5–15%
Extended interaction oscillator (EIO)	Oscillator	220	1 kW/10 W @ 95 GHz	n/a	0.2% (elec.) 4% (mech.)
Extended interaction Klystron (EIK)	Amplifier	140	1 kW/10 W @ 95 GHz	40–50 dB	0.5–1%
Klystron	Amplifier	35	50 kW/5 kW @ X-band	30–60 dB	0.1–2% (inst.) 1–10% (mech.)
Crossed-field amplifier (CFA)	Amplifier	18	500 kW/1 kW @ X-band	10–20 dB	5–15%
Solid state					
Silicon BJT	Amplifier	5	300 W/30 W @1 GHz	5–10 dB	10–25%
GaAs FET	Amplifier	30	10 W/3 W @ X-band	5–10 dB	5–20%
Impatt diode	Oscillator	140	30 W/10 W @ X-band	n/a	Fixed–5%

Source: Tracy V. Wallace, Georgia Tech Research Institute, Atlanta, Georgia.

trend is toward programmable digital signal synthesis because of the associated flexibility and performance stability.

Signal and Data Processing

Digital processing is generally divided between two processing subsystems according to the algorithm structure and throughput demands. Signal processing includes pulse compression, Doppler filtering, and detection threshold estimation and testing. Data processing includes track filtering, user interface support, and such specialized functions as electronic counter-counter measures (ECCM) and built-in test (BIT), as well as the resource management process required to control the radar system.

The signal processor is often optimized to perform the repetitive complex multiply-and-add operations associated with the fast Fourier transform (FFT). FFT processing is used for implementing **pulse compression** via fast convolution and for Doppler filtering. Fast convolution consists of taking the FFT of the digitized receiver output, multiplying it by the stored FFT of the desired filter function, and then taking the inverse FFT of the resulting product. Fast convolution results in significant computational saving over performing the time-domain convolution of returns with the filter function corresponding to the matched filter. The signal processor output can be characterized in terms of range gates and Doppler filters corresponding approximately to the range and Doppler resolution, respectively.

In contrast, the radar data processor typically consists of a general-purpose computer with a real-time operating system. Fielded radar data processors range from microcomputers to mainframe computers, depending on the requirements of the radar system. Data processor software and hardware requirements are significantly mitigated by off loading timing and control functions to specialized hardware. This timing and control subsystem typically functions as the two-way interface between the data processor and the other radar subsystems. The increasing inclusion of BIT and built-in calibration capability in timing and control subsystem designs promises to result in significant improvement in fielded system performance.

Radar Performance Prediction

Radar Line-of-Sight

With the exception over-the-horizon (OTH) radar systems, which exploit either sky-wave bounce or ground-wave propagation modes and sporadic ducting effects at higher frequencies, surface and airborne platform radar operation is limited to the refraction-constrained line of sight. Atmospheric refraction effects can be closely approximated by setting the earth's radius to 4/3 its nominal value in estimating horizon-limited range. The resulting line-of-sight range is depicted in Fig. 39.3 for a surface-based radar, an airborne surveillance radar, and a space-based radar.

As evident in the plot, airborne and space-based surveillance radar systems offer significant advantages in the detection of low-altitude targets that would otherwise be masked by earth curvature and terrain features from surface-based radars. However, efficient clutter rejection techniques must be used in order to detect targets since surface clutter returns will be present at almost all ranges of interest.

Radar Range Equation

The radar range equation is commonly used to estimate radar system performance, given that line-of-sight conditions are satisfied. This formulation essentially computes the signal-to-noise ratio (S/N) at the output of the radar signal processor. In turn, S/N is used to provide estimates of radar detection and position measurement performance as described in the subsections "Detection and Search" and "Estimation and Tracking." S/N can be calculated in terms of the number of pulses coherently integrated over a single coherent processing interval (CPI) using the radar range equation such that

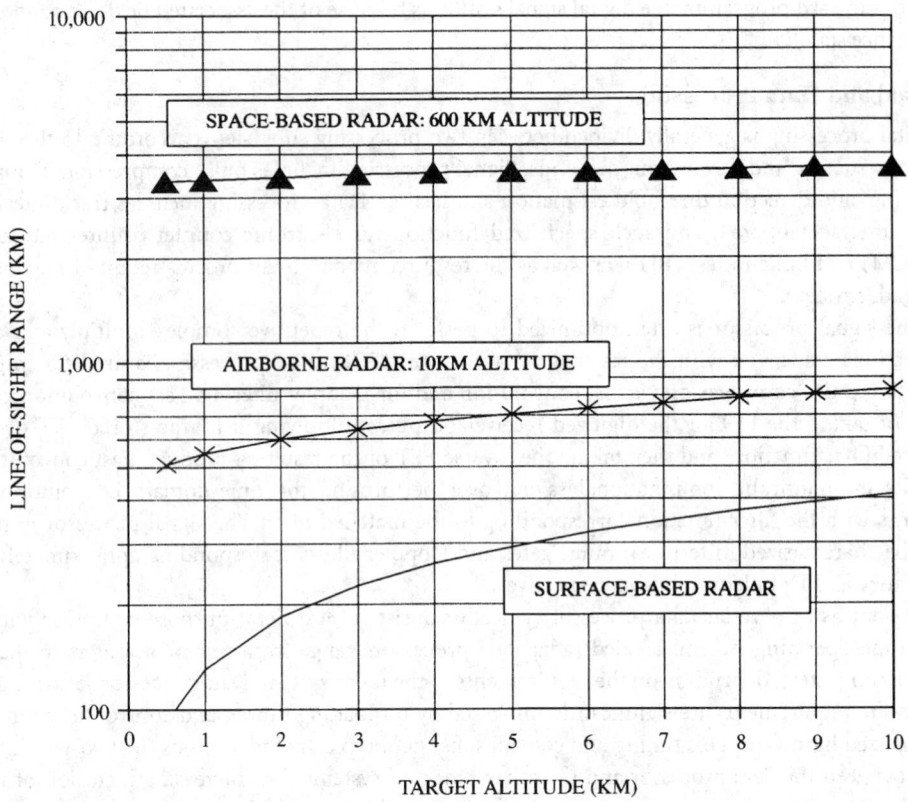

FIGURE 39.3 Maximum line-of-sight range for surface-based radar, an airborne surveillance radar, and a space-based radar.

$$S/N = \frac{PDAT_pN_p\sigma}{(4\pi)^2 R^4 L_t L_{rn} L_{sp} k T_s} \tag{39.1}$$

where P is peak transmitter power output, D is directivity of the transmit antenna, A is effective aperture area of the receive antenna in meters squared, T_p is pulse duration, σ is **radar cross section** in square meters, N_p is the number of coherently integrated pulses within the coherent processing interval, R is range to target in meters, L_t is system ohmic and nonohmic transmit losses, L_{rn} is system nonohmic receive losses, L_{sp} is signal processing losses, k is Boltzmann's constant (1.38×10^{-23} K), and T_s is system noise temperature, including receive ohmic losses (kelvin).

At X-band and above it may also be necessary to include propagation loss due to atmospheric absorption [Blake, 1986]. This form of the radar range equation is applicable to radar systems using pulse compression or pulse Doppler waveforms as well as the unmodulated single-pulse case. In many applications, average power is a better measure of system performance than peak power since it indicates the S/N improvement achievable with pulse integration over a given interval of time. Hence, the radar range equation can be modified such that

$$S/N = \frac{P_a DAT_c\sigma}{(4\pi)^2 R^4 L_t L_{rn} L_{sp} k T_s} \tag{39.2}$$

where P_a is average transmitter power and T_c is coherent processing interval (CPI).

The portion of time over which the transmitter is in operation is referred to as the radar duty cycle. The average transmitter power is the product of duty cycle and peak transmitter power. Duty cycle ranges from less than 1% for typical **noncoherent** pulse radars to somewhat less than 50% for high pulse repetition frequency (PRF) pulse Doppler radar systems. High PRF systems are sometimes referred to as interrupted continuous wave (ICW) systems because they operate essentially as a CW radar system with transmitter and receiver alternately turned on and off.

The CPI is the period over which returns are collected for **coherent** processing functions such as integration and Doppler filtering. The CPI can be estimated as the product of the number of coherently integrated pulses and the interval between pulses. Noncoherent integration is less efficient and alters the statistical character of the signal and interference.

Antenna Directivity and Aperture Area

The directivity of the antenna is

$$D = \frac{4\pi A \eta}{\lambda^2} \tag{39.3}$$

where η is aperture efficiency and λ is radar carrier wavelength. Aperture inefficiency is due to the antenna illumination factor.

The common form of the radar range equation uses power gain rather than directivity. Antenna gain is equal to the directivity divided by the antenna losses. In the design and analysis of modern radars, directivity is a more convenient measure of performance because it permits designs with distributed active elements, such as solid-state phased arrays, to be assessed to permit direct comparison with passive antenna systems. Beamwidth and directivity are inversely related; a highly directive antenna will have a narrow beamwidth. For typical design parameters,

$$D = \frac{10^7}{\theta_{az}\theta_{el}} \tag{39.4}$$

where θ_{az} and θ_{el} are the radar azimuth and elevation beamwidths, respectively, in milliradians.

Radar Cross Section

In practice, the radar cross section (RCS) of a realistic target must be considered a random variable with an associated correlation interval. Targets are composed of multiple interacting scatters so that the composite return varies in magnitude with the constructive and destructive interference of the contributing returns. The target RCS is typically estimated as the mean or median of the target RCS. The associated correlation interval indicates the rate at which the target RCS varies over time. RCS fluctuation degrades target detection performance at moderate to high probability of detection.

The median RCS of typical targets is given in Table 39.3. The composite RCS measured by a radar system may be composed of multiple individual targets in the case of closely spaced targets such as a bird flock.

Loss and System Temperature Estimation

Sources of S/N loss include ohmic and nonohmic (mismatch) loss in the antenna and other radio frequency components, propagation effects, signal processing deviations from matched filter operation, detection thresholding, and search losses. Scan loss in phased array radars is due to the combined effects of the decrease in projected antenna area and element mismatch with increasing scan angle.

Search operations impose additional losses due to target position uncertainty. Because the target position is unknown before detection, the beam, range gate, and Doppler filter will not be centered on the target return. Hence, straddling loss will occur as the target effectively straddles adjacent resolution cells in range and Doppler. Beamshape loss is a consequence of the radar beam not being

Table 39.3 Median Target RCS (m²)

Carrier Frequency, GHz	1–2	3	5	10	17
Aircraft (nose/tail avg.)					
Small propeller	2	3	2.5		
Small jet (Lear)	1	1.5	1	1.2	
T38-twin jet, F5	2	2–3	2	1–2/6	
T39-Sabreliner	2.5		10/8	9	
F4, large fighter	5–8/5	4–20/10	4	4	
737, DC9, MD80	10	10	10	10	10
727, 707, DC8-type	22–40/15	40	30	30	
DC-10-type, 747	70	70	70	70	
Ryan drone				2/1	
Standing man (180 lb)	0.3	0.5	0.6	0.7	0.7
Automobiles	100	100	100	100	100
Ships-incoming (×10⁴ m²)					
4K tons	1.6	2.3	3.0	4.0	5.4
16K tons	13	18	24	32	43
Birds					
Sea birds	0.002	0.001–0.004	0.004		
Sparrow, starling, etc.	0.001	0.001	0.001	0.001	0.001

Slash marks indicate different set.
Source: F. E. Nathanson, *Radar Design Principles,* 2nd ed., New York: McGraw-Hill, 1991. With permission.

pointed directly at the target so that there is a loss in both transmit and receive antenna gain. In addition, detection threshold loss associated with radar system adaptation to interference must be included [Nathanson, 1991]).

System noise temperature estimation corresponds to assessing the system thermal noise floor referenced to the antenna output. Assuming the receiver hardware is at ambient temperature, the system noise temperature can be estimated as

$$T_s = T_a + 290 \, (L_{ro}F - 1) \tag{39.5}$$

where T_a is the antenna noise temperature, L_{ro} is receive ohmic losses, and F is the receiver noise figure.

In phased array radars, the thermodynamic temperature of the antenna receive beam-former may be significantly higher than ambient, so a more complete analysis is required. The antenna noise temperature is determined by the external noise received by the antenna from solar, atmospheric, earth surface, and other sources.

Table 39.4 provides typical loss and noise temperature budgets for several major radar classes. In general, loss increases with the complexity of the radar hardware between the transmitter/receiver and the antenna radiator. Reflector antennas and active phased arrays impose relatively low loss, while passive array antennas impose relatively high loss.

Resolution and Accuracy

The fundamental resolution capabilities of a radar system are summarized in Table 39.5. In general, there is a trade-off between mainlobe resolution corresponding to the nominal range, Doppler, and angle resolution, and effective dynamic range corresponding to suppression of sidelobe components. This is evident in the use of weighting to suppress Doppler sidebands and angle sidelobes at the expense of broadening the mainlobe and S/N loss.

Cross range denotes either of the two dimensions orthogonal to the radar line of sight. Cross-range resolution in real-aperture antenna systems is closely approximated by the product of target

Table 39.4 Typical Microwave Loss and System Temperature Budgets

	Mechanically Scanned		Electronically Scanned
	Reflector Antenna	Slotted Array	Solid-State Phased Array
Nominal losses			
Transmit loss, L_t (dB)	1	1.5	0.5
Nonohmic receiver loss, L_r (dB)	0.5	0.5	0.1
Signal processing loss, L_{sp} (dB)	1.4	1.4	1.4
Scan loss (dB)	N/A	N/A	30 log [cos (scan angle)]
Search losses, L_{DS}			
Beam shape (dB)	3	3	3
Range gate straddle (dB)	0.5	0.5	0.5
Doppler filter straddle (dB)	0.5	0.5	0.5
Detection thresholding (dB)	1	1	1
System noise temperature (kelvin)	500	600	400

Table 39.5 Resolution and Accuracy

Dimension	Nominal Resolution	Noise-Limited Accuracy
Angle	$\dfrac{\alpha\lambda}{d}$	$\dfrac{\alpha\lambda}{dK_m\sqrt{2S/N}}$
Range	$\dfrac{\alpha C}{2B}$	$\dfrac{\alpha C}{2B\,K_i\sqrt{2S/N}}$
Doppler	$\dfrac{\alpha}{\text{CPI}}$	$\dfrac{\alpha}{\text{CPI}\,K_i\sqrt{2S/N}}$
SAR/ISAR	$\dfrac{\alpha\lambda}{2\Delta\theta}$	$\dfrac{\alpha\lambda}{2\Delta\theta\,K_i\sqrt{2S/N}}$

α, taper broadening factor, typically ranging from 0.89 (unweighted) to 1.3 (Hamming); d, antenna extent in azimuth/elevation; B, waveform bandwidth; K_m, monopulse slope factor, typically on the order of 1.5; K_i, interpolation factor, typically on the order of 1.8; $\Delta\theta$, line-of-site rotation of target relative to radar over CPI.

range and radar beamwidth in radians. Attainment of the nominal ISAR/SAR cross-range resolution generally requires complex signal processing to generate a focused image, including correction for scatterer change in range over the CPI.

The best accuracy performance occurs for the case of thermal noise-limited error. The resulting accuracy is the resolution of the radar divided by the square root of the S/N and an appropriate monopulse or interpolation factor. In this formulation, the single-pulse S/N has been multiplied by the number of pulses integrated within the CPI as indicated in Eqs. (39.1) and (39.2).

In practice, accuracy is also constrained by environmental effects, target characteristics, and instrumentation error as well as the available S/N. Environmental effects include multipath and refraction. Target glint is characterized by an apparent wandering of the target position because of coherent interference effects associated with the composite return from the individual scattering centers on the target. Instrumentation error is minimized with alignment and calibration but may significantly constrain track filter performance as a result of the relatively long correlation interval of some error sources.

Radar Range Equation for Search and Track

The radar range equation can be modified to directly address performance in the two primary radar missions: search and track.

Search performance is basically determined by the capability of the radar system to detect a target of specific RCS at a given maximum detection range while scanning a given solid angle extent within a specified period of time. S/N can be set equal to the minimum value required for a given detection performance, $S/N|r$, while R can be set to the maximum required target detection range, R_{max}. Manipulation of the radar range equation results in the following expression:

$$\frac{P_a A}{L_t L_r L_{sp} L_{os} T_s} \geq \left(\frac{S}{N}\right)_r \frac{R_{max}^4 \Omega}{\sigma \, T_{fs}} \cdot 16k \qquad (39.6)$$

where Ω is the solid angle over which search must be performed (steradians), T_{fs} is the time allowed to search Ω by operational requirements, and L_{os} is the composite incremental loss associated with search.

The left-hand side of the equation contains radar design parameters, while the right-hand side is determined by target characteristics and operational requirements. The right-hand side of the equation is evaluated to determine radar requirements. The left-hand side of the equation is evaluated to determine if the radar design meets the requirements.

The track radar range equation is conditioned on noise-limited angle accuracy as this measure stresses radar capabilities significantly more than range accuracy in almost all cases of interest. The operational requirement is to maintain a given data rate track providing a specified single-measurement angle accuracy for a given number of targets with specified RCS and range. Antenna beamwidth, which is proportional to the radar carrier wavelength divided by antenna extent, impacts track performance since the degree of S/N required for a given measurement accuracy decreases as the beamwidth decreases. Track performance requirements can be bounded as

$$\frac{P_a A^3}{\lambda^4 L_t L_r L_{sp} T_s} k_m^2 \eta^2 \geq 5k \frac{r N_t R^4}{\sigma \sigma_\theta^2} \qquad (39.7)$$

where r is the single-target track rate, N_t is the number of targets under track in different beams, σ_θ is the required angle accuracy standard deviation (radians), and σ is the RCS. In general, a phased array radar antenna is required to support multiple target tracking when $N_t > 1$.

Incremental search losses are suppressed during single-target-per-beam tracking. The beam is pointed as closely as possible to the target to suppress beamshape loss. The tracking loop centers the range gate and Doppler filter on the return. Detection thresholding loss is minimal since the track range window is small.

Radar Waveforms

Pulse Compression

Typical pulse radar waveforms are summarized in Table 39.6. In most cases, the signal processor is designed to closely approximate a matched filter. As indicated in Table 39.5, the range and Doppler/resolution of any match-filtered waveform are inversely proportional to the waveform bandwidth and duration, respectively. Pulse compression, using modulated waveforms, is attractive since S/N is proportional to pulse duration rather than bandwidth in matched filter implementations. Ideally, the intrapulse modulation is chosen to attain adequate range resolution and range sidelobe suppression performance while the pulse duration is chosen to provide the required sensitivity. Pulse compression waveforms are characterized as having a time bandwidth product

Table 39.6 Selected Waveform Characteristics

	Comments	Time Bandwidth Product	Range Sidelobes (dB)	S/N Loss (dB)	Range/Doppler Coupling	ECM/EMI Robustness
Unmodulated	No pulse compression	~1	Not applicable	0	No	Poor
Linear frequency modulation	Linearly swept over bandwidth	>10	Unweighted: −13.5 Weighted: >−40[a]	0 0.7–1.4	Yes	Poor
Nonlinear FM	Multiple variants	Waveform specific	Waveform specific	0	Waveform specific	Fair
Barker	N-bit biphase	≤ 13 (N)	−20 log(N)	0	No	Fair
LRS	N-bit biphase	~N; >64/pulse[a]	~−10 log (N)	0	No	Good
Frank	N-bit polyphase (N = integer²)	~N	~−10 log (π²N)	0	Limited	Good
Frequency coding	N subpulses noncoincidental in time and frequency	~N²	Waveform specific • Periodic • Pseudorandom	0.7–1.40 0	Waveform specific	Good

[a]Constraint due to typical technology limitations rather than fundamental waveform characteristics.

(TBP) significantly greater than unity, in contrast to an unmodulated pulse, which has a TBP of approximately unity.

Pulse Repetition Frequency

The radar system pulse repetition frequency (PRF) determines its ability to unambiguously measure target range and range rate in a single CPI as well as determining the inherent clutter rejection capabilities of the radar system. In order to obtain an unambiguous measurement of target range, the interval between radar pulses (1/PRF) must be greater than the time required for a single pulse to propagate to a target at a given range and back. The maximum unambiguous range is then given by $C/(2 \cdot PRF)$ where C is the velocity of electromagnetic propagation.

Returns from moving targets and clutter sources are offset from the radar carrier frequency by the associated Doppler frequency. As a function of range rate, $\dot{R}$, the Doppler frequency, f_D, is given by $2\dot{R}/\lambda$. A coherent pulse train samples the returns' Doppler modulation at the PRF. Most radar systems employ parallel sampling in the in-phase and quadrature baseband channels so that the effective sampling rate is twice the PRF. The target's return is folded in frequency if the PRF is less than the target Doppler.

Clutter returns are primarily from stationary or near-stationary surfaces such as terrain. In contrast, targets of interest often have a significant range rate relative to the radar clutter. Doppler filtering can suppress returns from clutter. With the exception of frequency ambiguity, the Doppler filtering techniques used to implement pulse Doppler filtering are quite similar to those described for CW radar in Section 39.2. Ambiguous measurements can be resolved over multiple CPIs by using a sequence of slightly different PRFs and correlating detections among the CPIs [Morris, 1988].

Detection and Search

Detection processing consists of comparing the amplitude of each range gate/Doppler filter output with a threshold. A detection is reported if the amplitude exceeds that threshold. A false alarm occurs when noise or other interference produces an output of sufficient magnitude to exceed the detection threshold. As the detection threshold is decreased, both the detection probability and the false alarm probability increase. *S/N* must be increased to enhance detection probability while maintaining a constant false alarm probability.

As noted in the subsection "Radar Cross Section," RCS fluctuation effects must be considered in assessing detection performance. The Swerling models which use chi-square probability density functions (PDFs) of 2 and 4 degrees of freedom (DOF) are commonly used for this purpose [Nathanson, 1991]. The Swerling 1 and 2 models are based on the 2 DOF PDF and can be derived by modeling the target as an ensemble of independent scatterers of comparable magnitude. This model is considered representative of complex targets such as aircraft. The Swerling 2 and 4 models use the 4 DOF PDF and correspond to a target with a single dominant scatterer and an ensemble of lesser scatterers. Missiles are sometimes represented by Swerling 2 and 4 models. The Swerling 1 and 3 models presuppose slow fluctuation such that the target RCS is constant from pulse to pulse within a scan. In contrast, the RCS of Swerling 2 and 4 targets is modeled as independent on a pulse to pulse basis.

Single-pulse detection probabilities for nonfluctuating, Swerling 1/2, and Swerling 3/4 targets are depicted in Fig. 39.4. This curve is based on a typical false alarm number corresponding approxi-

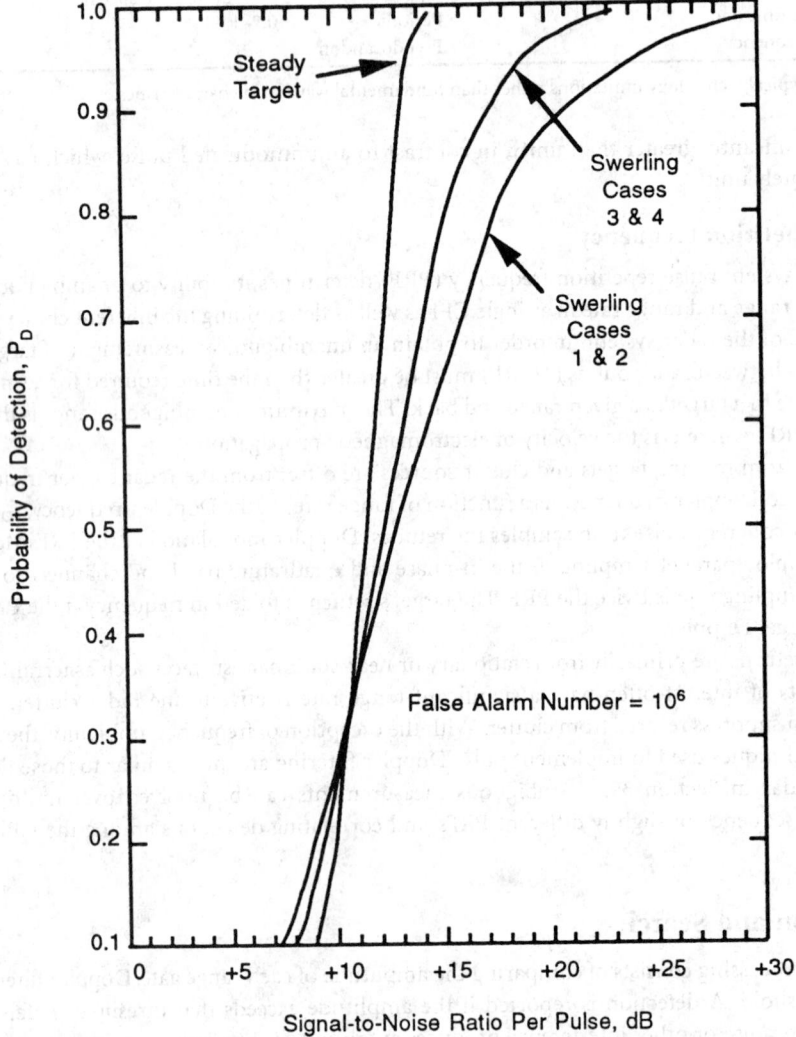

FIGURE 39.4 Detection probabilities for various target fluctuation models. (*Source:* F. E. Nathanson, *Radar Design Principles,* 2nd ed., New York: McGraw-Hill, 1991, p. 91. With permission.)

mately to a false alarm probability of 10^{-6}. The difference in S/N required for a given detection probability for a fluctuating target relative to the nonfluctuating case is termed the fluctuation loss.

The detection curves presented here and in most other references presuppose noise-limited operation. In many cases, the composite interference present at the radar system output will be dominated by clutter returns or electromagnetic interference such as that imposed by hostile electronic countermeasures. The standard textbook detection curves cannot be applied in these situations unless the composite interference is statistically similar to thermal noise with a Gaussian PDF and a white power spectral density. The presence of non-Gaussian interference is generally characterized by an elevated false alarm probability. Adaptive detection threshold estimation techniques are often required to search for targets in environments characterized by such interference.

Estimation and Tracking

Measurement Error Sources

Radars measure target range and angle position and, potentially, Doppler frequency. Angle measurement performance is emphasized here since the corresponding cross-range error dominates range error for most practical applications. Target returns are generally smoothed in a tracking filter, but tracking performance is largely determined by the measurement accuracy of the subject radar system. Radar measurement error can be characterized as indicated in Table 39.7.

The radar design and the alignment and calibration process development must consider the characteristics and interaction of these error components. Integration of automated techniques to support alignment and calibration is an area of strong effort in modern radar design that can lead to significant performance improvement in fielded systems.

As indicated previously, angle measurement generally is the limiting factor in measurement accuracy. Target azimuth and elevation position is primarily measured by a monopulse technique in modern radars though early systems used sequential lobing and conical scanning. Specialized monopulse tracking radars utilizing reflectors have achieved instrumentation and S/N angle residual systematic error as low as 50 µrad. Phased array antennas have achieved random error of 60 µrad, but the composite systematic residual errors remain to be measured. The limitations are primarily in the tolerance on the phase and amplitude of the antenna illumination function.

Figure 39.5 shows the monopulse beam patterns. The first is the received sum pattern that is generated by a feed that provides the energy from the reflector or phased array antenna through two ports in equal amounts and summed in phase in a monopulse comparator shown in Fig. 39.6. The second is the difference pattern generated by providing the energy through the same two ports in equal amounts but taken out with a phase difference of π radians, giving a null at the center. A target located at the center of the same beam would receive a strong signal from the sum pattern with which the target could be detected and ranged. The received difference pattern would produce a null return, indicating the target was at the center of the beam. If the target were off the null, the signal output would be almost linear proportional to the distance off the center (off-axis), as shown

Table 39.7 Radar Measurement Error

Random errors	Those errors that cannot be predicted except on a statistical basis. The magnitude of the random error can be termed the *precision* and is an indication of the repeatability of a measurement.
Bias errors	A systematic error whether due to instrumentation or propagation conditions. A nonzero mean value of a random error.
Systematic error	An error whose quantity can be measured and reduced by calibration.
Residual systematic error	Those errors remaining after measurement and calibration. A function of the systematic and random errors in the calibration process.
Accuracy	The magnitude of the rms value of the residual systematic and random errors.

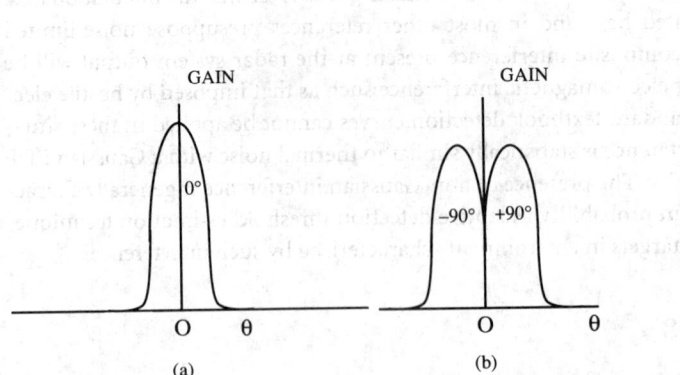

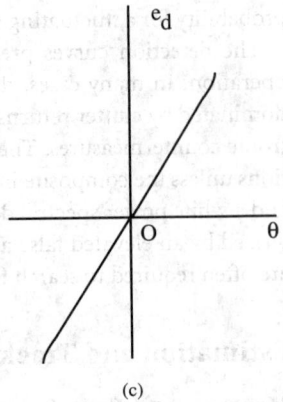

FIGURE 39.5 Monopulse beam patterns and difference voltage: (a) sum (Σ); (b) difference (Δ); (c) difference voltage.

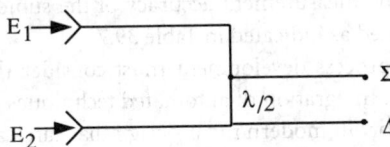

FIGURE 39.6 Monopulse comparator.

in the monopulse slope output as shown in the figure. This output is the dot product of the sum and the difference signals divided by the absolute magnitude of the sum signal squared, i.e.,

$$e_d = \frac{\Sigma \cdot \Delta}{|\Sigma|^2} \tag{39.8}$$

The random instrumentation measurement errors in the angle estimator are caused by phase and amplitude errors of the antenna illumination function. In reflector systems, such errors occur because of the position of the feedhorn, differences in electrical length between the feed and the monopulse comparator, mechanical precision of the reflector, and its mechanical rotation. In phased array radars, these errors are a function of the phase shifters, time delay units, and combines between the antenna elements and the monopulse comparator as well as the precision of the array. Although these errors are random, they may have correlation intervals considerably longer than the white noise considered in the thermal-noise random error and may depend upon the flight path of the target. For a target headed radially from or toward the radar, the correlation period of angle-measurement instrumental errors is essentially the tracking period. For crossing targets, the correlation interval may be pulse to pulse.

As in the estimate of range, the propagation effects of refraction and multipath also enter into the tracking error. The bias error in range and elevation angle by refraction can be estimated as

$$\Delta R = 0.007 \, N_s \, \text{cosecant } E_o \text{ (meters)}$$
$$\Delta E_o = N_s \cot E_o \, (\mu\text{rad}) \tag{39.9}$$

where N_s is the surface refractivity and E_o is the elevation angle [Barton and Ward, 1984].

One can calculate the average error in multipath. However, one cannot correct for it as in refraction since the direction of the error cannot be known in advance unless there are controlled conditions such as in a carefully controlled experiment. Hence, the general approach is to design the antenna sidelobes to be as low as feasible and accept the multipath error that occurs when tracking close to the horizon. There has been considerable research to find means to reduce the impact, including using very wide bandwidths to separate the direct path from the multipath return.

Tracking Filter Performance

Target tracking based on processing returns from multiple CPIs generally provides a target position and velocity estimate of greater accuracy than the single-CPI measurement accuracy delineated in Table 39.5. In principle, the error variance of the estimated target position with the target moving at a constant velocity is approximately $4/n \cdot \sigma_m^2$ where n is the number of independent measurements processed by the track filter and σ_m is the single measurement accuracy. In practice, the variance reduction factor afforded by a track filter is often limited to about an order of magnitude because of the reasons summarized in the following paragraphs.

Track filtering generally provides smoothing and prediction of target position and velocity via a recursive prediction-correction process. The filter predicts the target's position at the time of the next measurement based on the current smoothed estimates of position, velocity, and possibly acceleration. The subsequent difference between the measured position at this time and the predicted position is used to update the smoothed estimates. The update process incorporates a weighting vector that determines the relative significance given the track filter prediction versus the new measurement in updating the smoothed estimate.

Target model fidelity and adaptivity are fundamental issues in track filter mechanization. Independent one-dimensional tracking loops may be implemented to control pulse-to-pulse range gate positioning and antenna pointing. The performance of one-dimensional polynomial algorithms, such as the alpha-beta filter, to track targets from one pulse to the next and provide modest smoothing is generally adequate. However, one-dimensional closed-loop tracking ignores knowledge of the equations of motion governing the target so that their smoothing and long-term prediction performance is relatively poor for targets with known equations of motion. In addition, simple one-dimensional tracking-loop filters do not incorporate any adaptivity or measure of estimation quality.

Kalman filtering addresses these shortcomings at the cost of significantly greater computational complexity. Target equations of motion are modeled explicitly such that the position, velocity, and potentially higher-order derivatives of each measurement dimension are estimated by the track filter as a state vector. The error associated with the estimated state vector is modeled via a covariance matrix that is also updated with each iteration of the track filter. The covariance matrix determines the weight vector used to update the smoothed state vector in order to incorporate such factors as measurement S/N and dynamic target maneuvering.

Smoothing performance is constrained by the degree of *a priori* knowledge of the target's kinematic motion characteristics. For example, Kalman filtering can achieve significantly better error reduction against ballistic or orbital targets than against maneuvering aircraft. In the former case the equations of motion are explicitly known, while the latter case imposes motion model error because of the presence of unpredictable pilot or guidance system commands. Similar considerations apply to the fidelity of the track filter's model of radar measurement error. Failure to consider the impact of correlated measurement errors may result in underestimating track error when designing the system.

Defining Terms

Coherent: Integration where magnitude and phase of received signals are preserved in summation.

Noncoherent: Integration where only the magnitude of received signals is summed.

Phased array: Antenna composed of an aperture of individual radiating elements. Beam scanning is implemented by imposing a phase taper across the aperture to collimate signals received from a given angle of arrival.

Pulse compression: The processing of a wideband, coded signal pulse, of initially long time duration and low-range resolution, to result in an output pulse of time duration corresponding to the reciprocal of the bandwidth.

Radar cross section (RCS): A measure of the reflective strength of a radar target; usually represented by the symbol *s*, measured in square meters, and defined as 4π times the ratio of the power per unit solid angle scattered in a specified direction of the power unit area in a plane wave incident on the scatterer from a specified direction.

References

D. K. Barton and H.R. Ward, *Handbook of Radar Measurement,* Dedham, Mass.: Artech, 1984.

L. V. Blake, *Radar Range-Performance Analysis,* Dedham, Mass.: Artech, 1986.

J. L. Eaves and E. K. Reedy, Eds., *Principles of Modern Radar,* New York: Van Nostrand, 1987.

G.V. Morris, *Airborne Pulsed Doppler Radar,* Dedham, Mass.: Artech, 1988.

F. E. Nathanson, *Radar Design Principles,* 2nd ed., New York: McGraw-Hill, 1991.

Further Information

M. I. Skolnik, Ed., *Radar Handbook,* 2nd ed., New York: McGraw-Hill, 1990.

IEEE Standard Radar Definitions, IEEE Standard 686-1990, April 20, 1990.

39.2 Continuous Wave Radar

James C. Wiltse

Continuous wave (CW) radar employs a transmitter which is on all or most of the time. Unmodulated CW radar is very simple and is able to detect the **Doppler-frequency shift** in the return signal from a target which has a component of motion toward or away from the transmitter. While such a radar cannot measure range, it is used widely in applications such as police radars, motion detectors, proximity fuzes for projectiles or missiles, illuminators for semiactive missile guidance systems (such as the Hawk surface-to-air missile), and scatterometers (used to measure the scattering properties of targets or clutter such as terrain surfaces) [Nathanson, 1991; Saunders, 1990; Ulaby and Elachi, 1990].

Modulated versions include frequency-modulated (FM/CW), interrupted frequency-modulated (IFM/CW), and phase-modulated. Typical waveforms are indicated in Fig. 39.7. Such systems are used in altimeters, Doppler navigators, proximity fuzes, over-the-horizon radar, and active seekers for terminal guidance of air-to-surface missiles. The term *continuous* is often used to indicate a relatively long waveform (as contrasted to pulse radar using short pulses) or a radar with a high duty cycle (for instance, 50% or greater, as contrasted with the typical duty cycle of less than 1% for the usual pulse radar). As an example of a long waveform, planetary radars may transmit for up to 10 hours and are thus considered to be CW [Freiley *et al.,* 1992]. Another example is interrupted CW (or **pulse-Doppler**) radar, where the transmitter is pulsed at a high rate for 10 to 60% of the total time [Nathanson, 1991]. All of these modulated CW radars are able to measure range.

The first portion of this section discusses concepts, principles of operation, and limitations. The latter portion describes various applications. In general, CW radars have several potential advantages over pulse radars. Advantages include simplicity and the facts that the transmitter leakage is used as the local oscillator, transmitter spectral spread is minimal (not true for wide-deviation FM/CW), and peak power is the same as (or only a little greater than) the average power. This latter situation means that the radar is less detectable by intercepting equipment.

The largest disadvantage for CW radars is the need to provide antenna isolation (reduce spillover) so that the transmitted signal does not interfere with the receiver. In a pulse radar, the transmitter is off before the receiver is enabled (by means of a duplexer and/or receiver-protector switch). Isolation is frequently obtained in the CW case by employing two antennas, one for transmit and one for reception. When this is done, there is also a reduction of close-in clutter return from rain or terrain. A second disadvantage is the existence of noise sidebands on the transmitter signal which reduce sensitivity because the Doppler frequencies are relatively close to the carrier. This is considered in more detail below.

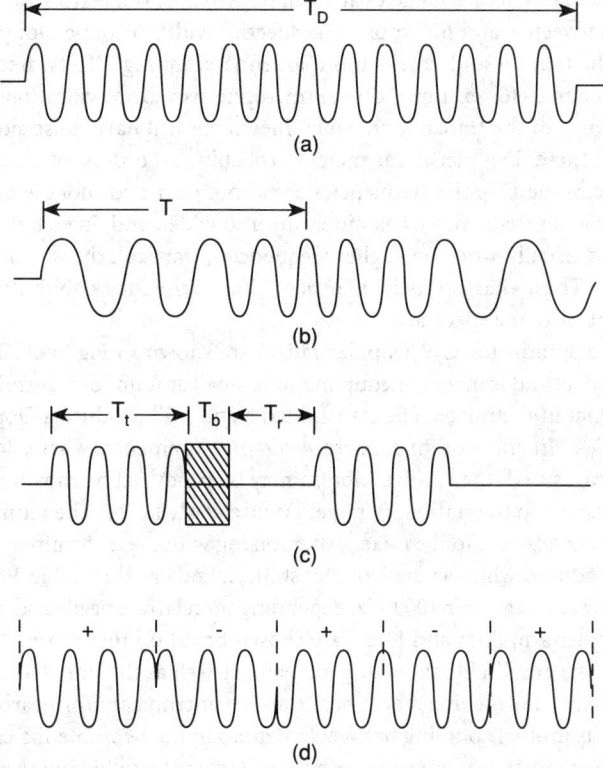

FIGURE 39.7 Waveforms for the general class of CW radar: (a) continuous sine wave CW; (b) frequency modulated CW; (c) interrupted CW; (d) binary phase-coded CW. (*Source:* F. E. Nathanson, *Radar Design Principles*, New York: McGraw-Hill, 1991, p. 450. With permission.)

CW Doppler Radar

If a sine wave signal were transmitted, the return from a moving target would be Doppler-shifted in frequency by an amount given by the following equation:

$$f_d = \frac{2v_r f_T}{c} = \text{Doppler frequency} \tag{39.10}$$

where f_T = transmitted frequency; c = velocity of propagation, 3×10^8 m/s; and v_r = radial component of velocity between radar and target.

Using Eq. (39.10) the Doppler frequencies have been calculated for several speeds and are given in Table 39.8.

Table 39.8 Doppler Frequencies for Several Transmitted Frequencies and Various Relative Speeds

Microwave Frequency—f_T	Relative Speed			
	1 m/s	300 m/s	1 mph	600 mph
3 GHz	20 Hz	6 kHz	8.9 Hz	5.4 kHz
10 GHz	67 Hz	20 kHz	30 Hz	17.9 kHz
35 GHz	233 Hz	70 kHz	104 Hz	63 kHz
95 GHz	633 Hz	190 kHz	283 Hz	170 kHz

As may be seen, the Doppler frequencies at 10 GHz (X-band) range from 30 Hz to about 18 kHz for a speed range between 1 and 600 mph. The spectral width of these Doppler frequencies will depend on target fluctuation and acceleration, antenna scanning effects, frequency variation in oscillators or components (for example, due to microphonism from vibrations), but most significantly, by the spectrum of the transmitter, which inevitably will have noise sidebands that extend much higher than these Doppler frequencies, probably by orders of magnitude. At higher microwave frequencies the Doppler frequencies are also higher and more widely spread. In addition, the spectra of higher frequency transmitters are also wider, and, in fact, the transmitter noise-sideband problem is usually worse at higher frequencies, particularly at millimeter wavelengths (i.e., above 30 GHz). These characteristics may necessitate frequency stabilization or phase locking of transmitters to improve the spectra.

Simplified block diagrams for CW Doppler radars are shown in Fig. 39.8. The transmitter is a single-frequency source, and leakage (or coupling) of a small amount of transmitter power serves as a local oscillator signal in the mixer. The transmitted signal will produce a Doppler-shifted return from a moving target. In the case of scatterometer measurements, where, for example, terrain reflectivity is to be measured, the relative motion may be produced by moving the radar (perhaps on a vehicle) with respect to the stationary target [Wiltse *et al.*, 1957]. The return signal is collected by the antenna and then also fed to the mixer. After mixing with the transmitter leakage, a difference frequency will be produced which is the Doppler shift. As indicated in Table 39.8, this difference is apt to range from low audio to over 100 kHz, depending on relative speeds and choice of microwave frequency. The Doppler amplifier and filters are chosen based on the information to be obtained, and this determines the amplifier bandwidth and gain, as well as the filter bandwidth and spacing. The transmitter leakage may include reflections from the antenna and/or nearby clutter in front of the antenna, as well as mutual coupling between antennas in the two-antenna case.

The detection range for such a radar can be obtained from the following [Nathanson, 1991]:

$$R^4 = \frac{\bar{P}_T G_T L_T A_e L_R L_p L_a L_s \delta_T}{(4\pi)^2 k T_s b(S/N)} \tag{39.11}$$

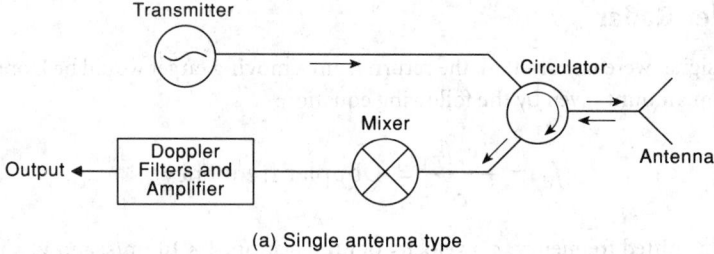

(a) Single antenna type

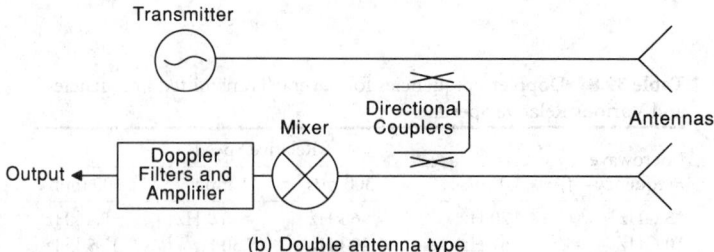

(b) Double antenna type

FIGURE 39.8 Block diagrams of CW-Doppler radar systems: (a) single antenna type; (b) double antenna type.

where $R =$ the detection range of the desired target.

$\bar{P}_T =$ the average power during the pulse.

$G_T =$ the transmit power gain of the antenna with respect to an omnidirectional radiator.

$L_T =$ the losses between the transmitter output and free space including power dividers, waveguide or coax, radomes, and any other losses not included in A_e.

$A_e =$ the effective aperture of the antenna, which is equal to the projected area in the direction of the target times the efficiency.

$L_R =$ the receive antenna losses defined in a manner similar to the transmit losses.

$L_p =$ the beam shape and scanning and pattern factor losses.

$L_a =$ the two-way-pattern propagation losses of the medium; often expressed as $\exp(-2 \propto R)$, where $\propto$ is the attenuation constant of the medium and the factor 2 is for a two-way path.

$L_s =$ signal-processing losses that occur for virtually every waveform and implementation.

$\delta_T =$ the radar cross-sectional area of the object that is being detected.

$k =$ Boltzmann's constant (1.38×10^{-23} W-s/K).

$T_s =$ system noise temperature.

$b =$ Doppler filter or *speedgate* bandwidth.

$S/N =$ signal-to-noise ratio.

$S_{\min} =$ the minimum detectable target-signal power that, with a given probability of success, the radar can be said to *detect, acquire,* or *track* in the presence of its own thermal noise or some external interference. Since all these factors (including the target return itself) are generally noiselike, the criterion for a detection can be described only by some form of probability distribution with an associated probability of detection P_D and a probability that, in the absence of a target signal, one or more noise or interference samples will be mistaken for the target of interest.

While the Doppler filter should be a matched filter, it usually is wider because it must include the target spectral width. There is usually some compensation for the loss in detectability by the use of post-detection filtering or integration. The S/N ratio for a CW radar must be at least 6 dB, compared with the value of 13 dB required with pulse radars when detecting steady targets [Nathanson, 1991, p. 449].

The Doppler system discussed above has a maximum detection range based on signal strength and other factors, but it cannot measure range. The rate of change in signal strength as a function of range has sometimes been used in fuzes to estimate range closure and firing point, but this is a relative measure.

FM/CW Radar

The most common technique for determining target range is the use of frequency modulation. Typical modulation waveforms include sinusoidal, linear sawtooth, or triangular, as illustrated in Fig. 39.9. For a linear sawtooth, a frequency increasing with time may be transmitted. Upon being reflected from a stationary point target, the same linear frequency change is reflected back to the receiver, except it has a time delay which is related to the range to the target. The time is $T = (2R)/c$, where R is the range. The received signal is mixed with the transmit signal, and the difference or beat frequency (F_b) is obtained. (The sum frequency is much higher and is rejected by filtering.) For a stationary target this is given by

$$F_b = \frac{4R}{c} \cdot \Delta F \cdot F_m \tag{39.12}$$

where $\Delta F =$ frequency deviation and $F_m =$ modulation rate.

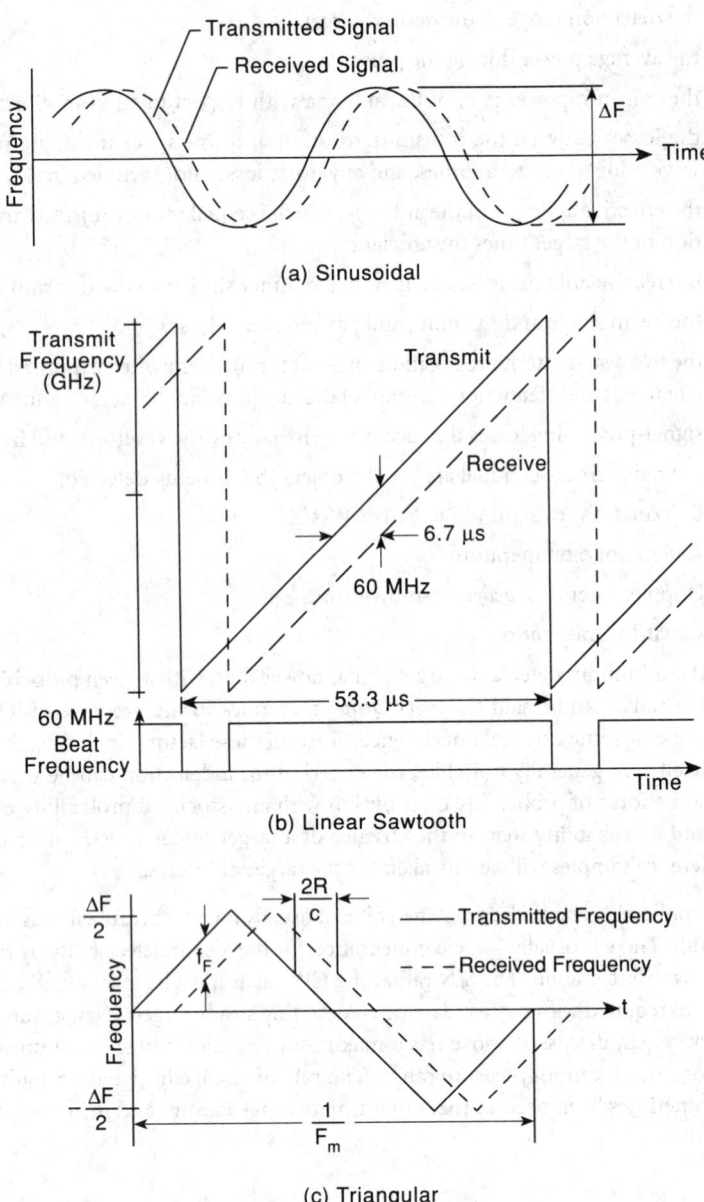

FIGURE 39.9   Frequency vs. time waveforms for FM/CW radar: (a) sinusoidal, (b) linear sawtooth, (c) triangular modulations.

The beat frequency is constant except near the turn-around region of the sawtooth, but, of course, it is different for targets at different ranges. (If it is desired to have a constant intermediate frequency for different ranges, which is a convenience in receiver design, then the modulation rate or the frequency deviation must be adjusted.) Multiple targets at a variety of ranges will produce multiple-frequency outputs from the mixer and frequently are handled in the receiver by using multiple range-bin filters.

If the target is moving with a component of velocity toward (or away) from the radar, then there will be a Doppler frequency component added to (or subtracted from) the difference frequency

(F_b), and the Doppler will be slightly higher at the upper end of the sweep range than at the lower end. This will introduce an uncertainty or ambiguity in the measurement of range, which may or may not be significant, depending on the parameters chosen and the application. For example, if the Doppler frequency is low (as in an altimeter) and/or the difference frequency is high, the error in range measurement may be tolerable. For the symmetrical triangular waveform, a Doppler less than F_b averages out, since it is higher on one-half of a cycle and lower on the other half. With a sawtooth modulation, only a decrease or increase is noted, since the frequencies produced in the transient during a rapid flyback are out of the receiver passband. Exact analyses of triangular, sawtooth, dual triangular, dual sawtooth, and combinations of these with noise have been carried out by Tozzi [1972]. Specific design parameters are given later in this chapter for an application utilizing sawtooth modulation in a **missile terminal guidance seeker**.

For the case of sinusoidal frequency modulation the spectrum consists of a series of lines spaced away from the carrier by the modulating frequency or its harmonics. The amplitudes of the carrier and these sidebands are proportional to the values of the Bessel functions of the first kind (J_n, $n = 0, \ldots 1, \ldots 2, \ldots 3, \ldots$), whose argument is a function of the modulating frequency and range. By choosing a particular modulating frequency, the values of the Bessel functions and thus the characteristics of the spectral components can be influenced. For instance, the signal variation with range at selected ranges can be optimized, which is important in fuzes. A short-range dependence that produces a rapid increase in signal, greater than that corresponding to the normal range variation, is beneficial in producing well-defined firing signals. This can be accomplished by proper choice of modulating frequency and filtering to obtain the signal spectral components corresponding to the appropriate order of the Bessel function. In a similar fashion, spillover and/or reflections from close-in objects can be reduced by filtering to pass only certain harmonics of the modulating frequency (F_m). Receiving only frequencies near $3F_m$ results in considerable spillover rejection, but at a penalty of 4 to 10 dB in signal-to-noise [Nathanson, 1991].

For the sinusoidal modulation case, Doppler frequency contributions complicate the analysis considerably. For details of this analysis the reader is referred to Saunders [1990] or Nathanson [1991].

Interrupted Frequency-Modulated CW (IFM/CW)

To improve isolation during reception, the IFM/CW format involves preventing transmission for a portion of the time during the frequency change. Thus, there are frequency gaps, or interruptions, as illustrated in Fig. 39.10. This shows a case where the transmit time equals the round-trip propagation time, followed by an equal time for reception. This duty factor of 0.5 for the waveform reduces the average transmitted power by 3 dB relative to using an interrupted transmitter. However, the improvement in the isolation should reduce the system noise by more than 3 dB, thus improving the signal-to-noise ratio [Piper, 1987]. For operation at short range, Piper points out that a high-speed switch is required [1987]. He also points out that the ratio of frequency deviation to beat frequency should be an even integer and that the minimum ratio is typically 6, which produces an out-of-band loss of 0.8 dB.

IFM/CW may be compared with pulse compression radar if both use a wide bandwidth. Pulse compression employs a "long" pulse (i.e., relatively long for a pulse radar) with a large frequency deviation or "chirp." A long pulse is often used when a transmitter is peak-power limited, because the longer pulse produces more energy and gives more range to targets. The frequency deviation is controlled in a predetermined way (frequently a linear sweep) so that a matched filter can be used in the receiver. The large time-bandwidth product permits the received pulse to be compressed in time to a short pulse in order to make an accurate range measurement. A linear-sawtooth IFM/CW having similar pulse length, frequency deviation, and pulse repetition rate would thus appear similar, although arrived at from different points of view.

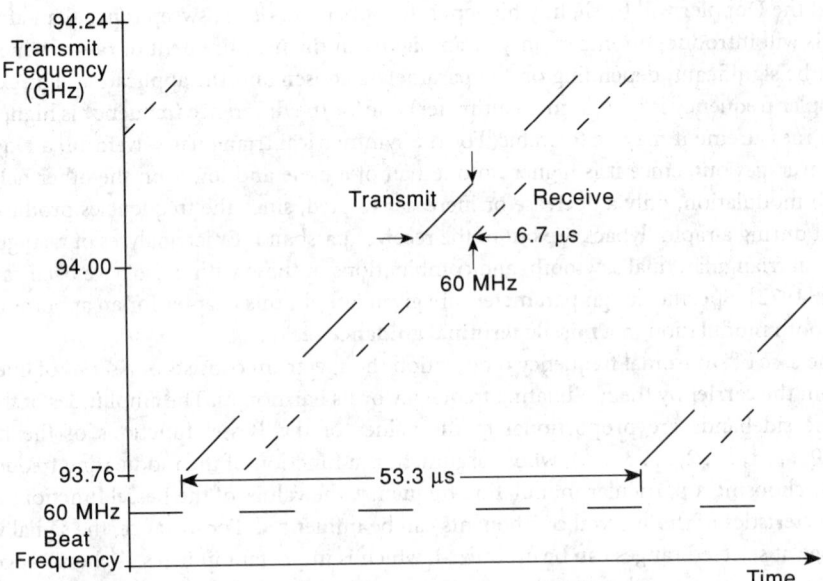

FIGURE 39.10 Interrupted FM/CW waveform. (*Source:* S.O. Piper, "MMW seekers," in *Principles and Applications of Millimeter Wave Radar*, N. Currie and C. E. Brown, Eds., Norwood, Mass.: Artech House, 1987, p. 683. With permission.)

Applications

Space does not permit giving a full description of the many applications mentioned at the beginning of this chapter, but several will be discussed.

Radar Proximity Fuzes

Projectiles or missiles designed to be aimed at ships or surface land targets often need a height-of-burst (HOB) sensor (or target detection device) to fire or fuze the warhead at a height of a few meters. There are two primary generic methods of sensing or measuring height to generate the warhead fire signal. The most obvious, and potentially the most accurate, is to measure target round trip propagation delay employing conventional radar ranging techniques. The second method employs a simple CW Doppler radar or variation thereof, with loop gain calibrated in a manner that permits sensing the desired burst height by measurement of target return signal amplitude and/or rate of change. Often the mission requirements do not justify the complexity and cost of the radar ranging approach. Viable candidates are thus narrowed down to variations on the CW doppler fuze.

In its simplest form, the CW Doppler fuze consists of a fractional watt RF oscillator, homodyne detector, Doppler amplifier, Doppler envelope detector, and threshold circuit. When the Doppler envelope amplitude derived from the returned signal reaches the preset threshold, a fire signal is generated. The height at which the fire signal occurs depends on the radar loop gain, threshold level, and target reflectivity. Fuze gain is designed to produce the desired height of burst under nominal trajectory angle and target reflectivity conditions, which may have large fluctuations due to glint effects, and deviations from the desired height due to antenna gain variations with angle, target reflectivity, and fuze gain tolerances are accepted. A loop gain change of 6 dB (2 to 1 in voltage), whether due to a change in target reflection coefficient, antenna gain, or whatever, will result in a 2 to 1 HOB change.

HOB sensitivity to loop gain factors can be reduced by utilizing the slope of the increasing return signal, or so-called rate-of-rise. Deriving HOB solely from the rate-of-rise has the disadvantage of rendering the fuze sensitive to fluctuating signal levels such as might result from a scintillating target. The use of logarithmic amplifiers decreases the HOB sensitivity to the reflectivity range. An early (excessively high) fire signal can occur if the slope of the signal fluctuations equals the rate-of-rise threshold of the fuze. In practice a compromise is generally made in which Doppler envelope amplitude and rate-of-rise contribute in some proportion of HOB.

Another method sometimes employed to reduce HOB sensitivity to fuze loop gain factors and angle of fall is the use of FM sinusoidal modulation of suitable deviation to produce a range correlation function comprising the zero order of a Bessel function of the first kind. The subject of sinusoidal modulation is quite complex, but has been treated in detail by Saunders [1990, pp. 1422–1446 and 144.41]. The most important aspects of fuze design have to do with practical problems such as low cost, small size, ability to stand very high-g accelerations, long life in storage, and countermeasures susceptibility.

Police Radars

Down-the-road police radars, which are of the CW Doppler type, operate at 10.525 or 24.150 GHz, frequencies approved in the United States by the Federal Communications Commission. Half-power beamwidths are typically in the 0.21 to 0.31 radian range. The sensitivity is usually good enough to provide a range exceeding 800 meters. Target size has a dynamic range of 30 dB (from smallest cars or motorcycles to large trucks). This means that a large target can be seen well outside the antenna 3-dB point at a range exceeding the range of a smaller target near the center of the beam. Thus there can be uncertainty about which vehicle is the target. Fisher [1992] has given a discussion of a number of the limitations of these systems, but in spite of these factors probably a hundred thousand have been built.

The designs typically have three amplifier gains for detection of short, medium, or maximum range targets, plus a squelch circuit so that sudden spurious signals will not be counted. The Doppler signal is integrated and this direct current provides a speed readout. Provision is made for calibration to assure the accuracy of the readings.

Altimeters

A very detailed discussion of FM/CW altimeters has been given by Saunders [1990, pp. 14.34–14.36], in which he has described modern commercial products built by Bendix and Collins. The parameters will be summarized below and if more information is needed, the reader may want to turn to other references [Saunders, 1990; Bendix Corp., 1982; and Maoz *et al.*, 1991]. In his material, Saunders gives a general overview of modern altimeters, all of which use wide-deviation FM at a low modulation frequency. He discusses the limitations on narrowing the antenna pattern, which must be wide enough to accommodate attitude changes of the aircraft. Triangular modulation is used, since for this waveform the Doppler averages out, and dual antennas are employed. There may be a step error or quantization in height (which could be a problem at low altitudes), due to the limitation of counting zero crossings. A difference of one zero crossing (i.e., 1/2 Hz) corresponds to 3/4 meter for a frequency deviation of 100 MHz. Irregularities are not often seen, however, since meter response is slow. Also, if terrain is rough, there will be actual physical altitude fluctuations. Table 39.9 shows some of the altimeters' parameters. These altimeters are not acceptable for military aircraft, because their relatively wide-open front ends make them potentially vulnerable to electronic countermeasures. A French design has some advantages in this respect by using a variable frequency deviation, a difference frequency that is essentially constant with altitude, and a narrowband front-end amplifier [Saunders, 1990].

Doppler Navigators

These systems are mainly sinusoidally modulated FM/CW radars employing four separate downward looking beams aimed at about 15 degrees off the vertical. Because commercial airlines have shifted to nonradar forms of navigation, these units are designed principally for helicopters. Saun-

Table 39.9 Parameters for Two Commercial Altimeters

Modulation Frequency	Frequency Deviation	Prime Power	Weight (pounds)	Radiated Power
Bendix ALA-52A	150 Hz	130 MHz	30 W	11*
Collins ALT-55	100 kHz	100 MHz	8	350 mW

* Not including antenna and indicator.

ders [1990] cites a particular example of a commercial unit operating at 13.3 GHz, employing a Gunn oscillator as the transmitter, with an output power of 50 mW, and utilizing a 30-kHz modulation frequency. A single microstrip antenna is used. A low-altitude equipment (below 15,000 feet), the unit weighs less than 12 pounds. A second unit cited has an output power of 300 mW, dual antennas, dual modulating frequencies, and an altitude capability of 40,000 feet.

Millimeter-Wave Seeker for Terminal Guidance Missile

Terminal guidance for short-range (less than 2 km) air-to-surface missiles has seen extensive development in the last decade. Targets such as tanks are frequently immersed in a clutter background which may give a radar return that is comparable to that of the target. To reduce the clutter return in the antenna footprint, the antenna beamwidth is reduced by going to millimeter wavelengths. For a variety of reasons the choice is usually a frequency near 35 or 90 GHz. Antenna beamwidth is inversely proportional to frequency, so in order to get a reduced beamwidth we would normally choose 90 GHz; however, more deleterious effects at 90 GHz due to atmospheric absorption and scattering can modify that choice. In spite of small beamwidths, the clutter is a significant problem, and in most cases signal-to-clutter is a more limiting condition than signal-to-noise in determining range performance. Piper [1987] has done an excellent job of analyzing the situation for 35- and 90-GHz pulse radar seekers and comparing those with a 90-GHz FM/CW seeker. His FM/CW results will be summarized below.

In his approach to the problem, Piper gives a summary of the advantages and disadvantages of a pulse system compared to the FM/CW approach. Most of these have already been covered in earlier sections, but one difficulty for the FM/CW can be emphasized again. That is the need for a highly linear sweep, and, because of the desire for the wide bandwidth, this requirement is accentuated. The wide bandwidth is desired in order to average the clutter return and to smooth the glint effects. In particular, glint occurs from a complex target because of the vector addition of coherent signals scattered back to the receiver from various reflecting surfaces. At some angles the vectors may add in phase (constructively) and at others they may cancel, and the effect is specifically dependent on wavelength. For a narrowband system, glint may provide a very large signal change over a small variation of angle, but, of course, at another wavelength it would be different. Thus, very wide bandwidth is desirable from this smoothing point of view, and typical numbers used in millimeter-wave radars are in the 450- to 650-MHz range. Piper chose 480 MHz.

Another tradeoff involves the choice of FM waveform. Here the use of a triangular waveform is undesirable because the Doppler frequency averages out and Doppler compensation is then required. Thus the sawtooth version is chosen, but because of the large frequency deviation desired, the difficulty of linearizing the frequency sweep is made greater. In fact many components must be extremely wideband, and this generally increases cost and may adversely affect performance. On the other hand, the difference frequency (F_b) and/or the intermediate frequency (F_{IF}) will be higher and thus further from the carrier, so the phase noise will be lower. After discussing the other tradeoffs, Piper chose 60 MHz for the beat frequency.

With a linear FM/CW waveform, the inverse of the frequency deviation provides the theoretical time resolution, which is 2.1 ns for 480 MHz (or range resolution of 0.3 meter). For an RF sweep linearity of 300 kHz, the range resolution is actually 5 meters at the 1000-meter nominal search range. (The system has a mechanically scanned antenna.) An average transmitting power of 25 mW was chosen, which was equal to the average power of the 5-W peak IMPATT assumed for the pulse system. The antenna diameter was 15 cm. For a target radar cross section of 20 m² and assumed weather conditions, the signal-to-clutter and signal-to-noise ratios were calculated and plotted for ranges out to 2 km and for clear weather or 4 mm per hour rainfall. The results show that for 1 km range the target-to-clutter ratios are higher for the FM/CW case than the pulse system in clear weather or in rain, and target-to-clutter is the determining factor.

Summary Comments

From this brief review it is clear that there are many uses for CW radars, and various types (such as fuzes) have been produced in large quantities. Because of their relative simplicity, today there are continuing trends toward the use of digital processing and integrated circuits. In fact, this is exemplified in an article describing an FM/CW radar built on a single microwave integrated circuit chip [Maoz *et al.*, 1991].

Defining Terms

Doppler-frequency shift: The observed frequency change between the transmitted and received signal produced by motion along a line between the transmitter/receiver and the target. The frequency increases if the two are closing and decreases if they are receding.

Missile terminal guidance seeker: Located in the nose of a missile, a small radar with short-range capability which scans the area ahead of the missile and guides it during the terminal phase toward a target such as a tank.

Pulse Doppler: A coherent radar, usually having high pulse repetition rate and duty cycle and capable of measuring the Doppler frequency from a moving target. Has good clutter suppression and thus can see a moving target in spite of background reflections.

References

Bendix Corporation, Service Manual for ALA-52A Altimeter; Design Summary for the ALA-52A, Bendix Corporation, Ft. Lauderdale, Fla., May 1982.

Collins (Rockwell International), ALT-55 Radio Altimeter System; Instruction Book, Cedar Rapids, Iowa, October 1984.

P. D. Fisher, "Improving on police radar," *IEEE Spectrum*, vol. 29, pp. 38–43, July 1992.

A. J. Freiley, B. L. Conroy, D. J. Hoppe, and A. M. Bhanji, "Design concepts of a 1-MW CW X-band transmit/receive system for planetary radar," *IEEE Transactions on Microwave Theory and Techniques*, vol. 40, pp. 1047–1055, June 1992.

B. Maoz, L. R. Reynolds, A. Oki, and M. Kumar, "FM-CW radar on a single GaAs/AlGaAs HBT MMIC chip," *IEEE Microwave and Millimeter-Wave Monolithic Circuits Symposium Digest*, pp. 3–6, June 1991.

F. E. Nathanson, *Radar Design Principles*, New York: McGraw-Hill, 1991, pp. 448–467.

S. O. Piper, "MMW seekers," in *Principles and Applications of Millimeter Wave Radar*, N. C. Currie and C. E. Brown, Eds., Norwood, Mass.: Artech House, 1987, chap. 14.

W. K. Saunders, "CW and FM radar," in *Radar Handbook*, M. I. Skolnik, Ed., New York: McGraw-Hill, 1990, chap. 14.

L. M. Tozzi, "Resolution in frequency-modulated radars," Ph.D. thesis, University of Maryland, College Park, 1972.

F. T. Ulaby and C. Elachi, *Radar Polarimetry for Geoscience Applications*, Norwood, Mass.: Artech House, 1990, pp. 193–200.

J. C. Wiltse, S. P. Schlesinger, and C. M. Johnson, "Back-scattering characteristics of the sea in the region from 10 to 50 GHz," *Proceedings of the IRE*, vol. 45, pp. 220–228, February 1957.

Further Information

For a general treatment, including analysis of clutter effects, Nathanson's [1991] book is very good and generally easy to read. For extensive detail and specific numbers in various actual cases, Saunders [1990] gives good coverage. The treatment of millimeter-wave seekers by Piper [1987] is excellent, both comprehensive and easy to read.

Maxwell in 1855 as a student at Cambridge University, England. (Courtesy of Burndy Library.)

40

Lightwave

Samuel O. Agbo
California Polytechnic State University

Allen H. Cherin
AT&T Bell Laboratories

Basant K. Tariyal
AT&T

40.1 Lightwave Waveguides 975
Ray Theory • Wave Equation for Dielectric Materials • Modes in Slab Waveguides • Fields in Cylindrical Fibers • Modes in Step-Index Fibers • Modes in Graded-Index Fibers • Attenuation • Dispersion and Pulse Spreading

40.2 Optical Fibers and Cables 987
Introduction • Classification of Optical Fibers and Attractive Features • Fiber Transmission Characteristics • Optical Fiber Cable Manufacturing

40.1 Lightwave Waveguides

Samuel O. Agbo

Lightwave waveguides fall into two broad categories: dielectric slab waveguides and optical fibers. As illustrated in Fig. 40.1, slab waveguides generally consist of a middle layer (the film) of **refractive index** n_1 and lower and upper layers of refractive indices n_2 and n_3, respectively.

Optical fibers are slender glass or plastic cylinders with annular cross sections. The core has a refractive index, n_1, which is greater than the refractive index, n_2, of the annular region (the cladding). Light propagation is confined to the core by total internal reflection, even when the fiber is bent into curves and loops. Optical fibers fall into two main categories: step-index and graded-index (GRIN) fibers. For step-index fibers, the refractive index is constant within the core. For GRIN fibers, the refractive index is a function of radius r given by

$$n(r) = \begin{cases} n_1\left[1 - 2\Delta\left(\dfrac{r}{a}\right)^{\alpha}\right]^{1/2} & ; r < a \\ n_1(1 - 2\Delta)^{1/2} = n_2 & ; a < r \end{cases} \qquad (40.1)$$

In Eq. (40.1), Δ is the **relative refractive index difference**, a is the core radius, and α defines the type of graded-index profile. For triangular, parabolic, and step-index profiles, α is, respectively, 1, 2, and ∞. Figure 40.2 shows the raypaths in step-index and graded-index fibers and the cylindrical coordinate system used in the analysis of lightwave propagation through fibers. Because rays propagating within the core in a GRIN fiber undergo progressive refraction, the raypaths are curved (sinusoidal in the case of parabolic profile).

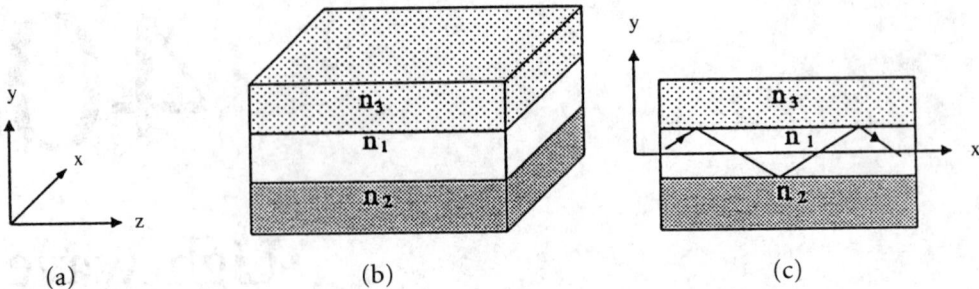

FIGURE 40.1 Dielectric slab waveguide: (a) the Cartesian coordinates used in analysis of slab waveguides; (b) the slab waveguide; (c) light guiding in a slab waveguide.

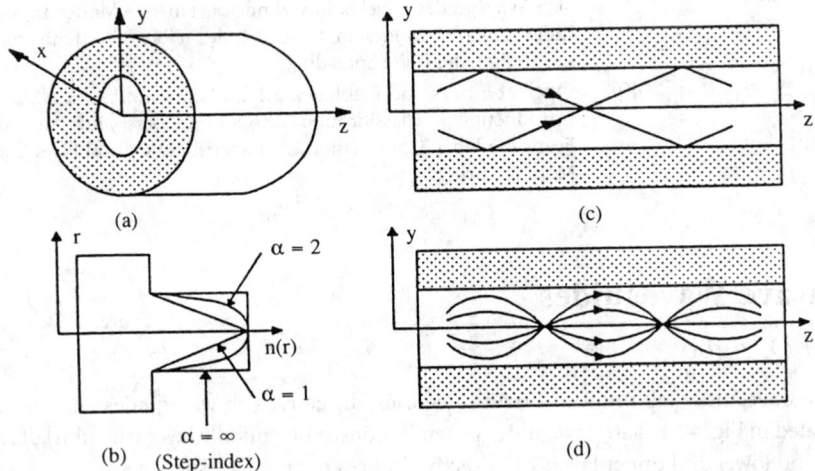

FIGURE 40.2 The optical fiber: (a) the cylindrical coordinate system used in analysis of optical fibers; (b) some graded-index profiles; (c) raypaths in step-index fiber; (d) raypaths in graded-index fiber.

Ray Theory

Consider Fig 40.3, which shows possible raypaths for light coupled from air (refractive index n_0) into the film of a slab waveguide or the core of a step-index fiber. At each interface, the transmitted raypath is governed by Snell's law. As θ_0 (the acceptance angle from air into the waveguide) decreases, the angle of incidence θ_i increases until it equals the critical angle, θ_c, making θ_0 equal to the maximum acceptance angle, θ_a. According to ray theory, all rays with acceptance angles less than θ_a propagate in the waveguide by total internal reflections. Hence, the numerical aperture (NA) for the waveguide, a measure of its light-gathering ability, is given by

$$NA = n_0 \sin \theta a = n_1 \sin\left(\frac{\pi}{2} - \theta_c\right) \tag{40.2}$$

By Snell's law, $\sin \theta_c = n_2/n_1$. Hence,

$$NA = \left[n_1^2 - n_2^2\right]^{1/2} \tag{40.3}$$

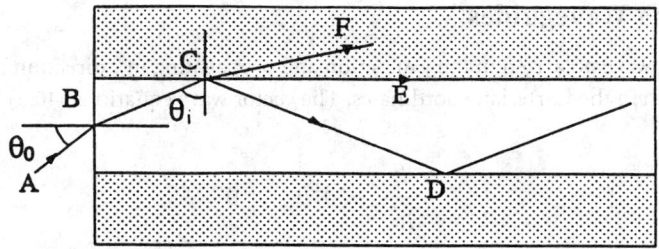

FIGURE 40.3 Possible raypaths for light coupled from air into a slab waveguide or a step-index fiber.

For step-index fibers, the preceding analysis applies to **meridional rays**. Skew (nonmeridional) rays have larger maximum acceptance angles, θ_{as}, given by

$$\sin \theta_{as} = \frac{NA}{\cos \gamma} \tag{40.4}$$

where NA is the numerical aperture for meridional rays and γ is the angle between the core radius and the projection of the ray onto a plane normal to the fiber axis.

Wave Equation for Dielectric Materials

Only certain discrete angles, instead of all acceptance angles less than the maximum acceptance angle, lead to guided propagation in lightwave waveguides. Hence, ray theory is inadequate, and wave theory is necessary, for analysis of light propagation in optical waveguides.

For lightwave propagation in an unbounded dielectric medium, the assumption of a linear, homogeneous, charge-free, and nonconducting medium is appropriate. Assuming also sinusoidal time dependence of the fields, the applicable Maxwell's equations are

$$\nabla \times \mathbf{E} = -j\omega\mu\mathbf{H} \tag{40.5a}$$

$$\nabla \times \mathbf{H} = j\omega\epsilon\mathbf{E} \tag{40.5b}$$

$$\nabla \times \mathbf{E} = 0 \tag{40.5c}$$

$$\nabla \times \mathbf{H} = 0 \tag{40.5d}$$

The resulting wave equations are

$$\nabla^2\mathbf{E} - \gamma^2\mathbf{E} = 0 \tag{40.6a}$$

$$\nabla^2\mathbf{H} - \gamma^2\mathbf{H} = 0 \tag{40.6b}$$

where

$$\gamma^2 = \omega^2\mu\epsilon = (j\kappa)^2 \tag{40.7}$$

and

$$\kappa = n\kappa_0 = \omega\sqrt{\mu\epsilon} = \frac{\omega}{v} \tag{40.8}$$

In Eq. (40.8) κ is the phase propagation constant and n is the refractive index for the medium, while κ_0 is the phase propagation constant for free space. The velocity of propagation in the medium is $v = 1/\sqrt{\mu\epsilon}$.

Modes in Slab Waveguides

Consider a plane wave polarized in the y direction and propagating in z direction in an unbounded dielectric medium in the Cartesian coordinates. The vector wave equations (40.6) lead to the scalar equations:

$$\frac{\partial^2 E_y}{\partial z} - \partial^2 E_y = 0 \tag{40.9a}$$

$$\frac{\partial^2 H_x}{\partial z} - \partial^2 H_x = 0 \tag{40.9b}$$

The solutions are

$$E_y = A e^{j(\omega t - \kappa z)} \tag{40.10a}$$

$$H_x = \frac{-E_y}{\eta} = \frac{A}{\eta} e^{j(\omega t - \kappa z)} \tag{40.10b}$$

where A is a constant and $\eta = \sqrt{\mu \epsilon}$ is the intrinsic impedance of the medium.

Because the film is bounded by the upper and lower layers, the rays follow the zigzag paths as shown in Fig. 40.3. The upward and downward traveling waves interfere to create a standing wave pattern. Within the film, the fields transverse to the z axis, which have even and odd symmetry about the x axis, are given, respectively, by

$$E_y = A \cos(hy) e^{j(\omega t - \beta z)} \tag{40.11a}$$

$$E_y = A \sin(hy) e^{j(\omega t - \beta z)} \tag{40.11b}$$

where β and h are the components of κ parallel to and normal to the z axis, respectively. The fields in the upper and lower layers are evanescent fields decaying rapidly with attenuation factors α_3 and α_2, respectively, and are given by

$$E_y = A_3 e^{-\alpha_3 \left(y - \frac{d}{2}\right)} e^{j(\omega t - \beta z)} \tag{40.12a}$$

$$E_y = A_2 e^{-\alpha_2 \left(y + \frac{d}{2}\right)} e^{j(\omega t - \beta z)} \tag{40.12b}$$

Only waves with raypaths for which the total phase change for a complete (up and down) zigzag path is an integral multiple of 2π undergo constructive interference, resulting in guided modes. Waves with raypaths not satisfying this mode condition interfere destructively and die out rapidly. In terms of a raypath with an angle of incidence $\theta_i = \theta$ in Fig. 40.3, the mode conditions [Haus, 1984] for fields transverse to the z axis and with even and odd symmetry about the x axis are given, respectively, by

$$\tan\left(\frac{hd}{2}\right) = \frac{1}{n_1 \cos \theta} \left[n_1^2 \sin^2 \theta - n_2^2\right]^{1/2} \tag{40.13a}$$

$$\tan\left(\frac{hd}{2} - \frac{\pi}{2}\right) = \frac{1}{n_1 \cos \theta} \left[n_1^2 \sin^2 \theta - n_2^2\right]^{1/2} \tag{40.13b}$$

where $h = \kappa \cos \theta = (2\pi n_1 / \lambda) \cos \theta$ and λ is the free space wavelength.

Equations (40.13a) and (40.13b) are transcendental, have multiple solutions, and are better solved graphically. Let $(d/\lambda)_0$ denote the smallest value of d/λ, the film thickness normalized with respect to the wavelength, satisfying Eqs. (40.13a) and (40.13b). Other solutions for both even and odd modes are given by

$$\left(\frac{d}{\lambda}\right)_m = \left(\frac{d}{\lambda}\right)_0 + \frac{m}{2n_1 \cos \theta} \tag{40.14}$$

where m is a nonnegative integer denoting the order of the mode.

Figure 40.4 [Palais, 1992] shows a **mode chart** for a symmetrical slab waveguide obtained by solving Eqs. (40.13a) and (40.13b). For the TE_m modes, the E field is transverse to the direction (z) of propagation, while the H field lies in a plane parallel to the z axis. For the TM_m modes, the reverse is the case. The highest-order mode that can propagate has a value m given by the integer part of

$$m = \frac{2d}{\lambda} \left[n_1^2 - n_2^2 \right]^{1/2} \tag{40.15}$$

To obtain a single-mode waveguide, d/λ should be smaller than the value required for $m = 1$, so that only the $m = 0$ mode is supported. To obtain a multimode waveguide, d/λ should be large enough to support many modes.

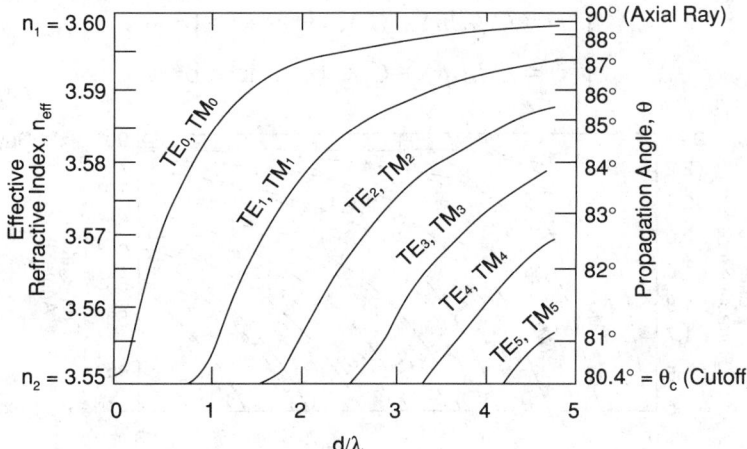

FIGURE 40.4 Mode chart for the symmetric slab waveguide with $n_1 = 3.6$, $n_2 = 3.55$. (*Source:* J.C. Palais, *Fiber Optic Communications*, Englewood Cliffs, N.J.: Prentice-Hall, 1992, p. 84. With permission.)

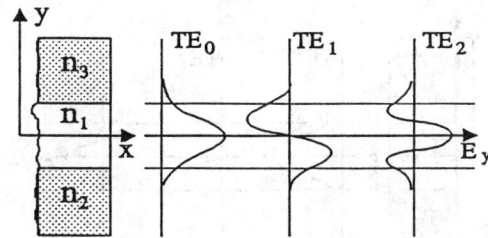

FIGURE 40.5 Transverse mode field patterns in the symmetric slab waveguide.

Shown in Fig. 40.5 are transverse mode patterns for the electric field in a symmetrical slab waveguide. These are graphical illustrations of the fields given by Eqs. (40.11) and (40.12). Note that, for TE_m, the field has m zeros in the film, and the evanescent field penetrates more deeply into the upper and lower layers for high-order modes.

For asymmetric slab waveguides, the equations and their solutions are more complex than those for symmetric slab waveguides. Shown in Fig. 40.6 [Palais, 1992] is the mode chart for the asymmetric slab waveguide. Note that the TE_m and TM_m modes in this case have different propagation constants and do not overlap. By contrast, for the symmetric case, TE_m and TM_m modes are degenerate, having the same propagation constant and forming effectively one mode for each value of m.

Figure 40.7 shows typical mode patterns in the asymmetric slab waveguide. Note that the asymmetry causes the evanescent fields to have unequal amplitudes at the two boundaries and to decay at different rates in the two outer layers.

Fields in Cylindrical Fibers

Let ψ represent E_z or H_z and β be the component of κ in z direction. In the cylindrical coordinates of Fig. 40.2, with wave propagation along the z axis, the wave equations (40.6) correspond to the scalar equation

$$\frac{\partial^2 \psi}{\partial r^2} + \frac{1}{r}\frac{\partial \psi}{\partial r} + \frac{1}{r^2}\frac{\partial^2 \psi}{\partial \Phi^2} + (\kappa^2 - \beta^2)\psi = 0 \qquad (40.16)$$

The general solution to the preceding equation is

$$\psi(r) = C_1 J_\ell(hr) + C_2 Y_\ell(hr); \quad \kappa^2 > \beta^2 \qquad (40.17a)$$

$$\psi(r) = C_1 I_\ell(qr) + C_2 K_\ell(qr); \quad \kappa^2 < \beta^2 \qquad (40.17b)$$

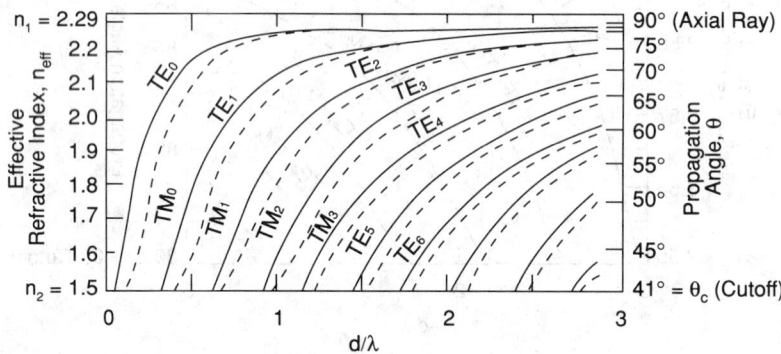

FIGURE 40.6 Mode chart for the asymmetric slab waveguide with $n_1 = 2.29$, $n_2 = 1.5$, and $n_3 = 1.0$. (*Source:* J. C. Palais, *Fiber Optic Communications*, Englewood Cliffs, N.J.: Prentice-Hall, 1992, p. 88. With permission.)

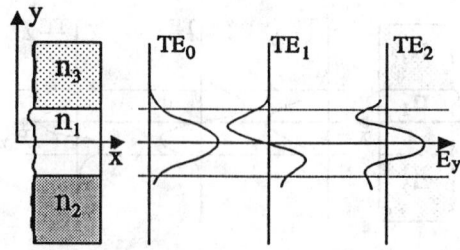

FIGURE 40.7 Transverse mode field patterns in the asymmetric slab waveguide.

In Eqs. (40.17) and (40.17b), J_ℓ and Y_ℓ are Bessel functions of the first kind and second kind, respectively, of order ℓ; I_ℓ and K_ℓ are modified Bessel functions of the first kind and second kind, respectively, of order ℓ; C_1 and C_2 are constants; $h^2 = \kappa^2 - \beta^2$ and $q^2 = \beta^2 - \kappa^2$.

E_z and H_z in a fiber core are given by Eq. (40.17a) or (40.17b), depending on the sign of $\kappa^2 - \beta^2$. For guided propagation in the core, this sign is negative to ensure that the field is evanescent in the cladding. One of the coefficients vanishes because of asymptotic behavior of the respective Bessel functions in the core or cladding. Thus, with A_1 and A_2 as arbitrary constants, the fields in the core and cladding are given, respectively, by

$$\psi(r) = A_1 J_\ell(hr) \tag{40.18a}$$

$$\psi(r) = A_2 \kappa_\ell(hr) \tag{40.18b}$$

Because of the cylindrical symmetry,

$$\psi(r,t) = \psi(r,\phi)e^{j(\omega t - \beta z)} \tag{40.19}$$

Thus, the usual approach is to solve for E_z and H_z and then express E_r, E_ϕ, H_r, and H_ϕ in terms of E_z and H_z.

Modes in Step-Index Fibers

Derivation of the exact modal field relations for optical fibers is complex. Fortunately, fibers used in optical communication satisfy the weekly guiding approximation in which the relative index difference, ∇, is much less than unity. In this approximation, application of the requirement for continuity of transverse and tangential electric field components at the core-cladding interface (at $r = a$) to Eqs. (40.18a) and (40.18b) results in the following eigenvalue equation [Snyder, 1969]:

$$haJ_{\ell \pm 1} \frac{(ha)}{J_\ell(ha)} = \pm \frac{qa\kappa_{\ell \pm 1}(qa)}{\kappa_\ell(qa)} \tag{40.20}$$

Let the normalized frequency V be defined as

$$V = a(q^2 + h^2)^{1/2} = a\kappa_0 \left(n_1^2 - n_2^2 \right)^{1/2} = \frac{2\pi}{\lambda} a(NA) \tag{40.21}$$

Solving Eq. (40.20) allows β to be calculated as a function of V. Guided modes propagating within the core correspond to $n_2\kappa_0 \le \beta \le n_1\kappa$. The normalized frequency V corresponding to $\beta = n_1\kappa$ is the cut-off frequency for the mode.

As with planar waveguides, TE ($E_z = 0$) and TM ($H_z = 0$) modes corresponding to meridional rays exist in the fiber. They are denoted by EH or HE modes, depending on which component, E or H, is stronger in the plane transverse to the direction or propagation. Because the cylindrical fiber is bounded in two dimensions rather than one, two integers, ℓ and m, are needed to specify the modes, unlike one integer, m, required for planar waveguides. The exact modes, $TE_{\ell m}$, $TM_{\ell m}$, $EH_{\ell m}$, and $HE_{\ell m}$, may be given by two linearly polarized modes, $LP_{\ell m}$. The subscript ℓ is now such that $LP_{\ell m}$ corresponds to $HE_{\ell + 1,m}$, $EH_{\ell - 1,m}$, $TE_{\ell - 1,m}$, and $TM_{\ell - 1,m}$. In general, there are 2ℓ field maxima around the fiber core circumference and m field maxima along a radius vector. Figure 40.8 illustrates the correspondence between the exact modes and the LP modes and their field configurations for the three lowest LP modes.

Figure 40.9 gives the mode chart for step-index fiber on a plot of the refractive index, β/κ_0, against the normalized frequency. Note that for a single-mode (LP_{01} or HE_{11}) fiber, $V < 2.405$. The number of modes supported as a function of V is given by

$$N = \frac{V^2}{2} \tag{40.22}$$

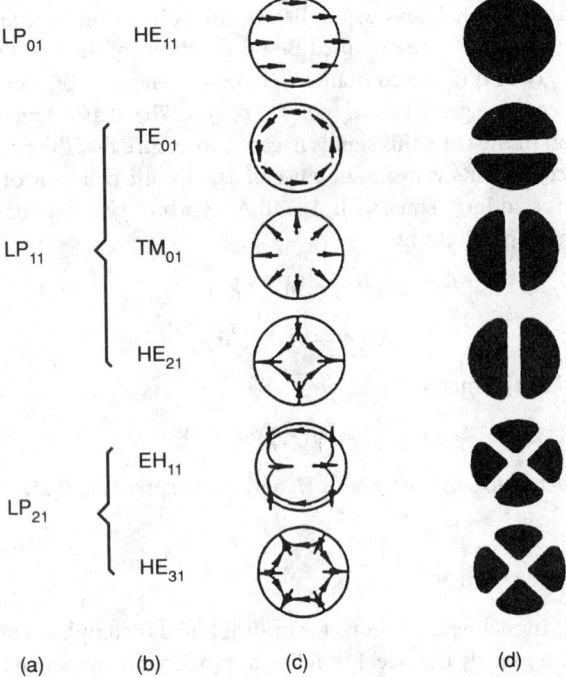

FIGURE 40.8 Transverse electric field patterns and field intensity distributions for the three lowest LP modes in a step-index fiber: (a) mode designations; (b) electric field patterns; (c) intensity distribution. (*Source:* J. M. Senior, *Optical Fiber Communications: Principles and Practice*, Englewood Cliffs, N.J.: Prentice-Hall, 1985, p. 36. With permission.)

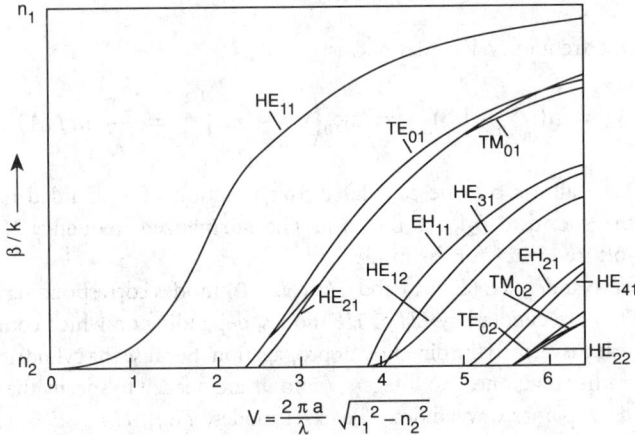

FIGURE 40.9 Mode chart for step-index fibers: $b = (\beta/\kappa_0 - n_2)/(n_1 - n_2)$ is the normalized propagation constant. (*Source:* D. B. Keck, *Fundamentals of Optical Fiber Communications*, M. K. Barnoski, Ed., New York: Academic Press, 1981, p. 13. With permission.)

Modes in Graded-Index Fibers

A rigorous modal analysis for optical fibers based on the solution of Maxwell's equations is possible only for step-index fiber. For graded-index fibers, approximate methods are used. The most widely used approximation is the WKB (Wenzel, Kramers, and Brillouin) method [Marcuse, 1982]. This method gives good modal solutions for graded-index fiber with arbitrary profiles, when the refractive index does not change appreciably over distances comparable to the guided wavelength [Yariv, 1991]. In this method, the transverse components of the fields are expressed as

$$E_t = \psi(r)e^{j\ell\phi}\,e^{j(\omega t - \beta z)} \tag{40.23}$$

$$H_t = \frac{\beta}{\omega\mu}\,E_t \tag{40.24}$$

In Eq. (40.23), ℓ is an integer. Equation (40.16), the scalar wave equation in cylindrical coordinates can now be written with $\kappa = n(r)\,\kappa_0$ as

$$\left[\frac{d^2}{dr^2} + \frac{1}{2}\frac{d}{dr} + p^2(r)\right]\psi(r) = 0 \tag{40.25}$$

where

$$p^2(r) = n^2(r)\kappa_0^2 - \frac{\ell^2}{r^2} - \beta^2 \tag{40.26}$$

Let r_1 and r_2 be roots of $p^2(r) = 0$ such that $r_1 < r_2$. A ray propagating in the core does not necessarily reach the core-cladding interface or the fiber axis. In general, it is confined to an annular cylinder bounded by the two caustic surfaces defined by r_1 and r_2. As illustrated in Fig. 40.10, the field is oscillatory within this annular cylinder and evanescent outside it. The fields obtained as solutions to Eq. (40.25) are

$$\psi(r) = \frac{A}{[rp(r)]^{1/2}}\,\exp\left[-\int_r^{r_1}|p(r)|\,dr\right];\, r < r_1 \tag{40.27a}$$

$$\psi(r) = \frac{B}{[rp(r)]^{1/2}}\,\sin\left[\int_{r_1}^r p(r)dr + \frac{\pi}{4}\right];\, r_1 < r \tag{40.27b}$$

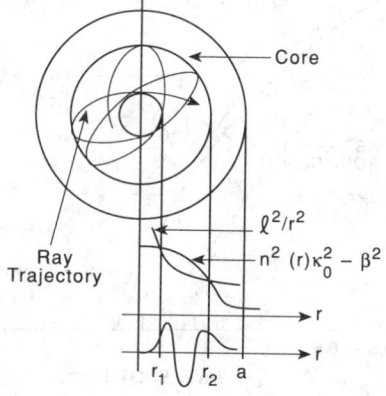

FIGURE 40.10 End view of a skew ray in a graded-index fiber, its graphical solution in the WKB method, and the resulting field that is oscillatory between r_1 and r_2 and evanescent outside that region.

$$\psi(r) = \frac{C}{[rp(r)]^{1/2}} \sin\left[\int_r^{r_2} p(r)dr + \frac{\pi}{4}\right]; r < r_2 \qquad (40.27c)$$

$$\psi(r) = \frac{D}{[rp(r)]^{1/2}} \exp\left[-\int_{r_2}^r |p(r)| dr\right]; r_2 < r \qquad (40.27d)$$

Equations (40.27b) and (40.27c) represent fields in the same region. Equating them leads to the mode condition:

$$\int_{r_1}^{r_2}\left[n^2(r)\kappa_0^2 - \frac{\ell^2}{r^2} - \beta^2\right]^{1/2} dr = (2M + 1)\frac{\pi}{2} \qquad (40.28)$$

In Eq. (40.28) ℓ and m are the integers denoting the modes. A closed analytical solution of this equation for β is possible only for a few simple graded-index profiles. For other cases, numerical or approximate methods are used.

Attenuation

The assumption of a nonconducting medium for dielectric waveguides led to solutions to the wave equation with no attenuation component. In practice, various mechanisms give rise to losses in lightwave waveguides. These mechanisms contribute a loss factor of $e^{-\alpha z}$ to Eq. (40.10) and comparable field expressions, where α is the attenuation coefficient. The attenuation due to these mechanisms and the resulting total attenuation as a function of wavelength is shown in Fig. 40.11 [Osanai *et al.*, 1976]. Note that the range of wavelengths (0.8 to 1.6 µm) in which communication fibers are usually operated corresponds to a region of low overall attenuation. Brief discussions follow of the mechanisms responsible for the various types of attenuation shown in Fig. 40.11.

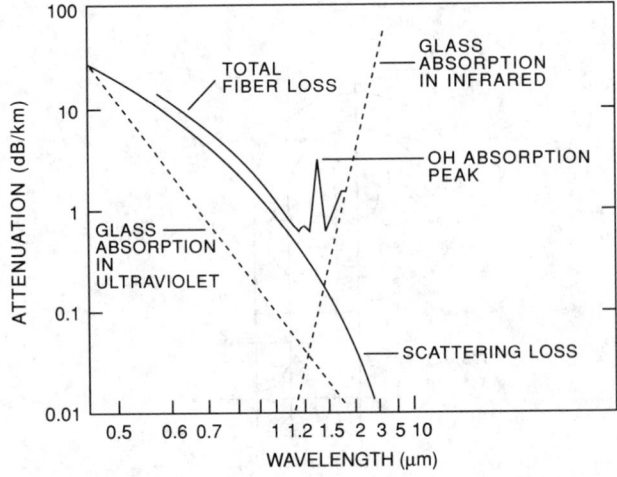

FIGURE 40.11   Attenuation of a germanium-doped low-loss silica glass fiber. (*Source:* H. Osanai *et al.*, "Effects of dopants on transmission loss of low-OH content optical fibers," *Electronic Letters*, vol. 12, no. 21, p. 550, 1976. With permission.)

Intrinsic Absorption

Intrinsic absorption is a natural property of glass. In the ultraviolet region, it is due to strong electronic and molecular transition bands. In the infrared region, it is caused by thermal vibration of chemical bonds.

Extrinsic Absorption

Extrinsic absorption is caused by metal (Cu, Fe, Ni, Mn, Co, V, Cr) ion impurities and hydroxyl (OH) ion impurity. Metal ion absorption involves electron transition from lower to higher energy states. OH absorption is caused by thermal vibration of the hydroxyl ion. Extrinsic absorption is strong in the range of normal fiber operation. Thus, it is important that impurity level be limited.

Rayleigh Scattering

Rayleigh scattering is caused by localized variations in refractive index in the dielectric medium, which are small relative to the optic wavelength. It is strong in the ultraviolet region. It increases with decreasing wavelength, being proportional to λ^{-4}. It contributes a loss factor of $\exp(-\alpha_R z)$. The Rayleigh scattering coefficient, α_R, is given by

$$\alpha_R = \left(\frac{8\pi^3}{3\lambda^4}\right)(\delta n^2)^2 \,\delta V \tag{40.29}$$

where δn^2 is the mean-square fluctuation in refractive index and V is the volume associated with this index difference.

Mie Scattering

Mie scattering is caused by inhomogeneities in the medium, with dimensions comparable to the guided wavelength. It is independent of wavelength.

Dispersion and Pulse Spreading

Dispersion refers to the variation of velocity with frequency or wavelength. Dispersion causes pulse spreading, but other nonwavelength-dependent mechanisms also contribute to pulse spreading in optical waveguides. The mechanisms responsible for pulse spreading in optical waveguides include material dispersion, waveguide dispersion, and multimode pulse spreading.

Material Dispersion

In material dispersion, the velocity variation is caused by some property of the medium. In glass, it is caused by the wavelength dependence of refractive index. For a given pulse, the resulting pulse spread per unit length is the difference between the travel times of the slowest and fastest wavelengths in the pulse. It is given by

$$\Delta\tau = \frac{-\lambda}{c}\, n''\Delta\lambda = -M\Delta\lambda \tag{40.30}$$

In Eq. 40.30, n'' is the second derivative of the refractive index with respect to λ, $M = (\lambda/c)n''$ is the material dispersion, and $\Delta\lambda$ is the **linewidth** of the pulse. Figure 40.12 shows the wavelength dependence of material dispersion [Wemple, 1979]. Note that for silica, zero dispersion occurs around 1.3 μm, and material dispersion is small in the wavelength range of small fiber attenuation.

Waveguide Dispersion

The effective refractive index for any mode varies with wavelength for a fixed film thickness, for a slab waveguide, or a fixed core radius, for an optical fiber. This variation causes pulse spreading, which is termed waveguide dispersion. The resulting pulse spread is given by

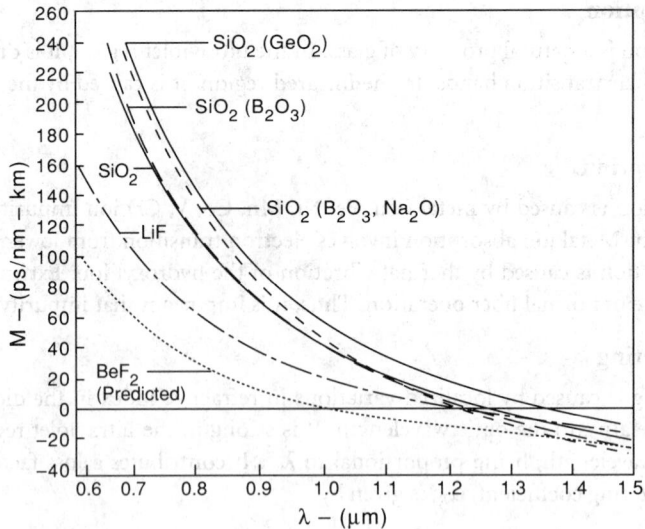

FIGURE 40.12 Material dispersion as a function of wavelength for silica and several solids. (*Source:* S.H. Wemple, "Material dispersion in optical fibers," *Applied Optics*, vol. 18, no. 1, p. 33, 1979. With permission.)

$$\Delta\tau = \frac{-\lambda}{c}\, n''_{\text{eff}}\Delta\lambda = -M_G\Delta\lambda \tag{40.31}$$

where $M_G = (\lambda/c)\,n''_{\text{eff}}$ is the waveguide dispersion.

Multimode Pulse Spreading

In a multimode waveguide, different modes travel different path lengths. This results in different travel times and, hence, in pulse spreading. Because this pulse spreading is not wavelength dependent, it is not usually referred to as dispersion. Multimode pulse spreads are given, respectively, for a slab waveguide, a step-index fiber, and a parabolic graded-index fiber by the following equations:

$$\Delta\tau_{\text{mod}} = \frac{n_1(n_1 - n_2)}{cn_2} \quad \text{(slab waveguide)} \tag{40.32}$$

$$\Delta\tau_{\text{mod}} = \frac{n_1\Delta}{c} \quad \text{(step-index fiber)} \tag{40.33}$$

$$\Delta\tau_{\text{mod}} = \frac{n_1\Delta^2}{8c} \quad \text{(GRIN fiber)} \tag{40.34}$$

Total Pulse Spread

Total pulse spread is the overall effect of material dispersion, waveguide dispersion, and multimode pulse spread. It is given by

$$\Delta\tau_T^2 = \Delta\tau_{\text{mod}}^2 + \Delta\tau_{\text{dis}}^2 \tag{40.35}$$

where

$$\Delta\tau_{\text{dis}} = \text{total dispersion} = -(M + M_G)\Delta\lambda$$

In a multimode waveguide, multimode pulse spread dominates, and dispersion can often be ignored. In a single-mode waveguide, only material and waveguide dispersion exist; material dispersion dominates, and waveguide dispersion can often be ignored.

Total pulse spread imposes an upper limit on the bandwidth of an optical fiber. This upper limit is equal to $1/(2\Delta\tau_T)Hz$.

Defining Terms

Linewidth: The range of wavelengths emitted by a source or present in a pulse.

Meridional ray: A ray that is contained in a plane passing through the fiber axis.

Mode chart: A graphical illustration of the variation of effective refractive index (or, equivalently, propagation angle θ) with normalized thickness d/λ for a slab waveguide or normalized frequency V for an optical fiber.

Refractive index: The ratio of the velocity of light in free space to the velocity of light in a given medium.

Relative refractive index difference: The ratio $(n_1^2 - n_2^2)/2n_1^2 \approx (n_1 - n_2)/n_1$, where $n_1 > n_2$ and n_1 and n_2 are refractive indices.

References

H. A. Haus, *Waves and Fields in Optoelectronics*, Englewood Cliffs, N.J.: Prentice-Hall, 1984.

D. B. Keck, "Optical fiber waveguides," in *Fundamentals of Optical Fiber Communications*, 2nd ed., M. K. Barnoski, Ed., New York: Academic Press, 1981.

D. Marcuse, *Light Transmission Optics*, 2nd ed., New York: Van Nostrand Reinhold, 1982.

H. Osanai *et al.*, "Effects of dopants on transmission loss of low-OH-content optical fibers," *Electronic Letters*, vol. 12, no. 21, 1976.

J. C. Palais, *Fiber Optic Communications*, Englewood Cliffs, N.J.: Prentice-Hall, 1992.

J. M. Senior, *Optical Fiber Communications: Principles and Practice*, Englewood Cliffs, N.J.: Prentice-Hall, 1985.

J. M. Snyder, "Asymptotic expressions for eigenfunctions and eigenvalues of a dielectric or optical waveguide," *Trans. IEEE Microwave Theory Tech.*, vol. MTT-17, pp. 1130–1138, 1969.

S. H. Wemple, "Material dispersion in optical fibers," *Applied Optics*, vol. 18, no. 1, p. 33, 1979.

A. Yariv, *Optical Electronics*, 4th ed., Philadelphia: Saunders College Publishing, 1991.

Further Information

IEEE Journal of Lightwave Technology, a bimonthly publication of the IEEE, New York.

IEEE Lightwave Telecommunications Systems, a quarterly magazine of the IEEE, New York.

Applied Optics, a biweekly publication of the Optical Society of America, 2010 Massachusetts Avenue NW, Washington, DC 20036.

D. Macruse, *Theory of Optical Waveguides*, 2nd ed., Boston: Academica Press, 1991.

40.2 Optical Fibers and Cables*

Allen H. Cherin and Basant K. Tariyal

Communications using light as a signal carrier and optical fibers as transmission media are termed optical fiber communications. The applications of optical fiber communications have increased at

---

* This section, including all illustrations, is modified from A. H. Cherin and B. K. Tariyal, "Optical fiber communication," in *Encyclopedia of Telecommunications*, R. A. Meyers, Ed., San Diego: Academic Press, 1988. With permission.

a rapid rate, since the first commercial installation of a fiber-optic system in 1977. Today every major telecommunication company is spending millions of dollars on optical fiber communication systems. In an optical fiber communication system voice, video, or data are converted to a coded pulse stream of light using a suitable light source. This pulse stream is carried by optical fibers to a regenerating or receiving station. At the final receiving station the light pulses are converted to electric signals, decoded, and converted into the form of the original information. Optical fiber communications arc currently used for telecommunications, data communications, military applications, industrial controls, medical applications, and CATV.

Introduction

Since ancient times humans have used light as a vehicle to carry information. Lanterns on ships and smoke signals or flashing mirrors on land are early examples of uses of how humans used light to communicate. It was just over a hundred years ago that Alexander Graham Bell (1880) transmitted a telephone signal a distance greater than 200 m using light as the signal carrier. Bell called his invention a "photophone" and obtained a patent for it. Bell, however, wisely gave up the photophone in favor of the electric telephone. Photophone at the time of its invention could not be exploited commercially because of two basic drawbacks: (1) the lack of a reliable light source and (2) the lack of a dependable transmission medium.

The invention of the laser in 1960 gave a new impetus to the idea of lightwave communications (as scientists realized the potential of the dazzling information-carrying capacity of these lasers). Much research was undertaken by different laboratories around the world during the early 1960s on optical devices and transmission media. The transmission media, however, remained the main problem, until K. C. Kao and G. A. Hockham in 1966 proposed that glass fibers with a sufficiently high-purity **core** surrounded by a lower refractive index **cladding** could be used for transmitting light over long distances. At the time, available glasses had losses of several thousand decibels per kilometer. In 1970, Robert Maurer of Corning Glass Works was able to produce a fiber with a loss of 20 dB/km. Tremendous progress in the production of low-loss optical fibers has been made since then in the various laboratories in the United States, Japan, and Europe, and today optical fiber communication is one of the fastest growing industries. Optical fiber communication is being used to transmit voice, video, and data over long distance as well as within a local network.

Fiber optics appears to be the future method of choice for many communications applications. The biggest advantage of a lightwave system is its tremendous information-carrying capacity. There are already systems that can carry several thousand simultaneous conversations over a pair of optical fibers thinner than human hair. In addition to this extremely high capacity, the lightguide cables are light weight, they are immune to electromagnetic interference, and they are potentially very inexpensive.

A lightwave communication system (Fig. 40.13) consists of a transmitter, a transmission medium, and a receiver. The transmitter takes the coded electronic signal (voice, video, or data) and converts it to the light signal, which is then carried by the transmission medium (an optical fiber cable) to either a **repeater** or the receiver. At the receiving end the signal is detected, converted to electrical pulses, and decoded to the proper output. This article provides a brief overview of the different components used in an optical fiber system, along with examples of various applications of optical fiber systems.

Classification of Optical Fibers and Attractive Features

Fibers that are used for optical communication are waveguides made of transparent dielectrics whose function is to guide light over long distances. An optical fiber consists of an inner cylinder of glass called the core, surrounded by a cylindrical shell of glass of lower refractive index, called the cladding. Optical fibers (lightguides) may be classified in terms of the refractive index profile of the

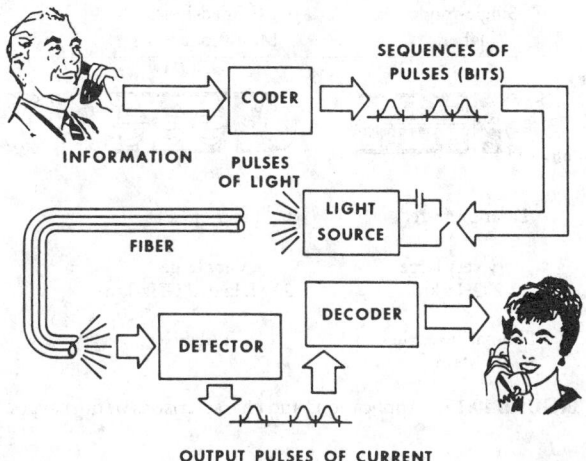

FIGURE 40.13 Schematic diagram of a lightwave communications system.

core and whether one **mode** (single-mode fiber) or many modes (multimode fiber) are propagating in the guide (Fig. 40.14). If the core, which is typically made of a high-silica-content glass or a multicomponent glass, has a uniform refractive index n_1, it is called a *step-index fiber*. If the core has a nonuniform refractive index that gradually decreases from the center toward the core-cladding interface, the fiber is called a *graded-index fiber*. The cladding surrounding the core has a uniform refractive index n_2 that is slightly lower than the refractive index of the core region. The cladding of the fiber is made of a high-silica-content glass or a multicomponent glass. Figure 40.14 shows the dimensions and refractive indexes for commonly used telecommunication fibers. Figure 40.15 enumerates some of the advantages, constraints, and applications of the different types of fibers. In general, when the transmission medium must have a very high **bandwidth**—for example, in an undersea or long-distance terrestrial system—a single-mode fiber is used. For intermediate

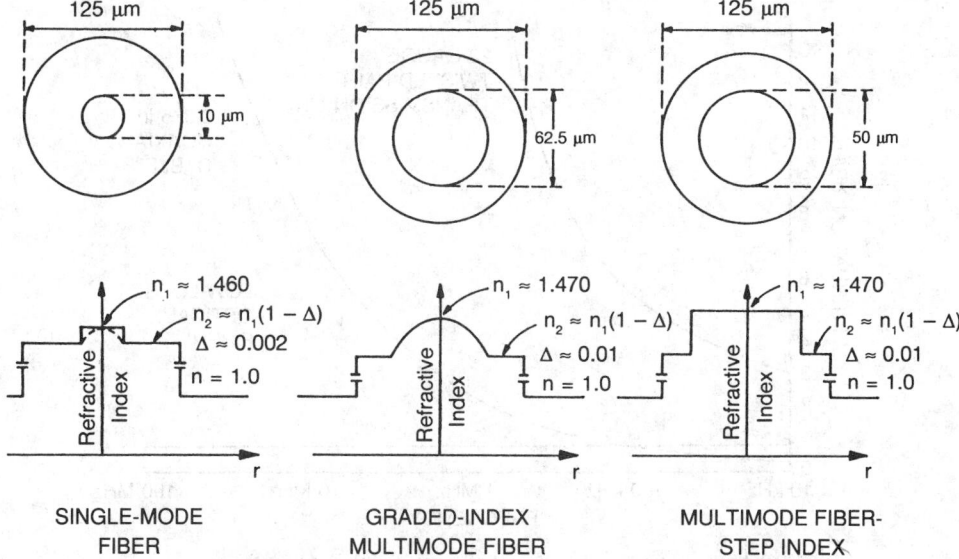

FIGURE 40.14 Geometry of single-mode and multimode fibers.

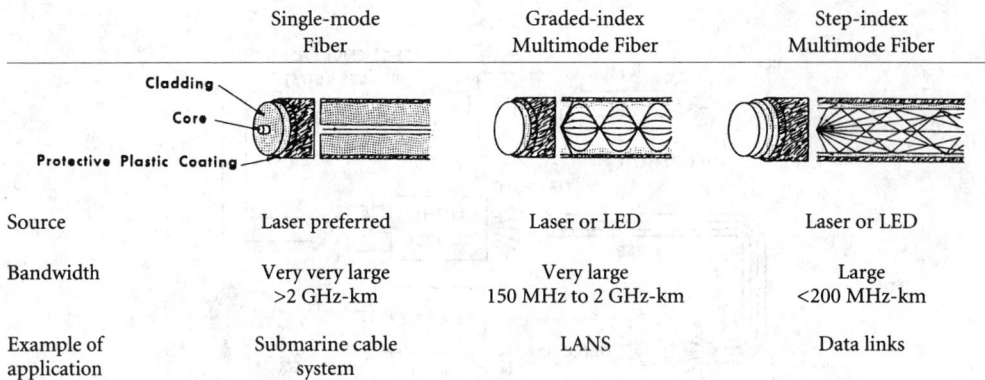

	Single-mode Fiber	Graded-index Multimode Fiber	Step-index Multimode Fiber
Source	Laser preferred	Laser or LED	Laser or LED
Bandwidth	Very very large >2 GHz-km	Very large 150 MHz to 2 GHz-km	Large <200 MHz-km
Example of application	Submarine cable system	LANS	Data links

FIGURE 40.15 Applications and characteristics of fiber types.

system bandwidth requirements between 200 MHz-km and 2 GHz-km, such as found in local-area networks, either a single-mode or graded-index multimode fiber would be the choice. For applications such as short data links where lower bandwidth requirements are placed on the transmission medium, either a graded-index or a step-index multimode fiber may be used.

Because or their low loss and wide bandwidth capabilities, optical fibers have the potential for being used wherever twisted wire pairs or coaxial cables are used as the transmission medium in a communication system. If an engineer were interested in choosing a transmission medium for a given transmission objective, he or she would tabulate the required and desired features of alternate technologies that may be available for use in the applications. With that process in mind, a summary of the attractive features and the advantages of optical fiber transmission will be given. Some of these advantages include (a) low loss and high bandwidth; (b) small size and bending radius; (c) nonconductive, nonradiative, and noninductive; (d) light weight; and (e) providing natural growth capability.

To appreciate the low loss and wide bandwidth capabilities of optical fibers, consider the curves of signal **attenuation** versus frequency for three different transmission media shown in Fig. 40.16. Optical fibers have a "flat" transfer function well beyond 100 MHz. When compared with wire pairs

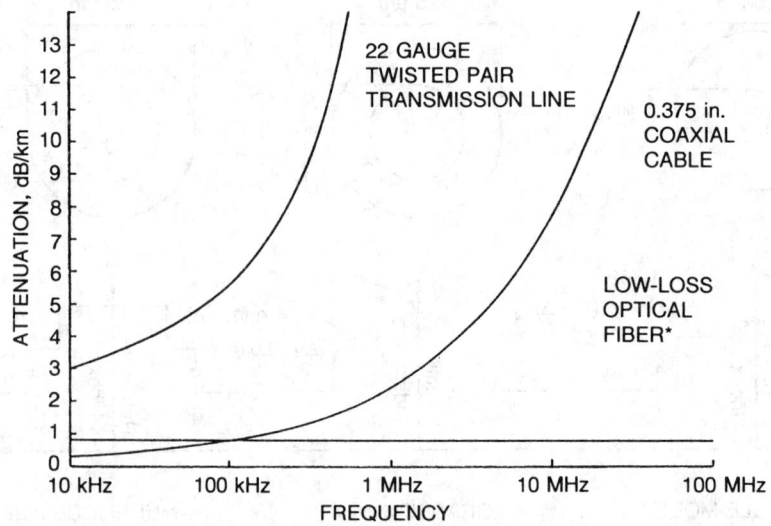

FIGURE 40.16 Attenuation versus frequency for three different transmission media. Asterisk indicates fiber loss at a carrier wavelength of 1.3 μm.

of coaxial cables, optical fibers have far less loss for signal frequencies above a few megahertz. This is an important characteristic that strongly influences system economics, since it allows the system designer to increase the distance between regenerators (amplifiers) in a communication system.

The small size, small bending radius (a few centimeters), and light weight of optical fibers and cables are very important where space is at a premium, such as in aircraft, on ships, and in crowded ducts under city streets.

Because optical fibers are dielectric waveguides, they avoid many problems such as radiative interference, ground loops, and, when installed in a cable without metal, lightning-induced damage that exists in other transmission media.

Finally, the engineer using optical fibers has a great deal of flexibility. He or she can install an optical fiber cable and use it initially in a low-capacity (low-bit-rate) system. As the system needs grow, the engineer can take advantage of the broadband capabilities of optical fibers and convert to a high-capacity (high-bit-rate) system by simply changing the terminal electronics.

Fiber Transmission Characteristics

The proper design and operation of an optical communication system using optical fibers as the transmission medium requires a knowledge of the transmission characteristics of the optical sources, fibers, and interconnection devices (connectors, couplers, and splices) used to join lengths of fibers together. The transmission criteria that affect the choice of the fiber type used in a system are signal attenuation, information transmission capacity (bandwidth), and source coupling and interconnection efficiency. Signal attenuation is due to a number of loss mechanisms within the fiber, as shown in Table 40.1, and due to the losses occurring in splices and connectors. The information transmission capacity of a fiber is limited by **dispersion**, a phenomenon that causes light that is originally concentrated into a short pulse to spread out into a broader pulse as it travels along an optical fiber. Source and interconnection efficiency depends on the fiber's core diameter and its **numerical aperture**, a measure of the angle over which light is accepted in the fiber. Absorption and scattering of light traveling through a fiber leads to signal attenuation, the rate of which is measured in decibels per kilometer (dB/km). As can be seen in Fig. 40.17, for both multimode and single-mode fibers, attenuation depends strongly on wavelength. The decrease in scattering losses with increasing wavelength is offset by an increase in material absorption such that attenuation is lowest near 1.55 μm (1550 nm).

Table 40.1 Loss Mechanisms

Intrinsic material absorption loss
  Ultraviolet absorption tail
  Infrared absorption tail
Absorption loss due to impurity ions
Rayleigh scattering loss
Waveguide scattering loss
Microbending loss

The measured values given in Table 40.2 are probably close to the lower bounds for the attenuation of optical fibers. In addition to intrinsic fiber losses, extrinsic loss mechanisms, such as absorption due to impurity ions, and **microbending** loss due to jacketing and cabling can add loss to a fiber.

The bandwidth or information-carrying capacity of a fiber is inversely related to its total dispersion. The total dispersion in a fiber is a combination of three components: intermodal dispersion (modal delay distortion), material dispersion, and waveguide dispersion.

Table 40.2 Best Attenuation Results (dB/km) in Ge-P-SiO_2 Core Fibers

Wavelength (nm)	$\Delta \approx 0.2\%$ (Single-mode Fibers)	$\Delta \approx 1.0\%$ (Graded-index Multimode Fibers)
850	2.1	2.20
1300	0.27	0.44
1500	0.16	0.23

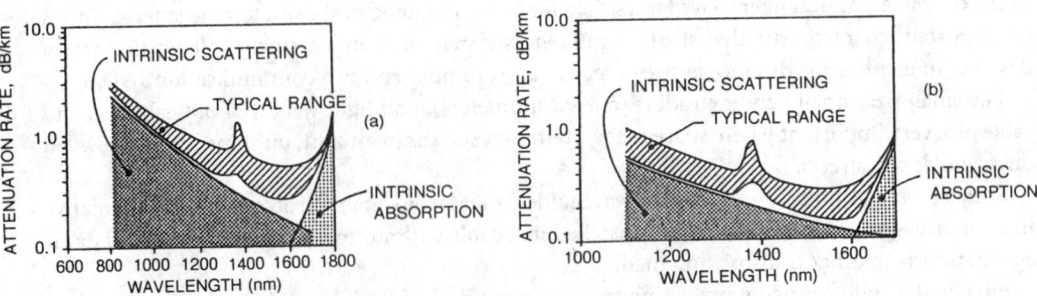

FIGURE 40.17 Spectral attenuation rate. (a) Graded-index multimode fibers. (b) Single-mode fibers.

Intermodal dispersion occurs in multimode fibers because rays associated with different modes travel different effective distances through the optical fiber. This causes light in the different modes to spread out temporally as it travels along the fiber. Modal delay distortion can severely limit the bandwidth of a step-index multimode fiber to the order of 20 MHz-km. To reduce modal delay distortion in multimode fibers, the core is carefully doped to create a graded (approximately parabolic shaped) refractive index profile. By carefully designing this index profile, the group velocities of the propagating modes are nearly equalized. Bandwidths of 1.0 GHz-km are readily attainable in commercially available graded-index multimode fibers. The most effective way of eliminating intermodal dispersion is to use a single-mode fiber. Since only one mode propagates in a single-mode fiber, modal delay distortion between modes does not exist and very high bandwidths are possible. The bandwidth of a single-mode fiber, as mentioned previously, is limited by the combination of material and waveguide dispersion. As shown in Fig. 40.18, both material and waveguide dispersion are dependent on wavelength.

Material dispersion is caused by the variation of the refractive index of the glass with wavelength and the spectral width of the system source. Waveguide dispersion occurs because light travels in both the core and cladding of a single-mode fiber at an effective velocity between that of the core and cladding materials. The waveguide dispersion arises because the effective velocity, the waveguide dispersion, changes with wavelength. The amount of waveguide dispersion depends on the design of the waveguide structure as well as on the fiber material. Both material and waveguide dispersion are measured in picoseconds (of pulse spreading) per nanometer (of source spectral width) per kilometer (of fiber length), reflecting both the increases in magnitude in source linewidth and the increase in dispersion with fiber length.

Material and waveguide dispersion can have different signs and effectively cancel each other's dispersive effect on the total dispersion in a single-mode fiber. In conventional germanium-doped silica fibers, the "zero-dispersion" wavelength at which the waveguide and material dispersion effects cancel each other out occurs near 1.30 μm. The zero-dispersion wavelength can be shifted to

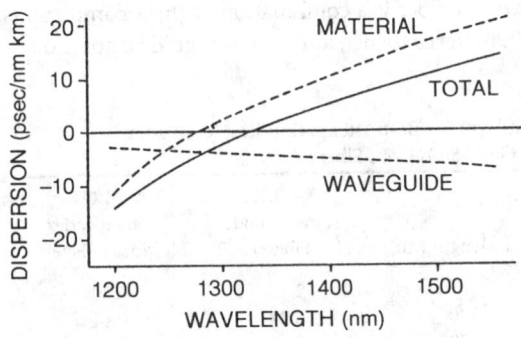

FIGURE 40.18 Single-mode step-index dispersion curve.

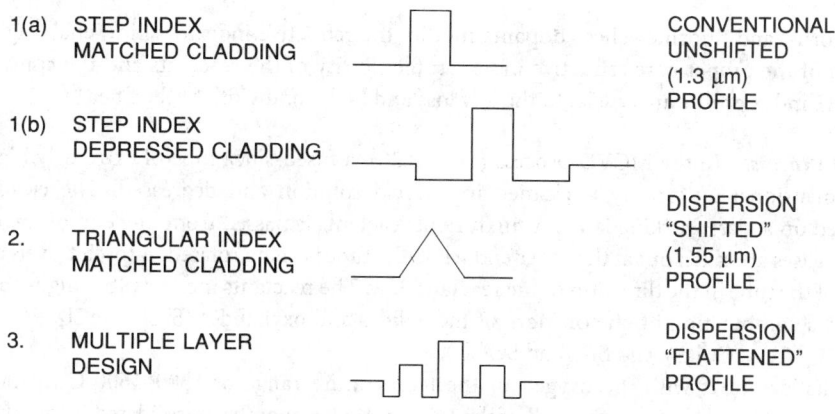

FIGURE 40.19 Single-mode refractive index profiles.

1.55 μm, or the low-dispersion characteristics of a fiber can be broadened by modifying the refractive index profile shape of a single-mode fiber. This profile shape modification alters the waveguide dispersion characteristics of the fiber and changes the wavelength region in which waveguide and material dispersion effects cancel each other. Figure 40.19 illustrates the profile shapes of "conventional," "dispersion-shifted," and "dispersion-flattened" single-mode fibers. Single-mode fibers operating in their zero-dispersion region with system sources of finite spectral width do not have infinite bandwidth but have bandwidths that are high enough to satisfy all current high-capacity system requirements.

Optical Fiber Cable Manufacturing

Optical fiber cables should have low loss and high bandwidth and should maintain these characteristics while in service in extreme environments. In addition, they should be strong enough to survive the stresses encountered during manufacture, installation, and service in a hostile environment. The manufacturing process used to fabricate the optical fiber cables can be divided into four steps: (1) preform fabrication, (2) fiber drawing and coating, (3) fiber measurement, and (4) fiber packaging.

Preform Fabrication

The first step in the fabrication of optical fiber is the creation of a glass preform. A preform is a large blank of glass several millimeters in diameter and several centimeters in length. The preform has all the desired properties (e.g., geometrical ratios and chemical composition) necessary to yield a high-quality fiber. The preform is subsequently drawn into a multi-kilometer-long hair-thin fiber. Four different preform manufacturing processes are currently in commercial use.

The most widely used process is the modified **chemical vapor deposition** (MCVD) process invented at the AT&T Bell Laboratories. Outside vapor deposition process (OVD) is used by Corning Glass Works and some of its joint ventures in Europe. Vapor axial deposition (VAD) process is the process used most widely in Japan. Philips, in Eindhoven, Netherlands, uses a low-temperature plasma chemical vapor deposition (PCVD) process.

In addition to the above four major processes, other processes are under development in different laboratories. Plasma MCVD is under development at Bell Laboratories, hybrid OVD-VAD processes are being developed in Japan, and Sol-Gel processes are being developed in several laboratories. The first four processes are the established commercial processes and are producing fiber economically. The new processes are aimed at greatly increasing the manufacturing productivity of preforms, and thereby reducing their cost.

All the above processes produce high-silica fibers using different dopants, such as germanium, phosphorus, and fluorine. These dopants modify the refractive index of silica, enabling the production of the proper core refractive index profile. Purity of the reactants and the control of the refractive index profile are crucial to the low loss and high bandwidth of the fiber.

MCVD Process. In the MCVD process (Fig. 40.20), a fused-silica tube of extremely high purity and dimensional uniformity is cleaned in an acid solution and degreased. The clean tube is mounted on a glass working lathe. A mixture of reactants is passed from one end of the tube and exhaust gases are taken out at the other end while the tube is being rotated. A torch travels along the length of the tube in the direction of the reactant flow. The reactants include ultra-high-purity oxygen and a combination of one or more of the halides and oxyhalides ($SiCl_4$, $GeCl_4$, $POCl_3$, BCl_3, BBr_3, SiF_4, CCl_4, CCl_2F_2, Cl_2, SF_6, and $SOCl_2$).

The halides react with the oxygen in the temperature range of 1300–1600°C to form oxide particles, which are driven to the wall of the tube and subsequently consolidated into a glassy layer as the hottest part of the flame passes over. After the completion of one pass, the torch travels back and the next pass is begun. Depending on the type of fiber (i.e., multimode or single-mode), a **barrier layer** or a cladding consisting of many thin layers is first deposited on the inside surface of the tube. The compositions may include B_2O_3-P_2O_5-SiO_2 or F-P_2O_5-SiO_2 for barrier layers and SiO_2, F-SiO_2, F-P_2O_5-SiO_2, or F-GeO_2-SiO_2-P_2O_5 for cladding layers. After the required number of barrier or cladding layers has been deposited the core is deposited. The core compositions depend on whether the fiber is single-mode, multimode, step-index, or multimode graded-index. In the case of graded-index multimode fibers, the dopant level changes with every layer, to provide a refractive index profile that yields the maximum bandwidth.

After the deposition is complete, the reactant flow is stopped except for a small flow of oxygen, and the temperature is raised by reducing the torch speed and increasing the flows of oxygen and hydrogen through the torch. Usually the exhaust end of the tube is closed first and a small positive pressure is maintained inside the deposited tube while the torch travels backward. The higher temperatures cause the glass viscosity to decrease, and the surface tension causes the tube to contract inward. The complete collapse of the tube into a solid preform is achieved in several passes. The speed of the collapse, the rotation of the tube, the temperature of collapse, and the positive pressure of oxygen inside the tube are all accurately controlled to predetermined values in order to produce a straight and bubble-free preform with minimum ovality. The complete preform is then taken off the lathe. After an inspection to assure that the preform is free of defects, the preform is ready to be drawn into a thin fiber.

The control of the refractive index profile along the cross section of the deposited portion of the

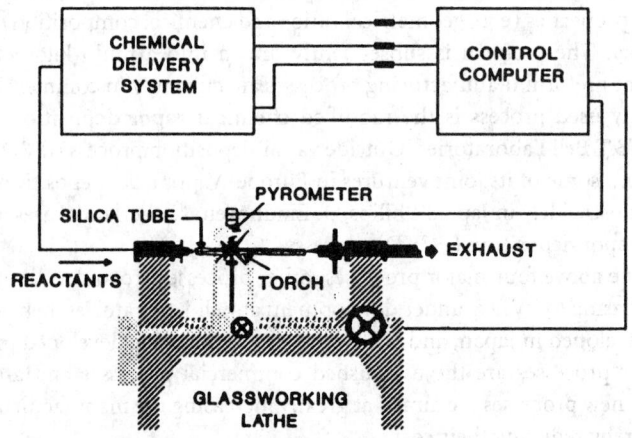

FIGURE 40.20 Schematic diagram of the MCVD process.

preform is achieved through a vapor delivery system. In this system, liquids are vaporized by passing a carrier gas (pure O_2) through the bubblers, made of fused silica. Accurate flows are achieved with precision flow controllers that maintain accurate carrier gas flows and extremely accurate temperatures within the bubblers. Microprocessors are used to automate the complete deposition process, including the torch travel and composition changes throughout the process. Impurities are reduced to very low levels by starting with pure chemicals, and there is further reducing of the impurities with in-house purification of these chemicals. Ultra-pure oxygen and a completely sealed vapor-delivery system are used to avoid chemical contamination. Transition-metal ion impurities of well below 1 ppb and OH^- ion impurities of less than 1 ppm are typically maintained to produce high-quality fiber.

The PCVD Process.   The PCVD process (Fig. 40.21) also uses a starting tube, and the deposition takes place inside the tube. Here, however, the tube is either stationary or oscillating and the pressure is kept at 10–15 torr. Reactants are fed inside the tube, and the reaction is accomplished by a traveling microwave plasma inside the tube. The entire tube is maintained at approximately 1200°C. The plasma causes the heterogeneous depositions of glass on the tube wall, and the deposition efficiency is very high. After the required depositions of the cladding and core are complete, the tube is taken out and collapsed on a separate equipment. Extreme care is required to prevent impurities from getting into the tube during the transport and collapse procedure. The PCVD process has the advantages of high efficiency, no tube distortion because of the lower temperature, and very accurate profile control because of the large number of layers deposited in a short time. However, going to higher rates of flow presents some difficulties, because of a need to maintain the low pressure.

The PMCVD Process.   The PMCVD is an enhancement of the MCVD process. Very high rates of deposition (up to 10 g/min, compared to 2 g/min for MCVD) are achieved by using larger diameter tubes and an RF plasma for reaction (Fig. 40.22). Because of the very high temperature of the plasma, water cooling is essential. An oxyhydrogen torch follows the plasma and sinters the deposition. The high rates of deposition are achieved because of very high thermal gradients from the

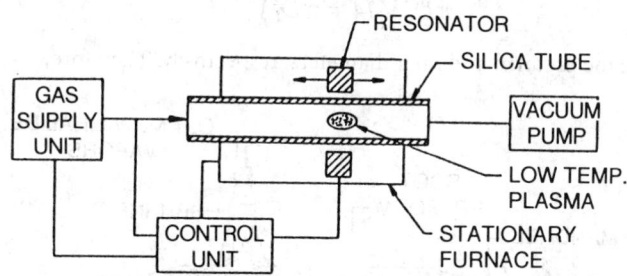

FIGURE 40.21   Schematic diagram of the PCVD process.

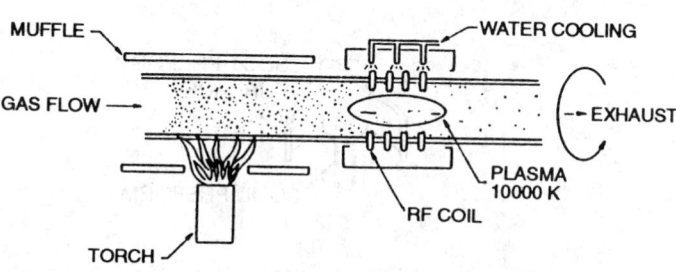

FIGURE 40.22   Schematic diagram of the PMCVD process.

center of the tube to the wall and the resulting high thermophoretic driving force. The PMCVD process is still in the development stage and has not been commercialized.

The OVD Process. The OVD process does not use a starting tube; instead, a stream of soot particles of desired composition is deposited on a bait rod (Fig. 40.23). The soot particles are produced by the reaction of reactants in a fuel gas-oxygen flame. A cylindrical porous soot preform is built layer by layer. After the deposition of the core and cladding is complete, the bait rod is removed. The porous preform is then sintered and dried in a furnace at 1400–1600°C to form a clear bubble-free preform under a controlled environment. The central hold left by the blank may or may not be closed, depending on the type of preform. The preform is now ready for inspection and drawing.

The VAD Process. The process is very similar to the OVD process. However, the soot deposition is done axially instead of radially. The soot is deposited at the end of a starting silica-glass rod (Fig. 40.24). A special torch using several annular holes is used to direct a stream of soot at the deposition surface. The reactant vapors, hydrogen gas, argon gas, and oxygen gas flow through different annular openings. Normally the core is deposited and the rotating speed is gradually withdrawn as the deposition proceeds at the end. The index profile is controlled by the composition of the gases flowing through the torch and the temperature distribution at the deposition surface. The porous preform is consolidated and dehydrated as it passes through a carbon-ring furnace in a controlled environment. $SOCl_2$ and Cl are used to dehydrate the preform. Because of the axial deposition, this process is semicontinuous and is capable of producing very large preforms.

Fiber Drawing

After a preform has been inspected for various defects such as bubbles, ovality, and straightness, it is taken to a fiber drawing station. A large-scale fiber drawing process must repeatedly maintain the optical quality of the preform and produce a dimensionally uniform fiber with high strength.

Draw Process. During fiber drawing, the inspected preform is lowered into a hot zone at a certain feed rate V_p, and the fiber is pulled from the softened neck-down region (Fig. 40.25) at a rate V_f. At steady state,

$$\pi D_p^2 V_p/4 = \pi D_f^2 V_f/4 \tag{40.36}$$

where D_p and D_f are the preform and fiber diameters, respectively. Therefore,

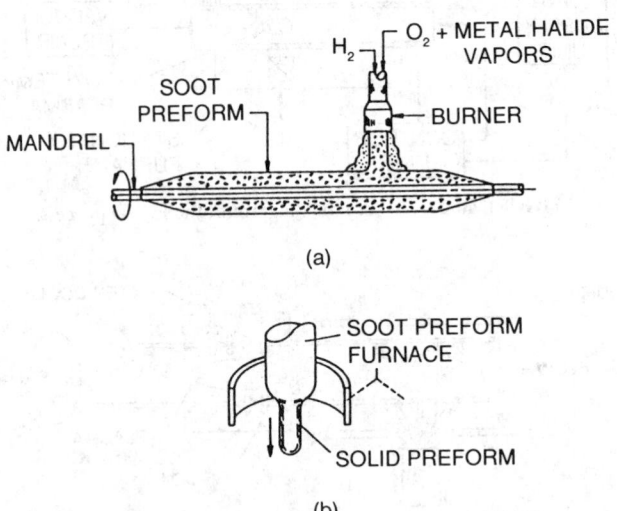

(a)

(b)

FIGURE 40.23 Schematic diagram of the outside vapor deposition. (a) Soot deposition. (b) Consolidation.

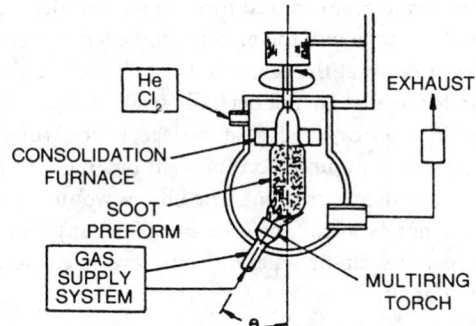

FIGURE 40.24 Schematic diagram of the vapor axial deposition.

$$V_f = (D_p^2/D_f^2)V_p \qquad (40.37)$$

A draw machine, therefore, consists of a preform feed mechanism, a heat source, a pulling device, a coating device, and a control system to accurately maintain the fiber diameter and the furnace temperature.

Heat Source. The heat source should provide sufficient energy to soften the glass for pulling the fiber without causing excessive tension and without creating turbulence in the neck-down region. A proper heat source will yield a fiber with uniform diameter and high strength. Oxyhydrogen torches, CO_2 lasers, resistance furnaces, and induction furnaces have been used to draw fibers. An oxyhydrogen torch, although a clean source of heat, suffers from turbulence due to flame. A CO_2 laser is too expensive a heat source to be considered for the large-scale manufacture of fibers. Graphite resistance furnaces and zirconia induction furnaces are the most widely used heat sources for fiber drawing. In the graphite resistance furnace, a graphite resistive element produces the required heat. Because graphite reacts with oxygen at high temperatures, an inert environment (e.g., carbon) is maintained inside the furnace. The zirconia induction furnace does not require inert environment. It is extremely important that the furnace environment be clean in order to produce high-strength fibers. A zirconia induction furnace, when properly designed and used, has produced very-high-strength long-length fibers (over 2.0 GPa) in lengths of several kilometers.

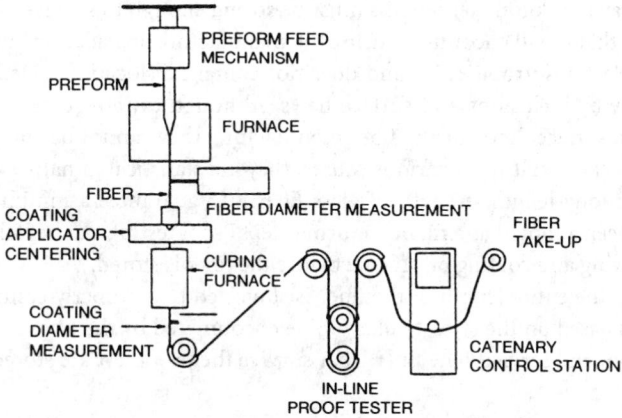

FIGURE 40.25 The fiber drawing process.

Mechanical Systems.   An accurate preform feed mechanism and drive capstan form the basis of fiber speed control. The mechanism allows the preform to be fed at a constant speed into the hot zone, while maintaining the preform at the center of the furnace opening at the top. A centering device is used to position preforms that are not perfectly straight. The preform is usually held with a collet-type chuck mounted in a vertically movable carriage, which is driven by a lead screw. A precision stainless-steel drive capstan is mounted on the shaft of a high-performance dc servomotor. The fiber is taken up on a proper-diameter spool. The fiber is wound on the spool at close to zero tension with the help of a catenary control. In some cases fiber is proof-tested in-line before it is wound on a spool. The proof stress can be set at different levels depending on the application for which the fiber is being manufactured.

Fiber Coating System.   The glass fiber coming out of the furnace has a highly polished pristine surface and the theoretical strength of such a fiber is in the range of 15–20 GPa. Strengths in the range of 4.5–5.5 GPa are routinely measured on short fiber lengths. To preserve this high strength, polymeric coatings are applied immediately after the drawing. The coating must be applied without damaging the fiber, it must solidify before reaching the capstan, and it should not cause microbending loss. To satisfy all these requirements, usually two layers of coatings are applied: a soft inner coating adjacent to the fiber to avoid microbending loss and a hard outer coating to resist abrasion. The coatings are a combination of ultraviolet- (UV) curable acrylates, UV-curable silicones, hot melts, heat-curable silicones, and nylons. When dual coatings are applied, the coated fiber diameter is typically 235–250 μm. The nylon-jacketed fiber typically used in Japan has an outside diameter of 900 μm. All coating materials are usually filtered to remove particles that may damage the fiber. Coatings are usually applied by passing the fiber through a coating cup and then curing the coating before the fiber is taken up by the capstan. The method of application, the coating material, the temperature, and the draw speed affect the proper application or a well-centered, bubble-free coating.

Fiber drawing facilities are usually located in a clean room where the air is maintained at class 10,000. The region of the preform and fiber from the coating cup to the top of the preform is maintained at class 100 or better. A class 100 environment means that there are no more than 100 particles of size greater than 0.5 μm in 1 ft³ of air. A clean environment, proper centering of the preform in the furnace and fiber in the coating cup, and proper alignment of the whole draw tower ensure a scratch-free fiber of a very high tensile strength. A control unit regulates the draw speed, preform feed speed, preform centering, fiber diameter, furnace temperature, and draw tension.

The coated fiber wound on a spool is next taken to the fiber measurement area to assure proper quality control.

Proof Testing of Fibers.   Mechanical failure is one of the major concerns in the reliability of optical fibers. Fiber drawn in kilometer lengths must be strong enough to survive all of the short- and long-term stresses that it will encounter during the manufacture, installation, and long service life. Glass is an ideal elastic isotropic solid and does not contain dislocations. Hence, the strength is determined mainly by inclusions and surface flaws. Although extreme care is taken to avoid inhomogeneities and surface flaws during fiber manufacture, they cannot be completely eliminated. Since surface flaws can result from various causes, they are statistical in nature and it is very difficult to predict the long-length strength of glass fibers. To guarantee a minimum fiber strength, proof testing has been adopted as a manufacturing step. Proof testing can be done in-line immediately after the drawing and coating or off-line before the fiber is stored.

In proof testing, the entire length of the fiber is subjected to a properly controlled proof stress. The proof stress is based on the stresses likely to be encountered by the fiber during manufacture, storage, installation, and service. The fibers that survive the proof test are stored for further packaging into cables.

Proof testing not only guarantees that the fiber will survive short-term stresses but also guarantees that the fiber will survive a lower residual stress that it may be subjected to during its long service life. It is well known that glass, when used in a humid environment, can fail under a long-term stress well below its instantaneous strength. This phenomenon is called static fatigue. Several models have been proposed to quantitatively describe the relationship between residual stress and the life of optical fibers. Use is made of the most conservative of these models, and the proof stress is determined by a consideration of the maximum possible residual stress in service and the required service life.

Fiber Packaging

In order to efficiently use one or more fibers, they need to be packaged so that they can be handled, transported, and installed without damage. Optical fibers can be used in a variety of applications, and hence the way they are packaged or cabled will also vary. There are numerous cable designs that are used by different cable manufacturers. All these designs, however, must meet certain criteria. A primary consideration in a cable design is to assure that the fibers in the cables maintain their optical properties (attenuation and dispersion) during their service life under different environmental conditions. The design, therefore, must minimize microbending effects. This usually means letting the fiber take a minimum energy position at all times in the cable structure. Proper selection of cabling materials so as to minimize differential thermal expansion or contraction during temperature extremes is important in minimizing microbending loss. The cable structure must be such that the fibers carry a load well below the proof-test level at all times, and especially while using conventional installation equipment. The cables must provide adequate protection to the fibers under all adverse environmental conditions during their entire service life, which may be as long as 40 years. Finally, the cable designs should be cost effective and easily connectorized or spliced.

Five different types (Fig. 40.26) of basic cable designs are currently in use: (a) loose tube, (b) fluted, (c) ribbon, (d) stranded, and (e) Lightpack Cable. The loose tube design was pioneered by Siemens in Germany. Up to 10 fibers are enclosed in a loose tube, which is filled with a soft filling compound. Since the fibers are relatively free to take the minimum energy configuration, the microbending losses are avoided. Several of these buffered loose tube units are stranded around a central glass-resin support member. Aramid yarns are stranded on the cable core to provide strength members (for pulling through ducts), with a final polyethylene sheath on the outside. The stranding lay length and pitch radius are calculated to permit tensile strain on the cable up to the rated force and to permit cooling down to the rated low temperature without affecting the fiber attenuation.

In the fluted designs, fibers are laid in the grooves of plastic central members and are relatively free to move. The shape and size of the grooves vary with the design. The grooved core may also contain a central strength member. A sheath is formed over the grooved core, and this essentially forms a unit. Several units may then be stranded around a central strength member to form a cable core of desired size, over which different types of sheaths are formed. Fluted designs have been pioneered in France and Canada.

The ribbon design was invented at AT&T Bell Laboratories and consists of a linear array of 12 fibers sandwiched between two polyester tapes with pressure-sensitive adhesive on the fiber side. The spacing and the back tension on the fibers is accurately maintained. The ribbons are typically 2.5 mm in width. Up to 12 ribbons can be stacked to give a cable core consisting of 144 fibers. The core is twisted to some lay length and enclosed in a polyethylene tube. Several combinations of protective plastic and metallic layers along with metallic or nonmetallic strength members are then applied around the core to give the final cable its required mechanical and environmental characteristics needed for use in specified conditions. The ribbon design offers the most efficient and economic packaging of fibers for high-fiber-count cables. It also lends the cable to preconnectorization and makes it extremely convenient for installation and splicing.

The tight-bound stranded designs were pioneered by Japanese and are used in the United States for several applications. In this design, several coated fibers are stranded around a central support member. The central support member may also serve as a strength member, and it may be metallic

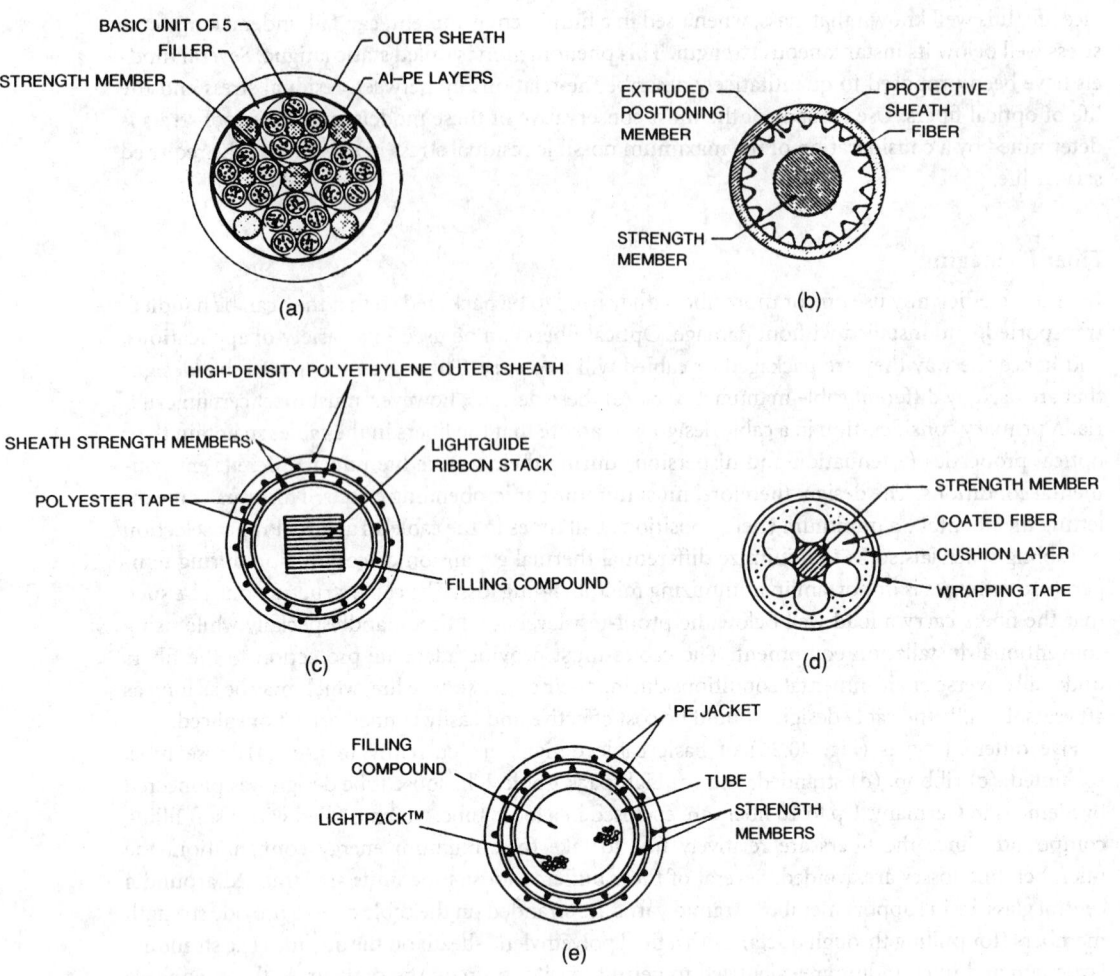

FIGURE 40.26   Fiber cable designs. (a) Loose tube design. (b) Slotted design. (c) Ribbon design. (d) Stranded unit. (e) Lightpack™ Cable design.

or nonmetallic. The stranded unit can have up to 18 fibers. The unit is contained within a plastic tube filled with a water-blocking compound. The final cable consists of several of these units stranded around a central member and protected on the outside with various sheath combinations.

The Lightpack Cable design, pioneered by AT&T, is one of the simplest designs. Several fibers are held together with a binder to form a unit. One or more units are laid inside a large tube, which is filled with a water-blocking compound. This design has the advantage of the loose tube design in that the fibers are free of strain, but is more compact. The tube-containing units can then be projected with various sheath options and strength members to provide the final cable.

The final step in cabling is the sheathing operation. After the fibers have been made into identifiable units, one or more of the units (as discussed earlier) form a core which is then covered with a combination of sheathing layers. The number and combination of the sheathing layers depend on the intended use. Typically, a polyethylene sheath is extruded over the filled cable core. In a typical cross-ply design, metallic or nonmetallic strength members are applied over the first sheath layer, followed by another polyethylene sheath, over which another layer of strength members is applied. The direction of lay of the two layers of the strength members is opposite to each other. A final sheath is applied and the cable is ready for the final inspection, preconnectorization, and shipment.

Metallic vapor barriers and lightning- and rodent-protection sheath options are also available. Further armoring is applied to cables made for submarine application.

In addition to the above cable designs, there are numerous other cable designs used for specific applications, such as fire-resistant cables, military tactical cables, cables for missile guidance systems, cables for field communications established by air-drop operations, air deployment cables, and cables for industrial controls. All these applications have unique requirements, such as ruggedness, low loss, and repeaterless spans, and the cable designs are accordingly selected. However, all these cable designs still rely on the basic unit designs discussed above.

Defining Terms

Attenuation: Decrease of average optical power as light travels along the length or an optical fiber.

Bandwidth: Measure of the information-carrying capacity of the fiber. The greater the bandwidth, the greater the information-carrying capacity.

Barrier layer: Layer of deposited glass adjacent to the inner tube surface to create a barrier against OH diffusion.

Chemical vapor deposition: Process in which products of a heterogeneous gas-liquid or gas-solid reaction are deposited on the surface of a substrate.

Cladding: Low refractive index material that surrounds the fiber core.

Core: Central portion of a fiber through which light is transmitted.

Cut-off wavelength: Wavelength greater than which a particular mode ceases to be a bound mode.

Dispersion: Cause of distortion of the signal due to different propagation characteristics of different modes, leading to bandwidth limitations.

Graded-index profile: Any refractive index profile that varies with radius in the core.

Microbending: Sharp curvatures involving local fiber axis displacements of a few micrometers and spatial wavelengths of a few millimeters. Microbending causes significant losses.

Mode: Permitted electromagnetic field pattern within an optical fiber.

Numerical aperture: Acceptance angle of the fiber.

Optical repeater: Optoelectric device that receives a signal and amplifies it and retransmits it. In digital systems the signal is regenerated.

References

M. K. Barnoski, Ed., *Fundamentals of Optical Fiber Communications*, New York: Academic Press, 1976.

B. Bendow and S. M. Shashanka. Eds., *Fiber Optics: Advances in Research and Development*, New York: Plenum Press, 1979.

A. H. Cherin, *Introduction to Optical Fibers*, New York: McGraw-Hill, 1983.

T. Li, Ed., *Optical Fiber Communications*, New York: Academic Press, 1985.

J. E. Midwinter, *Optical Fibers for Transmission*, New York: Wiley, 1979.

S. E. Miller and A. G. Chynoweth, Eds., *Optical Fiber Telecommunications*, New York: Academic Press, 1979.

Y. Suematsu and I. Ken-ichi, *Introduction to Optical Fiber Communication*, New York: Wiley, 1982.

MAGNETISM WITHOUT MAGNETS

Modern power and communications systems are designed according to laws governing forces between electric charges and magnetic poles. Without those laws, electrical engineering as we know it today would not exist.

The earliest researchers and experimenters confused the sources of electricity and magnetism. Rumors circulated that magnetic power was lost when the amber or glass rods were rubbed with garlic and regained when they were smeared with goat's blood. In 1600, William Gilbert, a physician to England's first Queen Elizabeth, cleared up much of the confusion in his book, *de Magnete*, in which he described his work on magnets.

In Germany around 1641, Otto von Guericke cast a sphere of sulphur and charged it by rubbing it with his bare hands while it was rotating. It would then attract pieces of lint or straw. In the most popular theory of the day, electricity was a fluid with properties that let it penetrate solids.

Charles Dufay in Paris in 1733 inferred the existence of two kinds of electricity—vitreous (from glass and silk) and resinous (from charged amber and wool).

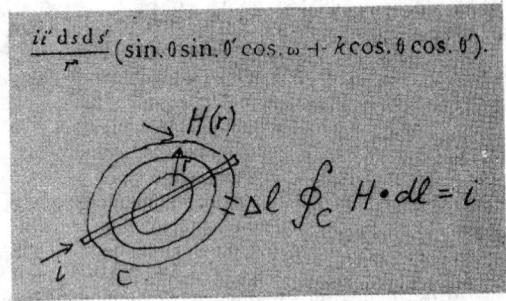

These early studies had only static charges to work with, and the inner electrode was charged by successive applications of a rubbed glass or resinous rod.

Despite this handicap, Charles Coulomb, in 1785, stated a quantitative law of electricity. He verified that the force of either attraction or repulsion between two small charged spheres is inversely proportional to the square of the distance between them.

It was not until 1800, when Alessandro Volta in Italy announced the primary electric battery, that continuous electric current could be made to flow at an experimenter's will. Volta found zinc and silver to work best when separated by a disk of cardboard moistened in brine. Layers of this sandwich made up Volta's pile.

Volta's invention of the battery opened up a whole new field for study. Humphrey Davy, in England, for one, used current from a voltaic pile in 1807 to isolate sodium and potassium by electrolysis.

In parallel with these discoveries experimenters were speculating that electricity and magnetism were related. The evidence in favor of a relationship kept mounting. But the precise nature of the link between electricity and magnetism remained unknown.

The availability of continuous current from Volta's pile set off further studies of the link in Europe. Hans Christian Oersted, a professor of natural philosophy (physics) at the University of Copenhagen, when lecturing in 1819, passed a current through a wire lying on top of a compass. As the current started to flow, he noticed that the needle of the compass moved. When the current ceased, the needle returned to its original position. The effect was found to be circular around the wire.

The Dane experimented for several months and in 1820 sent a paper to major journals in Europe, announcing he had observed the generation of a magnetic field by a current. His finding set off a frenzy of activity aimed at applying the action.

The news of Oersted's discovery reached Paris in September 1820, astounding a meeting of the Académie Royale des Sciences. Many members refused to believe the announcement—but not André Marie Ampère.

Within two weeks Ampère showed that the deflection of the magnetic needle could be predicted by what today is called the right-hand rule. He also observed that a coil of wire carrying a current developed magnetic poles and acted as a bar magnet without the presence of iron. This device he called a solenoid.

Ampère soon became the leader of electrodynamics, the name he coined for the newly discovered phenomena. His was the first experiment in magnetism to be performed without a magnet, and his work made electrodynamics a precise and mathematical subject. The current-field relationship is known as Ampère's law and is given in its original form (courtesy of the Burndy Library, Norwalk, Conn.) and in modern notation in the illustration. In the modern version, H is the magnetic field in amperes per meter, dl is a vector element of distance, C is the integration contour, and i is the current passing through the area bounded by C. Magnetic fields encircle currents.

Source: Adapted from J.D. Ryder, *IEEE Spectrum,* p. 28, October 1991. © 1991 IEEE.

41

Solid State Circuits

41.1 Introduction 1004
41.2 Amplifiers 1005
41.3 Oscillators 1009
41.4 Multipliers 1010
41.5 Mixers 1010
41.6 Control Circuits 1011
41.7 Summary and Future Trends 1013

I. J. Bahl
ITT Gallium Arsenide
Technology Center

41.1 Introduction

Over the past two decades, microwave active circuits have evolved from individual solid state transistors and passive elements housed in conventional waveguides and/or coaxial lines to fully integrated planar assemblies, including active and passive components and interconnections, generically referred to as a microwave integrated circuit (MIC). The **hybrid microwave integrated circuit** (HMIC) consists of an interconnect pattern and distributed circuit components printed on a suitable substrate, with active and lumped circuit components (in packaged or chip form) attached individually to the printed interconnect circuit by the use of soldering and wire bonding techniques. The solid state active elements are either silicon or gallium arsenide (or other III–V compound) devices. More recently, the solid state **monolithic microwave integrated circuit** (MMIC) approach has become commonplace. In MMICs, all interconnections and components, both active and passive, are fabricated simultaneously on a semi-insulating semiconductor substrate (usually gallium arsenide, GaAs) using deposition and etching processes, thereby eliminating discrete components and wire bond interconnects. The term MMIC is used for circuits operating in the millimeter wave (30–300 GHz) region of the frequency spectrum as well as the microwave (1–30 GHz) region. Major advantages of MMICs include low cost, small size, low weight, circuit design flexibility, broadband performance, elimination of circuit tweaking, high-volume manufacturing capability, package simplification, improved reproducibility, improved reliability, and multifunction performance on a single chip.

Microwave circuits use two types of active devices: two-terminal devices, referred to as diodes, such as Schottky, Gunn, tunnel, impact avalanche and transit time (IMPATT), varactor, and PIN, and three-terminal devices, referred to as transistors, such as bipolar junction transistor (BJT), metal semiconductor field effect transistor (MESFET), high electron mobility transistor (HEMT), and heterojunction bipolar transistor (HBT). Microwave circuits using these devices include amplifiers, oscillators, multipliers, mixers, switches, phase shifters, attenuators, modulators, and many others used for receiver or transmitter applications covering microwave and millimeter wave frequency bands. New devices, microwave computer-aided design (CAD) tools, and automated testing have played a significant role in the advancement of these circuits during the past decade. The theory and performance of most of these circuits have been well documented [1–10]. Solid state circuits are extensively used in such applications as radar, communication, navigation, electronic warfare (EW), smart weapons, consumer electronics, and microwave instruments and equipment. This section will briefly describe the performance status of amplifiers, oscillators, multipliers, mixers, and microwave control circuits.

41.2 Amplifiers

Amplifier circuits have received maximum attention in solid state circuits development. The two-terminal diode device amplifiers, such as parametric, tunnel, Gunn, and IMPATT, are normally called *reflection-type circuits*, or *negative resistance amplifiers*. A diagram for these amplifiers is shown in Fig. 41.1(a). Parametric amplifiers are narrowband (<10%) and have very good noise figure. Tunnel-diode amplifiers are high-gain, low-noise figure, and low-power circuits. Octave bandwidth of such amplifiers is possible. The performance of Gunn-diode amplifiers is quite similar to tunnel-diode amplifiers. IMPATT-diode amplifiers are high power and high efficiency. They are moderately noisy, and bandwidths up to an octave are possible.

The basic circuit configuration for three-terminal transistor device amplifiers is shown in Fig. 41.1(b). Several different types of amplifiers developed using transistors are low noise, power, high linearity, broadband, high efficiency, logarithmic, limiting, transimpedance, and variable gain. The silicon bipolar transistor performs very well up to about 4 GHz, with reliable performance, high power, high gain, and low cost. The GaAs MESFETs perform better than the bipolar transistors above 4 GHz. They are broadband, have a wide dynamic range, are highly reliable, and are low cost. Both low-noise and medium-power MESFET amplifiers are available. They compete with uncooled parametric amplifiers as well as moderate-power IMPATTs. HEMTs find a niche in low-noise and high-frequency applications. The noise figure of HEMT amplifiers is better than that of uncooled parametric amplifiers up to 100 GHz, as shown in Fig. 41.2.

Various techniques are used to realize small signal or low-power broadband amplifiers. Five of them are shown in Fig. 41.3. The distributed approach provides the unique capability of excellent gain-bandwidth product, low VSWR (voltage standing wave ratio), and moderately low noise figure. This technique has been successfully used in monolithic ultrabroadband amplifiers. The performance of such amplifiers using various transistor devices is given in Table 41.1.

The performance of solid state power amplifiers is shown in Fig. 41.4. Currently, IMPATT and Gunn diodes provide maximum power above 10 GHz, whereas bipolar junction transistor and

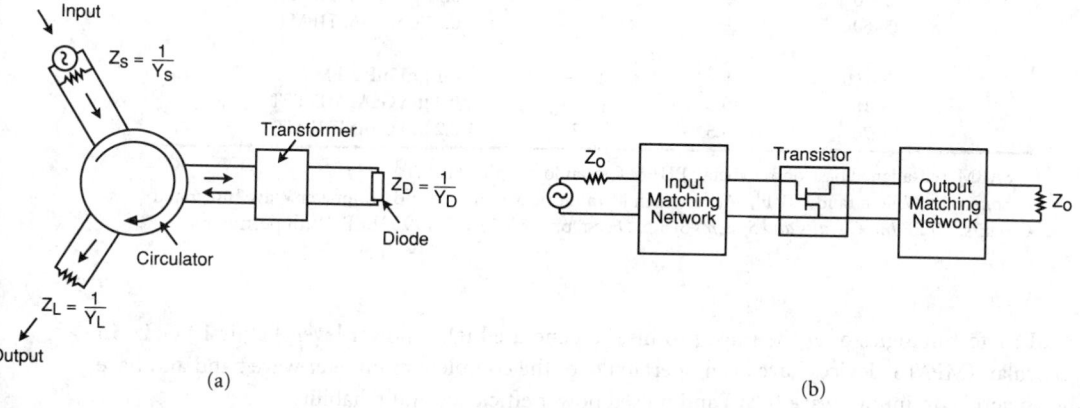

(a)   (b)

FIGURE 41.1 Amplifier circuits configurations. (a) Two-terminal negative resistance type requires a circulator to isolate the input and output ports. (b) Three-terminal transistor type requires input and output matching networks.

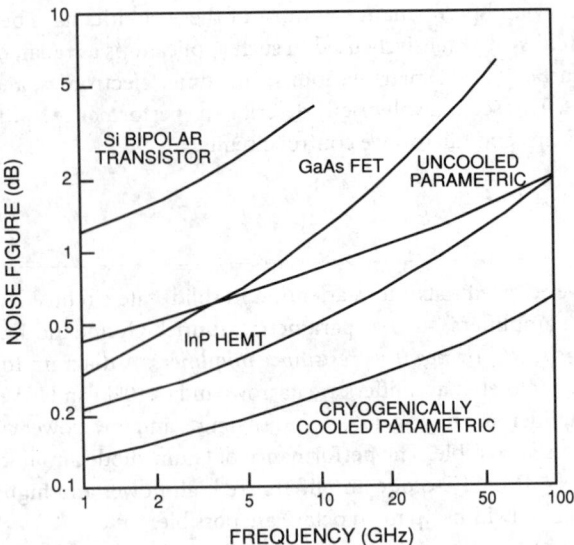

FIGURE 41.2 Comparison of noise performance of various solid state amplifiers; the InP HEMT LNA, which is also compatible with MMIC technology, is a clear choice for receiver applications where cryogenic cooling is precluded. (*Source:* D. Willems and I. Bahl, "Advances in Monolithic Microwave and Millimeter Wave Integrated Circuits," *IEEE Int. Circuits and Systems Symp. Digest,* pp. 783–786. © 1992 IEEE. With permission.)

Table 41.1   Broadband Single-Chip Distributed MMIC Amplifier Performance

Frequency Range (GHz)	Gain (dB)	Noise Figure (dB)	Device Used
0.5–26.5	6	5.2	0.32 μm GaAs HEMT
0.5–50	6	—	0.32 μm GaAs HEMT
2–18	9	5.7	0.5 μm dual gate FET
2–20	9.5	3.5	0.2 μm GaAs HEMT
2–24	6	—	2 μm SABM GaAs HBT
5–40	9	4	0.25 μm GaAs HEMT
5–60	8	—	0.25 μm GaAs HEMT
5–100	5	—	0.1 μm InP HEMT
6–18	10.5	—	0.4 μm GaAs MESFET
9–70	3.5	7	0.2 μm GaAs PHEMT

SABM, self-aligned base ohmic metal; PHEMT, pseudomorphic HEMT.
Source: D. Willems and I. Bahl, "Advances in Monolithic Microwave and Millimeter Wave Integrated Circuits," *IEEE Int. Circuits and Systems Symp. Digest,* pp. 783–786. © 1992 IEEE. With permission.

MESFET technologies offer the most promise to generate higher power levels below 10 GHz. In particular, IMPATT devices have been operated over the complete millimeter wave band and have shown good continuous wave (CW) and pulsed power efficiency and reliability.

During the past decade significant progress has been made in monolithic power amplifiers operating over both the narrowband and broadband [11, 12]. Power levels as high as 12 W from a single

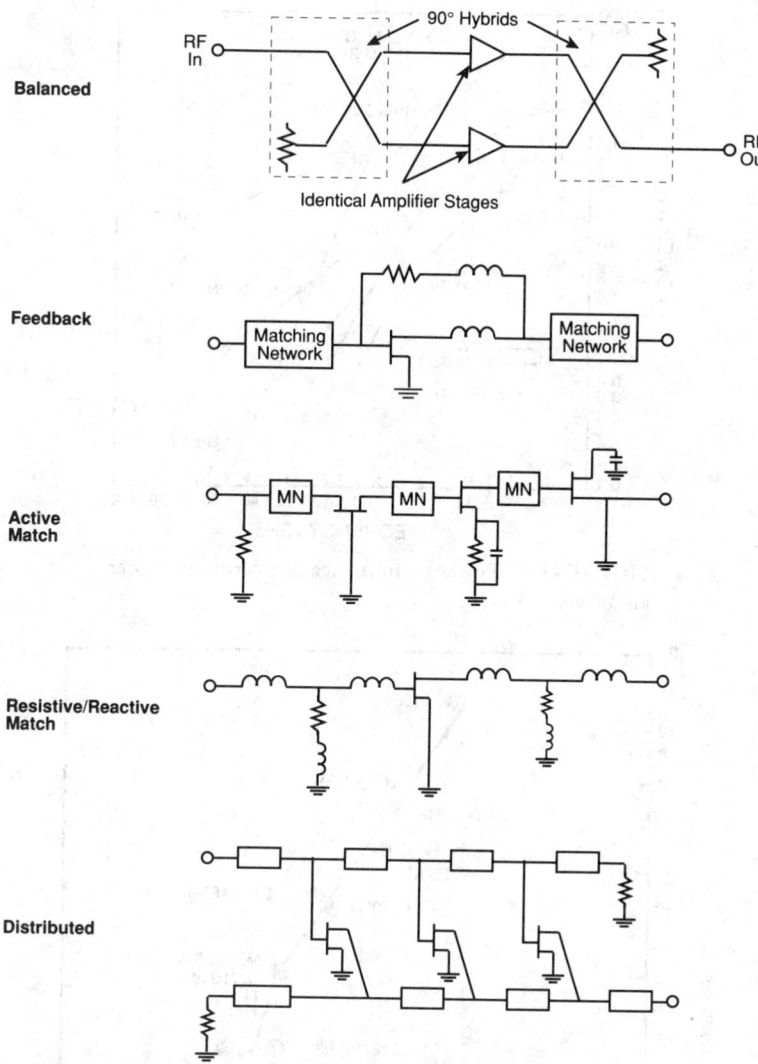

FIGURE 41.3 Broadband amplifier configurations. *Balanced* has low noise figure and better cascadability, *feedback* has small size, *active match* is more suitable for monolithic approach, and *distributed* is good for multioctave bandwidths. (*Source:* I. J. Bahl and P. Bhartia, *Microwave Solid State Circuit Design*, New York: John Wiley, 1988. Reproduced by permission of John Wiley & Sons Limited.)

MMIC chip at C-band with 36% power-added efficiency (PAE) and 0.65 W/mm^2 power density for the chip area (output power per unit area on GaAs) have been demonstrated. A 6-W MMIC chip has been developed at X-band. A 2-W power output has been obtained at 30 GHz. In the high-efficiency area, a C-band MMIC amplifier with 70% PAE, 8-dB gain, and 1.7-W power output has been demonstrated. For broadband amplifiers having an octave or more bandwidth, MMIC technology has been exclusively used and is quite promising. Figure 41.5 depicts power performance for single-chip MMIC amplifiers spanning microwave and millimeter wave frequencies. The state of the art in high efficiency and broadband power MMIC amplifiers is summarized in Tables 41.2 and 41.3, respectively. Note that the high-efficiency examples included in Table 41.2 all exceed 40% PAE.

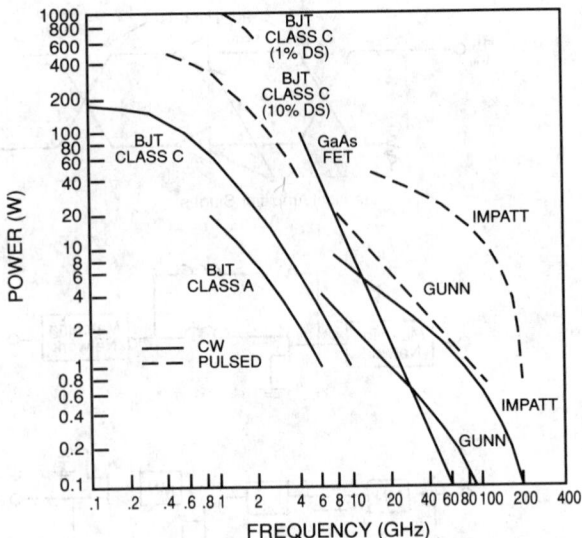

FIGURE 41.4 Power performance of microwave power amplifiers.

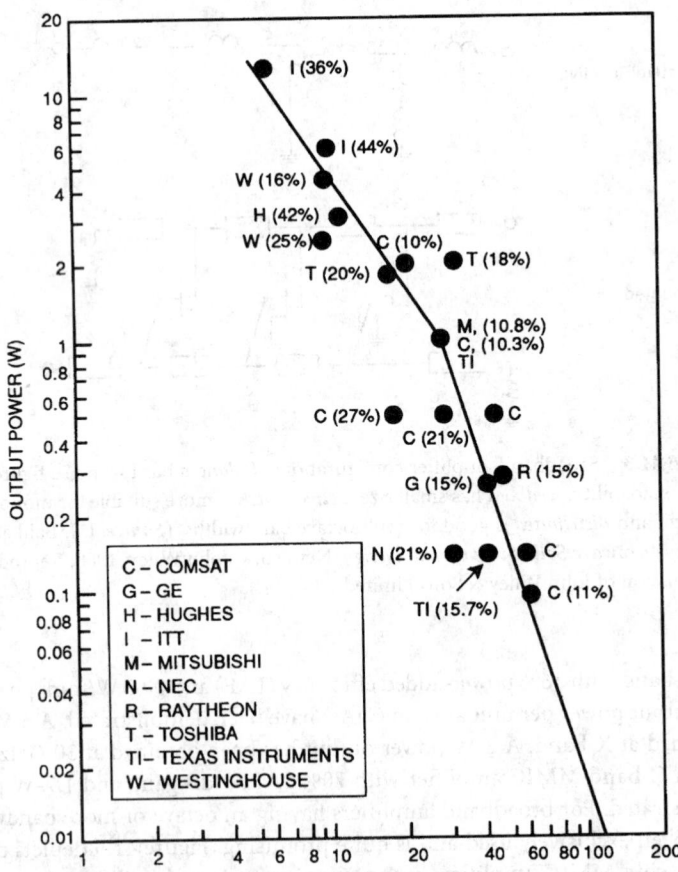

FIGURE 41.5 State of the art in single-chip power MMIC amplifiers. (*Source:* D. Willems and I. Bahl, "Advances in Monolithic Microwave and Millimeter Wave Integrated Circuits," *IEEE Int. Circuits and Systems Symp. Digest,* pp. 783–786. © 1992 IEEE. With permission.)

Table 41.2 Single-Chip High-Efficiency Power MMIC Performance

Frequency (GHz)	No. of Stages	P_O (W)	PAE (%)	Gain (dB)
5.0	1	5.0	60	9
5.5	1	1.7	70	8
8.5	2	3.2	52	—
10.0	1	5.0	48	7
10.0	1	6.0	44	6
11.5	2	3.0	42	12

Source: D. Willems and I. Bahl, "Advances in Monolithic Microwave and Millimeter Wave Integrated Circuits," *IEEE Int. Circuits and Systems Symp. Digest*, pp. 783–786. © 1992 IEEE. With permission.

Table 41.3 Single-Chip Broadband Power MMIC Performance

Frequency (GHz)	Configuration	No. of Stages	Gain (dB)	P_O (W)	PAE (%)
1.5–9.0	Reactive match	2	5	0.5	14
2.0–8.0	Distributed	1	5	1.0	—
2.0–20.0	Distributed	1	4	0.8	15
3.5–8.0	Reactive match	2	10	2.0	20
6–17	Distributed/reactive	4	16	0.8	11
6–20	Distributed	1	11	0.25	—
7–10.5	Reactive match	2	12.5	3.0	35
7.7–12.2	Reactive match	2	8.0	3.0	14
12–16	Reactive match	3	18	1.8	18
14–33	Distributed	1	4	0.1	—

Source: D. Willems and I. Bahl, "Advances in Monolithic Microwave and Millimeter Wave Integrated Circuits," *IEEE Int. Circuits and Systems Symp. Digest*, pp. 783–786. © 1992 IEEE. With permission.

41.3 Oscillators

Solid state oscillators represent the basic microwave energy source and have the advantages of light weight and small size compared with microwave tubes. As shown in Fig. 41.6, a typical microwave oscillator consists of a MESFET as an active device (a diode can also be used) and a passive frequency-determining resonant element, such as a microstrip, surface acoustic wave (SAW), cavity resonator, or dielectric resonator for fixed tuned oscillators and a varactor or a yttrium iron garnet (YIG) sphere for tunable oscillators. These oscillators have the capability of temperature stabilization and phase locking. Dielectric resonator oscillators provide stable operation from 1 to 100 GHz as fixed frequency sources. In addition to their good frequency

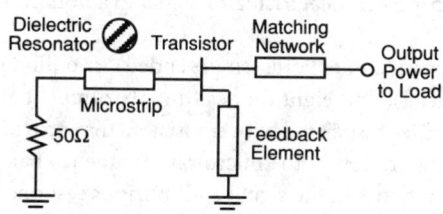

FIGURE 41.6 Basic configuration of a dielectric resonator oscillator. The feedback element is used to make the active device unstable, the matching network allows transfer of maximum power to the load, and the dielectric resonator provides frequency stability.

stability, they are simple in design, have high efficiency, and are compatible with MMIC technology. Gunn and IMPATT oscillators provide higher power levels and cover microwave and millimeter wave bands. The transistor oscillators using MESFETs, HEMTS, and HBTs provide highly cost-effective, miniature, reliable, and low-noise sources for use up to the millimeter wave frequency range, while BJT oscillators reach only 20 GHz. Compared to a GaAs MESFET oscillator, a BJT or a HBT oscillator typically has 6 to 10 dB lower phase noise very close to the carrier. Figure 41.7 shows the performance of various solid state oscillators. Higher power levels for oscillators are obtained by connecting high-power amplifiers at the output of medium-power oscillators.

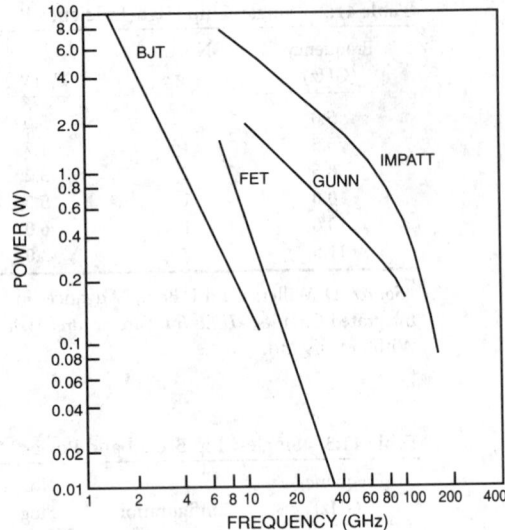

FIGURE 41.7 Maximum CW power available from solid state microwave oscillators.

41.4 Multipliers

Microwave frequency multipliers are used to generate microwave power at levels above those obtainable with fundamental frequency oscillators. Several different nonlinear phenomena can be used to achieve frequency multiplication, e.g., nonlinear reactance in varactors and step-recovery diodes and nonlinear resistance in Schottky barrier diodes and three-terminal devices (BJT, MESFET, HEMT, HBT).

Varactor multipliers offer the best frequency multiplier performance. Varactor multipliers (pulsed) have achieved power output in excess of 100 and 10 W at 4 and 10 GHz, respectively [5]. Table 41.4 shows the best performance measured in the millimeter wave range and above.

41.5 Mixers

Mixers convert (heterodyne) the input frequency to a new frequency, where filtering and/or gain is easier to implement, in contrast to detectors, which are used to provide an output signal that contains the amplitude or amplitude variation information of the input signal. A mixer is basically a multiplier, which requires two signals and uses any solid state device that exhibits nonlinear properties. Mixing is achieved by applying an RF and a high-power local oscillator signal to a nonlinear element, which can be a diode or a transistor.

As illustrated in Fig. 41.8, there are many types of mixers: one diode (single ended), two diodes (balanced or antiparallel), four diodes (double balanced), and eight diodes (double-double balanced). Mixers can also be realized using the nonlinearities associated with transistors that provide conversion gain. The most commonly used mixer configuration in the microwave frequency band is the double-balanced mixer, which has better isolation between the signal and spurious response. However, the single and balanced mixers place lower power requirements on the local oscillator and have lower conversion loss.

Subharmonic mixing (where the local oscillator frequency is approximately half that needed in conventional mixers) has been extensively used at millimeter wave frequencies. This technique is quite useful when reliable stable local oscillators are either unavailable or prohibitively expensive at high frequencies. Figure 41.9 gives the performance of millimeter wave mixers.

Table 41.4 Summary of State-of-the-Art Performance for Millimeter Wave Frequency Multipliers

Mount Type	Tunable Output Operating Band (GHz)	Minimum Output		Maximum Output			Maximum Pump Power (mW)	Notes[a]
		Effic. (%)	Power (mW)	Effic. (%)	Power (mW)	Freq. (GHz)		
Doubler	80–120	9.5	18	14.0	26.6	88 and 105	190	2, 3, 9
	80–120	10.7	16	15.5	23.2	100	150	1, 2, 3
	80–120	10	7	16	11	104	70	1, 4, 3
	100	—	—	25	20	100	80	6, 4
	110–170	10	8	15	12.0	120	80	1, 2, 3
	140–150	10	8	22	17.6	145	80	1, 2, 3, 5
	190–260	10	8	27	21.5	215	80	1, 2, 3
	200	—	—	19	18	200	150	6, 4
	400	—	—	8.5	0.44	300	5.1	1, 2, 3, 7
	500–600	7	0.7	—	—	—	10	1, 2, 8
Tripler	85–115	4	1.2	8	2.4	106	28	1, 2, 8
	96–120	1.8	1.8	8.2	8.2	110	100	1, 2, 3
	105	—	—	25	18	105	72	6, 4
	200–290	2.5	2.0	7.5	6	225	80	1, 2, 3
	190–240	1	0.3	10	3	230	30	1, 2, 8
	260–350	1.8	1.5	3.75	3.0	340	80	2, 3, 6
	300	—	—	2	2	300	100	6, 4
	450	—	—	1	0.079	450	6.3	1, 2, 3, 7
× 6 balanced doubler/ tripler	310–350	0.3	0.6	0.4	0.75	345	190	1, 2, 3, 6, 9

[a] 1, Crossed waveguide mount; 2, tuning and bias optimized at each operating frequency; 3, microstrip low-pass filter; 4, fixed tuning and bias; 5, narrowbanded version of NRAO 110- to 170-GHz doubler; 6, quasi-optical mount; 7, limited pump power available; 8, coaxial low-pass filter; 9, two-diode balanced cross guide mounts.

Source: K. Chang (Ed.), *Handbook of Microwave and Optical Components*, vol. 2, New York: John Wiley, 1990. Reproduced by permission of John Wiley & Sons Limited.

41.6 Control Circuits

Control components are widely used in communication, radar, EW, instrument, and other systems for controlling the signal flow or to adjust the phase and amplitude of the signal [5, 8, 13, 14]. PIN diodes and MESFETs are extensively used in HMICs and MMICs, respectively, for microwave control circuits, such as switches, phase shifters, attenuators, and limiters. PIN diode circuits have low loss and can handle higher power levels than do MESFET components; conversely, the latter have great flexibility in the design of integrated subsystems, consume negligible power, and are low cost.

Figure 41.10 shows various control configurations being developed using PIN and MESFET devices. Either device can be used in these circuits.

The most commonly used configuration for microwave switches is the single-pole double throw (SPDT) as shown in Fig. 41.10(a), which requires a minimum of two switching devices (diodes or transistors). Table 41.5 provides typical performance for broadband SPDT switches developed using GaAs MESFET monolithic technology. Table 41.5 also summarizes performance for phase shifters and attenuators, which are described briefly below.

There are four main types of solid state digitally controlled phase shifters: switched line, reflection, loaded line, and low-pass/high-pass, as shown in Fig. 41.10(b). The switched-line and low-pass/high-pass configurations, which are most suitable for broadband applications and compact size, are not suitable for analog operation. Reflection and loaded-line phase shifters are inherently narrowband; however, the loaded-line small bit phase shifters, 22.5 degrees or less, can be designed

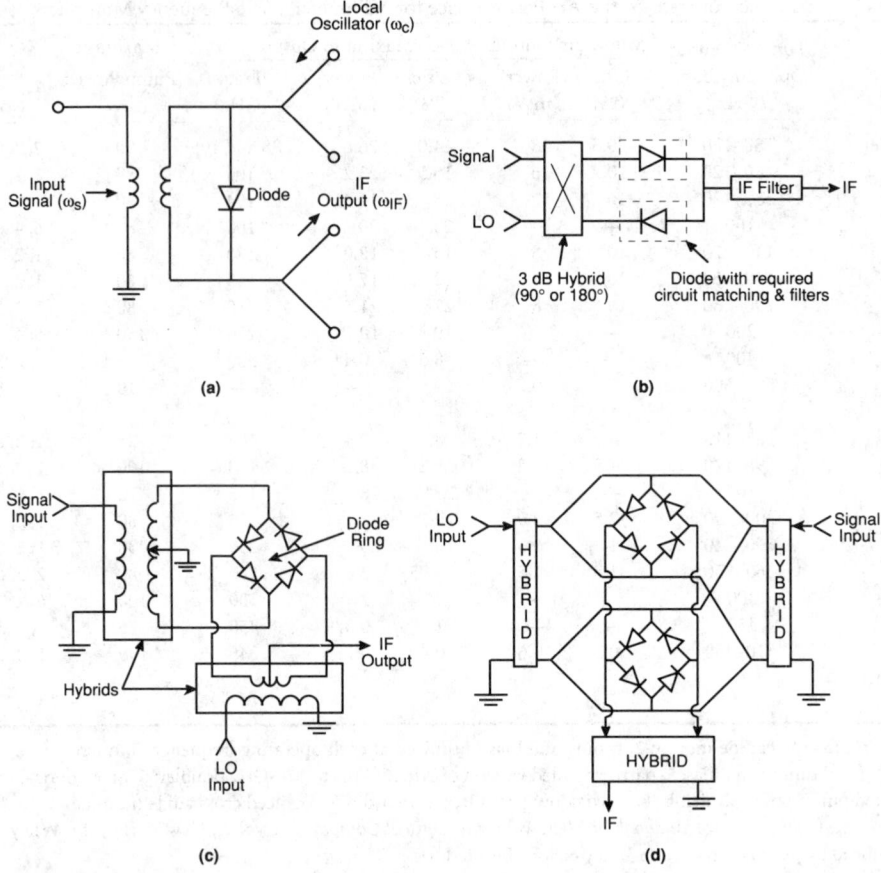

FIGURE 41.8 Basic mixer configurations: (a) single ended, (b) balanced, (c) double balanced, and (d) double-double balanced.

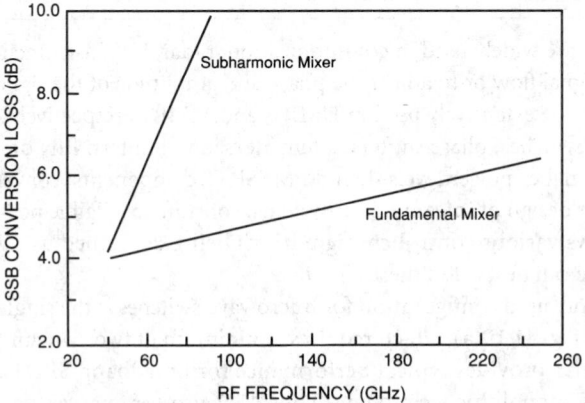

FIGURE 41.9 Single-sideband (SSB) conversion loss of millimeter wave mixers. Subharmonic type mixers have higher conversion loss but are generally less expensive. (*Source:* P. Bhartia and I. J. Bahl, *Millimeter Wave Engineering and Applications,* New York: John Wiley, 1984. Reproduced by permission of John Wiley & Sons Limited.)

Table 41.5 Performance of GaAs FET Control Monolithic ICs

Circuit Function	Frequency (GHz)	Maximum Insertion Loss (dB)	Minimum Return Loss (dB)	RMS Error (deg.)	Minimum Isolation (dB)	Minimum Dynamic Range (dB)	Maximum Power (dBm)	Minimum Switching Speed (ns)	Size (mm²)
Broadband SPDT switch	DC-20	2.0	7	—	24	—	25	0.4	1.27 × 1.27
	DC-40	3.0	7	—	23	—	25	0.8	0.84 × 1.27
2 × 2 Switch matrix	DC-20	3.5	10	—	30	—	—	—	2.03 × 1.78
6-Bit phase shifter	5–6	6.0	10	±2	—	—	—	5.0	9.43 × 4.20
3-Bit phase shifter	6–18	11.5	13	±8.5	—	—	—	—	2.2 × 1.25
Variable attenuator	DC-50	4.2	9	—	—	30	—	1.0	1.52 × 0.65
Multibit attenuator	DC-20	5.0	12	—	—	75	26	—	2.55 × 1.57

to have up to an octave bandwidth. Phase shifters using the vector-modulator concept have also been developed in monolithic form.

Voltage-controlled variable attenuators are important control elements and are widely used for automatic gain control circuits. They are indispensable for temperature compensation of gain variation in broadband amplifiers. Both PIN diodes and GaAs MESFET devices are used for variable attenuators. Figure 41.10(c) shows a variable attenuator configuration using MESFETs in the passive mode. Apart from the use of MESFETs in the passive mode, active MESFET amplifier circuits may also be used for variable attenuation circuits. Dual-gate MESFET amplifiers with controlled voltage applied to the second gate are ideal for this purpose and provide a lower noise figure than the passive attenuators.

A *limiter,* whose basic configuration is shown in Fig. 41.10(d), is an important control component used at microwave frequencies. An ideal limiter has no attenuation when low power is incident upon it but has an attenuation that increases with increasing power (above a threshold level) to maintain constant output power. Limiters are also used to protect the receivers from other nearby radar transmitters. Schottky and PIN diodes are commonly used to realize these components. Because they are used at the front ends of receivers, low loss is one of the basic requirements.

41.7 Summary and Future Trends

In this section we have briefly described the performance of microwave and millimeter wave solid state circuits. Currently the discrete silicon bipolar transistors, IMPATT diodes, and GaAs MESFETs are most widely used in solid state circuits and bipolar transistors, and IMPATT diodes are still the most powerful solid state sources below S-band and 10 to 300 GHz, respectively. The MESFET is the workhorse for microwave circuits up to 40 GHz. The HEMT applications are similar to the MESFET, including amplifiers, oscillators, frequency multipliers, mixers, and control circuits, and above 50 GHz, this device is exclusively used for power and low-noise amplifiers. The HBTs applications include high-efficiency amplifiers, ultrabroadband dc amplifiers, low-noise oscillators, frequency multipliers, and mixers.

During the past decade, GaAs monolithic technology has made tremendous progress in producing single and multifunction microwave and millimeter wave circuits using MESFETs, HEMTs, and HBTs. This technology is exerting a profound impact in producing low-cost and high-volume solid state circuits. Since monolithic circuits have the advantage of bandwidth and reproducibility over discrete devices because of wire bond elimination, the growth in millimeter wave solid state circuits will be based on this technology, and a cost reduction by a factor greater than 10 is expected in the near future. Wherever possible, the two-terminal devices such as IMPATT and Gunn diodes will be replaced by the transistors. Future trends in solid state circuits also include their optical control, tuning, and stabilization.

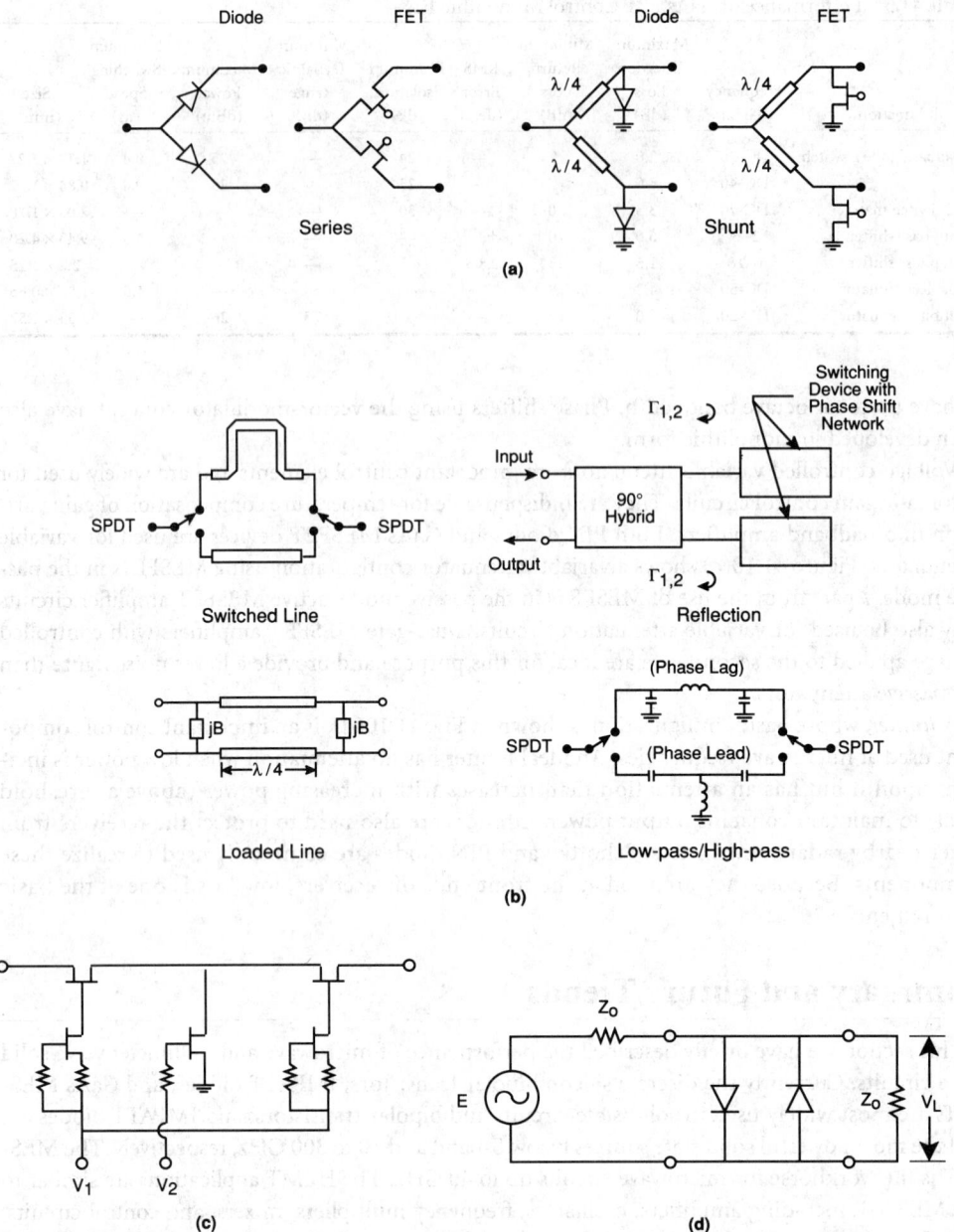

FIGURE 41.10 Microwave control circuits: (a) SPDT switch configurations, (b) basic phase shifter types, (c) schematic of a MESFET attenuator, and (d) basic limiter circuit using two rectifying diodes.

Defining Terms

Hybrid microwave integrated circuit (HMIC): A planar assembly that combines different circuit functions formed by strip or microstrip transmission lines printed onto a dielectric substrate and incorporating discrete semiconductor solid state devices and passive distributed or lumped circuit elements, interconnected with wire bonds.

Monolithic microwave integrated circuit (MMIC): An MMIC is formed by fabricating all active and passive circuit elements or components and interconnections onto or into the surface of a semi-insulating semiconducting substrate by deposition and etching schemes such as epitaxy, ion implantation, sputtering, evaporation, and/or diffusion, and utilizing photolithographic processes for pattern definition, thus eliminating the need for internal wire bond interconnects.

References

1. E. L. Kollberg (Ed.), Microwave and Millimeter-Wave Mixers, New York: IEEE Press, 1984.
2. P. Bhartia and I. J. Bahl, *Millimeter Wave Engineering and Applications*, New York: John Wiley, 1984.
3. R. A. Pucel (Ed.), *Monolithic Microwave Integrated Circuits*, New York: IEEE Press, 1985.
4. S. A. Maas, *Microwave Mixers*, Norwood, Mass.: Artech House, 1986.
5. I. J. Bahl and P. Bhartia, *Microwave Solid State Circuit Design*, New York: John Wiley, 1988.
6. R. Goyal (Ed.), *Monolithic Microwave Integrated Circuits: Technology and Design*, Norwood, Mass.: Artech House, 1989.
7. F. Ali, I. Bahl, and A. Gupta (Eds.), *Microwave and Millimeter-Wave Heterostructure Transistors and Their Applications*, Norwood, Mass.: Artech House, 1989.
8. K. Chang (Ed.), *Handbook of Microwave and Optical Components*, vol. 2, New York: John Wiley, 1990.
9. G. D. Vendelin, A. M. Pavio, and U. L. Rohde, *Microwave Circuit Design Using Linear and Nonlinear Techniques*, New York: John Wiley, 1990.
10. F. Ali and A. Gupta (Eds.), *HEMTs and HBTs: Devices, Fabrication and Circuits*, Norwood, Mass.: Artech House, 1991.
11. D. Willems and I. Bahl, "Advances in monolithic microwave and millimeter wave integrated circuits," *IEEE Int. Circuits and System Symp. Digest*, pp. 783–786, 1992.
12. H. Q. Tserng and P. Saunier, "Advances in power MMIC amplifier technology in space communications," *Proc. SPIE-Monolithic Microwave Integrated Circuits for Sensors, Radar, and Communications Systems*, pp. 74–85, 1991.
13. A. K. Sharma, "Solid-state control devices: State of the art," *Microwave Journal*, 1989 State of the Art Reference, pp. 95–112, Sept. 1989.
14. V. Sokolov, "Phase shifters technology assessment: Prospects and applications," *Proc. SPIE-Monolithic Microwave Integrated Circuits for Sensors, Radar and Communications Systems*, vol. 1475, pp. 228–332, 1991.

Further Information

The monthly journal *IEEE Transactions on Microwave Theory and Techniques* routinely publishes articles on the design and performance of solid state circuits. Special issues published in July 1982, January and February 1983, March 1984, and September 1989 exclusively deal with this topic.

IEEE Microwave and Millimeter-Wave Monolith Circuits Symposium Digests, published every year since 1982, include comprehensive information on the design and performance of monolithic microwave and millimeter-wave solid state circuits.

Books included in the references of this chapter discuss thoroughly the design, circuit implementation, and performance of solid state circuits.

THE RAD LAB

On January 4, 1941, an experimental microwave radar set at the recently organized Radiation Laboratory at the Massachusetts Institute of Technology picked up its first echoes reflected from buildings in the vicinity. In the fall of 1940, a British technical mission, headed by Sir Henry Tizard, had brought a resonant-cavity magnetron capable of high-power microwave generation to the United States. British disclosure of the revolutionary device led to the formation of the Rad Lab where an elite group of scientists and engineers devoted five years to the development of magnetrons and radar applications of them. In the process, sophisticated theories of magnetrons and other microwave devices and systems were formulated. The Rad Lab staff also contributed comprehensive design principles and developed novel methods of fabrication and testing. This body of knowledge was disseminated after the War in the well-known Rad Lab Series of 28 volumes.

Lee A. Dubridge from the physics department of the University of Rochester served as Director of the Rad Lab that began operation with a small staff of physicists in a room at MIT in November 1940. The facility included a "magnetron safe"

and the executive secretary, Edythe W. Baker, had the room windows painted "to keep outsiders from spying on the magnetrons." The number of personnel increased rapidly to about 2,000 in 1942 and led to organizational changes as the Rad Lab was divided into 11 Divisions. The Divisions were further divided into Groups. For example, Division 5 was responsible for transmitter components and it had 7 groups, including Group 52, known as the "Magnetron Group." The number of people employed at the Rad Lab peaked at 3,897 in August 1945. This number consisted of 1,189 technical staff, 1,301 nonstaff men, and 1,407 nonstaff women. At least 6,446 people spent at least some time at the Rad Lab by the time of its final termination. Nonstaff men worked as technicians, mechanics, draftsmen, and guards, while nonstaff women generally were assigned to secretarial and clerical work.

The authors of the book *Five Years*[1] stated that for most of the staff members, the Rad Lab became "a sort of second college." They described the staff as being a "heterogeneous, noisy, brawling, affectionate mob" who "were the Brooklyn Dodgers of the OSRD league." Although a major-

---

[1] *Five Years at the Radiation Laboratory,* a 1947 MIT publication somewhat like a college yearbook.

ity of the technical staff came from universities and 421 had graduate degrees, their backgrounds were diverse with degrees that included anthropology, music, biology, and political science as well as physics, engineering, and mathematics. Reportedly, the Rad Lab "seethed with sociability" with many parties and picnics. There were frequent weddings and one Division had its own orchestra. Historian Henry Guerlac wrote about the Rad Lab "that one of its outstanding merits, in the eyes of its own management, was that it was a physicist's world, run for, and as completely as possible by, physicists."

As part of an on-going research project on the Rad Lab, I have examined the activities of the Magnetron Group in some detail and plan to extend the analysis to other units. The authors of *Five Years* described this group as being "the heart of microwave radar development" and "U.S. magnetron headquarters." I have identified 117 people who spent at least some time in this group, 33 of them being women. Of the 56 who were technical staff members, four were women. Sixteen staff members held the doctoral degree with 14 in physics, one in physical chemistry and one in anthropology (but none in engineering). Monica Healea, the only woman among this group of Ph.D.'s, had received her doctorate from Radcliffe in 1936 and left the Rad Lab in 1944 to become head of the physics department at Vassar.

John Longyear had received a doctorate in anthropology from Harvard in 1940 and left in 1945 to join an archaeological expedition to Honduras. Eleven of these Ph.D.'s later joined universities while four joined industrial corporations. Thirty six of the 51 with college degrees were in physics. The dominance by physicists mentioned by Guerlac certainly seemed to apply at least in the Magnetron Group. It was staffed much like a large university physics department and the seven Ph.D.'s from Harvard and MIT made up a large segment of the intellectual leadership of Group 52.

The Rad Lab existed for only five years but became a paradigm case of what highly educated and motivated people from diverse backgrounds could achieve in a short time. It was a prototype of the type of institution that can facilitate rapid technological change. Its "graduates" carried the doctrine elsewhere when it closed in 1946. John Rigden and I.I. Rabi observed that the "Rad Lab experience had a profound effect on the Rad Lab physicists themselves and, subsequently, on physics." They noted that its legacy ranged "from the modern kitchen to the police cruiser, from the ocean liner to jet aircraft, from military defense warning systems to meteorological warning systems." Rigden and Rabi credited the Lab staff with having established "an entire field of science and engineering."

Source: Adapted from J.E. Brittain, *Proc. IEEE,* vol. 79, no. 1, p. 92, January 1991. © 1991 IEEE.

42

Three-Dimensional Analysis

42.1 Introduction 1018
42.2 The Field Equations 1018
    Low Frequency Fields • Statics Limit
42.3 Numerical Methods 1020
    Finite Elements • Edge Elements • Integral Methods
42.4 The Modern Design Environment 1023

C. W. Trowbridge
Vector Fields, Inc.

42.1 Introduction

The three-dimensional analysis of electromagnetic fields requires the use of numerical techniques exploiting the best available computer systems. The well-found laboratory will have at its disposal a range of machines that allow interactive data processing with access to software that provides geometric modeling tools and has sufficient central processing unit (CPU) power to solve the large (>10,000) set of algebraic equations involved.

42.2 The Field Equations

The classical equations governing the physical behavior of electromagnetic fields over the frequency range dc to light are **Maxwell's equations**. These equations relate the magnetic flux density (**B**), the electric field intensity (**E**), the magnetic field intensity (**H**), and the electric field displacement (**D**) with the electric charge density (ρ) and electric current density (**J**). The field vectors are not independent since they are further related by the material constitutive properties: $\mathbf{B} = \mu\mathbf{H}$, $\mathbf{D} = \epsilon\mathbf{E}$, and $\mathbf{J} = \sigma\mathbf{E}$ where μ, ϵ, and σ are the material permeability, permittivity, and conductivity, respectively. In practice these quantities may often be field dependent, and furthermore, some materials will exhibit both anisotropic and hysteretic effects. It must be strongly stated that accurate knowledge of the material properties is the single most important factor in obtaining reliable simulations.

Because the flux density vector satisfies a zero divergence condition (*div* **B** = 0), it can be expressed in terms of a magnetic vector potential **A**, i.e., **B** = *curl* **A**, and it follows from Faraday's law that $\mathbf{E} = -(\partial\mathbf{A}/\partial t + \nabla V)$, where V is the electric scalar potential. Neither **A** nor V is completely defined since the gradient of an arbitrary scalar function can be added to **A** and the time derivative of the same function can be subtracted from V without affecting the physical quantities **E** and **B**. These changes to **A** and V are the *gauge* transformations, and uniqueness is usually ensured by specifying the divergence of **A** and sufficient boundary conditions. If $\nabla \cdot \mathbf{A} = -(\mu\sigma V + \mu\epsilon\partial V/\partial t)$ (Lorentz gauge) is selected, then the field equations in terms of **A** and V are:

$$\nabla \times \frac{1}{\mu} \nabla \times \mathbf{A} + \sigma \frac{\partial \mathbf{A}}{\partial t} + \epsilon \frac{\partial^2 \mathbf{A}}{\partial t^2} = \nabla \left[\frac{1}{\mu} \nabla \cdot \mathbf{A} \right]$$

$$\mu\epsilon \frac{\partial^2 V}{\partial t^2} + \mu\sigma \frac{\partial V}{\partial t} = \nabla \cdot \nabla V$$

(42.1)

where σ has been assumed piecewise constant. This choice of gauge decouples the vector potential from the scalar potential. For the important class of two-dimensional problems there will be only one component of **A** parallel to the excitation current density. For fields involving time, at least two types can be distinguished: the time harmonic (ac) case in which the fields are periodic at a given frequency ω, i.e., $\mathbf{A} = \mathbf{A}_o \exp(j\omega t)$, and the transient case in which the time dependence is arbitrary.

Low Frequency Fields

In the important class of problems belonging to the low frequency limit, i.e., eddy current effects at power frequencies, the second derivative terms with respect to time (wave terms) in Eq. (42.1) vanish. This approximation is valid if the dimensions of the material regions are small compared with the wavelength of the prescribed fields. In such circumstances the displacement current term in Maxwell's equations is small compared to the free current density and there will be no radiation [Stratton, 1941]. In this case, while a full vector field solution is necessary in the conducting regions, in free space regions, where $\sigma = 0$ and *curl* $\mathbf{H} = \mathbf{J}_s$, Eqs. (42.1) can be replaced by $\nabla^2 \psi = 0$, where ψ is a scalar potential defined by $\mathbf{H} = -\nabla\psi$. The scalar and vector field regions are coupled together by the standard interface conditions of continuity of normal flux (**B**) and tangential field (**H**).

Statics Limit

In the statics limit (dc) the time-dependent terms in Eq. (42.1) vanish, and the field can be described entirely by the Poisson equation in terms of a single component scalar potential, which will be economic from the numerical point of view. In this case the defining equation is

$$\nabla \cdot \mu\nabla\phi = \nabla \cdot \mu\mathbf{H}_s$$

(42.2)

where ϕ is known as the reduced magnetic scalar potential with $\mathbf{H} = \mathbf{H}_s - \nabla\phi$, and $\mathbf{H}_s$ the source field given by the **Biot Savart law**. Some care is needed in solving Eq. (42.2) numerically, in practice, as $\mathbf{H}_s$ will often be calculated to a higher accuracy than ϕ. For instance, in regions with high permeability (e.g., ferromagnetic materials), the total field intensity **H** tends to a small quantity which can lead to significant errors due to cancellation between grad ϕ and $\mathbf{H}_s$, depending upon how the computations are carried out. One approach that avoids this difficulty is to use the total scalar potential ψ in regions that have zero current density [Simkin and Trowbridge, 1979], i.e., where $\mathbf{H} = -\nabla\psi$ and $\mathbf{H}_c$ is the coercive field for the material ψ satisfies

$$\nabla \cdot \mu\nabla\psi = \nabla \cdot \mu\mathbf{H}_c$$

(42.3)

Again, the two regions are coupled together by the standard interface condition that results, in this case, in a potential jump obtained by integrating the tangential continuity condition, i.e.,

$$\phi = \psi + \int_0^\Gamma \mathbf{H}_s d\Gamma$$

(42.4)

where Γ is the contour delineating the two regions that must not intersect a current-carrying region; otherwise the definition of ψ will be violated.

42.3 Numerical Methods

Numerical solutions for the field equations are now routine for a large number of problems encountered in magnet design; these include, for example, two-dimensional models taking into account nonlinear, anisotropic, and even hysteretic effects. Their use for complete three-dimensional models is not so widespread because of the escalating computer resources needed as the problem size increases. Nevertheless, solutions for nonlinear statics devices are regularly obtained in industry, and time-dependent solutions are rapidly becoming cost effective as computer hardware architectures develop.

Finite Elements

This increasing use of computer-based solutions has come about largely because of the generality of the finite element method (FEM). In this method, the problem space is subdivided (discretized) into finite regions (elements) over which the solution is assumed to follow a simple local approximating trial function (shape functions). In the simplest situation, a particular element could be a plane hexahedra defined by its eight vertices or nodes and a solution of Eq. (42.3) may be approximated by

$$\psi \approx u = \alpha_1 + \alpha_2 x + \alpha_3 y + \alpha_4 z + \alpha_5 xy + \alpha_6 yz + \alpha_7 zx + \alpha_8 xyz = \sum N_i U_i \quad (42.5)$$

Because a hexahedra has eight nodes it is natural to select a bilinear trial function with eight parameters; see Fig. 42.1 for other examples. The functions N_i are called the local shape functions and the parameters U_i are the solution values at the nodes. The finite elements can be integrated into a numerical model for the whole problem space either by (a) the variational method in which the total energy of the system is expressed in terms of the finite element trial functions and then minimized to determine the best solution or (b) the weighted residual method in which the formal error (residual), arising by substituting the trial functions into the defining equation, is weighted by a suitably chosen function and then integrated over the problem domain. The best fit for the trial function parameters can then be obtained by equating the integral to zero. Both methods lead to a set of algebraic equations and are equivalent if the weighting functions are chosen to be the trial functions (Galerkin's method [Zienkiewicz, 1990]). At the element level, the residual R_i is given by

$$R_i = \left[\int_{\text{elem}} \nabla N_i \mu \nabla N_j d\Omega \right] U_j + \int_{\text{elem}} N_i Q d\Omega \quad (42.6)$$

where Q (RHS) denotes the sources of Eqs. (42.2) or (42.3). The integrals can be readily evaluated and assembled for the whole problem by superposition, taking account of the boundary conditions and removing the redundancy at shared nodes. At interelement boundaries in a region of particular potential [reduced Eq. (42.2) or total Eq. (42.3)] the solution is forced to be continuous, but the normal flux (i.e., $\mu \partial U/\partial n$) will only be continuous in a weak sense, that is to say the discontinuity will depend upon the mesh refinement.

The FEM provides a systematic technique for replacing the continuum field equations with a set of discrete algebraic equations that can be solved by standard procedures. In Fig. 42.2 a typical field map is shown for a permanent magnet machine modeled by a computer simulator that can take

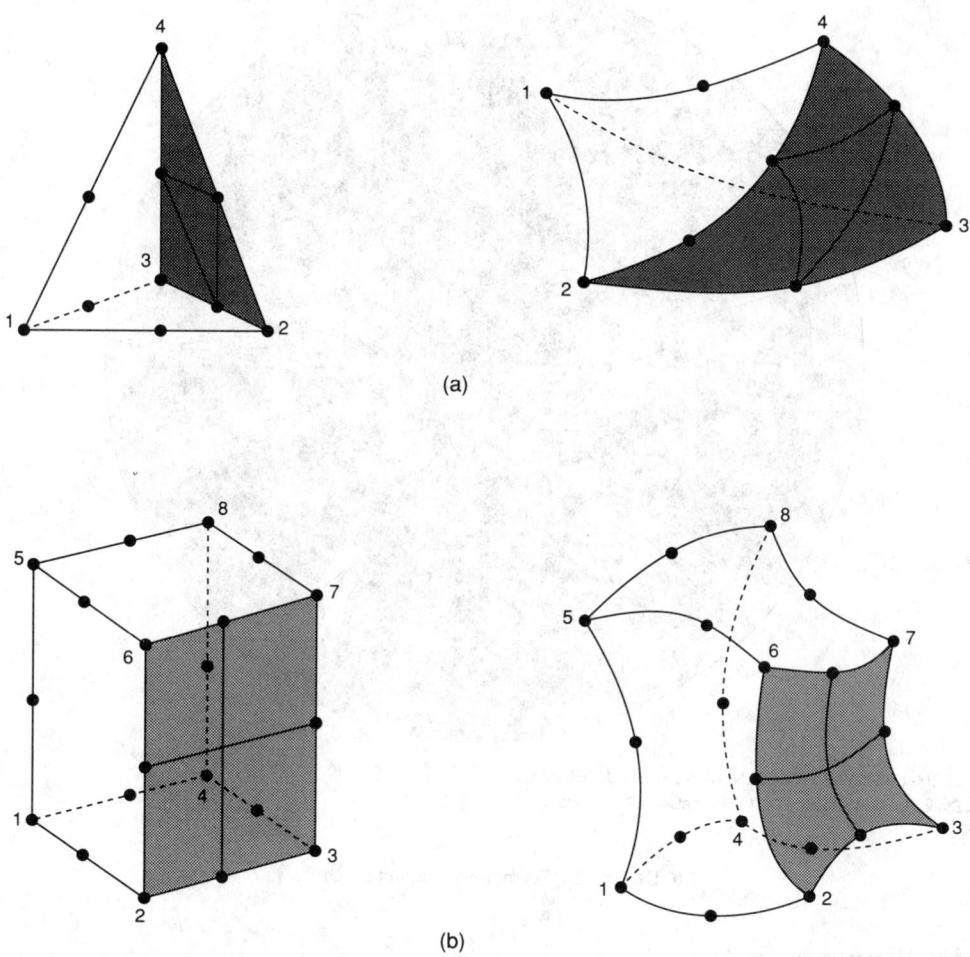

FIGURE 42.1 Three-dimensional second-order ISO plus parametric finite elements. (a) *Left*, master tetrahedron, 10 nodes in local space (ξ, η, ζ); *right*, actual tetrahedron, 10 nodes in global space (x, y, z). (b) *Left*, master hexahedron, 20 nodes in local space (ξ, η, ζ); *right*, actual hexahedron, 10 nodes in global space (x, y, z).

into account nonlinearity and permanently magnetized materials. Although hysteresis effects can be included, the computational resources required can be prohibitive because of the vector nature of magnetization. The magnetic material must be characterized by a large number of measurements to take account of the minor loops, and from these the convolution integrals necessary to obtain the constitutive relationships can be evaluated [Mayergoyz, 1990]. These characteristics must then be followed through time; this can be implemented by solving at a discrete set of time points, given the initial conditions in the material.

Although the FEM is widely used by industry for electromagnetic problems covering the entire frequency range, there are many situations where special methods are more effective. This is particularly the case for high-frequency problems, e.g., millimeter and microwave integrated circuit structures where integral equation techniques and such procedures as transmission line modeling (TLM), spectral domain approach, method of lines, and wire grid methods are often preferred [Itoh, 1989] (see Chapter 43).

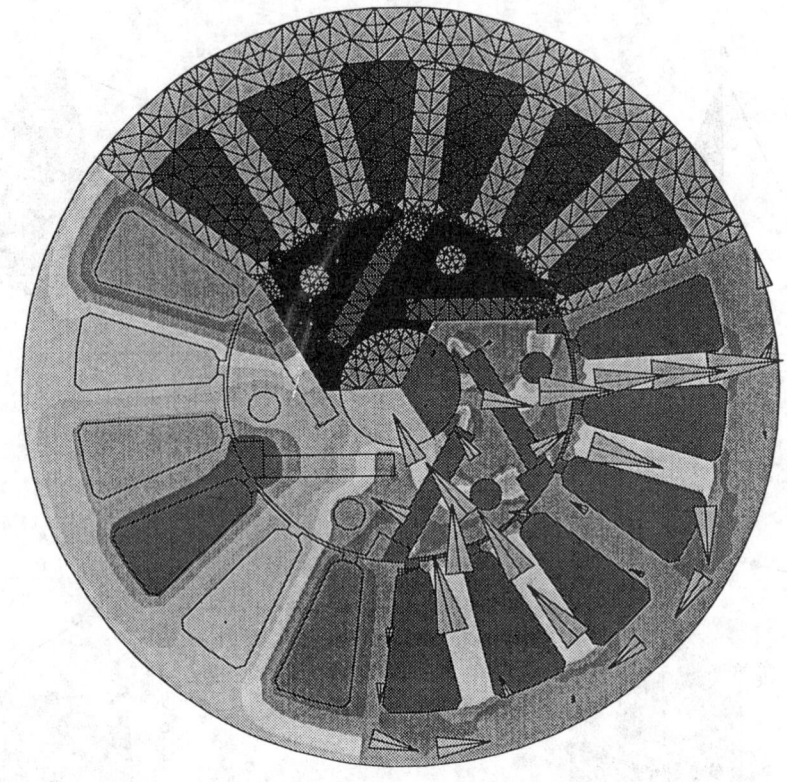

ELEM=LINE SYMM=XY SOLN=AT SCALE=1.0 FIEL=MAGN
Static Solution   Mesh   6610 Elements   42 Regions

FIGURE 42.2   Permanent magnet machine.

Edge Elements

Using potentials and nodal finite elements (see Fig. 42.1) rather than field components directly has the advantage that difficulties arising from field discontinuities at material interfaces can be avoided. However, if the element basis functions [see Eq. (42.5)] are expressed in terms of the field (**H**, say) constrained along an element edge, then tangential field continuity is enforced [Bossavit, 1988]. The field equations [Eq. (42.1)] in terms of the field intensity for the low frequency limit reduce to

$$\nabla \times \nabla \times \mathbf{H} + \sigma \frac{\partial (\mu \mathbf{H})}{\partial t} = 0 \tag{42.7}$$

and a suitable edge variable basis function form for solving this equation by finite elements using tetrahedral elements is

$$\mathbf{h}(\mathbf{r}) = \mathbf{a} + \mathbf{b} \times \mathbf{r} \tag{42.8}$$

where **r** is the position vector and **a** and **b**, respectively, are vectors dependent on the geometry of the element. The basis function expansion is given by

$$\mathbf{H} = \sum \mathbf{h}_e(\mathbf{r})\mathbf{H}_e \tag{42.9}$$

where $\mathbf{h}_e$ is the vector basis function for edge **e**, and $\mathbf{H}_e$ is the value of the field along an element edge (see Fig. 42.3). The functions, Eqs. (42.8) and (42.9), have the property of being divergence free, and most important they ensure that the tangential component of **H** is continuous while allowing for the possibility of a discontinuity in the normal component. In nonconducting regions where the field can be economically modeled by a scalar potential, standard nodal elements can be used. At the interface the edge elements couple exactly with the nodal elements.

Integral Methods

An alternative procedure is to solve the field equations in their integral form; for example, in magnetostatics, the magnetization vector **M** given by $\mathbf{M} = (\mu - 1)\mathbf{H}$ can be used instead of **H** to derive an integral equation over all ferromagnetic domains of the problem, i.e.,

$$\mathbf{M(r)} = (\mu - 1)\left[\mathbf{H}_s(\mathbf{r}') - \frac{1}{4\pi}\nabla\int_\Omega \mathbf{M}\cdot\nabla\left(\frac{1}{R}\right)d\Omega\right] \tag{42.10}$$

where R is the distance between the source and field point, respectively. For problems with linear materials Eq. (42.10) reduces to integrations over the bounding surfaces of materials in terms of the magnetic scalar potential, i.e.,

$$4\pi\phi = -\int_\Gamma \left(\frac{1}{R}\frac{\partial\phi}{\partial n} - \phi\frac{\partial 1/R}{\partial n}\right)d\Gamma \tag{42.11}$$

Equation (42.11) can be solved numerically by the boundary element method (BEM) in which the active surfaces are discretized into elements. The advantages of integral formulations compared to the standard differential approach using finite elements are (a) only active regions need to be discretized, (b) the far field boundary condition is automatically taken into account, and (c) the fields recovered from the solution are usually very smooth. Unfortunately, the computational costs rapidly escalate as the problem size increases because of the complexity of the system coefficients and because the resulting matrix is fully populated, whereas in the differential approach the coefficients are simple and the matrix is sparse, allowing the exploitation of fast equation solution methods.

42.4 The Modern Design Environment

The most common system used in software packages is one in which the preprocessor includes data input, model building, and mesh (element) generation. Although fully automated meshing is now a practical possibility, it needs to be combined with error estimation in order to allow the generation of optimal meshes. This approach is now becoming quite common for 2-D systems and can be expected in 3-D systems before long. Figure 42.4 illustrates such a field simulation environment in

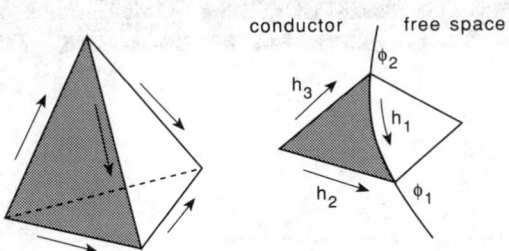

FIGURE 42.3 Edge variables for a tetrahedron element, $\mathbf{h}_1 = (\Phi_2 - \Phi_1)/l$.

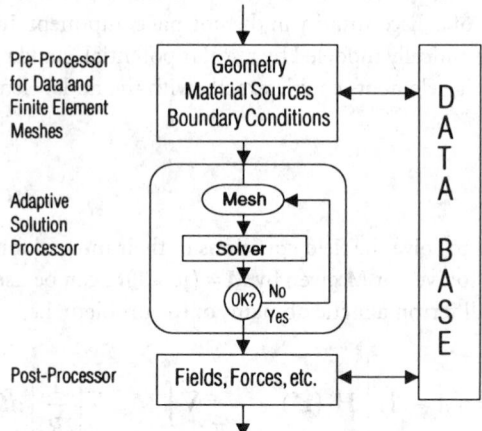

FIGURE 42.4   Computing environment.

which the solution processor includes an adaptive mesh generator controlled by *a posteriori* error estimation. This avoids the costly and essentially heuristic task of mesh generation, which in the past had to be performed by the designer.

An example of a finite element solution displayed on a modern workstation for a 3-D eddy current problem with a moving conductor retarded by a c-core electromagnet is shown in Fig. 42.5.

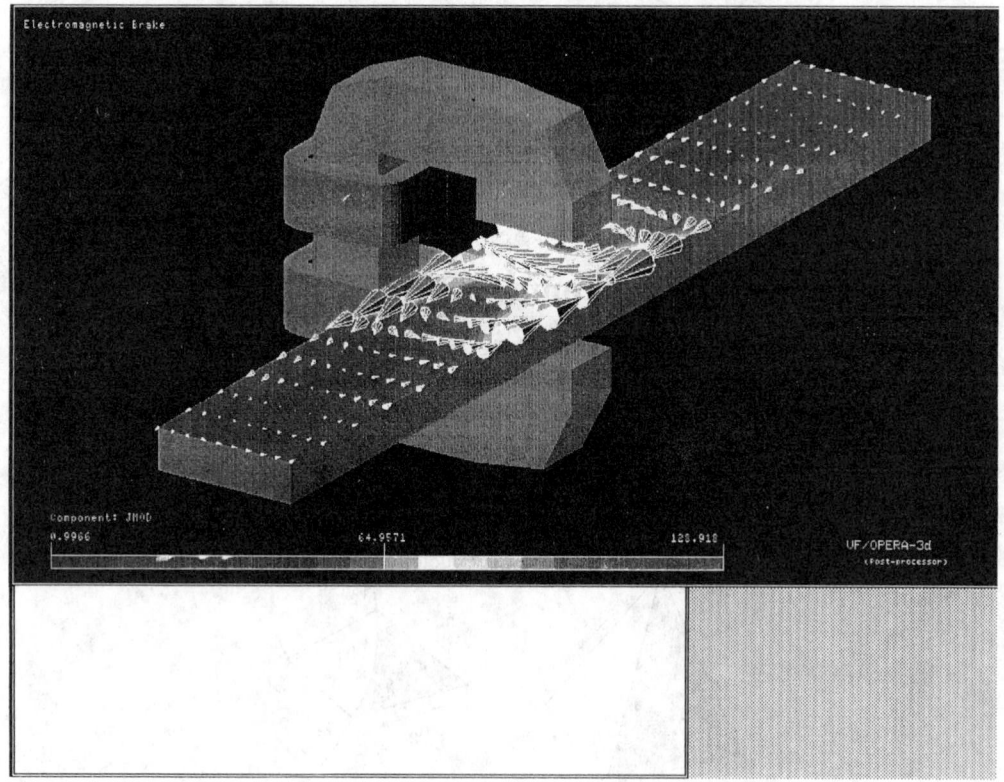

FIGURE 42.5   Moving copper strip, retarded by an electromagnet.

The postprocessor indicates the magnitude and direction of the induced eddy currents in the moving copper strip by the solid cones and the magnetic field by the gray scaled contours.

Defining Terms

Biot Savart law:

$$\mathbf{H}_s = \frac{1}{4\pi} \int_\Omega \mathbf{J}_s \times \nabla \frac{1}{R} \, d\Omega$$

where $\mathbf{R}$ is the distance from the source point to the field point.

Interface conditions:

$(\mathbf{B}_2 - \mathbf{B}_1) \cdot \mathbf{n} = 0$
$(\mathbf{D}_2 - \mathbf{D}_1) \cdot \mathbf{n} = \omega$
$(\mathbf{H}_2 - \mathbf{H}_1) \times \mathbf{n} = \mathbf{K}$
$(\mathbf{E}_2 - \mathbf{E}_1) \times \mathbf{n} = 0$

where $\mathbf{K}$ and ω are the surface current and charge densities, respectively.

Maxwell's equations:

$\nabla \cdot \mathbf{D} = \rho$ (Gauss's law)
$\nabla \cdot \mathbf{B} = 0$

$$\nabla \times \mathbf{E} = -\frac{\partial \mathbf{B}}{\partial t} \quad \text{(Faraday's law)}$$

$$\nabla \times \mathbf{H} = \mathbf{J} + \frac{\partial \mathbf{D}}{\partial t} \quad \text{(Ampere's law + displacement current)}$$

References

A. Bossavit, "Rationale for 'edge elements' in 3-D field computation," *IEEE Trans. on Magnetics*, vol. 24, no. 1, p. 74, 1988.

T. Itoh (ed.), *Numerical Techniques for Microwave and Millimeter-Wave Passive Structures*, New York: John Wiley, 1989.

I. Mayergoyz, *Mathematical Models of Hysteresis*, New York: Springer-Verlag, 1990.

P. P. Silvester and R. L. Ferrari, *Finite Elements for Electrical Engineers*, 2nd ed., Cambridge: Cambridge University Press, 1990.

J. Simkin and C. W. Trowbridge, "On the use of the total scalar potential in the numerical solution of field problems in electromagnetics," *IJNME*, vol. 14, p. 423, 1979.

J. A. Stratton, *Electromagnetic Theory*, New York: McGraw Hill, 1941.

O. C. Zienkiewicz, *The Finite Element Method*, 3rd ed., New York: McGraw Hill, 1990.

Further Information

Conferences on Computation of Electromagnetic Fields *Compumag* Proceedings:

Oxford, UK, 1976 (Ed. J. Simkin) Rutherford Appleton Laboratory, Chilton, Oxon, UK.
Grenoble, France, 1979 (Ed. J. C. Sabonnadiere) ERA 524 CNRS, Grenoble, France.
Chicago, USA, 1981 (Ed. L. Turner) *IEEE Trans. Mag.* 18 (2) 1982.
Genoa, Italy, 1983 (Ed. G. Molinari) *IEEE Trans. Mag.* 19 (6) 1983.
Fort Collins, USA, 1985 (Ed. W. Lord) *IEEE Trans. Mag.* 21 (6) 1985.
Graz, Austria, 1987 (Ed. K. Richter) *IEEE Trans. Mag.* 24 (1) 1988.
Tokyo, Japan, 1989 (Ed. K. Miya) *IEEE Trans. Mag.* 26 (2) 1990.
Sorrento, Italy, 1991 (Ed. R. Martone) *IEEE Trans. Mag.* 28 (2).

MAKING WAVES AT 100

A century after the publication of his historic paper, Heinrich Hertz's accomplishments were celebrated

In 1888, Heinrich Rudolph Hertz published his paper, "On Electric Radiation." The paper detailed the experiments through which he verified the theory of electromagnetism. The principles Hertz demonstrated with his experiments became the foundation of electromagnetic science and are seminal to all electronics.

In 1864, Maxwell proposed that energy can be transported through dielectrics, including empty space, at a finite velocity by electric and magnetic fields traveling together in space perpendicular to each other and both perpendicular to the direction of propagation. Maxwell offered a series of mathematical equations explaining the theory, but could not prove it empirically.

Hertz proved Maxwell's theory when he was confronted with a bipartite task: before he could develop a way to detect electromagnetic waves, he had to invent a way to generate them. The result was the first radio frequency oscillator-transmitter, built in 1886. About 2 meters long overall, the device generated 6-meter waves at a frequency of 50 MHz. It had a resonant circuit consisting of a balanced half-wavelength dipole, capacitively loaded by two 30-centimeter diameter spheres made of zinc sheeting. At the center of the dipole was a spark gap that acted as a switch. When current was applied, energy moved in turn from the electric field to the magnetic field, and the charge transferred from one end of the dipole to the other and back, flowing in simple harmonic motion. A portion of the energy present radiated at radio frequency.

Hertz soon turned to the subject of the velocity of electromagnetic wave propagation, which Maxwell had theorized would be finite and a product of wavelength and frequency. Hertz set up a 5-meter long, 30-cm diameter wire-over-ground-plane transmission line.

The signal source was a variation of his original transmitter, this time with square plates instead of spheres. Interference between incident waves and reflected waves traveling along the line resulted in a standing wave. To find the nodes in the standing wave, Hertz made a detector by mounting onto a wooden handle a helix of copper wire bent almost into a full circle, leaving only a spark gap.

When he inserted the detector between the outer wires and moved it along the transmission line, a continuous spark was produced across the spark gap until a node was encountered, where the spark would dwindle and cease. The distance between nodes determined the wavelength, and with the frequency calculated from estimates of the capacitance and the inductance of the dipole resonator,

In 1913, Hertz's original apparatus was collected at the Bavarian Academy of Science in Munich for archiving. (Unidentified man is thought to be a museum employee.) (Photo courtesy of the Museum of Science and Industry, Chicago.)

Hertz deduced the velocity of propagation, and so validated that aspect of Maxwell's theory.

Next, Hertz built the apparatus for a 70-centimeter wavelength at 450 MHz to address the next step of Maxwell's theory, proving that electromagnetic waves behaved similarly to light—then called optic waves—a novel notion to anyone unfamiliar with his or Maxwell's work.

Apparatus for Hertz's focused-beam experiments consisted of the 70-cm half-wavelength dipole transmitter, a receiver-detector, a cylindri-cal parabolic reflector, a wooden frame with parallel wires for polarization experiments, a stack of wedge-shaped wooden boxes to hold dielectric materials for refraction experiments, and a power supply. He generated electromagnetic waves, reflected them, refracted them, and polarized them, just as though they were a beam of light.

All of these experiments proving Maxwell's theory were completed by 1888, when Hertz was only in his early 30s. He died in 1894 from blood poisoning from an infected jaw; he was 37.

Source: Adapted from B. Santo, *IEEE Spectrum,* p. 58, May 1988. © 1988 IEEE.

43

Computational Electromagnetics

43.1 Introduction 1028
43.2 Background Discussion 1029
     Modeling as a Transfer Function • Some Issues Involved in
     Developing a Computer Model
43.3 Analytical Issues in Developing a Computer Model 1031
     Selection of Solution Domain • Selection of Field Propagator
43.4 Numerical Issues in Developing a Computer Model 1035
     Sampling Functions • The Method of Moments
43.5 Some Practical Considerations 1041
     Integral Equation Modeling • Differential Equation Modeling
     • Discussion • Sampling Requirements
43.6 Ways of Decreasing Computer Time 1044
43.7 Validation, Error Checking, and Error Analysis 1046
     Modeling Uncertainties • Validation and Error Checking
43.8 Concluding Remarks 1047

E. K. Miller
Los Alamos National Laboratory

43.1 Introduction

The continuing growth of computing resources is changing how we think about, formulate, solve, and interpret problems. In electromagnetics as elsewhere, computational techniques are complementing the more traditional approaches of measurement and analysis to vastly broaden the breadth and depth of problems that are now quantifiable. Computational electromagnetics (CEM) may be broadly defined as that branch of electromagnetics that intrinsically and routinely involves using a digital computer to obtain numerical results. With the evolutionary development of CEM during the past 20-plus years, the third tool of computational methods has been added to the two classical tools of experimental observation and mathematical analysis.

This discussion reviews some of the basic issues involved in CEM and includes only the detail needed to illustrate the central ideas involved. The underlying principles that unify the various modeling approaches used in electromagnetics are emphasized while avoiding most of the detail that makes them different. Listed throughout are representative, but not exhaustive, numbers of references that deal with various specialty aspects of CEM. For readers interested in broader, more general expositions, the well-known book on the moment method by Harrington [1968]; the

This chapter is excerpted from E. K. Miller, "A selective survey of computational electromagnetics," *IEEE Trans. Antennas Propagat.*, vol. AP-36, pp. 1281–1305, ©1988 IEEE.

books edited by Mittra [1973, 1975], Uslenghi [1978], and Strait [1980]; the monographs by Stutzman and Thiele [1981], Popovic, *et al.* [1982], Moore and Pizer [1984], and Wang [1991]; and an IEEE Press reprint volume on the topic edited by Miller *et al.* [1991] are recommended, as is the article by Miller [1988] from which this material is excerpted.

43.2 Background Discussion

Electromagnetics is the scientific discipline that deals with electric and magnetic sources and the fields these sources produce in specified environments. Maxwell's equations provide the starting point for the study of electromagnetic problems, together with certain principles and theorems such as superposition, reciprocity, equivalence, induction, duality, linearity, and uniqueness, derived therefrom [Stratton, 1941; Harrington, 1961]. While a variety of specialized problems can be identified, a common ingredient of essentially all of them is that of establishing a quantitative relationship between a cause (forcing function or input) and its effect (the response or output), a relationship which we refer to as a **field propagator**, the computational characteristics of which are determined by the mathematical form used to describe it.

Modeling as a Transfer Function

The foregoing relationship may be viewed as a generalized transfer function (see Fig. 43.1) in which two basic problem types become apparent. For the analysis or the direct problem, the input is known and the transfer function is derivable from the problem specification, with the output or response to be determined. For the case of the synthesis or inverse problem, two problem classes may be identified. The easier synthesis problem involves finding the input, given the output and transfer function, an example of which is that of determining the source voltages that produce an observed pattern for a known antenna array. The more difficult synthesis problem itself separates into two problems. One is that of finding the transfer function, given the input and output, an example of which is that of finding a source distribution that produces a given far field. The other and still more difficult is that of finding the object geometry that produces an observed scattered field from a known exciting field. The latter problem is the most difficult of the three synthesis problems to solve because it is intrinsically transcendental and nonlinear.

Electromagnetic propagators are derived from a particular solution of Maxwell's equations, as the cause mentioned above normally involves some specified or known excitation whose effect is to induce some to-be-determined response (e.g., a radar cross section, antenna radiation pattern). It therefore follows that the essence of electromagnetics is the study and determination of field propagators to thereby obtain an input–output transfer function for the problem of interest, and it follows that this is also the goal of CEM.

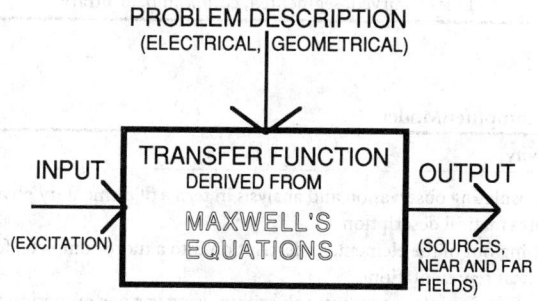

FIGURE 43.1 The electromagnetic transfer function relates the input, output, and problem.

Some Issues Involved in Developing a Computer Model

We briefly consider here a classification of model types, the steps involved in developing a **computer model**, the desirable attributes of a computer model, and finally the role of approximation throughout the modeling process.

Classification of Model Types

It is convenient to classify solution techniques for electromagnetic modeling in terms of the field propagator that might be used, the anticipated application, and the problem type for which the model is intended to be used, as is outlined in Table 43.1. Selection of a field propagator in the form, for example, of the Maxwell curl equations, a Green's function, modal or spectral expansions, or an optical description is a necessary first step in developing a solution to any electromagnetic problem.

Development of a Computer Model

Development of a computer model in electromagnetics or literally any other disciplinary activity can be decomposed into a small number of basic, generic steps. These steps might be described by different names but would include at a minimum those outlined in Table 43.2. Note that by its nature, validation is an open-ended process that cumulatively can absorb more effort than all the other steps together. The primary focus of the following discussion is on the issue of numerical implementation.

Table 43.1 Classification of Model Types in CEM

Field Propagator	Description Based on
Integral operator	Green's function for infinite medium or special boundaries
Differential operator	Maxwell curl equations or their integral counterparts
Modal expansions	Solutions of Maxwell's equations in particular coordinate system and expansion
Optical description	Rays and diffraction coefficients
Application	**Requires**
Radiation	Determining the originating sources of a field and patterns they produce
Propagation	Obtaining the fields distant from a known source
Scattering	Determining the perturbing effects of medium inhomogeneities
Problem type	**Characterized by**
Solution domain	Time or frequency
Solution space	Configuration or wave number
Dimensionality	1D, 2D, 3D
Electrical properties of medium and/or boundary	Dielectric, lossy, perfectly conducting, anisotropic, inhomogeneous, nonlinear
Boundary geometry	Linear, curved, segmented, compound, arbitrary

Table 43.2 Steps in Developing a Computer Model

Step	Activity
Conceptualization	Encapsulating observation and analysis in terms of elementary physical principles and their mathematical description
Formulation	Fleshing out of the elementary description into a more complete, formally solved, mathematical representation
Numerical implementation	Transforming into a computer algorithm using various numerical techniques
Computation	Obtaining quantitative results
Validation	Determining the numerical and physical credibility of the computed results

Desirable Attributes of a Computer Model

A computer model must have some minimum set of basic properties to be useful. From the long list of attributes that might be desired, we consider: (1) accuracy, (2) efficiency, and (3) utility the three most important as summarized in Table 43.3. Accuracy is put foremost because results of insufficient or unknown accuracy have uncertain value and may even be harmful. On the other hand, a code that produces accurate results but at unacceptable cost will have hardly any more value. Finally, a code's applicability in terms of the depth and breadth of the problems for which it can be used determines its utility.

The Role of Approximation

As approximation is an intrinsic part of each step involved in developing a computer model, we summarize some of the more commonly used approximations in Table 43.4. We note that the distinction between an approximation at the conceptualization step and during the formulation is somewhat arbitrary, but choose to use the former category for those approximations that occur before the formulation itself.

43.3 Analytical Issues in Developing a Computer Model

Attention here is limited primarily to propagators that use either the Maxwell curl equations or source integrals which employ a Green's function, although for completeness we briefly discuss modal and optical techniques as well.

Selection of Solution Domain

Either the **integral equation** (IE) or differential equation (DE) propagator can be formulated in the time domain, where time is treated as an independent variable, or in the frequency domain, where the harmonic time variation $\exp(j\omega t)$ is assumed. Whatever propagator and domain are chosen, the analytically formal solution can be numerically quantified via the **method of moments** (MoM) [Harrington, 1968], leading ultimately to a linear system of equations as a result of developing a discretized and sampled approximation to the continuous (generally) physical reality being modeled. Developing the approach that may be best suited to a particular problem involves making trade-offs among a variety of choices throughout the analytical formulation and numerical implementation, some aspects of which are now considered.

Selection of Field Propagator

We briefly discuss and compare the characteristics of the various propagator-based models in terms of their development and applicability.

Table 43.3 Desirable Attributes in a Computer Model

Attribute	Description
Accuracy	The quantitative degree to which the computed results conform to the mathematical and physical reality being modeled. Accuracy, preferably of known and, better yet, selectable value, is the single most important model attribute. It is determined by the physical modeling error (ε_P) and numerical modeling error (ε_N).
Efficiency	The relative cost of obtaining the *needed* results. It is determined by the *human* effort required to develop the computer input and interpret the output and by the associated *computer* cost of running the model.
Utility	The applicability of the computer model in terms of problem size and complexity. Utility also relates to ease of use, reliability of results obtained, etc.

Table 43.4 Representative Approximations that Arise in Model Development

Approximation	Implementation/Implications
Conceptualization	
Physical optics	Surface sources given by tangential components of incident field, with fields subsequently propagated via a Green's function. Best for backscatter and main-lobe region of reflector antennas, from resonance region ($ka > 1$) and up in frequency.
Physical theory of diffraction	Combines aspects of physical optics and geometrical theory of diffraction, primarily via use of edge-current corrections to utilize best features of each.
Geometrical theory diffraction	Fields propagated via a divergence factor with amplitude obtained from diffraction coefficient. Generally applicable for $ka > 2$–5. Can involve complicated ray tracing.
Geometrical optics	Ray tracing without diffraction. Improves with increasing frequency.
Compensation theorem	Solution obtained in terms of perturbation from a reference, known solution.
Born–Rytov	Approach used for low-contrast, penetrable objects where sources are estimated from incident field.
Rayleigh	Fields at surface of object represented in terms of only outward propagating components in a modal expansion.
Formulation	
Surface impedance	Reduces number of field quantities by assuming an impedance relation between tangential E and H at surface of penetrable object. May be used in connection with physical optics.
Thin-wire	Reduces surface integral on thin, wirelike object to a line integral by ignoring circumferential current and circumferential variation of longitudinal current, which is represented as a filament. Generally limited to $ka < 1$ where a is the wire radius.
Numerical Implementation	
$\partial f/\partial x \rightarrow (f_+ - f_-)/(x_+ - x_-)$ $\int f(x)\,dx \rightarrow \sum f(x_i)\Delta x_i$	Differentiation and integration of continuous functions represented in terms of analytic operations on sampled approximations, for which polynomial or trigonometric functions are often used. Inherently a discretizing operation, for which typically $\Delta x < \lambda/2\pi$ for acceptable accuracy.
Computation	
Deviation of numerical model from physical reality	Affects solution accuracy and relatability to physical problem in ways that are difficult to predict and quantify.
Nonconverged solution	Discretized solutions usually converge globally in proportion to $\exp(-AN_x)$ with A determined by the problem. At least two solutions using different numbers of unknowns N_x are needed to estimate A.

Integral Equation Model

The basic starting point for developing an IE model in electromagnetics is selection of a Green's function appropriate for the problem class of interest. While there are a variety of Green's functions from which to choose, a typical starting point for most IE MoM models is that for an infinite medium. One of the more straightforward is based on the scalar Green's function and Green's theorem. This leads to the Kirchhoff integrals [Stratton, 1941, p. 464 *et seq.*], from which the fields in a given contiguous volume of space can be written in terms of integrals over the surfaces that bound it and volume integrals over those sources located within it.

Analytical manipulation of a source integral that incorporates the selected Green's function as part of its kernel function then follows, with the specific details depending on the particular formulation being used. Perhaps the simplest is that of boundary-condition matching wherein the behavior required of the electric and/or magnetic fields at specified surfaces that define the problem geometry is explicitly imposed. Alternative formulations, for example, the Rayleigh–Ritz variational method and Rumsey's reaction concept, might be used instead, but as pointed out by Harrington [in Miller *et al.*, 1991], from the viewpoint of a numerical implementation any of these approaches lead to formally equivalent models.

This analytical formulation leads to an integral operator, whose kernel can include differential operators as well, which acts on the unknown source or field. Although it would be more accurate to refer to this as an integrodifferential equation, it is usually called simply an integral equation. Two general kinds of integral equations are obtained. In the frequency domain, representative forms for a perfect electric conductor are

$$\mathbf{n} \times \mathbf{E}^{\text{inc}}(\mathbf{r}) = \frac{1}{4\pi} \mathbf{n} \times \int_{S} \{j\omega\mu[\mathbf{n}' \times \mathbf{H}(\mathbf{r}')]\varphi(\mathbf{r}, \mathbf{r}')$$
$$-[\mathbf{n}' \cdot \mathbf{E}(\mathbf{r}, \mathbf{r}')\nabla'\varphi(\mathbf{r}, \mathbf{r}')\}ds'; \quad \mathbf{r} \in S \tag{43.1a}$$

$$\mathbf{n} \times \mathbf{H}(\mathbf{r}) = 2\mathbf{n} \times \mathbf{H}^{\text{inc}}(\mathbf{r}) + \frac{1}{2\pi} \mathbf{n} \times \int_{S} [\mathbf{n}' \times \mathbf{H}(\mathbf{r}')] \times \nabla'\varphi(\mathbf{r}, \mathbf{r}')\}ds'; \quad \mathbf{r} \in S \tag{43.1b}$$

where $\mathbf{E}$ and $\mathbf{H}$ are the electric and magnetic fields, respectively, $\mathbf{r}$, $\mathbf{r}'$ are the spatial coordinate of the observation and source points, the superscript inc denotes incident-field quantities, and $\varphi(\mathbf{r}, \mathbf{r}') = \exp[-jk|\mathbf{r} - \mathbf{r}'|]/|\mathbf{r} - \mathbf{r}'|$ is the free-space Green's function. These equations are known respectively as Fredholm integral equations of the first and second kinds, differing by whether the unknown appears only under the integral or outside it as well [Poggio and Miller in Mittra, 1973].

Differential Equation Model

A DE MoM model, being based on the defining Maxwell's equations, requires intrinsically less analytical manipulation than does derivation of an IE model. Numerical implementation of a DE model, however, can differ significantly from that used for an IE formulation in a number of ways for several reasons:

1. The differential operator is a local rather than global one in contrast to the Green's function upon which the integral operator is based. This means that the spatial variation of the fields must be developed from **sampling** in as many dimensions as possessed by the problem, rather than one less as the IE model permits if an appropriate Green's function is available.
2. The integral operator includes an explicit radiation condition, whereas the DE does not.
3. The differential operator includes a capability to treat medium inhomogeneities, non-linearities, and time variations in a more straightforward manner than does the integral operator, for which an appropriate Green's function may not be available.

These and other differences between development of IE and DE models are summarized in Table 43.5, with their modeling applicability compared in Table 43.6.

Modal-Expansion Model

Modal expansions are useful for propagating electromagnetic fields because the source-field relationship can be expressed in terms of well-known analytical functions as an alternate way of writing a Green's function for special distributions of point sources. In two dimensions, for example, the propagator can be written in terms of circular harmonics and cylindrical Hankel functions. Corresponding expressions in three dimensions might involve spherical harmonics, spherical Hankel functions, and Legendre polynomials. Expansion in terms of analytical solutions to the wave equation in other coordinate systems can also be used but requires computation of special functions that are generally less easily evaluated, such as Mathieu functions for the two-dimensional solution in elliptical coordinates and spheroidal functions for the three-dimensional solution in oblate or prolate spheroidal coordinates.

One implementation of modal propagators for numerical modeling is that due to Waterman [in Mittra, 1973], whose approach uses the extended boundary condition (EBC) whereby the required field behavior is satisfied away from the boundary surface on which the sources are located. This

Table 43.5 Comparison of IE- and DE-Field Propagators and Their Numerical Treatment

	Differential Form	Integral Form
Field propagator	Maxwell curl equations	Green's function
Boundary treatment		
At infinity (radiation condition)	Local or global "lookback" to approximate outward propagating wave	Green's function
On object	Appropriate field values specified on mesh boundaries to obtain stairstep, piecewise linear, or other approximation to the boundary	Appropriate field values specified on object contour which can in principle be a general, curvilinear surface, although this possibility seems to be seldom used
Sampling requirements		
No. of space samples	$N_x \propto (L/\Delta L)^D$	$N_x \propto (L/\Delta L)^{D-1}$
No. of time steps	$N_t \propto (L/\Delta L) \approx cT/\delta t$	$N_t \propto (L/\Delta L) \approx cT/\delta t$
No. of excitations (right-hand sides)	$N_{\text{rhs}} \propto (L/\Delta L)$	$N_{\text{rhs}} \propto (L/\Delta L)$
Linear system	Sparse, but larger	Dense, but smaller. In this comparison, note that we assume the IE permits a sampling of order one less than the problem dimension, i.e., inhomogeneous problems are excluded.
L is problem size		
D is no. of problem dimensions (1, 2, 3)		
T is observation time		
ΔL is spatial resolution		
δt is time resolution		
Dependence of solution time on highest-order term in $(L/\Delta L)$		
Frequency domain	$T_\omega \propto N_x^{2(D-1)/D+1}$ $= (L/\Delta L)^{3D-2}$	$T_\omega \propto N_x^3 = (L/\Delta L)^{3(D-1)}$
Time domain		
Explicit	$T_t \propto N_x N_t N_{\text{rhs}} = (L/\Delta L)^{D+1+r}$	$T_t \propto N_x^2 N_t N_{\text{rhs}} = (L/\Delta L)^{2D-1+r}; 0 \le r \le 1$
Implicit	$T_t \propto N_x^{2(D-1)/D+1} = (L/\Delta L)^{3D-2}, D = 2, 3;$ $\propto N_x N_t N_{\text{rhs}} = (L/\Delta L)^{2+r}, D = 1; 0 \le r \le 1$	$T_t \propto N_x^3 = (L/\Delta L)^{3(D-1)}$

Note that D is the number of *spatial* dimensions in the problem and is not necessarily the *sampling* dimensionality d. The distinction is important because when an appropriate Green's function is available, the source integrals are usually one dimension less than the problem dimension, i.e., $d = D - 1$. An exception is an inhomogeneous, penetrable body where $d = D$ when using an IE. We also assume for simplicity that matrix solution is achieved via factorization rather than iteration but that banded matrices are exploited for the DE approach where feasible. The solution-time dependencies given can thus be regarded as upper-bound estimates. See Table 43.10 for further discussion of linear-system solutions.

procedure, widely known as the *T-matrix* approach, has evidently been more widely used in optics and acoustics than in electromagnetics. In what amounts to a reciprocal application of EBC, the sources can be removed from the boundary surface on which the field-boundary conditions are applied. These modal techniques seem to offer some computational advantages for certain kinds of problems and might be regarded as using entire-domain basis and testing functions but nevertheless lead to linear systems of equations whose numerical solution is required. Fourier transform solution techniques might also be included in this category since they do involve modal expansions, but that is a specialized area that we do not pursue further here.

Geometrical Optics Model

Geometrical optics and the geometrical theory of diffraction (GTD) are high-frequency asymptotic techniques wherein the fields are propagated using such optical concepts as shadowing, ray tubes, and refraction and diffraction. Although conceptually straightforward, optical techniques are limited analytically by the unavailability of diffraction coefficients for various geometries and material bodies and numerically by the need to trace rays over complex surfaces. There is a vast literature on geometrical optics and GTD, as may be ascertained by examining the yearly and cumulative indexes of such publications as the *Transactions of the IEEE Antennas and Propagation Society*.

Table 43.6 Relative Applicability of IE- and DE-Based Computer Models

Time Domain			Frequency Domain	
DE	IE	Issue	DE	IE
		Medium		
√	√	Linear	√	√
~	x	Dispersive	√	√
√	x	Lossy	√	√
√	~	Anisotropic	√	√
√	x	Inhomogeneous	√	x
√	x	Nonlinear	x	x
√	x	Time-varying	x	x
		Object		
~	√	Wire	~	√
√	√	Closed surface	√	√
√	√	Penetrable volume	√	√
~	√	Open surface	~	√
		Boundary Conditions		
√	√	Interior problem	√	√
~	√	Exterior problem	~	√
√	√	Linear	√	√
√	√	Nonlinear	x	x
√	√	Time-varying	x	x
~	x	Halfspace	~	√
		Other Aspects		
~	~	Symmetry exploitation	√	√
~	√	Far-field evaluation	~	√
x	~	Number of unknowns	~	√
√	~	Length of code	~	x
		Suitability for Hybridizing with Other:		
~	√	Numerical procedures	√	√
x	~	Analytical procedures	~	√
x	~	GTD	x	√

√ signifies highly suited or most advantageous.
~ signifies moderately suited or neutral.
x signifies unsuited or least advantageous.

43.4 Numerical Issues in Developing a Computer Model

Sampling Functions

At the core of numerical analysis is the idea of polynomial approximation, an observation made by Arden and Astill [1970] in facetiously using the subtitle "Numerical Analysis or 1001 Applications of Taylor's Series." The basic idea is to approximate quantities of interest in terms of sampling functions, often polynomials, that are then substituted for these quantities in various analytical operations. Thus, integral operators are replaced by finite sums, and differential operators are similarly replaced by generalized finite differences. For example, use of a first-order difference to approximate a derivative of the function $F(x)$ in terms of samples $F(x_+)$ and $F(x_-)$ leads to

$$\frac{dF(x)}{dx} \approx \frac{F(x_+) - F(x_-)}{h}; \quad x_- \leq x \leq x_+ \tag{43.2a}$$

and implies a linear variation for $F(x)$ between x_+ and x_- as does use of the trapezoidal rule

$$\int_{x_-}^{x_+} F(x)dx \approx \frac{h}{2} \left[F(x_+) + F(x_-) \right] \tag{43.2b}$$

to approximate the integral of $F(x)$, where $h = x_+ - x_-$. The central-difference approximation for the second derivative,

$$\frac{d^2 F(x)}{dx^2} \approx \frac{[F(x_+) - 2F(x_0) + F(x_-)]}{h^2} \tag{43.2c}$$

similarly implies a quadratic variation for $F(x)$ around $x_0 = x_+ - h/2 = x_- + h/2$, as does use of Simpson's rule

$$\int_{x_-}^{x_+} F(x)dx \approx \frac{h}{6} \left[F(x_+) + 4F(x_0) + F(x_-) \right] \tag{43.2d}$$

to approximate the integral. Other kinds of polynomials and function sampling can be used, as discussed in a large volume of literature, some examples of which are Abramowitz and Stegun [1964], Acton [1970], and Press *et al.* [1986]. It is interesting to see that numerical differentiation and integration can be accomplished using the same set of function samples and spacings, differing only in the signs and values of some of the associated weights. Note also that the added degrees of freedom that arise when the function samples can be unevenly spaced, as in Gaussian quadrature, produce a generally more accurate result (for well-behaved functions) for a given number of samples. This suggests the benefits that might be derived from using unequal sample sizes in MoM modeling should a systematic way of determining the best nonuniform sampling scheme be developed.

The Method of Moments

Numerical implementation of the moment method is a relatively straightforward, and an intuitively logical, extension of these basic elements of numerical analysis, as described in the book by Harrington [1968] and discussed and used extensively in CEM [see, for example, Mittra, 1973, 1975; Strait, 1980; Poggio and Miller, 1988]. Whether it is an integral equation, a differential equation, or another approach that is being used for the numerical model, three essential sampling operations are involved in reducing the analytical formulation via the moment method to a computer algorithm as outlined in Table 43.7. We note that operator sampling can ultimately determine the sampling density needed to achieve a desired accuracy in the source–field relationships involving integral operators, especially at and near the "self term," where the observation and source points become coincident or nearly so and the integral becomes nearly singular. Whatever the method used for these sampling operations, they lead to a linear system of equations or matrix approximation of the original integral or differential operators. Because the operations and choices involved in developing this matrix description are common to all moment-method models, we shall discuss them in somewhat more detail.

When using IE techniques, the coefficient matrix in the linear system of equations that results is most often referred to as an impedance matrix because in the case of the E-field form, its multiplication of the vector of unknown currents equals a vector of electric fields or voltages. The inverse matrix similarly is often called an admittance matrix because its multiplication of the electric-field or voltage vector yields the unknown-current vector. In this discussion we instead use the terms *direct matrix* and *solution matrix* because they are more generic descriptions whatever the forms of the originating integral or differential equations. As illustrated in the following, development of the

Table 43.7 Sampling Operations Involved in MoM Modeling

Equation	DE Model $L(s')f(s') = g(s')$	IE Model $L(s,s')f(s') = g(s)$
Sampling of:		
Unknown via basis-functions $b_j(s')$ using $f(s') \approx \sum a_j b_j(s')$	Subdomain bases usually of low order are used. Known as FD procedure when pulse basis is used, and as FE approach when bases are linear.	Can use either subdomain or entire-domain bases. Use of latter is generally confined to bodies of rotation. Former is usually of low order, with piecewise linear or sinusoidal being the maximum variation used.
Equation via weight functions $w_i(s)$ $<w_i(s), L(s,s') \sum a_j b_j(s')>$ $= <w_i(s), g(s)>$ to get $Z_{ij} a_j = g_i$	Pointwise matching is commonly employed, using a delta function. Pulse and linear matching are also used.	Pointwise matching is commonly employed, using a delta function. For wires, pulse, linear, and sinusoidal testing is also used. Linear and sinusoidal testing is also used for surfaces.
Operator	Operator sampling for DE models is entwined with sampling the unknown in terms of the difference operators used.	Sampling needed depends on the nature of the integral operator $L(s,s')$. An important consideration whenever the field integrals cannot be evaluated in closed form.
Solution of: $Z_{ij} a_j = g_i$ for the a_j	Interaction matrix is sparse. Time-domain approach may be explicit or implicit. In frequency domain, banded-matrix technique usually used.	Interaction matrix is full. Solution via factorization or iteration.

direct matrix and solution matrix dominates both the computer time and storage requirements of numerical modeling.

In the particular case of an IE model, the coefficients of the direct or original matrix are the mutual impedances of the multiport representation which approximates the problem being modeled, and the coefficients of its solution matrix (or equivalent thereof) are the mutual admittances. Depending on whether a subdomain or entire-domain basis has been used (see Basic Function Selection), these impedances and admittances represent either spatial or modal interactions among the *N* ports of the numerical model. In either case, these coefficients possess a physical relatability to the problem being modeled and ultimately provide all the information available concerning any electromagnetic observables that are subsequently obtained.

Similar observations might also be made regarding the coefficients of the DE models but whose multiport representations describe local rather than global interactions. Because the DE model almost always leads to a larger, albeit less dense, direct matrix, its inverse (or equivalent) is rarely computed. It is worth noting that there are two widely used approaches for DE modeling, finite-difference (FD) and finite-element (FE) methods. They differ primarily in how the differential operators are approximated and the differential equations are satisfied, i.e., in the order of the basis and weight functions, although the FE method commonly starts from a variational viewpoint, while the FD approach begins from the defining differential equations. The FE method is generally better suited for modeling problems with complicated boundaries to which it provides a piecewise linear approximation as opposed to the cruder stairstep approximation of FD.

Factors Involved in Choosing Basis and Weight Functions

Basis and weight function selection plays a critical role in determining the accuracy and efficiency of the resulting computer model. One goal of the basis and weight function selection is to minimize computer time while maximizing accuracy for the problem set to which the model is to be applied. Another, possibly conflicting, goal might be that of maximizing the collection of problem sets to which the model is applicable. A third might be to replicate the problem's physical behavior with as few samples as possible. Some of the generic combinations of bases and weights that are used for MoM models are listed in Table 43.8 [Poggio and Miller from Mittra, 1973].

Basis Function Selection.   We note that there are two classes of bases used in MoM modeling, subdomain and entire-domain functions. The former involves the use of bases that are applied in a repetitive fashion over subdomains or sections (segments for wires, patches for surfaces, cells for volumes) of the object being modeled. The simplest example of a subdomain basis is the single-term basis given by the *pulse* or stairstep function, which leads to a single, unknown constant for each subdomain. Multiterm bases involving two or more functions on each subdomain and an equivalent number of unknowns are more often used for subdomain expansions.

The entire-domain basis, on the other hand, uses multiterm expansions extending over the entire object, for example, a circular harmonic expansion in azimuth for a body of revolution. As for subdomain expansions, an unknown is associated with each term in the expansion. Examples of hybrid bases can also be found, where subdomain and entire-domain bases are used on different parts of an object.

Although subdomain bases are probably more flexible in terms of their applicability, they have a disadvantage generally not exhibited by the entire-domain form, which is the discontinuity that occurs at the domain boundaries. This discontinuity arises because an n_s-term subdomain function can provide at most $n_s - 1$ degrees of continuity to an adjacent basis of the unknown it represents, assuming one of the n_s constants is reserved for the unknown itself. For example, the three-term or sinusoidal subdomain basis $a_i + b_i \sin(ks) + c_i \cos(ks)$ used for wire modeling can represent a current continuous at most up to its first derivative. This provides continuous charge density but produces a discontinuous first derivative in charge equivalent to a tripole charge at each junction.

As additional terms are used to develop a subdomain basis, higher-order continuity can be achieved in the unknown that the basis represents, assuming still that one unknown is reserved for the amplitude of the multiterm basis function. In the general case of the n_s-term subdomain basis, up to $n_s - 1$ constants can be determined from continuity conditions, with the remainder reserved

Table 43.8   Examples of Generic Basis/Weight-Function Combinations

Method	jth Term of Basis	ith Term of Weight
Galerkin	$a_j b_j(\mathbf{r}')$	$w_i(\mathbf{r}) = b_j(\mathbf{r})$
Least square	$a_j b_j(\mathbf{r}')$	$Q(\mathbf{r})\partial\varepsilon(\mathbf{r})/\partial a_i$
Point matching	$a_j \delta(\mathbf{r} - \mathbf{r}_j)$	$\delta(\mathbf{r} - \mathbf{r}_i)$
General collocation	$a_j b_j(\mathbf{r}')$	$\delta(\mathbf{r} - \mathbf{r}_i)$
Subsectional collocation	$U(\mathbf{r}_j)\sum a_{jk} b_k(\mathbf{r}')$	$\delta(\mathbf{r} - \mathbf{r}_i)$
Subsectional Galerkin	$U(\mathbf{r}_j)\sum a_{jk} b_k(\mathbf{r}')$	$U(\mathbf{r}_i)\sum b_i(\mathbf{r})$

   $\mathbf{r}'$ and $\mathbf{r}$ denote source and observation points respectively; a_j, a_{jk} are unknown constants associated with the jth basis function (entire domain) or the kth basis function of the jth subsection (subdomain); $U(\mathbf{r}_k)$ is the unit sampling function which equals 1 on the kth subdomain and is 0 elsewhere; $b_j(\mathbf{r}')$ is the jth basis function; $w_i(\mathbf{r})$ is the ith testing function; $\delta(\mathbf{r} - \mathbf{r}_i)$ is the Dirac delta function; $Q(\mathbf{r})$ is a positive-definite function of position; and $\varepsilon(\mathbf{r})$ is the residual or equation error [from Poggio and Miller in Mitra (1973)].

for the unknown. The kind of basis function employed ultimately determines the degree of fit that the numerical result can provide to the true behavior of the unknown for a given order of matrix. An important factor that should influence basis-function selection, then, is how closely a candidate function might resemble the physical behavior of the unknown it represents. Another consideration is whether a system of equations that is numerically easier to solve might result from a particular choice of basis and weight function, for example, by increasing its diagonal dominance so that an iterative technique will converge more rapidly.

Weight Function Selection. The simplest weight that might be used is a delta function which leads to a point-sampled system of equations, but point sampling of the field operators can be sensitive to any numerical anomalies that might arise as a result of basis function discontinuities. Distributed, multiterm weight functions can also be used on either a subdomain or an entire-domain basis to provide a further smoothing of the final equations to be solved. One example of this is the special case where the same functions are used for both the bases and weights, a procedure known as Galerkin's method. The kind of testing function used ultimately determines the degree to which the equations can be matched for a given basis function and number of unknowns. Some specific examples of basis and weight function combinations used in electromagnetics are summarized in Table 43.9.

Computing the Direct Matrix

We observe that obtaining the coefficients of the direct matrix in IE modeling is generally a two-step process. The first step is that of evaluating the defining integral operator in which the unknown is replaced by the basis functions selected. The second step involves integration of this result multiplied by the weight function selected. When using delta-function weights, this second step is numerically trivial, but when using nondelta weights, such as the case in a Galerkin approach, this second step can be analytically and numerically challenging.

Among the factors affecting the choice of the basis and weight functions, therefore, one of the most important is that of reducing the computational effort needed to obtain the coefficients of the direct matrix. This is one of the reasons, aside from their physical appeal, why sinusoidal bases are often used for wire problems. In this case, where piecewise linear, filamentary current sources are most often used in connection with the thin-wire approximation, field expressions are avail-

Table 43.9 Examples of Specific Basis/Weight-Function Combinations

Application	jth Term of Basis	ith Term of Weight
1D/wires	Constant—$a_j U_j(s)$	Delta function—$\delta(s - s_i)$
1D/wires	Piecewise linear—$a_{j1}(s - s_j - \delta_j/2)$ $+ a_{j2}(s - s_j + \delta_j/2)$	Piecewise linear—$(s - s_i - \delta_i/2)$ $+ (s - s_i + \delta_i/2)$
1D/wires	3-term sinusoidal—$a_{j1} + a_{j2}\sin[k(s - s_j)]$ $+ a_{j3}\cos[k(s - s_j)]$	Delta function—$\delta(s - s_i)$
1D/wires	Piecewise sinusoidal—$a_{j1}\sin[k(s - s_j - \delta_j/2)]$ $+ a_{j2}\sin[k(s - s_j + \delta_j/2)]$	Piecewise sinusoidal— $\sin[k(s - s_i - \delta_i/2)] + \sin[k(s - s_i + \delta_i/2)]$
2D/surfaces	Weighted delta function—$a_j\delta(s - s_j)\Delta_j$	Delta function— $\delta(s - s_i)$
2D/rotational surfaces	Piecewise linear axially, and $\exp(in\phi)$ azimuthally	Same (Galerkin's method)
2D/surfaces	Piecewise linear	Same (Galerkin's method)
2D/surfaces	Piecewise linear subdomain/Fourier series entire domain	Same (Galerkin's method)
3D/volumes	Piecewise linear	Same (Galerkin's method)

δ_k is the length of wire segment k; Δ_k is the area of surface patch k.

able in easily evaluated, analytical expressions. This is the case as well where Galerkin's method is used.

Aside from such special cases, however, numerical evaluation of the direct-matrix coefficients will involve the equivalent of point sampling of whatever order is needed to achieve the desired accuracy as illustrated below. Using a wirelike one-dimensional problem to illustrate this point, we observe that at its most elementary level evaluation of the *ij*th matrix coefficient then involves evaluating integrals of the form

$$
\begin{aligned}
Z_{i,j} &= \int_{C(\mathbf{r})} w_i(s) \int_{C(\mathbf{r})} [b_j(s')K(s, s')ds']ds \\
&\approx \sum_{m=1}^{M(i,j)} \sum_{n=1}^{N(i,j)} p_m q_n w_i(s_n) b_j(s'_m) K(s_n, s'_m) \qquad (43.3) \\
&= \sum_{m=1}^{M(i,j)} \sum_{n=1}^{N(i,j)} p_m q_n z(i, j, m, n); \quad i, j = 1, \ldots, N
\end{aligned}
$$

where $K(s,s')$ is the IE kernel function, and s_n and s'_m are the nth and mth locations of the observation and source integration samples. Thus, the final, direct-matrix coefficients can be seen to be constructed from sums of the more elementary coefficients $z(i,j,m,n)$ weighted by the quadrature coefficients p_m and q_n used in the numerical integration, which will be the case whenever analytical expressions are not available for the $Z_{i,j}$. These elementary coefficients, given by $w_i(s_n)b_j(s'_m)K(s_n,s'_m)$, can in turn be seen to be simply products of samples of the IE kernel or operator and sampled basis and testing functions. It should be apparent from this expanded expression for the direct-matrix coefficients that interchanging the basis and weight functions leaves the final problem description unchanged, although the added observation that two different IEs can yield identical matrices when using equivalent numerical treatments is less obvious.

Computing the Solution Matrix

Once the direct matrix has been computed, the solution can be obtained numerically using various approaches, ranging from inversion of the direct matrix to developing a solution via iteration as summarized in Table 43.10. A precautionary comment is in order with respect to the accuracy with which the solution matrix might be obtained. As computer speed and storage have increased, the number of unknowns used in modeling has also increased, from a few tens in earlier years to thousands now when using IE techniques. The increasing number of operations involved in solving these larger matrices increases sensitivity of the results to roundoff errors. This is especially the case when the direct matrix is not well conditioned. It is therefore advisable to perform some sensitivity analyses to determine the direct-matrix condition number and to ascertain the possible need for performing some of the computations in double precision.

Obtaining the Solution

When a solution matrix has been developed using inversion or factorization, subsequently obtaining the solution (most often a current) is computationally straightforward, involving multiplication of the right-hand side (RHS) source vector by the solution matrix. When an iterative approach is used, a solution matrix is not computed, but the solution is instead developed from RHS-dependent manipulation of the direct matrix. Motivation for the latter comes from the possibility of reducing the N_x^3 dependency of the direct procedure. As problem size increases, the computation cost will be increasingly dominated by the solution time.

Table 43.10 Summary of Operation Count for Solution of General Direct Matrix Having N_x Unknowns

Method	To Obtain Solution Matrix	To Obtain Solution	Comments
Cramer's rule	Expand in co-factors leading to →	$\sim N_x!$	Not an advisable procedure but useful to illustrate just how bad the problem could be
Inversion	N_x^3	N_x^2	Provides RHS-independent solution matrix
Factorization	$N_x^3/3$	N_x^2	RHS-independent solution matrix
Iteration			
General	—	$N_x^2 - N_x^3$	Each RHS requires separate solution
With FFT	—	$N_x - N_x^2$	Same, plus applicability to arbitrary problems uncertain
Symmetry			
Reflection	$(1 \text{ to } 2^p) \times (N_x/2^p)^3$	$N_x^2/2^p$	For $p = 1$ to 3 reflection planes
Translation (Toeplitz)	$n_x^3[t(\log_2 t)^2]$	N_x^2	For n_x unknowns per t sections of translation
Rotation (circulant)	$\log_2(N_x)N_x^3/n^2$	N_x	For n rotation sectors and a complete solution
	$m\log_2(N_x)(N_x/n)^3$	m	For $m = 1$ to n modes
Banded			
General	$N_x W^2$	$N_x W$	For a bandwidth of W coefficients
Toeplitz	$N_x \log_2 N_x$	W^2	For a bandwidth of W coefficients

43.5 Some Practical Considerations

Although the overall solution effort has various cost components, perhaps the one most considered is the computer time and storage required to obtain the numerical results desired. With the increasing computer memories becoming available, where even microcomputers and workstations can directly address gigabytes, the memory costs of modeling are becoming generally less important than the time cost, with which we are primarily concerned here. For each model class considered, the computer-time dependence on the number of unknowns is presented in a generic formula followed by the highest-order $(L/\Delta L)$ term in that formula to demonstrate how computer time grows with increasing problem size.

Integral Equation Modeling

Frequency Domain

If we consider an IE model specifically, we can show that, in general, the computer time associated with its application is dependent on the number of unknowns N_x in the frequency domain as

$$T_{\text{IE},\omega} \approx A_{\text{fill}}N_x^2 + A_{\text{solve}}N_x^3 + A_{\text{source}}N_x^2 N_{\text{rhs}} + A_{\text{field}}N_x N_{\text{rhs}}N_{\text{fields}}$$

$$\sim (L/\Delta L)^{3(D-1)}$$

(43.4a)

where the A's are computer- and algorithm-dependent coefficients that account for computation of A_{fill}, the direct (impedance) matrix; A_{solve}, the solution (admittance) matrix (assuming inversion or factorization); A_{source}, the source response (currents and charges) for one of N_{rhs} different excitations or right-hand sides (the g term of Table 43.7); and A_{field}, one of N_{field} fields, where $A_{\text{field}} \leq A_{\text{fill}}$, depending on whether a near-field (=) or far-field (<) value is obtained. D is the problem dimensionality (for a wire IE model, $D = 2$ except when used for wire-mesh approximations of surfaces in which case $D = 3$); L is a characteristic length of the object being modeled; and ΔL is the spatial resolution required, being proportional to the wavelength.

Time Domain

A similar relationship holds for a time-domain IE model which uses N_t time steps,

$$T_{\text{IE},t} \approx A_{\text{fill}}N_x^2 + A_{\text{solve}}N_x^3 + A_{\text{source}}N_x^2 N_t N_{\text{rhs}} + A_{\text{field}}N_x N_t N_{\text{rhs}}N_{\text{fields}}$$

$$\sim (L/\Delta L)^{2(D-1)+1+r}, \text{ explicit approach, } 0 \le r \le 1 \qquad (43.4b)$$

$$\sim (L/\Delta L)^{3(D-1)}, \qquad \text{implicit approach}$$

with the A's accounting for computation of the time-domain terms equivalent to their frequency-domain counterparts (with different numerical values), and r accounting for the RHS dependency. Although a direct matrix may require solution initially before time-stepping the model, that is normally avoided by using $\delta t \le \Delta x/c$, which yields an explicit solution, in which case $A_{\text{solve}} = 0$. As can be appreciated from these expressions, the number of unknowns that is required for these computations to be acceptably accurate has a strong influence on the computer time eventually needed.

Differential Equation Modeling

Frequency Domain

DE modeling is less commonly used in the frequency domain primarily because the order of the matrix that results depends on $(L/\Delta L)^D$ rather than the usual $(L/\Delta L)^{D-1}$ dependency of an IE model. On the other hand, the matrix coefficients require less computation whether the DE model is based on a finite-difference or finite-element treatment. Furthermore, the matrix is very sparse because a differential operator is a local rather than a global one, as is the integral operator. Matrix fill time is therefore generally not of concern, and the overall computer time is given approximately by

$$T_{\text{DE},\omega} \approx A_{\text{solve}}N_x W^2 + A_{\text{source}}N_x W N_{\text{rhs}} + A_{\text{field}}N_x^{(D-1)/D}N_{\text{rhs}}N_{\text{fields}}$$

$$\sim (L/\Delta L)^{3D-2} \qquad (43.4c)$$

which exhibits a dominance by the matrix-solution term. Note that the banded nature of the DE direct matrix has been taken into account where the bandwidth W varies as N_x^0, $N_x^{1/2}$, and $N_x^{2/3}$ respectively ($N_x^{[(D-1)/D]}$), in one, two, and three dimensions.

Time Domain

Time-domain DE modeling can use either implicit or explicit solution methods for developing the time variation of the solution. An explicit technique is one whereby the update at each time step is given in terms of solved-for past values of the unknowns and the present excitation, with no interaction permitted between unknowns within the same time step, and is the approach used in a technique known as finite-difference time domain (FDTD). An implicit technique, on the other hand, does allow for interaction of unknowns within the same time step but can therefore require the solution of a matrix equation. In spite of this disadvantage, implicit techniques are important because they are not subject to Courant instability when $c\delta t > \Delta\chi$ as is an explicit approach.

The solution time for the explicit case is approximated by

$$T_{\text{DE},t} \approx A_{\text{source}}N_x N_t N_{\text{rhs}} + A_{\text{field}}N_x^{(D-1)/D}N_{\text{rhs}}N_{\text{fields}}$$

$$\sim (L/\Delta L)^{D+1+r}, \text{ explicit approach; } 0 \le r \le 1 \qquad (43.4d)$$

$$\approx A_{\text{solve}}N_x W^2 + A_{\text{source}}N_x W N_t N_{\text{rhs}} + A_{\text{field}}N_x^{(D-1)/D}N_{\text{rhs}}N_{\text{fields}}$$

$$\sim (L/\Delta L)^{3D-2}, \text{ for } D = 2, 3$$

and

$$\sim (L/\Delta L)^{2+r}, \text{ for } D = 1, \text{ implicit approach; } 0 \leq r \leq 1$$

assuming a banded matrix is used to solve the implicit direct matrix.

Discussion

It should be recognized that the above computer-time estimates assume solutions are obtained via matrix factorization, an N^3 process, and that iterative techniques when applicable should be expected to reduce the maximum order of the $(L/\Delta L)$ dependency but at the cost, however, of requiring the computation to be repeated for each RHS. We also emphasize that these comparisons consider only problems involving homogeneous objects, thereby providing a more favorable situation for IE models because their sampling dimensionality $d = D - 1$ for a problem dimensionality of D but which increases to $d = D$ when an inhomogeneous object is modeled. Because of these and other factors that can lead to many different combinations of formulation and numerical treatment, the foregoing results should be viewed as only generic guidelines, with the computational characteristics of each specific model requiring individual analysis to obtain numerical values for the various A_x coefficients and their $(L/\Delta L)$ dependency. It is relevant to observe that the lowest-order size dependency for three-dimensional problems is exhibited by the DE explicit time-domain model which is on the order of $(L/\Delta L)^4$.

An additional factor that should be considered when choosing among computer models is the information needed for a particular application relative to the information provided by the model. A time-domain model, for example, can intrinsically provide a frequency response over a band of frequencies from a single calculation, whereas a frequency-domain model requires repeated evaluation at each of the frequencies required to define the wideband response. Iterative solution of the direct matrix may be preferable for problems involving only one, or a few, excitations such as is the case for antenna modeling, to avoid computing all N_x^2 admittances of the solution matrix when only a single column of that matrix is needed. A DE-based model necessarily provides the "near" fields throughout the region being modeled, while an IE-based model requires additional computations essentially the same as those done in filling the impedance matrix once the sources have been obtained to evaluate the near fields. For applications that require modest computer time and storage, these considerations may be relatively less important than those that strain available computer resources. Clearly, the overall objective from an applications viewpoint is to obtain the needed information at the required level of accuracy for the minimum overall cost.

Sampling Requirements

We may estimate the number of samples needed to adequately model the spatial, temporal, and angular variation of the various quantities of interest in terms of an object characteristic length L and sampling dimension d. This may be done from knowledge of the typical spatial and temporal densities determined from computer experiments and/or from invocation of Nyquist-like sampling rates for field variations in angle as a function of aperture size. The resulting estimates are summarized in Table 43.11 and apply to both IE and DE models.

These may be regarded as *wavelength-driven* sampling rates, in contrast with the *geometry-driven* sampling rates that can arise because of problem variations that are small in scale compared with λ. Geometry-driven sampling would affect primarily N_x, resulting in larger values than those indicated above.

We note that the computer time is eventually dominated by computation of the solution matrix and can grow as $(L/\Delta L)^3$, $(L/\Delta L)^6$, and $(L/\Delta L)^9$ (or f^3, f^6, and f^9), respectively, for wire, surface, and volume objects modeled using integral equations and matrix factorization or inversion. Thus, in spite of the fact that mainframe computer power has grown by a factor of about 10^6 from the UNIVAC-1 to the CRAY2, a growth that is anticipated to continue during the near future as shown

Table 43.11 Nominal Sampling Requirements for Various Field Quantities

Quantity	Value
N_x, total number of spatial samples (per scalar unknown)	$\sim (L/\Delta L)^d = (2\pi L/\lambda)^d$
N_t, number of time steps for time-domain model	$\sim (L/\Delta L) = (2\pi L/\lambda)$
N_f, number of frequency steps to characterize spectral response from frequency-domain model	$\sim (L/2\Delta L) = N_t/2$
N_{rhs}, number of excitation sources for monostatic radar cross section in one plane[a]	$\sim (4L/\Delta L) = 8\pi L/\lambda$
N_{fields}, number of far fields needed for bistatic pattern in one observation plane[a]	$\sim N_{rhs} = (4L/\Delta L)$

λ is the wavelength at the highest frequency of interest; ΔL is the spatial resolution being sought; L is object maximum object dimension or dimension in observation plane; d is the number of *spatial* dimensions being sampled and is not necessarily the *problem* dimensionality D. The distinction is important because when an appropriate Green's function is available, the source integrals are usually one dimension less than the problem dimension, i.e., $d = D - 1$. An exception is an inhomogeneous, penetrable body where $d = D$ when using an integral equation.

[a]Assuming ~6 samples per lobe of the scattering pattern are needed.

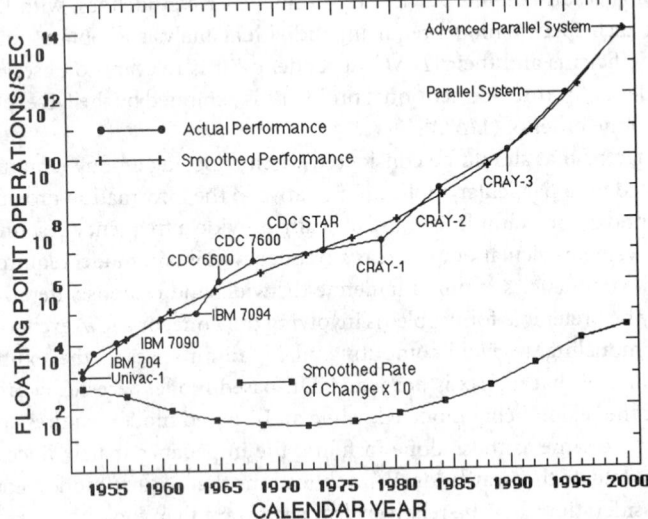

FIGURE 43.2 Raw and smoothed FLOP (floating-point operation) rates of mainframe computers and smoothed rate-of-change in speed, at year of introduction from the UNIVAC-1 to the projected performance of an advanced parallel system at year 2000. Future growth is increasingly dependent on computer architecture, requiring increasing parallelism as improvements due to component performance reach physical speed limits.

in Fig. 43.2, the growth in problem size is much less impressive, as illustrated by Fig. 43.3. The curves on this graph demonstrate emphatically the need for finding faster ways of performing the model computations, a point further emphasized by the results shown in Fig. 43.4 where the computer time required to solve a reference problem using various standard models is plotted as a function of frequency.

43.6 Ways of Decreasing Computer Time

The obvious drawback of direct moment-method models as N_x increases with increasing problem size and/or complexity suggests the need for less computationally intensive alternatives. There are various alternatives for decreasing the computer cost associated with solving electromagnetic prob-

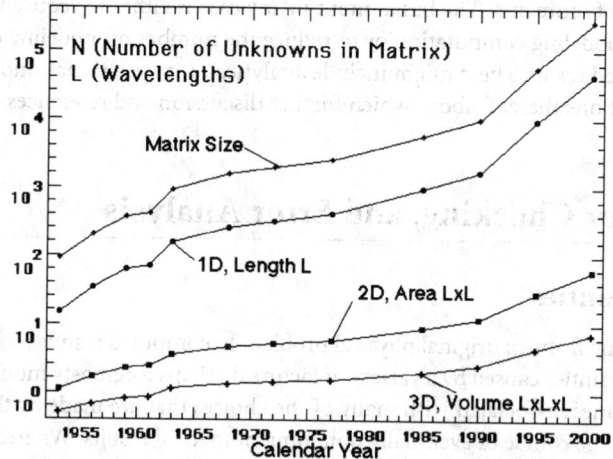

FIGURE 43.3 Time development of IE-based modeling capability for one-dimensional (e.g., a wire), two-dimensional (e.g., a plate), and three-dimensional (e.g., a penetrable, inhomogeneous cube) sampling of a problem of characteristic dimension L in wavelengths and matrix order N solvable in 1 h of computer time using mainframe computers introduced in the years indicated. Linear-systems solution using LU decomposition (an N^3 dependency) is assumed with number of unknowns proportional to L, L^2 and L^3, respectively, without any problem symmetry being exploited. These results should be viewed as upper bounds on solution time and might be substantially reduced by advances in linear-system solution procedures.

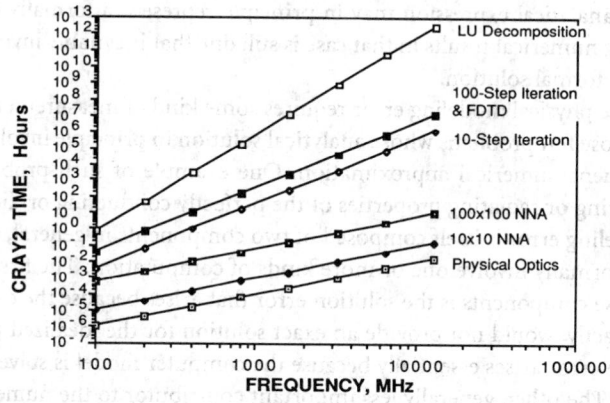

FIGURE 43.4 An illustration of the frequency dependence of the CRAY2 computer time required for some standard computer models applied to the reference problem of a perfectly conducting, space-shuttle-sized object having a surface area of 540 m^2 [Miller, 1988]. At a sampling density of $100/\lambda^2$, a total of 6,000 surface samples is assumed for an IE model at 100 MHz for which LU decomposition of the direct (impedance) matrix requires about 1 h of CRAY2 time. The top, LU, curve has a slope of f^6 as discussed in the text. The next two curves have slopes of f^4, the upper corresponding to use of a TD DE model (FDTD) as well as an iterative solution of the direct IE matrix, assuming acceptable convergence occurs in 100 iteration steps. The third curve is for a 10-step iterative solution of the IE matrix. The bottom three curves have f^2 slopes. The upper two of these are for 100- and 10-step iterative solutions used in connection with a near-neighbor approximation (NNA), wherein only the 100 and 10 largest interaction coefficients are retained in the matrix, respectively. The bottom curve is for the physical-optics approximation, in which the induced current is computed from the incident magnetic field. The effects of these different frequency slopes on the computer time can be seen to be extreme, emphasizing the need for developing more efficient solution procedures.

lems using the method of moments. The basic intent in any case is either to reduce the direct cost of performing a given modeling computation or to reduce the number of modeling computations needed to obtain a desired result. These might include analytical, computational, and experimental approaches or combinations thereof, about which further discussion and references may be found in Miller [1988].

43.7 Validation, Error Checking, and Error Analysis

Modeling Uncertainties

The process of proceeding from an original physical problem to computed results is one that is subject to numerous uncertainties caused by a variety of factors. Perhaps foremost among these factors is the degree of arbitrariness associated with many of the choices that are made by the code developer and/or modeler in the course of eventually obtaining numerical results. Whereas the numerical evaluation of classical boundary-value problems such as scattering from a sphere is numerically robust in the sense that different workers using different computers and different software can obtain results in agreement to essentially as many significant figures as they wish, the same observation cannot be made for moment-method modeling.

Modeling uncertainties can be assigned to two basic error categories, a physical **modeling error** ε_P and a numerical modeling error ε_N as outlined in Table 43.12. The former is due to the fact that for most problems of practical interest varying degrees of approximation are needed in developing a simplified or idealized problem representation that will be compatible with the computer code to be used for the modeling computations. The latter is due to the fact that the numerical results obtained are almost invariably only approximate solutions to that idealized representation. We note that although an analytical expression may in principle represent a formally exact solution, the process of obtaining numerical results in that case is still one that inevitably involves finite-precision evaluation of the formal solution.

By its very nature, the physical modeling error requires some kind of measurement for its determination, except for those few problems whose analytical solution in principle involves no physical idealization or subsequent numerical approximation. One example of such problems is that of determining the scattering or radiating properties of the perfectly conducting or dielectric sphere.

The numerical modeling error is itself composed of two components in general, the determination of which would normally involve one or more kinds of computation. The first and generally more important of these components is the solution error that arises because the computer model used, even if solved exactly, would not provide an exact solution for the idealized problem representation. The solution error arises essentially because the computer model is solved using a finite number of unknowns. The other, generally less important contributor to the numerical modeling error is the equation error that arises because the numerical results obtained from the computer

Table 43.12 Error Types that Occur in Computational Electromagnetics

Category	Definition
Physical modeling error, ε_P	Arises because the numerical model used is normally an idealized mathematical representation of the actual physical reality
Numerical modeling error, ε_N	Arises because the numerical results obtained are only approximate solutions to that idealized representation and consists of two components: (1) Solution error—The difference that can exist between the computed results and an exact solution even were the linear system of equations to be solved exactly, due to using a finite number of unknowns (2) Equation error—The equation mismatch that can occur in the numerical solution because of roundoff due to finite-precision computations or when using an iterative technique because of limited solution convergence

model used may not numerically satisfy the modeling equations. The equation error may be caused both by roundoff due to the computer word size as well as the solution algorithm used, as in the case of iteration, for example. The impact of equation error can be expected to increase with increasing condition number of the direct matrix.

Validation and Error Checking

One of the most time consuming and long lasting of the tasks associated with any model development is that of validation. Long after work on the model has been completed, questions will continue to arise about whether a given result is valid or whether the model can be applied to a given problem. There are essentially two kinds of validation procedures that can be considered to answer such questions: (1) internal validation, a check that can be made concerning solution validity within the model itself; and (2) external validation, a check that utilizes information from other sources which could be analytical, experimental, or numerical.

Existing computer models often do not perform internal checks on the results they produce but instead leave that as an exercise for the user. It would be of extremely great potential value if a variety of such checks could be built into the code and exercised as desired by the modeler. The topic of error checking and validation is an active one in CEM and receives a great deal of ongoing attention, for which the technical literature provides a good point of departure for the reader interested in more detail.

43.8 Concluding Remarks

In the preceding discussion we have presented a selective survey of computational electromagnetics. Attention has been directed to radiation and scattering problems formulated as integral equations and solved using the method of moments. Beginning from the viewpoint of electromagnetics as a transfer-function process, we concluded that the basic problem is one of developing source-field relationships or field propagators. Of the various ways by which these propagators might be expressed, we briefly discussed the Maxwell curl equations and Green's-function source integrals as providing the analytical basis for moment-method computer models. We then considered at more length some of the numerical issues involved in developing a computer model, including the idea of sampling functions used both to represent the unknowns to be solved for and to approximate the equations that they must satisfy. Some of the factors involved in choosing these sampling functions and their influence on the computational requirements were examined. Next, we discussed some ways of decreasing the needed computer time based on either analytical or numerical approaches. Some closing comments were directed to the important problems of validation, error checking, and error analysis. Throughout our discussion, emphasis has been given to implementation issues involved in developing and using computer models as opposed to exploring analytical details.

Defining Terms

Computer model: Based on a numerical solution of some appropriate analytical formulation that describes a physical phenomenon of interest. The model is realized in an *algorithm* or computer *code* that reduces the formulation to a series of operations suitable for computer solution.

Field propagator: The analytical description of how electromagnetic fields are related to the sources that cause them. Common field propagators in electromagnetics are the defining Maxwell equations that lead to differential equation models, Green's functions that produce integral equation models, optical propagators that lead to optics models, and multipole expansions that lead to modal models.

Integral equation: An analytical relationship in which the quantity whose solution is sought (the unknown) appears under an integral sign. When this is the only place that the unknown appears, the integral equation is commonly called a first-kind equation, while if the unknown also appears outside the integral, it is a second-kind integral equation.

Method of moments: A general technique for reducing integral, differential (including partial), and integrodifferential equations to a linear system of equations or matrix. The moment method involves discretizing, sampling, and approximating the defining equations using *basis* or *expansion* functions to replace the unknown and *testing* or *weighting* functions to satisfy the defining equations. The matrix that results may be full (all coefficients nonzero) or sparse (only a few per row are nonzero), depending on whether the model is an integral or differential equation.

Modeling errors: In essentially all computer modeling, there are two basic kinds of errors. One, the *physical modeling error*, arises from replacing some real-world physical problems with some idealized mathematical representation. The other, the *numerical modeling error*, comes from obtaining only approximate solution to that idealized representation. Usually, the numerical modeling error can be reduced below the physical modeling error if enough unknowns, i.e., a large enough matrix, are used to model the problem of interest.

Sampling: The process of replacing a continuous physical quantity by some sequence of sampled values. These values are associated with the analytical function used to approximate the behavior of the physical quantity whose solution is sought and are the unknowns of the moment-method matrix. Sampling is also involved in determining how well the defining equations are to be satisfied. A common approach for equation sampling is *point* sampling, where the equations are explicitly satisfied at a series of discrete points in some prescribed region of space. Unknown sampling can involve localized basis functions, an approach called *subdomain* sampling, while if the basis functions reside over the entire region occupied by the unknown, the approach is called *entire-domain* sampling.

Solution domain: Electromagnetic fields can be represented as a function of time, or a *time-domain* description, or as a function of frequency using a (usually) Fourier transform, which produces a *frequency-domain* description.

References

M. Abramowitz and I. A. Stegun, *Handbook of Mathematical Functions*, Applied Mathematics Series, vol. 55, Washington, D.C.: National Bureau of Standards, 1964.

F. S. Acton, *Numerical Methods that Work*, New York: Harper and Row, 1970.

B. W. Arden and K. N. Astill, *Numerical Algorithms: Origins and Applications*, Reading, Mass.: Addison-Wesley, 1970.

R. F. Harrington, *Time-Harmonic Electromagnetic Fields*, New York: McGraw-Hill, 1961.

R. F. Harrington, *Field Computation by Moment Methods*, New York: Macmillan, 1968.

E. K. Miller, "A selective survey of computational electromagnetics," *IEEE Trans. Antennas Propagat.*, vol. AP-36, pp. 1281–1305, 1988.

E. K. Miller, "Solving bigger problems—by decreasing the operation count and increasing the computation bandwidth," invited article in special issue of *IEEE Proc. Electromagnets*, vol. 79, no. 10, pp. 1493–1504, 1991.

E. K. Miller, L. Medgyesi-Mitschang, and E. H. Newman, *Computational Electromagnetics: Frequency-Domain Method of Moments*, New York: IEEE Press, 1991.

R. Mittra, ed., *Computer Techniques for Electromagnetics*, New York: Pergamon Press, 1973.

R. Mittra, ed., *Numerical and Asymptotic Techniques in Electromagnetics*, New York: Springer-Verlag, 1975.

J. Moore and R. Pizer, *Moment Methods in Electromagnetics: Techniques and Applications*, New York: Wiley, 1984.

A. J. Poggio and E. K. Miller, "Low frequency analytical and numerical methods for antennas," in *Antenna Handbook*, Y. T. Lo and S. W. Lee, eds., New York: Van Nostrand Reinhold, 1988.

B. D. Popovic, M. B. Dragovic, and A. R. Djordjevic, *Analysis and Synthesis of Wire Antennas*, Letchworth, Hertfordshire, England: Research Studies Press, 1982.

W. H. Press, B. R. Flannery, S. A. Teukolsky, and W. T. Vettering, *Numerical Recipes*, London: Cambridge University Press, 1986.

B. J. Strait, ed., *Applications of the Method of Moments to Electromagnetic Fields*, St. Cloud, Fla.: SCEEE Press, 1980.

J. A. Stratton, *Electromagnetic Theory*, New York: McGraw-Hill, 1941.

W. L. Stutzman and G. A. Thiele, *Antenna Theory and Design*, New York: John Wiley, 1981.

P. L. E. Uslenghi, ed., *Electromagnetic Scattering*, New York: Academic Press, 1978.

J. H. Wang, *Generalized Moment Methods in Electromagnetics*, New York: Wiley Interscience, 1991.

Further Information

The *International Journal of Numerical Modeling*, published by Wiley four times per year, includes numerous articles on modeling electronic networks, devices, and fields. Information concerning subscriptions should be addressed to Subscription Department, John Wiley & Sons Ltd., Baffins Lane, Chichester, Sussex PO19 1UD, England.

The *Journal of the Acoustical Society of America* is published by the American Institute of Physics on a monthly basis. Most issues contain articles about the numerical solution of acoustics problems which have much in common with problems in electromagnetics. Information about the society and journal can be obtained from Acoustical Society of America, 500 Sunnyside Blvd., Woodbury, NY 11797.

The *Journal of the Applied Computational Electromagnetics Society* is published two or three times a year, accompanied by a newsletter published about four times per year. The focus of the society and journal is the application of computer models, their validation, information about available software, etc. Membership and subscription information can be obtained from Dr. R.W. Adler, Secretary, Applied Computational Electromagnetics Society, Naval Postgraduate School, Code ECAB, Monterey, CA 93943.

The *Journal of Electromagnetic Waves and Applications* is published by VNU Science Press. It contains numerous articles dealing with the numerical solution of electromagnetic problems. Information about the journal can be obtained from its editor-in-chief, Professor J.A. Kong, Department of Electrical Engineering and Computer Science, Massachusetts Institute of Technology, Cambridge, MA 02139.

The *Proceedings of the IEEE, Transactions on Microwave Theory and Techniques of the IEEE, Transactions on Antennas and Propagation of the IEEE*, and *Transactions on Electromagnetic Compatibility of the IEEE* all are periodicals published by the Institute of Electrical and Electronics Engineers, about which information can be obtained from IEEE Service Center, 445 Hoes Lane, PO Box 1331, Piscataway, NJ 08855-1331.

Hewlett-Packard Company's HCPL-7800 isolation amplifier is one of the smallest current-sensing devices available. Shown is a typical Hall-effect sensor (top) and a typical isolation-amplifier module (bottom left) compared to the dual-inline packaged HCPL-7800. (Photo courtesy of Hewlett-Packard Company.)

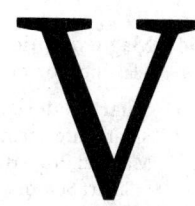

Electrical Effects and Devices

Lyle D. Feisel
State University of New York, Binghamton

44 Electroacoustic Devices *P. Rogers* 1055
Transduction Mechanisms • Sensitivity and Source Level • Reciprocity • Canonical Equations and Electroacoustic Coupling • Radiation Impedance • Directivity

45 Surface Acoustic Wave Filters *D. Malocha* 1062
SAW Material Properties • Basic Filter Specifications • SAW Transducer Modeling • Distortion and Second-Order Effects • Bidirectional Filter Response • Multiphase Unidirectional Transducers • Single-Phase Unidirectional Transducers • Dispersive Filters • Coded SAW Filters • Resonators

46 Ultrasound *G. Farnell* 1077
Propagation in Solids • Piezoelectric Excitation • One-Dimensional Propagation • Transducers

47 Ferroelectric and Piezoelectric Materials *K. Etzold* 1087
Mechanical Characteristics • Ferroelectric Materials

48 Piezoresistivity *A. Amin* 1099
Equation of State • Effect of Crystal Point Group on Π_{ijkl} • Geometric Corrections and Elastoresistance Tensor • Multivalley Semiconductors • Semiconducting (PTCR) Perovskites • Thick Film Resistors

49 The Hall Effect *A. Ehrlich* 1106
Theoretical Background • Relation to the Electronic Structure—(i) $\omega_c\tau \ll 1$ • Relation to the Electronic Structure—(ii) $\omega_c\tau \gg 1$

50 Superconductivity *K. Delin, T. Orlando* 1114
General Electromagnetic Properties • Superconducting Electronics • Types of Superconductors

51 Pyroelectric Materials and Devices *R. Whatmore* 1126
Polar Dielectrics • The Pyroelectric Effect • Pyroelectric Materials and Their Selection

52 Dielectrics and Insulators *R. Bartnikas* 1132
Dielectric Losses • Dielectric Breakdown • Insulation Aging • Dielectric Materials

53 Sensors *R. Smith* 1152
Physical Sensors • Chemical Sensors • Biosensors • Microsensors

54 **Magnetooptics**   *D. Young, C. Wang, Y. Pu* 1162
      Classification of Magnetooptic Effects • Applications of Magnetooptic Effects

55 **Smart Materials**   *P. Neelakanta* 1173
      Smart/Intelligent Structures • Objective-Based Classification of Smart/Intelligent Materials •
      Material Properties Conducive to Smart Material Applications • State-of-the-Art Smart Materials
      • Smart Sensors • Examples of Smart/Intelligent Systems • High-Tech Application Potentials

E VERY HIGH SCHOOL STUDENT who takes a course in physics or even general science is—or
      at least should be—familiar with the first-order, linear electrical effects such as resistance, induc-
tance, capacitance, etc. The more esoteric effects, however, are often neglected, even in otherwise
comprehensive undergraduate electrical engineering curricula. These effects, though, are not only
fascinating in their manifestations but are also potentially—and in some cases, currently—exceedingly useful in application. This section will describe many of these higher-order electrical and
magnetic effects and some of the devices that are based upon them. Readers are invited not only
to study the current applications but to let their imaginations extrapolate to other uses as yet
unproposed.

A number of phenomena are related to the interaction of mechanical energy with electrical
energy. The field of *acoustics* deals with those situations where that mechanical energy takes the
form of sound waves. Acoustic applications have been particularly fruitful, especially during the
last two decades. Surface acoustic wave (SAW) filters are among the more useful applications.
These elegant devices are a marriage of sophisticated signal theory and piezoelectricity, consum-
mated on the bed of thin-film technology. Unlike some elegant devices, they have been commercially successful as well.

A special class of acoustoelectric devices deals with acoustic frequencies beyond the range of
human hearing. The field of *ultrasonics* and its related devices and systems are finding broad application in the area of nondestructive testing. Of course, one of the testing applications where the
nondestructive property is especially important is in investigating the human body. Medical imaging has provided considerable impetus for advances in ultrasonics in the last few years.

Most people know that if a sample of certain types of material (e.g., iron) is subjected to a magnetic field, it will exhibit a retained magnetic behavior. Few, however, realize that some materials
exhibit a similar retention effect when an electric field is applied. *Ferroelectricity* is the phenomenon in which certain crystalline or polycrystalline materials retain electric polarization after
an external electric field has been applied and removed. Since the direction of the polarization
depends upon the direction of the applied field and since the polarization is quite persistent, memory devices can be based on this effect. Other applications have also been suggested.

For decades, the frequencies of radio transmitters have been stabilized with "crystals." In recent
years, the effect called *piezoelectricity*—in which a mechanical strain induces an electric field and
vice versa—has found many other applications. Like ferroelectrics, piezoelectric materials can be
either crystalline or polycrystalline and can be fabricated in a variety of shapes.

If an electric charge is moved with a velocity at some angle to a magnetic field, the charge will
experience a force at right angles to both the charge velocity and the magnetic field. If the charge is
inside a solid material, a charge inhomogeneity is created and an electric field results. This is the
well-known *Hall effect*, which finds practical application in such devices as magnetic field meters
and in more basic uses as measuring and understanding the properties of semiconductors.

Probably the second electrical phenomenon observed by humans (lightning was probably the first), *ferromagnetism* deals with the interaction of molecular magnetic dipoles with external and internal magnetic fields. Ferromagnetic materials retain some polarization after an external field is removed—a desirable property if the application is a permanent magnet or a recording device—but one which causes losses in a transformer. These materials have improved as the demands of magnetic recording have increased.

If certain materials get cold enough, their resistivity goes to zero—not to some very small value but, as nearly as we can tell, zero. *Superconductivity* has been known as an interesting phenomenon for many years, but applications have been limited because the phenomenon only occurred at temperatures within a few degrees of absolute zero. Recent advances, however, have produced materials which exhibit superconductive behavior at substantially higher temperatures, and there is renewed interest in developing applications. This is certainly an area to watch in the next few years.

Some very elegant devices have been developed to exploit the interactions between electric fields and photons or optical waves. *Electrooptics* is the key to many of the recent and, indeed, future advances in optical communication. The phenomena are generally higher-order, nonintuitive, and exceedingly interesting, and the devices are generally quite elegant but simple.

We have come a long way since the first Atlantic Cable was fabricated using gutta-percha, tarred hemp, and pitch for insulation. *Dielectrics and insulators* are now better understood and controlled for a wide variety of applications. At one time the only property of real interest was dielectric strength, the insulator's ability to stand up to high voltage. Today, many other properties, as well as ease and economy of fabrication, are at least as important.

The word *application* appears many times in the preceding paragraphs. What are these applications? Many of the practical uses of the phenomena described in this section are in measuring the variables that define the phenomena. Thus, *sensors* constitute a primary application. For instance, the Hall effect can be used to measure magnetic fields, and mechanical strain can be measured using the phenomenon of piezoelectricity.

Just as photons will interact with electric fields, so, too, will they affect and be affected by magnetic fields. *Magnetooptics* is the study and application of these interactions. As with electrooptics, the increased activity in optical communications has provided renewed interest in this field.

The use of *smart materials* may solve a variety of engineering problems. In general, these are materials which change their properties to adapt to their environments, thereby doing their jobs better. This promises to be an area of increased activity in the future.

Again, the reader is admonished not only to understand the applications presented in the following chapters but to understand, at least at the phenomenological level, the phenomena upon which the applications are based. Such understanding is likely to lead to even broader applications in the future.

Nomenclature

Symbol	Quantity	Unit	Symbol	Quantity	Unit
α	attenuation constant	Np/m	η	emissivity	
χ_o	magnetic susceptibility of free space		k	quantum mechanical wave factor	m^{-1}
D	diffraction constant		k^2	SAW coupling factor	
E	transducer efficiency		K	thermal conductivity of pyroelectric	$W/m^2/K$
ε	dielectric constant				
ϵ	complex permittivity	F/m	m	molar mass	kg
G_T	thermal conductance	W/K	R	Hall coefficient	m^3/C
η	viscosity	Poise	S	strain	

Symbol	Quantity	Unit	Symbol	Quantity	Unit
σ	conductivity	S/m	V	phase velocity	m/s
T	stress	N/m^2	V	Verdet constant	
τ_T	thermal time constant of element	s	W	electromagnetic energy density	W/m^2
θ_f	Faraday rotation coefficient		Z_R	radiation impedance	Ω

44

Electroacoustic Devices

44.1 Introduction 1055
44.2 Transduction Mechanisms 1055
    Piezoelectricity • Magnetostriction • Electrodynamic • Electrostatic
    • Magnetic • Hydraulic • Fiber Optic • Parametric Transducers
    • Carbon Microphones
44.3 Sensitivity and Source Level 1057
44.4 Reciprocity 1058
44.5 Canonical Equations and Electroacoustic Coupling 1058
44.6 Radiation Impedance 1059
44.7 Directivity 1060

Peter H. Rogers
Georgia Institute of Technology

44.1 Introduction

Electroacoustics is concerned with the transduction of acoustical to electrical energy and vice versa. Devices which convert acoustical signals into electrical signals are referred to as "**microphones**" or "hydrophones" depending on whether the acoustic medium is air or water. Devices which convert electrical signals into acoustical waves are referred to as "loudspeakers" (or earphones) in air and "projectors" in water.

44.2 Transduction Mechanisms

Piezoelectricity

Certain crystals produce charge on their surfaces when strained or conversely become strained when placed in an electric field. Important piezoelectric crystals include quartz, ADP, lithium sulphate, rochelle salt, and tourmaline. Lithium sulphate and tourmaline are "volume expanders," that is, their volume changes when subjected to an electric field in the proper direction. Such crystals can detect hydrostatic pressure directly. Crystals which are not volume expanders must have one or more surfaces shielded from the pressure field in order to convert the pressure to a uniaxial strain which can be detected. Tourmaline is relatively insensitive and used primarily in blast gauges, while quartz is used principally in high Q ultrasonic transducers.

Certain ceramics such as lead zirconate titanate (PZT), barium titanate, and lead metaniobate become piezoelectric when polarized. They exhibit relatively high electromechanical coupling, are

This chapter is reproduced from R. M. Besançon, *Encyclopedia of Physics*, 3rd ed., New York: Van Nostrand Reinhold, 1985, pp. 337–341. With permission.

capable of producing very large forces, and are used extensively as sources and receivers for underwater sound. PZT and barium titanate have only a small volume sensitivity; hence they must have one or more surfaces shielded in order to detect sound efficiently. Piezoelectric ceramics have extraordinarily high dielectric coefficients and hence high capacitance, and they are thus capable of driving long cables without preamplifiers.

Recently, it has been discovered that certain polymers, notably polyvinylidene fluoride, are piezoelectric when stretched. Such piezoelectric polymers are finding use in directional microphones and ultrasonic hydrophones.

Magnetostriction

Some ferromagnetic materials become strained when subjected to a magnetic field. The effect is quadratic in the field, so a bias field or dc current is required for linear operation. Important magnetostrictive metals and alloys include nickel and permendur. At one time, magnetostrictive transducers were used extensively in active sonars but have now been largely replaced by ceramic transducers. Magnetostrictive transducers are rugged and reliable but inefficient and configurationally awkward. Recently, it has been discovered that certain rare earth iron alloys such as terbium-dysprosium-iron possess extremely large magnetostrictions (as much as 100 times that of nickel). They have relatively low eddy current losses but require large bias fields, are fragile, and have yet to find significant applications. Metallic glasses have also recently been considered for magnetostrictive transducers.

Electrodynamic

Electrodynamic transducers exploit the forces produced on a current-carrying conductor in a magnetic field and, conversely, the currents produced by a conductor moving in a magnetic field. Direct radiation moving coil transducers dominate the loudspeaker field. Prototypes of high-power underwater projectors have been constructed using superconducting magnets. Electrodynamic microphones, particularly the directional ribbon microphones, are also common.

Electrostatic

Electrostatic sources utilize the force of attraction between charged capacitor plates. The force is independent of the sign of the voltage, so a bias voltage is necessary for linear operation. Because the forces are relatively weak, a large area is needed to obtain significant acoustic output. The effect is reciprocal, with the change in the separation of the plates (i.e., the capacitance) produced by an incident acoustic pressure generating a voltage. The impedance of a condenser microphone, however, is high, so a preamplifier located close to the sensor is required. Condenser microphones are very flat and extremely sensitive. The change in capacitance induced by an acoustic field can also be detected by making the capacitor a part of a bridge circuit or, alternatively, a part of an oscillator circuit. The acoustic signal will then appear as either an amplitude or frequency modulation of some ac carrier. The charge storage properties of electrets have been exploited to produce electrostatic microphones which do not require a bias voltage.

Magnetic

Magnetic transducers utilize the force of attraction between magnetic poles and, reciprocally, the voltages produced when the reluctance of a magnetic circuit is changed. Magnetic speakers are used extensively in telephone receivers.

Hydraulic

Nonreversible, low-frequency, high-power underwater projectors can be constructed utilizing hydraulic forces acting to move large pistons. Electroacoustic transduction is achieved by modulating the hydraulic pressure with a spool valve actuated by an electrostrictive (PZT) stack.

Fiber Optic

An acoustic field acting on an optical fiber will change the optical path length by changing the length and index of refraction of the fiber. Extremely sensitive hydrophones and microphones can be made by using a fiber exposed to an acoustic field as one leg of an optical interferometer. Path length changes of the order of 10^{-6} optical wavelengths can be detected. The principal advantages of such sensors are their configurational flexibility, their sensitivity, and their suitability for use with fiber optic cables. Fiber optic sensors which utilize amplitude modulation of the light (microbend transducers) are also being developed.

Parametric Transducers

The nonlinear interaction of sound waves can be used to produce highly directional sound sources with no side lobes and small physical apertures. In spite of their inherent inefficiency, substantial source levels can be achieved and such "parametric sonars" have found a number of underwater applications. Parametric receivers have also been investigated but practical applications have yet to be found.

Carbon Microphones

Carbon microphones utilize a change in electrical resistance with pressure and are used extensively in telephones.

44.3 Sensitivity and Source Level

A microphone or hydrophone is characterized by its free-field voltage sensitivity, M, which is defined as the ratio of the output voltage, E, to the free-field amplitude of an incident plane acoustic wave. That is, for an incident wave which *in the absence of the transducer* is given by

$$P = P_0 \cos(\mathbf{k} \cdot \mathbf{R} - \omega t) \tag{44.1}$$

M is defined by

$$M = E/P_0 \tag{44.2}$$

In general, M will be a function of frequency and the orientation of the transducer with respect to the wave vector $\mathbf{k}$ (i.e., the direction of incidence of the wave). Thus, for a given frequency, M is proportional to the directivity of the transducer. It is usually desirable for a microphone or hydrophone to have a flat (i.e., frequency independent) free-field voltage sensitivity over the broadest possible range of frequencies to assure fidelity of the output electrical signal.

A loudspeaker or projector is characterized in a similar manner by its transmitting current response, S, which is defined as the ratio of the acoustic source level to the driving current, I. In the farfield of a transducer the acoustic pressure is a spherical wave which can be expressed as

$$P(R) = P_s(\theta, \phi)(R_0/R) \cos(kR - \omega t) \tag{44.3}$$

where θ and ϕ are elevation and azimuth angles and R_0 an arbitrary reference distance (usually 1 meter). $P_s(\theta, \phi)$ is defined as the source level. Thus S is given by

$$S = P_s(\theta, \phi)/I \qquad (44.4)$$

which is a function of θ and ϕ and the frequency ω. For high-fidelity sound reproduction S should be as flat as possible over the broadest possible bandwidth. For some purposes, however, such as ultrasonic cleaning or long-range underwater acoustic propagation, fidelity is unnecessary and high Q resonant transducers are employed to produce high-intensity sound over a narrow bandwidth.

44.4 Reciprocity

Most conventional transducers are *reversible*, that is, they can be used as either sources or receivers of sound (a carbon microphone and a fiber optic hydrophone are examples of transducers which are *not* reversible). A transducer is said to be *linear* if the input and output variables are linearly proportional (hot-wire microphones and unbiased magnetostrictive transducers are examples of *nonlinear* transducers). A transducer is said to be *passive* if the only source of energy is the input electrical or acoustical signal (a microphone with a built-in preamplifier and a parametric projector are examples of *nonpassive* transducers). Most transducers which are linear, passive, and reversible exhibit a remarkable property called *reciprocity*. For a *reciprocal* transducer of any kind (moving coil, piezoelectric, magnetostrictive, electrostatic, magnetic, etc.) the ratio of the free-field voltage sensitivity to the transmitting current response is equal to the reciprocity factor J which is independent of the geometry and construction of the transducer. That is:

$$\frac{M(\omega, \theta, \phi)}{S(\omega, \theta, \phi)} = J(\omega) = \frac{4\pi R_0}{\rho_0 \omega} \qquad (44.5)$$

where ρ_0 is the density of the medium and R_0 is the reference distance used in defining the source level. Equation (44.5) has a number of useful consequences: (1) the receiving and transmitting beam patterns of a reciprocal transducer are identical, (2) a transducer cannot be simultaneously flat as a receiver and transmitter since S has an additional factor of ω, and (3) Eq. (44.5) provides the basis for the three-transducer reciprocity calibration technique whereby an absolute calibration of a hydrophone or microphone can be obtained from purely electrical measurements.

44.5 Canonical Equations and Electroacoustic Coupling

Simple acoustic transducers can be characterized by the following canonical equations:

$$E = Z_e I + T_{em} V \qquad (44.6)$$

$$F = T_{me} I + Z_m V \qquad (44.7)$$

where V is the velocity of the radiating or receiving surface, F is the total force acting on the surface (including acoustic reaction forces), Z_e is the blocked ($V = 0$) electrical impedance, Z_m is the open-circuit mechanical impedance, and T_{em} and T_{me} are the electromechanical coupling coefficients. For reciprocal transducers $T_{em} = \pm T_{me}$. For example, for a moving coil transducer where the "motor" is coil in a radial magnetic field, B,

$$T_{em} = -T_{me} = BL \qquad (44.8)$$

where L is the length of the wire in the coil and the electrical impedance Z_e is largely inductive. For a piston transducer with a piezoelectric "motor"

$$T_{me} = T_{em} = -id_{33}/(\epsilon^T s\omega) \qquad (44.9)$$

where d_{33} is the piezoelectric strain coefficient, s is the compliance, ϵ^T is the permittivity at constant stress, and the electrical impedance Z_e is largely capacitive.

If a piston transducer is placed in an acoustic field such that the average pressure over the surface of the piston is P_B, then $F = P_B A$, where A is the area of the piston, and for a receiver $I = 0$, so

$$E = (T_{em} A / Z_m) P_B \tag{44.10}$$

If the transducer is small compared with an acoustic wavelength $P_E \approx P_0$ (in general $P_B = D P_0$ where D is the diffraction constant) and the free-field voltage sensitivity is given by

$$M = T_{em} A / Z_m \tag{44.11}$$

From Eq. (44.5) the transmitting current response is

$$S = \frac{\rho_0 \omega T_{em} A}{4 \pi R_0 Z_m} \tag{44.12}$$

From these simple considerations a number of principles of practical transducer design can be deduced. The mechanical impedance Z_m is in general given by

$$Z_m = \frac{K_m}{i\omega} + i\omega M + R_m \tag{44.13}$$

where K_m is an effective spring constant, M the mass, and R_m the mechanical resistance. For a piezoelectric transducer [Eq. (44.9)] T_{em} is inversely proportional to frequency; hence from Eqs. (44.10) and (44.11) we see that a piezoelectric transducer will have a flat receiving sensitivity below resonance (i.e., where its behavior is controlled by stiffness). On the other hand, a moving coil microphone must have a resistive mechanical impedance to have a flat response. From Eq. (44.12) we derive the fundamental tenet of loudspeaker design, that a moving coil loudspeaker will have a flat transmitting current response above resonance (i.e., where it is mass controlled). Accordingly, moving coil loudspeakers are designed to have the lowest possible resonant frequency (by means of a high compliance since the output is inversely proportional to the mass) and piezoelectric hydrophones are designed to have the highest possible resonant frequency.

An interesting and important consequence of electromechanical coupling is the effect of the motion of the transducer on the electrical impedance. In the absence of external forces (including radiation reactance) from Eqs. (44.6) and (44.7)

$$E = \left(Z_e - \frac{T_{em} T_{me}}{Z_m} \right) I \tag{44.14}$$

That is, the electrical impedance has a "motional" component given by $T_{em} T_{me} / Z_m$. The motional component can be quite significant near resonance where Z_m is small. This effect is the basis of crystal-controlled oscillators.

44.6 Radiation Impedance

An oscillating surface produces a reaction force F_R on its surface given by

$$F_R = -Z_R V \tag{44.15}$$

where Z_R is the radiation impedance. We can thus rewrite Eq. (44.7) as

$$F_{ext} = T_{em}I + (Z_R + Z_m)V \tag{44.16}$$

where F_{ext} now includes only external forces. For an acoustically small baffled circular piston of radius a,

$$Z_R = \pi a^4 \rho_0 \omega^2/2c - i(8/3)\omega\rho_0 a^3 \tag{44.17}$$

The radiation impedance thus has a mass-like reactance with an equivalent "radiation mass" of $(8/3)\rho_0 a^3$ and a small resistive component proportional to ω^2 responsible for the radiated power. A transducer will thus have a lower resonant frequency when operated underwater than when operated in air or vacuum. The total radiated power of the piston transducer is given by

$$\pi = \mathrm{Re}\, Z_r |V|^2 = (\pi a^4 \rho_0 \omega^2/2c)\, V^2 \tag{44.18}$$

Most transducers are displacement limited, so for a direct-radiating transducer V in Eq. (44.18) is limited. To obtain the most output power the piston should have the largest possible surface area consistent with keeping the transducer omnidirectional (the transducer will become directional when $a \geq \lambda$). This is easy to do in air but difficult in water since it is hard to make pistons which are both lightweight and stiff enough to hold their shape in water. Alternatively, the driver can be placed at the apex of a horn. For a conical horn, the fluid velocity at the end of the horn (where the radius is a_e) will be reduced to $V(a/a_e)$ but the radiating piston will now have an effective radius of a_e so the radiated power will increase by a factor of $(a_e/a)^2$. For high-power operation at a single frequency, the driver can be placed at the end of a quarter wave resonator.

44.7 Directivity

It is often desirable for transducers to be directional. Directional sound sources are needed in diagnostic and therapeutic medical ultrasonics, for acoustic depth sounders; and to reduce the power requirements and reverberation in active sonars, etc. Directional microphones are useful to reduce unwanted noise (e.g., to pick up the voice of a speaker and not the audience); directional hydrophones or hydrophone arrays increase signal-to-noise and aid in target localization. One way to achieve directionality is to make the radiating surface large. A baffled circular piston has a directivity given by

$$D_e = 2J_1(ka\ \sin\theta)/ka \sin\theta \tag{44.19}$$

D_e equals unity for $\theta = 0$ and $1/2$ when $ka \sin\theta = 2.2$. For small values of ka, D_e is near unity for all angles.

Some transducers respond to the gradient of the acoustic pressure rather than pressure, for example, the ribbon microphone which works by detecting the motion of a thin conducting strip orthogonal to a magnetic field. Such transducers have a directivity which is dipole in nature, i.e.,

$$D_e = \cos\theta \tag{44.20}$$

Note that since the force in this case is proportional not to P_0 but to kP_0, a ribbon microphone (which like a moving coil microphone is electrodynamic) will have flat receiving sensitivity when its impedance is mass controlled. By combining a dipole receiver with a monopole receiver one obtains a unidirectional cardioid receiver with

$$D_e = (1 + \cos\theta) \tag{44.21}$$

Defining Terms

Electroacoustics: Concerned with the transduction of acoustical to electrical energy and vice versa.

Microphones: Devices which convert acoustical signals into electrical signals.

References

J.A. Bucaro, H.D. Dardy, and E.F. Carome, "Fiber optic hydrophone," *J. Acoust. Soc. Am.*, vol. 62, p. 1302, 1977.

R.J. Bobber, "New types of transducer," in *Underwater Acoustics and Signal Processing*, L. Bjorno (Ed.), Dordrecht, Holland: D. Riedel, 1981.

R.J. Bobber, Underwater Electroacoustic Measurements, Washington, D.C.: Government Printing Office, 1969.

J.V. Bouyoucos, "Hydroacoustic transduction," *J. Acoust. Soc. Am.*, vol. 57, p. 1341, 1975.

F.V. Hunt, *Electroacoustics*, Cambridge: Harvard University Press, and New York: Wiley, 1954.

S.W. Meeks and R.W. Timme, "Rare earth iron magnetostrictive underwater sound transducer," *J. Acoust. Soc. Am.*, vol. 62, p. 1158, 1977.

M.B. Moffett and R.M. Mellon, "Model for parametric acoustic sources," *J. Acoust. Soc. Am.*, vol. 61, p. 325, 1977.

D. Ricketts, "Electroacoustic sensitivity of piezoelectric polymer cylinders," *J. Acoust. Soc. Am.*, vol. 68, p. 1025, 1980.

G.M. Sessler and J.E. West, "Applications," in *Electrets*, G.M. Sessler (Ed.), New York: Springer-Verlag, 1980.

Further Information

IEEE Transactions on Acoustics, Speech, and Signal Processing.

45

Surface Acoustic Wave Filters

45.1	Introduction	1062
45.2	SAW Material Properties	1063
45.3	Basic Filter Specifications	1064
45.4	SAW Transducer Modeling	1066
	The SAW Superposition Impulse Response Transducer Model · Apodized SAW Transducers	
45.5	Distortion and Second-Order Effects	1070
45.6	Bidirectional Filter Response	1070
45.7	Multiphase Unidirectional Transducers	1071
45.8	Single-Phase Unidirectional Transducers	1073
45.9	Dispersive Filters	1073
45.10	Coded SAW Filters	1073
45.11	Resonators	1074

Donald C. Malocha
University of Central Florida

45.1 Introduction

A **surface acoustic wave (SAW)**, also called a Rayleigh wave, is composed of a coupled compressional and shear wave in which the SAW energy is confined near the surface. There is also an associated electrostatic wave for a SAW on a piezoelectric substrate which allows electroacoustic coupling via a transducer. SAW technology's two key advantages are its ability to electroacoustically access and tap the wave at the crystal surface and that the wave velocity is approximately 10,000 times slower than an electromagnetic wave. Assuming an electromagnetic wave velocity of 3×10^8 m/s and an acoustic wave velocity of 3×10^3 m/s, Table 45.1 compares relative dimensions versus frequency and delay. The SAW wavelength is on the same order of magnitude as line dimensions which can be photolithographically produced and the lengths for both small and long delays are achievable on reasonable size substrates. The corresponding E&M transmission lines or waveguides would be impractical at these frequencies.

Because of SAWs' relatively high operating frequency, linear delay, and tap weight (or sampling) control, they are able to provide a broad range of signal processing capabilities. Some of these include linear and dispersive filtering, coding, frequency selection, convolution, delay line, time impulse response shaping, and others. There are a very broad range of commercial and military system applications which include components for radars, front-end and IF filters, CATV and VCR components, cellular radio and pagers, synthesizers and analyzers, navigation, computer clocks, tags, and many, many others [Campbell, 1989; Matthews, 1977].

Table 45.1 Comparison of SAW and E&M Dimensions versus Frequency and Delay, Where Assumed Velocities are $v_{SAW} = 3000$ m/s and $v_{EM} = 3 \times 10^8$ m/s

Parameter	SAW	E&M
$F_0 = 10$ MHz	$\lambda_{SAW} = 300$ μm	$\lambda_{EM} = 30$ m
$F_0 = 2$ GHz	$\lambda_{SAW} = 1.5$ μm	$\lambda_{EM} = 0.15$ m
Delay = 1 ns	$L_{SAW} = 3$ μm	$L_{EM} = 0.3$ m
Delay = 10 μs	$L_{SAW} = 30$ mm	$L_{EM} = 3000$ m

There are four principal SAW properties: transduction, reflection, regeneration and nonlinearities. Nonlinear elastic properties are principally used for convolvers and will not be discussed. The other three properties are present, to some degree, in all SAW devices, and these properties must be understood and controlled to meet device specifications.

A finite-impulse response (FIR) or transversal filter is composed of a series of cascaded time delay elements which are sampled or "tapped" along the delay line path. The sampled and delayed signal is summed at a junction which yields the output signal. The output time signal is finite in length and has no feedback. A schematic of an FIR filter is shown in Fig. 45.1.

A SAW transducer is able to implement an FIR filter. The electrodes or fingers provide the ability to sample or "tap" the SAW and the distance between electrodes provides the relative delay. For a uniformly sampled SAW transducer, the delay between samples, Δt, is given by $\Delta t = \Delta L / v_a$, where ΔL is the electrode period and v_a is the acoustic velocity. The typical means for providing attenuation or weighting is to vary the overlap between adjacent electrodes which provides a spatially weighted sampling of a uniform wave. Figure 45.1 shows a typical FIR time response and its equivalent SAW transducer implementation. A SAW filter is composed of a minimum of two transducers and possibly other SAW components. A schematic of a simple SAW bidirectional filter is shown in Fig. 45.2. A **bidirectional transducer** radiates energy equally from each side of the transducer (or port). Energy not being received is absorbed to eliminate spurious reflections.

45.2 SAW Material Properties

There are a large number of materials which are currently being used for SAW devices. The most popular single-crystal piezoelectric materials are quartz, lithium niobate (LiNbO₃), and lithium

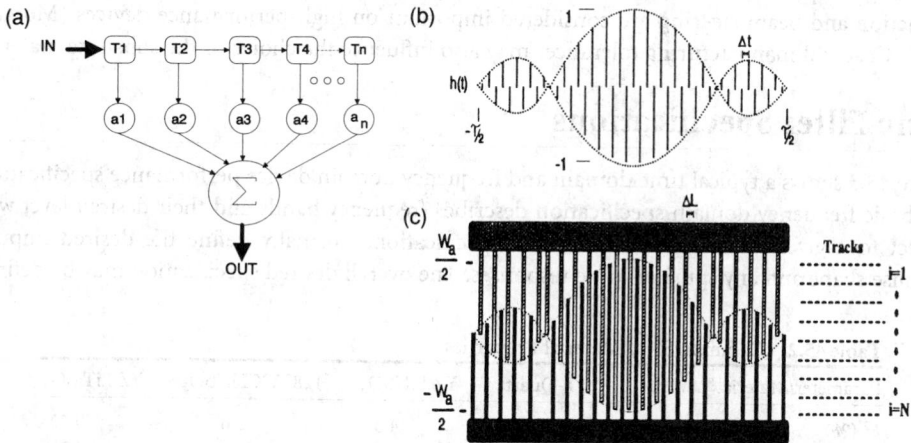

FIGURE 45.1 (a) Schematic of a finite-impulse response (FIR) filter. (b) An example of a sampled time function; the envelope is shown in the dotted lines. (c) A SAW transducer implementation of the time function $h(t)$.

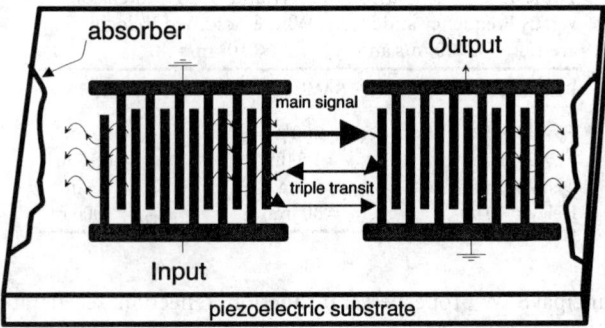

FIGURE 45.2 Schematic diagram of a typical SAW bidirectional filter consisting of two interdigital transducers. The transducers need not be identical. The input transducer launches waves in either direction and the output transducer converts the acoustic energy back to an electrical signal. The device exhibits a minimum 6-dB insertion loss. Acoustic absorber damps unwanted SAW energy to eliminate spurious reflections which could cause distortions.

tantalate ($LiTa_2O_5$). The materials are anisotropic, which will yield different material properties versus the cut of the material and the direction of propagation. There are many parameters which must be considered when choosing a given material for a given device application. Table 45.2 shows some important material parameters for consideration for four of the most popular SAW materials [Datta, 1986; Morgan, 1985].

The coupling coefficient, k^2, determines the electroacoustic coupling efficiency. This determines the fractional bandwidth versus minimum insertion loss for a given material and filter. The static capacitance is a function of the transducer electrode structure and the dielectric properties of the substrate. The values given in the table correspond to the capacitance per pair of electrodes having quarter wavelength width and one-half wavelength period. The free surface velocity, v_0, is a function of the material, cut angle, and propagation direction. The temperature coefficient of delay (TCD) is an indication of the frequency shift expected for a transducer due to a change of temperature and is also a function of cut angle and propagation direction.

The substrate is chosen based on the device design specifications and includes consideration of operating temperature, fractional bandwidth, and insertion loss. Second-order effects such as diffraction and beam steering are considered important on high-performance devices [Morgan, 1985]. Cost and manufacturing tolerances may also influence the choice of the substrate material.

45.3 Basic Filter Specifications

Figure 45.3 shows a typical time domain and frequency domain device performance specification. The basic frequency domain specification describes frequency bands and their desired level with respect to a given reference. Time domain specifications normally define the desired impulse response shape and any spurious time responses. The overall desired specification may be defined

Table 45.2 Common SAW Material Properties

Parameter/Material	ST-Quartz	YZ LiNbO$_3$	128° YX LiNbO$_3$	YZ LiTa$_2$O$_3$
k^2 (%)	0.16	4.8	5.6	0.72
C_S (pf/cm-pair)	0.05	4.6	5.4	4.5
v_0 (m/s)	3,159	3,488	3,992	3,230
Temp. coeff. of delay (ppm/°C)	0	94	76	35

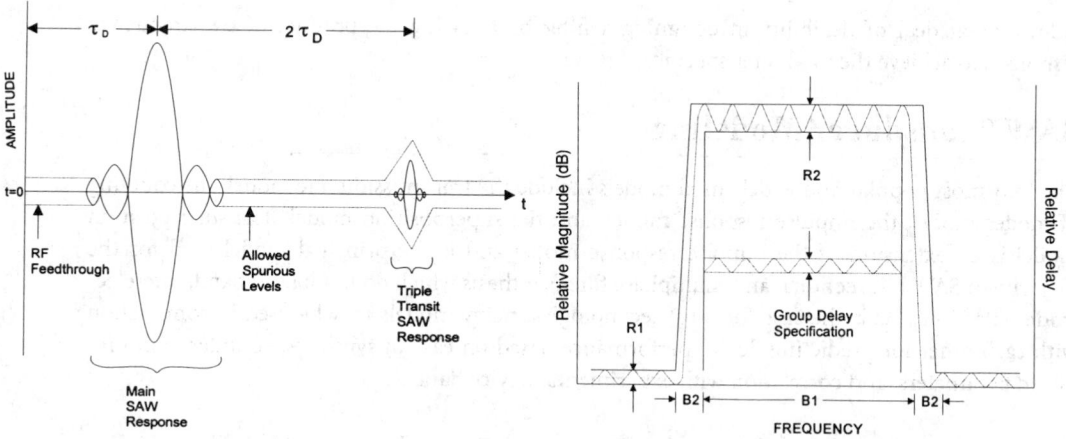

FIGURE 45.3 Typical time and frequency domain specification for a SAW filter. The filter bandwidth is B_1, the transition bandwidth is B_2, the inband ripple is R_2 and the out-of-band sidelobe level is R_1.

by combinations of both time and frequency domain specifications. Since time, $h(t)$, and frequency, $H(\omega)$, domain responses form unique Fourier transform pairs, given by

$$h(t) = 1 / 2\pi \int_{-\infty}^{\infty} H(\omega)e^{j\omega t}d\omega \tag{45.1}$$

$$H(\omega) = \int_{-\infty}^{\infty} h(t)e^{-j\omega t}dt \tag{45.2}$$

it is important that combinations of time and frequency domain specifications be self-consistent.

The electrodes of a SAW transducer act as sampling points for both transduction and reception. Given the desired modulated time response, it is necessary to sample the time waveform. For symmetrical frequency responses, sampling at twice the center frequency, $f_s = 2f_0$, is sufficient, while nonsymmetric frequency responses require sampling at twice the highest frequency of interest. A very popular approach is to sample at $f_s = 4f_0$. The SAW frequency response obtained is the convolution of the desired frequency response with a series of impulses, separated by f_s, in the frequency domain. The net effect of sampling is to produce a continuous set of harmonics in the frequency domain in addition to the desired response at f_0. This periodic, time-sampled function can be written as

$$g(t_n) = \sum_{-N/2}^{N/2} a_n \cdot \delta(t - t_n) \tag{45.3}$$

where a_n represents the sample values, $t_n = n\Delta t$, $n =$ nth sample, and $\Delta t =$ time sample separation. The corresponding frequency response is given by

$$G(f) = \sum_{-N/2}^{N/2} g(t_n)e^{-j2\pi f t_n} = \sum_{-N/2}^{N/2} g(t_n)e^{-j2\pi nf/f_s} \tag{45.4}$$

where $f_s = 1/\Delta t$. The effect of sampling in the time domain can be seen by letting $f = f + mf_s$, where m is an integer, which yields $G(f + mf_s) = G(f)$ which verifies the periodic harmonic frequency response.

Before leaving filter design, it is worth noting that a SAW filter is composed of two transducers which may have different center frequencies, bandwidth, and other filter specifications. This provides a great deal of flexibility in designing a filter by allowing the product of two frequency responses to achieve the total filter specification.

45.4 SAW Transducer Modeling

The four most popular and widely used models include the transmission line model, the coupling of modes model, the impulse response model, and the superposition model. The superposition model is an extension of the impulse response model and is the principal model used for the majority of SAW bidirectional and multiphase filter synthesis which do not have inband, interelectrode reflections. As is the case for most technologies, many models may be used in conjunction with each other for predicting device performance based on ease of synthesis, confidence in predicted parameters, and correlation with experimental device data.

The SAW Superposition Impulse Response Transducer Model

The impulse response model was first presented by Hartmann *et al.* [1973] to describe SAW filter design and synthesis. For a linear causal system, the Fourier transform of the device's frequency response is the device impulse time response. Hartmann showed that the time response of a SAW transducer is given by

$$h(t) = 4k\sqrt{C_s}\, f_i^{3/2}(t)\sin[\theta(t)] \quad \text{where } \theta(t) = 2\pi\int_0^t f_i(\tau)d\tau \qquad (45.5)$$

and where the following definitions are k^2 = SAW coupling coefficient, C_s = electrode pair capacitance per unit length (pf/cm-pair), and $f_i(t)$ = instantaneous frequency at a time, t. This is the general form for a uniform beam transducer with arbitrary electrode spacing. For a uniform beam transducer with periodic electrode spacing, $f_i(t) = f_0$ and $\sin\theta(t) = \sin\omega t$. This expression relates a time response to the physical device parameters of the material coupling coefficient and the electrode capacitance.

Given the form of the time response, energy arguments are used to determine the device equivalent circuit parameters. Assume a delta function voltage input, $v_{in}(t) = \delta(t)$, then $V_{in}(\omega) = 1$. Given $h(t)$, $H(\omega)$ is known and the energy launched as a function of frequency is given by $E(\omega) = 2\cdot|H(\omega)|^2$. Then

$$E(\omega) = V_{in}^2(\omega) \cdot G_a(\omega) = 1 \cdot G_a(\omega) \qquad (45.6)$$

or

$$G_a(\omega) = 2 \cdot |H(\omega)|^2 \qquad (45.7)$$

There is a direct relationship between the transducer frequency transfer function and the transducer conductance. Consider an **interdigital transducer (IDT)** with uniform overlap electrodes having N_p interaction pairs. Each gap between alternating polarity electrodes is considered a localized SAW source. The SAW impulse response at the fundamental frequency will be continuous and of duration τ, where $\tau = N \cdot \Delta t$, and $h(t)$ is given by

$$h(t) = \kappa \cdot \cos(\omega_0 t) \cdot rect(t/\tau) \qquad (45.8)$$

where $\kappa = 4k\sqrt{C_s}\, f_0^{3/2}$ and f_0 is the carrier frequency. The corresponding frequency response is given by

$$H(\omega) = \frac{\kappa\tau}{2}\left\{\frac{\sin(x_1)}{x_1} + \frac{\sin(x_2)}{x_2}\right\} \qquad (45.9)$$

where $x_1 = (\omega - \omega_0) \cdot \tau/2$ and $x_2 = (\omega + \omega_0) \cdot \tau/2$.

This represents the ideal SAW continuous response in both time and frequency. This can be related to the sampled response by a few substitutions of variables. Let

$$\Delta t = \frac{1}{2 \cdot f_0}, \qquad t_n = n \cdot \Delta t, \qquad N \cdot \Delta t = \tau, \qquad N_p \cdot \Delta t = \tau / 2 \qquad (45.10)$$

Assuming a frequency bandlimited response, the negative frequency component centered around $-f_0$ can be ignored. Then the frequency response, using Eq. (45.9), is given by

$$H(\omega) = \kappa \left\{ \frac{\pi N_p}{\omega_0} \right\} \cdot \frac{\sin(x_n)}{x_n} \qquad (45.11)$$

where

$$x_n = \frac{(\omega - \omega_0)}{\omega_0} \pi N_p = \frac{(f - f_0)}{f_0} \pi N_p$$

The conductance, given using Eqs. (45.6) and (45.10), is

$$G_a(f) = 2\kappa^2 \left\{ \frac{\pi N_p}{2\pi f_0} \right\}^2 \frac{\sin^2(x_n)}{x_n^2} = 8k^2 f_0 C_s N_p^2 \cdot \frac{\sin^2(x_n)}{x_n^2} \qquad (45.12)$$

This yields the frequency-dependent conductance per unit width of the transducer. Given a uniform transducer of width, W_a, the total transducer conductance is obtained by multiplying Eq. (45.12) by W_a. Defining the center frequency conductance as

$$G_a(f_0) = G_0 = 8k^2 f_0 C_s W_a N_p^2 \qquad (45.13)$$

the transducer conductance is

$$G_a(f_0) = G_0 \cdot \frac{\sin^2(x_n)}{x_n^2} \qquad (45.14)$$

The transducer electrode capacitance is given as

$$C_e = C_s W_a N_p \qquad (45.15)$$

Finally, the last term of the SAW transducer's equivalent circuit is the frequency-dependent susceptance. Given any system where the frequency-dependent real part is known, there is an associated imaginary part which must exist for the system to be real and causal. This is given by the Hilbert transform susceptance, defined as B_a, where [Datta, 1986]

$$B_a(\omega) = \frac{1}{\pi} \int_{-\infty}^{\infty} \frac{G_a(u)}{(u - \omega)} \, du = G_a(\omega) * 1/\omega \qquad (45.16)$$

where "$*$" indicates convolution.

These three elements compose a SAW transducer equivalent circuit. The equivalent circuit, shown in Fig. 45.4, is composed of one lumped element and two frequency-dependent terms which are related to the substrate material parameters, transducer electrode number, and the transducer configuration. Figure 45.5 shows the time and frequency response for a uniform transducer and the

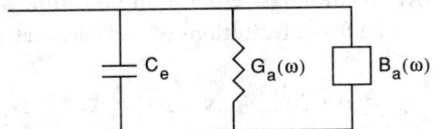

FIGURE 45.4   Electrical equivalent circuit model.

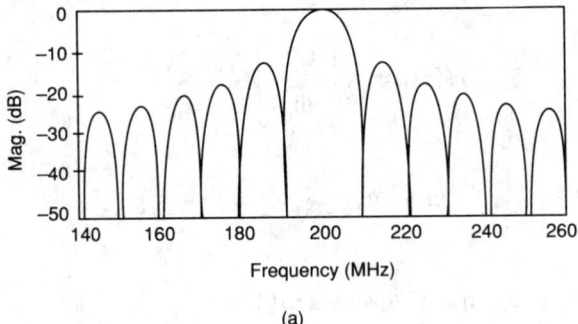

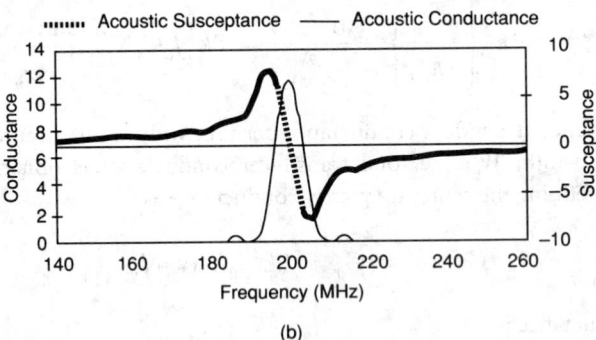

FIGURE 45.5   (a) Theoretical frequency response of a $rect(t/\tau)$ time function having a time length of 0.1 µs and a 200-MHz carrier frequency. (b) Theoretical conductance and susceptance for a SAW transducer implementing the frequency response. The conductance and susceptance are relative and are given in millisiemens.

associated frequency-dependent conductance and Hilbert transform susceptance. The simple impulse model treats each electrode as an ideal impulse; however, the electrodes have a finite width which distorts the ideal impulse response. The actual SAW potential has been shown to be closely related to the electrostatic charge induced on the transducer by the input voltage. The problem is solved assuming a quasi-static and electrostatic charge distribution, assuming a semi-infinite array of electrodes, solving for a single element, and then using superposition and convolution. The charge distribution solution for a single electrode with all others grounded is defined as the basic charge distribution function (BCDF). The result of a series of arbitrary voltages placed on a series of electrodes is the summation of scaled, time-shifted BCDFs. The identical result is obtained if an array factor, $a(x)$, defined as the ideal impulses localized at the center of the electrode or gap, is convolved with the BCDF, often called the element factor. This is very similar to the analysis of antenna arrays. Therefore, the ideal frequency transfer function and conductance given by the impulse response model need only be modified by multiplying the frequency-dependent element factor. The analytic solution to the BCDF is given in Datta [1986] and Morgan [1985], and is shown to place a small perturbation in the form of a slope or dip over the normal bandwidths of interest. The BCDF also predicts the expected harmonic frequency responses.

Apodized SAW Transducers

Apodization is the most widely used method for weighting a SAW transducer. The desired time-sampled impulse response is implemented by assigning the overlap of opposite polarity electrodes at a given position to a normalized sample weight at a given time. A tap having a weight of unity has an overlap across the entire beamwidth while a small tap will have a small overlap of adjacent electrodes. The time impulse response can be broken into tracks which have uniform height but whose time length and impulse response may vary. Each of these time tracks is implemented spatially across the transducer's beamwidth by overlapped electrode sections at the proper positions. This is shown in Fig. 45.1. The smaller the width of the tracks, the more exact the approximation of uniform time samples. There are many different ways to implement the time-to-spatial transformation; Fig. 45.1 shows just one such implementation.

The impulse response can be represented, to any required accuracy, as the summation of uniform samples located at the proper positions in time in a given track. Mathematically this is given by

$$h(t) = \sum_{i=1}^{I} h_i(t) \tag{45.17}$$

and

$$H(\omega) = \sum_{i=1}^{I} H_i(\omega) = \sum_{i=1}^{I} \left\{ \int_{-\tau/2}^{\tau/2} h_i(t) e^{-j\omega t} dt \right\} \tag{45.18}$$

The frequency response is the summation of the individual frequency responses in each track, which may be widely varying depending on the required impulse response. This spatial weighting complicates the calculations of the equivalent circuit for the transducer. Each track must be evaluated separately for its acoustic conductance, acoustic capacitance, and acoustic susceptance. The transducer elements are then obtained by summing the individual track values yielding the final transducer equivalent circuit parameters. These parameters can be solved analytically for simple impulse response shapes (such as the rect, triangle, cosine, etc.) but are usually solved numerically on a computer [Richie *et al.*, 1988].

There is also a secondary effect of apodization when attempting to extract energy. Not all of the power of a nonuniform SAW beam can be extracted by an a uniform transducer, and reciprocally, not all of the energy of a uniform SAW beam can be extracted by an apodized transducer. The transducer efficiency is calculated at center frequency as

$$E = \frac{\left| \sum_{i=1}^{I} H(\omega_0) \right|^2}{I \cdot \sum_{i=1}^{I} H^2(\omega_0)} \tag{45.19}$$

The apodization loss is defined as

$$\text{apodization loss} = 10 \cdot \log(E) \tag{45.20}$$

Typical apodization loss for common SAW transducers is 1 dB or less.

Finally, because an apodized transducer radiates a nonuniform beam profile, the response of two cascaded apodized transducers is not the product of each transducer's individual frequency responses, but rather is given by

$$H_{12}(\omega) = \sum_{i=1}^{I} H_{1i}(\omega) \cdot H_{2i}(\omega) \neq \sum_{i=1}^{I} H_{1i}(\omega) \cdot \sum_{i=1}^{I} H_{2i}(\omega) \qquad (45.21)$$

In general, filters are normally designed with one apodized and one uniform transducer or with two apodized transducers coupled with a spatial-to-amplitude acoustic conversion component, such as a multistrip coupler [Datta, 1986].

45.5 Distortion and Second-Order Effects

In SAW devices there are a number of effects which can distort the desired response from the ideal response. The most significant distortion in SAW transducers is called the **triple transit echo** (**TTE**) which causes a delayed signal in time and an inband ripple in the amplitude and delay of the filter. The TTE is primarily due to an electrically regenerated SAW at the output transducer which travels back to the input transducer, where it induces a voltage across the electrodes which in turn regenerates another SAW which arrives back at the output transducer. This is illustrated schematically in Fig. 45.2. Properly designed and matched **unidirectional transducers** have acceptably low levels of TTE due to their design. Bidirectional transducers, however, must be mismatched in order to achieve acceptable TTE levels. To first order, the TTE for a bidirectional two-transducer filter is given as

$$\text{TTE} \approx 2 \cdot IL + 6 \text{ dB} \qquad (45.22)$$

where IL = filter insertion loss, in dB [Matthews, 1977]. As examples, the result of TTE is to cause a ghost in a video response and intersymbol interference in data transmission.

Another distortion effect is electromagnetic feedthrough which is due to direct coupling between the input and output ports of the device, bypassing any acoustic response. This effect is minimized by proper device design, mounting, bonding, and packaging.

In addition to generating a SAW, other spurious acoustic modes may be generated. Bulk acoustic waves (BAW) may be both generated and received, which causes passband distortion and loss of out-of-band rejection. BAW generation is minimized by proper choice of material, roughening of the crystal backside to scatter BAWs, and use of a SAW track changer, such as a multistrip coupler.

Any plane wave which is generated from a finite aperture will begin to diffract. This is exactly analogous to light diffracting through a slit. Diffraction's principal effect is to cause effective shifts in the filter's tap weights and phase which results in increased sidelobe levels in the measured frequency response. Diffraction is minimized by proper choice of substrate and filter design.

Transducer electrodes are fabricated from thin film metal, usually aluminum, and are finite in width. This metal can cause discontinuities to the surface wave which cause velocity shifts and frequency-dependent reflections. In addition, the films have a given sheet resistance which gives rise to a parasitic electrode resistance loss. The electrodes are designed to minimize these distortions in the device.

45.6 Bidirectional Filter Response

A SAW filter is composed of two cascaded transducers. In addition, the overall filter function is the product of two acoustic transfer functions, two electrical transfer functions, and a delay line function, as illustrated in Fig. 45.6. The acoustic filter functions are as designed by each SAW transducer. The delay line function is dependent on several parameters, the most important being frequency and transducer separation. The propagation path transfer function, $D(\omega)$, is normally assumed unity, although this may not be true for high frequencies ($f > 500$ MHz) or if there are

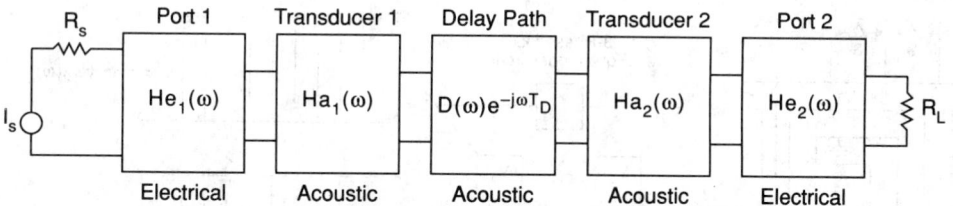

FIGURE 45.6 Complete transfer function of a SAW filter including the acoustic, electrical, and delay line transfer functions. The current generator is I_s, and R_s and R_L are the source and generator resistances, respectively.

films in the propagation path. The electrical networks may cause distortion of the acoustic response and are typically compensated in the initial SAW transducer's design.

The SAW electrical network is analyzed using the SAW equivalent circuit model plus the addition of packaging parasitics and any tuning or matching networks. Figure 45.7 shows a typical electrical network which is computer analyzed to yield the overall transfer function for one port of the two-port SAW filter [Morgan, 1985]. The second port is analyzed in a similar manner and the overall transfer function is obtained as the product of the electrical, acoustic, and propagation delay line effects.

45.7 Multiphase Unidirectional Transducers

The simplest SAW transducers are single-phase bidirectional transducers. Because of their symmetrical nature, SAW energy is launched equally in both directions from the transducer. In a two-transducer configuration, half the energy (3 dB) is lost at the transmitter, and reciprocally, only half the energy can be received at the receiver. This yields a net 6-dB loss in a filter. However, by adding nonsymmetry into the transducer, either by electrical multiphases or nonsymmetry in reflection and regeneration, energy can be unidirectionally directed yielding a theoretical minimum 0-dB loss.

The most common SAW UDTs are called the three-phase UDT (3PUDT) and the group type UDT (GUDT). The 3PUDT has the broadest bandwidth and requires multilevel metal structures with crossovers. The GUDT uses a single-level metal but has a narrower unidirectional bandwidth due to its structure. In addition, there are other UDT or equivalent embodiments which can be implemented but will not be discussed [Morgan, 1985]. The basic structure of a 3PUDT is shown in Fig. 45.8. A unit cell consists of three electrodes, each connected to a separate bus bar, where the electrode period is $\lambda_0/3$. One bus bar is grounded and the other two bus bars will be driven by an electrical network where $V_1 = V_2 \angle 60°$. The transducer analysis can be accomplished similar to a

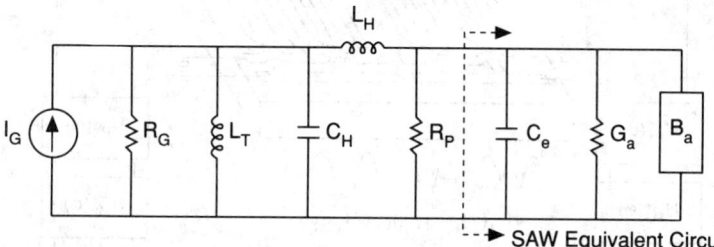

FIGURE 45.7 Electrical network analysis for a SAW transducer. I_G and R_G represent the generator source and impedance, L_T is a tuning inductor, C_H and L_H are due to the package capacitance and bond wire, respectively, and R_P represents a parasitic resistance due to the electrode transducer resistance. The entire network, including the frequency-dependent SAW network, is solved to yield the single-port transfer function.

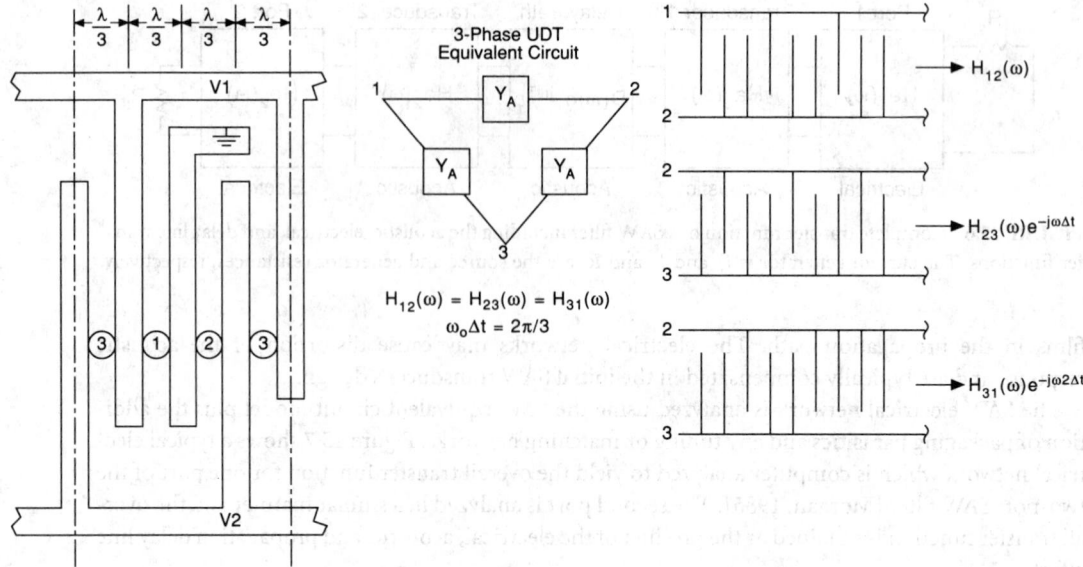

FIGURE 45.8 Schematic of a unit cell of a 3PUDT and the basic equivalent circuit. The 3PUDT can be analyzed as three collinear transducers with a spatial offset.

simple IDT by considering the 3PUDT as three collinear IDTs with a spatial phase shift, as shown in Fig. 45.8. The electrical phasing network, typically consisting of one or two reactive elements, in conjunction with the spatial offset results in energy being launched in only one direction from the SAW transducer. The transducer can then be matched to the required load impedance with one or two additional reactive elements. The effective unidirectional bandwidth of the 3PUDT is typically 20% or less, beyond which the transducer behaves as a normal bidirectional transducer. Figure 45.9 shows a 3PUDT filter schematic consisting of two transducers and their associated matching and phasing networks. The overall filter must be analyzed with all external electrical components in place for accurate prediction of performance. The external components can be miniaturized and

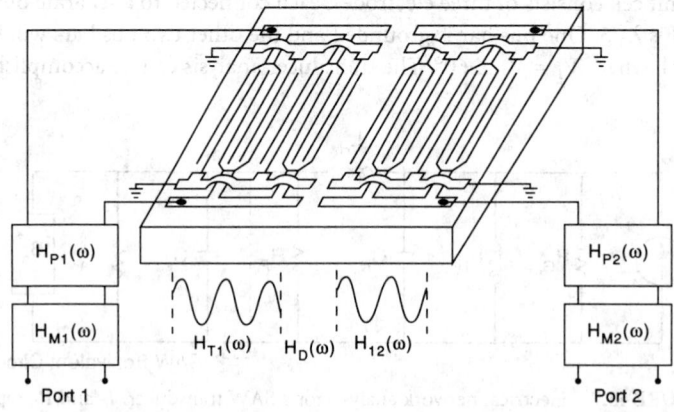

$$H_{FILTER}(\omega) = H_{M1}(\omega) \cdot H_{P1}(\omega) \cdot H_{T2}(\omega) \cdot H_D(\omega) \cdot H_{T2}(\omega) \cdot H_{P2}(\omega) \cdot H_{M2}(\omega)$$

FIGURE 45.9 Schematic diagram of a 3PUDT which requires the analysis of both the acoustic transducer responses as well as electrical phasing and matching networks.

may be fabricated using only printed circuit board material and area. This type of device has demonstrated as low as 2 dB insertion loss.

45.8 Single-Phase Unidirectional Transducers

Single-phase unidirectional transducers (SPUDT) use spatial offsets between mechanical electrode reflections and electrical regeneration to launch a SAW in one direction. A reflecting structure may be made of metal electrodes, dielectric strips, or grooved reflectors which are properly placed within a transduction structure. Under proper design and electrical matching conditions, the mechanical reflections can exactly cancel the electrical regeneration in one direction of the wave over a moderate band of frequencies. This is schematically illustrated in Fig. 45.10 which shows a reflector structure and a transduction structure merged to form a SPUDT. The transducer needs to be properly matched to the load for optimum operation. The mechanical reflections can be controlled by modifying the width, position, or height of the individual reflector. The regenerated SAW is primarily controlled by the electrical matching to the load of the transduction structure. SPUDT filters have exhibited as low as 3 dB loss over fractional bandwidths of 5% or less and have the advantage of not needing phasing networks when compared to the multiphase UDTs.

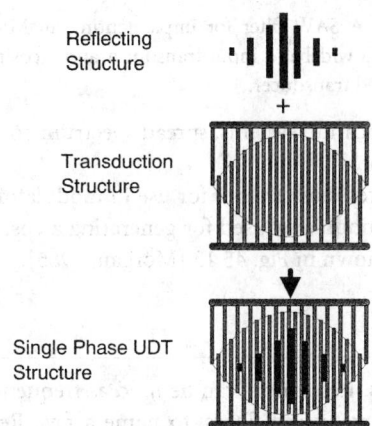

FIGURE 45.10 Schematic representation of a SPUDT which is a combination of transduction and reflecting structures to launch a SAW in one direction over moderate bandwidths.

45.9 Dispersive Filters

SAW filters can also be designed and fabricated using nonuniformly spaced electrodes in the transducer. The distance between adjacent electrodes determines the "local" generated frequency. As the spacing between the electrodes changes, the frequency is slowly changed either up (decreasing electrode spacing) or down (increasing electrode spacing) as the position progresses along the transducer. This slow frequency change with time is often called a "chirp." Figure 45.11 shows a typical dispersive filter consisting of a chirped transducer in cascade with a uniform transducer. Filters can be designed with either one or two chirped transducers and the rate of the chirp is variable within the design. These devices have found wide application in radar systems due to their small size, reproducibility, and large time bandwidth product.

FIGURE 45.11 A SAW dispersive filter consisting of a uniform transducer and a "down chirp" dispersive transducer. The high frequencies have a shorter delay than the low frequencies in this example.

45.10 Coded SAW Filters

Because of the ability to control the amplitude and phase of the individual electrodes or taps, it is easy to implement coding in a SAW filter. Figure 45.12 shows an example of a coded SAW filter implementation. By changing the phase of the taps, it is possible to generate an arbitrary code

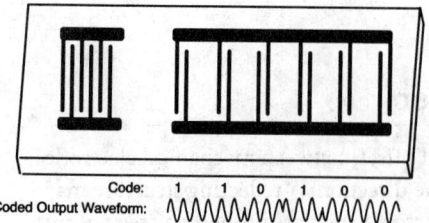

Code: 1   1   0   1   0   0
Coded Output Waveform: 〜〜〜〜〜〜〜〜〜〜〜〜〜〜

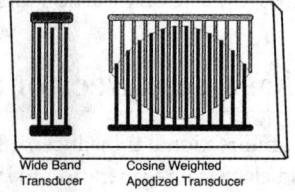

Wide Band   Cosine Weighted
Transducer   Apodized Transducer

FIGURE 45.12   Example of a coded SAW tapped delay line.

FIGURE 45.13   A SAW filter for implementing an MSK waveform using a wideband input transducer and a cosine envelope apodized transducer.

sequence. These types of filters are used in secure communication systems, spread spectrum communications, and tagging, to name a few [Matthews, 1977].

SAW devices can also be used to produce time impulse response shapes for use in modulators, equalizers, and other applications. An example of a SAW modulator used for generating a cosine envelope for a minimum shift keyed (MSK) modulator is shown in Fig. 45.13 [Morgan, 1985].

45.11 Resonators

Another very important class of devices is SAW resonators. Resonators can be used as frequency control elements in oscillators, as notch filters, and as narrowband filters, to name a few. Resonators are typically fabricated on piezoelectric quartz substrates due to its low TCD which yields temperature-stable devices. A resonator uses one or two transducers for coupling energy in/out of the device and one or more distributed reflector arrays to store energy in the device. This is analogous to an optical cavity with the distributed reflector arrays acting as the mirrors. A localized acoustic mirror, such as a cleaved edge, is not practical for SAW because of spurious mode coupling at edge discontinuities which causes significant losses.

A distributive reflective array is typically composed of a series of shorted metal electrodes, etched grooves in the substrate, or dielectric strips. There is a physical discontinuity on the substrate surface due to the individual reflectors. Each reflector is one-quarter wavelength wide and the periodicity of the array is one-half wavelength. This is shown schematically in Fig. 45.14. The net reflections from all the individual array elements add synchronously at center frequency, resulting in a very efficient reflector. The reflection from each array element is small and very little spurious mode coupling results.

Figure 45.14 shows a typical single-pole, single-cavity, two-port SAW resonator. Resonators can be made multipole by addition of multiple cavities, which can be accomplished by inline acoustic coupling, transverse acoustic coupling, and by electrical coupling. The equivalent circuit for SAW

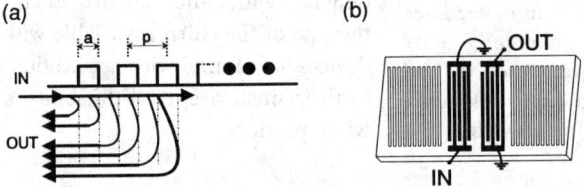

FIGURE 45.14   (a) SAW reflector array illustrating synchronous distributed reflections at center frequency. Individual electrode width (a) is 1/4 wavelength and the array period is 1/2 wavelength at center frequency. (b) A schematic of a simple single-pole, single-cavity two-port SAW resonator.

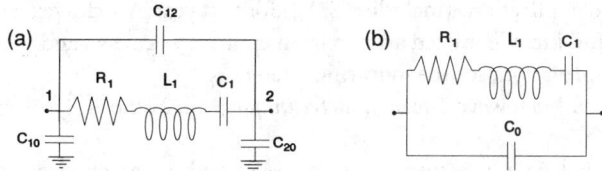

FIGURE 45.15 (a) Two-port resonator equivalent circuit and (b) one-port resonator equivalent circuit.

two-port and one-port resonators is shown in Fig. 45.15. SAW resonators have low insertion loss and high electrical Q's of several thousand [Campbell, 1989; Datta, 1986; Morgan, 1985].

Defining Terms

Bidirectional transducer: A SAW transducer which launches energy from both acoustic ports which are located at either end of the transducer structure.

Interdigital transducer: A series of collinear electrodes placed on a piezoelectric substrate for the purpose of launching a surface acoustic wave.

Surface acoustic wave (SAW): A surface acoustic wave (also known as a Rayleigh wave) is composed of a coupled compressional and shear wave. On a piezoelectric substrate there is also an electrostatic wave which allows electroacoustic coupling. The wave is confined at or near the surface and decays away rapidly from the surface.

Triple transit echo (TTE): A multiple transit echo received at three times the main SAW signal delay time. This echo is caused due to the bidirectional nature of SAW transducers and the electrical and/or acoustic mismatch at the respective ports. This is a primary delayed signal distortion which can cause filter distortion, especially in bidirectional transducers and filters.

Unidirectional transducer (UDT): A transducer which is capable of launching energy from primarily one acoustic port over a desired bandwidth of interest.

References

C. Campbell, *Surface Acoustic Wave Devices and their Signal Processing Applications*, San Diego, Calif.: Academic Press, 1989.

S. Datta, *Surface Acoustic Wave Devices*, Englewood Cliffs, N.J.: Prentice-Hall, 1986.

C. S. Hartmann, D. T. Bell, and R. C. Rosenfeld, "Impulse model design of acoustic surface wave filters," *IEEE Transactions on Microwave Theory and Techniques*, vol. 21, pp. 162–175, 1973.

H. Matthews, *Surface Wave Filters*, New York: Wiley Interscience, 1977.

D. P. Morgan, *Surface Wave Devices for Signal Processing*, New York: Elsevier, 1985.

S. M. Richie, B. P. Abbott, and D. C. Malocha, "Description and development of a SAW filter CAD system," *IEEE Transactions on Microwave Theory and Techniques*, vol. 36, no. 2, 1988.

Further Information

The *IEEE Transactions on Ultrasonics, Ferroelectrics, and Frequency Control* provides excellent information and detailed articles on SAW technology.

The *IEEE Ultrasonics Symposium Proceeding* provides information on ultrasonic devices, systems, and applications for that year. Articles present the latest research and developments and include invited articles from eminent engineers and scientists.

The *IEEE Frequency Control Symposium Proceedings* provides information on frequency control devices, systems, and applications (including SAW) for that year. Articles present the latest research and developments and include invited articles from eminent engineers and scientists.

For additional information, see the following references:

IEEE Transaction on Microwave Theory and Techniques, vol. 21, no. 4, 1973, special issue on SAW technology.

IEEE Proceedings, vol. 64, no. 5, special issue on SAW devices and applications.

Joint Special Issue of *IEEE Transaction on Microwave Theory and Techniques* and *IEEE Transactions on Sonics and Ultrasonics,* MTT-vol. 29, no. 5, 1981, on SAW device systems.

M. Feldmann and J. Henaff, *Surface Acoustic Waves for Signal Processing,* Norwood, Mass.: Artech House, 1989.

B. A. Auld, *Acoustic Fields and Waves in Solids,* New York: Wiley, 1973.

V. M. Ristic, *Principles of Acoustic Devices,* New York: Wiley, 1983.

A. Oliner, *Surface Acoustic Waves,* New York: Springer-Verlag, 1978.

46

Ultrasound

46.1 Introduction 1077
46.2 Propagation in Solids 1077
46.3 Piezoelectric Excitation 1080
46.4 One-Dimensional Propagation 1081
46.5 Transducers 1082

Gerald W. Farnell
McGill University

46.1 Introduction

In electrical engineering, the term *ultrasonics* usually refers to the study and use of waves of mechanical vibrations propagating in solids or liquids with frequencies in the megahertz or low gigahertz ranges. Such waves in these frequency ranges have wavelengths on the order of micrometers and thus can be electrically generated, directed, and detected with transducers of reasonable size. These ultrasonic devices are used for signal processing directly in such applications as filtering and pulse compression and indirectly in acousto-optic processors; for flaw detection in optically opaque materials; for resonant circuits in frequency control applications; and for medical imaging of human organs, tissue, and blood flow.

46.2 Propagation in Solids

If the solid under consideration is elastic (linear), homogeneous, and nonpiezoelectric, the components, u_i, of the displacement of an infinitesimal region of the material measured along a set of Cartesian axes, x_i, are interrelated by an equation of motion:

$$\rho \frac{\partial^2 u_i}{\partial t^2} = \sum_j \sum_k \sum_l c_{ijkl} \frac{\partial^2 u_j}{\partial x_k \partial x_l}, \qquad \text{Form:} \quad \rho \frac{\partial^2 u}{\partial t^2} = c \frac{\partial^2 u}{\partial x^2} \qquad (46.1)$$

where ρ is the mass density of the material and c_{ijkl} ($i, j, k, l = 1, 2, 3$) is called the stiffness tensor. It is the set of proportionality constants between the components of the stress tensor T and the strain tensor S in a three-dimensional Hooke's law (form: $T = cS$ with $S = \partial u/\partial x$). In Eq. (46.1) and in the subsequent equations the **form** of the equation is shown without the clutter of the many subscripts. The form is useful for discussion purposes; moreover, it gives the complete equation for cases in which the propagation can be treated as one dimensional, i.e., with variations in only one direction, one component of displacement, and one relevant c.

In an infinite medium, the simplest solutions of Eq. (46.1) are plane waves given by the real part of

$$u_i = U_i e^{-jk\left(\sum_j L_j x_j - Vt\right)} \qquad \text{Form:} \quad u = U e^{j(\omega t - kx)} \qquad (46.2)$$

where the polarization vector has components U_i along the axes. The **phase velocity** of the wave V is measured along the propagation vector **k** whose direction cosines with respect to these axes are given by L_i. Substituting the assumed solutions of Eq. (46.2) into Eq. (46.1) gives the third-order eigenvalue equations, usually known as the Christoffel equations:

$$\sum_j \sum_k \sum_l L_k L_l c_{ijkl} U_j = \rho V^2 U_i, \qquad \text{Form: } (c - \rho V^2) U = 0 \qquad (46.3)$$

The three eigenvalues in Eq. (46.3) give three values of ρV^2 and hence the phase velocities of three waves propagating in the direction of positive **k** and three propagating in the negative **k** direction. The eigenvectors of the three forward solutions give the polarization vector for each, and they form a mutually perpendicular triad. The polarization vector of one of the plane waves will be parallel, or almost parallel, to the **k** vector, and it is called the **longitudinal wave,** or quasi-longitudinal if the displacement is not exactly parallel to **k**. The other two waves will have mutually perpendicular polarization vectors, which will each be perpendicular, or almost perpendicular, to the **k** vector. If the polarization is perpendicular, the wave is called a transverse or **shear wave;** if almost perpendicular, it is called quasi-shear. The three waves propagate independently through the solid, and their respective amplitudes depend on the exciting source.

In an isotropic medium where there are only two independent values of c_{ijkl} in Eq. (46.1), there are one longitudinal wave and two degenerate shear waves. The phase velocities of these waves are independent of the direction of propagation and are given by

$$V_1 = \sqrt{\frac{c_{1111}}{\rho}} \quad \text{and} \quad V_s = \sqrt{\frac{c_{1212}}{\rho}} \qquad (46.4)$$

The phase velocities in isotropic solids are often expressed in terms of the so-called Lame constants defined by $\mu = c_{1212}$ and $\lambda = c_{1111} - 2c_{1212}$. The longitudinal velocity is larger than the shear velocity. Exact velocity values depend on fabrication procedures and purity, but Table 46.1 gives typical values for some materials important in ultrasonics.

In signal processing applications of ultrasonics, the propagating medium is often a single crystal, and thus a larger number of independent stiffness constants is required to describe the mechanical properties of the medium, e.g., three in a cubic crystal, five in a hexagonal, and six in a trigonal. Note that while the number of independent constants is relatively small, a large number of the c_{ijkl} are nonzero but are related to each other by the symmetry characteristics of the crystal. The phase velocities of each of the three independent plane waves in an anisotropic medium depend on the direction of propagation. Rather than plotting V as a function of angle of propagation, it is more common to use a **slowness surface** giving the reciprocal of V (or $k = \omega/V$ for a given ω) as a function of the direction of **k**. Usually planar cuts of such slowness surfaces are plotted as shown in Figs. 46.1(a) and (b).

In anisotropic materials the direction of energy flow (the ultrasonic equivalent of the electromagnetic Poynting vector) in a plane wave is not parallel to **k**. Thus the direction of **k** is set by the transducer but the energy flow or beam direction is normal to the tangent to the slowness surface at the point corresponding to **k**. The direction of propagation (of **k**) in Fig. 46.1 lies in the basal plane of a cubic crystal, here silicon. At each angle there are three waves—one is pure shear polarized perpendicular to this plane, one is quasilongitudinal for most angles, while the third is quasi-shear. For the latter two, the tangent to the slowness curves at an arbitrary angle is not normal to the radius vector, and thus there is an appreciable angle between the direction of energy flow and the direction of **k**. This angle is shown on the diagram by the typical **k** and **P** vectors, the latter being the direction of energy flow in an acoustic beam with this **k**. Along the cubic axes in a cubic crystal, the two shear waves are degenerate, and for all three waves the energy flow is parallel to **k**. When the particle displacement of a mode is either parallel to the propagation vector or perpen-

Table 46.1 Typical Acoustic Properties

Material	Velocity (km/s)		Impedance (kg/m² s×10⁶)		Density (kg/m³×10³)	Comments
	Longitudinal	Shear	Longitudinal	Shear		
Alcohol, methanol	1.103		0.872		0.791	Liq. 25°C
Aluminum, rolled	6.42	3.04	17.33	8.21	2.70	Isot.
Brass, 70% Cu, 30% Zn	4.70	2.10	40.6	18.14	8.64	Isot.
Cadmium sulphide	4.46	1.76	21.5	8.5	4.82	Piez crys Z-dir
Castor oil	1.507		1.42		0.942	Liq. 20°C
Chromium	6.65	4.03	46.6	28.21	7.0	Isot.
Copper, rolled	5.01	2.27	44.6	20.2	8.93	Isot.
Ethylene glycol	1.658		1.845		1.113	Liq. 25°C
Fused quartz	5.96	3.76	13.1	8.26	2.20	Isot.
Glass, crown	5.1	2.8	11.4	6.26	2.24	Isot.
Gold, hard drawn	3.24	1.20	63.8	23.6	19.7	Isot.
Iron, cast	5.9	3.2	46.4	24.6	7.69	Isot.
Lead	2.2	0.7	24.6	7.83	11.2	Isot.
Lithium niobate, LiNbO₃	6.57	4.08	30.9	19.17	4.70	Piez crys X-dir
		4.79		22.53		
Nickel	5.6	3.0	49.5	26.5	8.84	Isot.
Polystyrene, styron	2.40	1.15	2.52	1.21	1.05	Isot.
PZT-5H	4.60	1.75	34.5	13.1	7.50	Piez ceram Z
Quartz	5.74	3.3	15.2	8.7	2.65	Piez crys X-dir
		5.1		13.5		
Sapphire Al₂O₃	11.1	6.04	44.3	25.2	3.99	Cryst. Z-axis
Silver	3.6	1.6	38.0	16.9	10.6	Isot.
Steel, mild	5.9	3.2	46.0	24.9	7.80	Isot.
Tin	3.3	1.7	24.2	12.5	7.3	Isot.
Titanium	6.1	3.1	27.3	13.9	4.48	Isot.
Water	1.48		1.48		1.00	Liq. 20°C
YAG Y₃Al₁₅O₁₂	8.57	5.03	39.0	22.9	4.55	Cryst. Z-axis
Zinc	4.2	2.4	29.6	16.9	7.0	Isot.
Zinc oxide	6.37	2.73	36.1	15.47	5.67	Piez crys Z-dir

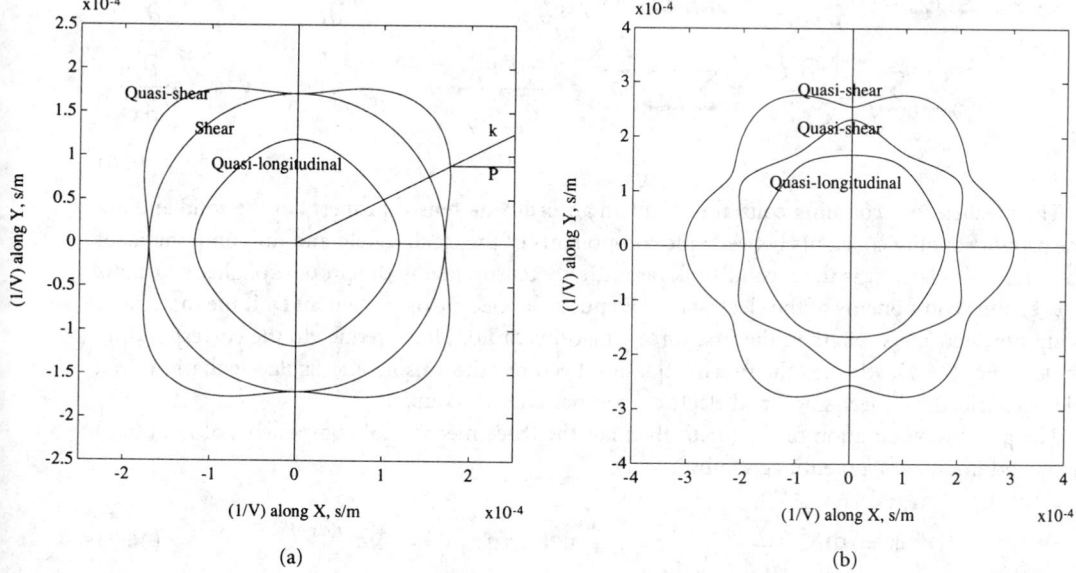

FIGURE 46.1 (a) Slowness curves, basal plane, cubic crystal, silicon. (b) Slowness curves, basal plane, trigonal crystal, quartz.

dicular to it and the energy flow is parallel to **k,** the mode is called a **pure mode.** The propagation vector in Fig. 46.1(b) lies in the basal plane of a trigonal crystal, quartz.

When ultrasonic waves propagate in a solid, there are various losses that attenuate the wave. Usually the attenuation per wavelength is small enough that one can neglect the losses in the initial calculation of the propagation characteristics of the material and the excitation, and then multiply the resulting propagating wave by a factor of the form exp[$-\alpha x$] where x is in the direction of **k** and α is called the *attenuation constant*. One loss mechanism is the viscosity of the material and due to it the attenuation constant is

$$\alpha = \eta \frac{\omega^2}{2V^3\rho} \tag{46.5}$$

in which η is the coefficient of viscosity. It should be noted that the attenuation constant for viscous loss increases as the square of the frequency. In polycrystalline solids there is also loss due to scattering from dislocation and grain structure; thus, for the same material the loss at high frequencies is much higher in a polycrystalline form than in a crystalline one. As a result, in high-frequency applications of ultrasound, such as for signal processing, the propagation material is usually in single-crystal form.

46.3 Piezoelectric Excitation

When a piezoelectric material is stressed, an electric field is generated in the stressed region; similarly, if an electric field is applied, there will be an induced stress on the material in the region of the field. Thus, there is a coupling between mechanical motion and time-varying electric fields. Analysis of wave propagation in piezoelectric solids should thus include the coupling of the mechanical equations such as Eq. (46.1) with Maxwell's equations. In most ultrasonic problems, however, the velocity of the mechanical wave solutions is slow enough that the electric fields can be described by a scalar potential ϕ. This is called the *quasi-static approximation*. Within this approximation, the equations of motion in a piezoelectric solid become

$$\rho\frac{\partial^2 u_i}{\partial t^2} - \sum_j\sum_k\sum_l c_{ijkl}\frac{\partial^2 u_j}{\partial x_k \partial x_l} = \sum_j\sum_k e_{ijk}\frac{\partial^2 \phi}{\partial x_j \partial x_k} \qquad \text{Form:} \quad \rho\frac{\partial^2 u}{\partial t^2} - c\frac{\partial^2 u}{\partial x^2} = e\frac{\partial^2 \phi}{\partial x^2}$$

$$\sum_i\sum_j \epsilon_{ij}\frac{\partial^2 \phi}{\partial x_i \partial x_j} = \sum_i\sum_j\sum_k e_{ijk}\frac{\partial^2 u_j}{\partial x_i \partial x_k} \qquad\qquad \epsilon\nabla^2\phi = e\frac{\partial^2 u}{\partial x^2}$$

$$\tag{46.6}$$

The piezoelectric coupling constants e_{ijk} form a third-rank tensor property of the solid and are the proportionality constants between the components of the electric field and the components of the stress. Similarly ϵ_{ij} is the second-rank permittivity tensor, giving the proportionality constants between the components of the electric field **E** and of the electric displacement **D**. If the material is nonpiezoelectric $e_{ijk} = 0$, then the first three equations of Eq. (46.6) reduce to the corresponding three of Eq. (46.1), whereas the fourth equation becomes the anisotropic Laplace equation. In a piezoelectric, these mechanical and electrical components are coupled.

The plane wave solution of Eq. (46.6) then has the three mechanical components of Eq. (46.2) and in addition has a potential given by

$$\phi = \Phi e^{-jk\left(\sum_j L_j x_j - Vt\right)} \qquad \text{Form:} \quad \phi = \Phi e^{\,j(\omega t - kx)} \tag{46.7}$$

Thus, for the quasi-static approximation there is a wave of potential that propagates with an acoustic phase velocity V in synchronism with the mechanical variations. As will be seen in Section 46.5, it is possible to use the corresponding electric field, $-\nabla\phi$, to couple to electrode configurations and thus excite or detect the ultrasonic wave from external electric circuits.

Rather than substituting Eq. (46.7) and Eq. (46.2) into Eq. (46.6) to obtain a set of four equations similar to Eq. (46.3), it is frequently more convenient to substitute Eq. (46.7) into the fourth equation in the set of Eq. (46.6). Because there are no time derivatives involved, this substitution gives the potential as a linear combination of the components of the mechanical displacement:

$$\Phi = \frac{\sum_i \sum_j \sum_k e_{ijk} L_i L_k U_j}{\sum_i \sum_j \epsilon_{ij} L_i L_j} \quad \text{Form:} \quad \Phi = \frac{e}{\epsilon} U \tag{46.8}$$

When this combination is substituted into the first three equations of Eq. (46.6) and terms gathered, they become identical to Eq. (46.1) but with each c_{ijkl} replaced by

$$\bar{c}_{ijkl} = c_{ijkl} + \frac{\sum_m \sum_n e_{mij} e_{nkl} L_m L_n}{\sum_m \sum_n \epsilon_{mn} L_m L_n} \quad \text{Form:} \quad \bar{c} = c\,(1 + K^2) \quad \text{with} \quad K^2 = \frac{e^2}{c\epsilon} \tag{46.9}$$

Using these so-called stiffened elastic constants, we obtain the same third-order eigenvalue equation, Eq. (46.3), and hence the velocities of each of the three modes and the corresponding mechanical displacement components. The potential is obtained from Eq. (46.8). The velocities obtained for the piezoelectric material are usually at most a few percent higher than would be obtained with the piezoelectricity ignored. The parameter K in Eq. (46.9) is called the electromechanical coupling constant.

46.4 One-Dimensional Propagation

If an acoustic plane wave as in Eq. (46.2) propagating within one medium strikes an interface with another medium, there will be reflection and transmission, much as in the corresponding case in optics. To satisfy the boundary conditions at the interface, it will be necessary in general to generate three transmitted modes and three reflected modes. Thus, the concepts of reflection and transmission coefficients for planar interfaces between anisotropic media are complicated. In many propagation and excitation geometries, however, one can consider only one independent pure mode with energy flow parallel to **k** and particle displacement polarized along **k** or perpendicular to it. This mode (plane wave) then propagates along the axis or its negative in Eq. (46.2). Discussion of the generation, propagation, and reflection of this wave is greatly assisted by considering analogies to the one-dimensional electrical transmission line.

With the transmission line model operating in the sinusoidal steady state, the particle displacement u_i of Eq. (46.2) is represented by a phasor, u. The time derivative of the particle displacement is the particle velocity and is represented by a phasor, $v = j\omega u$, which is taken as analogous to the current on the one-dimensional electrical transmission line. The negative of the stress, or the force per unit area, caused by the particle displacement is represented by a phasor, $(-T) = jkcu$, which is taken as analogous to the voltage on the transmission line. Here c is the appropriate stiffened elastic constant for the mode in question in Eq. (46.3). With these definitions, the general impedance, the **characteristic impedance**, the phase velocity, and the wave vector, respectively, of the equivalent line are given by

$$Z = \frac{(-T)}{v} \qquad Z_0 = \sqrt{\rho c} \qquad V = \sqrt{\frac{c}{\rho}} \qquad k = \frac{\omega}{V} \tag{46.10}$$

Some typical values of the characteristic impedance of acoustic media are given in Table 46.1. The characteristic impedance corresponding to a mode is given by the product of the density and the phase velocity, ρV, even in the anisotropic case where the effective stiffness c in Eq. (46.10) is difficult to determine.

As an example of the use of the transmission line model, consider a pure longitudinal wave propagating in an isotropic solid and incident normally on the interface with a second isotropic solid. There would be one reflected wave and one transmitted wave, both longitudinally polarized. The relative amplitudes of the stresses in these waves would be given, with direct use transmission line concepts, by the voltage reflection and transmission coefficients

$$\Gamma_R = \frac{Z_{02} - Z_{01}}{Z_{02} + Z_{01}} \quad \text{and} \quad \Gamma_T = \frac{2Z_{02}}{Z_{02} + Z_{01}} \qquad (46.11)$$

When an acoustic wave meets a discontinuity or a mismatch, part of the wave is reflected. For an incident mode, an interface represents a lumped impedance. If the medium on the other side of the interface is infinitely deep, that lumped impedance is the characteristic impedance of the second medium. However, if the second medium is of finite depth h in the direction of propagation and it in turn is terminated by a lumped impedance Z_{L2} the impedance seen by the incident wave at the interface is given, as in transmission line theory, by

$$Z_{in} = Z_{02} \frac{Z_{L2} \cos k_2 h + jZ_{02} \sin k_2 h}{Z_{02} \cos k_2 h + jZ_{L2} \sin k_2 h} \qquad (46.12)$$

Thus, as with transmission lines, an intervening layer can be used to match from one transmitting medium to another. For example, if the medium following the layer is infinite and of characteristic impedance Z_{03}, i.e., $Z_{L2} = Z_{03}$, the interface will look like Z_{01} to the incident wave if $kh = \pi/2$, quarter-wave thickness, and the layer characteristic impedance is $Z_{02}^2 = Z_{01}Z_{03}$. This matching, which provides complete power transfer from medium 1 to medium 3, is valid only at the frequency for which $kh = \pi/2$. For matching over a band of frequencies, multiple matching layers are required.

46.5 Transducers

Electrical energy is converted to acoustic waves in ultrasonic applications by means of electro-acoustic transducers. Most transducers are reciprocal in that they will also convert the mechanical energy in acoustic waves into electrical energy. The form of the transducer is very application dependent. Categories of applications include imaging, wherein one transducer is used to create an acoustic beam, discontinuities in the propagating medium scatter this beam, and the scattered energy is captured by the same or another transducer [see Fig. 46.4(b)]. From the changes of the scattered energy as the beam is moved, characteristics of the scatterer are determined. This is the process in the use of ultrasonics for nondestructive evaluation (NDE), flaw detection, for example, and in ultrasonic images for medical diagnosis. These are radar-like applications and are practical at reasonable frequencies because most solids and liquids support acoustic waves with tolerable losses and the wavelength is short enough that the resolution is adequate for practical targets. By recording both the amplitude and phase of the scattered signal as the transmitter-receiver combination is rotated about a target, one can generate tomographic-type images of the target.

A second category of transducer provides large acoustic standing waves at a particular frequency and, as a result, has a resonant electrical input impedance at this frequency and can be used as a narrowband filter in electrical circuits. In a third category of transducer, the object is to provide an acoustic beam that distorts the medium, as it passes through, in a manner periodic in space and time, and

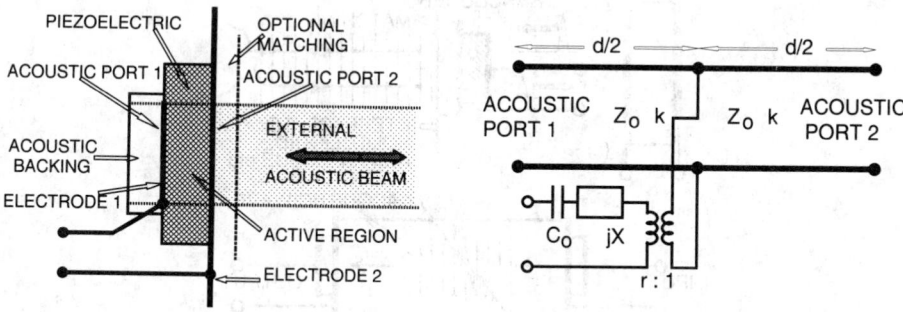

FIGURE 46.2   Prototype transducer geometry.   **FIGURE 46.3**   Model of active region.

thus provides a dynamic diffraction grating that will deflect or modulate an optical beam that is passed through it [see Fig. 46.4(c)]. Such acousto-optic devices are used in broadband signal processing.

Another category of transducer uses variation of the shape of the electrodes and the geometry of the electroacoustic coupling region so that the transfer function between a transmitting and a receiving transducer is made to have a prescribed frequency response. Such geometries find wide application in filtering and pulse compression applications in the frequency range up to a few gigahertz. Because of the ease of fabrication of complicated electrode geometries, special forms of the solution of the wave equation, Eq. (46.1), called surface acoustic waves (SAW) are dominant in such applications. Because surface acoustic waves are discussed in another section of this handbook, here we will confine the discussion to transducers that generate or detect acoustic waves that are almost plane and usually single mode, the so-called bulk modes.

The prototype geometry for a bulk-mode transducer is shown in Fig. 46.2. The active region is the portion of the piezoelectric slab between the thin metal electrodes, which can be assumed to be circular or rectangular in shape. Connections to these electrodes form the electrical port for the transducer and the voltage between them creates a spatially uniform electric field in the active region, and this time-varying electric field couples to the acoustic waves propagating between the electrodes. If the planar electrodes are many wavelengths in transverse dimensions and the active region is much thinner, and if the axial direction is a pure mode direction for the piezoelectric, the waves in the active region can be considered as plane waves. We then have the one-dimensional geometry considered earlier. The transducer may be in contact with another elastic medium on either side, as indicated in Fig. 46.2, so that the plane waves propagate in and out of the active regions in the cross-sectional region shown. Thus, the transducer has in general two acoustic ports for coupling to the outside world as well as the electrical port.

In the absence of piezoelectric coupling, the active region could be represented by a one-dimensional transmission line as discussed in the previous section and as indicated by the heavy lines in Fig. 46.3. With piezoelectricity there will be the stiffening of the appropriate stiffness constants as discussed in Eq. (46.9) with the concomitant perturbation of the characteristic impedance Z_{0p} and the phase velocity V_p, but more important there will also be coupling to the electrical port. One model including the latter coupling is shown in Fig. 46.3 in which the parameters are defined by

$$C_0 = \frac{\epsilon A}{d}; \quad jX = \frac{j}{\omega C_0} K^2 \frac{\sin(\pi\omega/\omega_0)}{\pi\omega/\omega_0} \quad r = \frac{2e/\epsilon}{\omega A Z_0} \sin(\pi\omega/2\omega_0) \quad (46.13)$$

Here C_0 is the capacity that would be measured between the electrodes if there were no mechanical strain on the piezoelectric, A is the cross-sectional area of the active region, and X is an effective reactance. The quantity r is the transformer ratio (with dimensions) of an ideal transformer coupling the electrical port to the center of the acoustic transmission line. K is the electromechanical

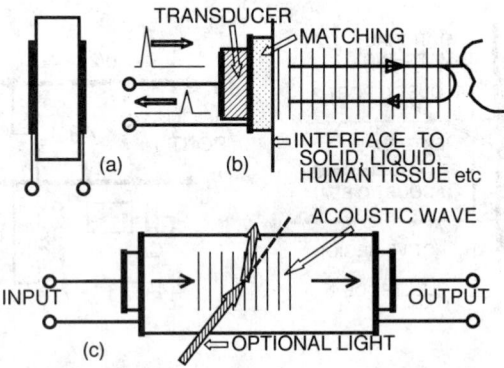

FIGURE 46.4 (a) Resonator structure; (b) acoustic probe; (c) acoustic delay line or optical modulator.

coupling constant for the material as defined in Eq. (46.9). The so-called resonant frequency ω_0 is that angular frequency at which the length d of the active region is one-half of the stiffened wavelength, $\omega_0 = \pi V/d$. In the physical configuration of Fig. 46.4(a), the transducer has zero stress on the surfaces of the active region and hence both acoustic ports of Fig. 46.3 are terminated in short circuits and the line is mechanically resonant at the angular frequency ω_0. At this frequency the secondary of the transformer of Fig. 46.3 is open circuited if there are no losses, and thus the electrical input impedance is infinite at this frequency and behaves like a parallel resonant circuit for neighboring frequencies. This configuration can be used as a high-Q resonant circuit if the mechanical losses can be kept low, as they are in single crystals of such piezoelectric materials as quartz. It should be noted, however, that the behavior is not as simple as that of a simple L-C parallel resonant circuit, primarily because of the frequency dependence of the effective reactance X and of the transformer ratio in the equivalent circuit. The electrical input impedance is given by

$$Z_{in} = \frac{1}{j\omega C_0}\left(1 - K^2\,\frac{\tan kd/2}{kd/2}\right) \tag{46.14}$$

Thus, while the input impedance is infinite as in a parallel resonant circuit at ω_0, it is zero as in a series resonant circuit at a slightly lower frequency where the bracketed term in Eq. (46.14) is zero. When losses are present or there is radiation out of an acoustic port, a resistive term is included in the reactive expression of Eq. (46.14).

Behavior analogous to that of coupled tuned electrical circuits for multipole filters can be achieved by subdividing the electrodes of Fig. 46.4(a) into different areas, each of which will act separately as a tuned circuit, but if they are close enough together there will be acoustic coupling between the different radiators. By controlling this coupling, narrowband filters of very high Q and of somewhat tailored frequency response can be built in the megahertz and low gigahertz range.

The basic geometry of Fig. 46.4(c) gives an electric-to-electric delay line whose delay is given by the length of the medium between the transducers divided by the phase velocity of the acoustic wave and would be on the order of 2 µs/cm. Since the solid has little dispersion, the bandwidth of the delay line is determined by that of the transducers. Here it is necessary to choose the characteristic impedances and thicknesses of the backing and matching layers in Fig. 46.2 in such a manner that the conversion of the electrical energy incident on the electrical port to the acoustic energy out of acoustic port 2 of Fig. 46.3 is independent of frequency over a large range about the resonant frequency of the **piezoelectric transducer** itself. Varying the matching and backing layers is equivalent to varying the terminating impedances on the acoustic line of Fig. 46.3. The matching is often assisted by lumped elements in the external electrical circuit.

The geometry of Fig. 46.4(c) is also the prototype form for acousto-optic interactions. Here the second transducer is not relevant and can be replaced by an acoustic absorber so that there is no reflected wave present in the active region. An optical wave coming into the crystal as shown in Fig. 46.4(c) sees a propagating periodic perturbation of the medium, and if the photoelastic coefficients of the solid are large, the wave sees appreciable variations in the refractive index and hence a moving diffraction grating. The angle of deflection of the output optical beam and its frequency as produced by the grating depend on the amplitude of the various frequency components in the acoustic beam when the optical beam traversed it. Thus, for example, the intensity versus angular position of the emerging optical beam is a measure of the frequency spectrum of any information modulated on the acoustic beam.

As noted previously, ultrasonic waves are often used as probes when the wavelength and attenuation are appropriate. For these radar-like applications, the acoustic beam is generated by a transducer and propagates in the medium containing the scatterer to be investigated as shown in Fig. 46.4(b). The acoustic wave is scattered by any discontinuity in the medium, and energy is returned to the same or to another transducer. If the outgoing signal is pulsed, then the delay for the received pulse is a measure of the distance to the scatterer. If the transducer is displaced or rotated, the change in delay of the echo gives a measure of the shape of the scatterer. Any movement of the scatterer, for example, flowing blood in an artery, causes a Doppler shift of the echo, and this shift, along with the known direction of the returned beam, gives a map of the flow pattern. Phasing techniques with multiple transducers or multiple areas of one transducer can be used to produce focused beams or beams electrically swept in space by differential variation of the phases of the excitation of the component areas of the transducer.

Defining Terms

Characteristic impedance: Ratio of the negative of the stress to the particle velocity in an ultrasonic plane wave.

Form: Term used to indicate the structure and dimensions of a multiterm equation without details within component terms.

Phase velocity: Velocity of propagation of planes of constant phase.

Piezoelectric transducers: Devices that convert electric signals to ultrasonic waves, and vice versa, by means of the piezoelectric effect in solids.

Pure longitudinal and shear waves (modes): Ultrasonic plane waves in which the particle motion is parallel or perpendicular, respectively, to the wave vector and for which energy flow is parallel to the wave vector.

Slowness surface: A plot of the reciprocal of the phase velocity as a function of direction in an anisotropic crystal.

References

B.A. Auld, *Acoustic Fields and Waves in Solids*, 2nd ed., Melbourne, Fla.: Robert E. Krieger, 1990.

E.A. Gerber and A. Ballato, *Precision Frequency Control*, vol.1, *Acoustic Resonators and Filters*, Orlando, Fla.: Academic Press, 1985.

G.S. Kino, *Acoustic Waves: Devices Imaging and Analog Signal Processing*, Englewood Cliffs, N.J.: Prentice-Hall, 1987.

Landolt-Bornstein, *Numerical Data and Functional Relationships in Science and Technology: Gp III Crystal and Solid State Physics*, vol. 11, *Elastic, Piezoelectric, Pyroelectric and Piezooptic Constants of Crystals*, Berlin: Springer-Verlag, 1979.

W.P. Mason and R.N. Thurston (Eds.), *Physical Acoustics, Principles and Methods*, multivolume series, New York: Academic Press.

J.F. Rosenbaum, *Bulk Acoustic Wave Theory and Devices*, Boston: Artech House, 1988.

Further Information

The main conferences in the ultrasonics area are the annual Ultrasonics Symposium sponsored by the IEEE Ultrasonics, Ferroelectrics and Frequency Control Society and the biannual Ultrasonics International Conference organized by the journal *Ultrasonics*, both of which publish proceedings. The periodicals include the *Transactions of the IEEE Ultrasonics, Ferroelectrics and Frequency Control Society*, the journal *Ultrasonics* published by Butterworth & Co., and the *Journal of the Acoustical Society of America*. The books by Kino and by Rosenbaum in the References provide general overviews of the field.

Transmission electron micrograph (TEM) of a PZT thin film prepared by the Sol-Gel method. Thin-film ferroelectric materials, such as PZT, have recently been investigated for their potential as a storage medium in nonvolatile memories (NV-RAM) and as a high-permittivity medium for integrated capacitors. Nonvolatile memories allow retention of data in the absence of power and thin-film dielectric films allow for denser and more efficient packaging of electronic circuits. Areas of light and dark shading are the domains with a different orientation of the polarization (see Chapter 47 for definitions of domains and polarization). Note that the direction of the strips varies according to which grain they are located in. Note also that in this example the domains do not cross grain boundaries.

0.25 µm

Courtesy of C. C. Hsueh, National Semiconductor Corporation.

Ferroelectric and Piezoelectric Materials

47.1 Introduction 1087
47.2 Mechanical Characteristics 1088
    Applications • Structure of Ferroelectric and Piezoelectric Materials
47.3 Ferroelectric Materials 1092
    Electrical Characteristics

K. F. Etzold
IBM T. J. Watson Research Center

47.1 Introduction

Piezoelectric materials have been used extensively in actuator and ultrasonic receiver applications, while **ferroelectric** materials have recently received much attention for their potential use in non-volatile (NV) memory applications. We will discuss the basic concepts in the use of these materials, highlight their applications, and describe the constraints limiting their uses. This chapter emphasizes properties which need to be understood for the effective use of these materials but are often very difficult to research. Among the properties which are discussed are **hysteresis** and **domains**.

Ferroelectric and piezoelectric materials derive their properties from a combination of structural and electrical properties. As the name implies, both types of materials have electric attributes. A large number of materials which are ferroelectric are also piezoelectric. However, the converse is not true. Pyroelectricity is closely related to ferroelectric and piezoelectric properties via the symmetry properties of the crystals.

Examples of the classes of materials that are technologically important are given in Table 47.1. It is apparent that many materials exhibit electric phenomena which can be attributed to ferroelectric, piezoelectric, and **electret** materials. It is also clear that vastly different materials (organic and inorganic) can exhibit ferroelectricity or piezoelectricity, and many have actually been commercially exploited for these properties.

As shown in Table 47.1, there are two dominant classes of ferroelectric materials, ceramics and organics. Both classes have important applications of their piezoelectric properties. To exploit the ferroelectric property, recently a large effort has been devoted to producing thin films of **PZT** (lead [Pb] zirconate titanate) on various substrates for silicon-based memory chips for nonvolatile storage. In these devices, data is retained in the absence of external power as positive and negative **polarization**. Organic materials have not been used for their ferroelectric properties. Liquid crystals in display applications are used for their ability to rotate the plane of polarization of light and not their ferroelectric attribute.

It should be noted that the prefix *ferro* refers to the permanent nature of the electric polarization in analogy with the magnetization in the magnetic case. It does not imply the presence of iron, even though the root of the word means iron. The root of the word piezo means pressure; hence the

Table 47.1 Ferroelectric, Piezoelectric, and Electrostrictive Materials

Type	Material Class	Example	Applications
Electret	Organic	Waxes	No recent
Electret	Organic	Fluorine based	Microphones
Ferroelectric	Organic	PVF2	No known
Ferroelectric	Organic	Liquid crystals	Displays
Ferroelectric	Ceramic	PZT thin film	NV-memory
Piezoelectric	Organic	PVF2	Transducer
Piezoelectric	Ceramic	PZT	Transducer
Piezoelectric	Ceramic	PLZT	Optical
Piezoelectric	Single crystal	Quartz	Freq. control
Piezoelectric	Single crystal	LiNbO$_3$	SAW devices
Electrostrictive	Ceramic	PMN	Actuators

original meaning of the word piezoelectric implied "pressure electricity"—the generation of electric field from applied pressure. This definition ignores the fact that these materials are reversible, allowing the generation of mechanical motion by applying a field.

47.2 Mechanical Characteristics

Materials are acted on by forces (stresses) and the resulting deformations are called strains. An example of a strain due to a force to the material is the change of dimension parallel and perpendicular to the applied force. It is useful to introduce the coordinate system and the numbering conventions which are used when discussing these materials. Subscripts 1, 2, and 3 refer to the *x*, *y*, and *z* directions, respectively. Displacements have single indices associated with their direction. If the material has a preferred axis, such as the poling direction in PZT, the axis is designated the *z* or 3 axis. Stresses and strains require double indices such as *xx* or *xy*. To make the notation less cluttered and confusing, contracted notation has been defined. The following mnemonic rule is used to reduce the double index to a single index:

$$
\begin{array}{ccc}
1 & 6 & 5 \\
xx & xy & xz \\
 & 2 & 4 \\
 & yy & yz \\
 & & 3 \\
 & & zz
\end{array}
$$

This rule can be thought of as a matrix with the diagonal elements having repeated indices in the expected order, then continuing the count in a counterclockwise direction. Note that $xy = yx$, etc. so that subscript 6 applies equally to *xy* and *yx*.

Any mechanical object is governed by the well-known relationship between stress and strain,

$$\mathbf{S} = \mathbf{sT} \tag{47.1}$$

where **S** is the strain (relative elongation), **T** is the stress (force per unit area), and **s** contains the coefficients connecting the two. All quantities are tensors; **S** and **T** are second rank, and **s** is fourth rank. Note, however, that usually contracted notation is used so that the full complement of subscripts is not visible. PZT converts electrical fields into mechanical displacements and vice versa. The connection between the two is via the *d* and *g* coefficients. The *d* coefficients give the displacement when a field is applied (transmitter), while the *g* coefficients give the field across the device when a stress is applied (receiver). The electrical effects are added to the basic Eq. (47.1) such that

$$\mathbf{S} = \mathbf{sT} + \mathbf{dE} \tag{47.2}$$

Table 47.2 Properties of Well-Known PZT Formulations (Based on the Original Navy Designations and Now Used by Commercial Vendor Vernitron)

	Units	PZT4	PZT5A	PZT5H	PZT8
ε_{33}	—	1300	1700	3400	1000
d_{33}	10^{-2} Å/V	289	374	593	225
d_{13}	10^{-2} Å/V	−123	−171	−274	−97
d_{15}	10^{-2} Å/V	496	584	741	330
g_{33}	10^{-3} Vm/N	26.1	24.8	19.7	25.4
k_{33}	—	70	0.705	0.752	0.64
T_{Θ}	°C	328	365	193	300
Q	—	500	75	65	1000
ρ	g/cm³	7.5	7.75	7.5	7.6
Application	—	High signal	Medium signal	Receiver	Highest signal

where **E** is the electric field and **d** is the tensor which contains the coupling coefficients. The latter parameters are reported in Table 47.2 for representative materials. One can write the matrix equation [Eq. (47.2)],

$$
\begin{bmatrix} S_1 \\ S_2 \\ S_3 \\ S_4 \\ S_5 \\ S_6 \end{bmatrix} = \begin{bmatrix} s_{11} & s_{12} & s_{13} & & & \\ s_{12} & s_{11} & s_{13} & & 0 & \\ s_{13} & s_{13} & s_{33} & & & \\ & & & s_{44} & & \\ & 0 & & & s_{44} & \\ & & & & & 2(s_{11} - s_{12}) \end{bmatrix} \begin{bmatrix} T_1 \\ T_2 \\ T_3 \\ T_4 \\ T_5 \\ T_6 \end{bmatrix} + \begin{bmatrix} 0 & 0 & d_{13} \\ 0 & 0 & d_{13} \\ 0 & 0 & d_{33} \\ 0 & d_{15} & 0 \\ d_{15} & 0 & 0 \\ 0 & 0 & 0 \end{bmatrix} \begin{bmatrix} E_1 \\ E_2 \\ E_3 \end{bmatrix} \quad (47.3)
$$

Note that **T** and **E** are shown as column vectors for typographical reasons; they are in fact row vectors. This equation shows explicitly the stress-strain relation and the effect of the electromechanical conversion.

A similar equation applies when the material is used as a receiver:

$$
\mathbf{E} = -\mathbf{gT} + (\varepsilon^{\mathrm{T}})^{-1}\mathbf{D} \quad (47.4)
$$

where T is the transpose and **D** the electric displacement. For all materials the matrices are not fully populated. Whether a coefficient is nonzero depends on the symmetry. For PZT, a ceramic which is given a preferred direction by the poling operation (the z-axis), only d_{33}, d_{13}, and d_{15} are nonzero. Also, again by symmetry, $d_{13} = d_{23}$ and $d_{15} = d_{25}$.

Applications

Historically the material which was used earliest for its piezoelectric properties was single-crystal quartz. Crude sonar devices were built by Langevin using quartz transducers, but the most important application was, and still is, frequency control. Crystal oscillators are today at the heart of every clock that does not derive its frequency reference from the ac power line. They are also used in every color television set and personal computer. In these applications at least one (or more) "quartz crystal" controls frequency or time. This explains the label "quartz" which appears on many clocks and watches. The use of quartz resonators for frequency control relies on another unique property. Not only is the material piezoelectric (which allows one to excite mechanical vibrations), but the material has also a very high mechanical "Q" or quality factor ($Q > 100,000$). The actual value depends on the mounting details, whether the crystal is in a vacuum, and other details. Compare this value to a Q for PZT between 75 and 1000. The Q factor is a measure of the rate of decay and thus the mechanical losses of an excitation with no external drive. A high Q leads to a very sharp resonance and thus tight frequency control. For frequency control it has been possible to find orientations of cuts of quartz which reduce the influence of temperature on the vibration frequency.

Ceramic materials of the PZT family have also found increasingly important applications. The piezoelectric but not the ferroelectric property of these materials is made use of in transducer applications. PZT has a very high efficiency (electric energy to mechanical energy coupling factor k) and can generate high-amplitude ultrasonic waves in water or solids. The coupling factor is defined by

$$k^2 = \frac{\text{energy stored mechanically}}{\text{total energy stored electrically}} \tag{47.5}$$

Typical values of k_{33} are 0.7 for PZT 4 and 0.09 for quartz, showing that PZT is a much more efficient transducer material than quartz. Note that the energy is a scalar; the subscripts are assigned by finding the energy conversion coefficient for a specific vibrational mode and field direction and selecting the subscripts accordingly. Thus k_{33} refers to the coupling factor for a longitudinal mode driven by a longitudinal field.

Probably the most important applications of PZT today are based on ultrasonic echo ranging. Sonar uses the conversion of electrical signals to mechanical displacement as well as the reverse transducer property, which is to generate electrical signals in response to a stress wave. Medical diagnostic ultrasound and nondestructive testing systems devices rely on the same properties. Actuators have also been built but a major obstacle is the small displacement which can conveniently be generated. Even then, the required voltages are typically hundreds of volts and the displacements are only a few hundred angstroms. For PZT the strain in the z-direction due to an applied field in the z-direction is (no stress, $\mathbf{T} = 0$)

$$s_3 = d_{33}E_3 \tag{47.6}$$

or

$$s_3 = \frac{\Delta d}{d} = d_{33}\frac{V}{d} \tag{47.7}$$

where s is the strain, E the electric field, and V the potential; d_{33} is the coupling coefficient which connects the two. Thus

$$\Delta d = d_{33}V \tag{47.8}$$

Note that this expression is independent of the thickness d of the material but this is true only when the applied field is parallel to the displacement. Let the applied voltage be 100 V and let us use PZT8 for which d_{33} is 225 (from Table 47.2). Hence $\Delta d = 225$ Å or 2.25 Å/V, a small displacement indeed. We also note that Eq. (47.6) is a special case of Eq. (47.2) with the stress equal to zero. This is the situation when an actuator is used in a force-free environment, for example, as a mirror driver. This arrangement results in the maximum displacement. Any forces which tend to oppose the free motion of the PZT will subtract from the available displacement with the reduction given by the normal stress-strain relation, Eq. (47.1).

It is possible to obtain larger displacements with mechanisms which exhibit mechanical gain, such as laminated strips (similar to bimetallic strips). The motion then is typically up to about 1 millimeter but at a cost of a reduced available force. An example of such an application is the video head translating device to provide tracking in VCRs.

There is another class of ceramic materials which recently has become important. **PMN** (lead [Pb], magnesium niobate), typically doped with $\approx 10\%$ lead titanate) is an **electrostrictive** material which has seen applications where the absence of hysteresis is important. For example, deformable mirrors require repositioning of the reflecting surface to a defined location regardless of whether the old position was above or below the original position.

Electrostrictive materials exhibit a strain which is quadratic as a function of the applied field. Producing a displacement requires an internal polarization. Because the latter polarization is induced by the applied field and is not permanent, as it is in the ferroelectric materials, electrostrictive materials have essentially no hysteresis. Unlike PZT, electrostrictive materials are not reversible; PZT will change shape on application of a field and generate a field when a strain is induced. Electrostrictive materials only change shape on application of a field and, therefore, cannot be used as receivers. PZT has inherently large hysteresis because of the domain nature of the polarization.

Organic electrets have important applications in self-polarized condenser (or capacitor) microphones where the required electric bias field in the gap is generated by the diaphragm material rather than by an external power supply.

Structure of Ferroelectric and Piezoelectric Materials

Ferroelectric materials have, as their basic building block, atomic groups which have an associated electric field, either as a result of their structure or as result of distortion of the charge clouds which make up the groups. In the first case, the field arises from an asymmetric placement of the individual ions in the group (these groupings are called unit cells). In the second case, the electronic cloud is moved with respect to the ionic core. If the group is distorted permanently, then a permanent electric field can be associated with each group. We can think of these distorted groups as represented by electric dipoles, defined as two equal but opposite charges which are separated by a small distance. Electric dipoles are similar to magnetic dipoles which have the familiar north and south poles. The external manifestation of a magnetic dipole is a magnetic field and that of an electric dipole an electric field.

Figure 47.1(a) represents a hypothetical slab of material in which the dipoles are perfectly arranged. In actual materials the atoms are not as uniformly arranged, but, nevertheless, from this model there would be a very strong field emanating from the surface of the crystal. The common observation, however, is that the fields are either absent or weak. This effective charge neutrality arises from the fact that there are free, mobile charges available which can be attracted to the surfaces. The polarity of the mobile charges is opposite to the charge of the free dipole end. The added charges on the two surfaces generate their own field, equal and opposite to the field due to the internal dipoles. Thus the effect of the internal field is canceled and the external field is zero, as if no charges were present at all [Fig. 47.1(b)].

In ferroelectric materials a crystalline asymmetry exists which allows electric dipoles to form. In their absence the dipoles are absent and the internal field disappears. Consider an imaginary horizontal line drawn through the middle of a dipole. We can see readily that the dipole is not symmetric about that line. The asymmetry thus requires that there be no center of inversion when the material is in the ferroelectric state.

All ferroelectric and piezoelectric materials have phase transitions at which the material changes crystalline symmetry. For example, in PZT there is a change from tetragonal or rhombohedral

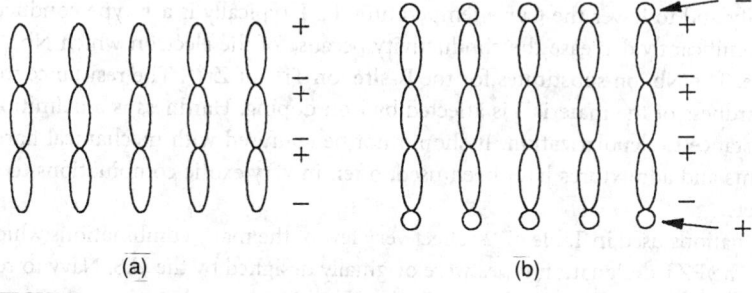

(a)   (b)

FIGURE 47.1 Charge configurations in ferroelectric model materials: (a) uncompensated and (b) compensated dipole arrays.

symmetry to cubic as the temperature is increased. The temperature at which the material changes **crystalline phases** is called the **Curie temperature**, T_Θ. For typical PZT compositions the Curie temperature is between 250 and 450°C.

A consequence of a phase transition is that a rearrangement of the lattice takes place when the material is cooled through the transition. Intuitively we would expect that the entire crystal assumes the same orientation throughout as we pass through the transition. By orientation we mean the direction of the preferred axis (say the tetragonal axis). Experimentally it is found, however, that the material breaks up into smaller regions in which the preferred direction and thus the polarization is uniform. Note that cubic materials have no preferred direction. In tetragonal crystals the polarization points along the c-axis (the longer axis) whereas in rhombohedral lattices the polarization is along the body diagonal. The volume in which the preferred axis is pointing in the same direction is called a domain and the border between the regions is called a domain wall. The energy of the multidomain state is slightly lower than the single-domain state and is thus the preferred configuration. The direction of the polarization changes by either 90° or 180° as we pass from one uniform region to another. Thus the domains are called 90° and 180° domains. Whether an individual crystallite or grain consists of a single domain depends on the size of the crystallite and external parameters such as strain gradients, impurities, etc. It is also possible that the domain extend beyond the grain boundary and encompasses two or more grains of the crystal.

Real materials consist of large numbers of unit cells, and the manifestation of the individual charged groups is an internal and an external electric field when the material is stressed. Internal and external refer to inside and outside of the material. The interaction of an external electric field with a charged group causes a displacement of certain atoms in the group. The macroscopic manifestation of this is a displacement of the surfaces of the material. This motion is called the piezoelectric effect, the conversion of an applied field into a corresponding displacement.

47.3 Ferroelectric Materials

PZT ($PbZr_xTi_{(1-x)}O_3$) is an example of a ceramic material which is ferroelectric. We will use PZT as a prototype system for many of the ferroelectric attributes to be discussed. The concepts, of course, have general validity. The structure of this material is ABO_3 where A is lead and B is one or the other atoms, Ti or Zr. This material consists of many randomly oriented crystallites which vary in size between approximately 10 nm and several microns. The crystalline symmetry of the material is determined by the magnitude of the parameter x. The material changes from rhombohedral to tetragonal symmetry when $x > 0.48$. This transition is almost independent of temperature. The line which divides the two phases is called a **morphotropic phase boundary** (change of symmetry as a function of composition only). Commercial materials are made with $x \approx 0.48$, where the d and g sensitivity of the material is maximum. It is clear from Table 47.2 that there are other parameters which can be influenced as well. Doping the material with donors or acceptors often changes the properties dramatically. Thus niobium is important to obtain higher sensitivity and resistivity and to lower the Curie temperature. PZT typically is a p-type conductor and niobium will significantly decrease the conductivity because of the electron which Nb^{5+} contributes to the lattice. The Nb ion substitutes for the **B-site** ion Ti^{4+} or Zr^{4+}. The resistance to depolarization (the hardness of the material) is affected by iron doping. Hardness is a definition giving the relative resistance to depolarization. It should not be confused with mechanical hardness. Many other dopants and admixtures have been used, often in very exotic combinations to affect aging, sensitivity, etc.

The designations used in Table 47.2 reflect very few of the many combinations which have been developed. The PZT designation types were originally designed by the U.S. Navy to reflect certain property combinations. These can be obtained with different combinations of compositions and dopants. The examples given in the table are representative of typical PZT materials, but today

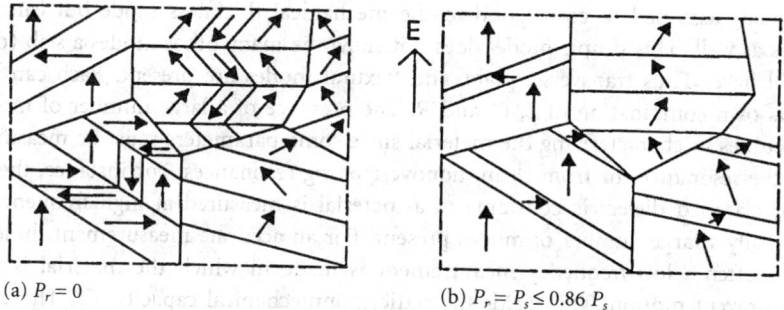

FIGURE 47.2 Domains in PZT, as prepared (a) and poled (b).

essentially all applications have their own custom formulation. The name PZT has become generic for the lead zirconate titanates and does not reflect Navy or proprietary designations.

When PZT ceramic material is prepared, the crystallites and domains are randomly oriented, and therefore the material does not exhibit any piezoelectric behavior [Fig. 47.2(a)]. The random nature of the displacements for the individual crystallites causes the net displacement to average to zero when an external field is applied. The tetragonal axis has three equivalent directions 90° apart and the material can be poled by reorienting the polarization of the domains into a direction nearest the applied field. When a sufficiently high field is applied, some but not all of the domains will be rotated toward the electric field through the allowed angle 90° or 180°. If the field is raised further, eventually all domains will be oriented as close as possible to the direction of the field. Note however, that the polarization will not point exactly in the direction of the field [Fig. 47.2(b)]. At this point, no further domain motion is possible and the material is saturated. As the field is reduced, the majority of domains retain the orientation they had with the field on leaving the material in an oriented state which now has a net polarization. Poling is accomplished for commercial PZT by raising the temperature to about 150°C (to lower the **coercive field**, E_c) and applying a field of about 30–60 kV/cm for several minutes. The temperature is then lowered but it is not necessary to keep the field on during cooling because the domains will not spontaneously rerandomize.

Electrical Characteristics

Before considering the dielectric properties, we will consider the equivalent circuit for a slab of ferroelectric material. In Fig. 47.3 the circuit shows a mechanical (acoustic) component and the static or clamped capacity C_o (and the dielectric loss R_d) which are connected in parallel. The acoustic components are due to their motional or mechanical equivalents, the compliance (capacity, C) and the mass (inductance, L). There will be mechanical losses, which are indicated in the mechanical branch by R. The electrical branch has the clamped capacity C_o and a dielectric loss (R_d), distinct from the mechanical losses. This configuration will have a resonance

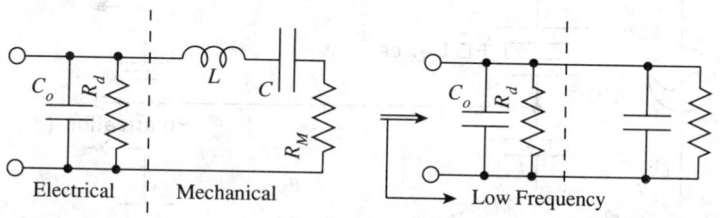

FIGURE 47.3 Equivalent circuit for a piezoelectric resonator. The reduction of the equivalent circuit at low frequencies is shown on the right.

which is usually assumed to correspond to the mechanical thickness mode but can represent other modes as well. This simple model does not show the many other modes a slab (or rod) of material will have. Thus transverse, plate, and flexural modes are present. Each can be represented by its own combination of L, C, and R. The presence of a large number of modes often causes difficulties in characterizing the material since some parameters must be measured either away from the resonances or from clean, nonoverlapping resonances. For instance, the clamped capacity (or clamped dielectric constant) of a material is measured at high frequencies where there are usually a large number of modes present. For an accurate measurement these must be avoided and often a low-frequency measurement is made in which the material is physically clamped to prevent motion. This yields the static, nonmechanical capacity, C_o. The circuit can be approximated at low frequencies by ignoring the inductor and redefining R and C. Thus, the coupling constant can be extracted from the value of C and C_o. From the previous definition of k we find

$$k^2 = \frac{\text{energy stored mechanically}}{\text{total energy stored electrically}} = \frac{CV^2/2}{(C + C_o)V^2/2} = \frac{1}{\frac{C_o}{C} + 1} \qquad (47.9)$$

It requires charge to rotate or flip a domain. Thus, there is charge flow associated with the rearrangement of the polarization in the ferroelectric material. If a bipolar, repetitive signal is applied to a ferroelectric material, its hysteresis loop is traced out and the charge in the circuit can be measured using the Sawyer Tower circuit (Fig. 47.4). In some cases the drive signal to the material is not repetitive and only a single cycle is used. In that case the starting point and the end point do not have the same polarization value and the hysteresis curve will not close on itself.

The charge flow through the sample is due to the rearrangement of the polarization vectors in the domains (the polarization) and contributions from the static capacity and losses (C_o and R_d in Fig. 47.3). The charge is integrated by the measuring capacitor which is in series with the sample. The measuring capacitor is sufficiently large to avoid a significant voltage loss. The polarization is plotted on a X-Y scope or plotter against the applied voltage and therefore the applied field.

Ferroelectric and piezoelectric materials are lossy. This will distort the shape of the hysteresis loop and can even lead to incorrect identification of materials as ferroelectric when they merely have nonlinear conduction characteristics. A resistive component (from R_d in Fig. 47.3) will introduce a phase shift in the polarization signal. Thus the display has an elliptical component, which looks like the beginnings of the opening of a hysteresis loop. However, if the horizontal signal has the same phase shift, the influence of this lossy component is eliminated, because it is in effect subtracted. Obtaining the exact match is the function of the optional phase shifter, and in the original circuits a bridge was constructed which had a second measuring capacitor in the comparison arm

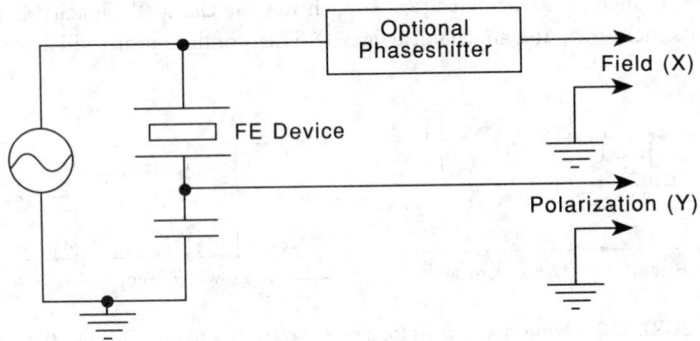

FIGURE 47.4 Sawyer Tower circuit.

(identical to the one in series with the sample). The phase was then matched with adjustable high-voltage components which match C_o and R_d.

This design is inconvenient to implement and modern Sawyer Tower circuits have the capability to shift the reference phase either electronically or digitally to compensate for the loss and static components. A contemporary version, which has compensation and no voltage loss across the integrating capacitor, is shown in Fig. 47.5. The op-amp integrator provides a virtual ground at the input, reducing the voltage loss to negligible values. The output from this circuit is the sum of the polarization and the capacitive and loss components. These contributions can be canceled using a purely real (resistive) and a purely imaginary (capacitive, 90° phaseshift) compensation component proportional to the drive across the sample. Both need to be scaled (magnitude adjustments) to match them to the device being measured and then have to be subtracted (adding negatively) from the output of the op amp. The remainder is the polarization. The hysteresis for typical ferroelectrics is frequency dependent and traditionally the reported values of the polarization are measured at 50 or 60 Hz.

The improved version of the Sawyer Tower (Fig. 47.5) circuit allows us to cancel C_o and R_d and the losses, thus determining the active component. This is important in the development of materials for ferroelectric memory applications. It is far easier to judge the squareness of the loop when the inactive components are canceled. Also, by calibrating the "magnitude controls" the value of the inactive components can be read off directly. In typical measurements the resonance is far above the frequencies used, so ignoring the inductance in the equivalent circuit is justified.

The measurement of the dielectric constant and the losses is usually very straightforward. A slab with a circular or other well-defined cross section is prepared, electrodes are applied, and the capacity and loss are measured (usually as a function of frequency). The dielectric constant is found from

$$C = \varepsilon_o \varepsilon \frac{A}{t} \qquad (47.10)$$

where A is the area of the device and t the thickness. In this definition (also used in Table 47.2) ε is the relative dielectric constant and ε_o is the permittivity of vacuum. Until recently, the dielectric constant, like the polarization, was measured at 50 or 60 Hz. Today the dielectric parameters are typically specified at 1 MHz or other high frequencies, which is possible because impedance analyzers with high-frequency capability are readily available. This avoids low-frequency anomalies which are often observed. These are not included in the equivalent circuit (Fig. 47.3) and are often due to interface layers. These anomalies will cause both the resistive and the reactive components to rise at low frequencies.

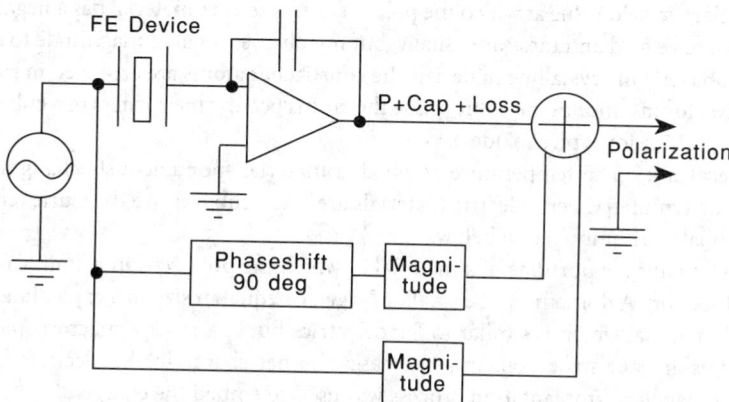

FIGURE 47.5 Modern hysteresis circuit. An op amp is used to integrate the charge; loss and static capacitance compensation are included.

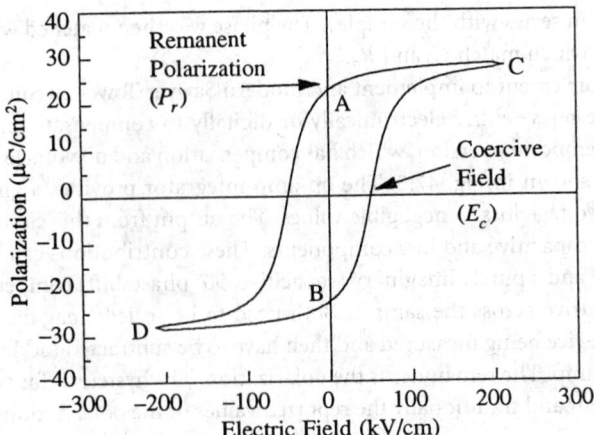

FIGURE 47.6 Idealized hysteresis curve for typical PZT materials. Many PZT materials display offsets from the origin and have asymmetries with respect to the origin. The curve shows how the remanent polarization ($\vec{P}_r$) and the coercive field ($\vec{E}_c$) are defined. While the loop is idealized, the values given for the polarization and field are realistic for typical PZT materials.

A piezoelectric component often has a very simple geometric shape, especially when it is prepared for measurement purposes. There will be mechanical resonances associated with the major dimensions of a sample piece. The resonance spectrum will be more or less complicated, depending on the shape of a sample piece. If the object has a simple shape, then some of the resonances will be well separated from each other and can be associated with specific vibrations and dimensions (modes). Each of these resonances has an electrical equivalent, and inspection of the equivalent circuit shows that there will be a resonance (maximum impedance) and an antiresonance (minimum impedance). Thus an impedance plot can be used to determine the frequencies and also the coupling constants and mechanical parameters for the various modes.

Defining Terms

A-site: Many ferroelectric materials are oxides with a chemical formula ABO_3. The A-site is the crystalline location of the A atom.

B-site: Analogous to the definition of the A-site.

Coercive field: When a ferroelectric material is cycled through the hysteresis loop the coercive field is the electric field value at which the polarization is zero. A material has a negative and a positive coercive field and these are usually, but not always, equal in magnitude to each other.

Crystalline phase: In crystalline materials the constituent atoms are arranged in regular geometric ways; for instance in the cubic phase the atoms occupy the corners of a cube (edge dimensions ≈ 2–15 Å for typical oxides).

Curie temperature: The temperature at which a material spontaneously changes its crystalline phase or symmetry. Ferroelectric materials are often cubic above the Curie temperature and tetragonal or rhombohedral below.

Domain: Domains are portions of a material in which the polarization is uniform in magnitude and direction. A domain can be smaller, larger, or equal in size to a crystalline grain.

Electret: A material which is similar to ferroelectrics but charges are macroscopically separated and thus are not structural. In some cases the net charge in the electrets is not zero, for instance when an implantation process was used to embed the charge.

Electrostriction: The change in size of a nonpolarized, dielectric material when it is placed in an electric field.

Ferroelectric: A material with permanent charge dipoles which arise from asymmetries in the crystal structure. The electric field due to these dipoles can be observed external to the material when certain conditions are satisfied (ordered material and no charge on the surfaces).

Hysteresis: When the electric field is raised across a ferroelectric material the polarization lags behind. When the field is cycled across the material the hysteresis loop is traced out by the polarization.

Morphotropic phase boundary (MPB): Materials which have a MPB assume a different crystalline phase depending on the composition of the material. The MPB is sharp (a few percent in composition) and separates the phases of a material. It is approximately independent of temperature in PZT.

Piezoelectric: A material which exhibits an external electric field when a stress is applied to the material and a charge flow proportional to the strain is observed when a closed circuit is attached to electrodes on the surface of the material.

PLZT: A PZT material with a lanthanum doping or admixture (up to approximately 15% concentration). The lanthanum occupies the A-site.

PMN: Generic name for electrostrictive materials of the lead (Pb) magnesium niobate family.

Polarization: The polarization is the amount of charge associated with the dipolar or free charge in a ferroelectric or an electret, respectively. For dipoles the direction of the polarization is the direction of the dipole. The polarization is equal to the external charge which must be supplied to the material to produce a polarized state from a random state (twice that amount is necessary to reverse the polarization). The statement is rigorously true if all movable charges in the material are reoriented (i.e., saturation can be achieved).

PVF2: An organic polymer which can be ferroelectric. The name is an abbreviation for polyvinyledene difluoride.

PZT: Generic name for piezoelectric materials of the lead (Pb) zirconate titanate family.

Remanent polarization: The residual or remanent polarization of a material after an applied field is reduced to zero. If the material was saturated, the remanent value is usually referred to as the polarization, although even at smaller fields a (smaller) polarization remains.

References

J. C. Burfoot and G. W. Taylor, *Polar Dielectrics and Their Applications*, Berkeley: University of California Press, 1979.

H. Diamant, K. Drenck, and R. Pepinsky, *Rev. Sci. Instrum.*, vol. 28, p. 30, 1957.

T. Hueter and R. Bolt, *Sonics*, New York: John Wiley and Sons, 1954.

B. Jaffe, W. Cook, and H. Jaffe, *Piezoelectric Ceramics*, London: Academic Press, 1971.

M. E. Lines and A. M. Glass, *Principles and Applications of Ferroelectric Materials*, Oxford: Clarendon Press, 1977.

C. B. Sawyer and C. H. Tower, *Phys. Rev.*, vol. 35, p. 269, 1930.

Z. Surowiak, J. Brodacki, and H. Zajosz, *Rev. Sci. Instrum.*, vol. 49, p. 1351, 1978.

Further Information

IEEE Transactions on Ultrasonics, Ferroelectrics, and Frequency Control (UFFC).

IEEE Proceedings of International Symposium on the Application of Ferroelectrics (ISAF) (these symposia are held at irregular intervals).

Materials Research Society, Symposium Proceedings, vols. 191, 200, and 243 (this society holds symposia on ferroelectric materials at irregular intervals).

K.-H. Hellwege, Ed., *Landolt-Bornstein: Numerical Data and Functional Relationships in Science and Technology*, New Series, Gruppe III, vols. 11 and 18, Berlin: Springer-Verlag, 1979 and 1984 (these volumes have elastic and other data on piezoelectric materials).

American Institute of Physics Handbook, 3rd ed., New York: McGraw-Hill, 1972.

THE FARADAY BICENTENNIAL

Michael Faraday (1791–1867), the eminent British natural philosopher, made numerous contributions to electrical science including the discovery of electromagnetic induction and the introduction of the concept of the electromagnetic field.

Born in Newington, near London, England, Faraday was the son of a blacksmith. His formal education was quite limited, but he took advantage of opportunities to attend public lectures and became an avid reader. Beginning when he was thirteen, he served an apprenticeship to a bookbinder. In 1813 he began his long career in science, when Sir Humphry Davy, the renowned chemist and lecturer at the Royal Institution in London, hired him as a laboratory assistant. Davy introduced Faraday to leading scientists on the continent during an extended tour that ended in 1815.

In 1821, Faraday demonstrated an electric motor effect where the pole of a permanent magnet fixed at one end rotated around a current-carrying conductor. In 1825, he was appointed director of the Royal Institution laboratory and soon initiated a lecture series that helped to popularize science and increase scientific literacy. For many years, he presented an annual series of lectures to children during the Christmas season on topics such as the chemical history of a candle.

In 1831, Faraday made a major discovery by showing that relative motion between a magnet and a nearby conductor produced an electromotive force in the conductor. This principle led to the development of magneto-electric generators and, in the late 1860's, to the dynamo. An extensive series of experiments on dielectrics led Faraday to the idea of specific inductive capacity (dielectric constant). This research was the basis for the adoption of the farad as the unit of capacitance at an international electrical congress in 1891.

Faraday also did important investigation in the field of electrochemistry, discovering what became known as Faraday's Law of Electrolysis. He introduced such familiar terms to the electrical vocabulary as electrode, electrolyte, anode, cathode, and ion.

During the 1840's, Faraday discovered that a magnet could cause rotation of the plane of polarization of a beam of light and discussed a theory of the electromagnetic field that later was to be given mathematical form by James C. Maxwell. Underlying Faraday's experimental program was his general belief in the unity and convertibility of forces, including gravitational as well as electric and magnetic. His friend and colleague, John Tyndall, wrote that Faraday "was incessantly theorizing" and that "theoretic ideas were the very sap of his intellect—the source from which all his strength as an experimenter was derived." Many of Faraday's more than 150 technical papers were included in his classic three-volume work, *Experimental Researches in Electricity,* which inspired Thomas A. Edison among others.

The year 1991 was recognized as the Faraday bicentennial, which stimulated numerous special events and observances. The British Post Office issued a postage stamp in his honor, and a new 20-pound bank note featured a portrait of Faraday. The British Science Museum created a major exhibition on his work which opened in the summer of 1991 and ran for six months. A meeting of the American Chemical Society held in Atlanta, Georgia, in April 1991 included a symposium with several papers on Faraday as a chemist and popular lecturer. A bicentenary symposium on both Faraday and Charles Babbage was held at Cambridge University in July 1991. The Royal Institution, the Royal Society of Chemistry, the Institute of Physics, and the Institution of Electrical Engineers (IEE) had also planned special meetings in celebration of the bicentennial.

Since the power and communication systems of today are based to a considerable degree on Faraday's work, it was most appropriate that electrical engineers and scientists celebrated the bicentennial of his birth and reflected on his legacy.

Source: Adapted from J.E. Brittain, *Proc. IEEE,* vol. 79, no. 9, p. 1342, September 1991. © 1991 IEEE.

48

Piezoresistivity

48.1 Introduction 1099
48.2 Equation of State 1099
48.3 Effect of Crystal Point Group on Π_{ijkl} 1100
48.4 Geometric Corrections and Elastoresistance Tensor 1101
48.5 Multivalley Semiconductors 1102
48.6 Semiconducting (PTCR) Perovskites 1104
48.7 Thick Film Resistors 1104

Ahmed Amin
Texas Instruments, Inc.

48.1 Introduction

Piezoresistivity is a linear coupling between mechanical stress X_{kl} and electrical resistivity ρ_{ij}. Hence, it is represented by a fourth rank polar tensor Π_{ijkl}. The piezoresistance properties of semiconducting silicon and germanium were discovered by Smith [1953] when he was verifying the form of their energy surfaces. Piezoresistance measurements can provide valuable insights concerning the conduction mechanisms in solids such as strain-induced carrier repopulation and intervalley scattering in multivalley semiconductors [Herring and Vogt, 1956], barrier tunneling in thick film resistors [Canali *et al.*, 1980] and barrier raising in semiconducting positive temperature coefficient of resistivity (PTCR) perovskites [Amin, 1989]. Piezoresistivity has also been investigated in compound semiconductors, thin metal films [Rajanna *et al.*, 1990], polycrystalline silicon and germanium thin films [Onuma and Kamimura, 1988], heterogeneous solids [Carmona *et al.*, 1987], and high T_c superconductors [Kennedy *et al.*, 1989]. Several sensors that utilize this phenomenon are commercially available.

48.2 Equation of State

The equation of state of a crystal subjected to a stress X_{kl} and an electric field E_i is conveniently formulated in the isothermal representation. The difference between isothermal and adiabatic changes, however, is negligible [Keyes, 1960]. Considering only infinitesimal deformations, where the linear theory of elasticity is valid, the electric field E_i is expressed in terms of the current density I_j and applied stress X_{kl} as [Mason and Thurston, 1957].

$$E_i = E_i\,(I_j, X_{kl}) \qquad i,j,k,l = 1,2,3 \tag{48.1}$$

In what follows the summation convention over repeated indices in the same term is implied, and the letter subscripts assume the values 1, 2, and 3 unless stated otherwise. Expanding in a McLaurin's series about the origin (state of zero current and stress)

$$dE_i = (\partial E_i/\partial I_j) \, dI_j + (\partial E_i/\partial X_{kl}) \, dX_{kl}$$
$$+ (1/2!) \, [(\partial^2 E_i/\partial I_j \partial I_m) \, dI_j \, dI_m$$
$$+ (\partial^2 E_i/\partial X_{kl} \partial X_{no}) \, dX_{kl} \, dX_{no} \tag{48.2}$$
$$+ 2 \, (\partial^2 E_i/\partial X_{kl} \partial I_j) \, dX_{kl} \, dI_j] \; + \ldots \text{H.O.T}$$

The partial derivatives in the expansion Eq. (48.2) have the following meanings: $(\partial E_i/\partial I_j) = \rho_{ij}$ (electric resistivity tensor); $(\partial E_i/\partial X_{kl}) = d_{ikl}$ (converse piezoelectric tensor); $(\partial^2 E_i/\partial X_{kl} \partial I_j) = (\partial/\partial X_{kl}) \, (\partial E_i/\partial I_j) = \Pi_{ijkl}$ (piezoresistivity tensor); $(\partial^2 E_i/\partial I_j \partial I_m) = \rho_{ijm}$ (nonlinear resistivity tensor); $(\partial^2 E_i/\partial X_{kl} \partial X_{no}) = \delta_{iklno}$ (nonlinear piezoelectric tensor).

Replacing the differentials in Eq. (48.2) by the components themselves, we get

$$E_i = \rho_{ij} I_j + d_{ikl} X_{kl} + 1/2 \, \rho_{ijm} I_j I_m + 1/2 \, \delta_{iklno} X_{kl} X_{no} + \Pi_{ijkl} X_{kl} I_j \tag{48.3}$$

Most of the technologically important piezoresistive materials, e.g., silicon, germanium, and polycrystalline films, are centrosymmetric. The effect of center of symmetry (i.e., the inversion operator) on Eq. (48.3) is to force all odd rank tensor coefficients to zero; hence, the only contribution to the resistivity change under stress will result from the piezoresistive term. Therefore, Eq. (48.3) takes the form

$$E_i = \Sigma_j \rho_{ij} I_j + \Sigma_j \Sigma_k \Sigma_l \Pi_{ijkl} X_{kl} I_j \tag{48.4}$$

taking the partial derivatives of Eq. (48.4) with respect to the current density I_j and rearranging

$$\partial E_i/\partial I_j = \rho_{ij}(X) - \rho_{ij}(0) = \Sigma_k \Sigma_l \Pi_{ijkl} X_{kl}$$

Thus, the specific change in resistivity with stress is given by

$$(\delta \rho_{ij}/\rho_0) = \Pi_{ijkl} X_{kl} \tag{48.5}$$

the piezoresistivity tensor Π_{ijkl} in Eq. (48.5) has the dimensions of reciprocal stress (square meters per newton in the MKS system of units). The effects of the intrinsic symmetry of the piezoresistivity tensor and the crystal point group are discussed next.

48.3 Effect of Crystal Point Group on Π_{ijkl}

The transformation law of Π_{ijkl} (a fourth rank polar tensor) is as follows:

$$\Pi'_{ijkl} = (\partial x'_i/\partial x_m)(\partial x'_j/\partial x_n)(\partial x'_k/\partial x_o)(\partial x'_l/\partial x_p)\Pi_{mnop} \tag{48.6}$$

where the primed and unprimed components refer to the new and old coordinate systems, respectively, and the determinants of the form $\|\partial x'_i/\partial x_m\|, \ldots$ etc. are the Jacobian of the transformation. A general fourth rank tensor has 81 independent components. The piezoresistivity tensor Π_{ijkl} has the following internal symmetry:

$$\Pi_{ijkl} = \Pi_{ijlk} = \Pi_{jilk} = \Pi_{jikl} \tag{48.7}$$

which reduces the number of independent tensor components from 81 to 36 for the most general triclinic point group $C_1(1)$. It is convenient to use the reduced (two subscript) matrix notation

$$\Pi_{ijkl} = \Pi_{mn} \tag{48.8}$$

where $m, n = 1, 2, 3, \ldots, 6$. The relation between the subscripts in both notations is

$$
\begin{array}{lcccccc}
\text{Tensor:} & 11 & 22 & 33 & 23, 32 & 13, 31 & 12, 21 \\
\text{Matrix:} & 1 & 2 & 3 & 4 & 5 & 6
\end{array}
$$

and

$$\Pi_{mn} = 2\Pi_{ijkl}, \text{ for } m \text{ and/or } n = 4, 5, 6$$

Thus, for example, $\Pi_{1111} = \Pi_{11}$, $\Pi_{1122} = \Pi_{12}$, $2\Pi_{2323} = \Pi_{44}$, $2\Pi_{1212} = \Pi_{66}$, and $2\Pi_{1112} = \Pi_{16}$. Hence, Eq. (48.5) takes the form

$$(\delta\rho_i/\rho_0) = \Pi_{ij} X_j, \qquad (i, j = 1, 2, \ldots, 6) \qquad (48.9)$$

Further reduction of the remaining 36 piezoresistivity tensor components is obtained by applying the generating elements of the point group to the piezoresistivity tensor transformation law Eq. (48.6) and demanding invariance. The following are two commonly encountered piezoresistivity matrices:

1. Cubic $O_h(m3m)$: single crystal silicon and germanium

$$
\begin{bmatrix}
\Pi_{11} & \Pi_{12} & \Pi_{12} & 0 & 0 & 0 \\
 & \Pi_{11} & \Pi_{12} & 0 & 0 & 0 \\
 & & \Pi_{11} & 0 & 0 & 0 \\
 & & & \Pi_{44} & 0 & 0 \\
 & & & & \Pi_{44} & 0 \\
 & & & & & \Pi'_{44}
\end{bmatrix}
$$

2. Spherical ($\infty \infty \, mmm$): polycrystalline silicon and germanium and films

$$
\begin{bmatrix}
\Pi_{11} & \Pi_{12} & \Pi_{12} & 0 & 0 & 0 \\
 & \Pi_{11} & \Pi_{12} & 0 & 0 & 0 \\
 & & \Pi_{11} & 0 & 0 & 0 \\
 & & & \Pi_{44} & 0 & 0 \\
 & & & & \Pi_{44} & 0 \\
 & & & & & \Pi_{44}
\end{bmatrix}
$$

where $\Pi_{44} = 2(\Pi_{11} - \Pi_{12})$. Thus, three coefficients Π_{11}, Π_{12}, and Π_{44} are required to completely specify the piezoresistivity tensor for silicon and germanium single crystals, and only two, Π_{11} and Π_{12}, for polycrystalline films. Under hydrostatic pressure conditions, the piezoresistivity coefficient Π_h for the preceding two symmetry groups is a linear combination of the longitudinal Π_{11} and transverse Π_{12} components, $\Pi_h = \Pi_{11} + 2\Pi_{12}$. Unlike the elastic stiffness c_{ij} (a fourth rank polar tensor), the piezoresistivity tensor Π_{mn} is not symmetric, i.e., $\Pi_{mn} \# \Pi_{nm}$, except for the following point groups, $C_{\infty v}(\infty \infty \, mmm)$, $O_h(m3m)$, $T_d(\overline{4}3m)$, and $O(432)$.

48.4 Geometric Corrections and Elastoresistance Tensor

The experimentally derived quantity is the piezoresistance coefficient $1/R_0(\partial R/\partial X)$. This must be corrected for the dimensional changes to obtain the piezoresistivity coefficient $1/\rho_0(\partial\rho/\partial X)$ as follows:

Table 48.1 Numerical Values of Π_{ij} and M_{ij} for Selected Materials

Material	Resistivity (Unstrained and RT)	$(10^{-11}$ m²/N)			Dimensionless		
		Π_{11}	Π_{12}	Π_{44}	M_{11}	M_{12}	M_{44}
Silicon							
n-type	11.7 (Ω-cm)	−102.2	53.4	−13.6	−72.6	86.4	−10.8
p-type	7.8 (Ω-cm)	6.6	−1.1	138.1	10.5	2.7	110
Ba$_{.648}$Sr$_{.35}$							
La$_{.002}$TiO$_3$	≈100 (Ω-cm)	250	250				
Thin films							
Si					15		
Ge					30		
Mn	160 (μΩ-cm)				3		
Thick film resistors							
DP 1351, main constituent Bi$_2$Ru$_2$O$_7$							
	100 (KΩ/□)				13.5		
ESL 2900	100 (KΩ/□)				13.8	11.6	

1. Uniaxial tensile stress parallel to current flow

$$1/R_0(\partial R/\partial X) - (s_{11} - 2s_{12}) = 1/\rho_0(\partial\rho/\partial x) = \Pi_{11} \tag{48.10}$$

2. Uniaxial tensile stress perpendicular to current flow

$$1/R_0(\partial R/\partial X) + s_{11} = 1/\rho_0(\partial\rho/\partial X) = \Pi_{12} \tag{48.11}$$

3. Hydrostatic pressure

$$1/R_0(\partial R/\partial p) - (s_{11} + 2s_{12}) = 1/\rho_0(\partial\rho/\partial p) = \Pi_h \tag{48.12}$$

where s_{11} and s_{12} are the elastic compliances that appear in the linear elasticity equation $x_{ij} = s_{ijkl} X_{kl}$, with x_{ij} the infinitesimal strain components. Details on the different geometries and methods of measuring the piezoresistance effect can be found in the References. Equation (48.9) could be written in terms of the strain conjugate x_o as follows

$$(\delta\rho_i/\rho_0) = M_{io} x_o \tag{48.13}$$

the dimensionless quantity M_{io} is the elastoresistance tensor (known as the gage factor in the sensors literature). It is related to the piezoresistivity Π_{ik} and the elastic stiffness c_{ko} tensors by

$$M_{io} = \Pi_{ik} c_{ko} \tag{48.14}$$

48.5 Multivalley Semiconductors

For a multivalley semiconductor, e.g., *n*-type silicon, the energy minima (ellipsoids of revolutions) of the unstrained state in momentum space are along the six <100> cubic symmetry directions; they possess the symmetry group $O_h(m3m)$. A tensile stress in the *x*-direction, for example, will strain the lattice in the *xy*-plane and destroy the three-fold symmetry, thereby lifting the degeneracy of the energy minima. However, the four-fold symmetry along the *x*-direction will be preserved. Thus, the two valleys along the direction of stress will be shifted relative to the four valleys in the perpendicular directions.

According to the deformation potential theory, the strain will shift the energy of all the states in a given band extremum by the same amount, i.e., the valley moves along the energy scale as a whole

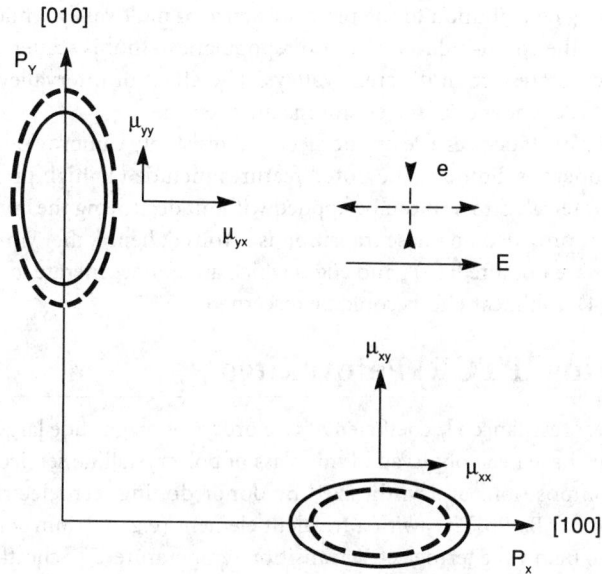

FIGURE 48.1 Two-dimensional representation of the constant energy surfaces in momentum space of a multivalley semiconductor (e.g., *n*-type silicon) showing only one quadrant, point group symmetry $C_{4v}(4mm)$. (*Source:* C.S. Smith, Piezoresistance effect in germanium and silicon, *Phys. Rev.*, vol. 94, p. 42, 1953. With permission.)

by an amount (the deformation potential constant) which is linearly proportional to the strain. Let's assume that the energy of those on the *y*- and *z*-axes are lowered with respect to those on the *x*-axis. This effect is represented by dashed lines in Fig. 48.1. As a result, there will be electron transfer from the high to low energy valleys. The components of the mobility tensor μ_{xy} $(= e \tau / m_{xy}$, where *e* is the electron charge, τ is the relaxation time, and m_{xy} is the effective mass) are illustrated by arrows in Fig. 48.1. The mobility anisotropy is due to the curvature of the conduction band near the bottom. The effective mass is inversely proportional to this curvature $(1/m_{xy} = (h/2\pi)^{-2}$ $(\partial^2 E/\partial k_x \partial k_y)$, which is larger for a direction perpendicular to the valley. For an applied *E* field parallel to the stress, the conductivity will increase (i.e., the resistivity decreases) relative to the unstressed state because of the increase in the number of electrons in the four valleys (*yz*-plane) for which the mobility is large in the field direction. If the field is perpendicular to the stress, the conductivity will decrease (i.e., the resistivity increases) with stress. Therefore, the piezoresistivity components Π_{11} and Π_{12} have opposite signs.

A shear stress about the crystallographic axes will not lift the degeneracy; hence, $\Pi_{44} = 0$. Similarly, a tensile stress along the <111> does not destroy the three-fold symmetry, and the degeneracy will not be lifted; thus, no piezoresistance should be there. Calculations based on the deformation potential model show that $\Pi_{11} = -2 \Pi_{12}$, and $\Pi_{44} = 0$.

Information concerning the symmetry properties of the valleys can be derived from the representation surface of the longitudinal Π_{11} piezoresistance component. This surface can be constructed by measuring the dependence of Π_{11} on the crystallographic direction. Smith showed that Π_{11} is maximum in the <001> directions of *n*-type silicon and not quite zero in the <111> directions. Reasons for the deviation from the deformation potential model of piezoresistivity in multivalley semiconductors are discussed in Keyes [1960]. For *n*-type germanium Π_{11} is maximum in the <111> directions. This is consistent with the loci of the valleys in these two materials. Qualitatively, a piezoresistance effect is produced whenever the stress destroys the symmetry elements that are responsible for the degeneracy of the valleys.

Intervalley scattering contribution to the piezoresistance of multivalley semiconductors may be comparable to that of the strain-induced electron repopulation. In this scattering process, the initial and final electron states are in different valleys. The effect of intervalley scattering can be deduced from the T^{-1} dependence of the elastoresistance tensor.

The influence of hydrostatic pressure on the electrical resistivity can provide additional insights on the transport properties. Some of the noted features include (1) high pressures (in the GPa range, versus MPa for tensile stresses) can be applied without destroying the crystal; (2) it does not destroy the symmetry, provided no phase transition is involved; hence, the symmetry degeneracies in the band structure are not lifted; (3) band edges which are not degenerate for symmetry reasons will be shifted; and (4) nonlinear effects could be discerned.

48.6 Semiconducting (PTCR) Perovskites

Large hydrostatic piezoresistance Π_h coefficients (two orders of magnitude larger than those of silicon and germanium) have been observed in this class of polycrystalline semiconductors [Sauer *et al.*, 1959]. PTCR compositions are synthesized by donor doping ferroelectric barium titanate $BaTiO_3$, $(Ba,Sr)TiO_3$, or $(Ba,Pb)TiO_3$ with a trivalent element (e.g., yttrium) or a pentavalent element (e.g., niobium). Below the ferroelectric transition temperature T_c, Schottky barriers between the conductive ceramic grains are neutralized by the spontaneous polarization P_s associated with the ferroelectric phase transition. Above T_c the barrier height increases rapidly with temperature (hence the electrical resistivity) because of the disappearance of P_s and the decrease of the paraelectric state dielectric constant. Analytic expressions that permit the computation of barrier heights under different elastic and thermal boundary conditions have been developed [Amin, 1989].

48.7 Thick Film Resistors

Thick film resistors consist of a conductive phase, e.g., rutile (RuO_2), perovskite ($BaRuO_3$), or pyrochlore ($Pb_2Ru_2O_{7-x}$), and an insulating phase (e.g., lead borosilicate) dispersed in an organic vehicle. They are formed by screen printing on a substrate, usually alumina, followed by sintering at $\approx 850°C$ for 10 min.

The increase of the piezoresistance properties of a commercial thick film resistor (ESL 2900 series) with sheet resistivity is illustrated in Fig. 48.2. The experimentally observed properties such as the resistance increase and decrease with tensile and compressive strains, respectively, and the increase of the elastoresistance tensor with sheet resistivity seem to support a barrier tunneling model [Canali *et al.*, 1980].

Defining Terms

ρ_{ij}:   Electric resistivity tensor.
d_{ikl}:   Converse piezoelectric tensor
Π_{ijkl}:   Piezoresistivity tensor
ρ_{ijm}:   Nonlinear resistivity tensor
δ_{iklno}:   Nonlinear piezoelectric tensor

References

A. Amin, "Numerical computation of the piezoresistivity matrix elements for semiconducting perovskite ferroelectrics," *Phys. Rev. B*, 40, 11603, 1989.

C. Canali, D. Malavasi, B. Morten, M. Prudenziati, and A. Taroni, "Piezoresistive effect in thick-film resistors," *J. Appl. Phys.*, 51, 3282, 1980.

F. Carmona, R. Canet, and P. Delhaes, "Piezoresistivity in heterogeneous solids," *J. Appl. Phys.*, 61, 2550, 1987.

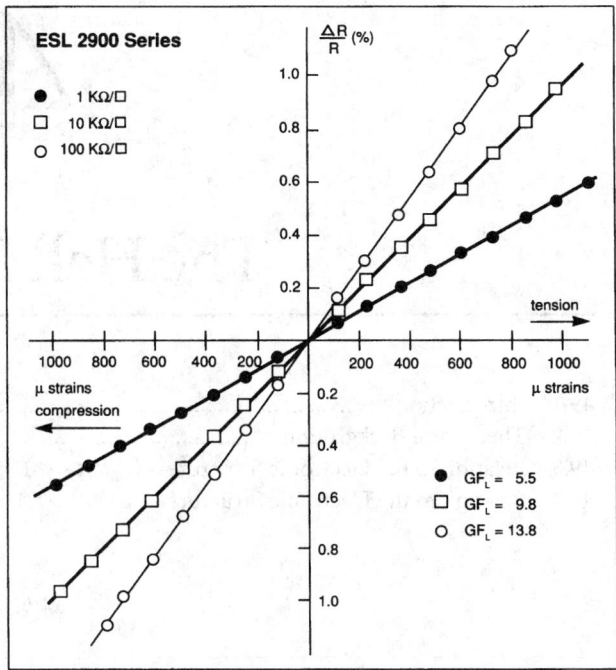

FIGURE 48.2 Relative changes of resistance for compressive and tensile strain applied parallel to the current direction. Note the increase of gage factor with sheet resistivity. (*Source:* C. Canali *et. al.,* Piezoresistive effect in thick film resistors, *J. Appl. Phys.,* vol. 51, p. 3282, 1980. With permission.)

C. Herring and E. Vogt, "Transport and deformation-potential theory for many valley semiconductors with anisotropic scattering," *Phys. Rev.,* 101, 944, 1956.

R. J. Kennedy, W. G. Jenks, and L. R. Testardi, "Piezoresistance measurements of $YBa_2Cu_3O_{7-x}$ showing large magnitude temporal anomalies between 100 and 300 K," *Phys. Rev. B,* 40, 11313, 1989.

R. W. Keyes, "The effects of elastic deformation on the electrical conductivity of semiconductors," *Solid State Phys.,* 11, 149, 1960.

W. P. Mason and R. N. Thurston, "Use of piezoresistive materials in the measurement of displacement, force, and torque," *J. Acoust. Soc. Am.,* 10, 1096, 1957.

Y. Onuma and K. K. Kamimura, "Piezoresistive elements of polycrystalline semiconductor thin films," *Sensors and Actuators,* 13, 71, 1988.

K. Rajanna, S. Mohan, M. M. Nayak, and N. Gunasekaran, "Thin film pressure transducer with manganese film as the strain gauge," *Sensor and Actuators,* A 24, 35, 1990.

H. A. Sauer, S. S. Flaschen, and D. C. Hoestery, "Piezoresistance and piezocapacitance effect in barium strontium titanate ceramics," *J. Am. Ceram. Soc.,* 42, 363, 1959.

C. S. Smith, "Piezoresistance effect in germanium and silicon," *Phys. Rev.,* 94, 42, 1953.

Further Information

M. Neuberger and S. J. Welles, *Silicon,* Electronic Properties Information Center, Hughes Aircraft Co., Culver City, Calif., 1969. This reference contains a useful compilation of the piezoresistance properties of silicon.

Electronic databases such as *Chemical Abstracts* will provide an update on the current research on piezoresistance materials and properties.

49

The Hall Effect

Alexander C. Ehrlich
U.S. Naval Research Laboratory

49.1 Introduction 1106
49.2 Theoretical Background 1107
49.3 Relation to the Electronic Structure—(i) $\omega_c\tau \ll 1$ 1108
49.4 Relation to the Electronic Structure—(ii) $\omega_c\tau \gg 1$ 1110

49.1 Introduction

The Hall effect is a phenomenon that arises when an electric current and magnetic field are simultaneously imposed on a conducting material. Specifically, in a flat plate conductor, if a current density, J_x, is applied in the x direction and (a component of) a magnetic field, B_z, in the z direction, then the resulting electric field, E_y, transverse to J_x and B_z is known as the Hall electric field E_H (see Fig. 49.1) and is given by

$$E_H = RJ_xB_z \tag{49.1}$$

where R is known as the Hall coefficient. The Hall coefficient can be related to the electronic structure and properties of the **conduction bands** in metals and semiconductors and historically has probably been the most important single parameter in the characterization of the latter. Some authors choose to discuss the Hall effect in terms of the Hall angle, ϕ, shown in Fig. 49.1, which is the angle between the net electric field and the imposed current. Thus,

$$\tan\phi = E_H/E_x \tag{49.2}$$

For the vast majority of Hall effect studies that have been carried out, the origin of E_H is the Lorentz force, F_L, that is exerted on a charged particle as it moves in a magnetic field. For an electron of charge e with velocity v, F_L is proportional to the vector product of $\mathbf{v}$ and $\mathbf{B}$; that is,

$$F_L = e\mathbf{v}\mathbf{X}\mathbf{B} \tag{49.3}$$

In these circumstances a semiclassical description of the phenomenon is usually adequate. This description combines the classical Boltzmann transport equation with the Fermi–Dirac distribution function for the charge carriers (electrons) [Ziman, 1960], and this is the point of view that will be taken in this chapter. Examples of Hall effect that cannot be treated semiclassically are the spontaneous (or extraordinary) Hall effect that occurs in ferromagnetic conductors [Berger and Bergmann, 1980], the quantum Hall effect [Prange and Girvin, 1990], and the Hall effect that arises in conjuction with hopping conductivity [Emin, 1977].

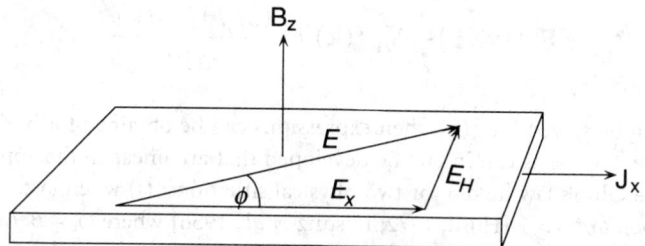

FIGURE 49.1 Typical Hall effect experimental arrangement in a flat plate conductor with current J_x and magnetic field B_z. The Hall electric field E_H arises because of the Lorentz force on the conducting charges and is of just such a magnitude that in combination with the Lorentz force there is no net current in the y direction. The angle ϕ between the current and net electric field is called the Hall angle.

In addition to its use as an important tool in the study of the nature of electrically conducting materials, the Hall effect has a number of direct practical applications. For example, the sensor in some commercial devices for measuring the magnitude and orientation of magnetic fields is a Hall sensor. The spontaneous Hall effect has been used as a nondestructive method for exploring the presence of defects in steel structures. The quantum Hall effect has been used to refine our knowledge of the magnitudes of certain fundamental constants such as the ratio of e^2/h where h is Planck's constant.

49.2 Theoretical Background

The Boltzmann equation for an electron gas in a homogeneous, isothermal material that is subject to constant electric and magnetic fields is [Ziman, 1960]

$$e[\mathbf{E} + \mathbf{v}\mathbf{X}\mathbf{B}]\left(\frac{1}{\hbar}\right)\nabla_{\mathbf{k}}f(\mathbf{k}) - \left(\frac{\partial f}{\partial t}\right)_c = 0 \tag{49.4}$$

Here $\mathbf{k}$ is the quantum mechanical wave vector, $\hbar$ is Planck's constant divided by 2π, t is the time, and f is the electron distribution function. In static equilibrium ($\mathbf{E} = 0$, $\mathbf{B} = 0$) f is equal to f_0 and f_0 is the Fermi–Dirac distribution function

$$f_0 = \frac{1}{e^{(\mathscr{E}(\mathbf{k})-\zeta)/KT} + 1} \tag{49.5}$$

where $\mathscr{E}(\mathbf{k})$ is the energy, ζ is the chemical potential, K is Boltzmann's constant, and T is the temperature. Each term in Eq. (49.4) represents a time rate of change of f and in dynamic equilibrium their sum has to be zero. The last term represents the effect of collisions of the electrons with any obstructions to their free movement such as lattice vibrations, crystallographic imperfections, and impurities. These collisions are usually assumed to be representable by a relaxation time, $\tau(\mathbf{k})$, that is

$$\left(\frac{\partial f}{\partial t}\right)_c = \frac{-(f - f_0)}{\tau(\mathbf{k})} = \frac{(\partial f_0/\partial \mathscr{E})g(\mathbf{k})}{\tau(\mathbf{k})} \tag{49.6}$$

where $f - f_0$ is written as $(\partial f_0/\partial \varepsilon)g(\mathbf{k})$, which is essentially the first term in an expansion of the deviation of f from its equilibrium value, f_0. Eqs. (49.6) and (49.4) can be combined to give

$$e[\mathbf{E} + \mathbf{v}\mathbf{X}\mathbf{B}] \frac{1}{\hbar} \nabla_\mathbf{k} f(\mathbf{k}) = \frac{(\partial f_0/\partial \mathcal{E})g(\mathbf{k})}{\tau(\mathbf{k})} \tag{49.7}$$

If Eq. (49.7) can be solved for $g(\mathbf{k})$, then expressions can be obtained for both the E_H and the magnetoresistance. Solutions can in fact be developed that are linear in the applied electric field (the regime where Ohm's law holds) for two physical situations: (i) when $\omega_c \tau \ll 1$ [Hurd, 1972, p. 69] and (ii) when $\omega_c \tau \gg 1$ [Hurd, 1972; Lifshitz *et al.*, 1956] where $\omega_c = Be/m$ is the cyclotron frequency. Situation (ii) means the electron is able to complete many cyclotron orbits under the influence of **B** in the time between scatterings and is called the high (magnetic) field limit. Conversely, situation (i) is obtained when the electron is scattered in a short time compared to the time necessary to complete one cyclotron orbit and is known as the low field limit. In effect, the solution to Eq. (49.7) is obtained by expanding $g(k)$ in a power series in $\omega_c \tau$ or $1/\omega_c \tau$ for (i) and (ii), respectively. Given $g(\mathbf{k})$ the current vector, $J_l(l = x,y,z)$ can be calculated from [Blatt, 1957]

$$J_l = \left(\frac{e}{4\pi^3}\right) \int v_l(\mathbf{k})g(\mathbf{k})\,(\partial f_0/\partial \mathcal{E})\,d^3k \tag{49.8}$$

where $v_l(\mathbf{k})$ is the velocity of the electron with wave vector **k**. Every term in the series defining J_l is linear in the applied electric field, **E**, so that the conductivity tensor σ_{lm} is readily obtained from $J_l = \sigma_{lm}E_m$ [Hurd, 1972, p. 9] This matrix equation can be inverted to give $E_l = \rho_{lm}J_m$. For the same geometry used in defining Eq. (49.1)

$$E_y = E_H = \rho_{21}J_x \tag{49.9}$$

where ρ_{21} is a component of the resistivity tensor sometimes called the Hall resistivity. Comparing Eqs. (49.1) and (49.9) it is clear that the B dependence of E_H is contained in ρ_{12}. However, nothing in the derivation of ρ_{12} excludes the possibility of terms to the second or higher powers in B. Although these are usually small, this is one of the reasons that experimentally one usually obtains R from the measured transverse voltage by reversing magnetic fields and averaging the measured E_H by calculating $(1/2)[E_H(\mathbf{B}) - E_H(-\mathbf{B})]$. This eliminates the second-order term in B and in fact all even power terms contributing to the E_H. Using the Onsager relation [Smith and Jensen, 1989, p. 60] $\rho_{12}(\mathbf{B}) = \rho_{21}(-\mathbf{B})$, it is also easy to show that in terms of the Hall resistivity

$$R = \frac{1}{2}\frac{1}{B}[\rho_{12}(\mathbf{B}) + \rho_{21}(\mathbf{B})] \tag{49.10}$$

Strictly speaking, in a single crystal the electric field resulting from an applied electric current and magnetic field, both of arbitrary direction relative to crystal axes and each other, cannot be fully described in terms of a second-order resistivity tensor [Hurd, 1972, p. 71]. On the other hand, Eqs. (49.1), (49.9), and (49.10) do define the Hall coefficient in terms of a second-order resistivity tensor for a polycrystalline (assumed isotropic) sample or for a cubic single crystal or for a lower symmetry crystal when the applied fields are oriented along major symmetry directions. In real world applications the Hall effect is always treated in this manner.

49.3 Relation to the Electronic Structure—(i) $\omega_c \tau \ll 1$

General expressions for R in terms of the parameters that describe the electronic structure can be obtained using Eqs. (49.7)–(49.10) and have been given by Blatt [Blatt, 1957] for the case of crystals having cubic symmetry. An even more general treatment has been given by McClure [McClure, 1956]. Here the discussion of specific results will be restricted to the free electron model wherein the material is assumed to have one or more conducting bands, each of which has a quadratic dispersion relationship connecting $\mathscr{E}$ and $\mathbf{k}$; that is

$$\mathscr{E}_i = \frac{\hbar^2 k_i^2}{2m_i} \tag{49.11}$$

where the subscript specifies the band number and m_i, the **effective mass** for each band. These masses need not be equal nor the same as the free electron mass. In effect, some of the features lost in the free electron approximation are recovered by allowing the masses to vary. The **relaxation times**, τ_i, will also be taken to be isotropic (not $\mathbf{k}$ dependent) within each band but can be different from band to band. Although extreme, these approximations are often qualitatively correct, particularly in polycrystalline materials, which are macroscopically isotropic. Further, in semiconductors these results will be strictly applicable only if τ_i is energy independent as well as isotropic.

For a single spherical band, R_H is a direct measure of the number of current carriers and turns out to be given by [Blatt, 1957]

$$R_H = \frac{1}{ne} \tag{49.12}$$

where n is the number of conduction carriers/volume. R_H depends on the sign of the charge of the current carriers being negative for electrons and positive for **holes**. This identification of the carrier sign is itself a matter of great importance, particularly in semiconductor physics. If more than one band is involved in electrical conduction, then by imposing the boundary condition required for the geometry of Fig. 49.1 that the total current in the y direction from all bands must vanish, $J_y = 0$, it is easy to show that [Wilson, 1958]

$$R_H = (1/\sigma)^2 \sum [\sigma_i^2 R_i] \tag{49.13}$$

where R_i and σ_i are the Hall coefficient and electrical conductivity, respectively, for the ith band ($\sigma_i = n_i e^2 \tau_i / m_i$), $\sigma = \Sigma \sigma_i$ is the total conductivity of the material, and the summation is taken over all bands. Using Eq. (49.12), Eq. (49.13) can also be written

$$R_H = \frac{1}{en_{\text{eff}}} = \frac{1}{e} \sum \left[\frac{1}{n_i} \left(\frac{\sigma_i}{\sigma} \right)^2 \right] \tag{49.14}$$

where n_{eff} is the effective or apparent number of electrons determined by a Hall effect experiment. (Note that some workers prefer representing Eqs. (49.12) and (49.13) in terms of the current carrier mobility for each band, μ_i, defined by $\sigma_i = n_i e \mu_i$.)

The most commonly used version of Eq. (49.14) is the so-called two-band model, which assumes that there are two spherical bands with one composed of electrons and the other of holes. Eq. (49.14) then takes the form

$$R_H = \frac{1}{e} \left[\frac{1}{n_e} \left(\frac{\sigma_e}{\sigma} \right)^2 - \frac{1}{n_h} \left(\frac{\sigma_h}{\sigma} \right)^2 \right] \tag{49.15}$$

From Eq. (49.14) or (49.15) it is clear that the Hall effect is dominated by the most highly conducting band. Although for fundamental reasons it is often the case that $n_e = n_h$ (a so-called compensated material), R_H would rarely vanish since the conductivities of the two bands would rarely be identical. It is also clear from any of Eqs. (49.12), (49.14), or (49.15) that, in general, the Hall effect in semiconductors will be orders of magnitude larger than that in metals.

49.4 Relation to the Electronic Structure—(ii) $\omega_c \tau \gg 1$

The high field limit can be achieved in metals only in pure, crystallographically well-ordered materials and at low temperatures, which circumstances limit the electron scattering rate from impurities, crystallographic imperfections, and lattice vibrations, respectively. In semiconductors, the much longer relaxation time and smaller effective mass of the electrons makes it much easier to achieve the high field limit. In this limit the result analogous to Eq. (49.15) is [Blatt, 1968, p. 290]

$$R_H = \frac{1}{e} \frac{1}{n_e - n_h} \tag{49.16}$$

Note that the individual band conductivities do not enter in Eq. (49.16). Eq. (49.16) is valid provided the cyclotron orbits of the electrons are closed for the particular direction of **B** used. It is not necessary that the bands be spherical or the τ's isotropic. Also, for more than two bands R_H depends only on the net difference between the number of electrons and the number of holes. For the case where $n_e = n_h$, in general, the lowest order dependence of the Hall electric field on B is B^2 and there is no simple relationship of R_H to the number of current carriers. For the special case of the two-band model, however, R_H is a constant and is of the same form as Eq. (49.15) [Fawcett, 1964].

Metals can have geometrically complicated Fermi surfaces wherein the Fermi surface contacts the Brillouin zone boundary as well as encloses the center of the zone. This leads to the possibility of open electron orbits in place of the closed cyclotron orbits for certain orientations of **B**. In these circumstances R can have a variety of dependencies on the magnitude of B and in single crystals will generally be dependent on the exact orientation of **B** relative to the crystalline axes [Hurd, 1972, p. 51; Fawcett, 1964]. R will not, however, have any simple relationship to the number of current carriers in the material.

Semiconductors have too few electrons to have open orbits but can manifest complicated behavior of their Hall coefficient as a function of the magnitude of B. This occurs because of the relative ease with which one can pass from the low field limit to the high field limit and even on to the so-called quantum limit with currently attainable magnetic fields. (The latter has not been discussed here.) In general, these different regimes of B will not occur at the same magnitude of B for all the bands in a given semiconductor, further complicating the dependence of R on B.

Defining Terms

Conducting band: The band in which the electrons primarily responsible for the electric current are found.

Effective mass: An electron in a lattice responds differently to applied fields than would a free electron or a classical particle. One can, however, often describe a particular response using classical equations by defining an effective mass whose value differs from the actual mass.

Electron band: A range or band of energies in which there is a continuum (rather than a discrete set as in, for example, the hydrogen atom) of allowed quantum mechanical states partially or fully occupied by electrons. It is the continuous nature of these states that permits them to respond almost classically to an applied electric field.

Hole or hole state: When a conducting band, which can hold two electrons/atom, is more than half full, the remaining unfilled states are called *holes*. Such a band responds to electric and magnetic fields as if it contained positively charged carriers equal in number to the number of holes in the band.

Relaxation time: The time for a distribution of particles, out of equilibrium by a measure Φ, to return exponentially toward equilibrium to Φ/e out of equilibrium when the disequilibrating fields are removed (e is the natural logarithm base).

References

L. Berger and G. Bergmann, in *The Hall Effect and Its Applications,* C. L. Chien and C. R. Westlake, Eds., New York: Plenum Press, 1980, p. 55.

F. L. Blatt in *Solid State Physics,* vol. 4, F. Seitz and D. Turnbull, Eds., New York: Academic Press, 1957, p. 199.

F. L. Blatt, *Physics of Electronic Conduction in Solids,* New York: McGraw-Hill, 1968, p. 290.

D. Emin, *Phil. Mag.,* vol. 35, p. 1189, 1977.

E. Fawcett, *Adv. Phys.* vol. 13, p. 139, 1964.

C. M. Hurd, *The Hall Effect in Metals and Alloys,* New York: Plenum Press, 1972, p. 69.

I. M. Lifshitz, M. I. Azbel, and M. I. Kaganov, *Zh. Eksp. Teor. Fiz.,* vol. 31, p. 63, 1956 [*Soviet Phys. JETP* (Engl. Trans.), vol. 4, p. 41, 1956].

J. W. McClure, *Phys. Rev.,* vol. 101, p. 1642, 1956.

R. E. Prange and S. M. Girvin, Eds., *The Quantum Hall Effect,* New York: Springer-Verlag, 1990.

H. Smith, and H. H. Jensen, *Transport Phenomena,* Oxford: Oxford University Press, 1989, p. 60.

A. H. Wilson, *The Theory of Metals,* London: Cambridge University Press, 1958, p. 212.

J. M. Ziman, *Electrons and Phonons,* London: Oxford University Press, 1960.

Further Information

In addition to the texts and review article cited in the references, an older but still valid article by J. P. Jan, in *Solid State Physics* (edited by F. Seitz and D. Turnbull, New York: Academic Press, 1957, p. 1) can provide a background in the various thermomagnetic and galvanomagnetic properties in metals. A parallel background for semiconductors can be found in the monograph by E. H. Putley, *The Hall Effect and Related Phenomena* (Boston: Butterworths, 1960).

Examples of applications of the Hall effect can be found in the book *Hall Generators and Magnetoresistors,* by H. H. Wieder, edited by H. J. Goldsmid (London: Pion Limited, 1971).

An index to the most recent work on or using any aspect of the Hall effect reported in the major technical journals can be found in *Physics Abstracts* (Science Abstracts Series A).

CARLSON'S DRY PRINTER

Before the Xerox Copier

As a poor, hardworking boy, who at the age of 12 assured his cousin that "someday I'm going to make a big invention," Carlson epitomizes the increasingly rare breed of the lone inventor who struggles for years to develop an idea and then spends even more years persuading someone to back it.

As usually told, however, the birth of xerography ends with the creation of that famous first image: 10-22-38 ASTORIA [see photo inset]. This image commemorated xerography's birth date and place, a small room in Astoria, N.Y., above a bar and grill and next door to a beauty parlor. The message was written in india ink on a 5-by-7.5-centimeter glass slide, transferred to a zinc plate dusted with sulfur and an electrically charged developing powder known as lycopodium, and finally "fixed" on heated wax paper.

Carlson was aided in his experiments by an Austrian refugee and electrical engineer, Otto Kornei, who on Oct. 1, 1938, began working for him part time—all Carlson could afford. And how three years before, when Carlson was 29 and working in P.R. Mallory's patent department, he had thought "how nice it would be if one could have a machine in the office to which one could take a document or a letter and put it in a slot and push the button and get a copy of it."

Such a machine was in fact conceived on that famous October Saturday, according to Carlson's notebooks, which are now in the New York Public Library. One of the entries made on Oct. 22 was a sketch of a Continuous Process Electro-

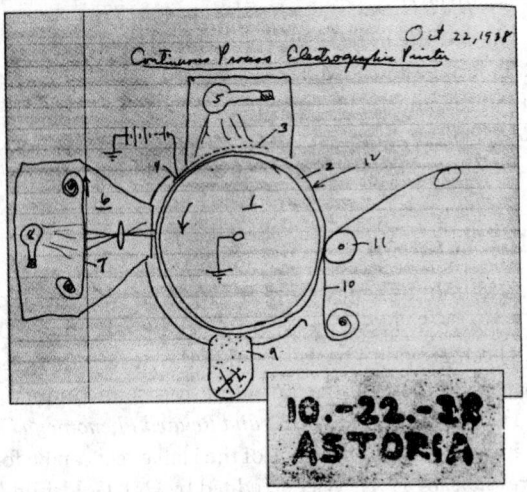

This Continuous Process Electrographic Printer, which Chester Carlson sketched in his notebook on Oct. 22, 1938, was a precursor of the first automatic office copier, introduced by Haloid-Xerox in 1959. On the same October 1938 day Chester Carlson and his assistant Otto Kornei made their first xerographic image (inset).

graphic Printer [see illustration]. This was a precursor of the Xerox 914, the first production-line automatic office copier, introduced in 1959.

In the notebook Carlson explained his Oct. 22 drawing as showing a sulfur-coated metal cylinder charged from two circular electrodes (3 and 4). An "exposing machine" (6) moved an image from microfilm (7) onto the sulfur layer where it was "developed" by lycopodium or a similar powder in a chamber (9). The powder image was then to be transferred to treated paper (10) passing under a roller (11).

On Dec. 3, 1938, Carlson sketched a more sophisticated paper feed. Then, on Jan. 14, 1939, he devoted five notebook pages to the advantages and disadvantages of making offset lithography masters with his process and with the conventional Multilith duplicating process. Carlson, who as a boy had kept a notebook of every penny he spent, even compared the cost of the services of the expert required to prepare the special sheets used on Multilith machines ($0.60 for 35 minutes) with the cost of the ordinary office worker able to use electrophotography ($0.06 for 5 minutes). He concluded his process would halve copying costs.

Carlson now realized he would need a working model to demonstrate to potential corporate sponsors. Without Kornei, however, who left him in 1939 for more remunerative employment, he was unable to get one built successfully. Instead, armed only with a demonstration kit and the faith that his invention would go of its own momentum "once it got started," he set forth to interest America's largest business-equipment firms.

Carlson approached 20 companies during the next several years, including IBM Corp. and Eastman Kodak Co., as well as the U.S. Army and the National Inventors Council. But not until he met a Battelle Memorial Institute engineer in 1944 did he arouse any real interest.

After that it took four years of intense scientific and engineering work by Battelle and a little Rochester, N.Y. firm, The Haloid Co., before xerography was ready for its unveiling on Oct. 22, 1948, at the Optical Society of America's annual meeting.

Another decade passed before the now renamed Haloid-Xerox (later Xerox Corp.) could deliver a practical, rotary-drum automatic copier for the office, as Carlson had envisioned 22 years earlier.

In all this there was a lesson for inventors everywhere. The incredibly persistent Chester Carlson summed it up in a 1964 lecture: "First of all you find a need that isn't supplied and then you immerse yourself completely in it for ten years or so; and just keep working and apply everything you see and read to the problem at hand and perhaps you will succeed."

Source: Adapted from M.F. Wolff, *IEEE Spectrum*, p. 44, December 1989. © 1989 IEEE.

50

Superconductivity

Kevin A. Delin
Massachusetts Institute of Technology

Terry P. Orlando
Massachusetts Institute of Technology

50.1 Introduction 1114
50.2 General Electromagnetic Properties 1115
50.3 Superconducting Electronics 1118
50.4 Types of Superconductors 1120

50.1 Introduction

The fundamental idea behind all of a superconductor's unique properties is that **superconductivity** is a quantum mechanical phenomenon on a macroscopic scale created when the motions of individual electrons are correlated. According to the theory developed by John Bardeen, Leon Cooper, and Robert Schrieffer (BCS theory), this correlation takes place when two electrons couple to form a Cooper pair. For our purposes, we may therefore consider the electrical charge carriers in a superconductor to be Cooper pairs (or more colloquially, superelectrons) with a mass m^* and charge q^* twice those of normal electrons. The average distance between the two electrons in a Cooper pair is known as the coherence length, ξ. Both the coherence length and the binding energy of two electrons in a Cooper pair, 2Δ, depend upon the particular superconducting material. Typically, the coherence length is many times larger than the interatomic spacing of a solid, and so we should not think of Cooper pairs as tightly bound electron molecules. Instead, there are many other electrons between those of a specific Cooper pair allowing for the paired electrons to change partners on a time scale of $h/(2\Delta)$ where h is Planck's constant.

If we prevent the Cooper pairs from forming by ensuring that all the electrons are at an energy greater than the binding energy, we can destroy the superconducting phenomenon. This can be accomplished, for example, with thermal energy. In fact, according to the BCS theory, the critical temperature, T_c, associated with this energy is

$$\frac{2\Delta}{k_B T_c} \approx 3.5 \tag{50.1}$$

where k_B is Boltzmann's constant. For low critical temperature (conventional) superconductors, 2Δ is typically on the order of 1 meV, and we see that these materials must be kept below temperatures of about 10 K to exhibit their unique behavior. High critical temperature superconductors, in contrast, will superconduct up to temperatures of about 100 K, which is attractive from a practical view because the materials can be cooled cheaply using liquid nitrogen. A second way of increasing the energy of the electrons is electrically driving them. In other words, if the critical current density, J_c, of a superconductor is exceeded, the electrons have sufficient kinetic energy to prevent the formation of Cooper pairs. The necessary kinetic energy can also be generated through the induced currents created by an external magnetic field. As a result, if a superconductor is placed in a magnetic field larger than its critical field, H_c, it will return to its normal metallic state. To summarize, a superconductor must be maintained under the appropriate temperature, electrical current density, and magnetic field conditions to exhibit its special properties. An example of this phase space is shown in Fig. 50.1.

50.2 General Electromagnetic Properties

The hallmark electromagnetic properties of a superconductor are its ability to carry a static current without any resistance and its ability to exclude a static magnetic flux from its interior. It is this second property, known as the Meissner effect, that distinguishes a superconductor from merely being a perfect conductor (which conserves the magnetic flux in its interior). Although superconductivity is a manifestly quantum mechanical phenomenon, a useful classical model can be constructed around these two properties. In this section, we will outline the rationale for this classical model, which is useful in engineering applications such as waveguides and high-field magnets.

The zero dc resistance criterion implies that the superelectrons move unimpeded. The electromagnetic energy density, w, stored in a superconductor is therefore

$$w = \frac{1}{2}\varepsilon \mathbf{E}^2 + \frac{1}{2}\mu_o \mathbf{H}^2 + \frac{n^*}{2}m^*\mathbf{v}_s^2 \tag{50.2}$$

where the first two terms are the familiar electric and magnetic energy densities, respectively. (Our electromagnetic notation is standard: ε is the permittivity, μ_o is the permeability, $\mathbf{E}$ is the electric field, and the magnetic flux density, $\mathbf{B}$, is related to the magnetic field, $\mathbf{H}$, via the constitutive law $\mathbf{B} = \mu_o\mathbf{H}$.) The last term represents the kinetic energy associated with the undamped superelectrons' motion (n^* and $\mathbf{v}_s$ are the superelectrons' density and velocity, respectively). Because the supercurrent density, $\mathbf{J}_s$, is related to the superelectron velocity by $\mathbf{J}_s = n^*q^*\mathbf{v}_s$, the kinetic energy term can be rewritten

$$n^*\left(\frac{1}{2}m^*\mathbf{v}_s^2\right) = \frac{1}{2}\Lambda \mathbf{J}_s^2 \tag{50.3}$$

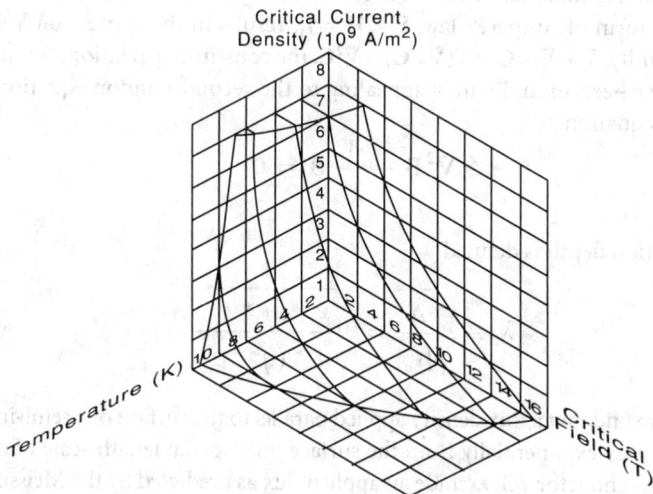

Critical Current
Density (10^9 A/m^2)

Temperature (K)

Critical Field (T)

FIGURE 50.1 The phase space for the superconducting alloy niobium–titanium. The material is superconducting inside the volume of phase space indicated. (*Source:* T.P. Orlando and K.A. Delin, Foundations of Applied Superconductivity, Reading, Mass.: Addison-Wesley, 1991, p.10. With permission. [As adapted from Wilson, 1983.])

where Λ is defined as

$$\Lambda = \frac{m^*}{n^* (q^*)^2} \tag{50.4}$$

Assuming that all the charge carriers are superelectrons, there is no power dissipation inside the superconductor, and so Poynting's theorem over a volume V may be written

$$-\int_V \nabla \cdot (\mathbf{E} \times \mathbf{H}) \, dv = \int_V \frac{\partial w}{\partial t} \, dv \tag{50.5}$$

where the left side of the expression is the power flowing into the region. By taking the time derivative of the energy density and appealing to Faraday's and Ampère's laws to find the time derivatives of the field quantities, we find that the only way for Poynting's theorem to be satisfied is if

$$\mathbf{E} = \frac{\partial}{\partial t} (\Lambda \mathbf{J}_s) \tag{50.6}$$

This relation, known as the first London equation (after the London brothers, Heinz and Fritz), is thus necessary if the superelectrons have no resistance to their motion.

Equation (50.6) also reveals that the superelectrons' inertia creates a lag between their motion and that of the electric field. As a result, a superconductor can support a time-varying voltage drop across itself. The impedance associated with the supercurrent, therefore, is an inductor, and it will be useful to think of Λ as an inductance created by the correlated motion of the superelectrons.

If the first London equation is substituted into Faraday's law, $\nabla \times \mathbf{E} = -(\partial \mathbf{B}/\partial t)$, and integrated with respect to time, the second London equation results:

$$\nabla \times (\Lambda \mathbf{J}_s) = -\mathbf{B} \tag{50.7}$$

where the constant of integration has been defined to be zero. This choice is made so that the second London equation is consistent with the Meissner effect as we now demonstrate. Taking the curl of the quasi-static form of Ampère's law, $\nabla \times \mathbf{H} = \mathbf{J}_s$, results in the expression $\nabla^2 \mathbf{B} = -\mu_o \nabla \times \mathbf{J}_s$, where a vector identity, $\nabla \times \nabla \times \mathbf{C} = \nabla(\nabla \cdot \mathbf{C}) - \nabla^2 \mathbf{C}$; the constitutive relation, $\mathbf{B} = \mu_o \mathbf{H}$; and Gauss's law, $\nabla \cdot \mathbf{B} = 0$, have been used. By now appealing to the second London equation, we obtain the vector Helmholtz equation

$$\nabla^2 \mathbf{B} - \frac{1}{\lambda^2} \mathbf{B} = 0 \tag{50.8}$$

where the penetration depth is defined

$$\lambda \equiv \sqrt{\frac{\Lambda}{\mu_o}} = \sqrt{\frac{m^*}{n^* (q^*)^2 \mu_o}} \tag{50.9}$$

From Eq. (50.8), we find that a flux density applied parallel to the surface of a semi-infinite superconductor will decay away exponentially from the surface on a spatial length scale of order λ. In other words, a bulk superconductor will exclude an applied flux as predicted by the Meissner effect.

The London equations reveal that there is a characteristic length λ over which electromagnetic fields can change inside a superconductor. This penetration depth is different from the more familiar skin depth of electromagnetic theory, the latter being a frequency-dependent quantity. Indeed, the penetration depth at zero temperature is a distinct material property of a particular superconductor.

Notice that λ is sensitive to the number of correlated electrons (the superelectrons) in the material. As previously discussed, this number is a function of temperature and so only at $T = 0$ do *all* the electrons that usually conduct ohmically participate in the Cooper pairing. For intermediate temperatures, $0 < T < T_c$, there are actually two sets of interpenetrating electron fluids: the uncorrelated electrons providing ohmic conduction and the correlated ones creating supercurrents. This two-fluid model is a useful way to build temperature effects into the London relations.

Under the two-fluid model, the electrical current density, J, is carried by both the uncorrelated (normal) electrons and the superelectrons: $J = J_n + J_s$ where J_n is the normal current density. The two channels are modeled in a circuit as shown in Fig. 50.2 by a parallel combination of a resistor (representing the ohmic channel) and an inductor (representing the superconducting channel). To a good approximation, the respective temperature dependences of the conductor and inductor are

$$\tilde{\sigma}_o(T) = \sigma_o(T_c)\left(\frac{T}{T_c}\right)^4 \qquad \text{for } T \leq T_c \qquad (50.10)$$

and

$$\Lambda(T) = \Lambda(0)\left(\frac{1}{1 - (T/T_c)^4}\right) \qquad \text{for } T \leq T_c \qquad (50.11)$$

where σ_o is the dc conductance of the normal channel. (Strictly speaking, the normal channel should also contain an inductance representing the inertia of the normal electrons, but typically such an inductor contributes negligibly to the overall electrical response.) Since the temperature-dependent penetration depth is defined as $\lambda(T) = \sqrt{\Lambda(T)/\mu_o}$, the effective conductance of a superconductor in the sinusoidal steady state is

$$\sigma = \tilde{\sigma}_o + \frac{1}{j\omega\mu_o\lambda^2} \qquad (50.12)$$

where the explicit temperature dependence notation has been suppressed.

Most of the important physics associated with the classical model is embedded in Eq. (50.12). As is clear from the lumped element model, the relative importance of the normal and superconducting channels is a function not only of temperature but also of frequency. The familiar L/R time constant, here equal to $\Lambda\tilde{\sigma}_o$, delineates the frequency regimes where most of the total current is carried by J_n (if $\omega\Lambda\tilde{\sigma}_o \gg 1$) or J_s (if $\omega\Lambda\tilde{\sigma}_o \ll 1$). This same result can also be obtained by comparing the skin depth associated with the normal channel, $\delta = \sqrt{2/(\omega\mu_o\tilde{\sigma}_o)}$, to the penetration depth to see which channel provides more field screening. In addition, it is straightforward to use Eq. (50.12) to rederive Poynting's theorem for systems that involve superconducting materials:

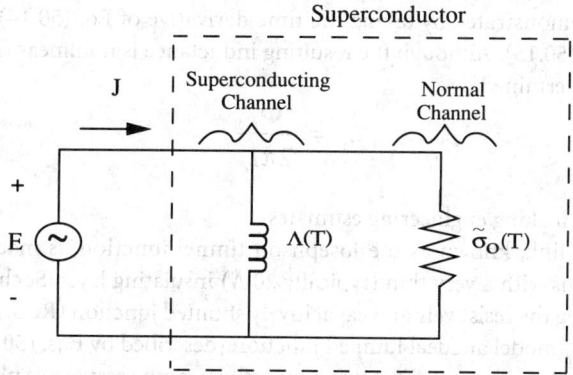

FIGURE 50.2 A lumped element model of a superconductor.

$$-\int_V \nabla \cdot (\mathbf{E} \times \mathbf{H})\, dv = \frac{d}{dt} \int_V \left(\frac{1}{2} \varepsilon \mathbf{E}^2 + \frac{1}{2} \mu_o \mathbf{H}^2 + \frac{1}{2} \Lambda(T) J_s^2 \right) dv$$

$$+ \int_V \frac{1}{\tilde{\sigma}_o(T)} J_n^2 dv \tag{50.13}$$

Using this expression, it is possible to apply the usual electromagnetic analysis to find the inductance (L_o), capacitance (C_o), and resistance (R_o) per unit length along a parallel plate transmission line. The results of such analysis for typical cases are summarized in Table 50.1.

50.3 Superconducting Electronics

The macroscopic quantum nature of superconductivity can be usefully exploited to create a new type of electronic device. Because all the superelectrons exhibit correlated motion, the usual wave–particle duality normally associated with a single quantum particle can now be applied to the entire ensemble of superelectrons. Thus, there is a spatiotemporal phase associated with the ensemble that characterizes the supercurrent flowing in the material.

Naturally, if the overall electron correlation is broken, this phase is lost and the material is no longer a superconductor. There is a broad class of structures, however, known as weak links, where the correlation is merely perturbed locally in space rather than outright destroyed. Coloquially, we say that the phase "slips" across the weak link to acknowledge the perturbation.

The unusual properties of this phase slippage were first investigated by Brian Josephson and constitute the central principles behind superconducting electronics. Josephson found that the phase slippage could be defined as the difference between the macroscopic phases on either side of the weak link. This phase difference, denoted as ϕ, determined the supercurrent, i_s, through and voltage, v, across the weak link according to the Josephson equations,

$$i_s = I_c \sin \phi \tag{50.14}$$

and

$$v = \frac{\Phi_o}{2\pi} \frac{\partial \phi}{\partial t} \tag{50.15}$$

where I_c is the critical (maximum) current of the junction and Φ_o is the quantum unit of flux. (The flux quantum has a precise definition in terms of Planck's constant, h, and the electron charge, e: $\Phi_o \equiv h/(2e) \approx 2.068 \times 10^{-15}$ Wb). As in the previous section, the correlated motion of the electrons, here represented by the superelectron phase, manifests itself through an inductance. This is straightforwardly demonstrated by taking the time derivative of Eq. (50.14) and combining this expression with Eq. (50.15). Although the resulting inductance is nonlinear (it depends on $\cos \phi$), its relative scale is determined by

$$L_j = \frac{\Phi_o}{2\pi I_c} \tag{50.16}$$

a useful quantity for making engineering estimates.

A common weak link, known as the Josephson tunnel junction, is made by separating two superconducting films with a very thin (typically 20 Å) insulating layer. Such a structure is conveniently analyzed using the resistively and capacitively shunted junction (RCSJ) model shown in Fig. 50.3. Under the RCSJ model an ideal lumped junction [described by Eqs. (50.14) and (50.15)] and a resistor R_j represent how the weak link structure influences the respective phases of the super and normal electrons, and a capacitor C_j represents the physical capacitance of the sandwich structure.

Table 50.1 Lumped Circuit Element Parameters Per Unit Length
for Typical Transverse Electromagnetic Parallel Plate Waveguides*

Transmission Line Geometry	L_o	C_o	R_o
Two identical, thin ($\lambda \gg b$) superconducting plates	$\dfrac{\mu_t h}{d} + \dfrac{2\mu_o \lambda^2}{db}$	$\dfrac{\varepsilon_t d}{h}$	$\dfrac{8}{db\tilde{\sigma}_o}\left(\dfrac{\lambda}{\delta}\right)^4$
Two identical, thick ($\lambda \ll b$) superconducting plates	$\dfrac{\mu_t h}{d} + \dfrac{2\mu_o \lambda}{d}$	$\dfrac{\varepsilon_t d}{h}$	$\dfrac{4}{d\delta\tilde{\sigma}_o}\left(\dfrac{\lambda}{\delta}\right)^3$
One thick ($\lambda \ll b$) superconducting plate and one thick ($\lambda \ll b$) ohmic plate	$\dfrac{\mu_t h}{d} + \dfrac{\mu_o \lambda}{d} + \dfrac{\mu_n \delta_n}{2d}$	$\dfrac{\varepsilon_t d}{h}$	$\dfrac{1}{d\delta_n \sigma_{o,n}}$

*The subscript n refers to parameters associated with a normal (ohmic) plate. Using these expressions, line input impedance, attenuation, and wave velocity can be calculated.

Source: T.P. Orlando and K.A. Delin, *Foundations of Applied Superconductivity*, Reading, Mass.: Addison-Wesley, 1991, p. 171. With permission.

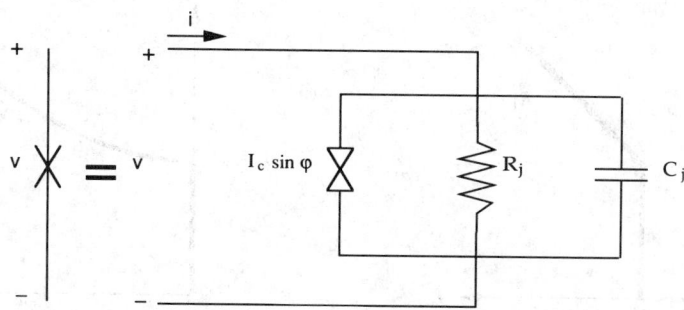

FIGURE 50.3 A real Josephson tunnel junction can be modeled using ideal lumped circuit elements.

If the ideal lumped junction portion of the circuit is treated as an inductor-like element, many Josephson tunnel junction properties can be calculated with the familiar circuit time constants associated with the model. For example, the quality factor Q of the RCSJ circuit can be expressed as

$$Q^2 = \frac{R_j C_j}{L_j / R_j} = \frac{2\pi I_c R_j^2 C_j}{\Phi_o} \equiv \beta \tag{50.17}$$

where β is known as the Stewart-McCumber parameter. Clearly, if $\beta \gg 1$ the ideal lumped junction element is underdamped in that the capacitor readily charges up, dominates the overall response of the circuit, and therefore creates a hysteretic *i-v* curve as shown in Fig. 50.4. In the case when the bias current is raised from zero, no time-averaged voltage is created until the critical current is exceeded. At this point, the junction switches to the voltage $2\Delta/e$ with a time constant $\sqrt{L_j C_j}$. Once the junction has latched into the voltage state, however, the bias current must be lowered to zero before it can again be steered through the superconducting path. Conversely, $\beta \ll 1$ implies that the L_j/R_j time constant dominates the circuit response, so that the capacitor does not charge up and the *i-v* curve is not hysteretic [Fig. 50.4(b)].

Just as bulk superconductors are influenced by the presence of a magnetic field, so too are weak links. This fact has been used to make extremely sensitive magnetometers known as superconducting quantum interference devices (SQUIDs). A SQUID is simply two weak links connected in parallel and biased with some current. Because the phase differences in the weak links are related to the flux, Φ, threading the loop between them by

$$\phi_2 - \phi_1 = (2\pi\Phi)/\Phi_o \tag{50.18}$$

it is possible to measure flux with a sensitivity on the order of the flux quantum.

50.4 Types of Superconductors

The macroscopic quantum nature of superconductivity also affects the general electromagnetic properties previously discussed. This is most clearly illustrated by the interplay of the characteristic lengths ξ, representing the scale of quantum correlations, and λ, representing the scale of electromagnetic screening. Consider the scenario where a magnetic field, H, is applied parallel to the surface of a semi-infinite superconductor. The correlations of the electrons in the superconductor must lower the overall energy of the system or else the material would not be superconducting in the first place. Because the critical magnetic field H_c destroys all the correlations, it is convenient to define the energy density gained by the system in the superconducting state as $(\frac{1}{2})\mu_o H_c^2$. The elec-

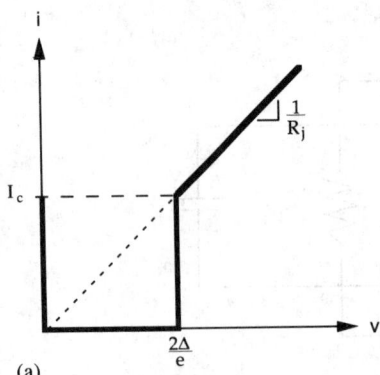

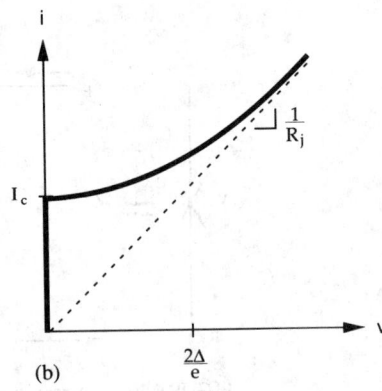

(a)   (b)

FIGURE 50.4   The *i-v* curves for a Josephson junction: (a) $\beta \gg 1$, and (b) $\beta \ll 1$.

trons in a Cooper pair are separated on a length scale of ξ, however, and so the correlations cannot be fully achieved until a distance roughly ξ from the boundary of the superconductor. There is thus an energy per unit area, $(1/2)\mu_o H_c^2 \xi$, that is lost because of the presence of the boundary. Now consider the effects of the applied magnetic field on this system. It costs the superconductor energy to maintain the Meissner effect, $\mathbf{B} = 0$, in its bulk; in fact the energy density required is $(1/2)\mu_o H^2$. However, since the field can penetrate the superconductor a distance roughly λ, the system need not expend an energy per unit area of $(1/2)\mu_o H^2 \lambda$ to screen over this volume. To summarize, more than a distance ξ from the boundary, the energy of the material is lowered (because it is superconducting), and more than a distance λ from the boundary the energy of the material is raised (to shield the applied field).

Now, if $\lambda < \xi$, the region of superconducting material greater than λ from the boundary but less than ξ will be higher in energy than that in the bulk of the material. Thus, the surface energy of the boundary is positive and so costs the total system some energy. This class of superconductors is known as type I. Most elemental superconductors, such as aluminum, tin, and lead, are type I. In addition to having $\lambda < \xi$, type I superconductors are generally characterized by low critical temperatures (~ 5 K) and critical fields (~ 0.05 T). Typical type I superconductors and their properties are listed in Table 50.2.

Conversely, if $\lambda > \xi$, the surface energy associated with the boundary is negative and lowers the total system energy. It is therefore thermodynamically favorable for a normal–superconducting interface to form inside these type II materials. Consequently, this class of superconductors does not exhibit the simple Meissner effect as do type I materials. Instead, there are now two critical fields: for applied fields below the lower critical field, H_{c1}, a type II superconductor is in the Meissner state, and for applied fields greater than the upper critical field, H_{c2}, superconductivity is destroyed. The three critical field are related to each other by $H_c \approx \sqrt{H_{c1}H_{c2}}$.

In the range $H_{c1} < H < H_{c2}$, a type II superconductor is said to be in the vortex state because now the applied field can enter the bulk superconductor. Because flux exists in the material, however, the superconductivity is destroyed locally, creating normal regions. Recall that for type II materials the boundary between the normal and superconducting regions lowers the overall energy of the system. Therefore, the flux in the superconductor creates as many normal–superconducting interfaces as possible without violating quantum criteria. The net result is that flux enters a type II superconductor in quantized bundles of magnitude Φ_o known as vortices or fluxons (the former name derives from the fact that current flows around each quantized bundle in the same manner as a fluid vortex circulates around a drain). The central portion of a vortex, known as the core, is a normal region with an approximate radius of ξ. If a defect-free superconductor is placed in a magnetic field, the individual vortices, whose cores essentially follow the local average field lines, form an ordered triangular array, or flux lattice. As the applied field is raised beyond H_{c1} (where the first vortex enters the superconductor), the distance between adjacent vortex cores decreases to maintain the appropriate flux density in the material. Finally, the upper critical field is reached when the

Table 50.2 Material Parameters for Type I Superconductors*

Material	T_c (K)	λ_o (nm)	ξ_o (nm)	Δ_o (meV)	$\mu_0 H_{co}$ (mT)
Al	1.18	50	1600	0.18	10.5
In	3.41	65	360	0.54	23.0
Sn	3.72	50	230	0.59	30.5
Pb	7.20	40	90	1.35	80.0
Nb	9.25	85	40	1.50	198.0

*The penetration depth λ_o is given at zero temperature, as are the coherence length ξ_o, the thermodynamic critical field H_{co}, and the energy gap Δ_o.

Source: R.J. Donnelly, "Cryogenics," in *Physics Vade Mecum*, H.L. Anderson, Ed., New York: American Institute of Physics, 1981. With permission.

normal cores overlap and the material is no longer superconducting. Indeed, a precise calculation of H_{c2} using the phenomenological theory developed by Vitaly Ginzburg and Lev Landau yields

$$H_{c2} = \frac{\Phi_o}{2\pi\mu_o\xi^2} \tag{50.19}$$

which verifies our simple picture. The values of typical type II material parameters are listed in Tables 50.3 and 50.4.

Type II superconductors are of great technical importance because typical H_{c2} values are at least an order of magnitude greater than the typical H_c values of type I materials. It is therefore possible to use type II materials to make high-field magnet wire. Unfortunately, when current is applied to the wire, there is a Lorentz-like force on the vortices, causing them to move. Because the moving vortices carry flux, their motion creates a static voltage drop along the superconducting wire by Faraday's law. As a result, the wire no longer has a zero dc resistance, even though the material is still superconducting. To fix this problem, type II superconductors are usually fabricated with intentional defects, such as impurities or grain boundaries, in their crystalline structure to pin the vortices and prevent vortex motion. The pinning is created because the defect locally weakens the superconductivity in the material, and it is thus energetically favorable for the normal core of the vortex to overlap the nonsuperconducting region in the material. Critical current densities usually

Table 50.3   Material Parameters for Conventional Type II Superconductors*

Material	T_c(K)	$\lambda_{GL}(0)$ (nm)	$\xi_{GL}(0)$ (nm)	Δ_o (meV)	$\mu_0 H_{c2,o}$ (T)
Pb-In	7.0	150	30	1.2	0.2
Pb-Bi	8.3	200	20	1.7	0.5
Nb-Ti	9.5	300	4	1.5	13
Nb-N	16	200	5	2.4	15
$PbMo_6S_8$	15	200	2	2.4	60
V_3Ga	15	90	2–3	2.3	23
V_3Si	16	60	3	2.3	20
Nb_3Sn	18	65	3	3.4	23
Nb_3Ge	23	90	3	3.7	38

*The values are only representative because the parameters for alloys and compounds depend on how the material is fabricated. The penetration depth $\lambda_{GL}(0)$ is given as the coefficient of the Ginzburg–Landau temperature dependence as $\lambda_{GL}(T) = \lambda_{GL}(0)(1 - T/T_c)^{-1/2}$; likewise for the coherence length where $\xi_{GL}(T) = \xi_{GL}(0)(1 - T/T_c)^{-1/2}$. The upper critical field $H_{c2,o}$ is given at zero temperature as well as the energy gap Δ_o.

Source: R.J. Donnelly, "Cryogenics," in *Physics Vade Mecum*, H.L. Anderson, Ed., New York: American Institute of Physics, 1981. With permission.

Table 50.4   Material Parameters for High-Temperature Type II Superconductors[†]

Material	T_c(K)	$\lambda_{a,b}$(nm)	λ_c(nm)	$\xi_{a,b}$(nm)	ξ_c(nm)
$YBa_2Cu_3O_7$	95	150	750	2	0.3
$Bi_2Sr_2CaCu_2O_8$	85	300	1500	3	0.2
$Bi_2Sr_2Ca_2Cu_3O_{10}$	110				
$Tl_2Ba_2CaCu_2O_8$	108				
$Tl_2Ba_2Ca_2Cu_3O_{10}$	125	200	1000		

[†]These values are only approximate because the parameters for high-temperature superconductors have not all been well established. The penetration depth and coherence length are estimated at zero temperature. However, since these materials are anisotropic, these lengths are given along the principal axes. The directions $\hat{a}$ and $\hat{b}$ are taken to lie in the plane of the Cu-O planes and $\hat{c}$ is taken to be perpendicular to that plane. An anisotropy of 5 was used, but this factor can be higher depending on the material preparation.

quoted for practical type II materials, therefore, really represent the depinning critical current density where the Lorentz-like force can overcome the pinning force. (The depinning critical current density should not be confused with the depairing critical current density, which represents the current when the Cooper pairs have enough kinetic energy to overcome their correlation. The depinning critical current density is typically an order of magnitude less than the depairing critical current density, the latter of which represents the theoretical maximum for J_c.)

By careful manufacturing, it is possible to make superconducting wire with tremendous amounts of current-carrying capacity. For example, standard copper wire used in homes will carry about 10^7 A/m^2, whereas a practical type II superconductor like niobium–titanium can carry current densities of 10^{10} A/m^2 or higher even in fields of several teslas. This property, more than a zero dc resistance, is what makes superconducting wire so desirable.

Defining Terms

Superconductivity: A state of matter whereby the correlation of conduction electrons allows a static current to pass without resistance and a static magnetic flux to be excluded from the bulk of the material.

References

R. J. Donnelly, "Cryogenics," in *Physics Vade Mecum*, H.L. Anderson, Ed., New York: American Institute of Physics, 1981.

S. Foner and B. B. Schwartz, *Superconducting Machines and Devices*, New York: Plenum Press, 1974.

S. Foner and B. B. Schwartz, *Superconducting Materials Science*, New York: Plenum Press, 1981.

T. P. Orlando and K. A. Delin, *Foundations of Applied Superconductivity*, Reading, Mass.: Addison-Wesley, 1991.

S. T. Ruggiero and D. A. Rudman, *Superconducting Devices*, Boston: Academic Press, 1990.

B. B. Schwartz and S. Foner, *Superconducting Applications: SQUIDs and Machines*, New York: Plenum Press, 1977.

T. Van Duzer and C. W. Turner, *Principles of Superconductive Devices and Circuits*, New York: Elsevier North Holland, 1981.

M. N. Wilson, *Superconducting Magnets*, Oxford: Oxford University Press, 1983.

Further Information

Every two years an Applied Superconductivity Conference is held devoted to practical technological issues. The proceedings of these conferences have been published every other year from 1977 to 1991 in the *IEEE Transactions on Magnetics*.

In 1991, the *IEEE Transactions on Applied Superconductivity* began publication. This quarterly journal focuses on both the science and the technology of superconductors and their applications, including materials issues, analog and digital circuits, and power systems. The proceedings of the Applied Superconductivity Conference now appear in this journal.

JAN. 28, 1958: A LASER IS BORN

Only a few decades old, the laser—a device for light amplification by stimulated emission of radiation—has revolutionized many industries. Its predecessor was the maser, which dealt in microwave amplification and so emitted at longer wavelengths than light.

The maser was conceived in 1951 by Charles H. Townes, then a professor at Columbia University, New York City, and first demonstrated by him in 1952. The steps that led from the maser to the laser owed much to his working relationship with one of his research associates at Columbia, Arthur Leonard Schawlow, a Canadian physicist who was studying the application of microwave spectroscopy to organic chemistry. Though Schawlow did not work on the maser, he kept an eye on it through his close link with Townes, with whom he was writing a book on microwave spectroscopy. Later, at Bell Telephone Laboratories in Murray Hill, NJ, Schawlow had begun by the summer of 1957 to think of developing a concentrated beam of far-infrared (not visible) light.

By this time, Townes was a consultant for Bell Labs, often conferring with Schawlow. It was at one of their meetings that the two first seriously debated how to develop some kind of infrared or even visible-light laser.

Schawlow opted to excite the atoms of potassium vapor, because potassium was the only substance he knew to have the first and second lines of its spectrum in the visible range.

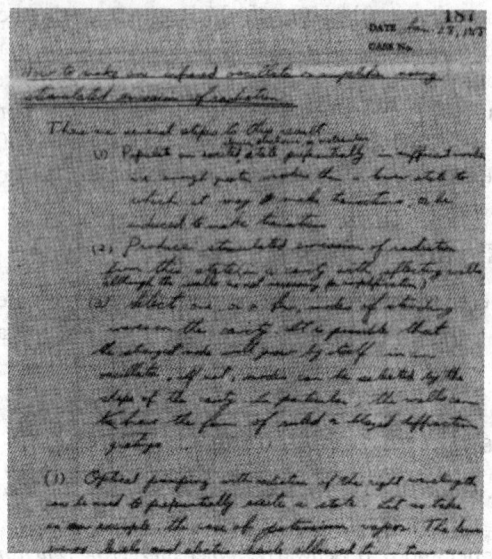

Arthur Schawlow's notebook entry dated Jan. 28, 1958, describes how to induce mode selection in a resonator.

Mode selection proved one of the biggest hurdles. It involved designing the resonator cavity so that it would reinforce only the waves of the desired frequency. Townes downplayed the problem, believing that once the atoms got excited, one or more modes would dominate over the others. Schawlow, though, felt just one mode should be selected because otherwise the light might emerge in all directions and over the whole bandwidth of the amplifying spectral line.

Of the many schemes Schawlow concocted to select a mode, one of particular interest is his notebook entry on Jan. 28, 1958. The entry describes how mode selection could be achieved through the shape of the resonator's cavity. It underscores the finding that mode selectivity could be increased by using diffraction gratings for the resonator walls (this was done nine years later in the first wavelength-tunable dye laser, built by Bernard H. Soffer and B.B. McFarland at Korad). Because the resonator cavity also provides stronger coupling between the light waves and atoms, the effects of stimulated emission are enhanced, the entry adds.

Schawlow later realized that to select one mode, all that was necessary was to select a direction for the waves. The two mirrors on opposite ends of the resonator would then cause the waves to go back and forth a number of times and generate a high mode selectivity. So he got rid of most of the resonator, keeping only its two ends.

At this point, Schawlow and Townes believed they had thoroughly analyzed the theory behind the laser and presented their ideas in their paper, "Infrared and Optical Masers," which was published in *Physical Review* (Vol. 112, 1958, pp. 1940–49).

In 1958 Schawlow discovered that the satellite lines at 701 and 704 nm, which accompany the R (red) line in the spectrum of concentrated ruby, were caused by the presence of chromium ion pairs. He also found that their lower energy levels were enough above the ground level that ions excited to those levels could be made to fall to a lower level by cryogenic cooling. With these lower levels essentially empty, he suggested there would be no absorption at these wavelengths, and so any ion pairs excited to the upper state of either satellite line would produce some amplification by stimulated emission—in this case by cooling. (In contrast, the R lines start out with nearly all of the chromium ions in the lower energy level, so that more than half of them would have to be excited before any optical amplification could begin.) An attempt with a low-power flash lamp failed.

After another pioneer in the field, Theodore H. Maiman, a researcher at the Hughes Research Laboratories, then in Culver City, Calif., demonstrated the first ruby laser on May 16, 1960, Schawlow attempted his cold ruby experiment with a bigger lamp.

That proved to be the solution. As Schawlow had predicted, he and co-worker George Devlin found that, at liquid-nitrogen temperature or below, these two satellite lines did indeed have a lower threshold for laser action than the R line in a ruby crystal with 0.5 percent chromium. The optical maser that Schawlow had analyzed in 1958 finally worked.

On March 22, 1960, the two were issued U.S. patent No. 2 929 922. Soon after, several companies produced the first lasers. For their contributions, Townes shared the Nobel Prize in Physics in 1969, and Schawlow shared it in 1981.

Source: Adapted from G. Likourezos, *IEEE Spectrum,* p. 43, May 1992. © 1992 IEEE.

51

Pyroelectric Materials and Devices

Roger W. Whatmore
GEC-Marconi Materials Technology

51.1 Introduction 1126
51.2 Polar Dielectrics 1126
51.3 The Pyroelectric Effect 1127
51.4 Pyroelectric Materials and Their Selection 1129

51.1 Introduction

It was known over 2000 years ago that certain minerals such as tourmaline would attract small objects when heated. It was understood over 200 years ago that this attraction was a manifestation of the appearance of electrical charges on the surface as a consequence of the temperature change. This is called the **pyroelectric** effect and over the last 15 years has become the basis of a major worldwide industry manufacturing detectors of infrared radiation. These are exploited in such devices as "people detectors" for intruder alarms and energy conservation systems, fire and flame detectors, spectroscopic gas analyzers—especially looking for pollutants from car exhausts—and, more recently, devices for thermal imaging. Such thermal imagers can be used for night vision and, by exploiting the smoke-penetrating properties of long-wavelength infrared radiation, in devices to assist firefighters in smoke-filled spaces. The major advantages of the devices in comparison with the competing infrared detectors that exploit narrow bandgap semiconductors are that no cooling is necessary and that they are cheap and consume little power.

The pyroelectric effect appears in any material which possesses a polar symmetry axis. This chapter describes the basic effect, gives a brief account of how it can be used in radiation detection, and discusses the criteria by which materials can be selected for use in this application, concluding with a comparison of the properties of several of the most commonly used materials.

51.2 Polar Dielectrics

A polar material is one whose crystal structure contains a unique axis, along which an electric dipole moment will exist. There are 10 polar crystal classes:

- Triclinic   1
- Tetragonal   4, 4mm
- Hexagonal   6, 6mm

- Monoclinic   2, m
- Orthorhombic   mm2
- Trigonal   3, 3m

All crystals whose structures possess one of these symmetry groups will exhibit both pyroelectric and **piezoelectric** characteristics. In **ferroelectrics,** which are a subset of the set of pyroelectrics, the orientation of the polar axis can be changed by application of an electric field of sufficient magnitude. The original and final states of the crystal are symmetrically related. It is important to note that

1. Not all polar materials are ferroelectric.
2. There is a set of point groups which lack a center of symmetry, without possessing a polar axis. The crystals belonging to these groups (222, $\bar{4}$, 422, $\bar{4}$2m, 32, $\bar{6}$, $\bar{6}$m2, 23, and $\bar{4}$3m) are piezoelectric without being pyroelectric. (432 is a noncentrosymmetric, nonpiezoelectric class.)

A very wide range of materials exhibit ferroelectric, and thus pyroelectric, behavior. These range from crystals, such as potassium dihydrogen phosphate and triglycine sulphate, to polymers, such as polyvinylidene fluoride, and liquid crystals and ceramics, such as barium titanate and lead zirconate titanate.

The fact that the orientation of the polar axis in ferroelectrics can be changed by the application of a field has a very important consequence for ceramic materials. If a polycrystalline body is made of a polar material, then the crystal axes will, in general, be randomly oriented. It cannot therefore show pyroelectricity. However, if an electric field greater than the **coercive field** (E_c) is applied to a ferroelectric ceramic, then the polar axes within the grains will tend to be reoriented so that they each give a component along the direction of the applied field. This process is called "poling." The resulting ceramic is polar (with a point symmetry ∞m) and will show both piezoelectricity and pyroelectricity.

95.3 The Pyroelectric Effect

The pyroelectric effect is described by:

$$P_i = p_i \Delta T \tag{51.1}$$

where P_i is the polarization change due to a change in temperature ΔT and p_i is the pyroelectric coefficient, which is a vector. The effect and its applications have been extensively reviewed in Whatmore [1986]. The effect of a temperature change on a pyroelectric material is to cause a current, i_p, to flow in an external circuit, such that

$$i_p = Ap\,dT/dt \tag{51.2}$$

where A is the electroded area of the material, p the component of the pyroelectric coefficient normal to the electrodes, and dT/dt the rate of change of temperature with time.

Pyroelectric devices detect changes in temperature in the sensitive material and as such are detectors of supplied energy. It can be seen that the pyroelectric current is proportional to the rate of change of the material with time and that in order to obtain a measurable signal, it is necessary to modulate the source of energy. As energy detectors, they are most frequently applied to the detection of incident electromagnetic energy, particularly in the infrared wavebands. Such devices are used for applications such as intruder detection, fire prevention, energy conservation, pollution monitoring, and thermal imaging.

Typically, a pyroelectric detector element will consist of a thin chip of the pyroelectric material cut perpendicular to the polar axis of the material, electroded with a conducting material such as an evaporated metal and connected to a low-noise, high-input impedance amplifier—for example, a junction field-effect transistor (JFET) or metal-oxide gate transistor (MOSFET)—as shown in Fig. 51.1. In some devices, the radiation is absorbed directly in the element. In this case the front electrode will be a thin metal layer matched to the permittivity of free space with an electrical resistivity of 367 Ω/square. However, in most high-performance devices, the element is coated with a layer designed to absorb the radiation of interest. The element itself must be thin to minimize the thermal mass and, in most cases, well isolated thermally from its environment. These measures are

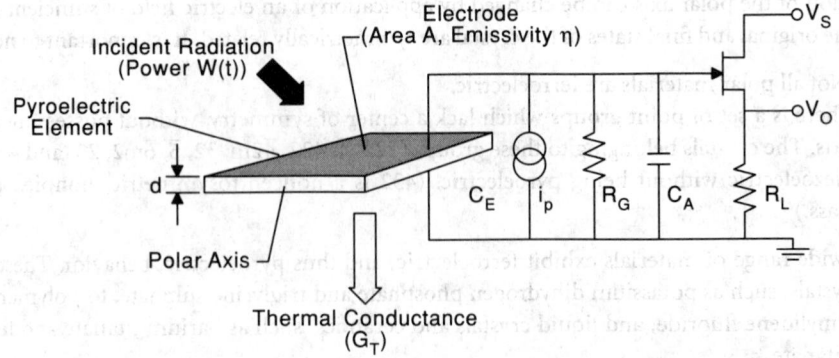

FIGURE 51.1   Pyroelectric detector with FET amplifier.

designed to increase the temperature change for a given amount of energy absorbed and thus the electrical signal generated. The necessary modulation of the radiation flux can be achieved either by deliberately only "looking" for moving objects or other radiation sources (e.g., flickering flames for a flame detector) or by interposing a mechanical radiation "chopper" such as a rotating blade.

The voltage responsivity of a device such as this is defined as $R_v = V_o/W$, where V_o is the output voltage and W is the input radiation power. For radiation sinusoidally modulated at a frequency ω, R_v is given by

$$R_v = \frac{R_G \eta p A \omega}{G_T (1 + \omega^2 \tau_T^2)^{1/2} (1 + \omega^2 \tau_E^2)^{1/2}} \tag{51.3}$$

where G_T is the thermal conductance from the element to the environment, τ_T is the thermal time constant of the element, τ_E is the electrical time constant of the element, R_G is the electrical resistance across the element, η is the emissivity of the element for the radiation being detected, and A is the sensitive area of the element.

It is easy to show that the response of a pyroelectric device maximizes at a frequency equal to the inverse of the geometric mean of the two time constants and that above and below the two frequencies given by τ_T^{-1} and τ_E^{-1}, R_v falls as ω^{-1}. The consequence of this is that pyroelectric detectors have their sensitivities maximized by having fairly long electrical time constants (0.1 to 10 s) and that such detectors thus work best at low frequencies (0.1 to 100 Hz). However, if high sensitivity is not required, extremely large bandwidths with little sensitivity variation can be obtained by shortening these time constants (making R_G and C_E low and G_T high). In this way, detectors have been made which give picosecond time responses for tracking fast laser pulses.

There are several noise sources in a pyroelectric device. These are discussed in detail in Whatmore [1986]. In many cases of interest, the dominant noise source is the Johnson noise generated by the ac conductance in the capacitance of the detector element. This noise is given by ΔV_j, where

$$\Delta V_j = \left\{ 4kT \frac{\tan\delta}{C_E} \right\}^{1/2} \omega^{-1/2} \quad \text{for } C_E \gg C_A \tag{51.4}$$

where k is Boltzmann's constant, T is the absolute temperature, $\tan\delta$ is the dielectric loss tangent of the detector material, C_E is the electrical capacitance of the element, and C_A is the input capacitance of the detector element amplifier.

The input radiation power required to give an output equal to the noise at a given frequency in unity bandwidth is known as the noise equivalent power (NEP). This is given by

$$NEP = V_n/R_v \tag{51.5}$$

A performance figure of merit frequently used when discussing infrared detectors is the detectivity, usually designated as D^*. This is given by

$$D^* = A^{1/2}/NEP \qquad (51.6)$$

Thus, the detectivity of a pyroelectric detector can be derived from Eqs. (51.3) to (51.6) and is given by

$$D^* = \frac{\eta d}{(4kT)^{1/2}} \cdot \frac{p}{c'(\epsilon \epsilon_o \tan\delta)^{1/2}} \cdot \frac{1}{\omega^{1/2}} \qquad (51.7)$$

where c' is the volume specific heat, ϵ is the dielectric constant of the pyroelectric, and d is the thickness of the pyroelectric element. The roll-off in D^* at high frequencies is thus $1/\omega^{1/2}$.

Pyroelectric single-element IR detectors come in many different varieties. A typical commercial device will have the sensitive element made from a material of the type discussed in the next section, such as a piece of lithium tantalate crystal or a ferroelectric ceramic. The element size will be a few millimeters square. Typical performance figures at about 10 Hz would be a responsivity of a few hundred volts per watt of input radiation, a noise equivalent power of about 8×10^{-9} W/Hz$^{1/2}$, and a detectivity of about 2×10^8 cm Hz$^{1/2}$ W^{-1} for unity bandwidth. The detector can be fitted with a wide variety of windows, depending upon the wavelength of the radiation to be detected.

Pyroelectric devices have also been used for thermal imaging. In this application, their main advantage when compared with photon detector materials such as mercury cadmium telluride (CMT) (which are more sensitive) is that they can be used at room temperature. All the photon detectors require cooling, typically to 77 K. A very successful device for pyroelectric thermal imaging is the pyroelectric vidicon which uses a thin plate of pyroelectric material contained in a vacuum tube. The thermal image is focused onto the surface of the material using a germanium lens. This causes the formation of a pattern of pyroelectric charges, which are "read" by means of an electron beam. Typical sensitivities for such devices are between 0.50 and 1 K temperature differences in the scene for an f/1 lens. This compares with <0.10 K for a cooled CMT detector-based imager. Recently, a solid-state approach to pyroelectric thermal imaging has been developed. In this, an array of many thousands of very small identical detectors, each between 50 and 100 μm square, depending on the array design, are linked to a silicon amplifier/multiplexer circuit which allows the signals from all the elements to be read onto a single output line. These devices have been primarily developed for thermal imaging applications and excellent sensitivities (close to those achieved by many cooled systems) have been demonstrated.

51.4 Pyroelectric Materials and Their Selection

There are many different types of pyroelectrics and the selection of a material depends strongly upon the application. It is possible to formulate from the given equations a number of figures of merit which describe the contribution of the physical properties of a material to the performance of a device. For example, the current responsivity is proportional to F_i:

$$F_i = p/c' \qquad (51.8)$$

The voltage response for a pyroelectric element feeding into a high-input impedance, unity gain amplifier (such as a source follower FET) as shown in Fig. 51.1 is proportional to F_v:

$$F_v = p/c' \epsilon \epsilon_o \qquad (51.9)$$

The detectivity is proportional to F_D:

$$F_D = p/\{c'(\epsilon \epsilon_o \tan\delta)^{1/2}\} \qquad (51.10)$$

Table 51.1 Pyroelectric Properties of Selected Materials

Material (Temperature)	Pyroelectric Coefficient P 10^{-4} cm^{-2} K^{-1}	Dielectric Properties (1 kHz)		Volume-Specific Heat c' 10^6 Jm^{-3} K^{-1}	Thermal Conductivity K 10^{-7} m^2 s^{-1}	F_v m^2 C^{-1}	F_D 10^{-5} Pa$^{-1/2}$	F_{vid} 10^6 sC^{-1}
		ϵ	tanδ					
TGS (35°C)	5.5	55	0.025	2.6	3.3	0.43	6.1	1.3
DTGS (40°C)	5.5	43	0.020	2.4	3.3	0.60	8.3	1.8
PVDF polymer	0.27	12	0.015	2.43	0.62	0.10	0.88	1.6
LiTaO$_3$ crystal	2.3	47	0.005	3.2	13.0	0.17	4.9	0.13
Modified PZ ceramic	3.8	290	0.003	2.5		0.06	5.8	
Modified PT ceramic	3.8	220	0.011	2.5		0.08	3.3	

PZ = PbZrO$_3$, PT = PbTiO$_3$.

For the pyroelectric vidicon, thermal spreading of the pattern on the target is important and the relevant figure of merit is F_{vid}:

$$F_{vid} = F_v/K \tag{51.11}$$

where K is the thermal conductivity of the pyroelectric. It should be noted that the use of these merit figures must be tempered with a knowledge of the type of detector the material is to be used in. It is necessary, if possible, to match the capacitance of the detector to the input capacitance of the amplifier. Hence, low-permittivity materials are better suited to large-area detectors, and conversely arrays of small-area detectors are better served by materials with a high permittivities.

Table 51.1 lists the pyroelectric properties of several different materials, single crystals, ceramics, and polymers. It can be seen that triglycine sulphate (TGS) and its deuterated isomorph (DTGS) exhibit the highest value of F_v and are frequently used for high-performance single-element detectors. These are the preferred materials for pyroelectric vidicon targets. However, they are water soluble, difficult to handle, and show poor long-term stability, both chemically and electrically, because of their low **Curie temperatures.** Furthermore, their dielectric loss is rather high, so that the F_D figures are not so favorable. Lithium tantalate, on the other hand, is an oxide single-crystal material which possesses a relatively low value of F_v but a very low loss so that F_D is favorable. The material is very stable and is now widely used for single-element detectors. Its thermal conductivity is quite high so that it is not a good material for the pyroelectric vidicon. The ferroelectric polymers possess relatively low pyroelectric coefficients and low dielectric constants with high losses, so that their figures of merit are also quite low. Their low thermal conductivities make them quite favorable for use in the pyroelectric vidicon and the fact that they are commercially available in thin sections (down to 6 μm) at low cost, removing any requirement for expensive lapping and polishing, makes them attractive for some low-cost detectors. Their low permittivities make them particularly well suited to large-area detectors.

The ceramic materials modified lead zirconate and modified lead titanate are interesting in that they possess high pyroelectric coefficients with relatively high permittivities and low losses. The modified lead zirconate is a solid solution of lead zirconate with lead iron niobate and lead titanate, with small additions of uranium as a stabilizing dopant. The use of uranium in this material minimizes the dielectric constant and loss (thus maximizing F_D) while also permitting control over the electrical resistivity, allowing the gate bias resistor in Fig. 51.1 to be designed into the sensor element. The modified lead titanate is doped with calcium titanate and lead cobalt tungstate. The use of hot pressing in ceramic manufacture permits the fabrication of very low porosity material, which can be lapped and polished to very thin sections (as low as 20 μm) while being mechanically strong enough to be placed on a mount which provides support only over a small area, permitting the fabrication of detectors with maximum sensitivity. While the F_v values are relatively small in these materials, the F_D values are as good as most of the single-crystal materials. They are very well suited to small-area detectors because their high dielectric constants enable the element capacitance to be matched to that of the amplifier. Pyroelectric ceramics are now finding use in a wide range of the infrared detector market, from low-cost intruder alarms to high-value imaging arrays.

Recently, a new class of pyroelectric materials which use the effect in the region of T_c have been developed [Whatmore, 1991]. In these materials a bias field must be applied to stabilize the effect, but F_D values as high as 10 to 15×10^{-5} Pa$^{-1/2}$ have been recorded in such materials as barium strontium titanate or lead scandium tantalate, both perovskite ceramics. This mode of operation is usually called "dielectric bolometer."

Defining Terms

Coercive field (E_c):   The field required to switch a sufficient proportion of the polarization of a body of a ferroelectric such that the net measurable external dipole moment is zero.

Curie temperature (T_c):   The temperature at which the spontaneous polarization of a ferroelectric goes to zero.

Ferroelectric: A polar dielectric in which the crystallographic orientation of the internal dipole moment can be changed by the application of an electric field.

Paraelectric: The nonpolar phase into which the ferroelectric transforms above T_c, frequently called the paraelectric phase.

Piezoelectric: A material which possesses a noncentrosymmetric crystal structure which will generate charge on the application of a mechanical stress. As in the case of a pyroelectric, this can be detected as either a potential difference or as a charge flowing in an external circuit.

Pyroelectric: A polar dielectric material in which the internal dipole moment is temperature dependent. This leads to a change in the charge balance at the surface of the material which can be detected as either a potential difference or as a charge flowing in an external circuit.

Remanent polarization: The value to which the externally measured polarization of a ferroelectric body relaxes after it has been subjected to an electric field much greater than the coercive field, which is then removed.

Saturation polarization: The value to which the externally measured electrical dipole moment of a ferroelectric body tends when subjected to an external electrical field greater than the coercive field.

Spontaneous polarization: The internal electrical dipole moment of a ferroelectric crystal.

References

R. W. Whatmore, "Pyroelectric devices and materials," *Rep. Prog. Phys.*, vol. 49, pp. 1335–1386, 1986.

R. W. Whatmore, "Pyroelectric ceramics and devices for thermal infrared detection and imaging," *Ferroelectrics*, vol. 118, pp. 241–259, 1991.

52

Dielectrics and Insulators

52.1 Introduction 1132
52.2 Dielectric Losses 1133
52.3 Dielectric Breakdown 1137
52.4 Insulation Aging 1140
52.5 Dielectric Materials 1143
       Gases • Insulating Liquids • Solid Insulating Materials • Solid-Liquid Insulating Systems

R. Bartnikas
Institut de Recherche d'Hydro-Québec

52.1 Introduction

Dielectrics are materials that are used primarily to isolate components electrically from each other or ground or to act as capacitive elements in devices, circuits, and systems. Their insulating properties are directly attributable to their large energy gap between the highest filled valence band and the conduction band. The number of electrons in the conduction band is extremely low, because the energy gap of a dielectric (5 to 7 eV) is sufficiently large to maintain most of the electrons trapped in the lower band. As a consequence, a dielectric, subjected to an electric field, will evince only an extremely small conduction or loss current; this current will be caused by the finite number of free electrons available in addition to other free charge carriers (ions) associated usually with contamination by electrolytic impurities as well as dipole orientation losses arising with polar molecules under ac conditions. Often the two latter effects will tend to obscure the miniscule contribution of the relatively few free electrons available. Unlike solids and liquids, vacuum and gases (in their nonionized state) approach the conditions of a perfect insulator—i.e., they exhibit virtually no detectable loss or leakage current.

Two fundamental parameters that characterize a dielectric material are its **conductivity** σ and the value of the real permittivity or **dielectric constant** ε'. By definition, σ is equal to the ratio of the leakage current density J_l to the applied electric field E,

$$\sigma = \frac{J_l}{E} \qquad (52.1)$$

Since J_l is in A cm^{-2} and E in V cm^{-1}, the corresponding units of σ are in S cm^{-1} or Ω^{-1} cm^{-1}. Alternatively, when only mobile charge carriers of charge e and mobility μ, in cm^2 V^{-1} s^{-1}, with a concentration of n per cm^3 are involved, the conductivity may be expressed as

$$\sigma = e\mu n \tag{52.2}$$

The conductivity is usually determined in terms of the measured insulation resistance R in Ω; it is then given by $\sigma = d/RA$, where d is the insulation thickness in cm and A the surface area in cm^2. Most practical insulating materials have conductivities ranging from 10^{-6} to 10^{-20} S cm^{-1}. Often dielectrics may be classified in terms of their resistivity value ρ, which by definition is equal to the reciprocal of σ.

The real value of the permittivity or dielectric constant ε' is determined from the ratio

$$\varepsilon' = \frac{C}{C_o} \tag{52.3}$$

where C represents the measured capacitance in F and C_o is the equivalent capacitance *in vacuo*, which is calculated for the same specimen geometry from $C_o = \varepsilon_o A/d$; here ε_o denotes the permittivity *in vacuo* and is equal to 8.854×10^{-14} F cm^{-1} (8.854×10^{-12} F m^{-1} in SI units) or more conveniently to unity in the Gaussian CGS system. In practice, the value of ε_o in free space is essentially the same as that for a gas (e.g., for air, $\varepsilon_o = 1.000536$). The majority of liquid and solid dielectric materials, presently in use, have dielectric constants extending from approximately 2 to 10.

52.2 Dielectric Losses

Under ac conditions **dielectric losses** arise mainly from the movement of free charge carriers (electrons and ions), space charge polarization, and dipole orientation [1]. Ionic, space charge, and dipole losses are temperature- and frequency-dependent, a dependency which is reflected in the measured values of σ and ε'. This necessitates the introduction of a complex permittivity ε defined by

$$\varepsilon = \varepsilon' - j\varepsilon'' \tag{52.4}$$

where ε'' is the imaginary value of the permittivity, which is equal to σ/ω. Note that the conductivity σ determined under ac conditions may include the contributions of the dipole orientation, space charge, and ionic polarization losses in addition to that of the drift of free charge carriers (ions and electrons) which determine its dc value.

The complex permittivity, ε, is equal to the ratio of the dielectric displacement vector $\bar{D}$ to the electric field vector $\bar{E}$, i.e., $\varepsilon = \bar{D}/\bar{E}$. Since under ac conditions the appearance of a loss or leakage current is manifest as a phase angle difference δ between the $\bar{D}$ and $\bar{E}$ vectors, then in complex notation $\bar{D}$ and $\bar{E}$ may be expressed as $D_o \exp[j(\omega t - \delta)]$ and $E_o \exp[j\omega t]$, respectively, where ω is the radial frequency term, t the time, and D_o and E_o the respective magnitudes of the two vectors. From the relationship between $\bar{D}$ and $\bar{E}$, it follows that

$$\varepsilon' = \frac{D_o}{E_o} \cos \delta \tag{52.5}$$

and

$$\varepsilon'' = \frac{D_o}{E_o} \sin \delta \tag{52.6}$$

It is customary under ac conditions to assess the magnitude of loss of a given material in terms of the value of its **dissipation factor,** $\tan\delta$; it is apparent from Eqs. (52.5) and (52.6), that

$$\tan\delta = \frac{\varepsilon''}{\varepsilon'} = \frac{\sigma}{\omega\varepsilon'} \tag{52.7}$$

Examination of Eq. (52.7) suggests that the behavior of a dielectric material may also be described by means of an equivalent electrical circuit. It is most commonplace and expedient to use a parallel circuit representation, consisting of a capacitance C in parallel with a large resistance R as delineated in Fig. 52.1. Here C represents the capacitance and R the resistance of the dielectric. For an applied voltage V across the dielectric, the leakage current is $\bar{I}_l = \bar{V}/R$ and the displacement current is $\bar{I}_C = j\omega C\bar{V}$; since $\tan\delta = \bar{I}_l/\bar{I}_C$, then

$$\tan\delta = \frac{1}{\omega RC} \tag{52.8}$$

It is to be emphasized that in Eq. (52.8), the quantities R and C are functions of temperature, frequency, and voltage. The equivalence between Eqs. (52.7) and (52.8) becomes more palpable if $\bar{I}_l$ and $\bar{I}_C$ are expressed as $\omega\varepsilon''C_o\bar{V}$ and $j\omega\varepsilon'C_o\bar{V}$, respectively.

Every loss mechanism will exhibit its own characteristic $\tan\delta$ loss peak, centered at a particular absorption frequency, ω_o for a given test temperature. The loss behavior will be contingent upon the molecular structure of the material, its thickness, and homogeneity, and the temperature, frequency, and electric field range over which the measurements are performed [1]. For example, dipole orientation losses will be manifested only if the material contains permanent molecular or side-link dipoles; a considerable overlap may occur between the permanent dipole and ionic relaxation regions. Ionic relaxation losses occur in dielectric structures where ions are able to execute short-range jumps between two or more equilibrium positions. Interfacial or space charge polarization will arise with insulations of multilayered structures where the conductivity and permittivity is different for the individual strata or where one dielectric phase is interspersed in the matrix of another dielectric. Alternatively, space charge losses will occur with mobile charge carriers whose movement becomes limited at the electrodes. This type of mechanism takes place often in thin-film dielectrics and exhibits a pronounced thickness effect. If the various losses are considered schematically on a logarithmic frequency scale at a given temperature, then the $\tan\delta$ and ε' values will appear as functions of frequency as delineated schematically in Fig. 52.2. For many materials the dipole and ionic relaxation losses tend to predominate over the frequency range extending from about 0.5 to 300 MHz, depending upon the molecular structure of the dielectric and temperature. For example, the absorption peak of an oil may occur at 1 MHz, while that of a much lower viscosity fluid such as water may appear at approximately 100 MHz. There is considerable overlap between the dipole and ionic relaxation loses, because the ionic jump distances are ordinarily of the same order of magnitude as the radii of the permanent dipoles. Space charge polarization losses manifest themselves normally over the low-frequency region extending from 10^{-6} Hz to 1 MHz and are characterized by very broad and intense peaks; this behavior is apparent from Eq. (52.7), which indicates that even small conductivities may lead to very large $\tan\delta$ values at very low frequencies. The nonrelaxation-type electronic conduction losses are readily perceptible over the low-frequency spectrum and decrease monotonically with frequency.

The dielectric loss behavior may be phenomenologically described by the Pellat-Debye equations, relating the imaginary and real values of the permittivity to the relaxation time, τ, of the loss process (i.e., the frequency at which the ε'' peak appears: $f_o = 1/2\,\pi\tau$), the low-frequency or static value of the real permittivity, ε_s, and the high- or optical-frequency value of the real permittivity, ε_∞,

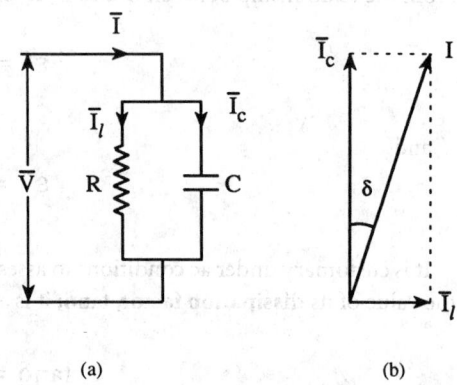

(a)   (b)

FIGURE 52.1 (a) Parallel equivalent RC circuit and (b) corresponding phasor diagram.

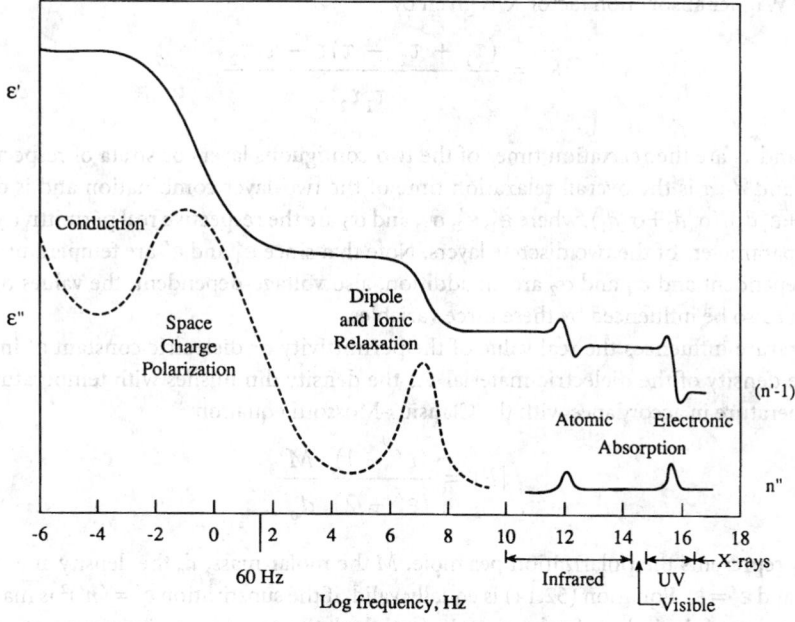

FIGURE 52.2 Schematic representation of various absorption regions [7].

$$\varepsilon' = \varepsilon_\infty + \frac{\varepsilon_s - \varepsilon_\infty}{1 + \omega^2 \tau^2} \qquad (52.9)$$

and

$$\varepsilon'' = \frac{(\varepsilon_s - \varepsilon_\infty)\omega\tau}{1 + \omega^2 \tau^2} \qquad (52.10)$$

Since the relaxation processes are thermally activated, an increase in temperature will cause a displacement of the loss peak to higher frequencies. In the case of ionic and dipole relaxation, the relaxation time may be described by the relation

$$\tau = \frac{h}{kT} \exp\left[\frac{\Delta H}{RT}\right] \exp\left[-\frac{\Delta S}{R}\right] \qquad (52.11)$$

where h is the Planck constant (6.624×10^{-34} J s^{-1}), k the Boltzmann constant (1.38×10^{-23} J K^{-1}), ΔH the activation energy of the relaxation process, R the universal gas constant (8.314×10^3 J K^{-1} kmol^{-1}), and ΔS the entropy of activation. For the ionic relaxation process, τ may alternatively be taken as equal to $1/2\Gamma$, where Γ denotes the ion jump probability between two equilibrium positions. Also for dipole orientation in liquids, τ may be approximately equated to $\eta/4\pi r^3 T$, where η represents the macroscopic viscosity of the liquid and r is the dipole radius [2]. With interfacial or space charge polarization, which may arise due to a pile-up of charges at the interface of two contiguous dielectrics of different conductivity and permittivity, Eq. (52.10) must be rewritten as [3]

$$\varepsilon'' = \varepsilon_\infty \left(\frac{\tau}{\omega \tau_1 \tau_2} + \frac{K\omega\tau}{1 + \omega^2 \tau^2} \right) \qquad (52.12)$$

where the Wagner absorption factor K is given by

$$K = \frac{(\tau_1 + \tau_2 - \tau)\tau - \tau_1\tau_2}{\tau_1\tau_2} \qquad (52.13)$$

where τ_1 and τ_2 are the relaxation times of the two contiguous layers or strata of respective thicknesses d_1 and d_2; τ is the overall relaxation time of the two-layer combination and is defined by $\tau = (\varepsilon_1'd_2 + \varepsilon_2'd_1)/(\sigma_1d_2 + \sigma_2d_1)$, where ε_1', ε_2', σ_1, and σ_2 are the respective real permittivity and conductivity parameters of the two discrete layers. Note that since ε_1' and ε_2' are temperature- and frequency-dependent and σ_1 and σ_2 are, in addition, also voltage-dependent, the values of τ and ε'' will in turn also be influenced by these three variables.

Temperature influences the real value of the permittivity or dielectric constant ε' insofar as it affects the density of the dielectric material. As the density diminishes with temperature, ε' falls with temperature in accordance with the Clausius-Mossotti equation

$$[P] = \frac{(\varepsilon' - 1)}{(\varepsilon' + 2)} \frac{M}{d_o} \qquad (52.14)$$

where $[P]$ represents the polarization per mole, M the molar mass, d_o the density at a given temperature, and $\varepsilon' = \varepsilon_s$. Equation (52.14) is equally valid, if the substitution $\varepsilon' = (n')^2$ is made; here n' is the real value of the index of refraction. In fact, the latter provides a direct connection with the dielectric behavior at optical frequencies. In analogy with the complex permittivity, the index of refraction is also a complex quantity, and its imaginary value n'' exhibits a loss peak at the absorption frequencies; in contrast with the ε' value which can only fall with frequency, the real index of refraction n' exhibits an inflection-like behavior at the absorption frequency. This is illustrated schematically in Fig. 52.2, which depicts the kn' or n'' and $n' - 1$ values as a function of frequency over the optical frequency regime. The absorption in the infrared results from atomic resonance that arises from a displacement and vibration of atoms relative to each other, while an electronic resonance absorption effect occurs over the ultraviolet frequencies as a consequence of the electrons being forced to execute vibrations at the frequency of the external field.

The characterization of dielectric materials must be carried out in order to determine their properties for various applications over different parts of the electromagnetic frequency spectrum. There are many techniques and methods available for this purpose that are too numerous and detailed to attempt to present here even in a cursory manner. However, Fig. 52.3 portrays schematically the different test methods that are commonly used to carry out the characterization over the different frequencies up to and including the optical regime.

The frequency response of dielectrics at the more elevated frequencies is primarily of interest in the electrical communications field. In contradistinction for the electrical power generation, transmission, and distribution, it is the low-frequency spectrum that constitutes the area of application. Also, the use of higher voltages in the electrical power area necessarily requires detailed knowledge of how the electrical losses vary as a function of the electrical field. Since most electrical power apparatus operates at a fixed frequency of 50 or 60 Hz, the main variable apart from the temperature is the applied or operating voltage. At power frequencies the dipole losses are generally very small and invariant with voltage up to the saturation fields which exceed substantially the operating fields, being in the order of 10^7 kV cm^{-1} or more. However, both the space charge polarization and ionic losses are highly field-dependent. As the electrical field is increased, ions of opposite sign are increasingly segregated; this hinders their recombination and, in effect, enhances the ion charge carrier concentration. As the dissociation rate of the ionic impurities is further augmented by temperature increases, combined rises in temperature and field may lead to appreciable dielectric loss. Thus, for example, for a thin liquid film bounded by two solids, tanδ increases with voltage until at some upper voltage value the physical boundaries begin to finally limit the amplitude of the ion

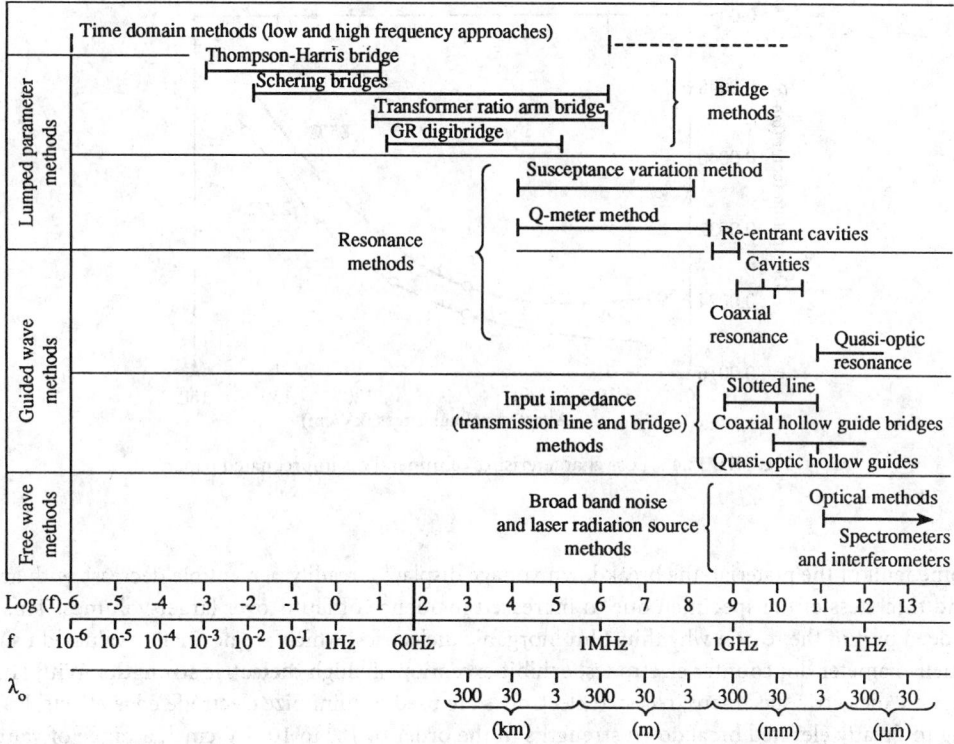

FIGURE 52.3 Frequency ranges of various dielectric test methods [7].

excursions, at which point tanδ commences a downward trend with voltage (Böning–Garton effect). The interfacial or space charge polarization losses may evince a rather intricate field dependence, depending upon the manner in which the discrete conductivities of the contiguous media change with applied voltage and temperature [as is apparent from the nature of Eqs. (52.12) and (52.13)]. The exact frequency value at which the space charge loss exhibits its maximum is contingent upon the value of the relaxation time τ. Figure 52.4 depicts typical tanδ versus applied voltage characteristics for an oil-impregnated paper-insulated cable model at two different temperatures, in which the loss behavior is primarily governed by ionic conduction and space charge effects. The monotonically rising dissipation factor with increasing applied voltage at room temperature is indicative of the predominating ionic loss mechanism, while at 85°C, an incipient decrease in tanδ is suggestive of space charge effects [4].

52.3 Dielectric Breakdown

As the voltage is increased across a dielectric material, a point is ultimately reached beyond which the insulation will no longer be capable of sustaining any further rise in voltage and breakdown will ensue, causing a short to develop between the electrodes. If the dielectric consists of a gas or liquid medium, the breakdown will be self-healing in the sense that the gas or liquid will support anew a reapplication of voltage until another breakdown recurs. In a solid dielectric, however, the initial breakdown will result in a formation of a permanent conductive channel, which cannot support a reapplication of voltage. The dielectric breakdown processes are distinctly different for the three states of matter.

In the case of solid dielectrics the breakdown is dependent not only upon the molecular structure and morphology of the solid but also upon extraneous variables such as the geometry of the material, the temperature, and the ambient environment. Since breakdown often occurs along

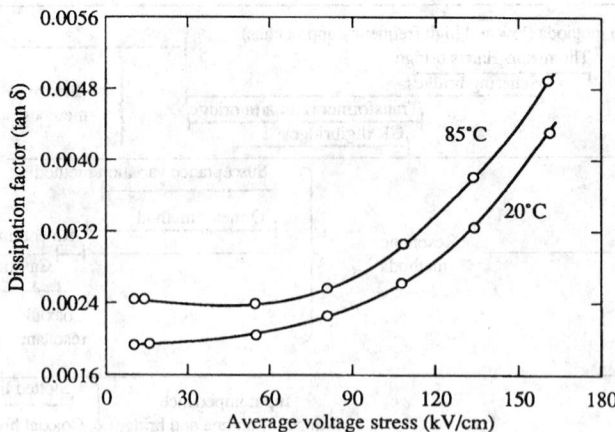

FIGURE 52.4 Loss characteristics of mineral oil-impregnated paper.

some fault of the material, the breakdown voltage displays a readily perceptible decrease with area and thickness of the specimen due to increased incidence of faults over larger volumes. This is indeed part of the reason why thin-film inorganic dielectrics, which are normally evaluated using small-diameter dot counter electrodes, exhibit exceptionally high **dielectric strengths.** With large organic dielectric specimens, recessed electrodes are used to minimize electrode edge effects, leading to greatly elevated breakdown strengths in the order of 10^6 to 10^7 kV cm^{-1}, a range of values which is considered to represent the ultimate breakdown strength of the material or its intrinsic breakdown strength; as the intrinsic breakdown occurs in approximately 10^{-8} to 10^{-6} s, an electronic mechanism is implicated.

The breakdown strength under dc and impulse conditions tends to exceed that at ac fields, thereby suggesting the ac breakdown process may be partially of a thermal nature. An additional factor, which may lower the ac breakdown strength, is that associated with the occurrence of partial discharges either in void inclusions or at the electrode edges; this leads to breakdown values very much less than the intrinsic value. In practice, the breakdowns are generally of an extrinsic nature, and the intrinsic values are useful conceptually insofar as they provide an idea of an upper value that can be attained only under ideal conditions. The intrinsic breakdown theories were essentially developed for crystalline dielectrics, for which it was assumed that a very small number of thermally activated electrons can be thermally excited to move from the valence to the conduction band and that under the influence of an external field they will be impelled to move in the direction of the field, colliding with the lattice of the crystalline dielectric and dissipating their energy by phonon interactions [1]. Accordingly, breakdown is said to occur when the average rate of energy gain by the electrons, $A(E, T, T_e, \xi)$, exceeds that lost in collisions with the lattice, $B(T, T_e, \xi)$. Hence, the breakdown criterion can be stated as

$$A(E, T, T_e, \xi) = B(T, T_e, \xi) \tag{52.15}$$

where E is the applied field, T the lattice temperature, T_e the electron temperature, and ξ an energy distribution constant. Thus in qualitative terms as the temperature is increased gradually, the breakdown voltage rises because the interaction between the electrons and the lattice is enhanced as a result of the increased thermal vibrations of the lattice. Ultimately, a critical temperature is attained where the electron–electron interactions surpass in importance those between the electrons and the lattice, and the breakdown strength commences a monotonic decline with temperature; this behavior is borne out in NaCl crystals, as is apparent from Fig. 52.5 [5]. However, with amorphous or partially crystalline polymers, as for example with polyethylene, the maximum in

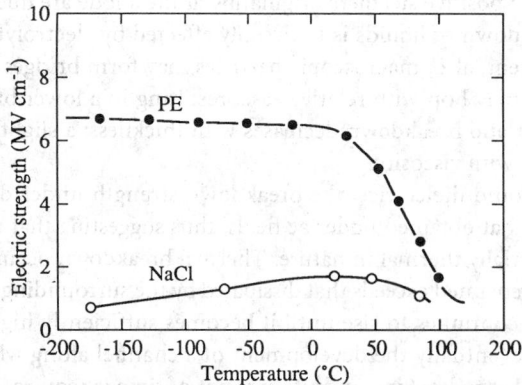

FIGURE 52.5 Dielectric breakdown characteristics of sodium chloride [5] and polyethylene [6].

breakdown strength is seen to be absent and only a decrease is observed [6]; as the crystalline content is increased in amorphous-crystalline solids, the breakdown strength is reduced.

The electron avalanche concept has also been applied to explain breakdown in solids, in particular to account for the observed decrease in breakdown strength with insulation thickness. Since breakdown due to electron avalanches results necessarily in the formation of a positive ion space charge, the space charge will tend to modify the conditions for breakdown.

The breakdown process in gases is relatively well understood and is explained in terms of the avalanche theory. A free electron, occurring in a gas due to cosmic radiation, will be accelerated in a field and upon collision with neutral molecules in its trajectory will eject, if its energy is sufficient, other electrons that will in turn undergo additional collisions resulting in a production of more free electrons. If the electric field is sufficiently high, the number of free electrons will increase exponentially along the collision route until ultimately an electron avalanche will form. As the fast-moving electrons in the gap disappear into the anode, they leave behind the slower-moving ions, which gradually drift to the cathode where they liberate further electrons with a probability γ. When the height of the positive ion avalanche becomes sufficiently large to lead to a regeneration of a starting electron, the discharge mechanism becomes self-sustaining and a spark bridges the two electrodes. The condition for the Townsend breakdown in a short gap is given by

$$\gamma[\exp(\alpha d) - 1] = 1 \tag{52.16}$$

where d is the distance between the electrodes and α represents the number of ionizing impacts per electron per unit distance. The value of γ is also enhanced by photoemission at the cathode and photon radiation in the gas volume by the metastable and excited gas atoms or molecules. In fact, in large gaps the breakdown is governed by steamer formation in which photon emission from the avalanches plays a dominant role. Breakdown characteristics of gases are represented graphically in terms of the Paschen curves, which are plots of the breakdown voltage as a function of the product of gas pressure p and the electrode separation d. Each gas is characterized by a well-defined minimum breakdown voltage at one particular value of the pd product.

The breakdown process in liquids is perhaps the least understood due to a lack of a satisfactory theory on the liquid state. The avalanche theory has been applied with limited success to explain the breakdown in liquids, by assuming that electrons injected from an electrode surface exchange energy with the atoms or molecules of the liquid, ultimately causing the atoms and molecules to ionize and thus precipitating breakdown. Recent investigations, utilizing electro-optical techniques, have demonstrated that breakdown involves steamers with tree- or bushlike structures that propagate from the electrodes [2]. The negative streamers emerging from the cathode form due to electron emission, while positive steamers originating at the anode are due to free electrons in the liquid itself. The breakdown of liquids is noticeably affected by electrolytic impurities as well as water and oxygen content; also, macroscopic particles may form bridges between the electrodes along which electrons may hop with relative ease, resulting in a lower breakdown. As in solids, there is a volume effect and breakdown decreases with thickness; a slight increase in breakdown voltage is also observed with viscosity.

In both solid and liquid dielectrics, the breakdown strength under dc and impulse fields is markedly greater than that obtained under ac fields, thus suggesting that under ac conditions the breakdown may be partially thermal in nature. Thermal breakdown occurs at localized hot spots where the rate of heat generated exceeds that dissipated by the surrounding medium. The temperature at such hot spots continues to rise until it becomes sufficiently high to induce fusion and vaporization, causing eventually the development of a channel along which breakdown ensues between the opposite electrodes. Since a finite amount of time is required for the heat buildup to occur to lead to the thermal instability, thermally induced breakdown is contingent upon the time of the alternating voltage application and is thus implicated as the leading cause of breakdown in many dielectrics under long-term operating conditions. However, under some circumstances thermal instability may develop over a very short time; for example, some materials have been found to undergo thermal breakdown when subjected to very short repetitive voltage pulses. In low-loss dielectrics, such as polyethylene, the occurrence of thermal breakdown is highly improbable under low operating temperatures, while glasses with significant ionic content are more likely to fail thermally, particularly at higher frequencies.

The condition for thermal breakdown may be stated as

$$KA \, \Delta T/l = \omega \varepsilon' E^2 \tan\delta \qquad (52.17)$$

where the left-hand side represents the heat transfer in J s^{-1} along a length l (cm) of sectional area A (cm^2) of the dielectric surface in the direction of the temperature gradient due to the temperature difference ΔT, in °C, such that the units of the thermal conductivity constant K are in J °C^{-1} cm^{-1} s^{-1}. The right-hand side of Eq. (52.17) is equal to the dielectric loss dissipated in the dielectric in J s^{-1}, where E is the external field, ε' the real value of the permittivity, and $\tan\delta$ the dissipation factor at the radial frequency ω.

Other causes of extrinsic breakdown are associated with particular defects in the dielectric or with the environmental conditions under which the dielectric material is employed. For example, some dielectrics may contain gas-filled cavities that are inherent with the porous structure of the dielectric or that may be inadvertently introduced either during the manufacturing process or created under load cycling. If the operating electrical field is sufficiently elevated to cause the gas within the cavities to undergo discharge, the dielectric will be subjected to both physical and chemical degradation by the partial discharges; should the discharge process be sustained over a sufficiently long period, breakdown will eventually ensue.

With overhead line insulators or bushings of electrical equipment, breakdown may occur along the surface rather than in the bulk of the material. Insulator surfaces may become contaminated by either industrial pollutants or salt spray near coastal areas, leading to surface tracking and breakdown below the normal flashover voltage. Surface tracking is enhanced in the presence of moisture, which increases the surface conductivity [7]. The latter is measured in S or Ω^{-1} and must be distinguished from the volume conductivity whose units are stated in S cm^{-1} or Ω^{-1} cm^{-1}. Surface tracking may be prevented by cleaning the surface and by applying silicone greases.

52.4 Insulation Aging

All insulating materials will undergo varying degrees of aging or deteriorating under normal operating conditions. The rate of aging will be contingent upon the magnitude of the electrical, thermal, and mechanical stresses to which the material is subjected; it will also be influenced by the composition and molecular structure of the material itself as well as the chemical, physical, and radiation environment under which the material must perform. The useful life of an insulating system will thus be determined by a given set and subset of aging variables. For example, the subset of variables in the voltage stress variable are the average and maximum values of the applied voltage, its frequency, and the recurrence rate of superposed impulse or transient voltage surges. For the thermal stress, the upper and lower ambient temperatures, the temperature gradient in the insulation, and the maximum permissible operating temperature constitute the subvariable set. Also, the character of the mechanical stress will differ, depending upon whether torsion, compression, or tension and bending are involved. Furthermore, the aging rate will be differently affected if all stresses (electrical, thermal, and mechanical) act simultaneously, separately, or in some predetermined sequence. The influence exerted on the aging rate by the environment will depend on whether the insulation system will be subjected to corrosive chemicals, petroleum fluids, water or high humidity, air or oxygen, ultraviolet radiation from the sun, and nuclear radiation. Organic insulations, in particular, may experience chemical degradation in the presence of oxygen. For example, polyethylene under temperature cycle will undergo both physical and chemical changes. These effects will be particularly acute at the emergency operating temperatures (90–130°C); at these temperatures partial or complete melting of the polymer will occur and the increased diffusion rate will permit the oxygen to migrate to a greater depth into the polymer. Ultimately the antioxidant will be consumed, resulting in an embrittlement of polymer and in extreme cases in the formation of macroscopic cracks. Subjection of the polymer to many repeated overload cycles will be accompanied by repeated melting and recrystallization of the polymer—a process that will inevitably cause the formation of cavities, which, when subjected to sufficiently high voltages, will undergo discharge, leading eventually to electrical breakdown.

There is a general consensus that electrically induced aging involves the mechanisms of treeing, partial discharge, and dielectric heating. Dielectric heating failures are more characteristic of lossy insulations or when highly conductive contaminants are involved. In the treeing mechanism, a distinction must be made between electrical and water trees [1]. The former refers to growth, resembling a tree, that occurs under dry conditions in the presence of an electric field; its branches or channels are hollow. In contrast, water trees require the presence of moisture, and their branches consist of fine filamentary channels, joining small cavities, all of which contain water; when placed in a dry environment, they eventually disappear. In a translucent dielectric, water trees are invisible and are rendered visible only when stained with a die, whereas electrical trees once formed remain readily discernible. Water trees are intrinsic to solid polymeric insulation, while electrical trees may occur in both solid and impregnated insulating systems. The actual failure path, when the breakdown current is limited to prevent destruction of evidence, invariably, even in liquids, consists of an electrical treelike structure.

Both electrical and water trees tend to propagate from electrical stress enhancement points, with electrical trees requiring appreciably higher stresses for their inception. Whereas the electrical tree mechanism usually involves partial discharges, the occurrence of the latter is conspicuously absent in water trees. The fact that water trees may bridge the opposite electrodes without precipitating failure infers a nonconductive nature of the filamentary channels or branches. Yet water trees are implicated in the overall degradation and aging process of the dielectric because of their presence in areas of failure as well as because of the often-observed phenomenon of an electrical tree emerging from some point along a water tree and causing abrupt failure.

The presence of voids or cavities within a solid or solid-liquid insulating system will almost invariably lead to eventual failure, with the proviso that the electrical field is sufficiently elevated to induce them to undergo either continuous or regularly recurring, though at times intermittent, discharge. Partial discharges give rise to both chemical and physical degradation of the insulation adjacent to the cavities. Electrical trees may readily ensue from discharging void inclusions. The deterioration rate due to partial discharges is proportional to the power dissipated P by the discharges, which may be expressed as [8]

$$P = \sum_{i=1}^{l} \sum_{j=1}^{m} n_{ij} \Delta Q_{ij} V_{ij} \qquad (52.18)$$

where n_{ij} is the recurrence rate of the jth discharge in the ith cavity and ΔQ_{ij} is its corresponding charge release at an applied voltage V_{ij}. Under ac conditions the discharges will tend to recur regularly in each cycle due to the capacitative voltage division across the void. Under dc conditions the discharge rate will be controlled by the time constant required to recharge the cavity following a discharge. The physical damage arising from discharges consists of surface erosion and pitting and is caused by the ion and electron bombardment incident on the void's walls at a given discharge site. Chemical degradation results from molecular chain scission due to particle bombardment of the surface and the reactions between the ambient ionized gases and the gases released due to the molecular chain scission processes. The final chemical composition of the reaction product is generally varied, depending primarily on the molecular structure of the dielectric materials involved and the composition of the ambient ionized gases; one discharge degradation product common to many polymers exposed to discharges in air is that of oxalic acid. Oxalic acid, as a result of its elevated conductivity, when deposited upon the cavity's walls may change the nature of the discharge (e.g., the discharge may change from a spark to a glow type) or may even ultimately extinguish the discharge (i.e., replace the discharge loss by an I^2R-type loss along the cavity's walls).

A number of aging models have been propounded to predict insulation aging under different types of stress. However, there are essentially only two models whose usefulness has been substantiated in practice and which have therefore gained wide acceptance. One of these is Dakin's classical thermal degradation model, which is based upon the approach used in chemical reaction rate theory [7],

$$t = A \exp[\Delta H/kT] \qquad (52.19)$$

where t represents the time to breakdown, A is a constant, ΔH is the activation energy of the aging process, T is the absolute temperature, and k is the Boltzmann constant (1.38×10^{-23} J K^{-1}). If the log t versus $1/T$ plot represents a straight line that is obtained when the insulation aging is accelerated at various temperatures above the operating temperature, extrapolation of the line to the operating temperature may yield a rudimentary estimate of the aging time or the service life that can be anticipated from the insulation system when operated under normal temperature and load conditions. Deviations from straight-line behavior are indicative of more than one thermal aging mechanism; for example, a polymeric insulation will exhibit such deviations when thermally stressed beyond its melting or phase transition temperature.

Another extremely useful model, applicable to accelerated aging studies under electrical stress, is the so-called inverse power law relationship given by [7]

$$t = BE^{-n} \qquad (52.20)$$

where t is the time to breakdown under an electric stress E, B is a constant, and n is an exponent parameter. The relationship is essentially empirical in nature, and its proof of validity rests primarily on experiment and observation. Again it is found that a simple type of electrically induced aging process will generally result in a straight-line relationship between log t and log E. Consequently, aging data obtained at much higher electrical stresses with correspondingly shorter times to breakdown when extrapolated to longer times at stresses in the vicinity of the operating stress should yield the value of the effective service life under the operating stress. The slope of the line determines the exponent n, which constitutes an approximate indicator of the type of aging involved. For example, with polymer insulation water treeing ordinarily results in n values less than 4, while under conditions that may involve discharges and electrical trees, the n values may approach 10 or greater.

52.5 Dielectric Materials

Dielectric materials comprise a variety of solids, liquids, and gases. The breakdown strength generally increases with the density of the material so that dielectric solids tend to have higher dielectric strengths than gases. The same tendency is observed also with the dielectric loss and permittivity; for example, the dielectric losses in gases are virtually too small to be measurable and their dielectric constant in most practical applications can be considered as unity. In this section we will describe a number of the more common insulating materials in use.

Gases

The 60-Hz breakdown strength of a 1-cm gap of air at 25°C at atmospheric pressure is 31.7 kV cm^{-1} [9]. Although this is a relatively low value, air is a most useful insulating medium for large electrode separations as is the case for overhead transmission lines. The only dielectric losses in the overhead lines are those due to corona discharges at the line conductor surfaces and leakage losses over the insulator surfaces. In addition, the highly reduced capacitance between the conductors of the lines ensures a small capacitance per unit length, thus rendering overhead lines an efficient means for transmitting large amounts of power over long distances.

Nitrogen, which has a slightly higher breakdown strength (33.4 kV cm^{-1}) than air, has been used when compressed in low-loss gas-insulated capacitors. However, as air, which was also used in its compressed form in circuit breakers, it has been replaced by sulfur hexafluoride, SF$_6$. The breakdown strength of SF$_6$ is ca. 79.3 kV cm^{-1}. Since the outside metal casing of metal-clad circuit breakers is ordinarily grounded, it has become common practice to have much of the substation interconnected bus and interrupting switching equipment insulated with SF$_6$. To achieve higher breakdown strengths, the SF$_6$ gas is compressed at pressures on the order of 6 atm. The use of SF$_6$ in substation equipment has resulted in the saving of space and the elimination of line insulator pollution problems.

The breakdown strength of gases is little affected with increasing frequency until the period of the wave becomes comparable to the transit time of the ions and, finally, electrons across the gap. At this point a substantial reduction in the breakdown strength is observed.

Table 52.1 Electrical Properties of a Number of Representative Insulating Liquids [2, 4, 10]

Liquid	Viscosity cSt (37.8°C)	Dielectric Constant (at 60 Hz, 25°C)	Dissipation Factor (at 60 Hz, 100°C)	Breakdown Strength, (kV cm^{-1})
Capacitor oil	21	2.2	0.001	>118
Pipe cable oil	170	2.15	0.001	>118
Self-contained cable oil	49.7	2.3	0.001	>118
Heavy cable oil	2365	2.23	0.001	>118
Transformer oil	9.75	2.25	0.001	128
Alkyl benzene	6.0	2.1	0.0004	>138
Polybutene pipe cable oil	110 (SUS)	2.14 (at 1 MHz)	0.0003	>138
Polybutene capacitor oil	2200 (SUS at 100°C)	2.22 (at 1 MHz)	0.0005	>138
Silicone fluid	50	2.7	0.00015	>138
Castor oil	98 (100°C)	3.74	0.06	>138
C$_8$F$_{16}$O fluorocarbon	0.64	1.86	<0.0005	>138

Insulating Liquids

Insulating liquids are rarely used by themselves and are intended for use mainly as impregnants with cellulose or synthetic papers. The 60-Hz breakdown strength of practical insulating liquids exceeds that of gases and for a 1-cm gap separation it is of the order of about 100 kV cm^{-1}. However, since the breakdown strength increases with decreasing gap length and as the oils are normally evaluated using a gap separation of 0.254 cm, the breakdown strengths normally cited range from approximately 138 to 240 kV cm^{-1} (cf. Table 52.1). The breakdown values are more influenced by the moisture and particle contents of the fluids than by their molecular structure.

Mineral oils have been extensively used in high-voltage electrical apparatus since the turn of the century. They constitute a category of hydrocarbon liquids that are obtained by refining petroleum crudes. Their composition consists of paraffinic, naphthenic, and aromatic constituents and is dependent upon the source of the crude as well as the refining procedure followed. The inclusion of the aromatic constituents is desirable because of their gas absorption and oxidation characteristics. Mineral oils used for cable and transformer applications have low polar molecule contents and are characterized by dielectric constants extending from about 2.10 to 2.25 with dissipation factors generally between 2×10^{-5} and 6×10^{-5} at room temperature, depending upon their viscosity and molecular weight. Their dissipation factors increase appreciably at higher temperatures when the viscosities are reduced. Oils may deteriorate in service due to oxidation and moisture absorption. Filtering and treatment with Fullers' earth may improve their properties, but special care must be taken to ensure that the treatment process does not remove the aromatic constituents which are essential to maintaining the gas-absorption characteristics of the oil.

Alkyl benzenes are used as impregnants in high-voltage cables, often as substitutes of the low-viscosity mineral oils in self-contained oil-filled cables. They consist of alkyl chains attached to a benzene ring having the general formula $C_6H_5(CH_2)_nCH_3$; however, branched alkyl benzenes are also employed. Their electrical properties are comparable to those of mineral oils, and they exhibit good gas inhibition characteristics. Due to their detergent character, alkyl benzenes tend to be more susceptible to contamination than mineral oils.

Polybutenes are synthetic oils that are derived from the polymerization of olefins. Their long chains, with isobutene as the base unit, have methyl group side chains with molecular weights in the range from 300 to 1350. Their electrical properties are comparable to those of mineral oils; due to their low cost, they have been used as pipe cable filling oils. Higher viscosity polybutenes have been used as capacitor impregnants. Mixtures of polybutenes and alkyl benzenes have been used to obtain higher ac breakdown strength with impregnated paper systems. They are also compatible and miscible with mineral oils.

Since the discontinued use of the nonflammable polychlorinated biphenyls (PCBs), a number of unsaturated synthetic liquids have been developed for application to capacitors, where due to high stresses evolved gases may readily undergo partial discharge. Most of these new synthetic capacitor fluids are thus gas-absorbing low-molecular-weight derivatives of benzene, with permittivities ranging from 2.66 to 5.25 at room temperature (compared to 3.5 for PCBs). None of these fluids have the nonflammable characteristics of the PCBs; however, they do have high boiling points [2].

Halogenated aliphatic hydrocarbons are derived by replacing the hydrogens by either chlorine or fluorine or both; they may also contain nitrogen and oxygen in their molecular structure. Their dielectric constants range from 1.8 to 3.0, the higher value reflecting some polarity due to molecular asymmetry as a result of branching. They have superior thermal properties to mineral oils and are highly flame-resistant. Fluorocarbons have been used in large power transformers, where both flammability and heat removal are of prime concern.

Silicone liquids consist of polymeric chains of silicon atoms alternating with oxygen atoms, with methyl side groups. For electrical applications, polydimethylsiloxane fluids are used, primarily for transformers as substitutes for the PCBs due to their inherently high flash and flammability points and reduced environmental concerns. They have lower tanδ values than mineral oils but somewhat higher dielectric constants because of their modestly polar nature. The viscosity of silicone fluids exhibits relatively little change with temperature, which is attributed to the ease of rotation about the Si–O–Si bond, thereby overcoming close packing of molecules and reducing intermolecular forces.

There are a large number of organic esters, but only a few are suitable for electrical applications. Their properties are adversely affected by hydrolysis, oxidation, and water content. Due to their reduced dielectric losses at elevated frequencies, they have been used in high-frequency capacitors. Castor oil has found specialized application in energy storage capacitors due to its exceptional resistance to partial discharges. The dielectric constants of esters are substantially higher than those for mineral oils.

Solid Insulating Materials

Solid insulating materials may be classified into two main categories, organic and inorganic. There are an extremely large number of solid insulants available, but in this section only the more commonly representative solid insulants will be considered.

Inorganic Solids

Below are described a number of the more prevalent inorganic dielectrics in use; their electrical and physical properties are listed in Table 52.2.

Alumina (Al_2O_3) is produced by heating aluminum hydroxide or oxyhydroxide; it is widely used as a filler for ceramic insulators. Further heating yields the corundum structure, which in its sapphire form is used for dielectric substrates in microcircuit applications.

Barium titanate ($BaTiO_3$) is an extraordinary dielectric in that below 120°C it behaves as a ferroelectric. That is, the electric displacement is both a function of the field as well as its previous history. Due to spontaneous polarization of the crystal, a dielectric hysteresis loop is generated. The dielectric constant is different in the x and z axis of the crystal (e.g., at 20°C, $\varepsilon' > 4000$ perpendicular to the z axis and $\varepsilon' < 300$ in the x-axis direction.

Porcelain is a multiphase ceramic material that is obtained by heating aluminum silicates until a mullite ($3Al_2O_3 \cdot 2SiO_2$) phase is formed. Since mullite is porous, its surface must be glazed with a high-melting-point glass to render it smooth and impervious for use in overhead line insulators. For high-frequency applications, low-loss single-phase ceramics, such as steatite ($3MgO \cdot 4SiO_2 \cdot H_2O$), are preferred.

Magnesium oxide (MgO) is a common inorganic insulating material, which due to its relatively high thermal conductivity is utilized for insulating heating elements in ovens. The resistance wire elements are placed concentrically within stainless steel tubes, with magnesium oxide packed around them to provide the insulation.

Electrical-grade glasses consist principally of SiO_2, B_2O_3, and P_2O_5 structures that are relatively open to permit ionic diffusion and migration. Consequently, glasses tend to be relatively lossy at high temperatures, though at low temperatures they are suitable for use in overhead line insulators and in transformer, capacitor, and circuit breaker bushings. At high temperatures, their main application lies with incandescent and fluorescent lamps as well as electronic tube envelopes.

Most of the mica used in electrical applications is of the muscovite [$KAl_2(OH)_2Si_3AlO_{10}$] type. Mica is a layer-type dielectric, and mica films are obtained by the splitting of mica blocks. The extended two-dimensionally layered strata of mica prevent the formation of conductive pathways across the mica, resulting in a high dielectric strength. It has excellent thermal stability and due to its inorganic nature is highly resistant to partial discharges. It is used in sheet, plate, and tape form in rotating machines and transformer coils. For example, a mica-epoxy composite is employed in stator bar insulation of rotating machines.

In metal-oxide-silicon (MOS) devices, the semiconductor surface is passivated by thermally growing a silicon dioxide, SiO_2, film (about 5000 Å) with the semiconductor silicon wafer exposed to an oxygen ambient at 1200°C. The resulting SiO_2 dielectric film has good adhesion properties, but due to its open glassy structure is not impervious to ionic impurities (primarily sodium). Accordingly, a denser film structure of silicon nitride, Si_3N_4, is formed in a reaction between silane and ammonia and is pyrolytically deposited on the SiO_2 layer. The thin film of Si_3N_4 is characterized by extremely low losses, and its relatively closed structure does not provide any latitude for free sodium movement, thereby providing complete passivation of the semiconductor device. The high dielectric strength of the double film layer of SiO_2 and Si_3N_4 renders it dielectrically effective in field-effect transistor (FET) applications.

In integrated circuit devices, a number of materials are suitable for thin-film capacitor applications. In addition to Al_2O_3, tantalum pentoxide, Ta_2O_5, has been extensively utilized. It is characterized by high-temperature stability and is resistant to acids with the exception of hydrofluoric acid (HF). The high dielectric constant material hafnia (HfO_2) has also been used in thin-film capacitors.

Organic Solids

Solid organic dielectrics consist of large polymer molecules, which generally have molecular weights in excess of 600. Primarily, with the exception of paper, which consists of cellulose that is comprised of a series of glucose units, organic dielectric materials are synthetically derived.

Polyethylene (PE) is perhaps one of the most common solid dielectrics, which is extensively used as a solid dielectric extruded insulant in power and communication cables. Linear PE is classified as a low- (0.910–0.925), medium- (0.926–0.940), or high- (0.941–0.965) density polymer (cf. Table 52.2). Since PE is essentially a long-chain hydrocarbon material in which the repeat unit is $-CH_2-CH_2-$, a low-density PE necessarily implies a high degree of branching. Decreased branching increases the crystallinity as molecules undergo internal folding, which leads to improved stiffness, tear strength, hardness, and chemical resistance. Cross linking of PE produces a thermosetting polymer with a superior temperature rating, improved tensile strength, and an enhanced resistance to partial discharges. Most of the PE used on extruded cables is of the cross-linked polyethylene (XLPE) type.

Ethylene-propylene rubber (EPR) is an amorphous elastomer, which is synthesized from ethylene and propylene. It is used as an extrudent on cables where its composition has a filler content that usually exceeds 50%, comprising primarily clay, with smaller amounts of added silicate and carbon black. The dielectric losses are appreciably enhanced by the fillers, and, consequently, EPR is not suitable for extra-high-voltage applications, with its use being confined to intermediate voltages (≤69 kV) and also where high cable flexibility due to its rubber properties may be additionally desired.

Polypropylene has a structure related to that of ethylene with one added methyl group and is a thermoplastic material having properties similar to high-density PE, though due to its lower density it has also a lower dielectric constant. It has many electrical applications both in bulk form as in molded and extruded insulations as well as in film form in taped capacitor, transformer, and cable insulations.

Polytetrafluoroethylene (PTFE) or Teflon is a fully fluorinated version of PE, having a repeat unit of $[-CF_2-CF_2-]$. It is characterized by a low dielectric constant , extremely low losses, and has excellent temperature stability and is resistive to chemical degradation. It has been extensively used in specialized applications on insulators, wires and cables, transformers, motors, and generators. Its relatively high cost is attributable to both the higher cost of the fluorinated monomers as well as the specialized fabrication techniques required.

Polyesters are obtained most commonly by reacting a dialcohol with a diester; they may be either thermosetting or thermoplastic. The former are usually employed in glass laminates and glass-fiber-reinforced moldings, while thermoplastic polyesters are used for injection-molding applications. They are used in small and large electrical apparatus as well as in electronic applications.

Polyimides, as nylons (polyamides), have nitrogen in their molecular structure. They constitute a class of high-temperature thermoplastics that may be exposed to continuous operation at 480°F.

Table 52.2 Electrical and Physical Properties of Some Common Solid Insulating Materials [1, 4, 10]

Material	Specific Gravity	Maximum Operating Temperature (°C)	Dielectric Constant 20°C			Dissipation Factor 20°C			AC Dielectric Strength (kV cm⁻¹)
			60 Hz	1 kHz	1 MHz	60 Hz	1 kHz	1 MHz	
Alumina (Al_2O_3)	3.1–3.9	1950	8.5	8.5	8.5	1×10^{-3}	1×10^{-3}	1×10^{-3}	98–157
Porcelain (mullite)	2.3–2.5	1000	8.2	8.2	8.2	1.4×10^{-3}	5.7×10^{-4}	2×10^{-4}	94–157
Steatite $3MgO \cdot 4SiO_2 \cdot H_2O$	2.7–2.9	1000–1100	5.5	5.0	5.0	1.3×10^{-3}	4.5×10^{-4}	3.7×10^{-4}	200
Magnesium oxide (MgO)	3.57	<2800	9.65	9.65	9.69	$<3 \times 10^{-4}$	$<3 \times 10^{-4}$	$<3 \times 10^{-4}$	>2000
Glass (soda lime)	2.47	110–460	6.25	6.16	6.00	5.0×10^{-3}	4.2×10^{-3}	2.7×10^{-3}	4500
Mica $(KAl_2(OH)_2Si_3AlO_{10})$	2.7–3.1	550	6.9	6.9	5.4	1.5×10^{-3}	2.0×10^{-4}	3.5×10^{-4}	3000–8200
SiO_2 film		<900		3.9			7×10^{-4}		1000–10,000
Si_3N_4		<1000		12.7			$<1 \times 10^{-4}$		1000–10,000
Ta_2O_5	8.2	<1800		28			1×10^{-2}		
HfO_2		4700°F		35			1×10^{-2}		
Low-density PE	(density: 0.910–0.925 g cm⁻³)	70	2.3	2.3	2.3	2×10^{-4}	2×10^{-4}	2×10^{-4}	181–276
Medium-density PE	(density: 0.926–0.940 g cm⁻³)	70	2.3	2.3	2.3	2×10^{-4}	2×10^{-4}	2×10^{-4}	197–295
High-density PE	(density: 0.941–0.965 g cm⁻³)	70	2.35	2.35	2.35	2×10^{-4}	2×10^{-4}	2×10^{-4}	177–197
XLPE	(density: 0.92 g cm⁻³)	90	2.3		2.28	3×10^{-4}		4×10^{-4}	217

Table 52.2 Electrical and Physical Properties of Some Common Solid Insulating Materials [1, 4, 10] (continued)

Material	Specific Gravity	Maximum Operating Temperature (°C)	Dielectric Constant 20°C			Dissipation Factor 20°C			AC Dielectric Strength (kV cm^{-1})
			60 Hz	1 kHz	1 MHz	60 Hz	1 kHz	1 MHz	
EPR	0.86	300–350°F		3.0–3.5		4×10^{-3}			354–413
Polypropylene	0.90	128–186	2.22–2.28	2.22–2.28	2.22–2.28	2–3×10^{-4}	2.5–3.0×10^{-4}	4.6×10^{-4}	295–314
PTFE	2.13–2.20	<327	2.0	2.0	2.0	$<2 \times 10^{-4}$	$<2 \times 10^{-4}$	$<2 \times 10^{-4}$	189
Glass-reinforced polyester premix	1.8–2.3	265	5.3–7.3		5.0–6.4	1–4×10^{-2}		0.8–2.2×10^{-2}	90.6–158
Thermoplastic polyester	1.31–1.58	250	3.3–3.8 (100 Hz)			1.5–2.0×10^{-3}			232–295
Polyimide polyester	1.43–1.49	480°F		3.4 (100 kHz)			1–5×10^{-3} (100 kHz)		220
Polycarbonate	1.20	215	3.17		2.96	9×10^{-4}		1×10^{-2}	157
Epoxy (with mineral filler)	1.6–1.9	200 (decomposition temperature)	4.4–5.6	4.2–4.9	4.1–4.6	1.1–8.3×10^{-2}	0.19–1.4×10^{-1}	0.13–1.4×10^{-1}	98.4–158
Epoxy (with silica filler)	1.6–2.0	200 (decomposition temperature)	3.2–4.5	3.2–4.0	3.0–3.8	0.8–3.0×10^{-2}	0.8–3.0×10^{-2}	2–4×10^{-2}	158–217
Silicone rubber	1.1–1.5	700°F	3.3–4.0		3.1–3.7	1.5–3.0×10^{-2}		3.0–5.0×10^{-3}	158–197

When glass-reinforced, they may be exposed to temperatures as high as 700°F; they are used both in molded and film form.

Polycarbonates are thermoplastics that are closely related to polyesters. They are primarily employed in the insulation of electrically powered tools and in the casings of electrical appliances. Polycarbonates may be either compression- or injection-molded and extruded as films or sheets.

Epoxy resins are prepared from an epoxide monomer. The first step involves a reaction between two comonomers, and in the subsequent step the prepolymer is cured by means of a cross-linking agent. Epoxy resins are characterized by low shrinkage and high mechanical strength; they may be reinforced with glass fibers and mixed with mica flakes. Epoxy resins have many applications such as, for example, for insulation of bars in the stators of rotating machines, solid-type transformers, and spacers for compressed-gas-insulated busbars and cables.

Silicone rubber is classified as an organic-inorganic elastomer, which is obtained from the polymerization of organic siloxanes. They are composed of dimethyl-siloxane repeat units, $(CH_3)_2SiO-$, with the side groups being methyl units. Fillers are added to obtain the desired silicone rubber compounds; cross linking is carried out with peroxides. Since no softeners and plasticizers are required, silicone rubbers are resistant to embrittlement and may be employed for low-temperature applications down to −120°F. Continuous operation is possible up to 500°F, with intermittent usage as high as 700°F.

Solid-Liquid Insulating Systems

Impregnated-paper insulation constitutes one of the earliest insulating systems employed in electrical power apparatus and cables. Although in some applications alternate solid- or compressed-gas insulating systems are now being used, the impregnated-paper system still constitutes one of the most reliable insulating systems available. Proper impregnation of the paper results in a cavity-free insulating system, thereby eliminating the occurrence of partial discharges that inevitably lead to deterioration and breakdown of the insulating system. The cellulose structure of paper has a finite acidity content as well as a residual colloidal or bound water, which is held by hydrogen bonds. Consequently, impregnated cellulose base papers are characterized by somewhat more elevated tanδ values in the order of 2×10^{-3} at 30 kV cm^{-1}. The liquid impregnants employed are either mineral oils or synthetic fluids. Since the dielectric constant of these fluids is normally about 2.2 and that of dried cellulose about 6.5–10, the resulting dielectric constant of the impregnated paper is approximately 3.1–3.5.

Lower-density cellulose papers have slightly lower dielectric losses, but the dielectric breakdown strength is also reduced. The converse is true for impregnated systems utilizing higher-density papers. The general chemical formula of cellulose paper is $C_{12}H_{20}O_{10}$. If the paper is heated beyond 200°C, the chemical structure of the paper breaks down even in the absence of external oxygen, since the latter is readily available from within the cellulose molecule. To avert this process from occurring, cellulose papers are ordinarily not used beyond 100°C.

Table 52.3 Electrical Properties of Taped Solid-Liquid Insulations [4]

Tape	Impregnating Liquid	Average Voltage Stress (kV cm^{-1})	tanδ at Room Temperature	tanδ at Operating Temperature
Kraft paper	Mineral oil	180	3.8×10^{-3} at 23°C	5.7×10^{-3} at 85°C
Kraft paper	Silicone liquid	180	2.7×10^{-3} at 23°C	3.1×10^{-3} at 85°C
Paper-polypropylene-paper (PPP)	Dodecyl benzene	180	9.8×10^{-4} at 18°C	9.9×10^{-4} at 100°C
Kraft paper	Polybutene	180	2.0×10^{-3} at 25°C	2.0×10^{-3} at 85°C

In an attempt to reduce the dielectric losses in solid-liquid systems, cellulose papers have been substituted in some applications by synthetic papers (cf. Table 52.3). For example in extra-high-voltage cables, cellulose paper–polypropylene composite tapes have been employed. A partial paper content in the composite tapes is necessary both to retain some of the impregnation capability of a porous cellulose paper medium and to maintain the relative ease of cellulose-to-cellulose tape sliding capability upon bending. In transformers the synthetic nylon or polyamide paper nomex has been used both in film and board form. It may be continuously operated at temperatures up to 220°C.

Defining Terms

Conductivity σ: Represents the ratio of the leakage current density to the applied electric field density. In general its ac and dc values differ because the mechanisms for establishing the leakage current in the two cases are not necessarily identical.

Dielectric: A material in which nearly all or a large portion of the energy required for its charging can be recovered when the external electric field is removed.

Dielectric constant ε':   A quantity that determines the amount of electrostatic energy which can be stored per unit volume per unit potential gradient. It is a real quantity and in Gaussian–CGS units is numerically equal to the ratio of the measured capacitance of the specimen, C, to the equivalent geometrical capacitance *in vacuo*, C_o. It is also commonly referred to as the real value of the permittivity. Note that when the SI system of units is employed, the ratio C/C_o defines the real value of the *relative permittivity* ε'_r.

Dielectric loss: The rate at which the electrical energy supplied to a dielectric material by an alternating electrical field is changed to heat.

Dielectric strength: Represents the value of the externally applied electric field at which breakdown or failure of the dielectric takes place. Unless a completely uniform field gradient can be assured across the dielectric specimen, the resulting breakdown value will be a function of the specimen thickness and the test electrode geometry; this value will be substantially below that of the intrinsic breakdown strength.

Dissipation factor (tan δ):   Equal to the tangent of the loss angle δ, which is the phase angle between the external electric field vector $\bar{E}$ and the resulting displacement vector $\bar{D}$. It is numerically equal to the ratio of the imaginary permittivity ε'' to the real permittivity, ε'; alternatively, it is defined by the ratio of the leakage current to the displacement (charging or capacitive) current.

References

1. R. Bartnikas and R. M. Eichhorn (eds.), *Engineering Dielectrics,* Vol. II A, *Electrical Properties of Solid Insulating Materials: Molecular Structure and Electrical Behavior,* STP 783, Philadelphia: ASTM, 1983.
2. R. Bartnikas (ed.), *Engineering, Dielectrics,* Vol. III, *Electrical Insulating Liquids,* Monograph 2, Philadelphia: ASTM, 1993 (in print).
3. A. von Hippel, *Dielectrics and Waves,* New York: Wiley, 1954.
4. R. Bartnikas and K.D. Srivastava (eds.), *Power Cable Engineering,* Waterloo, Ontario: Sandford Educational Press, 1987.
5. A. von Hippel and G.M. Lee. *Phys. Rev.,* vol. 59, pp. 824–826, 1941.
6. W.G. Oakes, *Proc. IEE,* vol. 90(I), pp. 37–43, 1949.
7. R. Bartnikas (ed.), *Engineering Dielectrics,* Vol. II B, *Measurement Techniques,* STP 926, Philadelphia: ASTM, 1987.
8. R. Bartnikas and E.J. McMahon (eds.), *Engineering Dielectrics,* Vol. I, *Corona Measurement and Interpretation,* STP 669, Philadelphia: ASTM, 1979.

9. J. D. Cobine, *Gaseous Conductors,* New York: McGraw-Hill, 1941.
10. Encyclopedia Issue, *Insul. Circuits,* June/July 1972.

Further Information

The IEEE Dielectrics and Electrical Insulation Society publishes regularly its *IEEE Transactions on Electrical Insulation,* wherein many new developments in the field of dielectrics are recorded in permanent form. It also sponsors on either an annual or a biennial basis a number of conferences, which provide a forum for rapid dissemination of both the applied and fundamental work carried out on dielectrics and electrical insulating systems. The reader may wish to consult the IEEE Conference Records on the Annual Report, Conference on Electrical Insulation and Dielectric Phenomena, the IEEE International Symposium on Electrical Insulation, and the Electrical/Electronics Insulation Conference. Also, a description of the different test methods on dielectric materials may be found in the *ASTM Book on Standards.*

Photograph taken in the General Electric Research Laboratories in 1923, showing, from the left, Irving Langmuir, Joseph J. Thomson, and William D. Coolidge. They are inspecting an early pliotron, a high-vacuum tube that was used both as a powerful amplifier and as an oscillator. Thomson was the discoverer of the electron, Langmuir was one of the pioneers in the development of thermionic-emission electron tubes, and Coolidge was a pioneer in the development of high-intensity x-ray tubes. (Courtesy of General Electric Company.)

53

Sensors

53.1 Introduction 1152
53.2 Physical Sensors 1155
    Temperature Sensors • Displacement and Force • Optical Radiation
53.3 Chemical Sensors 1157
    Ion-Selective Electrode • Gas Chromatograph
53.4 Biosensors 1158
    Immunosensor • Enzyme Sensor
53.5 Microsensors 1159

Rosemary L. Smith
University of California, Davis

53.1 Introduction

Sensors are critical components in all measurement and control systems. The need for computer-compatible sensors closely followed the advent of the microprocessor. Together with the always-present need for sensors in science and medicine, the demand for sensors in automated manufacturing and processing is rapidly growing. In addition, small, inexpensive sensors are finding their way into all sorts of consumer products, from childrens' toys to dishwashers to automobiles. Because of the vast variety of useful things to be sensed and sensor applications, sensor engineering is a multidisciplinary and interdisciplinary field of endeavor. This chapter introduces some basic definitions, concepts, and features of sensors and illustrates them with several examples. The reader is directed to the references and the sources listed under "Further Information" for more details and examples.

There are many terms which are often used synonymously for sensor, including transducer, meter, detector, and gage. Defining the term sensor is not an easy task; however the most widely used definition is that which has been applied to electrical transducers by the Instrument Society of America (ANSI MC6.1, 1975): *Transducer—A device which provides a usable output in response to a specified measurand.* A transducer is more generally defined as a device which converts energy from one form to another. A usable ouput refers to an optical, electronic, or mechanical signal. In the context of electrical engineering, however, a usable output refers to an electronic output signal. Therefore, the following discussion will be limited to electronic sensors. The measurand can be a physical, chemical, or biological property or condition to be measured.

Most, but not all, sensors are transducers, employing one or more transduction mechanisms to produce an electronic output signal. Sometimes sensors are classified as direct and indirect sensors according to how many transduction mechanisms are used. For example, a mercury thermometer produces a change in volume of mercury in response to a temperature change via thermal expansion, but the output is a mechanical displacement and not electrical. Another transduction mechanism is required. A thermometer is still a useful sensor since humans can read the change in

Table 53.1 Physical and Chemical Transduction Principles

Primary Signal	Secondary Signal					
	Mechanical	Thermal	Electrical	Magnetic	Radiant	Chemical
Mechanical	(Fluid) mechanical and acoustic effects (e.g., diaphragm, gravity balance, echo sounder)	Friction effects (e.g., friction calorimeter) Cooling effects (e.g., thermal flow meters)	Piezoelectricity Piezoresistivity Resistive, capacitive, and inductive effects	Magneto-mechanical effects (e.g., piezomagnetic effect)	Photoelastic systems (stress-induced birefringence) Interferometers Sagnac effect Doppler effect	
Thermal	Thermal expansion (bimetal strip, liquid-in-glass and gas thermometers, resonant frequency) Radiometer effect (light mill)		Seebeck effect Thermoresistance Pyroelectricity Thermal (Johnson) noise		Thermooptical effects (e.g., in liquid crystals) Radiant emission	Reaction activation (e.g., thermal dissociation)
Electrical	Electrokinetic and electro-mechanical effects (e.g., piezoelectricity, electro-meter, Ampere's law)	Joule (resistive) heating Peltier effect	Charge collectors Langmuir probe	Biot-Savart's law	Electrooptical effects (e.g., Kerr effect) Pockel's effect Electroluminescence	Electrolysis Electromigration
Magnetic	Magnetomechanical effects (e.g., magnetorestriction, magnetometer)	Thermomagnetic effects (e.g., Righi-Leduc effect) Galvanomagnetic effects (e.g., Ettingshausen effect)	Thermomagnetic effects (e.g., Ettingshausen-Nernst effect) Galvanomagnetic effects (e.g., Hall effect, magnetoresistance)		Magnetooptical effects (e.g., Faraday effect) Cotton-Mouton effect	
Radiant	Radiation pressure	Bolometer thermopile	Photoelectric effects (e.g., photovoltaic effect, photoconductive effect)		Photorefractive effects Optical bistability	Photosynthesis, -dissociation
Chemical	Hygrometer Electrodeposition cell Photoacoustic effect	Calorimeter Thermal conductivity cell	Potentiometry Conductimetry Amperometry Flame ionization Volta effect Gas-sensitive field effect	Nuclear magnetic resonance	(Emission and absorption) spectroscopy Chemiluminiscence	

Source: T. Grandke and J. Hesse, Introduction, Vol. 1: *Fundamentals and General Aspects, Sensors: A Comprehensive Survey*, W. Gopel, J. Hesse, and J. H. Zemel, Eds., Weinheim, Germany: VCH, 1989. With permission.

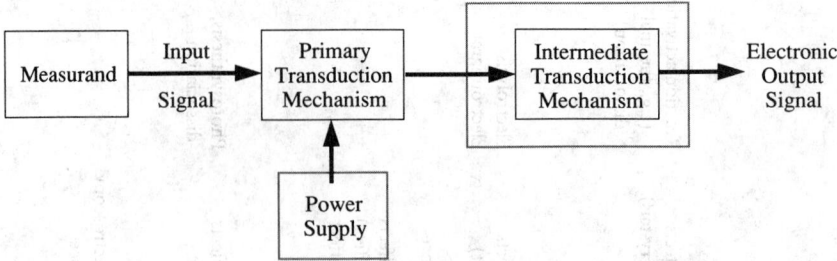

FIGURE 53.1 Sensor block diagram. Active sensors require input power to accomplish transduction. Many sensors employ multiple transduction mechanisms in order to produce an electronic output in response to the measurand.

mercury height using their eyes as the second transducing element. However, in order to produce an electrical output for use in a control loop, the height of the mercury would have to be converted to an electrical signal. This could be accomplished using capacitive effects. However, there are more direct temperature sensing methods, i.e., one where an electrical output is produced in response to a change in temperature. An example is given in the next section on physical sensors. Figure 53.1 depicts a sensor block diagram identifying the measurand and associated input signal, the primary and intermediate transduction mechanisms, and the electronic output signal. Active sensors require an external power source in order to produce a usable output signal, e.g., the piezoresistor. Table 53.1 is a 6×6 matrix of the more commonly employed physical and chemical transduction mechanisms. Many of the effects listed are described in more detail elsewhere in this handbook. (Detailed descriptions of the others can be found in most college-level physics textbooks.)

In choosing a particular sensor for a given application, there are many factors to be considered. These deciding factors or specifications can be divided into three major categories: environmental factors, economic factors, and the sensor characteristics. The most commonly encountered factors are listed in Table 53.2, although not all of these factors may be pertinent to a particular application. Most of the environmental factors determine the packaging of the sensor, with packaging meaning the encapsulation or insulation which provides protection and isolation and the input/output leads or connections and cabling. The economic factors determine the type of manufacturing and materials used in the sensor and to some extent the quality of the materials (with respect to lifetime). For example, a very expensive sensor may be cost effective if it is used repeatedly or for very long periods of time. On the other hand, a disposable sensor, such as is desired in many medical applications, should be inexpensive. The sensor characteristics of the sensor are usually the specifications of primary concern. The most important parameters are **sensitivity, stability,** and **repeatability.** Normally, a sensor is only useful if all three of these parameters are tightly specified for a given range of measurand and time of operation. For example, a highly sensitive device is not useful if its output signal drifts greatly during the measurement time and the data

Table 53.2

Environmental Factors	Economic Factors	Sensor Characteristics
Temperature range	Cost	Sensitivity
Humidity effects	Availability	Range
Corrosion	Lifetime	Stability
Size		Repeatability
Overrange protection		Linearity
Susceptibility to EM interferences		Error
Ruggedness		Response time
Power consumption		Frequency response
Self-test capability		

obtained is not reliable if the measurement is not repeatable. Other output characteristics, such as selectivity and linearity, can often be compensated for by using additional, independent sensor input or with signal conditioning circuits. In fact, most sensors have a response to temperature, since most tranducing effects are temperature dependent.

Sensors are most often classified by the type of measurand, i.e., physical, chemical, or biological. This is a much simpler means of classification than by transduction mechanism or output signal (e.g., digital or analog), since many sensors use multiple transduction mechanisms and the output signal can always be processed, conditioned, or converted by a circuit so as to cloud the definition of output. A description of each class and examples are given in the following sections. The last section introduces microsensors and gives some examples.

53.2 Physical Sensors

Physical measurands include temperature, strain, force, pressure, displacement, position, velocity, acceleration, optical radiation, sound, flow rate, viscosity, and electromagnetic fields. Referring to Table 53.1, all but those transduction mechanisms listed in the chemical column are used in the design of physical sensors. Clearly, they comprise a very large proportion of all sensors. It is impossible to illustrate all of them, but three measurands stand out in terms of their widespread application: temperature, displacement (or associated force), and optical radiation.

Temperature Sensors

Temperature is an important parameter in many control systems, most familiarly in environmental control systems. Several distinctly different transduction mechanisms have been employed. The mercury thermometer was mentioned in the Introduction as a nonelectronic sensor. The most commonly used electronic temperature sensors are thermocouples, thermistors, and resistance thermometers. Thermocouples employ the Seebeck effect, which occurs at the junction of two dissimilar metal wires. A voltage difference is generated at the hot junction due to the difference in the energy distribution of thermally energized electrons in each metal. This voltage is measured across the cool ends of the two wires and changes linearly with temperature over a given range, depending on the choice of metals. To minimize measurement error the cool end of the couple must be kept at a constant temperature, and the voltmeter must have a high input impedance.

The resistance thermometer relies on the increase in resistance of a metal wire with increasing temperature. As the electrons in the metal gain thermal energy, they move about more rapidly and undergo more frequent collisions with each other and the atomic nuclei. These scattering events reduce the mobility of the electrons, and since resistance is inversely proportional to mobility, the resistance increases. Resistance thermometers consist of a coil of fine metal wire. Platinum wire gives the largest linear range of operation. To determine the resistance indirectly, a constant current is supplied and the voltage is measured. A direct measurement can be made by placing the resistor in the sensing arm of a Wheatstone bridge and adjusting the opposing resistor to "balance" the bridge, which produces a null output. A measure of the sensitivity of a resistance thermometer is its temperature coefficient of resistance: $TCR = (\Delta R/R)(1/\Delta T)$ in units of % resistance per degree of temperature.

Thermistors are resistive elements made of semiconductor materials and have a negative coefficient of resistance. The mechanism governing the resistance change of a thermistor is the increase in the number of conducting electrons with an increase in temperature due to thermal generation, i.e., the electrons which are tightly bound to the nucleus by Coulombic attraction gain sufficient thermal energy to break away from the nucleus and become influenced by external fields. Thermistors can be measured in the same manner as resistance thermometers, but thermistors have up to 100 times higher TCR values.

Displacement and Force

Many types of forces are sensed by the displacements they create. For example, the force due to acceleration of a mass at the end of a spring will cause the spring to stretch and the mass to move. Its displacement from the zero acceleration position is governed by the force generated by the acceleration ($F = m \cdot a$) and the restoring force of the spring. Another example is the displacement of the center of a deformable membrane due to a difference in pressure across it. Both of these examples use multiple transduction mechanisms to produce an electronic output: a primary mechanism which converts force to displacement (mechanical to mechanical) and then an intermediate mechanism to convert displacement to an electrical signal (mechanical to electrical). Displacement can be measured by the associated capacitance. For example, the capacitance associated with a gap which is changing in length is given by $C =$ area $\times$ dielectric constant/gap length. The gap must be very small compared to the surface area of the capacitor, since most dielectric constants are of the order of 1×10^{-15} farads/cm and with present methods, capacitance is readily resolvable to only about 10^{-10} farads. This is because measurement leads and contacts create parasitic capacitances of about 10^{-11} farads. If the capacitance is measured at the generated site by an integrated circuit (see Section 53.5), capacitances as small as 10^{-15} farads can be measured. Displacement is also commonly measured by the movement of a ferromagnetic core inside of an inductor coil. The displacement produces a change in inductance which can be measured by placing the inductor in an oscillator circuit and measuring the change in frequency of oscillation.

The most commonly used force sensor is the strain gage. It consists of metal wires which are fixed to an immobile structure at one end and to a deformable element at the other. The resistance of the wire changes as it undergoes strain, i.e., a change in length, since the resistance of a wire is R = resistivity $\times$ length/cross-sectional area. The wire's resistivity is a bulk property of the metal which is a constant for constant temperature. For example, a strain gage can be used to measure acceleration by attaching both ends of the wire to a cantilever beam, with one end of the wire at the attached beam end and the other at the free end. The cantilever beam free end moves in response to an applied force, such as the force due to acceleration which produces strain in the wire and a subsequent change in resistance. The sensitivity of a strain gage is described by the unitless gage factor, $G = (\Delta R/R)/(\Delta L/L)$. For metal wires, gage factors typically range from 2 to 3. Semiconductors are known to exhibit piezoresistivity, which is a change in resistance in response to strain which involves a large change in resistivity in addition to the change in linear dimension. Piezoresistors have gage factors as high as 130. Piezoresistive strain gages are frequently used in **microsensors,** described in Section 53.5.

Optical Radiation

The intensity and frequency of optical radiation are parameters of growing interest and utility in consumer products such as the video camera and home security systems and in optical communications systems. The conversion of optical energy to electronic signals can be accomplished by several mechanisms (see radiant to electronic transduction in Table 53.1); however, the most commonly used is the photogeneration of carriers in semiconductors. The most often-used device is the *p-n* junction photodiode. The construction of this device is very similar to the diodes used in electronic circuits as rectifiers. The diode is operated in reverse bias, where very little current normally flows. When light is incident on the structure and is absorbed in the semiconductor, energetic electrons are produced. These electrons flow in response to the electric field sustained internally across the junction, producing an externally measurable current. The current magnitude is proportional to the light intensity and also depends on the frequency of the light. Figure 53.2 shows the effects of varying incident optical intensity on the terminal current versus voltage behavior of a *p-n* junction. Note that for zero applied voltage, a net negative current flows when the junction is illuminated. This device can therefore also be a source of power (a solar cell).

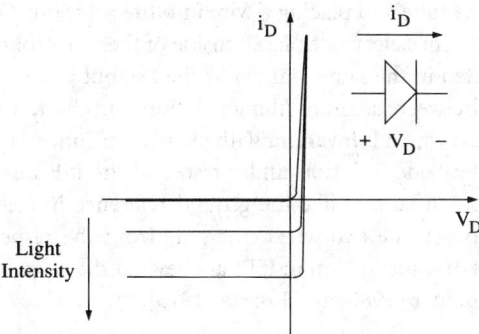

FIGURE 53.2 Sketch of the variation of current versus voltage characteristics of a *p-n* photodiode with incident light intensity.

53.3 Chemical Sensors

Chemical measurands include ion concentration, chemical composition, rate of reactions, reduction-oxidation potentials, and gas concentration. The last column of Table 53.1 lists some of the transduction mechanisms that have been, or could be, employed in chemical sensing. Two examples of chemical sensors are described here: the ion-selective electrode (ISE) and the gas chromatograph. They were chosen because of their general use and availability and because they illustrate the use of a primary (ISE) versus a primary plus intermediate (gas chromatograph) transduction mechanism.

Ion-Selective Electrode

As the name implies, ISEs are used to measure the concentration of a specific ion concentration in a solution of many ions. To accomplish this, a membrane which selectively generates a potential which is dependent on the concentration of the ion of interest is used. The generated potential is usually an equilibrium potential, called the Nernst potential, and develops across the interface of the membrane with the solution. This potential is generated by the initial net flow of ions (charge) across the membrane in response to a concentration gradient, and from thence forth the diffusional force is balanced by the generated electric force and equilibrium is established. This is very similar to the so-called built-in potential of a *p-n* junction diode. The ion-selective membrane acts in such a way as to ensure that the generated potential is dependent mostly on the ion of interest and negligibly on any other ions in solution. This is done by enhancing the exchange rate of the ion of interest across the membrane, so it is the fastest moving and, therefore, the species which generates and maintains the potential.

The most familiar ISE is the pH electrode. In this device the membrane is a sodium glass which possesses a high exchange rate for H^+. The generated Nernst potential, E, is given by the expression: $E = E_0 + (RT/F) \ln[H^+]$, where E_0 is a constant for constant temperature, R is the gas constant, and F is the Faraday constant. pH is defined as the negative of the $\log(H^+)$; therefore $pH = (E_0 - E)(2.03)F/RT$ and one pH unit change corresponds to a tenfold change in the molar concentration of H^+. Other ISEs have the same type of response, but specific to a different ion, depending on the choice of membrane. Many ISEs employ ionophores trapped inside of a polymeric membrane. An ionophore is a molecule which selectively and reversibly binds with an ion and thereby creates a high exchange rate for that particular ion.

The ISE consists of a glass tube with the ion-selective membrane closing that end of the tube which is immersed into the test solution. The Nernst potential is measured by making electrical contact to each side of the membrane. This is done by placing a fixed concentration, conductive filling solution inside of the tube and placing a wire into the solution. The other side of the membrane is contacted by a reference electrode placed inside of the same solution under test. The reference electrode is constructed in the same manner as the ISE but it has a porous membrane which creates a liquid junction between its inner filling solution and the test solution. That junction is designed to have a potential which is invariant with changes in concentration of any ion in the test solution. The reference electrode, solution under test, and the ISE form an electrochemical cell. The reference electrode potential acts like the ground reference in electric circuits, and the ISE potential is measured between the two wires emerging from the respective two electrodes. The details of the mechanisms of transduction in ISEs are beyond the scope of this chapter. The reader is referred to Bard and Faulkner [1980] and Koryta [1975].

Gas Chromatograph

Molecules in gases have thermal conductivities which are dependent on their masses; therefore, a pure gas can be identified by its thermal conductivity. One way to determine the composition of a gas is to first separate it into its components and then measure the thermal conductivity of each. A gas chromatograph does exactly that. The gas flows through a long narrow column, which is packed with an adsorbant solid (for gas–solid chromatography) wherein the gases are separated according to the retentive properties of the packing material for each gas. As the individual gases exit the end of the tube one at a time, they flow over a heated wire. The amount of heat transferred to the gas depends on its thermal conductivity. The gas temperature is measured a short distance downstream and compared to a known gas flowing in a separate sensing tube. The temperature is related to the amount of heat transferred and can be used to derive the thermal conductivity according to thermodynamic theory and empirical data. This sensor required two transductions: a chemical to thermal energy transduction followed by a thermal to electrical transduction.

53.4 Biosensors

Biological measurands are biologically produced substances, such as antibodies, glucose, hormones, and enzymes. Biosensors are not the same as biomedical sensors, which are any sensors used in biomedical applications, such as blood pressure sensors, or electrocardiogram electrodes. Many biosensors are biomedical sensors; however, they are also used in industrial applications, e.g., the monitoring and control of fermentation reactions. Table 53.1 does not include biological signals as a primary signal because they can be classified as either chemical or physical in nature. Biosensors are of special interest because of the very high selectivity of biological reactions and binding. However, the detection of that reaction or binding is often elusive. A very familiar commercial biosensor is the in-home pregnancy test sensor, which detects the presence of human growth factor in urine. That device is a nonelectronic sensor since the output is a color change which the eye senses. In fact, most biosensors require multiple transduction mechanisms to arrive at an electrical output signal. Two examples are given below: an immunosensor and an enzyme sensor. Rather than examine a specific species, the examples describe a general type of sensor and transduction mechanism, since the same principles can be applied to a very large number of biological species of the same type.

Immunosensor

Commercial techniques for detecting antibody-antigen binding utilize optical or x-radiation detection. An optically fluorescent molecule or radioisotope is nonspecifically attached to the species of interest in solution. The complementary binding species is chemically attached to a glass substrate or glass beads which are packed into a column. The tagged solution containing the species of interest, say the antibody, is passed over the antigen-coated surface, where the two selectively bind. After the specific binding occurs, the nonbound fluorescent molecules or radioisotopes are washed away, and the antibody concentration is determined by fluorescence spectroscopy or with a scintillation counter, respectively. These sensing techniques are quite costly and bulky, and therefore other biosensing mechanisms are rapidly being developed. One experimental technique uses the change in the mechanical properties of the bound antibody-antigen complex in comparison to an unbound surface layer of antigen. It uses a sheer mode, surface acoustic wave (SAW) device to sense this change as a change in the propagation time of the wave between the generating electrodes and the pick-up electrodes some distance away on the same piezoelectric substrate. The substrate surface is coated with the antigen and it is theorized that upon selectively binding with the antibody, this layer stiffens, changing the mechanical properties of the interface and therefore the velocity of the wave. The advantages of this device are that the SAW device produces an electrical signal (a change in oscillation frequency when the device is used in the feedback loop of an oscillator circuit) which is dependent on the amount of bound antibody; it requires only a very small amount of the antigen which can be very costly; the entire device is small, robust and portable; and the detection and readout method is inexpensive. There are numerous problems which currently preclude its commercial use, specifically a large temperature sensitivity and responses to nonspecific adsorption, i.e., by species other than the desired antibody.

Enzyme Sensor

Enzymes selectively react with a chemical substance to modify it, usually as the first step in a chain of reactions to release energy (metabolism). A well-known example is the selective reaction of glucose oxidase (enzyme) with glucose to produce gluconic acid and peroxide, according to

$$C_6H_{12}O_6 + O_2 \xrightarrow{\text{glucose oxidase}} \text{gluconic acid} + H_2O_2 + 80 \text{ kilojoules heat}$$

An enzymatic reaction can be sensed by measuring the rise in temperature associated with the heat of reaction or by the detection and measurement of byproducts. In the glucose example, the reaction can be sensed by measuring the local dissolved peroxide concentration. This is done via an electrochemical analysis technique called amperometry [Bard and Faulkner, 1980]. In this method, a potential is placed across two inert metal wire electrodes immersed in the test solution and the current which is generated by the reduction/oxidation reaction of the species of interest is measured. The current is proportional to the concentration of the reducing/oxidizing species. A selective response is obtained if no other available species has a lower redox potential. Because the selectivity of peroxide over oxygen is poor, some glucose sensing schemes employ a second enzyme called catalase which converts peroxide to oxygen and hydroxyl ions. The latter produces a change in the local pH. As described earlier, an ISE can then be used to convert the pH to a measurable voltage. In this latter example, glucose sensing involves two chemical-to-chemical transductions followed by a chemical-to-electrical transduction mechanism.

53.5 Microsensors

Microsensors are sensors that are manufactured using integrated circuit fabrication technologies and/or **micromachining.** Integrated circuits are fabricated using a series of process steps which are done in batch fashion, meaning that thousands of circuits are processed together at the same time in the same way. The patterns which define the components of the circuit are photolithographically transferred from a template to a semiconducting substrate using a photosensitive organic coating. The coating pattern is then transferred into the substrate or into a solid-state thin film coating through an etching or deposition process. Each template, called a mask, can contain thousands of identical sets of patterns, with each set representing a circuit. This "batch" method of manufacturing is what makes integrated circuits so reproducible and inexpensive. In addition, photoreduction enables one to make extremely small features, on the order of microns, which is why this collection of process steps is referred to as microfabrication. The resulting integrated circuit is contained in only the top few microns of the semiconductor substrate and the submicron thin films on its surface. Hence, integrated circuit technology is said to consist of a set of planar, microfabrication processes. Micromachining refers to the set of processes which produce three-dimensional microstructures using the same photolithographic techniques and batch processing as for integrated circuits. Here, the third dimension refers to the height above the substrate of the deposited layer or the depth into the substrate of an etched structure. Micromachining produces third dimensions in the range of 1–500 μm (typically). The use of microfabrication to manufacture sensors produces the same benefits as it does for circuits: low cost per sensor, small size, and highly reproducible behavior. It also enables the integration of signal conditioning, compensation circuits and actuators, i.e., entire sensing and control systems, which can dramatically improve sensor performance for very little increase in cost. For these reasons, there is a great deal of research and development activity in microsensors.

The first microsensors were integrated circuit components, such as semiconductor resistors and *p-n* junction diodes. The piezoresistivity of semiconductors and optical sensing by the photodiode were already discussed. Diodes are also used as temperature-sensing devices. When forward-biased with a constant diode current, the resulting diode voltage increases approximately linearly with increasing temperature. The first micromachined microsensor to be commercially produced was the silicon pressure sensor. It was invented in the mid-to-late 1950s at Bell Labs and commercialized in the 1960s by General Electric, Endevco, and Fairchild Control Division (now Foxboro/ICT, Inc.). This device contains a thin silicon diaphragm (≈10 μm) which deforms in response to a pressure difference across it (Fig. 53.3). The deformation produces two effects: a position-dependent displacement which is maximum at the diaphragm center and position-dependent strain which is maximum near the diaphragm edge. Both of these effects have been used in microsensors to produce an electrical output which is proportional to differential pressure. The membrane displacement is

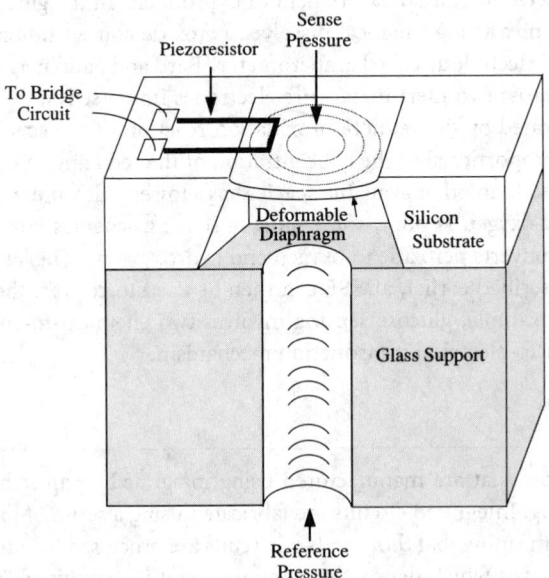

FIGURE 53.3 Schematic cross section of a silicon piezoresistive pressure sensor. A differential pressure deforms the silicon diaphragm, producing strain in the integrated piezoresistor. The change in resistance is measured via a Wheatstone bridge.

sensed capacitively as previously described in one type of pressure sensor. The strain is sensed in another by placing a piezoresistor, fabricated in the same silicon substrate, along one edge of the diaphragm. The two leads of the piezoresistor are connected to a Wheatstone bridge. The latter type of sensor is called a piezoresistive pressure sensor and is the commercially more common type of pressure microsensor. Pressure microsensors constituted about 5% of the total U.S. consumption of pressure sensors in 1991. Most of them are used in the medical industry as disposables due to their low cost and small, rugged construction. Many other types of microsensors are commercially under development, including accelerometers, mass flow rate sensors, and biosensors.

Defining Terms

Micromachining: The set of processes which produce three-dimensional microstructures using sequential photolithographic pattern transfer and etching or deposition in a batch processing method.

Microsensor: A sensor which is fabricated using integrated circuit and micromachining technologies.

Repeatability: The ability of a sensor to reproduce output readings for the same value of measurand, when applied consecutively and under the same conditions.

Sensitivity: The ratio of the change in sensor output to a change in the value of the measurand.

Sensor: A device which produces a usable output in response to a specified measurand.

Stability: The ability of a sensor to retain its characterisctics over a relatively long period of time.

References

[ANSI], "Electrical Transducer Nomenclature and Terminology," ANSI Standard MC6.1-1975 (ISA S37.1), Research Triangle Park, N.C.: Instrument Society of America, 1975.

A. J. Bard and L. R. Faulkner, *Electrochemical Methods: Fundamentals and Applications,* New York: John Wiley & Sons, 1980.

R. S. C. Cobbold, *Transducers for Biomedical Measurements: Principles and Applications,* New York: John Wiley & Sons, 1974.

W. Göpel, J. Hesse, and J. N. Zemel, Eds., *Sensors: A Comprehensive Survey,* vol. 1, *Fundamentals and General Aspects,* T. Grandke and W. H. Ko, Eds., Weinheim, Germany: VCH, 1989.

J. Koryta, *Ion-Selective Electrodes,* New York: Cambridge University Press, 1975.

H. N. Norton, *Handbook of Transducers,* Englewood Cliffs, N.J.: Prentice-Hall, 1989.

S. Wolf, "Guide to electronic measurements and laboratory practice," in *Prentice-Hall Electrical Engineering Series,* Englewood Cliffs, N.J.: Prentice-Hall, 1973.

Further Information

Sensors: A Comprehensive Survey, W. Gopel, J. Hesse, and J. N. Zemel, editors. VCH, Weinheim.

Vol. 1: *Fundamentals and General Aspects,* 1989

Vol. 2, 3: *Mechanical Sensors,* 1991

Vol. 4: *Thermal Sensors,* 1990

Vol. 5: *Magnetic Sensors,* 1989

Vol. 6: *Optical Sensors,* 1991

Vol. 7, 8: *Chemical and Biochemical Sensors,* 1990

Sensors and Actuators is a technical journal devoted to solid-state sensors and actuators, which is published bimonthly by Elsevier Press in two volumes: Vol. A: *Physical Sensors* and Vol. B: *Chemical Sensors.*

The International Conference on Solid-State Sensors and Actuators is held every 2 years, hosted in rotation by the U.S., Japan, and Europe. It is sponsored in part by IEEE in the U.S. and a digest of technical papers is published and available through IEEE. The most recent was *Transducers '91,* San Francisco.

54

Magnetooptics

David Young
Rockwell International

Chih-Lin Wang
Electro-Optek Corporation

Yuan Pu
University of California, Irvine

54.1 Introduction 1162
54.2 Classification of Magnetooptic Effects 1163
    Faraday Rotation or Magnetic Circular Birefringence • Cotton-Mouton Effect or Magnetic Linear Birefringence • Kerr Effects
54.3 Applications of Magnetooptic Effects 1166
    Optical Isolator and Circulator • MSW-Based Guided-Wave Magnetooptic Bragg Cell • Magnetooptic Recording

54.1 Introduction

When a magnetic field **H** is applied to a magnetic medium (crystal), a change in the magnetization **M** within the medium will occur as described by the constitution relation of the Maxwell equations $\mathbf{M} = \overleftrightarrow{\chi} \cdot \mathbf{H}$ where $\overleftrightarrow{\chi}$ is the magnetic susceptibility tensor of the medium. The change in magnetization can in turn induce a perturbation in the complex optical permittivity tensor $\overleftrightarrow{\varepsilon}$. This phenomenon is called the magnetooptic effect. Mathematically, the magnetooptic effect can be described by expanding the permittivity tensor as a series in increasing powers of the magnetization [Torfeh *et al.*, 1977] as follows:

$$\overleftrightarrow{\varepsilon} = \varepsilon_0[\varepsilon_{ij}] \tag{54.1}$$

where

$$\varepsilon_{ij}(\mathbf{M}) = \varepsilon_r\delta_{ij} + jf_1\, e_{ijk}M_k + f_{ijkl}M_kM_l$$

Here j is the imaginary number. M_1, M_2, and M_3 are the magnetization components along the principal crystal axes X, Y, and Z, respectively. ε_0 is the permittivity of free space. ε_r is the relative permittivity of the medium in the paramagnetic state (i.e., $\mathbf{M} = 0$), f_1 is the first-order magnetooptic scalar factor, f_{ijkl} is the second-order magnetooptic tensor factor, δ_{ij} is the Kronecker delta, and e_{ijk} is the antisymmetric alternate index of the third order. Here we have used Einstein notation of repeated indices and have assumed that the medium is quasi-transparent so that $\overleftrightarrow{\varepsilon}$ is a Hermitian tensor. Moreover, we have also invoked the Onsager relation in thermodynamical statistics, i.e., $\varepsilon_{ij}(\mathbf{M}) = \varepsilon_{ji}(-\mathbf{M})$. The consequences of Hermiticity and Onsager relation are that the real part of the permittivity tensor is an even function of **M** whereas the imaginary part is an odd function of **M**. For a cubic crystal, such as YIG (yttrium-iron-garnet), the tensor f_{ijkl} reduces to only three independent terms. In terms of Voigt notation, they are f_{11}, f_{12}, and f_{44}. In a principal coordinate system, the tensor can be expressed as

$$f_{ijkl} = f_{12}\delta_{ij}\delta_{kl} + f_{44}\,(\delta_{il}\delta_{kj} + \delta_{ik}\delta_{lj}) + \Delta f\delta_{kl}\delta_{ij}\delta_{jk} \tag{54.2}$$

where $\Delta f = f_{11} - f_{12} - 2f_{44}$.

In the principal crystal axes [100] coordinate system, the magnetooptic permittivity reduces to the following forms:

$$\overset{\leftrightarrow}{\varepsilon} = \varepsilon_0 \begin{bmatrix} \varepsilon_{11} & \varepsilon_{12} & \varepsilon_{13} \\ \varepsilon_{12}^* & \varepsilon_{22} & \varepsilon_{23} \\ \varepsilon_{13}^* & \varepsilon_{23}^* & \varepsilon_{33} \end{bmatrix}$$

where * denotes complex conjugate operation. The elements are given by

paramagnetic state

$$\overset{\leftrightarrow}{\varepsilon} = \varepsilon_0 \begin{bmatrix} \varepsilon_r & 0 & 0 \\ 0 & \varepsilon_r & 0 \\ 0 & 0 & \varepsilon_r \end{bmatrix}$$

Faraday rotation

$$+ \ \varepsilon_0 \begin{bmatrix} 0 & +jf_1M_3 & -jf_1M_2 \\ -jf_1M_3 & 0 & +jf_1M_1 \\ +jf_1M_2 & -jf_1M_1 & 0 \end{bmatrix}$$

Cotton-Mouton effect

$$+ \ \varepsilon_0 \begin{bmatrix} f_{11}M_1^2 + f_{12}M_2^2 + f_{12}M_3^2 & 2f_{44}M_1M_2 & 2f_{44}M_1M_3 \\ 2f_{44}M_1M_2 & f_{12}M_1^2 + f_{11}M_2^2 + f_{12}M_3^2 & 2f_{44}M_2M_3 \\ 2f_{44}M_1M_3 & 2f_{44}M_2M_3 & f_{12}M_1^2 + f_{12}M_2^2 + f_{11}M_3^2 \end{bmatrix}$$

$$(54.3)$$

In order to keep the discussion simple, analytic complexities due to optical absorption of the magnetic medium have been ignored. Such absorption can give rise to magnetic circular dichroism (MCD) and magnetic linear dichroism (MLD). Interested readers can refer to Hellwege [1978] and Arecchi and Schulz-DuBois [1972] for more in-depth discussions on MCD and MLD.

54.2 Classification of Magnetooptic Effects

Faraday Rotation or Magnetic Circular Birefringence

The classic Faraday rotation takes place in a cubic or isotropic transparent medium where the propagation direction of transmitted light is parallel to the direction of applied magnetization within the medium. For example, if the direction of magnetization and the propagation of light is taken as *Z*, the permittivity tensor becomes (assuming second-order effect is insignificantly small):

$$\overset{\leftrightarrow}{\varepsilon} \cong \varepsilon_0 \begin{bmatrix} \varepsilon_r & jf_1M_3 & 0 \\ -jf_1M_3 & \varepsilon_r & 0 \\ 0 & 0 & \varepsilon_r \end{bmatrix} \qquad (54.4)$$

The two eigenmodes of light propagation through the magnetooptic medium can be expressed as right circular polarized (RCP) light wave

$$\tilde{E}_1(Z) = \begin{bmatrix} 1 \\ j \\ 0 \end{bmatrix} \exp\left[j\left(\omega t - \frac{2\pi n_+}{\lambda_0} Z \right) \right] \tag{54.5a}$$

and left circular polarized (LCP) light wave

$$\tilde{E}_2(Z) = \begin{bmatrix} 1 \\ -j \\ 0 \end{bmatrix} \exp\left[j\left(\omega t - \frac{2\pi n_-}{\lambda_0} Z \right) \right] \tag{54.5b}$$

where $n_{\pm}^2 \cong \varepsilon_r \pm f_1 M_3$; ω and λ_0 are the angular frequency and the wavelength of the incident light, respectively. n_+ and n_- are the refractive indices of the RCP and LCP modes, respectively. These modes correspond to two counterrotating circularly polarized light waves. The superposition of these two waves produces a linearly polarized wave. The plane of polarization of the resultant wave rotates as one circular wave overtakes the other. The rate of rotation is given by

$$\theta_F \cong \frac{\pi f_1 M_3}{\lambda_0 \sqrt{\varepsilon_r}} \quad \text{rad/m}$$

$$= \frac{1.8 f_1 M_3}{\lambda_0 \sqrt{\varepsilon_r}} \quad \text{degree/cm} \tag{54.6}$$

θ_F is known as the Faraday rotation (FR) coefficient. When the direction of the magnetization is reversed, the angle of rotation changes its sign. Since two counterrotating circular polarized optical waves are used to explain FR, the effect is thus also known as optical magnetic circular birefringence (MCB). Furthermore, since the senses of polarization rotation of forward traveling and backward traveling light waves are opposite, FR is a nonreciprocal optical effect. Optical devices such as **optical isolator** and **optical circulator** use Faraday effect to achieve their nonreciprocal functions. For ferromagnetic and ferrimagnetic medium, the FR is characterized under magnetically saturated condition, i.e., $M_3 = M_S$, the saturation magnetization of the medium. For paramagnetic or diamagnetic materials, the magnetization is proportional to the external applied magnetic field H_0. Therefore, the FR is proportional to the external field or $\theta_F = V H_0$ where $V = \chi_0 f_1 \pi / (\lambda_0 \sqrt{\varepsilon_r})$ is called the Verdet constant and χ_0 is the magnetic susceptibility of free space.

Cotton-Mouton Effect or Magnetic Linear Birefringence

When a transmitted light is propagating perpendicular to the magnetization direction, first-order isotropic FR effect will vanish and the second-order anisotropic Cotton-Mouton (CM) effect will occur. For example, if the direction of magnetization is along the Z axis and light wave is propagating along the X axis, the permittivity tensor becomes

$$\overset{\leftrightarrow}{\varepsilon} = \varepsilon_0 \begin{bmatrix} \varepsilon_r + f_{12} M_3^2 & 0 & 0 \\ 0 & \varepsilon_r + f_{12} M_3^2 & 0 \\ 0 & 0 & \varepsilon_r + f_{11} M_3^2 \end{bmatrix} \tag{54.7}$$

The eigenmodes are two linearly polarized light waves polarized along and perpendicular to the magnetization direction:

$$\tilde{E}_{//}(x) = \begin{bmatrix} 0 \\ 0 \\ 1 \end{bmatrix} \exp\left[j\left(\omega t - \frac{2\pi}{\lambda_0} n_{//} x \right) \right] \tag{54.8a}$$

$$\tilde{E}_{\perp}(x) = \begin{bmatrix} 0 \\ 1 \\ 0 \end{bmatrix} \exp\left[j\left(\omega t - \frac{2\pi}{\lambda_0} n_{\perp} x \right) \right] \tag{54.8b}$$

with $n_{//}^2 = \varepsilon_r + f_{11} M_3^2$ and $n_{\perp}^2 = \varepsilon_r + f_{12} M_3^2$; $n_{//}$ and $n_{\perp}$ are the refractive indices of the parallel and perpendicular linearly polarized modes, respectively. The difference in phase velocities between these two waves gives rise to a magnetic linear birefringence (MLB) of light which is also known as the CM or Voigt effect. In this case, the light transmitted through the crystal has elliptic polarization. The degree of ellipticity depends on the difference $n_{//} - n_{\perp}$. The phase shift or retardation can be found by the following expression:

$$\psi_{cm} \cong \frac{\pi(f_{11} - f_{12})M_3^2}{\lambda_0 \sqrt{\varepsilon_r}} \quad \text{rad/m}$$

or

$$\frac{1.8(f_{11} - f_{12})M_3^2}{\lambda_0 \sqrt{\varepsilon_r}} \quad \text{degree/cm} \tag{54.9}$$

Since the sense of this phase shift is unchanged when the direction of light propagation is reversed, the CM effect is a reciprocal effect.

Kerr Effects

Kerr effects occur when a light beam is reflected from a magnetooptic medium. There are three distinct types of Kerr effects, namely, polar, longitudinal (or meridional), and transverse (or equatorial). Figure 54.1 shows the configurations of these Kerr effects. Reflectivity tensor relation between the incident light and the reflected light can be used to explain the phenomena as follows:

$$\begin{bmatrix} E_{r\perp} \\ E_{r//} \end{bmatrix} = \begin{bmatrix} r_{11} & r_{12} \\ r_{21} & r_{22} \end{bmatrix} \begin{bmatrix} E_{i\perp} \\ E_{i//} \end{bmatrix} \tag{54.10}$$

where r_{ij} is the reflectance matrix. $E_{i\perp}$ and $E_{i//}$ are, respectively, the perpendicular (TE) and parallel (TM) electric field components of the incident light waves (with respect to the plane of incidence). $E_{r\perp}$ and $E_{r//}$ are, respectively, the perpendicular and parallel electric field components of the reflected light waves.

The diagonal elements r_{11} and r_{22} can be calculated by Fresnel reflection coefficients and Snell's law. The off-diagonal elements r_{12} and r_{21} can be derived from the magnetooptic permittivity tensor, the applied magnetization and Maxwell equations with the use of appropriate boundary conditions [Arecchi and Schulz-DuBois, 1972]. It is important to note that all the elements of the

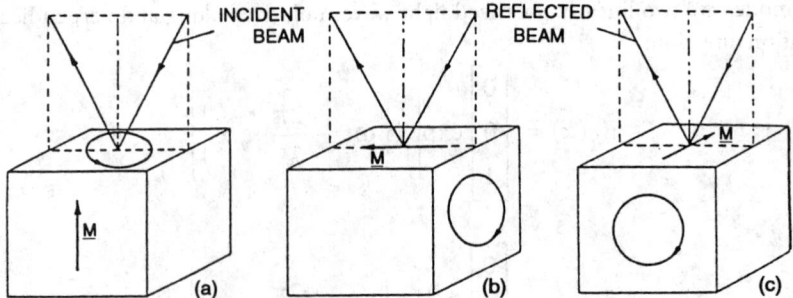

FIGURE 54.1 Kerr magnetooptic effect. The magnetization vector is represented by M while the plane of incidence is shown dotted. (a) Polar; (b) longitudinal; (c) transverse. (*Source:* A.V. Sokolov, *Optical Properties of Metals*, London: Blackie, 1967. With permission.)

reflectance matrix r_{ij} are dependent on the angle of incidence between the incident light and the magnetooptic film surface.

Polar Kerr Effect

The polar Kerr effect takes place when the magnetization is perpendicular to the plane of the material. A pair of orthogonal linearly polarized reflected light modes will be induced and the total reflected light becomes elliptically polarized. The orientation of the major axis of the elliptic polarization of the reflected light is the same for both TE ($E_{i\perp}$) or TM ($E_{i///}$) linearly polarized incident lights since $r_{12} = r_{21}$.

Longitudinal or Meridional Kerr Effect

The longitudinal Kerr effect takes place when the magnetization is in the plane of the material and parallel to the plane of incidence. Again, an elliptically polarized reflected light beam will be induced, but the orientation of the major axis of the elliptic polarization of the reflected light is opposite to each other for TE ($E_{i\perp}$) and TM ($E_{i///}$) linearly polarized incident lights since $r_{12} = -r_{21}$.

Transverse or Equatorial Kerr Effect

This effect is also known as the equatorial Kerr effect. The magnetization in this case is in the plane of the material and perpendicular to the plane of incidence. The reflected light does not undergo a change in its polarization since $r_{12} = r_{21} = 0$. However, the intensity of the TM ($E_{r///}$) reflected light will be changed if the direction of the magnetic field is suddenly reversed. For TE ($E_{r\perp}$) reflected light, this modulation effect is at least two orders of magnitude smaller and is usually ignored.

54.3 Applications of Magnetooptic Effects

Optical Isolator and Circulator

In fiber-optic-based communication systems with gigahertz bandwidth or coherent detection, it is often essential to eliminate back reflections from the fiber ends and other surfaces or discontinuities because they can cause amplitude fluctuations, frequency instabilities, limitation on modulation bandwidth, noise or even damage to the lasers. An optical isolator permits the forward transmission of light while simultaneously preventing reverse transmission with a high degree of extinction. The schematic configuration of a conventional optical isolator utilizing bulk rotator and permanent magnet [Johnson, 1966] is shown in Fig. 54.2. It consists of a 45-degree polarization rotator which is nonreciprocal so that back-reflected light is rotated by exactly 90 degrees and can therefore be excluded from the laser. The nonreciprocity is furnished by the Faraday effect. The

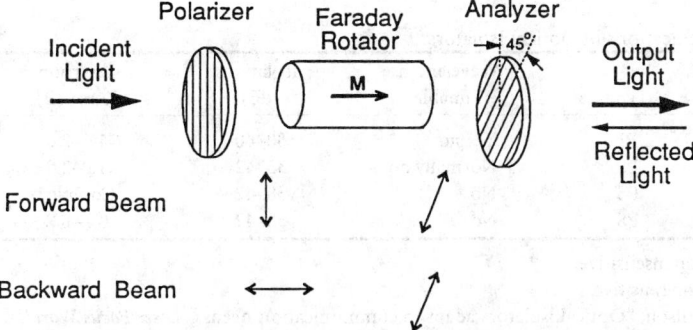

FIGURE 54.2 Schematic of an optical isolator. The polarization directions of forward and backward beams are shown below the schematic.

basic operation principle is as follows: A Faraday isolator consists of rotator material immersed in a longitudinal magnetic field between two polarizers. Light emitted by the laser passes through the second polarizer being oriented at 45 degrees relative to the transmission axis of the first polarizer. Any subsequently reflected light is then returned through the second polarizer, rotated by another 45 degrees before being extinguished by the first polarizer—thus optical isolation is achieved.

The major characteristics of an optical isolator include isolation level, insertion loss, temperature dependence, and size of the device. These characteristics are mainly determined by the material used in the rotator. Rotating materials generally fall into three categories: the paramagnetics (such as terbium-doped borosilicate glass), the diamagnetic (such as zinc selenide), and the ferromagnetic (such as rare-earth garnets). The first two kinds have small Verdet constants and mostly work in the visible or shorter optical wavelength range. Isolators for use with the InGaAsP semiconductor diode lasers ($\lambda_0 = 1100-1600$ nm), which serve as the essential light source in optical communication, utilize the third kind, especially the yttrium-iron-garnet (YIG) crystal. A newly available ferromagnetic crystal, epitaxially grown bismuth-substituted yttrium-iron-garnet (BIG), has an order-of-magnitude stronger Faraday rotation than pure YIG and magnetic saturation occurred with a smaller field [Matsuda *et al.*, 1987]. The typical parameters with YIG and BIG are shown in Table 54.1. As the major user of optical isolator, fiber optic communication system requires different input-output packaging for the isolators. Table 54.2 lists the characteristics of the isolators according to specific applications [Wilson, 1991].

Table 54.1   Characteristics of YIG and BIG Faraday Rotators

	YIG	BIG
Verdet constant (min/cm-Gauss)		
1300 nm	10.5	−806
1550 nm	9.2	−600
Saturated magnetooptic rotation (degree/mm)		
1300 nm	20.6	−136.4
1550 nm	18.5	−93.8
Thickness for 45-degree rotation (mm)		
1300 nm	2.14	0.33
1550 nm	2.43	0.48
Typical insertion loss (dB)	>0.4	<0.1
Typical reverse isolation (dB)	30–35	40
Required magnetic field (Gauss)	>1600	120
Magnetically tunable	No	Yes

Source: D.K. Wilson, "Optical isolators adapt to communication needs," *Laser Focus World*, p. 175, April 1991. With permission.

Table 54.2   Applications of Optical Isolators

Application	Type	Wavelength Tunable	Isolation (dB)	Insertion Loss (dB)	Return Loss (dB)
Fiber to fiber	PI	Yes/no	30–40	1.0–2.0	> 60
Fiber to fiber	PS	Normally no	33–42	1.0–2.0	> 60
Single fiber	PS	No	38–42	Complex	Complex
Bulk optics	PS	No	38–42	0.1–0.2	

PI = polarization insensitive.
PS = polarization sensitive.
Source: D.K. Wilson, "Optical isolators adapt to communication needs," *Laser Focus World*, p. 175, April 1991. With permission.

For the purpose of integrating the optical isolator component into the same substrate with the semiconductor laser to facilitate monolithic fabrication, integrated waveguide optical isolator becomes one of the most exciting areas for research and development. In a waveguide isolator, the rotation of the polarization is accomplished in a planar or channel waveguide. The waveguide is usually made of a magnetooptic thin film, such as YIG or BIG film, liquid phase epitaxially grown on a substrate, typically gadolinium-gallium-garnet (GGG) crystal. Among the many approaches in achieving the polarization rotation, such as the 45-degree rotation type or the unidirectional TE-TM mode converter type, the common point is the conversion or coupling process between the TE and TM modes of the waveguide. Although very good results have been obtained in some specific characteristics, for example, 60-dB isolation [Wolfe *et al.*, 1990], the waveguide optical isolator is still very much in the research and development stage.

Usually, precise wavelength of any given semiconductor diode is uncertain. Deviation from a specified wavelength could degrade isolator performance by 1 dB/nm, and an uncertainty of 10 nm

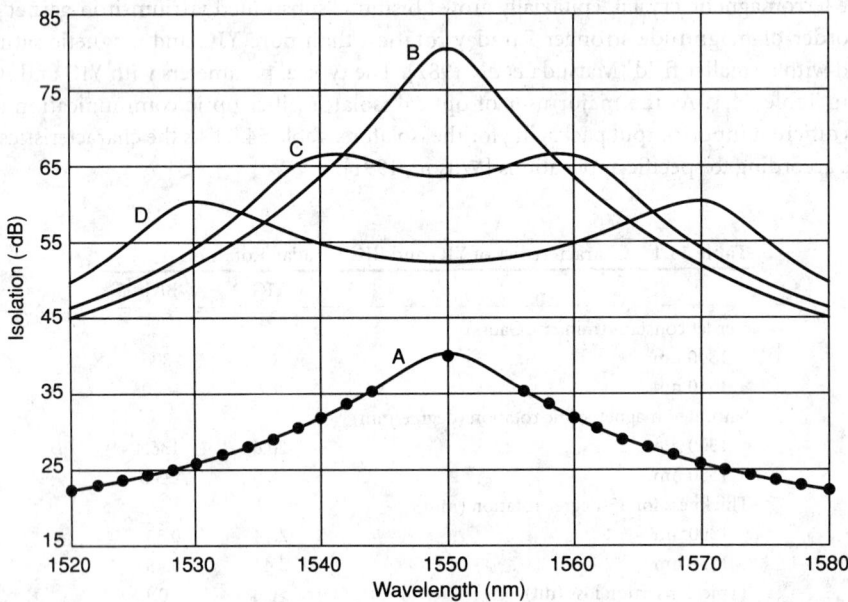

FIGURE 54.3   Isolation performance of four isolators centered around 1550 nm shows the effects of different configurations. A single-stage isolator (curve A) reaches about –40 dB isolation, and two cascaded single-wavelength isolators (curve B) hit –80 dB. Wavelength broadening (curves C and D) can be tailored by cascading isolators tuned to different wavelengths. (*Source:* D.K. Wilson, "Optical isolators adapt to communication needs," *Laser Focus World*, April 1991. With permission.)

can reduce isolation by 10 dB. Therefore, a tunable optical isolator is highly desirable. A typical approach is to simply place two isolators, tuned to different wavelengths, in tandem to provide a broadband response. Curves C and D of Fig. 54.3 show that isolation and bandwidth are a function of the proximity of the wavelength peak positions. This combination of nontunable isolators has sufficiently wide spectral bandwidth to accommodate normal wavelength variations found in typical diode lasers. In addition, because the laser wavelength depends on its operating temperature, this broadened spectral bandwidth widens the operating temperature range without decreasing isolation.

The factors that limit isolation are found in both the polarizers and the Faraday rotator materials. Intrinsic strain, inclusions, and surface reflections contribute to a reduction in the purity of polarization which affects isolation. About 40 dB is the average isolation for today's materials in a single-isolator stage. If two isolators are cascaded in tandem, it is possible to double the isolation value.

Finally, an optical circulator [Fletcher and Weisman, 1965] can be designed by replacing the polarizers in a conventional isolator configuration with a pair of calcite polarizing prisms. A laser beam is directed through a calcite prism, then through a Faraday rotator material which rotates the polarization plane by 45 degrees, then through a second calcite prism set to pass polarization at 45 degrees. Any reflection beyond this second calcite prism returns through the second prism, is rotated by another 45 degrees through the Faraday material, and, because its polarization is now 90 degrees from the incident beam, is deflected by the first calcite prism. The four ports of the circulator then are found as follows: (1) the incident beam, (2) the exit beam, (3) the deflected beam from the first calcite prism, and (4) the deflected beam from the second calcite prism.

MSW-Based Guided-Wave Magnetooptic Bragg Cell

When a ferrimagnet is placed in a sufficiently large externally applied dc magnetic field, H_0, the ferrimagnetic materials become magnetically saturated to produce a saturation magnetization $4\pi M_S$. Under this condition, each individual magnetic dipole will precess in resonance with frequency f_{res} = γH_0 where γ is the gyromagnetic ratio (γ = 2.8 MHz/Oe). However, due to the dipole–dipole coupling and quantum mechanical exchange coupling, the collective interactions among neighboring magnetic dipole moments produce a continuum spectrum of precession modes or spin waves at frequency bands near f_{res}. Exchange-free spin wave spectra obtained under the magnetostatic approximation are known as magnetostatic waves (MSWs) [Ishak, 1988]. In essence, MSWs are relatively slow propagating, dispersive, magnetically dominated electromagnetic (EM) waves which exist in biased ferrites at microwave frequencies (2–20 GHz). In a ferrimagnetic film with a finite thickness, such as a YIG thin film epitaxially grown on a nonmagnetic substrate such as GGG, MSW modes are classified into three types: magnetostatic surface wave (MSSW), magnetostatic forward volume wave (MSFVW), and magnetostatic backward volume wave (MSBVW), depending on the orientation of the dc magnetic field with respect to the film plane and the propagation direction of the MSW. At a constant dc magnetic field, each type of mode only exists in a certain frequency band. An important feature of MSW is that these frequency bands can be simply tuned by changing the dc magnetic field.

As a result of the Faraday rotation effect and Cotton-Mouton effect, the magnetization associated with MSWs will induce a perturbation in the dielectric tensor. When MSW propagates in the YIG film, it induces a moving optical grating which facilitates the diffraction of an incident guided light beam. If the so-called Bragg condition is satisfied between the incident guided light and the MSW-induced optical grating, Bragg diffraction takes place. An optical device built based on this principle is called the **magnetooptic Bragg cell** [Tsai and Young, 1990].

A typical MSFVW-based noncollinear coplanar guided-wave magnetooptic Bragg cell is schematically shown in Fig. 54.4. Here a homogeneous dc bias magnetic field is applied along the

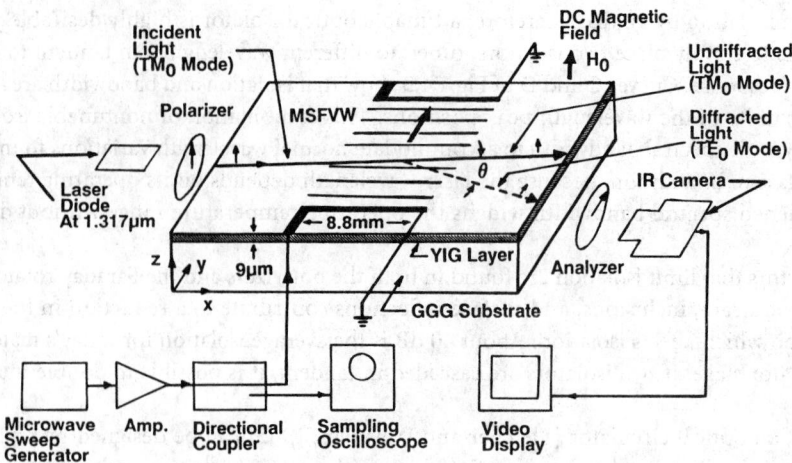

FIGURE 54.4 Experimental arrangement for scanning of guided-light beam in YIG-GGG waveguide using magnetostatic forward waves.

Z axis to excite a Y-propagating MSFVW generated by a microstrip line transducer. With a guided lightwave coupled into the YIG waveguide and propagating along the X axis, a portion of the lightwave is Bragg-diffracted and mode-converted (TE to TM mode and vice versa). The Bragg-diffracted light is scanned in the waveguide plane as the frequency of the MSFVW is tuned. Figure 54.5 shows the scanned light spots by tuning the frequency at a constant dc magnetic field.

MSW-based guided-wave magnetooptic Bragg cell is analogous to surface acoustic wave (SAW)-based guided-wave acoustooptic (AO) Bragg cell and has the potential to significantly enhance a wide variety of integrated optical applications which had previously been implemented with SAW. These include TE-TM mode converter, spectrum analyzer, convolvers/correlators, optical frequency shifters, tunable narrowband optical filters, and optical beam scanners/switches [Young, 1989]. In comparison to their AO counterparts, the MSW-based magnetooptic Bragg cell modules may possess the following unique advantages: (1) A much larger range of tunable carrier frequencies (2–20 GHz, for example) may be readily obtained by varying a dc magnetic field. Such high and tunable carrier frequencies with the magnetooptic device modules allow direct processing at the carrier frequency of wide-band RF signals rather than indirect processing via frequency down-conversion, as is required with the AO device modules. (2) A large magnetooptic bandwidth may be realized by means of a simpler transducer geometry. (3) Much higher and electronically tunable modulation/switching and scanning speeds are possible as the velocity of propagation for the MSW is higher than that of SAW by one to two orders of magnitude, depending upon the dc magnetic field and the carrier frequency.

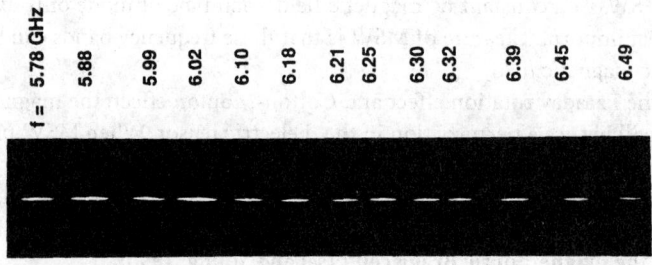

FIGURE 54.5 Deflected light spots obtained by varying the carrier frequency of MSFVW around 6 GHz.

Magnetooptic Recording

The write/erase mechanism of the **magnetooptical (MO) recording system** is based on a thermo-magnetic process in a perpendicularly magnetized magnetooptic film. A high-power pulsed laser is focused to heat up a small area on the magnetooptic medium. The coercive force of the MO layer at room temperature is much greater than that of a conventional non-MO magnetic recording medium. However, this coercive force is greatly reduced when the heated spot is raised to a critical temperature. Application of a bias magnetic field can then easily reverse the polarization direction of the MO layer within the heated spot. As a result, a very small magnetic domain with magnetization opposite to that of the surrounding area is generated. This opposite magnetic domain will persist when the temperature of the medium is lowered. The magnetization-reversed spot represents one bit of stored data. To erase data, the same thermal process can be applied while reversing the direction of the bias magnetic field.

To read the stored information optically, the Kerr effect is used to detect the presence of these very small magnetic domains within the MO layer. When a low-power polarized laser beam is reflected by the perpendicularly oriented MO medium, the polarization angle is twisted through a small angle θ_k, the Kerr rotation. Furthermore, the direction of this twist is either clockwise or counterclockwise, depending on the orientation of the perpendicular magnetic moment, which is either upward or downward. Therefore, as the read beam scans across an oppositely magnetized domain from the surrounding medium, there is a total change of $2\theta_k$ in the polarization directions from the reflected beam coming from the two distinct regions. Reading is done by detecting this phase difference.

The MO recording medium is one of the most important elements in a high-performance MO data-storage system. In order to achieve fast writing and erasing functions, a large Kerr rotation is required to produce an acceptable carrier-to-noise (C/N) ratio. In general, a high-speed MO layer has a read signal with a poor C/N ratio, while a good read-performance MO layer is slow in write/erase sensitivity.

Defining Terms

Cotton-Mouton effect: Second-order anisotropic reciprocal magnetooptic effect which causes a linearly polarized incident light to transmit through as an elliptically polarized output light wave when the propagation direction of the incident light is perpendicular to the direction of the applied magnetization of the magnetooptic medium. It is also known as magnetic linear birefringence (MLB).

Faraday rotation: First-order isotropic nonreciprocal magnetooptic effect which causes the polarization direction of a linearly polarized transmitted light to rotate when the propagation direction of the incident light wave is parallel to the direction of the applied magnetization of the magnetooptic medium. It is also known as magnetic circular birefringence (MCB).

Kerr effects: Reflected light from a magnetooptic medium can be described by the optical Kerr effects. There are three types of Kerr effects: polar, longitudinal, and transverse, depending on the directions of the magnetization with respect to the plane of incidence and the reflecting film surface.

Magnetooptic Bragg cell: A magnetically tunable microwave signal processing device which uses optical Bragg diffraction of light from a moving magnetooptic grating generated by the propagation of magnetostatic waves within the magnetic medium.

Magnetooptic recording system: A read/write data recording system based on a thermomagnetic process to write oppositely magnetized domains onto a magnetooptic medium by means of high-power laser heating. Magnetooptic Kerr effect is then employed to read the data by using a low-power laser as a probe beam to sense the presence of these domains.

Optical circulator: A four-ports optical device that can be used to monitor or sample incident light (input port) as well as reflected light (output port) with the two other unidirectional coupling ports.

Optical isolator: A unidirectional optical device which only permits the transmission of light in the forward direction. Any reflected light from the output port is blocked by the device from returning to the input port with very high extinction ratio.

References

F. T. Arecchi and E. O. Schulz-DuBois, *Laser Handbook*, D4, Amsterdam: North-Holland, 1972, pp. 1009–1027.

P. C. Fletcher and D. L. Weisman, "Circulators for optical radar systems," *Applied Optics*, vol. 4, pp. 867–873, 1965.

K. H. Hellwege, *Landolt-Bornstein Numerical Data and Functional Relationships in Science and Technology, New Series*, vols. 4 and 12, New York: Springer-Verlag, 1978.

W. S. Ishak, "Magnetostatic wave technology: A review," *Proc. IEEE*, vol. 76, pp. 171–187, 1988.

B. Johnson, "The Faraday effect at near infrared wavelength in rare-earth garnet," *Brit. J. Appl. Phys.*, vol. 17, p. 1441, 1966.

K. Matsuda, H. Minemoto, O. Kamada, and S. Isbizuka, "Bi-substituted, rare-earth iron garnet composite film with temperature independent Faraday rotation for optical isolators," *IEEE Trans. Mag.*, vol. MAG-23, p. 3479, 1987.

M. Torfeh, L. Courtois, L. Smoczynski, H. Le Gall, and J.M. Desvignes, "Coupling and phase matching coefficients in a magneto-optical TE-TM mode converter," *Physica*, vol. 89B, pp. 255–259, 1977.

C. S. Tsai and D. Young, "Magnetostatic-forward-volume-wave-based guided-waveless magneto-optic Bragg cells and applications to communications and signal processing," *IEEE Trans. on Microwave Theory and Technology*, vol. MTT-38(5), pp. 560–570, 1990.

D. K. Wilson, "Optical isolators adapt to communication needs," *Laser Focus World*, p. 175, April 1991.

R. Wolfe *et al.*, "Edge tuned ridge waveguide magneto-optic isolator," *Applied Physics Letters*, vol. 56, p. 426, 1990.

D. Young, "Guided Wave Magneto-Optic Bragg Cells Using Yttrium Iron Garnet-Gadolinium Gallium Garnet (YIG-GGG) Thin Film at Microwave Frequencies," Ph.D. Dissertation, University of California, Irvine, 1989.

Further Information

Current publications on magnetooptics can be found in the following journals: *Intermag Conference Proceedings* published by the IEEE and *International Symposium Proceedings on Magneto-Optics*.

For in-depth discussion on magnetic bubble devices, please see, for example, *Magnetic-Bubble Memory Technology* by Hsu Chang, published by Marcel Dekker, 1978.

An excellent source of information on garnet materials can be found in *Magnetic Garnet* by Gerhard Winkler, published by Friedr. Vieweg & Sohn, 1981.

Numerous excellent reviews on the subject of magnetooptics have been published over the years, for example, J.F. Dillon, Jr., "Magnetooptics and its uses," *Journal of Magnetism and Magnetic Materials*, vol. 31–34, pp. 1–9, 1983; M.J. Freiser, "A survey of magnetooptic effects," *IEEE Transactions on Magnetics*, vol. MAG-4, pp. 152–161, 1968; G. A. Smolenskii, R.V. Pisarev, and I.G. Sinii, "Birefringence of light in magnetically ordered crystals," *Sov. Phys. Usp.*, vol. 18, pp. 410–429, 1976; and A.M. Prokhorov, G.A. Smolenskii, and A.N. Ageev, "Optical phenomena in thin-film magnetic waveguides and their technical application," *Sov. Phys. Usp.*, vol. 27, pp. 339–362, 1984.

55

Smart Materials

55.1 Introduction 1173
55.2 Smart/Intelligent Structures 1174
55.3 Objective-Based Classification of Smart/Intelligent Materials 1175
Smart Structural Materials • Smart Thermal Materials • Smart Acoustical Materials • Smart Electromagnetic Materials • Pyrosensitive Smart Materials
55.4 Material Properties Conducive for Smart Material Applications 1176
Piezoelectric Effect • Magnetostrictive Effect • Electroplastic Effect • Shape-Memory Effects • Electrorheological Property • Nonlinear Electro-optic Properties • Nonlinear Electroacoustic Properties • Pyrosensitive Properties • Nonlinear Electromagnetic Properties
55.5 State-of-the-Art Smart Materials 1180
Piezoelectric Smart Materials • Magnetostrictive Smart Materials • Electroplastic Smart Materials • Shape-Memory Smart Materials • Electrorheological Smart Fluids • Electro-optic Smart Materials • Electroacoustic Smart Materials • Electromagnetic Smart Materials • Pyrosensitive Smart Materials
55.6 Smart Sensors 1182
Fiber-Optic-Based Sensors • Piezoelectric-Based Sensors • Magnetostriction-Based Sensors • Shape-Memory Effects-Based Sensors • Electromagnetics-Based Sensors • Electroacoustic Smart Sensors
55.7 Examples of Smart/Intelligent Systems 1183
Structural Engineering Applications • Electromagnetic Applications
55.8 High-Tech Application Potentials 1186
55.9 Conclusions 1188

P. S. Neelakanta
Florida Atlantic University

55.1 Introduction

Smart materials are a class of materials and/or composite media having inherent intelligence together with self-adaptive capabilities to external stimuli. Also known as **intelligent materials,** they constitute a few subsets of the material family that "manifest their own functions intelligently depending on environmental changes" [Rogers and Rogers, 1992].

Classically, such intelligent material systems have been conceived in the development of mechanical structures that contain their own sensors, actuators and self-assessing computational feasibilities in order to modify their structural (elastic) behavior via feedback control capabilities. The relevant concepts have stemmed from intelligent forms of natural (material) systems, namely, living organisms; hence, in modern concepts smart or intelligent materials and systems are conceived as those that mimic the life functions of sensing, actuation, control, and intelligence.

The inherent intelligence and self-adaptable control of artificial smart materials should be programmable in terms of the constituent processing, microstructural characteristics, and defects to permit the self-conditionings to adapt in a controlled manner to various types of stimuli. The dividing line between smart materials and the so-called **intelligent structures** is not, however, distinct. In simple terms, intelligent material systems are constructed of smart materials with a dedicated, discrete set of integrated actuators, sensors, and so on, and smart materials contain largely a built-in or embedded set of distributed sensors. In general, the term *smart materials* usually connotes the structural constituent in which the discrete functions of sensing, actuation, signal processing and control are tangibly integrated. Intelligent structures, as an extension, are constructed with smart materials to respond to the environment around them in a predetermined, desired manner.

Intelligent or smart materials that manifest their own functions intelligently vis-à-vis the changes in their surroundings are capable of performing, in general (Chong *et al.*, 1990):

- Primary functions specifying the adaptive roles of the sensor, the effector and processor capabilities (including the memory functions)

- Macroscopic functions that enclave the extensive or global aspects of the intelligence inherent in the materials

- Built-in social utility aspects with an instilled human-like intelligence with hyper-performance capabilities

55.2 Smart/Intelligent Structures

The framework of intelligent structures as a subset in the gamut of conventional material-based systems is illustrated in Fig. 55.1. This general classification of material structures refer to [Chong *et al.*, 1990]:

- Sensory structures, "which possess sensors that enable the determination or monitoring of system states or characteristics" [Chong *et al.*, 1990]

- Adaptive structures, which possess actuators that facilitate the alteration of system-states or characteristics in a controlled manner

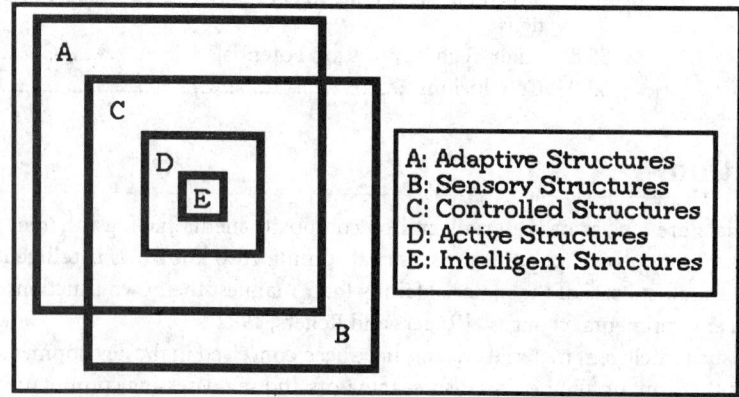

FIGURE 55.1   Set of structures. (Adapted from B. K. Wada, J. L. Fanson, and E. F. Crawley, "Adaptive structures," *J. Intell. Mat. Syst. Struct.*, 1, 1990.)

- Sensory systems, which may contain sensors, but no actuators
- Adaptive systems, which contain actuators, but no sensors

Referring to Fig. 55.1, the intersection of sensory versus adaptive structures depicts the controlled structures with a feedback architecture. That is, the active structure has an integrated controlled unit with sensors and/or actuators that have structural as well as control functionality. Hence, the logical subset that defines an intelligent structure is a highly integrated unit (with controlled logic, electronics, etc.) that provides the cognitive element of a distributed or a hierarchic controlled structure.

55.3 Objective-Based Classification of Smart/Intelligent Materials

Smart Structural Materials

Intelligent structural engineering materials are the classical versions of smart systems in which the mechanical (elastic) properties of a structure can be modified adaptively by means of an imbedded distribution of smart material(s), and an associated (integral) set of sensors and actuators together with an external control system to facilitate adaptive changes in the elastic behavior of structures so that motion, vibration, strength, stiffness, redistribution of load path in response to damage, etc. are controlled.

Smart Thermal Materials

A **smart thermal material,** in response to environmental demands, can self-adaptively influence its thermal states (temperature or such thermal properties as conductivity, diffusivity, absorptivity), by means of an integrated conglomeration of thermal sensors, heaters, or actuators with an associated control system.

Smart Acoustical Materials

Smart acoustical materials can be classified as those that have self-adaptive characteristics on their acoustical behavior (such as transmission, reflection, and absorption of acoustical energy) by means of sensors that assess the acoustical states (intensity, frequency, response, etc.), along with a set of actuators (dampers, exciters) with an associated control system. Again, the self-adaptive behavior of these materials is in response to ambient acoustical changes.

Smart Electromagnetic Materials

Smart Magnetic Shielding Materials

As warranted by the surroundings, the self-adaptive shielding effectiveness to magnetic fields at low frequencies (power frequencies such as 60 or 50 Hz) can be achieved by means of an integrated set of magnetic field sensors and actuators (magnetic biasing, current elements, etc.) plus a control system arrangement [Neelakanta and Subramaniam, 1992].

High-Frequency Smart Shielding Materials

Corresponding to radio and higher frequency environments, the shielding requirement warrants curtailing both electric and magnetic fields. Hence, the relevant self-adaptive intelligent shielding system would consist of an array of distributed electromagnetic sensors with appropriate elements (actuators) and a control system.

Smart Radar-Absorbing Materials

Absorption of microwave/millimeter wave energy at radar frequency is useful in radar stealth applications. Adaptively controllable smart radar-absorbing materials (smart RAMs) can be synthesized with integrated distribution of electromagnetic detectors (sensors) with appropriate actuators and control system [Neelakanta et al., 1992].

Smart Optical Surface Materials

Smart optical surface materials can be envisioned as those in which the surface optical properties (hue, intensity, etc.) can be adaptively controlled by means of an intelligent sensor/actuator combinational control system.

Pyrosensitive Smart Materials

Electromagnetic active surfaces constituted by pyrosensitive inclusions have been successfully developed to manage the electromagnetic reflection and/or absorption characteristics from the active surface by means of thermal actuation of the pyrosensitive nodes imbedded in the medium [Neelakanta et al., 1992]. With the inclusion of a feedback systems, smart operation in adaptively manipulating the active surface characteristics can be achieved.

55.4 Material Properties Conducive for Smart Material Applications

Certain specific characteristics of materials make them suitable for smart material applications. These properties are:

1. Piezoelectric effect
2. Magnetostrictive effect
3. Electroplastic effect
4. Shape-memory effects
5. Electrorheological properties
6. Nonlinear electro-optic properties
7. Nonlinear electroacoustic properties
8. Nonlinear electromagnetic properties
9. Pyrosensitive properties

Piezoelectric Effect

Piezoelectric property of a material refers to the ability to induce opposite charges at two faces (correspondingly, to exhibit a voltage difference between the faces) of the material as a result of the strain due to mechanical force (either tension or compression) applied across the surfaces. This process is also reversible in the sense that a mechanical strain would be experienced in the material when subjected to opposite electric charging at the two faces by means of an applied potential.

In the event of such an applied voltage being alternating, the material specimen will experience vibrations. Likewise, an applied vibration on the specimen would induce an alternating potential change between the two faces. The most commonly known materials that exhibit piezoelectric properties are natural materials like quartz and a number of crystalline and polycrystalline compounds.

The strain versus the electric phenomenon perceived in piezoelectric materials is dictated by a coefficient that has components referred to a set of orthogonal coordinate axes (which are correlated to standard crystallographic axes). For example, denoting the piezoelectric coefficient (ratio between piezoelectric strain component to applied electric field component at a constant mechanical stress or vice versa as d_{mn}, the subscript n (1 to 3) refers to the three euclidian orthogonal axes, and $m = 1$ to 6 specifies the mechanical stress-strain components. The unit for d_{mn} is meter/volt which is the same as coulomb/newton.

In the piezoelectric phenomenon, there is an electromechanical synergism expressed as a coupling factor K defined by K^2, which quantifies the ratio of mechanical energy converted into electric charges to the mechanical energy impressed on the material. Being a reversible process, a relevant inverse ratio is also applicable.

Magnetostrictive Effect

Magnetostrictive effect refers to the structural strain experienced in a material subjected to a polarizing magnetic flux. A static strain of $\Delta l/l$ is produced by a dc polarizing magnetic flux density B_o such that $\Delta l/l = CB_o^2$, where C is a material constant expressed in (meter^4/weber^2) taking the units for B_o as weber/meter^2.

The magnetic stress constant (Λ) in (newton/weber) is given by $\Lambda = 2CB_oY_o$ where Y_o refers to the Young's modulus of a linearly strained free bar. The coefficient (Λ) could be both positive or negative. For example, nickel contracts with increasing B_o, whereas magnetic alloys such as 45 Permalloy (45% Ni + 55% Fe), Alfer (13% Al, 87% Fe) exhibit positive magnetostrictive coefficient [Reed, 1988].

Electroplastic Effect

The **electroplastic effect** (EPE) refers to the plastic deformation of metals with the application of high-density electric current with an enhanced deformation rate (that persists in addition to that caused by the side effects of the current such as joule-heating and the magnetic pinch effect). The plastic strain rate resulting from a current pulse is given by $\varepsilon_I/\varepsilon_A = \alpha\, J^2 \exp(\beta J)$ where ε_I is the strain rate occurring during the current pulse, ε_A is the strain rate in the absence of the current pulse, J is the current density and α and β are material constants. Typically the EPE has been observed in zinc, niobium, titanium, etc.

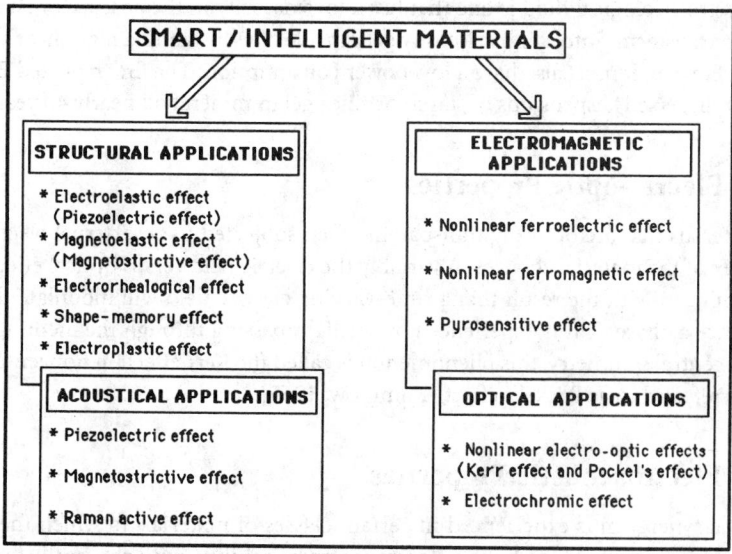

FIGURE 55.2 Application-specific classification of smart/intelligent materials.

Shape-Memory Effects

The mechanism by which a plastically deformed object in the low-temperature martensitic condition regains its original shape when the external stress is removed and heat is applied is referred to as the **shape-memory effect** (SME) [Jackson *et al.*, 1972]. It is a memory mechanism that is the result of a martensitic transformation taking place during heating.

Alhough the exact mechanism by which the shape-memory effect occurs is still under study, the process by which the original shape is regained is associated with a reverse transformation of the deformed martensitic phase to the higher temperature austenite phase. A group of nickel-titanium alloys (referred to as Nitinol) of proper composition exhibit the shape-memory property and are widely used in smart material applications [Jackson *et al.*, 1972].

Electrorheological Property

Electrorheological property is the property exhibited by certain fluids that are capable of altering their flow characteristics depending on an external applied electric field. These fluids have a fast response time, only a few milliseconds. Once the external field is applied, there is a form of progressive gelling of the fluid proportional to the applied field strength. Without the applied field, the fluid flows freely. If the electrified electrorheological (ER) fluid is sheared by an applied force larger than a certain critical value, it flows. Below this critical value of applied shear force, the electrified fluid remains in the gel phase [Gandhi and Thompson, 1989].

An electrorheological fluid requires particles (1 to 100 mm in diameter) dispersed in a carrier fluid. Sometimes a surfactant is also added to help the dispersion of particles in the fluid. The surfactant is used to prevent particle interaction that could otherwise result in a tendency for the particulates to clump together when the fluid is allowed to stand still over a stretch of time. The tendency of the particles to clump together is referred to as settling.

The applied electric field to perceive the electrorheological phenomenon is usually in the order of 4 kV/mm. When the electric field is applied, the positive and negative charges on the suspended particles are separated, forming a dipole of charges. These dipoles then align (polarize) themselves by mutual forces of attraction and repulsion to other similar dipoles, resulting in unique flow characteristics. In the absence of an electric field, there is no dipole separation of charges, and hence the fluid returns to its normal flow.

An ideal electrorheological fluid is one that has a low viscosity in the absence of an applied field and that which transforms into a high-viscosity gel capable of withstanding high shear stresses when the field is on. Further, it must also have a low power consumption. The first reported ER fluid consisted of finely dispersed suspensions of starch or silica gel in mineral oil nearly 40 years ago.

Nonlinear Electro-optic Properties

In certain materials that are optically transparent when subjected to an external electric field, the refractive index of the material changes. Invariably the electric field versus optical effect thus experienced is nonlinear, with the result that a time-varying electric field will modulate the refractive index, and hence a phase shift is experienced by the light passing through the medium. In materials that have a central symmetry, this phenomenon is called the Kerr effect; in noncentrosymmetric materials, it is referred to as Pockel's effect [Kaminow, 1965].

Nonlinear Electroacoustic Properties

Electroacoustic synergism is experienced in certain classes of materials in which the mechanical atomic vibrations are influenced by the electronic polarizability, with the result that nonlinear interaction between the atomic displacements versus the electric field causes modulation effects resulting in the generation of new sideband frequencies. Such sidebands (labeled Raman frequencies) and the response function of a Raman active medium have the form

$$H(\omega) = A_1 E(\omega) + A_2 E^2(\omega) + A_3 E^3(\omega) + \cdots$$

Pyrosensitive Properties

The **pyrosensitive property** is governed by a class of materials known as solid electrolytes. On thermally energizing such materials, they exhibit superionic electric conduction (also known as fast ion conduction), with the result that the medium, which is dielectric under cold conditions, becomes conducting at elevated temperatures. Correspondingly, the media that are embedded with solid electrolytes show different extents of electromagnetic reflection/transmission characteristics at low and high temperatures and hence can be manipulated thermally [Neelakanta *et al.*, 1992].

Typical solid electrolytes that can be adopted for such pyrosensitive applications are, for example, AgI and RbAg$_4$I$_5$. The materials like β-AgI and β-alumina show increasing conductivity with increasing temperature. The compound β-AgI exhibits superionic conductivity, with an abrupt transition at a temperature close to 147°C. This transition is known as the β- to α-phase transition, and there are a host of other materials that exhibit this phenomenon. For example, the material RbAg$_4$I$_5$ has a high electrical conductivity even at room temperature. It has also been observed that solid electrolytes provide sufficiently high electrical conductivity in the α-phase even when included in low volume fractions in a mixture with a nonsolid-electrolyte host [Neelakanta *et al.*, 1992].

Nonlinear Electromagnetic Properties

Basically, the nonlinear electromagnetic properties can manifest as two subsets of material characteristics, namely, **nonlinear dielectric properties** and **nonlinear magnetic properties**.

Nonlinear Dielectric Properties

Dielectric materials whose permittivity has a distinct dependence on the intensity of the applied electric field are referred to as active or nonlinear dielectrics. Such materials demonstrate very high values of permittivity (in the order of several thousand), pronounced dependence of dielectric parameters on the temperature, and a loop of electric hysteresis under the action of an alternating voltage.

Ferroelectrics are the most typical example of nonlinear dielectrics. Rochelle's salt (potassium sodium tartrate) was the first substance in which nonlinearity was discovered. All ferroelectrics, however, possess nonlinear properties only within a definite temperature range. The temperature transition points over which the ferroelectric materials gain or lose their ferroelectric properties are referred to as Curie points. The arsenates and dihydrogen phosphates of alkali metals are also examples of ferroelectric materials.

Piezoelectrics also fall under the category of active dielectrics. Electrets, which are capable of preserving an electric charge for a long period of time (hence regarded analogous to permanent magnets), exhibit highly nonlinear dielectric properties.

Nonlinear Magnetic Properties

Ferromagnetic materials are materials in which the permanent magnetic dipoles align themselves parallel to each other. These materials have a characteristic temperature below and above which their properties differ greatly. This temperature is referred to as the Curie temperature. Above the Curie temperature they behave as paramagnetic materials, while below it they exhibit the well known hysteresis B versus H curves. Examples of such ferromagnetic materials are iron, Mu-metal, and Supermalloy. Ferrimagnetic materials are similar in their hysteresis properties to ferromagnetic materials but differ from them in that their magnetic dipoles align themselves antiparallel to each other. Ferrites are the most popular ferrimagnetic materials, and they are of the greatest interest in electrical engineering applications.

55.5 State-of-the-Art Smart Materials

Piezoelectric Smart Materials

Piezoelectric smart materials find applications primarily in intelligent structures deploying electroelastic synergism, and a class of ceramics (popularly known as ferroelectric ceramics) have emerged in recent times for such applications. Typically, such ceramics include the base polycrystalline piezoelectrics such as $BaTiO_3$, $CdTiO_3$, $PbZrO_3$, and $PbTiO_3$, formulated with various stoichiometric proportions. Another class of piezoelectric flexible composite that has the potential for smart applications is a compound consisting of $PbTiO_3$ and chloroprene rubber. A set of glass ceramic composites containing the crystalline phases of Li_2SiO_3, $Li_2Si_2O_5$, $Ba_2TiSi_2O_8$, $Ba_2TiGe_2O_8$, $Li_2B_4O_7$, etc. are also emerging samples in smart material engineering [Chong *et al.*, 1990].

Piezoelectric smart materials can also be made from the family of polymers, namely, polyvinylidene fluoride (PVDF). The main advantages of using this polymer are that it can be formed into very thin sheets and has excellent mechanical strength combined with high sensitivity to pressure changes.

Another piezoelectric material recently developed in the NTK Research facility in Japan is a kind of rubber-based material referred to as piezoelectric rubber. This material is composed of a base material of synthetic rubber, namely, chloroben, dispersed with fine particles of a popular piezoelectric ceramic, called PZT (lead zirconium titanate). Piezoelectric rubber combines the favorable properties of PZT, namely, high sensitivity, chemical inertness, linearity, and simplicity, with that of the rubber base, namely, flexibility. The main drawback with the piezoelectric rubber is in making an electrical contact with it. This problem has been circumvented by the development of a coaxial cable connection that is easier to use [Ting, 1990].

Magnetostrictive Smart Materials

Materials with a high degree of magnetostriction are deployed in modern intelligent structures. Typically, the amount of strain inducible with intelligent materials in the current state of the art is 2000 ppm. These are alloys made with iron and rare earth materials such as terbium (Te), dysprosium (Dy), and niobium (Nb). A commercially known material of this category is Terfenol [Reed, 1988]. Magnetostrictive transducers for smart applications have also been developed with a certain class of metallic glass materials.

Electroplastic Smart Materials

Electroplastic materials are useful as smart elastic media inasmuch as the stimulus that modifies the elastic deformation is the electric current that can be controlled externally. The usefulness of these materials for smart systems under room temperature conditions is still under investigation.

Shape-Memory Smart Materials

Shape-memory smart materials include three categories, namely shape-memory alloys (SMA), shape-memory hybrid composites (SMHC), and shape-memory polymers (SMP).

Nickel–titanium (Nitinol) alloys of proper composition exhibit unique memory, or shape-restoration force characteristics, and are the most popular shape-memory alloys. When the material is plastically deformed in its low-temperature phase and then heated above its characteristic transition temperature, the original configuration or shape is restored. Deformations up to 6–8% can be completely restored by heating the material. It is this property that is used in smart electromechanical actuations.

Shape-memory hybrid composites are composite materials that contain SMA fibers or films in such a way that they can be mechanically controlled by heat. These materials can be heated by passing a current through the fibers. SMHCs offer a wide scope of applications in material–structure interaction. The fibers used in these composites are also made of Nitinol alloys.

The third form of shape-memory materials are the shape-memory polymers. These materials have an elastic memory, meaning that a large reversible change in the elastic modulus exists across the glass-transition temperature. In other words, across the glass-transition temperature, the material can change from a glass to rubbery state, allowing significant deformation in response to temperature changes. Shape-memory polymers, in general, are durable, lightweight, and transparent. Nippon Zeon Company and Mitsubishi Company have developed high-performance SMPs in the recent past [Chong *et al.*, 1990]. While the SMP of Nippon Zeon Company is polynorborene based, Mitsubishi's SMP is polyurethane based, which overcomes crucial weaknesses such as poor processability and limited-temperature operating range. In their applications SMPs can be used either as an elastic memory material or a shape-memory material. Depending on which of these possibilities are used, the range of applications differs.

Electrorheological Smart Fluids

Current research on electrorheological fluids is focused toward development of carrier–particle combinations that result in the desirable characteristics to achieve smart elastic behavior [Gandhi and Thompson, 1989]. The earlier versions of electrorheological fluids contained adsorbed water, which limited their operating temperature change (up to 80°C). Particles in the newer electrorheological fluids are, however, based on polymers, minerals, and ceramics, which have a higher operating range (200°C). Also, the increase in power consumption is less with temperature increments in the recent anhydrous systems. The most commonly used carrier fluids are silicone oil, mineral oil, and chlorinated paraffin, which offer good insulation and compatibility for particulate dispersion.

Electro-optic Smart Materials

Typically potassium dihydrogen phosphate (KDP) exhibits electro-optic behavior. Synthetic materials that have the ability to alter their refractive index (and hence the optical transmission and reflection characteristics) in the presence of an electric stimulus can be comprehended as viable **smart sensor** applications.

Electroacoustic Smart Materials

Although classically the nonlinear interaction of a vibrational (acoustic) wave and an electromagnetic wave has been studied in reference to Raman active media, relevant concepts can be exercised for smart engineering applications using those materials that exhibit strong vibrational versus piezoelectric characteristics. The NTK piezorubber, PZT ceramics, $LiNBO_3$, PZT with donor additives, insolvent additives, etc. are viable candidates for smart applications in addition to piezoelectric polymers.

Electromagnetic Smart Materials

In recent times a number of materials that possess ferroelectric properties have been discovered, the most popular of which is barium titanate ($BaTiO_3$). Barium titanate has an excellent prospect as a smart material because of the several advantages it offers, such as high mechanical strength, resistance to heat and moisture, and ease of manufacturing. $BaTiO_3$ and other similar materials are frequently referred to as ferroelectric ceramics. Also, electrets such as polymethylmethacrylate offer promise for smart applications.

Among the nonlinear magnetic materials, ferromagnetic materials such as Alnico V, platinum–cobalt, and a variety of ferrites are possible smart materials.

Pyrosensitive Smart Materials

Pyrosensitive smart materials are useful in realizing intelligent electromagnetic active surfaces, radar-absorbing materials, electromagnetic shielding, and so on. For example, it has been demonstrated [Neelakanta *et al.*, 1992] that the microwave reflection characteristics at a surface of a composite medium comprised of thermally controllable, solid-electrolytic zones (made of AgI pellets) show broadband microwave absorption/reflection characteristics under elevated temperatures. This principle can be adopted in conjunction with an electromagnetic sensor to provide a controllable feedback for thermal activation of fast-ion zones reconfigurably in order to acheive smart active-surface characteristics. Exclusive for this application, depending on the temperature limited conditions, the solid electrolyte can be chosen on the basis of its α- to β-phase transition characteristics. In order to keep the cost of the system low, a mixture phase can also be adopted, in which, commensurate with the elevated temperature operation, the host medium of the mixture could be a ceramic (dielectric).

55.6 Smart Sensors

Fiber-Optic-Based Sensors

The field of sensing technology has been revolutionized in the past decade by the entry of fiber optics. The properties of fiber optics that have made the technology suitable for communications are responsible for it being successful as a sensor as well. Fiber-optic sensors are of two types, namely, extrinsic and intrinsic. In the extrinsic type, the fiber itself acts only as a transmitter and does no part of the sensing. In an intrinsic type, however, the fiber acts as a sensor by using one of its intrinsic properties, such as induced birefringence or electrochromatism, to detect a phenomenon or quantify a measurement. Relevant to smart systems, the use of fiber optics in conjunction with optical (sensors) is based on changes in optical effects such as refractive index, optical absorption, luminescence, and chromic properties due to alterations in the environment in which the fiber is imbedded. Such alterations refer to strain or other elastic characteristics and thermal and/or electromagnetic properties [Claus, 1991]. Surfaces located with smart fiber sensors are known as smart skins.

Piezoelectric-Based Sensors

The most conventional form of sensing technology is that of piezoelectric materials, which generate an electrical response to a stimulus. In recent times piezoelectric materials have been greatly improved in mechanical strength and sensitivity. Pressure and vibration can be directly sensed as a one-to-one transduction effect resulting from the elastic-to-piezoelectric effect. Bending, on the other hand, can be sensed via piezoabsorption characteristics.

Magnetostriction-Based Sensors

The use of metallic glass as a distributive magnetostrictive sensor has been studied. Typically, in the imbedded smart sensing applications using the magnetostrictive property, the magnetic field is in the submicrogauss regime, and the nonlinearity associated with the hysteresis of magnetostriction provides a detectable sensor signal. Pressure and force, which cause static or quasi-static magnetic fields, as well as vibrations, which induce alternating magnetic fields, can be regarded as direct magnetostrictive sensor responses. In the bending mode, corresponding magnetostrictive absorption can also be sensed via reduction in the Q-factor due to absorption losses in a magnetostrictively tunable system.

Shape-Memory Effects-Based Sensors

The latest form of sensing technology utilizes shape-memory materials, namely, Nitinol alloys. The Nitinol sensors are used to measure strain and consist of superelastic Nitinol wires. The basic concept is to measure the change in resistance of a Nitinol wire used as an unbalanced arm of a Wheatstone bridge as a function of the strain. The desirable properties of Nitinol in such a sensing application are its high sensitivity and superelastic nature (which permits strains up to 6% to be accurately and repeatedly measured). The piezoelectric and Nitinol sensing materials can also be used for actuation applications.

Electromagnetics-Based Sensors

Smart electromagnetic sensors are simple deviations of classic electric/magnetic probes, more properly known as antennas or pickups. Depending on changes in the surroundings vis-à-vis the electromagnetic characteristics, these sensors respond and yield a corresponding signal. Again, the environmental changes refer to possible alterations caused by elastic, thermal, optical, magnetic, electric, and/or chemical influences.

Electroacoustic Smart Sensors

Electroacoustic smart sensors are embedded acoustic (vibration) sensors (similiar to a microphone) that adaptively yield a signal proportional to the acoustic input. Such inputs could result from changes in the alterations in the surroundings caused by elastic or thermal effects.

As far as smart sensor technology is concerned, in fact, all the synergistic responses and effects between the electric and nonelectric phenomena just discussed can be judiciously adopted. Considering the state-of-the-art technology and practical considerations, however, the existing smart sensors are limited to the aforesaid versions. Future trends could, however, include other possible electric to nonelectric synergistic responses.

55.7 Examples of Smart/Intelligent Systems

The method of synthesizing a smart/intelligent system is illustrated in Fig. 55.3. The output response under a given set of input condition(s) of a parent test system is normally decided by the properties of the constituent (conventional) materials. If the system-states (changes) under the influence of external inputs are sensed, however, an appropriate feedback control can be used to actuate an imbedded smart material in the parent unit, so that output will track adaptively a desired response. The feedback path may include relevant electronic hardware (such as microprocessors) for on-line processing of the feedback signal to optimize the system performance.

Essentially, the smart materials can be adopted in two regimes of the system shown in Fig. 55.3. The *sensing unit* can be zones of an integrated set of smart material that senses the response of the parent system on a real time basis. (Sometimes, conventional sensors/tranducers can serve this purpose, as well.) The *actuating unit*, built-in as a part of the parent structure, consists of a smart material, which upon receiving the electric signal from the feedback loop modifies the response of the

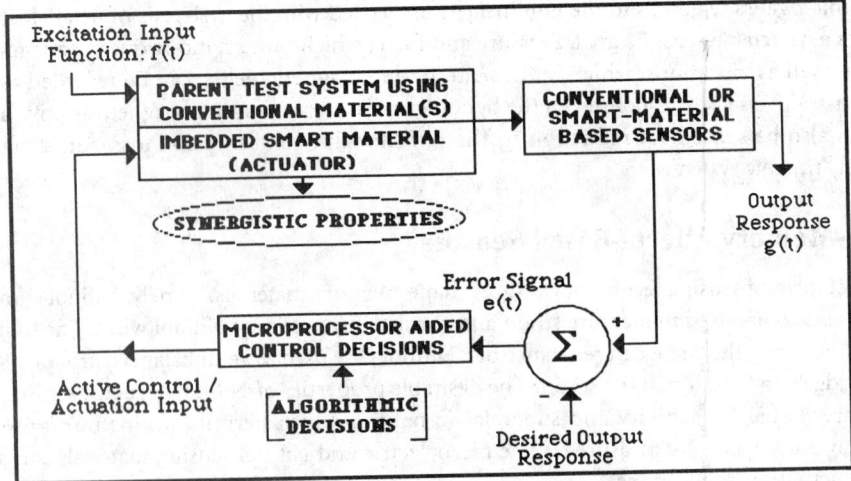

FIGURE 55.3   Schematic of a smart system.

parent system, as dictated by the input signal. Thus, the actuation is based on the synergism between the electric input to the corresponding material property of the parent structure being altered.

The feedback control unit may consist of decision logic, which can relatively modify the error signal being fed to the actuator. The decision logic refers to, for example, response linearization, time-averaged smoothing, amplitude-limiting, and bandwidth control. On the basis of the general schematic depicted in Fig. 55.3, the following discusses a few examples of application-specific intelligent systems using smart materials.

Structural Engineering Applications

Figure 55.4 illustrates a smart vibration control strategy in structural beams. Normally, the parent beam is made of conventional materials, and its vibrational characteristics are decided by the elastic behavior of the constituent materials. Suppose a smart material is imbedded in the test beam. This material could be one of the types indicated in Fig. 55.2. A vibration sensor yields an electric

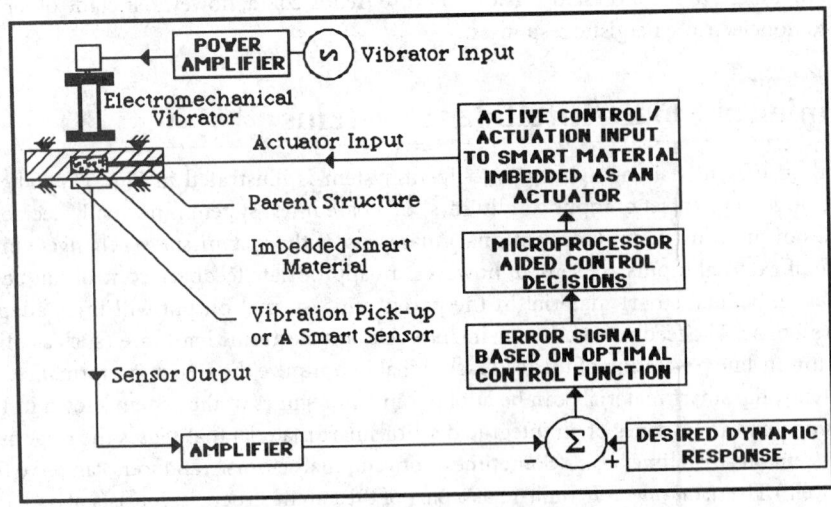

FIGURE 55.4   Active control of vibrating beams.

output proportional to the vibration. Suppose the dynamic response of the beam (as observed at the output of the sensor) deviates from the desired characteristics. Then an error signal can be generated, which in turn can be used to develop an optimal control signal and this control signal can be fed back to the smart material whose elastic behavior is then altered as a function of the control input. As a result, the vibration characteristics of the entire (parent) structure are modified, or the system is dynamically tuned in an adaptive manner.

The vibration sensor used can be either a conventional transducer (such as resistive, capacitive, inductive, or optical displacement versions) or it can be a smart sensor by itself. For example, an optical fiber with a leaky sheath (which permits the light energy to leak from the core to the outside surface) can be imbedded in the parent structure. When the structure is deformed, the extent of light leakage from the fiber to the surrounding material will modify proportionately. Hence, the detected light signal from the fiber optics, when detected, delivers information on the deformation or the dynamic structural characteristics of the test beam. This sensor can be made smart by integrating a distributed set of fibers that can sense strain, vibration, temperature, if needed, and so on, so that the network implemented with appropriate algorithms will provide exhaustive data for a comprehensive adaptive feedback control strategy.

Although the scheme illustrated in Fig. 55.4 refers to vibration control (or damping) in structures, judicious choice of subsystems and materials will permit adaptive control over other structural aspects also, namely, strain, bending moment, and redistribution of load path in response to failures.

Electromagnetic Applications

Smart material/structural techniques can be adopted in electromagnetic systems. The following are possible applications:

- Smart low-frequency magnetic shields
- Smart high-frequency electromagnetic shields
- Smart electrostatic dissipative/conductive surfaces
- Smart radar-absorbing materials (smart RAMs)
- Smart linear and aperture antennas

In all the preceding applications, the basic consideration is that the relevant structure can smartly and adaptively change its electromagnetic properties (normally specified via dielectric permittivity, magnetic permeability, and electrical conductivity parameters) so that the desired electromagnetic performance is acheived. Two typical systems are detailed next.

Electromagnetic Active Surface Embedded with Ferroelectric Inclusions

Figure 55.5 illustrates the concept of a smart electromagnetic active surface. The surface is made of a mixture of polyacrylamide, ferrite, and barium titanate on a ceramic substrate. This skin material, which represents a lossy, nonlinear electromagnetic medium with anisotropic ferroelectric and ferromagnetic properties, offers different extents of surface impedance, in the presence and absence of an electric voltage stimulus applied to it. Hence, the reflection coefficient of this material to electromagnetic energy can be altered via electric stimulus. Relevant feedback can facilitate adaptive smart responsiveness of the system as illustrated [Neelakanta et al., 1992].

Smart Electromagnetic Aperture

The aperture-radiation of microwaves can be smartly controlled by using a pyrosensitive material as illustrated in Fig. 55.6. A set of solid-electrolyte (AgI) pellets interconnected via nichrome heating elements is placed at the aperture of a microwave horn. At room temperature, the pellets behave as dielectrics (β-phase AgI). When heated, however, the β-phase AgI changes to a highly conducting medium (α-phase), which would mask a part of the aperture, thus modifying the radi-

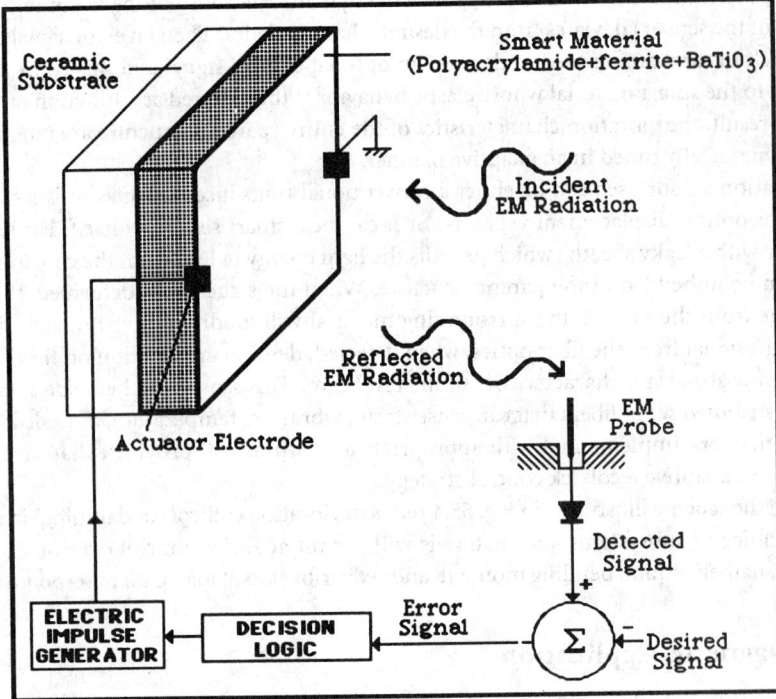

FIGURE 55.5   Smart electromagnetic active surface.

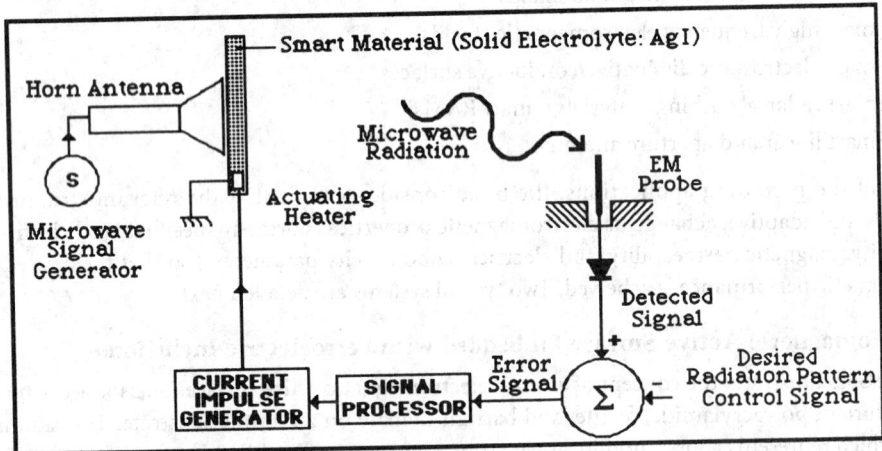

FIGURE 55.6   Smart electromagnetic aperture radiation control.

ation pattern of the horn antenna. Again, an appropriate feedback loop would render the functioning of this system intelligent [Neelakanta *et al.,* 1992].

55.8 High-Tech Application Potentials

Although smart material technology is in its infancy pending significant efforts to make it usable on a wide scale, the existing results and ongoing research have confirmed the usability of these materials in several avenues of modern high-technology systems.

Currently imaginable enclaves for the use of intelligent materials not only include structural engineering but also such areas as electromagnetics and biomedical, optical, and biological techniques. Relevant research has also been focused heavily in aerospace, aeronautics, marine vessel, and robotic applications.

Adaptive, self-monitoring of well-being by a system that has an integrated set of smart devices to self-assess its performance, diagnosing any malfunctions and failures and able to change its system characteristics vis-à-vis the environment, has been the objective of the relevant seed research pursued until now. For example, self-health checks by aircraft via a network of smart-skin sensors offer real-time monitoring of the structural well-being of tomorrow's aircraft [Claus, 1991]. The protocols in such efforts include self-diagnosis, prediction and notification, and self-repair strategies relevant to mechanical structures (such as aircraft bodies).

Another domain of smart material application is in self-induced morphologies in the infrastructure of the material with self-adaptive adjustments to the surroundings. Examples of this category include materials usable over a wide range of temperatures (as in space shuttles), with a smart adaptibility to transform according to the environment. Similarly, in radar stealth applications, the target skin could offer variable electromagnetic absorption over a broad band of radar frequencies.

Extensions of smart material concepts can cover selective acoustical absorptions and adaptive chromic controls in glasses, mirrors, etc. In short, viable smart systems can be conceived with various combinations of material characteristics discussed earlier together with the advent of new conventional materials, innovative sensors, advances in microcomputers, artificial intelligence, neural networking, and other upcoming technologies. Currently imaginable outlets for smart materials are summarized in the following list.

1. Structural/mechanical engineering
   - Airborne/space-borne systems with smart skins for adaptive self-health check feasibilities
   - Earthquake-resistant intelligent buildings
   - Large deployable space structures
   - Nondestructive evaluation of large structures
2. Thermal engineering
   - Adaptive heat transfer and heat-resistant structures (space shuttles, etc.)
3. Optical engineering
   - Adaptive hue, optical transparency, reflection, opaqueness control in glasses and mirrors
4. Electromagnetic engineering
   - Magnetic and electrostatic shielding
   - High-frequency shielding
   - Radar-absorbing materials
   - Active surfaces
   - Adaptive scattering/radiation control
5. Acoustical engineering
   - Active absorption/reflection of sonar radiation
   - Adaptive anechoic chambers
6. Chemical engineering
   - Materials with adaptive adsorption characteristics
   - Adaptive corrosion-resistant materials
7. Biomedical engineering
   - Materials with smart structural properties usable as artificial limbs
   - Materials with adaptive biochemical properties
8. Warfare systems
   - Smart shelters
   - Shock-resistant structures

55.9 Conclusions

The quest for new materials in scientific endeavors and engineering applications is everlasting. The emergence of the smart material concept has set a trend that science and technology in the coming years will rely to a large extent on the development of exotic materials, with intelligent materials being the leading candidates. Such materials will be hyper-functional with unstereotyped purposiveness responses to novel and changing situations.

Defining Terms

Electroacoustic smart materials: Materials that have self-adaptive characteristics in their acoustical behavior (such as transmission, reflection, and absorption of acoustical energy) in response to an external stimulus applied as a function of the sensed acoustical response.

Electromagnetic smart materials: Materials such as shielding materials, radar-absorbing materials (RAMs), and electromagnetic surface materials, in all of which some electromagnetic properties can be adaptively controlled by means of an external stimulus dictated by the sensed electromagnetic response.

Electro-optic smart materials: Materials in which optical properties are changed self-adaptively with an external electric stimulus proportional to the sensed optical characteristics.

Electroplastic effect: Plastic deformation of metals with the application of high-density electric current.

Electroplastic smart materials: Materials with smart properties of elastic deformation changes proportional to a controlled electric current applied in proportion to the sensed deformation.

Electrorheological property: Property exhibited by some fluids that are capable of altering their flow characteristics depending on an externally applied electric field.

Electrorheological smart fluids: Fluids with smart flow characteristics dictated to change self-adaptively by means of an electric field applied in proportion to the sensed flow parameters.

Intelligent structures: Structures constructed of smart materials with a dedicated, discrete set of integrated actuators, sensors, etc., in order to respond to the environment around them in a predetermined (desired) manner.

Magnetostrictive effect: Structural strain experienced in a material subjected to a polarizing magnetic flux, or reversibly, experiencing magnetic property changes to external mechanical stresses.

Magnetostrictive smart materials: A class of materials with self-adaptively modifiable elastic properties in response to a magnetic field applied in proportion to sensed stress–strain information.

Nonlinear dielectric property: The distinct dependence of the electric permittivity of certain dielectric materials on the intensity of an applied electric field.

Nonlinear electroacoustic property: Nonlinear interaction between the atomic displacement and the electric field experienced in certain materials that would cause modulation effects resulting in the generation of new sideband frequencies (called Raman frequencies).

Nonlinear electro-optic property: Nonlinear changes in the refractive index of certain optically transparent materials with change(s) in the externally applied electric field.

Nonlinear magnetic property: Nonlinear dependence of the magnetic susceptibility of certain materials on the intensity of an applied magnetic field.

Piezoelectric property: Ability of a material to induce opposite charges at two faces (correspondingly to exhibit a voltage difference between the faces) of the material as a result of the strain due to a mechanical force applied across the faces; reversibly, application of a potential across the faces would induce a mechanical strain.

Piezoelectric smart materials: Materials capable of changing their elastic characteristics (by virtue of their piezoelectric property) self-adaptively in response to an externally applied electric potential proportional to the observed elastic behavior.

Pyrosensitive properties: Exhibited by materials known as solid electrolytes whose electromagnetic properties can be altered by temperature.

Pyrosensitive smart materials: Materials that self-adaptively (smartly) manage the electromagnetic surface characteristics of active surfaces constituted by pyrosensitive inclusions, in response to an external temperature-inducing stimulus applied per the feedback information on electromagnetic characteristics.

Shape-memory effects: Mechanism by which a plastically deformed object in the low-temperature martensitic condition regains its original shape when the external stress is removed and heat is applied.

Shape-memory smart materials: Materials that smartly change their elastic characteristics by virtue of their shape-restoration characteristics achieved by means of an external stimulus in proportion to the magnitude of sensed shape changes.

Smart, or intelligent, materials: A class of materials and/or composite media having inherent intelligence together with self-adaptive capabilities to external stimuli applied in proportion to a sensed material response.

Smart sensors: Sensors with inherent intelligence via bulit-in electronics.

Smart structural materials: Materials in which the mechanical (elastic) properties can be modified adaptively through the application of external stimuli.

Smart thermal materials: Materials that can influence their thermal states (temperature or thermal properties such as conductivity) self-adaptively by means of an external control in response to environmental demands.

References

K.P. Chong, S.C. Liu, and J.C. Li (Eds.), *Intelligent Structures,* London and New York: Elsevier, 1990.

R.O. Claus, "Fiber sensors as nerves for smart materials," *Photonics Spectra,* vol. 25, no. 4, p. 75, 1991.

M.V. Gandhi and B.S. Thompson, "A new generation of revolutionary ultra-advanced intelligent materials featuring electrorheological fluids," in *Smart Materials, Structures, and Mathematical Issues,* Lancaster, Pa.: Technomic Publishing, 1989, pp. 63–68.

C.M. Jackson, H.J. Wagner, and R.J. Wasilewski, *55-Nitinol—The Alloy with a Memory: Its Physical Metallurgy, Properties and Application,* NASA-SP-5110, 1972.

I.P. Kaminow, "Parametric principles in optics," *IEEE Spectrum,* vol. 2, p. 40, 1965.

P.S. Neelakanta, J. Abello, and C. Gu, "Microwave reflection at an active surface imbedded with fast-ion conductors," *IEEE Trans. Microwave Theory Tech.,* vol. MTT-40, no. 5, pp. 28–30, 1992.

P.S. Neelakanta and K. Subramaniam, "Controlling the properties of electromagnetic composites," *Adv. Materials and Process,* vol. 141, no. 3, pp. 20–25, 1992.

R.S. Reed, "Shock isolation using an active magnetostrictive element," in *Proc. 59th Shock and Vibration Symp.* vol. I, Albuquerque, New Mex., Oct. 18–20, 1988.

C.A. Rogers and R.C. Rogers, *Recent Advances in Adaptive and Sensory Materials and Their Applications,* Lancaster, Pa.: Technomic Publishing, 1992.

R. Ting, "The hydroacoustic behavior of piezoelectric composite materials," *Ferroelectrics,* vol. 102, pp. 215–224, 1990.

Further Information

Intelligent Structures, edited by K.P Chong, S.C. Liu, and J.C. Li, contains the papers presented in an international workshop on intelligent structures held on 23–26 July 1990 in Taipei, Taiwan, Elsevier Science Publishers, 1990.

Another source is the *Proceedings of the International Workshop on Intelligent Materials,* The Society of Non-Traditional Technology, 1989.

Recent Advances in Adaptive and Sensory Materials and Their Applications, by C.A. Rogers and R.C. Rogers, Lancaster, Pa.: Technomic Publishing Co., Inc., 1992.

The author's publication with K. Subramaniam, "Controlling the Properties of Electromagnetic Composites" in *Advanced Materials and Processes,* vol. 141, no. 3, pp. 20–25, 1992.

These wind turbines are part of the largest wind farm in the world located in Altamont, California. Currently the United States has the bulk of the world's wind-generated power capacity, accounting for 68% of all installed capacity in 1992. Experts project that by the end of the decade, Europe will be the dominant force in wind power, accounting for 62% of the total installed wind power capacity worldwide. Aggressive government programs that set national goals for wind turbine installation and subsidize wind power projects are the driving force behind Europe's increased commitment to wind power. (Photo courtesy of American Wind Energy Association.)

VI

Energy

William H. Kersting
New Mexico State University

56 **Conventional Power Generation** *G. Karady* 1193
Fossil Power Plants • Nuclear Power Plants • Geothermal Power Plants • Hydroelectric Power Plants

57 **Distributed Power Generation** *R. Ramakumar* 1207
Photovoltaics • Wind-Electric Conversion • Hydro • Geothermal • Tidal Energy • Fuel Cells • Solar-Thermal-Electric Conversion • Biomass Energy • Thermoelectrics • Thermionics • Integrated System Concepts

58 **Transmission** *M. Chen, K. Lai, R. Thallam, M. El-Hawary, C. Gross, A. Phadke, R. Gungor, J. Glover* 1217
Alternating Current Overhead: Line Parameters, Models, Standard Voltages, Insulators • Alternating Current Underground: Line Parameters, Models, Standard Voltages, Cables • High-Voltage Direct-Current Transmission • Compensation • Fault Analysis in Power Systems • Protection • Transient Operation • Planning

59 **Power Transformers** *C. Gross, W. Feaster* 1296
Power Transformer Fundamentals • Transformer Construction • Transformer Performance • Transformers in Three-Phase Connections • Autotransformers

60 **Energy Distribution** *G. Karady* 1310
Primary Distribution System • Secondary Distribution System • Radial Distribution System • Secondary Networks • Load Characteristics • Voltage Regulation • Capacitors and Voltage Regulators

61 **Electrical Machines** *C. Liu, K. Vu, Y. Yu, D. Galler* 1321
Generators • Motors

62 **Energy Management** *K. Stanton, J. Giri, A. Bose* 1344
Power System Data Acquisition and Control • Automatic Generation Control • Load Management • Energy Management • Security Control • Operator Training Simulator

T HE GENERATION, TRANSMISSION, AND DISTRIBUTION of electrical energy remains one of the most exciting and challenging areas of electrical engineering. Without a safe, reliable, and economic supply of electrical energy, all industry would come to a grinding halt.

This section will present chapters discussing the theory and methods for the generation, transmission, and distribution of electrical energy. While the fundamentals have been around a long time, the application of the fundamentals will continue to take on many new forms.

The great majority of electrical energy continues to be generated by large conventional power plants, which are discussed in Chapter 56. As Chapter 57 will explain, fuel cells, wind, and solar are becoming important components of the overall generation picture.

The transmission of electrical energy over long distances and at increasingly higher voltages has become an ever more important component as "open access" of these facilities becomes a reality. Chapter 58 will present the theory of transmission, including alternating current and direct current transmission, both overhead and underground. In addition, the discussion will include the protection of these facilities, transient operation, and planning.

A modern power system operates at many different voltage levels. Because of this, transformers play a key role. The theory is the same for all voltage levels and will be presented in Chapter 59. The chapter will also discuss application of the theory for different types of transformers commonly used.

The final component in bringing electrical energy to the ultimate user is the distribution system. For many years the stepchild to the more costly generation and transmission components, distribution are now playing an important role in increasing reliability and service at a reduced cost. Chapter 60 presents an overview of this key component.

Generators and motors are still the primary devices for converting energy from mechanical to electrical and vice versa. Chapter 61 is devoted to the theory of ac/dc motors and generators.

The automatic control of the total power system is presented in Chapter 62. This is one area of power systems that has changed dramatically and continues to change on almost a daily basis. In many ways, the total field of electrical engineering is applied in the control of a modern power system.

Nomenclature

Symbol	Quantity	Unit	Symbol	Quantity	Unit
D	damping coefficient		n	Steinmetz constant	
DF	demand factor		N	number of turns	
δ	power angle	degree	N_p	number of stator pole pairs	
δ	torque angle	degree	ω_s	slip frequency radian	
f_s	slip frequency	Hz	P	real power	W
ϕ	core flux	Wb	P_e	eddy current loss	W
ϕ_f	magnetic field flux	Wb	P_h	hysteresis power loss	W
I_{ac}	rms ac current	A	Q	reactive power	
I_f	field current	A	r	radius of conductor	m
I_r	rotor current	A	ρ	resistivity of conductor	Ωm
J	moment of inertia of rotor	kg-m^2/rad	s	slip of an induction motor	
K_a	armature constant of a dc machine		$\overline{S}$	complex power	
			T	shaft torque	N-m
K_t	torque constant of a dc machine		θ	rotor angle	degree
			θ	power factor angle	degree
LSF	loss factor		V_a	armature back emf	V
λ	magnetic flux linkages	Wb/m	w	shaft speed	rad/s
n	rotor speed	rpm			

56

Conventional Power Generation

56.1 Introduction.. 1198
56.2 Fossil Power Plants 1194
    Fuel Handling • Boiler • Turbine • Generator • Electric System
    • Condenser • Stack and Ash Handling • Cooling and Feedwater
    System
56.3 Nuclear Power Plants... 1200
    Pressurized Water Reactor • Boiling-Water Reactor
56.4 Geothermal Power Plants..................................... 1202
56.5 Hydroelectric Power Plants.................................. 1203

George G. Karady
Arizona State University

56.1 Introduction

The electric energy demand of the world is continuously increasing, and most of the energy is generated by conventional power plants, which remain the only cost-effective method for generating large quantities of energy.

Power plants utilize energy stored in the earth and convert it to electrical energy that is distributed and used by customers. This process converts most of the energy into heat, which increases the entropy of the earth. In this sense, power plants deplete the earth's energy supply. Efficient operation becomes increasingly important to conserve energy.

Typical energy sources used by power plants include fossil fuel (gas, oil, and coal), nuclear fuel (uranium), geothermal energy (hot water, steam), and hydro energy (water falling through a head).

Around the turn of the century, the first fossil power plants used steam engines as the prime mover. These plants were evolved to an 8- to 10-MW capacity, but increasing power demands resulted in the replacement by a more efficient steam boiler–turbine arrangement. The first commercial steam turbine was introduced by DeLaval in 1882. The boilers were developed from heating furnaces. Oil was the preferred and most widely used fuel in the beginning. The oil shortage promoted coal-fired plants, but the adverse environmental effects (sulfur dioxide generation, acid rain, dust pollution, etc.) curtailed their use in the late seventies. Presently the most acceptable fuel is natural gas, which minimizes pollution and is available in large quantities. During the next two decades, gas-fired power plants will dominate the electric industry.

The hydro plants' ancestors are water wheels used for pumping stations, mill driving, etc. Water-driven turbines were developed in the last century and used for generation of electricity since the beginning of their commercial use. However, most of the sites that can be developed economically are currently being utilized. No significant new development is expected in the United States in the near future.

Nuclear power plants appeared after the Second World War. The major development occurred during the sixties; however, by the eighties environmental considerations stopped plant development in the United States and slowed it down all over the world. Presently, the future of nuclear power generation is unclear, but the abundance of nuclear fuel and the expected energy shortage in the early part of the next century may rejuvenate nuclear development if safety issues can be resolved.

Geothermal power plants are the product of the clean energy concept, although the small-scale, local application of geothermal energy has a long history. Presently only a few plants are in operation. The potential for further development is limited because of the unavailability of geothermal energy sites that can be developed economically.

Typical technical data for different power plants is shown in Table 56.1.

56.2 Fossil Power Plants

The operational concept and major components of a fossil power plant are shown in Fig. 56.1.

Fuel Handling

The most frequently used **fuels** are oil, natural gas, and coal. Oil and gas are transported by rail, on ships, or through pipelines. In the former case the gas is liquefied. Coal is transported by rail or ships if the plant is near a river or the sea. The power plant requires several days of fuel reserve. Oil and gas are stored in large metal tanks, and coal is kept in open yards. The temperature of the coal layer must be monitored to avoid self-ignition.

Oil is pumped and gas is fed to the burners of the **boiler.** Coal is pulverized in large mills, and the powder is mixed with air and transported by air pressure, through pipes, to the burners. The coal transport from the yard to the mills requires automated transporter belts, hoppers, and sometimes manually operated bulldozers.

Boiler

Two types of boilers are used in modern power plants: subcritical water-tube drum-type and supercritical once-through type. The former operates around 2500 psi, which is under the water critical pressure of 3208.2 psi. The latter operates above that pressure, at around 3500 psi. The superheated steam temperature is about 1000°F (540°C) because of turbine temperature limitations.

A typical subcritical water-tube drum-type boiler has an inverted-U shape. On the bottom of the rising part is the furnace where the fuel is burned. The walls of the furnace are covered by water pipes. The drum and the **superheater** are at the top of the boiler. The falling part of the U houses the reheaters, **economizer** (water heater), and air preheater, which is supplied by the forced-draft

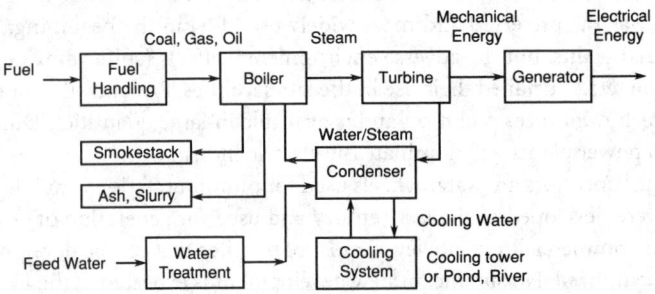

FIGURE 56.1 Major components of a fossil power plant.

Table 56.1 Power Plant Technical Data

Generation Type	Typical MW Size	Capitalized Plant Cost, $/kW	Construction Lead Time, Years	Heat Rate, Btu/kWh	Fuel Cost, $/MBtu	Fuel Type	Equivalent Forced Outage Rate	Equivalent Scheduled Outage Rate	O&M Fixed, $/kW/year	Cost Variable $/MWh
Nuclear	1200	2400	10	10,400	1.25	Uranium	20	15	25	8
Pulverized coal steam	500	1400	6	9,900	2.25	Coal	12	12	20	5
Atmospheric fluidized bed	400	1400	6	9,800	2.25	Coal	14	12	17	6
Gas turbine	100	350	2	11,200	4.00	Nat. gas	7	7	1	5
Combined-cycle	300	600	4	7,800	4.00	Nat. gas	8	8	9	3
Coal-gasification combined-cycle	300	1500	6	9,500	2.25	Coal	12	10	25	4
Pumped storage hydro	300	1200	6	—	—		5	5	5	2
Conventional hydro	300	1700	6	—	—		3	4	5	2

Source: H. G. Stoll. *Least-Cost Electric Utility Planning.* © 1989 John Wiley & Sons. Reprinted by permission of John Wiley & Sons, Inc.

fan. The induced-draft fan forces the flue gases out of the system and sends them up the stack, which is located behind the boiler. A flow diagram of the drum-type boiler is shown in Fig. 56.2. The steam generator has three major systems: fuel, air-flue gas, and water-steam.

Fuel System. Fuel is mixed with air and injected into the furnace through burners. The burners are equipped with nozzles, which are supplied by preheated air and carefully designed to assure the optimum air-fuel mix. The fuel mix is ignited by oil or gas torches. The furnace temperature is around 3000°F.

Air-Flue Gas System. Ambient air is driven by the forced-draft fan through the air preheater, which is heated by the high-temperature (600°F) flue gases. The air is mixed with fuel in the burners and enters into the furnace, where it supports the fuel burning. The hot combustion flue gas generates steam and flows through the boiler to heat the superheater, reheaters, economizer, etc. Induced-draft fans, located between the boiler and the stack, increase the flow and send the 300°F flue gases to the atmosphere through the stack.

Water-Steam System. Large pumps drive the feedwater through the high-pressure heaters and the economizer, which further increases the water temperature (400–500°F). The former is heated by steam removed from the turbine; the latter is heated by the flue gases. The preheated water is fed to the steam drum. Insulated tubes, called downcomers, are located outside the furnace and lead the water to a header. The header distributes the hot water among the risers. These are water tubes that line the furnace walls. The water tubes are heated by the combustion gases through both convection and radiation. The steam generated in these tubes flows to the drum, where it is separated from the water. Circulation is maintained by the density difference between the water in the downcomer and the water tubes. Saturated steam, collected in the drum, flows through the superheater. The superheater increases the steam temperature to about 1000°F. Dry superheated steam drives the high-pressure turbine. The exhaust from the high-pressure turbine goes to the reheater, which again increases the steam temperature. The reheated steam drives the low-pressure turbine.

The typical supercritical once-through-type boiler concept is shown in Fig. 56.3.

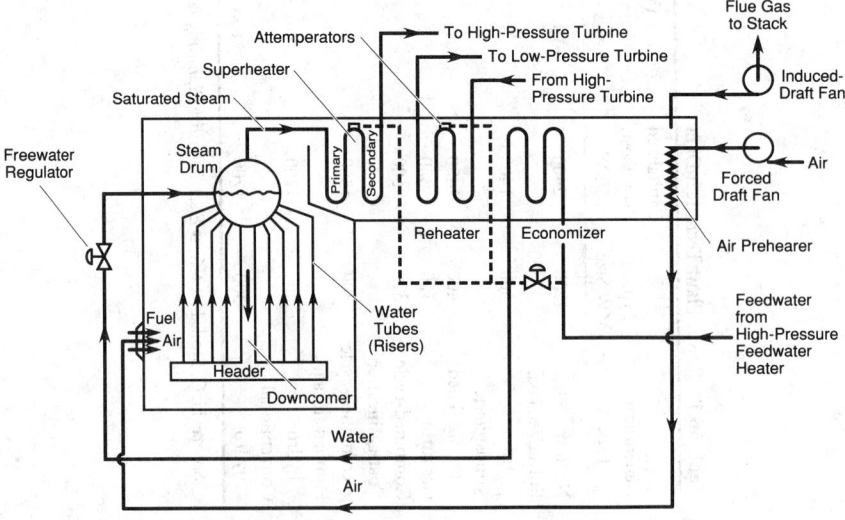

FIGURE 56.2 Flow diagram of a typical drum-type steam boiler. (*Source:* M.M. El-Wakil, *Power Plant Technology,* New York: McGraw-Hill, 1984, p. 210. With permission.)

The feedwater enters through the economizer to the boiler, which consists of riser tubes that line the furnace wall. All the water is converted to steam and fed directly to the superheater. The latter increases the steam temperature above the critical temperature of the water and drives the turbine. The construction of these steam generators is more expensive than the drum-type units but has a higher operating efficiency.

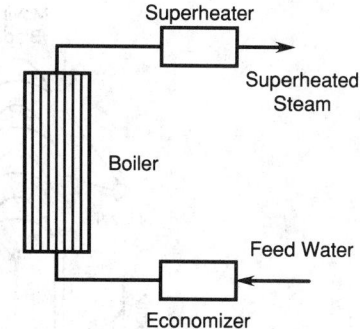

FIGURE 56.3 Concept of once-through-type steam generator.

Turbine

The turbine converts the heat energy of the steam into mechanical energy. Modern power plants usually use one high-pressure and one or two lower-pressure turbines. A typical turbine arrangement is shown in Fig. 56.4.

The figure shows that only one bearing is between each of the machines. The shafts are connected to form a tandem compound steam turbine unit. High-pressure steam enters the high-pressure turbine to flow through and drive the turbine. The exhaust is reheated in the boiler and returned to the lower-pressure units. Both the rotor and the stationary part of the turbine have blades. The length of the blades increases from the steam entrance to the exhaust.

Figure 56.5 shows the blade arrangement of an *impulse*-type turbine. Steam enters through nozzles and flows through the first set of moving rotor blades. The following stationary blades change the direction of the flow and direct the steam into the next set of moving blades. The nozzles increase the steam speed and reduce pressure, as shown in the figure. The impact of the high-speed steam, generated by the change of direction and speed in the moving blades, drives the turbine.

The *reaction*-type turbine has nonsymmetrical blades arranged like those shown in Fig. 56.5. The blade shape assures that the pressure continually drops through all rows of blades, but steam velocity decreases in the moving blades and increases in the stationary blades.

Generator

The generator converts mechanical energy from the turbines into electrical energy. The major components of the generator are the frame, stator core and winding, rotor and winding, bearings, and cooling system. Figure 56.6 shows the cross section of a modern hydrogen-cooled generator.

The stator has a laminated and slotted silicon steel iron core. The stacked core is clamped and held together by insulated axial through bolts. The stator winding is placed in the slots and consists of a copper-strand configuration with woven glass insulation between the strands and mica flakes, mica mat, or mica paper ground-wall insulation. To avoid insulation damage caused by vibration,

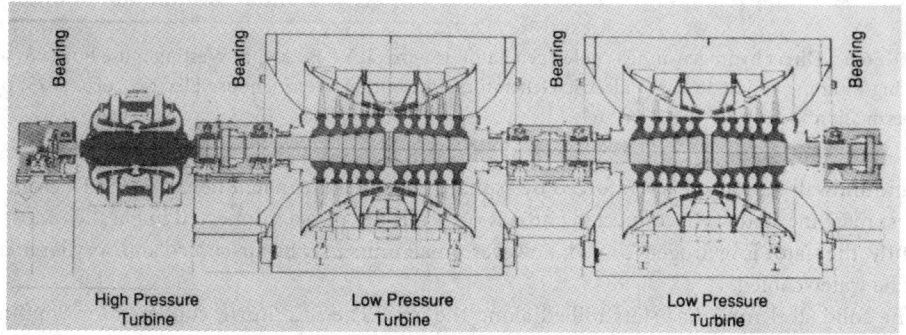

FIGURE 56.4 Large tandem compound steam turbine. (*Source:* M.M. El-Wakil, *Power Plant Technology,* New York: McGraw-Hill, 1984, p. 210. With permission.)

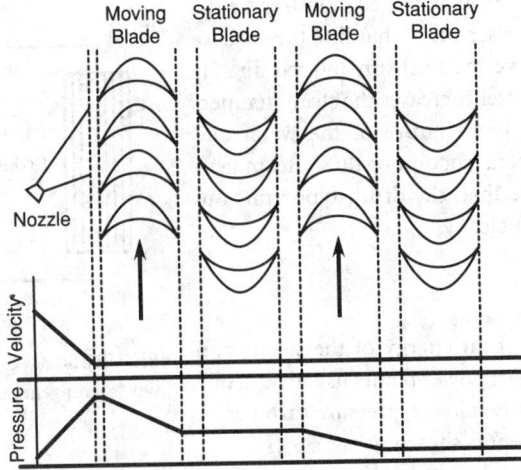

FIGURE 56.5   Velocity and pressure variation in an impulse turbine.

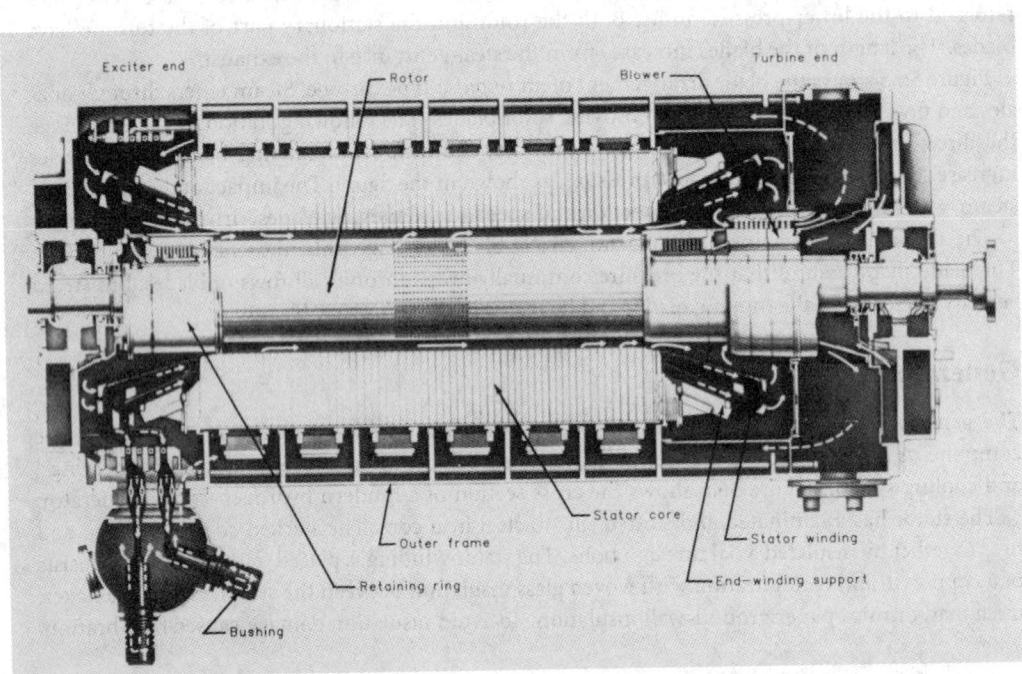

FIGURE 56.6   Direct hydrogen-inner-cooled generator. (*Source:* R.W. Beckwith, Westinghouse Power Systems Marketing Training Guide on Large Electric Generators, Pittsburgh: Westinghouse Electric Corp. 1979, p. 54. With permission.)

the groundwall insulation is reinforced by asphalt, epoxy-impregnated fiberglass, or Dacron. The largest machine stator is Y-connected and has two coils per phase, connected in parallel. Most frequently, the stator is hydrogen-cooled; however, small units may be air-cooled and very large units may be water-cooled.

The solid steel rotor has slots milled along the axis. The multiturn, copper rotor winding is placed in the slots and cooled by hydrogen. Cooling is enhanced by subslots and axial cooling passages. The rotor winding is restrained by wedges inserted in the slots.

The rotor winding is supplied by dc current, either directly by a brushless excitation system or through collector rings. The rotor is supported by bearings at both ends. The non-drive-end bearing is insulated to avoid shaft current generated by stray magnetic fields. The hydrogen is cooled by a hydrogen-to-water heat exchanger mounted on the generator or installed in a closed-loop cooling system.

The dc current of the rotor generates a rotating magnetic field that induces an ac voltage in the stator winding. This voltage drives current through the load and supplies the electrical energy.

Electric System

Energy generated by the power plant supplies the electric network through transmission lines. The power plant operation requires auxiliary power to operate mills, pumps, etc. The auxiliary power requirement is approximately 10 to 15%.

Smaller generators are directly connected in parallel using a busbar. Each generator is protected by a circuit breaker. The power plant auxiliary system is supplied from the same busbar. The transmission lines are connected to the generator bus, either directly or through a transformer.

The larger generators are unit-connected. In this arrangement the generator is directly connected, without a circuit breaker, to the main transformer. A conceptual one-line diagram is shown in Fig. 56.7. The generator supplies main and auxiliary transformers without circuit breakers. The units are connected in parallel at the high-voltage side of the main transformers by a busbar. The transmission lines are also supplied from this bus. Circuit breakers are installed at the secondary side of the main and auxiliary transformer. The application of generator circuit breaker is not economical in the case of large generators. Because of the generator's large short-circuit current, special expensive circuit breakers are required. However, the transformers reduce the short-circuit current and permit the use of standard circuit breakers at the secondary side. The disconnect switches permit visual observation of the off state and are needed for maintenance of the circuit breakers.

Condenser

The condenser condenses turbine exhaust steam to water, which is pumped back to the steam generator through various water heaters. The condensation produces a vacuum, which is necessary to exhaust the steam from the turbine. The condenser is a shell-and-tube heat exchanger, where steam condenses on water-cooled tubes. Cold water is obtained from the cooling towers or other cooling systems. The condensed water is fed through a deaerator, which removes absorbed gases from the water. Next, the gas-free water is mixed with the feedwater and returned to the boiler. The gases absorbed in the water may cause corrosion (oxygen) and increase condenser pressure, adversely affecting efficiency. Older plants use a separate deaerator heater, while deaerators in modern plants are usually integrated in the condensor, where injected steam jets produce pressure drop and remove absorbed gases.

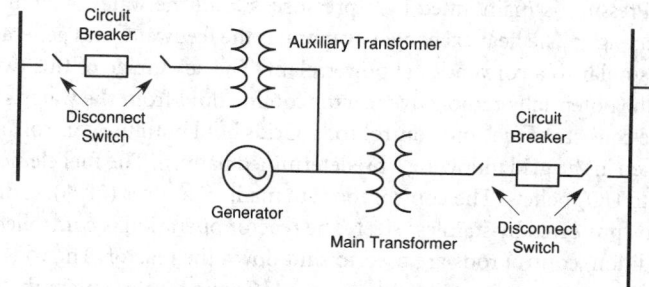

FIGURE 56.7   Conceptual one-line diagram for a unit-connected generator.

Stack and Ash Handling

The stack is designed to disperse gases into the atmosphere without disturbing the environment. This requires sufficient stack height, which assists the fans in removing gases from the boiler through natural convection. The gases contain both solid particles and harmful chemicals. Solid particles, like dust, are removed from the flue gas by electrostatic precipitators or bag-house filters. Harmful sulfur dioxide is eliminated by scrubbers. The most common is the lime/limestone scrubbing process.

Coal-fired power plants generate a significant amount of ash. The disposition of the ash causes environmental problems. Several systems have been developed in past decades. Large ash particles are collected by a water-filled ash hopper, located at the bottom of the furnace. Fly ash is removed by filters, then mixed with water. Both systems produce sludge that is pumped to a clay-lined pond where water evaporates and the ash fills disposal sites. The clay lining prevents intrusion of groundwater into the pond.

Cooling and Feedwater System

The condenser is cooled by cold water. The open-loop system obtains the water from a river or sea, if the power plant location permits it. The closed-loop system utilizes cooling towers, spray ponds, or spray canals. In the case of spray ponds or canals, the water is pumped through nozzles, which generate fine sprays. Evaporation cools the water sprays as they fall back into the pond. Several different types of cooling towers have been developed. The most frequently used is the wet cooling tower, where the hot water is sprayed on top of a latticework of horizontal bars. The water drifts downward and is cooled, through evaporation, by the air, which is forced by fans or natural draft upward.

The power plant loses a small fraction of the water through leakage. The feedwater system replaces this lost water. Replacement water has to be free from absorbed gases, chemicals, etc., because the impurities cause severe corrosion in the turbines and boiler. The water treatment system purifies replacement water by pretreatment, which includes filtering, chlorination, demineralization, condensation, polishing. These complicated chemical processes result in a corrosion-free high-quality feedwater.

56.3 Nuclear Power Plants

More than 500 nuclear power plants operate around the world. Close to 300 operate pressurized water reactors (PWRs), more than 100 are built with boiling-water reactors (BWRs), about 50 use gas-cooled reactors, and the rest are heavy-water reactors. In addition a few fast breeder reactors are in operation. These reactors are built for better utilization of uranium fuel. The modern nuclear plant size varies from 100 to 1200 MW.

Pressurized Water Reactor

The general arrangement of a power plant with a PWR is shown in Fig. 56.8(A).

The **reactor** heats the water from about 550 to about 650°F. High pressure, at about 2235 psi, prevents boiling. Pressure is maintained by a pressurizer, and the water is circulated by a pump through a heat exchanger. The heat exchanger evaporates the feedwater and generates steam, which supplies a system similar to a conventional power plant. The advantage of this two-loop system is the separation of the potentially radioactive reactor cooling fluid from the water-steam system.

The reactor core consists of fuel and control rods. Grids hold both the control and fuel rods. The fuel rods are inserted in the grid following a predetermined pattern. The fuel elements are Zircaloy-clad rods filled with UO_2 pellets. The control rods are made of a silver (80%), cadmium (5%), and indium (15%) alloy protected by stainless steel. The reactor operation is controlled by the position of the rods. In addition, control rods are used to shut down the reactor. The rods are released and fall in the core when emergency shutdown is required. Cooling water enters the reactor from the bottom, flows through the core, and is heated by nuclear fission.

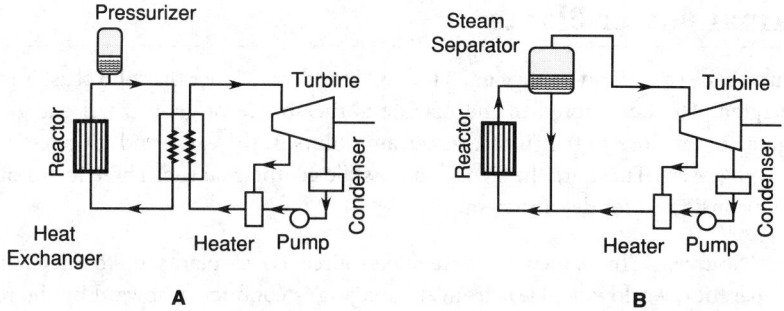

FIGURE 56.8 (A) Power plant with PWR; (B) power plant with BWR.

Boiling-Water Reactor

In the BWR shown in Fig. 56.8(B), the pressure is low, about 1000 psi. The nuclear reaction heats the water directly to evaporate it and produce wet steam at about 545°F. The remaining water is recirculated and mixed with feedwater. The steam drives a turbine that typically rotates at 1800 rpm. The rest of the plant is similar to a conventional power plant. A typical reactor arrangement is shown in Fig. 56.9. The figure shows all the major components of a reactor. The fuel and control rod assembly is located in the lower part. The steam separators are above the core, and the steam dryers are at the top of the reactor. The reactor is enclosed by a concrete dome.

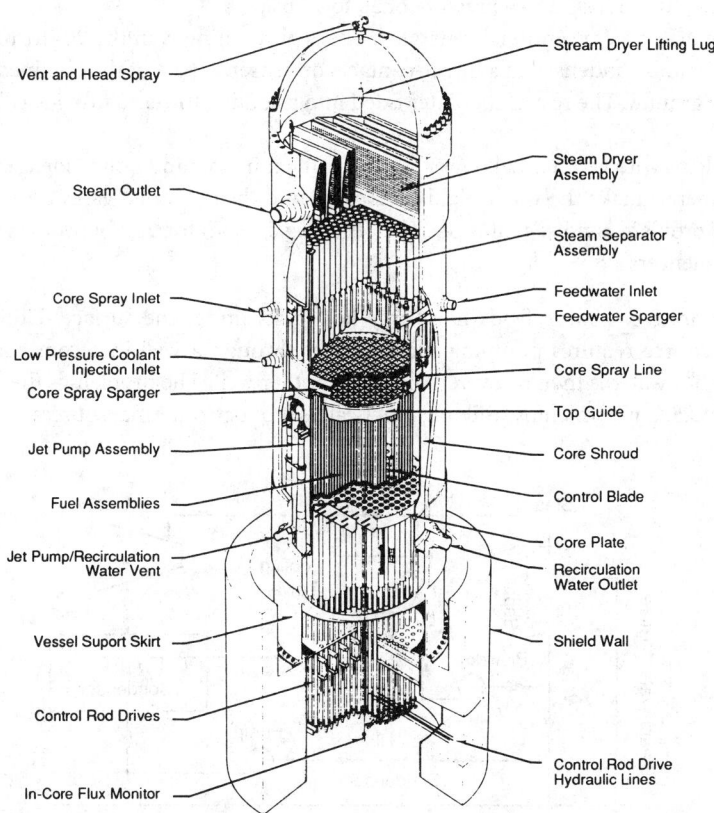

FIGURE 56.9 Typical BWR reactor arrangement. (*Source:* Courtesy of General Electric Company.)

56.4 Geothermal Power Plants

The solid crust of the earth is an average of 20 mil (32 km) deep. Under the solid crust is the molten mass, the magma. The heat stored in the magma is the source of geothermal energy. The hot molten magma comes close to the surface at certain points in the earth and produces volcanoes, hot springs, and geysers. These are the signs of a possible geothermal site. Three forms of geothermal energy are considered for development.

Hydrothermal Source. This is the most developed source. Power plants, up to a capacity of 2000 MW, are in operation worldwide. Heat from the magma is conducted upward by the rocks. The groundwater drifts down through the cracks and fissures to form reservoirs when water-impermeable solid rock bed is present. The water in this reservoir is heated by the heat from the magma. Depending on the distance from the magma and rock configuration, steam, hot pressurized water, or the mixture of the two are generated. Signs of these underwater reservoirs include hot springs and geysers. The reservoir is tapped by a well, which brings the steam-water mixture to the surface to produce energy. The geothermal power plant concept is illustrated in Fig. 56.10.

The hot water and steam mixture is fed into a separator. If the steam content is high, a centrifugal separator is used to remove the water and other particles. The obtained steam drives a turbine. The typical pressure is around 100 psi and the temperature is around 400°F (200°C). If the water content is high, the water-steam mixture is led through a flashed-steam system where the expansion generates a better quality of steam and separates the steam from the water. The water is returned to the ground, the steam drives the turbine. Typically the steam entering the turbine has a temperature of 120 to 150°C and a pressure of 30 to 40 psi.

The turbine drives a conventional generator. The typical rating is in the 20- to 100-MW range. The exhaust steam is condensed in a direct-contact condenser. A part of the obtained water is reinjected into the ground. The rest of the water is fed into a cooling tower to provide cold water to the condenser.

Major problems with geothermal power plants are the minerals and noncondensable gases in the water. The minerals make the water highly corrosive, and the separated gases cause air pollution. An additional problem is noise pollution. The centrifugal separator and blowdowns require noise dampers and silencers.

Petrothermal Source. Some fields have only hot rocks under the surface. Utilization of this petrothermal source requires pumping surface water through a well in a constructed hole to a reservoir. The hot water is then recovered through another well. The problem is the formation of a reservoir. The U.S. government is studying practical uses of petrothermal sources.

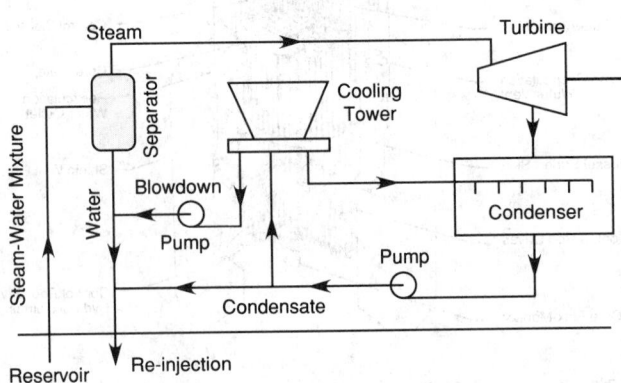

FIGURE 56.10 Concept of a geothermal power plant.

Geopressured Source. In deep underground holes (8000 to 30,000 ft) a mixture of pressurized water and natural gas, like methane, may sometimes be found. These geopressured sources promise power generation through the combustion of methane and the direct recovery of heat from the water. The geopressured method is currently in an experimental stage, with operating pilot plants.

56.5 Hydroelectric Power Plants

Hydroelectric power plants convert energy produced by a water head into electric energy. A typical hydroelectric power plant arrangement is shown in Fig. 56.11.

The head is produced by building a dam across a river, which forms the upper-level reservoir. In the case of low head, the water forming the reservoir is fed to the turbine through the intake channel or the turbine is integrated in the dam. The latter arrangement is shown in Fig. 56.11(A). **Penstock** tubes or tunnels are used for medium- [Fig. 56.11(B)] and high-head plants (Fig. 56.12). The spillway regulates the excess water by opening gates at the bottom of the dam or permitting overflow on the spillway section of the dam. The water discharged from the turbine flows to the lower or tail water reservoir, which is usually a continuation of the original water channel.

High-Head Plants. High-head plants (Fig. 56.12) are built with impulse turbines, where the head-generated water pressure is converted into velocity by nozzles and the high-velocity water jets drive the turbine runner.

Low- and Medium-Head Plants. Low- and medium-head installations (Fig. 56.11) are built with reaction-type turbines, where the water pressure is mostly converted to velocity in the turbine. The two basic classes of reaction turbines are the propeller or Kaplan type, mostly used for low-head plants, and the Francis type, mostly used for medium-head plants. The cross section of a typical low-head Kaplan turbine is shown in Fig. 56.13.

The vertical shaft turbine and generator are supported by a thrust bearing immersed in oil. The generator is in the upper, watertight chamber. The turbine runner has 4 to 10 propeller types, and adjustable pitch blades. The blades are regulated from 5 to 35 degrees by an oil-pressure-operated servo mechanism. The water is evenly distributed along the periphery of the runner by a concrete spiral case and regulated by adjustable wicket blades. The water is discharged from the turbine

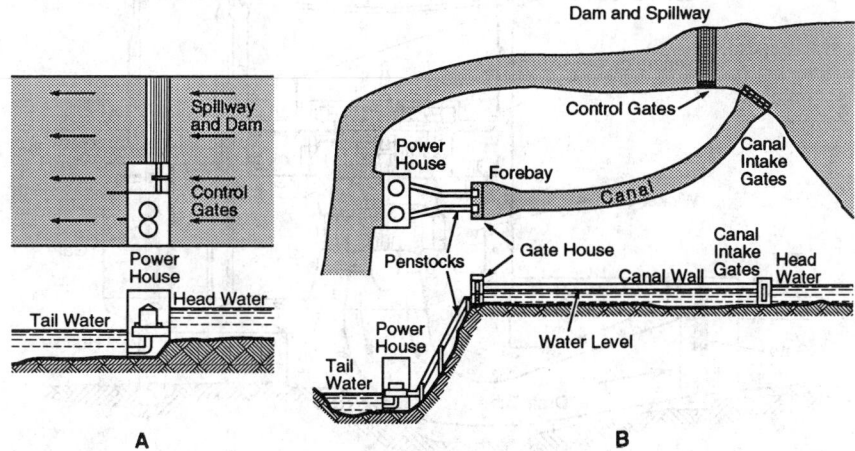

FIGURE 56.11 Hydroelectric power plant arrangement. (A) Low-head plant, (B) medium-head plant. (*Source:* D.G. Fink, *Standard Handbook for Electrical Engineers,* New York: McGraw-Hill, 1978. With permission.)

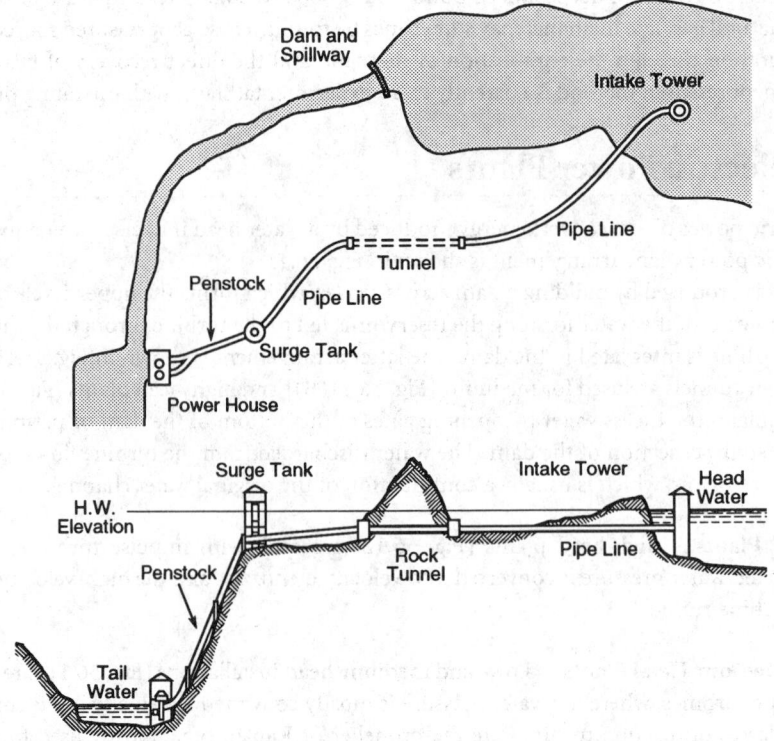

FIGURE 56.12 Hydroelectric power plant arrangement, high-head plant. (*Source:* D.G. Fink, *Standard Handbook for Electrical Engineers,* New York: McGraw-Hill, 1978. With permission.)

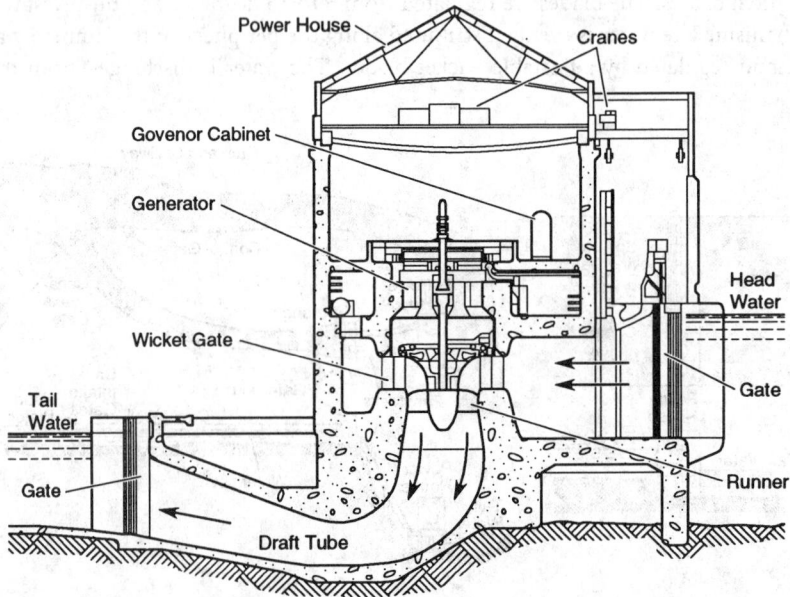

FIGURE 56.13 Typical low-head hydroplant with Kaplan turbine. (*Source:* D. G. Fink, *Standard Handbook for Electrical Engineers,* New York: McGraw-Hill, 1978. With permission.)

through an elbow-shaped draft tube. The conical profile of the tube reduces the water speed from the discharge speed of 10–30 ft/s to 1 ft/s to increase turbine efficiency.

Hydrogenerators. The hydrogenerator is a low-speed (100 to 360 rpm) salient-pole machine with a vertical shaft. A typical number of poles is from 20 to 72. They are mounted on a pole spider, which is a welded, spoked wheel. The spider is mounted on the forged steel shaft. The poles are built with a laminated iron core and stranded copper winding. Damper bars are built in the pole faces. The stator is built with slotted, laminated iron core that is supported by a welded steel frame. Windings are made of stranded conductors insulated between the turns by fiberglass or Dacron-glass. The ground insulation is multiple layers of mica tape impregnated by epoxy or polyester resins. The older machines use asphalt and mica tape insulation, which is sensitive to corona-discharge-caused insulation deterioration. Direct water cooling is used for very large machines, while the smaller ones are air- or hydrogen-cooled. Some machines use forced-air cooling with an air-to-water heat exchanger. A braking system is installed in larger machines to stop the generator rapidly and to avoid damage to the thrust.

Defining Terms

Boiler: A steam generator which converts the chemical energy stored in the fuel (coal, gas, etc.) to thermal energy by burning. The heat evaporates the feedwater and generates high-pressure steam.

Economizer: A heat exchanger which increases the feedwater temperature. It is heated by the flue gases.

Fuel: Thermal power plants use coal, natural gas, and oil as a fuel, which is burned in the boiler. Nuclear power plants use uranium as a fuel.

Penstock: A water tube which feeds the turbine. It is used when the slope is too steep for using an open canal.

Reactor: A container where the nuclear reaction takes place. The reactor converts the nuclear energy to heat.

Superheater: A heat exchanger which increases the steam temperature to about 1000°F. It is heated by the flue gases.

Surge tank: An empty vessel which is located at the top of the penstock. It is used to store water surge when the turbine valve is suddenly closed.

References

A.J. Ellis, "Using geothermal energy for power," *Power*, 123(10), October 1979.

M.M. El-Wakil, *Power Plant Technology*, New York: McGraw-Hill, 1984.

A.V. Nero, *A Guidebook to Nuclear Reactors*, Berkeley: University of California Press Ltd., 1979.

J. Weisman and L.E. Eckart, *Modern Power Plant Engineering*, Englewood Cliffs, N.J.: Prentice-Hall, 1985.

Further Information

Other recommended publications include the "Power Plant Electrical References Series," published by EPRI, which consists of several books dealing with power plant electrical system design. A good source of information on the latest developments is *Power* magazine, which regularly publishes articles on power plants.

Additional books include the following:

S. Glasstone and M.C. Edlund, *The Elements of Nuclear Reactor Theory*, New York: Van Nostrand, 1952, p. 416.

G. Murphy, *Elements of Nuclear Engineering,* New York: Wiley, 1961.

M. A. Schultz, *Control of Nuclear Reactors and Power Plants,* New York: McGraw-Hill, 1955.

R. H. Shannon, *Handbook of Coal-Based Electric Power Generation,* Park Ridge, N.J.: Noyes, 1982, p. 372.

E. J. G. Singer, *Combustion: Fossil Power Systems,* Windsor, Conn.: Combustion Engineering, Inc., 1981.

B. G. A. Skrotzki and W. A. Vopat, *Power Station Engineering and Economy,* New York: McGraw-Hill, 1960.

M. J. Steinberg and T. H. Smith, *Economy Loading of Power Plants and Electric Systems,* New York: Wiley, 1943, p. 203.

Various, *Electric Generation: Steam Stations,* B. G. A. Skrotzki, Ed., New York: McGraw-Hill, 1970, p. 403.

Various, *Steam,* New York: Babcock & Wilcox, 1972.

Various, *Steam: Its Generation and Use,* New York: Babcock & Wilcox, 1978.

Dynamo Room at the Pearl Street Station in New York City. This was Edison's first central station for incandescent electric lighting. It began operation in 1882. (Courtesy of General Electric Company.)

57

Distributed Power Generation

57.1 Introduction.. 1207
General Features • Potential and Future • DG Technologies
57.2 Photovoltaics.. 1208
57.3 Wind-Electric Conversion.. 1210
57.4 Hydro... 1211
57.5 Geothermal 1211
57.6 Tidal Energy 1212
57.7 Fuel Cells 1212
57.8 Solar-Thermal-Electric Conversion................................ 1213
57.9 Biomass Energy.. 1213
57.10 Thermoelectrics 1214
57.11 Thermionics... 1214
57.12 Integrated System Concepts... 1215

R. Ramakumar
Oklahoma State University

57.1 Introduction

Distributed generation (DG) refers to small (a few watts up to 1 MW) power plants at or near the loads, operating in a stand-alone mode or connected to a grid at the distribution or subtransmission level, and geographically scattered throughout the service area. Typically they harness unconventional energy resources such as **insolation,** wind, **biomass,** tides and waves, and **geothermal.** Small plants powered by site-specific conventional energy resources such as low-head and small hydro and natural gas are also included in this general group.

Interest in DG has been growing steadily since the dramatic oil embargo of 1973. In addition to the obvious advantages realized by the development of renewable energy sources, DG is ideally suited to power small remote loads, located far from the grid. An entire family of small power sources has been developed and employed for space, underwater, and biomedical applications. Another niche for these systems is in energizing remote rural areas of developing countries. It is estimated that there are more than one million remote villages in the world with no grid connection and minimally sustained by locally available energy sources. Integrated renewable energy systems (**IRES**), a special subset of DG, are ideally suited for these situations.

General Features

DG will have one or more of the following features:

- Small size
- Intermittent input resource

1207

- Stand-alone or interface at the distribution or subtransmission level
- Extremely site-specific inputs
- Located near the loads
- Remoteness from conventional grid supply
- Availability of energy storage and reconversion for later use

Potential and Future

Globally, the potential for DG is vast. Even extremely site-specific resources such as tides, geothermal, and small hydro are available in significant quantities. Assessments of the future for various DG technologies vary, depending on the enthusiasm of the estimator. However, in almost all cases, the limitations are economic rather than technical. Concerns over the unrestricted use of depletable energy resources and the ensuing environmental problems such as the greenhouse effect and global warming are providing the impetus necessary for the continued development of technologies for DG.

DG Technologies

Many technologies have been proposed and employed for DG. Power ratings of DG systems vary from milliwatts to megawatts, depending on the application. A listing of the technologies is given below.

- **Photovoltaics** (PV)
- **Wind-electric conversion** systems
- Mini and micro hydro
- Geothermal plants
- Tidal and wave energy conversion
- **Fuel cells**
- **Solar-thermal-electric conversion**
- Biomass utilization
- **Thermoelectrics**
- **Thermionics**
- Small cogeneration plants powered by natural gas and supplying electrical and thermal energies

The technology involved in the last item above is mature and very similar to that of conventional thermal power plants and therefore will not be considered in this section.

57.2 Photovoltaics

PV refers to the direct conversion of insolation (incident solar radiation) to electricity. A PV cell (also known as a solar cell) is simply a large-area semiconductor *pn* junction diode with the junction positioned very close to the top surface. Typically, a metallic grid structure on the top and a sheet structure in the bottom collect the minority carriers crossing the junction and serve as terminals. The minority carriers are generated by the incident photons with energies greater than or equal to the energy gap of the semiconductor material.

Since the output of an individual cell is rather low (1 or 2 W at a fraction of a volt), several (30 to 60) cells are combined to form a module. Typical module ratings range from 40 to 50 W at 15 to 17 V. PV modules are progressively put together to form panels, arrays (strings or trackers), groups, segments (subfields), and ultimately a PV plant consisting of several segments. Plants rated at several MW have been built and operated successfully.

Advantages of PV include demonstrated low operation and maintenance costs, no moving parts, silent and simple operation, almost unlimited lifetime if properly cared for, no recurring fuel costs, modularity, and minimal environmental effects. The disadvantages are its cost, need for large collector areas due to the diluteness of insolation, and the diurnal and seasonal variability of the output.

PV systems can be flat-plate or concentrating type. While flat-plate systems utilize the global (direct and diffuse) radiation, concentrator systems harness only the direct or beam radiation. As such, concentrating systems must track (one axis or two axis) the sun. Flat-plate systems may or may not be mounted on trackers.

By 1990, efficiencies of flat-plate crystalline and thin-film cells had reached 23 and 15%, respectively. Efficiencies as high as 34% were recorded for concentrator cells. Single-crystal and amorphous PV module efficiencies of 12 and 5% were achieved by the early '90s. For an average module efficiency of 10% and an insolation of 1 kW/m² on a clear afternoon, 10 m² of collector area is required for each kW of output.

The output of a PV system is dc and inversion is required for supplying ac loads or for utility-interactive operation. While the required fuel input to a conventional power plant depends on its output, the input to a PV system is determined by external factors such as cloud cover, time of day, season of the year, geographic location, orientation, and geometry of the collector. Therefore, PV systems are operated, as far as possible, at or near their maximum outputs. Also, PV plants have inertialess generation and are subject to rapid changes in their outputs due to moving clouds.

The current-voltage (*IV*) characteristic of an illuminated solar cell is given as

$$I = I_s - I_o \left[\exp\left(\frac{eV}{kT} \right) - 1 \right]$$

where I_o and I_s are the dark and source currents, respectively, k is the Boltzmann constant (1.38×10^{-23} J/K), T is the temperature in K, and e is the electronic charge. Under ideal conditions (identical cells), for a PV module with a series-parallel arrangement of cells, the *IV* characteristic will be similar, except that the current scale should be multiplied by the number of parallel branches and the voltage scale by the number of cells in series in the module. The source current varies linearly with insolation. The dark current increases as the cell operating temperature increases. Also, the larger the energy gap of the material, the smaller the dark current. The ratio of source current to dark current should be made as large as possible for improved operation.

Single-crystal silicon is still the dominant technology for fabricating PV devices. Polycrystalline, semicrystalline, and amorphous silicon technologies are developing rapidly to challenge this. Highly innovative technologies such as spheral cells are being introduced to reduce costs. Concentrator systems typically employ gallium arsenide or multiple junction cells. Many other materials and thin-film technologies are under investigation as potential candidates.

PV applications range from milliwatts (consumer electronics) to megawatts (central station plants). They are suitable for portable, remote, stand-alone, and utility-interactive applications. PV systems should be considered as energy sources and their design should maximize the conversion of insolation into useable electrical form. Power requirements of practical loads are met using an energy storage and reconversion system or utility interconnection. Concentrating systems have been designed and operated to provide both electrical and low-grade thermal outputs with combined peak utilization efficiencies approaching 60%.

The vigorous growth of PV technology is manifested by a doubling of world PV module shipments in five years — from 26 MW in 1986 to 53 MW in 1991. Tens of thousands of small (<1 kW) systems are in operation around the world. Thousands of kilowatt-size systems (1 to 10s of kW) also have been installed and are in operation. Many intermediate-scale systems (10 to 100s of kW) and large-scale systems (1 MW or larger) are being installed by utility- and government-sponsored programs as proof-of-concept experiments and to glean valuable operational data.

By 1988, nearly 11 MW of PV was interconnected to the utility system in the United States alone. Most were in the 1- to 5-kW range. The two major exceptions are the 1-MW Hesperia-Lugo project installed in 1982 and the 6.5-MW Carrisa Plains project installed in 1984, both in California. In Germany, a 340-kW system began operation in 1988 as part of a large program. Switzerland has a plan to install 1 MW of PV in 333 roof-mounted units of 3 kW each. By 1990, the installed capacity of PV in Italy exceeded 3 MW. Many nations have recognized the vast potential of PV and have established their own PV programs within the past decade.

From a capital cost of $7000/kW in 1988 with an associated levelized energy cost of 32¢/kWh, even with a business-as-usual scenario, a twofold reduction to $3500/kW by 2000 and an additional 3-to-1 reduction to $1175/kW by 2030 are being projected. The corresponding energy costs are 15 and 5¢/kWh, respectively. These estimates put the cost of energy from PV in par with the cost of energy from conventional plants in the early part of the twenty-first century.

57.3 Wind-Electric Conversion

Wind energy is intermittent, highly variable, and site-specific, exists in three dimensions, and is the least dependent upon latitude among all renewable resources. The power density (in W/unit area) in moving air (wind) is a cubic function of wind speed and therefore even small increases in average wind speeds can lead to significant increases in the capturable energy. Wind sites are typically classified as good, excellent, or outstanding, with associated mean wind speeds of 13, 16, and 19 mph, respectively.

Aeroturbines employ lift and/or drag forces to convert wind energy to rotary mechanical energy, which is then converted to electrical energy by coupling a suitable generator. The power coefficient C_p of an aeroturbine is the fraction of the incident power converted to mechanical shaft power, and it is a function of the tip speed-to-wind speed ratio λ. For a given propeller configuration, at any given wind speed, there is an optimum tip speed that maximizes C_p.

Several types of aeroturbines are available. They can have horizontal or vertical axes, number of blades ranging from one to several, mounted upwind or downwind, and fixed- or variable-pitch blades with full blade control or tip control. Vertical-axis (Darrieus) turbines are not self-starting and require a starting mechanism. Today, horizontal-axis turbines with two or more blades are the most prevalent, and considerable work is underway to develop advanced versions of these.

The electrical output P_e of a wind-electric conversion system (WECS) is given as

$$P_e = \eta_g \eta_m A C_p K v^3$$

where η_g and η_m are the efficiencies of the electrical generator and mechanical interface, respectively, A is the swept area, K is a constant, and v is the wind speed incident on the aeroturbine.

There are two basic options for wind-electric conversion. With varying wind speeds, the aeroturbine can be operated at a constant speed by blade-pitch control, and a conventional synchronous machine is then employed to generate constant-frequency ac. Alternatively, the aeroturbine rotational speed can be allowed to vary with wind to maintain a constant and optimum tip speed ratio, and then a combination of special energy converters and power electronics is employed to obtain utility-grade ac. The variable-speed option allows optimum efficiency operation of the turbine over a wide range of wind speeds, resulting in increased outputs with lower structural loads and stresses. All future utility-grade advanced turbines are expected to operate in the variable-speed mode and use power electronics to convert the variable-frequency output to constant frequency with minimal harmonic distortion.

Large-scale harnessing of wind energy will require hundreds or even thousands of WECS arranged in a wind farm with spacings of about 2 to 3 diameters crosswind and about 10 diameters apart downwind. The power output of an individual WECS will fluctuate over a wide range, and its statistics strongly depend on the wind statistics. When many WECS are used in a wind farm, some smoothing of the total power output will result, depending on the statistical independence of the outputs of individual WECS. This is desirable, especially with high (>20%) penetration of WECS in the generation mix. While the output of WECS is not dispatchable, with large wind farms the possibility of assigning some capacity credit to the overall output significantly improves.

Although wind-electric conversion has overall minimum environmental impacts, the large rotating structures involved do generate some noise and introduce visual aesthetics problems. By locating wind energy systems sufficiently far from centers of population, these effects can be minimized. The envisaged potential for bird kills turned out to be not a serious problem. Wind energy systems occupy only a very small fraction of the land. However, the area surrounding them can be used only for activities such as farming and livestock grazing. Thus, there is some negative impact on land use.

Today, the cost of energy delivered by wind plants rivals those obtained from some nonrenewable sources. By 1990, wind became the most utilized and competitive option among all the solar energy technologies for the bulk power market at a cost of generation of about 8¢/kWh (or roughly 7¢/kWh in 1987 dollars). Ongoing research and development work in new design tools, advanced airfoils, site tailoring, operating strategies, array spacing, and improved reliability and manufacturability is expected to bring the cost of energy further down by a factor of 2 to 3.

At around 1600 MW, nearly 90% of all the WECS installed in the world are in California. They are expected to generate nearly 3 billion kWh of electricity per year to the state's utilities to which they are interconnected. Although their lack of control and the intermittent nature of wind-derived energy are not embraced enthusiastically by electric utilities, this gap is expected to be bridged very soon with appropriate computer controls and operating strategies. Wind energy is already an economical option for remote areas endowed with good wind regimes. The modularity of WECS, coupled with the associated environmental benefits, potential for providing jobs, and economic viability point to a major role for wind energy in the generation mix of the world in the decades to come.

57.4 Hydro

Hydropower is a mature but neglected and one of the most promising renewable energy technologies. In the context of DG, small (less than 15 MW), mini (less than 1 MW), and micro (less than 100 kW) hydroelectric plants are of interest. The source of hydropower is the hydrologic cycle driven by the energy from the sun. Most of the sites for DG hydro are either low-head (2 to 20 m) or medium-head (20 to 150 m). The global hydroelectric potential is vast. One estimate puts it at 31 GW for Indonesia alone! The installed capacity of small hydro in the People's Republic of China was exceeding 7 GW by 1980.

Both impulse and reaction turbines have been employed for small-scale hydro for DG. Several standardized units are available in the market. Most of the units are operated at constant speed with governor control and are coupled to synchronous machines to generate utility-grade ac. If the water source is highly variable, it may be necessary to employ variable-speed operation. If the speed variations are not large, induction generators can be used. Special variable-speed constant-frequency (VSCF) generation schemes may be needed if the range of speed variations is large (> ±10%). Permanent magnet generators provide another alternative, especially if the output is to be rectified and stored for later use in the case of very small units.

57.5 Geothermal

Geothermal plants exploit the heat stored in the form of hot water and steam in the earth's crust at depths of 2000 to 8000 ft. By nature, these resources are extremely site-specific and slowly run down (depletable) over a period of years. For electric power generation, the resource should be at least around 250°C. Depending on the temperature and makeup, dry steam, flash steam, or binary technology can be employed. Of these, dry natural steam is the best since it eliminates the need for a boiler.

The three basic components of a geothermal plant are (1) a production well to bring the resource to the surface, (2) a turbine generator system for energy conversion, and (3) an injection well to recycle the spent geothermal fluids back into the reservoir.

Worldwide deployment of geothermal plants reached 5000 MW by 1987 in 17 countries. Nearly one-half of this was in the United States. The Geysers plant north of San Francisco is the largest in the world with an installed capacity of 516 MW. In some developing countries, the Philippines for example, geothermal plants supply nearly 20% of their electrical needs.

57.6 Tidal Energy

The origin of **tidal energy** is the upward-acting gravitational force of the moon, which results in a cyclic variation in the potential energy of water at a point on the earth's surface. These variations are amplified by topographical features such as the shape and size of estuaries. The ratio between maximum spring tide and minimum at neap can be as much as 3 to 1. In estuaries, the tidal range can be as large as 10 to 15 m.

Power can be generated from a tidal estuary in two basic ways. A single basin can be used with a barrage at a strategic point along the estuary. By installing turbines at this point, electricity can be generated both when the tide is ebbing or flooding. In the two-basin scheme, generation can be time-shifted to coincide with hours of peak demand by using the basins alternately.

As can be expected, tidal energy conversion is very site-specific. The largest tidal power plant is the single-basin scheme at La Rance in Brittany, France. It is rated at 240 MW and employs 24 vane-type horizontal turbines and alternator motors, each rated at 10 MVA. The plant has been in operation since 1966 with good technical and economic results. It has generated, on the average, around 500 GWh of net energy per year. The Severn estuary in the southwest of England and the Bay of Fundy in the border between the United States and Canada with the highest known tidal range of 17 m have been extensively studied for tidal power generation. There are several other possible sites around the world, but the massive capital costs required have delayed their exploitation.

57.7 Fuel Cells

A fuel cell is a simple static device that converts the chemical energy in a fuel directly, isothermally, and continuously into electrical energy. Fuel and oxidant (typically oxygen in air) are fed to the device in which an electrochemical reaction takes place that oxidizes the fuel, reduces the oxidant, and releases energy. The energy released is in both electrical and thermal forms. The electrical part provides the required output. Since a fuel cell completely bypasses the thermal-to-mechanical conversion involved in a conventional power plant and since its operation is isothermal, fuel cells are not Carnot-limited. Efficiencies in the range of 43 to 55% are forecasted for modular dispersed generators featuring fuel cells.

The low (< 0.05 lb/MWh) airborne emissions of fuel cell plants make them prime candidates for siting in urban areas. The possibility of using fuel cells in combined heat and power (CHP) units provides the cleanest and most efficient energy system option utilizing valuable (or imported) natural gas resources.

Hydrocarbon fuel (natural gas or LNG) or gasified coal is reformed first to produce hydrogen-rich (and sulphur-free) gas that enters the fuel cell stack where it is electrochemically "burned" to produce electrical and thermal outputs. The electrical output of a fuel cell is low-voltage high-current dc. By utilizing a properly organized stack of cells and an inverter, utility-grade ac output is obtained.

Early MW-scale demonstration plants employed phosphoric acid fuel cells. Molten carbonate fuel cell systems have shown considerable promise in recent years with demonstrated efficiencies in the 50 to 55% range based on the higher heating value. Another competitor in the long range is the solid oxide fuel cell that can be intergrated with a coal gasifier and a steam bottoming cycle.

57.8 Solar-Thermal-Electric Conversion

The quality of thermal energy needed for DG employing solar-thermal-electric conversion necessitates concentrated sunlight. Parabolic troughs, parabolic dishes, and central receivers are used to generate temperatures in the range of 400 to 500, 800 to 900, and >500°C, respectively.

Technical feasibility of the central receiver system was demonstrated in the early '80s by the 10-MWe Solar One system in Barstow, California. Over a six-year period, this system delivered 37 GWh of net energy to the Southern California Edison's grid with an overall system efficiency in the range of 7 to 8%. With improvements in heliostat and receiver technologies, annual system efficiencies of 14 to 15% and generation cost of 8 to 12¢/kWh have been projected.

Parabolic-dish electric-transport technology for DG was under active development at the Jet Propulsion Laboratory (JPL) in Pasadena, California, in the late '70s and early '80s. Prototype modules with Stirling engines reached a record 29% overall efficiency of conversion from insolation to electrical output. Earlier parabolic-dish designs collected and transported thermal energy to a central location for conversion to electricity. Advanced designs such as the one developed at JPL employed engine generators at the focal points of the dishes, and energy was collected and transported in electrical form.

By far the largest installed capacity (nearly 400 MW) of solar-thermal-electric DG employs parabolic-trough collectors and oil to transport the thermal energy to a central location for conversion to electricity via a steam-Rankine cycle. With the addition of a natural gas burner for hybrid operation, this technology, developed by LUZ under the code name SEGS (solar electric generating system), accounts for more than 90% of the world's solar electric capacity, all located in Daggett, Kramer Junction, and Harper Lake in California. Generation costs of around 8 to 9¢/kWh have been realized with SEGS. This technology uses natural gas to compensate for the temporal variations of insolation and firms up the power delivered by the system. This compensation may come during 7 to 11 P.M. in summer and during 8 A.M. to 5 P.M. in winter. SEGS will require about 5 acres/MW or can deliver 130 MW/mi^2 of land area.

57.9 Biomass Energy

Biological sources provide a wide array of materials that have been and continue to be used as energy sources. Wood, wood wastes, and residue from wood processing industries, sewage or municipal solid waste, cultivated herbaceous and other energy crops, waste from food processing industries, and animal wastes are lumped together by the term *biomass*. The most compelling argument for the use of biomass technologies is the inherent recycling of the carbon by photosynthesis. In addition to the obvious method of burning biomass, conversion to liquid and gaseous fuels is possible, thus expanding the application possibilities.

In the context of electric power generation, the role of biomass is expected to be for repowering old units and for use in small (20 to 50 MW) new plants. Several new high-efficiency conversion technologies are either already available or under development for the utilization of biomass. The technologies and their overall conversion efficiencies are listed below.

- FBC (fluidized-bed combustor), 36–38%
- EPS (energy performance system) combustor, 34–36%
- BIG/STIG (biomass-integrated gasifier/steam-injected gas turbine), 38–47%

Acid or enzymatic hydrolysis, gasification, and aqueous pyrolysis are some of the other technology options available for biomass utilization.

Anaerobic digestion of animal wastes is being used extensively in developing countries to produce biogas, which is utilized directly as a fuel in burners and for lighting. An 80–20 mixture of biogas and diesel has been used effectively in biogas engines to generate electricity in small quantities.

Biomass-fueled power plants are best suited in small (<100 MW) sizes for DG to serve base load and intermediate loads in the eastern United States and in many other parts of the world. This contribution is clean, renewable, and reduces CO_2 emissions. Since biomass fuels are sulphur-free, these plants can be used to offset CO_2 and SO_2 emissions from new fossil power plants. Ash from biomass plants can be recycled and used as fertilizer. A carefully planned and well-managed SRWC (short-rotation woody crop) plantation program with yields in the range of 6 to 12 dry tons/acre/year can be effectively used to mitigate greenhouse gases and contribute thousands of MW of DG to the U.S. grid by the turn of the century.

57.10 Thermoelectrics

Thermal energy can be directly converted to electrical energy by using the thermoelectric effects in materials. Semiconductors offer the best option as thermocouples since thermojunctions can be constructed using a *p*-type and an *n*-type material to cumulate the effects around a thermoelectric circuit. Moreover, by using solid solutions of tellurides and selenides doped to result in a low density of charge carriers, relatively moderate thermal conductivities and reasonably good electrical conductivities can be achieved.

In a thermoelectric generator, the Seebeck voltage generated under a temperature difference drives the current through the load circuit. Even though there is no mechanical conversion, the process is still Carnot-limited since it operates over a temperature difference. In practice, several couples are assembled in a series-parallel configuration to provide dc output power at the required voltage.

Typical thermoelectric generators employ radioisotope or nuclear reactor or hydrocarbon burner as the heat source. They are custom-made for space missions as exemplified by the SNAP (systems for nuclear auxiliary power) series and the RTG (radioisotope thermoelectric generator) used by the Apollo astronauts. Maximum performance over a large temperature range is achieved by cascading stages. Each stage consists of thermocouples electrically in series and thermally in parallel. The stages themselves are thermally in series and electrically in parallel.

Tellurides and selenides are used for power generation up to 600°C. Silicon germanium alloys turn out better performance above this up to 1000°C. With the materials available at present, conversion efficiencies in the 5 to 10% range can be expected. Whenever small amounts of silent reliable power is needed for long periods of time, thermoelectrics offer a viable option. Space, underwater, biomedical, and remote terrestrial power such as cathodic protection of pipelines fall into this category.

57.11 Thermionics

Direct conversion of thermal energy into electrical energy can be achieved by employing the Edison effect—the release of electrons from a hot body, also known as thermionic emission. The thermal input imparts sufficient energy (≥ work function) to a few electrons in the emitter (cathode), which helps them escape. If these electrons are collected using a collector (anode) and a closed path through a load is established for them to complete the circuit back to the cathode, then electrical output is obtained. Thermionic converters are heat engines with electrons as the working fluid and, as such, are subject to Carnot limitations.

Converters filled with ionizable gases such as cesium vapor in the interelectrode space yield higher power densities due to space charge neutralization. Barrier index is a parameter that signifies the closeness to ideal performance with no space charge effects. As this index is reduced, more applications become feasible.

A typical example of developments in thermionics is the TFE (thermionic fuel element) that integrates the converter and nuclear fuel for space nuclear power in the kW to MW level for very long (7 to 10 years) duration missions. Another niche is the thermionic cogeneration burner module, a high-temperature burner equipped with thermionic converters. Electrical outputs of 50 kW/MW of thermal output have been achieved. High (600 to 650°C) heat rejection temperatures of thermionic converters are ideally suited for producing flue gas in the 500 to 550°C range for industrial processes. A long-range goal is to use thermionic converters as toppers for conventional power plants. Such concepts are not economical at present.

57.12 Integrated System Concepts

DG technologies offer many possibilities for integrated operation. Integrated systems may be stand-alone with energy storage and reconversion or include grid connection. Also, both renewable and conventional systems can be integrated to achieve the required operational characteristics. Integrated renewable energy systems (**IRES**) that harness several manifestations of solar energy to supply a variety of energy and other needs have many advantages and applications worldwide. The complementary nature of some of the resources (insolation and wind, for example) over the annual cycle can be exploited by IRES to decrease the amount of energy storage necessary and lower the overall cost of energy.

Defining Terms

Biomass: General term used for wood, wood wastes, sewage, cultivated herbaceous and other energy crops, and animal wastes.

Distributed generation: Small power plants at or near loads and scattered throughout the service area.

Fuel cell: Device that converts the chemical energy in a fuel directly and isothermally into electrical energy.

Geothermal energy: Thermal energy in the form of hot water and steam in the earth's crust.

Hydropower: Conversion of potential energy of water into electricity using generators coupled to impulse or reaction water turbines.

Insolation: Incident solar radiation.

IRES: Acronym for integrated renewable energy system, a collection of devices that harness several manifestations of solar energy to supply a variety of energy and other needs.

Photovoltaics: Conversion of insolation into dc electricity by means of solid state *pn* junction diodes.

Solar-thermal-electric conversion: Collection of solar energy in thermal form using flat-plate or concentrating collectors and its conversion to electrical form.

Thermionics: Direct conversion of thermal energy into electrical energy by using the Edison effect (thermionic emission).

Thermoelectrics: Direct conversion of thermal energy into electrical energy using the thermoelectric effects in materials, typically semiconductors.

Tidal energy: The energy contained in the varying water level in oceans and estuaries, originated by lunar gravitational force.

Wind-electric conversion: The generation of electrical energy using electromechanical energy converters driven by aeroturbines.

References

S.W. Angrist, *Direct Energy Conversion*, 4th ed., Boston, Mass.: Allyn and Bacon, 1982.

R. C. Dorf, *Energy, Resources, & Policy*, Reading, Mass.: Addison-Wesley, 1978.

J. J. Fritz, *Small and Mini Hydropower Systems*, New York: McGraw-Hill, 1984.

J. F. Kreider and F. Kreith (eds.), *Solar Energy Handbook*, New York: McGraw-Hill, 1981.

T. Moore, "On-site utility applications for photovoltaics," *EPRI J.*, p. 27, 1991.

R. Ramakumar, "Renewable energy sources and developing countries," *IEEE Transactions on Power Apparatus and Systems,* vol. PAS-102, no. 2 , pp. 502–510, 1983.

R. Ramakumar, "Wind-electric conversion utilizing field modulated generator systems," *Solar Energy,* vol. 20, no. 1, pp. 109–117, 1978.

R. Ramakumar, H. J. Allison, and W. L. Hughes, "Solar energy conversion and storage systems for the future," *IEEE Transactions on Power Apparatus and Systems,* vol. PAS-94, no. 6, pp. 1926–1934, 1975.

R. H. Taylor, *Alternative Energy Sources for the Centralised Generation of Electricity,* Bristol, U.K.: Hilger, 1983.

The Potential of Renewable Energy, An interlaboratory white paper, prepared for the U.S. Department of Energy, Solar Energy Research Institute, Golden, Colo., 1990.

58

Transmission

Mo-Shing Chen
University of Texas at Arlington

K.C. Lai
University of Texas at Arlington

Rao S. Thallam
Salt River Project, Phoenix

Mohamed E. El-Hawary
Technical University of Nova Scotia

Charles Gross
Auburn University

Arun G. Phadke
Virginia Polytechnic Institute and State University

R.B. Gungor
University of South Alabama

J. Duncan Glover
FaAA Electrical Corporation

58.1 Alternating Current Overhead: Line Parameters, Models, Standard Voltages, Insulators 1217
Line Parameters • Models • Standard Voltages • Insulators

58.2 Alternating Current Underground: Line Parameters, Models, Standard Voltages, Cables 1223
Cable Parameters • Models • Standard Voltages • Cable Standards

58.3 High-Voltage Direct-Current Transmission 1227
Configurations of DC Transmission • Economic Comparison of AC and DC Transmission • Principles of Converter Operation • Converter Control • Developments

58.4 Compensation 1242
Series Capacitors • Synchronous Compensators • Shunt Capacitors • Shunt Reactors • Static VAR Compensators (SVC)

58.5 Fault Analysis in Power Systems 1252
Simplifications in the System Model • The Four Basic Fault Types • An Example Fault Study • Further Considerations

58.6 Protection 1268
Fundamental Principles of Protection • Overcurrent Protection • Distance Protection • Pilot Protection • Computer Relaying

58.7 Transient Operation 1279
Stable Operation of Power Systems

58.8 Planning 1287
Planning Tools • Basic Planning Principles • Equipment Ratings • Planning Criteria • Value-Based Transmission Planning

58.1 Alternating Current Overhead: Line Parameters, Models, Standard Voltages, Insulators

Mo-Shing Chen

The most common element of a three-phase power system is the overhead transmission line. The interconnection of these elements forms the major part of the power system network. The basic overhead transmission lines consist of a group of phase conductors that transmit the electrical energy, the earth return, and usually one or more neutral conductors (Fig. 58.1).

Line Parameters

The transmission line parameters can be divided into two parts: series impedance and shunt admittance. Since these values are subject to installation and utilization, e.g., operation frequency

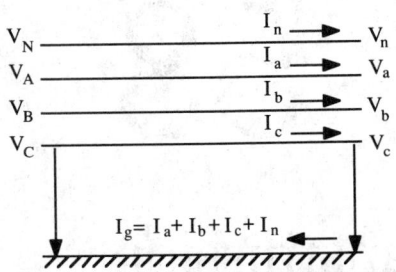

FIGURE 58.1 A three-phase transmission line with one neutral wire.

and distance between cables, the manufacturers are often unable to provide these data. The most accurate values are through measuring in the field, but it has been done only occasionally.

Though the symmetrical component method has been used to simplify many of the problems in power system analysis, the following paragraphs, which describe the formulas in the calculation of the line parameters, are much more general and are not limited to the application of symmetrical components. The sequence impedances and admittances used in the symmetrical components method can be easily calculated by a matrix transformation [Chen and Dillon, 1974]. A detailed discussion of symmetrical components can be found in Clarke [1943].

Series Impedance

The network equation of a three-phase transmission line with one neutral wire (as given in Fig. 58.1) in which only series impedances are considered is given as follows:

$$
\begin{bmatrix} V_A \\ V_B \\ V_C \\ V_N \end{bmatrix} = \begin{bmatrix} Z_{aa-g} & Z_{ab-g} & Z_{ac-g} & Z_{an-g} \\ Z_{ba-g} & Z_{bb-g} & Z_{bc-g} & Z_{bn-g} \\ Z_{ca-g} & Z_{cb-g} & Z_{cc-g} & Z_{cn-g} \\ Z_{na-g} & Z_{nb-g} & Z_{nc-g} & Z_{nn-g} \end{bmatrix} \begin{bmatrix} I_a \\ I_b \\ I_c \\ I_n \end{bmatrix} + \begin{bmatrix} V_a \\ V_b \\ V_c \\ V_n \end{bmatrix} \tag{58.1}
$$

where Z_{ii-g} = self-impedance of phase i conductor and Z_{ij-g} = mutual impedance between phase i conductor and phase j conductor.

The subscript g indicates a ground return. Formulas for calculating Z_{ii-g} and Z_{ij-g} were developed by J. R. Carson based on an earth of uniform conductivity and semi-infinite in extent [Carson, 1926]. For two conductors a and b with earth return, as shown in Fig. 58.2, the self- and mutual impedances in ohms per mile are

$$
Z'_{aa-g} = z_a + j\omega \frac{\mu_0}{2\pi} \ln \frac{2h_a}{r_a} + \omega \frac{\mu_0}{\pi} (P + jQ) \tag{58.2}
$$

$$
Z'_{ab-g} = j\omega \frac{\mu_0}{2\pi} \ln \frac{S_{ab}}{d_{ab}} + \omega \frac{\mu_0}{\pi} (P + jQ) \tag{58.3}
$$

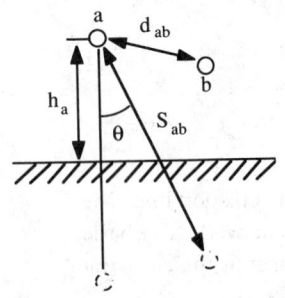

FIGURE 58.2 Geometric diagram of conductors a and b.

where the "prime" is used to indicate distributed parameters in per-unit length; $z_a = r_c + jx_i$ = conductor a internal impedance, Ω/mi; h_a = height of conductor a, ft; r_a = radius of conductor a, ft; d_{ab} = distance between conductors a and b, ft; S_{ab} = distance from one conductor to image of other, ft; $\omega = 2\pi f$; f = frequency, cycles/s; μ_0 = the magnetic permeability of free space, $\mu_0 = 4\pi \times 10^{-7} \times 1609.34$ H/mi; and P, Q are the correction terms for earth return effect and are given later.

The conductor internal impedance consists of the effective resistance and the internal reactance. The effective resistance is affected by three factors: temperature, frequency, and current density. In coping with the temperature effect on the resistance, a correction can be applied.

$$
R_{new} = R_{20°}[1 + \alpha (T_{new} - 20)] \tag{58.4}
$$

where R_{new} = resistance at new temperature, T_{new} = new temperature in °C, $R_{20°}$ = resistance at 20°C (Table 58.1), and α = temperature coefficient of resistance (Table 58.1).

Table 58.1 Electrical Properties of Metals Used in Transmission Lines

Metal	Relative Conductivity (Copper = 100)	Electrical Resistivity at 20°C, $\Omega \cdot m$ (10^{-8})	Temperature Coefficient of Resistance (per °C)
Copper (HC, annealed)	100	1.724	0.0039
Copper (HC, hard-drawn)	97	1.777	0.0039
Aluminum (EC grade, 1/2 H-H)	61	2.826	0.0040
Mild steel	12	13.80	0.0045
Lead	8	21.4	0.0040

An increase in frequency causes nonuniform current density. This phenomenon is called *skin effect*. Skin effect increases the effective ac resistance of a conductor and decreases its internal inductance. The internal impedance of a solid round conductor in ohms per meter considering the skin effect is calculated by

$$z = \frac{\rho m}{2\pi r} \frac{I_0(mr)}{I_1(mr)} \tag{58.5}$$

where ρ = resistivity of conductor, $\Omega \cdot m$; r = radius of conductor, m; I_0 = modified Bessel function of the first kind of order 0; I_1 = modified Bessel function of the first kind of order 1; and $m = \sqrt{j\omega\mu/\rho}$ = reciprocal of complex depth of penetration.

The ratios of effective ac resistance to dc resistance for commonly used conductors are given in many handbooks [such as *Electrical Transmission and Distribution Reference Book* and *Aluminum Electrical Conductor Handbook*]. A simplified formula is also given in Clarke [1943].

P and *Q* are the correction terms for earth return effect. For perfectly conducting ground, they are zero. The determination of *P* and *Q* requires the evaluation of an infinite integral. Since the series converge fast at power frequency or less, they can be calculated by the following equations:

$$P = \frac{\pi}{8} - \frac{1}{3\sqrt{2}} k \cos \theta + \frac{k^2}{16}\left[\left(0.6728 + \ln\frac{2}{k}\right)\cos 2\theta + \theta \sin 2\theta\right]$$
$$+ \frac{k^3 \cos 3\theta}{45\sqrt{2}} - \frac{\pi k^4 \cos 4\theta}{1536} \tag{58.6}$$

$$Q = -0.0386 + \frac{1}{2}\ln\frac{2}{k} + \frac{1}{3\sqrt{2}} k \cos \theta - \frac{\pi k^2 \cos 2\theta}{64} + \frac{k^3 \cos 3\theta}{45\sqrt{2}}$$
$$- \frac{k^4}{384}\left[\left(\ln\frac{2}{k} + 1.0895\right)\cos 4\theta + \theta \sin 4\theta\right] \tag{58.7}$$

with

$$k = 8.565 \times 10^{-4} D\sqrt{\frac{f}{\rho}}$$

where $D = 2h_i$ (ft), $\theta = 0$, for self-impedance; $D = S_{ij}$ (ft), for mutual impedance (see Fig. 58.2 for θ); and ρ = earth resistivity, Ω/m^3.

Shunt Admittance

The shunt admittance consists of the conductance and the capacitive susceptance. The conductance of a transmission line is usually very small and is neglected in steady-state studies. A capacitance matrix related to phase voltages and charges of a three-phase transmission line is

$$Qabc = Cabc \cdot Vabc \quad \text{or} \quad \begin{bmatrix} Q_a \\ Q_b \\ Q_c \end{bmatrix} = \begin{bmatrix} C_{aa} & -C_{ab} & -C_{ac} \\ -C_{ba} & C_{bb} & -C_{bc} \\ -C_{ca} & -C_{cb} & C_{cc} \end{bmatrix} \begin{bmatrix} V_a \\ V_b \\ V_c \end{bmatrix} \tag{58.8}$$

The capacitance matrix can be calculated by inverting a potential coefficient matrix.

$$Qabc = Pabc^{-1} \cdot Vabc \quad \text{or} \quad Vabc = Pabc \cdot Qabc$$

or

$$\begin{bmatrix} V_a \\ V_b \\ V_c \end{bmatrix} = \begin{bmatrix} P_{aa} & P_{ab} & P_{ac} \\ P_{ba} & P_{bb} & P_{bc} \\ P_{ca} & P_{cb} & P_{cc} \end{bmatrix} \begin{bmatrix} Q_a \\ Q_b \\ Q_c \end{bmatrix} \tag{58.9}$$

$$P_{ii} = \frac{l}{2\pi\varepsilon} \ln \frac{2h_i}{r_i} \tag{58.10}$$

$$P_{ij} = \frac{l}{2\pi\varepsilon} \ln \frac{S_{ij}}{d_{ij}} \tag{58.11}$$

where d_{ij} = distance between conductors i and j, h_i = height of conductor i, S_{ij} = distance from one conductor to the image of the other, r_i = radius of conductor i, ε = permittivity of the medium surrounding the conductor, and l = length of conductor.

Though most of the overhead lines are bare conductors, aerial cables may consist of cable with shielding tape or sheath. For a single-core conductor with its sheath grounded, the capacitance C_{ii} in per-unit length can be easily calculated by Eq. (58.12), and all C_{ij}'s are equal to zero.

$$C = \frac{2\pi\varepsilon_0\varepsilon_r}{\ln(r_2/r_1)} \tag{58.12}$$

where ε_0 = absolute permittivity (dielectric constant of free space), ε_r = relative permittivity of cable insulation, r_1 = outside radius of conductor core, and r_2 = inside radius of conductor sheath.

Models

In steady-state problems, three-phase transmission lines are represented by lumped-π equivalent networks, series resistances and inductances between buses are lumped in the middle, and shunt capacitances of the transmission lines are divided into two halves and lumped at buses connecting the lines (Fig. 58.3). More discussion on the transmission line models can be found in El-Hawary [1983].

Standard Voltages

Standard transmission voltages are established in the United States by the American National Standards Institute (ANSI). There is no clear delineation between distribution, subtransmission, and transmission voltage levels. Table 58.2 shows the standard voltages listed in ANSI Standard C84 and C92.2, all of which are in use at present.

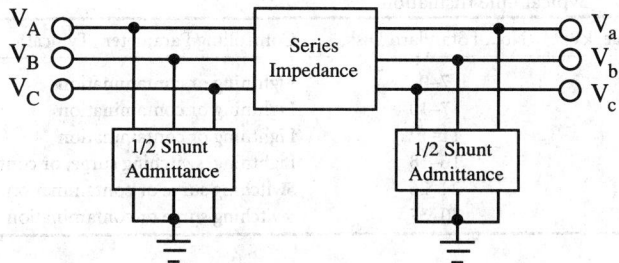

FIGURE 58.3 Generalized conductor model.

Table 58.2 Standard System Voltage, kV

Category	Rating	
	Nominal	Maximum
	34.5	36.5
	46	48.3
	69	72.5
	115	121
	138	145
	161	169
	230	242
Extra-high voltage (EHV)	345	362
	400 (principally in Europe)	
	500	550
	765	800
Ultra-high voltage (UHV)	1100	1200

Insulators

The electrical operating performance of a transmission line depends primarily on the insulation. Insulators not only must have sufficient mechanical strength to support the greatest loads of ice and wind that may be reasonably expected, with an ample margin, but must be so designed to withstand severe mechanical abuse, lightning, and power arcs without mechanically failing. They must prevent a flashover for practically any power-frequency operation condition and many transient voltage conditions, under any conditions of humidity, temperature, rain, or snow, and with accumulations of dirt, salt, and other contaminants which are not periodically washed off by rains.

The majority of present insulators are made of glazed porcelain. Porcelain is a ceramic product obtained by the high-temperature vitrification of clay, finely ground feldspar, and silica. Porcelain insulators for transmission may be disks, posts, or long-rod types. Glass insulators have been used on a significant proportion of transmission lines. These are made from toughened glass and are usually clear and colorless or light green. For transmission voltages they are available only as disk types. Synthetic insulators are usually manufactured as long-rod or post types. Use of synthetic insulators on transmission lines is relatively recent, and a few questions about their use are still under study. Improvements in design and manufacture in recent years have made synthetic insulators increasingly attractive since the strength-to-weight ratio is significantly higher than that of porcelain and can result in reduced tower costs, especially on EHV and UHV transmission lines.

NEMA Publication "High Voltage Insulator Standard" and AIEE Standard 41 have been combined in ANSI Standards C29.1 through C29.9. Standard C29.1 covers all electrical and mechanical tests for all types of insulators. The standards for the various insulators covering flashover voltages (wet, dry, and impulse; radio influence; leakage distance; standard dimensions; and mechanical-

Table 58.3   Typical Line Insulation

Line Voltage, kV	No. of Standard Disks	Controlling Parameter (Typical)
115	7–9	Lightning or contamination
138	7–10	Lightning or contamination
230	11–12	Lightning or contamination
345	16–18	Lightning, switching surge, or contamination
500	24–26	Switching surge or contamination
765	30–37	Switching surge or contamination

strength characteristics) are addressed. These standards should be consulted when specifying or purchasing insulators.

The electrical strength of line insulation may be determined by power frequency, switching surge, or lightning performance requirements. At different line voltages, different parameters tend to dominate. Table 58.3 shows typical line insulation levels and the controlling parameter.

Defining Term

Surge impedance loading (SIL):   The surge impedance of a transmission line is the characteristic impedance with resistance set to zero (resistance is assumed small compared to reactance). The power that flows in a lossless transmission line terminated in a resistive load equal to the line's surge impedance is denoted as the surge impedance loading of the line.

References

Aluminum Electrical Conductor Handbook, 2nd ed., Aluminum Association, 1982.

J. R. Carson, "Wave propagation in overhead wires with ground return," *Bell System Tech. J.,* vol. 5, pp. 539–554, 1926.

M. S. Chen and W. E. Dillon, "Power system modeling," *Proc. IEEE,* vol. 93, no. 7, pp. 901–915, 1974.

E. Clarke, *Circuit Analysis of A-C Power Systems,* vols. 1 and 2, New York: Wiley, 1943.

Electrical Transmission and Distribution Reference Book, Central Station Engineers of the Westinghouse Electric Corporation, East Pittsburgh, Pa.

M. E. El-Hawary, *Electric Power Systems: Design and Analysis,* Englewood Cliffs, N.J.: Prentice-Hall, 1983.

Further Information

Other recommended publications regarding EHV transmission lines include *Transmission Line Reference Book, 345 kV and Above,* 2nd ed., 1982, from Electric Power Research Institute, Palo Alto, Calif., and the IEEE Working Group on Insulator Contamination publication "Application guide for insulators in a contaminated environment," *IEEE Trans. Power Appar. Syst.,* September/October 1979.

Research on higher voltage levels has been conducted by several institutes: Electric Power Research Institute, Bonneville Power Administration, and others. The use of more than three phases for electric power transmission has been studied intensively by sponsors such as the U.S. Department of Energy.

58.2 Alternating Current Underground: Line Parameters, Models, Standard Voltages, Cables

Mo-Shing Chen and K. C. Lai

Although the capital costs of an underground power cable are usually several times those of an overhead line of equal capacity, installation of underground cable is continuously increasing for reasons of safety, security, reliability, aesthetics, or availability of right-of-way. In heavily populated urban areas, underground cable systems are mostly preferred.

Two types of cables are commonly used at the transmission voltage level: pipe-type cables and self-contained oil-filled cables. The selection depends on voltage, power requirements, length, cost, and reliability. In the United States, over 90% of underground cables are pipe-type design.

Cable Parameters

A general formulation of impedance and admittance of single-core coaxial and pipe-type cables was proposed by Prof. Akihiro Ametani of Doshisha University in Kyoto, Japan [Ametani, 1980]. The impedance and admittance of a cable system are defined in the two matrix equations

$$\frac{d(V)}{dx} = -[Z] \cdot (I) \tag{58.13}$$

$$\frac{d(I)}{dx} = -[Y] \cdot (V) \tag{58.14}$$

where (V) and (I) are vectors of the voltages and currents at a distance x along the cable and $[Z]$ and $[Y]$ are square matrices of the impedance and admittance. For a pipe-type cable, shown in Fig. 58.4, the impedance and admittance matrices can be written as Eqs. (58.15) and (58.16) by assuming:

1. The displacement currents and dielectric losses are negligible.
2. Each conducting medium of a cable has constant permeability.
3. The pipe thickness is greater than the penetration depth of the pipe wall.

$$[Z] = [Z_i] + [Z_p] \tag{58.15}$$

$$[Y] = j\omega[P]^{-1} \tag{58.16}$$

$$[P] = [P_i] + [P_p]$$

FIGURE 58.4 A pipe-type cable system.

where $[P]$ is a potential coefficient matrix.

$[Z_i]$ = single-core cable internal impedance matrix

$$= \begin{bmatrix} [Z_{i1}] & [0] & \cdots & [0] \\ [0] & [Z_{i2}] & \cdots & [0] \\ \vdots & \vdots & \ddots & \vdots \\ [0] & [0] & \cdots & [Z_{in}] \end{bmatrix} \tag{58.17}$$

$[Z_p]$ = pipe internal impedance matrix

$$= \begin{bmatrix} [Z_{p11}] & [Z_{p12}] & \cdots & [Z_{p1n}] \\ [Z_{p12}] & [Z_{p22}] & \cdots & [Z_{p2n}] \\ \vdots & \vdots & \ddots & \vdots \\ [Z_{p1n}] & [Z_{p2n}] & \cdots & [Z_{pnn}] \end{bmatrix} \tag{58.18}$$

The diagonal submatrix in $[Z_i]$ expresses the self-impedance matrix of a single-core cable. When a single-core cable consists of a core and sheath (Fig. 58.5), the self-impedance matrix is given by

$$[Z_{ij}] = \begin{bmatrix} Z_{ccj} & Z_{csj} \\ Z_{csj} & Z_{ssj} \end{bmatrix} \tag{58.19}$$

where

Z_{ssj} = sheath self-impedance

$$= Z_{\text{sheath-out}} + Z_{\text{sheath/pipe-insulation}} \tag{58.20}$$

Z_{csj} = mutual impedance between the core and sheath

$$= Z_{ssj} - Z_{\text{sheath-mutual}} \tag{58.21}$$

Z_{ccj} = core self-impedance

$$= (Z_{\text{core}} + Z_{\text{core/sheath-insulation}} + Z_{\text{sheath-in}}) + Z_{csj} - Z_{\text{sheath-mutual}} \tag{58.22}$$

where

$$Z_{\text{core}} = \frac{\rho m}{2\pi r_1} \frac{I_0(mr_1)}{I_1(mr_1)} \tag{58.23}$$

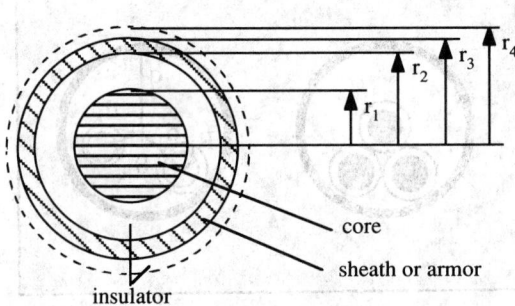

core

sheath or armor

insulator

FIGURE 58.5   A single-core cable cross section.

$$Z_{\text{core/sheath-insulation}} = \frac{j\omega\mu_1}{2\pi} \ln\frac{r_2}{r_1} \tag{58.24}$$

$$Z_{\text{sheath-in}} = \frac{\rho m}{2\pi r_2 D} [I_0(mr_2)K_1(mr_3) + K_0(mr_2)I_1(mr_3)] \tag{58.25}$$

$$Z_{\text{sheath-mutual}} = \frac{\rho}{2\pi r_2 r_3 D} \tag{58.26}$$

$$Z_{\text{sheath-out}} = \frac{\rho m}{2\pi r_3 D} [I_0(mr_3)K_1(mr_2) + K_0(mr_3)I_1(mr_2)] \tag{58.27}$$

$$Z_{\text{sheath/pipe-insulation}} = \frac{j\omega\mu_0}{2\pi} \cosh^{-1}\left(\frac{q^2 + R_i^2 - d_i^2}{2qR_i}\right) \tag{58.28}$$

where ρ = resistivity of conductor, $D = I_1(mr_3)K_1(mr_2) - I_1(mr_2)K_1(mr_3)$, γ = Euler's constant = 1.7811, I_i = modified Bessel function of the first kind of order i, K_i = modified Bessel function of the second kind of order i, and $m = \sqrt{j\omega\mu/\rho}$ = reciprocal of the complex depth of penetration.

A submatrix of $[Z_p]$ is given in the following form:

$$[Z_{pjk}] = \begin{bmatrix} Z_{pjk} & Z_{pjk} \\ Z_{pjk} & Z_{pjk} \end{bmatrix} \tag{58.29}$$

Z_{pjk} in Eq. (58.29) is the impedance between the jth and kth inner conductors with respect to the pipe inner surface. When $j = k$, $Z_{pjk} = Z_{\text{pipe-in}}$; otherwise Z_{pjk} is given in Eq. (58.31).

$$Z_{\text{pipe-in}} = \frac{\rho m}{2\pi q} \frac{K_0(mq)}{K_1(mq)} + \frac{j\omega\mu}{\pi} \sum_{n=1}^{\infty}\left[\left(\frac{d_i}{q}\right)^{2n} \frac{K_n(mq)}{n\mu_r K_n(mq) - mqK_n'(mq)}\right] \tag{58.30}$$

$$Z_{pjk} = \frac{j\omega\mu_0}{2\pi}\left\{ \begin{array}{l} \ln\dfrac{q}{S_{jk}} + \dfrac{\mu_r}{mq}\dfrac{K_0(mq)}{K_1(mq)} \\[2mm] + \displaystyle\sum_{n=1}^{\infty}\left(\dfrac{d_j d_k}{q^2}\right)^{n}\cos(n\theta_{jk})\left[2\mu_r \dfrac{K_n(mq)}{n\mu_r K_n(mq) - mqK_n'(mq)} - \dfrac{1}{n}\right] \end{array}\right\} \tag{58.31}$$

where q is the inside radius of the pipe (Fig. 58.4).

The formulation of the potential coefficient matrix of a pipe-type cable is similar to the impedance matrix.

$$[P_i] = \begin{bmatrix} [P_{i1}] & [0] & \cdots & [0] \\ [0] & [P_{i2}] & \cdots & [0] \\ \vdots & \vdots & \ddots & \vdots \\ [0] & [0] & \cdots & [P_{in}] \end{bmatrix} \quad (58.32)$$

$$[P_p] = \begin{bmatrix} [P_{p11}] & [P_{p12}] & \cdots & [P_{p1n}] \\ [P_{p12}] & [P_{p22}] & \cdots & [P_{p2n}] \\ \vdots & \vdots & \ddots & \vdots \\ [P_{p1n}] & [P_{p2n}] & \cdots & [P_{pnn}] \end{bmatrix} \quad (58.33)$$

The diagonal submatrix in $[P_i]$ expresses the potential coefficient matrix of a single-core cable. When a single-core cable consists of a core and sheath (Fig. 58.5), the submatrix is given by

$$[P_{ij}] = \begin{bmatrix} P_{cj} + P_{sj} & P_{sj} \\ P_{sj} & P_{sj} \end{bmatrix} \quad (58.34)$$

where

$$P_{sj} = \left(\frac{1}{2} \pi \varepsilon_0 \varepsilon_{sj} \right) \ln \frac{r_4}{r_3} \quad (58.35)$$

$$P_{cj} = \left(\frac{1}{2} \pi \varepsilon_0 \varepsilon_{cj} \right) \ln \frac{r_2}{r_1} \quad (58.36)$$

ε_0 = absolute permittivity of free space, ε_{sj} = relative permittivity of insulation outside sheath, and ε_{cj} = relative permittivity of insulation outside core.

Submatrix $[P_{pjk}]$ of $[P_p]$ is given by

$$[P_{pjk}] = \begin{bmatrix} P_{pjk} & P_{pjk} \\ P_{pjk} & P_{pjk} \end{bmatrix} \quad (58.37)$$

P_{pjk} in Eq. (58.37) is the potential coefficient between the jth and kth inner conductors with respect to the pipe inner surface. When $j = k$, $P_{pjk} = P_{\text{pipe-in}}$; otherwise P_{pjk} is given in Eq. (58.39).

$$P_{\text{pipe-in}} = \frac{\ln\left\{ \frac{q}{R_i} \left[1 - \left(\frac{d_i}{q} \right)^2 \right] \right\}}{2\pi \varepsilon_p \varepsilon_0} \quad (58.38)$$

$$P_{pjk} = \frac{1}{2\pi \varepsilon_p \varepsilon_0} \left[\ln \frac{q}{S_{jk}} - \sum_{n=1}^{\infty} \left(\frac{d_j d_k}{q^2} \right)^n \cdot \frac{\cos(n\theta_{jk})}{n} \right] \quad (58.39)$$

where ε_p is the relative permittivity of insulation inside the pipe; R_i is the outer radius of cable i; and d_i, d_j, and d_k are the inner radii of cables i, j, and k.

Models

Refer to "Models" in Section 58.1.

Standard Voltages

In the United States, the underground transmission cables are rated 69 to 345 kV (refer to Table 58.2 in Section 58.1). Cables rated 550 kV are used commercially in Japan. In the United States, cables installed at the 550-kV level are used in relatively short distances, for example, at the Grand Coulee Dam.

Cable Standards

The most universal standardizing authority for cables is the International Electrotechnical Commission (IEC). The IEC standards cater to a large variety of permissible options and serve mainly as a basis for the preparation of national standards. In the United States, in addition to national standards for materials and components, there are cable standards in widespread use by industry issued by four bodies: Underwriter's Laboratories (UL), Association of Edison Illuminating Companies (AEIC), and jointly by the Insulated Power Cables Engineers Association and the National Electrical Manufacturers' Association (IPCEA/NEMA).

References

A. Ametani, "A general formulation of impedance and admittance of cables," *IEEE Trans. Power Syst.,* vol. PAS-99, no. 3, pp. 902–910, 1980.

P. Graneau, *Underground Power Transmission,* New York: Wiley, 1979.

D. McAllister, *Electric Cables Handbook,* New York: Granada Technical Books, 1982.

B. M. Weedy, *Underground Transmission of Electric Power,* New York: Wiley, 1980.

Further Information

The development of advanced cable systems is continuously supported by government and utilities. Information and reports regarding these activities are available from two principal funding agencies, the Electric Power Research Institute (EPRI) and the U.S. Department of Energy.

58.3 High-Voltage Direct-Current Transmission

Rao S. Thallam

The first commercial high-voltage direct-current (HVDC) power transmission system was commissioned in 1954, with an interconnection between the island of Gotland and the Swedish mainland. It was an undersea cable, 96 km long, with ratings of 100 kV and 20 MW. There are now more than 50 systems operating throughout the world, and several more are in the planning, design, and construction stages. HVDC transmission has become acceptable as an economical and reliable method of power transmission and interconnection. It offers advantages over alternating current (ac) for long-distance power transmission and as asynchronous interconnection between two ac systems and offers the ability to precisely control the power flow without inadvertent loop flows in an interconnected ac system. Table 58.4 lists the HVDC projects to date (1992), their ratings, year commissioned (or the expected year of commissioning), and other details. The largest system in operation, Itaipu HVDC transmission, consists of two ±600-kV, 3150-MW-rated bipoles, transmitting a total of 6300 MW power from the Itaipu generating station to the Ibiuna (formerly Sao Roque) converter station in southeastern Brazil over a distance of 800 km.

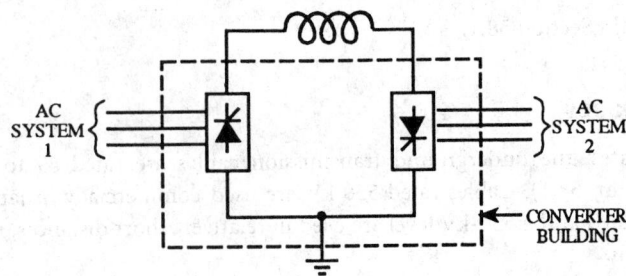

FIGURE 58.6 Back-to-back dc system.

Configurations of DC Transmission

HVDC transmission systems can be classified into three categories:

1. Back-to-back systems
2. Two-terminal, or point-to-point, systems
3. Multiterminal systems

These will be briefly described here.

Back-to-Back DC System

In a back-to-back dc system (Fig. 58.6), both rectifier and inverter are located in the same station, usually in the same building. The rectifier and inverter are usually tied with a reactor, which is generally of outdoor, air-core design. A back-to-back dc system is used to tie two **asynchronous ac systems** (systems that are not in synchronism). The two ac systems can be of different operating frequencies, for example, one 50 Hz and the other 60 Hz. Examples are the Sakuma and Shin-Shinano converter stations in Japan. Both are used to link the 50- and 60-Hz ac systems. The Acaray station in Paraguay links the Paraguay system (50 Hz) with the Brazilian system, which is 60 Hz. Back-to-back dc links are also used to interconnect two ac systems that are of the same frequency but are not operating in synchronism. In North America, eastern and western systems are not synchronized, and Quebec and Texas are not synchronized with their neighboring systems. A dc link offers a practical solution as a tie between nonsynchronous systems. Thus to date, there are 10 back-to-back dc links in operation interconnecting such systems in North America. Similarly, in Europe, eastern and western systems are not synchronized, and dc offers the practical choice for interconnection between them.

Two-Terminal, or Point-to-Point, DC Transmission

Two-terminal dc systems can be either **bipolar** or monopolar. Bipolar configuration, shown in Fig. 58.7, is the commonly used arrangement for systems with overhead lines. In this system, there will be two conductors, one for each polarity (positive and negative) carrying nearly equal currents. Only the difference of these currents, which is usually small, flows through ground return.

A monopolar system will have one conductor, either positive or negative polarity with current returning through either ground or another metallic return conductor. The monopolar ground return current configuration, shown in Fig. 58.8, has been used for undersea cable systems, where current returns through the sea. This configuration can also be used for short-term emergency operation for a two-terminal dc line system in the event of a pole outage. However, concerns for corrosion of underground metallic structures and interference with telephone and other utilities will restrict the duration of such operation. The total ampere-hour operation per year is usually the restricting criterion.

Table 58.4 HVDC Projects Data

	HVDC Supplier†	Year Commissioned	Power Rating, MW	DC Volts, kV	Line/Cable, km	Location
Mercury Arc Valves						
Gotland I[a]	A	1954	20	±100	96	Sweden
English Channel	A	1961	160	±100	64	England-France
Volgograd-Donbass[b]	F	1965	720	±400	470	Russia
Inter-Island	A	1965	600	±250	609	New Zealand
Konti-Skan I	A	1965	250	250	180	Denmark-Sweden
Sakuma	A	1965	300	2125	B-B[f]	Japan
Sardinia	I	1967	200	200	413	Italy
Vancouver I	A	1968	312	260	69	Canada
Pacific Intertie	JV	1970	1440	±400	1362	USA
		1982	1600			
Nelson River I[c]	I	1972	1620	±450	892	Canada
Kingsnorth	I	1975	640	±266	82	England
Thyristor Valves						
Gotland Extension	A	1970	30	±150	96	Sweden
Eel River	C	1972	320	2 × 80	B-B	Canada
Skagerrak I	A	1976	250	250	240	Norway-Denmark
Skagerrak II	A	1977	500	±250	240	Norway-Denmark
Skagerrak III	A	1993	440	±350	240	Norway-Denmark
Vancouver II	C	1977	370	−280	77	Canada
Shin-Shinano	D	1977	300	2 × 125	B-B	Japan
		1992	600	3 × 125		
Square Butte	C	1977	500	±250	749	USA
David A. Hamil	C	1977	100	50	B-B	USA
Cahora Bassa	J	1978	1920	±533	1360	Mozambique-S. Africa
Nelson River II	J	1978	900	±250	930	Canada
		1985	1800	±500		
C-U	A	1979	1000	±400	710	USA
Hokkaido-Honshu	E	1979	150	125	168	Japan
	E	1980	300	250		
		1993	600	±250		
Acaray	G	1981	50	25.6	B-B	Paraguay
Vyborg	F	1981	355	1 × 170 (±85)	B-B	Russia (tie with Finland)
	F	1982	710	2 × 170		
			1065	3 × 170		
Duernrohr	J	1983	550	145	B-B	Austria
Gotland II	A	1983	130	150	100	Sweden
Gotland III	A	1987	260	±150	103	Sweden
Eddy County	C	1983	200	82	B-B	USA
Chateauguay	J	1984	1000	2 × 140	B-B	Canada
Oklaunion	C	1984	200	82	B-B	USA
Itaipu	A	1984	1575	±300	785	Brazil
	A	1985	2383			
	A	1986	3150	±600		
	A	1987	6300	2 × ±600		
Inga-Shaba	A	1982	560	±500	1700	Zaire
Pac Intertie Upgrade	A	1984	2000	±500	1362	USA
Blackwater	B	1985	200	57	B-B	USA
Highgate	A	1985	200	±56	B-B	USA
Madawaska	C	1985	350	140	B-B	Canada
Miles City	C	1985	200	±82	B-B	USA
Broken Hill	A	1986	40	2 × 17 (±8.33)	B-B	Australia
Intermountain	A	1986	1920	±500	784	USA

Table 58.4 HVDC Projects Data (continued)

	HVDC Supplier†	Year Commissioned	Power Rating, MW	DC Volts, kv	Line/Cable, km	Location
			Thyristor Valves (continued)			
Cross-Channel						
Les Mandarins	H	1986	2000	±270	72	France
Sellindge	I	1986	2000	±270	72	England
Descantons-Comerford	C	1986	690	±450	172	Canada-USA
SACOI[d]	H	1986	200	200	415	Corsica Island
SACOI[e]		1992	300			Italy
Urguaiana Freq. Conv.	D	1987	53.7	17.9	B-B	Brazil (tie with Uruguay)
Virginia Smith (Sidney)	G	1988	200	55.5	B-B	USA
Gezhouba-Shanghai	B+G	1989	600	500	1000	China
		1990	1200	±500		
Konti-Skan II	A	1988	300	285	150	Sweden-Denmark
Vindhyachal	A	1989	500	2 × 69.7	B-B	India
Pac Intertie Expansion	B	1989	1100	±500	1362	USA
McNeill	I	1989	150	42	B-B	Canada
Fenno-Skan	A	1989	500	400	200	Finland-Sweden
Sileru-Barsoor	K	1989	100	+100	196	India
			200	+200		
			400	±200		
Rihand-Delhi	A	1991	750	+500	910	India
		1991	1500	±500		
Hydro Quebec-New Eng.	A	1990	2000[g]	±450	1500	Canada-USA
Welch-Monticello		1995	300		B-B	USA
		1998	600			
Etzenricht		1993	600	160	B-B	Germany (tie with Czech)
Vienna South-East	G	1993	550	145	B-B	Austria (tie with Hungary)
DC Hybrid Link	AB	1992	992	+270/−350	617	New Zealand
Chandrapur-Padghe		1996	1500	±500	900	India
Chandrapur-Ramagundam		1995	1000	2×	B-B	India
Gazuwaka-Jeypore		1997	500		B-B	India
Leyte-Luzun		1997	1000	350	440	Philippines
Haenam-Cheju I		1993	300	±180	100	South Korea
Baltic Cable Project		1994	500	450		Sweden-Germany
Victoria-Tasmania		1995	300	300		Australia
Kontek HVDC Intercon		1995	600	600		Denmark
Scotland-N. Ireland		1995	250	250		United Kingdom
Greece-Italy		1997	500			Italy
Tsq-Beijao		1997	1800	500		China

†A–ASEA;   H–CGEE Alsthom;
  B–Brown Boveri;   I–GEC (formerly Eng. Elec.);
  C–General Electric;   J–HVDC W.G. (AEG, BBC, Siemens);
  D–Toshiba;   K–(Independent);
  E–Hitachi;   AB–ABB Brown Boveri;
  F–Russian;   JV–Joint Venture (GE and ASEA).
  G–Siemens;

[a]Retired from service.
[b]2 valve groups replaced with thyristors in 1977.
[c]2 valve groups in Pole 1 replaced with thyristors by GEC in 1991.
[d]50-MW thyristor tap.
[e]Uprate with thyristor valves.
[f]Back-to-back HVDC system.
[g]Multiterminal system. Largest terminal is rated 2250 MW.

Source: Data compiled by D. J. Melvold, Los Angeles Department of Water and Power.

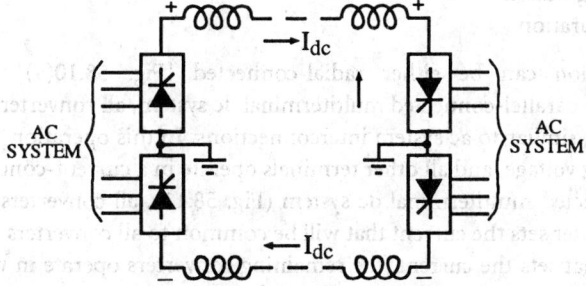

FIGURE 58.7 Bipolar dc system.

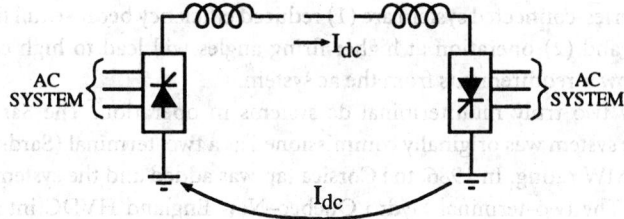

FIGURE 58.8 Monopolar ground return dc system.

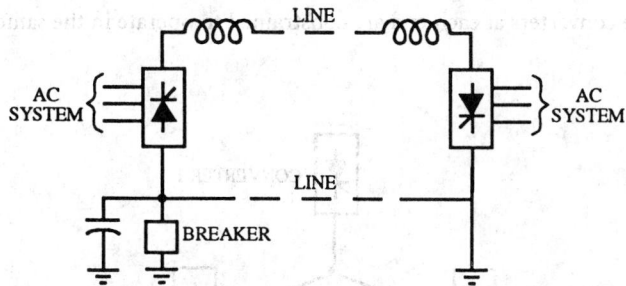

FIGURE 58.9 Monopolar metallic return dc system.

In a monopolar metallic return system, shown in Fig. 58.9, return current flows through a conductor, thus avoiding problems associated with ground return current. This method is generally used as a contingency mode of operation for a normal bipolar transmission system in the event of a partial converter (one-pole equipment) outage. In the case of outage of a one-pole converter, the conductor of the affected pole will be used as the returning conductor. A metallic return transfer breaker will be opened, diverting the return current from the ground path and into the pole conductor. This conductor will be grounded at one end and will be insulated at the other end. This system can transmit half the power of the normal bipolar system capacity (and can be increased if overload capacity is available). However, the percentage of losses will be doubled compared to the normal bipolar operation.

Multiterminal DC Systems

There are two basic configurations in which the dc systems can be operated as multiterminal systems:

1. Parallel configuration
2. Series configuration

Parallel configuration can be either radial-connected [Fig. 58.10(a)] or mesh-connected [Fig. 58.10(b)]. In a parallel-connected multiterminal dc system, all converters operate at the same nominal dc voltage, similar to ac system interconnections. In this operation, one converter determines the operating voltage, and all other terminals operate in a current-controlling mode.

In a series-connected multiterminal dc system (Fig. 58.11), all converters operate at the same current. One converter sets the current that will be common to all converters in the system. Except for the converter that sets the current, the remaining converters operate in voltage control mode (constant **firing angle** or constant **extinction angle**). The converters operate almost independently without requirement for high-speed communication between them. The power output of a non-current-controlling converter is varied by varying its voltage. At all times, the sum of the voltages across the rectifier stations must be larger than the sum of voltages across the inverter stations. Disadvantages of a series-connected system are (1) reduced efficiency because full line insulation is not used at all times and (2) operation at higher firing angles will lead to high converter losses and higher reactive power requirements from the ac system.

There are now two truly multiterminal dc systems in operation. The Sardinia–Corsica–Italy three-terminal dc system was originally commissioned as a two-terminal (Sardinia–Italy) system in 1967 with a 200-MW rating. In 1986, the Corsica tap was added and the system was upgraded to a 300-MW rating. The two-terminal Hydro Quebec–New England HVDC interconnection (commissioned in 1985) was extended to a five-terminal system and commissioned in 1990 (see Table 58.4). The largest terminal of this system at Radisson station in Quebec is rated at 2250 MW. Two more systems, the Nelson River system in Canada and the Pacific NW-SW Intertie in the United States, also operate as multiterminal systems. Each of these systems has two converters at each end of the line, but the converters at each end are constrained to operate in the same mode, either rectifier or inverter.

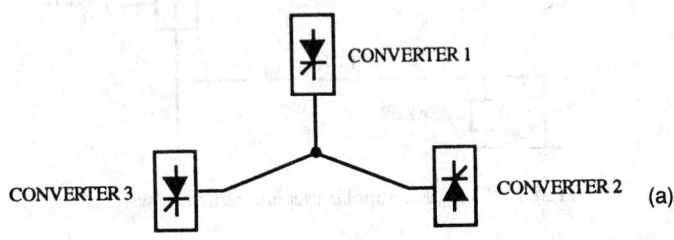

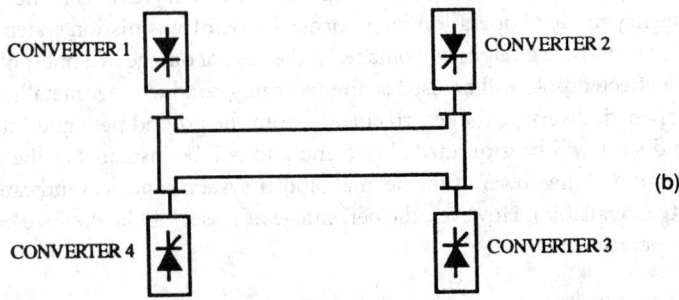

FIGURE 58.10   (a) Parallel-connected radial MTDC system; (b) parallel-connected mesh-type MTDC system.

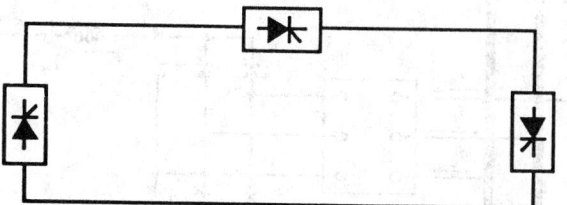

FIGURE 58.11　Series-connected MTDC system.

Economic Comparison of AC and DC Transmission

In cases where HVDC is selected on technical considerations, it may be the only practical option, as in the case of an asynchronous interconnection. However, for long-distance power transmission, where both ac and HVDC are practical, the final decision is dependent on total costs of each alternative. Total cost of a transmission system includes the line costs (conductors, insulators, and towers) plus the right-of-way (R-o-W) costs. A dc line with two conductors can carry almost the same amount of power as the three-phase ac line with the same size of line conductors. However, dc towers with only two conductors are simpler and cheaper than three-phase ac towers. Hence the per-mile costs of line and R-o-W will be lower for a dc line. Power losses in the dc line are also lower than for ac for the same power transmitted. However, the HVDC system requires converters at the two ends of the line; hence the terminal costs for dc are higher than for ac. Variation of total costs for ac and dc as a function of line length are shown in Fig. 58.12. There is a break-even distance above which the total costs of dc option will be lower than the ac transmission option. This is in the range of 500 to 800 km for overhead lines but much shorter for cables. It is between 20 and 50 km for submarine cables and twice as far for underground cables.

Principles of Converter Operation

Converter Circuit

Since the generation and most of the transmission and utilization is alternating current, HVDC transmission requires conversion from ac to dc (called rectification) at the sending end and conversion back from dc to ac (called inversion) at the receiving end. In HVDC transmission, the basic

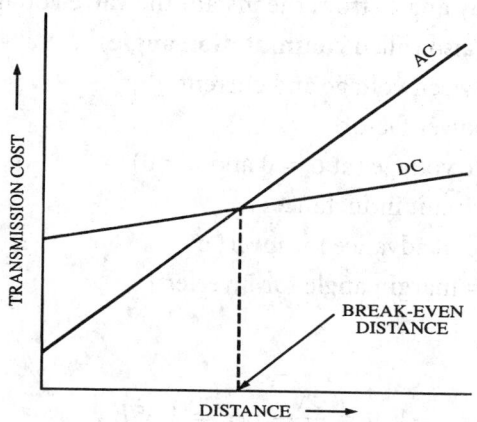

FIGURE 58.12　Transmission cost as function of line length.

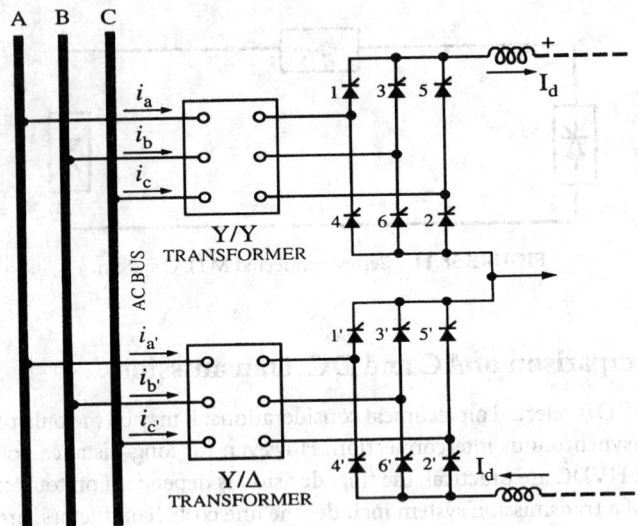

FIGURE 58.13 Basic circuit of a 12-pulse HVDC converter.

device used for conversion from ac to dc and from dc to ac is a three-phase full-wave bridge converter, which is also known as a Graetz circuit. This is a three-phase six-pulse converter. A three-phase twelve-pulse converter will be composed of two three-phase six-pulse converters, supplied with voltages differing in phase by 30 degrees (Fig. 58.13). The phase difference of 30 degrees is obtained by supplying one six-pulse bridge with a Y/Y transformer and the other by Y/Δ transformer.

Relationships between AC and DC Quantities

Voltages and currents on ac and dc sides of the converter are related and are functions of several converter parameters including the converter transformer. The following equations are provided here for easy reference. Detailed derivations are given in Kimbark [1971].

E_{LL} = rms line-to-line voltage of the converter ac bus

I_1 = rms value of fundamental frequency component of the converter ac current

h = harmonic number

α = valve firing delay angle (from the instant the valve voltage is positive)

u = **overlap angle** (also called **commutation angle**)

ϕ = phase angle between voltage and current

$\cos \phi$ = displacement power factor

V_{d0} = ideal no-load dc voltage (at $\alpha = 0$ and $u = 0$)

L_c = commutating circuit inductance

$\beta = 180 - \alpha$ = angle of advance for inverter

$\gamma = 180 - (\alpha + u)$ = margin angle for inverter

with $\alpha = 0$, $u = 0$,

$$V_{d0} = \frac{3\sqrt{2}}{\pi} E_{LL} = 1.35 E_{LL} \tag{58.40}$$

With $\alpha > 0$, and $u = 0$

$$V_d = V_{d0} \cos \alpha \qquad (58.41)$$

Theoretically α can vary from 0 to 180 degrees (with $u = 0$); hence V_d can vary from $+V_{d0}$ to $-V_{d0}$. Since the valves conduct current in only one direction, variation of dc voltage from V_{d0} to $-V_{d0}$ means reversal of power flow direction and the converter mode of operation changing from rectifier to inverter.

$$I_1 = \frac{\sqrt{6}}{\pi} I_d = 0.78 I_d \qquad (58.42)$$

$$\cos \phi = \cos \alpha = \frac{V_d}{V_{d0}} \qquad (58.43)$$

With $\alpha > 0$ and $0 < u > 60°$,

$$V_d = V_{d0} \frac{\cos \alpha + \cos(\alpha + u)}{2} \qquad (58.44)$$

$$V_d = \frac{3\sqrt{2}}{\pi} E_{LL} \frac{\cos \alpha + \cos(\alpha + u)}{2} \qquad (58.45)$$

$$I_1 \approx \frac{\sqrt{6}}{\pi} I_d \qquad (58.46)$$

The error in Eq. (58.46) is only 4.3% at $u = 60$ degrees (maximum overlap angle for normal steady-state operation), and it will be even lower (1.1%) for most practical cases when u is 30 degrees or less. It can be seen from Eqs. (58.45) and (58.46) that the ratio between ac and dc currents is almost fixed, but the ratio between ac and dc voltages varies as a function of α and u. Hence the HVDC converter can be viewed as a variable-ratio voltage transformer, with almost fixed current ratio.

$$P_{dc} = V_d I_d \qquad (58.47)$$

$$P_{ac} = \sqrt{3} E_{LL} I_1 \cos \phi \qquad (58.48)$$

Substituting for V_d and I_d in (58.47) and comparing with (58.48),

$$\cos \phi \approx \frac{\cos \alpha + \cos(\alpha + u)}{2} \qquad (58.49)$$

From Eqs. (58.44) and (58.49),

$$\cos \phi \approx \frac{V_d}{V_{d0}} \qquad (58.50)$$

From Eqs. (58.40), (58.44), and (58.49),

$$V_d \approx 1.35 E_{LL} \cos \phi \qquad (58.51)$$

AC Current Harmonics

The HVDC converter is a harmonic current source on the ac side. Fourier analysis of an ac current waveform, shown in Fig. 58.14, shows that it contains the fundamental and harmonics of the order 5, 7, 11, 13, 17, 19, etc. The current for zero degree overlap angle can be expressed as

$$i(t) = \frac{2\sqrt{3}}{\pi} I_d \left(\begin{array}{c} \cos \omega t - \dfrac{1}{5} \cos 5\omega t + \dfrac{1}{7} \cos 7\omega t \\[2mm] - \dfrac{1}{11} \cos 11\omega t + \dfrac{1}{13} \cos 13\omega t + \cdots \end{array} \right) \tag{58.52}$$

and

$$I_{h0} = \frac{I_{10}}{h} \tag{58.53}$$

where I_{10} and I_{h0} are the fundamental and harmonic currents, respectively, at $\alpha = 0$ and $u = 0$.

Equation (58.53) indicates that the magnitudes of harmonics are inversely proportional to their order.

Converter ac current waveform $i_{a'}$ for phase α with a Y/Δ transformer is also shown in Fig. 58.14. Fourier analysis of this current shows that the fundamental and harmonic components will have the same magnitude as in the case of the Y/Y transformer. However, harmonics of order 5, 7, 17, 19, etc. are in phase opposition, whereas harmonics of order 11, 13, 23, 25, etc. are in phase with the Y/Y transformer case. Hence harmonics of order 5, 7, 17, 19, etc. will be canceled in a 12-pulse converter and do not appear in the ac system. In practice they will not be canceled completely because of imbalances in converter and transformer parameters.

Effect of Overlap. The effect of overlap due to commutation angle is to decrease the amplitude of harmonics from the case with zero overlap. Magnitudes of harmonics for a general case with finite firing angle (α) and overlap angle (u) are given by

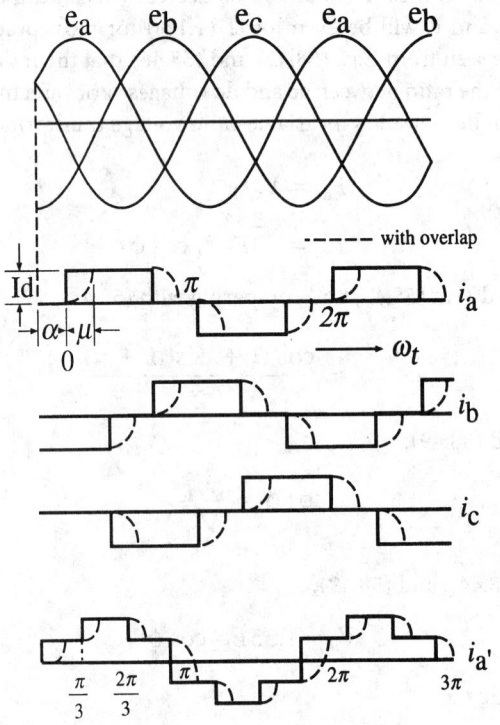

FIGURE 58.14 AC line current waveforms, i_a, i_b, i_c with Y/Y transformer and $i_{a'}$ with Y/Δ transformer.

$$\frac{I_h}{I_{h0}} = \frac{1}{x}\left[A^2 + B^2 - 2AB\cos(2\alpha + u)\right]^{1/2} \tag{58.54}$$

where

$$A = \frac{\sin(h+1)\dfrac{u}{2}}{h+1} \qquad B = \frac{\sin(h-1)\dfrac{u}{2}}{h-1}$$

$$x = \cos\alpha - \cos(\alpha + u)$$

Noncharacteristic Harmonics. In addition to the harmonics described above, converters also generate other harmonics due to "nonideal" conditions of converter operation. Examples of the nonideal conditions are converter ac bus voltage imbalance, perturbation of valve firing pulses, distortion of ac bus voltages, and unbalanced converter transformer impedances. Harmonics generated due to these causes are called *noncharacteristic* harmonics. These are usually smaller in magnitude compared to characteristic harmonics but can create problems if resonances exist in the ac system at these frequencies. In several instances additional filters were installed at the converter ac bus to reduce levels of these harmonics flowing into the ac system.

Converter Control

The static characteristic of a HVDC converter is shown in Fig. 58.15. There are three distinct features of this characteristic.

Constant Firing Angle Characteristic **(A–B).** If the converter is operating under constant firing angle control, the converter characteristic can be described by the equation

$$V_d = V_{d0}\cos\alpha - \frac{3\omega L_c}{\pi}I_d \tag{58.55}$$

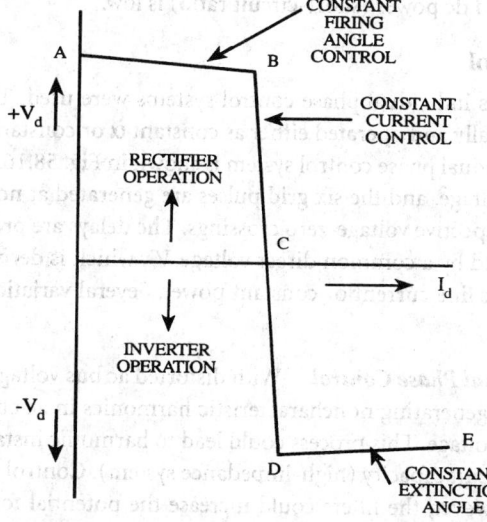

FIGURE 58.15 HVDC converter static characteristic.

When the ordered current is too high for the converter to deliver, it will operate at the minimum firing angle (usually 5 degrees). Then the current will be determined by the voltage V_d and the load. This is also referred to as the natural voltage characteristic. The converter in this mode is equivalent to a dc voltage source with internal resistance R_c, where

$$R_c = \frac{3\omega L_c}{\pi} \tag{58.56}$$

Constant Current Control. This is the usual mode of operation of the rectifier. When the converter is operating in constant current control mode, the firing angle is adjusted to maintain dc current at the ordered value. If the load current goes higher than the ordered current for any reason, control increases the firing angle to reduce dc voltage and the converter operation moves in the direction from B to C. At point C, the firing angle reaches 90 degrees (neglecting overlap angle), the voltage changes polarity, and the converter becomes an inverter. From C to D, the converter works as an inverter.

Constant Extinction Angle Control. At point D, the inverter firing angle has increased to a point where further increase can cause commutation failure. The inverter for its safe operation must be operated with sufficient angle of advance β, such that under all operating conditions the extinction angle γ is greater than the valve deionization angle. The deionization angle is defined as the time in electrical degrees from the instant current reaches zero in a particular valve to the time the valve can withstand the application of positive voltage. Typical minimum values of γ are 15 to 20 degrees for mercury arc valves and slightly less for thyristor valves. During the range D to E, the increase of load current increases the overlap angle, which reduces the dc voltage. This is the negative resistance characteristic of the inverter.

The functional requirements for HVDC converter control are:

1. Minimize the generation of noncharacteristic harmonics.
2. Safe inverter operation with fewest possible commutation failures even with distorted ac voltages.
3. Lowest possible consumption of reactive power. This requires operation with smallest possible delay angle α and extinction angle γ without increased risk of commutation failures.
4. Smooth transition from current control to extinction angle control.
5. Sufficient stability margins and response time when the ratio of the ac system short-circuit strength and the rated dc power (short-circuit ratio) is low.

Individual Phase Control

In the early HVDC systems individual phase control systems were used. The firing angle of each valve is calculated individually and operated either as constant α or constant γ control.

A schematic of the individual phase control system is shown in Fig. 58.16. Six timing voltages are derived from the ac bus voltage, and the six grid pulses are generated at nominally identical delay times subsequent to the respective voltage-zero crossings. The delays are produced by independent delay circuits and controlled by a common direct voltage V_c, which is derived through a feedback loop to control constant dc line current or constant power. Several variations of this control were used until the late 1960s.

Disadvantages of Individual Phase Control. With distorted ac bus voltages, the firing pulses will be unequally spaced, thus generating noncharacteristic harmonics in ac current. This in turn will further distort the ac bus voltage. This process could lead to harmonic instability, particularly with ac systems of low short-circuit capacity (high-impedance system). Control system filters were tried to solve this problem. However, the filters could increase the potential for commutation failures and also reduce the speed of control system response for faults or disturbances in the ac system.

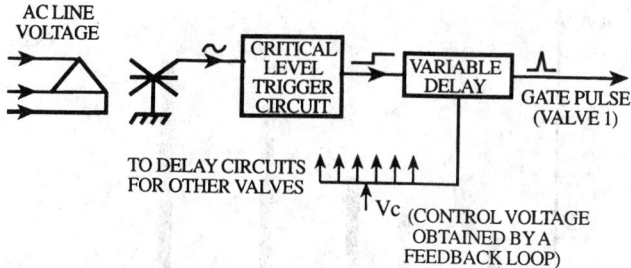

FIGURE 58.16   Constant α control.

Equal Pulse Spacing Control

A control system based on the principle of equal spacing of firing pulses at intervals of 60 degrees (electrical) independent of ac bus voltages was developed in the late 1960s. The basic components of this system, shown in Fig. 58.17, consist of a voltage-controlled oscillator and ring counter. The frequency of the oscillator is directly proportional to the dc control voltage V_c. Under steady-state conditions, pulse frequency is precisely $6f$, where f is the ac system frequency. The phase of each grid pulse will have some arbitrary value relative to the ac bus voltage. If the three-phase ac bus voltages are symmetrical sine waves with no distortion, then α is the same for all valves. The oscillator will be phase-locked with the ac system frequency to avoid drifting. The dc control voltage V_c is derived from a feedback loop for constant current, constant power, or constant extinction angle γ.

The control systems used in recent projects are digital-based and much more sophisticated than the earlier versions.

Developments

During the last two or three decades, several developments in HVDC technology have taken place that improved viability of the HVDC transmission. Prior to 1970 mercury arc valves were used for converting from ac to dc and dc to ac. They had several operational problems including frequent arcbacks. Arcback is a random phenomenon that results in failure of a valve to block conduction in the reverse direction. This is most common in the rectifier mode of operation. In rectifier operation, the valve is exposed to inverse voltage for approximately two-thirds of each cycle. Arcbacks result in line-to-line short circuits, and sometimes in three-phase short circuits, which subject the converter transformer and valves to severe stresses.

Thyristors

Thyristor valves were first used for HVDC transmission in the early 1970s, and since then have completely replaced mercury arc valves. The term thyristor valve, carried over from mercury arc valve, is used to refer to an assembly of series and parallel connection of several thyristors to make up the required voltage and current ratings of one arm of the converter. The first test thyristor valve

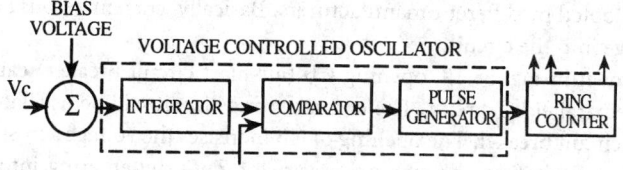

FIGURE 58.17   Equal pulse spacing control.

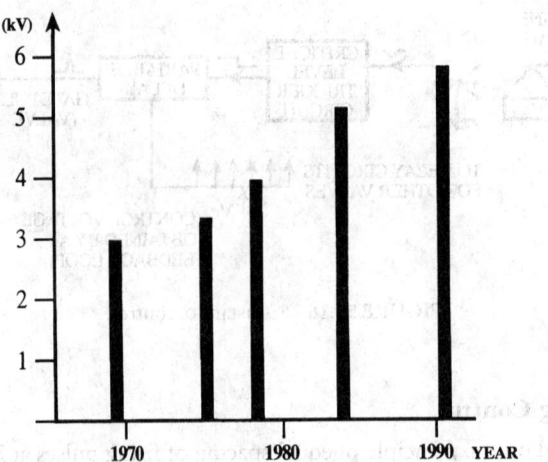

FIGURE 58.18   Maximum blocking voltage development.

in a HVDC converter station was installed in 1967, replacing a mercury arc valve in the Ygne converter station on the island of Gotland (see Gotland I in Table 58.4). The Eel River back-to-back station in New Brunswick, Canada, commissioned in 1972, was the first all-thyristor HVDC converter station. The voltage and current ratings of thyristors have increased steadily over the last two decades. Figure 58.18 shows the maximum blocking voltage of thyristors from the late 1960s to date. The current ratings have also increased in this period from 1 to 4 kA. Some of the increased current ratings were achieved with large-diameter silicon wafers (presently 100-mm diameter) and with improved cooling systems. Earlier projects used air-cooled thyristors. Water-cooled thyristors are used for all the recent projects.

Other recent developments include direct light-triggered thyristors and gate turn-off (GTO) thyristors. GTO thyristors have some advantages for HVDC converters connected to weak ac systems. They are now available in ratings up to 4.5 kV and 4 kA but have not yet been applied in HVDC systems.

DC Circuit Breakers

Interrupting the current in ac systems is aided by the fact that ac current goes through zero every half-cycle or approximately every 8 ms in a 60-Hz system. The absence of natural current zero in dc makes it difficult to develop a dc circuit breaker. There are three principal problems in designing a dc circuit breaker:

1. Forcing current zero in the interrupting element
2. Controlling the overvoltages caused by large di/dt in a highly inductive circuit
3. Dissipating large amounts of energy (tens of megajoules)

The second and third problems are solved by the application of zinc oxide varistors connected line to ground and across the breaking element. The first is the major problem, and several different solutions are adopted by different manufacturers. Basically, current zero is achieved by inserting a counter voltage into the circuit.

In the circuit shown in Fig. 58.19, opening CB (air-blast circuit breaker) causes current to be commutated to the parallel LC circuit. The commutating circuit will be oscillatory, which creates current zero in the circuit breaker. The opening of CB increases the voltage across the commutating circuit, which will be limited by the zinc oxide varistor ZnO_1 by entering into conduction. The resistance R is the closing resistor in series with switch S.

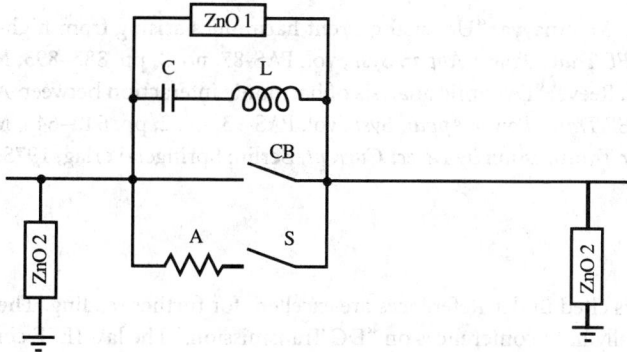

FIGURE 58.19 DC circuit breaker (one module).

It should be noted that a two-terminal dc system does not need a dc breaker since the fast converter control response can bring the current quickly to zero. In multiterminal systems, dc breakers can provide additional flexibility of operation. The multiterminal dc systems commissioned so far have not employed dc breakers.

Defining Terms

Asynchronous ac systems: AC systems with either different operating frequencies or that are not in synchronism.

Bipole: DC system with two conductors, one positive and the other negative polarity. Rated voltage of a bipole is expressed as ±100 kV, for example.

Commutation: Process of transferring current from one valve to another.

Commutation angle (overlap angle):   Time in electrical degrees from the start to the completion of the commutation process.

Extinction angle: Time in electrical degrees from the instant the current in a valve reaches zero (end of conduction) to the time the valve voltage changes sign and becomes positive.

Firing angle (delay angle):   Time in electrical degrees from the instant the valve voltage is positive to the application of firing pulse to the valve (start of conduction).

Pulse number of a converter: Number of ripples in dc voltage per cycle of ac voltage. A three-phase two-way bridge is a six-pulse converter.

Thyristor valve: Assembly of series and parallel connection of several thyristors to make up the required current and voltage ratings of one arm of the converter.

References

J. D. Ainsworth, "The phase locked oscillator: A new control system for controlled static converters," *IEEE Trans. Power Appar. Syst.*, vol. PAS-87, pp. 859–865, March 1968.

A. Ekstrom and L. Eklund, "HVDC thyristor valve development," in Proceedings of the International Conference on DC Power Transmission, Montreal, pp. 220–227, 1984.

A. Ekstrom and G. Liss, "A refined HVDC control system," *IEEE Trans. Power Appar. Syst.*, vol. PAS-89, no. 5, pp. 723–732, May/June 1970.

E. W. Kimbark, *Direct Current Transmission*, vol. I, New York: Wiley-Interscience, 1971.

W. F. Long *et al.*, "Considerations for implementing multiterminal dc systems," *IEEE Trans. Power Appar. Syst.*, pp. 2521–2530, September 1985.

K. R. Padiyar, *HVDC Power Transmission Systems—Technology and System Interactions*, New Delhi: Wiley Eastern Limited, 1990.

J. Reeve and P. C. S. Krishnayya, "Unusual current harmonics arising from high-voltage dc transmission," *IEEE Trans. Power Appar. Syst.*, vol. PAS-87, no. 3, pp. 883–893, March 1968.

R. S. Thallam and J. Reeve, "Dynamic analysis of harmonic interaction between AC and DC power systems," *IEEE Trans. Power Appar. Syst.*, vol. PAS-93, no. 2, pp. 640–646, March/April 1974.

E. Uhlmann, *Power Transmission by Direct Current*, Berlin: Springer-Verlag, 1975.

Further Information

The three textbooks cited under References are excellent for further reading. The IEEE (USA) and IEE (UK) periodically hold conferences on "DC Transmission." The last IEEE conference was held in 1984 in Montreal, and the IEE conference was held in 1991 (conf. publ. no. 345) in London. Proceedings can be ordered from these organizations.

58.4 Compensation

Mohamed E. El-Hawary

The term *compensation* is used to describe the intentional insertion of reactive power-producing devices, either capacitive or inductive, to achieve a desired effect in the electric power system. The effects include improved voltage profiles, enhanced stability performance, and improved transmission capacity. The devices are connected either in series or in **shunt** (parallel) at a particular point in the power circuit.

For illustration purposes, we consider the circuit of Fig. 58.20, where the link has an impedance of $R + jX$, and it is assumed that $V_1 > V_2$ and V_1 leads V_2. The corresponding phasor diagram for zero R and lagging load current I is shown in Fig. 58.21. The approximate relationship between the scalar voltage difference between two nodes in a network and the flow of reactive power Q can be shown to be [Weedy, 1972]

$$\Delta V = \frac{RP_2 + XQ_2}{V_2} \tag{58.57}$$

In most power circuits, $X \gg R$ and the voltage difference ΔV determines Q.

The flow of power and reactive power is from A to B when $V_1 > V_2$ and V_1 leads V_2. Q is determined mainly by $V_1 - V_2$. The direction of reactive power can be reversed by making $V_2 > V_1$. It can thus be seen that if a scalar voltage difference exists across a largely reactive link, the reactive power flows toward the node of lower voltage. Looked at from an alternative point of view, if there is a reactive power deficit at a point in an electric network, this deficit has to be supplied from the rest of the circuit and hence the voltage at that point falls. Of course, a surplus of reactive power generated will cause a voltage rise. This can be interpreted as providing voltage support by supplying reactive power at that point.

Assuming that the link is reactive, i.e., with $R = 0$, then $P_1 = P_2 = P$. In this case, the active power transferred from point A to point B can be shown to be given by

$$P = P_{max} \sin \delta \tag{58.58}$$

The maximum power transfer P_{max} is given by

$$P_{max} = \frac{V_1 V_2}{X} \tag{58.59}$$

It is clear that the power transfer capacity defined by Eq. (58.59) is improved if V_2 is increased.

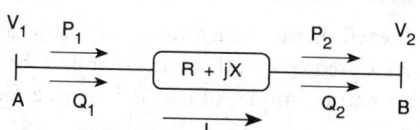

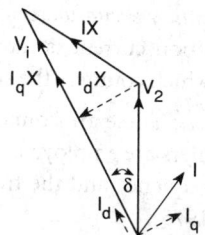

FIGURE 58.20 Two nodes connected by a link.

FIGURE 58.21 Phasor diagram for system shown in Fig. 58.20.

Series Capacitors

Series capacitors are employed to neutralize part of the inductive reactance of a power circuit, as shown in Fig. 58.22. From the phasor diagram of Fig. 58.23 we see that the load voltage is higher with the capacitor inserted than without the capacitor.

Introducing series capacitors is associated with an increase in the circuit's transmission capacity [from (58.59) with a net reduction in X] and enhanced stability performance as well as improved voltage conditions on the circuit. They are also valuable in other aspects such as:

- Controlling reactive power balance
- Load distribution and control of overall transmission losses

Series-capacitor compensation delays investments in additional overhead lines for added transmission capacity, which is advantageous from an environmental point of view.

The first worldwide series-capacitor installation was a 33-kV 1.25-MVAR bank on the New York Power & Light system, which was put in service in 1928. Since then, many higher-capacity, higher-voltage installations have been installed in the United States, Canada, Sweden, Brazil, and other countries.

The reduction in a circuit's inductive reactance increases the short-circuit current levels over those for the noncompensated circuit. Care must be taken to avoid exposing series capacitors to such large short-circuit currents, since this causes excessive voltage rise as well as heating that can damage the capacitors. Specially calibrated spark gaps and short-circuiting switches are deployed within a predetermined time interval to avoid damage to the capacitors.

The interaction between a series-capacitor-compensated ac transmission system in electrical **resonance** and a turbine-generator mechanical system in torsional mechanical resonance results in the phenomenon of **subsynchronous resonance** (SSR). Energy is exchanged between the electrical and mechanical systems at one or more natural frequencies of the combined system below the synchronous frequency of the system. The resulting mechanical oscillations can increase until mechanical failure takes place.

Techniques to counteract SSR include the following:

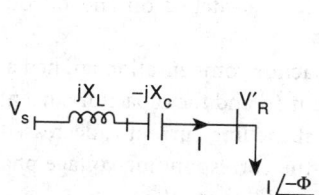

FIGURE 58.22 Line with series capacitor.

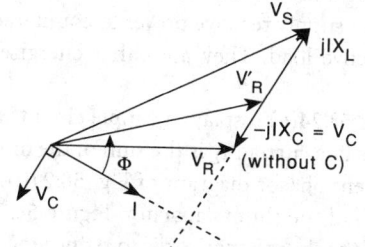

FIGURE 58.23 Phasor diagram corresponding to Fig. 58.22.

- *Supplementary excitation control:* The subsynchronous current and/or voltage is detected and the excitation current is modulated using high-gain feedback to vary the generator output voltage, which counters the subsynchronous oscillations [see El-Serafi and Sheltout, 1979].

- *Static filters:* These are connected in series with each phase of each main generator. Step-up transformers are employed. The filters are tuned to frequencies that correspond to the power system frequency and the troublesome machine natural modes of oscillations [see Tice and Bowler, 1975].

- *Dynamic filters:* In a manner similar to that of excitation control, the subsynchronous oscillation is detected, and a counter emf is generated by a thyristor cycloconverter or a similar device and injected in the power line through a series transformer [see Kilgore *et al.,* 1975].

- *Bypassing series capacitors:* To limit transient torque buildup, complete or partial bypass with the aid of low set gaps.

- Amortisseur windings on the pole faces of the generator rotors can be employed to improve damping.

- A more recent damping scheme [see Hingorani, 1981] is based on measuring the half-cycle period of the series-capacitor voltage, and if this period exceeds a preset value, the capacitor's charge is dissipated into a resistor shunting the capacitor through two antiparallel thyristors.

- A passive SSR countermeasure scheme [see Edris, 1990] involves using three different combinations of inductive and capacitive elements on the three phases. The combinations will exhibit the required equal degree of capacitive compensation in the three phases at power frequency. At any other frequency, the three combinations will appear as unequal reactances in the three phases. In this manner, asynchronous oscillations will drive unsymmetrical three-phase currents in the generator's armature windings. This creates an mmf with a circular component of a lower magnitude, compared with the corresponding component if the currents were symmetrical. The developed interacting electromagnetic torque will be lower.

Synchronous Compensators

A synchronous compensator is a synchronous motor running without a mechanical load. Depending on the value of excitation, it can absorb or generate reactive power. The losses are considerable compared with static capacitors. When used with a voltage regulator, the compensator can run automatically overexcited at high-load current and underexcited at low-load current. The cost of installation of synchronous compensators is high relative to capacitors.

Shunt Capacitors

Shunt capacitors are used to supply capacitive kVAR to the system at the point where they are connected, with the same effect as an overexcited synchronous condenser, generator, or motor. Shunt capacitors supply reactive power to counteract the out-of-phase component of current required by an inductive load. They are either energized continuously or switched on and off during load cycles.

Figure 58.24(a) displays a simple circuit with shunt capacitor compensation applied at the load side. The line current I_L is the sum of the motor load current I_M and the capacitor current I_c. From the current phasor diagram of Fig. 58.24(b), it is clear that the line current is decreased with the insertion of the shunt capacitor. Figure 58.24(c) displays the corresponding voltage phasors. The effect of the shunt capacitor is to reduce the source voltage to V_{s1} from V_{s0}.

From the above considerations, it is clear that shunt capacitors applied at a point in a circuit supplying a load of lagging power factor have the following effects:

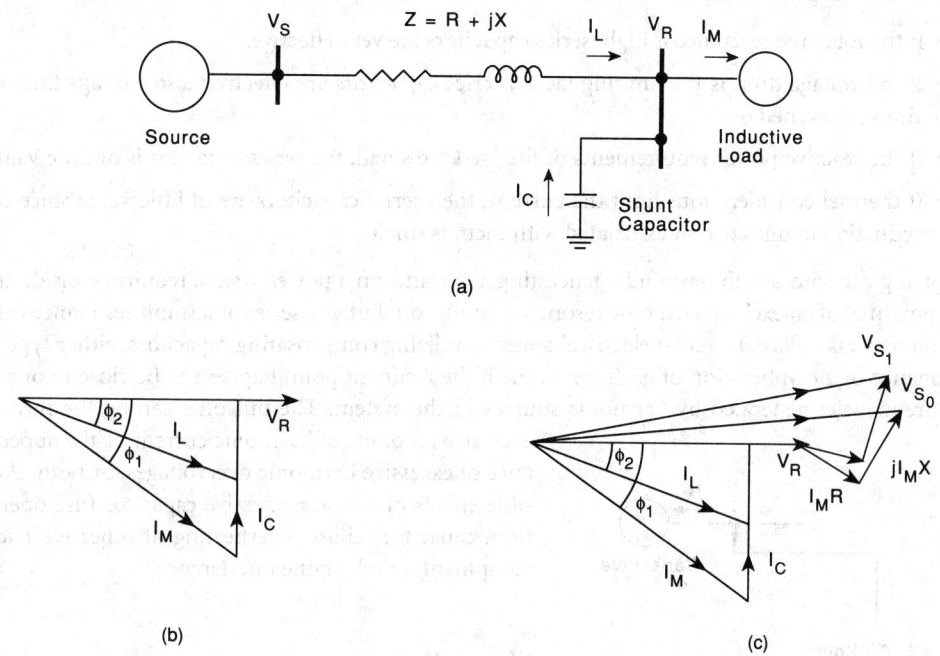

FIGURE 58.24 (a) Shunt-capacitor-compensated load; (b) current phasor diagram; (c) voltage phasor diagram.

- Increase voltage level at the load
- Improve voltage regulation if the capacitor units are properly switched
- Reduce I^2R power loss and I^2X kVAR loss in the system because of reduction in current
- Increase power factor of the source generators
- Decrease kVA loading on the source generators and circuits to relieve an overloaded condition or release capacity for additional load growth
- By reducing kVA load on the source generators, additional active power loading may be placed on the generators if turbine capacity is available
- Reduce demand kVA where power is purchased
- Reduce investment in system facilities per kW of load supplied

To reduce high inrush currents in starting large motors, a capacitor starting system is employed. This maintains acceptable voltage levels throughout the system. The high inductive component of normal reactive starting current is offset by the addition, during the starting period only, of capacitors to the motor bus. This differs from applying capacitors for motor power factor correction.

When used for voltage control, the action of shunt capacitors is different from that of synchronous condensers, since their reactive power varies as the square of the voltage, whereas the synchronous machine maintains approximately constant kVA for sudden voltage changes. The synchronous condenser has a greater stabilizing effect upon system voltages. The losses of the synchronous condenser are much greater than those of capacitors.

Note that in determining the amount of shunt capacitor kVAR required, since a voltage rise increases the lagging kVAR in the exciting currents of transformer and motors, some additional capacitor kVAR above that based on initial conditions without capacitors may be required to get the desired correction. If the load includes synchronous motors, it may be desirable, if possible, to increase the field currents to these motors.

The following are the relative merits of shunt and series capacitors:

- If the total line reactance is high, series capacitors are very effective.
- If the voltage drop is the limiting factor, series capacitors are effective; also, voltage fluctuations are evened out.
- If the reactive power requirements of the load are small, the series capacitor is of little value.
- If thermal considerations limit the current, then series capacitors are of little value since the reduction in line current associated with them is small.

Applying capacitors with harmonic-generating apparatus on a power system requires considering the potential of an excited harmonic resonance condition. Either a series or a shunt resonance condition may take place. In actual electrical systems utilizing compensating capacitors, either type of resonance or a combination of both can occur if the resonant point happens to be close to one of the frequencies generated by harmonic sources in the system. The outcome can be the flow of excessive amounts of harmonic current or the appearance of excessive harmonic overvoltages, or both. Possible effects of this are excessive capacitor fuse operation, capacitor failure, overheating of other electrical equipment, or telephone interference.

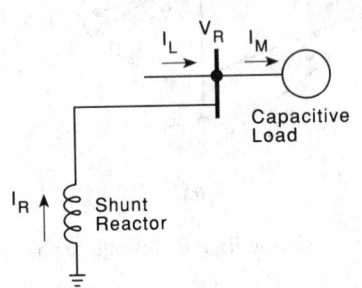

FIGURE 58.25   Shunt-reactor-compensated load.

Shunt Reactors

Shunt reactor compensation is usually required under conditions that are the opposite of those requiring shunt capacitor compensation (see Fig. 58.25). Shunt reactors are installed to remedy the following situations:

- Overvoltages that occur during low load periods at stations served by long lines as a result of the line's capacitance (Ferranti effect).
- Leading power factors at generating plants resulting in lower transient and steady-state stability limits, caused by reduced field current and the machine's internal voltage. In this case, shunt reactors are usually installed at either side of the generator's step-up transformers.
- Open-circuit line charging kVA requirements in extra-high-voltage systems that exceed the available generation capabilities.

Coupling from nearby energized lines can cause severe resonant overvoltages across the shunt reactors of unenergized compensated lines.

Static VAR Compensators (SVC)

Advances in **thyristor** technology for power systems applications have lead to the development of the static VAR compensators (SVC). These devices contain standard shunt elements (**reactors, capacitors**) but are controlled by thyristors.

Static VAR compensators provide solutions to two types of compensation problems normally encountered in practical power systems [Gyugyi *et al.*, 1978]. The first is load compensation, where the requirements are usually to reduce or cancel the reactive power demand of large and fluctuating industrial loads, such as electric arc furnaces and rolling mills, and to balance the real power drawn from the ac supply lines. These types of heavy industrial loads are normally concentrated in one plant and served from one network terminal, and thus can be handled by a local compensator con-

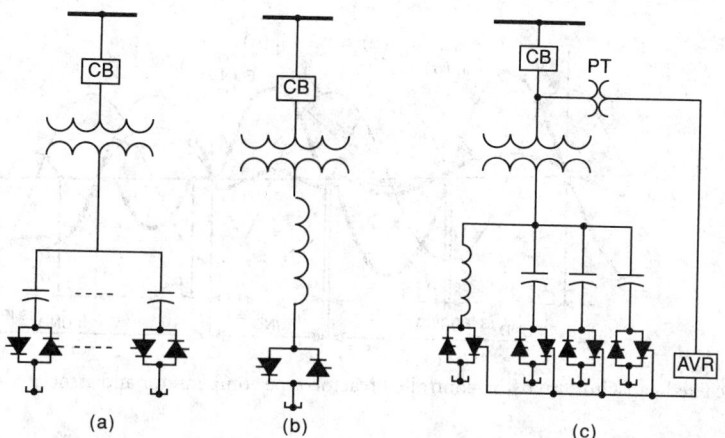

FIGURE 58.26 Basic static VAR compensator configurations. (a) Thyristor-switched shunt capacitors (TSC); (b) thyristor-switched shunt reactors (TCR); (c) combined TSC/TCR.

nected to the same terminal. The second type of compensation is related to voltage support of transmission lines at a given terminal in response to disturbances of both load and generation. The voltage support is achieved by rapid control of the SVC reactance and thus its reactive power output. The main objectives of dynamic VAR compensation are to increase the stability limit of the ac power system, to decrease terminal voltage fluctuations during load variations, and to limit overvoltages subsequent to large disturbances. SVCs are essentially thyristor-controlled reactive power devices.

The two fundamental thyristor-controlled reactive power device configurations are [Olwegard *et al.,* 1981]:

- Thyristor-switched shunt capacitors (TSC): The idea is to split a **capacitor bank** into sufficiently small capacitor steps and switch those steps on and off individually. Figure 58.26(a) shows the concept of the TSC. It offers stepwise control, virtually no transients, and no harmonic generation. The average delay for executing a command from the regulator is half a cycle.
- Thyristor-switched shunt reactors (TCR): In this scheme the fundamental frequency current component through the reactor is controlled by delaying the closing of the thyristor switch with respect to the natural zero crossings of the current. Figure 58.26(b) shows the concept of the TCR. Harmonic currents are generated from the phase-angle-controlled reactor.

The magnitude of the harmonics can be reduced using two methods. In the first, the reactor is split into smaller steps, while only one step is phase-angle controlled. The other reactor steps are either on or off. This decreases the magnitude of all harmonics. The second method involves the 12-pulse arrangement, where two identical connected thyristor-controlled reactors are used, one operated from a wye-connected secondary winding, the other from a delta-connected winding of a step-up transformer. TCR units are characterized by continuous control, and there is a maximum of one half-cycle delay for executing a command from the regulator.

In many applications, the arrangement of an SVC consists of a few large steps of thyristor-switched capacitor and one or two thyristor-controlled reactors, as shown in Fig. 58.26(c). The following are some practical schemes.

Fixed-Capacitor, Thyristor-Controlled Reactor (FC-TCR) Scheme

This scheme was originally developed for industrial applications, such as arc furnace "flicker" control [Gyugyi and Taylor, 1980]. It is essentially a TCR (controlled by a delay angle α) in paral-

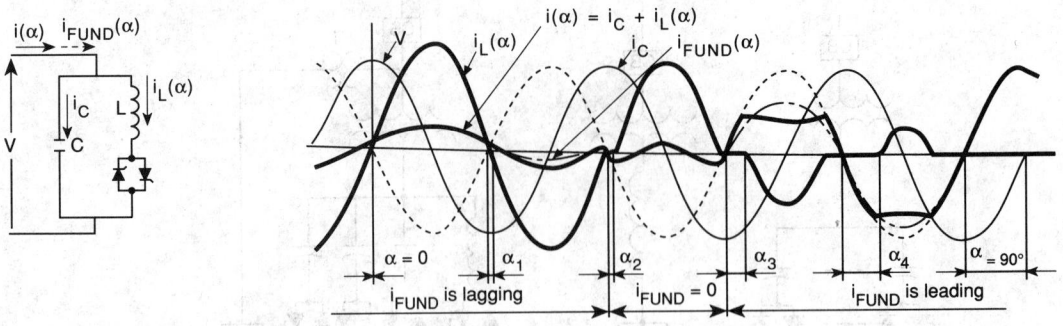

FIGURE 58.27   Basic fixed-capacitor, thyristor-controlled reactor-type compensator and associated waveforms.

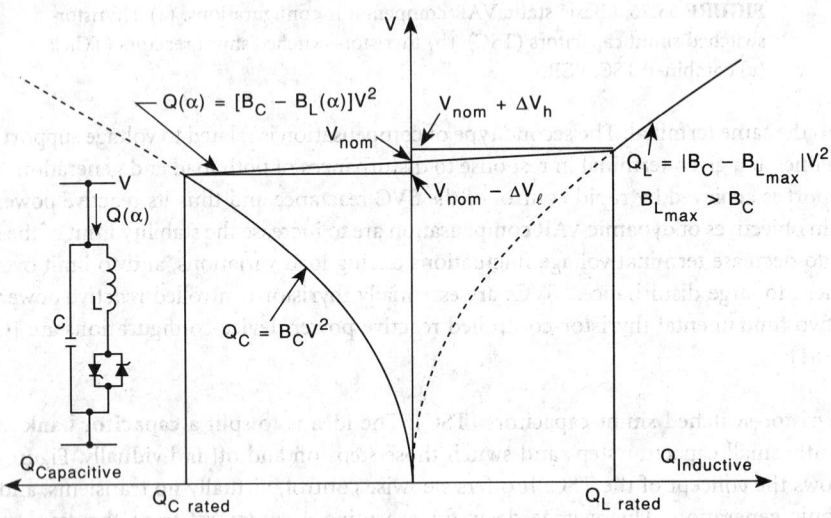

FIGURE 58.28   The steady-state reactive power versus terminal voltage characteristics of
a static VAR compensator.

lel with a fixed capacitor. Figure 58.27 shows a basic fixed-capacitor, thyristor-controlled reactor-type compensator and associated waveforms. Figure 58.28 displays the steady-state reactive power versus terminal voltage characteristics of a static VAR compensator. In the figure, B_C is the imaginary part of the admittance of the capacitor C, and B_L is the imaginary part of the equivalent admittance of the reactor L at delay angle α. The relation between the output VARs and the applied voltage is linear over the voltage band of regulation. In practice, the fixed capacitor is usually replaced by a filter network that has the required capacitive reactance at the power system frequency but exhibits a low impedance at selected frequencies to absorb troublesome harmonics.

The behavior and response of the FC-TCR type of compensator under large disturbances is uncontrollable, at least during the first few cycles following the disturbance. The resulting voltage transients are essentially determined by the fixed capacitor and the power system impedance. This can lead to overvoltage and resonance problems.

At zero VAR demand, the capacitive and reactive VARs cancel out, but the capacitor bank's current is circulated through the reactor bank via the thyristor switch. As a result, this configuration suffers from no load (standby) losses. The losses decrease with increasing the capacitive VAR output and, conversely, increase with increasing the inductive VAR output.

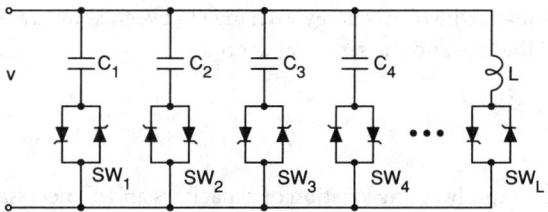

FIGURE 58.29 Basic thyristor-switched capacitor, thyristor-controlled reactor-type compensator.

Thyristor-Switched Capacitor, Thyristor-Controlled Reactor (TSC-TCR) Scheme

This hybrid compensator was developed specifically for utility applications to overcome the disadvantages of the FC-TCR compensators (behavior under large disturbances and loss characteristic). Figure 58.29 shows a basic circuit of this compensator. It consists in general of a thyristor-controlled reactor bank (or banks) and a number of capacitor banks, each in series with a solid-state switch, which is composed of either a reverse-parallel-connected thyristor pair or a thyristor in reverse parallel with a diode. The reactor's switch is composed of a reverse-parallel-connected thyristor pair that is capable of continuously controlling the current in the reactor from zero to maximum rated current.

The total capacitive range is divided into n operating intervals, where n is the number of capacitor banks in the compensator. In the first interval one capacitor bank is switched in, and at the same time the current in the TCR bank is adjusted so that the resultant VAR output from capacitor and reactor matches the VAR demand. In the ith interval the output is controllable in the range $[(i-1)\mathrm{VAR_{max}}/n]$ to $(i\,\mathrm{VAR_{max}}/n)$ by switching in the ith capacitor bank and using the TCR bank to absorb the surplus capacitive VARs. This scheme can be considered as a conventional FC-TCR, where the rating of the reactor bank is kept relatively small ($1/n$ times the maximum VAR output) and the value of the capacitor bank is changed in discrete steps so as to keep the operation of the reactor bank within its normal control range.

The losses of the TSC-TCR compensator at zero VARs output are inherently low, and they increase in proportion to the VAR output.

The mechanism by which SVCs introduce damping into the system can be explained as a result of the change in system voltage due to switching of a capacitor/reactor. The electrical power output of the generators is changed immediately due to the change in power transfer capability and the change in load power requirements.

Among the early applications of SVC for power system damping is the application to the Scandinavian system as discussed in Olwegard *et al.* [1981]. More recently, SVC control for damping of system oscillations based on local measurements has been proposed. The scheme uses phase-angle estimates based on voltage and power measurements at the SVC location as the control signal [see Lerch *et al.*, 1991].

For a general mathematical model of an SVC and an analysis of its stabilizing effects, see Hammad [1986]. Representing the SVC in transient analysis programs is an important consideration [see Gole and Sood, 1990; Lefebvre and Gerin-Lajoie, 1992].

It is important to recognize that applying static VAR compensators to series-compensated ac transmission lines results in three distinct resonant modes [Larsen *et al.*, 1990]:

- Shunt-capacitance resonance involves energy exchange between the shunt capacitance (line charging plus any power factor correction or SVCs) and the series inductance of the lines and the generator.

- Series-line resonance involves energy exchange between the series capacitor and the series inductance of the lines, transformers, and generators. The resonant frequency will depend on the level of series compensation.

- Shunt-reactor resonance involves energy exchange between shunt reactors at the intermediate substations of the line and the series capacitors.

Defining Terms

Capacitor bank: An assembly at one location of capacitors and all necessary accessories, such as switching equipment, protective equipment, and controls, required for a complete operating installation.

Reactor: A device whose primary purpose is to introduce reactance into a circuit. Inductive reactance is frequently abbreviated inductor.

Resonance: The enhancement of the response of a physical system to a periodic excitation when the excitation frequency is equal to a natural frequency of the system.

Shunt: A device having appreciable impedance connected in parallel across other devices or apparatus and diverting some of the current from it. Appreciable voltage exists across the shunted device or apparatus, and an appreciable current may exist in it.

Shunt reactor: A reactor intended for connection in shunt to an electric system to draw inductive current.

Subsynchronous resonance: An electric power system condition where the electric network exchanges energy with a turbine generator at one or more of the natural frequencies of the combined system below the synchronous frequency of the system.

Thyristor: A bistable semiconductor device comprising three or more junctions that can be switched from the off state to the on state, or vice versa, such switching occurring within at least one quadrant of the principal voltage-current characteristic.

References

I. S. Benko, B. Bhargava, and W. N. Rothenbuhler, "Prototype NGH subsynchronous resonance damping scheme, part II—Switching and short circuit tests," *IEEE Trans. Power Syst.*, vol. 2, pp. 1040–1049, 1987.

L. E. Bock and G. R. Mitchell, "Higher line loadings with series capacitors," *Transmission Magazine*, March 1973.

E. W. Bogins and H. T. Trojan, "Application and design of EHV shunt reactors," *Transmission Magazine*, March 1973.

C. E. Bowler, D. N. Ewart, and C. Concordia, "Self excited torsional frequency oscillations with series capacitors," *IEEE Trans. Power Appar. Syst.*, vol. 93, pp. 1688–1695, 1973.

G. D. Brewer, H. M. Rustebakke, R. A. Gibley, and H. O. Simmons, "The use of series capacitors to obtain maximum EHV transmission capability," *IEEE Trans. Power Appar. Syst.*, vol. 83, pp. 1090–1102, 1964.

C. Concordia, "System compensation, an overview," *Transmission Magazine*, March 1973.

S. E. M. de Oliveira, I. Gardos, and E. P. Fonseca, "Representation of series capacitors in electric power system stability studies," *IEEE Trans. Power Syst.*, vol. 6, no. 3, pp. 1119–1125, 1991.

A. A. Edris, "Series compensation schemes reducing the potential of subsynchronous resonance," *IEEE Trans. Power Syst.*, vol. 5, no. 1, pp. 219–226, 1990.

A. M. El-Serafi and A. A. Shaltout, "Damping of SSR Oscillations by Excitation Control," *IEEE PES Summer Meeting*, Vancouver, 1979.

A. M. Gole and V. K. Sood, "A static compensator model for use with electromagnetic transients simulation programs," *IEEE Trans. Power Delivery*, vol. 5, pp. 1398–1407, 1990.

L. Gyugyi, R. A. Otto, and T. H. Putman, "Principles and applications of static thyristor-controlled shunt compensators," *IEEE Trans. Power Appar. Syst.*, vol. PAS-97, pp. 1935–1945, 1978.

L. Gyugyi and E. R. Taylor, Jr., "Characteristics of static thyristor-controlled shunt compensators for power transmission system applications," *IEEE Trans. Power Appar. Syst.*, vol. PAS-99, pp. 1795–1804, 1980.

A. E. Hammad, "Analysis of power system stability enhancement by static VAR compensators," *IEEE Trans. Power Syst.*, vol. 1, pp. 222–227, 1986.

J. F. Hauer, "Robust damping controls for large power systems," *IEEE Control Systems Magazine*, pp. 12–18, January 1989.

R. A. Hedin, K. B. Stump, and N. G. Hingorani, "A new scheme for subsynchronous resonance damping of torsional oscillations and transient torque—Part II," *IEEE Trans. Power Appar. Syst.*, vol. PAS-100, pp. 1856–1863, 1981.

N. G. Hingorani, "A new scheme for subsynchronous resonance damping of torsional oscillations and transient torque—Part I," *IEEE Trans. Power Appar. Syst.*, vol. PAS-100, pp. 1852–1855, 1981.

N. G. Hingorani, B. Bhargava, G. F. Garrigue, and G. D. Rodriguez, "Prototype NGH subsynchronous resonance damping scheme, part I—Field installation and operating experience," *IEEE Trans. Power Syst.*, vol. 2, pp. 1034–1039, 1987.

IEEE Subsynchronous Resonance Working Group, "Proposed terms and definitions for subsynchronous oscillations," *IEEE Trans. Power Appar. Syst.*, vol. PAS-99, pp. 506–511, 1980.

IEEE Subsynchronous Resonance Working Group, "Countermeasures to subsynchronous resonance problems," *IEEE Trans. Power Appar. Syst.*, vol. PAS-99, pp. 1810–1818, 1980.

IEEE Subsynchronous Resonance Working Group, "Series capacitor controls and settings as countermeasures to subsynchronous resonance," *IEEE Trans. Power Appar. Syst.*, vol. PAS-101, pp. 1281–1287, June 1982.

G. Jancke, N. Fahlen, and O. Nerf, "Series capacitors in power systems," *IEEE Transactions on Power Appar. Syst.*, vol. PAS-94, pp. 915–925, May/June 1975.

L. A. Kilgore, D. G. Ramey, and W. H. South, "Dynamic filter and other solutions to the subsynchronous resonance problem," *Proceedings of the American Power Conference*, vol. 37, p. 923, 1975.

E. W. Kimbark, *Power System Stability*, vol. I, *Elements of Stability Calculations*, New York: Wiley, 1948.

E. W. Kimbark, "Improvement of system stability by switched series capacitors," *IEEE Trans. Power Appar. Syst.*, vol. 85, pp. 180–188, February 1966.

J. J. LaForest, K. W. Priest, Ramirez, and H. Nowak, "Resonant voltages on reactor compensated extra-high-voltage lines," *IEEE Trans. Power Appar. Syst.*, vol. PAS-91, pp. 2528–2536, November/December 1972.

E. V. Larsen, D. H. Baker, A. F. Imece, L. Gerin-Lajoie, and G. Scott, "Basic aspects of applying SVC's to series-compensated ac transmission lines," *IEEE Trans. Power Delivery*, vol. 5, pp. 1466–1472, July 1990.

S. Lefebvre and L. Gerin-Lajoie, "A static compensator model for the EMTP," *IEEE Trans. Power Systems*, vol. 7, no. 2, pp. 477–486, May 1992.

E. Lerch, D. Povh, and L. Xu, "Advanced SVC control for damping power system oscillations," *IEEE Trans. Power Syst.*, vol. 6, pp. 524–531, May 1991.

S. M. Merry and E. R. Taylor, "Overvoltages and harmonics on EHV systems," *IEEE Trans. Power Appar. Syst.*, vol. PAS-91, pp. 2537–2544, November/December 1972.

A. Olwegard, K. Walve, G. Waglund, H. Frank, and S. Torseng, "Improvement of transmission capacity by thyristor controlled reactive power," *IEEE Trans. Power Appar. Syst.*, vol. PAS-100, pp. 3930–3939, 1981.

J. B. Tice and C. E. J. Bowler, "Control of phenomenon of subsynchronous resonance," *Proceedings of the American Power Conference*, vol. 37, pp. 916–922, 1975.

B. M. Weedy, *Electric Power Systems*, London: Wiley, 1972.

Further Information

An excellent source of information on the application of capacitors on power systems is the Westinghouse *Transmission and Distribution* book, published in 1964. A most readable treatment of improving system stability by series capacitors is given by Kimbark's paper [1966]. Jancke *et al.* [1975] give a detailed discussion of experience with the 400-kV series-capacitor compensation installations on the Swedish system and aspects of the protection system. Hauer [1989] presents a discussion of practical stability controllers that manipulate series and/or shunt reactance.

An excellent summary of the state of the art in static VAR compensators is the record of the IEEE Working Group symposium conducted in 1987 on the subject (see IEEE Publication 87TH0187-5-PWR, Application of Static VAR Systems for System Dynamic Performance).

For state-of-the-art coverage of subsynchronous resonance and countermeasures, two symposia are available: IEEE Publication 79TH0059-6-PWR, State-of-the-Art Symposium—Turbine Generator Shaft Torsionals, and IEEE Publication 81TH0086-9-PWR, Symposium on Countermeasures for Subsynchronous Resonance.

58.5 Fault Analysis in Power Systems

Charles Gross

A **fault** in an electrical power system is the unintentional and undesirable creation of a conducting path (a *short circuit*) or a blockage of current (an *open circuit*). The short-circuit fault is typically the most common and is usually implied when most people use the term *fault*. We restrict our comments to the short-circuit fault.

The causes of faults include lightning, wind damage, trees falling across lines, vehicles colliding with towers or poles, birds shorting out lines, aircraft colliding with lines, vandalism, small animals entering switchgear, and line breaks due to excessive ice loading. Power system faults may be categorized as one of four types: single line-to-ground, line-to-line, double line-to-ground, and balanced three-phase. The first three types constitute severe unbalanced operating conditions.

It is important to determine the values of system voltages and currents during faulted conditions so that protective devices may be set to detect and minimize their harmful effects. The time constants of the associated transients are such that sinusoidal steady-state methods may still be used. The method of symmetrical components is particularly suited to fault analysis.

Our objective is to understand how symmetrical components may be applied specifically to the four general fault types mentioned and how the method can be extended to any unbalanced three-phase system problem.

Note that phase values are indicated by subscripts, *a, b, c*; sequence (symmetrical component) values are indicated by subscripts 0, 1, 2. The transformation is defined by

$$\begin{bmatrix} \overline{V}_a \\ \overline{V}_b \\ \overline{V}_c \end{bmatrix} = \begin{bmatrix} 1 & 1 & 1 \\ 1 & a^2 & a \\ 1 & a & a^2 \end{bmatrix} \begin{bmatrix} \overline{V}_0 \\ \overline{V}_1 \\ \overline{V}_2 \end{bmatrix} = [T] \begin{bmatrix} \overline{V}_0 \\ \overline{V}_1 \\ \overline{V}_2 \end{bmatrix}$$

$$\begin{bmatrix} \overline{V}_0 \\ \overline{V}_1 \\ \overline{V}_2 \end{bmatrix} = \frac{1}{3} \begin{bmatrix} 1 & 1 & 1 \\ 1 & a & a^2 \\ 1 & a^2 & a \end{bmatrix} \begin{bmatrix} \overline{V}_0 \\ \overline{V}_1 \\ \overline{V}_2 \end{bmatrix} = [T]^{-1} \begin{bmatrix} \overline{V}_a \\ \overline{V}_b \\ \overline{V}_c \end{bmatrix}$$

Simplifications in the System Model

Certain simplifications are possible and usually employed in fault analysis.

- Transformer magnetizing current and core loss will be neglected.
- Line shunt capacitance is neglected.
- Sinusoidal steady-state circuit analysis techniques are used. The so-called **dc offset** is accounted for by using correction factors.
- Prefault voltage is assumed to be $1/\underline{0°}$ per-unit. One per-unit voltage is at its nominal value prior to the application of a fault, which is reasonable. The selection of zero phase is arbitrary and convenient. Prefault load current is neglected.

In addition to the above, for hand calculations, neglect series resistance (this approximation will not be necessary for a computer solution). Also, the only difference in the positive and negative sequence networks is introduced by the machine impedances. If we select the subtransient reactance X_d'' for the positive sequence reactance, the difference is slight (in fact, the two are identical for nonsalient machines). The simplification is important, since it reduces computer storage requirements by roughly one-third. Circuit models for generators, lines, and transformers are shown in Figs. 58.30, 58.31, and 58.32, respectively.

Our basic approach to the problem is to consider the general situation suggested in Fig. 58.33(a). The general terminals brought out are for purposes of external connections that will simulate faults. Note carefully the positive assignments of phase quantities. Particularly note that the currents flow *out of* the system. We can construct general *sequence* equivalent circuits for the system, and such circuits are indicated in Fig. 58.33(b). The ports indicated correspond to the general three-phase entry port of Fig. 58.33(a). The positive sense of sequence values is compatible with that used for phase values.

The Four Basic Fault Types

The Balanced Three-Phase Fault

Imagine the general three-phase access port terminated in a fault impedance ($\overline{Z}_f$) as shown in Fig. 58.34(a). The terminal conditions are

$$\begin{bmatrix} \overline{V}_a \\ \overline{V}_b \\ \overline{V}_c \end{bmatrix} = \begin{bmatrix} \overline{Z}_f & 0 & 0 \\ 0 & \overline{Z}_f & 0 \\ 0 & 0 & \overline{Z}_f \end{bmatrix} \begin{bmatrix} \overline{I}_a \\ \overline{I}_b \\ \overline{I}_c \end{bmatrix}$$

Transforming to $[Z_{012}]$,

$$[Z_{012}] = [T]^{-1} \begin{bmatrix} \overline{Z}_f & 0 & 0 \\ 0 & \overline{Z}_f & 0 \\ 0 & 0 & \overline{Z}_f \end{bmatrix} [T] = \begin{bmatrix} \overline{Z}_f & 0 & 0 \\ 0 & \overline{Z}_f & 0 \\ 0 & 0 & \overline{Z}_f \end{bmatrix}$$

The corresponding network connections are given in Fig. 58.35(a). Since the zero and negative sequence networks are passive, only the positive sequence network is nontrivial.

$$\overline{V}_0 = \overline{V}_2 = 0 \tag{58.60}$$

$$\overline{I}_0 = \overline{I}_2 = 0 \tag{58.61}$$

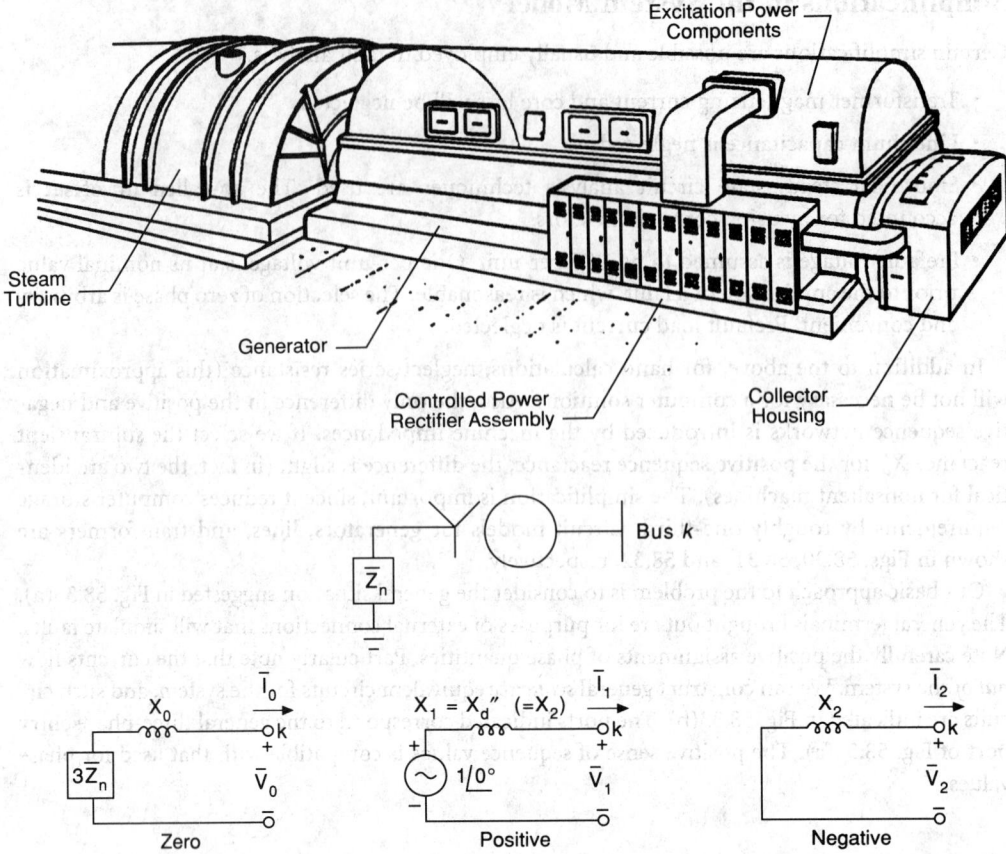

FIGURE 58.30 Generator sequence circuit models.

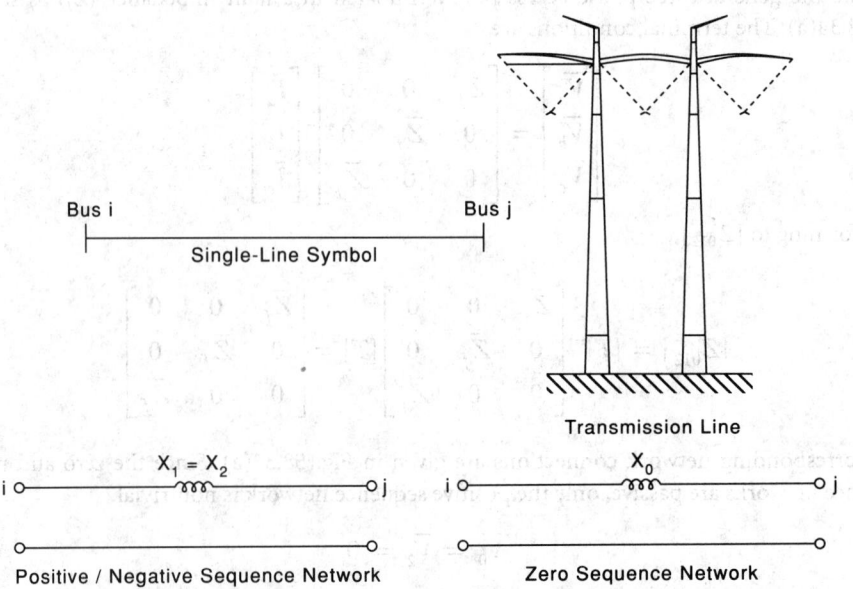

FIGURE 58.31 Line sequence circuit models.

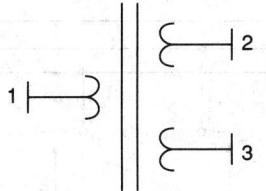

Single-Line Symbol

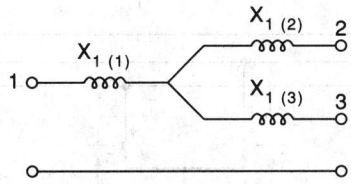

Positive / Negative Sequence Network

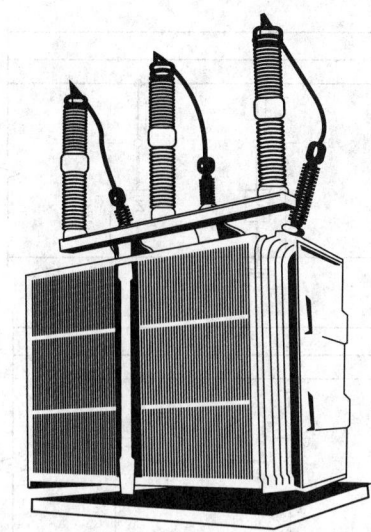

Zero Sequence Network

The "gaps" are connected to model Y–Δ connections:

FIGURE 58.32 Transformer sequence circuit models.

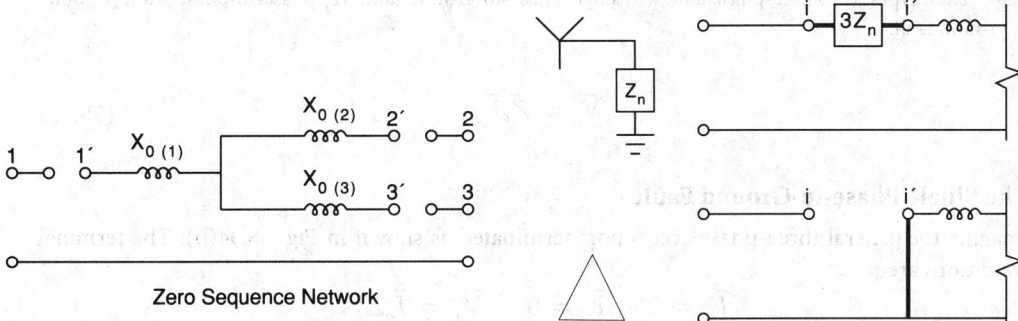

(a)

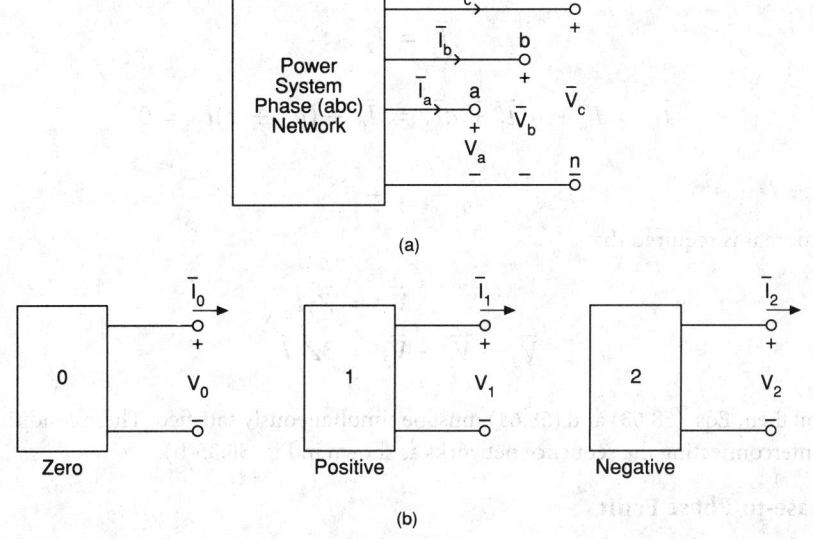

(b)

FIGURE 58.33 General fault port in an electric power system. (a) General fault port in phase (*abc*) coordinates; (b) corresponding fault ports in sequence (012) coordinates.

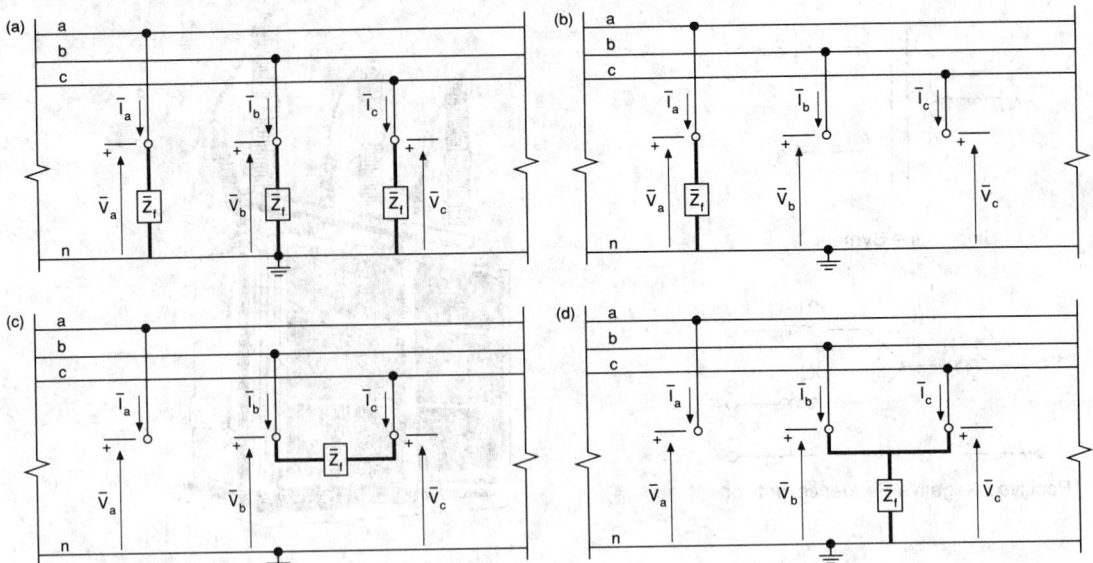

FIGURE 58.34 Fault types. (a) Three-phase fault; (b) single phase-to-ground fault; (c) phase-to-phase fault; (d) double phase-to-ground fault.

$$\overline{V}_1 = \overline{Z}_f \overline{I}_1 \tag{58.62}$$

The Single Phase-to-Ground Fault

Imagine the general three-phase access port terminated as shown in Fig. 58.34(b). The terminal conditions are

$$\overline{I}_b = 0 \quad \overline{I}_c = 0 \quad \overline{V}_a = \overline{I}_a \overline{Z}_f$$

Therefore

$$\overline{I}_0 + a^2 \overline{I}_1 + a \overline{I}_2 = \overline{I}_0 + a \overline{I}_1 + a^2 \overline{I}_2 = 0$$

or

$$\overline{I}_1 = \overline{I}_2$$

Also

$$\overline{I}_b = \overline{I}_0 + a^2 \overline{I}_1 + a \overline{I}_2 = \overline{I}_0 + (a^2 + a) \overline{I}_1 = 0$$

or

$$\overline{I}_0 = \overline{I}_1 = \overline{I}_2 \tag{58.63}$$

Furthermore it is required that

$$\overline{V}_a = \overline{Z}_f \overline{I}_a$$
$$\overline{V}_0 + \overline{V}_1 + \overline{V}_2 = 3 \overline{Z}_f \overline{I}_1 \tag{58.64}$$

In general then, Eqs. (58.63) and (58.64) must be simultaneously satisfied. These conditions can be met by interconnecting the sequence networks as shown in Fig. 58.35(b).

The Phase-to-Phase Fault

Imagine the general three-phase access port terminated as shown in Fig. 58.34(c). The terminal conditions are such that we may write

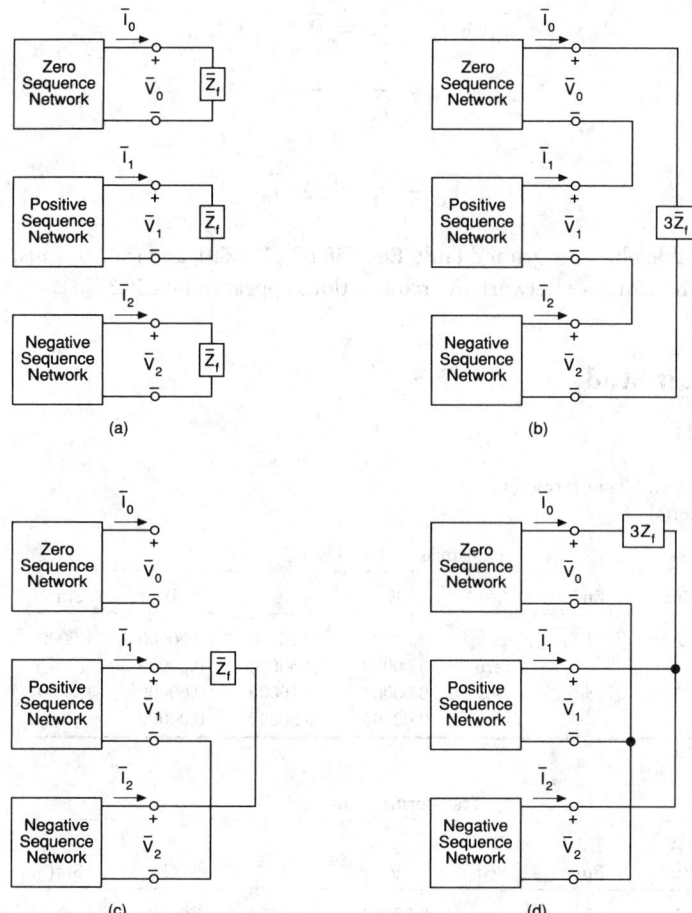

FIGURE 58.35 Sequence network terminations for fault types. (a) Balanced three-phase fault; (b) single phase-to-ground fault; (c) phase-to-phase fault; (d) double phase-to-ground fault.

$$\bar{I}_0 = 0 \qquad \bar{I}_b = -\bar{I}_c \qquad \bar{V}_b = \bar{Z}_f \bar{I}_b + \bar{V}_c$$

It follows that

$$\bar{I}_0 + \bar{I}_1 + \bar{I}_2 = 0 \tag{58.65}$$

$$\bar{I}_0 = 0 \tag{58.66}$$

$$\bar{I}_1 = -\bar{I}_2 \tag{58.67}$$

In general then, Eqs. (58.65), (58.66), and (58.67) must be simultaneously satisfied. The proper interconnection between sequence networks appears in Fig. 58.35(c).

The Double Phase-to-Ground Fault

Consider the general three-phase access port terminated as shown in Fig. 58.34(d). The terminal conditions indicate

$$\bar{I}_a = 0 \qquad \bar{V}_b = \bar{V}_c \qquad \bar{V}_b = (\bar{I}_b + \bar{I}_c)\bar{Z}_f$$

It follows that

$$\bar{I}_0 + \bar{I}_1 + \bar{I}_2 = \bar{0} \tag{58.68}$$

$$\bar{V}_1 = \bar{V}_2 \tag{58.69}$$

and

$$\bar{V}_0 - \bar{V}_1 = 3\bar{Z}_f \bar{I}_0 \tag{58.70}$$

For the general double phase-to-ground fault, Eqs. (58.68), (58.69), and (58.70) must be simultaneously satisfied. The sequence network interconnections appear in Fig. 58.35(d).

An Example Fault Study

Case: EXAMPLE SYSTEM
Run :
System has data for 2 Line(s); 2 Transformer(s);
4 Bus(es); and 2 Generator(s)

Transmission Line Data

Line	Bus	Bus	Seq	R	X	B	Srat
1	2	3	pos	0.00000	0.16000	0.00000	1.0000
			zero	0.00000	0.50000	0.00000	
2	2	3	pos	0.00000	0.16000	0.00000	1.0000
			zero	0.00000	0.50000	0.00000	

Transformer Data

Trans-former	HV Bus	LV Bus	Seq	R	X	C	Srat
1	2	1	pos	0.00000	0.05000	1.00000	1.0000
	Y	Y	zero	0.00000	0.05000		
2	3	4	pos	0.00000	0.05000	1.00000	1.0000
	Y	D	zero	0.00000	0.05000		

Generator Data

No.	Bus	Srated	Ra	Xd″	Xo	Rn	Xn	Con
1	1	1.0000	0.0000	0.200	0.0500	0.0000	0.0400	Y
2	4	1.0000	0.0000	0.200	0.0500	0.0000	0.0400	Y

Zero Sequence [Z] Matrix

0.0 + j(0.1144)	0.0 + j(0.0981)	0.0 + j(0.0163)	0.0 + j(0.0000)
0.0 + j(0.0981)	0.0 + j(0.1269)	0.0 + j(0.0212)	0.0 + j(0.0000)
0.0 + j(0.0163)	0.0 + j(0.0212)	0.0 + j(0.0452)	0.0 + j(0.0000)
0.0 + j(0.0000)	0.0 + j(0.0000)	0.0 + j(0.0000)	0.0 + j(0.1700)

Positive Sequence [Z] Matrix

0.0 + j(0.1310)	0.0 + j(0.1138)	0.0 + j(0.0862)	0.0 + j(0.0690)
0.0 + j(0.1138)	0.0 + j(0.1422)	0.0 + j(0.1078)	0.0 + j(0.0862)
0.0 + j(0.0862)	0.0 + j(0.1078)	0.0 + j(0.1422)	0.0 + j(0.1138)
0.0 + j(0.0690)	0.0 + j(0.0862)	0.0 + j(0.1138)	0.0 + j(0.1310)

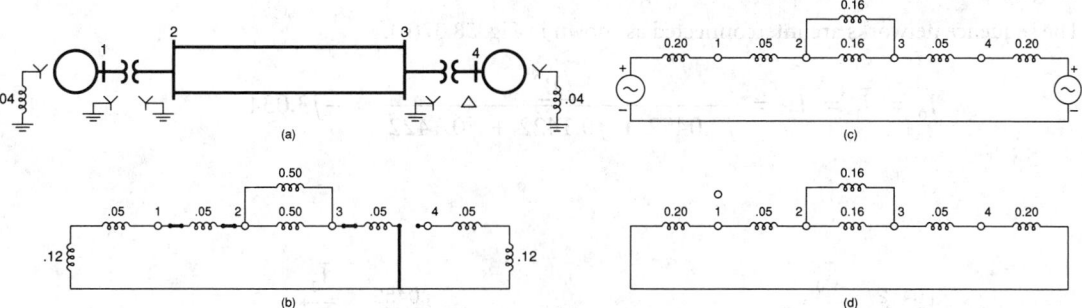

FIGURE 58.36   Example system. (a) Single-line diagram; (b) zero sequence network; (c) positive sequence network; (d) negative sequence network.

The single-line diagram and sequence networks are presented in Fig. 58.36.

Suppose bus 3 in the example system represents the fault location and $\bar{Z}_f = 0$. The positive sequence circuit can be reduced to its Thévenin equivalent at bus 3:

$$E_{T1} = 1.0 \underline{/0°} \qquad \bar{Z}_{T1} = j0.1422$$

Similarly, the negative and zero sequence Thévenin elements are

$$\bar{E}_{T2} = 0 \qquad \bar{Z}_{T2} = j0.1422$$
$$\bar{E}_{T0} = 0 \qquad \bar{Z}_{T0} = j0.0452$$

The network interconnections for the four fault types are shown in Fig. 58.37. For each of the fault types, compute the currents and voltages at the faulted bus.

Balanced Three-Phase Fault

The sequence networks are shown in Fig. 58.37(a). Obviously,

$$\bar{V}_0 = \bar{I}_0 = \bar{V}_2 = \bar{I}_2 = 0$$
$$\bar{I}_1 = \frac{1\underline{/0°}}{j0.1422} = -j7.032; \quad \text{also } \bar{V}_1 = 0$$

To compute the phase values,

$$\begin{bmatrix} \bar{I}_a \\ \bar{I}_b \\ \bar{I}_c \end{bmatrix} = [T] \begin{bmatrix} \bar{I}_0 \\ \bar{I}_1 \\ \bar{I}_2 \end{bmatrix} = \begin{bmatrix} 1 & 1 & 1 \\ 1 & a^2 & a \\ 1 & a & a^2 \end{bmatrix} \begin{bmatrix} 0 \\ -j7.032 \\ 0 \end{bmatrix} = \begin{bmatrix} 7.032 \underline{/-90°} \\ 7.032 \underline{/150°} \\ 7.032 \underline{/30°} \end{bmatrix}$$

$$\begin{bmatrix} \bar{V}_a \\ \bar{V}_b \\ \bar{V}_c \end{bmatrix} = [T] \begin{bmatrix} 0 \\ 0 \\ 0 \end{bmatrix} = \begin{bmatrix} 0 \\ 0 \\ 0 \end{bmatrix}$$

Single Phase-to-Ground Fault

The sequence networks are interconnected as shown in Fig. 58.37(b).

$$\bar{I}_0 = \bar{I}_1 = \bar{I}_2 = \frac{1 \underline{/0°}}{j0.0452 + j0.1422 + j0.1422} = -j3.034$$

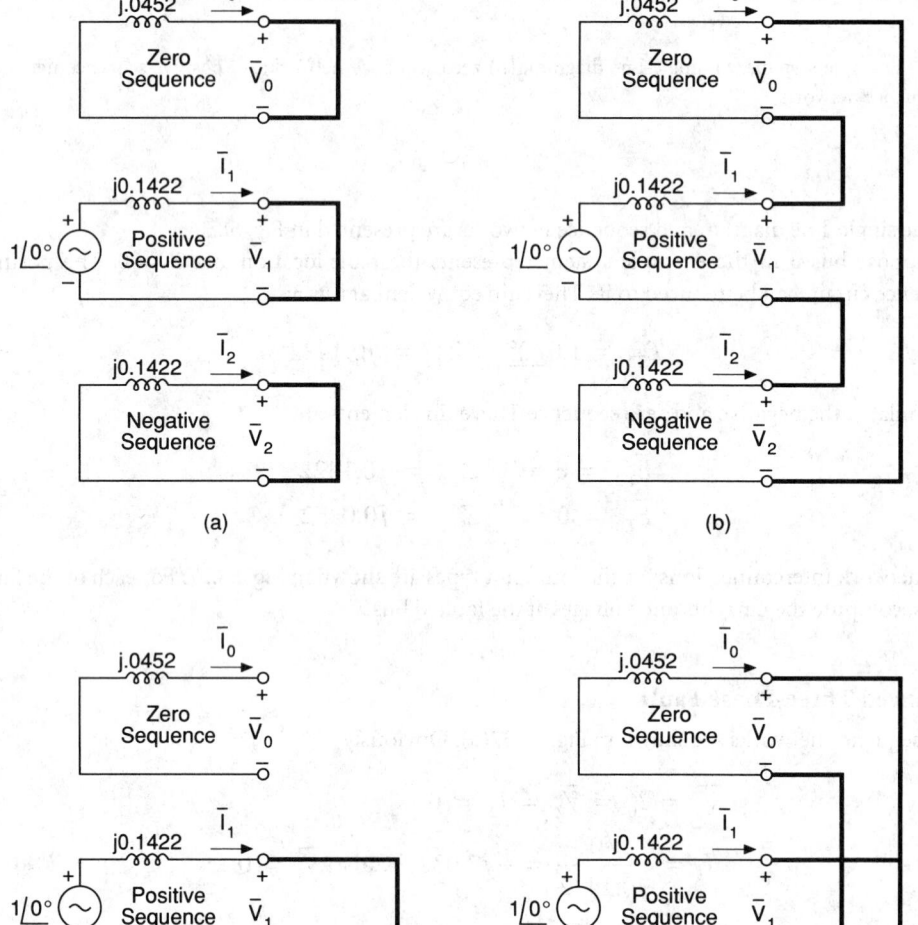

FIGURE 58.37 Example system faults at bus 3. (a) Balanced three-phase; (b) single phase-to-ground; (c) phase-to-phase; (d) double phase-to-ground.

$$\begin{bmatrix} \bar{I}_a \\ \bar{I}_b \\ \bar{I}_c \end{bmatrix} = \begin{bmatrix} 1 & 1 & 1 \\ 1 & a^2 & a \\ 1 & a & a^2 \end{bmatrix} \begin{bmatrix} -j3.034 \\ -j3.034 \\ -j3.034 \end{bmatrix} = \begin{bmatrix} -j9.102 \\ 0 \\ 0 \end{bmatrix}$$

The sequence voltages are

$$\bar{V}_0 = -j0.0452(-j3.034) = -1371$$

$$\bar{V}_1 = 1.0 - j0.1422(-j3.034) = 0.5685$$

$$\bar{V}_2 = -j0.1422(-j3.034) = -0.4314$$

The phase voltages are

$$\begin{bmatrix} \bar{V}_a \\ \bar{V}_b \\ \bar{V}_c \end{bmatrix} = \begin{bmatrix} 1 & 1 & 1 \\ 1 & a^2 & a \\ 1 & a & a^2 \end{bmatrix} \begin{bmatrix} -0.1371 \\ 0.5685 \\ -0.4314 \end{bmatrix} = \begin{bmatrix} 0 \\ 0.8901 \underline{/-103.4°} \\ 0.8901 \underline{/-103.4°} \end{bmatrix}$$

Phase-to-phase and double phase-to-ground fault values are calculated from the appropriate networks [Figs. 58.37(c) and (d)]. Complete results are provided.

Faulted Bus	Phase a	Phase b	Phase c
3	G	G	G

		Sequence Voltages				
Bus	V0		V1		V2	
1	0.0000/	0.0	0.3939/	0.0	0.0000/	0.0
2	0.0000/	0.0	0.2424/	0.0	0.0000/	0.0
3	0.0000/	0.0	0.0000/	0.0	0.0000/	0.0
4	0.0000/	0.0	0.2000/	−30.0	0.0000/	30.0

		Phase Voltages				
Bus	Va		Vb		Vc	
1	0.3939/	0.0	0.3939/	−120.0	0.3939/	120.0
2	0.2424/	0.0	0.2424/	−120.0	0.2424/	120.0
3	0.0000/	6.5	0.0000/	−151.2	0.0000/	133.8
4	0.2000/	−30.0	0.2000/	−150.0	0.2000/	90.0

			Sequence Currents				
Bus to Bus		I0		I1		I2	
1	2	0.0000/	167.8	3.0303/	−90.0	0.0000/	90.0
1	0	0.0000/	−12.2	3.0303/	90.0	0.0000/	−90.0
2	3	0.0000/	167.8	1.5152/	−90.0	0.0000/	90.0
2	3	0.0000/	167.8	1.5152/	−90.0	0.0000/	90.0
2	1	0.0000/	−12.2	3.0303/	90.0	0.0000/	−90.0
3	2	0.0000/	−12.2	1.5152/	90.0	0.0000/	−90.0
3	2	0.0000/	−12.2	1.5152/	90.0	0.0000/	−90.0
3	4	0.0000/	−12.2	4.0000/	90.0	0.0000/	−90.0
4	3	0.0000/	0.0	4.0000/	−120.0	0.0000/	120.0
4	0	0.0000/	0.0	4.0000/	60.0	0.0000/	−60.0

Faulted Bus	Phase a	Phase b	Phase c
3	G	G	G

Phase Currents

Bus to Bus		Ia		Ib		Ic	
1	2	3.0303/	−90.0	3.0303/	150.0	3.0303/	30.0
1	0	3.0303/	90.0	3.0303/	−30.0	3.0303/	−150.0
2	3	1.5151/	−90.0	1.5151/	150.0	1.5151/	30.0
2	3	1.5151/	−90.0	1.5151/	150.0	1.5151/	30.0
2	1	3.0303/	90.0	3.0303/	−30.0	3.0303/	−150.0
3	2	1.5151/	90.0	1.5151/	−30.0	1.5151/	−150.0
3	2	1.5151/	90.0	1.5151/	−30.0	1.5151/	−150.0
3	4	4.0000/	90.0	4.0000/	−30.0	4.0000/	−150.0
4	3	4.0000/	−120.0	4.0000/	120.0	4.0000/	−0.0
4	0	4.0000/	60.0	4.0000/	−60.0	4.0000/	−180.0

Faulted Bus	Phase a	Phase b	Phase c
3	G	0	0

Sequence Voltages

Bus		V0		V1		V2	
1		0.0496/	180.0	0.7385/	0.0	0.2615/	180.0
2		0.0642/	180.0	0.6731/	0.0	0.3269/	180.0
3		0.1371/	180.0	0.5685/	0.0	0.4315/	180.0
4		0.0000/	0.0	0.6548/	−30.0	0.3452/	210.0

Phase Voltages

Bus		Va		Vb		Vc	
1		0.4274/	0.0	0.9127/	−108.4	0.9127/	108.4
2		0.2821/	0.0	0.8979/	−105.3	0.8979/	105.3
3		0.0000/	89.2	0.8901/	−103.4	0.8901/	103.4
4		0.5674/	−61.8	0.5674/	−118.2	1.0000/	90.0

Sequence Currents

Bus to Bus		I0		I1		I2	
1	2	0.2917/	−90.0	1.3075/	−90.0	1.3075/	−90.0
1	0	0.2917/	90.0	1.3075/	90.0	1.3075/	90.0
2	3	0.1458/	−90.0	0.6537/	−90.0	0.6537/	−90.0
2	3	0.1458/	−90.0	0.6537/	−90.0	0.6537/	−90.0
2	1	0.2917/	90.0	1.3075/	90.0	1.3075/	90.0
3	2	0.1458/	90.0	0.6537/	90.0	0.6537/	90.0
3	2	0.1458/	90.0	0.6537/	90.0	0.6537/	90.0
3	4	2.7416/	90.0	1.7258/	90.0	1.7258/	90.0
4	3	0.0000/	0.0	1.7258/	−120.0	1.7258/	−60.0
4	0	0.0000/	90.0	1.7258/	60.0	1.7258/	120.0

Faulted Bus	Phase a	Phase b	Phase c
3	G	0	0

Phase Currents

Bus to Bus		Ia		Ib		Ic	
1	2	2.9066/	−90.0	1.0158/	90.0	1.0158/	90.0
1	0	2.9066/	90.0	1.0158/	−90.0	1.0158/	−90.0
2	3	1.4533/	−90.0	0.5079/	90.0	0.5079/	90.0
2	3	1.4533/	−90.0	0.5079/	90.0	0.5079/	90.0
2	1	2.9066/	90.0	1.0158/	−90.0	1.0158/	−90.0
3	2	1.4533/	90.0	0.5079/	−90.0	0.5079/	−90 0
3	2	1.4533/	90.0	0.5079/	−90.0	0.5079/	−90 0
3	4	6.1933/	90.0	1.0158/	90.0	1.0158/	90.0
4	3	2.9892/	−90.0	2.9892/	90.0	0.0000/	−90.0
4	0	2.9892/	90.0	2.9892/	−90.0	0.0000/	90.0

Faulted Bus	Phase a	Phase b	Phase c
3	0	C	B

Sequence Voltages

Bus	V0		V1		V2	
1	0.0000/	0.0	0.6970/	0.0	0.3030/	0.0
2	0.0000/	0.0	0.6212/	0.0	0.3788/	0.0
3	0.0000/	0.0	0.5000/	0.0	0.5000/	0.0
4	0.0000/	0.0	0.6000/	−30.0	0.4000/	30.0

Phase Voltages

Bus	Va		Vb		Vc	
1	1.0000/	0.0	0.6053/	−145.7	0.6053/	145.7
2	1.0000/	0.0	0.5423/	−157.2	0.5423/	157.2
3	1.0000/	0.0	0.5000/	−180.0	0.5000/	−180.0
4	0.8718/	−6.6	0.8718/	−173.4	0.2000/	90.0

Sequence Currents

Bus to Bus		I0		I1		I2	
1	2	0.0000/	−61.0	1.5152/	−90.0	1.5152/	90.0
1	0	0.0000/	119.0	1.5152/	90.0	1.5152/	−90.0
2	3	0.0000/	−61.0	0.7576/	−90.0	0.7576/	90.0
2	3	0.0000/	−61.0	0.7576/	−90.0	0.7576/	90.0
2	1	0.0000/	119.0	1.5152/	90.0	1.5152/	−90.0
3	2	0.0000/	119.0	0.7576/	90.0	0.7576/	−90.0
3	2	0.0000/	119.0	0.7576/	90.0	0.7576/	−90.0
3	4	0.0000/	119.0	2.0000/	90.0	2.0000/	−90.0
4	3	0.0000/	0.0	2.0000/	−120.0	2.0000/	120.0
4	0	0.0000/	90.0	2.0000/	60.0	2.0000/	−60.0

Faulted Bus	Phase a	Phase b	Phase c
3	0	C	B

Phase Currents

Bus to Bus		Ia		Ib		Ic	
1	2	0.0000/	180.0	2.6243/	180.0	2.6243/	0.0
1	0	0.0000/	180.0	2.6243/	0.0	2.6243/	180.0
2	3	0.0000/	−180.0	1.3122/	180.0	1.3122/	0.0
2	3	0.0000/	−180.0	1.3122/	180.0	1.3122/	0.0
2	1	0.0000/	180.0	2.6243/	0.0	2.6243/	180.0
3	2	0.0000/	−180.0	1.3122/	0.0	1.3122/	180.0
3	2	0.0000/	−180.0	1.3122/	0.0	1.3122/	180.0
3	4	0.0000/	−180.0	3.4641/	0.0	3.4641/	180.0
4	3	2.0000/	−180.0	2.0000/	180.0	4.0000/	0.0
4	0	2.0000/	0.0	2.0000/	0.0	4.0000/	−180.0

Faulted Bus	Phase a	Phase b	Phase c
3	0	G	G

Sequence Voltages

Bus	V0		V1		V2	
1	0.0703/	0.0	0.5117/	0.0	0.1177/	0.0
2	0.0909/	0.0	0.3896/	0.0	0.1472/	0.0
3	0.1943/	−0.0	0.1943/	0.0	0.1943/	0.0
4	0.0000/	0.0	0.3554/	−30.0	0.1554/	30.0

Phase Voltages

Bus	Va		Vb		Vc	
1	0.6997/	0.0	0.4197/	−125.6	0.4197/	125.6
2	0.6277/	0.0	0.2749/	−130.2	0.2749/	130.2
3	0.5828/	0.0	0.0000/	−30.7	0.0000/	−139.6
4	0.4536/	−12.7	0.4536/	−167.3	0.2000/	90.0

Sequence Currents

Bus to Bus		I0		I1		I2	
1	2	0.4133/	90.0	2.4416/	−90.0	0.5887/	90.0
1	0	0.4133/	−90.0	2.4416/	90.0	0.5887/	−90.0
2	3	0.2067/	90.0	1.2208/	−90.0	0.2943/	90.0
2	3	0.2067/	90.0	1.2208/	−90.0	0.2943/	90.0
2	1	0.4133/	−90.0	2.4416/	90.0	0.5887/	−90.0
3	2	0.2067/	−90.0	1.2208/	90.0	0.2943/	−90.0
3	2	0.2067/	−90.0	1.2208/	90.0	0.2943/	−90.0
3	4	3.8854/	−90.0	3.2229/	90.0	0.7771/	−90.0
4	3	0.0000/	0.0	3.2229/	−120.0	0.7771/	120.0
4	0	0.0000/	−90.0	3.2229/	60.0	0.7771/	−60.0

Faulted Bus	Phase a	Phase b	Phase c
3	0	G	G

		Phase Currents		
Bus to Bus		Ia	Ib	Ic
1	2	1.4396/ −90.0	2.9465/ 153.0	2.9465/ 27.0
1	0	1.4396/ 90.0	2.9465/ −27.0	2.9465/ −153.0
2	3	0.7198/ −90.0	1.4733/ 153.0	1.4733/ 27.0
2	3	0.7198/ −90.0	1.4733/ 153.0	1.4733/ 27.0
2	1	1.4396/ 90.0	2.9465/ −27.0	2.9465/ −153.0
3	2	0.7198/ 90.0	1.4733/ −27.0	1.4733/ −153.0
3	2	0.7198/ 90.0	1.4733/ −27.0	1.4733/ −153.0
3	4	1.4396/ −90.0	6.1721/ −55.9	6.1721/ −124.1
4	3	2.9132/ −133.4	2.9132/ 133.4	4.0000/ −0.0
4	0	2.9132/ 46.6	2.9132/ −46.6	4.0000/ −180.0

Further Considerations

Generators are not the only sources in the system. All rotating machines are capable of contributing to fault current, at least momentarily. Synchronous and induction motors will continue to rotate due to inertia and function as sources of fault current. The impedance used for such machines is usually the transient reactance X'_d or the subtransient X''_d, depending on protective equipment and speed of response. Frequently motors smaller than 50 hp are neglected. Connecting systems are modeled with their Thévenin equivalents.

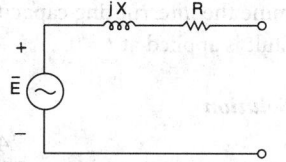

FIGURE 58.38 Positive sequence circuit looking back into faulted bus.

Although we have used ac circuit techniques to calculate faults, the problem is fundamentally transient since it involves sudden switching actions. Consider the so-called dc offset current. We model the system by determining its positive sequence Thévenin equivalent circuit, looking back into the positive sequence network at the fault, as shown in Fig. 58.38. The transient fault current is

$$i(t) = I_{ac}\sqrt{2}\,\cos(\omega t - \beta) + I_{dc}e^{-t/\tau}$$

This is a first-order approximation and strictly applies only to the three-phase or phase-to-phase fault. Ground faults would involve the zero sequence network also.

$$I_{ac} = \frac{E}{\sqrt{R^2 + X^2}} = \text{rms ac current}$$

$$I_{dc}(t) = I_{dc}e^{-t/\tau} = \text{dc offset current}$$

The maximum initial dc offset possible would be

$$\text{Max } I_{dc} = I_{max} = \sqrt{2}I_{ac}$$

The dc offset will exponentially decay with time constant τ, where

$$\tau = \frac{L}{R} = \frac{X}{\omega R}$$

The maximum dc offset current would be $I_{dc}(t)$

$$I_{dc}(t) = I_{dc}e^{-t/\tau} = \sqrt{2}I_{ac}e^{-t/\tau}$$

The *transient rms* current $I(t)$, accounting for both the ac and dc terms, would be

$$I(t) = \sqrt{I_{ac}^2 + I_{dc}^2(t)} = I_{ac}\sqrt{1 + 2e^{-2t/\tau}}$$

Define a multiplying factor k_i such that I_{ac} is to be multiplied by k_i to estimate the interrupting capacity of a breaker which operates in time T_{op}. Therefore,

$$k_i = \frac{I(T_{op})}{I_{ac}} = \sqrt{1 + 2e^{-2T_{op}/\tau}}$$

Observe that the maximum possible value for k_i is $\sqrt{3}$.

Example

In the circuit of Fig. 58.38, $E = 2400$ V, $X = 2\ \Omega$, $R = 0.1\ \Omega$, and $f = 60$ Hz. Compute k_i and determine the interrupting capacity for the circuit breaker if it is designed to operate in two cycles. The fault is applied at $t = 0$.

Solution

$$I_{ac} \cong \frac{2400}{2} = 1200\ \text{A}$$

$$T_{op} = \frac{2}{60} = 0.0333\ \text{s}$$

$$\tau = \frac{X}{\omega R} = \frac{2}{37.7} = 0.053$$

$$k_i = \sqrt{1 + 2e^{-2T_{op}/\tau}} = \sqrt{1 + 2e^{-0.0067/0.053}} = 1.252$$

Therefore

$$I = k_i I_{ac} = 1.252(1200) = 1503\ \text{A}$$

The Thévenin equivalent at the fault point is determined by normal sinusoidal steady-state methods, resulting in a first-order circuit as shown in Fig. 58.38. While this provides satisfactory results for the steady-state component I_{ac}, the X/R value so obtained can be in serious error when compared with the rate of decay of $I(t)$ as measured by oscillographs on an actual faulted system. The major reasons for the discrepancy are, first of all, that the system, for transient analysis purposes, is actually high-order, and second, the generators do not hold constant impedance as the transient decays.

Summary

Computation of fault currents in power systems is best done by computer. The major steps are summarized below:

- Collect, read in, and store machine, transformer, and line data in per-unit on common bases.
- Formulate the sequence impedance matrices.
- Define the faulted bus and Z_f. Specify type of fault to be analyzed.
- Compute the sequence voltages.
- Compute the sequence currents.
- Correct for wye-delta connections.
- Transform to phase currents and voltages.

Computer formulation of the sequence impedance matrices is required. Refer to Further Information for more detail. The procedure is followed "in triplicate," constructing the zero, positive, and negative sequence impedance matrices. Zero sequence networks for lines in close proximity to each other (on a common right-of-way) will be mutually coupled. If we are willing to use the same values for positive and negative sequence machine impedances,

$$[Z_1] = [Z_2]$$

Therefore, it is unnecessary to store these values in separate arrays, simplifying the program and reducing the computer storage requirements significantly. The error introduced by this approximation is usually not important. The methods previously discussed neglect the prefault, or load, component of current; that is, the usual assumption is that currents throughout the system were zero prior to the fault. This is almost never strictly true; however, the error produced is small since the fault currents are generally much larger than the load currents. Also, the load currents and fault currents are out of phase with each other, making their sum more nearly equal to the larger components than would have been the case if the currents were in phase. In addition, selection of precise values for prefault currents is somewhat speculative, since there is no way of predicting what the loaded state of the system is when a fault occurs. When it is important to consider load currents, a power flow study is made to calculate currents throughout the system, and these values are superimposed on (added to) results from the fault study.

A term which has wide industrial use and acceptance is the *fault level* or **fault MVA** at a bus. It relates to the amount of current that can be expected to flow out of a bus into a three-phase fault. As such, it is an alternate way of providing positive sequence impedance information. We define

$$\text{Fault level in MVA at bus } i = V_{i_{\text{pu nominal}}} \, I_{i_{\text{pu fault}}} \, S_{3\phi\,\text{base}}$$

$$= (1) \frac{1}{Z_{ii}^1} S_{3\phi\,\text{base}} = \frac{S_{3\phi\,\text{base}}}{Z_{ii}^1}$$

Fault study results may be further refined by approximating the effect of dc offset.

The basic reason for making fault studies is to provide data that can be used to size and set protective devices. The role of such protective devices is to detect and remove faults to prevent or minimize damage to the power system.

Defining Terms

DC offset: The natural response component of the transient fault current, usually approximated with a first-order exponential expression.

Fault: An unintentional and undesirable conducting path in an electrical power system.

Fault MVA: At a specific location in a system, the initial symmetrical fault current multiplied by the prefault nominal line-to-neutral voltage ($\times 3$ for a three-phase system).

Sequence (012) quantities: Symmetrical components computed from phase (*abc*) quantities. Can be voltages, currents, and/or impedances.

References

P. M. Anderson, *Analysis of Faulted Power Systems,* Ames: Iowa State Press, 1973.

M. E. El-Hawary, *Electric Power Systems: Design and Analysis,* Reston, Va.: Reston Publishing, 1983.

O. I. Elgerd, *Electric Energy Systems Theory: An Introduction,* 2nd ed., New York: McGraw-Hill, 1982.

General Electric, *Short-Circuit Current Calculations for Industrial and Commercial Power Systems,* Publication GET-3550.

C. A. Gross, *Power System Analysis,* 2nd ed., New York: Wiley, 1986.

I. Lazar, *Electrical Systems Analysis and Design for Industrial Plants,* New York: McGraw-Hill, 1980.

C. R. Mason, *The Art and Science of Protective Relaying,* New York: Wiley, 1956.

J. R. Neuenswander, *Modern Power Systems,* Scranton, Pa.: International Textbook, 1971.

G. Stagg and A. H. El-Abiad, *Computer Methods in Power System Analysis,* New York: McGraw-Hill, 1968.

Westinghouse Electric Corporation, *Applied Protective Relaying,* Relay-Instrument Division, Newark, N.J., 1976.

Further Information

For a comprehensive coverage of general fault analysis, see Paul M. Anderson, *Analysis of Faulted Power Systems,* Ames, Iowa: Iowa State Press, 1973. Also see Chapters 9 and 10 of *Power System Analysis* by C. A. Gross, New York: Wiley, 1986.

58.6 Protection

Arun G. Phadke

Fundamental Principles of Protection

Protective equipment—**relays**—is designed to respond to system abnormalities (faults) such as short circuits. When faults occur, the relays must signal the appropriate circuit breakers to trip and isolate the faulted equipment. The protection systems not only protect the faulty equipment from more serious damage, they also protect the power system from the consequences of having faults remain on the system for too long. In modern high-voltage systems, the potential for damage to the power system—rather than to the individual equipment—is often far more serious, and power system security considerations dictate the design of the protective system. The protective system con-

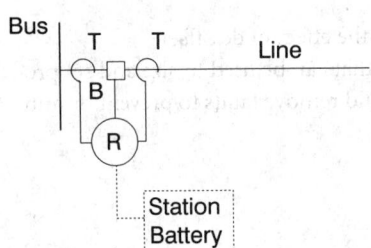

FIGURE 58.39 Elements of a protection system.

sists of four major subsystems as shown in Fig. 58.39. The **transducers** (T) are current and voltage transformers, which transform high voltages and currents to a more manageable level. In the United States, the most common standard for current transformers is a secondary current of 5 A (or less) for steady-state conditions. In Europe, and in some other foreign countries, a 1-A standard is also common. The voltage transformer standard is 69.3 V line-to-neutral or 120 V line-to-line on the transformer secondary side. Standardization of the secondary current and voltage ratings of the transducers has permitted independent development of the transducers and relays. The power handling capability of the transducers is expressed in terms of the volt-ampere burden, which they can supply without significant waveform distortion. In general, the transient response of the transducers is much more critical in relaying applications.

The second element of the protection system is the relay (R). This is the device that, using the current, voltage, and other inputs, can determine if a fault exists on the system, for which action on the part of the relay is needed. We will discuss relays in greater detail in the following. The third element of the protection chain is the circuit breaker (B), which does the actual job of interrupting the flow of current to the fault. Modern high-voltage circuit breakers are capable of interrupting currents of up to 100,000 A, against system voltages of up to 800,000 V, in about 15 to 30 ms. Lower-voltage circuit breakers are generally slower in operating speed. The last element of the protection chain is the station battery, which powers the relays and circuit breakers. The battery voltage has also been standardized at 125 V, although some other voltage levels may prevail in generating stations and in older substations.

The relays and circuit breakers must remove the faulted equipment from the system as quickly as possible. Also, if there are many alternative ways of deenergizing the faulty equipment, the protection system must choose a strategy that will remove from service the minimum amount of equipment. These ideas are embodied in the concepts of zones of protection, relay speed, and reliability of protection.

Zones of Protection

To make sure that a protection system removes the minimum amount of equipment from the power system during its operation, the power system is divided into zones of protection. Each zone has its associated protection system. A fault inside the zone causes the associated protection system to operate. A fault in any other zone must not cause an operation. A zone of protection usually covers one piece of equipment, such as a transmission line. The zone boundary is defined by the location of transducers (usually current transformers) and also by circuit breakers that will operate to isolate the zone. A set of zones of protection is shown in Fig. 58.40. Note that all zones are shown to overlap with their neighbors. This is to ensure that no point on the system is left unprotected. Occasionally, a circuit breaker may not exist at a zone boundary. In such cases, the tripping must be done at some other remote circuit breakers. For example, consider protection zone A in Fig. 58.40. A fault in that zone must be

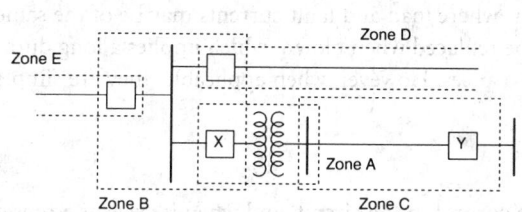

FIGURE 58.40 Zones of protection for a power system. Zones overlap; most zones are bounded by breakers.

isolated by tripping circuit breakers X and Y. While the breaker X is near the transformer and can be tripped locally, Y is remote from the station, and some form of communication channel must be used to transfer the trip command to Y. Although most zones of protection have a precise extent, there are some zones that have a loosely defined reach. These are known as *open* zones and are most often encountered in transmission line protection.

Speed of Protection

The faster the operation of a protection function, the quicker is the prospect of removing a fault from the system. Thus, all protection systems are made as fast as possible. However, there are considerations that dictate against making the protection faster than a minimum limit. Also, occasionally, it may be necessary to slow down a protection system in order to satisfy some specific system need. In general, the fastest protection available operates in about 5 to 10 ms after the inception of a fault [Thorp et al., 1979]. If the protection is made faster than this, it is likely to become "trigger happy" and operate falsely when it should not. When a protection system is intended as a backup system for some other protection, it is necessary to deliberately slow it down so that the primary protection may operate in its own time before the backup system will operate. This calls for a deliberate slowing of the backup protection. Depending upon the type of backup system being considered, the protection may sometimes be slowed down to operate in up to several seconds.

Reliability of Protection

In the field of relaying, **reliability** implies certain very specific concepts [Mason, 1956]. A reliable protection system has two attributes: *dependability* and *security*. A dependable relay is one that always operates for conditions for which it is designed to operate. A secure relay is one that will not operate for conditions for which it is not intended to operate. In modern power systems, the failure to operate when a fault occurs—lack of dependability—has very serious consequences for the power system. Therefore, most protective systems are made secure by duplicating relaying equipment, duplicating relaying functions, and providing several levels of backup protection. Thus modern systems tend to be very dependable, i.e., every fault is cleared, perhaps by more than one relay. As a consequence, security is somewhat degraded: modern protection systems will, occasionally, act and trip equipment falsely. Such occurrences are rare, but not uncommon. As power systems become leaner, i.e., they have insufficient margins of reserve generation and transmission, lack of security can be quite damaging. This has led to recent reevaluation of the proper balance between security and dependability of the protection systems.

Overcurrent Protection

The simplest fault detector is a sensor that measures the increase in current caused by the fault. The fuse is the simplest overcurrent protection; in fact, it is the complete protection chain—sensor, relay, and circuit breaker—in one package. Fuses are used in lower-voltage (distribution) circuits. They are difficult to set in high-voltage circuits, where load and fault currents may be of the same order of magnitude. Furthermore, they must be replaced when blown, which implies a long duration outage. They may also lead to system unbalances. However, when applicable, they are simple and inexpensive.

Inverse-Time Characteristic

Overcurrent relays sense the magnitude of the current in the circuit, and when it exceeds a preset value (known as the *pickup setting* of the relay), the relay closes its output contact, energizing the trip coil of the appropriate circuit breakers. The pickup setting must be set above the largest load current that the circuit may carry and must be smaller than the smallest fault current for which the relay must operate. A margin factor of 2 to 3 between the maximum load on the one hand and the minimum fault current on the other and the pickup setting of the relay is considered to be desirable. The overcurrent relays usually have an *inverse-time* characteristic as shown in Fig. 58.41. When the current exceeds the pickup setting, the relay operating time decreases in inverse proportion to the current magnitude. Besides this built-in feature in the relay mechanism, the relay also has a *time-dial* setting, which shifts the inverse-time curve vertically, allowing for more flexibility in setting the relays. The time dial has 11 discrete settings, usually labeled 1/2, 1, 2, . . . , 10, the lowest setting providing the fastest operation. The inverse-time characteristic offers an ideal relay for providing primary and backup protection in one package.

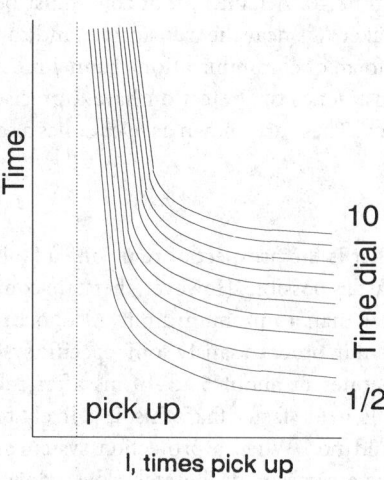

FIGURE 58.41 Inverse-time relay characteristic.

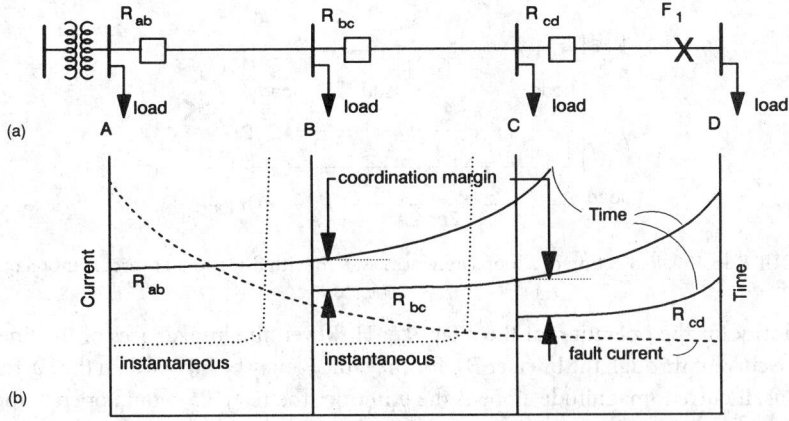

FIGURE 58.42 Coordination of inverse-time overcurrent and instantaneous relays for a radial system.

Coordination Principles

Consider the radial transmission system shown in Fig. 58.42. The transformer supplies power to the feeder, which has four loads at buses A, B, C, and D. For a fault at F_1, the relay R_{cd} must operate to open the circuit breaker B_{cd}. The relay R_{bc} is responsible for a zone of protection, which includes the entire zone of R_{cd}. This constitutes a remote backup for the protection at bus C. The backup relay (R_{bc}) must be slower than the primary relay (R_{cd}), its associated circuit breaker, with a safety margin. This delay in operating of the backup relay is known as the *coordination delay* and is usually about 0.3 s. In a similar fashion, R_{ab} backs up R_{bc}. The magnitude of the fault current varies as shown in Fig. 58.42(b), as the location of the fault is moved along the length of the feeder. We may plot the inverse time characteristic of the relay with the fault location as the abscissa, recalling that a smaller current magnitude gives rise to a longer operating time for the relay. The coordinating time delay between the primary and backup relays is also shown. It can be seen that, as we move from the far end of the feeder toward the source, the fault clearing time becomes progressively longer. The coordination is achieved by selecting relays with a time dial setting that will provide the proper separation in operating times.

The effect of cumulative coordination-time delays is slowest clearing of faults with the largest fault currents. This is not entirely satisfactory from the system point of view, and wherever possible, the inverse-time relays are supplemented by *instantaneous* overcurrent relays. These relays, as the name implies, have no intentional time delays and operate in less than one cycle. However, they cannot coordinate with the downstream relays and therefore must not operate ("see") for faults into the protection zone of the downstream relay. This criterion is not always possible to meet. However, whenever it can be met, instantaneous relays are used and provide a preferable compromise between fast fault clearing and coordinated backup protection.

Directional Overcurrent Relays

When power systems become meshed, as for most subtransmission and high-voltage transmission networks, inverse time overcurrent relays do not provide adequate protection under all conditions. The problem arises because the fault current can now be supplied from either end of the transmission line, and discrimination between faults inside and outside the zone of protection is not always possible. Consider the loop system shown in Fig. 58.43. Notice that in this system there must be a circuit breaker at each end of the line, as a fault on the line cannot be interrupted by opening one end alone. Zone A is the zone of protection for the line A–D. A fault at F_1 must be detected by the relays R_{ad} and R_{da}. The current through the circuit breaker B_{da} for the fault F_1 must be the deter-

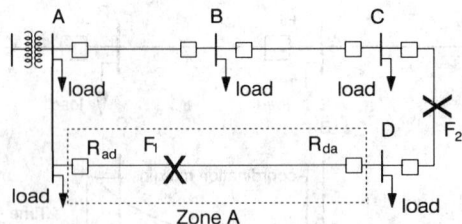

FIGURE 58.43   Protection of a loop (network) system with directional overcurrent relays.

mining quantity for the operation of the relay R_{da}. However, the impedances of the lines may be such that the current through the breaker B_{da} for the fault F_2 may be higher than the current for the fault F_1. Thus, if current magnitude alone is the criterion, the relay R_{da} would operate for fault F_2, as well as for the fault F_1. Of course, operation of R_{da} for F_2 is inappropriate, as it is outside its zone of protection, zone A. This problem is solved by making the overcurrent relays directional. By this is meant that the relays will respond as overcurrent relays only if the fault is in the forward direction from the relays, i.e., in the direction in which their zone of protection extends. The directionality is provided by making the relay sensitive to the phase angle between the fault current and a reference quantity, such as the line voltage at the relay location. Other reference sources are also possible, including currents in the neutral of a transformer bank at the substation.

Distance Protection

As the power networks become more complex, protection with directional overcurrent relays becomes even more difficult, if not impossible. Recall that the pickup setting of the relays must be set above the maximum load which the line is expected to carry. However, a network system has so many probable configurations due to various circuit breaker operations that the maximum load becomes difficult to define. For the same reason, the minimum fault current—the other defining parameter for the pickup setting—also becomes uncertain. Under these circumstances, the setting of the pickup of the overcurrent relays, and their reach, which will satisfy all the constraints, becomes impossible. Distance relays solve this problem.

Distance relays respond to a ratio of the voltage and current at the relay location. The ratio has the dimensions of an impedance, and the impedance between the relay location and fault point is proportional to the distance of the fault. As the zone boundary is related to the distance between the sending end and the receiving end of the transmission line, the distance to the fault forms an ideal relaying parameter. The distance is also a unique parameter in that it is independent of the current magnitude. It is thus free from most of the difficulties associated with the directional overcurrent relays mentioned above.

In a three-phase power system, 10 types of faults are possible: three single phase-to-ground faults, three phase-to-phase faults, three double phase-to-ground faults, and one three-phase fault. It turns out that relays responsive to the ratio of delta voltages and delta currents measure the correct distance to all multiphase faults. The delta quantities are defined as the difference between any two phase quantities; for example, $E_a - E_b$ is the delta voltage between a and b phases. Thus for a multiphase fault between phases x and y,

$$\frac{E_x - E_y}{I_x - I_y} = Z_1$$

where x and y can be a, b, or c and Z_1 is the positive sequence impedance between the relay location and the fault. For ground distance relays, the faulted phase voltage, and a compensated faulted phase current must be used

$$\frac{E_x}{I_x + mI_0} = Z_1$$

where m is a constant depending upon the line impedances and I_0 is the zero sequence current in the transmission line. A full complement of relays consists of three phase distance relays and three ground distance relays. As explained before, the phase relays are energized by the delta quantities, while the ground distance relays are energized by each of the phase voltages and the corresponding compensated phase currents. In many instances, ground distance protection is not preferred, and the time overcurrent relays may be used for ground fault protection.

Step-Distance Protection

The principle of distance measurement for faults is explained above. A relaying system utilizing that principle must take into account several features of the measurement principle and develop a complete protection scheme. Consider the system shown in Fig. 58.44. The distance relay R_{ab} must protect line AB, with its zone of protection as indicated by the dashed line. However, the distance calculation made by the relay is not precise enough for it to be able to distinguish between a fault just inside the zone and a fault just outside the zone, near bus B. This problem is solved by providing a two-zone scheme, such that if a fault is detected to be in zone 1, the relay trips instantaneously, and if the fault is detected to be inside zone 2, the relay trips with a time delay of about 0.3 s. Thus for faults near the zone boundary, the fault is cleared with this time delay, while for near faults, the clearing is instantaneous. This arrangement is referred to as a *step-distance* protection scheme, consisting of an underreaching zone (zone 1), and an overreaching zone (zone 2). The relays of the neighboring line (BC) can also be backed up by a third zone of the relay, which reaches beyond the zone of protection of relay R_{bc}. Zone 3 operation is delayed further to allow the zone 1 or zone 2 of R_{bc} to operate and clear the fault on line BC.

The distance relays may be further subdivided into categories depending upon the shape of their protection characteristics. The most commonly used relays have a directional distance, or a mho characteristic. The two characteristics are shown in Fig. 58.45. The directional impedance relay consists of two functions, a directional detection function and a distance measurement function. The mho characteristic is inherently directional, as the mho circle, by relay design, passes through the origin of the RX plane. Figure 58.45 also shows the multiple zones of the step distance protection.

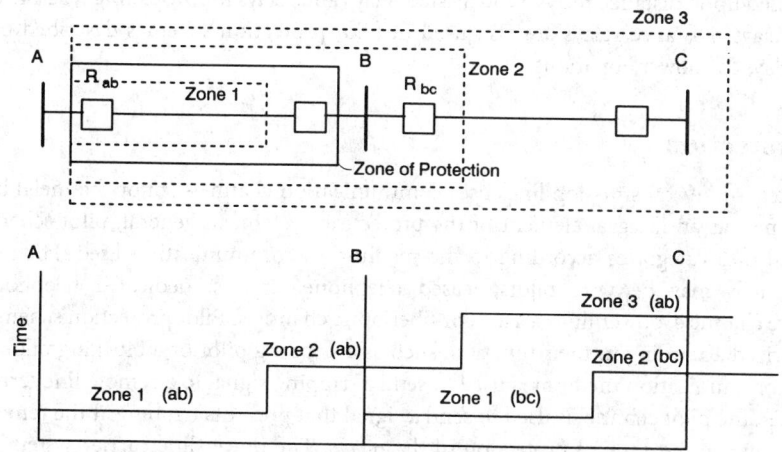

FIGURE 58.44 Zones of protection in a step-distance protection scheme. Zone 3 provides backup for the downstream line relays.

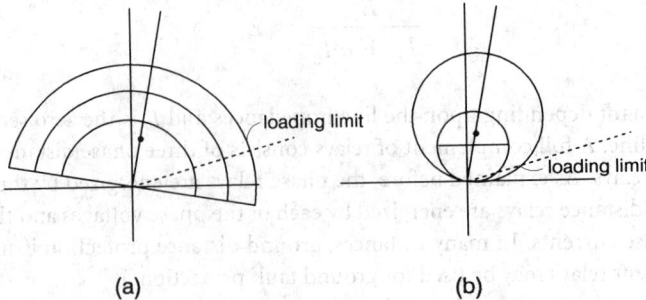

FIGURE 58.45 (a) Directional impedance characteristic. (b) Mho characteristic. Loadability limits as shown.

Loadability of Distance Relays

The load carried by a transmission line translates into an apparent impedance as seen by the relay, given by

$$Z_{app} = \frac{|E|^2}{P - jQ}$$

where $P - jQ$ is the load complex power and E is the voltage at the bus where a distance relay is connected. This impedance maps into the RX plane, as do all other apparent impedances, and hence the question arises whether this apparent load impedance could be mistaken for a fault by the distance relay. Clearly, this depends upon the shape of the distance relay characteristic employed. The loadability of a distance relay refers to the maximum load power (minimum apparent impedance) that the line can carry before a protective zone of a distance relay is penetrated by the apparent impedance. A typical load line is shown in Fig. 58.45. It is clear from this figure that the mho characteristic has a higher loadability than the directional impedance relay. In fact, other relay characteristics can be designed so that the loadability of a relay is increased even further.

Other Uses of Distance Relays

Although the primary use of distance relays is in protecting transmission lines, some other protection tasks can also be served by distance relays. For example, loss-of-field protection of generators is often based upon distance relays. Out-of-step relays and relays for protecting reactors may also be distance relays. Distance relays are also used in pilot protection schemes described next, and as backup relays for power apparatus.

Pilot Protection

Pilot protection of transmission lines uses communication channels (pilot channels) between the line terminals as an integral element of the protection system. In general, pilot schemes may be subdivided into categories according to the medium of communication used. For example, the pilot channels may be wire pilots, leased telephone circuits, dedicated telephone circuits, microwave channels, power line carriers, or fiber optic channels. Pilot protection schemes may also be categorized according to their function, such as a tripping pilot or a blocking pilot. In the former, the communication medium is used to send a tripping signal to a remote line terminal, while in the latter, the pilot channel is used to send a signal that prevents tripping at the remote terminal for faults outside the zone of protection of the relays. The power line carrier system is the most common system used in the United States. It uses a communication channel with a carrier signal frequency ranging between 30 and 300 kHz, the most common bands being around 100 kHz. The

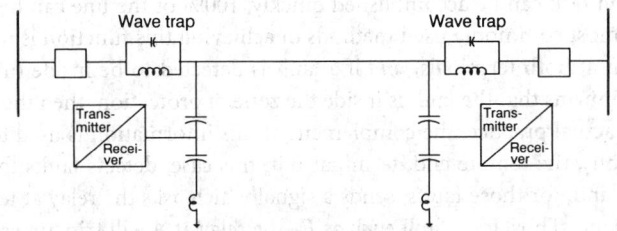

FIGURE 58.46 Carrier system for pilot protection of lines. Transmitter and receiver are connected to relays.

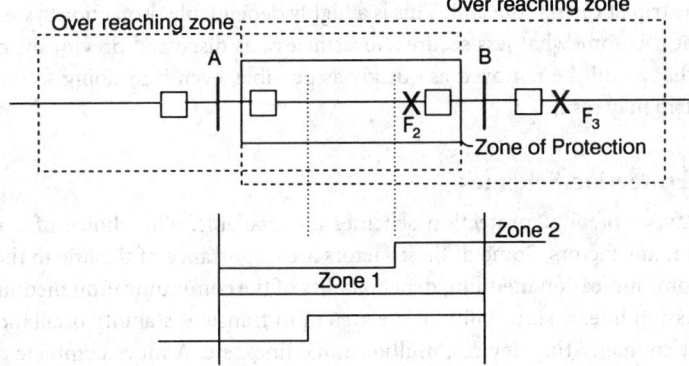

FIGURE 58.47 Pilot protection with overreaching zones of protection. This is most commonly used in a directional comparison blocking scheme.

modulated carrier signal is coupled into one or more phases of the power line through coupling capacitors. In almost all cases, the capacitors of the capacitive-coupled voltage transformers are used for this function (see Fig. 58.46). The carrier signal is received at both the sending and the receiving ends of the transmission line by tuned receivers. The carrier signal is blocked from flowing into the rest of the power system by blocking filters, which are parallel resonant circuits, known as *wave traps*.

Coverage of 100% of Transmission Line

The step-distance scheme utilizes the zone 1 and zone 2 combination to protect 100% of the transmission line. The middle portion of the transmission line, which lies in zone 1 of relays at the two ends of the line, is protected at high speed from both ends. However, for faults in the remaining portion of the line, the near end clears the fault at high speed, i.e., in zone 1 time, while the remote end clears the fault in zone 2 time. In effect, such faults remain on the system for zone 2 time, which may be of the order 0.3 to 0.5 s. This becomes undesirable in modern power systems where the margin of stability may be quite limited. In any case, it is good protection practice to protect the entire line with high-speed clearing of all internal faults from both ends of the transmission line. Pilot protection accomplishes this task.

Directional Comparison Blocking Scheme

Consider the fault at F_2 shown in Fig. 58.47. As discussed above, this fault will be cleared in zone 1 time by the step-distance relay at bus B, while the relay at bus A will clear the fault in zone 2 time. Since the relays at bus B can determine, with a high degree of certainty, that a fault such as F_2 is indeed inside the zone of protection of the relays, one could communicate this knowledge to terminal A, which can then cause the local circuit breaker to trip for the fault F_2. If the entire relaying and communication task can be accomplished quickly, 100% of the line can be protected at high speed. One of the most commonly used methods of achieving this function is to use overreaching zones of protection at both terminals, *and* if a fault is detected to be inside this zone, and if the remote terminal confirms that the fault is inside the zone of protection, then the local relay may be allowed to trip. In actual practice, the complement of this information is used to block the trip at the remote end. Thus, the remote end, terminal B in this case, detects faults that are outside the zone of protection and, for those faults, sends a signal which asks the relay at terminal A to block the tripping command. Thus, for a fault such as F_3, the relay at A will trip, unless the communication is received from terminal B that this particular fault is outside the zone of protection—as indeed fault F_3 happens to be. This mode, known as a blocking carrier, is preferred, since a loss of the carrier signal created by an internal fault, or due to causes that are unrelated to the fault, will not prevent the trip at the remote end. This is a highly dependable protection system, and precisely because of that it is somewhat less secure. Nevertheless, as discussed previously, most power systems require that a fault be removed as quickly as possible, even if in doing so for a few faults an unwarranted trip may result.

Other Pilot Protection Schemes

Several other types of pilot protection schemes are available. The choice of a specific scheme depends upon many factors. Some of these factors are importance of the line to the power system, the available communication medium, dependability of the communication medium, loading level of the transmission line, susceptibility of the system to transient stability oscillations, presence of series or shunt compensating devices, multiterminal lines, etc. A more complete discussion of all these issues will be found in the references [Westinghouse, 1982; Blackburn, 1987; Horowitz and Phadke, 1992].

Computer Relaying

Relaying with computers began to be discussed in technical literature in the mid-1960s. Initially, this was an academic exercise, as neither the computer speeds nor the computer costs could justify the use of computers for relaying. However, with the advent of high-performance microprocessors, computer relaying has become a very practical and attractive field of research and development. All major manufacturers of electric power equipment have computer relays to meet all the needs of power system engineers. Computer relaying is also being taught in several universities and has provided a very fertile field of research for graduate students. Computer relaying has also uncovered new ways of measuring power system parameters and may influence future development of power system monitoring and control functions.

Incentives for Computer Relaying

The acceptance of **computer relays** has been due to economic factors which have made microcomputers relatively inexpensive and computationally powerful. In addition to this economic advantage, the computer relays are also far more versatile. Through their self-diagnostic capability, they provide an assurance of availability. Thus, even if they should suffer the same (or even greater) number of failures in the field as traditional relays, their failures could be communicated to control centers and a maintenance crew called to repair the failures immediately. This type of diagnostic capability was lacking in traditional protection systems and often led to failures of relays, which went undetected for extended periods. Such hidden failures have been identified as one of the main sources of power system blackouts.

The computing power available with computer relays has also given rise to newer and better protection functions in several instances. Improved protection of transformers, multiterminal lines,

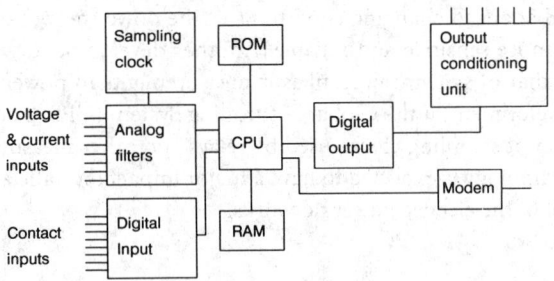

FIGURE 58.48 Block diagram of a computer relay architecture.

fault location, and reclosing are a few of the protection functions where computer relaying is likely to have a significant impact. Very significant developments in the computer relaying field are likely to occur in the coming years.

Architecture for a Computer Relay

There are many ways of implementing computer-based relays. Figure 58.48 is a fairly typical block diagram of a computer relay architecture. The input signals consisting of voltage and currents and contact status are filtered to remove undesired frequency components and potentially damaging surges. These signals are sampled by the CPU under the control of a sampling clock. Typical sampling frequency used in a modern digital relay varies between 4 and 32 times the nominal power system frequency. The sampled data is processed by the CPU with a digital filtering algorithm, which estimates the appropriate relaying quantity. A typical relaying quantity may be the rms value of a current, the voltage or current phasor, or the apparent impedance. The estimated parameters are then compared with prestored relay characteristics, and the appropriate control action is initiated. The decision of the relay is communicated to the substation equipment, such as the circuit breaker, through the output ports. These outputs must also be filtered to block any surges from entering the relay through the output lines. In most cases, the relay can also communicate with the outside world through a modem. The data created by a fault is usually saved by the relaying computer and can be used for fault analysis or for sequence-of-event analysis following a power system disturbance. The user may interface with the relay through a keyboard, a control panel, or a communication port. In any case, provision must be made to enter relay settings in the relay and to save these settings in case the station power supply fails. Although the block diagram in Fig. 58.48 shows different individual subsystems, the actual hardware composition of the subsystems is dependent on the computer manufacturer. Thus, we may find several microprocessors in a given implementation, each controlling one or more subsystems. Also, the hardware technology is in a state of flux, and in a few years, we may see an entirely different realization of the computer relays.

Experience and Future Trends

Field experience with the computer relays has been excellent so far. The manufacturers of traditional relays have adopted this technology in a big way. As more experience is gained with the special requirements of computer relays, it is likely that other—nontraditional—relay manufacturers will enter the field.

It seems clear that in computer relaying, power system engineers have obtained a tool with exciting new possibilities. Computers, with the communication networks now being developed, can lead to improved monitoring, protection, and control of power systems. An entirely new field, adaptive relaying, has been introduced recently [Phadke and Horowitz, 1990]. The idea is that protection systems should adapt to changing conditions of the power networks. In doing so, protection systems become more sensitive and reliable. Another development, which can be traced to computer relaying, is that of synchronized phasor measurements in power systems [Phadke and Thorp, 1991]. The development of the Global Positioning System (GPS) satellites has made possible the synchronization of sampling clocks used by relays and other measuring devices across the power system. This technology is expected to have a major impact on static and dynamic state estimation and on control of the electric power networks.

Defining Terms

Computer relays: Relays that use digital computers as their logic elements.

Distance protection: Relaying principle based upon estimating fault location (distance) and providing a response based upon the distance to the fault.

Electromechanical relays: Relays that use electromechanical logic elements.

Pilot: A communication medium used by relays to help reach a reliable diagnosis of certain faults.

Relays: Devices that detect faults on power equipment and systems and take appropriate control actions to deenergize the faulty equipment.

Reliability: For relays, reliability implies *dependability*, i.e., certainty of operating when it is supposed to, and *security*, certainty of not operating when it is not supposed to.

Solid state relays: Relays that use solid state analog components in their logic elements.

Transducers: Current and voltage transformers that reduce high-magnitude signals to standardized low-magnitude signals which relays can use.

References

J.L. Blackburn, "Protective relaying," Marcel Dekker, 1987.

S.H. Horowitz and A.G. Phadke, *Power System Relaying*, Research Studies Press, New York: Wiley & Sons, 1992.

C.R. Mason, *The Art and Science of Protective Relaying*, New York: Wiley & Sons, 1956.

A.G. Phadke and S.H. Horowitz, "Adaptive relaying," *IEEE Computer Applications in Power*, vol. 3, no. 3, pp. 47–51, July 1990.

A.G. Phadke and J.S. Thorp, "Improved control and protection of power systems through synchronized phasor measurements," in *Analysis and Control System Techniques for Electric Power Systems*, part 3, C.T. Leondes, Ed., San Diego: Academic Press, pp. 335–376, 1991.

J.S. Thorp, A.G. Phadke, S.H. Horowitz, and J.E. Beehler, "Limits to impedance relaying," *IEEE Trans. PAS*, vol. 98, no. 1, pp. 246–260, January/February 1979.

Westinghouse Electric Corporation, "Applied Protective Relaying," 1982.

Further Information

In addition to the references provided, papers sponsored by the Power System Relaying Committee of the IEEE and published in the *IEEE Transactions on Power Delivery* contain a wealth of information about protective relaying practices and systems. Publications of CIGRÉ also contain papers on relaying, through their Study Committee 34 on protection. Relays and relaying systems usually follow standards, issued by IEEE in this country, and by such international bodies as the IEC in Europe. The field of computer relaying has been covered in *Computer Relaying for Power Systems*, by A.G. Phadke and J.S. Thorp (New York: Wiley, 1988).

58.7 Transient Operation

R. B. Gungor

Stable operations of power transmission systems have been a great concern of utilities since the beginning of early power distribution networks. The transient operation and the stability under transient operation are studied for existing systems, as well as the systems designed for future operations.

Power systems must be stable while operating normally at steady state for slow system changes under switching operations, as well as under emergency conditions, such as lightning strikes, loss of some generation, or loss of some transmission lines due to **faults**.

The tendency of a power system (or a part of it) to develop torques to maintain its stable operation is known as **stability**. The determination of the stability of a system then is based on the static and dynamic characteristics of its synchronous generators. Although large induction machines may contribute energy to the system during the *subtransient* period that lasts one or two cycles at the start of the **disturbance**, in general, induction machine loads are treated as static loads for **transient stability** calculations. This is one of the simplification considerations, among others.

The per-phase model of an ideal synchronous generator with nonlinearities and the stator resistance neglected is shown in Fig. 58.49, where E_g is the generated (excitation) voltage and X_s is the steady-state direct axis *synchronous reactance*. In the calculation of transient and subtransient currents, X_s is replaced by *transient reactance X_s'* and *subtransient reactance X_s''*, respectively.

Per-phase electrical power output of the generator for this model is given by Eq. (58.71).

$$P_e = \frac{E_g V_t}{X_s} \sin \delta = P_{max} \sin \delta \qquad (58.71)$$

where δ is the **power angle**, the angle between the generated voltage and the terminal voltage.

The simple power-angle relation of Eq. (58.71) can be used for real power flow between any two voltages separated by a reactance. For the synchronous machine, the total real power is three times the value calculated by Eq. (58.71), when voltages in volts and the reactance in ohms are used. On the other hand, Eq. (58.71) gives *per-unit* power when per-unit voltages and reactance are used.

Figure 58.50 shows a sketch of the power-angle relation of Eq. (58.71). Here the power P_1 is carried by the machine under δ_1, and P_2 under δ_2. For gradual changes in the output power up to P_{max} for $\delta = 90°$, the machine will be stable. So we can define the **steady-state stability** limit as

$$\delta \leq 90° \qquad \frac{\partial P}{\partial \delta} > 0 \qquad (58.72)$$

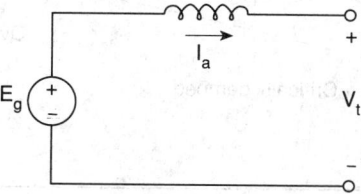

FIGURE 58.49 Per-phase model of an ideal synchronous generator. (*Source:* R. B. Gungor, *Power Systems,* San Diego: Harcourt Brace Jovanovich, 1988, chap. 11. With permission.)

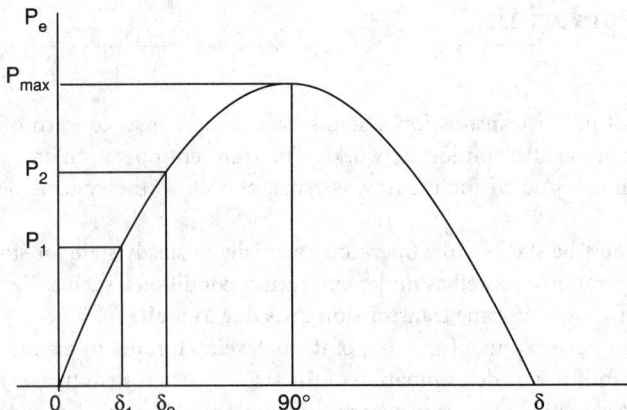

FIGURE 58.50 Power-angle characteristics of ideal synchronous generator. (*Source:* R. B. Gungor, *Power Systems,* San Diego: Harcourt Brace Jovanovich, 1988, chap. 11. With permission.)

A sudden change in the load of the generator, e.g., from P_1 to P_2, will cause the rotor to slow down so that the power angle δ is increased to supply the additional power to the load. However, the deceleration of the rotor cannot stop instantaneously. Hence, although at δ_2 the developed power is sufficient to supply the load, the rotor will overshoot δ_2 until a large enough opposite torque is built up to stop deceleration. Now the excess energy will start accelerating the rotor to decrease δ. Depending on the inertia and damping, these oscillations will die out or the machine will become unstable and lose its synchronism to drop out of the system. This is the basic **transient operation** of a synchronous generator. Note that during this operation it may be possible for δ to become larger than 90° and the machine still stay stable. Thus $\delta = 90°$ is not the transient stability limit.

Figure 58.51 shows typical power-angle versus time relations.

Stable Operation of Power Systems

Figure 58.52 shows an *N*-bus power system with *G* generators.

To study the stability of multimachine transmission systems, the resistances of the transmission lines and transformers are neglected and the reactive networks are reduced down to the generator

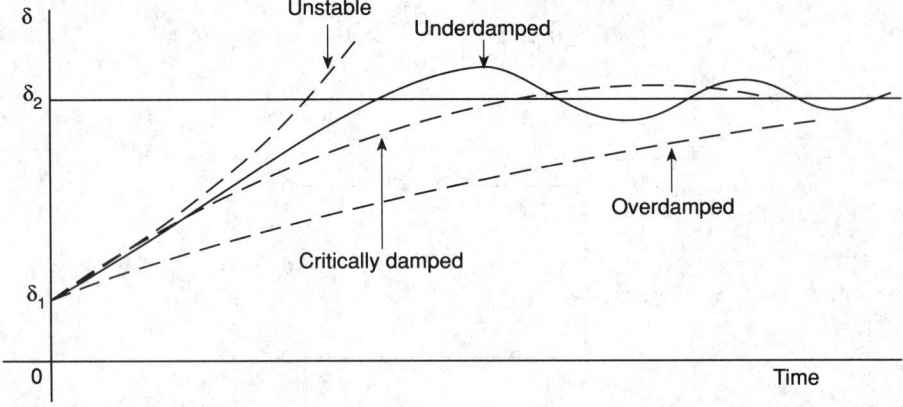

Power-angle swing curves

FIGURE 58.51 Typical power angle–time relations. (*Source:* R. B. Gungor, *Power Systems,* San Diego: Harcourt Brace Jovanovich, 1988, chap. 11. With permission.)

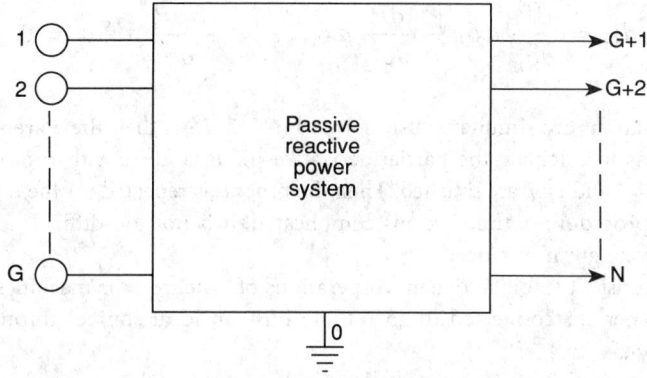

FIGURE 58.52 A multimachine reactive power system. (*Source:* R. B. Gungor, *Power Systems,* San Diego: Harcourt Brace Jovanovich, 1988, chap. 11. With permission.)

internal voltages by dropping the loads and eliminating the load buses. One such reduced network is sketched in Fig. 58.53.

The power flow through the reactances of a reduced network are

$$P_{ij} = \frac{E_i E_j}{X_{ij}} \sin \delta_{ij} \qquad i, j = 1, 2, \ldots, G \tag{58.73}$$

The generator powers are

$$P_i = \sum_{k=1}^{G} P_{ik} \tag{58.74}$$

The system will stay stable for

$$\frac{\partial P_i}{\partial \delta_{ij}} > 0 \qquad i = 1, 2, \ldots, G \tag{58.75}$$

Equation (58.75) is observed for two machines at a time by considering all but two (say k and n) of the powers in Eq. (58.74) as constants. Since the variations of all powers but k and n are zero, we have

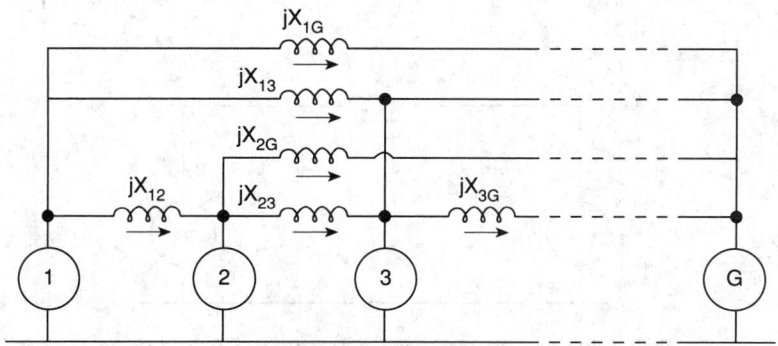

FIGURE 58.53 Multiport reduced reactive network. (*Source:* R. B. Gungor, *Power Systems,* San Diego: Harcourt Brace Jovanovich, 1988, chap. 11. With permission.)

$$dP_i = \frac{\partial P_i}{\partial \delta_{i1}} d\delta_{i1} + \frac{\partial P_i}{\partial \delta_{i2}} d\delta_{i2} + \cdots + \frac{\partial P_i}{\partial \delta_{iG}} d\delta_{iG} = 0 \qquad (58.76)$$

These G-2 equations are simultaneously solved for G-2 $d\delta_{ij}$s, then these are substituted in dP_k and dP_n equations to calculate the partial derivatives of P_k and P_n with respect to δ_{kn} to see if Eqs. (58.75) for $i=k$ and $i=n$ are satisfied. Then the process is repeated for the remaining pairs.

Although the procedure outlined seems complicated, it is not too difficult to produce a computer algorithm for a given system.

To study the transient stability, dynamic operations of synchronous machines must be considered. An ideal generator connected to an infinite bus (an ideal source) through a reactance is sketched in Fig. 58.54.

The so-called *swing equation* relating the accelerating (or decelerating) power (difference between shaft power and electrical power as a function of δ) to the second derivative of the power angle is given in Eq. (58.77).

$$P_a = P_s - P_e$$
$$M \frac{d^2\delta}{dt^2} = P_s - \frac{E_g E_i}{X} \sin \delta \qquad (58.77)$$

where $M = HS/180f$ (MJ/electrical degree); H is the inertia constant (MJ/MVA); S is the machine rating (MVA); f is the frequency (Hz); P_s is the shaft power (MW).

For a system of G machines, a set of G swing equations as given in Eq. (58.78) must be solved simultaneously.

$$M_i \frac{d^2\delta_i}{dt^2} = P_{s_i} - P_{max_i} \sin \delta_i \qquad i = 1, 2, \ldots, G \qquad (58.78)$$

The swing equation of the single-machine system of Fig. 58.54 can be solved either graphically or analytically. For graphical integration, which is called *equal-area criterion*, we represent the machine by its subtransient reactance, assuming that electrical power can be calculated by Eq. (58.71), and during the transients the shaft power P_s remains constant. Then, using the power-angle curve(s), we sketch the locus of operating point on the curve(s) and equate the areas for stability. Figure 58.55 shows an example for which the shaft power of the machine is suddenly increased from the initial value of P_o to P_s.

The excess energy (area A_1) will start to accelerate the rotor to increase δ from δ_o to δ_m for which the area (A_2) above P_s equals the area below. These areas are

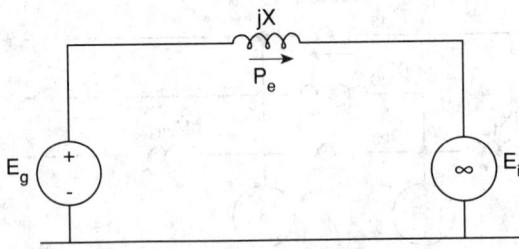

FIGURE 58.54 An ideal generator connected to an infinite bus. (*Source:* R. B. Gungor, *Power Systems,* San Diego: Harcourt Brace Jovanovich, 1988, chap. 11. With permission.)

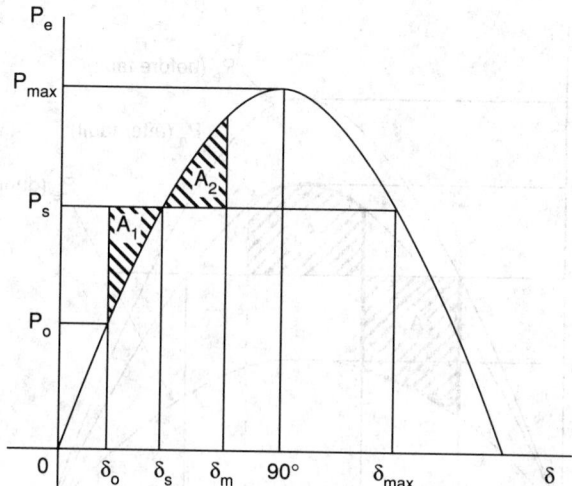

FIGURE 58.55 A sudden loading of a synchronous generator. (*Source:* R. B. Gungor, *Power Systems,* San Diego: Harcourt Brace Jovanovich, 1988, chap. 11. With permission.)

$$A_1 = P_s(\delta_s - \delta_o) - \int_{\delta_o}^{\delta_s} P_{max} \sin \delta \, d\delta$$

$$A_2 = \int_{\delta_s}^{\delta_m} P_{max} \sin \delta \, d\delta - P_s(\delta_m - \delta_s)$$

(58.79)

Substituting, the values of P_o, P_s, δ_o, and δ_s, δ_m can be calculated.

Figure 58.56 illustrates another example, where a three-phase fault reduces the power transfer to infinite bus to zero. δ_{cc} is the **critical clearing angle** beyond which the machine will not stay stable.

The third example, shown in Fig. 58.57, indicates that the power transfers before, during, and after the fault are different. Here the system is stable as long as $\delta_m \le \delta_{max}$.

For the analytical solution of the swing equation a *numerical integration* technique is used (Euler's method, modified Euler's method, Runge-Kutta method, etc.). The latter is most commonly used for computer algorithms.

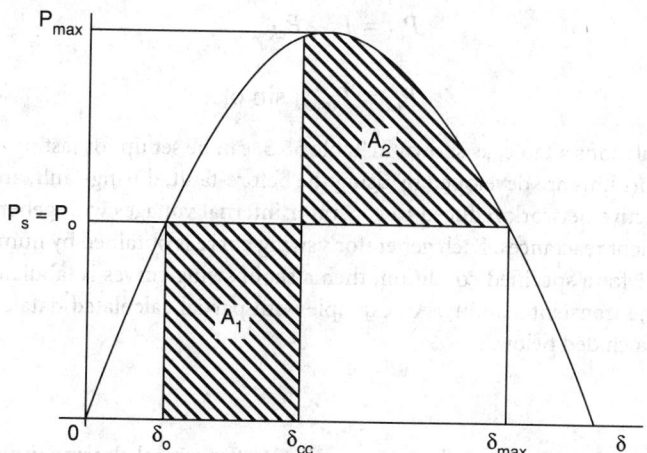

FIGURE 58.56 Critical clearing angle for stability. (*Source:* R. B. Gungor, *Power Systems,* San Diego: Harcourt Brace Jovanovich, 1988, chap. 11. With permission.)

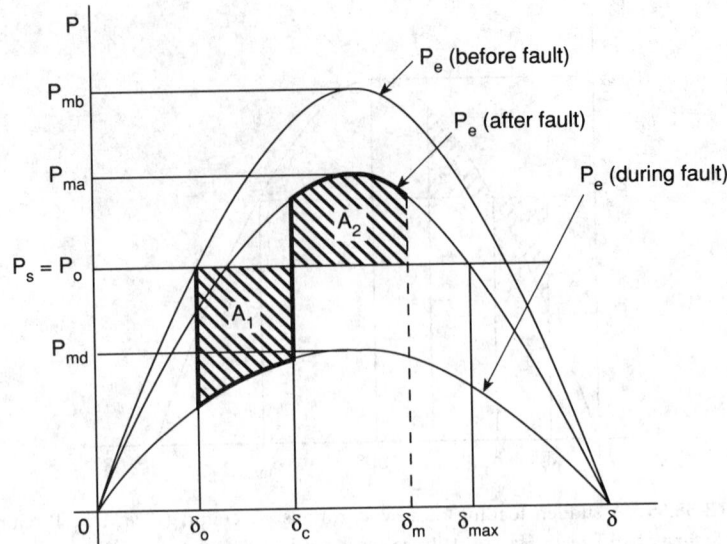

FIGURE 58.57 Power-angle relation for power transfer during fault. (*Source:* R. B. Gungor, *Power Systems,* San Diego: Harcourt Brace Jovanovich, 1988, chap. 11. With permission.)

The solution methods developed are based on various assumptions. As before, machines are represented by subtransient reactances, electrical powers can be calculated by Eq. (58.71), and the shaft power does not change during transients. In addition, the velocity increments are assumed to start at the beginning of time increments, and acceleration increments start at the middle of time increments; finally, an average acceleration can be used where acceleration is discontinuous (e.g., where circuit breakers open or close).

Figure 58.58 shows a sketch of angle, velocity, and acceleration changes related to time as outlined above. Under these assumptions the next value of the angle δ can be obtained from the previous value as

$$\delta_{k+1} = \delta_k + \Delta_{k+1}\delta = \delta_k + \Delta_k\delta + \frac{(\Delta t)^2}{M} P_{ak} \qquad (58.80)$$

where the accelerating power is

$$P_{ak} = P_s - P_{ek}$$

and

$$P_{ek} = P_{maxk} \sin \delta_k$$

For hand calculations a table, as shown in Table 58.5, can be set up for fast processing.

Computer algorithms are developed by using the before-fault, during-fault, and after-fault Z_{BUS} matrix of the reactive network reduced to generator internal voltages with generators represented by their subtransient reactances. Each generator's swing curve is obtained by numerical integration of its power angle for a specified condition, then a set of swing curves is tabulated or graphed for observation of the transient stability. An example with partial calculated data and a line plot for such a study are included below.

Defining Terms

Critical clearing angle: Power angle corresponding to the critical clearing time.

Critical clearing time: The maximum time at which a fault must be cleared for the system to stay transiently stable.

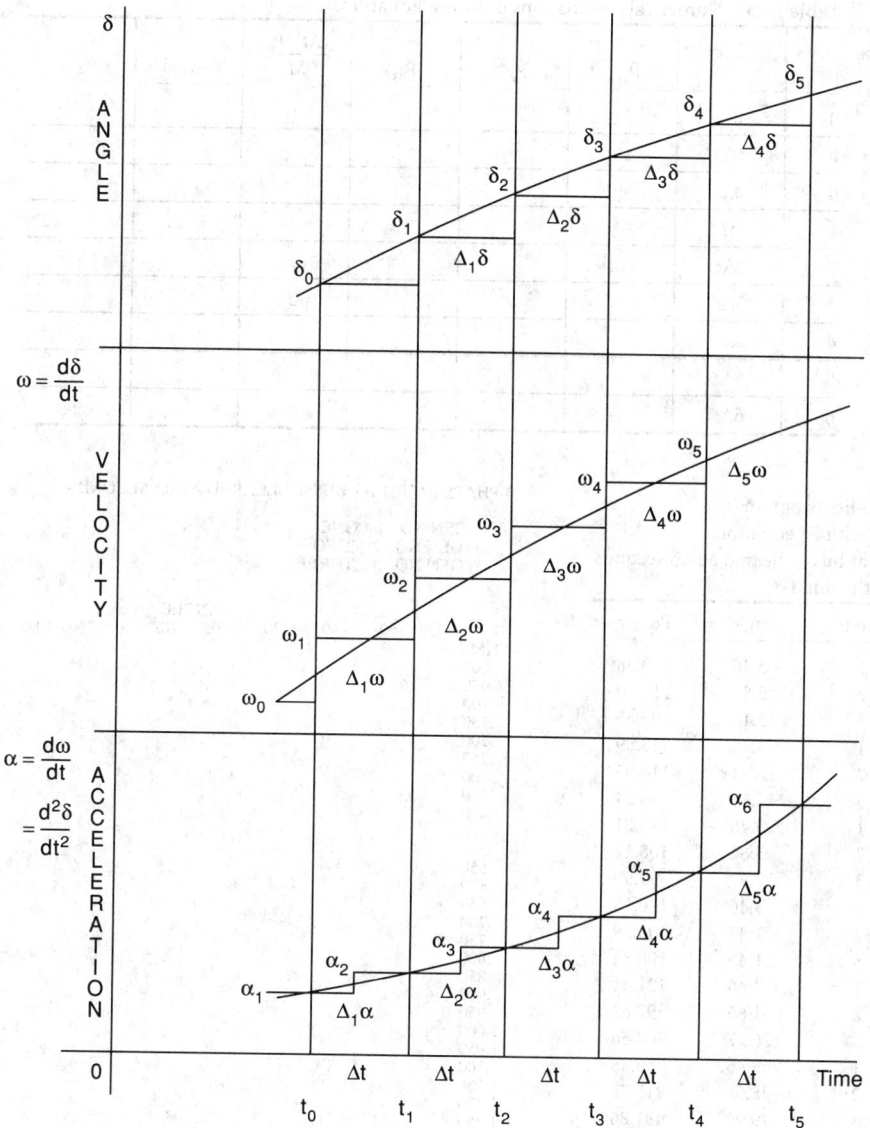

FIGURE 58.58 Incremental angle, velocity, and acceleration changes versus time. (*Source:* R. B. Gungor, *Power Systems,* San Diego: Harcourt Brace Jovanovich, 1988, chap. 11. With permission.)

Disturbance (fault): A sudden change or a sequence of changes in the components or the formation of a power system.

Large disturbance: A disturbance for which the equations for dynamic operation cannot be linearized for analysis.

Power angle: The electrical angle between the generated and terminal voltages of a synchronous generator.

Small disturbance: A disturbance for which the equations for dynamic operation can be linearized for analysis.

Stability: The tendency of a power system (or a part of it) to develop torques to maintain its stable operation for a disturbance.

Table 58.5 Numerical Calculations of Swing Equations

n	t	P_{max}	P_e	P_{ak}	$\dfrac{(\Delta t)^2 P_a}{M}$	$\Delta_{k+1}\delta$	δ_k
0	0_-						
0	0_+						
0	0_{av}						
1	Δt						
2	$2\Delta t$						
3	$3\Delta t$						
4	$4\Delta t$						
5	$5\Delta t$						
6	$6\Delta t$						

Transient stability program
7-Bus system with 3 generators
3-Phase fault at bus 6, cleared at 0.5 seconds
by removing the line 1-6

Time	Gen	Angle	Power
.000	1	3.46	118.38
.000	2	5.80	111.90
.000	3	15.16	95.65
.050	1	3.46	118.59
.050	2	5.31	110.95
.050	3	15.86	96.24
.100	1	3.46	119.21
.100	2	3.84	108.12
.100	3	17.97	97.99
.150	1	3.46	120.15
.150	2	1.47	103.59
.150	3	21.45	100.83
.200	1	3.46	121.31
.200	2	−1.66	97.62
.200	3	26.27	104.66
.500	1	3.46	55.48
.500	2	−26.55	−215.72
.500	3	79.92	481.86
.900	1	3.46	−198.56
.900	2	100.99	458.41
.900	3	49.43	72.78
1.950	1	3.46	125.86
1.950	2	−30.18	−216.29
1.950	3	41.40	425.31
2.000	1	3.46	125.86
2.000	2	−34.60	−216.29
2.000	3	57.78	425.31

3-PHASE FAULT AT BUS 6, CLEARED AT 0.5 SECONDS

. GEN NO. 1   ONE
+ GEN NO. 2   TWO
* GEN NO. 3   THREE

Steady-state stability: A power system is steady-state stable if it reaches another steady-state operating point after a small disturbance.

Transient operation: A power system operating under abnormal conditions because of a disturbance.

Transient stability: A power system is transiently stable if it reaches a steady-state operating point after a large disturbance.

References

J. Arrillaga, C. P. Arnold, and B. J. Harker, *Computer Modeling of Electrical Power Systems*, New York: Wiley, 1983.

A. R. Bergen, *Power System Analysis*, Englewood Cliffs, N.J.: Prentice-Hall, 1986.

H. E. Brown, *Solution of Large Networks by Matrix Methods*, New York: Wiley, 1985.

A. A. Fouad and V. Vittal, *Power System Transient Stability Analysis*, Englewood Cliffs, N.J.: Prentice-Hall, 1992.

J. D. Glover and M. Sarma, *Power System Analysis and Design*, Boston: PWS Publishers, 1987.

C. A. Gross, *Power System Analysis*, 2nd ed., New York: Wiley, 1986.

R. B. Gungor, *Power Systems*, San Diego: Harcourt Brace Jovanovich, 1988.

G. T. Heydt, *Computer Analysis Methods for Power Systems*, New York: Macmillan, 1986.

W. D. Stevenson, *Elements of Power System Analysis*, 4th ed., New York: McGraw-Hill, 1982.

Y. Wallach, *Calculations & Programs for Power System Networks*, Englewood Cliffs, N.J.: Prentice-Hall, 1986.

Further Information

In addition to the references listed above, further and more recent information can be found in IEEE publications, such as *IEEE Transactions on Power Systems*, *IEEE Transactions on Power Delivery*, *IEEE Transactions on Energy Conversion*, and *IEEE Transactions on Automatic Control*.

Power Engineering Review and *Computer Applications in Power* of the IEEE are good sources for paper summaries.

Finally, *IEEE Transactions on Power Apparatus and Systems* dating back to the 1950s can be consulted.

58.8 Planning

J. Duncan Glover

An electric utility transmission system performs three basic functions: delivers outputs from generators to the system, supplies power to the distribution system, and provides for power interchange with other utilities. The electric utility industry has developed planning principles and criteria to ensure that the transmission system reliably performs these basic functions.

The North American Electric Reliability Council (NERC) has provided definitions of the terms **reliability, adequacy,** and **security** (see Defining Terms at the end of this section).

System reliability may be viewed from two perspectives: short-term reliability and long-term reliability. The system operator is primarily concerned with real-time security aspects in the short term, that is, supplying steady, uninterrupted service under existing operating conditions and as they occur over the next few minutes, hours, days, or months. The transmission planning engineer, however, is concerned not only with security aspects in the short term but also adequacy and security aspects in the long term, as many as 25 or more years into the future.

The actual construction of a major transmission facility requires three to five years or more, depending largely on the siting and certification process. As such, the planning process requires up to ten years prior to operation of these facilities to ensure that they are available when required. The long lead times, environmental impacts, and high costs required for new transmission facilities require careful, near-optimal planning. Future changes in system operating conditions, such as changes in spatial load and generation patterns, create uncertainties that challenge the transmission planning engineer to select the best technical solution among several alternatives with due consideration of nontechnical factors. Transmission planning strives to maintain an optimal balance between system reliability, environmental impacts, and cost under future uncertainties.

Before transmission planning is started, long-term load forecasting and generation planning are completed. In long-term load forecasting, peak and off-peak loads in each area of the system under study are projected, year by year, from the present up to 25 years into the future. Such forecasts are based on present and past load trends, population growth patterns, and economic indicators. In generation planning, generation resources are selected with sufficient generation reserve margins to meet projected customer loads with adequate quality and reliability in an economic manner. New generating units both at new plant sites and at existing plants are selected, and construction schedules are established to ensure that new generation goes on-line in time to meet projected loads.

The results of long-term load forecasting and generation planning are used by transmission planning engineers to design the future transmission system so that it performs its basic functions. The following are selected during the transmission planning process.

- Routes for new lines
- Number of circuits for each route or right-of-way
- EHV versus HVDC lines
- Overhead versus underground line construction
- Types of towers for overhead lines
- Voltage levels
- Line ratings
- Shunt reactive and series capacitive line compensation
- Number and locations of substations
- Bus and circuit breaker configurations at substations
- Circuit breaker ratings
- Number, location, and ratings of bulk-power-system transformers
- Number, location, and ratings of voltage-regulating transformers and phase-shifting transformers
- Number, location, and ratings of static VAR systems, synchronous condensers, and shunt capacitor banks for voltage control
- Basic insulation levels (BILs)
- Surge arrester locations and ratings
- Protective relaying schemes
- Communications facilities
- Upgrades of existing circuits
- Reinforcements of system interconnections

Planning Tools

As electric utilities have grown in size and the number of interconnections has increased, making the above selections during the planning process has become increasingly complex. The increasing cost of additions and modifications has made it imperative that planning engineers consider a wide range of design options and perform detailed studies on the effects on the system of each option based on a number of assumptions: normal and emergency operating conditions, peak and off-peak loadings, and present and future years of operation. A large volume of network data must be collected and accurately handled. To assist the planning engineer, the following digital computer programs are used [Glover and Sarma, 1987]:

1. *Power-flow programs.* Power-flow (also called load-flow) programs compute voltage magnitudes, phase angles, and transmission line power flows for a power system network under

steady-state operating conditions. Other output results, including transformer tap settings, equipment losses, and reactive power outputs of generators and other devices, are also computed. To do this, the locations, sizes, and operating characteristics of all loads and generation resources of the system are specified as inputs. Other inputs include the network configuration as well as ratings and other characteristics of transmission lines, transformers, and other equipment. Today's computers have sufficient storage and speed to compute in less than 1 min power-flow solutions for networks with more than 2000 buses and 2500 transmission lines. High-speed printers then print out the complete solution in tabular form for analysis by the planning engineer. Also available are interactive power-flow programs, whereby power-flow results are displayed on computer screens in the form of single-line diagrams; the engineer uses these to modify the network from a keyboard or with a mouse and can readily visualize the results. Spreadsheet analyses are also used. The computer's large storage and high-speed capabilities allow the engineer to run the many different cases necessary for planning.

2. *Transient stability programs.* Transient stability programs are used to study power systems under disturbance conditions to predict whether synchronous generators remain in synchronism and system stability is maintained. System disturbances can be caused by the sudden loss of a generator or a transmission line, by sudden load increases or decreases, and by short circuits and switching operations. The stability program combines power-flow equations and generator dynamic equations to compute the angular swings of machines during disturbances. The program also computes critical clearing times for network faults and allows the planning engineer to investigate the effects of various network modifications, machine parameters, disturbance types, and control schemes.

3. *Short-circuits programs.* Short-circuits programs compute three-phase and line-to-ground fault currents in power system networks in order to evaluate circuit breakers and relays that detect faults and control circuit breakers. Minimum and maximum short-circuit currents are computed for each circuit breaker and relay location under various system operating conditions, such as lines or generating units out of service, in order to specify circuit breaker ratings and protective relay schemes.

4. *Transients programs.* Transients programs compute the magnitudes and shapes of transient overvoltages and currents that result from switching operations and lightning strikes. Planning engineers use the results of transients programs to specify BILs for transmission lines, transformers, and other equipment and to select surge arresters that protect equipment against transient overvoltages.

Research efforts aimed at developing computerized, automated transmission planning tools are ongoing. Examples and references are given in Back *et al.* [1989] and Smolleck *et al.* [1989]. Other programs for transmission planning include production-cost, investment-cost, relay-coordination, power-system database management, transformer thermal analysis, and transmission line design programs. Some of the vendors that offer software packages for transmission planning are given as follows:

- ABB Network Control Ltd., Switzerland
- CYME International, Burlington, Mass.
- EDSDA Micro Corporation, Bloomfield, Mich.
- Electric Power Consultants, Inc., Scotia, N.Y.
- Electrocon International, Inc., Ann Arbor, Mich.
- Power Technologies, Inc., Schenectady, N.Y.
- Operation Technology, Inc., Irvine, Calif.

Basic Planning Principles

The electric utility industry has established basic planning principles intended to provide a balance among all power system components so as not to place too much dependence on any one component or group of components. Transmission planning criteria are developed from these principles along with actual system operating history and reasonable contingencies. These planning principles are given as follows:

1. Maintain a balance among power system components based on size of load, size of generating units and power plants, the amount of power transfer on any transmission line or group of lines, and the strength of interconnections with other utilities. In particular:
    a. Avoid excessive generating capacity at one unit, at one plant, or in one area.
    b. Avoid excessive power transfer through any single transformer, through any transmission line, circuit, tower, or right-of-way, or though any substation.
    c. Provide interconnection capacity to neighboring utilities that is commensurate with the size of generating units, power plants, and system load.
2. Provide transmission capability with ample margin above that required for normal power transfer from generators to loads in order to maintain a high degree of flexibility in operation and to meet a wide range of contingencies.
3. Provide for power system operation such that all equipment loadings remain within design capabilities.
4. Utilize switching arrangements, associated relay schemes, and controls that permit:
    a. Effective operation and maintenance of equipment without excessive risk of uncontrolled power interruptions.
    b. Prompt removal and isolation of faulted components.
    c. Prompt restoration in the event of loss of any part of the system.

Equipment Ratings

Transmission system loading criteria used by planning engineers are based on equipment ratings. Both normal and various emergency ratings are specified. Emergency ratings are typically based on the time required for either emergency operator actions or equipment repair times. For example, up to 2 h may be required following a major event such as loss of a large generating unit or a critical transmission facility in order to bring other generating resources on-line and to perform appropriate line-switching operations. The time to repair a failed transmission line typically varies from 2 to 10 days, depending on the type of line (overhead, underground cable in conduit, or pipe-type cable). The time required to replace a failed bulk-power-system transformer is typically 30 days. As such, ratings of each transmission line or transformer may include normal, 2-h emergency, 2- to 10-day emergency, and in some cases 30-day emergency ratings.

The rating of an overhead transmission line is based on the maximum temperature of the conductors. Conductor temperature affects the conductor sag between towers and the loss of conductor tensile strength due to annealing. If the temperature is too high, proscribed conductor-to-ground clearances [ANSI, 1990] may not be met, or the elastic limit of the conductor may be exceeded such that it cannot shrink to its original length when cooled. Conductor temperature depends on the current magnitude and its time duration, as well as on ambient temperature, wind velocity, solar radiation, and conductor surface conditions. Standard assumptions on ambient temperature, wind velocity, etc., are selected, often conservatively, to calculate overhead transmission line ratings [ANSI/IEEE Std. 738–85, 1985]. It is common practice to have summer and winter normal line ratings, based on seasonal ambient temperature differences. Also, in locations with higher prevailing winds, such as coastal areas, larger normal line ratings may be selected. Emergency line ratings typically vary from 110 to 120% of normal ratings. Recently, real-time monitoring of actual conductor temperatures along a transmission line has been used for on-line dynamic transmission line ratings [Henke and Sciacca, 1989].

Normal ratings of bulk-power-system transformers are determined by manufacturers' nameplate ratings. Nameplate ratings are based on the following ANSI/IEEE standard conditions: (1) continuous loading at nameplate output; (2) 30°C average ambient temperature (never exceeding 40°C); and (3) 110°C average hot-spot conductor temperature (never exceeding 120°C) for 65°C-average-winding-rise transformers [ANSI/IEEE C57.92-1981, 1990]. For 55°C-average-winding-rise transformers, the hot-spot temperature limit is 95°C average (never exceeding 105°C). The actual output that a bulk-power-system transformer can deliver at any time with normal life expectancy may be more or less than the nameplate rating, depending on the ambient temperature and actual temperature rise of the windings. Emergency transformer ratings typically vary from 130 to 150% of nameplate ratings.

Planning Criteria

Transmission system planning criteria have been developed from the above planning principles and equipment ratings as well as from actual system operating data, probable operating modes, and equipment failure rates. These criteria are used to plan and build the transmission network with adequate margins to ensure a reliable supply of power to customers under reasonable equipment-outage contingencies. The transmission system should perform its basic functions under a wide range of operating conditions. Transmission planning criteria include equipment loading criteria, transmission voltage criteria, stability criteria, and regional planning criteria.

Equipment Loading Criteria

Typical equipment loading criteria are given in Table 58.6. With no equipment outages, transmission equipment loadings should not exceed normal ratings for all realistic combinations of generation and interchange. Operation of all generating units including base-loaded and peaking units during peak load periods as well as operation of various combinations of generation and interchange during off-peak periods should be considered. Also, normal ratings should not be exceeded with all transmission lines and transformers in service and with any generating unit out of service.

With any single-contingency outage, emergency ratings should not be exceeded. One loading criterion is not to exceed 2-h emergency ratings when any transmission line or transformer is out of service. This gives time to perform switching operations and change generation levels, including use of peaking units, to return to normal loadings.

With some of the likely double-contingency outages, the transmission system should supply all system load without exceeding emergency ratings. One criterion is not to exceed 2- to 10-day emergency ratings when any line and any transformer are out of service or when any line and any generator are out of service. This gives time to repair the line. With the outage of any transformer and any generator, 30-day emergency ratings should not be exceeded, which gives time to install a spare transformer.

The loading criteria in Table 58.6 do not include all types of double-contingency outages. For example, the outage of a double-circuit transmission line or two transmission lines in the same right-of-way is not included. Also, the loss of two transformers in the same load area is not included. Under these double-contingency outages, it may be necessary to shed load at some locations during heavy load periods. Although experience has shown that these outages are relatively unlikely, their consequences should be evaluated in specific situations. Factors to be evaluated include the size of load served, the degree of risk, and the cost of reinforcement.

Specific loading criteria may also be required for equipment serving critical loads and critical load areas. One criterion is to maintain service to critical loads under a double-contingency outage with the prior outage of any generator.

Transmission Voltage Criteria

Transmission voltages should be maintained within suitable ranges for both normal and reasonable emergency conditions. Abnormal transmission voltages can cause damage or malfunction of transmission equipment such as circuit breakers or transformers and adversely affect many customers. Low transmission voltages tend to cause low distribution voltages, which in turn cause increased distribution losses as well as higher motor currents at customer loads and at power plant auxiliaries. Transmission voltage planning criteria are intended to be conservative.

Maximum planned transmission voltage is typically 105% of rated nominal voltage for both normal and reasonable emergency conditions. Typical minimum planned transmission voltages are given in Table 58.7. System conditions in Table 58.7 correspond to equipment out of service in Table 58.6. Single-contingency outages correspond to the loss of any line, any transformer, or any generator. Double-contingency outages correspond to the loss of any transmission line and transformer, any transmission line and generator, any transformer and generator, or any two generators.

Typical planned minimum voltage criteria shown in Table 58.7 for EHV (345 kV and higher) substations and for generator substations are selected to maintain adequate voltage levels at interconnections, at power plant auxiliary buses, and on the lower-voltage transmission systems. Typical planned minimum voltage criteria for lower HV (such as 138 kV, 230 kV) transmission substations vary from 95 to 97.5% of nominal voltage under normal system conditions to as low as 92.5% of nominal under double-contingency outages.

Equipment used to control transmission voltages includes voltage regulators at generating units (excitation control), tap-changing transformers, regulating transformers, synchronous condensers, shunt reactors, shunt capacitor banks, and static VAR devices. When upgrades are selected during the planning process to meet planned transmission voltage criteria, some of this equipment should be assumed out of service.

Stability Criteria

System stability is the ability of all synchronous generators in operation to stay in synchronism with each other while moving from one operating condition to another. Steady-state stability refers to

Table 58.6   Typical Transmission Equipment Loading Criteria

Equipment Out of Service	Rating Not to Be Exceeded	Comment
None	Normal	
Any generator	Normal	
Any line or any transformer	2-h emergency	Before switching.
Any line and any transformer*	2- to 10-day emergency	After switching required for both outages. Line repair time.
Any line and any generator*	2- to 10-day emergency	After switching required for both outages. Line repair time.
Any transformer and any generator*	30-day emergency	After switching required for both outages. Install spare transformer.

*Some utilities do not include double-contingency outages in transmission system loading criteria.

Table 58.7   Typical Minimum Transmission Voltage Criteria

	Planned Minimum Transmission Voltage at Substations, % of Nominal		
System Condition	Generator Station	EHV Station	HV Station
Normal	102	98	95–97.5
Single-contingency outage	100	96	92.5–95
Double-contingency outage*	98	94	92.5

*Some utilities do not include double-contingency outages in planned minimum transmission voltage criteria.

small changes in operating conditions, such as normal load changes. Transient stability refers to larger, abrupt changes, such as the loss of the largest generator or a short circuit followed by circuit breakers opening, where synchronism or loss of synchronism occurs within a few seconds. Dynamic stability refers to longer time periods, from minutes up to a half hour following a large, abrupt change, where steam generators (boilers), automatic generation control, and system operator actions affect stability.

In the planning process, steady-state stability is evaluated via power-flow programs by the system's ability to meet equipment loading criteria and transmission voltage criteria under steady-state conditions. Transient stability is evaluated via stability programs by simulating system transient response for various types of disturbances, including short circuits and other abrupt network changes. The planning engineer designs the system to remain stable for the following typical disturbances:

1. With all transmission lines in service, a permanent three-phase fault (short circuit) occurs on any transmission line, on both transmission lines on any double-circuit tower, or at any bus; the fault is successfully cleared by primary relaying.
2. With any one transmission line out of service, a permanent three-phase fault occurs on any other transmission line; the fault is successfully cleared by primary relaying.
3. With all transmission lines in service, a permanent three-phase fault occurs on any transmission line; backup relaying clears the fault after a time delay, due to a circuit breaker failure.

Regional Planning Criteria

The North American Electric Reliability Council (NERC) defines nine geographical regions in North America, as shown in Fig. 58.59 [NERC, 1988]. Transmission planning studies are performed at two levels: (1) individual electric utility companies separately perform planning studies of their internal systems and (2) companies jointly participate in NERC committees or working groups to perform regional and interregional planning studies. The purpose of regional planning studies is to evaluate the transfer capabilities between interconnected utilities and the impact of severe disturbances.

One typical regional criterion is that the incremental power transfer capability, in addition to scheduled interchange, should provide a reasonable generation reserve margin under the following conditions: peak load, the most critical transmission line out of service, no component overloaded.

Another criterion is that severe disturbances to the interconnected transmission network should not result in system instability, widespread cascading outages, voltage collapse, or system blackouts. [NERC, 1988, 1989, and 1991]. Severe disturbances include the following:

1. With any three generating units or any combination of units up to 30% of system load out of service in an area, a sudden outage of any transmission line or any transformer occurs.
2. With any two generating units or any combination of units up to 20% of system load out of service in an area, a sudden outage of any generator or any double-circuit transmission line occurs.
3. With any transmission line or transformer out of service in an area, a sudden outage of any other transmission line or transformer occurs.
4. With any transmission line or transformer out of service in an area as well as any two generating units or any combination of units up to 20% of system load, a sudden outage of a transmission line occurs.
5. A sudden outage of all generating units at a power plant occurs.
6. A sudden outage of either a transmission substation or all transmission lines on a common right-of-way occurs.
7. A sudden outage of a large load or a major load center occurs.

When evaluating the impacts of the above severe disturbances, regional planning studies should consider steady-state stability, transient stability, and dynamic stability. These studies should also

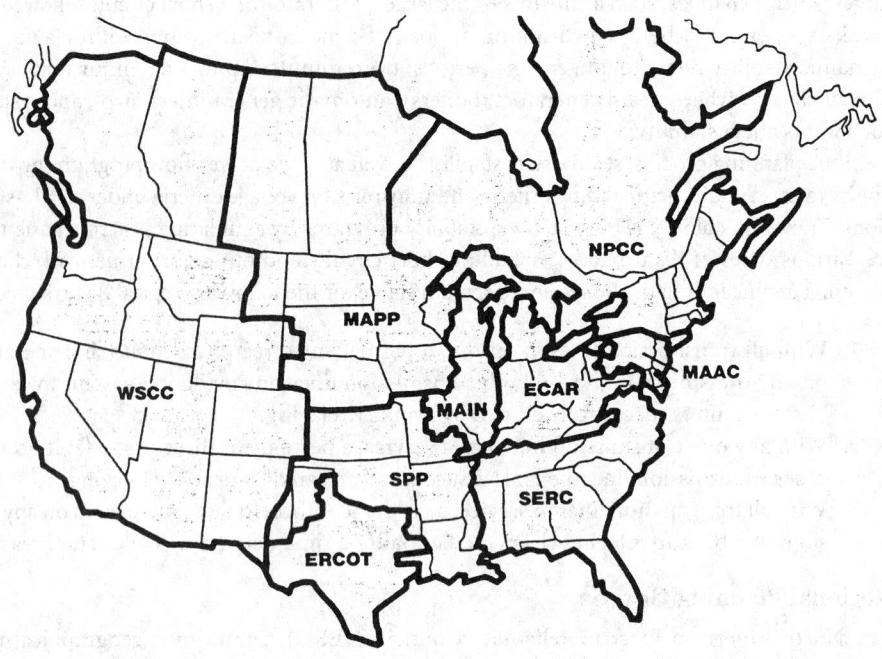

ECAR
East Central Area Reliability
Coordination Agreement

ERCOT
Electric Reliability Council of Texas

MAAC
Mid-Atlantic Area Council

MAIN
Mid-America Interpool Network

MAPP
Mid-continent Area Power Pool

NPCC
Northeast Power Coordinating Council

SERC
Southeastern Electric Reliability Council

SPP
Southwest Power Pool

WSCC
Western Systems Coordinating Council

FIGURE 58.59   Nine regional reliability councils established by NERC. (*Source: 1985 Electric Power Supply & Demand*, Princeton, N.J.: North American Electric Reliability Council, 1985. With permission.)

consider the effects of three-phase faults and slow fault clearing due to improper relaying or failure of a circuit breaker to open, as well as the anticipated load range and various operating conditions.

Value-Based Transmission Planning

Recently some utilities have begun to use a value-of-service concept in transmission planning [EPRI, 1986]. This concept establishes a method of assigning a dollar value to various levels of reliability in order to balance reliability and cost. For each particular outage, the amount and dollar value of unserved energy are determined. Dollar value of unserved energy is based on rate surveys of various types of customers. If the cost of the transmission project required to eliminate the outage exceeds the value of service, then that project is given a lower priority. As such, reliability is quantified, and benefit-to-cost ratios are used to compare and prioritize planning options.

Defining Terms

The North American Electric Reliability Council (NERC) defines *reliability* and the related terms *adequacy* and *security* as follows [NERC, 1988]:

Adequacy: The ability of the bulk-power electric system to supply the aggregate electric power and energy requirements of the consumers at all times, taking into account scheduled and unscheduled outages of system components.

Reliability: In a bulk-power electric system, reliability is the degree to which the performance of the elements of that system results in power being delivered to consumers within accepted standards and in the amount desired. The degree of reliability may be measured by the frequency, duration, and magnitude of adverse effects on consumer service.

Security: The ability of the bulk-power electric system to withstand sudden disturbances such as electric short circuits or unanticipated loss of system components.

References

ANSI C2-1990, National Electrical Safety Code, 1990 Edition, Piscataway, N.J.: IEEE, 1990.

ANSI/IEEE C57.92-1981, IEEE Guide for Loading Mineral-Oil Immersed Power Transformers Up to and Including 100 MVA with 55°C or 65°C Average Winding Rise, Piscataway, N.J.: IEEE, 1990.

ANSI/IEEE Std. 738-1985, Calculation of Bare Overhead Conductor Temperature and Ampacity under Steady-State Conditions, Piscataway, N.J.: IEEE, 1985.

H. Back *et al.*, "PLATINE—A new computerized system to help in planning the power transmission networks," *IEEE Trans. Power Systems*, vol. 4, no. 1, pp. 242–247, 1989.

Electric Power Research Institute (EPRI), Value of Service Reliability to Consumers, Report EA-4494, Palo Alto, Calif.: EPRI, March 1986.

J. D. Glover and M. S. Sarma, *Power System Analysis and Design with Personal Computer Applications*, Boston: PWS-Kent, 1987.

R. K. Henke and S. C. Sciacca, "Dynamic thermal rating of critical lines—A study of real-time interface requirements," *IEEE Computer Applications in Power*, pp. 46–51, July 1989.

NERC, *Reliability Concepts*, Princeton, N.J.: North American Electric Reliability Council, February 1985.

NERC, *Overview of Planning Reliability Criteria*, Princeton, N.J.: North American Electric Reliability Council, April 1988.

NERC, *Electricity Transfers and Reliability*, Princeton, N.J.: North American Electric Reliability Council, October 1989.

NERC, *A Survey of the Voltage Collapse Phenomenon*, Princeton, N.J.: North American Electric Reliability Council, 1991.

H. A. Smolleck *et al.*, "Translation of large data-bases for microcomputer-based application software: Methodology and a case study," *IEEE Comput. Appl. Power*, pp. 40–45, July 1989.

Further Information

The North American Electric Reliability Council (NERC) was formed in 1968, in the aftermath of the November 9, 1965, northeast blackout, to promote the reliability of bulk-electric-power systems of North America. Transmission planning criteria presented here are partially based on NERC criteria as well as on specific criteria used by transmission planning departments from three electric utility companies: American Electric Power Service Corporation, Commonwealth Edison Company, and Pacific Gas & Electric Company. NERC's publications, developed by utility experts, have become standards for the industry. In most cases, these publications are available at no charge from NERC, Princeton, N.J.

59

Power Transformers

59.1 Power Transformer Fundamentals... 1296
The Three-Winding Ideal Transformer Equivalent Circuit • A
Practical Three-Winding Transformer Equivalent Circuit • The
Two-Winding Transformer

59.2 Transformer Construction 1300
The Transformer Core • Core and Shell Types • Transformer
Windings

59.3 Transformer Performance 1302

59.4 Transformers in Three-Phase Connections............................ 1303
Phase Shift in Y–Δ Connections • The Three-Phase Transformer •
Determining Per-Phase Equivalent Circuit Values for Power
Transformers: An Example

59.5 Autotransformers.. 1307

Charles A. Gross
Auburn University

William M. Feaster
Auburn University

59.1 Power Transformer Fundamentals

Charles A. Gross

The electric power **transformer** is a major power system component which provides the capability of reliably and efficiently changing (transforming) voltage and current at high power levels. Because electrical power is proportional to the product of voltage and current, for a specified power level, low current levels can exist only at high voltage, and vice versa.

The Three-Winding Ideal Transformer Equivalent Circuit

To understand the basics of transformer operation, consider the three coils wrapped on a common core as shown in Fig. 59.1(a). Consider two idealizations: imagine the core permeability (μ) to be infinite, and the windings to be made of material of infinite conductivity (σ). As a result of the latter, all windings have zero resistance. Therefore, from Faraday's law:

$$v_1 = N_1 \frac{d\phi}{dt} \quad v_2 = N_2 \frac{d\phi}{dt} \quad v_3 = N_3 \frac{d\phi}{dt} \tag{59.1}$$

where ϕ is the core flux. This produces:

$$\frac{v_1}{v_2} = \frac{N_1}{N_2} \quad \frac{v_2}{v_3} = \frac{N_2}{N_3} \quad \frac{v_3}{v_1} = \frac{N_3}{N_1} \tag{59.2}$$

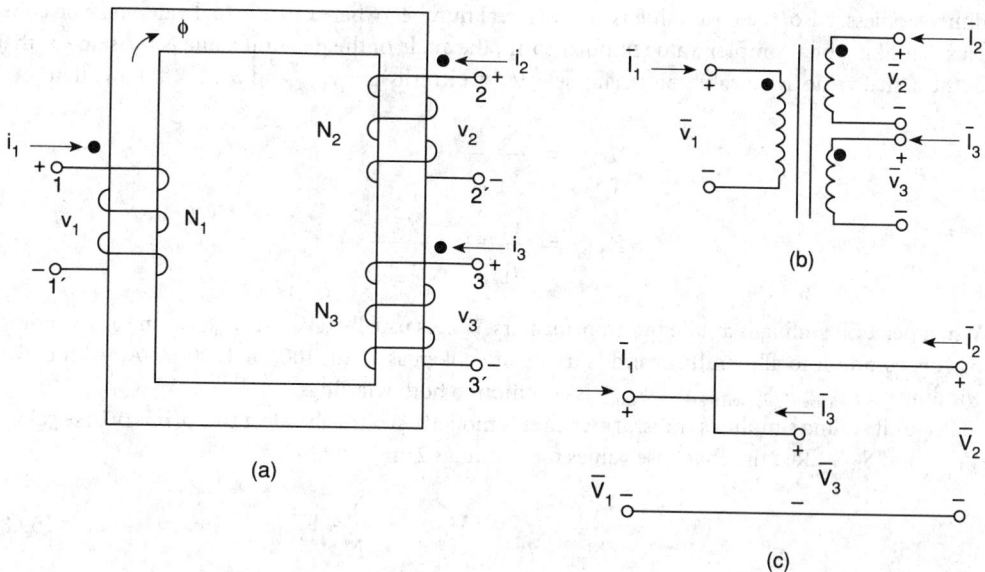

FIGURE 59.1   Ideal three-winding transformer. (a) Ideal three-winding transformer; (b) schematic symbol; (c) per-unit equivalent circuit.

For sinusoidal steady state performance, cross-multiply by the denominator voltages and transform into phasor values, producing:

$$\overline{V}_1 = \frac{N_1}{N_2}\overline{V}_2 \quad \overline{V}_2 = \frac{N_2}{N_3}\overline{V}_3 \quad \overline{V}_3 = \frac{N_3}{N_1}\overline{V}_1 \tag{59.3}$$

The circuit symbol to be used is shown in Fig. 59.1(b). Ampere's law requires that

$$\oint \hat{H} \cdot \hat{dl} = i_{\text{enclosed}} = 0 \tag{59.4}$$

$$0 = N_1 i_1 + N_2 i_2 + N_3 i_3 \tag{59.5}$$

Transform Eq. (59.5) into phasor notation:

$$N_1 \overline{I}_1 + N_2 \overline{I}_2 + N_3 \overline{I}_3 = 0 \tag{59.6}$$

Equations (59.3) and (59.6) are basic to understanding transformer operation. Consider Eq. (59.3). Also note that $\overline{V}_1$, $\overline{V}_2$, and $\overline{V}_3$ must be in phase, with dotted terminals defined positive. Now consider the total input complex power $\overline{S}$.

$$\overline{S} = \overline{V}_1 \overline{I}_1^* + \overline{V}_2 \overline{I}_2^* + \overline{V}_3 \overline{I}_3^* = 0 \tag{59.7}$$

Hence, ideal transformers can absorb neither real nor reactive power.

It is customary to scale system quantities (V, I, S, Z) into dimensionless quantities called per-unit values. The basic per-unit scaling equation is

$$\text{Per-unit value} = \frac{\text{actual value}}{\text{base value}}$$

The base value always carries the same units as the actual value, forcing the per-unit value to be dimensionless. Also, the base value is always a real number, whereas the actual value may be complex. Thinking of a complex value in polar form, the angle of the per-unit value is the same as that of the actual value. Base values normally selected arbitrarily are V_{base} and S_{base}. It follows that:

$$I_{base} = \frac{S_{base}}{V_{base}}$$

$$Z_{base} = \frac{V_{base}}{I_{base}} = \frac{V_{base}^2}{S_{base}}$$

When per-unit scaling is applied to transformers V_{base} is usually taken as V_{rated} as in each winding. S_{base} is common to all windings and is frequently taken as 1, 10, 100, or 1000 MVA; for the two-winding case S_{base} is S_{rated}, since S_{rated} is common to both windings.

Per-unit scaling simplifies transformer circuit models. Arbitrarily select two primary base values, $V_{1_{base}}$ and $S_{1_{base}}$. Require that base values for windings 2 and 3 be

$$V_{2_{base}} = \frac{N_2}{N_1} V_{1_{base}} \qquad V_{3_{base}} = \frac{N_3}{N_1} V_{1_{base}} \qquad (59.8)$$

and

$$S_{1_{base}} = S_{2_{base}} = S_{3_{base}} = S_{base} \qquad (59.9)$$

By definition:

$$I_{1_{base}} = \frac{S_{base}}{V_{1_{base}}} \qquad I_{2_{base}} = \frac{S_{base}}{V_{2_{base}}} \qquad I_{3_{base}} = \frac{S_{base}}{V_{3_{base}}} \qquad (59.10)$$

It follows that

$$I_{2_{base}} = \frac{N_1}{N_2} I_{1_{base}} \qquad I_{3_{base}} = \frac{N_1}{N_3} I_{1_{base}} \qquad (59.11)$$

Thus, Eqs. (59.3) and (59.6) scaled into per-unit become:

$$\overline{V}_{1_{pu}} = \overline{V}_{2_{pu}} = \overline{V}_{3_{pu}} \qquad (59.12)$$

$$\overline{I}_{1_{pu}} + \overline{I}_{2_{pu}} + \overline{I}_{3_{pu}} = 0 \qquad (59.13)$$

The basic per-unit equivalent circuit is shown in Fig. 59.1(c). Note that some circuit models and corresponding equations are correct *only if the parameters and variables are in per-unit*. The extension to the *n*-winding case is clear.

A Practical Three-Winding Transformer Equivalent Circuit

The circuit of Fig. 59.1(c) is reasonable for some power system applications, since the core and windings of actual transformers are constructed of materials of high μ and σ, respectively, though of course not infinite. However, for other studies, discrepancies between the performance of actual and ideal transformers are too great to be overlooked. The circuit of Fig. 59.1(c) may be modified into that of Fig. 59.2 to account for the most important discrepancies. Note:

R_1, R_2, R_3 —Since the winding conductors cannot be made of material of infinite conductivity, the windings must have some resistance.

X_1, X_2, X_3 —Since the core permeability is not infinite, not all of the flux created by a given winding current will be confined to the core. The part that escapes the core and seeks out parallel paths in surrounding structures and air is referred to as *leakage* flux. This effect may be modeled by the insertion of a linear inductive reactance as shown.

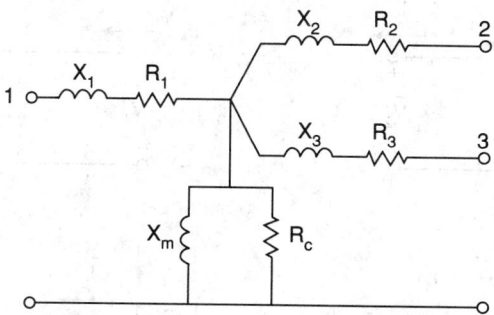

FIGURE 59.2 A practical equivalent circuit.

R_c, X_m —Also, since the core permeability is not infinite, the magnetic field intensity inside the core is not zero. Therefore, some current flow is necessary to provide this small H. The path provided in the circuit for this "magnetizing" current is through X_m. The core has internal power losses, referred to as *core loss*, due to hystereses and eddy current phenomena. The effect is accounted for in the resistance R_c. Sometimes R_c and X_m are neglected.

The circuit of Fig. 59.2 is a refinement on that of Fig. 59.1(c). The values $R_1, R_2, R_3, X_1, X_2, X_3$ are all small (less than 0.05 per-unit) and R_c, X_m, large (greater than 10 per-unit). The circuit of Fig. 59.2 requires that all values be in per-unit. Circuit data are available from the manufacturer or obtained from conventional tests. It must be noted that although the circuit of Fig. 59.2 is commonly used, it is not rigorously correct because it does not properly account for the mutual couplings between windings.

A comment about the terms **primary** and **secondary** is in order. The terms refer to source and load sides, respectively (i.e., energy flows from primary to secondary). However, in many applications energy can flow either way, in which case the distinction is meaningless. Also, the presence of a third winding (tertiary) confuses the issue. Here the terms (along with subscripts 1, 2, 3) simply distinguish between windings. The terms *step up* and *step down* refer to what the transformer does to the voltage from source to load. ANSI standards require that for a two-winding transformer the high-voltage and low-voltage terminals be marked as H1-H2 and X1-X2, respectively, with H1 and X1 markings having the same significance as *dots* for **polarity** markings. [Refer to ANSI C57 for comprehensive information.] *Additive* and *subtractive transformer polarity* refer to the physical positioning of high-voltage, low-voltage *dotted* terminals as shown in Fig. 59.3. If the dotted termi-

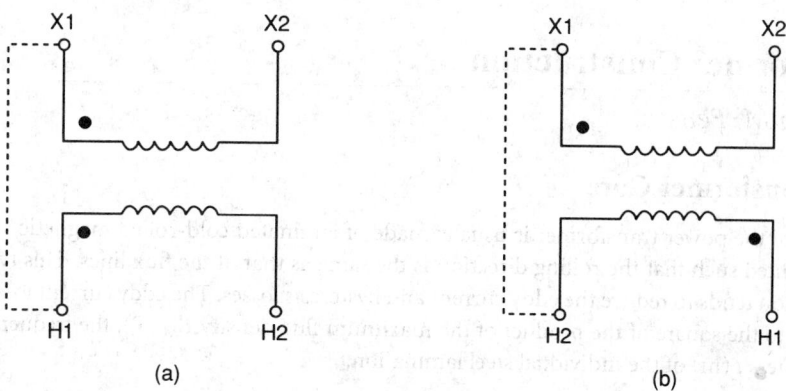

FIGURE 59.3 Transformer polarity terminology: (a) subtractive; (b) additive.

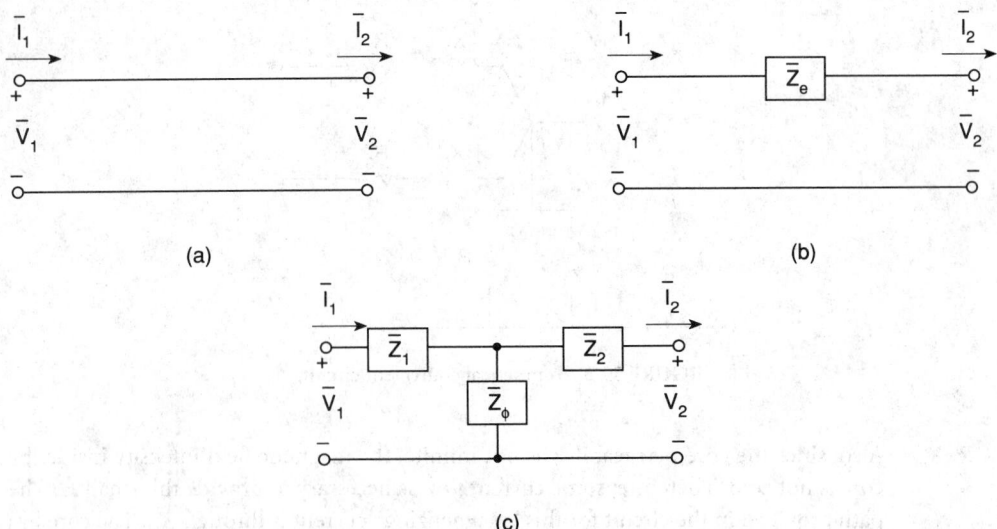

FIGURE 59.4 Two-winding transformer-equivalent circuits. All values in per-unit. (a) Ideal case; (b) no load current negligible; (c) precise model.

nals are adjacent, then the transformer is said to be *subtractive*, because if these adjacent terminals (H1-X1) are connected together, the voltage between H2 and X2 is the *difference* between primary and secondary. Similarly, if adjacent terminals X1 and H2 are connected, the voltage (H1-X2) is the *sum* of primary and secondary values.

The Two-Winding Transformer

The device can be simplified to two windings. Common two-winding transformer circuit models are shown in Fig. 59.4.

$$\bar{Z}_e = \bar{Z}_1 + \bar{Z}_2 \tag{59.14}$$

$$\bar{Z}_m = \frac{R_c(jX_m)}{R_c + jX_m} \tag{59.15}$$

Circuits (a) and (b) are appropriate when $\bar{Z}_m$ is large enough that magnetizing current and core loss is negligible.

59.2 Transformer Construction

William M. Feaster

The Transformer Core

The core of the power transformer is usually made of laminated cold-rolled magnetic steel that is grain oriented such that the rolling direction is the same as that of the flux lines. This type of core construction tends to reduce the eddy current and hysteresis losses. The eddy current loss P_e is proportional to the square of the product of the maximum flux density B_M (T), the frequency f (Hz), and thickness t (m) of the individual steel lamination.

$$P_e = K_e(B_M t f)^2 \quad (\text{W}) \tag{59.16}$$

K_e is dependent upon the core dimensions, the specific resistance of a lamination sheet, and the mass of the core. Also,

$$P_h = K_h f B_M^n \quad (\text{W}) \tag{59.17}$$

In Eq. (59.17), P_h is the hysteresis power loss, n is the Steinmetz constant ($1.5 < n < 2.5$) and K_h is a constant dependent upon the nature of core material and varies from $3 \times 10^{-3} m$ to $20 \times 10^{-3} m$, where m = core mass in kilograms.

The core loss therefore is

$$P_e = P_e + P_h \tag{59.18}$$

Core and Shell Types

Transformers are constructed in either a shell or a core structure. The shell-type transformer is one where the windings are completely surrounded by transformer steel in the plane of the coil. Core-type transformers are those that are not shell type. A power transformer is shown in Fig. 59.5.

Multiwinding transformers, as well as polyphase transformers, can be made in either shell- or core-type designs.

Transformer Windings

The windings of the power transformer may be either copper or aluminum. These conductors are usually made of conductors having a circular cross section; however, larger cross-sectional area conductors may require a rectangular cross section for efficient use of winding space.

The life of a transformer insulation system depends, to a large extent, upon its temperature. The total temperature is the sum of the ambient and the temperature rise. The temperature rise in a

FIGURE 59.5 230kVY:17.1kVΔ 1153-MVA 3ϕ power transformer. (Photo courtesy of General Electric Company.)

transformer is intrinsic to that transformer at a fixed load. The ambient temperature is controlled by the environment the transformer is subjected to. The better the cooling system that is provided for the transformer, the higher the "kVA" rating for the same ambient. For example, the kVA rating for a transformer can be increased with forced air (fan) cooling. Forced oil and water cooling systems are also used. Also, the duration of operating time at high temperature directly affects insulation life.

Other factors that affect transformer insulation life are vibration or mechanical stress, repetitive expansion and contraction, exposure to moisture and other contaminants, and electrical and mechanical stress due to overvoltage and short-circuit currents.

Paper insulation is laid between adjacent winding layers. The thickness of this insulation is dependent on the expected electric field stress. In large transformers oil ducts are provided using paper insulation to allow a path for cooling oil to flow between coil elements.

The short-circuit current in a transformer creates enormous forces on the turns of the windings. The short-circuit currents in a large transformer are typically 8 to 10 times larger than rated and in a small transformer are 20 to 25 times rated. The forces on the windings due to the short-circuit current vary as the square of the current, so whereas the forces at rated current may be only a few newtons, under short-circuit conditions these forces can be tens of thousands of newtons. These mechanical and thermal stresses on the windings must be taken into consideration during the design of the transformer. The current-carrying components must be clamped firmly to limit movement. The solid insulation material should be precompressed and formed to avoid its collapse due to the thermal expansion of the windings.

59.3 Transformer Performance

Charles A. Gross

There is a need to assess the quality of a particular transformer design. The most important measure for performance is the concept of efficiency, defined as follows:

$$\eta = \frac{P_{out}}{P_{in}} \tag{59.19}$$

where P_{out} is output power in watts (kW, MW) and P_{in} is input power in watts (kW, MW).

The situation is clearest for the two-winding case where the output is clearly defined (i.e., the secondary winding), as is the input (i.e., the primary). Unless otherwise specified, the output is understood to be rated power at rated voltage at a user-specified power factor. Note that

$$\Sigma L = P_{in} - P_{out} = \text{sum of losses}$$

The transformer is frequently modeled with the circuit shown in Fig. 59.6. Transformer losses are made up of the following components:

Electrical losses: $\qquad I'^2_1 R_{eq} = I^2_1 R_1 + I^2_2 R_2 \qquad (59.20\text{a})$

$\qquad\qquad\qquad\qquad \text{Primary winding loss} = I^2_1 R_1 \qquad (59.20\text{b})$

$\qquad\qquad\qquad\qquad \text{Secondary winding loss} = I^2_2 R_2 \qquad (59.20\text{c})$

Magnetic (core) loss: $\quad P_c = P_e + P_h = V^2_1/R_c \qquad (59.21)$

$\qquad\qquad\qquad\qquad \text{Core eddy current loss} = P_e$

$\qquad\qquad\qquad\qquad \text{Core hysterisis loss} = P_h$

Hence:

$$\Sigma L = I'^2_1 R_{eq} + V^2_1/R_c \tag{59.22}$$

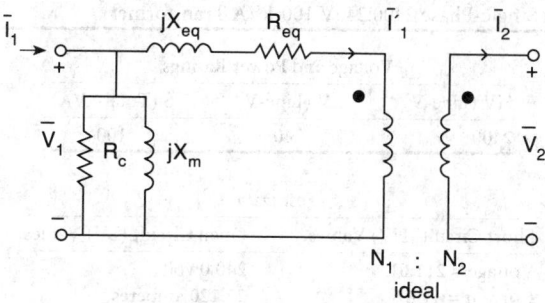

FIGURE 59.6   Transformer circuit model.

A second concern is fluctuation of secondary voltage with load. A measure of this situation is called *voltage regulation*, which is defined as follows:

$$\text{Voltage Regulation (VR)} = \frac{V_{2NL} - V_{2FL}}{V_{2FL}} \qquad (59.23)$$

where V_{2FL} = rated secondary voltage, with the transformer supplying rated load at a user-specified power factor, and V_{2NL} = secondary voltage with the load removed (set to zero), holding the primary voltage at the full load value.

A complete performance analysis of a 100 kVA 2400/240 V single-phase transformer is shown in Table 59.1.

59.4 Transformers in Three-Phase Connections

Charles A. Gross

Transformers are frequently used in three-phase connections. Recall that there are only two possible symmetric and balanced connections: the wye and the delta. If we use three identical three-winding transformers, we have a total of nine windings to account for. The three sets of windings may be individually connected in wye or delta in any combination. The symmetrical component transformation can be used to produce the sequence equivalent circuits shown in Fig. 59.7 which are essentially the circuits of Fig. 59.2 with R_c and X_m neglected.

The positive and negative sequence circuits are valid for both wye and delta connections. However, Y–Δ connections will produce a phase shift which is not accounted for in these circuits.

The zero sequence circuit requires special modification to account for wye, delta connections. Consider winding 1:

1. Solid grounded wye — short 1′ to 1″.
2. Ground wye through $\bar{Z}_n$ — connect 1′ to 1″ through $3\bar{Z}_n$.
3. Ungrounded wye — leave 1′ to 1″ open.
4. Delta — short 1″ to reference.

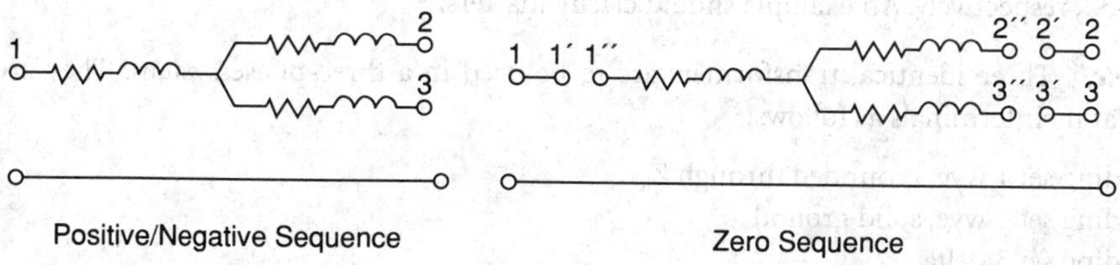

Positive/Negative Sequence   Zero Sequence

FIGURE 59.7   Sequence equivalent transformer circuits.

Table 59.1   Analysis of a Single-Phase 2400:240V 100-kVA Transformer

Voltage and Power Ratings

HV (Line-V)	LV (Line-V)	S (Total-kVA)
2400	240	100

Test Data

Short Circuit (HV) Values	Open Circuit (LV) Values
Voltage = 211.01	240.0 volts
Current = 41.67	22.120 amperes
Power = 1400.0	787.5 watts

Equivalent Circuit Values (in ohms)

Values referred to		HV Side	LV Side	Per-Unit
Series Resistance	=	0.8064	0.008064	0.01400
Series Reactance	=	4.9997	0.049997	0.08680
Shunt Magnetizing Reactance	=	1097.10	10.9714	19.05
Shunt Core Loss Resistance	=	7314.30	73.1429	126.98

Power Factor (—)	Efficiency (%)	Voltage Regulation (%)	Power Factor (—)	Efficiency (%)	Voltage Regulation (%)
0.0000 lead	0.00	−8.67	0.9000 lag	97.54	5.29
0.1000 lead	82.92	−8.47	0.8000 lag	97.21	6.50
0.2000 lead	90.65	−8.17	0.7000 lag	96.81	7.30
0.3000 lead	93.55	−7.78	0.6000 lag	96.28	7.86
0.4000 lead	95.06	−7.27	0.5000 lag	95.56	8.26
0.5000 lead	95.99	−6.65	0.4000 lag	94.50	8.54
0.6000 lead	96.62	−5.89	0.3000 lag	92.79	8.71
0.7000 lead	97.07	−4.96	0.2000 lag	89.56	8.79
0.8000 lead	97.41	−3.77	0.1000 lag	81.09	8.78
0.9000 lead	97.66	−2.16	0.0000 lag	0.00	8.69
1.0000 —	97.83	1.77			

Rated load performance at power factor = 0.866 lagging.

Secondary Quantities; LOW Voltage Side			Primary Quantities; HIGH Voltage Side		
	SI Units	Per-Unit		SI Units	Per-Unit
Voltage	240 volts	1.0000	Voltage	2539 volts	1.0577
Current	416.7 amperes	1.0000	Current	43.3 amperes	1.0386
Apparent power	100.0 kVA	1.0000	Apparent power	109.9 kVA	1.0985
Real power	86.6 kW	0.8660	Real power	88.9 kW	0.8888
Reactive power	50.0 kvar	0.5000	Reactive power	64.6 kvar	0.6456
Power factor	0.8660 lag	0.8660	Power factor	0.8091 lag	0.8091

Efficiency = 97.43%; voltage regulation = 5.77%.

Winding sets 2 and 3 interconnections produce similar connection constraints at terminals 2′–2″ and 3′–3″, respectively. An example should clarify matters.

Example.   Three identical transformers are to be used in a three-phase system. They are connected at their terminals as follows:

Winding set 1 wye, grounded through $\bar{Z}_n$
Winding set 2 wye, solid ground
Winding set 3 delta

Draw the zero sequence network.

Solution. The zero sequence network is as shown.

Phase Shift in Y–Δ Connections

The positive and negative sequence networks presented in Fig. 59.7 are misleading in one important detail. For Y–Y or Δ–Δ connections, it is always possible to label the phases in such a way that there is no phase shift between corresponding primary and secondary quantities. However, for Y–Δ or Δ–Y connections, it is impossible to label the phases in such a way that no phase shift between corresponding quantities is introduced. ANSI standard C57.12.10.17.3.2 is as follows:

> For either wye-delta or delta-wye connections, phases shall be labeled in such a way that positive sequence quantities on the high voltage side lead their corresponding positive sequence quantities on the low voltage side by 30°. The effect on negative sequence quantities is the reverse, i.e., HV values *lag* LV values by 30°.

This 30° phase shift is *not* accounted for in the sequence networks of Fig. 59.7. The effect only appears in the positive and negative sequence networks; the zero sequence network quantities are unaffected.

The Three-Phase Transformer

It is possible to construct a device (called a three-phase transformer) which allows the phase fluxes to share common magnetic paths. Such designs allow considerable savings in core material, and corresponding economies in cost, size, and weight. Positive and negative sequence impedances are equal; however, the zero sequence impedance is different. Otherwise the circuits of Fig. 59.7 apply as discussed previously.

Determining Per-Phase Equivalent Circuit Values for Power Transformers: An Example

One method of obtaining such data is through testing. Consider the problem of obtaining transformer equivalent circuit data from short-circuit tests. A numerical example will clarify per-unit scaling considerations.

The short-circuit test circuit arrangement is shown in Fig. 59.8. The objective is to derive equivalent circuit data from the test data provided in Fig. 59.8. Note that measurements are made in winding "i", with winding "j" shorted, and winding "k" left open. The short circuit impedance, looking into winding "i" with the transformer so terminated is designated as Z_{ij}. The indices i, j, and k, can be 1, 2, or 3.

The impedance calculations are done in per-unit, base values are provided in Fig. 59.8(c). The transformer ratings of the transformer of Fig. 59.1(a) would conventionally be provided as follows:

3φ 3W Transformer
15kVY/115kVY/4.157kVΔ
100/100/20 MVA

where 3φ means that the transformer is a three-phase piece of equipment (as opposed to an interconnection of three single-phase devices). 3W means three three-phase windings (actually nine windings). Usually the schematic is supplied also. The 15 kV rating is the *line* (phase-to-phase)

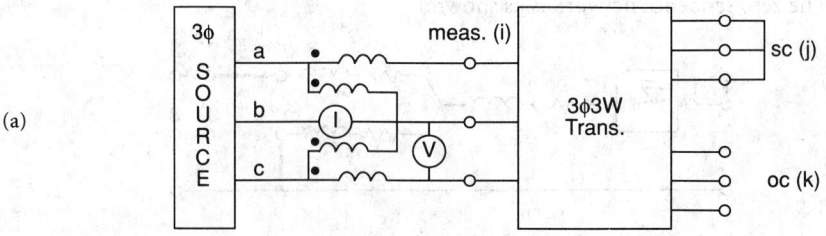

(b)

Line Voltage		3ph Power Ratings	S3ph Base = 100 MVA
Primary	15 kV	100 MVA	Zbase = 2.25
Secondary	115 kV	100 MVA	Zbase = 132.25
Tertiary	4.157 kV	20 MVA	Zbase = 0.1728

(c)

Voltage (line)	Current (line)	Power (3ph)	1	2	3
1200.0 V	3849.0 A	889.0 kW	meas	sc	oc
1840.0 V	100.0 A	100.0 kW	oc	meas	sc
33.0 V	2776.0 A	120.0 kW	sc	oc	meas

(d)

	R	X	Z		R	X	Z
Z12:	0.00889	0.07950	0.08000	Z1:	0.00686	0.20145	0.20157
Z23:	0.02520	0.07627	0.08033	Z2:	0.00203	−0.12195	0.12196
Z31:	0.03004	0.39967	0.40080	Z3:	0.02318	0.19822	0.19957

FIGURE 59.8 Transformer circuit data from short-circuit tests. (a) Setup for transformer short-circuit tests; (b) transformer data; (c) short-circuit test data; (d) short-circuit impedance values in per-unit.

value; three-phase apparatus is always rated in *line* values. "Y" means winding No. 1 is internally wye connected. 115kVY means that 115 kV is the line voltage rating, and winding No. 2 is wye connected. In 4.157kVΔ, again, "4.157kV" is the line voltage rating, and winding No. 3 is delta connected. 100/100/20 MVA are the *total* (3φ) power ratings for the primary, secondary, and tertiary winding, respectively; three-phase apparatus is always rated in three-phase terms.

The per-unit bases for $S_{3\phi_{base}}$ = 100 MVA are presented in Fig. 59.8(b). Calculating the short-circuit impedances from the test data in Fig. 59.8(c):

$$Z_{ij} = \frac{V_{i_{line}}/\sqrt{3}}{I_{i_{line}}}$$

$$R_{ij} = \frac{R_{3\phi}/3}{I_{1_{line}}^2}$$

$$X_{ij} = \sqrt{Z_{ij}^2 - R_{ij}^2}$$

Now calculate the transformer impedances from the short-circuit impedances:

$$\overline{Z}_1 = \frac{1}{2}\left(\overline{Z}_{12} - \overline{Z}_{23} + \overline{Z}_{31}\right)$$

$$\overline{Z}_2 = \frac{1}{2}\left(\overline{Z}_{23} - \overline{Z}_{13} + \overline{Z}_{12}\right)$$

$$\overline{Z}_3 = \frac{1}{2}\left(\overline{Z}_{31} - \overline{Z}_{12} + \overline{Z}_{23}\right)$$

Results are shown in Fig. 59.8(d). Observe that the Y–Δ winding connections had no impact on the calculations.

Another detail deserves mention. Although the real and reactive parts of the short-circuit impedances ($\overline{Z}_{12}, \overline{Z}_{23}, \overline{Z}_{31}$) will always be positive, this is not true for the transformer impedances ($\overline{Z}_1, \overline{Z}_2, \overline{Z}_3$). One or more of these can be, and frequently is, negative for actual short-circuit data. Negative values underscore that the circuit of Fig. 59.7 is a *port equivalent* circuit, producing correct values at the winding terminals. One should not attach physical significance to calculated *internal* voltages, currents, or powers.

59.5 Autotransformers

Charles A. Gross

Transformer windings, though magnetically coupled, are electrically isolated from each other. It is possible to enhance certain performance characteristics for transformers by electrically interconnecting primary and secondary windings. Such devices are called **autotransformers**. The benefits to be realized are lower cost, smaller size and weight, higher efficiency, and better voltage regulation. The basic connection is illustrated in Fig. 59.9. The issues will be demonstrated with an example.

Consider the conventional connection, shown in Fig. 59.9(a).

$$\overline{V}_2 = a\overline{V}_1$$

$$\overline{I}_2 = \frac{1}{a}\overline{I}_1$$

$$S_{\text{rating}} = V_1 I_1 = V_2 I_2 = S_{\text{load}}$$

Now for the autotransformer:

$$\overline{V}_2 = \overline{V}_1 + b\overline{V}_1 = (1 + b)\overline{V}_1$$

$$\overline{I}_1 = \overline{I}_2 + b\overline{I}_2 = (1 + b)\overline{I}_2$$

For the same effective ratio

$$1 + b = a$$

Therefore each winding rating is:

$$S_{\text{rated}} = S_{\text{load}}\left(\frac{b}{1+b}\right)$$

For example if $b = 1$ ($a = 2$)

$$S_{\text{rating}} = 1/2\, S_{\text{load}}$$

meaning that the transformer rating is only 50% of the load.

The principal advantage of the autotransformer is the increased power rating. Also, since the losses remain the same, expressed as a percentage of the new rating, they go down, and correspondingly, the efficiency goes up. The machine impedances in per unit drop for similar reasons. A disadvantage is the loss of electrical isolation between primary and secondary. Also, low impedance is not necessarily good, as we shall see when we study faults on power systems. Autotransformers are used in three-phase connections and in voltage control applications.

Defining Terms

Autotransformer: A transformer whose primary and secondary windings are electrically interconnected.

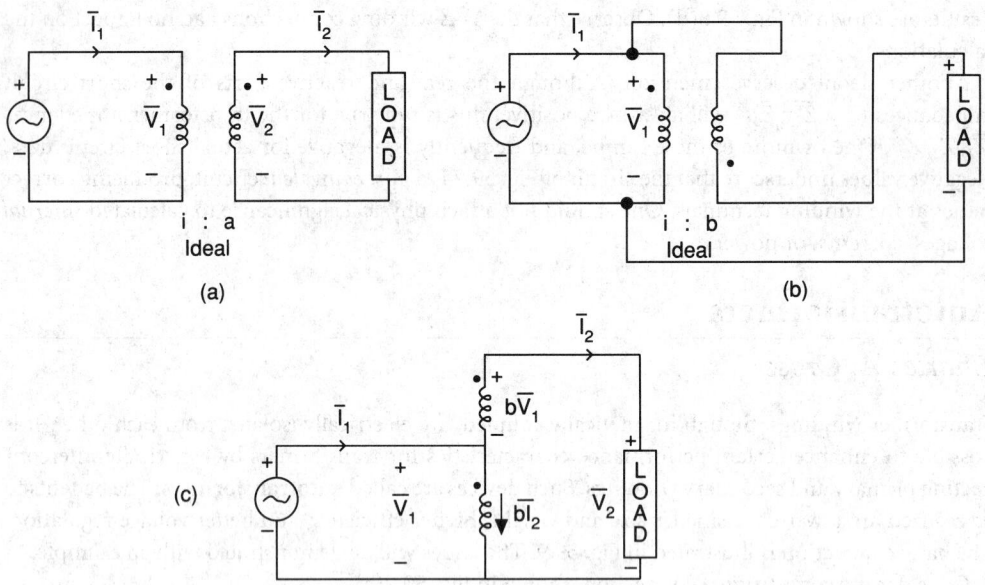

FIGURE 59.9 Autotransformer connection. (a) Conventional step-up connection; (b) autotransformer connection; (c) part (b) redrawn.

Polarity: Consideration of in-phase or out-of-phase relations of primary and secondary ac currents and voltages.

Primary: The source-side winding.

Secondary: The load-side winding.

Transformer: A device which converts ac voltage and current to different levels at essentially constant power and frequency.

References

1. ANSI Standard C57, New York: American National Standards Institute.
2. V. Del Toro, *Basic Electric Machines*, Englewood Cliffs, N.J.: Prentice-Hall, 1990.
3. M. E. El-Hawary, *Electric Power Systems: Design and Analysis*, Reston, Va.: Reston Publishing, 1983.
4. O. I. Elgerd, *Electric Energy Systems Theory: An Introduction*, 2nd ed., New York: McGraw-Hill, 1982.
5. A. E. Fitzgerald, C. Kingsley, and S. Umans, *Electric Machinery*, 5th ed., New York: McGraw-Hill, 1990.
6. C. A. Gross, *Power Systems Analysis*, 2nd ed., New York: Wiley, 1986.
7. N.N. Hancock, *Matrix Analysis of Electrical Machinery*, 2nd ed., Oxford: Pergamon, 1974.
8. G. McPherson, *An Introduction to Electrical Machines and Transformers*, New York: Wiley, 1981.
9. G. R. Slemon, *Magnetoelectric Devices*, New York: Wiley, 1966.
10. R. Stein and W. T. Hunts, Jr., *Electric Power System Components: Transformers and Rotating Machines*, New York: Van Nostrand Reinhold, 1979.

Further Information

For a comprehensive coverage of general transformer theory, see Chapter 2 of *Electric Machines* by G. R. Slemon and A. Straughen (Addison-Wesley, 1980). For transformer standards, see ANSI Standard C57. For a detailed explanation of transformer per-unit scaling, see Chapter 5 of *Power Systems Analysis* by C. A. Gross (John Wiley, 1986).

CHARLES E. L. BROWN AND POWER TRANSMISSION FROM LAUFFEN TO FRANKFURT IN 1891

In 1990 the IEEE Board of Directors voted to designate 1991 as the centennial year for the industrial use of alternating current power. Among the reasons for selecting 1991 was the successful and well-publicized transmission of polyphase power beginning August 24, 1891 from Lauffen, Germany, to the site of an international electrical exhibition in Frankfurt, a distance of about 175 km. This demonstration provided convincing evidence of the economic and technical feasibility of supplying power generated at remote locations to industrial centers.

The Lauffen-Frankfurt project was essentially a joint venture of a German electrical company, Allgemeine Elektrizitats Gesellschaft (AEG), and a Swiss company, Maschinenfabrik Oerlikon. Michael O. Dolivo-Dobrowolsky of AEG designed a polyphase motor that drove a pump supplying an artificial waterfall at the Frankfurt exhibition. An Oerlikon engineer, C. E. L. Brown, designed an innovative polyphase generator which was driven by a water turbine on the Neckar River in Lauffen. He also designed an oil-insulated transformer for the project. A portion of the power brought from Lauffen was used for an illuminated sign with 1,000 incandescent lamps.

The well-known British engineer, William E. Ayrton, who attended the Frankfurt exhibition, wrote that it had shown engineers that towns far from water power sources might become industrial centers. The demonstration convinced the city of Frankfurt to adopt alternating current for its municipal power plant which began operation in 1894 and also influenced the adoption of alternating current at the large hydroelectric plant at Niagara Falls, New York, which began operation in 1895.

C. E. L. Brown (1863–1924) was born in Winterthur, Switzerland. Brown was educated in Swiss schools and served an apprenticeship in a machine shop in Basel before joining the Oerlikon company in 1884. Two years later he became director of the electrical department at Oerlikon. He designed a variety of direct and alternating current machines and was awarded a grand prize for a dynamo design at the Paris exhibition in 1889.

In 1891, Brown and Walter Boveri, also an engineer with Oerlikon, decided to leave Oerlikon and form their own electrical manufacturing company, Brown, Boveri, and Company in Baden, Switzerland. The new firm received the contract to build the municipal power plant in Frankfurt and acquired rights to manufacture steam turbines covered by the patents of Charles Parsons. American versions of the Lauffen type alternator were introduced by General Electric in 1897 and later by the Westinghouse Company and Allis-Chalmers.

Brown-Boveri grew into one of the world's leading manufacturers of power machinery. In 1987, Brown-Boveri merged with Allmanna Svenska Elektriska Aktiebolaget (Asea) of Sweden to become reportedly "the world's largest electrotechnical concern, Asea Brown Boveri." In April 1988 Asea Brown Boveri and Westinghouse reached an agreement to engage in joint ventures in power generation and distribution. These developments have raised concern over the U.S. becoming increasingly dependent on non-U.S. firms for electrical power machinery and related apparatus. Similar concerns were voiced a century ago when Westinghouse was selected instead of Brown-Boveri to supply the generators for the first power plant at Niagara Falls.

Source: Adapted from J.E. Brittain, *Proc. IEEE*, vol. 79, no. 8, p. 1208, August 1991. © 1991 IEEE.

60

Energy Distribution

60.1   Introduction 1310
60.2   Primary Distribution System 1312
60.3   Secondary Distribution System 1314
60.4   Radial Distribution System 1314
60.5   Secondary Networks 1315
60.6   Load Characteristics 1316
60.7   Voltage Regulation 1317
60.8   Capacitors and Voltage Regulators 1318

George G. Karady
Arizona State University

60.1 Introduction

Distribution is the last section of the electrical power system. Figure 60.1 shows the major components of the electric power system. The power plants convert the energy stored in the fuel (coal, oil, gas, nuclear) or hydro into electric energy. The energy is supplied through step-up transformers to the electric network. To reduce energy transportation losses, step-up transformers increase the voltage and reduce the current. The high-voltage network, consisting of transmission lines, connects the power plants and high-voltage **substations** in parallel. The typical voltage of the high-voltage transmission network is between 240 and 765 kV. The high-voltage substations are located near the load centers, for example, outside a large town. This network permits load sharing among power plants and assures a high level of reliability. The failure of a line or power plant will not interrupt the energy supply.

The subtransmission system connects the high-voltage substations to the distribution substations. These stations are directly in the load centers. For example, in urban areas, the distance between the distribution stations is around 5 to 10 miles. The typical voltage of the subtransmission system is between 138 and 69 kV. In high load density areas, the subtransmission system uses a network configuration that is similar to the high-voltage network. In medium and low load density areas, the loop or radial connection is used. Figure 60.1 shows a typical radial connection.

The distribution system has two parts, primary and secondary. The primary distribution system consists of overhead lines or underground cables, which are called **feeders**. The feeders run along the streets and supply the distribution transformers that step the voltage down to the secondary level (120–480 V). The secondary distribution system contains overhead lines or underground cables supplying the consumers directly (houses, light industry, shops, etc.) by single- or three-phase power. Separate, dedicated primary feeders supply industrial customers requiring several megawatts of power. The subtransmission system directly supplies large factories consuming over 50 MW.

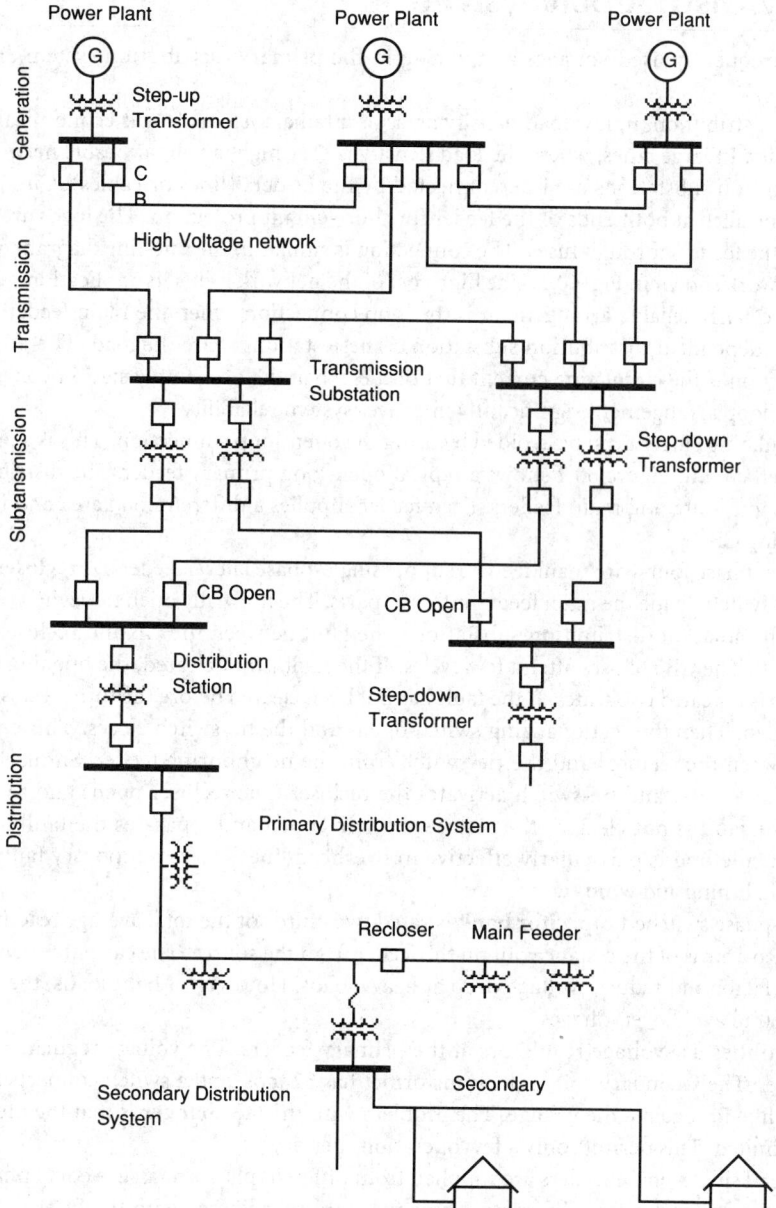

FIGURE 60.1 Electric energy system.

Table 60.1 Typical Primary Feeder Voltages

Class, kV	Voltage, kV	Wiring
2.5	2.4	3-wire delta
5	4.16	4-wire Y
8.66	7.2	4-wire Y
15	12.47	3-wire delta/4-wire Y
25	22.9	4-wire Y
35	34.5	4-wire Y

60.2 Primary Distribution System

The most frequently used voltages and wiring in the primary distribution system are listed in Table 60.1.

Primary distribution, in low load density areas, is a radial system. This is economical but yields low reliability. In large cities, where the load density is very high, a primary cable network is used. The distribution substations are interconnected by the feeders (lines or cables). Circuit breakers (CBs) are installed at both ends of the feeder for short-circuit protection. The loads are connected directly to the feeders through fuses. The connection is similar to the one-line diagram of the high-voltage network shown in Fig. 60.1. The high cost of the network limits its application. A more economical and fairly reliable arrangement is the loop connection, when the main feeder is supplied from two independent distribution substations. These stations share the load. The problem with this connection is the circulating current that occurs when the two supply station voltages are different. The loop arrangement significantly improves system reliability.

The circulating current can be avoided by using the open-loop connection. This is a popular, frequently used circuit. Figure 60.2 shows a typical open-loop primary feeder. The distribution substation has four outgoing main feeders. Each feeder supplies a different load area and is protected by a reclosing CB.

The three-phase four-wire main feeders supply single-phase lateral feeders. A **recloser** and a sectionalizing switch divide the main feeder into two parts. The normally open tie-switch connects the feeder to the adjacent distribution substation. The fault between the CB and recloser opens the reclosing CB. The CB recloses after a few cycles. If the fault is not cleared, the opening and reclosing process is repeated two times. If the fault has not been cleared before the third reclosing, the CB remains open. Then the sectionalizing switch opens and the tie-switch closes. This energizes the feeder between the recloser and the tie-switch from the neighboring feeder. Similarly, the fault between the recloser and tie-switch activates the recloser. The recloser opens and recloses three times. If the fault is not cleared, the recloser remains open and separates the faulty part of the feeder. This method is particularly effective in overhead lines where temporary faults are often caused by lightning and wind.

A three-phase switched **capacitor bank** is rated two-thirds of the total average reactive load and installed two-thirds of the distance out on the feeder from the source. The capacitor bank improves the power factor and reduces voltage drop at heavy loads. However, at light loads, the capacitor is switched off to avoid overvoltages.

Some utilities use voltage regulators at the primary feeders. The voltage regulator is an auto-transformer. The secondary coil of the transformer has 32 taps, and a switch connects the selected tap to the line to regulate the voltage. The problem with the **tap changer** is that the lifetime of the switch is limited. This permits only a few operations per day.

The lateral single-phase feeders are supplied from different phases to assure equal phase loading. Fuse cutouts protect the lateral feeders. These fuses are coordinated with the fuses protecting the distribution transformers. The fault in the distribution transformer melts the transformer fuse first. The lateral feeder fault operates the cutout fuse before the recloser or CB opens permanently.

A three-phase line supplies the larger loads. These loads are protected by CBs or high-power fuses.

Most primary feeders in rural areas are overhead lines using pole-mounted distribution transformers. The capacitor banks and the reclosing and sectionalizing switches are also pole-mounted. Overhead lines reduce the installation costs but reduce aesthetics.

In urban areas, an underground cable system is used. The switchgear and transformers are placed in underground vaults or ground-level cabinets. The underground system is not affected by weather and is highly reliable. Unfortunately, the initial cost of an underground cable is significantly higher than an overhead line with the same capacity. The high cost limits the underground system to high-density urban areas and housing developments.

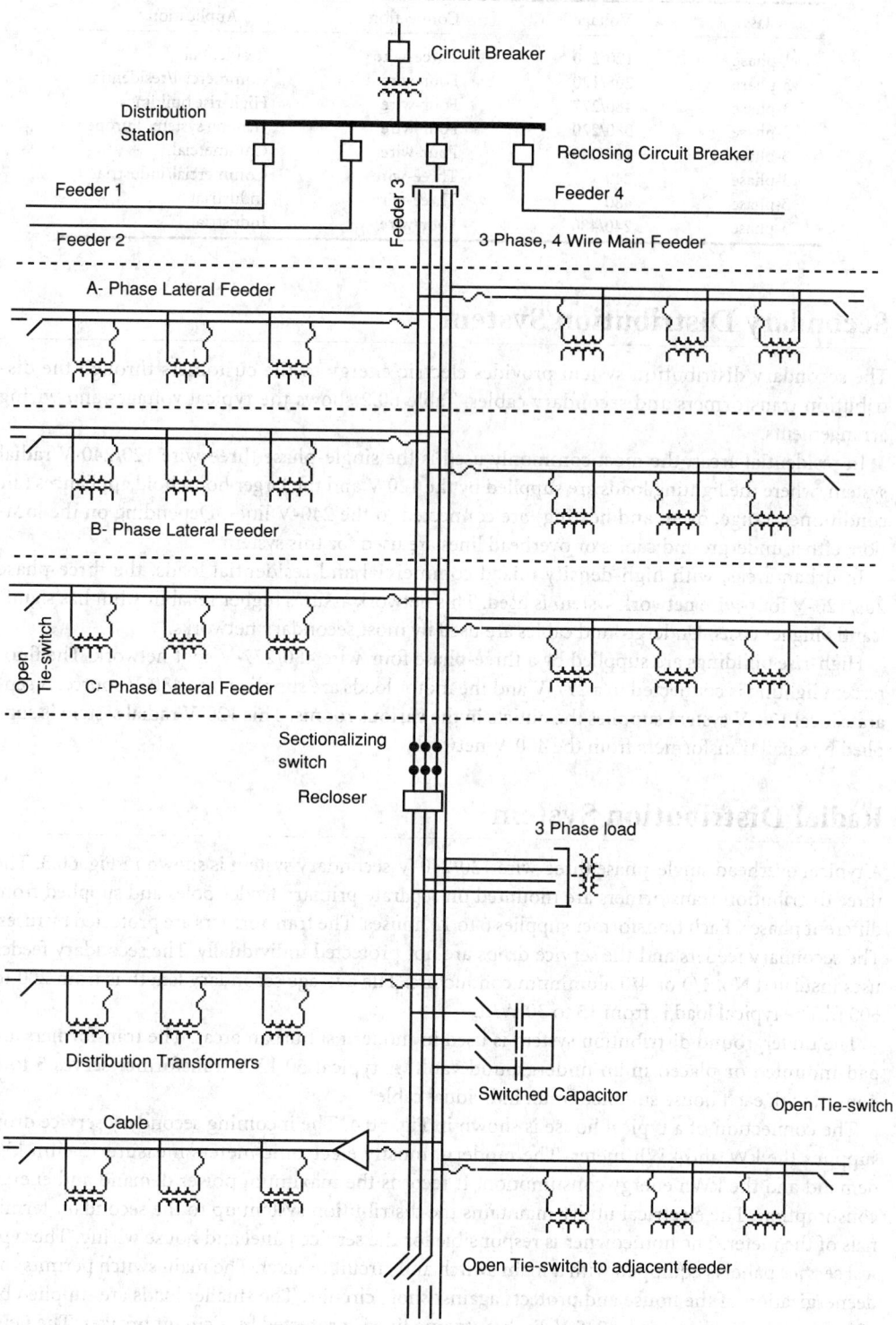

FIGURE 60.2 Radial primary distribution system.

Table 60.2 Secondary Voltages and Connections

Class	Voltage	Connection	Application
1-phase	120/240	Three-wire	Residential
3-phase	208/120	Four-wire	Commercial/residential
3-phase	480/277	Four-wire	High-rise buildings
3-phase	380/220	Four-wire	General system, Europe
3-phase	120/240	Four-wire	Commercial
3-phase	240	Three-wire	Commercial/industrial
3-phase	480	Three-wire	Industrial
3-phase	240/480	Four-wire	Industrial

60.3 Secondary Distribution System

The secondary distribution system provides electric energy to the customers through the distribution transformers and secondary cables. Table 60.2 shows the typical voltages and wiring arrangements.

In residential areas, the most commonly used is the single-phase three-wire 120/240-V radial system, where the lighting loads are supplied by the 120 V and the larger household appliances (air conditioner, range, oven, and heating) are connected to the 240-V lines. Depending on the location, either underground cables or overhead lines are used for this system.

In urban areas, with high-density mixed commercial and residential loads, the three-phase 208/120-V four-wire network system is used. This network assures higher reliability but has significantly higher costs. Underground cables are used by most secondary networks.

High-rise buildings are supplied by a three-phase four-wire 480/277-V spot network. The fluorescent lighting is connected to a 277-V and the motor loads are supplied by a 480-V source. A separate local 120-V system supplies the outlets in the various rooms. This 120-V radial system is supplied by small transformers from the 480-V network.

60.4 Radial Distribution System

A typical overhead single-phase three-wire 120/240-V secondary system is shown in Fig. 60.3. The three distribution transformers are mounted on separate primary feeder poles and supplied from different phases. Each transformer supplies 6 to 12 houses. The transformers are protected by fuses. The secondary feeders and the service drops are not protected individually. The secondary feeder uses insulated No. 1/0 or 4/0 aluminum conductors. The average secondary length is from 200 to 600 ft. The typical load is from 15 to 30 W/ft.

The underground distribution system is used in modern suburban areas. The transformers are pad-mounted or placed in an underground vault. A typical 50-kVA transformer serves 5 to 6 houses, with each house supplied by an individual cable.

The connection of a typical house is shown in Fig. 60.4. The incoming secondary service drop supplies the kW and kWh meter. The modern, mostly electronic meters measure 15-min kW demand and the kWh energy consumption. It records the maximum power demand and energy consumption. The electrical utility maintains the distribution system up to the secondary terminals of the meter. The homeowner is responsible for the service panel and house wiring. The typical service panel is equipped with a main switch and circuit breaker. The main switch permits the deenergization of the house and protects against short circuits. The smaller loads are supplied by 120 V and the larger loads by 240 V. Each outgoing line is protected by a circuit breaker. The neutral has to be grounded at the service panel, just past the meter. The water pipe was used for grounding in older houses. In new houses a metal rod, driven in the earth, provides proper grounding. In addition, a separate bare wire is connected to the ground. The ground wire con-

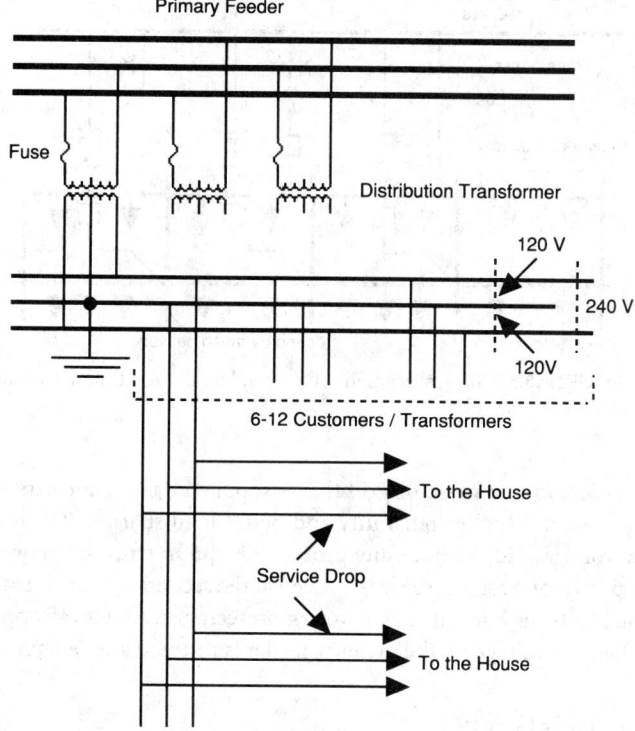

FIGURE 60.3 Typical 120/240-V radial secondary system.

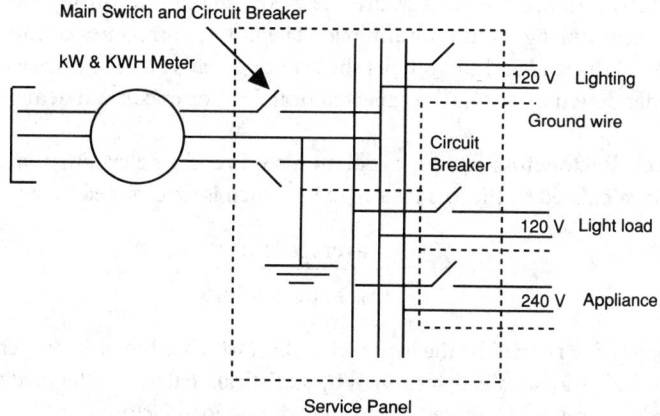

FIGURE 60.4 Residential electrical connection.

nects the metal parts of the appliances and service panel box together to protect against ground-fault-produced electric shocks.

60.5 Secondary Networks

The secondary network is used in urban areas with high load density. Figure 60.5 shows a segment of a typical secondary network.

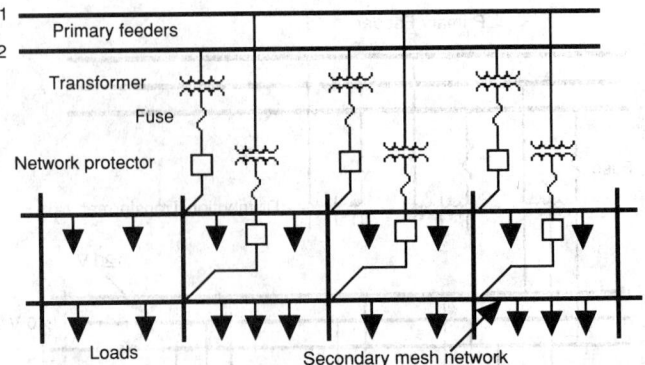

FIGURE 60.5   Typical segment of a secondary distribution network.

The secondary feeders form a mesh or grid that is supplied by transformers at the node points. The multiple supply assures higher reliability and better load sharing. The loads are connected directly to the low-voltage grid, without any protection equipment. The network is protected by fuses and network protector circuit breakers installed at the secondary transformers. A short circuit blows the fuses and limits the current. The network protectors automatically open on reverse current and reclose when the voltage on the primary feeder is restored after a fault.

60.6 Load Characteristics

The distribution system load varies during the day. The maximum load occurs in the early evening or late afternoon, and the minimum load occurs at night. The design of the distribution system requires both values, because the voltage drop is at the maximum during the peak load, and overvoltage may occur during the minimum load. The power companies continuously study the statistical variation of the load and can predict the expected loads on the primary feeders with high accuracy. The feeder design considers the expected peak load or maximum demand and the future load growth.

The economic conductor cross-section calculation requires the determination of average losses. The average loss is calculated by the loss factor (LSF), which is determined by statistical analyses of load variation.

$$LSF = \frac{\text{average loss}}{\text{loss at peak load}}$$

The average load is determined by the load factor (LF), which is the ratio of average load to peak load. The load factor for an area is determined by statistical analyses of the load variation in past years. The approximate relation between the loss factor and load factor is

$$LSF = 0.3LF + 0.7LF^2$$

This equation is useful because the load factor is measured continuously by utilities, and more accurate values are available for the load factor than for the loss factor. Typical values are given in Table 60.3.

The connected load or demand can be estimated accurately in residential and industrial areas. The connected load or demand is the sum of continuous ratings of apparatus connected to the system. However, not all equipment is used simultaneously. The actual load in a system is significantly lower than the connected load. The demand factor is used to estimate the actual or maximum demand. The demand factor (DF) is defined by

$$DF = \frac{\text{maximum demand}}{\text{total connected demand}}$$

The demand factor depends on the number of customers and the type of load. Typical demand factor values are given in Table 60.4.

60.7 Voltage Regulation

The voltage supplied to each customer should be within the ±5% limit, which, at 120 V, corresponds to 114 and 126-V. Figure 60.6 shows a typical voltage profile for a feeder at light and heavy load conditions. The figure shows that at heavy load, the voltage at the end of the line will be less than the allowable minimum voltage. However, at the light load condition the voltage supplied to each customer will be within the allowable limit. Calculation of the voltage profile, voltage drop, and feeder loss is one of the major tasks in distribution system design. The concept of voltage drop and loss calculation is demonstrated using the feeder shown in Fig. 60.6.

To calculate the voltage drop, the feeder is divided into sections. The sections are determined by the loads. Assuming a single-phase system, the load current is calculated by Eq. (60.1):

$$|I_i| = \frac{P_i}{V \cos \varphi_i}, \qquad I_i = |I_i| (\cos \varphi_i + \sin \varphi_i) \qquad (60.1)$$

where P is the power of the load, V is the rated voltage, and φ is the power factor.

The section current is the sum of the load currents. Equation (60.2) gives the section current between load i and $i-1$:

$$I_{(i, i-1)} = \sum_1^{i-1} I_i \qquad (60.2)$$

The electrical parameters of the overhead feeders are the resistance and reactance, which are given in Ω/mi. The underground feeders have significant capacitance in addition to the reactance and resistance. The capacitance is given in μF/mi. The actual values for overhead lines can be calculated using the conductor diameter and phase-to-phase and phase-to-ground distances [Fink and Beaty, 1978]. The residential underground system generally uses single-conductor cables with polyethylene insulation. The older systems use rubber insulation with neoprene jacket. Circuit parameters should be obtained from manufacturers. The distribution feeders are short transmission lines. Even the primary feeders are only a few miles long. This permits the calculation of the section resistance and reactance by multiplying the Ω/mi values by the length of the section. The length of the section in a single-phase two-wire system is two times the actual length. In a balanced three-phase system, it is the simple length. In a single-phase three-wire system the voltage drop on the neutral conductor must be calculated. Further information may be obtained from Pansini [1991].

Equation (60.3) gives the voltage drop, with a good approximation, for section i, $(i-1)$. The total voltage drop is the sum of the sections voltage drops.

Table 60.3 Typical Annual Load Factor Values

Type of Load	Load Factor
Residential	0.48
Commercial	0.66
Industrial	0.72

Table 60.4 Typical Demand Factors for Multifamily Dwellings

Number of Dwellings	Demand Factor, %
3 to 5	45
18 to 20	38
39 to 42	28
62 & over	23

Adapted from Article 220-32, Table 202-32, *National Electrical Code 1987*, Quincy, Mass.: National Fire Protection Association, 1986. With permission.

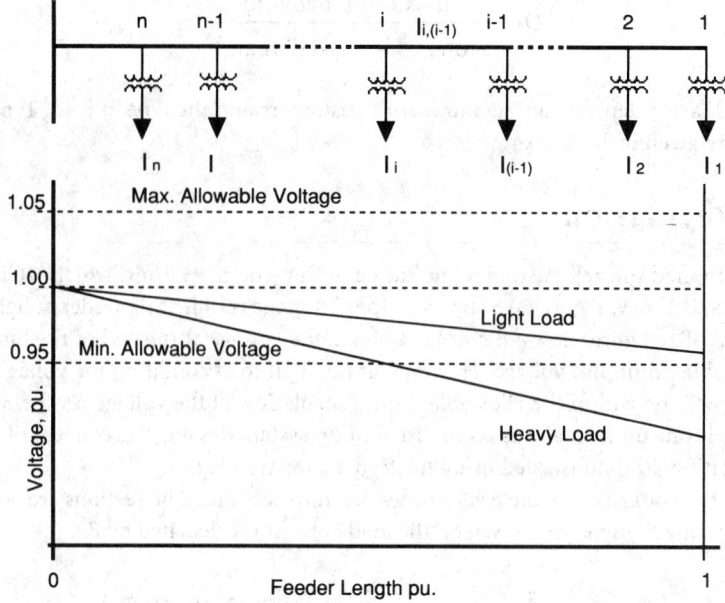

FIGURE 60.6   Feeder voltage profile.

$$e_{i,(i-1)} = |I_{i,(i-1)}|(R_{i,(i-1)} \cos \varphi_{i,(i-1)} + X_{i,(i-1)} \sin \varphi_{i,(i-1)}) \tag{60.3}$$

Equation (60.4) gives the losses on the line:

$$\text{Loss}_i = \sum_{1}^{i-1} (I_{i,(i-1)})^2 R_{i,(i-1)} \tag{60.4}$$

The presented calculation method describes the basic concept of feeder design; more details can be found in the literature.

60.8   Capacitors and Voltage Regulators

The voltage drop can be reduced by the application of a shunt capacitor. As shown in Fig. 60.7, a properly selected and located shunt capacitor assures that the voltage supplied to each of the customers will be within the allowable limit at the heavy load condition. However, at light load, the same capacitor will increase the voltage above the allowable limit. Most capacitors in the distribution system use switches. The capacitor is switched off during the night when the load is light and switched on when the load is heavy. The most frequent use of capacitors is on the primary feeders. In an overhead system, three-phase capacitor banks with vacuum switches are installed on the poles. Residential underground systems require less shunt capacitance for voltage control due to the reduced reactance. Even so, shunt capacitors are used for power factor correction and loss reduction.

The optimum number, size, and location of capacitor banks on a feeder is determined by detailed computer analyses. The concept of optimization includes the minimization of the operation, installation, and investment costs. The most important factor that affects the selection is the distribution and power factor of loads. In residential areas, the load is uniformly distributed. In this case the optimum location of the capacitor bank is around two-thirds of the length of the feeder.

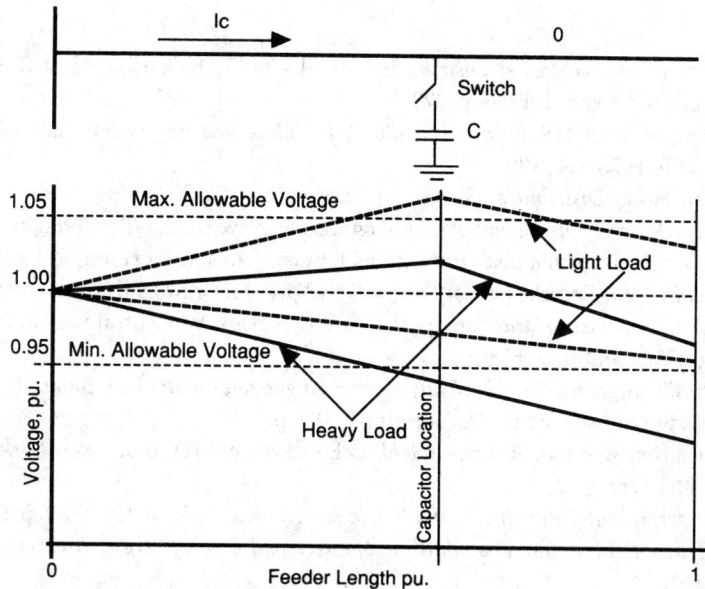

FIGURE 60.7   Capacitor effect on voltage profile.

The effect of capacitor bank can be studied by adding the capacitor current to the load current. The capacitor current flows between the supply and the capacitor as shown in Fig. 60.7. Its value can be calculated from Eq. (60.5) for a single-phase system:

$$I_c = j\omega CV, \qquad \omega = 2\pi f \tag{60.5}$$

where C is the capacitance, f is the frequency (60 Hz), and V is the voltage to ground.

The capacitive current is added to the inductive load current, reducing the total current, the voltage drop, and losses. The voltage drop and loss can be calculated from Eqs. (60.2) to (60.5).

The voltage regulator is a tap-changing transformer, which is located, in most cases, at the supply end of the feeder. The tap changer increases the supply voltage, which in turn increases the voltage above the allowable minimum at the last load. The tap changer transformer has two windings. The excitation winding is connected in parallel. The regulating winding is connected in series with the feeder. The latter has taps and a tap changer switch. The switch changes the tap position according to the required voltage. The tap changing requires the short interruption of load current. The frequent current interruptions reduce the lifetime of the tap changer switch. This problem limits the number of tap changer operations to between one to three per day.

Defining Terms

Capacitor bank: Consists of capacitors connected in parallel. Each capacitor is placed in a metal can and equipped with bushings.

Feeder: Overhead lines or cables which are used to distribute the load to the customers. They interconnect the distribution substations with the loads.

Recloser: A circuit breaker which is designed to interrupt short-circuit current and reclose the circuit after interruption.

Substation: A junction point in the electric network. The incoming and outgoing lines are connected to a busbar through circuit breakers.

Tap changer: A transformer. One of the windings is equipped with taps. The usual number of taps is 32. Each tap provides a 1% voltage regulation. A special circuit breaker is used to change the tap position.

References

D.F.S. Brass *et al.*, in *Electric Power Distribution, 415 V–33 kV*, E.O. Taylor and G.A. Boal (eds.), London: Edward Arnold, 1966, p. 272.

D.G. Fink and H.W. Beaty, *Standard Handbook for Electrical Engineers*, 11th ed., New York: McGraw-Hill, 1978, sec. 18.

T. Gönen, *Electric Power Distribution System Engineering*, New York: Wiley, 1986.

T. Gönen, *Electric Power Transmission System Engineering*, New York: Wiley, 1988, p. 723.

A.J. Pansini, *Power Transmission and Distribution*, Liburn, Ga.: The Fairmont Press, 1991.

E.P. Parker, *McGraw-Hill Encyclopedia of Energy*, New York: McGraw-Hill, 1981, p. 838.

Various, *Electrical Transmission and Distribution Reference Book*, W. Central Station Engineers, East Pittsburgh: Westinghouse Electric Corporation, 1950, p. 824.

Various, *Distribution Systems. Electric Utility Engineering Reference Books*, J. Billard (ed.), East Pittsburgh: Westinghouse Electric Corporation, 1965, p. 567.

Various, *EHV Transmission Line Reference Book*, G.E.C. Project EHV (ed.), New York: Edison Electric Institute, 1968, p. 309.

B.M. Weedy, *Underground Transmission of Electric Power*, New York: Wiley, 1980, p. 294.

W.L. Weeks, *Transmission and Distribution of Electrical Energy*, New York: Harper & Row, 1981, p. 302.

Further Information

Other recommended publications include J. M. Dukert, *A Short Energy History of the United States*, Edison Electric Institute, 1980. Also, the *IEEE Transactions on Power Delivery* publishes distribution papers sponsored by the Transmission and Distribution Committee. These papers deal with the latest development in the distribution area. Every-day problems are presented in two magazines: *Transmission & Distribution* and *Electrical World*.

61

Chen-Ching Liu
University of Washington

Khoi Tien Vu
Clemson University

Yixin Yu
Tianjing University

Donald Galler
FaAA Electrical Corporation

Electrical Machines

61.1 Generators 1321
  AC Generators • DC Generators
61.2 Motors 1333
  Motor Applications • Motor Analysis

61.1 Generators

Chen-Ching Liu, Khoi Tien Vu, and Yixin Yu

Electric generators are devices that convert energy from a mechanical form to an electrical form. This process, known as electromechanical energy conversion, involves magnetic fields that act as an intermediate *medium*. There are two types of generators: alternating current (ac) and direct current (dc). This section explains how these devices work and how they are modeled in analytical or numerical studies.

The input to the machine can be derived from a number of energy sources. For example, in the generation of large-scale electric power, coal can produce steam that drives the shaft of the machine. Typically, for such a thermal process, only about 1/3 of the raw energy (i.e., from coal) is converted into mechanical energy. The final step of the energy conversion is quite efficient, with an efficiency close to 100%.

The generator's operation is based on Faraday's law of electromagnetic induction. In brief, if a coil (or winding) is linked to a varying magnetic field, then an electromotive force, or voltage, emf, is induced across the coil. Thus, generators have two essential parts: one creates a magnetic field, and the other where the emf's are induced. The magnetic field is typically generated by electromagnets (thus, the field intensity can be adjusted for control purposes), whose windings are referred to as field windings or **field circuits**. The coils where the emf's are induced are called *armature* windings or **armature circuits**. One of these two components is stationary (stator), and the other is a rotational part (rotor) driven by an external torque. Conceptually, it is immaterial which of the two components is to rotate because, in either case, the armature circuits always "see" a varying magnetic field. However, practical considerations lead to the common design that for ac generators, the field windings are mounted on the rotor and the armature windings on the stator. In contrast, for dc generators, the field windings are on the stator and armature on the rotor.

AC Generators

Today, most electric power is produced by synchronous generators. Synchronous generators rotate at a constant speed, called **synchronous speed**. This speed is dictated by the operating frequency of the system and the machine structure. There are also ac generators that do not necessarily rotate at a fixed speed such as those found in windmills (induction generators); these generators, however, account for only a very small percentage of today's generated power.

1321

Synchronous Generators

Principle of Operation.   For an illustration of the steady-state operation, refer to Fig. 61.1 which shows a cross section of an ac machine. The rotor consists of a winding wrapped around a steel body. A dc current is made to flow in the rotor winding (or field winding), and this results in a mag-

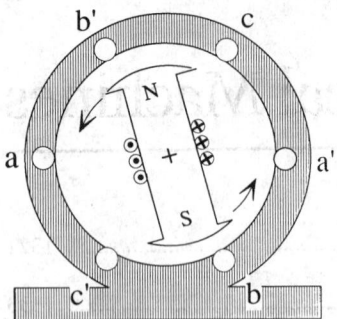

netic field (rotor field). When the rotor is made to rotate at a constant speed, the three stationary windings *aa'*, *bb'*, and *cc'* experience a periodically varying magnetic field. Thus, emf's are induced across these windings in accordance with Faraday's law. These emf's are ac and periodic; each period corresponds to one revolution of the rotor. Thus, for 60-Hz electricity, the rotor of Fig. 61.1 has to rotate at 3600 revolutions per minute (rpm); this is the synchronous speed of the given machine. Because the windings *aa'*, *bb'*, and *cc'* are displaced equally in space from each other (by 120 degrees), their emf waveforms are displaced in time by 1/3 of a period. In other words, the machine of Fig. 61.1 is capable of generating three-phase electricity. This machine has two poles since its rotor field resembles that of a bar magnet with a north pole and a south pole.

FIGURE 61.1 Cross section of a simple two-pole synchronous machine. The rotor body is salient. Current in rotor winding: ⊗ into the page, ⊙ out of the page.

When the stator windings are connected to an external (electrical) system to form a closed circuit, the steady-state currents in these windings are also periodic. These currents create magnetic fields of their own. Each of these fields is pulsating with time because the associated current is ac; however, the combination of the three fields is a **revolving field**. This revolving field arises from the space displacements of the windings and the phase differences of their currents. This combined magnetic field has two poles and rotates at the same speed and direction as the rotor. In summary, for a loaded synchronous (ac) generator operating in a steady state, there are two fields rotating at the same speed: one is due to the rotor winding and the other due to the stator windings. It is important to observe that the armature circuits are in fact exposed to two rotating fields, one of which, the armature field, is caused by and in fact tends to counter the effect of the other, the rotor field. The result is that the induced emf in the armature can be reduced when compared with an unloaded machine (i.e., open-circuited stator windings). This phenomenon is referred to as **armature reaction**.

It is possible to build a machine with *p* poles, where $p = 4, 6, 8, \ldots$ (even numbers). For example, the cross-sectional view of a four-pole machine is given in Fig. 61.2. For the specified direction of the (dc) current in the rotor windings, the rotor field has two pairs of north and south poles arranged as shown. The emf induced in a stator winding completes one period for every pair of north and south poles sweeping by; thus, each revolution of the rotor corresponds to two periods of the stator emf's. If the machine is to operate at 60 Hz then the rotor needs to rotate at 1800 rpm. In general, a *p*-pole machine operating at 60 Hz has a rotor speed of $3600/(p/2)$ rpm. That is, the lower the number of poles is, the higher the rotor speed has to be. In practice, the number of poles is dictated by the mechanical system (prime mover) that drives the rotor. Steam turbines operate best at a high speed; thus, two- or four-pole machines are suitable. Machines driven by hydro turbines usually have more poles.

Usually, the stator windings are arranged so that the resulting armature field has the same number of poles as the rotor field. In practice, there are many possible ways to arrange these windings; the essential idea, however, can be understood via the simple arrangement shown in Fig. 61.2. Each phase consists of a pair of windings (thus occupies four slots on the stator structure), e.g., those for phase *a* are labeled a_1a_1' and a_2a_2'. Geometry suggests that, at any time instant, equal emf's are induced across the windings of the same phase. If the individual windings are connected in series as shown in Fig. 61.2, their emf's add up to form the phase voltage.

Mathematical/Circuit Models. There are various models for synchronous machines, depending on how much detail one needs in an analysis. In the simplest model, the machine is equivalent to a constant voltage source in series with an impedance. In more complex models, numerous nonlinear differential equations are involved.

Steady-state model. When a machine is in a steady state, the model requires no differential equations. The representation, however, depends on the rotor structure: whether the rotor is cylindrical (round) or salient.

The rotors depicted in Figs. 61.1 and 61.2 are salient since the poles are protruding from the shaft. Such structures are mechanically weak, since at a high speed (3600 rpm and 1800 rpm, respectively) the centrifugal force becomes a serious problem. Practically, for high-speed turbines, round-rotor (or cylindrical-rotor) structures are preferred. The cross section of a two-pole, round-rotor machine is depicted in Fig. 61.3. From a practical viewpoint, salient rotors are easier to build because each pole and its winding can be manufactured separately and then mounted on the rotor shaft. For round rotors, slots need to be reserved in the rotor where the windings can be placed.

The mathematical model for round-rotor machines is much simpler than that for salient-rotor ones. This stems from the fact that the rotor body has a permeability much higher than that of the air. In a steady state, the stator field and the rotor body are at a standstill relative to each other. (They rotate at the same speed as discussed earlier.) If the rotor is salient, it is easier to establish the magnetic flux lines along the direction of the rotor body (when viewed from the cross section). Therefore, for the same set of stator currents, different positions of the rotor alter the stator field in different ways; this implies that the induced emf's are different. If the rotor is round, then the relative position of the rotor structure does not affect the stator field. Hence, the associated mathematical model is simplified.

In the following, the steady-state models of the round-rotor and salient-rotor generators are explained.

Refer to Fig. 61.3 which shows a two-pole round-rotor machine. Without loss of generality, one can select phase *a* (i.e., winding *aa'*) for the development of a mathematical model of the machine. As mentioned previously, the (armature or stator) winding of phase *a* is exposed to two magnetic fields: rotor field and stator field.

1. Rotor field. Its flux as seen by winding *aa'* varies with the rotor position; the flux linkage is largest when the N–S axis is perpendicular to the winding surface and minimum (zero) when this axis aligns with the surface. Thus, one can express the flux due to the rotor field as seen by winding *aa'* as $\lambda_1 = L(\theta)I_F$ where θ is to denote the angular position of the N–S axis (of the rotor field) relative to the surface of *aa'*, I_F is the rotor current (a dc current), and L is a periodic function of θ.

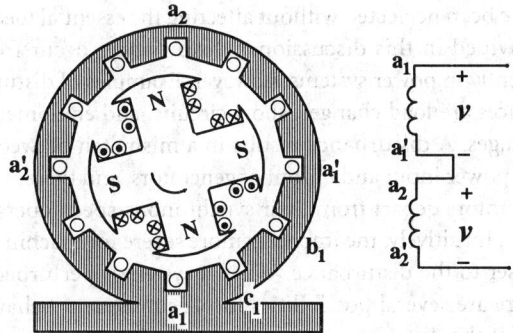

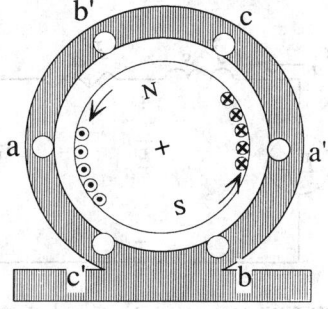

FIGURE 61.2 *Left*, cross section of a four-pole synchronous machine. Rotor has a salient pole structure. *Right*, schematic diagram for phase *a* windings.

FIGURE 61.3 Cross section of a two-pole round-rotor synchronous machine.

2. Stator field. Its flux as seen by winding aa' is a combination of three individual fields which are due to currents in the stator windings, i_a, i_b, and i_c. This flux can be expressed as $\lambda_2 = L_s i_a + L_m i_b + L_m i_c$, where L_s (L_m) is the self (mutual) inductance. Because the rotor is round, L_s and L_m are not dependent on θ, the relative position of the rotor and the winding. Typically, the sum of the stator currents $i_a + i_b + i_c$ is near zero; thus, one can write $\lambda_2 = (L_s - L_m)i_a$.

The total flux seen by winding aa' is $\lambda = \lambda_1 - \lambda_2 = L(\theta)I_F - (L_s - L_m)i_a$, where the minus sign in $\lambda_1 - \lambda_2$ is due to the fact that the stator field opposes the rotor field. The induced emf across the winding aa' is $d\lambda/dt$, the time derivative of λ:

$$e_a = \frac{d\lambda}{dt} = \frac{dL}{dt}I_F - (L_s - L_m)\frac{di_a}{dt} \triangleq e_F - (L_s - L_m)\frac{di_a}{dt}$$

The time-varying quantities are normally sinusoidal, and for practical purposes, can be represented by phasors. Thus the above expression becomes:

$$\overline{E}_a = \overline{E}_F - (L_s - L_m)j\omega_0\overline{I}_a \triangleq \overline{E}_F - jX_s\overline{I}_a$$

where ω_0 is the angular speed (rad/s) of the rotor in a steady state. This equation can be modeled as a voltage source $\overline{E}_F$ behind a reactance jX_s, as shown in Fig. 61.4; this reactance is usually referred to as *synchronous reactance*. The resistor R_a in the diagram represents the winding resistance, and V_t is the voltage measured across the winding.

As mentioned, the theory for salient-rotor machines is more complicated. In the equation $\lambda_2 = L_s i_a + L_m i_b + L_m i_c$, the terms L_s and L_m are now dependent on the (relative) position of the rotor. For example (refer to Fig. 61.1), L_s is maximum when the rotor is in a vertical position and minimum when the rotor is 90° away.

In the derivation of the mathematical/circuit model for salient-rotor machines, the stator field B_2 can be resolved into two components; when the rotor is viewed from a cross section, one component aligns along the rotor and the other is perpendicular to the rotor (Fig. 61.5). The component B_d, which directly opposes the rotor field, is said to belong to the *direct axis*; the other component, B_q, is weaker and belongs to the *quadrature axis*. The model for a salient-rotor machine consists of two circuits, direct-axis circuit and quadrature-axis circuit, each similar to Fig. 61.4. Any quantity of interest, such as I_a, the current in winding aa', is made up of two components, one from each circuit. The round-rotor machine can be viewed as a special case of the salient-pole theory where the corresponding parameters of the d-axis and q-axis circuits are equal.

Dynamic models. When a power system is in a steady state (i.e., operated at an equilibrium), the electrical output of each generator is equal to the power applied to the rotor shaft. (Various losses have been neglected without affecting the essential ideas provided in this discussion.) Disturbances occur frequently in power systems, however. Examples of disturbances are load changes, short circuits, and equipment outages. A disturbance results in a mismatch between the power input and output of generators, and therefore the rotors depart from their synchronous-speed operation. Intuitively, the impact is more severe for machines closer to the disturbance. When a system is perturbed, there are several possibilities for its subsequent behavior. If the disturbance is small, the machines may soon reach a new steady speed, which is close to or identical to their synchronous speed, in which case the system is said to be stable. It may also happen that some

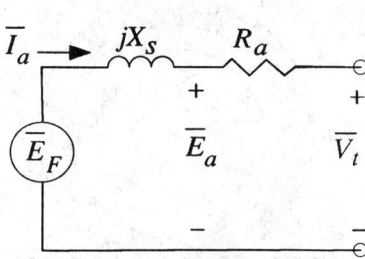

FIGURE 61.4 Per-phase equivalent circuit of round-rotor synchronous machines. $\overline{E}_F$ is the internal voltage (phasor form) and V_t is the terminal voltage.

machines speed up while others slow down. In a more complicated situation, a rotor may oscillate about its synchronous speed. This results in an unstable case. An unstable situation can result in abnormal changes in system frequency and voltage and, unless properly controlled, may lead to damage to machines (e.g., broken shafts). To study these phenomena, dynamic models are required. Details of a dynamic model depend on a number of factors such as location of disturbance and time duration of interest. An overview of dynamic generator models is given here. In essence, there are two aspects that need be modeled: electromechanical and electromagnetic.

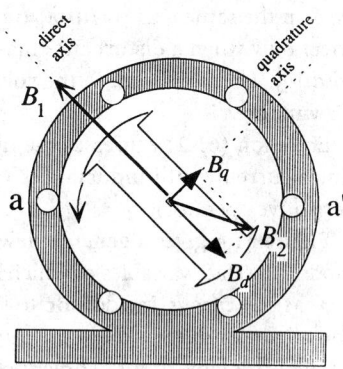

FIGURE 61.5 In the salient-pole theory, the stator field (represented by a single vector B_2) is decomposed into B_d and B_q. Note that $|B_d| > |B_q|$.

1. *Electromechanical equations.* Electromechanical equations are to model the effect of input–output imbalance on the rotor speed (and therefore on the operating frequency). The rotor of each machine can be described by the so-called **swing equation**,

$$M \frac{d^2\theta}{dt^2} + D \frac{d\theta}{dt} = P_{in} - P_{out}$$

where θ denotes the rotor position relative to a certain rotating frame, M the inertia of rotor, and D damping. The term $d\theta/dt$ represents the angular velocity and $d^2\theta/dt^2$ is the angular acceleration of the rotor. The preceding differential equation is derived from Newton's law for rotational motions and, in some respects, resembles the dynamical equation of a swinging pendulum (with $P_{in} \sim$ driving torque, and $P_{out} \sim$ restoring torque). The term P_{in}, which drives the rotor shaft, can be considered constant in many cases. The term P_{out}, the power sent out to the system, may behave in a very complicated way. Qualitatively, P_{out} tends to increase (respectively, decrease) as the rotor position moves forward (respectively, backward) relative to the synchronous rotating frame. However, such a stable operation can take place only when the system is capable of absorbing (respectively, providing) the extra power. In a multimachine system, conflict might arise when various machines compete with each other in sending out more (or sending out less) electrical power; as a result, the stabilizing effect might be reduced or even lost.

2. *Electromagnetic equations.* The (nonlinear) electromagnetic equations are derived from Faraday's law of electromagnetic induction—induced emf's are proportional to the rate of change of the magnetic fluxes. A general form is as follows:

$$\begin{cases} e_d = \dfrac{d}{dt}\lambda_d + \lambda_q \dfrac{d}{dt}\theta - ri_d \\[2mm] e_q = \dfrac{d}{dt}\lambda_q + \lambda_d \dfrac{d}{dt}\theta - ri_q \end{cases} \tag{61.1}$$

where

$$\begin{cases} \lambda_d = G(s)i_F - X_d(s)i_d \\[2mm] \lambda_q = -X_q(s)i_q \end{cases} \tag{61.2}$$

The true terminal voltage, e.g., e_a for phase a, can be obtained by combining the direct-axis and quadrature-axis components e_d and e_q, respectively, which are given in Eq. (61.1). On each line of Eq. (61.1), the induced emf is the combination of two sources: the first is the rate of change of the flux on the same axis [$(d/dt)\lambda_d$ on the first line, $(d/dt)\lambda_q$ on the second]; the second comes into effect only when a disturbance makes the rotor and stator fields depart from each other [given by $(d/dt)\theta$]. The third term in the voltage equation represents the ohmic loss associated with the stator winding.

Equation (61.2) expresses the fluxes in terms of relevant currents: flux is equal to inductance times current, with inductances $G(s)$, $X_d(s)$, $X_q(s)$ given in an operational form (s denotes the derivative operator).

Figure 61.6 gives a general view of the input–output state description of machine's dynamic model, the state variables of which appear in Eqs. (61.1) and (61.2).

3. *Miscellaneous.* In addition to the basic components of a synchronous generator (rotor, stator, and their windings), there are auxiliary devices which help maintain the machine's operation within acceptable limits. Three such devices are mentioned here: governor, damper windings, and excitation control system.

- Governor. This is to control the mechanical power input P_{in}. The control is via a feedback loop where the speed of the rotor is constantly monitored. For instance, if this speed falls behind the synchronous speed, the input is insufficient and has to be increased. This is done by opening up the valve to increase the steam for turbogenerators or the flow of water through the penstock for hydrogenerators. Governors are mechanical systems and therefore have some significant time lags (many seconds) compared to other electromagnetic phenomena associated with the machine. If the time duration of interest is short, the effect of governor can be ignored in the study; that is, P_{in} is treated as a constant.
- Damper windings (armortisseur windings). These are special conducting bars buried in notches on the rotor surface, and the rotor resembles that of a squirrel-cage-rotor induction machine (see Section 61.2). The damper windings provide an additional stabilizing force for the machine when it is perturbed from an equilibrium. As long as the machine is in a steady state, the stator field rotates at the same speed as the rotor, and no currents are induced in the damper windings. That is, these windings exhibit no effect on a steady-state machine. However, when the speeds of the stator field and the rotor become different (because of a disturbance), currents are induced in the damper windings in such a way as to keep, according to Lenz's law, the two speeds from separating.
- Excitation control system. Modern excitation systems are very fast and quite efficient. An excitation control system is a feedback loop that aims at keeping the voltage at machine terminals at a set level. To explain the main feature of the excitation system, it is sufficient to consider Fig. 61.4. Assume that a disturbance occurs in the system, and as a result, the machine's terminal voltage V_t drops. The excitation system boosts the internal voltage E_F; this action can increase the voltage V_t and also tends to increase the reactive power output.

From a system viewpoint, the two controllers of excitation and governor rely on local information (machine's terminal voltage and rotor speed). In other words, they are decentralized controls. For large-scale systems, such designs do not always guarantee a desired stable behavior since the effect of interconnection is not taken into account in detail.

Synchronous Machine Parameters.   When a disturbance, such as a short circuit at the machine terminals, takes place, the dynamics of a synchronous machine will be observed before a new steady state is reached. Such a process typically takes a few seconds and can be divided into subprocesses. The damper windings (armortisseur) exhibit their effect only during the first few cycles when the difference in speed between the rotor and the perturbed stator field is significant. This period is referred to as *subtransient*. The next and longer period, which is between the subtransient and the new steady state, is called *transient*.

Various parameters associated with the subprocesses can be visualized from an equivalent circuit. The *d*-axis and *q*-axis (dynamic) equivalent circuits of a synchronous generator consist of resistors, inductors, and voltage sources. In the sub-transient period, the equivalent of the damper windings needs to be considered. In the transient period, this equivalent can be ignored. When the new steady state is reached, the current in the rotor winding becomes a constant (dc); thus, one can further ignore the equivalent inductance of this winding. This approximate method results in three equivalent circuits, listed in order of complexity: subtransient, transient, and steady state. For each circuit, one can define parameters such as (effective) reactance and time constant. For example, the d-axis circuit for the transient period has an effective reactance X_d' and a time constant T_{do}' (computed from the R-L circuit) when open circuited. The parameters of a synchronous machine can be computed from experimental data and are used in

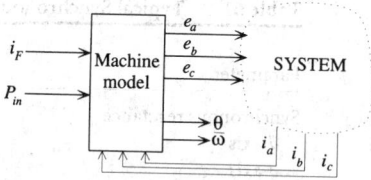

FIGURE 61.6 A block diagram depicting a qualitative relationship among various electrical and mechanical quantities of a synchronous machine. e_a, e_b, e_c are phase voltages; i_a, i_b, i_c phase currents; i_F rotor field current; θ relative position of rotor; ω deviation of rotor speed from synchronous speed; P_{in} mechanical power input. The state variables appear in Eqs. (61.1) and (61.2).

numerical studies. Typical values for these parameters are given in Table 61.1.

References on synchronous generators are numerous because of the historical importance of these machines in large-scale electric energy production. Reference 1 includes a derivation of the steady-state and dynamic models, dynamic performance, excitation, and trends in development of large generators. References 3 and 4 are among the basic sources of reference in electric machinery, where many practical aspects are given. An introductory discussion of power system stability as related to synchronous generators can be found in Reference 5. A number of handbooks that include subjects on ac as well as dc generators are also available in References 6, 7 and 8.

Superconducting Generators

The demand for electricity has increased steadily over the years. To satisfy the increasing demand, there has been a trend in the development of generators with very high power rating. This has been achieved, to a great extent, by improvement in materials and cooling techniques. Cooling is necessary because the loss dissipated as heat poses a serious problem for winding insulation. The progress in machine design based on conventional methods appears to reach a point where further increases in power ratings are becoming difficult. An alternative method involves the use of super-conductivity.

In a superconducting generator, the field winding is kept at a very low temperature so that it stays superconductive. An obvious advantage to this is that no resistive loss can take place in this winding, and therefore a very large current can flow. A large field current yields a very strong magnetic field, and this means that many issues considered important in the conventional design may no longer be critical. For example, the conventional design makes use of iron core for armature windings to achieve an appropriate level of magnetic flux for these windings; iron cores, however, contribute to heat loss—because of the effects of hysteresis and eddy currents—and therefore require appropriate designs for winding insulation. With the new design, there is no need for iron cores since the magnetic field can be made very strong; the absence of iron allows a simpler winding insulation, thereby accommodating additional armature windings.

There is, however, a limit to the field current increase. It is known that superconductivity and diamagnetism are closely related; that is, if a material is in the superconducting state, no magnetic lines of force can enter its interior. Increasing the current produces more and more magnetic lines of force, and this can continue until the dense magnetic field can penetrate the material. When this happens, the material fails to stay superconductive, and therefore resistive loss can take place. In other words, a material can stay superconductive until a certain *critical field strength* is reached. The critical field strength is dependent on the material and its temperature.

Table 61.1 Typical Synchronous Generator Parameters[a]

Parameter	Symbol	Round Rotor	Salient-Pole Rotor with Damper Windings
Synchronous reactance			
d-axis	X_d	1.0–2.5	1.0–2.0
q-axis	X_q	1.0–2.5	0.6–1.2
Transient reactance			
d-axis	X'_d	0.2–0.35	0.2–0.45
q-axis	X'_q	0.5–1.0	0.25–0.8
Subtransient reactance			
d-axis	X''_d	0.1–0.25	0.15–0.25
q-axis	X''_q	0.1–0.25	0.2–0.8
Time constants			
Transient			
Stator winding open-circuited	T'_{do}	4.5–13	3.0–8.0
Stator winding short-circuited	T'_d	1.0–1.5	1.5–2.0
Subtransient			
Stator winding short-circuited	T''_d	0.03–0.1	0.03–0.1

[a] Reactances are per unit, i.e., normalized quantities. Time constants are in seconds.
Source: M.A. Laughton and M.G. Say, eds., *Electrical Engineer's Reference Book*, Stoneham, Mass.: Butterworth, 1985.

A typical superconducting design of an ac generator, as in the conventional design, has the field winding mounted on the rotor and armature winding on the stator. The main differences between the two designs lie in the way cooling is done. The rotor has an inner body which is to support a winding cooled to a very low temperature by means of liquid helium. The liquid helium is fed to the winding along the rotor axis. To maintain the low temperature, thermal insulation is needed, and this can be achieved by means of a vacuum space and a radiation shield. The outer body of the rotor shields the rotor's winding from being penetrated by the armature fields so that the superconducting state will not be destroyed. The stator structure is made of nonmagnetic material, which must be mechanically strong. The stator windings (armature) are not superconducting and are typically cooled by water. The immediate surroundings of the machine must be shielded from the strong magnetic fields; this requirement, though not necessary for the machine's operation, can be satisfied by the use of a copper or laminated iron screen.

From a circuit viewpoint, superconducting machines have smaller internal impedance relative to the conventional ones (refer to equivalent circuit shown in Fig. 61.4). Recall that the reactance jX_s stems from the fact that the armature circuits give rise to a magnetic field that tends to counter the effect of the rotor winding. In the conventional design, such a magnetic field is enhanced because iron core is used for the rotor and stator structures; thus jX_s is large. In the superconducting design, the core is basically air; thus, jX_s is smaller. The difference is generally a ratio of 5:1 in magnitude. An implication is that, at the same level of output current I_a and terminal voltage V_t, it requires of the superconducting generator a smaller induced emf E_F or, equivalently, a smaller field current.

It is expected that the use of superconductivity adds another 0.4% to the efficiency of generators. This improvement might seem insignificant (compared to an already achieved figure of 98% by the conventional design) but proves considerable in the long run. It is estimated that given a frame size and weight, a superconducting generator's capacity is three times that of a conventional one. However, the new concept has to deal with such practical issues as reliability, availability, and costs before it can be put into large-scale operation.

Reference 2 provides more details on superconducting electric machines with issues such as design, performance, and application of such machines.

Induction Generators

Conceptually, a three-phase induction machine is similar to a synchronous machine, but the former has a much simpler rotor circuit. A typical design of the rotor is the squirrel-cage structure, where conducting bars are embedded in the rotor body and shorted out at the ends. When a set of three-phase currents (waveforms of equal amplitude, displaced in time by one-third of a period) is applied to the stator winding, a rotating magnetic field is produced. (See the discussion of a revolving magnetic field for synchronous generators in the section "Principle of Operation".) Currents are therefore induced in the bars, and their resulting magnetic field interacts with the stator field to make the rotor rotate in the same direction. In this case, the machine acts as a motor since, in order for the rotor to rotate, energy is drawn from the electric power source. When the machine acts as a motor, its rotor can never achieve the same speed as the rotating field (this is the synchronous speed) for that would imply no induced currents in the rotor bars. If an external mechanical torque is applied to the rotor to drive it beyond the synchronous speed, however, then electric energy is pumped to the power grid, and the machine will act as a generator.

An advantage of induction generators is their simplicity (no separate field circuit) and flexibility in speed. These features make induction machines attractive for applications such as windmills.

A disadvantage of induction generators is that they are highly inductive. Because the current and voltage have very large phase shifts, delivering a moderate amount of power requires an unnecessarily high current on the power line. This current can be reduced by connecting capacitors at the terminals of the machine. Capacitors have negative reactance; thus, the machine's inductive reactance can be compensated. Such a scheme is known as capacitive compensation. It is ideal to have a compensation in which the capacitor and equivalent inductor completely cancel the effect of each other. In windmill applications, for example, this faces a great challenge because the varying speed of the rotor (as a result of wind speed) implies a varying equivalent inductor. Fortunately, strategies for ideal compensation have been designed and put to commercial use.

In Reference 3, an analysis of induction generators and the effect of capacitive compensation on machine's performance are given.

DC Generators

To obtain dc electricity, one may prefer an available ac source with an electronic rectifier circuit. Another possibility is to generate dc electricity directly. Although the latter method is becoming obsolete, it is still important to understand how a dc generator works. This section provides a brief discussion of the basic issues associated with dc generators.

Principle of Operation

As in the case of ac generators, a basic design will be used to explain the essential ideas behind the operation of dc generators. Figure 61.7 is a schematic diagram showing an end of a simple dc machine.

The stator of the simple machine is a permanent magnet with two poles labeled N and S. The rotor is a cylindrical body and has two (insulated) conductors embedded in its surface. At one end of the rotor, as illustrated in Fig. 61.7, the two conductors are connected to a pair of copper segments; these semicircular segments, shown in the diagram, are mounted on the shaft of the rotor. Hence, they rotate together with the rotor. At the other end of the rotor, the two conductors are joined to form a coil.

Assume that an external torque is applied to the shaft so that the rotor rotates at a certain speed. The rotor winding formed by the two conductors experiences a periodically varying magnetic field, and hence an emf is induced across the winding. Note that this voltage periodically alternates in sign, and thus, the situation is conceptually the same as the one encountered in ac generators. To make the machine act as a dc source, viewed from the terminals, some form of rectification needs be introduced. This function is made possible with the use of copper segments and brushes.

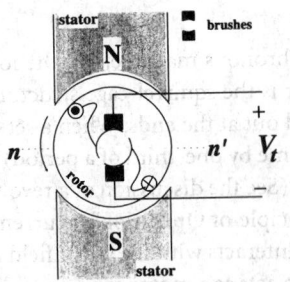

FIGURE 61.7 A basic two-pole dc generator. V_t is the voltage across the machine terminals. $\otimes$ and $\odot$ indicate the direction of currents (into or out of the page) that would flow if a closed circuit is made.

According to Fig. 61.7, each copper segment comes into contact with one brush half of the time during each rotor revolution. The placement of the (stationary) brushes guarantees that one brush always has positive potential relative to the other. For the chosen direction of rotation, the brush with higher potential is the one directly beneath the N-pole. (Should the rotor rotate in the reverse direction, the opposite is true.) Thus, the brushes can serve as the terminals of the dc source. In electric machinery, the rectifying action of the copper segments and brushes is referred to as **commutation**, and the machine is called a commutating machine.

A qualitative sketch of V_t, the voltage across terminals of an unloaded simple dc generator, as a function of time is given in Fig. 61.8. Note that this voltage is not a constant. A unidirectional current can flow when a resistor is connected across the terminals of the machine.

The pulsating voltage waveform generated by the simple dc machine usually cannot meet the requirement of practical applications. An improvement can be made with more pairs of conductors. These conductors are placed in slots that are made equidistant on the rotor surface. Each pair of conductors can generate a voltage waveform similar to the one in Fig. 61.8, but there are time shifts among these waveforms due to the spatial displacement among the conductor pairs. For instance, when an individual voltage is minimum (zero), other voltages are not. If these voltage waveforms are added, the result is a near constant voltage waveform. This improvement of the dc waveform requires many pairs of the copper segments and a pair of brushes.

When the generator is connected to an electrical load, load currents flow through the rotor conductors. Therefore, a magnetic field is set up in addition to that of the permanent magnet. This additional field generally weakens the magnetic flux seen by the rotor conductors. A direct consequence is that the induced emf's are less than those in an unloaded machine. Similar to the case of ac generators, this phenomenon is referred to as armature reaction, or flux-weakening effect.

The use of brushes in the design of dc generators can cause a serious problem in practice. Each time a brush comes into contact with two adjacent copper segments, the corresponding conductors are short-circuited. For a loaded generator, such an event occurs when the currents in these conductors are not zero, resulting in flashover at the brushes. This means that the life span of the brushes can be drastically reduced and that frequent maintenance is needed. A number of design techniques have been developed to mitigate this problem.

Mathematical/Circuit Model

The (no-load) terminal voltage V_t of a dc generator depends on several factors. First, it depends on the construction of the machine (e.g., the number of conductors). Second, the voltage magnitude depends on the magnetic field of the stator: the stronger the field is, the higher the voltage becomes. Third, since the induced emf is proportional to the rate of change of the magnetic flux (Faraday's law), the terminals have higher voltage with a higher machine speed. One can write

$$V_{t(\text{no load})} = K\lambda n$$

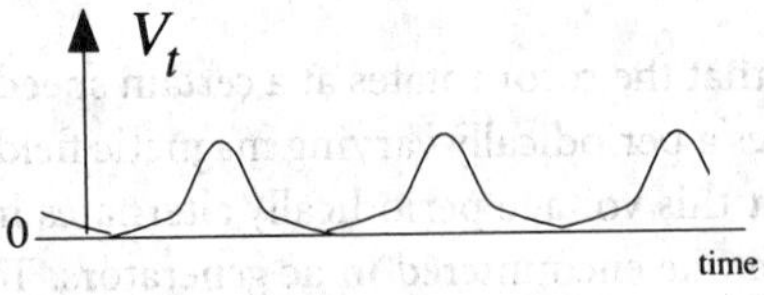

FIGURE 61.8 Open-circuited terminal voltage of the simple dc generator.

where K is a constant representing the first factor, λ is magnetic flux, and n is rotor speed. The foregoing equation provides some insights into the voltage control of dc generators. Among the three terms, it is impractical to modify K, which is determined by the machine design. Changing n over a wide range may not be feasible since this is limited by what drives the rotor.

Changing the magnetic flux λ can be done if the permanent magnet is replaced by an electromagnet, and this is how the voltage control is done in practice. The control of λ is made possible by adjusting the current fed to this electromagnet. Figure 61.9 shows the modified design of the simple dc generator. The stator winding is called the *field winding*, which produces excitation for the machine. The current in the field winding is adjusted by means of a variable resistor connected in series with this winding. It is also possible to use two field windings in order to have more flexibility in control.

The use of field winding(s) on the stator of the dc machine leads to a number of methods to produce the magnetic field. Depending on how the field winding(s) and the rotor winding are connected, one may have shunt excitation, series excitation, etc. Each connection yields a different terminal characteristic. The possible connections and the resulting current–voltage characteristics are given in Table 61.2.

References 3 and 7 provide more detailed discussions of dc generators. Specifically, Reference 3 shows how the characteristics are derived for various excitation methods.

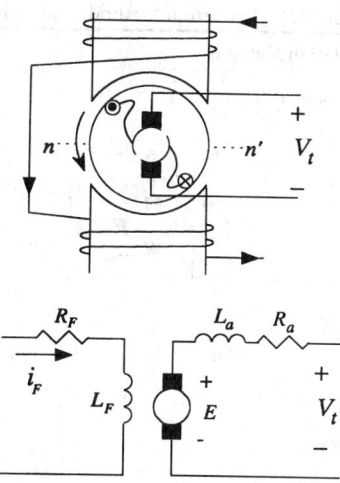

FIGURE 61.9   A simple two-pole dc generator with a stator winding to produce a magnetic field. *Top,* main components of the machine; *bottom,* coupled-circuit representation; the circuit on the left represents the field winding; the induced emf E is controlled by i_F.

Defining Terms

Armature circuit: A winding where the load current is carried.

Armature reaction: The phenomenon in which the magnetic field due to currents in the armature circuit counters the effect of the field circuit.

Commutation: A mechanical technique in which rectification can be achieved in dc machines.

Field circuit: A set of windings that produces a magnetic field so that the electromagnetic induction can take place in electric machines.

Revolving fields: A magnetic field created by multiphase currents on spatially displaced windings in rotating machines; the field revolves in the air gap.

Swing equation: A nonlinear differential equation describing the rotor dynamics of an ac synchronous machine.

Synchronous speed: A characteristic speed of synchronous and induction machines with a revolving field; it is determined by the rotor structure and the line frequency.

References

1. M. S. Sarma, *Synchronous Machines (Their Theory, Stability, and Excitation Systems),* New York: Gordon and Breach, 1979.
2. J. R. Bumby, *Superconducting Rotating Electrical Machines,* New York: Oxford University Press, 1983.
3. S. J. Chapman, *Electric Machinery Fundamentals,* New York: McGraw-Hill, 1991.
4. G. McPherson, *An Introduction to Electrical Machines and Transformers,* New York: Wiley, 1981.
5. A. R. Bergen, *Power Systems Analysis,* Englewood Cliffs, N.J.: Prentice-Hall, 1986.
6. M. A. Laughton and M. G. Say, eds., *Electrical Engineer's Reference Book,* Stoneham, Mass.: Butterworth, 1985.
7. D. G. Fink and H. W. Beaty, eds., *Standard Handbook for Electrical Engineers,* New York: McGraw-Hill, 1987.

Table 61.2   Excitation Methods and Voltage Current Characteristics for DC Generators

Excitation Methods	Characteristics

Separate

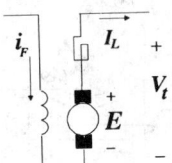

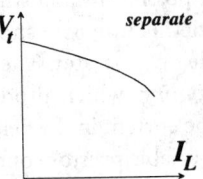

For low currents, the curve is nearly a straight line. As load current increases, the armature reaction becomes more severe and contributes to the nonlinear drop.

Series

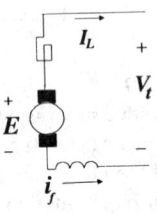

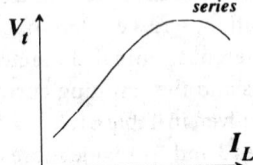

At no load, there is no field current, and voltage is due to the residual flux of the stator core. The voltage rises rapidly over the range of low currents, but the resistive drop soon becomes dominant.

Shunt

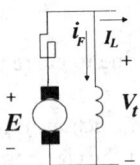

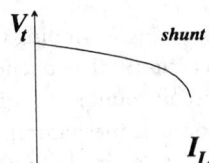

Voltage buildup depends on the residual flux. The shunt field resistance must be less than a critical value.

Compounded

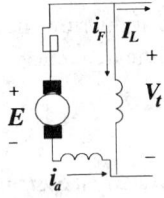

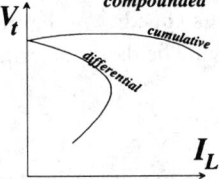

There are two field windings. Depending on how they are set up, one may have *cumulative* if the two fields are additive, *differential* if the two fields are subtractive.

Cumulative: An increase in load current increases the resistive drop, yet creates more flux. At high currents, however, resistive drop becomes dominant.
Differential: An increase in load current not only increases the resistive drop, but also reduces the net flux. Voltage drops drastically.

Source: S.J. Chapman, *Electric Machinery Fundamentals,* New York: McGraw-Hill, 1991.

8. S. S. L. Chang, ed., *Fundamentals Handbook of Electrical and Computer Engineering*, New York: Wiley, 1982.

Further Information

Several handbooks, e.g., References 6 and 7, give more details on the machine design. Reference 2 covers the subject of superconducting generators. Some textbooks in the area of rotating machines are listed as References 1, 3, and 4.

The quarterly journal *IEEE Transactions on Energy Conversion* covers the field of rotating machinery and power generation. Another IEEE quarterly journal, *IEEE Transactions on Power Systems,* is devoted to the general aspects of power system engineering and power engineering education.

The bimonthly journal *Electric Machines and Power Systems,* published by Hemisphere Publishing Corporation, covers the broad field of electromechanics, electric machines, and power systems.

61.2 Motors

Donald Galler

Electric motors are the most commonly used prime mover in industry. The classification of the types of ac and dc motors commonly used in industrial applications is shown in Fig. 61.10.

Motor Applications

DC Motors

Permanent magnet (PM) field motors occupy the low end of the horsepower (hp) range and are commercially available up to about 10 hp. Below 1 hp they are used for servo applications, such as in machine tools, for robotics, and in high-performance computer peripherals.

Wound field motors are used above about 10 hp and represent the highest horsepower range of **dc motor** application. They are commercially available up to several hundred horsepower and are commonly used in traction, hoisting, and other applications where a wide range of speed control is needed. The shunt wound dc motor is commonly found in industrial applications such as grinding

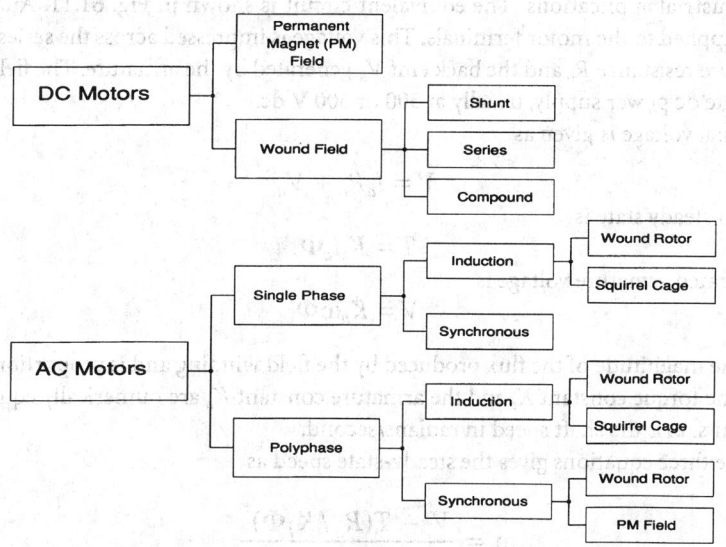

FIGURE 61.10 Classification of ac and dc motors for industrial applications.

and machine tools and in elevator and hoist applications. Compound wound motors have both a series and shunt field component to provide specific torque-speed characteristics. Propulsion motors for transit vehicles are usually compound wound dc motors.

AC Motors

Single-phase ac motors occupy the low end of the horsepower spectrum and are offered commercially up to about 5 hp. Single-phase **synchronous motors** are only used below about 1/10 of a horsepower. Typical applications are timing and motion control, where low torque is required at fixed speeds. Single-phase **induction motors** are used for operating household appliances and machinery from about 1/3 to 5 hp.

Polyphase ac motors are primarily three-phase and are by far the largest electric prime mover in all of industry. They are offered in ranges from 5 up to 50,000 hp and account for a large percentage of the total motor industry in the world. In number of units, the three-phase **squirrel cage induction motor** is the most common. It is commercially available from 1 hp up to several thousand horsepower and can be used on conventional ac power or in conjunction with adjustable speed ac drives. Fans, pumps, and material handling are the most common applications.

When the torque-speed characteristics of a conventional ac induction motor need to be modified, the **wound rotor induction motor** is used. These motors replace the squirrel cage rotor with a wound rotor and slip rings. External resistors are used to adjust the torque-speed characteristics for speed control in such applications as ac cranes, hoists, and elevators.

Three-phase synchronous motors can be purchased with PM fields up to about 5 hp and are used for applications such as processing lines and transporting film and sheet materials at precise speeds.

In the horsepower range above about 10,000 hp, three-phase synchronous motors with wound fields are used rather than large squirrel cage induction motors. Starting current and other characteristics can be controlled by the external field exciter. Three-phase synchronous motors with wound fields are available up to about 50,000 hp.

Motor Analysis

DC Motor Analysis

The **separately excited dc motor** is the simplest of all dc motors and is the one most commonly found in industrial applications. The equivalent circuit is shown in Fig. 61.11. An adjustable dc voltage V is applied to the motor terminals. This voltage is impressed across the series combination of the armature resistance R_a and the back emf V_a generated by the armature. The field is energized with a separate dc power supply, usually at 300 or 500 V dc.

The terminal voltage is given as

$$V = I_a R_a + V_a \qquad (61.3)$$

The torque in steady state is

$$T = K_t I_a \Phi \qquad (61.4)$$

and the generated armature voltage is

$$V = K_a \omega \Phi \qquad (61.5)$$

where Φ is the magnitude of the flux produced by the field winding and is proportional to the field current I_f. The torque constant K_t and the armature constant K_a are numerically equal in a consistent set of units. ω is the shaft speed in radians/second.

Solving the three equations gives the steady-state speed as

$$\omega = \frac{V - T(R_a / K_t \Phi)}{K_a \Phi} \qquad (61.6)$$

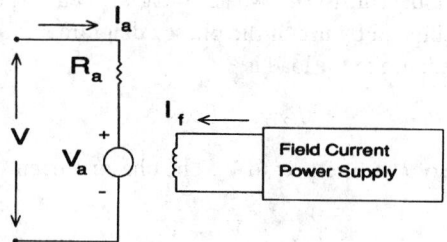

FIGURE 61.11   Equivalent circuit of separately excited dc motor.

The input power and output power are

$$P_{\text{in}} = I_a V \tag{61.7}$$

$$P_{\text{out}} = \omega T = I_a V - I_a^2 R_a \tag{61.8}$$

The efficiency (neglecting power loss in the field) is

$$\eta = \frac{P_{\text{out}}}{P_{\text{in}}}$$

$$= \frac{\omega T}{I_a V} \tag{61.9}$$

A simplified torque-speed curve is shown in Fig. 61.12. The torque capability is constant up to the base speed of the motor while the armature and field currents are held constant. The speed is controlled by armature voltage in this range. Operation above base speed is accomplished by reducing the field current. This is called *field weakening*. The motor operates at constant power in this range, and the torque reduces with increasing speed.

Synchronous Motor Analysis

Synchronous motor analysis may be conducted using either a round rotor or salient pole model for the motor. The round rotor model is used in the following discussion. The equivalent circuit is

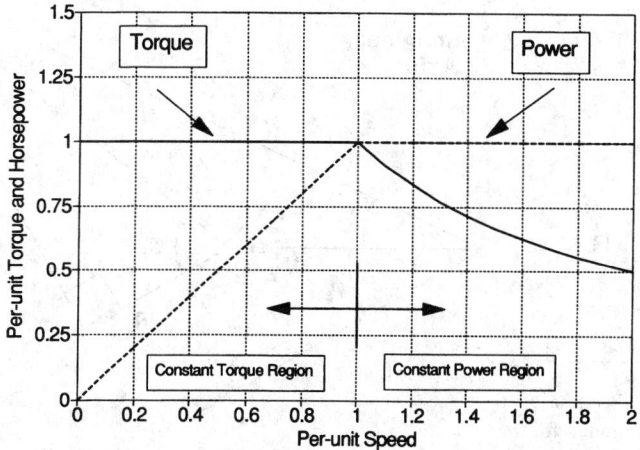

FIGURE 61.12   Torque-speed capability for the separately excited dc motor.

shown in Fig. 61.13. The model consists of two ac voltages V_1 and V_2 connected by an impedance $Z = R + jX$. Analysis is facilitated by use of the phasor diagram shown in Fig. 61.14. The power delivered through the impedance to the load is

$$P_2 = V_2 I \cos \phi_2 \tag{61.10}$$

where ϕ_2 is the phase angle of I with respect to V_2. The phasor current

$$I = \frac{V_1 - V_2}{Z} \tag{61.11}$$

is expressed in polar form as

$$I = \frac{V_1 \; \underline{/\delta} - V_2 \; \underline{/0°}}{Z \; \underline{/\phi_z}}$$

$$= \frac{V_1}{Z} \; \underline{/\delta - \phi_z} - \frac{V_2}{Z} \underline{/-\phi_z} \tag{61.12}$$

The equations make use of the fact that the three-phase operation is symmetrical and uses a "per-phase" equivalent circuit. This will also be true for the induction motor, which is analyzed in the following section.

The real part of I is

$$I \cos \phi_2 = \frac{V_1}{Z} \; \cos(\delta - \phi_z) - \frac{V_2}{Z} \; \cos(-\phi_z) \tag{61.13}$$

Using Eq. (61.13) in Eq. (61.10) gives

$$P_2 = \frac{V_1 V_2}{Z} \; \cos(\delta - \phi_z) - \frac{V_2^2 R}{Z^2} \tag{61.14}$$

Letting $\alpha = 90° - \phi_z = \arctan R/X$ gives the output power as

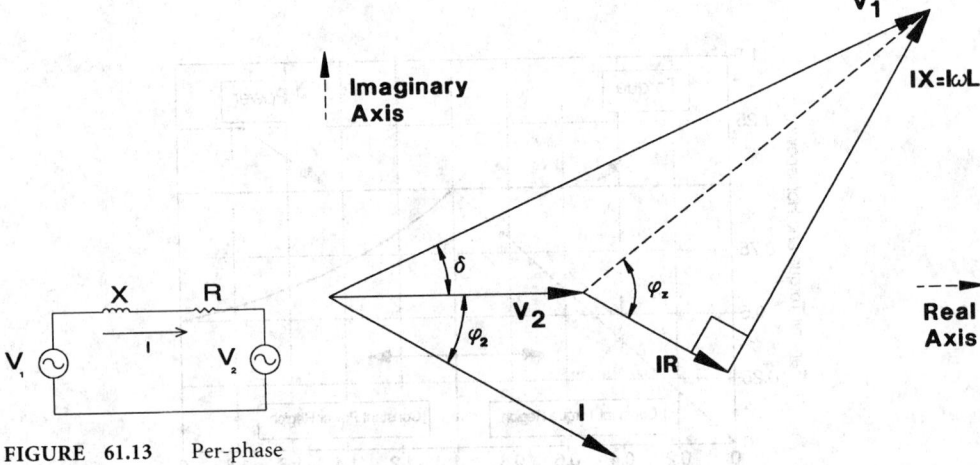

FIGURE 61.13   Per-phase equivalent circuit model for the synchonous motor (round rotor model).

FIGURE 61.14   Phasor diagram for the ac synchronous motor (round rotor model).

$$P_2 = \frac{V_1 V_2}{Z} \sin(\delta + \alpha) - \frac{V_2^2 R}{Z^2} \tag{61.15}$$

and the input power as

$$P_1 = \frac{V_1 V_2}{Z} \sin(\delta - \alpha) + \frac{V_1^2 R}{Z^2} \tag{61.16}$$

Usually R is neglected and

$$P_1 = P_2 = \frac{V_1 V_2}{X} \sin \delta \tag{61.17}$$

which shows that the power is maximum when $\delta = 90°$ and is

$$P_{MAX} = \frac{V_1 V_2}{X} \tag{61.18}$$

The current can be found from Eqs. (61.15) and (61.16) since the only loss occurs in R. Setting

$$I^2 R = P_2 - P_1 \tag{61.19}$$

and solving for I gives

$$I = \sqrt{(P_2 - P_1)/R} \tag{61.20}$$

which is the input line current.

The power factor is

$$\cos \theta = \frac{P_1}{V_1 I} \tag{61.21}$$

and $\theta = \delta + \phi_2$ as shown in Fig. 61.14.

All the foregoing values are per-phase values. The total input power is

$$P_{in} = 3 P_1 \tag{61.22}$$

The mechanical output power is

$$P_{out} = T\omega$$
$$= 3 \cdot P_2 \tag{61.23}$$

and the torque is

$$T = 3 \cdot P_{out} /\omega \tag{61.24}$$

where ω is the rotational speed of the motor expressed in radians per second.

Synchronous motor operation is determined by the torque angle δ and is illustrated in Fig. 61.15 for a typical motor. Input power, output power, and current are shown on a per-unit basis. Torque is not shown but is related to output power only by a constant.

Induction Motor Analysis

The characteristic algebraic equations for the steady-state power, torque, and efficiency of the ac induction motor are derived from the per-phase equivalent circuit of Fig. 61.16. All voltages and currents are in sinusoidal steady state. The derivation of the equations can be simplified by defining the complex motor impedance as

$$Z_m = \frac{\alpha}{\zeta} + j\frac{\beta}{\zeta} \tag{61.25}$$

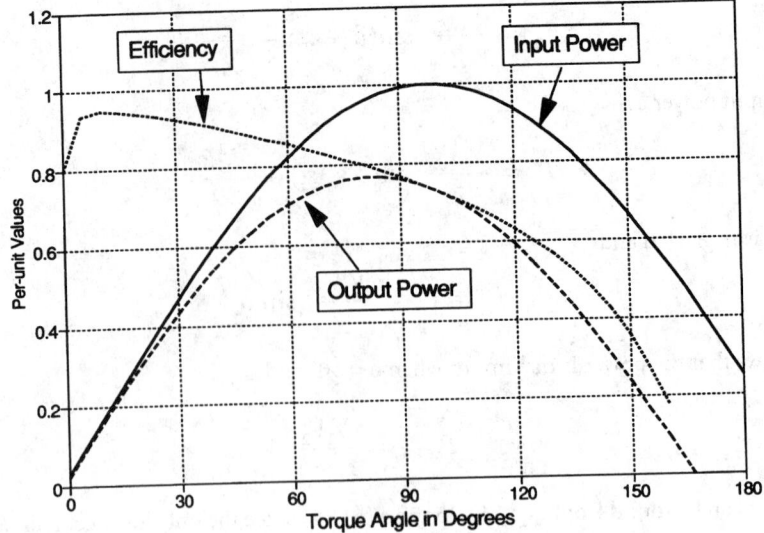

FIGURE 61.15 Synchronous motor performance.

By defining the following constants as

$$M_1 = R_1 R_2^2$$

$$M_2 = R_2 L_m^2$$

$$M_3 = L_2 + L_m \qquad (61.26)$$

$$M_4 = L_1 + L_m$$

$$M_5 = R_1 M_3^2 + M_2$$

the terms of Eq. (61.25) become

$$\zeta = R_2^2 + \omega_s^2 M_3^2 \qquad (61.27)$$

$$\alpha = \zeta R_1 + (\omega_m + \omega_s)\omega_s M_2 \qquad (61.28)$$

$$\beta = (\omega_m + \omega_s) [\zeta L_1 + L_M R_2^2 + \omega_s^2 M_3 L_2 L_m] \qquad (61.29)$$

The angular velocity ω_s is the slip frequency and is defined as follows:

$$\omega_s = \omega_f - \omega_m \qquad (61.30)$$

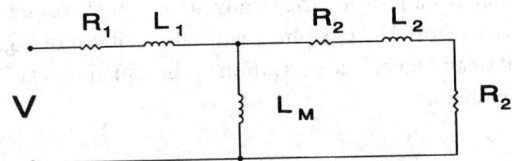

FIGURE 61.16 Equivalent circuit of ac induction motor.

where ω_f is the frequency applied to the **stator** and

$$\omega_m = \omega/N_p \qquad (61.31)$$

is the **rotor** angular velocity in terms of an equivalent stator frequency. N_p is the number of stator pole pairs. The average mechanical output power of the motor is the power in the resistance $R_2\omega_m/\omega_s$ and is given as

$$P_{out} = \frac{3V^2 \zeta \omega_m \omega_s M_2}{\alpha^2 + \beta^2} \qquad (61.32)$$

where V is the rms line-neutral voltage. Since

$$
\begin{aligned}
T &= \frac{P_{out}}{\omega} \\
&= \frac{P_{out} N_p}{\omega_m}
\end{aligned} \qquad (61.33)
$$

the torque becomes

$$T = \frac{3V^2 \zeta N_p \omega_s M_2}{\alpha^2 + \beta^2} \qquad (61.34)$$

The motor efficiency is defined as

$$\eta = \frac{P_{out}}{P_{in}} \qquad (61.35)$$

where the input power is

$$
\begin{aligned}
P_{in} &= \frac{3V_f^2 \cos\theta}{|Z_m|} \\
&= \frac{3V_f^2 \alpha}{|Z_m|^2 \zeta}
\end{aligned} \qquad (61.36)
$$

Using Eqs. (61.32) and (61.36), the efficiency becomes

$$\eta = \frac{\omega_m \omega_s M_2}{\alpha} \qquad (61.37)$$

Typical performance characteristics of the induction motor are shown in Fig. 61.17.

A simplified analysis [2] neglects R_1 and X_M. The slip, s, is defined as

$$s = \frac{\omega_f - \omega_m}{\omega_f} \qquad (61.38)$$

where ω_m is the equivalent mechanical frequency of the rotor, $\omega_m = \omega/N_p$, and ω_f is the angular velocity of the stator field in radians/second.

The output power is

$$P = I_2^2 R_2/s \qquad (61.39)$$

and is maximum when

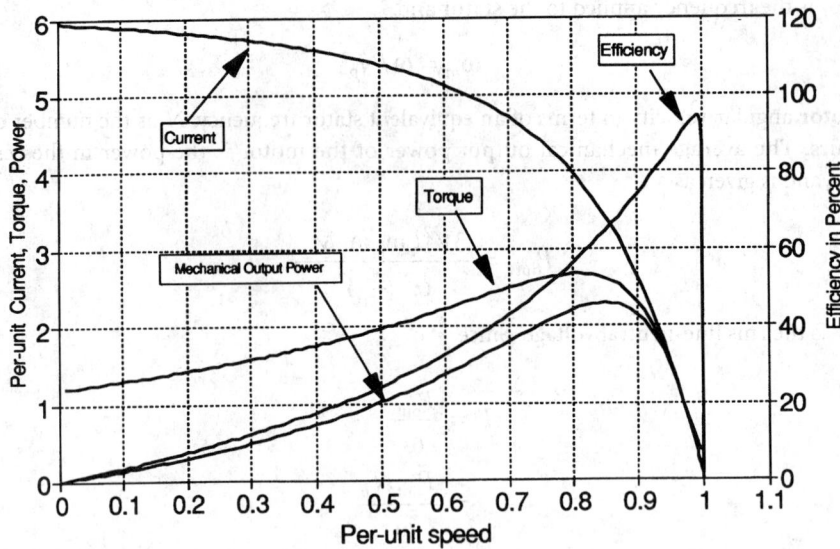

FIGURE 61.17   Induction motor operating characteristics, fixed voltage, and frequency.

$$R_2 = s(X_2 + X_1) \qquad (61.40)$$

The torque is given by

$$T = 2T_M \frac{ss_M}{s^2 + s_M^2} \qquad (61.41)$$

where T_M is the maximum torque and s_M is the slip at maximum torque.

AC and DC Motor Terms

General Terms

ω: Shaft angular velocity in radians/second

P_{out}:   Electrical output power

P_{in}:   Electrical input power

η: Efficiency

T: Shaft torque

DC Motor Terms

I_a: Armature current

I_f: Field current

V_a: Back emf generated by armature

V: Motor terminal voltage

R_a: Armature resistance

K_t: Torque constant

K_a: Armature constant

Φ: Field flux

AC Induction Motor Terms

L_1:   Stator winding inductance

R_1:   Stator winding resistance

L_2:   Rotor winding inductance

R_2:   Rotor winding resistance

L_M: Magnetizing inductance

N_p: Number of pole pairs in stator winding

ω_f: Frequency of voltage applied to stator

ω_m: Rotor equivalent mechanical frequency

ω_s: Slip frequency, $\omega_s = \omega_f - \omega_m$

s: Slip $s = (\omega_f - \omega_m)/\omega_f$

T_M: Maximum torque

s_M: Slip at maximum torque

AC Synchronous Motor Terms

V_1:	Terminal voltage	δ:	Torque angle (between V_1 and V_2)
V_2:	Back emf generated by rotor	ϕ_2:	Angle between I and V_2
R:	Rotor circuit resistance	ϕ_z:	Rotor circuit reactance angle $\phi_z = \tan^{-1} X/R$
X:	Rotor circuit reactance		
Z:	Rotor circuit impedance $Z = R + jX$	α:	$90° - \phi_z$
		θ:	Power factor angle $\theta = \delta + \phi_2$

Defining Terms

DC motor: A dc motor consists of a stationary active part, usually called the field structure, and a moving active part, usually called the armature. Both the field and armature carry dc.

Induction motor: An ac motor in which a primary winding on the stator is connected to the power source and polyphase secondary winding on the rotor carries induced current.

Permanent magnet dc motor: A dc motor in which the field flux is supplied by permanent magnets instead of a wound field.

Rotor: The rotating member of a motor including the shaft. It is commonly called the armature on most dc motors.

Separately excited dc motor: A dc motor in which the field current is derived from a circuit which is independent of the armature.

Squirrel cage induction motor: An induction motor in which the secondary circuit (on the rotor) consists of bars, short-circuited by end rings. This forms a squirrel cage conductor structure which is disposed in slots in the rotor core.

Stator: The portion of a motor that includes and supports the stationary active parts. The stator includes the stationary portions of the magnetic circuit and the associated windings and leads.

Synchronous motor: An ac motor in which the average speed of normal operation is exactly proportional to the frequency to which it is connected. A synchronous motor generally has rotating field poles which are excited by dc.

Wound rotor induction motor: An induction motor in which the secondary circuit consists of a polyphase winding or coils connected through a suitable circuit. When provided with slip rings, the term *slip-ring induction motor* is used.

References

1. A.E. Fitzgerald, C. Kingsley, Jr., and A. Umans, *Electric Machinery*, 4th ed., New York: McGraw-Hill, 1983.

2. D.R. Shoults, C.J. Rife, and T.C. Johnson, *Electric Motors In Industry*, New York: John Wiley, 1942.

3. D. Galler, "Energy efficient control of AC induction motor driven vehicles," *IEEE IAS Conference Proceedings*, 1980.

4. A. Kusko, *Solid-State DC Motor Drives*, Cambridge, Mass.: MIT Press, 1969.

Further Information

The theory of ac motor drive operation is covered in the collection of papers edited by Bimal K. Bose, *Adjustable Speed AC Drive Systems* (IEEE, 1981). The theory of operation of dc drives is covered in *Solid-State DC Motor Drives*, by Alexander Kusko. A good general text on motors of all kinds is *Electric Machinery*, by Fitzgerald, Kingsley, and Umans. Induction motors are described in detail in the book *The Nature of Polyphase Induction Machines*, by Alger. The analysis of synchronous machines is covered in the book *Alternating Current Machines*, by M.G. Say (Wiley, 1984). Standard ratings, characteristics, and terminology are included in the Standard MG-1, by The National Electrical Manufacturers Association.

VAN DE GRAAFF'S GENERATOR

Megavolt at $90

obert J. Van de Graaff, a young Rhodes Scholar back in 1925, was 24 when he went to Oxford that year. Soon afterward, he encountered a question in his physics text that "remained in my mind as a sort of challenge," he recalled later: why was it impossible to build an electrostatic machine with a large power output?

A memorandum he produced in 1933 reveals that his first major insight was that he could accelerate ions and electrons to "enormous" energies by generating a high voltage in a vacuum. At the same time, he recognized that direct current would be preferable to "the more usual sources, which have alternating, rippling, or impulsive characteristics."

Actual experimental work had to wait until September 1929, when the new Ph.D. arrived at Princeton University in New Jersey to work under physicist Karl T. Compton on a National Research Council (NRC) fellowship. Only a month later, Van de Graaff was able to show Compton the first model of his electrostatic generator. Its operating principle was explained formally the following March in a report that, as a fellow, Van de Graaff had to write for the NRC.

Referring to the drawing shown, he wrote: "A motor driven pulley P drives by means of the belt B a second pulley Q, which runs freely. The belt is of insulating material, but has at intervals carriers of conducting material. As these carriers pass under the inductor I, maintained at a considerable negative potential by an auxiliary Whimshurst machine, there is induced on each carrier a bound positive charge, while the free negative charge escapes to earth through the grounded metal pulley P. Thus the carriers move away toward the anode A carrying a positive charge, which they retain until they make contact with the metal pulley Q, situated in the interior of the anode and connected with it. At this contact the carrier gives up its entire charge to the surrounding anode and then returns uncharged to the first pulley P. In this way positive charge is continually brought up by the moving belt to the anode, so that its potential rises steadily, either until a constant potential is reached at which the leakage from the anode becomes equal to the input current, or until a spark passes."

The electrostatic machine was easy to build. The first model, dubbed "the tin can generator," had a belt made of silk ribbon bought at the local five-and-dime.

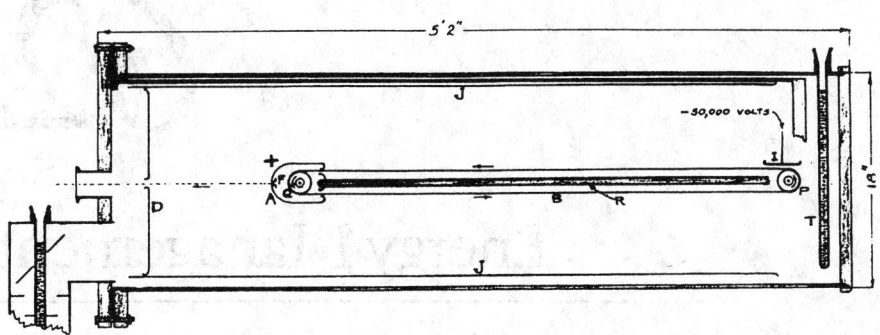

The small electrostatic generator shown in this March 1930 diagrammatic cross section marked the start of Robert J. Van de Graaff's search for an artificial source of copious high-energy radiation.

A few months later, Van de Graaff generated over 1 000 000 V between two 60-centimeter-diameter spheres on Pyrex columns 180 cm high. This version was published Oct. 28, 1931, in a three-page "Disclosure of invention," signed by Van de Graaff and two witnesses. The apparatus cost $90.

Van de Graaff's generator made headlines when it was demonstrated in New York City on Nov. 10, 1931, before a dinner meeting of the newly formed American Institute of Physics. Compton called it "the most important development that has ever taken place in the field of extremely high voltages."

Van de Graaff followed Compton to MIT in November, and he applied for a patent on Dec. 16, 1931 and assigned it to MIT in 1933 under an agreement that gave him 20 percent of the net income up to a maximum of $400 000.

On March 18, 1935, MIT vice president Vannevar Bush wrote Van de Graaff to inform him that patent 1 991 236 had been issued on Feb. 12. "It is a good looking patent, and I hope that our dreams in connection with it may some day come true," Bush added. In the light of the frontiers of nuclear research and cancer treatment the invention opened up in the ensuing years, it is likely those dreams were realized.

Source: Adapted from M.F. Wolff, *IEEE Spectrum,* p. 46, July 1990. © 1990 IEEE.

62

Energy Management

K. Neil Stanton
ESCA Corporation

Jay C. Giri
ESCA Corporation

Anjan Bose
Arizona State University

62.1 Introduction 1344
62.2 Power System Data Acquisition and Control 1345
62.3 Automatic Generation Control 1346
      Load Frequency Control • Economic Dispatch • Reserve Moni-
      toring • Interchange Transaction Scheduling
62.4 Load Management 1348
62.5 Energy Management 1349
62.6 Security Control 1350
62.7 Operator Training Simulator 1351
      Energy Control System • Power System Dynamic Simulation •
      Instructional System

62.1 Introduction

Energy management is the process of monitoring, coordinating, and controlling the generation, transmission, and distribution of electrical energy. The physical plant to be managed includes generating plants that produce energy fed through transformers to the high-voltage transmission network (grid), interconnecting generating plants and load centers. Transmission lines terminate at substations that perform switching, voltage transformation, measurement, and control. Substations at load centers transform to subtransmission and distribution levels. These lower-voltage circuits typically operate radially, i.e., no normally closed paths between substations through subtransmission or distribution circuits. (Underground cable networks in large cities are an exception.)

Since transmission systems provide negligible energy storage, supply and demand must be balanced by either generation or load. Production is controlled by turbine governors at generating plants, and automatic generation control is performed by control center computers remote from generating plants. Load management, sometimes called demand-side management, extends remote supervision and control to subtransmission and distribution circuits, including control of residential, commercial, and industrial loads.

Events such as lightning strikes, short circuits, equipment failure, or accidents may cause a system fault. Protective relays actuate rapid, local control through operation of circuit breakers before operators can respond. The goal is to maximize safety, minimize damage, and continue to supply load with the least inconvenience to customers. Data acquisition provides operators and computer control systems with status and measurement information needed to supervise overall operations. **Security** control analyzes the consequences of faults to establish operating conditions that are both robust and economical.

Energy management is performed at control centers (see Fig. 62.1), typically called system control centers, by computer systems called *energy management systems* (EMS). Data acquisition and remote control is performed by computer systems called *supervisory control and data acquisition* (SCADA) systems. These latter systems may be installed at a variety of sites including system control centers. An EMS typically includes a SCADA "front-end" through which it communicates with generating plants, substations, and other remote devices.

Figure 62.2 illustrates the **applications** layer of modern EMS as well as the underlying layers on which it is built: the operating system, a database manager, and a utilities/services layer.

FIGURE 62.1　Central dispatch operation arena of Gulf States Utilities Control Center (Beaumont, Texas) which includes a modern EMS.

62.2 Power System Data Acquisition and Control

A SCADA system consists of a master station that communicates with **remote terminal units** (RTUs) for the purpose of allowing operators to observe and control physical plants. Generating plants and transmission substations certainly justify RTUs, and their installation is becoming more common in distribution substations as costs decrease. RTUs transmit device status and measurements to, and receive control commands and setpoint data from, the master station. Communication is generally via dedicated circuits operating in the range of 600 to 4800 bits/s with the RTU responding to periodic requests initiated from the master station (polling) every 2 to 10 s, depending on the criticality of the data.

The traditional functions of SCADA systems are summarized:

- Data acquisition: Provides telemetered measurements and status information to operator.
- Supervisory control: Allows operator to remotely control devices, e.g., open and close circuit breakers. A "select before operate" procedure is used for greater safety.
- Tagging: Identifies a device as subject to specific operating restrictions and prevents unauthorized operation.
- Alarms: Informs operator of unplanned events and undesirable operating conditions. Alarms are sorted by criticality, area of responsibility, and chronology. Acknowledgment may be required.
- Logging: Logs all operator entry, all alarms, and selected information.
- Load shed: Provides both automatic and operator-initiated tripping of load in response to system emergencies.
- Trending: Plots measurements on selected time scales.

Since the master station is critical to power system operations, its functions are generally distributed among several computer systems depending on specific design. A dual computer system configured in primary and standby modes is most common. SCADA functions are listed below without stating which computer has specific responsibility.

- Manage communication circuit configuration
- Downline load RTU files
- Maintain scan tables and perform polling
- Check and correct message errors

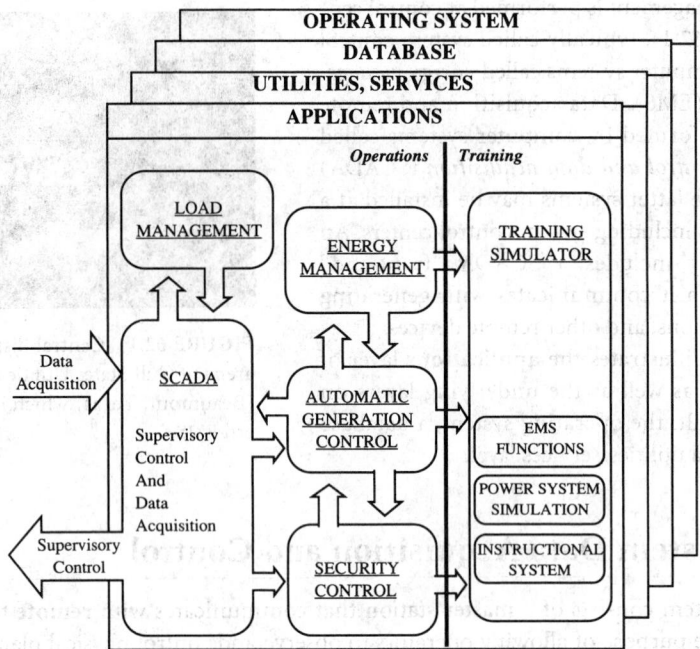

FIGURE 62.2   Layers of a modern EMS.

- Convert to engineering units
- Detect status and measurement changes
- Monitor abnormal and out-of-limit conditions
- Log and time-tag sequence of events
- Detect and annunciate alarms
- Respond to operator requests to:
    Display information
    Enter data
    Execute control action
    Acknowledge alarms
- Transmit control action to RTUs
- Inhibit unauthorized actions
- Maintain historical files
- Log events and prepare reports
- Perform load shedding

62.3 Automatic Generation Control

Automatic generation control (AGC) consists of two major and several minor functions that operate on-line in real time to adjust the generation against load at minimum cost. The major functions are load frequency control and economic **dispatch,** each of which is described below. The minor functions are reserve monitoring, which assures enough reserve on the system, **interchange** scheduling, which initiates and completes scheduled interchanges, and other similar monitoring and recording functions.

Load Frequency Control

Load frequency control (LFC) has to achieve three primary objectives which are stated below in priority order:

1. To maintain frequency at the scheduled value
2. To maintain net power interchanges with neighboring control areas at the scheduled values
3. To maintain power allocation among units at economically desired values

The first and second objectives are met by monitoring an error signal, called *area control error* (ACE), which is a combination of net interchange error and frequency error and represents the power imbalance between generation and load at any instant. This ACE must be filtered or smoothed such that excessive and random changes in ACE are not translated into control action. Since these excessive changes are different for different systems, the filter parameters have to be tuned specifically for each control area. The filtered ACE is then used to obtain the proportional plus integral control signal. This control signal is modified by limiters, deadbands, and gain constants that are tuned to the particular system. This control signal is then divided among the generating units under control by using participation factors to obtain *unit control errors* (UCE).

These participation factors may be proportional to the inverse of the second derivative of the cost of unit generation so that the units would be loaded according to their costs, thus meeting the third objective. However, cost may not be the only consideration because the different units may have different response rates and it may be necessary to move the faster generators more to obtain an acceptable response. The UCEs are then sent to the various units under control and the generating units monitored to see that the corrections take place. This control action is repeated every 2 to 6 s.

In spite of the integral control, errors in frequency and net interchange do tend to accumulate over time. These time errors and accumulated interchange errors have to be corrected by adjusting the controller settings according to procedures agreed upon by the whole interconnection. These accumulated errors as well as ACE serve as performance measures for LFC.

The main philosophy in the design of LFC is that each system should follow its own load very closely during normal operation, while during emergencies each system should contribute according to its relative size in the interconnection without regard to the locality of the emergency. Thus, the most important factor in obtaining good control of a system is its inherent capability of following its own load. This is guaranteed if the system has adequate regulation margin as well as adequate response capability. Systems that have mainly thermal generation often have difficulty in keeping up with the load because of the slow response of the units.

The design of the controller itself is an important factor, and proper tuning of the controller parameters is needed to obtain "good" control without "excessive" movement of units. Tuning is system-specific, and although system simulations are often used as aids, most of the parameter adjustments are made in the field using heuristic procedures.

Economic Dispatch

Since all the generating units that are on-line have different costs of generation, it is necessary to find the generation levels of each of these units that would meet the load at the minimum cost. This has to take into account the fact that the cost of generation in one generator is not proportional to its generation level but is a nonlinear function of it. In addition, since the system is geographically spread out, the transmission losses are dependent on the generation pattern and must be considered in obtaining the optimum pattern.

Certain other factors have to be considered when obtaining the optimum generation pattern. One is that the generation pattern provide adequate reserve margins. This is often done by constraining the generation level to a lower boundary than the generating capability. A more difficult set of constraints to consider are the transmission limits. Under certain real-time conditions it is possible that the most economic pattern may not be feasible because of unacceptable line flows or voltage conditions. The present-day economic dispatch (ED) algorithm cannot handle these security constraints. However, alternative methods based on optimal power flows have been suggested but have not yet been used for real-time dispatch.

The minimum cost dispatch occurs when the incremental cost of all the generators is equal. The cost functions of the generators are nonlinear and discontinuous. For the equal marginal cost algorithm to work it is necessary for them to be convex. These incremental cost curves are often represented as monotonically increasing piecewise-linear functions. A binary search for the optimal marginal cost is conducted by summing all the generation at a certain marginal cost and comparing it with the total power demand. If the demand is higher, a higher marginal cost is needed, and vice versa. This algorithm produces the ideal setpoints for all the generators for that particular demand, and this calculation is done every few minutes as the demand changes.

The losses in the power system are a function of the generation pattern, and they are taken into account by multiplying the generator incremental costs by the appropriate penalty factors. The penalty factor for each generator is a reflection of the sensitivity of that generator to system losses, and these sensitivities can be obtained from the transmission loss factors (Section 62.6).

This ED algorithm generally applies to only thermal generation units that have cost characteristics of the type discussed here. The hydro units have to be dispatched with different considerations. Although there is no cost for the water, the amount of water available is limited over a period, and the displacement of fossil fuel by this water determines its worth. Thus, if the water usage limitation over a period is known, say from a previously computed hydro optimization, the water worth can be used to dispatch the hydro units.

LFC and the ED functions both operate automatically in real time but with vastly different time periods. Both adjust generation levels, but LFC does it every few seconds to follow the load variation, while ED does it every few minutes to assure minimal cost. Conflicting control action is avoided by coordinating the control errors. If the unit control errors from LFC and ED are in the same direction, there is no conflict. Otherwise, a logic is set to either follow load (permissive control) or follow economics (mandatory control).

Reserve Monitoring

Maintaining enough reserve capacity is required in case generation is lost. Explicit formulas are followed to determine the spinning (already synchronized) and ready (10 min) reserves required. The availability can be assured by the operator manually, or, as mentioned previously, the ED can also reduce the upper dispatchable limits of the generators to keep such generation available.

Interchange Transaction Scheduling

The contractual exchange of power between utilities has to be taken into account by the LFC and ED functions. This is done by calculating the net interchange (sum of all the buy and sale agreements) and adding this to the generation needed in both the LFC and ED. Since most interchanges begin and end on the hour, the net interchange is ramped from one level to the new over a 10- or 20-min period straddling the hour. The programs achieve this automatically from the list of scheduled transactions.

62.4 Load Management

SCADA, with its relatively expensive RTUs installed at distribution substations, can provide status and measurements for distribution feeders at the substation. Distribution automation equipment is now available to measure and control at locations dispersed along distribution circuits. This equipment can monitor sectionalizing devices (switches, interruptors, fuses), operate switches for circuit reconfiguration, control voltage, read customers' meters, implement time-dependent pricing (on-peak, off-peak rates), and switch customer equipment to manage load. This equipment requires significantly increased functionality at distribution control centers.

Distribution control center functionality varies widely from company to company, and the following list is evolving rapidly.

- Data acquisition: Acquires data and gives the operator control over specific devices in the field. Includes data processing, quality checking, and storage.
- Feeder switch control: Provides remote control of feeder switches.
- Tagging and alarms: Provides features similar to SCADA.
- Diagrams and maps: Retrieves and displays distribution maps and drawings. Supports device selection from these displays. Overlays telemetered and operator-entered data on displays.
- Preparation of switching orders: Provides templates and information to facilitate preparation of instructions necessary to disconnect, isolate, reconnect, and reenergize equipment.
- Switching instructions: Guides operator through execution of previously prepared switching orders.
- Trouble analysis: Correlates data sources to assess scope of trouble reports and possible dispatch of work crews.
- Fault location: Analyzes available information to determine scope and location of fault.
- Service restoration: Determines the combination of remote control actions which will maximize restoration of service. Assists operator to dispatch work crews.
- Circuit continuity analysis: Analyzes circuit topology and device status to show electrically connected circuit segments (either energized or deenergized).
- Power factor and voltage control: Combines substation and feeder data with predetermined operating parameters to control distribution circuit power factor and voltage levels.
- Electrical circuit analysis: Performs circuit analysis, single-phase or three-phase, balanced or unbalanced.
- Load management: Controls customer loads directly through appliance switching (e.g., water heaters) and indirectly through voltage control.
- Meter reading: Reads customers' meters for billing, peak demand studies, time of use tariffs. Provides remote connect/disconnect.

62.5 Energy Management

Generation control and ED minimize the current cost of energy production and transmission within the range of available controls. Energy management is a supervisory layer responsible for economically scheduling production and transmission on a global basis and over time intervals consistent with cost optimization. For example, water stored in reservoirs of hydro plants is a resource that may be more valuable in the future and should, therefore, not be used now even though the cost of hydro energy is currently lower than thermal generation. The global consideration arises from the ability to buy and sell energy through the interconnected power system; it may be more economical to buy than to produce from plants under direct control. Energy accounting processes transaction information and energy measurements recorded during actual operation as the basis of payment for energy sales and purchases.

Energy management includes the following functions:

- System load forecast: Forecasts system energy demand each hour for a specified forecast period of 1 to 7 days.

- Unit commitment: Determines start-up and shut-down times for most economical operation of thermal generating units for each hour of a specified period of 1 to 7 days.
- Fuel scheduling: Determines the most economical choice of fuel consistent with plant requirements, fuel purchase contracts, and stockpiled fuel.
- Hydro-thermal scheduling: Determines the optimum schedule of thermal and hydro energy production for each hour of a study period up to 7 days while ensuring that hydro and thermal constraints are not violated.
- Transaction evaluation: Determines the optimal incremental and production costs for exchange (purchase and sale) of additional blocks of energy with neighboring companies.
- Transmission loss minimization: Recommends controller actions to be taken in order to minimize overall power system network losses.
- Security constrained dispatch: Determines optimal outputs of generating units to minimize production cost while ensuring that a network security constraint is not violated.
- Production cost calculation: Calculates actual and economical production costs for each generating unit on an hourly basis.

62.6 Security Control

Power systems are designed to survive all probable contingencies. A contingency is defined as an event that causes one or more important components such as transmission lines, generators, and transformers to be unexpectedly removed from service. Survival means the system stabilizes and continues to operate at acceptable voltage and frequency levels without loss of load. Operations must deal with a vast number of possible conditions experienced by the system, many of which are not anticipated in planning. Instead of dealing with the impossible task of analyzing all possible system states, security control starts with a specific state: the current state if executing the real-time network sequence; a postulated state if executing a study sequence. Sequence means sequential execution of programs that perform the following steps:

1. Determine the state of the system based on either current or postulated conditions.
2. Process a list of contingencies to determine the consequences of each contingency on the system in its specified state.
3. Determine preventive or corrective action for those contingencies which represent unacceptable risk.

Real-time and study network analysis sequences are diagramed in Fig. 62.3.

Security control requires topological processing to build network models and uses large-scale ac network analysis to determine system conditions. The required applications are grouped as a network subsystem which typically includes the following functions:

- Topology processor: Processes real-time status measurements to determine an electrical connectivity (bus) model of the power system network.
- State estimator: Uses real-time status and analog measurements to determine the "best" estimate of the state of the power system. It uses a redundant set of measurements; calculates voltages, phase angles, and power flows for all components in the system; and reports overload conditions.
- Power flow: Determines the steady-state conditions of the power system network for a specified generation and load pattern. Calculates voltages, phase angles, and flows across the entire system.
- Contingency analysis: Assesses the impact of a set of contingencies on the state of the power system and identifies potentially harmful contingencies that cause operating limit violations.

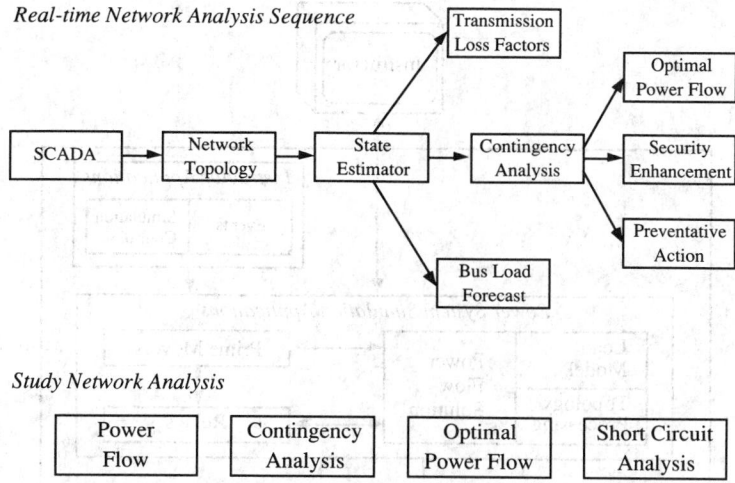

Real-time Network Analysis Sequence

Study Network Analysis

FIGURE 62.3 Real-time and study network analysis sequences.

- Optimal power flow: Recommends controller actions to optimize a specified objective function (such as system operating cost or losses) subject to a set of power system operating constraints.
- Security enhancement: Recommends corrective control actions to be taken to alleviate an existing or potential overload in the system while ensuring minimal operational cost.
- Preventive action: Recommends control actions to be taken in a "preventive" mode before a contingency occurs to preclude an overload situation if the contingency were to occur.
- Bus load forecasting: Uses real-time measurements to adaptively forecast loads for the electrical connectivity (bus) model of the power system network.
- Transmission loss factors: Determines incremental loss sensitivities for generating units; calculates the impact on losses if the output of a unit were to be increased by 1 MW.
- Short-circuit analysis: Determines fault currents for single-phase and three-phase faults for fault locations across the entire power system network.

62.7 Operator Training Simulator

Training simulators were originally created as generic systems for introducing operators to the electrical and dynamic behavior of power systems. Today, they model actual power systems with reasonable fidelity and are integrated with EMS to provide a realistic environment for operators and dispatchers to practice normal, every-day operating tasks and procedures as well as experience emergency operating situations. The various training activities can be safely and conveniently practiced with the simulator responding in a manner similar to the actual power system.

An operator training simulator (OTS) can be used in an investigatory manner to recreate past actual operational scenarios and to formulate system restoration procedures. Scenarios can be created, saved, and reused. The OTS can be used to evaluate the functionality and performance of new real-time EMS functions and also for tuning AGC in an off-line, secure environment.

The OTS has three main subsystems (Fig. 62.4).

Energy Control System

The energy control system (ECS) emulates normal EMS functions and is the only part of the OTS with which the trainee interacts. It consists of the supervisory control and data acquisition (SCADA) system, generation control system, and all other EMS functions.

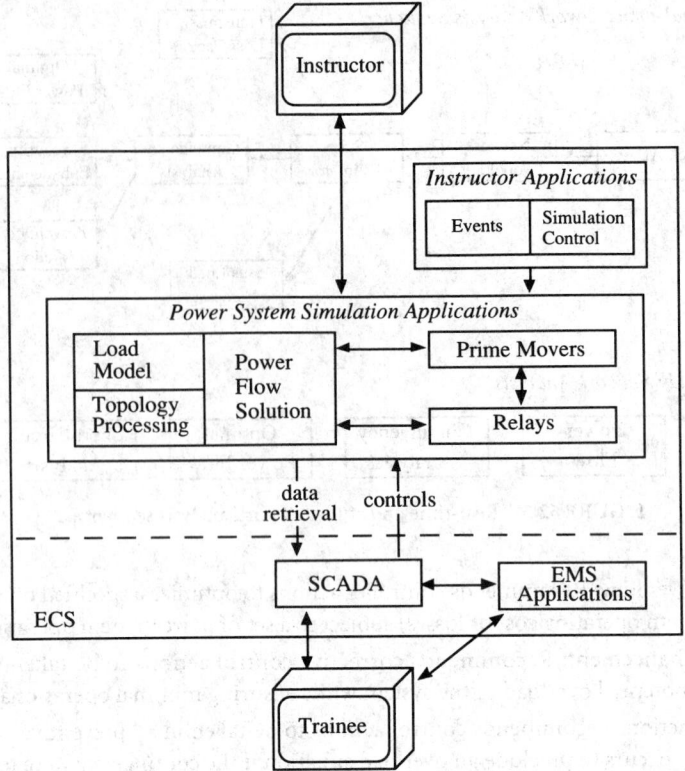

FIGURE 62.4 OTS block diagram.

Power System Dynamic Simulation

This subsystem simulates the dynamic behavior of the power system. System frequency is simulated using the "long-term dynamics" system model, where frequency of all units is assumed to be the same. The prime-mover dynamics are represented by models of the units, turbines, governors, boilers, and boiler auxiliaries. The network flows and states (bus voltages and angles, topology, transformer taps, etc.) are calculated at periodic intervals. Relays are modeled, and they emulate the behavior of the actual devices in the field.

Instructional System

This subsystem includes the capabilities to start, stop, restart, and control the simulation. It also includes making savecases, retrieving savecases, reinitializing to a new time, and initializing to a specific real-time situation.

It is also used to define event schedules. Events are associated with both the power system simulation and the ECS functions. Events may be deterministic (occur at a predefined time), conditional (based on a predefined set of power system conditions being met), or probabilistic (occur at random).

Defining Terms

Application: A software function within the energy management system which allows the operator to perform a specific set of tasks to meet a specific set of objectives.

Dispatch: The allocation of generation requirement to the various generating units that are available.

Distribution system: That part of the power system network which is connected to, and responsible for, the final delivery of power to the customer; typically the part of the network that operates at 33 kV and below, to 120 V.

Interchange or transaction: A negotiated purchase or sale of power between two companies.

Remote terminal unit (RTU): Hardware that gathers system-wide real-time data from various locations within substations and generating plants for telemetry to the energy management system.

Security: The ability of the power system to sustain and survive planned and unplanned events without violating operational constraints.

References

Application of Optimization Methods for Economy/Security Functions in Power System Operations, IEEE tutorial course, IEEE Publication 90EH0328-5-PWR, 1990.

Distribution Automation, IEEE Power Engineering Society, IEEE Publication EH0280-8-PBM, 1988.

Energy Control Center Design, IEEE tutorial course, IEEE Publication 77 TU0010-9 PWR, 1977.

Fundamentals of Load Management, IEEE Power Engineering Society, IEEE Publication EH0289-9-PBM, 1988.

Fundamentals of Supervisory Controls, IEEE tutorial course, IEEE Publication 91 EH0337-6 PWR, 1991.

"Special issue on computers in power system operations," *Proc. IEEE,* vol. 75, no. 12, 1987.

Further Information

Current innovations and applications of new technologies and algorithms are presented in the following publications:

- *IEEE Power Engineering Review* (monthly)
- *IEEE Transactions on Power Systems* (bimonthly)
- *Proceedings of the Power Industry Computer Application Conference* (biannual)

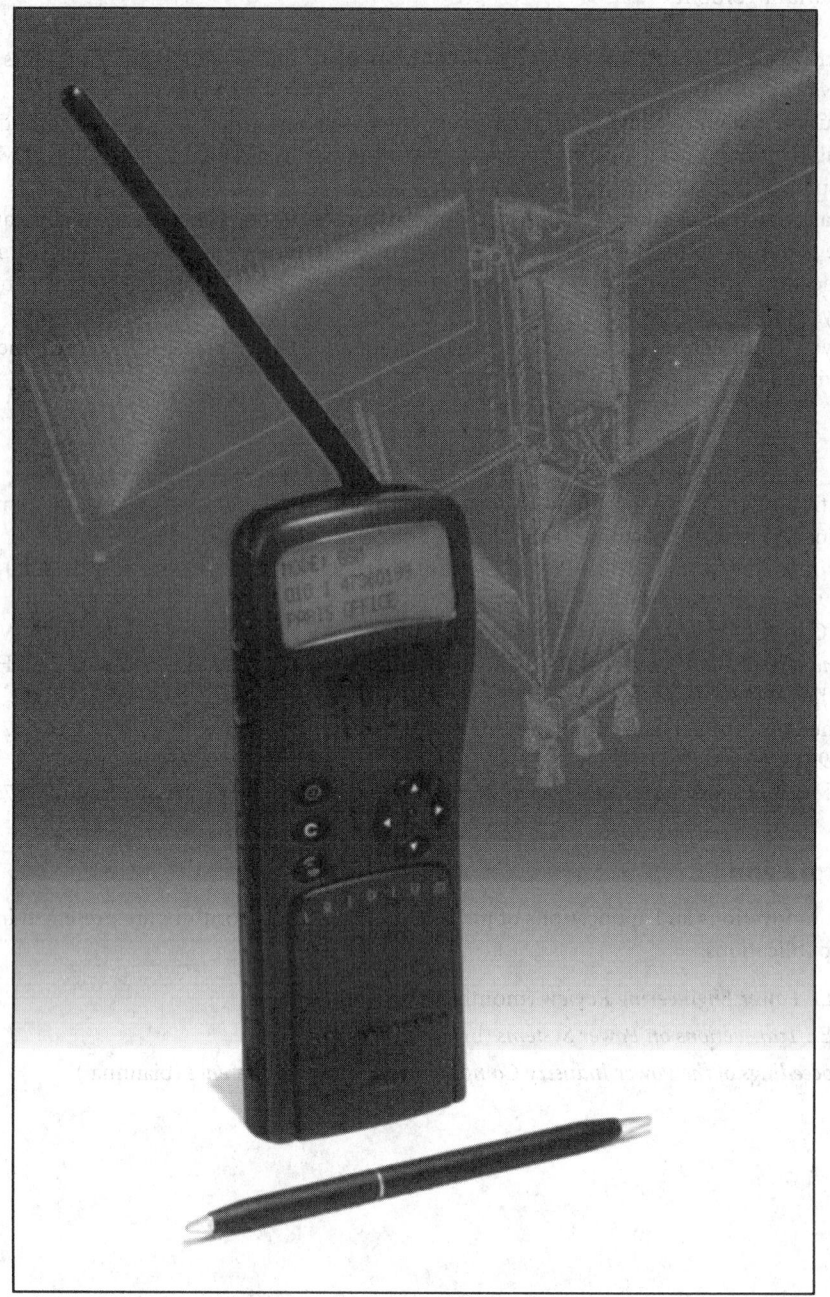

Motorola Satellite Communications is developing a personal communications system called Iridium, designed to provide global voice and data communications between virtually any locations in the world at any time. Subscribers will use small, hand-held, "pocketable" telephones (as shown) to communicate with any other telephone on earth through the use of low-orbiting satellites linked to each other and to terrestrial phone systems by gateway earth stations around the world. Motorola has received from the Federal Communications Commission an experimental license to construct and launch five satellites to demonstrate the feasibility of the system. The initial satellites are now planned for launch in 1996 and commencement of commercial service is planned for early 1998. (Photo provided by Planar Communications Corp., through the courtesy of Motorola, Inc.)

VII

Communications

Leonard Shaw
Polytechnic University, New York

63 **Broadcasting**   *R. Dorf, Z. Wan, J. Lindsey III, D. Doelitzsch, J. Whitaker, M. Roden, S. Salek, A. Clegg* 1359
Modulation and Demodulation • Radio • Television Systems • High-Definition Television • Digital Audio Broadcasting

64 **Digital Communication**   *R. Dorf, Z. Wan* 1405
Coding • Equalization

65 **Optical Communication**   *T. Darcie, J. Palais, I. Kaminow* 1417
Lightwave Technology for Video Transmission • Long Distance • Photonic Networks

66 **Networks**   *M. Huber, J. Daigle, J. Bannister, M. Gerla, R. Robrock II* 1441
B-ISDN • Computer Communication Networks • Local-Area Networks • The Intelligent Network

67 **Information Theory**   *H. Poor, C. Looney, R. Marks II, J. Thomas, T. Cover* 1478
Signal Detection • Noise • Stochastic Processes • The Sampling Theorem • Data Compression

68 **Satellites and Aerospace**   *D. DiFonzo* 1532
Satellite Applications • Satellite Functions • Satellite Orbits and Pointing Angles • Some Particular Orbits • Communications Link • System Noise Temperature and *G/T* • Digital Links • Interference and Total *C/N* • Access and Modulation • Frequency Allocations • Satellite Subsystems • Trends

69 **Personal and Office**   *W. Lee, R. Ziemer, M. Ovan* 1546
Mobile Radio and Cellular Communications • Facsimile • Wireless Local-Area Networks for the 1990s

70 **Phase-Locked Loop**   *S. Maddy* 1567
Loop Filter • Noise • PLL Design Procedures • Components • Applications

71 **Telemetry**   *C. Hoeppner* 1578
Measuring and Transmitting • Applications of Telemetry • Limitations of Telemetry • Transmitters and Batteries • Receivers and Discriminators • Antennas and Total System Operation • Calibration • Telemetry Frequency Allocations • Telemetry Antennas • Measuring and Transmitting • Modulating and Multiplexing • Passive Telemeters • The Receiving Station

72 **Computer-Aided Design and Analysis of Communication Systems**   *W. Tranter, K. Kosbar* 1593
The Role of Simulation • Motivation for the Use of Simulation • Limitations of Simulation • The Interdisciplinary Nature of Simulation • Model Design • Low-Pass Models • Pseudorandom Signal and Noise Generators • Transmitter, Channel, and Receiver Modeling • Symbol Error Rate Estimation • Validation of Simulation Results • A Simple Example Illustrating Simulation Products

ELECTRICAL TECHNOLOGY has been involved in aiding communication between a sender and a receiver of information since the advent of the electrical telegraph. The evolution of electrical communications technology has been influenced by both advances in devices for processing and transmitting electrical signals, as well as the growth and variety of communications applications that have become essential to modern society. A large fraction of electrical engineers are involved with some aspect of communications, as evidenced by the size of the IEEE Communications Society, which is second only to the Computer Society. In fact, communication between computers makes up a large part of communication system traffic, and communication technology is playing an increasing role *within* computers as they employ multiple processors and processors that are geographically distributed.

This section presents an overview of a variety of communication systems that have been developed to overcome the constraints of physical communication channels by exploiting the capabilities of the electronic and optoelectronic devices that are described elsewhere in this handbook. As a reflection of the dual influences of electrotechnology and user applications, some of the following chapters have application themes (broadcasting, satellite and aerospace, personal and office, and telemetry), while the rest have themes related to systems techniques (digital, optical, network, information theory, phase-locked loop, and computer-aided design).

The conventional radio station is a prototype of a broadcasting system in which a single transmitter sends the same message to multiple receivers. Chapter 63 reviews the basic notions of modulation needed to match the transmitted signal to the propagation and noise characteristics of the transmission medium and outlines recent developments in systems for high-definition television (HDTV) and digital audio broadcasting (DAB).

The chapter on digital techniques emphasizes the coding techniques used to detect and correct transmissions errors (which are inevitable even if systems can be designed to reduce their frequency of occurrence). Since the rate of pulse transmission over a channel can be maximized by having an accurate model for the channel, such systems are improved by continually readjusting the channel model as the characteristics change with time. This chapter also discusses adaptive equalizers that match electrical pulse shapes to changing channels.

The development of fiber-optic cables and efficient solid-state lasers has revolutionized telephone communications. Chapter 65 describes some of the related developments in signal design and transmission for optical systems that carry voice, video, and computer data messages.

Traditional telephone switching has evolved into a huge field of telecommunication networks, with the advent of new media such as fiber-optic cables and satellites and the rapidly growing digital traffic such as that between computers, electronic mail, etc. Chapter 66 describes switching and transmission protocols and other standards that are being developed to coordinate the design of equipment that sends and receives messages over the networks.

The chapter on information theory uses that term in a broad sense to describe mathematical models and techniques for describing and simplifying both deterministic and random signals. These techniques can be used for efficient communication by removing inessential information and by showing how a receiver can distinguish useful information from noise disturbances.

Satellite and aerospace applications, described in Chapter 68, provide dramatic examples of challenging communication environments where, due to equipment weight limitations and great distances, signals are weak compared to the associated noise, and propagation characteristics are nonlinear.

Personal and office innovations related to communication systems are as dramatic to the ordinary citizen as those in entertainment applications such as HDTV and digital audio. Chapter 69 describes how facsimile systems, which are especially useful for rapid transmission of graphical information, exploit standardized techniques for compressing black-and-white images. Also presented are new developments in modulation techniques and propagation modeling that have been stimulated by mobile telephone and wireless network applications.

Phase-locked loops are presented in Chapter 70 as good examples of electronic systems that are able to detect weak signals whose characteristics change with time in environments where there is strong interference from noise and from competing transmitters.

Telemetry systems are dedicated to collection and transmission of data from many sensors, often in hostile or distant environments. Chapter 71 describes how constraints on equipment size, weight and power often lead to novel methods for data multiplexing and transmission.

This section concludes with a chapter on computer-aided design methods that are being exploited to design communication systems more rapidly and effectively. Many of the problems, such as best location of a large number of nodes in a network where the construction costs and performance measures are a complex function of design parameters, are best solved by a designer who works interactively with computer algorithms.

Nomenclature

Symbol	Quantity	Unit	Symbol	Quantity	Unit
A_{eff}	effective area of antenna	m^2	M	detector gain	
B	bit rate	Mbytes/s	μ	rms modulation index	
B	channel bandwidth	Hz	n	effective input current noise density	
C	capacitance	F			
CIR	carrier-to-interference ratio		N	number of equalizer coefficients	
CNR	carrier-to-noise ratio		NF	noise ratio	
D	propagation delay	s	P	power density	W/m
$\Delta\lambda$	spectral width	Hz	P	probability of error	
$\Delta\tau$	pulse spread	s	PE	preemphasis factor	
E	electric field intensity	V/m	q	interference reduction factor	
f	carrier frequency	Hz	r	distance	m
F	noise figure		R_L	input impedance	Ω
$g(t)$	complex envelope		ρ	correlation coefficient	
G	power gain of antenna	dB	$s(t)$	modulated signal	
$H(x)$	entropy	bit	S	throughput	terabit/s
η	quantum efficiency		SNR	signal-to-noise ratio	
I	polarization isolation	dB	σ^2	variance of noise samples	
K	loop gain		t_R	rise time	s
$m(t)$	modulating signal		t_o	sample time	s
M	bit rate delay product		Z_{fs}	impedance of free space	$120\pi\ \Omega$

63

Broadcasting

Richard C. Dorf
University of California, Davis

Zhen Wan
University of California, Davis

Jefferson F. Lindsey III
Southern Illinois University at Carbondale

Dennis F. Doelitzsch
WDDD AM and FM

Jerry Whitaker
Technical Writer

Martin S. Roden
California State University

Stanley Salek
Hammett & Edison

Almon H. Clegg
CCi

63.1 Modulation and Demodulation 1359
Modulation • Superheterodyne Technique • Pulse-Code Modulation • Frequency-Shift Keying • *M*-ary Phase-Shift Keying • Quadrature Amplitude Modulation

63.2 Radio 1367
Standard Broadcasting (Amplitude Modulation) • Frequency Modulation

63.3 Television Systems 1379
Scanning Lines and Fields • Interlaced Scanning Fields • Synchronizing Video Signals • Television Industry Standards • Transmission Equipment • Television Reception

63.4 High-Definition Television 1394
Proposed Systems

63.5 Digital Audio Broadcasting 1397
The Need for DAB • DAB System Design Goals • Historical Background • Technical Overview of DAB • Audio Compression and Source Encoding • System Example: Eureka-147/DAB

63.1 Modulation and Demodulation

Richard C. Dorf and Zhen Wan

Modulation is the process of encoding the source information onto a bandpass signal with a carrier frequency f_c. This bandpass signal is called the modulated signal $s(t)$, and the baseband source signal is called the modulating signal $m(t)$. The modulated signal could be represented by

$$s(t) = \text{Re}\{g(t)e^{j\omega_c t}\} \qquad (63.1)$$

or, equivalently,

$$s(t) = R(t) \cos\left[\omega_c t + \theta(t)\right] \qquad (63.2)$$

and

$$s(t) = x(t) \cos \omega_c t - y(t) \sin \omega_c t \qquad (63.3)$$

where $\omega_c = 2\pi f_c$. The complex envelope is

$$g(t) = R(t)e^{j\theta(t)} = x(t) + jy(t) \qquad (63.4)$$

and $g(t)$ is a function of the modulating signal $m(t)$. That is,

$$g(t) = g[m(t)]$$

Thus $g[\cdot]$ performs a mapping operation on $m(t)$. The particular relationship that is chosen for $g(t)$ in terms of $m(t)$ defines the type of modulation used.

1359

In Table 63.1, examples of the mapping function $g(m)$ are given for the following types of modulation:

- AM: amplitude modulation
- DSB-SC: double-sideband suppressed-carrier modulation
- PM: phase modulation
- FM: frequency modulation
- SSB-AM-SC: single-sideband AM suppressed-carrier modulation
- SSB-PM: single-sideband PM
- SSB-FM: single-sideband FM
- SSB-EV: single-sideband envelope-detectable modulation
- SSB-SQ: single-sideband square-law-detectable modulation
- QM: quadrature modulation

Modulation

In Table 63.1, a generalized approach may be taken to obtain universal transmitter models that may be reduced to those used for a particular modulation type. We also see that there are equivalent models which correspond to difference circuit configurations, yet they may be used to produce the same type of modulated signal at their outputs. It is up to communications engineers to select an implementation method that will maximize performance, yet minimize cost based on the state of the art in circuit development.

There are two canonical forms for the generalized transmitter. Figure 63.1 is an AM-PM type circuit as described in Eq. (63.2). In this figure, the baseband signal processing circuit generates $R(t)$ and $\theta(t)$ from $m(t)$. The R and θ are functions of the modulating signal $m(t)$ as given in Table 63.1 for the particular modulation type desired.

Figure 63.2 illustrates the second canonical form for the generalized transmitter. This uses in-phase and quadrature-phase (IQ) processing. Similarly, the formulas relating $x(t)$ and $y(t)$ are shown in Table 63.1, and the baseband signal processing may be implemented by using either analog hardware or digital hardware with software. The remainder of the canonical form utilizes radio frequency (RF) circuits as indicated.

Any type of signal modulation (AM, FM, SSB, QPSK, etc.) may be generated by using either of these two canonical forms. Both of these forms conveniently separate baseband processing from RF processing.

Superheterodyne Technique

Most receivers employ the **superheterodyne receiving** technique (see Fig. 63.3). This technique consists of either down-converting or up-converting the input signal to some convenient frequency band, called the *intermediate frequency* (IF) band, and then extracting the information (or modulation) by using the appropriate detector. This basic receiver structure is used for the reception of all types of bandpass signals, such as television, FM, AM, satellite, and radar signals.

If the complex envelope $g(t)$ is desired for generalized signal detection or for optimum reception in digital systems, the $x(t)$ and $y(t)$ quadrature components, where $x(t) + jy(t) = g(t)$, may be obtained by using quadrature product detectors, as illustrated in Fig. 63.4. $x(t)$ and $y(t)$ could be fed into a signal processor to extract the modulation information. Disregarding the effects of noise, the signal processor could recover $m(t)$ from $x(t)$ and $y(t)$ (and, consequently, demodulate the IF signal) by using the inverse of the complex envelope generation functions given in Table 63.1.

Table 63.1 Complex Envelope Functions for Various Types of Modulation

Type of Modulation	Mapping Functions $g[m]$	Corresponding Quadrature Modulation		Corresponding Amplitude and Phase Modulation		Linearity	Remarks
		$x(t)$	$y(t)$	$R(t)$	$\theta(t)$		
AM	$1 + m(t)$	$1 + m(t)$	0	$\lvert 1 + m(t) \rvert$	$\begin{cases} 0, & m(t) > -1 \\ 180°, & m(t) < -1 \end{cases}$	L[b]	$m(t) > -1$ required for envelope detection.
DSB-SC	$m(t)$	$m(t)$	0	$\lvert m(t) \rvert$	$\begin{cases} 0, & m(t) > 0 \\ 180°, & m(t) < 0 \end{cases}$	L	Coherent detection required.
PM	$e^{jD_p m(t)}$	$\cos[D_p m(t)]$	$\sin[D_p m(t)]$	1	$D_p m(t)$	NL	D_p is the phase deviation constant (radian/volts).
FM	$e^{jD_f \int_{-\infty}^t m(\sigma)d\sigma}$	$\cos\left[D_f \int_{-\infty}^t m(\sigma)d\sigma\right]$	$\sin\left[D_f \int_{-\infty}^t m(\sigma)d\sigma\right]$	1	$D_f \int_{-\infty}^t m(\sigma)d\sigma$	NL	D_f is the frequency deviation constant (radian/volt-sec).
SSB-AM-SC[a]	$m(t) \pm j\hat{m}(t)$	$m(t)$	$\pm \hat{m}(t)$	$\sqrt{[m(t)]^2 + [\hat{m}(t)]^2}$	$\tan^{-1}[\pm \hat{m}(t)/m(t)]$	L	Coherent detection required.
SSB-PM[a]	$e^{jD_p[m(t)\pm j\hat{m}(t)]}$	$e^{\mp D_p \hat{m}(t)}\cos[D_p m(t)]$	$e^{\mp D_p \hat{m}(t)}\sin[D_p m(t)]$	$e^{\mp D_p \hat{m}(t)}$	$D_p m(t)$	NL	
SSB-FM[a]	$e^{jD_f \int_{-\infty}^t [m(\sigma)\pm j\hat{m}(\sigma)]d\sigma}$	$e^{\mp D_f \int_{-\infty}^t \hat{m}(\sigma)d\sigma}\cos\left[D_f \int_{-\infty}^t m(\sigma)d\sigma\right]$	$e^{\mp D_f \int_{-\infty}^t \hat{m}(\sigma)d\sigma}\sin\left[D_f \int_{-\infty}^t m(\sigma)d\sigma\right]$	$e^{\mp D_f \int_{-\infty}^t \hat{m}(\sigma)d\sigma}$	$D_f \int_{-\infty}^t m(\sigma)d\sigma$	NL	
SSB-EV[a]	$e^{\{\ln[1+m(t)]\pm j\hat{\ln}[1+m(t)]\}}$	$[1+m(t)]\cos\{\hat{\ln}[1+m(t)]\}$	$\pm[1+m(t)]\sin\{\hat{\ln}[1+m(t)]\}$	$1+m(t)$	$\pm\hat{\ln}[1+m(t)]$	NL	$m(t) > -1$ is required so that the ln will have a real value.
SSB-SQ[a]	$e^{(1/2)\{\ln[1+m(t)]\pm j\hat{\ln}[1+m(t)]\}}$	$\sqrt{1+m(t)}\cos\left\{\dfrac{1}{2}\hat{\ln}[1=m(t)]\right\}$	$\pm\sqrt{1+m(t)}\sin\left\{\dfrac{1}{2}\hat{\ln}[1+m(t)]\right\}$	$\sqrt{1+m(t)}$	$\pm\dfrac{1}{2}\hat{\ln}[1+m(t)]$	NL	$m(t) > -1$ is required so that the ln will have a real value.
QM	$m_1(t) + jm_2(t)$	$m_1(t)$	$m_2(t)$	$\sqrt{m_1^2(t) + m_2^2(t)}$	$\tan^{-1}[m_2(t)/m_1(t)]$	L	Used in NTSC color television: requires coherent detection.

L = linear, NL = nonlinear, $[\hat{\cdot}]$ is the Hilbert transform (i.e., −90° phase-shifted version) of $[\cdot]$. The Hilbert transform is

$$\hat{x}(t) \triangleq x(t) \star \frac{1}{\pi t} = \frac{1}{\pi} \int_{-\infty}^{\infty} \frac{x(\lambda)}{t - \lambda}\, d\lambda$$

[a] Use upper signs for upper sideband signals and lower signs for lower sideband signals.

[b] In the strict sense, AM signals are not linear because the carrier term does not satisfy the linearity (superposition) condition.

Source: L. W. Couch, *Digital and Analog Communication Systems*, New York: Macmillan, 1990. With permission.

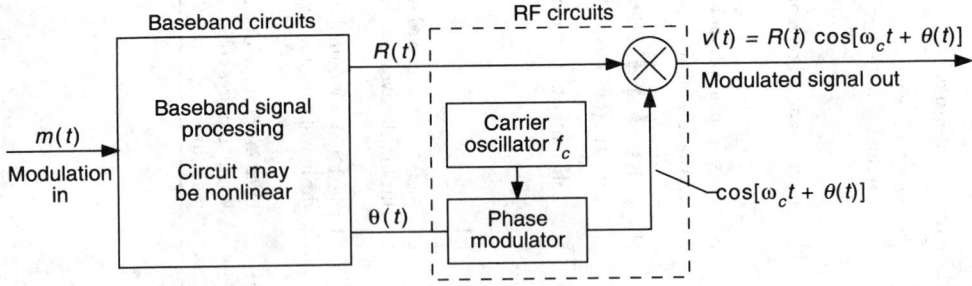

FIGURE 63.1 Generalized transmitter using the AM-PM generation technique. (*Source:* L. W. Couch, *Digital and Analog Communication Systems,* New York: Macmillan, 1990, p. 279. With permission.)

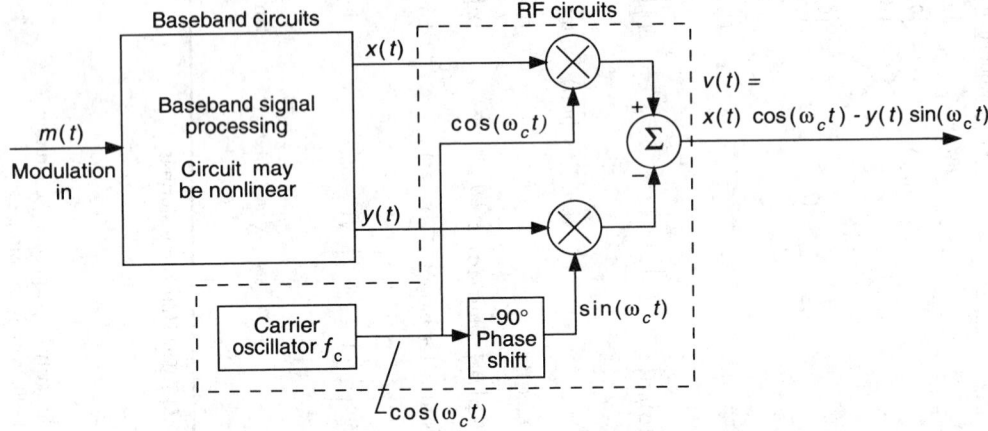

FIGURE 63.2 Generalized transmitter using the quadrature generation technique. (*Source:* L. W. Couch, *Digital and Analog Communication Systems,* New York: Macmillan, 1990, p. 280. With permission.)

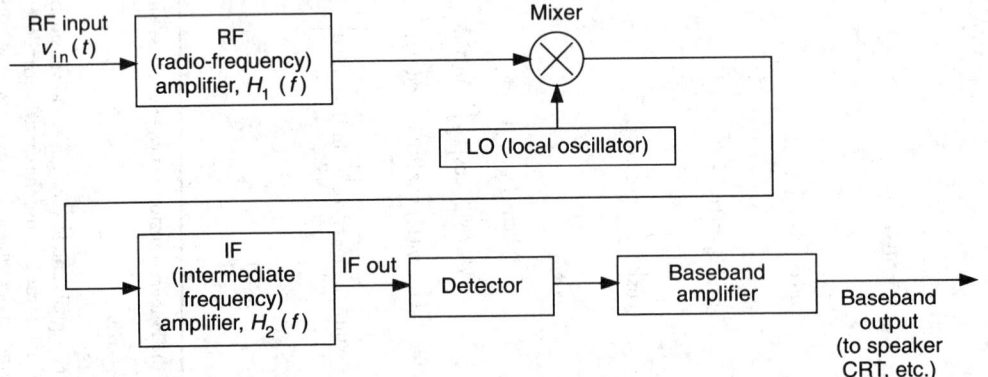

FIGURE 63.3 Superheterodyne receiver. (*Source:* L. W. Couch, *Digital and Analog Communication Systems,* New York: Macmillan, 1990, p. 281. With permission.)

The generalized modulation techniques are shown in Table 63.2. In digital communication systems, discrete modulation techniques are usually used to modulate the source information signal. Discrete modulation includes:

- PCM = pulse-code modulation
- DM = differential modulation

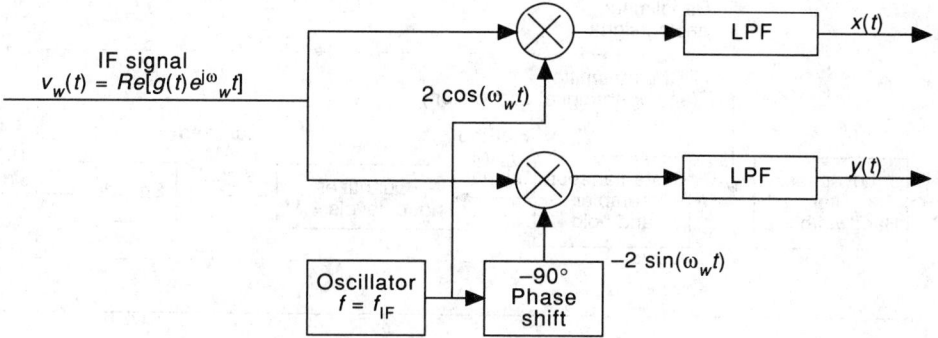

FIGURE 63.4 IQ (in-phase and quadrature-phase) detector. (*Source:* L. W. Couch, *Digital and Analog Communication Systems*, New York: Macmillan, 1990, p. 284. With permission.)

Table 63.2 Performance of a PCM System with Uniform Quantizing and No Channel Noise

Number of Quantizer Levels Used, M	Length of the PCM Word, n (bits)	Bandwidth of PCM Signal (First Null Bandwidth)[a]	Recovered Analog Signal Power-to-Quantizing Noise Power Ratios	
			$(S/N)_{pk\,out}$	$(S/N)_{out}$
2	1	$2B$	10.8	6.0
4	2	$4B$	16.8	12.0
8	3	$6B$	22.8	18.1
16	4	$8B$	28.9	24.1
32	5	$10B$	34.9	30.1
64	6	$12B$	40.9	36.1
128	7	$14B$	46.9	42.1
256	8	$16B$	52.9	48.2
512	9	$18B$	59.0	54.2
1024	10	$20B$	65.0	60.2

[a]B is the absolute bandwidth of the input analog signal.

- DPCM = differential pulse-code modulation
- FSK = frequency-shift keying
- PSK = phase-shift keying
- DPSK = differential phase-shift keying
- MPSK = M-ary phase-shift keying
- QAM = quadrature amplitude modulation

Pulse-Code Modulation

PCM is essentially analog-to-digital conversion of a special type, where the information contained in the instantaneous samples of an analog signal is represented by digital words in a serial bit stream. The PCM signal is generated by carrying out three basic operations: sampling, quantizing, and encoding (see Fig. 63.5). The sampling operation generates a flat-top phase amplitude modulation (PAM) signal. The quantizing converts the actual sampled value into the nearest of the M amplitude levels. The PCM signal is obtained from the quantized PAM signal by encoding each quantized sample value into a digital word.

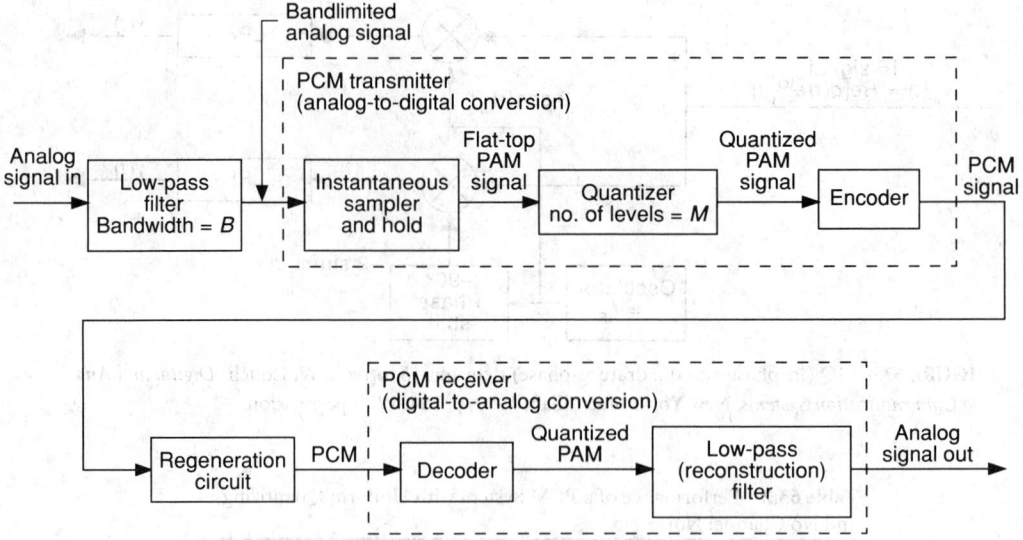

FIGURE 63.5   A PCM transmission system.

Frequency-Shift Keying

The FSK signal can be characterized as one of two different types. One type is called *discontinuous-phase* FSK since $\theta(t)$ is discontinuous at the switching times. The discontinuous-phase FSK signal is represented by

$$s(t) = \begin{cases} A_c \cos(\omega_1 t + \theta_1) & \text{for } t \text{ in time interval when a binary 1 is sent} \\ A_c \cos(\omega_2 t + \theta_2) & \text{for } t \text{ in time interval when a binary 0 is sent} \end{cases} \quad (63.5)$$

where f_1 is called the mark (binary 1) frequency and f_2 is called the space (binary 0) frequency. The other type is continuous-phase FSK. The continuous-phase FSK signal is generated by feeding the data signal into a frequency modulator, as shown in Fig. 63.6(b). This FSK signal is represented by

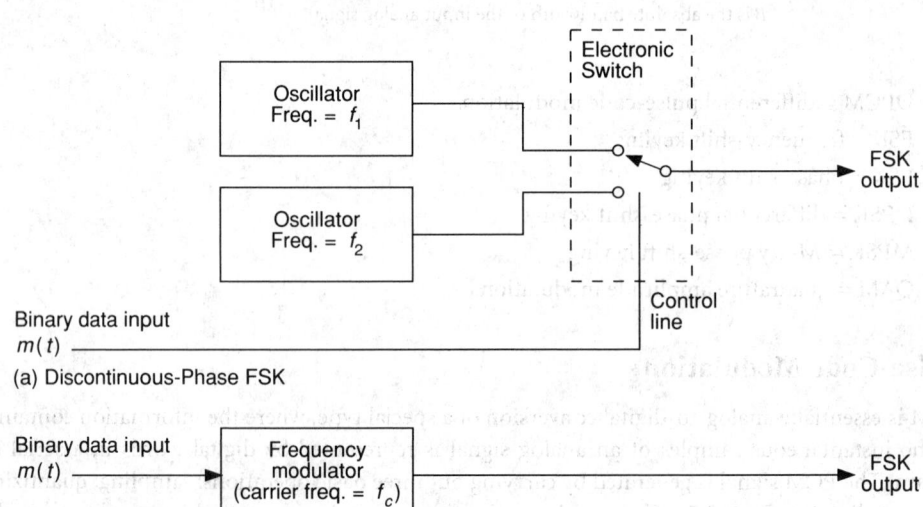

FIGURE 63.6   Generation of FSK. (*Source:* L. W. Couch, *Digital and Analog Communication Systems,* New York: Macmillan, 1990, p. 337. With permission.)

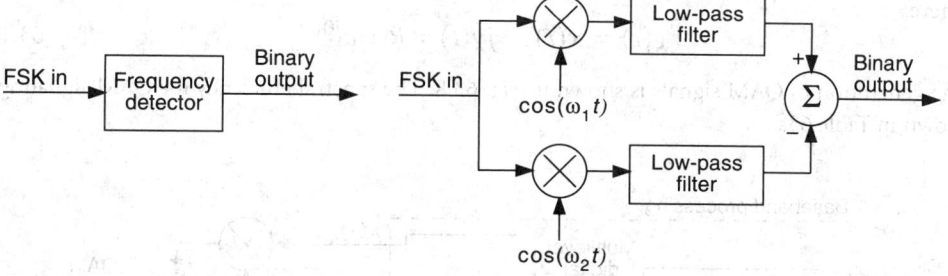

(a) Noncoherent Detection   (b) Coherent (Synchronous) Detection

FIGURE 63.7 Detection of FSK. (*Source:* L. W. Couch, *Digital and Analog Communication Systems,* New York: Macmillan, 1990, p. 344. With permission.)

$$s(t) = A_c \cos \omega_c t + D_f \int_{-\infty}^{t} m(\lambda)d\lambda$$

or

$$s(t) = \text{Re}\{g(t)e^{j\omega_c t}\} \tag{63.6}$$

where

$$g(t) = A_c^{j\theta(t)} \tag{63.7}$$

$$\theta(t) = D_f \int_{-\infty}^{t} m(\lambda)d\lambda \quad \text{for FSK} \tag{63.8}$$

Detection of FSK is illustrated in Fig. 63.7.

M-ary Phase-Shift Keying

If the transmitter is a PM transmitter with an *M*-level digital modulation signal, MPSK is generated at the transmitter output. A plot of the permitted values of the complex envelope, $A_c e^{j\theta(t)}$, would contain *M* points, one value of *g* (a complex number in general) for each of the *M* multilevel values, corresponding to the *M* phases that θ is permitted to have.

MPSK can also be generated using two quadrature carriers modulated by the *x* and *y* components of the complex envelope (instead of using a phase modulator)

$$g(t) = A_c e^{j\theta(t)} = x(t) + jy(t) \tag{63.9}$$

where the permitted values of *x* and *y* are

$$x_i = A_c \cos \theta_i \tag{63.10}$$

$$y_i = A_c \sin \theta_i \tag{63.11}$$

for the permitted phase angles θ_i, $i = 1, 2, \ldots, M$, of the MPSK signal. This is illustrated by Fig. 63.8, where the signal processing circuit implements Eqs. (63.10) and (63.11).

MPSK, where $M = 4$, is called quadrature-phase-shift-keyed (QPSK) signaling.

Quadrature Amplitude Modulation

Quadrature carrier signaling is called quadrature amplitude modulation (QAM). In general, QAM signal constellations are not restricted to having permitted signaling points only on a circle (of radius A_c, as was the case for MPSK). The general QAM signal is

$$s(t) = x(t) \cos \omega_c t - y(t) \sin \omega_c t \tag{63.12}$$

where
$$g(t) = x(t) + jy(t) = R(t)e^{j\theta(t)} \tag{63.13}$$

The generation of QAM signals is shown in Fig. 63.8. The spectral efficiency for QAM signaling is shown in Table 63.3.

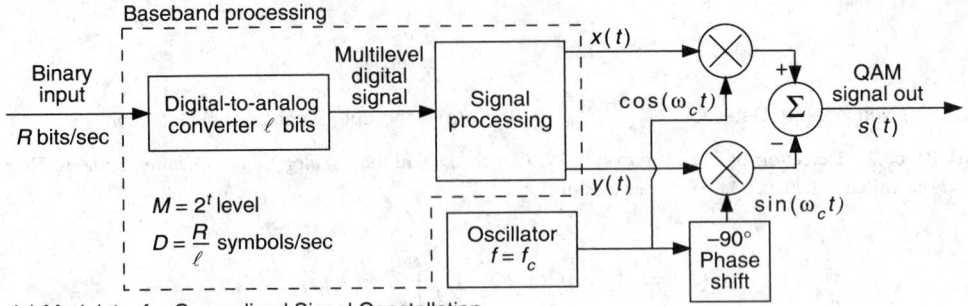

(a) Modulator for Generalized Signal Constellation

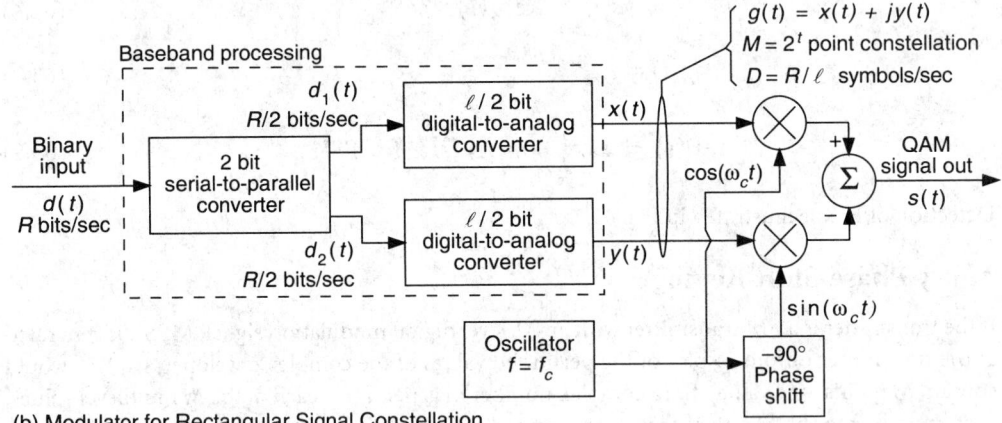

(b) Modulator for Rectangular Signal Constellation

FIGURE 63.8 Generation of QAM signals. (*Source:* L. W. Couch, *Digital and Analog Communication Systems*, New York: Macmillan, 1990, p. 346. With permission.)

Table 63.3 Spectral Efficiency for QAM Signaling with Raised Cosine-Roll-Off Pulse Shaping

		$\eta = \dfrac{R}{B_T} \dfrac{\text{bits/s}}{\text{Hz}}$					
Number of Levels, M (symbols)	Size of DAC, ℓ (bits)	$r = 0.0$	$r = 0.1$	$r = 0.25$	$r = 0.5$	$r = 0.75$	$r = 1.0$
2	1	1.00	0.909	0.800	0.667	0.571	0.500
4	2	2.00	1.82	1.60	1.33	1.14	1.00
8	3	3.00	2.73	2.40	2.00	1.71	1.50
16	4	4.00	3.64	3.20	2.67	2.29	2.00
32	5	5.00	4.55	4.0	3.33	2.86	2.50

DAC = digital-to-analog converter.
$\eta = R/B_T = \ell/2$ bits/s per hertz.
r is the roll-off factor of the filter characteristic.
Source: L. W. Couch, *Digital and Analog Communication Systems*, New York: Macmillan, 1990, p. 350. With permission.

Defining Terms

Modulation: The process of encoding the source information onto a bandpass signal with a carrier frequency f_c. It can be expressed as

$$s(t) = \text{Re}\{g(t)\, e^{j\omega_c t}\}$$

where $g(t)$ is a function of the modulating signal $m(t)$. That is,

$$g(t) = g[m(t)]$$

$g[\cdot]$ performs a mapping operation on $m(t)$. The particular relationship that is chosen for $g(t)$ in terms of $m(t)$ defines the type of modulation used.

Superheterodyne receiver: Most receivers employ the superheterodyne receiving technique, which consists of either down-converting or up-converting the input signal to some convenient frequency band, called the intermediate frequency band, and then extracting the information (or modulation) by using an appropriate detector. This basic receiver structure is used for the reception of all types of bandpass signals, such as television, FM, AM, satellite, and radar signals.

References

L. W. Couch, *Digital and Analog Communication Systems*, New York: Macmillan, 1990.

F. Dejager, "Delta modulation of PCM transmission using a 1-unit code," Phillips Res. Rep., no. 7, pp. 442–466, Dec. 1952.

J.H. Downing, *Modulation Systems and Noise*, Englewood Cliffs, N.J.: Prentice-Hall, 1964.

J. Dunlop and D.G. Smith, *Telecommunications Engineering*, London: Van Nostrand, 1989.

B.P. Lathi, *Modern Digital and Analog Communication Systems*, New York: CBS College, 1983.

J.H. Park, Jr., "On binary DPSK detection," *IEEE Trans. Commun.*, COM-26, pp. 484–486, 1978.

M. Schwartz, *Information Transmission, Modulation and Noise*, New York: McGraw-Hill, 1980.

Further Information

The monthly magazine *IEEE Transactions on Communications* describes telecommunication techniques. The performance of *M*-ary QAM schemes is evaluated in its March 1991 issue, pp. 405–408.

Another source is *IEEE Transactions on Broadcasting*, which is published quarterly by The Institute of Electrical and Electronics Engineers, Inc.

The biweekly magazine *Electronics Letters* investigates the error probability of coherent PSK and FSK systems with multiple co-channel interferences in its April 11, 1991 issue, pp. 640–642. Another relevant source regarding the coherent detection of MSK is described on pp. 623–625 of the same issue. All subscriptions inquiries and orders should be sent to IEE Publication Sales, P.O. Box 96, Stevenage, Herts, SG1 2SD, United Kingdom.

63.2 Radio

Jefferson F. Lindsey III and Dennis F. Doelitzsch

Standard Broadcasting (Amplitude Modulation)

Standard broadcasting refers to the transmission of voice and music received by the general public in the 535- to 1705-kHz frequency band. Amplitude modulation is used to provide service ranging from that needed for small communities to higher-power broadcast stations needed for larger regional areas. The *primary service area* is defined as the area in which the groundwave signal is not subject to objectionable interference or objectionable fading. The *secondary service area* refers to an area serviced by skywaves and not subject to objectionable interference. *Intermittent service area* refers to an area receiving service from either a groundwave or a skywave but beyond the primary service area and subject to some interference and fading.

Frequency Allocations

The carrier frequencies for standard broadcasting in the United States (referred to internationally as medium-wave broadcasting) are designated in the Federal Communications Commission (FCC) Rules and Regulations, Vol. III, Part 73. A total of 117 carrier frequencies are allocated from 540 to 1700 kHz in 10-kHz intervals. Each carrier frequency is required by the FCC rules to deviate no more than ±20 Hz from the allocated frequency, to minimize heterodyning from two or more interfering stations. Double-sideband full-carrier modulation, commonly called *amplitude modulation* (AM), is used in standard broadcasting for sound transmission. Typical modulation frequencies for voice and music range from 50 Hz to 10 kHz. Each channel is generally thought of as 10 kHz in width, and thus the frequency band is designated from 535 to 1705 kHz; however, when the modulation frequency exceeds 5 kHz, the radio frequency bandwidth of the channel exceeds 10 kHz and adjacent channel interference may occur. To improve the high-frequency performance of transmission and to compensate for the high-frequency roll-off of many consumer receivers, FCC rules require that stations boost the high-frequency amplitude of transmitted audio using pre-emphasis techniques. In addition stations may also use multiplexing to transmit stereophonic programming. The FCC does not specify the technique to be used, but the station must comply with the occupied bandwidth limitations and be compatible with AM receivers using envelope detectors and any applicable international agreements. The most common systems in use are Motorola's C-QUAM quadrature amplitude modulation and Kahn's independent sideband system. Approximately 700 AM stations transmit in stereo.

Channel and Station Classifications

In standard broadcast (AM), stations are classified according to their operating power, protection from interference, and hours of operation. A Class A station operates with 10 to 50 kW of power servicing a large area with primary, secondary, and intermittent coverage and is protected from interference both day and night. These stations are called "clear channel" stations because the channel is cleared of nighttime interference over a major portion of the country. Class B stations operate full time with transmitter powers of 0.25 to 50 kW and are designed to render primary service only over a principal center of population and the rural area contiguous thereto. While nearly all Class A stations operate with 50 kW, most Class B stations must restrict their power to 5 kW or less to avoid interfering with other stations. Class B stations operating in the 1605 to 1705 kHz band are restricted to a power level of 10 kW daytime and 1 kW nighttime. Class C stations operate on six designated channels (1230, 1240, 1340, 1400, 1450, and 1490) with a maximum power of 1 kW or less full time and render primarily local service to smaller communities. Class D stations operate on Class A or B frequencies with Class B transmitter powers during daytime, but nighttime operation, if permitted at all, must be at low power (less than 0.25 kW) with no protection from interference.

Although Class A stations cover large areas at night, approximately in a 1220-km (750-mi) radius, the nighttime coverage of Class B, C, and D stations is limited by interference from other stations, electrical devices, and atmospheric conditions to a relatively small area. Class C stations, for example, have an interference-free nighttime coverage radius of approximately 8 to 16 km. As a result, there may be large differences in the area that the station covers daytime versus nighttime. With over 5200 AM stations licensed for operation by the FCC, interference, both day and night, is a factor that significantly limits the service which stations may provide. In the absence of interference, a daytime signal strength of 2 mV/m is required for reception in populated areas of more than 2500, while a signal of 0.5 mV/m is generally acceptable in less populated areas. Secondary nighttime service is provided in areas receiving a 0.5-mV/m signal 50% or more of the time without objectionable interference. Table 63.4 indicates the daytime contour overlap limits. However, it should be noted that these limits apply to new stations and modifications to existing stations. Nearly every station on the air was allocated prior to the implementation of these rules when the interference criteria were less restrictive.

Field Strength

The field strength produced by a standard broadcast station is a key factor in determining the primary and secondary service areas and interference limitations of possible future radio stations. The field strength limitations are specified as field intensities by the FCC with the units volts per meter; however, measuring devices may read volts or decibels referenced to 1 mW (dBm), and a conversion may be needed to obtain the field intensity. The power received may be measured in dBm and converted to watts. Voltage readings may be converted to watts by squaring the root mean square (rms) voltage and dividing by the field strength meter input resistance, which is typically on the order of 50 or 75 Ω. Additional factors needed to determine **electric field intensity** are the power gain and losses of the field strength receiving antenna system. Once the power gain and losses are known, the effective area with loss compensation of the field strength receiver antenna may be obtained as

$$A_{\text{eff}} = G \frac{\lambda^2}{4\pi} L \tag{63.14}$$

where A_{eff} = effective area including loss compensation, m²; G = power gain of field strength antenna, W/W; λ = wavelength, m; and L = mismatch loss and cable loss factor, W/W.

From this calculation, the power density in watts per square meter may be obtained by dividing the received power by the effective area, and the electric field intensity may be calculated as

Table 63.4 Protected Service Signal Intensities for Standard Broadcasting (AM)

Class of Station	Power (kW)	Class of Channel Used	Signal Strength Contour of Area Protected from Objectionable Interference* (µV/m)		Permissible Interfering Signal	
			Day†	Night	Day†	Night‡
A	10–50	Clear	SC 100	SC 500 50% SW	SC 5	SC 25
			AC 500	AC 500 GW	AC 250	AC 250
B	0.25–50	Clear	500	2000†	25	25
		Regional			AC 250	250
C	0.25–1	Local	500	Not precise§	SC 25	Not precise
D	0.25–50	Clear	500	Not precise	SC 25	Not precise
		Regional			AC 250	

*When a station is already limited by interference from other stations to a contour of higher value than that normally protected for its class, this higher-value contour shall be the established protection standard for such station. Changes proposed by Class A and B stations shall be required to comply with the following restrictions. Those interferers that contribute to another station's RSS using the 50% exclusion method are required to reduce their contribution to that RSS by 10%. Those lesser interferers that contribute to a station's RSS using the 25% exclusion method but do not contribute to that station's RSS using the 50% exclusion method may make changes not to exceed their present contribution. Interferers not included in a station's RSS using the 25% exclusion method are permitted to increase radiation as long as the 25% exclusion threshold is not equaled or exceeded. In no case will a reduction be required that would result in a contributing value that is below the pertinent value specified in the table.

†Groundwave.

‡Skywave field strength for 10% or more of the time. For Alaska, Class SC is limited to 5 µV/m.

§During nighttime hours, Class C stations in the contiguous 48 states may treat all Class B stations assigned to 1230, 1240, 1340, 1400, 1450, and 1490 kHz in Alaska, Hawaii, Puerto Rico and the U.S. Virgin Islands as if they were Class C stations.

Note: SC = same channel; AC = adjacent channel; SW = skywave; GW = groundwave; RSS = root of sum squares.

Source: FCC Rules and Regulations, Revised 1991; vol. III, pt. 73.182(a).

$$E = \sqrt{\mathcal{P} Z_{fs}} \qquad (63.15)$$

where E = electric field intensity, V/m; $\mathcal{P}$ = power density, W/m²; and $Z_{fs} = 120\pi\ \Omega$, impedance of free space.

The protected service contours and permissible interference contours for standard broadcast stations shown in Table 63.4, along with a knowledge of the field strength of existing broadcast stations, may be used in determining the potential for establishing new standard broadcast stations.

Propagation

One of the major factors in the determination of field strength is the propagation characteristic that is described by the change in electric field intensity with an increase in distance from the broadcast station antenna. This variation depends on a number of factors including frequency, distance, surface dielectric constant, surface loss tangent, polarization, local topography, and time of day. Generally speaking, groundwave propagation occurs at shorter ranges both during day and night periods. Skywave propagation permits longer ranges and occurs during night periods, and thus some stations must either reduce power or cease to operate at night to avoid causing interference. Propagation curves in the broadcast industry are frequently referred to a reference level of 100 mV/m at 1 km; however, a more general expression of groundwave propagation may be obtained by using the Bremmer series [Bremmer, 1949]. A typical groundwave propagation curve for electric field strength as a function of distance is shown in Fig. 63.9 for an operating frequency of 770–810 kHz. The ground conductivity varies from 0.1 to 5000 mS/m, and the ground relative dielectric constant is 15.

The **effective radiated power** (ERP) refers to the effective power output from the antenna in a specified direction and includes the transmitter power output, transmission line losses, and antenna power gain. The ERP in most cases exceeds the transmitter output power, since that antenna power gain is normally 2 or more. For a hypothetical perfect isotropic radiator with a power gain of 1, the ERP is found to be

$$\text{ERP} = \frac{E^2 r^2}{30} \qquad (63.16)$$

where E is the electric field intensity, V/m, and r is the distance, m. For a distance of 1 km (1000 m), the ERP required to produce a field intensity of 100 mV/m is found to be 333.3 W. Since the field intensity is proportional to the square root of the power, field intensities may be determined at other powers.

Skywave propagation necessarily involves some fading and less predictable field intensities and is most appropriately described in terms of statistics or the percentage of time a particular field strength level is found. Figure 63.10 shows skywave propagation for a 100-mV/m field strength at a distance of 1 km for midpoint path latitudes of 35 to 50 degrees.

Transmitters

Standards that cover AM broadcast transmitters are given in the Electronic Industry Association (EIA) Standard TR-101A, "Electrical Performance Standard for Standard Broadcast Transmitters." Parameters and methods for measurement include the following: carrier output rating, carrier power output capability, carrier frequency range, carrier frequency stability, carrier shift, carrier noise level, magnitude of radio frequency (RF) harmonics, normal load, transmitter output circuit adjustment facilities, RF and audio interface definitions, modulation capability, audio input level for 100% modulation, audio frequency response, audio frequency harmonic distortion, rated power supply, power supply variation, operating temperature characteristics, and power input.

Standard AM broadcast transmitters range in power output from 5 W up to 50 kW units. While solid-state devices are used for many models (especially the lower-powered units), several manu-

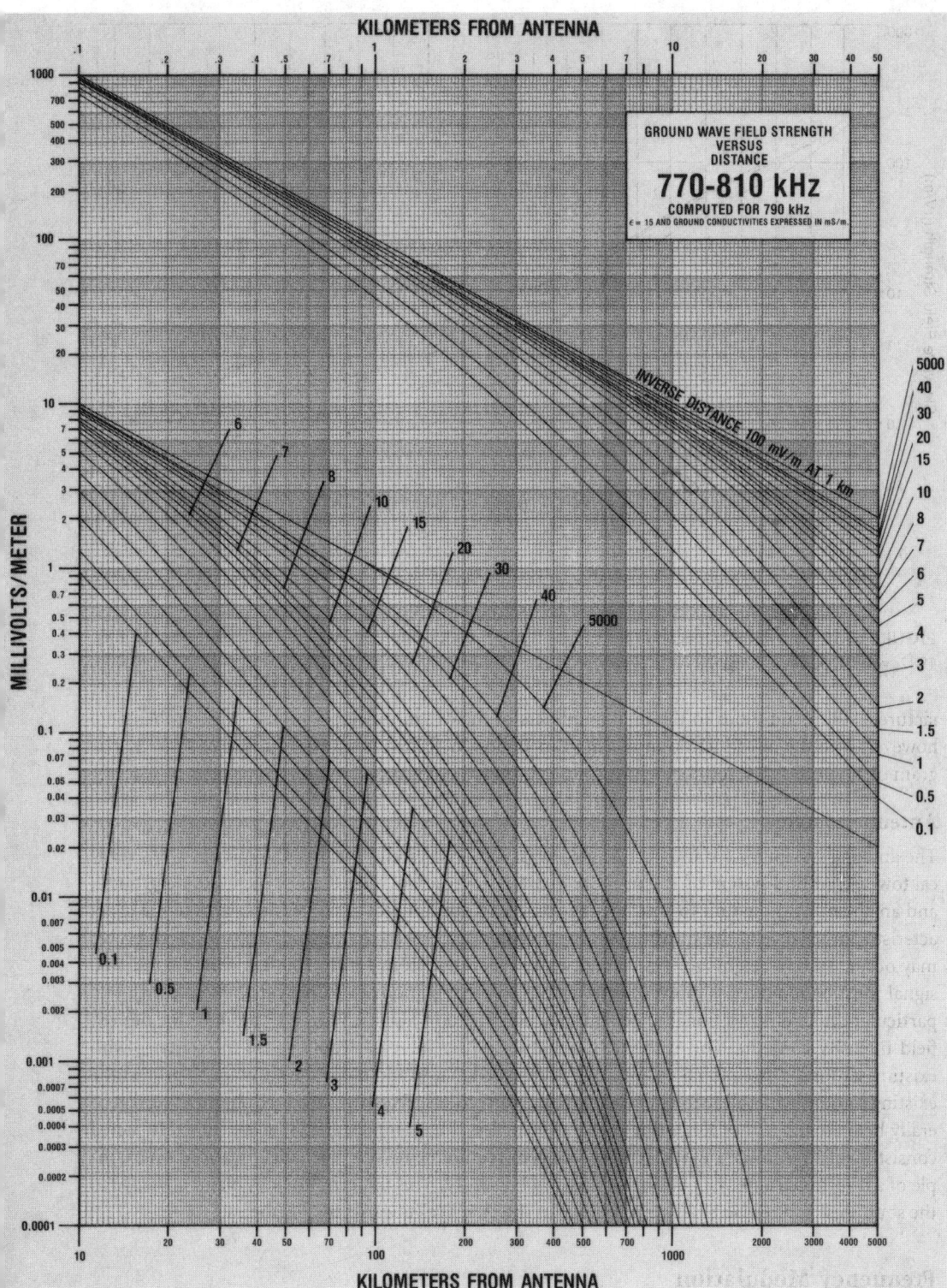

FIGURE 63.9 Typical groundwave propagation for standard AM broadcasting. (*Source:* 1986 National Association of Broadcasters.)

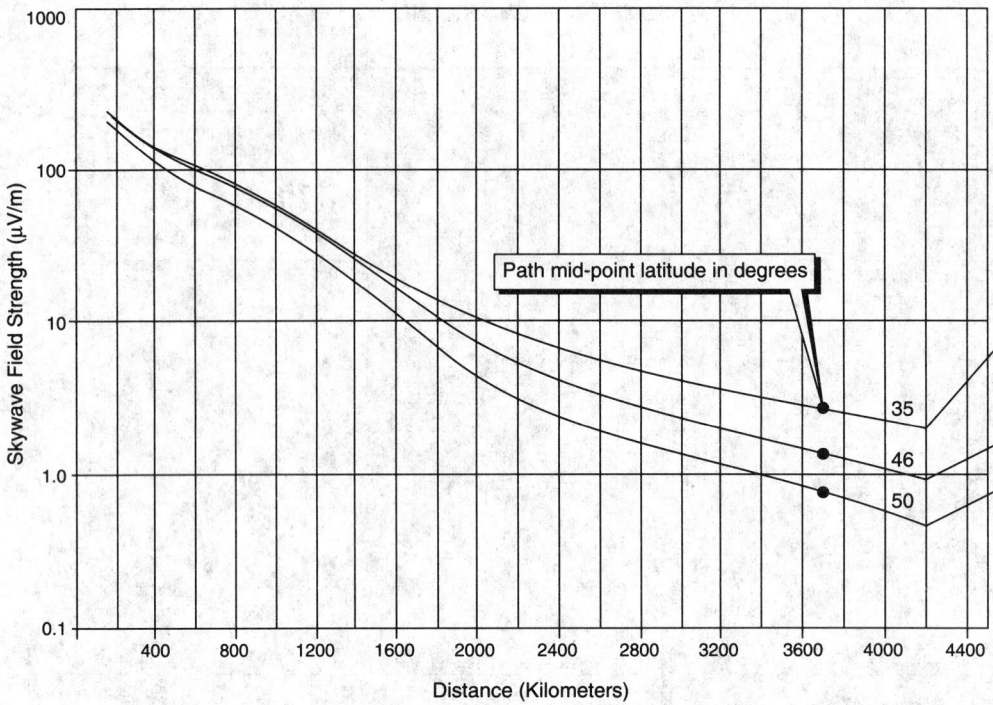

FIGURE 63.10 Skywave propagation for standard AM broadcasting. (*Source:* FCC Rules and Regulations, 1982, vol. III, pt. 73.190, fig. 2.)

facturers still retain tubes in the final amplifiers of their high-powered models. This is changing, however, with the introduction in recent years of 50-kW fully transistorized models. A block diagram of a typical 1-kW solid-state transmitter is shown in Fig. 63.11.

Antenna Systems

The antenna system for a standard AM broadcast station typically consists of a quarter-wave vertical tower, a ground system of 120 or more quarter-wave radials buried a few inches underground, and an antenna tuning unit to "match" the complex impedance of the antenna system to the characteristic impedance of the transmitter and transmission line so that maximum transfer of power may occur. Typical heights for AM broadcast towers range from 150 to 500 ft. When the radiated signal must be modified to prevent interference to other stations or to provide better service in a particular direction, additional towers may be combined in a phased array to produce the desired field intensity contours. For example, if a station power increase would cause interference with existing stations, a directional array could be designed that would tailor the coverage to protect the existing stations while allowing increases in other directions. The protection requirements can generally be met with arrays consisting of 4 towers or less, but complex arrays have been constructed consisting of 12 or more towers to meet stringent requirements at a particular location. An example of a directional antenna pattern is shown in Fig. 63.12. This pattern provides major coverage to the southwest and restricts radiation (and thus interference) towards the northeast.

Frequency Modulation

Frequency-modulation (FM) broadcasting refers to the transmission of voice and music received by the general public in the 88- to 108-MHz frequency band. FM is used to provide higher-fidelity reception than is available with standard broadcast AM. In 1961 stereophonic broadcasting was

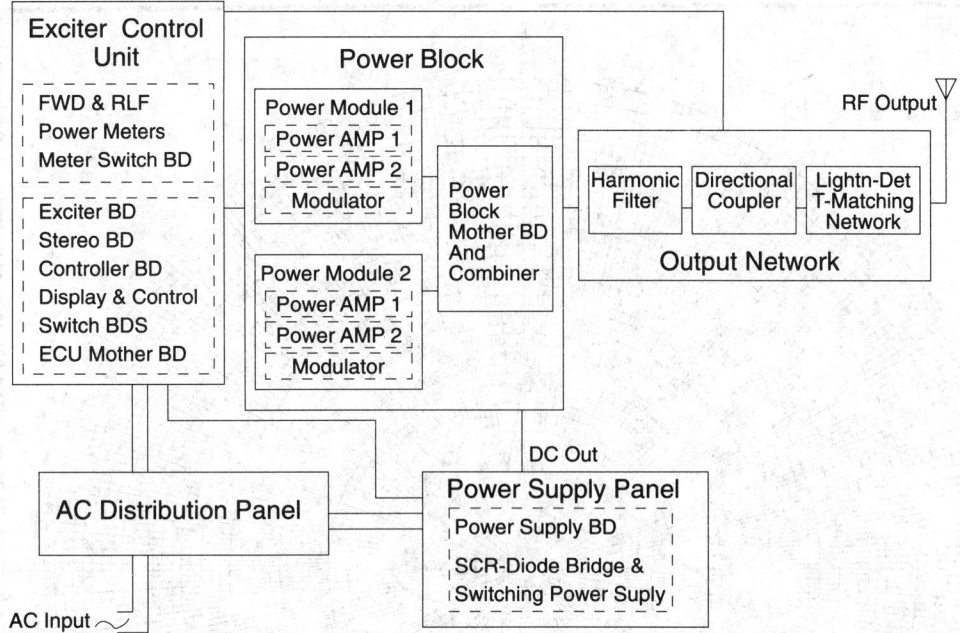

FIGURE 63.11   Block diagram of typical 1-kW solid-state AM transmitter. (*Source:* Broadcast Electronics Inc., Quincy, Ill. Reprinted with permission.)

introduced with the addition of a double-sideband suppressed carrier for transmission of a left-minus-right difference signal. The left-plus-right sum channel is sent with use of normal FM. Some FM broadcast systems also include a **subsidiary communications authorization (SCA)** subcarrier for private commercial uses. FM broadcast is typically limited to line-of-sight ranges. As a result, FM coverage is localized to a range of 75 mi (120 km) depending on the antenna height and ERP.

Frequency Allocations

The 100 carrier frequencies for FM broadcast range from 88.1 to 107.9 MHz and are equally spaced every 200 kHz. The channels from 88.1 to 91.9 MHz are reserved for educational and noncommercial broadcasting and those from 92.1 to 107.9 MHz for commercial broadcasting. Each channel has a 200-kHz bandwidth. The maximum frequency swing under normal conditions is ±75 kHz. Stations operating with an SCA may under certain conditions exceed this level, but in no event may exceed a frequency swing of ±82.5 kHz. The carrier frequency is required to be maintained within ±2000 Hz. The frequencies used for FM broadcasting limit the coverage to line-of-sight distances. The actual coverage area is determined by the ERP of the station and the height of the transmitting antenna above the average terrain in the area. Either increasing the power or raising the antenna will increase the coverage area.

Station Classifications

In FM broadcast, stations are classified according to their maximum allowable ERP and the transmitting antenna height above average terrain in their service area. Class A stations provide primary service to a radius of about 28 km with 6000 W of ERP at a maximum height of 100 m. The most powerful class, Class C, operates with maximums of 100,000 W of ERP and heights up to 600 m with a primary coverage radius of over 92 km. The powers and heights above average terrain (HAAT) for all of the classes are shown in Table 63.5. All classes may operate at antenna heights above those specified but must reduce the ERP accordingly. Stations may not exceed the maximum

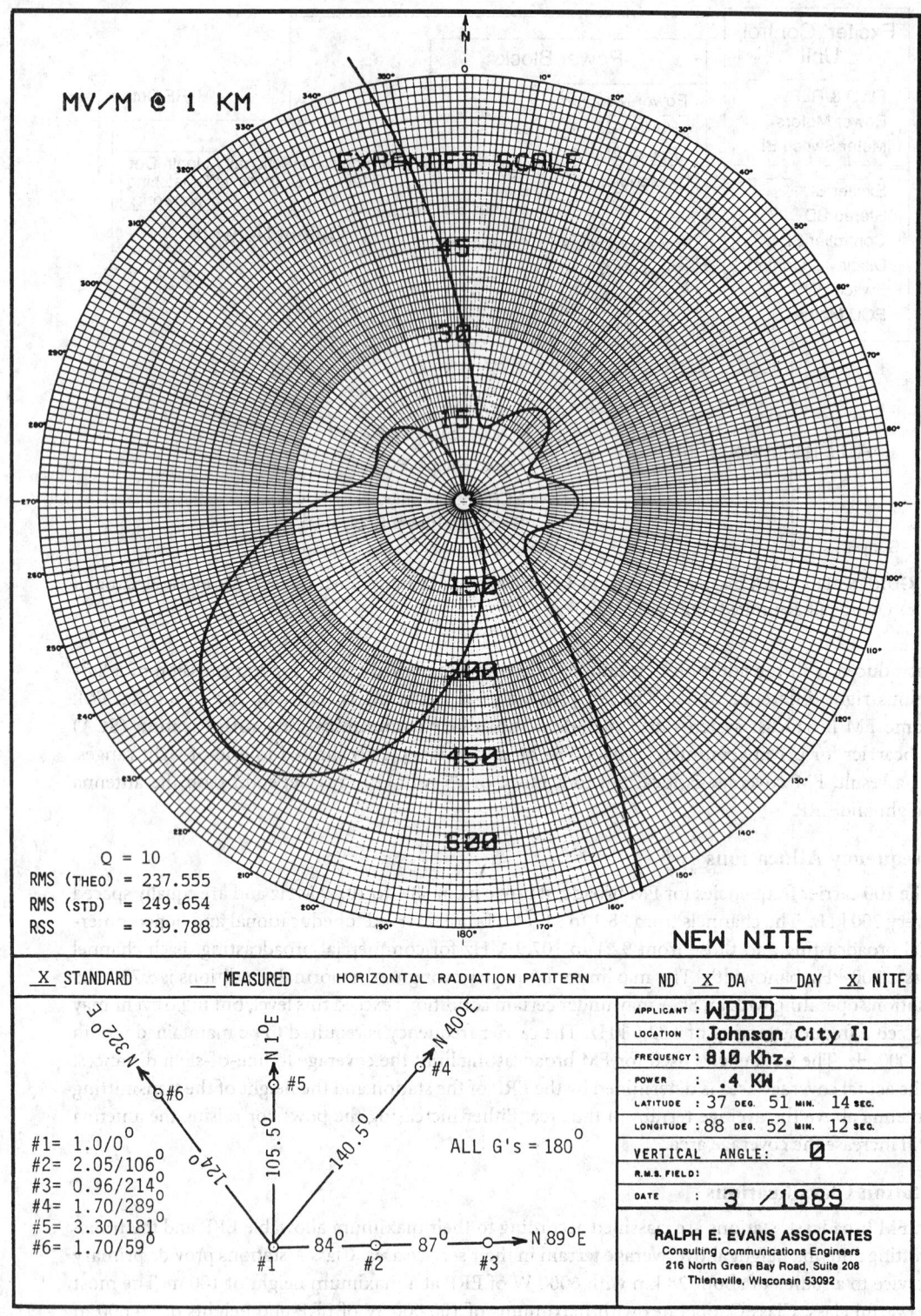

FIGURE 63.12 Directional AM antenna pattern for a six-element array. (*Source:* WDDD-AM, Marion, Ill., and Ralph Evans Associates.)

Table 63.5 FM Station Classifications, Powers, and Tower Heights

Station Class	Maximum ERP	HAAT, m (ft)	Distance, km
A	6 kW (7.8 dBk)	100 (328)	28
B1	25 kW (14.0 dBk)	100 (328)	39
B	50 kW (17.0 dBk)	150 (492)	52
C3	25 kW (14.0 dBk)	100 (328)	39
C2	50 kW (17.0 dBk)	150 (492)	52
C1	100 kW (20.0 dBk)	299 (981)	72
C	100 kW (20.0 dBk)	600 (1968)	92

Source: FCC Rules and Regulations, Revised 1991; vol. III, Part 73.211(b)(1).

power specified, even if antenna height is reduced. The classification of the station determines the allowable distance to other co-channel and adjacent channel stations.

Field Strength and Propagation

The field strength produced by an FM broadcast station depends on the ERP, antenna heights, local terrain, tropospheric scattering conditions, and other factors. From a statistical point of view, however, an estimate of the field intensity may be obtained from Fig. 63.13. A factor in the determination of new licenses for FM broadcast is the separation between allocated co-channel and adjacent channel stations, the class of station, and the antenna heights. The spacings are given in Table 63.6. The primary coverage of all classes of stations (except B and B1, which are 0.5 mV/m and 0.7 mV/m, respectively) is the 1.0 mV/m contour. The distance to the primary contour, as well as to the "city grade" or 3.16 mV/m contour may be estimated using Fig. 63.13. Although FM broadcast propagation is generally thought of as line-of-sight, larger ERPs along with the effects of diffraction, refraction, and tropospheric scatter allow coverage slightly greater than line-of-sight.

Transmitters

FM broadcast transmitters typically range in power output from 10 W to 50 kW. A block diagram of a dual FM transmitter is shown in Fig. 63.14. This system consists of two 25-kW transmitters that are operated in parallel and that provide increased reliability in the event of a failure in either the exciter or transmitter power amplifier. The highest-powered solid-state transmitters are currently 10 kW, but manufacturers are developing new devices that will make higher-power solid-state transmitters both cost-efficient and reliable.

Antenna Systems

FM broadcast antenna systems are required to have a horizontally polarized component. Most antenna systems, however, are circularly polarized, having both horizontal and vertical components. The antenna system, which usually consists of several individual radiating bays fed as a phased array, has a radiation characteristic that concentrates the transmitted energy in the horizontal plane toward the population to be served, minimizing the radiation out into space and down toward the ground. Thus, the ERP towards the horizon is increased with gains up to 10 dB. This means that a 5-kW transmitter coupled to an antenna system with a 10-dB gain would have an ERP of 50 kW. Directional antennas may be employed to avoid interference with other stations or to meet spacing requirements. Figure 63.15 is a plot of the horizontal and vertical components of a typical nondirectional circularly polarized FM broadcast antenna showing the effect upon the pattern caused by the supporting tower.

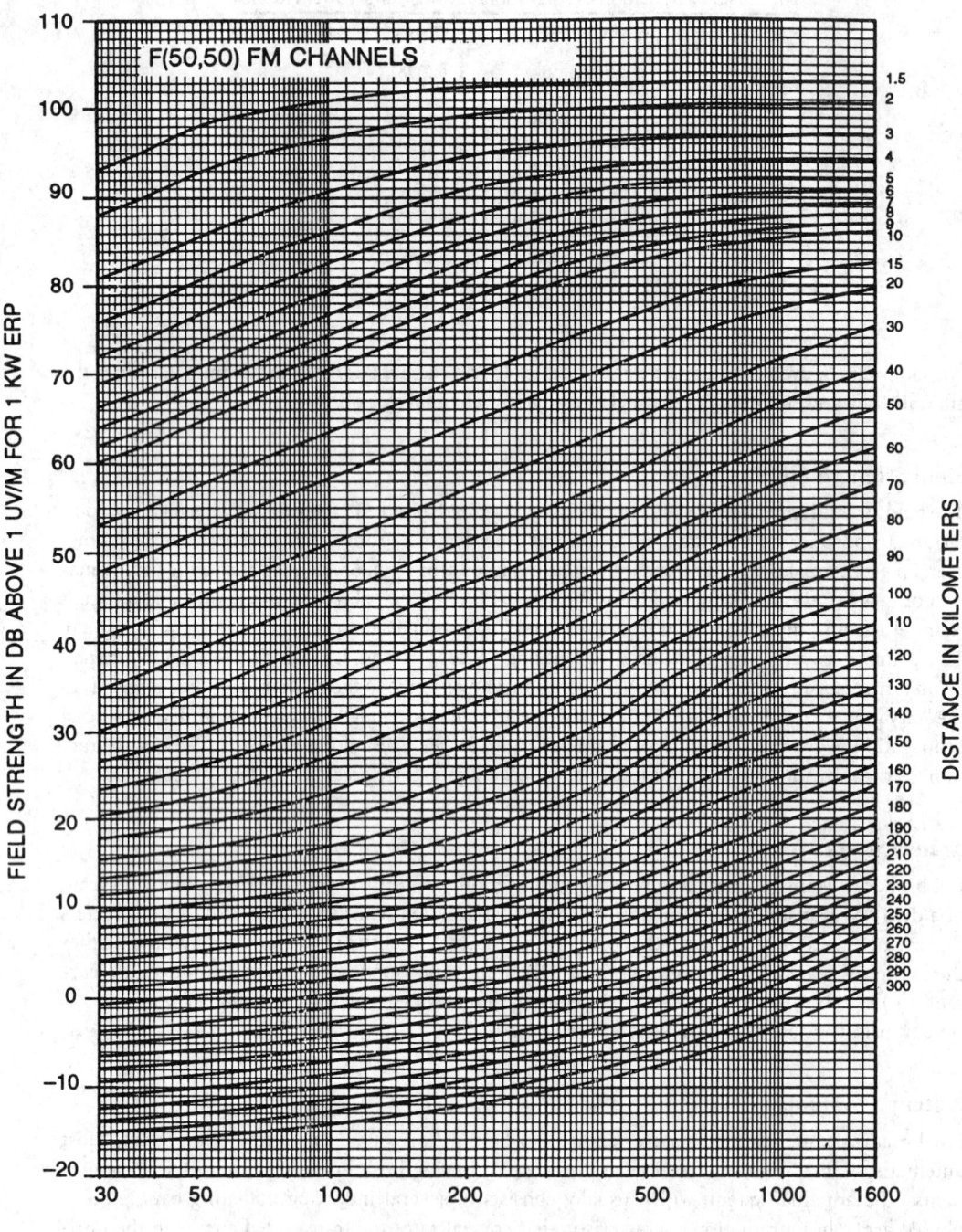

FM CHANNELS

Estimated Field Strength Exceeded at 50 Percent
of the Potential Receiver Locations for at Least 50 Percent
of the Time at a Receiving Antenna Height of 9 Meters

FIGURE 63.13 Propagation for FM broadcasting. (*Source:* FCC Rules and Regulations, Revised 1990; vol. III, pt. 73.333.)

Table 63.6 Distance Separation Requirement for FM Stations

Station Class Relation	Minimum Distance Separation Requirements, km (mi)			
	Co-Channel	200 kHz	400/600 kHz	10.6/10.8 MHz
A to A	115 (71)	72 (45)	31 (19)	10 (6)
A to B1	143 (89)	96 (60)	48 (30)	12 (7)
A to B	178 (111)	113 (70)	69 (43)	15 (9)
A to C3	142 (88)	89 (55)	42 (26)	12 (7)
A to C2	166 (103)	106 (66)	55 (34)	15 (9)
A to C1	200 (124)	133 (83)	75 (47)	22 (14)
A to C	226 (140)	165 (103)	95 (59)	29 (18)
B1 to B1	175 (109)	114 (71)	50 (31)	14 (9)
B1 to B	211 (131)	145 (90)	71 (44)	17 (11)
B1 to C3	175 (109)	114 (71)	50 (31)	14 (9)
B1 to C2	200 (124)	134 (83)	56 (35)	17 (11)
B1 to C1	233 (145)	161 (100)	77 (48)	24 (15)
B1 to C	259 (161)	193 (120)	105 (65)	31 (19)
B to B	241 (150)	169 (105)	74 (46)	20 (12)
B to C3	211 (131)	145 (90)	71 (44)	17 (11)
B to C2	211 (131)	145 (90)	71 (44)	17 (11)
B to C1	270 (168)	195 (121)	79 (49)	27 (17)
B to C	274 (170)	217 (135)	105 (65)	35 (22)
C3 to C3	153 (95)	99 (62)	43 (27)	14 (9)
C3 to C2	177 (110)	117 (73)	56 (35)	17 (11)
C3 to C1	211 (131)	144 (90)	76 (47)	24 (15)
C3 to C	237 (147)	176 (109)	96 (60)	31 (19)
C2 to C2	190 (118)	130 (81)	58 (36)	20 (12)
C2 to C1	224 (139)	158 (98)	79 (49)	27 (17)
C2 to C	237 (147)	176 (109)	96 (60)	31 (19)
C1 to C1	245 (152)	177 (110)	82 (51)	34 (21)
C1 to C	270 (168)	209 (130)	105 (65)	35 (22)
C to C	290 (180)	241 (150)	105 (65)	48 (30)

Source: FCC Rules and Regulations, Revised 1991; vol. III, pt. 73.207.

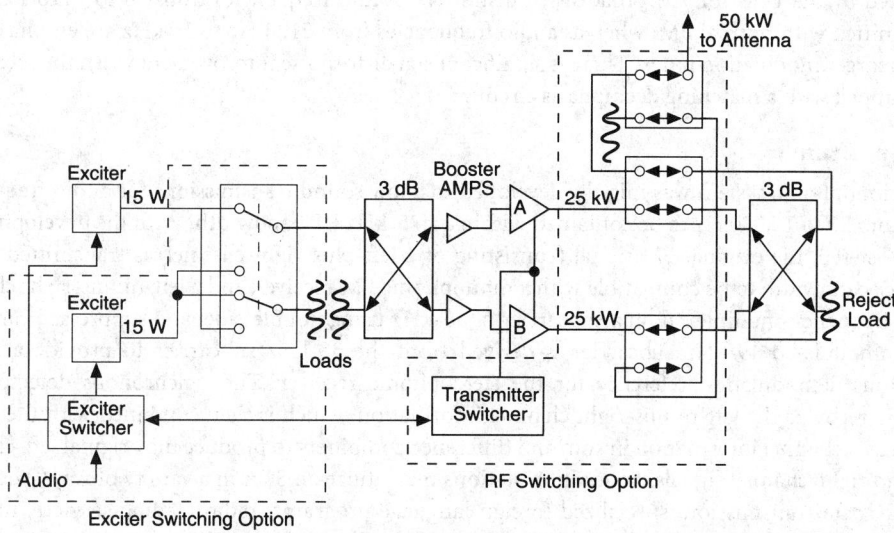

FIGURE 63.14 Block diagram of typical FM transmitter. (*Source:* Harris Corporation, Quincy, Ill.)

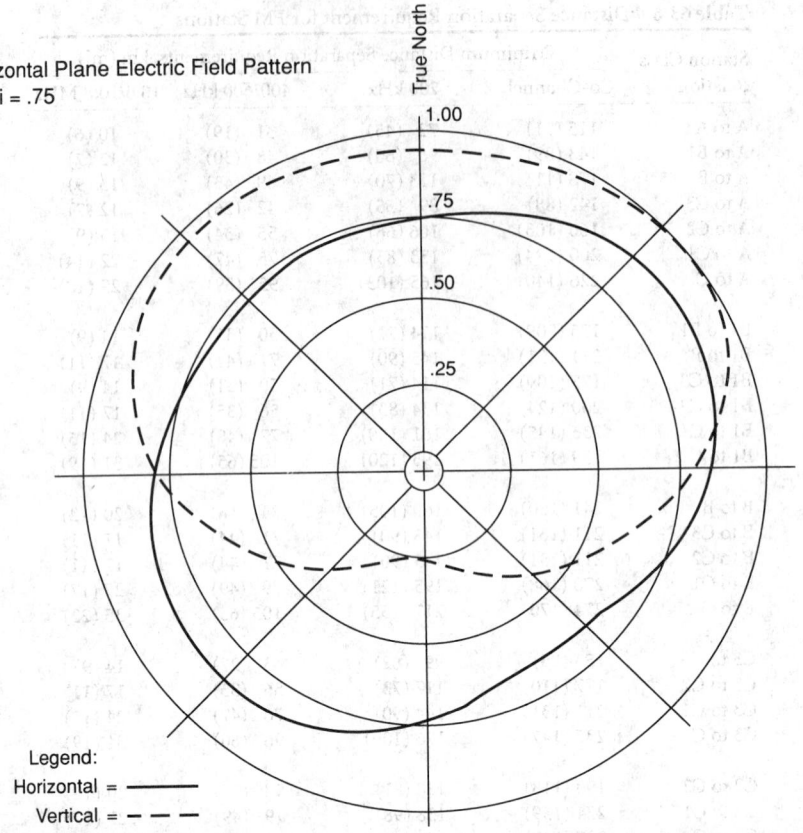

FIGURE 63.15 Typical nondirectional 92.5-MHz FM antenna characteristics showing the effect of the tower structure. (*Source:* Electronics Research, Inc., Newburgh, Ind.)

Preemphasis

Preemphasis is employed in an FM broadcast transmitter to improve the received signal-to-noise ratio. The preemphasis upper-frequency limit shown is based on a time constant of 75 µs as required by the FCC for FM broadcast transmitters. Audio frequencies from 50 to 2120 Hz are transmitted with normal FM, whereas audio frequencies from 2120 Hz to 15 kHz are emphasized with a larger modulation index. There is significant signal-to-noise improvement when the receiver is equipped with a matching deemphasis circuit.

FM Spectrum

The monophonic system was initially developed to allow sound transmissions for audio frequencies from 50 to 15,000 Hz to be contained within a ±75-kHz RF bandwidth. With the development of FM stereo, the original FM signal (consisting of a left-plus-right channel) is transmitted in a smaller bandwidth to be compatible with a monophonic FM receiver, and a left-minus-right channel is frequency-multiplexed on a subcarrier of 38-kHz using double-sideband suppressed carrier. An unmodulated 19-kHz subcarrier is derived from the 38-kHz subcarrier to provide a synchronous demodulation reference for the stereophonic receiver. The synchronous detector at 38 kHz recovers the left-minus-right channel information, which is then combined with the left-plus-right channel information in sum and difference combiners to produce the original left-channel and right-channel signals. In addition stations may utilize an SCA in a variety of ways, such as paging, data transmission, specialized foreign language programs, radio reading services, utility load management, and background music. An FM stereo station may utilize multiplex subcarriers within the range of 53 to 99 kHz with up to 20% modulation of the main carrier using any form of modulation. The only requirement is that the station does not exceed its occupied bandwidth limitations.

Defining Terms

Effective radiated power: Refers to the effective power output from an antenna in a specified direction and includes transmitter output power, transmission line loss and antenna power gain.

Electric field intensity: Measure of signal strength in volts per meter used to determine channel allocation criteria and interference considerations.

Primary service: Refers to areas in which the groundwave signal is not subject to objectionable interference or objectionable fading.

SCA: Subsidiary communications authorization for paging, data transmission, specialized foreign language programs, radio readings services, utility load management and background music using multiplexed subcarriers from 53–99 kHz in connection with broadcast FM.

Secondary service: Refers to areas serviced by skywaves and not subject to objectionable interference.

References

A. F. Barghausen, "Medium frequency sky wave propagation in middle and low latitudes," *IEEE Trans. Broadcast,* vol. 12, pp. 1–14, June 1966.

G.W. Bartlett, Ed., *National Association of Broadcasters Engineering Handbook*, 6th ed., Washington: The National Association of Broadcasters, 1975.

H. Bremmer, *Terrestrial Radio Waves: Theory of Propagation*, Amsterdam: Elsevier, 1949.

Electronic Industries Association, Standard TR-101A, *Electrical Performance Standards for AM Broadcast Transmitters*, 1948.

Federal Communications Commission, Rules and Regulations, vol. III, parts 73 and 74, October 1982.

Further Information

Pike & Fischer, Inc., in Bethesda, Md., offers an updated FCC rule service for a fee.

Several trade journals are good sources for up-to-date information such as *Broadcast Engineering*, Overland Park, Kan., and *Radio World*, Falls Church, Va.

Application-oriented computer software is available from R.F. Systems, Shawnee Mission, Kan.

The Society of Broadcast Engineers (SBE), Indianapolis, Ind., and the National Association of Broadcasters (NAB), Washington, D.C., are sources of further information.

63.3 Television Systems

Jerry Whitaker

The technology of television is based on the conversion of light rays from still or moving scenes and pictures into electronic signals for transmission or storage and subsequent reconversion into visual images on a screen. A similar function is provided in the production of motion picture film; however, where film records the brightness variations of a complete scene on a single frame in a short exposure no longer than a fraction of a second, the elements of a television picture must be scanned one piece at a time. In the television system, a scene is dissected into a **frame** composed of a mosaic of *picture elements* (pixels). A **pixel** is defined as the smallest area of a television image that can be transmitted within the parameters of the system. This process is accomplished by:

- Analyzing the image with a photoelectric device in a sequence of *horizontal **scans*** from the top to the bottom of the image to produce an electric signal in which the brightness and color values of the individual picture elements are represented as voltage levels of a video waveform
- Transmitting the values of the picture elements in sequence as voltage levels of a video signal
- Reproducing the image of the original scene in a video signal display of parallel scanning lines on a viewing screen

Scanning Lines and Fields

The image pattern of electrical charges on a camera tube target, corresponding to the brightness levels of a scene, are converted to a video signal in a sequential order of picture elements in the scanning process. At the end of each horizontal line sweep, the video signal is *blanked* while the beam returns rapidly to the left side of the scene to start scanning the next line. This process continues until the image has been scanned from top to bottom to complete one *field* scan.

After completion of this first field scan, at the midpoint of the last line, the beam again is blanked as it returns to the top center of the target where the process is repeated to provide a second field scan. The spot size of the beam as it impinges upon the target must be fine enough to leave unscanned areas between lines for the second scan. The pattern of scanning lines covering the area of the target, or the screen of a picture display, is called a **raster**.

Interlaced Scanning Fields

Because of the half-line offset for the start of the beam return to the top of the raster and for the start of the second field, the lines of the second field lie in between the lines of the first field. Thus, the lines of the two are **interlaced**. The two interlaced fields constitute a single television *frame*. Figure 63.16 shows a frame scan with interlacing of the lines of two fields.

Reproduction of the camera image on a cathode ray tube (CRT) is accomplished by an identical operation, with the scanning beam modulated in density by the video signal applied to an element of the electron gun. This control voltage to the CRT varies the brightness of each picture element on the phosphor screen.

Blanking of the scanning beam during the return trace is provided for in the video signal by a "blacker-than-black" pulse waveform. In addition, in most receivers and monitors another blanking pulse is generated from the horizontal and vertical scanning circuits and applied to the CRT ·electron gun to ensure a black screen during scanning retrace. The retrace lines are shown as diagonal dashed lines in Fig. 63.16.

The interlaced scanning format, standardized for monochrome and compatible color, was chosen primarily for two partially related and equally important reasons:

- To eliminate viewer perception of the intermittent presentation of images, known as *flicker*
- To reduce video bandwidth requirements for an acceptable flicker threshold level

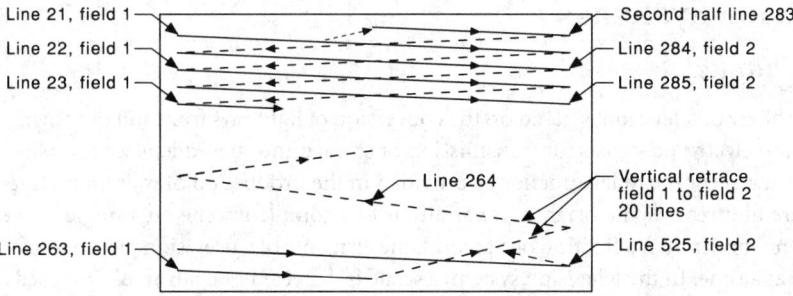

FIGURE 63.16   The interlaced scanning pattern (raster) of the television image. (*Source:* Electronic Industries Association.)

Perception of flicker is dependent primarily upon two conditions:

- The brightness level of an image
- The relative area of an image in a picture

The 30-Hz transmission rate for a full 525-line television frame is comparable to the highly successful 24-frame-per-second rate of motion-picture film. However, at the higher brightness levels produced on television screens, if all 483 lines (525 less blanking) of a television image were to be presented sequentially as single frames, viewers would observe a disturbing flicker in picture areas of high brightness. For a comparison, motion-picture theaters on average produce a screen brightness of 10 to 25 ft · L (footlambert), whereas a direct-view CRT may have a highlight brightness of 50 to 80 ft · L.

Through the use of interlaced scanning, single field images with one-half the vertical resolution capability of the 525-line system are provided at the high flicker-perception threshold rate of 60 Hz. Higher resolution of the full 490 lines (525 less vertical blanking) of vertical detail is provided at the lower flicker-perception threshold rate of 30 Hz. The result is a relatively flickerless picture display at a screen brightness of well over 50 to 75 ft · L, more than double that of motion-picture film projection. Both 60-Hz fields and 30-Hz frames have the same horizontal resolution capability.

The second advantage of interlaced scanning, compared to progressive scanning, is a reduction in video bandwidth for an equivalent flicker threshold level. Progressive scanning of 525 lines would have to be completed in 1/60 s to achieve an equivalent level of flicker perception. This would require a line scan to be completed in half the time of an interlaced scan. The bandwidth then would double for an equivalent number of pixels per line.

The standards adopted by the Federal Communications Commission (FCC) for monochrome television in the United States specified a system of 525 lines per frame, transmitted at a frame rate of 30 Hz, with each frame composed of two interlaced fields of horizontal lines. Initially in the development of television transmission standards, the 60-Hz power line waveform was chosen as a convenient reference for vertical scan. Furthermore, in the event of coupling of power line hum into the video signal or scanning/deflection circuits, the visible effects would be stationary and less objectionable than moving **hum bars** or distortion of horizontal-scanning geometry. In the United Kingdom and much of Europe, the 50-Hz interlaced system was chosen for many of the same reasons. With improvements in television receivers, the power line reference was replaced with a stable crystal oscillator.

The existing 525-line monochrome standards were retained for color in the recommendations of the National Television System Committee (NTSC) for compatible color television in the early 1950s. The NTSC system, adopted in 1953 by the FCC, specifies a scanning system of 525 horizontal lines per frame, with each frame consisting of two interlaced fields of 262.5 lines at a field rate of 59.94 Hz. Forty-two of the 525 lines in each frame are blanked as black picture signals and reserved for transmission of the vertical scanning synchronizing signal. This results in 483 visible lines of picture information.

Synchronizing Video Signals

In monochrome television transmission, two basic synchronizing signals are provided to control the timing of picture-scanning deflection:

- Horizontal sync pulses at the line rate.
- Vertical sync pulses at the field rate in the form of an interval of wide horizontal sync pulses at the field rate. Included in the interval are **equalizing pulses** at twice the line rate to preserve interlace in each frame between the even and odd fields (offset by a half line).

In color transmissions, a third synchronizing signal is added during horizontal scan blanking to provide a frequency and phase reference for color signal encoding circuits in cameras and decoding circuits in receivers. These synchronizing and reference signals are combined with the picture video signal to form a **composite video** waveform.

The scanning and color-decoding circuits in receivers must follow the frequency and phase of the synchronizing signals to produce a stable and geometrically accurate image of the proper color **hue** and **saturation.** Any change in timing of successive vertical scans can impair the interlace of the even and odd fields in a frame. Small errors in horizontal scan timing of lines in a field can result in a loss of resolution in vertical line structures. Periodic errors over several lines that may be out of the range of the horizontal scan automatic frequency control circuit in the receiver will be evident as jagged vertical lines.

Television Industry Standards

There are three primary color transmission standards in use today:

- *NTSC* (National Television Systems Committee): Used in the United States, Canada, Central America, most of South America, and Japan. In addition, NTSC is used in various countries or possessions heavily influenced by the United States.
- *PAL* (phase alternation each line): Used in England, most countries and possessions influenced by the British Commonwealth, many western European countries and China. Variation exists in PAL systems.
- *SECAM* (sequential color with [avec] memory): Used in France, countries and possessions influenced by France, Russia (generally the former Soviet Bloc nations, including East Germany), and other areas influenced by Russia.

The three standards are incompatible for a variety of reasons.

Television transmitters in the United States operate in three frequency bands:

- Low-band VHF (very high frequency), channels 2 through 6
- High-band VHF, channels 7 through 13
- UHF (ultra-high frequency), channels 14 through 83 (UHF channels 70 through 83 currently are assigned to mobile radio services)

Table 63.7 shows the frequency allocations for channels 2 through 83. Because of the wide variety of operating parameters for television stations outside the United States, this section will focus primarily on TV transmission as it relates to the Unites States.

Maximum power output limits are specified by the FCC for each type of service. The maximum **effective radiated power** (ERP) for low-band VHF is 100 kW; for high-band VHF it is 316 kW; and for UHF it is 5 MW. The ERP of a station is a function of transmitter power output (TPO) and antenna gain. ERP is determined by multiplying these two quantities together and subtracting transmission line loss.

The second major factor that affects the coverage area of a TV station is antenna height, known in the broadcast industry as *height above average terrain* (HAAT). HAAT takes into consideration the effects of the geography in the vicinity of the transmitting tower. The maximum HAAT permitted by the FCC for a low- or high-band VHF station is 1000 ft (305 m) east of the Mississippi River and 2000 ft (610 m) west of the Mississippi. UHF stations are permitted to operate with a maximum HAAT of 2000 ft (610 m) anywhere in the United States (including Alaska and Hawaii).

The ratio of visual output power to **aural** output power can vary from one installation to another; however, the aural is typically operated at between 10 and 20% of the visual power. This difference is the result of the reception characteristics of the two signals. Much greater signal strength is required at the consumer's receiver to recover the visual portion of the transmission than the aural portion. The aural power output is intended to be sufficient for good reception at the fringe of the station's coverage area but not beyond. It is of no use for a consumer to be able to receive a TV station's audio signal but not the video.

Table 63.7 Frequency Allocations for TV Channels 2 through 83 in the U.S.

Channel Designation	Frequency Band, MHz	Channel Designation	Frequency Band, MHz	Channel Designation	Frequency Band, MHz
2	54–60	30	566–572	58	734–740
3	60–66	31	572–578	59	740–746
4	66–72	32	578–584	60	746–752
5	76–82	33	584–590	61	752–758
6	82–88	34	590–596	62	758–764
7	174–180	35	596–602	63	764–770
8	180–186	36	602–608	64	770–776
9	186–192	37	608–614	65	776–782
10	192–198	38	614–620	66	782–788
11	198–204	39	620–626	67	788–794
12	204–210	40	626–632	68	794–800
13	210–216	41	632–638	69	800–806
14	470–476	42	638–644	70	806–812
15	476–482	43	644–650	71	812–818
16	482–488	44	650–656	72	818–824
17	488–494	45	656–662	73	824–830
18	494–500	46	662–668	74	830–836
19	500–506	47	668–674	75	836–842
20	506–512	48	674–680	76	842–848
21	512–518	49	680–686	77	848–854
22	518–524	50	686–692	78	854–860
23	524–530	51	692–698	79	860–866
24	530–536	52	698–704	80	866–872
25	536–542	53	704–710	81	872–878
26	542–548	54	710–716	82	878–884
27	548–554	55	716–722	83	884–890
28	554–560	56	722–728		
29	560–566	57	728–734		

Two classifications of low-power TV stations have been established by the FCC to meet certain community needs. They are:

- *Translator:* A low-power system that rebroadcasts the signal of another station on a different channel. Translators are designed to provide "fill-in" coverage for a station that cannot reach a particular community because of the local terrain. Translators operating in the VHF band are limited to 100 W power output (ERP), and UHF translators are limited to 1 kW.

- *Low-Power Television (LPTV):* A service recently established by the FCC designed to meet the special needs of particular communities. LPTV stations operating on VHF frequencies are limited to 100 W ERP, and UHF stations are limited to 1 kW. LPTV stations originate their own programming and can be assigned by the FCC to any channel, as long as sufficient protection against interference to a full-power station is afforded.

The composite video waveform is shown in Fig. 63.17. The actual radiated signal is inverted, with modulation extending from the synchronizing pulses at maximum carrier level (100%) to reference picture white at 7.5%. Because an increase in the amplitude of the radiated signal corresponds to a decrease in picture brightness, the polarity of modulation is termed *negative*.

Composite Video

The term *composite* is used to denote a video signal that contains:

- Picture luminance and chrominance information
- Timing information for synchronization of scanning and color signal processing circuits

The negative-going portion of the waveform shown in Fig. 63.17 is used to transmit information

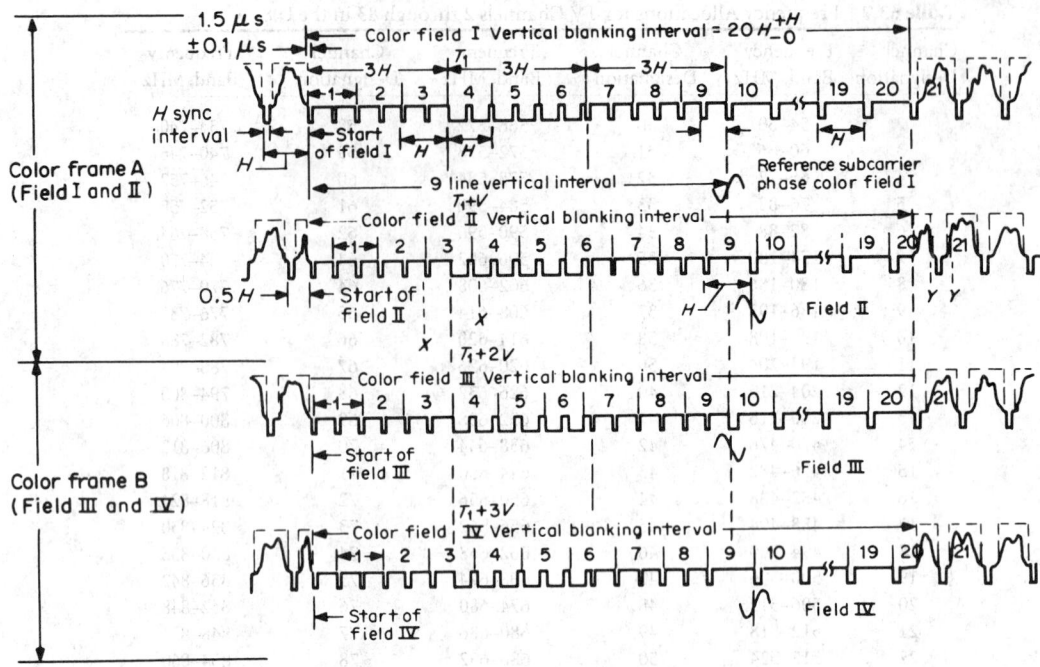

FIGURE 63.17 The principal components of the NTSC color television waveform. (*Source:* Electronic Industries Association.)

for synchronization of scanning circuits. The positive-going portion of the amplitude range is used to transmit luminance information representing brightness and, for color pictures, chrominance.

At the completion of each line scan in a receiver or monitor, a horizontal synchronizing (**H-sync**) pulse in the composite video signal triggers the scanning circuits to return the beam rapidly to the left of the screen for the start of the next line scan. During the return time, a horizontal blanking signal at a level lower than that corresponding to the blackest portion of the scene is added to avoid the visibility of the retrace lines. In a similar manner, after completion of each field, a vertical blanking signal blanks out the retrace portion of the scanning beam as it returns to the top of the picture to start the scan of the next field. The small-level difference between video reference black and blanking level is called **setup**. Setup is used as a guard band to ensure separation of the synchronizing and video-information functions and adequate blanking of the scanning retrace lines on receivers.

The waveforms of Fig. 63.18 shows the various reference levels of video and sync in the composite signal. The unit of measurement for video level was specified initially by the Institute of Radio Engineers (IRE). These **IRE** *units* are still used to quantify video signal levels. The primary IRE values are given in Table 63.8.

Color Signal Encoding

To facilitate an orderly introduction of color television broadcasting in the United States and other countries with existing monochrome services, it was essential that the new transmissions be compatible. In other words, color pictures would provide acceptable quality on unmodified monochrome receivers. In addition, because of the limited availability of the RF spectrum, another related requirement was the need to fit approximately 2-MHz bandwidth of color information into the 4.2-MHz video bandwidth of the existing 6-MHz broadcasting channels with little or no modification of existing transmitters. This is accomplished by using the band-sharing color signal system developed by the NTSC and by taking advantage of the fundamental characteristics of the eye regarding color sensitivity and resolution.

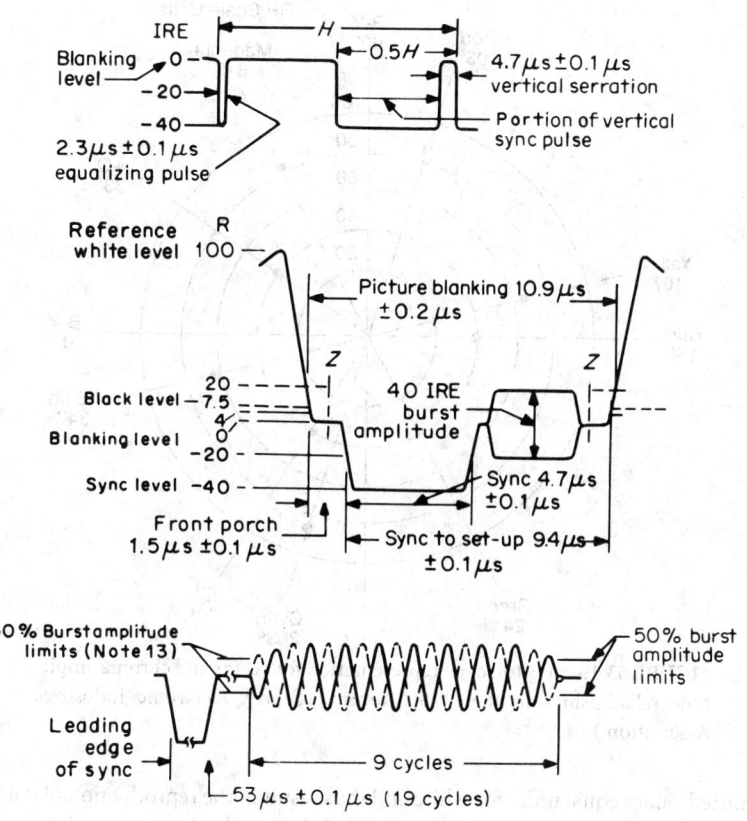

FIGURE 63.18 Sync pulse widths for the NTSC color system. (*Source:* Electronic Industries Association.)

The video-signal spectrum generated by scanning an image consists of energy concentrated near harmonics of the 15,734-Hz line scanning frequency. Additional lower-amplitude sideband components exist at multiples of 60 Hz (the field scan frequency) from each line scan harmonic. Substantially no energy exists halfway between the line scan harmonics, that is, at odd harmonics of one half line frequency. Thus, these blank spaces in the spectrum are available for the transmission of a signal for carrying color information and its sideband. In addition, a signal modulated with color information injected at this frequency is of relatively low visibility in the reproduced image because the odd harmonics are of opposite phase on successive scanning lines and in successive frames, requiring four fields to repeat. Furthermore, the visibility of the color video signal is reduced further by the use of a subcarrier frequency near the cutoff of the video bandpass.

In the NTSC system, color is conveyed using two elements:

- A luminance signal
- A chrominance signal

The luminance signal is derived from components of the three primary colors, red, green, and blue, in the proportions for *reference white*, E_y, as follows:

$$E_y = 0.3E_R + 0.59E_G + 0.11E_B$$

Table 63.8 Video and Sync Levels in IRE Units

Signal Level	IRE Level
Reference white	100
Blanking level width measurement	20
Color burst sine wave peak	+20 to −20
Reference black	7.5
Blanking	0
Sync pulse width measurement	−20
Sync level	−40

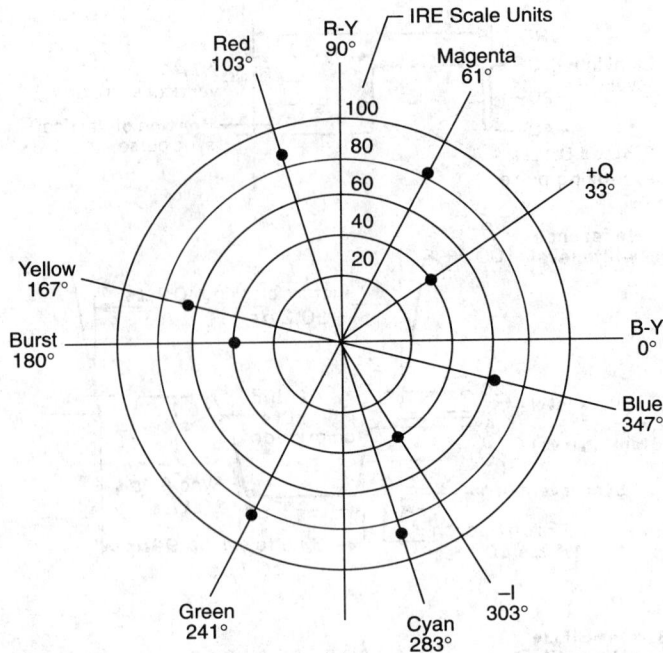

FIGURE 63.19   Vectorscope representation for vector and chroma amplitude relationships in the NTSC system. (*Source:* Electronic Industries Association.)

These transmitted values equal unity for white and thus result in the reproduction of colors on monochrome receivers at the proper luminance level. This is known as the *constant-luminance* principle.

The color signal consists of two chrominance components, I and Q, transmitted as amplitude-modulated sidebands of two 3.579545-MHz subcarriers in quadrature (differing in phase by 90°). The subcarriers are suppressed, leaving only the sidebands in the color signal. Suppression of the carriers permits demodulation of the color signal as two separate color signals in a receiver by reinsertion of a carrier of the phase corresponding to the desired color signal. This system for recovery of the color signals is called **synchronous demodulation.**

I and Q signals are composed of red, green, and blue primary color components produced by color cameras and other signal generators. The phase relationship among the I and Q signals, the derived primary and complementary colors, and the color synchronizing burst can be shown graphically on a **vectorscope** display. The horizontal and vertical sweep signals on a vectorscope are produced from R-Y and B-Y subcarrier sine waves in quadrature, producing a circular display. The chrominance signal controls the intensity of the display. A vectorscope display of an Electronic Industries Association (EIA) color bar signal is shown in Fig. 63.19.

Color-Signal Decoding

Each of the two chroma signal carriers can be recovered individually by means of synchronous detection. A reference subcarrier of the same phase as the desired chroma signal is applied as a gate to a balanced demodulator. Only the modulation of the signal in the same phase as the reference will be present in the output. A low-pass filter may be added to remove second harmonic components of the chroma signal generated in the process.

Transmission Equipment

Television transmitters are classified in terms of their operating band, power level, final-tube type, and cooling method. The transmitter is divided into two basic subsystems:

- The *visual* section, which accepts the video input, frequency modulates an RF carrier, and amplifies the signal to feed the antenna system
- The *aural* section, which accepts the audio input, frequency modulates a separate RF carrier, and amplifies the signal to feed the antenna system

The visual and aural signals are combined to feed a single radiating system.

Transmitter Design Considerations

Each manufacturer has a particular philosophy with regard to the design and construction of a broadcast TV transmitter. Some generalizations can, however, be made with respect to basic system design.

When the power output of a TV transmitter is discussed, the visual section is the primary consideration. Output power refers to the *peak power* of the visual section of the transmitter (*peak of sync*). The FCC-licensed ERP is equal to the transmitter power output minus feedline losses times the power gain of the antenna.

A low-band VHF station can achieve its maximum 100-kW power output through a wide range of transmitter and antenna combinations. A 35-kW transmitter coupled with a gain-of-4 antenna would work, as would a 10-kW transmitter feeding an antenna with a gain of 12. Reasonable pairings for a high-band VHF station would range from a transmitter with a power output of 50 kW feeding an antenna with a gain of 8 to a 30-kW transmitter connected to a gain-of-12 antenna. These combinations assume reasonable feedline losses. To reach the exact power level, minor adjustments are made to the power output of the transmitter, usually by a front panel power trim control.

UHF stations that want to achieve their maximum licensed power output are faced with installing a very high-power transmitter. Typical pairings include a transmitter rated for 220 kW and an antenna with a gain of 25, or a 110-kW transmitter and a gain-of-50 antenna. In the latter case, the antenna could pose a significant problem. UHF antennas with gains in the region of 50 are possible, but not advisable for most installations because of the coverage problems that can result. High-gain antennas have a narrow vertical radiation pattern that can reduce a station's coverage in areas near the transmitter site.

At first examination, it might seem reasonable and economical to achieve licensed ERP using the lowest transmitter power output possible and highest antenna gain. Other factors, however, come into play that make the most obvious solution not always the best solution. Factors that limit the use of high-gain antennas include:

- The effects of high-gain designs on coverage area and signal penetration
- Limitations on antenna size because of tower restrictions, such as available vertical space, weight, and windloading
- The cost of the antenna

The amount of output power required of a transmitter will have a fundamental effect on system design. Power levels dictate whether the unit will be of solid-state or vacuum-tube design; whether air, water, or vapor cooling must be used; the type of power supply required; the sophistication of the high-voltage control and supervisory circuitry; and many other parameters.

Tetrodes are generally used for VHF transmitters above 15 kW and for low-power UHF transmitters (below 10 kW). As solid-state technology advances, the power levels possible in a reasonable transmitter design steadily increase. As of this writing, all-solid-state VHF transmitters of 30 kW and greater have been produced.

In the realm of UHF transmitters, the **klystron** is the most common power output device. Klystrons use an *electron bunching* technique to generate high power (55 kW from a single tube is not uncommon) at microwave frequencies. They are currently the first choice for high-power service. The klystron, however, is relatively inefficient. A stock klystron with no efficiency-optimizing circuitry might be only 40 to 50% efficient, depending on the type of device used. Various schemes have been devised to improve klystron efficiency, the best known of which is **beam pulsing.** Two types of pulsing are in common use today:

- *Mod-anode pulsing*, a technique designed to reduce power consumption of the klystron during the color burst and video portion of the signal (and thereby improve overall system efficiency)
- *Annular control electrode* (ACE) pulsing, which accomplishes basically the same thing by incorporating the pulsing signal into a low-voltage stage of the transmitter, rather than a high-voltage stage (as with mod-anode pulsing).

Still another approach to improving UHF transmitter efficiency involves an entirely new class of vacuum tube: the **Klystrode** or the **multistage depressed collector (MSDC) klystron.** (The Klystrode is a registered trademark of Varian.) The Klystrode is a device that essentially combines the cathode/grid structure of the tetrode with the drift tube/collector structure of the klystron. The MSDC klystron incorporates a collector assembly that operates at progressively lower voltage levels. The net effect for the MSDC is to recover energy from the electron stream rather than dissipating the energy as heat.

Elements of the Transmitter

A television transmitter can be divided into four major subsystems:

- The exciter
- Intermediate power amplifier (IPA)
- Power amplifier (PA)
- High-voltage power supply

Figure 63.20 shows the audio, video, and RF paths for a typical television transmitter.

The modulated visual intermediate frequency (IF) signal is band-shaped in a vestigial sideband filter, typically a surface-acoustic-wave (SAW) filter. Envelope-delay correction is not required for the SAW filter because of the uniform delay characteristics of the device. Envelope-delay compensation may, however, be needed for other parts of the transmitter. The SAW filter provides many benefits to transmitter designers and operators. A SAW filter requires no adjustments and is stable with respect to temperature and time. A *color-notch filter* is required at the output of the transmitter because imperfect linearity of the IPA and PA stages introduces unwanted modulation products.

The power amplifier raises the output energy of the transmitter to the desired RF operating level. Tetrodes in television service are operated in the class B mode to obtain reasonable efficiency while maintaining a linear transfer characteristic. Class B amplifiers, when operated in tuned circuits, provide linear performance because of the flywheel effect of the resonance circuit. This allows a single tube to be used instead of two in push-pull fashion. The bias point of the linear amplifier is chosen so that the transfer characteristic at low modulation levels matches that at higher modulation levels. The plate (anode) circuit of a tetrode PA is usually built around a coaxial resonant cavity, which provides a stable and reliable tank circuit.

UHF transmitters using a klystron in the final output stage must operate class A, the most linear but also most inefficient operating mode for a vacuum tube. Two types of klystrons are presently in service: *integral cavity* and *external cavity* devices. The basic theory of operation is identical for each tube, but the mechanical approach is radically different. In the **integral cavity klystron,** the cavities are built into the device to form a single unit. In the **external cavity klystron,** the cavities are outside the vacuum envelope and are bolted around the tube when the klystron is installed in the transmitter. A number of factors come into play in a discussion of the relative merits of integral vs. external cavity designs. Primary considerations include operating efficiency, purchase price, and life expectancy.

The transmitter block diagram of Fig. 63.20 shows separate visual and aural PA stages. This con-

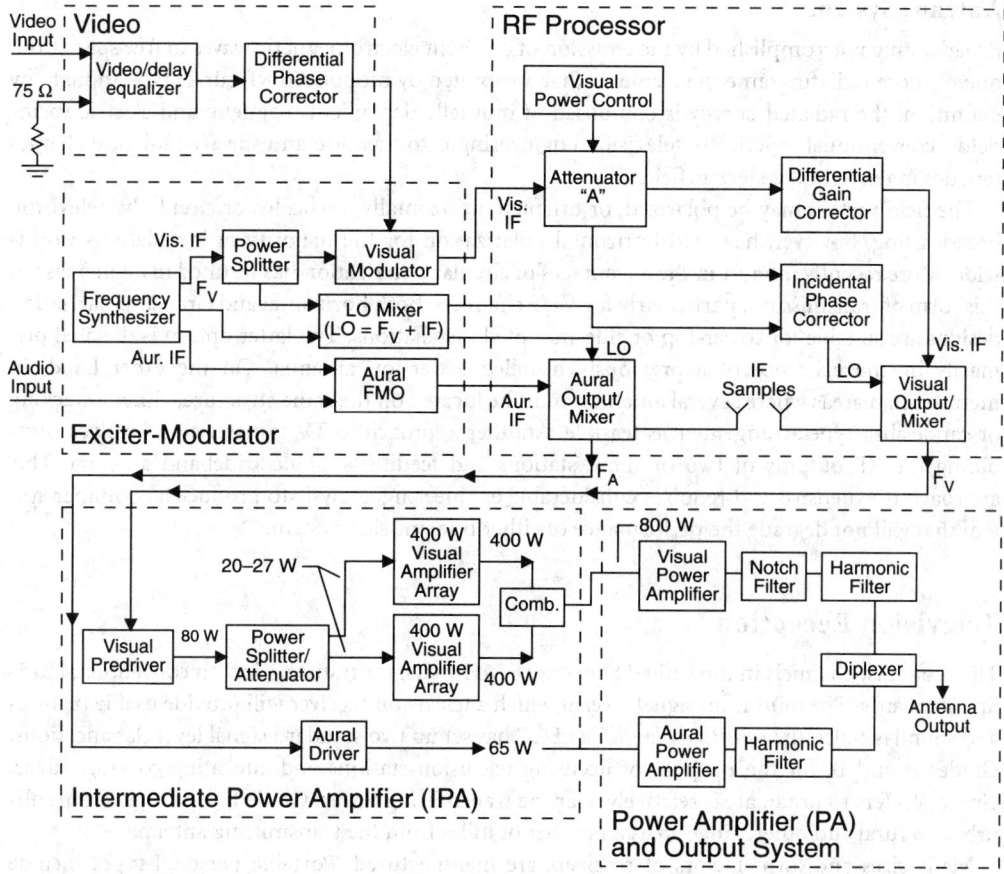

FIGURE 63.20 Simplified block diagram of a VHF television transmitter.

figuration is normally used for high-power transmitters. Low-power designs often use a combined mode in which the aural and visual signals are added prior to the PA. This approach offers a simplified system but at the cost of additional precorrection of the input video signal.

PA stages often are configured so that the circuitry of the visual and aural amplifiers is identical, providing backup protection in the event of a visual PA failure. The aural PA can then be reconfigured to amplify both the aural and the visual signals at reduced power.

The aural output stage of a television transmitter is similar in basic design to a frequency modulated (FM) broadcast transmitter. Tetrode output devices generally operate class C, providing good efficiency. Klystron-based aural PAs are used in UHF transmitters.

Harmonic filters are employed to attenuate out-of-band radiation of the aural and visual signals to ensure compliance with FCC requirements. Filter designs vary depending upon the manufacturer; however, most are of coaxial construction utilizing L and C components housed within a prepackaged assembly. Stub filters are also used, typically adjusted to provide maximum attenuation at the second harmonic of the operating frequency of the visual carrier and the aural carrier.

The filtered visual and aural outputs are fed to a hybrid diplexer where the two signals are combined to feed the antenna. For installations that require dual-antenna feedlines, a hybrid combiner with quadrature-phased outputs is used. Depending upon the design and operating power, the color-notch filter, aural and visual harmonic filters, and diplexer may be combined into a single mechanical unit.

Antenna System

Broadcasting is accomplished by the emission of coherent electromagnetic waves in free space from one or more radiating-antenna elements that are excited by modulated RF currents. Although, by definition, the radiated energy is composed of mutually dependent magnetic and electric vector fields, conventional practice in television engineering is to measure and specify radiation characteristics in terms of the electric field only.

The field vectors may be polarized, or oriented, horizontally, vertically, or circularly. Television broadcasting, however, has used horizontal polarization for the majority of installations worldwide. More recently interest in the advantages of circular polarization has resulted in an increase in this form of transmission, particularly for VHF channels. Both horizontal and circular polarization designs are suitable for tower-top or side-mounted installations. The latter option is dictated primarily by the existence of a previously installed tower-top antenna. On the other hand, in metropolitan areas where several antennas must be located on the same structure, either a stacking or candelabra-type arrangement is feasible. Another approach to TV transmission involves combining the RF outputs of two or more stations and feeding a single wideband antenna. This approach is expensive and requires considerable engineering analysis to produce a combiner system that will not degrade the performance of either transmission system.

Television Reception

The broadcast channels in the United States are 6 MHz wide for transmission on conventional 525-line standards. The minimum signal level at which a television receiver will provide usable pictures and sound is called the *sensitivity level*. The FCC has set up two standard signal level classifications, Grades A and B, for the purpose of licensing television stations and allocating coverage areas. Grade A refers to urban areas relatively near the transmitting tower; Grade B use ranges from suburban to rural and other fringe areas a number of miles from the transmitting antenna.

Many sizes and form factors of receivers are manufactured. Portable personal types include pocket-sized or hand-held models with picture sizes of 2 to 4 in. diagonal for monochrome and 5 to 6 in. for color. Large screen sizes are available in monochrome where low cost and light weight are prime requirements. However, except where portability is important, the majority of television program viewing is in color. The 19- and 27-in. sizes dominate the market.

Television receiver functions may be broken down into several interconnected blocks. With the increasing use of large-scale integrated circuits, the isolation of functions has become less obvious in the design and service of receivers. The typical functional configuration of a receiver using a tri-gun picture tube is shown in Fig. 63.21.

Display Systems

Color video displays may be classified under the following categories:

- Direct-view CRT
- Large-screen display, optically projected from a CRT
- Large-screen display, projected from a modulated light beam
- Large-area display of individually driven light-emitting CRTs or incandescent picture elements
- Flat-panel matrix of transmissive or reflective picture elements
- Flat-panel matrix of light-emitting picture elements

The CRT remains the dominant type of display for both consumer and professional 525-/625-line television applications. The Eidophor and light-valve systems using a modulated light source have found wide application for presentations to large audiences in theater environments, particularly

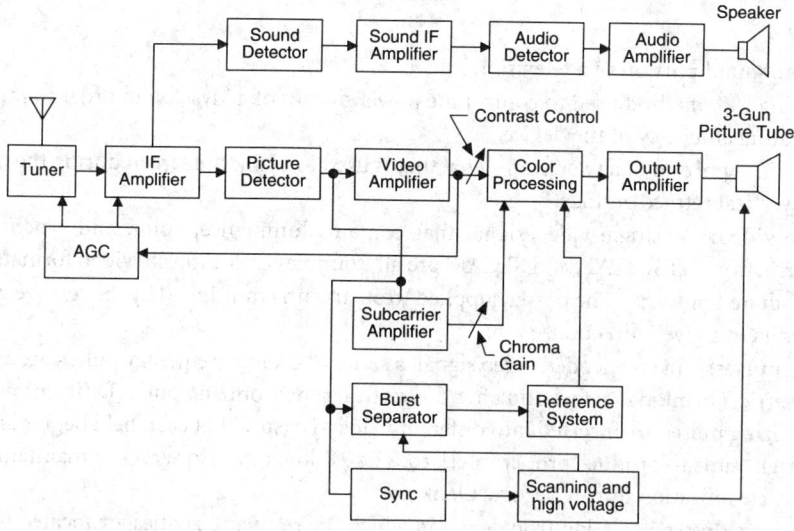

FIGURE 63.21 Simplified schematic block diagram of a color television receiver.

where high screen brightness is required. Matrix-driven flat-panel displays are used in increasing numbers for small-screen personal television receivers and for portable projector units.

Cathode Ray Tube Display

The direct-view CRT is the dominant display device in television. The attributes offered by CRTs include the following:

- High brightness
- High resolution
- Excellent gray-scale reproduction
- Low cost compared to other types of displays

From the standpoint of television receiver manufacturing simplicity and low cost, packaging of the display device as a single component is attractive. The tube itself is composed of only three basic parts: an electron gun, an envelope, and a shadow-mask phosphor screen. The luminance efficiency of the electron optical system and the phosphor screen is high. A peak beam current of under 1 μA in a 25-in. tube will produce a highlight brightness of up to 100 ft · L. The major drawback is the power required to drive the horizontal sweep circuit and the high accelerating voltage necessary for the electron beam. This requirement is partially offset through generation of the screen potential and other lower voltages by rectification of the scanning flyback voltage.

As consumer demands drive manufacturers to produce larger picture sizes, the weight and depth of the CRT and the higher power and voltage requirements become serious limitations. These are reflected in sharply increasing receiver costs. To withstand the atmospheric pressures on the evacuated glass envelope, CRT weight increases exponentially with the viewable diagonal. Nevertheless, manufacturers have continued to meet the demand for increased screen sizes with larger direct-view tubes. Improved versions of both tridot delta and in-line guns have been produced. The tridot gun provides small spot size at the expense of critical convergence adjustments for uniform resolution over the full-tube faceplate. In-line guns permit the use of a self-converging deflection yoke that will maintain dynamic horizontal convergence over the full face of the tube without the need for correction waveforms. The downside is slightly reduced resolution.

Defining Terms

Aural: The sound portion of a television signal.

Beam pulsing: A method used to control the power output of a klystron in order to improve the operating efficiency of the device.

Blanking: The portion of a television signal that is used to blank the screen during the horizontal and vertical retrace periods.

Composite video: A single video signal that contains luminance, color, and synchronization information. NTSC, PAL, and SECAM are all examples of composite video formats.

Effective radiated power: The power supplied to an antenna multiplied by the relative gain of the antenna in a given direction.

Equalizing pulses: In an encoded video signal, a series of 2X line frequency pulses occurring during vertical blanking, before and after the vertical synchronizing pulse. Different numbers of equalizing pulses are inserted into different fields to ensure that each field begins and ends at the right time to produce proper interlace. The 2X line rate also serves to maintain horizontal synchronization during vertical blanking.

External cavity klystron: A klystron device in which the resonant cavities are located outside the vacuum envelope of the tube.

Field: One of the two (or more) equal parts of information into which a frame is divided in interlace video scanning. In the NTSC system, the information for one picture is divided into two fields. Each field contains one-half the lines required to produce the entire picture. Adjacent lines in the picture are contained in alternate fields.

Frame: The information required for one complete picture in an interlaced video system. For the NTSC system, there are two fields per frame.

H (horizontal):   In television signals, H may refer to any of the following: the horizontal period or rate, horizontal line of video information, or horizontal sync pulse.

Hue: One of the characteristics that distinguishes one color from another. Hue defines color on the basis of its position in the spectrum (red, blue, green, yellow, etc.). Hue is one of the three characteristics of television color. Hue is often referred to as *tint*. In NTSC and PAL video signals, the hue information at any particular point in the picture is conveyed by the corresponding instantaneous phase of the active video subcarrier.

Hum bars: Horizontal black and white bars that extend over the entire TV picture and usually drift slowly through it. Hum bars are caused by an interfering power line frequency or one of its harmonics.

Integral cavity klystron: A klystron device in which the resonant cavities are located inside the vacuum envelope of the tube.

Interlaced: A shortened version of *interlaced scanning* (also called *line interlace*). Interlaced scanning is a system of video scanning whereby the odd- and even-numbered lines of a picture are transmitted consecutively as two separate interleaved fields.

IRE: A unit equal to 1/140 of the peak-to-peak amplitude of a video signal, which is typically 1 V. The 0 IRE point is at blanking level, with the sync tip at −40 IRE and white extending to +100 IRE. IRE stands for *Institute of Radio Engineers*, an organization preceding the IEEE, which defined the unit.

Klystrode: An amplifier device for UHF-TV signals that combines aspects of a tetrode (grid modulation) with a klystron (velocity modulation of an electron beam). The result is a more efficient, less expensive device for many applications. (Klystrode is a trademark of EIMAC, a division of Varian Associates.)

Klystron: An amplifier device for UHF and microwave signals based on velocity modulation of an electron beam. The beam is directed through an input cavity, where the input RF signal polarity initializes a *bunching effect* on electrons in the beam. The bunching effect excites subsequent cavities, which increase the bunching through an energy flywheel concept. Finally, the beam passes an output cavity that couples the amplified signal to the load (antenna system). The beam falls onto a collector element that forms the return path for the current and dissipates the heat resulting from electron beam bombardment.

Low-power TV (LPTV):   A television service authorized by the FCC to serve specific confined areas. An LPTV station may typically radiate between 100 and 1000 W of power, covering a geographic radius of 10 to 15 mi.

Multistage depressed collector (MSDC) klystron: A specially designed klystron in which decreasing voltage zones cause the electron beam to be reduced in velocity before striking the collector element. The effect is to reduce the amount of heat that must be dissipated by the device, improving operating efficiency.

Pixel: The smallest distinguishable and resolvable area in a video image. A pixel is a single point on the screen. The word pixel is derived from *picture element*.

Raster: A predetermined pattern of scanning the screen of a CRT. *Raster* may also refer to the illuminated area produced by scanning lines on a CRT when no video is present.

Saturation: The intensity of the colors in the active picture, the voltage levels of the colors. Saturation relates to the degree by which the eye perceives a color as departing from a gray or white scale of the same brightness. A 100% saturated color does not contain any white; adding white reduces saturation. In NTSC and PAL video signals, the color saturation at any particular instant in the picture is conveyed by the corresponding instantaneous amplitude of the active video subcarrier.

Scan: One sweep of the target area in a camera tube or of the screen in a picture tube.

Setup: A video term relating to the specified base of an active picture signal. In NTSC, the active picture signal is placed 7.5 IRE units above blanking (0 IRE). Setup is the separation in level between the *video blanking* and *reference black* levels.

Synchronous detection: A demodulation process in which the original signal is recovered by multiplying the modulated signal by the output of a synchronous oscillator locked to the carrier.

Translator: An unattended television or FM broadcast repeater that receives a distant signal and retransmits the picture and/or audio locally on another channel.

Vectorscope: An oscilloscope-type device used to display the color parameters of a video signal. A vectorscope decodes color information into R-Y and B-Y components, which are then used to drive the *X* and *Y* axis of the scope. The total lack of color in a video signal is displayed as a dot in the center of the vectorscope. The angle, distance around the circle, magnitude, and distance away from the center indicate the phase and amplitude of the color signal.

References

K. B. Benson and J. Whitaker, Eds., *Television Engineering Handbook,* rev. ed., New York: McGraw-Hill, 1991.

K. B. Benson and J. Whitaker, *Television and Audio Handbook for Technicians and Engineers*, New York: McGraw-Hill, 1990.

J. Whitaker, *Radio Frequency Transmission Systems: Design and Operation*, New York: McGraw-Hill, 1991.

J. Whitaker, *Maintaining Electronic Systems*, Boca Raton: CRC Press, 1991.

Further Information

Additional information on the topic of television system technology is available from the following sources:

Broadcast Engineering magazine, a monthly periodical dealing with television and radio technology. The magazine, published in Overland Park, Kan., is free to qualified subscribers.

The Society of Motion Picture and Television Engineers, which publishes a monthly journal and holds an annual conference and convention in the fall. The SMPTE is headquartered in White Plains, N.Y.

The Society of Broadcast Engineers, which holds an annual technical conference and convention in the fall. The SBE is located in Indianapolis.

The National Association of Broadcasters, which holds an annual engineering conference and trade show in the spring. The NAB is headquartered in Washington, D.C.

In addition, the following books are recommended:

K. B. Benson and J. Whitaker, Eds., *Television Engineering Handbook*, rev. ed., New York: McGraw-Hill, 1991.

K. B. Benson and J. Whitaker, Eds., *Television and Audio Handbook for Technicians and Engineers*, New York: McGraw-Hill, 1990.

National Association of Broadcasters Engineering Handbook, 8th ed., Washington, D.C.: NAB, 1992.

63.4 High-Definition Television

Martin S. Roden

When standards were developed for television, few people dreamed of its evolution into a type of universal communication terminal. While these traditional standards are acceptable for entertainment video, they are not adequate for many emerging applications, such as videotext. We must evolve into a high-resolution standard. High-definition TV (HDTV) is a term applied to a broad class of new systems whose developments have received worldwide attention.

We begin with a brief review of the current television standards. The reader is referred to Section 63.3 for a more detailed treatment of conventional television.

Japan and North America use the National Television Systems Committee (NTSC) standard that specifies 525 scanning lines per picture, a field rate of 59.94 per second (nominally 60 Hz), and 2:1 **interlaced scanning** (although there are about 60 fields per second, there are only 30 new frames per second). The **aspect ratio** (ratio of width to height) is 4:3. The bandwidth of the television signal is 6 MHz, including the sound signal. In Europe and some other countries, the phase-alternation line (PAL) or the sequential color and memory (SECAM) standard is used. This specifies 625 scanning lines per picture and a field rate of 50 per second. The bandwidth of this type of television signal is 8 MHz.

HDTV systems nominally double the number of scan lines in a frame and change the aspect ratio to 16:9. Of course, if we were willing to start from scratch and abandon all existing television systems, we could set the bandwidth of each channel to a number greater than 6 (or 8) MHz, thereby achieving higher resolution. The Japan Broadcasting Corporation (NHK) has done just this in their HDTV system. This system permits 1125 lines per frame with 30 frames per second and 60 fields per second (2:1 interlaced scanning). The aspect ratio is 16:9. The system is designed for a bandwidth of 10 MHz per channel. With the 1990 launching of the BS-3 satellite, two channels were devoted to this form of HDTV. To fit the channel within a 10-MHz bandwidth (instead of the approximately 50 MHz that would be needed to transmit using traditional techniques), bandwidth compression was required. It should be noted that the Japanese system is primarily analog frequency modulation (FM) (the sound is digital). The approach to decreasing bandwidth is multiple sub-Nyquist encoding (**MUSE**). The sampling below Nyquist lowers the bandwidth requirement, but moving images suffer from less resolution.

Europe began its HDTV project in mid-1986 with a joint initiative involving West Germany (Robert Bosch GmbH), the Netherlands (NV Phillips), France (Thomson SA), and the United Kingdom (Thorn/EMI Plc.). The system, termed **Eureka 95** or D2-MAC, has 1152 lines per frame, 50 fields per second, 2:1 interlaced scanning, and a 16:9 aspect ratio. A more recent European proposed standard is for 1250 scanning lines at 50 fields per second. This is known as the **Eureka EU95**.

It is significant to note that the number of lines specified by Eureka EU95 is exactly twice that of the PAL and SECAM standard currently in use. The field rate is the same, so it is possible to devise compatible systems that would permit reception of HDTV by current receivers (of course, with adapters and without enhanced definition). The HDTV signal requires nominally 30 MHz of bandwidth.

In the United States, the FCC has ruled that any new HDTV system must permit continuation of service to contemporary NTSC receivers. This significant constraint applies to terrestrial broadcasting (as opposed to videodisk, videotape, and cable television). One set of proposals involves sending the enhanced signals on **"taboo" channels,** those that are not used in metropolitan areas to provide adequate separation. Thus, these currently unused channels would be used for simulcast signals. The reason that the FCC has implemented these taboo channels is to relax the requirement for receiver tuner filter design. With better filters, there would be no problem with the use of taboo channels. However, since the proposals for HDTV in the United States concentrate on digital transmission, transmitter power can be less than that used for conventional television. Indeed, in heavily populated urban areas (where many stations are licensed for broadcast), the HDTV signals will have to be severely limited in power (less than 10% of that required for an NTSC transmitter). We will see soon that the proposed systems include error control coding because of anticipated low signal-to-noise ratios.

The proposals for U.S. HDTV have been predominantly digital. When a color television signal is converted from analog to digital (A/D), the luminance, hue, and saturation signals must each be digitized using 8 bits of A/D per sample. Digital transmission of conventional television therefore requires a nominal bit rate of about 216 megabits/s, while HDTV nominally requires about 1200 megabits/s. If we were to use a digital modulation system that transmits 1 bit per hertz of bandwidth, we see that the HDTV signal requires over 1 GHz of bandwidth. Clearly significant data compression is required!

Proposed Systems

As this volume goes to press, four digital HDTV approaches are being readied for FCC testing. The four are being proposed by General Instrument Corporation, the Advanced Television Research Consortium (composed of NBC, David Sarnoff Research Center, Philips Consumer Electronics, and Thomson Consumer Electronics, Inc.), Zenith Electronics in cooperation with AT&T Bell Labs and AT&T Microelectronics, and the American Television Alliance (General Instrument Corporation and MIT). There are many common aspects to the four proposals, and major differences exist in the data compression approaches. The data compression techniques can be viewed as two-dimensional extensions of techniques used in voice encoding. In the following, we describe a generic system. The reader is referred to the references for details of the four competing systems.

Figure 63.22 shows a general block diagram of a digital HDTV transmitter. Each frame from the camera is digitized, and the system has the capability of storing one entire frame. Thus the processor works with two inputs—the current frame (A) and the previous frame (B). The current frame and the previous frame are compared in a motion detector that generates coded motion information (C). Algorithms used for motion estimation attempt to produce three-dimensional parameters from sequential two-dimensional information. Parameters may include velocity estimates for blocks of the picture.

The parameters from the motion detector are processed along with the previous frame to produce a *prediction* of the current frame (D). Since the motion detector parameters are transmitted, the receiver can perform a similar prediction of the current frame.

The predicted current frame is compared to the actual current frame, and a difference signal (E) is generated. This difference signal will generally have a smaller dynamic range than the original signal. For example, if the television image is static (is not changing with time), the difference signal will be zero.

The difference signal is *compressed* to form the transmitted video signal (F). This compression

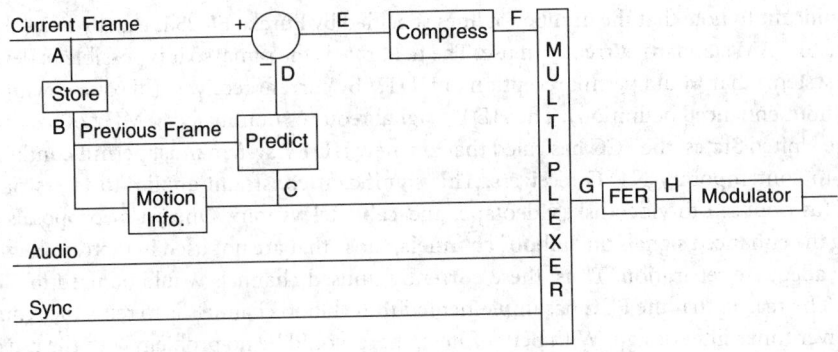

FIGURE 63.22   Block diagram of HDTV transmitter.

is performed both in the time and transform domains. **Entropy coding** of the type used in facsimile can be incorporated to take spatial continuity into account (i.e., a picture usually does not change over the span of a single picture element, so variations of "run length" coding can often compress the data). Several forms of transformation (e.g. *Hadamard* and *Karhunen-Loeve*) have potential for further lowering of the data rate by extracting essential parameters that describe the waveform.

Four data streams are asynchronously multiplexed to form the information to be transmitted (G). These four signals consist of the coded differential video, the motion detector parameters, the digital audio signal, and the synchronizing signals. Other information can be multiplexed, including various control signals that may be needed by cable operators.

Forward error correction is applied to the multiplexed digital signal to produce an encoded signal (H) that makes the transmission less susceptible to uncorrected bit errors. This is needed because of the anticipated low transmission power rates. Error control is also important because compression can amplify error effects—a single bit error can affect many picture elements. Proposals for error control include *BCH* and *Reed-Solomon* codes.

The encoded data signal forms the input to the modulator. To further conserve bandwidth, quadrature modulation is employed. This type of modulation is used in conventional NTSC color transmission where the two color difference signals are in phase quadrature with each other.

The corresponding receiver is shown in Fig. 63.23. The receiver simply forms the inverse of each transmitter operation. The received signal is first demodulated. The resulting data signal is decoded to remove the redundancy and correct errors. A demultiplexer separates the signal into the original four (or more) data signals. The audio and synchronization signals need no further processing.

The demultiplexed video signal is, hopefully, the same as the transmitted signal ("*F*"). (We use

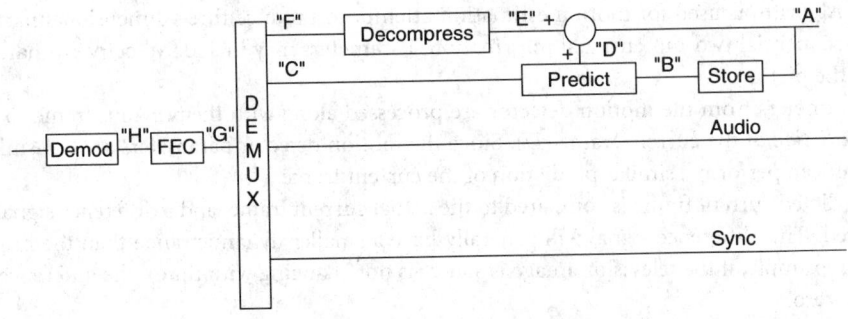

FIGURE 63.23   Block diagram of HDTV receiver.

letters with quotation marks to indicate that the signals are estimates of their transmitted counterpart.) This reproduced video signal is decompressed, using the inverse algorithm of that used in the transmitter, to yield an estimate of the original differential picture signal ("*E*"). The predict block in the receiver implements the same algorithm as that of the transmitter. Its inputs are the reconstructed motion signal ("*C*") and the previous reconstructed frame ("*B*"). When the predictor output ("*D*") is added to the reconstructed differential picture signal ("*E*"), the result is a reconstructed version of the current frame.

As this edition goes to press, none of the competing systems has received anything near uniform support. It is a situation where the technical aspects of the problem are merged with social and economic issues. In the United States, the FCC expects to make a decision regarding the many proposals it has by the spring of 1993. We can only hope that we do not end up with a multitude of incompatible systems.

Defining Terms

Aspect ratio: Ratio of frame width to height.

Entropy coding: A form of data compression that reduces a transmission to a shorter length by reducing signal redundancy.

Eureka 95 and EU95: European proposed HDTV systems.

Interlaced scanning: A bandwidth reduction technique wherein every other scan line is first transmitted followed by the "in between" lines.

MUSE: Multiple sub-Nyquist encoding, a technique used in Japanese HDTV system.

Taboo channels: Channels that the FCC does not currently assign in order to avoid interference from adjacent channels.

References

G.W. Beakley, "Channel coding for digital HDTV terrestrial broadcasting," *IEEE Transactions on Broadcasting*, vol. 37, no. 4, 1991.

R. Hopkins, "Digital HDTV broadcasting," *IEEE Transactions on Broadcasting*, vol. 37, no. 4, 1991.

R.K. Jurgen, Ed., "High-definition television update," *IEEE Spectrum*, April 1988.

R.K. Jurgen, Ed., "Consumer electronics," *IEEE Spectrum*, January 1989.

R.K. Jurgen, Ed., "The challenges of digital HDTV," *IEEE Spectrum*, April 1991.

J.C. McKinney, "HDTV approaches the end game," *IEEE Transactions on Broadcasting*, vol. 37, no. 4, 1991.

S. Prentiss, *HDTV*, Blue Ridge Summit, Pa.: TAB Books, 1990.

M.S. Roden, *Analog and Digital Communication Systems*, 3rd ed., Englewood Cliffs, N.J.: Prentice-Hall, 1991.

W.Y. Zou, "Digital HDTV compression techniques," *IEEE Transactions on Broadcasting*, vol. 37, no. 4, 1991.

63.5 Digital Audio Broadcasting

Stanley Salek and Almon H. Clegg

Digital audio broadcasting (DAB) is a developing technology that promises to give consumers a new and better aural broadcast system. DAB will offer dramatically better reception quality over existing AM and FM broadcasts by better audio quality and by superior resistance to interference in stationary and mobile/portable reception environments. Additionally, the availability of a digital data stream direct to consumers will open the prospects of providing extra services to augment basic sound delivery.

As of this writing, at least ten separate proponents have announced DAB transmission and reception systems. From the minimal data available describing these potential systems, it is clear that there is only partial agreement on which transmission method will provide the best operational balance. This chapter provides a general overview of the common aspects of DAB systems, as well as a description of one of the proposed transmission methods.

The Need for DAB

In the years since the early 1980s, the consumer marketplace has undergone a great shift toward digital electronic technology. The explosion of personal computer use has led to greater demands for information, including multimedia integration. Over the same time period, compact disc (CD) digital audio technology has overtaken long-playing records (and has nearly overtaken analog tape cassettes) as the consumer audio playback media of choice. Similar digital transcription methods and effects also have been incorporated into commonly available audio and video equipment. Additionally, it appears that the upcoming transition to a high-definition television broadcast system will incorporate full digital methods for video and audio transmission. Because of these market pressures, the radio broadcast industry has determined that the existing analog methods of broadcasting must be updated to keep pace with the advancing audio marketplace.

In addition to providing significantly enhanced audio quality, DAB systems are being developed to overcome the technical deficiencies of existing AM and FM analog broadcast systems. The foremost problem of current broadcast technology, as perceived by the industry, is its susceptibility to interference. AM medium-wave broadcasts, operating in the 530- to 1700-kHz frequency range, are prone to disruption by fluorescent lighting and by power system distribution networks, as well as by numerous other manufactured unintentional radiators, including computer and telephone systems. Additionally, natural effects, such as nighttime skywave propagation interference between stations and lightning, cause irritating service disruption to AM reception. FM broadcast transmissions in the 88- to 108-MHz band are much more resistant to these types of interference. However, multipath propagation and abrupt signal fading, especially found in urban and mountainous areas containing a large number of signal reflectors and shadowers (e.g., buildings and terrain), can seriously degrade FM reception, particularly in automobiles.

DAB System Design Goals

DAB systems are being designed with several technical goals in mind. The first goal is to create a service that delivers compact disc quality stereo sound for broadcast to consumers. The second is to overcome the interference problems of current AM and FM broadcasts, especially under portable and mobile reception conditions. Third, DAB must be spectrally efficient in that total bandwidth should be no greater than that currently used for FM broadcasts. Fourth, the DAB system should provide space in its data stream to allow for the addition of ancillary services, such as program textual information display or software downloading. Finally, DAB receivers must not be overly cumbersome, complex, or expensive, to foster rapid consumer acceptance.

In addition to these goals, desired features include the reduced RF transmission power requirements (when compared to AM and FM broadcast stations with the same signal coverage), the mechanism to seamlessly fill in coverage areas that are shadowed from the transmitted signal, and the ability to easily integrate DAB receivers into personal, home, and automotive sound systems.

Historical Background

DAB development work began in Europe in 1986, with the initial goal to provide high-quality audio services to consumers directly by satellite. Companion terrestrial systems were developed to evaluate the technology being considered, as well as to provide fill-in service in small areas where the satellite signals were shadowed. A consortium of European technical organizations known as Eureka-147/DAB demonstrated the first working terrestrial DAB system in Geneva in September 1988. Subsequent terrestrial demonstrations of the system followed in Canada in the summer of 1990, and in the United States in April and September of 1991.

For the demonstrations, VHF and UHF transmission frequencies between 200 and 900 MHz were used with satisfactory results. Because most VHF and UHF frequency bands suitable for DAB are already in use (or reserved for high-definition television and other new services), an additional Canadian study in 1991 evaluated frequencies near 1500 MHz (L-band) for use as a potential worldwide DAB allocation. This study concluded that L-band frequencies would support a DAB system such as Eureka-147, while continuing to meet the overall system design goals.

In early 1992, the World Administrative Radio Conference (WARC-92) was held, during which frequency allocations for many different radio systems were debated. As a result of WARC-92, a worldwide L-band standard of 1452 to 1492 MHz was designated for both satellite and terrestrial digital radio broadcasting. However, because of existing government and military uses of L-band, the United States was excluded from the standard. Instead, an S-band allocation of 2310 to 2360 MHz was substituted. Additionally, Asian nations including Japan, China, and CIS opted for an extra S-band allocation in the 2535- to 2655-MHz frequency range.

In mid-1991, because of uncertainty as to the suitability of using S-band frequencies for terrestrial broadcasting, most DAB system development work in the United States shifted from out-band (i.e., UHF, L-band, and S-band) to in-band. In-band terrestrial systems would merge DAB services with existing AM and FM broadcasts, using novel adjacent- and co-channel modulating schemes. In early 1992, one proponent demonstrated a proprietary method of extracting a compatible digital RF signal from a co-channel analog FM broadcast transmission [USA Digital Radio, 1992]. Thus, in-band DAB could permit a logical transition from analog to digital broadcasting for current broadcasters, within the current channel allocation scheme. As of July 1992, however, no in-band DAB system had been demonstrated under actual field conditions.

In 1991, a digital radio broadcasting standards committee was formed by the Electronic Industries Association (EIA). Present estimates are that the committee may complete its testing and evaluation of the various proposed systems by the mid 1990s.

Technical Overview of DAB

Regardless of the actual signal delivery system used, all DAB systems share a common overall topology. Figure 63.24 presents a block diagram of a typical DAB transmission system.

To maintain the highest possible audio quality, program material would be broadcast from digital sources, such as CD players and digital audio recorders, or digital audio feeds from network sources. Analog sources, such as microphones, are converted to a digital audio data stream using an analog-to-digital (A/D) converter, prior to switching or summation with the other digital sources.

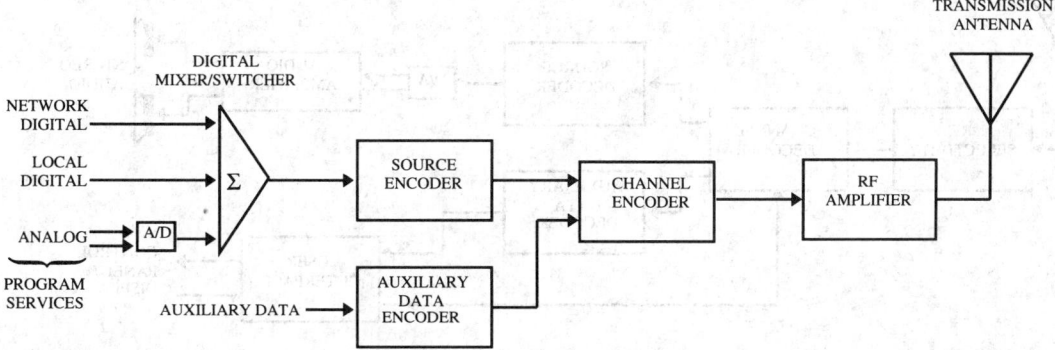

FIGURE 63.24 An example DAB transmission system. (*Source:* Hammett & Edison, Inc., Consulting Engineers.)

The linear digital audio data stream from the studio is then applied to the input of a **source encoder.** The purpose of this device is to reduce the required bandwidth of the audio information, helping to produce a spectrally efficient RF broadcast signal. For example, 16-bit linear digital audio sampled at 48 kHz (the standard professional rate) requires a data stream of 1.536 megabits/s to transmit a stereo program in a serial format. This output represents a bandwidth of approximately 1.5 MHz, much greater than that used by an equivalent analog audio modulating signal [Smyth, 1992]. Source encoders can reduce the data rate by factors of 8:1 or more, yielding a much more efficient modulating signal.

Following the source encoder, the resulting serial digital signal is applied to the input of the **channel encoder,** a device that modulates the transmitted RF wave with the reduced-rate audio information. Auxiliary serial data, such as program information and/or receiver control functions, also can be input to the channel encoder for simultaneous transmission.

The channel encoder uses sophisticated modulating techniques to accomplish the goals of interference cancellation and high spectral efficiency. Methods of interference cancellation include expansion of time and frequency diversity of the transmitted information, as well as the inclusion of error correction codes in the data stream. Time diversity involves transmitting the same information multiple times by using a predetermined time interval. Frequency diversity, such as that produced by spread-spectrum, multiple-carrier, or frequency-hopping systems, provides the means to transmit identical data on several different frequencies within the bandwidth of the system. At the receiver, real-time mathematical processes are used to locate the required data on a known frequency at a known time. If the initial information is found to be unusable because of signal interference, the receiver simply uses the same data found on another frequency and/or at another time, producing seamless demodulation.

Spectral efficiency is a function of the modulation system used. Among the modulation formats that have been proposed for DAB transmission are QPSK, M-ary QAM, and MSK [Springer, 1992]. Using these and other formats, digital transmission systems that use no more spectrum than their analog counterparts have been designed.

The RF output signal of the channel encoder is amplified to the appropriate power level for transmission. Because the carrier-to-noise (C/N) ratio of the modulated waveform is not generally so critical as that required for analog communications systems, relatively low transmission power often can be used. Depending on the sophistication of the data recovery circuits contained in the DAB receiver, the use of C/N ratios as low as 6 dB are possible, without causing a degradation to the received signal.

DAB reception is largely the inverse of the transmission process, with the inclusion of sophisticated error correction circuits. Fig. 63.25 shows a typical DAB receiver.

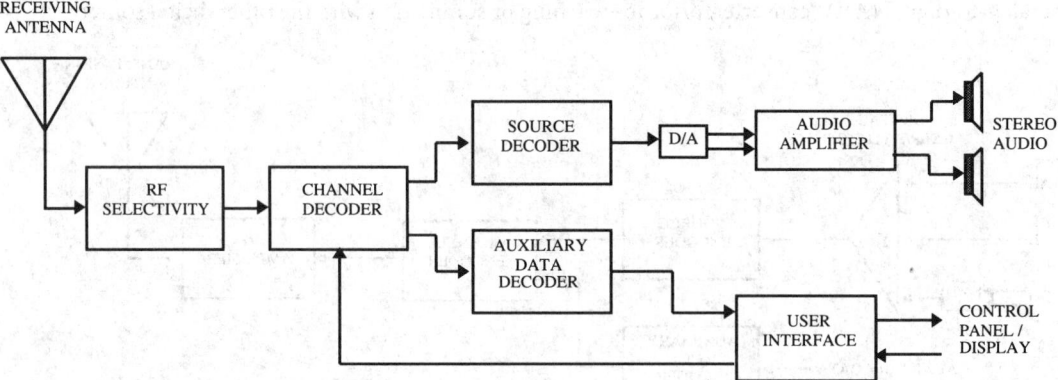

FIGURE 63.25   An example DAB receiver. (*Source:* Hammett & Edison, Inc., Consulting Engineers.)

DAB reception begins in a similar manner as is used in virtually all receivers. A receiving antenna feeds an appropriate stage of RF selectivity and amplification from which a sample of the coded DAB signal is derived. This signal then drives a channel decoder, which reconstructs the audio and auxiliary data streams. To accomplish this task, the channel decoder must demodulate and de-interleave the data contained on the RF carrier and then apply appropriate computational and statistical error correction functions.

The source decoder converts the reduced bit-rate audio stream back to pseudolinear at the original sampling rate. The decoder computationally expands the mathematically reduced data and fills the gaps left from the extraction of irrelevant audio information with averaged code or other masking data. The output of the source decoder feeds audio digital-to-analog (D/A) converters, and the resulting analog stereo audio signal is amplified for the listener.

In addition to audio extraction, DAB receivers likely will be capable of decoding auxiliary data. This data can be used in conjunction with the user interface to control receiver functions, or for a completely separate purpose. A typical user interface could contain a data display screen in addition to the usual receiver tuning and audio controls. This data screen could be used to obtain information about the programming, news reports, sports scores, advertising, or any other useful data sent by the station or an originating network. Also, external interfaces could be used to provide a software link to personal computer systems.

Audio Compression and Source Encoding

The development of digital audio encoding started with research into pulse-code modulation (PCM) in the late 1930s and evolved, shortly thereafter, to include work on the principles of digital PCM coding. Linear predictive coding (LPC) and adaptive delta pulse-code modulation (ADPCM) algorithms had evolved in the early 1970s and later were adopted into standards such as C.721 (published by the CCITT) and CD-I (Compact Disc-Interactive). At the same time, algorithms were being invented for use with phoneme-based speech coding. Phonetic coding, a first-generation "model-based" speech-coding algorithm, was mainly implemented for low bit-rate speech and text-to-speech applications. These classes of algorithms for speech further evolved to include both CELP (Code Excited Linear Predictive) and VSELP (Vector Selectable Excited Linear Predictive) algorithms by the mid-1980s. In the late 1980s, these classes of algorithms were also shown to be useful for high-quality audio music coding. These audio algorithms were put to commercial use from the late 1970s to the latter part of the 1980s.

Subband coders evolved from the early work on quadrature mirror filters in the mid-1970s and continued with polyphase filter-based schemes in the mid-1980s. Hybrid algorithms employing both subband and ADPCM coding were developed in the latter part of the 1970s and standardized (e.g., CCITT G.722) in the mid- to late 1980s. Adaptive transform coders for audio evolved in the mid-1980s from speech coding work done in the late 1970s.

By employing psychoacoustic noise-masking properties of the human ear, perceptual encoding evolved from early work of the 1970s and where high-quality speech coders were employed. Music quality bit-rate reduction schemes such as MPEG (Motion Picture Expert Group), MUSICAM (Masking pattern-adapted Universal Subband Integrated Coding And Multiplexing), PASC (Precision Adaptive Subband Coding), and ATRAC (Adaptive TRansform Acoustic Coding) have been recently developed, and by 1992 some commercial usage had been announced by manufacturers. Further refinements to the technology during the 1990s will focus attention on novel approaches such as wavelet-based coding and the use of entropy coding schemes.

However, progress has been significant, and the various audio coding schemes that have been demonstrated publicly over the time period from 1990 to 1992 have shown steady increases in compression ratios at given audio quality levels. Furthermore, with optimal use of preprocessing, psychoacoustic bit allocation, and postprocessing, industry consensus indicates that the compression ratios per given audio quality level will improve to more than about 2:1 over 1992's best results.

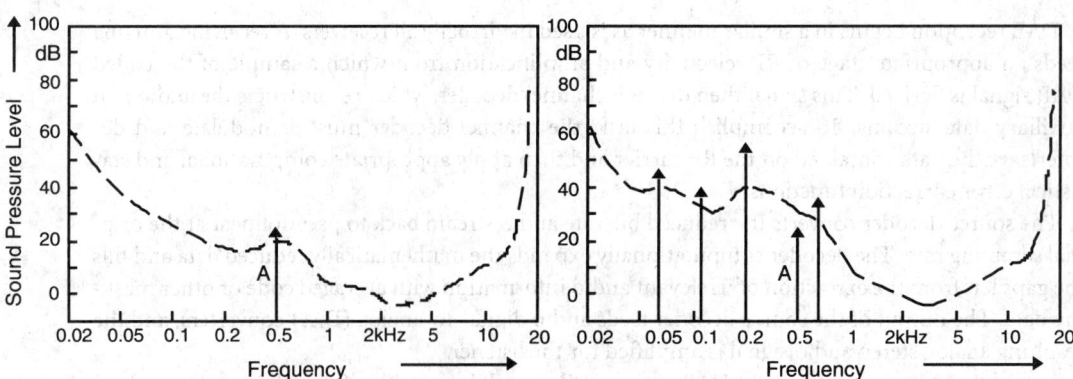

FIGURE 63.26 An example of the masking effect. Based on the hearing threshold of the human ear (dashed line), a 500-Hz sinusoidal acoustic waveform, shown at *A* on the left graph, is easily audible at relatively low levels. However, it can be masked by adding nearby higher-amplitude components, as shown on the right. (*Source:* CCi.)

Audio coding for digital broadcasting will likely use one of the many perceptual encoding schemes previously mentioned (such as MPEG/MUSICAM) or some variation thereof. Fundamentally, they all depend on two basic psychoacoustic phenomena: (1) the threshold of human hearing, and (2) masking of nearby frequency components. In the early days of hearing research, Harvey Fletcher, a researcher at Bell Laboratories, measured the hearing of many human beings and published the well-known Fletcher-Munson threshold-of-hearing chart. Basically it states that, depending on the frequency, audio sounds below certain levels cannot be heard by the human ear. Further, the masking effect, simply stated, is when two frequencies are very close to each other and one is a higher level than the other, the weaker of the two is masked and will not be heard. These two principles allow for as much as 80% of the data representing a musical signal to be discarded.

Figure 63.26 shows how introduction of frequency components affects the ear's threshold of hearing versus frequency. Figure 63.27 shows how the revised envelope of audibility results in the elimination of components that would not be heard.

The electronic implementation of these algorithms employs a digital filter that breaks the audio spectrum into many subbands, and various coefficient elements are built into the program to decide when it is permissible to remove one or more of the signal components. The details of how the bands are divided and how the coefficients are determined are usually proprietary to the individual system developers. Standardization groups have spent many worker-hours of evaluation attempting to determine the most accurate coding system.

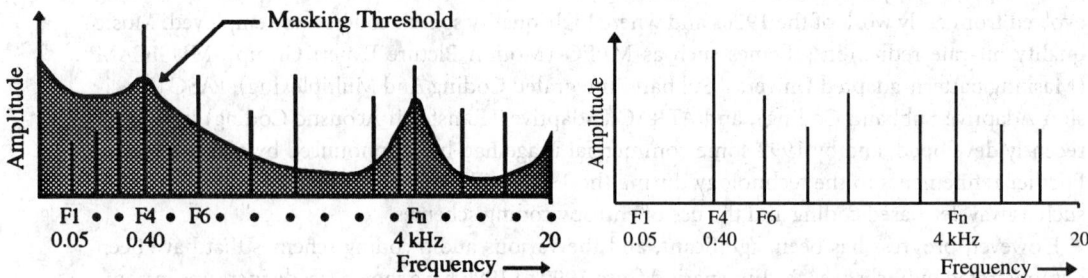

FIGURE 63.27 Source encoders use an empirically derived masking threshold to determine which audio components can be discarded (left). As shown on the right, only the audio components with amplitudes above the masking threshold are retained. (*Source:* CCi.)

System Example: Eureka-147/DAB

As of this writing, Eureka-147/DAB is the only fully developed DAB system that has demonstrated a capability to meet virtually all the described system goals. Developed by a European consortium, it is an out-band system in that its design is based on the use of a frequency spectrum outside the AM and FM radio broadcast bands. Out-band operation is required because the system packs up to 16 stereophonic broadcast channels (plus auxiliary data) into one contiguous band of frequencies, which can occupy a total bandwidth of up to 4 MHz. Thus, overall efficiency is maintained, with 16 digital program channels occupying about the same total bandwidth as 16 equivalent analog FM broadcast channels. System developers have promoted Eureka-147/DAB for satellite transmission, as well as for terrestrial applications in locations that have a suitable block of unused spectrum in the L-band frequency range or below.

In recent demonstrations, the MUSICAM source encoding/decoding system has been used. Originally developed by IRT (Institut für Rundfunktecknik) in Germany, the system works by dividing the original digital audio source into 32 subbands. As with the source encoders described earlier, each of the bands is digitally processed to remove redundant information and sounds that are not perceptible to the human ear. Using this technique, the original audio, sampled at a rate of 768 kilobits/s per channel, is reduced to 128 kilobits/s per channel, representing a compression ratio of 6:1.

The Eureka-147/DAB channel encoder operates by combining the transmitted program channels into a large number of adjacent narrowband RF carriers, which are each modulated using QPSK and grouped in a way that maximizes spectrum efficiency known as orthogonal frequency-division multiplex (OFDM). The information to be transmitted is distributed among the RF carriers and is also time-interleaved to reduce the effects of selective fading. A guard interval is inserted between blocks of transmitted data to improve system resistance to intersymbol interference caused by multipath propagation. Convolutional coding is used in conjunction with a Viterbi maximum-likelihood decoding algorithm at the receiver to make constructive use of echoed signals and to correct random errors [Alard and Lassalle, 1988].

RF power levels of just a few tens of watts per program channel have been used in system demonstrations, providing a relatively wide coverage area, depending on the height of the transmitting antenna above surrounding terrain. This low power level is possible because the system can operate at a C/N ratio of less than 10 dB, as opposed to the more than 30 dB that is required for high-fidelity demodulation of analog FM broadcasts.

Another demonstrated capability of the system is its ability to use "**gap filler**" transmitters to augment signal coverage in shadowed areas. A gap filler is simply a system that directly receives the DAB signal at an unobstructed location, provides RF amplification, and retransmits the signal, on the same channel, into the shadowed area. Because the system can make constructive use of signal reflections (within a time window defined by the guard interval and other factors), the demodulated signal is uninterrupted on a mobile receiver when it travels between an area served by the main signal into the service area of the gap filler.

Defining Terms

Channel encoder: A device that converts source-encoded digital information into an analog RF signal for transmission. The type of modulation used depends on the particular digital audio broadcasting (DAB) system, although most modulation techniques employ methods by which the transmitted signal can be made more resistant to frequency-selective signal fading and multipath distortion effects.

Gap filler: A low-power transmitter that boosts the strength of transmitted DAB RF signals in areas which normally would be shadowed due to terrain obstruction. Gap fillers can operate on the same frequency as DAB transmissions or on alternate channels that can be located by DAB receivers using automatic switching.

Source encoder: A device that substantially reduces the data rate of linearly digitized audio signals by taking advantage of the psychoacoustic properties of human hearing, eliminating redundant and subjectively irrelevant information from the output signal. Transform source encoders work entirely within the frequency domain, while time-domain source encoders work primarily in the time domain. Source decoders reverse the process, using various masking techniques to simulate the properties of the original linear data.

References

M. Alard and R. Lassalle, "Principles of modulation and channel coding for digital broadcasting for mobile receivers," in *Advanced Digital Techniques for UHF Satellite Sound Broadcasting* (collected papers), European Broadcasting Union, pp. 47–69, 1988.

R. Bruno, "Digital audio and video compression, present and future," presented to the Delphi Club, Tokyo, Japan, July 1992.

G. Chouinard and F. Conway, "Broadcasting systems concepts for digital sound," in *Proceedings of the 45th Annual Broadcast Engineering Conference*, National Association of Broadcasters, 1991, pp. 257–266.

F. Conway, R. Voyer, S. Edwards, and D. Tyrie, "Initial experimentation with DAB in Canada," in *Proceedings of the 45th Annual Broadcast Engineering Conference*, National Association of Broadcasters, 1991, pp. 281–290.

S. Kuh and J. Wang, "Communications systems engineering for digital audio broadcast," in *Proceedings of the 45th Annual Broadcast Engineering Conference*, National Association of Broadcasters, 1991, pp. 267–272.

P. H. Moose and J.M. Wozencraft, "Modulation and coding for DAB using multi-frequency modulation," in *Proceedings of the 45th Annual Broadcast Engineering Conference*, National Association of Broadcasters, 1991, pp. 405–410.

M. Rau, L. Claudy, and S. Salek, *Terrestrial Coverage Considerations for Digital Audio Broadcasting Systems*, National Association of Broadcasters, 1990.

S. Smyth, "Digital audio data compression," *Broadcast Engineering Magazine*, pp. 52–60, Feb. 1992.

K.D. Springer, *Interference Between FM and Digital M-PSK Signals in the FM Band*, National Association of Broadcasters, 1992.

USA Digital Radio, System description brochure, Annual Convention of the National Association of Broadcasters, April 1992.

Further Information

The National Association of Broadcasters publishes periodic reports on the technical, regulatory, and political status of DAB in the United States. Additionally, their Broadcast Engineering Conference proceedings published since 1990 contain a substantial amount of information on emerging DAB technologies.

IEEE Transactions on Broadcasting, published quarterly by the Institute of Electrical and Electronics Engineers, Inc., periodically includes papers on digital broadcasting.

Additionally, the biweekly newspaper publication *Radio World* provides continuous coverage of DAB technology, including proponent announcements, system descriptions, field test reports, and broadcast industry reactions.

64

Digital Communication

Richard C. Dorf
University of California, Davis

Zhen Wan
University of California, Davis

64.1 Coding 1405
Block Codes • Convolutional Codes • Code Performance • Trellis-Coded Modulation

64.2 Equalization 1410
Linear Transversal Equalizers • Nonlinear Equalizers • Linear Receivers • Nonlinear Receivers

64.1 Coding

Richard C. Dorf and Zhen Wan

Error correcting codes may be classified into two broad categories: **block codes** and **tree codes**. A block code is a mapping of k input binary symbols into n output binary symbols. Consequently, the block coder is a *memoryless* device. Since $n > k$, the code can be selected to provide redundancy, such as *parity bits*, which are used by the decoder to provide some error detection and error correction. The codes are denoted by (n, k), where the code rate R is defined by $R = k/n$. Practical values of R range from 1/4 to 7/8, and k ranges from 3 to several hundred [Clark and Cain, 1981]. Some properties of block codes are given in Table 64.1.

A tree code is produced by a coder that has *memory*. **Convolutional codes** are a subset of tree codes. The convolutional coder accepts k binary symbols at its input and produces n binary symbols at its output, where the n output symbols are affected by $v + k$ input symbols. Memory is incorporated since $v > 0$. The code rate is defined by $R = k/n$. Typical values for k and n range from 1 to 8, and the values for v range from 2 to 60. The range of R is between 1/4 and 7/8 [Clark and Cain, 1981].

Block Codes

In block code, the n code digits generated in a particular time unit depend only on the k message digits within that time unit. Some of the errors can be detected and corrected if $d \geq s + t + 1$, where s is the number of errors that can be detected, t is the number of errors that can be corrected, and d is the Hamming distance. Usually, $s \geq t$, thus, $d \geq 2t + 1$. A general code word can be expressed as a_1, $a_2, \ldots, a_k, c_1, c_2, \ldots, c_r$. k is the number of information bits and r is the number of check bits. Total word length is $n = k + r$.

In Fig. 64.1, the gain h_{ij} ($i = 1, 2, \ldots, r, j = 1, 2, \ldots, k$) are elements of the parity check matrix **H**. The k data bits are shifted in each time, while $k + r$ bits are simultaneously shifted out by the commutator.

Cyclic Codes

Cyclic codes are block codes such that another code word can be obtained by taking any one code word, shifting the bits to the right, and placing the dropped-off bits on the left. An encoding circuit with $(n - k)$ shift registers is shown in Fig. 64.2.

Table 64.1 Properties of Block Codes

Property	BCH	Reed–Solomon	Hamming	Maximal Length
Block length	$n = 2^m - 1$, $m = 3, 4, 5, \ldots$	$n = m(2^m - 1)$ bits	$n = 2^m - 1$	$n = 2^m - 1$
Number of parity bits		$r = m2t$ bits	$r = m$	
Minimum distance	$d \geq 2t + 1$	$d = m(2t + 1)$ bits	$d = 3$	$d = 2^m - 1$
Number of information bits	$k \geq n - mt$			$k = m$

[a] m is any positive integer unless otherwise indicated; n is the block length; k is the number of information bits; t is the number of errors that can be corrected; r is the number of parity bits; d is the distance.

Source: L. W. Couch, *Digital and Analog Communication Systems,* New York: Macmillan, 1990. With permission.

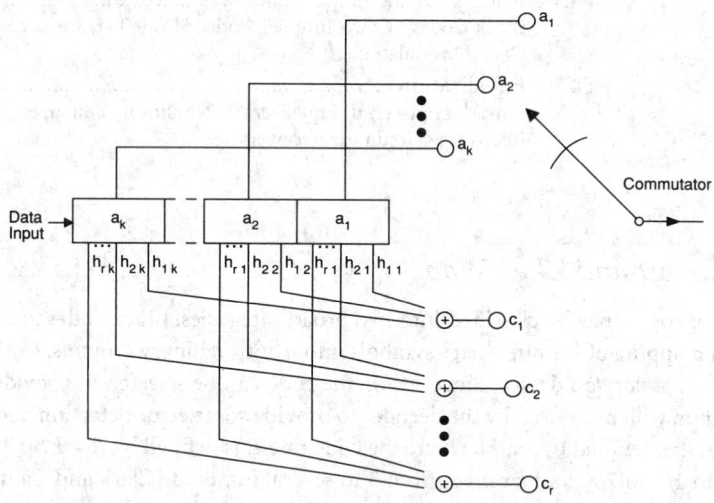

FIGURE 64.1 An encoding circuit of (n, k) block code.

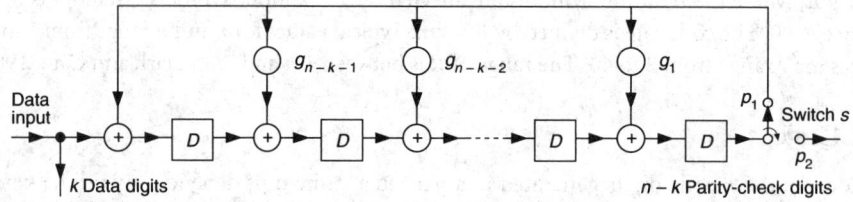

FIGURE 64.2 An encoder for systematic cyclic code. (*Source:* B.P. Lathi, *Modern Digital and Analog Communications,* New York: CBS College Publishing, 1983. With permission.)

In Fig. 64.2, the gain g_ks are the coefficients of the generator polynomial $g(x) = x^{n-k} + g_1 x^{n-k-1} + \cdots + g_{n-k-1} x + 1$. The gains g_k are either 0 or 1. The k data digits are shifted in one at a time at the input with the switch s held at position p_1. The symbol D represents a one-digit delay. As the data digits move through the encoder, they are also shifted out onto the output lines, because the first k digits of code word are the data digits themselves. As soon as the last (or kth) data digit clears the last $(n - k)$ register, all the registers contain the parity-check digits. The switch s is now thrown to position p_2, and the $n - k$ parity-check digits are shifted out one at a time onto the line.

Examples of cyclic and related codes are

1. Bose–Chaudhuri–Hocquenhem (BCH)
2. Reed–Solomon
3. Hamming
4. Maximal length
5. Reed–Muller
6. Golay codes

Convolutional Codes

In convolutional code, the block of n code digits generated by the encoder in a particular time unit depends not only on the block of k message digits within that time unit but also on the block of data digits within a previous span of $N - 1$ time units ($N > 1$). A convolutional encoder is illustrated in Fig. 64.3.

Here k bits (one input frame) are shifted in each time, and concurrently n bits (the output frame) are shifted out, where $n > k$. Thus, every k-bit input frame produces an n-bit output frame. Redundancy is provided in the output, since $n > k$. Also, there is memory in the coder, since the output frame depends on the previous K input frames where $K > 1$. The *code rate* is $R = k/n$, which is 3/4 in this illustration. The *constraint length*, K, is the number of input frames that are held in the kK bit shift register. Depending on the particular convolutional code that is to be generated, data from the kK stages of the shift register are added (modulo 2) and used to set the bits in the n-stage output register.

Code Performance

The improvement in the performance of a digital communication system that can be achieved by the use of coding is illustrated in Fig. 64.4. It is assumed that a digital signal plus channel noise is present at the receiver input. The performance of a system that uses binary-phase-shift-keyed (BPSK) signaling is shown both for the case when coding is used and for the case when there is no coding. For the BPSK no code case, $P_e = Q\left(\sqrt{2(E_b/N_o)}\right)$. For the coded case a (23,12) Golay code is used; P_e is the *probability of bit error*—also called the *bit error rate* (BER)—that is measured at the receiver output.

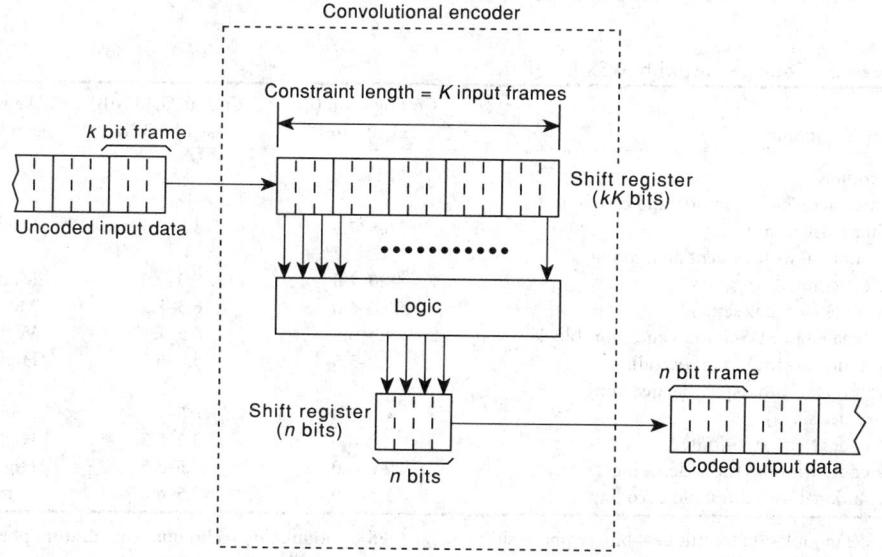

FIGURE 64.3 Convolutional encoding ($k = 3$, $n = 4$, $K = 5$, and $R = 3/4$). (*Source:* L. W. Couch, *Digital and Analog Communication Systems,* New York: Macmillan, 1990. With permission.)

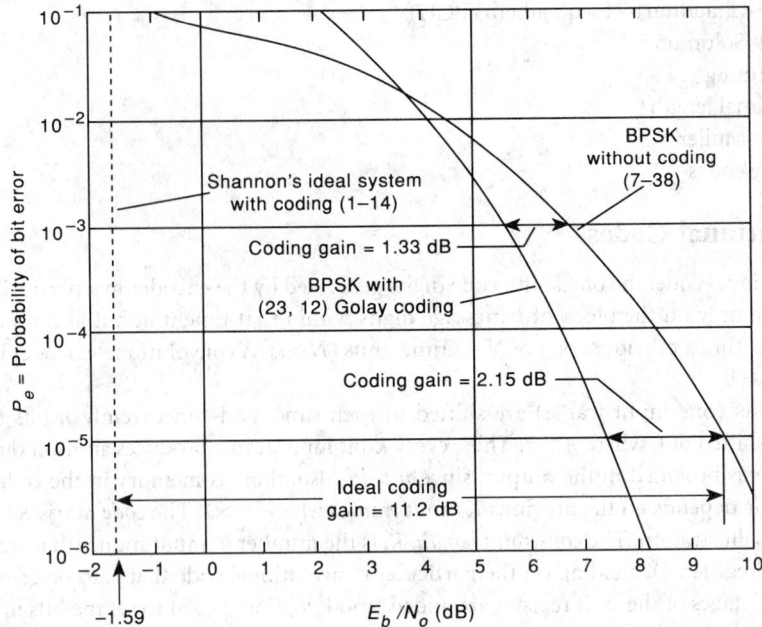

FIGURE 64.4 Performance of digital systems—with and without coding. E_b/N_o is the energy-per-bit to noise-density at the receiver input. The function $Q(x)$ is $Q(x) = (1/\sqrt{2\pi}x)e^{-x^2/2}$. (*Source:* L. W. Couch, *Digital and Analog Communication Systems*, New York: Macmillan, 1990. With permission.)

Trellis-Coded Modulation

Trellis-coded modulation (TCM) combines multilevel modulation and coding to achieve coding gain without bandwidth expansion [Ungerboeck, 1982, 1987]. TCM has been adopted for use in the new CCITT V.32 modem that allows an information data rate of 9600 b/s (bits per second) to be transmitted over VF (voice frequency) lines. The TCM has a coding gain of 4 dB [Wei, 1984].

Table 64.2 Coding Gains with BPSK or QPSK

Coding Technique Used	Coding Gain (dB) at 10^{-5} BER	Coding Gain (dB) at 10^{-8} BER	Data Rate Capability
Ideal coding	11.2	13.6	
Concatenated Reed–Solomon and convolution (Viterbi decoding)	6.5–7.5	8.5–9.5	Moderate
Convolutional with sequential decoding (soft decisions)	6.0–7.0	8.0–9.0	Moderate
Block codes (soft decisions)	5.0–6.0	6.5–7.5	Moderate
Concatenated Reed–Solomon and short block	4.5–5.5	6.5–7.5	Very high
Convolutional with Viterbi decoding	4.0–5.5	5.0–6.5	High
Convolutional with sequential decoding (hard decisions)	4.0–5.0	6.0–7.0	High
Block codes (hard decisions)	3.0–4.0	4.5–5.5	High
Block codes with threshold decoding	2.0–4.0	3.5–5.5	High
Convolutional with threshold decoding	1.5–3.0	2.5–4.0	Very high

BPSK: modulation technique—binary phase-shift keying; QPSK: modulation technique—quadrature phase-shift keying; BER: bit error rate.

Source: V.K. Bhargava, "Forward error correction schemes for digital communications," *IEEE Communication Magazine*, 21, 11–19, © 1983 IEEE. With permission.

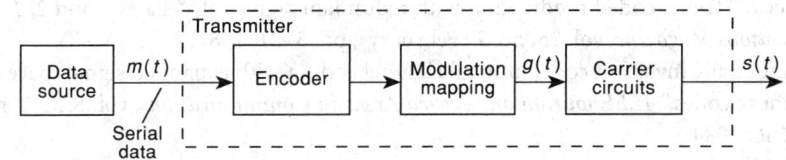

(a) Conventional Coding Technique

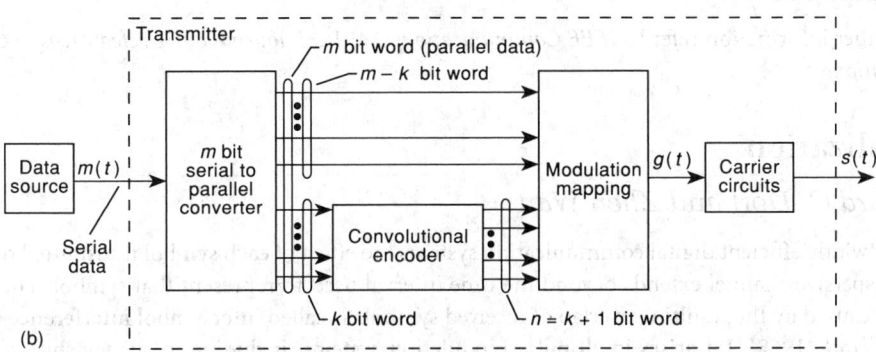

(b)

FIGURE 64.5 Transmitters for conventional coding and for TCM. (*Source:* L. W. Couch, *Digital and Analog Communication Systems,* New York: Macmillan, 1990. With permission.)

The combined modulation and coding operation of TCM is shown in Fig. 64.5(b). Here, the serial data from the source, $m(t)$, are converted into parallel (m-bit) data, which are partitioned into k-bit and ($m - k$)-bit words where $k \geq m$. The k-bit words (frames) are convolutionally encoded into ($n = k + 1$)-bit words so that the code rate is $R = k/(k + 1)$. The amplitude and phase are then set jointly on the basis of the coded n-bit word and the uncoded ($m - k$)-bit word. Almost 6 dB of coding gain can be realized if coders of constraint length 9 are used.

Defining Terms

Block code: A mapping of k input binary symbols into n output binary symbols.
Convolutional code: A subset of tree codes, accepting k binary symbols at its input and producing n binary symbols at its output.
Cyclic code: Block code such that another code word can be obtained by taking any one code word, shifting the bits to the right, and placing the dropped-off bits on the left.
Tree code: Produced by a coder that has memory.

References

V.K. Bhargava, "Forward error correction schemes for digital communications," *IEEE Communication Magazine,* 21, 1983.

G.C. Clark and J.B. Cain, *Error-Correction Coding for Digital Communications,* New York: Plenum, 1981.

L. W. Couch, *Digital and Analog Communication Systems,* New York: Macmillan, 1990.

B.P. Lathi, *Modern Digital and Analog Communication,* New York: CBS College Publishing, 1983.

G. Ungerboeck, "Channel coding with multilevel/phase signals," *IEEE Transactions on Information Theory,* vol. IT-28 (January), pp. 55–67, 1982.

G. Ungerboeck, "Trellis-coded modulation with redundant signal sets," Parts 1 and 2, *IEEE Communications Magazine*, vol. 25, no. 2 (February), pp. 5–21, 1987.

L. Wei, "Rotationally invariant convolutional channel coding with expanded signal space—Part II: Nonlinear codes," *IEEE Journal on Selected Areas in Communications*, vol. SAC-2, no. 2, pp. 672–686, 1984.

Further Information

For further information refer to *IEEE Communications* and *IEEE Journal on Selected Areas in Communications*.

64.2 Equalization

Richard C. Dorf and Zhen Wan

In bandwidth-efficient digital communication systems the effect of each symbol transmitted over a time dispersive channel extends beyond the time interval used to represent that symbol. The distortion caused by the resulting overlap of received symbols is called **intersymbol interference** (ISI) [Lucky *et al.*, 1968]. ISI arises in all pulse-modulation systems, including frequency-shift keying (FSK), phase-shift keying (PSK), and quadrature amplitude modulation (QAM) [Lucky *et al.*, 1968]. However, its effect can be most easily described for a baseband PAM system.

The purpose of an **equalizer,** placed in the path of the received signal, is to reduce the ISI as much as possible to maximize the probability of correct decisions.

Linear Transversal Equalizers

Among the many structures used for equalization, the simplest is the transversal (tapped delay line or nonrecursive) equalizer shown in Fig. 64.6. In such an equalizer the current and past values $r(t - nT)$ of the received signal are linearly weighted by equalizer coefficients (tap gains) c_n and summed to produce the output. In the commonly used digital implementation, samples of the received signal at the symbol rate are stored in a digital shift register (or memory), and the equalizer output samples (sums of products) $z(t_0 + kT)$ or z_k are computed digitally, once per symbol, according to

$$z_k = \sum_{n=0}^{N-1} c_n r(t_0 + kT - nt)$$

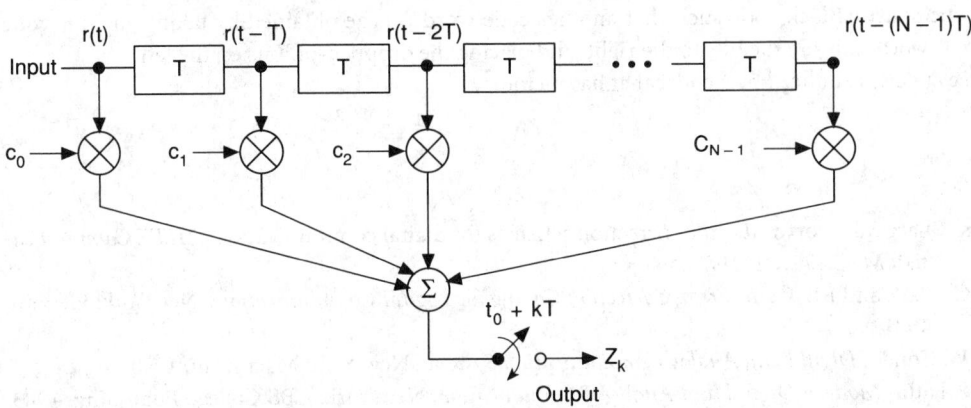

FIGURE 64.6 Linear transversal equalizer. (*Source*: K. Feher, *Advanced Digital Communications*, Englewood Cliffs, N.J.: Prentice-Hall, 1987, p. 648. With permission.)

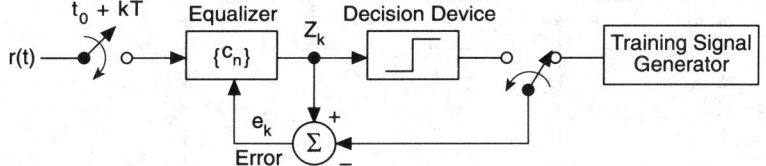

FIGURE 64.7 Automatic adaptive equalizer. (*Source:* K. Feher, *Advanced Digital Communications*, Englewood Cliffs, N.J.: Prentice-Hall, 1987, p. 651. With permission.)

where N is the number of equalizer coefficients and t_0 denotes sample timing.

The equalizer coefficients, c_n, $n = 0, 1, \ldots, N - 1$, may be chosen to force the samples of the combined channel and equalizer impulse response to zero at all but one of the NT-spaced instants in the span of the equalizer. Such an equalizer is called a *zero-forcing* (ZF) equalizer [Lucky, 1965].

If we let the number of coefficients of a ZF equalizer increase without bound, we would obtain an infinite-length equalizer with zero ISI at its output. An infinite-length zero-ISI equalizer is simply an inverse filter, which inverts the folded frequency response of the channel. Clearly, the ZF criterion neglects the effect of noise altogether. A finite-length ZF equalizer is approximately inverse to the folded frequency response of the channel. Also, a finite-length ZF equalizer is guaranteed to minimize the peak distortion or worst-case ISI only if the peak distortion before equalization is less than 100% [Lucky, 1965].

The *least-mean-squared* (LMS) equalizer [Lucky *et al.*, 1968] is more robust. Here the equalizer coefficients are chosen to minimize the mean squared error (MSE)—the sum of squares of all the ISI terms plus the noise power at the output of the equalizer. Therefore, the LMS equalizer maximizes the signal-to-distortion ratio (S/D) at its output within the constraints of the equalizer time span and the delay through the equalizer.

Automatic Synthesis

Before regular data transmission begins, automatic synthesis of the ZF or LMS equalizers for unknown channels may be carried out during a training period. During the training period, a known signal is transmitted and a synchronized version of this signal is generated in the receiver to acquire information about the channel characteristics. The automatic adaptive equalizer is shown in Fig. 64.7. A noisy but unbiased estimate:

$$\frac{\delta e_k^2}{\delta c_n(k)} = 2e_k r(t_0 + kT - nT)$$

is used. Thus, the tap gains are updated according to

$$c_n(k + 1) = c_n(k) - \Delta e_k r(t_0 + kT - nT), \qquad n = 0, 1, \ldots, N - 1$$

where $c_n(k)$ is the nth tap gain at time k, e_k is the error signal, and Δ is a positive adaptation constant or step size, error signals $e_k = z_k - q_k$ can be computed at the equalizer output and used to adjust the equalizer coefficients to reduce the sum of the squared errors. Note $q_k = \hat{x}_k$.

The most popular equalizer adjustment method involves updates to each tap gain during each symbol interval. The adjustment to each tap gain is in a direction opposite to an estimate of the gradient of the MSE with respect to that tap gain. The idea is to move the set of equalizer coefficients closer to the unique optimum set corresponding to the minimum MSE. This symbol-by-symbol procedure developed by Widrow and Hoff [Feher, 1987] is commonly referred to as the *stochastic gradient* method.

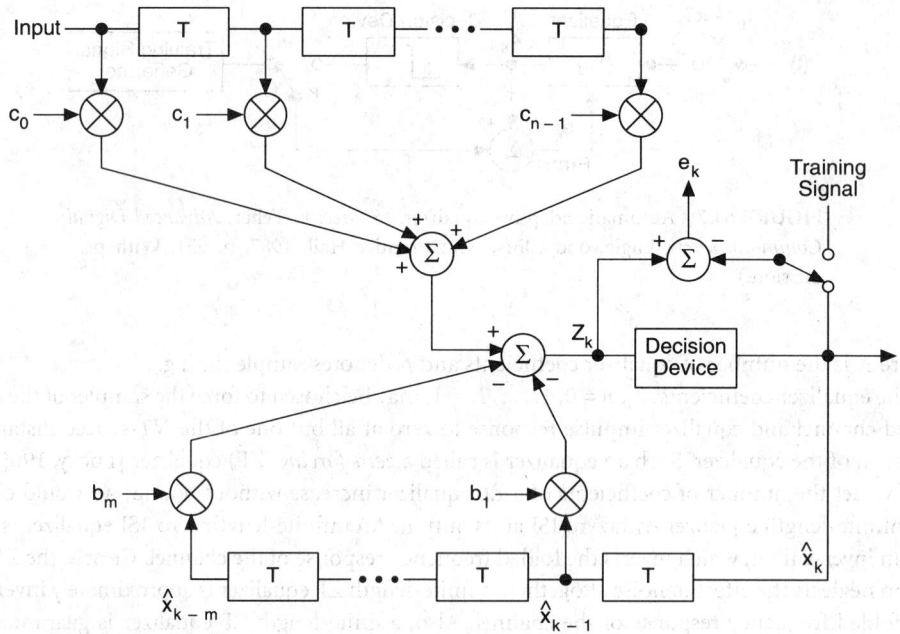

FIGURE 64.8   Decision-feedback equalizer. (*Source:* K. Feher, *Advanced Digital Communications,* Englewood Cliffs, N.J.: Prentice-Hall, 1987, p. 655. With permission.)

Adaptive Equalization

After the initial training period (if there is one), the coefficients of an adaptive equalizer may be continually adjusted in a *decision-directed* manner. In this mode the error signal $e_k = z_k - q_k$ is derived from the final (not necessarily correct) receiver estimate $\{q_k\}$ of the transmitted sequence $\{x_k\}$ where q_k is the estimate of x_k. In normal operation the receiver decisions are correct with high probability, so that the error estimates are correct often enough to allow the adaptive equalizer to maintain precise equalization. Moreover, a decision-directed adaptive equalizer can track slow variations in the channel characteristics or linear perturbations in the receiver front end, such as slow jitter in the sampler phase.

Nonlinear Equalizers

Decision-Feedback Equalizers

A decision-feedback equalizer (DFE) is a simple nonlinear equalizer [Monsen, 1971], which is particularly useful for channels with severe amplitude distortion and uses decision feedback to cancel the interference from symbols which have already been detected. Fig. 64.8 shows the diagram of the equalizer.

The equalized signal is the sum of the outputs of the forward and feedback parts of the equalizer. The forward part is like the linear transversal equalizer discussed earlier. Decisions made on the equalized signal are fed back via a second transversal filter. The basic idea is that if the values of the symbols already detected are known (past decisions are assumed to be correct), then the ISI contributed by these symbols can be canceled exactly, by subtracting past symbol values with appropriate weighting from the equalizer output.

The forward and feedback coefficients may be adjusted simultaneously to minimize the MSE. The update equation for the forward coefficients is the same as for the linear equalizer. The feedback coefficients are adjusted according to

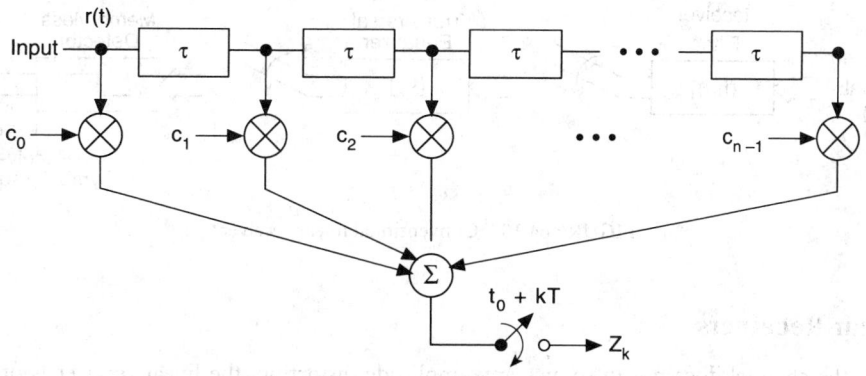

FIGURE 64.9 Fractionally spaced equalizer. (*Source:* K. Feher, *Advanced Digital Communications*, Englewood Cliffs, N.J.: Prentice-Hall, p. 656. With permission.)

$$b_m(k + 1) = b_m(k) + \Delta e_k \hat{x}_{k-m} \qquad m = 1, \ldots, M$$

where $\hat{x}_k$ is the kth symbol decision, $b_m(k)$ is the mth feedback coefficient at time k, and there are M feedback coefficients in all. The optimum LMS settings of b_m, $m = 1, \ldots, M$, are those that reduce the ISI to zero, within the span of the feedback part, in a manner similar to a ZF equalizer.

Fractionally Spaced Equalizers

The optimum receive filter in a linear modulation system is the cascade of a filter matched to the actual channel, with a transversal T-spaced equalizer [Forney, 1972]. The fractionally spaced equalizer (FSE), by virtue of its sampling rate, can synthesize the best combination of the characteristics of an adaptive matched filter and a T-spaced equalizer, within the constraints of its length and delay. A T-spaced equalizer, with symbol-rate sampling at its input, cannot perform matched filtering. A *fractionally spaced equalizer* can effectively compensate for more severe delay distortion and deal with amplitude distortion with less noise enhancement than a T-equalizer.

A fractionally spaced transversal equalizer [Monsen, 1971] is shown in Fig. 64.9. The delay-line taps of such an equalizer are spaced at an interval τ, which is less than, or a fraction of, the symbol interval T. The tap spacing τ is typically selected such that the bandwidth occupied by the signal at the equalizer input is $|f| < 1/2\tau$: that is, τ-spaced sampling satisfies the sampling theorem. In an analog implementation, there is no other restriction on τ, and the output of the equalizer can be sampled at the symbol rate. In a digital implementation τ must be KT/M, where K and M are integers and $M > K$. (In practice, it is convenient to choose $\tau = T/M$, where M is a small integer, e.g., 2.) The received signal is sampled and shifted into the equalizer delay line at a rate M/T, and one input is produced each symbol interval (for every M input sample). In general, the equalizer output is given by

$$z_k = \sum_{n=0}^{N-1} c_n r\left(t_0 + kT - \frac{nKT}{M}\right)$$

The coefficients of a KT/M equalizer may be updated once per symbol based on the error computed for that symbol according to

$$c_n(k + 1) = c_n(k) - \Delta e_k r\left(t_0 + kT - \frac{nKT}{M}\right), \qquad n = 0, 1, \ldots, N - 1$$

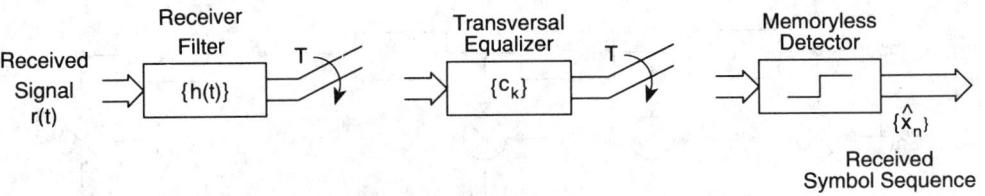

FIGURE 64.10   Conventional linear receiver.

Linear Receivers

When the channel does not introduce any amplitude distortion, the linear receiver is optimum with respect to the ultimate criterion of minimum probability of symbol error. The *conventional linear receiver* consists of a matched filter, a symbol-rate sampler, an infinite-length T-spaced equalizer, and a memoryless detector. The linear receiver structure is shown in Fig. 64.10.

In the conventional linear receiver, a memoryless threshold detector is sufficient to minimize the probability of error; the equalizer response is designed to satisfy the zero-ISI constraint, and the matched filter is designed to minimize the effect of the noise while maximizing the signal.

Matched Filter

The matched filter is the linear filter that maximizes $(S/N)_{out} = s_0^2(t)/n_0^2(t)$ of Fig. 64.11 and has a transfer function given by

$$H(f) = K \frac{S^*(f)}{P_n(f)} e^{-j\omega t_0}$$

where $S(f) = F[s(t)]$ is the Fourier transform of the known input signal $s(t)$ of duration T sec. $P_n(f)$ is the PSD of the input noise, t_0 is the sampling time when $(S/N)_{out}$ is evaluated, and K is an arbitrary real nonzero constant.

A general representation for a matched filter is illustrated in Fig. 64.11. The input signal is denoted by $s(t)$ and the output signal by $s_0(t)$. Similar notation is used for the noise.

Nonlinear Receivers

When amplitude distortion is present in the channel, a memoryless detector operating on the output of this receiver filter no longer minimizes symbol error probability. Recognizing this fact, several authors have investigated optimum or approximately optimum nonlinear receiver structures subject to a variety of criteria [Lucky, 1973].

Decision-Feedback Equalizers

A DFE takes advantage of the symbols that have already been detected (correctly with high probability) to cancel the ISI due to these symbols without noise enhancement. A DFE makes memory-

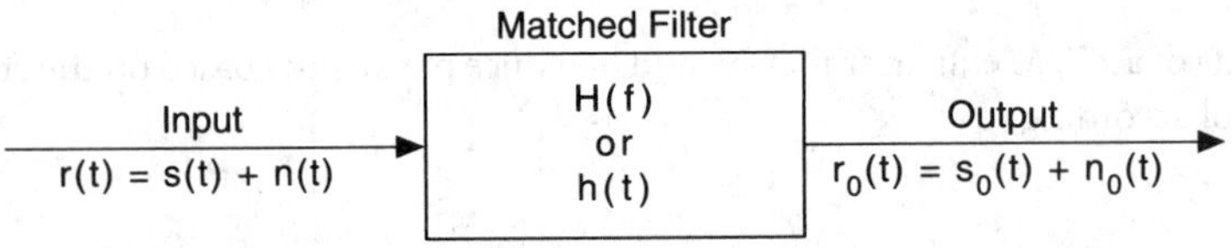

Matched Filter

Input
$r(t) = s(t) + n(t)$

$H(f)$
or
$h(t)$

Output
$r_0(t) = s_0(t) + n_0(t)$

FIGURE 64.11   Matched filter. (*Source:* L. W. Couch, *Digital and Analog Communication Systems*, New York: Macmillan, p. 497, 1990. With permission.)

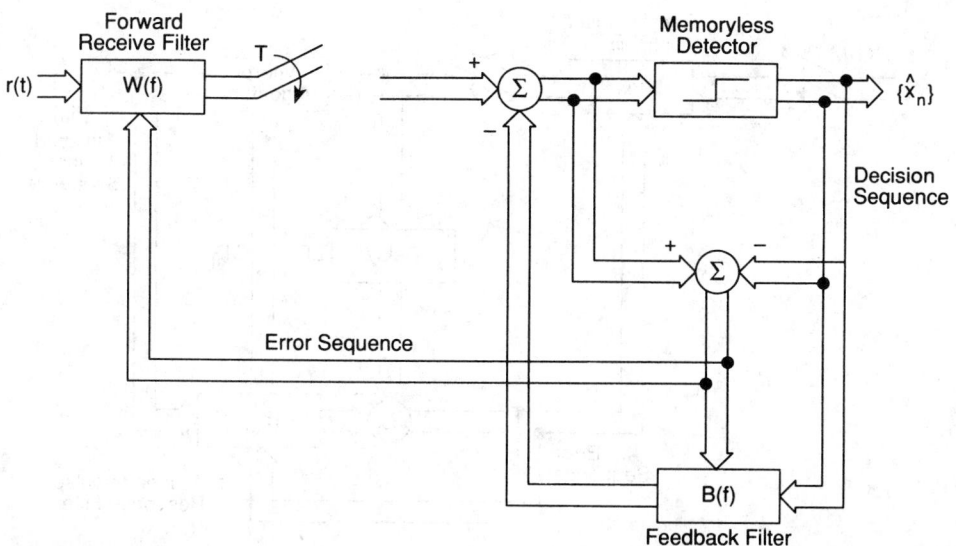

FIGURE 64.12   Conventional decision-feedback receiver. (*Source:* K. Feher, *Advanced Digital Communications*, Englewood Cliffs, N.J.: Prentice-Hall, 1987, p. 675. With permission.)

less decisions and cancels all trailing ISI terms. Even when the whitened matched filter (WMF) is used as the receive filter for the DFE, the DFE suffers from a reduced effective signal-to-noise ratio, and error propagation, due to its inability to defer decisions.

An infinite-length DFE receiver takes the general form (shown in Fig. 64.12) of a forward linear receive filter, symbol-rate sampler, canceler, and memoryless detector. The symbol-rate output of the detector is then used by the feedback filter to generate future outputs for cancellation.

Adaptive Filters for MLSE

For unknown and/or slowly time-varying channels, the receive filter must be adaptive in order to obtain the ultimate performance gain from MLSE (maximum-likelihood sequence estimation). Secondly, the complexity of the MLSE becomes prohibitive for practical channels with a large number of ISI terms. Therefore, in a practical receiver, an adaptive receive filter may be used prior to Viterbi detection to limit the time spread of the channel as well as to track slow time variation in the channel characteristics [Falconer and Magee, 1973].

Several adaptive receive filters are available that minimize the MSE at the input to the Viterbi algorithm. These methods differ in the form of constraint [Falconer and Magee, 1973] on the desired impulse response (DIR) which is necessary in this optimization process to exclude the selection of the null DIR corresponding to no transmission through the channel. The general form of such a receiver is shown in Fig. 64.13.

One such constraint is to restrict the DIR to be causal and to restrict the first coefficient of the DIR to be unity. In this case the delay (LT) in Fig. 64.13 is equal to the delay through the Viterbi algorithm and the first coefficient of $\{b_k\}$ is constrained to be unity.

The least restrictive constraint on the DIR is the unit energy constraint proposed by Falconer and Magee [1973]. This leads to yet another form of the receiver structure as shown in Fig. 64.13. However, the adaptation algorithm for updating the DIR coefficients $\{b_k\}$ is considerably more complicated [Falconer and Magee, 1973]. Note that the fixed predetermined WMF and *T*-spaced prefilter combination of Falconer and Magee [1973] has been replaced in Fig. 64.13 by a general fractionally spaced adaptive filter.

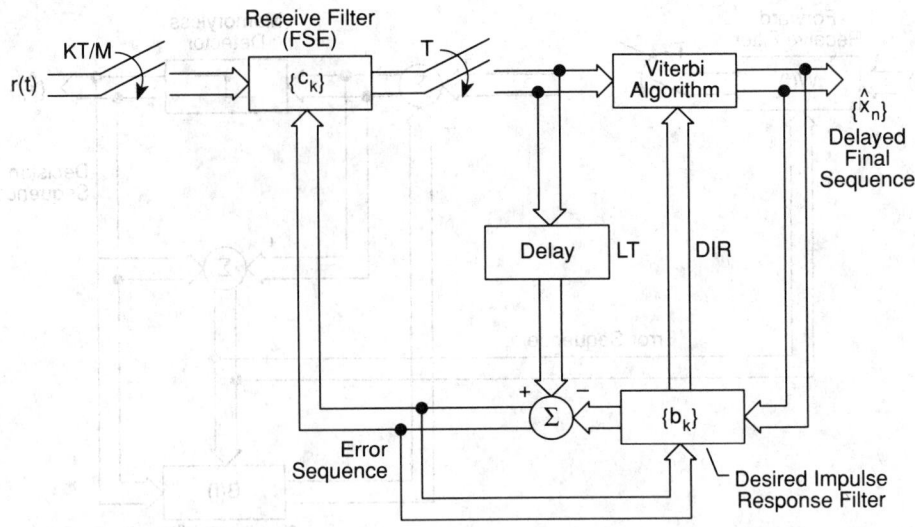

FIGURE 64.13 General form of adaptive MLSE receiver with finite-length DIR. (*Source:* K. Feher, *Advanced Digital Communications*, Englewood Cliffs, N.J.: Prentice-Hall, 1987, p. 684. With permission.)

Defining Terms

Equalizer: A filter used to reduce the effect of intersymbol interference.

Intersymbol interference: The distortion caused by the overlap (in time) of adjacent symbols.

References

L.W. Couch, *Digital and Analog Communication Systems,* New York: Macmillan, 1990.

D.D. Falconer and F.R. Magee, Jr., "Adaptive channel memory truncation for maximum likelihood sequence estimation," *Bell Syst. Technical Journal,* vol. 5, pp. 1541–1562, November 1973.

K. Feher, *Advanced Digital Communications,* Englewood Cliffs, N.J.: Prentice-Hall, 1987.

G.D. Forney, Jr., "Maximum-likelihood sequence estimation of digital sequences in the presence of intersymbol interference," *IEEE Trans. Information Theory,* vol. IT-88, pp. 363–378, May 1972.

R.W. Lucky, "Automatic equalization for digital communication," *Bell Syst. Tech. Journal,* vol. 44, pp. 547–588, April 1965.

R.W. Lucky, "A survey of the communication theory literature: 1968–1973," *IEEE Trans. Information Theory,* vol. 52, pp. 1483–1519, November 1973.

R.W. Lucky, J. Salz, and E.J. Weldon, Jr., *Principles of Data Communication,* New York: McGraw-Hill, 1968.

P. Monsen, "Feedback equalization for fading dispersive channels," *IEEE Trans. Information Theory,* vol. IT-17, pp. 56–64, January 1971.

65

Optical Communication

65.1 Lightwave Technology for Video Transmission 1417
Video Formats and Applications • Intensity Modulation • Noise
Limitations • Linearity Requirements • Laser Linearity • Clipping •
External Modulation • Miscellaneous Impairments • Summary

65.2 Long Distance 1427
Fiber • Modulator • Light Source • Source Coupler • Isolator
• Connectors and Splices • Optical Amplifier • Repeater
• Photodetector • Receiver • Other Components • System
Considerations • Error Rates and Signal-to-Noise Ratio • System
Design

65.3 Photonic Networks 1434
Data Links • Token Ring: FDDI, FFOL • Active Star Networks:
Ethernet, Datakit® • New Approaches to Optical Networks

T. E. Darcie
AT&T Bell Laboratories

Joseph C. Palais
Arizona State University

Ivan P. Kaminow
AT&T Bell Laboratories

65.1 Lightwave Technology for Video Transmission

T. E. Darcie

Lightwave technology has revolutionized the transmission of analog and, in particular, video information. Because the light output intensity from a semiconductor laser is linearly proportional to the injected current, and the current generated in a photodetector is linearly proportional to the incident optical intensity, analog information is transmitted as modulation of the optical intensity. The lightwave system is analogous to a **linear** electrical link, where current or voltage translates linearly into optical intensity. High-speed semiconductor lasers and photodetectors enable intensity-modulation bandwidths greater than 10 GHz. Hence, a wide variety of radio frequency (RF) and microwave applications have been developed [1].

Converting microwaves into intensity-modulated (IM) light allows the use of optical fiber for transmission in place of bulky inflexible coaxial cable or microwave waveguide. Since the fiber attenuation is 0.2–0.4 dB/km, compared with several decibels per meter for waveguide, entirely new applications and architectures are possible. In addition, the signal is confined tightly to the core of single-mode fiber, where it is immune to electromagnetic interference, cross talk, or spectral regulatory control.

To achieve these advantages, several limitations must be overcome. The conversion of current to light intensity must be linear. Several nonlinear mechanisms must be avoided by proper laser design or by the use of various linearization techniques. Also, because the photon energy is much larger than in microwave systems, the signal fidelity is limited by quantum or **shot noise**.

This section describes the basic technology for the transmission of various video formats. We begin by describing the most common video formats and defining transmission requirements for each. Sources of noise, including shot noise, **relative intensity noise** (RIN), and receiver noise are then quantified. Limitations imposed by source nonlinearity, for both **direct modulation** of the

1417

laser bias current and **external modulation** using an interferometric LiNbO$_3$ modulator, are compared. Finally, several other impairments caused by **fiber nonlinearity** or **fiber dispersion** are discussed.

Video Formats and Applications

Each video format represents a compromise between transmission bandwidth and robustness or immunity to impairment. With the exception of emerging digital formats, each is also an entrenched standard that often reflects the inefficiencies of outdated technology.

FM Video

Frequency-modulated (FM) video has served for decades as the basis for satellite video transmission [2], where high signal-to-noise ratios (SNRs) are difficult to achieve. Video information with a bandwidth of $B_v = 4.2$ MHz is used to FM modulate an RF carrier. The resulting channel bandwidth B is given by

$$B \sim \Delta f_{pp} + 2f_m \tag{65.1}$$

where Δf_{pp} is the frequency deviation (22.5 MHz) and f_m is the audio subcarrier frequency (6.8 MHz). As a result of this bandwidth expansion to typically 36 MHz, a high SNR can be obtained for the baseband video bandwidth B_v even if the received carrier-to-noise ratio (CNR) over the FM bandwidth B is small. The SNR is given by

$$SNR = CNR + 10 \log \left[\frac{3B}{2B_v} \left(\frac{\Delta f_{pp}}{B_v} \right) \right] + W + PE \tag{65.2}$$

where W is a weighting factor (13.8 dB) that accounts for the way the eye responds to noise in the video bandwidth, and PE is a pre-emphasis factor (0–5 dB) that is gained by emphasizing the high-frequency video components to improve the performance of the FM modulator. High-quality video (SNR = 55 dB) requires a CNR of only 16 dB. This is achieved easily in a lightwave transmission system.

Applications for lightwave FM video transmission include links to satellite transmission facilities, transport of video between cable television company head-ends (super-trunking), and perhaps delivery of video to subscribers over large fiber distribution networks [3, 4].

AM-VSB Video

The video format of choice, both for broadcast and cable television distribution, is AM-VSB. Each channel consists of an RF carrier that is amplitude modulated (AM) by video information. Single-sideband vestigial (VSB) filtering is used to minimize the bandwidth of the modulated spectrum. The resultant RF spectrum is dominated by the remaining RF carrier, which is reduced by typically 5.6 dB by the AM, and contains relatively low-level signal information, including audio and color subcarriers. An AM-VSB channel requires a bandwidth of only 6 MHz, but CNRs must be at least 50 dB.

For cable distribution, many channels are frequency-division multiplexed (FDM), separated nominally by 6 MHz (8 MHz in Europe), over the bandwidth supported by the coaxial cable. A typical 60-channel cable system operates between 55.25 and 439.25 MHz. Given the large dynamic range required to transmit both the remaining RF carrier and the low-level sidebands, transmission of this multichannel spectrum is a challenge for lightwave technology.

The need for such systems in cable television distribution systems has motivated the development of suitable high-performance lasers. Before the availability of lightwave AM-VSB systems, cable systems used long (up to 20 km) trunks of coaxial cable with dozens of cascaded electronic amplifiers to overcome cable loss. Accumulations of distortion and noise, as well as inherent reliability problems with long cascades, were serious limitations.

Fiber AM-VSB trunk systems can replace the long coaxial trunks so that head-end quality video can be delivered deep within the distribution network [5]. Inexpensive coaxial cable extends from the optical receivers at the ends of the fiber trunks to each home. Architectures in which the number of electronic amplifiers between each receiver and any home is approximately three or fewer offer a good compromise between cost and performance. The short spans of coaxial cable support bandwidths approaching 1 GHz, two or three times the bandwidth of the outdated long coaxial cable trunks. With fewer active components, reliability is improved. The cost of the lightwave components can be small compared to the overall system cost. These compelling technical and economic advantages resulted in the immediate demand for lightwave AM-VSB systems.

Compressed Digital Video

The next generation of video formats will be the product of compressed digital video (CDV) technology [6]. For years digital "NTSC-like" video required a bit rate of approximately 100 Mbps. CDV technology can reduce the required bit rate to less than 5 Mbps. This compression requires complex digital signal processing and large-scale circuit integration, but advances in chip and microprocessor design have made inexpensive implementation of the compression algorithms feasible.

Various levels of compression complexity can be used, depending on the ultimate bit rate and quality required. Each degree of complexity removes different types of redundancy from the video image. The image is broken into blocks of pixels, typically 8×8. By comparing different blocks and transmitting only the differences (DPCM), factors of 2 reduction in bit rate can be obtained. No degradation of quality need result. Much of the information within each block is imperceptible to the viewer. Vector quantization (VQ) or discrete-cosine transform (DCT) techniques can be used to eliminate bits corresponding to these imperceptible details. This intraframe coding can result in a factor of 20 reduction in the bit rate, although the evaluation of image quality becomes subjective. Finally, stationary images or moving objects need not require constant retransmission of every detail. Motion compression techniques have been developed to eliminate these interframe redundancies. Combinations of these techniques have resulted in coders that convert NTSC-like video (100 Mbps uncompressed) into a few megabits per second and HDTV images (1 Gbps uncompressed) into less than 20 Mbps.

CDV can be transmitted using time-division multiplexing (TDM) and digital lightwave systems or by using each channel to modulate an RF carrier and transmitting using analog lightwave systems. There are numerous applications for both alternatives. TDM systems for CDV are no different from any other digital transmission system and will not be discussed further.

Using RF techniques offers an additional level of RF compression, wherein advanced multilevel modulation formats are used to maximize the number of bits per hertz of bandwidth [7]. Quadrature-amplitude modulation (QAM) is one example of multilevel digital-to-RF conversion. For example, 64-QAM uses 8 amplitude and 8 phase levels and requires only 1 Hz for 5 bits of information. As the number of levels, hence the number of bits per hertz, increases, the CNR of the channel must increase to maintain error-free transmission. A 64-QAM channel requires a CNR of approximately 30 dB.

A synopsis of the bandwidth and CNR requirements for FM, AM-VSB, and CDV is shown in Fig. 65.1. AM-VSB requires high CNR but low bandwidth. FM is the opposite. Digital video can occupy a wide area, depending on the degree of digital and RF compression. The combination of CDV and QAM offers the possibility of squeezing a high-quality video channel into 1 MHz of bandwidth, with a required CNR of 30 dB. This drastic improvement over AM-VSB or FM could have tremendous impact on future video transmission systems.

Intensity Modulation

As mentioned in the introduction, the light output from the laser should be linearly proportional to the injected current. The laser is prebiased to an average output intensity L_0. Many video chan-

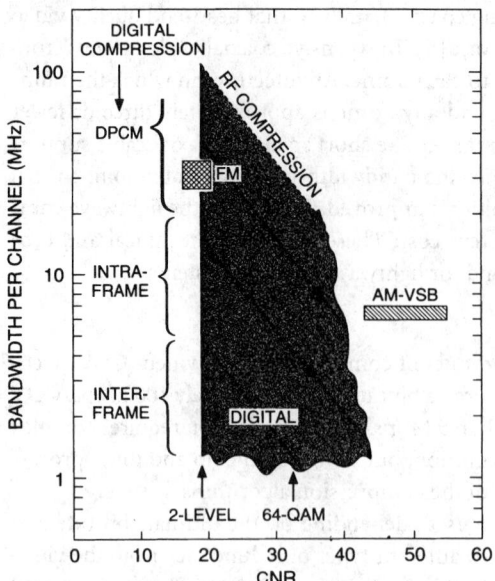

FIGURE 65.1 Bandwidth versus carrier-to-noise ratio (CNR) required for AM-VSB, FM, and digital video. Increasingly complex digital compression techniques reduce the bit rate required for NTSC-like video from 100 Mbps to less than 5 Mbps. Bandwidth efficient RF techniques like QAM minimize the bandwidth required for each bit rate but require greater CNRs.

nels are combined electronically, and the total RF signal is added directly to the laser current. The optical modulation depth (m) is defined as the ratio of the peak modulation L_0 for one channel, divided by L_0. For 60-channel AM-VSB systems, m is typically near 4%.

The laser (optical carrier) is modulated by the sum of the video channels that are combined to form the total RF signal spectrum. The resultant optical spectrum contains sidebands from the IM superimposed on unintentional frequency modulation, or **chirp**, that generally accompanies IM. This complex optical spectrum must by understood if certain subtle impairments are to be avoided.

A photodetector converts the incident optical power into current. Broadband InGaAs photodetectors with responsivities (R_0) of nearly 1.0 A/W and bandwidths greater than 10 GHz are available. The detector generates a dc current corresponding to the average received optical power L_r and the complete RF modulation spectrum that was applied at the transmitter. An ac-coupled electronic preamplifier is used to remove the dc component and boost the signal to usable levels.

Noise Limitations

The definition of CNR deserves clarification. Depending on the video format and RF modulation technique, the RF power spectrum of the modulated RF carrier varies widely. For AM-VSB video the remaining carrier is the dominant feature in the spectrum. It is thereby convenient to define the CNR as the ratio of the power remaining in the carrier to the integrated noise power in a 4-MHz bandwidth centered on the carrier frequency. For FM or digitally modulated carriers, the original carrier is not generally visible in the RF spectrum. It is then necessary to define the CNR as the ratio of the integrated signal power within the channel bandwidth to the integrated noise power.

Shot Noise

Shot noise is a consequence of the statistical nature of the photodetection process. It results in a noise power spectral density, or electrical noise power per unit bandwidth (dBm/Hz) that is proportional to the received photocurrent I_r ($= R_0 L_r$). The total shot noise power in a bandwidth B is given by

$$N_s = 2eI_rB \tag{65.3}$$

where e is the electronic charge.

With small m, the detected signal current is a small fraction of the total received current. The root mean square (rms) signal power for one channel is

$$P_s = \frac{1}{2}(mI_r)^2 \tag{65.4}$$

The total shot noise power then limits the CNR (P_s/N_s) to a level referred to as the quantum limit.

Received powers near 1 mW are required if CNRs greater than 50 dB are to be achieved for 40- to 80-channel AM-VSB systems.

Receiver Noise

Receiver noise is generated by the electronic amplifier used to boost the detected photocurrent to usable levels. The easiest receiver to build consists of a *pin* photodiode connected directly to a low-noise 50- to 75-Ω amplifier, as shown in Fig. 65.2(a). The effective input current noise density, (n), for this simple receiver is given by

$$n^2 = \frac{4kTF}{R_L} \tag{65.5}$$

where k is the Boltzmann constant, T is the absolute temperature, F is the **noise figure** of the amplifier, and R_L is the input impedance. For a 50-Ω input impedance and $F = 2$, $n = 20$ pA/$\sqrt{Hz}$.

A variety of more complicated receiver designs can reduce the noise current appreciably [8]. The example shown in Fig. 65.2(b) uses a high-speed FET. R_L can be increased to maximize the voltage developed by the signal current at the FET input. Input capacitance becomes a limitation by shunting high-frequency components of signal current. High-frequency signals are then reduced with respect to the noise generated in the FET, resulting in poor high-frequency performance. Various impedance matching techniques have been proposed to maximize the CNR for specific frequency ranges.

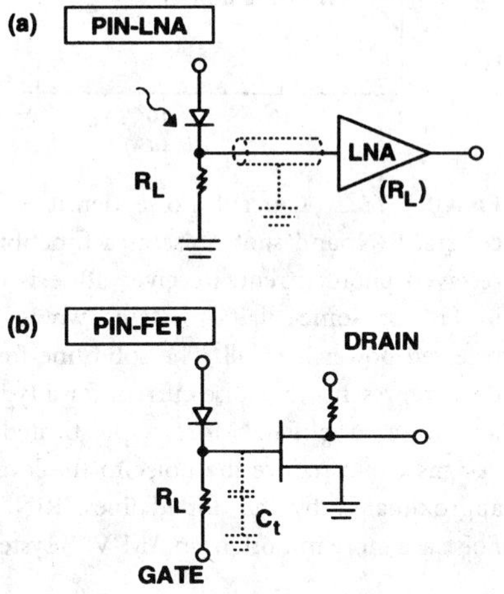

Relative Intensity Noise

Relative intensity noise (RIN) can originate from the laser or from reflections and **Rayleigh backscatter** in the fiber. In the laser, RIN is caused by spontaneous emission in the active layer. Spontaneous emission drives random fluctuations in the number of photons in the laser which appear as a random modulation of the output intensity, with frequency components extending to tens of gigahertz. The noise power spectral density from RIN is $I_r^2 \text{RIN}$, where RIN is expressed in decibels per hertz.

FIGURE 65.2 Receivers for broadband analog lightwave systems. Coupling a *pin* to a low-noise amplifier (a) is simple, but improved performance can be obtained using designs like the *pin* FET (b). C_t is the undesirable input capacitance.

RIN is also caused by component reflections and double-Rayleigh backscatter in the fiber, by a process called multipath interference. Twice-reflected signals arriving at the detector can interfere coherently with the unreflected signal. Depending on the modulated optical spectrum of the laser, this interference results in noise that can be significant [9].

The CNR, including all noise sources discussed, is given by

$$\text{CNR} = \frac{m^2 I_r^2}{2B[n^2 + 2eI_r + I_r^2 \text{RIN}]} \tag{65.6}$$

All sources of intensity noise are combined into RIN. Increasing m improves the CNR but increases the impairment caused by nonlinearity, as discussed in the following subsection. The optimum operating value for m is then a balance between noise and distortion.

Figure 65.3 shows the noise contributions from shot noise, receiver noise, and RIN. For FM or digital systems, the low CNR values required allow operation with small received optical powers.

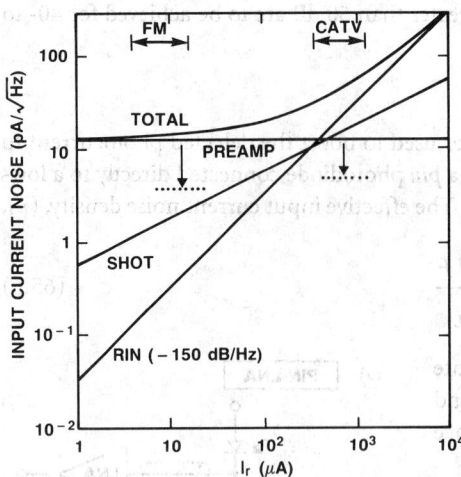

FIGURE 65.3 Current noise densities from receivers, RIN, and shot noise as a function of total received photocurrent. Receiver noise is dominant in FM or some digital systems where the total received power is small. The solid line for receiver noise represents the noise current for a typical 50-Ω low-noise amplifier. More sophisticated receiver designs could reduce the noise to the levels shown approximately by the dotted lines. RIN and shot noise are more important in AM-VSB systems.

Receiver noise is then generally the limiting factor. Much larger received powers are required if AM-VSB noise requirements are to be met. Although detecting more optical power helps to overcome shot and receiver noise, the ratio of signal to RIN remains constant. RIN can be dominant in high-CNR systems, when the received power is large. AM-VSB systems require special care to minimize all sources of RIN. The dominant noise source is then shot noise, with receiver noise and RIN combining to limit CNRs to within a few decibels of the quantum limit.

Linearity Requirements

Source linearity limits the depth of modulation that can be applied. Linearity, in this case, refers to the linearity of the current-to-light-intensity (*I-L*) conversion in the laser or voltage-to-light (*V-L*) transmission for an external modulator. Numerous nonlinear mechanisms must be considered for direct modulation, and no existing external modulator has a linear transfer function.

A Taylor-series expansion of the *I-L* or *V-L* characteristic, centered at the bias point, results in linear, quadratic, cubic, and higher-order terms. The linear term describes the efficiency with which the applied signal is converted to linear intensity modulation. The quadratic term results in second-order distortion, the cubic produces third-order distortion, and so on.

Requirements on linearity can be derived by considering the number and spectral distribution of the distortion products generated by the nonlinear mixing between carriers in the multichannel signal. Second-order nonlinearity results in sum and difference ($f_i \pm f_j$) mixing products for every combination of the two channels. This results in as many as 50 second-order products within a single channel, in a 60-channel AM-VSB system with the standard U.S. frequency plan. Similarly, for third-order distortion, products result from mixing among all combinations of three channels. However, since the number of combinations of three channels is much larger than for two, up to 1130 third-order products can interfere with one channel. The cable industry defines the **composite second-order** (**CSO**) distortion as the ratio of the carrier to the largest group of second-order products within each channel. For third-order distortion, the **composite triple beat** (**CTB**) is the ratio of the carrier to the total accumulation of third-order distortion at the carrier frequency in each channel.

The actual impairment from these distortion products depends on the spectrum of each RF channel and on the exact frequency plan used. A typical 42-channel AM-VSB frequency plan, with carrier frequencies shown as the vertical bars on Fig. 65.4, results in the distributions of second- and third-order products shown in Fig. 65.4(a) and (b), respectively. Since the remaining carrier is the dominant feature in the spectrum of each channel, the distortion products are dominated by the mixing between these carriers. Because high-quality video requires that the CSO is −60 dBc (dB relative to the carrier), each sum or difference product must be less than −73 dBc. Likewise, for the CTB to be less than 60 dB, each product must be less than approximately −90dB.

FM or CDV systems have much less restrictive linearity requirements, because of the reduced

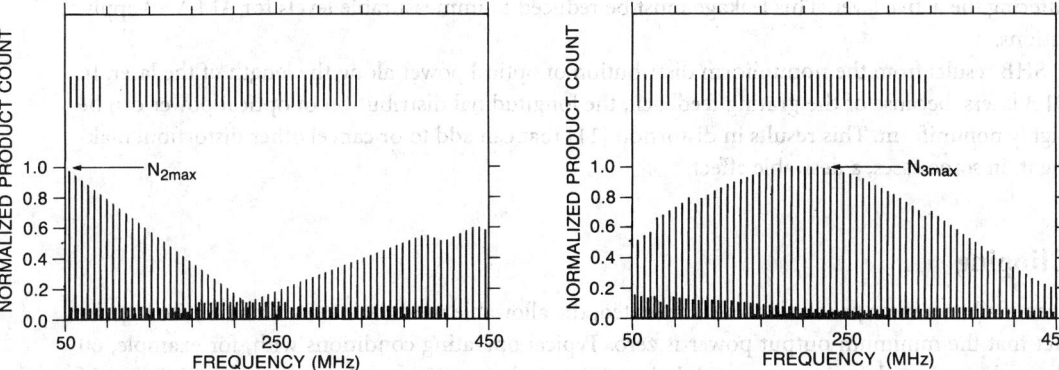

FIGURE 65.4 Second-order (a) and third-order (b) distortion products for 42-channel AM-VSB system. The maximum number of second-order products occurs at the lowest frequency channel, where 30 products contribute to the CSO. The maximum number of third-order products occurs near the center channel, where 530 products contribute to the CTB.

sensitivity to impairment. Distortion products must be counted, as with the AM-VSB example described previously, but each product is no longer dominated by the remaining carrier. Because the carrier is suppressed entirely by the modulation, each product is distributed over more than the bandwidth of each channel. The impairment resulting from the superposition of many uncorrelated distortion products resembles noise. Quantities analogous to the CSO and CTB can be defined for these systems.

Laser Linearity

Several factors limit the light-versus-current (*L-I*) linearity of directly modulated lasers. Early work on laser dynamics led to a complete understanding of resonance-enhanced distortion (RD). RD arises from the same carrier-photon interaction within the laser that is responsible for the relaxation-oscillation resonance.

The second-harmonic distortion ($2f_i$) and two-tone third-order distortion ($2f_i - f_j$) for a typical 1.3-μm wavelength directly modulated semiconductor laser are shown in Fig. 65.5 [10]. Both distortions are small at low frequencies but rise to maxima at half the relaxation resonance frequency. AM-VSB systems are feasible only within the low-frequency window. FM or uncompressed digital systems require enough bandwidth per channel that multichannel systems must operate in the region of large RD. Fortunately, the CNR requirements allow for the increased distortion. The large second-order RD can be avoided entirely by operating within a one-octave frequency band (e.g., 2–4 GHz), such that all second-order products are out of band.

Within the frequency range between 50 and 500 MHz, nonlinear gain and loss, intervalence-band absorption, and, more importantly, spatial-hole burning (SHB) and carrier leakage can all be significant. Carrier leakage prevents all of the current injected in the laser bond wire from

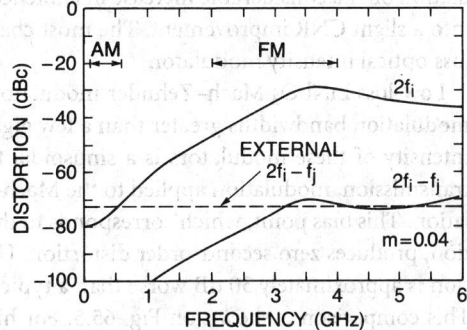

FIGURE 65.5 Resonance distortion for directly modulated laser with resonance frequency of 7 GHz. Both the second-harmonic $2f_i$ and two-tone third-order $2f_i \pm f_j$ distortion peak near half the resonance frequency and are small at low frequency. Also shown is the same third-order distortion for an external modulator biased at the point of zero second-order distortion.

entering the active layer. This leakage must be reduced to immeasurable levels for AM-VSB applications.

SHB results from the nonuniform distribution of optical power along the length of the laser. In DFB lasers, because of the grating feedback, the longitudinal distribution of optical power can be highly nonuniform. This results in distortion [11] that can add to or cancel other distortion, making it, in some cases, a desirable effect.

Clipping

Even if all nonlinear processes were eliminated, the allowable modulation would be limited by the fact that the minimum output power is zero. Typical operating conditions with, for example, 60 channels, each with an average modulation depth (m) near 4%, result in a peak modulation of 240%. Although improbable, modulations of more than 100% result in clipping.

The effects of clipping were first approximated by Saleh [12], who calculated the modulation level at which the total power contained in all orders of distortion became appreciable. Even for perfectly linear lasers, the modulation depth is bounded to values beyond which all orders of distortion increase rapidly. Assuming that half the total power in all orders of distortion generated by clipping is distributed evenly over each of N channels, clipping results in a carrier-to-interference ratio (CIR) given by

$$\mathrm{CIR} = \sqrt{2\pi}\, \frac{(1 + 6\mu^2)}{\mu^3}\, e^{1/2\mu^2} \tag{65.7}$$

where the rms modulation index μ is

$$\mu = m\,\sqrt{N/2} \tag{65.8}$$

External Modulation

Laser-diode-pumped YAG lasers with low RIN and output powers greater than 200 mW have been developed recently. Combined with linearized external LiNbO$_3$ modulators, these lasers have become high-performance competitors to directly modulated lasers. YAG lasers with external modulation offer a considerable increase in launched power, and the low RIN of the YAG laser translates into a slight CNR improvement. The most challenging technical hurdle is to develop a linear low-loss optical intensity modulator.

Low-loss LiNbO$_3$ Mach–Zehnder modulators are available with insertion losses less than 3 dB, modulation bandwidths greater than a few gigahertz, and switching voltages near 5 V. The output intensity of these modulators is a sinusoidal function of the bias voltage. By prebiasing to 50% transmission, modulation applied to the Mach–Zehnder results in the most linear intensity modulation. This bias point, which corresponds to the point of inflection in the sinusoidal transfer function, produces zero second-order distortion. Unfortunately, the corresponding third-order distortion is approximately 30 dB worse than a typical directly modulated DFB laser, at low frequencies. This comparison is shown on Fig. 65.5. For high-frequency applications where RD is important, external modulators can offer improved linearity. A means of linearizing the third-order nonlinearity is essential for AM-VSB applications.

Various linearization techniques have been explored. The two most popular approaches are feedforward and predistortion. Feedforward requires that a portion of the modulated output signal be detected and compared to the original applied voltage signal to provide an error signal. This error signal is then used to modulate a second laser, which is combined with the first laser such that the total instantaneous intensity of the two lasers is a replica of the applied voltage. In principle, this technique is capable of linearizing any order of distortion and correcting RIN from the laser.

Predistortion requires less circuit complexity than feedforward. A carefully designed nonlinear circuit is placed before the nonlinear modulator, such that the combined transfer function of the predistorter-modulator is linear. Various nonlinear electronic devices or circuits can act as second- or third-order predistorters. Difficulties include matching the frequency dependence of the predistorter with that of the modulator, hence achieving good linearity over a wide frequency range. Numerous circuit designs can provide reductions in third-order distortion by 15 dB.

Miscellaneous Impairments

Laser chirp can cause problems with direct laser modulation. Chirp is modulation of the laser frequency caused by modulation of the refractive index of the laser cavity in response to current modulation. The interaction of chirp and chromatic dispersion in the fiber can cause unacceptable CSO levels for AM-VSB systems as short as a few kilometers. Dispersion converts the FM into IM, which mixes with the signal IM to produce second-order distortion [13]. These systems must operate at wavelengths corresponding to low fiber dispersion, or corrective measures must be taken.

Chirp also causes problems with any optical component that has a transmission that is a function of optical frequency. This can occur if two optical reflections conspire to form a weak interferometer or in an **erbium-doped fiber amplifier** (EDFA) that has a frequency-dependent gain [14]. Once again, the chirp is converted to IM, which mixes with the signal IM to form second-order distortion.

Although externally modulated systems are immune to chirp-related problems, fiber nonlinearity, in the form of stimulated Brillouin scattering (SBS), places a limit on the launched power. SBS, in which light is scattered from acoustic phonons in the fiber, causes a rapid decrease in CNR for launched powers greater than approximately 10 mW [15]. Since the SBS process requires high optical powers within a narrow optical spectral width (20 MHz), it is a problem only in low-chirp externally modulated systems. Chirp in DFB systems broadens the optical spectrum so that SBS is unimportant.

Summary

A wide range of applications for transmission of video signals over optical fiber has been made possible by refinements in lightwave technology. Numerous technology options are available for each application, each with advantages or disadvantages that must be considered in context with specific system requirements. Evolution of these video systems continues to be driven by development of new and improved photonic devices.

Defining Terms

Chirp: Modulation of the optical frequency that occurs when a laser is intensity modulated.

Composite second order (CSO): Ratio of the power in the second-order distortion products to power in the carrier in a cable television channel.

Composite triple beat (CTB): Same as CSO but for third-order distortion.

Direct modulation: Modulation of the optical intensity output from a semiconductor diode laser by direct modulation of the bias current.

Erbium-doped fiber amplifier: Fiber doped with erbium that provides optical gain at wavelengths near 1.55 μm when pumped optically at 0.98 or 1.48 μm.

External modulation: Modulation of the optical intensity using an optical intensity modulator to modulate a constant power (cw) laser.

Fiber dispersion: Characteristic of optical fiber by which the propagation velocity depends on the optical wavelength.

Fiber nonlinearity: Properties of optical fibers by which the propagation velocity, or other characteristic, depends on the optical intensity.

Lightwave technology: Technology based on the use of optical signals and optical fiber for the transmission of information.

Linear: Said of any device for which the output is linearly proportional to the input.

Noise figure: Ratio of the output signal-to-noise ratio (SNR) to the input SNR in an amplifier.

Rayleigh backscatter: Optical power that is scattered in the backwards direction by microscopic inhomogeneities in the composition of optical fibers.

Relative intensity noise: Noise resulting from undesirable fluctuations of the optical power detected in an optical communication system.

Shot noise: Noise generated by the statistical nature of current flowing through a semiconductor *p-n* junction or photodetector.

References

1. T.E. Darcie, "Subcarrier multiplexing for lightwave networks and video distribution systems," *IEEE J. Selected Areas in Communications,* vol. 8, p. 1240, 1990.

2. T. Pratt and C.W. Bostian, *Satellite Communications,* New York: Wiley, 1986.

3. W. Way, C. Zah. C. Caneau, S. Menmocal, F. Favire, F. Shokoochi, N. Cheung, and T.P. Lee, "Multichannel FM video transmission using traveling wave amplifiers for subscriber distribution," *Electron. Lett.,* vol. 24, p. 1370, 1988.

4. R. Olshansky, V. Lanzisera, and P. Hill, "Design and performance of wideband subcarrier multiplexed lightwave systems," in *Proc. ECOC '88,* Brighton, U.K., Sept. 1988, pp. 143–146.

5. J.A. Chiddix, H. Laor, D.M. Pangrac, L.D. Williamson, and R.W. Wolfe, "AM video on fiber in CATV systems, need and implementation," *IEEE J. Selected Areas in Communications,* vol. 8, p. 1229, 1990.

6. A.N. Netravali and B.G. Haskel, *Digital Pictures,* New York: Plenum Press, 1988.

7. K. Feher, Ed., *Advanced Digital Communications,* Englewood Cliffs, N.J.: Prentice-Hall, 1987.

8. B.L. Kasper, "Receiver design," in *Optical Fiber Telecommunications II,* S.E. Miller and I.P. Kaminow, Eds., San Diego: Academic Press, 1988.

9. T.E. Darcie, G.E. Bodeep, and A.A.M. Saleh, "Fiber-reflection-induced impairments in lightwave AM-VSB CATV systems," *IEEE J. Lightwave Technol.,* vol. 9, no. 8, pp. 991–995, Aug. 1991.

10. T.E. Darcie, R.S. Tucker, and G.J. Sullivan, "Intermodulation and harmonic distortion in IaGaAsP lasers," *Electron. Lett.,* vol. 21, 665–666, erratum; vol 22, p. 619, 1986.

11. A. Takemoto, H. Watanabe, Y. Nakajima, Y. Sakakibara, S. Kakimoto, U. Yamashita, T. Hatta, and Y. Miyake, "Distributed feedback laser diode and module for CATV systems," *IEEE J. Selected Areas in Communications,* vol. 8, 1359, 1990.

12. A.A.M. Saleh, "Fundamental limit on number of channels in subcarrier mulitplexed lightwave CATV systems," *Electron. Lett.,* vol. 25, no. 12, pp. 776–777, 1989.

13. M.R. Phillips, T.E. Darcie, D. Marcuse, G.E. Bodeep, and N.J. Frigo, "Nonlinear distortion generated by dispersive transmission of chirped intensity-modulated signals," *IEEE Photonics Technol. Lett.,* vol. 3, no. 5, pp. 481–483, 1991.

14. C.Y. Kuo and E.E. Bergmann, "Erbium-doped fiber amplifier second-order distortion in analog links and electronic compensation," *IEEE Photonics Technol. Lett.,* vol. 3, p. 829, 1991.

15. X.P. Mao, G.E. Bodeep, R.W. Tkach, A.R. Chraplyvy, T.E. Darcie, and R.M. Derosier, "Brillouin scattering in lightwave AM-VSB CATV transmission systems," *IEEE Photonics Technol. Lett.,* vol. 4, no. 3, pp. 287–289, 1991.

Further Information

National Cable Television Association (NCTA), Proceedings from Technical Sessions, annual meetings, 1724 Massachusetts Ave. NW, Washington D.C., 20036, 1969.

Society of Cable Television Engineers (SCTE), Proceeding from Technical Sessions, biennial meetings, Exton Commons, Exton, Penn.

T.E. Darcie, "Subcarrier multiplexing for lightwave multiple-access lightwave networks," *J. Lightwave Technol.*, vol. LT-5, pp. 1103–1110, Aug. 1987.

T.E. Darcie and G.E. Bodeep, "Lightwave subcarrier CATV transmission systems," *IEEE Trans. Microwave Theory and Technol.*, vol. 38, no. 5, pp. 534–533, May 1990.

IEEE J. Lightwave Technol., Special Issue on "Broadband Analog Video Transmission Over Fibers," to be published Jan./Feb. 1993.

65.2 Long Distance

Joseph C. Palais

When the first laser was demonstrated in 1960, numerous applications of this magnificent new tool were anticipated. Some predicted that laser beams would transmit messages through the air at high data rates between distant stations. Although laser beams can indeed travel through the atmosphere, too many problems prevent this scheme from becoming practical. Included in the objections are the need for line-of-sight paths and the unpredictability of transmission through an atmosphere where weather variations randomly change path losses. Guided paths using optical fibers offer the only practical means of optical transmission over long distances.

Long-distance fiber systems tend to have the following operational characteristics: They are more than 10 km long, transmit digital signals (rather than analog), and operate at data rates above a few tens of megabits per second. This section primarily describes systems in this category.

Figure 65.6 illustrates the basic structure of a generalized long-distance fiber optic link. Each of the components will be described in the following paragraphs.

A useful figure of merit for these systems is the product of the system data rate and its length. This figure of merit is the well-known *rate-length product*. The bandwidth of the transmitting and receiving circuits (including the light source and photodetector) limits the achievable system data rate. The bandwidth of the fiber decreases with its length, so that the fiber itself limits the rate-length product. The losses in the system, including those in the fiber, also limit the path length. Systems are *bandwidth limited* if the rate-length figure is determined by bandwidth restraints and *loss limited* if determined by attenuation.

The first efficient fiber appeared in 1970, having a loss of 20 dB/km. Just 7 years later the first large-scale application, a link between two telephone exchanges in Chicago, was constructed. By

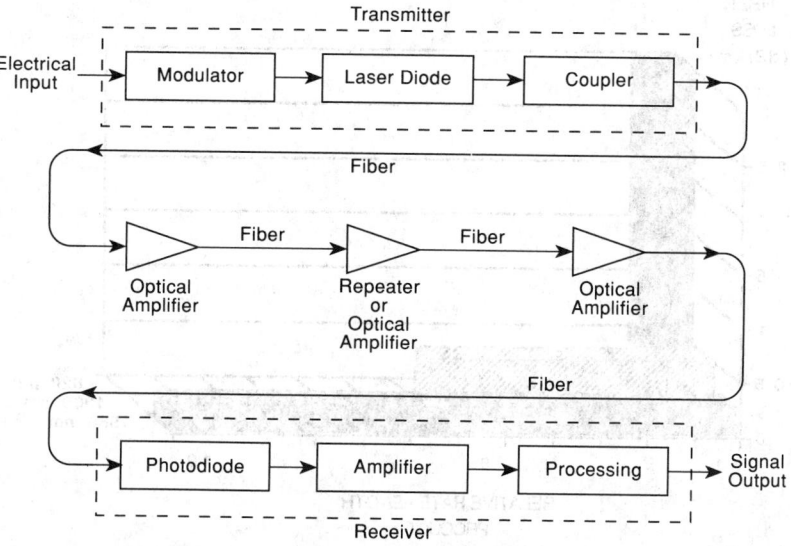

FIGURE 65.6 Long-distance fiber communication system.

this time the loss had been reduced to around 3 dB/km. The digital technology used could accommodate a rate of 45 Mbps over an unrepeatered length of 10 km and a total length of over 60 km with repeaters. The unrepeatered rate-length product for this initial system was a modest 0.5 Gbps × km. As fiber technology advanced, this figure steadily increased. Unrepeatered rate-length products have improved to 500 Gbps × km (e.g., 8 Gbps over a path of 60 km) and beyond. Allowing repeaters and/or optical amplifiers increases the net rate-length product considerably. Values beyond 70 Tbps × km (70,000 Gbps × km) are achievable with optical amplifiers. This latter figure allows construction of a transmission system operating at 5 Gbps over a 14,000-km path. The longest terrestrial paths are across the Atlantic and Pacific oceans, distances of about 6,000 and 9,000 km, respectively. Fibers are capable of spanning these distances with high-capacity links.

Fiber

All fibers used for long-distance communications are made of silica glass and allow only a single mode of propagation. The silica is doped with other materials to produce the required refractive index variations for the fiber core and cladding. The important fiber characteristics that limit system performance are its loss and its bandwidth. The loss limits the length of the link and the bandwidth limits the data rate.

Figure 65.7 shows the loss characteristics of single-mode silica fibers at the wavelengths of lowest attenuation. As indicated in the figure, there are three possible windows of operation. In the first window (around 820 nm), the loss is typically 3 dB/km. This is too high for long systems. In the second window (near 1300 nm), the loss is about 0.5 dB/km. In addition, the fiber bandwidth is quite high because of low pulse dispersion at this wavelength. The second window is a reasonable operating wavelength for high-capacity, long-distance systems. At 1550 nm (the third window) the loss is lowest, about 0.25 dB/km. This characteristic makes 1550 nm the optimum choice for the very longest links.

Dispersion refers to the spreading of a pulse as it travels along a **single-mode fiber**. It is due to material and waveguide effects. This spreading creates intersymbol interference if allowed to exceed about 70% of the original pulse width, causing receiver errors. The dispersion factor M is usually given in units of picoseconds of pulse spread per nanometer of spectral width of the light source and per kilometer of length of fiber.

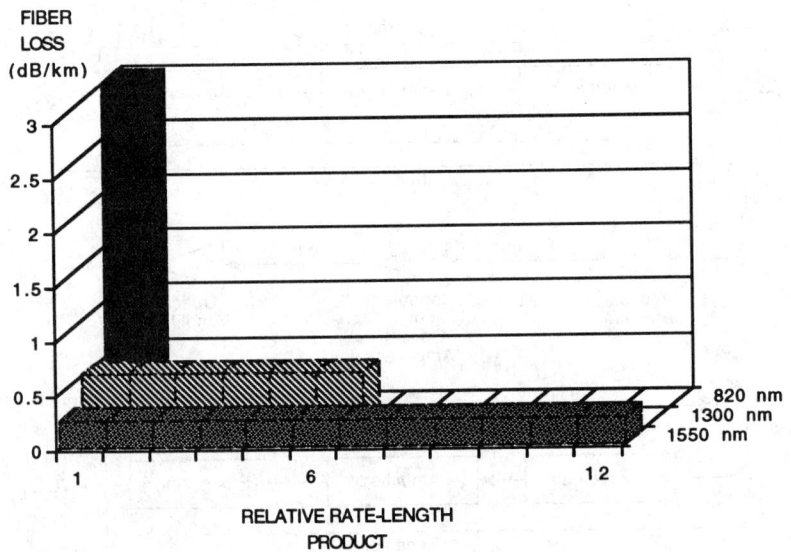

FIGURE 65.7 Fiber loss and relative unrepeatered, unamplified rate-length product.

In the range from 1200 to 1600 nm, the dispersion curve for silica can be approximated by the expression

$$M = \frac{M_0}{4}\left(\lambda - \frac{\lambda_0^4}{\lambda^3}\right) \tag{65.9}$$

where λ is the operating wavelength, λ_0 is the zero dispersion wavelength, and M_0 is the slope at the zero dispersion wavelength. M_0 is approximately 0.095 ps/(nm² × km). The pulse spread for a path length L, using a light source whose spectral width is $\Delta\lambda$, is then

$$\Delta\tau = ML\Delta\lambda \tag{65.10}$$

The zero dispersion wavelength, close to 1300 nm for silica fibers, makes this wavelength attractive for high-capacity links. The dispersion at 1550 nm is typically close to 20 ps/(nm × km). This is a moderate amount of dispersion. If a proposed 1550-nm system is bandwidth limited because of this spread, several alternatives are available. One solution is to use dispersion-shifted fiber, which is a special fiber with a refractive index profile designed to shift the zero dispersion wavelength from 1300 nm to 1550 nm. Another solution is to transmit soliton pulses, which use the nonlinearity of the fiber to maintain pulse shape during transmission.

Figure 65.7 includes relative unrepeatered, unamplified values of rate-length products in the three transmission windows. Because of high loss, the first window can be used only for moderate lengths (around 10 km). Because of high dispersion, the data rates are also limited in this region. In the second window, nearly zero dispersion allows high-rate transmission, but the losses limit the distance that can be covered (typically around 50 km). In the third window, the loss is about half the 1300-nm attenuation so that twice as much distance can be covered. Dispersion-shifted fiber allows the same high rates as does 1300-nm operation. Repeaters and amplifiers extend the useful distance of fiber links well beyond the distances listed here.

Modulator

A digital electrical signal modulates the light source. The driver circuit must be fast enough to operate at the system bit rate. As bit rates increase into the multigigabit per second range, this becomes increasingly difficult. Modulation can be done in the optical domain at very high speeds. In this case, the modulator follows the laser diode rather than preceding it. External modulation is usually accomplished using integrated-optic structures.

Light Source

Laser diodes or light-emitting diodes (LEDs) supply the optical carrier waves for most fiber links. LEDs cannot operate at speeds in the gigabit range, but laser diodes can. For this reason, laser diodes are normally required for high-rate, long-distance links. Laser diodes can be modulated at frequencies beyond 40 GHz.

Laser diodes emitting in the second and third fiber transmission windows are semiconductor heterojunctions made of InGaAsP. The exact emission wavelength is primarily determined by the proportions of the constituent atoms. Output powers are commonly on the order of a few milliwatts.

Typical laser diode **spectral widths** are between 1 and 5 nm when operating in more than one longitudinal mode. Single-mode laser diodes can have spectral widths of just a few tenths of a nanometer. As predicted by Eq. (65.10), narrow-spectral-width emitters minimize pulse spreading. Minimizing pulse spreading increases the fiber bandwidth and its data capacity.

Solid-state lasers other than semiconductor laser diodes may be useful in specific applications. Example of such lasers are the Nd:YAG laser and the erbium-doped fiber laser.

Source Coupler

The light emitted from the diode must be coupled as efficiently as possible into the fiber. Because the beam pattern emitted by a laser diode does not perfectly match the pattern of light propagating in the fiber, there is an inevitable mismatch loss. Good coupler designs, sometimes using miniature lenses, reduce this loss to about 3 dB when feeding a single-mode fiber.

Isolator

An optical isolator is a one-way transmission path. It allows power flow from the transmitter toward the receiver but blocks power flow in the opposite direction. It is used to protect the laser diode from back reflections, which tend to increase the laser noise.

Connectors and Splices

Connections between fibers and between the fiber and other components occur at numerous points in a long-distance link. Because there may be many splices in a long system, the loss of each one must be small. Fusion splices with an average loss of no more than 0.05 dB are often specified. Mechanical splices are also suitable. They often involve epoxy for fixing the connection. Connectors are used where remateable connections are required. Good fiber connectors introduce losses of just a few tenths of a decibel.

Optical Amplifier

Many fiber links are loss limited. One cause is the limited power available from the typical laser diode, which (together with the losses in the fiber and the other system components) restricts the length of fiber that can be used. The *fiber optic amplifier* increases the power level of the signal beam without conversion to the electrical domain. For example, gains of 30 dB are attainable at 1550 nm using the erbium-doped fiber amplifier (EDFA).

As indicated in Fig. 65.6, there are a number of possible locations for optical amplifiers in a system. An optical amplifier just following the transmitter increases the optical power traveling down the fiber. Amplifiers along the fiber path continually keep the power levels above the system noise. An amplifier located at the fiber end acts as a receiver preamplifier, enhancing its sensitivity. Many amplifiers can be placed in a fiber network, extending the total path length to thousands of kilometers.

Repeater

The repeater is a regenerator that detects the optical signal by converting it into electrical form. It then determines the content of the pulse stream and uses this information to generate a new optical signal and launch this improved pulse train into the fiber. The new optical pulse stream is identical to the one originally transmitted. The regenerated pulses are restored to their original shape and power level by the repeater.

Many repeaters may be placed in a fiber network, extending the total path length to thousands of kilometers. The advantage of the optical amplifier over the regenerator is its lower cost and improved efficiency. The greater cost of the regenerator arises from the complexity of conversion between the optical and electrical domains. The regenerator does have the advantage of restoring the signal pulse shape, which increases the system bandwidth. This advantage is negated by a system propagating soliton pulses, which do not degrade with propagation.

Photodetector

This device converts an incoming optical beam into an electrical current. In fiber receivers, the most commonly used photodetectors are semiconductor *pin* photodiodes and avalanche photodiodes (APD). Important detector characteristics are speed of response, spectral response, internal gain, and noise. Because avalanche photodiodes have internal gain, they are preferred for highly sensitive receivers. Both Ge and InGaAs photodiodes respond in the preferred second and third fiber windows. InGaAs performs better at low signal levels because it has smaller values of dark current (that is, it is less noisy).

The current produced by a photodetector in response to incident optical power P is

$$i = M\eta eP/hf \qquad (65.11)$$

where M is the detector's gain, η is its quantum efficiency (close to 0.9 for good photodiodes), h is Planck's constant (6.63×10^{-34} J s), e is the magnitude of the charge on an electron (1.6×10^{-19}), and f is the optical frequency. For *pin* photodiodes ($M = 1$), typical responses are on the order of 0.5 μA/μW.

Receiver

Because of the low power levels expected at the input to the receiver, an electronic amplifier is normally required following the photodetector. The remainder of the receiver includes such electronic elements as band-limiting filters, equalizers, decision-making circuitry, other amplification stages, switching networks, digital-to-analog converters, and output devices (e.g., telephones, video monitors, and computers).

Other Components

There are a number of other fiber components, not shown in Fig. 65.6, that can be found in some systems. These include passive couplers for tapping off some portion of the beam from the single fiber and wavelength-division multiplexers for coupling different optical carriers onto the transmission fiber.

System Considerations

Long-distance fiber links carry voice, video, and data information. Messages not already in digital form are first converted to it. A single voice channel is normally transmitted at a rate of 64,000 bits per second. Video requires a much higher rate. The rate could be as much as 90 Mbps or so, but video compression techniques can lower this rate significantly. Fiber systems for the telephone network operate at such high rates that many voice channels can be time-division multiplexed (TDM) onto the same fiber for simultaneous transmission. For example, a fiber operating at a rate of 2.3 Gbps could carry more than 30,000 digitized voice channels.

Several optical carriers can simultaneously propagate along the same fiber. Such wavelength-division multiplexed (WDM) links further increase the capacity of the system. Systems using two or three optical carriers are common. Adding more than a few channels puts unrealistic constraints on the multiplexers and light sources. In long systems wideband optical amplifiers are preferred over regenerators for WDM systems because a single amplifier can boost all the individual carriers simultaneously while separate regenerators are needed for each carrier wavelength.

Total cable capacity is also increased by placing numerous fibers inside the cable. This is a cost-effective strategy when installing long fiber cables. The added cost of the extra fibers is small compared to the costs of actually deploying the cable itself. Fiber counts above 100 are practical. Multi-fiber cables can have enormous total data capacities.

Still further capacity is possible using optical frequency-division multiplexing (OFDM). In this scheme, many optical carriers very closely spaced in wavelength (maybe a few tenths of a nanometer) operate as independent channels. Hundreds of channels can be visualized in each of the two low-loss fiber windows. Systems of this type require **coherent detection** receivers to separate the closely spaced carriers.

Error Rates and Signal-to-Noise Ratio

The signal-to-noise ratio is a measure of signal quality. It determines the error rate in a digital network. At the receiver, it is given by

$$\frac{S}{N} = \frac{(M\rho P)^2 R_L}{M^n 2eR_L B(I_D + \rho P) + 4kTB} \tag{65.12}$$

where P is the received optical power, ρ is the detector's unamplified responsivity, M is the detector gain if an APD is used, n accounts for the excess noise of the APD (usually between 2 and 3), B is the receiver's bandwidth, k is Boltzmann's constant ($k = 1.38 \times 10^{-23}$ J/K), e is the magnitude of the charge on an electron (1.6×10^{-19} coulomb), T is the receiver's temperature in degrees kelvin, I_D is the detector's dark current, and R_L is the resistance of the load resistor that follows the photodetector.

The first term in the denominator of Eq. (65.12) is caused by shot noise and the second term is attributed to thermal noise in the receiver. If the shot noise term dominates (and the APD excess loss and dark current are negligible), the system is shot-noise limited. In this case the probability of error has an upper bound given by:

$$P_e = e^{-n_s} \tag{65.13}$$

where n_s is the average number of photoelectrons generated by the signal during a single bit interval when a binary 1 is received. An error rate of 10^{-9} or better requires about 21 photoelectrons per bit. Shot noise depends on the optical signal level. Because the power level is normally low at the end of a long-distance system, the shot noise is small compared to the thermal noise. Avalanche photodiodes increase the shot noise compared to the thermal noise. With APD receivers, ideal shot-noise limited operation can be approached but (because of the APD excess noise and limited gain) not reached.

If the thermal noise dominates, the error probability is given by

$$P_e = 0.5 - 0.5 \text{ erf } (0.354 \sqrt{S/N}) \tag{65.14}$$

where erf is the error function. An error rate of 10^{-9} requires a signal-to-noise ratio of nearly 22 dB.

System Design

A major part of fiber system design involves the power budget and the bandwidth budget. The next few paragraphs describe these calculations.

In a fiber system, component losses (or gains) are normally given in decibels. The decibel is defined by

$$\text{dB} = 10 \log P_2/P_1 \tag{65.15}$$

where P_2 and P_1 are the output and input powers of the component. The decibel describes relative power levels. Similarly, dBm and dBμ describe absolute power levels. They are given by

$$\text{dBm} = 10 \log P \tag{65.16}$$

where P is in milliwatts and

$$\text{dBμ} = 10 \log P \tag{65.17}$$

where P is in microwatts.

Power budget calculations are illustrated in Table 65.1 for a system that includes an amplifier. A specific numerical example is found in the last two columns. The receiver sensitivity in dBm is subtracted from the power available from the light source in dBm. This difference is the loss budget (in decibels) for the system. All the system losses and gains are added together (keeping in mind that the losses are negative and the amplifier gains are positive). If the losses are more than the gains (as is usual), the system loss dB_{SL} will be a negative number. The loss margin is the sum of the loss budget and the system loss. It must be positive for the system to meet the receiver sensitivity

Table 65.1 Power Budget Calculations

Source power	dBm_s		3
Receiver sensitivity	dBm_r		−30
Loss budget: $dBm_s - dBm_r$		dB_{LB}	33
Component efficiencies			
Connectors	dB_c		−5
Splices	dB_s		−2
Source coupling loss	dB_{cl}		−5
Fiber loss	dB_f		−24
Isolator insertion loss	dB_i		−1
Amplifier gain	dB_a		10
Total system loss			
$dB_c + dB_s + dB_{cl} + dB_f + dB_i + dB_a$		dB_{SL}	−27
Loss margin: $dB_{LB} + dB_{SL}$		dB_{LM}	6

requirements. The system loss margin must be specified to account for component aging and other possible system degradations. A 6-dB margin was found for the system illustrated in the table. The fiber in the table has a total loss of 24 dB. If its attenuation is 0.25 dB/km, the total length of fiber allowed would be 24/0.25 = 96 km.

In addition to providing sufficient power to the receiver, the system must also satisfy the bandwidth requirements imposed by the rate at which data are transmitted. A convenient method of accounting for the bandwidth is to combine the rise times of the various system components and compare the result with the rise time needed for the given data rate and pulse coding scheme.

The system rise time is given in terms of the data rate by the expression

$$t = 0.7/R_{NRZ} \qquad (65.18)$$

for non-return-to-zero (NRZ) pulse codes and

$$t = 0.35/R_{RZ} \qquad (65.19)$$

for return-to-zero (RZ) codes.

An example of bandwidth budget calculations appears in Table 65.2. The calculations are based on the accumulated rise times of the various system components.

The system in Table 65.2 runs at 500 Mbps with NRZ coding for a 100-km length of fiber. Equation (65.18) yields a required system rise time no more than 1.4 ns. The transmitter is assumed to have a rise time of 0.8 ns. The receiver rise time, taken as 1 ns in the table, is a combination of the pho-todetector's rise time and that of the receiver's electronics.

Table 65.2 Bandwidth Budget Calculations[a]

Transmitter	t_t	0.8
Fiber	t_f	0.36
Receiver	t_r	1
System total: $\sqrt{t_t^2 + t_f^2 + t_r^2}$	t_s	1.33
System required	t	1.4

[a]All quantities in the table are rise time values in nanoseconds.

The fiber's rise time was calculated for a single-mode fiber operating at a wavelength of 1550 nm. Equation (65.9) shows that $M = 18$ ps/(nm × km) at 1550 nm. The light source was assumed to have a spectral width of 0.2 nm. Then, the pulse dispersion calculated from Eq. (65.10) yields a pulse spread of 0.36 ns. Because the fiber's rise time is close to its pulse spread, this value is placed in the table.

The total system rise time is the square root of the sum of the squares of the transmitter, fiber, and receiver rise times. That is:

$$t_s = \sqrt{t_t^2 + t_f^2 + t_r^2} \qquad (65.20)$$

In this example, the system meets the bandwidth requirements by providing a rise time of only 1.33 ns, where as much as 1.4 ns would have been sufficient.

Defining Terms

Coherent detection: The signal beam is mixed with a locally generated laser beam at the receiver. This results in improved receiver sensitivity and in improved receiver discrimination between closely spaced carriers.

Material dispersion: Wavelength dependence of the pulse velocity. It is caused by the refractive index variation with wavelength of glass.

Quantum efficiency: A photodiode's conversion efficiency from incident photons to generated free charges.

Single-mode fiber (SMF):   A fiber that can support only a single mode of propagation.

Spectral width: The range of wavelengths emitted by a light source.

References

E. E. Basch, Ed., *Optical-Fiber Transmission*, Indianapolis: Howard W. Sams & Co., 1987.

C. C. Chaffee, *The Rewiring of America*, San Diego: Academic Press, 1988.

R. J. Hoss, *Fiber Optic Communications Design Handbook*, Englewood Cliffs, N.J.: Prentice-Hall, 1990.

L. B. Jeunhomme, *Single-Mode Fiber Optics*, 2nd ed., New York: Marcel Dekker, 1990.

G. Keiser, *Optical Fiber Communications*, 2nd ed., New York: McGraw-Hill, 1991.

J. C. Palais, *Fiber Optic Communications*, 3rd ed., Englewood Cliffs, N.J.: Prentice-Hall, 1992.

A. Yariv, *Optical Electronics*, 4th ed., Philadelphia: Saunders College Publishing, 1991.

Further Information

Continuing information on the latest advances in long-distance fiber communications can be obtained from several professional society journals and several trade magazines including: *IEEE Journal of Lightwave Technology, IEEE Photonics Technology Letters, IEEE LTS: The Magazine of Lightwave Telecommunications, Lightwave*, and *Laser Focus World*.

65.3 Photonic Networks

Ivan P. Kaminow

Lightwave technology has been developed and widely utilized for local and long-distance transmission in the public telephone network (see Section 65.2 and Miller and Kaminow, 1988) and in modern CATV (cable TV) networks (Section 65.1). Computer communications have been provided utilizing copper transmission lines in private local-area networks (LAN) (Section 66.3) that cover short distances, $L < 10$ km, and involve low data throughputs $S < 10$ Mb/s. The throughput is defined as

$$S = NB$$

with N the number of simultaneous interconnections and B the communication bit-rate per user. At the higher ranges of the bit-rate-distance product, above $BL = (10$ Mb/s$) (10$ km$) = 100$ Mb/s $\cdot$ km or, equivalently, higher ranges of the bit-rate-delay product

$$M = BD = (BL)(n/c)$$

where D is the propagation delay, c/n is the (group) velocity of bits on the transmission line, and c is the velocity of light, optical fiber may be preferable to copper. For optical fibers, with refractive index $n = 1.5$, the delay is $n/c = 5 \times 10^{-9}$ s/m $= 5$ μs/km. Thus, with $BL = 100$ Mb/s $\cdot$ km, M is 500 bits, i.e., there are 500 bits in transit on the transmission line between transmitter and receiver.

As *M* gets larger, the performance of copper transmission lines—twisted pairs or coax—becomes unsatisfactory because of attenuation and pulse dispersion. The economic break-even value for *M*, where the added cost of lightwave technology is justified, though not precise, is in the neighborhood of 500 bits. In this section, we will cover aspects of lightwave data networks that utilize the lightwave technology discussed in Section 65.2 and some of the multiple-access methods for LANs discussed in Section 66.3. The latter section touches on commercial LAN standards that utilize optical data links for point-to-point transmission between nodes, often with multimode fiber. Here, we will first discuss some of the recent optical LAN standards and then briefly mention proposed approaches to photonic networks with terabit-per-second throughput and gigabit-per-second user access, and the novel optical components that are needed to realize this high performance. When such networks connect users separated by *L* ~ 1000s of kilometers, *M* ~ 10s of megabits may be in transit, requiring new approaches to congestion control for multiple access.

Data Links

A data link consists of a transmitter (T) that converts electrical pulses to optical pulses (E/O) and sends the optical pulses on an optical fiber to a receiver (R) which converts the optical pulses back to electrical pulses (O/E). The transmitter may use a light-emitting diode (LED) or a laser diode (LD) as the optical source. The LED is cheaper but has lower output power into the fiber (~10 μW vs. ~1 mW), lower modulation bandwidths (~100 Mb/s vs. ~1 Gb/s), and wider optical spectrum, leading to chromatic dispersion due to the variation of optical velocity in the fiber with wavelength. Pulse dispersion limits *BL* when the pulse spreading approaches a bit period. The receiver may employ a PIN (positive-intrinsic-negative) or APD (avalanche photodiode) photodetector. The former is cheaper and easier to bias but has poorer sensitivity by about 5 dB. The sensitivity of a good PIN receiver is about −50 dBm at 100 Mb/s and −35 dBm at 1 Gb/s for a bit-error-rate (BER) of 10^{-9}. Optical devices operating at a wavelength of ~0.87 μm use gallium-aluminum-arsenide materials and are less expensive than those operating at 1.3 or 1.5 μm and using indium-gallium-arsenide-phosphide materials. However, for 1.5 μm, the fiber attenuation and, for 1.3 μm, the chromatic dispersion is much less than for 0.87 μm. The fiber joining transmitter and receiver may be multimode or single mode. Multimode (with a typical core diameter of 62.5 μm) is cheaper, but since each mode travels at a different optical velocity, the modal dispersion further limits *BL*. The LED data links generally employ multimode fiber, and the combination of chromatic and modal dispersion limits *BL* to values below ~1 Gb/s · km. For an LD, single-mode fiber (core diameter ~10 μm) data link, *BL* of ~100 Gb/s · km is possible.

Optical data links are employed to connect electronic components of a LAN when copper is no longer feasible. However, because of its lower cost and the fact that it is often already installed, clever tricks are now being used to extend the utility of copper.

Token Ring: FDDI, FFOL

Figure 65.8 illustrates a ring network. The real topology may be a good deal more irregular than a circle, depending on the accessibility of stations. In its usual application, which uses a token-ring protocol for media access, an electronic repeater that operates at the aggregate network rate is required at each station.

A token—e.g., a "1" or a "0" bit, or a token packet—is propagated in one direction from station to station. When a station has a packet to send to another station, it adds the address of the receiving station in a header and holds the combined packet in a buffer. The sending station reads the tokens as they go by until it receives an empty token, a "0." It then converts the "0" to a "1," a busy token, and appends the packet.

Intermediate stations repeat the bits in the packet and also "listen" for their own addresses. If a station recognizes its address in the packet header, it copies the packet. When the packet returns to

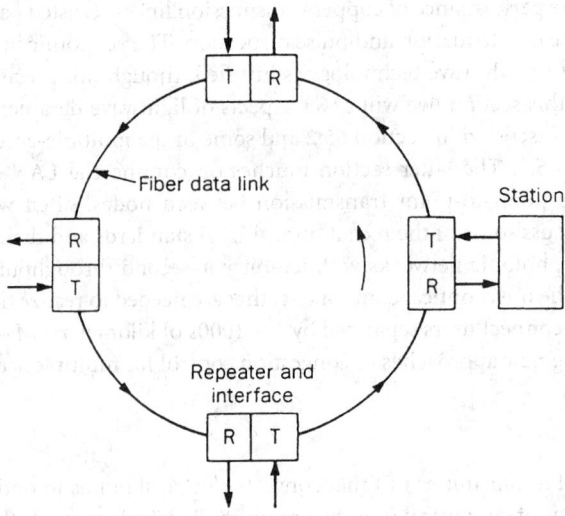

FIGURE 65.8 Undirectional ring network. R and T represent the receiver and transmitter functions, respectively.

the sender, it serves as an acknowledgment, and the sender removes it from the ring, after converting the token back to "0."

Commercial token rings use wire interconnections or optical data links to join stations at rates in the 10-Mb/s range. Actual network use is less than 10 Mb/s because of the time it takes an empty token to pass around the ring. This transit-time delay increases linearly with the number of stations. It includes propagation delay between stations and processing delay at each station, which must examine the header of every packet before repeating the bits to the next station.

A token-ring architecture is not especially attractive for a high-speed optical network (where $S \sim B$ is above 1 Gb/s) because of the cost of high-speed repeater optoelectronics at each station and the packet-processing delay. In addition, at high bit rates, the packet time may be much shorter than the propagation time around the ring, unless a packet contains an impractically large number of bits. Efficient use of the ring with short packets may call for multiple tokens, which can lead to complex protocols. Increasing the number of bits per packet increases the packet time but places added burden on the high-speed buffer at each station.

Reliability—if one station is disabled or if the fiber breaks—is a problem in both fiber and wire rings. To address these reliability problems, a double-ring optical network can provide for bypass of defective stations and loop back around a fiber break. Each station has two inputs and two outputs connected to two rings that operate in opposite directions; this, of course, increases the cost.

The fiber distributed data interface (FDDI) [Ross, 1986, 1989] is a standard proposed by the American National Standard Institute for a 100-Mb/s double-ring, time-division multiplexed (TDM) LAN that uses 1.3-μm multimode fiber and LED (or single-mode fiber and laser diode) data links between stations. This LAN is designed to provide both backbone services that interconnect lower speed LANs and back-end services that interconnect mainframe computers, mass storage systems, and other high-speed peripherals. FDDI provides datagram packet service with up to about 4,500 data bytes per frame. It employs 4B/5B coding so that the clock rate is 125 MHz for maximum S of 100 Mb/s. The FDDI network is designed to operate with low-cost components that were commercially available in 1986. The standard can provide both packet-switched and circuit-switched services. As many as 1000 stations can be connected, with a maximum of 2 km between stations and a maximum perimeter of 200 km.

An FDDI follow on LAN (FFOL) will operate with laser and single-mode fiber data links at bit rates corresponding to the synchronous optical network (SONET) or synchronous digital hierarchy (SDH) standards of 622 Mb/s and 2.5 Gb/s, possibly with ATM (asynchronous transfer mode) cells. The geographical size will also be increased.

Active Star Networks: Ethernet, Datakit®

Carrier sense multiple access with collision detection (CSMA/CD) Ethernet networks operating at 10 Mb/s connect users on a copper bus (Section 66.3). The length of the bus must be less than 1/2 the distance light propagates in the time required to transmit a packet frame. Thus, for speeds much greater than 10 Mb/s, where optics might be needed, the length of the bus will be limited unless the maximum frame contains an impractically large number of bits. Further, the number of stations that can be supported by an optical bus is limited by the nature of optical taps, as opposed to electrical taps [Kaminow, 1989]. Finally, the collision detection algorithm does not work well on an optical bus because the intensities of optical packets from two different stations may vary considerably along the bus. Thus, an active electronic star, as shown in Fig. 65.9, with optical data links from users is often employed.

The AT&T Datakit [Fraser, 1983] packet switch behaves as a virtual circuit switch (VCS) in that a reliable data path is set up for each session, and packet retransmission because of collisions is not required. Remote stations that may consist of mainframe computers, concentrators that bring together many terminals, or gateways to other networks are connected by 8-Mb/s fiber-optic data links to individual electronic modules at the central node as shown in Fig. 65.10 [Kaminow, 1988]. These modules plug into two electronic buses that are short (about 1 m) compared to a packet propagation length (16 bytes).

In the module, packets are formed and stored with a header that contains the source address. When the packet is complete (it has the full number of bytes, or a fixed waiting period for added bytes has passed), the module transmits its binary address on the contention bus while listening for

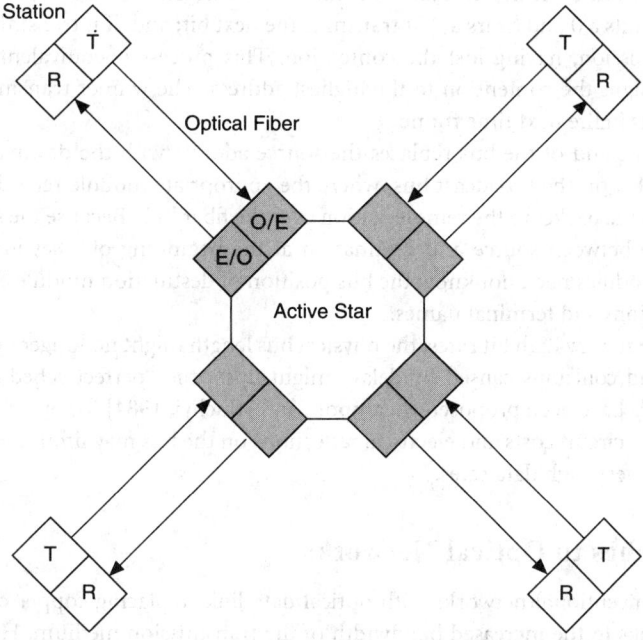

FIGURE 65.9 Active star network. Optical-to-electrical (O/E) and electrical-to-optical (E/O) converters must be provided at the star. R and T represent the receiver and transmitter functions, respectively.

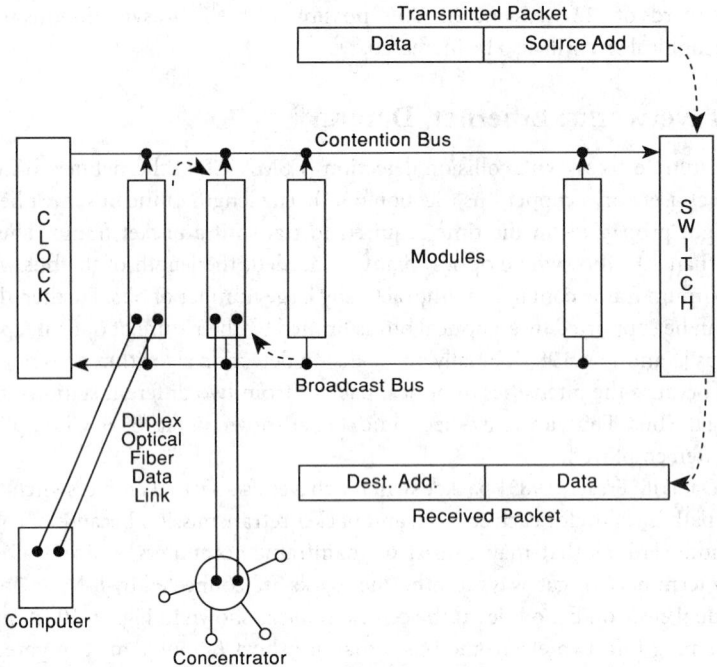

FIGURE 65.10   Datakit® VCS network. Remote stations are connected to the electrical node by 8-Mb/s data links. The length of the bus is much shorter than the propagation length of a packet.

bits transmitted by others. If the module transmits a 1 and hears a 1, it transmits the next address bit. But if it transmits a 0 and hears a 0, it transmits the next bit; and, if it transmits a 0 and hears a 1, it stops transmission, having lost the contention. This process is equivalent to a logical OR operation and assigns the contention to the highest address. The winner transmits the packet on the contention bus in the next time frame.

The switch at the end of the bus replaces the source address with the destination address and transmits the packet on the broadcast bus, where the appropriate module records the destination address and sends the packet to the remote station over the fiber link. Because the switch establishes a correspondence between source and destination at the beginning of a session (as in a circuit switch), source modules need not know the bus position of destination modules. The switch has a directory of positions and terminal names.

If we were to go to very high bit rates, the physical bus length might no longer be short compared with a packet, and collisions caused by delays might upset the "perfect scheduling" of packets. Although methods have been proposed [Acampora and Hluchyj, 1984] for overcoming this limitation, the electronic circuit costs and electrical reflections on the bus may limit the effectiveness of a centralized bus at very high data rates.

New Approaches to Optical Networks

The preceding conventional networks with optical data links replacing copper can improve their throughputs thanks to the increased bandwidth of the transmission medium. However, a revolutionary improvement in throughput to terabit-per-second levels with gigabit-per-second access requires entirely new approaches for the physical connectivity, architecture, and access protocols. We can use much of the photonic technology employed in long-haul lightwave systems to provide physical connectivity, but we also need devices with new functionality to realize proposed architectures, and, conversely, with new component functionality we can dream of new architectures.

We can provide connectivity among users in three dimensions: space, time, and optical frequency or wavelength, employing space-division multiplexing (SDM), optical time-division multiplexing (OTDM), and optical frequency-division multiplexing (OFDM) or wavelength-division multiplexing (WDM), respectively. To control the path routing we need optical switches for OTDM and frequency routing technology for OFDM. At present, network architectures and protocols are at the research stage. We mention some of these components and switches in the following paragraphs. More details can be found in the References [Miller and Kaminow, 1988; Special Issue, 1990].

A star topology seems most attractive for gigabit-per-second multiple-access photonic networks [Kaminow, 1989], as shown in Fig. 65.11. Each station has its own transmitter and receiver. For optical TDM, the connectivity can be provided by an $N \times N$ electrooptic switch and suitable controller, and for optical FDM, the connectivity is provided by a passive $N \times N$ star coupler. Electrooptic $N \times N$ switches based on integrated titanium-diffused lithium niobate waveguide elements [Korotky and Alferness, 1988] have been demonstrated with $N = 16$ and operating at $B = 2.5$ Gb/s for each input. The switch connections can be rearranged in a few nanoseconds. It is estimated that such switches could be interconnected to provide $N = 256$. Unlike electronic switches, electrooptic switches are transparent to the bit rate, i.e., they can connect any bit stream independent of B. The problems of suitable multiple-access protocols and controls have not yet been fully addressed.

The passive $N \times N$ star coupler [Kaminow, 1989] in Fig. 65.11 has N single-mode fiber inputs and N outputs. In an ideal passive star, a signal

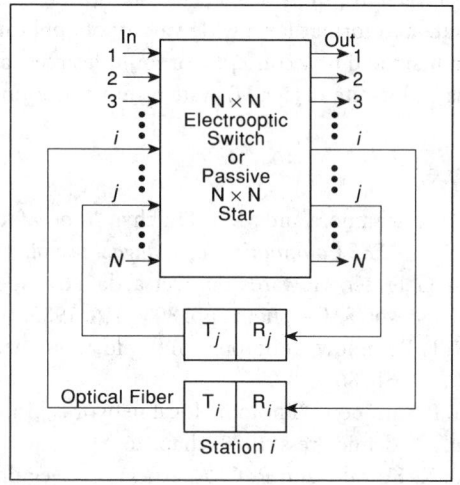

FIGURE 65.11 Electrooptic switch network for optical TDM or passive star network for optical FDM. The $N \times N$ switch or star have N single-mode optical fiber input ports and N optical fiber output ports. As in the active star (Fig. 65.9), two fibers connect a remote station with the star. R_x and T_x represent the receiver and transmitter functions, respectively, for station x.

incident on any input is divided equally among all the outputs, i.e., the star broadcasts every input to every output. Unlike the OTDM case, each transmitter uses a different optical frequency and each receiver must tune to the frequency of the channel intended for it, as illustrated in Fig. 65.12. Alter-

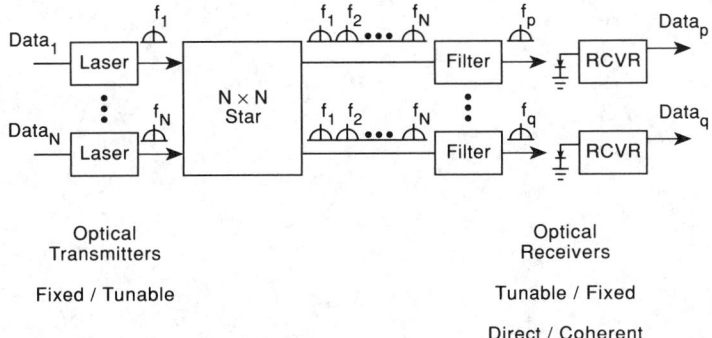

FIGURE 65.12 An optical FDM network with passive star distribution. Optical transmitter frequencies $f_1 \ldots f_N$ are modulated with data at the transmitter and selected by a filter at the receiver.

natively, the receiver frequencies may be fixed and the transmitters tunable. Thus, the control can be distributed in the terminals. Calculations indicate that such a network can support throughputs of several terabits per second. Current research [Special Issue, 1990] is aimed at devising multiple-access protocols and demonstrating the novel devices needed for optical frequency routing. These include fast tunable lasers and receivers that can cover many channels (switching speeds of ~10 ns at 2.5 Gb/s with ~50 channels appear feasible), optical frequency translators for frequency reuse, and integrated star couplers and integrated optical frequency routers.

One challenge in photonic network design is to make them "all-optical," as nearly as possible, in order to avoid throughput bottlenecks by electronic components and the expense of O/E and E/O conversions. In principle, clear all-optical channels would offer connectivity independent of data-rate and format for a wide variety of applications. However, many physical technology problems remain and new concepts for multple access and congestion control [Special Issue, 1991] suited to large bit-rate-delay (M) systems must be found.

References

A. S. Acampora and M. G. Hluchyj, "A new local area network architecture using a centralized bus," *IEEE Communications Magazine*, vol. 22, no. 8, pp. 12–21, 1984.

A. G. Fraser, "Towards a universal data transport system," *IEEE J. Selected Areas in Communications*, vol. SAC-1, no. 5, pp. 803–816, 1983.

I. P. Kaminow, "Photonic multiple access networks," *AT&T Technical Journal*, vol. 68, no. 2, pp. 61–86, 1989.

I. P. Kaminow, "Photonic local networks," in *Optical Fiber Telecommunications, II*, New York: Academic Press, 1988, chap. 26.

S. K. Korotky and R. C. Alferness, "Waveguide electrooptic devices for optical fiber communication," in *Optical Fiber Telecommunications, II*, New York: Academic Press, 1988, chap. 11.

S. E. Miller and I. P. Kaminow, Eds., *Optical Fiber Telecommunications, II*, New York: Academic Press, 1988.

F. E. Ross, "FDDI—A tutorial," *IEEE Communications Magazine*, vol. 24, no. 5, pp. 10–17, 1986.

F. E. Ross, "An overview of FDDI—The fiber distributed data interface," *IEEE J. Selected Areas in Communications*, vol. 7, no. 7, pp. 1043–1051, 1989.

Special Issue, "Congestion control in high speed networks," *IEEE Communications Magazine*, vol. 29, no. 10, 1991.

Special Issue "Dense wavelength division multiplexing techniques for high capacity and multiple access communications systems," *IEEE J. Selected Areas in Communication*, vol. 8, no. 6, 1990.

66

Networks

Manfred N. Huber
Siemens

J. N. Daigle
Mitre Corporation

Joseph Bannister
The Aerospace Corporation

Mario Gerla
*University of California,
Los Angeles*

Richard B. Robrock II
Bell Communications Research

66.1 B-ISDN 1441
      B-ISDN Services and Applications • Asynchronous Transfer Mode
      • Transmission of B-ISDN Signals • ATM Adaptation Layer
      • B-ISDN Signaling
66.2 Computer Communication Networks 1447
      General Networking Concepts • Computer Communication
      Network Architecture • Local-Area Networks and Internets • Some
      Additional Recent Developments
66.3 Local-Area Networks 1460
      The LAN Service Model • Other Features • The Importance of
      LAN Standards
66.4 The Intelligent Network 1468
      A History of Intelligence in the Network • The Intelligent
      Network • Intelligent Network Systems • The CCS7 Network •
      The Service Control Point • Data Base 800 Service • Alternate
      Billing Services • Other Services • The Advanced Intelligent
      Network • Back to the Future

66.1 B-ISDN

Manfred N. Huber

Since the mid-1980s the idea of the **integrated services digital network** (ISDN) has become reality. In ISDN voice services with supplementary features and data services with a bit rate of up to 64 kbit/s are integrated in one network. For voice communication and many text and data applications the 64-kbit/s ISDN will be sufficient. Although it is minor as yet, there exists already a growing demand for **broadband** communication with bit rates from some megabits per second up to approximately 130 Mbit/s [Wiest, 1990] (e.g., high-speed data communication, video communication, high-resolution graphics).

In order to provide the same advantages of ISDN to broadband communication users, network operators, and service providers, the development of an intelligent broadband-ISDN (B-ISDN) is necessary. The future B-ISDN will become the universal network integrating different kinds of services with their individual features and requirements. B-ISDN will support switched, semipermanent and permanent, point-to-point, and point-to-multipoint connections and provide on-demand, reserved, and permanent services. B-ISDN connections support packet mode and circuit mode services of mono- and/or multimedia type of a connection-oriented or connectionless nature in a unidirectional or bidirectional configuration [Händel and Huber, 1991b].

B-ISDN Services and Applications

As already mentioned, there exists some demand for broadband communication which originates from business customers as well as residential customers. In the residential area, on the one hand, people are interested in video distribution services for entertainment purposes, like television and high-definition TV; on the other hand, they will use video telephony with acceptable quality. Over the long term, video mail services and video retrieval services will become more important.

Voice and text are no longer sufficient for business customers. In the offices and factories of tomorrow, interactive broadband services will be required. Handling complex tasks in the future demands comprehensive support by services for voice, text, data, graphics, video, and documents. In addition to the individual services, the multimedia services and the simultaneous or alternating use of several services with multifunction workstations will gain importance [Armbrüster, 1990].

Interconnection of local-area networks (LANs) or large computers, computer-aided design, and computer-aided manufacturing will become important data applications. The first video services will be video telephony and video conferencing (studio-to-studio and workstation video conferencing). Initially these services may have diminished quality, but for the long term TV quality can be expected.

The bit rates of all services mentioned above are in the range of 2 to 130 Mbit/s (depending on the individual application). Taking into account that in the future more enhanced video coding mechanisms will be available, the required bit rates for video services will become lower without influencing quality significantly.

Asynchronous Transfer Mode

In today's public switched networks the synchronous transfer mode (STM) predominates. Applying STM technology, for the duration of a connection a synchronous channel with constant bit rate is allocated to that connection. STM does not fit very well for the integration of services with bit rates from some kilobits per second to 130 Mbit/s. Therefore, in B-ISDN a new transfer mode called **asynchronous transfer mode** (ATM) is used.

In ATM all kinds of information is transported in **cells**. A cell is a block of fixed length, which consists of a 5-octet cell header and a 48-octet cell payload (see Fig. 66.1). The cell header contains all necessary information for transferring the cell through the network and the cell payload includes the user information. The cell rate of a connection is proportional to the service bit rate. Only if information is available is a cell used by the connection. By having different routing labels,

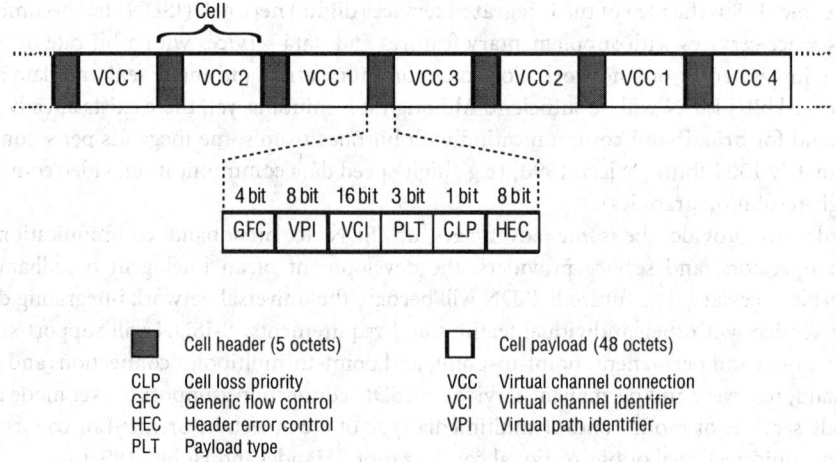

FIGURE 66.1   ATM principle.

cells of different connections can be transported on the same transmission line (cell multiplexing). If no connection has information ready to transport, idle cells will be inserted. Idle cells do not belong to any connection; they are identified by a standardized cell header.

ATM uses only cells; multiplexing and switching of cells is independent of the applications and of the bit rates of the individual connections. Applying ATM technology, the idea of one universal integrated network becomes a reality. However, the ATM technology also causes some problems. Because of the asynchronous multiplexing buffers are necessary, which results in cell delay, cell delay variation, and cell loss. In order to compensate for these effects additional measures have to be provided.

Figure 66.1 also shows the individual subfields of the cell header. The first field, called generic flow control (GFC), is only available at the user-network interface (UNI). Its main purpose is media access control in shared medium configurations (LAN-like configurations) within the customer premises [Göldner and Huber, 1991]. The proposed GFC procedures are based either on the distributed queue algorithm or the reset timer control mechanism [Händel and Huber, 1991a]. At the network-node interface (NNI) these bits are part of the virtual path identifier (VPI).

The VPI together with the virtual channel identifier (VCI) form the routing label (identifier of the connection). The VPI itself marks only the virtual path (VP). The VP concept allows the flexible configuration of individual subnetworks (e.g., **signaling** network or virtual private network), which can be independent of the underlying transmission network. VP networks are under the control of network management. The bandwidth of a VP will be allocated according to its requirements. Within the VP network the individual connections are established and cleared down dynamically (by signaling).

The payload type field in the cell header differentiates the information in the cell payload of one connection (e.g., user information, operation and maintenance information for ATM). The value of the cell loss priority bit distinguishes cells that can be discarded under some exceptional network conditions without disturbing the quality significantly from those cells that may not be discarded. The last field of the cell header forms the header error control field. The cell header is protected against errors with a mechanism that allows the correction of a single bit error and the detection of multibit errors.

The high transmission speeds for ATM cell transfer require very high-performance switching nodes. Therefore, the switching networks (SNs) have to be implemented in fast hardware. Within the SN the self-routing principle will be applied [Schaffer, 1990]. At the inlet of the SN the cell is extended by an SN-internal header. It is evident that the SN-internal operational speed has to be increased. When passing the individual switching elements, for the processing of the SN-internal header only simple hard-wired logic is necessary. This reduces the control complexity and provides a better failure behavior. When starting several years ago with the implementation of the ATM technology, only the emitter coupled logic (ECL) was available. Nowadays, the complementary metal-oxide semiconductor (CMOS) technology with its low power consumption is used [Fischer *et al.*, 1991].

Transmission of B-ISDN Signals

Transmission systems at the UNI provide bit rates of around 150 and 622 Mbit/s. In addition to these rates, at the NNI around 2.5 Gbit/s and up to 10 Gbit/s will be used in the future [Baur, 1991]. In addition to the high-capacity switching and multiplexing technology, high-speed transmission systems are required. Optical fibers are especially suitable for this purpose; however, for the lower bit rates coaxial cables can be used. Optical transmission uses optical fibers as the transmission medium in low-diameter and low-weight cables to provide large transmission capacities over long distances without the need for repeaters. Optical transmission equipment currently tends to monomode fiber and laser diodes with wavelengths of around 1310 nm. For both directions in a transmission system either two separate fibers or one common fiber with wavelength division multi-

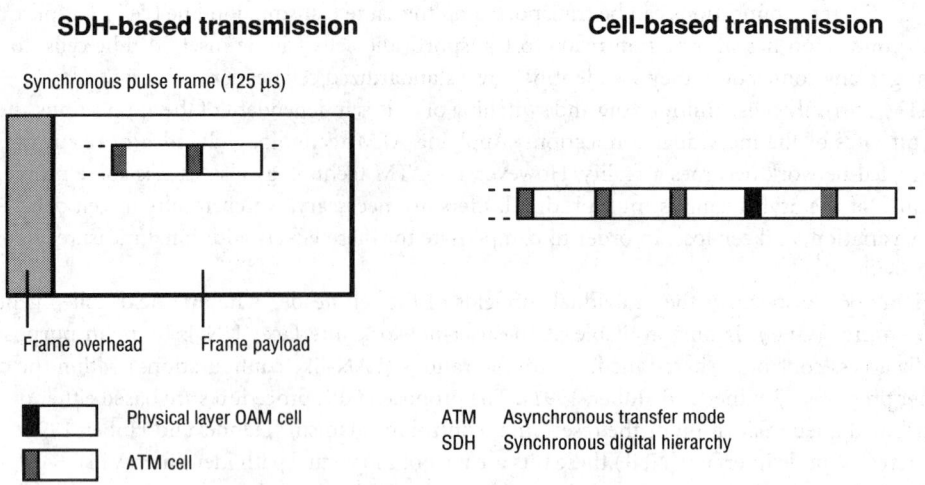

FIGURE 66.2   Transmission principles for B-ISDN.

plexing can be used. The second solution may be a good alternative for subscriber lines and short trunk lines [Bauch, 1991].

For ATM cell transmission, two possibilities exist, which are shown in Fig. 66.2: synchronous pulse frame or continuous cell stream (cell-based). The basis for the pulse frame concept is the existing **synchronous digital hierarchy** (SDH). In SDH the cells are transported within the SDH payload; the frame overhead includes operation and maintenance (OAM) of the transmission system. In the cell-based system the OAM for the transmission system is transported within cells. The SDH solution is already defined, whereas for cell-based transmission some problems remain to be solved (e.g., OAM is not yet fully defined).

ATM Adaptation Layer

The **ATM adaptation layer** (AAL) is between the ATM layer and higher layers. Its basic function is the enhanced adaptation of the services provided by ATM to the requirements of the layers above. In order to minimize the number of AAL protocols, the service classification shown in Fig. 66.3 was defined. This classification was made with respect to timing relation, bit rate, and connection mode.

The AAL protocols are subdivided into two parts. The lower part performs, at the sending side,

	Class A	Class B	Class C	Class D
Timing relation between source and destination	Required		Not required	
Bitrate	Constant	Variable		
Connection mode	Connection oriented			Connection-less

FIGURE 66.3   AAL service classification.

the segmentation of long messages into the cell payload and, at the receiving side, reassembly into long messages. The upper part is service dependent and provides the AAL service to the higher layer.

B-ISDN Signaling

For signaling in B-ISDN, existing protocols and infrastructure will be reused as much as possible. Figure 66.4 shows the protocol stacks for UNI and NNI. The upper part concerns signaling applications and the lower part signaling transfer.

For the introduction of simple switched services in B-ISDN, at UNI and NNI, existing signaling application protocols will be reused. The 64-kbit/s ISDN-specific information elements will be removed and new B-ISDN-specific information elements will be added. Right from the beginning these protocols will provide means that allow smooth migration toward future applications, which will include highly sophisticated features like multimedia services [Huber *et al.*, 1992]. This approach guarantees compatibility for future protocol versions.

At the NNI the existing signaling system no. 7 (SS7) can be reused (see right part of the NNI protocol stack in Fig. 66.4). SS7 is a powerful and widespread network that will continue to be applied for rather a long period until ATM penetration has been reached. For the middle term, however, a fully ATM-based network will be available which also carries signaling messages (see left part of the NNI protocol stack in Fig. 66.4). ATM-based signaling at the NNI needs a suitable AAL which provides the services of the existing message transfer part level 2.

At the UNI, right from the beginning, all kinds of traffic (including signaling) is carried within cells. An AAL for signaling at the UNI is also required. This AAL has to provide the services of the existing layer 2 UNI protocol. The AAL for signaling at UNI and NNI will be common as much as possible. In contrast to the NNI, at the UNI meta-signaling is necessary. Meta-signaling establishes, checks, and removes the signaling channels between customer equipment and the central office in a dynamic way. The signaling channels at the NNI are semipermanent and, therefore, meta-signaling is not required.

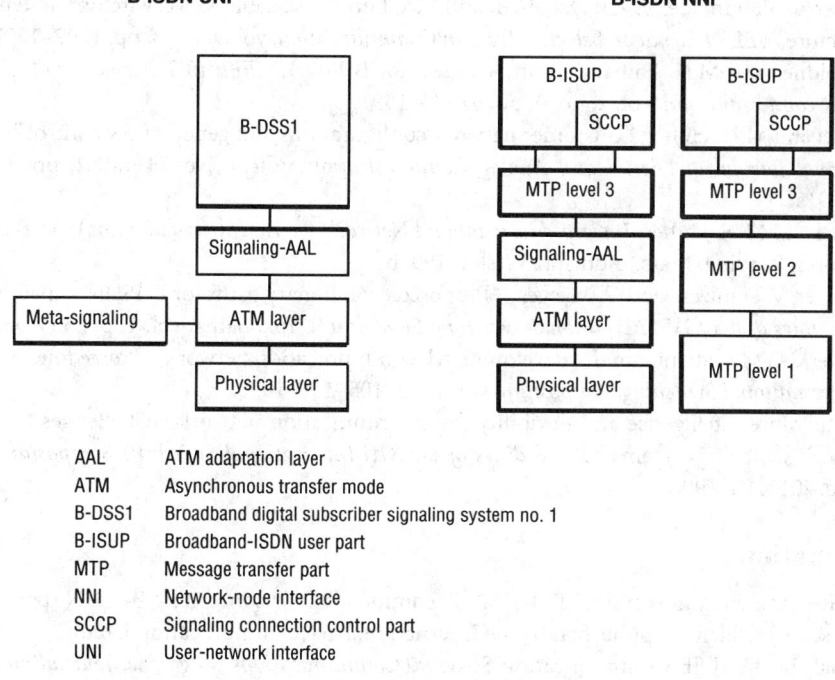

AAL	ATM adaptation layer
ATM	Asynchronous transfer mode
B-DSS1	Broadband digital subscriber signaling system no. 1
B-ISUP	Broadband-ISDN user part
MTP	Message transfer part
NNI	Network-node interface
SCCP	Signaling connection control part
UNI	User-network interface

FIGURE 66.4   Protocol stacks for B-ISDN signaling.

Defining Terms

Asynchronous transfer mode: A transfer mode in which the information is organized into cells; it is asynchronous in the sense that the recurrence of cells containing information from an individual user is not necessarily periodic.

ATM adaptation layer: A layer which provides the adaptation of higher layers to ATM.

Broadband: A service or system requiring transmission channels capable of supporting bit rates greater than 2 Mbit/s.

Cell: A block of fixed length which is subdivided into a cell header and an information field. The cell header contains a label which allows the clear allocation of a cell to a connection.

Integrated services digital network: A network which provides end-to-end digital connectivity to support a wide range of services, including voice and nonvoice services, to which users have access by a limited set of standard multipurpose user-network interfaces.

Signaling: Procedures which are used to control (set up and clear down) calls and connections within a telecommunication network.

Synchronous digital hierarchy: A standard for optical transmission which provides transmission facilities with flexible add/drop capabilities to allow simple multiplexing and demultiplexing of signals.

References

H. Armbrüster, "Blueprint for future telecommunications," *Telcom Report International,* vol. 13, no. 1, pp. 5–8, 1990.

H. Bauch, "Transmission systems for B-ISDN," *IEEE LTS, Magazine of Lightwave Telecommunication,* vol. 2, no. 3, pp. 31–36, 1991.

H. Baur, "Technological perspective of telecommunications for the nineties," *Integration, Interoperation and Interconnection: This Way to Global Services, Proceedings of the Technical Symposium,* Geneva, part 2, vol. 1 paper 1.1, 1991.

W. Fischer, O. Fundneider, E.-H. Goeldner, and K.A. Lutz, "A scalable ATM switching system architecture," *IEEE Journal on Selected Areas in Communication,* vol. 9, no. 8, pp. 1299–1307, 1991.

E.-H. Göldner and M.N. Huber, "Multiple access for B-ISDN," *IEEE LTS, Magazine of Lightwave Telecommunication,* vol. 2, no. 3, pp. 37–43, 1991.

R. Händel and M.N. Huber, "Customer network configurations and generic flow control," *International Journal of Digital and Analog Communication Systems,* vol. 4, no. 2, pp. 117–122, 1991a.

R. Händel and M.N. Huber, *Integrated Broadband Networks — An Introduction to ATM-Based Networks,* Reading, Mass.: Addison-Wesley, 1991b.

M.N. Huber, V. Frantzen, and G. Maegerl, "Proposed evolutionary paths for B-ISDN signalling," *Proceedings of the XIV International Switching Symposium,* Yokohama, vol. 1, pp. 334–338, 1992.

B. Schaffer, "ATM switching in the developing telecommunication network," *Proceedings of the XIII International Switching Symposium,* vol. 1, pp. 105–110, 1990.

G. Wiest, "More intelligence and flexibility for communication network—Challenges for tomorrow's switching systems," *Proceedings of the XIII International Switching Symposium,* vol. 5, pp. 201–204, 1990.

Further Information

CCITT Recommendations and CCITT Draft Recommendations concerning B-ISDN (parts of F, G, I and Q series), which are published by the International Telecommunication Union.

Journals of the IEEE Communication Society (*Communications Magazine, Journal on Selected Areas in Communications, LTS: Magazine of Lightwave Telecommunication, Networks, Transactions on Communications*), which are published by the Institute of Electrical and Electronics Engineers, Inc.

International Journal of Digital and Analog Communication System, which is published by John Wiley & Sons, Ltd.

Proceedings of international conferences such as GLOBECOM, INFOCOM, International Conference on Communications, International Conference on Computer Communication, International Switching Symposium, International Symposium on Subscriber Loops and Services, and International Teletraffic Congress.

A detailed description of ISDN is given in *ISDN—The Integrated Services Digital Network—Concepts, Methods, Systems*, by P. Bocker, published by Springer-Verlag.

66.2 Computer Communication Networks

J. N. Daigle

A **computer communication network** is a collection of applications hosted on different machines and interconnected by an infrastructure that provides communications among the communicating entities. While the applications are generally understood to be computer programs, the generic model includes the human being as an application. In fact, one or all of the "applications" that are communicating may be human beings.

This section summarizes the major characteristics of computer communication networks. The objective is to provide a concise introduction that will allow the reader to gain an understanding of the key distinguishing characteristics of the major classes of networks that exist today and some of the issues involved in the introduction of emerging technologies.

There are a significant number of well-recognized books in this area. Among these are the excellent texts by Schwartz [1987], Tanenbaum [1988], and Spragins [1991], which have enjoyed wide acceptance by both students and practicing engineers and cover most of the general aspects of computer communication networks. Stallings [1990a, 1990b, 1990c] covers a broad array of standards in this area. Other books that have been found to be especially useful by practitioners are those by Rose [1990] and Black [1992].

The latest developments are, of course, covered in the current literature, conference proceedings, and the notes of standards meetings. A pedagogically oriented magazine that specializes in computer communications networks is *IEEE Network*, but *IEEE Communications* and *IEEE Computer* often also contain interesting articles in this area. *ACM Communications Review*, in addition to presenting pedagogically oriented articles, often presents very useful summaries of the latest standards activities. Major conferences that specialize in computer communications include the IEEE INFOCOM and ACM SIGCOMM series, which are held annually.

We will begin our discussion with a brief statement of how computer networking came about and a capsule description of the networks that resulted from the early efforts. Networks of this generic class, called **wide-area networks (WANs),** are broadly deployed today, and there are still a large number of unanswered questions with respect to their design. The issues involved in the design of those networks are basic to the design of most networks, whether wide area or otherwise. In the process of introducing these early systems, we will describe and contrast three basic types of communication switching: circuit, message, and packet.

We will next turn to a discussion of computer communication **architecture,** which describes the structure of communication-oriented processing software within a communication processing system. Our discussion is limited to the **International Standards Organization/Open Systems Interconnection** (ISO/OSI) **reference model** (ISORM) because it provides a framework for discussion of some of the modern developments in communications in general and communication networking in particular. This discussion is necessarily simplified in the extreme, thorough coverage requiring on the order of several hundred pages, but we hope our brief description will enable the reader to appreciate some of the issues.

Having introduced the basic architectural structure of communication networks, we will next turn to a discussion of an important variation on this architectural scheme: the **local-area network** (LAN). Discussion of this topic is important because it helps to illustrate what the reference model is and what it is not. In particular, the architecture of LANs illustrates how the ISORM can be adapted for specialized purposes. Specifically, early network architectures anticipate networks in which individual node pairs are interconnected via a single link, and connections through the network are formed by concatenating node-to-node connections.

LAN architectures, on the other hand, anticipate all nodes being interconnected in some fashion over the same communication link (or medium). This, then, introduces the concept of adaption layers in a natural way. It also illustrates that if the services provided by an architectural layer are carefully defined, then the services can be used to implement virtually any service desired by the user, possibly at the price of some inefficiency.

After discussing LANs, we will conclude our article with a discussion of two of the variants in packet switching transmission technology: frame relay and a recent development in basic transmission technology called the **asynchronous transfer mode**, which is a part of the larger **broadband integrated services digital network** effort. These technologies are likely to be important building blocks for the computer communication networks of the future.

General Networking Concepts

Data communication networks have existed since about 1950. The early networks existed primarily for the purpose of connecting users of a large computer to the computer itself, with additional capability to provide communications between computers of the same variety and having the same operating software. The lessons learned during the first twenty or so years of operation of these types of networks have been valuable in preparing the way for modern networks. For the purposes of our current discussion, however, we will think of communication networks as being networks whose purpose is to interconnect a set of applications that are implemented on hosts manufactured by possibly different vendors and managed by a variety of operating systems. Networking capability is provided by software systems that implement standardized interfaces specifically designed for the exchange of information among heterogeneous computers.

During the late 1960s, many forward-looking thinkers began to recognize that significant computing resources (that is, supercomputers) would be expensive and unlikely to be affordable by many of the researchers needing this kind of computer power. In addition, they realized that significant computing resources would not be needed all of the time by those having local access. If the computing resource could be shared by a number of research sites, then the cost of the resource could be shared by its users.

Many researchers at this time had computing resources available under the scenario described in the first paragraph above. The idea of interconnecting the computers to extend the reach of these researchers to other computers developed. In addition, the interconnection of the computers would provide for communication among the researchers themselves. In order to investigate the feasibility of providing the interconnectivity anticipated for the future using a new technology called **packet switching,** the Advanced Research Projects Agency (ARPA) of the Department of the Army sponsored a networking effort, which resulted in the computer communication network called the ARPANET.

The end results of the ARPA networking effort, its derivatives, and the early initiatives of many companies such as AT&T, DATAPOINT, DEC, IBM, and NCR have been far-reaching in the extreme. Any finitely delimited discussion of the accomplishments of those efforts would appear to underestimate their impact on our lives. We will concentrate on the most visible product of these efforts, which is a collection of programs that allows applications running in different computers to intercommunicate. Before turning to our discussion of the software, however, we will provide a brief description of a generic computer communication network.

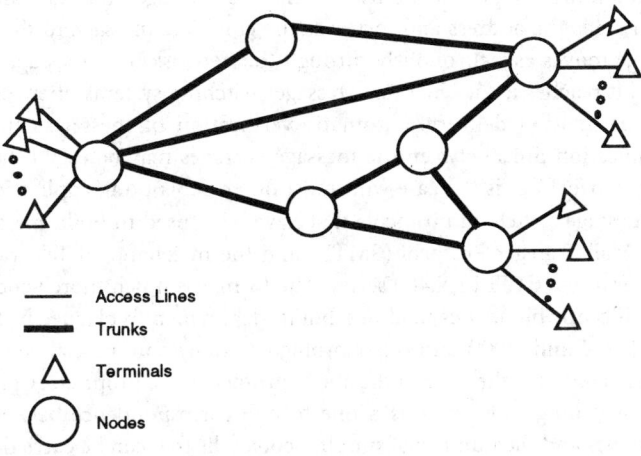

FIGURE 66.5   Generic computer communication network.

Figure 66.5 shows a diagram of a generic computer communication network. The most visible components of the network are the terminals, the **access lines,** the **trunks,** and the **switching nodes.** Work is accomplished when the users of the network, the terminals, exchange messages over the network.

The terminals represent the set of communication terminating equipment communicating over the network. Equipment in this class includes, but is not limited to, user terminals, general-purpose computers, and database systems. This equipment, either through software or through human interaction, provides the functions required for information exchange between pairs of application programs or between application programs and people. The functions include, but are not limited to, call set-up, session management, and message transmission control. Examples of applications include electronic mail transfer, terminal-to-computer connection for time sharing or other purposes, and terminal-to-database connections.

Access lines provide for data transmission between the terminals and the network switching nodes. These connections may be set up on a permanent basis or they may be switched connections, and there are numerous transmission schemes and protocols available to manage these connections. The essence of these connections, however, from our point of view is a channel that provides data transmission at some number of bits per second (bps), called the channel capacity, C. The access line capacities may range from a few hundred bits per second to in excess of millions of bits per second, and they are usually not the same for all terminating equipments of a given network. The actual information-carrying capacity of the link depends upon the protocols employed to effect the transfer; the interested reader is referred to Bertsekas and Gallagher [1987], especially Chapter 2, for a general discussion of the issues involved in transmission of data over communication links.

Trunks, or internodal trunks, are the transmission facilities that provide for transmission of data between pairs of communication switches. These are analogous to access lines, and, from our point of view, they simply provide a communication path at some capacity, specified in bits per second.

There are three basic switching paradigms: circuit, message, and packet switching. **Circuit switching** and packet switching are transmission technologies while message switching is a service technology. In circuit switching, a call connection between two terminating equipments corresponds to the allocation of a prescribed set of physical facilities that provide a transmission path of a certain bandwidth or transmission capacity. These facilities are dedicated to the users for the duration of the call. The primary performance issues, other than those related to quality of transmission, are related to whether or not a transmission path is available at call set-up time and how calls are handled if facilities are not available.

Message switching is similar in concept to the postal system. When a user wants to send a message to one or more recipients, the user forms the message and addresses it. The message switching system reads the address and forwards the complete message to the next switch in the path. The message moves asynchronously through the network on a message switch-to-message switch basis until it reaches its destination. Message switching systems offer services such as mail boxes, multiple destination delivery, automatic verification of message delivery, and bulletin boards. Communication links between the message switches may be established using circuit or packet switching networks as is the case with most other networking applications.

Examples of message switching protocols that have been used to build message switching systems are Simple Mail Transfer Protocol (SMTP) and the International Telegraph and Telephone Consultative Committee (CCITT) X.400 series. The former is much more widely deployed, while the latter has significantly broader capabilities, but its deployment is plagued by having two incompatible versions (1984 and 1988) and other problems. Many commercial vendors offer message switching services based on either one of the above protocols or a proprietary protocol.

In the circuit switching case, there is a one-to-one correspondence between the number of trunks between nodes and the number of simultaneous calls that can be carried. That is, a trunk is a facility between two switches that can service exactly one call, and it does not matter how this transmission facility is derived. Major design issues include the specification of the number of trunks between node pairs and the routing strategy used to determine the path through a network in order to achieve a given call blocking probability. When blocked calls are queued, the number of calls that may be queued is also a design question.

A packet-switched communication system exchanges messages among users by transmitting sequences of packets which comprise the messages. That is, the sending terminal equipment partitions a message into a sequence of packets, the packets are transmitted across the network, and the receiving terminal equipment reassembles the packets into messages. The transmission facility interconnecting a given node pair is viewed as a single trunk, and the transmission capacity of this trunk is shared among all users whose packets traverse both nodes. While the trunk capacity is specified in bits per second, the packet handling capacity of a node pair depends both upon the trunk capacity and the nodal processing power.

In many packet-switched networks, the path traversed by a packet through the network is established during a call set-up procedure, and the network is referred to as a virtual circuit packet switching network. Other networks provide datagram service, a service that allows users to transmit individually addressed packets without the need for call set-up. Datagram networks have the advantage of not having to establish connections before communication takes place, but they have the disadvantage that every packet must contain complete addressing information. Virtual circuit networks have the advantage that addressing information is not required in each packet, but have the disadvantage that a call set-up must take place before communication can occur. Datagram is an example of **connectionless service** while virtual circuit is an example of **connection-oriented service.**

Prior to the late 1970s, signaling for circuit establishment was in-band. That is, in order to set up a call through the network, the call set-up information was sent sequentially from switch to switch using the actual circuit that would eventually become the circuit used to connect the end users. In an extreme case, this amounted to trying to find a path through a maze, sometimes having to retrace one's steps before finally emerging at the destination or just simply giving up when no path could be found. This had two negative characteristics: first, the rate of signaling information transfer was limited to the circuit speed, and second, the circuits that could have been used for accomplishing the end objective were being consumed simply to find a path between the end-points. This resulted in tremendous bottlenecks on major holidays, which were solved by virtually disallowing alternate routes through the toll switching network.

An alternate out-of-band signaling system, usually called **common-channel interoffice signaling** (CCIS), was developed primarily to solve this problem. Signaling now takes place over a signaling network that is partitioned from the network that carries the user traffic. This principle is incorporated into the concept of integrated services digital networks (ISDNs), which is described thoroughly in

Helgert [1991]. The basic idea of ISDN is to offer to the user some number of 64-kbps access lines plus a 16-kbps access line through which the user can describe to an ISDN how the user wishes to use each of the 64-kbps circuits at any given time. The channels formed by concatenating the access lines with the network interswitch trunks having the requested characteristics are established using an out-of-band signaling system, the most modern of which is signaling system #7 (SS#7).

In either virtual circuit or datagram networks, packets from a large number of users may simultaneously need transmission services between nodes. Packets arrive at a given node at random times. The switching node determines the next node in the transmission path, and then places the packet in a queue for transmission over a trunk facility to the next node. Packet arrival processes tend to be bursty, that is, the number of packet arrivals over fixed-length intervals of time has a large variance. Because of the burstiness of the arrival process, packets may experience significant delays at the trunks. Queues may also build due to the difference in transmission capacities of the various trunks and access lines. Combining of packets that arrive at random times from different users onto the same line, in this case a trunk, is called statistical multiplexing.

In addition to the delays experienced at the input to trunks, packets may also experience queueing delays within the switching nodes. In particular, the functions required for packet switching are effected by executing various software processes within the nodes, and packets must queue while awaiting execution of the various processes on their behalf.

Both transmission capacities and nodal processing capabilities are available over a wide range of values. If the trunk capacities are relatively low compared to nodal processing capability, then delays at switching nodes may be relatively small. If line capacities are large compared to nodal processing capabilities, however, delays due to nodal processing may be significant. In the general case, all possible sources of delay should be examined to determine where bottlenecks, and consequently delays, occur.

It is often the case that a particular point in the communication network, either a processing node or a trunk, is the primary source of delay. In this case, this point is usually singled out for analysis, and a simple model is invoked to analyze the performance at that point. The results of this analysis, combined with results of other analyses, result in a profile of overall system performance. In this case, the key aspect of the analysis is to choose an appropriate model for the isolated analysis. In this way, a simplified analysis leading to useful results can be performed, and this can lead to an improved network design.

Protocol design and performance issues are frequent topics of discussion at both general conferences in communications and those specialized to networking. The reader is encouraged to consult the proceedings of the conferences mentioned earlier for a better appreciation of the range of issues and the diversity of the proposed solutions to the issues.

Computer Communication Network Architecture

In this section, we will begin with a brief, high-level definition of the ISORM. The reference model has seven layers, none of which can be bypassed conceptually. In general, a layer is defined by the types of services it provides to its users and the quality of those services. For each layer in the ISO/OSI architecture, the user of a layer is the next layer up in the hierarchy, except for the highest layer for which the user is an application. Clearly, when a layered architecture is implemented under this philosophy, then the quality of service obtained by the end user, the application, is a function of the quality of service provided by all of the layers. In order to clarify the communications strategy of the ISO/OSI architecture, we will provide a discussion of the layer 2 services in some detail.

There is significant debate over whether the efforts of the ISO/OSI community are leading to the best standards (or even standards that have any merit whatever!). Limiting our discussion to the ISORM is, by no means, an endorsement of the actual protocols that have been developed in the ISO arena; there are actually more widely deployed and successful standards in other arenas. On the

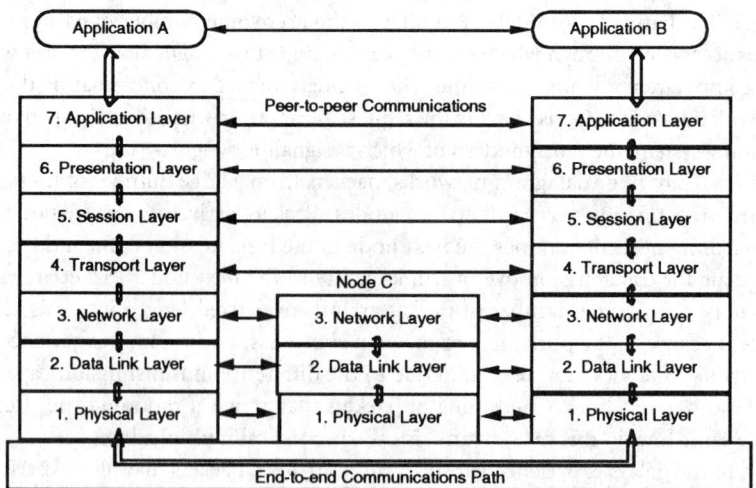

FIGURE 66.6   Layered architecture for ISO/OSI reference model.

other hand, the ISORM is very useful for discussing network architecture principles, and these principles apply across the board. Thus, we choose to base our discussion on the ISORM.

Figure 66.6, adopted from Spragins [1991], shows the basic structure of the OSI architecture and how this architecture is envisaged to provide for exchange of information between applications. As shown in the figure, there are seven layers: application, presentation, session, transport, network, data link, and physical. Brief definitions of the layers follow, but the reader should bear in mind that substantial further study will be required to develop an understanding of the practical implications of the definitions:

- *Physical layer:*   Provides electrical, functional, and procedural characteristics to activate, maintain, and deactivate physical data links that transparently pass the bit stream for communication between data link **entities.**

- *Data link layer:*   Provides functional and procedural means to transfer data between network entities; provides for activation, maintenance, and deactivation of data link connections, character and frame synchronization, grouping of bits into characters and frames, error control, media access control, and flow control.

- *Network layer:*   Provides switching and routing functions to establish, maintain, and terminate network layer connections, and transfer data between transport layers.

- *Transport layer:*   Provides host-to-host, cost-effective, transparent transfer of data, end-to-end flow control, and end-to-end quality of service as required by applications.

- *Session layer:*   Provides mechanisms for organizing and structuring dialogues between application processes.

- *Presentation layer:*   Provides for independent data representation and syntax selection by each communicating application and conversion between selected contexts and the internal architecture standard.

- *Application layer:*   Provides applications with access to the ISO/OSI communication stack and certain distributed information services.

As we have mentioned previously, a layer is defined by the types of services it provides to its users. In the case of a request or a response, these services are provided via invocation of **service primitives** of the layer in question by the layer that wants the service performed. In the case of an indication or a confirm, these services are provided via invocation of service primitives of the layer in question by the same layer that wants the service performed.

This process is not unlike a user of a programming system calling a subroutine from a scientific subroutine package in order to obtain a service, say, matrix inversion or memory allocation. For example, a request is analogous to a CALL statement in a FORTRAN program, and a response is analogous to the RETURN statement in the subroutine that has been CALLed. The requests for services are generated asynchronously by all of the users of all of the services and these join (typically prioritized) queues along with other requests and responses while awaiting servicing by the processor or other resource such as a transmission line.

The service primitives fall into four basic types: request, indication, response, and confirm. These types are defined as follows:

- *Request:* A primitive sent by layer $(N + 1)$ to layer N to request a service.
- *Indication:* A primitive sent by layer N to layer $(N + 1)$ to indicate that a service has been requested of layer N by a different layer $(N + 1)$ entity.
- *Response:* A primitive sent by layer $(N + 1)$ to layer N in response to an *indication* primitive.
- *Confirm:* A primitive sent by layer N to layer $(N + 1)$ to indicate that a response to an earlier *request* primitive has been received.

In order to be more specific about how communication takes place, we will now turn to a brief discussion of layer 2, the data link layer. The primitives provided by the ISO data link (DL) layer are as follows [Stallings, 1990a]:

DL_CONNECT.request	DL_RESET.request
DL_CONNECT.indication	DL_RESET.indication
DL_CONNECT.response	DL_RESET.response
DL_CONNECT.confirm	DL_RESET.confirm
DL_DATA.request	DL_DISCONNECT.request
DL_DATA.indication	DL_DISCONNECT.indication
DL_DATA.response	DL_UNITDATA.request
DL_DATA.confirm	DL_UNITDATA.indication

Each primitive has a set of **formal parameters,** which are analogous to the formal parameters of a procedure in a programming language. For example, the parameters for the DL_CONNECT.request primitive are the Called Address, the Calling Address, and the Quality of Service Parameter Set. The four primitives are used in the establishment of data link connections. The called address and the calling address are analogous to the telephone numbers of two parties of a telephone call, while the quality of service parameter set allows for the negotiation of various agreements such as throughput measured in bits per second.

All four DL_CONNECT primitives are used to establish a data link. An analogy to an ordinary phone call can better illustrate the basic idea of the primitives. The DL_CONNECT.request is equivalent to picking up the phone and dialing. The phone ringing at the called party's end is represented by the DL_CONNECT.indication. DL_CONNECT.response is equivalent to the called party lifting the receiver and answering, and DL_CONNECT.confirm is equivalent to the calling party hearing the response of the called party.

In general, communication takes place between peer layer protocols by the exchange of **protocol data units (PDUs)**, which contain all of the information required for the receiving protocol entity to provide the required service. In order to exchange PDUs, entities at a given layer use the services of the next lower layer. The data link primitives listed above include both connection-mode primitives and connectionless-mode primitives. For connection-mode communications, a connection must be established between two peer entities before they can exchange PDUs.

For example, suppose a network layer entity in Host A wishes to be connected to a network layer entity in Host B, as shown in Figure 66.6. Then the connection would be accomplished by the concatenation of two data link connections: one between A and C, and one between C and B. In order to establish the connection, the network layer entity in Host A would issue a DL_CON-NECT.request to its associated data link entity, providing the required parameters. This data link entity would then transmit this request to a data link entity in C, which would issue a DL_CON-NECT.indication to a network entity in C. The network entity in C would then analyze the parameters of the DL_CONNECT.indication and realize that the target destination is B. This network layer entity would then reissue the DL_CONNECT.request to its data link entity, which would transmit the request to a data link entity in B. The data link entity in B would send a DL_CONNECT.indication to a network layer entity in B, and this entity would issue a DL_CONNECT.response back to the data link entity in B. This DL_CONNECT.response would be relayed back to the data link entity in A following the same sequence of events as in the forward path. Eventually, this DL_CON-NECT.response would be converted to a DL_CONNECT.confirm by the data link entity in A and passed to the network entity in A, thus completing the connection.

Once the connection is established, data exchange between the two network layer entities can take place; that is, the entities can exchange PDUs. For example, if a network layer entity in Host A wishes to send a PDU to a network layer entity in Host B, the network layer entity in Host A would issue a DL_DATA.request to the appropriate data link layer entity in Host A. This entity would package the PDU together with appropriate control information into a data link service data unit (DLSDU) and send it to its peer at C. The peer at C would deliver it to the network entity at C, which would forward it to the data link entity in C providing the connection to Host B. This entity would then send the DLSDU to its peer in Host B, and this data link entity would pass the PDU to Host B network entity via a DL_DATA.indication.

Network layer PDUs are called packets and data link layer PDUs are called frames. The data link layer does not know that the information it is transmitting is a packet; to the data link layer entity, the packet is simply user information. From the perspective of a data link entity, it is not necessary to have a network layer. The network layer exists to add value for the user of the network layer to the services provided by the data link layer. In the example above, value was added by the network layer by providing a relaying capability since Hosts A and C were not directly connected. Similarly, the data link layer functions on a hop-by-hop basis, each hop being completely unaware that there are any other hops involved in the communication. We will see later that the data link need not be limited to a single physical connection.

The philosophy of the ISO/OSI architecture is that in addition to the software being layered, implementations are not allowed to bypass entire layers; that is, every layer must appear in the implementation. This approach was developed after the approach defined for the ARPANET project, which is hierarchical, was fully developed. In the hierarchical approach, the layer interfaces are carefully designed, but any number of layers of software can be bypassed by any application (or other higher-layer protocol) that provides the appropriate functionality. These two approaches have been hotly debated for a number of years, but as the years pass, the approaches are actually beginning to look more and more alike for a variety of reasons that will not be discussed here.

The ISO/OSI layered architecture described above would appear to be very rigid, not allowing for any variations in underlying topology or variations in link reliability. However, as we shall see, this is not necessarily the case. As an example, ISO 8348, which developed as a result of the X.25 project, provides only connection-oriented service, and it was originally intended as the only network layer standard for ISO/OSI. However, ISO 8473, or ISO-IP, which is virtually identical to the Department of Defense (DoD) internet protocol (DoD-IP) developed in the ARPANET project, has since been added to the protocol suite to provide connectionless service as well as internet service. An interesting aside is that because of the addressing limitations of DoD-IP, the Internet Administrative Board (IAB) has recently recommended replacement of the DoD-IP protocol by the ISO-IP protocol, thus bringing the process full circle.

The ISO/OSI protocol suite is in a constant state of revision as new experience reveals the need for additional capabilities and flexibility. Some of this additional flexibility and functionality is being provided through the use of so-called **adaption sublayers**, which enhance the capabilities of a given layer so that it can use the services of a lower layer with which it was not specifically designed for compatibility.

Interestingly, the use of adaption sublayers is only a short step away from using adaption layers that would allow applications to directly interface with any ISO layer. This would result in a hierarchical rather than layered architecture; to wit: ISORM becomes DoDRM. Indeed, fundamental changes in the national (and worldwide) communications infrastructure appear to be leading naturally in the hierarchical direction. Of course, the indiscriminate use of such adaptions would lead back to the proliferation of incompatible protocols and interfaces, the frustration that led to the current twenty-year standardization crusade! It is refreshing to note that a return to our former state does not appear to be around the corner; most standardization work is actually headed in the direction of allowing open systems to intercommunicate.

Local-Area Networks and Internets

We will now turn to a discussion of LANs, which have inherent properties that make the use of sublayers particularly attractive. In this section, we will discuss the organization of communications software for LANs. In addition, we will introduce the idea of **internets**, which were brought about to a large extent by the advent of LANs. We will discuss the types of networks only briefly and refer the reader to the many excellent texts on the subject. Layers 4 and above for local-area communications networks are identical to those of wide-area networks. However, because the hosts communicating over a LAN share a single physical transmission facility, the routing functions provided by the network layer, layer 3, are not necessary. Thus, the functionality of a layer 3 in a LAN can be substantially simplified without loss of utility. On the other hand, a data link layer entity must now manage many simultaneous data link layer connections because all connections entering and leaving a host on a single LAN do so over a single physical link. Thus, in the case of connection-oriented communications, the software must manage several virtual connections over a single physical link.

There were several basic types of transmission schemes in use in early LANs. Three of these received serious consideration for standardization: the **token ring, token bus,** and **carrier-sense multiple access** (CSMA). In a token ring network, the stations are configured on a physical ring around the medium. A token rotates around this physical ring, visiting each host (or station) in turn. A station wishing to transmit data must wait until the token is available to that station. In a token bus LAN, the situation is the same, except that the stations share a common bus and the ring is logical rather than physical. In a CSMA network, the stations are bus connected, and a station may transmit whenever other stations are not currently transmitting. That is, a station wishing to transmit senses the channel, and if there is no activity, the station may transmit. Of course, the actual access protocol is significantly more complicated than this.

In the early 1980s, there was significant debate over which LAN connection arrangement was superior, a single choice being viewed as necessary. This debate centered on such issues as cost, network throughput, network delay, and growth potential. Performance evaluation based on queueing theory played a major role in putting these issues in perspective. For thorough descriptions of LAN protocols and queueing models used to evaluate their performance, the interested reader is referred to Hammond and O'Reilly [1986].

All three of the access methods mentioned above became IEEE standards (IEEE 802) and eventually became ISO standards (ISO 8802 series) because all merited standardization. On the other hand, all existed for the express purpose of exchanging information among peers, and it was recognized at the outset that the upper end of the data link layer could be shared by all three access techniques. On the other hand, the lower-level functions of the layer deal with interfacing to the physical media. Here, drastic differences in the way the protocol had to interface with the media were recognized. Thus, a different **media-access control sublayer** (MAC) was needed for each of the access techniques.

The decision to use a common logical link control (LLC) sublayer for all of the LAN protocols apparently ushered in the idea of adaption sublayers. The reason for splitting the layer is simple: a user of the data link control (DLC) layer need not know what kind of medium provides the communications; all that is necessary is that the user understand the interface to the DLC layer.

On the other hand, the media of the three types of access protocols provide transmission service in different ways, so software is needed to bridge the gap between what the user of the service needs, which is provided by the LLC, and how the LLC uses the media to provide the required service. Thus, the MAC sublayer was born.

This idea has proven to be valuable as new types of technologies have become available. For example, the new fiber-distributed digital interface (FDDI) uses the LLC of all other LAN protocols, but its MAC is completely different from the token ring MAC even though FDDI is a token ring protocol. Reasons for needing a new MAC for LLC are provided in Stallings [1990b].

One of the more interesting consequences of the advent of local-area networking is that many traditional computer communication networks became internets overnight. LAN technology was used to connect stations to a host computer, and these host computers were already on a WAN. It was then a simple matter to provide a relaying, or bridging, service at the host in order to provide wide-area interconnection of stations to LANs to each other. In short, the previously established WANs became networks for interconnection of LANs; that is, they were interconnecting networks rather than stations. Internet performance suddenly became a primary concern in the design of networks.

More recently, FDDI is being thought of as a mechanism to provide LAN interconnection on a site basis, and a new type of network, the **metropolitan-area network** (MAN) has been under study for the interconnection of LANs within a metropolitan area. The primary media configuration for MANs is a dual bus configuration and it is implemented via the distributed queue, dual bus (DQDB) protocol, also known as IEEE 802.6. The net effect of this protocol is to use the dual bus configuration to provide service approaching the first-come–first-served service discipline to the traffic entering the FDDI network, which is remarkable considering that the LANs being interconnected are geographically dispersed. Interestingly, DQDB concepts have recently been adapted to provide wide-area communications. Specifically, structures have been defined for transmitting DQDB frames over standard DS-1 (1.544 megabits per second [Mb/s]) and DS-3 (6.312 Mb/s) facilities, and these have been used as the basis for a new service offering called switched multimegabit data services (SMDS).

As of this writing, advances in LANs design and new forms of LANs are emerging. One example is wireless LANs, which are LANs in which radio or photonic links serve as cable replacements. Wireless LAN technology is viewed by many as crucial to the evolution of personal communication networks. Another example is the asynchronous transfer mode-based LAN, which is mentioned in the next section following a general discussion of asynchronous transfer mode.

Some Additional Recent Developments

In this subsection, we will describe two recent developments of significant interest in communication networking: **fast packet networks** and transmission using the *asynchronous transfer mode* (ATM), which is a part of the larger *broadband integrated services digital network* (B-ISDN) effort.

As we have mentioned previously, there is really no requirement that the physical media between two adjacent data link layers be composed of a single link. In fact, if a path through the network is initially established between two data link entities, there is no reason that DLC protocols need to be executed at intermediate nodes. Figure 66.7, adapted from Bhargava and Hluchyj [1990] shows how the end-to-end connection might be implemented. A network implemented in the fashion indicated in Fig. 66.7 is called a fast packet network (FPN).

From Fig. 66.7, it is seen that the data link layer is partitioned into three sublayers: the data link control sublayer (which parallels the LLC layer of LANs), the fast packet adaption (FPA) sublayer, and the fast packet relay (FPR) sublayer. The function of the fast packet adaption sublayer is to segment the layer-2 PDU, the frame, into smaller units, called fast packets, for transmission over the

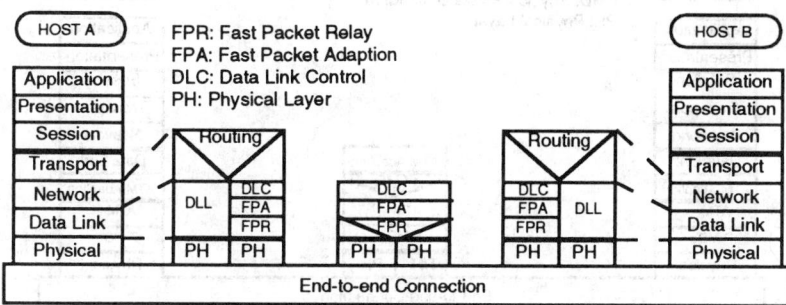

FIGURE 66.7   Fast packet switched layered architecture.

FPN. These fast packets contain information that identifies the source and destination node names and the frame to which they belong so that they can be routed through the network and reassembled at the destination.

The fast packets are statistically multiplexed onto a common physical link by the FPR sublayer for transmission. At intermediate nodes, minor error checking, fast packet framing, fast packet switching, and queueing takes place. If errors are found, then the fast packet is dropped. When the fast packets reach their destination, they are reassembled into a frame by the FPA sublayer and passed on to the DLC sublayer where normal DLC functions are performed.

The motivation for FPNs is that since link transmission is becoming more reliable, extensive error checking and flow control are not needed across individual links; an end-to-end check should be sufficient. Meanwhile, the savings in processing due to not processing at the network layer can be applied to frame processing, which allows interconnection of the switches at higher line speeds.

Since bits-per-second costs decrease with increased line speed, service providers can offer savings to their customers through FPNs. Significant issues are fast packet loss probability and retransmission delay. Such factors will determine the retransmission strategy deployed in the network. Of course, the goal is to improve network efficiency, so a significant issue is whether FPNs are better than ordinary packet networks and, if so, by how much.

Another recent innovation is the ATM, usually associated with B-ISDN. The idea of ATM is to partition a user's data into many small segments, called cells, for transmission over the network. Independent of the data's origin, the cell size is 53 octets, of which 5 octets are for use by the network itself for routing and error control. Users of the ATM are responsible for segmentation and reassembly of their data. Any control information required for this purpose must be included in the 48 octets of user information in each cell. In the usual case, these cells would be transmitted over networks that would provide users with 135 Mb/s and above data transmission capacity (with user overhead included in the capacity).

The segmentation of units of data into cells introduces tremendous flexibility for handling different types of information, such as voice, data, image, and video, over a single transmission facility. As a result, LANs, WANs, and MANs based on the ATM paradigm are being designed, and indeed deployed. A significant portion of the deployment activity is a national testbed program, which involves industrial/academia cooperation, under joint sponsorship of the National Science Foundation (NSF) and the Defense Advanced Research Projects Agency (DARPA). There is also significant private investment in developing this technology; for example, experimental ATM-based LANs are already in place at the Digital Equipment Corporation (DEC) research facility in Palo Alto, California. There is some possibility that LANs of this type, rather than of the FDDI type, will be the dominant means of providing high-speed LAN and LAN-interconnect services.

There are numerous possibilities for connection of hosts to ATM networks, but they all share a common architecture, which consists of three sublayers: the ATM adaption layer (AAL), the ATM layer, and the physical media-dependent (PMD) layer. Connection of hosts to ATM at a given layer

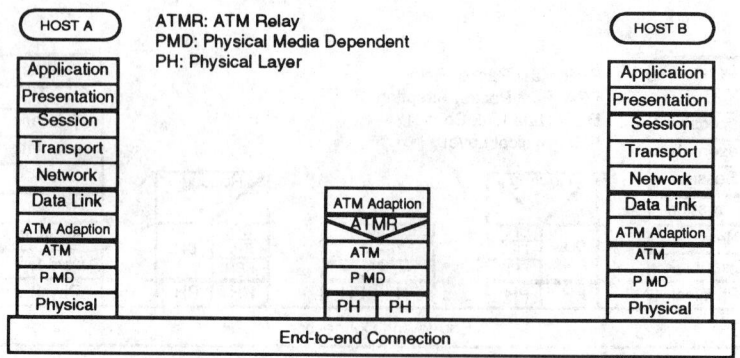

FIGURE 66.8   Asynchronous transfer mode layered architecture.

is achieved through developing an AAL for the layer in question. For example, one might decide to adapt to ATM at the network layer. In that case, the transport layer would operate as usual, and the AAL would be designed to process data structures from the transport layer to produce data structures for use by the ATM layer and vice versa. Of course, all hosts communicating with each other in this way would use the same AAL.

Figure 66.8 shows an example of how an ISO/OSI host might connect to an ATM network. Below the data link layer is the ATM adaption layer (AAL), which provides for call control across the ATM network and for segmentation and reassembly of frames from the data link layer. The current estimate for the amount of overhead needed per cell for AAL purposes is 4 octets, leaving 44 octets for user information.

At the present time, end-to-end connections at the ATM level are expected to be connection oriented. As cells traverse the network, they are switched on a one-by-one basis, using information contained in the five ATM overhead octets to follow the virtual path established during the ATM call set-up. Typically, cells outbound on a common link are statistically multiplexed, and if buffers are full, cells are dropped. In addition, if one or more errors are found in a cell, then the cell is dropped.

In the case of data transmission, a lost cell will result in an unusable frame unless the data is encoded to guard against cell loss prior to transmission. Coding might be provided by the AAL, for example. The trade-offs involved in coding and retransmission and their impact upon network throughput, delay and complexity are not well understood at the time of this writing. Part of the reason for this is that cell loss probability and the types of traffic that are likely to use the network are not thoroughly understood. Resolution of these issues accounts for a significant portion of the research activity in computer communication networking at this time. The relevant American National Standards Institute and CCITT documents are frequently updated to include the results.

Defining Terms

Access line: A communication line that connects a user's terminal equipment to a switching node.

Adaption sublayer: Software that is added between two protocol layers to allow the upper layer to take advantage of the services offered by the lower layer in situations where the upper layer is not specifically designed to interface directly to the lower layer.

Architecture: The set of protocols defining a computer communication network.

Asynchronous transfer mode (ATM):   A mode of communication in which communication takes place through the exchange of tiny units of information called cells.

Broadband integrated services digital network (B-ISDN):   A generic term that generally refers to the future network infrastructure that will provide ubiquitous availability of integrated voice, data, imagery, and video services.

Carrier-sense multiple access: A random-access method of sharing a bus-type communications medium in which a potential user of the medium listens before beginning to transmit.

Circuit switching: A method of communication in which a physical circuit is established between two terminating equipments before communication begins to take place. This is analogous to an ordinary phone call.

Common-channel interoffice signaling: The use of a special network, dedicated to signaling, to establish a path through a communication network, which is dedicated to the transfer of user information.

Computer communication network: Collection of applications hosted on different machines and interconnected by an infrastructure that provides intercommunications.

Connection-oriented service: A mode of packet switching in which a call is established prior to any information exchange taking place. This is analogous to an ordinary phone call, except that no physical resources need to be allocated.

Connectionless service: A mode of packet switching in which packets are exchanged without first establishing a connection. Conceptually, this is very close to message switching, except that if the destination node is not active, then the packet is lost.

Entity: A software process that implements a part of a protocol in a computer communication network.

Fast packet networks: Networks in which packets are transferred by switching at the frame layer rather than the packet layer. Such networks are sometimes called frame relay networks. At this time, it is becoming in vogue to think of frame relay as a service, rather than transmission, technology.

Formal parameters: The parameters passed during the invocation of a service primitive; similar to the arguments passed in a subroutine call in a computer program.

International Standards Organization reference model: A model, established by ISO, that organizes the functions required by a complete communication network into seven layers.

Internet: A network formed by the interconnection of networks.

Local-area network: A computer communication network spanning a limited geographic area, such as a building or college campus.

Media-access control: A sublayer of the link layer protocol whose implementation is specific to the type of physical medium over which communication takes place and which controls access to that medium.

Message switching: A service-oriented class of communication in which messages are exchanged among terminating equipments by traversing a set of switching nodes in a store-and-forward manner. This is analogous to an ordinary postal system. The destination terminal need not be active at the same time as the originator in order that the message exchange take place.

Metropolitan-area network: A computer communication network spanning a limited geographic area, such as a city; sometimes features interconnection of LANs.

Packet switching: A method of communication in which messages are exchanged between terminating equipments via the exchange of a sequence of fragments of the message called packets.

Protocol data unit (PDU): The unit of exchange of protocol information between entities. Typically, a PDU is analogous to a structure in C or a record in Pascal; the protocol is executed by processing a sequence of PDUs.

Service primitive: The name of a procedure that provides a service; similar to the name of a subroutine or procedure in a scientific subroutine library.

Switching node: A computer or computing equipment that provides access to networking services.

Token bus: A method of sharing a bus-type communications medium that uses a token to schedule access to the medium. When a particular station has completed its use of the token, it broadcasts the token on the bus, and the station to which it is addressed takes control of the medium.

Token ring: A method of sharing a ring-type communications medium that uses a token to schedule access to the medium. When a particular station has completed its use of the token, it transmits the token on the ring, and the station that is physically next on the ring takes control.

Trunk: A communication line between two switching nodes.

Wide-area network: A computer communication network spanning a broad geographic area, such as a state or country.

References

D. Bertsekas and R. Gallagher, *Data Networks*, Englewood Cliffs, N.J.: Prentice-Hall, 1987.

A. Bhargava and M. G. Hluchyj, "Frame losses due to buffer overflow in fast packet networks," *Proc. IEEE INFOCOM '90*, San Francisco, 1990, pp. 132–139.

U. Black, *TCP/IP and Related Protocols*, New York: McGraw-Hill, 1992.

J. L. Hammond and P. J. P. O'Reilly, *Performance Analysis of Local Computer Networks*, Reading, Mass.: Addison-Wesley, 1986.

H. J. Helgert, *Integrated Services Digital Networks*. Reading, Mass.: Addison-Wesley, 1991.

M. Rose, *The Open Book: A Practical Perspective on OSF*, Englewood Cliffs, N.J.: Prentice-Hall, 1990.

M. Schwartz, *Telecommunications Networks: Protocols, Modeling and Analysis*, Reading, Mass.: Addison-Wesley, 1987.

J. D. Spragins, *Telecommunications: Protocols and Design*, Reading, Mass.: Addison-Wesley, 1991.

W. Stallings, *Handbook of Computer-Communications Standards: The Open Systems Interconnection (OSI) Model and OSI-Related Standards*, New York: Macmillan, 1990a.

W. Stallings, *Handbook of Computer-Communications Standards: Local Network Standards*, New York: Macmillan, 1990b.

W. Stallings, *Handbook of Computer-Communications Standards: Department of Defense (DOD) Protocol Standards*, New York: Macmillan, 1990c.

A. S. Tanenbaum, *Computer Networks*, 2nd ed., Englewood Cliffs, N.J.: Prentice-Hall, 1988.

Further Information

There are many conferences and workshops that provide up-to-date coverage in the computer communications area. Among these are the IEEE INFOCOM and ACM SIGCOMM conferences and the IEEE Computer Communications Workshop, which specialize in computer communications and are held annually. In addition, IEEE GLOBCOM (annual), IEEE ICC (annual), IFIPS ICCC (biannual), and the International Telecommunications Congress (biannual) regularly feature a substantial number of paper and panel sessions in networking.

The *ACM Communications Review,* a quarterly, specializes in computer communications and often presents summaries of the latest standards activities. *IEEE Network,* a bimonthly, contains tutorial articles on all aspects of computer communications and includes a regular column on books related to the discipline. Additionally, *IEEE Communications* and *IEEE Computer,* monthly magazines, frequently have articles on specific aspects of networking.

For those who wish to be involved in the most up-to-date activities, there are many interest groups on the Internet, a worldwide TCP/IP-based network, that specialize in some aspect of networking. *The User's Directory of Computer Networks* (Digital Press, T. L. LaQuey, Ed.) provides an excellent introduction to the activities surrounding internetworking and how to obtain timely information.

66.3 Local-Area Networks

Joseph Bannister and Mario Gerla

The local-area network (LAN) is a communication network that interconnects computers and computer-based devices, such as file servers, printers, and graphics terminals. The LAN is characterized as being contained completely within the premises of a single business entity—which almost always owns and operates the network—and this distinguishes the LAN from public-domain networks such as metropolitan- or wide-area networks. The LAN, then, is normally restricted to a few hundred stations (i.e., devices that attach directly to the LAN) that span a limited geographical area, so that no two connected stations are separated by a distance of more than a few kilometers. Moreover, the LAN can be distinguished from the computer or backplane bus, which interconnects components, boards, or devices that comprise a single computer. The LAN uses serial—rather than parallel—transmission, which also differentiates it from the computer bus. In contrast to today's wide-area networks, information is transmitted over LANs at high speeds and with very low error rates.

The LAN often employs fully broadcast media, or **physical media** that allow each station's transmissions to be received by all other stations. Thus, a broadcast capability is often an integral feature of the LAN. Frequently, the LAN also provides for multicasting, a generalization of broadcasting in which a specified subset of stations receives a transmission.

LANs are based on a variety of technologies that include twisted copper-wire pairs, coaxial cable, optical fibers, wireless infrared and radio for signal transport, as well as several integrated circuit families for transmitters, receivers, and the implementation of low-level protocols.

The **topology** of a LAN refers to the physical layout of the transmission media and the logical arrangement of the stations on those media. Four topologies are commonly used in LANs: the bus, ring, star, and tree topologies, which are illustrated in Fig. 66.9.

The LAN Service Model

Within the scope of the well-known Open Systems Interconnection seven-layer reference model, the LAN occupies the two bottom layers, namely, the physical and data link layers, as shown in Fig.

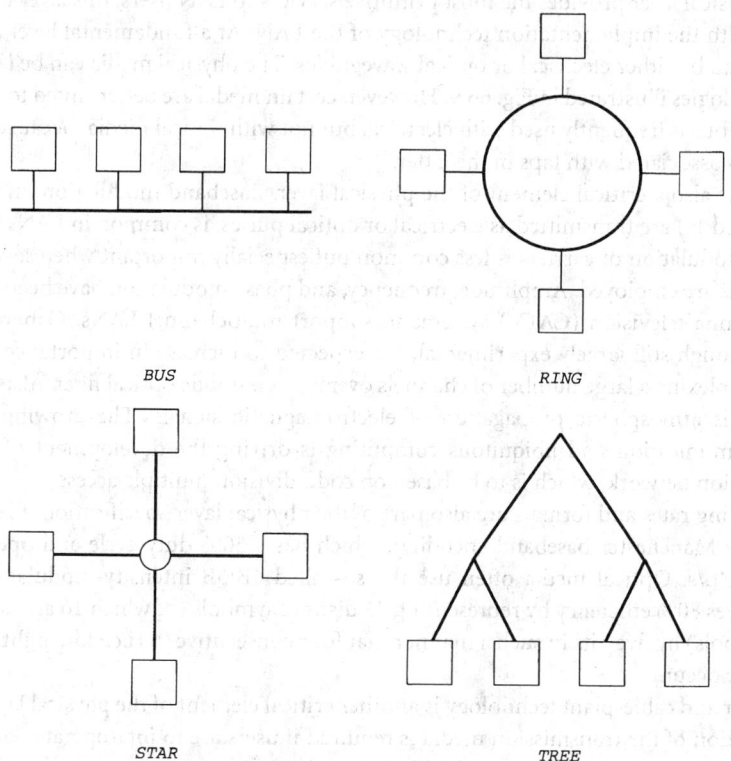

FIGURE 66.9 LAN topologies.

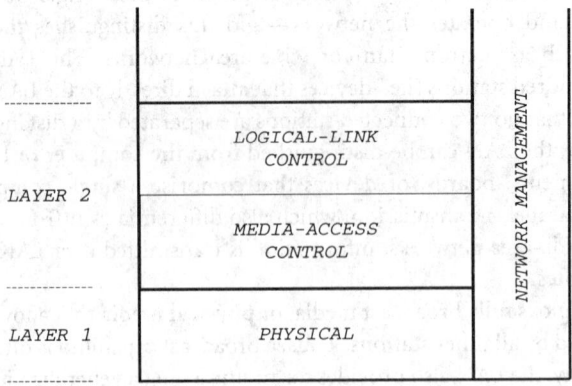

FIGURE 66.10   LAN service model.

66.10. The physical layer specifies the most primitive services of the LAN, e.g., media characteristics, signal formats, waveforms, signaling rates, timing, and mechanical aspects of connectors, etc. The data link layer uses the services of the physical layer to provide multiple access for stations sharing the media. Station or network management, which is shown as a vertical "layer" in Fig. 66.10, is responsible for maintaining a necessary level of performance, fault detection and recovery, and configuration and security functions.

The Physical Layer

Since the physical layer provides the most primitive services to LAN users, this layer is most closely associated with the implementation technology of the LAN. At a fundamental level the transmission media can be either electrical or optical waveguides. The physical media can be laid out as one of those topologies illustrated in Fig. 66.9. However, certain media are better suited to some topologies, e.g., the bus is frequently used with electrical but not with optical media, because there is high insertion loss associated with taps in the latter.

Signaling is also a critical element of the physical layer. Baseband modulation, in which digital signals (0s and 1s) are transmitted as electrical or optical pulses, is common in LANs because of its simplicity. Modulation of carriers is less common but especially important when several independent channels are employed. Amplitude, frequency, and phase modulation have been used in community antenna television (CATV) systems to support multichannel LANs. Coherent lightwave systems, although still largely experimental, are expected to increase in importance because they permit multiplexing a large number of channels over a single-mode optical fiber. Also of increasing importance is atmospheric propagation of electromagnetic signals. The growing demand for mobile communication and ubiquitous computing is driving the development of the personal communication network, which is to be based on code-division multiple access.

The signaling rates and formats are also part of the physical layer specification. Electrical media generally use Manchester baseband encoding, which has a 50% duty cycle and operates at rates below 100 Mb/s. Optical media often use the so-called 4B/5B intensity-modulation encoding, which achieves 80% efficiency by representing 27 distinct symbols (of which 16 are data and 11 are control symbols) by five bits in such a manner that four consecutive 0s (i.e., low-light power levels) should never occur.

Connector and cable-plant technology is another critical element of the physical layer. Thorough characterization of the transmission media is required if users are to interoperate with each other. The type of cable—e.g., shielded or unshielded twisted copper-wire pairs, coaxial cable, and single-mode or multimode optical fiber—must be specified. Furthermore, the connectors between stations and the cable plant are defined as part of the physical layer. Stations can attach via passive taps or can actively repeat signals; in the latter case a bypass switch is usually provided as the station's interface to the cable plant.

The Data Link Layer

The data link layer is often divided into two sublayers, i.e., the media-access and logical-link control (MAC and LLC) sublayers, as shown in Fig. 66.10. The LLC sublayer [see *Logical Link Control*] uses the services of the MAC sublayer to provide to its user a connection-oriented service between stations that includes flow and error control or a connectionless service that does little more than multiplex upper-layer connections. The connection-oriented LLC protocol gives the service user the illusion of having a dedicated point-to-point link between a pair of communicating stations.

The MAC sublayer specifies the **media access protocol** that stations use to share the media. In fully broadcast media no more than one station may transmit at a time, so the MAC sublayer manages exclusive access to the broadcast media. The ring topology is well suited to a token-passing MAC protocol, which gives transmission rights to the station holding the token. The token is represented by a special packet that is passed sequentially from station to station. When a station recognizes the token, it seizes it and begins transmitting buffered packets, or passes the token to the next station if it has no packet to transmit. To limit the time that a station can hold the token, the MAC protocol can implement one of several disciplines:

- One-shot service, in which the station releases the token when it has transmitted one packet

- Exhaustive service, in which the station releases the token when it has no more packets to transmit

- Gated service, in which the station releases the token when all packets that were buffered at token-acquisition time have been transmitted

- Token-timing service, in which the station releases the token at the expiration of a timer

The IEEE 802.5 token ring standard specifies a token-timing service discipline that requires transmissions to be completed within a fixed time after the token is seized, but implementations sometimes use the simpler one-shot service discipline. The ANSI X3T9.5 fiber distributed data interface (FDDI) standard uses an adaptive token-timing service discipline that is intended to guarantee a minimum amount of (synchronous) bandwidth to each station.

A variation of the token ring protocol is the token bus protocol, which allows the token to be passed in a specified order. In the token bus protocol, which is often used with the bus or tree topologies, a station broadcasts the token, specifying the successor station in an address field of the token packet. Although all stations receive the token, only the addressed successor station can seize it.

A MAC scheme that is widely used with the bus topology is carrier-sense multiple access (CSMA). A contention protocol, CSMA operates by allowing any station to transmit a buffered packet if it senses that the bus is idle. If two stations are ready to transmit their packets at nearly the same time, they will both sense that the bus is idle and their transmissions will collide, i.e., the superimposed bits of the packets will be garbled. The propagation delay—or time it takes for the packet to travel from one station to the other—dictates the window of vulnerability for CSMA; the larger the window, the more collisions are likely. To overcome the problem of collisions, CSMA is often enhanced with collision detection (CSMA/CD) by enabling stations to monitor their transmissions for the garbled bits associated with collisions. When a collision is detected, the station aborts its transmission and reschedules it by backing off for a period of time. The binary exponential backoff algorithm specifies that the random backoff time is drawn uniformly from the interval between 0 and $2^n - 1$ time units, where n is the number of times the packet has collided.

A time-slotted bus maintains on the bus a continuous stream of short, fixed-length frames that are initially empty but can be filled with data as they pass stations with waiting packets. The distributed queue, dual bus (DQDB) local- and metropolitan-area network uses two-directional buses so that a station can reserve on its downstream bus a frame for its upstream-destined packets.

In the star topology stations are homed into a central hub which can manage their access to the media. Active hubs physically control media access, while passive hubs merely broadcast incoming packets to specific output ports. Linear combiner/dividers based on lithium niobate technology allow incoming optical signals to be combined and distributed to output ports according to electronically programmed combining and dividing ratios. A common scheme is to use time-division multiplexing with the star topology. The hub can serve as the central controller, allocating time slots to individual stations, or reservations can be used in the manner of a satellite-based network.

The Management Layer

LAN-specific network management functions are referred to as station management. Station management covers five areas—configuration, performance, fault, accounting, and security management.

Monitoring and controlling the LAN are essential elements of station management. By monitoring the media, stations maintain a record of important measurements, such as the number of a specific type of packet transmitted or received, the number of different kinds of errors, and the source addresses of received packets. Such measurements are made available to an application in the station or to a management center. Thus are applications able to monitor and collect, correlate, and act upon key LAN statistics. Likewise, designated applications are able to effect changes in the LAN by writing to specific variables within stations, which collectively comprise the so-called management information base. For instance, station management informs the MAC sublayer of its unique LAN address by writing the value to a special MAC register.

Some management functions are distributed across the LAN and are implemented at a low level. To recover after the failure of a dual-fiber cable, stations automatically enter into a procedure to reconfigure around the failure and reestablish connectivity. Although such procedures can be viewed as station management, they are sometimes specified as part of the MAC sublayer, because they are so tightly integrated with media access.

Other Features

The basic features of media access are often augmented to provide specialized services and features.

Specialized LAN Services

LAN users have special communication requirements that must be supported by the physical and data link layers. In particular, the MAC sublayer is responsible for providing specialized services. Although all MAC sublayers support asynchronous traffic by providing for the simple, best-effort delivery of packets, some MAC sublayers also support other classes of traffic. To synchronous traffic, which requires a set amount of preallocated bandwidth, the properly designed MAC sublayer guarantees a maximum packet response time. The adaptively timed token-passing protocol of FDDI is capable of supporting synchronous traffic, i.e., at token-capture time the station has a fixed amount of time during which to transmit synchronous packets, and the token is guaranteed to return within a certain amount of time. Isochronous traffic, which requires a fixed amount of traffic to be periodically delivered, is also accommodated by some MAC sublayers. DQDB uses preallocated time slots to provide isochronous service.

Priorities are also important in LANs. Therefore many LANs transmit queued packets in accordance with priorities assigned to the packets. Prioritization can be on a LAN-wide basis or merely within the station. Most LANs offer some method for prioritizing the transmission of packets.

Reliability and Availability

Being a shared resource, the LAN should have a high degree of reliability and availability. The media should not be a single point of failure, and no individual station should be able to prevent—maliciously or otherwise—the delivery of service to other stations. LANs are designed to withstand both transient and permanent failures of the media and stations.

Transmitted information is subject to short bursts of errors and must also be protected. The connection-oriented service at the LLC sublayer is intended to recover from errors—such as garbled, dropped, or out-of-sequence packets—by positively acknowledging packets and retransmitting packets not acknowledged within the timeout window. The MAC sublayer usually provides error-detecting codes that can recognize an error burst of several consecutive bits (a favorite is the 32-bit cyclic redundancy code, which is easily implemented as a linear feedback shift register). Errors can also be recognized at the physical layer when they cause code violations, e.g., the absence of transitions in the Manchester or 4B/5B codes. Some LANs even use error-correcting codes for protecting time-sensitive information.

Other protection mechanisms are used to tolerate cable breaks and station malfunctions. The use of fully broadcast media makes a LAN vulnerable to media failure, since this effectively partitions the stations into noncommunicating groups. To cope with this problem, redundant cables are provided and a mechanism for reconfiguring from the bad to the good cable is built into the LAN protocols. A popular approach for the token ring can be seen in the counter-rotating dual-ring scheme, which is illustrated in Fig. 66.11. If a cable segment or an active station fails, the stations

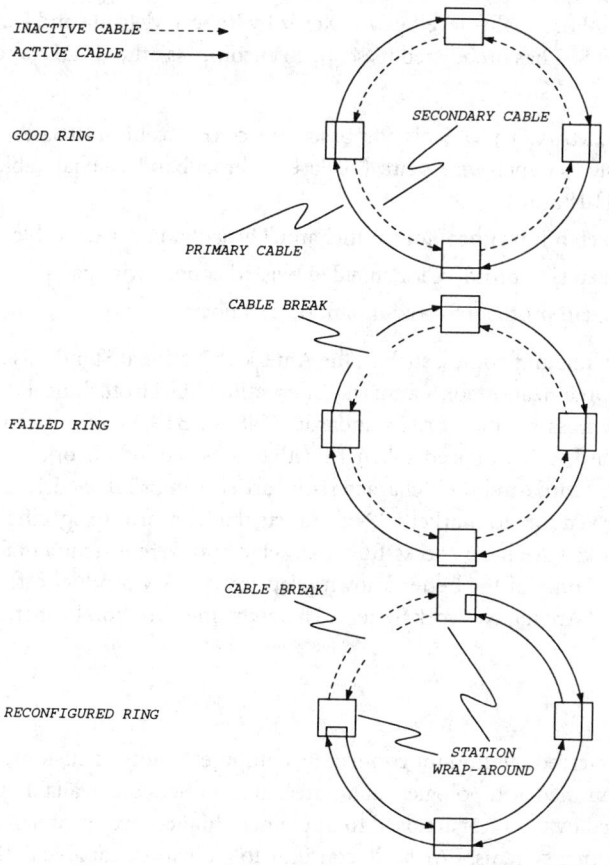

FIGURE 66.11 Reconfiguration of dual counter-rotating rings.

adjacent to the failure can reconfigure the ring by executing "wrap-around" operations. The new configuration uses the spare cable in conjunction with the original cable to form a new ring. Given the complexity of such a reconfiguration procedure, it is usually necessary for station management to coordinate the actions of the stations.

Special mechanisms for adding and removing stations to and from the LAN might also be required. Since the physical addition or removal of a station can disrupt the transmission of data, protocols for reestablishing a lost token could also be necessary.

The Importance of LAN Standards

LAN standards play a central role in promoting the goal of universal connectivity among a community of users. The standardization of communication services and protocols allows all conforming implementations to exchange information. Consequently, the importance of LAN standards has grown steadily. Currently, several LAN standards have been established to support the different communication requirements of users.

The first LANs—developed in the early 1970s—were proprietary products meant to interconnect one vendor's computer products. By 1980, however, Project 802 of the Institute of Electrical and Electronics Engineers (IEEE) had recognized the need for publicly disseminated LAN standards and eventually published a specification of the CSMA/CD protocol that any vendor may implement. Furthermore, the definition of the standard was sanctioned by companies that participated in the IEEE Working Group's balloting process, so that the standard was viewed as an open, nonproprietary solution. The IEEE 802.3 Working Group chose a protocol that was based closely on the Ethernet LAN originally developed at Xerox by Robert Metcalfe and David Boggs.

The IEEE Project 802 has broadened its scope to encompass other LAN standards. These include the following:

- 802.3: The CSMA/CD protocols for baseband coaxial cable (10Base5 and 10Base2), unshielded twisted copper-wire pairs (10BaseT), broadband coaxial cable (10Broad36), and optical fiber (10BaseF)
- 802.4: The token bus protocol for multichannel broadband coaxial cable
- 802.5: The token ring protocol for shielded twisted copper-wire pairs
- 802.6: The DQDB protocol for redundant optical fibers

Other standards-making bodies, such as the American National Standards Institute (ANSI) and the International Organization for Standards/International Electrotechnical Committee (ISO/IEC) have developed or cross-adopted LAN standards. ANSI's X3T9.5 committee is responsible for the FDDI LAN standard, a high-speed token ring that uses redundant optical fibers. Some of the important LAN standards and their characteristics are shown in Table 66.1.

The trend is for vendors to market LAN products that conform to specific standards. However, proprietary networks have been successfully marketed and were instrumental in the development of LAN standards. Some of the better known proprietary-LAN product offerings were the Xerox Ethernet, Datapoint Arcnet, Network Systems Hyperchannel, Proteon Pronet, Sytek System 20, and AT&T DATAKIT.

Summary

The LAN is the preferred method for connecting computers within a customer's premises. A number of transmission media, topologies, data rates, and services are available to meet users' needs. The services offered by the LAN are used to implement higher-layer protocols that are required by distributed computing systems. LANs will continue to grow more capable in the data and bit-error rates they achieve, the functionality they provide, and the number and geographical span of the stations they support.

Table 66.1   Characteristics of Standard LANs

	CSMA/CD	Token Ring	Token Bus	FDDI	DQDB
Standard	IEEE 802.3	IEEE 802.5	IEEE 802.4	ANS X3T9.5	IEEE 802.6
Topology	Bus, tree, star	Ring	Tree	Ring	Pseudobus
Media	Coax, UTP, MMF	STP	Coax	MMF, SMF	SMF
Encoding	MC, FSK, AM/PSK	DMC	FSK, AM/PSK	4B/5B	4B/5B
Data rate	10 Mb/s	4 Mb/s 16 Mb/s	1 Mb/s 5 Mb/s 10 Mb/s	100 Mb/s	34 Mb/s 45 Mb/s 140 Mb/s 155 Mb/s
Features		Priorities	Priorities, ST, multichannel	Priorities, ST, dual ring	Priorities, IT, dual bus

AM/PSK = amplitude modulation/phase-shift keying	IT = isochronous traffic
ANS = American National Standard	MC = Manchester coding
CSMA/CD = carrier-sense multiple-access with collision detection	MMF = multimode fiber
DMC = differential Manchester coding	SMF = single-mode fiber
DQDB = distributed queue, dual bus	ST = synchronous traffic
FDDI = fiber distributed data interface	STP = shielded twisted pair
FSK = frequency-shift keying	UTP = unshielded twisted pair
IEEE = Institute of Electrical and Electronics Engineers	

Defining Terms

Media-access protocol: The protocol that permits one of a group of contending stations to access the media exclusively. Media-access protocols are generally based on token passing or carrier sense.

Physical media: The communication channel over which signals are transmitted. Broadcast media, in which all stations receive each transmission, are primarily used in local-area networks. Common media are optical fibers, coaxial cable, twisted copper-wire pairs, and airwaves.

Topology: The paths and switches of a local-area network that provide the physical interconnection among stations. The most common topologies are the bus, ring, tree, and star.

References

American National Standard for Information Systems—Fiber Distributed Data Interface (FDDI), ANSI Standards X3.139, X3.148, X3.166, X3.184.

Carrier Sense Multiple Access with Collision Detection (CSMA/CD) Access Method and Physical Layer Specifications, ANSI/IEEE Standard 802.3, ISO/IEC Standard 8802/3.

Distributed Queue Dual Bus (DQDB) Metropolitan Area Network (MAN), Proposed IEEE Standard 802.6.

Logical Link Control, ANSI/IEEE Standard 802.2, ISO/IEC Standard 8802/2.

Token-Passing Bus Access Method and Physical Layer Specifications, ANSI/IEEE Standard 802.4, ISO/IEC Standard 8802/4.

Token Ring Access Method and Physical Layer Specifications, ANSI/IEEE Standard 802.5, ISO/IEC Standard 8802/5.

Further Information

A popular, frequently updated textbook on LANs is W. Stallings, *Local Networks,* 3rd ed., New York: Macmillan, 1990.

Leading journals that publish research articles on LANs include:
- *IEEE Transactions on Communications*
- *IEEE Transactions on Networking*
- *Computer Networks and ISDN Systems*
- *IEEE Network*

Four annual conferences that cover the topic of LANs are:
- The IEEE INFOCOM Conference on Computer Communications
- The IEEE Conference on Local Computer Networks
- The ACM SIGCOMM Conference on Communications Architectures and Protocols
- The EFOC/LAN European Fibre Optic Communications and Local Area Networks Conference

66.4 The Intelligent Network

Richard B. Robrock II

The term *intelligent network* refers to the concept of deploying centralized databases in the telecommunications network and querying those databases to provide a wide variety of network services such as 800 Service (toll-free service) and credit card calling. The first use of these centralized databases was in AT&T's network in 1981 where they were used to facilitate the setup of telephone calls charged to a calling card. Today such databases are widely deployed throughout the United States and support the handling of billions of calls per year; they represent an infrastructure on which to build new services.

A History of Intelligence in the Network

The first "intelligence" in the telephone network took the form of rows of human telephone operators, sitting side by side, plugging cords into jacks to facilitate the handling of calls. These operators established calls to far-away points, selected the best routes and provided billing information. They were also an information source—providing time or weather or perhaps disseminating the local news. Moreover, they had the opportunity to demonstrate a kind of heroism—gathering volunteers to save a house from fire, helping to catch a prowler, locating a lost child, and on and on. In the early years of telephony, the feats of the telephone operator were indeed legendary.

In the 1920s, however, technology became available that allowed automatic switching of telephone calls through the use of sophisticated electromechanical switching systems. Initially, these switches served as an aid to operators; ultimately, they led to the replacement of operators. The combination of the rotary telephone dial and the electromechanical switch allowed customers to directly dial calls without the assistance of operators. This led to a reduction of human intelligence in the network.

Another dramatic change took place in the telephone network in 1965; it was called software. It came with the marriage of the computer and the telephone switching system in the first stored-program control switch. With the introduction of switching software came a family of Custom Calling services (speed calling, call waiting, call forwarding, and three-way calling) for residential customers, and a robust set of Centrex features (station attendant, call transfer, abbreviated dialing, etc.) for business customers. The first software programs for these stored-program control switches contained approximately 100,000 lines of code; by 1990 some of these switching systems became enormously complex, containing 10 million lines of code and offering hundreds of different services to telephone users.

During the 1980s, a new architectural concept was introduced; it came to be called the intelligent network. It allowed new telecommunications services to be introduced rapidly and in a ubiquitous and uniform fashion. Feature and service availability in the network ceased to be solely dependent upon the hardware and software in stored-program control switches. Rather some new intelligence was centralized in databases which were accessed using packet switching techniques. Most significantly, the intelligent network started to provide some of the capabilities that operators had made available in the early years of telephony. The remaining sections of this chapter describe the intelligent network, its characteristics, and its services. They also provide a description of the advanced intelligent network, which dramatically broadens the participation in the creation of new services.

The Intelligent Network

The intelligent network architecture is illustrated in Fig. 66.12; its primary elements are a switching system, a signaling network, a centralized database, and an operations support system which supports the database. The architectural concept is a simple one. When a customer places a telephone call which requires special handling, such as a toll-free call (800 Service) or credit card call, that call is intercepted by the switching system which suspends call processing while it launches a query through a signaling network to a centralized database. The database, in turn, retrieves the necessary information to handle the call and returns that information through the signaling network to the switch so that the call can be completed. The role of the operations support system is to administer the appropriate network and customer information that resides in the database.

It is conceivable that the database in this architecture could reside in the switching system, and the signaling network in this instance would not be required. However, that would magnify the task of administering the customer information, since that information would be contained in thousands of switches instead of dozens of centralized databases. In addition, even more importantly, there are two shortcomings associated with basing many of the potential new services in switches, rather than utilizing centralized databases to provide information for the switches. The first is a deployment problem. As of 1990 there were approximately 15,000 switches in the United States, and a single switch can cost millions of dollars. To introduce a new service in local switches and to make it widely available generally requires some not-so-simple changes in those switches or, in some cases, replacement of certain switch types altogether. These switch modifications typically take years to implement and require a tremendous capital investment. As a result, ten years after introduction, Custom Calling services were available to fewer than 1% of the residential customers in the United States.

A second problem with switch-based services has been that a single service sometimes behaves differently in different switch types. For example, the speed calling access patterns are different in various stored-program control switches. The public is not particularly sensitive to this fact,

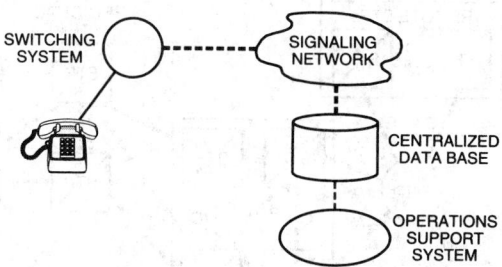

FIGURE 66.12 Intelligent network architecture—telephone calls which require special handling are intercepted in a switching system which launches queries through a signaling network to a centralized database. (*Source:* R. B. Robrock II, "The intelligent network—Changing the face of telecommunications," *Proc. IEEE*, vol. 79, no. 1, pp. 7–20, January 1991. © 1991 IEEE.)

because speed calling is not associated with an individual but rather an individual's station set. People live in a mobile society, however, and they want to have their services available from any station set and have them behave the same from any station set.

The intelligent network architecture has been the key to solving both the deployment problem and service uniformity problem associated with switch-based services. Services deployed using an intelligent network centralized database are immediately ubiquitous and uniform throughout a company's serving area.

Intelligent Network Systems

In 1981, AT&T introduced into the Bell System a set of centralized databases called network control points; they supported two applications—the Billing Validation Application for Calling Card Service (credit card calling) and the INWATS database used to support 800 Service. Queries were launched to these databases through AT&T's common-channel interoffice signaling (CCIS) network.

In 1984, following the divestiture of the Regional Bell Operating Companies from AT&T, the regional companies began planning to deploy their own **common-channel signaling (CCS)** networks and their own centralized databases. They selected the **signaling system 7** protocol for use in their signaling networks, called CCS7 networks, and they named their databases **service control points (SCPs).**

The CCS7 Network

A general architecture for a regional signaling network is shown in Fig. 66.13. The network is made up of **signal transfer points (STPs),** which are very reliable, high-capacity packet switches that route signaling messages between network access nodes such as switches and SCPs. To perform these routing functions, the STPs each possess a large routing database containing translation data.

The CCS7 network in Fig. 66.13 contains both local STPs and regional STPs. The STPs are typically deployed in geographically separated pairs so that in the event of a natural disaster at one site, such as an earthquake, flood, or fire, the total traffic volume can be handled by the second site. Indeed, redundancy is provided at all key points so that no single failure can isolate a node.

As illustrated in Fig. 66.13, the following link types have been designated:

- A-links connect an access node, such as a switching system or SCP, to both members of an STP pair.

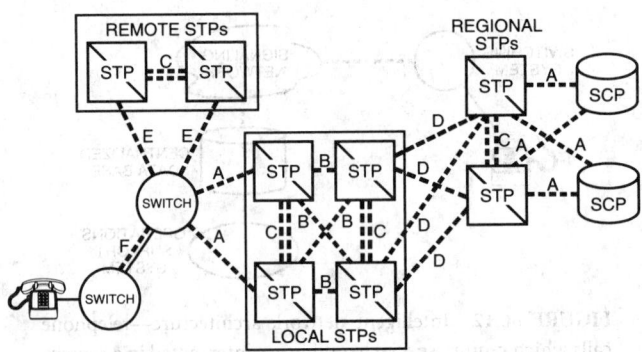

FIGURE 66.13 Link arrangements in a CCS7 signaling network. (*Source:* R. B. Robrock II, "The intelligent network—Changing the face of telecommunications," *Proc. IEEE,* vol. 79, no. 1, pp. 7–20, January 1991. © 1991 IEEE.)

- B-links interconnect two STP pairs forming a "quad" of four signaling links where each STP independently connects to each member of the other pair.

- C-links are the high-capacity connections between the geographically separated members of an STP pair.

- D-links connect one STP pair to a second STP pair at another level in the signaling hierarchy or to another carrier.

- E-links connect an access node to a remote STP pair in the signaling network and are rarely used.

- F-links directly interconnect two access nodes without the use of an STP; they are nonredundant.

The CCS7 links normally function at 56 kb/s in North America while links operating at 64 kb/s are common in Europe.

The CCS7 signaling network provides the underlying foundation for the intelligent network, and the regional telephone companies in the United States began wide-scale deployment of these networks in 1986; several large independent telephone companies and interexchange carriers (ICs) soon followed. They used these networks for both trunk signaling between switches as well as for direct signaling from a switch to a database.

The Service Control Point

The "brains" of the intelligent network is the SCP. It is an on-line, fault-tolerant, transaction-processing database which provides call handling information in response to network queries. The SCP deployed for 800 Service is a high-capacity system capable of handling more than 300 queries per second or 1 million per hour. It is a real-time system with a response time of less than one half second, and it is a high-availability system with a downtime of less than 3 minutes per year for a mated SCP pair. The SCP is also designed to accommodate growth, which means that processing power or memory can be added to an in-service system without interrupting service. In addition, it is designed to accommodate graceful retrofit, which means that a new software program can be loaded into an in-service SCP without disrupting service.

Data Base 800 Service

SCPs have been deployed throughout the United States in support of the Data Base 800 Service mandated by the Federal Communications Commission. This service provides its subscribers with number portability so that a single 800 number can be used with different carriers. The Data Base 800 Service architecture is shown in Fig. 66.14. With this architecture, 800-number calls are routed from an end office to a service switching point (SSP) which launches queries through a CCS7 signaling network to the SCP. The SCP identifies the appropriate carrier, as specified by the 800 Service subscriber, and then, if appropriate, translates the 800 number to a plain old telephone (POTS) number. This information is subsequently returned to the SSP so that the call can be routed through the network by handing the call off to the appropriate carrier. This technology allows subscribers to select the carrier and the POTS number as a function of criteria such as time of day, day of week, percent allocation, and the location of the calling station. Thus the SCP provides two customer-specified routing information functions: a carrier identification function and an address translation function.

The SCP 800 Service database is administered by a single national **service management system (SMS)**. The SMS is an interactive operations support system that is used to process and update customer records. It is the interface between the customer and the SCP. It translates a language which is friendly to a customer into a language which is friendly to on-line, real-time databases. Along the way, it validates the customer input.

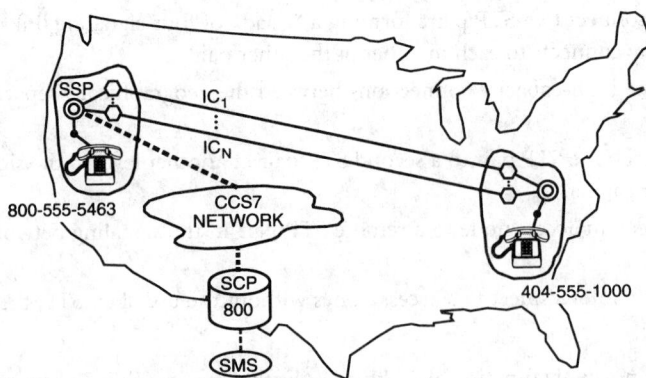

FIGURE 66.14 Data Base 800 Service—800-number calls are routed to an SSP which launches queries through a CCS7 network to an SCP containing the 800 database. In this example, the SCP translates the 800 Service number of 800-555-5463 into the POTS number of 404-555-1000. (*Source:* R. B. Robrock II, "The intelligent network—Changing the face of telecommunications," *Proc. IEEE,* vol. 79, no. 1. pp. 7–20, January 1991. © 1991 IEEE.)

Alternate Billing Services

Alternate billing services (ABS) have also been implemented using the intelligent network architecture. Alternate billing is an umbrella title which includes Calling Card Service, collect calling, and bill-to-third-number calling. The network configuration supporting ABS is shown in Fig. 66.15.

With this architecture, when a customer places a Calling Card call, the call is routed to an operator services system (OSS) which suspends call processing and launches a query through a CCS7 signaling network. The query is delivered to an SCP which contains the **line information database (LIDB)** application software. The LIDB application can provide routing information, such as identifying the customer-specified carrier which is to handle the call, as well as provide screening functions, such as the Calling Card validation used to authorize a call. The LIDB then returns the appropriate information to the OSS so that the call can be completed. The LIDBs are supported by the

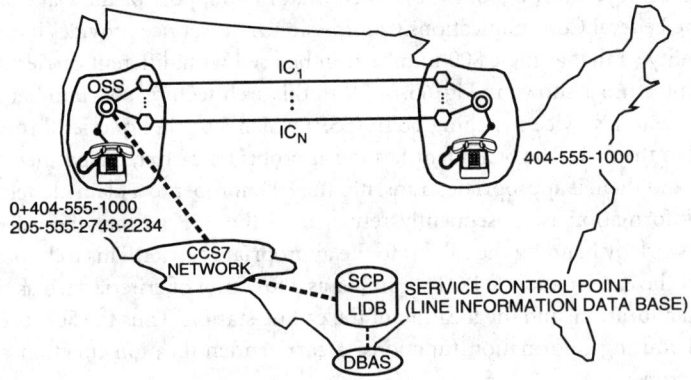

FIGURE 66.15 Alternate billing services—calls are routed to an OSS which launches queries through the CCS7 network to SCPs containing the LIDB application. (*Source:* R. B. Robrock II, "The intelligent network—Changing the face of telecommunications," *Proc. IEEE,* vol. 79, no. 1. pp. 7–20, January 1991. © 1991 IEEE.)

database administration system (DBAS), which is an operations support system that processes updates for Calling Card Service as well as other services. Multiple DBAS systems typically support each LIDB.

During 1991, each of the Regional Bell Operating Companies and a number of large independent telephone companies interconnected their CCS7 networks, mostly through STP hubs, to create a national signaling network; it was a process called LIDB interconnect. When it was finished, it meant that a person carrying a particular company's Calling Card could, from anywhere in the United States, query the LIDB containing the associated Calling Card number.

Other Services

For alternate billing services, the SCP is essentially designed to perform two functions: carrier identification and billing authorization. For 800 Service, the SCP provides carrier identification and address translation. These basic functions of authorization, address translation and carrier identification can be used again and again in many different ways. For example, the intelligent network has been used to support private virtual networks (PVNs). PVNs make use of the public telephone network but, by means of software control, appear to have the characteristics of private networks. A PVN serves a closed-user group, and a caller requires authorization to gain access to the network. This screening function on originating calls uses an authorization function. Second, a PVN may offer an abbreviated dialing plan, for example, four-digit dialing. In this instance, the SCP performs an address translation function, converting a four-digit number to a ten-digit POTS number. There may also be a customer-specified routing information function which involves selecting from a hierarchy of facilities; this can be accomplished through use of the SCP carrier identification function.

The SCP in the intelligent network can support a vast number of services ranging from Calling Name Delivery service to messaging service. With Calling Name Delivery, a switch sends a query to the SCP with the ten-digit calling party number; the response is the calling party name which is then forwarded by the switch to a display unit attached to the called party station set. In support of messaging services, the address translation capability of the SCP can be used to translate a person's telephone number to an electronic-mail address. As a result, the sender of electronic mail need only know a person's telephone number.

The Advanced Intelligent Network

The intelligent network architecture discussed thus far is often referred to in the literature as Intelligent Network/1; this architecture has addressed the deployment problem and the service uniformity problem. The next phase in the evolution of this network has come to be called the advanced intelligent network (AIN).

The concept of AIN is that new services can be developed and introduced into the network without requiring carriers to wait for switch generics to be upgraded. Some AIN applications introduce powerful service-creation capabilities which allow nonprogrammers to invoke basic functions offered in the network and stitch together those functions, as illustrated in Fig. 66.16, to constitute a new service. As a result, AIN promises to dramatically shorten the interval required to develop new services. Perhaps of greater significance, it promises to broaden the participation in service creation. In addition, it offers the opportunity to personalize or customize services. The silicon revolution has driven the cost of memory down to the point where it is economically viable to have enough memory in the network to store the service scripts or call processing scenarios that are unique to individuals.

Many people think of the AIN as a collection of network elements, network systems and operations systems; this view might be called a technologist's view. Perhaps a better representation is shown in Fig. 66.17; it shows a collection of people—people empowered to create services.

Historically, the creation of new services provided by the telephone network has been the sole

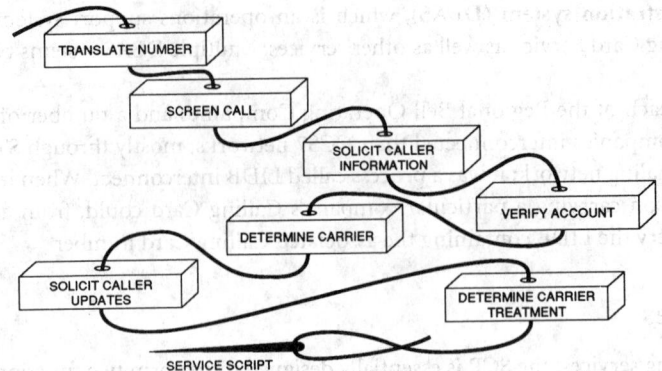

FIGURE 66.16   Creating the service script or scenario for a call by stitching together functional blocks. (*Source:* R. B. Robrock II, "The intelligent network—Changing the face of telecommunications," *Proc. IEEE,* vol. 79, no. 1. pp. 7–20, January 1991. © 1991 IEEE.)

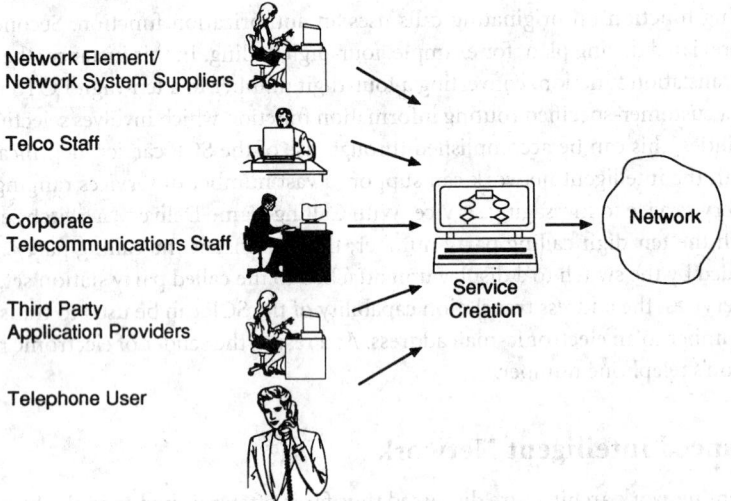

FIGURE 66.17   The advanced intelligent network—a business perspective. (*Source:* R. B. Robrock II, "Putting the Telephone User in the Driver's Seat," International Council for Computer Communication Conference on Intelligent Networks, pp. 144–150, May 1992.)

domain of the network element and network system suppliers. There is perhaps a good analogy with the automobile industry. A market study in the early 1900s predicted that 200,000 was the maximum number of cars that could ever be sold in a single year in the United States; the reasoning was that 200,000 was the maximum number of chauffeurs that could enter the workforce in a single year. In the telecommunications business, the network element and network system suppliers have been the chauffeurs of the network services business.

The service-creation tools offered by the AIN, however, empower telephone company staff to create new services. Moreover, similar tools may well be used by the telecommunications staff of large corporations, or by third-party application providers or even by some segment of the telephone user population. As a result, we may see an explosion in the number of network services.

The AIN introduces very powerful service-creation tools which are used to produce service-logic scripts (programs). In one arrangement, the service creation is done by assembling service-logic graphs from graphical icons that represent functional components of services. The completed graph is then validated with an expert system and tested off-line by executing every leg of the service-logic graph. At this point the service-logic program can be downloaded into the service control point so that it is ready for execution.

To make use of the new service, it is then necessary to set "triggers" in the appropriate service switching point. These triggers can be set for both originating and terminating calls, and they represent events which, should they occur, indicate the need for the switch to launch a query to the SCP for information the switch needs to process the call. The AIN switch generics, which are presently deployed, support several triggers such as "immediate off hook" or "called address." Future AIN switch generics are expected to support several dozen triggers. The first phase of the AIN, called AIN 0, became reality in late 1991 when friendly user trials began in two of the Regional Bell Operating Companies. The rollout has begun.

Back to the Future

The intelligent network, with its centralized databases, has offered a means to rapidly introduce new services in a ubiquitous fashion and with operational uniformity as seen by the end user. The advanced intelligent network has gone on to provide a service-independent architecture, and, with its powerful service-creation capabilities, has empowered nonprogrammers to participate in the development of new services. In many ways, as we go into the future, we are going back to a time when operators were the "human intelligence" in the network. The human intelligence was all but eliminated with the introduction of switching systems, but now the intelligent network is working to put the intelligence of the human operator back into the network.

Defining Terms

Common-channel signaling (CCS): A technique for routing signaling information through a packet-switched network.

Database administration systems (DBAS): An operations support system that administers updates for the line information database.

Line information database (LIDB): An application running on the service control point that contains information on telephone lines and Calling Cards.

Service control point (SCP): An on-line, real-time, fault-tolerant, transaction-processing database which provides call-handling information in response to network queries.

Service management system (SMS): An operations support system which administers customer records for the service control point.

Signal transfer point (STP): A packet switch found in the common-channel signaling network; it is used to route signaling messages between network access nodes such as switches and SCPs.

Signaling system 7 (SS7): A communications protocol used in common-channel signaling networks.

References

AT&T Bell Laboratories, "Common channel signaling," *The Bell System Tech. J.,* vol. 57, no. 2, pp. 221–477, February 1978.

AT&T Bell Laboratories, "Stored program controlled network," *The Bell System Tech. J.,* vol. 61, no. 7, part 3, pp. 1573–1815, September 1982.

Globecom '86: The Global Telecommunications Conference, *Conference Record,* vol. 3, pp. 1311–1335, December 1986.

R.J. Hass and R.W. Humes, "Intelligent network/2: A network architecture concept for the 1990s," International Switching Symposium, *Conference Record,* vol. 4, pp. 944–951, March 1987.

R.B. Robrock, II, "The intelligent network—Changing the face of telecommunications," *Proc. IEEE,* vol. 79, no. 1. pp. 7–20, January 1991.

R.B. Robrock, II, "Putting the telephone user in the driver's seat," International Council for Computer Communication Intelligent Networks Conference, pp. 144–150, May 1992.

R.B. Robrock, II, "The many faces of the LIDB data base," International Conference on Communications, *Conference Record,* June 1992.

Further Information

The bimonthly magazine *Bellcore Exchange* has numerous articles on the intelligent network, particularly in the following issues: July/August 1986, November/December 1987, July/August 1988, and March/April 1989. Articles on AIN service creation appear in the January/February 1992 issue. Subscriptions or single copies are available from the Bellcore Exchange Circulation Manager, 60 New England Avenue, Piscataway, NJ 08854-4196.

The monthly publication *IEEE Communications Magazine* contains numerous articles on the intelligent network. A special issue on the subject was published in January 1992. Copies are available from the IEEE Service Center, 445 Hoes Lane, Piscataway, NJ 08854-4150.

The monthly publication *The Bellcore Digest* lists recent Bellcore publications. There are a series of technical advisories, technical requirements, and special reports that have been issued on the intelligent network. Copies are available by contacting Bellcore Customer Service Toll-Free 1-800-521-CORE (2673).

The bimonthly publication *The AT&T Technical Journal* contains numerous articles on the intelligent network. The advanced intelligent network is the subject of a special issue: Summer 1991, vol. 70, nos. 3–4. Current or recent issues may be obtained from the AT&T Customer Information Center, P.O. Box 19901, Indianapolis, IN 46219.

WALTER R. G. BAKER AND THE ADVENT OF COMMERCIAL TELEVISION IN 1941

It was over fifty years ago that commercial monochrome television broadcasting was introduced in the United States. On 1 July 1941, station WNBT transmitted the first officially sanctioned commercial. It took the form of an image of a Bulova clock that remained on the screen for sixty seconds and cost the sponsor four dollars.

Walter R. G. Baker, who worked for the General Electric Company and served as chairman of the National Television System Committee (NTSC), played a major role in the resolution of conflicting views that enabled the television industry to reach this important stage of development. The NTSC standards that have lasted to the present day included a 525 line picture with interlaced scanning at 30 frames per second and a channel bandwidth of 6 MHz. The Federal Communications Commission (FCC) had approved the recommended standards and issued specific rules for operation in April 1941. Baker presided over the NTSC from July 1940 to March 1941 and also served as chairman of a second NTSC that formulated standards for color television in the early 1950s.

Baker was born in Lockport, New York, in 1892 and graduated in electrical engineering from Union College in Schenectady, New York, in 1916. He received a Master's degree in EE, also from Union College, in 1919. He took a job with General Electric (GE) in 1916 and was a member of the radio engineering department at GE during the early 1920s. From 1924–1929, he had administrative responsibility for radio products manufactured by GE. In 1929 he left GE but returned in 1935 as the manager of GE's radio-television facility in Bridgeport, Connecticut. He was named a GE vice president in 1941.

Baker also served as director of the engineering department of the Radio Manufacturers Association (RMA) in the 1930s. The RMA established two television committees to prepare reports on standards and frequency allocation for FCC hearings held in June 1936. Donald G. Fink has described the work of these two committees as being "the first major step in television standardization in the United States."

In the face of disagreement among RMA and its competitors, the FCC in May 1940 announced that it would be prepared to authorize full commercialization of television "as soon as the engineering opinion of the industry is prepared to approve any one of the competing systems of broadcasting as the standard system." After a meeting with the FCC Chairman, James L. Fly, Baker agreed to organize and chair the NTSC in an effort to reach a consensus. Over the next several months, 168 members devoted approximately 4,000 person hours to meetings. At its final meeting on March 8, 1941, the NTSC changed its recommendation on the number of lines from 441 to 525. The NTSC standards became the basis for an explosive growth in the industry after WW II.

Baker served as president of the Institute of Radio Engineers in 1947 and then chaired the second NTSC from January 1950 until its final meeting in July 1953. The FCC adopted this committee's recommended standards for compatible color television and authorized commercial broadcasting under these standards beginning in January 1954. Baker retired from GE in 1957 and died in 1960.

A process is currently underway to formulate standards for a proposed high definition television (HDTV) service that is quite reminiscent of the process followed by Baker and the NTSC committees. Recent reports indicate that a digital HDTV system is a strong contender for a proposed simulcast HDTV standard mandated by the FCC. FCC testing of several HDTV systems is now underway, although a decision on an HDTV standard is not anticipated before 1993.

Source: Adapted from J. E. Brittain, *Proc. IEEE*, vol. 79, no. 7, pp. 1085–1086, July 1991. © 1991 IEEE.

67

Information Theory

67.1 Signal Detection 1478
General Considerations • Detection of Known Signals • Detection
of Parametrized Signals • Detection of Random Signals • Deciding
among Multiple Signals • Detection of Signals in More General
Noise Processes • Robust and Nonparametric Detection •
Distributed and Sequential Detection • Detection with Continuous-Time Measurements

67.2 Noise 1488
Statistics of Noise • Noise Power • Effect of Linear Transformations
on Autocorrelation and Power Spectral Density • White, Gaussian,
and Pink Noise Models • Thermal Noise as Gaussian White Noise •
Some Examples • Measuring Thermal Noise • Effective Noise and
Antenna Noise • Noise Factor and Noise Ratio • Equivalent Input
Noise • Other Electrical Noise • Measurement and Quantization
Noise • Coping with Noise •

67.3 Stochastic Processes 1499
Introduction to Random Variables • Stochastic Processes • Classifi-
cations of Stochastic Processes • Stationarity of Processes • Gaussian
and Markov Processes • Examples of Stochastic Processes • Linear
Filtering of Weakly Stationary Processes • Cross-Correlation of
Processes • Coherence • Ergodicity

67.4 The Sampling Theorem 1510
The Cardinal Series • Proof of the Sampling Theorem • The Time-
Bandwidth Product • Sources of Error • Generalizations of the
Sampling Theorem

67.5 Data Compression 1517
Entropy • The Huffman Algorithm • Entropy Rate • Arithmetic
Coding • Lempel–Ziv Coding • Rate Distortion Theory • Quanti-
zation and Vector Quantization • Kolmogorov Complexity • Data
Compression in Practice

H. Vincent Poor
Princeton University

Carl G. Looney
University of Nevada

R. J. Marks II
University of Washington

Joy A. Thomas
IBM

Thomas M. Cover
Stanford University

67.1 Signal Detection

H. Vincent Poor

The field of signal detection and estimation is concerned with the processing of information-bearing signals for the purpose of extracting the information they contain. The applications of this methodology are quite broad, ranging from areas of electrical engineering such as automatic control, digital communications, image processing, and remote sensing, into other engineering disciplines and the physical, biological, and social sciences.

There are two basic types of problems of interest in this context. *Signal detection* problems are concerned primarily with situations in which the information to be extracted from a signal is discrete in nature. That is, signal detection procedures are techniques for deciding among a discrete (usually finite) number of possible alternatives. An example of such a problem is the demodulation of a digital communication signal, in which the task of interest is to decide which of several possible transmitted symbols has elicited a given received signal. *Estimation* problems, on the other hand, deal with the determination of some numerical quantity taking values in a continuum. An example of an estimation problem is that of determining the phase or frequency of the carrier underlying a communication signal.

Although signal detection and estimation is an area of considerable current research activity, the fundamental principles are quite well developed. These principles, which are based on the theory of statistical inference, explain and motivate most of the basic signal detection and estimation procedures used in practice. In this section, we will give a brief overview of the basic principles underlying the field of signal detection. Estimation is treated elsewhere this volume, notably in Section 15.2. A more complete introduction to these subjects is found in Poor [1988].

General Considerations

The basic principles of signal detection can be conveniently discussed in the context of decision-making between two possible statistical models for a set of real-valued measurements, $Y_1, Y_2, \ldots, Y_n$. In particular, on observing $Y_1, Y_2, \ldots, Y_n$, we wish to decide whether these measurements are most consistent with the model

$$Y_k = N_k, \quad k = 1, 2, \ldots, n \tag{67.1}$$

or with the model

$$Y_k = N_k + S_k, \quad k = 1, 2, \ldots, n \tag{67.2}$$

where $N_1, N_2, \ldots, N_n$ is a random sequence representing noise, and where $S_1, S_2, \ldots, S_n$ is a sequence representing a (possibly random) signal.

In deciding between Eqs. (67.1) and (67.2), there are two types of errors possible: a *false alarm,* in which (67.2) is falsely chosen, and a *miss,* in which (67.1) is falsely chosen. The probabilities of these two types of errors can be used as performance indices in the optimization of rules for deciding between (67.1) and (67.2). Obviously, it is desirable to minimize both of these probabilities to the extent possible. However, the minimization of the **false-alarm probability** and the minimization of the **miss probability** are opposing criteria. So, it is necessary to effect a trade-off between them in order to design a signal detection procedure. There are several ways of trading off the probabilities of miss and false alarm: the **Bayesian detector** minimizes an average of the two probabilities taken with respect to prior probabilities of the two conditions (67.1) and (67.2), the *minimax* detector minimizes the maximum of the two error probabilities, and the **Neyman-Pearson detector** minimizes the miss probability under an upper-bound constraint on the false-alarm probability.

If the statistics of noise and signal are known, the Bayesian, minimax, and Neyman-Pearson detectors are all of the same form. Namely, they reduce the measurements to a single number by computing the **likelihood ratio**

$$L(Y_1, Y_2, \ldots, Y_n) \triangleq \frac{p_{S+N}(Y_1, Y_2, \ldots, Y_n)}{p_N(Y_1, Y_2, \ldots, Y_n)} \tag{67.3}$$

where p_{S+N} and p_N denote the probability density functions of the measurements under signal-plus-noise (67.2) and noise-only (67.1) conditions, respectively. The likelihood ratio is then compared to a *decision threshold,* with the signal-present model (67.2) being chosen if the threshold is exceeded, and the signal-absent model (67.1) being chosen otherwise. Choice of the decision threshold determines a trade-off of the two error probabilities, and the optimum procedures for the three criteria mentioned above differ only in this choice.

There are several basic signal detection structures that can be derived from Eqs. (67.1) to (67.3) under the assumption that the noise sequence consists of a set of independent and identically distributed (i.i.d.) Gaussian random variables with zero means. Such a sequence is known as **discrete-time white Gaussian noise.** Thus, until further notice, we will make this assumption about the noise.

Detection of Known Signals

If the signal sequence $S_1, S_2, \ldots, S_n$ is known to be given by a specific sequence, say $s_1, s_2, \ldots, s_n$ (a situation known as *coherent detection*), then the likelihood ratio (67.3) is given in the white Gaussian noise case by

$$\exp\left\{\left(\sum_{k=1}^{n} s_k Y_k - \frac{1}{2}\sum_{k=1}^{n} s_k^2\right) / \sigma^2\right\} \tag{67.4}$$

where σ^2 is the variance of the noise samples. The only part of (67.4) that depends on the measurements is the term $\sum_{k=1}^{n} s_k Y_k$ and the likelihood ratio is a monotonically increasing function of this quantity. Thus, optimum detection of a coherent signal can be accomplished via a **correlation detector**, which operates by comparing the quantity

$$\sum_{k=1}^{n} s_k Y_k \tag{67.5}$$

to a threshold, announcing signal presence when the threshold is exceeded.

Note that this detector works on the principle that the signal will correlate well with itself, yielding a large value of (67.5) when present, whereas the random noise will tend to average out in the sum (67.5), yielding a relatively small value when the signal is absent. This detector is illustrated in Fig. 67.1.

Detection of Parametrized Signals

The correlation detector cannot usually be used directly unless the signal is known exactly. If, alternatively, the signal is known up to a short vector $\boldsymbol{\theta}$ of random parameters (such as frequencies or phases) that are independent of the noise, then an optimum test can be implemented by threshold comparison of the quantity

$$\int_{\Lambda} \exp\left\{\left(\sum_{k=1}^{n} s_k(\boldsymbol{\theta}) Y_k - \frac{1}{2}\sum_{k=1}^{n} \left[s_k(\boldsymbol{\theta})\right]^2\right) / \sigma^2\right\} p(\boldsymbol{\theta}) d\boldsymbol{\theta} \tag{67.6}$$

where we have written $S_k = s_k(\boldsymbol{\theta})$ to indicate the functional dependence of the signal on the parameters, and where Λ and $p(\boldsymbol{\theta})$ denote the range and probability density function, respectively, of the parameters.

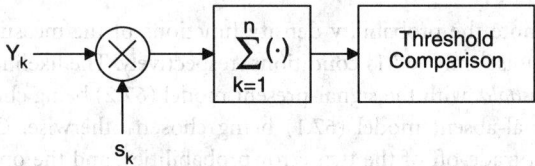

FIGURE 67.1 Correlation detector for a coherent signal in additive white Gaussian noise.

The most important example of such a parametrized signal is that in which the signal is a modulated sinusoid with random phase, i.e.,

$$S_k = a_k \cos(\omega_c k + \theta), \quad k = 1, 2, \ldots, n \tag{67.7}$$

where $a_1, a_2, \ldots, a_n$ is a known amplitude modulation sequence, ω_c is a known (discrete-time) carrier frequency, and the random phase θ is uniformly distributed in the interval $[-\pi, \pi]$. In this case, the likelihood ratio is a monotonically increasing function of the quantity

$$\left[\sum_{k=1}^{n} a_k \cos(\omega_c k) Y_k \right]^2 + \left[\sum_{k=1}^{n} a_k \sin(\omega_c k) Y_k \right]^2 \tag{67.8}$$

Thus, optimum detection can be implemented via comparison of (67.8) with a threshold, a structure known as an **envelope detector.** Note that this detector correlates the measurements with two orthogonal components of the signal, $a_k \cos(\omega_c k)$ and $a_k \sin(\omega_c k)$. These two correlations, known as the *in-phase* and *quadrature* components of the measurements, respectively, capture all of the energy in the signal, regardless of the value of θ. Since θ is unknown, however, these two correlations cannot be combined coherently, and thus they are combined *noncoherently* via (67.8), before the result is compared with a threshold. This detector is illustrated in Fig. 67.2.

Parametrized signals also arise in situations in which it is not appropriate to model the unknown parameters as random variables with a known distribution. In such cases, it is not possible to compute the likelihood ratio (67.6) so an alternative to the likelihood ratio detector must then be used. (An exception is that in which the likelihood ratio detector is invariant to the unknown parameters—a case known as *uniformly most powerful detection.*) Several alternatives to the likelihood ratio detector exist for these cases.

One useful such procedure is to test for the signal's presence by threshold comparison of the *generalized likelihood ratio,* given by

$$\max_{\theta \in \Lambda} L_{\theta}(Y_1, Y_2, \ldots, Y_n) \tag{67.9}$$

where L_{θ} denotes the likelihood ratio for Eqs. (67.1) and (67.2) for the known-signal problem with the parameter vector fixed at θ. In the case of white Gaussian noise, we have

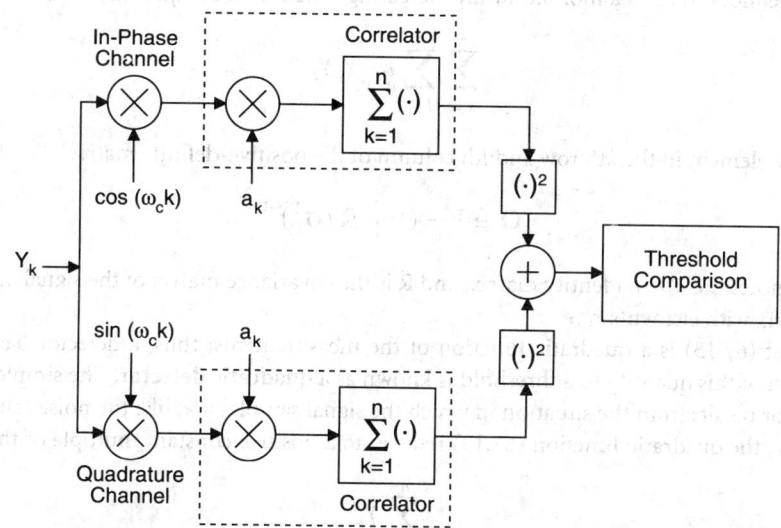

FIGURE 67.2 Envelope detector for a noncoherent signal in additive white Gaussian noise.

$$L_{\boldsymbol{\theta}}(Y_1, Y_2, \ldots, Y_n) = \exp\left\{ \left(\sum_{k=1}^{n} s_k(\boldsymbol{\theta})Y_k - \frac{1}{2} \sum_{k=1}^{n} \left[s_k(\boldsymbol{\theta}) \right]^2 \right) / \sigma^2 \right\} \qquad (67.10)$$

(It should be noted that this formulation is also valid if the statistics of the noise have unknown parameters, e.g., the noise variance in the white Gaussian case.)

One common application in which the generalized likelihood ratio detector is useful is that of detecting a signal that is known except for its time of arrival. That is, we are often interested in signals parametrized as

$$s_k(\boldsymbol{\theta}) = a_{k-\theta} \qquad (67.11)$$

where $\{a_k\}$ is a known finite-duration signal sequence and where θ ranges over the integers. Assuming white Gaussian noise and an observation interval much longer than the duration of $\{a_k\}$, the generalized likelihood ratio detector in this case announces the presence of the signal if the quantity

$$\max_{\theta} \sum_{k} a_{k-\theta} Y_k \qquad (67.12)$$

exceeds a fixed threshold. This type of detector is known as a *matched filter*, since it can be implemented by filtering the measurements with a digital filter whose pulse response is a time-reversed version of the known signal $\{a_k\}$ (hence it is "matched" to the signal), and announcing the signal's presence if the filter output exceeds the decision threshold at any time.

Detection of Random Signals

In some applications, particularly in remote sensing applications such as sonar and radio astronomy, it is appropriate to consider the signal sequence $S_1, S_2, \ldots, S_n$ itself to be a random sequence, statistically independent of the noise. In such cases, the likelihood ratio formula of (67.6) is still valid with the parameter vector $\boldsymbol{\theta}$ simply taken to be the signal itself. However, for long measurement records (i.e., large n), (67.6) is not a very practical formula except in some specific cases, the most important of which is the case in which the signal is Gaussian.

In particular, if the signal is Gaussian with zero-mean and autocorrelation sequence $r_{k,l} \overset{\Delta}{=} E\{S_k S_l\}$, then the likelihood ratio is a monotonically increasing function of the quantity

$$\sum_{k=1}^{n} \sum_{l=1}^{n} q_{k,l} Y_k Y_l \qquad (67.13)$$

with $q_{k,l}$ the element in the kth row and lth column of the positive-definite matrix

$$\mathbf{Q} \overset{\Delta}{=} \mathbf{I} - (\mathbf{I} + \mathbf{R}/\sigma^2)^{-1} \qquad (67.14)$$

where $\mathbf{I}$ denotes the $n \times n$ identity matrix, and $\mathbf{R}$ is the covariance matrix of the signal, i.e., it is the $n \times n$ matrix with elements $r_{k,l}$.

Note that (67.13) is a quadratic function of the measurements; thus, a detector based on the comparison of this quantity to a threshold is known as a **quadratic detector**. The simplest form of this detector results from the situation in which the signal samples are, like the noise samples, i.i.d. In this case, the quadratic function (67.13) reduces to a positive constant multiple of the quantity

$$\sum_{k=1}^{n} Y_k^2 \qquad (67.15)$$

A detector based on (67.15) simply measures the energy in the measurements and then announces the presence of the signal if this energy is large enough. This type of detector is known as a *radiometer*.

Thus, radiometry is optimum in the case in which both signal and noise are i.i.d. Gaussian sequences with zero means. Since in this case the presence of the signal is manifested only by an increase in energy level, it is intuitively obvious that radiometry is the only way of detecting the signal's presence. More generally, when the signal is correlated, the quadratic function (67.13) exploits both the increased energy level and the correlation structure introduced by the presence of the signal. For example, if the signal is a narrowband Gaussian process, then the quadratic function (67.13) acts as a narrowband radiometer with bandpass characteristic that approximately matches that of the signal. In general, the quadratic detector will make use of whatever spectral properties the signal exhibits.

If the signal is random but not Gaussian, then its optimum detection [described by (67.6)] typically requires more complicated nonlinear processing than the quadratic processing of (67.13) in order to exploit the distributional differences between signal and noise. This type of processing is often not practical for implementation, and thus approximations to the optimum detector are typically used. An interesting family of such detectors uses cubic or quartic functions of the measurements, which exploit the higher-order spectral properties of the signal [Mendel, 1991]. As with deterministic signals, random signals can be parametrized. In this case, however, it is the distribution of the signal that is parametrized. For example, the power spectrum of the signal of interest may be known only up to a set of unknown parameters. Generalized likelihood ratio detectors (67.9) are often used to detect such signals.

Deciding among Multiple Signals

The preceding results have been developed under the model (67.1)–(67.2) that there is a single signal that is either present or absent. In digital communications applications, it is more common to have the situation in which we wish to decide between the presence of two (or more) possible signals in a given set of measurements. The foregoing results can be adapted straightforwardly to such problems. This can be seen most easily in the case of deciding among known signals. In particular, consider the problem of deciding between two alternatives:

$$Y_k = N_k + s_k^{(0)}, \quad k = 1, 2, \ldots, n \tag{67.16}$$

and

$$Y_k = N_k + s_k^{(1)}, \quad k = 1, 2, \ldots, n \tag{67.17}$$

where $s_1^{(0)}, s_2^{(0)}, \ldots, s_n^{(0)}$ and $s_1^{(1)}, s_2^{(1)}, \ldots, s_n^{(1)}$ are two known signals. Such problems arise in data transmission problems, in which the two signals $\mathbf{s}^{(0)}$ and $\mathbf{s}^{(1)}$ correspond to the waveforms received after transmission of a logical "zero" and "one," respectively. In such problems, we are generally interested in minimizing the *average probability of error*, which is the average of the two error probabilities weighted by the prior probabilities of occurrence of the two signals. This is a Bayesian performance criterion, and the optimum decision rule is a straightforward extension of the correlation detector based on (67.5). In particular, under the assumptions that the two signals are equally likely to occur prior to measurement, and that the noise is white and Gaussian, the optimum decision between (67.16) and (67.17) is to choose the model (67.16) if $\sum_{k=1}^{n} s_k^{(0)} Y_k$ is larger than $\sum_{k=1}^{n} s_k^{(1)} Y_k$, and to choose the model (67.17) otherwise.

More generally, many problems in digital communications involve deciding among M equally likely signals with $M > 2$. In this case, again assuming white Gaussian noise, the decision rule that minimizes the error probability is to choose the signal $s_1^{(j)}, s_2^{(j)}, \ldots, s_n^{(j)}$, where j is a solution of the maximization problem

$$\sum_{k=1}^{n} s_k^{(j)} Y_k = \max_{0 \le m \le M-1} \sum_{k=1}^{n} s_k^{(m)} Y_k \qquad (67.18)$$

There are two basic types of digital communications applications in which the problem (67.18) arises. One is in *M-ary data transmission,* in which a symbol alphabet with M elements is used to transmit data, and a decision among these M symbols must be made in each symbol interval [Proakis, 1983]. The other type of application in which (67.18) arises is that in which data symbols are correlated in some way because of intersymbol interference, coding, or multiuser transmission. In such cases, each of the M possible signals represents a frame of data symbols, and a joint decision must be made about the entire frame since individual symbol decisions cannot be decoupled. Within this latter framework, the problem (67.18) is known as *sequence detection.* The basic distinction between M-ary transmission and sequence detection is one of degree. In typical M-ary transmission, the number of elements in the signaling alphabet is typically a small power of 2 (say 8 or 32), whereas the number of symbols in a frame of data could be on the order of thousands. Thus, solution of (67.18) by exhaustive search is prohibitive for sequence detection, and less complex algorithms must be used. Typical digital communications applications in which sequence detection is necessary admit dynamic programming solutions to (67.18) [see, e.g., Verdú, 1993].

Detection of Signals in More General Noise Processes

In the foregoing paragraphs, we have described three basic detection procedures: correlation detection of signals that are completely known, envelope detection of signals that are known except for a random phase, and quadratic detection for Gaussian random signals. These detectors were all derived under an assumption of white Gaussian noise. This assumption provides an accurate model for the dominant noise arising in many communication channels. For example, the thermal noise generated in signal processing electronics is adequately described as being white and Gaussian. However, there are also many channels in which the statistical behavior of the noise is not well described in this way, particularly when the dominant noise is produced in the physical channel rather than in the receiver electronics.

One type of noise that often arises is noise that is Gaussian but not white. In this case, the detection problem (67.1)–(67.2) can be converted to an equivalent problem with white noise by applying a linear filtering process known as *prewhitening* to the measurements. In particular, on denoting the noise covariance matrix by Σ, we can write

$$\Sigma = \mathbf{C}\mathbf{C}^T \qquad (67.19)$$

where $\mathbf{C}$ is an $n \times n$ invertible, lower-triangular matrix and where the superscript T denotes matrix transposition. The representation (67.19) is known as the *Cholesky decomposition.* On multiplying the measurement vector $\mathbf{Y} \triangleq (Y_1, Y_2, \ldots, Y_n)^T$ satisfying (67.1)–(67.2) with noise covariance Σ, by $\mathbf{C}^{-1}$, we produce an equivalent (in terms of information content) measurement vector that satisfies the model (67.1)–(67.2) with white Gaussian noise and with the signal conformally transformed. This model can then be treated using the methods described previously.

In other channels, the noise can be modeled as being i.i.d. but with an amplitude distribution that is not Gaussian. This type of model arises, for example, in channels dominated by impulsive phenomena, such as urban radio channels. In the non-Gaussian case the procedures discussed previously lose their optimality as defined in terms of the error probabilities. These procedures can still be used, and they will work well under many conditions; however, there will be a resulting performance penalty with respect to optimum procedures based on the likelihood ratio. Generally speaking, likelihood-ratio-based procedures for non-Gaussian noise channels involve more complex nonlinear processing of the measurements than is required in the standard detectors, although the retention of the i.i.d. assumption greatly simplifies this problem. A treatment of methods for such channels can be found in Kassam [1988].

When the noise is both non-Gaussian and dependent, the methodology is less well developed, although some techniques are available in these cases. An overview can be found in Poor and Thomas [1993].

Robust and Nonparametric Detection

All of the procedures outlined in the preceding subsection are based on the assumption of a known (possibly up to a set of unknown parameters) statistical model for signals and noise. In many practical situations it is not possible to specify accurate statistical models for signals or noise, and so it is of interest to design detection procedures that do not rely heavily on such models. Of course, the parametrized models described in the foregoing paragraphs allow for uncertainty in the statistics of the observations. Such models are known as *parametric* models, because the set of possible distributions can be parametrized by a finite set of real parameters.

While parametric models can be used to describe many types of modeling uncertainty, composite models in which the set of possible distributions is much broader than a parametric model would allow are sometimed more realistic in practice. Such models are termed *nonparametric models*. For example, one might be able to assume only some very coarse model for the noise, such as that it is symmetrically distributed. A wide variety of useful and powerful detectors have been developed for signal-detection problems that cannot be parametrized. These are basically of two types: *robust* and *nonparametric*. Robust detectors are those designed to perform well despite small, but potentially damaging, nonparametric deviations from a nominal parametric model, whereas nonparametric detectors are designed to achieve constant false-alarm probability over very wide classes of noise statistics.

Robustness problems are usually treated analytically via minimax formulations that seek best worst-case performance as the design objective. This formulation has proven to be very useful in the design and characterization of robust detectors for a wide variety of detection problems. Solutions typically call for the introduction of light limiting to prevent extremes of gain dictated by an (unrealistic) nominal model. For example, the correlation detector of Fig. 67.1 can be made robust against deviations from the Gaussian noise model by introducing a soft-limiter between the multiplier and the accumulator.

Nonparametric detection is usually based on relatively coarse information about the observations, such as the algebraic signs or the ranks of the observations. One such test is the *sign test*, which bases its decisions on the number of positive observations obtained. This test is nonparametric for the model in which the noise samples are i.i.d. with zero median and is reasonably powerful against alternatives such as the presence of a positive constant signal in such noise. More powerful tests for such problems can be achieved at the expense of complexity by incorporating rank information into the test statistic.

Distributed and Sequential Detection

The detection procedures discussed in the preceding paragraphs are based on the assumption that all measurements can and should be used in the detection of the signal, and moreover that no constraints exist on how measurements can be combined. There are a number of applications, however, in which constraints apply to the information pattern of the measurements.

One type of constrained information pattern that is of interest in a number of applications is a network consisting of a number of distributed or local decision makers, each of which processes a subset of the measurements, and a *fusion center,* which combines the outputs of the distributed decision makers to produce a global detection decision. The communication between the distributed decision makers and the fusion center is constrained, so that each local decision maker must reduce its subset of measurements to a summarizing local decision to be transmitted to the fusion center. Such structures arise in applications such as the testing of large-scale integrated circuits, in which data collection is decentralized, or in detection problems involving very large data sets, in which it is desirable to distribute the computational work of the detection algorithm. Such problems lie in the field of *distributed detection*. Except in some trivial special cases, the constraints imposed by distributing the detection algorithm introduce a further level of difficulty into the design of optimum detection systems. Nevertheless, considerable progress has been made on this problem, a survey of which can be found in Tsitsiklis [1993].

Another type of nonstandard information pattern that arises is that in which the number of measurements is potentially infinite, but in which there is a cost associated with taking each measurement. This type of model arises in applications such as the synchronization of wideband communication signals. In such situations, the error probabilities alone do not completely characterize the performance of a detection system, since consideration must also be given to the cost of sampling. The field of *sequential detection* deals with the optimization of detection systems within such constraints. In sequential detectors, the number of measurements taken becomes a random variable depending on the measurements themselves. A typical performance criterion for optimizing such a system is to seek a detector that minimizes the expected number of measurements for given levels of miss and false-alarm probabilities.

The most commonly used sequential detection procedure is the *sequential probability ratio test*, which operates by recursive comparison of the likelihood ratio (67.3) to two thresholds. In this detector, if the likelihood ratio for a given number of samples exceeds the larger of the two thresholds, then the signal's presence is announced and the test terminates. Alternatively, if the likelihood ratio falls below the smaller of the two thresholds, the signal's absence is announced and the test terminates. However, if neither of the two thresholds is crossed, then another measurement is taken and the test is repeated.

Detection with Continuous-Time Measurements

Note that all of the preceding formulations have involved the assumption of discrete-time (i.e., sampled-data) measurements. From a practical point of view, this is the most natural framework within which to consider these problems, since implementations most often involve digital hardware. However, the procedures discussed in this section all have continuous-time counterparts, which are of both theoretical and practical interest. Mathematically, continuous-time detection problems are more difficult than discrete-time ones, because they involve probabilistic analysis on function spaces. The theory of such problems is quite elegant, and the interested reader is referred to Poor [1988] or Grenander [1981] for more detailed exposition.

Continuous-time models are of primary interest in the front-end stages of radio frequency or optical communication receivers. At radio frequencies, continuous-time versions of the models described in the preceding paragraphs can be used. For example, one may consider the detection of signals in continuous-time Gaussian white noise. At optical wavelengths, one may consider either continuous models (such as Gaussian processes) or point-process models (such as Poisson counting processes), depending on the type of detection used [see, e.g., Snyder and Miller, 1991]. In the most fundamental analyses of optical detection problems, it is sometimes desirable to consider the quantum mechanical nature of the measurements [Helstrom, 1976].

Defining Terms

Bayesian detector: A detector that minimizes the average of the false-alarm and miss probabilities, weighted with respect to prior probabilities of signal-absent and signal-present conditions.

Correlation detector: The optimum structure for detecting coherent signals in the presence of additive white Gaussian noise.

Discrete-time white Gaussian noise: Noise samples modeled as independent and identically distributed Gaussian random variables.

Envelope detector: The optimum structure for detecting a modulated sinusoid with random phase in the presence of additive white Gaussian noise.

False-alarm probability: The probability of falsely announcing the presence of a signal.

Likelihood ratio: The optimum processor for reducing a set of signal-detection measurements to a single number for subsequent threshold comparison.

Miss probability: The probability of falsely announcing the absence of a signal.

Neyman-Pearson detector: A detector that minimizes the miss probability within an upper-bound constraint on the false-alarm probability.

Quadratic detector: A detector that makes use of the second-order statistical structure (e.g., the spectral characteristics) of the measurements. The optimum structure for detecting a zero-mean Gaussian signal in the presence of additive Gaussian noise is of this form.

References

U. Grenander, *Abstract Inference*, New York: Wiley, 1981.

C. W. Helstrom, *Quantum Detection and Estimation Theory*, New York: Academic Press, 1976.

S. A. Kassam, *Signal Detection in Non-Gaussian Noise*, New York: Springer-Verlag, 1988.

J. M. Mendel, "Tutorial on higher-order statistics (spectra) in signal processing and systems theory: Theoretical results and some applications," *Proc. IEEE*, vol. 79, pp. 278–305, 1991.

H. V. Poor, *An Introduction to Signal Detection and Estimation*, New York: Springer-Verlag, 1988.

H. V. Poor and J. B. Thomas, "Signal detection in dependent non-Gaussian noise," in *Advances in Statistical Signal Processing*, vol. 2, Signal Detection, H. V. Poor and J. B. Thomas, Eds., Greenwich, Conn.: JAI Press, 1993.

J. G. Proakis, *Digital Communications*, New York: McGraw-Hill, 1983.

D. L. Snyder and M. I. Miller, *Random Point Processes in Time and Space*, New York: Springer-Verlag, 1991.

J. Tsitsiklis, "Distributed detection," in *Advances in Statistical Signal Processing*, vol. 2, Signal Detection, H. V. Poor and J. B. Thomas, Eds., Greenwich, Conn.: JAI Press, 1993.

S. Verdú, "Multiuser detection," in *Advances in Statistical Signal Processing*, vol. 2, Signal Detection, H. V. Poor and J. B. Thomas, Eds., Greenwich, Conn.: JAI Press, 1993.

Further Information

Except as otherwise noted in the accompanying text, further details on the topics introduced in this section can be found in the textbook:

Poor, H. V. *An Introduction to Signal Detection and Estimation*, New York: Springer-Verlag, 1988.

The bimonthly journal, *IEEE Transactions on Information Theory*, publishes recent advances in the theory of signal detection. It is available from the Institute of Electrical and Electronics Engineers, Inc., 345 East 47th Street, New York, NY 10017.

Papers describing applications of signal detection are published in a number of journals, including the monthly journals *IEEE Transactions on Communications*, *IEEE Transactions on Signal Processing*, and the *Journal of the Acoustical Society of America*. The IEEE journals are available from the IEEE, as above. The *Journal of the Acoustical Society of America* is available from the American Institute of Physics, 335 East 45th Street, New York, NY 10017.

67.2 Noise

Carl G. Looney

Every information signal $s(t)$ is corrupted to some extent by the superimposition of extra-signal fluctuations that assume unpredictable values at each time instant t. Such undesirable signals were called **noise** due to early measurements with sensitive audio amplifiers.

Noise sources are (1) *intrinsic*, (2) *external*, or (3) *process induced*. Intrinsic noise in conductors comes from thermal agitation of molecularly bound ions and electrons, from microboundaries of impurities and grains with varying potential, and from transistor junction areas that become temporarily depleted of electrons/holes. External electromagnetic interference sources include airport radar, x-rays, power and telephone lines, communications transmissions, gasoline engines and electric motors, computers and other electronic devices; and also include lightning, cosmic rays, plasmas (charged particles) in space, and solar/stellar radiation (conductors act as antennas). Reflective objects and other macroboundaries cause multiple paths of signals. Process-induced errors include measurement, quantization, truncation, and signal generation errors. These also corrupt the signal with noise power and loss of resolution.

Statistics of Noise

Statistics allow us to analyze the spectra of noise. We model a noise signal by a **random** (or *stochastic*) **process** $N(t)$, a function whose realized value $N(t) = x_t$ at any time t is chosen by the outcome of the random variable $N_t = N(t)$. $N(t)$ has a probability distribution for the values x it can assume. Any particular trajectory $\{(t, x_t)\}$ of outcomes is called a **realization** of the noise process. The *first-order statistic* of $N(t)$ is the *expected value* $\mu_t = E[N(t)]$. The *second-order statistic* is the *autocorrelation function* $R_{NN}(t, t+\tau) = E[N(t)N(t+\tau)]$, where $E[-]$ is the expected value operator. **Autocorrelation** measures the extent to which noise random variables $N_1 = N(t_1)$ and $N_2 = N(t_2)$ at times t_1 and t_2 depend on each other in an average sense.

When the first- and second-order statistics do not change over time, we call the noise a **weakly** (or *wide-sense*) **stationary process**. This means that: (1) $E[N(t)] = \mu_t = \mu$ is constant for all t, and (2) $R_{NN}(t, t+\tau) = E[N(t)N(t+\tau)] = E[N(0)N(\tau)] = R_{NN}(\tau)$ for all t [see Brown, 1983, p. 82; Gardner, 1990, p. 108; or Peebles, 1987, p. 153 for properties of $R_{NN}(\tau)$]. In this case the autocorrelation function depends only on the *offset* τ. We assume hereafter that $\mu = 0$ (we can subtract μ, which does not change the autocorrelation). When $\tau = 0$, $R_{NN}(0) = E[N(t)N(t+0)] = E[(N(t))^2] = \sigma_N^2$, which is the fixed variance of each random variable N_t for all t. Ws processes are the most commonly encountered cases and are the ones considered here. *Evolutionary* processes have statistics that change over time and are difficult to analyze.

Figure 67.3 shows a realization of a noise process $N(t)$, where at any particular time t, the probability density function is shown coming out of the page in a third dimension. For a ws noise, the distributions are the same for each t. The most mathematically tractable noises are *Gaussian* ws processes, where at each time t the probability distribution for the random variable $N_t = N(t)$ is Gaussian (also called *normal*). The first- and second-order statistics completely determine Gaussian distributions, and so ws makes their statistics of all orders stationary over time also. It is well known [see Brown, 1983, p. 39] that linear transformations of Gaussian random variables are also Gaussian random variables. The probability density function for a Gaussian random variable N_t is $f_N(x) = \{1/[2\pi\sigma_N^2]^{1/2}\exp[-(x - \mu_N)^2/2\sigma_N^2]$, which is the familiar bell-shaped curve centered on $x = \mu_N$. The standard Gaussian probability table [Peebles, 1987, p. 314] is useful, e.g., $\Pr[-\sigma_N < N_t < \sigma_N) = 2\Pr[0 < N_t < \sigma_N) = 0.8413$ from the table.

Noise Power

The noise signal $N(t)$ represents voltage, so the autocorrelation function at offset 0, $R_{NN}(0) = E[N(t)N(t)]$ represents expected power in volts squared, or watts per ohm. When $R = 1\ \Omega$, then

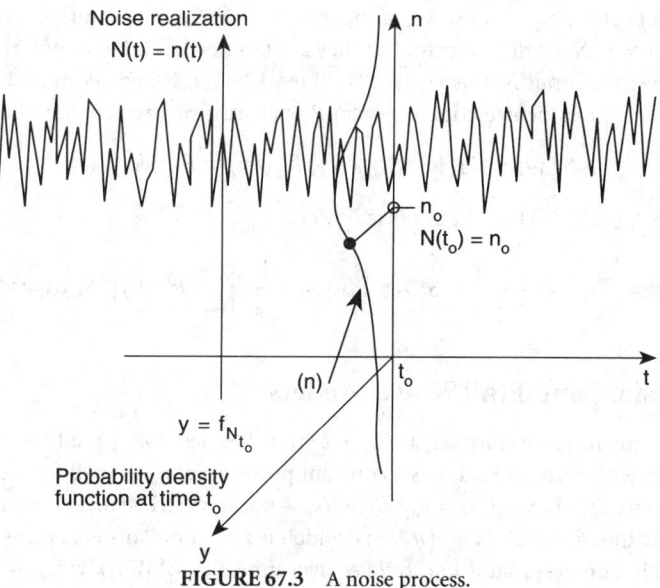

FIGURE 67.3 A noise process.

$N(t)N(t) = N(t)[N(t)/R] = N(t)I(t)$ volt-amperes = watts (where $I(t)$ is the current in a 1-Ω resistor). The Fourier transform $\mathbf{F}[R_{NN}(\tau)]$ of the autocorrelation function $R_{NN}(\tau)$ is the power spectrum, called the **power spectral density function** (psdf), $S_{NN}(w)$ in W/(rad/s). Then

$$S_{NN}(w) = \int_{-\infty}^{\infty} R_{NN}(\tau)e^{-jws}d\tau = \mathbf{F}[R_{NN}(\tau)]$$

$$R_{NN}(\tau) = \frac{1}{2\pi}\int_{-\infty}^{\infty} S_{NN}(w)e^{jws}dw = \mathbf{F}^{-1}[S_{NN}(w)]$$

$$(67.20)$$

The psdf at frequency f is defined to be the expected power that the voltage $N(t)$, bandlimited to an incremental band df centered at f, would dissipate in a 1-Ω resistance divided by df.

Equations (67.20) are known as the *Wiener-Khinchin* relations that establish that $S_{NN}(w)$ and $R_{NN}(\tau)$ are a Fourier transform pair for ws random processes [Brown, 1983; Gardner, 1990, p. 230; Peebles, 1987]. The psdf $S_{NN}(w)$ has units of W/(rad/s), whereas the autocorrelation function $R_{NN}(\tau)$ has units of watts. When $\tau = 0$ in the second integral of Eq. (67.20), the exponential becomes $e^0 = 1$, so that $R_{NN}(0)$ $(= E[N(t)^2] = \sigma_N^2)$ is the integral of the psdf $S_{NN}(w)$ over all radian frequencies, $-\infty < w < \infty$. The rms (root-mean-square) voltage is $N_{\text{rms}} = \sigma_N$ (the *standard deviation*). The power spectrum in W/(rad/s) is a density that is summed up via an integral over the radian frequency band w_1 to w_2 to obtain the total power over that band.

$$P_{NN}(w_1, w_2) = \frac{1}{2\pi}\int_{w1}^{w2} S_{NN}(w) \cdot dw \quad \text{watts}$$

$$\sigma_N^2 = E[N(t)^2] = \frac{1}{2\pi}\int_{-\infty}^{\infty} S_{NN}(w) \cdot dw \quad \text{watts}$$

$$(67.21)$$

The variance $\sigma_N^2 = R_{NN}(0)$ is the mean instantaneous power over all frequencies at any time t.

Effect of Linear Transformations on Autocorrelation and Power Spectral Density

Let $h(t)$ be the impulse response function of a time-invariant linear system L and $H(w) = \mathbf{F}[h(t)]$ be its transfer function. Let an input noise signal $N(t)$ have autocorrelation function $R_{NN}(\tau)$ and psdf

$S_{NN}(w)$. We denote the output noise signal by $Y(t) = L[N(t)]$. The Fourier transforms $Y(w) \equiv \mathbf{F}[Y(t)]$ and $N(w) \equiv \mathbf{F}[N(t)]$ do not exist, but they are not needed. The output $Y(t)$ of a linear system is ws whenever the input $N(t)$ is ws [see Gardner, 1990, p. 195; or Peebles, 1987, p. 215]. The output psdf $S_{YY}(w)$ and autocorrelation function $R_{YY}(\tau)$ are given by, respectively,

$$S_{YY}(w) = |H(w)|^2 S_{NN}(w), \ R_{YY}(\tau) = \mathbf{F}^{-1}[S_{YY}(w)] \qquad (67.22)$$

[see Gardner, 1990, p. 223]. The output noise power is

$$\sigma_Y^2 = P_{YY} = \frac{1}{2\pi} \int_{-\infty}^{\infty} S_{YY}(w)dw = \frac{1}{2\pi} \int_{-\infty}^{\infty} |H(w)|^2 S_{NN}(w)dw \qquad (67.23)$$

White, Gaussian, and Pink Noise Models

White noise [see Brown, 1983; Gardner, 1990, p. 234; or Peebles, 1987] is a theoretical model $W(t)$ of noise that is ws with zero mean. It has a constant power level n_o over all frequencies (analogous to white light), so its psdf is $S_{WW}(w) = n_o$ W/(rad/s), $-\infty < w < \infty$. The inverse Fourier transform of this is the impulse function $R_{WW}(\tau) = (n_o)\delta(\tau)$, which is zero for all offsets except $\tau = 0$. Therefore, white noise $W(t)$ is a process that is *uncorrelated* over time, i.e., $E[W(t_1)W(t_2)] = 0$ for t_1 not equal to t_2. Figure 67.4(a) shows the autocorrelation and psdf for white noise with offset $s = \tau$. A *Gaussian white noise* is white noise such that the probability distribution of values that each random variable $W_t = W(t)$ may assume must be Gaussian. When two Gaussian random variables W_1 and W_2 are uncorrelated, i.e., $E[W_1 W_2] = 0$, they are independent [see Gardner, 1990, p. 37]. We use Gaussian models because of the *central limit theorem* that states that the sum of a number of random variables is approximately Gaussian.

Actual circuits attenuate signals above cut-off frequencies, and also the power must be finite. However, for white noise, $P_{WW} = R_{NN}(0) = \infty$, so we often truncate the white noise spectral density (psdf) at cut-offs $-w_c$ to w_c. The result is known as *pink noise*, $P(t)$, and is usually taken to be Gaussian because linear filtering of any white noise (through the effect of the central limit theorem) tends to make the noise Gaussian [see Gardner, 1990, p. 241]. Figure 67.4(b) shows the sinc function $R_{PP}(s) = \mathbf{F}^{-1}[S_{PP}(w)]$ for pink noise. Random variables P_1 and P_2 at times t_1 and t_2 are correlated only for t_1 and t_2 close.

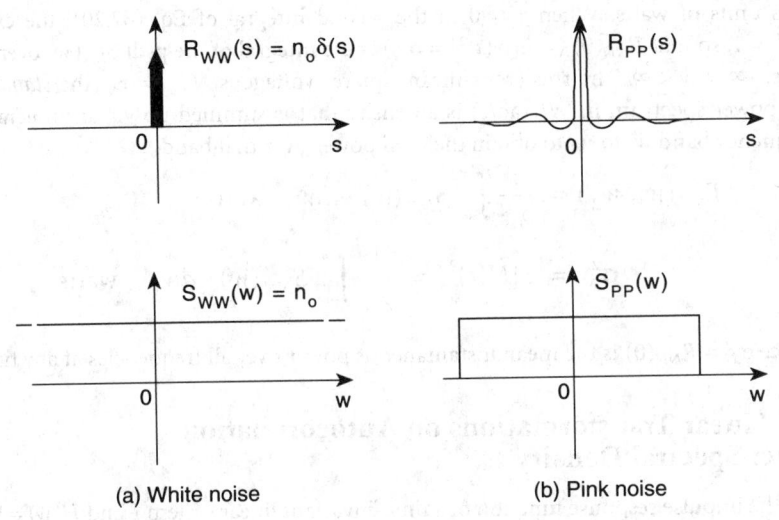

(a) White noise   (b) Pink noise

FIGURE 67.4 Power transform pairs for white and pink noise.

Thermal Noise as Gaussian White Noise

Brown observed in 1828 that pollen and dust particles moved randomly when suspended in liquid. In 1906, Einstein analyzed such motion based on the random walk model. Perrin confirmed in 1908 that the thermal activity of molecules in a liquid caused irregular bombardment of the much larger particles. It was predicted that charges bound to thermally vibrating molecules would generate electromotive force (emf) at the open terminals of a conductor, and that this placed a limit on the sensitivity of galvanometers. Thermal noise (also called *Johnson noise*) was first observed by J. B. Johnson at Bell Laboratories in 1927. Figure 67.5 displays white noise as seen in the laboratory on an oscilloscope.

The voltage $N(t)$ generated thermally between two points in an open circuit conductor is the sum of an extremely large number of superimposed, independent electronically and ionically induced microvoltages at all frequencies up to $f_c = 6,000$ GHz at room temperature [see Gardner 1990, p. 235], near infrared. The mean relaxation time of free electrons is $1/f_c = 0.5 \times 10^{-10}/T$ s, so at room temperature of $T = 290$K, it is 0.17 ps (1 picosecond $= 10^{-12}$ s). The values of $N(t)$ at different times are uncorrelated for time differences (offsets) greater than $\tau_c = 1/f_c$. The expected value of $N(t)$ is zero. The power is fairly constant across a broad spectrum, and we cannot sample signals at picosecond periods, so we model Johnson noise $N(t)$ with Gaussian white noise $W(t)$. Although $\mu = E[W(t)] = 0$, the average power is positive at temperatures above 0K, and is $\sigma_W^2 = R_{WW}(0)$ [see the right side of Eq. (67.21)]. A disadvantage of the white noise model is its infinite power, i.e., $R_{WW}(0) = \sigma_W^2 = \infty$, but it is valid over a limited bandwidth of B Hz, in which case its power is finite.

In 1927, Nyquist [1928] theoretically derived thermal noise power in a resistor to be

$$P_{WW}(B) = 4kTRB \quad \text{(watts)} \qquad (67.24)$$

where R is resistance (ohms), B is the frequency bandwidth of measurement in Hz (all emf fluctuations outside of B are ignored), $P_{WW}(B)$ is the mean power over B (see Eq. 67.21), and Boltzmann's constant is $k = 1.38 \times 10^{-23}$ J/K [see Ott, 1976; Gardner, 1990, p. 288; or Peebles, 1987, p. 227]. Under external emf, the thermally induced collisions are the main source of resistance in conductors (electrons pulled into motion by an external emf at 0K meet no resistance). The rms voltage is $W_{\text{rms}} = \sigma_W = [(4kTRB)]^{1/2}$ V over a bandwidth of B Hz.

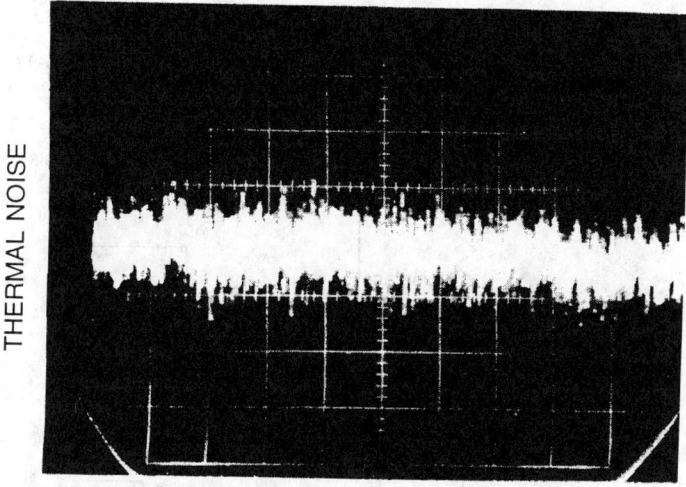

TIME, 200 MICROSECONDS PER DIVISION

FIGURE 67.5   Thermal noise in a resistor. (*Source:* H. W. Ott, *Noise Reduction Techniques in Electronic Systems*, New York: Wiley-Interscience, 1976, p. 203. With permission.)

Planck's radiation law is $S_{NN}(w) = (2h|f|)/[\exp(h|f|/kT) - 1]$, where $h = 6.63 \times 10^{-34}$ J/s is Planck's constant, and f is the frequency [see Gardner, 1990, p. 234]. For $|f|$ much smaller than $kT/h = 6.04 \times 10^{12}$ Hz $\approx 6,000$ GHz, the exponential above can be approximated by $\exp(h|f|/kT) = 1 + h|f|/kT$. The denominator of $S_{NN}(w)$ becomes $h|f|/kT$, so $S_{NN}(w) = (2h|f|)/(h|f|/kT) = 2kT$ W/Hz in a 1-Ω resistor. Over a resistance of $R\,\Omega$ and a bandwidth of B Hz (positive frequencies), this yields the total power $P_{WW}(B) = 2BRS_{NN}(w) = 4kTRB$ W over the two-sided frequency spectrum. This is Nyquist's result.

Thermal noise is the same in a 1000-Ω carbon resistor as it is in a 1000-Ω tantalum thin-film resistor [see Ott, 1976]. While the intrinsic noise may never be less, it may be higher because of other superimposed noise (described in later sections). We model the thermal noise in a resistor by an internal source (generator), as shown in Fig. 67.6. Capacitance cannot be ignored at high f, but pure reactance (C or L) cannot dissipate energy, and so cannot generate thermal noise. The white noise model $W(t)$ for thermal noise $N(t)$ has a constant psdf $S_{WW}(w) = n_o$ W/(rad/s) for $-\infty < w < \infty$. By Eq. 67.21, the white noise mean power over the frequency bandwidth B is

$$P_{WW}(B) = \frac{1}{2\pi} \int_{-2\pi B}^{2\pi B} S_{WW}(w)\,dw = n_o(4\pi B/2\pi) = 2n_o B \qquad (67.25)$$

Solving for the constant n_o, we obtain $n_o = P_{WW}(B)/2B$, which we put into Eq. (67.20) to get the spectral density as a function of temperature and resistance using Nyquist's result above.

$$S_{WW}(w) = n_o = P_{WW}(B)/4\pi B = 4kTR2\pi B/4\pi B = 2kTR \quad \text{watts}/(\text{rad/s}) \qquad (67.26)$$

Some Examples

The parasitic capacitance in the terminals of a resistor may cause a roll-off of about 20 dB/octave in actual resistors [Brown, 1983, p. 139]. At 290K (room temperature), we have $2kT = 2 \times 1.38 \times 10^{-23} \times 290 = 0.8 \times 10^{-20}$ W/Hz due to each ohm [see Ott, 1976]. For $R = 1$ MΩ (10^6 Ω), $S_{WW}(w) = 0.8 \times 10^{-14}$. Over a band of 10^8 Hz, we have $P_{WW}(B) = S_{WW}(w)B = 0.8 \times 10^{-14} \times 10^8 = 0.8 \times 10^{-6}$ W $= 0.8$ μW by Eqs. (67.24) and (67.26). In practice, parasitic capacitance causes thermal noise to be bandlimited (pink noise). Now consider Fig. 67.6(b) and let the temperature be 300K, $R = 10^6$ Ω, $C = 1$ pf (1 picofarad $= 10^{-12}$ farads), and assume L is 0 H. By Eq. (67.26), the thermal noise power is

$$S_{WW}(w) = 2kTR = 2 \times 1.38 \times 10^{-23} \times 300 \times 10^6 = 828 \times 10^{-17} \text{ W/Hz}$$

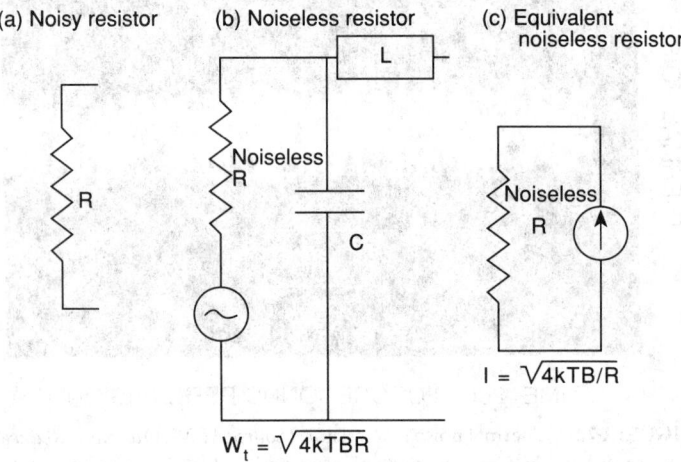

(a) Noisy resistor   (b) Noiseless resistor   (c) Equivalent noiseless resistor

$W_t = \sqrt{4kTBR}$

$I = \sqrt{4kTB/R}$

FIGURE 67.6   Thermal noise in a resistor.

The power across a bandwidth $B = 10^6$ is $P_{WW}(B) = S_{WW}(w)B = 8280 \times 10^{-12}$ W, so the rms voltage is $W_{rms} = [P_{WW}(B)]^{1/2} = 91$ μV.

Now let $Y(t)$ be the output voltage across the capacitor. The transfer function can be seen to be $H(w) = \{I(w)(1/jwC)\}/\{I(w)[R + (1/jwC)]\} = (1/jwC)/[R + 1/jwC] = 1/[1 + jwRC]$ (where $I(w)$ is the Fourier transform of the current). The output psdf [see Eq. (67.22)] is

$$S_{YY}(w) = |H(w)|^2 S_{WW}(w) = (1/[1 + w^2R^2C^2])S_{WW}(w)$$

Integrating $S_{YY}(w) = (1/[1 + w^2R^2C^2])S_{WW}(w)$ over all radian frequencies $w = 2\pi f$ [see Eq. (67.21)], we obtain the antiderivative $(828 \times 10^{-17})(1/RC)\text{atan}(RCw)/2\pi$. Upon substituting the limits $w = \pm\infty$, this becomes $828 \times 10^{-17}[\pi/2 + \pi/2]/2\pi RC = 414 \times 10^{-17}(1/2RC) = 207 \times 10^{-17} \times 10^6 = 2070 \times 10^{-12}$ W/Hz. Then $\sigma_Y^2 = E[Y(t)^2] = P_{YY}(-\infty,\infty) = 2070 \times 10^{-12}$ W, so $Y_{rms}(t) = \sigma_Y = [P_{YY}(-\infty,\infty)]^{1/2} = 45.5$ μV. The half-power (cut-off) radian frequency is $w_c = 1/RC = 10^6$ rad/s, or $f_c = w_c/2\pi = 159.2$ kHz. Approximating $S_{YY}(w)$ by the rectangular spectrum $S_{YY}(w) = n_o$, $-10^6 < w < 10^6$ rad/s (0 elsewhere), we have that $R_{YY}(\tau) = (w_c/\pi)\text{sinc}(w_c\tau)$, which has the first zeros at $|w_c\tau| = \pi$, that is $|\tau| = 1/(2f_c)$ [see Fig. 67.4(b)]. We approximate the autocorrelation by $R_{YY}(\tau) = 0$ for $|s| \geq 1/2f_c$.

Measuring Thermal Noise

In Fig. 67.7, the thermal noise from a noisy resistor R is to be measured, where R_L is the measurement load. The incremental noise power in R over an incremental frequency band of width df is $P_{WW}(df) = 4kTRdf$ W, by Eq. (67.24). $P_{YY}(df)$ is the integral of $S_{YY}(w)$ over df by Eqs. (67.21), where $S_{YY}(w) = |H(w)|^2 S_{WW}(w)$, by Eq. (67.22). In this case, the transfer function $H(w)$ is nonreactive and does not depend upon the radian frequency (we can factor it out of the integral). Thus,

$$P_{YY}(df) = \int_{-df}^{df} |H(f)|^2 (2kTR)df = \{R_L/(R + R_L)^2\}(4kTRdf)$$

To maximize the power measured, let $R_L = R$. The *incremental available power* measured is then $P_{YY}(df) = 4kTR^2df/(4R^2) = kTdf$ [see Ott, 1976, p. 201; Gardner, 1990, p. 288; or Peebles, 1987, p. 227]. Thus, we have the result that incremental available power over bandwidth df depends only on the temperature T.

$$P_{YY}(df) = kTdf \text{ (output power over } df) \tag{67.27}$$

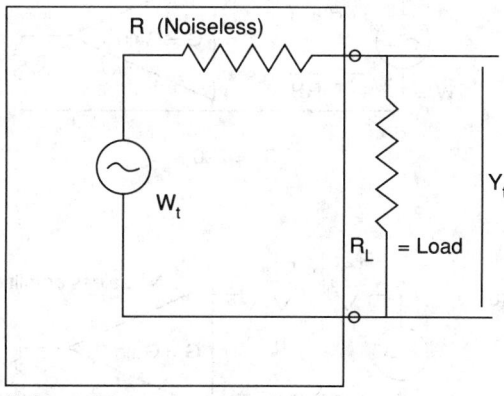

FIGURE 67.7 Measuring thermal noise voltage.

Albert Einstein used statistical mechanics in 1906 to postulate that the mean kinetic energy per degree of freedom of a particle, $(1/2)mE[v^2(t)]$, is equal to $(1/2)kT$, where m is the mass of the particle, $v(t)$ is its instantaneous velocity in a single dimension, k is Boltzmann's constant, and T is the temperature in kelvin. A shunt capacitor C is charged by the thermal noise in the resistor [see Fig. 67.6(b), where L is taken to be zero]. The average potential energy stored is $(1/2)CE[W(t)^2]$. Equating this to $1/2kT$ and solving, we obtain the mean square power

$$E[W(t)^2] = kT/C \qquad (67.28)$$

For example, let $T = 300K$ and $C = 50$ pf, and recall that $k = 1.38 \times 10^{-23}$ J/K. Then $E[W(t)^2] = kT/C = 82.8 \times 10^{-12}$, so that the input rms voltage is $\{E[W(t)^2]\}^{1/2} = 9.09$ μV.

Effective Noise and Antenna Noise

Let two series resistors R_1 and R_2 have respective temperatures of T_1 and T_2. The total noise power over an incremental frequency band df is $P_{\text{Total}}(df) = P_{11}(df) + P_{22}(df) = 4kT_1R_1df + 4kT_2R_2df = 4k(T_1R_1 + T_2R_2)df$. By putting

$$T_E = (T_1R_1 + T_2R_2)/(R_1 + R_2) \qquad (67.29)$$

we can write $P_{\text{Total}}(df) = 4kT_E(R_1 + R_2)df$. T_E is called the *effective noise temperature* [see Gardner, 1990, p. 289; or Peebles, 1987, p. 228]. An antenna receives noise from various sources of electromagnetic radiation, such as radio transmissions and harmonics, switching equipment (such as computers, electrical motor controllers), thermal (blackbody) radiation of the atmosphere and other matter, solar radiation, stellar radiation, and galaxial radiation (the ambient noise of the universe). To account for noise at the antenna output, we model the noise with an equivalent thermal noise using an effective noise temperature T_E. The incremental available power (output) over an incremental frequency band df is $P_{YY}(df) = kT_Edf$, from Eq. (67.27). T_E is often called *antenna temperature*, denoted by T_A. Although it varies with the frequency band, it is usually virtually constant over a small bandwidth.

Noise Factor and Noise Ratio

In reference to Fig. 67.8(a), we define the *noise factor F = (noise power output of actual device)/(noise power output of ideal device)*, where (noise power output of ideal device) = (power output due to

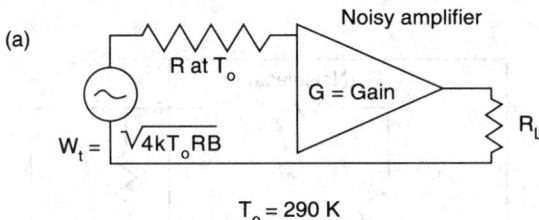

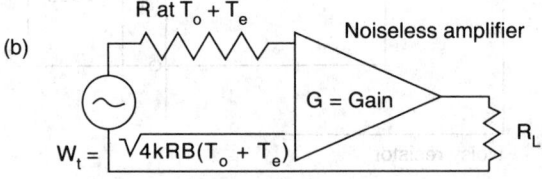

FIGURE 67.8 Equivalent input noise and noise factor.

thermal noise source). The noise source is taken to be a noisy resistor R at a temperature T, and all output noise measurements must be taken over a resistive load R_L (reactance is ignored). Letting $P_{WW}(B) = 4kTRB$ be the open circuit thermal noise power of the source resistor over a frequency bandwidth B, and noting that the gain of the device is G, the output power due to the resistive noise source becomes $G^2 P_{WW}(B) = 4kTRBG^2/R_L$. Now let $Y(t)$ be the output voltage measured at the output across R_L. Then the noise factor is

$$F = (P_{YY}(B)/R_L)/(G^2 P_{WW}(B)/R_L) = (P_{YY}(B))/(4kTRBG^2) \qquad (67.30)$$

F is seen to be independent of R_L, but not R. To compare two noise factors, the same source must be used. In the ideal noiseless case, $F = 1$, but as the noise level in the device increases, F increases. Because this is a power ratio, we may take the logarithm, called the *noise ratio*, which is

$$N_F = 10 \log_{10}(F) = 10 \log_{10}(P_{YY}(B)) - 10 \log_{10}(4kTRBG^2) \qquad (67.31)$$

The noise power output $P_{YY}(B)$ of an actual device is a superposition of the amplified source thermal noise $G^2 P_{WW}(B)$ and the device noise, i.e., $P_{YY}(B) = G^2 P_{WW}(B) + $ (device noise). The output noise across R_L can be measured by putting a single frequency (in the passband) source generator $S(t)$ as input. First, $S(t)$ is turned off, and the output rms voltage $Y(t)$ is measured and the output power $P_{Y(w)}(B)$ is recorded. This is the sum of the thermal available power and the device noise. Next, $S(t)$ is turned on and adjusted until the output power doubles, i.e., until the output power $P_{Y(W)}(B) + P_{Y(S)}(B) = 2P_{Y(W)}(B)$. This $P_{SS}(B)$ is recorded. Solving for $P_{Y(S)}(B) = P_{Y(W)}(B)$, we substitute this in $F = P_{Y(W)}(B)/(G^2 P_{WW}(B))$ to obtain

$$F = P_{Y(S)}(B)/(G^2 \cdot P_{WW}(B)) = (G^2 P_{SS}(B))/(G^2 4kTRB) = P_{SS}(B)/4kTRB \qquad (67.32)$$

A better way is to input white noise $W(t)$ in place of $S(t)$ (a noise diode may be used). The disadvantages of noise factors are (1) when the device has low noise relative to thermal noise, the noise factor has value close to 1; (2) a low resistance causes high values; and (3) increasing the source resistance decreases the noise factor while increasing the total noise in the circuit [Ott, 1976, p. 216]. Thus, accuracy is not good. For cascaded devices, the noise factors can be conveniently computed [see Buckingham, 1985, p. 67; or Ott, 1976, p. 228].

Equivalent Input Noise

Shot noise (see below) and other noise can be modeled by equivalent thermal noise that would be generated in an input resistor by increased temperature. Recall that the (maximum) incremental available power (output) in a frequency bandwidth df is $P_{WW}(df) = kTdf$ from Eq. (67.27). Figure 67.8(b) presents the situation. Let the resistor be the noise source at temperature T_o with thermal noise $W(t)$. Then $E[W(t)^2] = 4kT_o Rdf$, by Eq. (67.24) (Nyquist's result). Let the open circuit output noise power at R_L be $E[Y(t)^2]$. The incremental available noise power $P_{YY}(df)$ at the output ($R_L = R$) can be considered to be due to the resistor R having a higher temperature and an ideal (noiseless) device, usually an amplifier. We must find a temperature T_e at which a pseudothermal noise power $E[W_e(t)^2] = 4kT_e Rdf$ yields the extra "input" noise power. Let $V(t) = W(t) + W_e(t)$. Then $P_{VV}(df) = 4kT_o Rdf + 4kT_e Rdf = 4k(T_o + T_e)Rdf$ W, from Eq. (67.24). T_e is called the *equivalent input noise temperature*. It is related to the noise factor F by $T_e = 290(F - 1)$. In cascaded amplifiers with gains $G_1, G_2, \ldots$ and equivalent input noise temperatures $T_{e1}, T_{e2}, \ldots$, the total equivalent input noise temperature is

$$T_{e(\text{Total})} = T_{e1} + T_{e2}/G_1 + T_{e3}/G_1 G_2 + \cdots \qquad (67.33)$$

[see Gardner, 1990, p. 289].

Other Electrical Noise

Thermal noise and shot noise (which can be modeled by thermal noise with equivalent input noise) are the main noise sources. Other noises are discussed in the following paragraphs.

Shot Noise

In a conductor under an external emf, there is an average flow of electrons, holes, photons, etc. In addition to this induced net flow and thermal noise, there is another effect. The potential differs across the boundaries of metallic grains and particles of impurities, and when the kinetic energy of electrons exceeds this potential, electrons jump across the barrier. This summed random flow is known as *shot noise* [see Gardner, 1990, p. 239; Ott, 1976, p. 208]. The shot effect was analyzed by Schottky in 1918 as $I_{sh} = (2qI_{dc}B)^{1/2}$, where $q = 1.6 \times 10^{-19}$ coulombs per electron, I_{dc} = average dc current in amperes, and B = noise bandwidth (Hz).

Partition Noise

Partition noise is caused by a parting of the flow of electrons to different electrodes into streams of randomly varying density. Suppose that electrons from some source S flow to destination electrodes A and B. Let $n(A)$ and $n(B)$ be the average numbers of electrons per second that go to nodes A and B respectively, so that $n(S) = n(A) + n(B)$ is the average total number of electrons emitted per second. It is a success when an electron goes to A, and the probability of success on a single trial is p, where

$$p = n(A)/n(S), \ 1 - p = n(B)/n(S) \tag{67.34}$$

The current to the respective destinations is $I(A) = n(A)q$, $I(B) = n(B)q$, where q is the charge of an electron, so that $I(A)/I(S) = p$ and $I(B)/I(S) = 1 - p$. Using the binomial model, the average numbers of successes are $E[n(A)] = n(S)p$ and $E[n(B)] = n(S)(1 - p)$. The variance is $\text{Var}(n(A)) = n(S)p(1 - p) = \text{Var}(n(B))$ (from the binomial formula for variance). Therefore, substitution yields

$$\text{Var}(I(A)) = q^2[n(S)p(1 - p)] = q^2 n(S)\{I(A)I(B)/[I(A) + I(B)]\} \tag{67.35}$$

Partition noise applies to pentodes, where the source is the cathode, A is the anode (success), and B is the grid. For transistors, the source is the emitter, A is the collector, and B represents recombination in the base. In photo devices, a photoelectron is absorbed, and either an electron is emitted (a success) or not. Even a partially silvered mirror can be considered to be a partitioner: the passing of a photon is a success and reflection is a failure. While the binomial model applies to partitions with destinations A and B, multinomial models are analogous for more than two destinations.

Flicker, Contact, and Burst Noise

J. B. Johnson first noticed in 1925 that noise across thermionic gates exceeded the expected shot noise at lower frequencies. It is most noticeable up to about 2 kHz. The psdf of the extra noise, called *flicker noise*, is

$$S(f) = I^2/\alpha f, \ f > 0 \tag{67.36}$$

where I is the dc current flowing through the device and f is the positive frequency. Empirical values of α are about 1 to 1.6 for different sources. These sources vary but include the irregularity of the size of macro regions of the cathode surface, impurities in the conducting channel, and generation and recombination noise in transistors. In the early days of transistors, this generation-recombination was of great concern because the materials were not of high purity. Flicker noise occurs in thin layers of metallic or semiconducting material, solid state devices, carbon resistors, and vacuum tubes [see Buckingham, 1985, p. 143]. It includes *contact noise* because it is caused by fluctuating conductivity due to imperfect contact between two surfaces, especially in switches and relays. Flicker noise may be high at low frequencies.

Burst noise is also called *popcorn noise:* audio amplifiers sound like popcorn popping in a frying pan background (thermal noise). Its characteristic is $1/f^n$ (usually $n = 2$), so its power density falls off rapidly, where f is frequency. It may be problematic at low frequencies. The cause is manufacturing defects in the junction of transistors (usually a metallic impurity).

Barkhousen and Other Noise

Barkhousen noise is due to the variations in size and orientation of small regions of ferromagnetic material and is especially noticeable in the steeply rising region of the hysteresis loop. There is also secondary emission, photo and collision ionization, etc.

Measurement and Quantization Noise

Measurement Error

The measurement X_t of a signal $X(t)$ at any t results in a measured value $X = x_e$ that contains error, and so is not equal to the true value $X = x_T$. The probability is higher that the magnitude of $e = (x_e - x_T)$ is closer to zero. The bell-shaped Gaussian probability density $f(e) = [1/(2\pi\sigma^2)]^{1/2}\exp(-e^2/2\pi\sigma)$ fits the error well. This noise process is stationary over time. The expected value is $\mu_e = 0$, the mean-square error is σ_e^2, and the rms error is σ_e. Its instantaneous power at time t is σ_e^2. To see this, the error signal $e(t) = (x_e - x_T)$ has instantaneous power per Ω

$$P_i = e(t)i(t) = e(t)[e(t)/R] = e^2(t) \tag{67.37}$$

where $R = 1\ \Omega$ and $i(t)$ is the current. The average power is the summed instantaneous power over a period of time T, divided by the time, taken in the limit as $T \to \infty$, i.e.,

$$P_{\text{ave}} = \lim_{T\to\infty}(1/T)\int_0^T e^2(t)dt$$

This average power can be determined by sampling on known signal values and then computing the sample variance (assuming ergodicity: see Gardner [1990, p. 163]). The error and signal are probabilistically independent (unless the error depends on the values of X). The signal-to-noise power ratio is computed by $S/N = P_{\text{signal}}/P_{\text{ave}}$.

Quantization Noise

Quantization noise is due to the digitization of an exact signal value $v_t = v(t)$ captured at sampling time t by an A/D converter. The binary representation is $b_{n-1}b_{n-2}\ldots b_1b_0$ (an n-bit word). The n-bit digitization has 2^n different values possible, from 0 to $2^n - 1$. Let the voltage range be R. The *resolution* is $dv = R/2^n$. Any voltage v_t is coded into the nearest lower binary value x_b, where the error $e = x_t - x_b$ satisfies $0 \le e \le dv$. Thus, the errors e are distributed over the interval $[0,dv]$ in an equally likely fashion that implies the uniform distribution on $[0, dv]$. The expected value of $e = e_t = e(t)$ at any time is $\mu_e = dv/2$, and the variance is $\mu_e^2 = dv^2/12$ (the variance of a uniform distribution on an interval $[a,b]$ is $\sigma = (b - a)^2/12$). Thus the noise is ws and the power of quantization noise is

$$\sigma_e^2 = \int_0^{dv} (e - dv/2)^2 (1/dv)de$$
$$= (e - dv/2)^3/3dv \Big|_0^{dv} = [(dv)^3 + (dv)^3]/24dv = dv^2/12 \tag{67.38}$$

We can find the signal-to-noise voltage ratio for the total range R via $R/(dv/(12)^{1/2}) = 2^n dv/(dv/(12)^{1/2}) = 2^n(12)^{1/2}$. The power ratio is the square of this, which is $(2^{2n})(12)$. In decibels this becomes $(S/N)_{\text{dB}} = 10\log_{10}(2^{2n} \cdot 12) = 10\log_{10}(12) + 20n\log_{10}(2) = 10.8 + 6.02n$. Thus, quantization *S/N* power ratio depends directly upon the number of bits *n* in that the higher *S/N* power ratio is better, just as we would have expected.

Coping with Noise

External interference is ubiquitous. Intrinsic noise is present up to the incremental available power at temperatures above absolute zero, and other intrinsic noises depend on material purity and connection integrity. Processing error is always introduced in some form.

External Sources

Standard defenses are (1) shielding of lines and circuits, (2) twisted wire pairs or coaxial cables, (3) short lines and leads, (4) digital regeneration at waypoints of digital signals, (5) narrowband signals, (6) correlation of received signals with multipaths, and (7) adaptive notch filtering to eliminate interference at known frequencies; e.g., the second harmonic of 60-Hz ac power lines may interfere with biological microvoltage measurements but could be eliminated via adaptive notch filtering. Ferrite beads can dampen interference [Barnes, 1987]. Digital signal processing, spectral shaping filters [see Brown, 1983], and frequency-shift filters [see Gardner, 1990, p. 400] can be used to lower noise power. Kalman filtering is a powerful estimation method, and frequency-shift filtering is a newer technique for discriminating against both measurement error (e.g., in system identification applications) and extrinsic sources of both noise and interference [Gardner, 1990, p. 400].

Intrinsic Sources

Strategies for minimizing intrinsic noise are (a) small bandwidth *B*, (b) small resistances *R*, (c) low temperature *T* (higher temperatures can be devastating), (d) low voltage and currents (CMOS transistors), (e) modern materials of high purity, (f) wrapped wire resistors (thermal noise is the same, but other noise will be less), (g) fewer and better connections (of gold), (h) smaller circuits of lower power, and (i) shunt capacitors to reduce noise bandwidth. Greater purity of integrated circuit materials nowadays essentially reduces intrinsic noise to thermal noise. Better design and materials are the keys to lower noise.

Processing Sources

Processing errors can be reduced by using higher resolution of analog-to-digital converters, i.e., more bits to represent each value. This lowers the quantization error power. Measurement error can be reduced while using the same instruments by taking multiple measurements and averaging. Other estimation/correlation can yield better values (e.g., the Global Positioning System location determination can be reduced from a meter to a few centimeters by multiple measurement estimation).

Defining Terms

Autocorrelation: A function associated with a random signal $X(t)$ that is defined on pairs of time instants t_1 and t_2 and whose value is the expected value of the product of the random variables $X(t_1)$ and $X(t_2)$, i.e., $R_{XX}(t_1,t_2) = E[X(t_1)X(t_2)]$. For weakly stationary random signals, it depends only on the offset $\tau = t_2 - t_1$, so we write $R_{XX}(\tau) = E[X(t)X(t+\tau)]$.

Noise: A signal $N(t)$ whose value at any time t is randomly selected by events beyond our control. At any time instant t, $N(t)$ is a random variable N_t with a probability distribution that determines the relative frequencies at which N_t assumes values. The statistics of the family of random variables $\{N_t\}$ may be constant (stationary) over time (the usual case) or may vary.

Power spectral density: The Fourier transform of the power $X^2(t)$ does not necessarily exist, but it does for $X_T^2(t)/2T$ ($X_T(t) = 0$ for $|t| > T$, $= X(t)$ elsewhere), for any $T > 0$. Letting $T \to \infty$, the expected value of the Fourier transforms $E[\mathbf{F}[X_T^2(t)/2T] = \mathbf{F}[E[X_T^2(t)]]/2T$ goes to the limit of the average power in $X(t)$ over $-T$ to T, known as the power spectral density function $S_{xx(w)}$. Summed up over all frequencies, it gives the total power in the signal $X(t)$.

Random process: (signal): A signal that is either a noise, an interfering signal $s(t)$, or a sum of these such as $X(t) = s_1(t) + \cdots + s_m(t) + N_1(t) + \cdots + N_n(t)$.

Realization: A trajectory $\{(t, x_t): X(t) = x_t\}$ determined by the actual outcomes $\{x_t\}$ of values from a random signal $X(t)$, where $X(t) = x_t$ at each instant t. A trajectory is also called a *sample function* of $X(t)$.

Weakly stationary (ws) random process (signal): A random signal whose first- and second-order statistics remain stationary (fixed) over time.

References

J. R. Barnes, *Electronic System Design: Interference and Noise Control,* Englewood Cliffs, N.J.: Prentice-Hall, 1987.

R. G. Brown, *Introduction to Random Signal Analysis and Kalman Filtering,* New York: Wiley, 1983.

M. J. Buckingham, *Noise in Electronic Devices and Systems,* New York: Halstead Press, 1985.

W. A. Gardner, *Introduction to Random Processes,* 2nd ed., New York: McGraw-Hill, 1990.

J. B. Johnson, "Thermal agitation of electricity in conductors," *Phys. Rev.,* vol. 29, pp. 367–368, 1927.

J. B. Johnson, "Thermal agitation of electricity in conductors," *Phys. Rev.,* vol. 32, pp. 97–109, 1928.

H. Nyquist, "Thermal agitation of electric charge in conductors," *Phys. Rev.,* vol. 32, pp. 110–113, 1928.

H. W. Ott, *Noise Reduction Techniques in Electronic Systems,* New York: Wiley-Interscience, 1976.

P. Z. Peebles, Jr., *Probability, Random Variables, and Random Signal Principles,* 2nd ed., New York: McGraw-Hill, 1987.

Further Information

The IEEE Individual Learning Program, *Random Signal Analysis with Random Processes and Kalman Filtering,* prepared by Carl G. Looney (IEEE Educational Activities Board, PO Box 1331, Piscataway, NJ 08855-1331, 1989) contains a gentle introduction to estimation and Kalman filtering.

Also see H. M. Denny, *Getting Rid of Interference,* IEEE Video Conference, Educational Activities Dept., Piscataway, NJ, 08855-1331, 1992.

67.3 Stochastic Processes

Carl G. Looney

Introduction to Random Variables

A *random variable* (rv) A is specified by its *probability density function* (pdf)

$$f_A(a) = \lim_{\varepsilon \to 0} (1/\varepsilon) P[a - (\varepsilon/2) < A \leq a + (\varepsilon/2)]$$

In other words, the rectangular area $\varepsilon \cdot f_A(a)$ approximates the probability $P[(A \leq a + (\varepsilon/2)] - P[a - (\varepsilon/2) < A]$. The joint pdf of two rv's A and B is specified by

$$f_{AB}(a,b) = \lim_{\varepsilon \to 0} (1/\varepsilon^2) P[a - \varepsilon < A \leq a + (\varepsilon/2) \text{ and } b - \varepsilon < B \leq b + (\varepsilon/2)]$$

A similar definition holds for any finite number of rv's.

The *expected value* $E[A]$, or *mean* μ_A, of a rv A is the first moment of the pdf, and the *variance* of A is the second centralized moment, defined respectively by

$$\mu_A = E[A] \equiv \int_{-\infty}^{\infty} a f_A(a) da \qquad (67.39a)$$

$$\sigma_A^2 = E[(A - \mu_A)^2] \equiv \int_{-\infty}^{\infty} (a - \mu_A)^2 f_A(a) da \qquad (67.39b)$$

The square root of the variance is the *standard deviation*, which is also called the *root mean square* (rms) error. The *covariance* of two rv's A and B is the second-order centralized joint moment

$$\sigma_{AB} = E[(A - \mu_A)(B - \mu_B)] \equiv \int_{-\infty}^{\infty} \int_{-\infty}^{\infty} (a - \mu_A)(b - \mu_B) f_{AB}(a,b) da\, db \qquad (67.40)$$

The noncentralized second moments are the *mean-square value* and the *correlation*, respectively,

$$E[A^2] = \int_{-\infty}^{\infty} a^2 f_A(a) da = \sigma_A^2 + \mu_A^2, \quad E[AB] = \int_{-\infty}^{\infty} \int_{-\infty}^{\infty} ab f_{AB}(a,b) da\, db = \sigma_{AB} + \mu_A \mu_B$$

A set of rv's A, B, and C is defined to be *independent* whenever their joint pdf factors as

$$f_{ABC}(a,b,c) = f_A(a) f_B(b) f_C(c) \qquad (67.41)$$

for all a, b, and c, and similarly for any finite set of rv's. A *weak independence* holds when the second moment of the joint pdf, the correlation, factors as $E[AB] = E[A]E[B]$, so that $\sigma_{AB} = 0$, in which case the rv's are said to be *uncorrelated*. The covariance of A and B is a measure of how often A and B vary together (have the same sign) over trials of outcomes, and of how often they vary oppositely (different signs) and by how much, on the average. To standardize so that units do not influence the measure of dependence, we often use the *correlation coefficient*

$$\rho_{AB} \equiv \sigma_{AB} / \sigma_A \sigma_B$$

The accuracy of approximating a rv A as a linear function of another rv B, $A \approx cB + d$, for real coefficients c and d, is found by minimizing the mean-square error $\varepsilon = E\{[A - (cB + d)]^2\}$. Upon squaring and taking the expected values, we can obtain $\varepsilon_{min} = \sigma_A^2(1 - |\rho_{AB}|^2)$, which shows $|\rho_{AB}|$ to be a measure of the degree of linear relationship between A and B. Because $\varepsilon_{min} \geq 0$, this shows that $|\rho_{AB}| \leq 1$, which demonstrates the Cauchy-Schwarz inequality

$$|E[AB]| \leq \{E[A^2]E[B^2]\}^{1/2} \qquad (67.42)$$

When $|\rho_{AB}| = 1$, then knowledge of one of A or B completely determines the other ($c \neq 0$), and so A and B are completely dependent, while $|\rho_{AB}| = 0$ indicates there is no linear relationship, i.e., that A and B are uncorrelated.

An important result is the *fundamental theorem of expectation*: if $g(\cdot)$ is any real function, then the expected value of the rv $B = g(A)$ is given by

$$E[B] = E[g(A)] = \int_{-\infty}^{\infty} g(a) f_A(a) da \qquad (67.43)$$

Stochastic Processes

A **stochastic** (or *random*) **process** is a collection of random variables $\{X_t : t \in T\}$, indexed on an ordered set T that is usually a subset of the real numbers or integers. Examples are the Dow-Jones averages $D(t)$ at each time t, the pressure $R(t)$ in a pipe at distance t, or a noise voltage $N(t)$ at time t. A process is thus a *random function* $X(t)$ of t whose value at each t is drawn randomly from a range of outcomes for the rv $X_t = X(t)$ according to a probability distribution for X_t. A trajectory

$\{x_t: t \in T\}$ of outcomes over all $t \in T$, where $X_t = x_t$ at each t, is called a **sample function** (or *realization*) of the process. A stochastic process $X(t)$ has mean value $E[X(t)] = \mu(t)$ at time t, and **autocorrelation function** $R_{XX}(t, t + \tau) = E[X(t)X(t + \tau)]$ at times t and $t + \tau$, the correlation of two rv's at two times offset by τ. When $\mu(t) = 0$ for all t, the autocorrelation function equals the *autocovariance function* $C_{XX}(t, t + \tau) = E[(X(t) - \mu(t))(X(t + \tau) - \mu(t + \tau))]$.

A process $X(t)$ is completely determined by its joint pdf's $f_{X(t(1)) \ldots X(t(n))}(x(t_1), \ldots, x(t_n))$ for all time combinations $t_1, \ldots, t_n$ and all positive integers n (where $t(j) = t_j$). When the rv's $X(t)$ are *iid* (independent, identically distributed), then knowledge of one pdf yields the knowledge of all joint pdf's. This is because we can construct the joint pdf by factorization, per Eq. (67.41).

Classifications of Stochastic Processes

The ordered set T can be continuous or discrete, and the values that $X(t)$ assumes at each t may also be continuous or discrete, as shown in Table 67.1.

In another classification, a stochastic process $X(t)$ is *deterministic* whenever an entire sample function can be determined from an initial segment $\{x_t: t \leq t_1\}$ of $X(t)$. Otherwise, it is *nondeterministic* [see Brown, 1983, p. 79; or Gardner, 1990, p. 304].

Stationarity of Processes

A stochastic process is *nth order (strongly) stationary* whenever all joint pdf's of n and fewer rv's are independent of all translations of times $t_1, \ldots, t_n$ to times $\tau + t_1, \ldots, \tau + t_n$. The case of $n = 2$ is very useful. Another type of process is called **weakly stationary** (ws), or *wide-sense stationary*, and is defined to have first- and second-order moments that are independent of time (see Section 67.2 on noise). These satisfy (1) $\mu(t) = \mu$ (constant) for all t, and (2) $R_{XX}(t, t + \tau) = R_{XX}(t + s, t + s + \tau)$ for all values of s. For $s = -t$, this yields $R_{XX}(t, t + \tau) = R_{XX}(0, 0 + \tau)$, which is abbreviated to $R_{XX}(\tau)$. $X(t)$ is *uncorrelated* whenever $C_{XX}(\tau) = 0$ for τ not zero [we say $X(t)$ has *no memory*]. If $X(t)$ is correlated, then $X(t_1)$ depends on values $X(t)$ for $t \neq t_1$ [$X(t)$ has *memory*].

Some properties of autocorrelation functions for ws processes follow. First, $|R_{XX}(\tau)| \leq R_{XX}(0)$, $-\infty < \tau < \infty$, as can be seen from Eq. (67.42) with $|R_{XX}(\tau)|^2 = E[X(0)X(\tau)] \leq E[X(0)^2]E[X(\tau)^2] = R_{XX}(0)R_{XX}(\tau)$. Next, $R_{XX}(\tau)$ is real and even, i.e., $R_{XX}(-\tau) = R_{XX}(\tau)$, which is evident from substituting $s = t - \tau$ and using time independence. If $X(t)$ has a periodic component, then $R_{XX}(\tau)$ will have that same periodic component, which follows from the definition. Finally, if $X(t)$ has a nonzero mean μ and no periodic components, then the variance goes to zero (the memory fades) and so $\lim_{\tau \to \infty} R_{XX}(\tau) \to 0 + \mu^2 = \mu^2$.

Gaussian and Markov Processes

A process $X(t)$ is defined to be *Gaussian* if for every possible finite set $\{t_1, \ldots, t_n\}$ of times, the rv's $X(t_1), \ldots, X(t_n)$ are *jointly Gaussian*, which means that every linear combination $Z = a_1 X(t_1) + \cdots + a_n X(t_n)$ is a Gaussian rv, defined by the Gaussian pdf

Table 67.1 Continuous/Discrete Classification of Stochastic Processes

T Values	X Values	
	Continuous	Discrete
Continuous	Continuous stochastic processes	Discrete valued stochastic processes
Discrete	Continuous random sequences	Discrete valued random sequences

$$f_Z(z) = \left[1/(\sigma_Z\sqrt{2\pi})\right]\exp\{-(z-\mu_Z)^2/2\sigma_Z^2\} \tag{67.44}$$

In case the n rv's are *linearly independent*, i.e., $Z = 0$ only if $a_1 = \cdots = a_n = 0$, the joint pdf has the Gaussian form [see Gardner, 1990, pp. 39–40]

$$f_{X(t(1))\ldots X(t(n))}(x_1,\ldots,x_n) = \left[1/(2\pi)^{n/2}|C|^{1/2}\right]\cdot\exp\{-(\mathbf{x}-\boldsymbol{\mu})^t C^{-1}(\mathbf{x}-\boldsymbol{\mu})\} \tag{67.45}$$

where $\mathbf{x} = (x_1,\ldots,x_n)$ is a column vector, $\mathbf{x}^t$ is its transpose, $\boldsymbol{\mu} = (\mu_1,\ldots,\mu_n)$ is the mean vector, C is the *covariance matrix*

$$C = \begin{pmatrix} \sigma_1^2 & \cdots & \sigma_{1n} \\ \vdots & \vdots & \\ \sigma_{n1} & \cdots & \sigma_n^2 \end{pmatrix} \tag{67.46}$$

and $|C|$ is the determinant of C. If $X(t_1),\ldots,X(t_n)$ are linearly dependent, then the joint pdf takes on a form similar to Eq. (67.45), but contains impulses [see Gardner, 1990, p. 40].

A weakly stationary Gaussian process is strongly stationary to all orders n: all Gaussian joint pdf's are completely determined by their first and second moments by Eq. (67.45), and those moments are time independent by weak stationarity, and so all joint pdf's are also. Every second-order strongly stationary stochastic process $X(t)$ is also weakly stationary because the time translation independence of the joint pdf's determines the first and second moments to have the same property. However, non-Gaussian weakly stationary processes need not be strongly second-order stationary.

Rather than with pdf's, a process $X(t)$ may be specified in terms of conditional pdf's

$$f_{X(t(1))\ldots X(t(n))}(x_1,\ldots,x_n) = f_{X(t(n))|X(t(n-1))}(x_n|x_{n-1})\cdot\cdots\cdot f_{X(t(2))|X(t(1))}(x_2|x_1)f_{X(t(1))}(x_1)$$

by successive applications of Bayes' law, for $t_1 < t_2 < \cdots < t_n$. The conditional pdf's satisfy

$$f_{A|B}(a|b) = f_{AB}(a,b)/f_B(b) \tag{67.47}$$

The conditional factorization property may satisfy

$$f_{X(t(n))\,|\,X(t(n-1))\ldots X(t(1))}(x_n\,|\,x_{n-1},\ldots,x_1) = f_{X(t(n))\,|\,X(t(n-1))}(x_n\,|\,x_{n-1}) \tag{67.48}$$

which indicates that the pdf of the process at any time t_n, given values of the process at any number of previous times $t_{n-1},\ldots,t_1$, is the same as the pdf at t_n given the value of the process at the most recent time t_{n-1}. Such an $X(t)$ is called a *first-order Markov process*, in which case we say the process remembers only the previous value (the previous value has influence). In general, an *nth-order Markov* process remembers only the n most recent previous values. A first-order Markov process can be fully specified in terms of its first-order conditional pdf's $f_{X(t)|X(s)}(x_t,x_s)$ and its unconditional first-order pdf at some initial time t_0, $f_{X(t(0))}(x_0)$.

Examples of Stochastic Processes

Figure 67.9 shows two sample functions of nonstationary processes. Now consider the example $X(k) = A$, for all $k \geq 0$, where A is a rv (a *random initial condition*) that assumes a value 1 or -1 with respective probabilities p and $1 - p$ at $k = 0$. This value does not change, once the initial random draw is done at $k = 0$. This stochastic sequence has two sample functions only, the constant

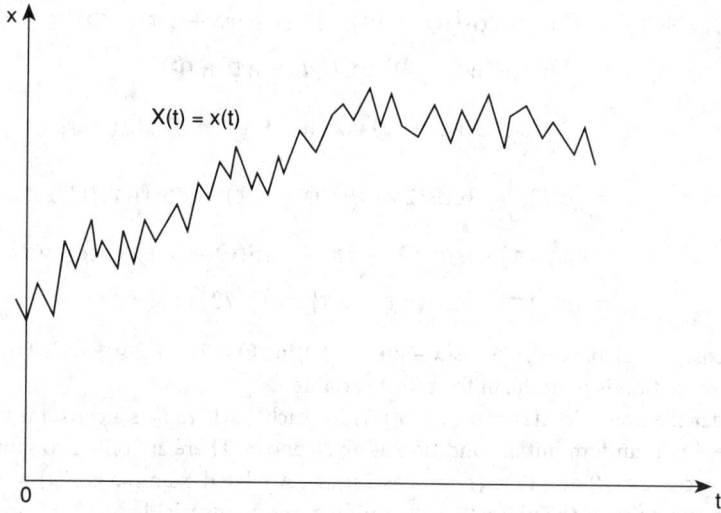

(a) Sample function with nonstationary mean and stationary variance

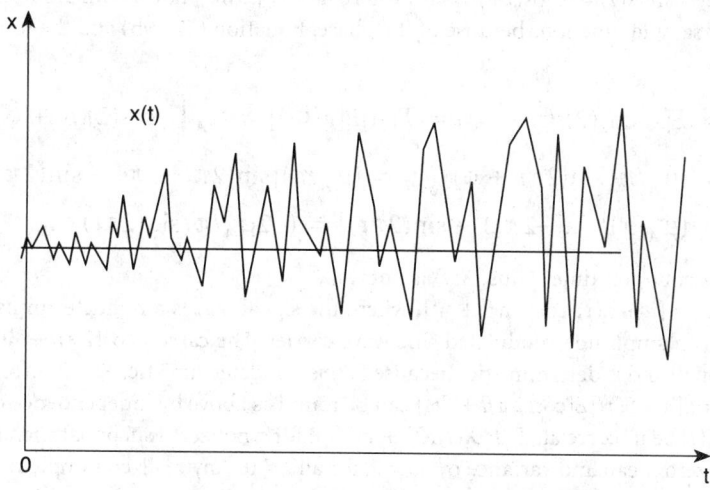

(b) Sample function with stationary mean and nonstationary (increasing) variance

FIGURE 67.9   Examples of nonstationary processes.

sequences $\{-1\}$ and $\{1\}$. The expected value of $X(k)$ at any time k is $E[X(k)] = E[A] = p \cdot 1 + (1 - p) \cdot (-1) = 2p - 1$, which is independent of k. The autocorrelation function is, by definition, $E[X(k)X(k + m)] = E[A \cdot A] = E[A^2] = p \cdot 1^2 + (1 - p) \cdot (-1)^2 = 1$ which is also independent of time k. Thus $X(k)$ is perfectly correlated for all time (the process has *infinite memory*). This process is deterministic.

For another example, put $X(t) = (c) \cdot \cos(wt + \Phi)$, where Φ is the uniform rv on $(-\pi, \pi)$. Then $X(t)$ is a function of the rv Φ (as well as t), so by use of Eq. (67.39a), we obtain

$$E[X(t) = c \cdot \int_{-\pi}^{\pi} \cos(wt + \phi)f_\Phi(\phi)d\phi = (c/2\pi) \sin(wt + \phi)\Big|_{-\pi}^{\pi} = 0$$

Therefore, the mean does not vary with time t. The autocorrelation is

$$R_{XX}(t, t + \tau) = E[(c) \cdot \cos(wt + \Phi)(c) \cdot \cos(wt + w\tau + \Phi)]$$

$$= c^2 E[\cos(wt + \Phi) \cos(wt + w\tau + \Phi)]$$

$$= c^2 \int_{-\pi}^{\pi} \cos(wt + \phi) \cos(wt + w\tau + \Phi) f_\Phi(\phi) d\phi$$

$$= (c^2/2) \int_{-\pi}^{\pi} \{\cos(2wt + 2\phi + w\tau) + \cos(w\tau)\}(1/2\pi) d\phi$$

$$= (c^2/4\pi) \cdot \{\sin(\Theta + 2\pi) - \sin(\Theta - 2\pi) + \cos(w\tau) \cdot 2\pi\}$$

$$= (c^2/4\pi) \cdot \{\cos(w\tau) \cdot 2\pi\} = (c^2/2) \cos(w\tau)$$

[using $\cos(x)\cos(y) = \frac{1}{2}\{\cos(x + y) + \cos(x - y)\}$ and letting $\Theta = 2wt + 2\Phi + w\tau$]. Therefore, $X(t)$ is ws. The autocorrelation is periodic in the offset variable τ.

Now consider the example $X(t) = A\cos(2\pi f_0 t)$ for each t, where f_0 is a constant frequency, and the amplitude A is a random initial condition as given above. There are only two sample functions here: (1) $x(t) = \cos(2\pi f_0 t)$ and (2) $x(t) = -\cos(2\pi f_0 t)$. A related example is $X(t) = A\cos(2\pi f_0 t + \Phi)$, where A is given above, the phase Φ is the uniform random variable on $[0,\pi]$, and A and Φ are independent. Again, Φ does not depend on time (an initial random condition). Thus, the sample functions for $X(t)$ are $x(t) = \pm\cos(2\pi ft + \phi)$, where $\Phi = \phi$ is the value assumed initially. There are infinitely many sample functions because of the phase. Equation (67.39b) and the independence of A and Φ yield

$$E[X(t)] = E[A\cos(2\pi ft + \Phi)] = E[A]E[g(\Phi)] = \mu_A \int_0^\pi \cos(2\pi ft + \phi)(1/\pi) d\phi$$

$$= (\mu_A/\pi) \sin(2\pi t + \phi)\big|_{\phi=0}^\pi = (\mu_A/\pi)[\sin(2\pi t + \pi) - \sin(2\pi t)]$$

$$= (\mu_A/\pi)[\sin(-2\pi t) - \sin(2\pi t)] = (-2\mu_A/\pi)\sin(2\pi t)$$

which is dependent upon time. Thus, $X(t)$ is not ws.

Next, let $X(t) = [a + S(t)]\cos[2\pi ft + \Phi]$, where the signal $S(t)$ is a nondeterministic stochastic process. This is an amplitude-modulated sine wave carrier. The carrier $\cos[2\pi ft + \Phi]$ has random initial condition Φ and is deterministic. Because $S(t)$ is nondeterministic, $X(t)$ is also. The expected value $E[X(t)] = E[a + S(t)]E[\cos(2\pi ft + \Phi)]$ can be found as above by independence of $S(t)$ and Φ.

Finally, let $X(t)$ be uncorrelated ($E[X(t)X(t + \tau)] = 0$ for τ not zero) such that each rv $X(t) = X_t$ is Gaussian with zero mean and variance $\sigma^2(t) = t$, for all $t > 0$. Any realized sample function $x(t)$ of $X(t)$ cannot be predicted in any average sense based on past values (uncorrelated Gaussian random variables are independent). The variance grows in an unbounded manner over time, so $X(t)$ is neither stationary nor deterministic. This is called the *Wiener* process.

A useful model of a ws process is that for which $\mu = 0$ and $R_{XX}(\tau) = \sigma_X^2 \exp(-\alpha|\tau|)$. If this process is also Gaussian, then it is strongly stationary and all of its joint pdf's are fully specified by $R_{XX}(\tau)$. In this case it is also a first-order Markov process and is called the *Ornstein-Uhlenbeck* process [see Gardner, 1990, p. 102]. Unlike white noise, many real-world ws stochastic processes are correlated ($|R_{XX}(t, t + \tau)| > 0$) for $|\tau| > 0$. The autocorrelation either goes to zero as τ goes to infinity, or else it has periodic or other nondecaying memory. We consider ws processes henceforth [for nonstationary processes, see Gardner, 1990]. We will also assume without loss of generality that $\mu = 0$.

Linear Filtering of Weakly Stationary Processes

Let the ws stochastic process $X(t)$ be the input to a linear time-invariant stable filter with impulse response function $h(t)$. The output of the filter is also a ws stochastic process and is given by the convolution

$$Y(t) = h(t) * X(t) = \int_{-\infty}^{\infty} h(s)X(t - s)ds \tag{67.49}$$

The mean of the output process is obtained by using the linearity of the expectation operator [see Gardner, 1990, p. 32]

$$\mu_Y = E[Y(t)] = E\left[\int_{-\infty}^{\infty} h(s)X(t-s)ds\right] = \int_{-\infty}^{\infty} h(s)E[X(t-s)]ds = \int_{-\infty}^{\infty} h(s)\mu_X ds$$

$$= \mu_X \int_{-\infty}^{\infty} h(s)ds = \mu_X \cdot H(0)$$

(67.50)

where $H(f) = \int_{-\infty}^{\infty} h(t)e^{-j2\pi ft}dt$ is the filter transfer function and $H(0)$ is the dc gain of the filter. The autocorrelation of the output process, obtained by using the linearity of $E[\cdot]$, is

$$R_{YY}(\tau) = E[Y(t)Y(t+\tau)] = E\left[\int_{-\infty}^{\infty} h(v)X(t-v)dv \int_{-\infty}^{\infty} h(u)X(t+\tau-u)du\right]$$

$$= \int_{-\infty}^{\infty}\int_{-\infty}^{\infty} E[X(t-v)X(t+\tau-u)]h(v)h(u)dvdu$$

$$= \int_{-\infty}^{\infty}\left\{\int_{-\infty}^{\infty} R_{XX}(\tau-u+v)h(u)du\right\}h(v)dv$$

$$= \int_{-\infty}^{\infty}\left\{\int_{-\infty}^{\infty} R_{XX}([\tau-(-v)]-u)h(u)du\right\}h(-v)(dv)$$

(67.51)

$$= \int_{-\infty}^{\infty}\left\{R_{XX}(\tau+v)*h(\tau+v)\right\}h(-v)dv$$

$$= [R_{XX}(\tau)*h(\tau)]*h(-\tau) = R_{XX}(\tau)*[h(\tau)*h(-\tau)] = R_{XX}(\tau)*r_h(\tau)$$

where $r_h(\tau) = \int_{-\infty}^{\infty} h(\tau+u)h(u)du$. However, $r_h(\tau)$ has Fourier transform $H(f)H*(f) = |H(f)|^2$, because the Fourier transform of the convolution of two functions is the product of their Fourier transforms, and the Fourier transform of $h(-\tau)$ is the complex conjugate $H*(f)$ of the Fourier transform $H(f)$ of $h(\tau)$. Thus, the Fourier transform of $R_{YY}(\tau)$, denoted by $\mathbf{F}\{R_{YY}(\tau)\}$, is

$$\mathbf{F}\{R_{YY}(\tau)\} = \mathbf{F}\{R_{XX}(\tau)*h(\tau)*h(-\tau)\} = \mathbf{F}\{R_{XX}(\tau)\} \cdot H(f)H*(f) = \mathbf{F}\{R_{XX}(\tau)\} \cdot |H(f)|^2$$

Upon defining

$$S_{XX}(f) \equiv \mathbf{F}\{R_{XX}(\tau)\}, \quad S_{YY}(f) \equiv \mathbf{F}\{R_{YY}(\tau)\}$$

(67.52)

we can also determine $R_{YY}(\tau)$ via the two steps

$$S_{YY}(f) = S_{XX}(f) \cdot |H(f)|^2$$

(67.53)

$$R_{YY}(\tau) = \mathbf{F}^{-1}\{S_{YY}(f)\} = \int_{-\infty}^{\infty} S_{YY}(f)e^{j2\pi f\tau}df$$

(67.54)

Equations (67.52) define the *power spectral density functions* (psdf's) $S_{XX}(f)$ for $X(t)$ and $S_{YY}(f)$ for $Y(t)$. Thus, $R_{XX}(\tau)$ and $S_{XX}(f)$ are Fourier transform pairs, as are $R_{YY}(\tau)$ and $S_{YY}(f)$. Further, the psdf $S_{XX}(f)$ of $X(t)$ is a power spectrum (in an average sense). If $X(t)$ is the voltage dropped across a 1-Ω resistor, then $X^2(t)$ is the instantaneous power dissipation in the resistance. Consequently, $R_{XX}(0) = E[X^2(t)]$ is the expected power dissipation over all frequencies, i.e., by Eq. (67.54) with $\tau = 0$, we have

$$R_{XX}(0) = \int_{-\infty}^{\infty} S_{XX}(f)df$$

We want to show that when we pass $X(t)$ through a narrow bandpass filter with a bandwidth δ centered at the frequency $\pm f_0$, the expected power at the output terminals, divided by the bandwidth δ, is $S_{XX}(f_0)$ in the limit as $\delta \to 0$. This shows that $S_{XX}(f)$ is a density function (whose area is the total expected power over all frequencies, just as the area under a pdf is the total probability). This result that $R_{XX}(\tau)$ and $S_{XX}(f)$ are a Fourier transform pair is known as the *Wiener-Khinchin* relation [see Gardner, 1990, p. 230].

To verify this relation, let $H(f)$ be the transfer function of an ideal bandpass filter, where

$$H(f) = 1, \|f\| - f_0\| < \delta/2; \qquad H(f) = 0, \text{ otherwise}$$

Let Y(t) be the output of the filter. Then Eqs. (67.54) and (67.53) provide

$$E[Y^2(t)] = R_{YY}(0) = \int_{-\infty}^{\infty} S_{YY}(f)df = \int_{-\infty}^{\infty} S_{XX}(f)\,|H(f)|^2\,df$$
$$= \int_{f_0-\delta/2}^{f_0+\delta/2} S_{XX}(f)df + \int_{-f_0-\delta/2}^{-f_0+\delta/2} S_{XX}(f)df$$

Dividing by 2δ and taking the limit as $\delta \to 0$ yields $(1/2)S_{XX}(f_0) + (1/2)S_{XX}(-f_0)$, which becomes $S_{XX}(f_0)$ when we use the fact that psdf's are even and real functions (because they are the Fourier transforms of autocorrelation functions, which are even and real).

For example, let $X(t)$ be white noise, with $S_{XX}(f) = N_0$, being put through a first-order linear time-invariant system with respective impulse response and transfer functions

$$h(t) = \exp\{-\alpha t\}, t \geq 0; \ h(t) = 0, t < 0 \qquad H(f) = 1/[\alpha + j2\pi f], \text{ all } f$$

The temporal correlation of $h(t)$ with itself is $r_h(\tau) = (1/2\alpha)\exp\{-\alpha|\tau|\}$, so the power transfer function is $|H(f)|^2 = 1/[\alpha^2 + (2\pi f)^2]$. The autocorrelation for the input $X(t)$ is then

$$R_{XX}(\tau) = \int_{-\infty}^{\infty} N_0 e^{j2\pi f\tau} df = N_0 \delta(\tau)$$

which is an impulse. It follows that the output $Y(t)$ has respective autocorrelation and psdf

$$R_{YY}(\tau) = [N_0\delta(\tau)]*[(1/2\alpha)e^{-\alpha|\tau|}] = (N_0/2\alpha)e^{-\alpha|\tau|}, \quad S_{YY}(f) = N_0/[\alpha^2 + (2\pi f)^2]$$

The output expected power $E[Y^2(t)]$ can be found from either one of

$$E[Y^2(t)] = R_{YY}(0) = N_0/2\alpha \qquad \text{or} \qquad E[Y^2(t)] = \int_{-\infty}^{\infty} S_{YY}(f)df = N_0/2\alpha$$

If the input $X(t)$ to a linear system is Gaussian, then the output will also be Gaussian [see Brown, 1983; or Gardner, 1990]. Thus, the output of a first-order linear time-invariant system driven by Gaussian white noise is the Ornstein–Uhlenbeck process, which is also a first-order Markov process.

For another example, let $X(t) = A \cos(w_o t + \Theta)$, where the random amplitude A has zero mean, the random phase Θ is uniform on $[-\pi, \pi]$, and A and Θ are independent. As before, we obtain $R_{XX}(\tau) = \sigma_A^2 \cos(w_o\tau)$, from which it follows that $S_{XX}(f) = (\sigma_A^2/2)[\delta(f - w_o/2\pi) + \delta(f + w_o/2\pi)]$. These impulses in the psdf, called *spectral lines*, represent positive amounts of power at discrete frequencies.

Cross-Correlation of Processes

The *cross-correlation function* for two random processes $X(t)$ and $Y(t)$ is defined via

$$R_{XY}(t, t + \tau) \equiv E[X(t)Y(t + \tau)] \tag{67.55}$$

Let both processes be ws with zero means, so the covariance coincides with the correlation function. We say that two ws processes $X(t)$ and $Y(t)$ are *jointly ws* whenever $R_{XY}(t, t + \tau) = R_{XY}(\tau)$. In this case we can find the cross-correlation $R_{XY}(\tau)$ via

$$
\begin{aligned}
R_{XY}(\tau) &= E[X(t)Y(t + \tau)] = E[X(t)\int_{-\infty}^{\infty} h(u)X(t + \tau - u)du] \\
&= \int_{-\infty}^{\infty} h(u)E[X(t)X(t + \tau - u)]\,du \\
&= \int_{-\infty}^{\infty} h(u)R_{XX}(\tau - u)du = R_{XX}(\tau) * h(\tau)
\end{aligned}
\tag{67.56}
$$

Cross-correlation functions of ws processes satisfy (1) $R_{XY}(-\tau) = R_{YX}(\tau)$, (2) $|R_{XY}(\tau)|^2 \leq R_{XX}(0)R_{YY}(0)$, and (3) $|R_{XY}(\tau)| \leq (1/2)[R_{XX}(0) + R_{YY}(0)]$. The first follows from the definition, while the second comes from expanding $E[\{Y(t + \tau) - \alpha X(t)\}^2] \geq 0$. The third comes from the fact that the geometric mean cannot exceed the arithmetic mean [see Peebles, 1987, p. 154].

Taking the Fourier transform of the leftmost and rightmost sides of Eqs. (67.56) yields

$$
S_{XY}(f) = S_{XX}(f)H(f)
\tag{67.57}
$$

The Fourier transform of the cross-correlation function is the *cross-spectral density function*

$$
S_{XY}(f) = \int_{-\infty}^{\infty} R_{XY}(\tau)e^{-j2\pi f\tau}d\tau
\tag{67.58}
$$

According to Gardner [1990, p. 228], this is a *spectral correlation density function* that does not represent power in any sense.

Equation (67.57) suggests a method for identifying a linear time-invariant system. If the system is subjected to a ws input $X(t)$ and the power spectral density of $X(t)$ and the cross-spectral density of $X(t)$ and the output $Y(t)$ are measured, then the ratio yields the system transfer function

$$
H(f) = S_{XY}(f)/S_{XX}(f)
\tag{67.59}
$$

In fact, it can be shown that this method gives the best linear time-invariant model of the (possibly time varying and nonlinear) system in the sense that the time-averaged mean-square error between the outputs of the actual system and of the model, when both are subjected to the same input, is minimized [see Gardner, 1990, pp. 282–286].

As an application, suppose that an undersea sonar-based device is to find the range to a target, as shown in Fig. 67.10, by transmitting a sonar signal $X(t)$ and receiving the reflected signal $Y(t)$. If v is the velocity of the sonar signal, and τ_o is the offset that maximizes the cross-correlation $R_{XY}(\tau)$, then the range (distance) d can be determined from $d = v\tau_o/2$ (note that the signal travels twice the range d).

Coherence

When $X(t)$ and $Y(t)$ have no spectral lines at f, the finite spectral correlation $S_{XY}(f)$ is actually a spectral covariance and the two associated variances are $S_{XX}(f)$ and $S_{YY}(f)$. We can normalize $S_{XY}(f)$ to obtain a *spectral correlation coefficient* $\Upsilon_{XY}(f)$ defined by

$$
\Upsilon_{XY}(f)^2 = |S_{XY}(f)|^2/S_{XX}(f)S_{YY}(f)
\tag{67.60}
$$

We call $\Upsilon_{XY}(f)$ the *coherence function*. It is a measure of the power correlation of $X(t)$ and $Y(t)$ at each frequency f. When $Y(t) = X(t)*h(t)$, it has a maximum: by Eqs. (67.53), (67.59), and (67.60), $|\Upsilon_{XY}(f)|^2 = |S_{XX}(f) \cdot H(f)|^2/[S_{XX}(f) \cdot S_{XX}(f) \cdot |H(f)|^2] = 1$. In the general case we have

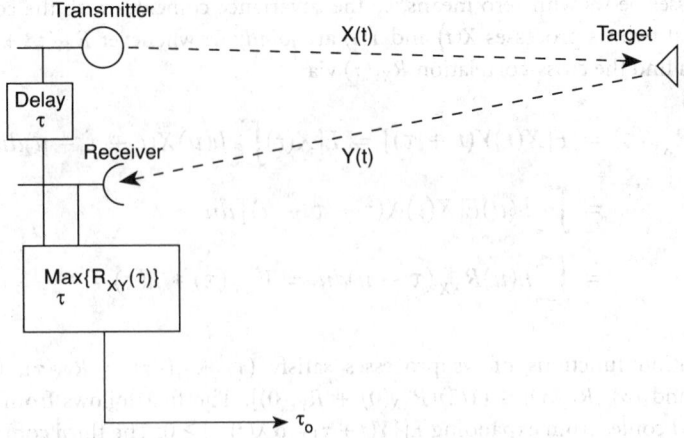

FIGURE 67.10   A sonar range finder.

$$|S_{XY}(f)| \le [S_{XX}(f)S_{YY}(f)]^{1/2} \tag{67.61}$$

Upon minimizing the mean-square error $\varepsilon = E[(Y(t) - X(t)*h(t))^2]$ over all possible impulse response functions $h(t)$, the optimal one, $h_o(t)$, has transfer function

$$H_o(f) = S_{XY}(f)/S_{XX}(f) \tag{67.62}$$

Further, the resultant minimum value is given by

$$\varepsilon_{min} = \int_{-\infty}^{\infty} S_{YY}(f)[1 - |\Upsilon_{XY}(f)|^2]\, df$$

[see Gardner, 1990, pp. 434–436; or Bendat and Piersol, 1986]. At frequencies f where $|\Upsilon_{XY}(f)| \approx 1$, $\varepsilon_{min} \approx 0$. Thus $1 - |\Upsilon_{XY}(f)|^2$ is the mean-square proportion of $Y(t)$ not accounted for by $X(t)$, while $|\Upsilon_{XY}(f)|^2$ is the proportion due to $X(t)$. When $Y(t) = X(t)*h(t)$, $\varepsilon_{min} = 0$.

The optimum system $H_o(f)$ of Eq. (67.62) is known as the *Wiener filter* for minimum mean-square error estimation of one process $Y(t)$ using a filtered version of another process $X(t)$ [see Gardner, 1990; or Peebles, 1987, p. 262].

Ergodicity

When the **time average**

$$\lim_{T \to \infty} (1/T) \int_{-T/2}^{T/2} X(t)dt$$

exists and equals the corresponding expected value $E[X(t)]$, then the process $X(t)$ is said to possess an *ergodic property associated with the mean*. There are ergodic properties associated with the mean, autocorrelation (and power spectral density), and all finite-order joint moments, as well as finite-order joint pdf's. If a process has all possible ergodic properties, it is said to be an *ergodic process*.

Let $Y(t) = g[X(t + t_1), \ldots, X(t + t_n)]$, where $g[\cdot]$ is any nonrandom real function, so that $Y(t)$ is a function of a finite number of time samples of a strongly stationary process. For example, let (1) $Y(t) = X(t + t_1)X(t + t_2)$, $E[Y(t)] = R_{XX}(t_1 - t_2)$ and (2) $Y(t) = 1$ if $X(t) < t$, $Y(t) = 0$, otherwise, so that

$$E[Y(t)] = 1 \cdot P(X(t) < x) + 0 \cdot P(X(t) \ge x) = P(X(t) < x) = \int_{-\infty}^{1} f_{X(t)}(z)dz$$

We want to know under what conditions the mean-square error between the time average

$$\langle Y(t) \rangle_T \equiv (1/T) \int_{-T/2}^{T/2} Y(t) dt$$

and the expected value $E[Y(t)]$ will converge to zero. It can be shown that a necessary and sufficient condition for the mean-square ergodic property

$$\lim_{T \to \infty} E[\{\langle Y(t) \rangle_T - E[Y(t)]\}^2] = 0 \qquad (67.63)$$

to hold is that

$$\lim_{T \to \infty} (1/T) \int_0^T C_{YY}(\tau) d\tau = 0 \qquad (67.64)$$

For example, if $C_{YY}(\tau) \to 0$ as $\tau \to \infty$, then Eq. (67.64) will hold, and thus Eq. (67.63) will also, where $C_{YY}(\tau)$ is the covariance function of $Y(t)$. As long as the two sets of rv's $\{X(t + t_1), \ldots, X(t + t_n)\}$ and $\{X(t + t_1 + \tau), \ldots, X(t + t_n + \tau)\}$ become independent of each other as $\tau \to \infty$, the above condition holds, so Eq. (67.63) holds [see Gardner, 1990, pp. 163–174].

In practice, if $X(t)$ exhibits ergodicity associated with the autocorrelation, then we can estimate $R_{XX}(\tau)$ using the time average

$$\langle X(t)X(t + \tau) \rangle_T \equiv (1/T) \int_{-T/2}^{T/2} X(t)X(t + \tau) dt \qquad (67.65)$$

In this case the mean-square estimation error $E[\{\langle X(t)X(t + \tau) \rangle_T - R_{XX}(\tau)\}^2]$ will converge to zero as T increases to infinity, and the power spectral density $S_{XX}(f)$ can also be estimated via time averaging [see Gardner, 1990, pp. 230–231].

Defining Terms

Autocorrelation function: A function $R_{XX}(t, t + \tau) = E[X(t)X(t + \tau)]$ that measures the degree to which any two rv's $X(t)$ and $X(t + \tau)$, at times t and $t + \tau$, are correlated.

Sample function: A real-valued function $x(t)$ of t where at each time t the value $x(t)$ at the argument t was determined by the outcome of a rv $X_t = x(t)$.

Stochastic process: A collection of rv's $\{X_t: t \in T\}$, where T is an ordered set such as the real numbers or integers [also called a random function, $X(t)$, on the domain T].

Time average: Any real function $g(t)$ of time has average value g_{ave} on the interval $[a,b]$ such that the rectangular area $g_{ave}(b - a)$ is equal to the area under the curve between a and b, i.e., $g_{ave} = [1/(b - a)] \int_a^b g(t) dt$. The time average of a sample function $x(t)$ is the limit of its average value over $[0,T]$ as T goes to infinity.

Weakly stationary: The property of a stochastic process $X(t)$ whose mean $E[X(t)] = \mu(t)$ is a fixed constant μ over all time t, and whose autocorrelation is also independent of time in that $R_{XX}(t, t + \tau) = R_{XX}(s + t, s + t + \tau)$ for any s. Thus, $R_{XX}(t, t + \tau) = R_{XX}(0, \tau) = R_{XX}(\tau)$.

References

The author is grateful to William Gardner of the University of California, Davis for making substantial suggestions.

J. S. Bendat and A. G. Piersol, *Random Data: Analysis and Measurement*, 2nd ed., New York: Wiley-Interscience, 1986.

R. G. Brown, *Introduction to Random Signal Analysis and Kalman Filtering*, New York: Wiley, 1983.

W. A. Gardner, *Introduction to Random Processes,* 2nd ed., New York: McGraw-Hill, 1990.

P. Z. Peebles, Jr., *Probability, Random Variables, and Random Signal Principles,* 2nd ed., New York: McGraw-Hill, 1987.

Further Information

The IEEE Individual Learning Package, *Random Signal Analysis with Random Processes and Kalman Filtering,* prepared for the IEEE in 1989 by Carl G. Looney, IEEE Educational Activities Board, PO Box 1331, Piscataway, NJ 08855-1331.

R. Iranpour and P. Chacon, *Basic Stochastic Processes: The Mark Kac Lectures,* New York: Macmillan, 1988.

A. Papoulis, *Probability, Random Variables, and Stochastic Processes,* 3rd ed., New York: Macmillan, 1991.

67.4 The Sampling Theorem

R. J. Marks II

Much of that which is ordinal is modeled as analog. Most computational engines, on the other hand, are digital. Transforming from analog to digital is straightforward: we simply sample. Regaining the original signal from these samples and assessing the information lost in the sampling process are the fundamental questions addressed by the **sampling theorem**.

The fundamental result of the sampling theorem is, remarkably, that a bandlimited signal is uniquely specified by its sufficiently close equally spaced samples. Indeed, the sampling theorem illustrates how the original signal can be regained from knowledge of the samples and the sampling rate at which they were taken.

Popularization of the sampling theorem is credited to Shannon [1948] who, in 1948, used it to show the equivalence of the information content of a bandlimited signal and a sequence of discrete numbers. Shannon was aware of the pioneering work of Whittaker [1915] and Whittaker's son [1929] in formulating the sampling theorem. Kotel'nikov's [1933] independent discovery in the then Soviet Union deserves mention. Higgins [1985] credits Borel [1897] with first recognizing that a signal could be recovered from its samples.

Surveys of sampling theory are in the widely cited paper of Jerri [1977] and in two books by the author [1991, 1993]. Marvasti [1987] has written a book devoted to nonuniform sampling.

The Cardinal Series

If a signal has finite energy, the minimum **sampling rate** is equal to two samples per period of the highest frequency component of the signal. Specifically, if the highest frequency component of the signal is B Hz, then the signal, $x(t)$, can be recovered from the samples by

$$x(t) = \frac{1}{\pi} \sum_{n=-\infty}^{\infty} x\left(\frac{n}{2B}\right) \frac{\sin[\pi(2Bt - n)]}{2Bt - n} \tag{67.66}$$

The frequency B is also referred to as the signal's bandwidth and, if B is finite, $x(t)$ is said to be bandlimited. The signal, $x(t)$, is here being sampled at a rate of $2B$ samples per second. If sampling were done at a lower rate, the replications would overlap and the information about $X(\omega)$ [and thus $x(t)$] is irretrievably lost. Undersampling results in *aliased* data. The minimum sampling rate at which **aliasing** does not occur is referred to as the **Nyquist rate** which, in our example, is $2B$. Eq. (67.66) was dubbed the **cardinal series** by the junior Whittaker [1929].

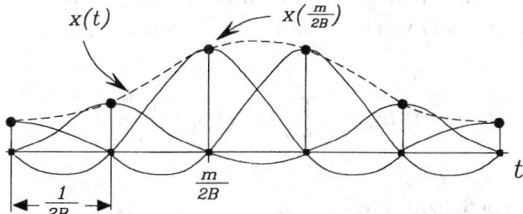

FIGURE 67.11 Illustration of the interpolation that results from the cardinal series. A sinc function, weighted by the sample, is placed at each sample bottom. The sum of the sincs exactly generates the original bandlimited function from which the samples were taken.

A signal is bandlimited in the low-pass sense if there is a $B > 0$ such that

$$X(\omega) = X(\omega)\Pi\left(\frac{\omega}{4\pi B}\right) \tag{67.67}$$

where the gate function $\Pi(\xi)$ is one for $\xi \le 1/2$ and is otherwise zero, and

$$X(\omega) = \int_{-\infty}^{\infty} x(t)e^{-j\omega t}\,dt \tag{67.68}$$

is the **Fourier transform** of $x(t)$. That is, the spectrum is identically zero for $|\omega| > 2\pi B$. The B parameter is referred to as the **signal's bandwidth**. The inverse Fourier transform is

$$x(t) = \frac{1}{2\pi}\int_{-\infty}^{\infty} X(\omega)\,e^{j\omega t}\,d\omega \tag{67.69}$$

The sampling theorem reduces the normally continuum infinity of ordered pairs required to specify a function to a countable—although still infinite—set. Remarkably, these elements are obtained directly by sampling.

How can the cardinal series interpolate uniquely the bandlimited signal from which the samples were taken? Could not the same samples be generated from another bandlimited signal? The answer is no. Bandlimited functions are smooth. Any behavior deviating from smooth would result in high-frequency components which in turn invalidates the required property of being bandlimited. The smoothness of the signal between samples precludes arbitrary variation of the signal there.

Let's examine the cardinal series more closely. Evaluate Eq. (67.74) at $t = m/2B$. Since sinc(n) is one for $n = 0$ and is otherwise zero, only the sample at $t = m/2B$ contributes to the interpolation at that point. This is illustrated in Fig. 67.11, where the reconstruction of a signal from its samples using the cardinal series is shown. The value of $x(t)$ at a point other than a sample location [e.g., $t = (m + \frac{1}{2})/2B$] is determined by all of the sample values.

Proof of the Sampling Theorem

Borel [1897] and Shannon [1948] both discussed the sampling theorem as the Fourier transform dual of the Fourier series. Let $x(t)$ have a bandwidth of B. Consider the periodic signal

$$Y(\omega) = \sum_{n=-\infty}^{\infty} X(\omega - 4\pi nB) \tag{67.70}$$

The function $Y(\omega)$ is a periodic function with period $4\pi B$. From Eq. (67.67) $X(\omega)$ is zero for

$\omega > 2\pi B$ and is thus finite in extent. The terms in Eq. (67.70) therefore do not overlap. Periodic functions can be expressed as a Fourier series.

$$Y(\omega) = \sum_{n=-\infty}^{\infty} \alpha_n \exp\left(\frac{-jn\omega}{2B}\right) \tag{67.71}$$

where the Fourier series coefficients are

$$\alpha_n = \frac{1}{4\pi B} \int_{-2\pi B}^{2\pi B} Y(\omega) \exp\left(\frac{jn\omega}{2B}\right) d\omega$$

or

$$\alpha_n = \frac{1}{2B} x\left(\frac{n}{2B}\right) \tag{67.72}$$

where we have used the inverse Fourier transform in Eq. (67.69). Substituting into the Fourier series in Eq. (67.71) gives

$$Y(\omega) = \frac{1}{2B} \sum_{n=-\infty}^{\infty} x\left(\frac{n}{2B}\right) \exp\left(\frac{-jn\omega}{2B}\right) \tag{67.73}$$

Since a period of $Y(\omega)$ is $X(\omega)$, we can get back the original spectrum by

$$X(\omega) = Y(\omega) \Pi\left(\frac{\omega}{4\pi B}\right)$$

Substitute Eq. (67.73) and inverse transforming gives, using Eq. (67.69),

$$x(t) = \frac{1}{4\pi B} \int_{-2\pi B}^{2\pi B} \sum_{n=-\infty}^{\infty} x\left(\frac{n}{2B}\right) \exp\left(\frac{-jn\omega}{2B}\right) e^{j\omega t} d\omega$$

or

$$x(t) = \sum_{n=-\infty}^{\infty} x\left(\frac{n}{2B}\right) \text{sinc}\,(2Bt - n) \tag{67.74}$$

where

$$\text{sinc}\,(t) = \frac{\sin \pi t}{\pi t}$$

is the inverse Fourier transform of $\Pi\,(\omega/2\pi)$. Eq. (67.74) is, of course, the cardinal series.
   The sampling theorem generally converges uniformly, in the sense that

$$\lim_{N \to \infty} |\,x(t) - x_N(t)\,|^2 = 0$$

where the truncated cardinal series is

$$x_N(t) = \sum_{n=-N}^{N} x\left(\frac{n}{2B}\right) \text{sinc}\,(2Bt - n) \tag{67.75}$$

Sufficient conditions for uniform convergence are [Marks, 1991]

1. the signal, $x(t)$, has finite energy, E,

$$E = \int_{-\infty}^{\infty} |x(t)|^2 \, dt < \infty$$

2. or $X(\omega)$ has finite area,

$$A = \int_{-\infty}^{\infty} |X(\omega)| \, d\omega < \infty$$

Care must be taken in the second case, though, when singularities exist at $\omega = \pm 2\pi B$. Here, sampling may be required to be strictly greater than $2B$. Such is the case, for example, for the signal, $x(t) = \sin(2\pi Bt)$. Although the signal is bandlimited, and although its Fourier transform has finite area, all of the samples of $x(t)$ taken at $t = n/2B$ are zero. The cardinal series in Eq. (67.74) will thus interpolate to zero everywhere. If the sampling rate is a bit greater than $2B$, however, the samples are not zero and the cardinal series will uniformly converge to the proper answer.

The Time-Bandwidth Product

The cardinal series requires knowledge of an infinite number of samples. In practice, only a finite number of samples are required. If most of the energy of a signal exists in the interval $0 \leq t \leq T$, and we sample at the Nyquist rate of $2B$ samples per second, then a total of $S = \langle 2BT \rangle$ samples are taken. ($\langle \theta \rangle$ denotes the largest number not exceeding θ.) The number S is a measure of the degrees of freedom of the signal and is referred to as its **time-bandwidth product.** A 5-min single-track audio recording requiring fidelity up to 20,000 Hz, for example, requires a minimum of $S = 2 \times 20,000 \times 5 \times 60 = 12$ million samples. In practice, audio sampling is performed well above the Nyquist rate.

Sources of Error

Exact interpolation using the cardinal series assumes that (1) the values of the samples are known exactly, (2) the sample locations are known exactly, and (3) an infinite number of terms are used in the series. Deviation from these requirements results in interpolation error due to (1) data noise, (2) jitter, and (3) truncation, respectively. The effect of data error on the restoration can be significant. Some innocently appearing sampling theorem generalizations, when subjected to performance analysis in the presence of data error, are revealed as ill-posed. In other words, a bounded error on the data can result in unbounded error on the restoration [Marks, 1991].

Data Noise

The source of data noise can be the signal from which samples are taken, or from round-off error due to finite sampling precision. If the noise is additive and random, instead of the samples

$$x\left(\frac{n}{2B}\right)$$

we must deal with the samples

$$x\left(\frac{n}{2B}\right) + \xi\left(\frac{n}{2B}\right)$$

where $\xi(t)$ is a stochastic process. If these noisy samples are used in the cardinal series, the interpolation, instead of simple $x(t)$, is

$$x(t) + \eta(t)$$

where the interpolation noise is

$$\eta(t) = \sum_{n=-\infty}^{\infty} \xi\left(\frac{n}{2B}\right) \mathrm{sinc}\,(2Bt - n)$$

If $\xi(t)$ is a zero mean process, then so is the interpolation noise. Thus, the noisy interpolation is an unbiased version of $x(t)$. More remarkably, if $\xi(t)$ is a zero-mean (wide-sense) stationary process with uncertainty (variance) σ^2, then so is $\eta(t)$. In other words, *the uncertainty at the sample point locations is the same as at all points of interpolation* [Marks, 1991].

Truncation

The truncated cardinal series is in Eq. (67.75). A signal cannot be both bandlimited and of finite duration. Indeed, a bandlimited function cannot be identically zero over any finite interval. Thus, other than the rare case where an infinite number of the signal's zero crossings coincide with the sample locations, truncation will result in an error.

The magnitude of this **truncation error** can be estimated through the use of Parseval's theorem for the cardinal series that states

$$E = \int_{-\infty}^{\infty} |x(t)|^2 dt$$

$$= \frac{1}{2B} \sum_{-\infty}^{\infty} \left| x\left(\frac{n}{2B}\right) \right|^2 \qquad (67.76)$$

The energy of a signal can thus be determined directly from either the signals or the samples. The energy associated with the truncated signal is

$$E_N = \frac{1}{2B} \sum_{-N}^{N} \left| x\left(\frac{n}{2B}\right) \right|^2$$

If $E - E_N \ll E$, then the truncation error is small.

Jitter

Jitter occurs when samples are taken near to but not exactly at the desired sample locations. Instead of the samples $x\,(n/2W)$, we have the samples

$$x\left(\frac{n}{2W} - \sigma_n\right)$$

where σ_n is the jitter offset of the nth sample. For jitter, the σ_n's are not known. If they were, an appropriate nonuniform sampling theorem [Marks, 1993; Marvasti, 1987] could be used to interpolate the signal.

Using the jittered samples in the cardinal series results in an interpolation that is not an unbiased estimate of $x(t)$. Indeed, if the probability density function of the jitter is the same at all sample locations, the expected value of the jittered interpolation is the convolution of $x(t)$ with the probability density function of the jitter. This bias can be removed by inverse filtering at a cost of decreasing the signal-to-noise ratio of the interpolation [Marks, 1993].

Generalizations of the Sampling Theorem

There exist numerous generalizations of the sampling theorem [Marks, 1991; Marks, 1993].

1. **Stochastic processes.** A wide-sense stationary stochastic process, $\chi(t)$, is said to be bandlimited if its autocorrelation, $R_\chi(t)$, is a bandlimited function. The cardinal series

$$\hat{\chi}(t) = \sum_{n=-\infty}^{\infty} \chi\left(\frac{n}{2B}\right) \operatorname{sinc}(2Bt - n)$$

   converges to $\chi(t)$ in the sense that

$$E\left[\left|\hat{\chi}(t) - \chi(t)\right|^2\right] = 0$$

   where E denotes expectation.

2. **Nonuniform sampling.** There exist numerous scenarios wherein interpolation can be performed from samples that are not spaced uniformly. Marvasti [1987] devotes a book to the topic.

3. **Kramer's generalization.** Kramer generalized the sampling theorem to integral transforms other than Fourier, for example, to Legendre and Laguerre transforms.

4. **Papoulis' generalization.** Shannon noted that a bandlimited signal could be restored when sampling was performed at half the Nyquist rate if, at every sample location, a sample of the signal's derivative were also taken. Recurrent nonuniform sampling is where P samples are spaced the same in every P Nyquist intervals. Another sampling scenario is when a signal and its Hilbert transform are both sampled at half their respective Nyquist rates. Restoration of the signal from these and numerous other sampling scenarios are subsumed in an eloquent generalization of the sampling theorem by Papoulis.

5. **Lagrangian interpolation.** Lagrangian interpolation is a topic familiar in numerical analysis. An Nth order polynomial is fit to $N + 1$ arbitrarily spaced sample points. If an infinite number of samples are equally spaced, then Lagrangian interpolation is equivalent to the cardinal series.

6. **Trigonometric polynomials.** All periodic bandlimited signals can be expressed as trigonometric polynomials (i.e., a Fourier series with a finite number of terms). If the series has M terms, then the signal has M degrees of freedom which can be determined from M samples taken within a single period.

7. **Multidimensional sampling theorems.** Multidimensional signals, such as images, require dimensional extensions of the sampling theorem. The sampling of the signal now requires geometrical interpretation. Uniform sampling of an image, for example, can either be done on a rectangular or hexagonal grid. The minimum sampling density for one geometry may differ from that of another. The smallest sampling density that does not result in aliasing can be achieved, in many cases, with a number of different uniform sampling geometries and is referred to as the Nyquist density. Interestingly, sampling can sometimes be performed below the Nyquist density with nonuniform sampling geometries such that the multidimensional signal can be restored. Such is not the case for one dimension.

8. **Continuous sampling.** When a signal is known on one or more disjoint intervals, it is said to have been continuously sampled. Divide the time line into intervals of T. Periodic continuous sampling assumes that the signal is known on each interval over an interval of αT where α is the duty cycle. Continuously sampled signals can be accurately interpolated even in the presence of aliasing. Other continuously sampled cases, each of which can be considered as a limiting case of continuously periodically sampled restoration, include

(a) **Interpolation.** The tails of a signal are known and we wish to restore the middle.

(b) **Extrapolation.** We wish to generate the tails of a function with knowledge of the middle.

(c) **Prediction.** A signal for $t > 0$ is to be estimated from knowledge of the signal for $t < 0$.

Final Remarks

Since its popularization in the late 1940s, the sampling theorem has been studied in depth. More than 1000 papers have been generated on the topic [Marks, 1993]. Its understanding is fundamental in matching the largely continuous world to digital computation engines.

Defining Terms

Aliasing: A phenomenon that occurs when a signal is undersampled. High-frequency information about the signal is lost.

Cardinal series: The formula by which samples of a bandlimited signal are interpolated to form a continuous time signal.

Fourier transform: The mathematical operation that converts a time-domain signal into the frequency domain.

Jitter: A sample is temporally displaced by an unknown, usually small, interval.

Kramer's generalization: A sampling theory based on other than Fourier transforms and frequency.

Lagrangian interpolation: A classic interpolation procedure used in numerical analysis. The sampling theorem is a special case.

Nyquist rate: The minimum sampling rate that does not result in aliasing.

Papoulis' generalization: A sampling theory applicable to many cases wherein signal samples are obtained either nonuniformly and/or indirectly.

Sampling rate: The number of samples per second.

Sampling theorem: Samples of a bandlimited signal, if taken close enough together, exactly specify the continuous time signal from which the samples were taken.

Signal bandwidth: The maximum frequency component of a signal.

Time bandwidth product: The product of a signal's duration and bandwidth approximates the number of samples required to characterize the signal.

Truncation error: The error that occurs when a finite number of samples are used to interpolate a continuous time signal.

References

E. Borel, "Sur l'interpolation," *C.R. Acad. Sci. Paris*, vol. 124, pp. 673–676, 1897.

J. R. Higgins, "Five short stories about the cardinal series," *Bull. Am. Math. Soc.*, vol. 12, pp. 45–89, 1985.

A. J. Jerri, "The Shannon sampling theorem—its various extension and applications: a tutorial review," *Proc. IEEE*, vol. 65, pp. 1565–1596, 1977.

V. A. Kotel'nikov, "On the transmission capacity of 'ether' and wire in electrocommunications," *Izd. Red. Upr. Svyazi RKKA* (Moscow), 1933.

R. J. Marks II, *Introduction to Shannon Sampling and Interpolation Theory*, New York: Springer-Verlag, 1991.

R. J. Marks II, Ed., *Advanced Topics in Shannon Sampling and Interpolation Theory*, New York: Springer-Verlag, 1993.

F. A. Marvasti, *A Unified Approach to Zero-Crossing and Nonuniform Sampling*, Oak Park, Ill.: Nonuniform, 1987.

C. Shannon, "A mathematical theory of communication," *Bell System Technical Journal,* vol. 27, pp. 379, 623, 1948.

E. T. Whittaker, "On the functions which are represented by the expansions of the interpolation theory," *Proc. Royal Society of Edinburgh,* vol. 35, pp. 181–194, 1915.

J. M. Whittaker, "The Fourier theory of the cardinal functions," *Proc. Math. Soc. Edinburgh,* vol. 1, pp. 169–176, 1929.

Further Information

An in-depth study of the sample theorem and its numerous variations is provided in R. J. Marks II, Ed., *Introduction to Shannon Sampling and Interpolation Theory,* New York: Springer-Verlag, 1991.

In-depth studies of modern sampling theory with over 1000 references are available in R. J. Marks II, Ed., *Advanced Topics in Shannon Sampling and Interpolation Theory,* New York: Springer-Verlag, 1993.

The specific case of nonuniform sampling is treated in the monograph by F. A. Marvasti, *A Unified Approach to Zero-Crossing and Nonuniform Sampling,* Oak Park, Ill: Nonuniform, 1987.

The sampling theorem is treated generically in the *IEEE Transactions on Signal Processing.* For applications, topical journals are the best source of current literature.

67.5 Data Compression

Joy A. Thomas and Thomas M. Cover

Data compression is a process of finding the most efficient representation of an information source in order to minimize communication or storage. It often consists of two stages—the first is the choice of a (probabilistic) model for the source and the second is the design of an efficient coding system for the model. In this section, we will concentrate on the second aspect of the compression process, though we will touch on some common sources and models in the last subsection.

Shannon [1948] was the first to distinguish the probabilistic model that underlies an information source from the semantics of the information. An information source produces one of many possible messages; the goal of communication is to transmit an unambiguous specification of the message so that the receiver can reconstruct the original message. For example, the information to be sent may be the result of a horse race. If the recipient is assumed to know the names and numbers of the horses, then all that must be transmitted is the number of the horse that won. In a different context, the same number might mean something quite different, e.g., the price of a barrel of oil. The significant fact is that the difficulty in communication depends only on the length of the representation. Thus, finding the best (shortest) representation of an information source is critical to efficient communication.

When the possible messages are all equally likely, then it makes sense to represent them by strings of equal length. For example, if there are 32 possible equally likely messages, then each message can be represented by a binary string of 5 bits. However, if the messages are not equally likely, then it is more efficient on the average to allot short strings to the frequently occurring messages and longer strings to the rare messages. Thus, the Morse code allots the shortest string (a dot) to the most frequent letter (E) and allots long strings to the infrequent letters (e.g., dash, dash, dot, dash for Q). The minimum average length of the representation is a fundamental quantity called the entropy of the source, which is defined in the next subsection.

Entropy

An information source will be represented by a random variable X, which takes on one of a finite number of possibilities $i \in X$ with probability $p_i = \Pr(X = i)$. The **entropy** of the random variable X is defined as

$$H(X) = -\sum_{i \in X} p_i \log p_i \qquad (67.77)$$

where the log is to base 2 and the entropy is measured in *bits*. We will use logarithms to base 2 throughout this chapter.

Example 67.1   Let X be a random variable that takes on a value 1 with probability θ and takes on the value 0 with probability $1 - \theta$. Then $H(X) = -\theta \log \theta - (1 - \theta) \log (1 - \theta)$. In particular, the entropy of a fair coin toss is 1 bit.

This definition of entropy is related to the definition of entropy in thermodynamics. It is the fundamental lower bound on the average length of a code for the random variable.

A **code** for a random variable X is a mapping from X, the range of X, to the set of finite-length binary strings. We will denote the code word corresponding to i by $C(i)$, and the length of the code word by l_i. The average length of the code is then $L(C) = \sum_i p_i l_i$.

A code is said to be *instantaneous* or *prefix-free* if no code word is a prefix of any other code word. This condition is sufficient (but not necessary) to allow a sequence of received bits to be parsed unambiguously into a sequence of code words.

Example 67.2   Consider a random variable X taking on the values $\{1, 2, 3\}$ with probabilities $(0.5, 0.25, 0.25)$. An instantaneous code for this random variable might be $(0, 10, 11)$. Thus, a string 01001110 can be uniquely parsed into 0, 10, 0, 11, 10, which decodes to the string $x = (1, 2, 1, 3, 2)$. Note that the average length of the code is 1.5 bits, which is the same as the entropy of the source.

For any instantaneous code, the following property of binary trees called the *Kraft inequality*

$$\sum_i 2^{-l_i} \leq 1 \qquad (67.78)$$

must hold. Conversely, it can be shown that given a set of lengths that satisfies the Kraft inequality, we can find a set of prefix-free code words of those lengths.

The problem of finding the best source code then reduces to finding the optimal set of lengths that satisfies the Kraft inequality and minimizes the average length of the code. Simple calculus can then be used to show [Cover and Thomas, 1991] that the average length of any instantaneous code is larger than the entropy of the random variable, i.e., the minimum of $\sum p_i l_i$ over all l_i satisfying $\sum 2^{-l_i} \leq 1$ is $-\sum p_i \log p_i$. Also, if we take $l_i = \lceil \log 1/p_i \rceil$ (where $\lceil t \rceil$ denotes the smallest integer greater than or equal to t), we can verify that this choice of lengths satisfies the Kraft inequality and that

$$L(C) = \sum_i p_i \left\lceil \log \frac{1}{p_i} \right\rceil < \sum_i p_i \left(\log \frac{1}{p_i} + 1 \right) = H(X) + 1 \qquad (67.79)$$

The optimal code can only have a shorter length, and therefore we have the following theorem:

Theorem 67.1   Let L^* be the average length of the optimal instantaneous code for a random variable X. Then

$$H(X) \leq L^* < H(X) + 1 \qquad (67.80)$$

This theorem is one of the fundamental theorems of information theory. It identifies the entropy as the fundamental limit for the average length of the representation of a discrete information source and shows that we can find representations with average length within one bit of the entropy.

The Huffman Algorithm

The choice of code word lengths $l_i = \lceil \log 1/p_i \rceil$ (called the *Shannon code lengths*) is close to optimal, but not necessarily optimal, in terms of average code word length. We will now describe an algorithm (the **Huffman algorithm**) that produces an instantaneous code of minimal average length for a random variable with distribution $p_1, p_2, \ldots, p_m$. The algorithm is a greedy algorithm for building a tree from the bottom up.

Step 1. Arrange the probabilities in decreasing order so that $p_1 \geq p_2 \geq \ldots \geq p_m$.

Step 2. Form a subtree by combining the last two probabilities p_{m-1} and p_m to a single node of weight $p'_{m-1} = p_{m-1} + p_m$.

Step 3. Recursively execute Steps 1 and 2, decreasing the number of nodes each time, until a single node is obtained.

Step 4. Use the tree constructed above to allot code words.

The algorithm for tree construction is illustrated for a source with distribution $(0.5, 0.2, 0.2, 0.1)$ in Fig. 67.12. After constructing the tree, the leaves of the tree (which correspond to the symbols of X) can be assigned code words that correspond to the paths from the root to the leaf. We will not give a proof of the optimality of the Huffman algorithm; the reader is referred to Gallager [1968] or Cover and Thomas [1991] for details.

Entropy Rate

The entropy of a sequence of random variables $X_1, X_2, \ldots, X_n$ with joint distribution $p(x_1, x_2, \ldots, x_n)$ is defined analogously to the entropy of a single random variable as

$$H(X_1, X_2, \ldots, X_n) = -\sum_{x_1} \sum_{x_2} \ldots \sum_{x_n} p(x_1, x_2, \ldots, x_n) \log p(x_1, x_2, \ldots, x_n) \quad (67.81)$$

For a stationary process $X_1, X_2, \ldots$, we define the *entropy rate* $\mathcal{H}(X)$ of the process as

$$\mathcal{H}(X) = \lim_{n \to \infty} \frac{H(X_1, X_2, \ldots, X_n)}{n} \quad (67.82)$$

It can be shown [Cover and Thomas, 1991] that the entropy rate is well defined for all stationary processes. In particular, if $X_1, X_2, \ldots, X_n$ is a sequence of independent and identically distributed (i.i.d.) random variables, then $H(X_1, X_2, \ldots, X_n) = nH(X_1)$, and $\mathcal{H}(X) = H(X_1)$.

In the previous subsection, we showed the existence of a prefix-free code having an average length within one bit of the entropy. Now instead of trying to represent one occurrence of the random variable, we can form a code to represent a block of n random variables. In this case, the aver-

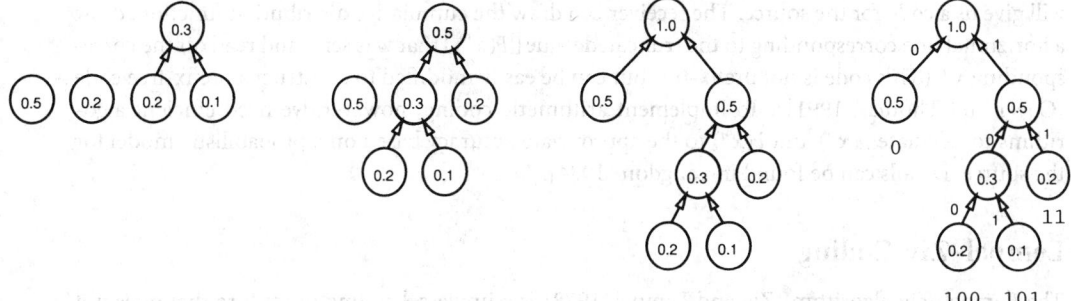

FIGURE 67.12 Example of the Huffman algorithm.

age code length is within one bit of $H(X_1, X_2, \ldots, X_n)$. Thus, the average length of the code per input symbol satisfies

$$\frac{H(X_1, X_2, \ldots, X_n)}{n} \leq \frac{L_n^*}{n} < \frac{H(X_1, X_2, \ldots, X_n)}{n} + \frac{1}{n} \qquad (67.83)$$

Since $[H(X_1, X_2, \ldots, X_n)]/n \to \mathcal{H}(X)$, we can get arbitrarily close to the entropy rate by using longer and longer block lengths. Thus, the entropy rate is the fundamental limit for data compression for stationary sources, and we can achieve rates arbitrarily close to this limit by using long blocks.

All the above assumes that we know the probability distribution that underlies the information source. In many practical examples, however, the distribution is unknown or too complex to be used for coding. There are various ways to handle this situation:

- Assume a simple distribution and design an appropriate code for it. Use this code on the real source. If an estimated distribution $\hat{p}$ is used when in fact the true distribution is p, then the average length of the code is lower bounded by $H(X) + \Sigma_x p(x) \log [p(x)/\hat{p}(x)]$. The second term, which is denoted $D(p\|\hat{p})$ is called the *relative entropy* or the *Kullback Leibler distance* between the two distributions.
- Estimate the distribution empirically from the source and adapt the code to the distribution. For example, with *adaptive Huffman coding*, the empirical distribution of the source symbols is used to design the Huffman code used for the source.
- Use a *universal coding algorithm* like the **Lempel–Ziv algorithm** (see the subsection "Lempel–Ziv Coding").

Arithmetic Coding

In the previous subsections, it was shown how we could construct a code for a source that achieves an average length within one bit of the entropy. For small source alphabets, however, we have efficient coding only if we use long blocks of source symbols. For example, if the source is binary, and we code each symbol separately, we must use 1 bit per symbol, irrespective of the entropy of the source. If we use long blocks, we can achieve an expected length per symbol close to the entropy rate of the source.

It is therefore desirable to have an efficient coding procedure that works for long blocks of source symbols. Huffman coding is not ideal for this situation, since it is a bottom-up procedure with a complexity that grows rapidly with the block length. Arithmetic coding is an incremental coding algorithm that works efficiently for long block lengths and achieves an average length within one bit of the entropy for the block.

The essential idea of arithmetic coding is to represent a sequence $x^n = x_1, x_2, \ldots, x_n$ by the cumulative distribution function $F(x^n) = \Sigma_{\tilde{x}^n \leq x^n} p(\tilde{x}^n)$ expressed to an appropriate accuracy. The cumulative distribution function for x^n is illustrated in Fig. 67.13. We can use any real number in the interval $[F(x^n) - p(x^n), F(x^n)]$ as the code for x^n. Expressing $F(x^n)$ to an accuracy of $\lceil \log 1/p(x^n) \rceil$ will give us a code for the source. The receiver can draw the cumulative distribution function, draw a horizontal line corresponding to the truncated value $\lfloor F(x^n) \rfloor$ that was sent, and read off the corresponding x^n. (This code is not prefix-free but can be easily modified to construct a prefix-free code [Cover and Thomas, 1991]). To implement arithmetic coding, however, we need efficient algorithms to calculate $p(x^n)$ and $F(x^n)$ to the appropriate accuracy based on a probabilistic model for the source. Details can be found in Langdon [1984].

Lempel–Ziv Coding

The Lempel–Ziv algorithm [Ziv and Lempel, 1978] is a universal coding procedure that does not require knowledge of the source statistics and yet is asymptotically optimal. The basic idea of

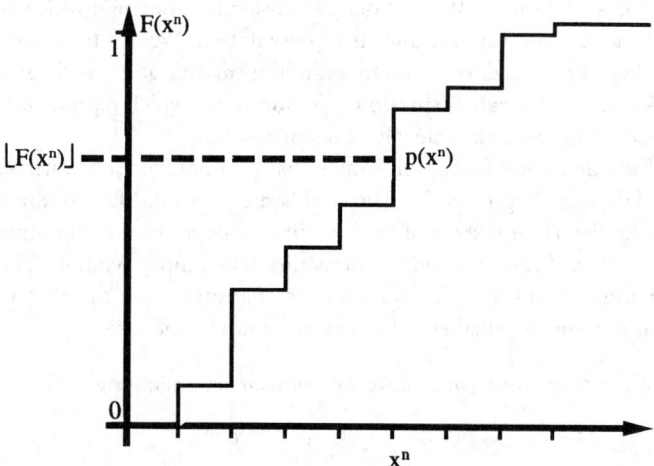

FIGURE 67.13 Cumulative distribution function for sequences x^n.

the algorithm is to construct a table or dictionary of frequently occurring strings and to represent new strings by pointing to their prefixes in the table. We first parse the string into sequences that have not appeared so far. For example, the binary string 11010011011100 is parsed into 1,10,100,11,0,111,00. Then instead of sending the bits of each phrase, we send a pointer to its prefix and the value of the last bit. Thus, if we use three bits for the pointer, we will represent this string by (000,1), (001,0), (010,0), (001,1), (000,0), (100,1), (101,0), etc. For this short example, the algorithm has not compressed the string—it has in fact expanded it.

The very surprising fact is that, as Lempel and Ziv have shown, the algorithm is asymptotically optimal for any stationary ergodic source. This is expressed in the following theorem [Ziv and Lempel, 1978; Cover and Thomas, 1991]:

Theorem 67.2 Let L_n be the length of the Lempel–Ziv code for n symbols drawn from a stationary ergodic process $X_1, X_2, \ldots, X_n$ with entropy rate $\mathcal{H}(X)$. Then

$$\frac{L_n}{n} \to \mathcal{H}(X) \quad \text{with probability 1} \tag{67.84}$$

Thus, for long enough block lengths, the Lempel–Ziv algorithm (which does not make any assumptions about the distribution of the source) does as well as if we knew the distribution in advance and designed the optimal code for this distribution.

The algorithm described above is only one of a large class of similar adaptive dictionary-based algorithms, which are all rather loosely called Lempel–Ziv. These algorithms are simple and fast and have been implemented in both software and hardware, e.g., in the *compress* command in UNIX and the *PKZIP* command on PCs. On ASCII text files, the Lempel–Ziv algorithm achieves compressions on the order of 50%. It has also been implemented in hardware and has been used to "double" the capacity of data storage media or to "double" the effective transmission rate of a modem. Many variations on the basic algorithm can be found in Bell *et al.* [1990].

Rate Distortion Theory

An infinite number of bits are required to describe an arbitrary real number, and therefore it is not possible to perfectly represent a continuous random variable with a finite number of bits. How "good" can the representation be? We first define a distortion measure, which is a measure of the distance between the random variable and its representation. We can then consider the trade-off between the number of bits used to represent a random variable and the distortion incurred. This trade-off is represented by the **rate distortion function** $R(D)$, which represents the minimum rate required to represent a random variable with a distortion D.

We will consider a discrete information source that produces random variables $X_1, X_2, \ldots, X_n$ that are drawn i.i.d. according to $p(x)$. (The results are also valid for continuous sources.) The encoder of the rate distortion system of rate R will encode a block of n outputs X^n as an index $f(X^n) \in \{1, 2, \ldots, \lfloor 2^{nR} \rfloor\}$. (Thus, the index will require R bits/input symbol.) The decoder will calculate a representation $\hat{X}^n(f(X^n))$ of X^n. Normally, the representation alphabet $\hat{X}$ of the representation is the same as the source alphabet X, but that need not be the case.

Definition: A *distortion function* or *distortion measure* is a mapping

$$d : X \times \hat{X} \rightarrow R^+ \tag{67.85}$$

from the set of source alphabet–reproduction alphabet pairs into the set of nonnegative real numbers. The distortion $d(x, \hat{x})$ is a measure of the cost of representing the symbol x by the symbol $\hat{x}$.

Examples of common distortion functions are

- *Hamming (probability of error) distortion.* The Hamming distortion is given by

$$d(x, \hat{x}) = \begin{cases} 0 \text{ if } x = \hat{x} \\ 1 \text{ if } x \neq \hat{x} \end{cases} \tag{67.86}$$

  and thus $Ed(X, \hat{X}) = \Pr(X \neq \hat{X})$.
- *Squared error distortion.* The squared error distortion

$$d(x, \hat{x}) = (x - \hat{x})^2 \tag{67.87}$$

is the most popular distortion measure used for continuous alphabets. Its advantages are its simplicity and its relationship to least squares prediction. However, for information sources such as images and speech, the squared error is not an appropriate measure for distortion as perceived by a human observer.

The distortion between sequences x^n and $\hat{x}^n$ of length n is defined by

$$d(x^n, \hat{x}^n) = \frac{1}{n} \sum_{i=1}^{n} d(x_i, \hat{x}_i) \tag{67.88}$$

For a rate distortion system, the expected distortion D is defined as

$$D = Ed(X^n, \hat{X}^n(f(X^n))) = \sum_{x^n} p(x^n) d(x^n, \hat{X}^n(f(x^n))) \tag{67.89}$$

Definition: The rate distortion pair (R,D) is said to be achievable if there exists a rate distortion code of rate R with expected distortion D. The *rate distortion function* $R(D)$ is the infimum of rates R such that (R,D) is achievable for a given D.

Definition: The *mutual information* $I(X, \hat{X})$ between random variables X and $\hat{X}$, with joint probability mass function $p(x, \hat{x})$ and marginal probability mass functions $p(x)$ and $p(\hat{x})$ is defined as

$$I(X; \hat{X}) = \sum_{x \in X} \sum_{\hat{x} \in \hat{X}} p(x, \hat{x}) \log \frac{p(x, \hat{x})}{p(x)p(\hat{x})} \qquad (67.90)$$

The mutual information is a measure of the amount of information that one random variable carries about another.

The main result of rate distortion theory is contained in the following theorem, which provides a characterization of the rate distortion function in terms of the mutual information of joint distributions that satisfy the expected distortion constraint:

Theorem 67.3 The rate distortion function for an i.i.d. source X with distribution $p(x)$ and distortion function $d(x, \hat{x})$ is

$$R(D) = \min_{p(\hat{x}|x):\Sigma_{(x,\hat{x})}p(x)p(\hat{x}|x)d(x,\hat{x}) \leq D} I(X; \hat{X}) \qquad (67.91)$$

We can construct rate distortion codes that can achieve distortion D at any rate greater than $R(D)$, and we cannot construct such codes at any rate below $R(D)$.

The proof of this theorem uses ideas of random coding and long block lengths as in the proof of the channel capacity theorem. The basic idea is to generate a code book of 2^{nR} reproduction code words $\hat{X}^n$ at random and show that for long block lengths, for any source sequence, it is very likely that there is at least one code word in this code book that is within distortion D of that source sequence. See Gallager [1968] or Cover and Thomas [1991] for details of the proof.

Example 67.3 (*Binary source*) The rate distortion function for a Bernoulli (p) source (a random variable that takes on values $\{0, 1\}$ with probabilities $p, 1 - p$) with Hamming distortion is given by

$$R(D) = \begin{cases} H(p) - H(D), & 0 \leq D \leq \min\{p, 1-p\} \\ 0, & D > \min\{p, 1-p\} \end{cases} \qquad (67.92)$$

where $H(p) = -p \log p - (1 - p)\log (1 - p)$ is the binary entropy function.

Example 67.4 (*Gaussian source*) The rate distortion function for a Gaussian random variable with variance σ^2 and squared error distortion is

$$R(D) = \begin{cases} \frac{1}{2} \log \frac{\sigma^2}{D}, & 0 \leq D \leq \sigma^2 \\ 0, & D > \sigma^2 \end{cases} \qquad (67.93)$$

Thus, with nR bits, we can describe n i.i.d. Gaussian random variables $X_1, X_2, \ldots, X_n \sim \mathcal{N}(0, \sigma^2)$ with a distortion of $\sigma^2 2^{-2R}$ per symbol.

Quantization and Vector Quantization

The rate distortion function represents the lower bound on the rate that is needed to represent a source with a particular distortion. We now consider simple algorithms that represent a continuous random variable with a few bits. Suppose we want to represent a single sample from a continuous source. Let the random variable to be represented be X and let the representation of X be denoted as $\hat{X}(X)$. If we are given R bits to represent X, then the function $\hat{X}$ can take on 2^R values. The problem of optimum **quantization** is to find the optimum set of values for $\hat{X}$ (called the reproduction points or code points) and the regions that are associated with each value $\hat{X}$ in order to minimize the expected distortion.

For example, let X be a Gaussian random variable with mean 0 and variance σ^2, and assume a squared error distortion measure. In this case, we wish to find the function $\hat{X}(X)$ such that $\hat{X}$ takes on at most 2^R values and minimizes $E(X - \hat{X}(X))^2$. If we are given 1 bit to represent X, it is clear that the bit should distinguish whether $X > 0$ or not. To minimize squared error, each reproduced symbol should be at the conditional mean of its region. If we are given 2 bits to represent the sample, the situation is not as simple. Clearly, we want to divide the real line into four regions and use a point within each region to represent the samples within that region. We can state two simple properties of optimal regions and reconstruction points for the quantization of a single random variable:

- Given a set of reconstruction points, the distortion is minimized by mapping a source random variable X to the representation $\hat{X}(w)$ that is closest to it (in distortion). The set of regions defined by this mapping is called a Voronoi or Dirichlet partition defined by the reconstruction points.

- The reconstruction points should minimize the conditional expected distortion over their respective assignment regions.

These two properties enable us to construct a simple algorithm to find a "good" quantizer: we start with a set of reconstruction points, find the optimal set of reconstruction regions (which are the nearest neighbor regions with respect to the distortion measure), then find the optimal reconstruction points for these regions (the centroids of these regions if the distortion measure is squared error), and then repeat the iteration for this new set of reconstruction points. The expected distortion is decreased at each stage in the algorithm, so the algorithm will converge to a local minimum of the distortion. This algorithm is called the *Lloyd algorithm*.

It follows from the arguments of rate distortion theory that we will do better if we encode long blocks of source symbols rather than encoding each symbol individually. In this case, we will consider a block of n symbols from the source as a vector-valued random variable, and we will represent these n-dimensional vectors by a set of 2^{nR} code words. This process is called **vector quantization** (VQ). We can apply the Lloyd algorithm to design a set of representation vectors (the code book) and the corresponding nearest neighbor regions. Instead of using the probability distribution for the source to calculate the centroids of the regions, we can use the empirical distribution from a training sequence. Many variations of the basic vector quantization algorithm are described in Gersho and Gray [1992].

Common information sources like speech produce continuous waveforms, not discrete sequences of random variables as in the models we have been considering so far. By sampling the signal at twice the maximum frequency present (the Nyquist rate), however, we convert the continuous time signal into a set of discrete samples from which the original signal can be recovered (the sampling theorem). We can then apply the theory of rate distortion and vector quantization to such waveform sources as well.

Kolmogorov Complexity

In the 1960s, the Russian mathematician Kolmogorov considered the question "What is the intrinsic descriptive complexity of a binary string?" From the preceding discussion, it follows that if the binary string were a sequence of i.i.d. random variables $X_1, X_2, \ldots, X_n$, then on the average it would take $nH(X)$ bits to represent the sequence. But what if the bits were the first million bits of the binary expansion of π? In that case, the string appears random but can be generated by a simple computer program. So if we wanted to send these million bits to another location which has a computer, we could instead send the program and ask the computer to generate these million bits. Thus, the descriptive complexity of π is quite small.

Motivated by such considerations, Kolmogorov defined the complexity of a binary string to be the length of the shortest program for a universal computer that generates that string. (This concept was also proposed independently and at about the same time by Chaitin and Solomonoff.)

Definition: The ***Kolmogorov complexity*** $K_u(x)$ of a string x with respect to a universal computer $\mathcal{U}$ is defined as

$$K_u(x) = \min_{p:\mathcal{U}(p)=x} l(p) \qquad (67.94)$$

the minimum length over all programs that print x and halt. Thus $K_u(x)$ is the shortest description length of x over all descriptions interpreted by computer $\mathcal{U}$.

A universal computer can be thought of as a Turing machine that can simulate any other universal computer. At first sight, the definition of Kolmogorov complexity seems to be useless, since it depends on the particular computer that we are talking about. But using the fact that any universal computer can simulate any other universal computer, any program for one computer can be converted to a program for another computer by adding a constant length "simulation program" as a prefix. Thus, we can show [Cover and Thomas, 1991] that for any two universal computers, $\mathcal{U}$ and $\mathcal{A}$,

$$\left| K_u(x) - K_{\mathcal{A}}(x) \right| < c \qquad (67.95)$$

where the constant c, though large, does not depend on the string x under consideration. Thus, Kolmogorov complexity is universal in that it does not depend on the computer (up to a constant additive factor).

Kolmogorov complexity provides a unified way to think about problems of data compression. It is also the basis of principles of inference (Occam's razor: "The simplest explanation is the best") and is closely tied with the theory of computability.

Data Compression in Practice

The previous subsections discussed the fundamental limits to compression for a stochastic source. We will now consider the application of these algorithms to some practical sources, namely, text, speech, images, and video. In real applications, the sources may not be stationary or ergodic, and the distributions underlying the source are often unknown. Also, in addition to the efficiency of the algorithm, important considerations in practical applications include the computational speed and memory requirements of the algorithm, the perceptual quality of the reproductions to a human observer, etc. A considerable amount of research and engineering has gone into the development of these algorithms, and many issues are only now being explored. We will not go into the details but simply list some popular algorithms for the different sources.

Text

English text is normally represented in ASCII, which uses 8 bits/character. There is considerable redundancy in this representation (the entropy rate of English is about 1.3 bits/character). Popular compression algorithms include variants of the Lempel–Ziv algorithm, which compress text files by about 50% (to about 4 bits/character).

Speech

Telephone quality speech is normally sampled at 8 kHz and quantized at 8 bits/sample (a rate of 64 kbits/s) for uncompressed speech. Simple compression algorithms like adaptive differential pulse code modulation (ADPCM) [Jayant and Noll, 1984] use the correlation between adjacent samples to reduce the number of bits used by a factor of two to four or more with almost imperceptible distortion. Much higher compression ratios can be obtained with algorithms like linear predictive coding (LPC), which model speech as an autoregressive process, and send the parameters of the process as opposed to sending the speech itself. With LPC-based methods, it is possible to code speech at less than 4 kbits/s. At very low bit rates, however, the reproduced speech sounds synthetic.

Images

A single high-quality color image of 1024 by 1024 pixels with 24 bits per pixel represents about 3 MB of storage in an uncompressed form, which will take more than 40 minutes to transmit over a 9600-baud modem. It is therefore very important to use compression to save storage and communication capacity for images. Many different algorithms have been proposed for image compression, and standards are still being developed for compression of images. For example, the popular GIF standard uses Lempel–Ziv coding, and the JPEG standard being developed by the Joint Photographic Experts Group uses an 8 by 8 discrete cosine transform (DCT) followed by quantization (the quality of which can be chosen by the user) and Huffman coding. The compression ratios achieved by these algorithms are very dependent on the image being coded. The lossless compression methods achieve compression ratios of up to about 3:1, whereas lossy compression methods achieve ratios up to 50:1 with very little perceptible loss of quality.

Video

Video compression methods exploit the correlation in both space and time of the sequence of images to improve compression. There is a very high correlation between successive frames of a video signal, and this can be exploited along with methods similar to those used for coding images to achieve compression ratios up to 200:1 for high-quality lossy compression. A standard for full-motion video and audio compression is being developed by the Moving Pictures Experts Group (MPEG). Applications of video compression techniques include videoconferencing, multimedia CD-ROMs, and high-definition TV.

A fascinating and very readable introduction to different sources of information, their entropy rates, and different compression algorithms can be found in the book by Lucky [1989]. Implementations of popular data compression algorithms including adaptive Huffman coding, arithmetic coding, Lempel–Ziv and the JPEG algorithm can be found in Nelson [1991].

Defining Terms

Code: A mapping from a set of messages into binary strings.

Entropy: A measure of the average uncertainty of a random variable. For a random variable with probability distribution $p(x)$, the entropy $H(X)$ is defined as $\Sigma_x - p(x) \log p(x)$.

Huffman coding: A procedure that constructs the code of minimum average length for a random variable.

Kolmogorov complexity: The minimum length description of a binary string that would enable a universal computer to reconstruct the string.

Lempel-Ziv coding: A dictionary-based procedure for coding that does not use the probability distribution of the source and is nonetheless asymptotically optimal.

Quantization: A process by which the output of a continuous source is represented by one of a set of discrete points.

Rate distortion function: The minimum rate at which a source can be described to within a given average distortion.

Vector quantization: Quantization applied to vectors or blocks of outputs of a continuous source.

References

T. Bell, J. Cleary, and I. Witten, *Text Compression*, Englewood Cliffs, N.J.: Prentice-Hall, 1990.

T. M. Cover and J. A. Thomas, *Elements of Information Theory*, New York: Wiley, 1991.

R. Gallager, *Information Theory and Reliable Communication*, New York: Wiley, 1968.

A. Gersho and R. Gray, *Vector Quantization and Source Coding*, Boston: Kluwer Academic, 1992.

N. Jayant and P. Noll, *Digital Coding of Waveforms: Principles and Applications to Speech and Video*, Englewood Cliffs, N.J.: Prentice-Hall, 1984.

G. Langdon, "An introduction to arithmetic coding," *IBM Journal of Research and Development*, vol. 28, pp. 135–149, 1984.

R. Lucky, *Silicon Dreams: Information, Man and Machine*, New York: St. Martin's Press, 1989.

M. Nelson, *The Data Compression Book*, San Mateo, Calif.: M & T Books, 1991.

C. E. Shannon, "A mathematical theory of communication," *Bell Sys. Tech. Journal*, vol. 27, pp. 379–423, 623–656, 1948.

J. Ziv and A. Lempel, "Compression of individual sequences by variable rate coding," *IEEE Trans. Inform. Theory*, vol. IT–24, pp. 530–536, 1978.

Further Information

Discussion of various data compression algorithms for sources like speech and images can be found in the *IEEE Transactions on Communications* and the *IEEE Transactions on Signal Processing*, while the theoretical underpinnings of compression algorithms are discussed in the *IEEE Transactions on Information Theory*.

Some of the latest developments in the areas of speech and image coding are described in a special issue of the *IEEE Journal on Selected Areas in Communications*, June 1992. It includes an excellent survey by N.S. Jayant of current work on signal compression, including various data compression standards.

CLAUDE E. SHANNON

His colleagues label him inventor, tinkerer, puzzle-solver, prankster, and the father of information theory

Who is the real Claude Shannon? A visitor to Entropy House, the stuccoed mansion outside Boston where Shannon and his wife Betty have lived for more than 30 years, might reach different conclusions in different rooms. One room, prim and tidy, is lined with plaques that solemnly testify to Shannon's numerous honors, including the National Medal of Science, which he received in 1966; the Kyoto Prize, Japan's equivalent of the Nobel; and the IEEE Medal of Honor.

That room enshrines the Shannon whose work Robert Lucky, the executive director of research for AT&T Bell Laboratories, has called the greatest "in the annals of technological thought," and whose "pioneering insight" IBM Fellow Rolf W. Landauer has equated with Einstein's. That Shannon is the one who, as a young engineer at Bell Laboratories in 1948, defined the field of information theory. With a brilliant paper in the *Bell System Technical Journal*, he established the intellectual framework for the efficient packaging and transmission of electronic data. The paper, entitled "The Mathematical Theory of Communication," still stands as the Magna Carta of the communications age.

But showing a recent visitor his awards, Shannon, at 75, seemed almost embarrassed. After a fidgety minute, he bolted into the room next door. This room has framed certificates, too, including one certifying Shannon as a "doctor of juggling." But it is also lined with tables heaped with all kinds of gadgets.

Some of these treasures—such as the talking chess-playing machine, the hundred-bladed jack-knife, the motorized pogo-stick, and the countless musical instruments—Shannon has collected through the years. Others he has built himself: a miniature stage with three juggling clowns, a mechanical mouse that finds its way out of a maze, a juggling mannekin of the comedian W. C. Fields, and a computer called Throbac (Thrifty Roman Numeral Backward Computer) that calculates in Roman numerals.

This roomful of gadgets reveals the other Shannon, the one who rode through the halls of Bell Laboratories on a unicycle while simultaneously juggling four balls, invented a rocket-powered frisbee, and designed a "mind-reading" machine.

Shannon makes no apologies. "I've always pursued my interests without much regard for financial value or value to the world," he said cheerfully. "I've spent lots of time on totally useless things."

Shannon's delight in both mathematical abstractions and gadgetry emerged early on. Shannon played with radio kits and erector sets supplied by his father, a probate judge. He also enjoyed solving mathematical puzzles given to him by his sister, Catherine, who eventually became a professor of mathematics.

As an undergraduate at the University of Michigan in Ann Arbor, Shannon majored in both mathematics and electrical engineering. His familiarity with the two fields helped him notch his first big success as a graduate student at the Massachusetts Institute of Technology (MIT) in Cambridge. Following a discussion of complex telephone switching circuits with Amos Joel, famed Bell Laboratories expert in the topic, in his master's thesis Shannon showed how an algebra invented by the British mathematician George Boole in the mid-1800s—which deals with such concepts as "if X or Y happens but not Z, then Q results"—could represent the workings of switches and relays in electronic circuits.

The implications of the paper by the 22-year-old student were profound: circuit designs could be tested mathematically, before they were built, rather than through tedious trial and error. Engineers now routinely design computer hardware and software, telephone networks, and other complex systems with the aid of Boolean algebra.

Shannon's paper has been called "possibly the

most important master's thesis in the century," but Shannon, typically, downplays it. "It just happened that no one else was familiar with both those fields at the same time," he said. After a moment's reflection, he added, "I've always loved that word, 'Boolean.'"

Receiving his doctorate from MIT in 1940 (his Ph.D. thesis addressed the mathematics of genetic transmission), Shannon then spent a year at the Institute for Advanced Study in Princeton, N.J. Lowering his voice dramatically, Shannon recalled how he was giving a talk at the institute when suddenly the legendary Einstein entered a door at the rear of the room. Einstein looked at Shannon, whispered something to another scientist, and departed. After his talk, Shannon rushed over to the scientist and asked him what Einstein had said. The scientist gravely told him that the great physicist had "wanted to know where the tea was," Shannon said, and burst into laughter.

HOW DO YOU SPELL 'EUREKA'? Shannon went to Bell Laboratories in 1941 and remained there for 15 years. During World War II, he was part of a group that developed digital encryption systems, including one that Churchill and Roosevelt used for transoceanic conferences.

It was this work, Shannon said, that led him to develop his theory of communication. He realized that, just as digital codes could protect information from prying eyes, so could they shield it from the ravages of static or other forms of interference. The codes could also be used to package information more efficiently, so that more of it could be carried over a given channel.

"My first thinking about [information theory]," Shannon said, "was how you best improve information transmission over a noisy channel. This was a specific problem, where you're thinking about a telegraph system or a telephone system. But when you get to thinking about that, you begin to generalize in your head about all these broader applications." Asked whether at any point he had a "Eureka!"-style flash of insight, Shannon deflected the simplistic question with: "I would have, but I didn't know how to spell the word."

The definition of information set forth in Shannon's 1948 paper is crucial to his theory of communication. Shannon demonstrated that informa-

tion is a measurable commodity. The amount of information in a given message, he showed, is determined by the probability that—out of all the messages that could be sent—that particular message would be selected.

He defined the overall potential for information in a system as its "entropy," which in thermodynamics denotes the randomness—or "shuffledness," as one physicist has put it—of a system. (The great mathematician and computer theoretician John von Neumann persuaded Shannon to use the word entropy. The fact that no one knows what entropy really is, von Neumann argued, would give Shannon an edge in debates over information theory.)

Shannon defined the basic unit of information, which John Tukey of Bell Laboratories dubbed a binary unit and then a bit, as a message representing one of two states.

Building on this mathematical foundation, Shannon then showed that any given communications channel has a maximum capacity for reliably transmitting information. Actually, he showed that although one can approach this maximum through clever coding, one can never quite reach it. The maximum has come to be known as the Shannon limit.

Shannon's 1948 paper showed how to calculate the Shannon limit. The first step was to eliminate redundancy from the message. Just as a laconic Romeo can get his message across with a mere "i lv u," a good code first compresses information to its most efficient form.

A so-called error-correction code then adds just enough redundancy to ensure that the stripped-down message is not obscured by noise. For example, an error-correction code processing a stream of numbers might add a polynomial equation on whose graph the numbers fall. The decoder on the receiving end knows that any numbers diverging from the graph have been altered in transmission.

Aaron D. Wyner, the head of the Communications Analysis Research Department at AT&T Bell Laboratories, Murray Hill, noted that some scientific discoveries seem in retrospect to be inevitably products of their times—but not Shannon's.

In fact, Shannon's ideas were almost too prescient to have an immediate impact. "A lot of practical people around the labs thought it was an interesting theory, but not very useful," said Edgar Gilbert, who went to Bell Labs in 1948—in part to work alongside Shannon. Vacuum-tube circuits simply could not handle the complex codes needed to approach the Shannon limit, Gilbert explained. Shannon's paper even received a negative review from J. L. Doov, a prominent mathematician at the University of Illinois in Urbana-Champaign. Historian William Aspray also noted that in any event the conceptual framework was not in place to permit the application of information theory at the time.

Not until the early 1970s—with the advent of high-speed integrated circuits—did engineers begin to fully exploit information theory. Today Shannon's insights help shape virtually all systems that store, process, or transmit information in digital form—from compact discs to super-computers, from facsimile machines to deep-space probes such as Voyager.

INFORMATION THEORY AND RELIGION. Especially early on, however, information theory captivated an audience much larger than the one for which it was intended. People in linguistics, psychology, economics, biology, even music and the arts sought to fuse information theory to their disciplines.

John R. Pierce, a former co-worker of Shannon who is now a professor emeritus at California's Stanford University, has compared the "widespread abuse" of information theory to that inflicted on two other profound and much misunderstood scientific ideas: Heisenberg's uncertainty principle and Einstein's theory of relativity.

Some physicists went to extraordinary lengths to prove that the entropy of information theory was mathematically equivalent to the entropy of thermodynamics. That turned out to be true but of little consequence, according to Bell Labs veteran David Slepian, a former colleague of Shannon at the laboratories who was also a seminal figure in information coding. Many engineers, too, "jumped on the bandwagon without really understanding" the theory, Slepian said. Shannon's work inspired the formation of the IEEE Information Theory Society in 1956, and subgroups dedicated to economics, biological systems, and other applications

soon formed. In the early 1970s, the *IEEE Transactions on Information Theory* published an editorial, titled "Information Theory, Photosynthesis, and Religion," deploring the over-extension of Shannon's theory.

Shannon, while also skeptical of some of the uses of his theory, was rather free-ranging in his own investigations. In the 1950s, he did living-room experiments on the redundancy of language with his wife, Betty, who was a Bell computer scientist, Bernard Oliver, another Bell scientist (and a former IEEE president), and Oliver's wife. One person would offer the first few letters of a word, or words in a sentence, and the others would try to guess what came next. Shannon also directed an experiment at Bell Labs in which workers counted the number of times various letters appeared in a written text, and their order of appearance.

Indeed, Shannon's work on information theory and his love of gadgets led to a precocious fascination with intelligent machines. Shannon was one of the first scientists to propose that a computer could compete with humans in chess; in 1950 he wrote an article for *Scientific American* explaining how that task might be accomplished.

Shannon did not restrict himself to chess. He built a "mind-reading" machine that played the game of penny-matching, in which one person tries to guess whether the other has chosen heads or tails. A colleague at Bell Laboratories, David W. Hagelbarger, built the prototype; the machine recorded and analyzed its opponents' past choices, looking for patterns that would foretell the next choice. Because humans almost invariably fall into such patterns, the machine won more than 50 percent of the time. Shannon then built his own version of the machine and challenged Hagelbarger to a now legendary duel.

In 1950 Shannon created a mechanical mouse that could learn to find its way through a maze to a chunk of brass cheese, seemingly unassisted. Shannon named the mouse Theseus, after the mythical Greek hero who slew the Minotaur and found his way out of the dreaded labyrinth. Actually the "brains" of the mouse were contained in a bulky set of vacuum-tube circuitry under the floor of the maze; the circuits controlled the movement of a magnet which in turn controlled the mouse.

When in 1977 the editor of *IEEE Spectrum* challenged readers to create a "micromouse" whose "brains" were self-contained, who could through trial and error solve a maze, and who could then learn through its mistakes and get through the maze in a repeat attempt without error, a former colleague of Shannon called *Spectrum* and insisted that Shannon had built such a micromouse two decades earlier.

Knowing the technology of the '50s would not have permitted it, the editor nevertheless called Shannon, who laughed, saying he had fooled a lot of people as he took his "smart" mouse around the country. The drapes around the table that hid the vacuum tubes and lead-screw machinery were vital, he chuckled.

Asked about prospects for artificial intelligence, Shannon noted that current computers, in spite of their extraordinary power, are still "not up to the human level yet" in terms of raw information processing. Simply replicating human vision in a machine, he points out, remains a formidable task. But he added that "it is certainly plausible to me that in a few decades machines will be beyond humans."

UNIFIED FIELD THEORY OF JUGGLING. In 1956 Shannon left his permanent position at Bell Labs (he remained affiliated for more than a decade) to become a professor of communications science at MIT. In recent years, his great obsession has been juggling. He has built several juggling machines and devised what may be the unified field theory of juggling: if B equals the number of balls, H the number of hands, D the time each ball spends in a hand, F the time of flight of each ball, and E the time each hand is empty, then $B/H = (D + F)/(D + E)$.

Shannon has also developed various mathematical models to predict stock performance and has tested them—successfully, he said—on his own portfolio.

After the late 1950s, Shannon published little on information theory. Some former Bell colleagues suggested that by the time he went to MIT Shannon had "burned out" and tired of the field he had created.

But Shannon denied that claim. He said he continued to work on various problems in information theory through the 1960s, and even published a few papers, though he did not consider most of his investigations then important enough to publish. "Most great mathematicians have done their finest work when they were young," he observed.

In the l960s Shannon also stopped attending meetings dedicated to the field he had created. Berlekamp offered one possible explanation. In 1973, he recalled, he persuaded Shannon to give the first annual Shannon lecture at the International Information Theory Symposium, but Shannon almost backed out at the last minute. "I never saw a guy with so much stagefright," Berlekamp said. "In this crowd, he was viewed as a godlike figure, and I guess he was worried he wouldn't live up to his reputation."

Berlekamp said Shannon eventually gave an inspiring speech, which anticipated ideas on the universality of feedback and self-referentiality in nature.

Shannon nevertheless fell out of sight once again. But in recent years, encouraged by his wife, he has begun to drop in on small meetings and to visit various laboratories where his work is carried on.

In 1985 he made an unexpected appearance at the International Information Theory Symposium in Brighton, England. The meeting was proceeding smoothly, if uneventfully, when news raced through the halls and lecture rooms that the snowy-haired man with the shy grin who was wandering in and out of the sessions was none other than Claude Shannon. Some of those at the conference had not even known he was still alive.

At the banquet, the meeting's organizers somehow persuaded Shannon to address the audience. He spoke for a few minutes and then—fearing that he was boring his audience, he recalled later—pulled three balls out of his pockets and began juggling. The audience cheered and lined up for autographs. Said Robert J. McEliece, a professor of electrical engineering at the California Institute of Technology and chairman of the symposium: "It was as if Newton had showed up at a physics conference."

Source: Adapted from J. Horgan, *IEEE Spectrum*, pp. 72–75, April 1992. © 1992 IEEE.

68

Satellites and Aerospace

68.1	Introduction	1532
68.2	Satellite Applications	1532
68.3	Satellite Functions	1533
68.4	Satellite Orbits and Pointing Angles	1535
68.5	Some Particular Orbits	1537
68.6	Communications Link	1537
68.7	System Noise Temperature and G/T	1538
68.8	Digital Links	1540
68.9	Interference and Total C/N	1540
68.10	Access and Modulation	1541
68.11	Frequency Allocations	1542
68.12	Satellite Subsystems	1542
68.13	Trends	1544

Daniel F. DiFonzo
Planar Communications
Corporation

68.1 Introduction

The impact of satellites on world communications since commercial operations began in the mid-1960s is such that we now take for granted many services that were not available a few decades ago: worldwide TV, reliable communications with ships and aircraft, wide-area data networks, communications to remote areas, direct TV broadcast to homes, position determination, and earth observation (weather and mapping). Future satellite-based global personal communications to hand-held portable telephones may usher in yet another new era.

Satellites function as line-of-sight microwave relays in orbits high above the earth which can "see" large areas of the earth's surface. This unique feature ensures satellites' continued growth even as fiber-optic cables capture a larger market share of high-density point-to-point traffic. Satellites provide cost-effective access for areas with low (thin-route) communications traffic because earth terminals can be installed in locations where the high investment cost of terrestrial facilities might not be warranted. Satellites are particularly well suited to wide-area coverage for broadcasting, mobile communications, and point-to-multipoint communications.

68.2 Satellite Applications

Figure 68.1 depicts several kinds of satellite links and orbits. The geostationary earth orbit (GEO) is in the equatorial plane at an altitude of 36,000 km with a period of one sidereal day (23h 56m 4.09s). GEO satellites appear to be almost stationary from the ground (subject to small perturbations) and the earth antennas pointing to these satellites may need only limited or no tracking

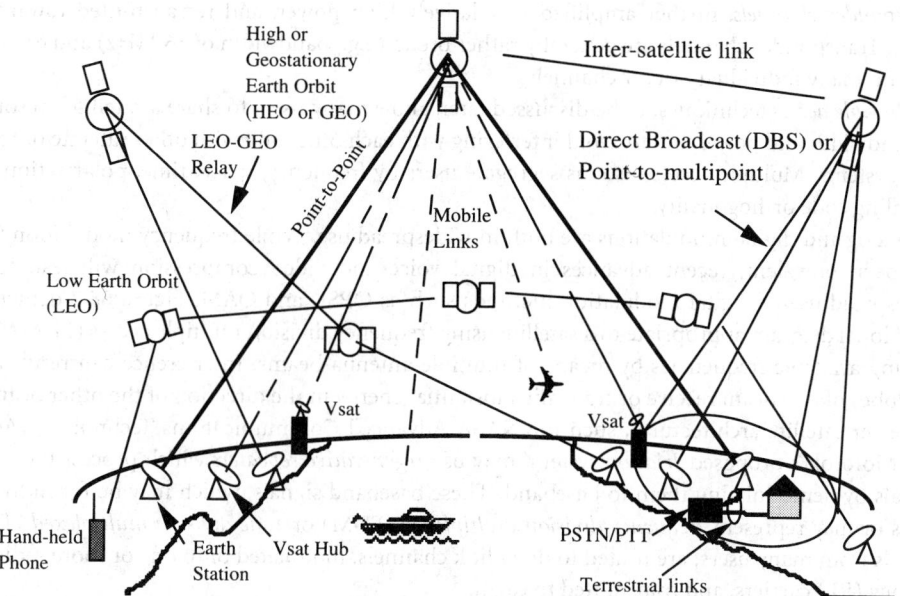

FIGURE 68.1 Several types of satellite links. Illustrated are point-to-point, point-to-multipoint, VSAT, direct broadcast, mobile, personal communications, and intersatellite links.

capability. The orbits for which the highest altitude (apogee) is at or greater than GEO are sometimes referred to as high earth orbits (HEO). Low earth orbits (LEO) typically range from a few hundred kilometers to about 1000 km and medium earth orbits (MEO) are at intermediate altitudes.

Initially, satellites were used primarily for point-to-point traffic in the GEO fixed satellite service (FSS), e.g., for telephony across the oceans and for point-to-multipoint TV distribution to cable *head end* stations. Large earth station antennas with high-gain narrow beams and high uplink powers were needed to compensate for limited satellite power. This type of system, exemplified by the early global network of the International Telecommunications Satellite Consortium (Intelsat), used "Standard A" earth antennas with 30-m diameters. Since the start of Intelsat, many other satellite organizations and consortia have been formed around the world to provide international, regional, and domestic services [Rees, 1990].

As satellites have grown in power and sophistication, the average size of the earth terminals has been reduced. High-gain satellite antennas and relatively high-power satellite transmitters have led to *very small aperture* earth *terminals* (VSAT) with diameters of less than 2 m and modest powers of less than 10 W [Gagliardi, 1991]. As depicted in Fig. 68.1, VSAT terminals may be placed atop urban office buildings, permitting private networks of hundreds or thousands of terminals which bypass terrestrial lines. VSATs are usually incorporated into *star* networks where the small terminals communicate through the satellite with a larger *Hub* terminal. The Hub retransmits through the satellite to another small terminal. Therefore, VSAT-to-VSAT links require two *hops* with attendant time delays. With high-gain satellite antennas and relatively narrowband digital signals (e.g., compressed voice at ≤8 kbps), direct single-hop *mesh* interconnections of VSATs may be used.

68.3 Satellite Functions

The traditional function of a satellite is that of a "bent pipe" quasi-linear repeater in space. As shown in Fig. 68.2, *uplink* signals from earth terminals directed at the satellite are received by the satellite's antennas, amplified, translated to a different *downlink* frequency band, channelized into *transponder channels*, further amplified to relatively high power, and retransmitted toward the earth. Transponder channels are generally rather broad (e.g., bandwidth of 36 MHz) and each may contain many individual or user channels.

Multiple access techniques, to be discussed later, allow many users to share a satellite's resources of bandwidth and power and to avoid interfering with each other and with other satellite or terrestrial systems. Multiple access systems segregate users by frequency, space, time, polarization, and signaling code orthogonality.

Analog and digital modulations are both in widespread use. While frequency modulation (FM) has been prevalent, recent advances in digital voice and video compression will lead to the widespread use of digital modulation methods such as QPSK and QAM. Figure 68.2 depicts the functional diagram appropriate to a satellite using frequency division multiple access (FDMA) and reusing available frequencies by means of multiple antenna beams. Interference can result if the sidelobes of one beam receive or transmit substantial energy in the direction of the other beam.

Newer satellite architectures, such the NASA Advanced Communications Technology (ACTS) and Motorola's proposed *Iridium* system, may use *regenerative repeaters* which process the uplink signals by demodulating them to baseband. These baseband signals, which may be for individual users or may represent *frequency division multiplexed* (FDM) or *time division multiplexed* (TDM) signals from many users, are routed to downlink channels, modulated onto one or more radio frequency (RF) carriers, and transmitted to earth.

High-power *direct broadcast satellites* (DBS) operating at Ku-band (around 12 GHz) deliver TV directly to home receivers having antennas less than 1 m in size. Such systems using analog FM are operational in Japan and Europe. In the United States, DBS with digital modulation and compressed video will provide more than four NTSC TV channels per 24-MHz transponder channel. For the United States, where each DBS orbital location is allocated 32 transponder channels of 24 MHz each, more than 128 conventional TV channels can be provided from a single DBS orbital location. DBS is seen as an attractive medium for delivery of high-definition TV (HDTV) to a large number of homes.

Mobile satellite services (MSS) operating at L-band around 1.6 GHz have revolutionized communications with ships and, more recently, with aircraft which would normally be out of reliable

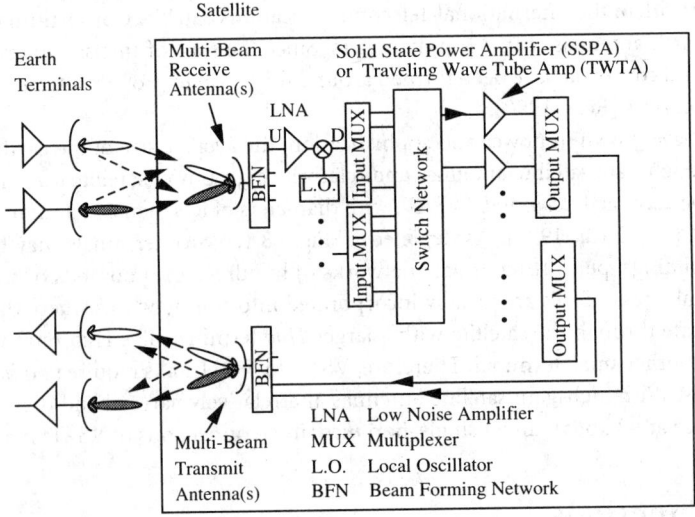

FIGURE 68.2 A satellite repeater receives uplink signals (U), translates them to a downlink frequency band (D), channelizes, amplifies to high power, and retransmits to earth. Multiple beams allow reuse of the available band. Interference (dashed lines) can limit performance.

communications range of terrestrial radio signals. The International Maritime Satellite Consortium (Inmarsat) operates the dominant system of this type.

Links between LEO satellites (or the NASA shuttle) and GEO satellites are used for data relay, e.g., via the NASA Tracking and Data Relay Satellite System (TDRSS). Some systems will use intersatellite links (ISL) to improve the interconnectivity of a wide-area network. ISL systems would typically operate at frequencies above 20 GHz or even use optical links.

An exciting new development is the prospective use of L-band frequencies with a large number (12 to 66) of LEO satellites for personal communications systems (PCS) directly with small handheld portable telephones anywhere in the world.

68.4 Satellite Orbits and Pointing Angles

Reliable communication to and from a satellite requires a knowledge of its position and velocity relative to a location on the earth. Details of the relevant astrodynamics formulas for satellite orbits are given in Griffin and French [1991] and Morgan and Gordon [1989].

A satellite, having mass m, in orbit around the earth, having mass M_e, traverses an elliptical path such that the centrifugal force due to its acceleration is balanced by the earth's gravitational attraction, leading to the equation of motion for two bodies :

$$d^2\mathbf{r}/dt^2 + \mu/r^3\mathbf{r} = 0 \tag{68.1}$$

where $\mathbf{r}$ is the radius vector joining the earth's center and the satellite and $\mu = G(M_e + m) \approx GM_e = 398,600.5$ km^3/s^2 is the product of the gravitational constant and the mass of the earth. Because $m \approx M$, the center of rotation of the two bodies may be taken as the earth's center, which is at one of the focal points of the orbit ellipse.

Figure 68.3 depicts the orbital elements for a geocentric right-handed coordinate system where the x axis points to the "first point of Aries", i.e., the fixed position against the stars where the sun's apparent path around the earth crosses the earth's equatorial plane while traveling from the southern toward the northern hemisphere at the vernal equinox. The z axis points to the north and the y axis is in the equatorial plane and points to the winter solstice. The elements shown are longitude or right ascension of the ascending node, Ω, measured in the equatorial plane; the orbit's inclination angle, i, relative to the equatorial plane; the ellipse semimajor axis length, a; the ellipse eccentricity, e; the argument (angle) of perigee, ω, measured in the orbit plane from the ascending node to the satellite's closest approach to the earth; and the true anomaly (angle) in the orbit plane from the perigee to the satellite, ν.

The mean anomaly, M, is the angle from perigee that would be traversed by a satellite moving at its mean angular velocity, n. Given an initial value, M_0, usually taken as 0 for a particular epoch (time) at perigee, the mean anomaly at time t is $M = M_0 + n(t - t_0)$, where $n = \mu^{1/2}/a^{3/2}$. The eccentric anomaly, E, may then be found from Kepler's transcendental equation $M = E - e \sin E$ which must be solved numerically by, for example, guessing an initial value for E and using a root finding method. For small eccentricities, the series approximation $E \approx M + e \sin M + (e^2/2)\sin 2M + (e^3/8)(3 \sin 3M - \sin M)$ yields good accuracy [Morgan and Gordon, 1989, p. 806]. Other useful quantities include the orbit radius, r; the period, P, of the orbit (i.e., for $n(t - t_0) = 2\pi$); the velocity, V; and the radial velocity, V_v:

$$r = a(1 - e \cos E) \tag{68.2}$$

$$P = 2\pi\sqrt{a^3/\mu} \tag{68.3}$$

$$V = \mu\left(\frac{2}{r} - \frac{1}{a}\right) \tag{68.4}$$

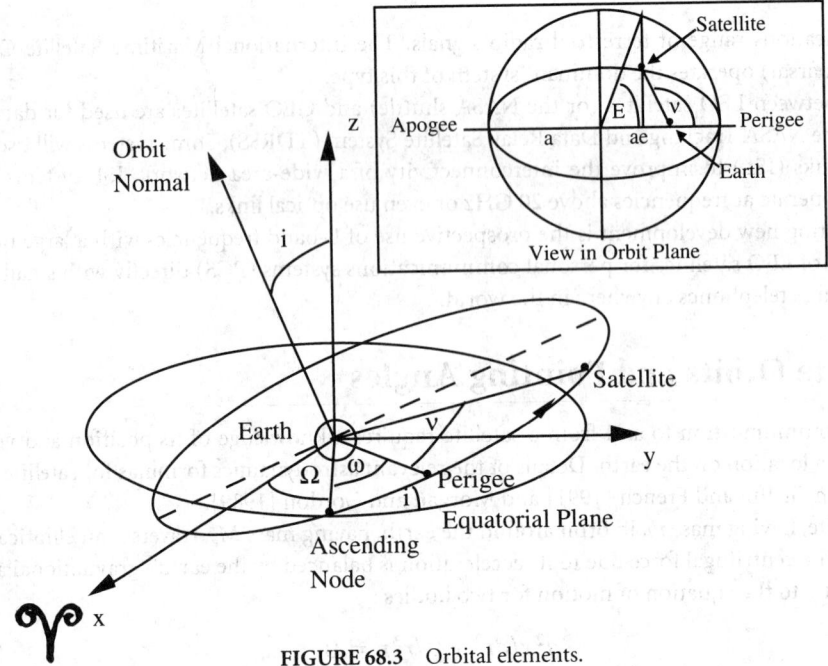

FIGURE 68.3   Orbital elements.

$$V_v = \frac{e(\mu a)^{1/2} \sin E}{a(1 - e \cos E)} \tag{68.5}$$

Figure 68.4 depicts quantities useful for communications links in the plane formed by the satellite, a point on the earth's surface and the earth's center. For a satellite at altitude h and earth radius $r_e = 6378.14$ km, the slant range, r_s, elevation angle to the satellite from the local horizon, el, and the satellite's nadir angle, θ, are related by simple spherical trigonometry formulas. Note that $\theta + (el) + \gamma = 90°$, where γ is the earth central (core) angle and the ground range from the subsatellite point is γr_e. Then,

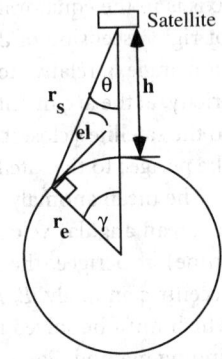

$$k = \frac{r_e + h}{r_e} = \frac{\cos(el)}{\sin \theta} \tag{68.6}$$

$$\tan(el) = \frac{(\cos \gamma - 1/k)}{\sin \gamma} \tag{68.7}$$

$$r_s = \sqrt{1 + k^2 - 2k \cos \gamma} \tag{68.8}$$

FIGURE 68.4 Geometry for a satellite in the plane defined by the satellite, the center of the earth and a point on the earth's surface. The elevation angle, *el*, is the angle from the local horizon to the satellite.

The earth station azimuth angle to the satellite measured clockwise from north in the horizon plane is given in terms of the satellite's declination, δ, the observer's latitude, ϕ, and the difference of the east longitudes of observer and satellite, $\Delta\lambda$. Then:

$$\tan A = \frac{\sin \Delta\lambda}{\cos \phi \tan \delta - \sin \phi \cos \Delta\lambda} \tag{68.9}$$

Table 68.1 Comparison of Orbit Parameters

Orbit Parameter	GEO	Tundra	Molniya
Perigee radius, km	42,164	30,841	7,161
Apogee radius, km	42,164	53,474	45,951
Eccentricity, e	0	0.27	0.73
Inclination, i, degrees	0	63.4	63.4
Period, h	23.93	23.93	11.97
Apogee latitude, degrees	0	63.4	63.4

Source: M. J. Lynch, Alternative Orbits for Commercial Communication Networks, AIAA-92-1989-CP, p. 1370.

taking due account of the sign of the denominator to ascertain the quadrant. The fraction of the earth's surface area covered by the satellite $= (1 - \cos \gamma)/2$.

68.5 Some Particular Orbits

Table 68.1 compares several orbits. For GEO, $e = 0$, $i = 0$, $k = (r_e + h)/r_e = 6.61$, and $h = 35,786$ km. When $el = 0$ the maximum nadir angle $\theta = 8.7°$, the maximum slant range is 41,680 km, and, from Eq. (68.7), $\gamma = 81.3°$. Therefore, a GEO satellite cannot "see" the earth above 81.3° latitude. A geosynchronous orbit has a period which is a multiple of the earth's rotation period, but it is not necessarily circular and it may be inclined. Therefore, GEO is a special case of a geosynchronous orbit. Molniya and Tundra orbits are critically inclined ($i = 63.4°$) elliptical orbits (HIEO) to cause the satellite's subsatellite ground trace to dwell at apogee at the same place each day. Such orbits whose subsatellite paths trace a repetitive loop (LOOPUS) allow several satellites to be phased to offer quasi-stationary satellite service at high latitudes. For full earth coverage from a constellation of LEO satellites, circular polar constellations [Adams and Rider, 1987] and constellations of orbit planes with different inclinations, e.g., "Walker orbits" [Walker, 1977], have received attention. Intersatellite links among the differently inclined planes of a Walker orbit are difficult.

68.6 Communications Link

Figure 68.5 illustrates the elements of the RF link between a satellite and earth terminals. The overall link performance is determined by computing the link equation for the uplink and downlink separately and then combining the results along with interference and intermodulation effects.

The uplink received carrier-to-noise ratio at the satellite is

$$(C/N)_u = EIRP_u + (G_{su} - 10 \log T_{su}) + 228.6$$
$$-10 \log(4\pi r_{su}^2) - 10 \log(4\pi / \lambda^2) - A_u - \Gamma_u - B \quad \text{(dB)} \quad (68.10)$$

where each term is a quantity expressed in dB, e.g., P (dBW) $= 10 \log$ (power in watts). The downlink equation has identical form with appropriate downlink quantities substituted in Eq. (68.10). The relevant quantities (without subscripts) are described below.

The carrier-to-noise ratio (C/N) is the dB *difference* between carrier and noise powers expressed in dBW, i.e., $P = 10 \log$ (power in watts). EIRP $= P + G$ dBW is the equivalent isotropically radiated power where the antenna gain, G in dBi (dB relative to isotropic), is the antenna gain *in the direction of the link*, i.e., it is not necessarily the antenna's peak gain. The received noise power is $N = kTB$ W where $k = 1.38 \times 10^{-23}$ J/K is Boltzmann's constant and $10 \log(k) = -228.6$ dBW/K/Hz, T is the system noise temperature in kelvins (K), and B is the bandwidth in dB Hz. Then, $(G - 10 \log T)$

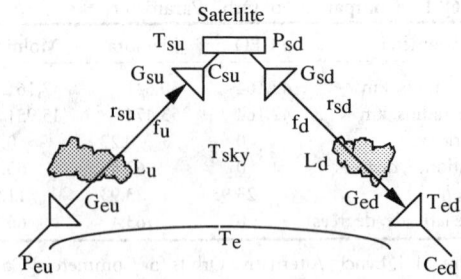

FIGURE 68.5   Quantities for a satellite RF link. P = transmit power (dBW). G = antenna gain (dBi). C = received carrier power (dBW). T = noise temperature (K). L = dissipative loss (dB). r = slant range (m). f = frequency (Hz). u = uplink. d = downlink. e = earth. s = satellite.

dB/K is a figure of merit for the satellite or earth receiving system. It is usually written as G/T and read as "gee over tee." The antenna gain and the noise temperature must be defined at the same reference point, e.g., at the receiver's input port.

The spreading factor, $4\pi r_s^2$, depends *only* on the slant range distance r_{su} or r_{sd}. The gain of a 1-m^2 antenna is $10\ \log(4\pi/\lambda^2)$, where the wavelength, $\lambda = c/f$, f is the frequency in Hz, and $c = 2.9979 \times 10^8$ m/s is the velocity of light. The dB sum of the spreading factor and the gain of a 1-m^2 antenna is the frequency-dependent "path loss." A is the signal attenuation due to dissipative losses in the propagation medium, Γ is the polarization mismatch loss between the incident wave and the receive antenna $(0 \le \Gamma \le \infty)$, and B is the bandwidth in dB Hz, i.e., $B = 10\ \log$ (bandwidth in Hz).

The polarization coupling factor, Γ, is given

$$\Gamma = \frac{(1 + r_w^2)(1 + r_A^2) \pm 4r_w r_A(1 - r_w^2)(1 - r_A^2)\cos 2\delta}{2(1 + r_w^2)(1 + r_A^2)} \qquad (0 \le \Gamma \le 1) \qquad (68.10a)$$

where r_w and r_A are the voltage axial ratios of the wave and antenna, respectively, and δ is the angle between the major axes of the polarization ellipses. The minus sign is used if the wave and antenna have opposite polarization senses.

68.7 System Noise Temperature and G/T

The system noise temperature, T, incorporates contributions to the noise power radiated into the antenna from the sky, ground, and galaxy, as well as the noise temperature due to circuit and propagation losses, and the noise figure of the receiver. The clear sky antenna temperature for an earth station depends on the elevation angle since the antenna's sidelobes will receive a small fraction of the thermal noise power radiated by the earth which has a noise temperature $T_{\text{earth}} \approx 290$ K. At 11 GHz the clear sky *antenna* noise temperature, T_{aclear}, ranges from 5 to 10 K at zenith $(el = 90°)$ to more than 50 K at $el = 5°$ [Pratt and Bostian, 1986].

As shown in Fig. 68.6, the system noise temperature is developed from the standard formula for the equivalent temperature of tandem elements including the antenna in clear sky, propagation (rain) loss of A dB, circuit losses between the aperture and receiver of L_c dB, and receiver noise figure of F dB (corresponding to receiver noise temperature T_r, K). The system noise temperature referred to the receiver input is approximated by Eq. (68.11) where $T_{\text{rain}} \approx 270$ K is a reasonable approximation for the physical temperature of the rain.

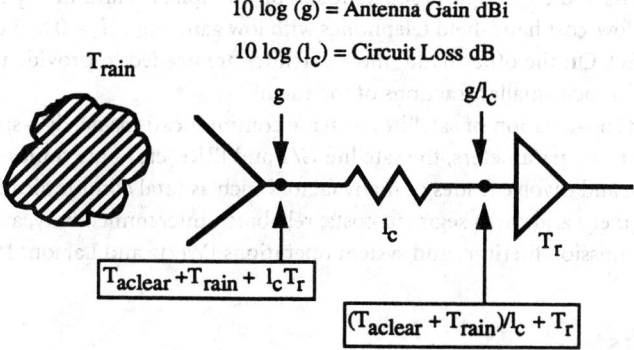

FIGURE 68.6 Tandem connection of antenna, loss elements such as waveguide, and receiver front end. The noise temperature depends on the reference plane but the G/T is the same for both points shown.

$$
\begin{aligned}
T_{ed} = {} & [T_{aclear} + 270(1 - 10^{-A/10})]/10^{L_c/10} \\
& + 290(1 - 10^{-L_c/10}) + 290(10^{F/10} - 1) \quad \text{K}
\end{aligned}
\tag{68.11}
$$

The system noise temperature is defined at a specific reference point such as the input to the receiver. However, G/T is *independent* of the reference point when G correctly accounts for circuit losses. The satellite's noise temperature is generally higher than an earth terminal's under clear sky conditions because the satellite antenna "sees" a warm earth temperature of $\approx$150 to 300 K, depending on the proportion of clouds, oceans, and land in the satellite antenna's beam, whereas an earth antenna generally "sees" cold sky. Furthermore, a satellite receiving system generally has higher circuit losses due to beam forming networks, protection circuitry, and extra components for redundancy.

Figure 68.7 illustrates the link loss factors, maximum nadir angle, and earth central angle as a function of satellite altitude. The path losses are shown for several satellite frequencies in use. The

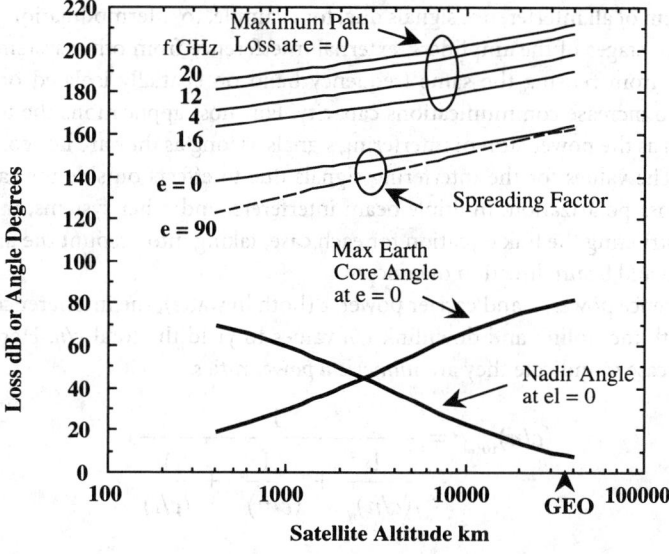

FIGURE 68.7 Satellite link losses, spreading factors, maximum nadir angle, θ ($el = 0$), and earth central angle, γ, vs. satellite altitude, h (km).

variation in path loss and core angle is substantial. For example, L-band LEO personal communications systems to low-cost hand-held telephones with low gain (e.g., $G_u \approx 0$ to 3 dBi) need less link power than for GEO. On the other hand, more satellites are needed to provide full coverage since LEO satellites "see" much smaller fractions of the earth.

The design for a constellation of satellites to serve communications needs—such as the number of satellites, their orbital parameters, the satellite G/T and EIRP, etc.—are topics related to mission analysis and design and involve trades of many factors such as total communications capacity, link margins, space segment and earth segment costs, reliability, interconnectivity, availability and cost of launch vehicles, mission lifetime, and system operations [Wertz and Larson, 1991].

68.8 Digital Links

For digital modulation systems, bit error rate (BER) is related to the dimensionless ratio (dB difference) of energy per bit, E_b dB J, to the noise power density, $N_o = 10 \log(kT)$ dB J.

$$\left(\frac{E_b}{N_o}\right) = \left(\frac{C}{N}\right) + B - R$$

$$= \left(\frac{C}{N_o}\right) - R \qquad \text{dB}$$

(68.12)

where R is 10 log (transmission bit rate in bit/s), B is the bandwidth (dB Hz), and (C/N_o) is the *carrier-to-noise density ratio*, i.e., C/N normalized to unit bandwidth. Curves relating the communications performance measure of BER vs. E_b/N_o for different modulations may be found elsewhere in this handbook.

68.9 Interference and Total *C/N*

The *C/N* for a complete transponder link combines the contributions of the uplink, downlink, and also the power sum of all interference signals due, for example, to intermodulation products generated in the output stages of the amplifiers, external interference from other systems, and intrasystem interference from reusing the same frequency band on spatially isolated or dual polarized antenna beams to increase communications capacity. For most applications the total interference power may taken as the power sum of interfering signals as long as they are not correlated with the desired carrier. The values for the interfering signals due to effects outside the satellite, e.g., **frequency reuse** cross-polarization, multiple beam interferers, and other systems, must be obtained by carefully constructing the link equation for each case, taking into account the antenna gains for each polarization and beam direction of concern.

For an interference power, i, and carrier power, c (both in watts), the interference ratio, c/i, must be combined with the uplink and downlink c/n values to yield the total c/n. Here, the ratios are written in lower case to indicate they are *numerical power ratios*.

$$(c/n)_{\text{total}} = \cfrac{1}{\cfrac{1}{(c/n)_u} + \cfrac{1}{(c/n)_d} + \cfrac{1}{(c/i)}}$$

(68.13)

Equation (68.10) applies to a "bent pipe" satellite. If on-board signal regeneration is used for digital transmission, the uplink signal is demodulated and a "clean" set of baseband bits is remodulated. This has the effect of separating the accumulation of uplink and downlink noise contributions by causing the uplink noise to be effectively modulated onto the downlink carrier with the desired signal [Gagliardi, 1991]. Remodulation is also useful for intersatellite links. In each case, a savings in power or antenna size may be obtained at the expense of circuit and processing complexity.

For a system employing frequency reuse via dual polarizations, the polarization coupling factor, Γ, from Eq. (68.10), between a wave and antenna, determines the interference power. The ratio of desired (copolarized) receive power and undesired (cross-polarized) powers is called the **polarization isolation**. For nearly circularly polarized systems, the isolation can be approximated as:

$$I \approx 10 \log (R_w^2 + R_A^2 + 2R_w R_A \cos \delta) - 24.8 \qquad \text{dB} \qquad (68.14)$$

where R_w and R_A are the axial ratios in dB of the wave and antenna, respectively, and d is the relative tilt angles of the polarization ellipses. In this equation, the isolation is the Γ of Eq. (68.10) for the case where the wave and antenna have opposite polarization sense. For ideal antennas and no *depolarization* due to precipitation [Gagliardi, 1991; Pratt and Bostian, 1986; Pritchard and Sciulli, 1986] the coupling would be zero or $-\infty$ dB. The copolarized coupling factor would normally be ≈ 1 or 0 dB.

68.10 Access and Modulation

Satellites act as central relay nodes which are visible to a large number of users who must efficiently use the limited power and bandwidth resources. For detailed discussions of access issues see Gagliardi [1991], Pritchard and Sciulli [1986], Miya [1985], and Shimbo [1988]. A brief summary of issues specific to satellite systems is given below.

Frequency division multiple access (FDMA) has been the most prevalent access for satellite systems until recently. Individual users assigned a particular frequency band may communicate at any time. Satellite filters subdivide a broad frequency band into a number of *transponder channels*, e.g., the 500-MHz uplink FSS band from 5.925 to 6.425 GHz may be divided into 12 transponder channels of 36 MHz bandwidth plus guard bands. This limits the interference among adjacent channels in the corresponding downlink band of 3.7–4.2 GHz.

FDMA implies that several individual carriers coexist in the transmit amplifiers. In order to operate the amplifiers in a quasi-linear region relative to their saturated output power to limit intermodulation products, the amplifiers, must be operated in a **backed off** condition. For example, in order to limit third-order intermodulation power for two carriers in a conventional TWT (traveling wave tube) amplifier to ≈ -20 dBc, its input power must be reduced (*input backoff*) by about 10 dB relative to the power that would drive it to saturation. The output power of the carriers is reduced by about 4 to 5 dB (*output backoff*). Amplifiers with fixed bias levels will consume power even if no carrier is present. Therefore, dc-to-RF efficiency degrades as the operating point is backed off. For amplifiers with many carriers, the intermodulation products have a noise-like spectrum and the *noise power ratio* is a better measure of multicarrier performance.

Time division multiple access (TDMA) users share a common frequency band and are each assigned a unique time slot for their digital transmissions. At any instant there is only one carrier in the transmit amplifier, requiring little or no backoff from saturation. The dc-RF efficiency is high. A drawback is the system complexity required to synchronize widely dispersed users in order to avoid intersymbol interference caused by more than one user signal appearing in a given time slot. Also, the total transmission rate in a TDMA satellite channel must be essentially the sum of the users' rates, including overhead bits such as for framing, synchronization and clock recovery, and source coding. At the present state of the art, earth terminal hardware costs may be higher than for FDMA. Nevertheless, TDMA systems are gaining acceptance for some applications.

Code division multiple access (CDMA) modulates each carrier with a unique pseudo random code, usually by means of either a direct sequence or frequency hopping spread spectrum modulation. As the CDMA users occupy the same frequency band at the same time, the aggregate signal in the satellite amplifier is noise-like. Individual signals are extracted at the receiver by correlation processes. CDMA tolerates noise-like interference but does not tolerate large deviations from average loading conditions. One or more very strong carriers could violate the noise-like interference condition and generate strong intermodulation signals.

User access is via assignments of a frequency, time slot, or code. Fixed assigned channels allow a user unlimited access. However, this may result in poor utilization efficiency for the satellite resources and may imply higher user costs (analogous to a leased terrestrial line). Other assignment schemes include *demand assigned multiple access* (DAMA) and *random access* (e.g., for the Aloha concept). DAMA systems require the user to first send a channel request over a common control channel. The network controller (at another earth station) seeks an empty channel and instructs the sending and receiving units to tune to it (either in frequency or time slot). A link is maintained for the call duration and then released to the system for other users to request. Random access is economical for lightly used burst traffic such as data. It relies on random time of arrival of data packets and protocols are in place for repeat requests in the event of collisions [Gagliardi, 1991].

In practice, combinations of multiplexing and access techniques may be used. A broad band may be channelized or *frequency division multiplexed* (FDM) and FDMA may be used in each subband, e.g., FDM/FDMA.

The traditional satellite modulation format has been FM. However, recent trends indicate that digital modulations such as M-ary PSK and QAM will become more prevalent for nearly all applications including voice, data, and TV. The efficiencies afforded by digital modulations arise partly because they allow signal processing for bandwidth compression. Compressed digital TV transmission allows a significant improvement in the capacity compared with FM.

68.11 Frequency Allocations

Table 68.2 contains a partial list of frequency allocations for satellite communications. The World Administrative Radio Conference, WARC-92, allocated new L-band frequencies for LEO personal communications services and for LEO small satellite data relay. Most of the other bands have been in force for years.

68.12 Satellite Subsystems

The major satellite subsystems are described in Griffin and French [1991]. They are propulsion, power, antenna, communications repeater, structures, thermal, **attitude** determination and control, telemetry, tracking, and command.

The satellite *antennas* typically are offset-fed paraboloids. Typical sizes are constrained by launch vehicles and have ranged from <1 m to >5 m for some applications. The Intelsat 6 satellite used a 3.2-m antenna at 4 GHz. Ku-band satellites may use a diameter >2 m (i.e., $D > 80\,\lambda$). Multiple feeds in the focal region each produce a narrow "component beam" whose beamwidth is $\approx 65\,\lambda/D$ and whose directions are established by the displacement of the feed from the focal point. These beams are combined to produce shaped beam with relatively high gain over a geographical region. Multiple beams are also used to reuse frequencies on the satellite. Figure 68.2 suggests that a satellite may have several beams for frequency reuse. In that case, the carriers occupying the same frequencies must be isolated from each other by either polarization orthogonality or antenna sidelobe suppression. As long as the sidelobes of one beam do not radiate strongly in the direction of another, both may use the same frequency band and increase the satellite's capacity.

Table 68.2   Partial List of Satellite Frequency Allocations (Frequencies in GHz)

Band	Uplink	Downlink	Satellite Service
VHF		0.137–0.138	Mobile
VHF	0.3120–0.315	0.387–0.390	Mobile
	0.8210–0.825	0.866–0.870	Mobile
	0.8450–0.851	0.890–0.896	Mobile
L-band	1.610–1.6138		Mobile, Radio Astronomy
	1.6138–1.6265	1.6138–1.6265	Mobile LEO
	1.6265–1.6605	1.525–1.535	Mobile
		1.575	Global Positioning System
		1.227	GPS
S-band	2.110–2.120	2.290–2.300	Deep-space research
		2.4835–2.500	Mobile
C-band	5.9–6.4	3.7–4.2	Fixed (FSS)
X-band	7.9–9.4	7.25–7.75	Military (U.S.)
Ku-band	14.0–14.5	11.7–12.2	FSS
		12.2–12.7	Direct Broadcast (BSS)
Ka-band	27–31	17–21	Unassigned
Q	50–51	40–41	Fixed
V		54–58 and 59–64	Intersatellite

Sources: R.M. Gagliardi, *Satellite Communications,* New York: Van Nostrand Reinhold, 1991; Final Acts of the World Administrative Radio Conference (WARC-92), Malaga-Torremolinos, 1992.

The *repeaters* include the following main elements (see Fig. 68.2). A low-noise amplifier (LNA) amplifies the received signal and establishes the uplink noise. The *G/T* of the satellite receiver includes the effect of losses in the satellite antenna, the noise figure of the LNA, and the noise temperature of the earth seen from space (from 150 to 290 K depending on the percentage of the beam area over oceans and clouds). In a conventional repeater, the overall frequency band is down-converted by a local oscillator (L.O.) and mixer from the uplink band to the downlink band. It is channelized by an input multiplexer into a number (e.g., 12) of transponder channels. These channelized signals each are amplified by a separate high-power amplifier. Typically, a TWT amplifier is used with powers from a few watts to >200 W for a DBS. Solid-state amplifiers can now provide more than 15 W at C- and Ku-bands.

The *attitude determination and control system* (ADCS) must maintain the proper angular orientation of the satellite in its orbit, in order to keep the antennas pointed to the earth and the solar arrays aimed toward the sun (for example). The two prevalent stabilization methods are spin stabilization and body stabilization. In the former, the satellite body spins and the angular momentum maintains a gyroscopic stiffness. The latter uses momentum wheels to keep the spacecraft body orientation fixed. Components of this subsystem include the momentum wheels, torquers (which interact with the earth's magnetic field), gyros, sun and earth sensors, and thrusters to maintain orientation.

The *telemetry tracking and command* (TT&C) subsystem receives data from the ground and enables functions on the satellite to be activated by appropriate codes transmitted from the ground. This system operates with low data rates and requires omnidirectional antennas to maintain ground contact in the event the satellite loses its orientation.

The *power* subsystem comprises batteries and a solar array. The solar array must provide enough power to drive the communications electronics as well as the housekeeping functions, and it must also have enough capacity to charge the batteries which power the satellite during eclipse, i.e., when it is shadowed and receives no power from the sun. Typical battery technology uses nickel-hydrogen cells which can provide a power density of more than 50 W-h/kg. Silicon solar cells can yield more than 170 W/m^2 at a satellite's beginning of life (BOL). Gallium-arsenide solar cells (GaAs) yield more than 210 W/m^2. However, they are more expensive than silicon cells.

The space environment including radiation, thermal, and debris issues are described in Wertz and Larson [1991] and Griffin and French [1991]. The structure must support all the functional components and withstand the rigors of the launch environment. The thermal subsystem must control the radiation of heat to maintain a required operating temperature for critical electronics.

68.13 Trends

Satellite communications have approached a mature stage of development and their competitiveness for point-to-point voice traffic, compared with fiber, has been questioned. However, as mentioned, satellites will continue to exploit their unique wide view of the earth for such applications as broadcast and personal communications. The satellite industry's maturity also presents another challenge. To date, satellite construction has resembled a craft industry with extensive custom design, long lead times, long test programs, and high cost. Satellites will benefit from modern "lean production" and "design-to-cost" concepts that could lead to systems having lower cost per unit of capacity and higher reliability. Technology advances that are being pursued include development of light-weight "lightsats" for economical provision of services at low cost, more sophisticated on-board processing to improve interconnectivity, intersatellite links, improved components such as batteries, and even such speculative concepts as providing satellite power from the ground via high-power laser beams using adaptive optics [Landis and Westerlund, 1992].

Defining Terms

Attitude: The angular orientation of a satellite in its orbit, characterized by roll (R), pitch (P), and yaw (Y). The roll axis points in the direction of flight, the yaw axis points toward the earth's center, and pitch axis is perpendicular to the orbit plane such that $R \times P \rightarrow Y$.

Backoff: Amplifiers are not linear devices when operated near saturation. To reduce intermodulation products for multiple carriers, the drive signal is reduced or backed off. Input backoff is the dB difference between the input power required for saturation and that employed. Output backoff refers to the reduction in output power relative to saturation.

Bus: The satellite bus is the ensemble of all the subsystems that support the antennas and payload electronics. It includes subsystems for electrical power, attitude control, thermal control, TT&C, and structures,

Frequency reuse: A way to increase the effective bandwidth of a satellite system when available spectrum is limited. Dual polarizations and multiple beams pointing to different earth regions may utilize the same frequencies as long as, for example, the gain of one beam or polarization in the directions of the other beams or polarization (and vice versa) is low enough. Isolations of 27–35 dB are typical for reuse systems.

Polarization isolation: Frequency reuse allocates the same bands to several independent satellite transponder channels. The only way these signals can be kept separate is to isolate the antenna response for one reuse channel in the direction or polarization of another. The beam isolation is the coupling factor for each interfering path (ideally 0 or $-\infty$ dB).

References

W. S. Adams and L. Rider. "Circular polar constellations providing continuous single or multiple coverage above a specified latitude," *Journal of the Astronautical Sciences*, vol. 35, no. 2, pp. 155–192, April–June 1987.

K. Feher, *Digital Communications: Satellite/Earth Station Engineering*, Englewood Cliffs, N.J.: Prentice-Hall, 1983.

R. M. Gagliardi, *Satellite Communications,* New York: Van Nostrand Reinhold, 1991.

M. D. Griffin and J. R. French, *Space Vehicle Design,* Washington, D.C.: American Institute of Aeronautics and Astronautics, 1991.

G. A. Landis and L. H. Westerlund, Laser Beamed Power: Satellite Demonstration Applications, 43rd Congress of the International Astronautical Federation, Washington, D.C., August 28–September 5, 1992.

M. Long, *World Satellite Almanac,* Indianapolis: Howard W. Sams & Co., 1987.

M. Long, *The 1990 World Satellite Annual,* Winter Beach, Fla.: MLE, Inc. 1990.

K. Miya (Ed.), *Satellite Communications Technology,* Tokyo: KDD Engineering and Consulting, Inc., 1985.

W. L. Morgan and G. D. Gordon, *Communications Satellite Handbook,* New York: John Wiley & Sons, 1989.

J. J. Pocha, *An Introduction to Mission Design for Geostationary Satellites,* Dordrecht, The Netherlands: D. Reidel, 1987.

T. Pratt and C. W. Bostian, *Satellite Communications,* New York: John Wiley & Sons, 1986.

W. L. Pritchard and J. A. Sciulli, *Satellite Communications Systems Engineering,* Englewood Cliffs, N.J.: Prentice-Hall, 1986.

D. Rees, *Satellite Communications: The First Quarter Century of Service,* New York: John Wiley & Sons, 1990.

D. Roddy, *Satellite Communications,* Englewood Cliffs, N.J.: Prentice-Hall, 1989.

O. Shimbo, *Transmission Analysis in Communications Systems,* vol. 1 & 2, New York: Computer Science Press, 1988.

J. G. Walker, Continuous Whole-Earth Coverage by Circular Orbit Satellite Patterns, Technical Report 77044, Royal Aircraft Establishment, Farnborough, Hants, U.K., 1977.

J. R. Wertz and W. J. Larson (Eds.), *Space Mission Analysis and Design,* Dordrecht, The Netherlands: Kluwer Academic Publishers, 1991.

Further Information

For a brief history of satellite communications see *Satellite Communications: The First Quarter Century of Service,* by D. Rees (Wiley, 1990). Propagation issues are summarized in *Propagation Effects Handbook for Satellite Systems Design,* 1983, NASA Reference Publication 1082(03), November 1990; descriptions of the proposed LEO personal communications systems are in the FCC filings for *Iridium* (Motorola), *Globalstar* (SS/Loral), *Odyssey* (TRW), *Ellipso* (Ellipsat), and *Aries* (Constellation Communications), 1991 and 1992. For a discussion of the trends in satellite communications see *An Assessment of the Status and Trends in Satellite Communications 1986–2000,* NASA Technical Memorandum 88867, NASA Lewis Research Center, Cleveland Ohio, November 1986.

69

Personal and Office

William C. Y. Lee
PacTel Corporation

Rodger E. Ziemer
University of Colorado at Colorado Springs

Mil Ovan
Motorola, Inc.

69.1 Mobile Radio and Cellular Communications 1546
The Difference between Fixed-to-Fixed Radio Communication and Mobile Communication • Natural Problems in Mobile Radio Communications • Description of Mobile Radio Systems • Mobile Data Systems • Personal Communication Service Systems

69.2 Facsimile 1554
Scanning • Encoding • Modulation and Transmission • Demodulation and Decoding • Recoding • Personal Computer Facsimile

69.3 Wireless Local-Area Networks for the 1990s 1557
The Wireless In-Building Vision • Market Research • LAN Market Factors • Cabling Problems • User Requirements Environment • Product Requirements: End User Reaction • Technology Alternatives in Meeting Customer Requirements

69.1 Mobile Radio and Cellular Communications

William C. Y. Lee

The Difference between Fixed-to-Fixed Radio Communication and Mobile Communication

In fixed-to-fixed radio communications, the transmitter power, antenna location, antenna height, and antenna gain can be determined after calculating the link budget. Also, depending on the frequency range of the carrier affected on the atmospheric variation, a different "margin" value will be put in the budget calculation. The fixed-to-fixed radio links are usually 10 miles or longer and high above the ground. The signal variation over the link is due mostly to atmospheric changes. Satellite communications, microwave links, troposcatter, etc. are fixed-to-fixed radio communications. In mobile radio communications, the parameters such as transmitter power, antenna location, antenna height, and antenna gain are determined by covering an area or cell. In mobile radio communications, at least one end is in motion. The sizes of cells in urban and suburban areas are less than 10 miles. In mobile radio communications, the design of cell coverage is based on the average power. No "margin is applied in calculating the cell coverage."

Natural Problems in Mobile Radio Communications

In mobile radio communications, there are many problems which never occur in fixed-to-fixed radio communication system:

1. *Excessive pathloss:* Vehicles are referred to as mobile units. The antenna height of the mobile unit is very close to the ground. Therefore, the average signal strength received at the mobile unit has two components, a direct wave and a ground-reflected wave. These two waves act in canceling their average signal strengths and result in excessive pathloss at the receiver.
2. *Multipath fading:* Due to the human-made environment in which mobile units travel, the instantaneous signal sent from the base station is reflected back and forth from the buildings before arriving at the mobile unit and causes signal fading received in time domain. These signal fadings cause an increase in the bit error rate (BER) and in the degradation of voice quality.
3. *Human-made noise:* The antenna height of mobile units is usually low. Therefore, human-made industrial noise, automotive ignition noise, etc. are very easily received by the mobile unit. These noises will raise the noise floor and impact system performance.
4. *Dispersive medium:* Due to the human-made environment and the low antenna height of the mobile unit, the signal after bouncing back and forth from the human-made structures becomes multiple reflected waves which arrive at the mobile unit at different times. One impulse sent from the base station propagating through the medium becomes multiple reflected impulses received at different times at the mobile unit. This medium is called a dispersive medium. First the dispersive medium does not affect the analog voice channel, but does affect the data channels. Second, the medium becomes effective depending on the transmission symbol rate of the system. The dispersive medium will impact the reception performance when the transmission rate is over 20 kbps. Third, the dispersive medium becomes more effective in urban areas than in suburban areas.

Description of Mobile Radio Systems

There are two basic systems. One is called trunked systems and the other is called **cellular systems**.

Trunked Systems

A trunked system is assigned a channel over a number of available channels to a user. The user is never assigned to a fixed channel.

1. Specialized mobile radio (SMR) is a trunked system. The SMR operator is licensed by the FCC to a group of 10 or 20 channels within 14 MHz of the spectrum between 800 and 900 MHz.
   - Loading requirement: A minimum of 70 mobile units per channel is required. SMR can offer privacy, speedier channel access, and efficient services. It can serve up to 125–150 mobile units per channel.
   - Channel spacing: 25 kHz.
   - Channel allocation: FCC allocates a spectrum of either 500 kHz or 1 MHz to a SMR operator who will serve 10 or 20 paired transmit-receiver voice channels.
   - Coverage: Coverage is about 25 miles in radius since SMR uses only one high-power transmitting tower covering a large area.
   - Telephone interconnect: Public service telephone network (PSTN) extends mobile telephone service to SMR users.
   - Roaming: Mobile units are equipped with software that allows the radio to roam to any SMR system in the network.
   - No handoff: No tower-to-tower **handoff** capability, the channel frequency does not change as the unit moves from one cell to another.
2. ESMR (enhanced SMR): A system used to enhance the SMR system.
   - The SMR band.

- **TDMA** (time division multiple access) digital technology, the same digital TDMA standard adopted by the cellular industry.
- A network of low-power cells.
- Handoff, providing cell-to-cell handoffs through a centralized switching facility.
- A spectrum average of 7–8 MHz in each market. The spectrum is not continuous.
- A channel bandwidth of 25 kHz. Three time slots per channel.
- Modulation 16 QAM.
- No equalizer.

Cellular Systems

The cellular system [Lee, 1989] is a high-capacity system that uses the frequency reuse concept. The same frequency is used over and over again in different geographical locations. In large cities, the same frequency can be reused over 30 times.

Key Elements: There are several key elements in the cellular system.

- **Cochannel interference reduction factor** q (see Fig. 69.1):   Two cells using the same frequency channels are called cochannel cells. The required distance between two cochannel cells in order to receive the accepted voice quality is D_s, and the radius of the cell is R. Then the cochannel interference reduction factor q is

$$q = D_s/R$$

The value of q is different in different kinds of cellular systems such as analog, TDMA, and **CDMA** (code division multiple access).

- **Handoff:**   Handoff is a feature implemented in cellular systems to handoff a frequency of a cell while the mobile unit changes to another frequency of another cell while the vehicle is entering. The handoff is handled by the system and the user does not notice the handoff occurrences.

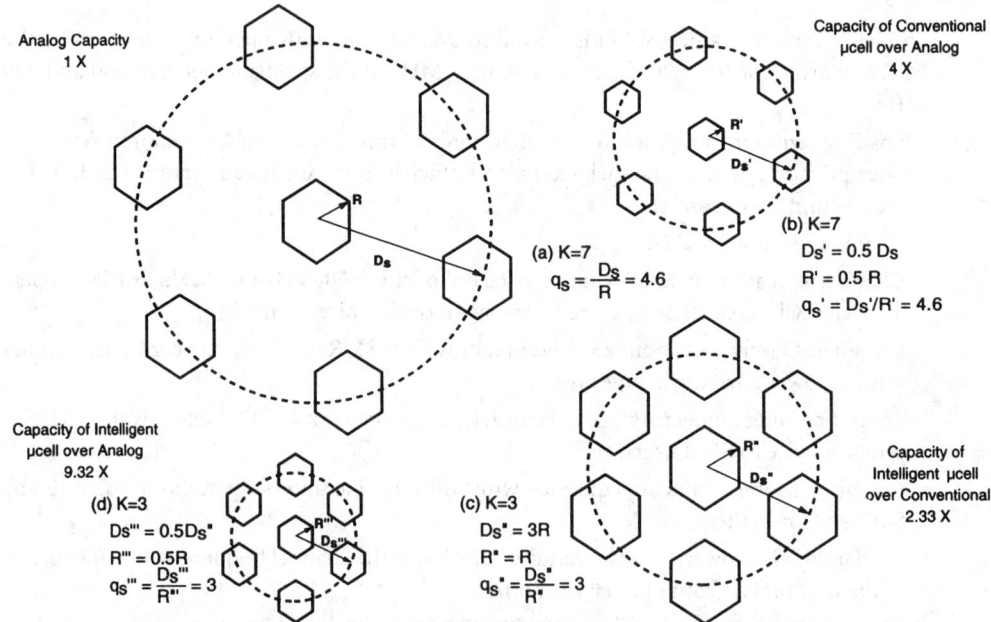

FIGURE 69.1   Four cases of expression of cochannel interference reduction factor.

- **Cell splitting:** When a cell provides a maximum of 60 radio channels and all are used during busy hours, the cell has to be split into smaller cells in order to provide more radio channels, normally reducing the cell by using a half radius. As a result a cell will be covered by four subcells. Each subcell provides 60 channels. The total area of an original cell will provide 240 radio channels which is four times higher in capacity as compared with the original cell capacity before splitting.

Spectrum Allocation in the United States, Europe, and Japan: In the United States there are 50 MHz allocated to cellular within 800–900 MHz. Based on duopoly, each city has two licensed operators. Each one operates on a 25-MHz band. There are two bands, Band A and Band B. Each band consists of 416 channels. The channel bandwidth is 30 kHz. Among 416 channels, 21 channels are used for setting up and 395 for voice channels.

- *Analog:* The frequency management of both Band A and Band B is shown in Table 69.1.
- *Digital:* There are two potential systems, TDMA and CDMA.

TDMA	
Bandwidth per channel	30 kHz
Time slots	3
Modulation	π/4-DQPSK
Speech coder	8 kbps—VCELP (vector code excited LPC)
Channel coding	1/2 convolutional (13 kbps)
Total transmit rate	48 kbps per channel
Equalizer	Up to 40 μs

CDMA	
Bandwidth per channel	1.23 MHz
Speech coder	8 kbps (max.)—a variable rate vocoder
Forward radio channels	Pilot (1) sync (1), paging (7), traffic channels (55), total 64 channels
Reverse radio channels	Access (9), traffic channels (55)
Power control	Forward, reverse
Diversity	Rake receiver

In Europe spectrum allocation is

Analog

	England	Scandinavia	West Germany
System	TACS	NMT	C450
Transmission frequency (kHz)			
Base station	935–960	463–467.5	461.3–465.74
Mobile station	890–915	453–457.5	451.3–455.74
Spacing between transmission and receiving frequencies (MHz)	45	10	10
Spacing between channels (kHz)	25	25	20
Number of channels	1000 (control channel 21 × 2); interleave used	180	222
Coverage radius (km)	2–20	1.8–40	5–30
Audio signal			
Type of modulation	FM	FM	FM
Frequency deviation (kHz)	±9.5	±5	±4
Control signal			
Type of modulation	FSK	FSK	FSK
Frequency deviation (kHz)	±6.4	±3.5	±2.5
Data transmission rate (kbps)	8	1.2	5.28
Message protection	Principle of majority decision is employed	Receiving steps are predetermined according to the content of the message	Message is sent again when an error is detected

GSM European Standard

GSM Characteristics

- TDMA: 8 slots/radio carrier
- 124 radio carriers (200 kHz/carrier) 935–960 MHz, 890–915 MHz
- GMSK modulation
- Slow frequency hopping (FH) (217 HOPS/s)
- Block and convolutional channel coding
- Synchronization (up to 233 μs absolute delay)
- Equalization (16 μs dispersion)
- TDMA structure: one frame (8 slots) 4.615 ms; each slot 0.557 ms
- Radio transmission rate: 270.833 kbps

GSM Physical Channels

- RACH: random-access control channel
- BCCH: broadcast common control channel (system parameters, sync.)
- PCH: paging channel
- SDCCH: stand-alone dedicated control channel (for transmit user's data)
- FACCH: fast associate control channel (for handoff)
- SACCH: slow associate control channel (for signaling)
- TCH: traffic channel
    Full rate: use full rate speech code
    Half rate

In Japan spectrum allocation is

Analog

System	NTT
Transmission frequency (kHz)	
Base station	870–885
Mobile station	925–940
Spacing between transmission and receiving frequencies (MHz)	55
Spacing between channels (kHz)	25
Number of channels	600
Coverage radius (km)	5 (urban area)
	10 (suburbs)
Audio signal	
Type of modulation	FM
Frequency deviation (kHz)	±5
Control signal	
Type of modulation	FSK
Frequency deviation (kHz)	±4.5
Data transmission rate (kbps)	0.3
Message protection	Transmitted signal is checked when it is sent back to the sender by the receiver

Digital

System	PHP (Japan)
Frequency band	1.9 GHz
Access method	TDMA/TDD (MC)
Traffic channels/RF carrier	1 (or 8 channels at half rate)
Modulation	π/4-QPSK
Voice codec	32 kbit/s ADPCM
Output power (PS)	10 mW
Radio transmission rate	384 kpbs
Carrier spacing	300 kHz

Table 69.1   New Frequency Management (Full Spectrum)

Block A

1A	2A	3A	4A	5A	6A	7A	1B	2B	3B	4B	5B	6B	7B	1C	2C	3C	4C	5C	6C	7C
1	2	3	4	5	6	7	8	9	10	11	12	13	14	15	16	17	18	19	20	21
22	23	24	25	26	27	28	29	30	31	32	33	34	35	36	37	38	39	40	41	42
43	44	45	46	47	48	49	50	51	52	53	54	55	56	57	58	59	60	61	62	63
64	65	66	67	68	69	70	71	72	73	74	75	76	77	78	79	80	81	82	83	84
85	86	87	88	89	90	91	92	93	94	95	96	97	98	99	100	101	102	103	104	105
106	107	108	109	110	111	112	113	114	115	116	117	118	119	120	121	122	123	124	125	126
127	128	129	130	131	132	133	134	135	136	137	138	139	140	141	142	143	144	145	146	147
148	149	150	151	152	153	154	155	156	157	158	159	160	161	162	163	164	165	166	167	168
169	170	171	172	173	174	175	176	177	178	179	180	181	182	183	184	185	186	187	188	189
190	191	192	193	194	195	196	197	198	199	200	201	202	203	204	205	206	207	208	209	210
211	212	213	214	215	216	217	218	219	220	221	222	223	224	225	226	227	228	229	230	231
232	233	234	235	236	237	238	239	240	241	242	243	244	245	246	247	248	249	250	251	252
253	254	255	256	257	258	259	260	261	262	263	264	265	266	267	268	269	270	271	272	273
274	275	276	277	278	279	280	281	282	283	284	285	286	287	288	289	290	291	292	293	294
295	296	297	298	299	300	301	302	303	304	305	306	307	308	309	310	311	312	667	668	669
670	671	672	673	674	675	676	677	678	679	680	681	682	683	684	685	686	687	688	689	690
691	692	693	694	695	696	697	698	699	700	701	702	703	704	705	706	707	708	709	710	711
712	713	714	715	716	X	X	X	X	991	992	993	994	995	996	997	998	999	1000	1001	1002
1003	1004	1005	1006	1007	1008	1009	1010	1011	1012	1013	1014	1015	1016	1017	1018	1019	1020	1021	1022	1023
313*	**314**	**315**	**316**	**317**	**318**	**319**	**320**	**321**	**322**	**323**	**324**	**325**	**326**	**327**	**328**	**329**	**330**	**331**	**332**	**333**

Block B

1A	2A	3A	4A	5A	6A	7A	1B	2B	3B	4B	5B	6B	7B	1C	2C	3C	4C	5C	6C	7C
334*	**335**	**336**	**337**	**338**	**339**	**340**	**341**	**342**	**343**	**344**	**345**	**346**	**347**	**348**	**349**	**350**	**351**	**352**	**353**	**354**
355	356	357	358	359	360	361	362	363	364	365	366	367	368	369	370	371	372	373	374	375
376	377	378	379	380	381	382	383	384	385	386	387	388	389	390	391	392	393	394	395	396
397	398	399	400	401	402	403	404	405	406	407	408	409	410	411	412	413	414	415	416	417
418	419	420	421	422	423	424	425	426	427	428	429	430	431	432	433	434	435	436	437	438
439	440	441	442	443	444	445	446	447	448	449	450	451	452	453	454	455	456	457	458	459
460	461	462	463	464	465	466	467	468	469	470	471	472	473	474	475	476	477	478	479	480
481	482	483	484	485	486	487	488	489	490	491	492	493	494	495	496	497	498	499	500	501
502	503	504	505	506	507	508	509	510	511	512	513	514	515	516	517	518	519	520	521	522
523	524	525	526	527	528	529	530	531	532	533	534	535	536	537	538	539	540	541	542	543
544	545	546	547	548	549	550	551	552	553	554	555	556	557	558	559	560	561	562	563	564
565	566	567	568	569	570	571	572	573	574	575	576	577	578	579	580	581	582	583	584	585
586	587	588	589	590	591	592	593	594	595	596	597	598	599	600	601	602	603	604	605	606
607	608	609	610	611	612	613	614	615	616	617	618	619	620	621	622	623	624	625	626	627
628	629	630	631	632	633	634	635	636	637	638	639	640	641	642	643	644	645	646	647	648
649	650	651	652	653	654	655	656	657	658	659	660	661	662	663	664	665	666	X	X	X
X	X	X	X	X	717	718	719	720	721	722	723	724	725	726	727	728	729	730	731	732
733	734	735	736	737	738	739	740	741	742	743	744	745	746	747	748	749	750	751	752	753
754	755	756	757	758	759	760	761	762	763	764	765	766	767	768	769	770	771	772	773	774
775	776	777	778	779	780	781	782	783	784	785	786	787	788	789	790	791	792	793	794	795
796	797	798	799																	

*Boldface numbers indicate 21 control channels for Block A and Block B, respectively.

Mobile Data Systems

The design aspect of developing a mobile data system is different from that of developing a cellular voice system, although the mobile radio environment is the same. The quality of a voice channel has to be determined based on a subjective test. The quality of a data transmission is based on an objective test. In a data transmission, the bit error rate and the word error rate are the parameters to be used to measure the performance at any given carrier-to-interference ratio (C/I). The burst errors caused by the multipath fading and the intersymbol interference caused by the time delay spread are the major concerns in receiving the mobile data. The burst errors can be reduced by interleaving and coding. The intersymbol interference can be reduced by using equalizers or lowering the transmission rate or applying diversity.

The wireless data transmission can be sent via a circuit switched network or a packet switched network. Also, the mobile data transmission can be implemented on cellular systems or on a stand-alone system.

ARDIS	
Transmission rate	4.8 kbps and 19.2 kbps
Transmit power	1 W
Channel	Packet radio
Vendors	IBM/Motorola

RAM	
Transmission rate	8 kbps
Transmit power	4 W
Channel	Packet radio
Vendor	Ericsson

Cellular Plan II	
Transmission rate	19.2 kbps
Transmit power	0.6–1.2 W
Channel	Packet cellular
Vendor	IBM/PCSI

Cellular Modems	
Transmission rate	38.4 kbps
Transmission	3 W
Channel	Circuit cellular, carry data over cellular voice channels
Modem vendor	AT&T, PowerTek, Vital

Personal Communication Service Systems

In June 1990, the FCC started to ask the wireless communication industry to study the development of future personal communication service (PCS) systems. The PCS systems need to have more capacity. The technologies of increasing the capacity apply to CDMA (code division multiple access) and the new microcell system.

CDMA

A San Diego trial held in 1991 showed that a cellular CDMA scheme can provide higher capacity than cellular TDMA (time division multiple access). A cellular CDMA system [Lee, May 1991] does not need to apply a frequency reuse scheme. All the cellular CDMA cells share the same radio channel. Therefore, the capacity of a cellular CDMA system is higher than either cellular **FDMA** (frequency division multiple access) or cellular TDMA systems.

Assume that a spectral bandwidth of 1.2 MHz can be divided into 120 radio channels with a channel bandwidth of 10 kHz. This is an FDMA scheme. A spectral bandwidth of 1.2 MHz can also be divided into 40 radio channels with a radio channel bandwidth of 30 kHz but each radio channel carries three time slots. Therefore, a total of 120 time-slot channels is obtained. This is a TDMA scheme. A spectral bandwidth of 1.2 MHz can also be used as one radio channel but provide 40 code-sequence traffic channels for each sector of a cell. A cell of three sectors will have a total of 120 traffic channels. This is a CDMA scheme. Now we can visualize that as far as channel efficiency is concerned, TDMA, FDMA, and CDMA provide the same number of traffic channels. However, in FDMA or TDMA, the frequency reuse has to be applied. Let the **frequency reuse factor** $K = 7$ maintain a required C/I $\geq$ 18 dB; then the total channels will be divided by 7 as:

$$\frac{120}{7} = 17 \text{ channels/cell (in TDMA or FDMA)}$$

In CDMA no frequency reuse is needed. Therefore, every cell can have the same 120 channels: 120 channels/cell (in CDMA). In cellular, because the factor of frequency reuse is applied on FDMA and TDMA schemes but not on CDMA, therefore, cellular CDMA has a greater spectrum efficiency than cellular FDMA or TDMA [Lee, May 1991].

New Microcell System

The conventional microcell system [Lee, Nov. 1991, 1993] reduces the transmit power and makes a cell less than 1 km in radius. The concept of using cell splitting is to increase capacity. Furthermore, the new microcell system is to make a microcell intelligently. The conventional microcell does not have the intelligence to know where the mobile or portable units are located within the cell. Therefore, the cell site has to cover the signal strength over the whole cell or whole sector. The more unnecessary signal power transmitted, the more interference will be caused in the system and less capacity will be achieved. In this new intelligent microcell system, each cell is an intelligent cell. In a new microcell, there are three or more zones. The cell will know which zone a particular mobile unit is in. Then a small amount of power will be needed to deliver to that zone. The cochannel interference reduction factor (CIRF) now will be measured from two cochannel zones instead of two cochannel cells. Then the two cochannel cells can be located much closer. In this new microcell system, the frequency reuse factor K becomes $K = 3$. As compared to the conventional microcell $K = 7$, the new microcell system has a capacity increase of 2.33 (= 7/3) times. These two techniques can be used in buildings and outside buildings.

Defining Terms

CDMA: A multiple access scheme by using code sequences as traffic channels in a comom radio channel.

Cell splitting: A method of increasing capacity by reducing the size of the cell.

Cochannel interference reduction factor (CIRF): A key factor used to design a cellular system to avoid the cochannel interference.

FDMA: A multiple access scheme by dividing an allocated spectrum into different radio channels.

Frequency reuse factor (K): A number based on frequency reuse to determine how many channels per cell.

Handoff: A frequency channel will be changed to a new frequency channel as the vehicle moves from one cell to another cell without the user's intervention.

Mobile cellular systems: A high-capacity system operating at 800–900 MHz using a frequency reuse scheme for vehicle and portable telephone communications.

TDMA: A multiple access scheme by dividing a radio channel into many time slots where each slot carries a traffic channel.

References

W. C. Y. Lee, *Mobile Cellular Telecommunication Systems,* New York: McGraw Hill, 1989.

W. C. Y. Lee, "Overview cellular CDMA," *IEEE Trans. on Veh. Tech.,* vol. 40, pp. 290–302, May 1991.

W. C. Y. Lee, "Microcell architecture—Smaller cells for greater performance," *IEEE Magazine,* vol. 29, pp. 19–23, Nov. 1991.

W. C. Y. Lee, *Mobile Communications Design Fundamentals,* 2nd ed., New York: Wiley, 1993.

69.2 Facsimile

Rodger E. Ziemer

Facsimile combines copying with data transmission to produce an image of a **subject copy** at another location, either nearby or distant. Although the Latin phrase *fac simile* means to "make similar," the compressed phrase facsimile has been taken to mean "exact copy of a transmission" since 1815 [Quinn, 1989]. The image of the subject copy is referred to as a *facsimile copy*, or **record copy**. Often the abbreviated reference "fax" is used in place of the longer term *facsimile*.

Facsimile was invented by Alexander Bain in 1842; Bain's system used a synchronized pendulum arrangement to send a facsimile of dot patterns and record them on electrosensitive paper. Over the years, much technological development has taken place to make facsimile a practical and affordable document transmission process. An equally important role in the wide acceptance of facsimile for image transmission has been the adoption of standards by the Consultative Committee on International Telephone and Telegraph (CCITT). The advent of a nationwide dial telephone network in the 1960s provided impetus for the rebirth of facsimile after television put the damper on early facsimile use. Group 2 fax machines which appeared in the mid-1970s were capable of transmitting a page within a couple of minutes. These machines, based on analog transmission methods, were developed by Graphic Sciences and 3M. The Group 3 fax machines, developed in the mid-1970s by the Japanese, are based on digital transmission technology and are capable of transmitting a page in 20 seconds or less. They can automatically switch to an analog mode to communicate with the older Group 1 and 2 fax machines. Group 4 fax units offer the highest resolution at the fastest rates but rely on digital telephone lines which are just now becoming widely available [Quinn, 1989]. Group 3 facsimile will be featured in the remainder of this article. Group 3 facsimile refers to apparatus which is capable of transmitting an 8.5 × 11-inch page over telephone-type circuits in one minute or less. Detailed standards for Group 3 equipment may be found in Recommendation T.4 of CCITT, Vol. VII.

Facsimile transmission involves the separate processes of *scanning, encoding, modulation, transmission, demodulation, decoding,* and *recording* [Stamps, 1982]. Each of these will be described in greater detail below.

Scanning

Before transmission of the facsimile signal, the subject copy must be **scanned**. This involves the sensing of the diffuse reflectances of light from the elemental areas making up the subject copy. For CCITT Group 3 high-resolution facsimile, these elemental areas are rectangles 1/208 inch wide by 1/196 inch high. The signal corresponding to an elemental area is called a **pixel** which stands for picture element. For pixels that can assume only one of two possible states (i.e., white on black or vice versa), the term used is a **pel**. Various arrangements of illuminating sources, light-sensing transducers, and mechanical scanning methods can be employed [Stamps, 1982]. For more than six sweeps per second across the subject copy, electronic scanning utilizing a cathode-ray tube or photosensitive arrays or laser sources with polygon mirrors are utilized. A photosensitive array arrangement for scanning a flood-illuminated subject copy is illustrated in Fig. 69.2. This is the most often encountered scanning mechanism for modern facsimile scanners, and the sensors are typically silicon photosensitive devices. Two photosensor arrays in common use are photodiode arrays and charge-coupled device linear image sensors. For digital facsimile, the array is composed of 1728 sensors in a row 1.02 inches long with the optics designed so that an 8.5 inch subject copy can be scanned.

Encoding

The output of the photosensor array for one scan or row of the subject copy consists of 1728 pels (1s or 0s) since Group 3 facsimile recognizes only black or white. Typically, facsimile subject copy is

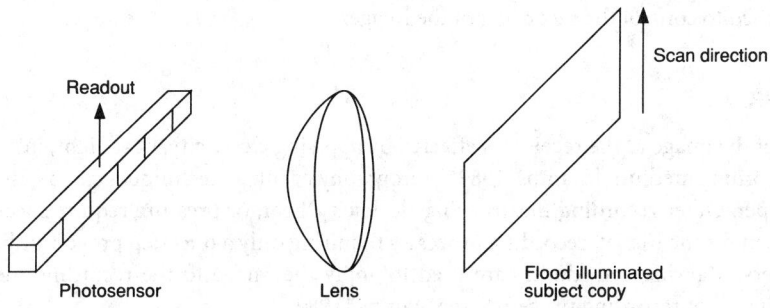

FIGURE 69.2 Arrangement for scanning by means of a linear photosensitive array.

85% white. The data from scanning the subject copy is reduced through **run-length encoding**. In the encoding process, it is assumed that a white pel (0) always occurs first. A white run is the number of 0s until the first 1 is encountered (the run length is 0 if the first pel is a 1); after a white run, a black run must follow with length equal to the number of 1s until the first 0 is encountered. All possible run lengths of white and black are then encoded into a binary code using a modified Huffman encoding technique [Jayant and Noll, 1984]. On the average, fewer binary symbols are needed to encode run lengths of the subject copy than if the binary values of the pels themselves were transmitted. For Group 3 facsimile, compression is optionally extended to the vertical dimension through employment of a READ (relative element address designate) code.

Run lengths from 0 to 63 are encoded by *terminating codes*, and run lengths in equal multiples of 64 from 64 to 1728 are encoded by *makeup codes*. Thus any run length up to 1728 can be described by a makeup code plus an appropriate terminating code. Additional makeup codewords are available for equipment that accommodates wider paper while maintaining the same resolution [Stamps, 1982]. Tables of modified Huffman run-length terminating and makeup codes are given in Stamps [1982].

Modulation and Transmission

Transmission of the encoded facsimile signal makes use of modem signaling techniques based on CCITT recommendations V.27 (standard) and V.29 (optional addition). The former utilizes 8-phase modulation at 4800 bits per second (bps), and the latter employs 16-QAM (quadrature amplitude modulation) at 9600 bps with adaptive, linear equalization. A facsimile telephone call consists of five phases, labeled A through E [Stamps, 1982]. In phase A, the telephone call is placed, with a training sequence sent consisting of signals to establish carrier detection, AGC, timing synchronization, and adjust equalizer tap settings. Phase B consists of the called station responding with a confirmation to receive (CFR) signal. The response is a 300 bps binary coded frequency-shift keyed signal (1 = 1650 ± 6 Hz and 0 = 1850 ± 6 Hz), except for the equalizer training sequence which is at the fast data rate of the digital modem. In phase C the encoded facsimile image is transmitted. Phase D consists of the end-of-transmission signal consisting of six consecutive end-of-lines (EOLs), with receipt required from the receiver. If no more images are to be sent or received, phase E (going on-hook) is effected at both terminals.

Demodulation and Decoding

Demodulation consists of the inverse of the modulation process. Standard techniques are used to demodulate the phase-modulated or QAM signals. Also included in the demodulation process is equalization. The decoding process converts the run-length encoded information to a series of 1s and 0s corresponding to the black and white pels of the image. The demodulated and decoded signal is then used to control the recording of the image.

Recording

Recording of the image at the receiver is effected by applying electricity, heat, light, ink jet, or pressure to a recording medium [Stamps, 1982]. Xerography or ink jet techniques can be used to record on plain paper. Other recording means using electricity, heat, or pressure require specially coated papers. Except for the ink jet, recording processes requiring only a one-step process utilize specially coated papers. Marking transducers are used to apply the image to the recording medium. For more description of recording processes, see Stamps [1982].

Personal Computer Facsimile

Whereas character-oriented text is readily transmitted between personal computers by means of teletex or computer mail, facsimile transmission in conjunction with personal computers extends this capability to images [Hayashi and Motegi, 1989].

Defining Terms

Facsimile: The process of making an exact copy of a document through scanning of the subject copy, electronic transmission of the resultant signals modulated by the subject copy, and making a record copy at a remote location.

Pel: A picture element which has been encoded as black or white, with no gray scale in between.

Pixel: A picture element of a subject or record copy that is represented in shades of gray.

Record copy: The copy of the document made at the receiving end of a facsimile system.

Run-length encoding: The assignment of a codeword to each possible run of 0s (white pel sequence) or run of 1s (black pel sequence) in a scan of the subject copy.

Scanning: The process of scanning the subject copy in a facsimile transmission from left to right and from top to bottom.

Subject copy: The document that is scanned and transmitted in a facsimile system.

References

K. Hayashi and C. Motegi, "Personal computer image communications using facsimile," *IEEE Journal on Selected Areas in Communications,* vol. 7, pp. 276–282, Feb. 1989.

N. S. Jayant and P. Noll, *Digital Coding of Waveforms,* Englewood Cliffs, N.J.: Prentice Hall, 1984, chap. 10.

G. V. Quinn, *The FAX Handbook,* Blue Ridge Summit, Pa.: Tab Books, 1989.

G. M. Stamps, "Facsimile systems," in *Electronic Engineers' Handbook,* D. G. Fink, Ed., New York: McGraw-Hill, 1982, pp. 20-87–20-107.

Further Information

D. Cannon and G. Luecke, *Understanding Communications Systems,* Dallas: Texas Instruments Incorporated, 1980, chap. 9.

C. Chamzas and D. L. Duttweiler, "Encoding facsimilie images for packet-switched networks," *IEEE Journal on Selected Areas in Communications,* vol. 7, pp. 857–864, June 1989.

D. M. Costigan, *Electronic Delivery of Documents and Graphics,* New York: Van Nostrand Reinhold, 1978.

K. McConnell, D. Bodson, and R. Schaphorst, *FAX: Digital Facsimilie Technology and Applications,* Boston: Artech House, 1992.

Y. Yasuda, Y. Yamazaki, T. Kamae, and K. Kobayashi, "Advances in FAX," *IEEE Proceedings,* vol. 73, pp. 706–730, April 1985.

69.3 Wireless Local-Area Networks for the 1990s

Mil Ovan

Wireless local-area networks (LANs) represent a new form of communications among personal computers inside buildings. To better understand its applicability, this paper defines the customer challenges in networking personal computers as well as specific product requirements for a wireless LAN. These insights were gained through market studies of over 1000 corporate and government entities surveyed through different marketing research techniques.

The Wireless In-Building Vision

To date, the evolution of wireless communications has been exemplified by the dramatic growth in **cellular communications**. Cellular has enabled customers to transcend the constraints of fixed telephony in communicating outside of buildings with portable and now personal communications devices.

There has been significant interest and publicity regarding wireless in-building communications lately, both for data and voice. Throughout the 1980s, we have seen the development of a significant range of in-building business communications problems that have been caused by changes in the technological, business, and regulatory environments. Because of these developments, buyers of telecommunications and data communications systems increasingly are having to face significant time, cost, and logistical problems associated with the installation, movement, and management of computing and communications equipment in dynamic office environments.

Over the next 20 years, society will witness a significant "wireless evolution" in both personal and professional communications and change the way we conduct our lives at home, on the road, and at work (see Fig. 69.3). New forms of wireless communications will free us from the "bonds" of wire that today restrict our movements or interaction.

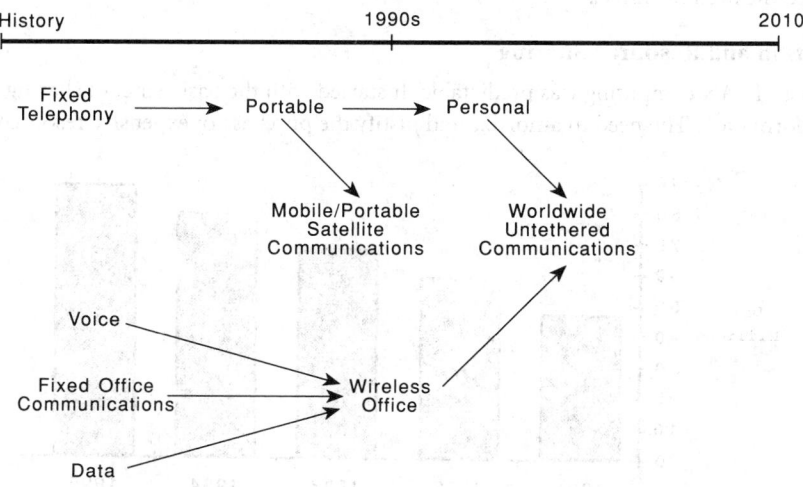

FIGURE 69.3 Evolution of wireless communications. (Courtesy of Motorola, Inc.)

Market Research

Beginning in the middle to late 1980s, a systematic evaluation of the technological and environmental attributes necessary to anticipate and define wireless in-building communications was undertaken. This included a comprehensive marketing needs assessment and research program. The overriding objective was to anticipate and identify customer needs and trends; that is, "What are the specific needs of various customer groups, and what type of product attributes will satisfy their needs?"

To determine answers to these and a whole host of other questions, a multiphased marketing research program was conducted. The overall aim of the program was to anticipate and ascertain the customer need, where this need existed currently, what were the market and customer environmental characteristics, and what product characteristics would be needed to provide an optimal wireless solution.

The remainder of this paper describes a higher-level overview of the results from these market research phases. This includes an overview of market needs, the problems/difficulties with current cabling methods, and a description of market requirements.

LAN Market Factors

Personal Computer Explosion

The move from mainframe and central information processing of the 1960s and 1970s provided an opportunity for minicomputers to enter the market. The minicomputer provided greater computer and applications access by employees. Throughout the 1980s the move to more intelligent desktop devices like personal computers was just that—personal. Organizations, in an effort to empower the worker, provided all types of applications, software, and hardware to the worker. The decremental costs of technology facilitated the distribution of personal computers. More importantly, projections state that business personal computer growth will continue its aggressive pace (see Fig. 69.4).

However, the growth of decentralized storage and computing created yet another problem—work groups needed to share information—and much of this information resided in individual hard disks. Furthermore, despite the declining costs of personal computers and associated technology, it was and still is considerably expensive to "fully load" the workforce with all of the applications it needs. The ability to share applications became desirable. It was these two trends which highlighted the need for LANs.

Information and Resource Sharing

The success of LAN computing was predictable. It started with the basic tenet of sharing resources and/or information. The need to amortize and justify the purchase of expensive resources, such as

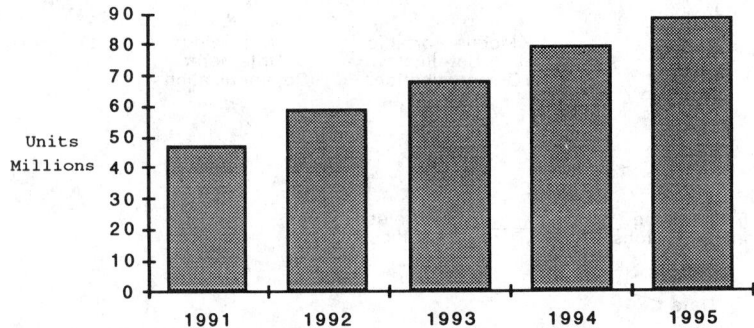

FIGURE 69.4 Worldwide business personal computer-installed base. (*Source:* International Data Corporation.)

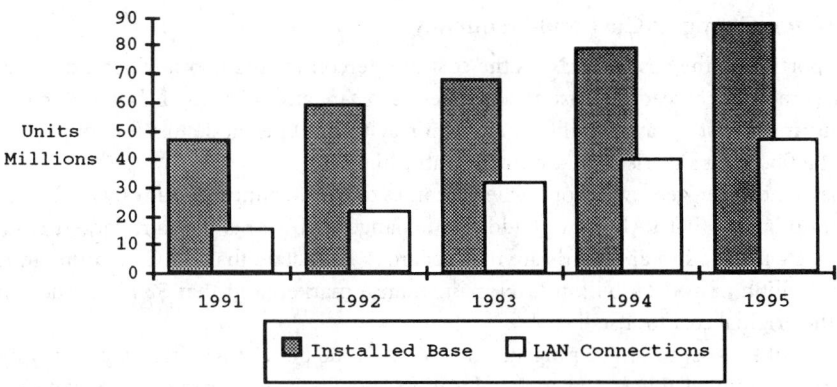

FIGURE 69.5   Worldwide business PCs and those PCs that are LAN connected. (*Source: International Data Corporation.*)

printers and storage, was an obvious factor which supported LAN growth. The need for knowledge workers to exchange data was and is imperative. Furthermore, the ability to share applications supported the growth of network computing.

LAN Growth

The success of LANs throughout the 1980s has been phenomenal. However, the projected growth throughout the 1990s is equally as impressive (see Fig. 69.5). This can be attributed not only to new installations of LANs, but also to the physical and logical segmentation of LANs as traffic and throughput degradations are observed.

Moves/Adds/Changes and Increasing Mobility

The world economies will continue to develop interdependencies and, likewise, global competition. The increasing competitive environment will demand greater worker mobility, changing assignments and reassignments, changing work groups, and mission mobility. The demand to have information how we want it, when we want it, and where we want it will be a strategic and competitive weapon. The need to improve efficiency and the growing need for information will accelerate the adoption of wireless communications.

Today's wired network, for all its great strides, is very restrictive. The cost to deploy and redeploy personnel and workgroups is time consuming and expensive. Cabling in today's environment inhibits the ability to attain efficiency and competitive advantages. The next section will highlight some of the author's market research findings.

Cabling Problems

As each phase of market investigation was conducted, several problems with today's wired networks were uncovered. Whether **copper twisted pair, coax, or optical fiber,** hard wiring for telecommunications and data communications systems within a building environment is expensive and troublesome to install, maintain, and, especially, change. Beneath today's increasingly dense office electronic environment lies a tangled, confusing, unmanageable maze of wiring.

What appeared to be very significant in the focus group research was how quickly the respondents stated the problems they have with wiring. Among the majority of respondents, the most favorable solution was to free themselves of all wiring. Therefore, their first choice solution would be a wireless system, minimizing the time and effort of implementing a move, add, or change.

Moves/Adds/Changes: Cost and Frequency

A major portion of the cost of LANs is the cost of interconnecting them, which experts acknowledge can sometimes exceed the cost of computer hardware and software. Labor and material costs for wiring are almost always significant and can reach $1000 per node just for copper wire. Coax and optical fiber, not surprisingly, are considerably higher.

The news, however, gets even worse when it comes to maintenance. A study by the Frost and Sullivan[1] group quotes that LAN moves, adds, and changes (MACs) is the third largest cost component for LAN installation and hardware maintenance. They state that MACs account annually for almost $2 billion of a $12.2 billion LAN maintenance market, and that $2 billion does not even include the original cost to install cable.

Estimates of the cost to rewire range from $200 to $1000 per change. In fact, a survey by KPMG Peat Marwick[2] quoted that the average relocation cost for just rewiring a LAN station averages $300 per node. But those are just the direct costs; the time to effect the wired change is a significant problem as well. Moves, for example, often take weeks or longer to coordinate in addition to the time to actually make the physical wiring change.

Most of the research respondents were asked what proportion of their company's staff was involved in some kind of a move involving wiring or rewiring. The majority of the respondents, almost 80%, had some type of relocation or addition of personnel over the last year surveyed. Their responses ranged from as few as 20% per year up to as much as 200% annually. Furthermore, according to the KPMG Peat Marwick study, the average company moves its employees approximately 50% annually. Telecommunications consultant Richard Kuehn states that data terminals are moved as often as 1.5 to 3 times per year. The combined problems of the actual hard relocation costs, however, are just the beginning. Soft, or hidden, costs further exacerbate the cabling dilemma.

Hidden Costs

Significant problems arise when these moves or changes are implemented. There is always the disruption of the workers involved in the move or change, not to mention the loss in productivity. The problems, however, become much more involved when dealing with whole departments and more complex user equipment. In fact, surveyed firms responded that when a relocation takes place, over 60% of the time it involves the movement of an entire department.

The toll of wait time and down time on productivity varies greatly and is difficult to quantify, but certainly is significant and costly. In today's increasingly mobile working environment, it is likely to grow. The situation is exacerbated by relocations and additions which require reconstructions, thereby continuing to add to the effective cost of a move, add, or change.

Costs to rewire rise enormously with the age and complexity of the building. The majority of high-rise office space in large metropolitan areas presents major problems and expense for tenants trying to install, add, or move network wiring. Buildings more than 30 or 40 year old, with designs and construction that did not consider today's electronic office, poorly accommodate communications wiring. If asbestos insulation exists in the building, as it does even in many pre–health-safety regulated buildings, rewiring costs can take on huge proportions.

The coordination of personnel and the moving of one group out to prepare for the new group moving in is a very costly and labor intensive ordeal. In some cases, wiring had to be installed, or different cabling may have been needed to accommodate new or different types of users' equipment.

Cable Is Not Business Friendly

Although office planners, building managers, and network operators are well aware of the problems with wire, the limitations and huge costs of wire have not generated focused attention outside of this community. The general business world seems to accept wire as inevitable. Perhaps that is

---

[1] *PC Week Magazine*, "Maintenance Costs of LANs Keep Soaring," Frost & Sullivan, Inc.

[2] KPMG Peat Marwick Study, January 1991.

because there have been no real alternatives. Yet, as computing and telecommunications power continues to proliferate and becomes more widely distributed to the "knowledge worker," the problem will increase. Easy, quick, efficient movement of "people assets" within the working environment is also increasingly being recognized as essential to the productivity and competitiveness of a business. Wiring severely inhibits that movement.

The research indicated a need for a flexible, compatible, cost-effective, yet high-performance wireless alternative to extend and complement, if not replace, the capabilities of wire, cable, and fiber for in-building communications networks. More specifically, it is the convenience and flexibility that users need. In fact, the aggregate need for flexibility and convenience was found to be twice that of the perceived benefit for cost savings.

When research respondents were asked how they could improve upon their experiences when implementing a move, add, or change, many solutions were offered. These solutions ranged from having more compatibility among different vendors' equipment, to providing a better way to organize all the different cabling.

Structured Distribution Systems

A number of firms in the research study had deployed a **structured distribution system (SDS)**, a topology which advocates cabling saturation of a desired environment to accommodate all potential personnel movements and reconstructions within that office. SDS requires firms to invest large sums of capital initially on the assumption of not knowing how many telecommunications devices may be employed or where the devices are to be located. Consequently SDS usually plans for worst-case conditions, meaning that some or much of wiring systems capability may never be utilized.

However, many firms which have an SDS deployed also expressed those problems which stress their SDS investment. Some of the most frequently mentioned include:

- High equipment addition/relocations exceeding 40% annually
- Expansion and contraction of their workforce
- Changing technology and business support
- Continued investment and vigilance to maintaining the SDS and its intrinsic advantage
- Continued departmental LAN growth requirements

In short, the latter group of SDS respondents provided some notable requirements. A wireless system must:

- Extend the capabilities of their SDS system
- Facilitate the inherent advantages of the SDS
- Offer enhanced flexibility to nonserviced SDS portions of their building or occupancy

These points indicate that even in SDS environments, there is an opportunity to employ wireless devices. Wiring—the expense, time, and inflexibility of installing, moving, and changing—limits the way companies can productively use networks. To stay productive, these LANs have to move and change with the workforce they support. Therefore, a wireless offering must be a *complementary* solution for buildings with an SDS, in bringing wireless flexibility and extensibility to today's networks.

User Requirements Environment

Office Friendly

Several notable conclusions were derived from the marketing research. **Secondary market research** suggests that over 70% of LAN node installations were estimated to reside in an office environment (as opposed to factories and warehouses). Therefore, as an office-oriented offering, a wireless system would have to be, by definition, office friendly. A traditional office is composed of hard offices with opening and closing doors, furniture and personnel movement, cubicles, conference rooms, and walls of varying thickness and substance. Therefore, a wireless system must continually adapt to different and changing conditions and office layouts.

Optimized Service Area

The second wireless in-building need expressed by the office market is the manageability and reuse of any potential system. Unlike the signal propagation characteristics of many lower-frequency radio products, LAN administrators desired the ability to control or, more aptly, contain the coverage of a potential wireless system. The reasons were twofold:

- LAN managers wanted the ability to add different services to a new group of users. In fact these new users may very well be physically adjacent to another system, wireless or wired.
- These same managers wanted the flexibility to connect a new or existing user group to either a **backbone** or to create a stand-alone LAN.

LAN Workgroup Sizes

Respondents were asked as to where a wireless offering might be first installed. The market research indicated that approximately 70% of the installations would contain less than 30 users (see Fig. 69.6). Furthermore, the average LAN appeared to be in the 12- to-15-node range. This is further corroborated by the KPMG Peat Marwick study which found that the average LAN size is about 15 users per LAN.

Furthermore, the system must have the flexibility to manage the service area. That ability, to either incrementally add systems whether on the backbone or in a stand-alone configuration, must accommodate scalability within an organization.

It is interesting to note that these figures are consistent with good LAN administration practices for purposes of maintaining high throughput and fault isolation. As LANs become larger and traffic more intensive, there is a natural inclination to begin segmenting LANs into more logical and defined user areas/groups.

Coverage Area

To satisfy the majority of requirements, we determined that approximately 70% of LANs would be deployed in areas of less than 5000 ft^2 (see Fig. 69.7). This must take into consideration the fairly dense environment, made up of cubicles and apportioned hallway space. The market investigations indicated that a wireless offering must accommodate, at the least, 150 ft^2 per user. This is equivalent to 32 users/system in a 5,000 ft^2 area.

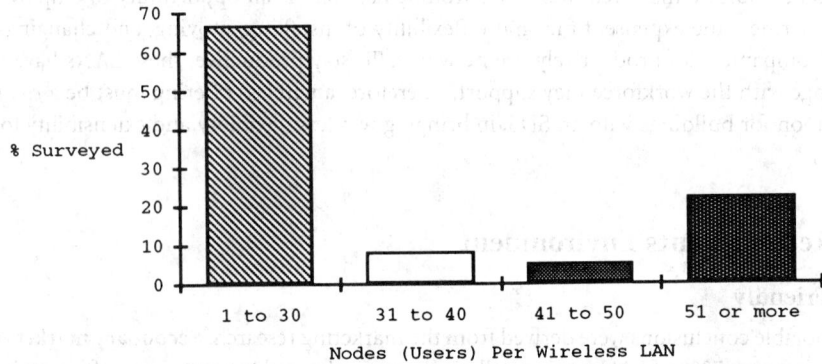

FIGURE 69.6   Survey results: forecast users per wireless LAN. (Courtesy of Motorola, Inc.)

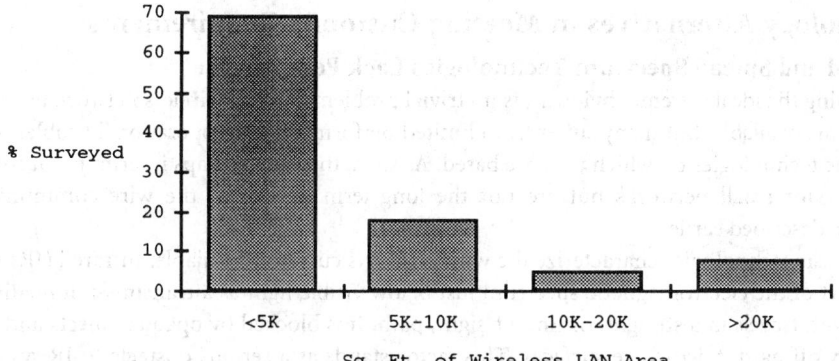

FIGURE 69.7 Survey results: forecast office area of wireless LAN. (Courtesy of Motorola, Inc.)

Product Requirements: End User Reaction

Transparency, Compatibility, and Performance

To justify the expense of a wireless system to end users, a wireless offering would have to provide reliable performance, as well as be practical and cost effective. Our market research indicated that the ideal system should be:

- Easy both to install and move, preferably by the user
- Able to coexist with both existing wire and cable, as well as with future optical fiber
- Easy to operate, virtually transparent to the user
- Almost universally applicable, suitable to replace any LAN cable or wire, in any office environment
- Secure, absolutely reliable, and cost effective

A wireless system must be totally transparent. If customers are to enjoy the attributes of wireless, the respondents indicated that the wireless implementation must not require the users to change the way they operate or interface with their personal computers. Also, a wireless offering must provide true compatibility. The wireless connection must be compatible with standards-based components such as operating systems and applications, LAN cards and other devices, as well as LAN wire that is already in place.

Security and Reliability

In addition, the market mandated that a wireless product offering be absolutely secure and reliable. Security was a requirement across several dimensions. To provide sufficient data security a wireless system should first prevent the effective capture of data by a receiver outside of the wireless system and, second, prevent capture of the data by unauthorized wireless hardware within the system. A wireless product must be secure from eavesdropping, either accidental or intentional.

Reliability was another important attribute. The users required absolute reliability. That is, users wanted a guaranteed packet delivery from the entry/exit wireline points—and they wanted it at least as error free as their current cabled environment.

Cost Effective

Finally, most businesses will place any capital or expense under rigorous financial analysis. As such, the acceptance of a new technology/application must pass the payback test for that business. Therefore, demonstrable payback and justification is needed to facilitate an organization's evaluation of any potential wireless offering.

Technology Alternatives in Meeting Customer Requirements

Infrared and Spread Spectrum Technologies Lack Performance

Developing the ideal system, obviously, is no trivial problem. Several wireless network products, to be sure, are available, but many suffer from limited performance and operational problems inherent to the technologies on which they are based. As such, they are perhaps interim point solutions, primarily for small networks but are not the long-term answer to the wire communications dilemma described earlier.

Two basic technologies characterize the wireless LANs currently available. Infrared (IR) systems use a part of the electromagnetic spectrum just below visible light as a transmission medium. IR, being light, travels in a straight, or line-of-sight, path. It is blocked by opaque objects and reflects well only off hard, mirror-like surfaces. This factor stands as a serious obstacle to IR systems for applications other than in open working environments.

Radio technologies form the platform for other wireless LAN products. Many of the radio LANs are based on spread spectrum technology. This technology, developed by the U.S. military, uses a combination of several small, narrow bands within a general region of this band as carrier frequencies. However, the commercial products operate with fewer frequencies than are available for military systems; hence interference rejection and performance are lower. (In radio transmission, the wider the bandwidth or available frequencies on which to encode data, the higher the achievable total data rate.)

Another important issue in this ultrahigh frequency (UHF) environment is that radio frequency (RF) energy at these frequencies tends to propagate through and around obstacles, reaching beyond the confines of the network it is serving. That property makes this RF band suitable for receiving commercial broadcast signals from distant stations through the walls of buildings to receivers inside. It also makes it suitable for mobile cellular telephones, but it cannot be well contained within the confines of the **microcell** described earlier, which limits **spectrum reuse** and overall network capacity requirements absolutely essential in a viable wireless network communications system. Current UHF spread spectrum wireless in-building communications systems, then, suffer from critical bandwidth and spectrum reuse shortcomings that seem likely to prevent them from expanding beyond limited applications.

The 18- to 19-GHz Radio Band: An Ideal Choice for Wireless In-Building Microcellular Networks

A region of the electromagnetic spectrum above the kilohertz and megahertz radio bands, yet below the extremely high frequencies of the infrared band, offers two very compelling advantages. Specifically, the 18- to 19-GHz portion of this band fulfills the key requirements of spectrum reuse and bandwidth availability that eliminates most other frequencies from consideration.

Properties Right for Both Microcell Coverage and Confinement

The first major advantage of the 18-GHz band is its excellent propagation characteristics for a microcellular network. Indeed, the behavior and properties of these higher frequencies that are disadvantages for traditional long-range broadcast applications become critical advantages for wireless microcellular network applications. Propagation characteristics of 18-GHz radio waves make them well suited to diffuse thoroughly through a network microcell using only a minimum of transmitted power, yet still stay confined within it so that the same frequencies can be reused by another system within as little as 120 feet or so, or even on the other side of a dense, continuous barrier such as a cement floor. Typical microcells might encompass, for example, a level or floor, or portions of a floor, in a standard office building.

As one might expect, 18-GHz radio waves exhibit a blend of the characteristics of UHF frequencies below them and IR light above them. For example, 18-GHz waves act like light and unlike radio in that they are blocked and reflected by large structures such as concrete and steel. Reflecting back and forth would allow them to fill an area defined by concrete floors and walls with only very small amounts of transmitted power, yet not pass beyond. They also refract like light, penetrating tiny holes and cracks such as closed doors to diffuse and spread through the space beyond. What little radio signal that might escape the microcell would be rapidly dissipated.

Also, unlike lower frequency radio, 18-GHz radio is of a high enough frequency that not only office equipment but even high-energy factory equipment and processes do not interfere with it. Likewise, with its high-frequency and low-required transmitted power, the 18-GHz signals themselves from such a system would not interfere with other electronic systems or equipment. On the other hand, like radio and unlike light, 18-GHz signals still can pass through less dense materials such as drywall and interior office separators and, combined with their reflectivity, are thus not subject to "line-of-sight" limitations. They can also be modulated to carry information just as traditional radio signals are. Finally, since antenna size and design are largely a function of wavelength, which decreases as frequency increases, the antennas for the 16-mm wavelengths of an 18-GHz radio system would be relatively small and compact.

Plenty of Bandwidth in an Otherwise Crowded Spectrum

The second major advantage of 18-GHz radio is its available bandwidth. Few other areas of the electromagnetic spectrum are as interference-free, clear, and available as this band, certainly not the VHF and UHF bands, which must accommodate television, FM radio, cellular telephone, baby monitors, and more.

The reason for this clear band is largely that these higher frequencies have been difficult to work with. The particular technical properties of 18-GHz frequencies and the expense, size, and complexity of the equipment to use them have prevented them from being an attractive option for many commercial applications. As a result, the military has been the primary developer and user of the 15- to 300-GHz band, and the few commercial uses that have emerged (weather, aircraft and police radar, point-to-point telecom transmission, etc.) use expensive technology pioneered by military-funded research.

Developing a comprehensive in-building radio system, however, had never been done until Motorola recently developed the Altair™ wireless ethernet network. Such an application required the creation of new, improved performance data handling and signal processing hardware and software, as well as a radio antenna system that could transmit and capture these data speeds on 18-GHz frequencies in an in-building environment.

Summary

The numerous problems with wiring will become even more acute in the office of the 1990s. This environment will be characterized by:

- The proliferation of decentralized computing resources
- Increased number of telephones and personal computers as an outgrowth of a country's economic shift toward service industries

As the penetration of personal computers nears a one-to-one relationship with phones in the office workplace, the limitations of separate voice and data networks will become even more evident. If these problems are not addressed, an organization's flexibility in redeploying "people assets" and ultimately competitiveness will be seriously hindered. The time it takes to move/add/change equipment and reconfigure communications wires will be the limiting factor in rapidly reorganizing workgroups and responding to new assignments. Wireless LANs will become an attractive solution in the office of the 1990s, interconnecting personal computers and offering data communications capabilities without the need for elaborate cabling methodologies. The obvious and inherent flexibility offered by wireless LANs is the obvious primary benefit. However, the ability to retrieve that investment, never retrievable until now, clearly presents a significant economical benefit.

Defining Terms

Backbone: Wiring which runs within and between floors of a building and connects local-area network segments together.

Cellular communications: Traditionally an outside-of-building radio telephone system that allows users to communicate from their car or from their portable telephone.

Copper twisted pair, coax, and optical fiber: Wired media which connects telephone and computer equipment.

Microcell   A low-power radio network which transmits its signal over a confined distance.

Secondary marketing research: Market research conducted by other organizations.

Spectrum reuse: Reusing frequencies over and over again in a confined area, resulting in more efficient utilization and higher radio network capacity.

Structured distribution systems (SDS):   A topology which advocates cabling saturation of a desired environment to accommodate all potential personnel movements and reconstructions within that office.

Wireless local-area networks: A method of connecting personal computers together without extensive cabling, allowing communications among these devices in an area such as a department or floor of a building.

70

Phase-Locked Loop

Steven L. Maddy
RLM Research

70.1 Introduction 1567
70.2 Loop Filter 1568
70.3 Noise 1570
70.4 PLL Design Procedures 1572
70.5 Components 1572
70.6 Applications 1574

70.1 Introduction

A *phase-locked loop* (PLL) is a system that uses feedback to maintain an output signal in a specific phase relationship with a reference signal. PLLs are used in many areas of electronics to control the frequency and/or phase of a signal. These applications include frequency synthesizers, analog and digital modulators and demodulators, and clock recovery circuits. Figure 70.1 shows the block diagram of a basic PLL system. The *phase detector* consists of a device that produces an output voltage proportional to the phase difference of the two input signals. The *VCO* (voltage-controlled oscillator) is a circuit that produces an ac output signal whose frequency is proportional to the input control voltage. The *divide by N* is a device that produces an output signal whose frequency is an integer (denoted by N) division of the input signal frequency. The **loop filter** is a circuit that is used to control the PLL dynamics and therefore the performance of the system. The $F(s)$ term is used to denote the Laplace transfer function of this filter.

Servo theory can now be used to derive the equations for the output signal phase relative to the reference input signal phase. Because the VCO control voltage sets the frequency of the oscillation (rather than the phase), this will produce a pure integration when writing this expression. Several of the components of the PLL have a fixed gain associated with them. These are the **VCO** control voltage to output frequency conversion gain (K_v), the **phase detector** input signal phase difference to output voltage conversion gain (K_ϕ), and the feedback division ratio (N). These gains can be combined into a single factor called the **loop gain** (K). This loop gain is calculated using Eq. (70.1) and is then used in the following equations to calculate the loop transfer function.

$$K = \frac{K_\phi \times K_v}{N} \tag{70.1}$$

The closed-loop transfer function [$H(s)$] can now be written and is shown in Eq. (70.2). This function is typically used to examine the frequency or time-domain response of a PLL and defines the relationship of the phase of the VCO output signal (θ_o) to the phase of the reference input (θ_i). It also describes the relationship of a change in the output frequency to a change in the input frequency. This function is low-pass in nature.

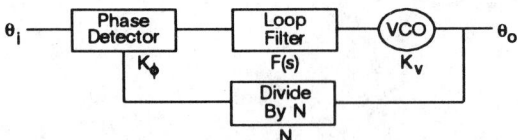

FIGURE 70.1   PLL block diagram.

$$H(s) = \frac{\theta_o(s)}{\theta_i(s)} = \frac{KF(s)}{s + KF(s)} \tag{70.2}$$

The loop error function, shown in Eq. (70.3), describes the difference between the VCO phase and the reference phase and is typically used to examine the performance of PLLs that are modulated. This function is high-pass in nature.

$$\frac{\theta_i(s) - \theta_o(s)}{\theta_i(s)} = \frac{\theta_e(s)}{\theta_i(s)} = \frac{s}{s + KF(s)} \tag{70.3}$$

The open-loop transfer function $[G(s)]$ is shown in Eq. (70.4). This function describes the operation of the loop before the feedback path is completed. It is useful during the design of the system in determining the gain and phase margin of the PLL. These are indications of the stability of a PLL when the feedback loop is connected.

$$G(s) = \frac{KF(s)}{s} \tag{70.4}$$

These functions describe the performance of the basic PLL and can now be used to derive synthesis equations. The synthesis equations will be used to calculate circuit components that will give a desired performance characteristic. These characteristics usually involve the low-pass corner frequency and shape of the closed-loop response characteristic [Eq. (70.2)] and determine such things as the loop lock-up time, the ability to track the input signal, and the output signal noise characteristics.

70.2 Loop Filter

The loop filter is used to shape the overall response of the PLL to meet the design goals of the system. There are two implementations of the loop filter that are used in the vast majority of PLLs: the passive lag circuit shown in Fig. 70.2 and the active circuit shown in Fig. 70.3. These two circuits both produce a PLL with a second-order response characteristic.

The transfer functions of these loop filter circuits may now be derived and are shown in Eqs. (70.5) for the passive circuit (Fig. 70.2) and (70.6) for the active circuit (Fig. 70.3).

$$F_p(s) = \frac{sC_1R_2 + 1}{s(R_1 + R_2)C_1 + 1} \tag{70.5}$$

$$F_a(s) = \frac{sR_2C_1 + 1}{sR_1C_1} \tag{70.6}$$

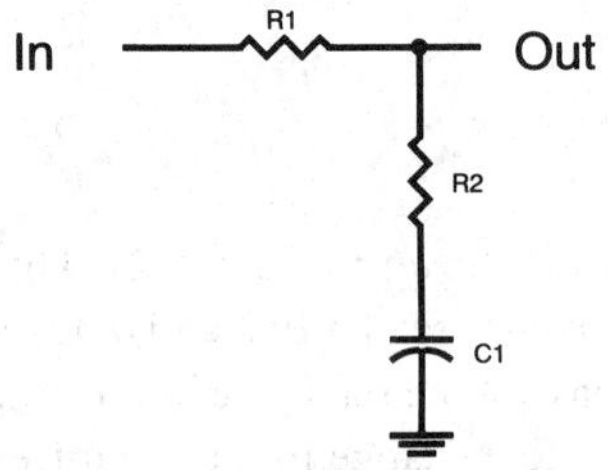

FIGURE 70.2   Passive loop filter.

These loop filter equations may now be substituted into Eq. (70.2) to form the closed-loop transfer functions of the PLL. These are shown as Eqs. (70.7) for the case of the passive filter and (70.8) for the active.

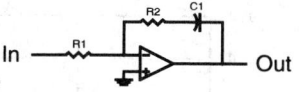

FIGURE 70.3 Active loop filter.

$$H_p(s) = \frac{s\dfrac{KR_2}{R_1 + R_2} + \dfrac{K}{(R_1 + R_2)C_1}}{s^2 + s\left[\dfrac{1}{(R_1 + R_2)C_1} + \dfrac{KR_2}{R_1 + R_2}\right] + \dfrac{K}{(R_1 + R_2)C_1}} \tag{70.7}$$

$$H_a(s) = \frac{s\dfrac{KR_2}{R_1} + \dfrac{K}{R_1 C_1}}{s^2 + s\dfrac{KR_2}{R_1} + \dfrac{K}{R_1 C_1}} \tag{70.8}$$

These closed-loop equations can also be written in the forms shown below to place the function in terms of the **damping factor** (ζ) and the loop **natural frequency** (ω_n). It will be shown later that these are very useful parameters in specifying PLL performance. Equation (70.9) is the form used for the PLL with a passive loop filter, and Eq. (70.10) is used for the active loop filter case.

$$H_p(s) = \frac{s[2\zeta\omega_n - (\omega_n^2/K)] + \omega_n^2}{s^2 + s2\zeta\omega_n + \omega_n^2} \tag{70.9}$$

$$H_a(s) = \frac{s2\zeta\omega_n + \omega_n^2}{s^2 + s2\zeta\omega_n + \omega_n^2} \tag{70.10}$$

Solving Eqs. (70.7) and (70.9) for R_1 and R_2 in terms of the loop parameters ζ and ω_n, we now obtain the synthesis equations for a PLL with a passive loop filter. These are shown as Eqs. (70.11) and (70.12).

$$R_2 = \frac{2\zeta}{\omega_n C} - \frac{1}{KC} \tag{70.11}$$

$$R_1 = \frac{2\zeta}{\omega_n C} - R_2 \tag{70.12}$$

To maintain resistor values that are positive the passive loop filter PLL must meet the constraint shown in Eq. (70.13).

$$\zeta > \frac{\omega_n}{2K} \tag{70.13}$$

For the active loop filter case Eqs. (70.8) and (70.10) are solved and yield the synthesis equations shown in Eqs. (70.14) and (70.15). It can be seen that no constraints on the loop damping factor exist in this case.

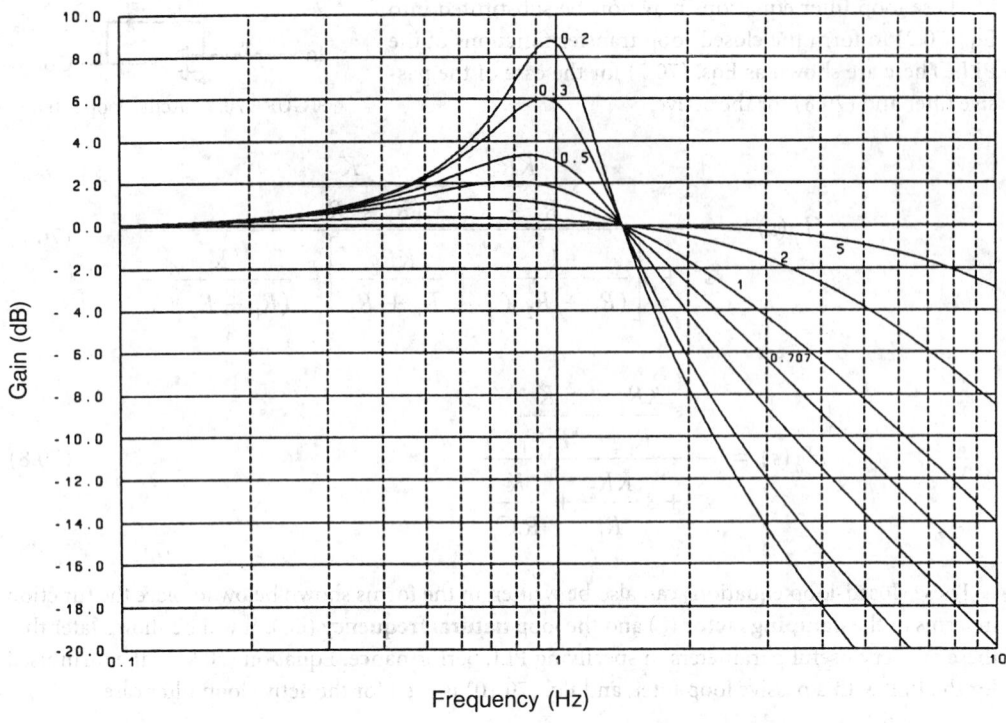

FIGURE 70.4   Closed-loop second-order type-2 PLL error response for various damping factors.

$$R_1 = \frac{K}{\omega_n^2 C} \tag{70.14}$$

$$R_2 = \frac{2\zeta}{\omega_n C} \tag{70.15}$$

A typical design procedure for these loop filters would be, first, to select the loop damping factor and natural frequency based on the system requirements. Next, all the loop gain parameters are determined. A convenient capacitor value may then be selected. The remaining resistors can now be computed from the synthesis equations presented above.

Figure 70.4 shows the closed-loop frequency response of a PLL with an active loop filter [Eq. (70.10)] for various values of damping factor. The loop natural frequency has been normalized to 1 Hz for all cases.

Substituting Eq. (70.6) into (70.3) will give the loop error response in terms of damping factor. This function is shown plotted in Fig. 70.5. These plots may be used to select the PLL performance parameters that will give a desired frequency response shape.

The time response of a PLL with an active loop filter to a step in input phase was also computed and is shown plotted in Fig. 70.6.

70.3 Noise

An important design aspect of a PLL is the noise content of the output. The dominant resultant noise will appear as phase noise (jitter) on the output signal from the VCO. Due to the dynamics of

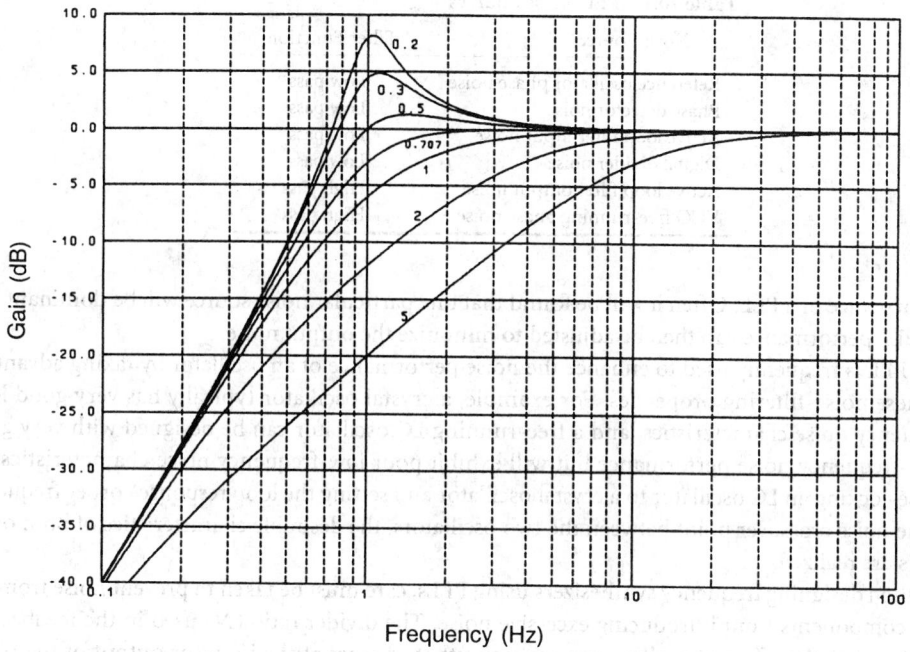

FIGURE 70.5 Closed-loop second-order type-2 PLL step response for various damping factors.

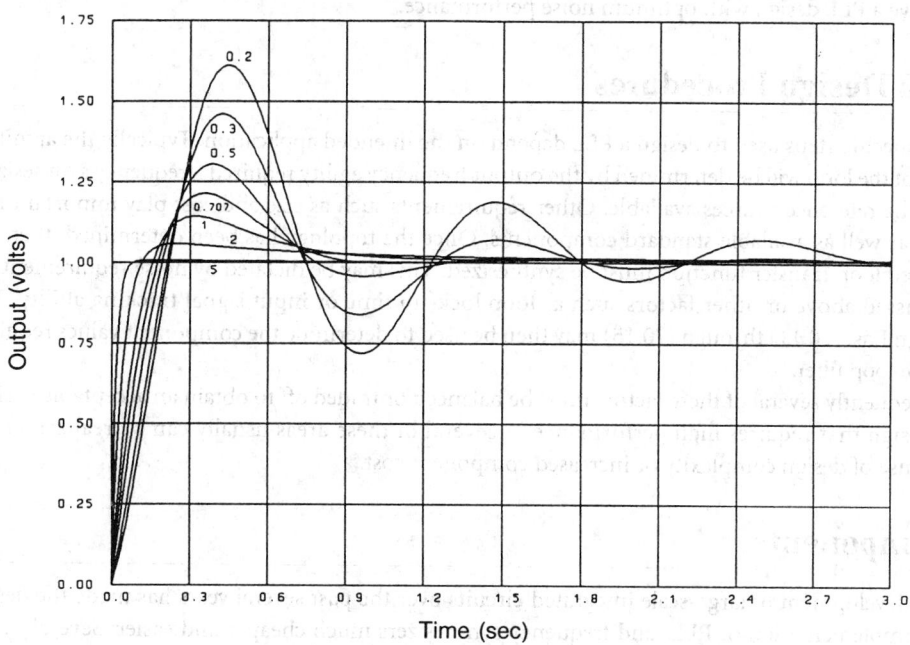

FIGURE 70.6 Closed-loop PLL response for various damping factors.

the PLL some of these noise sources will be filtered by the loop transfer function [Eq. (70.2)] that is a low-pass characteristic. Others will be processed by the loop error function [Eq. (70.3)] that is a high-pass characteristic. Table 70.1 shows the major sources of noise in a PLL and the effect of the loop dynamics on this noise. All these factors must be combined to evaluate the complete noise

Table 70.1 PLL Noise Sources

Noise Source	Filter Function
Reference oscillator phase noise	Low pass
Phase detector noise	Low pass
Active loop filter input noise	Low pass
Digital divider noise	Low pass
Active loop filter output noise	High pass
VCO free-running phase noise	High pass

performance of a PLL. Often it will be found that one particular noise source will be dominant and the PLL performance can then be adjusted to minimize the output noise.

A PLL is frequently used to enhance the noise performance of an oscillator by taking advantage of these noise-filtering properties. For example, a crystal oscillator typically has very good low-frequency noise characteristics, and a free-running LC oscillator can be designed with very good high-frequency noise performance but will exhibit poor low-frequency noise characteristics. By phase-locking an LC oscillator to a crystal oscillator and setting the loop response corner frequency to the noise crossover point between the two oscillators, the desirable characteristics of both oscillators are realized.

When designing frequency synthesizers using PLLs, care must be taken to prevent noise from the PLL components from introducing excessive noise. The divider ratio (N) used in the feedback of the loop has the effect of multiplying any noise that appears at the input or output of the phase detector by this factor. Frequently, a large value of N is required to achieve the desired output frequencies. This can cause excessive output noise. All these effects must be taken into account to achieve a PLL design with optimum noise performance.

70.4 PLL Design Procedures

The specific steps used to design a PLL depend on the intended application. Typically the architecture of the loop will be determined by the output frequency agility required (frequency synthesizer) and the reference sources available. Other requirements such as size and cost play important factors, as well as available standard components. Once the topology has been determined, then the desired loop transfer function must be synthesized. This may be dictated by noise requirements as discussed above or other factors such as loop lock-up time or input signal tracking ability. The design Eqs. (70.11) through (70.15) may then be used to determine the component values required in the loop filter.

Frequently several of these factors must be balanced or traded off to obtain an acceptable design. A design that requires high performance in several of these areas usually can be realized at the expense of design complexity or increased component cost.

70.5 Components

The development of large-scale integrated circuits over the past several years has made the design and implementation of PLLs and frequency synthesizers much cheaper and easier. Several major manufacturers (Motorola, Signetics, National, Plessey, etc.) currently supply a wide range of components for PLL implementation. The most complex of these are the synthesizer circuits that provide a programmable reference divider, programmable divide by N, and a phase detector. Several configurations of these circuits are available to suit most applications. Integrated circuits are also available to implement most of the individual blocks shown in Fig. 70.1.

A wide variety of phase detector circuits are available, and the optimum type will depend on the circuit requirements. An analog multiplier (or mixer) may be used and is most common in applications where the comparison frequency must be very high. This type of phase detector produces an output that is the multiplication of the two input signals. If the inputs are sine waves, the output will consist of a double-frequency component as well as a dc component that is proportional to the cosine of the input phase difference. The double-frequency component can be removed with a **low-pass filter**, leaving only the dc component. The analog multiplier has a somewhat limited phase range of ±90 degrees. The remainder of the phase detector types discussed here are digital in nature and operate using digital edges or transitions of the signals to be compared.

The sample-and-hold phase detector is widely used where optimum noise performance is required. This circuit operates by using one of the phase detector inputs to sample the voltage on the other input. This latter input is usually converted to a triangle wave to give a linear phase detector characteristic. Once the input is sampled, its voltage is held using a capacitor. The good noise performance is achieved since most of the time the phase detector output is simply a stored charge on this capacitor. The phase range of the sample-and-hold phase detector depends on the type of waveform shaping used and can range from ±90 to ±180 degrees.

One of the simplest types of phase detectors to implement uses an exclusive OR gate to digitally multiply the two signals together. The output must then be low-pass filtered to extract only the dc component. The main drawback to this circuit is the large component that exists in the output at twice the input frequency. This requires a large amount of low-pass filtering and may restrict the PLL design. The phase range of this type of circuit is ±90 degrees.

One of the main drawbacks of all the above types of phase detectors is that they only provide an output that is proportional to phase and not to a frequency difference in the input signals. For many applications the PLL input signals are initially not on the same frequency. Several techniques have been used in the past to resolve this such as sweeping the VCO or using separate circuitry to first acquire the input frequency. The sequential (sometimes called phase/frequency) phase detector has become the most commonly used solution due to its wide availability in integrated form. This type of phase detector produces pulses with the width of the pulses indicating the phase difference of the inputs. It also has the characteristic of providing the correct output to steer the VCO to the correct frequency. The noise characteristic of this type of phase detector is also quite good since either no or very narrow pulses are produced when the inputs are in phase with each other. The phase range of this type of circuit is ±360 degrees.

Digital dividers are widely available and may either have programmable or fixed division ratios depending on the application. For optimum noise performance a synchronous type of divider should be used. When a programmable divider is required to operate at a high frequency (>50 MHz), a dual modulus circuit is normally used. This circuit uses a technique called *pulse swallowing* to extend the range of normal programmable divider integrated circuits by using a dual modulus prescaler (usually ECL). The dual modulus prescaler is a high-frequency divider that can be programmed to divide by only two sequential values. A second programmable divider section is then used to control the prescaler. Further details of this type of divider are available from component manufacturers' data sheets as well as in the references.

The voltage-controlled oscillator is typically the most critical circuit in determining the overall noise performance of a PLL. For this reason it is often implemented using discrete components, especially at the higher frequencies. Some digital integrated circuits exist for lower-frequency VCOs, and microwave integrated circuit VCOs are now available for use to several gigahertz. The major design parameters for a VCO include the operating frequency, tuning range, tuning linearity, and phase noise performance. Further information on the design of VCOs is contained in the references.

Loop filters used in PLLs may be either active or passive depending on the specific application. Active filters are normally used in more critical applications when superior control of loop parameters and reference frequency suppression is required. The loop filter is typically followed by a low-pass filter to remove any residual reference frequency component from the phase detector. This low-pass filter will affect the calculated loop response and will typically appear to reduce the loop damping factor as its corner frequency is brought closer to the loop natural frequency. To avoid this degradation the corner frequency of this filter should be approximately one order of magnitude greater than the loop natural frequency. In some cases a notch filter may be used to reduce the reference frequency when it is close to the reference frequency.

70.6 Applications

Phase-locked loops are used in many applications including frequency synthesis, modulation, demodulation, and clock recovery. A frequency synthesizer is a PLL that uses a programmable divider in the feedback. By selecting various values of division ratio, several output frequencies may be obtained that are integer multiples of the reference frequency (Fref). Frequency synthesizers are widely used in radio communications equipment to obtain a stable frequency source that may be tuned to a desired radio channel. Since the output frequency is an integer multiple of the reference frequency, this will determine the channel spacing obtained. The main design parameters for a synthesizer are typically determined by the required channel change time and output noise.

Transmitting equipment for radio communications frequently uses PLLs to obtain frequency modulation (FM) or phase modulation (PM). A PLL is first designed to generate a radio frequency signal. The modulation signal (i.e., voice) is then applied to the loop. For FM the modulating signal is added to the output of the loop filter. The PLL will maintain the center frequency of the VCO, while the modulation will vary the VCO frequency about this center. The frequency response of the FM input will exhibit a high-pass response and is described by the error function shown in Eq. (70.3). Phase modulation is obtained by adding the modulation signal to the input of the loop filter. The modulation will then vary the phase of the VCO output signal. The frequency response of the PM input will be a low-pass characteristic described by the closed-loop transfer function shown in Eq. (70.2).

A communications receiver must extract the modulation from a radio frequency carrier. A PLL may be used by phase locking a VCO to the received input signal. The loop filter output will then contain the extracted FM signal, and the loop filter input will contain the PM signal. In this case the frequency response of the FM output will be a low-pass function described by the closed-loop transfer function and the PM output response will be a high-pass function described by the error function.

In digital communications (modems) it is frequently necessary to extract a coherent clock signal from an input data stream. A PLL is often used for this task by locking a VCO to the input data. Depending on the type of data encoding that is used, the data may first need to be processed before connecting the PLL. The VCO output is then used as the clock to extract the data bits from the input signal.

Defining Terms

Capture range: The range of input frequencies over which the PLL can acquire phase lock.

Damping factor: A measure of the ability of the PLL to track an input signal step. Usually used to indicate the amount of overshoot present in the output to a step perturbation in the input.

Free-run frequency: The frequency at which the VCO will oscillate when no input signal is presented to the PLL. Sometimes referred to as the rest frequency.

Lock range: The range of input frequencies over which the PLL will remain in phase lock once acquisition has occurred.

Loop filter: The filter function that follows the phase detector and determines the system dynamic performance.

Loop gain: The combination of all dc gains in the PLL.

Low-pass filter: A filter that usually follows the loop filter and is used to remove the reference frequency components generated by the phase detector.

Natural frequency: The characteristic frequency of the PLL dynamic performance. The frequency of the closed-loop transfer function dominant pole.

Phase detector gain: The ratio of the dc output voltage of the phase detector to the input phase difference. This is usually expressed in units of volts/radian.

VCO gain: The ratio of the VCO output frequency to the dc control input level. This is usually expressed in units of radians/second/volt.

References

AFDPLUS Reference Manual, Boulder, Colo.: RLM Research, 1991 (software used to generate the graphs in this section).

R. G. Best, *Phase-Locked Loops—Theory, Design & Applications*, New York: McGraw-Hill, 1984.

A. Blanchard, *Phase-Locked Loops, Application to Coherent Receiver Design*, New York: Wiley Interscience, 1976.

W. F. Egan, *Frequency Synthesis by Phase Lock*, New York: Wiley Interscience, 1981.

F. M. Gardner, *Phaselock Techniques*, New York: Wiley, 1979.

J. Gorski-Popiel, *Frequency Synthesis; Techniques & Applications*, Piscataway, N.J.: IEEE Press, 1975.

W.C. Lindsey and M.K. Simon, *Phase-Locked Loops & Their Applications*, Piscataway, N.J.: IEEE Press, 1978.

V. Manassewtsch, *Frequency Synthesizers: Theory and Design*, New York: Wiley Interscience, 1980.

U. L. Rhode, *Digital PLL Frequency Synthesizers Theory and Design*, Englewood Cliffs, N.J.: Prentice-Hall, 1983.

Further Information

Recommended periodicals that cover the subject of PLLs include *IEEE Transactions on Communications, IEEE Transactions on Circuits and Systems*, and *IEEE Transactions on Signal Processing*. Occasionally articles dealing with PLLs may also be found in *EDN, Electronic Design, RF Design*, and *Microwaves and RF Magazine*. A four-part PLL tutorial article titled *PLL Primer*, by Andrzej B. Przedpelski, appeared in *RF Design Magazine* in the March/April 1983, May/June 1983, July/August 1983, and November 1987 issues.

Another good source of general PLL design information can be obtained from application notes available from various PLL component manufacturers. *Phase-Locked Loop Design Fundamentals*, by Garth Nash, is available from Motorola, Inc. as AN-535 and gives an excellent step-by-step synthesizer design procedure.

THE LEGACY OF EDWIN HOWARD ARMSTRONG

Edwin Armstrong is widely regarded as one of the foremost contributors to the field of radio-electronics and was inducted into the National Inventors Hall of Fame in 1980. Among his principal contributions were regenerative feedback circuits, the superheterodyne radio receiver, and a frequency-modulation radio broadcasting system. As an independent inventor whose patented inventions served as important pawns in an arena of corporate conflicts over dominance in the radio industry, he devoted much money, time, and energy to efforts to establish his priority both in the legal sense and in the eyes of his peers. The frustrations that he experienced in these efforts caused his final years to unfold like a Shakespearean tragedy.

Armstrong's inventive style was somewhat unusual in that he expressed almost a phobic distrust of mathematical analysis, and his well-crafted technical papers rarely contained an equation. Instead he employed circuit diagrams, oscillograms, and graphical tube characteristics to explain his discoveries. He was like an artist except that his medium was three-dimensional combinations of vacuum tubes, resistors, inductors, and capacitors. He exhibited an intuitive grasp of the effect of changing circuit parameters and could isolate the cause of unexpected phenomena that baffled less perceptive experimenters.

Armstrong graduated with a degree in electrical engineering from Columbia University in 1913 and observed the phenomenon of regenerative feedback in vacuum-tube circuits while still an undergraduate. At Columbia, he came under the influence of the legendary professor-inventor, Michael I. Pupin, who had become wealthy from selling his loading-coil patents to the Bell Telephone Company. Pupin served as a role model for Armstrong and became an effective promoter of the young inventor. In 1915 Armstrong presented an influential paper on regenerative amplifiers and oscillators to the Institute of Radio Engineers (IRE). The paper provoked a dispute with Lee de Forest, who questioned Armstrong's interpretation and his priority in the discovery. Subsequently, regenerative feedback was incorporated into a comprehensive engineering science developed by Harold Black, Harry Nyquist, Hendrik Bode, and others in the period between 1915 and 1940. The methods of analysis and design that they developed proved highly applicable to the related field of servomechanisms during World War II. Bode characterized this convergence of control and communications as having been a "sort of shotgun marriage."

Armstrong conceived the superheterodyne receiver principle in 1918 while serving in the Army Signal Corps in France. As with regenerative feedback, there were other claimants to priority, but he played a key role in the commercialization of the invention during the early 1920s. The Radio Corporation of America (RCA) used his superheterodyne patent to monopolize the market for this type of receiver until 1930. The superheterodyne eventually extended its domain far beyond commercial broadcast receivers and, for example, proved ideal for microwave radar receivers developed during World War II.

In 1933, Armstrong was granted patents on a frequency-modulation system that he promoted as a superior alternative to the established amplitude-modulation broadcasting service. Wideband FM became almost a paradigm of how a generation of engineers may sometimes be so misled by a valid but limited mathematical theory as to overlook a major innovation. Armstrong drew on his own financial resources in a determined effort to establish FM broadcasting until the War intervened. His crusade for FM received additional setbacks in the post-war years from certain regulatory decisions by the Federal Communications Commission. In 1948 he brought suit against RCA for patent infringement, and the litigation process placed great stress on his resources. The case was not resolved until after his tragic demise early in 1954. His inventive creativity had flourished in an earlier era of radio mania, but he seemed unable to adapt to the new environment of government regulation and protracted litigation after the war.

Armstrong was highly esteemed by a generation of radio-electronics engineers and amateur radio enthusiasts. He received the first medal of honor of the IRE in 1918. The importance of his legacy might be appreciated by means of a thought experiment of imagining what the field of telecommunication would be like without regenerative feedback, the superheterodyne receiver, and FM.

Source: Adapted from J. E. Brittain, *Proc. IEEE*, vol. 79, no. 2, p. 248, February 1991. © 1991 IEEE.

71

Telemetry

71.1 Introduction to Telemetry 1578
71.2 Measuring and Transmitting 1582
71.3 Applications of Telemetry 1583
        Power Sources • Power Plants
71.4 Limitations of Telemetry 1585
71.5 Transmitters and Batteries 1586
71.6 Receivers and Discriminators 1586
71.7 Antennas and Total System Operation 1586
71.8 Calibration 1587
71.9 Telemetry Frequency Allocations 1587
71.10 Telemetry Antennas 1588
71.11 Measuring and Transmitting 1589
71.12 Modulating and Multiplexing 1589
71.13 Passive Telemeters 1590
71.14 The Receiving Station 1590

Conrad H. Hoeppner
The Johns Hopkins University

71.1 Introduction to Telemetry

Telemetry, or measurement at a distance, takes many and varied forms. It may use the principles of radio, electricity, optics, mechanics, or hydraulics to convey measurements made at one place to indicators, actuators, recorders, or computers at another. By far the most popular telemetry systems are electrical and use radio or wire links to convey information. In this respect, all of the considerations in the foregoing chapters on communications apply, as well as considerations of antennas, power supplies and convertors, heat removal, and radio frequency interference. Additional considerations that are unique to telemetry are treated here.

The deeper an instrumented vehicle probes into the remote reaches of outer space, the more technologically spectacular seem the achievements of telemetry. There is still something exciting and uncanny about performing measurements of a physical quantity at a distant location and precisely reproducing them at a more convenient place for reading or recording them.

Yet the vast distances spanned by telemetry signals are less challenging technically than the stubborn problems of almost sheer inaccessibility in some industrial applications to the quantities being measured. Signals from a missile-launched space probe soaring toward the sun are often easier to obtain than measurements from inside a stolid, earthbound motor only a foot or two away. To find the temperature of the spinning rotor, housed in a steel casing and surrounded by a strong alternating magnetic field, may require more ingenuity to transcend the operating environment than taking measurements from the most distant instrument payload speeding through the unaccommodating environment of space.

The technology that has produced missile and space telemetry is also spawning new forms of industrial radio telemetry, capitalizing on the development of new transducers, powerful miniature radio transmitters, improved self-contained power sources, and better techniques of environmental protection.

Simply enough, to telemeter is to measure at a distance. First, at the remote point, is needed a transducer, a device that converts the physical quantity being measured into a signal, usually an electrical one, so that it can be more conveniently transmitted. Then a connecting link between the location where the measurement is being made and the point where one can read or record the signal being sent is required. This link can be either an electrical circuit—there have been wired telemetry systems since long before the turn of the century—or pneumatic or hydraulic lines, a beam of light, or now, more practically, a radio carrier.

A radio telemetry system comprises (1) transducers that convert measurements into electrical systems, (2) a subcarrier oscillator modulated by the transducers, (3) a radio transmitter modulated by the subcarrier, (4) a transmitting antenna, (5) a receiving antenna, (6) a radio receiver, and (7) a subcarrier discriminator. The radio link can transmit an analog of the continuous variable being measured, or, with pulse-code methods, it sends the measurement data digitally as a finite number of symbols representing a finite number of possible values of the measurement signal at the time it is sampled. The range of a radio link is limited by the power radiated toward the receiver from the transmitter and by the sensitivity of the receiver. The wider the bandwidth, the more the effect from noise, and therefore the more transmitted power required for a detectable signal.

Optical, mechanical, and hydraulic telemeters represent a smaller segment of the telemetry field than do electrical and radio telemeters; they will be given only brief treatment here.

Optical telemeters use light transmitted through space or through optical fibers, the light being modulated by the measurement signal. The modulation may be produced either electrically or mechanically. Electrically, light-emitting diodes, lasers, or electroluminescent material are used to convert the electricity to light. The light may then be modulated with the measurement information by modulating the electricity that produces the light or it may be polarized and rotated by Kerr or Pockels cells, absorbed by electrically activated chromofors, or converted to another wavelength by electrically controlled nonlinear elements, all of which are activated by the modulating signal.

Hydraulic telemeters are generally used in conjunction with hydraulic sensors and hydraulic displays, such as pressure gauges. They are immune to all electrical and optical interference and hence find application in unfavorable electrical and optical environments. A typical hydraulic telemeter is used to measure load lifted and boom angle on a crane. Here a hydraulic piston is activated by the tension of a lifting rope pushed sidewise by the piston roller. An increase in tension produced by the load tends to straighten the deflected rope and press the piston into its cylinder. The change in pressure is communicated through fluid in a tube to a remote indicator. The hydraulic telemeter measures the boom angle by simply placing a fluid reservoir on the boom, which when it is raised provides increased pressure through its tube to a second remote indicator. In this way, with remote indicators, the operator monitors the crane to prevent overloading and/or overturning.

Electrical telemeters proliferate through (1) space, (2) battlefields, and (3) industrial sites, varying in size, configuration, and information-carrying capacity with their various applications. Space research and missile development used the first significant multichannel telemeters. Telemeters developed at the Naval Research Laboratories and built by the Raytheon Company were first used to explore outside the earth's atmosphere in German V-2 rockets launched at the White Sands Proving Ground. These telemeters used 1000-MHz pulse position modulated signals at ranges greater than 100 miles. Conrad H. Hoeppner designed the equipment and managed the installations and operation. From 1945 onward, telemetry developed rapidly and found its way to the various missile ranges also being developed.

To permit tests to be made interchangeably at all ranges, it was necessary to standardize types of telemeters at the ranges. To this end the Department of Defense Research and Development Board formed the Guided Missiles Committee, which in turn formed the Working Group on Telemetry. This later became the Inter-Range Instrumentation Group (IRIG), which has published telemetry standards that are widely accepted.

Meanwhile, industrial telemetry has developed along different lines, producing miniaturized complete capsules for applications to process control, detection of defects, and machine design. Medical science is currently using telemetry in experimental, clinical, and diagnostic applications. Some of the particular body characteristics telemetered include heartbeat, brain waves, blood pressure, temperature, voice patterns, heart sounds, respiration sounds, and muscle tensions. Similar studies are being pursued in the biological and psychological fields, where more experimental latitude permits embedding of transmitters within living animals.

The basic telemetry system consists of three building blocks. These are (1) input transducer, (2) the transmitter, and (3) the receiving station. Transducers convert the measured physical quantity into a usable form for transmission. The conversion of the desired information into a form capable of being transmitted to the receiver is a function of the type of transducer employed. Transducers convert the physical quantities to be measured into electrical, light, pneumatic, or hydraulic energy. The type of energy conversion is determined by the type of transmission desired. In a radio telemetry system, the transmitter and receiver have much in common with communications equipment. The transducers, however, are unique to telemetry and will be described in some detail here.

One of the most common types of transducers generates electrical signals as a function of the changing physical quantity, and one of the most common varieties of this type is the resistance wire strain gauge. In this transducer, the ability of the wire to change its dimension as it is stressed causes a corresponding change in its electrical resistance. A decrease in wire diameter generally results in greater resistance to the flow of electricity. Similarly, temperature-sensitive materials that have electrical characteristics changing with temperature make temperature detection possible.

In most transducers, the electrical output is varied as a function of changes in the physical parameter. These electrical changes can be transmitted by wire direct to a control center, data display area, or to a data analysis section for evaluation. The difficulties with the use of wire in many applications have given rise to wireless telemetry. In order to transmit the transducer information through the air, it is necessary to apply this information to a high-frequency electrical carrier, as is commonly done in radio. Application of the transducer information to a high-frequency carrier is commonly called modulation. High-frequency or rapidly changing electricity has the capability of being propagated through space, whereas low-frequency or battery, nonchanging voltage does not possess this ability.

The technique used for applying or modulating the high-frequency carrier by the transducer output involves any one of three different methods. It is possible to modulate a carrier by a change in amplitude, a change in frequency, or a change in the carrier phase. The last technique is similar to the modulation used in transmitting color by television. In color TV the brightness signal is transmitted as amplitude modulation (AM), the sound as frequency modulation (FM), and the color as phase modulation (PM), or pulse coding. Pulse coding is used to modulate the radio frequency carrier in either AM, FM, or PM.

A common and extremely useful technique for increasing the information-carrying capability of a single transmitting telemetry line is called multiplexing. When it is desirable to monitor different physical parameters, such as temperature and pressure, it may be wasteful to have duplicating telemetry transmission lines. Multiplexing techniques can usually be considered to be of two types: frequency division multiplexing and time division multiplexing. In the frequency division multiplexing system, different subcarrier frequencies are modulated by their respective changing physical parameter; these subcarrier frequencies are then used to modulate the carrier frequency, enabling the transmission of all desired channels of information, simultaneously by one carrier. At the receiver, these subcarrier frequencies must be individually removed. This is accomplished by filters that allow any one of the respective subcarrier frequencies to pass. Each subcarrier frequency is then converted back to a voltage by the discriminator. The discriminator voltages can be used to actuate recorders and/or similar devices. Time division telemetry systems may use pulse modulation or pulse code modulation. In these systems the information signal is applied, in time sequence, to modulate the radio carrier. The characteristics of a pulse signal can be affected by modulating its amplitude, frequency, or phase.

Telemetry began as a wire communication technique between two remotely located stations. As science extends its domains into the realm of space, telemetry will be the essential communicating link among satellites, spaceships, robots, and other scientific devices yet to be designed.

The range of a radio link is limited by the strength of the signal radiated by the transmitter toward the receiver and by the sensitivity of that receiver. A 10-microwatt (μW) output will transmit data easily 100 feet with a bandwidth of 100 kHz. The wider the bandwidth, the more the effect from noise, and therefore the more transmitting power required for an acceptable signal.

At the receiving station, there are usually no space restrictions in accommodating large antennas, sensitive radio tuners and recorders, and an ample power supply, but the transmitting station often must be small, possibly doughnut-size, but sometimes no bigger than a pea, and must be self-sufficient, carrying its own power or perhaps receiving it by radio.

On the surface, industrial radio telemetry seems to be simply a matter of hardware. It almost is, except that the functional requirements are a lot different from those in missile and space telemetry. Distances are much shorter, a matter of a few feet to a few hundred yards; signal power can be radiated directly from the transmitter circuitry or from an antenna as simple as an inch or two of wire. Most tests are repeatable—no missile blowing up on the pad here, taking with it valuable instruments and invaluable records of the events leading up to that failure.

Quantities can be measured one or two at a time, rather than requiring an enormous amount of information to be transmitted at once. This results in relatively inefficient use of the radio link but enables simpler circuitry at both the transmitting and receiving ends.

Surprisingly, environment plays the most critical role in industrial telemetry. It makes by far the largest difference between telemetry operations from missiles and spacecraft and those used in industrial remote measurement. While missile telemetry equipment is expected to withstand accelerations of 10 to 20 *g*, the rotating applications of telemetry in industry, such as the embedding of a transducer in a spinning shaft, require immunity to 10,000 or 20,000 *g* centrifugal accelerations.

The environmental extremes under which industrial telemeters must work are considered normal operating conditions by their users. Unlike missile telemetry equipment, which is shielded and insulated against extremes of temperature, shock, and vibrations and which is carefully calibrated for weeks before it is used only once in an actual shot, industrial telemeters must operate repeatedly without adjustment and calibration. Used outdoors, they are often subjected to a temperature range of −40 to +140°F. They must operate when immersed in hot or cold fluids, and thus it is almost mandatory that they be completely encapsulated to be impervious not only to humidity and water but to many other chemical fluids and fumes. Many lubricating oils operate at temperatures of 300 or 350°F.

We know that missile telemetry components must be small and light, yet an order of magnitude reduction in size and weight has been necessary to make telemetry suitable for high-speed rotating shafts or for biological implants. They must be so reliable that no maintenance is required, for there are no service centers set up to handle this kind of equipment, and it must work without failure to continue to gain industrial acceptance.

Information theory has been used extensively to develop space telemetry for the most efficient data transmission over a maximum distance with a minimum of transmitted power. Inefficiencies, being of no real consequence in industrial telemetry, make for less elaborate, less costly equipment. Radio channels are used in a relatively inefficient manner, and the distances between transmitter and receiver are usually so short that there are few problems of weak signals. In many cases, measurement and testing via telemetry links takes place in completely shielded buildings or in metal housings.

Although telemetry is usually defined as measurement at a distance, it has also gradually begun to embody the concept of control from a distance. In telemetry—the transmission of the value of a quantity from a remote point—it may serve merely to communicate the reading on an instrument at a distance. The output of the instrument can also be fed into a control mechanism, however, such as a relay or an alarm, so that the telemetered signal can activate, stop, or otherwise regulate a process. Measurement may be taken at one location, indication provided at a second location, and the remote control function initiated at one of those two locations or at a third point.

For example, a motor might be pumping oil from one location while oil pressure is being measured at another. When the pressure reading is telemetered to a control station, a decision can be made there to reduce pump motor speed when the pressure is too high, or a valve can be opened at still another location to direct the oil to flow in another path. The decision-making controller may be an experienced pipeline dispatcher or an automatic device.

71.2 Measuring and Transmitting

Telemetry, then, really begins with measurement. A physical quantity is converted to a signal for transmission to another point. The transducer that converts this physical quantity into an electrical signal may be a piezoelectric crystal, a variable resistance, or perhaps an accelerometer. Telemetered information need be no less accurate than that obtained directly under laboratory conditions. For instance, in telemetering strain measurements, it is possible to achieve accuracies of a few microinches per inch or greater. The only limitation is usually the degree of stability in the bond of the strain gauge to the specimen, and not the strain gauge itself.

If great accuracy in temperature measurement is desired, it can be attained by choosing a transducer that provides a large variation of output signal over a small range of process property variation. The resolution which this provides may be translated into true accuracy by careful transducer calibration. Accuracy is reduced, of course, if a wider range of temperature needs to be detected. Typical single-channel analog telemetry links maintain a measurement accuracy of 1–5%. This is not a limitation of the total system, however, since 1% of a 100° temperature change would only be 1°, so several telemetry channels can easily share the total temperature range to be measured, say a 100°F range divided into four 25°F ranges, to produce an accuracy of one-fourth of a degree.

Special temperature probes have been produced for the range of 70 to 400°F and higher to maximize the stability and accuracy of temperature telemetry. These probes, when used with the proper choice of transmitters and receivers, can provide temperature measurements to closer than 0.05°F.

One of the limitations to accuracy and to repeatability in telemetry is the output level of the transducer. The low electrical levels produced by thermocouples and strain gauges (millivolts) are much more difficult to telemeter than a higher-voltage level of, say, 5 V. At low signal levels, extraneous electrical noises produce great degradation. This noise may be thermally generated, caused by atmospheric effects, or generated by nearby electrical equipment. When low-level transducers are used, stable amplifiers are required to raise the signal voltage to useful modulation levels.

There may be great variations in the strength of the radio signal received because of variations in distance between transmitter and receiver or because of the interposition of metallic objects. In industrial radio telemetry transmission, these effects can be prevented from disturbing the data by resorting to FM of both the subcarrier and the carrier so that the telemetered signal is unchanged by undesirable amplitude variations. This method is called FM/FM telemetry.

If FM is used in the subcarrier of the transmitter, the transducer signal modulates the frequency of the subcarrier oscillator. This can be done by a simple resonant circuit that produces a given frequency in the audio range, say 1,000 Hz, which is varied above or below by the signal from the transducer as it responds to the variable it is measuring. If the signal were fed to a loudspeaker, a rising or falling tone would be heard. The subcarrier oscillator then modulates a radio frequency carrier, varying its frequency (FM) or its amplitude (AM) in accordance with the subcarrier signal. The radio frequency in FM industrial radio telemetry links is usually in the 88- to 108-MHz band. At the receiving end of the link, the radio receiver demodulates the signal, removing the carrier and feeding the subcarrier to a special discriminator circuit that removes the modulation and precisely reproduces the original measurement signal for calibrated indication or recording.

Multiple measurements can also be transmitted over the carrier by sampling the output of several transducers in rapid sequence, a technique called time-division multiplexing. This technique has been employed to handle as many as a million samples per second. It provides for simple data displays and easier separation of channels for recording or analysis, and it is free of cross talk. If possible, it is advantageous to use no multiplexing at all for concurrent data taking, but to use separate radio carriers for each measurement being transmitted.

Many and varied kinds of modulation have been used in telemetry systems. All have general usefulness, with cost and application being the drivers. Synchronized modulation is generally used with other systems, being synchronized to them to give additional information such as range. Typically, a command control uplink to an aircraft is used to synchronize a telemetry downlink with the delay being proportional to distance or the length of the link. Signal-to-noise advantages are also achieved. Another example is one in which telemetry is tacked on to a radar transponder to add additional pulses to indicate altitude, heading identification, or other conditions of the vehicle carrying the transponder. This is usually accomplished with pulse position modulation.

In many instances, a reconnaissance vehicle will carry a television camera. Its signals may be recorded on board but are often telemetered for real-time observations. Data may be placed on the same carrier using a few lines of the TV picture or an additional subcarrier. The much greater bandwidth of the TV signal seriously compromises the combination of range, transmitter power, and antenna directivity, and typically signal-to-noise ratio is reduced as much as 30 dB.

71.3 Applications of Telemetry

High-voltage transmission lines are an excellent example of how inaccessible an object of measurement may be. These lines vibrate in the wind, and the stresses and strains require measurement under the dynamic conditions that contribute to fatigue failure. Strain tests to determine fatigue will show quickly whether the endurance limit of the line has been exceeded, and only if it is exceeded need we be concerned about fatigue failure. Therefore, it is necessary to measure the number and magnitude of the strain reversals in order to predict the time of failure. Telemetry techniques permit dynamic testing under actual service conditions rather than by simulated laboratory conditions or static tests.

While the transducer that produces an electrical signal proportional to strain may have an output of 0.01 V, the live transmission line to which it is attached may be at a potential of several hundred thousand volts. The problem is to detect this hundredth of a volt in the presence of a very large signal. In the language of the telemetry engineer, this is rejection of a common mode voltage on the order of 10^8 to 1. Then why not deenergize the line? It's a simple matter of economics—an idle line transmits no power, and the wind forces that cause the line to vibrate are neither predictable nor constant. So, weeks or months may be spent in gathering measurements for a particular set of spans. However, a radio telemetry link makes it possible to transmit the strain signal even while power is being carried.

A self-contained FM radio transmitter is attached to the transmission line at a point adjacent to a strain gauge. All remain at the same electrical potential as the line, much like a bird sitting safely on the wire, transmitting the strain gauge output to a radio receiver and recorder located at some convenient point on the ground, where vibration analysis can be made. As a result, armor rods may be placed around the line at the vulnerable points, or vibration absorbers of the correct resonant frequency can be installed at the proper points on the line.

More down to earth, but equally inaccessible to measurement, is strain on the chain belt of an earth mover. Too light a chain will quickly fail from fatigue caused by the alternating stresses imposed by the full and empty buckets it transports. Measurements made under actual operating conditions of the earth hoist mean attaching strain gauges to a chain traveling at 500 ft per minute, subjecting them to violent shock and vibration. On this kind of moving equipment, slip rings and wire-link remote measurements will not work. Here again, radio telemetry is now providing the dynamic measurements needed to test the earth-moving equipment at work. A transducer and a small, rugged transmitter are attached to points along the chain—strain varies from link to link, depending on the proximity to the bucket—until the most vulnerable part of the chain is found. It is preferable to use several transducers and multiple-channel telemetry equipment for such measurements to simplify correlation between load and the resulting strain at various links.

Telemetry can also determine water levels and flow rates of rivers to provide vital data for flood control or for efficient hydroelectric power generation. Data on the potential watershed into rivers can be obtained by analyzing the water content of snow that would eventually melt and feed them. One requirement is to measure the depth and water content of snow in the mountains, then transmit these data from remote points to a central receiving station. The snow-measuring transducer may consist of a radioactive source atop a tall pole and a radiation intensity meter on the ground beneath the snow. The gamma-ray intensity reaching the meter is a function of the height and water content of the intervening snow. Both the meter and the transmitting equipment can be powered by a storage battery and controlled by a clock timer that sets the time of transmission to a few seconds per day.

In the design of machinery, one of the most difficult factors to cope with is alternating fatigue-producing stresses that occur at some parts of the machine. It has long been the custom to measure stress in equipment with bonded strain gauges to predict the failure limits before actual failure occurs. This had only been possible on those portions of the equipment that could be connected by wires. With radio telemetry, it is not possible on all members. Costly fatigue failures are now avoidable through installation of miniature telemetry components that are reliable, rugged, and accurate in heretofore inaccessible locations and environments. Industrial uses are virtually limitless; systems can be built to specifications and encapsulated to withstand the most adverse conditions. Low-cost measurement and telemetry systems have been applied to read internal vibrations and strain in rotating equipment, chains, vehicles, and projectiles—eliminating slip rings and wires. Measurement can be made under operating conditions of vibration, acceleration, strain, temperature, pressure, magnetic fields, electrical current, and voltage, under such adverse conditions as in a field of high electrical potential, in fluids, in steam, or in high-velocity gases.

Power Sources

Power sources for the transmitter in industrial telemetry applications are seldom a problem. Batteries can be used for temporary applications and at temperatures below 200°F. Small and light, rechargeable and expendable batteries are available solidly encapsulated in epoxy resin to withstand almost as rugged environments as the telemeter itself.

In a moving or rotating application, stationary magnets can be placed so that they generate electricity in a moving coil and are used to provide automatic power generation. If this method is not feasible, a stationary coil can be placed in the vicinity of the transmitter and fed electrical energy at a high frequency, so that its field can easily couple into a moving coil in almost any environment. The stationary coil ring may be large, even encompassing a whole room; usually only one turn of wire is necessary. The stationary coil may also be made extremely small, ¼ to ½ in. in diameter, and coupled to the end of a rotating shaft. These power supplies and coil configurations are standard available units.

Power Plants

In power plants, coal is fed in turn to a number of hoppers by conveyor belt. A tripper on the conveyor belt diverts the coal into a particular hopper until it is full. Either an operator or a mechanical sensing device determines when the hopper is full, and a signal is transmitted to the conveyor to move onto the next hopper. Before telemetering equipment was in use, costly accidents could occur if the operator should be away momentarily or if the sensor failed to function. As much as six tons of coal a minute could overflow onto the power station floor.

To prevent this, pressure switches are installed in the tripper chute to activate a radio transmitter if coal backs up into the tripper. The transmitter sends its signal to a receiver located at the conveyor belt and sounds an alarm. This type of control is difficult if not impossible to achieve by wired power connections because the tripper is moving and because the corrosive coal dust atmosphere attacks the wires. For this reason, a radio transmitter equipped with long-life batteries is mounted on the tripper. The receiver at the control end is powered by ac. Subcarrier tone (frequency) coding is used to eliminate the effects of interference and noise, giving positive protection at all times.

71.4 Limitations of Telemetry

The preceding paragraphs describe a number of the requirements placed upon telemetry systems by the transducers and quantities being measured. Unfortunately, the development of telemetry has not been such as to satisfy all requirements, and in many cases the telemetry system seriously limits the measurement. A compromise is therefore required between telemetry capabilities and the requirements of measurement. The shortcomings and limitations of the telemetry system place restrictions upon measurements above and beyond those encountered in the laboratory when the telemeter is not used. In the first place, an electrical output from the measuring device is required in order that the measurement may be placed on a radio link. Consequently, transducers that produce an electrical output on one form or another are necessary. Also, the telemetry system may not be perfectly stable down to zero frequency (dc), and transducers and methods of measurement must be chosen to minimize the effects of drift. Overmodulating the subcarrier, or the time-division multiplexer, may also affect adjacent channels, as well as produce erroneous data in its own channel. If various measuring devices are switched, the switching transients must be minimized, or the accuracy of the telemetry system may be impaired. When mechanical commutators or time multiplexers are used, the measurement of the time occurrence of the event, such as the impact of cosmic particles or the receipt of a guidance pulse, is made more difficult and the time ambiguity of the multiplexed system is a serious limitation.

The measurement of a large number of parameters requires extensive and bulky equipment, unless the parameters can be combined in groups of similar inputs to minimize the signal conditioning required. This fact generally dictates a relatively standard transducer rather than an optimum one for each particular measurement.

The bandwidth of the measurement, or the frequency with which the measured quantity changes, is also seriously limited by the telemeter. In the FM/FM telemeter, the permissible bandwidth varies from a relatively low value on the lower-frequency subcarriers to a reasonably high value on the high-frequency subcarriers. The bandwidth of the measurement must not exceed the subcarrier bandwidth limitations, or sidebands will be generated in adjacent channels, thereby reducing the accuracy of other measurements (if multiplexed), or interference with adjacent RF signals will be caused.

In a time-multiplexed system, the problem of "folded data" is present whenever the rate of data change is faster than one-half the sampling rate. When this occurs, it is not known whether the measured quantity has reversed itself several times between samples or if there has been no reversal at all. It is considered desirable to limit the bandwidth of the data so that this ambiguity is not present; however, with refined techniques of analysis, this is not a rigid requirement. The form in which the data is displayed or recorded is also a limitation on measurement. In general, time-history plots of the measured quantity are desired. In this case, the speed at which the recording medium moves is often a severe limitation. If sampling is not regular, demultiplexing difficulties are magnified.

71.5 Transmitters and Batteries

The transmitter is made up of two components: the subcarrier oscillator and the radio frequency oscillator. The subcarrier can be bridge controlled (BCO) or voltage controlled (VCO). The sub-carrier center frequency is 4,000 Hz, which can be modulated ±400 Hz by the strain or voltage being measured. Using BCOs, a strain as large as 2,500 microinches per inch (μin./in.) and as small as 2 μin./in. can be measured and transmitted. The temperature measurement range of the VCO is from −200 to 4,000°F. With a copper-constantan thermocouple, a temperature change as small as 2°F can be sensed and transmitted.

The single-resistance strain gauge transmitter does not have a subcarrier oscillator and can be used from −40 to 212°F. It has only a radio frequency oscillator, which is modulated by the sensor signal. For this reason, it is not suitable for static strain measurements and must be used for dynamic strain measurements only. It has a frequency response to 25,000 Hz or greater. A static strain signal transmitted by this device will drift. It is provided with self-contained rechargeable nickel-cadmium batteries. Pins protruding through the epoxy case are used for all electrical connections. Only one screw adjustment is provided, and this is used to set the radio frequency.

Rechargeable nickel-cadmium batteries are used with the BCO and the VCO. The BCO batteries have useful lives of 4 and 9 hr. A VCO battery has 40 hr useful life. The single-resistance strain gauge transmitter has a built-in nickel-cadmium battery with a life of 4 hr.

71.6 Receivers and Discriminators

A typical industrial receiver has a tuning range of 88 to 108 MHz. When the transmitter is used in its greatest sensitivity mode, the output of the discriminator is approximately 1 V for a 25-μin./in. strain with a single active gauge in the bridge. At the most insensitive mode 1 V is obtained for an approximately 500-μin./in. strain. The discriminator can withstand a 500% overload, which means that a 5-V signal will be obtained from a strain of 125-μin./in. at the maximum sensitivity and from 2,500-μin./in. at the minimum sensitivity.

71.7 Antennas and Total System Operation

A nickel-cadmium battery supplies the power to the transmitter. For the BCO, the resistance change of the strain gauge changes the frequency of the subcarrier. In the case of the VCO, the millivolt output of the thermocouple changes the frequency of the subcarrier. This change modulates the radio frequency transmitted by the antenna. The receiving antenna picks up the signal and conducts it by wire link to the radio receiver, which is tuned to the transmitting frequency. The radio receiver demodulates the FM carrier to reproduce the subcarrier signal. The subcarrier signal is then fed to the discriminator, which demodulates this signal to obtain a dc voltage, which is then amplified by the dc amplifier and recorded on the oscillograph. The oscillograph record, properly calibrated, is then a display of the strain in microinches per inch for the BCOs, or the temperature in degrees for the VCO. At the same time the dc signal can be read on a VTVM and can be used as a check on the oscillograph.

The transmitter subcarrier oscillators are factory set to operate at a center frequency of about 4,000 Hz. They have a frequency range of ±400 Hz about the 4,000-Hz center frequency. The center frequency is set with a counter at the time of testing. The change of ±400 Hz is the information frequency change brought about by the change in strain or temperature measured by the sensor. It is this information frequency change that the discriminators isolate as a dc voltage change, which is proportional to the measured strain and is recorded on the oscillograph.

71.8 Calibration

Batteries are calibrated under simulated service conditions for voltage drop versus time. Bridge-controlled transmitters are calibrated for strain subcarrier frequency change using a cantilever beam instrumented with resistance strain gauges. The beam is calibrated for load versus strain using a strain indicator. It is then used to calibrate the bridge-controlled transmitters statically, by measuring the subcarrier frequency change as a function of strain. A dynamic calibration can also be made by using a second cantilever beam driven by a vibration generator. Two resistance strain gauges are mounted back-to-back on the second beam and calibrated. One of the gauges is monitored through the telemetry system and the other by wire link to the oscillograph, and the two signals are then compared. The single-resistance strain gauge transmitter is similarly calibrated, but in this case, the beam is fixed in a fatigue machine operating at 30 Hz. Calibrations are performed at various strain levels. Again, two calibrated gauges are monitored and compared, one using the telemetry systems and the other using wire link.

The effect of temperature on a transmitter and battery is measured at temperatures from 65 to 135°F by placing both in an air-circulating oven, with the receiving equipment and the calibration beams to room temperature outside the oven.

The voltage-controlled transmitter is calibrated for temperature subcarrier frequency change from 78 to 640°F. Two calibrated thermocouples, welded next to one another on a piece of stainless steel, are heated simultaneously. After determining by wire link instrumentation that both thermocouples are indicating the same temperature, the millivolt output of one is fed into the transmitter, and the output of the other is fed by wire into a precision potentiometer. The subcarrier frequency change is determined as a function of temperature, and the radio signal is recorded on the calibrated oscillograph, with a galvanometer determining its deflection as a function of temperature. The data obtained by wire link and radio are then compared to establish the calibration. The effect of thermocouple lengths can also be investigated in the same test setup. The receivers and discriminators are calibrated before these tests.

Cold junction compensation may be investigated from −40 to 258°F by cooling the transmitter and battery, with leads shorted and with a 20-mV input, in a cold chamber below room temperature and by heating in an air-circulating oven to above room temperature. The 20-mV input is imposed with a dc power supply kept outside the temperature chamber.

The discriminators are calibrated with the transmitters. The subcarrier frequency, which is the input to the discriminator, is monitored with a digital counter as the calibration beam is loaded. The voltage output corresponding to the frequency change can be monitored with a vacuum tube voltmeter. The strain, subcarrier frequency change, and the voltage output of the discriminator are then correlated. A digital frequency counter is used to set the transmitter center frequency.

71.9 Telemetry Frequency Allocations

Frequency bands for telemetry have been allocated as follows:

88–108 MHz	Low power, noninterference
216–260 MHz	General telemetry
400–475 MHz	Command destruct
1435–1540 MHz	General telemetry
1710–1850 MHz	Video telemetry
2.2–2.3 GHz	General telemetry

The low-power, 88 to 108 MHz, band is shared with FM broadcast stations. Telemetry is allowed to be used, but it must not interfere with broadcast reception. Transmitter power and antennas are limited to provide a signal strength no greater than 50 μV/m at 50 ft from the antenna and/or transmitter. In use, the telemetry transmitters are generally tuned to operate at frequencies between local FM broadcast stations.

The remaining frequency bands are used mainly with aircraft, unmanned vehicles, space vehicles, and for military applications. Equipment for these applications is rigidly constrained for stability, low spurious emissions, low cross talk, good linearity, etc.

71.10 Telemetry Antennas

When transmitter and receiver are stationery, antenna considerations for telemetry are no different than for communications. The usual case, however, is that the transmitter is moving, both rotating and translating and often obscuring the transmission with reflective material. This poses problems in both receiving the signal and tracking the moving transmitter with a directive receiving antenna. In many cases it is necessary to have two or more receiving antennas to receive from a single transmitter.

Transmitting antennas may be either conformal or protruding. The protruding antennas are usually cheaper and simpler. The radiation pattern must include downward directivity if an aircraft or space vehicle will fly directly overhead. A vertical whip antenna does not provide this coverage. A simple choice to provide smooth coverage from the nadir to the horizon is circular polarization at the nadir and elliptic polarization in between. A circular polarized receiving antenna is used to receive the signals over the complete pattern. It must be polarized in the same circular direction as the transmitting antenna. A spiral or helix antenna is usually used on the ground. Two complete radio frequency systems are generally used to receive signals of any polarization. One system uses a circular polarized right-hand antenna and the other uses a circular polarized left-hand antenna. In this manner, as the vehicle tilts or spins, signals are received continuously unless the transmitting antenna is occluded. To receive occluded signals, diversity reception is required. Each receiving station is located such that one fills in the occluded pattern of the other.

In short distance telemetry the same problems are encountered but from a different cause. If the transmitter is occluded at one position by the shaft of a rotating machine, it would be expected that reflections from nearby objects or walls would fill the gap. In practice it is the usual occurrence that a gap is generated once per revolution. This is caused by multiple signal cancellation. There are two methods of overcoming this signal drop-out: (1) with diverse polarization and (2) by locating the receiving antenna close to the transmitting antenna and effectively surrounding the shaft with it.

While information theory has been used extensively to develop space telemetry for the most efficient data transmission over a maximum distance with a minimum of transmitted power, the very inefficiencies permitted in industrial telemetry make for less elaborate, less costly equipment.

Radio channels are used in a relatively inefficient manner, and the distances between transmitter and receiver are usually so short that there are few problems of weak signals. In many cases, measurement and testing via telemetry links take place in completely shielded buildings or in metal housings.

Although telemetry is usually defined as measurement at a distance, it has also gradually begun to embody the concept of control from a distance. In a telemeter—the transmission of the value of a quantity from a remote point—it may only be necessary to observe the reading of an instrument to determine the temperature, pressure, or vibration of a distant or inaccessible object. One can also feed the output of the instrument into a control mechanism, however, such as a relay or an alarm device, so that the telemetered signal may activate or stop a controllable process. Measurement may be performed at one location, indication provided at a second location, and the remote control function initiated at one of the first two locations or even at a third point.

Take, for example, an oil pipeline in which a motor is pumping oil from one location and oil pressure is being measured at a second location. The pressure reading is telemetered to a station where a decision can be made to reduce the speed of the pump motor when the pressure is too high, or a valve may be opened at still another location to cause the oil flow in another path. The decision-making element may be human, an experienced pipeline dispatcher, or an automatic controller. Human or automatic device—either one telemeters a command to the control points.

71.11 Measuring and Transmitting

Telemetry, then, really begins with measurement. A physical quantity is converted to a signal for transmission to another point. The transducer that converts the physical quantity into an electric signal typically may be a piezoelectric crystal, a variable resistance, or perhaps an accelerometer.

Telemetering the measurement signal of the best transducers in no way degrades the measurement below accuracies attainable under laboratory conditions. For instance, in strain measurement it is possible to achieve accuracies of a few microinches per inch or greater, but the limitation is usually the degree of stability in the bond of the strain gauge to the specimen.

If one wants accuracy in temperature measurement, it can be attained by choosing a transducer that provides a large variation in output signal over a small range of temperature. The resolution that this provides may be translated to true accuracy by careful transducer calibration. Typical analog telemetry links maintain a measurement accuracy on a single channel to 1%. This is not a limitation of the total system, since 1% of a 100-degree temperature change would only be 1 degree, so several telemetry channels can easily share the total temperature range to be measured, say, a 100 range divided into four 25 ranges to produce an accuracy of ¼ degree.

One of the limitations to accuracy and repeatability in telemetry is the output level of the transducer. The low electrical levels produced by thermocouples and strain gauges (0.010 V) are more difficult to telemeter than high-voltage levels of 5 V. At low signal levels, extraneous electrical noises produce greater degradation. These may be thermally generated or caused by atmospheric effects or generated by nearby electrical equipment. When low-level transducers are used, stable amplifiers are required to raise their signal voltage to useful modulation levels.

71.12 Modulating and Multiplexing

The transducer signal modulates the frequency of the subcarrier oscillator. This is simply a resonant circuit that produces a given frequency in the audio range, say 100 Hz, and is varied plus or minus this center frequency by the signal from the transducer as it responds to the variable that it is measuring. When the signal is fed to a loudspeaker, one actually hears a rising or falling tone. The subcarrier oscillator modulates a radio frequency carrier, varying its frequency in accordance with the subcarrier voltage signal. The radio frequency in FM industrial radio telemetry links is usually in the 88- to 108-MHz band, permitting the use of high-grade radio tuners already mass produced for the high-fidelity market. The radio receiver demodulates the signal, removing the carrier and feeding it to a special discriminator circuit that removes the double modulation and reproduces an analog of the original measurement signal for calibrated indication or recording.

There can be great variations in the strength of the radio signal received because of variations in distance between transmitter and receiver or because of the interposition of metallic objects. In industrial radio telemetry transmission, these effects are prevented from disturbing the data by resorting to FM of both the subcarrier and the carrier so that the telemetered signal is unchanged by undesirable amplitude variations. This method is called FM/FM telemetry. There are other methods of carrier modulation, such as pulse amplitude modulation, phase modulation, and pulse duration modulation. Each has its proper place in missile and space telemetry, where great distances must be spanned with a maximum of data over crowded and often noisy communication channels. Pulse code modulation of an FM link, however, may be expected to become more widespread in industrial telemetry.

Particularly in missile telemetry, it is important that multiple measurements be transmitted over a single carrier to save power and minimize electronic equipment and antennas. Such simultaneous transmission of signals over a common path, called multiplexing, is sometimes used in industrial telemetry. When concurrent data about several simultaneous events are transmitted by several subcarriers, each subcarrier oscillator has a distinctive reference frequency and swings from this center frequency toward arbitrary maximum and minimum frequencies in response to signals from a corresponding transducer. Thus a number of separate audio frequency bands are sent over the radio frequency carrier. This is called frequency division multiplexing. The frequency division multiplex requires careful adjustment of subcarrier frequencies and the corresponding filters at the receiver and strong suppression of harmonics to avoid cross talk or interaction between channels.

Multiple measurements may also be transmitted over the carrier by sampling the output of each transducer in rapid sequence, a technique called time-division multiplexing. The technique has been used to handle as many as a million samples per second. It provides for very simple data displays, easier separation of channels for recording or analysis, and is free of cross talk. If possible, though, it is advantageous to use no multiplexing at all for concurrent data talking but to use separate radio carriers for each measurement being transmitted. The multiplex telemeter requires careful adjustment of subcarrier frequencies and precisely tuned filters to separate them at the receiver. This adds to the cost of the equipment and requires considerable experience of an operator.

The telemetry data received may be recorded in a number of ways, but such records must preserve the accuracy of the entire system. For example, if one is monitoring a 1% system and can distinguish 1/64th in. on a paper graph, the minimum graph size for full scale should be approximately 2 in. Similarly, numeric data should be printed to enough decimal places to preserve the accuracy of the system.

A single channel of industrial FM/FM telemetry equipment may cost between $1000 and $2000, depending on the flexibility required and the measurements being made. It buys everything needed for a given remote measurement—transducer, radio link, power supply, and simple indicator.

71.13 Passive Telemeters

Passively powered telemeters offer some interesting advantages. When a number of telemeters are used, they can be powered in sequence or only when measurements are required, thus preventing radio frequency congestion. In medical applications of telemetry, passive devices eliminate the danger involved in swallowing or implanting batteries.

Most passive telemeters are essentially an inductance and a capacitance coupled as a resonant circuit. Either of these components may be pressure sensitive or temperature sensitive. A nearby magnetic coil coupled to this circuit can, by means of a varying frequency, determine the resonant point of the telemeter, which can then be a function of the temperature or pressure being measured.

71.14 The Receiving Station

The industrial telemetry receiving station differs vastly in purpose and principle from the transmitting station. Its usual environment is no more difficult to cope with, in terms of ambient temperature, shock, and vibrations, than an automobile radio. It receives signals over relatively short distances in which the subcarrier frequencies are so widely spaced that harmonics and drift are no problem.

In the FM broadcast band, available professional-grade high-fidelity tuners have 1 or 2 μV sensitivity for 30 dB of quieting of extraneous noises and automatic frequency control circuits, which compensate for both transmitter and receiver drift. They feed telemetry phase-lock discriminators, which lock the receiver into the frequency and phase of the incoming signal.

The transmitters usually radiate from the resonant elements themselves, avoiding elaborate antennas that might be required for longer distance transmission; the receivers use simple dipole or commercial TV antennas. Thus, industrial radio telemetry has become a carefully engineered blend of the borrowed and the new.

The September 1899 yacht race for the America's Cup was unique in one respect: New York newspapers carried up-to-the-minute reports on its progress at sea. Guglielmo Marconi had installed his primitive transmitters on two ships, and the success of this demonstration of wireless telegraphy aroused worldwide interest and brought the financial support he needed for his first transatlantic communication in 1901. By that time, six telegraph cables had been laid under the Atlantic and the demand for rapid point-to-point communication was well established.

John Fleming, a consultant with the Marconi International Marine Communication Company, soon invented his diode detector, and in 1906 Lee De Forest's triode provided sensitive detection and amplification of radio waves. But Marconi's spark-gap transmitter was an inefficient device. When high voltage was applied through an inductor, the gap broke down, the air was ionized, and the energy in the inductor was dissipated in a series of highly damped oscillatory currents coupled into the aerial-to-ground circuit. The hertzian waves radiated could be

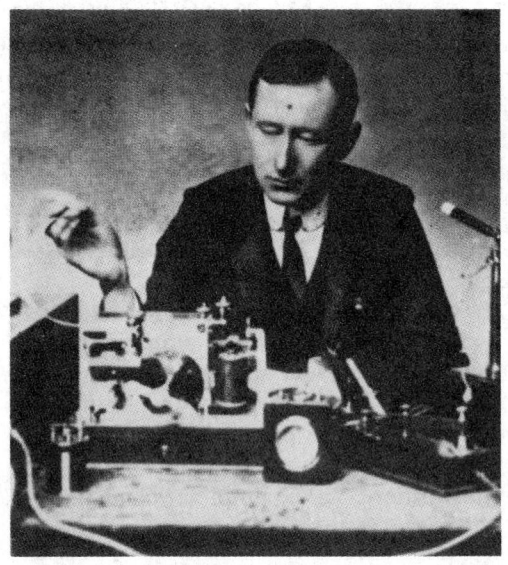

Marconi with the receiving apparatus used at Signal Hill, 1901. (Courtesy of the IEEE Center for the History of Electrical Engineering.)

interrupted into dots and dashes, but the transmission of voice signals would require a *continuous wave* as a carrier.

72

Computer-Aided Design and Analysis of Communication Systems

72.1   Introduction 1593
72.2   The Role of Simulation 1594
72.3   Motivation for the Use of Simulation 1595
72.4   Limitations of Simulation 1595
72.5   Simulation Structure 1596
72.6   The Interdisciplinary Nature of Simulation 1597
72.7   Model Design 1597
72.8   Low-Pass Models 1598
72.9   Pseudorandom Signal and Noise Generators 1599
72.10   Transmitter, Channel, and Receiver Modeling 1601
72.11   Symbol Error Rate Estimation 1602
72.12   Validation of Simulation Results 1604
72.13   A Simple Example Illustrating Simulation Products 1604
72.14   Conclusions 1607

William H. Tranter
University of Missouri–Rolla

Kurt L. Kosbar
University of Missouri–Rolla

72.1 Introduction

It should be clear from the preceding chapters that communication systems exist to perform a wide variety of tasks. The demands placed on today's communication systems necessitate higher data rates, greater flexibility, and increased reliability. Communication systems are therefore becoming increasingly complex, and the resulting systems cannot usually be analyzed using traditional (pencil and paper) analysis techniques. In addition, communication systems often operate in complicated environments that are not analytically tractable. Examples include channels that exhibit severe bandlimiting, multipath, fading, interference, non-Gaussian noise, and perhaps even burst noise. The combination of a complex system and a complex environment makes the design and analysis of these communication systems a formidable task. Some level of computer assistance must usually be invoked in both the design and analysis process. The appropriate level of computer assistance can range from simply using numerical techniques to solve a differential equation defining an element or subsystem to developing a **computer simulation** of the end-to-end communication system.

There is another important reason for the current popularity of computer-aided analysis and simulation techniques. It is now practical to make extensive use of these techniques. The

1593

computing power of many personal computers and workstations available today exceeds the capabilities of many large mainframe computers of only a decade ago. The low cost of these computing resources make them widely available. As a result, significant computing resources are available to the communications engineer within the office or even the home environment. Personal computers and workstations tend to be resources dedicated to a specific individual or project. Since the communications engineer working at his or her desk has control over the computing resource, lengthy simulations can be performed without interfering with the work of others. Over the past few years a number of software packages have been developed that allow complex communication systems to be simulated with relative ease [1]. The best of these packages contains a wide variety of subsystem models as well as integrated graphics packages that allow waveforms, spectra, histograms, and performance characteristics to be displayed without leaving the simulation environment. For those motivated to generate their own simulation code, the widespread availability of high-quality C, Pascal, and FORTRAN compilers makes it possible for large application-specific simulation programs to be developed for personal computers and workstations. When computing tools are both available and convenient to use, they will be employed in the day-to-day efforts of system analysts and designers.

The purpose of this chapter is to provide a brief introduction to the subject of **computer-aided design and analysis** of communication systems. Since computer-aided design and analysis almost always involves some level of simulation, we focus our discussion on the important subject of the simulation of communication systems.

Computer simulations can, of course, never replace a skilled engineer, although they can be a tremendous help in both the design and analysis process. The most powerful simulation program cannot solve all the problems that arise, and the process of making trade-off decisions will always be based on experience. In addition, evaluating and interpreting the results of a complex simulation require considerable skill and insight. While these remarks seem obvious, as computer-aided techniques become more powerful, one is tempted to replace experience and insight with computing power.

72.2 The Role of Simulation

The main purposes of simulation are to help us understand the operation of a complex communication system, to determine acceptable or optimum parameters for implementation of a system, and to determine the performance of a communication system. There are basically two types of systems in which communication engineers have interest: **communication links** and communication networks.

A communication link is usually a single source, a single user, and the components and channel between source and user. A typical link architecture is shown in Fig. 72.1. The important performance parameter in a digital communication link is typically the reliability of the communication link as measured by the symbol or bit error rate (BER). In an analog communication link the performance parameter of interest is typically the signal-to-noise ratio (SNR) at the receiver input or the mean-square error of the receiver output. The simulation is usually performed to determine the effect of system parameters, such as filter bandwidths or code rate, or to determine the effect of environmental parameters, such as noise levels, noise statistics, or power spectral densities.

A communication network is a collection of communication links with many signal sources and many users. Computer simulation programs for networks often deal with problems of routing, flow and congestion control, and the network delay. While this chapter deals with the communica-

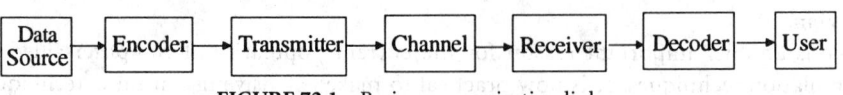

FIGURE 72.1   Basic communication link.

tion link, the reader is reminded that network simulation is also an important area of study. The simulation methodologies used for communication networks are different from those used on links because, in a communication link simulation, each waveform present in the system is sampled using a constant sampling frequency. In contrast, network simulations are event-driven, with the important events being such quantities as the time of arrival of a message.

Simulations can be developed to investigate either transient phenomena or steady-state properties of a system. The study of the acquisition time of a phase-lock loop receiver is an example of a transient phenomenon. Simulations that are performed to study transient behavior often focus on a single subsystem such as a receiver synchronization system. Simulations that are developed to study steady-state behavior often model the entire system. An example is a simulation to determine the BER of a system.

72.3 Motivation for the Use of Simulation

As mentioned previously, simulation is a reasonable approach to many design and analysis problems because complex problems demand that computer-based techniques be used to support traditional analytical approaches. There are many other motivations for making use of simulation.

A carefully developed simulation is much like having a breadboard implementation of the communication system available for study. Experiments can be performed using the simulation much like experiments can be performed using hardware. System parameters can be easily changed, and the impact of these changes can be evaluated. By continuing this process, parameteric studies can easily be conducted and acceptable, or perhaps even optimum, parameter values can be determined. By changing parameters, or even the system topology, one can play "what if" games much more quickly and economically using a simulation than with a system realized in hardware.

It is often overlooked that simulation can be used to support analysis. Many people incorrectly view simulation as a tool to be used only when a system becomes too complex to be analyzed using traditional analysis techniques. Used properly, simulation goes hand in hand with traditional techniques in that simulation can often be used to guide analysis. A properly developed simulation provides insight into system operation. As an example, if a system has many parameters, these can be varied in a way that allows the most important parameters, in terms of system performance, to be identified. The least important parameters can then often be discarded, with the result being a simpler system that is more tractable analytically. Analysis also aids simulation. The development of an accurate and efficient simulation is often dependent upon a careful analysis of various portions of the system.

72.4 Limitations of Simulation

Simulation, useful as it is, does have limitations. It must be remembered that a system simulation is an approximation to the actual system under study. The nature of the approximations must be understood if one is to have confidence in the simulation results. The accuracy of the simulation is limited by the accuracy to which the various components and subsystems within the system are modeled. It is often necessary to collect extensive experimental data on system components to ensure that simulation models accurately reflect the behavior of the components. Even if this step is done with care, one can only trust the simulation model over the range of values consistent with the previously collected experimental data. A main source of error in a simulation results because models are used at operating points beyond which the models are valid.

In addition to modeling difficulties, it should be realized that the digital simulation of a system can seldom be made perfectly consistent with the actual system under study. The simulation is affected by phenomena not present in the actual system. Examples are the aliasing errors resulting from the sampling operation and the finite word length (quantization) effects present in the simulation. Practical communication systems use a number of filters, and modeling the analog filters present in the actual system by the digital filters required by the simulation involves a number of approximations. The assumptions and approximations used in modeling an analog filter using impulse-invariant digital filter synthesis techniques are quite different from the assumptions and approximations used in bilinear z-transform techniques. Determining the appropriate modeling technique requires careful thought.

Another limitation of simulation lies in the excessive computer run time that is often necessary for estimating performance parameters. An example is the estimation of the system BER for systems having very low nominal bit error rates. We will expand on this topic later in this chapter.

72.5 Simulation Structure

As illustrated in Fig. 72.1, a communication system is a collection of subsystems such that the overall system provides a reliable path for information flow from source to user. In a computer simulation of the system, the individual subsystems must first be accurately modeled by signal processing operations. The overall simulation program is a collection of these signal processing operations and must accurately model the overall communication system. The important subject of subsystem modeling will be treated in a following section.

The first step in the development of a simulation program is to define the topology of the system, which specifies the manner in which the individual subsystems are connected. The subsystem models must then be defined by specifying the signal processing operation to be performed by each of the various subsystems. A simulation structure may be either fixed topology or free topology. In a fixed topology simulation, the basic structure shown in Fig. 72.1 is modeled. Various subsystems can be bypassed if desired by setting switches, but the basic topology cannot be modified. In a free topology structure, subsystems can be interconnected in any way desired and new additional subsystems can be added at will.

A simulation program for a communication system is a collection of at least three operations, shown in Fig. 72.2, although in a well-integrated simulation these operations tend to merge together. The first operation, sometimes referred to as the *preprocessor*, defines the parameters of each subsystem and the intrinsic parameters that control the operation of the simulation. The second operation is the *simulation exercisor*, which is the simulation program actually executed on the computer. The third operation performed in a simulation program is that of *postprocessing*. This is a collection of routines that format the simulation output in a way which provides insight into system operations and allows the performance of the communication system under study to be evaluated. A postprocessor usually consists of a number of graphics-based routines, allowing the user to view waveforms and other displays generated by the simulation. The postprocessor also consists of a number of routines that allow estimation of the bit error rate, signal-to-noise ratios, histograms, and power spectral densities.

When faced with the problem of developing a simulation of a communication system, the first fundamental choice is whether to develop a custom simulation using a general-purpose high-level language or to use one of the many special-purpose communication system simulation languages available. If the decision is made to develop a dedicated simulation using a general-purpose language, a number of resources are needed beyond a quality compiler and a mathematics library. Also needed are libraries for filtering routines, software models for each of the subsystems contained in the overall system, channel models, and the waveform display and data analysis routines needed for

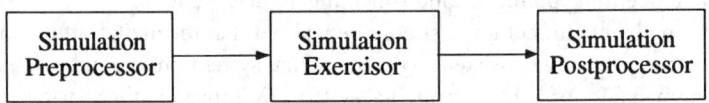

FIGURE 72.2   Typical structure of a simulation program.

the analysis of the simulation results (postprocessing). While at least some of the required software will have to be developed at the time the simulation is being written, many of the required routines can probably be obtained from digital signal processing (DSP) programs and other available sources. As more simulation projects are completed, the database of available routines becomes larger.

The other alternative is to use a **dedicated simulation language**, which makes it possible for one who does not have the necessary skills to create a custom simulation using a high-level language to develop a communication system simulation. Many simulation languages are available for both personal computers and workstations [1]. While the use of these resources can speed simulation development, the user must ensure that the assumptions used in developing the models are well understood and applicable to the problem of interest. In choosing a dedicated language from among those that are available, one should select a language that has an extensive model library, an integrated postprocessor with a wide variety of data analysis routines, on-line help and documentation capabilities, and extensive error-checking routines.

72.6 The Interdisciplinary Nature of Simulation

The subject of computer-aided design and analysis of communication systems is very much interdisciplinary in nature. The major disciplines that bear on the subject are communication theory, DSP, numerical analysis, and stochastic process theory. The roles played by these subjects is clear. The simulation user must have knowledge of the behavior of communication theory if the simulation results are to be understood. The analysis techniques of communication theory allow simulation results to be verified. Since each subsystem in the overall communication system is a signal processing operation, the tools of DSP provide the algorithms to realize filters and other subsystems. Numerical analysis techniques are used extensively in the development of signal processing algorithms. Since communication systems involve random data signals, as well as noise and other disturbances, the concepts of stochastic process theory are important in developing models of these quantities and also for determining performance estimates.

72.7 Model Design

Practicing engineers frequently use models to investigate the behavior of complex systems. Traditionally, models have been physical devices or a set of mathematical expressions. The widespread use of powerful digital computers now allows one to generate computer programs that model physical systems. Although the detailed development and use of computer models differs significantly from their physical and mathematical counterparts, the computer models share many of the same design constraints and trade-offs. For any model to be useful one must guarantee that the response of the model to stimuli will closely match the response of the target system, the model must be designed and fabricated in much less time and at significantly less expense than the target system, and the model must be reasonably easy to validate and modify. In addition to these constraints, designers of computer models must assure that the amount of processor time required to execute the model is not excessive. The optimal model is the one that appropriately balances these conflicting requirements. Figure 72.3 describes the typical design trade-off faced when developing computer models. A somewhat surprising observation is that the optimal model is often not the one that most closely approximates the target system. A highly detailed model will typically require a tremendous amount of time to develop, will be difficult to validate and modify, and may require prohibitive processor time to execute. Selecting a model that achieves a good balance between these constraints is as much an art as a science. Being aware of the trade-offs which exist, and must be addressed, is the first step toward mastering the art of modeling.

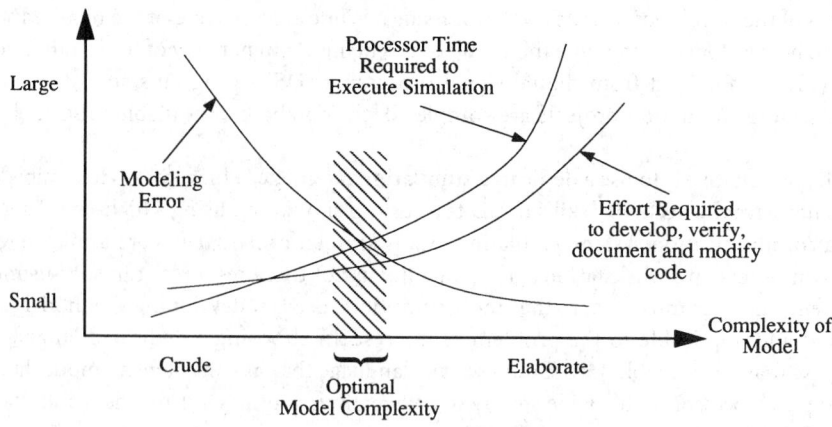

FIGURE 72.3   Design constraints and trade-offs.

72.8 Low-Pass Models

In most cases of practical interest the physical layer of the communication system will use continuous time (CT) signals, while the simulation will operate in discrete time (DT). For the simulation to be useful, one must develop DT signals and systems that closely match their CT counterparts. This topic is discussed at length in introductory DSP texts. A prominent result in this field is the Nyquist sampling theorem, which states that if a CT signal has no energy above frequency f_h Hz, one can create a DT signal that contains *exactly* the same information by sampling the CT signal at any rate in excess of $2 f_h$ samples per second. Since the execution time of the simulation is proportional to the number of samples it must process, one naturally uses the lowest sampling rate possible. While the Nyquist theorem should not be violated for arbitrary signals, when the CT signal is bandpass one can use low-pass equivalent (LPE) waveforms that contain all the information of the CT signal but can be sampled slower than $2 f_h$.

Assume the energy in a bandpass signal is centered about a carrier frequency of f_c Hz and ranges from f_l to f_h Hz, resulting in a bandwidth of $f_h - f_l = W$ Hz, as in Fig. 72.4. It is not unusual for W to be many orders of magnitude less than f_c. The bandpass waveform $x(t)$ can be expressed as a function of two low-pass signals. Two essentially equivalent LPE expansions are known as the envelope/phase representation [2],

$$x(t) = A(t) \cos\left[2\pi f_c t + \theta(t)\right] \tag{72.1}$$

and the quadrature representation,

$$x(t) = x_c(t) \cos(2\pi f_c t) - x_s(t) \sin(2\pi f_c t) \tag{72.2}$$

All four real signals $A(t)$, $\theta(t)$, $x_c(t)$, and $x_s(t)$ are low pass and have zero energy above $W/2$ Hz. A computer simulation that replaces $x(t)$ with a pair of LPE signals will require far less processor time since the LPE waveforms can be sampled at W as opposed to $2 f_h$ samples per second. It is cumber-

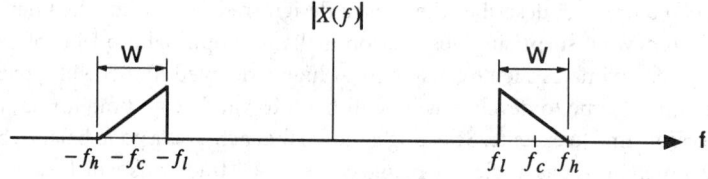

FIGURE 72.4   Amplitude spectrum of a bandpass signal.

some to work with two signals rather than one signal. A more mathematically elegant LPE expansion is

$$x(t) = \text{Re}\{v(t)e^{j2\pi f_c t}\} \tag{72.3}$$

where $v(t)$ is a low-pass, *complex-time domain signal* that has no energy above $W/2$ Hz. Signal $v(t)$ is known as the complex envelope of $x(t)$ [3]. It contains all the information of $x(t)$ and can be sampled at W samples per second without aliasing. This notation is disturbing to engineers accustomed to viewing all time domain signals as real. However, a complete theory exists for complex time domain signals, and with surprisingly little effort one can define convolution, Fourier transforms, analog-to-digital and digital-to-analog conversions, and many other signal processing algorithms for complex signals. If f_c and W are known, the LPE mapping is one-to-one so that $x(t)$ can be completely recovered from $v(t)$. While it is conceptually simpler to sample the CT signals at a rate in excess of $2f_h$ and avoid the mathematical difficulties of the LPE representation, the tremendous difference between f_c and W makes the LPE far more efficient for computer simulation. This type of trade-off frequently occurs in computer simulation. A careful mathematical analysis of the modeling problem conducted *before* any computer code is generated can yield substantial performance improvements over a conceptually simpler, but numerically inefficient approach.

The fundamental reason the LPE representation outlined above is popular in simulation is that one can easily generate **LPE models** of linear time-invariant bandpass filters. The LPE of the output of a bandpass filter is merely the convolution of the LPE of the input signal and the LPE of the impulse response of the filter. It is far more difficult to determine a LPE model for nonlinear and time-varying systems. There are numerous approaches that trade off flexibility and simplicity. If the system is nonlinear and time invariant, a Volterra series can be used. While this series will exactly represent the nonlinear device, it is often analytically intractable and numerically inefficient. For nonlinear devices with a limited amount of memory the AM/AM, AM/PM [4] LPE model is useful. This model accurately describes the response of many microwave amplifiers including traveling-wave tubes, solid-state limiting amplifiers, and, under certain conditions, devices which exhibit hysteresis. The Chebyshev transform [5] is useful for memoryless nonlinearities such as hard and soft limiters. If the nonlinear device is so complex that none of the conventional LPE models can be used, one may need to convert the LPE signal back to its bandpass representation, route the bandpass signal through a model of the nonlinear device, and then reconvert the output to a LPE signal for further processing. If this must be done, one has the choice of increasing the sampling rate for the entire simulation or using different sampling rates for various sections of the simulation. The second of these approaches is known as a *multirate simulation* [6]. The interpolation and decimation operations required to convert between sampling rates can consume significant amounts of processor time. One must carefully examine this trade-off to determine if a multirate simulation will substantially reduce the execution time over a single, high sampling rate simulation. Efficient and flexible modeling of nonlinear devices is in general a difficult task and continues to be an area of active research.

72.9 Pseudorandom Signal and Noise Generators

The preceding discussion was motivated by the desire to efficiently model filters and nonlinear amplifiers. Since these devices often consume the majority of the processor time, they are given high priority. However, there are a number of other subsystems that do not resemble filters. One example is the data source that generates the message or waveform which must be transmitted. While signal sources may be analog or digital in nature, we will focus exclusively on binary digital sources. The two basic categories of signals produced by these devices are known as *deterministic* and *random*. When performing worst-case analysis, one will typically produce known, repetitive signal patterns designed to stress a particular subsystem within the overall communication system. For example, a signal with few transitions may stress the symbol synchronization loops, while a signal with many regularly spaced transitions may generate unusually wide bandwidth signals. The generation of this type of signal is straightforward and highly application dependent. To test the nominal system performance one typically uses a random data sequence. While generation of a truly random signal is arguably impossible [7], one can easily generate pseudorandom (PN) sequences. PN sequence generators have been extensively studied since they are used in Monte Carlo integration and simulation [8] programs and in a variety of wideband and secure communication systems. The two basic structures for generating PN sequences are binary shift registers (BSRs) and linear congruential algorithms (LCAs).

Digital data sources typically use BSRs, while noise generators often use LCAs. A logic diagram for a simple BSR is shown in Fig. 72.5. This BSR consists of a clock, six D-type flip-flops (F/F), and an exclusive OR gate denoted by a modulo-two adder. If all the F/F are initialized to 1, the output of the device is the waveform shown in Fig. 72.6. Notice that the waveform is periodic with period 63 $= 2^6 - 1$, but within one cycle the output has many of the properties of a random sequence. This demonstrates all the properties of the BSR, LCA, and more advanced PN sequence generators. All PN generators have memory and must therefore be initialized by the user before the first sample is generated. The initialization data is typically called the seed. One must choose this seed carefully to ensure the output will have the desired properties (in this example, one must avoid setting all F/F to zero). All PN sequence generators will produce periodic sequences. This may or may not be a problem. If it is a concern, one should ensure that one period of the PN sequence generator is longer than the total execution time of the simulation. This is usually not a significant problem, since one can easily construct BSRs that have periods greater than 10^{27} clock cycles. The final concern is how closely the behavior of the PN sequence generator matches a truly random sequence. Standard statistical analysis algorithms have been applied to many of these generators to validate their performance.

Many digital communication systems use m bit (M-ary) sources where $m > 1$. Figure 72.7 depicts a simple algorithm for generating a M-ary random sequence from a binary sequence. The clock must now cycle through m cycles for every generated symbol, and the period of the generator has been reduced by a factor of m. This may force the use of a longer-period BSR. Another common application of PN sequence generators is to produce samples of a continuous stochastic process, such as Gaussian noise. A structure for producing these samples is shown in Fig. 72.8. In this case the BSR has been replaced by an LCA [7]. The LCA is very similar to BSR in that it requires a seed value, is clocked once for each symbol generated, and will generate a periodic sequence. One can generate a white noise process with an arbitrary first-order probability density function (pdf) by

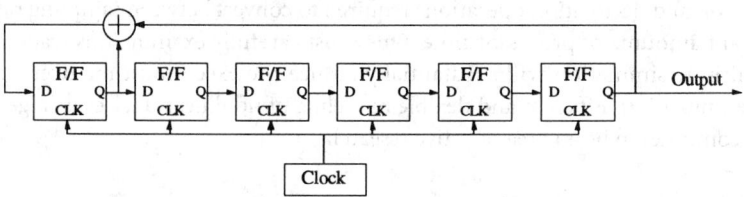

FIGURE 72.5   Six-stage binary shift register PN generator.

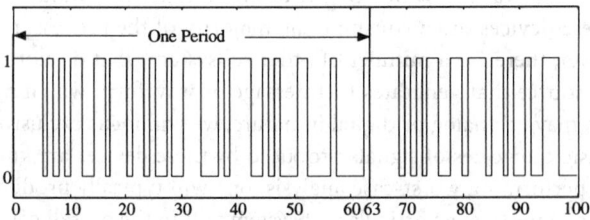

FIGURE 72.6   Output of a six-stage maximal length BSR.

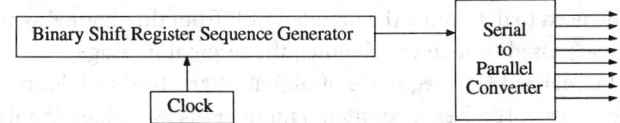

FIGURE 72.7   *M*-ary PN sequence generator.

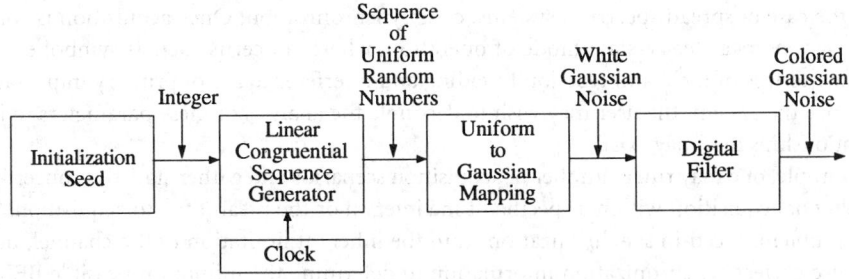

FIGURE 72.8   Generation of Gaussian noise.

passing the output of the LCA through an appropriately designed nonlinear, memoryless mapping. Simple and well-documented algorithms exist for the uniform to Gaussian mapping. If one wishes to generate a nonwhite process, the output can be passed through the appropriate filter. Generation of a wide-sense stationary Gaussian stochastic process with a specified power spectral density is a well-understood and -documented problem. It is also straightforward to generate a white sequence with an arbitrary first-order pdf or to generate a specified power spectral density if one does not attempt to control the pdf. However, the problem of generating a noise source with an arbitrary pdf *and* an arbitrary power spectral density is a significant challenge [9].

72.10 Transmitter, Channel, and Receiver Modeling

Most elements of transmitters, channels, and receivers are implemented using standard DSP techniques. Effects that are difficult to characterize using mathematical analysis can often be included in the simulation with little additional effort. Common examples include gain and phase imbalance in quadrature circuits, nonlinear amplifiers, oscillator instabilities, and antenna platform motion. One can typically use LPE waveforms and devices to avoid translating the modulator output to the carrier frequency. Signal levels in physical systems often vary by many orders of magnitude, with the output of the transmitters being extremely high energy signals and the input to receivers at very low energies. To reduce execution time and avoid working with extremely large and small signal level simulations, one often omits the effects of linear amplifiers and attenuators and uses normalized signals. Since the performance of most systems is a function of the signal-to-noise ratio, and not of absolute signal level, normalization will have no effect on the measured performance. One must be careful to document the normalizing constants so that the original signal levels can be reconstructed if needed. Even some rather complex functions, such as error detecting and correcting codes, can be handled in this manner. If one knows the uncoded error rate for a system, the coded error rate can often be closely approximated by applying a mathematical mapping. As will be pointed out below, the amount of processor time required to produce a meaningful error rate estimate is often inversely proportional to the error rate. While an uncoded error rate may be easy to measure, the coded error rate is usually so small that it would be impractical to execute a simulation to measure this quantity directly. The performance of a coded communication system is most often determined by first executing a simulation to establish the channel **symbol error rate**. An analytical mapping can then be used to determine the decoded BER from the channel symbol error rate.

Once the signal has passed though the channel, the original message is recovered by a receiver. This can typically be realized by a sequence of digital filters, feedback loops, and appropriately selected nonlinear devices. A receiver encounters a number of clearly identifiable problems that one may wish to address independently. For example, receivers must initially synchronize themselves to the incoming signal. This may involve detecting that an input signal is present, acquiring an estimate of the carrier amplitude, frequency, phase, symbol synchronization, frame synchronization, and, in the case of spread spectrum systems, code synchronization. Once acquisition is complete, the receiver enters a steady-state mode of operation, where concerns such as symbol error rate, mean time to loss of lock, and reaction to fading and interference are of primary importance. To characterize the system, the user may wish to decouple the analysis of these parameters to investigate relationships that may exist.

For example, one may run a number of acquisition scenarios and gather statistics concerning the probability of acquisition within a specified time interval or the mean time to acquisition. To isolate the problems faced in synchronization from the inherent limitation of the channel, one may wish to use perfect synchronization information to determine the minimum possible BER. Then the symbol or carrier synchronization can be held at fixed errors to determine sensitivity to these parameters and to investigate worst-case performance. Noise processes can be used to vary these parameters to investigate more typical performance. The designer may also wish to investigate the performance of the synchronization system to various data patterns or the robustness of the synchronization system in the face of interference. The ability to measure the system response to one parameter while a wide range of other parameters are held fixed and the ability to quickly generate a wide variety of environments are some of the more significant advantages that simulation enjoys over more conventional hardware and analytical models.

72.11 Symbol Error Rate Estimation

One of the most fundamental parameters to measure in a digital communication system is the steady-state BER. The simplest method for estimating the BER is to perform a **Monte Carlo (MC) simulation**. The simulation conducts the same test one would perform on the physical system. All data sources and noise sources produce typical waveforms. The output of the demodulator is compared to the output of the message source, and the BER is estimated by dividing the number of observed errors by the number of bits transmitted. This is a simple technique that will work with any system that has ergodic [10] noise processes. The downside of this approach is that one must often pass a very large number of samples through the system to produce a reliable estimate of the BER. The question of how many samples must be collected can be answered using confidence intervals. The confidence interval gives a measure of how close the true BER will be to the estimate produced by the MC simulation. A typical confidence interval curve is shown in Fig. 72.9. The ratio of the size of the confidence interval to the size of the estimate is a function of the number of errors observed. Convenient rules of thumb for this work are that after one error is observed the point estimate is accurate to within 3 orders of magnitude, after 10 errors the estimate is accurate to within a factor of 2, and after 100 errors the point estimate will be accurate to a factor of 1.3. This requirement for tens or hundreds of errors to occur frequently limits the usefulness of MC simulations for systems that have low error rates and has motivated research into more efficient methods of estimating BER.

Perhaps the fastest method of BER estimation is the semi-analytic (SA) or quasi-analytic technique [11]. This technique is useful for systems that resemble Fig. 72.10. In this case the mean of the decision metric is a function of the transmitted data pattern and is independent of the noise. All other parameters of the pdf of the decision metric are a function of the noise and are independent

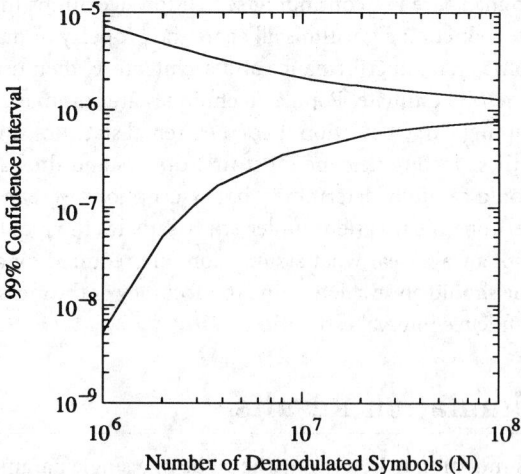

FIGURE 72.9 Typical confidence interval (BER) point estimate = 10^{-6}.

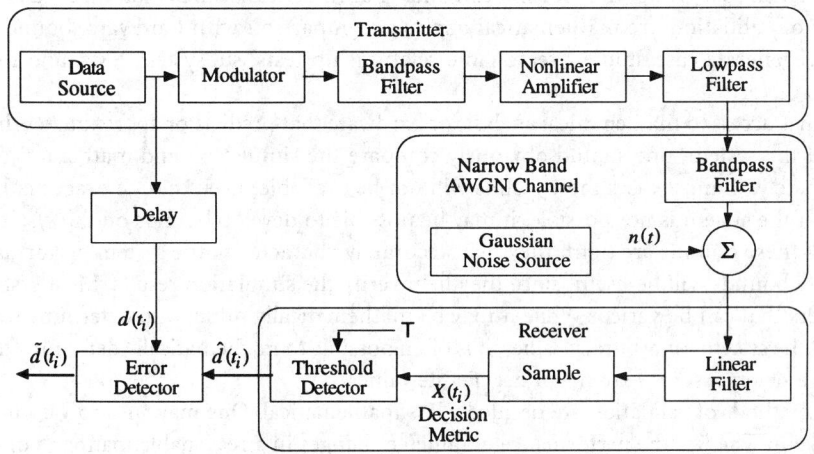

FIGURE 72.10 Typical digital communication system.

of the data. This means that one can analytically determine the conditional pdf of the decision metric given the transmitted data pattern. By using total probability one can then determine the unconditional error rate. The problem with conventional mathematical analysis is that when the channel has a significant amount of memory or the nonlinearity is rather complex, one must compute a large number of conditional density functions. Simulation can easily solve this problem for most practical systems. A noise-free simulation is executed, and the value of the decision metric is recorded in a data file. Once the simulation is complete, this information can be used to reconstruct the conditional and ultimately the unconditional error rate. This method generates highly accurate estimates of the BER and makes very efficient use of computer resources, but can only be used in the special cases where one can analytically determine the conditional pdf.

The MC and SA techniques fall at the two extremes of BER estimation. MC simulations require no *a priori* information concerning the system performance or architecture but may require tremendous amounts of computer time to execute. SA techniques require an almost trivial amount of computer time for many cases but require the analyst to have a considerable amount of information concerning the system. There is a continuing search for algorithms that fall in between these extremes. These variance reduction algorithms all share the property of making a limited number of assumptions concerning system performance and architecture, then using this information to reduce the variance of the MC estimate. Popular techniques are summarized in Reference 12 and include importance sampling, large deviation theory, extremal statistics, and tail extrapolation. To successfully use one of these techniques one must first understand the basic concept behind the technique. Then one should carefully determine what assumptions were made concerning the system architecture to determine if the system under study satisfies the requirements. This can be a difficult task since it is not always clear what assumptions are required for a specified technique to be applicable. Finally, one should always determine the accuracy of the measurement through some technique similar to confidence interval estimation.

72.12 Validation of Simulation Results

One often constructs a simulation to determine the value of a single parameter, such as the system BER. However the estimate of this parameter has little or no value unless one can ensure that the simulation model closely resembles the physical system. A number of methods can be used to validate a simulation. Individually, none of them will guarantee that the simulation results are accurate, but taken as a group, they form a convincing argument that the results are realistic. Seven methods of validation are mathematical analysis, comparison with hardware, bounding techniques, degenerate case studies, reasonable relationship tests, subsystem tests, and redundant simulation efforts.

If one has access to mathematical analysis or hardware that predicts or approximates the performance of the system, one should obviously compare the simulation and mathematical results. Unfortunately, in most cases these results will not be available. Even though exact mathematical analysis of the system is not possible, it may be possible to develop bounds on the system performance. If these bounds are tight, they may accurately characterize the system performance, but even loose bounds will be useful since they help verify the simulation results. Most systems have parameters that can be varied. While it may be mathematically difficult to determine the performance of the system for arbitrary values, it is often possible to mathematically determine the results when parameters assume extreme or degenerate values.

Other methods of validation are decidedly less mathematical. One may wish to vary parameters and ascertain whether the performance parameter changes in a reasonable manner. For example, small changes in SNR rarely cause dramatic changes in system performance. When constructing a simulation, each subsystem, such as filters, nonlinear amplifiers, and noise and data sources, should be thoroughly tested before being included in a larger simulation. Be aware, however, that correct operation of all the various subsystems that make up a communication system does not imply that the overall system performs correctly. If one is writing his or her own code, one must verify that there are no software bugs or fundamental design errors. Even if one purchases a commercial software package, there is no guarantee that the designer of the software models made the same assumptions the user will make when using the model. In most cases it will be far easier to test a module before it is inserted into a simulation than it will be to isolate a problem in a complex piece of code. The final check one may wish to perform is a redundant simulation. There are many methods of simulating a system. One may wish to have two teams investigate a problem or have a single team implement a simulation using two different techniques to verify that the results are reasonable.

72.13 A Simple Example Illustrating Simulation Products

To illustrate the output that is typically generated by a communication system simulation, a simple example is considered. The system is that considered in Fig. 72.10. An OQPSK (offset quadrature

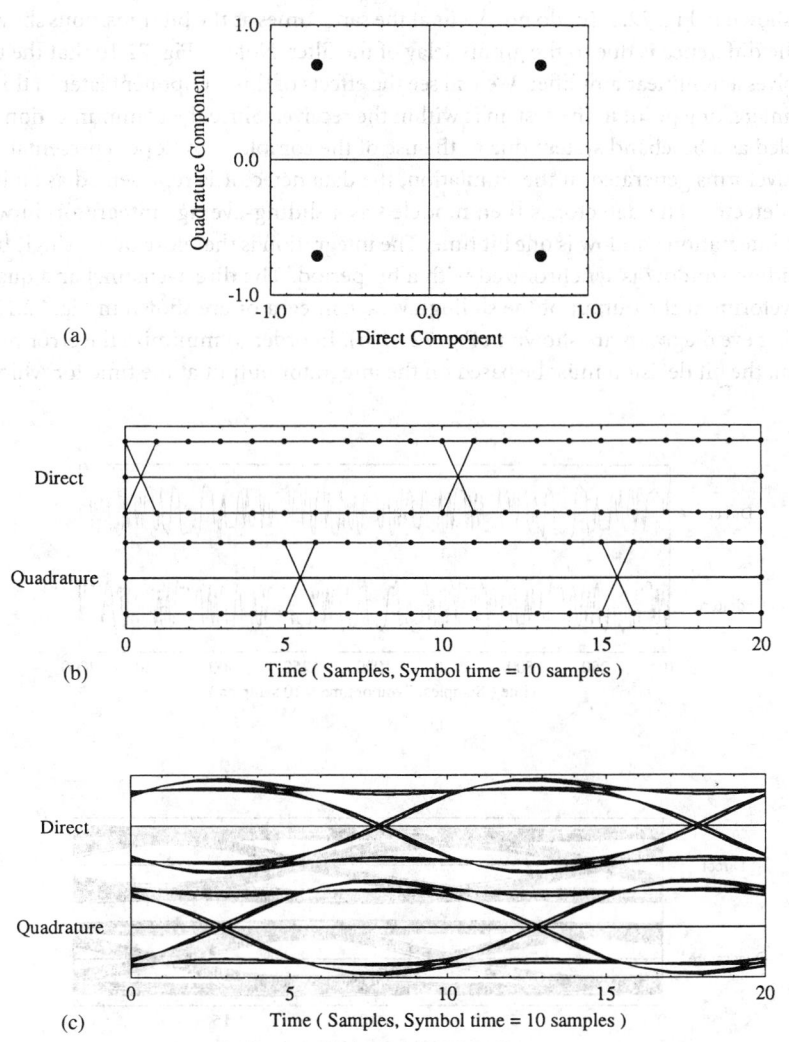

FIGURE 72.11 Transmitter signal constellation and eye diagrams: (a) OQPSK signal constellation; (b) eye diagram of modulator output; (c) eye diagram of filtered modulator output.

phase-shift keyed) modulation format is assumed so that one of four waveforms is transmitted during each symbol period. The data source may be viewed as a single binary source, in which the source symbols are taken two at a time when mapped onto a transmitted waveform, or as two parallel data sources, with one source providing the direct channel modulation and the second source providing the quadrature channel modulation. The signal constellation at the modulator output appears as shown in Fig. 72.11(a), with the corresponding eye diagram appearing as shown in Fig. 72.11(b). The eye diagram is formed by overlaying successive time intervals of a time domain waveform onto a single graph, much as would be done with a common oscilloscope. Since the simulation sampling frequency used in generating Fig. 72.11(b) was 10 samples per data symbol, it is easily seen that the eye diagram was generated by retracing every 2 data symbols or 20 simulation samples. Since Fig. 72.11(a) and (b) correspond to the modulator output, which has not yet been filtered, the transitions between binary states occur in one simulation step. After filtering, the eye diagram appears as shown in Fig. 72.11(c). A seventh-order Butterworth bilinear z-transform digital filter was assumed with a 3-dB bandwidth equal to the bit rate. It should be noted that the bit transitions shown in Fig. 72.11(c) do not occur at the same times as the bit transitions shown in Fig. 72.11(b). The difference is due to the group delay of the filter. Note in Fig. 72.10 that the transmitter also involves a nonlinear amplifier. We will see the effects of this component later in this section.

Another interesting point in the system is within the receiver. Since the communication system is being modeled as a baseband system due to the use of the complex-envelope representation of the bandpass waveforms generated in the simulation, the data detector is represented as an integrate-and-dump detector. The detector is then modeled as a sliding-average integrator, in which the width of the integration window is one bit time. The integration is therefore over a single bit period when the sliding window is synchronized with a bit period. The direct-channel and quadrature-channel waveforms at the output of the sliding-average integrator are shown in Fig. 72.12(a). The corresponding eye diagrams are shown in Fig. 72.12(b). In order to minimize the error probability of the system, the bit decision must be based on the integrator output at the time for which the eye

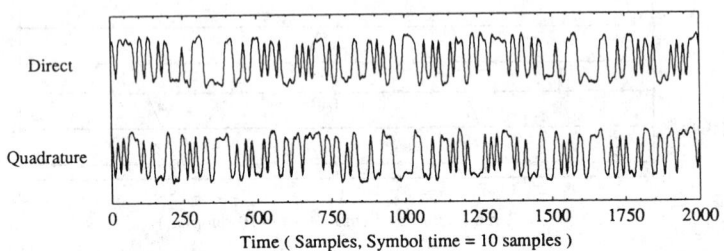

(a)

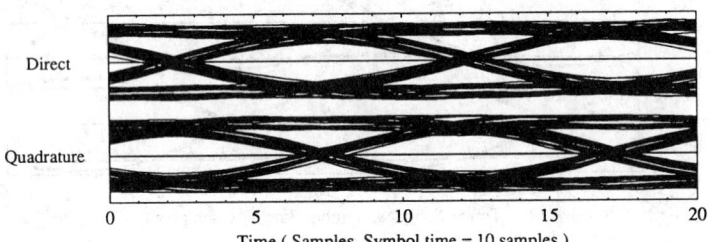

(b)

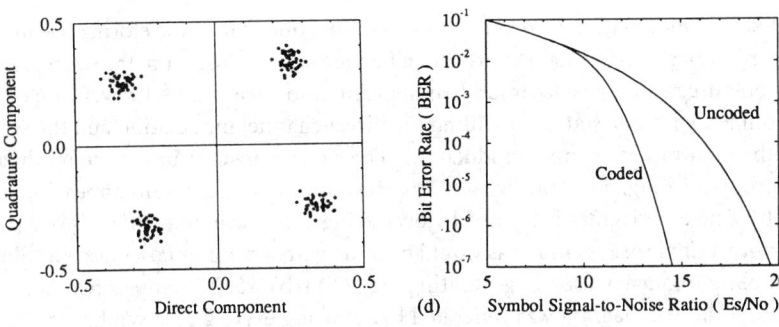

(c)   (d)

FIGURE 72.12   Integrator output signals and system error probability: (a) sliding integrator output signals; (b) sliding integrator output eye diagram; (c) sliding integrator output signal constellation; (d) error probability.

opening is greatest. Thus the eye diagram provides important information concerning the sensitivity of the system to timing errors.

The signal constellation at the sliding integrator output is shown in Fig. 72.12(c) and should be carefully compared to the signal constellation shown in Fig. 72.11(a) for the modulator output. Three effects are apparent. First, the signal points exhibit some scatter, which, in this case, is due to intersymbol interference resulting from the transmitter filter and additive noise. It is also clear that the signal is both compressed and rotated. These effects are due to the nonlinear amplifier that was mentioned previously. For this example simulation the nonlinear amplifier is operating near the saturation point, and the compression of the signal constellation is due to the AM/AM characteristic of the nonlinearity and the rotation is due to the AM/PM characteristic of the nonlinearity.

The performance of the overall communication system is illustrated in Fig. 72.12(d). The error probability curve is perhaps the most important simulation product. Note that both uncoded and coded results are shown. The coded results were calculated analytically from the uncoded results assuming a (63, 55) Reed–Solomon code. It should be mentioned that **semi-analytic simulation** was used in this example since, as can be seen in Fig. 72.10, the noise is injected into the system on the receiver side of the nonlinearity so that linear analysis may be used to determine the effects of the noise on the system performance.

This simple example serves to illustrate only a few of the possible simulation products. There are many other possibilities including histograms, correlation functions, estimates of statistical moments, estimates of the power spectral density, and estimates of the signal-to-noise ratio at various points in the system.

A word is in order regarding spectral estimation techniques. Two basic techniques can be used for spectral estimation: Fourier techniques and model-based techniques. In most simulation problems one is blessed with a tremendous amount of data concerning sampled waveforms but does not have a simple model describing how these waveforms are produced. For this reason model-based spectral estimation is typically not used. The most common form of spectral estimation used in simulation is the Welch periodogram. While this approach is straightforward, the effects of windowing the data sequence must be carefully considered, and tens or even hundreds of data windows must be averaged to achieve an accurate estimate of the power spectral density.

72.14 Conclusions

We have seen that the analysis and design of today's complex communication systems often requires the use of computer-aided techniques. These techniques allow the solution of problems that are otherwise not tractable and provide considerable insight into the operating characteristics of the communication system.

Defining Terms

Communication link: A point-to-point communication system that typically involves a single information source and a single user. This is in contrast to a communications network, which usually involves many sources and many users.

Computer-aided design and analysis: The process of using computer assistance in the design and analysis of complex systems. In our context, the design and analysis of communication systems, computer-aided design and analysis often involves the extensive use of simulation techniques. Computer-aided techniques often allow one to address design and analysis problems that are not tractable analytically.

Computer simulation: A set of computer programs which allows one to imitate the important aspects of the behavior of the specific system under study. Simulation can aid the design process by, for example, allowing one to determine appropriate system design parameters or aid the analysis process by, for example, allowing one to estimate the end-to-end performance of the system under study.

Dedicated simulation language: A computer language, either text based or graphics based, specifically developed to facilitate the simulation of the various systems under study, such as communication systems.

Low-pass equivalent (LPE) model: A method of representing bandpass signals and systems by low-pass signals and systems. This technique is extremely useful when developing discrete time models of bandpass continuous-time systems. It can substantially reduce the sampling rate required to prevent aliasing and does not result in any loss of information. This in turn reduces the execution time required for the simulation. This modeling technique is closely related to the quadrature representation of bandpass signals.

Monte Carlo simulation: A technique for simulating systems that contain signal sources producing stochastic or random signals. The signal sources are modeled by pseudorandom generators. Performance measures, such as the symbol error rate, are then estimated by time averages. This is a general-purpose technique that can be applied to an extremely wide range of systems. It can, however, require large amounts of computer time to generate accurate estimates.

Pseudorandom generator: An algorithm or device that generates deterministic waveforms which in many ways resemble stochastic or random waveforms. The power spectral density, auto-correlation, and other time averages of pseudorandom signals can closely match the time and ensemble averages of stochastic processes. These generators are useful in computer simulation where one may be unable to generate a truly random process, and they have the added benefit of providing reproducible signals.

Semi-analytic simulation: A numerical analysis technique that can be used to efficiently determine the symbol error rate of digital communication systems. It can be applied whenever one can analytically determine the probability of demodulation error given a particular transmitted data pattern. Although this technique can only be applied to a restricted class of systems, in these cases it is far more efficient, in terms of computer execution time, than Monte Carlo simulations.

Simulation validation: The process of certifying that simulation results are reasonable and can be used with confidence in the design or analysis process.

Symbol error rate: A fundamental performance measure for digital communication systems. The symbol error rate is estimated as the number of errors divided by the total number of demodulated symbols. When the communication system is ergodic, this is equivalent to the probability of making a demodulation error on any symbol.

References

1. K. Shanmugan, "An update on software packages for simulation of communication systems (links)," *IEEE J. Selected Areas Commun.*, no. 1, 1988.

2. W. Davenport and W. Root, *An Introduction to the Theory of Random Signals and Noise*, New York: McGraw-Hill, 1958.

3. S. Haykin, *Communication Systems*, New York: Wiley, 1983.

4. O. Shimbo, "Effects of intermodulation, AM-PM conversion, and additive noise in multi-carrier TWT systems," *Proc. IEEE*, no. 2, 1971.

5. N. Blachman, "Bandpass nonlinearities," *IEEE Trans. Inf. Theory*, no. 2, 1964.

6. R. Cochiere and L. Rabiner, *Multirate Digital Signal Processing*, Englewood Cliffs, N.J.: Prentice-Hall, 1983.

7. D. Knuth, *The Art of Computer Programming*, vol. 2, *Seminumerical Algorithms*, 2nd ed., Reading, Mass.: Addison-Wesley, 1981.

8. R. Rubinstein, *Simulation and the Monte Carlo Method*, New York: Wiley, 1981.

9. M. Sondhi, "Random processes with specified spectral density and first-order probability density," *Bell Syst. Tech. J.*, vol. 62, 1983.

10. A. Papoulis, *Probability, Random Variables, and Stochastic Processes*, New York: McGraw-Hill, 1965.

11. M. Jeruchim, P. Balaban, and K. Shanmugan, *Simulation of Communication Systems*, New York: Plenum, 1992.

12. M. Jeruchim, "Techniques for estimating the bit error rate in the simulation of digital communication systems," *IEEE J. Selected Areas Commun.*, no. 1, January 1984.

13. P. Bratley, B. L. Fox, and L. E. Schrage, *A Guide to Simulation*, New York: Springer-Verlag, 1987.

14. P. Balaban, K. S. Shanmugan, and B. W. Stuck (eds.), "Special issue on computer-aided modeling, analysis and design of communication systems," *IEEE J. Selected Areas Commun.*, no. 1, 1984.

15. P. Balaban, E. Biglieri, M. C. Jeruchim, H. T. Mouftah, C. H. Sauer, and K. S. Shanmugan (eds.), "Computer-aided modeling, analysis and design of communication systems II," *IEEE J. Selected Areas Commun.*, no. 1, 1988.

16. H. T. Mouftah, J. F. Kurose, and M. A. Marsan (eds.), "Computer-aided modeling, analysis and design of communication networks I," *IEEE J. Selected Areas Commun.*, no. 9, 1990.

17. H. T. Mouftah, J. F. Kurose, and M. A. Marsan (eds.), "Computer-aided modeling, analysis and design of communication networks II," *IEEE J. Selected Areas Commun.*, no. 1, 1991.

Further Information

Until recently the subject of computer-aided analysis and simulation of communication systems was a very difficult research area. There were no textbooks devoted to the subject, and the fundamental papers were scattered over a large number of technical journals. While a number of excellent books treated the subject of simulation of systems in which random signals and noise are present [8, 13], none of these books specifically focused on communication systems.

Starting in 1984, the *IEEE Journal on Selected Areas in Communications (JSAC)* initiated the publication of a sequence of issues devoted specifically to the subject of computer-aided design and analysis of communication systems. A brief study of the contents of these issues tells much about the rapid development of the discipline. The first issue, published in January 1984 [14], emphasizes communication links, although there are a number of papers devoted to networks. The portion devoted to links contained a large collection of papers devoted to simulation packages.

The second issue of the series was published in 1988 and is roughly evenly split between links and networks [15]. In this issue the emphasis is much more on techniques than on simulation packages. The third part of the series is a two-volume issue devoted exclusively to networks [16, 17].

As of this writing, the book by Jeruchim *et al.* is the only comprehensive treatment of the simulation of communication links [11]. It treats the component and channel modeling problem and the problems associated with using simulation techniques for estimating the performance of communication systems in considerable detail. This textbook, together with the previously cited *JSAC* issues, gives a good overview of the area.

The HDSP-253X from Hewlett-Packard Company is one of four new families of alphanumeric LED displays incorporating CMOS decoder and driver circuitry. This display is available in red, orange, green, yellow, and super-bright AlGaAs red for visibility under full sunlight conditions. (Photo courtesy of Hewlett-Packard Company.)

VIII

Digital Devices

Richard S. Sandige
University of Wyoming

73 **Logic Elements** *G. Moss, P. Graham, R. Sandige, H. Hinton* 1613
IC Logic Family Operation and Characteristics • Logic Gates (IC) • Bistable Devices • Optical Devices

74 **Memory Devices** *W. Pricer, R. Katz, P. Lee, M. Mansuripur* 1651
Integrated Circuits (RAM, ROM) • Basic Disk System Architectures • Magnetic Tape • Magneto-Optical Disk Data Storage

75 **Logical Devices** *F. Preparata, R. Sandige, B. Bannister, D. Whitehead, M. Bolton, B. Carroll* 1695
Combinational Networks and Switching Algebra • Logic Circuits • Registers • Programmable Arrays • Arithmetic Logic Units

76 **Microprocessors** *J. Staudhammer, S. Chen, P. Windley, J. Frenzel* 1748
Practical Microprocessors • Applications

77 **Displays** *J. Morris, A. Martin, L. Weber* 1763
Light-Emitting Diodes • Liquid-Crystal Displays • The Cathode Ray Tube • Plasma Displays

78 **Data Acquisition** *D. Kurumbalapitiya, S. Hoole* 1799
The Analog and Digital Signal Interface • Analog Signal Conditioning • Sample-and-Hold and A/D Techniques in Data Acquisition • The Communication Interface of a Data Acquisition System • Data Recording • Software Aspects

79 **Testing** *M. Serra, B. Dervisoglu* 1808
Digital IC Testing • Design for Test

E LECTRONIC DESIGNERS have placed increasing significance on digital devices since the late 1960s. This is due primarily to the greater reliability and improved accuracy gained when using electronic devices in a two-level mode (binary mode) as compared to using electronic devices in a continuous mode (analog mode). As silicon integrated circuits (ICs) became denser and more consistently reproducible over the past few decades, so did digital electronic devices. Today digital circuits and digital systems produced from digital devices can be found in every walk of life ranging from children's toys, kitchen appliances, laboratory instruments, personal and workstation computers to space shuttle and satellite applications.

The intent of this section is to present topics related to the utilization and application of digital devices. Chapter 73 establishes the foundation for digital logic elements beginning with switching logic (IC), logic gates, bistable devices, and optical devices. Discussed in the next chapter are memory devices, which include integrated circuits (RAM, ROM), disk systems, magnetic tape, and optical disks. Chapter 75 on logical devices discusses Boolean algebra, logic circuits, registers, programmable arrays (PAL, FPGA), and arithmetic units.

The next chapter explains the microprocessor, perhaps the best-known digital device. The topics covered include programming the microprocessor, practical microprocessors, and microprocessor applications. Chapter 77 on optical displays includes the light-emitting diode, liquid-crystal display, the cathode ray tube, and the plasma panel. The gathering of digital information, referred to as data acquisition, is discussed next. Finally, no digital system is released to production without extensive testing. The last chapter of this section presents methods of testing and design for testing.

The combination of topics presented in this section should provide readers with a contemporary overview of digital devices. To obtain additional information, the reader may refer to the References and Further Information in each chapter.

Nomenclature

Symbol	Quantity	Unit	Symbol	Quantity	Unit
A	area	m^2	λ	radiation wavelength	nm
α	average absorption coefficient		m	magnification factor	
B	luminance off the projection screen		μ_n	electron mobility	
			n	aperture	
C	brightness contrast		ν	photon frequency	Hz
CMRR	common-mode rejection ratio		p	photon momentum	kg · m/s
			R	recombination rate	ns
C_R	contrast ratio		R_i	reflectivity	
d	diameter	m	S	emitting screen surface	m^2
E_g	band gap energy	eV	SNR	signal-to-noise ratio	
ε	screen efficiency	lumen/W	t_{add}	add time	ns
f	focal length	m	t_h	hold time	ns
F	maximum flux	lumen	t_{pd}	propagation delay time	ns
h	Planck's constant	6.626×10^{-34} J · s	T	transmission ratio	
			T	transmission of faceplate	
η	quantum efficiency		τ	lifetime	ns
			θ_c	critical angle	degree
I	beam current	amp	V_B	accelerating voltage	V
L	luminance	cd/m^2	V_s	screen voltage	V
L	raster luminance				

73

Logic Elements

Gregory L. Moss
Purdue University

Peter Graham
Florida Atlantic University
(Retired)

Richard S. Sandige
University of Wyoming

H. S. Hinton
McGill University

73.1 IC Logic Family Operation and Characteristics..................... 1613
   IC Logic Families and Subfamilies • TTL Logic Family • CMOS
   Logic Family • ECL Logic Family • Logic Family Circuit Parameters
   • Interfacing Between Logic Families
73.2 Logic Gates (IC)... 1622
   Gate Specification Parameters • Bipolar Transistor Gates • Com-
   plementary Metal-Oxide Semiconductor (CMOS) Logic •
   Choosing a Logic Family
73.3 Bistable Devices 1635
   Basic Latches • Gated Latches • Flip-Flops • Edge-Triggered
   Flip-Flops • Special Notes on Using Latches and Flip-Flops
73.4 Optical Devices 1641
   All-Optical Devices • Optoelectronic Devices • Limitations

73.1 IC Logic Family Operation and Characteristics

Gregory L. Moss

Digital logic circuits can be classified as belonging to one of two categories, either combinational (also called combinatorial) or sequential logic circuits. The output logic level of a combinatorial circuit depends only on the current logic levels present at the circuit's inputs. Sequential logic circuits, on the other hand, have a memory characteristic so the sequential circuit's output is dependent not only on the current input conditions but also on the current output state of the circuit. The primary building block in combinational circuits is the logic gate. The three simplest logic gate functions are the inverter (or NOT), AND, and OR. Other common basic logic functions are derived from these three. Table 73.1 gives **truth table** definitions of the various types of logic gates. The memory elements used to construct sequential logic circuits are called latches and flip-flops.

The integrated circuit switching logic used in modern digital systems will generally be from one of three families: transistor-transistor logic (TTL), complementary metal-oxide semiconductor logic (CMOS), or emitter-coupled logic (ECL). Each of the logic families has its advantages and disadvantages. The three major families are also divided into various subfamilies derived from performance improvements in integrated circuit (IC) design technology. Bipolar transistors provide the switching action in both TTL and ECL families, while enhancement-mode MOS transistors are the basis for the CMOS family. Recent improvements in switching circuit performance are also attained using BiCMOS technology, the merging of bipolar and CMOS technologies on a single chip. A particular logic family is usually selected by digital designers based on such criteria as

Table 73.1 Defining Truth Tables for Logic Gates

1-Input Function		2-Input Functions							
Input	Output	Inputs		Output Functions					
A	NOT	A	B	AND	OR	NAND	NOR	XOR	XNOR
0	1	0	0	0	0	1	1	0	1
1	0	0	1	0	1	1	0	1	0
		1	0	0	1	1	0	1	0
		1	1	1	1	0	0	0	1

1. Switching speed
2. Power dissipation
3. PC board area requirements (levels of integration)
4. Output drive capability (**fan-out**)
5. Noise immunity characteristics
6. Product breadth
7. Sourcing of components

IC Logic Families and Subfamilies

The integrated circuit logic families actually consist of several subfamilies of ICs that differ in various performance characteristics. The TTL logic family has been the most widely used family type for applications that employ small-scale integration (SSI) or medium-scale integration (MSI) integrated circuits. Lower power consumption and higher levels of integration are the principal advantages of the CMOS family. The ECL family is generally used in applications that require high-speed switching logic. Today, the most common device numbering system used in the TTL and CMOS families has a prefix of 54 (generally used in military applications and having an operating temperature range of –55 to 125°C) and 74 (generally used in industrial/commercial applications and hav-

Table 73.2 Logic Families and Subfamilies

Family and Subfamily	Description
TTL	Transistor-transistor logic
74xx	Standard TTL
74Lxx	Low-power TTL
74Hxx	High-speed TTL
74Sxx	Schottky TTL
74LSxx	Low-power Schottky TTL
74ASxx	Advanced Schottky TTL
74ALSxx	Advanced low-power Schottky TTL
74Fxx	Fast TTL
CMOS	Complementary metal-oxide semiconductor
4xxx	Standard CMOS
74Cxx	Standard CMOS using TTL numbering system
74HCxx	High-speed CMOS
74HCTxx	High-speed CMOS—TTL compatible
74FCTxx	Fast CMOS—TTL compatible
74ACxx	Advanced CMOS
74ACTxx	Advanced CMOS—TTL compatible
ECL (or CML)	Emitter-coupled (current-mode) logic
10xxx	Standard ECL
10Hxxx	High-speed ECL

ing an operating temperature range of 0 to 70°C). Table 73.2 identifies various logic families and subfamilies.

TTL Logic Family

The TTL family has been the most widely used logic family for many years in applications that use SSI and MSI. It is relatively fast and offers a great variety of standard chips.

The active switching element used in all TTL family circuits is the *npn* bipolar junction transistor (BJT). The transistor is turned on when the base is approximately 0.7 V more positive than the emitter and there is a sufficient amount of base current flowing. The turned on transistor is said to be in saturation and, ideally, acts like a closed switch between the collector and emitter terminals. The transistor is turned off when the base is not biased with a high enough voltage (with respect to the emitter). Under this condition, the transistor acts like an open switch between the collector and emitter terminals.

Figure 73.1 illustrates the transistor circuit blocks used in a standard TTL inverter. Four transistors are used to achieve the inverter function. The input to the gate connects to the emitter of transistor Q1, the input coupling transistor. A clamping diode on the input prevents negative input voltage spikes from damaging Q1. The collector voltage (and current) of Q1 controls Q2, the phase splitter transistor. Q2, in turn, controls the Q3 and Q4 transistors forming the output circuit, which is called a totem-pole arrangement. Q4 serves as a pull-up transistor to pull the output high when it is turned on. Q3 does just the opposite to the output and serves as a pull-down transistor. Q3 pulls the output low when it is turned on. Only one of the two transistors in the totem pole may be turned on at a time, which is the function of the phase splitter transistor Q2.

When a high **logic level** is applied to the inverter's input, Q1's base-emitter junction will be reverse biased and the base-collector junction will be forward biased. This circuit condition will allow Q1 collector current to flow into the base of Q2, saturating Q2 and thereby providing base current into Q3, turning it on also. The collector voltage of Q2 is too low to turn on Q4 so that it appears as an open in the top part of the totem pole. A diode between the two totem-pole transistors provides an extra voltage drop in series with the base-emitter junction of Q4 to ensure that Q4

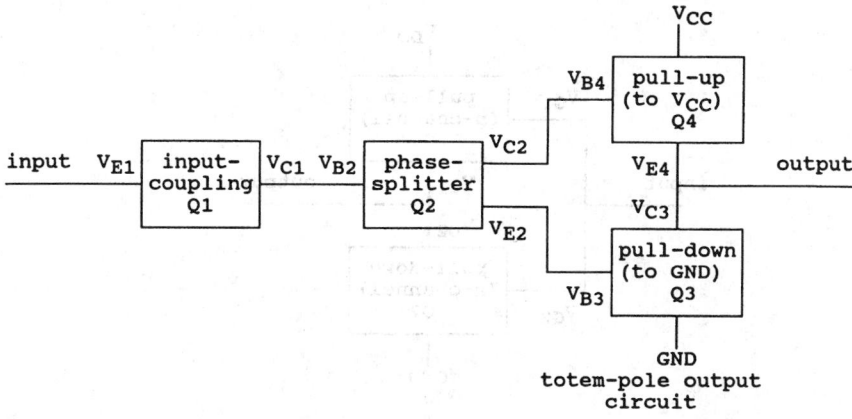

input	V_{C1}	Q2	V_{C2}	V_{E2}	Q3	V_{C3}	Q4	V_{E4}	output
hi	hi	on	low	hi	on	low	off	open	low
low	low	off	hi	low	off	open	on	hi	hi

FIGURE 73.1 TTL inverter circuit block diagram and operation.

will be turned off when Q2 is turned on. The saturated Q3 transistor brings the output near ground potential, producing a low output result for a high input into the inverter.

When a low logic level is applied to the inverter's input, Q1's base-emitter junction will be forward biased and the base-collector junction will be reverse biased. This circuit condition will turn on Q1 so that the collector terminal is shorted to the emitter and, therefore, to ground (low level). This low voltage is also on the base of Q2 and turns Q2 off. With Q2 off, there will be insufficient base current into Q3, turning it off also. Q2 leakage current is shunted to ground with a resistor to prevent the partial turning on of Q3. The collector voltage of Q2 is pulled to a high potential with another resistor and, as a result, turns on Q4 so that it appears as a short in the top part of the totem pole. The saturated Q4 transistor provides a low resistance path from V_{CC} to the output, producing a high output result for a low input into the inverter.

A TTL NAND gate is very similar to the inverter circuit, with the exception that the input coupling transistor Q1 is constructed with multiple emitter-base junctions and each input to the NAND is connected to a separate emitter terminal. Any of the transistor's multiple emitters can be used to turn on Q1. The TTL NAND gate thus functions in the same manner as the inverter in that if any of the NAND gate inputs are low, the same circuit action will take place as with a low input to the inverter. Therefore, any time a low input is applied to the NAND gate it will produce a high ouput. Only if all of the NAND gate inputs are simultaneously high will it then produce the same circuit action as the inverter with its single input high, and the resultant output will be low. Input coupling transistors with up to eight emitter-base junctions, and therefore, eight input NAND gates, are constructed.

CMOS Logic Family

The active switching element used in all CMOS family circuits is the metal-oxide semiconductor field-effect transistor (MOSFET). CMOS stands for complementary MOS transistors and refers to the use of both types of MOSFET transistors, *n*-channel and *p*-channel, in the design of this type of switching circuit. While the physical construction and the internal physics of a MOSFET is quite different from that of the BJT, the circuit switching action of the two transistor types is quite similar. The MOSFET switch is essentially turned off and has a very high channel resistance by applying the same potential to the gate terminal as the source. An *n*-channel MOSFET is turned on and has

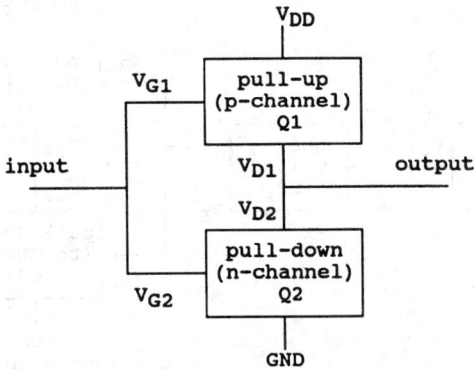

input	Q1	V_{D1}	Q2	V_{D2}	output
hi	off	open	on	low	low
low	on	hi	off	open	hi

FIGURE 73.2   CMOS inverter circuit block diagram and operation.

a very low channel resistance when a high voltage with respect to the source is applied to the gate. A *p*-channel MOSFET operates in the same fashion but with opposite polarities; the gate must be more negative than the source to turn on the transistor.

A block diagram for a CMOS inverter circuit is shown in Fig. 73.2. Note that it is a simpler and much more compact circuit design than that for the TTL inverter. That fact is a major reason why MOSFET integrated circuits have a much higher circuit density than BJT integrated circuits and is one advantage that MOSFET ICs have over BJT ICs. As a result, CMOS is used in all levels of integration, from SSI through VLSI (very large scale integration).

When a high logic level is applied to the inverter's input, the *p*-channel MOSFET Q1 will be turned off and the *n*-channel MOSFET Q2 will be turned on. This will cause the output to be shorted to ground through the low resistance path of Q2's channel. The turned off Q1 has a very high channel resistance and acts nearly like an open.

When a low logic level is applied to the inverter's input, the *p*-channel MOSFET Q1 will be turned on and the *n*-channel MOSFET Q2 will be turned off. This will cause the output to be shorted to V_{DD} through the low resistance path of Q1's channel. The turned off Q2 has a very high channel resistance and acts nearly like an open.

CMOS NAND gates are constructed by paralleling *p*-channel MOSFETs, one for each input, and putting in series an *n*-channel MOSFET for each input, as shown in the block diagram of Fig. 73.3. The NAND gate will produce a low output only when both Q3 and Q4 are turned on, creating a low resistance path from the output to ground through the two series channels. This can be accomplished by having a high on both input A and input B. This input condition will also turn off Q1 and Q2. If either input A or input B or both is low, the respective parallel MOSFET will be turned on, providing a low resistance path for the output to V_{DD}. This will also turn off at least one of the series MOSFETs, resulting in a high resistance path for the output to ground.

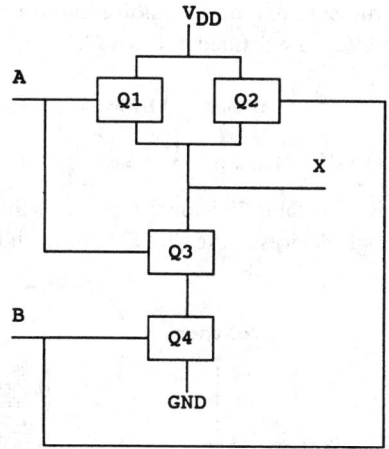

| inputs | | Q1 | Q2 | Q3 | Q4 | output |
A	B					X
low	low	on	on	off	off	hi
low	hi	on	off	off	on	hi
hi	low	off	on	on	off	hi
hi	hi	off	off	on	on	low

FIGURE 73.3 CMOS two-input NAND circuit block diagram and operation.

ECL Logic Family

ECL is a higher-speed logic family. While it does not offer as large a variety of IC chips as are available in the TTL family, it is quite popular for logic applications requiring high-speed switching.

The active switching element used in the ECL family circuits is also the *npn* BJT. Unlike the TTL family, however, which switches the transistors into saturation when turning them on, ECL switching is designed to prevent driving the transistors into saturation. Whenever bipolar transistors are driven into saturation, their switching speed will be limited by the charge carrier storage delay, a transistor operational characteristic. Thus, the switching speed of ECL circuits will be significantly higher than for TTL circuits. ECL operation is based on switching a fixed amount of bias current that is less than the saturation amount between two different transistors. The basic circuit found in the ECL family is the differential amplifier. One side of the differential amplifier is controlled by a bias circuit and the other is controlled by the logic inputs to the gate. This logic family is also referred to as current-mode logic (CML) because of its current switching operation.

Logic Family Circuit Parameters

Digital circuits and systems operate with only two states, logic 1 and 0, usually represented by two different voltage levels, a *high* and a *low*. The two logic levels actually consist of a range of values with the numerical quantities dependent upon the specific family that is used. Minimum high logic levels and maximum low logic levels are established by specifications for each family. Minimum device output levels for a logic high are called $V_{OH(min)}$ and minimum input levels are called $V_{IH(min)}$. The abbreviations for maximum output and input low logic levels are $V_{OL(max)}$ and $V_{IL(max)}$, respectively. Figure 73.4 shows the relationships between these parameters. Logic voltage level parameters are illustrated for selected prominent logic subfamilies in Table 73.3. As seen in this illustration, there are many operational incompatibilities between major logic family types.

Noise margin is a quantitative measure of a device's **noise immunity.** High-level noise margin (V_{NH}) and low-level noise margin (V_{NL}) are defined in Eqs. (73.1) and (73.2).

$$V_{NH} = V_{OH(min)} - V_{IH(min)} \tag{73.1}$$

$$V_{NL} = V_{IL(max)} - V_{OL(max)} \tag{73.2}$$

Using the logic voltage values given in Table 73.3 for the selected subfamilies reveals that highest noise immunity is obtained with logic devices in the CMOS family, while lowest noise immunity is endemic to the ECL family.

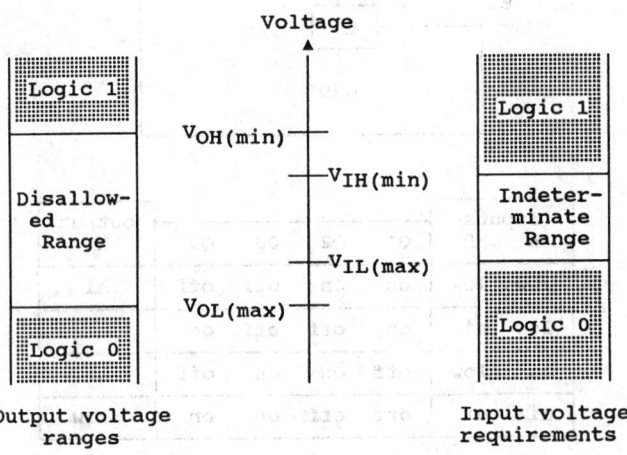

FIGURE 73.4 Switching device logic levels.

Table 73.3 Logic Signal Voltage Parameters for Selected Logic Subfamilies (in Volts)

Subfamily	$V_{OH(min)}$	$V_{OL(max)}$	$V_{IH(min)}$	$V_{IL(max)}$
74xx	2.4	0.4	2.0	0.8
74LSxx	2.7	0.5	2.0	0.8
74ASxx	2.5	0.5	2.0	0.8
74ALSxx	2.5	0.4	2.0	0.8
74HCxx	4.9	0.1	3.15	0.9
74HCTxx	4.9	0.1	2.0	0.8
74ACxx	3.8	0.4	3.15	1.35
74ACTxx	3.8	0.4	2.0	0.8
10xxx	−0.96	−1.65	−1.105	−1.475
10Hxxx	−0.98	−1.63	−1.13	−1.48

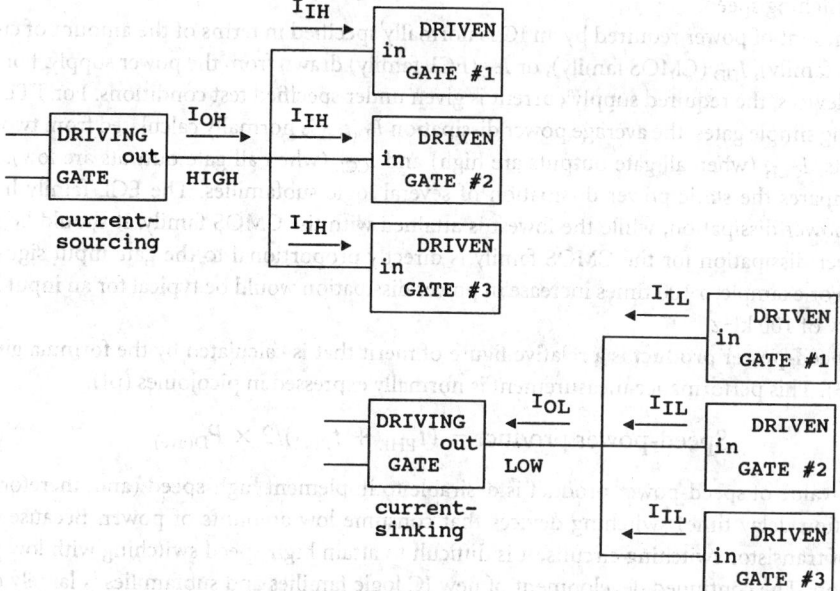

FIGURE 73.5 Current loading of driving gates.

Table 73.4 Worst Case Current Parameters for Selected Logic Subfamilies

Subfamily	$I_{OH(max)}$	$I_{OL(max)}$	$I_{IH(max)}$	$I_{IL(max)}$
74xx	−400 μA	16 mA	40 μA	−1.6 μA
74LSxx	−400 μA	8 mA	20 μA	−400 μA
74ASxx	−2 mA	20 mA	200 μA	−2 mA
74ALSxx	−400 μA	8 mA	20 μA	−100 μA
74HCxx	−4 mA	4 mA	1 μA	−1 μA
74HCTxx	−4 mA	4 mA	1 μA	−1 μA
74ACxx	−24 mA	24 mA	1 μA	−1 μA
74ACTxx	−24 mA	24 mA	1 μA	−1 μA
10xxx	50 mA	−50 mA	−265 μA	500 nA
10Hxxx	50 mA	−50 mA	−265 μA	500 nA

Switching circuit outputs are loaded by the inputs of the devices that they are driving, as illustrated in Fig. 73.5. Worst case input loading current levels and output driving current capabilities are listed in Table 73.4 for various logic subfamilies. The fan-out of a driving device is the ratio between its output current capabilities at each logic level and the corresponding gate input current loading value. Switching circuits based on bipolar transistors have fan-out limited primarily by the current-sinking and current-sourcing capabilities of the driving device.

CMOS switching circuits are limited by the charging and discharging times associated with the output resistance of the driving gate and the input capacitance of the load gates. Thus, CMOS fan-out depends on the frequency of switching. With fewer (capacitive) loading inputs to drive, the maximum switching frequency of CMOS devices will increase.

The switching speed of logic devices is dependent of the device's **propagation delay time.** The propagation delay of a logic device limits the frequency at which it can be operated. There are two propagation delay times specified for logic gates: t_{PHL}, delay time for the output to change from high to low, and t_{PLH}, delay time for the output to change from low to high. Average typical propagation delay times are listed for several logic subfamilies in Table 73.5. The ECL family has the fastest switching speed.

The amount of power required by an IC is normally specified in terms of the amount of current I_{CC} (TTL family), I_{DD} (CMOS family), or I_{EE} (ECL family) drawn from the power supply. For complex IC devices, the required supply current is given under specified test conditions. For TTL chips containing simple gates, the average power dissipation $P_{D(ave)}$ is normally calculated from two measurements, I_{CCH} (when all gate outputs are high) and I_{CCL} (when all gate outputs are low). Table 73.5 compares the static power dissipation of several logic subfamilies. The ECL family has the highest power dissipation, while the lowest is attained with the CMOS family. It should be noted that power dissipation for the CMOS family is directly proportional to the gate input signal frequency. For example, a 100 times increase in power dissipation would be typical for an input signal frequency of 100 kHz.

The **speed-power product** is a relative figure of merit that is calculated by the formula given in Eq. (73.3). This performance measurement is normally expressed in picojoules (pJ).

$$\text{Speed-power product} = (t_{PHL} + t_{PLH})/2 \times P_{D(ave)} \qquad (73.3)$$

A low value of speed-power product is desirable to implement high-speed (and, therefore, low propagation delay time) switching devices that consume low amounts of power. Because of the nature of transistor switching circuits, it is difficult to attain high-speed switching with low power dissipation. The continued development of new IC logic families and subfamilies is largely due to the trade-offs between these two device switching parameters. The speed-power product for various subfamilies is also compared in Table 73.5.

Table 73.5   Speed-Power Comparison for Selected Logic Subfamilies

Subfamily	Propagation Delay Time, ns (ave.)	Static Power Dissipation, mW (per gate)	Speed-Power Product, pJ
74xx	10	10	100
74LSxx	9.5	2	19
74ASxx	1.5	2	13
74ALSxx	4	1.2	5
74HCxx	8	0.003	24×10^{-3}
74HCTxx	14	0.003	42×10^{-3}
74ACxx	5	0.010	50×10^{-3}
74ACTxx	5	0.010	50×10^{-3}
10xxx	2	25	50
10Hxxx	1	25	25

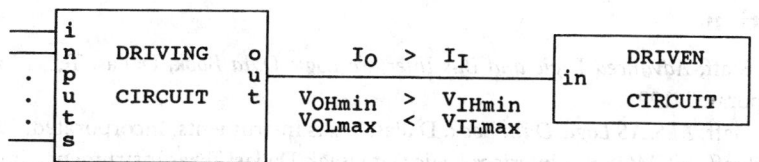

FIGURE 73.6   Circuit interfacing requirements.

Interfacing Between Logic Families

The interconnection of logic chips requires that input and output specifications be satisfied. Figure 73.6 illustrates voltage and current requirements. The driving chip's $V_{OH(min)}$ must be greater than the driven circuit's $V_{IH(min)}$, and the driver's $V_{OL(max)}$ must be less than $V_{IL(max)}$ for the loading circuit. Voltage level shifters must be used to interface the circuits together if these voltage requirements are not met. Of course, a driving circuit's output must not exceed the maximum and minimum allowable input voltages for the driven circuit. Also, the current sinking and sourcing ability of the driver circuit's output must be greater than the total current requirements for the loading circuit. Buffer gates or stages must be used if current requirements are not satisfied. All chips within a single logic family are designed to be compatible with other chips in the same family. Mixing chips from multiple subfamilies together within a single digital circuit can have adverse effects on the overall circuit's switching speed and noise immunity.

Defining Terms

Fan-out: The specification used to identify the limit to the number of loading inputs that can be reliably driven by a driving device's output.

Logic level: The high or low value of a voltage variable that is assigned to be a 1 or a 0 state.

Noise immunity: A logic device's ability to tolerate input voltage fluctuation caused by noise without changing its output state.

Propagation delay time: The time delay from when the input logic level to a device is changed until the resultant output change is produced by that device.

Speed-power product: An overall performance measurement that is used to compare the various logic families and subfamilies.

Truth table: A listing of the relationship of a circuit's output that is produced for various combinations of logic levels at the inputs.

References

D. J. Comer, *Digital Logic and State Machine Design,* 2nd ed., Philadelphia: Saunders College Publishing, 1990.

T. L. Floyd, *Digital Fundamentals,* 4th ed., Columbus, Ohio: Merrill, 1990.

E. J. McCluskey, *Logic Design Principles,* Englewood Cliffs, N.J.: Prentice-Hall, 1986.

R. J. Prestopnik, *Digital Electronics: Concepts and Applications for Digital Design,* Philadelphia: Saunders College Publishing, 1990.

R. S. Sandige, *Modern Digital Design,* New York: McGraw-Hill, 1990.

S. G. Shiva, *Introduction to Logic Design,* Glenview, Ill.: Scott, Foresman, 1988.

R. J. Tocci, *Digital Systems: Principles and Applications,* 5th ed., Englewood Cliffs, N.J.: Prentice-Hall, 1991.

S. H. Unger, *The Essence of Logic Circuits,* Englewood Cliffs, N.J.: Prentice-Hall, 1990.

J. F. Wakerly, *Digital Design: Principles and Practices,* Englewood Cliffs, N.J.: Prentice-Hall, 1990.

Further Information

Engineering Staff, *Advanced Logic and Bus Interface Logic Data Book,* Dallas: Texas Instruments, Incorporated, 1991.

Engineering Staff, *ALS/AS Logic Data Book,* Dallas: Texas Instruments, Incorporated, 1986.

Engineering Staff, *BiCMOS Bus Interface Logic Data Book,* Dallas: Texas Instruments, Incorporated, 1989.

Engineering Staff, *FACT Data,* Phoenix: Motorola, Incorporated, 1989.

Engineering Staff, *High-Speed CMOS Logic Data Book,* Dallas: Texas Instruments, Incorporated, 1991.

Engineering Staff, *MECL Device Data,* Phoenix: Motorola, Incorporated, 1989.

Engineering Staff, *TTL Logic Data Book,* Dallas: Texas Instruments, Incorporated, 1988.

73.2 Logic Gates (IC)*

Peter Graham

This section introduces and analyzes the electronic circuit realizations of the basic gates of the three technologies: transistor-transistor logic (TTL), emitter-coupled logic (ECL), and complementary metal-oxide semiconductor (CMOS) logic. These circuits are commercially available on small-scale integration chips and are also the building blocks for more elaborate logic systems. The three technologies are compared with regard to speed, power consumption, and noise immunity, and parameters are defined which facilitate these comparisons. Also included are recommendations which are useful in choosing and using these technologies.

Gate Specification Parameters

Theoretically almost any logic device or system could be constructed by wiring together the appropriate configuration of the basic gates of the selected technology. In practice, however, the gates are interconnected during the fabrication process to produce a desired system on a single chip. The circuit complexity of a given chip is described by one of the following four rather broad classifications:

- **Small-Scale Integration (SSI).** The inputs and outputs of every gate are available for external connection at the chip pins (with the exception that exclusive OR and AND-OR gates are considered SSI).
- **Medium-Scale Integration (MSI).** Several gates are interconnected to perform somewhat more elaborate logic functions such as flip-flops, counters, multiplexers, etc.
- **Large-Scale Integration (LSI).** Several of the more elaborate circuits associated with MSI are interconnected within the integrated circuit to form a logic system on a single chip. Chips such as calculators, digital clocks, and small microprocessors are examples of LSI.
- **Very-Large-Scale Integration (VLSI).** This designation is usually reserved for chips having a very high density, 1000 or more gates per chip. These include the large single-chip memories, gate arrays, and microcomputers.

Specifications of logic speed require definitions of switching times. These definitions can be found in the introductory pages of most data manuals. Four of them pertain directly to gate circuits. These are (see also Fig. 73.7):

---

*Based on P. Graham, "Gates," in *Handbook of Modern Electronics and Electrical Engineering,* C. Belove, Ed., New York: Wiley-Interscience, 1986, pp. 864–876. With permission.

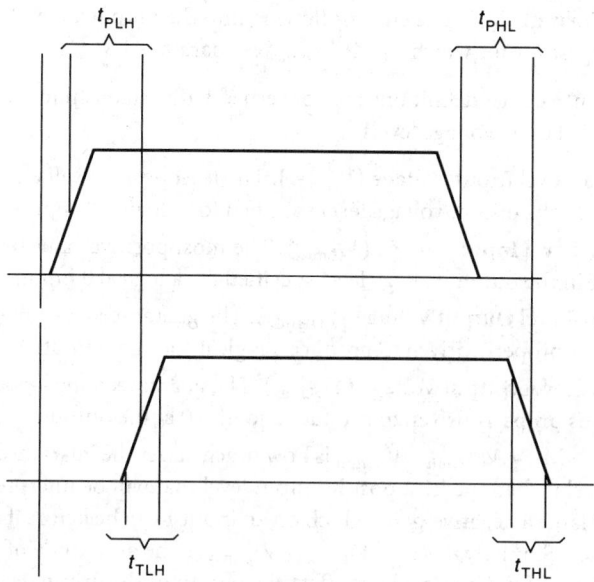

FIGURE 73.7 Definitions of switching times. (*Source:* P. Graham, "Gates," in *Handbook of Modern Electronics and Electrical Engineering,* C. Belove, Ed., New York: Wiley-Interscience, 1986, p. 865. With permission.)

- **LOW-to-HIGH Propagation Delay Time (t_{PLH}).** The time between specified reference points on the input and output voltage waveforms when the output is changing from low to high.
- **HIGH-to-LOW Propagation Delay Tune (t_{PHL}).** The time between specified reference points on the input and output voltage waveforms when the output is changing from high to low.
- **Propagation Delay Time (t_{PD}).** The average of the two propagation delay times: $t_{PD} = (t_{PD} + t_{PHL})/2$.
- **LOW-to-HIGH Transition Time (t_{TLH}).** The rise time between specified reference points on the LOW-to-HIGH shift of the output waveform.
- **HIGH-to-LOW Transition Time (t_{THL}).** The fall time between specified reference points on the HIGH-to-LOW shift of the output waveform. The reference points usually are 10 and 90% of the voltage level difference in each case.

Power consumption, driving capability, and effective loading of gates are defined in terms of currents.

- **Supply Current, Outputs High (I_{xxH}).** The current delivered to the chip by the power supply when all outputs are open and at the logical 1 level. The xx subscript depends on the technology.
- **Supply Current, Outputs Low (I_{xxL}).** The current delivered to the chip by the supply when all outputs are open and at the logical 0 level.
- **Supply Current, Worst Case (I_{xx}).** When the output level is unspecified, the input conditions are assumed to correspond to maximum supply current.
- **Input HIGH Current (I_{IH}).** The current flowing into an input when the specified HIGH voltage is applied.
- **Input LOW Current (I_{IL}).** The current flowing into an input when the specified LOW voltage is applied.
- **Output HIGH Current (I_{OH}).** The current flowing into the output when it is in the HIGH state. I_{OHmax} is the largest I_{OH} for which $V_{OH} \geq V_{OHmin}$ is guaranteed.

- **Output LOW Current (I_{OL}).** The current flowing into the output when it is in the LOW state. I_{OLmax} is the largest I_{OL} for which $V_{OL} \geq V_{OLmax}$ is guaranteed.

The most important voltage definitions are concerned with establishing ranges on the logical 1 (HIGH) and logical 0 (LOW) voltage levels.

- **Minimum High-Level Input Voltage (V_{IHmin}).** The least positive value of input voltage guaranteed to result in the output voltage level specified for a logical 1 input.
- **Maximum Low-Level Input Voltage (V_{ILmax}).** The most positive value of input voltage guaranteed to result in the output voltage level specified for a logical 0 input.
- **Minimum High-Level Output Voltage (V_{OHmin}).** The guaranteed least positive output voltage when the input is properly driven to produce a logical 1 at the output.
- **Maximum Low-Level Output Voltage (V_{OLmax}).** The guaranteed most positive output voltage when the input is properly driven to produce a logical 0 at the output.
- **Noise Margins.** $NM_H = V_{OHmin} - V_{IHmin}$ is how much larger the guaranteed least positive output logical 1 level is than the least positive input level that will be interpreted as a logical 1. It represents how large a negative-going glitch on an input 1 can be before it affects the output of the driven device. Similarly, $NM_L = V_{ILmax} - V_{OLmax}$ is the amplitude of the largest positive-going glitch on an input 0 that will not affect the output of the driven device.

Finally, three important definitions are associated with specifying the load that can be driven by a gate. Since in most cases the load on a gate output will be the sum of inputs of other gates, the first definition characterizes the relative current requirements of gate inputs.

- **Load Factor (LF).** Each logic family has a reference gate, each of whose inputs is defined to be a unit load in both the HIGH and the LOW conditions. The respective ratios of the input currents I_{IH} and I_{IL} of a given input to the corresponding I_{IH} and I_{IL} of the reference gate define the HIGH and LOW load factors of that input.
- **Drive Factor (DF).** A device output has drive factors for both the HIGH and the LOW output conditions. These factors are defined as the respective ratios of I_{OHmax} and I_{OLmax} of the gate to I_{OHmax} and I_{OLmax} of the reference gate.
- **Fan-Out.** For a given gate the fan-out is defined as the maximum number of inputs of the same type of gate that can be properly driven by that gate output. When gates of different load and drive factors are interconnected, fan-out must be adjusted accordingly.

Bipolar Transistor Gates

A logic circuit using bipolar junction transistors (BJTs) can be classified either as saturated or as nonsaturated logic. A saturated logic circuit contains at least one BJT that is saturated in one of the stable modes of the circuit. In nonsaturated logic circuits none of the transistors is allowed to saturate. Since bringing a BJT out of saturation requires a few additional nanoseconds (called the storage time), nonsaturated logic is faster. The fastest circuits available at this time are emitter-coupled logic (ECL), with transistor-transistor logic (TTL) having Schottky diodes connected to prevent the transistors from saturating (Schottky TTL) being a fairly close second. Both of these families are nonsaturated logic. All TTL families other than Schottky are saturated logic.

Transistor-Transistor Logic

TTL evolved from resistor-transistor logic (RTL) through the intermediate step of diode-transistor logic (DTL). All three families are catalogued in data books published in 1968, but of the three only TTL is still available.

The basic circuit of the standard TTL family is typified by the two-input NAND gate shown in Fig. 73.8(a). To estimate the operating levels of voltage and current in this circuit, assume that any transistor in saturation has $V_{CE} = 0.2$ and $V_{BE} = 0.75$ V. Let drops across conducting diodes also be 0.75 V and transistor current gains (when nonsaturated) be about 50. As a starting point, let the voltage levels at both inputs A and B be high enough that T_1 operates in the reversed mode. In this case the emitter currents of T_1 are negligible, and the current into the base of T_1 goes out the collector to become the base current of T_2. This current is readily calculated by observing that the base of T_1 is at $3 \times 0.75 = 2.25$ V so there is a 2.75-V drop across the 4-kΩ resistor. Thus $I_{BI} = I_{B2} = 0.7$ mA, and it follows that T_2 is saturated. With T_2 saturated, the base of T_3 is at $V_C + V_{BE4} = 0.95$ V. If T_4 is also saturated, the emitter of T_3 will be at $V_{D3} + V_{CE4} = 0.95$ V, and T_3 will be cut off. The voltage across the 1.6-kΩ resistor is $5 - 0.95 = 4.05$ V, so the collector current of T_2 is about 2.5 mA. This means the emitter current of T_2 is 3.2 mA. Of this, 0.75 mA goes through the 1-kΩ resistor, leaving 2.45 mA as the base current of T_4. Since the current gain of T_4 is about 50, it will be well into saturation for any collector current less than 100 mA, and the output at C is a logic 0. The corresponding minimum voltage levels required at the inputs are estimated from $V_{BE4} + V_{ECI}$, or about 1.7 V.

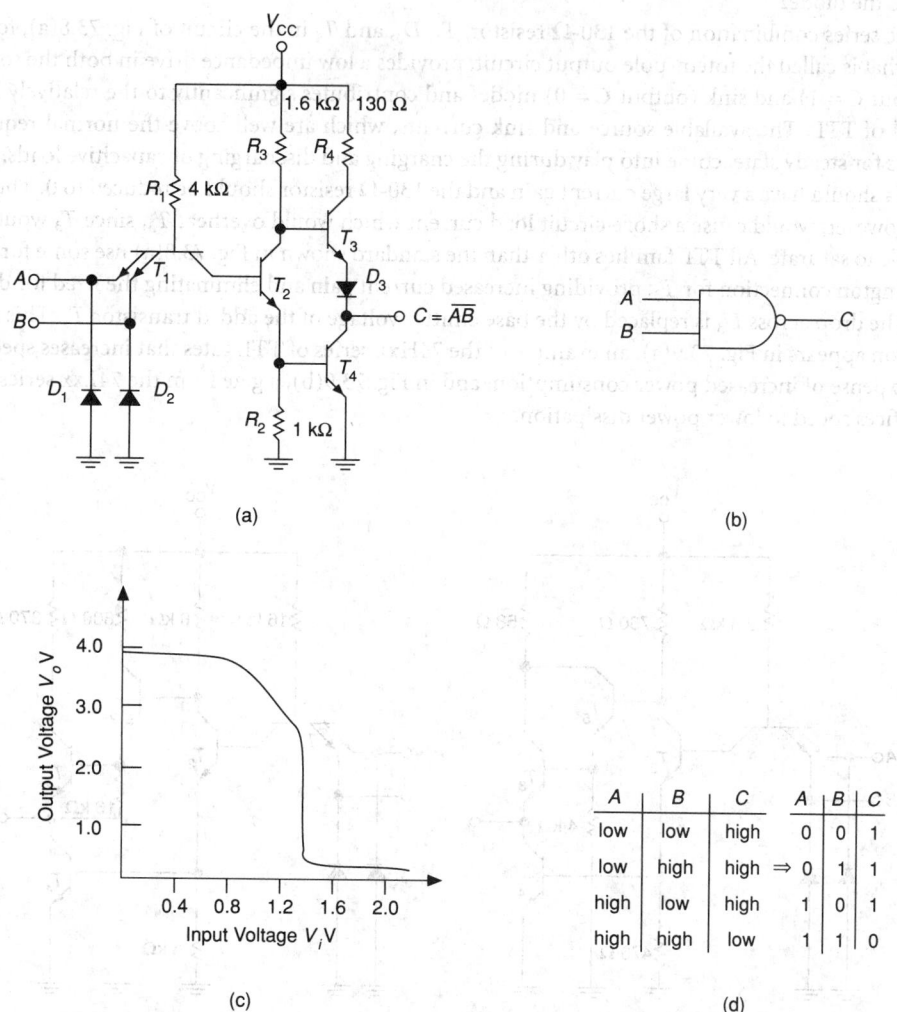

(a)

(b)

(c)

(d)

A	B	C		A	B	C
low	low	high		0	0	1
low	high	high	$\Rightarrow$	0	1	1
high	low	high		1	0	1
high	high	low		1	1	0

FIGURE 73.8 Two-input transistor-transistor logic (TTL) NAND gate type 7400: (a) circuit, (b) symbol, (c) voltage transfer characteristic (V_i to both inputs), (d) truth table. (*Source:* P. Graham, "Gates," in *Handbook of Modern Electronics and Electrical Engineering*, C. Belove, Ed., New York: Wiley-Interscience, 1986, p. 867. With permission.)

Now let either or both of the inputs be dropped to 0.2 V. T_1 is then biased to saturation in the normal mode, so the collector current of T_1 extracts the charge from the base region of T_2. With T_2 cut off, the base of T_4 is at 0 V and T_4 is cut off. T_3 will be biased by the current through the 1.6-kΩ resistor (R_3) to a degree regulated by the current demand at the output C. The drop across R_3 is quite small for light loads, so the output level at C will be $V_{CC} - V_{BE3} - V_{D3}$, which will be about 3.5 V corresponding to the logical 1.

The operation is summarized in the truth table in Fig. 73.8(d), identifying the circuit as a two-input NAND gate. The derivation of the input-output voltage transfer characteristic [Fig. 73.8(c)], where V_i is applied to inputs A and B simultaneously, can be found in most digital circuit textbooks. The sloping portion of the characteristic between $V_i = 0.55$ and 1.2 V corresponds to T_2 passing through the active region in going from cutoff to saturation.

Diodes D_1 and D_2 are present to damp out "ringing" that can occur, for example, when fast voltage level shifts are propagated down an appreciable length (20 cm or more) of microstripline formed by printed circuit board interconnections. Negative overshoots are clamped to the 0.7 V across the diode.

The series combination of the 130-Ω resistor, T_3, D_3, and T_4 in the circuit of Fig. 73.8(a), forming what is called the totem-pole output circuit, provides a low impedance drive in both the source (output $C = 1$) and sink (output $C = 0$) modes and contributes significantly to the relatively high speed of TTL. The available source and sink currents, which are well above the normal requirements for steady state, come into play during the charging and discharging of capacitive loads. Ideally T_3 should have a very large current gain and the 130-Ω resistor should be reduced to 0. The latter, however, would cause a short-circuit load current which would overheat T_3, since T_3 would be unable to saturate. All TTL families other than the standard shown in Fig. 73.8(a) use some form of Darlington connection for T_3, providing increased current gain and eliminating the need for diode D_3. The drop across D_3 is replaced by the base emitter voltage of the added transistor T_5. This connection appears in Fig. 73.9(a), an example of the 74Hxx series of TTL gates that increases speed at the expense of increased power consumption, and in Fig. 73.9(b), a gate from the 74Lxx series that sacrifices speed to lower power dissipation.

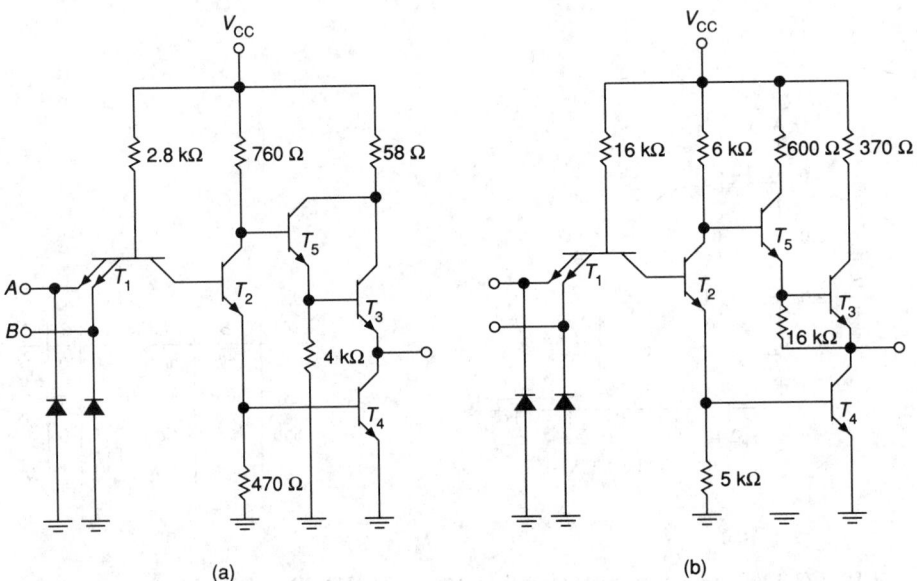

FIGURE 73.9 Modified transistor-transistor logic (TTL) two-input NAND states: (a) type 74Hxx, (b) type 74L00. (*Source:* P. Graham, "Gates," in *Handbook of Modern Electronics and Electrical Engineering*, C. Belove, Ed., New York: Wiley-Interscience, 1986, p. 868. With permission.)

A number of TTL logic function implementations are available with open collector outputs. For example, the 7403 two-input NAND gate shown in Fig. 73.10 is the open collector version of Fig. 73.8(a). The open collector output has some useful applications. The current in an external load connected between the open collector and V_{CC} can be switched on and off in response to the input combinations. This load, for example, might be a relay, an indicator light, or an LED display. Also, two or more open collector gates can share a common load, resulting in the anding together of the individual gate functions. This is called a "wired-AND connection." In any application, there must be some form of load or the device will not function. There is a lower limit to the resistance of this load which is determined by the current rating of the open collector transistor. For wired-AND applications the resistance range depends on how many outputs are being wired and on the load being driven by the wired outputs. Formulas are given in the data books. Since the open collector configuration does not have the speed enhancement associated with an active pull-up, the low to high propagation delay (t_{PLH}) is about double that of the totem-pole output. It should be observed that totem-pole outputs should not be wired, since excessive currents in the active pull-up circuit could result.

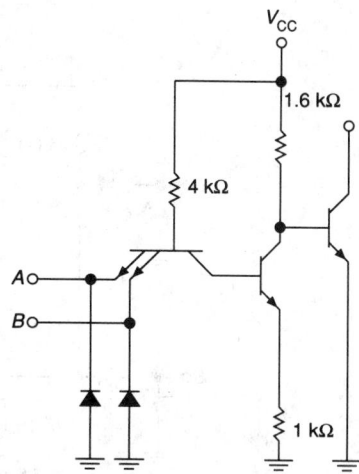

FIGURE 73.10 Open collector two-input NAND gate. (*Source:* P. Graham, "Gates," in *Handbook of Modern Electronics and Electrical Engineering,* C. Belove, Ed., New York: Wiley-Interscience, 1986, p. 868. With permission.)

Nonsaturated TTL. Two TTL families, the Schottky (74Sxx) and the low-power Schottky (74LSxx), can be classified as nonsaturating logic. The transistors in these circuits are kept out of saturation by the connection of Schottky diodes, with the anode to the base and the cathode to the collector.

Schottky diodes are formed from junctions of metal and an *n*-type semiconductor, the metal fulfilling the role of the *p*-region. Since there are thus no minority carriers in the region of the forward-biased junction, the storage time required to bring a *pn* junction out of saturation is eliminated. The forward-biased drop across a Schottky diode is around 0.3 V. This clamps the collector at 0.3 V less than the base, thus maintaining V_{CE} above the 0.3-V saturation threshold. Circuits for the two-input NAND gates 74LS00 and 74S00 are given in Fig. 73.11(a) and (b). The special transistor symbol is a short-form notation indicating the presence of the Schottky diode, as illustrated in Fig. 73.11(c).

Note that both of these circuits have an active pull-down transistor T_6 replacing the pull-down resistance connected to the emitter of T_2 in Fig. 73.9. The addition of T_6 decreases the turn-on and turn-off times of T_4. In addition, the transfer characteristic for these devices is improved by the squaring off of the sloping region between $V_i = 0.55$ and 1.2 V [see Fig. 73.8(c)]. This happens because T_2 cannot become active until T_6 turns on, which requires at least 1.2 V at the input.

The diode AND circuit of the 74LS00 in place of the multi-emitter transistor will permit maximum input levels substantially higher than the 5.5-V limit set for all other TTL families. Input leakage currents for 74LSxx are specified at $V_i = 10$ V, and input voltage levels up to 15 V are allowed. The 74LSxx has the additional feature of the Schottky diode D_1 in series with the 100-Ω output resistor. This allows the output to be pulled up to 10 V without causing a reverse breakdown of T_5. The relative characteristics of the several versions of the TTL two-input NAND gate are compared in Table 73.6. The 74F00 represents one of the new technologies that have introduced improved Schottky TTL in recent years.

TTL Design Considerations. Before undertaking construction of a logic system, the wise designer consults the information and recommendations provided in the data books of most manufacturers. Some of the more significant tips are provided here for easy reference.

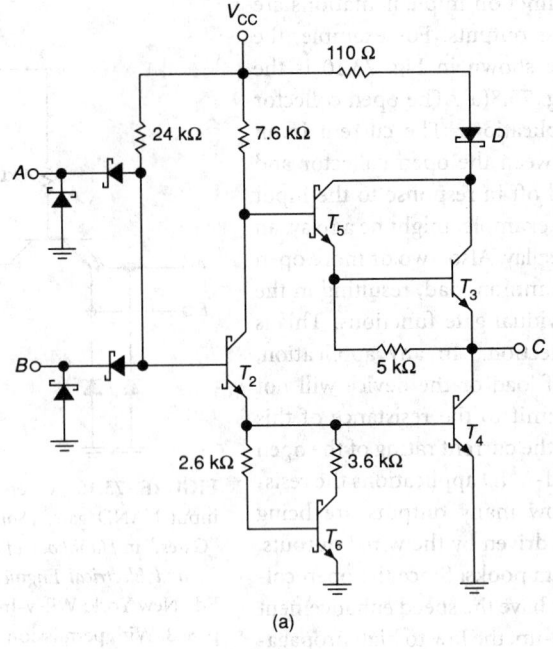

(a)

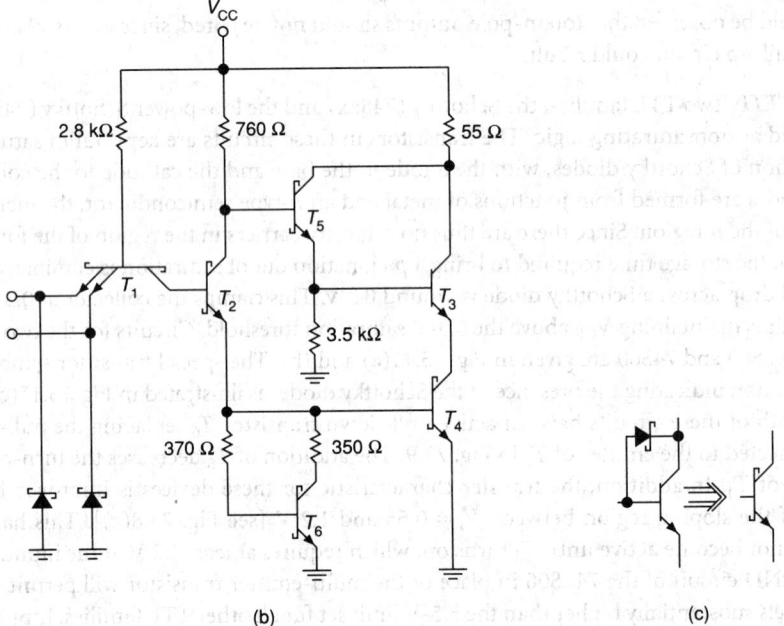

(b)

(c)

FIGURE 73.11 Transistor-transistor logic (TTL) nonsaturated logic. (a) Type 74LS00 two-input NAND gate, (b) type 74S00 two-input NAND gate, (c) significance of the Schottky transistor symbol. (*Source:* P. Graham, "Gates," in *Handbook of Modern Electronics and Electrical Engineering,* C. Belove, Ed., New York: Wiley-Interscience, 1986, p. 870. With permission.)

1. **Power supply, decoupling, and grounding.** The power supply voltage should be 5 V with less than 5% ripple factor and better than 5% regulation. When packages on the same printed circuit board are supplied by a bus there should be a 0.05-μF decoupling capacitor between the bus and the ground for every five to ten packages. If a ground bus is used, it should be as

Table 73.6 Comparison of TTL Two-Input NAND Gates

TTL Type	Supply Current		Propagation Delay Time		Noise Margins		Load Factor, H/L	Drive Factor, H/L	Fan-out
	I_{CCH}[a] (mA)	I_{CCL} (mA)	t_{PLH} (ns)	t_{PHL} (ns)	NM_H (V)	NM_L (V)			
74F00	2.8	10.2	2.9	2.6	0.7	0.3	0.5/0.375	25/12.5	33
74S00	10	20	3	3	0.7	0.3	1.25/1.25	25/12.5	10
74H00	10	26	5.9	6.2	0.4	0.4	1.25/1.25	12.5/12.5	10
74LS00	0.8	2.4	9	10	0.7	0.3	0.5/0.25	10/5	20
7400	4	12	11	7	0.4	0.4	1/1	20/10	10
74L00	0.44	1.16	31	31	0.4	0.5	0.24/0.1125	5/2.25	20

[a]See text for explanation of abbreviations.

Source: P. Graham, "Gates," in *Handbook of Modern Electronics and Electrical Engineering,* C. Belove, Ed., New York: Wiley-Interscience, 1986, p. 871. With permission.

wide as possible, and should surround all the packages on the board. Whenever possible, use a ground plane. If a long ground bus is used, both ends must be tied to the common system ground point.

2. **Unused gates and inputs.** If a gate on a package is not used, its inputs should be tied either high or low, whichever results in the least supply current. For example, the 7400 draws three times the current with the output low as with the output high, so the inputs of an unused 7400 gate should be grounded. An unused input of a gate, however, must be connected so as not to affect the function of the active inputs. For a 7400 NAND gate, such an input must either be tied high or paralleled with a used input. It must be recognized that paralleled inputs count as two when determining the fan-out. Inputs that are tied high can be connected either to V_{CC} through a 1-kΩ or more resistance (for protection from supply voltage surges) or to the output of an unused gate whose input will establish a permanent output high. Several inputs can share a common protective resistance. Unused inputs of low-power Schottky TTL can be tied directly to V_{CC}, since 74LSxx inputs tolerate up to 15 V without breakdown. If inputs of low-power Schottky are connected in parallel and driven as a single input, the switching speed is decreased, in contrast to the situation with other TTL families.

3. **Interconnection.** Use of line lengths of up to 10 in. (5 in. for 74S) requires no particular precautions, except that in some critical situations lines cannot run side by side for an appreciable distance without causing cross talk due to capacitive coupling between them. For transmission line connections, a gate should drive only one line, and a line should be terminated in only one gate input. If overshoots are a problem, a 25- to 50-Ω resistor should be used in series with the driving gate input and the receiving gate input should be pulled up to 5 V through a 1-kΩ resistor. Driving and receiving gates should have their own decoupling capacitors between the V_{CC} and ground pins. Parallel lines should have a grounded line separating them to avoid cross talk.

4. **Mixing TTL subfamilies.** Even synchronous sequential systems often have asynchronous features such as reset, preset, load, and so on. Mixing high-speed 74S TTL with lower speed TTL (74LS for example) in some applications can cause timing problems resulting in anomalous behavior. Such mixing is to be avoided, with rare exceptions which must be carefully analyzed.

Emitter-Coupled Logic

ECL is a nonsaturated logic family where saturation is avoided by operating the transistors in the common collector configuration. This feature, in combination with a smaller difference between the HIGH and LOW voltage levels (less than 1 V) than other logic families, makes ECL the fastest logic available at this time. The circuit diagram of a widely used version of the basic two-input ECL gate is given in Fig. 73.12. The power supply terminals V_{CC1}, V_{CC2}, V_{EE}, and V_{TT} are available for flexibility in biasing. In normal operation, V_{CC1} and V_{CC2} are connected to a common ground, V_{EE} is biased to –5.2 V, and V_{TT} is biased to –2 V. With these values the nominal voltage for the logical 0 and 1 are, respectively, –1.75 and –0.9 V. Operation with the V_{CC} terminals grounded maximizes the immunity from noise interference.

A brief description of the operation of the circuit will verify that none of the transistors saturates. For the following discussion, V_{CC1} and V_{CC2} are grounded, V_{EE} is –5.2 V, and V_{TT} is –2 V. Diode drops and base-emitter voltages of active transistors are 0.8 V.

First, observe that the resistor-diode (D_1 and D_2) voltage divider establishes a reference voltage of –0.55 V at the base of T_3, which translates to –1.35 V at the base of T_2. When either or both of the inputs A and B are at the logical 1 level of –0.9 V, the emitters of T_{1A}, T_{1B}, and T_2 will be 0.8 V lower, at –1.7 V. This establishes the base-emitter voltage of T_2 at $-1.35 - (-1.7) = 0.35$ V, so T_2 is cut off. With T_2 off, T_4 is biased into the active region, and its emitter will be at about –0.9 V, corresponding to a logical 1 at the $(A + B)$ output. Most of the current through the 365-Ω emitter resistor, which is $[-1.7 - (-5.2)]/0.365 = 9.6$ mA, flows through the 100-Ω collector resistor, dropping the base voltage of T_5 to –0.96 V. Thus the voltage level at the output terminal designated $(\overline{A + B})$ is –1.76 V, corresponding to a logical 0.

When both A and B inputs are at the LOW level of –1.75 V, T_2 will be active, with its emitter voltage at $-1.35 - 0.8 = -2.15$ V. The current through the 365-Ω resistor becomes $[-2.15 - (-5.2)]/0.365 = 8.2$ mA. This current flows through the 112-Ω resistor pulling the base of T_4 down to –0.94 V, so that the $(A + B)$ output will be at the LOW level of –1.75 V. With T_{1A} and T_{1B} cut off, the base of T_5 is close to 0.0 V, and the $(\overline{A + B})$ output will therefore be at the nominal HIGH level of –0.9 V.

Observe that the output transistors T_4 and T_5 are always active and function as emitter followers, providing the low-output impedances required for driving capacitive loads. As T_{1A} and/or T_{1B} turn on, and T_2 turns off as a consequence, the transition is accomplished with very little current change in the 365-Ω emitter resistor. It follows that the supply current from V_{EE} does not undergo the sudden increases and decreases prevalent in TTL, thus eliminating the need for decoupling capacitors. This is a major reason why ECL can be operated successfully with the low noise margins which are inherent in logic having a relatively small voltage difference between the HIGH and LOW voltage levels (see Table 73.7). The small level shifts between LOW and HIGH also permit low propagation

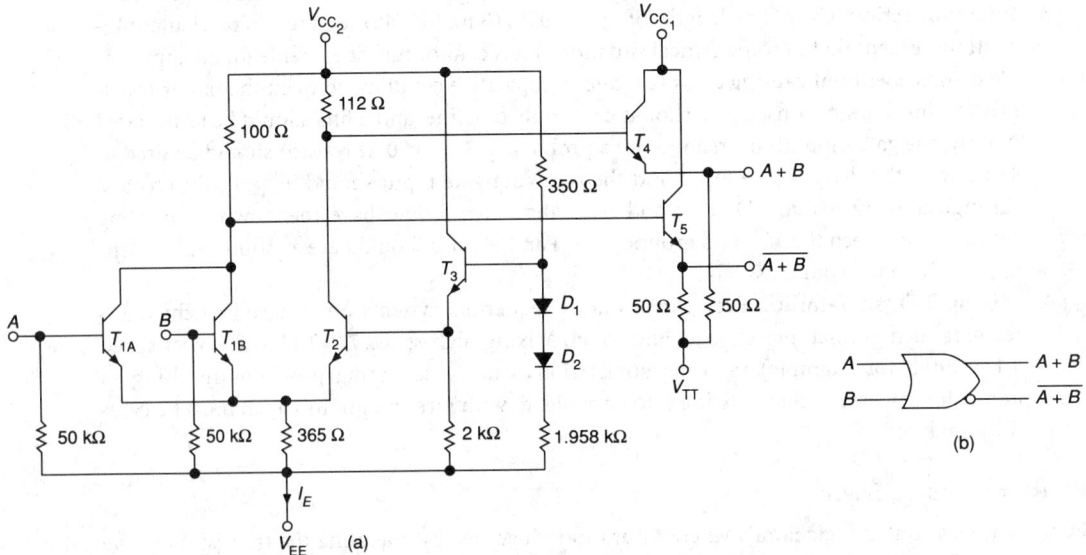

FIGURE 73.12   Emitter-coupled logic basic gate (ECL 10102): (a) circuit, (b) symbol. (*Source:* P. Graham, "Gates," in *Handbook of Modern Electronics and Electrical Engineering*, C. Belove, Ed., New York: Wiley-Interscience, 1986, p. 872. With permission.)

Table 73.7 Comparison of ECL Quad Two-Input NOR Gates ($V_{TT} = V_{EE} = 5.2$ V, $V_{CC1} = 0$ V)

ECL Type	Power Supply Terminal V_{EE} (V)	Power Supply Current I_E (mA)	Propagation Delay Time		Transition Time		Noise Margins		Test Load
			$t_{PLH}{}^a$ (ns)	t_{PHL} (ns)	$t_{TLH}{}^b$ (ns)	$t_{THL}{}^b$ (ns)	NM_H (V)	NM_L (V)	
ECL II									
1012	−5.2	18^c	5	4.5	4	6	0.175	0.175	Fan-out of 3
95102	−5.2	11	2	2	2	2	0.14	0.145	50 Ω
10102	−5.2	20	2	2	2.2	2.2	0.135	0.175	50 Ω
ECL III									
1662	−5.2	56^c	1	1.1	1.4	1.2	0.125	0.125	50 Ω
100102^d	−4.5	55	0.75	0.75	0.7	0.7	0.14	0.145	50 Ω
11001^e	−5.2	24	0.7	0.7	0.7	0.7	0.145	0.175	50 Ω

[a]See text for explanation of abbreviations.
[b]20 to 80% levels.
[c]Maximum value (all other typical).
[d]Quint 2-input NOR/OR gate.
[e]Dual 5/4-input NOR/OR gate.

Source: P. Graham, "Gates," in *Handbook of Modern Electronics and Electrical Engineering*, C. Belove, Ed., New York: Wiley-Interscience, 1986, p. 873. With permission.

times without excessively fast rise and fall times. This reduces the effects of residual capacitive coupling between gates, thereby lessening the required noise margin. For this reason the faster ECL (100xxx) should not be used where the speed of the 10xxx series is sufficient. A comparison of three ECL series is given in Table 73.7. The propagation times t_{PLH} and t_{PHL} and transition times t_{TLH} and t_{THL} are defined in Fig. 73.7. Transitions are between the 20 and 80% levels.

The 50-Ω pull-down resistors shown in Fig. 73.12 are connected externally. The outputs of several gates can therefore share a common pull-down resistor to form a wired-OR connection. The open emitter outputs also provide flexibility for driving transmission lines, the use of which in most cases is mandatory for interconnecting this high-speed logic. A twisted pair interconnection can be driven using the complementary outputs $(A + B)$ and $(A + B)$ as a differential output. Such a line should be terminated in an ECL line receiver (10114).

Since ECL is used in high-speed applications, special techniques must be applied in the layout and interconnection of chips on circuit boards. Users should consult design handbooks published by the suppliers before undertaking the construction of an ECL logic system.

While ECL is not compatible with any other logic family, interfacing buffers, called translators, are available. In particular, the 10124 converts TTL output levels to ECL complementary levels, and the 10125 converts either single-ended or differential ECL outputs to TTL levels. Among other applications of these translators, they allow the use of ECL for the highest speed requirements of a system while the rest of the system uses the more rugged TTL. Another translator is the 10177, which converts the ECL output levels to n-channel metal-oxide semiconductor (NMOS) levels. This is designed for interfacing ECL with n-channel memory systems.

Complementary Metal-Oxide Semiconductor (CMOS) Logic

Metal-oxide semiconductor (MOS) technology is prevalent in LSI systems due to the high circuit densities possible with these devices. p-Channel MOS was used in the first LSI systems, and it still is the cheapest to produce because of the higher yields achieved due to the longer experience with PMOS technology. PMOS, however, is largely being replaced by NMOS (n-channel MOS), which has the advantages of being faster (since electrons have greater mobility than holes) and having TTL compatibility. In addition, NMOS has a higher function/chip area density than PMOS, the highest density in fact of any of the current technologies. Use of NMOS and PMOS, however, is limited to LSI and VLSI fabrications. The only MOS logic available as SSI and MSI is CMOS (complementary MOS).

CMOS is faster than NMOS and PMOS, and it uses less power per function than any other logic. While it is suitable for LSI, it is more expensive and requires somewhat more chip area than NMOS or PMOS. In many respects it is unsurpassed for SSI and MSI applications. Standard CMOS (the 4000 series) is as fast as low-power TTL (74Lxx) and has the largest noise margin of any logic type.

A unique advantage of CMOS is that for all input combinations the steady-state current from V_{DD} to V_{SS} is almost zero because at least one of the series FETs is open. Since CMOS circuits of any complexity are interconnections of the basic gates, the quiescent currents for these circuits are extremely small, an obvious advantage which becomes a necessity for the practicality of digital watches, for example, and one which alleviates heat dissipation problems in high-density chips. Also a noteworthy feature of CMOS digital circuits is the absence of components other than FETs. This attribute, which is shared by PMOS and NMOS, accounts for the much higher function/chip area density than is possible with TTL or ECL. During the time the output of a CMOS gate is switching there will be current flow from V_{DD} to V_{SS}, partly due to the charging of junction capacitances and partly because the path between V_{DD} and V_{SS} closes momentarily as the FETs turn on and off. This causes the dc supply current to increase in proportion to the switching frequency in a CMOS circuit. Manufacturers specify that the supply voltage for standard CMOS can range over $3\ V \le V_{DD} - V_{SS} \le 18\ V$, but switching speeds are slower at the lower voltages, mainly due to the increased resistances of the "on" transistors. The output switches between low and high when the input is midway between V_{DD} and V_{SS}, and the output logical 1 level will be V_{DD} and the logical 0 level V_{SS} [Fig. 73.13(c)]. If CMOS is operated with $V_{DD} = 5\ V$ and $V_{SS} = 0\ V$, the V_{DD} and V_{SS} levels will be almost compatible with TTL except that the TTL totem-pole output high of 3.4 V is marginal as a logical 1 for CMOS. To alleviate this, when CMOS is driven with TTL a 3.3-kΩ pull-up resistor between the TTL output and the common V_{CC}, V_{DD} supply terminal should be used. This raises V_{OH} of the TTL output to 5 V.

All CMOS inputs are diode protected to prevent static charge from accumulating on the FET gates and causing punch-through of the oxide insulating layer. A typical configuration is illustrated in Fig. 73.14. Diodes D_1 and D_2 clamp the transistor gates between V_{DD} and V_{SS}. Care must be taken to avoid input voltages that would cause excessive diode currents. For this reason manufacturers specify an input voltage constraint from $V_{SS} - 0.5\ V$ to $V_{DD} + 0.5\ V$. The resistance R_s helps protect the diodes from excessive currents but is introduced at the expense of switching speed, which is deteriorated by the time constant of this resistance and the junction capacitances.

Advanced versions of CMOS have been developed which are faster than standard CMOS. The

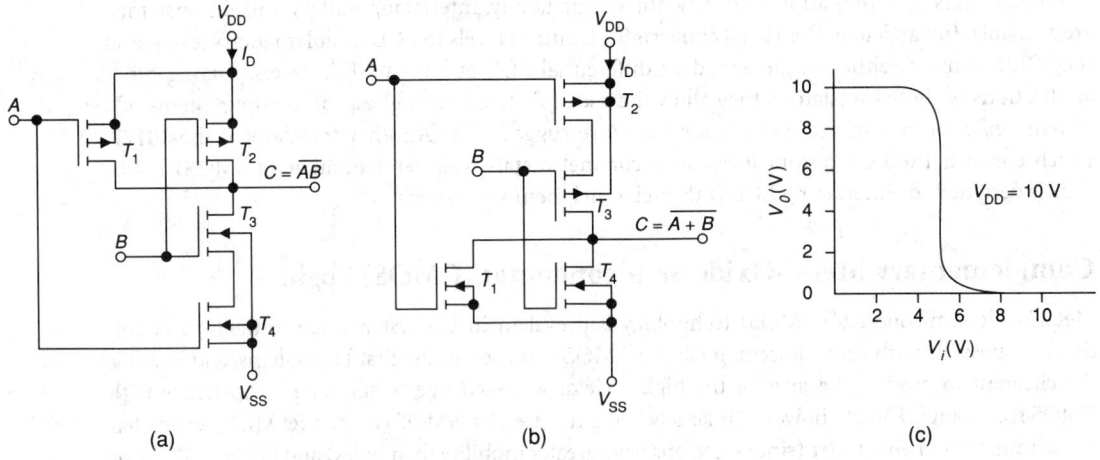

FIGURE 73.13 (a) Complementary metal-oxide semiconductor (CMOS) NAND gate, (b) NOR gate, and (C) inverter transfer characteristic. (*Source:* P. Graham, "Gates," in *Handbook of Modern Electronics and Electrical Engineering,* C. Belove, Ed., New York: Wiley-Interscience, 1986, p. 874. With permission.)

first of these to appear were designated 74HCxx and 74HCTxx. The supply voltage range for this series is limited to $2\text{ V} \le V_{DD} - V_{SS} \le 6\text{ V}$. The pin numbering of a given chip is the same as its correspondingly numbered TTL device. Furthermore, gates with the HCT code have skewed transfer characteristics which match those of its TTL cousin, so that these chips can be directly interchanged with low-power Schottky TTL.

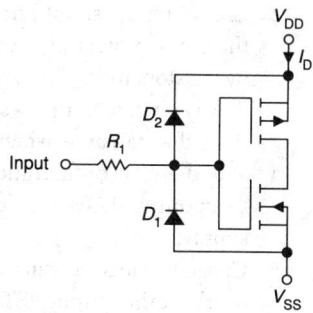

More recently, a much faster CMOS has appeared and carries the designations 74ACxx and 74ACTxx. These operate in the same supply voltage range and bear the same relationship with TTL as the HCMOS. The driving capabilities (characterized by I_{OH} and I_{OL}) of this series are much greater, such that they can be fanned out to 10 low-power Schottky inputs.

The three types of CMOS are compared in Table 73.8. The relative speeds of these technologies are best illustrated by including in the table the maximum clock frequencies for D flip-flops. In each case, the frequency given is the maximum for which the device is guaranteed to work. It is worth noting that a typical maximum clocking of 160 MHz is claimed for the 74ACT374 D flip-flop.

FIGURE 73.14 Diode protection of input transistor gates. $200\ \Omega < R_s < 1.5\ \text{k}\Omega$. (*Source:* P. Graham, "Gates," in *Handbook of Modern Electronics and Electrical Engineering*, C. Belove, Ed., New York: Wiley-Interscience, 1986, p. 875. With permission.)

CMOS Design Considerations

Design and handling recommendations for CMOS, which are included in several of the data books, should be consulted by the designer using this technology. A few selected recommendations are included here to illustrate the importance of such information.

1. All unused CMOS inputs should be tied either to V_{DD} or V_{SS}, whichever is appropriate for proper operation of the gate. This rule applies even to inputs of unused gates, not only to protect the inputs from possible static charge buildup, but to avoid unnecessary supply current drain. Floating gate inputs will cause all the FETs to be conducting, wasting power and heating the chip unnecessarily.

Table 73.8 Comparison of Standard, High-Speed, and Advanced High-Speed CMOS

Parameter	Symbol	Unit	Standard CMOS NOR Gates		High-Speed CMOS Inverter		Advanced CMOS Inverter	
			4001B	4011UB	74HC04	74HCT04	74AC04	74ACT04
Supply voltage	V_{DD}-V_{SS}	V	15	15	6	5.5	5.5	5.5
Input voltage	V_{IHmin}	V	11	12.5	4.2	2	3.85	2
thresholds	V_{ILmax}	V	4	2.5	1.8	0.8	1.65	0.8
Guaranteed output	V_{OHmin}	V	13.5	13.5	5.9	4.5	4.86	4.76
levels at maximum IO	V_{OLmax}	V	1.5	1.5	0.1	0.26	0.32	0.37
Maximum	I_{OH}	mA	−8.8	−3.5	−4	−4	−24	−24
output currents	I_{OL}	mA	8.8	8.8	4	4	24	24
Noise	NM_L	V	2.5	2.5	1.7	0.54	1.33	.43
margins	NM_H	V	2.5	2.5	1.7	2.5	1.01	1.24
Propagation	t_{PLH}	ns	40	40	16	15	4	4.3
times	t_{PHL}	ns	40	40	16	17	3.5	3.9
Max input current leakage	I_{INmax}	μA	0.1	0.1	0.1	0.1	0.1	0.1
D-flip-flop			4013B		74HC374	74HCT374A	74AC374	74ACT374
max frequency (guaranteed minimum)	f_{max}	MHz	7.0	N.A.	35	30	100	100

2. CMOS inputs should never be driven when the supply voltage V_{DD} is off, since damage to the input-protecting diodes could result. Inputs wired to edge connectors should be shunted by resistors to V_{DD} or V_{SS} to guard against this possibility.

3. Slowly changing inputs should be conditioned using Schmitt trigger buffers to avoid oscillations that can arise when a gate input voltage is in the transition region.

4. Wired-AND configurations cannot be used with CMOS gates, since wiring an output HIGH to an output LOW would place two series FETs in the "on" condition directly across the chip supply.

5. Capacitive loads greater than 5000 pF across CMOS gate outputs act as short circuits and can overheat the output FETs at higher frequencies.

6. Designs should be used that avoid the possibility of having low impedances (such as generator outputs) connected to CMOS inputs prior to power-up of the CMOS chip. The resulting current surge when V_{DD} is turned on can damage the input diodes.

While this list of recommendations is incomplete, it should alert the CMOS designer to the value of the information supplied by the manufacturers.

Choosing a Logic Family

A logic designer planning a system using SSI and MSI chips will find that an extensive variety of circuits is available in all three technologies: TTL, ECL, and CMOS. The choice of which technology will dominate the system is governed by what are often conflicting needs, namely, speed, power consumption, noise immunity, cost, availability, and the ease of interfacing. Sometimes the decision is easy. If the need for a low static power drain is paramount, CMOS is the only choice. It used to be the case that speed would dictate the selection; ECL was high speed, TTL was moderate, and CMOS low. With the advent of advanced TTL and, especially, advanced CMOS the choice is no longer clear-cut. All three will work at 100 MHz or more. ECL might be used since it generates the least noise because the transitions are small, yet for that same reason it is more susceptible to externally generated noise. Perhaps TTL might be the best compromise between noise generation and susceptibility. Advanced CMOS is the noisiest because of its rapid rise and fall times, but the designer might opt to cope with the noise problems to take advantage of the low standby power requirements.

A good rule is to use devices which are no faster than the application requires and which consume the least power consistent with the needed driving capability. The information published in the manufacturers' data books and designer handbooks is very helpful when choice is in doubt.

Defining Term

Logic gate: Basic building block for logic systems that controls the flow of pulses.

References

Advanced CMOS Logic Designers Handbook, Dallas: Texas Instruments, Inc., 1987.

C. Belove and D. Schilling, *Electronic Circuits, Discrete and Integrated,* 2nd ed., New York: McGraw-Hill, 1979.

FACT Data, Phoenix: Motorola Semiconductor Products, Inc., 1989.

Fairchild Advanced Schottky TTL, Calif.: Fairchild Camera and Instrument Corporation, 1980.

W. I. Fletcher, *An Engineering Approach to Digital Design,* Englewood Cliffs, N.J.: Prentice-Hall, 1980.

High Speed CMOS Logic Data, Phoenix: Motorola Semiconductor Products, Inc., 1989.

P. Horowitz and W. Hill, *The Art of Electronics,* 2nd ed., New York: Cambridge University Press, 1990.

MECL System Design Handbook, Phoenix: Motorola Semiconductor Products, Inc., 1988.

H. Taub and D. Schilling, *Digital Integrated Electronics,* New York: McGraw-Hill, 1977.

The TTL Data Book for Design Engineers, Dallas: Texas Instruments, Inc., 1990.

Further Information

An excellent presentation of the practical design of logic systems using SSI and MSI devices is developed in the referenced book *An Engineering Approach to Digital Design* by William I. Fletcher. The author pays particular attention to the importance of device speed and timing.

The Art of Electronics by Horowitz and Hill is particularly helpful for its practical approach to interfacing digital with analog.

Everything one needs to know about digital devices and their interconnection can be found somewhere in the data manuals, design handbooks, and application notes published by the device manufacturers. Unfortunately, no single publication has it all, so the serious user should acquire as large a collection of these sources as possible.

73.3 Bistable Devices

Richard S. Sandige

This section deals with bistable devices which are also commonly referred to as **bistables, latches,** or **flip-flops.** Bistable devices are **memory elements.** Each bistable provides storage for only 1-bit, i.e., it can store a 1 or a 0. Figure 73.15 shows a graphic classification of bistable devices.

Manufacturers supply integrated circuit (IC) packages containing several bistable devices. One data book for the transistor-transistor logic (TTL) circuit technology lists 4-, 8-, 9-, and 10-bit latches in one IC package. The same data book lists 2-, 4-, 6-, 8-, 9-, and 10-bit flip-flops in one IC package. While a 1-bit bistable can only store 1 bit of information, 8-bit bistables are capable of storing 8 bits of information. Bistable devices implemented with logic gates are **volatile devices.** When power is first applied the first stored value of the bistable is random (it can store a 1 or a 0), and when power is removed the bistable loses its storage capability. Certain memories (also called stores) are nonvolatile and therefore retain their data when power is removed. These devices will not be discussed in this section.

Basic Latches

A latch can be either basic or gated. Figure 73.16 is an example of a basic *S-R* NOR latch implementation using two cross-coupled NOR gates. The logic symbol recommended for the *S-R* NOR latch

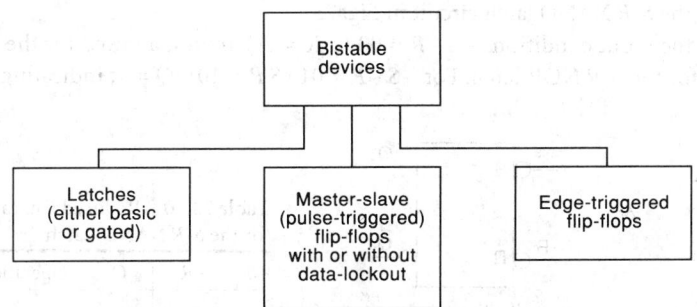

FIGURE 73.15 Graphic classification of bistable devices. (*Source:* Modified from R. S. Sandige, *Modern Digital Design,* New York: McGraw-Hill, 1990, p. 467. With permission.)

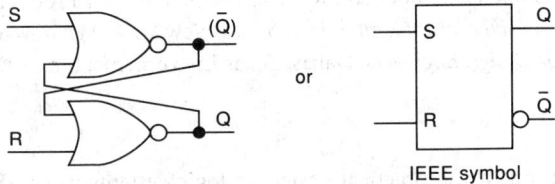

FIGURE 73.16 Basic *S-R* NOR latch implementation. (*Source:* Modified from R. S. Sandige, *Modern Digital Design,* New York: McGraw-Hill, 1990, p. 448. With permission.)

Table 73.9 Reduced Characteristic Table for the *S-R* NOR Latch

S	R	Q	Operation
0	0	Q_0	no change
0	1	0	reset
1	0	1	set
1	1	0	not normally allowed

by the Institute of Electrical and Electronics Engineers (IEEE) is shown to the right of the logic circuit implementation.

The input signal named *S* stands for set while the input signal named *R* stands for reset. Manufacturers often select *Q* as the output signal name for bistable devices in their data books. The *Q*s on the outputs are added for clarity and are not part of the IEEE symbol. The *S-R* NOR latch shown in Fig. 73.16 is a basic latch circuit since the *S* and *R* inputs are not gated with a control signal. The **reduced characteristic table** in Table 73.9 shows the operation of the *S-R* NOR latch circuit.

For *S R* = 00, *Q* = Q_0, illustrating that the output for the next state *Q* is the same as the present state output Q_0. For *S R* = 01, *Q* = 0, specifying that the output for the next state is reset. For *S R* = 10, *Q* = 1, indicating that the output for the next state is set. In most cases the input conditions *S R* = 11 are not allowed for two reasons. If *S R* = 11, then the alternate output for the bistable, shown in parentheses as $\bar{Q}$ in Fig. 73.16, is not logically correct as it is for all other input combinations. The second reason is more subtle since the next state of the bistable can be set or reset due to a **critical race** condition when the inputs are changed from 11 to 00. Such unpredictability is not desirable and therefore the *S R* = 11 condition is generally not allowed. Latches and flip-flops that contain a *Q* and a $\bar{Q}$ output (complementary outputs) provide double-rail outputs.

The *S-R* NAND latch implementation shown in Fig. 73.17 uses two cross-coupled NAND gates. The tildes shown in the logic circuit diagram preceding *S* and *R* represent inline symbols for the logical complements of *S* and *R*, respectively, as recommended by the IEEE. Data books usually refer to the *S-R* NAND latch as the $\bar{S}$-$\bar{R}$ latch. The logic symbol recommended for the *S-R* NAND latch by IEEE is shown to the right of the logic circuit diagram.

The ~*S* and ~*R* on the inputs and *Q*s on the outputs of the IEEE symbol are added for clarity and are not part of the IEEE symbol. The reduced characteristic table illustrated in Table 73.10 shows the operation of the *S-R* NAND latch circuit in Fig. 73.17.

In most cases the input conditions ~*S* ~*R* = 00 (*S R* = 11) are not allowed for the same reasons provided above for the *S-R* NOR latch. For ~*S* ~*R* = 01 (*S R* = 10), *Q* = 1, indicating that the out-

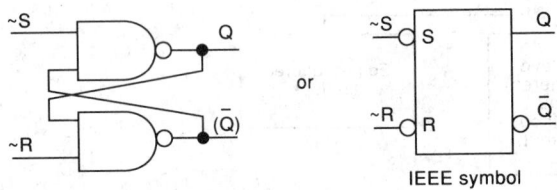

FIGURE 73.17 Basic *S-R* NAND latch implementation. (*Source:* Modified from R. S. Sandige. *Modern Digital Design,* New York: McGraw-Hill, 1990, p. 449. With permission.)

Table 73.10 Reduced Characteristic Table for the *S-R* NAND Latch

~S	~R	Q	Operation
0	0	1	not normally allowed
0	1	1	set
1	0	0	reset
1	1	Q_0	no change

put for the next state is set. For $\sim S \sim R = 10$ ($S R = 01$), $Q = 0$, specifying that the output for the next state is reset. For $\sim S \sim R = 11$ ($S R = 00$), $Q = Q_0$, illustrating that the output for the next state Q is the same as the present state output Q_0.

Gated Latches

All other gate level latches and flip-flops are functionally equivalent to either the configuration of the cross-coupled NOR latch circuit or the cross-coupled AND latch circuit. A gated S-R NOR latch circuit and a gated S-R AND latch are illustrated in Fig. 73.18 along with the recommended IEEE symbol. The reduced characteristic table for both of these circuits is provided in Table 73.11.

In each circuit both the S and R inputs are gated with a control signal C. Notice in the reduced characteristic table that the S and R inputs are only enabled, and thus have an effect on the output, when $C = 1$ **(transparent mode).**

Whatever value the output has when C goes to 0 is latched, captured, or stored (memory mode). Like the basic latches, the input conditions for $S R = 11$ are not generally allowed for the gated S-R latches when C goes to 1. The gated D latch circuit is perhaps the most used latch circuit since the added Inverter shown in the circuit diagram in Fig. 73.19 ensures that the input conditions for $S R = 11$ cannot occur when C goes to 1. The reduced characteristic table for the gated D latch circuit is shown in Table 73.12.

Flip-Flops

We will use the term *flip-flop* to distinguish between the bistable device called a latch and the bistable device that allows feed-back without oscillation. Early types of flip-flops were of the master-slave (pulse-triggered) variety that had no data-lockout circuitry and caused a storage error if improperly used due to

Table 73.11 Reduced Characteristic Table for the Gated S-R Latches

C	S	R	Q	Operation
0	0	0	Q_0	no change
0	0	1	Q_0	no change
0	1	0	Q_0	no change
0	1	1	Q_0	no change
1	0	0	Q_0	no change
1	0	1	0	reset
1	1	0	1	set
1	1	1	0,1	reset (S-R NOR), set (S-R NAND)

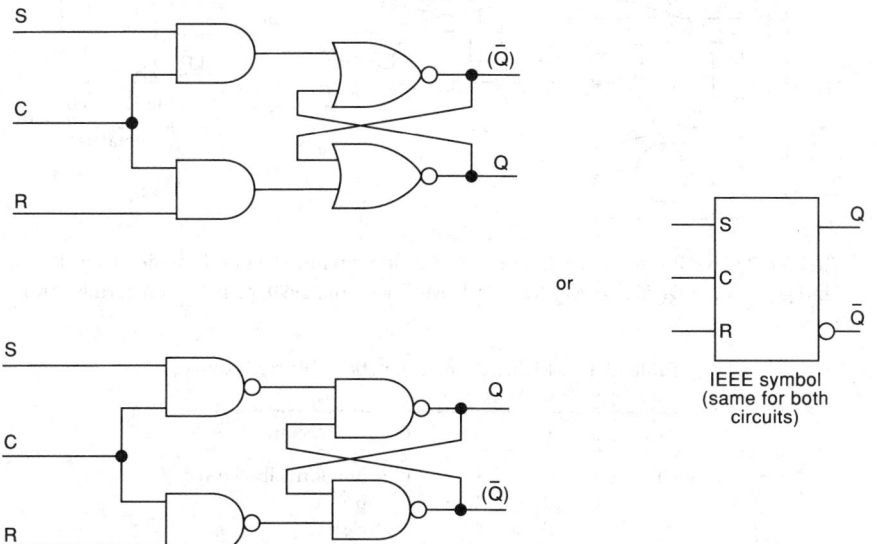

FIGURE 73.18 Gated S-R NOR and gated S-R NAND latch circuit. (*Source:* Modified from R. S. Sandige. *Modern Digital Design,* New York: McGraw-Hill, 1990, p. 468. With permission.)

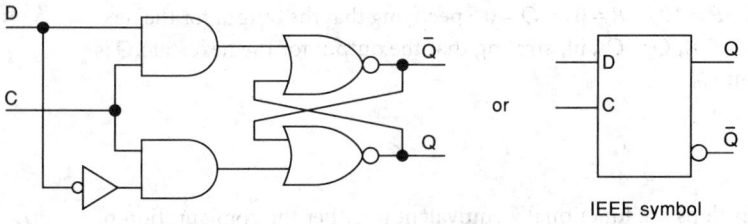

Table 73.12 Reduced Characteristic Table for the Gated D Latch

C	D	Q	Operation
0	0	Q_0	no change
0	1	Q_0	no change
1	0	0	reset
1	1	1	set

FIGURE 73.19 Gated D latch circuit. (*Source:* Modified from R. S. Sandige, *Modern Digital Design,* New York: McGraw-Hill, 1990, p. 470. With permission.)

1s and 0s catching. To prevent 1s and 0s catching, data-lockout (also called variable-skew) circuitry was added to a few master-slave flip-flop types. Due to the better design features and popularity of **edge-triggered** flip-flops, master-slave flip-flops are not recommended for newer designs and in some cases have been made obsolete by manufacturers, making them difficult to obtain even for repair parts. For this reason only edge-triggered flip-flops will be discussed.

Edge-Triggered Flip-Flops

Two types of edge-triggered flip-flops are predominantly used in modern designs. These are the D type and J-K type. The D type is perhaps the most used because its circuitry generally takes up less real estate on an IC chip and because most engineers consider it an easier device with which to design. An example of a positive edge-triggered D flip-flop circuit is shown in Fig. 73.20. The reduced characteristic table illustrating the operation of this flip-flop is shown in Table 73.13.

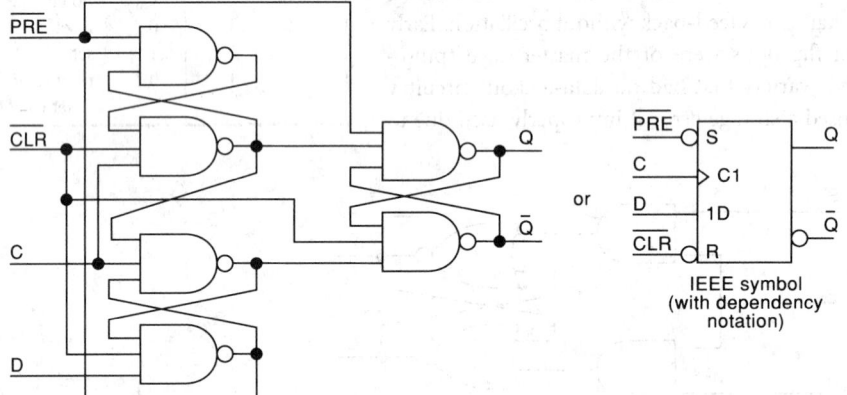

FIGURE 73.20 Positive edge-triggered D flip-flop circuit. (*Source:* Modified from R. S. Sandige, *Modern Digital Design,* New York: McGraw-Hill, 1990, p. 490. With permission.)

Table 73.13 Reduced Characteristic Table for Positive Edge-Triggered D Flip-Flop

$\overline{PRE}$	$\overline{CLR}$	C	D	Q	Operation
0	0	X	X	1	not normally allowed
0	1	X	X	1	preset
1	0	X	X	0	clear
1	1	↑	1	1	set
1	1	↑	0	0	reset
1	1	0	X	Q_0	no charge

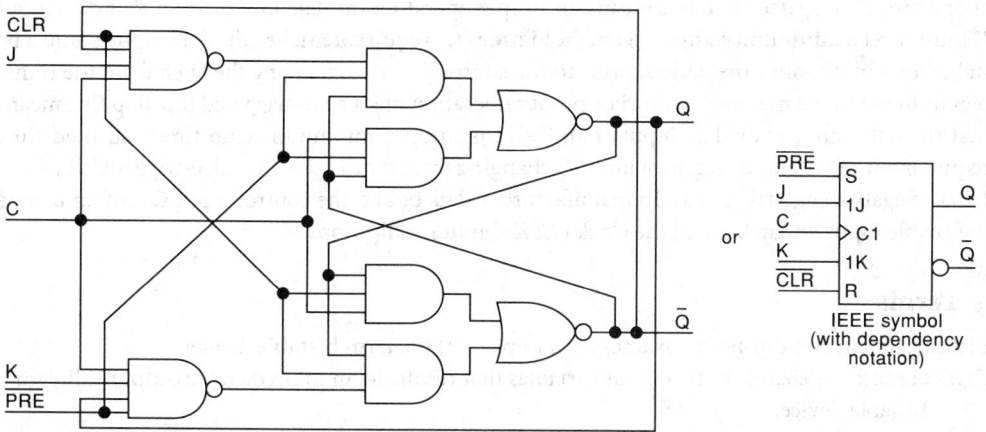

FIGURE 73.21 Negative edge-triggered *J-K* flip-flop circuit. (*Source:* Modified from R. S. Sandige. *Modern Digital Design,* New York: McGraw-Hill, 1990, p. 493. With permission.)

The main difference between a latch and an edge-triggered flip-flop is the question of transparency. The gated *D* latch is transparent (the *Q* output follows the *D* input when the control input $C = 1$) and it latches, captures, or stores the value at the *D* input at the time the control input *C* goes to 0. The *positive edge-triggered D* flip-flop is never transparent from its data input *D* to its output *Q*. When the control input *C* is 0 the output *Q* does not follow the *D* input and remains unchanged; however, the value at the *D* input is latched, captured, or stored at the time the *control input C makes a 0 to 1 transition*. The characteristic that makes edge-triggered flip-flops desirable for feedback applications is that, due to their nontransparent property, their outputs can be fed back as inputs to the device without causing oscillation. This is true for all types of edge-triggered flip-flops. A negative edge-triggered *J-K* flip-flop circuit is shown in the circuit diagram in Fig. 73.21 with its corresponding IEEE symbol. Notice that the *J-K* flip-flop requires eight logic gates compared to only six logic gates for the *D* flip-flop in Fig. 73.20. The reduced characteristic table for this negative edge-triggered flip-flop is shown in Table 73.14.

Notice in the reduced characteristic table (Table 73.14 for the *J-K* flip-flop) when the *J* and *K* inputs are both 1 and the control input *C* makes a 1 to 0 transition, the flip-flop **toggles,** i.e., the next state output *Q* changes to the complement of the present state output Q_0. By simply connecting *J* and *K* together and renaming it *T* for toggle, one can obtain a negative edge-triggered *T* flip-flop.

Special Notes on Using Latches and Flip-Flops

Since bistable devices are asynchronous **fundamental mode** sequential logic circuits, only one input is allowed to change at a time. This means that for proper operation for a basic latch, only one of the data inputs *S* or *R* for a *S-R* NOR latch (~*S* or ~*R* for a *S-R* NAND latch) may be changed

Table 73.14 Reduced Characteristic Table for Negative Edge-Triggered *J-K* Flip-Flop

$\overline{PRE}$	$\overline{CLR}$	C	J	K	Q	Operation
0	0	X	X	X	1	not normally allowed
0	1	X	X	X	1	preset
1	0	X	X	X	0	clear
1	1	↓	0	0	Q_0	no change
1	1	↓	1	0	1	set
1	1	↓	0	1	0	reset
1	1	↓	X	X	$\overline{Q}_0$	toggle
1	1	1	X	X	Q_0	no change

at one time. For a gated latch this means for proper operation the data inputs S and R or data input D must meet a **minimum setup** (t_{su}) **and hold time** (t_h) **requirement,** i.e., the data input(s) must be stable for a minimum time period, prior to the control input C changing the latch from the transparent mode to the memory mode. For proper operation of an edge-triggered flip-flop this means that the data input D or data inputs J and K must meet a minimum setup time and hold time requirement relative to the control input C changing from 0 to 1 (positive edge-triggered) or from 1 to 0 (negative edge-triggered). In manufacturers' data books, the control input C is often named the enable input for latches and the clock (CLK) input for flip-flops.

Defining Terms

Bistable, latch, and flip-flop: Names used in place of the term **bistable device.**

Critical race: A change in two input variables that results in an unpredictable output value for a bistable device.

Edge-triggered: Term used to describe the edge of a positive or negative pulse applied to the control input of a nontransparent bistable device to latch, capture, or store the value indicated by the data input(s).

Fundamental mode: Operating mode of a circuit that allows only one input to change at a time.

Memory element: A bistable device or element that provides data storage for a logic 1 or a logic 0.

Reduced characteristic table: A tabular representation used to illustrate the operation of various bistable devices.

Setup and hold time requirement: Setup time (hold time) is the time required for the data input(s) to be held stable prior to (or after) the control input C changes to latch, capture, or store the value indicated by the data input(s).

Toggle: Change of state from logic 0 to logic 1 or from logic 1 to logic 0 in a bistable device.

Transparent mode: Mode of a bistable device where an output responds to data input signal changes.

Volatile device: A memory or storage device that loses its storage capability when power is removed.

References

ANSI/IEEE Std 91-1984, *IEEE Standard Graphic Symbols for Logic Functions,* New York: Institute of Electrical and Electronics Engineers.

ANSI/IEEE Std 991-1986, *IEEE Standard for Logic Circuit Diagrams,* New York: Institute of Electrical and Electronics Engineers.

D. L. Dietmeyer, *Logic Design of Digital Systems,* 2nd ed., Boston: Allyn and Bacon, 1988.

F. J. Hill and G. R. Peterson, *Introduction to Switching Theory & Logical Design,* 3rd ed., New York: John Wiley, 1981.

E. L. Johnson and M. A. Karim, *Digital Design Pragmatic Approach,* Boston: Prindle, Weber & Schmidt Publishers, 1987.

I. Kampel, *A Practical Introduction to the New Logic Symbols,* 2nd ed., London: Butterworths, 1986.

C. H. Roth, Jr., *Fundamentals of Logic Design,* 4th ed., St. Paul: West Publishing, 1992.

R. S. Sandige, *Modern Digital Design,* New York: McGraw-Hill, 1990.

Texas Instruments, *The TTL Data Book,* vol. 3, Advanced Low-Power Schottky, Advanced Schottky, Dallas: Texas Instruments, 1984.

Further Information

The monthly magazine *IEEE Transactions on Computers* presents papers discussing bistable devices, for example, "A Simulation-Based Method for Generating Tests for Sequential Circuits" in its December 1990 issue, pp. 1456–1463.

Another monthly magazine, *IEEE Transactions on Computer-Aided Design*, sometimes presents papers discussing bistable devices, for example, "Schematic Generation with an Expert System" in its December 1990 issue, pp. 1289-1306.

73.4 Optical Devices

H. S. Hinton

Since the first demonstration of optical logic devices in the late 1970s, there have been many different experimental devices reported. Figure 73.22 categorizes optical logic devices into four main classes. The first division is between all-optical and optoelectronic devices. All-optical devices are devices that do not use electrical currents to create the nonlinearity required by digital devices. These devices can be either single-pass devices (light passes through the nonlinear material once) or they can use a resonant cavity to further enhance the optical nonlinearity (multiple passes through the same nonlinear material). Optoelectronic devices, on the other hand, use electrical currents and electronic devices to process a signal that has gone through an optical-to-electrical conversion process. The output of these devices is either provided by electrically driving an optical source such as a laser or LED (detect/emit) or by modulating some external light source (detect/modulate). Below each of these categories are listed some of the devices that have been experimentally demonstrated.

All-Optical Devices

To create an all-optical logic device requires a medium that will allow one beam of light to affect another. This phenomenon can arise from the cubic response to the applied field. These third-order processes can lead to purely dielectric phenomena, such as irradiance-dependent refractive indices. By exploiting purely dielectric third-order nonlinearities, such as the optical Kerr effect, changes can be induced in the optical constants of the medium which can be read out directly at the same wavelength as that inducing them. This then opens up the possibilities for digital optical circuitry based on cascadable all-optical logic gates. Although there have been many different all-optical gates demonstrated, this section will only briefly review the **soliton** gate (single-pass) and one example of the **nonlinear Fabry-Perot** structures (cavity-based).

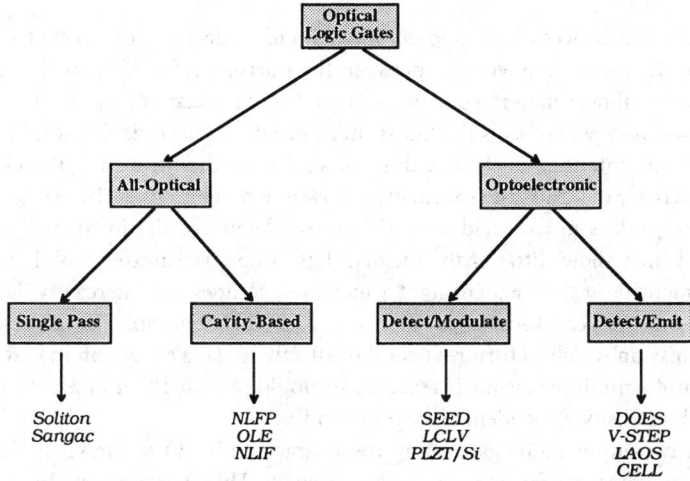

FIGURE 73.22 Classification of optical logic devices.

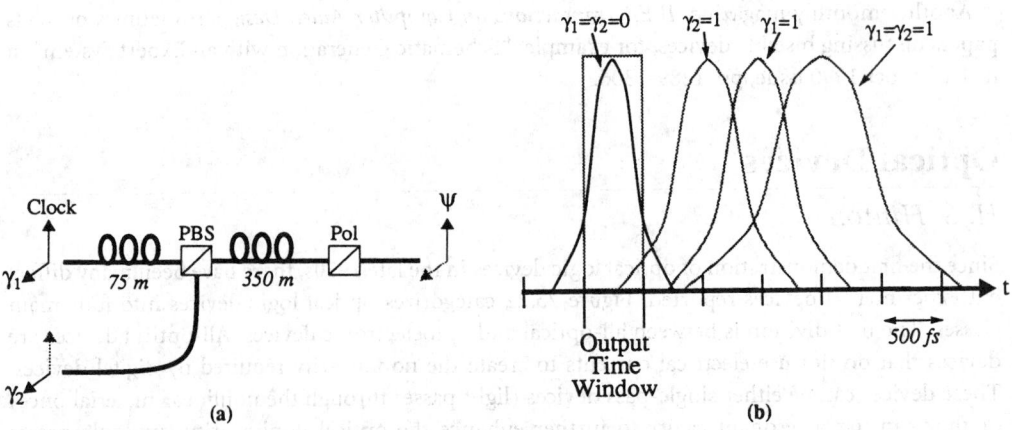

FIGURE 73.23 Soliton NOR gate: (a) physical implementation, (b) timing diagram.

Single-Pass Devices

An example of an all-optical single-pass optical logic gate is the soliton NOR gate. It is an all-fiber logic gate based on time shifts resulting from soliton dragging. A NOR gate consists of two birefringent fibers connected through a polarizing beamsplitter with the output filtered by a polarizer as shown in Fig. 73.23. The clock pulse, which provides both gain and logic level restoration, propagates along one principal axis in both fibers. For the NOR gate the fiber length is trimmed so that in the absence of any signal the entering clock pulse will arrive within the output time window corresponding to a "1." When either or both of the input signals are incident, they interact with the clock pulse through soliton dragging and shift the clock pulse out of the allowed output time window creating a "0" output. In soliton dragging two temporally coincident, orthogonally polarized pulses interact in the fiber through cross-phase modulation and shift each other's velocities. This velocity shift converts into a time shift after propagating some distance in the fiber. To implement the device, the two input signal pulses γ_1 and γ_2 are polarized orthogonal to the clock. The signals are timed so that γ_1 and the clock pulse coincide at the input to the first fiber and γ_2 and the clock pulse coincide (in the absence of γ_1) at the input to the second fiber. At the output the two input signals are blocked by the polarizer, allowing only the temporally modified clock pulse to pass. In a prototyped demonstration this all-optical NOR gate required 5.8 pJ of signal energy and provided an effective gain of 6.

Cavity-Based Devices

Cavity-based optical logic devices are composed of two highly reflective mirrors that are separated by a distance d [Fig. 73.24(a)]. The volume between the mirrors, referred to as the cavity of the etalon, is filled with a nonlinear material possessing an index of refraction that varies with intensity according to $n_c = n_0 + n_2 \gamma_c$ where n_0 is the linear index of refraction, n_2 is the nonlinear index of refraction, and γ_c is the intensity of light within the cavity. In the ideal case, the characteristic response of the reflectivity of a Fabry-Perot cavity, R_{fp}, is shown in Fig. 73.24(b). At low intensities, the cavity resonance peak is not coincident with the wavelength of the incident light; thus the reflectivity is high, which allows little of the incident light to be transmitted [solid curves in Fig. 73.24(b)]. As the intensity of the incident light γ increases, so does the intercavity light intensity which shifts the resonance peak [dotted curve in Fig. 73.24(b)]. This shift in the resonant peak increases the transmission which in turn reduces the reflectivity. This reduction in ψ will continue with increasing γ until a minimum value is reached. It should be noted that in practice all systems of interest have both intensity-dependent absorption and n_2.

To implement a two-input NOR gate using the characteristic curve shown in Fig. 73.24(c) requires a third input which is referred to as the *bias beam*, γ_b. This energy source biases the etalon at a point on its operating curve such that any other input will exceed the nonlinear portion of the curve moving the etalon from the high reflection state. This is illustrated in Fig. 73.24(c) where the γ_b combines with the inputs γ_1 and γ_2 to exceed the threshold of the nonlinear characteristic curve.

The first etalon-based optical logic device was in the form of a nonlinear interference filter (NLIF). A simple interference filter has a general form similar to a Fabry-Perot etalon, being constructed by depositing a series of thin layers of transparent material of various refractive indices on a transparent substrate. The first several layers deposited form a stack of alternating high and low refractive indices, all of optical thickness equal to one quarter of the operating wavelength. The next layer is a low integer (1–20) number of half wavelengths thick and finally a further stack is deposited to form the filter. The two outer stacks have the property of high reflectivity at one wavelength, thus playing the role of mirrors forming a cavity. A high finesse cavity is usually formed when both mirrors are identical, i.e., of equal reflectivity. However, unlike a Fabry-Perot etalon with a nonabsorptive material in the cavity, matched (equal) stack reflectivities do not give the optimum cavity design to minimize switch power because of the absorption in the spacer (which may be necessary to induce nonlinearity). A balanced design which takes into account the effective decrease in back mirror reflectivity due to the double pass through the absorbing cavity is preferable and also results in greater contrast between bistable states. The balanced design is easily achieved by varying one or all of the available parameters: number of periods, thickness and refractive index of each layer within either stack.

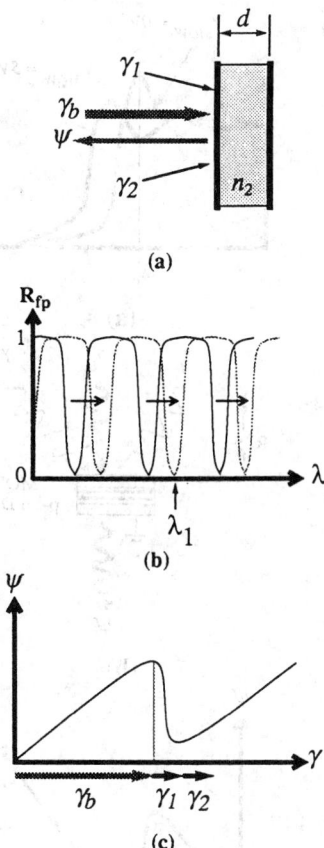

FIGURE 73.24 (a) Nonlinear Fabry-Perot etalon, (b) reflection peaks of NLFP, and (c) NLFP in reflection (NOR).

Optoelectronic Devices

Optoelectronic devices take advantage of both the digital processing capabilities of electronics and communications capabilities of the optical domain. This section will review both the SEED-based optical logic gates and the *pnpn* structures that have demonstrated optical logic.

Detect/Modulate Devices

In the most general terms the self-electro-optic effect device **(SEED) technology** corresponds to any device based on **multiple quantum well** (MQW) modulators. The basic physical mechanism used by this technology is the quantum confined Stark effect. This mechanism creates a shift in the bandedge of a semiconductor with an applied voltage. This is illustrated in Fig. 73.25(a). This shift in the bandedge is then used to vary the absorption of incident light on the MQW material. The characteristic curve shown in Fig. 73.25(c) results when this MQW material is placed in the intrinsic region of a *pin* diode and electrically connected to a resistor as shown in Fig. 73.25(b). When the incident intensity, γ_i, is low there is no current flowing through the *pin* diode or resistor; thus the majority of the voltage is across the *pin* diode. If the device is operating at the wavelength λ_0, the device will be in a low absorptive state. As the incident intensity increases so does the current flowing in the *pin* diode; this in turn reduces the voltage across the diode which increases the absorption and current flow. This state of increasing absorption creates the nonlinearity in the output signal, ψ, shown in Fig. 73.25(c). Optical logic gates can be formed by optically biasing the R-SEED close to the nonlinearity, γ_b, and then applying lower level data signals γ_1 and γ_2 to the device.

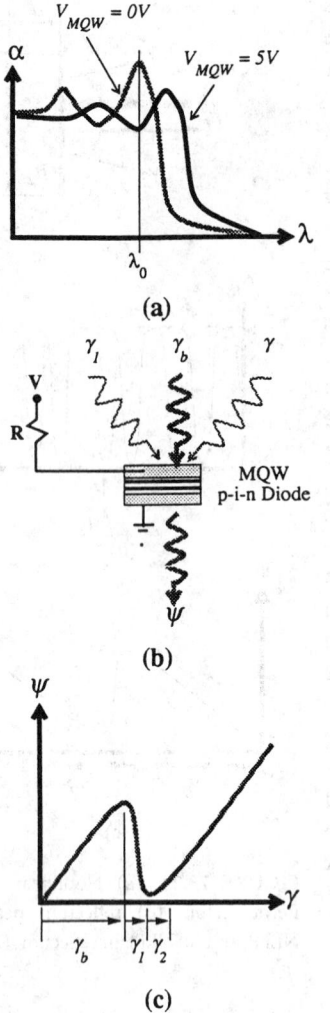

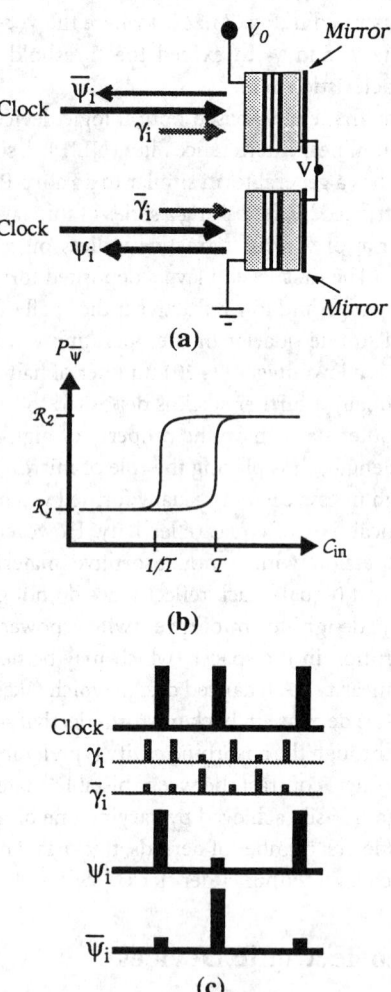

FIGURE 73.25 (a) Absorption spectra of MQW material for both 0 and 5 V, (b) schematic of MQW *pin* diode, (c) input/output characteristics of MQW *pin* diode.

FIGURE 73.26 Symmetric self-electro-optic effect device (S-SEED). (a) S-SEED with inputs and outputs, (b) power transfer characteristics, (c) optically enabled S-SEED, and (d) electrically enabled S-SEED.

The S-SEED, which behaves like an optical inverting *S-R* latch, is composed of two electrically connected MQW *pin* diodes as illustrated in Fig. 73.25(a). In this figure, the device inputs include the signal, γ_i (Set), and its complement, $\bar{\gamma}_i$ (Reset), and a clock signal. To operate the S-SEED the γ_i and $\bar{\gamma}_i$ inputs are also separated in time from the clock inputs as shown in Fig. 73.25(c) and (d). The γ_i and $\bar{\gamma}_i$ inputs, which represent the incoming data and its complement, are used to set the state of the device. When $\bar{\gamma}_i > \gamma_i$, the S-SEED will enter a state where the upper MQW *pin* diode will be reflective, forcing the lower diode to be absorptive. When $\bar{\gamma}_i > \bar{\gamma}_i$ the opposite condition will occur. Low switching intensities are able to change the device's state when the clock signals are not present. After the device has been put into its proper state, the clock beams are applied to both inputs. The ratio of the power between the two clock beams should be approximately one, which will prevent the device from changing states. These higher energy clock pulses, on reflection, will transmit the state of the device to the next stage of the system. Since the inputs γ_i and $\bar{\gamma}_i$ are low-intensity pulses and the clock signals are high-intensity pulses, a large differential gain may be achieved. This type of gain is referred to as time-sequential gain.

The operation of an S-SEED is determined by the power transfer characteristic shown in Fig. 73.26(b). The optical power reflected by the ψ_i window, when the clock signal is applied, is plotted against the ratio of the total optical signal power impinging on the γ_i and $\bar{\gamma}_i$ windows (when the clock signal is not applied). Assuming the clock power incident on both signal windows, γ_i and $\bar{\gamma}_i$, the output power is proportional to the reflectivity, R_i. The ratio of the input signal powers is defined as the input contrast ratio $C_{in} = P_\gamma/P_{\bar{\gamma}_i}$. As C_{in} is increased from zero, the reflectivity of the ψ_i window switches from a low value, R_1, to a high value, R_2, at a C_{in} value approximately equal to the ratio of the absorbances of the two optical windows: $T = (1 - R_1)/(1 - R_2)$. Simultaneously, the reflectivity of the other window (ψ_i) switches from R_2 to R_1. The return transition point (ideally) occurs when $C_{in} = (1 - R_2)/(1 - R_1) = 1/T$. The ratio of the two reflectivities, R_2/R_1, is the output contrast, C_{out}. Typical measured values of the preceding parameters include $C_{out} = 3.2$, $T = 1.4$, $R_2 = 50\%$ and $R_1 = 15\%$. The switching energy for these devices has been measured at ~7 fJ/μm^2.

The S-SEED is also capable of performing optical logic functions such as NOR, OR, NAND, and AND. The inputs will also be differential, thus still avoiding any critical biasing of the device. A method of achieving logic gate operation is shown in Fig. 73.27. The logic level of the inputs will be defined as the ratio of the optical power on the two optical windows. When the power of the signal incident on the γ_i input is greater than the power of the signal on the $\bar{\gamma}_i$ input, a logic "1" will be present on the input. On the other hand, when the power of the signal incident on the γ_i input is less than the power of the signal on the $\bar{\gamma}_i$ input, a logic "0" will be incident on the input.

For the noninverting gates, OR and AND, we can represent the output logic level by the power of the signal coming from the ψ output relative to the power of the signal coming from the $\bar{\psi}$ output. As before, when the power of the signal leaving the ψ output is greater than the power of the signal leaving the $\bar{\psi}$ output, a logic "1" will be represented on the output. To achieve AND operation, the device is initially set to its "off" or logic "0" state (i.e., ψ low and $\bar{\psi}$ high) with preset pulse, *Preset*$_\psi$ incident on only one *pin* diode as shown in Fig. 73.27. If both input signals have logic levels of "1" (i.e., set = 1, reset = 0), then the S-SEED AND gate is set to its "on" state. For any other input combination, there is no change of state, resulting in AND operation. After the signal beams determine the state of the device, the clock beams are then set high to read out the state of the AND gate. For NAND operation, the logic level is represented by the power of the $\bar{\psi}$ output signal relative to the power of the ψ output signal. That is, when the power of the signal leaving the $\bar{\psi}$ output is greater than the power of the signal leaving the ψ output, a logic "1" is present on the output. The

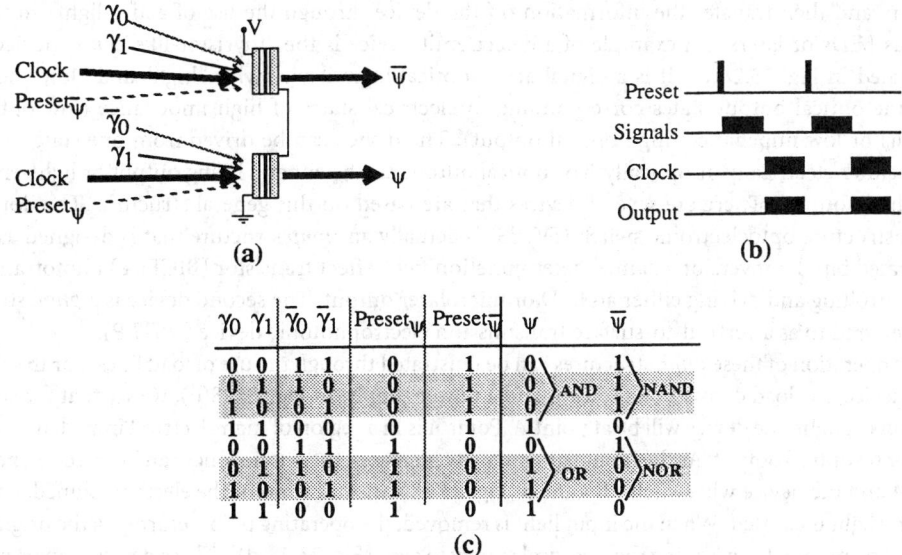

γ_0 γ_1	$\bar{\gamma}_0$ $\bar{\gamma}_1$	Preset$_\psi$	Preset$_{\bar{\psi}}$	ψ	$\bar{\psi}$	
0 0	1 1	0	1	0	1	
0 1	1 0	0	1	0	1	AND / NAND
1 0	0 1	0	1	0	1	
1 1	0 0	0	1	1	0	
0 0	1 1	1	0	0	1	
0 1	1 0	1	0	1	0	OR / NOR
1 0	0 1	1	0	1	0	
1 1	0 0	1	0	1	0	

(c)

FIGURE 73.27 Logic using S-SEED devices.

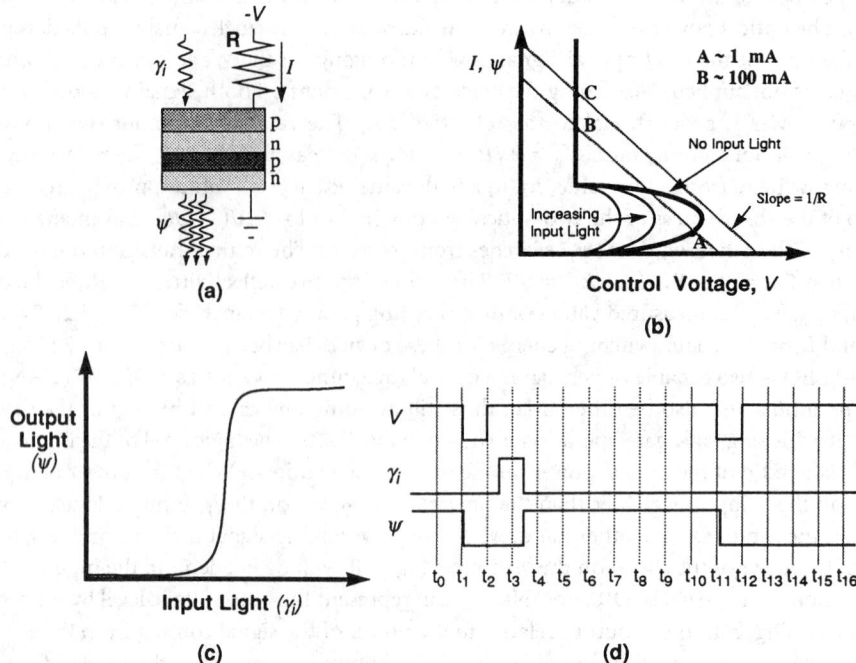

FIGURE 73.28 *pnpn* devices: (a) basic structure, (b) voltage/output characteristics, (c) input/output characteristics, and (d) timing diagram of device operation.

operation of the OR and NOR gates is identical to the AND and NAND gates, except that preset pulse *Preset*$_\psi$ is used instead of the preset pulse *Preset* $_{\overline{\psi}}$. Thus, a single array of devices can perform any or all of the four logic functions and memory functions with the proper optical interconnections and preset pulse routing.

Detect/Emit Devices

Detect/emit devices are optoelectronic structures that detect the incoming signal, process the information, and then transfer the information off the device through the use of active light emitters such as LEDs or lasers. An example of a detect/emit device is the "thyristor-like" *pnpn* device as illustrated in Fig. 73.28(a). It is a digital active optical logic device with "high" and "low" light-emitting optical output states corresponding to electrical states of high impedance (low optical output) or low impedance (high optical output). The device can be driven from one state to the other either electrically or optically. The optical output can be either a lasing output or light-emitting diode output. There are several devices that are based on this general structure. The double heterostructure optoelectronic switch (DOES) is actually an *npnp* structure that is designed as an integrated bipolar inversion channel heterojunction field-effect transistor (BICFET) phototransistor controlling and driving either an LED or microlaser output. The second device is a *pnpn* structure referred to as a vertical-to-surface transmission electrophotonic device (VSTEP).

The operation of these *pnpn* structures can be illustrated through the use of load lines. For the simplest device, the load consists of a resistor and a power supply. In Fig. 73.28(b), we see that for small amounts of light, the device will be at point *A*. Point *A* is in a region of high electrical impedance with little or no optical output. As the input light intensity increases, there is no longer an intersection point near *A* and the device will switch to point *B* [Fig. 73.28(c)]. At this point the electrical impedance is low and light is emitted. When the input light is removed, the operating point returns via the origin to point *A* by momentarily setting the electrical supply to zero [Fig. 73.28(d)]. These devices can be used as either optical OR or AND gates using a bias beam and several other optical inputs.

The device can also be electrically switched. Assuming no input intensity, the initial operating point is at point *A*. By increasing the power supply voltage, the device will switch to point *C*. Point *C* like point *B* is in the region of light emission and low impedance. To turn off the device, the power supply must then be reduced to zero, after which it may be increased up to some voltage where switching occurs.

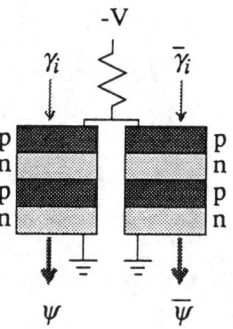

FIGURE 73.29 Differential *pnpn* device.

A differential *pnpn* device made by simply connecting two *pnpn* devices in parallel and connecting that combination in series with a resistive load is illustrated in Fig. 73.29. The operation of the device can be described as follows. When the device is biased below threshold, that is, with the device unilluminated, both optical switches are "off." When the device is illuminated, the one with the highest power is switched "on." The increase in current leads to a voltage drop across the resistor which in turn leads to a lowering of the voltage across both optical switches. Therefore, the one with the lower input cannot be switched "on." Unless both inputs were illuminated with precisely the same power and both devices had identical characteristics (both of these are impossible), only one of the two optical switches will emit light.

The required input optical switching energy density can be quite low if the device without light is biased critically just below threshold. Since incoherent light from an LED cannot be effectively collected from small devices or focused onto small devices, a lasing *pnpn* is needed. Microlaser-based structures are also required to reduce the total power dissipation to acceptable levels. Surface-emitting microlasers provide an ideal laser because of their small size, single-mode operation, and low thresholds. The surface-emitting microlasers consist of two AlAs/AlGaAs dielectric mirrors with a thin active layer in between. This active layer typically consists of one or a few MQWs. The material can be etched vertically into small posts, typically 1–5 μm in diameter. Thresholds are typically on the order of milliwatts.

The switching speed of these devices is limited by the time it takes the photogenerated carriers to diffuse into the light-emitting region. Optical turn-off times are also limited by the RC time constant. For devices made so far, the RC time constants are in the range of 1–10 ns, and optical switch-on times were ~10 ns. Performance of the devices is expected to improve as the areas are reduced; switching times comparable to the best electronic devices (~10 ps) are possible, although the optical turn-on times of at least the surface-emitting LED devices will continue to be slower since this time is determined by diffusion effects and not device capacitance and resistance. Lasing devices should offer improved optical turn-on times.

Another approach to active devices is to combine lasers/modulators with electronics and photodiodes as has been proposed for optical interconnections of electronic circuits. Since the logic function is implemented with electronic circuitry, any relatively complex functionality can be achieved. Several examples of logic gates have been made using GaAs circuitry and light-emitting diodes. Again surface-emitting microlasers provide an ideal emitter for this purpose, because of their small size and low threshold current. However, the integration of these lasers with the required electrical components has yet to be demonstrated.

Limitations

In the normal operating regions of most devices, a fixed amount of energy, the switching energy, is required to make them change states. This switching energy can be used to establish a relationship between both the switching speed and the power required to change the state of the device. Since the power required to switch the device is equal to the switching energy divided by the switching time, then a shorter switching time will require more power. As an example, for a photonic device with an area of 100 μm^2 and a switching energy of 1 fJ/μm^2 to change states in 1 ps requires 100 mW of power instead of the 100 μW that would be required if the device were to switch at 1 ns. Thus, for high power signals the device will change states rapidly, while low power signals yield a slow switching response.

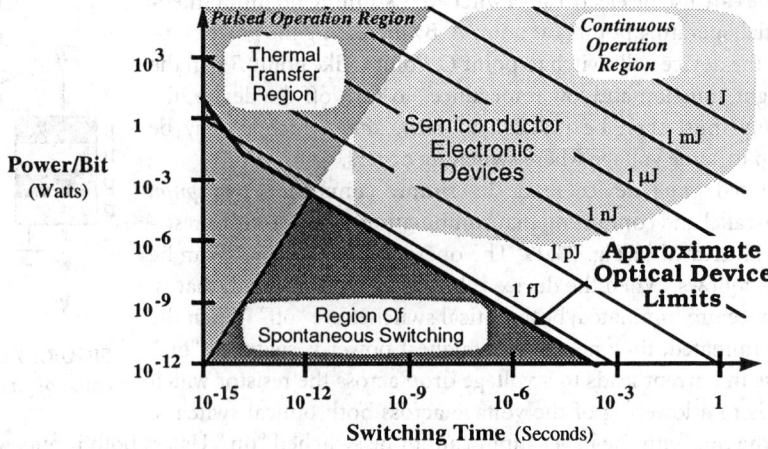

FIGURE 73.30 Fundamental limitations of optical logic devices.

Some approximate limits on the possible switching times of a given device, whether optical or electrical, are illustrated in Fig. 73.30. In this figure the time required to switch the state of a device is on the abscissa while the power/bit required to switch the state of a device is on the ordinate. The region of spontaneous switching is the result of the background thermal energy that is present in a device. If the switching energy for the device is too low, the background thermal energy will cause the device to change states spontaneously. To prevent these random transitions in the state of a device, the switching energy required by the device must be much larger than the background thermal energy. To be able to differentiate statistically between two states, this figure assumes that each bit should be composed of at least 1000 photons. Thus, the total energy of 1000 photons sets the approximate boundary for this region of spontaneous switching. For a wavelength of 850 nm, this implies a minimum switching energy on the order of 0.2 fJ. For the thermal transfer region, it is assumed that for continuous operation, the thermal energy present in the device cannot be removed any faster than 100 W/cm^2 (1 μW/μm^2). There has been some work done to indicate that this value could be as large as 1000 W/cm^2. This region also assumes that there will be no more than an increase of 20°C in the temperature of the device. Devices can be operated in this region using a pulsed rather than continuous mode of operation. Thus, high energy pulses can be used if sufficient time is allowed between pulses to allow the absorbed energy to be removed from the devices. The cloud represents the performance capabilities of current electronic devices. This figure illustrates that optical devices will not be able to switch states orders of magnitude faster than electronic devices when the system is in a continuous rather than a pulsed mode of operation. There are, however, other considerations in the use of photonic switching devices than how fast a single device can change states. Assume that several physically small devices need to be interconnected so that the state information of one device can be used to control the state of another device. To communicate this information, there needs to be some type of interconnection with a large bandwidth that will allow short pulses to travel between the separated devices. Fortunately, the optical domain can support the bandwidth necessary to allow bit rates in excess of 100 Gb/s, which will allow high-speed communication between these individual switching devices. In the electrical domain, the communications bandwidth between two or more devices is limited by the resistance, capacitance, and inductance of the path between the different devices. Therefore, even though photonic devices cannot switch orders of magnitude faster than their electronic counterparts, the communications capability or transmission bandwidth present in the optical domain should allow higher data rate systems than are possible in the electrical domain.

Defining Terms

Light-amplifying optical switch (LAOS): Vertically integrated heterojunction phototransistor and light-emitting diode which has latching thyristor-type current-voltage characteristics.

Liquid-crystal light valve (LCLV): Optical controlled spatial light modulator based on liquid crystals.

Multiple quantum well (MQW): Collection of alternating thin layers of semiconductors (e.g., GaAs and AlGaAs) that results in strong peaks in the absorption spectrum which can be shifted with an applied voltage.

Nonlinear Fabry-Perot (NLFP): Fabry-Perot etalon or interferometer that has an optically nonlinear medium in its cavity.

Optical logic etalon (OLE): Pulsed nonlinear Fabry-Perot etalon that requires two wavelengths (λ_1 = signal, λ_2 = clock).

PLZT/Si: Technology based on conventional silicon electronics using silicon detectors for the device inputs and PLZT modulators for the outputs.

Sagnac logic gate: An all-optical gate based on a Sagnac interferometer. A Sagnac interferometer is composed of two coils of optical fiber arranged so that light from a single source travels clockwise in one and counterclockwise in the other.

SEED technology: Any device based on multiple quantum well (MQW) modulators.

Soliton: Any isolated wave that propagates without dispersion of energy.

Surface-emitting laser logic (CELL): Device that integrates a phototransistor with a low-threshold vertical-cavity surface-emitting laser.

References

H. S. Hinton, "Architectural consideration for photonic switching networks," *IEEE Journal on Selected Areas in Communications,* 6, 1988.

M. N. Islam *et al.,* "Ultrafast all-optical fiber-soliton gates," in *Proceedings on Photonic Switching,* vol. 8, H. S. Hinton and J. W. Goodman, eds., Washington, D.C.: Optical Society of America, 1991, pp. 98–104.

J. L. Jewell *et al.,* "Use of a single nonlinear Fabry-Perot etalon as optical logic gates," *Applied Physics Letters,* 44, 1984.

K. Kasahara *et al.,* "Double heterostructure optoelectronic switch as a dynamic memory with low-power consumption," *Applied Physics Letters,* 52, 1988.

A. L. Lentine *et al.,* "Symmetric self-electrooptic effect device: Optical set-reset latch, differential logic gate, and differential modulator/detector," *IEEE Journal of Quantum Electronics,* 25, 1989.

J. E. Midwinter, "Digital optics, optical logic or smart interconnect or optical logic," *Physics in Technology,* 19, 1988.

D. A. B. Miller, "Quantum well self-electro-optic effect devices," *Optical and Quantum Electronics,* 22, 1990.

P. W. Smith, "On the physical limits of digital optical switching and logic elements," *Bell System Technical Journal,* 61, 1982.

S. D. Smith, "Optical bistability, photonic logic, and optical computation," *Applied Optics,* 25, 1986.

G. W. Taylor *et al.,* "A new double heterostructure optoelectronic device using molecular beam epitaxy," *Journal of Applied Physics,* 59, 1986.

Further Information

Books which cover this material in more detail include:

H. H. Arsenault, T. Szoplik, and B. Macukow, *Optical Processing and Computing,* New York: Academic Press, 1989.

H. M. Gibbs, *Optical Bistability: Controlling Light with Light*, New York: Academic Press, 1985.

M. N. Islam, *Ultrafast Fiber Switching Devices and Systems*, London: Cambridge University Press, 1992.

A. D. McAulay, *Optical Computer Architectures*, New York: John Wiley, 1991.

B. S. Wherrett and F. A. P. Tooley, *Optical Computing*, Scottish Universities Summer School in Physics, 1989.

In the eighteenth century, French artisans had become highly skilled in manufacturing intricately designed fabrics on looms. The machinery for weaving the designs was directed by a method perfected by Joseph Jacquard (1752–1834) early in the nineteenth century. Jacquard used cards punched with holes to position threads for the weaving process. A hole allowed a hooded wire, containing a thread, to be inserted into the pattern. If no hole was present,

Hollerith equipment. *Courtesy of IBM Corporation.*

Jacquard's loom. Program cards, attached to each other in belt fashion, fed from the floor into the loom. *Courtesy of IBM Corporation.*

no wire emerged and no colored thread was allowed into the pattern during that operation of the loom. For each operation, a card was provided; the whole collection of cards made up a program which directed the weaving.

The punched card appealed to manufacturers of calculating equipment. In 1890, the United States Census was compiled with the aid of Hollerith computing machines, named for Herman Hollerith, an inventor who had adapted the punched card system to the special needs of census-taking. Hollerith had added electrical sensing equipment to take advantage of the information holes in the punched cards. Hollerith's company was to become the International Business Machines Corporation.

74

Memory Devices

W. David Pricer
IBM

Randy H. Katz
*University of California,
Berkeley*

Peter A. Lee
*Department of Trade and
Industry, London*

M. Mansuripur
University of Arizona, Tucson

74.1 Integrated Circuits (RAM, ROM).. 1651
      Dynamic RAMs (DRAMs) • Static RAMs (SRAMs) • Nonvolatile
      Programmable Memories • Read-Only Memories (ROMs)
74.2 Basic Disk System Architectures 1658
      Basic Magnetic Disk System Architecture • Characterization of I/O
      Workloads • Extensions to Conventional Disk Architectures
74.3 Magnetic Tape.. 1670
      A Brief Historical Review • Introduction • Magnetic Tape • Tape
      Format • Recording Modes
74.4 Magneto-Optical Disk Data Storage.. 1675
      Preliminaries and Basic Definitions • The Optical Path • Automatic
      Focusing • Automatic Tracking • Thermomagnetic Recording
      Process • Magneto-Optical Readout • Materials of Magneto-
      Optical Data Storage

74.1 Integrated Circuits (RAM, ROM)

W. David Pricer

The major forms of semiconductor memory in descending order of present economic importance are

1. Dynamic Random-Access Memories (DRAMs)
2. Static Random-Access Memories (SRAMs)
3. Nonvolatile Programmable Memories (PROMs, EEPROMs, EAROMs, EPROMs)
4. Read-Only Memories (ROMs)

DRAMs and SRAMs differ little in their applications. DRAMs are distinguished from SRAMs in that no bistable electronic circuit internal to the storage cell maintains the information. Instead DRAM information is stored "dynamically" as charge on a capacitor. All modern designs feature one field-effect transistor (FET) to access the information for both reading and writing and a thin film capacitor for information storage. SRAMs maintain their bistability, so long as power is applied, by a cross-coupled pair of inverters within each storage cell. Almost always two additional transistors serve to access the internal nodes for reading and writing. Most modern cell designs are CMOS, with two P-channel and four N-channel FETs.

Programmable memories operate much like read-only memories with the important attribute that they can be programmed at least once, and some can be reprogrammed a million times or more. Storage is almost always by means of a floating-gate FET. Information in such storage cells is not indefinitely nonvolatile. The discharge time constant is on the order of ten years. ROMs are generally programmed by a custom information mask within the fabrication sequence. As the name implies, information thence can only be read. The information thus stored is truly non-volatile, even when power is removed. This is the most dense form of semiconductor storage (and the least flexible). Other forms of semiconductor memories, such as associative memories and charge-coupled devices, are used rarely.

Dynamic RAMs (DRAMs)

The universally used storage cell circuit of one transistor and one capacitor has remained unchanged for over 20 years. The physical implementation, however, has undergone much diversity and many refinements. The innovation in physical implementation is driven primarily by the need to maintain a nearly constant value of capacitance while the surface area of the cell has decreased. A nearly fixed value of capacitance is needed to meet two important design goals. The cell has no internal amplification. Once the information is accessed, the stored voltage is vastly attenuated by the much larger bit line capacitance (see Fig. 74.1). The resulting signal must be kept larger than the resolution limits of the sensing amplifier. DRAMs in particular are also sensitive to a problem called soft errors. These are typically initiated by atomic events such as the incidence of a single alpha particle. An alpha particle can cause a spurious signal of 50,000 electrons or more. All modern DRAM designs resolve this problem by constructing the capacitor in space out of the plane of the transistors (see Fig. 74.2 for examples). Placing the capacitor in space unusable for transistor fabrication has allowed great strides in DRAM density, generally at the expense of fabrication complexity. DRAM chip capacity has increased by about a factor of four every three years.

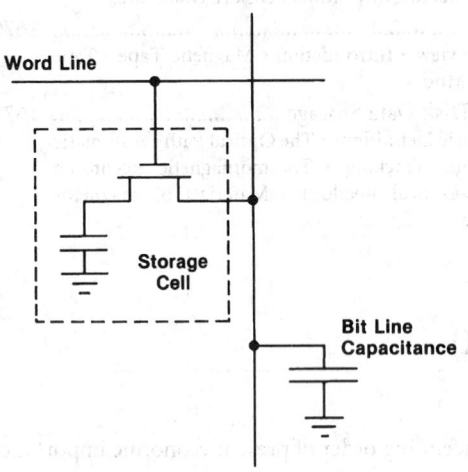

FIGURE 74.1   Cell and bit line capacitance.

DRAMs are somewhat slower than SRAMs. This relationship derives directly from the smaller signal available from DRAMs and from certain constraints put on the support circuitry by the DRAM array. DRAMs also require periodic intervals to "refresh" lost charge from the capacitor. This charge is lost primarily across the semiconductor junctions and must be replenished every few milliseconds. The manufacturer usually supplies these "housekeeping" functions with on-chip circuitry.

Signal detection and amplification remain a critical focus of good DRAM design. Figure 74.3 illustrates an arrangement called a "folded bit line." This design cancels many of the noise sources originating in the array and decreases circuit sensitivity to manufacturing process variations. It also achieves a high ratio of storage cells per sense amplifier. Note the presence of the dummy cells, which create a reference signal midway between a "one" and a "zero" for the convenience of the sense amplifier. The stored reference voltage in this case is created by shorting two driven bit lines after one of the storage cells has been written.

Large DRAM integrated circuit chips frequently provide other features that users may find useful. Faster access is provided between certain adjacent addresses, usually along a common word line. Some designs feature on-chip buffer memories, low standby power modes, or error correction circuitry. A few DRAM chips are designed to mesh with the constraints of particular applications such as image support for CRT displays. Some on-chip features are effectively hidden from the user. These may include redundant memory addresses which the maker activates by laser to improve manufacturing yield.

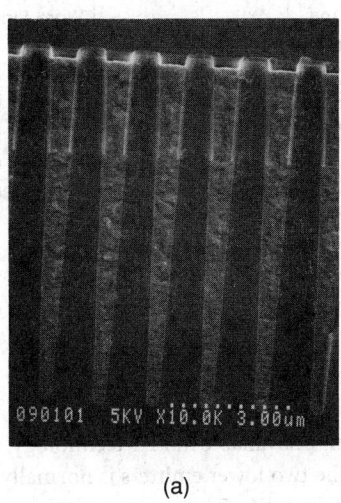

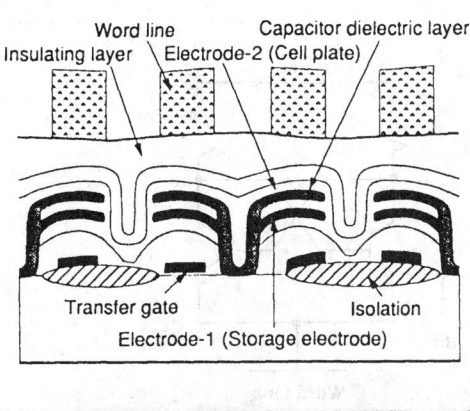

Word line / Insulating layer / Electrode-2 (Cell plate) / Capacitor dielectric layer

Transfer gate / Isolation / Electrode-1 (Storage electrode)

(a)

(b)

FIGURE 74.2 (a) Cross section of "trench capacitors" etched vertically into the semiconductor surface of a DRAM integrated circuit. (Courtesy of IBM.) (b) Cross section of "stacked" capacitors fabricated above the semiconductor surface of a DRAM integrated circuit. (*Source:* M. Taguchi *et al.*, "A 40-ns 64-b parallel data bus architecture," *IEEE J. Solid State Circuits*, vol. 26, no. 11, p. 1495. © 1991 IEEE. With permission.)

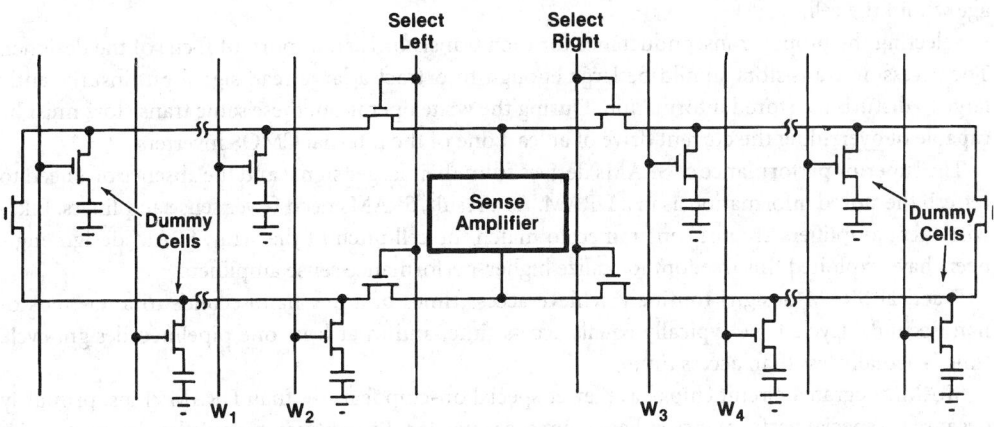

FIGURE 74.3 Folded bit line array.

Static RAMs (SRAMs)

The primary advantages of SRAMs as compared to DRAMs are high speed and ease of use. In addition, SRAMs fabricated in CMOS technology exhibit extremely low standby power. This later feature is effectively used in much portable equipment like pocket calculators. Bipolar SRAMs are generally faster but less dense than FET versions. Figure 74.4 illustrates two cells. SRAM performance is dominated by the speed of the support circuits, leading some manufacturers to design bipolar support circuits to FET arrays.

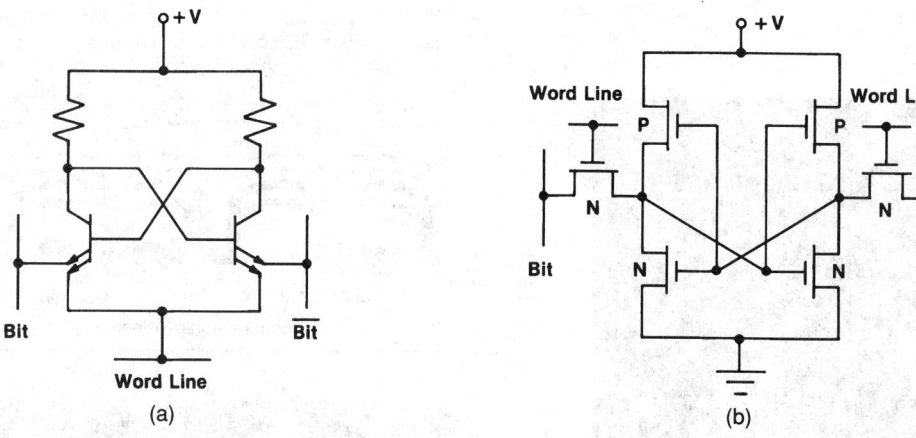

FIGURE 74.4 (a) Bipolar SRAM cell. (b) CMOS SRAM cell.

Bipolar designs frequently incorporate circuit consolidation unavailable in FET technology, such as the multi-emitter cell shown in Fig. 74.4(a). Here one of the two lower emitters is normally forward biased, turning one inverter on and the other off for bistability. The upper emitters can be used either to extract a differential signal or to discharge one collector towards ground in order to write the cell. The word line is pulsed positive to both read and write the cell.

A few RAMs use **polysilicon** load resistors of very high resistance value in place of the two P-channel transistors shown in Fig. 74.4(b). Most are full CMOS designs like the one shown. Sometimes the P-channel transistors are constructed by thin film techniques and are physically placed over the N-channel transistors to improve density. When both P- and N-channel transistors are fabricated in the same plane of the single-crystal semiconductor, the standby current can be extremely low. Typically this can be microamps for megabit chips. The low standby current is possible because each cell sources and sinks only that current needed to overcome the actual node leakage within the cell.

Selecting the proper transconductance for each transistor is an important focus of the designer. The accessing transistors should be large enough to extract a large read signal but insufficiently large to disturb the stored information. During the write operation, these same transistors must be capable of overriding the current drive of at least one of the internal CMOS inverters.

The superior performance of SRAMs derives from their larger signal and the absence of a need to refresh the stored information as in a DRAM. As a result, SRAMs need fewer sense amplifiers. Likewise these amplifiers are not constrained to match the cell pitch of the array. SRAM design engineers have exploited this freedom to realize higher-performance sense amplifiers.

Practical SRAM designs routinely achieve access times of a few nanoseconds to a few tens of nanoseconds. Cycle time typically equals access time, and in at least one pipelined design, cycle time is actually less than access time.

SRAM integrated circuit chips have fewer special on-chip features than DRAM chips, primarily because no special performance enhancements are needed. By contrast, many other integrated circuit chips feature on-chip SRAMs. For example, many **ASICs (application-specific integrated circuits)** feature on-chip RAMs because of their low power and ease of use.

Nonvolatile Programmable Memories

A few nonvolatile memories are programmable just once. These have arrays of diodes or transistors with fuses or **antifuses** in series with each semiconductor cross point. Aluminum, titanium, tungsten, platinum silicide, and polysilicon have all been successfully used as fuse technology (see Fig. 74.5).

Most nonvolatile cells rely on trapped charge stored on a floating gate in an FET. These can be rewritten many times. The trapped charge is subject to very long term leakage, on the order of ten years. The number of times the cell may be rewritten is limited by programming stress-induced degradation of the dielectric. Charge reaches the floating gate either by **tunneling** or by **avalanche injection** from a region near the drain. Both phenomena are induced by over-voltage conditions and hence the degradation after repeated erase/write cycles. Commercially available chips typically promise 100 to 100,000 write cycles. Erasure of charge from the floating gate may be by tunneling or by exposure to ultraviolet light. Asperities on the polysilicon gate and silicon-rich oxide have both been shown to enhance charging and discharging of the gate. The nomenclature used is not entirely consistent throughout the industry. However, EPROM is generally used to describe cells which are electronically written but UV erased. EEPROM is used to describe cells which are electronically both written and erased.

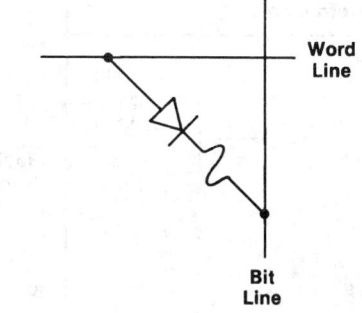

Cells are of either a two- or a one-transistor design. Where two transistors are used, the second transistor is a conventional **enhancement mode** transistor (see Fig. 74.6). The second transistor works to minimize the disturb of unselected cells. It also removes some constraints on the writing limits of the programmable transistor, which in one state may be **depletion mode**. The two transistors in series then assume the threshold of the second (enhancement) transistor, or a very high threshold as determined by the programmable transistor. Some designs are so cleverly integrated that the features of the two transistors are merged.

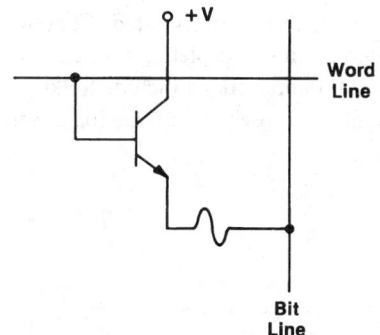

Flash EEPROMs describe a family of single-transistor cell EPPROMs. Cell sizes are about half that of two-transistor EEPROMs, an important economic consideration. Care must be taken that these cells are not programmed into the depletion mode. An array of depletion mode cells would confound the read operation by providing multiple signal paths. Programming to enhancement only thresholds can be accomplished by a sequence of

FIGURE 74.5   PROM cells.

partial program and then monitor subcycles, until the threshold is brought to compliance with specification limits. Flash EEPROMs require bulk erasure of large portions of the array.

NVRAM is a term used to describe a SRAM or DRAM with nonvolatile circuit elements. The cell is built to operate as a RAM with normal power applied. On command or with power failure imminent, the EEPROM elements can be activated to capture the last state of the RAM cell. The nonvolatile information is restored to a SRAM cell by normal internal cell regeneration when power is restored.

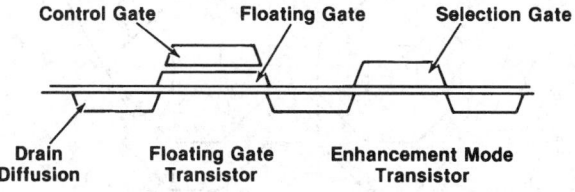

FIGURE 74.6   Cross section of two-transistor EEPROM cells.

Read-Only Memories (ROMs)

ROMs are the only form of semiconductor storage which is permanently nonvolatile. Information is retained without power applied, and there is not even very gradual information loss as in EEPROMs. It is also the most dense form of semiconductor storage. ROMs are, however, less used than RAMs or EEPROMs. ROMs must be personalized by a mask in the fabrication process. This method is cumbersome and expensive unless many identical parts are to be made. Furthermore it seems much "permanent" information is not really permanent and must be occasionally updated.

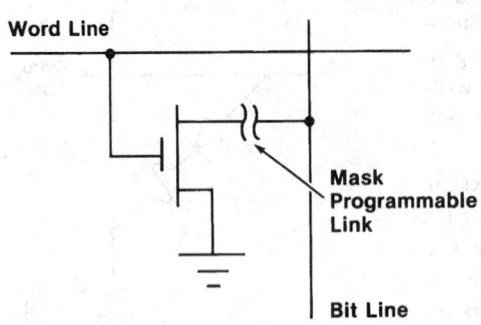

FIGURE 74.7 ROM cell.

ROM cells can be formed as diodes or transistors at every intersection of the word and bit lines of a ROM array (see Fig. 74.7). One of the masks in the chip fabrication process programs which of these devices will be active. Clever layout and circuit techniques may be used to obtain further density. Two such techniques are illustrated in Figs. 74.8 and 74.9. The X array shares bit and virtual ground lines. The AND array places many ROM cells in series. Each of these series AND ROM cells is either an enhancement or a depletion channel of an FET. Sensing is accomplished by pulsing the gates of all series cells positive except the gate which is to be interrogated. Current will flow through all series channels only if the interrogated channel is depletion mode.

ROM applications include look-up tables, machine-level instruction code for computers, and small arrays used to perform logic (see PAL in Section 75.4 of this handbook).

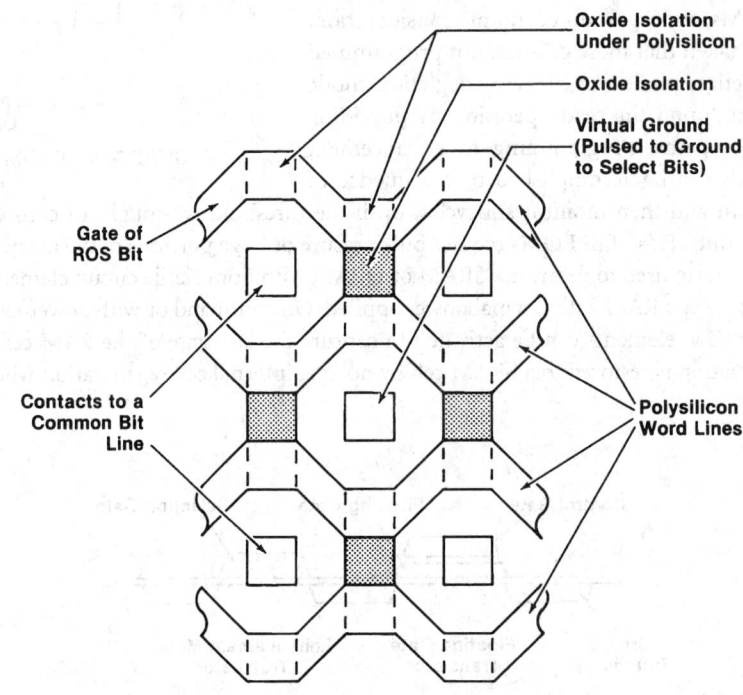

FIGURE 74.8 Layout of ROS X array.

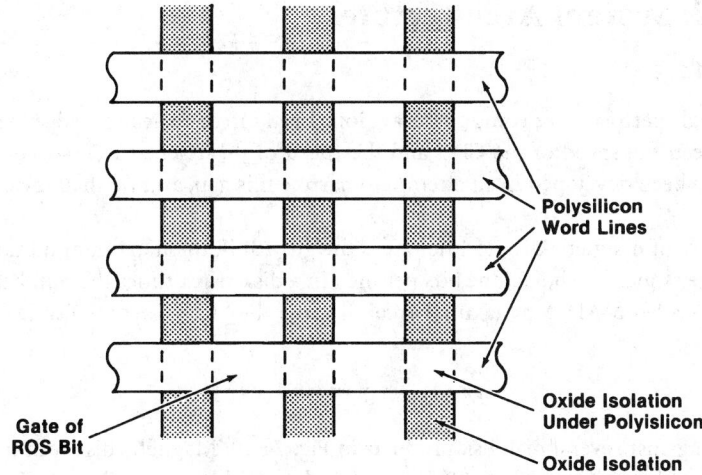

FIGURE 74.9 Layout of ROS AND array.

Defining Terms

Antifuse: A fuse-like device which when activated becomes low impedance.

Application-specific integrated circuits (ASICs): Integrated circuits specifically designed for one particular application.

Avalanche injection: The physics whereby electrons highly energized in avalanche current at a semiconductor junction can penetrate into a dielectric.

Depletion mode: An FET which is on when zero volts bias is applied from gate to source.

Enhancement mode: An FET which is off when zero volts bias is applied from gate to source.

Polysilicon: Silicon in polycrystalline form.

Tunneling: A physical phenomenon whereby an electron can move instantly through a thin dielectric.

References

H. Kalter *et al.*, "A 50 nsec 16 Mb DRAM with 10 nsec data rate and on-chip ECC," *IEEE Journal of Solid-State Circuits*, vol. SC 25, no. 5, 1990.

H. Kato, "A 9 nsec 4 Mb BiCMOS SRAM with 3.3 V operation," *Digest of Technical Papers ISSCC*, vol. 35, 1992.

H. Kawague, and N. Tsuji, "Minimum size ROM structure compatible with silicon-gate E/D MOS LSI," *IEEE Journal of Solid State Circuits*, vol. SC 11, no. 2, 1976.

Further Information

W. Donoghue *et al.*, "A 256K H CMOS ROM using a four state cell approach," *IEEE Journal of Solid-State Circuits*, vol. SC20, no. 2, 1985.

D. Frohmann-Bentchkowsky, "A fully decoded 2048 bit electronically programmable MOS-ROM," *Digest of Technical Papers ISSCC*, vol. 14, 1971.

L. A. Glasser and D. W. Dobberpuhl, *The Design and Analysis of VLSI Circuits*, Reading, Mass.: Addison-Wesley, 1985.

F. Masuoka, "Are you ready for the next generation dynamic RAM chips," *IEEE Spectrum Magazine*, vol. 27, no. 11, 1990.

R. D. Pashley and S. K. Lai, "Flash memories: The best of two worlds," *IEEE Spectrum Magazine*, vol. 26, no. 12, 1989.

74.2 Basic Disk System Architectures

Randy H. Katz

Architects of high-performance computers have long been forced to acknowledge the existence of a large gap between the speed of the CPU and the speed of its attached I/O devices. A number of techniques have been developed in an attempt to narrow this gap, and we shall review them in this chapter.

A key measure of magnetic disk technology is the growth in the maximum number of bits that can be stored per square inch, i.e., the bits per inch in a disk **track** times the number of tracks per inch of media. Called MAD, for **maximal areal density**, the "First Law in Disk Density" predicts [Frank, 1987]:

$$MAD = 10^{(\text{Year}-1971)/10} \qquad (74.1)$$

This is plotted against several real disk products in Fig. 74.10. Magnetic disk technology has doubled capacity and halved price every three years, in line with the growth rate of semiconductor memory. Between 1967 and 1979 the growth in disk capacity of the average IBM data processing system more than kept up with its growth in main memory, maintaining a ratio of 1000:1 between disk capacity and physical memory size [Stevens, 1981].

In contrast to primary memory technologies, the performance of conventional magnetic disks has improved only modestly. These *mechanical* devices, the elements of which are described in more detail in the next section, are dominated by seek and rotation delays: from 1971 to 1981, the raw seek time for a high-end IBM disk improved by only a factor of two while the rotation time did not change [Harker *et al.*, 1981]. Greater recording density translates into a higher transfer rate once the information is located, and extra positioning actuators for the read/write **heads** can reduce the average seek time, but the raw seek time only improved at a rate of 7% per year. This is to be compared to a doubling in processor power every year, a doubling in memory density every two years, and a doubling in disk density every three years. The gap between processor performance and disk speeds continues to widen, and there is no reason to expect a radical improvement in raw disk performance in the near future.

To maintain balance, computer systems have been using even larger main memories or solid-state disks to buffer some of the I/O activity. This may be an acceptable solution for applications

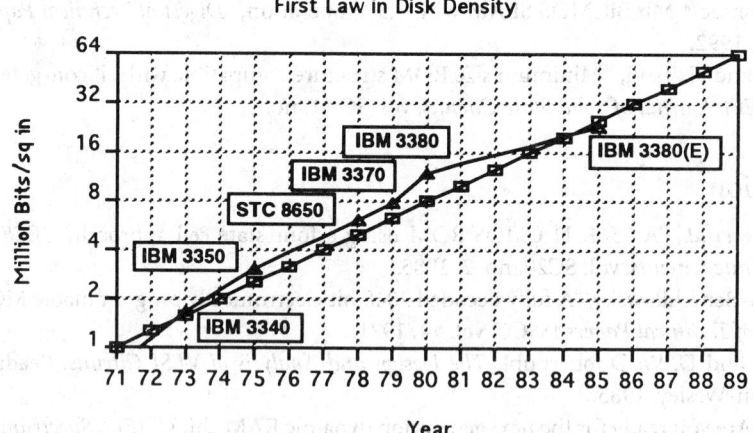

FIGURE 74.10 Maximal areal density law. Squares represent predicted density; triangles are the MAD reported for the indicated products.

whose I/O activity has locality of reference and for which volatility is not an issue, but applications dominated by a high rate of random requests for small pieces of data (e.g., transaction processing) or by a small number of sequential requests for massive amounts of data (e.g., supercomputer applications) face a serious performance limitation.

The rest of the chapter is organized as follows. In the next section, we will briefly review the fundamentals of disk system architecture. The third section describes the characteristics of the applications that demand high I/O system performance. Conventional ways to improve disk performance are discussed in the last section.

Basic Magnetic Disk System Architecture

We will review here the basic terminology of magnetic disk devices and controllers and then examine the disk subsystems of three manufacturers (IBM, Cray, and DEC). Throughout this section we are concerned with technologies that support random access, rather than sequential access (e.g., magnetic tape). A more detailed discussion, focusing on the structure of small dimension **disk drives,** can be found in Vasudeva [1988]. The basic concepts are illustrated in Fig. 74.11. A **spindle** consists of a collection of **platters.** Platters are metal disks covered with a magnetic material for recording information. Each platter contains a number of circular recording *tracks.* A **sector** is a unit of a track that is physically read or written at the same time. In traditional magnetic disks, the constant angular rotation of the platters dictates that sectors on inner tracks are recorded more densely than sectors on the outer tracks. Thus, the platter can spin at a constant rate and the same amount of data can be recorded on the inner and outer tracks.[1] Some modern disks use zone recording techniques to more densely record data on the outer tracks, but this requires more sophisticated read/write electronics.

The read/write *head* is an electromagnet that produces switchable magnetic fields to read and record bit streams on a platter's track. It is associated with a disk **arm,** attached to an actuator. The head "flies" close to, but never touches, the rotating platter (except perhaps when powered down). This is the classical definition of a **Winchester disk.** The actuator is a mechanical assembly that

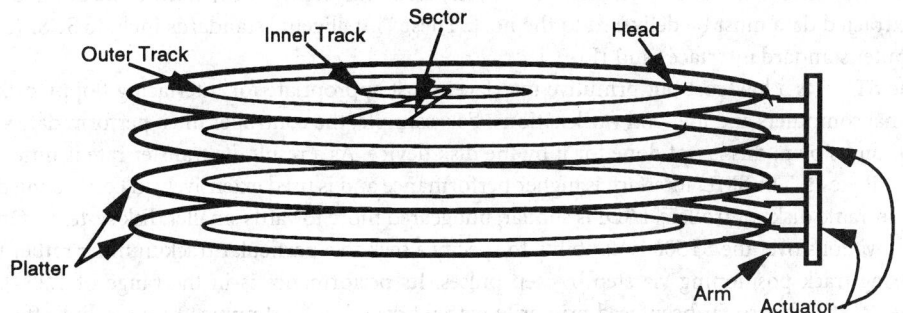

FIGURE 74.11 Disk terminology. Heads reside on arms which are positioned by actuators. Tracks are concentric rings on platters. A sector is the basic unit of read/write. A cylinder is a stack of tracks at one actuator position. An HDA is everything in the figure plus the air-tight casing. In some devices it is possible to transfer from multiple surfaces simultaneously. The collection of heads that participate in a single logical transfer that is spread over multiple surfaces is called a head group.

[1]Some optical disks use a technique called constant linear velocity (CLV), where the platter rotates at different speeds depending on the relative position of the track. This allows more data to be stored on the outer tracks than the inner tracks, but because it takes more delay to vary the speed of rotation, the technique is better suited to sequential rather than random access.

positions the head electronics over the appropriate track. It is possible to have multiple read/write mechanisms per surface, e.g., multiple heads per arm—at one extreme, one could have a head-per-track position, that is, the disk equivalent of a magnetic drum—or multiple arms per surface through multiple actuators. Due to costs and technical limitations, it is usually uneconomical to build a device with a large number of actuators and heads.

A **cylinder** is a stack of tracks at one actuator position. A **head disk assembly** (HDA) is the collection of platters, heads, arms, and actuators, plus the air-tight casing. A *disk drive* is an HDA plus all associated electronics. A *disk* might be a platter, an actuator, or a drive depending the context.

We can illustrate these concepts by describing two first-generation supercomputer disks, the Cray DD-19 and the CDC 819 [Bucher and Hayes, 1980]. These were state-of-the-art disks around 1980. Each disk has 40 recording surfaces (20 platters), 411 cylinders, and 18 (DD-19) or 20 (CDC 819) 512-byte sectors per track. Both disks possess a limited "parallel read-out" capability. A given data word is actually byte interleaved over four surfaces. Rather than a single set of read/write electronics for the actuator, these disks have four sets, so it is possible to read or write with four heads at a time. Four heads on adjacent arms are called a *head group*. A disk track is thus composed of the stacked recording tracks of four adjacent surfaces, and there are 10 tracks per cylinder, spread over 40 surfaces. The advances over the last decade can be illustrated by the Cray DD-49, which is a typical high-end supercomputer disk of today. It consists of 16 recording surfaces (9 platters), 886 cylinders, 42 4096-byte sectors per track, with 32 read/write heads organized into eight head groups, four groups on each of two independent actuators. Each actuator can sweep the entire range of tracks, and by "scheduling" the arms to position the actuator closest to the target track of the pending request, the average seek time can be reduced. The DD-49 has a capacity of 1.2 Gbytes of storage and can transfer at a sustained rate of 9.6 Mbytes/s.

A variety of standard and proprietary interfaces are defined for transferring the data recorded on the disk to or from the host. We concentrate on industry standards here. On the disk surface, information is represented as alternating polarities of magnetic fields. These signals need to be sensed, amplified, and decoded into synchronized pulses by the read electronics. For example, the pulse-level protocol ST506/412 standard describes the way pulses can be extracted from the alternating flux fields. The bit-level ESDI, SMD, and IPI-2 standards describe the bit encoding of signals. At the packet level, these bits must be aligned into bytes, error correcting codes need to be applied, and the extracted data must be delivered to the host. These "intelligent" standards include SCSI (small computer standard interface) and IPI-3.

The ST506 is a low-cost but primitive interface, most appropriate for interfacing floppy disks to personal computers and low-end workstations. For example, the controller must perform data separation on its own; this is not done for it by the disk device. As a result, its transfer rate is limited to 0.625 Mbytes/s. The SMD interface is higher performance and is used extensively in connecting disks to mainframe disk controllers. ESDI is similar, but geared more towards smaller disk systems. One of its innovations over the ST506 is its ability to specify a seek to a particular track number rather than requiring track positioning via step-by-step pulses. Its performance is in the range of 1.25–1.875 Mbytes/s. SCSI has so far been used primarily with workstations and minicomputers, but offers the highest degree of integration and intelligence. Implementations with performance at the level of 1.5–4 Mbytes/s are common. The newer IPI-3 standard has the advantages of SCSI, but provides even higher performance at a higher cost. It is beginning to make inroads into mainframe systems. However, because of the very widespread use of SCSI, many believe that SCSI-2, an extension of SCSI to wider signal paths, will become the de facto standard for high-performance small disks.

The connection pathway between the host and the disk device varies widely depending on the desired level of performance. A low-end workstation or personal computer would use a SCSI interface to directly connect the device to the host. A higher end file server or minicomputer would typically use a separate disk controller to manage several devices at the same time. These devices attach to the controller through SMD interfaces. It is the controller's responsibility to implement error checking and corrections and direct memory transfer to the host.

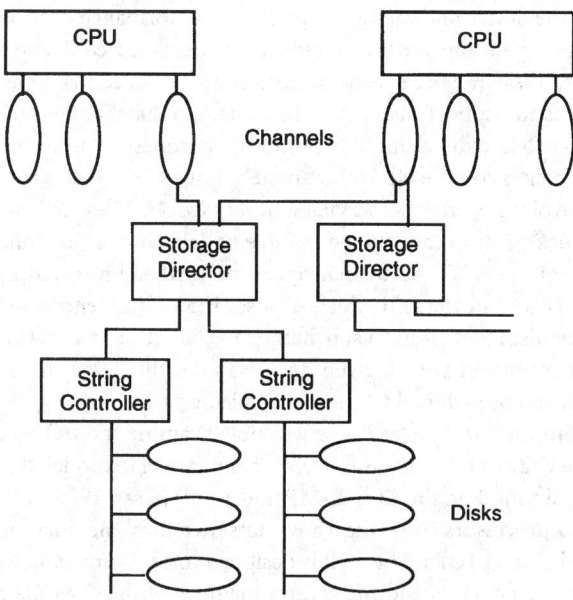

FIGURE 74.12 Host-to-device pathways. For large IBM mainframes, the connection between host and device must pass through a channel, storage director, and string controller. Note that multiple storage directors can be attached to a channel, multiple string controllers per storage director, and multiple devices per string controller. This multipathing approach makes it possible to share devices among hosts and to provide alternative pathways to better utilize the drives and controllers. While logically correct, the figure does not reflect the true physical components of high-end IBM systems (308X, 3090). The concept of channel has disappeared from these systems and has been replaced by a channel path.

Mainframes tend to have more devices and more complex interconnection schemes to access them. In IBM terminology [Buzen and Shum, 1986], the *channel path*, i.e., the set of cables and associated electronics that transfer data and control information between an I/O device and main memory, consists of a *channel*, a *storage director*, and a *head of string* (see Fig. 74.12). The collection of disks that share the same pathway to the head of string is called a *string*.

In earlier IBM systems, a channel path and channel are essentially the same thing. The channel processor is the hardware that executes channel programs, which are fetched from the host's memory. A *subchannel* is the execution environment of a channel program, similar to a process on a conventional CPU. Formerly, a subchannel was statically assigned for execution to a particular channel, but a major innovation in high-end IBM systems (308X and 3090) allows subchannels to be dynamically switched among channel paths. This is like allocating a process to a new processor within a multiprocessor system every time it is rescheduled for execution.

I/O program control statements, e.g., *transfer in channel*, are interpreted by the channel, while the storage director (also known as the *device controller* or *control unit*) handles seek and data-transfer requests. Besides these control functions, it may also perform certain datapath functions, such as error detection/correction and mapping between serial and parallel data. In response to requests from the storage director, the device will position the access mechanism, select the appropriate head, and perform the read or write. If the storage director is simply a control unit, then the datapath functions will be handled by the head of string (also known as a *string controller*).

To minimize the **latency** caused by copying into and out of buffers, the IBM I/O system uses little buffering between the device and memory.[2] In a high-performance environment, devices spend a good deal of time waiting for the pathway's resources to become free. These resources are used for time periods related to disk transfer speeds, measured in milliseconds. One possible method for improving utilization is to support disconnect/reconnect. A subchannel can connect to a device, issue a seek, disconnect to free the channel path for other requests, and reconnect later to perform the transfer when the seek is completed. Unfortunately, not all reconnects can be serviced immediately, because the control units are busy servicing other devices. These *RPS misses* (to be described in more detail in the next section) are a major source of delay in heavily utilized IBM storage subsystems [Buzen and Shum, 1987]. Performance can be further improved by providing multiple paths between memory and devices. To this purpose, IBM's high-end systems support *dynamic path reconnect*, a mechanism that allows a subchannel to change its channel path each time it cycles through a disconnect/reconnect with a given device. Rather than wait for its currently allocated path to become free, it can be assigned to another available path.

Turning to supercomputer I/O systems, we will now examine the I/O architecture of the Cray machines. Because the Cray I/O system (IOS) varies from model to model, the following discussion concentrates on the IOS found on the Cray X-MP and Y-MP [Cray, 1988]. In general, the IOS consists of two to four I/O processors (IOPs), each with its own local memory and sharing a common buffer memory with the other IOPs. The IOP is designed to be a simple, fast machine for controlling data transfers between devices and the central memory of the Cray main processors. Since it executes the control statements of an I/O program, it is not unlike the IBM channel processor in terms of its functionality, except that IO programs reside in its local memory rather than in the host's. An IOP's local memory is connected through a high-speed communications interface, called a *channel* in Cray terminology, to a disk control unit (DCU). A given port into the local memory can be time multiplexed among multiple channels. Data is transferred back and forth between devices and the main processors through the IOP's local memory, which is interfaced to central memory through a 100-Mbyte/s channel pair (one pathway for each direction of transfer).

The DCU provides the interface between the IOP and the disk drives and is similar in functionality to IBM's storage director. It oversees the data transfers between devices and the IOP's local memory, provides speed matching buffer storage, and transmits control signals and status information between the IOP and the devices. Disk storage units (DSUs) are attached to the DCU through point-to-point connections. The DSU contains the disk device and is responsible for dealing with its own defect management, by using a technique called sector slipping. Fig. 74.13 summarizes the elements of the Cray I/O system.

Digital Equipment Corporation's high-end I/O strategy is described in terms of the digital storage architecture (DSA) and is embodied in system configurations such as the VAXCluster shared disk system (see Fig. 74.14). The architecture provides a rigorous definition of how storage subsystems and host computers interact. It achieves this by defining a client/server message-based model for I/O interaction based on device-independent interfaces [Massiglia, 1986; Kronenberg *et al.*, 1986]. A *mass storage subsystem* is viewed at the architectural level as consisting of logical block machines capable of storing and retrieving fixed blocks of data, i.e., the I/O system supports the transfer of logical blocks between CPUs and devices given a logical block number. From the viewpoint of physical components, a subsystem consists of *controllers* which connect computers to *drives*.

The software architecture is divided into four levels: the *Operating System Client* (also called the Class Driver), the *Class Server* (Controller), the *Device Client* (Data Controller), and the *Device Server* (Device). The Disk Class Driver, resident on a host CPU, accepts requests for disk I/O service from applications, packages these requests into messages, and transmits them via a communications interface (such as the *Computer Interconnect* port driver) to the Disk Class Server resident within a controller in the I/O subsystem. The command set supported by the Class Server includes

---

[2]Only the most recent generation of storage directors (e.g., IBM 3880, 3990) incorporate disk caches, but care must be taken to avoid cache management-related delays [Buzen, 1982].

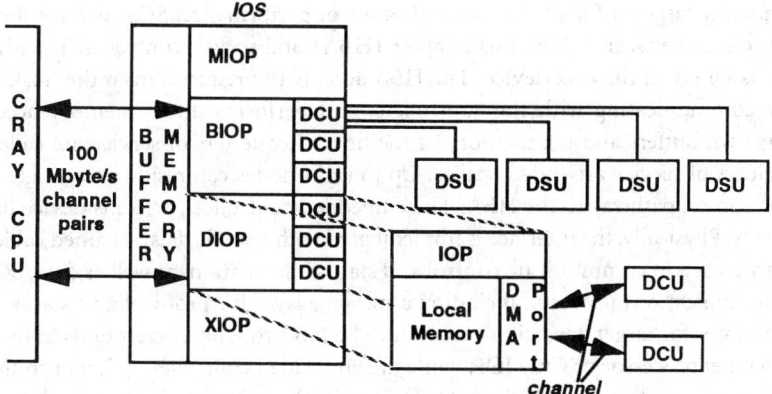

FIGURE 74.13 Elements of the Cray I/O system for the Y-MP. An IOS contains up to four IOPs. The MIOP connects to the operator workstation and performs mainly maintenance functions. The XIOP supports block multiplexing and is most appropriate for controlling relatively slow speed devices, such as tapes. The BIOP and DIOP are designed for controlling high-speed devices like disks. Up to four disk storage units (DSUs) can be attached through the disk control unit (DCU) to the IOP. Three DCUs can be connected to each of the BIOP and DIOP, leading to a total of 24 disks per IOS. The Y-MP can be configured with two IOSs, for a system total of 48 devices.

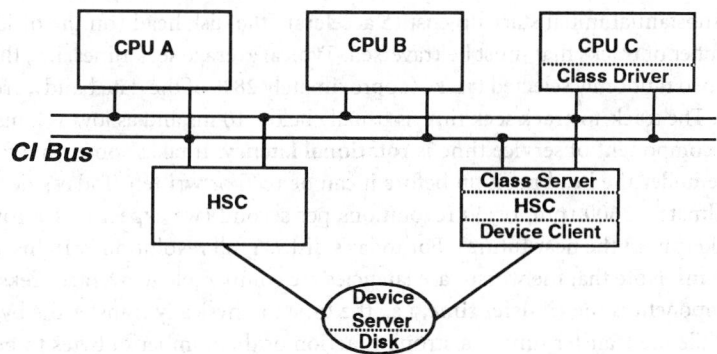

FIGURE 74.14 VAXCluster architecture. CPUs are connected to HSCs (hierarchical storage controllers) through a dual CI (computer interconnect) bus. Thirty-one hosts and 31 HSCs can be connected to a CI. Up to 32 disks can be connected to an HSC-70.

such relatively device-independent operations as read logical block, write logical block, bring online, and request status. The Disk Class Server[3] interprets the transmitted commands, handles the scheduling of command execution, tracks their progress, and reports status back to the Class Driver. Note the absence of seek or select head commands. This interface can be used equally well for solid-state disks as for conventional magnetic disks. Device-specific commands are issued at a lower level of the architecture, i.e., between the Device Client (disk controller) and Device Server (disk device). The former provides the path for moving commands and data between hosts and drives, and it is usually realized physically by a piece of hardware that corresponds to the device controller. The latter coincides with the physical drives used for storing and retrieving data.

It is interesting to contrast these proprietary approaches with an industry standard approach like

---

[3]Other kinds of class servers are also supported, such as for tape drives.

SCSI, admittedly targeted for the low to mid range of performance. SCSI defines the logical and physical interface between a host bus adapter (HBA) and a disk controller, usually embedded within the assembly of the disk device. The HBA accepts I/O requests from the host, initiates I/O actions by communicating with the controllers, and performs direct memory access transfers between its own buffers and the memory of the host. Requesters of service are called initiators, while providers of service are called targets. Up to eight nodes can reside on a single SCSI string, sharing a common pathway to the HBA. The embedded controller performs device handling and error recovery. Physically, the interface is implemented with a single daisy-chained cable, and the 8-bit datapath is used to communicate control and status information, as well as data. SCSI defines a layered communications protocol, including a message layer for protocol and status, and a command/status layer for target operation execution. The HBA roughly corresponds to the function of the IBM channel processor or Cray IOP, while the embedded controller is similar to the IBM storage director/string controller or the Cray DCU. Despite the differences in terminology, the systems we have surveyed exhibit significant commonality of function and similar approaches for partitioning these functions among hardware components.

Characterization of I/O Workloads

Before characterizing the I/O behavior of different workloads, it is necessary to first understand the elements of disk performance. Disk performance is a function of the service time, which consists of three main components: *seek time, rotational latency*, and *data transfer time*.[4] **Seek time** is the time needed to position the heads to the appropriate track position containing the desired data. It is a function of a substantial initial start-up cost to accelerate the disk head (on the order of 6 ms) as well as the number of tracks that must be traversed. Typical average seek times, i.e., the time to traverse between two randomly selected tracks (approximately 28% of the data band), are in the range of 10 to 20 ms. The track-to-track seek time is usually below 10 ms and as low as 2 ms.

The second component of service time is **rotational latency**. It takes some time for the desired sector to rotate under the head position before it can be read or written. Today's devices spin at a rate of approximately 3600 rpm, or 60 revolutions per second (we expect to see rotation speeds increase to 5400 rpm in the near future). For today's disks, a full revolution is 16 ms, and the average latency is 8 ms. Note that the worst-case latencies are comparable to average seeks.

The last component is the **transfer time,** i.e., the time to physically transfer the bytes from disk to the host. While the transfer time is a strong function of the number of bytes to be transferred, seek and rotational latencies times are independent of the transfer blocksize. If data is to be read or written in large chunks, it makes sense to choose a large blocksize, since the "fixed cost" of seek and latency is better amortized across a large data transfer.

A low-performance I/O system might dedicate the pathway between the host and the disk for the entire duration of the seek, rotate, and transfer times. Assuming small blocksizes, transfer time is a small component of the overall service time, and these pathways can be better utilized if they are shared among multiple devices. Thus, higher performance systems support independent seeks, in which a device can be directed to detach itself from the pathway while seeking to the desired track (recall the discussion of dynamic path reconnect in the previous section). The advantage is that multiple seeks can be overlapped, reducing overall I/O latency and better utilizing the available I/O **bandwidth**.

However, to make it possible for devices to reattach to the pathway, the I/O system must support a mechanism called **rotational position sensing,** i.e., the device interrupts the I/O controller when the desired sector is under the heads. If the pathway is currently in use, the device must pay a full

---

[4]In a heavily utilized system, delays waiting for a device can match actual disk service times, which in reality is composed of device queuing, controller overhead, seek, rotational latency, reconnect misses, error retries, and data transfer.

rotational delay before it can again attempt to transfer. These rotational positional reconnect miss delays (RPS delays) represent a major source of degradation in many existing I/O systems [Buzen and Shum, 1987]. This arises from the lack of device buffering and the real-time service requirements of magnetic disks. At the time that these architectures were established, buffer memories were expensive and the demands for high I/O performance were less pressing with slower speed CPUs. An alternative, made more attractive by today's relative costs of electronic and mechanical components, is to associate a **track buffer** with the device that can be filled immediately. This can then be used as the source of the transfer when the pathway becomes available [Houtekamer, 1985].

I/O intensive applications vary widely in the demand they place on the I/O system. They run the gamut from processing small numbers of bulk I/Os that must be handled with minimum delay (supercomputer I/O) to large numbers of simple tasks that touch small amounts of data (transaction processing). An important design challenge is to develop an I/O system that can handle the performance needs of these diverse workloads.

A given workload's demand for I/O service can be specified in terms of three metrics: *throughput*, *latency*, and *bandwidth*. **Throughput** refers to the number of requests for service made per unit time. Latency measures how long it takes to service an individual request. Bandwidth gauges the amount of data flowing between service requesters (i.e., applications) and service providers (i.e., devices).

As observed by Bucher and Hayes [1980], supercomputer I/O can be characterized almost entirely by sequential I/O. Typically, computation parameters are moved in bulk from disk to in-memory data structures, and results are periodically written back to disk. These workloads demand large bandwidth and minimum latency, but are characterized by low throughput. Contrast this with transaction processing, which is characterized by enormous numbers of random accesses, relatively small units of work, and a demand for moderate latency with very high throughput.

Figure 74.15 shows another way of thinking about the varying demands of I/O intensive applications. It shows the percentage of time different applications spend in the three components of I/O service time. Transaction processing systems spend the majority of their service time in seek and rotational latency; thus technological advances which reduce the transfer time will not affect their performance very much. On the other hand, scientific applications spend a more equal amount of time in seek and data transfer, and their performance is sensitive to any improvement in disk technology.

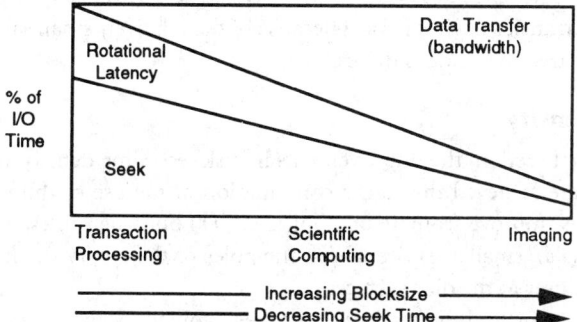

FIGURE 74.15 I/O system parameters as a function of application. Transaction processing applications are seek and rotational latency limited, since only small blocks are usually transferred from disk. Image-processing applications, on the other hand, transfer huge blocks and thus spend most of their I/O time in data transfer. Scientific computing applications tend to fall in between. (*Source:* I. Y. Bucher and A. H. Hayes, "I/O performance measurement on Cray-1 and CDC 7600 computers," *Proc. Cray Users Group Conference*, October 1980. With permission.)

Extensions to Conventional Disk Architectures

In this subsection, we will focus on techniques for improving the performance of conventional disk systems, i.e., methods which allow us to reduce the seek time, rotational latency, or transfer time of conventional disks. By reducing disk service times, we also decrease device queuing delays. These techniques include fixed-head disks, parallel transfer disks, increased disk density, solid-state disks, disk caches, and disk scheduling.

Fixed-Head Disk

The concept of a fixed-head disk is to place a read/write head at every track position. The need for positioning the heads is eliminated, thus eliminating the seek time altogether. The approach does not assist in reducing rotational latencies, nor does it lessen the transfer time. Fixed-head disks were often used in the early days of computing systems as a back-end store for virtual memory. However, since modern disks have hundreds of tracks per surface, placing a head at every position is no longer viewed as an economical solution.

Parallel Transfer Disks

Some high-performance disk drives make it possible to read or write from multiple disk surfaces at the same time. For example, the Cray DD-19 and DD-49 disks described in the second section have a parallel transfer capability. The advantage is that much higher transfer rates can be achieved, but no assistance is provided for seek or rotational latency. Thus transfer units are correspondingly larger in these systems.

A number of economic and technological issues limit the usefulness of parallel transfer disks. From the economic perspective, providing more than one set of read/write electronics *per actuator* is expensive. Further, current disks use sophisticated control systems to lock onto an individual track, and it is difficult to do this simultaneously across tracks within the same cylinder. Hence, the Cray strategy is limiting head groups to only four surfaces. There appears to be a fundamental trade-off between track density and the number of platters: as the track density increases, it becomes ever more difficult to lock onto tracks across many platters, and the number of surfaces that can participate in a parallel transfer is reduced. For example, current Cray track densities are around 980 tracks/inch, and require a rather sophisticated closed-loop track-following servo system to position the heads accurately with finely controlled voice coil actuators. A lower cost ($/megabyte) high-performance disk system can be constructed from several standard drives than from a single parallel transfer device, in part because of the relatively small sales volume of parallel transfer devices compared to standard drives.

Increasing Disk Density

As described in the first section, the improvements in disk recording density are likely to continue. Higher bit densities are achieved through a combination of the use of thinner films on the disk platters (e.g., densities improve from 16,000 bpi to 21,000 bpi when thick iron oxide is replaced with thin film materials), smaller gaps between the poles of the read/write head's electromagnet, and heads which fly closer to the disk surface.

While vertical recording techniques have long been touted as the technology of the future, advances in head technology make it possible to continue using conventional horizontal methods, but still keep disks on the MAD curve. These *magneto-resistive heads* employ noninductive methods for reading, which work well with dense horizontal recording fields. However, a more conventional head is needed for writing, but this dual-head organization permits separate optimizations for read and write.

Also, the choice of coding technique can have a significant effect on density. Standard modified frequency modulation techniques require approximately one flux change per bit, while more advanced run-length limited codes can increase density by an additional factor of 50%. Densities as high as 31,429 bpi can be attained with these techniques. As the recording densities increase, the transfer times decrease, as more bits transit beneath the heads per unit time. Of course, this approach provides no improvement in seek and latency times. Most of the increase in density comes from increases in the number of tracks per inch, which does not improve (and may actually reduce) performance.

Although increased densities are inevitable, the problem is primarily economic. Increasing the tracks per inch may make seeks slower as it becomes more time consuming for the heads to correctly "lock" onto the appropriate track. The sensing electronics get more complex and thus more expensive. Once again, it can be argued that higher capacity can be achieved at lower cost by using several smaller disks rather than one expensive high-density disk.

Solid-State Disks

Solid-state disks (SSD), constructed from relatively slow memory chips, can be viewed either as a kind of large and slow main memory or as a small and high-speed disk. When viewed as large main memory, the SSD is often called expanded storage (ES). The expanded storage found in the IBM 3090 class machines [Buzen and Shum, 1986] supports operations for paging data blocks from and to main memory. Usually, the expanded storage looks to the system more like memory than an I/O device: it is directly attached to main memory through a high-speed bus rather than an I/O controller. The maximum transfer bandwidth on the IBM 3090 between expanded store and memory is two orders of magnitude faster than conventional devices: approximately 216 Mbytes/s—one word each 18.5 ns!

Further, unlike conventional devices, a transfer between memory and expanded storage is performed synchronously with the CPU. This is viewed as acceptable, because the transfer requires so little time and does not involve the usual operating system overheads of I/O set-up and interrupts. Note that to transfer data from ES to disk requires the data to be first staged into main memory.

The Cray X-MP and Y-MP also support SSDs, which can come in configurations of up to 4096 Mbytes, approximately four times the capacity of the DD-49. The SSD has the potential for enormous bandwidth. It can be attached to the Cray IO system or directly to the CPU through up to two 1000-Mbyte/s channels. Access can be arranged in one of three ways [Reinhardt, 1988]. The first alternative is to treat the SSD as a logical disk, with users responsible for staging heavily accessed files to it. Unfortunately, this leads to the inevitable contention for SSD space. Further, the operating system's disk device drivers are not tuned for the special capabilities of SSDs, and some performance is lost. The second alternative is to use the SSD as an extended memory, in much the same manner as IBM's extended storage. Special system calls for accessing the SSD bypass the usual disk-handling code, and a 4096-byte sector can be accessed in 25 μs. The last alternative is to use the SSD as a logical device cache, i.e., as a second-level cache for multitrack chunks of files that resides between the system's in-main memory file cache and the physical disk devices. Cray engineers have observed workload speedups for their UNIX-like operating system of a factor of four over conventional disk when the cache is enabled. These results indicate that SSDs are most appropriate for containing "hot spot" data [Gawlick, 1987]. Conventional wisdom has it that 20% of the data receives 80% of the accesses, and this has been widely observed in transaction processing systems [Gawlick, 1987].

If SSDs are to be used to replace magnetic disks, then they must be made nonvolatile, and herein lies their greatest weakness. This can be achieved through battery back-up, but the technique is controversial. First, it is difficult to verify that the batteries will be fully charged when needed, i.e., when conventional power fails. Second, it is difficult to determine how long is long enough when powering the SSD with batteries. This should probably be long enough to off-load the disk's contents to magnetic media. Fortunately, low-power DRAM and wafer scale integration technology are making feasible longer battery hold times.

Another weakness is their cost. At the present time, there is more than a 10 to 20 times difference in price between the cost of a megabyte of magnetic disk memory and a megabyte of DRAM. While wafer scale integration may bring this price down in the future, for the near term SSDs will be limited to a staging or caching function.

Disk Caches

Disk caches place buffer memories between the host and the device. If disk data is likely to be re-referenced, caches can be effective in eliminating the seek and rotational latencies. Unfortunately, this effectiveness depends critically on the access behavior of the applications. Truly random access with little re-referencing cannot make effective use of disk caches. However, applications that exhibit a large degree of sequential access can use a cache to good purpose, because data can be staged into the cache before it is actually requested.

Disk caches can become even more useful if they are made nonvolatile using the battery back-up techniques described in the previous subsection (and with the same potential problems). A non-volatile cache will allow "fast writes": the application need not wait for the write I/O to actually complete before it is notified that it has completed. For some applications environments, disk caches have the beneficial effect of reducing the number of reads and thus the number of I/O requests seen by the disks. This has the interesting side effect of increasing the percentage of writes found in the I/O mix, and some observers believe that writes may dominate I/O performance in future systems.

As already mentioned, a disk cache can also lead to better utilization of the host-to-device pathways. A device can transfer data into a cache even if the pathway is in use by another device on the same string. Thus caches are effective in avoiding rotational position sensing misses.

Disk Scheduling

The mechanical delays as seen by a set of simultaneous I/O requests can be reduced through effective disk scheduling. For example, seek times can be reduced if a *shortest-seek-time-first* scheduling algorithm is used [Smith, 1981]. That is, among the queue of pending I/O requests, the one next selected for service is the one that requires the shortest seek time from the current location of the read/write heads. The literature on disk scheduling algorithms is vast, and the effectiveness of a particular scheduling approach depends critically on the workload. It has been observed that scheduling algorithms work best when there are long queues of pending requests; unfortunately, this situation seems to occur rarely in existing systems [Smith, 1981].

Disk Arrays

An alternative to the approaches just described is to exploit parallelism by grouping together a number of physical disks and making these appear to applications as a single logical disk. This has the advantage that the bandwidth of several disks can be harnessed to service a single logical I/O request or can support multiple independent I/Os in parallel. Further, arrays can be constructed using existing, widely available disk technology, rather than the more specialized and more expensive approaches described in the previous subsection. For example, Cray offers a device called the DS-40, which appears as a single logical disk device but which is actually implemented internally as four drives. A logical track is constructed from sectors across the four disks. The DS-40 can transfer at a peak rate of 20 Mbyte/s, with a sustained transfer rate of 9.6 Mbyte/s, and thus is strictly faster than the DD-49.

Defining Terms

Arm: A mechanical assembly that positions the head to the correct track for reading or writing.
Bandwidth: The amount of data per unit time flowing between host computers and storage devices.
Cylinder: A stack of tracks at one acuator position.
Disk drive: An HDA plus all associated electronics.

Head: An electromagnet that produces switchable magnetic fields to read and record bit streams on a platter's track.

Head disk assembly (HDA):   The collection of platters, heads, arms, and actuators, plus the airtight casing, that makes up the storage device. Basically, this is everything but the electronics for controlling the drive and interfacing it to a computer system.

Latency: How long it takes to service an individual request.

Maximal areal density (MAD):   The maximum number of bits that can be stored per square inch. Computed by multiplying the bits per inch in a disk track times the number of tracks per inch of media.

Platters: Metal disks covered with a magnetic material for recording information.

Rotational latency: The time it takes for the desired sector to rotate under the head position before it can be read or written.

Rotational position sensing: A storage device interrupts the I/O controller when the desired sector is under the heads.

Sector: A unit of a storage that is physically read or written at the same time.

Seek time: The time needed to position the heads to the appropriate track position containing the desired data.

Spindle: The collection of disk platters.

Track buffer: A memory buffer embedded in the disk drive. It can hold the contents of the current disk track.

Tracks: The circular recording regions on a platter.

Transfer time: The time taken to physically transfer the bytes from disk to the host.

Throughput: The number of requests for disk service per unit time.

Winchester disk: A magnetic disk in which the read/write heads fly above the recording surface on an air bearing. This is in contrast to contact recording, such as a floppy disk, in which the head and the magnetic media are actually touching.

References

I. Y. Bucher and A. H. Hayes, "I/O performance measurement on Cray-1 and CDC 7600 computers," *Proceedings of the Cray Users Group Conference,* October 1980.

J. Buzen, "BEST/1 analysis of the IBM 3880-13 cached storage controller," *Proc. CMG XIII Conference,* 1982.

J. P. Buzen and A. Shum, "I/O architecture in MVS/370 and MVS/XA," *CMG Transactions,* vol. 54, pp. 19–26, Fall 1986.

J. P. Buzen and A. Shum, "A unified operational treatment of RPS reconnect delays," *Proc. 1987 Sigmetrics Conference,* Performance Evaluation Review, vol. 15, no. 1, May 1987.

Cray Research, Inc., "CRAY Y-MP Computer Systems Functional Description Manual," HR-4001, January 1988.

P. D. Frank, "Advances in head technology," presentation at Challenges in Disk Technology Short Course, Institute for Information Storage Technology, University of Santa Clara, Santa Clara, Calif., December 12–15, 1987.

D. Gawlick, Private Communication, November 1987.

J. M. Harker *et al.,* "A quarter century of disk file innovation," *IBM Journal of Research and Development,* vol. 25, no. 5, pp. 677–689, September 1981.

G. Houtekamer, "The local disk controller," *Proc. 1985 Sigmetrics Conference,* August 1985.

N. P. Kronenberg, H. Levy, and W. D. Strecker, "VAXClusters: A closely-coupled distributed system," *ACM Trans. on Comp. Systems,* vol. 4, no. 2, pp. 130–146, May 1986.

P. Massiglia, *Digital Large System Mass Storage Handbook,* Colorado Springs, Col.: Digital Equipment Corporation, 1986.

S. Reinhardt, "A blueprint for the UNICOS operating system," *Cray Channels*, vol. 10, no. 3, pp. 20–24, Fall 1988.

A. J. Smith, "Input/output optimization and disk architectures: A survey," in *Performance and Evaluation 1*, North-Holland Publishing Company, 1981, pp. 104–117.

L. D. Stevens, "The evolution of magnetic storage," *IBM Journal of Research and Development*, vol. 25, no. 5, pp. 663–675, September 1981.

A. Vasudeva, "A case for disk array storage system," *Proc. Reliability Conference*, Santa Clara, Calif., 1988.

J. Voelcker, "Winchester disks reach for a gigabyte," *IEEE Spectrum*, pp. 64–67, February 1987.

Further Information

International Business Machines (IBM) Corporation developed the first rotating magnetic storage device in the mid-1950s and has always been an industry leader in the storage industry. In honor of the 25-year anniversary of the invention of the magnetic disk, *IBM's Journal of Research and Development* in September 1981 reviewed the development of the technology up to that time. Two particularly notable papers are

L. D. Stevens, "The evolution of magnetic storage," *IBM Journal of Research and Development*, vol. 25, no. 5, pp. 663–675, September 1981.

J. M. Harker *et al.,* "A quarter century of disk file innovation," *IBM Journal of Research and Development*, vol. 25, no. 5, pp. 677–689, September 1981.

For a more up-to-date review of progress in the disk drive industry, see:

J. Voelker, "Winchester disks reach for a gigabyte," *IEEE Spectrum*, pp. 64–67, February 1987.

74.3 Magnetic Tape*

Peter A. Lee

Computers depend on memory to execute programs and to store program code and data. They also need access to stored program code and data in a **nonvolatile memory** (i.e., a form in which the information is not lost when the power is removed from the computer system). Different types of memory have been developed for different tasks. This memory can be categorized according to its price per bit, **access time**, and other parameters. Table 74.1 shows a typical hierarchy for memory which places the smallest and fastest memory at the top in level 0 and in general the largest, slowest, and cheapest at the bottom in level 4 [Ciminiera and Valenzano, 1987]. Auxiliary (secondary or

Table 74.1 Memory Hierarchy

	Data	Code	MMU
Level 0	CPU register	Instruction registers	MMU registers
Level 1	Data cache	Instruction cache	MMU memory
Level 2		On-board cache	
Level 3		Main memory	
Level 4		Auxiliary memory	

Source: P. A. Lee, "Memory subsystems," in *Digital Systems Reference Book*, B. Holdsworth and G. R. Martin, Eds., Oxford: Butterworth-Heinemann, 1991, p. 2.6/3. With permission.

*Based on P. A. Lee, "Memory subsystems," in *Digital Systems Reference Book,* B. Holdworth and G. R. Martins, Eds., Oxford: Butterworth-Heinemann, 1991, chap. 2.6. With permission.

mass) memory of level 4 forms the large storage capacity for program and code that are not currently required by the CPU. This is usually nonvolatile and is at a low cost per bit. Computer **magnetic tape** falls within this category and is the subject of this section.

A Brief Historical Review

Probably the first recorded storage device, developed by Schickard in 1623, used mechanical positions of cogs and gears to work a semi-automatic calculator. Then came Pascal's calculating machine based on 10 digits per wheel. In 1812 punched cards were used in weaving looms to store patterns for woven material. Since that time there have been many mechanical and, latterly, electromechanical devices developed for memory and storage.

In 1948 at Manchester University in England the cathode ray tube (Williams) and the magnetic drum were developed. These consisted of 1024 bits and 1280 bits and a magnetic drum capacity of 120K bits. Cambridge University developed the mercury delay line in 1949, which represented the first fully operational delay line memory, consisting of 576 bits per tube with a total capacity of 18K bits and a circulation time of 1.1 ms.

The first commercial computer with a magnetic tape system was introduced in 1951. The UNIVAC I had a magnetic tape system of 1.44M bits on 150 feet of tape and was capable of storing 128 characters per inch. The tape could be read at a rate of 100 ips. Optical memories are now available as very fast storage devices and will replace magnetic storage in the next few years. At present these devices are expensive although it is envisaged that optical disks with large silicon caches will be the storage arrangement of the future where computer systems utilizing CAD software and image processing can take advantage of the large storage capacities with fast access times. In the future, semiconductor memories are likely to continue their advancing trend.

Introduction

Today's microprocessors are capable of addressing up to 16 Mbytes of main memory. To take advantage of this large capacity, it is usual to have several programs residing in memory at the same time. With intelligent memory management units (MMUs), the programs can be swopped in and out of the main memory to the auxiliary memory when required. For the system to keep pace with this program swopping, it must have a fast auxiliary memory to write to. In the past, most auxiliary systems like magnetic tape and disks have had slow access times, and this has meant that expensive systems have evolved to cater for this requirement. Now that auxiliary memory has improved, and access times are fast and the memory cheap, computer systems have been developed that provide memory swopping with large nonvolatile storage systems. Although the basic technology has not changed over the last 20 years, new materials and different approaches have meant that a new form of auxiliary memory has been brought to the market at a very cheap cost.

Magnetic Tape

Magnetic tape currently provides the cheapest form of storage for large quantities of computer data in a nonvolatile form. The tape is arranged on a reel and has several different packaging styles. It is made from a polyester transportation layer with a deposited layer of oxide having a property similar to a ferrite material with a large hysteresis. Magnetic tape is packaged either in a cartridge, on a reel, or in a cassette. The magnetic cartridge is manufactured in several tape lengths and cartridge sizes capable of storing up to 2 G (giga) bytes of data. These can be purchased in many popular preformatted styles.

The magnetic tape reel is usually ½ inch or 1 inch wide and has lengths of 600, 1200, and 2400 feet. Most reels can store data at rates from 800 bits per inch (bpi) up to 6250 bpi. The reel-to-reel magnetic tape reader is generally bulkier and more expensive than the cartridge readers due to the

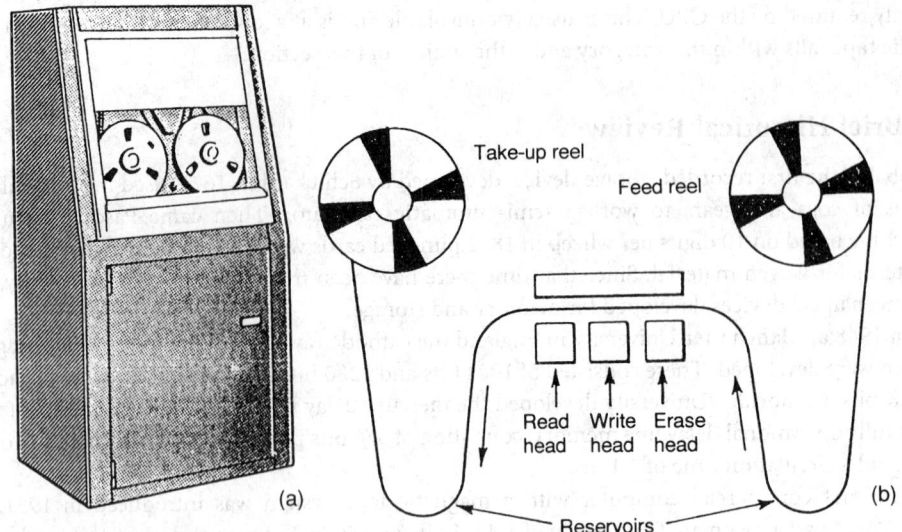

FIGURE 74.16 (a) Magnetic tape drive. (b) Magnetic tape reel arrangement. (*Source:* K. London, *Introduction to Computers*, London: Faber and Faber, 1986, p. 141. With permission.)

complicated pneumatic drive mechanisms, but it provides a large data storage capacity with high access speeds [Wiehler, 1974]. An example of a typical magnetic tape drive with the reel-to-reel arrangement is shown in Fig. 74.16.

A cheap storage medium is the magnetic cassette. Based on the audio cassette, this uses the normal audio cassette recorder for reading and writing data via the standard Kansas City interface through a serial computer I/O line. A logic data "1" is recorded by a high frequency and a logic data "0" by a lower frequency. High-density cassettes can store up to 60 Mbytes of data on each tape and are popular with the computer games market as a cheap storage medium for program distribution.

Both reel-to-reel and cartridge tapes are generally organized by using nine separate tracks across the tape as shown in Fig. 74.17(a).

Each track has its own read and write head operated independently from other tracks [see Fig. 74.17(b)]. Tracks 1 to 8 are used for data and track nine for the parity bit. Data is written on the tape in rows of magnetized islands, using for example EBCDIC (Extended Binary Coded Decimal Interchange Code).

Each read/write head is shaped from a **ferromagnetic material** with an air gap 1 μm wide as seen in Fig. 74.18. The writing head is concerned with converting an electrical pulse into a magnetic state and can be magnetized in one of two directions. This is done by passing a current through the magnetic coil which sets up a leakage field across the 1-μm gap. When the current is reversed the field across the gap is changed, reversing the polarity of the magnetic field on the tape. The head magnetizes the passing magnetic tape recording the state of the magnetic field in the air gap. A logic 1 is recorded as a change in polarity on the tape, and a logic 0 is recorded as no change in polarity, as seen in Fig. 74.19. Reading the magnetic tape states from the tape and converting them to electrical signals is done by the read head. The bit sequences in Fig. 74.19 show the change in magnetic states on the tape. When the tape is passed over the read head, it induces a voltage into the magnetic coil which is converted to digital levels to retrieve the original data.

Tape Format

Information is stored on magnetic tape in the form of a coherent sequence of rows forming a block. This usually corresponds to a page of computer memory and is the minimum amount of data written

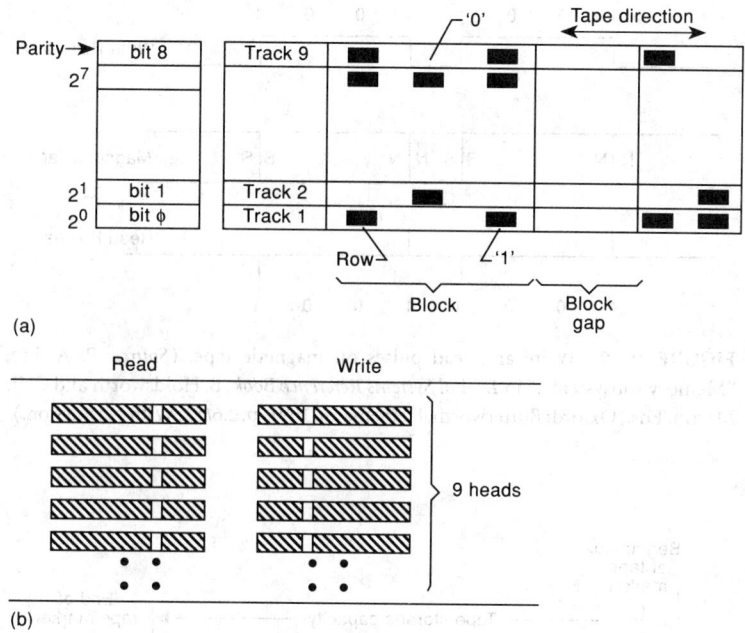

FIGURE 74.17   Magnetic tape format. (*Source:* P. A. Lee, "Memory subsystems," in *Digital Systems Reference Book,* B. Holdsworth and G. R. Martin, Eds., Oxford: Butterworth-Heinemann, 1991, p. 2.6/11. With permission.)

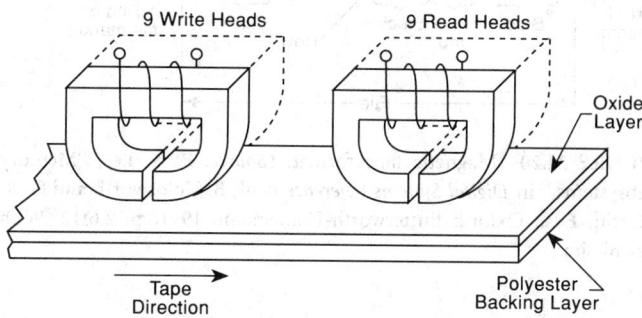

FIGURE 74.18   Read/write head layout. (*Source:* P. A. Lee, "Memory subsystems," in *Digital Systems Reference Book,* B. Holdsworth and G. R. Martin, Eds., Oxford: Butterworth-Heinemann, 1991, p. 2.6/12. With permission.)

to or read from magnetic tape with each program statement. Each block of data is separated by a block gap which is approximately 15 mm long and has no data stored in it. This is shown in Fig. 74.20.

Block gaps are used to allow the tape to accelerate to its operational speed and for the tape to decelerate when stopping at the end of a block. Block gaps use up to 50% of the tape space available for recording, although this may be reduced by making the block sizes larger but has the disadvantage of requiring larger memory buffers to accommodate the data.

A number of blocks make up a file identified by a tape file marker which is written to the tape by the tape controller. The entire length of tape is enclosed between the beginning and end of tape markers. These normally consist of a photosensitive material that triggers sensors on the read/write heads. When a new tape is loaded, it normally advances to the beginning of a tape marker and then it is ready for access by the CPU. The end of tape marker is used to prevent the tape from running off the end of the tape spool and indicates the limit of the storage length.

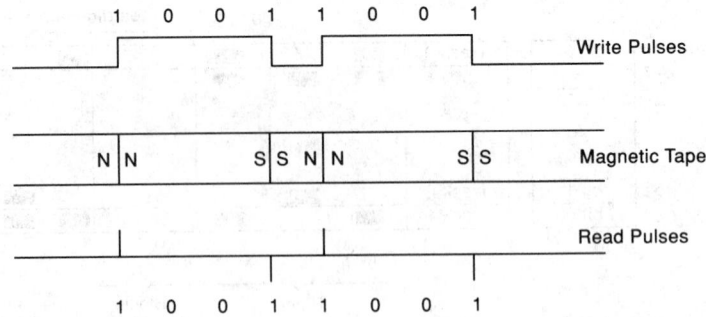

FIGURE 74.19 Write and read pulses on magnetic tape. (*Source:* P. A. Lee, "Memory subsystems," in *Digital Systems Reference Book,* B. Holdsworth and G. R. Martin, Eds., Oxford: Butterworth-Heinemann, 1991, p. 2.6/12. With permission.)

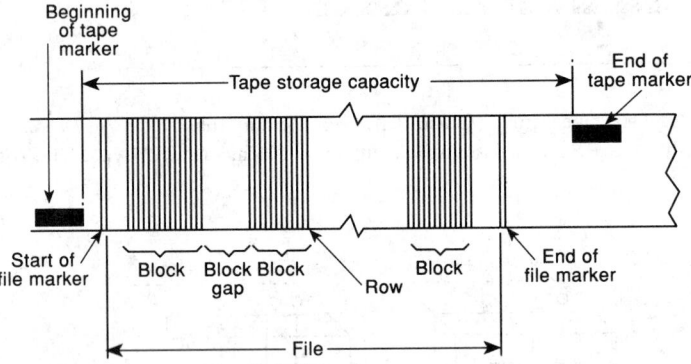

FIGURE 74.20 Magnetic tape format. (*Source:* P. A. Lee, "Memory subsystems," in *Digital Systems Reference Book,* B. Holdsworth and G. R. Martin, Eds., Oxford: Butterworth-Heinemann, 1991, p. 2.6/12. With permission.)

Recording Modes

Several recording modes are used with the express objective of storing data at the highest density and with the greatest reliability of noncorruption of retrieved data. Two popular but contrasting modes are the *non-return-to-zero* (NRZ) and *phase encoding* (PE) modes. These are incompatible although some magnetic tape drives have detectors to sense the mode and operate in a bimodal way. The NRZ technique is shown in Fig. 74.19, where only the 1 bit is displayed by a reversal of magnetization on the tape. The magnetic polarity remains unchanged for logic 0. An external clock track is also required for this mode because a pulse is not always generated for each row of data on the tape.

The PE technique allows both the 0 and 1 states to be displayed by changes of magnetization. A 1 bit is given by a north-to-north pole on the tape, and a 0 bit is given by a south-to-south pole on the tape. PE provides approximately double the recording density and processor speed of NRZ. PE tapes carry an identification mark called a *burst*, which consists of successive magnetization changes at the beginning of track 4. This allows the tape drive to recognize the tape mode and configure itself accordingly.

Defining Terms

Access time: The cycle time for the computer store to present information to the CPU. Access times vary from less than 40 ns for level 0 register storage up to tens of seconds for magnetic tape storage.

Auxiliary (secondary, mass, or backing) storage: Computer stores which have a capacity to store enormous amounts of information in a *nonvolatile* form. This type of memory has an access time usually greater than main memory and consists of magnetic tape drives, magnetic disk stores, and optical disk stores.

Ferromagnetic material: Materials that exhibit high magnetic properties. These include metals such as cobalt, iron, and some alloys.

Magnetic tape: A polyester film sheet coated with a *ferromagnetic* powder, which is used extensively in auxiliary memory. It is produced on a reel, in a cassette, or in a cartridge transportation medium.

Nonvolatile memory: The class of computer memory that retains its stored information when the power supply is cut off. It includes magnetic tape, magnetic disks, flash memory, and most types of ROM.

References

L. Ciminiera and A. Valenzano, *Advanced Microprocessor Architectures*, Reading, Mass.: Addison-Wesley, 1987.

B. Holdsworth and G. Martin, Eds, *Digital Systems Reference Book*, Oxford: Butterworth-Heinemann, 1991, pp 2.6/1–2.6/11.

R. Hyde, "Overview of memory management," *Byte*, pp. 219–225, April 1988.

J. Isailović, *Video Disc and Optical Memory Systems*, Englewood Cliffs, N.J.: Prentice-Hall, 1985.

K. London, *Introduction to Computers*, London: Faber and Faber Press, 1986, p. 141.

M. Mano, *Computer Systems Architecture*, Englewood Cliffs, N.J.: Prentice-Hall, 1982.

R. Matick, *Computer Storage Systems & Technology*, New York: John Wiley, 1977.

A. Tanenbaum, *Structured Computer Organisation*, Englewood Cliffs, N.J.: Prentice-Hall, 1990.

G. Wiehler, *Magnetic Peripheral Data Storage*, Heydon & Son, 1974.

Further Information

The *IEEE Transactions on Magnetics* is available from the IEEE Service Center, Customer Service Department, 445 Hoes Lane, Piscataway, NJ 08855-1331; 800-678-IEEE (outside the USA: 908-981-0060). An IEEE-sponsored Conference on Magnetism and Magnetic Materials was held in December 1992. The British Tape Industry Association (BTIA) has a computer media committee, and further information on standards, etc. can be obtained from British Tape Industry Association, Carolyn House, 22-26 Dingwall Road, Croydon CR0 9XF, England. The equivalent American Association also provides information on computer tape and can be contacted at International Tape Manufacturers' Association, 505 Eighth Avenue, New York, NY 10018.

74.4 Magneto-Optical Disk Data Storage

M. Mansuripur

Since the early 1940s, magnetic recording has been the mainstay of electronic information storage worldwide. Audio tapes provided the first major application for the storage of information on magnetic media. Magnetic tape has been used extensively in consumer products such as audio tapes and video cassette recorders (VCRs); it has also found application in backup/archival storage of computer files, satellite images, medical records, etc. Large volumetric capacity and low cost are the hallmarks of tape data storage, although sequential access to the recorded information is perhaps the main drawback of this technology. Magnetic hard disk drives have been used as mass storage devices in the computer industry ever since their inception in 1957. With an areal density that has doubled roughly every other year, hard disks have been and remain the medium of choice for secondary storage in computers.[1] Another magnetic data storage device, the floppy disk, has been successful in areas where compactness, removability, and fairly rapid access to the recorded information have been of prime concern. In addition to providing backup and safe storage, inexpensive floppies with their moderate capacities (2 Mbyte on a 3.5-in. diameter platter is typical nowadays) and reasonable transfer rates have provided the crucial function of file/data transfer between isolated machines. All in all, it has been a great half-century of progress and market dominance for magnetic recording devices, which are only now beginning to face a potentially serious challenge from the technology of optical recording.

Like magnetic recording, a major application area for optical data storage systems is the secondary storage of information for computers and computerized systems. Like the high-end magnetic media, optical disks can provide recording densities in the range of 10^7 bits/cm^2 and beyond. The added advantage of optical recording is that, like floppies, these disks can be removed from the drive and stored on the shelf. Thus the functions of the hard disk (i.e., high capacity, high data transfer rate, rapid access) may be combined with those of the floppy (i.e., backup storage, removable media) in a single optical disk drive. Applications of optical recording are not confined to computer data storage. The enormously successful audio **compact disk (CD)**, which was introduced in 1983 and has since become the de facto standard of the music industry, is but one example of the tremendous potentials of the optical technology.

A strength of optical recording is that, unlike its magnetic counterpart, it can support read-only, write-once, and erasable/rewritable modes of data storage. Consider, for example, the technology of optical audio/video disks. Here the information is recorded on a master disk which is then used as a stamper to transfer the embossed patterns to a plastic substrate for rapid, accurate, and inexpensive reproduction. The same process is employed in the mass production of read-only files (CD-ROM, O-ROM) which are now being used to distribute software, catalogues, and other large databases. Or consider the write-once read-many (WORM) technology, where one can permanently store massive amounts of information on a given medium and have rapid, random access to them afterwards. The optical drive can be designed to handle read-only, WORM, and erasable media all in one unit, thus combining their useful features without sacrificing performance and ease of use or occupying too much space. What is more, the media can contain regions with prerecorded information as well as regions for read/write/erase operations, both on the same platter. These possibilities open new vistas and offer opportunities for applications that have heretofore been unthinkable; the interactive video disk is perhaps a good example of such applications.

In this article we will lay out the conceptual basis for optical data storage systems; the emphasis will be on disk technology in general and magneto-optical disk in particular. The first section is devoted to a discussion of some elementary aspects of disk data storage including the concept of track and definition of the access time. The second section describes the basic elements of the optical path and its functions; included are the properties of the semiconductor laser diode, characteristics of the beamshaping optics, and certain features of the focusing objective lens. Because of the limited depth of focus of the objective and the eccentricity of tracks, optical disk systems must have a closed-loop feedback mechanism for maintaining the focused spot on the right track. These

---

[1]At the time of this writing, achievable densities on hard disks are in the range of 10^7 bits/cm^2. Random access to arbitrary blocks of data in these devices can take on the order of 10 ms, and individual read/write heads can transfer data at the rate of several megabits per second.

mechanisms are described in the third and fourth sections for automatic focusing and automatic track following, respectively. The physical process of thermomagnetic recording in magneto-optic (MO) media is described next, followed by a discussion of the MO readout process in the sixth section. The final section describes the properties of the MO media.

Preliminaries and Basic Definitions

A disk, whether magnetic or optical, consists of a number of **tracks** along which the information is recorded. These tracks may be concentric rings of a certain width, W_t, as shown in Fig. 74.21. Neighboring tracks may be separated from each other by a guard band whose width we shall denote by W_g. In the least sophisticated recording scheme imaginable, marks of length Δ_0 are recorded along these tracks. Now, if each mark can be in either one of two states, present or absent, it may be associated with a binary digit, 0 or 1. When the entire disk surface of radius R is covered with such marks, its capacity C_0 will be

$$C_0 = \frac{\pi R^2}{(W_t + W_g)\Delta_0} \quad \text{bits per surface} \tag{74.2}$$

Consider the parameter values typical of current optical disk technology: $R = 67$ mm corresponding to 5.25-in. diameter platters, $\Delta_0 = 0.5$ μm which is roughly determined by the wavelength of the read/write laser diodes, and $W_t + W_g = 1$ μm for the track pitch. The disk capacity will then be around 28×10^9 bits, or 3.5 gigabytes. This is a reasonable estimate and one that is fairly close to reality, despite the many simplifying assumptions made in its derivation. In the following paragraphs we examine some of these assumptions in more detail.

The disk was assumed to be fully covered with information-carrying marks. This is generally not the case in practice. Consider a disk rotating at $\mathcal{N}$ revolutions per second (rps). For reasons to be clarified later, this rotational speed should remain constant during the disk operation. Let the electronic circuitry have a fixed clock duration T_c. Then only pulses of length T_c (or an integer multiple thereof) may be used for writing. Now, a mark written along a track of radius r, with a pulse-width equal to T_c, will have length ℓ, where

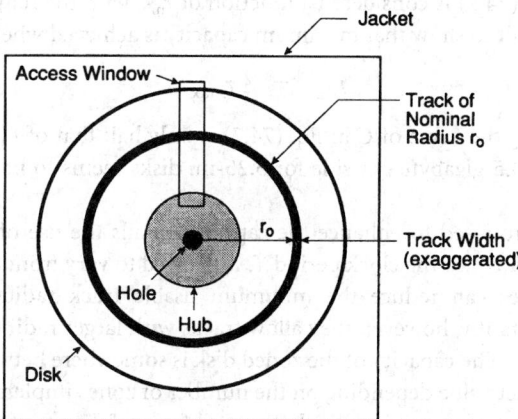

FIGURE 74.21 Physical appearance and general features of an optical disk. The read/write head gains access to the disk through a window in the jacket; the jacket itself is for protection purposes only. The hub is the mechanical interface with the drive for mounting and centering the disk on the spindle. The track shown at radius r_0 is of the concentric-ring type.

$$\ell = 2\pi \mathcal{N} r T_c \tag{74.3}$$

Thus for a given rotational speed $\mathcal{N}$ and a fixed clock cycle T_c, the minimum mark length ℓ is a linear function of track radius r, and ℓ decreases toward zero as r approaches zero. One must, therefore, pick a minimum usable track radius, r_{min}, where the spatial extent of the recorded marks is always greater than the minimum allowed mark length, Δ_0. Equation (74.3) yields

$$r_{min} = \frac{\Delta_0}{2\pi \mathcal{N} T_c} \tag{74.4}$$

One may also define a maximum usable track radius r_{max}, although for present purposes $r_{max} = R$ is a perfectly good choice. The region of the disk used for data storage is thus confined to the area between r_{min} and r_{max}. The total number N of tracks in this region is given by

$$N = \frac{r_{max} - r_{min}}{W_t + W_g} \tag{74.5}$$

The number of marks on any given track in this scheme is independent of the track radius; in fact, the number is the same for all tracks, since the period of revolution of the disk and the clock cycle uniquely determine the total number of marks on any individual track. Multiplying the number of usable tracks N with the capacity per track, we obtain for the usable disk capacity

$$C = \frac{N}{\mathcal{N} T_c} \tag{74.6}$$

Replacing for N from Eq. (74.5) and for $\mathcal{N} T_c$ from Eq. (74.4), we find,

$$C = \frac{2\pi r_{min}(r_{max} - r_{min})}{(W_t + W_g)\Delta_0} \tag{74.7}$$

If the capacity C in Eq. (74.7) is considered a function of r_{min} with the remaining parameters held constant, it is not difficult to show that maximum capacity is achieved when

$$r_{min} = \tfrac{1}{2} r_{max} \tag{74.8}$$

With this optimum r_{min}, the value of C in Eq. (74.7) is only half that of C_0 in Eq. (74.2). In other words, the estimate of 3.5 gigabyte per side for 5.25-in. disks seems to have been optimistic by a factor of two.

One scheme often proposed to enhance the capacity entails the use of multiple zones, where either the rotation speed $\mathcal{N}$ or the clock period T_c is allowed to vary from one zone to the next. In general, zoning schemes can reduce the minimum usable track radius below that given by Eq. (74.8). More importantly, however, they allow tracks with larger radii to store more data than tracks with smaller radii. The capacity of the zoned disk is somewhere between C of Eq. (74.7) and C_0 of Eq. (74.2), the exact value depending on the number of zones implemented.

A fraction of the disk surface area is usually reserved for **preformat** information and cannot be used for data storage. Also, prior to recording, additional bits are generally added to the data for **error correction coding** and other housekeeping chores. These constitute a certain amount of overhead on the user data and must be allowed for in determining the capacity. A good rule of thumb is that overhead consumes approximately 20% of the raw capacity of an optical disk, although the exact number may vary among the systems in use. Substrate defects and film contaminants during the deposition process can create bad **sectors** on the disk. These are typically identified during the certification process and are marked for elimination from the sector directory. Needless to say, bad sectors must be discounted when evaluating the capacity.

Modulation codes may be used to enhance the capacity beyond what has been described so far. Modulation coding does not modify the minimum mark length of Δ_0, but frees the longer marks from the constraint of being integer multiples of Δ_0. The use of this type of code results in more efficient data storage and an effective number of bits per Δ_0 that is greater than unity. For example, the popular (2, 7) modulation code has an effective bit density of 1.5 bits per Δ_0. This or any other modulation code can increase the disk capacity beyond the estimate of Eq. (74.7).

The Concept of Track

The information on magnetic and optical disks is recorded along tracks. Typically, a track is a narrow annulus at some distance r from the disk center. The width of the annulus is denoted by W_t, while the width of the guard band, if any, between adjacent tracks is denoted by W_g. The track pitch is the center-to-center distance between neighboring tracks and is therefore equal to $W_t + W_g$. A major difference between the magnetic floppy disk, the magnetic hard disk, and the optical disk is that their respective track pitches are presently of the order of 100, 10, and 1 μm. Tracks may be fictitious entities, in the sense that no independent existence outside the pattern of recorded marks may be ascribed to them. This is the case, for example, with the audio compact disk format where prerecorded marks simply define their own tracks and help guide the laser beam during readout. In the other extreme are tracks that are physically engraved on the disk surface before any data is ever recorded. Examples of this type of track are provided by pregrooved WORM and magneto-optical disks. Figure 74.22 shows micrographs from several recorded optical disk surfaces. The tracks along which the data are written are clearly visible in these pictures.

It is generally desired to keep the read/write head stationary while the disk spins and a given track is being read from or written onto. Thus, in an ideal situation, not only should the track be perfectly circular, but also the disk must be precisely centered on the spindle axis. In practical systems, however, tracks are neither precisely circular, nor are they concentric with the spindle axis. These eccentricity problems are solved in low-performance floppy drives by making tracks wide enough to provide tolerance for misregistrations and misalignments. Thus the head moves blindly to a radius where the track center is nominally expected to be and stays put until the reading or writing is over. By making the head narrower than the track pitch, the track center is allowed to wobble around its nominal position without significantly degrading the performance during the read/write operation. This kind of wobble, however, is unacceptable in optical disk systems, which have a very narrow track, about the same size as the focused beam spot. In a typical situation arising in practice, the eccentricity of a given track may be as much as ±50 μm while the track pitch is only about 1 μm, thus requiring active track-following procedures.

One method of defining tracks on an optical disk is by means of pregrooves that are either etched, stamped, or molded onto the substrate. In **grooved media of optical storage**, the space between neighboring grooves is the so-called land [see Fig. 74.23(a)]. Data may be written in the grooves with the land acting as a guard band. Alternatively, the land regions may be used for recording while the grooves separate adjacent tracks. The groove depth is optimized for generating an optical signal sensitive to the radial position of the read/write laser beam. For the push-pull method of track-error detection the groove depth is in the neighborhood of $\lambda/8$, where λ is the wavelength of the laser beam.

In digital data storage applications, each track is divided into small segments or sectors, intended for the storage of a single block of data (typically either 512 or 1024 bytes). The physical length of a sector is thus a few millimeters. Each sector is preceded by header information such as the identity of the sector, identity of the corresponding track, synchronization marks, etc. The header information may be preformatted onto the substrate, or it may be written on the storage layer prior to shipping the disk. Pregrooved tracks may be "carved" on the optical disk either as concentric rings or as a single continuous spiral. There are certain advantages to each format. A spiral track can contain a

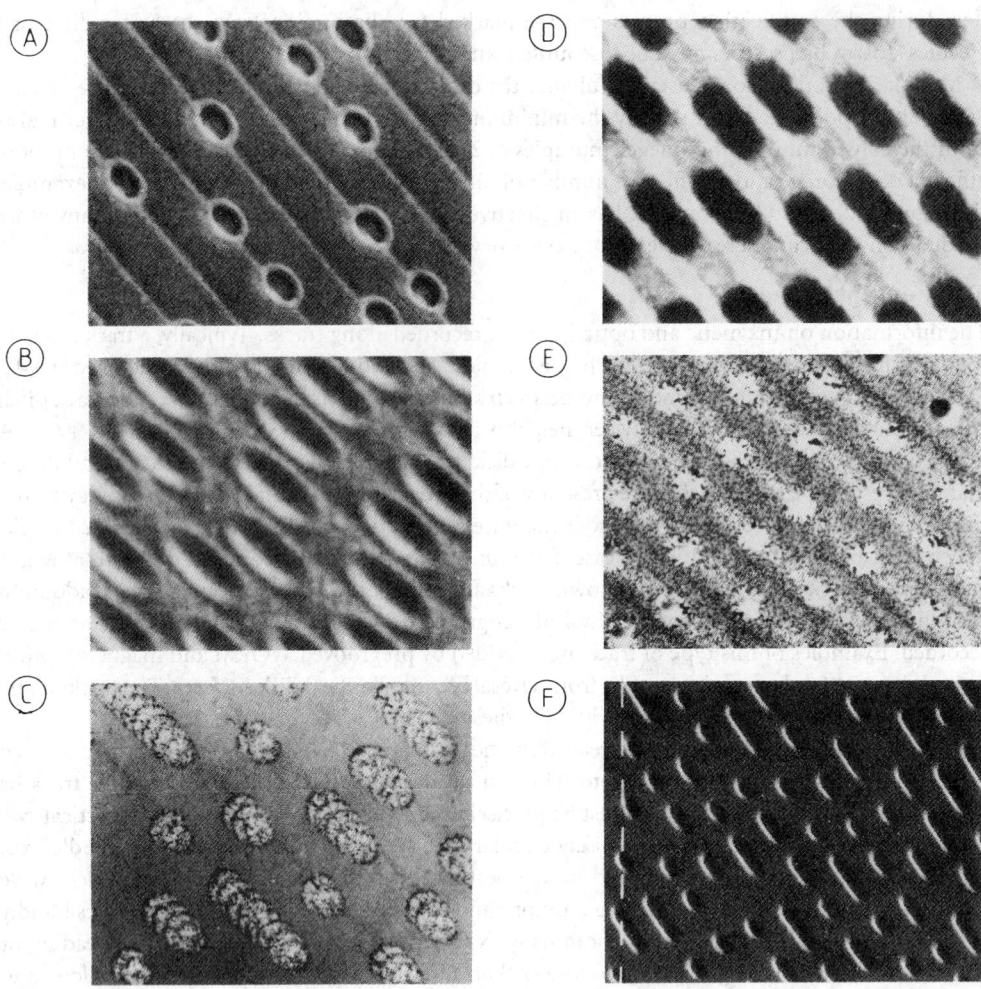

FIGURE 74.22 Micrographs of several types of optical storage media. The tracks are straight and narrow (track pitch = 1.6 μm), with an orientation angle of ≃ −45°. (A) Ablative, write-once tellurium alloy. (B) Ablative, write-once organic dye. (C) Amorphous-to-crystalline, write-once phase-change alloy GaSb. (D) Erasable, amorphous magneto-optic alloy GdTbFe. (E) Erasable, crystalline-to-amorphous phase-change tellurium alloy. (F) Read-only CD-audio, injection-molded from polycarbonate with a nickel stamper. (*Source: Ullmann's Encyclopedia of Industrial Chemistry*, 5th ed., vol. A14, Weinheim: VCH, 1989, p. 196. With permission.)

succession of sectors without interruption, whereas concentric rings may each end up with some empty space that is smaller than the required length for a sector. Also, large files may be written onto (and read from) spiral tracks without jumping to the next track, which occurs when concentric tracks are used. On the other hand, multiple-path operations such as write-and-verify or erase-and-write, which require two paths each for a given sector, or still-frame video are more conveniently handled on concentric-ring tracks.

Another track format used in practice is based on the sampled-servo concept. Here the tracks are identified by occasional marks placed permanently on the substrate at regular intervals, as shown in Fig. 74.23. Details of track following by the sampled-servo scheme will follow shortly; suffice it to say at this point that servo marks help the system identify the position of the focused spot relative to the track center. Once the position is determined it is fairly simple to steer the beam and adjust its position.

Disk Rotation Speed

When a disk rotates at a constant angular velocity ω, a track of radius r moves with the constant linear velocity $V = r\omega$. Ideally, one would like to have the same linear velocity for all the tracks, but this is impractical except in a limited number of situations. For instance, when the desired mode of access to the various tracks is sequential, such as in audio and video disk applications, it is possible to place the head in the beginning at the inner radius and move outward from the center thereafter while continuously decreasing the angular velocity. By keeping the product of r and ω constant, one can thus achieve constant linear velocity for all the tracks.[2] Sequential access mode, however, is the exception rather than the norm in data storage systems. In most applications, the tracks are accessed randomly with such rapidity that it becomes impossible to adjust the rotation speed for constant linear velocity. Under these circumstances, the angular velocity is best kept constant during the normal operation of the disk. Typical rotation speeds are 1200 and 1800 rpm for slower drives and 3600 rpm for the high data rate systems. Higher rotation rates (5000 rpm and beyond) are certainly feasible and will likely appear in future storage devices.

Access Time

The direct-access storage device or DASD, used in computer systems for the mass storage of digi-

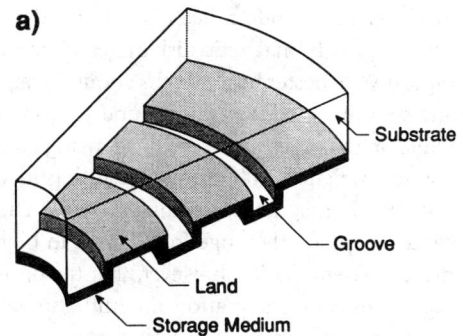

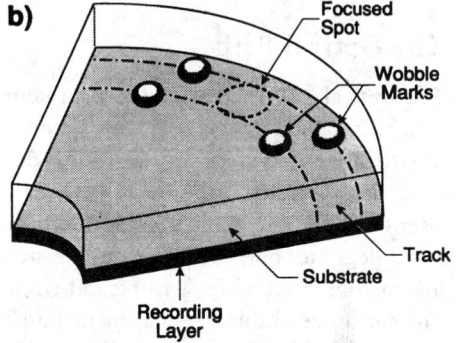

FIGURE 74.23 (a) Lands and grooves in an optical disk. The substrate is transparent, and the laser beam must pass through it before reaching the storage medium. (b) Sampled-servo marks in an optical disk. These marks which are offset from the track-center provide information regarding the position of focused spot.

tal information, is a disk drive capable of storing large quantities of data and accessing blocks of this data rapidly and in arbitrary order. In read/write operations it is often necessary to move the head to new locations in search of sectors containing specific data items. Such relocations are usually time-consuming and can become the factor that limits performance in certain applications. The access time τ_a is defined as the average time spent in going from one randomly selected spot on the disk to another. τ_a can be considered the sum of a seek time, τ_s, which is the average time needed to acquire the target track, and a latency, τ_l, which is the average time spent on the target track waiting for the desired sector. Thus,

$$\tau_a = \tau_s + \tau_l \qquad (74.9)$$

The latency is half the revolution period of the disk, since a randomly selected sector is, on the average, halfway along the track from the point where the head initially lands. Thus for a disk rotating at 1200 rpm $\tau_l = 25$ ms, while at 3600 rpm $\tau_l \simeq 8.3$ ms. The seek time, on the other hand, is independent of the rotation speed, but is determined by the traveling distance of the head during an average seek, as well as by the mechanism of head actuation. It can be shown that the average length

---

[2]In compact disk players the linear velocity is kept constant at 1.2 m/s. The starting position of the head is at the inner radius $r_{min} = 25$ mm, where the disk spins at 460 rpm. The spiral track ends at the outer radius $r_{max} = 58$ mm, where the disk's angular velocity is 200 rpm.

of travel in a random seek is one third of the full stroke. (In our notation the full stroke is $r_{max} - r_{min}$.) In magnetic disk drives where the head/actuator assembly is relatively light-weight (a typical Winchester head weighs about 5 grams) the acceleration and deceleration periods are short, and seek times are typically around 10 ms in small drives (i.e., 5.25 and 3.5 in.). In optical disk systems, on the other hand, the head, being an assembly of discrete elements, is fairly large and heavy (typical weight $\simeq 100$ grams), resulting in values of τ_s that are several times greater than those obtained in magnetic recording systems. The seek times reported for commercially available optical drives presently range from 20 ms in high-performance 3.5-in. drives to about 80 ms in larger drives. We emphasize, however, that the optical disk technology is still in its infancy; with the passage of time, the integration and miniaturization of the elements within the optical head will surely produce lightweight devices capable of achieving seek times of the order of a few milliseconds.

The Optical Path

The **optical path** begins at the light source which, in practically all laser disk systems in use today, is a semiconductor GaAs diode laser. Several unique features have made the laser diode indispensable in optical recording technology, not only for the readout of stored information but also for writing and erasure. The small size of this laser has made possible the construction of compact head assemblies, its coherence properties have enabled diffraction-limited focusing to extremely small spots, and its direct modulation capability has eliminated the need for external modulators. The laser beam is modulated by controlling the injection current; one applies pulses of variable duration to turn the laser on and off during the recording process. The pulse duration can be as short as a few nanoseconds, with rise and fall times typically less than 1 ns. Although readout can be accomplished at constant power level, i.e., in CW mode, it is customary for noise reduction purposes to modulate the laser at a high frequency (e.g., several hundred megahertz during readout).

Collimation and Beam Shaping

Since the cross-sectional area of the active region in a laser diode is only about one micrometer, diffraction effects cause the emerging beam to diverge rapidly. This phenomenon is depicted schematically in Fig. 74.24(a). In practical applications of the laser diode, the expansion of the emerging beam is arrested by a collimating lens, such as that shown in Fig. 74.24(b). If the beam happens

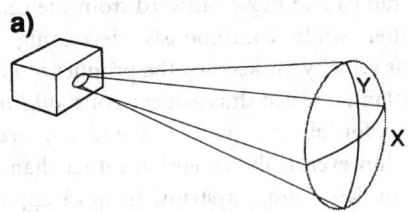

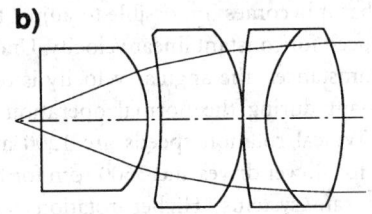

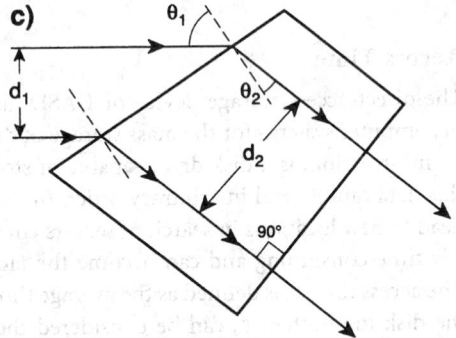

FIGURE 74.24 (a) Away from the facet, the output beam of a diode laser diverges rapidly. In general, the beam diameter along X is different from that along Y, which makes the cross section of the beam elliptical. Also, the radii of curvature R_x and R_y are not the same, thus creating a certain amount of astigmatism in the beam. (b) Multi-element collimator lens for laser diode applications. Aside from collimating, this lens also corrects astigmatic aberrations of the beam. (c) Beam shaping by deflection at a prism surface. θ_1 and θ_2 are related by the Snell's law, and the ratio d_2/d_1 is the same as $\cos \theta_2/\cos \theta_1$. Passage through the prism circularizes the elliptical cross section of the beam.

to have aberrations (astigmatism is particularly severe in diode lasers), then the collimating lens must be designed to correct this defect as well.

In optical recording it is most desirable to have a beam with circular cross section. The need for shaping the beam arises from the special geometry of the laser cavity with its rectangular cross section. Since the emerging beam has different dimensions in the directions parallel and perpendicular to the junction, its cross section at the collimator becomes elliptical, with the initially narrow dimension expanding more rapidly to become the major axis of the ellipse. The collimating lens thus produces a beam with elliptical cross section. Circularization may be achieved by bending various rays of the beam at a prism, as shown in Fig. 74.24(c). The bending changes the beam's diameter in the plane of incidence but leaves the diameter in the perpendicular direction intact.

Focusing by the Objective Lens

The collimated and circularized beam of the diode laser is focused on the surface of the disk using an **objective lens**. The objective is designed to be aberration-free, so that its focused spot size is limited only by the effects of diffraction. Figure 74.25(a) shows the design of a typical objective made from spherical optics. According to the classical theory of diffraction, the diameter of the beam, d, at the objective's focal plane is given by

$$d \simeq \frac{\lambda}{NA} \qquad (74.10)$$

where λ is the wavelength of light, and NA is the numerical aperture of the objective.[3]

In optical recording it is desired to achieve the smallest possible spot, since the size of the spot is directly related to the size of marks recorded on the medium. Also, in readout, the spot size determines the resolution of the system. According to Eq. (74.10) there are two ways to achieve a small spot: first by reducing the wavelength and, second, by increasing the numerical aperture of the objective. The wavelengths currently available from GaAs lasers are in the range of 670–840 nm. It is possible to use a nonlinear

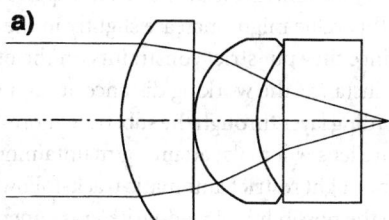

a)

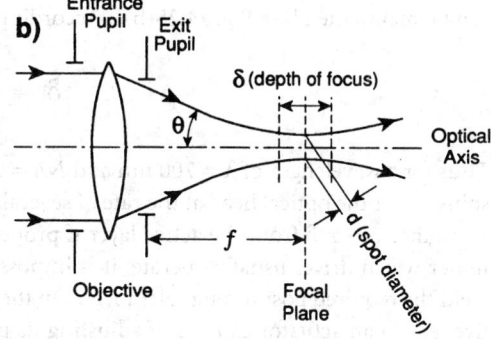

b)

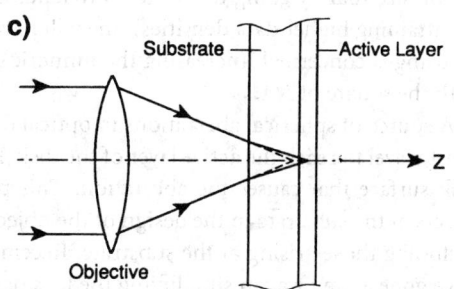

c)

FIGURE 74.25 (a) Multi-element lens design for a high numerical aperture video disk objective. (*Source:* D. Kuntz, "Specifying laser diode optics," *Laser Focus*, March 1984. With permission.) (b) Various parameters of the objective lens. The numerical aperture is $NA = \sin \theta$. The spot diameter d and the depth of focus δ are given by Eqs. (74.10) and (74.11), respectively. (c) Focusing through the substrate can cause spherical aberration at the active layer. The problem can be corrected if the substrate is taken into account while designing the objective.

[3]Numerical aperture is defined as $NA = n \sin \theta$, where n is the refractive index of the image space, and θ is the half-angle subtended by the exit pupil at the focal point. In optical recording systems the image space is air whose index is very nearly unity; thus for all practical purposes $NA = \sin \theta$.

optical device to double the frequency of these diode lasers, thus achieving blue light. Good efficiencies have been demonstrated by frequency doubling. Also recent developments in II–VI materials have improved the prospects for obtaining green and blue light directly from semiconductor lasers. Consequently, there is hope that in the near future optical storage systems will operate in the wavelength range of 400–500 nm. As for the numerical aperture, current practice is to use a lens with NA ≃0.5–0.6. Although this value might increase slightly in the coming years, much higher numerical apertures are unlikely, since they put strict constraints on the other characteristics of the system and limit the tolerances. For instance, the working distance at high numerical aperture is relatively short, making access to the recording layer through the substrate more difficult. The smaller depth of focus of a high numerical aperture lens will make attaining/maintaining proper focus more of a problem, while the limited field of view might restrict automatic track-following procedures. A small field of view also places constraints on the possibility of read/write/erase operations involving multiple beams.

The depth of focus of a lens, δ, is the distance away from the focal plane over which tight focus can be maintained [see Fig. 74.25(b)]. According to the classical diffraction theory

$$\delta \simeq \frac{\lambda}{NA^2} \tag{74.11}$$

Thus for a wavelength of $\lambda = 700$ nm and $NA = 0.6$, the depth of focus is about ±1 μm. As the disk spins under the optical head at the rate of several thousand rpm, the objective lens must stay within a distance of $f \pm \delta$ from the active layer if proper focus is to be maintained. Given the conditions under which drives usually operate, it is impossible to make rigid enough mechanical systems to yield the required positioning tolerances. On the other hand, it is fairly simple to mount the objective lens in an actuator capable of adjusting its position with the aid of closed-loop feedback control. We shall discuss the technique of **automatic focusing** in the next section. For now, let us emphasize that by going to shorter wavelengths and/or larger numerical apertures (as is required for attaining higher data densities) one will have to face a much stricter regime as far as automatic focusing is concerned. Increasing the numerical aperture is particularly worrisome, since δ drops with the square of *NA*.

A source of spherical aberrations in optical disk systems is the substrate through which the light must travel to reach the active layer of the disk. Figure 74.25(c) shows the bending of the rays at the disk surface that causes the aberration. This problem can be solved by taking into account the effects of the substrate in the design of the objective, so that the lens is corrected for all aberrations including those arising at the substrate. Recent developments in molding of aspheric glass lenses have gone a long way in simplifying the lens design problem. Figure 74.26 shows a pair of molded glass aspherics designed for optical disk system applications; both the collimator and the objective are single-element lenses and are corrected for aberrations.

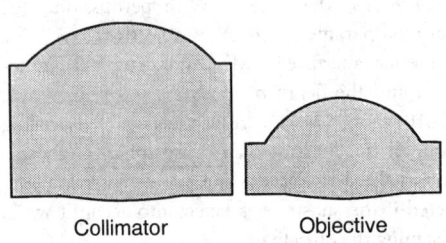

Collimator   Objective

FIGURE 74.26 Molded glass aspheric lens pair for optical disk applications. These singlets can replace the multi-element spherical lenses shown in Figs. 74.24(b) and 74.25(a).

Automatic Focusing

We mentioned in the preceding section that since the objective has a large numerical aperture ($NA \geq 0.5$), its depth of focus δ is rather shallow ($\delta \simeq \pm 1$ μm at $\lambda = 780$ nm). During all read/write/erase operations, therefore, the disk must remain within a fraction of a micrometer from the focal plane of the objective. In practice, however, the disks are not flat and they are not always mounted rigidly parallel to the focal plane, so that movements away from focus occur a few times during each revolution. The peak-to-peak movement in and out of focus may be as much as 100 μm. Without automatic focusing of the objective along the optical axis, this runout (or disk flutter) will be detrimental to the operation of the system. In practice, the objective is mounted on a small motor (usually a voice coil) and allowed to move back and forth in order to keep its distance within an acceptable range from the disk. The spindle turns at a few thousand rpm, which is a hundred or so revolutions per second. If the disk moves in and out of focus a few times during each revolution, then the voice coil must be fast enough to follow these movements in real time; in other words, its frequency response must extend to several kilohertz.

The signal that controls the voice coil is obtained from the light reflected from the disk. There are several techniques for deriving the focus error signal, one of which is depicted in Fig. 74.27(a). In this so-called obscuration method a secondary lens is placed in the path of the reflected light, one-half of its aperture is covered, and a split detector is placed at its focal plane. When the disk is in focus, the returning beam is collimated and the secondary lens will focus the beam at the center of the split detector, giving a difference signal ΔS equal to zero. If the disk now moves away from the objective, the returning beam will become converging, as in Fig. 74.27(b), sending all the light to detector #1. In this case ΔS will be positive and the voice coil will push the lens towards the disk. On the other hand, when the disk moves close to the objective, the returning beam becomes diverging and detector #2 receives the light [see Fig. 74.27(c)]. This results in a negative ΔS that forces the voice coil to pull back in order to return ΔS to zero. A given focus error detection scheme is generally characterized by the shape of its focus error signal ΔS versus the amount of defocus Δz; one such curve is shown in Fig. 74.27(d). The slope of the focus error signal (FES) curve near the origin is of particular importance, since it determines the overall performance and stability of the servo loop.

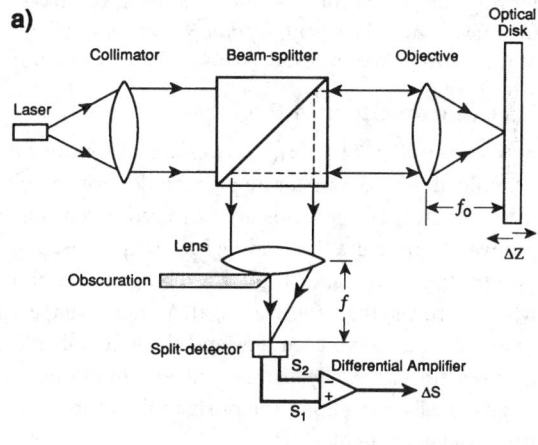

Automatic Tracking

Consider a track at a certain radial location, say r_0, and imagine viewing this track through the access window shown in Fig. 74.21. It is through this window that the head gains access to arbitrarily selected tracks. To a viewer looking through the window, a perfectly circular track centered on the spindle axis will look stationary, irrespective of the rotation rate. However, any eccentricity will cause an apparent radial motion of the track. The peak-to-peak distance traveled by a track (as seen through the window) depends on a number of factors including centering accuracy of the hub, deformability of the substrate, mechanical vibrations, manufacturing tolerances, etc. For a typical 3.5-in. disk, for example, this peak-to-peak motion can be as much as 100 μm during one revolution. Assuming a revolution rate of 3600 rpm, the apparent velocity of the track in the radial direction will be several millimeters

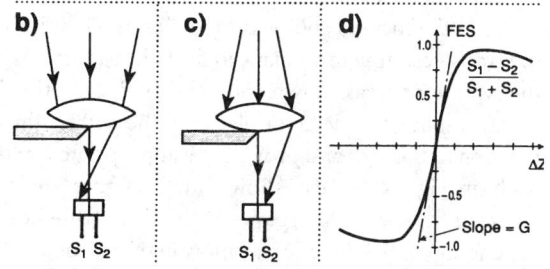

FIGURE 74.27 Focus error detection by the obscuration method. In (a) the disk is in focus, and the two halves of the split detector receive equal amounts of light. When the disk is too far from the objective (b) or too close to it (c), the balance of detector signals shifts to one side or the other. A plot of the focus error signal (FES) versus defocus is shown in (d), and its slope near the origin is identified as the FES gain, G.

per second. Now, if the focused spot remains stationary while trying to read from or write to this track, it is clear that the beam will miss the track for a good fraction of every revolution cycle.

Practical solutions to the above problem are provided by **automatic tracking** techniques. Here the objective is placed in a fine actuator, typically a voice coil, which is capable of moving the necessary radial distances and maintaining a lock on the desired track. The signal that controls the movement of this actuator is derived from the reflected light itself, which carries information about the position of the focused spot. There exist several mechanisms for extracting the track error signal (TES); all these methods require some sort of structure on the disk surface in order to identify the track. In the case of read-only disks (CD, CD-ROM, and video disk), the embossed pattern of data provides ample information for tracking purposes. In the case of write-once and erasable disks, tracking guides are "carved" on the substrate in the manufacturing process. As mentioned earlier, the two major formats for these tracking guides are pregrooves (for continuous tracking) and sampled-servo marks (for discrete tracking). A combination of the two schemes, known as continuous/composite format, is often used in practice. This scheme is depicted in Fig. 74.28 which shows a small section containing five tracks, each consisting of the tail end of a groove, synchronization marks, a mirror area used for adjusting focus/track offsets, a pair of wobble marks for sampled tracking, and header information for sector identification.

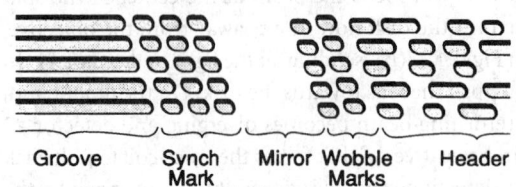

Groove   Synch   Mirror Wobble   Header
          Mark   Marks

FIGURE 74.28 Servo fields in continuous/composite format contain a mirror area and offset marks for tracking (*Source:* A. B. Marchant, *Optical Recording*, Reading, Mass.: Addison-Wesley, 1990, p. 264. With permission.)

Tracking on Grooved Regions

As shown in Fig. 74.23(a), grooves are continuous depressions that are either embossed or etched or molded onto the substrate prior to deposition of the storage medium. If the data is recorded on the grooves, then the lands are not used except for providing a guard band between neighboring grooves. Conversely, the land regions may be used to record the information, in which case grooves provide the guard band. Typical track widths are about one wavelength. The guard bands are somewhat narrower than the tracks, their exact shape and dimensions depending on the beam size, required track-servo accuracy, and the acceptable levels of cross-talk between adjacent tracks. The groove depth is usually around one-eighth of one wavelength ($\lambda/8$), since this depth can be shown to give the largest TES in the push-pull method. Cross sections of the grooves may be rectangular, trapezoidal, triangular, etc.

When the focused spot is centered on track, it is diffracted symmetrically from the two edges of the track, resulting in a balanced far field pattern. As soon as the spot moves away from the center, the symmetry breaks down and the light distribution in the far field tends to shift to one side or the other. A split photodetector placed in the path of the reflected light can therefore sense the relative position of the spot and provide the appropriate feedback signal. This strategy is depicted schematically in Fig. 74.29; also shown in the figure are intensity plots at the detector plane for light reflected from various regions of the disk. Note how the intensity shifts to one side or the other depending on the direction of motion of the spot.

Sampled Tracking

Since dynamic track runout is usually a slow and gradual process, there is actually no need for continuous tracking as done on grooved media. A pair of embedded marks, offset from the track center as in Fig. 74.23(b), can provide the necessary information for correcting the relative position of the focused spot. The reflected intensity will indicate the positions of the two servo marks as two successive short pulses. If the beam happens to be on track, the two pulses will have equal magnitudes and there will be no need for correction. If, on the other hand, the beam is off-track, one of

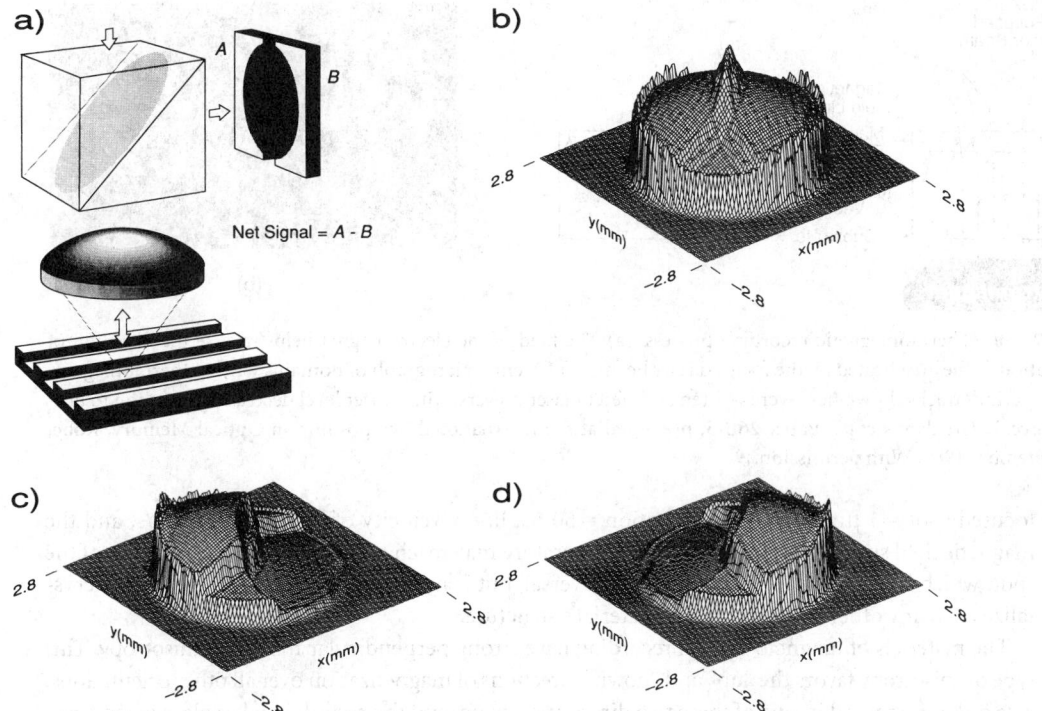

FIGURE 74.29 (a) Push-pull sensor for tracking on grooves. (*Source:* A. B. Marchant, *Optical Recording*, Reading, Mass.: Addison-Wesley, 1990, p. 175. With permission.) (b) Calculated distribution of light intensity at the detector plane when the disk is in focus and the beam is centered on track. (c) Calculated intensity distribution at the detector plane with disk in focus but the beam centered on the groove edge. (d) Same as (c) except for the spot being focused on the opposite edge of the groove.

the pulses will be stronger than the other. Depending on which pulse is the stronger, the system will recognize the direction in which it has to move and will correct the error accordingly. The servo marks must appear frequently enough along the track to ensure proper track following. In a typical application, the track might be divided into groups of 18 bytes, 2 bytes dedicated as servo offset areas and 16 bytes filled with other format information or left blank for user data.

Thermomagnetic Recording Process

Recording and erasure of information on a magneto-optical disk are both achieved by the **thermomagnetic process**. The essence of thermomagnetic recording is shown in Fig. 74.30. At the ambient temperature the film has a high magnetic coercivity[4] and therefore does not respond to the externally applied field. When a focused beam raises the local temperature of the film, the hot spot becomes magnetically soft (i.e., its coercivity drops). As the temperature rises, coercivity drops continuously until such time as the field of the electromagnet finally overcomes the material's resistance to reversal and switches its magnetization. Turning the laser off brings the temperatures back to normal, but the reverse-magnetized domain remains frozen in the film. In a typical situation in practice, the film thickness may be around 300 Å, laser power at the disk $\simeq 10$ mW, diameter of the

---

[4]Coercivity of a magnetic medium is a measure of its resistance to magnetization reversal. For example, consider a thin film with perpendicular magnetic moment saturated in the $+Z$ direction. A magnetic field applied along $-Z$ will succeed in reversing the direction of magnetization only if the field is stronger than the coercivity of the film.

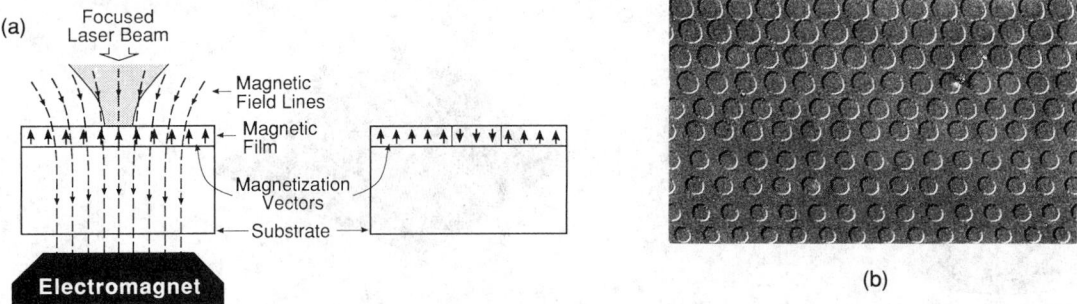

FIGURE 74.30 Thermomagnetic recording process. (a) The field of the electromagnet helps reverse the direction of magnetization in the area heated by the focused laser beam. (b) Lorentz micrograph of domains written thermomagnetically. The various tracks shown here were written at different laser powers, with power level decreasing from top to bottom. (*Source:* F. Greidanus *et al.*, Paper 26B-5, presented at the International Symposium on Optical Memory, Kobe, Japan, September 1989. With permission.)

focused spot $\simeq 1$ μm, laser pulse duration $\simeq 50$ ns, linear velocity of the track $\simeq 10$ m/s, and the magnetic field strength $\simeq 200$ gauss. The temperature may reach a peak of 500 K at the center of the spot, which is sufficient for magnetization reversal, but is not nearly high enough to melt or crystalize or in any other way modify the material's structure.

The materials of magneto-optical recording have strong perpendicular magnetic anisotropy. This type of anisotropy favors the "up" and "down" directions of magnetization over all other orientations. The disk is initialized in one of these two directions, say up, and the recording takes place when small regions are selectively reverse-magnetized by the thermomagnetic process. The resulting magnetization distribution then represents the pattern of recorded information. For instance, binary sequences may be represented by a mapping of zeros to up-magnetized regions and ones to down-magnetized regions (non-return to zero or NRZ). Alternatively, the NRZI scheme might be used, whereby transitions (up-to-down and down-to-up) are used to represent the ones in the bit-sequence.

Recording by Laser Power Modulation (LPM)

In this traditional approach to thermomagnetic recording, the electromagnet produces a constant field, while the information signal is used to modulate the power of the laser beam. As the disk rotates under the focused spot, the on/off laser pulses create a sequence of up/down domains along the track. The Lorentz electron micrograph in Fig. 74.30(b) shows a number of domains recorded by LPM. The domains are highly stable and may be read over and over again without significant degradation. If, however, the user decides to discard a recorded block and to use the space for new data, the LPM scheme does not allow direct overwrite; the system must erase the old data during one disk revolution cycle and record the new data in a subsequent revolution cycle.

During erasure, the direction of the external field is reversed, so that up-magnetized domains in Fig. 74.30(a) now become the favored ones. Whereas writing is achieved with a modulated laser beam, in erasure the laser stays on for a relatively long period of time, erasing an entire sector. Selective erasure of individual domains is not practical, nor is it desired, since mass data storage systems generally deal with data at the level of blocks, which are recorded onto and read from individual sectors. Note that at least one revolution period elapses between the erasure of an old block and its replacement by a new block. The electromagnet therefore need not be capable of rapid switchings. (When the disk rotates at 3600 rpm, for example, there is a period of 16 ms or so between successive switchings.) This kind of slow reversal allows the magnet to be large enough to cover all the tracks simultaneously, thereby eliminating the need for a moving magnet and an actuator. It also affords a relatively large gap between the disk and the magnet, which enables the use of double-sided disks and relaxes the mechanical tolerances of the system without overburdening the magnet's driver.

The obvious disadvantage of LPM is its lack of direct overwrite capability. A more subtle concern is that it is perhaps unsuitable for the PWM (pulse width modulation) scheme of representing binary waveforms. Due to fluctuations in the laser power, spatial variations of material properties, lack of perfect focusing and track following, etc., the length of a recorded domain along the track may fluctuate in small but unpredictable ways. If the information is to be encoded in the distance between adjacent domain walls (i.e., PWM), then the LPM scheme of thermomagnetic writing may suffer from excessive domain-wall jitter. Laser power modulation works well, however, when the information is encoded in the position of domain centers (i.e., pulse position modulation or PPM). In general, PWM is superior to PPM in terms of the recording density, and, therefore, recording techniques that allow PWM are preferred.

Recording by Magnetic Field Modulation

Another method of thermomagnetic recording is based on magnetic field modulation (MFM) and is depicted schematically in Fig. 74.31(a). Here the laser power may be kept constant while the information signal is used to modulate the magnetic field. Photomicrographs of typical domain patterns recorded in the MFM scheme are shown in Fig. 74.31(b). Crescent-shaped domains are

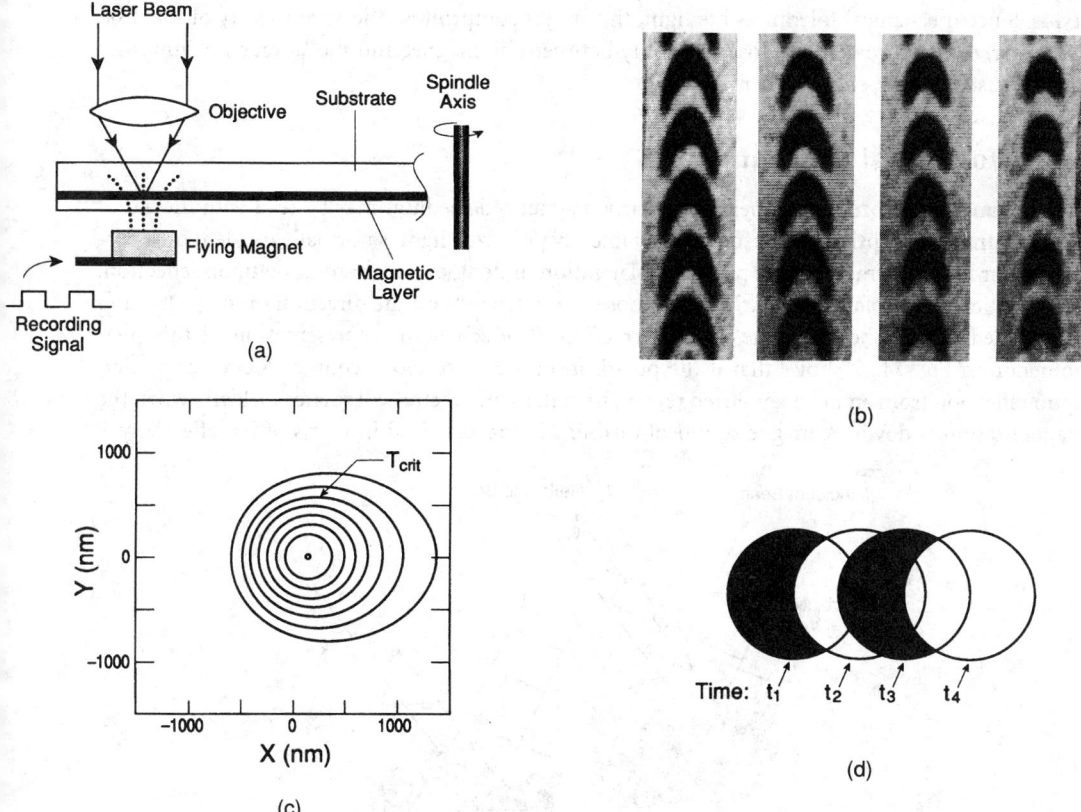

FIGURE 74.31 (a) Thermomagnetic recording by magnetic field modulation. The power of the beam is kept constant, while the magnetic field direction is switched by the data signal. (b) Polarized-light microphotograph of recorded domains. (c) Computed isotherms produced by a CW laser beam, focused on the magnetic layer of a disk. The disk moves with constant velocity under the beam. The region inside the isotherm marked as T_{crit} is above the critical temperature for writing, that is, its magnetization aligns with the direction of the applied field. (d) Magnetization within the heated region (above T_{crit}) follows the direction of the applied field, whose switchings occur at times t_n. The resulting domains are crescent-shaped.

the hallmark of the field modulation technique. If one assumes (using a much simplified model) that the magnetization aligns itself with the applied field within a region whose temperature has passed a certain critical value, T_{crit}, then one can explain the crescent shape of these domains in the following way: With the laser operating in the CW mode and the disk moving at constant velocity, temperature distribution in the magnetic medium assumes a steady-state profile, such as that shown in Fig. 74.31(c). Of course, relative to the laser beam, the temperature profile is stationary, but in the frame of reference of the disk the profile moves along the track with the linear track velocity. The isotherm corresponding to T_{crit} is identified as such in the figure; within this isotherm the magnetization aligns itself with the applied field. Figure 74.31(d) shows a succession of critical isotherms along the track, each obtained at the particular instant of time when the magnetic field switches direction. From this picture it is easy to infer how the crescent-shaped domains form and also understand the relation between the waveform that controls the magnet and the resulting domain pattern.

The advantages of magnetic field modulation recording are that (1) direct overwriting is possible and (2) domain-wall positions along the track, being rather insensitive to defocus and laser power fluctuations, are fairly accurately controlled by the timing of the magnetic field switchings. On the negative side, the magnet must now be small and fly close to the disk surface, if it is to produce rapidly switched fields with a magnitude of a hundred gauss or so. Systems that utilize magnetic field modulation often fly a small electromagnet on the opposite side of the disk from the optical stylus. Since mechanical tolerances are tight, this might compromise the removability of the disk. Moreover, the requirement of close proximity between the magnet and the storage medium dictates the use of single-sided disks in practice.

Magneto-Optical Readout

The information recorded on a perpendicularly magnetized medium may be read with the aid of the polar **magneto-optical Kerr effect**. When linearly polarized light is normally incident on a perpendicular magnetic medium, its plane of polarization undergoes a slight rotation upon reflection. This rotation of the plane of polarization, whose sense depends on the direction of magnetization in the medium, is known as the polar Kerr effect. The schematic representation of this phenomenon in Fig. 74.32 shows that if the polarization vector suffers a counterclockwise rotation upon reflection from an up-magnetized region, then the same vector will rotate clockwise when the magnetization is down. A magneto-optical medium is characterized in terms of its reflectivity R

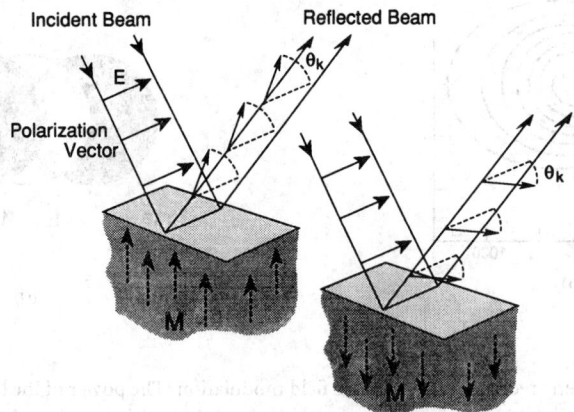

FIGURE 74.32 Schematic diagram describing the polar magneto-optical Kerr effect. Upon reflection from the surface of a perpendicularly magnetized medium, the polarization vector undergoes a rotation. The sense of rotation depends on the direction of magnetization, **M**, and switches sign when **M** is reversed.

and its Kerr rotation angle θ_k. R is a real number (between 0 and 1) that indicates the fraction of the incident power reflected back from the medium at normal incidence. θ_k is generally quoted as a positive number, but is understood to be positive or negative depending on the direction of magnetization; in MO readout, it is the sign of θ_k that carries the information about the state of magnetization, i.e., the recorded bit pattern.

The laser used for readout is usually the same as that used for recording, but its output power level is substantially reduced in order to avoid erasing (or otherwise obliterating) the previously recorded information. For instance, if the power of the write/erase beam is 20 mW, then for the read operation the beam is attenuated to about 3 or 4 mW. The same objective lens that focuses the write beam is now used to focus the read beam, creating a diffraction-limited spot for resolving the recorded marks. Whereas in writing the laser was pulsed to selectively reverse-magnetize small regions along the track, in readout it operates with constant power, i.e., in CW mode. Both up- and down-magnetized regions are read as the track passes under the focused spot. The reflected beam, which is now polarization-modulated, goes back through the objective and becomes collimated once again; its information content is subsequently decoded by polarization-sensitive optics, and the scanned pattern of magnetization is reproduced as an electronic signal.

Differential Detection

Figure 74.33 shows the differential detection system that is the basis of magneto-optical readout in practically all erasable optical storage systems in use today. The beam splitter (BS) diverts half of the reflected beam away from the laser and into the detection module.[5] The polarizing beam splitter (PBS) splits the beam into two parts, each carrying the projection of the incident polarization along one axis of the PBS, as shown in Fig. 74.33(b). The component of polarization along one of the axes goes straight through, while the component along the other axis splits off and branches to the side. The PBS is oriented such that in the absence of the Kerr effect its two branches will receive equal amounts of light. In other words, if the polarization, upon reflection from the disk, did not undergo any rotations whatsoever, then the beam entering the PBS would be polarized at 45° to the PBS axes, in which case it would split equally between the two branches. Under this condition, the two detectors generate identical signals and the differential signal ΔS will be zero. Now, if the beam returns from the disk with its polarization rotated clockwise (rotation angle = θ_k), then detector #1 will receive more light than detector #2, and the differential signal will be positive. Similarly, a counterclockwise rotation will generate a negative ΔS. Thus, as the disk rotates under the focused spot, the electronic signal ΔS reproduces the pattern of magnetization along the scanned track.

Materials of Magneto-Optical Data Storage

Amorphous rare earth transition metal alloys are presently the media of choice for erasable optical data storage applications. The general formula for the composition of the alloy may be written $(Tb_yGd_{1-y})_x(Fe_zCo_{1-z})_{1-x}$ where terbium and gadolinium are the rare earth (RE) elements, while iron and cobalt are the transition metals (TM). In practice, the transition metals constitute roughly 80 atomic percent of the alloy (i.e., $x \simeq 0.2$). In the transition metal subnetwork, the fraction of cobalt is usually small, typically around 10%, and iron is the dominant element ($z \simeq 0.9$). Similarly, in the rare earth subnetwork Tb is the main element ($y \simeq 0.9$) while the gadolinium content is small or it may even be absent in some cases. Since the rare earth elements are highly reactive to oxygen, RE-TM films tend to have poor corrosion resistance and, therefore, require protective coatings. In multilayer disk structures, the dielectric layers that enable optimization of the medium

---

[5]The use of an ordinary beam splitter is an inefficient way of separating the incoming and outgoing beams, since half the light is lost in each pass through the splitter. One can do much better by using a so-called "leaky" polarizing beam splitter.

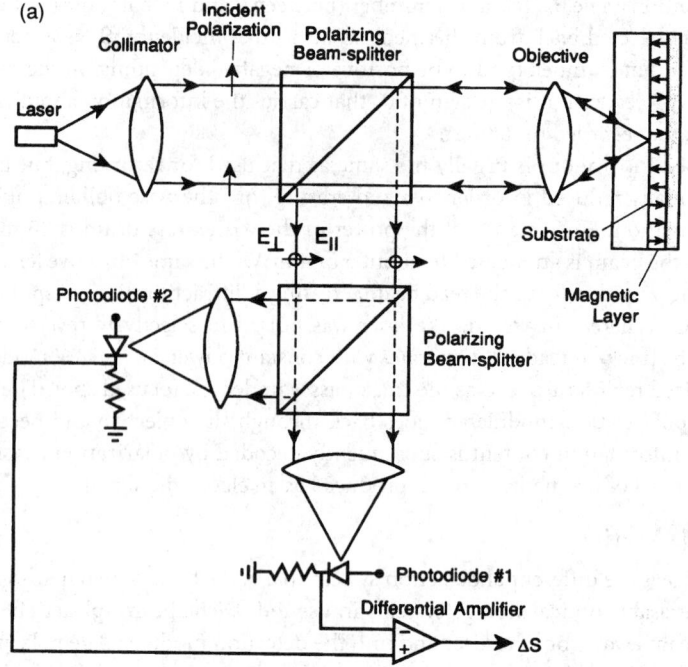

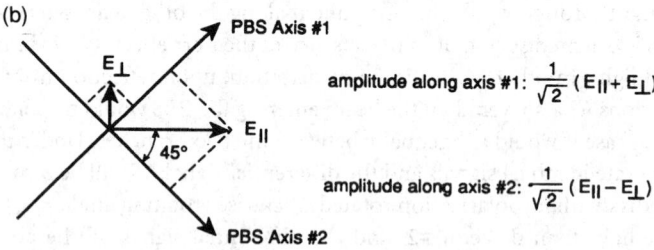

FIGURE 74.33 Differential detection scheme utilizes a polarizing beam splitter and two photodetectors in order to convert the rotation of polarization to an electronic signal. $E_{||}$ and $E_{\perp}$ are the reflected components of polarization; they are, respectively, parallel and perpendicular to the direction of incident polarization. The diagram in (b) shows the orientation of the PBS axes relative to the polarization vectors.

for the best optical/thermal behavior also perform the crucial function of protecting the MO layer from the environment.

The amorphous nature of the material allows its composition to be continuously varied until a number of desirable properties are achieved. In other words, the fractions x, y, z of the various elements are not constrained by the rules of stoichiometry. Disks with very large areas can be coated uniformly with thin films of these media, and, in contrast to polycrystalline films whose grains and grain boundaries scatter the beam and cause noise, amorphous films are continuous, smooth, and substantially free from noise. The films are deposited either by sputtering from an alloy target or by co-sputtering from multiple elemental targets. In the latter case, the substrate moves under the various targets and the fraction of a given element in the alloy is determined by the time spent under each target as well as the power applied to that target. During film deposition the substrate is kept at a low temperature (usually by chilled water) in order to reduce the mobility of deposited atoms and thus inhibit crystal growth. The type of the sputtering gas (argon, krypton, xenon, etc.) and its pressure during sputtering, the bias voltage applied to the substrate, deposition rate, nature of the substrate and its pretreatment, and temperature of the substrate all can have dramatic effects on the composition and short-range order of the deposited film. A comprehensive discussion of the factors that influence film properties will take us beyond the intended scope here; the interested reader may consult the vast literature of this field for further information.

Defining Terms

Automatic focusing: The process in which the distance of the disk from the objective's focal plane is continuously monitored and fed back to the system in order to keep the disk in focus at all times.

Automatic tracking: The process in which the distance of the focused spot from the track center is continuously monitored and the information fed back to the system in order to maintain the read/write beam on track at all times.

Compact disk (CD):   A plastic substrate embossed with a pattern of pits that encode audio signals in digital format. The disk is coated with a metallic layer (to enhance its reflectivity) and read in a drive (CD player) that employs a focused laser beam and monitors fluctuations of the reflected intensity in order to detect the pits.

Error correction coding (ECC):   Systematic addition of redundant bits to a block of binary data, as insurance against possible read/write errors. A given error-correcting code can recover the original data from a contaminated block, provided that the number of erroneous bits is less than the maximum number allowed by that particular code.

Grooved media of optical storage: A disk embossed with grooves of either the concentric-ring type or the spiral type. If grooves are used as tracks, then the lands (i.e., regions between adjacent grooves) are the guard bands. Alternatively, lands may be used as tracks, in which case the grooves act as guard bands. In a typical grooved optical disk in use today the track width is 1.1 μm, the width of the guard band is 0.5 μm, and the groove depth is 70 nm.

Magneto-optical Kerr effect: The rotation of the plane of polarization of a linearly polarized beam of light upon reflection from the surface of a perpendicularly magnetized medium.

Objective lens: A well-corrected lens of high numerical aperture, similar to a microscope objective, used to focus the beam of light onto the surface of the storage medium. The objective also collects and recollimates the light reflected from the medium.

Optical path: Optical elements in the path of the laser beam in an optical drive. The path begins at the laser itself and contains a collimating lens, beam shaping optics, beam splitters, polarization-sensitive elements, photodetectors, and an objective lens.

Preformat: Information such as sector address, synchronization marks, servo marks, etc., embossed permanently on the optical disk substrate.

Sector: A small section of track with the capacity to store one block of user data (typical blocks are either 512 or 1024 bytes). The surface of the disk is covered with tracks, and tracks are divided into contiguous sectors.

Thermomagnetic process: The process of recording and erasure in magneto-optical media, involving local heating of the medium by a focused laser beam, followed by the formation or annihilation of a reverse-magnetized domain. The successful completion of the process usually requires an external magnetic field to assist the reversal of the magnetization.

Track: A narrow annulus or ring-like region on a disk surface, scanned by the read/write head during one revolution of the spindle; the data bits of magnetic and optical disks are stored sequentially along these tracks. The disk is covered either with concentric rings of densely packed circular tracks or with one continuous, fine-pitched spiral track.

References

A. B. Marchant, *Optical Recording*, Reading, Mass.: Addison-Wesley, 1990.

P. Hansen and H. Heitman, "Media for erasable magneto-optic recording," *IEEE Trans. Mag.*, vol. 25, pp. 4390–4404, 1989.

M. H. Kryder, "Data-storage technologies for advanced computing," *Scientific American*, vol. 257, pp. 116–125, 1987.

G. Bouwhuis, J. Braat, A. Huijser, J. Pasman, G. Van Rosmalen, and K. S. Immink, *Principles of Optical Disk Systems*, Bristol: Adam Hilger Ltd., 1985, chap. 2 and 3.

Special issue of *Applied Optics* on video disks, July 1, 1978.

E. Wolf, "Electromagnetic diffraction in optical systems. I. An integral representation of the image field," *Proc. R. Soc. Ser. A*, vol. 253, pp. 349–357, 1959.

M. Mansuripur, "Certain computational aspects of vector diffraction problems," *J. Opt. Soc. Am. A*, vol. 6, pp. 786–806, 1989.

D. O. Smith, "Magneto-optical scattering from multilayer magnetic and dielectric films," *Opt. Acta*, vol. 12, p. 13, 1965.

P. S. Pershan, "Magneto-optic effects," *J. Appl. Phys.*, vol. 38, pp. 1482–1490, 1967.

K. Egashira and R. Yamada, "Kerr effect enhancement and improvement of readout characteristics in MnBi film memory," *J. Appl. Phys.*, vol. 45, pp. 3643–3648, 1974.

H. S. Carslaw and J. C. Jaeger, *Conduction of Heat in Solids*, London: Oxford University Press, 1954.

P. Kivits, R. deBont, and P. Zalm, "Superheating of thin films for optical recording," *Appl. Phys.*, vol. 24, pp. 273–278, 1981.

M. Mansuripur, G. A. N. Connell, and J. W. Goodman, "Laser-induced local heating of multi-layers," *Appl. Opt.*, vol. 21, p. 1106, 1982.

J. Heemskerk, "Noise in a video disk system: experiments with an (AlGa)As laser," *Appl. Opt.*, vol. 17, p. 2007, 1978.

A. Arimoto, M. Ojima, N. Chinone, A. Oishi, T. Gotoh, and N. Ohnuki, "Optimum conditions for the high frequency noise reduction method in optical video disk players," *Appl. Opt.*, vol. 25, p. 1398, 1986.

M. Mansuripur, G. A. N. Connell, and J. W. Goodman, "Signal and noise in magneto-optical read-out," *J. Appl. Phys.*, vol. 53, p. 4485, 1982.

J. W. Beck, "Noise considerations of optical beam recording," *Appl. Opt.*, vol. 9, p. 2559, 1970.

S. Chikazumi and S. H. Charap, *Physics of Magnetism*, New York: John Wiley, 1964.

B. G. Huth, "Calculation of stable domain radii produced by thermomagnetic writing," *IBM J. Res. Dev.*, pp. 100–109, 1974.

A. P. Malozemoff and J. C. Slonczewski, *Magnetic Domain Walls in Bubble Materials*, New York: Academic Press, 1979.

A. M. Patel, "Signal and error-control coding," in *Magnetic Recording*, vol. II, C. D. Mee and E. D. Daniel, Eds. New York: McGraw-Hill, 1988.

K. A. S. Immink, "Coding methods for high-density optical recording," *Philips J. Res.*, vol. 41, pp. 410–430, 1986.

L. I. Maissel and R. Glang, Eds., *Handbook of Thin Film Technology*, New York: McGraw-Hill, 1970.

G. L. Weissler and R. W. Carlson, Eds., *Vacuum Physics and Technology*, vol. 14 of *Methods of Experimental Physics*, New York: Academic Press, 1979.

Further Information

Proceedings of the *Optical Data Storage Conference* are published annually by SPIE, the International Society for Optical Engineering. These proceedings document the latest developments in the field of optical recording each year. Two other conferences in this field are the *International Symposium on Optical Memory* (ISOM), whose proceedings are published as a special issue of the *Japanese Journal of Applied Physics*, and the *Magneto-Optical Recording International Symposium* (MORIS), whose proceedings appear in a special issue of the *Journal of the Magnetics Society of Japan*.

75

Logical Devices

Franco P. Preparata
Brown University

Richard S. Sandige
University of Wyoming

B. R. Bannister
University of Hull

D. G. Whitehead
University of Hull

Martin Bolton
SGS-Thomson Microelectronics

Bill D. Carroll
University of Texas at Arlington

75.1 Combinational Networks and Switching Algebra 1695
Introduction to Binary Functions of Binary Variables • Switching
Functions of One and Two Variables • Networks and Expressions •
Switching Algebra • Boolean Expressions: Normal and Canonical
Forms • Other Important Boolean Connectives

75.2 Logic Circuits 1711
Combinational Logic Circuits • Sequential Logic Circuits •
Synchronous Sequential Logic Circuits • Asynchronous Sequential
Logic Circuits

75.3 Registers 1721
Gated Registers • Shift Registers • Register Transfer Language •
Input/Output Ports • Counters • Registered ASICs • Standard
Graphic Symbols

75.4 Programmable Arrays 1735
PLA-Structured Devices • Programmable Gated Arrays (FPGAs) •
Computer-Aided Design for Programmable Arrays

75.5 Arithmetic Logic Units 1741
Basic Adders and Subtracters • High-Speed Adders • Multifunction
Arithmetic Logic Units • Standard Integrated Circuit ALUs

75.1 Combinational Networks and Switching Algebra*

Franco P. Preparata

Introduction to Binary Functions of Binary Variables

A digital system can be analyzed as an interconnection of functional components of basically two types:

1. *Storage* components, called memory or registers, as appropriate, which store information
2. *Combinational* components, which do not have any memory capability and whose function is the implementation of the required information processing activities (computing)

In this section we shall be concerned with the study of combinational components. Consider, for example, a circuit (the *adder*) designed to compute the sum of two integers represented in binary (Fig. 75.1). The adder has two sets of input lines, one for each of the two operands A and B and one set of output lines for the sum. Each set of lines (input and output) consists of as many wires as there are bits in the number it is designed to carry.

*All material in this section is adapted from F. Preparata, *Introduction to Computer Engineering*, New York: John Wiley, 1985, chap. 3. With permission.

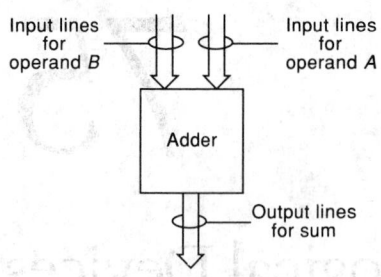

FIGURE 75.1 A binary adder.

Such an adder can be realized by means of a collection of simpler blocks, called *adder cells*, each of which is assigned to a fixed position of the operands (that is, the ith cell receives the ith bits of the addends). The adder is completed by connecting the carry-out output of the ith cell to the carry-in input of the $(i + 1)$th cell and setting c_0 permanently to 0 (see Fig. 75.2).

Thus, all we need do to design the adder is to design the adder cell, whose behavior is specified in Fig. 75.2(b). The adder cell receives three binary inputs A, B, and C, of which A and B are the operand bits and C is the carry-in bit, and produces two binary outputs C' and S, of which S is the sum bit and C' is the carry-out bit. A, B, and C (the input variables) are independent and therefore can appear in any one of the eight configurations shown in Fig. 75.2(b); C' and S, instead, depend entirely upon the binary values of A, B, and C. Specifically, the ordered pair (C', S) corresponding to a given triple (A, B, C) will represent in binary the number of 1's appearing in the binary string ABC; for example, in $ABC = 110$ there are two 1's, whence (C', S) will represent in binary the number 2, that is, $C'S = 10$ [see Fig. 75.2(b)].

We recognize a familiar notion; both C' and S are functions of (A, B, C), that is, for each of them we have a *domain* consisting of the eight possible triples of binary values for (A, B, C) [Fig. 75.2(b), three left columns] and a range consisting of the two values 0,1. Thus C' and S are each binary functions of binary variables; now, we have the following:

A binary function of binary variables is called a switching function. A combinational circuit or network is a digital subsystem which realizes a switching function.

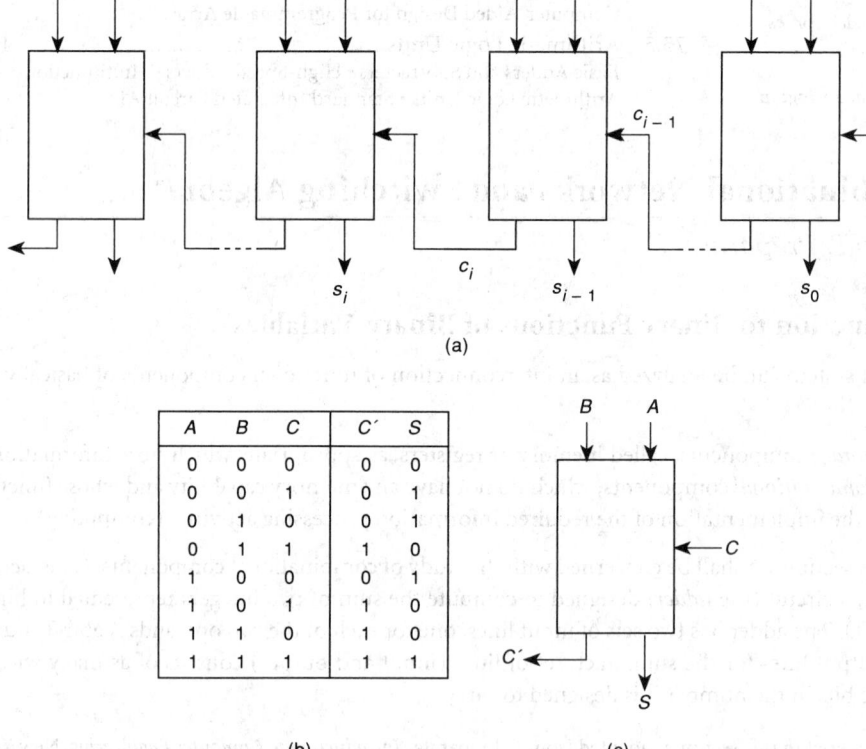

FIGURE 75.2 An adder (a); an adder cell (b); and its behavior (c).

The conventional way to display a switching function f is that shown in Fig. 75.2(b). Specifically, the combinations of 0's and 1's are ordered so that, when viewed as binary numbers, they are in natural order. Next to each combination, the value of the function is given: this table of function values is called the *truth table* of f. The reason for this name is that a binary variable is said to be *true* when equal to 1 and *false* otherwise. So, the function table gives the *truth* values of the function.

Our objective is to develop a methodology for the design of a combinational network that realizes a given switching function.

Switching Functions of One and Two Variables

To gain insight into the nature of switching functions, we begin by considering binary functions of one binary variable x. This variable can assume only two values, 0 and 1, which form the domain (see Fig. 75.3). All the functions of one variable are obtained by filling a two-place truth table in all possible ways, that is, in four ways, shown in Fig. 75.3.

We say that a function is *degenerate* if it does not depend upon all of its arguments and *nondegenerate* otherwise. So we see that function f_0 is degenerate: in fact, it is constant and equal to 0, so we will call it the

Functions Domain	f_0	f_1	f_2	f_3
0	0	0	1	1
1	0	1	0	1

FIGURE 75.3 Truth tables of all functions of one variable.

constant 0; similarly, f_3 will be called the constant 1. Instead, f_1 and f_2 are nondegenerate functions: notice that $f_1(0) = 0$ and $f_1(1) = 1$, thus $f_1(x) = x$, so f_1 will be called the identity; $f_2(0) = 1$ and $f_2(1) = 0$ (f_2 maps 0 to 1 and 1 to 0) and will be called the complement function and denoted by $f_2(x) = \bar{x}$.

We are now ready to consider the binary function of two binary variables x_2 and x_1. Here the pair (x_2, x_1) can assume four possible values (00, 01, 10, 11), which form the domain (see Fig. 75.4). All functions are now obtained by filling a truth table with four entries in all possible ways (Fig. 75.4). Obviously, this can be done in 16 ways; that is, we have 16 binary functions of 2 binary variables. Let us examine these functions $g_0, g_1, \ldots, g_{15}$. We realize that g_0 and g_{15} are, respectively, the constants 0 and 1; moreover, we notice that $g_3 = x_2$, $g_5 = x_1$, $g_{10} = \bar{x}_1$, and $g_{12} = \bar{x}_2$, that is, the latter are actually nondegenerate functions of *only one* variable. The remaining 10 functions $g_1, g_2, g_4, g_6, g_7, g_8, g_9, g_{11}, g_{13}, g_{14}$ are nondegenerate functions of two variables (i.e., *each* of them depends upon *both* variables). We could analyze all of them, but temporarily we content ourselves with the study of g_1 and g_7.

For convenience, we redraw the truth tables of functions g_1 and g_7 in Fig. 75.5. Function g_1 is equal to 1 only when both x_2 *and* x_1 are equal to 1; for this reason it is called the AND function. Function g_7 is equal to 1 when either x_2 or x_1, or both, are equal to 1; for this reason it is called the OR function.

We can now imagine that special devices are available for the realization of some of the functions we have considered. Specifically, we have a one-input–one-output device, called an inverter [in Fig. 75.6(a) we give the conventional symbol for this device], which realizes the function COMPLEMENT; a two-input–one-output device [Fig. 75.6(b)], called *AND gate*, which realizes the function AND, and an analogous device [Fig. 75.6(c)], called *OR gate*, which realizes the function OR.

x_1	x_2	g_0	g_1	g_2	g_3	g_4	g_5	g_6	g_7	g_8	g_9	g_{10}	g_{11}	g_{12}	g_{13}	g_{14}	g_{15}
0	0	0	0	0	0	0	0	0	0	1	1	1	1	1	1	1	1
0	1	0	0	0	0	1	1	1	0	0	0	0	0	1	1	1	1
1	0	0	0	1	1	0	0	1	1	0	0	1	1	0	0	1	1
1	1	0	1	0	1	0	1	0	1	0	1	0	1	0	1	0	1

FIGURE 75.4 Truth tables of all functions of two variables.

x_2	x_1	g_1	g_1
0	0	0	0
0	1	0	1
1	0	0	1
1	1	1	1

FIGURE 75.5 Study of the AND and OR functions.

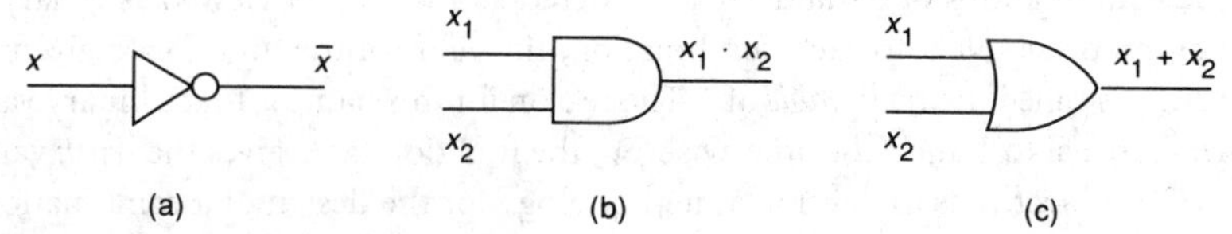

FIGURE 75.6 Symbols for inverter (a), AND gate (b), and OR gate (c).

The output of an AND gate with inputs x_1 and x_2 will be denoted by $x_1 \cdot x_2$ or simply $x_1 x_2$; analogously, the output of an OR gate with inputs x_1 and x_2 will be denoted by $x_1 + x_2$. (The context will avoid confusion with the symbols "$\cdot$" and "$+$" when used in ordinary arithmetic.)

Networks and Expressions

Consider an interconnection of the basic building blocks, AND gates, OR gates, and inverters, such as that shown in Fig. 75.7(a). Such an interconnection we call a network. Notice that each gate output, except the single-network output, feeds exactly one gate input and that there are no loops; that is, when tracing a path in the obvious way, in no case will this path traverse the same gate twice. The input terminals are all the unconnected gate inputs. We may think of constructing this network by connecting to the two input terminals of gate G_1 [see Fig. 75.7(b)] the output terminals of two smaller networks. In turn, the latter networks could be decomposed into even smaller networks, and so on until we reach the simplest networks of all: terminals. This analysis actually enables us to give an (inductive) definition of a network.

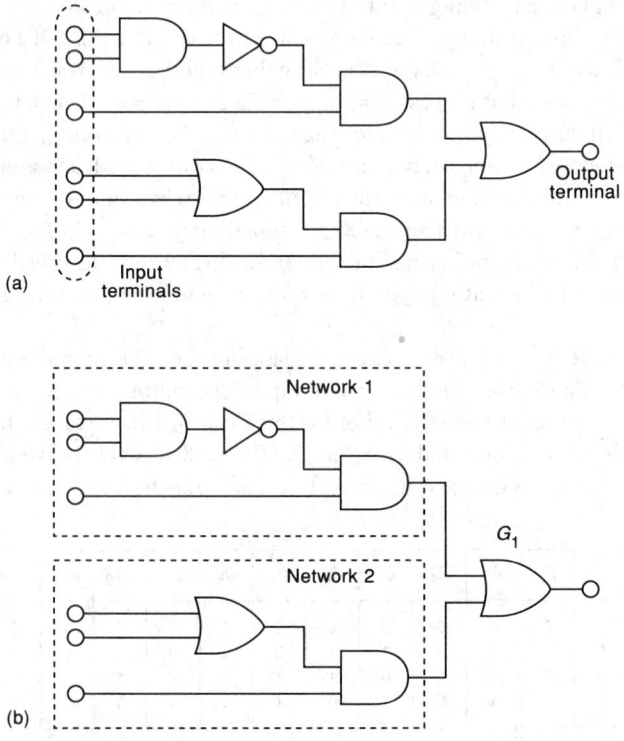

FIGURE 75.7 A combinatorial network.

Definition of Combinational Networks

1. Input terminals are networks (elementary networks).
2. If N_1 and N_2 are networks (represented as black boxes), so are the following:

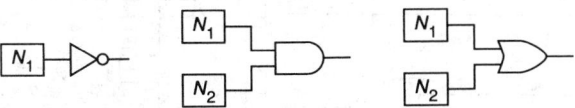

In analogy with the preceding definition, consider now the following definition of **Boolean expressions**.

Definition of Boolean Expressions

1. Variables (both complemented and uncomplemented) and constants are expressions (elementary expressions).
2. If E_1 and E_2 are expressions, so are $\overline{E}_1$, $E_1 \cdot E_2$, and $E_1 + E_2$.[1]

Example 1. $x_1, 0, x_2$ are elementary expressions; $[x_1 + (\overline{x}_2 x_3)] \cdot x_4$ is an expression; specifically, it is the AND of expressions $E_1 = x_1 + (\overline{x}_2 x_3)$ and $E_2 = x_4$.

Suppose that we now assign either a variable (complemented or otherwise) or a constant to each input terminal of a nonelementary network. Then, with the output of each gate whose inputs are connected to the input terminals, we can associate an expression, and so on downstream until we associate an expression with the output of the network [for example see Fig. 75.8(a)]. Therefore, we see that there is a one-to-one correspondence between expressions and networks whose inputs have been assigned (input-assigned network). Specifically we say that

An expression ↔ an input-assigned network

Normally we will drop the qualifier *input-assigned* whenever the context makes it obvious.

Consider the network of Fig. 75.8(a). We may now assign to each binary input variable, that is, to x_1, x_2, and x_3, one of the two possible values 0 or 1. Once this assignment has been made we can easily trace the network downstream and calculate a binary value on each internal wire of the network until we obtain a binary value at the output terminal [Fig. 75.8(b)]. Consider what we have done: We have chosen a set of binary values for (x_1, x_2, x_3) (in our example 001) and have obtained a binary value of u, and for each different choice of input values the network embodies a well-defined rule for obtaining a value of u. This means that u is a function of (x_1, x_2, x_3), from which we conclude that *any given switching network computes a switching function of its inputs.*

A most remarkable fact—to be shown later—is that the converse of the above statement is also true, that is, for any given switching function we can design a network that realizes it. The corresponding design techniques will be presented in the next subsections.

The notion of combinational networks can be slightly generalized to encompass the class of networks in which a gate output may feed *more than one* gate.[2] A gate output feeding two or more gate inputs is said to have *multiple fan-out*. Notice, however, that a network with multiple fan-out gates does not correspond to a single Boolean expression: indeed, if we want to be able to reconstruct the network from its description, we must have a distinct expression for each gate having a multiple fan-out.

Switching Algebra

The techniques for designing combinational networks rest on the properties of a fundamental formal system called *switching algebra* (which is a special case of more general systems called Boolean algebras, although frequently switching algebra is referred to as *Boolean algebra*).

---

[1]To avoid any ambiguity, the new expressions $E_1 \cdot E_2$ and $E_1 + E_2$ should be parenthesized as $(E_1 \cdot E_2)$ and $(E_1 + E_2)$. However, we shall conform here to the familiar rules for parentheses adopted in ordinary algebra for "+" and ".".

[2]The number of inputs driven by the output of a gate is called the *fan-out* of that gate. Also, the number of inputs of a gate is called the *fan-in* of the gate.

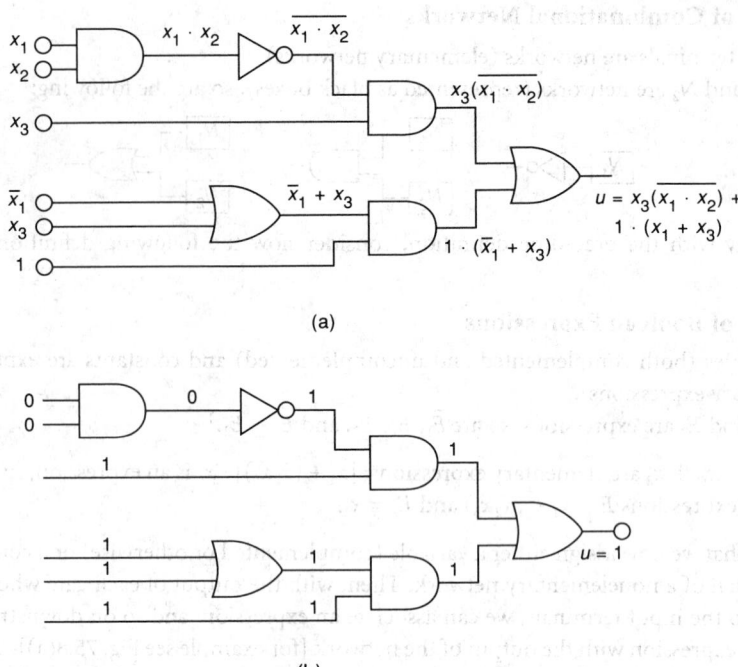

FIGURE 75.8 Determination of the output expression of a network.

The objects that switching algebra deals with are the (Boolean) expressions defined in the preceding subsection. The basic axiom is as follows.

Axiom. Each expression assumes either the value 0 or the value 1 for all assignments of values (0 or 1) to its variables.

We begin by regarding the function of one variable, COMPLEMENT, as a *unary* operation, that is, as an operation with *one* operand x, the function's argument, which is itself to be regarded as an expression and can only assume the two values 0 and 1. The table of the operation is repeated below.

$$\begin{array}{cc} \text{Operand} & \text{Complement} \\ 0 & 1 \\ 1 & 0 \end{array}$$

Notice that the complement of 0 is 1, $\overline{0} = 1$; similarly $\overline{1} = 0$. It follows that

$$(\overline{\overline{0}}) = \overline{1} = 0, (\overline{\overline{1}}) = \overline{0} = 1$$

This is summarized by the identity

$$\overline{\overline{x}} = x \qquad\qquad \text{Involution} \quad (75.1)$$

which describes a fundamental property of COMPLEMENT. Notice also that the constants 0 and 1 are mutually complementary.

We now regard the functions AND and OR of two variables x_1 and x_2 as *binary* operations, that is, as operations with two operands. Here again x_1 and x_2 are to be regarded as expressions, and by the axiom, each can only assume either the value 0 or the value 1. The transformation of each of the function tables to the corresponding operation table, shown in Fig. 75.9, should be self-explana-

x_2	x_1	AND
0	0	0
0	1	0
1	0	0
1	1	1

(a)

AND	x_2 0	1
x_1 0	0	0
1	0	1

x_2	x_1	OR
0	0	0
0	1	1
1	0	1
1	1	1

(b)

OR	x_2 0	1
x_1 0	0	1
1	1	1

FIGURE 75.9 Operational tables for (a) AND (b) OR.

tory. From the inspection of these operation tables, we can now deduce their characteristic properties. First of all, both tables are *symmetric* with respect to the main diagonal, that is, we can exchange the role of x_1 and x_2; we shall summarize this as follows:

$$x_1 \cdot x_2 = x_2 \cdot x_1 \qquad x_1 + x_2 = x_2 + x_1 \qquad \text{Commutativity} \qquad (75.2)$$

Next, we notice that $0 \cdot 0 = 0 + 0 = 0$ and $1 \cdot 1 = 1 + 1 = 1$, which leads to the property, for any expression x,

$$xx = x \qquad x + x = x \qquad \text{Idempotency} \qquad (75.3)$$

Since $0 \cdot 1 = 0$ and $1 \cdot 1 = 1$ we extract the rule $x \cdot 1 = x$; similarly $0 + 0 = 0$ and $1 + 0 = 1$ gives $x + 0 = x$, and we have the properties

$$x \cdot 1 = x \qquad x + 0 = x \qquad (75.4)$$

Also, $0 \cdot 0 = 0$ and $1 \cdot 0 = 0$ yields $x \cdot 0 = 0$; similarly $0 + 1 = 1$ and $1 + 1 = 1$ yields $x + 1 = 1$, thus the properties

$$x \cdot 0 = 0 \qquad x + 1 = 1 \qquad (75.5)$$

Finally, considering the off-diagonal elements in both operation tables, we see that $0 \cdot 1 = 1 \cdot 0 = 0$ and $0 + 1 = 1 + 0 = 1$; therefore

$$x \cdot \bar{x} = 0 \qquad x + \bar{x} = 1 \qquad \text{Complementarity} \qquad (75.6)$$

There is now a collection of additional properties that can be easily derived by means of a useful proof mechanism called *perfect induction*, which is stated as follows.

Perfect Induction. Two expressions E_1 and E_2 on the same set of variables are equivalent (denoted by $E_1 = E_2$) if, for all possible assignments of values to the variables, the values of E_1 and E_2 coincide.

Perfect induction will now be used to prove the following identities:

$$x \cdot (y + z) = xy + xz \qquad x + yz = (x + y)(x + z) \qquad \text{Distributivity} \qquad (75.7)$$

Indeed, the claim is proved by the tables in Fig. 75.10.

By perfect induction we can also prove the identities

$$x(x + y) = x \qquad\qquad x + xy = x \qquad\qquad \text{Absorption} \qquad (75.8)$$

$$(xy)z = x(yz) \qquad\qquad (x + y) + z = x + (y + z) \qquad\qquad \text{Associativity} \qquad (75.9)$$

$$\overline{xy} = \bar{x} + \bar{y} \qquad\qquad \overline{x + y} = \bar{x} \cdot \bar{y} \qquad\qquad \text{De Morgan's Law} \qquad (75.10)$$

(Notice that because associativity holds, we will omit parentheses when writing the AND or the OR of more than two variables.)

Consider now the identities (75.2)-(75.10) which we have established. They are offered in pairs such that one term of the pair is obtained from the other by interchanging AND and OR and by interchanging the constants 0 and 1. This fact is summarized as follows.

x	y	z	y + z	x(y + z)	xy	xz	xy + xz	yz	x + yz	x + y	x + z	(x + y)(x + z)
0	0	0	0	0	0	0	0	0	0	0	0	0
0	0	1	1	0	0	0	0	0	0	0	1	0
0	1	0	1	0	0	0	0	0	0	1	0	0
0	1	1	1	0	0	0	0	1	1	1	1	1
1	0	0	0	0	0	0	0	0	1	1	1	1
1	0	1	1	1	0	1	1	0	1	1	1	1
1	1	0	1	1	1	0	1	0	1	1	1	1
1	1	1	1	1	1	1	1	1	1	1	1	1

FIGURE 75.10   Perfect induction proofs of identities (75.7).

Principle of Duality.   Given a valid identity, we obtain another valid identity by:

1. Interchanging the operators AND and OR
2. Interchanging the constants 0 to 1

We can now concisely summarize the properties of switching algebra which we have just established. *Switching algebra* is a set $\mathfrak{B}$ of elements (Boolean expressions) containing the constants 0 and 1, with the following operations:

1. Two binary operations, AND and OR, which are commutative (75.2), associative (75.9), idempotent (75.3), absorptive (75.8), and mutually distributive (75.7).
2. A unary operation, COMPLEMENT (or NEGATION), with the properties of involution (75.1), complementarity (75.6), De Morgan's law (75.10).

The constants 0 and 1 have the following properties (75.4), (75.5):

$$\overline{1} = 0$$

$$x \cdot 1 = x \quad x + 0 = x$$

$$x \cdot 0 = 0 \quad x + 1 = 1$$

Identities (75.1)–(75.10) given above represent a set of rules—given in dual pairs—that can be applied to transform an expression into an equivalent expression. It can be shown that rules (75.1)–(75.10) are not independent and that we can select five of them and derive the others from these; this, however, is outside our present scope.

Example 2.   Prove the following identities, without using perfect induction and by transforming the left side to the right side.

$$a + \overline{a}b = a + b \tag{75.11}$$

$$
\begin{aligned}
a + \overline{a}b &= a \cdot 1 + \overline{a}b & &\text{[by (75.4)]} \\
&= a(b + \overline{b}) + \overline{a}b & &\text{[by (75.6)]} \\
&= ab + a\overline{b} + \overline{a}b & &\text{[by (75.7)]} \\
&= a\overline{b} + ab + \overline{a}b & &\text{[by (75.2)]} \\
&= a\overline{b} + ab + ab + \overline{a}b & &\text{[by (75.3)]} \\
&= a(\overline{b} + b) + (a + \overline{a})b & &\text{[by (75.7)]} \\
&= a \cdot 1 + 1 \cdot b & &\text{[by (75.6)]} \\
&= a + b & &\text{[by (75.4)]}
\end{aligned}
$$

An alternative and simpler proof of (75.11) runs as follows:

$$a + \bar{a}b = (a + \bar{a})(a + b) \qquad \text{[by (75.7)]}$$

$$= 1 \cdot (a + b) \qquad \text{[by (75.6)]}$$

$$= a + b \qquad \text{[by (75.4)]}$$

$$ab + bc + \bar{a}c = ab + \bar{a}c \qquad (75.12)$$

$$ab + bc + \bar{a}c = ab + 1 \cdot bc + \bar{a}c \qquad \text{[by (75.4)]}$$

$$= ab + (a + \bar{a})bc + \bar{a}c \qquad \text{[by (75.6)]}$$

$$= ab + abc + \bar{a}bc + \bar{a}c \qquad \text{[by (75.7)]}$$

$$= ab \cdot 1 + abc + \bar{a}bc + \bar{a}c \cdot 1 \qquad \text{[by (75.4)]}$$

$$= ab(1 + c) + \bar{a}c(b + 1) \qquad \text{[by (75.7)]}$$

$$= ab \cdot 1 + \bar{a}c \cdot 1 \qquad \text{[by (75.5)]}$$

$$= ab + \bar{a}c \qquad \text{[by (75.4)]}$$

Identities (75.11) and (75.12) are actually theorems that have been proved by using the valid identities (75.1)–(75.10); (75.11) is sometimes, but improperly, called *absorption* because of its similarity with (75.8), and (75.12) is known as the *consensus* identity. These two identities are quite convenient because they are relatively easy to memorize and can themselves be applied to accomplish transformations of Boolean expressions. We could continue deriving identities of this kind to be included in our bag of valid rules; however, the burden of memorization will rapidly reach the point of diminishing return.

Table 75.1 is a summary of the manipulative rules of switching algebra.

Boolean Expressions: Normal and Canonical Forms

We saw earlier that every Boolean expression involving n distinct variables describes a switching function of those n variables. We shall now show a very important fact, namely, the converse of the above statement: *given any binary function of n binary variables we can construct a Boolean expression describing that function*. Since every Boolean expression corresponds to a combinational circuit consisting of single fan-out gates, we obtain the far-reaching result that every switching function is realizable by means of a combinational network.

An expression involves variables in *uncomplemented or complemented* forms: we call *literal* any

Table 75.1   Switching Algebra Summary

(P1) $XY = YX$	(S1) $X + Y = Y + X$	Commutativity
(P2) $X(YZ) = (XY)Z$	(S2) $X + (Y + Z) = (X + Y) + Z$	Associativity
(P3) $XX = X$	(S3) $X + X = X$	Idempotency
(P4) $X(X + Y) = X$	(S4) $X + XY = X$	Absorption
(P5) $X(Y + Z) = XY + XZ$	(S5) $X + YZ = (X + Y)(X + Z)$	Distributivity
(P6) $X\bar{X} = 0$	(S6) $X + \bar{X} = 1$	Complementarity
(C1) $\bar{\bar{X}} = X$		Involution
(P7) $\overline{XY} = \bar{X} + \bar{Y}$	(S7) $\overline{X+Y} = \overline{X}\overline{Y}$	De Morgan's
(P8) $X(\bar{X} + Y) = XY$	(S8) $X + \bar{X}Y = X + Y$	
(B1) $\bar{1} = 0$		
(P10) $X \cdot 0 = 0$	(S10) $X + 1 = 1$	
(P11) $X \cdot 1 = X$	(S11) $X + 0 = X$	
(P13) $(X + Y)(Y + Z)(\bar{X} + Z) = (X + Y)(\bar{X} + Z)$	(S13) $XY + YZ + \bar{X}Z = XY + \bar{X}Z$	Consensus

occurrence of a variable in either form. For example, the expression $[x_1 = x_2\,(\overline{x_3 + x_4\overline{x_1}})]\overline{x}_3 + \overline{x}_2 x_4$ has four variables and six literals.

An expression is in *normal sum-of-products* (SOP) *form* when it is the OR (sum) of ANDs (products) of literals. We shall now describe how an arbitrary Boolean expression E can be transformed into an equivalent expression in normal SOP form.

Reduction to Normal SOP Form

Let expression $E = [x_1 + x_2\,(\overline{x_3 + x_4\overline{x}_1})]x_3 + \overline{x}_2 x_4$ be given.

1. Place all complements directly on variables (by using De Morgan's laws). In our example,

$$E = (x_1 + x_2 \cdot \overline{x}_3 \cdot \overline{x_4\overline{x}_1})x_3 + \overline{\overline{x}}_2 + \overline{x}_4$$

$$= (x_1 + x_2 \cdot \overline{x}_3(\overline{x}_4 + \overline{\overline{x}}_1))x_3 + x_2 + \overline{x}_4$$

$$= [x_1 + x_2\overline{x}_3(\overline{x}_4 + x_1)]x_3 + x_2 + \overline{x}_4$$

2. Apply the distributive law. In our example,

$$E = (x_1 + x_2\overline{x}_3\overline{x}_4 + x_1 x_2\overline{x}_3)x_3 + x_2 + \overline{x}_4$$

$$= x_1 x_3 + x_2\overline{x}_3 x_3\overline{x}_4 + x_1 x_2\overline{x}_3 x_3 + x_2 + \overline{x}_4$$

3. Eliminate redundant terms (using idempotency and complementarity). In our example notice that, by (75.6), $x_3\overline{x}_3 = 0$ and that, by (75.5), all product terms containing a factor 0 are 0 themselves, whereby

$$E = x_1 x_3 + 0 + 0 + x_2 + \overline{x}_4$$

$$= x_1 x_3 + x_2 + \overline{x}_4$$

The latter expression is in normal SOP form and is equivalent to the given expression.

With reference to expressions on n variables, a special type of product term (AND term) is one that contains as a factor each variable, either uncomplemented or complemented: these terms are called *fundamental products* or *minterms*. For example, for $n = 4$, $\overline{x}_1\overline{x}_2\overline{x}_3 x_4$ and $\overline{x}_1 x_2\overline{x}_3 x_4$ are minterms but $\overline{x}_1\overline{x}_2 x_4$ is not. A normal SOP expression is said to be in *canonical* (SOP) *form* if its product terms are all minterms. We shall now describe how to transform a normal form expression into a canonical form expression.

Transformation from Normal Form to Canonical Form

Let the normal form expression $x_1 x_3 + x_2 + \overline{x}_3$ be given.

1. If a product term contains neither x_i nor $\overline{x}_i$, "multiply" it by $(x_i + \overline{x}_i)$. [Notice that this transforms the product term into an equivalent expression since $x_i + \overline{x}_i = 1$ by (75.6).] In our example $x_1 x_3$ does not contain a literal with index 2: x_2 does not contain literals with indices 1 and 3; $\overline{x}_3$ does not contain literals with indices 1 and 2. Thus,

$$x_1 x_3 + x_2 + \overline{x}_3 = x_1 x_3(x_2 + \overline{x}_2) + x_2(x_1 + \overline{x}_1)(x_3 + \overline{x}_3) + x_3(x_1 + \overline{x}_1)(x_2 + \overline{x}_2)$$

2. Apply the distributive law. In our example,

$$E = x_1 x_3\overline{x}_2 + x_1 x_3 x_2 + x_2\overline{x}_1\overline{x}_3 + x_2\overline{x}_1 x_3 + x_2 x_1\overline{x}_3 + x_2 x_1 x_3$$

$$+ \overline{x}_3\overline{x}_1\overline{x}_2 + \overline{x}_3\overline{x}_1 x_2 + \overline{x}_3 x_1\overline{x}_2 + \overline{x}_3 x_1 x_2$$

3. Eliminate repeated product term using idempotency. In our example, the following sets of terms are sets of identical terms: (2nd, 6th) (3rd, 8th) (5th, 10th). Thus, after eliminating the

repeated terms and rearranging the order of the indices as $(3,2,1)$, we obtain the canonical expression for $x_1 x_3 + x_2 + \bar{x}_3$:

$$E = x_3 \bar{x}_2 x_1 + x_3 x_2 x_1 + \bar{x}_3 x_2 \bar{x}_1 + x_3 x_2 \bar{x}_1 + \bar{x}_3 x_2 x_1 + \bar{x}_3 \bar{x}_2 \bar{x}_1 + \bar{x}_3 \bar{x}_2 x_1$$

We begin by introducing a useful notation for minterms. Consider, for example, the minterm $\bar{x}_4 x_3 x_2 \bar{x}_1$; we associate with this minterm an ordered binary 4-tuple $(b_4 b_3 b_2 b_1)$, where b_i corresponds either to x_i or $\bar{x}_i$, and $b_i = 1$ if x_i is uncomplemented and is 0 otherwise. In our example, with $\bar{x}_4 x_3 x_2 \bar{x}_1$ we associate 0110: this string is the binary equivalent of the integer 6, so that we shall denote $\bar{x}_4 x_3 x_2 \bar{x}_1$ by m_6. Referring to the previous example, the expression E in this new notation becomes $m_5 + m_7 + m_2 + m_6 + m_3 + m_0 + m_1$, or equivalently, OR $(m_0, m_1, m_2, m_3, m_5, m_6, m_7)$.

Suppose now that we have combinational networks F and G, which respectively compute switching functions $f(x_1, \ldots, x_n)$ and $g(x_1, \ldots, x_n)$, and that we connect the outputs of these networks to the inputs of gates or inverters. Clearly, if f and g are fed, say, to an AND gate, then the function u at the output of this gate will be 1 only when both f and g are 1, and similarly for the other cases. Obviously u is a function of the same set of variables $\{x_1, \ldots, x_n\}$ as f and g, so we obtain the following simple rules for its truth table:

The truth table of $\bar{f}$ is the entry-by-entry (component-wise) complement of the truth table of f; the truth table of $f \cdot g$ [or $(f + g)$] is the component-wise AND (or OR) of the truth tables of f and g.

Given a minterm m of $x_1, x_2, \ldots, x_n$ (the minterm itself obviously describes a function of these variables), we recognize that $m = 1$ exactly when the following conditions hold: if x_i appears in m in uncomplemented form, then $x_i = 1$; if x_i appears in m in complemented form, then $\bar{x}_i = 1$, that is, $x_i = 0$. Thus $m = 1$ only when each variable attains a specific value, that is, $m = 1$ for a unique combination of the variables or, equivalently, *the truth table of a minterm has exactly one "1."* (Incidentally, this explains the denomination *minterm*: a nondegenerate function with the minimum number of 1's in its truth table.)

Example 3. For $n = 4$, $\bar{x}_4 \bar{x}_3 \bar{x}_2 x_1 = m_1$ is 1 when and only when $\bar{x}_4 = 1$, $\bar{x}_3 = 1$, $\bar{x}_2 = 1$, $x_1 = 1$, that is, when $(x_4 x_3 x_2 x_1) = (0001)$.

Now, let f be a switching function of arguments $x_1, \ldots, x_n$, given by means of its truth table (see Fig. 75.11 where $n = 3$). We may view the truth table of f as the OR of as many distinct truth tables as it has 1's, each of the latter truth tables having exactly a single 1 [Fig. 75.11(a)]. However, each such table is the table of a minterm! Moreover, each minterm is a product of literals, which for $i = 1, 2, \ldots, n$ contains either x_i or $\bar{x}_i$, depending upon whether in the combination corresponding to the single 1 in the table the x_i entry is 1 or 0. In Fig. 75.11, $f_1 = 1$ in correspondence to $(x_3 x_2 x_1) = (011)$, whence $f_1 = \bar{x}_3 x_2 x_1$. Similarly, we obtain $f_2 = x_3 \bar{x}_2 \bar{x}_1$ and $f_3 = x_3 x_2 \bar{x}_1$.

x_3	x_2	x_1	f	f_1	f_2	f_3
0	0	0	0	0	0	0
0	0	1	0	0	0	0
0	1	0	0	0	0	0
0	1	1	1	1	0	0
1	0	0	1	0	1	0
1	0	1	0	0	0	0
1	1	0	1	0	0	1
1	1	1	0	0	0	0

(a)

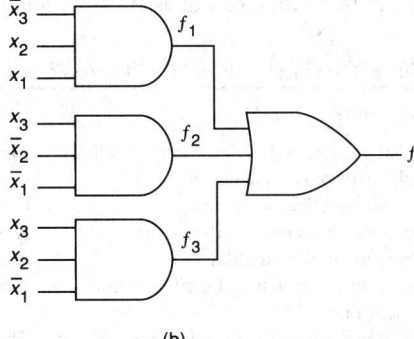

(b)

FIGURE 75.11 A switching function and its corresponding AND-to-OR network.

In conclusion, since $f = f_1 + f_2 + f_3$, we have

$$f = \bar{x}_3 x_2 x_1 + x_3 \bar{x}_2 \bar{x}_1 + x_3 x_2 \bar{x}_1$$

Notice that this is a most remarkable finding: given a function f by means of its truth table (i.e., as a binary function of binary variables), we have obtained an expression (actually, a canonical expression) describing that function!

Once we have an expression for the given function, we shall design the corresponding combinational network. Before proceeding, however, we recall that in the subsection "Switching Functions of One and Two Variables" we have introduced AND gates and OR gates as two-input–one-output devices; since we have proved [identity (75.9)] that the AND and OR operations are associative, we may think of using in our networks devices that realize the AND (or OR) *of more than two* inputs; this is indeed technically possible, although, for physical reasons, the number of inputs may not be too large. Therefore, we see that by using these newly introduced gates, we can construct the network of Fig. 75.11(b), which computes the given function. (This network is a collection of AND gates feeding a single OR gate and is therefore called an AND-to-OR network.) Notice that we have achieved the objective set forth at the end of the subsection "Networks and Expressions" and summarized below:

> Given a switching function f by means of its truth table (i.e., as a binary function of binary variables), we can construct a switching network that computes it.

In Table 75.2, left side, we summarize the important notions concerning SOP canonical expressions. A discussion analogous to the one just completed can be carried out with reference to *normal product-of-sums* (POS) expressions, that is, an AND of ORs of literals. Indeed, all we need in the preceding discussion is a set of substitutions as dictated by the principle of duality

$$\text{AND} \leftrightarrow \text{OR}$$

$$\text{Product} \leftrightarrow \text{Sum}$$

$$\text{Minterm} \leftrightarrow \text{Maxterm (see below)}$$

$$0 \leftrightarrow 1$$

The conclusions are summarized in Table 75.2, right side. Notice the perfect duality of corresponding statements in this table. The only novel term in this table is *maxterm,* the dual of minterm: the reason for the denomination is that a maxterm describes a nondegenerate function with the maximum number of 1's in its truth table. A maxterm is usually denoted by the symbol M_j, specifically $M_j = \bar{m}_j$, that is, $M_j = 0$ if and only if $m_j = 1$, and vice versa. For example, for variables x_3, x_2, x_1, $M_5 = \bar{m}_5 = \overline{x_3 \bar{x}_2 x_1} = (x_3 + x_2 + \bar{x}_1)$, that is, M_5 is the maxterm which is 0 exactly for $(x_3 x_2 x_1) = (101)$ and is 1 otherwise.

In conclusion, a **Boolean function** can be specified either as an *OR of minterms* (corresponding to the 1's in the truth table) or as an *AND of maxterms* (corresponding to the 0's in the truth table).

Table 75.2 Duality of Canonical SOP and POS Expressions

Canonical SOP Expressions	Canonical POS Expressions
Minterm—a *product* of literals that has as a "factor" each of the *n* variables either true or complemented	*Maxterm*—a *sum* of literals that has as an "addend" each of the *n* variables either true or complemented
Canonical SOP expressions—*OR* of *minterms*	Canonical *POS* expression—*AND* of *maxterms*
A *minterm* is a canonical *SOP* expression that is 1 for exactly one combination of the variables	A *maxterm* is a canonical *POS* expression that is 0 for exactly one combination of the variables
A *minterm* corresponds to one switching function whose truth table has exactly one "1"	A *maxterm* corresponds to one switching function whose truth table has exactly one "0"
There exists a one-to-one correspondence between switching functions and canonical *SOP* expressions	There exists a one-to-one correspondence between switching functions and canonical *POS* expressions

Example 4

Decimal equivalent	x_3	x_2	x_1	f
0	0	0	0	0
1	0	0	1	0
2	0	1	0	0
3	0	1	1	1
4	1	0	0	1
5	1	0	1	0
6	1	1	0	1
7	1	1	1	0

$$f = \text{OR}\ (m_3, m_4, m_6),\ \text{SOP}$$

$$f = \text{AND}\ (M_0, M_1, M_2, M_5, M_7),\ \text{POS}$$

Consider now a Boolean function $f(x_1, x_2, \ldots, x_n)$ expressed in canonical SOP form. Each minterm of f contains either $\bar{x}_n$ or x_n; therefore, we associate into two separate expressions, F_0 and F_1, the minterms of f depending upon whether they contain $\bar{x}_n$ or x_n, respectively, that is,

$$f = F_0 + F_1$$

Now from all terms of F_0 we can factor out $\bar{x}_n$, i.e., F_0 can be written as the AND of $\bar{x}_n$ and an expression f_0, which consists exactly of minterms over the variables $x_1, x_2, \ldots, x_{n-1}$, that is, $F_0 = \bar{x}_n f_0 (x_1, x_2, \ldots, x_{n-1})$. Similarly, we can express F_1 as $F_1 = x_n f_1 (x_1, x_2, \ldots, x_{n-1})$. It follows that f can be expressed as

$$f(x_1, \ldots, x_{n-1}, x_n) = \bar{x}_n f_0(x_1, \ldots, x_{n-1}) + x_n f_1(x_1, \ldots, x_{n-1}) \qquad (75.13)$$

In the above relation, we now set $x_n = 0$ and obtain

$$f(x_1, \ldots, x_{n-1}, 0) = 1 \cdot f_0(x_1, \ldots, x_{n-1}) + 0 \cdot f_1(x_1, \ldots, x_{n-1})$$

that is, $f_0(x_1, \ldots, x_{n-1}) = f(x_1, \ldots, x_{n-1}, 0)$. Similarly, if we set $x_n = 1$ in (75.13) we have

$$f_1(x_1, \ldots, x_{n-1}, 1) = 0 \cdot f_0(x_1, \ldots, x_{n-1}) + 1 \cdot f_1(x_1, \ldots, x_{n-1})$$

that is, $f_1(x_1, \ldots, x_{n-1}) = f_1(x_1, \ldots, x_{n-1}, 1)$. This result is called the fundamental theorem of Boolean algebra and can be stated as follows.

Fundamental Theorem of Boolean Algebra. Every function $f(x_1, \ldots, x_n)$ of $x_1, \ldots, x_n$, for any x_i can be expressed as

$$f = \bar{x}_i f_0 + x_i f_1$$

where $f_0 = f(x_1, \ldots, x_{i-1}, 0, x_{i+1}, \ldots, x_n)$ and $f_1 = f(x_1, \ldots, x_{i-1}, 1, x_{i+1}, \ldots, x_n)$ are both functions of the $(n-1)$ variables $x_1, \ldots, x_{i-1}, x_{i+1}, \ldots, x_n$.

Other Important Boolean Connectives

Although the operators AND, OR, and NOT are perfectly adequate for the realization of any combinational network, there are other connectives that are quite important and are now introduced.

The NAND and NOR Connectives

The first of these connectives, called NAND (AND followed by NOT), realizes the function $\overline{x \cdot y}$ of two variables x and y; its circuit symbol is given, for two inputs, in Fig. 75.12(a). (Note that, as already has been done for the AND and OR connectives, the function NAND can be generalized to any number of variables, as $\overline{x \cdot y \ldots w}$.) The connective NAND is interesting because alone it can be

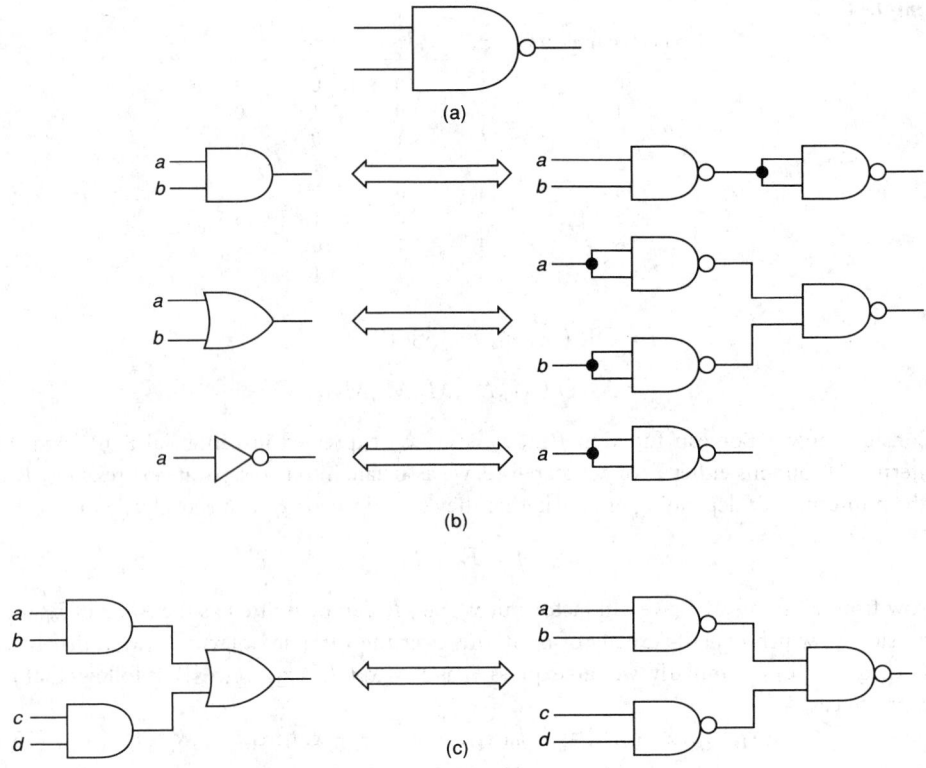

FIGURE 75.12   (a) Circuit symbol of NAND gate. (b) and (c) NAND gate realization of AND, OR, and NOT gates and SOP expression.

used to realize any combinational network. (We refer to this property by saying that NAND is *logically complete.*) Indeed, since we know that AND, OR, and NOT are adequate for realizing combinational networks, all we need to show is that each of these three connectives can, in turn, be realized by an expression involving only NAND. This is readily shown below:

(rules C1 and P3, Table 75.1)

$$a \cdot b = \overline{\overline{a \cdot b}} = \overline{\overline{a \cdot b} \cdot \overline{a \cdot b}} = \text{NAND}[\text{NAND}(a,b), \text{NAND}(a,b)]$$

(rules P7 and P3, Table 75.1)

$$a + b = \overline{\overline{a} \cdot \overline{b}} = \overline{\overline{aa} \cdot \overline{bb}} = \text{NAND}[\text{NAND}(a,a), \text{NAND}(b,b)]$$

(rule P3, Table 75.1)

$$\overline{a} = \overline{aa} = \text{NAND}(a,a)$$

These transformations are illustrated in Fig. 75.12(b). Thus, given a network consisting of AND, OR, and NOT gates, by using the above rules, one can transform it into an equivalent one consisting of NAND gates alone. Besides this rather cumbersome transformation, there is a more direct and useful correspondence between NAND expressions and SOP expressions, as shown below:

$$ab + cd = \overline{\overline{ab + cd}} = \overline{\overline{ab} \cdot \overline{cd}} = \text{NAND}[\text{NAND}(a,b), \text{NAND}(c,d)]$$

So we see that any SOP expression can be realized by a network consisting of NAND gates alone [see Fig. 75.12(c)] by simply replacing with NAND gates both the AND gates and the OR gate in the standard AND-to-OR realization of the given expression.

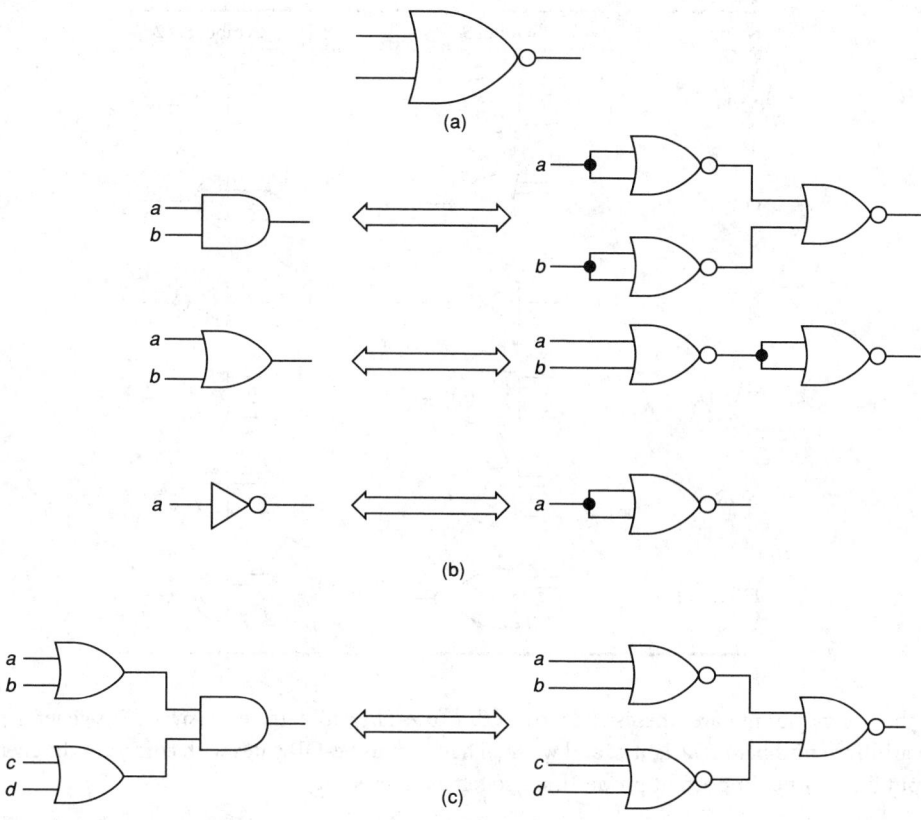

FIGURE 75.13 (a) Circuit symbol of NOR gate. (b) and (c) NOR gate realizations of AND, OR, and NOT gates and POS expression.

As we may expect from duality, there is another connective, NOR, which enjoys analogous properties. This connective NOR (OR followed by NOT) realizes the function $\overline{x + y}$ of x and y, and its symbol is given, for two inputs, in Fig. 75.13(a). Transformations analogous to those obtained above can be easily derived and the results are shown in Fig. 75.13(b) and 75.13(c). Notice that a two-level NAND network corresponds to an SOP expression, and a two-level NOR network corresponds to a POS expression. These properties make NOR and NAND gates very attractive and popular in digital design, since entire systems can be realized by using just one type of component.

The XOR Connective

Finally, we introduce the connective exclusive-OR (frequently abbreviated as XOR), which realizes the function $(x\overline{y} + \overline{x}y)$ of two variables x and y. The symbol used for this connective is $\oplus$, while the circuit symbol is given in Fig. 75.14. The reason for the name exclusive-OR is that $(x \oplus y)$ is equal to 1 if and only if either x or y, *but not both,* is equal to 1. (By contrast, the ordinary OR is correctly called *inclusive* OR, although the adjective inclusive is normally omitted.)

FIGURE 75.14 Symbol for the exclusive-OR gate.

The exclusive-OR has several interesting properties, whose proof is left as an exercise. First, exclusive-OR is associative

$$(x \oplus y) \oplus z = x \oplus (y \oplus z)$$

Table 75.3

Name	Symbol Set 1	Symbol Set 2
AND		
OR		
NOT		
NAND		
NOR		
EXCLUSIVE OR		

so that we may omit parentheses and write $x \oplus y \oplus z$. Therefore, the exclusive-OR is generalized to an arbitrary number of variables, and we shall have exclusive-OR gates with correspondingly many input lines. Other important properties, also left as an exercise, are

$$x(y \oplus z) = xy \oplus xz$$
$$x \oplus 1 = \bar{x}$$

It is appropriate to introduce at this point an alternative set of standard symbols for the logic gates discussed in this chapter. They are displayed in Table 75.3 vis-à-vis their by now familiar counterparts (symbol set 1) and deserve no additional comments.

Notes and References

Boolean algebra, which—as we saw—provides the formalism for the description of (binary) digital networks, was developed in the last century, originating with the English mathematician George Boole, who in 1854 published his fundamental work, *An Investigation of the Laws of Thought*. Boole's goal was essentially the development of a formalism to compute the truth or falsehood (i.e., the *truth value*) of complex compound statements from the truth values of their component statements. The discipline developed later into a more complex body of knowledge, known as symbolic logic.

Apparently, early in this century, more than one scientist perceived the applicability of Boolean algebra to the design of telephone circuits [Ehrenfest, 1910]. It was only in the thirties that the potential was fully realized, when C.E. Shannon [1938] published his paper "A Symbolic Analysis of Relay and Switching Circuits," which became the foundation of switching theory and logical design. Because of the context in which it was originally used (telephone networks, also called switching networks), the name *switching algebra* has become standard for the algebra of functions of two-valued variables. Although initially the interests of researchers focused on relay networks (also called contact networks), as mechanical devices were gradually replaced by electronic devices the techniques were tailored to gate networks of the type described above. The term *gate* was already in use in the forties to denote the logical elements discussed earlier.

There are very many good references on Boolean algebra and we may quote only a selected few of them. Suffice it to mention the texts by Hill and Peterson [1974], Kohavi [1978], and Hohn [1966]. These books give a sufficiently rigorous formulation of the subject, tailored to the analysis and the design of combinational networks. In addition, like most of the earlier books, Hohn's and Kohavi's texts also contain a discussion of the Boolean techniques used in connection with relay circuits. (Some of the more recent works completely omit this topic, which has been but totally overshadowed by the impressive development of electronic networks.) The reader interested in studying the relation of switching algebra to Boolean algebras in general is referred to Preparata and Yeh [1973] for an elementary introduction.

Defining Terms

Boolean algebra: The algebra of logical values enabling the logical designer to obtain expressions for digital circuits.

Boolean expressions: Expressions of logical variables constructed using the connectives *and, or,* and *not.*

Boolean functions: Common designations of binary functions of binary variables.

Combinational lock: Interconnections of memory-free digital elements.

Switching theory: The theory of digital circuits viewed as interconnections of elements whose output can switch between the logical values of 0 and 1.

References

G. Boole, *An Investigation of the Laws of Thought,* New York: Dover Publication, 1954.

F. J. Hill and G. R. Peterson, *Introduction to Switching Theory and Logical Design,* New York: Wiley, 1974.

F. E. Hohn, *Applied Boolean Algebra,* New York: Macmillan, 1966.

Z. Kohavi, *Switching and Finite Automata Theory,* New York: McGraw-Hill, 1978.

F. P. Preparata and R. T. Yeh, *Introduction to Discrete Structures,* Reading, Mass.: Addison-Wesley, 1973.

C. E. Shannon, "A symbolic analysis of relay and switching circuits," *Trans. AIEE,* vol. 57, pp. 713–723, 1938.

75.2 Logic Circuits

Richard S. Sandige

Section 75.2 deals with two-state (high or low, 1 or 0, or true or false) logic circuits. Two-state logic circuits can be broken down into two major types of circuits: **combinational logic circuits** and **sequential logic circuits.** By definition, the external output signals of combinational logic circuits are totally dependent on the external input signals applied to the circuit. In contrast, the present state output signals of sequential logic circuits are dependent on all or part of the present state output signals of the circuit fed back as input signals to the circuit as well as any external input signals if they should exist. Sequential logic circuits can be subdivided into **synchronous** or **clock-mode** circuits and asynchronous circuits. Asynchronous circuits can be further divided into fundamental-mode circuits and pulse-mode circuits. Fig. 75.15 is the graphic classification of logic circuits.

Combinational Logic Circuits

The block diagram in Fig. 75.16 illustrates the model for combinational logic circuits. The logic elements inside the block entitled *combinational logic circuit* can be any configuration of two-state logic elements such that the output signals are totally dependent on the input signals to the circuit as indicated by the functional relationships in the figure.

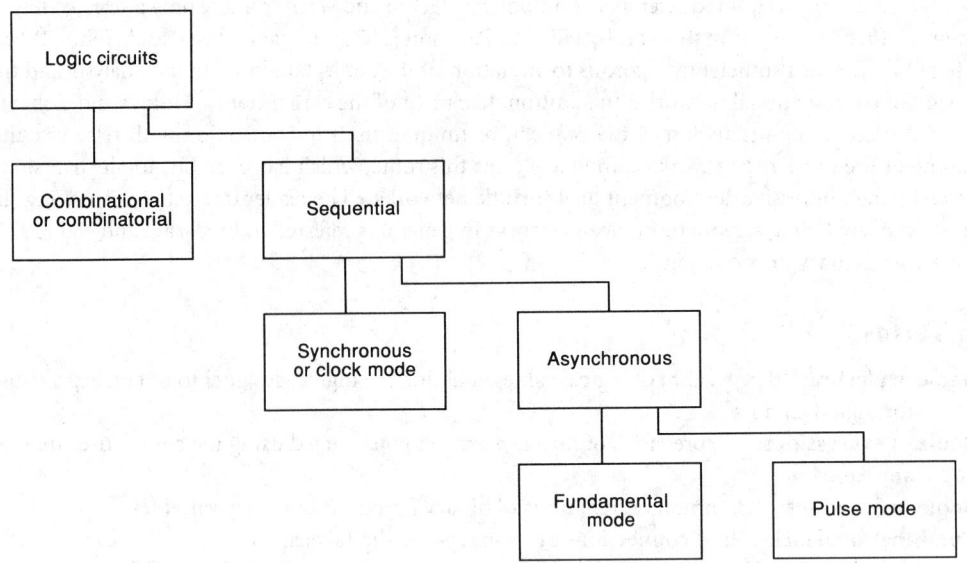

FIGURE 75.15 Graphic classification of logic circuits. (*Source:* R. S. Sandige, *Modern Digital Design,* New York: McGraw-Hill, 1990, p. 440. With permission.)

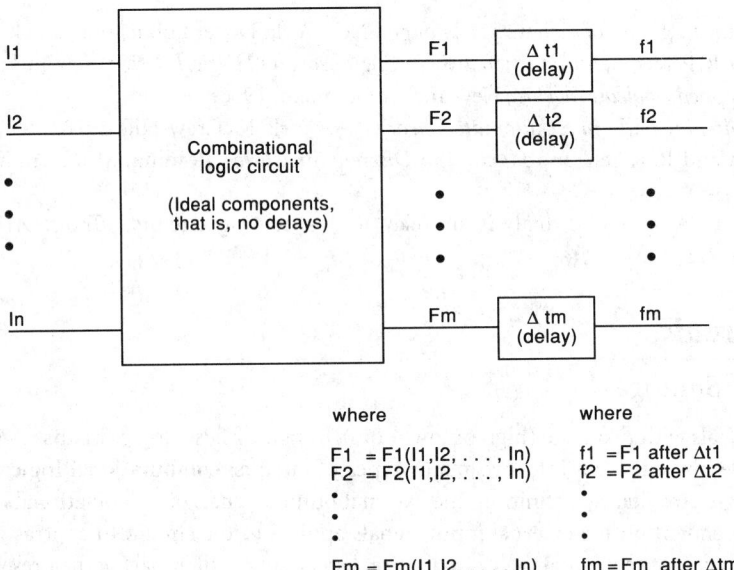

FIGURE 75.16 Block diagram model for combinational logic circuits. (*Source:* R. S. Sandige, *Modern Digital Design,* New York: McGraw-Hill, 1990, p. 440. With permission.)

The logic elements can be anything from relays with their slow on and off switching action to modern off-the-shelf integrated circuit (IC) transistor switches with their extremely fast switching action. Modern ICs exist in various technologies and circuit configurations such as transistor-transistor logic (TTL), complementary metal-oxide semiconductor (CMOS), emitter-coupled logic (ECL), and integrated injection logic (I^2L), just to name a few.

The delays in the outputs of the model in Fig. 75.16 represent lumped delays, that is, worst-case delays through the longest delay path from the inputs to each respective output of the combinational logic circuit. The lumped delays provide an approximate measure of circuit speed or settling time (the time it takes an output signal to become stable after the input signals have become stable).

Figure 75.17 illustrates the gate-level method (random logic method) of implementing a binary to seven-segment hexadecimal character generator suitable for driving a seven-segment common cathode LED display like the one in Fig. 75.18.

The propagation delays of logic circuits are seldom shown on logic circuit diagrams; however, these delays are inherent in each logic element and must be considered in systems designs. This gate-level combinational logic circuit converts the binary input code 0000 though 1111 represented

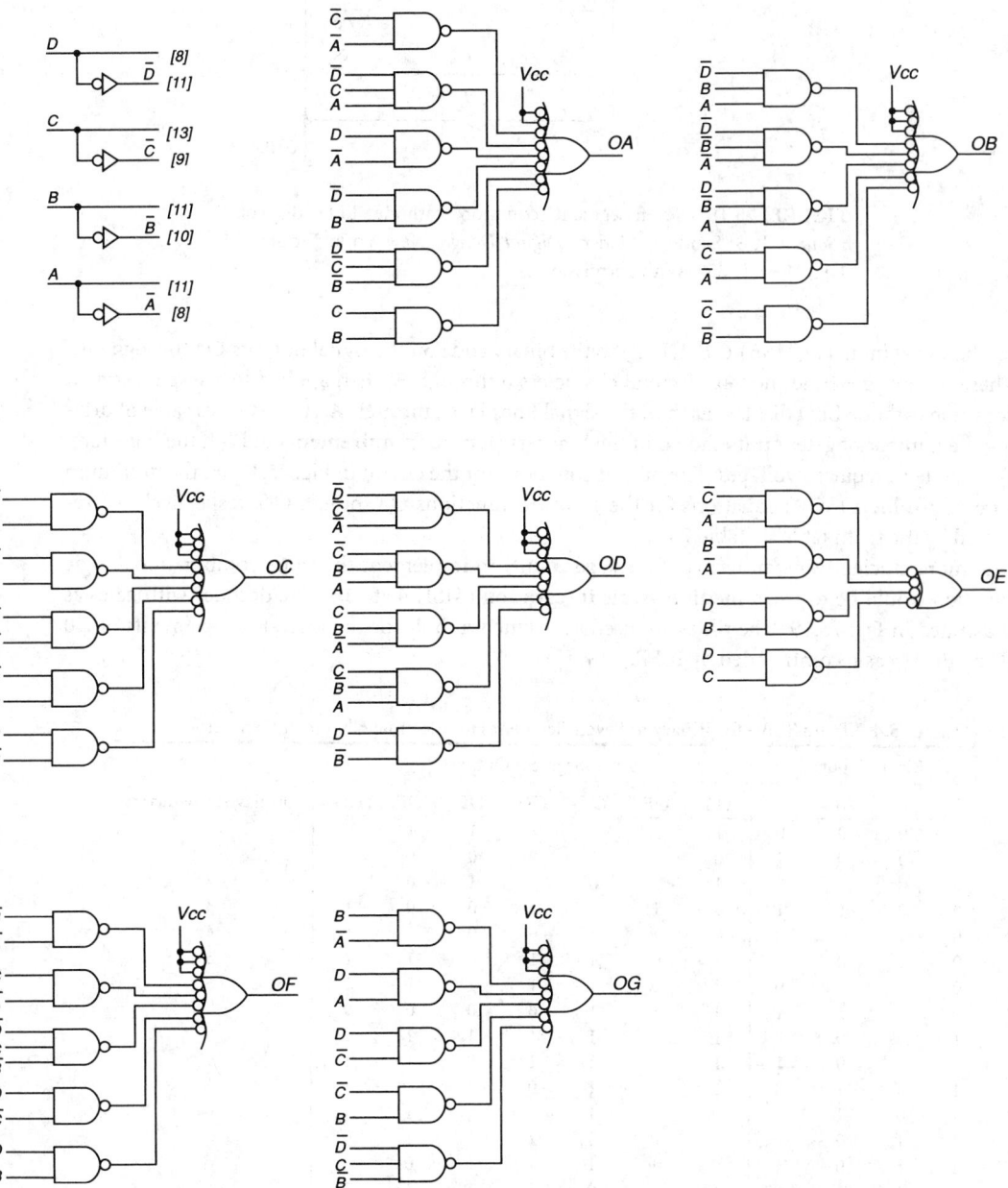

FIGURE 75.17 Gate-level logic circuit for binary to seven-segment hexadecimal character generator. (*Source:* R. S. Sandige, *Modern Digital Design*, New York: McGraw-Hill, 1990, pp. 258–259. With permission.)

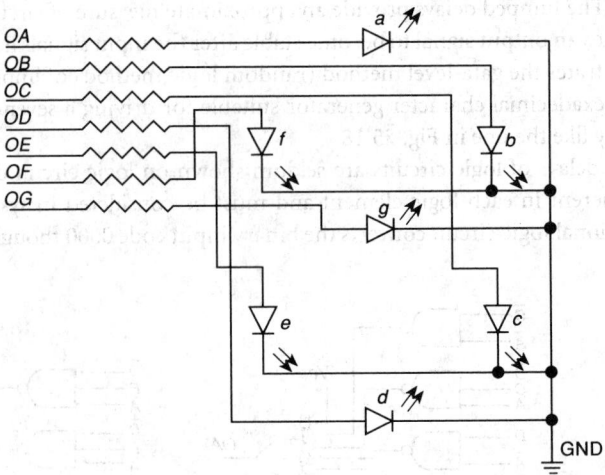

FIGURE 75.18 Seven-segment common cathode LED display. (*Source:* R. S. Sandige, *Modern Digital Design,* New York: McGraw-Hill, 1990, p. 255. With permission.)

on the signal inputs D(MSB) C B A(LSB) to the binary code on the signal outputs OA through OG. These outputs generate the hexadecimal characters 0 through F when applied to a seven-segment common cathode LED display. Each of the signal lines D, $\bar{D}$, through A, $\bar{A}$ must be capable of driving the number of gate inputs shown in the brackets (**fan-out requirement**) to both the high-level and low-level required voltages. The output equations for the circuit in Fig. 75.17 are the minimum **sum of products** (SOP) equations for the 1's of the functions OA though OG, respectively, represented by the truth table in Table 75.4.

A more efficient way (in terms of package count) to implement the same combinational logic function would be to use a **medium-scale integration** (MSI) 4- to 16-1ine decoder with gates as illustrated in Fig. 75.19. The tildes are used as in-line symbols for the logical complements of D0 through D15 as recommended by IEEE.

Table 75.4 Truth Table for Binary to Seven-Segment Hexadecimal Character Generator

Binary Inputs				Seven-Segment Outputs							Displayed Characters
D	C	B	A	OA	OB	OC	OD	OE	OF	OG	
0	0	0	0	1	1	1	1	1	1	0	0
0	0	0	1	0	1	1	0	0	0	0	1
0	0	1	0	1	1	0	1	1	0	1	2
0	0	1	1	1	1	1	1	0	0	1	3
0	1	0	0	0	1	1	0	0	1	1	4
0	1	0	1	1	0	1	1	0	1	1	5
0	1	1	0	1	0	1	1	1	1	1	6
0	1	1	1	1	1	1	0	0	0	0	7
1	0	0	0	1	1	1	1	1	1	1	8
1	0	0	1	1	1	1	1	0	1	1	9
1	0	1	0	1	1	1	0	1	1	1	A
1	0	1	1	0	0	1	1	1	1	1	b
1	1	0	0	1	0	0	1	1	1	0	C
1	1	0	1	0	1	1	1	1	0	1	d
1	1	1	0	1	0	0	1	1	1	1	E
1	1	1	1	1	0	0	0	1	1	1	F

Source: R. S. Sandige, *Modern Digital Design,* New York: McGraw-Hill, 1990, p. 252. With permission.

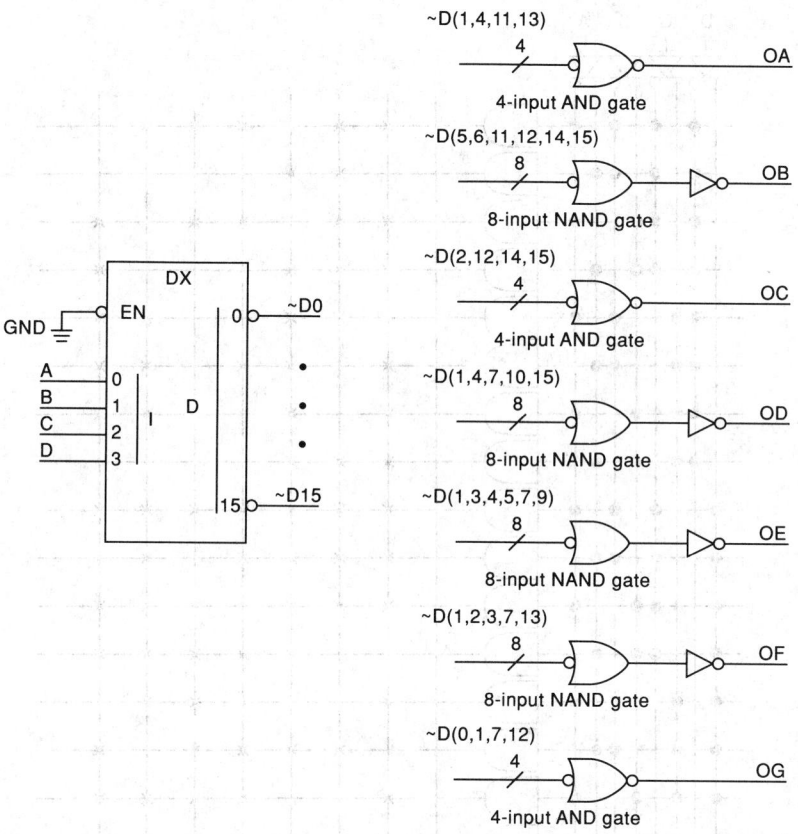

FIGURE 75.19 Decoder logic circuit for binary to seven-segment hexadecimal character generator. (*Source:* R. S. Sandige, *Modern Digital Design,* New York: McGraw-Hill, 1990, p. 359. With permission.)

The decoder circuit in Fig. 75.19 requires only 8 IC packages compared to the gate-level circuit in Fig. 75.17 which requires 18 IC packages. Functionally, both circuits perform the same. The output equations for the circuit in Fig. 75.19 are the canonical or standard SOP equations for the 0's of the functions *OA* though *OG*, respectively, represented by the truth table (Table 75.4). The gates shown in Fig. 75.19 with more than four inputs are eight-input NAND gates with each unused input tied to V_{CC} via a pull-up resistor (not shown on the logic diagram).

An even more efficient way to implement the same combinational logic function would be to utilize part of a simple programmable read only memory (PROM) circuit such as the 27S19 fuse programmable PROM in Fig. 75.20. An equivalent architectural gate structure for a portion of the PROM is shown in Fig. 75.21. The X's in Fig. 75.21 represent fuses that are left intact after programming the device. The code for programming the PROM or generating the truth table of the function

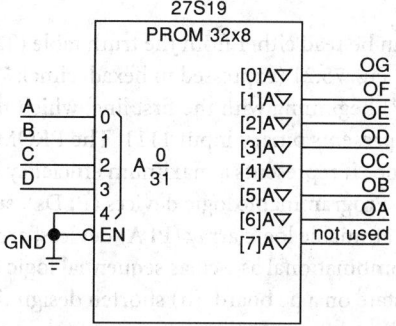

FIGURE 75.20 PROM implementation for binary to seven-segment hexadecimal character generator. (*Source:* R. S. Sandige, *Modern Digital Design,* New York: McGraw-Hill, 1990, p. 382. With permission.)

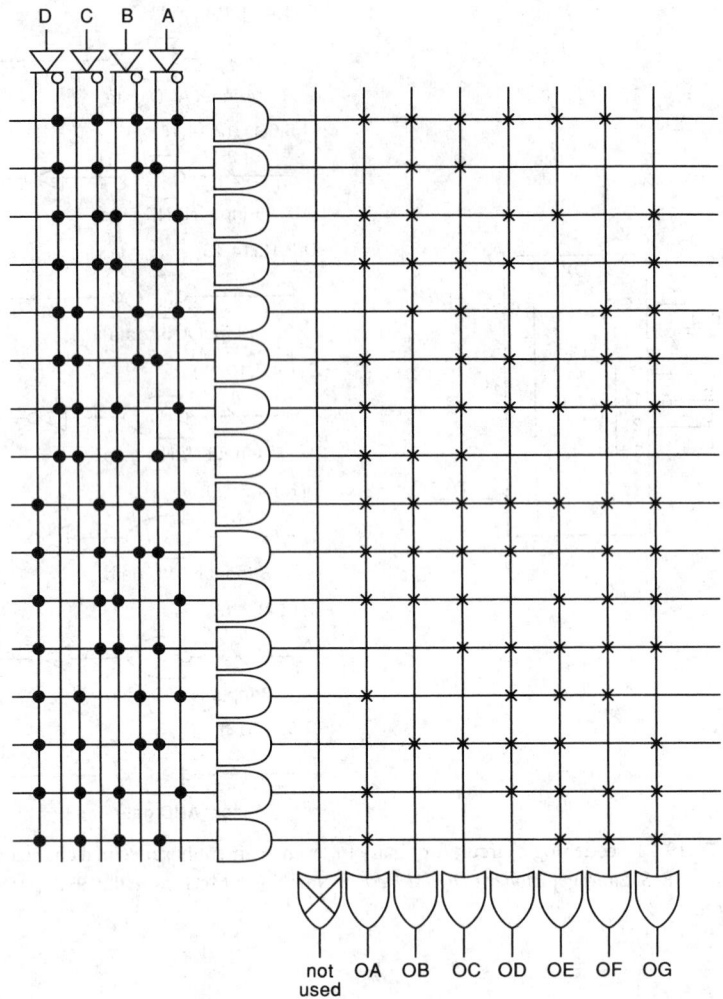

FIGURE 75.21 PROM logic circuit. (*Source:* R. S. Sandige, *Modern Digital Design,* New York: McGraw-Hill, 1990, p. 381. With permission.)

can be read either from the truth table (Table 75.4) or directly from each line of the circuit diagram in Fig. 75.21 (expressed in hexadecimal: 7E, 30, 6D, 79, 33, 5B, 5F, 70, 7F, 7B, 77, 1F, 4E, 3D, 4F, and 47) beginning with the first line, which represents binary input 0000, down to the last line, which represents binary input 1111. The PROM solution for the combinational logic circuit is optimum since it represents a maximum efficiency by requiring only a single IC package.

Programmable logic devices (PLDs), such as PROM, programmable array logic (PAL), and programmable logic array (PLA) devices, are fast becoming the preferred devices when implementing combinational as well as sequential logic circuits. This is true because these devices (a) use less real estate on a pc board, (b) shorten design time, (c) allow design changes to be made more easily, and (d) improve reliability because of fewer connections.

Figure 75.22 shows a PAL16L8 implementation for the binary to seven-segment hexadecimal character generator that also requires just a single IC package. The fuse map for this design was obtained using the software program PLDesigner. Karnaugh maps are handy tools that allow a designer to easily obtain minimum SOP equations for either the 1's or 0's of Boolean functions of

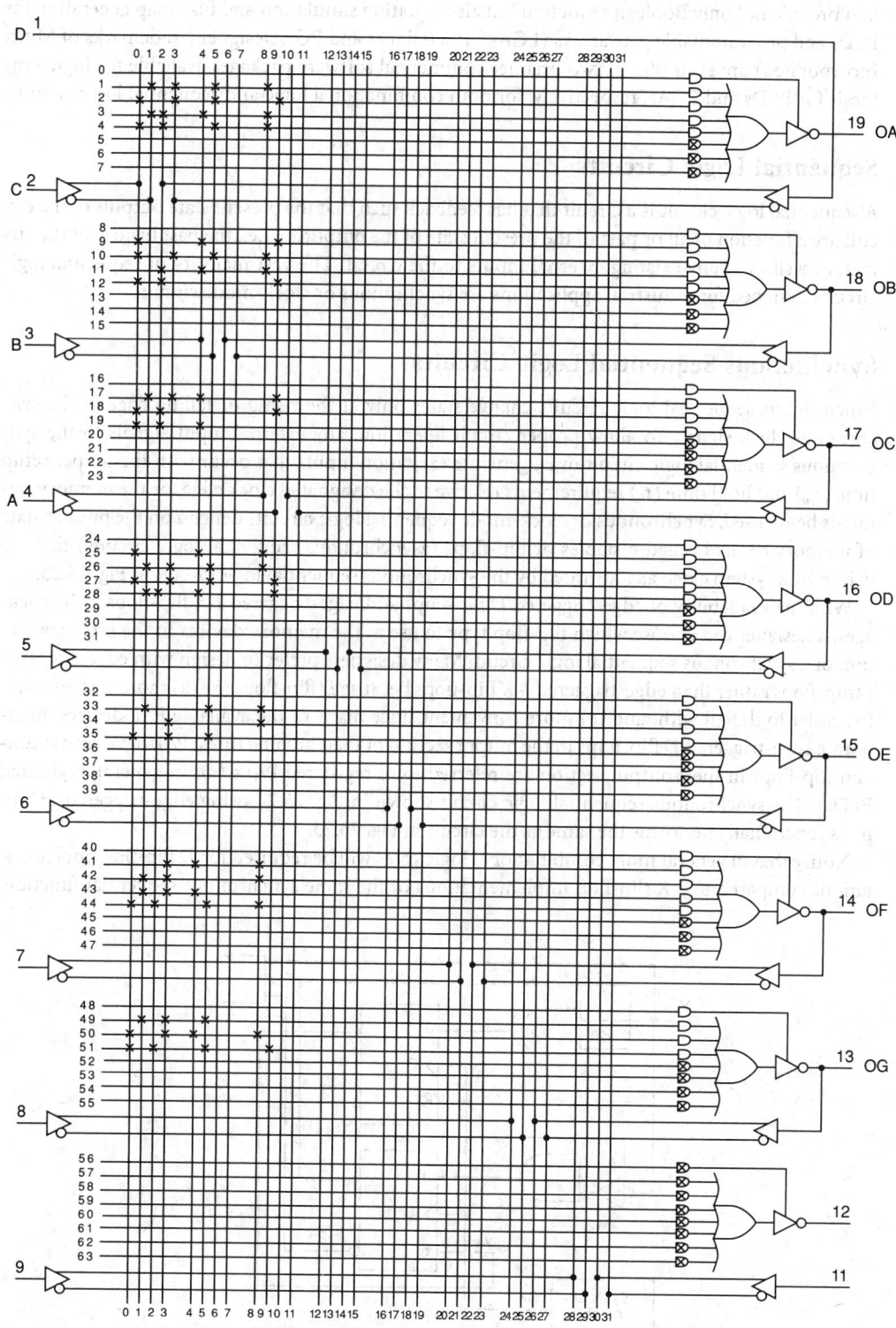

FIGURE 75.22 PAL16L8 implementation for the binary to seven-segment hexadecimal character generator. (*Source: PAL Device Data Book*, Advanced Micro Devices, Sunnyvale, Calif., 1988, p. 5–46.)

up to four or five variables; however, there are a host of commercially available software programs that provide not only Boolean reduction but also equation simulation and fuse map generation for PLDs and programmable gate arrays (PGAs). PLDesigner and PGADesigner (trademarks of Minc, Incorporated) are examples of two premier commercial software packages available for logic synthesis for PLDs and PGAs, respectively, for both combinational logic and sequential logic circuits.

Sequential Logic Circuits

A sequential logic circuit is a circuit that has feedback such that the present state outputs of the circuit are a function of all or part of the present state of the output(s), i.e., the past history of the circuit, as well as of any existing external inputs to the circuit. The vast majority of sequential logic circuits designed for industrial applications are synchronous or clock-mode circuits.

Synchronous Sequential Logic Circuits

Synchronous sequential logic circuits change states only at the rising or falling edge of the synchronous clock signal. To allow proper circuit operation, any external input signals to the synchronous sequential logic circuit must generate excitation inputs that occur with the proper setup time (t_{su}) and hold time (t_h) requirements relative to the designated clock edge for the memory elements being used. Synchronous or clock-mode sequential logic circuits depend on the present state of memory devices called bistables or flip-flops (asynchronous sequential logic circuits) that are driven by a system clock as illustrated by the synchronous sequential logic circuit in Fig. 75.23.

With the availability of edge-triggered D flip-flops and edge-triggered J-K flip-flops in IC packages, a designer can choose which flip-flop type to use as the memory devices in the memory section of a synchronous sequential logic circuit. Many designers prefer to design with edge-triggered D flip-flops rather than edge-triggered J-K flip-flops because D flip-flops are (a) more cost efficient, (b) easier to design with, and (c) more convenient since many of the available PAL devices incorporate edge-triggered D flip-flops in the output section of their architectures. PAL devices that contain flip-flops in their output section are referred to as registered PALs (or, in general, registered PLDs). The synchronous sequential logic circuit shown in Fig. 75.24 using edge-triggered D flip-flops functionally performs the same as the circuit in Fig. 75.23.

Notice that in general more combinational logic gates will be required for D flip-flop implementations compared to J-K flip-flop implementations of the same synchronous sequential function.

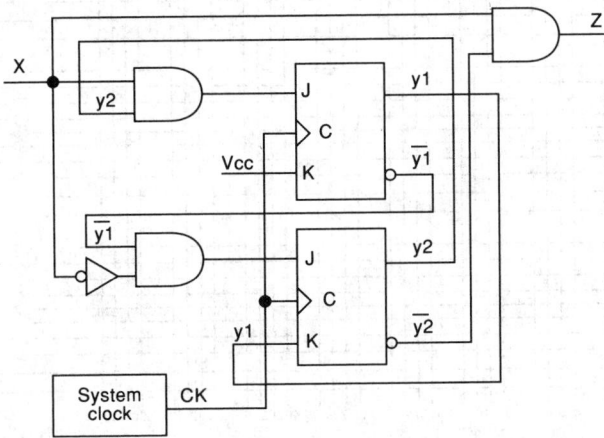

FIGURE 75.23   Synchronous sequential logic circuit using positive edge-triggered J-K flip-flops.

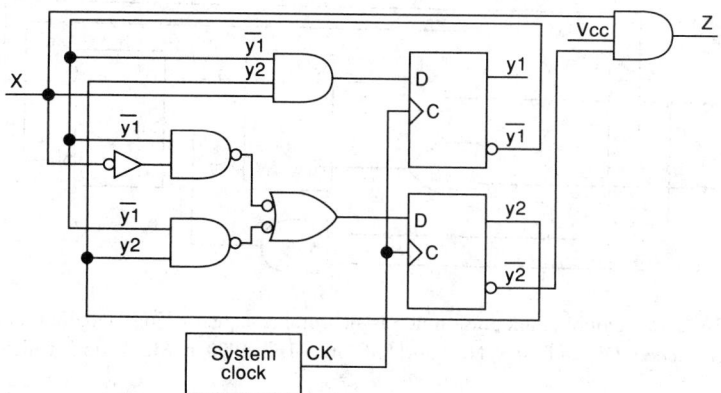

FIGURE 75.24 Synchronous sequential logic circuit using positive edge-triggered *D* flip-flops.

Using a registered PAL such as a PAL16RP4A would only require one IC package to implement the circuit in Fig. 75.24. The PAL16RP4A has four edge-triggered *D* flip-flops in its output section, of which only two are required for this design.

Generally speaking, synchronous sequential logic circuits can be designed much more easily (considering design time as the criteria) than fundamental-mode asynchronous sequential logic circuits. With a system clock and edge-triggered flip-flops, a designer does not have to worry about **hazards** or **glitches** (momentary error conditions that occur at the outputs of combinational logic circuits), since outputs are allowed to become stable before the next clock edge occurs. Thus, sequential logic circuit designs allow the use of combinational hazardous circuits as well as the use of arbitrary state assignments, provided the resulting combinational logic gate count or package count is acceptable.

Asynchronous Sequential Logic Circuits

Asynchronous sequential logic circuits change states any time a level change occurs at one of their inputs (fundamental mode) or any time a pulse occurs at one of their inputs (pulse mode). Latches and edge-triggered flip-flops are asynchronous sequential logic circuits and must be designed with care by utilizing hazard-free combinational logic circuits and race-free or critical **race-free state assignments.** Both hazards and race conditions interfere with the proper operation of asynchronous logic circuits. The gated *D* latch circuit illustrated in Fig. 75.25 is an example of a fundamental-mode asynchronous sequential logic circuit that is used extensively in microprocessor systems for the temporary storage of data.

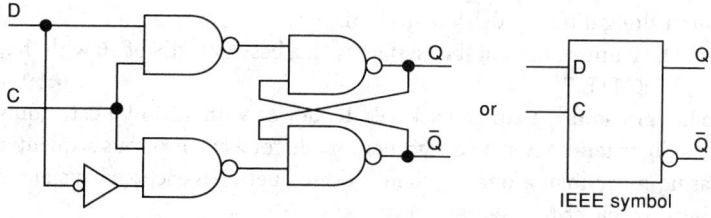

FIGURE 75.25 Fundamental-mode asynchronous sequential logic circuit. (*Source:* R. S. Sandige, *Modern Digital Design,* New York: McGraw-Hill, 1990, p. 470. With permission.)

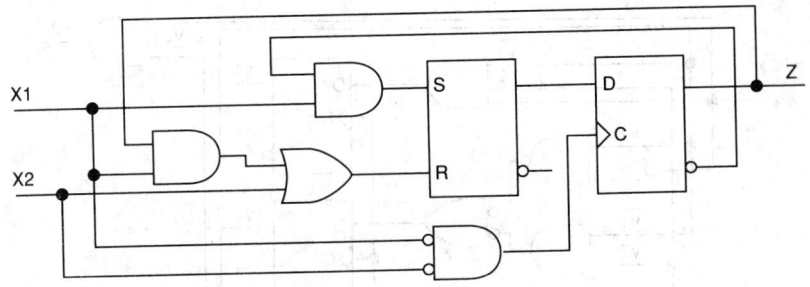

FIGURE 75.26   Double-rank pulse-mode asynchronous sequential logic circuit. (*Source:* R. S. Sandige, *Modern Digital Design*, New York: McGraw-Hill, 1990, p. 615. With permission.)

Quad, octal, 9-bit, and 10-bit transparent latches are readily available as off-the-shelf IC devices for these types of applications. For proper asynchronous circuit operation, the signal applied to the data input D of the fundamental-mode circuit in Fig. 75.25 must meet a minimum setup time and hold time requirement relative to the control input C, changing the latch to the memory mode when C goes to 0. This is a basic requirement for asynchronous circuits with level inputs, i.e., only one input signal is allowed to change at one time. Another restriction requires letting the circuit reach a stable state before allowing the next input signal to change.

An example of a reliable pulse-mode asynchronous sequential logic circuit is shown in Fig. 75.26. While the inputs to asynchronous fundamental-mode circuits are logic levels, the inputs to asynchronous pulse-mode circuits are pulses. Pulse-mode circuits have the restriction that the maximum pulse width of any input pulse must be sufficiently narrow such that an input pulse is no longer present when the new present state output signal becomes available. The purpose of the double-rank circuit in Fig. 75.26 is to ensure that the maximum pulse width requirement is easily met, since the output is not fed back until the input pulse is removed, i.e., goes low or goes to logic 0. The input signals to pulse-mode circuits must also meet the following restrictions: (a) only one input pulse may be applied at one time, (b) the circuit must be allowed to reach a new stable state before applying the next input pulse, and (c) the minimum pulse width of an input pulse is determined by the time it takes to change the slowest flip-flop used in the circuit to a new stable state.

Defining Terms

Asynchronous circuit: A sequential logic circuit without a system clock.

Combinational logic circuit: A circuit with external output signal(s) that are totally dependent on the external input signals applied to the circuit.

Fan-out requirement: The maximum number of loads a device output can drive and still provide dependable 1 and 0 logic levels.

Hazard or glitch: A momentary output error that occurs in a logic circuit because of input signal propagation along different delay paths in the circuit.

Hexadecimal: The name of the number system with a base or radix of 16 with the usual symbols of 0 . . . 9,A,B,C,D,E,F.

Medium-scale integration: A single packaged IC device with 12 to 99 gate-equivalent circuits.

Race-free state assignment: A state assignment made for asynchronous sequential logic circuits such that no more than a one-bit change occurs between each stable state transition, thus preventing possible critical races.

Sequential logic circuit: A circuit with the present state output signals dependent on all or part of the present state output signals fed back as input signals as well as any external input signals if they should exist.

Sum of products (SOP): A standard form for writing a Boolean equation that contains product terms (input variables or signal names either complemented or uncomplemented ANDed together) that are logically summed (ORed together).

Synchronous or clock-mode circuit: A sequential logic circuit that is synchronized with a system clock.

References

Advanced Micro Devices, *PAL Device Data Book,* Sunnyvale, Calif.: Advanced Micro Devices, Inc., 1988.

ANSI/IEEE Std 91-1984, *IEEE Standard Graphic Symbols for Logic Functions,* New York: The Institute of Electrical and Electronics Engineers, 1984.

ANSI/IEEE Std 991-1986, *IEEE Standard for Logic Circuit Diagrams,* New York: The Institute of Electrical and Electronics Engineers, 1986.

K. J. Breeding, *Digital Design Fundamentals,* 2nd ed., Englewood Cliffs, N.J.: Prentice-Hall, 1992.

F. J. Hill and G. R. Peterson, *Introduction to Switching Theory & Logical Design,* 3rd ed., New York: John Wiley, 1981.

M. M. Mano, *Digital Design,* 2nd ed., Englewood Cliffs, N.J.: Prentice-Hall, 1991.

E. J. McCluskey, *Logic Design Principles,* Englewood Cliffs, N.J.: Prentice-Hall, 1986.

Minc, *PLDesigner User's Manual (Systems 200, 300, 500, Version 2.1),* Colorado Springs: Minc, Incorporated, 1991.

R. S. Sandige, *Modern Digital Design,* New York: McGraw-Hill, 1990.

Further Information

The monthly magazine *IEEE Journal on Solid-State Circuits* presents papers discussing logic circuits, for example, "Automating the Design of Asynchronous Sequential Logic Circuits," in its March 1991 issue, pp. 364–370.

The monthly magazine *IEEE Transactions on Computers* presents papers discussing logic circuits, for example, "Concurrent Logic Programming as a Hardware Descriptive Tool," in its January 1990 issue, pp. 72–88.

Also, the monthly magazine *Electronics and Wireless World* presents articles discussing logic circuits, for example, "DIY PLD," in its June 1989 issue, pp. 578–581.

75.3 Registers

B. R. Bannister and D. G. Whitehead

The basic building block of any **register** is the flip-flop, but, just as there are several types of flip-flop, there are many different register arrangements, and an idea of the vast range and their interrelationships is given in Fig. 75.27.

The simplest type of flip-flop is the set-reset flip-flop which can be constructed simply by cross-connecting two NAND/NOR gates. This forms an *asynchronous* flip-flop in which the set or reset signal determines both *what* the flip-flop is to do and *when* it is to operate. In fact, if a state change is required, the flip-flop begins to change state as soon as the input change is detected. This flip-flop is therefore useful as a *latch* which is used to detect when some event has occurred, and is often referred to as a *flag* since it indicates to other circuitry that the event has occurred and remains set until the controlling circuitry responds by resetting it.

Flags are widely used in digital systems to indicate a change of state and all microprocessors have a set of flags which, among other things, are used in deciding whether a program branch should or should not be made. Thus the 8086 family of microprocessors [Intel, 1989], for example, has a

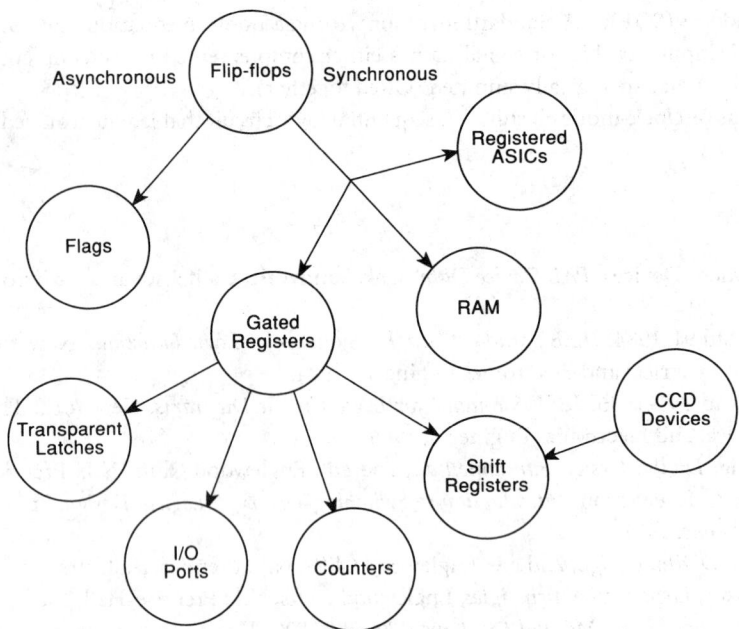

FIGURE 75.27   The register family.

group of nine flags—three control flags used to control particular modes of operation of the processor and six status flags indicating whether certain conditions have resulted from the most recent arithmetic or logical instruction: zero, carry, auxiliary carry, overflow, sign and parity. For convenience, although they all act independently, these flags are grouped together into what is known as the *flag register* or *program status word register*.

Gated Registers

The more conventional meaning of register applies to a collection of identical flip-flops which are activated as a set rather than individually. They are, in general, available as four-bit or eight-bit and are used in multiples of eight bits in most cases. It is the number of flip-flops in each register that determines the width of the data bus in a microprocessor or other bus-based system and that is used to describe the microprocessor. The Z80, for instance, is said to be an eight-bit processor, indicating that the working registers are eight bits wide. The bit values in the register may represent a numerical value in standard fixed point binary, floating point, or some other coded form. Alternatively, they may indicate some logical pattern such as the settings of switches used in an industrial controller.

In order to control *when* the flip-flops set or reset we must make use of *synchronous* flip-flops, leaving the *D* or *J-K* inputs to determine *what* the flip-flop is to do logically, that is to set or reset. In all bus-organized systems it is necessary to control when the data held in the register is fed on to the output bus. This is usually achieved by means of three-state (3S) gates at the register outputs which are disabled, that is, set to their high-impedance state, until the data is required.

A multi-register, bus-structured digital system will have the *n*th bit of each register connected to bit *n* of the data bus at both the input and output of the register. In order to transfer data from one register to another, or, more strictly, to transfer a copy of the contents of one register to another register, the output gates of the source register must be enabled so that the data is fed on to the bus. This data becomes available at the inputs of all registers and is latched in under the control of the appropriate input signals. It is important in the design of the sequencing circuitry that only one set of register output gates can be enabled at any time, although the data can be latched into as many registers as required.

The signal controlling the input to the register is applied to all flip-flops simultaneously and its action depends on the type of flip-flop used. Edge-triggered flip-flops set or reset according to the value on the data inputs at the time the control signal changes. These registers are sometimes known as *staticizers*. After the few nanoseconds required for the flip-flops to settle to their new values, the register content is available at the output gating. The correct operation of the circuitry depends upon certain timing criteria being satisfied and minimum values are quoted by the manufacturers. Each is the smallest time above which the device is guaranteed to operate correctly, but in practice the device probably functions satisfactorily with smaller time intervals on at least some of the parameters. The main timing constraints occur at the inputs to the flip-flops and are illustrated in Fig. 75.28. The interval preceding the active transition of the control input is the setup time, t_{su}, during which the data signal must be held steady; t_h is the hold time and is the interval during which the data signal must be retained following the active transition of the control input; t_w is a minimum pulse width indication which applies to the control inputs such as the clock, reset, and clear. The clock pulse width is usually quoted both for the high state and for the low and is related to the maximum clocking frequency of the flip-flops used in the register.

The 74LS374, Fig. 75.29(a), is a typical eight-bit register using positive edge-triggering at the inputs. It includes 3S output gates designed specifically for driving highly capacitive loads, such as are found in bus-organized systems, and which respond to an output control signal operating quite independently of the flip-flops.

An alternative flip-flop is the transparent latch, an interesting development of the simple latch. When enabled by the control signal, the latch becomes a transparent section of the data path and data values at the input simply reappear at the output. When the control signal disables the latch, however, the last value applied to the latch is "frozen" and held until the enabling signal is applied again. The 74LS374, Fig. 75.29(b), is a transparent version of the 74LS373 register, consisting of eight transparent latches with a common ENABLE input. Typical minimum timing figures for the 74LS373 and 74LS374 are shown in Fig. 75.29(c) and the waveforms occurring for the two different types of register are illustrated in Fig. 75.29(d).

An extended form of transparent operation is provided in addressable latches such as the eight-bit 74LS259. As well as being able to store successive bits arriving at a single input, *D*, in the eight addressable latches, using a three-bit address, any latch can be selected for output so that the device can also act as a 1-of-8 decoder or demultiplexer. Four modes of operation are possible under the control of the *enable* and *clear* inputs. In the addressable latch mode the single-addressed latch acts transparently, with all other latches retaining their previous states. When in the memory mode all latches retain their previous states and are unaffected by address or data inputs. In the decode mode the output of the addressed latch follows the level at the *D* input, and in the clear mode all outputs are set low.

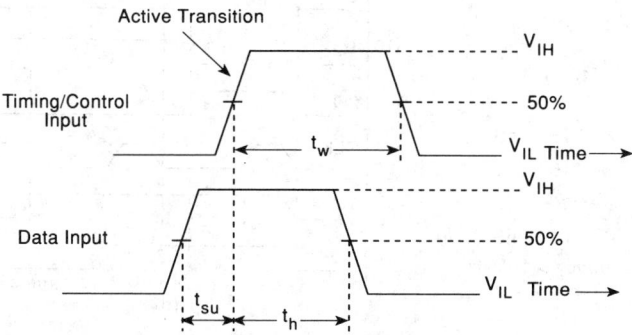

FIGURE 75.28 Control timing parameters.

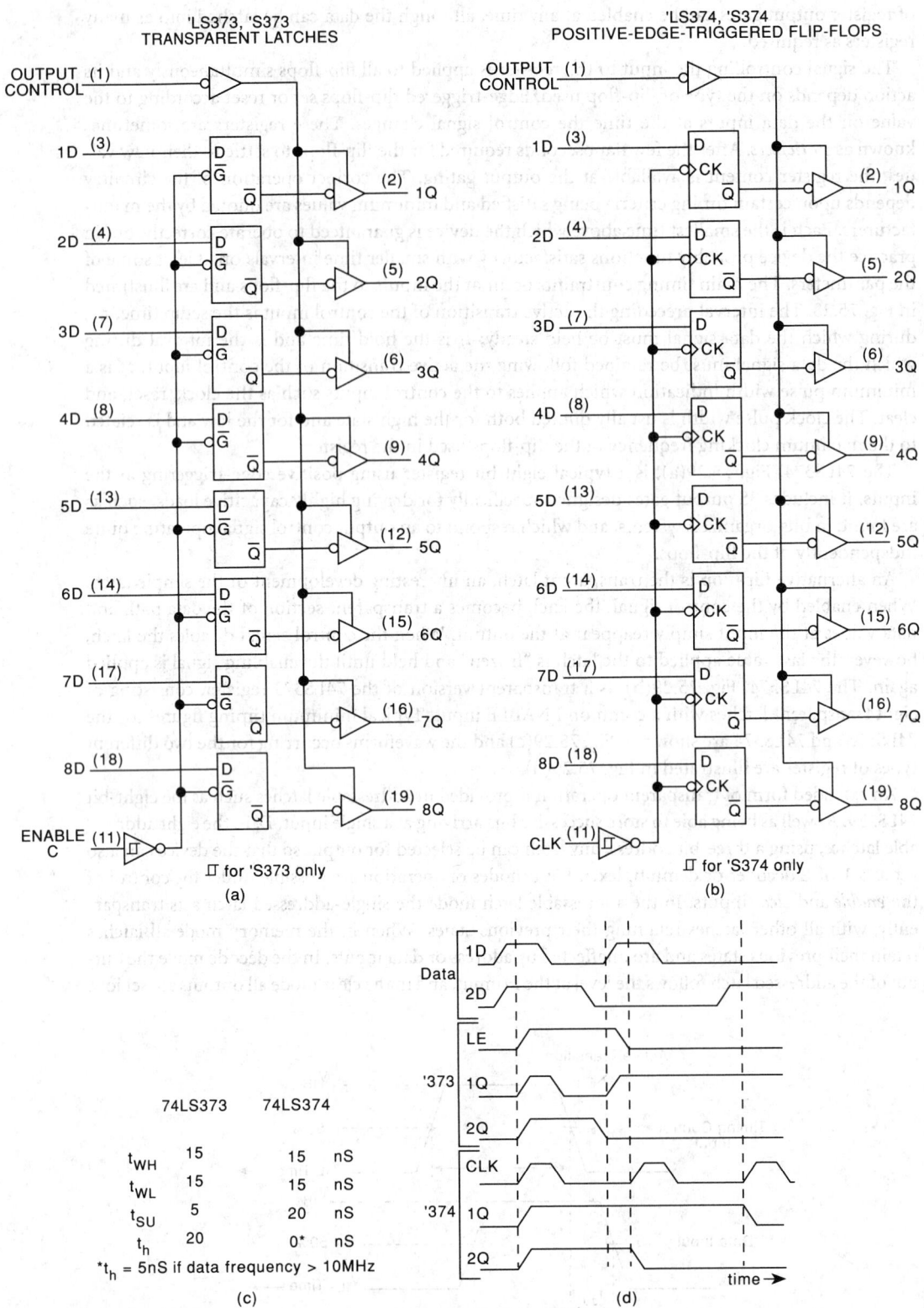

FIGURE 75.29 (a) 74LS373 and (b) 74LS374. (*Source: TTL Data Book*, Vol. 1, Texas Instruments, Inc.) (c) Typical minimum timing values and (d) I/O waveforms for the 'LS373/374.

Shift Registers

There are essentially two modes of operation for a register, either *serial* or *parallel,* and those we have considered so far have operated in parallel mode. **Parallel operation** affects the entire group of bits held in the register during a single clock pulse. In **serial operation** data bits are inputted (or outputted) sequentially to (or from) the register, one bit for every clock pulse.

A register which has the facility to move the stored bits one place at a time left or right under the control of the clock pulse is called a *shift register* (Fig. 75.30).

Shift registers are normally implemented by means of *D, S-R* or *J-K* flip-flops. As an example, the 74LS165A is shown (Fig. 75.31), consisting of eight *S-R*-type flip-flops with clock, clock inhibit, and shift/load control inputs.

Data presented to the eight separate inputs is loaded into the register in parallel when the shift/load input is taken low. Shifting occurs when the shift/load input is high and the clock pulse is applied, the action taking place on the low-to-high transition of the clock pulse. Registers are available which switch on the other clock edge. For example, the 74LS295A, which is a four-bit shift register with serial and parallel operating modes, carries out all data transfers and shifting operations on the high-to-low clock transition. This device also provides 3S operation. Selection of the mode of operation is carried out by suitable combinations of the MODE SELECT inputs tabulated in Fig. 75.32.

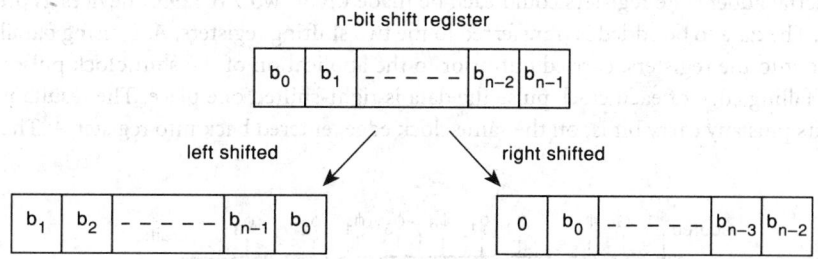

FIGURE 75.30 Shift register operation.

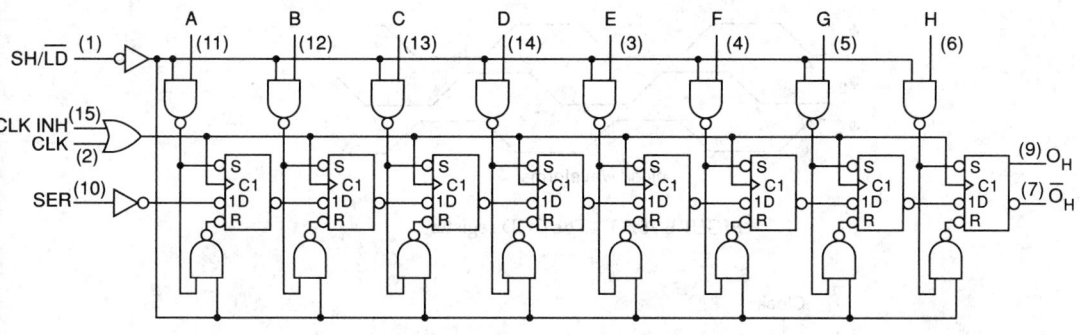

FIGURE 75.31 The 74LS165A shift register. (*Source: TTL Data Book,* Vol. 1, Texas Instruments, Inc.)

Operating Mode	INPUTS				OUTPUTS			
	PE	$\overline{CP}$	Ds	Pn	Q_0	Q_1	Q_2	Q_3
Shift Right	0	⌐_	0	X	0	q_0	q_1	q_2
	0	⌐_	1	X	1	q_0	q_1	q_2
Parallel Load	1	⌐_	X	Pn	P_0	P_1	P_2	P_3

FIGURE 75.32 Operating modes for 74LS295A.

Large-capacity shift registers make use of charge-coupled devices (CCD). These are MOS devices in which data bits are stored *dynamically* as charge between gate and substrate on what is effectively a distributed multi-gate MOS transistor. Consider the diagram (Fig. 75.33) depicting a section through part of an *n*-type substrate which has a series of very closely spaced gate electrodes separating the drain and source of a "stretched" MOS transistor. Using gate "G" and the first storage gate a charge packet of electrons can be introduced into the structure. The overlapping clock pulses allow this charge packet to be moved along the array. At the drain the presence of a charge under the final storage gate is detected by a change in current. Steady-state operation of this type of register is not possible since thermally generated carriers (leakage current) will ultimately cause stationary charge, held beneath a storage gate, to leak away. The result is a minimum operating *shift* frequency of around 20 kHz. The maximum length of the CCD shift register is limited by the charge transfer efficiency; each time the charge packet is transferred between storage gates a fraction is lost. The transfer efficiency also reduces as the frequency increases, limiting the maximum shift frequency to typically 10 MHz for a 256-bit register. The CCD register is structurally a very simple device and large storage arrays are possible. Devices are available operating from two-, three- and four-phase clocks.

Transfer of data between registers is an important operation in all digital systems and sometimes the data may be modified during the transfer. For example, in an addition routine the contents of one register will be added to the contents of another, the resulting sum then being returned to one of the two registers. Parallel or serial addition could be used, but the example shown in Fig. 75.34 is that of a serial adder. The registers could each be made up of two 74LS295A devices as previously described. The data to be added is transferred to the two shifting registers, *A, B*, using parallel loading of data into the registers, carried out prior to the application of the shift clock pulses shown.

On the falling edge of each clock pulse the data is right-shifted one place. The resultant sum of the two bits plus any carry bit is, on the same clock edge, entered back into register *A*. The *D*-type

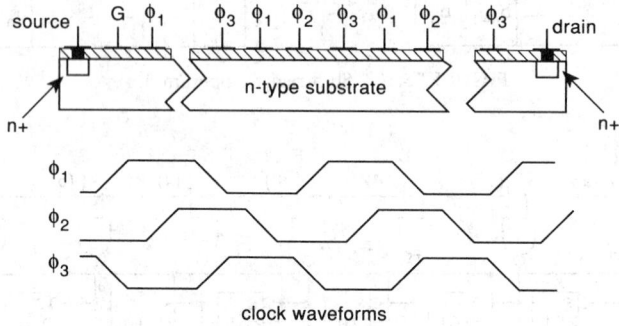

FIGURE 75.33   The CCD register.

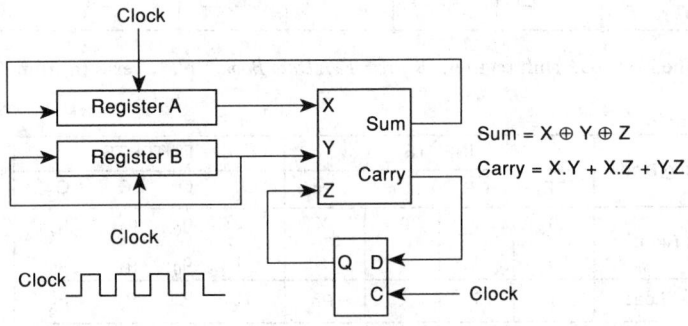

$$Sum = X \oplus Y \oplus Z$$

$$Carry = X.Y + X.Z + Y.Z$$

FIGURE 75.34   The serial adder using shifting registers.

flip-flop is used to delay the carry bit until the next add time: Data entered at D does not appear at Q until the falling edge of the clock pulse has occurred, and at such time it is, therefore, too late to modify the previous addition. At the end of the addition process, when all the data bits have been shifted through the registers, register A contains the sum, and register B is unchanged.

A single shift register can be arranged to provide its own input by means of feedback circuits, and its action then becomes **autonomous**, since the only external signal required is the clock signal. There are only a finite number of states of the feedback shift register (FSR), and the output sequence from the register will, therefore, repeat with a cycle length not greater than 2^n bits, where n is the number of flip-flops in the register. This property can be used to create a counter known as a Johnson counter in which the shift register has the J and K inputs of the first stage fed directly from the Q' *and* Q outputs, respectively, of the last stage. This simple form of feedback leads to the name *twisted-ring counter,* and the result is the generation of a *creeping* or *stepping* code with $2n$ different states. This form of counter is convenient only when the count is small, as the number of flip-flops quickly becomes excessive, but is ideal for a simple decade counter. Unlike standard binary decade counters, the Johnson decade counter requires five flip-flops but no additional feedback circuitry. Gating needed to detect specific settings of the counter is also very simple [Bannister and Whitehead, 1987].

Another set of sequences is obtained if the feedback arrangements are restricted to the use of exclusive-OR, that is modulo-2 functions, and, by correct choice of function, the **linear feedback shift register (LFSR)** so formed generates a *maximal length sequence,* or *m-sequence*. A maximal length sequence has a length of $2^n - 1$ bits (the all-zeros state is not included, since the mod-2 feedback would not allow any escape from that state, so the sequence has a 0 missing) with useful properties of repeatable randomness and is, therefore, described as a *pseudorandom binary sequence* (PRBS). The number of maximal length sequences for a register of length n, and the feedback arrangements to achieve them, are not at all obvious, but have been worked out for a large number of cases [Messina, 1972]. A 4-bit LFSR will produce only one maximal length *sequence,* but a 10-bit register can produce 30 distinct *m*-sequences, and a 30-bit register produces no less than 8,910,000 distinct sequences!

Shift registers can be used in parallel to form a *first-in, first-out (FIFO)* memory. These are typically 128×8-bit register memories with independent input and output buses. At the input port, data is controlled by a shift-in clock operating in conjunction with an input ready signal which indicates whether the memory is able to accept further words or is now full. The data entered is automatically shifted in parallel to the adjacent memory location if it is empty and as this continues the data words stack up at the output end of the memory. At the output port, data transfers are controlled by a shift-out clock and its associated output ready signal. The output ready signal indicates either that a data word is ready to be shifted out or that the memory is now empty. FIFOs can easily be cascaded to any desired depth and operated in parallel to give any required word length. This type of memory is widely used in controlling transfers of data between digital subsystems which operate at different clock rates and is often known as an *elastic buffer*.

Register Transfer Language

The transfer of data between registers is described using a simple notation termed the register transfer language or RTL. For data transferred from register A to register B we write: $B \leftarrow A$. The symbol $\leftarrow$ is called the *transfer operator*. Note that this statement does not indicate how many bits are to be transferred. To define the size of the register we declare the size thus: $A[8]$, $B[16]$, here defining an 8-bit register and a 16-bit register. If the action to be taken is the transfer of the most significant bit (7th bit) of register A to the least significant bit (bit 0) of register B, then we write: $B[0] \leftarrow A[7]$. Usually data is transferred by the enable control signal or a clock pulse. If such a signal is designated "C," then we would describe the action by $C: B \leftarrow A$.

Returning to the serial adder circuit shown earlier, we could describe the register transfers thus:

$$A[8], B[8], D[1]$$

$$C: A[7] \leftarrow A[0] \oplus B[0] \oplus D[0], \; B[7] \leftarrow B[0], \; D[0] \leftarrow \text{Carry}$$

Here in the declaration statement we refer to the D-type flip-flop as a single-bit register. Simultaneous processes are separated by a comma; sequential processes would be separated by a semicolon. The symbol "$\oplus$" is the exclusive-or (XOR) operator. Other logical operations include NOT, AND, OR. The AND operation is also called the *masking* operation because it can be used to remove (or select) specific sections of data from a register. Thus the operation $A[8] \leftarrow A[8] \; \&3CH$ will result in the most significant two bits and the least significant two bits of the eight-bit register A being set to zero. Note that 3CH refers to the hexadecimal number 3C, i.e., 00111100 in binary. Some other terms commonly used are as follows:

$$D \leftarrow A' \qquad \text{transfer the complement of } A \text{ to } D$$

$$A \leftarrow A + 1 \qquad \text{increment } A$$

$$A[8{:}15] \leftarrow B[8{:}15] \qquad \text{transfer bits 8 through 15 from } B \text{ to } A$$

In order to differentiate between arithmetic and logical operations it is usual to represent OR and AND by $\vee$ and $\wedge$. Table 75.5 lists some typical RTL examples that include arithmetic, bit-by-bit logic, shift, rotate, scale and conditional operations. It is assumed that the three registers are set initially to $A = 10110$, $B = 11000$ and $C = 00001$.

Input/Output Ports

The working registers provided in microprocessors may be thought of as high-speed extensions to the memories used for storing programs and data. The random access memories (RAM) themselves are also arrays of registers, though the form of circuit used differs considerably from the more conventional register. The need to transfer data in and out of the system has led manufacturers to produce special registers which are further extensions to the internal memory and are known as **input and output ports.** One of the simplest of input/output ports is the Intel 8212 (Fig. 75.35). This has two modes of operation selected by the mode input, *MD*. With *MD* at 0 the device acts as an input port and a peripheral unit can enter data on the *DI* lines by sending an active high strobe signal, STB. When the central processor is ready for the data it selects the port by setting the correct address bits on the *device select* inputs. This enables the 3S output buffers and data is routed to the processor data bus via the *DO* lines. This device also includes a service request flip-flop to generate an interrupt signal to the processor when the data is ready. In the alternative mode of use, with the mode input at 1, the device select logic routes data from the processor, now connected to the *DI* inputs, so the 8212 acts as an output port. The data is immediately available to the peripheral unit on the *DO* lines, as the 3S output buffers are permanently enabled. A more sophisticated range of input and output facilities is provided by most microprocessor manufacturers in the form of programmable input/output ports or peripheral interfaces. These are special registers with appropriate buffers and additional built-in control and status registers to facilitate proper system operation.

The majority of input/output devices use 3S bidirectional buffers which switch to the high-impedance state when not enabled. Some input/output ports, however, such as the 8051 family of microcontrollers and derivatives, are provided with *quasi-bidirectional* ports. In this construction, each bit has an internal pull-up transistor, as shown in Fig. 75.36. For the bit to be operative as an input the port latch must contain a 1, so that the output FET driver is turned off. (All port latches in the 8051 are set by the reset function.) Under this condition, the pin voltage is pulled high but can be taken low by an external signal when required. These inputs can, therefore, be driven in a

Table 75.5 Typical RTL Examples

Type of Operation	Meaning	Register Bits after Operation
General		
$A_3 \leftarrow A_2$	Bit 2 of A to bit 3 of A	$A = 11110$
$A_3 \leftarrow B_4$	Bit 4 of B to bit 3 of A	$A = 11110$
$A_{1-3} \leftarrow B_{1-3}$	Bits 1 through 3 of B to bits 1 through 3 of A	$A = 11000$
$A_{1,4} \leftarrow B_{1,4}$	Bits 1 and 4 of B to bits 1 and 4 of A	$A = 10100$
$A_{1-3} \leftarrow B_z$	Groups of bit Z of B to bits 1 through 3 of A	$A = 11000$
Arithmetic		
$B \leftarrow 0$	Clear B	$B = 00000$
$A \leftarrow B_2 + C$	Sum of B and C to A	$A = 11001$
$A \leftarrow B - C$	Difference $B - C$ to A	$A = 10111$
$C \leftarrow C + 1$	Increment C by 1	$C = 00010$
Logic		
$A \leftarrow B \wedge C$	Bit-by-bit AND result of B and C to A	$A = 00000$
$A \leftarrow B \vee C_4$	OR operation result of B with bit 4 of C to A	$A = 11000$
$C \leftarrow \bar{C}$	Complement C	$C = 11110$
$B \leftarrow \bar{B} + 1$	2's complement of B	$B = 01000$
$B \leftarrow A \oplus C$	XOR operation result of A and C to B	$B = 10111$
Serial		
$B \leftarrow \text{sr } B$	Shift right B one bit	$B = 01100$
$B \leftarrow \text{sl } B$	Shift right B one bit	$B = 00110$
$B \leftarrow \text{sr2 } B$	Shift right B two bits	$B = 00110$
$B \leftarrow \text{rr } B$	Rotate right B one bit	$B = 01100$
$B \leftarrow \text{rl2 } B$	Rotate left two bits	$B = 00011$
$B \leftarrow \text{scr } B$	Scale B one bit (shift right with sign bit unchanged)	$B = 11100$
$B \leftarrow \text{scl } B$	Scale B one bit (shift left with sign bit unchanged)	$B = 10000$
$B, C \leftarrow \text{sr2 } B, C$	Shift right concatenated B and C two bits	$B, C = 0011000000$
Conditional		
IF $(B_4 = 1)$ $C \leftarrow 0$	If bit 4 of B is a 1, then C is cleared	$C = 00000$
IF $(B \geq C)$ $B \leftarrow 0, C_1 \leftarrow 1$	If B is greater than or equal to C, then B is cleared and C is set to 1	$B = 00000$ $C = 00011$

Initial values: $A = 10110$, $B = 1\underbrace{1000}_{z}$ and $C = 00001$.

Source: E. L. Johnson and M. A. Karim, *Digital Design*, Boston: PWS Publishers, 1987. With permission.

normal way by TTL and MOS circuits and can also cope with open-collector or open-drain circuits using the pull-up transistors as load resistors.

Yet more flexible capabilities are provided in the Rockwell 6522 versatile interface adapter (VIA), which, in addition to two 8-bit bidirectional input/output ports, contains two 16-bit programmable timer/counters and a serial data port using a serial-parallel, parallel-serial shift register. Each line of the input/output ports, A and B, can be individually programmed as an input or an output by means of *data direction registers*, DDRA and DDRB, respectively, which "mirror" the ports. That is, a bit in the data direction register set at 1 causes the corresponding bit of the port to act as an output, with a value determined by the output latch setting. If the bit in the data direction register is set at 0, this causes the corresponding bit of the port to act as an input, and incoming data may be latched into an internal register under the control of a special peripheral control line, $CA1$ or $CB1$. Data may be written into register bits which are programmed as inputs but the output signal is unaffected. The input latching may be enabled or disabled. When reading a peripheral port with the latching disabled, the value read is the current value on the input pin; with the latching enabled, the value read is the value which existed at the input pin when the active transition occurred at the control input.

The interval timer/counters can operate in a variety of modes and are retriggerable so that an

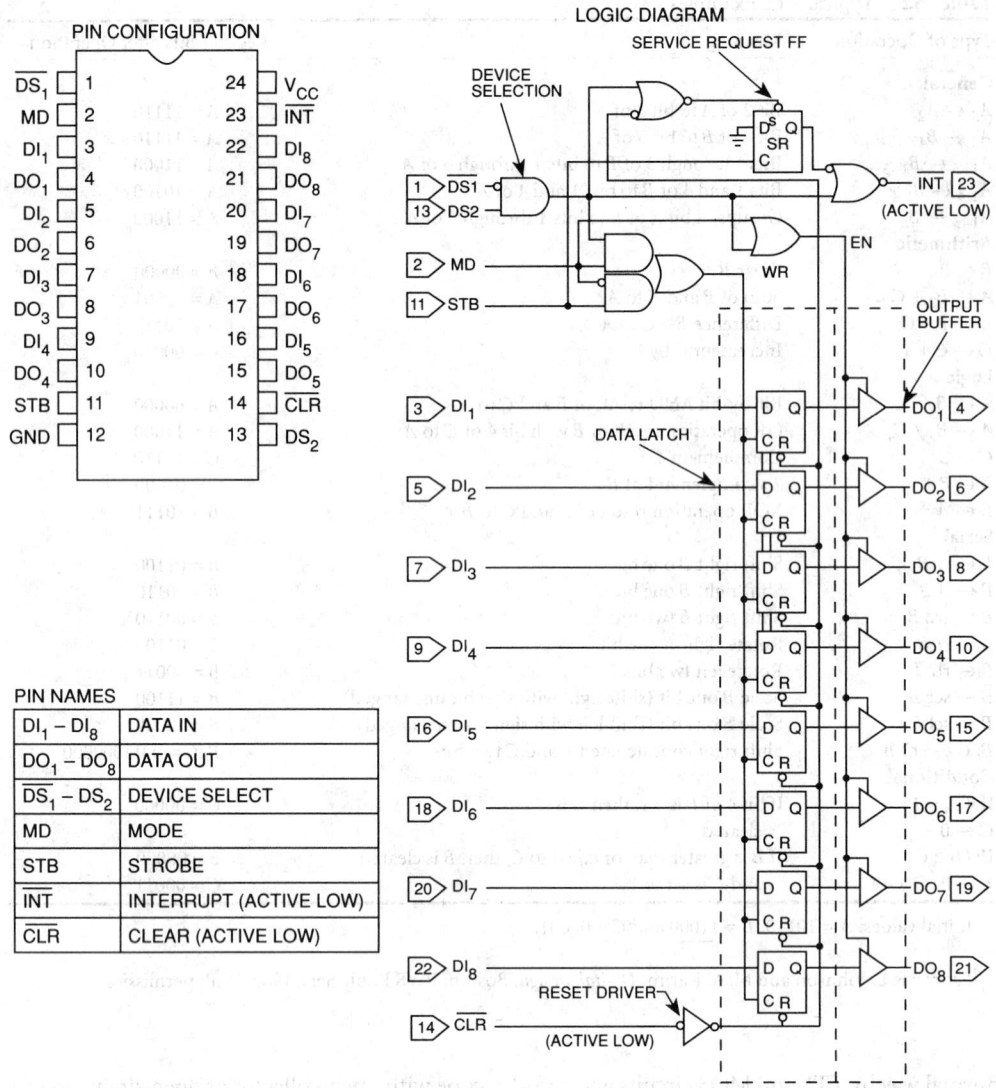

FIGURE 75.35   The Intel 8212. (*Source: Microprocessor Component Handbook,* Intel Corp.)

interrupt signal is generated either once, when the interval expires, or continually, when subsequent intervals expire. If an external output is required, it can be either a one-shot output or a squarewave, with the output changing level each time the interval count is completed. The main timer/counter consists of two 8-bit latches operating in conjunction with a 16-bit counter. The latches store the data which is to be loaded into the counter and are loaded sequentially, since 16 bits are required for the counter but the data bus is only 8 bits wide. When loading the counter with a specific value, the low-order byte is actually routed to the low-order 8-bit latch. Then, when the high-order byte is supplied, it is simultaneously written into both the high-order 8-bit latch and the high-order counter byte, and the low-order byte is also transferred from the latch to the counter. Countdown then begins on the next clock pulse. This method ensures that the correct 16-bit value is loaded, but it also means that the current count value can be modified, during counting if required, by writing to the counter, or the *subsequent* counts can be modified by writing to the latches.

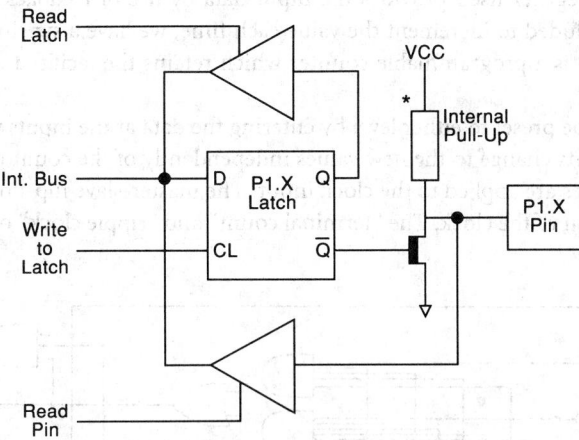

FIGURE 75.36 Intel 8051 quasi-bidirectional port.

Register Number	RS Coding				Register Desig.	Description	
	RS3	RS2	RS1	RS0		Write	Read
0	0	0	0	0	ORB/IRB	Output Register "B"	Input Register "B"
1	0	0	0	1	ORA/IRA	Output Register "A"	Input Register "A"
2	0	0	1	0	DDRB	Data Direction Register "B"	
3	0	0	1	1	DDRA	Data Direction Register "A"	
4	0	1	0	0	T1C-L	T1 Low-Order Latches	T1 Low-Order Counter
5	0	1	0	1	T1C-H	T1 High-Order Counter	
6	0	1	1	0	T1L-L	T1 Low-Order Latches	
7	0	1	1	1	T1L-H	T1 High-Order Latches	
8	1	0	0	0	T2C-L	T2 Low-Order Latches	T2 Low-Order Counter
9	1	0	0	1	T2C-H	T2 High-Order Latches	
10	1	0	1	0	SR	Shift Register	
11	1	0	1	1	ACR	Auxiliary Control Register	
12	1	1	0	0	PCR	Peripheral Control Register	
13	1	1	0	1	IFR	Interrupt Flag Register	
14	1	1	1	0	IER	Interrupt Enable Register	
15	1	1	1	1	ORA/IRA	Same as Reg 1 Except No "Handshake"	

FIGURE 75.37 Internal register summary. (*Source: Synertek Data Book.*)

The shift register performs serial data transfers in and out of a given pin under the control of an internal modulo-8 counter. When all the necessary control registers and interrupt handling registers are included with the data direction registers, together with the ports and counters of direct interest to the user, the device requires a total of 16 registers. These are individually addressed using four register select pins, *RS0–RS3*.

Counters

A register can be loaded with any combination by applying the correct bit pattern to the input data lines and activating the control line. As with the feedback shift registers, it is then only a small step to arrange that the register itself provides the input data by use of feedback connections and, if other circuitry is included to increment the value each time, we have a synchronous counter. The 74L5191 (Fig. 75.38) is a programmable counter which retains the facility for parallel loading of external data.

Each output may be preset to either level by entering the data at the inputs while the LOAD signal is low. The outputs change to the new values independently of the count pulses, and counting continues when pulses are applied to the clock input. The master-slave flip-flops are triggered by a low-to-high transition of the clock. The "terminal count" and "ripple clock" outputs facilitate cas-

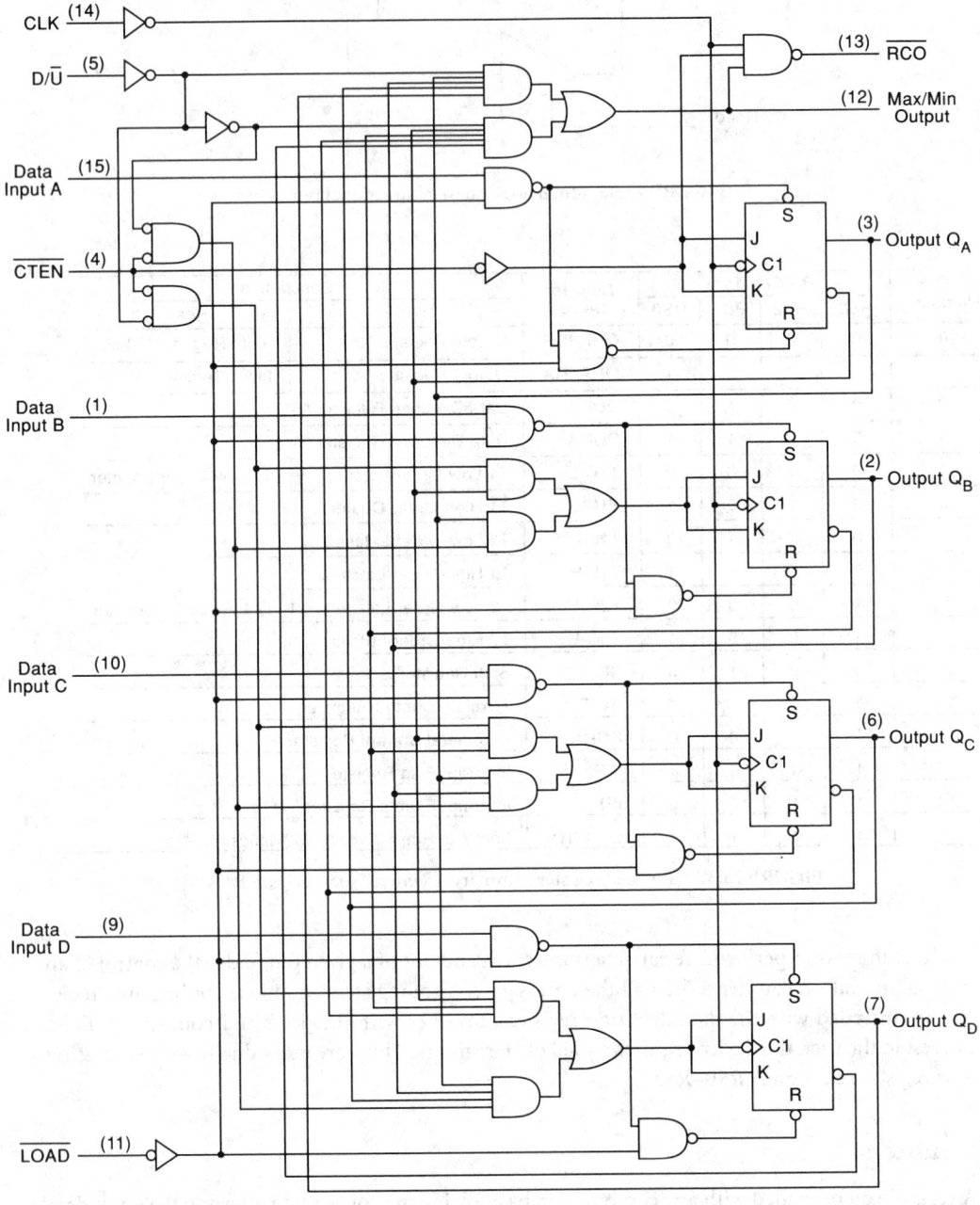

FIGURE 75.38 The 74LS191 programmable counter. (*Source: TTL Data Book,* Vol. 1, Texas Instruments, Inc.)

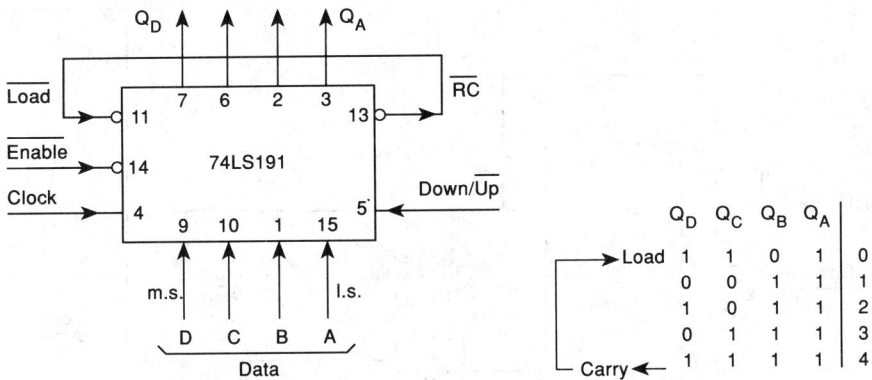

FIGURE 75.39   Programmable counter giving modulo-5.

cading of several counters. The ripple clock carry/borrow output signal is a pulse equal in length to the clock pulse when the counter overflows or underflows, that is, when it is incremented from 1111 or decremented from 0000. By using this signal to reload the value at the data inputs we create a counter of modulus less than 16. Figure 75.39, for example, shows the arrangement to give a modulo-5 count.

Registered ASICs

Developments in application-specific integrated circuits (ASICs) and field-programmable gate arrays (FPGAs) over recent years have provided digital system designers with a wide range of flexible devices which can be programmed for the specific job in hand. The vast majority of digital subsystems involve sequential logic to a greater or lesser extent, and the array manufactures have provided most devices with registers at either input or output and, sometimes, at both. The Altera EP512 is a good example, and its input structure is shown in Fig. 75.40. The device has eight programmable inputs in which each may be configured as any one of the following:

Synchronous D-type flip-flop (register)
Asynchronous D-type flip-flop (register)
Synchronous latch
Asynchronous latch
Flowthrough latch (transparent latch)

The *internal latch enable* (ILE) input carries the clock when synchronous mode is selected, whereas asynchronous mode control signals are generated internally. The EP512 contains 12 macrocells and the input/output architecture provides each macrocell with over 50 possible configurations. Each I/O unit can be individually configured for combinatorial or registered output, with the output polarity also programmable. Four different types of flip-flops ($D, T, J\text{-}K, S\text{-}R$) can be implemented in each I/O unit, which either synchronous or asynchronous operation.

Standard Graphic Symbols

The use of standardized graphical symbols is becoming widespread and the family of registers have their own coherent set of symbols. Two representative examples are given in Fig. 75.41. As shown, the eight-bit shift register is designated SGR8. The direction of shift is given by the arrow. The "1D" refers to the single data input to the "D" input of the first stage of the register. The four-bit register, denoted RG4, has four "D" inputs, but only one is shown. The "1" relates to the clock input, signi-

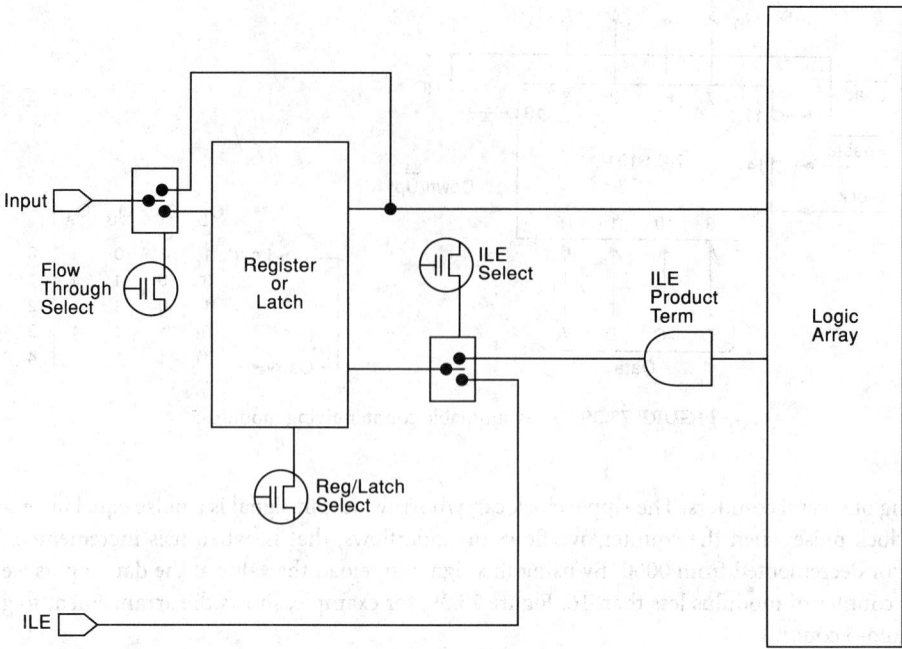

FIGURE 75.40   Input arrangements of the Altera EP512. (*Source: Altera Data Book,* 1988.)

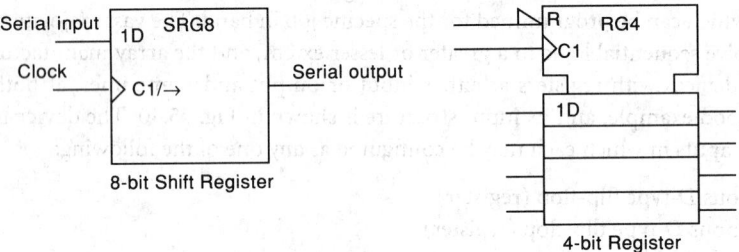

FIGURE 75.41   Standard graphic symbols. (*Source:* ANSI/IEEE Std. 91-1984.)

fying that data is clocked in on the "number one" clock. The reset "R" and the clock are common to all units and are shown as inputs to the common block The reset operates in negative logic. For further details see the References at the end of this section.

Defining Terms

Autonomous operation: Operation of a sequential circuit in which no external signals, other than clock signals, are applied. The necessary logic inputs are derived internally using feedback circuits.

Input/output port: A form of register designed specifically for data input-output purposes in a bus-oriented system.

Linear feedback shift register (LFSR):   An autonomous feedback shift register in which the feedback function involves only exclusive-OR operations.

Parallel operation: Data bits on separate lines (often in multiples of eight) are transferred simultaneously under control of signals common to all lines.

Register: A circuit formed from several identical gated flip-flops or latches and capable of storing several bits of data.

Serial operation: Data bits on a single line are transferred sequentially under the control of a single signal.

References

Altera, *Data Book,* 1988.

B. R. Bannister and D. G. Whitehead, *Fundamentals of Modern Digital Systems,* London: Macmillan, 1987.

IEEE, Standard Graphic Symbols for Logic Functions, ANSI/IEEE Std. 91-1984, New York, 1984.

Intel Corporation, *8086/8088 User's Manual.*

Intel Corporation, *Microprocessor and Peripheral Handbook.*

E. L. Johnson and M. A. Karim, *Digital Design,* Boston: PWS Engineering, 1987.

I. Kampel, *A Practical Introduction to the New Logic Symbols,* Boston: Butterworth, 1985.

A. Messina, "Considerations for non-binary counter applications," *Computer Design,* vol. 11, no. 11, Nov. 1972.

Synertek, *Data Book.*

Texas Instruments, Inc., *TTL Data Book.*

Further Information

The monthly journal *IEEE Transactions on Computers* regularly has articles involving the design and application of registers and associated systems. Further information can be obtained from IEEE Service Center, 445 Hoes Lane, P.O. Box 1331, Piscataway, NJ 08855-1331.

The *IEE Proceedings-E, Computers and Digital Techniques,* published bi-monthly by the Institution of Electrical Engineers (Michael Faraday House, Six Hills Way, Stevenage, Herts. SG1 2AY UK), is also a useful source of information on the application of register devices.

75.4 Programmable Arrays

Martin Bolton

Programmable arrays or **programmable logic devices** (PLDs) are general-purpose combinational or sequential digital components whose ultimate function is determined by the designer. They leave the manufacturer in an unprogrammed state. The configurations of internal switches are fixed after the particular logic function for the PLD has been prepared and checked using a computer-aided design package appropriate for the PLD family used. PLDs belong to the family of application-specific integrated circuits (ASICs).

PLDs are manufactured today in most digital integrated circuit technologies—principally CMOS and bipolar silicon, and gallium arsenide. The programmable switches themselves can be fuses, antifuses, floating-gate MOSFETs, and RAM cells. The floating-gate devices can usually be erased and reprogrammed, while the RAM-based devices can be reconfigured dynamically. Most PLDs have to be programmed with the aid of a programmer, a unit which is able to deliver the appropriate sequences of programming pulses which configure the PLD's arrays of switches in the pattern specified by the user. Some PLDs can be programmed by sending a data stream to the device in its application environment. This is *in-system programming.*

There are today two major classes of PLDs—those based on the **programmable logic array** (PLA) and **field-programmable gate arrays** (FPGAs). PLA technology is the oldest and is now restricted to the less complex circuits. FPGAs, on the other hand, are a more recent technology and are able to implement complex systems equivalent to networks of several thousand logic gates. The first part of this section will explain the principles of PLA-based PLDs; the second will introduce the concepts of FPGAs. A final section will briefly cover the requirements of computer-aided design for programmable logic.

PLA-Structured Devices

It is possible to represent any Boolean function in the form of a *sum of products*. For example, the expressions for the outputs of a full adder can be represented as:

$$\text{SUM} = A\overline{BC} + ABC + \overline{A}B\overline{C} + \overline{AB}C \tag{75.14}$$

$$\text{CARRY} = AC + AB + BC \tag{75.15}$$

where *A*, *B*, and *C* are the inputs. The first equation has four *product terms,* or *products;* the second has three.

A PLA enables functions expressed in this form to be directly implemented. Each product term is generated by a gate which can be programmed to form the AND of any subset of the inputs and their complements, while subsets of products can be summed in a set of programmable OR gates. The programmable gates are constructed in the form of arrays, with the input lines being orthogonal to the product lines, which are themselves orthogonal to the output lines, as shown in Fig. 75.42.

This PLA has three inputs, five product terms, and two outputs. Each input is fed into the first array, the *AND array,* in true and complement form. This enables any product to be formed. The products are all fed into the *OR array.* A cross at an intersection indicates a connection. Notice that some products are used to contribute to both outputs. The ability to share product terms in this way is an important feature of the PLA. The product terms which make up the inputs to the OR gate for the SUM output are those given in Eq. (75.14). The equation for the CARRY output has been modified for use in this PLA which also has programmable output inversions, indicated by the two exclusive-OR gates. The modified equation for the inverse of CARRY, which now shares some products with the first function, is

$$\overline{\text{CARRY}} = A\overline{BC} + \overline{A}B\overline{C} + \overline{AB}C + \overline{ABC} \tag{75.16}$$

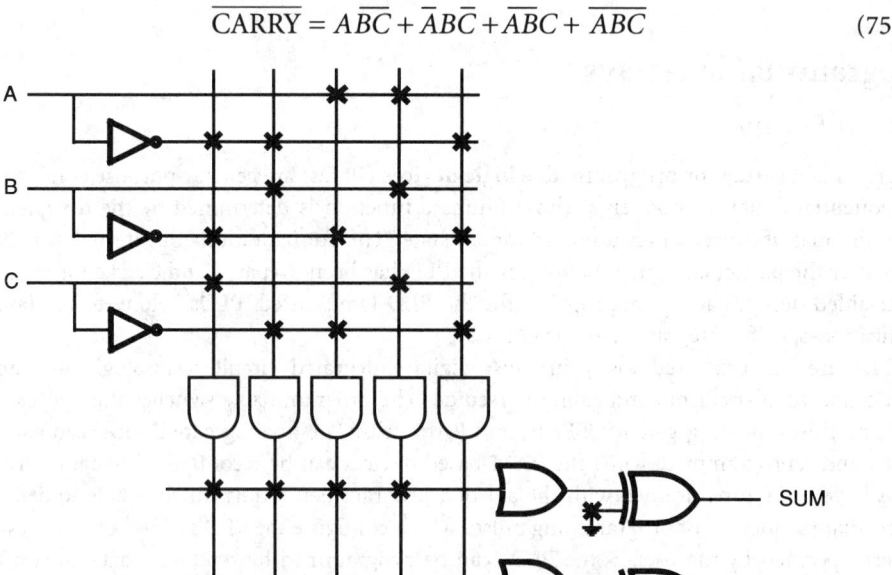

FIGURE 75.42   A full adder implemented in a PLA with programmable output inversion.

Only one new product is now required instead of the three which would have been required without the double negation of the CARRY function.

A PLA is able to implement any set of combinational logic functions, limited only by the number of inputs, number of outputs, and number of product terms. A memory also has this ability, but since in this type of device every combination of inputs is decoded and mapped to a unique set of outputs (via a memory word), the number of inputs handled cannot be as great as with a PLA of similar physical size, since a PLA does not decode every input combination. However, where the set of functions to be implemented would require too many products for a practical PLA, a memory, or a multiple-level network, would have to be used.

An important economy is possible by making use of the fact that for many sets of functions the OR array is not needed for the sharing of products between outputs. Figure 75.43 shows an example of such a function, a multiplexer. Such a function can be implemented by a PLA with only an AND array; the OR array has degenerated into a set of OR gates. The drawback is that the PLA is no longer universal, because a fixed allocation of products to OR gates has to be chosen. PLAs of this structure have become known as PAL devices or just *PALs*. (The term "PAL," which stands for **"programmable array logic,"** is a trademark of Advanced Micro Devices, Inc.). There are many applications where the product term sharing capability of the full PLA is wasted, and a PAL-based solution is more economical. Also, because there is only one programmable array, the propagation delay will generally be smaller. Address decoding in microprocessor systems is a very common application of combinational PAL devices.

PAL devices are available in a wide variety of sizes. Because there is a fixed allocation of products to outputs in a simple PAL device, more different types are needed than in the case of PLAs. This limitation has lately been overcome to some extent in some devices by allowing a limited allocation of products between adjacent outputs.

The flexibility of PAL-structured PLAs can be enhanced by adding controllable three-state output drivers, as shown in Fig. 75.44. The drivers, controlled by additional product terms (or *control terms*), allow those pins to which they are connected to act as either inputs, outputs, or both. This feature allows a single device to be used in a wider range of applications than would be possible with a device with a fixed allocation of inputs and outputs. Note also the polarity control, similar to that in the PLA of Fig. 75.42.

The array structures introduced above can be extended by adding clocked flip-flops to the outputs of the arrays to create general-purpose sequential circuits. A PLA extended in this way is known as a **sequencer**, whose generic structure is shown in Fig. 75.45.

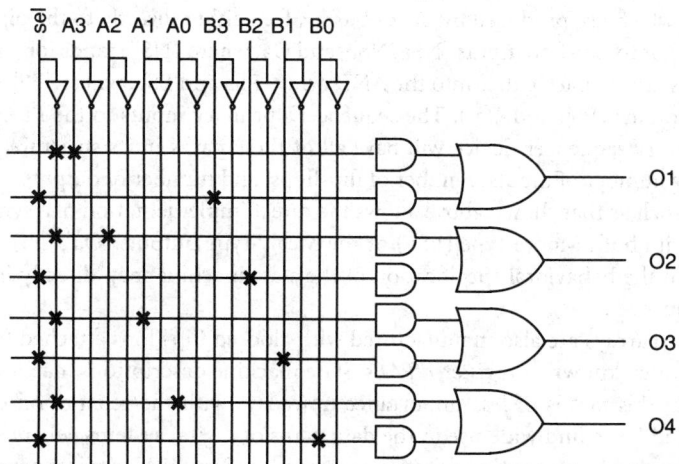

FIGURE 75.43   A four-bit multiplexer implemented in a PAL-structured programmable array.

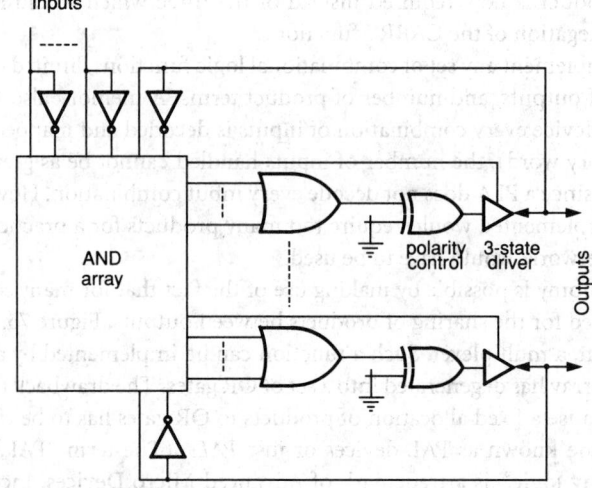

FIGURE 75.44   PAL-structured programmable array with output controls.

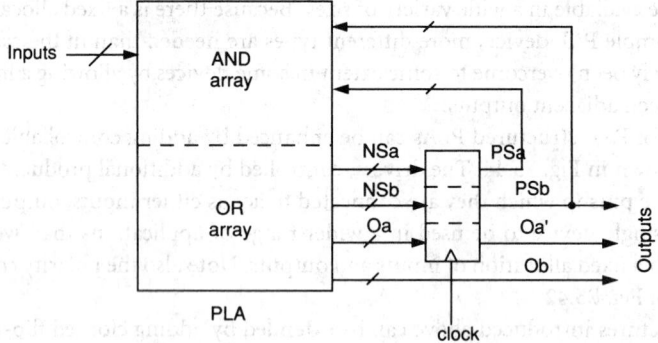

FIGURE 75.45   The structure of a sequencer.

The outputs, which are produced by the OR array, feed either directly to the pins (indicated by Ob in the figure) or to flip-flop inputs (NSa, NSb, and Oa, where "NS" stands for "next state"). The flip-flop outputs are fed back either into the AND array (PSa and PSb, where "PS" stands for "present state") or to pins (PSb and Oa'). The sequencer's primary inputs to the PLA AND array are labeled I. Not every sequencer device will have all of these paths in its structure. Sequencers are characterized by number of inputs, number of flip-flops, and number of outputs.

A sequencer such as that shown above allows the direct implementation of a synchronous finite state machine with both Moore-type (PSb) or the Mealy-type outputs (Oa', Ob). The transitions between states in the behavioral specification of the state machine map directly into the product terms of the sequencer.

PAL-structured arrays are also manufactured with clocked flip-flops attached to the array outputs. These are often known as *registered PALs*. State machine descriptions map less naturally into these arrays, but this fact is of less importance nowadays with the reliance on computer-aided design. Registered PALs find wide use in the data paths of digital systems, where special-purpose registers, counters, and data routing functions are needed. Some PAL devices have enhancements such as the exclusive-ORing of outputs to make them more useful in these applications.

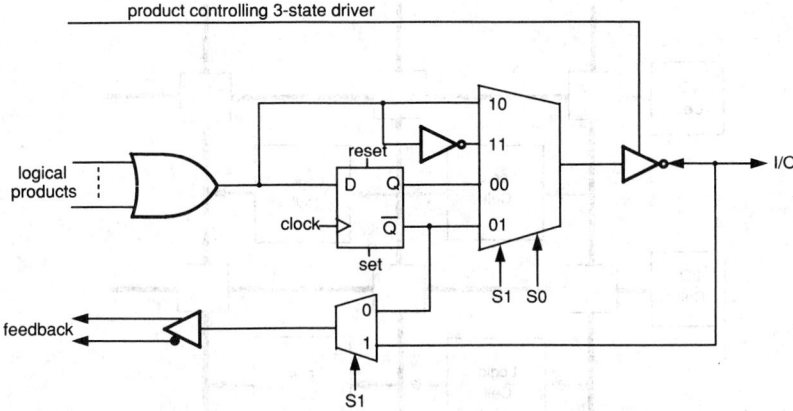

FIGURE 75.46 The output macrocell of the 22V10 PAL device.

Just as a combinational PAL device can be made more general purpose by the addition of a controlled output, a sequential one can be generalized by adding programmable *macrocells* to the array outputs. The output macrocell of the 22V10 device, the first of the *generic* PAL devices to be introduced, is shown in Fig. 75.46.

The modes of this macrocell are controllable by two dedicated bits S1 and S0, programmed into the device, and by a product term controlling the three-state driver. The four possible configurations determined by the programming of S1 and S0 are:

00: registered output, active low, fed back into array
01: registered output, active high, fed back into array
10: combinational input/output, active low
11: combinational input/output, active high

This form of PAL device is now very widely used and has the advantage that many fewer device types are needed to handle a range of applications. The fastest devices have combinational propagation delays in the 5- to 10-ns range and maximum clock frequencies in excess of 100 MHz.

The single-array devices described so far cannot be extended in size indefinitely; it is difficult to efficiently use large arrays, and the long internal lines required cannot be driven fast enough while maintaining a reasonable power consumption. For these reasons many PLD architectures based on partitioned arrays and switching matrices have been introduced. There is not space to elaborate on these here. A number of these designs are described in Moore and Luk [1991]. However, the most important high-complexity array structure, the FPGA, is introduced in the next section.

Programmable Gate Arrays (FPGAs)

Gate arrays are semicustom devices based on an array of simple cells selected from a library surrounded by an interconnection network. In conventional gate array technology, the interconnection pattern is defined by metallization layers applied at the final stage of manufacture. FPGAs dispense with this final stage by possessing a fixed interconnection network which includes programmable crosspoint switches, as shown in Fig. 75.47. The cells, instead of being selected from a library, are generic and have programmable function. The price for doing this is a lower logic density and higher resistance interconnections, with concomitant effect on speed of operation. Nevertheless, this disadvantage is more than overcome by the short design turnaround times achievable. A major tradeoff in gate array architecture is between cell complexity and interconnection channel capacity. This remains the case with FPGAs.

Each logic cell in an FPGA is a small programmable logic block which will usually contain one or more flip-flops. Typically these cells will have up to ten inputs and a smaller number of outputs.

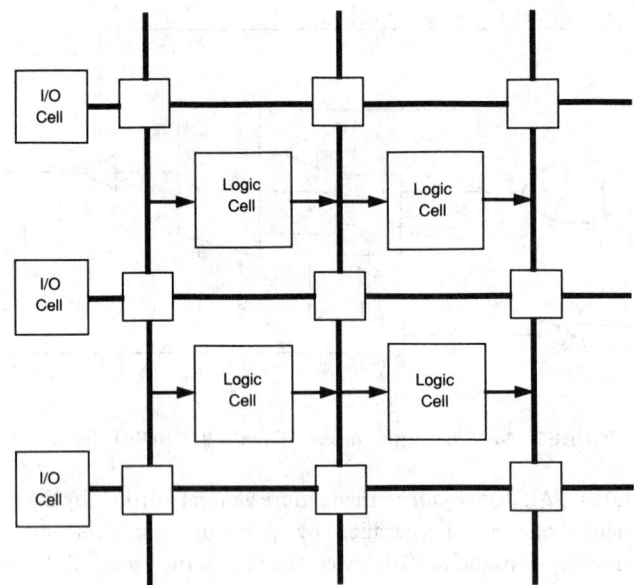

FIGURE 75.47   Interconnection structure in an FPGA.

Computer-Aided Design for Programmable Arrays

As programmable devices become more complex, the capabilities of the computer-aided design support become more and more important. Small PLA and PAL devices can be programmed from a manually created table, but this method is error-prone and cannot be recommended. The earliest widely available design aid for programmable logic was the compiler known as PALASM. This was able to translate the functional specification of a device, expressed in the form of Boolean equations, into the device programming specification for a PAL device. It also allowed the function of the device to be simulated. Other manufacturer-specific compilers for the simpler PLDs have been produced, but nowadays these needs can be served by one of the very capable Boolean language-based universal compilers.

Recognizing the need for design entry based on schematics, the manufacturers of the more complex PLDs and FPGAs began to offer this alternative; in fact for FPGAs this is an easier option for the design tool writer. Schematics for a PLD describe the interconnection of a collection of notional primitive cells.

Converting a language-based description into the layout of an FPGA is an application of *logic synthesis,* a technology which has evolved independently of programmable logic design tools, but which has now converged with them. However, programmable devices offer new problems to the designer of general-purpose design tools due to the great variety of architectures now existing for the complex devices. Programmable logic design tools are now becoming more integrated into mainstream computer-aided design systems and are starting to adopt the same standards, for example use of the VHDL language for specification.

Defining Terms

Field-programmable gate array (FPGA):   A PLD which consists of a matrix of programmable cells embedded in a programmable routing mesh. The combined programming of the cell functions and routing network define the function of the device.

Programmable array logic (PAL) device: A PLA with no OR array, but instead a fixed set of OR gates into which are fed sets of product terms. ("PAL" is a trademark of Advanced Micro Devices, Inc.)

Programmable logic array (PLA):   A PLD which consists of an AND array forming logical products of the input literals and an OR array which sums these products to form a set of output functions.

Programmable logic device (PLD):   An integrated circuit which is able to implement combinational and/or sequential digital functions which are defined by the designer and programmed into the device.

Sequencer: A PLA which has a set of flip-flops for storage of outputs which can be fed back into the PLA as inputs, enabling the implementation of a finite state machine.

References

R. C. Alford, *Programmable Logic Designer's Guide*, Indianapolis: H. W. Sams, 1989.

J. Birkner and V. Coli, *PAL Handbook*, 2nd ed., New York: McGraw-Hill, 1981.

M. J. P. Bolton, *Digital Systems Design with Programmable Logic*, Wokingham, England: Addison-Wesley, 1990.

G. Bostock, *Programmable Logic Handbook*, London: Collins, 1987 (also published as *Programmable Logic Devices: Technology and Applications*, New York: McGraw-Hill, 1988).

J. D. Broesch, *Practical Programmable Circuits: A Guide to PLDs, State Machines and Microcontrollers*, San Diego: Academic Press, 1991.

P. K. Lala, *Digital Systems Design with Programmable Logic Devices*, Englewood Cliffs, N.J.: Prentice-Hall, 1990.

W. R. Moore and W. Luk, Eds., *FPGAs*, Abingdon, England: Abingdon EE&CS Books, 1991.

D. Pellerin and M. Holley, *Practical Design Using Programmable Logic*, Englewood Cliffs, N.J.: Prentice-Hall, 1991.

Further Information

Programmable logic is a fast-moving field, with many vendors continually introducing new devices and improved versions of existing architectures. The magazines *Electronic Design, EDN,* and *Computer Design* carry regular announcements of new programmable logic hardware and software products and often print articles illustrating applications and design methods. All of the vendors publish application notes which are the best source of device-specific design information.

75.5 Arithmetic Logic Units

Bill D. Carroll

Arithmetic logic units (ALUs) are combinational logic circuits that can perform basic arithmetic (addition or subtraction) or logical (AND, OR, NOT, etc.) operations on two m-bit operands. ALUs may be constructed from standard integrated circuits or programmable logic devices and are available as single-chip medium-scale integrated circuits. Integrated ALUs may be cascaded to form longer word lengths than are available in a single device.

This section covers the design of arithmetic and logic circuits in sufficient detail for the reader to design and implement basic arithmetic logic units and to understand the operation and utilization of commercial ALU chips. The reader wanting more details or more in-depth discussion of the subject is referred to the References and other sources given at the end of the section.

In the material that follows, it is assumed that operands are signed n-bit binary numbers with the left-most bit representing the sign (0 for positive and 1 for negative) when discussing arithmetic operations/circuits. Negative numbers will be represented in two's complement form. Recall

that the two's complement of an n-bit number A is $A' + 1$ where A' represents the bit-wise complement of A. Unsigned n-bit binary numbers are assumed for logic operations/circuits.

Basic Adders and Subtracters

The basic building block for most arithmetic circuits is the **full adder**. A full adder is a logic circuit
that produces the two-bit sum (S and C) of three one-bit binary numbers (X, Y, and Z). Table 75.6
shows the truth table and logic equations of a full adder. A logic symbol and a gate-level realization
of a full adder are shown in Fig. 75.48.

The addition of two n-bit binary numbers ($X = x_{n-1} \ldots x_1 x_0$ and $Y = Y_{n-1} \ldots Y_1 Y_0$) can be accomplished with n full adders cascaded as shown in Fig. 75.49. Such a circuit is called a **ripple-carry
adder** since carries produced by lower-order stages must propagate or ripple through the higher-
order stages before the addition operation is complete.

Ripple-carry adders are simple in both operation and structure but are slow since in the worst
case ($X = 1 \ldots 11$ and $Y = 0 \ldots 01$) a carry produced in the least significant full adder must propagate through all the more significant ones. The worst-case add time, t_{add}, is given below where t_{pd} is
the propagation delay introduced at each stage.

$$t_{add} = n t_{pd}$$

Table 75.6 Full Adder Truth Table and Logic Equations

X	Y	Z	S	C	
0	0	0	0	0	
0	0	1	1	0	$S = XYZ + XY'Z' + X'YZ' + X'Y'Z$
0	1	0	1	0	
0	1	1	0	1	
1	0	0	1	0	$C = XY + XZ = YZ$
1	0	1	0	1	
1	1	0	0	1	
1	1	1	1	1	

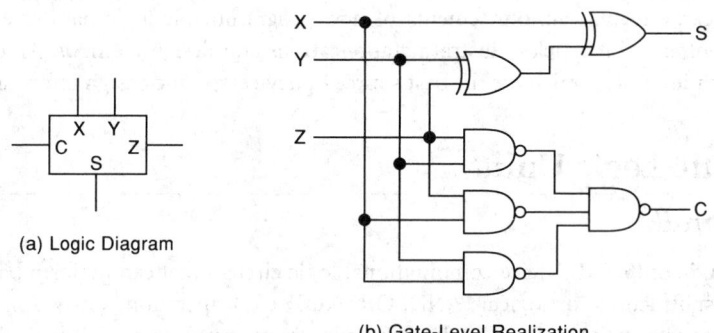

(a) Logic Diagram

(b) Gate-Level Realization

FIGURE 75.48 Full adder.

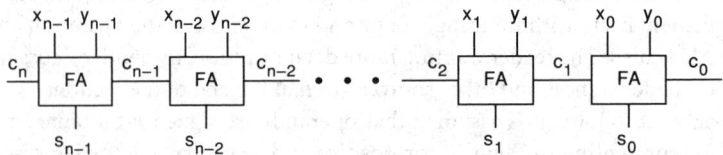

FIGURE 75.49 Ripple-carry adder for two n-bit binary numbers.

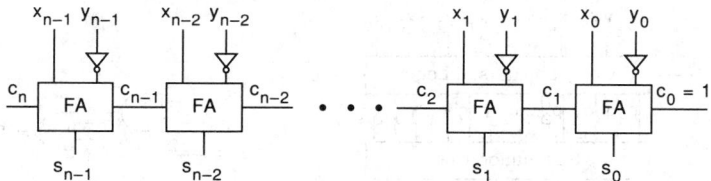

FIGURE 75.50   Two's complement subtracter.

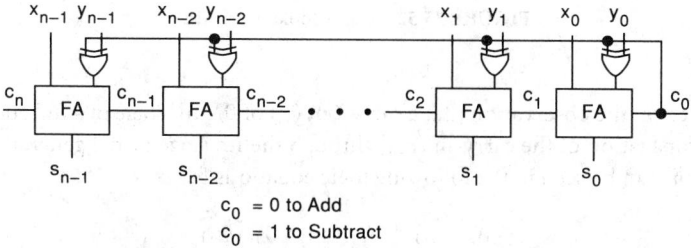

$$c_0 = 0 \text{ to Add}$$
$$c_0 = 1 \text{ to Subtract}$$

FIGURE 75.51   Two's complement adder/subtracter.

This assumes that all addend bits are presented to the adder simultaneously. It is important to note that in the least significant full adder, t_{pd} represents the time to compute c_1 from x_0 and y_0 and in the most significant full adder the time to compute s_{n-1} after c_{n-1} is received. In the intermediate stages, t_{pd} is the time needed to compute c_{i+1} from c_i. The propagation delay is approximately equal to that of a three-level logic circuit which is consistent with the realization of a full adder given in Fig. 75.48.

Subtraction can easily be performed by adding the minuend to the negative of the subtrahend. In a two's complement number system, $X - Y$ can thus be obtained by computing $X + Y' + 1$. The ripple-carry adder described above can be easily modified to perform this computation by placing inverters on the Y inputs of each full adder and by making the carry-in (c_0) equal to 1. The resulting two's complement subtracter is shown in Fig. 75.50.

A device that can perform either addition or subtraction can be built by replacing the inverters in the subtracter with exclusive-OR gates and using the carry-in (c_0) as a control signal. The resulting two's complement adder/subtracter is shown in Fig. 75.51. The device will function as a ripple-carry adder when $c_0 = 0$ and as a two's complement subtracter when $c_0 = 1$.

High-Speed Adders

Several different adder designs have been developed for performing high-speed addition. These include **carry lookahead adders (CLAs)**, carry-completion adders, conditional-sum adders, and carry-select adders. Carry lookahead adders have gained wide acceptance in the design of ALUs due to the speed obtained and because they can be conveniently implemented in integrated circuit form.

This material covers only the carry lookahead approach. However, before beginning that discussion, let's briefly explain why fully parallel adders are not feasible. Addition is a combinational process so it is theoretically possible to construct a $2n$-bit "full-adder" that can be realized by a three-level combinational logic circuit and that can perform addition of two n-bit numbers in the time equal to the delay of the circuit. However, such circuits are too costly in terms of gate fan-in to be implemented for reasonable values of n. Carry lookahead is a practical and effective compromise between fully parallel adders and ripple-carry adders. The block diagram of a four-bit CLA is shown in Fig. 75.52(a).

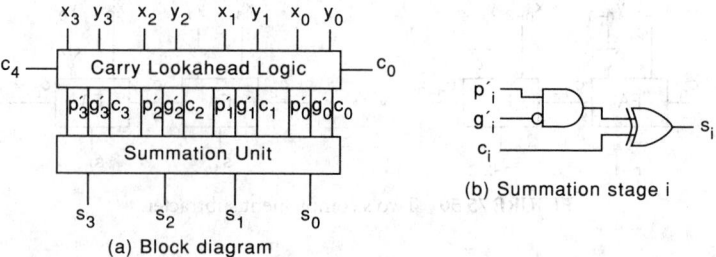

FIGURE 75.52   Carry lookahead adder.

CLAs are based on the observation that a carry-out (c_i) of the ith stage of a full adder is produced by either the propagation of the carry-in (c_{i-1}) through the ith stage or the generation of a carry in the ith stage. This can be seen in the following logic equations for c_i:

$$c_i = x_{i-1}y_{i-1} + x_{i-1}c_{i-1} + y_{i-1}c_{i-1}$$

$$= x_{i-1}y_{i-1} + (x_{i-1} + y_{i-1})c_{i-1}$$

$$= g_{i-1} + p_{i-1}c_{i-1}$$

where $g_i = x_i y_i$ and $p_i = x_i + y_i$ are the generate and propagate terms, respectively, for stage i for $i = 0$ to $n - 1$.

The carry equations for an n-bit adder can be derived by repeatedly applying the above equation. The following set of equations results for the $n = 4$ case:

$$c_1 = g_0 + p_0c_0$$

$$c_2 = g_1 + p_1g_0 + p_1p_0c_0$$

$$c_3 = g_2 + p_2g_1 + p_2p_1g_0 + p_2p_1p_0c_0$$

$$c_4 = g_3 + p_3g_2 + p_3p_2g_1 + p_3p_2p_1g_0 + p_3p_2p_1c_0$$

The carry equations can be realized by three-level combinational logic circuits to form the carry lookahead logic block shown in Fig. 75.52(a). The sum (s_i) bits for the ith stage of an adder can be written in terms of g_i, p_i, and c_i and generated by the logic circuit given in Fig. 75.52(b). This completes the description of the four-bit CLA.

Now let's examine the add-time, t_{add}, for a CLA. Assume that both addends are applied to the CLA simultaneously and that $c_0 = 0$. Also, let t_{pd} represent the propagation delay of a three-level logic circuit. There are two components that contribute to the add time. First, the three-level carry lookahead logic must produce the carries. This takes t_{pd}. Then, the summation unit must produce the final values of the sum bits. This step takes a time equal to the propagation delay of the exclusive-OR gate in the summation unit which is t_{pd} since an exclusive-OR gate can be realized as a three-level combinational logic circuit. Hence, the add time for a CLA is

$$t_{add} = 2t_{pd}$$

The above result indicates that the add time of a CLA is not only much faster than a ripple-carry adder but is also independent of the length (n) of the addends. Hence, one might conclude that CLAs are the final answer to the high-speed adder problem. However, a closer look at the set of carry equations given above quickly reveals that the equations become progressively more complex in the number of product terms and literals. Therefore, fan-in constraints will eventually limit the practicality of realizing the equations in three-level logic. The actual limit is technology dependent.

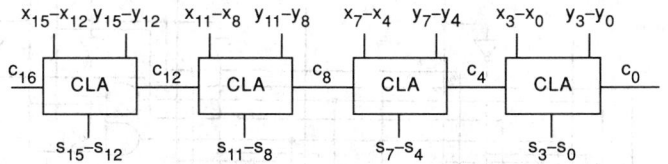

FIGURE 75.53 Cascaded carry lookahead adders.

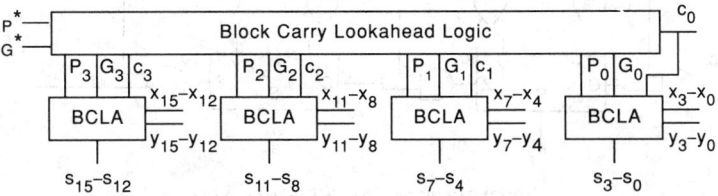

FIGURE 75.54 Block carry lookahead adders.

Standard single-chip medium-scale ALUs typically handle four-bit operands, although longer lengths are certainly feasible with today's technology.

CLAs may be cascaded to produce an adder for longer operands. Figure 75.53 shows a cascade of four 4-bit CLAs to produce a 16-bit adder. Carries are produced using carry lookahead logic within each CLA stage but must propagate between stages in a manner reminiscent of a ripple-carry adder. Hence the add time of cascaded CLAs is dependent on the number of stages in the cascade. The four-stage adder shown in Fig. 75.53 has a worst-case add time of $5t_{pd}$. In general, the add time of an m-stage cascade is $(m + 1)t_{pd}$.

The carry lookahead approach can be applied at a higher level to eliminate the propagation of carries between CLA stages or blocks. This approach uses **block carry lookahead adders** (BCLAs) and block carry lookahead (BCL) logic as shown in Fig. 75.54. A BCLA is a CLA modified to produce block carry propagate (P) and block carry generate (G) outputs instead of a carry-out. BLC logic is a combinational logic circuit that generates block carries (C_j) for each BCLA from the P and G outputs of lower-order BCLAs and c_0. Logic equations for the block carry logic can be derived by repeated application of the following equations for a typical block:

$$C_j = G_j + P_j C_{j-1}$$

where

$$G_j = [g_3 + p_3 g_2 + p_3 p_2 g_1 + p_3 p_2 p_1 g_0]_j$$

and

$$P_j = [p_3 p_2 p_1 p_0]_j$$

BCLAs and block carry logic units are available in standard medium-scale integrated circuits. Extension of the carry lookahead concept to k levels is possible in theory. However, more than three levels is usually not practical.

Multifunction Arithmetic Logic Units

Devices that can provide a variety of addition, subtraction, and logical operations can be easily designed around the adders/subtracters presented in the previous sections. The logic diagram of the first two stages of an n-bit multifunction ALU is given in Fig. 75.55. Operand inputs for the device are $X = x_{n-1} \ldots x_1 x_0$ and $Y = y_{n-1} \ldots y_1 y_0$ and the output is $S = s_{n-1} \ldots s_1 s_0$. The function performed on the operand(s) is determined by the values of the control inputs k_2, k_1, k_0, and c_{in} as shown in Table 75.7. The given realization is based on a ripple-carry adder for simplicity of presentation. However, the same design approach can be used with other adders such as carry lookahead.

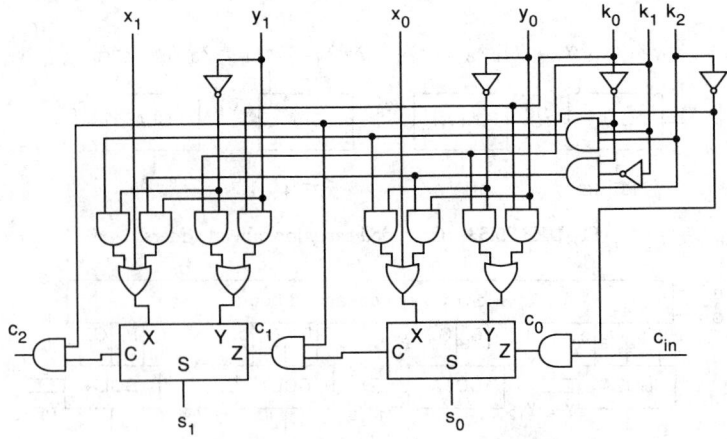

FIGURE 75.55   Multifunction ALU.

Table 75.7   Functions Performed by the Multifunction ALU

Control Inputs				Result	Function
k_2	k_1	k_0	c_{in}		
0	0	0	0	$S = X$	Transfer X
0	0	0	1	$S = X + 1$	Increment X
0	0	1	0	$S = X + Y$	Addition
0	0	1	1	$S = X + Y + 1$	Add with carry in
0	1	0	0	$S = X - Y - 1$	Subtract with borrow
0	1	0	1	$S = X - Y$	Subtraction
0	1	1	0	$S = X - 1$	Decrement X
0	1	1	1	$S = X$	Transfer X
1	0	0	...	$S = X$ OR Y	Logical OR
1	0	1	...	$S = X$ OR Y	Exclusive-OR
1	1	0	...	$S = X$ AND Y	Logical AND
1	1	1	...	$S =$ NOT X	Bit-wise complement

Table 75.8   Typical Integrated Circuit Arithmetic and Logic Devices

Part Number	Function	Features
74LS181	4-bit multifunction (16) ALU	BCL outputs
74LS182	Carry lookahead generator	Use with 74LS181 for BCL
74LS183	Full adder	Two per package
74LS283	4-bit binary adder	Internal CL
74LS381	4-bit multifunction (8) ALU	BCL outputs
74LS382	4-bit multifunction (8) ALU	Ripple-carry output

Standard Integrated Circuit ALUs

The devices described above are generic in nature but are similar in function and realization to many commercially available integrated circuit products. Representative products are summarized in Table 75.8.

Defining Terms

Arithmetic logic unit (ALU):   A combinational logic circuit that can perform basic arithmetic and logical operations on n-bit binary operands.

Block carry lookahead adder (BCLA): An adder that uses two levels of carry lookahead logic.

Carry lookahead adder (CLA): A high-speed adder that uses extra combinational logic to generate all carries in an m-bit block in parallel.

Full adder (FA): A combinational logic circuit that produces the two-bit sum of three one-bit binary numbers.

Ripple-carry adder (RCA): A basic n-bit adder that is characterized by the need for carries to propagate from lower- to higher-order stages.

References

J. Gosling, *Design of Arithmetic Units for Digital Computers,* New York: Springer-Verlag, 1980.

K. Hwang, *Computer Arithmetic: Principles, Architecture, and Design,* New York: John Wiley and Sons, 1979.

M. M. Mano, *Digital Logic and Computer Design,* Englewood Cliffs, N.J.: Prentice-Hall, 1979.

H. T. Nagle, B. D. Carroll, and J. D. Irwin, *An Introduction to Computer Logic,* Englewood Cliffs, N.J.: Prentice-Hall, 1975.

E. E. Swartzlander, Jr., Ed., *Computer Arithmetic, Volume I,* Los Alamitos, Calif.: IEEE Computer Society Press, 1980.

E. E. Swartzlander, Jr., Ed., *Computer Arithmetic, Volume II,* Los Alamitos, Calif.: IEEE Computer Society Press, 1990.

TTL Data Book, Texas Instruments, Inc., Dallas, Texas, 1988.

J. F. Wakerly, *Digital Design Principles and Practices,* Englewood Cliffs, N.J.: Prentice-Hall, 1990.

S. Waser and M. J. Flynn, *Introduction to Arithmetic for Digital Systems Designers,* New York: CBS College Publishing, 1982.

Further Information

The reader wanting information on the theoretical aspects of computer arithmetic is referred to the *IEEE Transactions on Computers,* a monthly publication of the Institute for Electrical and Electronics Engineers, Inc., 445 Hoes Lane, P.O. Box 1331, Piscataway, NJ 08855-1331.

More information on the specifications and applications of integrated circuits can be found in the data books and application notes published by electronics manufacturers such as Texas Instruments, Motorola, and National Semiconductor.

Discussions of multiplication, division, and floating-point arithmetic can be found in numerous textbooks on computer architecture.

Microprocessors

John Staudhammer
University of Florida

Sue-Ling Chen
Wavefront Technologies

Phillip J. Windley
University of Idaho

James F. Frenzel
University of Idaho

76.1 Practical Microprocessors 1748
  Types of Microprocessors and Microcontrollers • Software for
  µP/µC Systems • Packaging and Cost • Programming of µPs •
  Development Support • Comparison of µP/µC Chips • Trends in
  µP/µC Developments

76.2 Applications 1753
  Data Collection • Control • Computing

76.1 Practical Microprocessors

John Staudhammer and Sue-Ling Chen

A **microprocessor** (µP) is a semiconductor die containing the components of a computer central processor, complete with instruction processing unit, arithmetic, **interrupt**, and basic communication facilities. Such devices have been around since the early 1970s and have greatly benefitted from the continuing improvements in electronics. As microelectronic technology allows the **feature sizes** of components to be shrunk, more powerful µP systems are being put on single **dies**. One of the early microprocessors, the Intel 8080, is still a widely used controller; its understanding and study are recommended for all designers of microprocessor systems [see Gaonkar, 1984]. Early processors contained a few thousands of transistors, while the top-end microprocessor of 1992 has over 1.5 million. By the mid-1990s processors made with two million transistors will be in widespread use. The early processors were 4-bit machines, initially intended for hand-held calculator use; the largest chips today are intended for use as full central processing units of large computers. What makes these devices *micro* is that they are microelectronic devices. These devices are typically built of NMOS (N-type metal-oxide semiconductor) or CMOS (complementary metal-oxide semiconductor) circuits; the CMOS version is typically a bit more expensive and requires less power. An excellent overview of microprocessors and systems using them is found in Raffiquzzaman [1990].

Application of a microprocessor involves adding memory for program and data, and input/output circuitry, which may involve analog/digital and digital/analog converters. As feature sizes of electronic components shrink, the manufacturer of the chip has three options:

1. Reduce the chip size (the die size)—this may result in a cheaper device of the same capability.
2. Increase the processing power at the same cost.
3. Add **peripheral** circuits to the processor, thus putting on-chip devices normally added to the µP. Adding the right combination of peripherals has the greatest benefit for system cost.

A **microcontroller** (µC) is a microprocessor with peripherals on the same chip. These include various types of memory, interrupt structures, communication means, timing and data acquisition circuits.

All three chip developments listed above occur simultaneously, thus keeping a successful product line going for many years. For example, the Motorola 6800 processor was introduced about 20 years ago; today it is still available, but the manufacturer steers designers to the successor chips, the 6809 μP and the 68HC11 μC.

A μP chip communicates to its peripherals by means of three sets of lines: the bidirectional data bus, the memory address bus, and the control bus. In addition, the μP will also send/receive data on communication lines; the number and types of these vary greatly among chips. A μC will have on the same chip a number of memory elements, timers, communication ports and buffers, counters, and analog/digital converters in addition to a processor; communication to the peripheral circuits is through ports assigned to these devices, as well as any needed external memories. A discussion of general μC systems is given in Clements [1987], while Myers and Budde [1988] present the details of a very-high-performance μC.

All processors require an external clock, typically a crystal in the few-megahertz range. The processor internal circuits run at this speed, but external devices usually use a submultiple of this rate, the bus clock, which is usually 2 to 4 times slower. The advertised clock rate is usually the external clock rate, not the bus cycle speed.

There are well over a hundred different μP systems on the market. They each have peculiarities and may have some advantages. They differ by the kind of data they handle, the amount of processing they do, and the software support they enjoy. What makes μP practical is not so much its claimed prowess, typically stated in peak instruction execution capability (MIPS, millions of instructions per second), but rather its ease of use in a given application, determined to a large degree by the kind and amount of support software available from the manufacturer.

Types of Microprocessors and Microcontrollers

The yearly compendium of μP/μC chips [Markowitz, 1991] categorizes these chips by the width of the data path: 4, 8, 16, 32, and 64 bits. In addition, high-performance chips include bit/word slice chips—these are meant to implement the functions of a central processing unit for a limited number of bits (4 or 8 bits) and are meant to be concatenated for handling an entire computer word. For a discussion of these chips and their uses, see Mick and Brick [1980].

The most precious resource in a μP chip is the number of connection pins. Great efforts are made to utilize the ones used by various control lines and mode selections for the processor. Hence, a given μP chip may have four or more modes of operation: it is simply cheaper to build a flexible system, rather than several different ones.

The vast majority of μPs possess a richness of data access means (addressing modes); they support many different ways of working with memory and external data items. They are complex instruction set computers (CISC). A prime example is the Motorola 68000 with five data groupings and nine addressing modes.

Even the simplest 4-bit processor in wide use (National Semiconductor COP400) has a 10-bit address bus (and a 4-bit data bus). The device can have up to 1 K words (1024 bytes) of memory on the chip, which is often enough for a simple dedicated task. Thus, these chips often appear as single items in simple computer-controlled devices. Most of these chips are found in embedded applications, in kitchen appliances, and in toys.

By far the largest volume of μPs are 8-bit devices; they use 8-bit-wide (1 byte) data paths, but have address busses usually 16 (or more) bits. Often the data bus is time-multiplexed with the lower byte of the address bus. These types of devices can access as much as 65 K memory locations. The typical instruction execution time is 3 to 7 bus cycles. The most widely used 8-bit processor is the Zilog Z-80.

The 16-bit μP chips have 16 data and address lines. These are typically the μP chips at the heart of personal computer systems. The typical execution time is 2 to 5 bus cycles.

The 32-bit and 64-bit μP chips are the high end of these devices and find application in advanced personal computers, high-performance workstations, and digital controllers. They are characterized by high cost (compared to most μC chips) and extensive support circuitry. The typical execution time is 1 to 2 bus cycles. They represent the developing efforts in μP technology.

Microcontroller chips contain a μP and various items that make up a μP-based system: there is usually some random access memory (RAM) for holding volatile data, at least one kind of read only memory (ROM) for holding the control program, communication peripherals, including parallel interface(s), serial communication adapter(s), various counting and timing circuits for measuring input pulses, and analog/digital converters. Because each processor has an external clock, usually a crystal clock, timing intervals can be determined to great precision. Many procedures have been developed to take advantage of this precision: voltage-to-frequency converters are used externally to bring pulses to the μP system, which then proceeds to accurately count them. The timing and counting capabilities of μP systems gives them their ubiquitous applicability.

Software for μP/μC Systems

Manufacturers of μP/μC devices have gone to extraordinary efforts to make their devices attractive to system designers. Much software is offered for use with the chips and to support the design effort. ROM-based monitor programs are available in most chips so that normal communication tasks can be accomplished easily; programs for usual input/output tasks, data acquisition, timing, and program examples are distributed so that (relatively) error-free software is available for the designers. Most of these are distributed through public-access dial-up bulletin boards. Programming then becomes largely an adaptation of these programs to the tasks that the system is to perform. Of course, the system task analysis and program system design are still the designer's responsibilities, as is the conduct of a software validation.

The μP receives instructions, from its internal or external memory, and data, as a combination of zeros and ones; it is this machine code instruction that controls the operation of the system. The writing of such machine code is far too tedious. An English-like language, machine mnemonics, is used to describe each machine operation and this is then translated to the 0's and 1's which are stored in the memory. For example ADD 123 may stand for the machine code necessary to perform the addition of the number contained in memory location 123 to the content of the computer's arithmetic register, leaving the result in the arithmetic register. The manufacturers supply assemblers to accomplish the mnemonic language assembly into code that may be loaded into the μP memory. These assemblers typically will run on personal computers and various minicomputer systems. Programs that monitor the operation of an actual chip through the μP serial port(s) are effective development tools.

Hence, the typical software includes an assembler, a loader, and a monitor. More user-oriented software for these tasks is available from third-party vendors; many of these extra-cost software packages are highly cost effective. Interactive assemblers and debugging tools are particularly good investments.

Packaging and Cost

The simplest processors are housed in normal dual in-line packages (DIP) of 16 pins or more, spaced 0.3 inches wide. Thus, they resemble conventional transistor-transistor logic (TTL) packages. The mid-range ones are usually in large DIP packages, 0.8 inches wide, with 40 to 68 pins (2 to 4 inches long.) The high-performance ones come in multirow pin packages and may require 200 connections and may use unusual chip sockets. The high performance chips may require special cooling.

Virtually all processors are multisourced, i.e., they are available from more than one manufacturer. This is an important consideration for continued product support. The processors come in various speed grades; the higher speed ones may be 3 to 5 times as costly as the slow-speed versions.

The low-end processors, with no memory and a few communication ports, cost less than $1 in quantity (1992). The most widely used Z-80, developed about 15 years ago, costs about $1, while the high-end processors (i860, i486) cost around $1000, in low quantities.

Programming of μPs

Most μP systems are programmed in assembly language. Each of the instructions that the processor can execute is given an English-like name (i.e., ADD for the instruction to add two numbers) and translator programs are available from the μP manufacturers to translate the English-like statements (the source code) into the machine codes (which are streams of zeros/ones that the μP interprets). Such assemblers are available from public-access bulletin boards maintained by several vendors; however, convenient assemblers for specific μP systems are usually bought from third-party vendors. Almost all μP assemblers are meant to be used on personal computers; some are available for minicomputer systems. One should check with the chip manufacturer for current software.

High-level language compilers are also available for most μPs, typically from third-party vendors. These are usually C-language compilers, but Pascal compilers also may be available. Here again, the best guide is the chip manufacturer.

The programming of μP systems usually involves carefully tailored code to control signals that interact with the processor chip. Usually the functions that are required of the processor are time-critical real-time control actions. Programs for such applications can easily become highly intricate and require detailed knowledge of all the functions of the processor chip. For a discussion of these problems, see Chapters 3 and 4 of Peatman [1988].

Development Support

Complex μP/μC systems are designed top-down: the task statements are successively refined to a set of smaller tasks, until they can be implemented in relatively short subroutines. The design approach for micro-based systems is described and illustrated in Peatman [1988].

The most important support software for a designer is programs for checking the system, both for logical flow and for cycle-by-cycle activity. Often subtle errors and data dependencies will occur; finding them may be a daunting task. Simulators are the first-level checkout tools for programs. Simulators do not use the actual hardware; rather they use a software model of it. They calculate and show the contents of all computer registers, of ports and selected memory locations, so that an effective check of the internal operations can be made. Most simulators are from third-party vendors.

The actual operation of the μP/μC system may be checked with an in-circuit emulator (ICE), an expensive but effective tool that replaces the μP or μC pin-for-pin in the actual circuit. The ICE typically uses a more powerful computer to mimic the performance of the μP/μC being developed. The ICE tracks signals (including many transients) and can be used to effectively show the behavior of the system, as well as the expected response from the μP and its associated software.

Comparison of μP/μC Chips

As with all computer devices, the advertised speeds and performance figures must be carefully interpreted. These numbers tend to measure only the performance of the manufacturer's test cases and may not be directly comparable [see Hennessey and Patterson, 1990]. These numbers may not be applicable to any one user's requirements. Since the market for μP/μC chips is highly competitive, small differences become amplified in advertising.

Unfortunately, getting a valid number for comparison purposes requires an extensive effort at benchmarking. The advertised performance figures may be taken as a guide, if the task is similar to the benchmark programs. If the processor does not pass the comparison with a comfortable margin, however, one should opt for a higher-performance version of the same chip (if much programming already has been done) or select a clearly superior other candidate.

Trends in µP/µC Developments

An entire microprocessor may be used as a building block in an application-specific integrated circuit (ASIC), available from various vendors. ASIC and VLSI design tools may be used to design such systems, tailored to specific user applications. The µP manufacturers are developing µC chips extending the use of their basic processors. For any application, the best procedure is to invite several vendors to propose alternate systems. Current chips are merely indications of devices to come.

High-performance µP systems are becoming more RISC (reduced instruction set computer) oriented. The object is to execute one instruction per clock period (versus the prevailing 3 to 7 cycles) and to obtain a more regular processor structure. The most popular µP/µC systems are CISC processors. For a user it makes little difference what the internal structure of the processor is; the availability of user support software is far more critical. The trend is to make processors simpler, thus speeding up program execution, even if some programs may have to use more program steps in replacing many "convenience" instructions.

Nevertheless, the very high end µP systems have such a wide data path (32 or 64 bits) that more than one instruction may be accessed at one time. These machines are termed very long instruction word (VLIW) machines. They are able to execute more than one instruction per clock cycle, but they will be more complex internally. For users, both the simpler RISC and the complex VLIW machines will provide increased performance.

By the mid-1990s microelectronics will be able to produce 2 million transistor chips. While they will be expensive, they will replace virtually all of the peripheral chips: memory, timers, and communication channels can all be controlled from a single chip. However, the humble 8-bit chip will still be the most cost-effective workhorse of the bulk of µC applications.

Defining Terms

Die: The piece a silicon wafer containing all electronics.

Feature size: The characteristic size of electronic components on a die.

Microcontroller (µC): A microelectronic chip incorporating a µP and memory, communication, as well as other computer support functions.

Microprocessor (µP): A microelectronic chip that carries out all operations of a computer central processing unit: instruction fetch, execution, interrupt and management of address, data and control lines which are connected to the chip.

Peripheral: Device that supports the functions of the processor. The peripheral may be all electronic (a communications adapter) or may contain mechanical parts (a disk memory). To the processor a peripheral appears as electronics with timing constraints.

References

A. Clements, *Microprocessor System Design*, Boston: PWS Publishers, 1987.

R.S. Gaonkar, *Microprocessor Architecture, Programming, and Applications with the 8085/8080A*, 2nd ed., Columbus, Ohio: Merrill Publishing Company, 1984.

J.L. Hennessey and D.A. Patterson, *Computer Architecture: A Quantitative Approach*, Palo Alto, Calif.: Morgan Kaufmann Publishers, 1990.

M.C. Markowitz, "EDN's 18th annual µP/µC chip directory," *EDN Magazine*, November 21, 1991; see also *EDN Magazine*, November 26, 1992.

J. R. Mick and J. Brick, *Bit-Slice Microprocessor Design*, New York: McGraw-Hill, 1980.

G. J. Myers and D. L. Budde, *The 80960 Microprocessor Architecture*, New York: Wiley Interscience, 1988.

J. B. Peatman, *Design with Microcontrollers*, New York: McGraw-Hill, 1988.

M. Rafiquzzaman, *Microprocessors and Microcomputer-Based System Design*, Boca Raton, Fla.: CRC Press, 1990.

Further Information

The magazine *EDN* runs an annual microprocessor/microcontroller review, typically late in the year (i.e., November 23, 1989; November 22, 1990; November 21, 1991). These are handy compendia of characteristics.

The IEEE magazine *Micro* presents detailed articles on device development and applications of microprocessors and systems which embed them. At mid-year the magazine carries a set of articles based on the Hot Chips Symposium, presenting developments in μP and chip technology for high-performance workstations and systems.

Specifics of various microprocessor and microsystem chips are found in the respective manufacturer's reference literature; for any design one must become familiar with the applicable manual and design notes. Even for a modest chip the reference manual may run 300 pages; in addition, the manufacturer's free-of-charge support software is of comparable size. For example the Motorola 68HC11 reference manual is 512 pages, and there is over 1 megabyte of design support software available from the manufacturer. The Intel i860 data book is 150 pages, hardware and programmer reference manuals are over 300 pages, and the support software is several megabytes.

76.2 Applications

Phillip J. Windley and James F. Frenzel

Microprocessors are cheap, small, and consume little power. In addition, in recent years their performance has increased at a greater rate than the performance of larger computers. These factors have led to an explosion in the application of microprocessors. A short section could never do justice to every application; therefore, we will view representative applications in three broad areas:

- Data collection, where microprocessors are used to monitor sensors and either record the collected information or communicate the information to some other computer.
- Control, where microprocessors have largely replaced analog electronics for controlling everything from manufacturing robots to home appliances.
- Computing, where microprocessors have transformed the concept of computer and made parallel processing possible.

Admittedly, these categories are not strictly disjoint. They do, however, represent the most pervasive uses for microprocessors at an abstract level.

Data Collection

In data collection the microprocessor-based system serves primarily as a low-cost data recorder. Basic functions include the polling of sensors, acceptance of data, data storage, and data transmission or display. Additional features might include preprocessing of the raw data. Such a classification spans a broad range of applications, from automotive diagnostics to space-born monitoring stations.

Microprocessors are well suited as the controller for such tasks because of their cost and flexibility. Sufficient numbers of processors may be used to allow real-time data acquisition. Because the microprocessor is programmable, sensors may be added, removed, or rearranged without major system impact. Finally, because the microprocessor is a computational device, calculations may be performed on the recorded data to produce useful information, such as calculating speed from distance and time. In the next section we will examine the components of one such system, the retail point-of-sale terminal [Hordeski, 1984].

Point-of-Sale Terminal

The function of a point-of-sale (POS) terminal is characteristic of the applications under the category of data collection. The microprocessor is not being used for intensive computations, nor for controlling a complex process, but rather to collect data, perform some processing, and then pass the results on to a central collector. The cost and flexibility of the microprocessor make it an excellent choice over special-purpose hardware.

System Components.   In addition to the microprocessor and storage capability, the typical retail terminal has one or more input devices for entering prices (e.g., keyboard, bar code scanner) and one or more output devices for displaying totals (e.g., paper tape, display). Often these terminals are part of a large network of terminals and may support additional features beyond totaling purchases such as automated inventory control and credit checking. A complete system is shown in Fig. 76.1, including magnetic tape for storing transactions and a **universal asynchronous receiver/transmitter (UART)** for communication with a central processing facility. Because of the high unit volume, it is desirable to keep the cost and complexity low. Typically, each terminal will have limited storage capability, relying on a central processor for maintaining store inventory and credit checks. In order to reduce communication traffic with the central processor, however, each terminal generally has in storage the current price for all items.

Universal Product Code.   The use of the Universal Product Code (UPC) has enabled the development of intelligent POS terminals which can "read" the UPC symbol and determine the identity of the item. The UPC symbol consists of ten decimal digits, split into two fields of five digits each. Each digit is encoded using a 7-bit binary number, represented by a group of 7 dark (binary 1) and light (binary 0) bars. The five left-hand digits are encoded using odd parity and the right-hand digits are encoded using even parity. This allows correct recognition of the symbol, independent of its orientation.

For groceries, the first five decimal digits identify the manufacturer and the second group of five digits identify the specific product. There are additional codes in use as well, such as the National

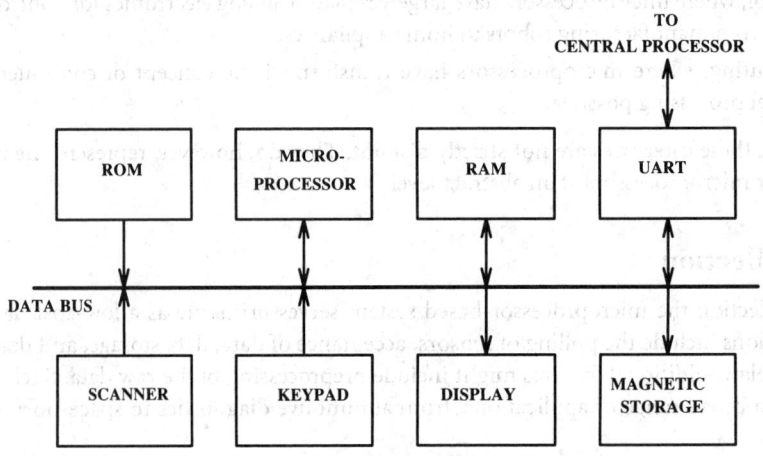

FIGURE 76.1   Point-of-sale terminal system.

Drug Code. By using a microprocessor-based system, a POS terminal can be quickly reconfigured to recognize a different code (or multiple codes) through a simple software change.

Operation. A typical sale might involve the following steps. The clerk inquires whether the sale is to be a cash purchase or charged to an account. If the latter, the clerk enters the necessary information and the terminal transmits a request to a central processor, inquiring as to the available credit. In the interim, items are entered, either through the bar code scanner or the keypad, and the price and running total are displayed. The identity of the items purchased is also stored for later transmission to the central processor responsible for inventory control. Finally, the terminal checks the available credit against the total and records the transaction for later transmission to the central processor.

Digital Tachometer

Another example of using a microprocessor for data collection is the implementation of a digital tachometer [Bonert, 1989]. The microprocessor samples the output of a shaft **encoder** and compares it with a reference signal to determine the rotational speed. The calculated value is passed to a digital-to-analog converter to generate an analog speed signal. The system is shown in Fig. 76.2.

Speed Evaluation Methods. Various methods may be used to evaluate the speed value, all of which involve some combination of pulse counting and time measurement. The constant elapsed time (CET) method provides a good compromise between measurement accuracy and response time. The CET method records the number of encoder pulses observed during a fixed time interval. The rotational speed, *n*, is then given by

$$n = C_p/(C_t m/T_c)$$

where C_p is the number of encoder pulses, C_t is the number of clock pulses, *m* is the number of encoder marks per turn, and T_c is the clock pulse period.

Implementation. Rather than continuously stopping and resetting external counters, it is possible to take advantage of features often found in modern **microcontrollers**, microprocessors containing additional interface circuitry. Microcontrollers often contain counters, timers, and **capture registers**. Capture registers allow the storing of timer or counter values triggered by an external signal. At the start of evaluation, the rising edge of the next encoder pulse triggers the capture of the timer count and the pulse count. After a minimum evaluation time has elapsed, the next encoder pulse again triggers the capture of the current counter values. The rotational speed can then be computed using the difference between the captured values. A flowchart of the algorithm is shown in Fig. 76.3.

Performance. Using an encoder with 1024 marks per revolution, a 2-MHz reference clock, and an evaluation period of 2.3 ms resulted in a measurable speed range of 25.5–4883 rpm. The maximum relative error was 0.123%, induced primarily by the encoder tolerance [Bonert, 1989].

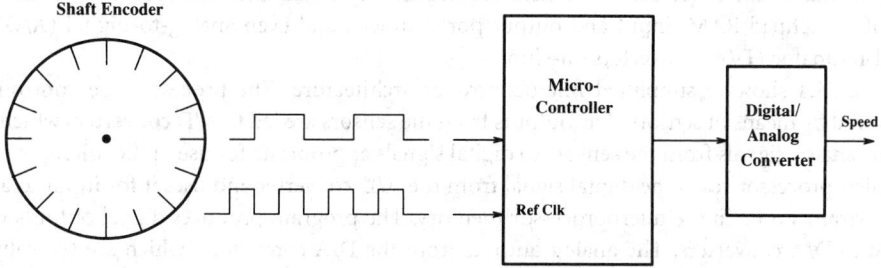

FIGURE 76.2 Digital tachometer.

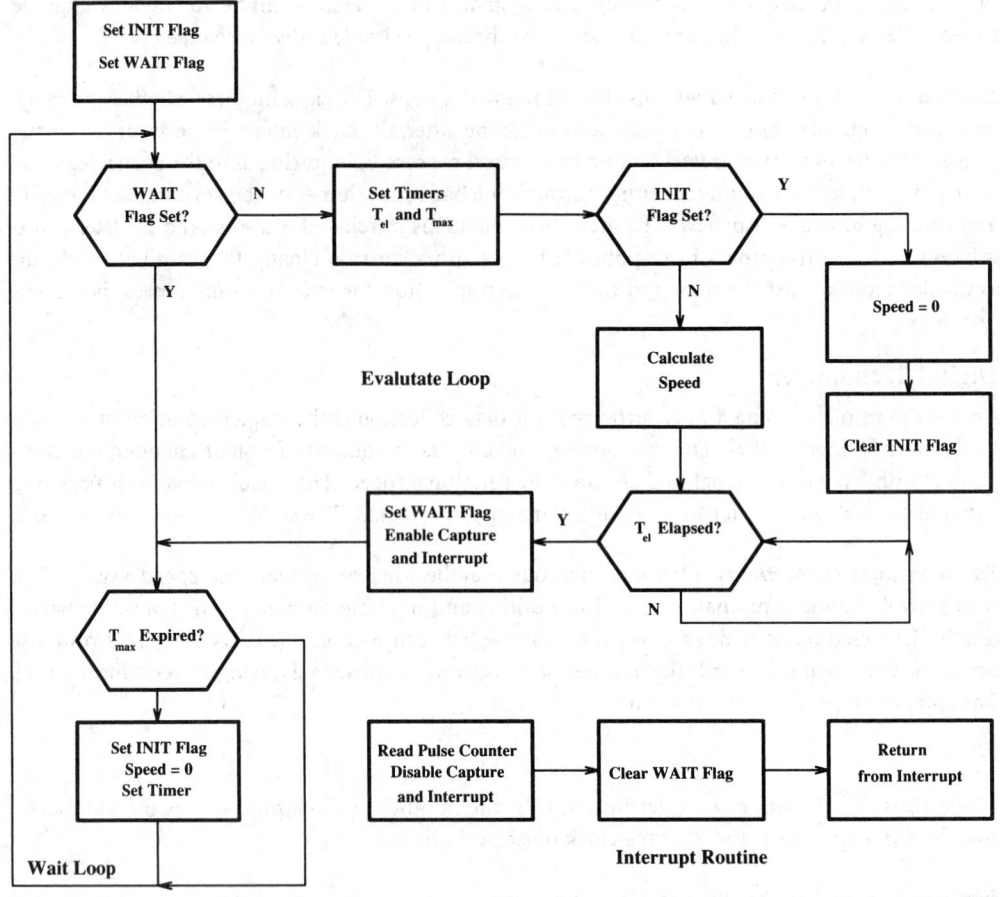

FIGURE 76.3   Tachometer program. (*Source:* Bonert, 1989.)

Control

Microprocessors are ubiquitous in control applications. While some custom analog controllers are still built, the advantages of cost and flexibility inherent in microprocessors make them a natural choice. The advantages of microprocessors are particularly obvious in mass-produced goods where time-to-market can be a significant driving force.

Microcontrollers

Microprocessors designed especially for use in control applications are called *microcontrollers*. Typically, the major difference between a microcontroller and a standard microprocessor is the presence of scratchpad RAM, input and output ports, timers, and even analog-to-digital (**A/D**) and digital-to-analog (**D/A**) converters on-chip.

Figure 76.4 shows a simplified microcontroller architecture. The process to be controlled is monitored by means of sensors. The outputs from the sensors are fed to A/D converters which convert the analog signals from the sensors to digital signals appropriate for use in the microprocessor. The microprocessor reads the digital signal from the A/D converter and uses it for input to a control program stored in the microprocessor memory. The program produces digital outputs which are fed to D/A converters. The analog outputs from the D/A converters (which are typically low power) are fed to amplifiers, and the amplified signal is used to control actuators that affect the process being controlled.

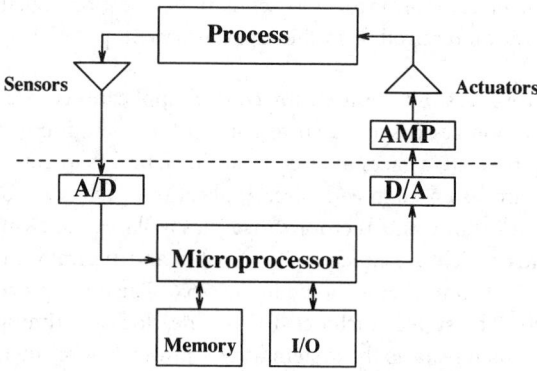

FIGURE 76.4 Typical microcontroller design.

Control Applications

Consumer Electronics. A survey of the typical home will show numerous microprocessors where 10 years ago, there were none. Microprocessors are used for controlling VCRs, TVs, stereo equipment, microwave ovens, sprinkler systems, telephone equipment, heating systems, and virtually every other appliance using electricity.

Manufacturing. Microprocessors have found numerous applications in manufacturing. Perhaps none is better known than the robot. Microprocessor technology has made the modern robot possible. Robot arms used in manufacturing typically have five or six joints. Current practice is to treat each joint in the robot arm as a separate servomechanism with its own control system. For example, the PUMA 560 robot arm, manufactured by Unimation, has six rotating joints. Each joint is controlled by an individual microcontroller system. Another computer calculates paths and sends individual joint motion information to the six joint servomechanisms [Fu *et al.*, 1987].

The servomechanism system shown in Fig. 76.5 consists of an 8-bit Rockwell 6503 microprocessor, a D/A converter, an amplifier, a joint motor, and an encoder. The 6503 microprocessor receives joint position information from the supervisory computer every 28 ms. The microprocessor calculates the joint error information by comparing the current position to the desired joint position using the PID (proportional-integral-derivative) control method. The error is converted to an analog signal by the D/A converter and amplified before going to the joint motor. The encoder is connected to the motor shaft and provides a digital signal to the microprocessor.

The microprocessor performs the following functions:

1. Receives the desired joint position from the supervisory computer every 28 ms
2. Reads the position signal from the encoder every 0.875 ms
3. Calculates the error every 0.875 ms
4. Sends the error to the D/A converter

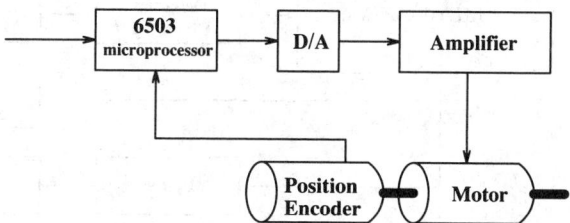

FIGURE 76.5 Microprocessor-controlled servomechanism from a PUMA 560 robot.

The microprocessor calculates joint error and sends the correction signal to the joint motor 32 times for every joint position received from the supervisory computer.

Transportation. Microprocessors are used for control applications in every facet of the transportation industry. Microprocessors are used to control the operation of the vehicles themselves such as controlling engines, air surfaces in aircraft, antilock brakes in automobiles, and rudders in ships. Microprocessors are also used in wide-area applications such as traffic control.

In controllers for motor traffic, the microprocessor has replaced hardwired logic and analog systems to provide systems which are much more capable and typically more reliable [Hordeski, 1984]. A typical traffic light controller is shown in Fig. 76.6. The microprocessor provides the CPU, memory, and I/O ports. The system includes a real-time clock for timing external events and a power-fail restart unit which restarts the system after a power failure (including restoring volatile data). The system monitors traffic at the intersection through the use of loop detectors and controls the traffic by changing the traffic lights. Other components of the system monitor and control pedestrian traffic and provide an interface to the system for human operators.

The loop detectors are paired coils of wire placed under the pavement. The impedance of the loop detectors changes in response to the presence of a car on the roadway. The change in impedance changes the frequency of an RC oscillator, which is converted to a digital signal reported to the microprocessor. Loop detectors can be used to monitor the presence of a car at a traffic light, the length of a line of cars, and the speed of traffic.

The function of the traffic controller is to optimize traffic flow. For example, during busy periods of the day, the goal may be to optimize flow through an intersection. Another goal may be to ensure that traffic flows smoothly in certain directions to effectively feed larger roads. Traffic lights can be synchronized to provide a highway through a busy network of roads by ensuring that a car that enters the roadway and maintains a recommended speed can travel along the entire length without stopping at a traffic light. On the other hand, during periods of low use, such as night and early morning, the system may monitor for the presence of a car at an intersection and immediately switch the light to let it pass.

Microprocessors offer advantages in traffic control situations in addition to optimized traffic flow. When properly designed, the system can provide a certain degree of fault tolerance. When a loop detector is giving a faulty value, the system can be programmed to ignore its value and use values from adjoining lanes. An error report can be forwarded to a central traffic facility and after repairs are made, the loop brought automatically on-line. The system can also monitor feedback information from the traffic light to ensure that the lights are actually lit. When a problem is detected, the system can enter an emergency mode and report the problem.

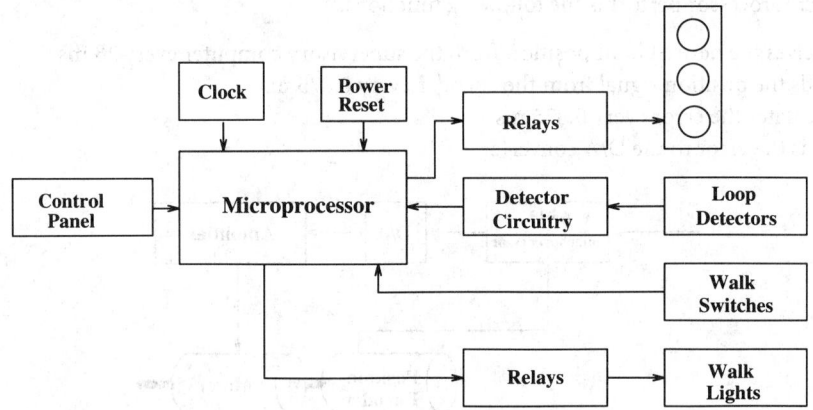

FIGURE 76.6 Traffic control system. (*Source:* M. Hordeski, *Microprocessors in Industry*, New York: Van Nostrand Reinhold, 1984, p. 398. With permission.)

Social Issues

The explosive growth in the use of microprocessors in control applications has caused discussion about the utility and safety of such devices.

An issue many people can identify with is feature overload. The advent of cheap microprocessors has turned design upside-down. Designers can add additional features for very little additional increased manufacturing cost. Competition spurs even more features until even the simplest of consumer items come with thick instruction manuals. Naturally, consumers become frustrated with features that are difficult to use.

Perhaps more important are the safety hazards that may be engendered by replacing analog control systems with digital control systems. Most analog systems are based on physical properties with continuous behavior. Digital systems, on the other hand, are discrete and are thus much more prone to problems where small errors can result in large changes in behavior due to the digital representation of value; a single bit change can result in a large change in magnitude. Digital control systems are becoming more and more prevalent in systems controlling aircraft, automobiles, nuclear power plants, and other safety-critical systems. Engineers who design the systems and officials charged with ensuring their safety are still coming to grips with the implications of this trend. New techniques for analyzing computer system designs for errors are being developed which promise to alleviate some of these concerns [Windley, 1990].

Computing

While microprocessors have been put to a plethora of interesting special-purpose uses such as data collection and control, perhaps the most visible use of microprocessors has been in the area of general-purpose computing.

Microcomputers

The advent of microprocessors has resulted in a personal computer on virtually every desktop. Even the slowest of these computers rival the performance of the largest computers available 15 years ago.

Figure 76.7 shows the major hardware components of a simple microcomputer. The central processing unit (CPU) is the execution engine of the microcomputer and is most often a microprocessor. One popular family of microprocessors used as the CPU in microcomputers is manufactured by Intel. These chips, with names such as the 8088, 80286, 80386, and 80486, are used in microcomputers such as the IBM personal computer. Another important family of microprocessors is the Motorola 68000 series, which is used in microcomputers manufactured by Apple Computer [Matloff, 1992].

In addition to the CPU, there are a number of other components in a microcomputer. General-purpose memory is not typically part of the microprocessor but must be added as a separate component. In simple microcomputers, the memory may be directly attached to the microprocessor. In more complex designs, the memory is attached to the microprocessor by a system bus that allows

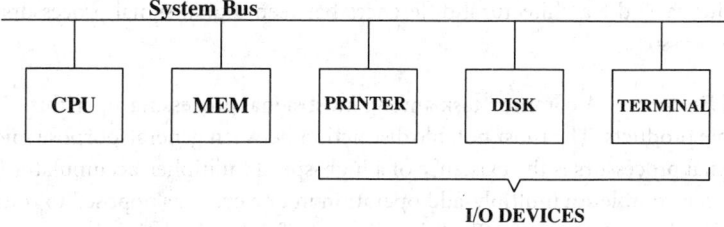

FIGURE 76.7 Major components of a simple microcomputer.

system components other than the microprocessor to access memory as well. In addition, the memory may have its own controller, called a memory management unit.

Other components in the system include input/output (I/O) interfaces to devices such as printers, terminals, disks, mice, and so on. The common feature of all of these devices is that they interface the microprocessor to the outside world. All of the components in the microcomputer are connected together by a system bus. The bus is a set of parallel wires that carry information from one component to another.

Multiprocessing

The desire for greatly increased computer performance has fueled research in using microprocessors as the computing engines in multiprocessors which would achieve performance gains over single-processor computers through the use of numerous low-cost microprocessors.

There are numerous multiprocessor architectures. An example architecture that is well suited to using large numbers of microprocessors is the hypercube. The hypercube architecture was originally developed by Charles Seitz and others at California Institute of Technology in the early 1980s. The hypercube depends on using large numbers of commodity microprocessors, each with private memory, in a hypercube network [Bell, 1989].

In a hypercube network, N microprocessors are arranged in an n-dimensional cube, where $N = 2^n$. Each processor is connected to n other processors and the longest communications path from any processor to any other is n links. For example, a three-dimensional hypercube contains eight processors and is arranged as a standard cube, where the nodes are the processors and the edges of the cube are the communication paths.

Figure 76.8 shows a four-dimensional hypercube represented as a tesseract. A four-dimensional hypercube has 16 processors, each is connected to 4 other processors, and the longest path between any two processors (shown in bold in Fig. 76.8) is 4. Thus, doubling the number of processors results in a unit increase in the communications path length. This logarithmic relationship results in the great advantage of the hypercube: it scales well. A system with 1024 processors has a maximum communications path length of just 10.

There are several manufacturers of hypercube systems including NCUBE and Intel. Most of these systems have between 32 and 1024 processors. NCUBE has a hypercube architecture with 8192 nodes operating at 2.4 megaflops each.

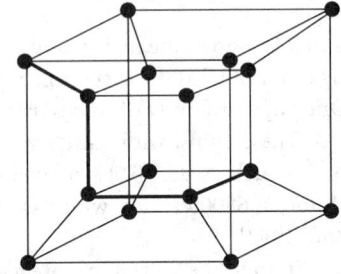

FIGURE 76.8   A four-dimensional hypercube.

Digital Signal Processing

Digital signal processing (DSP) may be considered a specific example belonging to the category of computation. Specialized microprocessors are finding widespread application in many areas of digital signal processing such as telecommunications, speech processing, medical imaging, and radar [Aliphas and Feldman, 1987]. These microprocessors are designed for very high data rates and contain specialized circuitry to accelerate computations that are specific to signal processing. Figure 76.9 illustrates the architectural differences between digital signal processors and conventional microprocessors.

Architectural Features.   A common task among most signal processing algorithms is the summation of multiple products. The most notable distinction between general-purpose microprocessors and digital signal processors is the existence of a high-speed multiplier-accumulator [Allen, 1985]. This circuitry can complete a multiply-add operation in one cycle, as opposed to roughly 25 cycles for a conventional microprocessor. Traditionally, only fixed-point arithmetic was available, but newer DSP chips provide floating-point arithmetic with 32 bits of precision.

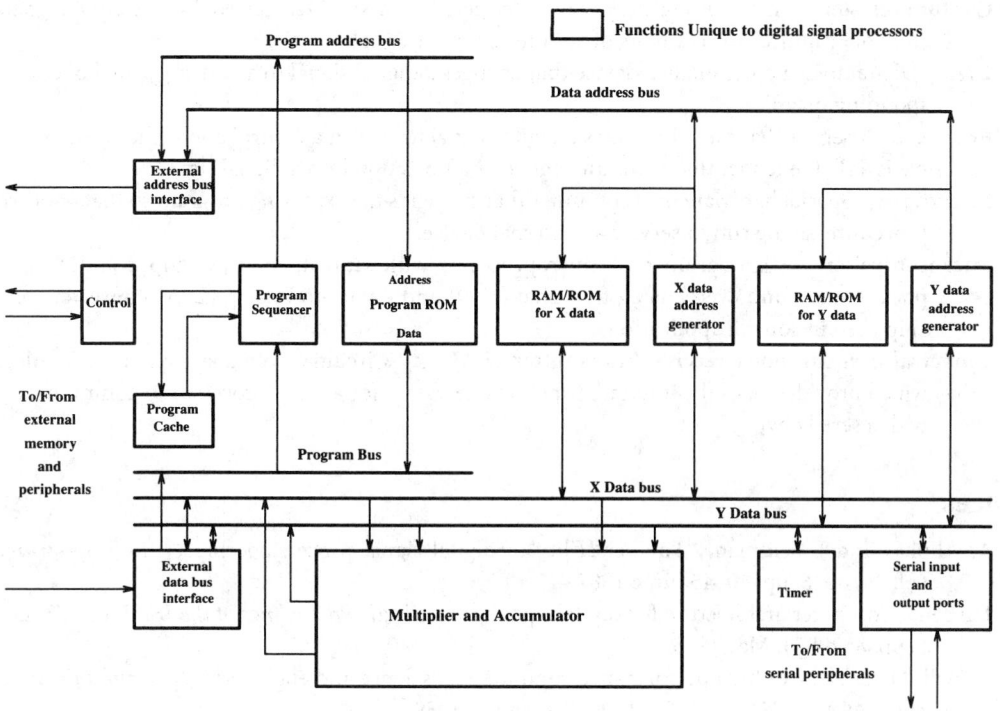

FIGURE 76.9 Digital signal processor architecture. (*Source:* Aliphas and Feldman, 1987.)

The second most noticeable feature on DSP chips is the existence of multiple data buses and memories. Many chips have two data memories, each with a data bus, allowing the simultaneous fetch of two operands for the multiply-accumulate operation. Furthermore, most chips use the Harvard architecture, characterized by separate program and data memories, so that instructions and data can be fetched simultaneously. Others use a modified Harvard architecture, where data can be stored in slower, cheaper program memory and moved to the faster data memory as needed.

Finally, DSP chips typically have separate arithmetic-logic units (ALU) for data arithmetic and address calculations. This serves two purposes: (1) data calculations can proceed unhindered by address calculations, maintaining a high throughput, and (2) each unit can be specialized for its particular task. For example, the data ALU may have additional circuitry to support saturation arithmetic, whereas the ALU used for address calculations may provide indexing, auto-increment, or even bit-reversal, an operation required for the fast Fourier transform (FFT).

Dedicated digital signal processors offer an excellent alternative or supplement to general-purpose microprocessors for signal processing applications. As a slave to a conventional processor, the DSP chip is freed from communicating with peripherals, increasing throughput. For additional performance, DSP chips may be operated in a multiprocessor configuration, controlled by a central processor. Such an arrangement would be appropriate for applications such as phased-array radar, where the volume of data and uniformity of the calculations lend themselves to distributed processing.

Defining Terms

A/D: Analog to digital. Usually a device that changes an analog signal to a digital signal of corresponding magnitude.

Capture registers: Internal registers which, triggered by a specified internal or external signal, store or "capture" the contents of an internal timer or counter.

D/A: Digital to analog. Usually a device that changes a digital signal to an analog signal of corresponding magnitude.

Encoder: A sensor that directly creates a digital signal for use in a control application. An example is a shaft encoder that turns an angular shaft position into a digital signal.

Interrupts: Special hardware on a computer that suspends the executing program so that another procedure can be run to service an external device.

Microcontroller: A special-purpose microprocessor with scratchpad RAM, input and output ports, timers, and even analog to digital (A/D) and digital-to-analog (D/A) converters on-chip used in control applications.

Universal asynchronous receiver/transmitter (UART):   Circuitry (often a separate module), which provides all of the interface functions necessary for a microprocessor to communicate with a serial device.

References

A. Aliphas and J. Feldman, "The versatility of digital signal processing chips," *IEEE Spectrum*, vol. 24, no. 6, pp. 40–45, June 1987.

J. Allen, "Computer architecture for digital signal processing," *Proceedings of the IEEE*, vol. 73, no. 5, pp. 852–873, May 1985.

G. Bell, "The future of high performance computers in science and engineering," *Communications of the ACM*, 32(9), pp. 1091–1099, September 1989.

R. Bonert, "Design of a high performance digital tachometer with a microcontroller," *IEEE Transactions on Instrumentation and Measurement*, vol. 38, no. 6, pp. 1104–1108, December 1989.

K.S. Fu, R.C. Gonzalez, and C.S.G. Lee, *Robotics: Control, Sensing, Vision, and Intelligence*, New York: McGraw-Hill, 1987.

M. Hordeski, *Microprocessors in Industry*, New York: Van Nostrand Reinhold, 1984.

N. S. Matloff, *IBM Microcomputer Architecture and Assembly Language*, Englewood Cliffs, N.J.: Prentice-Hall, 1992.

P. J. Windley, The Formal Verification of Generic Interpreters, Ph.D. Thesis, University of California, Davis, 1990.

Further Information

Byte magazine is a good resource for entry-level articles on microprocessor applications. For subscriptions contact: BYTE, One Phoenix Mill Lane, Petersborough, NH 03458.

The Institute of Electrical and Electronics Engineers (IEEE) publishes several magazines and journals that frequently contain articles concerning microprocessor applications. *IEEE Micro* is a bimonthly magazine which addresses the design and use of microprocessors and minicomputers. *IEEE Computer* is a monthly magazine covering all aspects of computing. Three pertinent journals published bimonthly by the IEEE are *Transactions on Industry Applications, Transactions on Industrial Electronics,* and *Transactions on Instrumentation and Measurement.* The address for the IEEE Service Center is 445 Hoes Lane, Piscataway, NJ 08855.

77

Displays

James E. Morris
State University of New York, Binghamton

André Martin
Hughes Display Products

Larry F. Weber
Plasmaco, Inc.

77.1 Light-Emitting Diodes.. 1763
   Semiconductor Device Principles • Semiconductor Materials •
   Device Efficiency • Interfacing
77.2 Liquid-Crystal Displays.. 1772
   Principle of Operation • Interfacing
77.3 The Cathode Ray Tube 1778
   Monochrome CRTs • Color CRTs • Contrast and Brightness •
   Measurements on CRTs • Projection Screen • Conclusion
77.4 Plasma Displays 1786
   Plasma Display Attributes • Gas Discharge Characteristics • Cur-
   rent Limiting for Plasma Displays • Plasma Display Products • Gray
   Scale • Color Plasma Displays

77.1 Light-Emitting Diodes

James E. Morris

The light-emitting diode (LED) has found a multitude of roles as the field of optoelectronics has bloomed. Infrared devices are used in conjunction with spectrally matched phototransistors in optoisolation couplers, hand-held remote controllers, interruptive, reflective and fiber-optic sensing techniques, etc. Visible spectrum applications include simple status indicators and dynamic power level bar graphs on a stereo or tape deck. This section will concentrate on digital display applications of visible output devices.

Semiconductor Device Principles

The operation of an LED is based on the recombination of electrons and holes in a semiconductor. As an electron carrier in the conduction band recombines with a hole in the valence band, it loses energy ΔE equal to the bandgap E_g with the emission of a photon of frequency

$$\upsilon = c/\lambda = \Delta E/h \tag{77.1}$$

where λ is the radiation wavelength and h is Planck's constant.

The incidence of recombination under equilibrium conditions is insufficient for practical applications but can be enhanced by increasing the minority carrier density. In an LED, this is accomplished by forward biasing the diode, the injected minority carriers recombining with the majority carriers within a few diffusion lengths of the junction edge. Figure 77.1 illustrates the process. The potential barrier eV_o is reduced by forward bias eV, leading to net forward current and the minority carrier distributions shown on either side of the depletion layer. As the carriers diffuse away

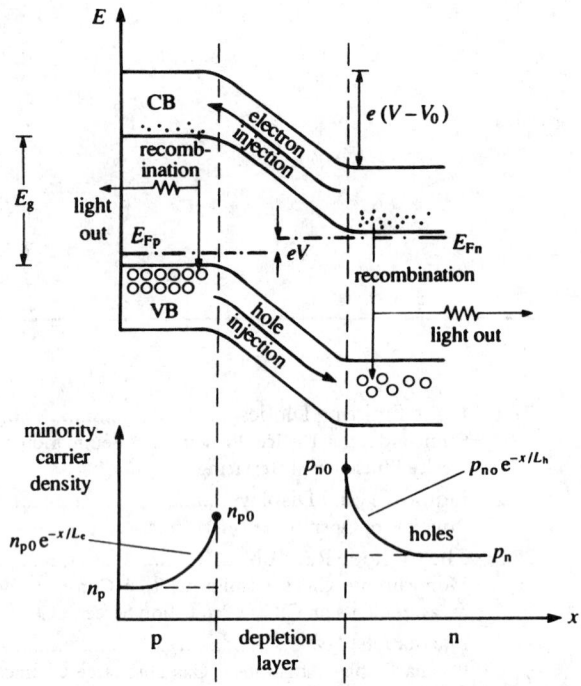

FIGURE 77.1 Light emission due to radiative recombination of injected carriers in a forward-biased *pn* junction. (*Source:* J. Allison, *Electronic Engineering Semiconductors and Devices*, 2nd ed., London: McGraw-Hill, 1990, p. 302. With permission.)

from the junction edges, these distributions decay exponentially because of recombination with the majority carriers. Each recombination event shown on either side of the junction gives off a photon. This process is called **injection electroluminescence.**

Equation (77.1) implies that the radiation emitted will be monochromatic, but in practice $\Delta E > E_g$, and there is a spectral distribution corresponding to the energy distributions of the carriers in the conduction and valence bands.

Semiconductor Materials

Silicon is the most common material used in current semiconductor technologies, but it is not at all suitable for an LED. The reason is that silicon has an indirect bandgap, and a direct bandgap is required for process efficiency. Direct and indirect bandgaps are compared in Fig. 77.2, where carrier energy is plotted versus momentum for both cases. The photon momentum

$$p = h\lambda = h\upsilon/c \tag{77.2}$$

(where c is the velocity of light) is very small, and conservation of momentum can be readily accommodated by small deviations from the vertical transition shown in Fig. 77.2(a). For the indirect case illustrated in Fig. 77.2(b), the energy change ΔE defines the photon energy and momentum, again according to Eqs. (77.1) and (77.2), but conservation of momentum additionally requires that the much greater electron momentum on the order of $h/2a$ be accounted for. For lattice dimensions, a, on the order of 10^{-10} m and wavelengths, λ, on the order of 10^{-6} m, it is clearly not possible for both conservation criteria to be met without the participation of a third body, i.e., a phonon. The two consequences of this result are that the indirect transition is inefficient (in that

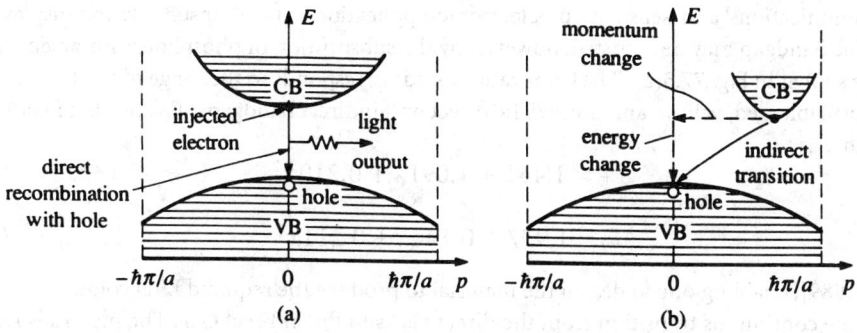

FIGURE 77.2 (a) Interband recombination in a direct-bandgap semiconductor; (b) recombination in an indirect-gap semiconductor also involves a momentum change. (*Source:* J. Allison, *Electronic Engineering Semiconductors and Devices*, 2nd ed., London: McGraw-Hill, 1990, p. 303. With permission.)

it must transfer momentum and hence thermal energy to the lattice) and less likely to occur than the direct transition (because of the requirement for all three particles to simultaneously meet the energy and momentum conditions). Indirect bandgaps therefore lead to long diffusion lengths and recombination times, which produce good transistors but poor LEDs.

The most common direct-bandgap semiconductor is GaAs, but the photon wavelength calculated for $E_g = E_D = 1.43$ eV as listed in Fig. 77.3(b) is in the infrared. Such a material may be ideal

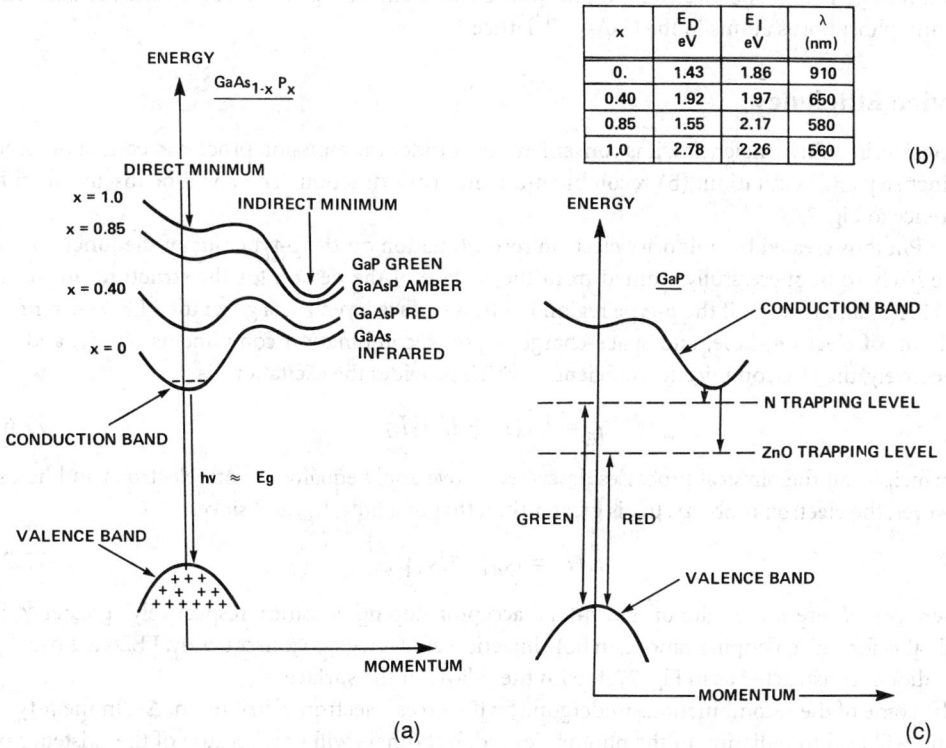

x	E_D eV	E_I eV	λ (nm)
0.	1.43	1.86	910
0.40	1.92	1.97	650
0.85	1.55	2.17	580
1.0	2.78	2.26	560

FIGURE 77.3 (a) Plot of momentum versus bandgap energy, and (b) corresponding semiconductor parameters for various compounds of the GaAs/GaP system; (c) plot of momentum versus bandgap energy for indirect GaP materials showing special trapping levels. (*Source:* S. Gage *et al.*, *Optoelectronics/Fiber-Optics Applications Manual*, 2nd ed., New York: Hewlett-Packard/McGraw-Hill, 1981, pp. 1.3–4. With permission.)

for communications and sensory optoelectronic applications but is unsuitable for display purposes. The bandgap may be adjusted, however, by the substitution of phosphorus for arsenic in the lattice as shown in Fig. 77.3(a). The color range listed corresponds to the range of LED colors commonly available: red, yellow, and green. The direct and indirect bandgaps, E_D and E_I, of $GaAs_{1-x}P_x$ vary with x as

$$E_D = 1.441 + 1.091x + 0.210x^2 \tag{77.3}$$

and

$$E_I = 1.977 + 0.144x + 0.211x^2 \tag{77.4}$$

[Wang, 1989], enabling one to design the material to produce the required LED color.

Note the continuous transition from the direct GaAs to the indirect GaP. The materials have an indirect bandgap for $x > 0.4$ and have the same problems as light emitters as silicon. The efficiency of an indirect-gap emitter can be greatly enhanced by the introduction of appropriate impurity recombination centers, as shown in Fig. 77.3(c). In the process shown, an injected minority carrier electron (in p-type material) is first trapped by the localized impurity (which is itself electrically neutral but which introduces a local potential to the lattice which attracts electrons). The center is then negatively charged and attracts a hole to complete the recombination process, which produces the photon. The recombination center solves the momentum transfer problem, because the trapped electron is localized to the impurity lattice site and has a momentum range according to the Heisenberg Uncertainty Principle of

$$\Delta p \sim h/2\pi a \tag{77.5}$$

that is, sufficient to include the processes shown in the diagram at $p \sim 0$. In the cases used as examples, a nitrogen atom substitutes for a phosphorus, or a zinc–oxygen pair substitutes for adjacent gallium–phosphorus atoms in the $GaAs_{1-x}P_x$ lattice.

Device Efficiency

In considering LED efficiencies, it is convenient to consider the emission process to consist of three distinct steps: (a) excitation, (b) recombination, and (c) extraction. These will be discussed with reference to Fig. 77.4.

(a) Photons created by minority electron recombination on the p-type side of the junction are more likely to be successfully emitted from the surface of the device, for the structure shown in Fig. 77.4(a) and (b), i.e., if the p-type region is a thin surface layer. For a given total LED current, I, made up of electron, hole, and space-charge region recombination components, I_n, I_p, and I_r, respectively, the electron injection efficiency (which provides the excitation) is

$$\gamma_n = I_n/(I_n + I_p + I_r) \tag{77.6}$$

In principle, all the physical processes described above apply equally to both electrons and holes. However, the electron mobility, μ_n, is greater than that of a hole, μ_p, and since

$$I_n/I_p = N_d\mu_n/N_a\mu_p \tag{77.7}$$

(where N_d, N_a are n-type donor and p-type acceptor doping densities, respectively) greater γ_n is attainable for a given doping ratio than hole injection efficiency, γ_p. Consequently, LEDs are usually p-n^+ diodes constructed as in Fig. 77.4, with the p-layer at the surface.

(b) Some of the recombinations undergone by the excess electron distribution, Δn, in the p-type region will lead to radiation of the photon desired, but others will not, because of the existence of doping and various impurity levels in the bandgap. The total recombination rate, R, can be written in terms of the radiative and nonradiative rates, R_r and R_{nr}, as

$$R = R_r + R_{nr} \tag{77.8}$$

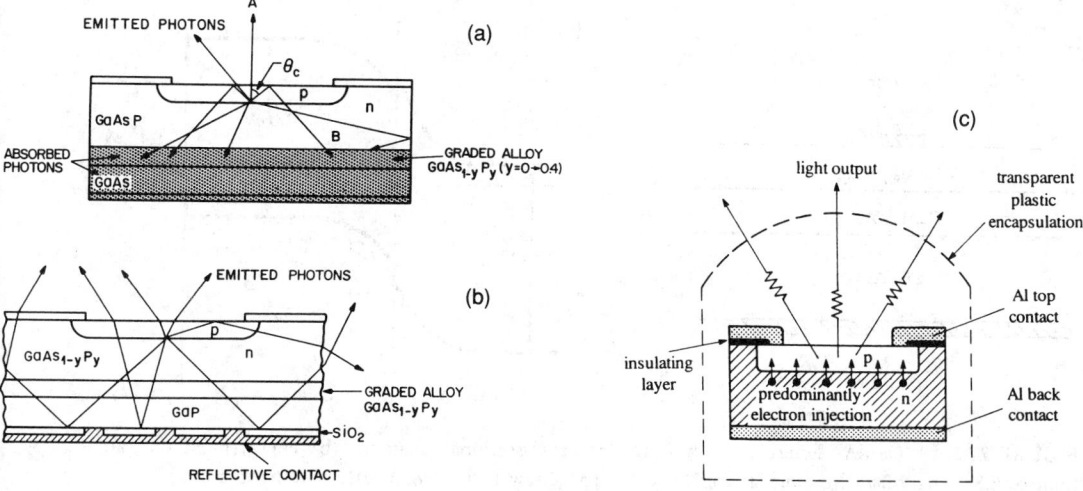

FIGURE 77.4 Effect of (a) opaque substrate, (b) transparent substrate, and (c) encapsulation on photons emitted at the *pn* junction. (*Source:* (a and b) S.M. Sze, *Semiconductor Devices: Physics and Technology,* New York: Wiley, 1985, p. 262. Reprinted by permission of John Wiley & Sons, Inc. (c) J. Allison, *Electronic Engineering Semiconductors and Devices,* 2nd ed., London: McGraw-Hill, 1990, p. 307. With permission.)

where

$$R_r = \Delta n/\tau_r, \quad R_{nr} = \Delta n/\tau_{nr}, \quad R = \Delta n/\tau \qquad (77.9)$$

and where τ_r and τ_{nr} are the minority carrier lifetimes associated with the radiative and nonradiative recombination processes, and τ is the effective lifetime. The radiative efficiency is defined as

$$\eta = R_r/(R_r + R_{nr}) = \tau/\tau_r \qquad (77.10)$$

and the **internal quantum efficiency** is

$$\eta_i = \eta\gamma \qquad (77.11)$$

(c) It is clear from Fig. 77.4 that many of the photons generated on either side of the junction will pass through sufficient bulk semiconductor to be reabsorbed. In fact the photon energy may be ideally suited to reabsorption if it exceeds the semiconductor direct bandgap. It is obvious, then, why GaAs is opaque and GaP transparent to photons from Ga(As:P) junctions. Clearly, a greater efficiency might be expected from the transparent substrate with reflecting contact [Fig. 77.4(b)].

The photon must strike the LED surface at an angle less than the critical angle for total internal reflection, θ_c, where

$$\sin \theta_c = n_{ext}/n_{LED} = 1/n \qquad (77.12)$$

and n_{ext}, n_{LED} are the external and internal refractive indices, respectively. For air, $n_{ext} = 1$, but critical angle loss can be reduced by encapsulating the device in an epoxy lens cap [Fig. 77.4(c)] to increase both $n_{ext} > 1$ and the angle of incidence at the air interface.

Even within angles less than θ_c, there is Fresnel loss, with transmission ratio

$$T = 4n/(1 + n)^2 \qquad (77.13)$$

The total **external quantum efficiency** is then the fraction of photons emitted [Neamen, 1992], given by [Yang, 1988]

$$\eta_e = 1/(1 + \alpha v_o/AT) \qquad (77.14)$$

where α is the average absorption coefficient, v_o is the LED volume, and A is the emitting area.

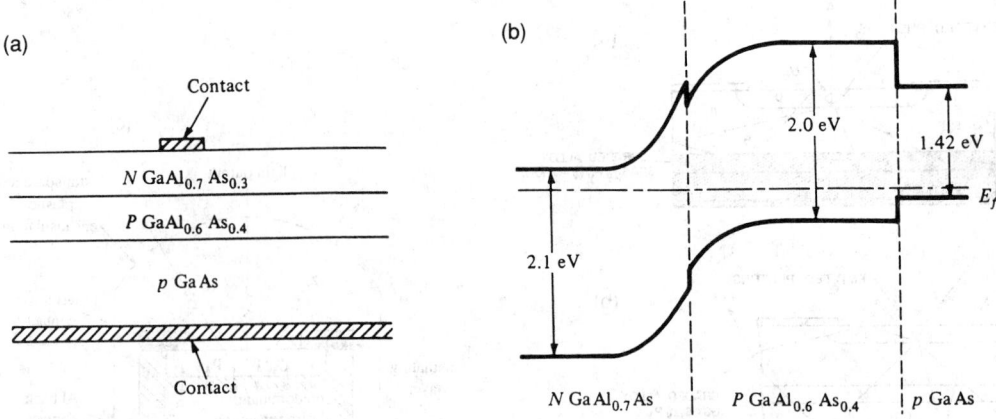

FIGURE 77.5 A GaAlAs heterojunction LED: (a) cross-sectional diagram; (b) energy-band diagram. (*Source:* E.S. Yang, *Microelectronic Devices*, New York: McGraw-Hill, 1988, p. 401. With permission.)

In considering LED effectiveness for display purposes, one must also include radiation wavelength in relation to the spectral response of the human eye [Sze, 1985]. Although the GaP green LED is intrinsically less efficient than the GaAsP red LED, the eye compensates for the deficiency with a greater sensitivity to green.

More recently developed heterojunction LEDs (Fig. 77.5) offer two mechanisms to improve LED efficiencies [Yang, 1988]. The electron injection efficiency can be enhanced, but, in addition, absorption losses through the wider 2.1-eV bandgap *n*-type layer are essentially eliminated for photons emitted by recombination in the lower 2.0-eV bandgap *p*-type region.

Interfacing

In circuit design applications, the LED may be treated much as a regular diode, but with a much greater forward voltage, V_F. Since one usually seeks maximum brightness from the device, it is usually conducting heavily and V_F approaches the contact potential. As one moves from GaAs to GaP [Fig. 77.3(a)], V_F varies from about 1.5 to around 2.0 V. The variation in V_F with temperature (at constant current) follows similar rules as apply to conventional diodes, but radiant power and wavelengths also change [Gage *et al.*, 1981].

Single LEDs are commonly driven by logic gates, perhaps as status indicators, and some of the simplest interface circuits are shown in Fig. 77.6. In many cases, the gate output will not be able to source or sink sufficient current for visibility, and an amplifier will be required, as in Fig. 77.7. Bar

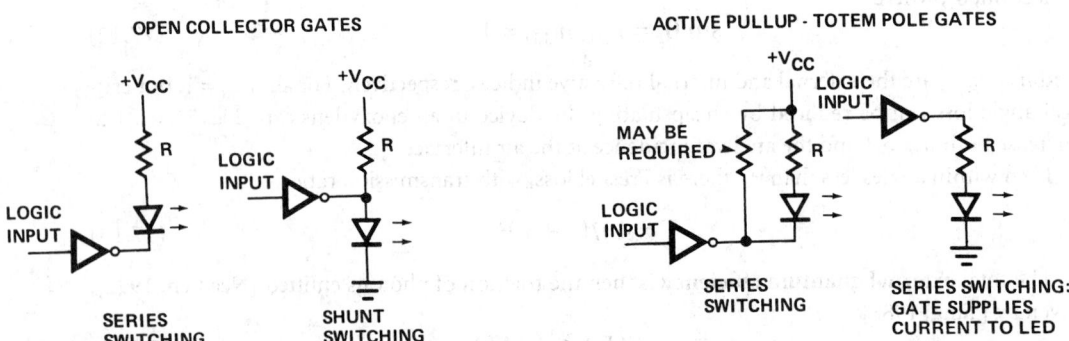

FIGURE 77.6 Digital logic can interface directly to LED lamps. (*Source:* S. Gage *et al.*, *Optoelectronics/Fiber-Optics Applications Manual*, 2nd ed., New York: Hewlett-Packard/McGraw-Hill, 1981, p. 2.20. With permission.)

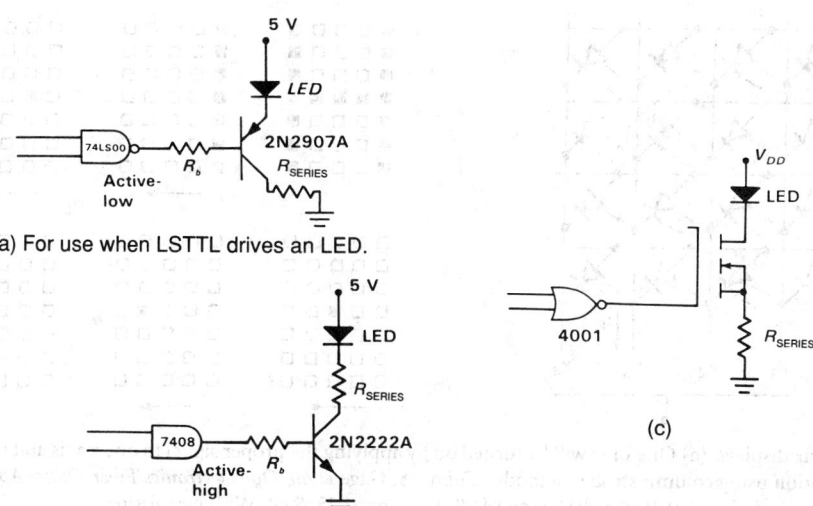

(a) For use when LSTTL drives an LED.

(b) For use when a logic high is needed to drive an LED.

(c)

FIGURE 77.7 LED interfacing for (a) low-power transistor-transistor logic, (b) logic high drive, and (c) CMOS. (*Source:* M. Forbes and B.B. Brey, *Digital Electronics,* Indianapolis: Bobbs-Merrill, 1985, p. 242. With permission.)

graph displays are commonly used to indicate signal level on audio equipment, with a modification of the position indicator seen in Fig. 77.8 to guide fine tuning. Matrix LED arrays can be used for flexible, high-density panel displays [Fig. 77.9(a)] and are conventionally controlled by row or column strobing [Fig. 77.9(b)] controlled by a microprocessor interface.

Multiple LEDs are commonly packaged together in a single integrated device, organized in one

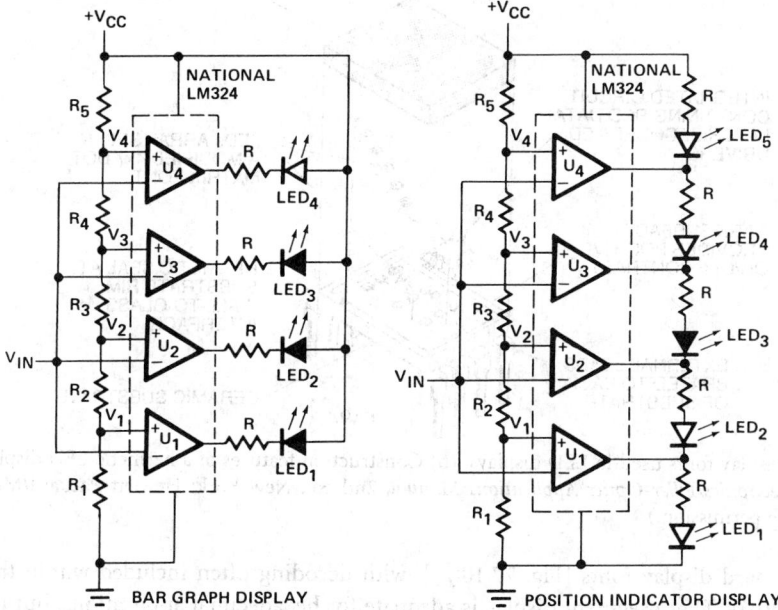

FIGURE 77.8 Operational amplifiers or voltage comparators used to decode an analog signal into a bar graph or position indicator display. (*Source:* S. Gage *et al., Optoelectronics/Fiber-Optics Applications Manual,* 2nd ed., New York: Hewlett-Packard/McGraw-Hill, 1981, p. 23.3. With permission.)

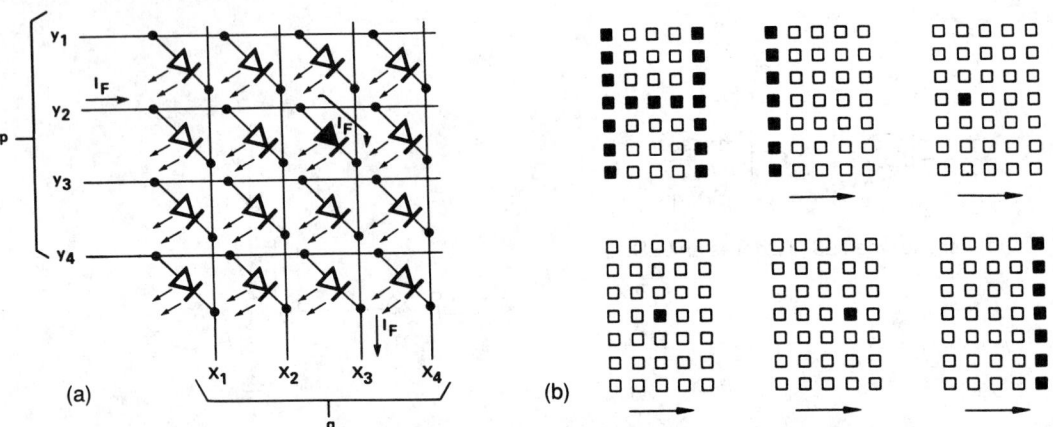

FIGURE 77.9 Matrix displays. (a) One LED will be turned on by applying the proper signal to one *x* axis and one *y* axis. (b) Character generation using column strobe methods. (*Source:* S. Gage *et al., Optoelectronics/Fiber-Optics Applications Manual,* 2nd ed., New York: Hewlett-Packard/McGraw-Hill, 1981, pp. 2.25, 5.44. With permission.)

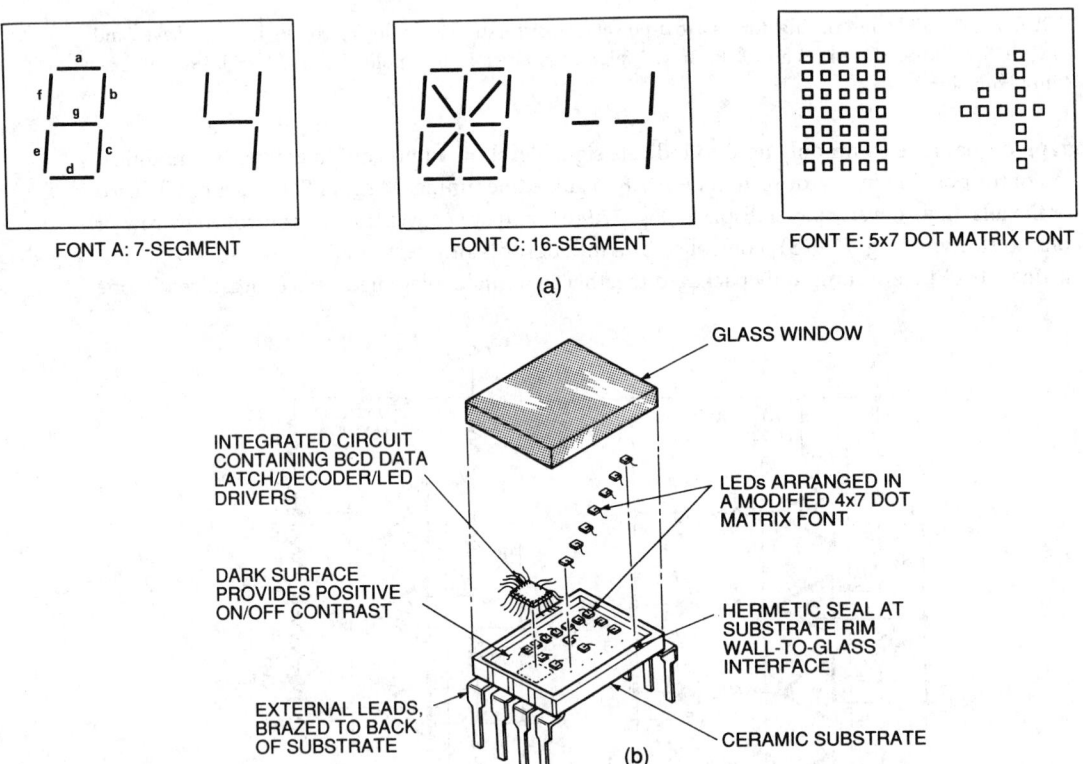

FIGURE 77.10 (a) Display fonts used in LED displays. (b) Construction features of a hermetic LED display. (*Source:* S. Gage *et al., Optoelectronics/Fiber-Optics Applications Manual,* 2nd ed., New York: Hewlett-Packard/McGraw-Hill, 1981, pp. 5.3, 5.6. With permission.)

of the standard display fonts [Fig. 77.10(a)], with decoding often included within the package [Fig. 77.10(b)]. The 7-segment display is adequate for hexadecimal applications, but the 16-segment is required for alphanumerics. To limit pin-out requirements, the LEDs of a single package are connected in either the common anode or common cathode configuration [Fig. 77.11(a)], with multiple display digits multiplexed as illustrated in Fig. 77.11(b).

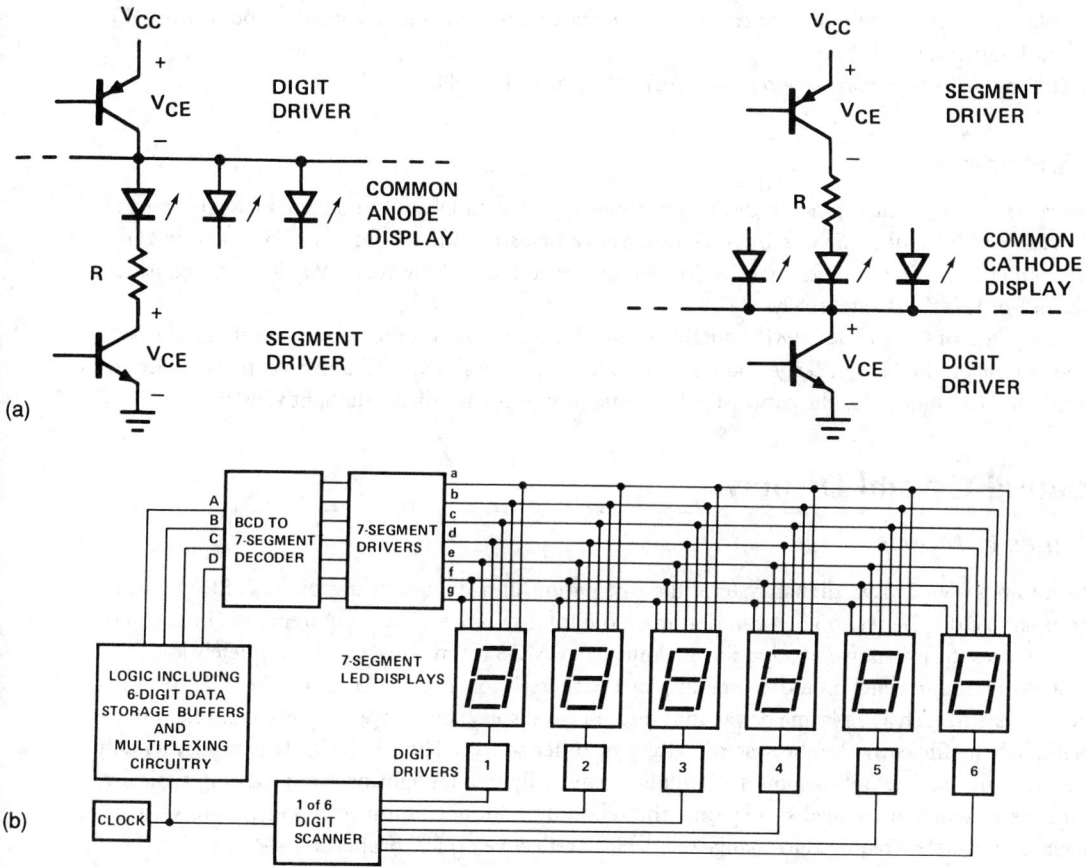

FIGURE 77.11   (a) Generalized drive circuits for strobed operation. (b) Block diagram of a strobed (multiplexed) six-digit LED display. (*Source*: S. Gage *et al.*, *Optoelectronics/Fiber-Optics Applications Manual*, 2nd ed., New York: Hewlett-Packard/McGraw-Hill, 1981, pp. 5.25, 5.23. With permission.)

Defining Terms

External quantum efficiency: The proportion of the photons emitted from the *pn* junction that escape the device structure (but sometimes alternatively defined as $\eta_i \eta_e$).

Injection electroluminescence: *Electroluminescence* is the general term for optical emission resulting from the passage of electric current; *injection electroluminescence* refers to the case where the mechanism involves the injection of carriers across a *pn* junction.

Internal quantum efficiency: The product of injection efficiency and radiative efficiency corresponds to the ratio of power radiated from the junction to electrical power supplied.

References

J. Allison, *Electronic Engineering Semiconductors and Devices*, 2nd ed., London: McGraw-Hill, 1990.

M. Forbes and B. B. Brey, *Digital Electronics*, Indianapolis: Bobbs-Merrill, 1990.

S. Gage, D. Evans, M. Hodapp, H. Sorensen, R. Jamison, and R. Krause, *Optoelectronics/Fiber-Optics Applications Manual*, 2nd ed., New York: Hewlett-Packard/McGraw-Hill, 1981.

D. A. Neamen, *Semiconductor Physics and Devices: Basic Principles*, Boston: Irwin, 1992.

S. M. Sze, *Semiconductor Devices: Physics and Technology*, New York: Wiley, 1985.

S. Wang, *Fundamentals of Semiconductor Theory and Device Physics*, Englewood Cliffs, N.J.: Prentice-Hall, 1989.

E. S. Yang, *Microelectronic Devices*, New York: McGraw-Hill, 1988.

Further Information

More extensive semiconductor device treatments of the LED are contained in *Semiconductor Devices and Integrated Circuits* by A. G. Milnes (Van Nostrand Reinhold, New York) and in *Introduction to Optical Electronics* by K. A. Jones (Harper and Row, New York). Wang [1989] considers second-order effects extensively.

Chapter 2 of Gage *et al.* [1981] contains detailed information on the optical and thermal design constraints on the LED package and on LED back-lit display systems. Chapter 6 considers filtering and other techniques for the contrast enhancement required for direct sunlight viewing.

77.2 Liquid-Crystal Displays

James E. Morris

In a low-power CMOS digital system, the dissipation of a light-emitting diode (LED) or other comparable display technology can dominate the total system's power requirements. In such circumstances the low-power dissipation advantage of CMOS technology can be completely lost. This is the situation in which liquid-crystal display (LCD) technology must be used. The LED (or other active system, such as a plasma or vacuum fluorescent display) emits optical power supplied (comparatively inefficiently) by the system battery or other source. The passive LCD is fundamentally different in that the optical power is supplied externally (by sunlight or room lighting typically) and the system source need supply only the relatively minute amount of power (microwatts per square centimeter) required to change the device's reflective optical properties.

Principle of Operation

Materials classed as liquid crystals are typically liquid at high temperatures and solid at low temperatures, but in the intermediate temperature range they display characteristics of both. Although there are many different types of liquid crystals used, we will concentrate here on the use of **nematic** crystals in **twisted nematic** devices, the most common by far.

The essential feature of a liquid crystal is the long rod-like molecule. In a nematic crystal, the molecules align as shown in Fig. 77.12. If the container surface is microscopically grooved, the interface molecules will be aligned by the grooves and intermolecular forces will maintain that orientation across the liquid crystal [Fig. 77.12(a)]. The molecules will align in an electric field, and beyond a critical value, the field may be sufficient to overcome the alignment with the grooves [Fig. 77.12(b)]. (In practice, the transition is not so abrupt, and groove alignment persists at the interface itself [Fig. 77.13].)

The process of alignment in the electric field is the result of the anisotropic dielectric constant characteristic of liquid crystals. For the electric field parallel to the molecular alignment, $\varepsilon_r = \varepsilon_\parallel$, and for a perpendicular field, $\varepsilon_r = \varepsilon_\perp$. In a "positive" liquid crystal, $\varepsilon_\parallel > \varepsilon_\perp$, and the molecules align parallel to the field as described above in order to minimize the system's potential energy.

The principle of the twisted nematic cell is illustrated in Fig. 77.14. The confining plates, typically 10 μm apart, are grooved orthogonally, forcing the molecular orientation to spiral through 90 degrees [Fig. 77.14(a)]. In the LCD, two polarizers and a mirror are added as shown in Fig. 77.14(b). Incident ambient light is polarized and enters the liquid-crystal cell with the plane of polarization parallel to the molecular orientation. As the light traverses the cell, the plane of polarization is rotated by the twist in the liquid crystal, so that it reaches the opposite face with a polar-

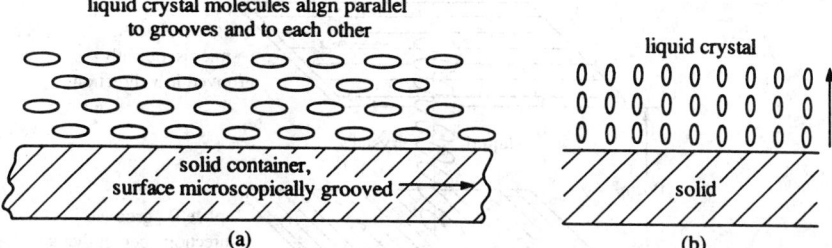

FIGURE 77.12 Liquid-crystal/grooved interface: (a) with no field applied, and (b) with an electric field ε > a critical value. (*Source:* J. Allison, *Electronic Engineering Semiconductors and Devices*, 2nd ed., London: McGraw-Hill, 1990, p. 308. With permission.)

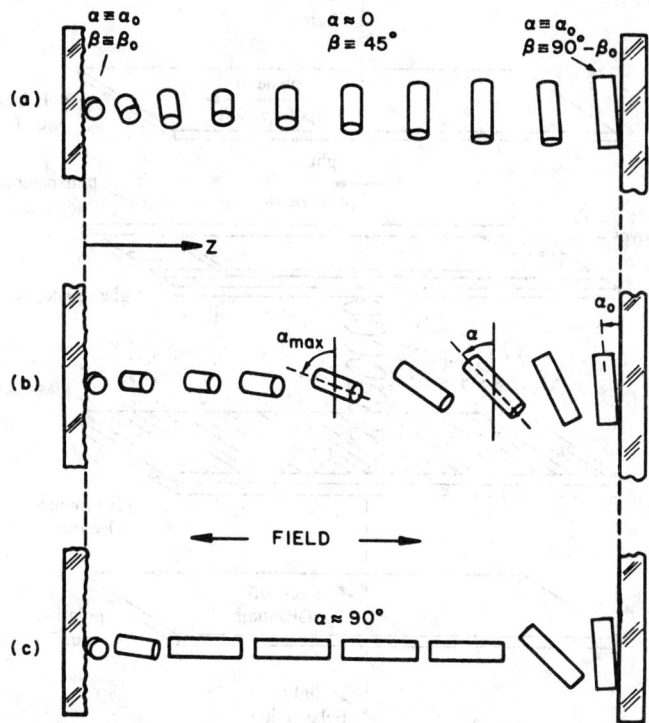

FIGURE 77.13 Diagram of the orientation of the liquid-crystal axis in a cell (a) with no applied field, (b) with about twice the critical field, and (c) with several times the critical field. Note slight permanent tilt (α_0) and turn (β_0) at the surfaces. (*Source:* G. Baur, in *The Physics and Chemistry of Liquid Crystal Devices*, G.J. Sprokel, Ed., New York: Plenum, 1980, p. 62. With permission.)

ization 90 degrees to the original direction, but parallel now to the direction of the second polarizer, through which it may therefore pass. The light is then reflected from the mirror and passes back through the cell, reversing the prior sequence.

When an electric field (greater than the critical field) is applied between the transparent electrodes, usually conductive **indium–tin oxide** (**ITO**) thin films, the 90-degree twist in the crystal is destroyed as the molecules align parallel to the field, so that the rotation of the light's plane of polarization cannot be sustained. Consequently, the crossed polarizers effectively block reflection of the incident light from the backing mirror, and the surface appears to be dark, with excellent contrast to the light gray color of the device in the reflecting mode. The contrast ratio can be fur-

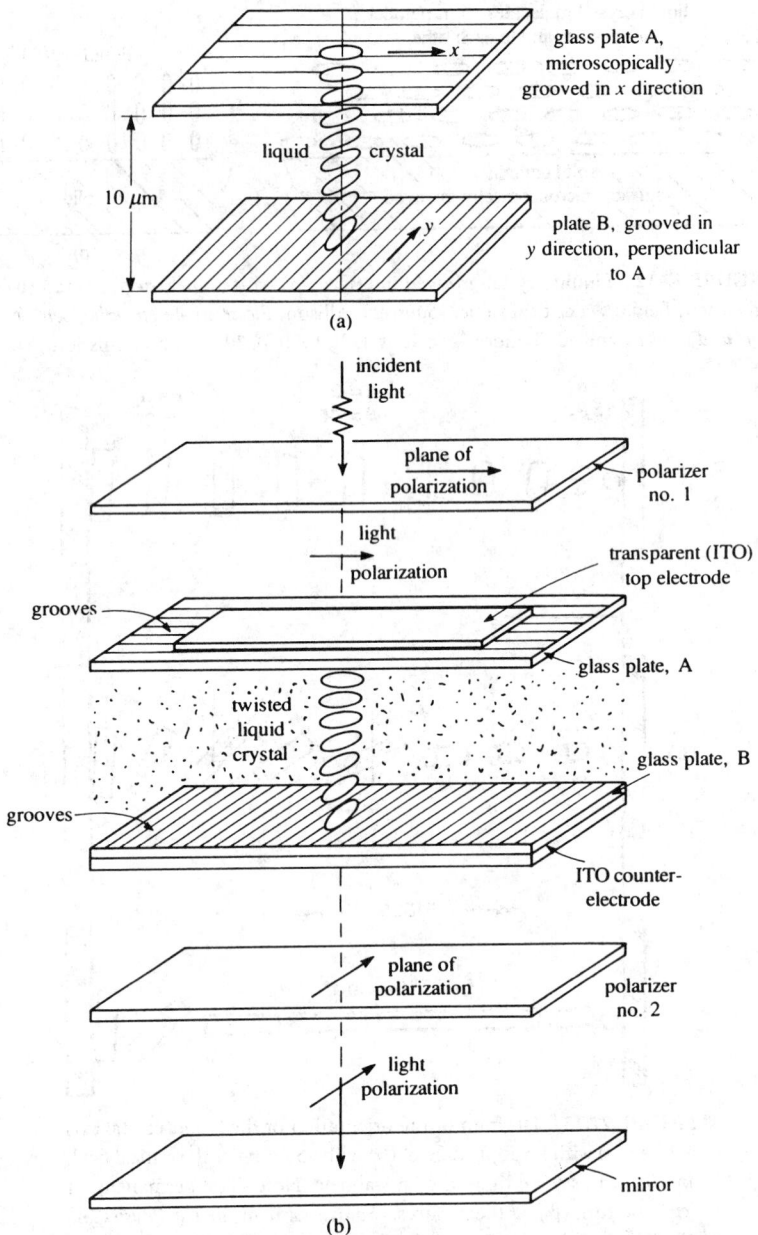

FIGURE 77.14   (a) Twisted nematic cell, $\varepsilon = 0$. (b) Liquid-crystal display element. (*Source:* J. Allison, *Electronic Engineering Semiconductors and Devices*, 2nd ed., London: McGraw-Hill, 1990, p. 309. With permission.)

ther enhanced by the use of the *super twisted nematic* crystal, where the molecular orientation is rotated through 270 degrees rather than 90 degrees.

Transmission LCDs function very similarly to the devices just described, but without the mirror, which is replaced by a powered backlighting source. Obviously, the low-power advantage of the passive device is lost in this active alternative, but monochromatic backlighting does provide one means of constructing displays with varied background colors.

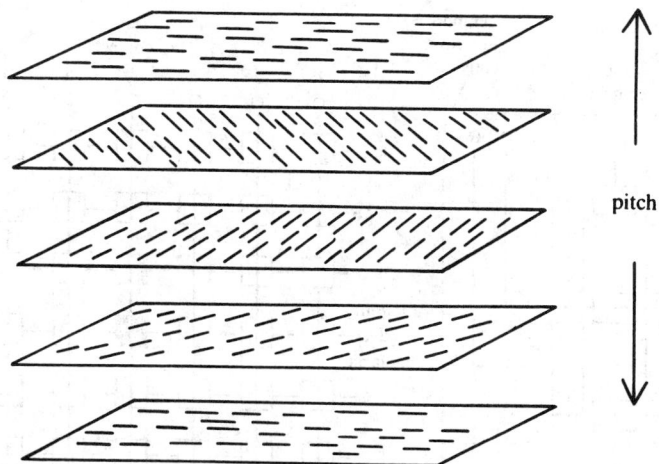

pitch

FIGURE 77.15 Cholesteric ordering: a large number of planes of nematic ordering are formed where the directors rotate as we move along a direction perpendicular to the planes. (*Source:* J. Wilson and J.F.B. Hawkes, *Optoelectronics: An Introduction,* London: Prentice-Hall, 1989, p. 145. With permission.)

Another form of color display is provided by **cholesteric** crystals. The three main types of liquid crystals, nematic, cholesteric, and smectic, are distinguished by the different types of molecular ordering they display. In the cholesteric crystal, the direction of molecular alignment rotates in each successive parallel plane (Fig. 77.15). The spatial period of the rotation, p, is called the pitch, and Bragg reflections occur when the wavelength of incident light meets the condition

$$\lambda = p/n \qquad (77.15)$$

where n is an integer. The liquid crystal can thus appear to be colored in incident white light. In practice, the color is strongly temperature dependent and the effect is more appropriate to temperature-sensing applications than to digital displays.

There is current interest in the development of liquid-crystal color switches where an electrical control signal would be able to change the device color, whether from white to monochromatic or continuously through the spectrum.

Interfacing

LCDs can be organized in all the ways available to competing technologies, e.g., LEDs (see Section 77.1), including seven-segment, alphanumeric, and dot matrix. The LCD differs from LED displays, where each pixel or segment must be a separate device, because the LCD segment or pixel areas are defined by transparent electrodes separated from a common overlapping backplane by a single liquid crystal [Fig. 77.16(a)]. In a large matrix array, it may take a significant period to scan all pixels, and the simple addressing scheme of Fig. 77.16(b) may lead to noticeable flicker. The high off resistance of the MOSFETs of Fig. 77.16(c) can reduce this problem by increasing the discharge time to hold the LCD on after the address pulse has gone. An amorphous silicon implementation of this active matrix technology is described in Shur [1990].

The interfacing requirements, which are otherwise similar in multiplexing techniques, etc., are complicated by the requirement for zero net dc bias across the cell in order to avoid electrochemical degradation of the material. LCDs require ac drive signals, and square waves of frequency

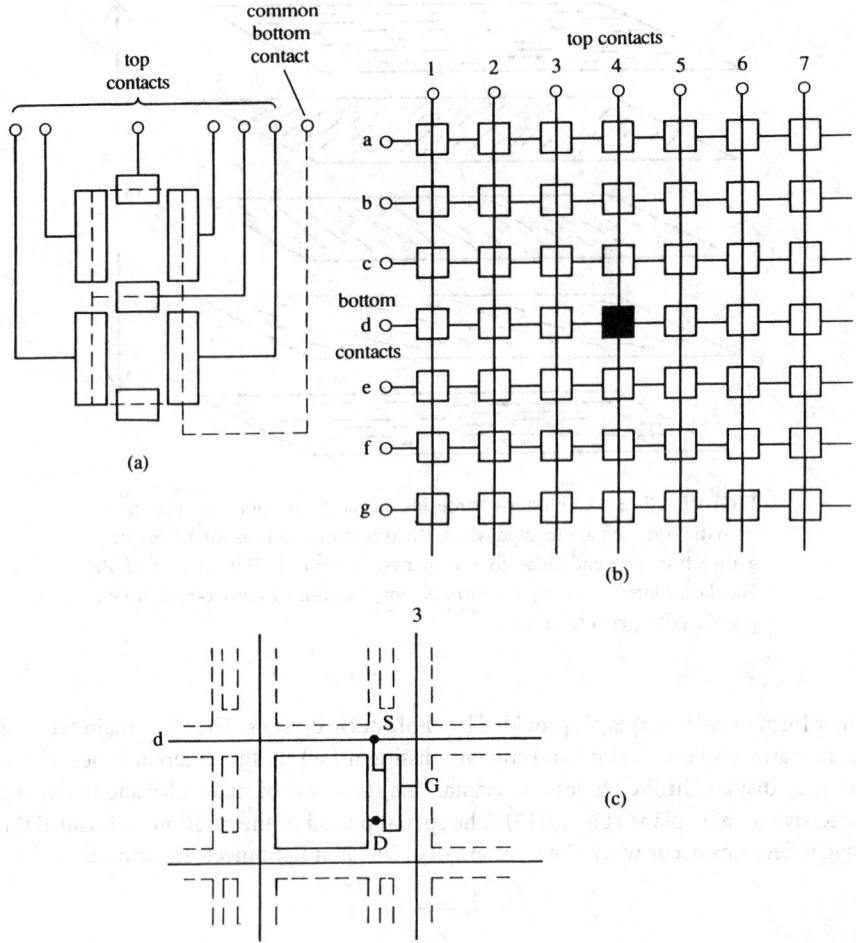

FIGURE 77.16 LCD addressing: (a) simple (seven-segment) addressing; (b) matrix addressing; and (c) matrix addressing with MOSFETs. (*Source:* J. Allison, *Electronic Engineering Semiconductors and Devices,* 2nd ed., London: McGraw-Hill, 1990, p. 312. With permission.)

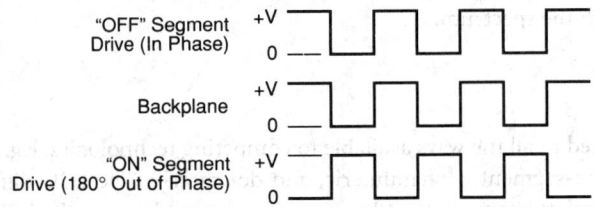

FIGURE 77.17 Drive signals from a direct connect LCD driver. (*Source:* R. Lutz, Application Note 350, in *Interface Databook,* Santa Clara, Calif.: National Semiconductor Corporation, 1990, p. 4–109. With permission.)

between 25 Hz and 1 kHz are typically used [Wilson and Hawkes, 1989]. A square wave is applied to the backplane, with in-phase and antiphase signals to the counter electrode determining whether the given pixel or segment is on or off. In practice, the state is determined by the root-mean-square (rms) value of the differential voltage applied.

Figure 77.18 illustrates the additional complexity that would be required by even a simple multi-

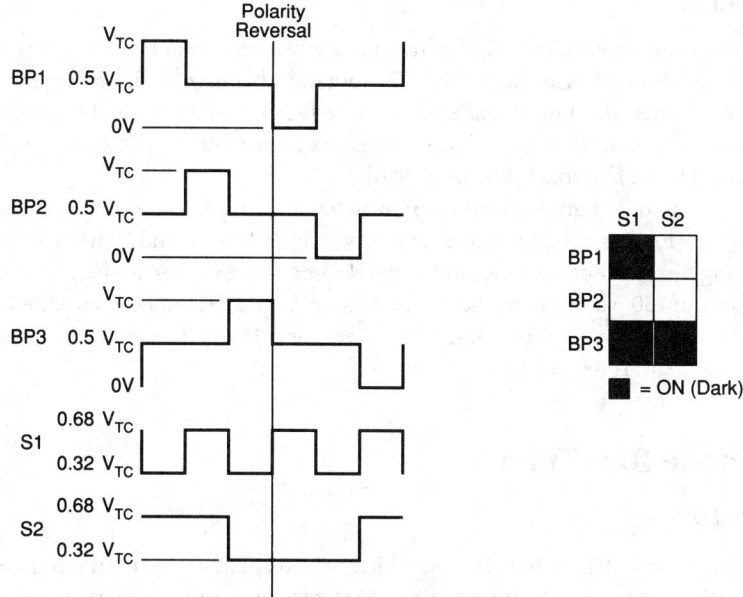

FIGURE 77.18 Example of backplane and segment patterns. (*Source*: R. Lutz, Application Note 350, in *Interface Databook*, Santa Clara, Calif.: National Semiconductor Corporation, 1990, p. 4–109. With permission.)

plexed addressing system. The backplane and segment drivers might correspond to rows and columns of a dot matrix, as implied in the diagram, or the backplanes may identify specific characters of an alphanumeric display. Calculating the rms values of the difference voltages shown gives 0.42 V_{tc} for the *on* pixels and 0.24 V_{tc} for *off*, from which V_{tc} can be calculated for reliable operation if the critical voltage is known for the LCD to be used.

Defining Terms

Cholesteric: In the cholesteric liquid crystal, successive layers of aligned molecules are rotated naturally.

Indium–tin oxide (ITO): A mixture of the semiconducting oxides SnO_2 and In_2O_3; the most common transparent conductor.

Nematic: The type of liquid crystal in which the molecular chains align; such alignment can be controlled across the liquid crystal if it can be constrained at the boundaries.

Twisted nematic: The alignments of the nematic planes are rotated through 90 degrees across the crystal by constraining alignments to be orthogonal at the boundaries.

References

J. Allison, *Electronic Engineering Semiconductors and Devices*, 2nd ed., London: McGraw-Hill, 1990.

G. Baur, "Optical characteristics of liquid crystal displays," in *The Physics and Chemistry of Liquid Crystal Devices*, G. J. Sprokel, Ed., New York: Plenum, 1980.

R. Lutz, "Designing an LCD dot matrix display interface, application note 350," in *Interface Databook*, Santa Clara, Calif.: National Semiconductor Corporation, 1990.

M. Shur, *Physics of Semiconductor Devices*, Englewood Cliffs, N.J.: Prentice-Hall, 1990.

J. Wilson and J. F. B. Hawkes, *Optoelectronics: An Introduction*, London: Prentice-Hall, 1989.

Further Information

Nematic liquid-crystal molecules typically incorporate two separated benzene rings in a complex chain molecule [Wilson and Hawkes, 1989]. The organic chemistry of liquid-crystal compounds will lie outside the interests of most readers but is briefly reviewed in "Liquid Crystal Materials for Display Devices," by J. A. Castellano and K. J. Harrison in *The Physics and Chemistry of Liquid Crystal Devices*, edited by G. J. Sprokel [Plenum, 1980].

One technique used in liquid-crystal color switches requires the use of electrically controlled birefringence. This topic is covered at an elementary level by Wilson and Hawkes [1989].

An interesting historical perspective on the development of LCD technology is provided by the extensive reviews of 150 patents in the field contained in *Liquid Crystal Devices*, edited by T. Kallard (*State of the Art Review*, Vol. 7, Optosonic Press, New York, 1973). The book also contains a bibliography of more than 1100 entries.

77.3 The Cathode Ray Tube

André Martin

The **cathode ray tube** (CRT) is the element which, in a display, converts an electrical signal into visual information using an electron beam adequately intensity modulated and deflected to impinge on a cathodoluminescent screen surface, in a glass envelope under vacuum.

Because of the extensive growth of electronic communication since the Second World War, information is very often presented on CRTs, mainly when the information content exceeds 100,000 picture elements (pixels).

Monochrome CRTs are widely used for computer terminals, radars, oscilloscopes, projection systems, etc., while color CRTs are preferred for complex imaging from computers, such as CAD-CAM systems and digital image processors.

In 1991, the worldwide market for CRT monitors was 37 million units [Stanford Resources, 1992a], of which 20 million units were high-resolution color monitors. These 37 million do not take into account the oscilloscope, radar, projection, and other special-purpose tubes that would add a few hundred thousand to the figures.

The importance of the CRT in the display world can be explained by two key factors:

- The CRT is using a single serial data input to generate a picture.
- The CRT is a very efficient light-emitting display.

For example, a typical high-resolution 20-inch diagonal color monitor requires 70 W from the main ac 50/60-Hz power supply and generates a picture visible in any office environment with a luminous efficiency of 6 to 8 lm/W. These two factors, high luminous efficiency and convenience of addressing, make the CRT difficult to replace by any other type of display for large image contents.

We will describe first the monochrome CRT and then discuss the color CRTs.

Monochrome CRTs

General

The monochrome CRT [Martin, 1986] is composed of:

- A glass envelope with the necessary glass-to-metal seals for anode and electron gun connections. This envelope is under vacuum (about 10^{-7} mm Hg).
- A cathodoluminescent screen deposited on the faceplate, usually aluminized to improve brightness and obtain good screen potential uniformity.

- An electron gun using a hot cathode to emit electrons that are accelerated toward the screen and deflected either by electrostatic plates or by an electromagnetic deflection coil.
- Various outside and inside conductive coatings to normalize the potentials.

The Electron Gun

A typical electron gun is composed of (Fig. 77.19):

- A hot cathode that emits electrons when heated to $\approx 800°C$ and a suitable potential is applied to the adjacent electrodes G_1 and G_2.
- An apertured grid No. 1, also called G_1, which is maintained negative with respect to the cathode and whose potential controls the flow of electrons from the cathode.
- An apertured grid No. 2 placed close to G_1 (usually a few thousandths of an inch) and set at a positive voltage of a few hundred volts with respect to the cathode. This G_2 attracts the electrons controlled by the G_1 aperture potential and shapes the beam.
- An anode composed of metal cylinders to accelerate the electron beam toward the focus electrode and a final anode to further accelerate the beam toward the screen, where it focuses into a spot.

Figure 77.19 describes the unipotential lens focus structure, also called an EINZEL lens design. Another structure, widely used in modern CRTs, is the bipotential lens focus structure represented in Fig. 77.20. The bipotential structure is theoretically a better performer than the unipotential focus structure, because the lenses have less curvature of the line forces and less spherical aberration.

When optimum resolution is required, electromagnetic focus is used instead of the electrostatic focus systems described previously. Because this magnetic focus lens just bends the electron trajectories without changing the electron's speed, and because of its large diameter, the spherical aberration is reduced and the spot size is optimized.

The Cathode

The cathode used on most CRTs is the oxide cathode, which consists typically of a heated nickel substrate coated with barium, strontium, and calcium oxides. This cathode works at a temperature

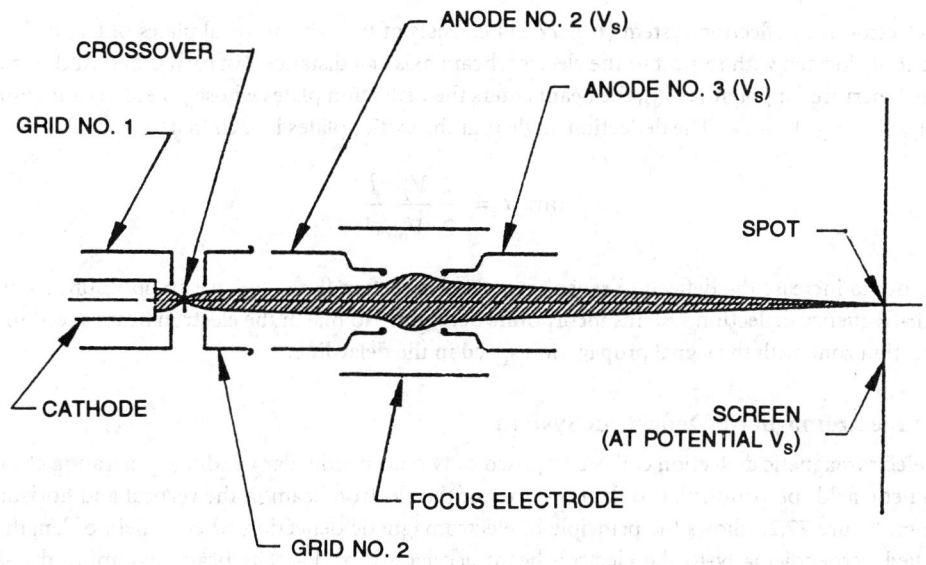

FIGURE 77.19 A typical electron gun with unipotential lens structure.

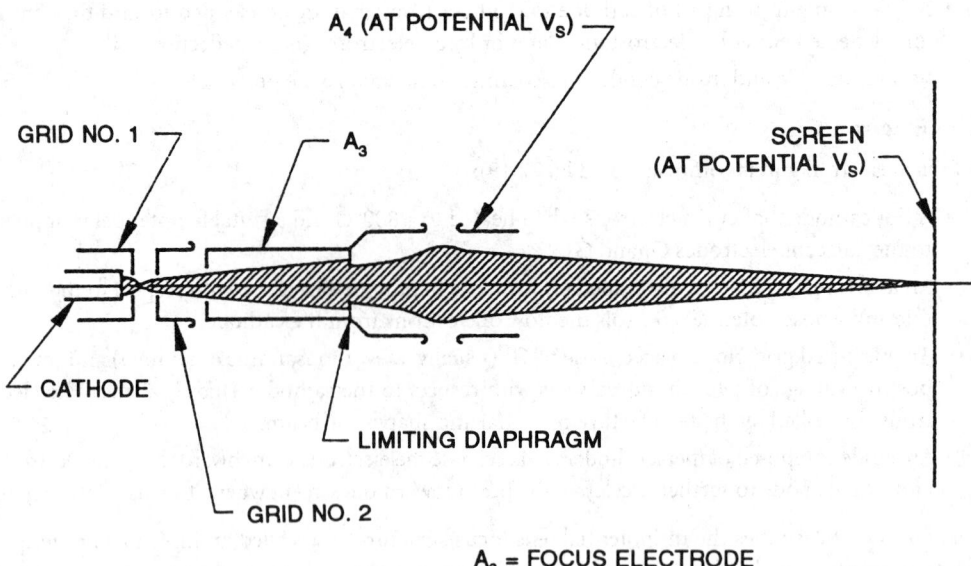

FIGURE 77.20   A typical electron gun with bipotential lens structure.

of $\approx$800°C and provides a dc emission density up to 0.2 amp/cm^2 and 2 amp/cm^2 peak current density. When higher current densities are required, a tungsten-impregnated cathode (porous tungsten matrix heated at $\approx$1000°C and impregnated with barium and calcium aluminates) can be used. The impregnated cathode operates at higher temperature than the oxide cathode and requires sophisticated materials and techniques for its processing. The impregnated cathode, also known as a dispenser cathode, is commonly used on projection tubes and on other tubes where high beam currents are required, typically above 1.5 mA.

The Electrostatic Deflection System

An electrostatic deflection system (Fig. 77.21) consists of two sets of metal plates of length l symmetrically located with respect to the electron beam axis at a distance d of each other. At the anode outlet aperture (at potential V_0), the beam enters the deflection plates whose potentials are, respectively, $V_0 + V_D$, $V_0 - V_D$. The deflection angle α at the exit of plates is such that

$$\tan \alpha = \frac{1}{2} \frac{V_D}{V_0} \frac{l}{d}$$

In order to increase the deflection sensitivity, plates are often flared to have an optimum contour. High-frequency deflection systems incorporate delay lines to match the electron beam speed in the deflection zone with the signal propagation speed in the delay line.

The Electromagnetic Deflection System

An electromagnetic deflection coil is composed of two perpendicular windings generating electromagnetic fields perpendicular to the trajectory of the electron beam in the vertical and horizontal planes. Figure 77.22 shows the principle of electromagnetic deflection where a field of length l is applied perpendicularly to the electron beam accelerated at V_B. The beam, assuming the field intensity is uniform and of length l, is deflected onto a circular path of radius r. The corresponding angle of deflection is θ such as:

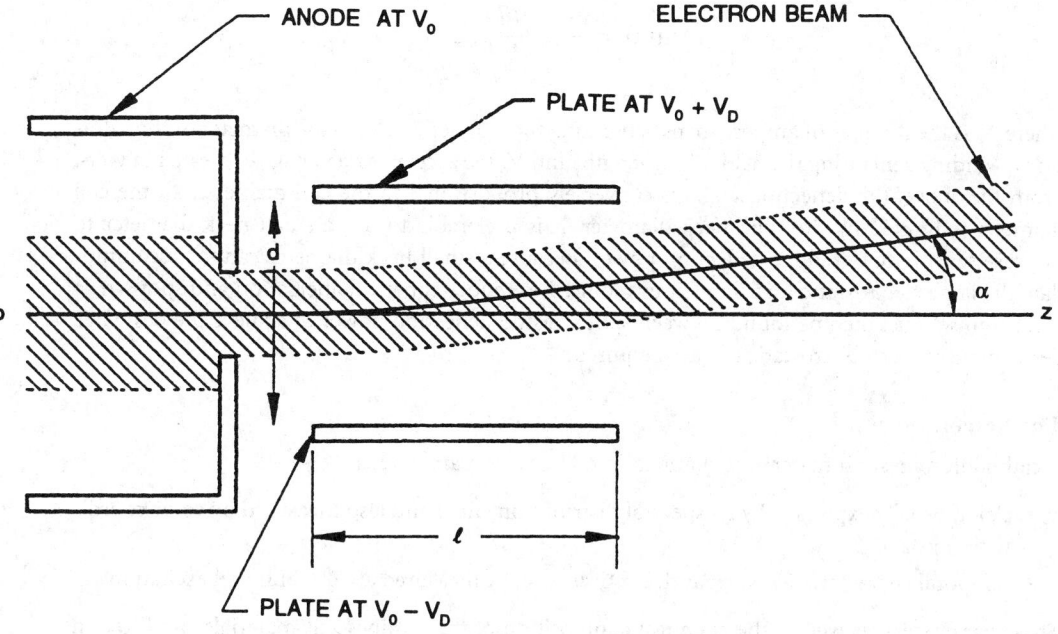

FIGURE 77.21 Principle of electrostatic deflection.

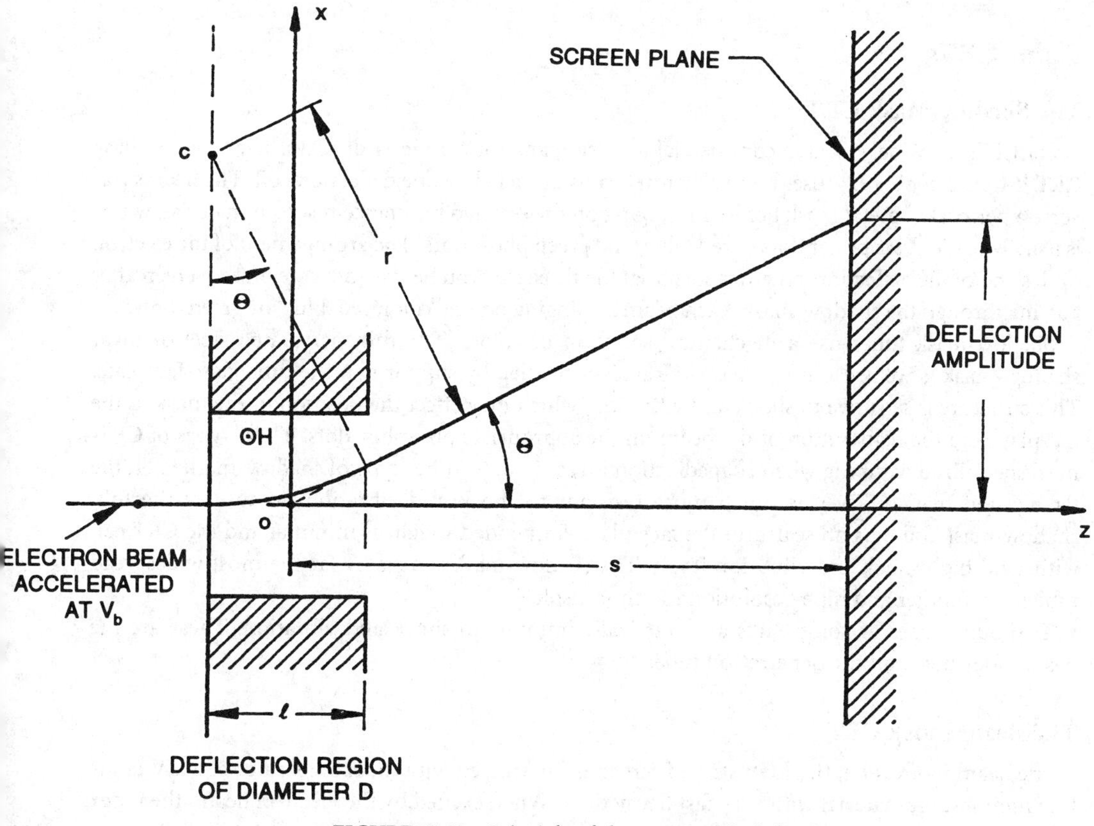

FIGURE 77.22 Principle of electromagnetic deflection.

$$\sin \theta = \frac{Nil}{2.68D\sqrt{V_B}}$$

where Ni is the number of ampere turns generating the magnetic field, D the diameter of the cylindrical winding generating the field, l is its length, and V_B the accelerating voltage expressed in volts. From the above, the deflection angle is conversely proportional to the coil diameter. As the coil diameter is limited by the tube neck diameter, it is preferable to use a small neck diameter to increase sensitivity to reduce deflection power. Because a small neck diameter cannot accommodate the large electrostatic focusing lenses required to reduce spherical aberration and spot size, a compromise must then be found between spot size and deflection power to achieve the best CRT performance when electrostatic focus is required.

The Screen

A cathodoluminescent material is characterized by several parameters:

- Color, usually expressed by its spectral distribution curve and also measured by color coordinates xy or $u'v'$
- Temporal characteristics, such as decay time, usually measured at 10% of initial excitation

These parameters, as well as the chemical composition of the luminescent materials, are listed in the E.I.A. Publication TEP 116 C [Publ. TEP].

Other parameters such as luminous efficiency (expressed in lumens/watt) or energy conversion efficiency (expressed in watts/watt) are also necessary and have to be required from manufacturers.

Color CRTs

The Shadow-Mask CRT

Color CRTs are widely used in commercial television and for computer displays. The shadow-mask CRT [Morrell *et al.*, 1974] uses three electron beams deflected by one deflection coil. The beams traverse a perforated metal mask before impinging on the selected luminescent screen material, which is usually made of stripes or dots of red, blue, and green phosphors. The arrangement of the electron optics and of the deflection system is such that the three electron beams converge on the screen after passing through the shadow-mask, each beam impinging on one color, red, blue, or green, only.

Shadow-mask tubes use a mechanical selection of colors. The thin perforated steel or invar shadow-mask is welded onto a metallic frame suspended by supports in the tube glass faceplate. This structure is sensitive to shock and vibration, which may affect the position of the mask in the faceplate and the registration of the beam on the appropriate phosphor dots. These types of CRTs need specialized damping when ruggedization is required. Another type of shadow-mask CRT, the flat tension mask (FTM), is much more resistant to shock and vibration because of the thin shadow-mask foil tension sealed to the face plate. Suspended weight is minimal and the CRT can withstand high shock and vibration levels. The shadow-mask tube is by far the most widespread tube for computer and high-resolution monitor displays.

Two other types of color CRTs are practically immune to shock and vibration. These are the beam index tube and the penetration tube.

The Beam Index CRT

In the beam index tube, the RGB striped screen is intermixed with indexing stripes of a UV emitting luminescent material with very fast decay time. When excited by the electron beam, the index stripes emit a light pulse that is detected by photosensors located on the transparent bulb. The signals are then digitally processed, permitting video signals to be fired at the correct position of the electron beam on the screen.

Beam index CRTs are basically used for avionic applications. A typical $150 \times 150 \text{ mm}^2$ (6×6 in.) CRT can display images with a white brightness around 3400 cd/m² (1000 fL).

The Penetration CRT

With the penetration CRT, operation is based on the variation in depth of penetration of an electron beam in successive layers of different luminescent materials, typically red and green emitting.

At low voltage, such as 9 kV, the first layer of luminescent material is excited and a red color is obtained. At high voltage, such as 18 kV, the electrons penetrate the red layer without losing much of their energy and excite the green layer to produce a slightly desaturated green. Typically, only four colors can be produced.

These tubes can be built in any size and can also use a variety of short and long persistence luminescent materials to achieve variable persistence. This is convenient for radar or other specific limited color applications.

Contrast and Brightness

Contrast ratio is defined as the ratio of the luminance L_1 of the picture element to the luminance L_2 of the background as follows:

$$C_R = \frac{L_1}{L_2}$$

Brightness contrast, important to the observer, is defined as the ratio of the luminance of the picture element plus background $L_1 + L_2$ to the background luminance L_2

$$C = \frac{L_1 + L_2}{L_2} = 1 + C_R$$

In order to improve contrast, the usual technique is to use an absorbing faceplate or spectrally matched filter to absorb the light emitted by the luminescent screen once and the ambient incident light twice. In addition, antireflection coatings and antiglare treatments can be applied to the CRT face to reduce the reflections from the front surface. Depending upon the ambient lighting conditions, a compromise must usually be found between light output required and the contrast need.

Measurements on CRTs

Line Width and Modulation Transfer Function

The emitted phosphor spot of a CRT usually presents a Gaussian energy distribution. The measurement is usually performed at the 50% height of the Gaussian curve and is called $L_{0.5}$. It can be done also at the σ point of the Gaussian curve and called $L_{0.6}$, with $L_{0.5} = 1.175\ L_{0.6} = 1.175\ L\sigma$.

Spot size is related to modulation transfer function (MTF) [E.I.A., 1986] by

$$L_{0.5} = \sqrt{\frac{1}{3.56(N)^2}\ L\ \frac{1}{MTF}} = \frac{0.53}{N}\sqrt{L\ \frac{1}{MTF}}$$

where MTF is fractional, L is the neperian logarithm, N is the number of cycles/millimeter, and $L_{0.5}$ is expressed in millimeters, and can be related to $L\sigma$ by the formula

$$L_{0.5} = 1.175 L\sigma$$

The spot image is normally measured using a microscope with a suitable detector such as a fiber-optic probe coupled to a photomultiplier or a CCD array.

Another method widely used is the shrinking **raster** technique. After a raster of n horizontal lines is scanned on the CRT screen, the vertical size of the screen is reduced until the line structure disappears and produces a uniform luminance to the observer. The number of raster lines is then divided into the raster height:

$$L_{SR} = \frac{\text{compressed raster height}}{n}$$

Comparing L_{SR} to $L\sigma$, authors find values of L_{SR} between 1.17 and 1.23 $L\sigma$.

Brightness

Brightness is measured as area brightness (raster luminance) L_R or peak line brightness L_p. Peak line brightness is an inverse function of the writing speed, a direct function of the refresh rate of the displayed line and, of course, of the beam current. Raster luminance L_R is related to the raster emitting surface (S), to the beam current (I), to the screen efficiency (ε), transmission (T) of the faceplate, and to the screen voltage (V_s) by the formula

$$L_R = \frac{V_s I}{\pi S} T\varepsilon$$

where ε is lumens/watt, I is amperes, S is in square meters, T is fractional, and $\pi = 3.1416$. There is no convenient relation between peak line brightness and raster brightness.

Illumination

Illumination is usually measured by illuminance in lux where: 1 lux = 1 lumen/m². The S.I. units are the candela/square meter or nit for luminance and the lux for illuminance. They are related to the commonly used footlambert (f L) and footcandle (f C) by the relations:

$$1 \text{ fL} = \frac{1}{\pi} \text{ cd/ft}^2 = 3.42 \text{ cd/m}^2$$

$$1 \text{ fC} = 1 \text{ lm/ft}^2 = 10.8 \text{ lux}$$

$$1 \text{ lux} = 1 \text{ lm/m}^2$$

The notion of illumination has to be differentiated from the notion of brightness. The footcandle is often used for illuminance and is produced approximately by a source having a luminance of 1 fL (when the reflectance of the surface is 1), which makes the contrast calculation easy. However, these units must not be confused.

Projection Screen

Very often a CRT is used to project an image onto a screen. For a lambertian screen, which rediffuses the incident light in all directions, we can write

$$B = L \frac{T}{4n^2(m+1)^2}$$

where B is the luminance off the projection screen, L is the luminance of the CRT, f is the focal length of the lens used, m is the magnification factor, d is lens diameter, n is aperture, f/d, and T is the transmission of the optics.

The maximum **flux** F (lumens) from the CRT can be expressed by

$$F = \pi\, LS$$

where L is the CRT screen brightness (candelas/square meters) and S the emitting screen surface (square meters), $\pi = 3.1416$.

Conclusion

The worldwide market for CRTs is predicted to grow in 1997 at a level of 54 million units from 38 million units in 1991 [Stanford Resources, 1992b]. It is now obvious that the competitive technologies have still to progress to be able to replace the CRT in many applications and that this progress is much slower than the sometimes optimistic predictions of the scientific and industrial community.

It is also important to note that the Electronics Industries Association is monitoring several standard committees for CRTs and that proposals for standards are submitted by the industry and by laboratories such as the National Information Display Laboratory (see Further Information). These standards are following the evolution of measurement techniques. The adoption of these standards by the industry will make it possible for display users to choose their equipment according to criteria uniformly approved around the world.

The CRT is widely used aboard commercial and military aircraft for cockpit displays, aboard surface ships and submarines, and aboard military and commercial vehicles. Because of its wide range of capabilities, versatility of use, brightness and contrast, image quality, and efficiency, the qualities of the CRT largely offset its bulkiness and weight for most applications. CRTs are still dominating high-resolution applications ranging from 12-mm helmet-mounted displays to 1000-mm diagonal 2000×2000 pixels color monitors and, according to P. Brody [1980], will continue to dominate the market for a few decades.

Defining Terms

Cathode ray tube: A vacuum tube which uses *cathode rays* to generate a picture on a fluorescent screen. These cathode rays are in fact the electron beam deflected and modulated, which impinges on a phosphor screen to generate a picture according to a repetitive pattern refreshed at a frequency usually between 25 and 72 Hz. The term cathode rays stems from the discovery by Plücker [1858] and Hittorf [1869] of a blue glow present at some spots on a glass tube when studying high-voltage discharge in a low-pressure gas. Crookes [1879] showed that these cathode rays were deflected by a magnetic field and were in fact electrons emitted by the negative electrode. The more poetic term *cathode rays* has been kept instead of electron beam for the cathode ray tube.

Contrast: For a cathode ray tube, contrast is the evaluation of the visibility of the picture presented on the phosphor screen in a given ambient lighting. Contrast is usually measured by the contrast ratio, which is the ratio of the luminance of the picture element under evaluation to the background luminance.

Flux: Also called *radiant flux*, the radiant power emitted by a source. It can be expressed in watts for a radiometric source or in lumens when the source spectral energy distribution is between 400 and 700 nm.

Illumination: The effect of a visible radiation flux received on a given surface. Illumination is measured by the illuminance, which is the luminous flux received by surface unit, usually expressed in lux. One lux equals 1 lumen/m^2.

Raster: Also called *television raster;* it is developed by a moving spot of light generated by an electron beam scanning a CRT phosphor screen in a predetermined and repetitive pattern. A picture is generated by modulating the beam intensity, hence the spot light output, when scanning the screen surface. Usually, horizontal lines are generated scanning in a left-to-right sequence and developed top-to-bottom of the image surface.

References

T. P. Brody, "When—if ever—will the CRT be replaced by a flat display panel?" *Microelectronics Journal*, vol. 11, pp. 5–9, 1980.

W. Crookes, "On the illumination of lines of molecular pressure and the trajectory of molecules," *Philos. Trans. R. Soc. London*, vol. 170, pp. 135–164, 1879.

E.I.A. J.T 20 Committee—Meeting #59 (1986)—Test Method—Measurement of M.T.F. for Monochrome CRTs by Fourier Transform.

W. Hittorf, "Über die Elektricitätsleitung der Base," *Ann. Phys. (Leipzig)* [2], vol. 136, pp. 1–31, 1869.

A. Martin, *Cathode Ray Tubes for Industrial and Military Applications*, vol. 67, New York: Academic Press, 1986, pp. 183–328.

A.M. Morrell, H.B. Law, E.G. Ramberg, and E.W. Herold, *Color Television Picture Tubes*, New York: Academic Press, 1974.

J. Plücker, "Über die Einwirkung der Magneten auf die elektrischen Entladungen in verdüunten Gasen," *Ann. Phys. (Leipzig)* [2], vol. 103, pp. 88–106, 1858.

Publication TEP 116C, Washington, D.C.: Electronic Industries Association.

Stanford Resources—Monitor Market Trends—1991, Menlo Park, Calif.: Information Associates, 1992a, 552 pp.

Stanford Resources—Electronic Display World, San Jose, Calif.: Stanford Resources, Inc., vol. 12, no. 3, p. 4, 1992b.

Further Information

Electronic Industries Association (E.I.A.)
2001 I Street, N.W.
Washington, D.C. 20006

E.I.A. prints a wide range of literature on electronics components, computers and industrial electronics, communications and services. E.I.A. also administers many committees for standardization, safety, etc., for cathode ray tubes and other components. E.I.A. offers a complete Electronics Technology Curriculum and technical training books, tapes, etc.

National Information Display Laboratory
David Sarnoff Research Center–CN 8619
Princeton, NJ 08543-8619

The NIDL's strategic objective is to provide direct support to government users while promoting the development and commercialization of advanced soft copy technologies. The NIDL participates in the elaboration of standards for soft copy, for example, for high-resolution monitors, in close relation with E.I.A.

77.4 Plasma Displays

Larry F. Weber

Plasma displays have emerged as one of the dominant flat-panel display technologies. The major market success has been the acceptance of plasma displays in portable computer terminals with worldwide sales of over a half million terminals per year. The users of these terminals have now recognized the familiar orange glow of the plasma panel as an attractive alternative to the liquid-crystal display (LCD). The utility of having a fast, bright emissive display that spreads light over a wide viewing angle has given some top-of-the-line portable personal computers very strong market success. Much of this success is due to the availability of relatively low-cost plasma displays and high-volume production capacity. While the excitement generated by the recent LCD research developments has distracted attention from plasma displays, the present plasma display market success is catching the eye of those who need high-performance flat-panel displays.

From the perspective of this section, written in 1992, the research and development activity on plasma displays is currently directed toward achieving lower-cost higher-performance computer displays and the ultimate dream of flat-panel color television. Plasma displays have a rather impressive list of research and development accomplishments. A number of full-color plasma displays have been demonstrated in laboratories, and these are steadily advancing toward the market. Many of these appear as attractive as the CRT. Plasma displays with multiple intensity levels (gray scale) are widely available. Plasma displays have achieved the largest size and the largest number of pixels of any of the flat-panel technologies. Plasma display products are available from Photonics having 2048×2048 pixels and measuring 1.06×1.06 meters or 1.5 meters diagonal. The very large size and very high resolution capability of these monochrome display products is motivating researchers at more than 15 institutions worldwide to develop large-sized full-color plasma displays for both high-definition and NTSC home entertainment television systems.

Plasma displays have been successful over a very wide range of products. One of the purposes of this section is to discuss some of these products as shown in the plasma display family tree of Fig. 77.23. Under each of these types of plasma display are the manufacturers that offer these plasma products. Included in this list are only those manufacturers that actually make plasma display devices. This list would be much larger if it included those companies that buy plasma devices from these manufacturers and then add electronics for further resale.

Plasma Display Attributes

Table 77.1 shows some of the attributes of plasma displays which have made them so successful. The following is a review of each attribute.

1. The plasma display has a very large nonlinearity due to the electrical characteristic of the gas discharge used in all plasma displays. This is an electrical nonlinearity which means that below a certain threshold voltage the gas discharge will emit no light. Of course above that threshold voltage the gas discharge fires and emits a pleasant glow. This very sharp nonlin-

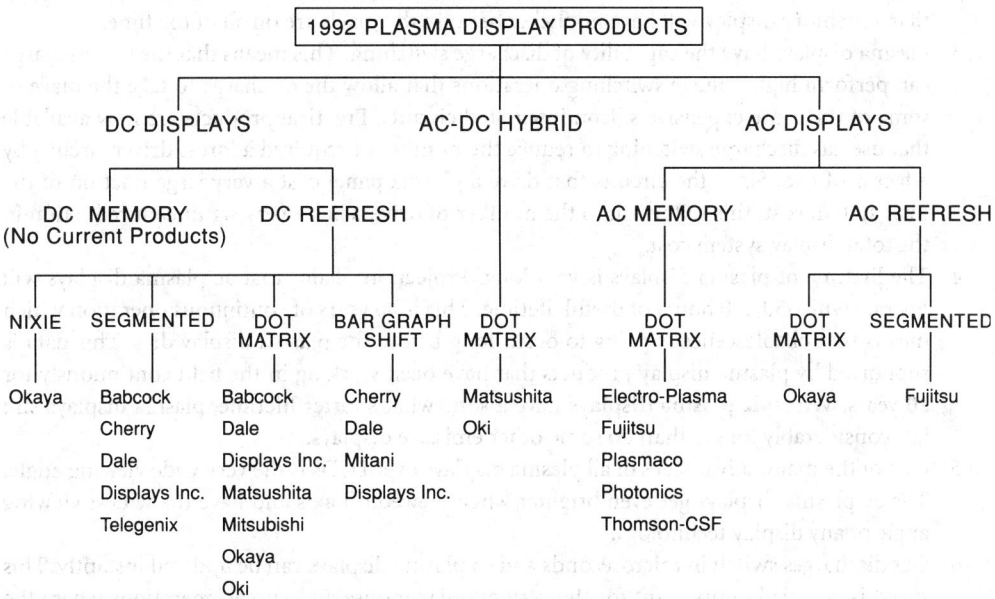

FIGURE 77.23 Family tree of plasma display products.

Table 77.1   Plasma Display Attributes

   1. Very strong nonlinearity
   2. Memory
   3. Discharge switching
   4. Long lifetime
   5. Very wide viewing angle
   6. Instant update time
   7. Good brightness and luminous efficiency
   8. Amorphous structure, ease of fabrication
   9. Rugged self-supporting structure
  10. High resolution and large size
  11. Transparent display medium with $n = 1$ does not scatter ambient light
  12. Tolerant to harsh environment and temperature extremes
  13. Reasonable impedance characteristics
  14. Diffuse glow
  15. Multicolor and gray-scale capability

earity allows plasma displays to be multiplexed without limit, which means that very large plasma displays are practical. This is demonstrated by the 1.5-meter diagonal 2048×2048 plasma display product mentioned in the introduction. This is a considerable advantage when compared to other display technologies such as the LCD. The LCD does not have a very good nonlinearity, and, therefore, some other nonlinear element such as a thin-film transistor is sometimes added in series with each liquid-crystal element to increase the display nonlinearity. This greatly complicates and adds cost to this active matrix liquid crystal.

2. Some plasma displays have inherent **memory**. This memory is stored directly in the glass plasma panel. Memory is very desirable for flat-panel displays because it allows the display to be very bright even for very large sizes, because a display with memory has a pixel duty cycle of 1. Displays without memory have a pixel duty cycle of 1 divided by the number of scanned lines. This means that as the nonmemory displays get bigger and the number of scanned lines increases, the duty cycle and therefore the brightness of the display decreases. The **ac plasma displays** achieve this memory function readily. An additional value of memory is that a memory display will have no flicker because the pixels are on all of the time.

3. Plasma displays have the capability of discharge switching. This means that the gas discharge can perform high-voltage switching operations that allow the discharge to take the place of some of the more expensive silicon integrated circuits. Practical products are now available that use gas discharge switching to reduce the number of required address driver circuits by a factor of two. Since the circuits that drive a plasma panel cost a very large fraction of the total system cost, this reduction in the number of drivers has a very significant reduction in the total display system cost.

4. The lifetime of plasma displays is very long. Projections claim that ac plasma displays will operate with 350,000 hours of useful lifetime. This is 40 years of continuous operation which means that obsolescence is going to occur long before the plasma display dies. This data is supported by plasma display products that have been working in the field continuously for 20 years. While **dc plasma displays** have a somewhat shorter lifetime, plasma displays still last considerably longer than do some other emissive displays.

5. One of the major advantages of all plasma displays over LCDs is the very wide viewing angle. The ac plasma displays get even brighter when viewed off axis and have the widest viewing angle of any display technology.

6. Gas discharges switch in microseconds and so plasma displays can be updated instantly. This speed is especially important for the very popular mouse and cursor operations where the cursor would disappear when moving on an LCD.

7. The brightness and **luminous efficiency** of plasma displays are good. Some other display technologies such as ac electroluminescence (EL) have a higher material luminous efficiency, but this fact must be tempered by the fact that the plasma panel has 1000 times less electrical capacitance than the ac electroluminescent devices. This means that the plasma panel frequently takes less power than the EL devices when the switching loss of the EL panel is considered. This favors plasma panels for larger numbers of scanned lines and favors EL panels for smaller numbers of scanned lines. The crossover point is somewhere in the region of a few hundred scanned lines.

8. Plasma displays have an amorphous glass structure that is easy to mass-produce.

9. The structure can withstand very high levels of shock and vibration when properly mounted. Military plasma displays have been designed for in excess of 150 g of shock.

10. Plasma displays have been made with sizes as great as 1.5 meters in diagonal and with over 4 million pixels. The highest-resolution plasma display products have achieved 0.2 pixels per millimeter.

11. The gas discharge does not scatter ambient light since it has an index of refraction of 1. This means that a lower luminance on a plasma panel may have a much higher contrast than a higher luminance on a CRT phosphor that scatters the ambient light.

12. Plasma displays can easily operate at both high- and low-temperature extremes. The ac plasma displays have a temperature limit dependent almost solely on the drive circuit characteristics, while dc plasma displays which use mercury should not be operated for long periods at low temperatures without an external heater.

13. Plasma displays have a high-input impedance characteristic that makes them easy to drive. The dielectric constant of the gas is equal to 1, meaning that plasma displays have virtually the lowest possible electrode capacitance. This is 1000 times smaller than electroluminescent displays and about 100 times smaller than LCDs. This translates to lower current requirements and therefore smaller drive circuit silicon area for the plasma displays. While plasma displays do require high-voltage drivers, it is frequently easier to design high-voltage circuits than high-current circuits.

14. The plasma displays have a diffuse glow that allows the gas discharge to spread around the electrode. This is a great advantage because low-resistance, opaque electrodes can be used and yet the light still escapes around the sides of the electrode. Transparent electrodes can still be used with plasma displays but they frequently have too much resistance, which gives too much voltage drop and power loss. While the opaque electrode shadows the gas discharge, it still allows virtually all of the light to get out of the panel. Note that many of the plasma panels are transparent, a characteristic which allows images to be back projected with maps or other images. In addition, a very pleasant display can be achieved by placing a green ac powder electroluminescent panel behind the orange plasma panel. This gives an attractive color contrast.

15. As will be covered later in this section, plasma displays are fully capable of full-color television with gray scale. This color capability coupled with the above attributes make plasma displays the current most promising technology for future large-area flat-panel displays for high-definition television.

Gas Discharge Characteristics

A brief account of the gas discharges used in plasma display follows, but a more detailed discussion of this material is presented in Weber [1985].

Figure 77.24 shows the important reactions that occur in a gas discharge. The reactions in the gas volume include ionization (I), excitation (E), metastable generation (M), and Penning ionization (P). The three surface reactions that occur at the cathode cause ejection of electrons from the cathode by a bombarding neon ion, a neon metastable atom, or a high-energy photon. The most

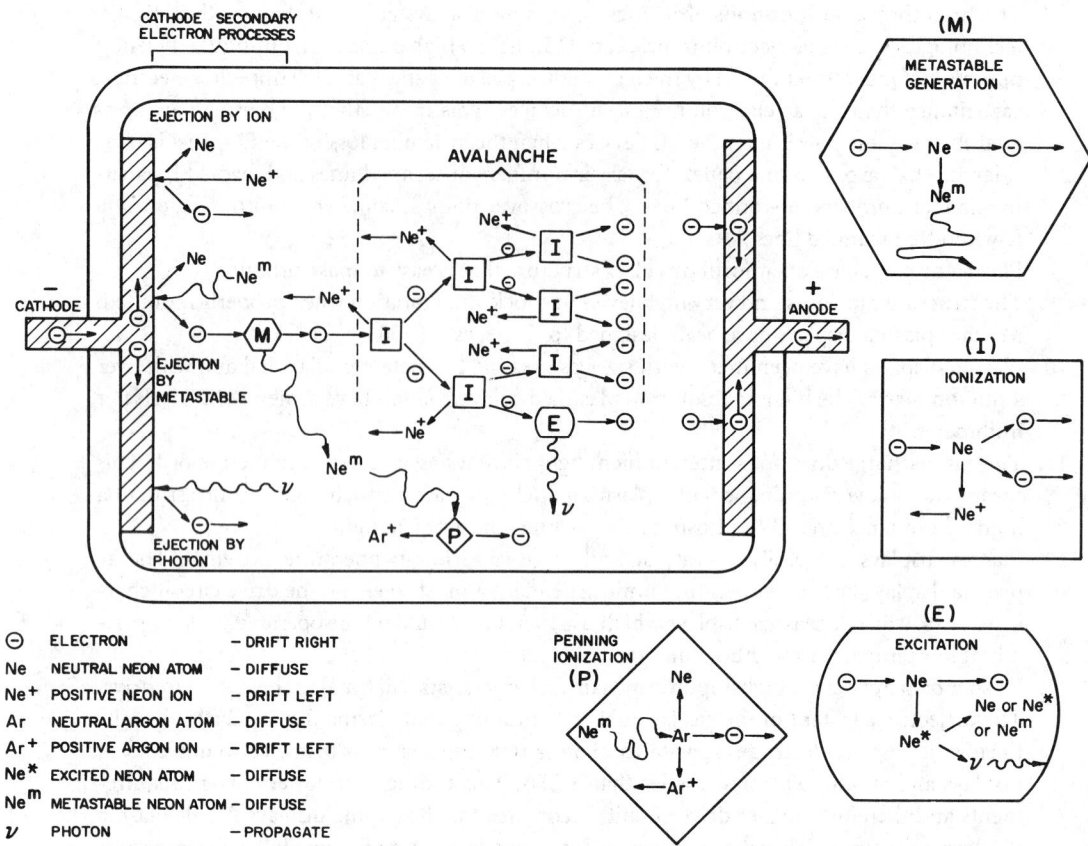

FIGURE 77.24   Model of important gas discharge reactions.

important volume reaction is ionization (I), which can cause the generation of an avalanche in the gas volume as shown in Fig. 77.24. This avalanche is started by an electron near the cathode and, as it grows toward the anode, the avalanche generates a large number of electron-ion pairs. The number of electron-ion pairs increases with increasing applied voltage across the gas. Ions, photons, or metastable atoms that are transported to the cathode can then eject electrons with a cathode surface-dependent probability and these ejected electrons will initiate further avalanches. These mechanisms act as a positive feedback system that becomes unstable when the loop gain is greater than 1. The onset of the unstable condition is defined as the gas firing voltage. Above this firing voltage the discharge current will continue to grow without bounds if the initial avalanche is primed with at least a single electron.

Figure 77.25 shows the I-V characteristic of a typical gas discharge found in plasma displays. Note that the current is plotted on a log scale over nine orders of magnitude. The most striking feature is the very strong nonlinearity at the firing voltage. This is a major attribute of gas discharges that allows matrix addressing. When the discharge current has sufficient magnitude, space charge distortion sets in and the characteristic achieves a negative resistance region. Most plasma displays operate near the junction of the normal and the abnormal glow regions of the characteristic.

Figure 77.26 shows the characteristics of the glow discharge commonly found in operating plasma displays. The light comes from two luminous regions: the negative glow and the positive column. All plasma displays on the market today use light from the negative glow, but a number of research displays have used the light from the positive column. These regions are caused by the space charge distribution of the electrons and ions that distort the electric field and voltage distribution.

Figure 77.27 shows the wavelength distribution found in plasma displays. This is the classic line spectrum of neon gas which appears as the familiar neon orange. Neon is the only luminous gas currently used in commercial plasma displays because of its high luminous efficiency.

Current Limiting for Plasma Displays

To avoid a catastrophic arc shown in the upper regions of Fig. 77.25, the current in a gas discharge must be limited by some means. There are a number of ways of accomplishing this, but only two, shown in Fig. 77.28, have achieved commercial success. The dc plasma displays use an external resistor to limit the current and have the electrodes in intimate contact with the gas discharge, while ac plasma displays limit the current with an internal glass dielectric that couples the electrodes capacitively to the gas discharge.

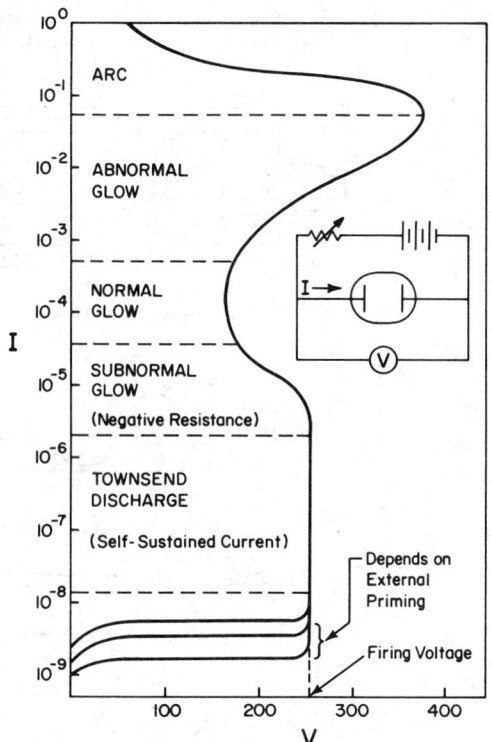

FIGURE 77.25 The I-V characteristic of a gas discharge.

Most of the commercially successful dc plasma displays have the resistors connected to a display electrode external to the panel. This allows only one discharge to be ignited along that electrode at any one time. This works well for scanned displays. Multiple discharges and dc memory require placing the resistor internal to the panel in series with each pixel. It is very difficult to fabricate internal high-voltage resistors with the required impedance (10k to 1 MΩ). Because of this there are currently no surviving dc memory products.

The ac plasma displays can achieve memory because it is easy to make a dielectric layer with the desired impedance so that a current-limiting capacitor can be placed in series with each pixel. When a voltage pulse is applied to an ac panel, the discharge deposits a charge on the wall that reduces the voltage across the gas. After a short time, the discharge will extinguish and the light output will end until the applied voltage reverses polarity and a new discharge pulse occurs. This wall charge allows the ac plasma displays to operate in a memory mode, which greatly increases the brightness of large displays.

Plasma Display Products

Returning now to the plasma display products family tree of Fig. 77.23, we see that plasma displays are divided into three major categories: dc plasma displays, ac plasma displays, and the ac-dc hybrid based on the current limiting technique just discussed. Each of the differing plasma display products is discussed in the following sections.

DC Plasma Displays

Some of the simplest dc plasma displays are the segmented displays shown in Fig. 77.29. The electrode patterns are typically etched transparent electrode anodes with screen-printed cathode electrodes. A large number of manufacturers supply these segmented displays, as shown in Fig. 77.23.

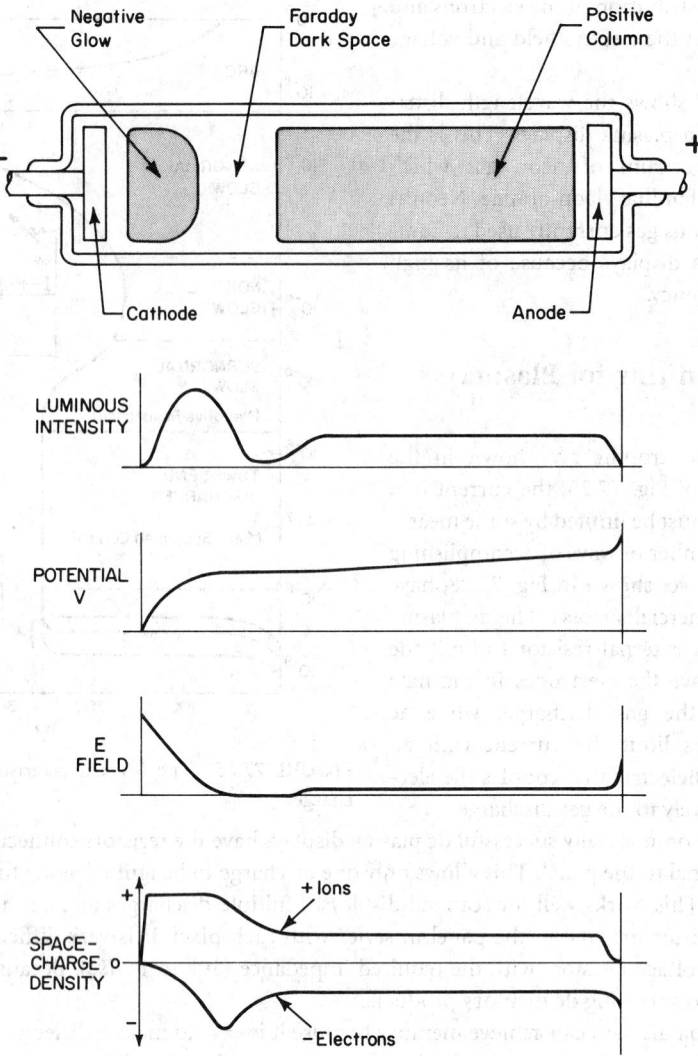

FIGURE 77.26   The luminous regions of a gas discharge.

It is important to note that most dc plasma displays have a small amount of mercury added to the neon gas. This serves to greatly extend the lifetime of the dc display by inhibiting the sputtering action of the gas discharge on the cathode. The ac plasma displays do not need mercury to inhibit the sputtering because sputtering does not significantly influence the refractory oxide, MgO, used for the ac dielectric cathode material. Care must be taken to prevent dc displays from being operated at too low a temperature or the mercury will freeze out and the dc displays will then have a very short lifetime.

The dc plasma display structure that has achieved the greatest commercial success for computer displays is shown in Fig. 77.30 and is primarily manufactured by Matsushita and Oki in Japan [Akutsu and Nakagawa, 1981, 1982]. It is similar to the dc segmented displays since it uses transparent indium–tin oxide anode electrodes and screen-printed nickel electrodes for the cathodes. Horizontal dielectric ribs are added to isolate the individual discharges and also to separate the top anode plate from the bottom cathode plate. External alumina substrates with thick-film resistors

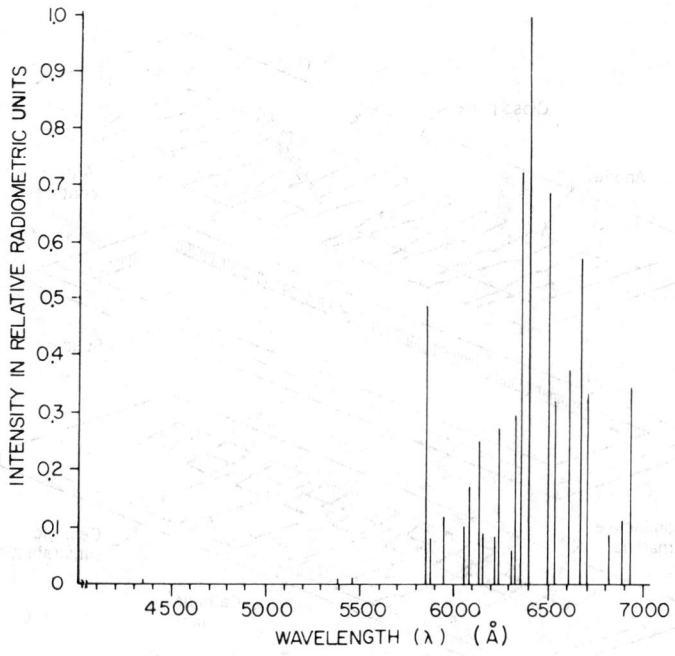

FIGURE 77.27 The light wavelength distribution for a plasma display having neon gas.

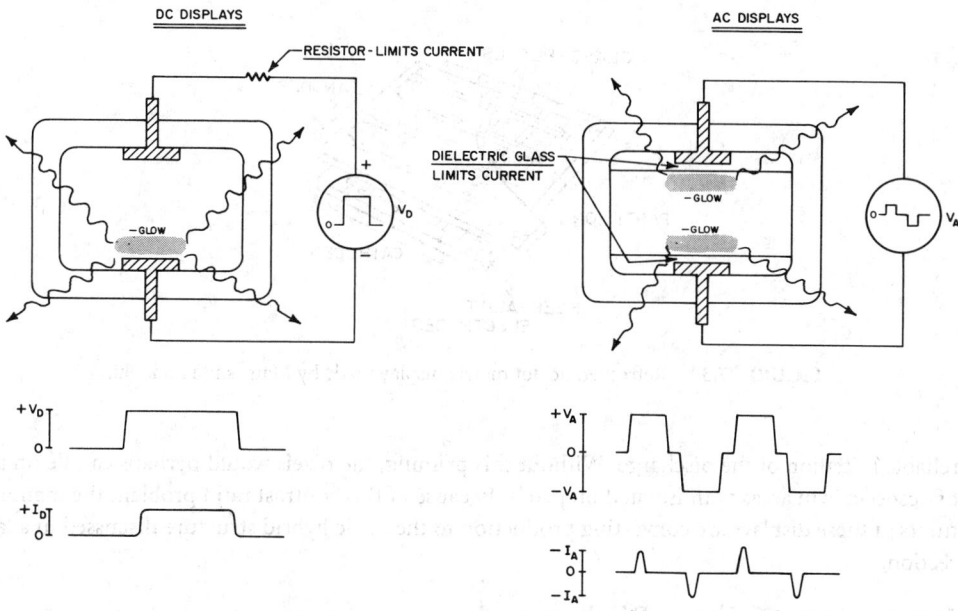

FIGURE 77.28 The two current-limiting techniques used in plasma display products.

are connected to the column electrodes to limit the current. An alternate technique is to use a special integrated circuit driver that has very high-impedance current-sourcing capabilities.

One characteristic of these dc displays is a background glow that limits the contrast ratio to about 6:1. This background glow is due to a short pulse of discharge activity applied to each off pixel during every refresh scan. This short discharge provides the priming particles necessary for

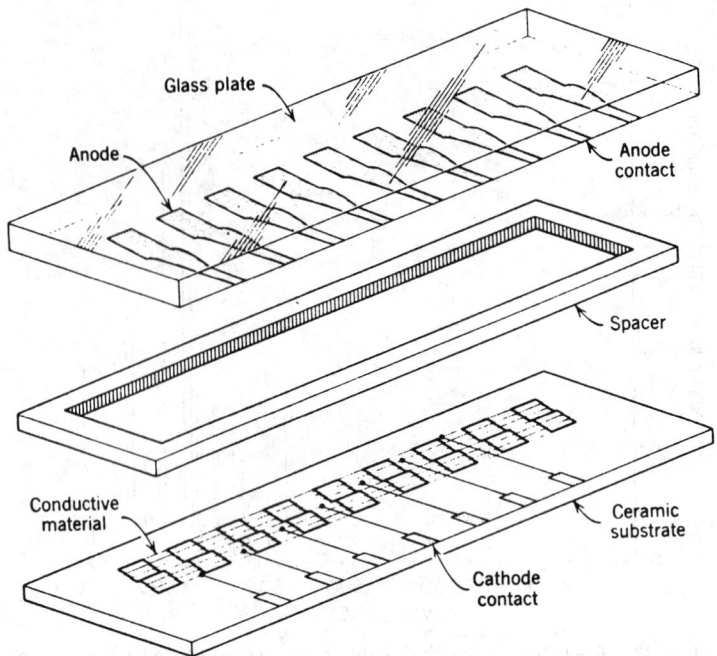

FIGURE 77.29   Segmented dc plasma display.

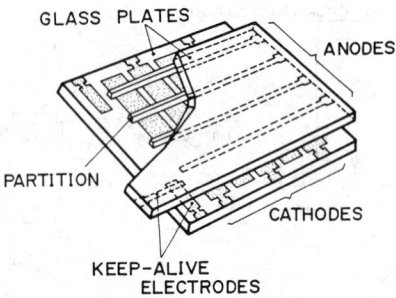

FIGURE 77.30   Refreshed dc dot matrix display made by Matsushita and Oki.

reliable initiation of the discharge. Without this priming, the pixels would perhaps candle on and off, especially in areas with isolated on pixels. Because of this contrast ratio problem the manufacturers of these displays are converting production to the ac-dc hybrid structure discussed in a later section.

Memory-Type AC Plasma Displays

A look at the plasma displays products family tree (Fig. 77.23) shows that ac displays are divided into the memory type and the refresh type which does not use the memory feature. This section will discuss only the memory type because of its much greater commercial success [Criscimagna and Pleshko, 1980].

Figure 77.31 shows the ac plasma display structure. These panels are made by depositing thin-film electrodes on the front and back substrates and then covering those electrodes with a thin dielectric glass. Recall from Fig. 77.28 that this dielectric glass makes a capacitor that is used to limit

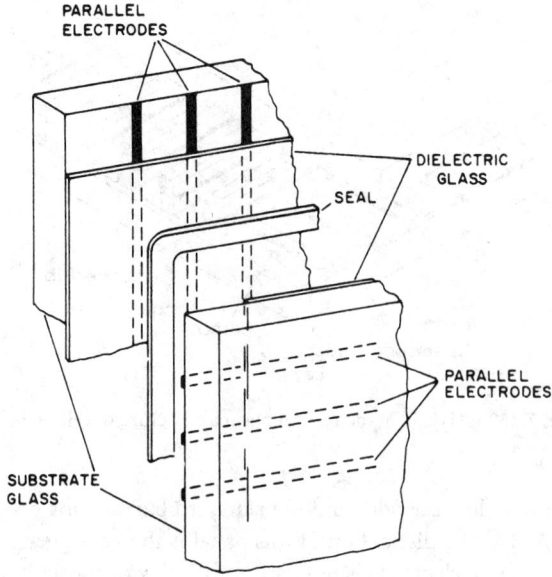

FIGURE 77.31   AC memory plasma display structure.

the discharge current. This dielectric also is used to store the charge that gives these panels inherent memory. The two substrates are then sealed together around the perimeter and filled with neon gas. There are no barrier ribs between the discharges as in the dc panels shown in Fig. 77.30. The ac panels have a simpler structure that allows the pixels to be isolated simply by the action of electric fields.

The ac plasma panels require that the inner surface of the dielectric that is in contact with the neon gas has a special coating of magnesium oxide. This MgO layer is necessary for the panel to have low operating voltages and long life. It acts similarly to the mercury in the dc displays as an antisputtering agent.

The ultimate limits of the size and the number of pixels are being explored by the impressive ac display having a diagonal of 1.5 meters and a 2048 × 2048 array of more than 4 million pixels [Wedding *et al.*, 1987]. This display operates at a very high update rate so that it will work with a standard NTSC video source. The memory feature allows this display to have the same luminance as the smaller page-sized displays.

The ac displays require that an ac signal called the sustain voltage be applied during operation, as shown in Fig. 77.28. When a pixel is discharging, charge collects on the dielectric glass walls and influences the voltage across the gas. The component of voltage due to this charge is called the wall voltage. When a pixel is on, the wall voltage changes for each polarity reversal of the sustain voltage. This change in wall voltage coincides with a pulse of light due to the gas discharge. When the pixel is off, there are no light pulses, and the wall voltage remains at a zero level.

Pixel addressing is achieved through a partial discharge introduced by an address pulse timed between the sustain voltage pulses. A write pulse causes the wall voltage to transit from zero volts to the final equilibrium wall voltage level. Likewise, an erase pulse causes the wall voltage to return to zero.

Hybrid AC-DC Plasma Display

Figure 77.32 shows the structure of the hybrid plasma display initially developed by Sony and manufactured for a number of years by DIXY [Amano *et al.*, 1982]. This design has emerged as the dominant plasma technology for VGA-sized displays. This is basically a dc plasma panel with the

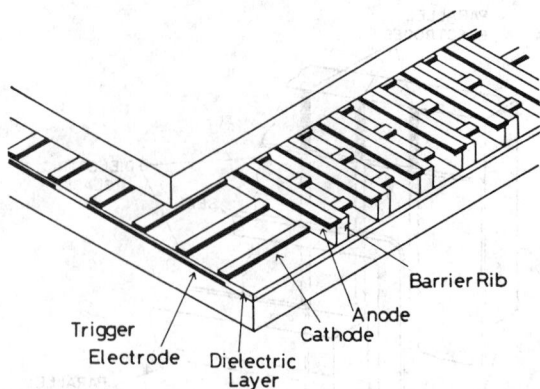

FIGURE 77.32   Hybrid ac-dc refresh plasma structure developed by Sony.

normal anode and cathode dc electrodes and the standard barrier ribs needed to isolate the discharges just as in Fig. 77.30. The distinction of this panel is the ac trigger electrode that is buried under the dc cathodes. This ac electrode is not used for display purposes but rather is used to create priming particles for the dc discharge.

The addressing sequence is initiated with a negative pulse applied to the ac trigger electrode. This creates the priming particles that eventually are used to ignite the normal dc discharge that is used for the display. These priming particles have two beneficial effects. First they allow use of lower voltages to drive the normal dc address electrodes. These lower voltages require lower-cost circuit drivers. The second advantage of this priming action is the shorter time delay between the start of the address pulse and the ignition of the discharge. This allows a very significant reduction in the pixel-to-pixel brightness variation due to the adverse statistical time delay. The major consequence of this design is a very low background glow when compared to the conventional dc displays of Fig. 77.30 since it is not necessary to partially discharge off pixels to such a high degree to generate priming particles.

Gray Scale

Gray-scale research for plasma displays was very active in the early 1970s. This research demonstrated the first high-quality flat-panel television, first on a Burroughs self-scan dc plasma panel and then on a memory ac plasma panel [Chodil, 1976]. This work created a great deal of excitement in the display community and it stimulated most of the subsequent flat-panel television research.

In addition to true gray scale, an attractive spatial gray-scale technique was developed called ordered dither that could be used with any bilevel display [Judice and Slusky, 1981]. Later research found more sophisticated adaptive spatial gray-scale techniques that are used today with the advent of powerful local microprocessors.

Gray-scale products are available for dc panels from Matsushita and Oki and for ac panels from Fujitsu, Thomson CSF, and Photonics. Currently the highest number of intensity levels available is 16. However, the combination of spatial techniques and true gray scale can be quite useful when more intensity levels are needed. The dc displays use pulse-width modulation to control the intensity of the pixels. Special-purpose column driver integrated circuits are needed to apply the varying-width pulses to the different columns during a single-row line time. The ac memory displays cannot use pulse-width modulation because the pixel is either on or off and a narrow pulse would probably have the undesirable effect of erasing the pixel. Instead, the ac memory displays modulate the time that the pixel is on in a given frame. This means that the pixels must be addressed multiple times per frame. A binary chop technique is used so that each pixel is addressed at most once per frame for each bit of gray-scale resolution. For instance, a 16-intensity gray scale would require a maximum of four addresses per frame.

Color Plasma Displays

Color is achieved by placing phosphors in the plasma panel and then exciting those phosphors with the ultraviolet light of the gas discharge. This is the same principle as used in the fluorescent lamp. As a practical product example, Fig. 77.33 shows the structure of a 21-inch diagonal 640×480 full-color ac plasma panel developed by Fujitsu [Yoshikawa *et al.*, 1992]. The phosphors are placed on the rear glass substrate of the panel and are excited by the ultraviolet light generated by the sustain electrodes on the front glass substrate. This electrode arrangement is not of the ordinary design for the ac panel shown in Fig. 77.31. In this case both sets of power-supplying sustain electrodes are on the front glass substrate. The ac sustaining voltage is applied to pairs of these sustain electrodes in the normal way and the fringing fields from these electrodes reach into the gas and create a discharge. This method is used to reduce the ionic bombardment of the phosphor that results in its degradation and short display lifetime. The ions usually bombard the cathode, which in this structure is always on the front glass substrate, and so the ions are unlikely to hit the phosphors on the rear glass substrate.

A major problem with color plasma displays is designing them to obtain high brightness at high luminous efficiency. Unfortunately, the high brightness and luminous efficiency of fluorescent lamps is not achieved when applied to the very small pixels in a plasma display. In addition, almost all plasma displays use light from the negative glow, which is known to be considerably less efficient than light from the positive column. There remains a considerable opportunity for future improvement of plasma panel luminous efficiency because current plasma panels are a factor of 100 less power efficient than common fluorescent lamps.

Research and development activity in full-color plasma displays is very active. In addition to the 21-inch ac display shown in Fig. 77.33, other developments include a 23-inch diagonal ac display by Thomson CSF, a 19-inch ac display by Photonics, and a 40-inch dc plasma display developed by NHK for high-definition television [Murakami *et al.*, 1991]. While some of these displays are of CRT-like quality, each has a different structure and there has not yet emerged a clear leader.

The greatest opportunity for full-color plasma seems to be in displays with diagonals greater than 15 to 20 inches. This is primarily because smaller diagonal full-color thin-film transistor LCDs are already commercially available. However, the very expensive LCD manufacturing plants have processing equipment limitations of roughly 15 inches, which greatly reduces the likelihood of

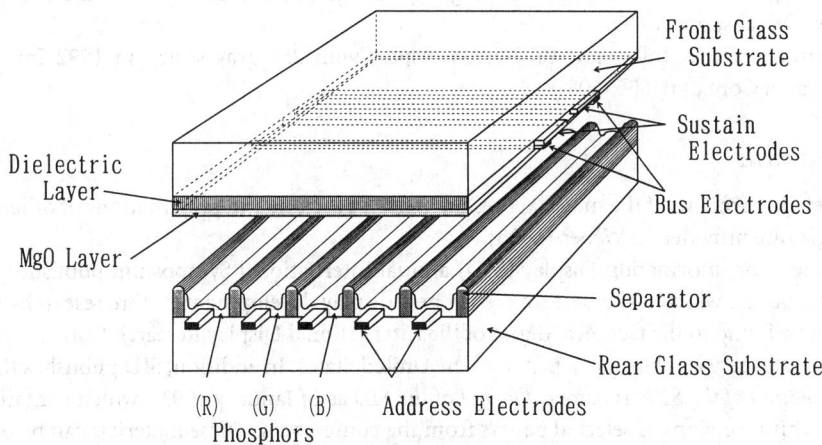

FIGURE 77.33 Full-color ac memory plasma structure developed by Fujitsu.

larger diagonal LCDs. The many attributes of plasma displays favoring large-screen displays make plasma a very attractive candidate for the large flat-panel television displays of the future.

Defining Terms

AC plasma displays: These employ an internal capacitive dielectric layer to limit the gas discharge current.

DC plasma displays: These employ an external resistor to limit the gas discharge current.

Luminous efficiency: The measure of the display output light luminance for a given input power, usually measured in lumens per watt, which is equivalent to the nit.

Memory: The property of a display pixel that allows it to remain stable in an initially established state of luminance. Memory gives a display high luminance and absence of flicker.

Plasma: The fourth state of matter comprised of positive ions and negative electrons of equal and sufficiently high density to nearly cancel out any applied electric field. Not to be confused with blood plasma.

References

H. Akutsu and Y. Nakagawa, "Scanning in a DC plasma panel," in 1981 SID Int. Symposium, New York, pp. 166–167.

H. Akutsu and Y. Nakagawa, "A DC plasma display panel unit with higher reliability and simpler construction," *Proc. SID*, vol. 23, pp. 61–65, 1982.

Y. Amano, K. Yoshida, and T. Shionoya, "High-resolution DC plasma display panel," in 1982 SID Int. Symposium, San Diego, pp. 160–161.

G. Chodil, "Gas discharge displays for flat-panel," *Proc. SID*, vol. 17, pp. 14–22, 1976.

T. N. Criscimagna and P. Pleshko, "AC plasma display," in *Topics in Applied Physics*, vol. 40, *Display Devices*, Berlin: Springer-Verlag, 1980, pp. 91–150.

C. N. Judice and R. D. Slusky, "Processing images for bilevel digital displays," in *Advances in Image Pickup and Display*, vol. 4, New York: Academic Press, 1981, pp. 157–229.

S. Kanagu *et al.*, "A 31-in. diagonal full-color surface-discharge AC plasma display panel," in 1992 SID Int. Symposium, Boston, pp. 713–716.

H. Murakami *et al.*, "A 33-in. diagonal HDTV display using gas discharge pulse memory technology," in 1991 SID Int. Symposium, Anaheim, pp. 713–716.

L. F. Weber, "Plasma displays," in *Flat-Panel Displays and CRTs*, L.E. Tannas Jr., Ed., New York: Van Nostrand Reinhold, 1985, pp. 332–414.

D. K. Wedding *et al.*, "A 1.5 m diagonal AC gas discharge display," in 1987 SID Int. Symposium, New Orleans, pp. 96–99.

K. Yoshikawa *et al.*, "A full color AC plasma display with 256 gray scale," in 1992 Int. Display Research Conf., pp. 605–608.

Further Information

A more detailed account of the material presented in this section and presentations of other display technologies are provided in Weber [1985].

The Society for Information Display (SID) annual International Symposium publishes a digest of technical papers which is the best source for new display developments. More research-oriented papers can be found in the technical digest of the International Display Research Conference which rotates annually between Europe, Japan, and the United States. In addition, SID publishes the quarterly *Proceedings of the SID* (renamed *Journal of the SID* as of January 1993), which contains more detailed archival versions of selected papers from the conferences. These materials can be obtained from the SID at 8055 West Manchester Ave., Playa del Rey, CA 90293.

Data Acquisition

Dhammika
Kurumbalapitiya
Harvey Mudd College

S. Ratnajeevan H. Hoole
Harvey Mudd College

78.1 Introduction 1799
78.2 The Analog and Digital Signal Interface 1801
78.3 Analog Signal Conditioning 1802
78.4 Sample-and-Hold and A/D Techniques in Data
      Acquisition 1804
78.5 The Communication Interface of a Data Acquisition
      System 1804
78.6 Data Recording 1806
78.7 Software Aspects 1806

78.1 Introduction

Data acquisition includes everything from gathering data, to transporting it, to storing it. The term *data acquisition* is described as the "phase of data handling that begins with sensing of variables and ends with a magnetic recording of raw data, may include a complete telemetering link" (McGraw-Hill, *Dictionary of Scientific and Technical Terms,* Second Edition, 1978). Here, the term *variables* refers to those physical quantities that are associated with a natural or artificial process. A data acquisition phase involves a real-time computing environment where the computer must be keyed to the time scale of the process. Figure 78.1 gives a simplified block diagram of a data acquisition system current in the early 1990s.

The path the data travels through the system is called the data acquisition channel. Data are first captured and subsequently translated into usable signals using transducers. In this discussion, usable signals are assumed to be electrical voltages, either unipolar (that is, single ended, with a common ground so that we need just one lead wire to carry the signal) or bipolar (that is, common mode, with the signal carried by a wire pair, so that the reference of the rest of the system is not part of the output). These voltages can be either analog or digital, depending on the nature of the mea-surand (the quantity being captured). When there is more than one analog input, they are subsequently sent to an analog **multiplexer** (MUX). Both the analog and the digital signals are then conditioned using signal conditioners. There are two additional steps for those conditioned analog signals. First they must be sampled (see Chapter 67.4) and next converted to digital data. This conversion is done by **analog-to-digital converters** (ADC) (see Chapter 31).

Once the analog-to-digital conversion is done, the rest of the steps have to deal with digital data only. The calendar/clock block shown in Fig 78.1 is used to add the time-of-date information, an important parameter of a real-time processing environment, into the half-processed data. The digital processor performs the overall system control tasks using a software program, which is usually called system software. These control tasks also include display, printer, data recorder, and commu-

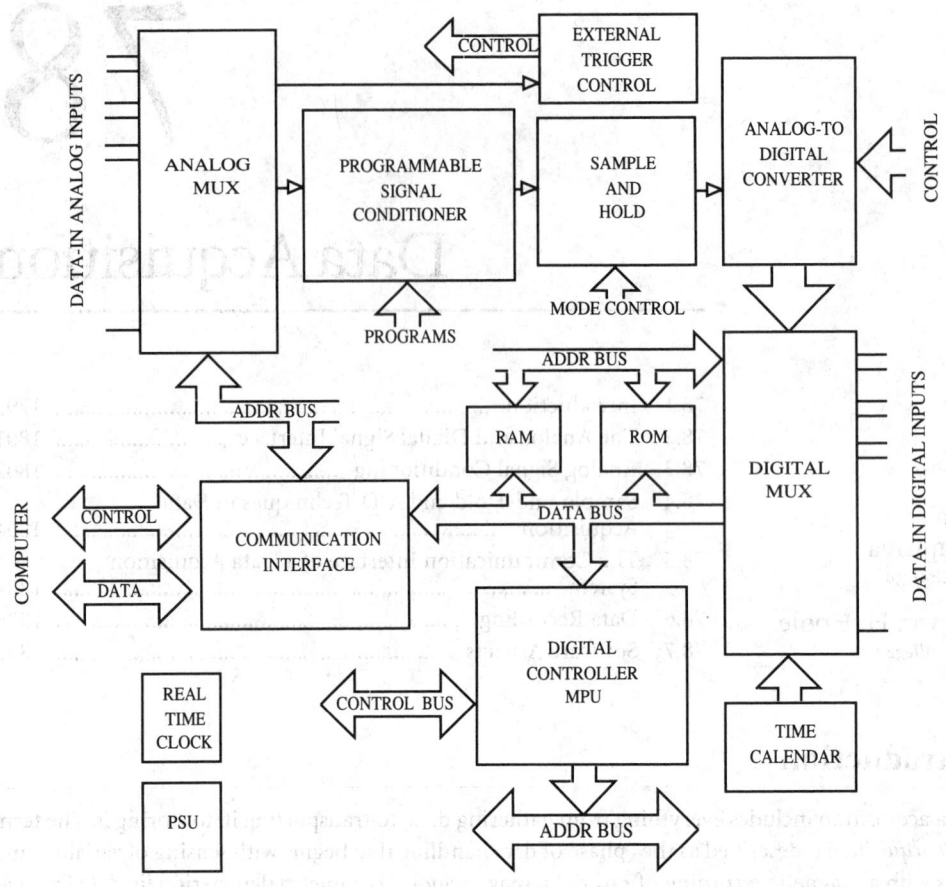

FIGURE 78.1   The block diagram of a data acquisition system.

nication interface management. A well-regulated **power supply unit** (PSU) and a stable clock are essential components in many data acquisition systems. There are systems where massive amounts of data points are produced within a very short period of time, and they are equipped with *on-board memory* so that a considerable amount of data points can be stored locally. Data are transmitted to the host computer once the local storage has reached its full capacity. Historically, data acquisition evolved in modular form, until monolithic silicon came along and reduced the size of the modules.

The analysis and design of data acquisition systems is a discipline that has roots in the following subject areas: signal theory, transducers, analog signal processing, noise, sampling theory, quantizing and encoding theory, analog-to-digital conversion theory, analog and digital electronics, data communication, and systems engineering. Cost, accuracy, bit resolution, speed of operation, on-board memory, power consumption, stability of operation under various operating conditions, number of input channels and their ranges, on-board space, supply voltage requirements, compatibility with existing bus interfaces, and the types of data recording instruments involved are some of the prime factors that must be considered when designing or buying a data acquisition system. Data acquisition systems are involved in a wide range of applications, such as machine control, robot control, medical and analytical instrumentation, vibration analysis, spectral analysis, correlation analysis, transient analysis, digital audio and video, seismic analysis, test equipment, machine monitoring, and environmental monitoring.

78.2 The Analog and Digital Signal Interface

The data acquisition system must be designed to match the process being measured as well as the end-user requirements. The nature of the process is mainly characterized by its speed and number of measuring points, whereas the end-user requirement is mainly the flexibility in control. Certain processes require data acquisition with no interruption where computers are used in controlling. On the other hand, there are cases where the acquisition starts at a certain instance and continues for a definite period. In this case the acquisition cycle is repeated in a periodic manner, and it can be controlled manually or by software. Controllers access the process via the analog and digital interface submodules, which are sometimes called analog and digital front ends.

Many applications require information capturing from more than one channel. The use of the analog MUX in Fig. 78.1 is to cater to multiple analog inputs. A detailed diagram of this input circuitry is shown in Fig. 78.2 and the functional description is as follows. When the MUX is addressed to select an input, say, $x_i(t)$, the same address will be decoded by the decoding logic to generate another address, which is used in addressing the programmable register. The programmable register contains further information regarding how to handle $x_i(t)$. The outcome of the register is then used in subsequent tuning of the signal conditioner. Complex programmable control tasks might include automatic gain selection for each channel, and hence the contents of this register are known as the channel gain list. The MUX address generator could be programmed in many ways, and one simple way is to scan the input channels in a cyclic fashion where the address can be generated by means of a binary counter. Microprocessors are also used in addressing MUXs in applications where complex channel selection tasks are involved. Multiplexers are available in integrated circuit form, though relay MUXs are widely used because they minimize errors due to cross talk and bias currents. Relay MUX modules are usually designed as plugged-in units and can be connected according to the requirements.

There are applications where the data acquisition cycle is triggered by the process itself. In this case an analog or digital trigger signal is sent to the unit by the process, and a separate external trigger interface circuitry is supplied. The internal controller assumes its duties once it has been triggered. It takes a finite time to settle the signal $x_i(t)$ through the MUX up to the signal conditioner once it is addressed. Therefore, it is possible to process $x_{i-1}(t)$ during the selection time of $x_i(t)$ for greater speeds. This function is known as pipelining and will be illustrated in Section 78.3.

In some data acquisition applications the data acquisition module is a plugged-in card in a computer, which is installed far away from the process. In such cases, transducers—the process sensing

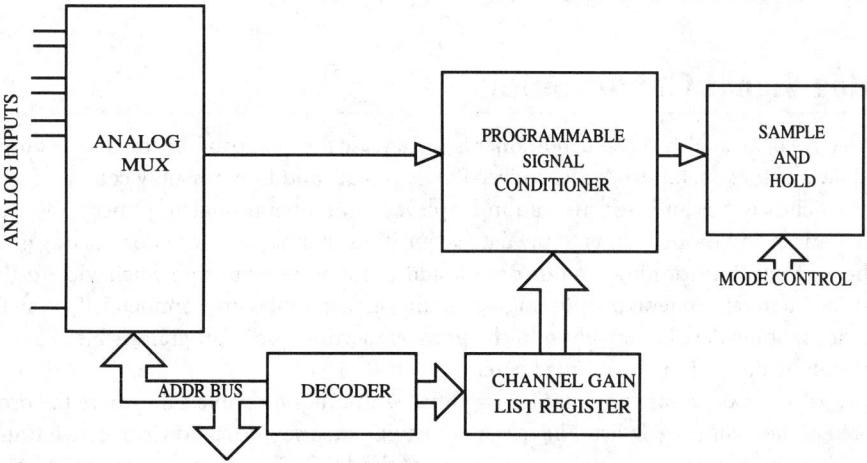

FIGURE 78.2 Analog input circuitry—the analog front end.

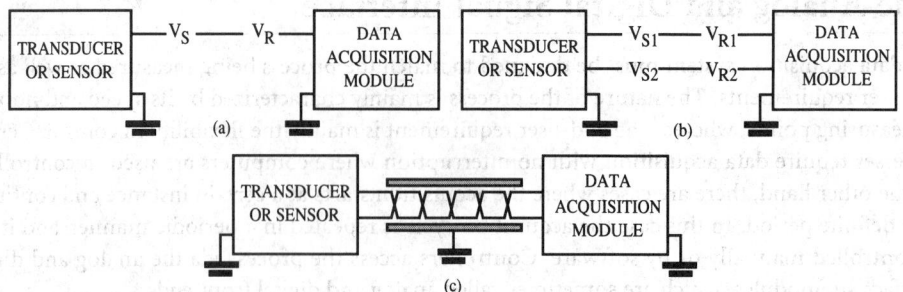

FIGURE 78.3   (a) Connecting transducers to the data acquisition unit, (b) single-ended (unipolar) output, and (c) common-mode (bipolar) output.

elements—are connected to the data acquisition module using transmission lines or a radio link. In the latter case a complete demodulating unit is required at the input. When transmission lines are used in the interconnection, care must be taken to minimize electromagnetic interference since transmission lines pick up noise easily. In the case of a single-ended transducer output configuration, a single wire is adequate for the signal transmission, but a common ground must be established between the two ends as given in Fig. 78.3(a). For the transducers that have common mode outputs, a shielded twisted pair of wires will carry the signal. In this case, the shield, the transducer's encasing chassis, and the data acquisition module's reference may be connected to the same ground as shown in Fig. 78.3(c). In high-speed applications the transmission line impedance should be matched with the output impedance of the transducer in order to prevent reflected traveling waves. If the transducer output is not strong enough to transmit for a long distance, then it is best to amplify it before transmission.

Transducers that produce digital outputs may be first connected to Schmitt trigger circuits for pulse shaping purposes, and this can be considered as a form of digital signal conditioning. This becomes an essential requirement when such inputs are connected through long transmission lines where the line capacitance significantly affects the rising and falling edges of the incoming wave. Opto-isolators are sometimes used in coupling when the voltage levels of the two sides of the transducer and the input circuit of the data acquisition unit do not match each other. Special kinds of connectors are designed and widely used in interconnecting transmission lines and data acquisition equipment in order to screen the signals from noise. Analog and digital signal grounds should be kept separate where possible to prevent digital signals from flowing in the analog ground circuit and including spurious analog signal noise.

78.3 Analog Signal Conditioning

The objective of an analog signal conditioner is to increase the quality of the transducer output to a desired level before analog-to-digital conversion. A signal conditioner mainly consist of a preamplifier, which is either an instrumentation amplifier or an operational amplifier and/or a filter. Coupling more and more circuits to the data acquisition channel has to be done taking great care that these signal conditioning circuits do not add more noise or unstable behavior to the data acquisition channel. General purpose signal conditioner modules are commercially available for applications. Some details were given in the previous section about programmable signal conditioners and the discussion is continued here.

Figure 78.4 shows an instrumentation amplifier with programmable gain where the programs are stored in the channel-gain list. The reason for having such sophistication is to match transducer outputs with the maximum allowable input range of the ADC. This is very important in improving accuracy in cases where transducer output voltage ranges are much smaller than the full-scale input range of an ADC, as is usually the case. Indeed, this is equally true for signals that are larger than the full-scale range, and in such cases the amplifier functions as an attenuator. Furthermore, the instrumentation amplifier converts a bipolar voltage signal into a unipolar voltage with respect to the system ground. This action will reduce a major control task as far as the ADC is concerned; that is, the ADC is always sent unipolar voltages, and hence it is possible to maintain unchanged the mode control input which toggles the ADC between the unipolar and bipolar modes of an ADC.

Values of the **signal-to-noise ratio**

$$\text{SNR} = \left[\frac{\text{RMS signal}}{\text{RMS noise}}\right]^2 \tag{78.1}$$

at the input and the output of the instrumentation amplifier are related to its **common-mode rejection ratio** (CMRR) given by

$$\text{CMRR} = \sqrt{\frac{\text{SNR}_{\text{output}}}{\text{SNR}_{\text{input}}}} \tag{78.2}$$

Hence, higher values of $\text{SNR}_{\text{output}}$ indicate low noise power. Therefore, instrumentation amplifiers are designed to have very high CMRR figures. The existence of noise will result in an error in the ADC output. The allowable error is normally expressed as a fraction of the **least significant bit** (LSB) of the code such as $\pm(1/X)$LSB. The amount of error voltage (V_{error}) corresponding to this figure can be found considering the bit resolution (N) and the ADC's maximum analog input voltage (V_{max}) as given in

$$V_{\text{error}} = \pm\left[\frac{V_{\text{max}}}{2^N - 1} \times \frac{1}{X}\right] \text{ volts} \tag{78.3}$$

Other specifications of amplifiers include the temperature dependence of the input offset voltage (V_{offset}, μV/°C) and the current (I_{offset}, pA/°C) associated with the operational amplifiers in use. High slew rate (V/μs) amplifiers are recommended in high-speed applications. Generally, the higher the bandwidth, the better the performance.

Cascading a filter with the preamplifier will result in better performance by eliminating noise. Active filters are commonly used because of their compact design, but passive filters are still in use. The cut-off frequency, f_c, is one of the important performance indices of a filter that has to be designed to match the channel's requirements. The value f_c is a function of the preamplifier bandwidth, its output SNR, and the output SNR of the filter.

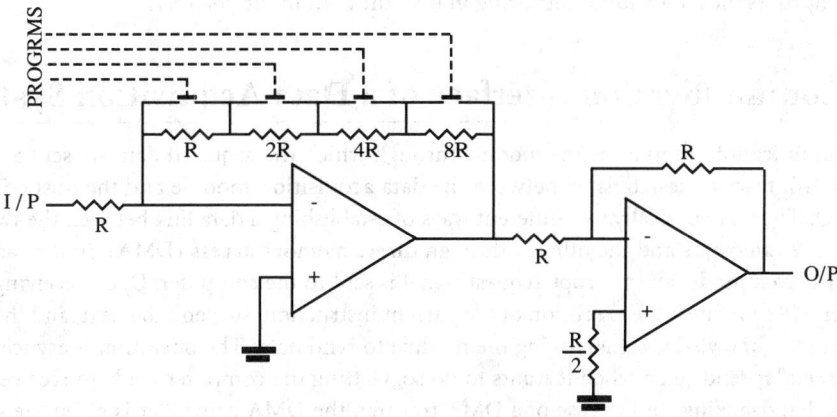

FIGURE 78.4 Programmable gain instrumentation amplifier.

78.4 Sample-and-Hold and A/D Techniques in Data Acquisition

Sample-and-hold systems are primarily used to maintain a constant magnitude representing the input, across the input of the ADC throughout a precisely known period of time. Such systems are called **sample-and-hold amplifiers** (SHA), and their characteristics are crucial to the overall system accuracy and reliability of digital data. The SHA is not an essential item in applications where the analog input does not vary more than $\pm(1/2)$LSB of voltage. As the name indicates, a SHA operates in two different modes, which are digitally controlled. In the sampling mode it acts as an input voltage follower, where, once it is triggered into its hold mode, it should ideally retain the signal voltage level at the time of the trigger. When it is brought back into the sampling mode, it instantly assumes the voltage level at the input.

Figure 78.5 shows the simplified circuit diagram of a monolithic sampling-and-hold circuit and the associated switching waveforms. The differential amplifiers function as input and output buffers, and the capacitor acts as the storage mechanism. When the mode control switch is at its *on* position, the two buffers are connected in series and the capacitor follows the input with minimum time delay, if it is small. Now, if the mode control is switched *off*, the feedback loop is interrupted, and the capacitor ideally retains its terminal voltage until the next sampling signal occurs. Leakage and bias currents usually cause the capacitor to discharge and or charge in the hold mode and the fluctuation of the hold voltage is called *droop,* which could be minimized by having a large capacitor. Therefore, the capacitance has to be selected such that the circuit performs well in both modes. Several time intervals are defined relative to the switching waveform of SHAs. The *acquisition time* (t_a) is the time taken by the device to reach its final value after the sample command has been given. The *setting time* (t_s) is the time taken to settle the output. The *aperture uncertainty* or *aperture jitter* (t_{us}) is the range of variation of the aperture time. It is important to note here that the sampling techniques have a well-formulated theoretical background.

ADCs perform a key function in the data acquisition process. The application of various ADC technologies in a data acquisition system depends mainly on the cost, bit resolution, and speed. Successive approximation types are more common at high resolution at moderate speeds (<1 MHz). This kind of ADC offers the best trade-offs among bit resolution, accuracy, speed, and cost. Flash converters, on the other hand, are best suited for high-speed applications. Integrating-type converters are suitable for high-resolution and -accuracy applications.

Many techniques have been developed in coupling sample-hold circuits and ADCs in data acquisition systems because no single ADC or sampling technology is able to satisfy the ever increasing requirements of data acquisition applications. Figure 78.6 illustrates the various sampling and ADC configurations used in practice. It can be seen that the sampling frequencies are increased because of pipelining, parallelism, or concurrent architecture. The increase in the sampling frequency improves the bandwidth, improving in turn the SNR in the channel.

78.5 The Communication Interface of a Data Acquisition System

The communication interface is the module through which the acquired data are sent as well as other control tasks are established between the data acquisition module and the host computer (Fig. 78.1). There are basically two different ways of establishing a data link between the two. One way is to use interrupts and the other is through **direct memory access** (DMA). In the case of an interrupt-driven mode, an interrupt-request signal is sent to the computer. Upon receiving it, the computer will first finish the execution of the current instruction, suspend the next, and then send an interrupt-acknowledge signal asking the module to send data. The operation is asynchronous since the sender sends data when it wants to do so. Getting the computer ready to receive data is known as handshaking. In the case of a DMA transfer, the DMA controller is given the starting address of the memory location where the data have to be written. The DMA controller asks the

INPUT
MODE
CONTROL
C
INPUT
OUTPUT
HOLD
SAMPLE
t_s
t_{us}
t_a

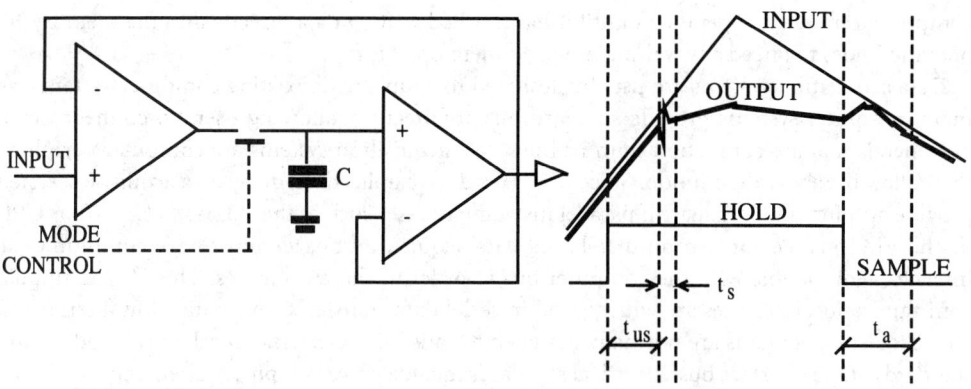

FIGURE 78.5 Sample-and-hold circuit diagram and switching waveforms.

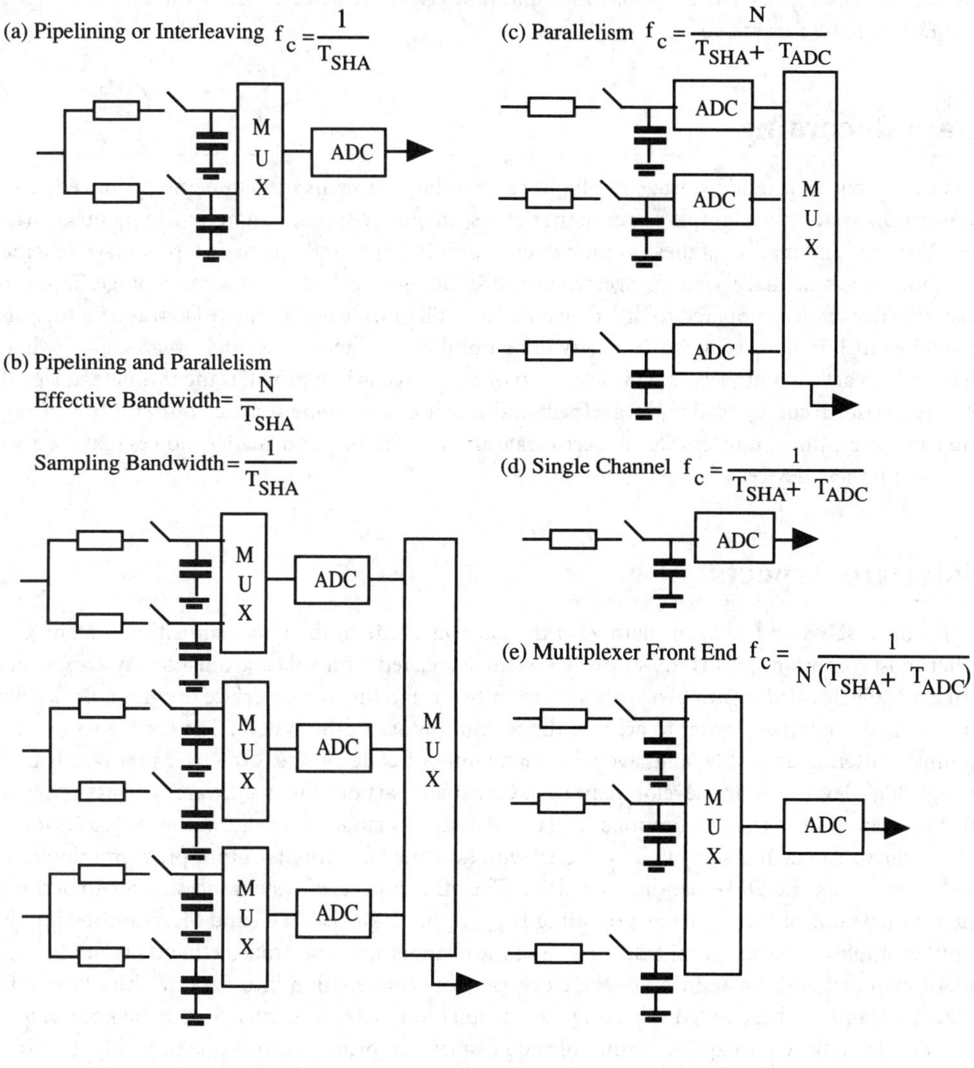

(a) Pipelining or Interleaving $f_c = \frac{1}{T_{SHA}}$
MUX
ADC
(b) Pipelining and Parallelism
Effective Bandwidth= $\frac{N}{T_{SHA}}$
Sampling Bandwidth= $\frac{1}{T_{SHA}}$
(c) Parallelism $f_c = \frac{N}{T_{SHA} + T_{ADC}}$
(d) Single Channel $f_c = \frac{1}{T_{SHA} + T_{ADC}}$
(e) Multiplexer Front End $f_c = \frac{1}{N(T_{SHA} + T_{ADC})}$

FIGURE 78.6 Coupling techniques for SHA and ADC systems.

computer to freeze its operations until it has finished writing data directly into the memory. The operation does not need any waiting time and therefore it is fast.

Data acquisition systems are usually designed to couple with existing computer systems, and many computer systems provide standard bus architecture, allowing users to connect various peripherals that are compatible with its bus. Data acquisition systems are computer peripherals that follow the above description. Since ADCs produce parallel data, many data acquisition systems provide outputs compatible with parallel instrument buses such as the IEEE-488 (HP-IB or GPIB) or the VMEbus. Personal computer-based data acquisition boards must have communication interfaces compatible with the computer bus in order to share resources. The RS-232 standard communication interfaces are widely used in serial data transfer. Communication interfaces for data acquisition systems are normally designed to satisfy the electrical, mechanical, and protocol standards of the interface bus. Electrical standards include power supply requirements, methods of supply, the data transfer rate (baud rate), the width of the address, and the line terminating impedance. Mechanical requirements are the type, size, and the pin assignments of the connectors. The data transfer protocol determines the procedure of data transfer between the two systems. A definition of the timing and input–output philosophy—whether the transfer is in synchronous, asynchronous, or quasi-synchronous mode and how errors are detected and handled—are important factors to be considered.

78.6 Data Recording

It is important to provide storage media to cater to large streams of data being produced. Data acquisition systems use graph paper, paper tapes, magnetic tapes, magnetic floppy disks, hard disks, or any combination of these as their data recorders. Paper and magnetic tape storage schemes are known as sequential access storage, whereas disk storage is called direct access storage. Tapes are cost-effective media compared to disk drives and are still in wide use. In many laboratory situations it will be much more cost effective to network a number of systems to a single, high-capacity hard drive, which acts as a file server. This adoption of digital recording provides the ultimate in signal-to-noise ratio, accuracy of signal waveform and freedom from tape transfer flutter. Data storage capacity, access time, transfer rate, and error rate are some of the performance indices that are associated with these devices.

78.7 Software Aspects

So far the discussion has been mainly on the hardware side of the data acquisition system. The other most important part is the software system associated with a data acquisition system, which can generally be divided into two—the system software and the user-interface program. Both must be designed properly in order to achieve the maximum use of the system. The system software is mainly written in assembly language with many lines of code, whereas the user interface is built using a high-level software development tool. One main part of system software is written to handle the input–output (I/O) operations. The use of assembly language results in the fast execution of I/O commands. The I/O software has to deal with how the basic input–output programming tasks such as interrupt and DMA handling are done. The other aspects of system software are to perform the internal control tasks such as providing trigger pulses for the ADC and SHA, addressing the input multiplexer, the accessing and editing of the channel-gain list, transferring data into the on-board memory, and the addition of the clock/calendar information into data. Multitasking software programs are best suited for many data acquisition systems because it may be necessary to read data from the data acquisition module and display and print it at the same time. Menu-driven user interfaces are common and have a variety of functions built into them.

Defining Terms

Analog-to-digital converter (ADC):   A device that converts analog input voltage signals into digital form.

Common-mode rejection ratio (CMRR):   A measure of quality of an amplifier with differential inputs and defined as the ratio between the common-mode gain and the differential gain.

Direct memory access (DMA):   The process of sending data from an external device into the computer memory with no involvement of the computer's central processing unit.

Least significant bit (LSB):   The 2^0th bit in a digital word.

Multiplexer (MUX):   A combinational logic device with many input channels and usually just one output. The function performed by the device is connecting one and only one input channel at a time to the output. The required input channel is selected by sending the channel address to the MUX.

Power supply unit (PSU):   The one that generates the necessary voltage levels required by a system.

Sample-and-hold amplifier (SHA):   A unity gain amplifier with a mode control switch where the input of the amplifier is connected to a time-varying voltage signal. A trigger pulse at the mode control switch causes it to read the input at the instance of the trigger and maintain that value until the next trigger pulse.

Signal-to-noise ratio (SNR):   The ratio between the signal power and the noise power at a point in the signal traveling path.

References

For further reading consult the following texts, which were used along with the authors' experience and other sources as a basis for this article:

Analog Devices, *Data Conversion Handbook*, Analog Devices, Inc., 1989/90.

R. Annino and R. Driver, *Scientific and Engineering Applications with Personal Computers*, New York: Wiley Interscience, 1986.

D. L. Feucht, *Handbook of Analog Circuit Design*, San Diego: Academic Press, 1990.

D. G. Fink and D. Christiansen (eds.), *Electronic Engineers' Handbook*, 3rd ed., New York: McGraw-Hill, 1989.

P. M. Garrett, *Analog Systems for Microprocessor and Minicomputers*, Reston, Va.: Reston Publishing Company, 1978.

P. Holloway, "Technology focus interview," *Electronic Engineering*, December 1990.

F. Jorgensen, *The Complete Handbook of Magnetic Recording*, 3rd ed., Blue Ridge Summit, Penn.: Tab Books, 1988.

F. F. Mazda, *Electronic Instruments and Measurement Techniques*, New York: Cambridge University Press, 1987.

D. A. Mellichamp (ed.), *Real-Time Computing With Applications to Data Acquisition and Control*, New York: Van Nostrand Reinhold, 1983.

M. Tatkow and J. Turner, "New techniques for high-speed data acquisition," *Electronic Engineering*, September 1990.

Further Information

To probe further in the subject area, refer to the *Data Acquisition Handbook*, published by Data Translation, Marlboro, Mass., 1990.

79

Testing

Micaela Serra
University of Victoria

Bulent I. Dervisoglu
Hewlett-Packard Company

79.1 Digital IC Testing 1808
    Taxonomy of Testing • Fault Models • Test Pattern
    Generation • Output Response Analysis
79.2 Design for Test 1816
    The Testability Problem • Design for Testability • Future for
    Design for Test

79.1 Digital IC Testing

Micaela Serra

In this section we give an overview of digital testing techniques with appropriate reference to material containing all details of the methodology and algorithms. First, we present a general introduction of terminology and a taxonomy of testing methods. Next, we present a definition of fault models, and finally we discuss the main approaches for test pattern generation and data compaction, respectively.

Taxonomy of Testing

The evaluation of the reliability and quality of a digital IC is commonly called *testing*, yet it comprises distinct phases that are mostly kept separate both in the research community and in industrial practice [Pradhan, 1986].

1. *Verification* is the initial phase in which the first prototype chips are "tested" to ensure that they match their functional specification, that is, to verify the correctness of the design. Verification checks that all design rules are adhered to, from layout to electrical parameters; more generally, this type of functional testing checks that the circuit (a) implements what it is supposed to do *and* (b) does not do what it is not supposed to do. Both conditions are necessary. This type of evaluation is done at the design stage and uses a variety of techniques, including logic verification with the use of hardware description languages, full functional simulation, and generation of functional test vectors. We do not discuss verification techniques here.

2. *Testing* correctly refers to the phase when one must ensure that only defect-free production chips are packaged and shipped and detect faults arising from manufacturing and/or wear-out. Testing methods must (a) be fast enough to be applied to large amounts of chips during production, (b) take into consideration whether the industry concerned has access to large expensive external tester machines, and (c) consider whether the implementation of **built-in self-test (BIST)** proves to be advantageous. In BIST, the circuit is designed to include its own self-testing extra circuitry and thus can signal directly, during testing, its possible failure

status. Of course, this involves a certain amount of overhead in area, and trade-offs must be considered. The development of appropriate testing algorithms and their tool support can require a large amount of engineering effort, but one must note that it may need to be done only once per design. The speed of application of the algorithm (applied to many copies of the chips) can be of more importance.

3. *Parametric testing* is done to ensure that components meet design specification for delays, voltages, power, etc. Lately much attention has been given to I_{DDq} **testing,** a parametric technique for CMOS testing. I_{DDq} testing monitors the current I_{DD} that a circuit draws when it is in a quiescent state. It is used to detect faults such as bridging faults, transistor stuck-open faults, or gate oxide leaks, which increase the normally low I_{DD} [Jacomino *et al.,* 1989].

The density of circuitry continues to increase, while the number of I/O pins remains small. This causes a serious escalation of complexity, and testing is becoming one of the major costs to industry (estimated up to 30%). ICs should be tested before and after packaging, after mounting on a board, and periodically during operation. Different methods may be necessary for each case. Thus by testing we imply the means by which some qualities or attributes are determined to be fault-free or faulty. The main purpose of testing is the detection of malfunctions (Go/NoGo test), and only subsequently one may be interested in the actual location of the malfunction; this is called *fault diagnosis* or *fault location.*

Most testing techniques are designed to be applied to combinational circuits only. While this may appear a strong restriction, in practice it is a realistic assumption based on the idea of designing a sequential circuit by partitioning the memory elements from the control functionality such that the circuit can be reconfigured as combinational at testing time. This general approach is one of the methods in *design for testability* (DFT) (see Section 79.2). DFT encompasses any design strategy aimed at enhancing the testability of a circuit. In particular, scan design is the best-known implementation for separating the latches from the combinational gates such that some of the latches can also be reconfigured and used as either tester units or as input generator units (essential for built-in testing).

Figure 79.1(a) shows the general division for algorithms in testing [McCluskey, 1986]. *Test pattern generation* implies a fair amount of work in generating an appropriate subset of all input combinations, such that a desired percentage of faults is activated and observed at the outputs. *Output response analysis* encompasses methods which capture only the output stream, with appropriate transformations, with the assumption that the circuit is stimulated by either an exhaustive or a random set of input combinations. Both methodologies are introduced below.

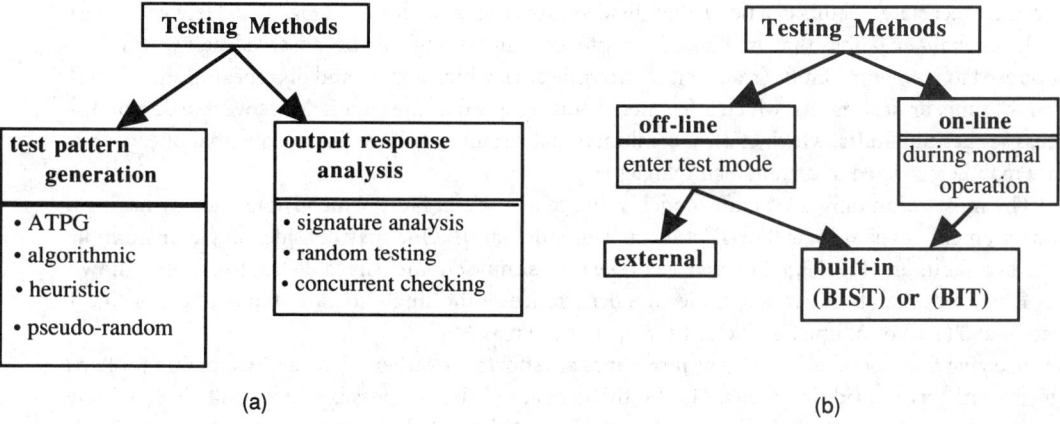

(a)   (b)

FIGURE 79.1 Taxonomy of testing methods. (a) Test pattern generation; (b) on-line and off-line methods.

Moreover a further division can be seen between *on-line* and *off-line* methods [see Fig. 79.1(b)]. In the former, each output word from the circuit is tested during normal operation. In the latter, the circuit must suspend normal operation and enter a "test mode," at which time the appropriate method of testing is applied. While **off-line testing** can be executed either through external testing (a tester machine external to the circuitry) or through the use of BIST, **on-line testing** (also called *concurrent checking*) usually implies that the circuit contains some coding scheme which has been previously embedded in the design of the circuitry.

Table 79.1 Examples of Defect Levels

Y	FC	DL
0.15	0.90	0.18
0.25	0.00	0.75
0.25	0.90	0.15

If many defects are present during the manufacturing process, the manufacturing yield is lowered, and testing becomes of paramount importance. Some estimation can be given about the relationship between manufacturing yield, effectiveness of testing and defect level remaining after test [Williams, 1986]. Let Y denote the yield, where Y is some value between 1 (100% defect-free production) and 0 (all circuits faulty after testing). Let FC be the **fault coverage**, calculated as the percentage of detected faults over the total number of detectable modeled faults (see below for fault models). The value of FC ranges from 1 (all possible faults detected) to 0 (no testing done). We are interested in the final defect level (DL), after test, defined as the probability of shipping a defective product. It has been shown that tests with high fault coverage (for certain fault models, see below) also have high defect coverage. The empirical equation is

$$DL = (1 - Y^{1-FC})\ 100\%$$

Plotting this equation gives interesting and practical results. Table 79.1 shows only a few examples of some practical values of Y and FC. The main conclusion to be drawn is that a very high fault coverage must be achieved to obtain any acceptable defect level value, and manufacturing yield must be continually improved to maintain reliability of shipped products.

Fault Models

At the defect level, an enormous number of different failures could be present, and it is totally infeasible to analyze them as such. Thus failures are grouped together with regards to their logical fault effect on the functionality of the circuit, and this leads to the construction of logical fault models as the basis for testing algorithms [Abramovici *et al.,* 1990]. More precisely, a *fault* denotes the physical failure mechanism, the *fault effect* denotes the logical effect of a fault on a signal-carrying net, and an *error* is defined as the condition (or state) of a system containing a fault (deviation from correct state). Faults can be further divided into classes, as shown in Fig. 79.2. Here we discuss only *permanent* faults, that is, faults in existence long enough to be observed at test time, as opposed to *temporary* faults (transient or intermittent), which appear and disappear in short intervals of time, or *delay* faults, which affect the operating speed of the circuit. Moreover we do not discuss **sequential faults**, which cause a combinational circuit to behave like a sequential one, as they are mainly restricted to certain technologies (e.g., CMOS).

The most commonly used fault model is that of a **stuck-at fault**, which is modeled by having a line segment stuck at logic 0 or 1 (stuck-at 1 or stuck-at 0). One may consider single or multiple stuck-at faults and Fig. 79.3 shows an example for a simple circuit. The fault-free function is shown as F, while the faulty functions, under the occurrence of the single stuck-at faults of either line 1 stuck-at 0 (1/0) or of line 2 stuck-at 1 (2/1), are shown as F^*.

Bridging faults occur when two or more lines are shorted together. There are two main problems in the analysis of bridging faults: (1) the theoretical number of possible such faults is extremely high and (2) the operational effect is of a wired logic AND or OR, depending on technology, and it can even have different effects in complex CMOS gates.

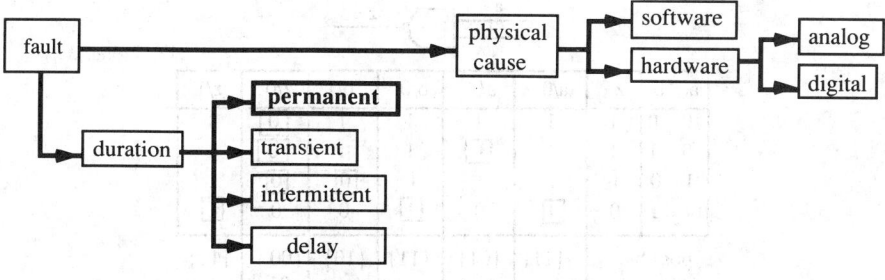

FIGURE 79.2 Fault characteristics.

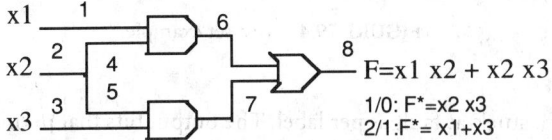

FIGURE 79.3 Single stuck-at fault example.

CMOS stuck-open faults have been examined recently, as they cannot be modeled from the more classical fault models and are restricted to the CMOS technology. They occur when the path through one of the *p*-channel or one of the *n*-channel transistors becomes an open circuit. The main difficulty in detecting this type of fault is that it changes the combinational behavior of a cell into a sequential one. Thus the logical effect is to retain, on a given line, the previous value, introducing a memory state. To detect such a fault, one must apply two stimuli: the first to set a line at a certain value and the second to try and change that value. This, of course, increases the complexity of fault detection.

Test Pattern Generation

Test pattern generation is the process of generating a (minimal) set of input patterns to stimulate the inputs of a circuit such that detectable faults can be exercised (if present) [Pradhan, 1986; Abramovici *et al.*, 1990]. The process can be divided in two distinct phases: (1) derivation of a test and (2) application of a test. For (1), one must first select appropriate models for the circuit (gate or transistor level) and for faults; one must construct the test such that the output signal from a faulty circuit is different from that of a good circuit. This can be computationally very expensive, but one must remember that the process is done only once at the end of the design stage. The generation of a test set can be obtained either by manual methods, by algorithmic methods (with or without heuristics), or by pseudo-random methods. On the other hand, for (2), a test is subsequently applied many times to each IC and thus must be efficient both in space (storage requirements for the patterns) and in time. Often such a set is not minimal, as near minimality may be sufficient. The main considerations in evaluating a test set are the time to construct a minimal test set; the size of the test pattern generator, i.e., the software or hardware module used to stimulate the circuit under test; the size of the test set itself; the time to load the test patterns; and the equipment required (if external) or the BIST overhead.

Most algorithmic test pattern generators are based on the concept of sensitized paths. Given a line in a circuit, one wants to find a *sensitized* path to take a possible error all the way to an observable output. For example, to sensitize a path that goes through one input of an AND gate, one must set all other inputs of the gate to logic 1 to permit the sensitized signal to carry through. Figure 79.4 summarizes the underlying principles of trying to construct a test set. Each column shows the expected output for each input combination of a NAND gate. Columns 3 to 8 show the output

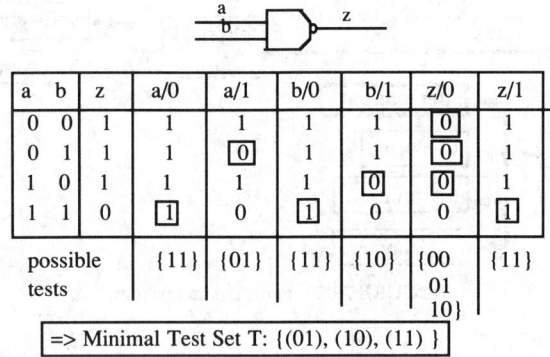

a	b	z	a/0	a/1	b/0	b/1	z/0	z/1
0	0	1	1	1	1	1	[0]	1
0	1	1	1	[0]	1	1	[0]	1
1	0	1	1	1	1	[0]	[0]	1
1	1	0	[1]	0	[1]	0	0	[1]
possible tests			{11}	{01}	{11}	{10}	{00 01 10}	{11}

=> Minimal Test Set T: {(01), (10), (11) }

FIGURE 79.4 Test set example.

under the presence of a stuck-at fault as per label. The output bits that permit detection of the corresponding fault are shown in a square, and thus at the bottom the minimal test set is listed, comprising the minimal number of distinct patterns necessary to detect all single stuck-at faults.

The best-known algorithms are the D-algorithm (precursor to all), PODEM, and FAN [Pradhan, 1986]. Three steps can be identified in most automatic test pattern generation (ATPG) programs: (1) listing the signals on the inputs of a gate controlling the line on which a fault should be detected, (2) determining the primary input conditions necessary to obtain these signals (back propagation) and sensitizing the path to the primary outputs such that the signals and fault can be observed, and (3) repeating this procedure until all detectable faults in a given fault set have been covered. PODEM and FAN introduce powerful heuristics to speed the three steps by aiding in the sequential selection of faults to be examined and by cutting the amount of back and forward propagation necessary.

Notwithstanding heuristics, algorithmic test pattern generation is very computationally expensive and can encounter numerous difficulties, especially in certain types of networks. Newer alternatives are based on **pseudo-random pattern generation** [Bardell *et al.*, 1987] and **fault simulation**. In this strategy, a large set of patterns is generated pseudo-randomly with the aid of an inexpensive (hardware or software) generator. Typical choices for these are linear feedback shift registers and linear cellular automata registers (see below). The pseudo-random set is used to stimulate a circuit, and, using a fault simulator, one can evaluate the number of faults that are covered by this set. An algorithmic test pattern generator is then applied to find coverage for the remaining faults (hopefully, a small number), and the pseudo-random set is thus augmented. The disadvantages are that the resulting set is very large and fault simulation is also computationally expensive. However, this method presents an alternative for circuits where the application of deterministic algorithms for all faults is infeasible.

Output Response Analysis

Especially when designing a circuit including some BIST, one must decide how to check the correctness of the circuit's responses [Bardell *et al.*, 1987]. It is infeasible to store on-chip all expected responses, and thus a common solution is to reduce the circuit responses to relatively short sequences: this process is called *data compaction* and the short, compacted resulting sequence is called a *signature*. The normal configuration for data compaction testing is shown in Fig. 79.5. The n-input circuit is stimulated by an input pattern generator (pseudo-random or exhaustive if $n < 20$); the resulting output vector(s), of length up to 2^n, is compacted to a very short signature of length $k << 2^n$ (usually k is around 16 to 32 bits). The signature is then compared to a known good value. The main advantages of this method are that (1) the testing can be done at circuit speed,

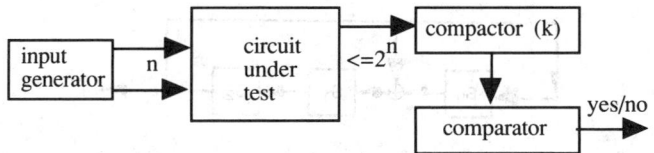

FIGURE 79.5 Data compaction testing.

(2) there is no need to generate test patterns, and (3) the testing circuitry involves a very small area, especially if the circuit has been designed using scan techniques (see Section 79.2). The issues revolve around designing very efficient input generators and compactors.

The main disadvantage of this method is the possibility of **aliasing**. When the short signature is formed, a loss of information occurs, and it can be the case that a faulty circuit produces the same signature of a fault-free circuit, thus remaining undetected. The design method for data compaction aims at minimizing the probability of aliasing. Using the compactors explained below, the probability of aliasing has been theoretically proven to be 2^{-k}, where k is the length of the compactor (and thus the length of the signature). It is important to note that (1) the result is asymptotically independent of the size and complexity of the circuit under test; (2) for $k = 16$, the probability of aliasing is only about 10^{-6} and thus quite acceptable; and (3) the empirical results show that in practice this method is even more effective. Most of all, this is the chosen methodology when BIST is required for its effectiveness, speed, and small area overhead.

A secondary issue in data compaction is in the determination of the expected "good" signature. The best way is to use fault-free simulation for both the circuit and the compactor, and then the appropriate comparator can be built as part of the testing circuitry [Bardell *et al.*, 1987; Abramovici, 1990].

The most important issue is in the choice of a compactor. Although no "perfect" compactor can be found, several have been shown to be very effective. Several compaction techniques have been researched: *counting techniques*, as in one's count, syndrome testing, transition count, and Walsh spectra coefficients; and *signature analysis techniques* based on linear feedback shift registers (**LFSRs**) or linear cellular automata registers (LCARs). Only these latter ones are discussed here. LFSRs and LCARs are also the preferred implementation for the input pattern generators.

LFSRs as Pseudo-Random Pattern Generators

An autonomous LFSR is a clocked synchronous shift register augmented with appropriate feedback taps and receiving no external input [Bardell *et al.*, 1987; Abramovici, 1990]. It is an example of a general linear finite state machine, where the memory cells are simple D flip-flops and the next state operations are implemented by EXOR gates only. Figure 79.6 shows an example of an autonomous LFSR of length $k = 3$. An LFSR of length k can be described by a polynomial with binary coefficients of degree k, where the nonzero coefficients of the polynomial denote the positions of the respective feedback taps. In Fig. 79.6, the high-order coefficient for x^3 is 1, and thus there is a feedback tap from the rightmost cell s_2; the coefficient for x^2 is 0, and thus no feedback tap exists after cell s_1; however, taps are present from cell s_0 and to the leftmost stage since x and x^0 have nonzero coefficients. Since this is an autonomous LFSR, there is no external input to the leftmost cell.

The state of the LFSR is denoted by the binary state of its cells. In Fig. 79.6, the next state of each cell is determined by the implementation given by its polynomial and can be summarized as follows: $s_0^+ = s_2, s_1^+ = s_0 \oplus s_2, s_2^+ = s_1$, where the s_i^+ denotes the next state of cell s_i at each clock cycle. If the LFSR is initialized in a nonzero state, it cycles through a sequence of states and eventually comes back to the initial state, following the functionality of the next-state rules implemented by its polynomial description. An LFSR that goes through all possible $2^k - 1$ nonzero states is said to be described by a *primitive* polynomial (see theory of Galois fields for the definition of primitive), and such polynomials can be found from tables [Bardell *et al.*, 1987].

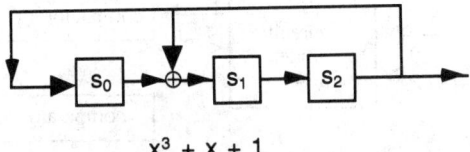

$$x^3 + x + 1$$

FIGURE 79.6   Autonomous LFSR.

By connecting the output of each cell to an input of a circuit under test, the LFSR implements an ideal input generator, as it is inexpensive in its implementation and it provides the stimuli in pseudo-random order for either exhaustive or pseudo-exhaustive testing.

LFSRs as Signature Analyzer

If the leftmost cell of an LFSR is connected to an external input, as shown in Fig. 79.7, the LFSR can be used as a compactor [Bardell *et al.,* 1987; Abramovici, 1990]. In general, the underlying operation of the LFSR is to compute polynomial division over a finite field, and the theoretical analysis of the effectiveness of **signature analysis** is based on this functionality. The polynomial describing the LFSR implementation is seen to be the divisor polynomial. The binary input stream can be seen to represent the coefficients (high order first) of a dividend polynomial. For example, if the input stream is 1001011 (bits are input left to right in time), the dividend polynomial is $x^6 + x^3 + x + 1$. After seven clock cycles for all the input bits to have entered the LFSR, the binary output stream exiting from the right denotes the quotient polynomial, while the last state of the cells in the LFSR denotes the remainder polynomial.

In the process of computing a signature for testing the circuit, the input stream to the LFSR used as a compactor is the output stream from the circuit under test. At the end of the testing cycles, only the last state of the LFSR is examined and considered to be the compacted signature of the circuit. In most real cases, circuits have many outputs, and the LFSR is converted into a multiple-input shift register (**MISR**). A MISR is constructed by adding EXOR gates to the input of some or all the flip-flop cells; the outputs of the circuit are then fed through these gates into the compactor. The probability of aliasing for a MISR is the same as that of an LFSR; however, some errors are missed due to cancellation. This is the case when an error in one output at time t is canceled by the EXOR operation with the error in another output at time $t + 1$. Given an equally likely probability of errors occurring, the probability of error cancellation has been shown to be 2^{1-m-N}, where m is the number of outputs compacted and N is the length of the output streams.

Given that the normal length of signatures used varies between $k = 16$ and $k = 32$, the probability of aliasing is minimal and considered acceptable in practice. In MISR, the length of the compactor also depends on the number of outputs tested. If the number of outputs is greater than the length of the MISR, algorithms or heuristics exist for combining outputs with EXOR trees before feeding them to the compactor. If the number of outputs is much smaller, various choices can be evaluated. The amount of aliasing that actually occurs in a particular circuit can be computed by full fault simulation, that is, by injecting each possible fault into a simulated circuit and computing

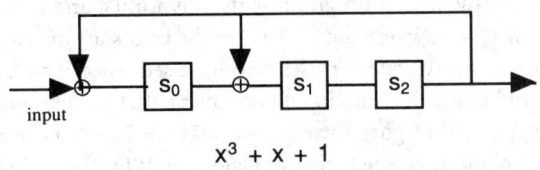

$$x^3 + x + 1$$

FIGURE 79.7   LFSR for signature analysis.

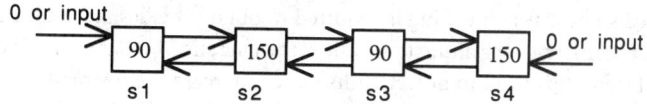

FIGURE 79.8   LCAR for signature analysis.

the resulting signature. Changes in aliasing can be achieved by changing the polynomial used to define the compactor. It has been shown that primitive polynomials, essential for the generation of exhaustive input generators (see above), also possess better aliasing characteristics.

Data Compaction with Linear Cellular Automata Registers

LCARs are one-dimensional arrays composed of two types of cells: rule 150 and rule 90 cells [Serra et al., 1990]. Each cell is composed of a flip-flop that saves the current state of the cell and an EXOR gate used to compute the next state of the cell. A rule 150 cell computes its next state as the EXOR of its present state and of the states of its two (left and right) neighbors. A rule 90 cell computes its next state as the EXOR of the states of its two neighbors only. As can be seen in Fig. 79.8, all connections in an LCAR are near-neighbor connections, thus saving routing area and delays (common for long LFSRs).

Up to two inputs can be trivially connected to an LCAR, or it can be easily converted to accept multiple inputs fed through the cell rules. There are some advantages of using LCARs instead of LFSRs: first, the localization of all connections, and second, and most importantly, it has been shown that LCARs are much "better" pseudo-random pattern generators when used in autonomous mode, as they do not show the correlation of bits due to the shifting of the LFSRs. Finally, the better pattern distribution provided by LCARs as input stimuli has been shown to provide better detection for delay faults and open faults, normally very difficult to test.

As for LFSRs, LCARs are fully described by a characteristic polynomial, and through it any linear finite state machine can be built either as an LFSR or as an LCAR. It is, however, more difficult, given a polynomial, to derive the corresponding LCAR, and tables are now used. The main disadvantage of LCARs is in the area overhead incurred by the extra EXOR gates necessary for the implementation of the cell rules. This is offset by their better performance. The corresponding multiple-output compactor is called a MICA.

Summary

Accessibility to internal dense circuitry is becoming a greater problem, and thus it is essential that a designer consider how the IC will be tested and incorporate structures in the design. Formal DFT techniques are concerned with providing access points for testing (see *controllability* and *observability* in Section 79.2). As test pattern generation becomes even more prohibitive, probabilistic solutions based on compaction and using fault simulation are more widespread, especially if they are supported by DFT techniques and they can avoid the major expense of dedicated external testers. However, any technique chosen must be incorporated within the framework of a powerful CAD system providing semiautomatic analysis and feedback, such that the *rule of ten* can be kept under control: if one does not find a failure at a particular stage, then detection at the next stage will cost 10 times as much!

Defining Terms

Aliasing: Whenever the faulty output produces the same signature as a fault-free output.
Built-in self-test (BIST):   The inclusion of on-chip circuitry to provide testing.
Fault coverage: The fraction of possible failures that the test technique can detect.

Fault simulation: An empirical method used to determine how faults affect the operation of the circuit and also how much testing is required to obtain the desired fault coverage.

I_{DDq} **testing:** A parametric technique to monitor the current I_{DD} that a circuit draws when it is in a quiescent state. It is used to detect faults which increase the normally low I_{DD}.

LFSR: A shift register formed by D flip-flops and EXOR gates, chained together, with a synchronous clock, used either as input pattern generator or as signature analyzer.

MISR: Multiple-input LFSR.

Off-line testing: Testing process carried out while the tested circuit is not in use.

On-line testing: Concurrent testing to detect errors while circuit is in operation.

Pseudo-random pattern generator: Generates a binary sequence of patterns where the bits appear to be random in the local sense (1 and 0 are equally likely), but they are repeatable (hence only pseudo-random).

Random testing: The process of testing using a set of pseudo-randomly generated patterns.

Sequential fault: A fault that causes a combinational circuit to behave like a sequential one.

Signature analysis: A test where the responses of a device over time are compacted into a characteristic value called a signature, which is then compared to a known good one.

Stuck-at fault: A fault model represented by a signal stuck at a fixed logic value (0 or 1).

Test pattern (test vector): Input vector such that the faulty output is different from the fault-free output (the fault is stimulated and detected).

References

M. Abramovici, M.A. Breuer and A.D. Friedman, *Digital Systems Testing and Testable Design,* Rockville, Md.: Computer Science Press, 1990.

P.H. Bardell, W.H. McAnney, and J. Savir, *Built-In Test for VLSI: Pseudorandom Techniques,* New York: John Wiley and Sons, 1987.

M. Jacomino, J.L. Rainard, and R. David, "Fault detection in CMOS circuits by consumption measurement," *IEEE Trans. Instrumentation and Measurement,* vol. 38, no. 3, pp. 773–778, June 1989.

E.J. McCluskey, *Logic Design Principles,* Englewood Cliffs, N.J.: Prentice-Hall, 1986.

D.K. Pradhan (Ed.), *Fault Tolerant Computing,* Englewood Cliffs, N.J.: Prentice-Hall, 1986.

M. Serra, T. Slater, J.C. Muzio, and D.M. Miller, "The analysis of one-dimensional linear cellular automata and their aliasing properties," *IEEE Trans. Computer Aided Design,* vol. 9, no. 7, pp. 767–778, 1990.

T.W. Williams (Ed.), *VLSI Testing,* Amsterdam: North-Holland, 1986.

Further Information

The author would like to recommend reading the book by Abramovici *et al.* [1990] that, at the present time, gives the most comprehensive view of testing methods and design for testability. More information on deterministic pattern generation can also be found in the book edited by D.K. Pradhan [1986], and for newer approaches of random testing the book by Bardell *et al.* contains basic information. The latest state-of-the-art research is to be found mainly in proceedings of the IEEE International Test Conference.

79.2 Design for Test

Bulent I. Dervisoglu

Testing of electronic circuits, which has long been pursued as an activity that follows the design and manufacture of (at least) the prototype product, has currently become a topic of up-front investigation and commitment. Today, it is not uncommon to list the *design for testability* (DFT) features of a product among the so-called *functional* requirements in the definition of a new product to be developed. Just how such a major transformation has occurred can be understood by examining the testability problems faced by manufacturing organizations and considering their impact on time to market (TTM).

The Testability Problem

The primary objective of testing digital circuits at chip, board, or system level is to detect the presence of hardware failures induced by faults in the manufacturing processes or by operating stress or wearout mechanisms. Furthermore, during manufacturing, a secondary but equally important objective is to accurately determine which component or physical element (e.g., connecting wire) is faulty so that quick diagnosis/repair of the product becomes possible. These objectives are necessary due to imperfections in the manufacturing processes used in building digital electronic components/systems. All digital circuits must undergo appropriate level testing to avoid shipping faulty components/systems to the customer. Analog circuits may have minimum and maximum allowable input signal values (e.g., input voltage) as well as infinitely many values in between these that the component has to be able to respond to. Testing of analog circuits is often achieved by checking the circuit response at the specified upper and lower bounds as well as observing/quantifying the change of the output response with varying input signal values. On the other hand, the behavior of a digital system is characterized by discrete (as opposed to continuous) responses to discrete operating state/input signal permutations such that testing of digital circuits may be achieved by checking their behavior under every operating mode and input signal permutation. In principle this approach is valid. However, in practice, most digital circuits are too complex to be tested using such a brute force technique. Instead, test methods have been developed to test digital circuits using only a fraction of all possible test conditions without sacrificing test coverage. Here, *test coverage* is used to refer to the ratio of faults that can be detected to all faults which are taken into consideration, expressed as a percentage. At the present time the most popular *fault model* is the so-called *stuck-at* fault model that refers to individual nets being considered to be fault-free (i.e., *good network*) or considered to be permanently stuck at either one of the logic 1 or logic 0 values. For example, if the *device under test* (DUT) contains several components (or building blocks), where the sum of all input and output terminals (*nodes*) of the components is k, there are said to be $2k$ possible stuck-at faults, corresponding to each of the circuit nodes being permanently stuck at one of the two possible logic states. In general, a larger number of possible stuck-at faults leads to increased difficulty of testing the digital circuit.

For the purpose of *test pattern* (i.e., input stimulus) generation it is often assumed that the *circuit under test* (CUT) is either fault-free or it contains only one node which is permanently stuck at a particular logic state. Thus, the most widely used fault model is the so-called *single stuck-at fault* model. Using this model each fault is tested by applying a specific test pattern that, in a good circuit, drives the particular node to the logic state which has the opposite value from the state of the fault assumed to be present in the faulty circuit. For example, to test if node v is stuck at logic state x (denoted by v/x or v-x), a test pattern must be used that would cause node v to be driven to the opposite of logic state x if the circuit is not faulty. Thus, the test pattern attempts to show that node v is not stuck at x by driving the node to a value other than x, which for a two-valued digital circuit must be the opposite of x (denoted by $\sim x$). This leads to the requirement that to detect any stuck-at fault v/x, it is necessary to be able to control the logic value at node v so that it can be set to $\sim v$. If the signal value at node v can be observed directly by connecting it to a test equipment, the particular fault v/x can be detected readily. However, in most cases, node v may be an *internal* node, which is inaccessible for direct observation from outside the component package. In that case, it is necessary to create a condition where the value of the signal on an externally observable node, say node t, will be different for each of the two possible values that node v can take on, that is, node t

shall be driven to logic state y or $\sim y$ depending upon whether node v is at logic state x or $\sim x$, respectively. Note that x and y may represent the same or different logic states.

The external pins of a component are the only means of applying the stimuli and observing the behavior of that component. During testing, a test pattern is used as the stimulus to detect the presence of a particular fault by causing at least one output pin of the component to take on a different value depending upon whether the targeted fault is present or not. Thus, a test pattern is used for *controlling* the circuit's nodes so that the presence of a fault on a circuit node can be *observed* on at least one of the circuit's external pins. Solving the dual problems of *controllability* and *observability* is the primary objective of all test methods. The *logic-to-pin ratio* of a digital circuit is a relative measure of the ratio of possible faults in the circuit to the number of signal pins (i.e., not including the constant power/ground pins) of that component. A large-value logic-to-pin ratio implies that logic states of a large number of circuit nodes must be controlled using a small number of external pins. As a result, conflicting requirements for controllability and observability become harder to satisfy, and the circuit is considered to be more difficult to test.

Consider Fig. 79.9, which depicts a single (hypothetical) *integrated circuit* (IC) component and shows its internal circuitry which uses four NAND gates. The nodes of the circuit are numbered 1 through 12 and the external pins of the component are labeled *A, B,* and *C.* To detect if node 7 is stuck at logic 0 (i.e., 7/0), a test pattern must be found that sets node 7 (and hence, node 5) to the logic 1 state. This can be achieved by setting either or both of the external pins *A* and *B* to the logic 0 state. Furthermore, to observe (or deduce) the value of node 7 at the only externally visible circuit pin, *C,* it is necessary to create a condition where the logic state of node 12 becomes dependent on the value of node 7. The only path from node 7 to node 12 passes through node 10, and since node 10 is the output of a NAND gate the second input to that gate (i.e., node 6) must be set to the logic 1 state by setting input pin *A* to the logic 1 state. Therefore, the only possible test pattern for 7/0 is *A* = 1 and *B* = 0. At this point, we must still continue the analysis to see if indeed node 12 will reflect the value of node 7. With input terminals *A* and *B* set to logic 1 and logic 0, respectively, node 9 will be set to logic 0, which causes node 11 to become logic 1. With these settings, the value at node 12 will be determined by the value at node 10 and the test pattern is valid. Table 79.2 shows the values of all circuit nodes when this test pattern is applied to the circuit of Fig. 79.9.

It should be evident from the simple example of a *combinational circuit* described above that test pattern generation for digital circuits can be very difficult and involved. The problem becomes much more complex when dealing with *sequential circuits,* where the *internal state variables* (i.e., bistable memory storage elements such as latches and flip-flops) must be treated as *pseudo-inputs* and *pseudo-outputs* that must be controlled and observed using the external pins of the component. In this case test patterns become *test sequences* that must be applied in precise order, and out-

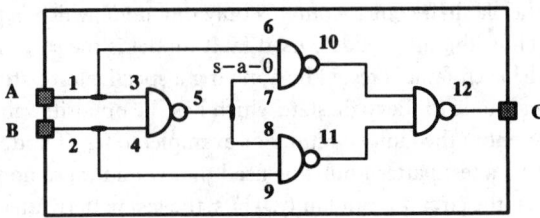

FIGURE 79.9   Example logic circuit with internal node 7 stuck at 0 (7/0).

Table 79.2   Test Pattern for Node 7/0 for the Circuit in Fig. 79.9

A	B	1	2	3	4	5	6	7	8	9	10	11	12	C
1	0	1	0	1	0	1	1	1	1	0	0	1	1	1 good circuit
1	0	1	0	1	0	1	1	**0**	1	0	1	1	**0**	**0** with fault 7/0

puts must be observed only at prescribed times. Thus, the testing of sequential circuits is much harder to achieve compared to the testing of combinational circuits. Computer programs, called automatic test pattern generation (ATPG) programs, have been developed for generating test patterns for combinational or sequential circuits. By far, the generation of test patterns for combinational circuits is better understood and automated than doing the same for sequential circuits.

Before discussing the various techniques that may be used to improve testability of digital circuits, it is necessary to mention the related problem of determining test effectiveness. A typical digital system contains a very large number of possible stuck-at faults. This and the logical complexity of the circuits make it unacceptable to "guess" how effective the test patterns (or the diagnostic program) will be in detecting all possible faults. This problem is often approached in a formal manner by using a class of test tool called a *fault simulator* program. A fault simulator uses the given set of test patterns to simulate the given circuit first when there are no faults assumed present (i.e., good circuit simulation). Next, the circuit is simulated with the same set of test patterns, but this time the effects of each possible stuck-at fault are considered one at a time. For a given test pattern, and given stuck-at-type fault, if the output of the good circuit simulation differs from the output obtained during fault simulation, then the given fault will be detected by the given test pattern. This way, it is possible to determine the percentage of all possible stuck-at faults that may be present in a digital circuit which will be covered by the given set of test patterns.

Most ATPG programs operate by picking a possible fault from among the possible faults, generating a specific test pattern that covers it, simulating the logic circuit with the newly generated test pattern to determine which other faults are incidentally covered by the same pattern, and continuing the process until all faults have been considered. Of the two related processes of *test pattern generation* and *fault simulation*, the latter is by far the more time-consuming one.

A different approach is taken in some testability analysis tools whereby rather than determining which faults are covered by a given test pattern, the analysis program assigns a numeric value to indicate the degree of difficulty of controlling and observing the digital circuit's nodes. This analysis, which can be done much more quickly compared to performing fault simulation, should be done prior to attempting to generate the test patterns for a circuit so that time will not be spent unnecessarily on digital circuits which are likely to present difficulties for the ATPG/fault-simulation process to deal with.

Design for Testability

Low-cost/high-volume manufacturing requires that product testability be considered up front since a product which is inherently hard to test will cost both time and money to achieve a desired level of quality. There are many steps that can be taken to improve the testability of digital circuits and systems. The following subsections describe some of the techniques that can be used.

Ad-Hoc Techniques [Abramovici *et al.*, 1990; Bardell *et al.*, 1978]

Circuit/System Reset Requirements. A simple and straightforward mechanism for resetting a digital circuit to a known state is an essential requirement for testability. It should be noted that the requirement is not only for having the reset function provided but further that it should be simple to execute. For example, applying a defined sequence of external signals to a circuit which must be synchronized with a free-running clock signal would not be considered a simple reset mechanism. Instead, keeping an external signal at some logic value for a minimum duration is a much more desirable approach. It is very desirable that the reset function be asynchronous (i.e., not require system clock pulses to execute) since during power-up a circuit may need to be reset even before free-running clock pulses can be started.

Clock Control Requirements. Another very important requirement for implementing DFT is the ability to control the clocking of the internal logic of the digital circuit. If the external clock signal is gated with some other signals such that it is necessary to determine how to set these other signals to their required values to allow the externally applied clock pulse to reach the internal flip-flop clock terminals, then the ATPG program has another level of constraints to resolve in generating the test patterns. Furthermore, some of these additional requirements may pose difficulties in satisfying them during component and/or system testing. Most ATPG programs assume that once the test pattern has been applied to the pins of the component, the system's response to that pattern can be captured by applying an external clock pulse which enables the internal flip-flops to respond to the test pattern. Thus, the ATPG programs assume that the internal flip-flop clock inputs are controlled directly from an external pin of the component. This very desirable characteristic is often expressed by stating that *externally applied clock pulses are not allowed to be gated by other signals before these reach the clock terminals of the internal flip-flops.* A side benefit of this design rule is that it prevents glitches (i.e., undesirable pulses) which might be generated at the flip-flop clock terminals due to changing the other inputs to the clock gating circuit while the clock pulse is present.

Managing "Unused" Inputs of Components. When designing digital systems from existing components there may be inputs of those components that, for the current implementation, are not needed. For example, if a two-input AND gate is needed to implement a logic circuit on a printed circuit board, it may be possible to use one of the unused three-input AND gate elements from an IC package already present on that board. In this case, the unused third input of that AND gate must be connected to the logic 1 level in order that a three-input AND function may be implemented using the other two inputs to that gate. Thus, the unused input to the AND gate may be connected directly to the V_{cc} (i.e, power supply) signal. Similarly, if a flip-flop contains unused *preset* or *clear* terminals, these may be tied off to their respective deasserted states. In many cases printed circuit boards are tested using an *in-circuit tester* which uses a *bed-of-nails* test fixture to make physical contact with selected nets on the board so that their values can be observed or controlled by the tester. For the in-circuit tester to control the value of a net it has to backdrive the output of the component which normally drives that net. Since IC components have limited output drive capabilities, the in-circuit tester can overcome the electrical drive from that component and can force that net to a value opposite the value which the driving IC is trying to achieve. By keeping such backdriving conditions to last only a very short period, damage to the opposing IC component is prevented. However, if the net is driven not by an IC but directly from the V_{cc} or ground (*Gnd*) signals, then the in-circuit tester may not be able to overcome their drive. Furthermore, backdriving the V_{cc} or *Gnd* levels would prevent the other IC components from being able to perform their normal functions. Instead, if the logic signals to such unused terminals are applied using *pull-up* or *pull-down* resistors when connecting these to the V_{cc} or *Gnd* levels, respectively, these signals may be controlled by the in-circuit tester. For example, this way it becomes possible to set/reset a flip-flop value by using the normally "unused" preset/clear terminal of that flip-flop. Note that if the flip-flop contains both a preset and a clear input which are unused, these must be pulled up (or pulled down) through separate resistors so that each can be controlled by the in-circuit tester independent of the other. This is illustrated in Fig. 79.10.

Synchronous versus Asynchronous Design Style. More than any other issue, discussions concerning synchronous versus asynchronous design style create the most disagreements concerning design for testability. Many logic designers who are experienced in using SSI and MSI IC chips have adapted a design style where synchronous (e.g., clocked) and asynchronous (e.g., self-timed) designs are freely mixed together. Using clocked flip-flops with asynchronous preset/clear inputs is a typical example of this design style. Similarly, building latches out of, say, cross-coupled NAND gates and using these as state variables in implementing finite-state machines used to be a very common technique. However, concerns about system initialization and pattern generation have made this style undesirable for implementing DFT. Indeed, most of the so-called *structured* design styles described below make it a requirement that all internal storage elements be constructed from

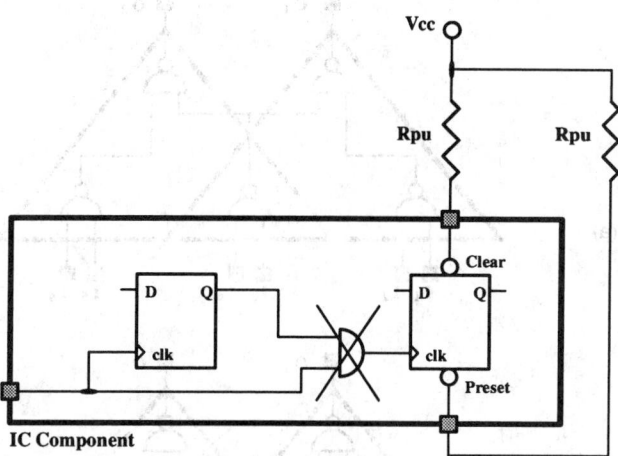

FIGURE 79.10   Using pull-up resistor to tie off unused preset/clear inputs of flip-flops.

clocked flip-flops, and feedback loops in combinational circuits are broken with the insertion of such flip-flops, along the feedback paths. Asynchronous circuits suffer from combinational circuit hazards that are glitches created as a result of delay differences along circuit paths. Some hazards may be prevented by constraining the manner (i.e., sequence) in which circuit inputs are allowed to be changed. Whereas such constraints may be met during regular system operation, often test pattern generation algorithms cannot take such constraints into account. Therefore, asynchronous logic may create severe problems during testing.

Avoiding Redundant Logic.   Technically speaking, redundancy is the only reason why a given stuck-at fault might not be detectable by any test. For example, if an INVERTER function is implemented by tying both inputs of a two-input NAND gate together, then a stuck-at 1 fault on either one of the inputs becomes undetectable since the output signal can still be determined correctly by the remaining nonfaulty input signal. This creates two problems. First, conventional ATPG programs might spend a lot of time trying to generate a test pattern for such a fault before they declare the fault untestable. Second, the presence of an undetectable fault can cause a detectable fault to become undetectable (it may also cause an undetectable fault to become detectable). For example, consider a parity checking circuit in which an existing stuck-at fault may cause the wrong parity to be generated, and the existence of a second fault may correct the parity and hence hide both failures. The remedy for these situations is to try to avoid redundancy in the first place, and when this is not possible provide additional circuit modes where the redundant circuits might be isolated. Alternately (or in addition) it may be useful to provide additional test points, as described below.

Providing Test Points.   A test point is an input or output signal to control or observe intermediate signals in a logic circuit. For example, if triple redundancy has been used to implement a fault-tolerant circuit, additional output signals might be provided so that signal values from the identical functional units become individually observable, improving the testability of the overall circuit. Similarly, control signals might be provided so that, during testing, outputs from some functional units may be forced into certain states which allow easier observation of the outputs from other circuits. Recommended sites for inserting test points include redundant nets, nets with large fan-outs, preset and clear inputs of flip-flops, nets that carry system clock signals, (at least some of the) inputs to logic circuit gates with large number of inputs (i.e., large fan-in), data and/or address lines of bus lines, as well as intermediate points in cascaded circuits (such as long ripple counters, shift registers).

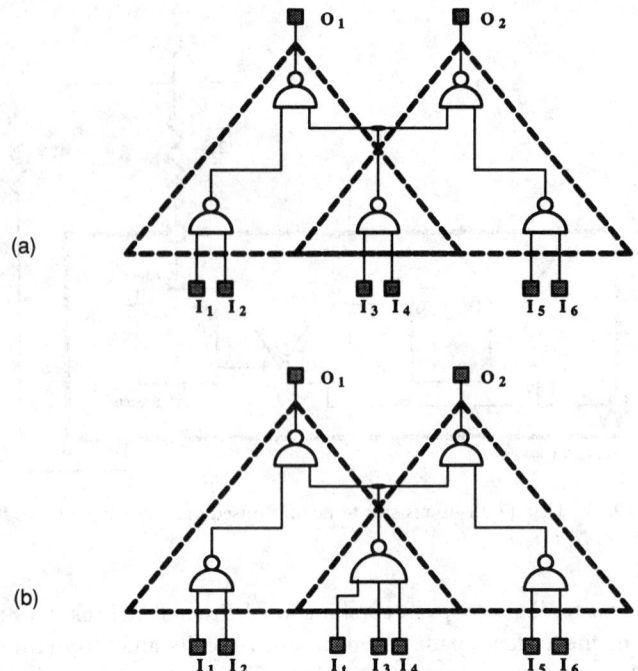

FIGURE 79.11   (a) Logic partitioning with overlapping logic cones.
(b) Adding an additional test point to reduce dependence on primary inputs.

Logic Partitioning.   Traditionally logic partitioning has been used as a strategy when the circuit is
too large/complex for the test generation tools to handle. Thus, its objective is to reduce the number of circuit nodes that must be considered jointly in order to generate test patterns. The partitioning process identifies the *logic cones*, which are sections of logic receiving inputs from multiple
input sources and generating a single output. Thus, a digital circuit would be broken into as many
individual logic cones as there are individually observable output signals. Obviously, the logic
cones may (and often do) overlap with each other since they share common input signals or intermediate signals generated from inside one partition and used in another partition. This is illustrated in Fig. 79.11(a), where two overlapping cones of logic are shown. Here, logic cones O_1 and
O_2 contain primary inputs I_1, I_2, I_3, I_4 and I_3, I_4, I_5, I_6, respectively. When either partition is dependent on more inputs than what the ATPG tools or the tester can accommodate, it is possible to
insert an additional gate, controlled by a tester input in order to test each partition independently
of the other. This is illustrated in Fig. 79.11(b), where an additional input pin I_t has been added
such that with I_t set to logic 0 by the tester, it is possible to test either partition without requiring to
control shared inputs I_3 or I_4. Logic partitioning has become more important as a result of
increased use of *pseudo-exhaustive testing* (to be described later).

Testing Embedded Memory Blocks.   A major testability problem arises when a regular-structure
memory block such as random-access memory (RAM) or read-only memory (ROM) is embedded
into a logic circuit. This creates three problems:

1. Testing logic that is downstream from the RAM block (i.e., output of RAM block drives the
   downstream logic) is difficult since this requires setting the test pattern at the RAM outputs.
   This problem is usually solved by providing a bypass mode where data inputs to the RAM (or
   ROM) block are channeled directly to the RAM (or ROM) outputs without (or in addition

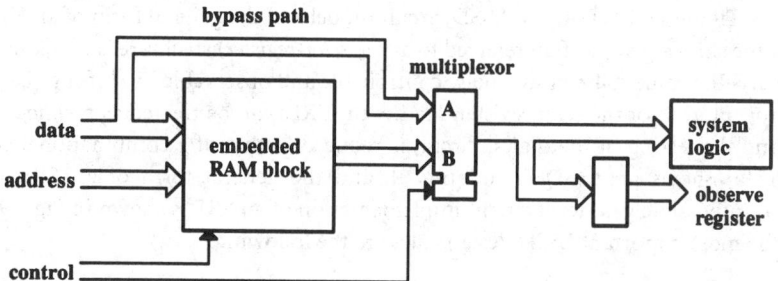

FIGURE 79.12   Providing testability in a design containing an embedded memory block.

to) being stored inside the RAM block. This way the RAM data outputs can be controlled by controlling the data inputs as desired.

2. Testing logic that is upstream from the RAM block (i.e., outputs from logic circuit are captured by the RAM block) is difficult since the observation point is the RAM block. That is, it is necessary to access the RAM block in order to observe the test results. This problem might be solved by improving the observability of the RAM inputs and/or making the RAM outputs more easily observable as well as providing the *bypass* capability. This way, inputs to the RAM might be bypassed directly to the RAM outputs where they may be observed. This may require adding an *observe-only* register to capture the RAM outputs.

3. Testing of the RAM block itself is difficult since controlling its inputs and observing its outputs require manipulating the upstream and downstream logic circuit blocks, which may be difficult to achieve. Solution to this problem involves providing adequate control of the RAM block inputs (data, address, and read/write control) as well as providing observability of the RAM outputs. In effect, the embedded RAM block can be made testable as if it was a standalone block where established memory test algorithms can be applied [Breuer and Friedman, 1976].

Figure 79.12 illustrates how to improve testability of an embedded RAM structure.

Structured Techniques

An alternate approach to improving the testability of digital circuits is to carry out the circuit design by following certain rules that, by construction, assure high testability of the resulting circuits. Since the main problem in achieving testability of a digital circuit is achieving adequate controllability/observability of its internal nodes, structured DFT approaches [Bardell and McAnney, 1978] follow strict design rules that are aimed at achieving this goal. Furthermore, most structured DFT approaches require/recommend additional design rules aimed at preventing incorrect circuit operation as a result of signal races and hazards.

Level-Sensitive Scan Design (LSSD).   Level-sensitive **scan design** [Eichelberger and Williams, 1978] imposes strict rules on clock signal usage and allows implementing sequential behavior to be implemented only using the shift-register latch (SRL). In the first place, by not allowing any feedback involving combinational circuit elements alone, the LSSD approach prevents timing failures that might be present in purely asynchronous designs. Furthermore, rigid clocking rules are stated in order to prevent SRL data inputs from changing while the clock pulse(s) is (are) transitioning. Hence, the digital circuit is separated into two sections: (1) a robust (i.e., level-sensitive) multi-input/multi-output combinational circuit and (2) a set of SRL elements with which sequential behavior is implemented. In addition to their normal system interconnections each SRL is also connected to its two neighboring SRLs to form a shift-register structure. The serial shift input and shift output signals are labeled *scan-in* and *scan-out,* respectively, and treated as primary input/output terminals. Figure 79.13 shows an LSSD circuit model and the general form of an SRL. The significance of the shift-register (often referred to as the *scan-register*) structure is that, during testing, it allows each SRL's value to be individually controllable and observable by shifting (i.e., scanning) a serial vector into/out of the scan register. Hence, the SRLs can be treated as *pseudo-input/output terminals,* and the testing of the digital circuit is reduced to that of a combinational circuit only. Figure 79.13(a) shows an LSSD circuit model, and the general form of an SRL is given in Fig. 79.13(b). A possible gate-level circuit implementation of an SRL is shown in Fig. 79.13(c).

Among the most important LSSD design rules are the following:

1. All internal storage is implemented using SRLs. Each SRL operates such that the L1 latch accepts one or the other of the system data-in or the scan-in data values depending upon whether the system clk or the scan-in clk clock pulse is applied, respectively. The L2 latch accepts the L1 latch value when the scan-out clk clock pulse is applied. The L1 and L2 latches are stable (i.e., cannot change) when the clocks are off.
2. The SRL clocks system clk, scan-in clk, and scan-out clk must be controlled from primary circuit terminals and must be operated in nonoverlapping fashion. This eliminates dependency on minimum circuit delay and assures hazard-free (i.e, level-sensitive) operation.
3. System data-out from SRL_1 may feed the system data-in terminal of SRL_2 only if the system clk which feeds SRL_1 does not overlap with the system clk which feeds SRL_2. This rule prevents the data input to a latch from changing while its clock signal is transitioning.

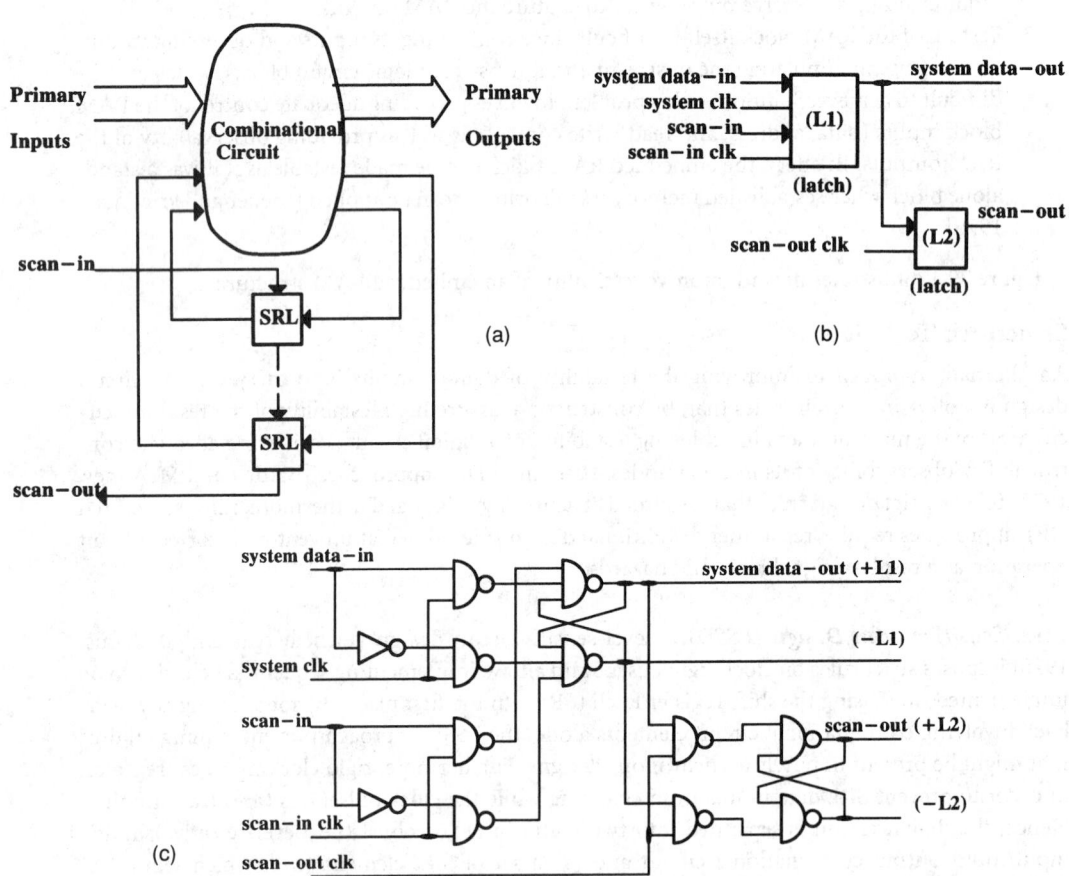

FIGURE 79.13 (a) LSSD circuit model. (b) SRL block diagram. (c) SRL logic diagram.

4. All SRLs are interconnected into one or multiple shift registers by connecting the scan-out terminal from one SRL to the scan-in terminal of the next one in series. If multiple shift registers are implemented, each must be capable of being shifted simultaneously with the others and must have its own scan-in and scan-out primary terminals.

Scan Path.   The *scan-path* [Funatsu *et al.*, 1975] approach can be seen as a generalization of the LSSD approach since it follows the same principles but uses standard *D*-type flip-flops as the storage elements instead of the SRLs. The scannable flip-flops can be implemented using dual-ported latches (similar to the L1 latch in the SRL) or using a multiplexor to select between the scan-in and system data-in signals to feed the *D* input of a standard *D*-type flip-flop, as shown in Fig. 79.14.

Scan/Set Logic.   Scan/set [Stewart, 1977] is another form of implementing scan technology whereby the sequential circuit structure is separated from its accompanying scan/set register. This is illustrated in Fig. 79.15. A variation on this scheme is the so-called shadow-register concept that has been implemented in some off-the-shelf IC components [AMDI, 1987].

Random-Access Scan.   Random-access scan [Ando, 1980] uses a technique akin to addressing locations in a memory (e.g., RAM) block in order to make the states of all storage elements controllable and observable from primary input/output terminals. Using this approach, each storage element is made individually addressable (i.e., accessible) so that in order to control and/or observe the value of an individual storage element it is not necessary to shift in/shift out all other storage elements as well. Figure 79.16(a) shows the general model of a digital circuit employing the random-access scan approach. A possible gate-level circuit implementation of an addressable latch is given in Fig. 79.16(b).

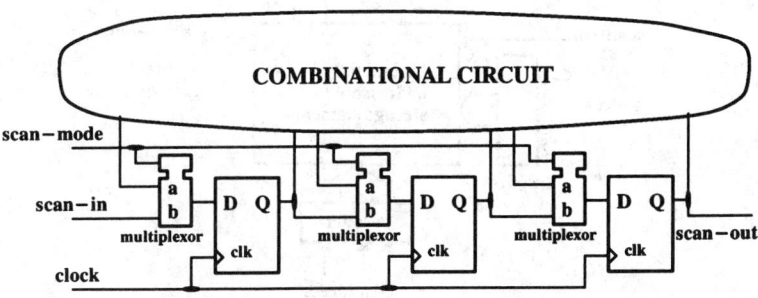

FIGURE 79.14   Model of a digital circuit with scan path.

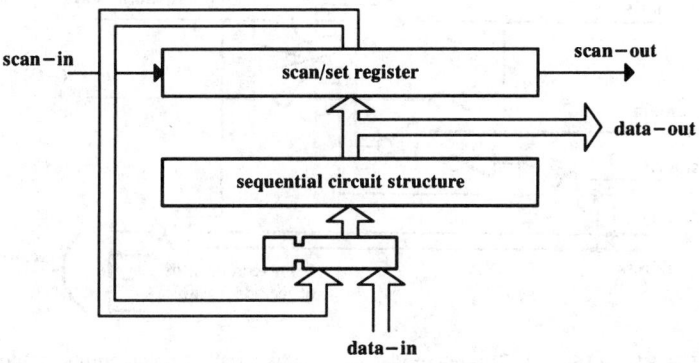

FIGURE 79.15   Generic scan/set circuit design.

Using this approach, each storage element in the circuit is given a unique *x/y* address and the decoded address signals are connected to the *x/y* address inputs of the latches. As seen in the circuit of Fig. 79.16(b), each latch can then be individually written into using the *scan-in* terminal or its output can be observed using the *scan-out* terminal, provided that the pair of *x/y* address lines connected to the current latch are both asserted (i.e., set to logic 1). Furthermore, whereas it is also necessary to apply the *scan-in clk* in order to write into the latch, no clock is necessary to observe the latch output. This is a convenient feature that allows the latch values to be selectively observable even while the regular system operations are being executed. The *scan-out* values from the individual latches are combined together into a single AND gate and brought out to a primary output terminal of the circuit. This arrangement works since for any given address only one of the addressable latches will be selected and the scan-out from all other latches will be forced to the logic 1 state. On the other hand, a disadvantage of this approach is that before addressing each latch its proper address must first be applied to the circuit.

Boundary Scan.   Unlike the other scan-based techniques described above, **boundary scan** [IEEE, 1990] is intended primarily for testing the board-level interconnections among the IC components on a printed circuit board (PCB). In effect, boundary scan is a special form of scan path that is

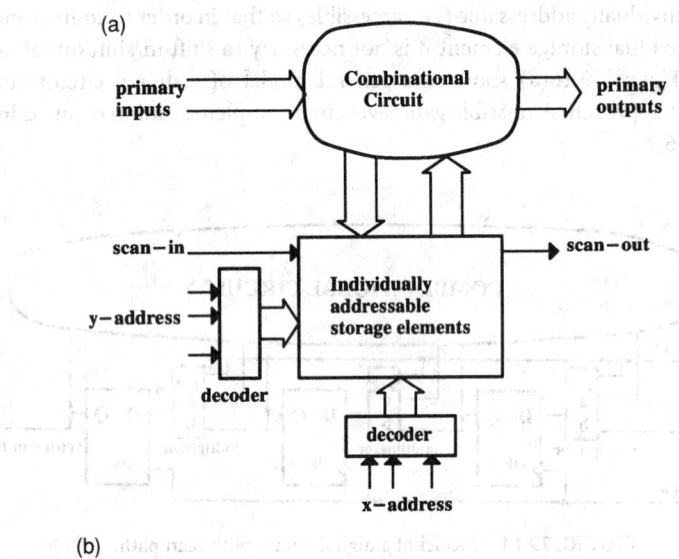

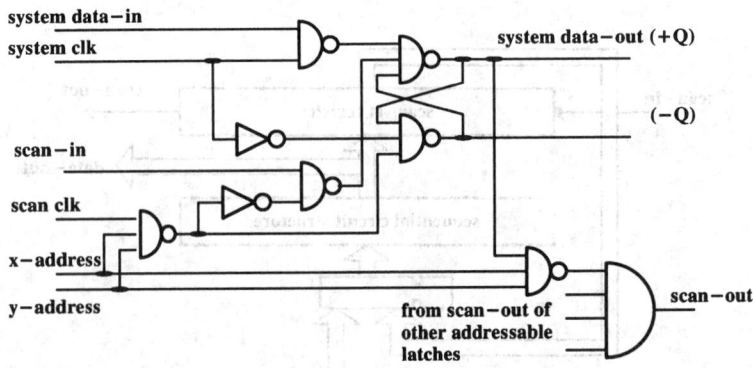

FIGURE 79.16   (a) General model for digital circuit implementing random-access scan. (b) Logic diagram for addressable latch.

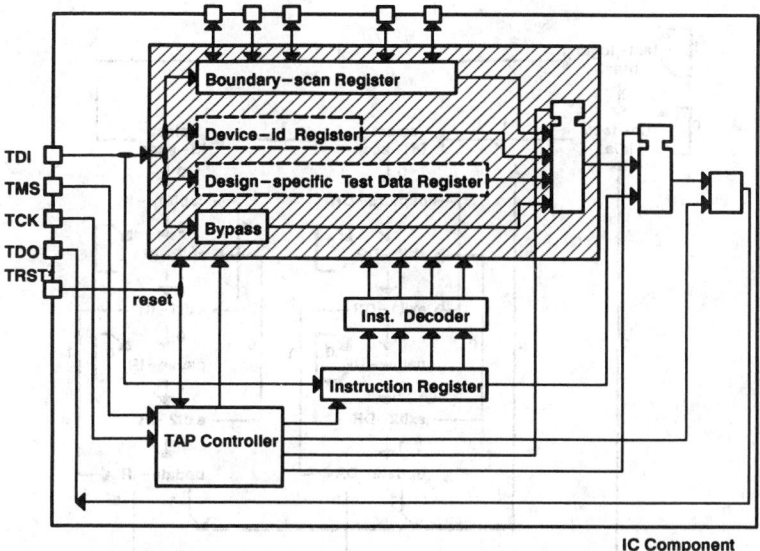

FIGURE 79.17 Architecture of IEEE 1149.1 boundary-scan standard.

implemented around every I/O pin of an IC component in order to provide controllability and observability of the I/O pin values during testing. Test control signals provided by an on-chip controller are used to disable the boundary-scan cells during regular system operation so that signal values can flow in/out of the IC component without interference from the test circuits. During testing, *output* pin values can be controlled using values preloaded into the boundary-scan register. Similarly, signal values received on the *input* pins can be captured into the boundary-scan register and subsequently shifted out to be observed on an external tester.

Boundary scan has become an important tool in achieving design for testability following the adoption of the IEEE 1149.1 Test Access Port and Boundary-Scan Architecture in 1990. The IEEE 1149.1 Standard defines a mandatory four-pin (plus an optional fifth pin) test access port (TAP) for providing the interface between the IC component and a digital tester. TAP signals comprise test data input (TDI), test data output (TDO), test clock (TCK), and test mode select (TMS) plus an optional asynchronous tap reset (TRST*) signal. The overall IEEE 1149.1 test architecture (see Fig. 79.17) includes:

- The TAP
- The TAP controller
- The instruction register (IR)
- A group of mandatory and optional test data registers (TDRs)

The TAP controller is characterized by a 16-state finite-state machine (FSM) whose behavior is defined by the IEEE 1149.1 Standard. State transitions of the TAP FSM are controlled by the TMS input line and the dedicated test clock, TCK. Figure 79.18 shows the state-transition diagram for the TAP FSM.

A most important test data register defined by the IEEE 1149.1 Standard is the boundary-scan register that has individual cells associated with each I/O pin of the IC component. Mandatory and permissible features of the boundary-scan register cells are defined by the standard. In addition, a special single-bit register called the BYPASS register has been provided to furnish a more efficient way to shift data through IC components when multiple ICs are chained together by connecting the TDO output from one component to the TDI input of another.

Another mandatory feature of the IEEE 1149.1 Standard is the instruction register and an asso-

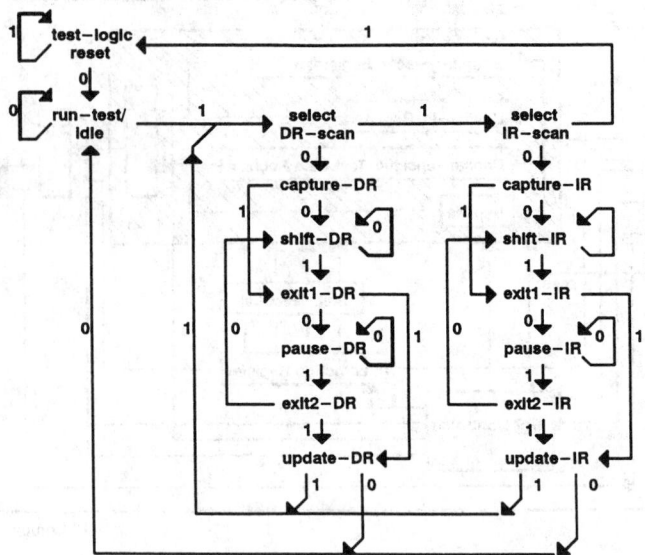

FIGURE 79.18   State-transition diagram for the TAP FSM.

ciated list of mandatory/permissible instructions that govern the behavior of the IC component during testing. The three mandatory instructions are called SAMPLE/PRELOAD, BYPASS, and EXTEST. SAMPLE allows taking a snapshot of the normal operation of the IC, whereas PRELOAD is used for shifting the captured values out while new values are loaded into the boundary-scan register. BYPASS allows shortening the (electrical) distance between the TDI and TDO pins by providing a single-bit register as a shortcut during scan operations involving multiple IC components that are connected in series. EXTEST is the "workhorse" instruction that allows driving the signal values on the component's output pads from the boundary register while capturing the input values into their respective cells in the boundary register. This is followed by shifting the captured values out (using the TDO output) while simultaneously shifting in the new driving values (using the TDI input).

An alternative to using boundary scan is to use a "traditional" in-circuit tester that uses a special "bed-of-nails" fixture. In this approach [Parker, 1987], every net on a PCB would be probed using a tester pin which comes in physical contact with that net such that the current signal value of the net can be observed by the tester. The tester can also be used to control the signal values of the individual nets by injecting appropriate currents through the tester pins. However, since each net is already connected to an output pin of a component on the PCB, this approach amounts to *backdriving* the output drivers of IC components and therefore poses a potential risk of damage to the IC components. This approach is becoming more difficult and/or costly to implement as the number of nets goes up and IC pin spacing is reduced. Furthermore, due to fixturing difficulties, double-sided PCBs cannot be tested in this manner. The IEEE 1149.1 boundary-scan standard [IEEE, 1990] helps solve these problems by providing convenient direct access to the I/O pins of an IC component without requiring the traditional bed-of-nails fixture.

The "CrossCheck" Technique.   The CrossCheck approach [Gheewala, 1989] uses cells with built-in test points to observe critical signal values. The test points are connected to an underlining grid structure using very small FETs called *cross-point switches*. An on-chip test control circuit generates the necessary signals to address the individual probe lines and capture the results in a *multi-input signature register* (MISR). Test patterns can be generated externally or by using an on-chip pattern

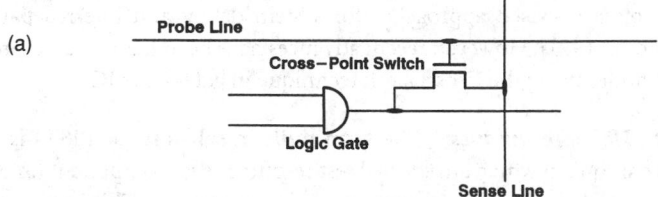

(a)

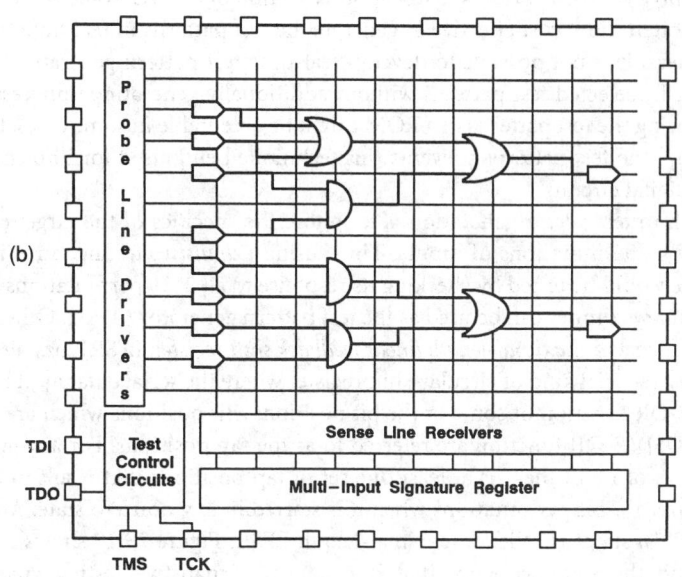

(b)

FIGURE 79.19 (a) Cross-point switch implementation. (b) Overview of the CrossCheck technique.

generator, and the final test signature (i.e., contents of MISR) can be accessed using dedicated test pins, such as by providing an IEEE 1149.1 TAP (see previous subsection). Figure 79.19 shows how the CrossCheck technique is implemented on an ASIC.

CrossCheck methodology provides a high degree of observability of the ASIC. Since it is not possible to provide observability of all signals of a design, careful analysis must be performed to determine the most effective points for inserting the cross-point switches. Similarly, the size of the grid structure for the probe lines might be chosen to be design-dependent. However, in many instances it may be better to implement the probe lines as part of the IC master slice in order to reduce the amount of customization to a minimum.

The benefit offered by the CrossCheck technique is due to the potential for the reduced number of test patterns necessary to test an ASIC. This is due to the fact that as observability of internal nodes is increased it becomes easier to generate efficient test patterns which can detect many faults simultaneously. Furthermore, increased observability of internal nodes also improves diagnosability and may help determine the root cause of a failure sooner. On the negative side, the CrossCheck technique does not help improve controllability of internal nodes as achieved using scan-path techniques. Also, a primary disadvantage of the CrossCheck methodology is area penalty due to routing channels that must be set aside for the grid structure. Furthermore, added capacitance of the cross-point switches may affect performance, especially in high-speed applications. In addition, since the technique offers very good observability but no controllability of the internal nodes, it lacks the advantage offered by scan-based approaches for system debug and internal **path-delay testing** [Dervisoglu and Stong, 1991]. However, recent advances have been made that improve the controllability of internal nodes using the CrossCheck technique in gate-array ICs.

Built-in Self-Test (BIST) Techniques.   The term **built-in self-test** (or BIST) is a generic name given to any test technique in which an external test resource (e.g., component tester) is not needed to apply test patterns and check a circuit's response to those patterns. This implies that the test patterns must be preloaded into the target device or be generated by the target device itself, in real time. For example, dedicating a section of an IC component for implementing a ROM-based sequencer to apply prestored patterns to test another section of that IC would be classified as a BIST technique. It is often more cost effective to generate the test patterns in real time (i.e., during testing), but in general it is not possible to develop real-time test pattern generation techniques that generate arbitrarily selected test patterns without additionally generating unnecessary ones. Note that whereas storing the test patterns in a ROM might be acceptable in some cases, the size of ROM necessary to store the test patterns prevents this technique being used for implementing BIST in large/complex digital circuits.

One approach to test vector generation is to ignore the specifics of the target circuit and enumerate all possible permutations of inputs. Thus, using *exhaustive* testing, an n-input combinational logic cone would be tested by checking its response to all $2^{**}n$ permutations of input values. In this case, a binary counter can be used as the test pattern generator (TPG). Other, more efficient counter forms (such as a *maximal-length linear feedback shift register,* LFSR) may also be used as the TPG. An LFSR is a special kind of circular-shift register where the serial data input is determined by an EXCLUSIVE-OR function of some of the bit positions. Bit positions which are included in the feedback EXCLUSIVE-OR function are referred to as the tap positions. For any given *degree* (i.e., number of bits) n of LFSR there is at least one set of tap positions that result in the LFSR going through all nonzero n-bit permutations when it is started in any nonzero state. An LFSR that can go through all $2^{**}n$ states is called a maximal-length LFSR. Figure 79.20 shows a 3-bit maximal-length LFSR and the state sequence that it produces. Exhaustive testing guarantees that all *detectable* faults which do not transform a combinational circuit into a sequential circuit will be detected. Depending upon the clock frequency, this approach becomes impractical to apply when the number of input variables goes up (usually above 22) [McCluskey, 1984].

In cases where the number of test patterns necessary to achieve exhaustive testing is too large to be applicable, a related technique, called **pseudo-random testing**, may be used. Pseudo-random testing achieves many of the benefits of exhaustive testing but requires much fewer test patterns. This is achieved by generating the test patterns in random fashion from among the $2^{**}n$ possible patterns. However, the random generation of test patterns is done using a deterministic algorithm that produces test patterns in repeatable sequence. Before pseudo-random testing is chosen, it is necessary to examine the pseudo-random test resistance of the circuit. For example, if 500,000 pseudo-random test patterns are applied to a 20-input AND gate, there is only a 0.00004% probability that an essential test pattern (which sets all 20 inputs to logic 1) will be included among them.

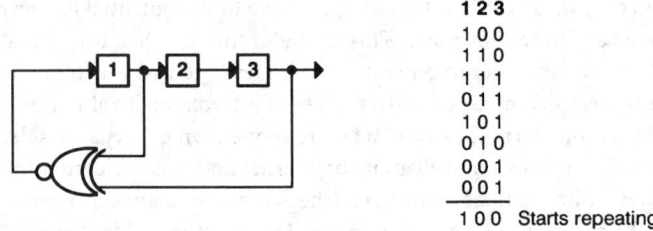

FIGURE 79.20   Three-bit maximal-length LFSR.

Yet another related technique is to use *pseudo-exhaustive* testing that aims at breaking a circuit into separate partitions and testing each partition exhaustively [Barzilai *et al.*, 1985; Dervisoglu, 1985; Bardell and McAnney, 1984]. Pseudo-exhaustive testing uses the same techniques used in exhaustive testing for testing the individual partitions without generating test patterns that cover the entire circuit. Mathematical considerations for pseudo-random/pseudo-exhaustive testing are too complex to describe here. The following example is presented for illustration purposes only. Figure 79.21 depicts the combinational portion of a digital circuit consisting of a number of overlapping logic cones that each produce a single output signal. All inputs are assumed to be connected to scannable flip-flops (i.e., pseudo-inputs) or to primary input pins of the component such that all inputs are 100% controllable either by controlling the values in the flip-flops or the primary input pins. All flip-flops are assumed to be scannable and are arranged into a single *scan path* such that the logic cones have n or fewer inputs all of which lie within k consecutive bits along the scan path. Outputs from the individual logic cones connect (not shown here) to the inputs of flip-flops and/or primary output pins. Thus, all logic cone outputs are also 100% observable. Now, assume that the serial output from the LFSR shown in Fig. 79.20 is connected as the "scan-in" input to the scan-path register shown in Fig. 79.21. In this case any *consecutive* 3-bit partition of the scan-path register will go through the same state sequence as the LFSR itself, delayed from it by the number of flip-flops between that partition and the output bit of the LFSR. For example, the third logic cone that has inputs from flip-flops 4, 5, and 6 will see all input permutations except the all-zeros case which can be applied separately as a special case. On the other hand, the first logic cone, with inputs from flip-flops 1, 2, and 4, will not receive all possible nonzero permutations of three input variables. This is because the first logic cone receives its three inputs from three *nonconsecutive* positions of the scan-path register. In this case only input permutations that have even parity across positions 1, 2, and 4 will be received by the first logic cone. Furthermore, the fourth logic cone that also receives inputs from three nonconsecutive bit positions which are 4 bits apart will receive all 3-bit nonzero input permutations. Analysis of which set of input permutations may be generated across nonconsecutive n bits of a scan-path register which receives the outputs from an mth degree ($m \geq n$) LFSR is based on *linear dependence* and is outside the scope of this section. However, the problem may also be approached statistically by choosing the degree of the LFSR to be higher than n but smaller than k which is the largest span of inputs to any logic cone. For example, in Fig. 79.21 the degree of the LFSR may be chosen as 4. In this case, the probability that a logic cone which has 4 or fewer inputs separated by k bits (here, $k = 5$) may be calculated [Lempel and Cohn, 1985]. It should be noted that a logic cone may be tested in full even when it has not received all $2^{**}n$ input permutations.

BIST also requires ability to capture the test results without the need for an external tester. This is often achieved by using a *multi-input signature register* (MISR) to capture individual test results and compress these into an overall value called the test *signature*. Figure 79.22 shows a sample sig-

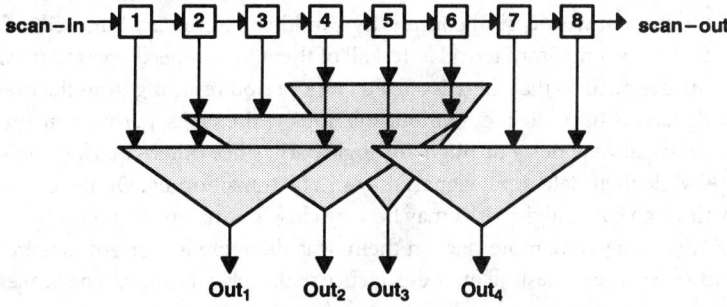

FIGURE 79.21 Overlapping logic cones connected to a common scan path.

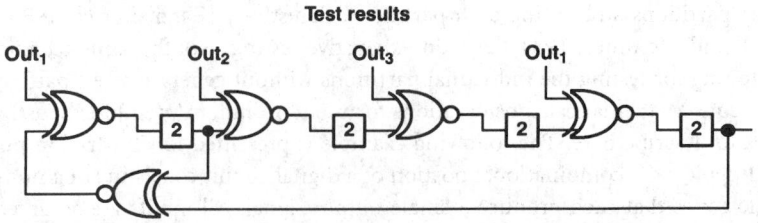

FIGURE 79.22   A four-bit parallel-input signature register.

nature register that can compress test results captured from four separate outputs into a single 4-bit signature. Provided that the test circuit has deterministic behavior, a signature register can be started in a given starting state, and its final value may be compared to a known good signature to determine pass/fail status. However, compressing test results into a single overall signature may prevent proper fault detection if multiple erroneous outputs (which may result from the same fault being detected on multiple test vectors) causes the final test signature to be correct even though interim signatures were wrong. The probability that a faulty circuit signature will be the same as the good circuit signature is known as aliasing probability. It can be shown that if the test length is sufficiently long, aliasing probability diminishes toward 2^{-t}, where t is the number of bits of the signature register [Dervisoglu, 1985].

The two constructs of LFSR and the MISR can be merged into a single multipurpose register in a *built-in logic block observation* (BILBO) approach [Konemann *et al.*, 1979] where each register can have multiple modes of operation including the LFSR mode, MISR mode, SCAN mode, and NORMAL mode. In this case an on-chip test-control circuit may be used to control the modes of operation of the BILBO registers so that, in turn, each register is used as a test pattern generator or signature register to test a digital component. Figure 79.23 illustrates how to use the BILBO scheme in a stepwise fashion to test a large digital circuit.

Path-Delay Testing

Path-delay testing is aimed at testing whether a given component/system operates at a specified performance level that is often measured as the maximum system clock frequency. For example, the lower bound for the maximum clock frequency which a microprocessor IC is specified that it can reach needs to be verified. However, due to the very large number of different operations that a microprocessor can perform it is not practical to verify correct behavior of such a component operating at maximum clock frequency for every possible single operation or sequence of operations that it is designed to perform. On the other hand, it may be possible to examine the structure of the design to discover its *logic paths* and verify that signals can be propagated along these paths within a specified propagational delay time between the initiation of a signal transition at the beginning of the path and the arrival of the final values at the end of that path. This is called *path-delay* testing. A modern IC component with typical complexity would contain many hundreds of thousands of logic paths, so that it becomes impractical to test all of them for at-speed operation. All *synchronous* digital circuits are designed so that there is a fixed clock period resulting from the use-constant frequency clock signals to time their operation. Obviously, the clock period constitutes an upper bound for the propagational delay through any logic path, since otherwise clock pulses may arrive at the flip-flops while their data input signals may still be transitioning. On the other hand, propagational delay through some logic paths may be very close to this upper bound (i.e., clock period) value whereas others may have more slack in them. It is therefore important to identify the *critical* paths and perform path-delay testing on these. Hence path-delay testing can be broken into the two phases of critical-path selection and path-delay test pattern generation.

Several different approaches can be used in identifying the critical paths, including:

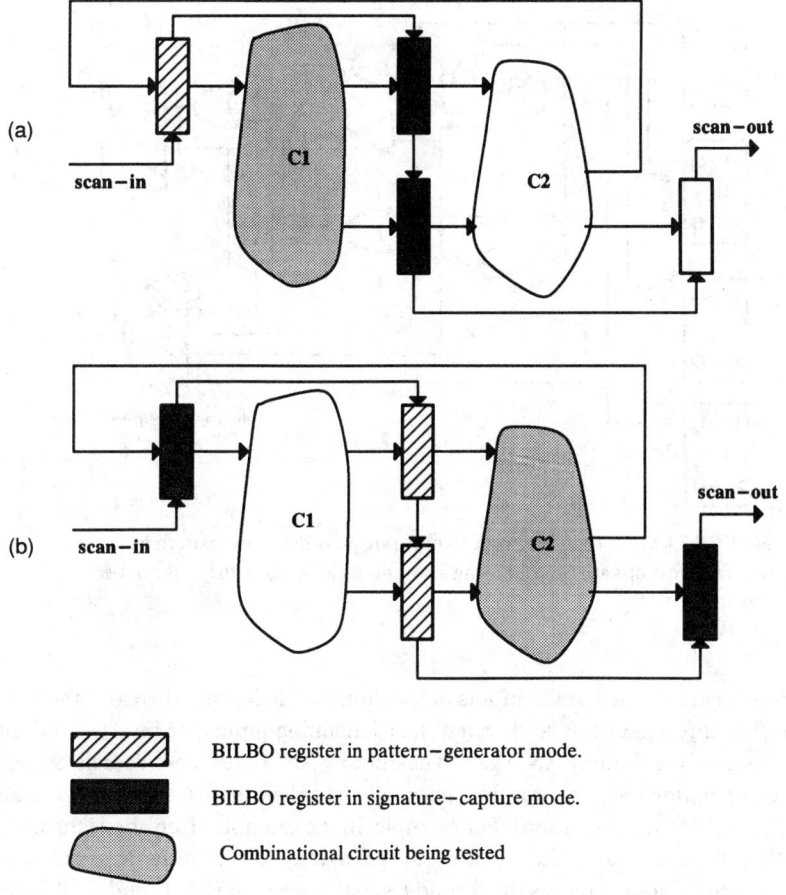

FIGURE 79.23 Using BILBO technique to partition and test a large circuit. (a) Testing combinatorial circuit C_1. (b) Testing combinatorial circuit C_2.

1. Select sufficiently large number of paths selected at random from a list of all logic paths.
2. Calculate worst-case timing for all logic paths and select a certain percentage of the slowest paths.
3. First identify certain key nodes and then select paths that pass through those nodes using either of the two approaches listed in (1) and (2) above.

The more challenging problem is to generate the test patterns to verify that none of the signal propagations along a given logic path require longer than the clock-period time to complete. A path-delay test pattern is a pair of patterns that generates the desired signal transition(s) and provides the sensitization of the signal paths whereby the generated transition(s) is (are) sensitized through the combinational circuit to the input of a flip-flop where it will be captured when the system clock is applied. For example, Fig. 79.24 shows a combinational circuit and identifies a specific signal path for which the path delay is to be measured. To determine the appropriate path-delay test patterns, a dummy AND gate is first added to the circuit as shown. An input to the AND gate is derived from the output of the combinational circuit through which the input signal transition is to be propagated. This signal is used in its true or complemented form depending upon whether the final value of the signal transition is a logic 1 or logic 0, respectively. Other inputs to the dummy AND gate come from all remaining inputs of gates through which the desired signal transitions must flow. If the desired signal transition is flowing through an AND or NAND gate, the remaining

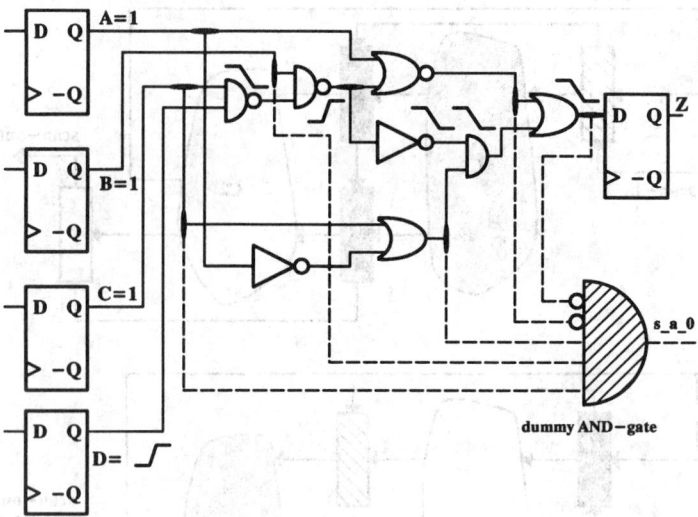

FIGURE 79.24 Circuit example to illustrate path-delay test pattern generation (all flip-flops are clocked using a common clock signal that has not been
shown).

inputs of these gates are also fed to the inputs of the dummy AND gate, whereas if the desired signal transitions flow through OR or NOR gates, their remaining inputs are inverted and then connected to the inputs of the dummy AND gate. The dummy AND gate is not actually implemented
as part of the combinational logic but rather acts as a convenient place to collect all the necessary
conditions for sensitizing the transitions. For example, in the example given above the first pattern
requires input flip-flops A, B, and C all to be set to the logic 1 value in order to sensitize a *low-to-
high* transition at the D input, whereas the second test pattern requires A, B, and C all to remain at
logic 1 while D is changed from logic 0 to the logic 1 value. This way the transitions created on
input D will travel through the identified signal path to reach the destination flip-flop Z.

Path-delay test patterns become much easier to generate and also apply to a circuit if the circuit
is designed using scannable flip-flops that are additionally capable of storing two arbitrarily
selected values in them. This can be done in such a fashion that the initial value available at the flip-
flop output will be replaced by the second value when a first clock pulse is applied, and the flip-flop
will revert to its normal mode of operation before the second clock pulse is applied. This way the
pair of test patterns that form a path-delay test are first loaded into the flip-flops (using scan) and
then two clock pulses are applied at speed. The final result captured by the second clock pulse is
then scanned out and examined to determine pass/fail status. It is also possible to get an actual
measurement of the path delays by repeating the same test over and over again while systematically
reducing the time distance between the two clock pulses to determine the minimum separation of
the two clock pulses required for proper operation.

Figure 79.25 shows a modified LSSD latch design [Malaiya and Narayanaswamy, 1983] that can
be used to enable path-delay testing as described above. Using this design, it is possible to load any
two arbitrary test vectors to the combinational circuit in rapid succession. First, test vector Q_1,
$Q_2,\ldots,Q_n$ would be scanned into the L1 latches outputs by using clocks C_3 and C_2. Next, the test
vector would be moved into the L2 latches by applying a single C clock. This way the flip-flop outputs would be set to their initial values defined by $Q_1, Q_2,\ldots,Q_n$. Following this, the second test vector $Y_1, Y_2,\ldots,Y_n$ would be scanned into the L1 latches using clock signals C_3 and C_2. Now applying
the C clock causes the first test vector (Q_i) to be replaced by the second test vector (Y_i), and if the C_1
clock is applied next, the response of the combinational circuit will be captured in the L1 latches.

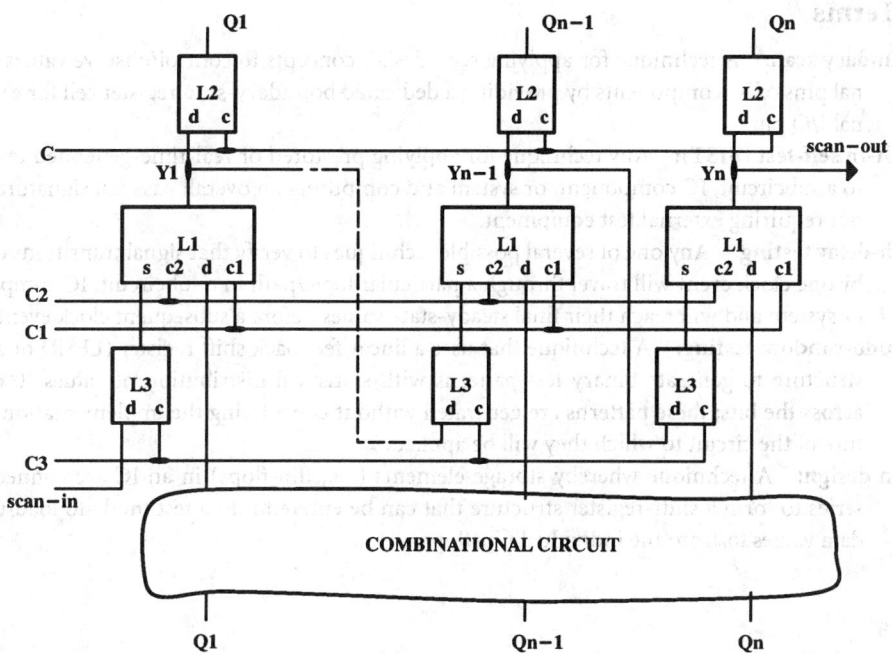

FIGURE 79.25 Using a three-latch flip-flop design to enable path-delay testing.

This way, the minimum delay between the clock signals C and C_1 that is necessary to allow the signals to propagate through the combinational circuit can be determined. Other flip-flop designs with built-in features to support *double-strobe* testing are also possible [Dervisoglu and Stong, 1991].

A different and more difficult-to-use approach for generating test patterns for path-delay measurement is to perform scan-in to load the internal flip-flops with a special pattern that prior circuit analysis will have determined will be transformed into the actually intended test pattern when the first functional clock pulse is applied. The circuit analysis required to use this approach amounts to performing simulation in reverse time flow to determine what state the device under test should be placed in (using scan) so that its next state corresponds to the desired test pattern.

Future for Design for Test

Present-day trends for striving to achieve shorter time to market while at the same time meeting competitive cost demands are going to continue into the foreseeable future. Design for testability is one of several areas that manufacturers from IC components to complete systems are paying increased emphasis to in order to meet their product goals. Twenty years ago some product managers considered testing as being necessary to weed out the bad from the good but did not consider DFT to be adding value to a product. However, since testing is essential, the value of DFT is seen in reducing the cost of an essential item. Hence DFT adds value to a product at least by an amount equal to the savings in test costs that it brings about. Furthermore, DFT improves time to market by making it possible to identify initial production problems at an earlier point in time. For example, initial productions of high-performance ASIC components may contain flaws that prevent their at-speed operation under certain circumstances. If these flaws are not discovered in a timely manner, they may turn into "showstopper" issues causing serious delays in revenue shipments of products. Whereas no "guaranteed" solutions exist to prevent and/or find a solution for all types of problems, design for testability is a rapidly maturing field of digital design.

Defining Terms

Boundary scan: A technique for applying scan design concepts to control/observe values of signal pins of IC components by providing a dedicated boundary-scan register cell for each signal I/O pin.

Built-in self-test (BIST): Any technique for applying prestored or real-time-generated test cases to a subcircuit, IC component, or system and computing an overall pass/fail signature without requiring external test equipment.

Path-delay testing: Any one of several possible techniques to verify that signal transitions created by one clock event will travel through a particular logic/path in a subcircuit, IC component, or system and will reach their final steady-state values before a subsequent clock event.

Pseudo-random testing: A technique that uses a linear feedback shift register (LFSR) or similar structure to generate binary test patterns with statistical distribution of values (0 and 1) across the bits; these patterns are generated without considering the implementation structure of the circuit to which they will be applied.

Scan design: A technique whereby storage elements (i.e., flip-flops) in an IC are connected in series to form a shift-register structure that can be entered into a test mode to load/unload data values to/from the individual flip-flops.

References

M. Abramovici, M. A. Breuer, and A. D. Friedman, *Digital Systems Testing and Testable Design,* Rockville, Md.: Computer Science Press, 1990.

Advanced Micro Devices Inc. [AMDI], "Am29C818 CMOS Pipeline Register with SSR Diagnostics," product specification, Bus Interface Products Data Book, 1987, pp. 47–55.

H. Ando, "Testing VLSI with random access scan," in digest of papers, COMPCON, February 1980, pp. 50–52.

P. H. Bardell and W. H. McAnney, "Parallel pseudorandom test sequences for built-in test," in Proc. International Test Conference, October 1984, pp. 302–308.

P. H. Bardell, W. H. McAnney, and J. Savir, *Built-In Test for VLSI. Pseudorandom Techniques,* New York: Wiley, 1978.

Z. Barzilai, D. Coppersmith, and A. L. Rosenberg, "Exhaustive generation of bit patterns with applications to VLSI self-testing," *IEEE Trans. on Computers,* vol. C-32, no. 2, pp. 190–194, February 1985.

M. A. Breuer and A. D. Friedman, *Diagnosis and Reliable Design of Digital Systems,* Rockville, Md.: Computer Science Press, 1976, pp. 139–146, 156–160.

B. I. Dervisoglu, "VLSI self-testing using exhaustive bit patterns," in Proc. IEEE International Conference on Computer Design, October 1985, pp. 558–561.

B. I. Dervisoglu and G. E. Stong, "Design for testability: Using scanpath techniques for path-delay test and measurement," in Proc. International Test Conference, October 1991, pp. 364–374.

E. B. Eichelberger and T. W. Williams, "A logic design structure for LSI testability," *Journal of Design Automation and Fault-Tolerant Computing,* vol. 2, no. 2, pp. 165–178, 1978.

S. Funatsu, N. Wakatsuki, and T. Arima, "Test generation systems in Japan," in Proc. 12th Design Automation Symposium, June 1975, pp. 114–122.

T. Gheewala, "CrossCheck: A cell based VLSI testability solution," in Proc. 26th Design Automation Conference, 1989, pp. 706–709.

"IEEE Standard Test Access Port and Boundary-Scan Architecture," IEEE Std. 1149.1-1990, May 1990.

B. Konemann, J. Mucha, and G. Zwiehoff, "Built-in logic block observation technique," in digest of papers, International Test Conference, October 1979, pp. 37–41.

A. Lempel and M. Cohn, "Design of universal test sequences for VLSI," *IEEE Trans. on Information Theory,* vol. IT-31, no. 1, pp. 10–17, 1985.

Y. K. Malaiya and R. Narayanaswamy, "Testing for timing faults in synchronous sequential integrated circuits," in Proc. International Test Conference, 1983, pp. 560–571.

E. J. McCluskey, "Verification testing. A pseudoexhaustive test technique," *IEEE Trans. on Computers,* vol. C-33, no. 6, pp. 541–546, June 1984.

K. P. Parker, *Integrating Design and Test,* New York: IEEE Computer Society Press, 1987.

J. H. Stewart, "Future testing of large LSI circuit cards," in Proc. Semiconductor Test Symposium, Cherry Hill, N.J., October 1977, pp. 6–15.

Further Information

An excellent treatment of design for testability topics is found in Abramovici *et al.* [1990]. Also, Breuer and Friedman [1976] provide a very good treatment of pseudo-random test topics.

C. M. Maunder and R. E. Tulloss (*The Test Access Port and Boundary-Scan Architecture,* IEEE Computer Society Press Tutorial, 1990) provide a user's guide for boundary-scan and the IEEE 1149.1 Standard.

B. I. Dervisoglu ("Using Scan Technology for Debug and Diagnostics in a Workstation Environment," in Proc. International Test Conference, 1988, pp. 976–986) provides a very good example of applying DFT techniques all the way from the IC component level to the system level. Also, B. I. Dervisoglu ("Scan-Path Architecture for Pseudorandom Testing," *IEEE Design & Test of Computers,* vol. 6, no. 4, pp. 32–48, August 1989) describes using pseudo-random testing at the system level. Similarly, P. H. Bardell and M. J. Lapointe ("Production Experience with Built-in Self-Test in the IBM ES/9000 System," in Proc. International Test Conference, October 1991, pp. 28–36) describe application of BIST for testing a commercial product at the system level.

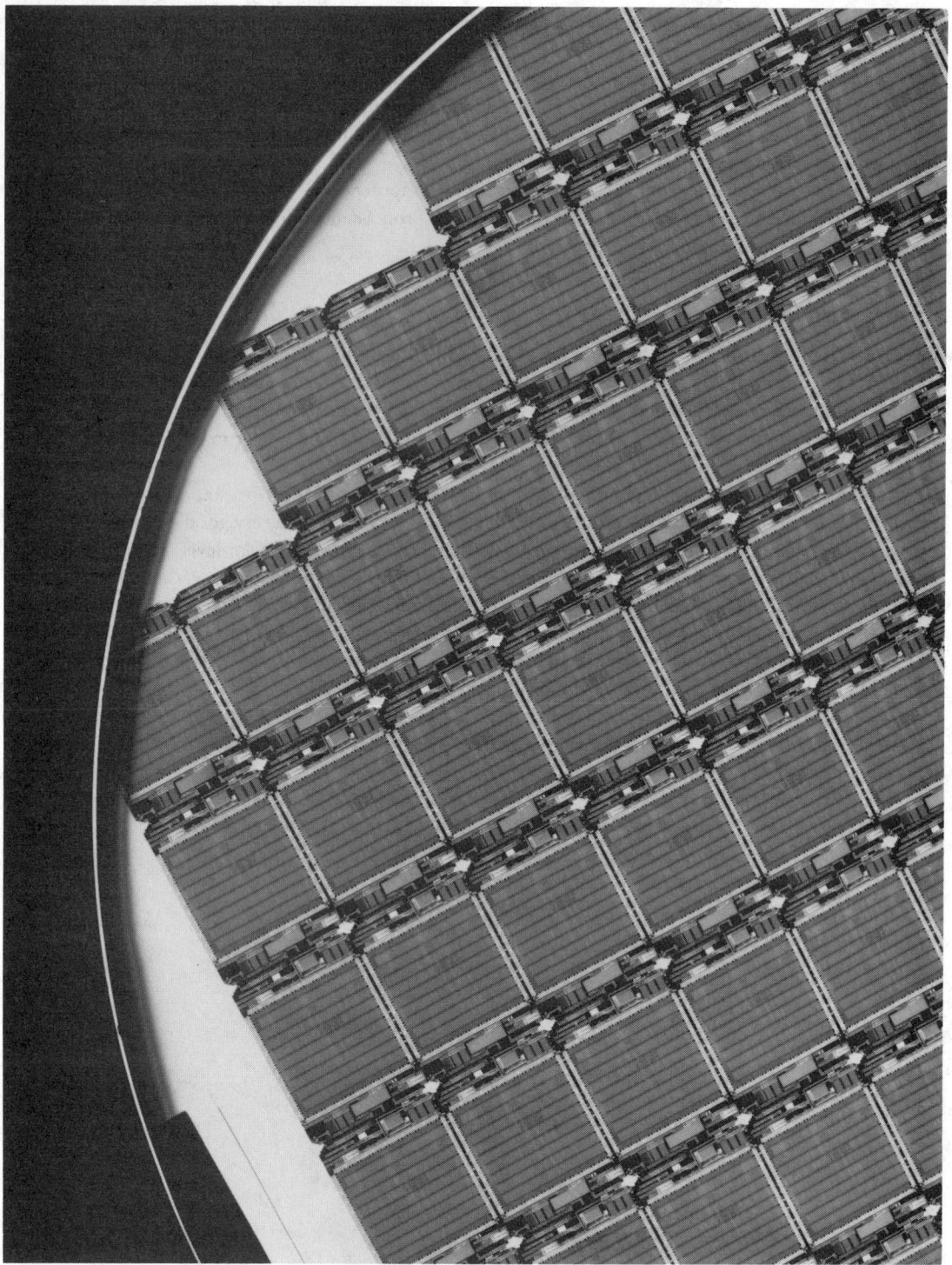

First logic chips on 8-inch wafers. Application-specific integrated circuit (ASIC) logic chips are shown on an 8-inch diameter silicon wafer. These 128,000-circuit, 0.8-micron, 5-volt logic chips are fabricated in an advanced complementary metal-oxide semiconductor (CMOS) technology process. The chips measuring 12.7 millimeters square are used in a variety of IBM information processing systems. (Photo courtesy of IBM Corp. Photographer: Tom Way.)

IX

Computer Engineering

John V. Oldfield
Syracuse University

80 Organization *R. Tinder, V. Oklobdzija, V. Hamacher, Z. Vranesic,*
S. Zaky, J. Raymond 1843
Number Systems • Computer Arithmetic • Architecture • Microprogramming

81 Programming *J. Feldman, E. Czeck, T. Lewis, J. Martin* 1878
Assembly Language • High-Level Languages • Data Types and Data Structures

82 Memory Systems *J. Goodman, G. Sohi* 1927
Parallel and Interleaved Memories • Memory Hierarchies • Virtual Memory

83 Input and Output *S. Sherr, R. Durbeck* 1938
Input Devices • Computer Output Printer Technologies

84 Software Engineering *C. Argila, C. Jones, J. Martin* 1976
Tools and Techniques • Testing, Debugging, and Verification • Programming Methodology

85 Computer Graphics *E. Rozanski* 2004
Graphics Hardware • Graphics Software

86 Computer Networks *T. Robertazzi* 2015
Local Area Networks • Metropolitan Area Networks • Wide Area Networks • The Future

87 Fault Tolerance *B. Johnson* 2020
Hardware Redundancy • Information Redundancy • Time Redundancy • Software Redundancy
• Dependability Evaluation

88 Knowledge Engineering *M. Abdelguerfi, J. Liebowitz* 2032
Databases • Rule-Based Expert Systems

89 Parallel Processors *T. Feng* 2052
Classifications • Types of Parallel Processors • System Utilization

90 Operating Systems *J. Boykin* 2061
Types of Operating Systems • Distributed Computing Systems • Fault-Tolerant Systems •
Parallel Processing • Real-Time Systems • Operating System Structure • Industry Standards

91 Computer Security *J. Cooper* 2072
Physical Security • Cryptology • Software Security • Hardware Security • Network Security •
Personnel Security

92 Computer Reliability *C. Guy* 2087
Definitions of Failure, Fault, and Error • Failure Rate and Reliability • Relationship Between
Reliability and Failure Rate • Mean Time to Failure • Mean Time to Repair • Mean Time
Between Failures • Availability • Calculation of Computer System Reliability • Markov
Modeling • Software Reliability • Reliability Calculations for Real Systems

COMPUTER ENGINEERING was originally coined as a term to describe the engineering processes required to develop, construct, and operate digital computers. As *digital,* or more correctly, *discrete* circuits and systems have become all-pervasive in computing, instrumentation, control, and measurement, it is appropriate to use it in a wider sense. Since the previous section of this handbook covered digital devices, this section will concentrate on system-level topics. No other field of engineering or human endeavor can match the rate of change of computer engineering, which has seen advances in performance and size of at least six orders of magnitude within the career lifetimes of some contemporary engineers.

Computer systems are often subdivided into software and hardware, corresponding to the Von Neumann model of a digital computer, i.e., of a control unit which fetches instructions from an addressed memory, decodes them, and directs a central processor capable of arithmetic and logical operations. This paradigm is still dominant and explains the sustained appeal of the computer as an adaptable information processing instrument applicable to a semi-infinite set of tasks, each described by specific software. In fact, the recent trend to reduced-instruction-set processors has had the effect of simplifying hardware to a small but extremely efficient processor, surrounded, so to speak, by layers of software, of which the application layer is outermost. As computer technology has benefited from very large scale integration (VLSI), we have seen alternative, more parallel architectures become feasible. The division between hardware and software is changing, as specialized and demanding tasks are carried out by dedicated hardware. Although Von Neumann–type computers will continue to improve in speed, major advances in computer performance will depend on more parallel architectures. At the same time, advances in digital techniques to specific application fields will blur the distinction between computers, controllers, and instruments.

Some of the technology-specific aspects of computer engineering were covered in Section VIII. This section, however, is concerned with higher-level aspects which are substantially independent of circuit technology. Chapter 80 reviews organizational matters which particularly affect computer processor design, such as the arithmetic and logical functions required. The next chapter considers the major topic of programming, which may be different in each "layer," using the previous analogy. Programming too has long been dominated by a particular paradigm, the so-called *imperative* model, in which the programmer expresses an algorithm, i.e., a process for solving a problem, as a sequence of instructions—either simple or complex, depending on the type of programming required. Recently others have emerged, such as rule-based programming, which has a *declarative* model, i.e., the user specifies the facts and rules of a situation and poses a question, leaving the computer ("knowledge-engine") to make its own inferences en route to finding a solution or set of solutions.

Computer memory systems are considered in Chapter 82. Early purists preferred the term *storage systems,* since the organization of a computer memory bears little resemblance to what we know of the organization of the human brain. For economic reasons, computer memories have been organized as a hierarchy of different technologies, with decreasing cost per bit as well as increased access times as one moves away from the central processor. The introduction of virtual memory in the Manchester Atlas project (c. 1964) was a major breakthrough in removing memory management from the tasks of the programmer, but recently the availability of vast quantities of semiconductor memory at ultralow prices has reduced the need for this technique.

The topic of Chapter 83 is the input and output of information. Early computers were confined almost exclusively to character information, but "input/output" now refers to any form of *transducer,* to choose an engineering term, which allows any form of information to be sensed whether in analog or digital form, entered into a computer system, and be output by it in a correspondingly useful form. Information may vary in time such as a temperature indication, in two dimensions such as the user's action in moving a mouse, or even in three dimensions, and output may be as simple as closing a contact or drawing a picture containing a vast range of colors.

Software engineering as discussed in Chapter 84 refers to the serious problem of managing the complexity of the layers of software. This problem has few parallels in other walks of life and is exacerbated by the rate of change in computing. It is dominated by the overall question "Is this computer system reliable?" which will be referred to in Chapter 92. Some parallels can be drawn with other complex human organizations, and, fortunately, the computer itself can be applied to the task.

Graphical input and output is the topic of Chapter 85. Early promise in the mid-1960s led to the pessimistic observation a decade later that this was "a solution looking for a problem," but as computer display technology improved in quality, speed, and, most importantly, cost, attention was focused on visualization algorithms, e.g., the task of producing a two-dimensional representation of a three-dimensional object. This is coupled with the need to provide a natural interface between the user and the computer and has led to the development of interactive graphical techniques for drawing, pointing, etc., as well as consideration of the human factors involved.

As computers have extended their scope, it has become necessary for a computer to communicate with other computers, whether nearby, such as a file server, or across a continent or ocean, such as in electronic mail. Chapter 86 reviews the major concepts of both local and wide area computer networks.

Many engineers were skeptical as to whether early computers would operate sufficiently long before a breakdown would prevent the production of useful results. Little recognition has been given to the pioneers of component and circuit reliability that have made digital systems virtually, but still not totally, fault-free. Critical systems, whether in medicine or national defense, must operate even if components and subsystems fail. The next chapter reviews the techniques employed to make computer systems fault-tolerant.

The idea of a rule-based system, referred to earlier, is covered in Chapter 88. Application software naturally reflects the nature of the application, and the term *knowledge engineering* has been coined to include languages and techniques for particularly demanding tasks, which cannot readily be expressed in a conventional scientific or business programming language.

Parallel systems are emerging as the power of computer systems is extended by using multiple units. The term *unit* may correspond to anything from a rudimentary processor, such as a "smart word" in a massively parallel "fine grain" architecture, to a full-scale computer, in a coarse-grain parallel system with a few tens of parallel units. Chapter 89 discusses the hardware and software approaches to a wide variety of parallel systems.

Operating systems, which are described in the next chapter, turn a "raw" computer into an instrument capable of performing useful, low-level tasks, such as creating a file or starting a process corresponding to an algorithm, or transferring its output to a device such as a printer, which may be busy with other tasks.

As society has become more dependent upon the computer and computer technology, it has become increasingly concerned with protecting the privacy of individuals and maintaining the integrity of computer systems against infiltration—by individuals, groups, and even on occasion by governments, as in the case of the computer virus said to have been introduced into the Iraqi defense system during the 1991 Gulf War. Techniques for protecting the security of a system and ensuring individual privacy are discussed in Chapter 91.

The final chapter discusses the overall reliability of computer systems, based on the inevitable limitations of both hardware and software mentioned earlier. Given the inevitability of failure, human or component, what can be said about the probability of a whole computer system failing? This may not be an academic issue for a passenger reading this section while flying in a modern jet airliner, which may spend over 95% of a flight under the control of an automatic pilot. He or she may be reassured to know, however, that the practitioners of reliability engineering have reduced the risk of system failure to truly negligible proportions.

Nomenclature

Symbol	Quantity	Unit
A	area	m^2
A_m	main amplifier gain	
A_p	preamplifier gain	
A_v	availability	
BW	bandwidth	Mbyte/s
C	capacitance	F
d	distance	m
E_L	illuminance	
f	proportionality factor	
h	Planck's constant	6.625×10^{-34} J · s
L	latency	ns
λ	failure rate	
μ_f	flip-flop sensitivity	
μ_p	photodetector sensitivity	

Symbol	Quantity	Unit
μ_s	Schmitt trigger sensitivity	
ν	hardware utilization	
ω	angular velocity	rad/s
P	parallelism	
P_c	character pitch	
q	drop charge	
R	wheel radius	
R_1	shaft radius	
S	sensitivity	fL
S	speed-up ratio	
t_L	optical loss	
V_b	band velocity	m/s
ξ	rotation angle	rad
$z(t)$	hazard rate	

80

Organization

Richard F. Tinder
Washington State University

Vojin G. Oklobdzija
University of California, Davis

V. Carl Hamacher
Queen's University, Canada

Zvonko G. Vranesic
University of Toronto

Safwat G. Zaky
University of Toronto

Jacques Raymond
University of Ottawa

80.1 Number Systems 1843
Positional and Polynomial Representations • Unsigned Binary
Number System • Unsigned Binary-Coded Decimal, Hexadecimal,
and Octal Systems • Conversion between Number Systems • Signed
Binary Numbers • Floating-Point Number Systems

80.2 Computer Arithmetic 1858
Number Representation • Algorithms for Basic Arithmetic
Operations • Floating-Point Representation

80.3 Architecture 1865
Functional Units • Basic Operational Concepts • Bus Structures •
Parallel and Distributed Computing

80.4 Microprogramming 1870
Levels of Programming • Microinstruction Structure • Microprogram Development • High-Level Languages for Microprogramming • Emulation • Applications of Microprogramming

80.1 Number Systems

Richard F. Tinder

Number systems provide the basis for conveying and quantifying information. Weather data, stocks, pagination of books, weights and measures—these are just a few examples of the use of numbers that affect our daily lives. For this purpose we find the decimal (or arabic) number system to be reliable and easy to use. This system evolved presumably because early humans were equipped with a crude type of calculator, their ten fingers. A number system that is appropriate for humans, however, may be intractable for use by a machine such as a computer. Likewise, a number system appropriate for a machine may not be suitable for human use.

Before concentrating on those number systems that are useful in computers, it will be helpful to review the characteristics that are desirable in any number system. There are *four* important characteristics in all:

- Distinguishability of symbols
- Arithmetic operations capability
- Error control capability
- Tractability and speed

To one degree or another the decimal system of numbers satisfies these characteristics for hard-copy transfer of information between humans. Roman numerals and **binary** are examples of number systems that do not satisfy all four characteristics for human use. On the other hand, the binary number system is preferable for use in digital computers. The reason is simply put: current digital electronic machines recognize only two identifiable states physically represented by a high voltage level and a low voltage level. These two physical states are logically interpreted as the binary symbols 1 and 0.

A fifth desirable characteristic of a number system to be used in a computer should be that it have a minimum number of easily identifiable states. The binary number system satisfies this condition. However, the digital computer must still interface with humankind. This is done by converting the binary data to a decimal and character-based form that can be readily understood by humans. A minimum number of identifiable characters (say 1 and 0, or true and false) is not practical or desirable for direct human use. If this is difficult to understand, imagine trying to complete a tax form in binary or in any number system other than decimal. On the other hand, use of a computer for this purpose would not only be practical but, in many cases, highly desirable.

Positional and Polynomial Representations

The *positional form* of a number is a set of side-by-side (juxtaposed) digits given generally in *fixed-point* form as

$$N_r = (a_{n-1} \ldots a_3 a_2 a_1 a_0 . a_{-1} a_{-2} a_{-3} \ldots a_{-m})_r \tag{80.1}$$

where the **radix** (or base) r is the total number of digits in the number system and a is a digit in the set defined for radix r. Here, the radix point separates n integer digits on the left from m fraction digits on the right. Notice that a_{n-1} is the most significant (highest-order) digit, called MSD, and that a_{-m} is the least significant (lowest-order) digit, denoted by LSD.

The *value* of the number in Eq. (80.1) is given in *polynomial form* by

$$
\begin{aligned}
N_r &= \sum_{i=-m}^{n-1} a_i r^i \\
&= \left(
\begin{aligned}
&a_{n-1} r^{n-1} + \cdots + a_2 r^2 + a_1 r^1 + a_0 r^0 \\
&+ a_{-1} r^{-1} + a_{-2} r^{-2} + \cdots + a_{-m} r^{-m}
\end{aligned}
\right)_r
\end{aligned}
\tag{80.2}
$$

where a_i is the digit in the ith position with a *weight* r^i.

Application of Eqs. (80.1) and (80.2) follows directly. For the decimal system $r = 10$, indicating that there are 10 distinguishable characters recognized as decimal numerals $0, 1, 2, \ldots, r - 1(= 9)$. Examples of the positional and polynomial representations for the decimal system are

$$N_{10} = (d_3 d_2 d_1 d_0 . d_{-1} d_{-2} d_{-3})_{10}$$
$$= 3017.528$$

and

$$
\begin{aligned}
N_{10} &= \sum_{i=-3}^{n-1} d_i 10^i \\
&= 3 \times 10^3 + 0 \times 10^2 + 1 \times 10^1 + 7 \times 10^0 + 5 \times 10^{-1} + 2 \times 10^{-2} + 8 \times 10^{-3} \\
&= 3000 + 10 + 7 + 0.5 + 0.02 + 0.008
\end{aligned}
$$

where d_i is the decimal digit in the ith position. Exclusive of possible leading and trailing zeros, the MSD and LSD for this number are 3 and 8, respectively. This number could have been written in a form such as $N_{10} = 03017.52800$ without altering its value but implying greater accuracy of the fraction portion.

Unsigned Binary Number System

Applying Eqs. (80.1) and (80.2) to the binary system requires that $r = 2$, indicating that there are two distinguishable characters, typically 0 and $(r - 1) = 1$, that are used. In positional representation these characters (numbers) are called *binary digits* or *bits*. Examples of the positional and polynomial notations for a binary number are

$$N_2 = (b_{n-1}\ldots b_3 b_2 b_1 b_0 \, . \, b_{-1} b_{-2} b_{-3} \ldots b_{-m})_2$$
$$= 101101.101_2$$

$$\text{MSB} \longrightarrow \qquad \text{LSB}$$

and

$$N = \sum_{i=-m}^{n-1} b_i 2^i$$
$$= 1 \times 2^5 + 0 \times 2^4 + 1 \times 2^3 + 1 \times 2^2 + 0 \times 2^1 + 1 \times 2^0 + 1 \times 2^{-1} + 0 \times 2^{-2} + 1 \times 2^{-3}$$
$$= 32 + 8 + 4 + 1 + 0.5 + 0.125$$
$$= 45.625_{10}$$

where b_i is the bit in the ith position. Thus, the bit positions are weighted $\ldots$, 16, 8, 4, 2, 1, ½, ¼, ⅛, $\ldots$ for any number consisting of integer and fraction portions. Binary numbers so represented are sometimes referred to as *natural* binary. In positional representation the bits on the extreme left and extreme right are called the MSB (most significant bit) and LSB (least significant bit), respectively. Notice that by obtaining the value of a binary number a conversion from binary to decimal has been performed. The subject of radix (base) conversion will be dealt with more extensively later.

For reference purposes Table 80.1 provides the binary-to-decimal conversion for two-, three-, four-, five-, and six-bit binary. The six-bit binary column is only halfway completed for brevity.

Table 80.1 Binary-to-Decimal Conversion

Two-Bit Binary	Decimal Value	Three-Bit Binary	Decimal Value	Four-Bit Binary	Decimal Value	Five-Bit Binary	Decimal Value	Six-Bit Binary	Decimal Value
00	0	000	0	0000	0	10000	16	100000	32
01	1	001	1	0001	1	10001	17	100001	33
10	2	010	2	0010	2	10010	18	100010	34
11	3	011	3	0011	3	10011	19	100011	35
		100	4	0100	4	10100	20	100100	36
		101	5	0101	5	10101	21	100101	37
		110	6	0110	6	10110	22	100110	38
		111	7	0111	7	10111	23	100111	39
				1000	8	11000	24	101000	40
				1001	9	11001	25	101001	41
				1010	10	11010	26	101010	42
				1011	11	11011	27	101011	43
				1100	12	11100	28	101100	44
				1101	13	11101	29	101101	45
				1110	14	11110	30	101110	46
				1111	15	11111	31	101111	47
								⋮	⋮

In the natural binary system the number of bits in a unit of data is commonly assigned a name. Examples are:

- 4-data-bit unit: nibble (or half-byte)
- 8-data-bit unit: byte
- 16-data-bit unit: two bytes (or half-word)
- 32-data-bit unit: word (or four bytes)
- 64-data-bit unit: double-word

    etc.

The word size for a computer is determined by the number of bits that can be manipulated and stored in registers. The foregoing list of names would be applicable to a 32-bit computer.

Unsigned Binary-Coded Decimal, Hexadecimal, and Octal Systems

While the binary system of numbers is most appropriate for use in computers, it has several disadvantages when used by humans who have become accustomed to the decimal system. For example, binary machine code is long, difficult to assimilate, and tedious to convert to decimal. However there exist simpler ways to represent binary numbers for conversion to decimal representation. Three examples, commonly used, are natural binary-coded decimal (NBCD), binary-coded **hexadecimal** (BCH), and binary-coded **octal** (BCO). These number systems are useful in applications where a digital device, such as a computer, must interface with humans. The NBCD code representation is also useful in carrying out computer arithmetic.

The NBCD Representation

The BCD system as used here is actually an 8, 4, 2, 1 weighted code called *natural* BCD or NBCD. This system uses patterns of four bits to represent each decimal position of a number and is one of several such weighted BCD code systems. The NBCD code is converted to its decimal equivalent by polynomials of the form

$$N_{10} = b_3 \times 2^3 + b_2 \times 2^2 + b_1 \times 2^1 + b_0 \times 2^0$$
$$= b_3 \times 8 + b_2 \times 4 + b_1 \times 2 + b_0 \times 1$$

for any $b_3 b_2 b_1 b_0$ code integer. Thus, decimal 6 is represented as $(0 \times 8) + (1 \times 4) + (1 \times 2) + (0 \times 1)$, or 0110 in NBCD code. Like natural binary, NBCD code is also called "natural" because its bit positional weights are derived from integer powers of 2^n. Table 80.2 shows the NBCD bit patterns for decimal integers 0 through 9.

Table 80.2   NBCD Bit Patterns and Decimal Equivalent

NBCD Bit Pattern	Decimal	NBCD Bit Pattern	Decimal
0000	0	1000	8
0001	1	1001	9
0010	2	1010	NA
0011	3	1011	NA
0100	4	1100	NA
0101	5	1101	NA
0110	6	1110	NA
0111	7	1111	NA

NA = not allowed.

The NBCD code is currently the most widely used of the BCD codes. There are many excellent sources of information on BCD codes. One, in particular, provides a fairly extensive coverage of both weighted and unweighted BCD codes [Tinder, 1991].

Decimal numbers greater than 9 or less than 1 can be represented by the NBCD code if each digit is given in that code and if the results are combined. For example, the number 63.98 is represented by (or converted to) NBCD code as

$$
\begin{array}{ccccc}
6 & 3 & . & 9 & 8 \\
\end{array}
$$
$$
63.98_{10} = 0110 \ \ 0011 \ . \ 1001 \ \ 1000)_{\text{NBCD}}
$$
$$
= 1100011.10011_{\text{NBCD}}
$$

Here, the code weights are 80, 40, 20, 10; 8, 4, 2, 1; 0.8, 0.4, 0.2, 0.1; and 0.08, 0.04, 0.02, 0.01 for the tens, units, tenths, and hundredths digits, respectively, representing four decades. Conversion between binary and NBCD requires conversion to decimal as an intermediate step. For example, to convert from NBCD to binary requires that groups of four bits be selected in both directions from the radix point to form the decimal number. If necessary, zeros are added to the leftmost or rightmost ends to complete the groups of four bits as in the above example. Negative NBCD numbers can be represented either in sign-magnitude notation or 1's or 2's **complement** notation as discussed later.

Another BCD code that is used for number representation and manipulation is called excess 3 BCD (or XS3 NBCD, or simply XS3). XS3 is an example of a *biased-weighted* code (a bias of 3). This code is formed by adding $0011_2 \ (= 3_{10})$ to the NBCD bit patterns in Table 80.2. Thus, to convert XS3 to NBCD code, 0011 must be subtracted from XS3 code. In four-bit quantities the XS3 code has the useful feature that when adding two numbers together in XS3 notation a carry will result and yield the correct value any time a carry results in decimal (i.e., when 9 is exceeded). This feature is not shared by either natural binary or NBCD addition.

The Hexadecimal and Octal Systems

The hexadecimal number system requires that $r = 16$ in Eqs. (80.1) and (80.2), indicating that there are 16 distinguishable characters in the system. By convention, the permissible hexadecimal digits are 0, 1, 2, 3, 4, 5, 6, 7, 8, 9, A, B, C, D, E, and F for decimals 0 through 15, respectively. Examples of the positional and polynomial representations for a hexadecimal number are

$$
N_{16} = (h_{n-1} \ldots h_3 h_2 h_1 h_0 \ . \ h_{-1} h_{-2} h_{-3} \ldots h_{-m})_{16}
$$
$$
= (AF3.C8)_{16}
$$

with a decimal value of

$$
N = \sum_{i=-m}^{n-1} h_i 16^i
$$
$$
= 10 \times 16^2 + 15 \times 16^1 + 3 \times 16^0 + 12 \times 16^{-1} + 8 \times 16^{-2}
$$
$$
= 2803.78125_{10}
$$

Here, it is seen that a hexadecimal number has been converted to decimal by using Eq. (80.2).

The octal number system requires that $r = 8$ in Eqs. (80.1) and (80.2), indicating that there are eight distinguishable characters in this system. The permissible octal digits are 0, 1, 2, 3, 4, 5, 6, and 7, as one might expect. Examples of the application of Eqs. (80.1) and (80.2) are

$$
N_8 = (o_{n-1} \ldots o_3 o_2 o_1 o_0 \ . \ o_{-1} o_{-2} o_{-3} \ldots o_{-m})_8
$$
$$
= 501.74_8
$$

Table 80.3 The BCH and BCO Number Systems

Binary	BCH	BCO	Decimal	Binary	BCH	BCO	Decimal
0000	0	0	0	1010	A	12	10
0001	1	1	1	1011	B	13	11
0010	2	2	2	1100	C	14	12
0011	3	3	3	1101	D	15	13
0100	4	4	4	1110	E	16	14
0101	5	5	5	1111	F	17	15
0110	6	6	6	10000	10	20	16
0111	7	7	7	11011	1B	33	27
1000	8	10	8	110001	31	61	49
1001	9	11	9	1001110	4E	116	78

with a decimal value of

$$N = \sum_{i=-m}^{n-1} o_i 8^i$$

$$= 5 \times 8^2 + 0 \times 8^1 + 1 \times 8^0 + 7 \times 8^{-1} + 4 \times 8^{-2}$$

$$= 321.9375_{10}$$

When the hexadecimal and octal number systems are used to represent bit patterns in binary, they are called binary-coded hexadecimal (BCH) and binary-coded octal (BCO), respectively. These two number systems are examples of *binary-derived radices*. Table 80.3 lists several selected examples showing the relationships between BCH, BCO, binary, and decimal.

What emerges on close inspection of Table 80.3 is that each hexadecimal digit corresponds to four binary digits and that each octal digit corresponds to three binary digits. The following example illustrates the relationships between these number systems:

$$10110111111.11011_2 = \overset{5}{0101} \; \overset{B}{1011} \; \overset{F}{1111} \; . \; \overset{D}{1101} \; \overset{8}{1000}$$

$$= 5BF.D8_{16}$$

$$= \overset{2}{010} \; \overset{6}{110} \; \overset{7}{111} \; \overset{7}{111} \; . \; \overset{6}{110} \; \overset{6}{110}$$

$$= 2677.66_8$$

$$= 1471.84375_{10}$$

To separate the binary digits into groups of four (for BCH) or groups of three (for BCO), counting must begin from the radix point and continue outward in both directions. Then, where needed, zeros are added to the leading and trailing ends of the binary representation to complete the MSDs and LSDs for the BCH and BCO forms.

Conversion between Number Systems

It is not the intent of this section to cover all methods for radix (base) conversion. Rather, the plan is to provide general approaches, separately applicable to the integer and fraction portions, followed by specific examples.

Conversion of Integers

Since the polynomial form of Eq. (80.2) is a geometrical progression, the integer portion can be represented in *nested radix* form. In source radix s, the nested representation is

$$N_s = \left(a_{n-1}s^{n-1} + a_{n-2}s^{n-2} + \cdots + a_1 s^1 + a_0 s^0\right)_s$$

$$= a_0 + s(a_1 + s(a_2 + \cdots + a_{n-1})))))_s$$

$$= a_0 + s\left(\sum_{i=1}^{n-1} a_i s^{i-1}\right) \tag{80.3}$$

for digits a_i having integer values from 0 to $s - 1$. The nested radix form not only suggests a conversion process but also forms the basis for computerized conversion.

Consider that the number in Eq. (80.3) is to be represented in nested radix r form

$$N_r = b_0 + r(b_1 + r(b_2 + \cdots + b_{m-1})))))_r$$

$$= b_0 + r\left(\sum_{i=1}^{m-1} b_i r^{i-1}\right) \tag{80.4}$$

where, in general, $m \neq n$. Then, if N_s is divided by r, the results are of the form

$$\frac{N_s}{r} = Q + \frac{R}{r} \tag{80.5}$$

where Q is the integer quotient rearranged as $Q_0 = b_1 + r(b_2 + \cdots + b_{m-1})))$, and R is the remainder $R_0 = b_0$. A second division by r yields $Q_0/r = Q_1 + R_1/r$, where Q_1 is arranged as $Q_1 = b_2 + r(b_3 + \cdots + b_{m-1}))$, and $R_1 = b_1$. Thus, by repeated division of the integer result Q_i by r, the remainders yield $(b_0, b_1, b_2, \ldots, b_{m-1})_r$ in that order.

The conversion method just described, called the *radix divide method*, can be used to convert between any two integers of different radices. However, the requirement is that *the arithmetic required by N_s/r must be carried out in source radix, s.* Except for source radices 10 and 2, this poses a severe problem for humans. Table 80.4 provides the recommended procedures for integer conversion. The radix divide method is suitable for computer conversion providing, of course, that the computer is programmed to carry out the arithmetic in different radices.

The integer conversion methods of Table 80.4 can be illustrated by the following simple examples:

Example 1. $139_{10} \rightarrow N_2$

N/r	Q	R
$139/2 =$	69	1
$69/2 =$	34	1
$34/2 =$	17	0
$17/2 =$	8	1
$8/2 =$	4	0
$4/2 =$	2	0
$2/2 =$	1	0
$1/2 =$	0	1

$139_{10} = 10001011_2$

Example 2. $10001011_2 \rightarrow N_{10}$. By positional weights,

$$N_{10} = 128 + 8 + 2 + 1 = 139_{10}$$

Table 80.4 Summary of Recommended Methods for
Integer Conversion by Noncomputer Means

Integer Conversion	Conversion Method
$N_{10} \rightarrow N_r$	Radix division by radix r using Eq. (80.5)
$N_s \rightarrow N_{10}$	Eq. (80.2) or Eq. (80.3)
$N_s)_{s \neq 10} \longrightarrow N_r)_{r \neq 10}$	$N_s \rightarrow N_{10}$ by Eq. (80.2) or (80.3)
	$N_{10} \rightarrow N_r$ radix division by r using Eq. (80.5)

Special Cases for Binary Forms

$N_2 \rightarrow N_{10}$	Positional weighting
$N_2 \rightarrow N_{BCH}$	Partition N_2 into groups of four bits starting from radix point, then apply Table 80.3
$N_2 \rightarrow N_{BCO}$	Partition N_2 into groups of three bits starting from radix point, then apply Table 80.3
$N_{BCH} \rightarrow N_2$	Reverse of $N_2 \rightarrow N_{BCH}$
$N_{BCO} \rightarrow N_2$	Reverse of $N_2 \rightarrow N_{BCO}$
$N_{BCH} \rightarrow N_{BCO}$	$N_{BCH} \rightarrow N_2 \rightarrow N_{BCO}$
$N_{BCO} \rightarrow N_{BCH}$	$N_{BCO} \rightarrow N_2 \rightarrow N_{BCH}$
$N_{NBCD} \rightarrow N_{XS3}$	Add $0011_2 (= 3_{10})$ to N_{NBCD}
$N_{XS3} \rightarrow N_{NBCD}$	Subtract $0011_2 (= 3_{10})$ from N_{NBCD}

Example 3. $139_{10} \rightarrow N_8$

$$
\begin{array}{rcl}
N/r & Q & R \\
139/8 = & 17 & 3 \\
17/8 = & 2 & 1 \\
2/8 = & 0 & 2 \quad\quad 139_{10} = 213_8
\end{array}
$$

Example 4. $10001011_2 \rightarrow N_{BCO}$

$$
\begin{array}{ccc}
2 & 1 & 3 \\
010 & 001 & 011 = 213_{BCO}
\end{array}
$$

Example 5. $213_{BCO} \rightarrow N_{BCH}$

$$
\begin{array}{cccccc}
2 & 1 & 3 & & 8 & B \\
213_{BCO} = 010 & 001 & 011 = 10001011_2 = 1000 & 1011 = 8B_{16}
\end{array}
$$

Example 6. $213_8 \rightarrow N_5$

$$213_8 = 2 \times 8^2 + 1 \times 8^1 + 3 \times 8^0 = 139_{10}$$

$$
\begin{array}{rcl}
N/r & Q & R \\
139/5 = & 27 & 4 \\
27/5 = & 5 & 2 \\
5/5 = & 1 & 0 \\
1/5 = & 0 & 1 \quad\quad 213_8 = 1024_5
\end{array}
$$

Check: $1 \times 5^3 + 2 \times 5^1 + 4 \times 5^0 = 125 + 10 + 4 = 139_{10}$

Conversion of Fractions

By extracting the fraction portion from Eq. (80.2) one can write

$$.N_s = (a_{-1}s^{-1} + a_{-2}s^{-2} + \cdots + a_{-m}s^{-m})_s$$
$$= s^{-1}(a_{-1} + s^{-1}(a_{-2} + \cdots + a_{-m})))))_s$$
$$= s^{-1}(a_{-1} + \sum_{i=2}^{m} a_{-i}s^{-i+1})_s \qquad (80.6)$$

in radix s. This is called the *nested inverse radix* form that provides the basis for computerized conversion.

If the fraction in Eq. (80.6) is represented in nested inverse radix r form, then

$$.N_r = r^{-1}(b^{-1} + r^{-1}(b^{-2} + \cdots + b^{-p})))))_r$$
$$= r^{-1}(b_{-1} + \sum_{i=2}^{p} b_{-i}r^{-i+1})_r \qquad (80.7)$$

for any fraction represented in radix r. Now, if N_s is multiplied by r, the result is of the form

$$.N_s \times r = I + F \qquad (80.8)$$

where I is the product integer, $I_1 = b_{-1}$, and F_0 is the product fraction arranged as $F_1 = r^{-1}(b_{-2} + r^{-1}(b_{-3} + \cdots + b_{-p})))_r$. By repeated multiplication by r of the remaining fractions F_i, the resulting integers yield $(b_{-1}, b_{-2}, b_{-3}, \ldots b_{-m})_r$, in that order.

The conversion just described is called the *radix multiply method* and is perfectly general for converting between fractions of different radices. However, as in the case of integer conversion, the requirement is that *the arithmetic required by* $.N_s \times r$ *must be carried out in source radix, s.* For noncomputer use by humans, this procedure is usually limited to fraction conversions $N_{10} \rightarrow N_r$, where the source radix is 10 (decimal). The recommended methods for converting between fractions of different radices are given in Table 80.5. The radix multiply method is well suited to computer use.

For any integer of radix s, there exists an exact representation in radix r. This is not the case for a fraction whose conversion is a geometrical progression that never converges. Terminating a fraction conversion at n digits (to the right of the radix point) results in an error or uncertainty. In decimal, this error is given by

Table 80.5 Summary of Recommended Methods for Fraction Conversion by Noncomputer Means

Fraction Conversion	Conversion Method
$.N_{10} \rightarrow .N_r$	Radix multiplication by using Eq. (80.8)
$.N_s \rightarrow .N_{10}$	Equation (80.2) or Eq. (80.6)
$.N_r)_{s \neq 10} \rightarrow .N_r)_{r \neq 10}$	$N_s \rightarrow N_{s10}$ by Eq. (80.2) or Eq. (80.6)
	$N_{10} \rightarrow N_r$ radix multiply by Eq. (80.8)

Special Cases for Binary Forms

$.N_2 \rightarrow .N_{BCH}$	Partition $.N_2$ into groups of four bits from radix point, then apply Table 80.3
$.N_2 \rightarrow .N_{BCO}$	Partition $.N_2$ into groups of three bits from radix point, then apply Table 80.3
$.N_{BCH} \rightarrow .N_2$	Reverse of $.N_2 \rightarrow .N_{BCH}$
$.N_{BCO} \rightarrow .N_2$	Reverse of $.N_2 \rightarrow .N_{BCO}$
$.N_{BCH} \rightarrow .N_{BCO}$	$.N_{BCH} \rightarrow .N_2 \rightarrow .N_{BCO}$
$.N_{BCO} \rightarrow .N_{BCH}$	$.N_{BCO} \rightarrow .N_2 \rightarrow .N_{BCH}$

$$\epsilon_{10} = a_{-n}r^{-n} + a_{-(n+1)}r^{-(n+1)} + a_{-(n+2)}r^{-(n+2)} + \cdots$$

$$= r^{-n}\left[a_{-n} + \sum_{i=1}^{\infty} a_{-(n+i)}r^{-(n+i)}\right]_r$$

where the quantity in brackets approaches the value of $a_{-n} + 1$. Therefore, terminating a fraction conversion at n digits from the radix point results in an error with bounds

$$0 < \epsilon_{10} \le r^{-n}(a_{-n} + 1) \tag{80.9}$$

in decimal. Equation (80.9) is useful in deciding when to terminate a fraction conversion.

Often, it is desirable to terminate a fraction conversion at $(n + 1)$ digits and then round off to n from the radix point. A suitable method for rounding to n digits in radix r is: Perform the fraction conversion to $(n + 1)$ digits from the radix point, then drop the $(n + 1)$ digit if $a_{-(n+1)} < r/2$, or add $r^{-(n-1)}$ to the result if $a_{-(n+1)} \ge r/2$.

After rounding off to n digits, the maximum error becomes the difference between the rounded result and the smallest value possible. By using Eq. (80.9), this difference is

$$\epsilon_{max} = r^{-n}(a_{-n} + 1) - r^{-n}(a_{-n} + a_{-(n+1)}/r)$$

$$= r^{-n}(1 - a_{-(n+1)}/r)$$

Then, by rounding to n digits, there results an error with bounds

$$0 < \epsilon_{10} \le r^{-n}(1 - a_{-(n+1)}/r) \tag{80.10}$$

in decimal. If $a_{-(n+1)} < r/2$ and the $(n + 1)$ digit is dropped, the maximum error is r^{-n}. Note that for $N_s \to N_{10} \to N_r$ type conversions, the bounds of errors aggregate.

The following examples illustrate the fraction conversion methods of Table 80.5.

Example 7. $0.654_{10} \to N_2$ rounded to eight bits

$.N_s \times r$	F	I	
0.654×2	0.308	1	
0.308×2	0.616	0	
0.616×2	0.232	1	
0.232×2	0.464	0	
0.464×2	0.928	0	
0.928×2	0.856	1	$0.654_{10} = 0.10100111_2$
0.856×2	0.712	1	
0.712×2	0.424	1	
0.424×2	0.848	0	$\epsilon_{max} = 2^{-8}$

Example 8. $0.654_{10} \to N_8$ terminated at four digits

$.N_s \times r$	F	I	
0.654×8	0.232	5	
0.232×8	0.856	1	$0.654_{10} = 5166_8$
0.856×8	0.848	6	with error bounds
0.848×8	0.784	6	$0 < \epsilon_{10} \le 7 \times 8^{-4} = 1.71 \times 10^{-3}$

Example 9. $0.5166_8 \to N_2$ rounded to eight bits and let $0.5166_8 \to N_{10}$ be rounded to four decimal places.

$$0.5166_8 = 5 \times 8^{-1} + 1 \times 8^{-2} + 6 \times 8^{-3} + 6 \times 8^{-4}$$
$$= 0.625000 + 0.015625 + 0.011718 + 0.001465$$
$$= 0.6538 \text{ rounded to four decimal places; } \epsilon_{10} \leq 10^{-4}$$

$.N_s \times r$	F	I
0.6538×2	0.3076	1
0.3076×2	0.6152	0
0.6152×2	0.2304	1
0.2304×2	0.4608	0
0.4608×2	0.9216	0
0.9216×2	0.8432	1
0.8432×2	0.6864	1
0.6864×2	0.3728	1
0.3728×2	0.7457	0

$0.5166_8 = 0.10100111_2$ (compare with Example 7)

$\epsilon_{10} \leq 10^{-4} + 2^{-8} = 0.0040$

Example 10. $0.10100111_2 \rightarrow N_{BCH}$

$$\begin{array}{cc} A & 7 \\ 0.10100111_2 = 0.1010 & 0111 = 0.A7_{BCH} \end{array}$$

Signed Binary Numbers

To this point only unsigned numbers (assumed to be positive) have been considered. However, both positive and negative numbers must be used in computers. Several schemes have been devised for dealing with negative numbers in computers, but only four are commonly used:

- Signed-magnitude representation
- Radix complement representation
- Diminished radix complement representation
- Excess (offset) code representation

Of these, the radix 2 complement representation, called 2's complement, is the most widely used system in computers.

Signed-Magnitude Representation

A signed-magnitude number consists of a magnitude together with a symbol indicating its sign (positive or negative). Such a number lies in the decimal range of $-(r^{n-1} - 1)$ through $+(r^{n-1} - 1)$ for n integer digits in radix r. A fraction portion, if present, would consist of m digits to the right of the radix point.

The most common examples of signed-magnitude numbers are those in the decimal and binary systems. The sign symbols for decimal (+ or −) are well known. In binary it is established practice to use 0 = plus and 1 = minus for the sign symbols and to place one of them in the MSB position for each number. Examples in eight-bit binary are

$$\begin{array}{ll} +45.5_{10} = 0 \; \overbrace{101101.1}^{\text{Magnitude}}_2 & +0_{10} = 0 \; 0000000_2 \\ \text{Sign bit} \nearrow \end{array}$$

$$\begin{array}{ll} -123_{10} = 1 \; \overbrace{1111011}^{\text{Magnitude}}_2 & -0_{10} = 1 \; 0000000_2 \\ \text{Sign bit} \nearrow \end{array}$$

Although the sign-magnitude system is used in computers, it has two drawbacks. There is no unique zero, as indicated by the examples, and addition and subtraction calculations require time-consuming decisions regarding operation and sign as, for example, (-7) minus (-4). Even so, the sign-magnitude representation is commonly used in **floating-point** number systems.

Radix Complement Representation

The *radix complement* of an n-digit number N_r is obtained by subtracting it from r^n, that is $r^n - N_r$. The operation $r^n - N_r$ is equivalent to complementing the number and adding 1 to the LSD. Thus, the radix complement is $\bar{N}_r + 1_{\text{LSD}}$ where $\bar{N}_r = r^n - 1 - N_r$ is the complement of a number in radix r. Therefore, one may write

$$\text{Radix complement of } N_r = r^n - N_r \hspace{3cm} (80.11)$$
$$= \bar{N}_r + 1$$

The complements $\bar{N}_r$ for digits in three commonly used number systems are given in Table 80.6. Notice that the complement of a binary number is formed simply by replacing the 1's with 0's and 0's with 1's as required by $2^n - 1 - N_2$.

With reference to Table 80.6 and Eq. (80.11), the following examples of radix complement representation are offered.

Example 11. The 10's complement of 47.83 is

$$\bar{N}_{10} + 1_{\text{LSD}} = 52.17$$

Example 12. The 2's complement of 0101101.101 is

$$\bar{N}_2 + 1_{\text{LSB}} = 1010010.011$$

Example 13. The 16's complement of A3D is

$$\bar{N}_{16} + 1_{\text{LSD}} = 5C2 + 1 = 5C3$$

The decimal value of Eq. (80.11) can be found from the polynomial expression

$$N_{\text{radix compl.}})_{10} = -(a_{n-1}r^{n-1}) + \sum_{i=-m}^{n-2} a_i r^i \hspace{2cm} (80.12)$$

for any n-digit number of radix r. In Eqs. (80.11) and (80.12) the MSD is taken to be the position of the sign symbol.

2's Complement Representation.   The radix complement for binary is the 2's complement representation. In 2's complement the MSB is the sign bit, 1 indicating a negative number or 0 if positive. The decimal range of representation for n-integer bits in 2's complement is from $-(2^{n-1})$ through $+(2^{n-1})$. From Eq. (80.11), the 2's complement is formed by

$$N_2)_{\text{2's compl.}} = 2^n - N_2 = \bar{N}_2 + 1 \hspace{3cm} (80.13)$$

A few examples in eight-bit binary are shown in Table 80.7. Notice that application of Eq. (80.13) changes the sign of the decimal value of a binary number (+ to −, and vice versa) and that only one zero representation exists.

Application of Eq. (80.12) gives the decimal value of any 2's complement number, including those containing a radix point. For example, the pattern $N_{\text{2's compl.}} = 11010010.011$ has a decimal value

Table 80.6 Complements for Three Commonly Used Number Systems

	Complement ($\bar{N}_r$)		
Digit	Binary	Decimal	Hexadecimal
0	1	9	F
1	0	8	E
2		7	D
3		6	C
4		5	B
5		4	A
6		3	9
7		2	8
8		1	7
9		0	6
A			5
B			4
C			3
D			2
E			1
F			0

Table 80.7 Examples of Eight-Bit 2's and 1's Complement Representations (MSB = Sign Bit)

Decimal Value	2's Complement	1's Complement
−128	10000000	
−127	10000001	10000000
−31	11100001	11100000
−16	11110000	11101111
−15	11110001	11110000
−3	11111101	11111100
−0	00000000	11111111
+0	00000000	00000000
+3	00000011	00000011
+15	00001111	00001111
+16	00010000	00010000
+31	00011111	00011111
+127	01111111	01111111
+128		

$$N_{2\text{'s compl.}})_{10} = -1 \times 2^7 + 1 \times 2^6 + 1 \times 2^4 + 1 \times 2^1 + 1 \times 2^{-2} + 1 \times 2^{-3}$$
$$= -128 + 64 + 16 + 2 + 0.25 + 0.125$$
$$= -45.625_{10}$$

The same result could have easily been obtained by first applying Eq. (80.13) to $N_{2\text{'s compl.}}$ followed by the use of positional weighting to obtain the decimal value. Thus,

$$N_{2\text{'s compl.}} = 00101101.101$$
$$= 32 + 8 + 5 + 0.5 + 0.125$$
$$= 45.625_{10}$$

which is known to be a negative number, -45.625_{10}.

Negative NBCD numbers can be represented in 2's complement. The foregoing discussion on 2's complement applies to NBCD with consideration of how NBCD is formed from binary. As an example, -59.24_{10} is represented by

$$0101\ 1001.0010\ 0100)_{\text{NBCD}} = 10100110.11011100)_{2\text{'s compl. NBCD}}$$

In a similar fashion, negative NBCD numbers can also be represented in 1's complement following the procedure given in the next paragraph. Sign-magnitude representation of a negative NBCD number simply requires the addition of a sign bit to the NBCD magnitude.

Diminished Radix Complement Representation

The diminished radix complement of a number is obtained by

$$N_r)_{\text{dim. rad. compl.}} = r^n - N_r - 1 \qquad (80.14)$$
$$= \bar{N}_r$$

Thus, the complement of a number is its diminished radix complement. It also follows that the radix complement of a number is the diminished radix complement with 1 added to the LSD as in Eq. (80.13). The range of representable numbers is $-(r^{n-1} - 1)$ through $+(r^{n-1} - 1)$ for radix r.

In the binary and decimal number systems, the diminished radix complement representations are the 1's complement and 9's complement, respectively. Examples of 1's complement are shown in Table 80.7 for comparison with those of 2's complement. Notice that in 1's complement there are two representations for zero, one for +0 and the other for −0. This fact limits the usefulness of the 1's complement representation for computer arithmetic.

Excess (Offset) Representations

Other systems for representing negative numbers use *excess* or *offset* codes. Here, a bias B is added to the true value N_r of the number to produce an excess number N_{xs} given by

$$N_{xs} = N_r + B \qquad (80.15)$$

When $B = r^{n-1}$ exceeds the usable bounds of negative numbers, N_{xs} remains positive. Perhaps the most common use of the excess representation is in floating-point number systems—the subject of the next section.

Two examples are given below in eight-bit excess 128 code.

Example 14.

$$
\begin{array}{rll}
-43_{10} & 11010101 & N_{2\text{'s compl.}} \\
+128_{10} & 10000000 & B \\
\hline
85_{10} & 01010101 & N_{xs} = -43_{10} \text{ in excess 128 code}
\end{array}
$$

Example 15.

$$
\begin{array}{rll}
27_{10} & 00011011 & N_{2\text{'s compl.}} \\
+128_{10} & 10000000 & B \\
\hline
155_{10} & 10011011 & N_{xs} = 27_{10} \text{ in excess 128 code}
\end{array}
$$

The representable decimal range for an excess 2^{n-1} number system is -2^{n-1} through $+(2^{n-1} - 1)$ for an n-bit binary number. However, if $N_2 + B > 2^{n-1} - 1$, *overflow* occurs and 2^{n-1} must be subtracted from $(N_2 + B)$ to give the correct result in excess 2^{n-1} code.

Floating-Point Number Systems

In fixed-point representation [Eq. (80.1)], the radix point is assumed to lie immediately to the right of the integer field and at the left end of the fraction field. The fixed-point system is the most commonly used system for representing bounded orders of magnitude. For example, with 32 bits a binary number could represent decimal numbers with upper and lower bounds of the order of $\pm 10^{10}$ and $\pm 10^{-10}$. However, for greatly expanded bounds of representation, as in scientific notation, the *floating-point* representation is needed.

A floating-point number (FPN) in radix r has the general form

$$\text{FPN})_r = F \times r^E \qquad (80.16)$$

where F is the *fraction* (or **mantissa**) and E is the *exponent*. Only fraction digits are used for the mantissa! Take, for example, Planck's constant $h = 6.625 \times 10^{-34}$ J·s. This number can be represented many different ways in floating point notation:

$$
\begin{aligned}
\text{Planck's constant } h &= 0.625 \times 10^{-33} \\
&= 0.0625 \times 10^{-32} \\
&= 0.00625 \times 10^{-31}
\end{aligned}
$$

All three adhere to the form of Eq. (80.16) and are, therefore, legitimate floating-point numbers in radix 10. Thus, as the radix point *floats* to the left, the exponent is *scaled* accordingly. The first form for *h* is said to be *normalized* because the MSD of *F* is nonzero, a means of standardizing the radix point position. Notice that the sign for *F* is positive while that for *E* is negative.

In computers the FPN is represented in binary where the normalized representation requires that the MSB for *F* always be 1. Thus, the range in *F* in decimal is

$$0.5 \leq F < 1$$

Also, the mantissa *F* is represented in sign-magnitude from. The normalized format for a 32-bit floating-point number in binary, which agrees with the IEEE standard [IEEE, 1985], is shown in Fig. 80.1. Here, the sign bit (1 if negative or 0 if positive) is placed at bit position 0 to indicate the sign of the fraction. Notice that the radix point is assumed to lie between bit positions 8 and 9 to separate the *E* bit-field from the *F* bit-field.

Before two FPNs can be added or subtracted in a computer, the *E* fields must be compared and equalized and the *F* fields adjusted. The decision-making process can be simplified if all exponents are converted to positive numbers by using the excess representation given by Eq. (80.15). For a *q*-digit number in radix *r*, the exponent in Eq. (80.16) becomes

$$E_{xs} = E_r + r^{q-1} \tag{80.17}$$

where *E* is the actual exponent augmented by a bias of $B = r^{q-1}$. The range in the actual exponent E_r is usually taken to be

$$-(r^{q-1} - 1) \leq E_r \leq +(r^{q-1} - 1)$$

In the binary system, required for computer calculations, Eq. (80.17) becomes

$$E_{xs} = E_2 + 2^{q-1} \tag{80.18}$$

with a range in actual exponent of $-(2^{q-1}-1) \leq E_2 \leq +(2^{q-1} - 1)$. In 32-bit normalized floating-point form, the exponent is stored in excess 128 code, while the number is stored in sign-magnitude form.

There still remains the question of how the number 0 is to be represented. If the *F* field is zero, then the exponent can be anything and the number will be zero. However, in computers the normalized FPN$_2$ limits *F* to $0.5 \leq F < 1$ since the MSB for *F* is always 1. The solution to this problem is to assume that the number is zero if the exponent bits are all zero regardless of the value of the mantissa. This leads, however, to a discontinuity in normalized FPN$_2$ representation at the low end.

The IEEE standard for normalized FPN$_2$ representation attempts to remove the problem just described. The IEEE system stores the exponent in excess $2^{q-1} - 1$ code and limits the decimal range of the actual exponent to

$$-(2^{q-1} - 2) \leq E_2 \leq +(2^{q-1} - 1)$$

For 32-bit FPN representation, the exponent is stored in excess 127 code as indicated in Fig. 80.1. Thus, the allowable range of representable exponents is from

$$-126_{10} = 00000001_2 \quad \text{through} \quad +127_{10} = 11111110_2$$

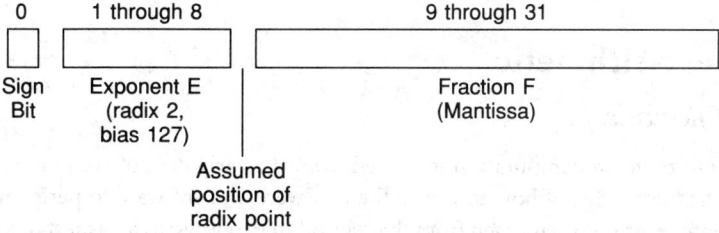

FIGURE 80.1 IEEE standard bit format for normalized floating-point representation.

This system reserves the use of all 0's or all 1's in the exponent for special conditions [IEEE, 1985; Pollard, 1990]. So that the F field magnitude can diminish linearly to zero when $E = -126$, the MSB $= 1$ for F is not specifically represented in the IEEE system but is implied.

The following example attempts to illustrate the somewhat confusing aspects of the IEEE normalized representation:

The number 101101.11001_2 is to be represented in IEEE normalized FPN$_2$ notation.

$$101101.11001_2 = .10110111001 \times 2^6$$

Sign bit $= 0$ (positive)

$$E_{xs} = 6 + 127 = 133_{10} = 10000101_2$$

$$F = 0110111001. . .00 \text{ (the MSB} = 1 \text{ is not shown)}$$

Therefore, the IEEE normalized FPN is

$$FPN_2 = 0 \ \ 10000101 \ \ 0110111001. . .0$$

Still other forms of FPNs are in use. In addition to the IEEE system, there are the IBM, Cray, and DEC systems of representation, each with its own single- and double-precision forms.

Defining Terms

Binary: Representation of quantities in base 2.
Complement: Opposite form of a number system.
Floating point: Similar to "scientific notation" except used to represent binary operations in a computer.
Hexadecimal: Base 16 number system.
Mantissa: Fraction portion of a floating-point number.
Octal: Base 8 number system.
Radix: Base to which numbers are represented.

References and Further Information

H. L. Garner, "Number systems and arithmetic," in *Advances in Computers*, vol. 6, F.L. Alt *et al.*, Eds., New York: Academic, 1965, pp. 131–194.

K. Hwang, *Computer Arithmetic*, New York: Wiley, 1978.

IEEE, *IEEE Standard for Binary Floating-Point Arithmetic*, ANSI/IEEE Std. 754–1985.

D. E. Knuth, *The Art of Computer Programming: Seminumerical Algorithms*, vol. 2, Reading, Mass: Addison-Wesley, 1969.

L. H. Pollard, *Computer Design and Architecture*, Englewood Cliffs, N.J.: Prentice-Hall, 1990.

R. F. Tinder, *Digital Engineering Design: A Modern Approach*, Englewood Cliffs, N.J.: Prentice-Hall, 1991.

C. Tung, "Arithmetic," in *Computer Science*, A. F. Cardenas *et al.*, Eds., New York: Wiley-Interscience, 1972, chap. 3.

80.2 Computer Arithmetic

Vojin G. Oklobdzija

As the ability to perform computation increased from the early days of computers up to the present, so has the knowledge of how to utilize the hardware and software to perform computation. Digital computer arithmetic emerged from that period in two ways: as an aspect of logic design and as a development of efficient **algorithms** to utilize the available hardware.

Given that numbers in a digital computer are represented as a string of zeros and ones and that hardware can perform only a relatively simple and primitive set of Boolean operations, all the arithmetic operations performed are based on a hierarchy of operations which are built upon the very simple ones.

What distinguishes computer arithmetic is its intrinsic relation to technology and the ways things are designed and implemented in a digital computer. This comes from the fact that the value of a particular way to compute, or a particular algorithm, is directly evaluated from the actual speed with which this computation is performed. Therefore, there is a very direct and strong relationship between the technology in which digital logic is implemented to compute and the way the computation is structured. This relationship is one of the guiding principles in the development of computer arithmetic.

The subject of computer arithmetic can be, for simpler treatment, divided into number representation; basic, arithmetic operations (such as addition, multiplication, and division); and evaluation of functions.

Number Representation

The only way to represent information in a digital computer is via a string of bits, i.e., zeros and ones. The number of bits being used depends on the length of the *computer word,* which is a quantity of bits on which hardware is capable of operating (sometimes also a quantity that is brought to the CPU from memory in a single access). The first question is what relationship to use in establishing correspondence between those bits and a number. Second, we need to make sure that certain properties which exist in the corresponding **number representation system** are satisfied and that they directly correspond to the operations being performed in hardware over the taken string of bits.

This relationship is defined by the rule that associates one numerical value designated as X (in the text we will use capital X for the numerical value) with the corresponding bit string designated as x.

$$x = \{x_{n-1}, x_{n-2}, \ldots, x_0\}$$

where

$$x_i \in 0, 1$$

In this case the associated word (the string of bits) is n bits long.

When for every value X there exists one and only one corresponding bit string x, we define the number system as **nonredundant.** If however, we could have more than one bit string x that represents the same value X, the number system is **redundant.**

Most commonly we are using numbers represented in a *weighted* number system, where a numerical value is associated with the bit string x according to the equation

$$x = \sum_{i=0}^{n-1} x_i w_i$$

where

$$w_0 = 1 \quad \text{and} \quad w_i = (w_i - 1)(r_i - 1)$$

The value r_i is an integer designated as the radix, and in a nonredundant number system it is an integer equal to the number of allowed values for x_i. In general x_i could consist of more than one bit. The numerical value associated with x is designated as the *explicit value* of x_e. In conventional number systems the radix r_i is the same positive integer for all the digit positions x_i and with the canonical set of digit values

$$\Sigma i = \{0, 1, 2, 3, \ldots, r_i - 1\} \quad \text{for } 0 \le i \le n - 1$$

An example of a weighted number system with a mixed radix would be the representation of time in weeks, days, hours, minutes, and seconds with a range for representing 100 weeks:

$$r = 10, 10, 7, 24, 60, 60$$

In digital computers the radices encountered are 2, 4, 10, and 16, with 2 being the most commonly used one.

The digit set x_i can be *redundant* and *nonredundant*. If the number of different values x_i can assume is $n_x \leq r$, then we have a *nonredundant* digit set. Otherwise, if $n_x > r$, we have a *redundant* digit set. Use of the *redundant* digit set has its advantages in efficient implementation of algorithms (multiplication and division in particular).

Other number representations of interest are *nonweighted* number systems, where the relative position of the digit does not affect the weight so that the appropriate interchange of any two digits will not change the value x. The best example of such a number system is the residue number system (RNS).

We also define **explicit value x_e** and **implicit value X_i** of a number represented by a bit string x. The *implicit value* is the only value of interest to the user, while the *explicit value* provides the most direct interpretation of the bit string x. Mapping of the *explicit value* to the *implicit value* is obtained by an arithmetic function that defines the number representation used. It is a task of the arithmetic designer to devise algorithms that result in the correct implicit value of the result for the operations on the operand digits representing the explicit values. In other words, the arithmetic algorithm needs to satisfy the *closure* property.

The relationship between the *implicit value* and the *explicit value* is best illustrated by Table 80.8.

Representation of Signed Integers

The two most common representations of signed integers are sign and magnitude (SM) representation and true and complement (TC) representation. While SM representation might be easier to understand and convert to and from, it has its own problems. Therefore, we will find TC representation to be more commonly used.

Sign and Magnitude Representation (SM). In SM representation signed integer X_i is represented by sign bit x_s and magnitude x_m (x_s, x_m). Usually 0 represents the positive sign (+) and 1 represents the negative sign (−). The magnitude of the number x_m can be represented in any way chosen for the representatation of positive integers. A disadvantage of SM representation is that two representations of zero exist, positive and negative zero: $x_s = 0$, $x_m = 0$ and $x_s = 1$, $x_m = 0$.

True and Complement Representation (TC). In TC representation there is no separate bit used to represent the sign. Mapping between the explicit and implicit value is defined as

$$X_i = \begin{cases} x_e & x_e < C/2 \\ x_e - C & x_e > C/2 \end{cases}$$

Table 80.8   The Relationship between the Implicit Value and the Explicit Value for $x = 11011$

Implied Attributes: Radix Point, Negative Number Representation, Others	Expression for Implicit Value X_i as a Function of Explicit Value x_e	Numerical Implicit Value X_i (in Decimal)
Integer magnitude	$X_i = x_e$	27
Integer, two's complement	$X_i = -2^5 + x_e$	−5
Integer, one's complement	$X_i = -(2^5 - 1) + x_e$	−4
Fraction, magnitude	$X_i = -2^{-5}x_e$	27/32
Fraction, two's complement	$X_i = -2^{-4}(2^{-5} + x_e)$	−5/16
Fraction, one's complement	$X_i = -2^{-4}(2^{-5} + 1 + x_e)$	−4/16

Source: A. Avizienis, "Digital computer arithmetic: A unified algorithmic specification," in *Symp. Computers and Automata,* Polytechnic Institute of Brooklyn, April 13–15, 1971.

Table 80.9   True and Complement Mapping

x_e	X_i
0	0
1	1
2	2
⋮	⋮
$C/2 - 1$	$C/2 - 1$
$C/2 + 1$	$-(C/2 + 1)$
⋮	⋮
$C - 2$	-2
$C - 1$	-1
C	0

Source: M. Ercegovac, *Digital Systems and Hardware/Firmware Algorithms,* New York: John Wiley, 1985, chap. 12. With permission.

Table 80.10   Mapping of the Explicit Value x_e into RC and DC Number Representations

x_e	X_i(RC)	X_i(RDC)
0	0	0
1	1	1
2	2	2
⋮	⋮	⋮
$\frac{1}{2}r^n - 1$	$\frac{1}{2}r^n - 1$	$\frac{1}{2}r^n - 1$
$\frac{1}{2}r^n$	$-\frac{1}{2}r^n$	$-(\frac{1}{2}r^n - 1)$
⋮	⋮	⋮
$r^n - 2$	-2	-1
$r^n - 1$	-1	0

The illustration of TC mapping is given in Table 80.9. In this representation positive integers are represented in the *true form,* while negative integers are represented in the *complement form.*

With respect to how the complementation constant C is chosen, we can further distinguish two representations within the TC system. If the complementation constant is chosen to be equal to the range of possible values taken by x_e, $C = r^n$ in a conventional number system where $0 \le x_e \le r^n - 1$, then we have defined the *range complement* (RC) system (also known as the *radix complement* system). If, on the other hand, the complementation constant is chosen to be $C = r^n - 1$, we have defined the *digit complement* (DC) system (also known as the *diminished radix complement* number system). Representations of the RC and DC number representation systems are shown in Table 80.10.

As can be seen from Table 80.10, the RC system provides for one unique representation of zero because the complementation constant $C = r^n$ falls ouside the range. There are two representations of zero in the DC system, $x_e = 0$ and $r^n - 1$. The RC representation is not symmetrical, and it is not a closed system under the change of sign operation. The range for RC is $[-\frac{1}{2}r^n, \frac{1}{2}r^n - 1]$. The DC is symmetrical and has the range of $[-(\frac{1}{2}r^n - 1), \frac{1}{2}r^n - 1]$.

For the radix $r = 2$, RC and DC number representations are commonly known as *two's complement* and *one's complement* number representation systems, respectively. Those two representations are illustrated by an example in Table 80.11 for the range of values $-(4 \le X_i \le 3)$.

Algorithms for Basic Arithmetic Operations

The algorithms for the arithmetic operation are dependent on the number representation system used. Therefore, their implementation should be examined for each number representation system

Table 80.11   Two's Complement and One's Complement Representation

X_i	Two's Complement,	$C = 8$	One's Complement,	$C = 7$
	x_e	X_i (2's complement)	x_e	X_i (1's complement)
3	3	011	3	011
2	2	010	2	010
1	1	001	1	001
0	0	000	0	000
−0	0	000	7	111
−1	7	111	6	110
−2	6	110	5	101
−3	5	101	4	100
−4	4	100	3	

separately, given that the complexity of the algorithm, as well as its hardware implementation, is dependent on it.

Addition and Subtraction in Sign and Magnitude System

In the *SM number system* addition/subtraction is performed on pairs (u_s, u_m) and (w_s, w_m) resulting in a sum (s_s, s_m), where u_s and w_s are sign bits and u_m and w_m are magnitudes. The algorithm is relatively complex because it requires comparisons of the signs and magnitudes as well. Extending the addition algorithm in order to perform subtraction is relatively easy because it only involves change of the sign of the operand being subtracted. Therefore, we will consider only the addition algorithm.

The algorithm can be described as

if $u_s = w_s$ (signs are equal) **then**
    $s_s = u_s$ and $s_m = u_m + w_m$ (operation includes checking for the overflow)
if $u_s \neq w_s$ **then**
    if $u_m > w_m$: $s_m = u_m - w_m$ $s_s = u_s$
    else: $s_m = w_m - u_m$ $s_s = w_s$

Addition and Subtraction in True and Complement System

Addition in the *TC system* is relatively simple. It is sufficient to perform modulo addition of the explicit values; therefore,

$$s_e = (u_e + w_e) \bmod C$$

Proof will be omitted.

In the *RC number system* this is equivalent to passing the operands through an adder and discarding the carry-out of the most significant position of the adder which is equivalent to performing the modulo addition (given that $C = r^n$).

In the *DC number system* the complementation constant is $C = r^n - 1$. Modulo addition in this case is performed by subtracting r^n and adding 1. It turns out that this operation can be performed by simply passing the operands through an adder and feeding the carry-out from the most significant digit position into the carry-in at least significant digit position. This is also called addition with *end-around carry*.

Subtracting two numbers is performed by simply changing the sign of the operand to be subtracted preceding the addition operation.

Change of Sign Operation.   The change of sign operation involves the following operation:

$$W_i = -Z_i$$
$$w_e = (-z_e) = (-z_e) \bmod C = C - Z_i \bmod C = C - z_e$$

which means that the change of sign operation consists of subtracting the operand z_e from the complementation constant C.

In the *DC system* complementation is performed by simply complementing each digit of the operand Z_i with respect to $r - 1$. In the case of $r = 2$, this results in a simple inversion of bits.

In the *RC system* the complementation is performed by complementing each digit of the operand Z_i with respect to $r - 1$ and adding 1 to the resulting z_e.

Multiplication Algorithm

The multiplication operation is performed in a variety of forms in hardware and software. In the beginning of computer development any complex operation was usually programmed in software or coded in the microcode of the machine. Some limited hardware assistance was provided. Today it is more likely to find full hardware implementation of the multiplication for reasons of speed and reduced cost of hardware. However, in all of them, multiplication shares the basic algorithm with some adaptations and modifications to the particular implementation and number system used. For simplicity we will describe a basic multiplication algorithm that operates on positive n-bit-long integers X and Y resulting in the product P, which is $2n$ bits long:

$$P = XY = X \times \sum_{i=0}^{n-1} y_i r^i = \sum_{i=0}^{n-1} X \times y_i r^i$$

This expression indicates that the multiplication process is performed by summing n terms of a *partial product*: $X \times y_i r^i$. This product indicates that the ith term is obtained by a simple arithmetic left shift of X for the i positions and multiplication by the single digit y_i. For the binary radix $r = 2$, y_i is 0 or 1 and multiplication by the digit y_i is very simple to perform. The addition of n terms can be performed at once, by passing the partial products through a network of adders (which is the case of full hardware multiplier), or sequentially, by passing the *partial product* through an adder n times. The algorithm to perform multiplication of X and Y can be described as [Ercegovac, 1985]

$$p^{(0)} = 0$$

$$p^{j+1} = 1/r(p^j + r^n X y_j) \qquad \text{for } j = 0, \ldots, n-1$$

It can be easily proved that this recurrence results in $p^{(n)} = XY$.

Various modifications of the multiplication algorithm exist; one of the most famous is the *modified Booth recoding algorithm* described by Booth in 1951. This algorithm allows for the reduction of the number of partial products, thus speeding up the multiplication process. Generally speaking, the Booth algorithm is a case of using the redundant number system with the radix higher than 2.

Division Algorithm

Division is a more complex process to implement because, unlike multiplication, it involves *guessing* the digits of the quotient. Here, we will consider an algorithm for division of two positive integers designated as *dividend Y* and *divisor X* resulting in a *quotient Q* and an integer *remainder Z* according to the relation given by

$$Y = XQ + Z$$

In this case the dividend contains $2n$ integers and the divisor has n digits in order to produce a quotient with n digits.

The algorithm for division is given with the following recurrence relationship [Ercegovac, 1985]:

$$z^{(0)} = Y$$

$$z^{(j+1)} = r z^{(j)} - X r^n Q_{n-1-j} \qquad \text{for } j = 0, \ldots, n-1$$

this recurrence relation yields

$$z^{(n)} = r^n (Y - XQ)$$

$$Y = XQ + z^{(n)} r^{-n}$$

which defines the division process with remainder $Z = z^{(n)} r^{-n}$.

The selection of the quotient digit is done by satisfying that $0 \leq Z < X$ at each step in the division process. This selection is a crucial part of the algorithm, and the best known are *restoring* and *non-restoring* division algorithms. In the former the value of the *tentative partial remainder* $z^{(j)}$ is restored after the wrong guess is made of the quotient digit q_j. In the latter this correction is not done in a separate step, but rather in the step following. The best-known division algorithm is the so-called SRT algorithm independently developed by Sweeney, Robertson, and Tocher. Algorithms for a higher radix were further developed by Robertson and his students, most notably Ercegovac.

Floating-Point Representation

Numbers represented as signed integers can only cover the range limited by the number of digits n and choice of the radix r. For the choice of radix r and n digits used, the maximum positive integer that can be represented is $r^n - 1$.

Often scientific computation requires the use of very small numbers as well as very large ones represented with some required precision. To satisfy those needs of the scientific or engineering computation, *floating-point* (FP) format is used to represent the numbers that are represented as

$$X = S \times B^{\mp \text{Exp}} \times \text{Fract}$$

where S is a sign bit (0, 1), B is a selected base, Exp is an exponent (which contains its own sign or is biased), and Fract is a fraction of the number. A typical FP number represented in a 32-bit word is shown:

0	1		8	9		31
S		Exp			Fraction	

Using this particular representation we can represent the range of the negative numbers from

$$-(1 - 2^{-24}) \times 2^{127} \leq X \leq -0.5 \times 2^{-128}$$

and

$$0.5 \times 2^{-128} \leq X \leq (1 - 2^{-24}) \times 2^{127}$$

for the positive numbers.

Different computer manufacturers such as IBM, DEC, and Intel have adopted their own standards for the floating-point number representation. This has led to an effort to introduce a standard for floating-point representation and computation resulting in IEEE Standard 754. More about floating-point computation and representation can be found in Waser and Flynn [1982].

Defining Terms

Algorithm: Decomposition of the computation into subcomputations with an associated precedence relation that determines the order in which these subcomputations are performed [Ercegovac, 1985].

Explicit value x_e: A value associated with the bit string according to the rule defined by the number representation system being used.

Implicit value X_i: The value obtained by applying the arithmetic function defined for the interpretation of the explicit value x_e.

Nonredundant number system: The system where for each bit string there is one and only one corresponding numerical value x_e.

Number representation system: A defined rule that associates one numerical value x_e with every *valid* bit string x.

Redundant number system: The system in which the numerical value x_e could be represented by more than one bit string.

References

A. Avizienis, "Digital computer arithmetic: A unified algorithmic specification," in *Symposium on Computers and Automata*, Polytechnic Institute of Brooklyn, April 13–15, 1971.

M. Ercegovac, *Digital Systems and Hardware/Firmware Algorithms*, New York: Wiley, 1985, chap. 12.

S. Waser and M. Flynn, *Introduction to Arithmetic for Digital Systems Designers*, New York: Holt, 1982.

Further Information

For more information about specific arithmetic algorithms and their implementation see K. Hwang, *Computer Arithmetic: Principles, Architecture and Design*, New York: Wiley, 1979 and also E. Swartzlander, *Computer Arithmetic*, vols. I and II, Los Alamitos, Calif.: IEEE Computer Society Press, 1980.

Publications in *IEEE Transactions on Electronic Computers* and *Proceedings of the Computer Arithmetic Symposia* by various authors, most notably by Milos Ercegovac, are very good sources for detailed information on a particular algorithm or implementation.

Good coverage of floating-point arithmetic can be found in the above book by Hwang and further details are provided in *IEEE Standard for Binary Floating Point Arithmetic*, ANSI/IEEE Standard 754 (1985) and its discussion in *IEEE Computer Magazine*, vol. 14, no. 3, pp. 51–62, 1981.

80.3 Architecture*

V. Carl Hamacher, Zvonko G. Vranesic, and Safwat G. Zaky

Computer architecture can be defined to mean the functional operation of the individual hardware units in a computer system and the flow of information and control among them. This is a somewhat more general definition than is sometimes used in the computer literature—for example, some articles and books refer to the instruction set architecture or the system bus architecture—but for an overview handbook article such as this, the more general definition is more appropriate [Hamacher *et al.*, 1990].

The main functional units of a single-processor system will be described. Some alternatives for connecting them together into an effective computing system will be given. Following this, a brief introduction to systems that have more than one processor will be provided.

Functional Units

Let us first define the term **digital computer**, or simply computer. In its simplest form, a contemporary computer is a fast electronic calculating machine that accepts digitized input information, processes it according to a list of internally stored instructions, and produces the resultant output information. The list of instructions is called a *program*, and internal storage is called *computer memory*.

A computer has five functionally independent main parts: input, memory, arithmetic and logic, output, and control. The input unit accepts coded information from human operators, from electromechanical devices such as a keyboard, or from other computers over digital communication lines. The information received is either stored in the memory for later reference or immediately used by the arithmetic and logic circuitry to perform the desired operations. The processing steps are determined by a program stored in the memory. Finally, the results are sent back to the outside world through the output unit. All these actions are coordinated by the control unit.

It has been traditional to refer to the arithmetic and logic circuits in conjunction with the main control circuits as a **central processing unit** (CPU), or simply a *processor*. The word *central* was originally used because most of the control functions in early computers were centralized in a sin-

*Adapted from V. C. Hamacher, Z. G. Vranesic, and S. G. Zaky, *Computer Organization*, 3rd ed., New York: McGraw-Hill, 1990. With permission.

gle processing unit. Modern systems often contain many processors as will be shown later, but the term CPU is still widely used.

Input and output equipment is usually combined under the term **input-output unit** (*I/O unit*). This is reasonable because some standard equipment provides both input and output functions. The simplest example of this is the video terminal consisting of a keyboard for input and a cathode-ray tube for output. We should emphasize that input and output functions are, of course, separated within the terminal. Thus, the computer recognizes two distinct devices, even though the human operator associates them as being part of the same physical unit.

The **memory unit** stores programs and data. There are two classes of memory devices called *primary* and *secondary* memory.

Primary storage, or main memory, is a fast electronic memory. Programs are stored in the main memory during the time that they are executed. Main memory consists of a large number of semiconductor storage cells, each capable of storing one bit of information. These cells are rarely read or written as individual cells but are instead processed in groups of fixed size called *words*. The main memory is organized so that the contents of one word can be stored or retrieved in one basic operation called a *memory cycle*.

To provide easy access to any word in the main memory, a distinct address is associated with each word location. Addresses are numbers that identify successive locations and a given word is accessed by specifying its address and issuing a control command that starts the storage or retrieval process.

The number of bits in a word is referred to as the *word length* of the computer. Word lengths vary from 8 to 32 bits in microcomputers and from 32 to 64 bits in large machines and supercomputers. See Vranesic and Zaky [1989] for information on microcomputer structures, and refer to Hayes [1988] for larger machines. Small machines may have only a few hundred thousand words in the main memory while larger machines have millions of words. The time required to access a word for reading or writing is in the range of about 50 to 500 ns for most modern computers. See Sedra and Smith [1991] for details of modern integrated-circuit memory design.

Although primary memory is essential, it tends to be expensive and volatile. Thus cheaper, more permanent, magnetic media secondary storage is used for files of information that contain either programs or data. A wide selection of suitable devices is available, including magnetic disks, drums, and tapes.

Execution of most operations within a computer takes place in the **arithmetic and logic unit** (ALU) of a processor. Consider a typical example. Suppose that two numbers located in the main memory are to be added. They are brought into the arithmetic unit where the addition is performed, and the sum may then be stored back into the main memory. Multiplication, division, or comparison of numbers are other typical operations. Processors almost always contain a number of high-speed storage elements called *registers*, which may be used for temporary storage of often-used operands. Each electronic register contains one word of data and its access time is 5 to 10 times faster than memory access time.

In concept it is reasonable to think of the *control unit* as a well-defined, physically separate unit that interacts with other parts of the machine, but in practice this is seldom the case. Much of the control circuitry is distributed throughout the machine. A large set of control lines (wires) carries the signals used for timing and synchronization of events in all units. It should also be mentioned that modern large-scale microelectronic fabrication techniques allow whole processors to be implemented on a single semiconductor chip containing more than a million transistors.

Basic Operational Concepts

To perform a given computational task, an appropriate program consisting of a set of instructions is stored in the main memory. Individual instructions are brought from the memory into the processor for execution. Data used as operands are also stored in the memory. A typical instruction may be

ADD   LOCA,R0

This instruction adds the operand at memory location LOCA to the operand in processor register R0, and then places the sum into R0. This instruction requires several individual steps to be performed. First, the instruction must be transferred from the main memory into the processor. Then the operand at location LOCA must be fetched into the processor and added to the contents of R0. Finally the resultant sum is stored into R0. Figure 80.2 shows how the connection between the main memory and the processor can be made. It also shows a few details of the processor that have not been discussed yet but are operationally essential.

The processor contains, in addition to the ALU and some storage registers that have already been mentioned, a number of registers that are needed to sequence the detailed processing steps described above. The instruction register (IR) contains the current instruction that is being executed, and its output is available to the control circuits. The control circuits generate the timing signals for controlling the processing circuits that execute the instruction. The program counter (PC) register contains the address of the instruction currently being executed. During the execution of an instruction the PC is updated to contain the address of the next instruction to be executed. The memory address register (MAR) holds the address of the memory location to or from which data are to be transferred and the memory data register (MDR) holds the data.

Instruction set design has been intensively studied in recent years to determine the effectiveness of the various alternatives. See Hennessey and Patterson [1990] for a thorough discussion.

Normal execution of programs may sometimes be preempted if some I/O device requires urgent control action or servicing. For example, a monitoring device in a computer-controlled industrial process may detect a dangerous condition that requires the execution of a special service program dedicated to the device. To cause this service program to be executed, the device activates an interrupt signal to the processor. The processor temporarily suspends the program it has been executing and executes the service routine. It then returns to the interrupted program. To appreciate the complexity of the computer system software programs needed to control such switching from one program task to another and to manage the general movement of programs and data between primary and secondary storage, consult Tanenbaum [1990].

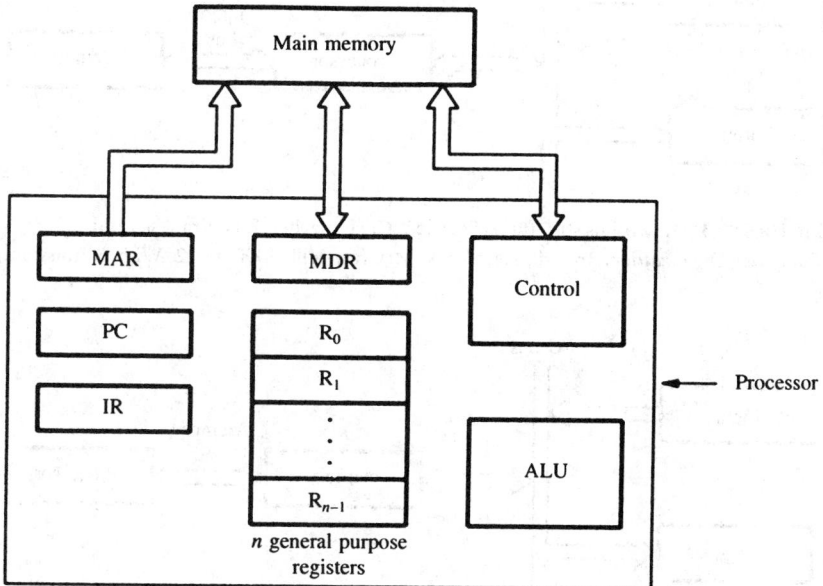

FIGURE 80.2 Connections between the processor and the main memory. (*Source:* V.C. Hamacher, Z.G. Vranesic, and S.G. Zaky, *Computer Organization,* 3rd ed., New York: McGraw-Hill, 1990, p. 10. With permission.)

Bus Structures

So far, we have discussed the functional characteristics of individual parts of a computer. To form an operational system, these parts must be connected together in an organized way. We will discuss a few standard techniques that have proven to be effective in practice. See Hayes [1988] for a more advanced treatment.

A collection of wires that enables computer system units to exchange a single word of data is called a **bus**. In addition to the data bits, the communicating devices, or a location in them, must be specified, and control signals are needed to sequence the required operations. Thus a bus consists of data, address, and control lines.

Figure 80.3 shows the simplest form of a computer system that contains two buses. The processor is central to the control of all transfers that are done with either memory or I/O devices. A somewhat more complex system that requires extra controls but leads to higher performance is shown in Fig. 80.4.

The scheme in Fig. 80.4 requires control circuitry (actually a special processor) that can handle the direct transfer of data between the memory and I/O devices, without the need for detailed control by the main processor itself. In this configuration, the main processor initiates I/O data transfer activity, then proceeds to the normal execution of programs while the data transfer takes place. The detailed control of the transfer is handled by a dedicated special-purpose processor called a *direct-memory access controller* (not shown in the figure). Assuming that accesses to the main memory from both I/O devices and the main processor can be appropriately interwoven, this configuration can potentially lead to more parallel activity, and thus higher performance, than that available in the scheme of Fig. 80.3.

The last bus scheme that must be mentioned is a single-bus system in which all units are connected to a single bus. This is the least costly possibility, but it also has lower performance because

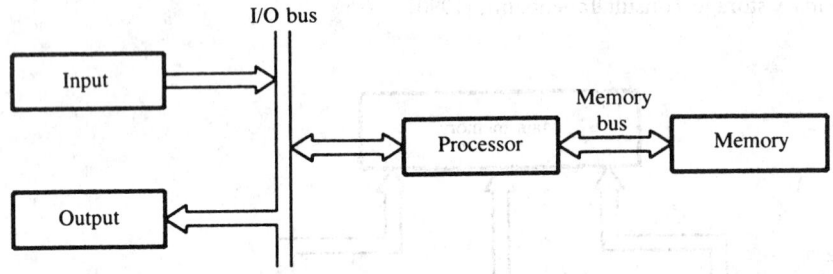

FIGURE 80.3   A two-bus structure. (*Source:* V.C. Hamacher, Z.G. Vranesic, and S.G. Zaky, *Computer Organization,* 3rd ed., New York: McGraw-Hill, 1990, p. 12. With permission.)

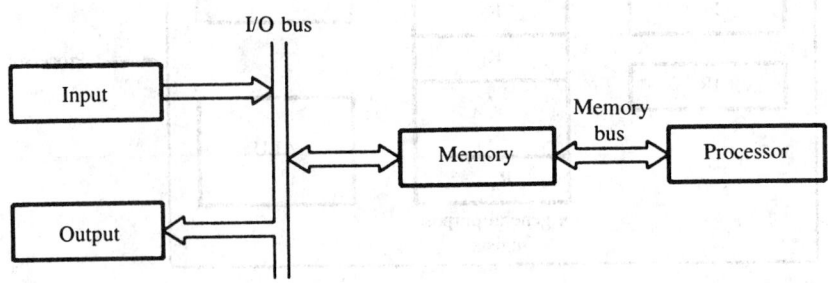

FIGURE 80.4   An alternative two-bus structure. (*Source:* V.C. Hamacher, Z.G. Vranesic, and S.G. Zaky, *Computer Organization,* 3rd ed., New York: McGraw-Hill, 1990, p. 13. With permission.)

it cannot handle as much parallel activity as the other two configurations. Nevertheless, single-bus systems have proven to be very effective because of their flexibility and have been used extensively in low-cost systems. Their flexibility is derived from the fact that all devices are attached to the single bus using the same type of control requirements, and new devices can be readily added.

Parallel and Distributed Computing

Computer systems have evolved from machines based on a single processing unit, as has been described so far, into configurations that contain a number of processors. The processors can serve different roles. An instruction set processor, for example, executes the computational steps of a program, whereas an I/O processor performs I/O tasks. It is also possible to use a number of identical processors to perform certain tasks in parallel. For example, large matrix calculations can be done on such **parallel processors.**

There are many ways in which processors can be interconnected via buses and communication lines with memory units and I/O devices to form complete computing systems. For example, a computer-aided design (CAD) system may consist of one large computer and a number of workstations. Each workstation is a computer in its own right with graphics input and output capability. A designer at one of these stations generates a description of a desired object using the graphics capabilities. The large computer provides simulation facilities that may require a large amount of computing power to test the designed object in terms of its predicted operation and effectiveness. The workstations might be dispersed throughout a building or plant site and connected to the large computer via a high-speed, bit-serial, local-area network consisting of coaxial or optical lines. Such systems have proven to be very cost-effective and are generally referred to as **distributed computing** systems.

Defining Terms

Arithmetic and logic unit: A collection of logic gates and register storage elements used to perform the basic operations of addition, subtraction, multiplication, and division of numeric operands and the comparison, shifting, and alignment operations on more general forms of numeric and nonnumeric data.

Bus: A collection of data, address, and control wires that enables the exchange of data, usually in word-size quantities, among the various computer system units. In practice, a large number of units can be connected to a single bus. These units contend in an orderly way for use of the bus for individual transfers.

Central processing unit (CPU):   The arithmetic and logic processing circuits of a digital computer, including the main control circuits needed to sequence the execution of instructions.

Computer architecture: The functional operation of the individual hardware units in a computer system and the flow of information and control among them.

Digital computer: A fast electronic calculating machine that accepts digitized input information, processes it according to a list of internally stored instructions, and produces the resultant output information. The list of instructions is called a program, and internal storage is called computer memory.

Distributed computing: A general form of using a number of processors, or complete computer systems, to accelerate the execution of some large computational task that has a number of subtasks. This situation is general in the sense that the computers do not need to be the same, and the subtasks do not need to be identical either. An example is a collection of interconnected workstations in a building or plant site, used to cooperatively process the subtasks of a large engineering design task. In such a network, all permanent program and data files may reside on a central computer, and some of the network nodes may provide high-performance numerical calculations on behalf of any individual workstation.

Input-output unit (I/O):   The equipment and controls necessary for a computer to interact with a human operator, to access mass storage devices such as disks and tapes, or to communicate with other computer systems over communication facilities.

Memory unit: Stores programs and data. The two main classes of memory devices are semiconductor chips consisting of millions of bit storage cells and magnetic disks and tapes that store orders of magnitude more bits. Semiconductor memories are called primary memory because programs and data must reside there when processing takes place; magnetic devices are called secondary memory because they contain permanent copies of program and data files when they are not being processed from primary memory.

Parallel processors: A collection of processors capable of operating in parallel on distinct operand pairs from some large vector or array of data so that total processing time for the full set of data can be reduced to the time needed to process an individual pair of values.

References

V. C. Hamacher, Z. G. Vranesic, and S. G. Zaky, *Computer Organization*, 3rd ed., New York: McGraw-Hill, 1990.

J. P. Hayes, *Computer Architecture and Organization*, 2nd ed., New York: McGraw-Hill, 1988.

J. Hennessey and D. Patterson, *Computer Architecture: A Quantitative Approach*, San Mateo, Calif.: Morgan Kaufman, 1990.

A. S. Sedra and K. C. Smith. *Microelectronic Circuits*, 3rd ed., Philadelphia: Saunders, 1991.

A. S. Tanenbaum, *Structured Computer Organization*, 3rd ed., Englewood Cliffs, N.J.: Prentice-Hall, 1990.

Z. G. Vranesic and S. G. Zaky, *Microcomputer Structures*, New York: Holt, 1989.

Further Information

The IEEE technical magazines *Computer, Micro,* and *Software* all have interesting articles on subjects either directly related to computer architecture or very closely related, including software aspects. Very readable but technically rigorous articles on computer architecture occasionally appear in *Scientific American.*

80.4 Microprogramming

Jacques Raymond

Since the 1950s when Wilkes *et. al.* [1958] defined the term and the concept, microprogramming has been used as a clean and systematic way to define the instruction set of a computer. It has also been used to define a virtual architecture out of a real hardwired one.

Levels of Programming

In Fig. 80.5, we see that a computer application is obtained by programming (generally) a given algorithm in a high-level language. A system offering a high-level language capability is implemented at the system level via a compiler. The operating system is (usually) implemented in low-level language (machine instructions). The machine instruction set can be hardwired (in a hardware implementation) or implemented via microprogramming.

Therefore, microprogramming is simply an extra level in the general structure. Since it is used to define the machine instruction set, it can be considered at the *hardware* level. Since this definition is done via a program at low level, but still eventually modifiable, it can also be considered to be at the *software* level. For these reasons, the term **firmware** has been coined to name sets of micropro-

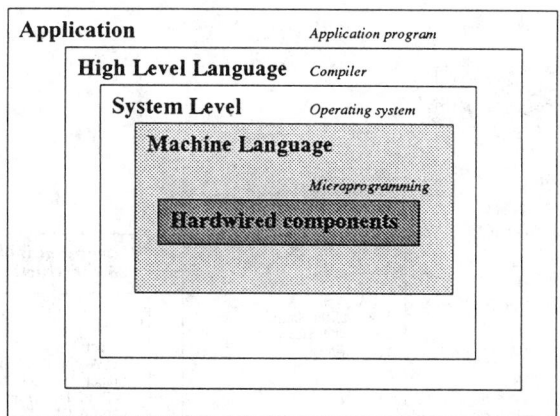

FIGURE 80.5 Levels of programming in a computer system.

grams. In short, **microinstructions** that specify hardware functions (*microoperations* such as Open a path, Select operation) are used to form a more complex instruction (Convert to binary, Add decimal). The machine instruction set is defined via a set of microprogram routines and a micropro-grammed instruction decoder.

In a microprogrammed machine, the hardware is designed in terms of its capabilities (ALU, data paths, I/O, processing units) with little concern over how these capabilities will have to be specified by the programmers. The microoperations are gated and decoded in microinstructions. The way programmers *view* the machine is defined at the microprogramming level.

This approach offers some differences over the hardwired approach. The advantages are that it is more systematic in implementation, modifiable, economical on most designs, and easier to debug. The disadvantages are that it is uneconomical on simple machines, slower, and needs support software.

Microinstruction Structure

On a given hardware, many processing functions are available. In general a subset O of these functions can be effected in parallel, for example, carrying on an addition between two registers while copying a register on an I/O bus. These functions are called **microcommands.**

Horizontal Microinstructions

Each of the fields **f** of a microinstruction specifies a microcommand. If the structure of the microinstruction is such that all possible microcommands can be specified, the instruction is called *horizontal.* Most of the time, it is wasteful in memory as, in a microprogram, not every possible microcommand is specified in each microinstruction. However, it permits the microprogrammer to fully take advantage of all possible parallelisms and to build faster machines.

For example, the horizontal specification of an ALU operation,

ALUOperation	SourcePathA	SourcePathB	ResultPath	CvtDecimal

specifies both operands, which register will contain the result, whether or not the result is to be converted to decimal, and the operation to be performed. If this instruction is, for example, part of a microprogram defining a 32-bit addition instruction, assuming a 16-bit path, it is wasteful to specify twice the source and result operands.

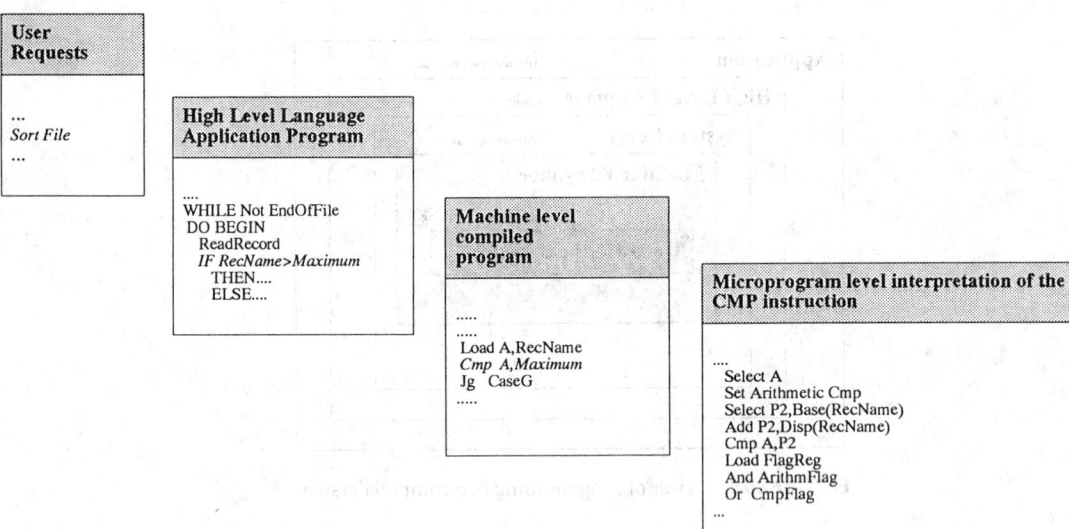

FIGURE 80.6   A view of computer system levels.

In some cases, it is possible to design a microinstruction that specifies more microcommands than can be executed in parallel. In that case, the execution of the microcommand is carried out in more than one clock cycle. For this reason they are called *polyphase* microinstructions (as opposed to *monophase*).

Vertical Microinstructions

At the other extreme, if the structure of the microinstruction allows only the specification of a single microcommand at a time, the instruction is then called *vertical*. In that case, only the necessary commands are specified, resulting in smaller **control memory** requirements. However, it is not possible to take advantage of possible parallelisms offered by the hardware, since only one microcommand is executed at a time. For example, the vertical specification of an ALU operation is as follows:

SourceA	Reg#	1st Operand
SourceB	Reg#	2nd Operand
Result	Reg#	Result
ALU	Op	Operation

Diagonal Microinstructions

Most cases fit in between these two extremes. Some parallelism is possible; however, microcommands pertaining to a given processing unit are regrouped. This results in shorter microprograms than in the vertical case and may still allow some optimization. For example, the diagonal specification of an ALU operation is as follows:

SelectSources	RegA#	RegB#	Select Operands
SelectResult	Reg#	Dec/Bin	Result place and format
Select ALU Operation			Perform the operation

Optimization

Time and space optimization studies can be performed before designing the microinstruction format. The reader is refered to Das *et al.* [1973] and Agerwala [1976] for details and more references.

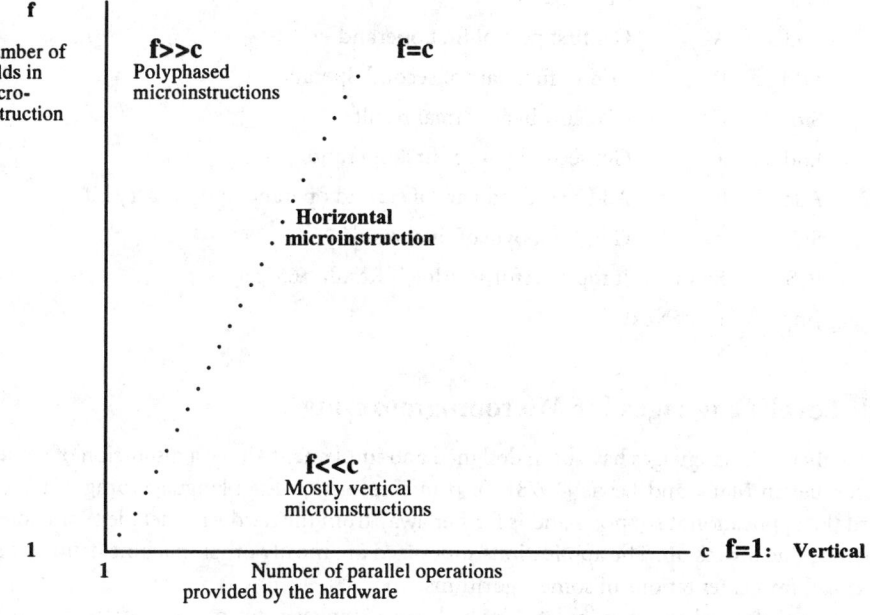

FIGURE 80.7   Microinstruction fields versus microcommands.

Microprogram Development

Microprogramming by Hand

At a very low level, and in quite special circumstances (for example, a quick patch while debugging), it might be necessary to enter some microinstructions in binary directly on some toggle switches. This is rare, and indeed it is very prone to errors with severe consequences. Reading microprograms in binary, or even in the next step higher, hexadecimal or octal, is more often done than one would wish. The abundance of high-level debugging tools has not yet eliminated completely "digging into the bits."

Microassemblers

The first level of sophistication in the specification of microinstructions is, just like its counterpart at the machine level, the assembler. Although the process and philosophy is exactly the same, it is traditionally called a microassembler. A microassembler is a software program (it is not relevant to know in which language it is written) whose function is to translate a source program into the binary code equivalent. Obviously, to write a source program, a language has to be designed. At assembly level, languages are usually very reminiscent of the hardware structure, and the objects defined are microregisters, gate level controls, and paths. Operations are the microoperations (sometimes slightly more sophisticated with a microassembler with macrofacilities).

This level provides a much better understanding and readable microprogram and does much to help avoid syntax errors. In binary, only the programmer's mind can catch a faulty 1 or 0; the microassembler can catch syntax errors or some faulty register specifications. No microassembler exists that can catch all logic errors in the implementation of a given instruction algorithm. It is still very easy to make mistakes. It should be noted that this level is a good compromise between convenience and cost.

The following is a typical example of a microprogram in the microassembler (it implements a 16-bit add on an 8-bit path and ALU):

CLC		Clear Carry
Lod	A	Get first part of first operand
Add	B	Add to first part of second operand
Sto	C	Give low byte of final result
Lod	a	Get second part of first operand
Adc	b	Add to second part of second operand and to carry bit
Sto	c	Give high byte of final result
JCS	Error	Jump to error routine if Result >65536
Jmp	FetchNext	

High-Level Languages for Microprogramming

Many higher-level languages have been designed and implemented; see a discussion of some design philosophies in Malik and Lewis [1978]. In principle, a high-level language program is oriented toward the application it supports and is farther away from the hardware-detailed implementation of the machine it runs on. The applications supported are mostly other machine definitions (**emulators**) and implementations of some algorithms.

The objects defined and manipulated by high-level languages for microprogramming are therefore the virtual components of virtual machines. They are usually much the same as their real counterparts: registers, paths, ALUs, microoperations, etc. Furthermore, writing a microprogram is usually defining a machine architecture. It involves a lot of intricate and complicated details, but the algorithms implemented are mostly quite simple. The advantages offered by high-level languages to write better algorithms without getting lost in implementation details are therefore not well used.

Firmware Implementation

Microprogramming usually requires the regular phases of design, coding, test, conformance acceptation, documentation, etc. It differs from other programming activities when it comes to the delivery. The usual final product takes the form of hardware, namely, a control memory, PROM, ROM, or other media, containing the bit patterns equivalent to the firmware. These implementation steps require special hardware and software support. They include a linker, loader, PAL programmer, or ROM burner; a test program in a control memory test bench is also necessary.

Supporting Software

It is advisable to test the microprogram before its actual hard implantation, if the implantation process is irreversible or too costly to repeat. Software simulators have been implemented to allow thorough testing of the newly developed microprogram. Needless to say these tools are very specialized to a given environment and therefore costly to develop, as their development cost cannot be distributed over many applications.

Emulation

Concept

In a microprogrammed environment, a computer is softly (or firmly) defined by the microprogram designed to implement its architecture, its data paths, and its machine-level instructions. It is easy to see that if one changes the microprogram for another one, then a new computer is defined. In this environment, the desired operation is simulated by the execution of the firmware, instead of being the result of action on real hardwired components.

Since the word *simulation* was already in use for simulation by software, the word *emulation* was chosen to mean simulation by firmware. Of course, "simulation" by hardware is not a simulation but the *real thing*.

The general structure of an emulator consists of the following pseudocode algorithm:

```
BEGIN
Initialize Machine Components
Repeat

Fetch Instruction
Emulate Operation of the current instruction
Process interrupts
Update instruction counter

Until MachineIsOff

Perform shutdown procedure
END
```

Many variations exist, in particular to process interrupts within the emulation of a lengthy operation or to optimize throughput, but the general principle and structure are fairly constant.

Emulation of CPU Operation

One of the advantages of microprogramming is that the designer can implement his or her *dream instructions* simply by emulating its operation. We have seen already the code for a typical 16-bit adder, but it is not difficult to code a parity code generator, a cyclic redundancy check calculator, or an instruction that returns the eigenvalues of an $n \times n$ matrix. This part is straight programming. One consideration is to make sure that the machine is still listening to the outside world (interruptions) or actively monitoring it (I/O flags) in order not to loose asynchronous data while looking for a particular pattern in a 1 megabyte string. Another consideration is to optimize memory usage by combining common processes for different operations. For example, emulating a 32-bit add instruction and emulating a 16-bit add instruction have common parts. This is, however, a daily programming concern.

I/O System and Interrupts

Programming support for I/Os and interrupts is more complicated than for straight machine instructions. This is due to the considerable speed differences between I/O devices and a CPU, the need for synchronization, the need for not losing any external event, and the concerns for optimizing processing time. Microprogramming offers considerable design flexibility, as these problems are more easily handled by programming than with hardware components.

Applications of Microprogramming

The main application of microprogramming is the emulation of a virtual machine architecture on a different host machine. It is, however, easy to see that the concept of emulation can be broadened to other functions than the traditional hardware operation.

It is mainly a pure matter of point of view. Emulation and simulation are essentially the same process but viewed from different levels. Realizing a 64-bit addition and implementing a communication controller are qualitatively the same type of task. Once this is considered, there are theoretically no limits to the uses of microprogramming.

From the actual programmer's point of view, programming is the activity to produce, in some language, an implementation of some algorithm. If the language is at the very lowest level, as is the case with microprogramming, and at the same time the algorithm is filled with intricate data structures and complex decisions, the task might be enormous, but nothing says it cannot be done (except, maybe, experience). With this perspective of the field, we now look at some existing applications of microprogramming.

Operating System Support

One of the first applications, besides emulation, was to support some operating system functions. Since microprograms are closer to the hardware and programming directly in microcode removes the overhead of decoding machine-level instructions, it was thought that directly coding operating system (OS) functions would improve their performance. Success was achieved in some areas, such as virtual memory. In general, people write most OS functions in assembly language, probably because the cost is not offset by the benefits, especially with rapidly changing OS versions. The problems raised by the human side of programming have changed the question "Should it be in microcode or in assembler?" to the question "Should it be in assembler or in C?" These are debates where psychology and economics have more impact than technology.

High-Level Languages Support

Early research was done also in the area of support for high-level languages. Support can be in the form of microprogrammed implementations of some language primitive (for example, the trigonometric functions) or support for the definition and processing of data structures (for example, trees and lists primitives). Many interesting research projects have led to esoteric laboratory machines. More common examples include the TRT and TR instructions (IBM 360 series), string searches and compares, or indexing multidimensional arrays.

Paging, Virtual Memory

An early and typical application of microprogramming is the implementation of the paging algorithm for a virtual memory system. It is a typical application since it is a low-level function that must be time optimized and is highly hardware dependent. Furthermore, the various maintenance functions which are required by the paging algorithms and the disk I/Os can be done during the idle time of the processing of other functions or during part of that processing in order to avoid I/O delays.

Diagnostics

Diagnostic functions have also been an early application of microprogramming. A firmware implementation is ideally suited to test the various components of a computer system, since the gates, paths, and units can be exercised in an isolated manner, therefore allowing one to precisely pinpoint the trouble area.

Controllers

Real-time controllers benefit from a microprogrammed implementation, due to the speed gained by programming only the required functions, therefore avoiding the overhead of general-purpose instructions. Since the microprogrammer can better make use of the available parallelisms in the machine, long processes can still support the asynchronous arrival of data by incorporating the interrupt polling at intervals in these processes.

High-Level Machines

Machines that directly implement the constructs of high-level languages can be easily implemented via microprogramming. For example, Prolog machines and Lisp machines have been tried. It is also possible to conceive an application directly microcoded. Although this could provide a very fast hardware, human errors and software engineering practice seem to make such a machine more of a curiosity than a maintainable system.

Defining Terms

Control memory: A memory containing a set of microinstructions (a microprogram) that defines the instruction set and operations of a CPU.

Emulator: The firmware that simulates a given machine architecture.

Firmware: Meant as an intermediate between software, which can be modified very easily, and hardware, which is practically unchangeable (once built); the word *firmware* was coined to represent the microprogram in control memory, i.e., the modifiable representation of the CPU instruction set.

High-level language for microprogramming: A high-level language more or less oriented toward the description of machine. Emulators can more easily be written in a high-level language; the source code is compiled into the microinstructions for actual implementation.

Horizontal microinstruction: Theoretically, a completely horizontal microinstruction is made up of all the possible microcommands available in a given CPU. In practice, some encoding is provided to reduce the length of the instruction.

Microcommand: An n (usually $n = 1$) bit field indicating if a gate is open or closed, if a function is enabled or not, if a control path is active or not, etc. A microcommand is therefore the specification of some action within the control structure of a CPU.

Microinstruction: The set of microcommands to be executed or not, enabled or not. Each field of a microinstruction is a microcommand. The instruction specifies the new state of the CPU.

Vertical microinstruction: A completely vertical microinstruction would contain one field and therefore would specify one microcommand. An Op code is used to specify which microcommand is specified. In practice, microinstructions that typically contain three or four fields are called vertical.

References

T. Agerwala, "Microprogram optimization: a survey," *IEEE Trans. Comput.*, vol. C25, no. 10, pp. 862–873, 1976.

J.D. Bagley, "Microprogrammable virtual machines," *Computer*, pp. 38–42, 1976.

D.K. Banerji and J. Raymond, *Elements of Microprogramming*, Englewood Cliffs, N.J.: Prentice-Hall, 1982.

G.F. Casaglia, "Nanoprogramming vs. microprogramming," *Computer*, pp. 54–58, 1976.

S.R. Das, D. K. Banerji, and A. Chattopadhyay, "On control memory minimization in microprogrammed digital computers," *IEEE Trans. Comput.*, vol. C22, no. 9, pp. 845–848, 1973.

L.H. Jones, "An annotated bibliography on microprogramming," *SIGMICRO Newsletter*, vol. 6, no. 2, pp. 8–31, 1975.

L.H. Jones, "Instruction sequencing in microprogrammed computers," *AFIPS Conf. Proc.*, vol. 44, pp. 91–98, 1975.

K. Malik and T.J. Lewis, "Design objectives for high level microprogramming languages," in *Proceedings of the 11th Annual Microprogramming Workshop*, Englewood Cliffs, N.J.: Prentice-Hall, 1978, pp. 154–160.

J. Raymond and D.K. Banerji, "Using a microprocessor in an intelligent graphics terminal," *Computer*, pp. 18–25, 1976.

M.V. Wilkes, W. Renwick, and D.J. Wheeler, "The design of the control unit of an electronic digital computer," *Proc. IEE*, pp. 121–128, 1958.

81
Programming

James M. Feldman
Northeastern University

Edward W. Czeck
Northeastern University

Ted G. Lewis
Oregon State University

Johannes J. Martin
University of New Orleans

81.1 Assembly Language 1878
   NumberCount( ) • Comparisons Down on the Factory Floor •
   Compiler Optimization and Assembly Language
81.2 High-Level Languages 1902
   What Is a HLL? • How High Is a HLL? • HLLs and Paradigms
81.3 Data Types and Data Structures 1915
   Abstract Data Types • Fundamental Data Types • Type Con-
   structors • Dynamic Types • More Dynamic Data Types • Object-
   Oriented Programming

81.1 Assembly Language

James M. Feldman and Edward W. Czeck

The true language of computers is a stream of 1s and 0s—bits. Everything in the computer, be it numbers or text or program, spreadsheet or database or 3-D rendering, is nothing but an array of bits. The meaning of the bits is in the "eye of the beholder"; it is determined entirely by context. Bits are not a useful medium for human consumption. Instead, we insist that what *we* read be formatted spatially and presented in a modest range of visually distinguishable characters. 0 and 1 arranged in a dense, page-filling array do not fulfill these requirements *in any way*. The several languages that are presented in this handbook are all intended to make something readable to two quite different readers. On the one hand, they serve the human reader with his/her requirements on symbols and layout; on the other, they provide a grammatically regular language for interpretation by a **compiler**. A compiler, of course, is normally a program running on a computer, but human beings can and sometimes do play both sides of this game. They want to play with both the input and output. Such accessibility requires that not only the input but the output of the compilation process be comfortably readable by humans. The language of the input is called a **high-level language** (**HLL**). Examples are C, Pascal, Ada and Modula II. They are designed to express both regularly and concisely the kinds of operations and the kinds of constructs that programmers manipulate. The output end of the compiler generates **object code**—a generally unreadable, binary representation of machine language, lacking only the services of a **linker** to turn it into true machine language. The language that has been constructed to represent object code for human consumption is *assembly language*. That is the subject of this section.

Some might object to our statement of purpose for assembly language. While few will contest the concept of assembly language as the readable form of object code, some see writing assembly code as the way to "get their hands on the inner workings of the machine." They see it as a "control" issue. Since most HLLs today give the user reasonably direct ways to access hardware, where does the "control" issue arise? What assembly proponents see as the essential reason for having an assembly language is the option to optimize the "important" sections of a program by doing a better job of machine code generation than the compiler does. This perspective was valid enough when compilers were mediocre optimizers. It was not unlike the old days when a car came with a complete set of tools because you needed them. The same thing that has happened to cars has happened to compilers. They are engineered to be "fuel efficient" and perform their assigned functions with remarkable ability. When the cars or compilers get good enough and complex enough, the tinkerer may do more harm than good. IBM's superscalar RISC computer—the RS6000—comes with superb compilers *and no **assembler** at all*. The Pentagon took a long look at their costs of programming their immense array of computers. Contrary to popular legend, they decided to save money. The first amendment not withstanding, their conclusion was: "Thou shalt not assemble."

The four principal reasons for *not* writing assembly language are

- Any sizable programming job gets done at least four times faster in a HLL.

- Most modern compilers are good optimizers of code; some are superb.

- Almost all important code goes through revisions—maintenance. Reworking old assembly code is similar to breaking good encryption; it takes forever.

- Most important of all is portability. To move any program to a different computer, you must generate machine code for that new platform. With a program in a HLL, a new platform is almost free; all it requires is another pass through the compiler for the target platform. With assembly code, you are back to square one. Assembly code is unique to the platform.

Given all of that, the question naturally arises: Why have an article on assembly language? We respond with two reasons, both of which we employ in our work as teachers and programmers:

- An essential ingredient in understanding computer hardware and in designing new computer systems and compilers is a detailed appreciation of the operations of central processing units (CPUs). These are best expressed in assembly language. Our undergraduate Computer Engineering courses include a healthy dose of assembly language programming for this specific reason.

- If you are concerned about either CPU design or compiler effectiveness, you have to be able to look in great detail at the interface between them—*machine language*. As we have said, the easiest way to read machine language is by translating it to assembly language. This is one way to get assembly language, not by writing in it as a source of code but by running the object code itself through a backward translator called a **disassembler**. While many compilers will oblige you by providing an assembly listing if asked, often that listing does not include optimizations that occur only when the several modules are linked together, providing opportunities for truly global optimization. Some compilers "help" the reader by using **macros** (names for predefined blocks of code) in place of the real machine instructions and register assignments. The absence of the optimizations and the inclusion of unexpected macros can make the assembly listing almost useless for obtaining insight into the program's fine detail. The compilers that we have used on the DECstations and SPARC machines do macro inclusion. To see what is really going on in these machines, you must disassemble the machine code. That is precisely what the Think C® compiler on the Macintosh does when you ask for machine code. It disassembles what it just did in compiling and linking the whole program. What you see is what is really there. The code we present for the 68000 was obtained in that way.

These are important applications. Even if most or all other programming needs can be better met in HLLs, these applications are sufficient reason for many engineers to want to know something about assembly language.

There are other applications of assembly language, but they tend to be specific to rather specialized and infrequent tasks. For example, the back end of most HLL compilers is a machine code generator. To write one of those, you certainly must know something about assembly language. On rare occasions, you may find some necessary machine-specific transaction which is not supported by the HLL of choice or which requires some special micro optimization. A "patch" of assembly code is a way to fit this inexpressible thought into the program's vocabulary. These are rare events. The reason why we recommend to you this section on assembly code is that it improves your under-standing of HLLs and of computer architecture.

We will take a single subroutine which we express in C and look at the machine code that is generated on two representative machines. The machines include two widely used *complex instruc-tion set computers* (**CISCs**) and one *reduced instruction set computer* (**RISC**). These are the 68000®, the VAX®, and a SPARC®. We will have two objectives:

- To see how a variety of paradigms in HLLs are translated (or, in other words, to see what is really going on when you ask for a particular HLL operation)

- To compare the several architectures to see how they are the same and how they differ

The routine attempts to get a count of the number of numbers which occur in a block of text. Since we are seeking numbers and not *digits*, the task is more complex than you might first assume. This is why we say "attempts." The function that we present below handles all of the normal text forms:

- Integers, such as 123 or –17

- Numbers written in a fixed-point format, such as 12.3 or 0.1738

- Numbers written in a floating-point format, such as –12.7e+19 or 6.781E2

If our program were to scan the indented block of code above, it would report finding six numbers. The symbols that the program recognizes as potentially part of a number include the digits 0 to 9 and the symbols 'e', 'E', ' . ', '–' and '+'. Now it is certainly possible to include other symbols in legitimate numbers, such as HEX numbers or the like, but this little routine will not properly deal with them. Our purpose was not to handle all comers but to provide a routine with some variety of expression and possible application. Let us begin.

NumberCount( )

We enter the program at the top with one pointer passed from the calling routine and a set of local variables comprising two integers and eight Boolean variables. Most of the Boolean variables will be used in pairs. The first element of a pair, for instance, *ees* of *ees* and *latche*, indicates that the cur-rent character is one of a particular class of non-numeric characters which might be found inside a number. If you consider that the number begins at the first digit, then these characters can occur legally only once within a given number. *ees* will be set TRUE if the current character is the *first instance* of either 'e' or 'E'. The paired variable, *latche*, is set TRUE if there has ever been one of those characters in the current number. The other pairs are *period* and *latchp* and *sign* and *latchs*.

There is also a pair of Booleans which indicate if the current character is a *digit* and if the scan-ner is currently *inside* a number. Were you to limit your numbers to integers, these two are the only Booleans which would be needed. At the top of the program, all Booleans are reset (made FALSE). Then we step through the block looking for numbers. The search stops when we encounter the first null [char(0)] or reach the end of the block. Try running through the routine with text containing the three forms of number. You will quickly convince yourself that the routine works with all nor-mal numbers. If someone writes "3..14" or "3.14ee6", the program will count 2 numbers. That is probably right in the first case. Who knows in the second?

Let us look at this short routine in C.

```c
int NumberCount(char block[])
{ int count=0,inside=0,digit;
  int ees=0, latche=0, latchp=0, period=0, latchs=0, sign=0;
  char *source;

  source = block;
  while (*source!='\0'){
     digit = (*source >= '0') && (*source <= '9');
     period = (*source=='.') && inside && !latchp;
     latchp = (latchp || period);
     ees = ((*source=='E') || (*source=='e')) && inside && !latche;
     latche = (latche || ees);
     sign = ((*source=='+') || (*source=='-')) && inside && latche && !latchs;
     latchs = (latchs || sign);
     if (inside) {
        if (!(digit || ees || period || sign)) inside=latchp=latche=latchs=0;
     }
     else if (digit) {
        count++;
        inside = 1;
     }
  source++;
  }
  return count;
}
```

To access values within the character array, the normal C paradigm is to step a pointer along the array. *Source* points at the current character in the array; *source is the character ("what *source* points at"). *source* is initialized at the top of the program before the loop (source = block;) and incremented (source++;) at the bottom of the loop. Note the many repetitions of *source*. Each one means the same current character. If you read that expression as *the character which source is pointing to*, it looks like an invitation to fetch the same character from memory eight times. A compiler that optimizes by removing *common subexpressions* should eliminate all but the first such fetch. This optimization is one of the things that we want to look for.

For those less familiar with C, the meanings of the less familiar symbols are:

==	equal (in the logical sense)
!	not
!=	not equal
&&	and
||	or
count++	increment *count* by 1 unit (after using it)

C uses 0 as FALSE and anything else as TRUE.

Comparisons Down on the Factory Floor

Now let us see what we can learn by running these two programs through their respective compilers on several quite different hosts. The items that we wish to examine include:

    I.   Subroutine operations comprising:
        A. Building the call block
        B. The call itself
        C. Obtaining memory space for local variables
        D. Accessing the call block
        E. Returning the function value
        F. Returning to the calling routine

    II. Data operations
       A. Load and store
       B. Arithmetic
       C. Logical
       D. Text
   III. Program control
       A. Looping
       B. if and the issue of multiple tests

Our objectives are to build three quite different pictures:

- An appreciation for the operations underlying the HLL statements
- An overview of the architectures of several important examples of CISC and RISC processors
- An appreciation for what a HLL optimizer should be doing for you

We will attempt to do all three all of the time.

Let us begin with the calling operations. Our first machine will be the MC68000, one of the classical and widely available CISC processors. It or one of its progeny is found in many machines and forms the heart of the Macintosh and the early Sun workstations. Programmatically, the 68000 family shares a great deal with the very popular VAX family of processors. Both of these CISC designs derive in rather linear fashion from DEC's PDP-11 machines that were so widely used in the 1970s. Comparisons to that style of machine will be done with the SPARC, a RISC processor found in Sun, Solbourne, and other workstations.

Memory and Registers

All computers will have data stored in memory and some space in the CPU for manipulating data. Memory can be considered to be a long list of bytes (8-bit data blocks) with *addresses* (locations in the list) spanning some large range of numbers from 0 to typically 4 billion (4 GB). The memory is constructed physically by grouping chips so that they appear to form enormously deep columns of bytes, as shown in Fig. 81.1. Since each column can deliver one byte on each request, the number of adjacent columns determines the number of bytes which may be obtained from a single request. Machines today have 1, 2, 4, or 8 such columns. (Some machines, the 68000 being our current example, have only 2 columns but arrange to have the CPU ask for two successive transfers to get a total of 4 bytes.) In general, the CPU may manipulate in a single step a datum as wide as the memory. For all of the machines which we will consider, that maximum datum size is 32 bits or 4 bytes. While convention would have us call this biggest datum a *word*, historical reason has led both the VAX and MC68000 to call it a *longword*. Then, 2 bytes is either a *halfword* or a *word*. We will use the VAX/68000 notation (longword, word, and byte) wherever possible to simplify the reading. To load data from memory, the CPU sends the address and the datum size to the memory and gets the datum as the reply. To store data, the address is sent and then the datum and datum size.

Some machines require that the datum be properly aligned with the stacking order of the memory columns in Fig. 81.1. Thus, on

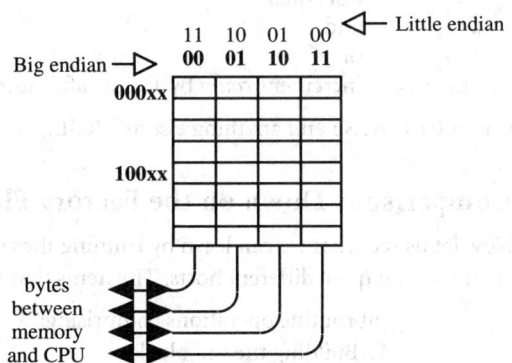

FIGURE 81.1 Memory arranged as 4 columns of bytes. The binary addresses are shown in the two formats widely used in computers. The illustration shows only 32 bytes in a 4 × 8 array, but a more realistic span would be 4 × 1,000,000 or 4 × 4,000,000 (4 MB to 16 MB).

the SPARC, a longword must have an address ending in 00 (xxx00 in Fig. 81.1), and a word address must end in 0. The programmer who arranges to violate this rule will be greeted with an **address error**. Since the MC68000 has only two columns, it complains only if you ask for words or long-words with odd addresses. Successor models of that chip (68020, 30, and 40), like the VAX, accept any address and have the CPU read two longwords and do the proper repacking.

The instruction explicitly specifies the size and indicates how the CPU should calculate the address. An instruction to load a byte, for example, is LB, MOVE.B, or MOVB on the SPARC, MC68000, and VAX, respectively. These are followed immediately by an expression which specifies an address. We will discuss how to specify an address later. First, we must introduce the concept of a register.

The space for holding data and working on it in the CPU is the *register set*. Registers are a very important resource. Bringing data in from memory is quite separate from any operations on that data. Data in memory must first be fetched, then acted upon. Data in registers can be acted on immediately. Thus, the availability of registers to store very active variables and intermediate results makes a processor inherently faster. In some machines, most or all of the registers are tied to specific uses. The most prevalent example would be Intel's 80x86 processors, which power the ubiquitous PC. Such architectures, however, are considered quite old-fashioned. All of the machines that we are considering are of a type called *general register machines* in that they have a large group of registers which may be used for any purpose. The machines that we include have either 16 or 32 registers, with only a few tied to specific machine operations.

Table 81.1 shows the general register resources in the three machines. The SPARC is a little strange. The machine provides eight *global* registers and then a *window blind* of 128 registers which sits behind a frame which exposes 24 of the 128. A program can ask the machine to raise or lower the blind by 16 registers. That leaves an overlap of eight between successive yanks or rewinds. This arrangement is called a *multiple overlapping register set* (MORS). If you think of starting with register r8 at the bottom and r31 at the top, a yank of 16 on the blind will now have r47 at the top and r24 at the bottom. r24 to r31 are shared between the old set and the new. To avoid having to keep track of which registers are showing, the set of 24 are divided into what came *in* from the last set, those that are only *local*, and those that will go *out* to the next set. These names apply to going toward increasing numbers. In going the other direction, the *ins* of the current set will become the

Table 81.1   General Registers in the Three Machines

	Reg	Special	Names	Comments
MC68000	16	1	D0..D7 A0..A7	A(ddress) register operations are 32 bits wide. Address generation uses A registers as bases. D (data) registers allow byte, word, and longword operations. A7 is SP.
VAX	16	4	r0..r11 AP, FP, SP, PC	AP,FP,SP and PC hold the addresses of the argument block, the frame, the top of the stack and the current place in the program, respectively. All data instructions can use any register.
SPARC	32 (136)	4	zero, g1..g7, i0..i5, FP, RA, l0..l7, o0..o5, SP, o7	The 4 groups of eight registers comprise: global (g), incoming parameters (i), local (l) and outgoing parameters (o). g0 is a hardwired 0 as a data source and a wastebasket as a destination. The registers are arranged as a window blind (see text) with the g's always visible and the others moveable in multiple overlapping frames of 24.

The special registers are within the set of general registers. Where a PC is not listed, it exists as a special register and can be used as an address when the program uses *program-relative* addressing.

outs of the next set. Almost all other machines keep their registers screwed down to the local masonry, but you will see in a moment how useful a MORS can be. (Like other useful but expensive accessories, the debate is always on whether it is worth it [Patterson and Hennessy, 1989].)

Stack.   Most subroutines define a number of local variables. NumberCount in C, for example, defines 10 local variables. While these local variables will often be created and kept in register, there is always some need for a bit of memory for each *invocation of* (call to) a subroutine. In the "good old days," this local storage was often tied to the block of code comprising the subroutine. However, such a fixed block means that a subroutine could never call itself or be called by something that it called. To avoid that problem (and for other purposes) a memory structure called a *stack* was invented which got its name because it behaved like the spring-loaded plate stack in a restaurant. Basically, it is a *last-in-first-out* (LIFO) structure whose top is defined by a pointer (address) which resides in a register commonly called the *stack pointer* or SP.

Heap.   When a subroutine needs space to store local variables, it acquires that space on the stack. When the subroutine finishes, it returns that stack space for use by other routines. Thus, local variable allocations live and die with their subroutines. It is often necessary to create a data structure which is passed to other routines whose lives are independent of the creating routine. This kind of storage must be independent of the creator. To meet this need, the *heap* was invented. This is an expandable storage area managed by the system. You get an allocation by asking for it [*malloc (structure_size)* in C]. You get back a pointer to the allocation and the routine can pass that pointer to any other routine and then go away. When it comes time to dispose of the allocation—that is, return the space for other uses—the program must do that actively by a deallocation call [*free(pointer)* in C]. Thus, one function can create a structure, several may use it, and another one can return the memory for other uses, all by passing the pointer to the structure from one to another.

  Both heap and stack provide a mechanism to obtain large (or small) amounts of storage dynamically. Thus, large structures which are created only at run time need not have static space stored for them in programs that are stored on disk nor need they occupy great chunks of memory when the program does not need them. Dynamic allocation is very useful and all modern HLLs provide for it.

  Since there are two types of dynamic storage, there must be some way to lay out memory so that unpredictable needs in either stack or heap can be met at all times. The mechanism is simplicity itself. The program is stuffed into low addresses in memory along with any static storage (e.g., globals) which are declared in the program. The entire remaining space is then devoted to dynamic storage. The heap starts right after the program and grows toward higher addresses; the stack goes in at the top of memory and grows down. The system is responsible to see that they never collide (a *stack crash*). When it all goes together, it looks like Fig. 81.2 [Aho *et al.*, 1986].

  There is one last tidbit that an assembly programmer must be aware of in looking at memory. Just as some human alphabets are written left to right and some right to left (not to mention top to bottom), computer manufacturers have chosen to disagree on how to arrange words in memory. The two schemes are called *big-endian* and *little-endian* (after which end of a number goes in the lowest-numbered byte and also after a marvelous episode in *Gulliver's Travels*). The easiest way to perceive how it is done in the two systems is to think of all numbers as being written in conventional order (left to right), but for big-endian you start counting on the upper left of the page and on little-endian you start counting on the

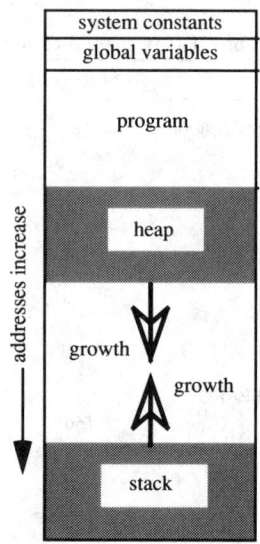

FIGURE 81.2   Layout of a program, static storage, and dynamic storage in memory.

upper right (see Fig. 81.1). Since each character in a text block is *a number* of length 1 byte, this easy description makes big-endian text read in normal order (left to right) but little-endian text reads from right to left. Figure 81.3 shows the sentence "This is a sentence" followed by the two hexadecimal (HEX) numbers 01020304 and 0A0B0C0D written to consecutive bytes in the two systems. Why must we bring this up? Because anyone working in assembly language must know how the bytes are arranged. Furthermore, two of the systems we are considering are big-endian and one (the VAX) is little-endian. Which is the better system? Either one. It is having both of them that is a nuisance.

As you look at Fig. 81.3, undoubtedly you will prefer big-endian, but that is only because it appeals to your prejudices. In truth, either works well. What is important is that you be able to direct your program to go fetch the item of choice. In both systems, you use the lowest-numbered byte to indicate the item of choice. Thus, for the number 01020304, the address will be 13. For the big-endian system, 13 will point to the byte containing 04 and for the little-endian system, it will point at the byte containing 01.

Figure 81.3 contains a problem for some computers which we alluded to in the discussion of Fig. 81.1. We have arranged the bytes to be four in a row as in Fig. 81.1. That is the way that the memory is arranged in two of our three machines. (In the 68000, there are only two columns.) A good way to look at the fetch operation is that the memory always delivers a whole row and then the processor must acquire the parts that it wants and then properly arrange them. (This is the effect if not always the method.) Some processors—the VAX being a conspicuous example—are willing to undertake getting a longword by fetching two longwords and then piecing together the parts that it wants. Others (in our case, the 68000 and the SPARC) are not so accommodating. Those machines opt for simplicity and speed and require that the program keep its data aligned. To use one of those machines, you (or the compiler or assembler) must rearrange Fig. 81.3 by inserting a null byte into Fig. 81.2. This modification is shown in Fig. 81.4. With this modification, all three machines could fetch the two numbers in one operation without rearrangement.

Look closely at the numbers 01020304 and 0A0B0C0D in Fig. 81.4. Notice that for both configurations, the numbers read from left to right and that (visually) they appear to be in the same place. Furthermore, as pointed out in the discussion of Fig. 81.3, the "beginning" or address of each of the numbers is identical. However, the byte that is pointed at by the address is not the same and the internal bytes do not have the same addresses. Getting big-endian and little-endian machines in a conversation is not easy. It proves to be even more muddled than these figures suggest. A delightful and cogent discussion of the whole issue is found in Cohen [1981].

The principal objective in this whole section has been accomplished if looking at Fig. 81.4 and given the command to load a byte from location 0000 0019, you get the number 0B in the big-endian machine and 0C in the little-endian machine.

00			03
T	h	i	s
i	s		
a		s	e
n	t	e	n
c	e	.	01
02	03	04	0A
0B	0C	0D	
18			1B

Big Endian
(SPARC, MC68000)

03			00
s	i	h	T
	s	i	
e	s		a
n	e	t	n
04	.	e	c
0D	01	02	03
	0A	0B	0C
1B			18

Little Endian
(VAX)

FIGURE 81.3 Byte numbering and number placement for big- and little-endian systems. Hexadecimal numbers are used for the memory addresses.

00			03
T	h	i	s
i	s		
a		s	e
n	t	e	n
c	e	.	00
01	02	03	04
0A	0B	0C	0D

18   1B

Big Endian

(SPARC, MC68000)

03			00
s	i	h	T
		s	i
e	s		a
n	e	t	n
00	.	e	c
01	02	03	04
0A	0B	0C	0D

1B   18

Little Endian

(VAX)

FIGURE 81.4 The same items as in Fig. 81.3, but with justification of the long integers to begin on a longword boundary.

If you are not already familiar with storing structures in memory, look at the string (sentence) and ask how those letters get in memory. To begin with, every typeable symbol and all of the unprintable actions such as tabbing and carriage returns have been assigned a numerical value from the *ASCII code*. Each assignment is a byte-long number. What "This" really looks like (HEX, left to right) is 54 68 69 73. The spaces are HEX 20; the period 2E. With the alignment null byte at the end, this list of characters forms a proper C string. It is a structure of 20 bytes. A structure of any number of bytes can be stored, but from the assembly point of view, it is all just a list of bytes. You may access them two at a time, four at a time, or one at a time. Any interpretation of those bytes is entirely up to the program. Unlike the HLL which requires that you tell it what each named variable is, assembly language knows only bytes and groups of bytes. In assembly language, the "T" can be thought of as a letter or the number 54 (HEX). Your choice. Or, more importantly, your program's choice.

Addressing

Now that we have both memory and addresses, we should next consider how these processors require that programmers specify the data that is to be acted upon by the instructions.

All of these machines have multiple modes of address. The VAX has the biggest vocabulary; the SPARC the smallest. Yet all can accomplish the same tasks. Four general types of address specification are quite common among assembly languages. These are shown in Table 81.2. They are spelled out in words in the table, but their usage is really developed in the examples which follow in this and the succeeding sections.

In Table 81.2, formats 1.4 and 1.5 and the entries in 4 require some expansion. The others will be clear in the examples we will present. Base-index addressing is the mechanism for dealing with subscripts. The base points at the starting point of a data structure, such as a string or a vector; the index measures the offset from the start of the structure to the element in question. For most machines, the index is simply a separate register which counts the bytes from the base to the item in question. If the items in the list are 4 bytes long, then to increment the index, you add 4. While that is not hard to remember, the VAX does its multiplication by the item length for you. Furthermore, it allows you to index any form of address that you can write. To show you what that means, consider expanding numbers stored in words into numbers stored in longwords. The extension is to preserve sign. The VAX provides specific instructions for conversions. If we were moving these words in one array to longwords in another array, we would write:

CVTWL (r4)[r5],(r6)[r5]   ;convert the words starting at (r4) to longwords starting at (r6)

Note that the same index, [r5], is used for both arrays. On the left, the contents of r5 are multiplied by 2 and added to r4 to get the address; on the right, the address is r5*4+r6. You would be saying: "Convert the 4th word to the 4th longword." This is undoubtedly compact and sometimes convenient. It is also unique to the VAX.

Table 81.2 Addressing Modes

1.	**Explicit addresses**	**Example**	
	1.1. Absolute addressing	765	The actual address written into the instruction.
	1.2. Register indirect	(r3)	Meaning "the address is in register 3."
	1.3. Base-displacement	−12(r3)	Meaning "12 bytes before the address in register 3."
	1.4. Base-index	(r3,r4)	Meaning make an address by adding the contents of r3 and r4. This mode has many variations which are discussed below.
	1.5. Double indirect	@5(r4)	Very uncommon! Means calculate an address as in 1.3, then fetch the longword there, and then use it as the address of what you really want.
2.	**Direct data specification**		
	2.1. Immediate/literal	#6 or 6	Meaning "use 6 as the datum." In machines which use #6, 6 without # means address 6. This is called "absolute addressing."
3.	**Program-relative**		
	3.1. Labels	loop:	The label (typically an alphanumeric ending in a colon) is a marker in the program which the assembler and linker keep track of. The common uses are to jump to a labeled spot or to load labeled constants stored with the program.
4.	**Address-modifying forms** (CISC only)		
	4.1. Postincrement	(sp)+	Same as 1.2 except that, after the address is used, it is incremented by the size of the datum in bytes and returned to the register from which it came.
	4.2. Predecrement	−(sp)	The value in SP is decremented by the size of the datum in bytes, used as the address and returned to the register from which it came.

For the 68000, the designers folded both base-displacement and base-index into one mode and made room for word or longword indices. It looks like:

```
add.1 64(A3,D2.w),D3        ;address = (A3+64) +sign-extended(D2)
```

The 68000 limits the displacement to a signed byte, but other than that, it is indeed a rather general indexing format. If you do not want the displacement, set it to 0.

For the powerful but simple SPARC, the simple base-index form shown in 1.4 is all that you have (or need).

The double-indirect format, 1.5, is so rarely used that it has been left out of almost all designs but the VAX. What makes it occasionally useful is that subroutines get pointers to "pass by pointer" variables. Thus, if you want to get the variable, first you must load the address and then the variable. The VAX allows you to do this in one instruction. While that sounds compact, it is expensive in memory cycles. If you want to use that pointer again, it pays to have it in register.

The two items under heading 4 are strange at first. Their principal function is adding items to and removing them from a dynamic stack, or for C, to execute the operation *X++ or *(−−X). The action may be viewed with the code below and the illustration of memory in Fig. 81.2:

```
movl r4, −(sp)        ;make room on the stack (subtract 4 from SP) and put
                      the contents of r4 in that spot

movl (sp)+, r4        ;take a longword off the stack, shorten the stack by 4
                      bytes, and put the longword in r4
```

RISCs abhor instructions which do two unrelated things. Instead of using a dynamic stack, they use a quasi-static stack. If a subroutine needs 12 bytes of stack space, it explicitly subtracts 12 from SP. Then it works from there with the base-displacement format (1.3) to reference any place in the block of bytes just defined. If you want to use a pointer and then increment the pointer, RISCs will do that as two independent instructions.

Let us consider one short section of MC68000 code from our sample program in C to see how these modes work and to sample some of the flavor of the language:

;ees = ((*source=='E') || (*source=='e')) && inside && !latche;

```
           CMPI.B    #$45,(A4)     ; 'E'      "compare immediate"        literal hex 45,
                                                                            what A4 points at
           BEQ       first         ;          "branch if equal"          to label first
           CMPI.B    #$65,(A4)     ; 'e'      "compare immediate"        literal hex 65, what
                                                                            A4 points at
           BNE       second        ;          "branch if not equal"      to label second
first:
           TST.W     D6            ;          "test word" (subtract 0)   D6 ('inside')
           BEQ       second        ;          "branch if equal"          to label second
           TST.W     D3            ;          "test word" (subtract 0)   D3 ('latche')
           BEQ       third         ;          "branch if equal"          to label third
second:
           MOVEQ     #00,D0        ;          "move quick"               literal 0 to D0
           BRA       fourth        ;          "branch always"            to label fourth
third:
           MOVEQ     #$01,D0       ;          "move quick"               literal 1 to D0
fourth:
           MOVE.W    D0,-6(A6)     ;          "move word"                from D0 to -6(FP)
```

There are all sorts of little details in this short example. For example, a common way to indicate a comment is to start with a ";". The assembler will ignore your comments. The "#" indicates a literal, and the "$" that the literal is written in hexadecimal notation. The VAX would use #^x to express the same idea. "Compare" means "subtract but save only the **condition codes** of the result" (*v* or *overflow*, *n* or *negative*, *z* or *zero*, and *c* or *carry*). Thus, the first two lines do a subtraction of whatever A4 is pointing at (*source) from the ASCII value for 'E' and then, if the two were equal (the result, zero), the program jumps to line 5. If *source is not 'E', then it simply goes to the next line, line 3. The instruction, TST.W D6, is quite equivalent to CMPI.W D6, #0, but the TST instruction is inherently shorter and faster. On a SPARC, where it would be neither shorter nor faster, TST does not exist.

Exactly what the assembler or linker does to replace the label references with proper addresses, while interesting, is not particularly germane to our current topic. Note that the range of the branch is somewhat limited. In the 68000, the maximum branch is ±32K and in the VAX a mere +127 to −128. If you need to go further, you must combine a branch with a jump. For example, if you were doing BEQ farlabel, you would instead do:

```
           BNE       nearlabel
           jmp       farlabel    ;    this instruction can go any distance
           nearlabel:
```

Follow through the example above until the short steps of logic and the addressing modes are clear. Then progress to the next section where we use the addressing modes to introduce the general topic of subroutine calling conventions.

Calling Conventions

Whenever you invoke a subroutine in a HLL, the calling routine (*caller*) must pass to the called routine (*callee*) the parameters that the subroutine requires. These parameters are defined at compile time to be either *pass-by-value* or *pass-by-pointer* (or *pass-by-reference*), and they are listed in some particular order. The convention for passing the parameters varies from architecture to architecture and HLL to HLL, but basically it always consists of building a *call block* which contains all of the parameters and which will be found where the recipient expects to find it.

Along with the passing of parameters, for each system, a convention is defined for register and stack use which establishes:

- Which registers must be returned from callee to caller with the same contents that the callee received (such registers are said to be *preserved across a call*)
- Which registers may be used without worrying about their contents (such registers are called *scratch registers*)
- Where the return address is to be found
- Where the value returned by a function will be found

The convention may be supported by hardware or simply a gentlemanly rule of the road. However the rules come into being, they define the steps which must be accomplished coming into and out of a subroutine. The whole collection of such rules forms the *calling convention* for that machine. In this section, we look at our three different machines to see how all accomplish the same tasks but by rather different mechanisms.

The two CISCs do almost all of their passing and saving on the stack. The call block will be built on the stack; the return address will be put on the stack; saved registers will be put on the stack. Only a few stack references are passed forward in register; the value returned by the function will be passed back in register.

How different is the SPARC! The parameters to be passed are placed in the *out* registers (six are available for this purpose). Only the overflow, if any, would go on the stack. In general, registers are saved by window-blinding rather than moving them to the stack. On return, data is returned in the *in* registers and the registers restored by reverse window-blinding.

MC68000 Call and Return. Let us look at the details for two of the machines. We start with the 68000, because that is the most open and "conventional." We continue with the function Number-Count. Only a single parameter must be passed—the pointer to the text block. The HLL callee sees NumberCount(block) as an integer (i.e., what will be returned), but the assembly program must do a call and then use the returned integer as instructed. A typical assembly routine would be:

```
MOVE.L  A2,-(SP)     ; move pointer to block onto the stack
JSR     NumberCount  ; save return address on the stack and start
                     ; executing NumberCount
                     ; do something with value returned in D0
```

The first instruction puts the pointer to the block, which is in A2, on the stack. It first must make room, so the "−" in −(A7) first subtracts 4 from A7 (making room for the longword) and then moves the longword into the space pointed to by the now-modified A7. The one instruction does two things: the decrementing of SP and the storing of the longword in memory.

```
MOVE.L  A2,-(A7)    A7 ⇐ A7-4   ;A7 = SP
                    M(A7) ⇐ A2  ;M(x) = memory(address x)
```

The next instruction, *jump subroutine* (JSR), does three things. It decrements SP (i.e., A7) by 4, stores the return address on the top of the stack, and puts the address of NumberCount in the *program counter*. We have just introduced two items which need specific definition:

Return address (RA):   This will always be the address of the instruction which the callee should return to. In the 68000 and the VAX (and all other CISCs), the RA points to the first instruction after the JSR. In the SPARC and almost any RISC, RA will point to the second instruction after JSR. That curious difference will be discussed later.

Program counter (PC):   This register (which is a *general register* on the VAX but a special register on the other machines) points to the place (memory location) in the machine language instruction stream where the program is currently operating. As each instruction is fetched, the PC is automatically incremented. The action of the JSR is to save the last version of the PC—the one for the next fetch—and replace it with the starting address of the routine to be jumped to.

Summing up these transactions in algebraic form:

```
JSR NumberCount  SP ⇐ SP-4   ;A7 = SP
                 M(SP) ⇐ PC   ;M(x) = memory(address x)
                 PC ⇐ address of NumberCount
```

Should you wonder how the address of NumberCount gets in there, the *linker*, which assigns each section of code to its proper place in memory and therefore knows where all the labels are, will insert the proper address in place of the name.

This completes the call as far as building the call block, doing the call itself, and picking up the result. Had there been more parameters to pass, that first instruction would have been replicated enough times to *push* all of the parameters, one at a time, onto the stack. Now let us look at the conventions from the point of view of the callee. The callee has more work.

When the callee picks up the action, the stack and registers are as shown in Figure 81.5. With the exception of D0 and A7, the callee has no registers . . . yet. The callee must make room for local variables in either register or memory. If it wants to use registers, it must save the user's data from

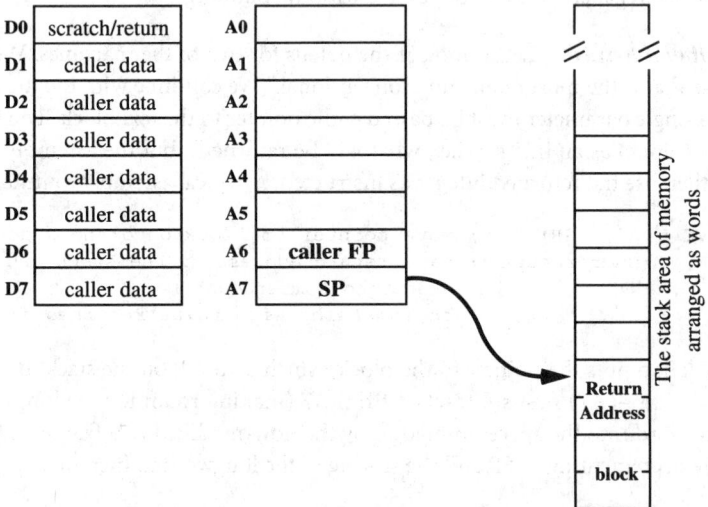

FIGURE 81.5 The stack area of the 68000's memory and the register assignments that the called subroutine sees as it is entered at the top. The registers all hold longwords, the size of an address. In typical PC/Macintosh compilers, integers are defined as 16-bit words. Accordingly, the stack area of memory is shown as words, or half the width of a register.

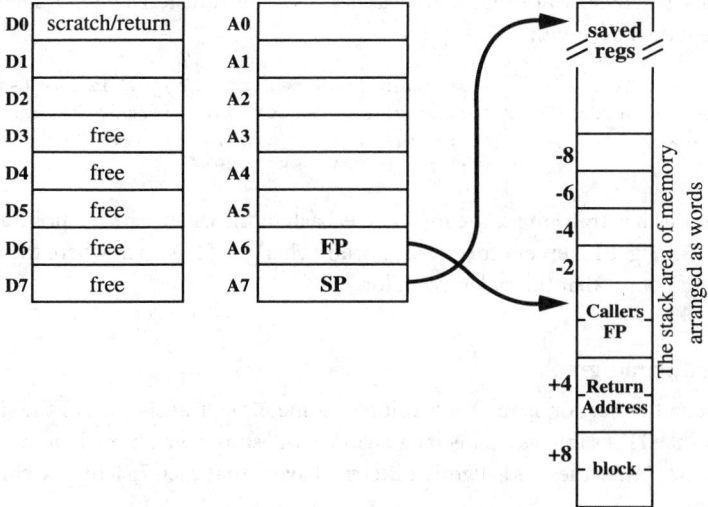

FIGURE 81.6   The stack area of the 68000's memory and the register situation just after MOVEM has been executed. The memory area between the two arrows is the subroutine's *frame.*

the registers. The subroutine can get whatever space it needs on the stack. Only after the setup will it get down to work. The entire section of stack used for local variables and saving registers is called the callee's frame. It is useful to have a pointer (FP) to the bottom of the frame to provide a static reference to the return address, the passed parameters, and the subroutine's local variable area on the stack. In the 68000, the convention is to use A6 as FP. When our routine, NumberCount, begins, the address in A6 points to the start of the caller's frame. The first thing the callee must do is to establish a local frame. It does that with the instruction LINK.

Typical of a CISC, each instruction does a large piece of the action. The whole entry operation for the 68000 is contained in two instructions:

```
    LINK      A6,#$FFF8
    MOVEM.L   D3-D7/A4,-(SP)
```

The first instruction does the frame making; the second does the saving of registers. There are multiple steps in each. Each double step of decrementing SP and moving a value onto the stack will be called a *push.* The steps are as follows:

```
    LINK      A6,#$FFF8        ;push A6   (A7 < A7-4, M(A7) < A6)
                               ;move A7 to A6 (SP to FP)
                               ;add FFF8 (-8) to SP (4 words for local variables)
    MOVEM.L   D3-D7/A4,-(A7)   ;push 5 data registers (3..7) and 1 address
                               ;register (A4)
```

At this point, the stack looks like Fig. 81.6.

The subroutine is prepared to proceed. How it uses those free registers and the working space set aside on the stack is the subject of the section on optimization in this chapter. For the moment, however, we simply assume that it will do its thing in exemplary fashion, get the count of the numbers, and return. We continue in this section by considering the rather simple transaction of getting back.

The callee is obliged to put the answer back where the caller expects to find it. Two paradigms for return are common. The one that our compiler uses is to put the answer in D0. The other common paradigm is to put the answer back on the stack. The user will have left enough room for that answer at FP+8, whether or not that space was also used for transferring parameters in. Using our paradigm, the return becomes:

```
MOVE.W  $FFFC(A6),D0      ;answer from callee's stack frame [-4(FP)] to D0
MOVEM.L  (A7)+,D3-D7/A4   ;registers restored to former values
UNLK    A6                ;SP < FP, FP < M(SP), SP < SP+4
RTS                       ;PC < M(SP), SP < SP+4
```

When all of this has transpired, the machine is back to the caller with SP pointing at *block*. The registers look like Fig. 81.5 except for two important changes. SP is back where the caller left it and D0 contains the answer that the caller asked for.

Transactional Paradigms

The final topic in this section is the description of some of the translations of the simple and ordinary phrases of the HLLs into assembly language. We will show some in each of our three machines to show both the similarities and slightly different flavors that each machine architecture gives to the translation.

The paradigms that we will discuss comprise:

- Arithmetic
- Replacement
- Testing and branching, particularly multiple Boolean expressions
- Stepping through a structure

Many studies have shown that most computer arithmetic is concerned with addressing, testing, and indexing. In NumberCount there are several examples of each. For example, near the bottom of the program, there are statements such as:

```
count++;
```

For all three machines, the basic translation is the same: Add an *immediate* (a constant stored right in the instruction) to a number in register. However, for the CISCs, one may also ask that the number be brought in and even put back in memory. The three translations of this pair comprise:

MC68000	VAX	SPARC
ADDQ.W #$1,$FFFE(A6)	INCL R0	add %o2,1,%o2

Typical of the VAX, it makes a special case out of adding 1. There is no essential difference in asking it to add 1 by saying "1," but if one has a special instruction, it saves a byte of program length. With today's inexpensive memories, a byte is no longer a big deal, but when the VAX first emerged (1978), they were delivered with less memory than a PC or Mac would have today. The VAX, of course, can say ADDL #1, r0 just like the 68000, and for any number other than 1 or 0, it would. Note also that the VAX compiler chose to keep *count* in register, while Think C® decided to put it on the stack (−2(SP)). A RISC has no choice. If you want arithmetic, your numbers must be in register. However, once again, we are really talking about the length of the code, not the speed of the transaction. *All* transactions take place from registers. The only issues are whether the programmer can see the registers and whether a single instruction can include both moving the operands and doing the operand arithmetic. The RISC separates the address arithmetic (e.g., −2(SP)) from the operand arithmetic, putting each in its own instruction. Both get the job done.

The next items we listed were *replacement* and *testing and branching*. We have both within the statement:

```
digit = (*source >= '0') && (*source <= '9');
```

The translation requires several statements:

MC68000	VAX	SPARC
MOVE.B (A4),D3	clrb r1	add %g0,0,%o1
CMPI.B #$30,D3	cmpb @4(ap),#48	ldsb [%o2],%o0
BLT ZERO	blss ZERO	subcc %o0, 47,%g0
CMPI.B #$39,D3	cmpb @4(ap),#57	ble ZERO
BLE ONE	bgtr ZERO	nop
	incb r1	subcc %o0,57,%g0
		bg ZERO
		nop
		add %g0, 1,%o1
		add %o1,0,%l3
ZERO:	**ZERO:**	**ZERO:**
MOVEQ #$00,D0		
BRA DONE		
ONE:		
MOVEQ #$01,D0		
DONE:		
MOVE.W D0,$FFF6(A6)		

To begin with, all three do roughly the same thing. The only noticeable difference in concept is that the SPARC compiler chose to compare the incoming character (*source) to 47 (the character before '0') and then branch if the result showed the letter to be "less than or equal," while the other two compared it to '0' as asked and then branched if the result was "less than." No big deal. But let us walk down the several columns to see the specific details. Prior to beginning, note that all three must bring in the character, run one or two tests, and then set an integer to either *zero* (false) or *not zero* (true). Also, let it be said that each snatch of code is purportedly optimized, but at least with the small sample that we have, it looks as if each could be better. We begin with three parallel walk-downs. Notes as needed are provided below.

MC68000	VAX	SPARC
character from M → D#	Set (byte) DIGIT to 0	Set (byte) DIGIT to 0
Is (D3-'0') <=0?	Is (*source-'0') <=0?	character from Mfi out1
If <, branch to label ZERO	If <, branch to ZERO	Is (*source-'/') <=0?
Is (D3-'9') <=0?	Is (*source-'9') <=0?	If <=, branch to ZERO
If <=. branch to label ONE	If neither, branch to ZERO	Is (*source-'9') <=0?
	Add (byte) 1 to DIGIT	If neither, branch to
ZERO:	**ZERO:**	Add 1 to DIGIT
Put a longword 0 in D0		**ZERO:**
Branch to label DONE		
ONE:		
Put a longword 1 in D0		
DONE:		
Put value in D0 into DIGIT		

Notes:

1. Moving the character into register to compare it with '0' and '9':
   a. The first 68000 line moves the next character *as a byte* into register D3. The other 3 bytes will be ignored in the byte operations. Remember that the program had already moved the pointer to the string into A4.
   b. The SPARC does the same sort of thing with a pointer in %o2, except with the difference that it sign-extends the byte to a longword. Sign extension simply copies the *sign-bit* into the higher-order bits, effectively making 3E into 0000 003E or C2 into FFFF FFC2. That is what the mnemonic means: "LoaD Signed Byte."
   c. The VAX compiler takes a totally different approach—a rather poor one, actually. It leaves not only the byte in memory but even the pointer to the byte. Thus, every time it wants to refer to the byte—and it does so numerous times—it must first fetch the address from memory and then fetch the byte itself. This double memory reference is what @4(ap) means: "At the address you will find at address 4(ap)." The only thing that makes all this

apparent coming and going even remotely acceptable is that the VAX will cache both the address and the character the first time that it gets them. Then, it can refer to them rapidly again. Cache references, however, are not as fast as register references.

2. Testing the character:
The next line (3rd for the SPARC) does a comparison between 48 (or 47) and the character. *Compare* is an instruction which subtracts one operand from the other, but instead of putting the results somewhere, it stores only the facts on whether the operation delivered a negative number or zero or resulted in either an overflow or a carry. These are stored in **flags**, single bits associated with the arithmetic unit. The bits can contain only one result at a time. The 68000 and VAX must test immediately after the comparison or they will lose the bits. The SPARC changes the bits only when the instruction says so (the CC on the instruction — "change condition codes"). Thus, the subtraction can be remote from the test-and-branch.

The SPARC is explicit about where to store the subtraction—in %g0. %g0 is a pseudo-register. It is always a 0 as a source and is a garbage can as a destination. The availability of the 0 and the garbage can serves all the same functions that the special instructions for zeros and comparisons do on the CISCs.

3. The differences in the algorithm to do the tests:
There are two different paradigms expressed in these three examples. One says: "Figure out which thing you want to do and then do that one thing." The other says: "First set the result false and then figure out if you should set it true." While the second version would seem to do a lot of unnecessary settings to zero, the other algorithm will execute one less branch. That would make it roughly equivalent. However, the 68000 algorithm is definitely longer—uses more memory for code. That is not really much of an issue, but why put the result first into a temporary register and then where you really want it?

Compiler Optimization and Assembly Language

Compiler Operations

To understand the optimizing features of compilers and their relation to assembly language, it is best to understand some of the chores for the compiler. This section examines variable allocation and how it can be optimized, and the optimization task of constant expression removal. Examples of how compilers perform these operations are taken from the various systems used in the article.

Variable Allocation. Variables in high-level languages are an abstraction of memory cells or locations. One of the compiler's tasks is to assign the abstract variables into physical locations—either registers within the processor or locations within memory. Assignment strategies vary, but an easy and often-used strategy is to place *all* variables in memory. Easy, indeed, but wasteful of execution time in that it requires memory fetches for all HLL variables. Another assignment strategy is to assign as many variables to the registers as possible and then assign any remaining variables to memory; this method is typically sufficient, except when there is a limited number of registers, such as in the 68000. In these cases, the best assignment strategy is to assign registers to the variables which have the greatest use and then assign any remaining variables to memory. In examining the compilers and architecture used in this article, we find examples of all these methods.

In the unoptimized mode, VAX and SPARC compilers are among the many which take the easy approach and assign variables only to memory locations. In Figs. 81.6 and 81.7, the variable assignments are presented for the unoptimized and optimized options. Note that only one or two registers are used, both as scratch pads, in the unoptimized option, whereas the optimization assigns registers to all variables. The expected execution time savings is approximately 42 of the 50 memory references per loop iteration. That does not include additional savings caused by compact code.

Detailed comparisons are not presented since the interpretation of architectural comparisons is highly subjective.

Unlike the VAX and SPARC compilers, the 68000 compiler assigns variables to registers in both the unoptimized and unoptimized options; these assignments are depicted in Figs. 81.7 and 81.8. Since there are only eight general-purpose data registers in the 68000 and two are assigned as scratch pads, only six of the program's ten variables can be assigned to registers. The question is how the 68000 compiler chose which variables to assign to registers and which to leave in memory. As might be expected, the compiler assigned registers based on their usage for the unoptimized option as well as the optimized. The exact counting strategy is unknown. However, a fair estimate, which yields the same assignment as the compiler, is to count only the variable's usage in the loop—the likely place for the program to spend most of its execution time. There are other ways to estimate the variable usage such as assigning weights to a reference based on its placement within a loop, its placement within a conditional (if-then-else) statement, etc. These estimates and how they are applied within a compiler can change the variable allocation as well as the efficiency of the code.

In the optimized case, a slightly different register assignment is used. This is because the optimizer created another character variable—*source—which it assigned to a register. The motivation for its creation and its assignment to a register is shown in the next section on constant expression removal.

Even though the assignment of variables to registers gives an improvement in performance, it is not always possible to assign a variable to a register. In C, one operation is to assign the address of a variable to a pointer-type variable (e.g., ip = & i). If i were assigned to a register, the operation would become invalid, because a register does not have a memory address. Although this example appears limited to the C language, the use of a variable's address is widespread when subroutine parameters are passed by reference. (Variables sent to a subroutine are either passed by reference, where the address of the variable is passed to the subroutine, allowing modifications to the original variable, or they are passed by value, where a copy of the variable is passed to the subroutine.) When a parameter is passed by reference its address must be obtained and passed to the subroutine,

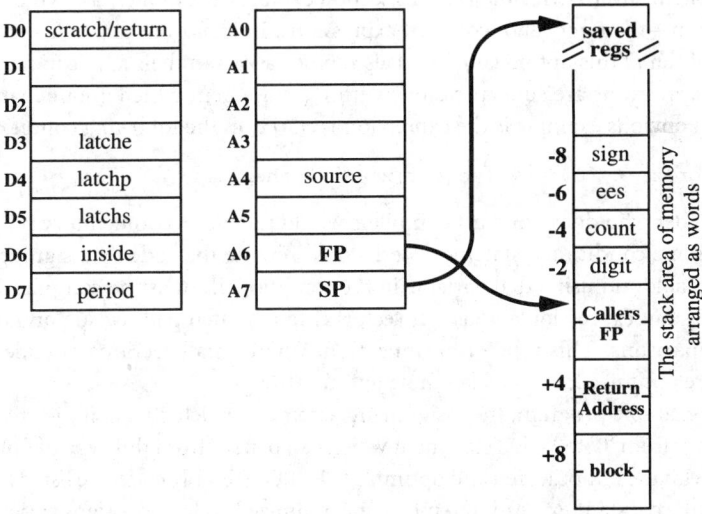

FIGURE 81.7 The stack area of the 68000's memory and the register assignments that the Think C® compiler made with global optimization *turned off.* The stack is shown just after the MOVEM instruction. The items in bold are as they would be after that instruction. While the registers all hold longwords, in typical PC/Macintosh compilers, integers are defined as words. This figure is the programmatic successor to Fig. 81.6.

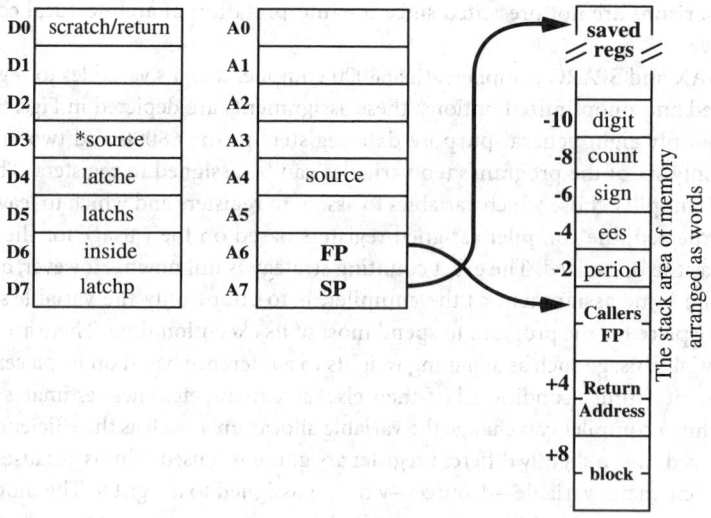

FIGURE 81.8 The stack area of the 68000's memory and the register assignments that the Think C® compiler made with global optimization *turned on*. The stack is shown just after the MOVEM instruction. The items in bold are as they would be after that instruction. This figure should be compared with Fig. 81.7.

an action commonly found in most languages. This action compounds the task of selecting candidate variables for register assignment.

Constant Expression Removal. The task of allocating program variables to physical locations is accomplished by all compilers; we have shown that there are many ways to achieve this goal with varying ease or run-time performance. This section explores a compiler task which is done strictly for optimization—the removal of constant expressions. In exploring this task, we show strategies for the recognition of this optimization and also some caveats in their application.

Constant expressions are subexpressions within a statement which are unchanged during its execution. An obvious example is the expression *vector[x]* in the following conditional statement.

```
if (vector[x] < 100) && (vector[x] > 0) then ...
```

An astute coder who does not trust compilers would not allow two memory references for vector[x] in the same conditional statement and would rewrite the code by assigning vector[x] to a temporary variable and using that variable in the conditional. An astute compiler would also recognize the constant expression and assign vector[x] to a scratch pad register and use this register for both comparisons. This type of optimization, where small sections of code (typically one source line) are optimized, is called key-hole optimization.

Within the example program, the assignment statement which checks if the character is a digit within the range from '0' to '9' is a statement which can benefit from this type of optimization. The C code lines, with the unoptimized and optimized SPARC assembly code, are listed below. Note that in addition to the constant expression removal the optimized code also assigns variables to registers.

```
digit = (*source >= '0') && (*source <= '9') ;
```

In translating this line on the SPARC, Fig. 81.9 shows the 32 registers visible at any moment in the window-blinding SPARC. The top 24 shift by 16 in a normal call. The eight globals remain the same. The shift of the registers is accompanied by copying SP to o6 and the call instruction puts the return address into o7. Accordingly, a call wipes out the caller's o7. Register g0 serves as a 0 (as a

SPARC GCC no optimization

```
L2:  add 0, %g0, %o0          ; put 0 in o0 — i.e., assume a false result
     ld [%fp-92], %o1         ; get source from memory
     ldsb [%o1], %o1          ; get *source
     add 47, %g0, %o2         ; move 47 into o2
     subcc %o1, %o2, %g0      ; compare *source to '0'
     ble L5                   ; branch if less then, digit is false
     nop
     ld [%fp-92], %o1         ; get source again
     ldsb [%o1], %o1          ; get *source again
     add 57, %g0, %o2         ;
     subcc %o1, %o2, %g0      ; compare to '9'
     bg L5
     nop
     add 1, %g0, %o0          ; results is a true, change temporary
L5:                           ; jump target if result is false
     st %o0, [%fp-36]         ; move the result to variable.
```

SPARC GCC optimization

```
L2:  add 0, %g0, %o1          ; assume a false result from statement.
     ldsb [%o2], %o0          ; get *source save in register o0
     addcc -47, %o0, %g0      ; compare to '0'
     ble L5
     nop
     addcc -57, %o0, %g0      ; reuse *source and compare to '9'
     bg L5
     nop
     add 1, %g0, %o1          ; results is a true, so change temporary
L5:                           ; jump target if result is false
     add %o1, %g0, %l3        ; move the result to variable.
```

i7	return address	r31
i6	frame pointer	r30
i5	param #5	r29
i4	param #4	r28
i3	param #3	r27
i2	param #2	r26
i1	param #1	r25
i0	param #0	r24
l7	local	r23
l6	local	r22
l5	local	r21
l4	local	r20
l3	local	r19
l2	local	r18
l1	local	r17
l0	local	r16
o7	scratch	r15
o6	stack pointer	r14
o5	param out #5	r13
o4	param out #4	r12
o3	param out #3	r11
o2	param out #2	r10
o1	param out #1	r9
o1	param out #1	r8
g7	global	r7
g6	global	r6
g5	global	r5
g4	global	r4
g3	global	r3
g2	global	r2
g1	global	r1
g0	src=0,dst=WB	r0

FIGURE 81.9 SPARC register assignments.

source) and as a *wastebasket* (as a destination). *ld* loads a longword, and *ldsb* sign-extends a byte into a longword. The instruction after a branch is executed whether the branch is taken or not (delayed branching). An instruction such as add 47, %g0, %o2 adds a constant to 0 and puts it in the register. This is equivalent to move.l #47, d4 on the 68000. An *add* or *sub* with *cc* appended changes the condition codes. To do a *compare*, one uses *addcc* or *subcc* and puts the result in g0 (the wastebasket).

The same type of constant expression can be found and removed with a global perspective, typically from within loops. A simple example is the best way to describe how they can be removed and to offer some caveats when the compiler cannot see the global picture. The following example code updates each element in the vector by adding the first element scaled by a constant y. An observation shows that the subexpression, vector[0] * y, is constant throughout all executions of the loop. An obvious improvement in code is to calculate the product of vector[0] * y outside of the loop and store its value in a temporary variable. This is done in the second example.

Constant expression present

```
for( i= 0 ; i < size; i++)
{ vector[i] = vector[i] + (vector[0] * y); }
```

Constant expression removed

```
temp = vector[0] * y ;
for( i= 0 ; i < size; i++) { vector[i] = vector[i] + temp ; }
```

Ideally, the compiler should find and remove these constant expressions from loops, but this is not as obvious as it may seem. Consider the above example if the following line were inserted in the loop:

```
y = vector[i-1] + vector[i+1]
```

If each source line is taken in isolation, y appears constant, but y is dependent on the loop index i. Hence before removing constant expressions, the compiler must map the dependencies of each variable on the other variables and the loop index. Additionally, other not-so-obvious dependencies—such as when two pointers modify the same structure—are difficult to map and can result in erroneous object code. This is one of the difficulties in optimizing compiler operation and why its extent is limited.

A subtle example for constant expression removal is found in our sample program in the reference to *source. In these statements, the character referenced (addressed) by source is obtained from memory. The pointer (address) *source* is changed only at the bottom of the loop and the memory contents addressed by *source* are static. A global optimization should obtain the character once at the top of each pass of the loop and save on subsequent memory references throughout. The 68000 C compiler with the optimization option determined *source to be constant throughout the loop and assigned register D3 to hold its contents. This saved seven of the eight memory accesses to *source in each loop pass. The unoptimized 68000 option, the SPARC, and the VAX compilers did not use global constant expression removal and must fetch the operand from memory before its use each time.

The Problems.   With optimization yielding more efficient code resulting in improved system performance, why would you not use it? Our favorite, among the several reasons, is the following quote from compiler documentation: "Compiling with optimization may produce incorrect object code." Problems are caused by assumptions used by the compiler which are not held by the programmer. For example, an optimization which assumes that memory contents do not change with time is erroneous for multi-tasking systems which share memory structures and also for memory-mapped I/O devices, where memory contents are changed by external events. For these cases, the data in register may not match the newer data in memory.

Additionally, HLL debuggers do not always work well with the optimization option since the one-to-one correspondence between HLL code and the object code may have been removed by optimization. Consider the reassignment of *source to a data register which is performed by the 68000 C compiler. If a debugger were to modify the contents of *source, then it would have to know about the two locations where it is stored: the memory and the register. Other types of optimizations which may cause problems are when unneeded variables are removed or when code is resequenced. If a HLL debugger tries to single-step through HLL code, there may not be corresponding assembly code, and its execution will appear erroneous.

High-Level Language and Assembly Language Relations

In comparing the various assembly languages from the compiler-generated code, we have not presented a full vocabulary of assembly languages or the minutiae of the underlying machines. Exploring only the code generated by the compilers may lead one to believe that all assembly languages and processor architectures are pretty much the same. This is not really the case. What we have shown is that compilers typically use the same assembly language instructions regardless of the underlying machines. The compiler writer's motivation for this apparent similarity is not because all architectures are the same, but because it is difficult—arguably even nonproductive—for the compiler to take advantage of the complex features which some CPU architectures offer. An argument may be made that compilers generate the best code by developing code rather independently of the underlying architecture. Only in the final stages of code generation is the underlying platform's hardware architecture specifically considered [see Aho *et al.*, 1986]. Differences in the architectures and assembly languages are plentiful; compilers typically do not and probably should not take advantage of such features.

The VAX is one of the best examples of an architecture having an almost extraordinary vocabulary, which is why it is often considered the prototypical *CISC* machine. What were the motivations for having this rich vocabulary if compilers simply ignore it? Early computer programming was accomplished through slightly alphabetized machine language—mnemonics for opcodes and sometimes for labels. Assembly language represented a vast improvement in readability, but even though FORTRAN, COBOL and Algol were extant at the same time as assembly language, their crude or absent optimization abilities led to the popular belief that *really good* programming was always done in assembly. This was accepted lore until early studies of optimization began to have an impact on commercial compilers. It is fair to say that this impact did not occur until the early 1980s. The VAX and the 68000 were products of the middle and late 1970s. It is no great surprise then to find that CISC computer architectures were designed to enable and assist the assembly language programmer. Such an objective promotes the inclusion of many complex features which the programmer might want to utilize. However, two facts emerged in the late 1970s which suggested that this rich vocabulary was provided at too high a cost for its benefit:

- It was widely observed that the generation, testing, and maintenance of large programs in assembly code was extremely expensive when compared to doing the same task in a HLL. Furthermore, the compilers were improving to the point where they competed quite well with typical assembly code.
- Although initially not widely accepted, it was observed that the rich vocabulary made it very difficult to set up an efficient processing pipeline for the instruction stream. In essence, the assembly line was forced to handle too many special cases and slowed down under the burden. When the analysis of compiled programs showed that only a limited span of instructions was being used, these prescient designers decided to include only the heavily used instructions and to restrict even these instructions so that they would flow in unblemished, uniform streams through the production line. Because this focus resulted in noticeably fewer instructions— though that was not the essential objective—the machines were called *RISC*, a sobriquet that came out of a VLSI design project at Berkeley [Patterson and Hennessy, 1989].

Even though RISC hardware designs have increased performance in essence by reducing the complexity and richness of the assembly language, back at the ranch the unrepentant assembly language programmer still desired complex features. Some of these features were included in the assembly languages not as native machine instructions but essentially as a *high-level* extension to assembly language. A universal extension is the inclusion of *macros*. In some sense, a macro looks like a subroutine, but instead of a call to a distant block of code, the macro results in an inline insertion of a predefined block of code. Formally, a macro is a name which identifies a particular sequence of assembly instructions. Then, wherever the name of the macro appears in the text, it is replaced by the lines of code in the macro definition. In some sense, macros make assembly language a little more like a HLL. It makes code more readable, makes code maintenance a little faster and more reliable (fix the macro definition and you fix all of the invocations of the macro), and it speeds up the programmer's work.

Another extension to some assembly languages is extended mnemonics. Here the coder places a mnemonic in place of specific assembly language instructions; during code assembly the mnemonic is automatically translated to an optimal and correct instruction or instruction sequence. These free the coder from the management of low-level details, leaving the task to a program where it is better suited. Examples of extended mnemonics include *get* and *put*, which generate memory transfers by selecting the addressing mode as well as the specific instructions based on the operand locations. An increasingly common feature of assembly languages is the inclusion of structured control statements which emulate high-level language control-flow constructs such as: if .. then .. else, for loops, while .. do loops, repeat .. until loops, break, and next. These features remove the tedium from the programmer's task, allow for a more readable code, and reduce the cost of code development. An amusing set of examples are found in the assemblers that we have used on the SPARC. Architecture not withstanding, the assembly programmers wanted VAX assembly code! In spite of the absence of such constructs in the SPARC architecture, you find expressions such as CMP (compare) and MOV. Since these are easily translated to single lines of real SPARC code, their only raison d'etre is to keep the old assembly language programmers happy. Presumably, those who knew not the VAX are not writing SPARC assembly code.

Summary

After all this fuss over compilers and how they generate assembly code, the obvious question is "Why bother to write any assembly code at all?" Three reasons why "some assembly may be required" follow.

- A human writing directly in assembly language is probably better than an optimizing compiler in extracting the last measure of resources (e.g., performance, hardware usage) from the machine. That inner loop—the code where the processor spends most of its execution cycles—may need to be hand-optimized to achieve acceptable performance. Real-time systems, where the expedient delivery of the data is as critical as its correctness, are another area where the required optimization may be greater than that achievable by compilers. The disparity in performance between human optimizers and their machine competitors comes from two special capabilities of the human programmer. These are the ability to know what the program will be doing—forward vision based on the program's intent rather than its structure—and the ability to take advantage of special quirks or tricks that have no general applicability. If you really need to extract this last full measure of performance, assembly language is the route. The cost of doing such hand-optimization is much greater than the hours spent in doing it and getting it debugged. Special quirks and tricks expressible only in assembly language will not translate to another machine and may disappear even in an "upgrade" of the intended processor.

- There is overhead in using HLL conventions, some of which can be eliminated by directly coding in assembly language. A typical embedded processor does not need the full span of HLL conventions and support, such as parameter passing or memory and stack allocation. One can *get away with* such dangerous things as global variables which do not have to be passed at all. By eliminating these conventions, increased performance is obtained. It should be pointed out that code written without such standard conventions is likely to be *very* peculiar, bug-prone, and hard to maintain.

- HLLs provide only limited access to certain hardware features of the underlying machine. Assembly language may be required to access these features. Again, this makes the code unportable and hard to maintain, but small stubs of assembly code may be required to invoke hardware actions which have no representation in a HLL. For example, setting or clearing certain bits in a special register may not be expressible in C. While any address can be explicitly included in C code, how do you reference a register which has no address? An example of such usage is writing or reading into or out of the status register. Some machines map these transactions into special addresses so that C could be used to access them, but for the majority of machines which do not provide this route to the hardware, the only way to accomplish these actions is with assembly code. To this end, some C compilers provide an inline assembler. You can insert a few lines of assembly language right in the C code, get your datum into or out of the special register, and move right back to HLL. Those compilers which provide this nonstandard extension also provide a rational paradigm for using HLL variable names in the assembly statements. Where necessary, the name gets expanded to allow the variable to be fetched and then used.

These reasons are special; they are not valid for most applications. Using assembly language loses development speed, loses portability, and increases the maintenance costs. While this caveat is well taken and widely accepted, at least for the present, few would deny the existence of situations where assembly language programming provides the best or only solution.

Defining Terms

Address error: An exception (error interrupt) caused by a program's attempt to access unaligned words or longwords on a processor which does not accommodate such requests. The address error is detected within the CPU. This contrasts with problems which arise in accessing the memory itself, where a logic circuit external to the CPU itself must detect and signal the error to cause the CPU to process the exception. Such external problems are called *bus errors.*

Assembler: A computer program (*application*) for translating an assembly-code text file to an *object* file suitable for linking to become an executable image (*application*) in machine language. Some HLL compilers include an inline assembler, allowing the programmer to drop into and out of assembly language in the midst of a HLL program.

CISC: *Complex instruction set computer,* a name to mean "not a RISC," but generally one that offers a very rich vocabulary of computer operations at a cost of making the processor which must handle this variety of operations more complex, expensive, and often slower than a RISC designed for the same task. One of the benefits of a CISC is that the code tends to be very compact. When memory was an expensive commodity, this was a substantial benefit. Today, speed of execution rather than compactness of code is the dominant force.

Compiler: A computer program (*application*) for translating a HLL text file to an *object* file suitable for linking to become an executable image (*application*) in machine language. Some compilers do both compilation and linking, so their output is an application.

Condition codes: Many computers provide a mechanism for saving the characteristics of results of a particular calculation. Such characteristics as *sign, zero result, carry* or *borrow,* and *overflow* are typical of integer operations. The program may reference these flags to determine whether to branch or not.

Disassembler: A computer program which can take an executable image and convert it back into assembly code. Such a reconstruction will be true to the machine language but normally loses much of the convenience factors, such as *macros* and name equivalencies, that an original assembly language program may contain.

Flags: See *Condition codes.*

High-level language (HLL): A computer programming language generally designed to be efficient and succinct in expressing human programming concepts and paradigms. To be contrasted with low-level programming languages such as *assembly language.*

Linker: A computer program which takes one or more object files, assembles them into blocks which are to fit in particular blocks in memory, and resolves all external (and possibly internal) references to other segments of a program and to libraries of precompiled subroutines. The output of the linker is a single file called an *executable image* which has all addresses and references resolved and which the operating system can load and run on request.

Macro: A single line of code-like text, defined by the programmer, which the assembler will then recognize and which will result in an inline insertion of a predefined block of code. In most cases, the assembler allows both hidden and visible local variables and local labels to be used within a macro. Macros also appear in some HLLs, such as C (the *define* paradigm).

Object code: A file comprising an intermediate description of a segment of a program. The object file contains binary data, machine language for program, tables of offsets with respect to the beginning of the segment for each label in the segment, and data that would be of use to debugger programs.

RISC: *Reduced instruction set computer,* a name coined by Patterson *et al.* at the University of California at Berkeley to describe a computer with an instruction set designed for maximum execution speed on a particular class of computer programs. Such designs are characterized by requiring separate instructions for load/store operations and arithmetic operations on data in registers. The earliest computers explicitly designed by these rules were designs by Seymour Cray at CDC in the 1960s. The earliest development of the RISC philosophy of design was given by John Cocke in the late 1970s at IBM. See *CISC* above for the contrasting approach.

References

A. V. Aho, R. Sethi, and J. D. Ullman, *Compiler Principles, Techniques and Tools*, Reading, Mass.: Addison-Wesley, 1986. A detailed text on the principles of compiler operations and tools to help you write a compiler. This text is good for those wishing to explore the intricacies of compiler operations.

D. Cohen, "On holy wars and a plea for peace," *Computer*, pp. 11–17, Sept. 1981. A delightful article on the comparisons and motivations of byte ordering in memory.

D. Patterson and J. Hennessy, *Computer Architecture, A Quantitative Approach*, San Mateo, Calif.: Morgan Kaufman, 1989. An excellent though rather sophisticated treatment of the subject. The appendices present a good summary of several seminal RISC designs.

81.2 High-Level Languages

Ted G. Lewis

High-level languages (**HLLs**), also known as higher-order languages (**HOLs**), have a rich history in the annals of computing. From their inception in the 1950s until advances in the 1970s, HLLs were thought of as simple mechanical levers for producing machine-level instructions (see Table 81.3). Removing the details of the underlying machine, and automatically converting from a HLL statement to an equivalent machine-level statement, releases the programmer from the drudgery of the computer, allowing one to concentrate on the solution to the problem at hand.

Over the years, HLLs evolved into a field of study of their own, finding useful applications in all areas of computing. Some HLLs are designed strictly for solving numerical problems, and some for symbolic problems. Other HLLs are designed to control the operation of the computer itself, and yet even more novel languages have been devised to describe the construction of computer hardware. The number of human-crafted languages has multiplied into the hundreds, leading to highly special-purpose HLLs.

This evolution is best characterized as a shift away from the mechanical lever view of a HLL toward HLLs as notations for encoding **abstractions**. An abstraction is a model of the real world whose purpose is to de-emphasize mundane details and highlight the important parts of a problem, system, or idea. Modern HLLs are best suited to expressing such abstractions with little concern for the underlying computer hardware.

Abstraction releases the HLL designer from the bounds of a physical machine. A HLL can adopt a metaphor or arbitrary model of the world. Such unfettered languages provide a new interface between human and computer, allowing the human to use the machine in novel and powerful ways. Abstractions rooted in logic, symbolic manipulation, database processing, or operating systems, instead of the instruction set of a central processing unit (CPU), open the engineering world to new horizons. Thus, the power of computers depends on the expressiveness of HLLs.

To illustrate the paradigm shifts brought on by HLLs over the past 30 years, consider PROLOG, LISP, SQL, C++, and various operating system command languages. PROLOG is based

Table 81.3 Each Statement of a HLL Translates into More than One Statement in a Machine-Level Language Such as Assembler

Language	Typical Number of Machine-Level Statements
FORTRAN	4–8
COBOL	3–6
Pascal	5–8
APL	12–15
C	3–5

on first-order logic. Instead of computing a numerical answer, PROLOG programs derive a conclusion. LISP is based on symbolic processing instead of numerical processing and is often used to symbolically solve problems in calculus, robotics, and artificial reasoning. SQL is a database language for manipulating large quantities of data without regard for whether it is numeric or symbolic. C++ is based on the **object-oriented paradigm**, a model of the world that is particularly powerful for engineering, scientific, and business problem solving.

None of these modern languages bear much resemblance to the machines they run on. The idea of a mechanical lever has been pushed aside by the more powerful idea of language as world builder. The kinds of worlds that can be constructed, manipulated, and studied are limited only by the HLL designer's formulation of the world as a paradigm.

In this section we answer some fundamental questions about HLLs: What are they? What do we mean by "high level"? What constitutes a paradigm? What are the advantages and disadvantages of HLLs? Who uses HLLs? What problems can be solved with these languages?

What Is a HLL?

At a rudimentary level, all languages, high and low, must obey a finite set of rules that specify both their syntax and semantics. **Syntax** specifies legal combinations of symbols that make up statements in the language. **Semantics** specifies the meanings attached to a syntactically correct statement in the language. To illustrate the difference between these two fundamental traits of all languages consider the statement, "The corners of the round table were sharp." This is syntactically correct according to the rules of English grammar, but what does it mean? Round tables do not have sharp corners, so this is a meaningless statement. We say it is semantically incorrect.

Statements of a language can be both syntactically and semantically correct and still be unsuitable for computer languages. For example, the phrase ". . . time flies . . ." has two meanings: one as an expression of clock speed, and the other as a reference to a species of insects. Therefore, we add one other requirement for computer languages: there must be only one meaning attached to each syntactically correct statement of the language. That is, the language must be *unambiguous*.

This definition of a computer language does not separate a HLL from all other computer languages. To understand the features of HLLs that make them different from other computer languages, we must understand the concepts of mechanical translation and abstraction. Furthermore, to understand the differences among HLLs, we must know how abstractions are used to change the computing paradigm. But first, what is a HLL in terms of translation and abstraction?

Defining the syntax of a HLL is easy. We simply write rules that define all legal combinations of the symbols used by the language. Thus, in FORTRAN, we know that arithmetic statements obey the rules of algebra, with some concessions to accommodate keyboards. A **metalanguage** is sometimes used as a kind of shorthand for defining the syntax of other languages, thus reducing the number of cases to be listed.

Defining the semantics of a language is more difficult because there is no universally accepted metalanguage for expressing semantics. Instead, semantics is usually defined by another program that translates from the HLL into some machine-level language. In a way, the semantics of a certain HLL is defined by writing a program that unambiguously maps each statement of the HLL into an equivalent sequence of machine-level statements. For example, the FORTRAN statement below is converted into an equivalent machine-level sequence of statements as shown to the right:

```
X = (B**2 – 4*A*C)    PUSH B
                      PUSH #2
                      POWER      //B**2
                      PUSH #4
                      PUSH A
                      PUSH C
```

```
MULT        //A*C
MULT        //4*(A*C)
SUB         //(B**2)–(4*(A*C))
POP  X      //X=
```

In this example, we assume the presence of a **pushdown stack** (see Fig. 81.10). The PUSH and POP operations are machine-level instructions for loading/storing the top element of the stack. The POWER, MULT, and SUB instructions take their arguments from the top of the stack and return the results of exponentiation, multiplication, and subtraction to the top of the stack. The symbolic expression of calculation in FORTRAN becomes a sequence of low-level machine instructions which often bear little resemblance to the HLL program.

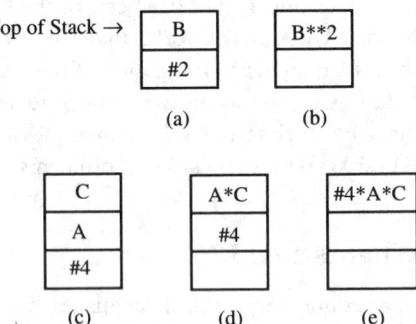

The foregoing example illustrates the mechanical advantage provided by FORTRAN because one FORTRAN statement is implemented by many machine-level statements. Furthermore, it is much easier for a human programmer to read and write FORTRAN than to read and write machine-level instructions. One major advantage of a HLL is the obvious improvement in program creation and, later on, its maintenance. As the size of the program increases, this advantage becomes larger as we con-

FIGURE 81.10   (a) The stack after PUSH #4 and PUSH B; (b) the stack after POWER; (c) the stack after PUSH #4, PUSH A, and PUSH C; (d) the stack after MULT; and (e) the stack after MULT a second time.

sider the total cost to design, code, test, and enhance an application program.

The FORTRAN program containing the example statement is treated like input data by the translating program which produces a machine-level sequence as output. In general, the input data is called the **source program**, and the resulting translated output is called the **object program**. There are two ways to obtain an object program from a source program: compiling and interpreting.

In most cases, FORTRAN is translated by a compiler program. The idea behind a **compiler** is that the translator converts the source program in its entirety before any part of the resulting object program is actually run on the computer. That is, compiling is a two-step process. In some HLLs, however, it is impossible to entirely convert a source program into an object program until the program executes.

Suppose the storage for A, B, and C in the previous example is not known at the time the program is compiled. We might want to allocate storage on-the-fly while the program is running, because we do not know in advance that the storage is needed. This is an example of **delayed binding** of a variable to its storage location in memory.

Powerful languages such as Pascal and C permit a limited amount of delayed binding, as illustrated in the following example written in Pascal. This example also illustrates a limited amount of abstraction introduced by the HLL.

```
type    rnumber = real;             {template}
        rptr = ^rnumber;            {pointer}
var     Aptr, Bptr, Cptr : rptr;    {instance}
...
{later in the program...}
new(Aptr); read( Aptr^);            {binding}
new(Bptr); read( Bptr^);
new(Cptr); read( Cptr^);
X := (Bptr^) * (Bptr^) – 4 * (Aptr^) * (Cptr^);
```

The **type** statement is an abstraction that defines a template and access mechanism for the variables A, B, and C that are to be created on-the-fly. The **var** statement is similar to the DIMENSION statement in that it tells the translator to allocate space for three pointers: Aptr, Bptr, and Cptr. Each of these allocations will point to the actual values of A, B, and C according to the previous **type** statement.

The actual allocation of storage is not known until the program executes the sequence of new() functions in the body of the program. Each new() function allocates space according to the **type** statement and returns a pointer to that space. To access the actual values in these storage spaces, the up arrow, ∧, is written following the variable name. Thus, the read() function gets a number from the keyboard and puts it in the space pointed to by the pointer variable. Similarly, the value of X is computed by indirect reference to each value stored at the newly allocated memory location.

The purpose of this example is to illustrate the use of delayed binding in a HLL. Languages such as LISP and C++ require even greater degrees of delayed binding because of the abstractions they support. When the amount of delayed binding becomes so great that very little of the program can be compiled, we say that the HLL is an *interpreted language*, and the translator becomes an **interpreter** rather than a compiler. This crossover is often obscure, so some HLLs are translated by both a compiler and an interpreter. BASIC is a classic example of a HLL that is both interpreted and compiled.

The purpose of delayed binding is to increase the level of a HLL by introducing abstraction. Abstraction is the major differentiating feature between HLLs and other computer languages. Without abstraction and delayed binding, most HLLs would be no more powerful than a macro **assembler** language. However, with abstraction, HLLs permit a programmer to express ideas that transcend the boundaries of the physical machine.

We can now define HLL based on the concept of abstraction. *A HLL is a set of symbols which obey unambiguous syntactic and semantic rules: the syntactic rules specify legal combinations of symbols, and the semantic rules specify legal meanings of syntactically correct statements relative to a collection of abstractions.*

The notion of abstraction is very important to understanding what a HLL is. The example above illustrates a simple abstraction, e.g., that of data structure abstraction, but other HLLs employ much more powerful abstraction mechanisms. In fact, the *level of abstraction* of a HLL defines how high a HLL is. But, how do we measure the level of a HLL? What constitutes a HLL's height?

How High Is a HLL?

There have been many attempts to quantify the level of a programming language. The major obstacle has been to find a suitable measure of level. This is further complicated by the fact that nearly all computer languages contain some use of abstraction, and therefore nearly all languages have a "level." Perhaps the most interesting approach comes from information theory.

Suppose a certain HLL program uses P operators and Q operands to express a solution to some problem. For example, a four-function pocket calculator uses $P = 4$ operators for addition, subtraction, multiplication, and division. The same calculator might permit $Q = 2$ operands by saving one number in a temporary memory and the other in the display register. In a HLL the number of operators and operands might number in the hundreds or thousands.

We can think of the set of P operators as a grab bag of symbols that a working programmer selects one at a time and places in a program. Suppose each symbol is selected with probability $1/P$, so the information content of the entire set is

$$-\sum_{1}^{P} \frac{1}{P} \log\left(\frac{1}{P}\right) = \log(P)$$

Assuming the set is not depleted, the programmer repeats this process P times, until all of the operators have been selected and placed in the program. The information content contributed by the operators is $P \log(P)$, and if we repeat the process for selecting and placing all Q operands, we get $Q\log(Q)$ steps again. The sum of these two processes yields $P\log(P) + Q\log(Q)$ symbols. This is known as Halstead's **metric** for *program length* [Halstead, 1977].

Similar arguments can be made to derive the volume of a program, V, level of program abstraction, L, and level of the HLL, λ, as follows.

P = Number of distinct operators appearing in the program

p = Total number of operators appearing in the program

Q = Number of distinct operands appearing in the program

q = Total number of operands appearing in the program

N = Number of operators and operands appearing in the program

V = Volume = $N\log_2(P + Q)$

L = Level of abstraction used to write the program $\approx (2/P)*(Q/q)$

λ = Level of the HLL used to write the program = L^2V

E = Mental effort to create the program = V/L

Halstead's theory has been applied to English (*Moby Dick*) and a number of programs written in both HLL and machine-level languages. A few results based on the values reported in Halstead [1977] are given in Table 81.4. This theory quantifies the level of a programming language: *PL/I* is higher level than Algol-58, but lower level than English.

In terms of the mental effort required to write the same program in different languages, Table 81.4 suggests that a HLL is about twice as high level as assembler language. That is, the level of abstraction of *PL/I* is more than double that of assembler. This abstraction is used to reduce mental effort and solve the problem faster.

Table 81.4 Comparison of Languages in Terms of Level, λ, and Programming Effort, E

Language	Level, λ	Effort, E
English	2.16	1.00
PL/I	1.53	2.00
Algol-58	1.21	3.19
FORTRAN	1.14	3.59
Assembler	0.88	6.02

HLLs and Paradigms

A programming **paradigm** is a way of viewing the world, e.g., an idealized model. HLLs depend on paradigms to guide their design and use. In fact, one might call HLL designers *paradigm engineers* because a good HLL starts with a strong model. Without such a model, the abstraction of a HLL is meaningless. In this section we examine the variety of paradigms embodied in a number of HLLs.

The **procedural paradigm** was the earliest programming paradigm. It is the basis of COBOL, FORTRAN, Pascal, C, BASIC, and most early languages. In this paradigm the world is modeled by an algorithm. Thus, an electrical circuit's behavior is modeled as a system of equations. The equations are solved for voltage, current, and so forth by writing an algorithmic procedure to numerically compute these quantities.

In the procedural paradigm a large system is composed of modules which encapsulate procedures which in turn implement algorithms. Hierarchical decomposition of a large problem into a collection of subordinate problems results in a hierarchical program structure. Hence, a large FORTRAN or C program is typically composed of a collection of procedures (subroutines in FORTRAN and functions in C) organized in layers, forming a tree structure, much like the organization chart of a large corporation (see Fig. 81.11).

Hierarchy is used in the procedural paradigm to encapsulate low-level algorithms, thus abstracting them away. That is, algorithm abstraction is the major contributor to leveling in a procedural HLL. Figure 81.11 illustrates this layering as a tree where each box is a procedure and subordinate boxes represent procedures called by parent boxes, the top-most box is the most abstract, and the lowest boxes in the tree are the most concrete.

Intellectual leverage is limited to control flow encapsulation in most procedural languages. Only the execution paths through the program are hidden in lower levels. While this is an improvement

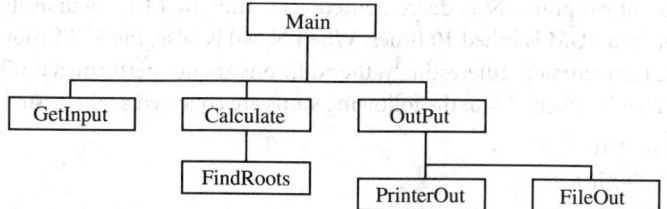

FIGURE 81.11 Hierarchical decomposition of procedural program.

over machine-level languages, it does not permit much flexibility. For example, algorithmic abstraction is not powerful enough to easily express non-numerical ideas. Thus, a C program is not able to easily model an electronic circuit as a diagram or object that can be reasoned about, symbolically.

One of the reasons procedural HLLs fail to fully hide all details of an abstraction is that they typically have weak models of data. Data is allowed to flow across many boundaries, which leads to problems with encapsulation. In FORTRAN, BASIC, Pascal, and C, for example, access to any data is given freely through globals, parameter passing, and files. This is called **coupling** and can have disastrous implications if not carefully controlled.

One way to reduce coupling in a procedural language is to eliminate side-effects caused by unruly access to data. Indeed, if procedures were prohibited from directly passing and accessing data altogether, many of the problems of procedural languages would go away. An alternative to the procedural paradigm is the **functional paradigm**. In this model of the world, everything is a function that returns a value. Data is totally abstracted away so that algorithms are totally encapsulated as a hierarchical collection of functions. LISP is the most popular example of a functional HLL [Winston and Horn, 1989].

A LISP statement that limits data access usually consists of a series of function calls; each function returns a single value which is used as an argument by another function and so on until the calculation is finished. For example, the FORTRAN statement X = (B**2 − 4*A*C) given earlier is written in functional form as follows:

ASSIGN(X, MINUS(SQUARE(B), TIMES(4, TIMES(A,C))))

This statement means to multiply A times C, then multiply the result returned by TIMES by 4, then subtract this from the result returned by SQUARE, and so forth. The final result is assigned to X.

One of the most difficult concepts to adjust to when using the procedural paradigm is the idea that all things are functions. The most significant implication of this kind of thinking is the replacement of loops with **recursion** and branches with guards. Recall that everything is a function that must return a value—even control structures. To illustrate, consider the functional (non-LISP) equivalent of the summation loop in FORTRAN, below.

```
S=0
DO 20 I=1,10              SUM( XList, N):
S=S+X(I)                  N>0 |
20 CONTINUE               N is N–1,
                          SUM is Head( XList ) + SUM (TAIL(Xlist), N)
```

The functional form will seem strange to a procedural programmer because it is higher level, e.g., more abstract. It hides more details and uses functional operators HEAD (for returning the first element of XList), TAIL (for returning the N–1 tail elements of XList), and **is** for binding a value to a name. Also, notice the disappearance of the loop. Recursion on SUM is used to run through the entire list, one element at a time. Finally, the guard N>0 prevents further recursion when N reaches zero.

In the functional program, N is decremented each time SUM is recursively called. Suppose N = 10, initially; then SUM is called 10 times. When N > 0 is false, the SUM routine does nothing, thus terminating the recursion. Interestingly, the additions are not performed until the final attempt to recurse fails. That is, when N = 0, the following sums are collected as the nested calls unwind:

```
SUM     : XList(10)
        : SUM+Xlist(9)
. . .     . . .
        : SUM+XList(1)
```

Functional HLLs are higher level than procedural languages because they reduce the number of symbols needed to encode a solution as a program. The problem with functional programs, however, is their high execution overhead caused by the delayed binding of their interpreters. This makes LISP and PROLOG, for example, excellent **prototyping** languages but expensive production languages. LISP has been confined to predominantly research use; few commercial products based on LISP have been successfully delivered without first rewriting them in a lower-level language such as C. Other functional languages such as PROLOG and STRAND88 have had only limited success as commercial languages.

Another alternative is the **declarative paradigm.** Declarative languages such as Prolog and STRAND88 are both functional and declarative. In the declarative paradigm, solutions are obtained as a byproduct of meeting limitations imposed by constraints. Think of the solution to a problem as the only (first) solution that satisfies all constraints declared by the program.

An example of the declarative paradigm is given by the simplified PROLOG program below for declaring an electrical circuit as a list of constraints. All of the constraints must be true for the circuit() constraint to be true. Thus, this program eliminates all possible R, L, C, V circuits from consideration, except the one displayed in Fig. 81.12. The declarations literally assert that Circuit(R, L, C, V) is a thing with "R connected to L, L connected to C, L connected to R, C connected to V, and V connected to R." This eliminates "V connected to L," for example, and leaves only the solution shown in Fig. 81.12.

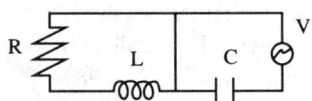

FIGURE 81.12   Solution to declaration for Circuit(R, L, C, V).

```
Circuit(R, L, C, V) :
        Connected(R, L)
        Connected(L, C)
        Connected(L, R)
        Connected(C, V)
        Connected(V, R)
```

One interesting feature of declarative languages is their ability to represent infinite calculations. A declaration might constrain a solution to be in an infinite series of numbers, but the series may not need to be fully computed to arrive at an answer.

Another feature of such languages is their ability to compute an answer when in fact there may be many answers that meet all of the constraints. In many engineering problems, the first answer is as good as any other answer.

The declarative paradigm is a very useful abstraction for unbounded problems. Adding abstraction to the functional paradigm elevates declarative languages even higher. Solutions in these languages are arrived at in the most abstract manner, leading to comparatively short, powerful programs.

Perhaps the most common use of declarative languages is for construction of expert systems [Smith, 1988]. These kinds of applications are typically diagnostic. That is, they derive a conclusion based on assertions of fact. An electrical circuit board might be diagnosed with an expert system that takes symptoms of the ailing board as its input and derives a conclusion based on rules of electronic circuits—human rules of thumb given it by an experienced technician—and declarative reasoning. In this example, the rules are constraints expressed as declarations. The expert system program may derive more than one solution to the problem because many solutions may fit the constraints.

Declarative languages have the same inefficiencies as functional languages. For this reason, expert system applications are usually developed in a specialized declarative system called an *expert system shell*. A shell extracts the declarative or constraint-based capability from functional languages such LISP and PROLOG to improve efficiency. Often it is possible to simplify the shell so that early binding is achieved, thus leading to compiling translators rather than interpreters. Very large and efficient expert systems have been developed for commercial use using this approach.

Yet another paradigm used as the basis of modern languages such as C++ and Object Pascal is the *object-oriented programming (OOP) paradigm* [Budd, 1991]. OOP merges data and procedural abstractions into a single concept. In OOP, an object has both storage capacity and algorithmic functionality. These two abstractions are encapsulated in a construct called a **class**. One or more objects can be created by cloning the class. Thus, an **object** is defined as an instance of a class [Lewis, 1991].

OOP actually represents a culmination of ideas of procedural programming that have evolved over the past three to four decades. It is a gross oversimplification to say that OOP is procedural programming, because it is not, but consider the following evolution.

Procedure = Algorithm + Data Structures
Abstract Data Structure = Implementation Part + Interface Part
Class = Abstract Data Structure + Functions
Object = Class + Inheritance

The first "equation" states that a procedure treats algorithms and data separately, but the programmer must understand both the data structure and the algorithms for manipulating the data structures of an application. This separation between algorithms and data is a key feature of the procedural paradigm. During the 1970s structured programming was introduced to control the complexity of the procedural paradigm. While only partially successful, structured programming limited procedures to less powerful control structures by eliminating the GOTO and programs with labels. However, structured programming did not go far enough.

The next improvement in procedural programming came in the form of increased abstraction, called **ADT** (abstract data structures). An ADT separates the interface part of a procedure from its implementation part. Modula II and Ada™ were designed to support ADTs in the form of modules and packages. The interface part is an abstraction that hides the details of the algorithm. Programming in this form of the procedural paradigm reduces complexity by elevating a programmer's thoughts to a higher level of abstraction, but it still does not remove the problem of how procedures are related to one another.

Classes group data together into clusters that contain all of the functions that are allowed to access and manipulate the data. The class concept is a powerful structuring concept because it isolates the data portion of a program, thus reducing coupling and change propagation.

The class construct invented by the designers of Simula67 enforced the separation of interface and implementation parts of a module, and in addition introduced a new concept. Inheritance is the ability to do what another module can do. Thus, inheritance relates modules by passing on the algorithmic portion of a module to other modules. Inheritance in a programming language like SmallTalk, Object Pascal, and C++ means even greater abstraction because code can be reused without being understood.

An object is an instance of a class. New objects inherit all of the functions defined on all of the classes used to define the parent of the object. This simple idea, copied from genetics, has a profound impact on both design and programming. It changes the way software is designed and constructed, i.e., it is a new paradigm.

Object-oriented thinking greatly differs from procedural thinking (see Table 81.5). In OOP a problem is decomposed into objects which mimic the real world. For example, the objects in Fig. 81.12 are resistor, inductor, capacitor, and voltage source. These objects have real-world fea-

Table 81.5   Procedural versus Object-Oriented Thinking

Procedural	Object-Oriented
Instructions and data are separated.	Objects consist of both data and instructions.
Software design is linear, e.g., it progresses from design through coding and testing. This means change is difficult to accommodate.	Software design is interactive with coding and testing. This means change is easier to accommodate.
Programs are top-down decompositions of procedures, e.g., trees.	Programs are networks of objects that send messages to one another without concern for tree structure.
Program components are the real world, thus making programming more of a magic art.	Program components have abstractions of correspondence with the real world, thus making programming more of a discipline.
New programs are mostly custom built with little reuse from earlier programs. This leads to high construction costs and errors.	New programs are mostly specializations of earlier programs through reuse of their components. This leads to low construction costs and higher quality.

tures (state) such as resistance, inductance, capacitance, and voltage. They also have behaviors defined by sinusoidal curves or phase shifts. In short, the real-world objects have both state and function. The state is represented in a program by storage and the function is represented by an algorithm. A resistor is a program module containing a variable to store the resistance and a function to model the behavior of the resistor when it is subjected to an input signal.

The objects in an object-oriented world use inheritance to relate to one another. That is, objects of the same class all inherit the same functions. These functions are called **methods** in SmallTalk and **member functions** in C++[Ellis and Stroustrup, 1990]. However, the state or storage attributes of objects cloned from the same class differ. The storage components of an object are called **instance variables** in SmallTalk and **member fields** in C++.

The wholism of combining data with instructions is known as *ADTs*; the concept of sending messages instead of calling procedures is the *message-passing paradigm*; the concept of interactive, nonlinear, and iterative development of a program is a consequence of an object's **interface specification** being separated from its **implementation part**; the notion of modeling the real world as a network of interacting objects is called *OOD* (object-oriented design); the concept of specialization and **reuse** is known as *inheritance*; and OOP is the act of writing a program in an object-oriented language while adhering to an object-oriented view of the world.

Perhaps an analogy will add a touch of concreteness to these vague concepts. Suppose automobiles were constructed using both technologies. Table 81.6 repeats the comparison of Table 81.5 using an automobile design and manufacturing analogy.

We illustrate these ideas with a simple C++ example. The following code declares a class and two subclasses which inherit some properties of the class. The code also shows how interface and implementation parts are separated and how to override unwanted methods. Figure 81.13 depicts the inheritance and class hierarchy for this example.

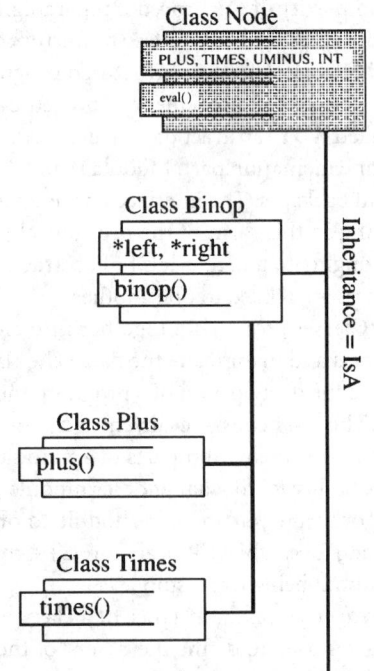

FIGURE 81.13   Partial class hierarchy for a C++ program that simulates a pocket calculator. The shaded Node class is an abstract class that all other classes use as a template.

Table 81.6 Analogy with an Automobile Manufacturer

Procedural	Object-Oriented
Vendors and Assemblers work from their own plans. There is little coordination between the two.	Vendors and Assemblers follow the same blueprints; thus the resulting parts are guaranteed to fit together.
Manufacturing and design are sequential processes. A change in the design causes everyone to wait while the change propagates from designers to workers on the production line.	Design interacts with production. Prototypes are made and discarded. Production workers are asked to give suggestions for improvement or on how to reduce costs.
Changes on the manufacturing floor are not easily reflected as improvements to the design on the drafting board.	Changes in implementation rarely affect the design as interfaces are separated from implementation. Thus, the materials may change, but not the need for the parts themselves.
New cars are designed and constructed from the ground up, much like the first car ever produced.	New cars are evolutionary improvement to existing base technology, plus specializations that improve over last year's model.

```
class Node{
  public:                          //The interface part...
        Node() {}                  //Constructor function
        virtual ~Node() {}         //Destructor function
        virtual int eval() { error(); return 0;}    //Override this function
}
```

The Node class consists of public functions that are to be overridden by descendants of the class. We know this because the functions are virtual, which in C++ means we expect to replace them later. Therefore we call this an **abstract class**. Also, Node() is the name of both the constructor and destructor member functions. A constructor is executed when a new object is created from the class, and the destructor is executed when the object is destroyed. These two functions take care of initialization and garbage collection which must be performed before and after dynamic binding of objects to memory space. Figure 81.13 shows this Node as an abstract class from which all other subclasses of this application are derived.

Now, we create a subclass that inherits the properties (interface) of Node() and adds a new property, e.g., Binop. Binop is an abstraction of the binary operators of a pocket calculator, which is the real-world object being simulated by this example program. The expression to be calculated is stored in a binary tree, and Binop sprouts a new left and right subtree when it is created and deletes this subtree when it is disposed.

```
class Binop : public Node {       //Derive Binop from Node
  public:
        Node *left, *right;        //Pointers to left and right subtrees
        ~Binop() { delete left; delete right;}  //Collect garbage
        Binop( Node *lptr, Node *rptr) {left = lptr; right=rptr;}
}
```

Next, we define further specializations of Binop: one for addition of two numbers, Plus(), and the other for multiplication, Times(). The reader can see how to extend this to other operators.

```
class Plus: public Binop {
  public:                         //Add member functions to Binop
        Plus( Node *lptr, Node *rptr) : Binop( lptr, rptr) {} //Use Binop
        int eval() { return left->eval()+right->eval();}     //Do Addition
};
```

```
class Times: public Binop {
  public:                          //Add member functions to Binop
        Times( Node *lptr, Node *rptr) : Binop( lptr, rptr) {}  //Use Binop
        int eval() { return left->eval()*right->eval();}        //Do Multiply
};
```

In each case, the special-purpose operations defined in Plus() and Times() reuse Binop's code to perform the pointer operations. Then they add a member function eval() to carry out the operation. This illustrates reuse and the value of inheritance.

At this point, we have a collection of abstractions in the form of C++ classes. An object, however, is an instance of a class. Where are the objects in this example? We must dynamically create the required objects using the new function of C++.

```
Node *ptr = new Plus ( lptr, rptr );     //Create an object and point to it
int result = ptr->eval();                //Add
delete ptr;
```

The foregoing code instantiates an object that ptr points to, sends a message to the object telling it to perform the eval() function, and then disposes of the object. This example assumes that lptr and rptr have already been defined elsewhere.

Clearly, the level of abstraction is greatly raised by OOP. Once a class hierarchy is established, the actual data processing is hidden or abstracted away. This is the power of the OOP paradigm.

The pure object-oriented languages such as SmallTalk80 and CLOS have achieved only modest success due to their unique syntax and heavy demands on computing power. Hybrid HLLs such as Object Pascal and C++ have become widely accepted because they retain the familiar syntax of procedural languages, and they place fewer demands on hardware.

Although OOP is an old technology (circa 1970), it began to gain widespread acceptance in the 1990s because of the growing power of workstations, the increased use of graphical user interfaces, and the invention of hybrid object-oriented languages such as C++. Typically, C++ adds 15–20% overhead to an application program due to delayed binding of objects to their methods. Given that the power of the hardware increases more than 20% per annum, this is an acceptable performance penalty. In addition, OOP is much more suitable for the design of graphical user-interface-intensive applications because the display objects correspond with programming objects, thus simplifying design and coding. Finally, if you know C, it is a small step to learn C++.

Summary and Conclusions

HLLs: What are they? What do we mean by "high level"? What constitutes a paradigm? What are the advantages and disadvantages of HLLs? Who uses HLLs? What problems can be solved with these languages?

HLLs are human inventions that allow humans to control and communicate with machines. They obey rules of syntax and unambiguous semantics which are combined to express abstract ideas. HLLs are called "high level" because they express abstractions.

We have chosen to define the level of a HLL in terms of the information content of its syntax and semantics. The Halstead measure of language level essentially says that the higher a HLL is, the fewer symbols are needed to express an idea. Thus, if language A is higher than language B, a certain program can be expressed more succinctly in A than B. This is clearly the case when comparing HLLs with various machine-level languages, where a single statement in the HLL requires many statements in the machine-level language.

HLLs differ from one another in the abstractions they support. Abstract views of the world are called paradigms, and the guiding principle of any HLL is its programming paradigm.

We have compared the following programming paradigms: procedural, functional, declarative, and object-oriented. Procedural programming has the longest history because the first HLLs were based on low-level abstractions that are procedural. FORTRAN, COBOL, C, and Pascal are classical examples of the procedural languages.

Functional and declarative languages employ higher levels of abstraction by restricting the world view to simple mechanisms: mathematical functions and constraints. It may seem odd that such restrictions increase the level of abstraction, but languages like LISP and PROLOG hide much of the detail found to be necessary in the procedural paradigm. This increases the measure of level defined in this section.

Object-oriented programming embraces a novel abstraction that seems to fit the world of computing: objects. In this paradigm, the world is modeled as a collection of objects that communicate by sending messages to one another. The objects are related to each other through an inheritance mechanism that passes on the algorithmic behavior from one class of objects to another class. Inheritance permits reuse and thus raises the programming abstraction to a level above previous paradigms.

The future of HLLs is uncertain and unpredictable. It is unlikely that anyone in 1970 would have predicted the acceptance of functional, declarative, or object-oriented paradigms in the 1990s. Therefore, it is unlikely that the following predictions bear much relationship to computing in the year 2000. However, it is instructive to project a few scenarios and explain their power.

Functional and declarative programming result in software that can be mathematically analyzed, thus leading to greater assurances that the software actually works. Currently these paradigms consume too much memory and machine cycles. However, in the year 2000, very high-speed machines will be commonplace. What will we use these powerful machines for? One answer is that we will no longer be concerned with the execution efficiency of a HLL. The drawbacks of functional and declarative languages will fade, to be replaced by concern for the correctness and expressiveness of the HLL. If this occurs, functional and declarative languages will be the preferred HLLs because of the elevated abstractions supported by the functional and declarative paradigms. Applications constructed from these HLLs will exhibit more sophisticated logic, communicate in non-numeric languages such as speech and graphics, and solve problems that are beyond the reach of current HLLs.

Object-orientation is an appealing idea whose time has come. OOP will be to the 1990s what structured programming was to the 1970s. Computer hardware is becoming increasingly distributed and remote. Networks of workstations routinely solve problems in concert rather than as stand-alone systems. This places greater demands on flexibility and functionality of applications. Consider the next step beyond objects—servers:

$$Server = Object + Process$$

A server is an object that is instantiated as an operating system process or task. The server sends messages to other servers to get work done. The servers "live" on any processor located anywhere on the network. Software is distributed and so is the work. OOP offers the greatest hope for distributing applications in this fashion without loss of control. Should this scenario come true, the OOP paradigm will not only be appropriate, but contribute to greater HLL leverage through reusable objects, distributed servers, and delayed binding of methods to these servers.

Object-oriented languages, databases, and operating systems are on the immediate horizon. Graphical user-interface servers such as X-Windows already exist, lending credibility to this scenario. At least for the near future, HLLs are most likely to become identical with the object-oriented paradigm.

Defining Terms

Abstract class: A class consisting of only an interface specification. The implementation part is unspecified, because the purpose of an abstract class is to establish an interface.

Abstraction: Abstraction in computer languages is a measure of the amount of separation between the hardware and an expression of a programming idea. The level of abstraction of a high-level language defines the level of that language.

ADT: An abstract data type (ADT) is a software module that encapsulates data and functions allowed to be performed on that data. ADTs also separate the interface specification of a module from the implementation part to minimize coupling among modules.

Assembler: A computer program for translating symbolic machine instructions into numerical machine instructions. Assemblers are considered low-level languages for programming a computer.

Class: A specification for one or more objects that defines state (data) and functions (algorithms) that all objects may inherit when created from the class. A class is a template for implementing objects.

Compiler: A computer program that translates the source program statements of a high-level language into lower-leveled object program statements. Compilers differ from interpreters in that they do not immediately perform the operations specified in the source program. Instead, a compiler produces an object program that in turn performs the intended operations when it is run.

Coupling: A measure of the amount of interaction between modules in a computer program. High coupling means that a change in one module is likely to affect another module. Low coupling means there is little impact on other modules whenever a change is made in one module.

Declarative paradigm: A programming paradigm in which the world is modeled as a collection of rules and constraints.

Delayed binding: The process of postponing the meaning of a programming object until the object is manipulated by the computer. Delayed binding is used by interpreters and compilers, but more often it is employed by interpreters.

Functional paradigm: A programming paradigm in which the world is modeled as a collection of mathematical functions.

HLL (also HOL): A HLL is a set of symbols which obey unambiguous syntactic and semantic rules: the syntactic rules specify legal combinations of symbols, and the semantic rules specify legal meanings of syntactically correct statements relative to a collection of abstractions.

Implementation part: The definition or algorithm for a programming module which gives the details of how the module works.

Instance variables: Data encapsulated by an object.

Interface specification: The definition of a programming module without any indication of how the module works.

Interpreter: A computer program that translates and performs the intended operations of the source statements of a high-level language program. Interpreters differ from compilers in that they immediately perform the intended operations specified in the source program, and they do not produce an object program.

Member fields: Instance variables of a C++ object.

Member functions: Methods defined on a C++ object.

Metalanguage: A formal language for defining other languages. A metalanguage is typically used to define the syntax of a high-level language.

Methods: Functions allowed to be performed on the data of an object.

Metric: A measure of a computer program's complexity, clarity, length, difficulty, etc.

Object: An instance of a class. Objects have state (data) and function (algorithms) that are allowed to manipulate the data.

Object-oriented paradigm: A programming paradigm in which the world is modeled as a collection of self-contained objects that interact by sending messages. Objects are modules that contain data and all functions that are allowed to be performed on the encapsulated data. In addition, objects are related to one another through an inheritance hierarchy.

Object program: Machine form of a computer program, which is the output from a translator.

Paradigm: An idealized model, typically used as a conceptual basis for software design. Programming paradigms dictate the approach taken by a programmer to organize, and then write, a computer program.

Procedural paradigm: A programming paradigm in which the world is modeled as a collection of procedures which in turn encapsulate algorithms.

Prototyping: A simplified version of a software system is a prototype. Prototyping is the process of designing a computer program through a series of versions; each version becomes a closer approximation to the final one.

Pushdown stack: A data structure containing a list of elements which are restricted to insertions and deletions at one end of the list, only. Insertion is called a push operation and deletion is called a pull operation.

Recursion: A procedure is called recursive if it calls itself.

Reuse: Programming modules are reused when they are copied from one application program and used in another. Reusability is a property of module design that permits reuse.

Semantics: The part of a formal definition of a language that specifies the meanings attached to a syntactically correct statement in the language.

Source program: Symbolic form of a computer program, which is the input to a translator.

Syntax: The part of a formal definition of a language that specifies legal combinations of symbols that make up statements in the language.

References

T. Budd, *Object-Oriented Programming,* Reading, Mass.: Addison-Wesley, 1991.

M. A. Ellis and B. Stroustrup, *The Annotated C++ Reference Manual,* Reading, Mass.: Addison-Wesley, 1990.

M. H. Halstead, *Elements of Software Science,* New York: Elsevier North-Holland, 1977.

T. G. Lewis, *CASE: Computer-Aided Software Engineering,* New York: Van Nostrand Reinhold, 1991.

P. Smith, *Expert System Development in Prolog and Turbo-Prolog,* New York: Halsted Press, 1988.

P. H. Winston and B. K. P. Horn, *LISP,* Reading, Mass.: Addison-Wesley, 1989.

81.3 Data Types and Data Structures

Johannes J. Martin

The study of *data types* and *data structures* is a part of the discipline of computer programming. The terms refer to the two aspects of data objects: their usage and their implementation, respectively. The study of data *types* deals with the *identification* of (abstract) data objects in the context of a programming project and with methods of their more or less formal *specification*; the study of data *structures*, on the other hand, is concerned with the *implementation* of such objects using already existing data objects as raw material.

Concretely, the area addresses a basic problem of programming: the reduction of complex objects, such as vectors, tensors, text, graphic images, sound, functions, directories, maps, corporate organizations, models of ecosystems or machinery, or anything else that a program may have to deal with, to the only native objects of digital computers: arrays of binary digits (bits). The fundamental problem of this reduction is managing program complexity. Two organizational tools are essential to its solution: abstraction and hierarchical structuring. Abstraction refers to the separation of *what* computational objects are used for from *how* they are reduced to (i.e., implemented by means of) simpler ones. Hierarchical structuring refers to breaking this reduction up into small manageable steps. Through several steps of abstraction more and more complex objects are constructed, each one reduced to the previous, simpler generation of objects. This process ends when the desired objects have been composed.

Abstract Data Types

An **abstract data type** is one or more *sets* of computational objects together with some basic operations on those objects. One of these sets is defined by the type either by enumeration or by generating operations and is called the *carrier set* of the type. Customarily it is given the same name as the type. All other sets are called auxiliary sets of the type. In exceptional cases a type may have more than one carrier set.

The heart of the specification of an abstract data type is the definition of its functions, their syntax and semantics. Their syntax is specified by their **functionalities** and their semantics by algebraic axioms. [For more details see, e.g., Martin, 1986].

With sets A and B, the expression $A \to B$ denotes the set of all functions that have the domain A and the codomain B. Functions $f \in A \to B$ (traditionally denoted by $f: A \to B$) are said to have the functionality $A \to B$.

The collection of basic operations does not need to be minimal. It should, however, be rich enough so that all other operations that one might wish to perform on the objects of the carrier set can be expressed exclusively by these basic operations. The type *Boolean*, for example, consists of the set of Boolean values, Boolean = {true, false}, with, e.g., the operations *not*, *and*, and *or*.

In general, things are not quite this simple. To be useful for programming purposes, even the type Boolean requires at least one additional function. This function, called a *conditional expression*, provides a choice of one of two given values depending on a given Boolean value. It has the form:

$$f: \text{Boolean} \times \text{SomeType} \times \text{SomeType} \to \text{SomeType}$$

and, with $a, b \in$ SomeType, is defined by:

$$f(\text{true}, a, b) = a \quad \text{and}$$
$$f(\text{false}, a, b) = b$$

The syntactical form of conditional expressions varies for different programming languages that provide this construct. For example, in the language C it has the form:

    Boolean ? SomeType : SomeType.   /* with the result type of SomeType */

The set SomeType is an auxiliary set of the type Boolean.

Fundamental Data Types

The fundamental types listed next are supported by almost all modern high-level programming languages (reference books on Pascal, Modula II, C, and Ada are listed among the references at the end of this section):

    Integer, Real (sometimes called Float), Character, and Boolean

Since their carrier sets are ordered (one of the operations of these types is $\leq$), these types are also called scalar types. All provide operations for comparing values; in addition, Integer and Real come equipped with the usual arithmetic operations ($+, -, *, /$) and Boolean with the basic logical operations (not, and, or). Most computers support these operations by hardware instructions. Thus, while bit arrays are the original native objects of a digital computer, the fundamental scalar types may be viewed as the given elementary building blocks for the construction of all other types.

Type Constructors

Enumerated Types

Beginning with Pascal, modern languages provide a rather useful constructor for scalar types, called enumerated types. Enumerated types have finite (small) carrier sets that the programmer defines by enumerating the constants of the type (specified as identifiers). For example, if the type Boolean were not available in Pascal, its carrier set could simply be defined by:

> **type**   Boolean = (false, true)

In Pascal, enumerated types are automatically equipped with operations for comparison as well as with the functions succ and pred, i.e., successor and predecessor. In the above example, succ(false) = true, pred(true) = false; succ(true) and pred(false) are not allowed.

Records

Values of scalar types can be arranged into tuples by a construct called a *record* (also called a *structure* in some languages). Into programming languages, records introduce the equivalent of the Cartesian product. Tuples can be viewed as abstract data types. Consider pairs as an example:

> The type *Pairs:*
>> Carrier set: *Pairs,*
>> Auxiliary sets: *A, B;*
>> Operations: pair $\in A \times B \to$ *Pairs*;
>>   first $\in$ *Pairs* $\to A$;
>>   scnd $\in$ *Pairs* $\to B$;
>> where $\forall\ a \in A, b \in B$
>>   first (pair (a, b)) = a; scnd (pair(a, b)) = b;

Using Pascal notation this type is defined by:

> **type** Pairs = **record** first: A; scnd: B **end**

By providing the so-called field names, first and scnd, the programmer implicitly chooses the names for the selector functions. Pascal does not provide the function "pair." Instead, it permits the declaration of variables of type Pairs whose component values may be set (by assignment) and retrieved:

> p: Pairs;   {declaration of p as a variable of type Pairs}
> p.first := a; p.scnd := b;   {p now has the value of "pair(a,b)" above}
> if p.first = x then . . . else . . .   {p.first is Pascals notation for "first(p)" }

The sets A and B can be of any type including records. Furthermore, since there is no restriction on the number of fields records may have, they can represent arbitrary tuples.

Arrays

Arrays permit the construction of simple function spaces, $I \to A$. The domain I, a scalar type—in some languages restricted to a subset $\{0, 1, \ldots, n\}$ of the integers—is called the index set of the array; A is an arbitrary type. In Pascal, the mathematical notation $f \in I \to A$ assumes the form:

> f: **array**[I] **of** A

I has to be a finite scalar type (e.g., 0 .. 40). The function f can now be defined by associating values of type A with the values of the domain I using the assignment operation:

> f[i] := a; where $i \in I$ and $a \in A$

Application of f to a value j ∈ I is expressed by f[j]. This expression has a value of type A and can be used as such.

As with records, Pascal allows the naming of the function space I → A by the definition:

type myFunctions = **array**[I] **of** A

and the subsequent declaration of a specific function (array) by:

f : myFunctions

Functions of several arguments are represented by so-called multidimensional arrays:

$$f \in I_1 \times \cdots \times I_n \to A$$

is defined by

f : **array**$[I_1, \ldots, I_n]$ **of** A

Variant Records

Variant records model *disjoint* (also called *tagged*) *unions*. In contrast to an ordinary union C = A ∪ B, a disjoint union D = A + B is formed by tagging the elements of A and B before forming D such that elements of D can be recognized as elements of A or of B. In programming, this amounts to creating variables that can be used to house values of both type A and type B. A tag field, (usually) part of the variable, is used to keep track of the type of the value currently stored in the variable. In Pascal, D = A + B is expressed by:

```
type    tagType       =    (inA, inB){an enumerated type};
        D             =    record
                           case kind: tagType of
                             inA:  (aValue: A);
                             inB:  (bValue:B);
                           end.
```

Variables of type D are now used as follows:

```
            mix:   D;               {mix is declared to be of type D}
mix.kind          :=    inA;
aValue            :=    a;
. . .
if mix.kind = inA
    then {do something with mix.aValue, which is of type A}
    else {do something with mix.bValue, which is of type B}
```

Conceptually, only one of the two fields, mix.aValue or mix.bValue, exists at any one time. The proper use of the tag is policed in some languages (e.g., Ada) and left to the programmer in others (e.g., Pascal).

An Example of a User-Defined Abstract Data Type

Most carrier sets are assumed to contain a distinguished value: *error. Error* is not a proper computational object: a function is considered to compute *error* if it does not return to the point of its invocation due to some error condition. Functions are called **strict** if they compute the value *error* whenever one or more of their arguments have the value *error.*

The following example models a cafeteria tray stack with the following operations:

1. create a new stack with n trays;
2. remove a tray from a stack;
3. add a tray to the stack;
4. check if the stack is empty;

Specification:

Cts (cafeteria tray stacks)		is the carrier set of the type;
Boolean and *Integer*		are auxiliary sets;
newStack, remove, add, isEmpty		are the operations of the type

where

newStack	$\in$ Integer	$\to$ Cts; {create a stack of n trays}
remove, add	$\in$ Cts	$\to$ Cts;
isEmpty	$\in$ Cts	$\to$ Boolean;

Axioms (logical expressions that describe the semantics of the operations): All functions are strict and, for all non-negative values of n,

1. remove (newStack(n)) = if n = 0 then *error* else newStack(n − 1);
2. add(newStack(n)) = newStack(n + 1);
3. isEmpty (newStack(n)) = (n = 0).

These axioms suffice to describe the desired behavior of Cts exactly, i.e., using these axioms, arbitrary expressions built with the above operations can be reduced to *error* or newStack(m) for some m.

Implementation:

For the representation of Cts (i.e. its data structure) we will choose the type Integer.

type Cts = integer;

function newStack(n: Integer):Integer;
    begin if n < 0   **then** error ('n must be >= 0')   **else** newStack := n **end**;

function remove (s: Cts):Cts;
    begin if s = 0   **then** error ('stack empty')   **else** remove := s - 1 **end**;

function add (s: Cts):Cts;   **begin** add := s + 1 **end**;

function isEmpty (s: Cts):Boolean;   **begin** isEmpty := (s = 0) **end**;

Above, "error" is a function that prints an error message and stops the program.

Dynamic Types

The carrier sets of dynamic types contain objects of vastly different size. For these types, variables (memory space) must be allocated dynamically, i.e., at run time, when the actual sizes of objects are known. Examples for dynamic types are character strings, lists, tree structures, sets, and graphs. A classical example of a dynamic type is a special type of a list: a queue. As the name suggests, a queue object is a sequence of other objects with the particular restrictive property that objects are inspected at and removed from its front and added to its rear.

Specification of Queues

Carrier set:	*Queues*	
Auxiliary sets:	*Boolean, A*	(*A* contains the items to be queued)
Operations:	newQueue, isEmpty, queue, pop, front	

1. newQueue	$\in$ *Queues*;		{a new, empty queue}
2. isEmpty	$\in$ *Queues*	$\to$ *Boolean*;	{check if a queue is empty}
3. queue	$\in A \times Queues$	$\to$ *Queues*;	{add an object to the rear of a queue}
4. front	$\in$ *Queues*	$\to$ *A*;	{return front element for inspection}
5. pop	$\in$ *Queues*	$\to$ *Queues*;	{remove front element from a queue}

Axioms: All functions are strict and, for a ∈ A and s ∈ Queues,

	isEmpty (newQueue);	(i.e. isEmpty (newQueue)	is true)
not	isEmpty (queue(a, s));	(i.e. isEmpty (queue(a, s))	is false)

$$\text{pop (newQueue)} = \text{error;}$$
$$\text{pop (queue(a, s))} = \text{if } s = \text{newQueue}$$
$$\text{then newQueue else queue(a, pop(s));}$$

$$\text{front(newQueue)} = \text{error;}$$
$$\text{front(queue(a, s))} = \text{if } s = \text{newQueue}$$
$$\text{then } a \text{ else front(s).}$$

Implementation of Queues

The following implementation represents queues of the form queue(a, s) by the ordered pair (a, s) and a new, empty queue by the null pair *nil*. For the moment we assume that a data type, *Pairs*, which provides pairs on demand at run time, already exists and is defined as follows:

Specification of Pairs

Carrier Set: *Pairs*
Auxiliary sets: *Boolean, A*
Operations: nil, isNil, pair, first, scnd

1. nil	∈ Pairs,		{a distinguished pair, the null pair}
2. isNil	∈ Pairs	→ Boolean,	{test for the null pair}
3. pair	∈ A × Pairs	→ Pairs,	{combine an item and a pair to a new pair}
4. first	∈ Pairs	→ A,	{the first component, i.e., the item}
5. scnd	∈ Pairs	→ Pairs,	{the second component, i.e., the pair}

Axioms: All functions are strict and, for a ∈ A and p ∈ Pairs,

	isNil(nil);	(i.e. isNil(nil)	is true)
not	isNil(pair(a,p));	(i.e. isNil(pair(a,p))	is false)

$$\text{first (nil)} = \text{error;}$$
$$\text{first(pair(a,p))} = a;$$
$$\text{scnd (nil)} = \text{error;}$$
$$\text{scnd(pair(a,p))} = p;$$

With pairs, queues may now be implemented as follows:

```
type  Queues = Pairs;
function newQueue    : Queues;              begin  newQueue  := nil         end;
function isEmpty (s : Queues)   : Boolean;  begin  isEmpty   := (s = nil)   end;
function queue (x : A;  s : Queues)  : Queues;  begin  queue     := pair(x, s)  end;
function pop (s : Queues) : Queues;
begin
   if isNil(s)
      then error ('cannot pop empty queue')
      else if scnd(s) = nil
           then pop := nil
           else pop := pair (first(s), pop (scnd(s)))
   end;

function front (s : Queues) : A;

begin

   if isNil(s)
```

       then error ('an empty queue does not have a front')
       else if scnd(s) = nil
           then front := first(s)
           else front := front(scnd(s))
    end;

The logic of these programs echoes the axioms. Such implementations are sometimes not very efficient but useful for prototype programs, since the probability of their correctness is very high. The queues behave as *values*, i.e., the functions queue(a,s) and pop(s) do not modify the queues s but compute new queues; after the execution of, e.g., s1 := pop(s) there are two independent queue values, s and s1. This is exactly the behavior postulated by the axioms. However, practical applications frequently deal with *mutable objects*, objects that can be modified. With mutable objects memory may often be used more efficiently, since it is easier to decide when a memory cell, used, e.g., for storing an ordered pair, is no longer needed and thus may be recycled. If queues are viewed as mutable objects, the operations *queue* and *pop* are implemented as procedures that modify a queue. In order to apply the style of axioms introduced above for the description of mutable objects, these objects are best viewed as containers of values. The *procedure* queue(a, qobj), for example, takes the queue value, e.g., s, out of the container qobj, applies the *function* queue(a, s) (described by the axioms), and puts the result back into qobj.

If more than one place in a program needs to maintain a reference to such an object, the object must be implemented using a *head cell*: a storage location that represents the object and that is not released. The following implementation uses a head cell with two fields, one pointing to the front and one to the rear of the queue. We assume that the type *Pairs* has two additional functions:

    6. pairH ∈ Pairs × Pairs → Pairs;   {create a pair with 2 pair fields}
    7. firstH ∈ Pairs   → Pairs;   {retrieve the first field of such a 2-pair cell}

and three procedures:

    8. setfirstH (p: Pairs; q: Pairs);   {change the firstH field of q to p}
    9. setscnd (p: Pairs; q: Pairs);   {change the scnd field of q to p}
    10. delete (s: Pairs)   {free the storage space occupied by s}

```
type    Queues = Pairs;
procedure newQueue(var q : Queues);    begin q        := pairH(nil, nil)    end;
function isEmpty (s : Queues) : Boolean;    begin isEmpty    := (firstH(s) = nil)    end;
procedure queue (x : A;  s : Queues);
    var temp : Pairs;
    begin temp := pair(x, nil);
        if isNil(firstH(s)) then setfirstH(temp, s) else setscnd(temp, scnd(s));
        setscnd(temp, s);
    end;
function pop (s : Queues) : Queues;
    var temp : Pairs;
    begin
        if isNil(firstH(s))
            then error ('cannot pop empty queue')
            else begin temp := firstH(s);  setfirstH(scnd(temp), s);  delete(temp) end
    end;
function front (s : Queues) : A;
    begin
        if isNil(firstH(s))
            then error ('an empty queue does not have a front')
            else front := first(firstH(s))
    end;
```

Compared to the value implementation given earlier, this implementation improves the performance of *front* and *pop* from O(n) to O(1).

An algorithm has O(f(n)) (pronounced: order f(n) or proportional to f(n)) time performance if there exists a constant c, such that, for arbitrary values of the input size n, the time that the algorithm needs for its computation is $t \leq c \cdot f(n)$.

Most modern programming languages support the implementation of the type *pairs* (*n-tuples*) whose instances can be created dynamically. It requires two operations, called *new* and *dispose* in Pascal, that dynamically allocate and deallocate variables, respectively. These operations depend on the concept of a *reference* (or *pointer*), which serves as a name for a variable. References always occupy the same storage space independently of the type of variable they refer to. The following implementation of *Pairs* explains the concept.

```
type  CellKind = (headCell, bodyCell);
      Pairs    = ^PairCell; {Pairs are references to PairCells}
      PairCell = record     tail: Pairs;
                     case kind: CellKind of
                         headCell: (frnt: Pairs);
                         bodyCell: (val: A)
                     end;
function pair(item: A; next:Pairs):Pairs;
    var p: Pairs;
    begin
        new(p, bodyCell); {a new "bodyCell" is created and accessible through p}
        p^.kind := bodyCell; p^.val := item; p^.tail := next; pair := p
    end;
function first(p: Pairs):A;
    begin
        if p = nil then error('...')
            else if p^.kind = bodyCell then first := p^.val else error('...')
    end;
procedure setfirstH(p, q:Pairs);
    begin
        if q = nil then error('...')
            else if q^.kind = headCell then q^.frnt := p else error('...')
    end;
```

(Note: The Pascal constant **nil** denotes the null pointer, a reference to nothing.)

```
function isNil(p: Pairs): Boolean; begin isNil := (p = nil) end;
```

The reader should have no difficulty filling in the rest of the implementation of Pairs. Most of the algorithms on dynamic data structures have been developed in the 1960s; still, an excellent reference is Knuth [1973].

More Dynamic Data Types

Stacks and Lists with a Point of Interest

A queue is the type of (linear) list used to realize first-come-first-served behavior. In contrast, another linear type, the *stack*, realizes last-come-first-served behavior. Sometimes it is necessary to scan a list object without dismantling it. This is accomplished by giving the list a point of interest (see Fig. 81.14). This requires four additional operations:

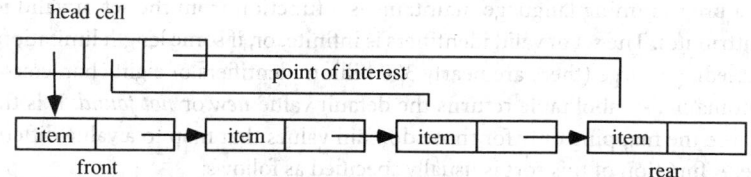

FIGURE 81.14 A list implementation with a point of interest and access to front and rear.

restart(l: List);	{moves point of interest to beginning of list l}
current(l:List):A;	{returns object at point of interest}
advance(l:List);	{advances point of interest by one toward end of list l}
endOfList(l : List): Boolean;	{true, if end of list has been reached}

The type can be extended further by allowing insertions and deletions at the point of interest.

N-ary, Binary, and General Trees

An **n-ary tree** is the smallest set containing the *empty tree* and all ordered n+1-tuples $t = (a, t_1, \ldots, t_n)$ where a is member of some auxiliary set and the t_i are n-ary trees. The element a is called the *root element* or simply the root of t and the t_i are called *subtrees* of t.

Note that in this sense, a list is a unary tree. Binary trees used as searchtrees access finite ordered sets.

A **binary searchtree** is a tree that accesses a set. A tree t *accesses* a set s if the root of t is some element a of s, and s_1, called the left subtree of t, accesses the subset $\{x|x \in s \text{ and } x < a\}$ and s_2, called the right subtree of t, accesses the subset $\{x|x \in s \text{ and } x > a\}$.

If the left and right subtrees of the above definition are of similar size, then the time for finding an element in the set is proportional to $\log(n)$ where n is the cardinality of the set.

Quaternary trees (usually called quad trees) and octonary trees (called oct trees) are used to access two-dimensionally and three-dimensionally organized data, respectively. As with lists, the implementation of n-ary trees is based on n+1-tuples. A minimal set of operations for binary trees includes:

nilTree	∈ Trees;		{the empty tree, represented by nil}
isNil	∈ Trees	→ Boolean;	{test if tree is empty}
tree	∈ A × Trees × Trees	→ Trees;	{build tree from an item and subtrees}
root	∈ Trees	→ A;	{retrieve root item of tree}
left	∈ Trees	→ Trees;	{retrieve left subtree}
right	∈ Trees	→ Trees;	{retrieve right subtree}

A **general tree** is the smallest set containing all order pairs $t = (a,s)$ where a is a member of some auxiliary set and s is a possibly empty list of general trees. The element a is called the root element or simply the root of t and the trees in s are called subtrees of t.

Note that there is no empty general tree; the simplest tree has a root and an empty list of subtrees. General trees are useful for the representation of hierarchical organizations such as the table of contents of a book or the organization of a corporation.

Functions, Sets, Relations, Graphs

Functions with reasonably small domains can be represented by arrays, as described earlier. Similarly, sets formed from a reasonably small universal set, relations on small domains, and graphs with not too many vertices can be represented by their characteristic functions implemented as bit arrays. In fact, Pascal provides a type constructor for sets that are derived from small universal sets.

Frequently domains are far too large for this approach. For example, the symbol table that a compiler of a programming language maintains is a function from the set of valid identifiers to some set of attributes. The set of valid identifiers is infinite, or, if some length limitation is imposed, finite but exceedingly large (there are nearly 300 billion identifiers of eight characters or less). For most of its domain a symbol table returns the default value *new* or *not found*. It is therefore economical to store the mapping only for those domain values that map to a value different from the default value. A function of this sort is usually specified as follows:

Specification of Functions:

      Carrier Set: *Functions*
      Auxiliary sets: *Dom, Cod*   (domain and codomain)
      Operations: newFun, apply, update

    1. newFun $\in$ Functions,   (returns default everywhere)
    2. apply $\in$ Functions $\times$ Dom $\to$Cod,
    3. update $\in$ Functions $\times$ Dom $\times$ Cod $\to$ Functions;

Axioms: All functions are strict and, for x,z $\in$ Dom, y $\in$ Cod and f $\in$ Functions,

    apply(newFun, x)   =   default;
    apply (update(f,x,y), z) = if x = z then y else apply (f, z);

An implementation based on these axioms amounts to representing a function as a list of those of its individual mappings that differ from *default* and leads to an **O(1) performance** for *update* and an O(n) performance of *apply*. Better is an implementation by binary searchtrees with a performance of O(log(n)) for both *apply* and *update*.

Hashing

The fastest method for the implementation of functions is *hash coding* or *hashing*. By means of a *hash function*, h $\in$ Dom $\to$ 0 .. k–1, the domain of the function is partitioned into k sections and each section is associated with an index. For each partition a simple list implementation is used and the lists are stored in an array A: array[1 .. k–1]. In order to compute apply(f,x) or update(f,x,y), the list at A[hash(x)] is searched or updated. If the hash function has been properly chosen and if k and the number of function values different from default are of similar size, then the individual lists can be expected to be very short and independent of the number of nondefault entries of the function; thus performance for apply and update is O(1).

The above discussion applies also to sets, relations, and graphs, since these objects can be represented by their characteristic functions.

Object-Oriented Programming

In languages that support object-oriented programming, *classes* (i.e., types) of objects are defined by specifying (1) the variables that each object will own as *instance variables* and (2) operations, called *methods*, applicable to the objects of the class. As a difference in style, these methods are not invoked like functions or procedures, but are *sent* to an object as a *message*. The expression [*window moveTo* : *x* : *y*] is an example of a message in the programming language *Objective C*, a dialect of C. Here the object *window*, which may represent a window on the screen, is instructed to apply to itself the method *moveTo* using the parameters *x* and *y*.

New objects of a class are created—usually dynamically—by *factory methods* addressed to the class itself. These methods allocate the equivalent of a record whose fields are the instance variables of the object and return a reference to this record, which represents the new object. After its creation an object can receive messages from other objects.

To data abstraction, object-oriented programming adds the concept of inheritance: from an existing class new (sub)classes can be derived by adding additional instance variables and/or methods. Each subclass inherits the instance variables and methods of its superclass. This encourages the use of existing code for new purposes. As an example, consider a class of a *list* objects. Each object has two instance variables pointing to the front and the rear of the list. In Objective C, the specification of the interface for this list class, i.e., the declaration of the instance variables and headers (functionalities) of the methods, has the following form:

```
@interface  MyLists : Object     /* Object is the universal (system) class from which
                                    all classes are derived directly or indirectly */
{                                /* declaration of the instance variables;
    listRef front;                 listRef is the type of a pointer to a list assumed
    listRef rear;                  to be defined elsewhere */

}
– initList;                      /* initializes instance variables with null pointers */
– (BOOL) isEmpty;                /* test for empty list; note: parameter list is implied */
– add : (item) theThing;         /* item is the type of things on the list */
– pop;
– (item) front;
@end
```

As a companion of the *interface* file there is also an *implementation* file that contains the executable code for the methods of the class. A list with a point of interest can be defined as a subclass of MyList as follows:

```
@interface   ScanList : MyList  /* ScanList is made a subclass of MyList */

{   listRef  pointOfInterest; }

– restart;
– (BOOL)endOfList;
– advance;
– (Item)current;

@end
```

If we also need a list that can add and delete at the point of interest, we define:

```
@interface       InsertionList : ScanList
{ }                                    /* there are no new instance variables */
– insert : (item) theThing;
– shrink;                              /* removes item at the point of interest */

@end
```

If a subclass defines a method already defined in the superclass, the new definition overrides the old one. Suppose we need a list where items are kept in ascending order:

```
@interface   SortedList : MyList
{ }
– add : (item) theThing;         /* this version of add inserts theThing at the proper
                                    place to keep the list sorted */
@end
```

Defining Terms

Abstract data type: One or more sets of computational objects together with some basic operations on those objects. One of these sets is defined by the type either by enumeration or by generating operations and is called the *carrier set* of the type. Customarily it is given the same name as the type. All other sets are called auxiliary sets of the type. In exceptional cases a type may have more than one carrier set.

Binary searchtree: A tree that accesses a set. A tree t accesses a set s if the root of t is some element a of s, and s_1, called the left subtree of t, accesses the subset $\{x \mid x \in s \text{ and } x < a\}$, and s_2, called the right subtree of t, accesses the subset $\{x \mid x \in s \text{ and } x > a\}$.

Functionality: With sets A and B, the expression $A \rightarrow B$ denotes the set of all functions that have the domain A and the codomain B. Functions $f \in A \rightarrow B$ (traditionally denoted by $f\colon A \rightarrow B$) are said to have the functionality $A \rightarrow B$.

General tree: The smallest set containing all ordered pairs $t = (a, s)$ where a is member of some auxiliary set and s is a possibly empty list of general trees. The element a is called the root element or simply the root of t and the trees in s are called subtrees of t.

n-ary tree: The smallest set containing the *empty tree* and all ordered n+1-tuples $t = (a, t_1, \ldots, t_n)$ where a is member of some auxiliary set and the t_i are n-ary trees. The element a is called the root element or simply the root of t and the t_i are called subtrees of t.

O(f(n)) performance: An algorithm has $O(f(n))$ (pronounced: order $f(n)$ or proportional to $f(n)$) time performance if there exists a constant c, such that, for arbitrary values of the input size n, the time that the algorithm needs for its computation is $t \leq c \cdot f(n)$.

Strictness: Most carrier sets are assumed to contain a distinguished value: *error*. *Error* is not a proper computational object: a function is considered to compute *error* if it does not return to the point of its invocation due to some error condition. Functions are called *strict* if they compute the value *error* whenever one or more of their arguments have the value *error*.

References

K. Jensen and N. Wirth, *Pascal: User Manual and Report,* Berlin, Springer-Verlag, 1974.

B. W. Kernighan and D. M. Ritchie, *The C Programming Language,* Englewood Cliffs, N.J.: Prentice-Hall, 1988.

D. E. Knuth, *The Art of Computer Programming,* vol. 1, Reading, Mass.: Addison-Wesley, 1973, chap. 2.

J. J. Martin, *Data Types and Data Structures,* C.A.R. Hoare, Series Ed., Englewood Cliffs, N.J.: Prentice-Hall International, 1986.

NeXT Step Concepts, Chapter 3, *Objective C,* NeXT Developers' Library, NeXT Inc.

B. Stroustrup, *The C++ Programming Language,* Reading, Mass.: Addison-Wesley, 1991.

United States Department of Defense, *Reference Manual for the Ada® Programming Language,* Washington D.C.: U.S. Government Printing Office, 1983.

N. Wirth, *Programming in Modula-2,* Berlin: Springer-Verlag, 1983.

Further Information

There is a wealth of textbooks on data structures. Papers on special aspects of data types and their relation to programming languages are found regularly in periodicals such as *ACM Transactions on Programming Languages and Systems, the Journal of the Association for Computing Machinery, IEEE Transactions on Computers, IEEE Transactions on Software Engineering,* or *Acta Informatica.*

82

Memory Systems

James R. Goodman
University of Wisconsin-Madison

Gurindar S. Sohi
University of Wisconsin-Madison

82.1 Introduction 1927
82.2 Parallel and Interleaved Memories 1928
82.3 Memory Hierarchies 1931
82.4 Virtual Memory 1934

82.1 Introduction

A *memory system* serves as a repository of information in a computer system. The processing unit [or processor or central processing unit (CPU)] accesses information from the memory system, operates on it, and stores it back. The memory system is a collection of storage locations. Each storage location, or *memory word*, has an *address*; a collection of storage locations form an *address space*. Figure 82.1 shows the essentials of how a processor is connected to a memory system via address, data, and control lines.

When a processor attempts to read a memory location, the request is very urgent. In virtually all computers, the work soon comes to a halt (in other words, the processor *stalls*) if the memory request is not serviced. High-performance computers may be able to continue briefly by overlapping memory requests, but even the most sophisticated computers will frequently exhaust their ability to process data and stall momentarily in the face of long memory delays. Thus, a key performance parameter in the design of any computer, fast or slow, is the speed of its memory.

Ideally, the memory system must be infinitely large so that it can contain an arbitrarily large amount of information and infinitely fast so that it does not limit the processing unit. Practically, however, this is not possible. There are three properties of memory that are inherently in conflict: speed, capacity, and cost. In general, technology tradeoffs can be employed to optimize any two of the three factors at the expense of the other. Thus it is possible to have memories that are (1) large and cheap, but not fast; (2) cheap and fast, but small; or (3) large and fast, but expensive. The last of the three is further limited by physical constraints. A large-capacity memory that is very fast is also physically large, and speed-of-light delays place a limit on the **latency** of such a memory system.

The *latency* (*L*) of the memory is the delay from when the processor first requests a word from memory until that word arrives and is available for use by the processor. The latency of a memory system is one attribute of performance; the other is **bandwidth.** The bandwidth (*BW*) is the rate at which information can be transferred from the memory system. The bandwidth and the latency are related. If *R* is the number of requests that the memory can service simultaneously, then:

$$BW = \frac{R}{L} \tag{82.1}$$

From Eq. (82.1) we see that a decrease in the latency will result in an increase in bandwidth, and vice

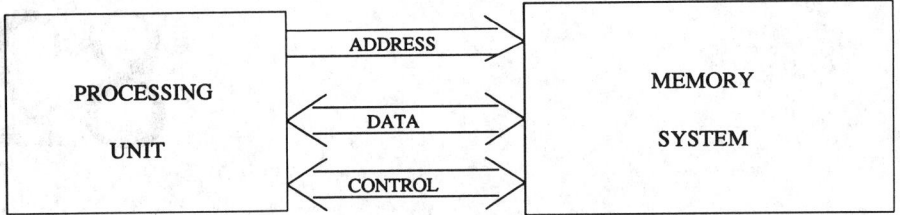

FIGURE 82.1   The memory interface.

versa, if R is unchanged. Also obvious is that the bandwidth can be increased by increasing R, if L does not increase proportionately. For example, we can build a memory system that takes 20 ns to service the access of a single 32-bit word. Its latency is 20 ns per 32-bit word, and its bandwidth is

$$\frac{32}{20 \times 10^{-9}} \frac{\text{bits}}{\text{sec}}$$

or 200 Mbytes/s. If the memory system is modified to accept a new request for a 32-bit word every 5 ns, then its bandwidth is

$$\frac{32}{5 \times 10^{-9}} \frac{\text{bits}}{\text{sec}}$$

or 800 Mbytes/s. Equation (82.1) also tells us that this memory system has to be designed to handle four requests at a given time.

The basic unit of construction of a semiconductor memory system is a *module* or *bank*. A memory bank, constructed from several memory chips, can service a single request at a time. The time that a bank is busy servicing a request is called the *bank busy time*. The bank busy time limits the bandwidth of a memory bank.

Since an ideal memory system (infinite capacity, zero latency and infinite bandwidth, with affordable cost) is not possible, the challenge is, given the cost and technology constraints, to engineer a memory system whose abilities match the abilities that the processor demands of it. That is, engineering a memory system that performs as close to an ideal memory system (for the given processing unit) as is possible. For a processor that stalls when it makes a memory request [most current (1992) microprocessors fall in this category], this implies the engineering of a memory system with the lowest possible latency; engineering a memory system that can handle more than one request at a time is unnecessary since the processor can never have more than one request outstanding. For vector processors which can have multiple memory requests outstanding, it is important not only to reduce latency, but also to increase bandwidth (over what is possible by latency reduction alone) by designing a memory system that is capable of servicing multiple requests simultaneously.

Parallel or **interleaved** memories are used to improve bandwidth, and memory hierarchies are used to decrease latency.

82.2 Parallel and Interleaved Memories

Multiple memory banks can be connected together to form an *interleaved* or *parallel* memory system. Since each bank can service a request, an interleaved memory system with K banks can service K requests simultaneously, increasing the peak bandwidth of the memory system to K times the bandwidth of a single bank. In most interleaved memory systems, the number of banks is a power of two, that is, $K = 2^k$. An n-bit memory word address is broken into two parts: a k-bit bank number and an m-bit address of a word within a bank. Though the k bits used to select a bank number could be any k bits of the n-bit word address, typical interleaved memory systems use the low-order

k address bits to select the bank number; the higher order $m = n - k$ bits of the word address are used to access a word in the selected bank. The reason for using the low-order k bits will be discussed shortly. An interleaved memory system which uses the low-order k bits to select the bank is referred to as a *low-order* or a *standard* interleaved memory.

There are two ways of connecting multiple memory banks: *simple interleaving* and *complex interleaving*. Sometimes simple interleaving is also referred to as *interleaving*, and complex interleaving as *banking*.

Figure 82.2 shows the structure of a simple interleaved memory system. m address bits are supplied to all the memory banks simultaneously. All banks are also connected to the same read/write control line (not shown in the figure). For a read operation, the banks start the read operation and deposit the data in their latches. Data can then be read from the latches, one by one, by setting the switch appropriately. Meanwhile, the banks could be accessed again, to carry out another read or write operation. For a write operation, the latches are loaded, one by one. When all the latches have been written, their contents can be written into the memory banks by supplying m bits of address (they will be written into the same word in each of the different banks). In a simple interleaved memory, all banks are cycled at the same time; each bank starts and completes its individual operations at the same time as every other bank; a new memory cycle can start (for all banks) once the previous cycle is complete. Details of the timing of accesses can be found in *The Architecture of Pipelined Computers*, by P. M. Kogge.

One use of a simple interleaved memory system is to back up a **cache memory** (discussed in the next section). To do so, the memory must be able to read blocks of contiguous words (a cache block) and supply them to the cache. If the low-order k bits of the address are used to select the bank number, then consecutive words of the block reside in different banks, and they can all be read in parallel, and supplied to the cache one by one. If some other address bits are used for bank selection, then multiple words from the block might fall in the same memory bank, requiring multiple accesses to the same bank to fetch the block.

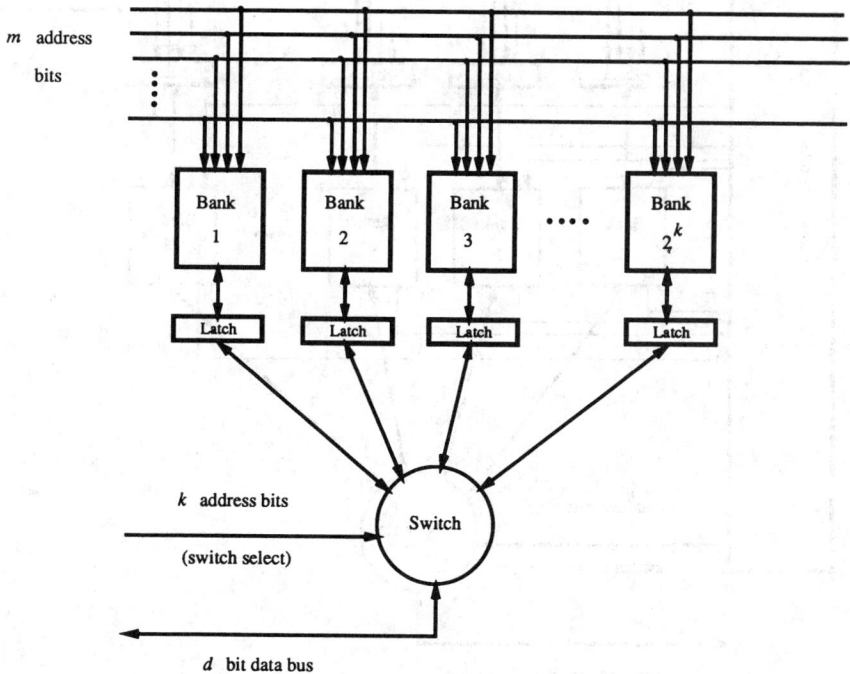

FIGURE 82.2 A simple interleaved memory system. (*Source:* Adapted from Kogge, 1981.)

Figure 82.3 shows the structure of a complex interleaved memory system. In such a system, each bank is set up to operate on its own, independent of the operation of the other banks. For example, bank 1 could be carrying out a read operation on a particular memory address, and bank 2 could be carrying out a write operation on a completely unrelated memory address. (Contrast this with the operation in a simple interleaved memory where all banks are carrying out the same operation, read or write, and the locations accessed within each bank represent a contiguous block of memory.) This is accomplished by providing an address latch and a read/write command line for each bank. The overall operation of the interleaved memory is controlled by the *memory controller*. The processing unit submits the memory request to the memory controller, which determines which bank needs to be accessed. It determines if the bank is busy (by monitoring a busy line for each bank). The request is held if the bank is busy, and submitted to the bank when it is available to access the request. When the bank responds to a read request, the switch is set by the controller to accept the request from the bank and forward it to the processing unit. Details of the timing of accesses can be found in *The Architecture of Pipelined Computers*, by P. M. Kogge.

A typical use of a complex interleaved memory system is in a vector processor. In a vector processor, the processing units operate on a vector, for example a portion of a row or a column of a matrix. If consecutive elements of a vector are present in different memory banks, then the memory system can sustain a bandwidth of one element per clock cycle. By arranging the data suitably in memory and using standard interleaving (for example, storing the matrix in row-major order will place consecutive elements in consecutive memory banks), the vector can be accessed at the rate of one element per clock cycle as long as the number of banks is greater than the bank busy time.

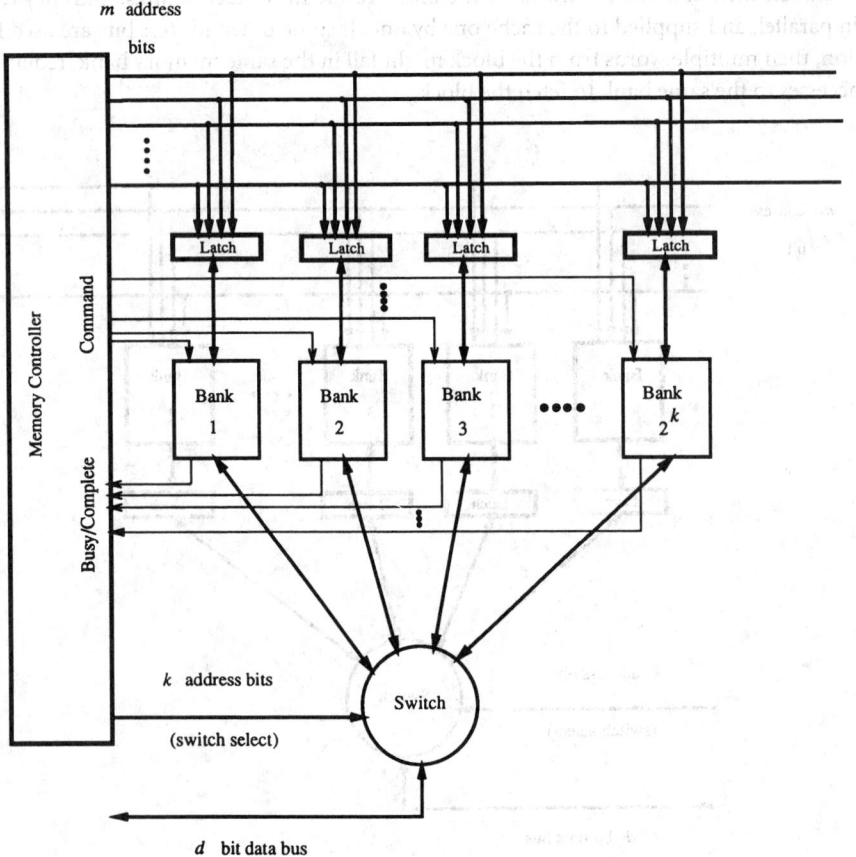

FIGURE 82.3   A complex interleaved memory system. (*Source:* Adapted from Kogge, 1981.)

Interleaved memory systems found in high-end *vector supercomputers* are slight variants on the basic complex interleaved memory system of Fig. 82.3. Such memory systems may have hundreds of banks, with multiple memory controllers which allow multiple independent memory requests to be made every clock cycle.

82.3 Memory Hierarchies

As mentioned above, technology does not permit memories that are cheap, large, and fast. By recognizing the nonrandom nature of memory requests, and emphasizing the *average* rather than worst case latency, it is possible to build a hierarchical memory system. A small amount of very fast memory, placed in front of a large, slow memory, can be designed to satisfy most requests at the speed of the small memory. This, in fact, is the primary motivation for the use of registers in the CPU: in this case, the programmer makes sure that the most commonly accessed variables are allocated to registers.

A variety of techniques, employing either hardware or software, or a combination of the two, can be employed to assure that most memory references are satisfied by the faster memory. The foremost of these techniques is the exploitation of the principle of *locality of reference*. This principle captures the fact that some memory locations are referenced much more frequently than others. *Spatial locality* is the property that an access to a given memory location greatly increases the probability that neighboring locations will be accessed immediately. This is largely, but not exclusively, a result of the tendency to access memory locations sequentially. *Temporal locality* is the property that access to a given memory location greatly increases the probability that the same location will be accessed again soon. This is largely, but not exclusively, a result of the frequency of looping behavior of programs. Particularly for temporal locality, a good predictor of the future is the past: the longer a variable has gone unreferenced, the less likely it is to be accessed next.

A commonly employed memory hierarchy includes a special, high-speed memory, known as a *cache,* in addition to the conventional memory, referred to as *main memory,* or *backing storage.* The hardware can dynamically allocate parts of this memory for addresses deemed most likely to be accessed soon. The cache contains only redundant copies of the address space, which is wholly contained in the main memory. The cache memory is *associative,* or *content-addressable,* which is to say, the address of a memory location is stored, along with its content. Rather than reading data directly from a memory location, the cache is given an address and responds by providing data which may or may not be that requested. The cache indicates that it is providing the correct data by indicating a *hit.* Otherwise it indicates a *miss.* On a miss, the memory access is then performed with respect to the backing storage, and the cache is updated to include the new data.

A useful performance parameter is the *effective latency.* If the latency L_{HIT} is known in the case of a hit and the latency in the case of a miss is L_{MISS}, the effective latency can be determined from the *hit ratio* (H), the portion of memory accesses that are hits:

$$L_{\text{average}} = L_{\text{HIT}} \cdot H + L_{\text{MISS}} \cdot (1 - H) \qquad (82.2)$$

The portion of memory accesses that miss is called the *miss ratio* ($M = 1 - H$). The hit ratio is strongly influenced by the program being executed, but is largely independent of the ratio of cache size to memory size. It is not uncommon for a cache containing a few thousand bytes capacity to exhibit a hit ratio greater than 90%.

The cache is intended to hold the most active portions of the memory, and the hardware dynamically selects portions of main memory to store in the cache. When the cache is full, bringing in new data must be matched by deleting old data. Thus a strategy for cache management is necessary. Cache management strategies exploit the principle of locality. Spatial locality is exploited in the choice of what is brought into the cache. Temporal locality is exploited in the choice of what gets deleted. When a cache miss occurs, hardware copies into the cache a large, contiguous block of

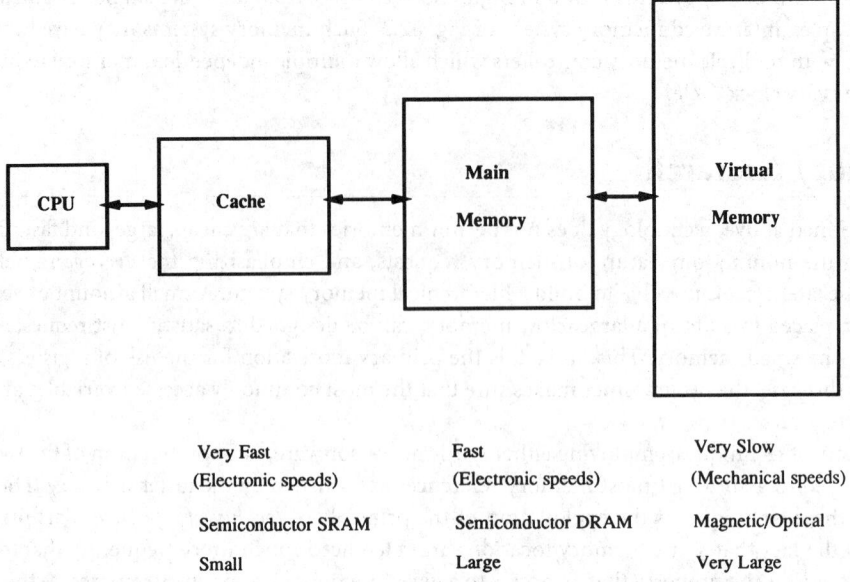

FIGURE 82.4   A memory hierarchy.

memory that includes the word requested. This fixed-size region of memory, known as a cache *line* or *block*, may be as small as a single word, or up to several hundred bytes. A block is a set of contiguous memory locations, usually a power of two. A block is said to be *aligned* if the lowest address in the block is exactly divisible by the block size. That is to say, for a block of size *B* beginning at location *A*, the block is aligned if

$$A \; modulo \; B = 0 \qquad\qquad (82.3)$$

Conventional caches require that all blocks be aligned.

When a block is brought into the cache, it is likely that another block will have to be evicted. The evicted block is selected based on some attempt to capture temporal locality. Since prescience is so difficult to achieve, other methods are generally used to predict future memory accesses. Least-recently-used (LRU) is often the basis for the choice. Other replacement policies are sometimes used, particularly because true LRU replacement requires extensive logic and bookkeeping.

The cache often comprises two conventional memories, one known as the data memory and one as the tag memory as shown in Fig. 82.5. The address of each cache line contained in the data memory is stored in the tag memory, as well as other information (*state* information), particularly the fact that a cache line is present (the cache is initially empty). Each line contained in the data memory is allocated a corresponding word in the tag memory to indicate the full address of the cache line.

The requirement that the cache memory be associative complicates the design. Addressing data by content is inherently more complicated than by its address. All the tags must be compared concurrently, of course, because the whole point of the cache is to achieve low latency. Thus the cache can be made simpler by introducing a mapping of memory locations to cache cells, limiting the number of possible cells where a given line can be held. The extreme case is known as *direct mapping*, where each memory location is mapped to a single location in the cache. This makes many aspects of the design easy, since there is no choice in where the line might reside, and no choice as to the line to be deleted, but can result in poor utilization of the cache when two memory locations are alternately accessed and must share a single cache cell.

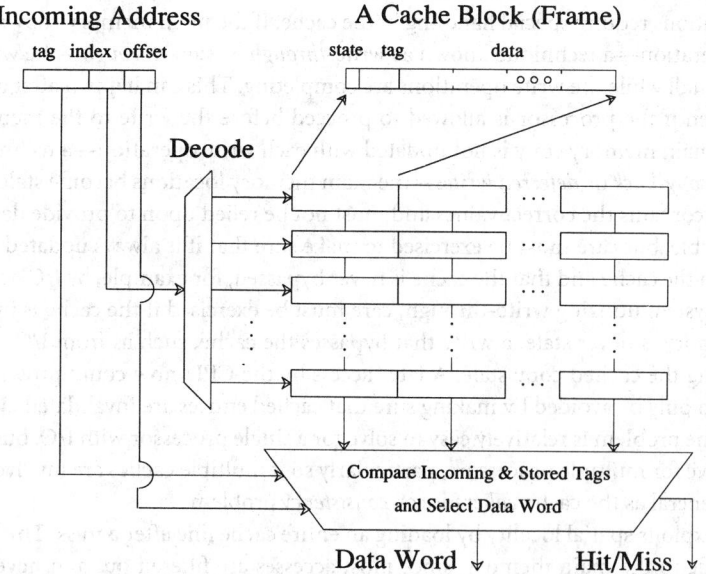

FIGURE 82.5 Components of a cache memory. (*Source:* Adapted from Hill, 1988.)

A hashing algorithm is used to determine the cache address from the memory address. The conventional mapping algorithm consists of a function of the form

$$A_{\text{cache}} = \frac{A_{\text{memory}} \text{ modulo } size\_of\_cache}{size\_of\_cache\_line} \tag{82.4}$$

where A_{cache} is the address within the cache for main memory location A_{memory}, *size_of_cache* is the capacity of the cache in addressable units (usually bytes), and *size_of_cache_line* is the size of the cache line in addressable units. Since the hashing function is simple bit selection, the tag memory need only contain the part of the address not implied by the hashing function. That is,

$$A_{\text{tag}} = A_{\text{memory}} \text{ div } size\_of\_cache \tag{82.5}$$

where A_{tag} is stored in the tag memory and div is the integer divide operation. In testing for a match, the complete address of a line stored in the cache can be inferred from the tag and its storage location within the cache.

A *two-way set-associative* cache maps each memory location into either of two locations in the cache and can be constructed essentially as two identical direct-mapped caches. However, both caches must be searched with each memory access and the appropriate data selected and multiplexed on a match. On a miss, a choice must be made between the two possible cache lines as to which is to be deleted. A single LRU bit can be saved for each such pair of lines to remember which line has been accessed more recently. This bit must be toggled to the current state each time either of the cache lines is accessed.

In the same way, an *M-way associative* cache maps each memory location into any of M memory locations in the cache and can be constructed from M identical direct-mapped caches. The problem of maintaining the LRU ordering of M cache lines quickly becomes hard, however, since there are M! possible orderings, and therefore it takes at least

$$\lceil \log_2 (M!) \rceil \tag{82.6}$$

bits to store the ordering. In practice, this requirement limits true LRU replacement to 3- or 4-way set associativity.

Write operations require special handling in the cache. If the main memory copy is updated with each write operation—a technique known as *write-through* or *store-through*—the writes may force operations to stall while the write operations are completing. This can happen after a series of write operations even if the processor is allowed to proceed before the write to the memory has completed. If the main memory copy is not updated with each write operation—a technique known as *write-back* or *copy-back* or *deferred writes*—the main memory locations become stale, that is, memory no longer contains the correct values and must not be relied upon to provide data. This is generally permissible, but care must be exercised to make sure that it is always updated before the line is purged from the cache and that the cache is never bypassed, for example, by I/O accesses (DMA).

Even for a system utilizing write-through, care must be exercised if the cache is bypassed. While the main memory is never stale, a write that bypasses the cache, such as from I/O, could have the effect of leaving the cached copy stale. A later access by the CPU now could provide an incorrect value. This can only be avoided by making sure that cached entries are invalidated even if the cache is bypassed. The problem is relatively easy to solve for a single processor with I/O, but becomes very difficult to solve for multiple processors, particularly so if multiple caches are involved as well. This is known in general as the cache *coherence* or *consistency* problem.

The cache exploits spatial locality by loading an entire cache line after a miss. This tends to result in bursty traffic to the main memory, since most accesses are filtered out and never reach it, but after a miss, the memory system must provide an entire line at once. Cache memory nicely complements an interleaved, high-bandwidth memory, since a cache line can be interleaved across many banks in a regular manner, avoiding memory conflicts, and loaded rapidly into the cache.

Conventional caches cannot accept requests when they are servicing a miss request. In other words, they *lock up* or *block* when they are servicing a miss. In some processing situations, it may be necessary for the memory system to continue accepting (and servicing) requests from the processor while a miss is being serviced. In fact, it might even be necessary to service multiple miss requests simultaneously. To allow this mode of operation, the cache has to be designed to be *lockup-free* or *non-blocking*.

The notion of a memory hierarchy can be applied at multiple levels. For example, a computer may have a small on-chip cache and a larger off-chip cache inserted between the CPU and the main memory. The technique is remarkably broad in its applicability and can also be applied at a different level between main memory and a disk. This approach is traditionally known as **virtual memory.**

82.4 Virtual Memory

Cache memory contains portions of the main memory in dynamically allocated cache lines. Since the data portion of the cache memory is itself a conventional memory, at any time a line present in the cache has two addresses associated with it: its main memory address and the cache address. Thus the main memory address of a word can be divorced from a particular storage location and thought of abstractly simply as an element in the address space. The use of a two-level hierarchy consisting of main memory and a slower, larger disk storage device evolved by making a clear distinction between the address space and the locations in memory. An address generated during the execution of a program is known as a *virtual address*, which must be translated to a *real address* before it can be accessed in main memory. The total address space is only an abstraction.

A virtual memory address is mapped to a real address, which indicates the location in main memory where the data actually reside. The mapping is maintained through a table, known as the *page table*, and maintained in software by the operating system. Like the tag memory of a cache memory, the page table is accessed through a virtual address to determine the main memory (real) address of the entry. Unlike the tag memory, however, the table is usually sorted by virtual addresses, making the translation process a simple matter of an extra access to determine the real address of the desired item. A system maintaining the page table in the way analogous to a cache tag memory is said to have *inverted page tables*. In addition to the real address mapped to a virtual page, and an indication of whether the page is present at all, a page table entry often contains other information, for example the location on the disk of the place where the page is stored when not present in main memory. The unit of transfer between the disk and main memory is known as a page and is generally somewhat larger than a cache line, often in the thousands of bytes.

The virtual memory can be thought of as a collection of blocks. These blocks are often of fixed size and aligned, in which case they are known as *pages*. A virtual address can be broken into two parts, a virtual page number and an offset. The page number specifies the page to be accessed, and the page offset indicates the offset from the beginning of the page to the indicated address.

A real address can also be broken into two parts, a real page number, also called a *page frame* number, and an offset. The mapping is done at the level of pages, so the page table can be indexed by means of the virtual page number. The page frame number is contained in the page table and is read out during the translation, along with other information about the page. In most implementations the page offset is the same for a virtual address and the real address to which it is mapped.

The virtual memory hierarchy is different than the cache/main memory hierarchy in a number of respects, resulting primarily from the fact that there is a much greater difference in latency between the disk and the main memory in the virtual memory hierarchy. While a typical latency ratio for cache and main memory is one order of magnitude (main memory has a latency ten times larger than the cache), the latency ratio between disk and main memory is often four orders of magnitude or more. This results from the fact that the disk is a mechanical device, and its latency is limited by velocity and inertia, whereas main memory is limited only by electronic and energy constraints. Because of this much larger penalty for a miss, many design decisions are affected by the need to minimize the frequency of misses. In addition, when a miss does occur, the processor would be idle for the period in which it could execute tens of thousands of instructions. Rather than stall during this time, as is done upon a cache miss, the processor invokes the operating system and may switch to a different task. Because the operating system is being invoked anyway, it is convenient to rely on the operating system to set up and maintain the page table, unlike cache memory, where it is done entirely in hardware.

Hardware support provided for a virtual memory system generally includes the ability to translate the virtual addresses provided by the processor into the real addresses needed to access main

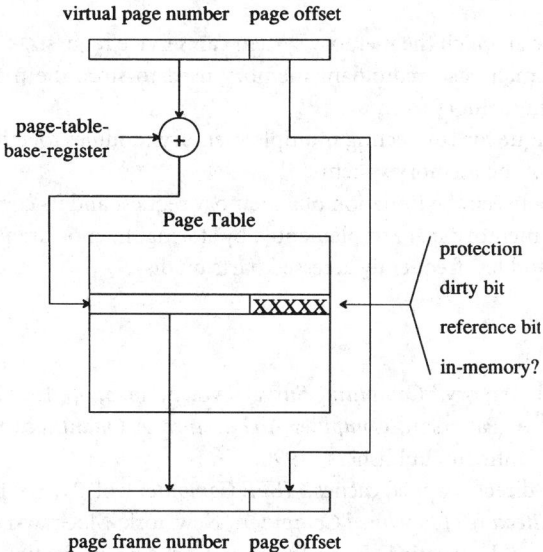

FIGURE 82.6 Virtual-to-real address translation.

memory. Thus only on a virtual address miss is the operating system invoked. An important aspect of a computer implementing virtual memory, however, is the necessity to freeze the processor at the point where a miss occurs, service the page table fault, and later return to continue the execution as if no page fault had occurred. This requirement means either that it must be possible to halt execution at any point—including possibly in the middle of a complex instruction—or it must be possible to guarantee that all memory accesses will be to pages resident in main memory.

As described above, virtual memory requires two memory accesses to fetch a single entry from memory, one into the page table to map the virtual address into the real address, and the second to fetch the data itself. This process can be sped up in a variety of ways. First, a special-purpose cache memory to store the active portion of the page table can be used to speed up the first access. This special-purpose cache is often called a *translation lookaside buffer* (TLB). Second, if the system also employs a cache memory, it may be possible to overlap the access of the cache memory with the access to the TLB, allowing the requested item to be accessed in a single cache access time in typical cases. The two accesses can be fully overlapped if the virtual address supplies sufficient information to fetch the data from the cache before the virtual-to-real address translation has been accomplished. This is true for an *M*-way set-associative cache of capacity *C* if the following relationship holds:

$$Page\_size \geq \frac{C}{M} \qquad (82.7)$$

For such a cache, the index into the cache can be determined strictly from the page offset. Since the virtual page offset is identical to the real page offset, no translation is necessary, and the cache can be accessed concurrently with the TLB. Of course the real address must be obtained before the tag can be compared.

An alternative method applicable to a system containing both virtual memory and a cache is to store in the tag memory the virtual address instead of the real address. This technique introduces consistency problems in virtual memory systems that permit more than a single address space or that allow a single physical page to be mapped to more than a single virtual page. This is known as the *aliasing* problem.

Defining Terms

Bandwidth: The rate at which the memory system can service requests.
Cache memory: A small, fast, redundant memory used to store the most frequently accessed parts of the main memory.
Interleaving: Technique for connecting multiple memory modules together in order to improve the bandwidth of the memory system.
Latency: The time between the initiation of a memory request and its completion.
Virtual memory: A memory space implemented by storing the most frequently accessed parts in main memory and less frequently accessed parts on disk.

References

P. J. Denning, "Virtual memory," *Computing Surveys*, vol. 2, no. 3, pp. 153–170, Sept. 1970.

J. L. Hennessy and D. A. Patterson, *Computer Architecture: A Quantitative Approach*, San Mateo, Calif.: Morgan Kaufmann Publishers, 1990.

M. D. Hill, "A case for direct-mapped caches," *IEEE Computer*, vol. 21, no. 12, Dec. 1988.

P. M. Kogge, *The Architecture of Pipelined Computers*, New York: McGraw-Hill, 1981.

D. Kroft, "Lockup-Free Instruction Fetch/Prefetch Cache Organization," in *Proceedings 8th Annual International Symposium on Computer Architecture*, May 1981, pp. 81–87.

A. J. Smith, "Bibliography and readings on CPU cache memories and related topics," *ACM SIGARCH Computer Architecture News*, vol. 14, no. 1, pp. 22–42, Jan. 1986.

A. J. Smith, "Second bibliography on cache memories," in *ACM SIGARCH Computer Architecture News*, vol. 19, no 4, pp. 154–182, June 1991.

Further Information

A. V. Pohm and O. P. Agrawal, *High-Speed Memory Systems*, Reston, Va.: Reston Publishing Company, 1983.

I n the mid-1940s, John Von Neumann, a brilliant mathematician at Princeton University, conceived a theoretical machine in which binary logic and arithmetic could work together in storing detailed programs and performing complex calculations. Von Neumann demonstrated that one could encode instructions to the machine in the same language used for the data it processed. This great advance meant that a computer could read instructions, accept data, perform calculations, and store results all in a single code.

These ideas pointed the way toward the design, construction, and operation of units that can be employed separately or combined for greater power and flexibility. They also focused attention on the newly developed integrated circuits in which very small, highly reliable components could store and process digital information. Finally, Von Neumann's concepts led to the idea of computers sending and receiv-

John Von Neumann with the IAS computer, completed in 1952 under his direction. (Courtesy of IEEE Center for the History of Electrical Engineering.)

ing information to and from other computers. The modern computer network incorporating many diverse computing elements is one outcome.

83

Input and Output

83.1	Input Devices	1938
	Keyboards • Light Pen • Data Tablet (Graphics, Digitizer) • Mouse • Trackball • Joystick • Touch Input • Scanners • Voice • Summary • Advantages and Disadvantages	
83.2	Computer Output Printer Technologies	1958
	Classification of Printer Technologies • Page Printer Technologies • Serial Nonimpact Printer Technologies • Impact Printer Technologies	

Solomon Sherr
Westland Electronics

Robert C. Durbeck
IBM Corporation (Retired)

83.1 Input Devices*

Solomon Sherr

Input devices are those portions of computer, data processing, and information systems that perform the essential function of providing some means for entering commands and data into the system. Therefore, input devices are found in all such systems, but are treated here as a separate equipment group, independent of the total system configuration. However, the place of input devices in a representative computer system may be clarified by reference to Fig. 83.1(a), which shows the interface of the main input device categories in relation to the portions of the generalized system that accept the inputs. The categories and the devices listed in Table 83.1 are the subject of this section.

Keyboards

Keyboards are essentially electromechanical devices, and are still so ubiquitous, in spite of the inroads of other input devices, that they deserve at least a limited treatment. The primary type of keyboard in use as an input device is the alphanumeric (A/N) form, well known in its typewriter application, but with various additions and expansions consisting of numeric and special function keys. This type of keyboard is shown in Fig. 83.1(b) with a standard QWERTY format, so named because of the layout of the top left alpha keys, for the A/N portion, a separate numeric set to the right, and a group of function keys at the top. Other layouts for the A/N portion have been proposed and at least one (Dvorak) accepted by the American National Standards Institute (ANSI), but it has not received much use in spite of its advantages in increased efficiency. At present, the overwhelming majority of system keyboards still use the QWERTY layout, and it is the only one considered here.

---

*The material contained in this section is a shortened version of that which appears in *Electronic Displays*, 2nd ed., by Sol Sherr, Chapter 6, Section 6.1, 1993 (in press), published by John Wiley & Sons, Inc., and is reprinted here by permission.

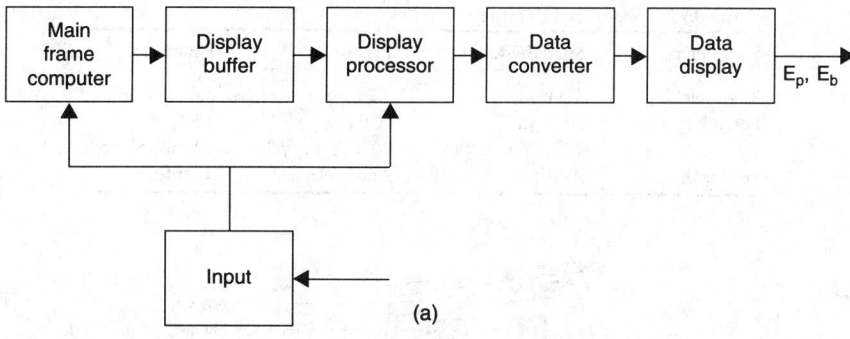

(a)

(b)

FIGURE 83.1 (a) Generalized display-system block diagram. (*Source:* After S. Sherr, *Electronic Displays,* New York: John Wiley & Sons, 1979. With permission.) (b) Alphanumeric keyboard. (Courtesy of Key tronic.)

As illustrated in Fig. 83.1, a keyboard consists of a number of keyswitches whose exact structure is of prime importance in keyboard design. The relevant characteristics of keyswitch operation are life, actuation force, travel distance, and feedback. Accepted values are shown in Table 83.2 for different keyswitch designs. The elastomer type is preferred to a limited extent over the other two when the electronic audio feedback is included. This indicates that some type of audio feedback is desirable. One form of keyswitch design using an elastomer or "molded boot" is shown in Fig. 83.2(a), in which the boot consists of two collapsible domes. In this design, the internal movement of the keyswitch is completely silent so that some source of sound must be added to achieve the desired audible feedback.

Table 83.1 List of Input Devices

Category	Designation	Operation Mode
Keyboard	Alphanumeric	Electromechanical
Keyboard	Function	Electromechanical
Pointing	Light pen	Screen pointing
Pointing	Touchscreen	Screen pointing
Pointing	Pen tablet	Tablet pointing
Coordinates	Digitizer	X-Y conversion
Coordinates	Data tablet	X-Y location
Cursor	Mouse	Movement
Cursor	Trackball	Movement
Cursor	Joystick	Movement
Image	Scanner	Conversion
Verbal	Voice	Conversion

Table 83.2   Keyboard Parameter Values

Parameter	Snap Switch	Elastomer	Foam Pad
Key travel	3.8 mm	3.2 mm	3.8 mm
Force	>60 gm	>50 gm	>30 gm
Life	10 million cycles	10 million cycles	10 million cycles
Feedback	Audio mechanical	Audio electric	Tactile

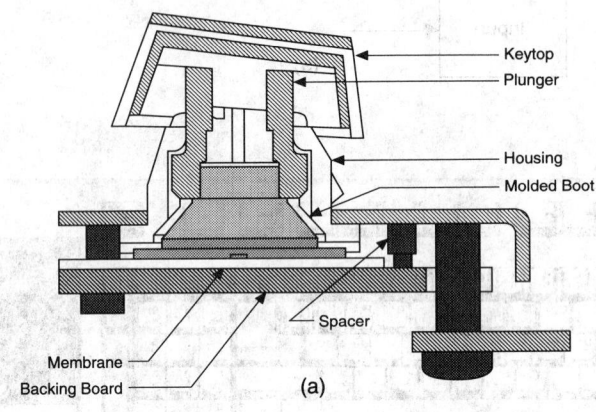

(a)

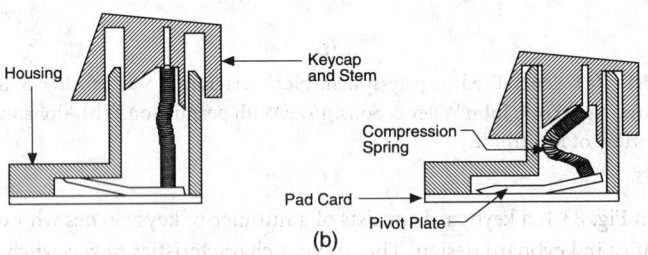

(b)

FIGURE 83.2   (a) Elastomer-type keyswitch. (b) Snap switch. (*Source:* After H. Brunner *et al.*, "Effects of key action design on keyboard preference and throughput performance," Micro Switch. With permission.)

The snap switch design shown in Fig. 83.2(b) has built-in sound and achieves a small reduction in insertion errors over the elastomer design with audio feedback.

The life requirement is estimated on the basis of workstation users operating at approximately half the accepted rate of 20 million actuations per key used for electronic typewriters. The actual layout and content of the keyboard may vary greatly, ranging from the standard typewriter arrangement, through different combinations of alphanumerics and symbols, to the special-function keyboards that contain legends and symbols specific to the particular application. However, the outputs of each type are the same in that they must contain coded signals that relate the action to be performed by the information system to that defined by the key being operated, in terms of the input code of the system. Thus, many of the keyboards output the ASCII code, and the system is usually designed so that it can accept this type of standard code. Incidentally, ASCII, the acronym for American Standard Code for Information Interchange, is the standard means for encoding alphanumerics and a group of selected symbols for transmission to a display system, among others. It is the standard code used in the United States and most other English-speaking countries and corresponds to the ISO seven-bit code. The seven-bit ASCII is usually used, and it should be noted that for serial data transmission an eighth bit is added for parity. Various keyboard arrangements are possible, and many variants are found in particular applications. The means for coding the key operation may be through magnetic reed relays, solid-state circuits, or more exotic devices such as Hall effect sensors. However, we do not pursue these means any further in order to avoid getting bogged down in a whole new class of device characteristics that are only incidental to the operation. Similarly, we do not discuss the human-factors aspects of keyboard design, not because they are not important, but because, apart from the visual considerations, the other factors have to do with tactile and physical features best left to others.

Light Pen

The light pen initially was a very popular means for accomplishing manual input to the random deflection information display systems, but fell out of favor when raster systems became more popular due to its being somewhat difficult to use with raster systems. This device goes by a misleading cognomen, as despite its name it does not emit light and is not a pen other than being somewhat similar to one in its physical appearance, as shown in Fig. 83.3(a). However, when we consider its functional characteristics, the validity of the term becomes apparent, since it is used to cause the electron beam to "write" patterns on the cathode ray tube (CRT) that are defined by the motion of the light pen on the CRT faceplate.

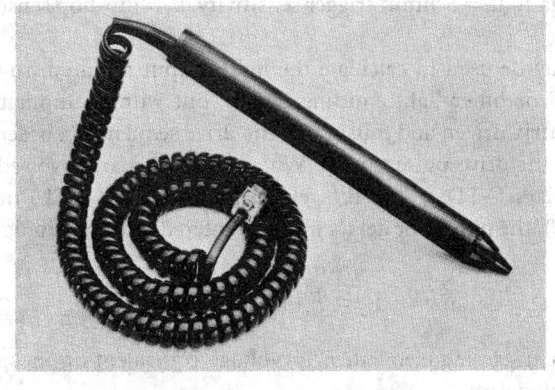

(a)

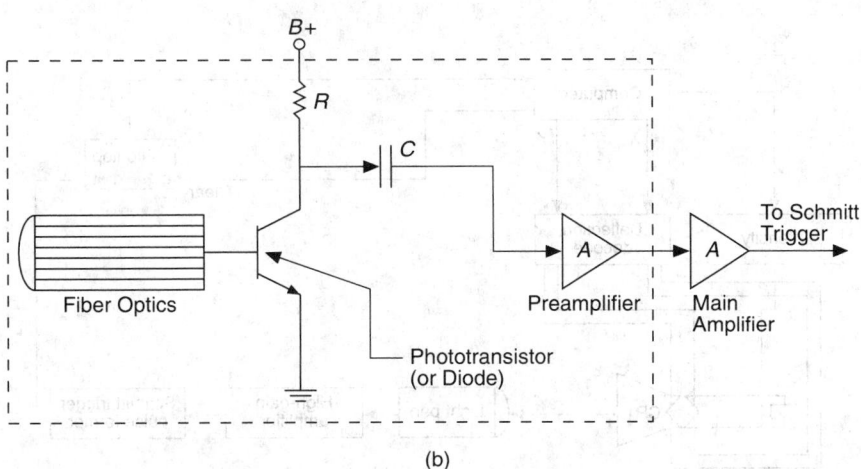

(b)

FIGURE 83.3 (a) Light pen. (Courtesy of FTG Data Systems.) (b) Light pen schematic. (*Source:* After S. Sherr, *Electronic Displays*, New York: John Wiley & Sons, 1979, p. 388. With permission.)

The light pen operates by sensing the existence or nonexistence of a pulse of light at the point on the screen of the CRT or surface of any other light-emitting device where the point of the pen is placed. This is accomplished by means of the circuit shown in Fig. 83.3(b), where the light pulse is collected and transmitted through the fiber optics to a light-sensitive device that converts the light pulse into an electrical pulse which is shaped by some form of electronics (of which a Schmitt trigger is one example). We need not concern ourselves with the exact form of the electronics except to note that this pulse is then sent to the computer, as shown in Fig. 83.4, and provides a complete, closed-loop system. As the electronic pulse occurs at the time when the light pulse passes under the light pen, the computer is informed of the location at which the designated operation is to be performed and may proceed accordingly. Thus, the light pen is a pointing device that designates a point on the display screen and can be used as an input device. Various light pen programs have been written to expand the capabilities of the original one, and it should be noted that the light pen is coming back into favor as improvements in accuracy, ease of operation, and reliability occur, and it is once more a valid input device.

There are two characteristics of light pen operation that affect the capabilities of this input device. The first is the sensitivity, given by

$$S = E_L \mu_p A_p A_m \mu_s \mu_f t_L \qquad (83.1)$$

where E_L = illuminance at photodetector, μ_p = photodetector sensitivity, A_p = preamplifier gain, A_m = main amplifier gain, μ_s = Schmitt trigger sensitivity, μ_f = flip-flop sensitivity, and t_L = optical loss.

Equation (83.1) may be used to calculate the light output required from the display surface, which may be a CRT or other light-emitting device, but with the limitation that most of the flat panel units are matrix driven and must track the drive sequence in order to know the location of the light pen from the drive pulse timing. When phosphors are involved as for the CRT, vacuum fluorescent displays (VFDs), thin-film electroluminescent (TFEL) units, and color liquid crystal displays (LCDs), the phosphor delays must be entered into the timing, and the total delay is given by

$$E_o = E_i(1 - e^{-t/\tau}) \qquad (83.2)$$

where E_o = voltage at triggering element, E_i = voltage equivalent of phosphor light output, t = time, and τ = sum of all delays.

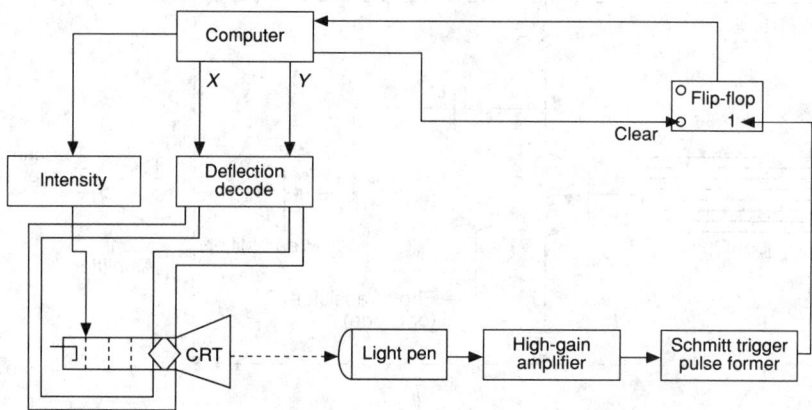

FIGURE 83.4 Block diagram of light pen computer system. (*Source:* S. Sherr, *Electronic Displays*, New York: John Wiley & Sons, 1979, p. 389. With permission.)

These delays set limits to the positional accuracy, as the computer tracking the signal will be in error by this amount. Other inaccuracies are due to the dimensions of the optical pickup surface, all of which somewhat negate the simplicity of operation. The result is the parameter values shown in Table 83.3.

Table 83.3 Light Pen Data

Field of View	Response Time	Sensitivity
0.02–0.08 in.	120–150 ns	0.02–0.04 ft·L

Data Tablet (Graphics, Digitizer)

A very convenient means for data entry, retaining some of the ease of operation of the light pen but with much better accuracy, are the various forms of data tablets available. These tablets differ from the light pen in another significant way in that they do not require a moving spot of light to detect the location of the beam or direct it to a new location. This need for a moving light spot made the light pen difficult to use with the data tablets initially designed to overcome this limitation while still using a device with a pen-like input. The first successful example was the Rand tablet, a digital device that used an X–Y assembly wire from which a wand placed above some point on the X–Y wire matrix would pick up the output from a pulse generator that fed X and Y electrical pulses into the matrix. By determining the number of pulses in a time period, the location of the wand is established. Another similar device used magnetostrictive rather than electrical signals to accomplish the same result, and this location is converted into display coordinates used to position a cursor on the CRT screen. The cursor may then be used as a visual feedback element so that the operator can correct the position of the wand until the cursor is properly placed. At this time the information from the tablet may also be transferred to either the host computer or the resident desktop or portable computer, as desired. Since the cursor is not used to signal its position to a pickup device, as is the case with the light pen, it may be used with any type of display system, including the non-light-emitting flat panel displays. Another advantage of the tablet is that it may be used to position cursors in the blank areas of the display, where no light pulses are available unless they are specially generated by the light pen.

Subsequent to the development of the first data tablet as exemplified by the Rand Tablet, there have been numerous improvements and new developments using a variety of technologies. These technologies include magnetostrictive, electromagnetic, electrostatic or capacitive, scanned X–Y grid, resistive, and sonic. Of these, electromagnetic tablets dominate the digitizer market, and sonic is of interest because it does not require a tablet, but most of the other technologies are essentially restricted to touch input devices covered later. As noted previously, electromagnetic is the most popular technology for high-performance digitizer tablets. Operation is based on transformer principles, whereby a conductor carrying ac creates a magnetic field around it that induces a current in a second conductor. The digitizer tablet uses the amplitude and phase of the induced current to determine digitizing data. The tablet contains an X–Y pattern of conductors beneath its surface, in a manner similar to the Rand Tablet, but instead of counting pulses in a time period a circular conductor is used as the pick-up element for the induced current. This coil is placed on the tablet surface, and its position is determined by measuring the phase and amplitude of the current in the coil. Its center is interpolated by sweeping through the X–Y grid lines and demodulating the signal in the coil to determine the phase reversal point, or by calculating this point using digitized data fed into a microprocessor. The X–Y coordinates may be resolved to better than 0.025 mm using either of these two techniques. Figure 83.5(a) is a photograph of a representative digitizer tablet.

Another digitizer technology which should be noted is that using the measurement of the time required for sound waves to travel from a source to movable microphone pickups. This sonic technology has the advantage that no special digitizing board is required, and either a stylus or a cursor can be used as the digitizer. Two sonic sources are contained in an L frame so that both X and Y coordinates can be determined by calculating the time it takes for the sound wave to reach the

(a)

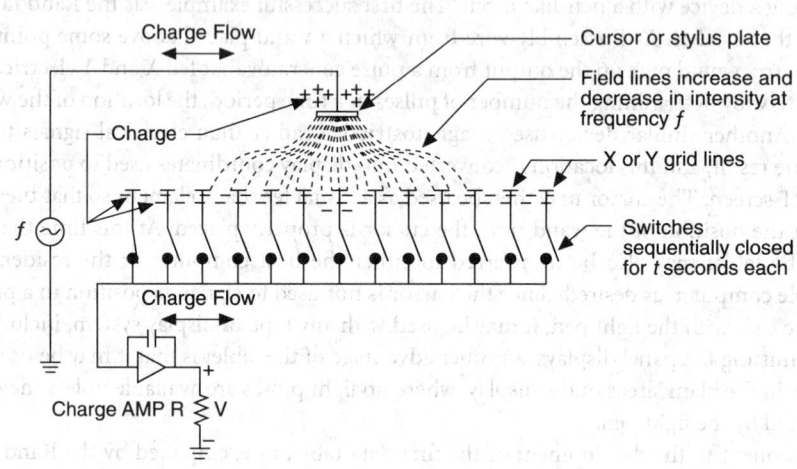

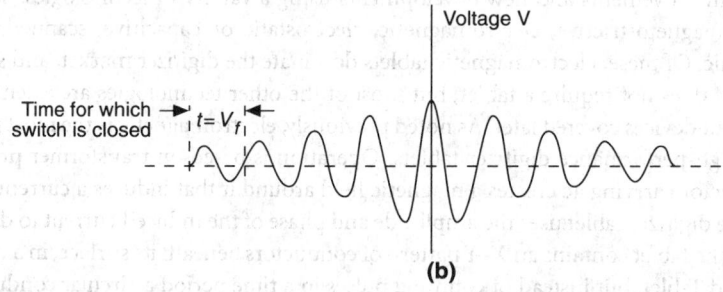

(b)

FIGURE 83.5 (a) Digitizer tablet. (Courtesy of Numonics.) (b) Capacitive technology. (*Source:* After T. E. Davies *et al.*, "Digitizers and input tablets," in *Input Devices*, S. Sherr, Ed., New York: Academic Press, 1988, p. 186. With permission.)

microphones contained in the pickup device. This calculation is made on the basis of sound traveling at 345 m/s at 20°C, and the accuracy is dependent on stable ambient conditions. This tends to limit the resolution to about 300 lpi, and the accuracy to ±0.1%. The device may also be implemented with a single sonic source as the digitizing means and a pair of microphones located outside the digitizing area. In this case the location of the transducer is calculated by triangulation and converted into Cartesian coordinates.

Digitizers are used primarily for inputting accurate coordinate data from maps and engineering drawings. Their high accuracy requirements have led to relatively high prices. Alternative means for inputting data are the data and graphics tablets that meet most input requirements at a lower cost and accuracy. The main technology is still electromagnetic, and the units are essentially the same as the digitizers, but with lower accuracies. However, several of the other technologies have also been used to achieve lower costs. Most successful among them are the capacitive and resistive versions, which may also be used as digitizers. The capacitive units, also termed electrostatic, use capacitive coupling where the coupling between the tablet and the cursor or stylus is determined by the capacitance made up of the tablet surface as one plate and the pickup element as the other. In this case, the capacitance is given by

$$C = f(\epsilon A/d) \tag{83.3}$$

where C = capacitance, ϵ = permittivity of dielectric, A = relative area of two plates, d = distance between plates, and f = proportionality factor.

A scanned grid approach is used to determine the location of the cursor. As in the electromagnetic tablet, an X–Y grid of conductors is embedded in the tablet, with semiconductor switches on each line providing contact on a scanned basis. The charge flowing from each capacitance is summed through a summing amplifier as shown in Fig. 83.5(b). The resultant voltage peaks twice, once for the X and once for the Y lines, as they are scanned. The peak positions are digitized by means of a counter that starts at the beginning of the scan, and runs at some multiple of the scan rate. The digital values represent the coordinates of the cursor location.

Mouse

The mouse has gone a long way from its original invention by Engelbart in 1965, through its redesign at Xerox and introduction by Apple as a main input device, and its general acceptance by computer users as an important addition to the group of input devices. It should be noted, in passing, that the mouse is essentially an upside-down trackball, although the latter is now being referred to as an upside-down mouse. However, the trackball has priority as is described further in the next section.

Mice contain motion-sensing elements and are operated by moving mechanical or optical elements. One form uses wheels and shafts to drive the sensing elements, as shown schematically in Fig. 83.6. The angular velocity (ω) of the wheel and shaft is given by

$$\omega = V_r/R \quad \text{rad/s} \tag{83.4}$$

where V_r = velocity of wheel and R = wheel radius.

The rotation angle (θ) is given by

$$(\theta) = X/R \quad \text{rad} \tag{83.5}$$

where X = distance moved.

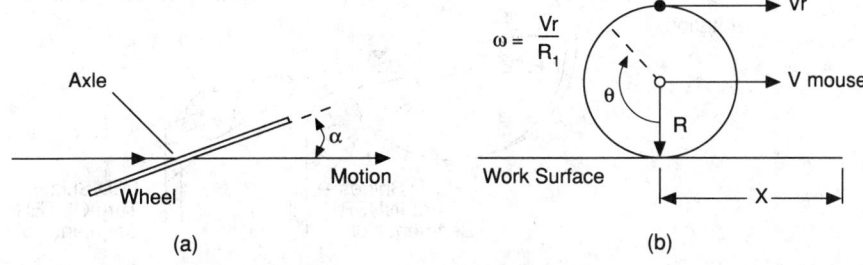

(a)   (b)

FIGURE 83.6 Wheel showing velocities and slip angle. (*Source:* After C. Goy, "Mice," in *Input Devices*, S. Sherr, Ed., New York: Academic Press, 1988, p. 225. With permission.)

This type of mouse has two sets of wheels and shafts, one for horizontal and the other for vertical motion.

A more popular type of mechanical mouse is the one that uses a ball for the motion sensing device, as shown in Fig. 83.7. Again, the velocity of the ball circumference equals the velocity of the mouse, and the angular velocity is given by

$$\omega = V/R_1 \text{ rad/s} \qquad (83.6)$$

where R_1 = shaft radius.

The smaller the shaft the more rapid its rotation for a given mouse velocity. Another form of the ball-and-shaft mouse is the one that uses an optical interrupter, as shown in Fig. 83.8. In this form, the light from the light-emitting diodes (LEDs) is interrupted by the coded disks that are rotated by the shafts, and is then picked up by the phototransistors and converted into the digital signal that represents the disk rotation. An optical interrupter is also used for the optomechanical mouse, and here the interrupter con-

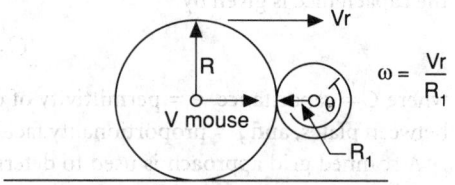

Work Surface

FIGURE 83.7 Ball and shaft. (*Source:* C. Goy, "Mice," in *Input Devices*, S. Sherr, Ed., New York: Academic Press, 1988. With permission.)

tains a set of slots; as the interrupter rotates quadrature signals are created that correspond to the shaft rotation.

In addition to the shaft and optomechanical mice, an early form of mouse used multiturn potentiometers connected to the wheels, and the output voltage that represented the motion varied in direct proportion to the mouse motion. The voltage was then converted by means of an analog-to-digital converter into digital form for input to the computer.

Finally, there are the true optical mice that use a special surface that is printed with a set of geometric shapes, usually a grid of lines or dots, that are illuminated and focused on a light detector. The most common form uses a grid made up of orthogonal lines, with the vertical and horizontal lines printed in different colors. These colors absorb light at different frequencies so that the optical detectors can differentiate between horizontal and vertical movement of the mouse. If such a structure is used as the mouse, then the photodetector will pick up a series of light-dark impulses consisting of the reflections from the mirror surface and the grid lines and convert them into square waves. A second LED and photodetector that is mounted orthogonally to the first is used to detect motion in the orthogonal direction, and the combination of the two inks avoids confusion between the two directions of motion. The system then counts the number of impulses created by

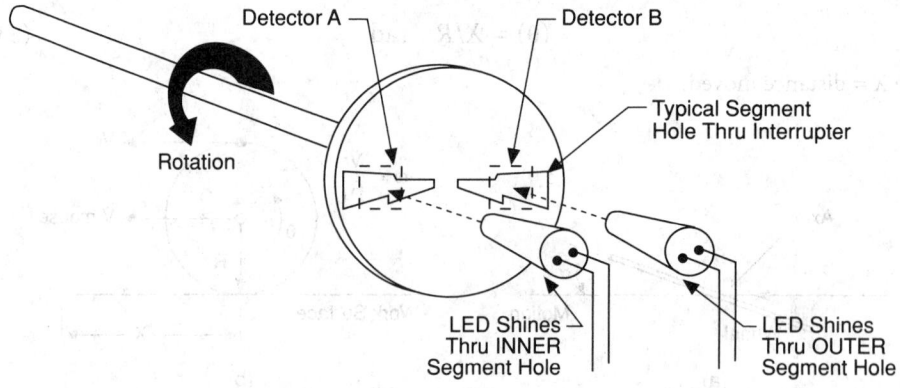

FIGURE 83.8 Optical interrupter. (*Source:* C. Goy, "Mice," in *Input Devices*, S. Sherr, Ed., New York: Academic Press, 1988, p. 229. With permission.)

the mouse motion and converts the result into motion information for the cursor. This type of mouse has the advantage that no mechanical elements are required.

Trackball

As noted previously, the trackball uses technology similar to the mouse, but preceded it as an input device. Thus, the comment that it is an upside-down mouse should be reversed. The movable element is housed in an assembly as is shown in Fig. 83.9, and the assembly remains stationary so that much less desk space is required than for the mouse. In addition, the trackball may be mounted on a keyboard so that very little additional desk space is needed. The movable element can be the same as used in the mouse, and the output can be a set of bits corresponding to the coordinates to which the cursor should be driven, or where the command should be carried out. The output format is essentially equivalent to that used for the mouse, and the same protocols are used.

The typical trackball has an X and Y optical encoder that generates a pulse for each 0.76 mm of incremental motion of the ball. This means that the pulse train may range from 10 to 2500 pulses per second (pps), depending on how fast the ball is rotated. This is much more rapid than required for satisfactory updates, which need not be greater than about 100 times per second. This can easily be accomodated by the RS-232 protocol using an eight-bit word. Thus, the trackball is an excellent alternative for the mouse, and is rapidly returning to a preferred position as an input device.

Joystick

The joystick has not achieved much acceptance as an input device for electronic display systems, except for video games, although it has been the preferred control for many types of aircraft. However, it can be used to some extent in display systems other than those used in video games, and therefore warrants inclusion in this section. There are two basic types of joysticks, termed "displacement" and "force-operated," respectively. A typical displacement joystick is shown in Fig. 83.10, and may have two or three degrees of freedom. The activating means may vary from as few as four switches mounted 90 degrees apart, to full potentiometers for analog output, and optical encoders for digital output. A third axis may be added by allowing the handle to rotate and drive a third potentiometer. Spring forces of 5 to 10 lbs. are usual for the other two axes, and displacements go from 6 to 30 degrees.

The force joystick operates by responding to pressure on the handle to generate the X–Y coordinates. It may be either a two- or three-dimensional version, with the same types of handles as for

FIGURE 83.9   Trackball. (Courtesy of CH Products, Vista, Calif.)

the displacement joysticks. However, it is difficult to use a rotating handle for the third dimension because some force is usually transmitted to the other dimensions causing crosstalk. Therefore, a separate lever is preferred. The force is detected by means of piezoelectric sensors that are bonded to the handle rod, and a voltage source is applied across the network, as shown in Fig. 83.11. The output is taken from the strain gauge and the analog voltage will be proportional to the amount of force. The same type of protocol and output circuitry may be used as for the displacement unit, and both can generate either position or rate data. An exponential curve with a dead zone threshold is preferred for pulse rates in order to avoid starting pulse rate uncertainties, with the first pulse starting as soon as the threshold is exceeded.

Touch Input

Touch input devices come in two basic forms, either placed directly on the display surface, or as a separate panel attached to the computer system. In its second form it is

FIGURE 83.10 Three-axis displacement joystick. (Courtesy of CH Products, Vista, Calif.)

essentially a data tablet and differs mainly in that it acts as another display unit with some form of a touch-sensitive surface. In this implementation it is the same as the *Touchscreen* input device, and this discussion concentrates on the technologies used for Touchscreens. There are five different technologies used for touch input devices, which are capacitive or resistive overlays, piezoelectric, light beam interruption, and surface acoustic wave. The system may be divided into the sensor unit, which senses the location of the pointing element, and the controller that interfaces with the sensor and communicates the location information to the system computer. Since the controller is an electronic device that does not use technology different from the computer it is not covered here. The main differences among the different touch input devices are due to the choice of sensor technology, and the discussion concentrates on these technologies.

Capacitive. Capacitive overlay technology is illustrated in Fig. 83.12 where a transparent metallic coating is placed over the display screen and the finger or stylus capacitance is sensed to determine the touch location. The overlay may consist of a group of separate sections etched into the surface with each separate section connected to the controller, or a continuous surface connected at the

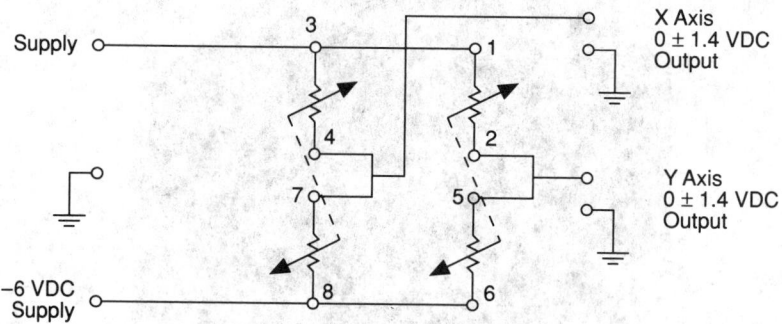

FIGURE 83.11 Schematic connections in a force joystick. (*Source:* After D. Doran, "Trackballs and joysticks," in *Input Devices*, S. Sherr, Ed., New York: Academic Press, 1988, p. 260. With permission.)

four corners. The first form is termed discrete capacitive, and touch location is determined by having each section sequentially connected to an oscillator circuit where the frequency of oscillations is affected by the pointing device. The oscillation frequency is measured and compared to a stored reference frequency. If the frequency difference is large enough then it is recognized as a touch at that location. It is a simple system, but suffers from low resolution and slow response so that it is only practical for menu selection.

The analog capacitive system uses the same metallic overlay, but the metallic surface is continuous rather than etched. The connections at the four ends are each connected to a separate oscillator, and the frequency of each is measured and stored. Then when the overlay is touched the change in capacitance will have a different effect on the frequency of each oscillator. These are measured and the differences are used to determine the coordinates of the touch by means of an algorithm. This technique is capable of much higher resolution (250 × 250) than the digital approach and is preferred for graphics or other high-density displays.

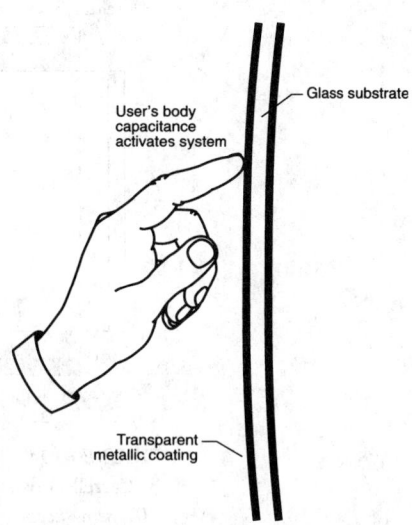

FIGURE 83.12 Capacitive overlay technology. (*Source:* After A. B. Carrell and J. Carstedt, "Touch input technology," *SID Sem. Lecture Notes,* p. 15.30, 1987. With permission.)

Resistive.   Resistive overlay technology requires a more complex assembly consisting of two layers, as illustrated in Fig. 83.13. The layers both contain transparent metallic surfaces and are separated by spacers so that an air gap exists between the layers in the absence of any pressure on the touch panel. The metallic layers face each other and when the outer panel is pressed the metallic layers make contact and form a conductive path at the point of contact. When a voltage is applied between the top of the outer layer and the bottom of the inner layer, the two layers act as a voltage divider, and the voltage at the point of contact may be measured in the X and Y directions by applying the voltage in first one and then the other direction. The measured voltages are then transmitted to the controller where they are converted into coordinates which are then sent to the computer.

The panel may be discrete, in which the conductive coating on the top layer is etched in one direction and that on the bottom layer in the other direction, or analog, where the conductive coatings in both layers are continuous. In the discrete case, the panel then acts as an X–Y matrix, and the resolution is determined by the number of etched lines. The analog configuration requires the addition of linearization networks on each edge of the panel so that a large-area resistor is created with a voltage drop in one direction. Other linearization techniques are also possible, but only the

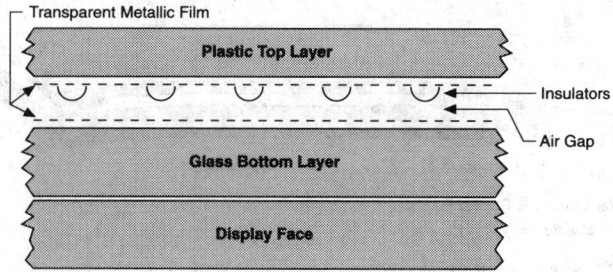

FIGURE 83.13   Resistive overlay technology. (*Source:* After A. B. Carrell and J. Carstedt, "Touch input technology," *SID Sem. Lecture Notes,* p. 15.31, 1987. With permission.)

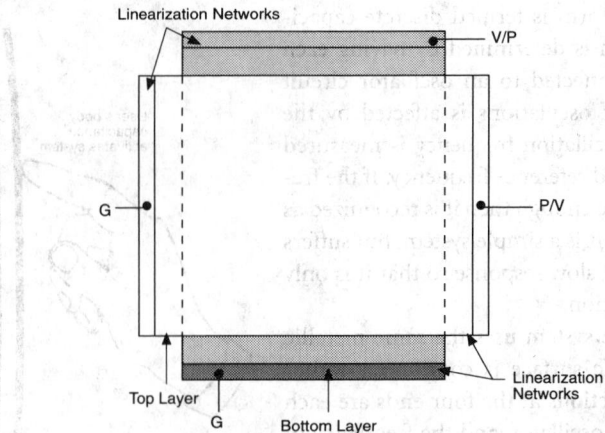

FIGURE 83.14 Four-wire analog resistive. (*Source:* A. B. Carrell and J. Carstedt, "Touch input technology," *SID Sem. Lecture Notes,* p. 15,32, 1987. With permission.)

four-element system is described here as shown in Fig. 83.14. In this arrangement, one of the layers acts as the large-area resistor and the other as a voltage probe where either can function in either role. For the Y coordinate value the top layer is the voltage probe, and the voltage is applied by the controller to the bottom layer. Similarly, the X coordinate is found by connecting the voltage to the top layer and making the bottom layer into the voltage probe. In either type of system, the resolution can be very high, but the transmissivity is reduced to under 80%.

Piezoelectric. The piezoelectric technology uses pressure-sensitive transducers as the means for determining the location of the touch, as shown in Fig. 83.15. The sensor is a glass plate with transducers connected to the four corners. Pressure on the plate causes readings to occur at each of the transducers, which depend on the location of the pressure. Thus, the controller can measure the readings and obtain the coordinates by means of a proper algorithm. This technique allows a high-transmissivity plate to be used that can be curved to follow the CRT face plate curvature, but it allows only a limited number of touch points to be used.

Light Beam Interruption. This is a fairly straightforward technology that requires a matrix of light sources and detectors facing each other in the X and Y directions. When the beams from the X and Y light sources are interrupted, this is sensed by the facing light detectors and the signals are sent to the controller. The light beams are turned on sequentially by pulsing the LEDs and thus create a full matrix of light beams without requiring each of them to be on continuously. This system does not

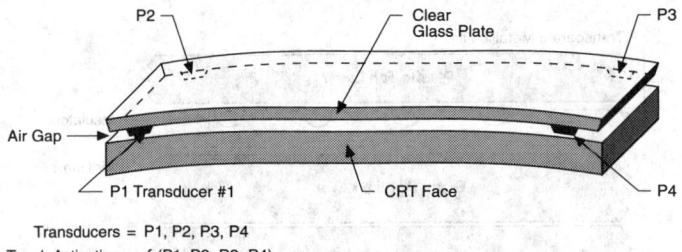

Transducers = P1, P2, P3, P4
Touch Activation = f (P1, P2, P3, P4)

FIGURE 83.15 Piezoelectric technology. (*Source:* A. B. Carrell and J. Carstedt, "Touch input technology," *SID Sem. Lecture Notes,* p. 15.34, 1987. With permission.)

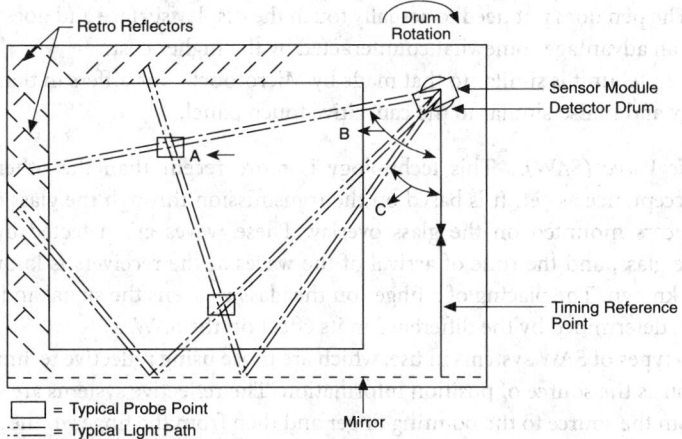

FIGURE 83.16   Rotating infrared beam technology. (*Source:* A. B. Carrell and J. Carstedt, "Touch input technology," *SID Sem. Lecture Notes,* p. 15.34, 1987. With permission.)

reduce the screen transmissivity as there is no obstruction of the screen output, but it is limited in resolution to the number of LED detector pairs that can be placed on the periphery of the screen.

Another approach to light interruption is to use a rotating beam of light, which has the advantage that only one light source and detector pair is required. This technology is depicted in Fig. 83.16 and consists of a LED and a light detector placed inside a rotating drum which has a slit that allows light to be transmitted outside the drum. The light is swept across the surface and strikes the retroreflectors that sends it back directly to the detector. The beam scan is sampled 256 times on each scan, and Fig. 83.16 shows how two angles of interruption are created, angle B by direct interruption, and angle C by mirror reflection interruption. The result is that the location of the interruption can be calculated by comparing the two angles. Again, there is no obstruction of the screen but a moving element must be added, and parallax errors may occur.

Pen-Based Computing.   This is an application for touch input devices that is growing at a rapid rate. The input device comes in several forms, each of which can recognize hand printing, or act as an input switch, with the special operating system and software recognizing this type of input. The pen-based input device comes in several forms, of which TouchPen™ is one type that can function both as a digitizer with a touch tablet, and as the touch input device with a touch input pen-based computer system. A second one is that developed by Wacom, Inc., primarily for the GO Systems computer, but used by other pen-based systems as well. Finally, a third unit is that made by Scriptel Corp. and used by Wang Laboratories in its system.

TouchPen™ was developed by Microtouch Systems, Inc., initially for use in GridPad made by the Grid Systems Corp. It is essentially a high-resolution digitizer consisting of an all-glass tablet that can be used with a number of stylus input operating systems to digitize handwriting. It is basically a touch input device using resistive techniques to digitize the handwriting appearing on the display surface of pen-based computer systems. The glass tablet is placed on the display surface and the system pen is used to transmit the digitized data to the computer. As noted previously, the tablet may also be used as a standard touch input device.

The second form of pen-based input device is one that uses electromagnetic technology and consists of a grid of wires that transmit radio waves that are picked up by a tuned circuit in the stylus. This circuit resonates at its own frequency and transmits that signal back to the wires at the grid location it is touching. The pen also transmits its signal to the computer, which turns off the grid transmission, and locates the position of the pen by determining which of the grid wires pick up the pen signal. The pen does not need to actually touch the display surface and does not require any power, which is an advantage somewhat counteracted by the higher cost.

Finally, the Scriptel unit is similar to that made by Microtouch, but differs in that it uses electrostatic technology and is also similar to the capacitive touch panel.

Surface Acoustic Wave (SAW). This technology is more recent than the others and has not received wide acceptance as yet. It is based on the transmission through the glass of SAWs generated by transducers mounted on the glass overlay. These waves are detected by receivers also mounted on the glass, and the time of arrival of the waves at the receivers is known because the wave velocity is known. The placing of a finger on the glass weakens the signal and the location of the finger can be determined by the difference in its effect on the SAW.

There are two types of SAW systems in use, which are those using reflective techniques and those using attenuation as the source of position information. The reflective systems are similar to sonar and the time from the source to the pointing finger and then from the finger to the receiver is measured to arrive at finger location. The attenuation technology is illustrated in Fig. 83.17 and consists of two transducers, two receivers, and four reflector strips, all mounted on a glass substrate. One transducer-receiver pair is used for X and the other for Y location. Figure 83.17 shows the X axis pair, and the transducer transmits a burst of acoustic energy in a horizontal wave. The wave is partially reflected by the top reflector strips and travels down to the bottom strip where the reflectors are at an angle such that it is reflected to the lower left corner receiver. The wave now has a long rectangular shape, and each point in time corresponds to a specific vertical path across the substrate. The Y axis is scanned in the same fashion after the X wave dies out. Then, when the finger touches the substrate, its water content absorbs some of the energy in the wave, and the wave is attenuated. The dip in the wave amplitude corresponds to the amount of absorbed energy, and the time of the lowest point can be determined, allowing the location of the finger to be calculated. Finally, in addition to the X and Y coordinates, a Z coordinate can be determined, depending on how hard the user presses. This depends on surface contact, which affects the amount of attenuation. The advantages of this system are high resolution, speed of transmission, and the availability of a Z axis component. Its main disadvantages are the variation in moisture content in fingers and

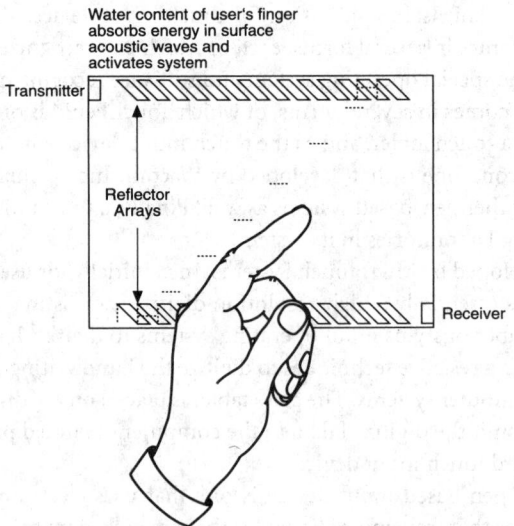

FIGURE 83.17 Attenuation SAW technology. (*Source:* A. B. Carrell and J. Carstedt, "Touch input technology," *SID Sem. Lecture Notes,* p. 15.35, 1987. With permission.)

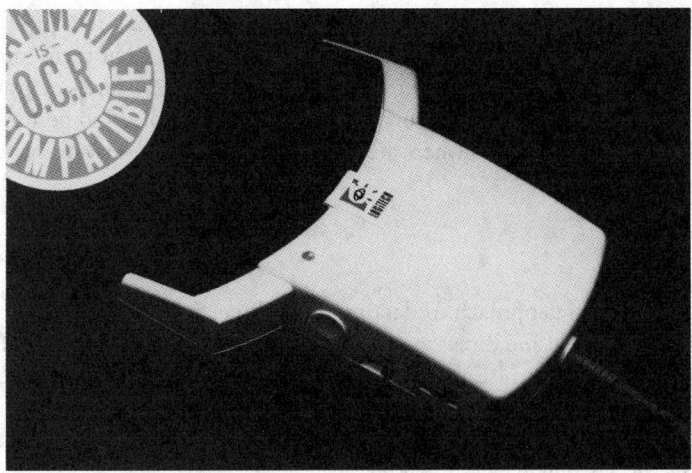

FIGURE 83.18 Hand-held scanner. (Courtesy of Logitech, Freemont, Calif.)

sensitivity to local moisture on the substrate. However, it is being used in developmental units and should be considered as another input device technology.

Scanners

Scanners are a means for inputting text and/or images directly into the computer system, thus avoiding the need for retyping and redrawing information contained in other sources. It is a relatively convenient way to avoid repetition if the data to be entered already exist in readable form. This is done by special image-recognition software that accompanies the scanning hardware, and can transfer an entire image containing both text and illustrations, but without the capability to modify the image. However, the addition of optical character recognition (OCR) software allows the entered text to be modified as if it were entered by typewriter. This can greatly simplify entering and editing text from some preexistent source and has resulted in a proliferation of devices that can perform this function.

These devices come in two main forms, that is *hand-held* and *page* scanners, with or without OCR software in addition to the standard image-recognition software. A typical hand-held scanner is shown in Fig. 83.18 and it consists of a light source, a light-sensitive device such as a charge-coupled device (CCD) array, and the electronics to actuate the elements of the array sequentially under software control. The scanner window is placed over the page, and is moved down or across the page so that the window covers as much of the page as falls within the capability of the software. The light source is reflected from the page to the CCD and the charge in the CCD is modified by the reflectivity of the printed material.

The window area ranges from 4 to 5 in. in width by 0.5 in. in height and may be moved through 14 to 20 in., so that a fairly large area may be covered in a single manual scan. Images wider than the maximum window may be scanned in two passes, and the OCR software can stitch the two scans together into a single image, although this procedure requires considerable care in scanning so that the scans line up properly. Therefore, when images wider than the window of the hand-held scanner are to be scanned, it is advisable to

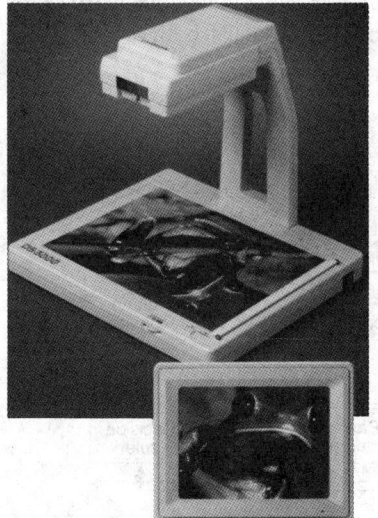

FIGURE 83.19 Page scanner. (Courtesy of Chinon, Torrance, Calif.)

use a *flatbed scanner* of the type shown in Fig. 83.19 which can handle a full 8.5 in. by 11 in. page, or some of the larger scanners than can accept large drawings and input them into the computer system. Resolutions of 400 dpi and higher are available with up to 250 levels of gray and 24 bits of color resolution available. Thus, scanners offer a wide variety of choice and performance capabilities, and are powerful input devices when prepared data in visual form is to be entered into the computer system.

Voice

Voice input is an intriguing approach to data input, with particular attractiveness to upper-level managers who want a simple and direct means for inputting data and commands. For many years, this technology tended to promise more than it could achieve, but recent developments have brought it to the point where it can be considered as a viable input means. This has been due to new developments in software that have made it possible to minimize the amount of training required and increased the success rate to close to 100%.

One basic approach to speech recognition is represented by the block diagram shown in Fig. 83.20. This is a system that is built around a special chip developed by Texas Instruments. This system uses templates and special algorithms for recognizing the input speech patterns. The system is speaker dependent, with the capability of storing up to 32 word templates and user-defined phrases. The output portion may be superfluous when the system is used only for inputting data and commands, but can be a useful adjunct to the visual response. Other chips are available from other manufacturers, and other techniques such as speaker-independent and phoneme-recognition systems are also available. Vocabularies range from 50 to 5000 active words, and both isolated and connected words can be recognized, although the larger numbers tend to be associated with isolated word systems. In general, it seems feasible that a combination of speech input and pen-based computing may find a viable market.

Summary

The multiplicity of input devices that are available makes it difficult to determine which is most suitable for any specific set of requirements. However, the limited functional comparison of the input devices covered in this section shown in Table 83.4 may be of some use, and in any event is a starting point in this evaluation. It should be noted that what appears best at one time may become unpopular or obsolete at a later time, as occurred for light pens and trackballs, both of which have come back into favor.

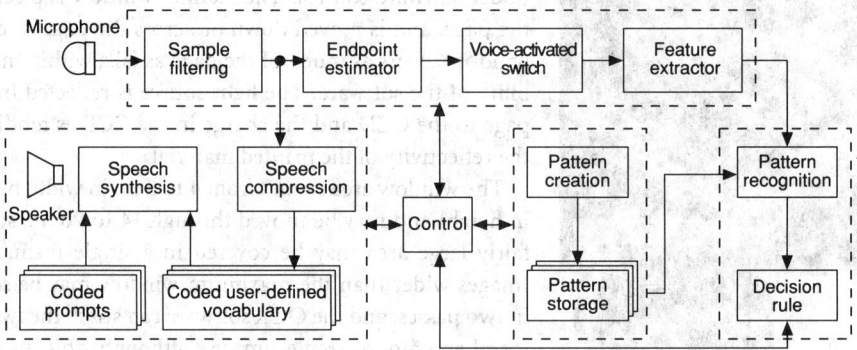

FIGURE 83.20 Block diagram of speech recognition and synthesis chip. (*Source:* After M. Leonard, "Speech poised to join man-machine interface," *Electronic Design*, pp. 43–48, September 26, 1991. With permission.)

Table 83.4 Input Device Functional Evaluation

| | | Function | | |
Input Device	Control	Data/Text	Data/Graphics	Total
Keyboard	E	E	P	9
Light Pen	G	G	E	10
Tablet	E	G	E	11
Mouse	E	G	E	11
Trackball	E	G	E	11
Joystick	F	F	G	5
Touchscreen	G	F	G	8
Scanner	F	E	G	9
Voice	G	F	P	6
Total	29	23	28	80

E = Excellent = 4; G = Good = 3; F = Fair = 3; P = Poor = 2

In addition to the generalized evaluation shown in Table 83.4, it is also of interest to examine representative performance parameters. These are shown in Table 83.5 and while representative do not necessarily cover the range of performance parameters offered. More data may be obtained from the vendors of these devices.

Table 83.5 Representative Performance Parameters

Input Device	Parameter	Value
Light pen	Response time	150–500 ns
	Spectral response	4200–9500 A
	Luminous sensitivity	0.03–0.7 nts
	Field of view	0.02–0.1 in.
	Ambient rejection	350 nts
Data tablet (digitizers)	Resolution (l/in.)	100–2000
	Accuracy (in.)	0.0005–0.02
	Active area (in.)	12×12–60×120
	Active height (in.)	0.02–2.5
	Digitizing rate (pps)	100–350
	Transducers	Stylus, puck, cursor
Mouse	Resolution	10–1000 dpi
	Speed	1–20 in./s
	Accuracy	25–1000 dpi
Trackball	Resolution	100–1000 cpi
	Speed	1200–9600 BPS
	Accuracy	100–1000 dpi
	Ball diameter	1.5–2.5 in.
Joystick	Travel	25–30°
	Accuracy	5–10%
	Repeatability	1%
Touchscreen	Resolution	256×256–4096×4096
	Transmissivity	60–100%
	Viewing area (in.)	3×4.5–15×20
	Speed	80–200 touch pts./s
Scanner	Resolution (dpi)	75–1600
	Scan rate (in./s)	0.5–2.0
	Scanning width (in.)	4.1–36 gray shades 32–256
	Scan time (s/page)	1–30
Voice	Active vocabulary	13–5000 words

Table 83.6   Input Devices—Advantages and Disadvantages

Device	Advantages	Disadvantages
Keyboard	Simple operation	Requires many keys
	Well known	Requires training
	Standard interface	No graphics
Light pen	Eye-hand coordination	Arm fatigue
	Low cost models	Limited resolution
	No desk space required	May block display
Graphic tablet	Natural hand movements	Eye-hand conflict
	Screen not blocked	Requires desk space
	No parallax	Breakable stylus
	Good for graphics	Poor for A/N entry
Mouse	Small space needed	Some space needed
	Low cost	Slow transmission
	Screen viewing	Low resolution
	Any surface may be used	Grid for optical
	(Optical) noiseless	Mechanical noise
Trackball	High resolution	Poor for A/N input
	Fixed desk space	Slow transmission
	Screen viewing	Mechanical noise
	Tactile feedback	3-D difficult
Joystick	Fixed desk space	Low accuracy
	Low fatigue	Low resolution
	Low cost	No A/N input
Touchscreen	Eye-hand coordination	Arm fatigue
	Minimal training	May block display
	Minimum input errors	Varied resolution
	User acceptance	Parallax
	No special commands	Slow data entry
Scanner	Full A/N page input	Hand scanner width
	Color scan input	High cost for color
	High resolution	Slow input
	OCR software	Compatibility
Voice	Ease of use	Limited words
	Minimal training	Machine training
	No special devices	Graphics difficult

Advantages and Disadvantages

Input devices make up one of the functional groups of the display systems, and their technical characteristics are covered in some detail at the beginning of this chapter, with performance information provided in Table 83.5 containing characteristic parameter values for each type, as available. The following material expands somewhat on that information by placing these devices in the context of a full graphics display system and evaluating the functions that the various types of input devices perform in that type of system in terms of their advantages and disadvantages. It is of some interest to compare the advantages and disadvantages of each type at this point, as listed in Table 83.6. This is an imposing list and may be used to aid in choosing the best input devices for specific applications. It also concludes this section on input devices.

Defining Terms

Data tablet/digitizer: A device consisting of a surface, usually flat, and incorporating means for selecting a specific location on the surface of the device and transmitting the coordinates of this location to a computer or other data processing unit that can use this information for moving a cursor on the screen of the display unit.

Joystick: An input device somewhat in the form of the navigation control device found in early aircraft and operating in a somewhat similar manner by generating series of pulses whose frequency or number depend on how far, with what force, and in what direction the control stick is moved from the central position.

Keyboards: Electromechanical devices consisting of sets of keys labeled with alphanumeric, numeric, and functional designations that enable the user to describe and define the operation to be performed.

Light pen: Neither a pen or a light source but rather an input device in the shape of a pen that operates by sensing the existence or nonexistence of light pulses at specific locations on the surface of a display device and uses this information to signal the computer as to the location of the pen.

Mouse: An input device based on a much older type known as a trackball and fancifully named because it bears only a casual resemblance to a mouse. It consists of a roller ball that is moved on a flat surface and causes orthogonal potentiometers or other types of X–Y-position signal generators to move and produce electrical signals defining the desired coordinates of the cursor on the screen so that the cursor can be moved to that position.

Scanners: Means for converting hard copy into electrical signals that can be entered into a computer or data processing system. The usual means for accomplishing such conversion is to move a light beam over the surface containing the data either by hand or automatically and using arrays of light-sensitive devices to convert the reflected light into electrical pulses.

Touch input: A means for selecting a location on the surface of the display unit using a variety of technologies that can respond to the placing of a finger or other pointing device on the surface. These are essentially data panels placed either on the display surface or between the user and the display surface.

Trackball: The earliest version of an input device using a roller ball, differing from the mouse in that the ball is contained in a unit that can remain in a fixed position while the ball is rotated. It is sometimes referred to as an upside-down mouse, but the reverse is more appropriate as the trackball came first.

Voice: Means for enabling a computer or data processing system to recognize spoken commands and input data and convert them into electrical signals that can be used to cause the system to carry out these commands or accept the data. Various types of algorithms and stored templates are used to achieve this recognition.

References

H. Brunner *et al.*, "Effects of key action design on keyboard preference and throughput performance," Micro Switch.

A.B. Carrell and J. Carstedt, "Touch input technology," *SID Sem. Lec. Notes*, pp. 15.30–15.35, 1987.

T.E. Davies *et al.*, "Digitizers and input tablets," in *Input Devices*, S. Sherr, Ed., New York: Academic Press, 1988, p. 186.

D. Doran, "Trackballs and joysticks," in *Input Devices*, S. Sherr, Ed., New York: Academic Press, 1988, pp. 251–262.

C. Goy, "Mice," in *Input Devices*, S. Sherr, Ed., New York: Academic Press, 1988, pp. 225–232.

M. Leonard, "Speech poised to join man-machine interface," *Elec. Des.*, pp. 43–48, Sept. 26, 1991.

S. Sherr, *Electronic Displays*, New York: Wiley, 1979, pp. 323, 388–389.

Further Information

Electronic Displays, 2nd ed., by Sol Sherr and published by John Wiley & Sons, Inc., contains an extensive and detailed discussion of other aspects of display systems and technology, as well as a somewhat expanded version of this section. In addition, *Input Devices*, edited by Sol Sherr, and *Output Hardcopy Devices*, edited by Robert C. Durbeck and Sol Sherr, both published by Academic Press, include extensive discussions of a wide variety of devices.

The Society for Information Display (SID) sponsors a yearly symposium at which a large amount of information on new developments in information display as well as tutorials and seminars on basic information display topics are presented and made available in published form. In addition, it publishes two journals, namely, *Proceedings of the Society for Information Display* and *Information Display*. Other relevant meetings and publications are those sponsored by the Computer Society and Electron Devices groups of the IEEE, the SIGGRAPH group of the Association for Computing Machines (ACM), and the National Computer Graphics Association (NCGA).

83.2 Computer Output Printer Technologies

Robert C. Durbeck

Electronic printers for computer output represent a very important part of the computer industry. They range from small, inexpensive printers for personal computers and workstations to very large and fast page printers used for bulk printing output for large-scale computer systems. The technologies employed for this wide scope of printing requirements are diverse: some based on "impact" methods to transfer ink from a sheet or ribbon to paper, others based on more sophisticated "nonimpact" methods. Today, there is no single technology which completely dominates. A wide range of user needs has led to the present proliferation of printer technologies. The most prevalent are discussed in the following.

Classification of Printer Technologies

Table 83.7 illustrates the two main classifications of printer technologies and also the wide range of extant technologies. Those technologies listed under *line impact* and *page printing* are used for large computer system printing; by far the most popular are *fully formed character* and *electrophotographic* (so-called laser printers). The most common printing technologies used for personal/workstation computer systems are *serial wire matrix, electrophotographic* and *ink-jet*. Emphasis in the following is directed to these favored technologies, with brief descriptions of the others included.

Page Printer Technologies

By far the most important page printer technology is electrophotography (EP). EP, as well as ionographic and magnetographic technologies, uses "powder toning" development of an intermediate "image" created in the process. Liquid toners are also used to develop electrostatic images. Thermal page printers employ a full-page-width linear array of thin-film heater elements to melt and transfer ink from a ribbon to the paper or to mark special thermal-sensitive paper.

Table 83.7 Types of Printer Technologies

Impact	Nonimpact
Line impact	**Page printing**
Fully formed character	Electrophotographic
Dot band matrix	Ionographic
Shuttle hammer matrix	Magnetographic
	Electrostatic
	Thermal
Serial impact	**Serial nonimpact**
Fully formed character	Ink-jet
Serial wire matrix	Continuous
	Piezoelectric/impulse
	Thermal/bubble-jet
	Thermal
	Direct (thermal paper)
	Thermal transfer
	Resistive ribbon

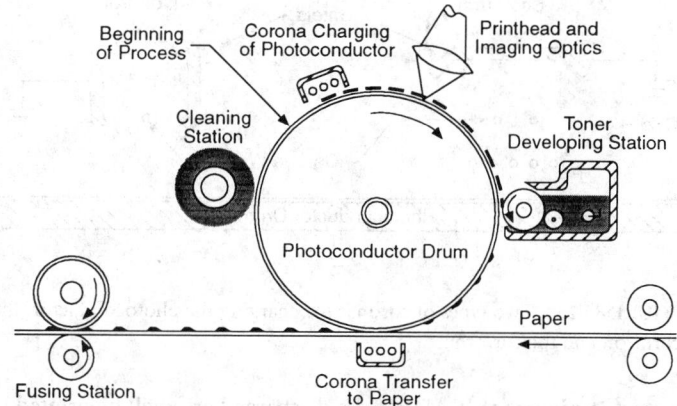

FIGURE 83.21 The six basic electrophotographic printer process steps: charge, image, develop, transfer, fuse, and clean.

Electrophotographic Printing

EP printers use essentially the same technology found in most "plain paper" copiers, the major exception being the printhead. Instead of using page- or line-imaging optics as in a copier, the printhead utilizes a solid-state laser (usually GaAlAs) or gas laser (typically HeNe) to scan across and expose a photoconductor drum or belt to create a "latent image" (see below). A few EP printers use stitched arrays of light-emitting $GaAs_{1-x}P_x$ diodes (LEDs) with Selfoc™ glass fiber optics or an array of liquid crystal shutters, the latter to modulate light from a bright line light source. Other possibilities are electroluminescent, magnetooptic or electrooptic arrays, but these have not been commercialized to any extent.

There are basically six major steps employed in the EP printing process (see Fig. 83.21): (1) charging the photoconductor (PC) electrostatically; (2) exposing the PC to the image light pattern, which results in selective discharge of the uniform area charge created in Step 1, creating an electrostatic image; (3) developing the PC by bringing electrostatically charged toner particles (black or colored) to the surface of the PC where they selectively adhere to appropriately charged regions; (4) electrostatically transferring the toned image from the PC to the final medium (usually paper); (5) thermal fusing of the toner to the paper; and (6) cleaning residual toner from the surface of the PC to allow reinitiation of the six step cycle.

Step 1—Charging the PC. The most common approach used is corona charging. One or more thin corona wires (typically tungsten) are supported directly above the PC and are energized to 5–8 kV. The resultant high electric field surrounding the wire causes electrons in the immediate region to be accelerated to energies sufficient to ionize local air molecules. Either positive or negative ions are then attracted to the outer surface of the PC (which, when unexposed to light, acts for a short period of time as an insulator) depending on the sign of the potential difference. At the same time, a counter-image charge is formed on the inner side of the PC. The two corona structures most commonly used are the corotron and the scorotron (see Fig. 83.22). The grid on the scorotron is used to more precisely control the resultant voltage charge level on the PC (approximates the grid voltage). Both dc and ac designs are used; the latter usually include a glass sleeve around the corona wire to reduce localized high-emission spots on the wire due to contaminants. To save on cost for very low-cost EP printers and to reduce corona by-products (e.g., ozone), a lower-voltage conductive elastomer charge roll in direct contact with the PC has also been used in place of the corona wire.

Step 2—Exposing the PC. The wavelength of the exposing light source must match the spectral sensitivity of the PC. If the PC is discharged in areas that will be printed white, the overall process

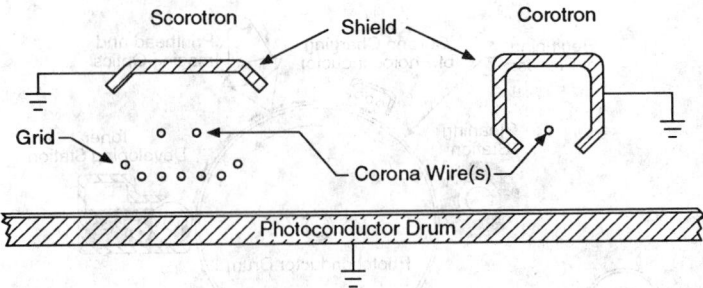

FIGURE 83.22 Two types of coronas for charging the photoconductor: the scorotron and the corotron.

is termed charge area development (CAD); if the discharged area will be printed black (or color), the process is called discharge area development (DAD). Both CAD and DAD processes are used in EP printers but CAD is the only common process used in copiers.

Selective discharge of the PC involves two steps: (1) photogeneration of electron-hole pairs and (2) transport of the electrons and holes in opposite directions under the influence of a high-dc bias field, locally dissipating the surface charges created in Step 1 (see Fig. 83.23).

Both organic and inorganic PC materials are used (see Fig. 83.24). A variety of charge-generation and charge-transport material systems have been developed for organic PCs; most use separate layers for charge generation and charge transport. Examples of efficient organic charge-generation materials sensitive at both GaAlAs (7800 Å) and HeNe (6328 Å) wavelengths include squarylium and thiopyridium dyes in an appropriate binder layer ($\approx$0.5 μm thick). The charge-transport layer (CTL) consists of a thicker layer (20–30 μm) of a charge-transport molecule dispersed in an inert binder, or simply a good charge-transporting polymer (e.g., polyvinylcarbazole). Since most polymer transport materials are essentially hole-transport materials, this requires that the PC surface charge produced in Step 1 be negative. The CTL must be transparent at the imaging wavelength.

Examples of inorganic PC materials include amorphous chalcogenide alloys such as a-Se, a-As_2Se_3 and Te-doped Se. Most do not have a high sensitivity at GaAlAs wavelength, but a-Si does. a-Si also offers other desirable properties (e.g., durability and lack of fatigue), but is quite expensive to produce.

The laser beam is scanned linearly over the PC in a direction orthogonal to the PC motion; the combined motions covering a "page". The most common gas laser scanning technique employs a high-speed rotating polygon mirror along with beam-expanding optics, an acoustic-optic modulator (not required for solid-state lasers) and an f-θ imaging lens. To produce a quality image, the multifaceted scanning mirror system must be essentially free of facet defects, up-and-down wobble, variations in polygon rotational velocity and lack of synchronization with the **pixel** clock.

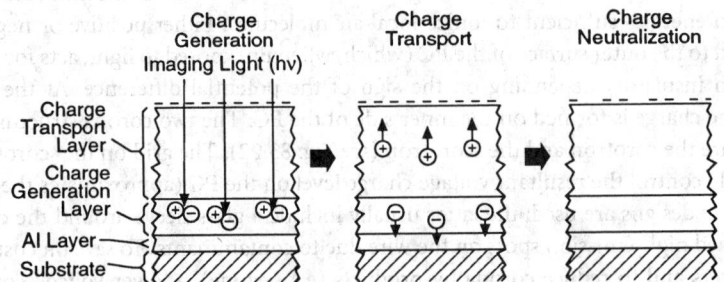

FIGURE 83.23 Light is used to selectively discharge the photoconductor. Electron-hole pairs are photogenerated in the charge-generation layer, followed by charge transport under a dc bias field and then selective neutralization of the surface charges, thereby creating an electrostatic image.

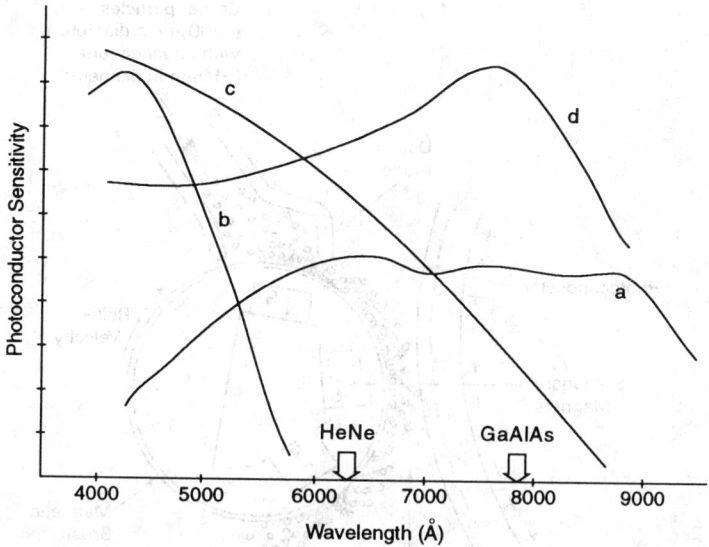

FIGURE 83.24 Spectral absorption characteristics of several photoconductor materials: (a) squarylium dye, (b) SeTe, (c) As_2Se_3, and (d) a-Si.

Some nonplanarity of the facet surfaces can be corrected with anamorphic optics. LED arrays and liquid crystal shutter systems do not have these same technical challenges but, so far, they are more expensive to produce.

Step 3—Developing the PC. There are basically three development techniques: dual component, monocomponent and liquid development. The first two use powder toner. Until the advent of EP printers for personal and workstation computers, the most common method of development was dual component, where polymer-coated magnetic carrier beads are mixed with the toner particles and development is done with a "magnetic brush." This technique is still prevalent in high-end printers. With this approach, the 5–20 μm toner particles (consisting mainly of resin plus carbon particles or colorant) are triboelectrically charged by repeated contacts during mixing with the much larger (60–250 μm) magnetic carrier beads. The toner particles then electrostatically adhere to the opposite-sign charged carrier beads. Charge control agents (e.g., complex organometallic salts) are often included in the toner composition to control the charge level, rate of charging, and consistency of charge. The mix is mechanically directed to a nonmagnetic rotating shell which has fixed magnets located within its core, adjacent to the gap between the PC and the shell (see Fig. 83.25). The gap is typically 0.5 to 6 mm, depending on the specific system. As the shell rotates, the mixture is carried to the gap, and chains of carrier beads (coated with toner—the "magnetic brush"!) form along the local magnetic field lines. These field lines are approximately perpendicular to the shell at the smallest gap. In addition, a development voltage (200–500 V) is applied between the PC and shell. This provides a high field in the gap whose local value is determined by the applied voltage, gap dimension and the electrical properties of the material mix in the gap. The electric field at the end of the last carrier bead (next to the PC) in a chain may be up to 50 times the nominal unfilled gap field, the value being greatest with uncoated carrier beads. It can be shown that for coated carrier beads, the mass of toner per unit area developed on the PC is approximately

$$\frac{m}{A} = \frac{\pi \epsilon_0 V}{2 L_g C_m} \left\{ r_c / [(r_t / \kappa) + \delta_c] \right\} \cdot \left| v_r / v_p \right| \tag{83.7}$$

where v_r/v_p is the surface speed ratio between the developer roll (shell) and the PC surface, V the voltage across the gap L_g, C_m the toner charge-to-mass ratio, ϵ_0 the permittivity of free space, r_c the

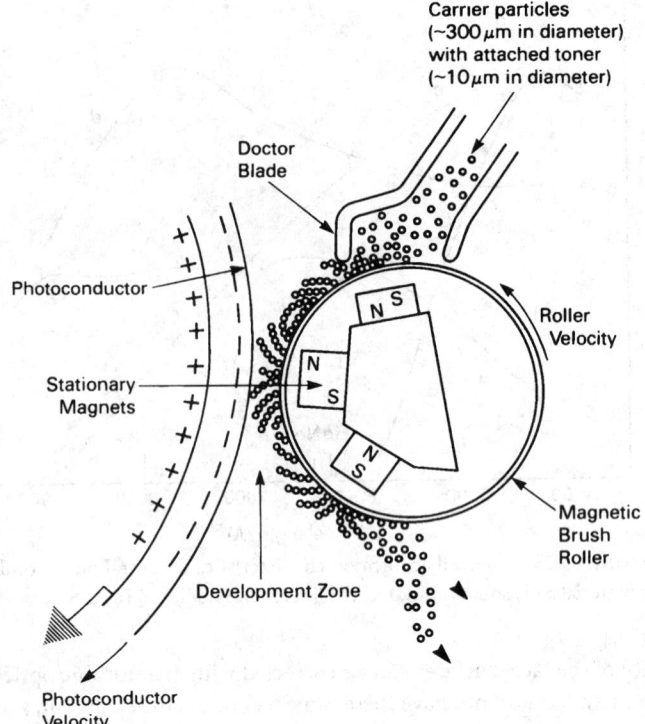

Carrier particles
(~300 μm in diameter)
with attached toner
(~10 μm in diameter)

Doctor
Blade

Photoconductor

Stationary
Magnets

Roller
Velocity

Magnetic
Brush
Roller

Development Zone

Photoconductor
Velocity

FIGURE 83.25   Dual-component magnetic brush developer. Toner particles, adhering electrostatically to much larger magnetic carrier beads, are transported into the photoconductor-developer gap to tone the electrostatic image. (*Source:* R. C. Durbeck and S. Sherr, Eds., *Output Hardcopy Devices,* San Diego, Calif.: Academic Press, 1988, p. 242. With permission.)

carrier bead radius, r_t the toner radius, κ the dielectric constant of the bulk toner, and δ_c the thickness of the carrier polymer coating.

Monocomponent development is used almost exclusively for low-end printers because this process does not require carrier beads, toner concentration sensors or toner replenishment hardware, resulting in much lower manufacturing costs. This approach has also allowed the use of replaceable toner/developer cartridges which, although more costly on a supplies cost-per-page basis, adds greatly to the user-perceived reliability of the system (if it fails—just replace the cartridge!). A rotating donor/development roll with appropriate charging properties is employed to charge the toner by touch-and-rubbing contacts (see Fig. 83.26). The toner electrostatically adheres to the donor roll and is transported to contact the PC at the nip. Here, in the presence of a development bias field, the toner is selectively transferred to those areas on the PC with opposite sign charge.

Liquid development employs a high-resistivity hydrocarbon dispersion of very fine toner particles (<1 μm) that are charged naturally in the solvate. Mechanical means are used to bring the liquid into contact with the PC, and the toner is then electrophoretically transferred to the latent image areas on the PC.

Color can be accomplished by using multiple development stations, one each for the subtractive colors (cyan, yellow and magenta) plus black. Toners are colored by either dyes or pigments. The four-colored images may be accumulated on the PC (see Fig. 83.27) or alternatively on an intermediate belt or drum, or even on the paper itself. A wide range of colors and the visual illusion of gray scale can be achieved by the use of laser pulse-width modulation and the use of "super-pixels" consisting of $N \times M$ arrays of binary pixels to provide digital half-toning.

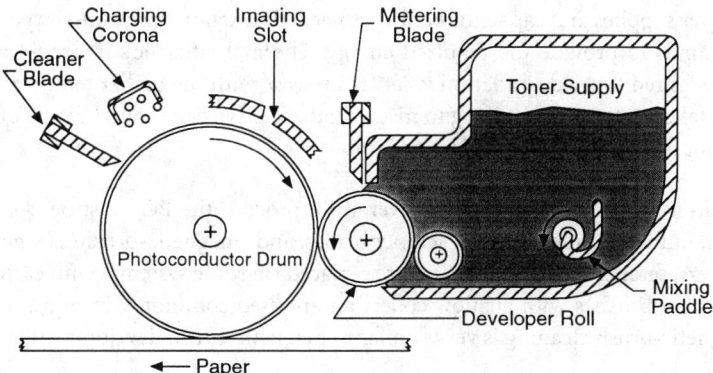

FIGURE 83.26 Monocomponent developer system.

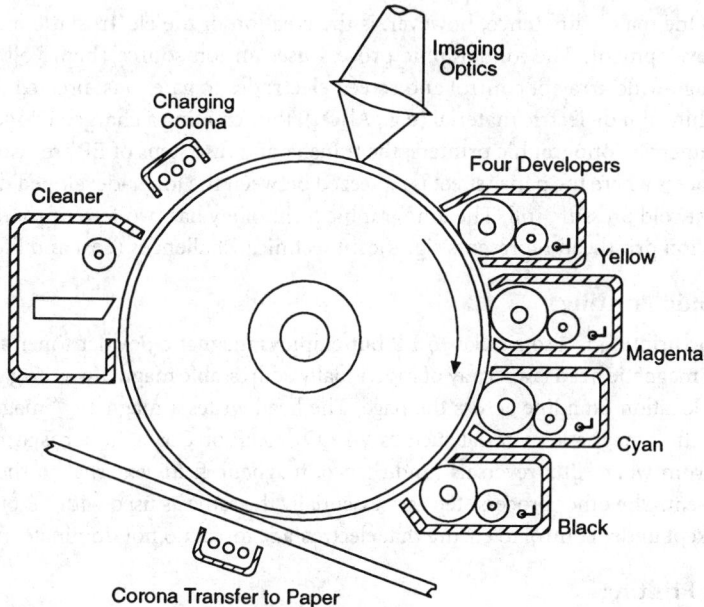

Corona Transfer to Paper

FIGURE 83.27 Four developers are used to produce color plus black. Four photoconductor drum rotations are needed to accumulate the four-color toner images.

Step 4—Transfer of Toner to Paper. Here a corona is used to charge the back side of the paper and the toner is transferred from the PC to the paper. Typically, some small fraction of the toner usually has a charge of the wrong sign. It is at this step where most of this wrong-sign toner is removed, minimizing "background" on the page, since primarily only toner with the correct sign gets transferred.

Step 5—Fusing Toner on the Paper. After transfer the toner is only loosely held (electrostatically) onto the paper; it must be fixed by fusing. There are several fusing techniques: (1) hot-roll using hot fuser rolls under pressure, (2) cold-roll, (3) solvent vapor, (4) flash lamp, and (5) radiant heating. Hot-roll fusing is predominantly used but all five techniques have been commercialized. The most efficient approach is hot-roll but all of the thermal approaches require 0.5–2.0 J/cm^2 energy at the paper surface. The energy required is determined by the contact duration (related to the nip compliance and roll speed), pressure, paper water content, and the melt-flow rheological characteristics of the toner. Flash fusing has only been used in mid-range to high-end printers because expensive power supplies and capacitor banks are needed. A xenon flash lamp is typically pulsed for a few milliseconds to provide the required energy. Thermal efficiency improves with very short pulses, but undesired toner degradation volatiles increase with the higher temperatures produced. Care must be taken with radiant fusing to insure that excessive paper overheating does not occur if the paper is stopped under the incandescent lamp.

Step 6—Cleaning the PC.   To restart the overall EP process, the PC must be cleaned of residual toner and contaminants. Fiber brushes, scraper blades, and "magnetic-brush" cleaners are used. For low-end printers, scraper blades suffice and are replaced (in some systems) with each new cartridge. Rotating soft-fiber brushes with air flow collection are used commonly in mid-range to high-end printers. Magnetic-brush cleaning is very similar to magnetic-brush development (see above).

Ionographic Printing

Ionographic printer devices also use powder toner technology and the overall technology is very similar to EP. One major difference, however, is the creation of the electrostatic image for subsequent toner development. The ionographic process uses an ion source (high-voltage drive electrode) and a page-wide array of control and screen electrodes to gate ions directed toward a drum coated with a thin-film dielectric material (e.g., Al_2O_3), thus creating a charged image on the drum. In present commercial ionographic printers, the transfer and fuse steps of EP are also replaced with a "transfix" process where the paper sheet is squeezed between the toner-developed dielectric drum and a compliant cold pressure roll. The ionographic technology has two fewer process steps (vis-à-vis EP) but the ion printhead represents significant technical challenges (cost and lifetime).

Magnetographic Printing

Magnetographic printing is also similar to EP but employs magnetic powder toner and a magnetic printhead. The magnetic head is an array of individually addressable magnetic write gaps, each representing a pixel location on a line across the page. The head writes a magnetic "image" on a belt or drum coated with a magnetic material such as γ-Fe_2O_3, Co:P or Co:Cr. Toner is attracted to those areas on the drum where flux reversals (and, hence, magnetic fields external to the the magnetic media) are present. The other process steps are essentially the same as used with EP but toner charge levels must be kept under control to ensure that electrostatic forces do not dominate magnetic forces.

Electrostatic Printing

There are two basic approaches that have been developed: (1) one using a special dielectric paper, and (2) the other using plain paper. In the first approach, a page-wide linear array of electrode discharge pins are independently pulsed to charge the surface of a moving, conductive-base, dielectric-coated paper. Both powder and liquid toners may be used to develop the image; thermal fusing of the image is required for powder toner. With the plain paper version, either a precharged or uncharged dielectric drum or belt can be used depending on whether the array of electrode pins are used to charge or discharge the media surface. The other process steps are essentially the same as with the nominal EP printing process. Cost savings can be realized by multiplexing the electrode driver lines because there is a process threshold of hundreds of volts. This requires, however, that segmented counter electrodes be positioned behind the receiving surface (drum, belt, or paper).

Thermal Page Printing

Thermal transfer printing is accomplished by using a stationary page-wide linear array of thermal heating elements coupled with a page-wide transfer ribbon roll (see Serial Thermal Printing below for discussion of thermal printhead technologies). The ribbon is positioned between the printhead and the receiving paper. The ribbon typically consists of a polycarbonate or polyester film substrate (10–20 μm) coated with a waxy ink layer. Heat from the thermal printhead elements must penetrate through the substrate to heat (melt) the dye/wax coating. The melted ink layer is pressed into the paper surface by printhead pressure, and after the paper and ribbon move together away from the printhead, the partially cooled ink layer adheres better to the paper than to the ribbon substrate. The ribbon is then peeled away from the paper, leaving the desired image on the paper. Considerable energy must be applied to melt the ink layer (e.g., 2–4 J/cm^2) which does restrict speed. Also, comparably smooth paper ($\leq$50–100 ml/min Sheffield roughness) is required for quality printing with minimal gaps or voids. Print resolution of better than 8 pixels/mm can be achieved with smooth paper.

Sublimable transfer dyes have been used for high-quality colored images since the amount of dye transferred can be controlled somewhat by the energy supplied to the head heater elements. Since the sublimable dyes tend to penetrate into the paper sizing and fibers, a less smooth paper can be used than with waxy materials. The energy required for dye transfer is similar to that required for waxy ribbons.

Serial Nonimpact Printer Technologies

Ink-jet and thermal printers represent the two major classes of serial nonimpact technologies used for personal and workstation computer printer output. As shown in Table 83.7, the ink-jet technologies can be subclassified as continuous, piezoelectric, and thermal/bubble-jet. The thermal technologies can be subdivided into direct thermal, thermal transfer, and resistive ribbon.

Serial Ink-Jet Printing

Ink-jet technologies have evolved over the last two decades with continuous ink-jet being the first to be developed, followed by piezoelectric and then thermal or bubble-jet technologies. All three have been commercialized, although bubble-jet is the most popular today and continuous ink-jet is used primarily for page printer or high-quality color graphics and image applications. Both piezoelectric and bubble-jet are "drop-on-demand" or "impulse" technologies, i.e., a drop is ejected from the printhead only when desired. With continuous ink-jet, a continuous stream of droplets is generated by the printhead and undesired drops are deflected electrostatically away from the paper (or vice versa). Ink requirements for these technologies are very demanding and the development of appropriate inks is as important (and difficult) as the development of the printer and printhead hardware. Demands on the ink are both extensive and conflicting, and include the requirements of nonclogging in the nozzle but fast dry time on the paper, water-based but water-resistant after drying on the paper, and quality printing on a wide range of papers (requires minimal feathering and controlled spot size). Nonaqueous solid inks have also been used; the ink is solid at room temperature but is liquid at an elevated head temperature.

Continuous Ink-Jet Printing. With this technology, ink is continuously jetted from a small-diameter nozzle under pressure (see Fig. 83.28). Although the resultant jet stream will naturally break into small drops, this phenomenon is assisted and stabilized by the inclusion of a piezoelectric perturbation transducer, driven at the desired drop rate. Lord Rayleigh was the first to determine that the *dimensionless instability factor*

$$\frac{g_r}{(\sigma/\rho d^3)^{1/2}} \tag{83.8}$$

(where g_r is the growth rate of jet instability, σ the surface tension, ρ the fluid density, and d the jet diameter) is maximized for $\lambda/d = 4.51$, where λ is the perturbation wavelength. Operation at this design point produces very consistent drop breakoff and diameter. Operation at frequencies in excess of 100 kHz is possible with drop velocities of typically 25 m/s with a 50-μm nozzle diameter. Because the drops must be electrostatically charged at the breakoff point, conductive ink must be

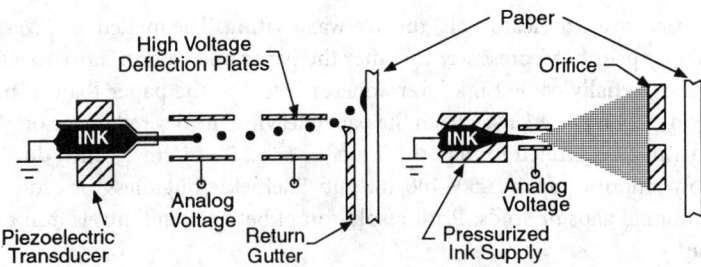

FIGURE 83.28   (a) Continuous drop ink-jet, and (b) continuous spray ink-jet technologies.

used. The ink source is typically grounded and a controllable voltage electrode is placed at the breakoff point (usually surrounding the jet stream). The charge level on each drop is then proportional to the applied voltage. Each charged drop can then be deflected by parallel downstream deflection plates with a field of typically 10 kV/cm. This deflection (assuming parallel plates and a uniform electric field between the plates) can be approximated by:

$$d_l = \frac{l_p}{2}\left(\frac{qE}{mv^2}\right) \cdot (2l_z - l_p) \qquad (83.9)$$

where d_l is the deflection length, q the drop charge, m the drop mass, E the electric field, v the drop velocity, l_p the deflection plate length, and l_z the distance from the upstream end of the deflection plates to the paper. Drop deflection can be either binary or analog. With the former, the drop either reaches the paper or is directed into a collection gutter. With analog deflection, the drop may be deflected linear (e.g., any position over the height of a printed character).

Smaller satellite drops may be also produced between the primary drops. To eliminate their effect, the excitation system is designed to produce forward-merging satellites as these charge simultaneously with the preceding drop; thus the forward-merging (through "drafting") and subsequent drop coalescence will not alter the charge-to-mass ratio of the augmented drop.

Compensation must be provided for both aerodynamic and electrostatic interactions. Two primary examples are that (1) the first drop in a sequence of drops encounters much greater aerodynamic drag than subsequent drops, and (2) the charge on a drop is influenced by the charge on the previous few drops. One approach to greatly reduce electrostatic drop interaction and to stabilize merging effects is to include noncharged drops between charged drops. This obviously reduces the effective drop rate by a factor of two and requires a design configuration where only charged drops reach the paper. Charging electrode voltage adjustment algorithms based on voltages applied to prior drops are also used to reduce the electrostatic interaction.

An alternate approach is the *continuous spray* design where a smaller nozzle (10–20 µm) is used. Much smaller drops are produced at higher velocities (≈40 m/s). It is often called the Hertz method [see Fig. 83.28(b)]. As with the *continuous drop* approach, a controlled voltage electrode is positioned around the breakoff point of the jet stream, a conductive ink is used, and the smaller droplets are charged proportionally to the applied voltage. The stream of droplets is directed to the paper when no voltage is applied. When a voltage is applied, the resulting electrostatic charge on the droplets produces strong mutual repulsion forces and the stream transforms into a spray, the cone angle of which is determined by the applied voltage. The spray is intercepted by a collecting surface surrounding the collection orifice which allows only uncharged or low-charged droplets through to the paper. With analog voltage control, the amount of spray that passes through the orifice can be varied, thus providing gray-scale capability. This approach (with multiple orifices) has been commercialized for very high-quality color image and graphics applications.

Piezoelectric Ink-Jet Printing. Piezoelectric ceramic transducers are employed with this technology. These materials (e.g., lead zirconate titanate and barium titanate), when polarized, change their physical dimensions when subject to an electric field—usually applied through surface electrodes. Deflections of several angstroms per volt are typical. When the transducer is pulsed with a voltage, the deflection generates a pressure wave in an adjacent ink chamber, resulting in the ejection of a single drop—hence the descriptors, "impulse" or "drop-on demand". Four implementations are shown in Fig. 83.29. Often arrays of these devices are integrated into a serial printhead which allows the printing of a one-character-high-per-head pass across the paper. Color printing can be accomplished by assigning one or more nozzles per color.

Only a very small (100–1000 Å) deflection of the piezoelectric transducer is needed to create the ink chamber pressure wave if the displacement is very rapid (10–100 μs). Hence, these devices are very efficient; only a few microjoules per drop are required for robust operation. Drop ejection rates of over 20 kHz have been demonstrated in the laboratory but commercial devices are typically designed for the 5–10 kHz range. Drops that produce a spot size of 150–200 μm on paper can be achieved with an orifice diameter of about 50 μm. Even greater device efficiency can be obtained by synchronizing the arrival of a direct wave and a reflected wave at the nozzle (e.g., a negatively reflected wave caused by an initial expansion pulse plus the direct wave from a following compression pulse). These two waves can be made to reinforce one another at the time and place of drop ejection.

Figure 83.29(a) shows a piezoelectric/membrane laminate design with a disk-shaped piezoelectric transducer which, upon being pulsed, deforms (assumes a finite radius of curvature) and thus creates a pressure wave in the ink chamber. Figure 83.29(b) illustrates an "oil-can" version of the laminate approach. In Figure 83.29(c), a tubular transducer is used to squeeze the ink chamber. Figure 83.29(d) represents a push-rod or piston design where the piezoelectric material is used in the extensional mode. Attempts have also been made to use modern semiconductor planar/etching processing techniques to create low-cost arrays of devices.

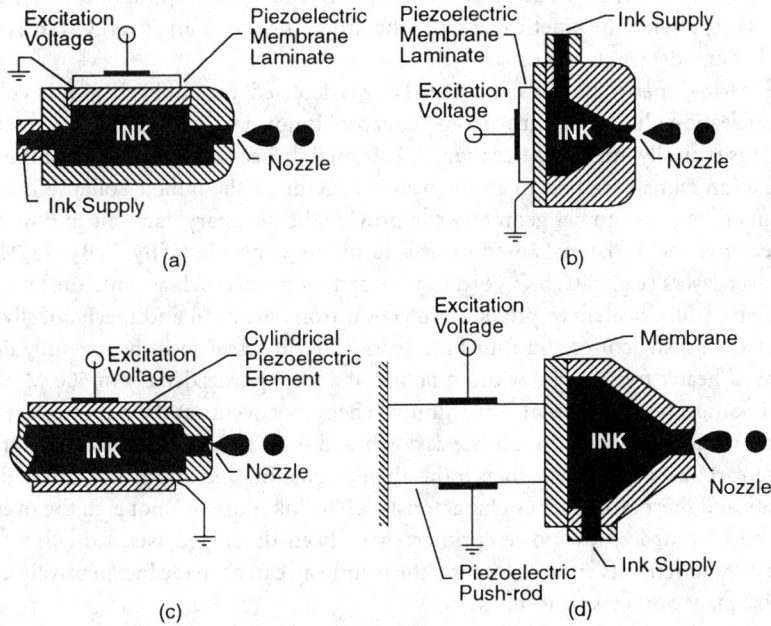

FIGURE 83.29 Four approaches to piezoelectric drop-on-demand ink-jet technology: (a) piezoelectric/membrane laminate design, (b) "oil-can" version of (a), (c) piezoelectric squeeze-tube approach, and (d) piezoelectric push-rod method.

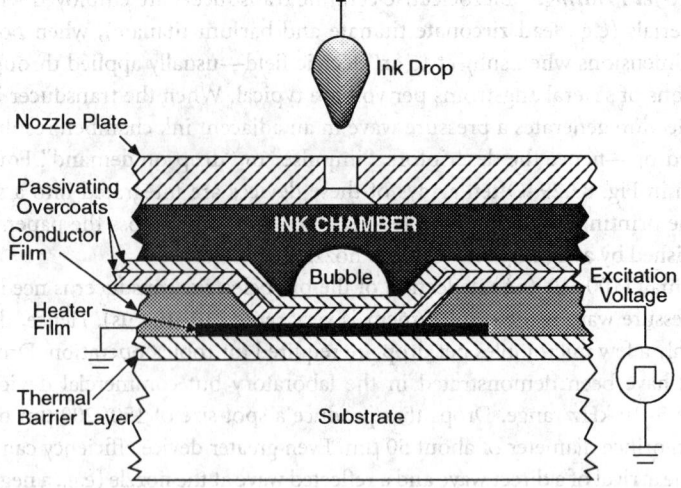

FIGURE 83.30 Thermal-jet device showing one heater-nozzle package. Joule heating of the thin-film heater causes nucleate boiling in the ink adjacent to the heater. A bubble forms and pushes a drop of ink out of the nozzle.

Bubble-Jet Printing.   This technology has also been called thermal-jet. With this approach, very small thermal resistors on the ink chamber wall are electrically pulsed. Joule heating of the resistor causes the temperature of the ink adjacent to the heater to rise to 350–400°C (see Fig. 83.30). Because the ink becomes locally superheated, nucleation of tiny bubbles takes place on the surface over the heater. These bubbles coalesce and very rapidly form a single expanding bubble which, by displacement (like a piston), propels a single drop of ink out at the orifice. The electrical pulse must be short (typically 3–6 μs) to insure low conductive heat losses; however, the power density is extremely high (~500 MW/m²). The energy applied per drop is 30–50 μJ but only a small fraction (a few percent) represents the kinetic energy of the drop. The remaining energy is thermally dissipated in the ink and device structure.

When the thermal energy in the superheated layer is depleted, the bubble begins to collapse. The total cycle (nucleation plus bubble growth and collapse) is normally complete in about 20 μs. Drop rate, however, is typically limited to less than 10 kHz, mainly because of the limits of thermal dissipation. Cavitation damage can occur to the heater structure if the bubble collapse is too violent. Proper design of the ink chamber geometry can provide the necessary damping and minimize this problem. Heater element materials used for this technology include HfB_2, ZrB_2, Ta_2Al and TaN. Passivation over-layers (e.g., SiC, SiO_2, Si_3N_4, plus certain metals such as tantalum) are deposited on top of the thin-film heaters to provide protection from chemical and mechanically enhanced corrosion. Materials, structures and thin-film deposition processes must be carefully designed to allow billions of heater pulse cycles without failure at a local power density of 500 MW/m². Also, very special ink must be used so that very minimal chemistry occurs at the hot interface layer, e.g., so that the chemicals in the ink do not break down and form a thick carbonaceous film over the heater. These carbonaceous films, when more than a few hundred angstroms thick, destroy the device velocity and thermal efficiency characteristics. The ink must also not etch the overcoats.

Very low-cost, compact, low-power printers have been developed based on this technology because energy requirements are very low and the printhead can be made inexpensively using semiconductor-like planar processing techniques.

Serial Thermal Printing

Thermal technologies have been used in many low-cost serial printer applications. There are three basic approaches: (1) use of a heat-sensitive special paper, (2) thermal transfer of ink from a ribbon to paper, and (3) a variant on (2) with a special ribbon structure and printhead that improve thermal efficiency and allow printing on a wide range of standard papers (including the relatively rough office bond papers).

Direct Thermal Printing. The key aspects to this technology are the thin-film resistive serial printhead and the special paper required. A typical array structure (see Fig. 83.31) consists of photolithographically defined and deposited heater material (e.g., Ta_2N or TaAl) on a contoured insulating substrate. One or more protective layers (e.g., SiO_2, Ta_2O_5 or SiC) are deposited on top of the heater layer for abrasion resistance and electrical insulation. Contours (raised areas or bumps) on the surface under the heater areas are constructed by a raised glass "glaze" of about 40 µm and provide both improved contact with the paper and short time constant thermal insulation to the substrate. This structure is often called the *thin-film head structure*. Alternates include (1) the *silicon mesa technology*, where a two-dimensional array of silicon mesas is fabricated from monolithic silicon, and (2) the *thick-film technology*. With the former, each mesa contains its own resistor-transistor/diode on the base of the silicon chip. Joule heating occurs when voltage is applied to the resistor via the transistor/diode. With the thick-film approach, a resistor paste is typically screened onto a ceramic substrate. Materials for the resistor paste include borosilicate glass and lanthanum glass-ruthenium oxide. The elements are typically 25 µm thick and slowly wear down by contact with the paper. Since, with each approach, the resistive elements must be allowed to cool before the head is ready to "write" the next pixel position on the paper, the heating and cooling time constants of the head primarily determine the attainable print speed.

Special thermal paper typically has a coating with a leuco dye plus a phenol developer in a polymer binder which react at the elevated temperature provided by the printhead to form a colored species. The resulting optical density is a function of the temperature reached so that some gray-scale capability is possible. Energy density requirements are in the range of 2–6 J/cm^2.

Thermal Transfer Printing. The desire to eliminate costly special thermal papers even for low-cost serial printers led to the development of the thermal transfer printing technology where a separate transfer ribbon is interposed between the thermal printhead and the paper (see Thermal Page Printing above for discussion of ribbon technology).

Resistive Ribbon Printing. This technology grew from the need to provide quality thermal printing at higher speeds and on a wide range of papers, including standard office bonds. With this tech-

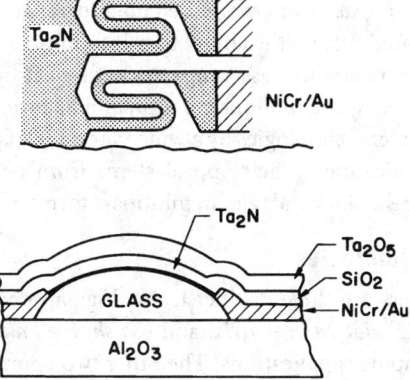

FIGURE 83.31 Thin-film thermal head structure. (*Source:* R.C. Durbeck and S. Sherr, Eds., *Output Hardcopy Devices*, San Diego, Calif.: Academic Press, 1988, p. 282. With permission.)

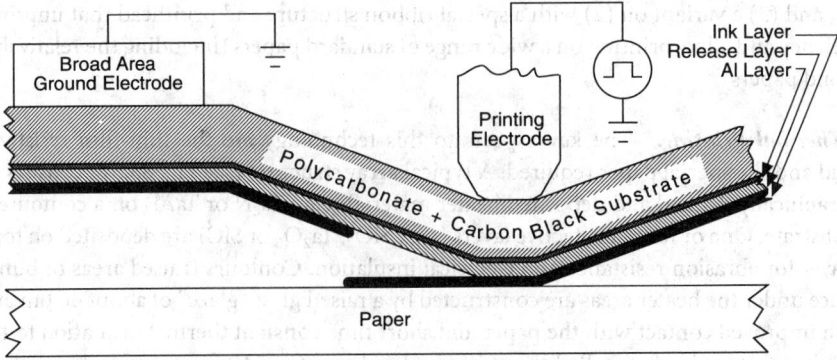

FIGURE 83.32   Resistive ribbon head and ribbon system.

nology, the heating function is repositioned from within the printhead into the ribbon itself. The ribbon (see Fig. 83.32) includes an additional aluminum conductive and heating thin-film layer, sandwiched between the "standard" thermoplastic ink layer and the ribbon substrate. The substrate is made conductive (350–900 $\Omega/\square$) by the incorporation of sufficient (20–30%) carbon black in the polycarbonate film. Joule heating occurs as current flows from the addressable head pixel-size electrodes through the aluminum layer to the large single-ground return electrode. Current density is maximum directly beneath the addressable electrodes so that temperatures there (but not elsewhere) are sufficient to cause ink layer melt and transfer. In addition, a thin release layer can be added between the ink layer and the aluminum heating layer to provide enhanced and smooth-edged pixels; this provides sharp character edges and allows more detail in images.

This technology allows (1) 2–3 speed increase (vis-à-vis conventional thermal transfer printing) because printhead thermal time constants are much less of a factor, (2) very high-resolution printing (up to 40 pixels/mm), and (3) printing on office-quality bond papers. Also, already deposited ink can be selectively lifted off the paper by reducing the printhead power and using a fresh ribbon area. Offsetting these advantages is the fact that the ribbon is more complex and costly than conventional transfer ribbons.

Impact Printer Technologies

The earliest electronic printing technologies were impact devices employed on the early plug-board programmable accounting machines that used punched-card input. Before that there were typewriters and teleprinters. *Line impact* printers are used on mid-range to high-end computer systems; their continuing appeal is based first on the ability to print multipart forms, and secondly on lower cost of printing and greater reliability (as compared with electrophotographic printers). *Serial impact* printers continue to sell by the millions per year, and historically represent the greatest sales volume for all computer printer technologies by a wide margin. By far the largest volume is for personal computer output applications. Their appeal stems from products offering very low cost (device and supplies), high reliability and, again, multipart forms.

Line Impact Printer Technologies

As shown in Table 83.7, there are three distinct technology approaches for line impact printers: (1) *fully formed character*, (2) *dot band matrix*, and (3) *shuttle hammer matrix*. The first method dominates in high-end computer applications. The other two approaches are primarily used with mid-range computer systems.

Fully Formed Character Printing.   Various line printer technologies have evolved over the past but the most popular extant technology makes use of a band of etched engraved characters (plus

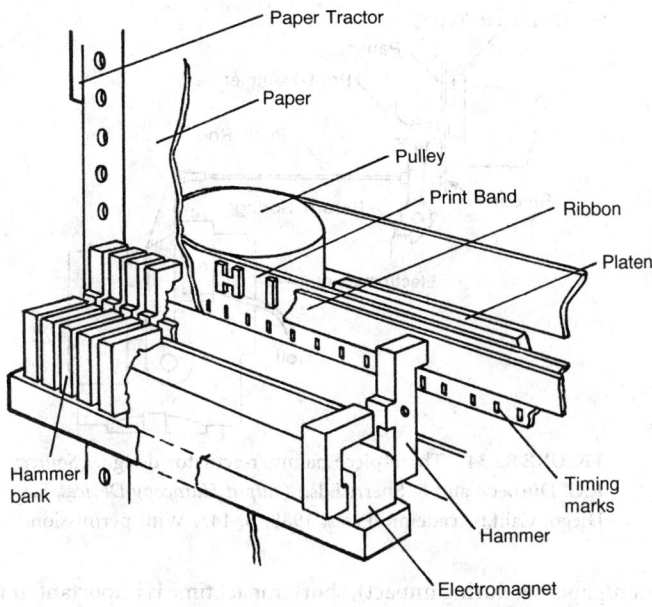

FIGURE 83.33 Line impact printer mechanism. (*Source:* R.C. Durbeck and S. Sherr, Eds., *Output Hardcopy Devices*, San Diego, Calif.: Academic Press, 1988, p. 130. With permission.)

timing and position marks) in a continuously moving loop configuration (see Fig. 83.33). This band is positioned between a page-wide bank (typically 132) of hammers-actuators and the paper/ribbon set. With timing and position marks, the location of the characters (replicated several times on the band loop) can be easily tracked and the appropriate hammers can be asynchronously fired to provide a full line of print in tens of milliseconds. A limit to the throughput P_r (lines per minute [lpm]) of such a printer may be approximated by:

$$P_r = \frac{V_b}{V_b T_i + P_c N_c} \tag{83.10}$$

where V_b is the band velocity (in./min), T_i the time to increment the paper to the next print line position (min), P_c the character pitch (in.), and N_c the number of characters in the character string. This relation assumes that the hammer settle-out time is less than the paper advance time, the latter normally not the limiting speed factor. The usual limiting speed factor is related to print quality, i.e., well-registered, sharp-edged, high-optical-density characters. In addition, because not all lines on a page are usually printed, with line skipping the actual throughput may be significantly faster than indicated above.

Several factors are important to achieve high-quality print characters. First, the hammer-actuator must provide enough impact force and energy to produce optically dense print. Second, it must impact the ribbon/paper set at just the right instant to provide good character registration. Third, it must have a short impact time. The impact force and energy are determined by the kinetic energy transferred by the hammer; this is the main factor in determining the amount of ink transfer to the paper. The hammer energy is typically 4–8 mJ. The total time from electronic print impulse to impact with the ribbon/paper set must be closely controlled to achieve good character registration, increasing in importance with band loop speed. The flight time variance must not be more than 1.7 μs to yield a print registration error of no greater than 0.05 mm with a loop speed of 30 m/s. This can best be done by automatically and periodically measuring the time to impact (using a piezoelectric impact bar) and then making microcode/electronic impulse delay adjustments. To minimize "slur" (i.e., the blurring of a character caused by lateral relative motion between the engraved char-

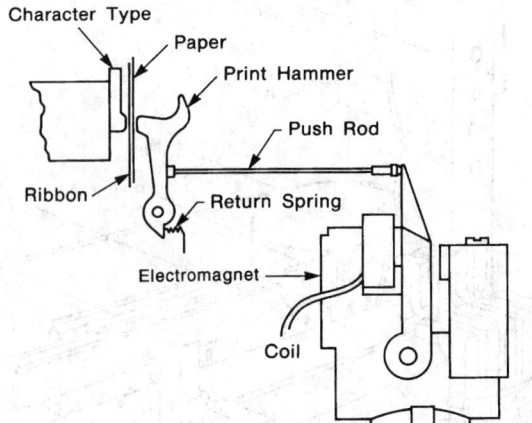

FIGURE 83.34   Three-piece hammer-actuator design. (*Source:*
R.C. Durbeck and S. Sherr, Eds., *Output Hardcopy Devices,* San
Diego, Calif.: Academic Press, 1988, p. 143. With permission.)

acter and the ribbon/paper set during impact), short impact time is important. It may be shown that
this impact time is inversely proportional to the square root of the hammer mass (for given hammer
kinetic energy), and inversely proportional to the hammer velocity (for given ribbon/paper set
thickness and compliance). For a given amount of acceptable slur, the hammer velocity must
increase proportional to the band loop velocity, and the hammer mass (for fixed kinetic energy)
must decrease—inversely proportional to the square of the band loop velocity.

Hammer-actuator systems have been designed with one, two and even three moving piece components. With the one-piece design, the mass impacting the paper is largest, limiting this design to
slower printers (≤650 lpm). To have less mass in the hammer impacting the ribbon/paper set, separate parts are used for the armature and the impacting hammer (two-piece design). In this case,
most of the kinetic energy from the pivoting armature is transferred to the hammer without the
large mass handicap of the ferromagnetic armature. Of course, the residual energy in the armature
must be absorbed and dissipated within the printhead. Further design improvements are possible
by interposing a push rod between the the armature and the hammer pieces (see Fig. 83.34). Since
the lengths of adjacent push rods can be made alternately short and long, close packing of these
actuator assemblies having a single pivot axis is possible. With the most efficient designs, printing
speeds of up to 5000 lpm are possible.

Both moving-coil and stored-energy actuators are used, the latter using a "bucking" coil and a
stored-energy flexible spring. The bucking coil, when energized, cancels the magnetic flux from a
permanent magnet, holding the actuator in the "cocked" position, and thus converts the stored
spring energy to kinetic energy.

Dot Band Matrix Printing.   This technology is employed on some printers used for mid-range
computer systems, and, in a sense, is a hybrid technology combining attributes of high-end line
band printers with features found in serial wire matrix printers (see below). Present designs also
use a moving metal band but incorporate pixel-size raised bumps instead of etched characters.
These bumps (typically 120 on a band) are positioned at the apex of chevron-shaped springs
etched into the band (see Fig. 83.35). Also, timing/position slots are etched into the band above the
spring slots. Instead of one full line of characters being printed in one cycle, one full line of dots is
printed with this technology. Also, each hammer covers typically three character positions across
the page; thus, 45 hammers can cover a page-width of 135 character positions. Hammer cycle time
is typically 1.2 ms and the band speed can be 0.28 m/s or higher. After a line of dots is printed, the
paper is advanced *N* times until a full character height is achieved.

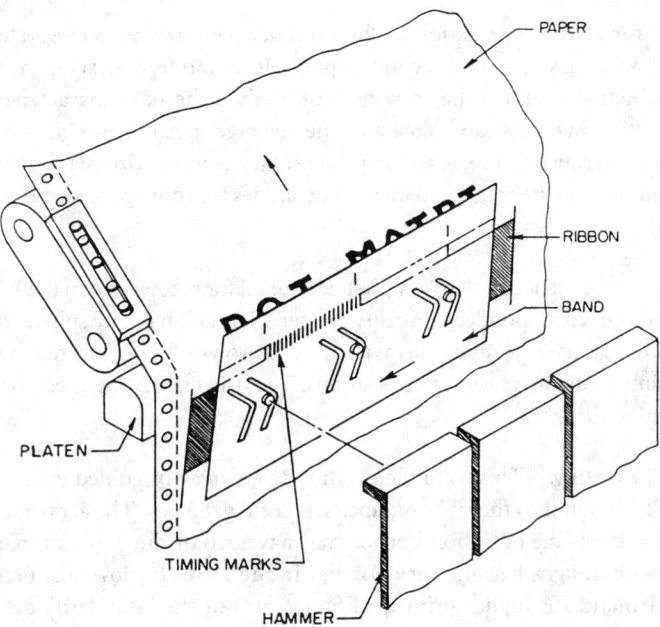

FIGURE 83.35 Dot band printer technology. (*Source:* R.C. Durbeck and S. Sherr, Eds., *Output Hardcopy Devices*, San Diego, Calif.: Academic Press, 1988, p. 198. With permission.)

This technology is much less expensive than typical engraved band technology, and allows the use of a faster draft mode using fewer dots per character; however, normal throughput (lpm) is generally much less.

Shuttle Hammer Matrix Printing. Transverse shuttling a reduced number of horizontally spaced hammers to cover a full print line is another way to implement matrix line printing. Here, a pixel-size raised bump is incorporated into the strike surface of each hammer. One design includes 33 hammers (each spaced 1 cm from its neighbor) covering 132 character positions. With an oscillatory shuttle motion of 1 cm, all line dot positions are covered, and the shuttle return time can be used for paper advance. Moderate printing speeds (600–900 lpm) can be accomplished with $1/4$-cm hammer spacing; 300 lpm is possible with 1-cm spacing. Opposite moving counterweights have been employed to reduce shuttle vibration.

Serial Impact Printing

There are two classes of technologies presently employed for serial impact printing: fully formed character and wire matrix. The former category includes both daisywheel and typeball technologies, the latter now in little use for computer output except for a few dedicated word processing applications. Daisywheel technology is used in many dedicated word processing systems and for some personal computer systems. With these two technologies, font change can be implemented by simply changing a wheel or ball, but all-points-addressable printing for image and graphics is basically not possible. Serial wire matrix technology has for years been the most popular output device for personal computers, but is now losing market share to low-end laser printers and ink-jet.

Daisywheel Printing. This technology makes use of a "petal-like" rotating print wheel with typically 96 spring-like fingers or "petals" radiating outward from a central core. An engraved raised character is present at the end of each petal. The wheel is rotated by a dc servomotor to the desired character. A single hammer-actuator then impacts the back side of the petal, forcing the raised character into the ribbon/paper set. The simplest designs have a constant carriage speed and a fixed time delay to allow for up to 180° rotation of the wheel between strokes. These systems are typically limited to about 30 characters per second (cps). More modern designs have incorporated microcode logic which slow down the carriage motor when the next character position is more than, for example, 30° away. This can increase the net average speed to as high as 60 cps.

Variants on this design use thimble- and cup-like rotary devices with petals bent parallel to the axis of rotation. This lowers the rotational inertia and allows for more petals (and hence more characters—up to 128).

Typeball Printing.   Typeball technology was first developed for typewriters in the 1950s, but has also been used for low-speed, correspondence-quality printers. The golf ball-size sphere, rotated and tilted to reach the desired character position, has typically four rows of 11 characters, repeated on both hemispheres, making a total of 88 character positions. The usual operating speed is 10 cps but up to 15 cps has been achieved.

Serial Wire Matrix Printing.   This technology employs an array of guided wires (often tungsten) that are individually driven into the ribbon/paper set (see Fig. 83.36). The array may form a single-row configuration at the plane of ribbon impact, may have two in-line rows, or may have two staggered rows. Nine wire designs became very popular in the 1980s for low-cost personal computer systems, but the demand for higher print quality has moved the "standard" design point to 24 wires. Early technology limited wire cycle repetition rates to about 500 Hz. More advanced designs can perform at 2500 Hz, but most available products operate in the 1000–1500 Hz range.

Actuators to drive these wires are quite robust and can print through many layers (up to 6) of a multipart form. The kinetic energy needed to print through 4–6 layers is normally about 0.5 mJ for 200-μm diameter wires, 1.0 mJ for 300-μm wires. For 2-layer forms, these energies drop to about 0.3 mJ and 0.45 mJ, respectively. There are two common approaches to driving the guided wires. One is a pivot-type actuator and the other is a direct solenoid design (see Fig. 83.37). The magnetic circuit in all cases must be designed to maximize either (1) the moving wire kinetic energy in the case of a free flight wire configuration or (2) the kinetic energy of the combined armature and moving wire in the case where these two components are either permanently connected or remain in contact during the drive cycle.

Other actuator designs use a stored energy approach where a preloaded leaf spring is held by a permanent magnet until released by a "bucking coil" to counteract the static flux. Also, experimental stacked piezoelectric transducers have been investigated using high lever ratios (e.g., 30:1) which

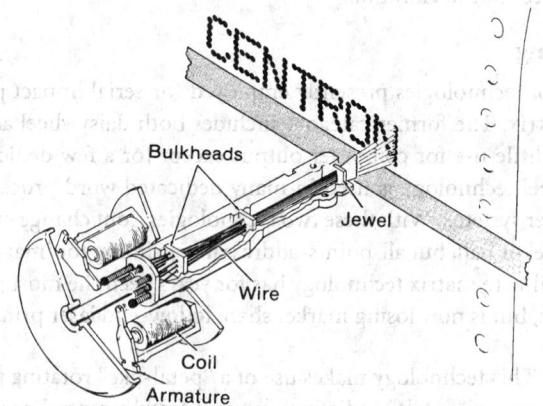

FIGURE 83.36   Wire matrix serial printhead. (*Source:* R.C. Durbeck and S. Sherr, Eds., *Output Hardcopy Devices*, San Diego, Calif.: Academic Press, 1988, p. 187. With permission.)

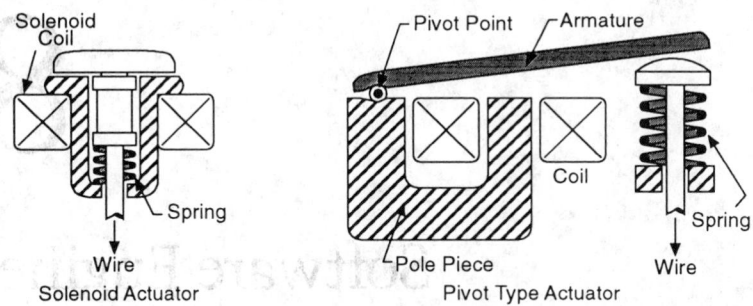

FIGURE 83.37 Solenoid and pivot-type actuators for serial wire matrix printing.

have operated at 3 kHz and above. Synthetic ruby or ceramic wire guide holes are sometimes used to combat wear. Several hundred million cycles lifetime operation for each wire is typically required for today's devices.

Higher draft-quality speed (e.g., 200 cps) can be achieved without decreasing wire cycle time by simply increasing the carriage speed. Higher than normal print quality can be accomplished by both slowing down the carriage and interlacing horizontal rows of dots. This near letter-quality printing is usually at greatly reduced speed (e.g., 48 cps). Most printers also print during the carriage return. Color can be produced by shifting four-color ribbons.

Defining Terms

Line printing: A printer prints one full line width of characters or dots at a time. The paper is then moved into the next print line position, ready for the next line of characters or dots. The printer may pause or stop between lines. Printing speed is often given in units of lines per minute (lpm).

Page printing: The information to be printed on a page is electronically composed and stored before shipping to the printer. The printer then prints the full page nonstop. Printing speed is usually given in units of pages per minute (ppm).

Pixel: The nominal printed spot area or "picture element" addressed by a particular printing device. It is sometimes called "pel."

Serial printing: Printing is done one character at a time. The print head must move across the entire page to print a line of characters. The printer may pause or stop between characters. Printing speed is usually given in units of characters per second (cps).

References

R.C. Durbeck and S. Sherr, Eds., *Output Hard Copy Devices*, San Diego, Calif.: Academic Press, 1988.

J. Heinzel and C.H. Hertz, *Advances in Electronics and Electron Physics*, vol. 65, P.W. Hawkes, Ed., San Diego, Calif.: Academic Press, 1985, pp. 91–285.

L.B. Schein, *Electrophotography and Development Physics*, Berlin: Springer-Verlag, 1988.

J.M. Sturge *et al.*, Eds., *Imaging Processes and Materials*, New York: Van Nostrand Reinhold, 1989.

D. Winkelmann *et al.*, *Ullmann's Encyclopedia of Industrial Chemistry*, vol. A13, Weinheim, Germany: VCH Publishers, 1989, pp. 571–660.

Further Information

Proceedings of the Seven SPSE International Congresses on Nonimpact Printing, published by the Society for Imaging Sciences & Technology, 7003 Kilworth Lane, Springfield, VA 22151.

Proceedings of the Society for Information Display, 1980–1992, published by The Society for Information Display, 8055 West Manchester Ave., Suite 615, Playa del Rey, CA 90203.

84

Software Engineering

Carl A. Argila
Software Engineering Consultant

Capers Jones
Software Productivity Research, Inc.

Johannes J. Martin
University of New Orleans

84.1 Tools and Techniques 1976
   Approach • Methods • Information Modeling • Essential Modeling • Implementation Modeling • CASE Tools
84.2 Testing, Debugging, and Verification 1985
   The Origins and Causes of Software Defects • The Taxonomy and Efficiency of Software Defect Removal • Pre-Test Defect Removal • Testing Software • Selecting an Optimal Series of Defect Prevention and Removal Operations • Post-Release Defect Removal
84.3 Programming Methodology 1996
   Analysis of Algorithms • Flow of Control • Abstraction • Modularity • Simple Hierarchical Structuring • Object-Oriented Programming • Program Testing

84.1 Tools and Techniques

Carl A. Argila

The last decade has seen a revolution in software engineering tools and techniques. This revolution has been fueled by the ever-increasing complexity of the software component of delivered systems. Although the software component of delivered systems may not be the most expensive component, it is usually, however, "in series" with the hardware component; if the software doesn't work, the hardware is useless.

Traditionally, software engineering has focused primarily on computer programming with ad hoc analysis and design techniques. Each software system was a unique piece of intellectual work; little emphasis was placed on architecture, interchangeability of parts, reusability, etc. These ad hoc software engineering methods resulted in the production of software systems which did not meet user requirements, were usually delivered over budget and beyond schedule, and were extraordinarily difficult to maintain and enhance.

In an attempt to find some solutions to the "software crisis," large governmental and private organizations motivated the development of so-called "waterfall" methods. These methods defined formal requirement definition and analysis phases, which had to be completed before commencing a formal design stage, which in turn had to be completed before beginning a formal implementation phase, etc. Although waterfall methods were usually superior to ad hoc methods, large and complex software systems were still being delivered over budget and beyond schedule, which did not meet user requirements. There were several reasons for this. First, waterfall methods focus on the generation of *work products* rather than "engineering." Simply put, writing documents is not the same as doing good engineering. Second, the waterfall methods do not support the *evolution* of system requirements throughout the development life cycle. Also, the prose English specifications produced within the waterfall methods are not well suited to describing the complex behaviors of software systems.

The basic, underlying philosophy of how software systems should be developed changed dramatically in 1978 when Tom DeMarco published his truly seminal book, *Structured Analysis and System Specification* [DeMarco, 1979]. DeMarco proposed that software systems should be developed like any large, complex engineering systems—by first building scale models of proposed systems so as to investigate their behavior. This *model-based software engineering* approach is analogous to that used by architects to specify and design large complex buildings (see Fig. 84.1). We build scale models of software systems for the same reason that architects build scale models of houses, so that users can visualize living with the systems of the future. These models serve as vehicles for communication and negotiation between users, developers, sponsors, builders, etc. Model-based software engineering holds considerable promise for enabling large, complex software systems to be developed on budget, within schedule, while meeting user requirements [see Harel, 1992].

As shown in Fig. 84.2, a number of specific software development models may be built as part of the software development process. These models may be built by different communities of users, developers, customers, etc. Most importantly, however, these models are built in an *iterative* fashion. Although work products (documents, milestone reviews, code releases, etc.) may be delivered chronologically, models are built iteratively throughout the software system's development life cycle.

In Fig. 84.3 we illustrate the distinction between *methodology, tool,* and *work product.* A number of differing software development methods have evolved, all based on the underlying model-based philosophy. Different methods may in fact be used for the requirements and analysis phases of project development than for design and implementation. These differing methods may or may not integrate well. Tools such as **CASE** may support all, or only a part, of a given method. Work products, such as document production or code generation, may be generated manually or by means of CASE tools.

This article will present a synopsis of various practical software engineering techniques which can be used to construct software development models; these techniques are illustrated within the context of a simple case study system.

Approach

One of the most widely accepted approaches in the software engineering industry is to build two software development models. An **essential model** captures the behavior of a proposed software system, independent of implementation specifics. An essential model of a software system is analogous to the scale model of a house built by an architect; this model is used to negotiate the *essential* requirements of a system between customers and developers. A second model, an **implementation model**, of a software system describes the technical aspects of a proposed system within a particular implementation environment. This model is analogous to the detailed blueprints created by an architect; it specifies the *implementation* aspects of a system to those who will do the construction. These models [described in Argila, 1992] are shown in Fig. 84.4. The essential and implementation models of a proposed software system are built in an iterative fashion.

Methods

The techniques used to build the essential and implementation models of a proposed software system are illustrated by means of a simple case study. The Radio Button System (RBS) is a component of a fully automated, digital automobile sound system. The RBS monitors a set of front-panel *station selection buttons* and performs station selection functions.

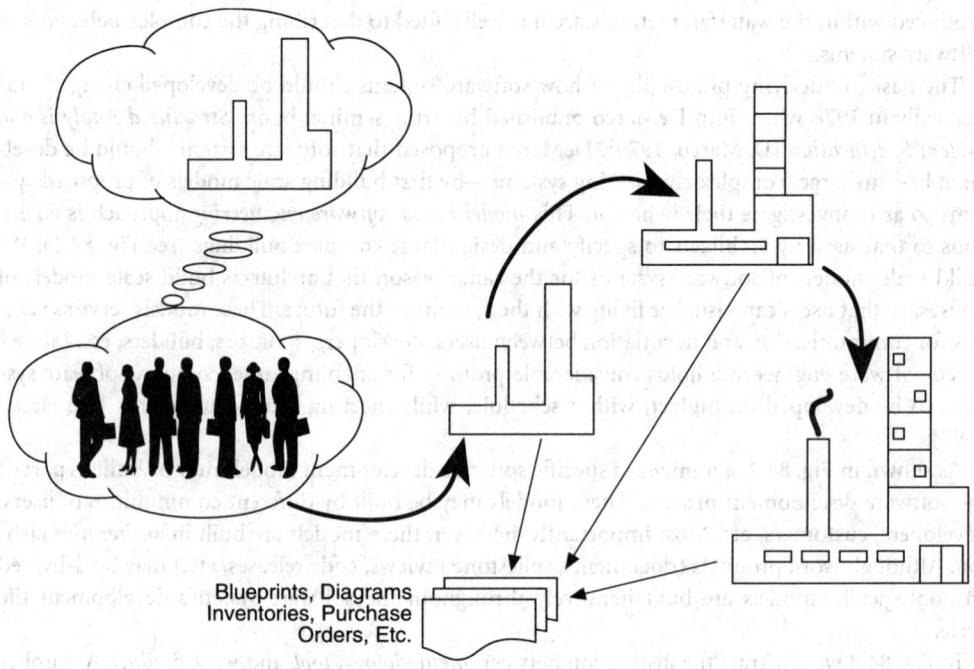

FIGURE 84.1 Model-based software engineering.

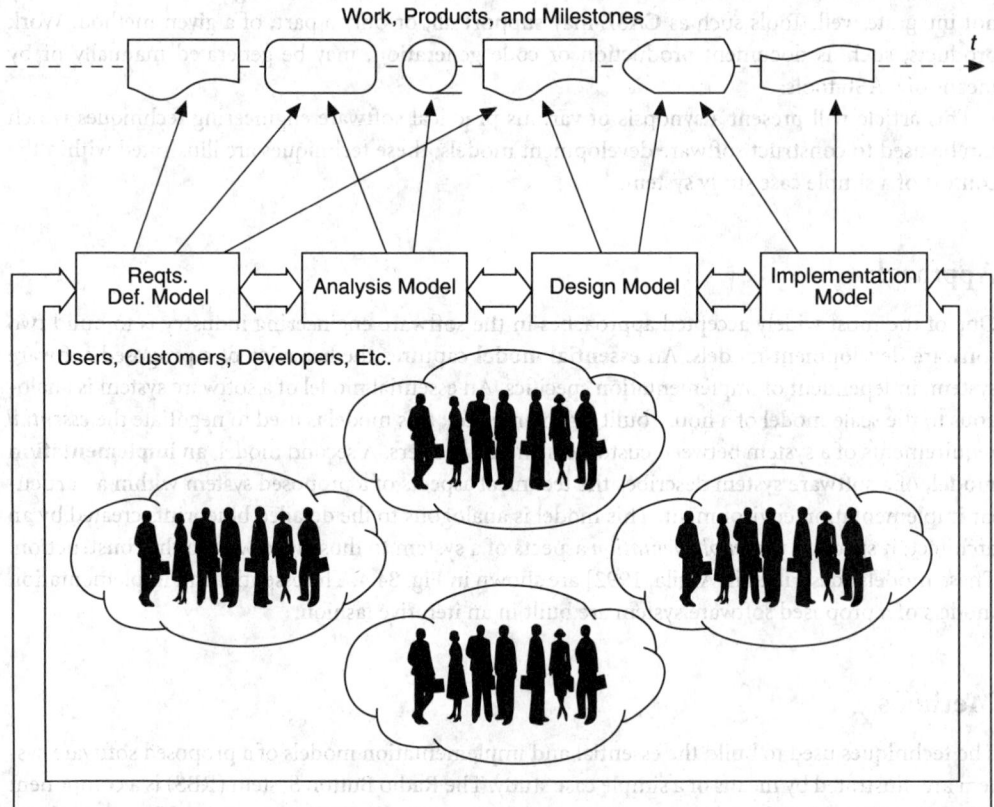

FIGURE 84.2 Modeling life cycle.

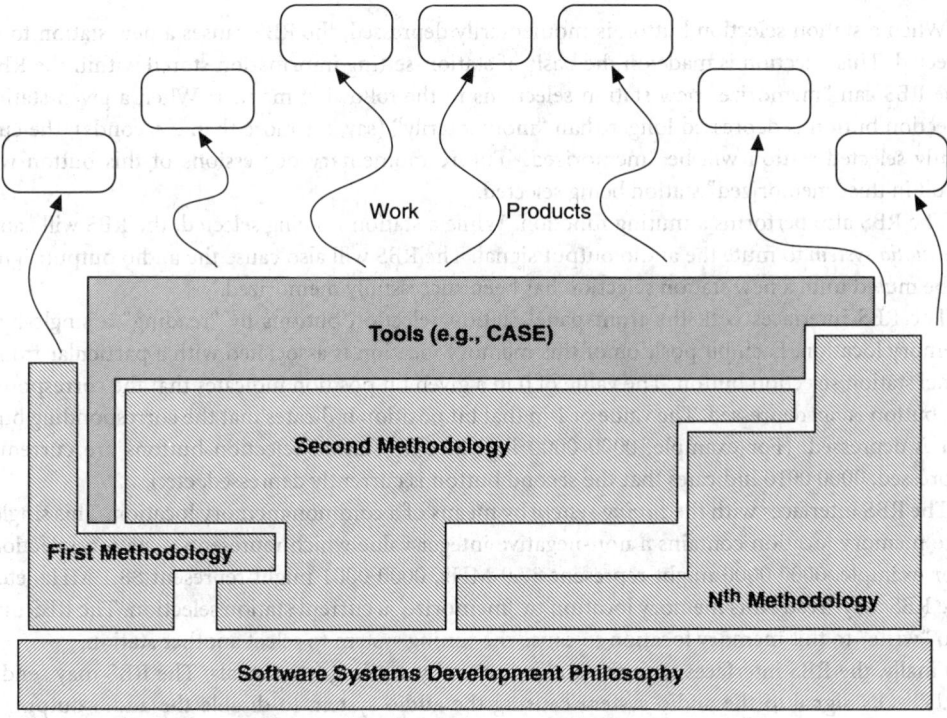

FIGURE 84.3 Methods, tools and work products.

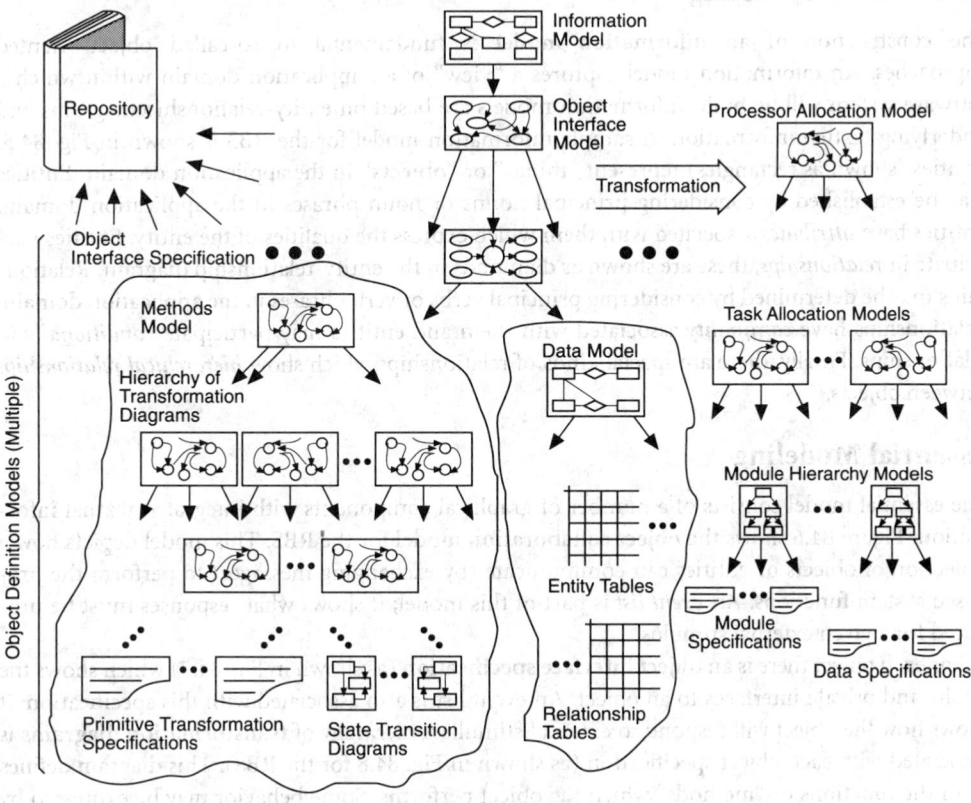

FIGURE 84.4 Software engineering methods overview.

When a station selection button is momentarily depressed, the RBS causes a new station to be selected. This selection is made on the basis of station-setting information stored within the RBS. The RBS can "memorize" new station selections in the following manner: When a given station selection button is depressed longer than "momentarily" (say, for more than 2 seconds), the currently selected station will be "memorized." Future momentary depressions of this button will result in this "memorized" station being selected.

The RBS also performs a muting function. While a station is being selected, the RBS will cause the *audio system* to mute the audio output signal. The RBS will also cause the audio output signal to be muted until a new station selection has been successfully memorized.

The RBS interfaces with the front-panel station selection buttons by "reading" a single-byte memory location. Each bit position of this memory location is associated with a particular front-panel station selection button. The value of 0 in a given bit position indicates that the corresponding button is *not* depressed. The value of 1 in that bit position indicates that the corresponding button *is* depressed. (For example, 0000 0000 indicates no station selection buttons are currently depressed; 0000 0010 indicates that the second button is currently depressed, etc.)

The RBS interfaces with the *tuning system* by means of a common memory location. This single-byte memory location contains a non-negative integer value which represents a station selection. (For example, 0000 0000 might represent 87.9 MHz, 0000 0001 might represent 88.1 MHz, etc.) The RBS may "read" this memory location to "memorize" a current station selection. The RBS may also "write" to this memory location to cause the tuning system to select another station.

Finally, the RBS interfaces with the audio system by sending two signals. The RBS may send a MUTE-ON signal to the audio system causing the audio system to disable the audio output. A MUTE-OFF signal would cause the audio system to enable the audio output.

Information Modeling

The construction of an **information model** is fundamental to so-called object-oriented approaches. An information model captures a "view" of an application domain within which a software system will be built. Information models are based on entity-relationship diagrams and underlying textual information. A sample information model for the RBS is shown in Fig. 84.5. Entities (shown as rectangles) represent "things" or **"objects"** in the application domain. Entities may be established by considering principal nouns or noun phrases in the application domain. Entities have *attributes* associated with them which express the qualities of the entity. Entities participate in *relationships;* these are shown as diamonds in the entity-relationship diagram. Relationships may be determined by considering principal verbs or verb phrases in the application domain. Relationships have *cardinality* associated with them and entities may participate *conditionally* in relationships. Finally, there are special kinds of relationships which show *hierarchical relationships* between objects.

Essential Modeling

The essential model consists of a number of graphical components with integrated textual information. Figure 84.6 shows the **object collaboration model** for the RBS. This model depicts how a collection of objects or entities can communicate (by exchanging messages) to perform the proposed system functions. An *event list* is part of this model; it shows what responses must be produced for a given external stimulus.

For each object there is an **object interface specification** (as shown in Fig. 84.7) which shows the public and private interfaces to an object. An event list is also associated with this specification; it shows how the object will respond to external stimuli. A hierarchy of **transformation diagrams** is associated with each object specification (as shown in Fig. 84.8 for the RBS). This diagram defines all of the functions or "methods" which the object performs. Some behavior may be expressed by means of a **state transition diagram** (Fig. 84.9).

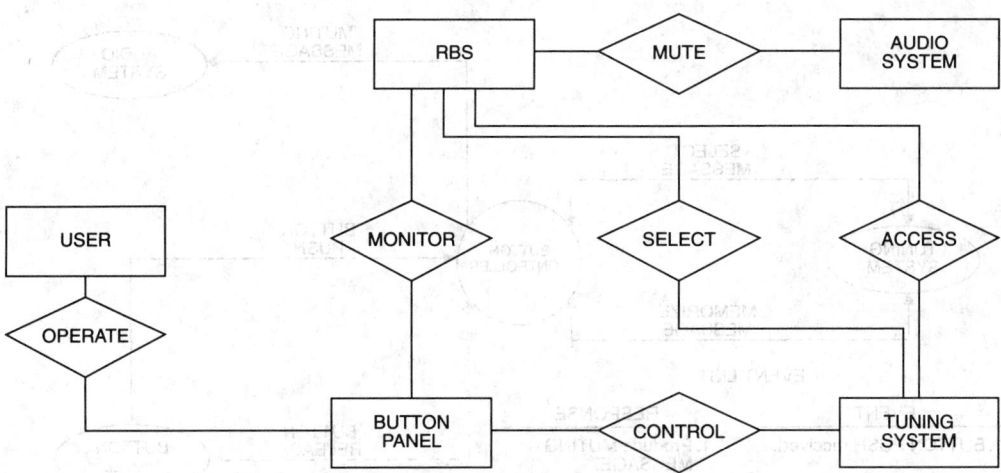

FIGURE 84.5 RBS information model.

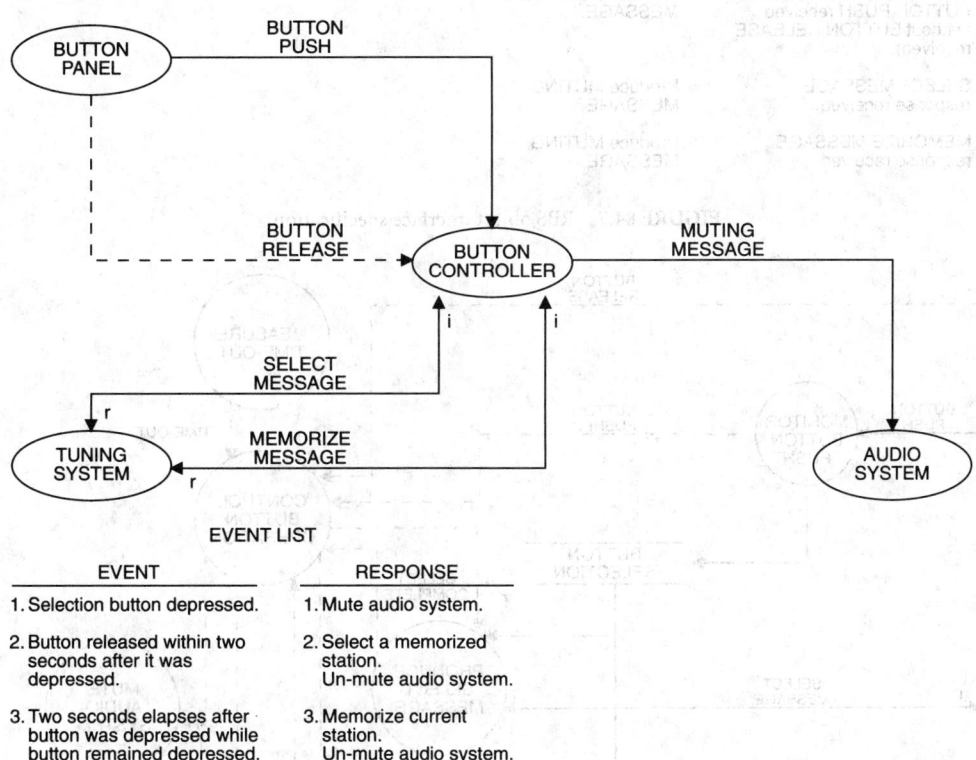

EVENT LIST

EVENT	RESPONSE
1. Selection button depressed.	1. Mute audio system.
2. Button released within two seconds after it was depressed.	2. Select a memorized station. Un-mute audio system.
3. Two seconds elapses after button was depressed while button remained depressed.	3. Memorize current station. Un-mute audio system.

FIGURE 84.6 RBS object collaboration model.

Implementation Modeling

Two principal activities must be accomplished in transitioning from the essential to the implementation model. First, all of the methods and data encapsulated by each object must be mapped to the implementation environment. This process is illustrated in Fig. 84.10. Second, all of the details which were ignored in the essential model (such as user interfaces, communication protocols, hardware limitations, etc.) must now be accounted for.

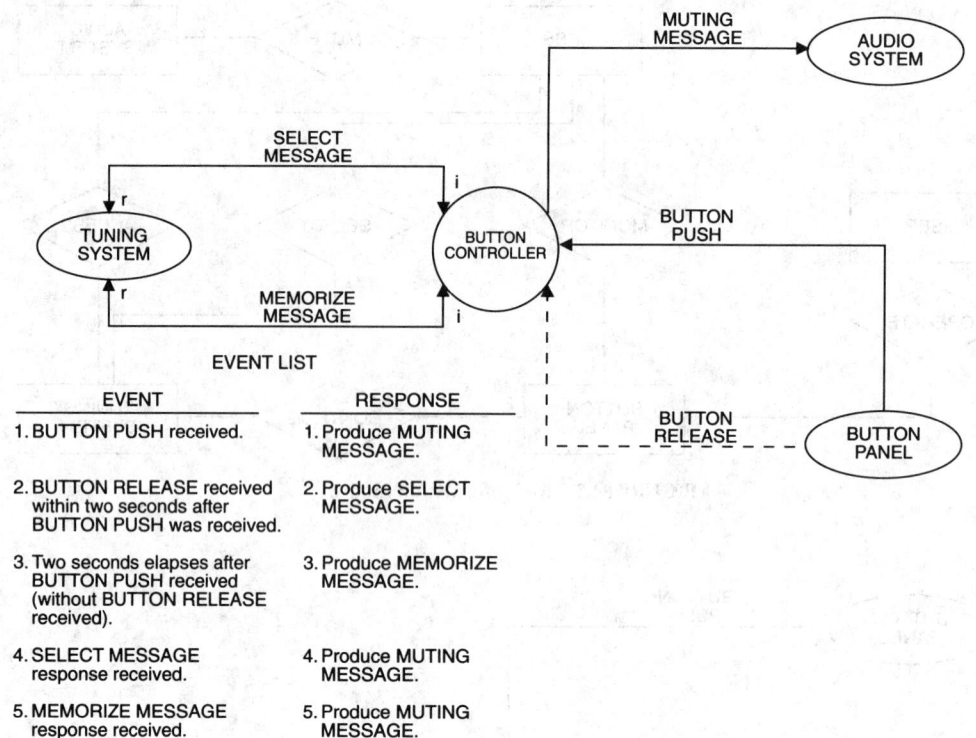

FIGURE 84.7   RBS object interface specification.

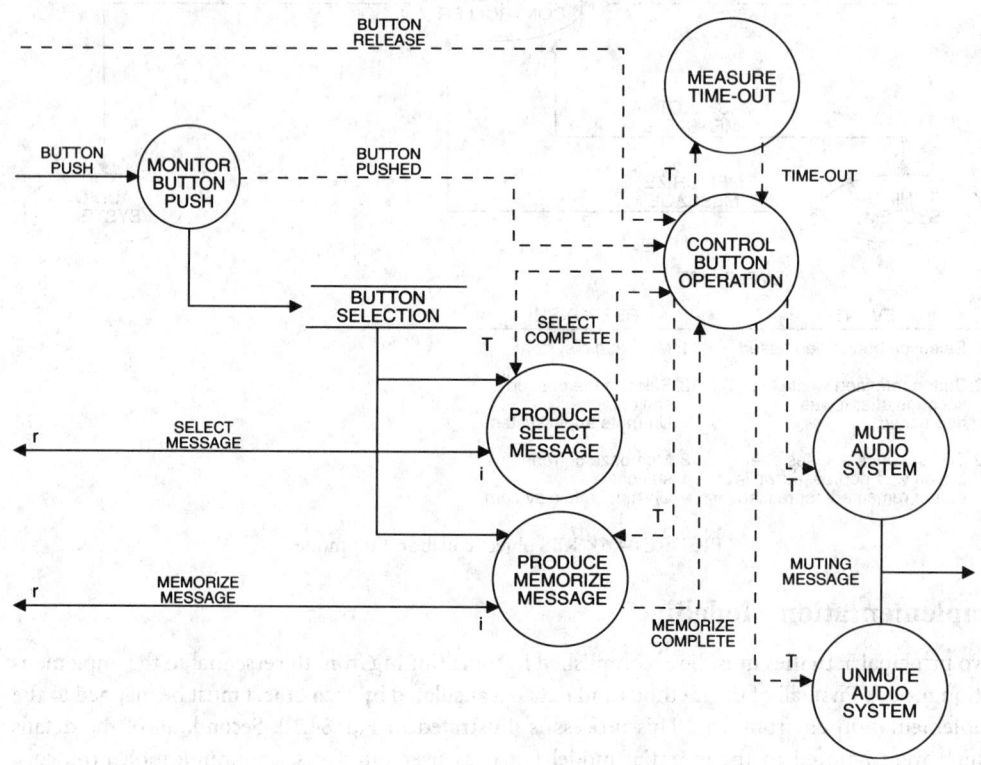

FIGURE 84.8   RBS transformation diagram.

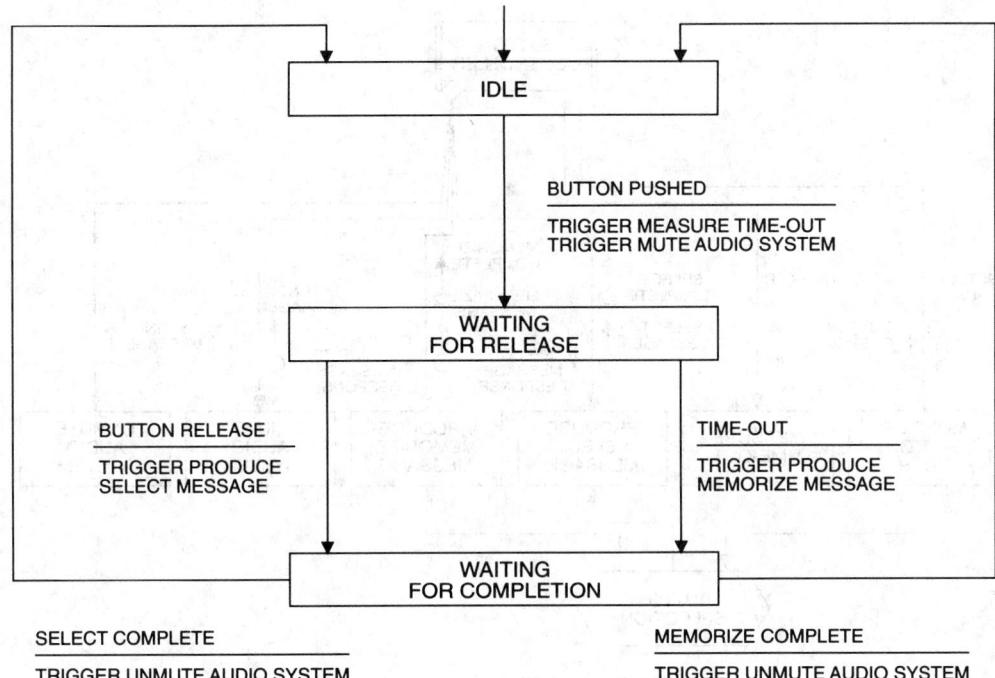

FIGURE 84.9   RBS state transition diagram.

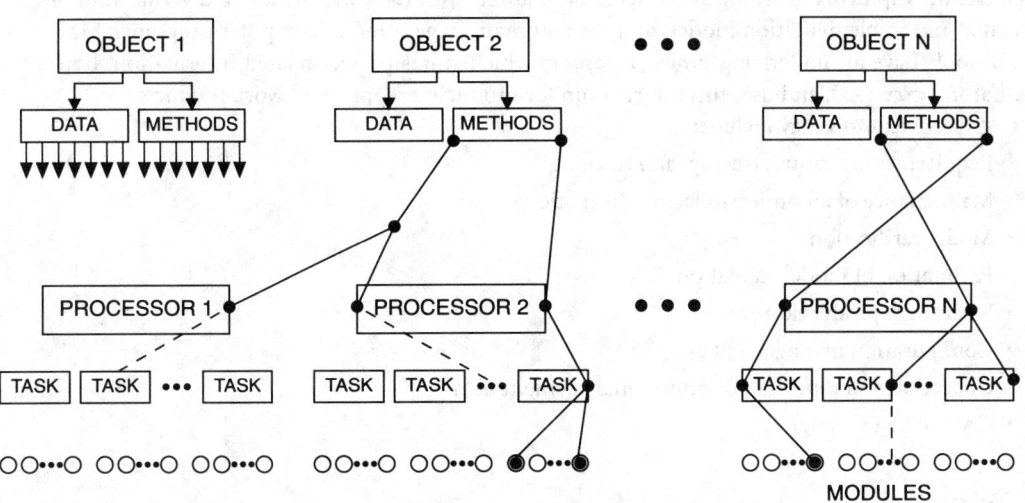

FIGURE 84.10   Implementation modeling.

Each component of the essential model must be allocated to hardware processors. Within each hardware processor, allocation must be continued to the *task* level. Within each task, the computer program controlling that task must be described. This latter description is accomplished by means of a **module structure chart.** As illustrated in Fig. 84.11 for one component of the RBS, the module structure chart is a formal description of each of the computer program units and their interfaces.

CASE Tools

The term *computer-aided software engineering* (CASE) is used to describe a collection of tools which automate all or some of various of the software engineering life cycle phases. These tools may

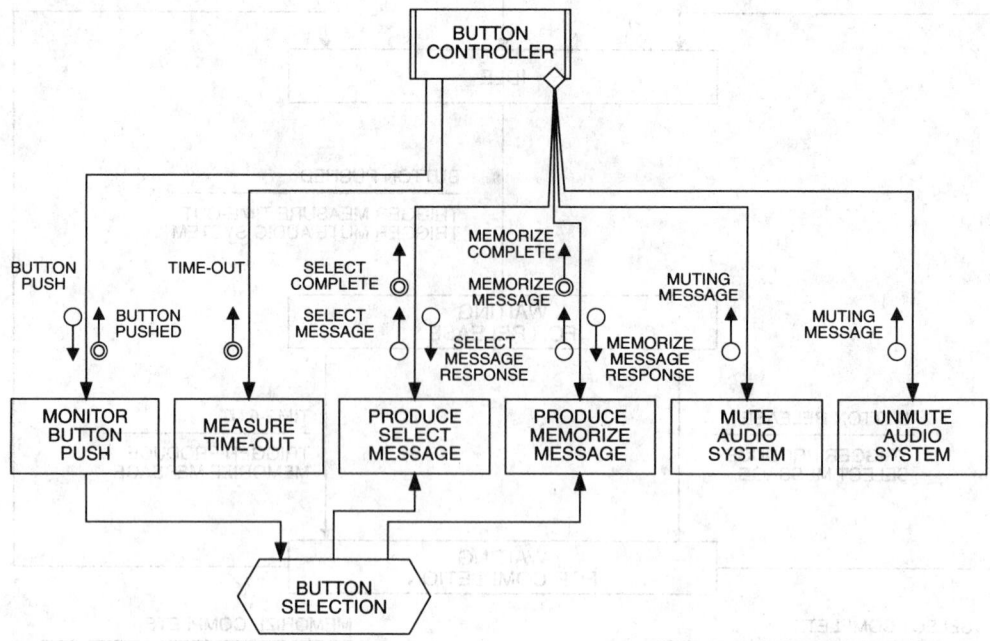

FIGURE 84.11   RBS module structure chart.

facilitate the capturing, tracking and tracing of requirements, the construction and verification of essential and implementation models and the automatic generation of computer programs. Most CASE tools have an underlying *project repository* which stores project-related information, both textual and graphical, and uses this information for producing reports and work products.

CASE tool features may include:

- Requirements capture, tracing, and tracking
- Maintenance of all project-related information
- Model verification
- Facilitation of model validation
- Document production
- Configuration management
- Collection and reporting of project management data
- CASE data exchange

Defining Terms

CASE: Computer-aided software engineering. A general term for tools which automate various of the software engineering life cycle phases.

Essential model: A software engineering model which describes the behavior of a proposed software system independent of implementation aspects.

Implementation model: A software engineering model which describes the technical aspects of a proposed system within a particular implementation environment.

Information model: A software engineering model which describes an application domain as a collection of objects and relationships between those objects.

Module structure chart: A component of the implementation model; it describes the architecture of a single computer program.

Object: An "entity" or "thing" within the application domain of a proposed software system.

Object collaboration model: A component of the essential model; it describes how objects exchange messages in order to perform the work specified for a proposed system.

Object interface specification: A component of the essential model; it describes all of the public and private interfaces to an object.

State transition diagram: A component of the essential model; it describes event-response behaviors.

Transformation diagram: A component of the essential model; it describes system functions or "methods."

References

C. Argila, "Object-oriented real-time systems development" (video course notes), Los Angeles: University of Southern California IITV, June 11, 1992.

G. Booch, *Object-Oriented Design with Applications*, Redwood City, Calif.: Benjamin/Cummings, 1991.

P. Coad and E. Yourdon, *Object-Oriented Analysis*, 2nd ed., New York: Prentice-Hall, 1991.

P. Coad and E. Yourdon, *Object-Oriented Design*, New York: Prentice-Hall, 1991.

T. DeMarco, *Structured Analysis and System Specification*, New York: Prentice-Hall, 1979.

D. Harel, "Biting the silver bullet," *Computer*, January 1992.

J. Rumbaugh *et al.*, *Object-Oriented Modeling and Design*, New York: Prentice-Hall, 1991.

S. Shlaer and S. Mellor, *Object-Oriented Systems Analysis: Modeling the World in Data*, New York: Prentice-Hall, 1988.

S. Shlaer and S. Mellor, *Object Life-Cycles: Modeling the World in States*, New York: Prentice-Hall, 1992.

P. Ward and S. Mellor, *Structured Development for Real-Time Systems*, New York: Prentice-Hall, vol. 1, 1985; vol. 2, 1985; vol. 3, 1986.

E. Yourdon and L. Constantine, *Structured Design*, 2nd ed., New York: Prentice-Hall, 1975, 1978.

Further Information

A video course presenting the software engineering techniques described here is available [see Argila, 1992]. The author may be contacted for additional information and comments at (800) 347-6903.

84.2 Testing, Debugging, and Verification

Capers Jones

Achieving acceptable levels of software quality has been one of the most troublesome areas of software engineering since the industry began. It has been so difficult to achieve error-free software that, historically, the cost of finding and fixing **"bugs"** has been the most time-consuming and expensive aspect of large-scale software development.

Software quality control is difficult to achieve, but the results of the careful application of **defect prevention** and **defect removal** techniques are quite good. The software producers who are most effective in quality control discover several derivative benefits as well: software with the highest quality levels also tends to be the least expensive to produce, tends to have minimum schedules during development, and also tends to have the highest levels of post-release user satisfaction.

The topic of defect prevention is outside the primary scope of this article. However, it is important to note that the set of technologies associated with defect prevention must be utilized concurrently with the technologies of defect removal in order to achieve high levels of final quality. The technologies which prevent defects are those concerned with optimizing both clarity and structure and with minimizing ambiguity. Joint application design (JAD) for preventing requirements defects; prototyping; reuse of certified material; clean-room development; any of the standard structured analysis, design, and coding techniques; and quality function deployment (QFD) for evaluating end-user quality demands are examples of the technologies associated with defect prevention. Many aspects of total quality management (TQM) programs are also associated with defect prevention.

The Origins and Causes of Software Defects

Before software defects can be prevented or removed effectively, it is necessary to know where defects originate and what causes them. There are five primary origin points for software defects and three primary causes. The five primary origin points are (1) requirements, (2) design, (3) code, (4) user documentation, and (5) bad fixes, or secondary defects that occur while attempting to repair a primary defect. The three primary causes of software defects are (1) errors of omission, (2) errors of commission, and (3) errors of clarity or ambiguity. Table 84.1 shows the interaction of software defect origins and defect types.

The phrase *errors of omission* refers to problems caused by a failure to include vital information. Such errors are frequently encountered in requirements, specifications, source code, and user documents. An example of such an error became highly visible on February 29, 1992, when it was discovered that the calendar routine for a particular application omitted leap year calculations, thus shutting down the system at midnight on the 28th. From 15% to more than 30% of the post-deployment problems encountered in software can be traced to errors of omission. The probability and frequency of such errors rises with the size of the system.

The phrase *errors of commission* refers to problems caused by an action that is not correct, such as looping through a counter routine one time too often or including conditions in a specification that are mutually contradictory. Such errors are frequently encountered in design, code, and in "bad fixes" or secondary defects created as a by-product of repairing prior defects. From 40 to 65% of the post-deployment problems encountered in software can be traced to errors of commission, thus making it the most common source of software problems.

The phrase *errors of clarity or ambiguity* refers to problems caused by two or more interpretations of the same information. For example, a requirement may contain a phrase that an application should provide "rapid response time" without actually defining what "rapid" means. To the client, sub-second response time may have been intended, but to the development team one-minute response time may have been their interpretation. Such errors are frequently encountered in all software deliverables based on natural language such as English. They are also frequently encountered in source code itself, especially so if the code were not well structured. Also, certain languages such as APL and those using nested parentheses are notorious for being ambiguous. From less than 5% to more than 10% of the post-deployment problems encountered in software can be traced to errors of clarity or ambiguity.

When considering the origins of software defects, it is significant that the requirements and specifications themselves may be the source of as many as 40% of the defects later encountered after

Table 84.1 Origins and Causes of Software Defects

	Defect Causes		
Defect Origins	Errors of Omission	Errors of Commission	Errors of Clarity or Ambiguity
Requirements defects	Frequent	Seldom	Frequent
Design defects	Frequent	Frequent	Frequent
Coding defects	Frequent	Frequent	Frequent
Document defects	Frequent	Seldom	Frequent
Bad fix defects	Seldom	Frequent	Seldom

Table 84.2 Software Defect Origins and Project Size

	Software Project Size (Statements in C Language)			
	1000	10,000	100,000	1,000,000
Requirements	5%	7%	10%	15%
Design	10%	15%	20%	25%
Code	70%	60%	50%	40%
Documents	5%	8%	10%	10%
Bad fixes	10%	10%	10%	10%
Total	100%	100%	100%	100%

deployment. This fact implies that some forms of verification and validation, which assume the requirements and specifications are complete and correct, have hidden logical flaws.

The distribution of defects among the common origin points varies with the size and complexity of the application. Table 84.2 shows the probable defect distributions for four size ranges of software projects, using the language C as the nominal coding language.

The sum of the five defect categories is termed the *defect potential* of a software program or system [Jones, 1986]. The defect potential constitutes the universe of all errors and bugs that might cause an application to either fail to operate or to produce erratic and unacceptable results while operating.

The defect potential of software tends to be alarmingly high. When expressed with the traditional metric "KLOC" (where K stands for 1000 and LOC stands for lines of code) the observed defect potentials have ranged from about 10 bugs per KLOC to more than 100 bugs per KLOC, assuming procedural languages such as C, Fortran, or Jovial. When expressed with the newer Function Point metric, the range of software defect potentials is from less than 2 to more than 10 bugs per Function Point.

When historical defect data is analyzed, it can easily be seen why defect prevention and defect removal are complementary and synergistic technologies. The defect prevention approaches are used to lower defect potentials, and the defect removal technologies are then used to eliminate residual defects.

Another dimension of concern is the *severity level* of software defects. The severity level scheme used by IBM, and widely adopted by other software producers, is based on a four-point scale. Severity 1 defects are those which stop the software completely or prevent it from being used at all. Severity 2 defects are those where major functions are disabled or unusable. Severity 3 defects are those that can be bypassed or which have a minor impact. Severity 4 defects are very minor problems which do not affect the operation of the software, for example, a spelling error in a text message.

Software **debugging,** testing, and verification methods should be effective against the entire defect potential of software, and not just against coding defects. Defect removal methods should approach 100% in efficiency against severity 1 and severity 2 defects. Further, defect removal methods should deal with errors of omission as well as errors of commission. Software operates at extremely high speed, so defect removal operations must deal with timing and performance problems as well as with structural and logical problems. Finally, and most difficult, defect removal methods should deal with errors of clarity or ambiguity. These are challenging tasks.

The Taxonomy and Efficiency of Software Defect Removal

There is no standard taxonomy that encompasses all of the many forms of defect removal that can be applied to software [Dunn, 1984]. For convenience, we will divide defect removal operations into three broad categories: pre-test defect removal, testing, and post-release defect removal.

Since the goal of defect removal is the elimination of bugs or defects, the primary figure of merit for a defect removal operation is its *efficiency* [Jones, 1991]. **Defect removal efficiency** is defined as the percent of latent defects which a given removal operation will detect. Cumulative defect removal efficiency is the overall efficiency of a complete series of pre-test removal activities combined with all test stages.

Calculating the efficiency of a defect removal operation, or a series, is necessarily a long-range operation. All defects encountered prior to release are enumerated and running totals are kept. After the first year of production, the client-reported defects and the pre-release defects are aggregated as a set, and the efficiencies of pre-release operations are then calibrated. Thus for large systems with multi-year development cycles, it may take up to five years before defect removal efficiency rates are fully known.

Major computer companies, telecommunications manufacturers, defense contractors, and some software vendors have been measuring defect removal efficiencies for more than 20 years. The body of empirical data is large enough to make general observations about the efficiencies of many different kinds of defect removal operation.

Empirical observations and long-range studies of commercial software have demonstrated that most forms of testing are less than 30% efficient. That is, any given stage of testing such as system testing or acceptance testing is likely to find less than 30% of the latent defects that are actually present. However, a carefully planned series of defect removal operations can achieve very respectable levels of cumulative defect removal efficiency. Computer manufacturers, commercial software producers, and defense software producers may utilize as many as 9 to 20 consecutive defect removal operations, and sometimes achieve cumulative defect removal efficiencies that approach the six sigma quality level, i.e., defect removal efficiency rates approaching 99.999999%.

Pre-Test Defect Removal

The set of defect removal activities for software carried out prior to the commencement of testing can be further subdivided into *manual defect removal* and *automated defect removal*.

A fundamental form of manual defect removal is termed *desk checking*, or the private review of a specification, code document, or document by the author. The efficiency of desk checking varies widely in response to both individual talents and to the clarity and structure of the materials being analyzed. However, most humans are not particularly efficient in finding their own mistakes, so the overall efficiency of desk checking is normally less than 35%.

The most widely utilized forms of manual defect removal are reviews, inspections, and walkthroughs. Unfortunately common usage blurs the distinction among these three forms of defect removal. All three are group activities where software deliverables such as the requirements, preliminary or detailed design, or a source code listing are discussed and evaluated by technical team members.

Casual reviews or informal walkthroughs seldom keep adequate records, and so their defect removal efficiencies are not precisely known. A few controlled studies carried out by large companies such as IBM indicate that informal reviews or walkthroughs may average less than 45% in defect removal efficiency.

The term *inspection* is often used to define a very rigorous form of manual analysis [Fagan, 1976]. Ideally, the inspection team will have received formal training before participating in their first inspection. Also, the inspection team will follow formal protocols in terms of preparation, execution, recording of defects encountered, and follow-up of the inspection session. The normal complement for a formal inspection includes a moderator, a recorder, a reader to paraphrase the material being inspected, the developer whose work is undergoing inspection, and often one or more additional inspectors.

Formal inspections of the kind just discussed have the highest measured efficiency of any kind of defect removal operation yet observed. For inspections of plans, specifications, and documents the defect removal efficiency of formal inspections can exceed 65%. Formal inspections of source code listings can exceed 60%.

Formal inspections are rather expensive, but extremely valuable and cost-effective. Not only do inspections achieve high rates of defect removal efficiency, but they are also effective against both errors of omission and errors of clarity or ambiguity, which are extremely difficult to find via testing. Formal inspections also operate as a defect prevention mechanism. Those who participate in the inspection process obviously learn a great deal about the kinds of defects encountered. These observations are kept in memory and usually lead to spontaneous reductions in potential defects within a few months of the adoption of formal inspections.

Military software projects in the United States are governed by various military standards such as DOD 2167A and DOD 1614. These standards call for an extensive series of reviews or inspections that are given the generic name of *verification and validation*. The word *verification* is generally defined as ensuring that each stage in the software process follows the logical intent of its predecessor. The term *validation* is generally defined as ensuring that each delivered feature or function can be traced back to a formal requirement.

U.S. military verification and validation has developed its own argot, and those interested in the full set of U.S. military verification and validation steps should refer to the military standards themselves or to some of the specialized books that deal with this topic. Examples of the forms of verification and validation used on U.S. military projects, and the three-letter abbreviations by which they are commonly known, include system requirements review (SRR), system design review (SDR), preliminary design review (PDR), critical design review (CDR).

The defect removal efficiencies of the various forms of military review have not been published, but there is no reason to doubt that they are equivalent or even superior to similar reviews in the civilian domain. The cumulative defect removal efficiency of the full series of U.S. military defect removal operations is rather good: from about 94% to well over 99% for critical weapons systems.

Large military software projects use a special kind of external review, termed *independent verification and validation* (IV&V). The IV&V activities are often performed by contracting organizations which specialize in this kind of work. The efficiency ranges of IV&V reviews have not been published, but there is no reason to doubt that they would achieve levels similar to those of civilian reviews and inspections, i.e., in the 40 to 60% range.

The most elaborate and formal type of manual defect removal methodology is that of *correctness proofs*. The technique of correctness proofs calls for using various mathematical and logical procedures to validate the assertions of software algorithms. Large-scale studies and data on the efficiency of correctness proofs have not been published, and anecdotal results indicate efficiency levels of less than 30%. To date there is no empirical evidence that software where correctness proofs were used actually achieves lower defect rates in the field than software not using such proofs. Also, the number of correctness proofs which are themselves in error appears to constitute an alarming percentage, perhaps more than half, of all proofs attempted.

Among commercial and military software producers, a very wide spectrum of manual defect removal activities are performed by *software quality assurance* (SQA) teams [Dunn and Ullman, 1982]. The full set of activities associated with formal software quality assurance is outside the primary scope of this article. However, defect prediction, defect measurement, moderating and participating in formal inspections, test planning, test case execution, and providing training in quality-related topics are all aspects of software quality assurance. Organizations that have formal software quality assurance teams will typically average 10 to 15% higher in cumulative defect removal efficiency than organizations which lack such teams.

The suite of automated pre-test tools that facilitate software defect removal has improved significantly in recent years, but still has a number of gaps that may be filled in the future.

For requirements, specifications, user documents, and other software deliverables based on natural language, the use of word processors and their built-in spelling checkers has minimized the presence of minor spelling errors. Also available, although not used as widely as spelling checkers, are a variety of automated grammar and syntax checkers and even textual complexity analyzers. Errors of clarity and ambiguity, long a bane of software, can be reduced significantly by judicious usage of such tools.

Several categories of specialized tools have recently become available for software projects. Many CASE (computer-aided software engineering) tool suites have integral support for both producing and verifying structural descriptions of software applications. There are also tools that can automatically generate test cases from specifications and tools that can trace test cases back to the original requirements. While such tools do not fully eliminate errors of commission, they do provide welcome assistance.

Errors of omission in requirements, specifications, and user documents are the most resistant to automatic detection and elimination. Studies carried out on operating systems and large switching systems have revealed that in spite of the enormous volume of software specifications (sometimes more than 100 English words per source statement, with many diagrams and tables also occurring) more than 50% of the functionality in the source code could not be found described in the specifications or requirements. Indeed, one of the unsolved challenges of software engineering is to determine if it is even theoretically possible to fully specify large and complex software systems. Assuming that full specification is possible, several practical issues are also unknown: (1) What combination of text, graphics, and other symbols is optimal for software specifications? (2) What is the optimum volume or size of the specifications for an application of any given size? (3) What will be the impact of multi-media software extensions on specifications, debugging, and testing?

The number and utility of debugging tools for programming and programmers have made enormous strides over the past few years, and continued progress can be expected indefinitely. Software development in the 1990s often takes place with the aid of what is termed a *programming environment*. The environment constitutes the set of tools and aids which support various programming activities. For debugging purposes, software syntax checkers, trace routines, trap routines, static analyzers, complexity analyzers, cross-reference analyzers, comparators, and various data recording capabilities are fairly standard. Not every language and not every vendor provides the same level of debugging support, but the best are very good indeed.

Testing Software

There is no standard taxonomy for discussing testing. For convenience, it is useful to consider test planning, test case construction, test case execution, test coverage analysis, and test library control as the major topics encompassing software testing.

For such a mainstream activity as test planning, both the literature and tool suites are surprisingly sparse. The standard reference is Myers [1979]. The basic aspects of test planning are to consider which of the many varieties of testing will be carried out and to specify the number and kind of test cases that will be performed at each step. U.S. military specifications are fairly thorough in defining the contents of test plans. There are also commercial books and courses available, but on the whole the topic of test planning is underreported in the software literature.

A new method for estimating the number of test cases required for software was developed in 1991 and is starting to produce excellent results. A metric termed *Function Points* was invented by A. J. Albrecht of IBM and placed in the public domain in 1978 [Albrecht, 1979]. This metric is derived from the weighted sums of five parameters: the numbers of inputs, outputs, inquiries, logical files, and interfaces that constitute a software application [Garmus, 1991]. The Function Point total of an application can be enumerated during the requirements and design phases.

It is obvious that testing must be carried out for each of the factors utilized by the Function Point metric. Empirical observations indicate that from two to four test cases per Function Point are the normal quantities created by commercial software vendors and computer manufacturers. Thus for an application of 1000 Function Points in size, from 2000 to 4000 test cases may be required to test the user-defined functionality of the application.

Testing can be approached from several directions. Testing which attempts to validate user requirements or the functionality of software, without regard for its inner structure of the application, is termed *black-box* testing.

Black-box test case construction for software has long been a largely manual operation that is both labor-intensive and unreliable. (Empirical observations of operating system test libraries revealed more errors in the test cases than in the product being tested.) Test case generators have been used experimentally since the 1970s and are starting to appear as both stand-alone products and as parts of CASE tool suites. However, in order for test case generation to work effectively, the specifications or written description of the software must be fairly complete, rigorous, and valid in its own right. It is to no purpose to generate automatic test cases for incorrect specifications.

Testing which attempts to exercise the structure, branching, and control flows of software is termed *white-box* or sometimes *glass-box* testing. In this domain, a fairly rich variety of tools has come into existence since 1985 that can analyze the structure and complexity of software. For certain languages such as COBOL and C, tools are available that not only analyze complexity but can restructure or simplify it. In addition, there are tools that can either create or aid in the creation of test cases. Quite a number of tools are available that can monitor the execution of test cases and identify portions of code which have or have not been reached by any given test run.

Although not full test case generators, many supplemental testing tools are available that provide services such as creating matrices of how test cases interact with functions and modules. Also widely used are *record and playback* tools which capture all events during testing. One of the more widely used testing tool classes is that of *test coverage analyzers*. Test coverage analyzers dynamically monitor the code that is being executed while test cases are being run, and then report on any code or paths that may have been missed. Test coverage is not the same as removal efficiency: it is possible to execute 100% of the instructions in a software application without finding 100% of the bugs. However, for areas of code that are not executed at all, the removal efficiency may be zero. Software *defect tracking systems* are exceptionally useful.

Once created, effective test cases will have residual value long after the first release of a software product. Therefore, formal test case libraries are highly desirable, to ensure that future changes to software do not cause regression or damage to existing functionality. Test library tools occur in several CASE tool suites, are available as stand-alone products, and are also often constructed as custom in-house tools by the larger software vendors and computer manufacturers.

Test case execution can be carried out either singly for individual test cases or for entire sets of related test cases using *test scripts*. It is obvious that manually running one test case at a time is too expensive for large software projects. Tools for multi-test execution support, sometimes called a *test harness*, are available in either stand-alone form or as part of several CASE tool suites.

There are no rigorous definitions or standard naming conventions for the kinds and varieties of testing that occur for software. Some of the more common forms of testing include the following.

The testing of an individual module or program by the programmer who created it is normally called *unit test*. Unit testing can include both black-box and white-box test cases. The efficiency of unit test varies with the skill of the programmer and the size and complexity of the unit being tested. However, the observed defect removal efficiency of unit testing seldom exceeds 50% and the average efficiency hovers around 25%. For small projects developed by a single programmer, unit test may be the only test step performed.

For large software projects involving multiple programmers, modules, or components a number of test stages will normally occur that deal with testing multiple facets of the application.

The phrase *new function test* is used to define the testing of capabilities being created or added to an evolving software project. New function testing may consist of testing the aggregated work of several programmers. New function testing may be carried out by the developers themselves, or it may be carried out by a team of testing specialists. The observed efficiency of new function testing is in the 30% range when carried out by developers and in the 35% range when carried out by testing specialists.

The phrase *regression test* is used to define the testing of an evolving software product to ensure that existing capabilities have not been damaged or degraded as a by-product of adding new capabilities. Regression testing is normally carried out using test cases created for earlier releases, or at least created earlier in the development cycle of the current release. The observed removal efficiency of regression testing against the specific class of errors that it targets may exceed 50% when carried out by sophisticated organizations such as computer manufacturers or defense contractors. However, efficiencies in the 20 to 25% range are more common.

The phrase *stress test* is used to define a special kind of testing often carried out for real-time or embedded software. With stress testing, the software is executed at maximum load and under maximum performance levels, to ensure that it can meet critical timing requirements. The observed removal efficiency of stress testing often exceeds 50% against timing and performance-related problems. Stress testing is normally carried out by testing and performance specialists who are supported by fairly sophisticated test tool suites.

The phrase *integration test* is used to define a recurring series of tests that occur when components or modules of an evolving software product are added to the fundamental system. The purpose of integration testing is to ensure that the newer portions of the product can interact and operate safely with the existing portions. The observed removal efficiency of integration testing is normally in the 20 to 35% range. When performed by commercial software producers, integration test is normally carried out by testing specialists supported by fairly powerful tools.

The phrase *system test* is used to define the final, internal testing stage of a complete product before it is released to customers. The test suites that are executed during system test will include both special tests created for system test and also regression tests drawn from the product's test library. The observed removal efficiency of system testing is in the 25 to 35% range when such testing is done by the developers themselves. When system testing is carried out by testing specialists for commercial software, its defect removal efficiency may achieve levels approaching 50%. However, high defect removal efficiency this late in a development cycle often leads to alarming schedule slippages.

The phrase *independent test* is used to define a form of testing by an independent testing contractor or an outside organization. Independent testing is standard for U.S. military software and sometimes used for commercial software as well. Defect removal efficiency of independent test may approach 50%.

The phrase *field test* is used to define testing by early customers of a software product, often under a special agreement with the software vendor. Field testing often uses live data and actual customer usage as the vehicle for finding bugs, although prepared test cases may also occur. The defect removal efficiency of field test fluctuates wildly from customer to customer and product to product, but seldom exceeds 30%.

The phrase *acceptance test* is used to define testing by a specific client, as a precursor to determining whether to accept a software product or not. Acceptance testing may be covered by an actual contract, and if so is a major business factor. The defect removal efficiency of acceptance testing fluctuates from client to client and product to product, but seldom exceeds 30%.

Sometimes the phrases *alpha test* and *beta test* are used as a linked pair of related terms. Alpha testing defines the set of tests run by a development organization prior to release to early customers. Beta testing defines the set of tests and actual usage experiences of a small group of early customers. Unfortunately the term alpha test is so ambiguous that it is not possible to assign a defect removal efficiency rating. Beta testing is better understood, and the observed removal efficiency is usually in the 25% range.

Selecting an Optimal Series of Defect Prevention and Removal Operations

Since most forms of defect removal are less than 60% efficient, it is obvious that more than a single defect removal activity will be necessary for any software project. For mission-critical software systems where it is imperative to achieve a cumulative defect removal efficiency higher than 99%, then at least 10 discrete defect removal activities should be utilized, and careful defect prevention activities should also be planned (i.e., prototyping, use of structured techniques, etc.).

An effective series of defect removal operations utilized by computer manufacturers, telecommunication manufacturers, and defense contractors who wish to approach the six-sigma quality level (99.999999% efficiency) will include from 16 to 20 steps and resemble the following, although there are variances from company to company and project to project:

Pre-Test Defect Removal Activities

1. Requirements inspection
2. Functional design inspection
3. Logical design inspection
4. Test plan inspection
5. User documentation inspection
6. Desk checking and debugging by developers
7. Code inspection
8. Software quality assurance review
9. Independent verification and validation (military projects)

Testing Activities

10. Unit testing by developers
11. New function testing
12. Regression testing
13. Integration testing
14. Stress testing
15. Independent testing (military projects)
16. System testing
17. Field testing
18. Acceptance testing

Post-Release Defect Removal Activities

19. Incoming defect report analysis
20. Defect removal efficiency calibration (after one year of deployment)

The series of defect removal operations just illustrated can achieve cumulative defect removal efficiencies well in excess of 99%. However, these levels of efficiency are normally encountered only for large mission-critical or important commercial-grade software systems, such as operating systems, defense systems, and telecommunication systems.

A rough rule of thumb can predict that number of defect removal operations that are normally carried out for software projects. Using the Function Point total of the application as the starting point, calculate the 0.3rd power, and express the result as an integer. Thus an application of 100 Function Points in size would normally employ a series of 4 defect removal operations. An application of 1000 Function Points in size would normally employ a series of 8 defect removal operations. An application of 10,000 Function Points in size would normally employ 16 defect removal operations.

Post-Release Defect Removal

Defect removal operations do not cease when a software project is released. Indeed, one of the embarrassing facts about software is that post-deployment defect removal must sometimes continue indefinitely.

One of the critical tasks of post-release defect removal is the retroactive calculation of defect removal efficiency levels after software has been in use for one year. For a software project of some nominal size, such as 1000 Function Points, assume that 2000 bugs or defects were found via inspections and tests prior to release. The first year total of user-reported defects might be 50. Dividing the pre-release defect total (2000) by the sum of all defects (2050) indicates a provisional defect removal efficiency of 97.56%.

The defect potential of the application can be retroactively calculated at the same time, and in this case is 2.05 defects per Function Point. The annual rate of incoming user-reported defects for the first year is 0.05 defects per Function Point per year.

In addition to calculating defect removal efficiency, it is also useful to trace each incoming user-reported defect back to its origin, i.e., whether the defect originated in requirements, design, code, documentation, or as a result of a "bad fix." Defect origin analysis also requires a sophisticated tracking system, and full configuration control and traceability tools are useful adjuncts.

Note that the use of Function Points rather than the traditional KLOC metric is preferred for both defect potential and for defect origin analysis. KLOC metrics produce invalid and paradoxical results when applied to requirements, specification, and documentation error classes. Since defects outside the code itself often constitute more than 50% of the total bugs discovered, KLOC metrics are harmful to the long-range understanding of software quality.

If additional bugs are reported during the second or subsequent years, then defect removal efficiency levels should be recalculated as necessary. Removal efficiency analysis should also be performed for each specific review, inspection, and test step utilized on the project.

Calculating post-release defect removal efficiency requires an accurate quality measurement system throughout the life cycle. In addition, rules for counting post-release defects must be established. For example, if the same bug is reported more than once by different users, it should still count as only a single bug. If upon investigation user bug reports should turn out to be usage errors, hardware errors, or errors against some other software package, such bug reports should not be considered in calculating defect removal efficiency rates. If bugs are found by project personnel, rather than users, after release to customers, those bugs should still be considered to be in the post-release category.

The leading computer manufacturers, telecommunication companies, defense contractors, and software vendors utilize powerful defect tracking systems for monitoring both pre-release and post-release software defects. These systems record all incoming defects and the symptoms which were present when the defect was discovered. They offer a variety of supplemental facilities as well. Obviously statistical analysis on defect quantities and severities is a by-product of defect tracking systems. Less obvious, but quite important, is support for routing defects to the appropriate repair center and for notifying users of anticipated repair schedules. The most powerful defect tracking systems can show defect trends by product, by development laboratory, by time period, by country, by state, by city, by industry, and even by specific customer.

Summary and Conclusions

Software has been prone to such a wide variety of error conditions that much of the cost and time associated with developing software projects is devoted to defect removal. In spite of the high expense levels and long schedules historically associated with defect removal, deployed software is often unstable and requires continuous maintenance.

Synergistic combinations of defect prevention and defect removal operations, accompanied by careful measurements and the long-range calibration of inspection and test efficiencies, can give software-producing organizations the ability to consistently create stable and reliable software packages. As the state of the art advances, both six-sigma quality levels or even zero-defect quality levels may be achievable. Achieving high quality levels also reduces development schedules and lowers development costs.

Defining Terms

Bug: The generic term for a defect or error in a software program. The term originated with Admiral Grace Hopper in the 1950s, who discovered an actual insect that was blocking a contact in an electromechanical device.

Debugging: The generic term for the process of eliminating bugs, i.e., errors, from software programs. The phrase is often used in a somewhat narrow sense to refer to the defect removal activities of individual programmers prior to the commencement of formal testing.

Defect prevention: The set of technologies which simplify complexity and reduce the probability of making errors when constructing software. Examples of defect prevention technologies include prototypes, structured methods, and reuse of certified components.

Defect removal: The set of activities concerned with finding and eliminating errors in software deliverables. This is a broad term which encompasses both manual and automated activities and which covers errors associated with requirement, design, documentation, code, and bad fixes or secondary defects created as a by-product of eliminating prior bugs.

Defect removal efficiency: The ratio of defects discovered and eliminated to defects present. Removal efficiency is normally expressed as a percent and is calculated based on all defects discovered prior to release and for the first year of deployment.

References

A.J. Albrecht, "Measuring application development productivity," *Proceedings of the Joint SHARE, GUIDE, IBM Application Development Conference,* October 1979, pp. 83–92.

R.H. Dunn, *Software Defect Removal,* New York: McGraw-Hill, 1984.

R.H. Dunn and R. Ullman, *Quality Assurance for Computer Software,* New York: McGraw-Hill, 1982.

M.E. Fagan, "Design and code inspections to reduce errors in program development," *IBM Systems Journal,* vol. 15, no. 3, pp. 182–211, 1976.

D. Garmus, Ed., *Function Point Counting Practices Manual,* Version 3.2. Westerville, Ohio: International Function Point Users Group (IFPUG), 1991.

C. Jones, *Programming Productivity,* New York: McGraw-Hill, 1986.

C. Jones, *Applied Software Measurement,* New York: McGraw-Hill, 1991.

G. J. Myers, *The Art of Software Testing,* New York: John Wiley, 1979.

Further Information

Both specialized and general-interest journals cover defect removal and software quality control. There are also frequent conferences, seminars, and workshops on these topics. Some of the journals include:

ACM Transactions on Software Engineering and Methodology (TOSEM). The Association for Computing Machinery, 11 West 42nd Street, New York, NY 10036.

American Programmer. American Programmer, Inc., 161 West 86th Street, New York, NY 10024-3411.

IEE Software Engineering Journal. Institution of Electrical Engineers and the British Computer Society, Michael Faraday House, Six Hills Way, Stevenage, Herts, SG1 2AY, United Kingdom.

IEEE Transactions on Software Engineering. Institute of Electrical and Electronics Engineers, 345 East 47th Street, New York, NY 10017.

ITEA Journal of Test and Evaluation. International Test and Evaluation Association, 4400 Fair Lakes Court, Fairfax, VA 22033.

Software Maintenance News. Software Maintenance Society, 141 Saint Marks Place, Suite 5F, Staten Island, NY 10301.

84.3 Programming Methodology

Johannes J. Martin

Programming methodology is concerned with the problem of producing and managing large software projects. It relates to programming as the theory of style relates to writing prose. Abiding by its rules does not, in itself, guarantee success, but ignoring them usually creates chaos. Many useful books have been written on the subject and there is a wealth of primary literature. Some of these books, that themselves contain numerous pointers to the primary literature, are listed at the end of this section.

The rules and recommendations of programming methodology are rather independent of particular programming languages; however, significant progress in the understanding of programming methods has always led to the design of new languages or the enhancement of existing ones. Because of the necessity of upward compatibility, enhanced old languages are, in comparison to new designs, less likely to realize progressive programming concepts satisfactorily. Consequently, the year of its design usually determines how well a language supports state-of-the-art programming methods.

Programming methodology promotes program correctness, maintainability, portability, and efficiency.

Program Correctness.   For some relatively simple programs specifications and even verifications can be formalized. Unfortunately, however, formal specification methods are not available or not sufficiently developed for most programs of practical interest, and the specifications, given as narratives, are most likely less than complete. For the situations not explicitly covered by the specifications, the resulting program may exhibit bizarre behavior or simply terminate. Furthermore, informal methods of demonstrating correctness may also miss points that were, indeed, addressed by the specifications. The costs of these problems may range from time wasted in the workplace to injuries and loss of lives.

Short of formal methods for proving **program correctness,** simplicity of design gives informal methods a chance to succeed.

Maintainability.   Since the programmer who will do the **program maintenance** is usually not the one who also did the original coding (or if he did he cannot be expected to remember the details), proper documentation and straightforward design and implementation of the pertinent algorithms is mandatory.

Portability.   A program is called (easily) portable if it can be adapted to a different computer system without much change. One step toward **program portability** is using a high-level language common to both computer systems. The second step is avoiding the use of idiosyncratic features of a system if at all possible or, if such a feature must be used, to isolate its use to small, well-designed program segments that can easily be rewritten for the new system.

Efficiency.   The costs of running a computer program are determined (1) by the time needed to compute the desired result and (2) by the amount of storage space the program requires to run. As the costs of computer hardware have declined dramatically over the past decade, the importance of program efficiency has similarly declined. Yet, there are different dimensions to efficiency. Choosing the right algorithm is still crucial, while local optimization should not be pursued, if it increases the complexity and decreases the clarity of a program and thereby jeopardizes its correctness and maintainability.

Analysis of Algorithms

The solution of a problem as a sequence of computational steps, each taking finite time, is called an algorithm. As an example consider the algorithm that finds the greatest element in a random list of *n* elements:

1. Give the name *A* to the first element of the list.
2. For each element *x* of the list beginning with the second and proceeding to the last, if $x > A$ then give the name *A* to *x*.
3. The element named *A* is the greatest element in the list.

Assume that examining the "next" element, comparing it with *A* and possibly reassigning the name *A* to the new element takes a fixed amount of time that does not depend on the number of elements in the list. Then the total time for finding the maximum increases proportional to the length *n* of the list. We say the *time complexity* of the algorithm is of *order n*, denoted by $O(n)$.

While there are not many algorithms for finding the maximum (or minimum) in an unordered list, other tasks may have many different solutions with distinctly different performances. The task of putting a list in ascending or descending order (sorting), for example, has many different solutions, with time complexities of $O(n^2)$ and $O(n \cdot \log(n))$. Compared with the $O(n \cdot \log(n))$ algorithms, the $O(n^2)$ algorithms are simpler and their individual steps take less time, so that they are the appropriate choice, if *n*, the number of items to be sorted, is relatively small. However, as *n* grows, the balance tips decidedly toward the $O(n \cdot \log(n))$ algorithms. In order to compute the break-even point suppose that the individual steps of an $O(n \cdot \log(n))$ and an $O(n^2)$ algorithm take $k \cdot t$ and t units of time, respectively. Thus, for the break-even point we have

$$t \cdot n^2 = k \cdot t \cdot n \cdot \log(n)$$

hence

$$k = n/\log(n)$$

With $k = 5$, for example, the break-even point is $n = 13$. The difference in performance of the two algorithms for large numbers of *n* is quite dramatic. Again with $k = 5$, the $O(n \cdot \log(n))$ algorithm is 217 times faster if $n = 10,000$, and it is 14,476 times faster if $n = 1,000,000$. If the faster algorithm would need, e.g., 10 minutes to sort a million items, the slower one would require about 100 days.

Consequently, while efficiency is no longer a predominant concern, analyzing algorithms and choosing the proper one is still a most important step in the process of program development.

Flow of Control

The term *flow of control* refers to the order in which individual instructions of a program are performed. For the specification of the flow of control, assembly (machine) languages (and older high-level languages) provide conditional and unconditional branch instructions. Programmers discovered that the indiscriminate use of these instructions frequently leads to flow structures too complex to be analyzed at reasonable costs. On the other hand, keeping the flow of control reasonably simple is, indeed, possible since three elementary control constructors are sufficient to express arbitrary algorithms. Using only these constructors contributes in an essential way to the manageability of programs. The basic rules are as follows:

1. A proper computational unit has exactly one entry point and one exit.
2. New computational units are formed from existing units A, B, . . . by
   a. sequencing (perform A, then B, and so forth)
   b. selection (perform A or B or . . . , depending on some condition)
   c. iteration (while some condition prevails, repeat A).

Figure 84.12 shows diagrams that illustrate these rules.

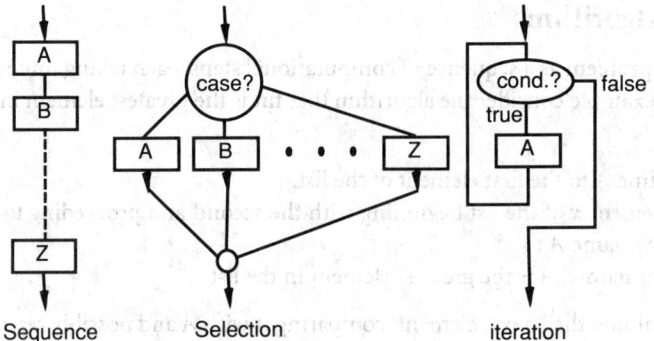

Sequence   Selection   iteration

FIGURE 84.12   Basic control constructors.

   High-level languages support these three constructors by (1) juxtaposition for sequencing, (2) if-then-else and case (switch) constructs for selection, and (3) while-do and for loops for iteration. The somewhat rigid restriction on loops sometimes requires the duplication of code. Some languages ease this problem by providing a *break* or *leave* statement that permits leaving a loop prematurely (Fig. 84.13). A typical example for a program that profits from this mechanism is a loop that reads and processes records and terminates when a record with some given property is found. In Fig. 84.13 block B reads and block A processes a record. With a strict while-loop the code of B must be duplicated.

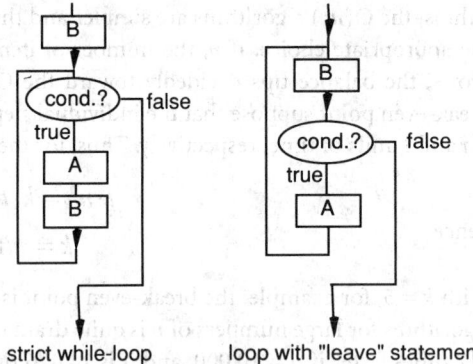

strict while-loop   loop with "leave" statement

FIGURE 84.13   The use of a leave statement.

Abstraction

In programming, abstraction refers to the separation of what needs to be done from the details of how to do it. As it facilitates the separation of unrelated issues, it is a most powerful tool for making complex structures comprehensible. There are two varieties: *functional (procedural) abstraction* and *data abstraction*.

   Almost all programming languages support *functional abstraction* by providing facilities for the definition of *subprograms*. The definition of a subprogram (procedure or function) associates a possibly complex sequence of instructions with an identifier and (usually) a list of parameters. The instruction sequence may then be performed by referring to its associated identifier and providing the required parameters. We could, for example, define a procedure *findRecord* that searches a given list of records for a record with a given key, e.g., an identification number. We could then use the statement

```
findRecord(theRecord, id_number, list);
```

instead of the instruction sequence that actually searches the list for the record. If this operation is needed more than once, then the obvious advantage is saving the replication of instructions. Much more important is a second advantage: When we study the text of a program that *uses findRecord*, we most likely wish to understand how this program processes and updates records or uses the information they contain. As this in itself may be quite complex, we do not wish to be burdened simultaneously with the details of how to *find* a record, since this second problem has absolutely nothing to do with the process that we are trying to understand. Functional abstraction allows us to separate these different concerns and, thus, make each comprehensible.

The essential property of a data object is not its representation (is it an array, a record or something else?) but the operations that can be applied to it. The separation of these (abstract) operations from the subprograms that implement them, as well as from the representation of the object, is called *data abstraction*. This separation is possible because only the implementations of the operations need to know about the actual representation of the object. A system of data objects and operations defined by data abstraction is called an *abstract data type*. See Chapter 81.3 on *data types and data structures* for more details.

Modularity

With the tool of abstraction programs are made modular, that is, they are broken into fairly small, manageable parts. Each part, called a module, consists of one or more procedures or functions and addresses a particular detail of a project. A module may consist of a sorting program, a package for matrix or time and date calculations, a team of programs that preprocess input or format output, and the like.

A program should be broken up into smaller parts for one of two reasons: (1) if it addresses more than one problem and (2) if it is physically too large.

First, a procedure (function) should address a single problem. If more than one problem is addressed by the same procedure, a reader of the program text experiences a distracting shift in subject, which makes the comprehension of the text unnecessarily difficult. For example, a program involved in interest calculations should not also solve calendar problems (such as how many days are between March 15 and June 27) but use a calendar module that provides the abstract data type *Date* to obtain the desired results. Similarly, a sorting procedure should not be burdened with the details of comparing records. Instead, a different function should be written that contains the details of the comparison, and this function should then be invoked by the sorting program. The program fragments in Example 1, written in the programming language C, illustrate the separation of these levels of concern.

Considering Example 2, the advantage of separating levels of concern may appear to be subtle. The reason is the small size of the original sort program. A quote by B. Stroustrup [1991] emphasizes this point, "You can make a small program (less than 1000 lines) work through brute force even when breaking every rule of good style. For a larger program, this is simply not so. . . ." Advantages that do not seem very significant for small programs become invaluable for large ones. Writing large programs is an inherently difficult activity that demands a high level of concentration, helped by separating and hindered by mixing unrelated levels of concern.

Second, as a rule of thumb, the text of a procedure should fit onto one page (or into a window). A single page of text can be studied exclusively by eye movement and, thus, comprehended more easily than text that spans over several pages. That is not to say that longer procedures are necessarily an indication of poor programming methods, but there should be compelling reasons if the limit of one page is to be exceeded.

Simple Hierarchical Structuring

The relation among modules should be hierarchical, i.e., modules should be the vertices of a directed acyclic graph whose edges describe how modules refer to other modules. In such a structure, called a top-down design, each module can be understood in terms of the modules to which it refers directly. Yet, hierarchical structuring by itself does not yield simple and comprehensible programs. In addition, interfaces between modules must be kept narrow, that is, (1) the use of nonlocal variables must be avoided and (2) the number of parameters passed to a subprogram should be held to a minimum.

First, the scope rules of many programming languages permit subprograms to access entities of the program in which they are defined (static scoping) or of the calling program (dynamic scop-

Example 1. A program that addresses more than one problem:

```
void sort (recType *table, int size)  /* recType is a structure with fields
                                           suitable for a personnel record  */
{
     int i, j, result;
     recType temp;
   for (i = 0; i < size-1; i++)
     for (j = i+1; j<size; j++){
         if ((result = strcmp(table[i].last , table[j].last)) == 0)
             if ((result = strcmp(table[i].first , table[j].first)) == 0)
                 if ((result = strcmp(table[i].midl , table[j].midl)) == 0)
                     if ((result = datecmp(table[i].birthdate , table[j].birthdate)) == 0)
                         result = addresscmp(table[i].address , table[j].address);
         if (result > 0){
             temp = table[i];
             table[i] = table[j];
             table[j] = temp;
         }
     }
}
```

Example 2. Programs tackling one problem at a time:

```
BOOL isGreater (recType a, recType b);
void swap (recType *a, recType *b);

void sort (recType *table, int size)
{
     int i, j;
   for (i = 0; i < size-1; i++)
     for (j = i+1; j<size; j++)
         if (isGreater(table[i] , table[j])) swap(&table[i], &table[j]);
}

BOOL isGreater (recType a, recType b)
{
     int result;
   if ((result = strcmp(a.last , b.last)) == 0)
       if ((result = strcmp(a.first , b.first)) == 0)
           if ((result = strcmp(a.midl , b.midl)) == 0)
               if ((result = datecmp(a.birthdate , b.birthdate)) == 0)
                   result = addresscmp(a.address , b.address);
   return result > 0;
}

void swap (recType *a, recType *b)
{
     recType temp;
   temp = * a;
   *a = *b;
   *b = temp;
}
```

ing). These entities are called nonlocal in contrast to those defined within the subprogram itself. *Nonvariable* entities can and should be broadcast by means of nonlocal, preferably global definitions (i.e., definitions available to all modules and subprograms within an entire project). For these entities—types, procedures, and constant values—global definitions are very beneficial, since changes that may become necessary need to be made only at the place of definition, not at every place of usage. Moreover, global definitions that replace numeric or alphanumeric constants by descriptive identifiers improve the readability of a program. Nonlocal *variables,* on the other hand, are to be avoided. For reasons of simplicity and clarity, the task performed by a subprogram should be determined *exclusively* by the subprogram itself and the values of its parameters. As a result, all information needed to understand why a subprogram may behave differently for different invocations is provided at each point of invocation. This rule, related to the concept of *referential transparency,* is clearly violated if a subprogram uses nonlocal variables: in order to understand the behavior of the subprogram, information contained in the nonlocal variables must be consulted. These variables may appear nowhere near the point of invocation.

Similarly, because their return values are clearly identified, functions that do not change their parameters are preferred over those that do and over procedures. As functions are not always appropriate, a programmer choosing to use a procedure should aid the reader's comprehension by adopting some documented standard of consistently changing either the procedure's first or last parameter(s).

Second, the number of parameters should be kept to a minimum. If more than four or five parameters seem to be required, programmers should seriously consider packaging some of them into a record or an array. Of course, only values that in some logical sense belong together should be packaged, while parameters that serve different functions should be kept separate. For example, suppose the operation *updateRecord* is to find a personnel record with a given identification number in a file or a list and then change certain fields such as the job title, rate of pay, etc. A programmer may be tempted to define

```
updateRecord (Rectype myRecord);
```

yet the form

```
updateRecord (Idtype idNumber, Packtype attributes)
```

is better, since it suggests that the record with the key *idNumber* is the one to be updated by modifying the *attributes.*

The term *top-down design* (also called step-wise refinement) is frequently misunderstood as exclusively describing the *method* of design rather than the *structure* of the *finished product.* However, the creative process of designing software (or anything else) frequently proceeds in a rather erratic fashion, and there is no fault in this as long as the final product through analyses and reorganizations has a simple hierarchical structure. This is not to say that the process of design should not be guided by a top-down analysis; however, neither should a programmer's creativity be hampered by the straightjacket of a formalistic design principle, nor can a poorly designed program be defended with reference to the superior (top-down design) method used for its creation.

Object-Oriented Programming

In languages that support object-oriented programming, *classes* (i.e., data types) of objects are defined by specifying (1) the variables that each object will own as *instance variables* and (2) operations, called *methods,* applicable to the objects of the class. As a difference in style, these methods are not invoked like functions or procedures, but are *sent* to an object as a *message.* The expression [*window moveTo* : x : y], for example, is a message in the programming language *Objective C,* a dialect of C. Here the object *window,* which may represent a window on the screen, is instructed to apply to itself the method *moveTo* using the parameters x and y.

New objects of a class are created—usually dynamically—by *factory methods* addressed to the *class* itself. These methods allocate the equivalent of a record whose fields are the instance variables of the object and return a reference to this record, which represents the new object. After its creation, an object can receive messages from other objects.

To data abstraction, object-oriented programming adds the concept of inheritance: From an existing class new (sub)classes can be derived by adding additional instance variables and/or methods. Each subclass inherits the instance variables and methods of its superclass. This encourages the use of existing code for new purposes.

With object-oriented programming, classes become the modules of a project. For the most part, the rules of hierarchical structuring apply, interfaces should be kept narrow, and nonlocal variables ought to be restricted to the instance variables of an object. There are, however, at least two situations, both occurring with modern graphical user interfaces, where the strict hierarchical structure is replaced by a structure of mutual referencing.

Objects—instances of classes—frequently represent visible objects on the screen such as windows, panels, menus, or parts thereof. It may now happen that actions (such as mouse clicks) applied to one object influence another object and vice versa. For example, the selection of an object, such as a button or a text field, may launch or modify an inspector panel that, in turn, allows the modification of the appearance or function of the button or the text field (mutual referencing).

In conventional programming, library routines are, with rare exceptions, at the bottom of the hierarchy, i.e., they are called by subprograms written by the user of the library (later referred to as the "user") but they do not call subprograms that the user has written.

With object-oriented systems, library programs may have user-written *delegates* whose methods are used by the library object. For example, a window—a library object—may send the message "windowWillClose" (written by the user) to its delegate in response to the operators clicking of the window's close button. In response, the delegate may now interrogate the window in order to determine whether its contents have been changed and possibly require saving (mutual referencing).

Furthermore, buttons, menus, and the like are library objects. Yet, when operated, they usually activate user methods. Again, the user program may then interrogate the button about its state or request that the button highlight itself, etc. (mutual referencing).

While in the cases discussed, mutual referencing seems to be natural and appropriate, it should, in general, be used very sparingly.

Program Testing

Aside from keeping the structure of a program as simple as possible, a top-down design can be tested in a divide-and-conquer fashion. There are basically two possible strategies: bottom-up testing and top-down testing.

Bottom-up testing proceeds by first testing all those modules that do not depend on other modules (except library modules). One then proceeds always testing those modules that use only modules already tested earlier. Testing of each (sub)program must exercise all statements of the program, and it must ensure that the program handles all exceptional responses from supporting subprograms correctly. For each module a driver program must be written that invokes the module's functions and procedures in an appropriate way. If a program is developed in a top-down fashion, this method cannot be used until a substantial part of the project is completed. Thus design flaws may not be detected until corrections have become difficult and costly.

Top-down testing begins with the modules that are not used by any other module. Since the supporting modules have not been tested yet (or do not even exist yet), simple stand-in programs must be written that simulate the actual programs. This can be done by replacing the computation of the actual program with the programmer, who enters the required results from the keyboard in response to the parameter values displayed on the screen. This method is nontrivial, especially if the supporting program deals with complex objects.

In practice both methods are being used. Frequently, developing the procedure for top-down testing by itself leads to the discovery of errors and can prevent major design flaws.

Defining Terms

Program correctness: A program's conformation with its specifications.

Program maintenance: Modifications and (late) corrections of programs already released for use.

Program portability: A program is called (easily) portable if it can be adapted to a different computer system without much change.

References

B.W. Kernighan and D.M. Ritchie, *The C Programming Language*, Englewood Cliffs, N.J.: Prentice-Hall, 1988.

J.J. Martin, *Data Types and Data Structures*, C.A.R. Hoare, Series Ed., New York: Prentice-Hall, 1986.

NeXT Step Concepts, *Objective C*, NeXT Developers' Library, NeXT, Inc., 1991, chap 3.

J.C. Reynolds, *The Craft of Programmings*, C.A.R. Hoare, Series Ed., New York: Prentice-Hall, 1983.

B. Stroustrup, *The C++ Programming Language*, Reading, Mass.: Addison-Wesley, 1991.

J. Welsh *et al., Sequential Program Structures*, C.A.R. Hoare, Series Ed., New York: Prentice-Hall, 1984.

N. Wirth, *Algorithms + Data Structures = Programs*, Berlin: Springer-Verlag, 1984.

Further Information

Textbooks on programming usually address the subject of good programming style in great detail. More information can be found in articles on object-oriented programming and design and on software engineering as published, for example, in the proceedings of the annual conferences on *Object Oriented Programming Systems, Languages and Applications (OOPSLA)* sponsored by the Association for Computing Machinery (ACM) and on *Computer Software and Applications (CompSac)* sponsored by the Institute of Electrical and Electronics Engineers, Inc. (IEEE). Articles on the subject are also found in periodicals such as *IEEE Transactions on Software Engineering, IEEE Transactions on Computers, ACM Transactions on Software Engineering and Methodology, Software Engineering Notes* published by the ACM Special Interest Group on Software Engineering, *ACM Transactions on Programming Languages and Systems*, the *Communications of the ACM*, or *Acta Informatica*.

85

Computer Graphics

85.1 Introduction 2004
85.2 Graphics Hardware 2005
     Hard Copy Technologies • Display Technologies • Standard
     CRT • Other Display Technologies
85.3 Graphics Software 2007
     Engineering Software Packages • General Purpose Libraries and
     Packages • Solid Modeling Packages • Object-Oriented Pro-
     gramming • Plotting and Page Description Languages • Inter-
     action
85.4 Conclusion 2012

Evelyn P. Rozanski
Rochester Institute of Technology

85.1 Introduction

The term **computer graphics** refers to the generation, representation, manipulation, processing, and display of data by a computer. Computer-generated images may be real or imagined, animated or still, two-dimensional (2-D) or three-dimensional (3-D). Today many computers, particularly those in the PC, microcomputer, or workstation categories, have graphics functionality. Their central components are a graphical display device, usually a cathode ray tube (CRT), and one or more input devices (e.g., keyboard, mouse, digitizer, data glove). Output devices include laser printers or video or such other displays as goggles or "eyephones" as in the case of some **virtual reality** systems.

Computer graphics encompasses a wide variety of applications. It has expanded its scope from the mundane business/presentation graphics to placing desktop publishing at everyone's fingertips. Highly interactive real-time systems are used in flight simulators where the display represents changes in the scene or landscape. In engineering, computer-aided design (CAD) systems allow users to create, store, manipulate, and test objects and designs. Fully integrated systems allow standard component parts libraries to be incorporated into a product. Product design and drafting information is fed into manufacturing operations via numerical control interfaces. Other engineering applications that make extensive use of graphics include very large scale integration (VLSI) and solid modeling.

In the sciences, graphics is used for visualization of physical phenomena and multidimensional data in order to aid in its understanding and interpretation [Purgathofer and Schonhut, 1989; Vince, 1990]. One application simulates laboratory testing of a new friction material for disc brakes and visualizes temperature distribution of the brakes' ability to conduct or absorb heat [Purgathofer and Schonhut, 1989]. In mathematics, B. B. Mandelbrot defined the geometry of **fractals**. Fractals, geometrical self-similar objects with fractional dimension, form a powerful tool for generating objects that resemble natural phenomena such as mountains, trees, and coastlines [de Ruiter, 1988; Mandelbrot, 1982].

In the world of animation, the computer has taken the drudgery out of transforming and redrawing objects. It has enhanced cell animation as well as produced glitzy Hollywood special effects such as morphing, a process of letting the computer transform one image to another by generating all the in-between images.

One of the most spectacular uses of graphics is in the area of virtual reality (VR). This technology, which uses high-resolution graphics terminals and head-mounted displays (HMD) or eyephones, provides the user with a stereo view of a virtual world and an ability to navigate through it. These systems have a tracking device to determine the position of the user and devices, such as data gloves, for inputting commands [Thomas and Stuart, 1992]. Applications include simulation and architecture.

Research in the area of computer graphics has centered on all aspects of hardware, software, and algorithm development. Some of these areas are

1. Object-oriented environments: Design of programming languages, tools, databases, user interfaces, and animation [Purgathofer and Schonhut, 1989; Cunningham *et al.*, 1992; de Ruiter, 1988].

2. Virtual reality: The design of system architecture, the creation and integration of component hardwares, the creation of software, the building of virtual environments, the development of real-world applications, and the study of philosophical and human perceptual issues [Stuart, 1992].

3. **Scientific visualization:** Graphics software solutions, practical implementations, user interfaces, high-resolution hard copy, data representation and metafiles [Purgathofer and Schonhut, 1989].

4. Algorithmic design: Ray tracing [Straber, 1987].

5. Hardware design: Workstation architectures, support for geometric modeling [Straber, 1987].

6. Color models and manipulation [Purgathofer and Schonhut, 1989].

7. Page description languages (PDLs): PostScript interpreters [Purgathofer and Schonhut, 1989].

8. CAD and solid modeling: VLSI, data exchange, geometric modeling [de Ruiter, 1988; Purgathofer and Schonhut, 1989].

85.2 Graphics Hardware

Computer graphics systems comprise several different output components in which to display computer-generated images. These components are classified into two groups: (1) hard copy technologies and (2) display technologies.

Hard Copy Technologies

Hard copy technologies include printers, pen plotters, electrostatic plotters, laser printers, ink-jet plotters, thermal transfer plotters, and film recorders [Foley *et al.*, 1990]. These devices use either a raster or vector style of drawing. The raster style uses discrete dots, and the vector style uses a continuous drawing motion. Each display device is distinguished by its dot size and the number of dots per inch, known as *addressability*. The closer the dots, the smoother the image. The smaller the dot, the finer the detail. *Resolution* is related to dot size and is the number of distinguishable lines per inch. This may vary in the horizontal and vertical directions. High-resolution devices have fine detail, smooth lines, and crisp images.

Poisson's Ratio vs. Treatment for Ductile Irons

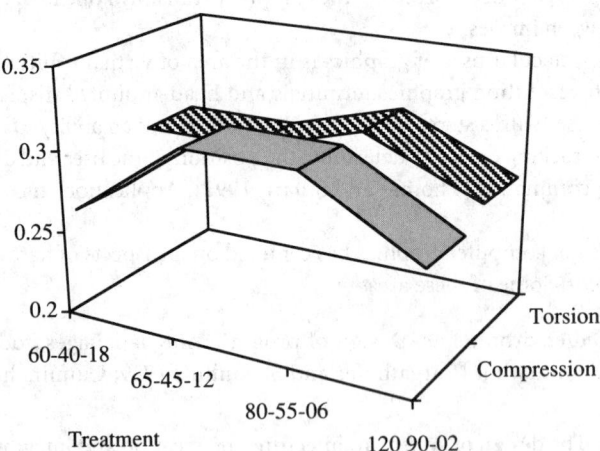

FIGURE 85.1 An example of a figure generated on the Macintosh with Microsoft Excel 3.0, showing the effect of four different treatments on two different measured variables. Although this information could be presented in two dimensions, the 3-D illustration can be more intuitive and interesting.

Color may be achieved in several ways, depending on the device. Some devices use multicolored ribbons with single print heads, multiple print heads with different ribbons, or overstriking to combine colors. Other devices use color pens, spray (e.g., ink jet), toner (e.g., laser printer, electrostatic plotters), or pigment from colored wax paper (e.g., thermal transfer).

The hard copy devices vary in color and intensity levels, addressability, dot size, cost, image quality, and speed. The laser printer is becoming the most common, high-quality output device in this category [Foley *et al.*, 1990].

Display Technologies

Displays are, for the most part, characterized by their responsiveness to a changing image. As with the hard copy technologies, display technologies vary greatly with respect to performance and cost. Guidelines for comparisons are based on the following characteristics: power consumption, screen size, depth, weight, ruggedness, brightness, addressability, contrast, intensity levels per dot, viewing angle, color capability, and relative cost.

Standard CRT

The most common component of graphics displays has been the CRT, which is used in televisions. The CRT is composed of five parts: (1) the electron gun, which when heated emits electrons at an appropriate rate; (2) the control grid, which regulates the flow of electrons; (3) the focusing system, which concentrates the beam into a fine point; (4) the deflection system, which directs the beam to the appropriate location; and (5) the phosphor screen, which glows when bombarded with the electron beam. The *persistence* of the phosphor is defined as the time from the removal of excitation to when the phosphorescence has decayed to 10% of the initial light output [Foley *et al.*, 1990]. Depending on the persistence of the phosphor used, the screen will need to be continually *refreshed*, or redrawn. Color is produced by laying triads of red-green-blue (RGB) phosphors on the screen and using three electron guns, one for each color, to excite the corresponding phosphor.

The raster CRT scans the image, one row at a time, from a matrix whose elements correspond to a *pixel*, or point on the screen. This matrix is referred to as the *frame buffer* and allows for a constant refresh rate, usually 60 times per second. Systems may also have more than one frame buffer to facilitate faster image generation. In vector CRT displays, the picture is generated in a continuous sweep, much like tracing an image on paper. The refresh rate on the vector displays is a function of the complexity of the image. The result may be a noticeable flicker on the screen.

Other Display Technologies

1. Direct view storage tubes (DVST): These devices were the primary displays used in earlier systems. These vector drawing devices stored their images on a grid, which was continually bombarded with electrons in order to transfer the image to the screen. The advantage was that once the image was drawn, the refresh process took place independently of the complexity of the image, thereby producing a constant image on the screen. The disadvantage of these systems was that no part of the image could be selectively erased without erasing the entire grid and resending the modified image to the display.

2. Liquid crystal display (LCD): This device uses matrix addressing and refreshes the display one row at a time. Appropriate voltages are applied to the crystals, causing them to line up. They remain polarized, not allowing light to pass through; light is absorbed, causing dark spots on the display. These devices are light in weight, rugged, and have a low power consumption, fair intensity, and low cost.

3. Plasma panels: These devices have an array of neon bulbs between glass plates, which may be turned on or off. While color is possible, it has not been commercially available. These devices excel in screen size, weight, ruggedness, and brightness characteristics but are generally high in cost.

4. Electroluminescent displays: These devices also use a grid-like structure for addressing elements. The light-emitting material, a zinc sulfide doped with manganese, is available in color. These devices have excellent brightness characteristics but are high in cost.

85.3 Graphics Software

Software for scientific and engineering applications has changed dramatically in the past several years. In the 1970s and early 1980s, there were few graphics software tools available. Most of the engineering packages were in the CAD area. Many specific engineering applications required users to develop and implement programs to solve their problems. These programs were written in the Fortran or C programming languages using low-level graphical commands or calls to some standard or quasi-standard (e.g., the CORE package) graphical routines. Most of these systems were developed for a mainframe computer environment. A trend begun in the late 1980s resulted in a change in computing hardware environments as well as in software approaches. Predominantly, the hardware platforms are PCs, microcomputers, and powerful Unix workstations, with most of these machines having excellent graphics capabilities. Software moved from code generation to customized stand-alone scientific and engineering software tools. Software development uses standard languages and graphical user interfaces for C, Fortran, and Pascal, as well as more sophisticated high-level, fourth-generation languages such as HyperText, Unix X.11, Microsoft's Windows, and PostScript. The technical community is relying more and more on the increased power of computers to easily support software packages that manipulate complex data and represent them in a visual manner.

Engineering Software Packages

Several commercial scientific and engineering software packages have graphics functionality. It is difficult to distinguish graphics or visualization capabilities without discussing some of these packages. An excellent reference is found in the *IEEE Spectrum Focus Report: Software*.

These graphical application software packages fall into five categories:

1. Logic simulation for application-specific integrated circuits (ASICs). Software in this area might display a schematic of a multigate ASIC from large functional building blocks. These blocks could represent a finite-state machine with several states and gates. Representative packages are Mentor Graphics' AutoLogic, Cadence Design Systems' HDL Synthesizer and Optimizer, and Teradyne's Frenchip. HDL is a hardware description language.

2. Electromagnetic design and simulation. Software in this area might simulate a printed-circuit board for a 32-bit-wide, 8-bit-byte reversal network. Multilayers of a board are displayed, with colors indicating current densities in lines. Representative systems are Hewlett-Packard's High Frequency Structure Simulator (HFSS), a finite-element-based product having animation of field plots and conductor loss and 3-D full-wave solution and S-parameter output; Sonnet Software's "em" package with animation of conductor currents; and Compact Software's Microwave Explorer with X-Windows and OSF Motif graphical interfaces.

3. Data acquisition, analysis, display, and technical reporting. Systems in this area have compute-intensive analysis routines and enhanced visualization of data which capitalize on sharper display resolutions. These packages could produce plots and graphs based on acquired data that are displayed in several windows at once; changes to one window could result in recalculation and updating of corresponding windows. Packages in this area frequently have support for standard languages and graphical user interfaces for C and Fortran as well as the Unix X.11 interface or Microsoft's Windows. Representative packages are HP's VEE-Test; Design Science's MathType; DSP Development's DADiSP; National Instruments' LabWindows; Speakeasy Computing's Speakeasy Zeta, which features user-tailored graphical user interface and PostScript output; and Mihalisin Associates' Temple-Graph, which produces a color PostScript output link to *Mathematica*.

4. Mathematical calculations and graphics for visualization. Applications for these packages would be curve fitting, evaluation of integrals, statistical analysis, signal processing, and numerical analysis. Features include programmability in languages such as C, Fortran, and Pascal and 2-D and 3-D representations. The leading package in this area is *Mathematica* by Wolfram Research, which is a general system and programming language for numerical, symbolic, and graphical computations in engineering, research, science, financial analysis, and education [Wolfram, 1991]. Other packages are Amtec Engineering's Tecplot, Integrated Systems' Xmath, MathWorks' Mathlab, Jandel Scientific's SigmaPlot, and NAG's Axiom.

5. Digital signal processors for embedded systems. The tools available in this area let the engineer focus on the application rather than the programming details. The Audio Frequency Fourier Analyzer by National Instruments is a combination of graphical programming with development software for the Macintosh environment. Signals can be analyzed, manipulated, and displayed using custom graphics software. Some packages allow for programming in C and user interfaces. Representative packages include Signal Technology's N!Power, which has **object-oriented programming** and linkage to X-Windows, and Bitware Research Systems' DsqHq with real-time graphics and algorithm design.

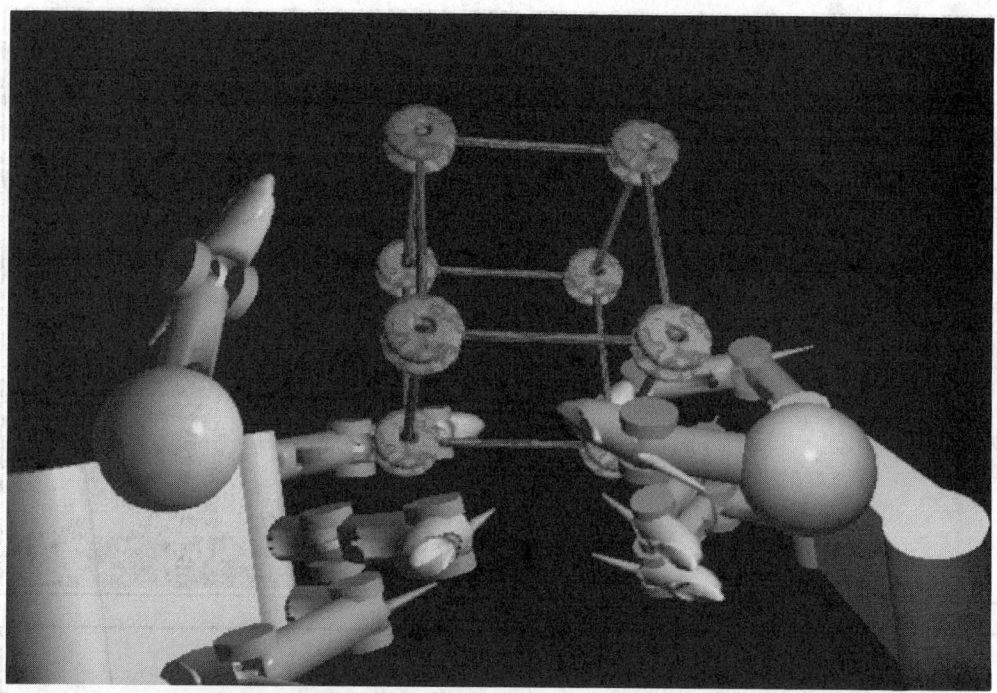

FIGURE 85.2 Examples of rendered 3-D figures generated on the Macintosh Quadra 700/8-bit color system with Infini-d software, showing the effects of different rendering, coloring, and lighting parameters on a 3-D scene. Computer artist: Kent Francis, M.F.A. student, Rochester Institute of Technology.

FIGURE 85.3   Examples of stills from a computer-animated short generated on the Macintosh system with Macro-Mind Director, Modern Artist, and MacPaint software. Computer artist: Silvina Manrique, M.F.A. student, Rochester Institute of Technology.

General Purpose Libraries and Packages

Traditionally, graphical software systems are developed as a result of programming in high-level languages with interfaces to standard or quasi-standard software packages. These packages attempt to address the issues of device independence and application program portability by allowing systems to drive a wide variety of display devices as well as isolating the programmer from machine-specific graphics commands. Portability of programs is enhanced by allowing the user to move an application from one system to another. The primary programming languages include C, Fortran, and Pascal.

The quasi-standard graphical package is ACM/SIGGRAPH's Core system developed in 1977 and revised in 1979. While it was not a formally recognized standard, it did fulfill a role as a baseline specification for graphical systems [Foley *et al.*, 1990]. The two official standards are GKS-3D, the 3-D Graphical Kernel System; and PHIGS and PHIGS+, the Programmer's Hierarchical Interactive Graphics System. Both systems support graphics primitives, such as lines, polygons, and character strings, and their attributes. The GKS system allows for groupings into segments with no nesting capabilities. PHIGS supports geometrical transformations (i.e., scaling, translating, and rotating) and a database structure that allows for selective editing and manipulation of the picture. PHIGS runs best when there is hardware support for the transformation, clipping, and **rendering** features.

In traditional graphical systems development, image data are stored either as Cartesian coordinates or as vectors. These data are manipulated through the geometrical transformations of scaling, translating, and rotating in a reference system known as the *world coordinate system* (WCS). The units of the WCS system might be inches, millimeters, or miles. Physical devices use their own coordinate systems known as *screen coordinate systems* (SCS). In order to ready the image for display, a *viewing transformation* takes place, which changes the image data in the WCS to its corresponding device-specific screen coordinates in SCS. A *window* or portion of the world picture is chosen to be shown in an area of the display known as the *viewport*. Because some of the data in the world could be outside the window, a *clipping* operation is necessary. Clipping will eliminate any data points outside the window. These values are then converted to an intermediate coordinate system known as the *normalized device coordinate system* (NDC). Values in this system are in the range of 0 to 1. Because a viewport may be any portion of the display area and the image could be displayed on more than one device, the NDC values are easily adjusted to screen coordinates. In 3-D, the viewing transformation must also take into account the *view reference point* (i.e., the position from which an object is to be viewed) and the *perspective* or *parallel projection* (i.e., the conversion from the object's 3-D coordinates to the screen's 2-D coordinates).

Solid Modeling Packages

Feature-based systems such as solid or geometric modeling rather than mathematical-based systems form the basis of some CAD systems. **Solid modeling** (SM) systems use constructive solid geometry to build complicated objects. These systems have a descriptive language which uses a database of 3-D primitive objects such as block, cylinder, sphere, wedge, cone, and torus. These solids are combined to form other solids using the set operators of union, intersection, and difference. The resultant object can then be named, saved, and positioned into a picture or drawing. Attributes stored with the objects allow them to be displayed in wire-frame format or as a completely rendered image. Representative SM systems are MAGI (Mathematical Applications Group, Inc.), Synthavision, PADL-2 (Production Automation Project), GM Solid (a proprietary package of General Motors), and McDonnell Douglas's UNISOLID [Teicholz, 1985].

Object-Oriented Programming

Object-oriented programming is a new paradigm for designing and implementing software and is particularly important in computer graphics. An engineering approach, these languages allow software to be constructed from reusable, interchangeable, and extensible parts [Cunningham *et al.*, 1992]. *Class* libraries of graphical objects are being developed. Classes of objects are defined in terms of what an object can do (i.e., what actions and reactions it might produce) and communicate via messages. Subclasses *inherit* actions or characteristics of the superclass. For example, a robot could be constructed from instances of such classes as legs, arms, and head. Each class would have actions defined for it (e.g., a head would be able to nod up and down or shake from side to side). An instance of a head in the object robot would preserve these characteristics. Representative object-oriented languages are Smalltalk, C++, Objective-C, Actor, and Object Pascal.

Plotting and Page Description Languages

Plotting packages, such as ISSCO's DISSPLA and Precision Visuals' DI-3000, consist of routines that are callable from a high-level program. These packages handle 2-D and 3-D images and generally display them in a wire-frame format.

Page description languages are desktop publishing formats that produce graphical output on a printer, display, or other output device. They are used in application programs such as composition systems and illustrators where text, graphical shapes, and sampled images are to be combined into a single document. The dominant language in this category is PostScript, which is a simple interpretive programming language with powerful graphics capabilities. It communicates the description of a document to a printing system in a high level, device-independent manner. PostScript features construction of arbitrary shapes, which may self-intersect, be painted, transformed, cropped, or rendered. The commands are embedded in a general purpose programming language. PostScript programs can be created, transmitted, and interpreted in the form of ASCII source text. The resultant representations will allow for document interchange [Adobe, 1990].

Interaction

The power of computer graphics is to be able to input commands or data in a manner that is appropriate for an application and to have the program react in a timely fashion. These interactions may involve typing words or labels, pointing to items or commands, specifying values or directions for movement, or choosing picture parts displayed on the screen.

Some of the input devices that are available include mouse, special purpose keyboards using buttons or dials, touch panels and screens, light pens, graphics tablets, joysticks, 3-D digitizers, trackballs, and voice systems. Each of these devices is capable of sending appropriate values to the graphics program for action [Hearn and Baker, 1986].

Graphics software packages categorize input devices as one of the following logical devices:

1. Locator: a device for specifying a coordinate position (x,y) or orientation (e.g., tablet)
2. Valuator: a device for specifying scalar values (e.g., dials)
3. Keyboard: a device for specifying text input
4. Pick: a device for selecting displayed entities (e.g., mouse)
5. Choice/button: a device for selecting among alternatives (e.g., function keys)

In some systems, an input device might be used for more than one operation. For example, in the Macintosh computer, the mouse is used as a locator, valuator, and pick device [Foley *et al.*, 1990].

85.4 Conclusion

The field of computer graphics has changed dramatically over the past decade. Scientific and engineering applications have expanded from the CAD systems to scientific visualization of complex systems, enhanced solid modeling systems, real-time animated simulations, and now to another dimension, virtual reality.

On the hardware side, we have seen a movement from large, costly systems to putting the power and speed of computers with advanced graphics capabilities on a desktop. PCs, microcomputers, and professional workstations have provided cost-effective platforms that are within the reach of every engineer.

Interaction with a system has been simplified. In most cases, the user has been relieved of the task of keying in and remembering commands. By merely pointing to menu items, the user is led through a system.

Advances in hardware have driven the software development side. Gone are the days of tediously programming and interfacing with low-level graphics commands. Off-the-shelf and vendor-supplied applications packages that incorporate sophisticated graphics abound. These systems are characterized by user-friendly interfaces and high-quality output capabilities.

When programming is necessary, high-level picture constructs through object-oriented environments make manipulation of graphical images more natural. Other support allows for high-level interfaces to X-Windows, Windows, and PostScript by providing the programmer with more graphical development tools.

Overall, scientists and engineers will find the visual dimension for their applications an integral and common component of their tool kit.

Defining Terms

Computer graphics: The generation, representation, manipulation, processing, and display of data by a computer.

Fractals: Geometrical self-similar objects with fractional dimension.

Object-oriented programming: An engineering approach that uses software constructs that are reusable, interchangeable, and extensible.

Rendering: The preparation of the representation of an image to include illumination, shading, depth cueing, coloring, texture, and reflection.

Scientific visualization: The use of computer graphics techniques to represent complex physical phenomena and multidimensional data in order to aid in its understanding and interpretation.

Solid modeling: The use of constructive geometry to build complicated 3-D objects.

Virtual reality: Three or more dimensionality of computer-generated images, which give the user a sense of presence (i.e., a first-person experience) in the scene.

References

Adobe Systems Incorporated, *PostScript Language Reference Manual*, 2nd ed., Reading, Mass.: Addison-Wesley, 1990.

L. Ammeraal, *Programming Principles in Computer Graphics*, New York: John Wiley, 1986.

M. Brown, *Understanding PHIGS*, TEMPLATE, San Diego: Megatek Corporation, 1985.

S. Cunningham, N.K. Craighill, M.W. Fong, and J.R. Brown, Eds., *Computer Graphics Using Object-Oriented Programming*, New York: John Wiley, 1992.

M.M. de Ruiter, Ed., *Advances in Computer Graphics III*, New York: Springer-Verlag, 1988.

J.D. Foley, A. van Dam, S.K. Feiner, and J.F. Hughes, *Computer Graphics: Principles and Practice*, 2nd ed., Reading, Mass.: Addison-Wesley, 1990.

S. Harrington, *Computer Graphics: A Programming Approach*, New York: McGraw-Hill, 1987.

D. Hearn and M.P. Baker, *Computer Graphics*, Englewood Cliffs, N.J.: Prentice-Hall, 1986.

IEEE Spectrum Focus Report: Software, vol. 28, no. 11, November 1991.

B.B. Mandelbrot, *The Fractal Geometry of Nature*, San Francisco: W.H. Freeman, 1982.

W. Purgathofer and J. Schonhut, Eds., *Advances in Computer Graphics V*, New York: Springer-Verlag, 1989.

W. Straber, Ed., *Advances in Computer Graphics Hardware I*, New York: Springer-Verlag, 1987.

R. Stuart, "Virtual reality: directions in research and development," Interactive Learning Int., vol, 8, pp. 95–100, 1992.

E. Teicholz, Ed., *CAD/CAM Handbook*, New York: McGraw-Hill, 1985.

J.C. Thomas and R. Stuart, "Virtual reality and human factors," Proc. Human Factors Society, 36th Annual Meeting, 1992.

J. Vince, *The Language of Computer Graphics,* New York: Van Nostrand Reinhold, 1990.

S. Wolfram, *Mathematica: A System for Doing Mathematics by Computer,* 2nd ed., Redwood City, Calif.: Addison-Wesley, 1991.

Further Information

Two professional computing organizations publish periodicals that are specifically devoted to the field of computer graphics and provide an excellent forum for current research and techniques. The Association for Computing Machinery (ACM) publishes *ACM Transactions on Graphics* and the IEEE publishes *IEEE Computer Graphics and Applications.*

SIGGRAPH, ACM's special interest group on graphics, sponsors an annual conference and exhibit as well as offering a variety of tutorials and course notes. Other major conferences are sponsored by the National Computer Graphics Association and Eurographics.

86

Computer Networks

86.1 Introduction 2015
86.2 Local Area Networks 2015
    Carrier Sense Buses • Token Ring • Token Bus • Private
    Branch Exchange
86.3 Metropolitan Area Networks 2017
    FDDI • DQDB
86.4 Wide Area Networks 2018
86.5 The Future 2018

Thomas G. Robertazzi
State University of New York,
Stony Brook

86.1 Introduction

Computer networks are geographically distributed collections of communication links and switching processors, the purpose of which is to transport data between computers, workstations, and terminals. In general the elements of a computer network must follow compatible rules of operation together to function effectively. These rules of operation are known as protocols.

There are three broad categories of computer networks, distinguished by geographical extent. Local **area networks** (LANs) connect computer equipment in a single building or floor of a building. Metropolitan area networks (MANs) interconnect network users over a campus or metropolitan-sized region. Finally, wide area networks (WANs) interconnect users on a national or an international scale. In the following, key features of these types of networks will be outlined.

86.2 Local Area Networks

There are four main types of local network architectures that have been commercially produced to date: carrier sense multiple-access buses with collision detection, token rings, token buses, and private branch exchanges. The first three have been standardized in the IEEE 802 series standards.

Carrier Sense Buses

The **IEEE 802.3 standard** deals with a network architecture and protocol first constructed at Xerox in the 1970s and termed *Ethernet*. All stations in an Ethernet are connected, through interfaces, to a **coaxial cable** that is usually run through the ceiling near each user's computer equipment.

The coaxial cable essentially acts as a private radio channel for the users. An interesting protocol called carrier sense multiple-access with collision detection (CSMA/CD) is used in such a network. Each station constantly monitors the cable and can detect when it is idle (no user transmitting), when one user is transmitting (successfully), or when more than one user is simultaneously transmitting (resulting in an unsuccessful collision on the channel). The cable basically acts as a broad-cast bus. Any station can transmit on the cable if the station detects it to be idle. Once a station transmits, other stations will not interrupt the transmission. As there is no central control in the network, occasionally two or more stations may attempt to transmit at about the same time. The transmissions will overlap and be unintelligible (collision). The transmitting stations will detect such a situation and each will retransmit at a randomly chosen later time.

Ethernet and 802.3 networks have raw speeds of up to 10 million bits per second (Mbps). Idle time and collisions, however, can reduce the useful information throughput significantly. **Fiber-optic cables** cannot be used in such networks because they are not suited for direct taps by stations, which is possible with an electrical coaxial cable. The maximum length of these networks is limited by signal propagation delay. An 802.3 coaxial bus differs from an internal computer bus in size and in the lack of a bus controller.

Token Ring

Token ring LANs were developed by IBM in the early 1980s. Topologically, stations are arranged in a circle with point-to-point links between neighbors. Transmissions flow in only one direction (clockwise or counterclockwise). A message transmitted is relayed over the point-to-point links to the receiving station and then forwarded around the rest of the ring and back to the sender to serve as an acknowledgment.

Only a station possessing a single digital code word known as a *token* may transmit. When a station is finished transmitting, it passes the token to its downstream neighbor. Thus, there are no collisions in a token ring, and utilization can approach 100% under heavy loads.

Because of the use of point-to-point links, token rings can use various transmission media such as twisted-pair wire or fiber-optic cables. The transmission speed of a token ring can range from 1 to 16 Mbps, depending on the type of point-to-point links used. Token rings are often wired in star configurations for ease of installation. Token rings are covered by the **IEEE 802.5 standard.**

Token Bus

A token bus uses a coaxial cable along with the token concept to produce a LAN with improved throughput compared to the 802.3 protocol. That is, stations pass a token from one to another to determine which station currently has permission to transmit. Also, in a token bus (and in a token

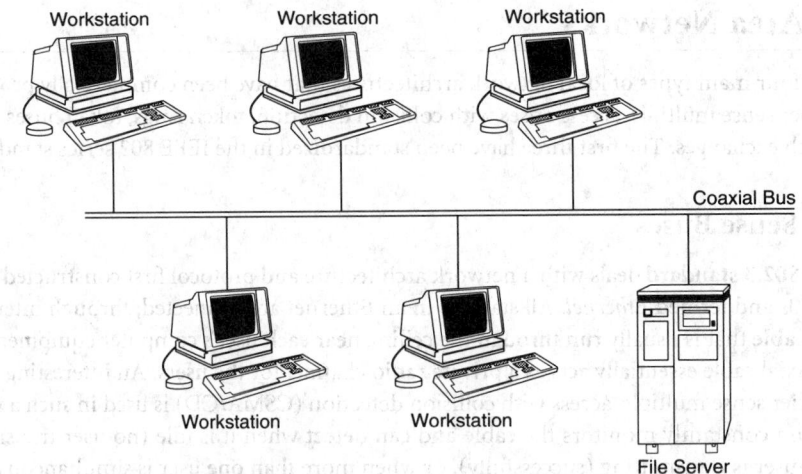

FIGURE 86.1   Bus-type local area network.

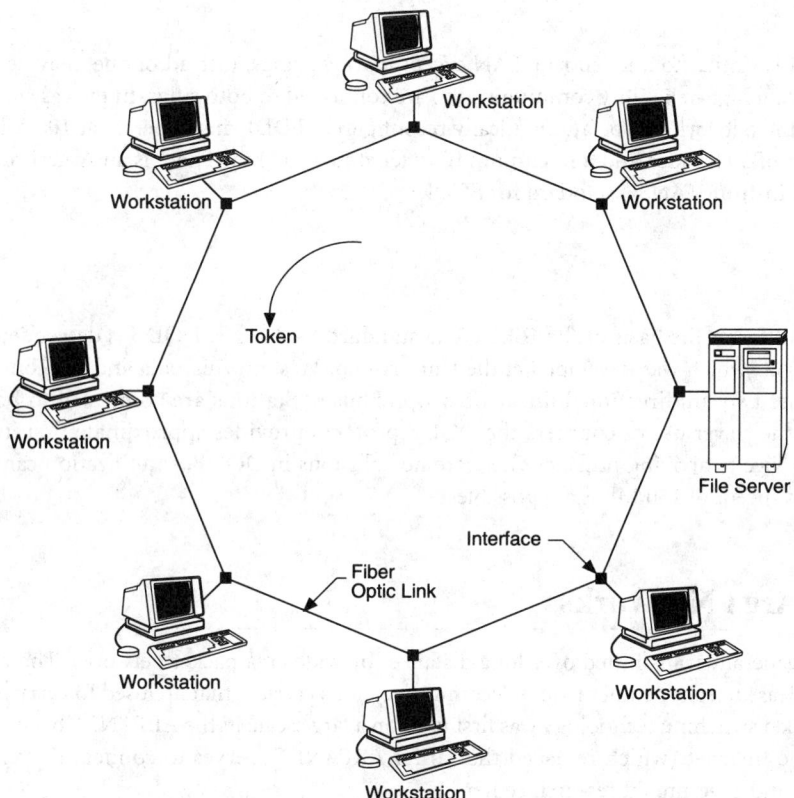

FIGURE 86.2   Token ring local area network.

ring), response times can be deterministically bounded. This is important in factory automation, where commands to machines must be received by set times. By way of comparison, response times in an Ethernet-like network can only be probabilistically defined. For this reason, General Motors' Manufacturing Automation Protocol makes use of the token bus. Token buses can operate at 1, 5 and 10 Mbps. Token bus operation is standardized in the **IEEE 802.4 standard.**

Private Branch Exchange

Historically, private branch exchanges (PBXs) were privately owned telephone switching computers that would be placed in the basement of a building and serve to interconnect phones in the building and provide access to outside lines provided by common carriers. However, PBXs are now available that offer both telephone and data service. In a typical system a phone may have a data socket for terminals or workstations. PBXs are wired in a star topology with the PBX at the center of the star and each user wired directly to it.

86.3 Metropolitan Area Networks

While several network architectures have been proposed for use as MANs, the two that are closest to widespread commercial implementation are fiber-distributed data interface (FDDI) and distributed queue dual bus (DQDB) interface. A key feature of a MAN is the ability to interconnect LANs. This is a problem because of the high data rates at which LANs operate.

FDDI

The FDDI is similar to a token ring LAN except that two rings, instead of one, may be used. Stations needing high-reliability communication are connected to both rings. In the case of a break in the rings the network can be automatically reconfigured. FDDI rings operate at 100 Mbps with a maximum of 500 nodes and a maximum fiber length of 200 km. There is an American National Standards Institute (ANSI) standard for FDDI.

DQDB

The DQDB forms the basis of the **IEEE 802.6 standard** for MANs. DQDB is descended from the earlier QPSX, which was developed at the University of Western Australia and Telecom Australia. DQDB uses two unidirectional linear fiber-optic buses. Stations are connected to both buses. Through the clever use of counters the DQDB protocol provides approximate first in, first out (FIFO) service to arriving packets. There are no collisions in DQDB, so utilization can approach 100%. Bus speeds of 150 Mbps are possible.

86.4 Wide Area Networks

Data are generally transmitted over long distances by wide area packet networks. These networks generally lease telephone lines from telecommunications carriers that are used to carry data exclusively. Packet switching technology was first used on a large scale in the ARPANET beginning in the 1960s. The Internet (which replaced the earlier ARPANET) serves to connect universities and industrial and government research centers.

One problem area unique to wide area packet networks is that of routing. Unlike the previously mentioned networks, there are usually multiple routes available between sources and destinations. Distributed routing algorithms have been developed that route based on current traffic conditions.

86.5 The Future

The future is likely to see an increase in data rates as fiber-optic cables are widely deployed. This will spur the development of faster switching nodes through the use of parallel processing and VLSI implementation. Protocols will have to be simplified to increase processor throughput. New forms of traffic such as video and graphics will become more important. Computer networks will proliferate throughout the world, making possible the ubiquitous transport of data between any two points. These networks are likely to consist of both private networks and new service offerings from telecommunications companies.

Defining Terms

Area networks: LAN, within single building; MAN, metropolitan-sized region; WAN, national/international region.

Coaxial cable: A shielded cable that conducts electrical signals and is used in bus-type local area networks.

Fiber-optic cable: A glass fiber cable that conducts light signals and can be used in token ring local area networks and metropolitan area networks. Fiber optics can provide higher data rates than coaxial cable. They are also immune to electrical interference.

IEEE standards: 802.3, CSMA/CD bus; 802.4, token bus; 802.5, token ring; 802.6, DQDB MAN.

References

U. Black, *Data Networks: Concepts, Theory and Practice*, Englewood Cliffs, N.J.: Prentice-Hall, 1989.

J. D. Spragins, J. L. Hammond, and K. Pawlikowski, *Telecommunications Protocols and Design*, Reading, Mass.: Addison-Wesley, 1991.

A. S. Tanenbaum, *Computer Networks*, 2nd ed., Englewood Cliffs, N.J.: Prentice-Hall, 1988.

J. Walrand, *Communication Networks: A First Course*, Boston, Mass. and Homewood, Ill.: Aksen Associates Incorporated Publishers and Richard D. Irwin, Inc., 1991.

The first electronic computer was built in 1946 by J. P. Eckert and J. W. Mauchly at the University of Pennsylvania. The nearly instantaneous working of electronic components made it possible for this machine, called ENIAC,* to multiply two ten-digit numerals in three thousandths of a second. ENIAC contained 18,000 vacuum tubes; it was huge, taking up the walls of a room 20 by 40 feet in size. Its use was limited to the special problems of ballistics. The designers avoided the mistakes that had caused Babbage to fail, among them those which come from trying to accomplish too much at one time.

Mark I. *Courtesy of IBM Corporation.*

A machine constructed at the University of Iowa by John Atanasoff between 1939 and 1942 also contends for honor as the first electronic computer.

*Electronic Numerical Integrator and Computer. The use of acronyms like ENIAC increased rapidly from this point, with engineers vying with each other to find names that would yield interesting or catchy acronyms. It was not long before a MANIAC appeared, for instance.

87

Fault Tolerance

87.1 Introduction 2020
87.2 Hardware Redundancy 2020
87.3 Information Redundancy 2021
87.4 Time Redundancy 2024
87.5 Software Redundancy 2024
87.6 Dependability Evaluation 2025

Barry W. Johnson
University of Virginia

87.1 Introduction

Fault tolerance is the ability of a system to continue correct performance of its tasks after the occurrence of hardware or software faults. A **fault** is simply any physical defect, imperfection, or flaw that occurs in hardware or software. Applications of fault-tolerant computing can be categorized broadly into four primary areas: long-life, critical computations, maintenance postponement, and high availability. The most common examples of long-life applications are unmanned space flight and satellites. Examples of critical-computation applications include aircraft flight control systems, military systems, and certain types of industrial controllers. Maintenance postponement applications appear most frequently when maintenance operations are extremely costly, inconvenient, or difficult to perform. Remote processing stations and certain space applications are good examples. Banking and other time-shared systems are good examples of high-availability applications. Fault tolerance can be achieved in systems by incorporating various forms of redundancy, including hardware, information, time, and software redundancy [Johnson, 1989].

87.2 Hardware Redundancy

The physical replication of hardware is perhaps the most common form of fault tolerance used in systems. As semiconductor components have become smaller and less expensive, the concept of hardware redundancy has become more common and more practical. There are three basic forms of hardware redundancy. First, *passive* techniques use the concept of fault masking to hide the occurrence of faults and prevent the faults from resulting in **errors**. Passive approaches are designed to achieve fault tolerance without requiring any action on the part of the system or an operator. Passive techniques, in their most basic form, do not provide for the detection of faults but simply mask the faults. An example of a passive approach is triple modular redundancy (TMR), which is illustrated in Fig. 87.1. In the TMR system three identical units perform identical functions, and a majority vote is performed on the output.

The second form of hardware redundancy is the *active* approach, which is sometimes called the *dynamic* method. Active methods achieve fault tolerance by detecting the existence of faults and performing some action to remove the faulty hardware from the system. In other words, active

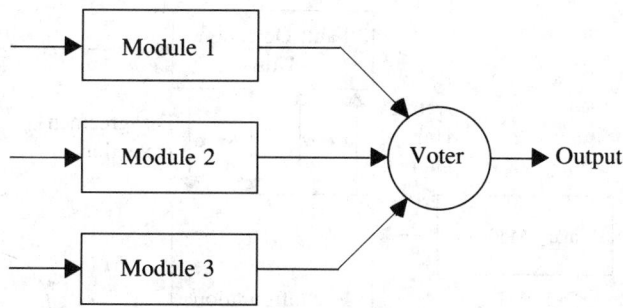

FIGURE 87.1 Fault masking using triple modular redundancy (TMR). (*Source:* B. W. Johnson, *Design and Analysis of Fault-Tolerant Digital Systems,* Reading, Mass.: Addison-Wesley, 1989, p. 52. With permission.)

techniques require that the system perform reconfiguration to tolerate faults. Active hardware redundancy uses fault detection, fault location, and fault recovery in an attempt to achieve fault tolerance. An example of an active approach to hardware redundancy is standby sparing, which is illustrated in Fig. 87.2. In standby sparing one or more units operate as spares and replace the primary unit when it fails.

The final form of hardware redundancy is the *hybrid* approach. Hybrid techniques combine the attractive features of both the passive and active approaches. Fault masking is used in hybrid systems to prevent erroneous results from being generated. Fault detection, fault location, and fault recovery are also used in the hybrid approaches to improve fault tolerance by removing faulty hardware and replacing it with spares. Providing spares is one form of providing redundancy in a system. Hybrid methods are most often used in the critical-computation applications where fault masking is required to prevent momentary errors, and high reliability must be achieved. The basic concept of the hybrid approach is illustrated in Fig. 87.3.

87.3 Information Redundancy

Another approach to fault tolerance is to employ redundancy of information. Information redundancy is simply the addition of redundant information to data to allow fault detection, fault mask-

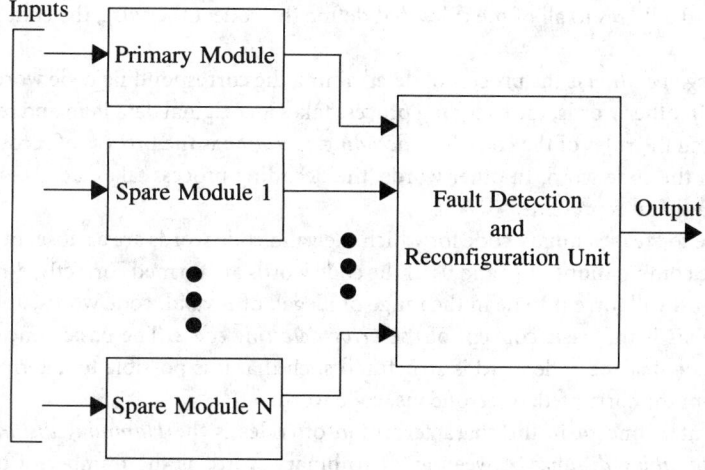

FIGURE 87.2 General concept of standby sparing.

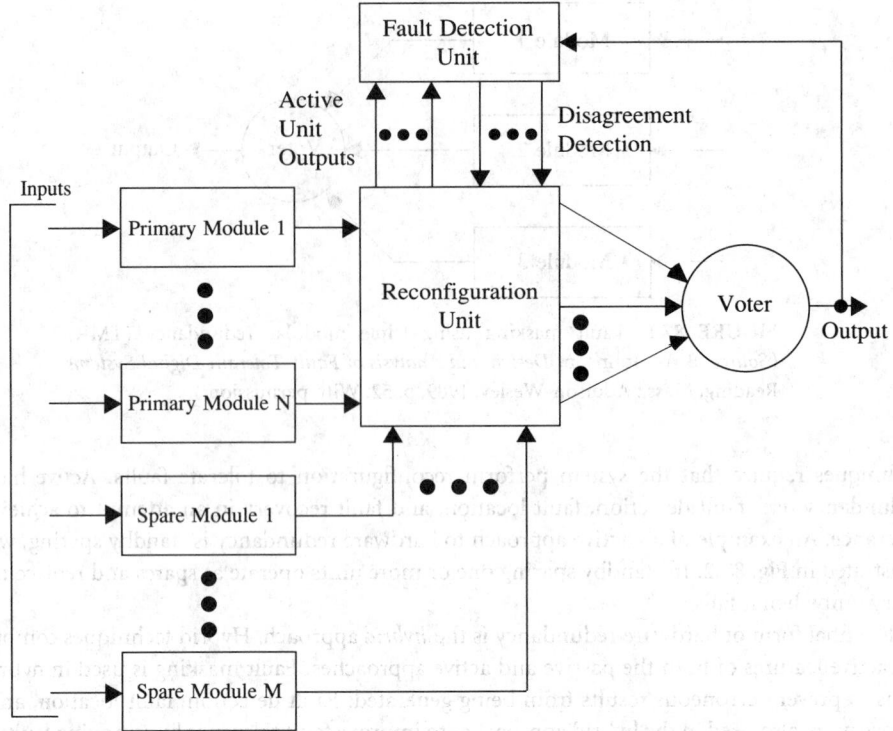

FIGURE 87.3 Hybrid redundancy approach. (*Source:* B. W. Johnson, *Design and Analysis of Fault-Tolerant Digital Systems,* Reading, Mass.: Addison-Wesley, 1989, p. 70. With permission.)

ing, or possibly fault tolerance. Good examples of information redundancy are error detecting and error correcting codes, formed by the addition of redundant information to data words or by the mapping of data words into new representations containing redundant information [Lin and Costello, 1983].

In general, a *code* is a means of representing information, or data, using a well-defined set of rules. A *code word* is a collection of symbols, often called digits if the symbols are numbers, used to represent a particular piece of data based upon a specified code. A *binary code* is one in which the symbols forming each code word consist of only the digits 0 and 1. A code word is said to be *valid* if the code word adheres to all of the rules that define the code; otherwise, the code word is said to be *invalid*.

The *encoding operation* is the process of determining the corresponding code word for a particular data item. In other words, the encoding process takes an original data item and represents it as a code word using the rules of the code. The *decoding operation* is the process of recovering the original data from the code word. In other words, the decoding process takes a code word and determines the data that it represents.

It is possible to create a binary code for which the valid code words are a subset of the total number of possible combinations of 1s and 0s. If the code words are formed correctly, errors introduced into a code word will force it to lie in the range of illegal, or invalid, code words, and the error can be detected. This is the basic concept of the *error detecting codes*. The basic concept of the *error correcting code* is that the code word is structured such that it is possible to determine the correct code word from the corrupted, or erroneous, code word.

A fundamental concept in the characterization of codes is the *Hamming distance* [Hamming, 1950]. The *Hamming distance* between any two binary words is the number of bit positions in which the two words differ. For example, the binary words 0000 and 0001 differ in only one position and therefore have a Hamming distance of 1. The binary words 0000 and 0101, however, differ in two positions; consequently, their Hamming distance is 2. Clearly, if two words have a Hamming distance of 1, it is possible to change one word into the other simply by modifying one bit in one of the words. If, however, two words differ in two bit positions, it is impossible to transform one word into the other by changing one bit in one of the words.

The Hamming distance gives insight into the requirements of error detecting codes and error correcting codes. We define the *distance* of a code as the minimum Hamming distance between any two valid code words. If a binary code has a distance of two, then any single-bit error introduced into a code word will result in the erroneous word being an invalid code word because all valid code words differ in at least two bit positions. If a code has a distance of 3, then any single-bit error or any double-bit error will result in the erroneous word being an invalid code word because all valid code words differ in at least three positions. However, a code distance of 3 allows any single-bit error to be corrected, if it is desired to do so, because the erroneous word with a single-bit error will be a Hamming distance of 1 from the correct code word and at least a Hamming distance of 2 from all others. Consequently, the correct code word can be identified from the corrupted code word.

In general, a binary code can correct up to c bit errors and detect an additional d bit errors if and only if

$$2c + d + 1 \leq H_d$$

where H_d is the distance of the code [Nelson and Carroll, 1986]. For example, a code with a distance of 2 cannot provide any error correction but can detect single-bit errors. Similarly, a code with a distance of 3 can correct single-bit errors or detect a double-bit error.

A second fundamental concept of codes is *separability*. A *separable code* is one in which the original information is appended with new information to form the code word, thus allowing the decoding process to consist of simply removing the additional information and keeping the original data. In other words, the original data is obtained from the code word by stripping away extra bits, called the code bits or check bits, and retaining only those associated with the original information. A *nonseparable code* does not possess the property of separability and, consequently, requires more complicated decoding procedures.

Perhaps the simplest form of a code is the parity code. The basic concept of parity is very straightforward, but there are variations on the fundamental idea. Single-bit parity codes require the addition of an extra bit to a binary word such that the resulting code word has either an even number of 1s or an odd number of 1s. If the extra bit results in the total number of 1s in the code word being odd, the code is referred to as *odd parity*. If the resulting number of 1s in the code word is even, the code is called *even parity*. If a code word with odd parity experiences a change in one of its bits, the parity will become even. Likewise, if a code word with even parity encounters a single-bit change, the parity will become odd. Consequently, a single-bit error can be detected by checking the number of 1s in the code words. The single-bit parity code (either odd or even) has a distance of 2, therefore allowing any single-bit error to be detected but not corrected. Figure 87.4 illustrates the use of parity coding in a simple memory application.

Arithmetic codes are very useful when it is desired to check arithmetic operations such as addition, multiplication, and division [Avizienis, 1971]. The basic concept is the same as all coding techniques. The data presented to the arithmetic operation is encoded before the operations are performed. After completing the arithmetic operations, the resulting code words are checked to make sure that they are valid code words. If the resulting code words are not valid, an error condition is signaled. An arithmetic code must be invariant to a set of arithmetic operations. An arithmetic code, A, has the property that $A(b*c) = A(b)*A(c)$, where b and c are operands, $*$ is some arithmetic operation, and $A(b)$ and $A(c)$ are the arithmetic code words for the operands b and c, respectively. Stated verbally, the performance of the arithmetic operation on two arithmetic code words will produce the arithmetic code word of the result of the arithmetic operation. To completely define an arithmetic code, the method of encoding and the arithmetic operations for which

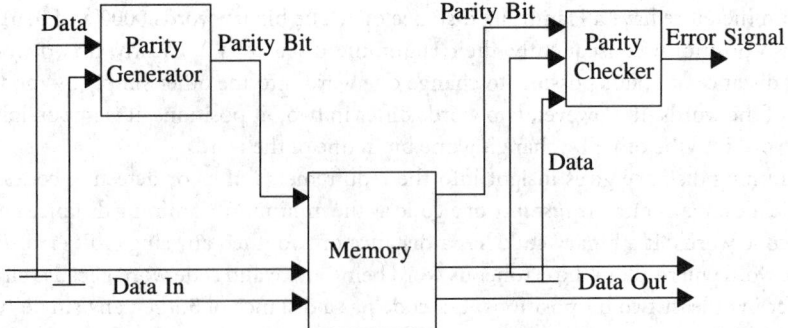

FIGURE 87.4   Use of parity coding in a memory application. (*Source:* B. W. Johnson, *Design and Analysis of Fault-Tolerant Digital Systems,* Reading, Mass.: Addison-Wesley, 1989, p. 85. With permission.)

the code is invariant must be specified. The most common examples of arithmetic codes are the *AN* codes, residue codes, and the inverse residue codes.

87.4 Time Redundancy

Time redundancy methods attempt to reduce the amount of extra hardware at the expense of using additional time. In many applications, the time is of much less importance than the hardware because hardware is a physical entity that impacts weight, size, power consumption, and cost. Time, on the other hand, may be readily available in some applications. The basic concept of time redundancy is the repetition of computations in ways that allow faults to be detected. Time redundancy can function in a system in several ways. The fundamental concept is to perform the same computation two or more times and compare the results to determine if a discrepancy exists. If an error is detected, the computations can be performed again to see if the disagreement remains or disappears. Such approaches are often good for detecting errors resulting from transient faults but cannot provide protection against errors resulting from permanent faults.

The main problem with many time redundancy techniques is assuring that the system has the same data to manipulate each time it redundantly performs a computation. If a transient fault has occurred, a system's data may be completely corrupted, making it difficult to repeat a given computation. Time redundancy has been used primarily to detect transients in systems. One of the biggest potentials of time redundancy, however, now appears to be the ability to detect permanent faults while using a minimum of extra hardware. The fundamental concept is illustrated in Fig. 87.5. During the first computation or transmission, the operands are used as presented and the results are stored in a register. Prior to the second computation or transmission, the operands are encoded in some fashion using an encoding function. After the operations have been performed on the encoded data, the results are then decoded and compared to those obtained during the first operation. The selection of the encoding function is made so as to allow faults in the hardware to be detected. Example encoding functions might include the complementation operator and an arithmetic shift.

87.5 Software Redundancy

Software faults are unusual entities. Software does not break as hardware does, but instead software faults are the result of incorrect software designs or coding mistakes. Therefore, any technique that detects faults in software must detect design flaws. A simple duplication and comparison procedure

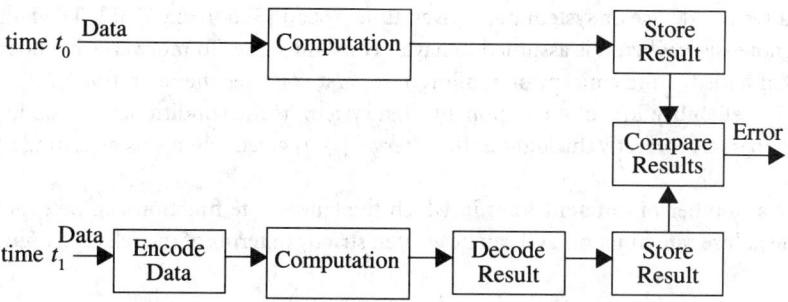

FIGURE 87.5   Time redundancy concept. (*Source:* B. W. Johnson, *Design and Analysis of Fault-Tolerant Digital Systems,* Reading, Mass.: Addison-Wesley, 1989, p. 137. With permission.)

will not detect software faults if the duplicated software modules are identical, because the design mistakes will appear in both modules.

The concept of N self-checking programming is to first write N unique versions of the program and to develop a set of acceptance tests for each version. The acceptance tests are essentially checks performed on the results produced by the program and may be created using consistency checks and capability checks, for example. Selection logic, which may be a program itself, chooses the results from one of the programs that passes the acceptance tests. This approach is analogous to the hardware technique known as hot standby sparing. Since each program is running simultaneously, the reconfiguration process can be very fast. Provided that the software faults in each version of the program are independent and the faults are detected as they occur by the acceptance tests, then this approach can tolerate $N - 1$ faults. It is important to note that the assumptions of fault independence and perfect fault coverage are very big assumptions to make in almost all applications.

The concept of N-version programming was developed to allow certain design flaws in software modules to be tolerated [Chen and Avizienis, 1978]. The basic concept of N-version programming is to design and code the software module N times and to vote on the N results produced by these modules. Each of the N modules is designed and coded by a separate group of programmers. Each group designs the software from the same set of specifications such that each of the N modules performs the same function. However, it is hoped that by performing the N designs independently, the same mistakes will not be made by the different groups. Therefore, when a fault occurs, the fault will either not occur in all modules or it will occur differently in each module, so that the results generated by the modules will differ. Assuming that the faults are independent the approach can tolerate $(N - 1)/2$ faults where N is odd.

The recovery block approach to software fault tolerance is analogous to the active approaches to hardware fault tolerance, specifically the cold standby sparing approach. N versions of a program are provided, and a single set of acceptance tests is used. One version of the program is designated as the primary version, and the remaining $N - 1$ versions are designated as spares, or secondary versions. The primary version of the software is always used unless it fails to pass the acceptance tests. If the acceptance tests are failed by the primary version, then the first secondary version is tried. This process continues until one version passes the acceptance tests or the system fails because none of the versions can pass the tests.

87.6 Dependability Evaluation

Dependability is defined as the quality of service provided by a system [Laprie, 1985]. Perhaps the most important measures of dependability are reliability and availability. Fundamental to reliability calculations is the concept of failure rate. Intuitively, the *failure rate is* the expected number of

failures of a type of device or system per a given time period [Shooman, 1968]. The failure rate is typically denoted as λ when it is assumed to have a constant value. To more clearly understand the mathematical basis for the concept of a failure rate, first consider the definition of the reliability function. The **reliability** $R(t)$ of a component, or a system, is the conditional probability that the component operates correctly throughout the interval $[t_0, t]$ given that it was operating correctly at the time t_0.

There are a number of different ways in which the failure rate function can be expressed. For example, the failure rate function $z(t)$ can be written strictly in terms of the reliability function $R(t)$ as

$$z(t) = \left(-\frac{dR(t)/dt}{R(t)} \right)$$

Similarly, $z(t)$ can be written in terms of the unreliability $Q(t)$ as

$$z(t) = -\frac{dR(t)/dt}{R(t)} = \frac{dQ(t)/dt}{1 - Q(t)}$$

where $Q(t) = 1 - R(t)$. The derivative of the unreliability, $dQ(t)/dt$, is called the *failure density function.*

The failure rate function is clearly dependent upon time; however, experience has shown that the failure rate function for electronic components does have a period where the value of $z(t)$ is approximately constant. The commonly accepted relationship between the failure rate function and time for electronic components is called the bathtub curve and is illustrated in Fig. 87.6. The bathtub curve assumes that during the early life of systems, failures occur frequently due to substandard or weak components. The decreasing part of the bathtub curve is called the early-life or infant mortality region. At the opposite end of the curve is the wear-out region where systems have been functional for a long period of time and are beginning to experience failures due to the physical wearing of electronic or mechanical components. During the intermediate region, the failure

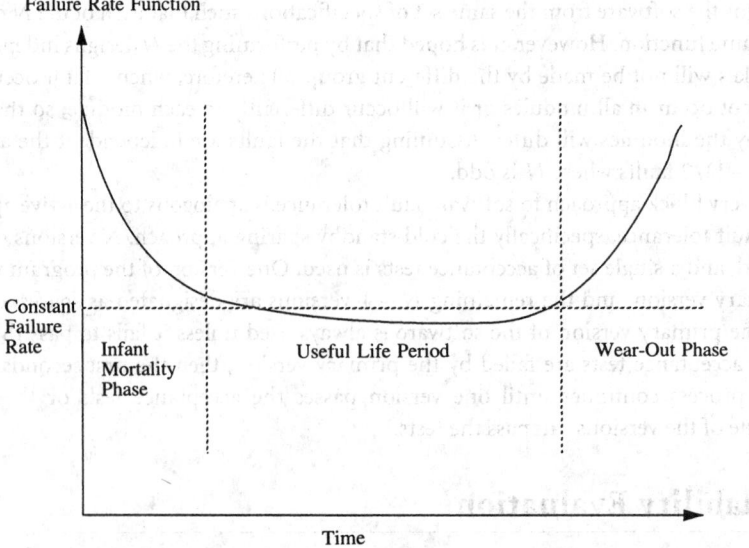

FIGURE 87.6   Bathtub curve relationship between the failure rate function and time. (*Source:* B. W. Johnson, *Design and Analysis of Fault-Tolerant Digital Systems,* Reading, Mass.: Addison-Wesley, 1989, p. 173. With permission.)

rate function is assumed to be a constant. The constant portion of the bathtub curve is called the useful-life phase of the system, and the failure rate function is assumed to have a value of λ during that period. λ is referred to as the failure rate and is normally expressed in units of failures per hour.

The reliability can be expressed in terms of the failure rate function as a differential equation of the form

$$\frac{dR(t)}{dt} = -z(t)R(t)$$

The general solution of this differential equation is given by

$$R(t) = e^{-\int z(t)dt}$$

If we assume that the system is in the useful-life stage where the failure rate function has a constant value of λ, the solution to the differential equation is an exponential function of the parameter λ given by

$$R(t) = e^{-\lambda t}$$

where λ is the constant failure rate. The exponential relationship between the reliability and time is known as the *exponential failure law*, which states that for a constant failure rate function, the reliability varies exponentially as a function of time.

In addition to the failure rate, the mean time to failure (MTTF) is a useful parameter to specify the quality of a system. The MTTF is the expected time that a system will operate before the first failure occurs. The MTTF can be calculated by finding the expected value of the time of failure.

From probability theory, we know that the expected value of a random variable, X, is

$$E[X] = \int_{-\infty}^{\infty} x f(x) dx$$

where $f(x)$ is the probability density function. In reliability analysis we are interested in the expected value of the time of failure (MTTF), so

$$MTTF = \int_{0}^{\infty} t f(t) dt$$

where $f(t)$ is the failure density function, and the integral runs from 0 to ∞ because the failure density function is undefined for times less than 0. We know, however, that the failure density function is

$$f(t) = \frac{dQ(t)}{dt}$$

so, the MTTF can be written as

$$MTTF = \int_{0}^{\infty} t \frac{dQ(t)}{dt} dt$$

Using integration by parts and the fact that $dQ(t)/dt = -dR(t)/dt$ we can show that

$$MTTF = \int_{0}^{\infty} t \frac{dQ(t)}{dt} dt = -\int_{0}^{\infty} t \frac{dR(t)}{dt} dt = \left[-tR(t) + \int R(t)dt \right]\Big|_{0}^{\infty} = \int_{0}^{\infty} R(t)dt$$

Consequently, the MTTF is defined in terms of the reliability function as

$$\text{MTTF} = \int_0^\infty R(t)\,dt$$

which is valid for any reliability function that satisfies $R(\infty) = 0$.

The mean time to repair (MTTR) is simply the average time required to repair a system. The MTTR is extremely difficult to estimate and is often determined experimentally by injecting a set of faults, one at a time, into a system and measuring the time required to repair the system in each case. The MTTR is normally specified in terms of a repair rate, μ, which is the average number of repairs that occur per time period. The units of the repair rate are normally number of repairs per hour. The MTTR and the rate, μ, are related by

$$\text{MTTR} = \frac{1}{\mu}$$

It is very important to understand the difference between the MTTF and the mean time between failure (MTBF). Unfortunately, these two terms are often used interchangeably. While the numerical difference is small in many cases, the conceptual difference is very important. The MTTF is the average time until the first failure of a system, while the MTBF is the average time between failures of a system. If we assume that all repairs to a system make the system perfect once again just as it was when it was new, the relationship between the MTTF and the MTBF can be determined easily. Once successfully placed into operation, a system will operate, on the average, a time corresponding to the MTTF before encountering the first failure. The system will then require some time, MTTR, to repair the system and place it back into operation once again. The system will then be perfect once again and will operate for a time corresponding to the MTTF before encountering its next failure. The time between the two failures is the sum of the MTTF and the MTTR and is the MTBF. Thus, the difference between the MTTF and the MTBF is the MTTR. Specifically, the MTBF is given by

$$\text{MTBF} = \text{MTTF} + \text{MTTR}$$

In most practical applications the MTTR is a small fraction of the MTTF, so the approximation that the MTBF and MTTF are equal is often quite good. Conceptually, however, it is crucial to understand the difference between the MTBF and the MTTF.

An extremely important parameter in the design and analysis of fault-tolerant systems is fault coverage. The fault coverage available in a system can have a tremendous impact on the reliability, safety, and other attributes of the system. Fault coverage is mathematically defined as the conditional probability that, given the existence of a fault, the system recovers [Bouricius *et al.*, 1969]. The fundamental problem with fault coverage is that it is extremely difficult to calculate. Probably the most common approach to estimating fault coverage is to develop a list all of the faults that can occur in a system and to form, from that list, a list of faults from which the system can recover. The fault coverage factor is then calculated appropriately.

Reliability is perhaps one of the most important attributes of systems. The reliability of a system is generally derived in terms of the reliabilities of the individual components of the system. The two models of systems that are most common in practice are the series and the parallel. In a series system, each element of the system is required to operate correctly for the system to operate correctly. In a parallel system, on the other hand, only one of several elements must be operational for the system to perform its functions correctly.

The series system is best thought of as a system that contains no redundancy; that is, each element of the system is needed to make the system function correctly. In general, a system may contain N elements, and in a series system each of the N elements is required for the system to function correctly. The reliability of the series system can be calculated as the probability that none of the elements will fail. Another way to look at this is that the reliability of the series system is the probability that all of the elements are working properly. The reliability of a series system is given by

$$R_{\text{series}}(t) = R_1(t)R_2(t)\ldots R_N(t)$$

or

$$R_{\text{series}}(t) = \prod_{i=1}^{N} R_i(t)$$

An interesting relationship exists in a series system if each individual component satisfies the exponential failure law. Suppose that we have a series system made up of N components, and each component, i, has a constant failure rate of λ_i. Also assume that each component satisfies the exponential failure law. The reliability of the series system is given by

$$R_{\text{series}}(t) = e^{-\lambda_1 t}e^{-\lambda_2 t}\ldots e^{-\lambda_N t}$$

$$R_{\text{series}}(t) = e^{-\sum\limits_{i=1}^{N}\lambda_i t}$$

The distinguishing feature of the basic parallel system is that only one of N identical elements is required for the system to function. The reliability of the parallel system can be written as

$$R_{\text{parallel}}(t) = 1.0 - Q_{\text{parallel}}(t) = 1.0 - \prod_{i=1}^{N} Q_i(t) = 1.0 - \prod_{i=1}^{N}(1.0 - R_i(t))$$

It should be noted that the equations for the parallel system assume that the failures of the individual elements that make up the parallel system are independent.

M-of-N systems are a generalization of the ideal parallel system. In the ideal parallel system, only one of N modules is required to work for the system to work. In the M-of-N system, however, M of the total of N identical modules are required to function for the system to function. A good example is the TMR configuration where two of the three modules must work for the majority voting mechanism to function properly. Therefore, the TMR system is a 2-of-3 system.

In general, if there are N identical modules and M of those are required for the system to function properly, then the system can tolerate $N - M$ module failures. The expression for the reliability of an M-of-N system can be written as

$$R_{M\text{-of-}N}(t) = \sum_{i=0}^{N-M} \binom{N}{i} R^{N-i}(t)(1.0 - R(t))^i$$

where

$$\binom{N}{i} = \frac{N!}{(N-i)!\,i!}$$

The **availability**, $A(t)$, of a system is defined as the probability that a system will be available to perform its tasks at the instant of time t. Intuitively, we can see that the availability can be approximated as the total time that a system has been operational divided by the total time elapsed since the system was initially placed into operation. In other words, the availability is the percentage of time that the system is available to perform its expected tasks. Suppose that we place a system into operation at time $t = 0$. As time moves along, the system will perform its functions, perhaps fail, and hopefully be repaired. At some time $t = t_{\text{current}}$, suppose that the system has operated correctly for a total of t_{op} hours and has been in the process of repair or waiting for repair to begin for a total of t_{repair} hours. The time t_{current} is then the sum of t_{op} and t_{repair}. The availability can be determined as

$$A(t_{\text{current}}) = \frac{t_{\text{op}}}{t_{\text{op}} + t_{\text{repair}}}$$

where $A(t_{\text{current}})$ is the availability at time t_{current}.

If the average system experiences N failures during its lifetime, the total time that the system will be operational is $N(\text{MTTF})$ hours. Likewise, the total time that the system is down for repairs is $N(\text{MTTR})$ hours. In other words, the operational time, t_{op}, is $N(\text{MTTF})$ hours and the downtime, t_{repair}, is $N(\text{MTTR})$ hours. The average, or steady-state, availability is

$$A_{SS} = \frac{N(\text{MTTF})}{N(\text{MTTF}) + N(\text{MTTR})}$$

We know, however, that the MTTF and the MTTR are related to the failure rate and the repair rate, respectively, for simplex systems, as

$$\text{MTTF} = \frac{1}{\lambda}$$

$$\text{MTTR} = \frac{1}{\mu}$$

Therefore, the steady-state availability is given by

$$A_{SS} = \frac{1/\lambda}{1/\lambda + 1/\mu} = \frac{1}{1 + \lambda/\mu}$$

Defining Terms

Availability, $A(t)$:   The probability that a system is operating correctly and is available to perform its functions at the instant of time t.

Dependability: The quality of service provided by a particular system.

Error: The occurrence of an incorrect value in some unit of information within a system.

Failure: A deviation in the expected performance of a system.

Fault: A physical defect, imperfection, or flaw that occurs in hardware or software.

Fault avoidance: A technique that attempts to prevent the occurrence of faults.

Fault tolerance: The ability to continue the correct performance of functions in the presence of faults.

Maintainability, $M(t)$:   The probability that an inoperable system will be restored to an operational state within the time t.

Performability, $P(L,t)$:   The probability that a system is performing at or above some level of performance, L, at the instant of time t.

Reliability, $R(t)$:   The conditional probability that a system has functioned correctly throughout an interval of time, $[t_0, t]$, given that the system was performing correctly at time t_0.

Safety, $S(t)$:   The probability that a system will either perform its functions correctly or will discontinue its functions in a well-defined, safe manner.

References

A. Avizienis, "Arithmetic error codes: Cost and effectiveness studies for application in digital system design," *IEEE Transactions on Computers,* vol. C-20, no. 11, pp. 1322–1331, November 1971.

W. G. Bouricius, W. C. Carter, and P. R. Schneider, "Reliability modeling techniques for self-repairing computer systems," in *Proceedings of the 24th ACM Annual Conference,* pp. 295–309, 1969.

L. Chen and A. Avizienis, "N-version programming: A fault tolerant approach to reliability of software operation," in *Proceedings of the International Symposium on Fault Tolerant Computing,* pp. 3–9, 1978.

R. W. Hamming, "Error detecting and error correcting codes," *Bell System Technical Journal,* vol. 26, no. 2, pp. 147–160, April 1950.

B. W. Johnson, *Design and Analysis of Fault-Tolerant Digital Systems,* Reading, Mass.: Addison-Wesley, 1989.

J-C. Laprie, "Dependable computing and fault tolerance: Concepts and terminology," in *Proceedings of the 15th Annual International Symposium on Fault-Tolerant Computing,* Ann Arbor, Mich.: pp. 2–11, June 19–21, 1985.

S. Lin and D. J. Costello, Jr., *Error Control Coding: Fundamentals and Applications,* Englewood Cliffs, N.J.: Prentice-Hall, 1983.

V. P. Nelson and B. D. Carroll, *Tutorial: Fault-Tolerant Computing,* Washington, D.C.: IEEE Computer Society Press, 1986.

M. L. Shooman, *Probabilistic Reliability: An Engineering Approach,* New York: McGraw-Hill, 1968.

Further Information

The *IEEE Transactions on Computers, IEEE Computer* magazine, and the *Proceedings of the IEEE* have published numerous special issues dealing exclusively with fault tolerance technology. Also, the IEEE International Symposium on Fault-Tolerant Computing has been held each year since 1971. Finally, the following textbooks are available, in addition to those referenced above:

P. K. Lala, *Fault Tolerant and Fault Testable Hardware,* Englewood Cliffs, N.J.: Prentice-Hall, 1985.

D. K. Pradhan, *Fault-Tolerant Computing: Theory and Techniques,* Englewood Cliffs, N.J.: Prentice-Hall, 1986.

D. P. Siewiorek and R. S. Swarz, *The Theory and Practice of Reliable Systems Design,* 2nd ed., Bedford, Mass.: Digital Press, 1992.

88

Knowledge Engineering

88.1 Databases 2032
     Database Abstraction • Data Models • Relational Databases • Hierarchical Databases • Network Databases • Architecture of a DBMS • Data Integrity and Security • Emerging Trends
88.2 Rule-Based Expert Systems 2048
     Problem Selection • Knowledge Acquisition • Knowledge Representation • Knowledge Encoding • Knowledge Testing and Evaluation • Implementation and Maintenance

M. Abdelguerfi
University of New Orleans

Jay Liebowitz
George Washington University

88.1 Databases

M. Abdelguerfi

In the past, file processing techniques were used to design information systems. These systems usually consist of a set of files and a collection of application programs. Permanent records are stored in the files, and application programs are used to update and query the files. The application programs were in general developed individually to meet the needs of different groups of users. In many cases, this approach leads to a duplication of data among the files of different users. Also, the lack of coordination between files belonging to different users often leads to a lack of data consistency. In addition, changes to the underlying data requirements usually necessitate major changes to existing application programs. Among other major problems that arise with the use of file processing techniques are lack of data sharing, reduced programming productivity, and increased program maintenance. Because of their inherent difficulties and lack of flexibility, file processing techniques have lost a great deal of their popularity and are being replaced by **database management systems (DBMS)**.

A DBMS is designed to efficiently manage a shared pool of interrelated data (**database**). This includes the existence of features such as a *data definition language* for the definition of the logical structure of the database (*database scheme*), a *data manipulation language* to query and update the database, a *concurrency control* mechanism to keep the database consistent when shared by several users, a *crash recovery* strategy to avoid any loss of information after a system crash, and *safety* mechanisms against any unauthorized access.

Database Abstraction

A DBMS is expected to provide for *data independence*, i.e., user requests are made at a *logical level* without any need for the knowledge of how the data is stored in actual files. This implies that the internal file structure could be modified without any change to the user's perception of the database. To achieve data independence, the Standards Planning and Requirements Committee (SPARC) of the American National Standards Institute (ANSI) in its 1977 report recommended

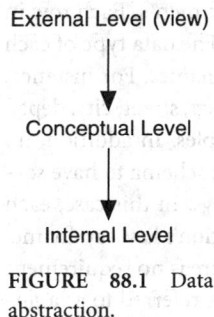

External Level (view)

Conceptual Level

Internal Level

FIGURE 88.1 Data abstraction.

three levels of database abstraction (see Fig. 88.1). The lowest level in the abstraction is the internal level. Here, the database is viewed as a collection of files organized according to one of several possible internal data organizations (e.g., B⁺-tree data organization). In the conceptual level, the database is viewed at an abstract level. The user at this level is shielded from the internal storage details. At the external level, each group of users has his own perception or *view* of the database. Each view is derived from the conceptual database and is designed to meet the needs of a particular group of users. Such a group can only have access to the data specified by its particular view. This, of course, ensures both privacy and security.

The mapping between the three levels of abstraction is the task of the DBMS. When changes to the internal level (such as a change in file organization) do not affect the conceptual and external levels, the system is said to provide for *physical data independence*. *Logical data independence* prevents changes to the conceptual level to affect users' views. Both types of data independence are desired features in a database system.

Data Models

A **data model** refers to an integrated set of tools used to describe the data and its structure, data relationships, and data constraints. Some data models provide a set of operators that is used to update and query the database. Data models can be classified in two main categories: *record-based* and *object-based*. Both classes are used to describe the database at the conceptual and external levels. With object-based data models, constraints on the data can be specified more explicitly.

There are three main record-based data models: the *relational, network,* and *hierarchical* data models. In the relational model, data at the conceptual level is represented as a collection of interrelated tables. These tables are *normalized* so as to minimize data redundancy and update anomalies. In this model, data relationships are implicit and are derived by matching columns in tables. In the hierarchical and network models, the data is represented as a collection of records and data relationships are explicit and are represented by *links*. The difference between the last two models is that in the hierarchical model, data is represented as a tree structure, while it is represented as a generalized graph in the network model.

In hierarchical and network models, the existence of physical pointers (links) to link related records allows an application program to retrieve a single record at a time by following the pointer's chain. The process of following the pointer's chain and selecting one record at a time is referred to as *navigation*. In nonnavigational models such as the relational, records are not related through pointer's chains, but relationships are established by matching columns in different tables.

The hierarchical and network models require the application programmer to be aware of the internal structure of the database. The relational model, on the other hand, allows for a high degree of physical and logical data independence. Earlier DBMSs were for the most part navigational systems. Because of its simplicity and strong theoretical foundations, the relational database model has since received wide acceptance. Today, most DBMSs are based on the relational paradigm.

Relational Databases

The relational database model was introduced by E. F. Codd [1970]. Since the theoretical underpinnings of the relational model have been well defined, it has become the focus of most commercial DBMSs.

In the relational model, the data is represented as a collection of relations. To a large extent, each relation can be thought of as a table. The example of Fig. 88.2 shows part of a university database composed of two relations. FAC_INFO gives personal information (last name, social security, street and city of residence, and department) of a faculty. DEP_CHAIR gives the last name of the chairman of each department. A faculty is not allowed to belong to two departments. Each row in a relation is referred to as a *tuple*. A column name is called an *attribute name*. The data type of each attribute name is known as its *domain*. A *relation scheme* is a set of attribute names. For instance, the relation scheme (or schema) of the relation FAC_INFO is (lname, social_sec#, street, city, dept). A *key* is a set of attribute names whose composite value is distinct for all tuples. In addition, no proper subset of the key is allowed to have this property. It is not unusual for a schema to have several possible keys. In FAC_INFO, both lname and social_sec# are possible keys. In this case, each possible key is known as a *candidate key*, and the one selected to act as the relation's key, say, lname, is referred to as the *primary key*. A *superkey* is a key with the exception that there is no requirement for minimality. In a relation, an attribute name (or a set of attribute names) is referred to as a *foreign key*, if it is the primary key of another relation. In FAC_INFO, the attribute name dept is a foreign key, since the same attribute is a key in DEP_CHAIR. Because of updates to the database, the content of a relation is dynamic. For this reason, the data in a relation at a given time instant is called *instance* of the relation.

There are three integrity constraints that are usually imposed on each instance of a relation: primary key integrity, entity integrity, and referential integrity. The key integrity constraint requires that no two tuples of a relation have the same key value. The entity integrity constraint specifies that the key value of each tuple should have a known value (i.e., no *null* values are allowed for primary keys). The referential integrity constraint specifies that if a relation r_1 contains a foreign key that matches the primary key of a relation r_2, then each value of the foreign key in r_1 must either match a value of the primary key in r_2 or must be null. For the database of Fig. 88.2 to be consistent, each value of dept in FAC_INFO must match a value of dept in DEP_CHAIR.

Relational Database Design

The relational database design [Maier, 1983] refers to the process of generating a set of relation schemes that minimizes data redundancy and removes update anomalies. One of the most popular approaches is the use of the *normalization theory*. The normalization theory is based on the notion of *functional dependencies*.

Functional dependencies are constraints imposed on the database. The notion of superkey, introduced in the previous section, can be formulated as follows: A subset of a relation schema is a superkey if, in any instance of the relation, no two distinct tuples have the same superkey value. If $r(R)$ is used to denote a relation r on a schema R, $K \subseteq R$ a superkey, and $t(k)$ the K-value of tuple t, then no two tuples t_1 and t_2 in $r(R)$ are such that $t_1(K) = t_2(K)$.

The notion of a functional dependency can be seen as a generalization of the notion of superkey. Let X and Y be two subsets of R; the functional dependency $X \rightarrow Y$ exists in $r(R)$ if whenever two tuples in $r(R)$ have the same X-value, their Y-value is also the same. That is, if $t_1(X) = t_2(X)$, then

```
FAC_INFO( lname,    social_sec#,  street,     city,        dept)
          Hosch     383909164     Esplanade   Kenner       CS
          Loggins   482233364     Bonnabel    Metairie     EE
          Martin    399254402     Williams    Kenner       CH
          Krad      100995678     Burbon      New-Orleans  ME
          Hosch     383988164     Esplanade   Kenner       CS
          Hanoura   400919945     Bonnabel    Metairie     CH
          Prados    388998800     Severn      Metairie     EE
          Abdel     383909164     St Charles  New Orleans  CS

DEP_CHAIR( dept,    chair)
           CS       Hosch
           EE       Prados
           CH       Martin
```

FIGURE 88.2   An example of two relations: FAC_INFO and DEP_CHAIR.

$t_1(Y) = t_2(Y)$. Using functional dependencies, one can define the notion of a key more precisely. A key k of a relation $r(R)$ is such that $k \to R$ and no proper subset of k has this property. Note that if the schema R is composed of attribute names $\{A_1, A_2, \ldots, A_n\}$, then each attribute name A_i is functionally determined by the key k, i.e., $k \to A_i$, $i = 1, \ldots, n$. An attribute name that is part of a key is referred to as a *prime attribute*. In the example of Fig. 88.2, both attribute names street and city are nonprime attributes.

The normalization process can be thought of as the process of decomposing a schema with update anomalies and data redundancy into smaller schemas in which these undesirable properties are to a large extent eliminated. Depending on the severity of these undesirable properties, schemas are classified into *normal forms*. Originally, Codd defined three normal forms: *first normal form* (1NF), *second normal form* (2NF), and *third normal form* (3NF). Thereafter, a stronger version of the 3NF, known as Boyce-Codd normal form (BCNF), was suggested. These four normal forms are based on the concept of functional dependencies.

The 1NF requires that attribute name values be *atomic*. That is, composite values for attribute names are not allowed. A 2NF schema is a 1NF schema in which all nonprime attributes are fully dependent on the key. Consider the relation of Fig. 88.3. Each tuple in PRODUCT gives the name of a supplier, a product name, its price, and the supplier's location. The schema (supplier_name, product_name, price, quantity) is in 1NF since each attribute name is atomic. It is assumed that many products can be supplied by a single supplier, that a given product can be supplied by more than one supplier, and that a supplier has only one location. So, (supplier_name, product_name) is the relation's key and the functional dependency supplier_name $\to$ location should hold for any instance of PRODUCT.

The structure of the relation of Fig. 88.3 does not allow a supplier to figure in the relation unless the supplier supplies at least one product. Even the use of null values is not of much help in this case as product_name is part of a key and therefore cannot be assigned a null value. Another anomaly can be encountered during the deletion process. For instance, deleting the last tuple in the relation results in the loss of the information that Rudd is a supplier located in Metairie. It is seen that the relation PRODUCT suffers from insertion and deletion anomalies.

Modifications can also be a problem in the relation PRODUCT. Suppose that the location of the supplier Martin is moved from Kenner to Slidell. In order not to violate the functional dependency supplier_name $\to$ location, the location attribute name of all tuples where the supplier is Martin needs to be changed from Kenner to Slidell. This modification anomaly has a negative effect on performance.

In addition, the relation PRODUCT suffers from data redundancy. For example, although Martin has only one location "Kenner", such a location appears in all three tuples where the supplier_name is Martin.

The update anomalies and data redundancy encountered in PRODUCT are all due to the functional dependency supplier_name $\to$ location. The right-hand side of this dependency "location" is a nonprime attribute, and the left-hand side represents part of the key. Therefore, we have a nonprime attribute that is only partially dependent on the key (supplier_name, product_name). As a consequence, the schema (supplier_name, product_name, price, location) is not in 2NF. The removal of the partial dependency supplier_name $\to$ location will eliminate all the above anomalies.

PRODUCT (	supplier_name,	Product_name,	price,	location)
	Martin	sofa	500.99	Kenner
	Martin	bed	100.95	Kenner
	Martin	desk	150.99	Kenner
	Evans	sofa	600.99	Metairie
	Evans	desk	250.99	Metairie
	Rudd	bed	110.95	Metairie

FIGURE 88.3 Instance of PRODUCT (supplier_name, product_name, price, quantity).

```
PRO_INFO( supplier_name,  product_name,   price)
          Martin          sofa            500.99
          Martin          bed             100.95
          Martin          desk            150.99
          Evans           sofa            600.99
          Evans           desk            250.99
          Rudd            bed             110.95

SUP_LOC( supplier_name,   location)
         Martin           Kenner
         Evans            Metairie
         Rudd             Metairie
```

FIGURE 88.4   Decomposition of PRODUCT into PRO_INFO and SUP_LOC.

The removal of the partial dependency is achieved by decomposing the schema (supplier_name, product_name, price, quantity) into two 2NF schemas: (supplier_name, product_name, price), and (supplier_name, location). This decomposition results in relations PRO_INFO and SUP_LOC shown in Fig. 88.4. The keys of PRO_INFO and SUP_LOC are (supplier_name, product_name), and supplier_name, respectively.

Normalizing schemas into 2NF removes all update anomalies due to nonprime attributes being partially dependent on keys. Anomalies of a different nature are still possible.

Update anomalies and data redundancy can originate from *transitive dependencies*. A nonprime attribute A_i is said to be transitively dependent on a key k via attribute name A_j, if $k \rightarrow A_j$, $A_j \rightarrow A_i$, and A_j does not functionally determine A_k. A 3NF is a 1NF where no nonprime attribute is transitively dependent on a key.

The relation of Fig. 88.5 highlights update anomalies and data redundancy due to the transitive dependency of a nonprime attribute on a key. The relation gives the name of a client (client_name), the corresponding supplier (supplier_name), and the supplier's location. Each client is assumed to have one supplier. The relation's key is client_name, and each supplier has only one location. A supplier and his location cannot be inserted in SUPPLIES unless the supplier has at least one client. In addition, the relation has a deletion anomaly since if Tillis is no longer a client of Rudd, the information about Rudd as a supplier and his location is lost. A change to a supplier's location may require updating the location attribute name of several tuples in the relation. Also, although each supplier has only one location, such a location is sometimes repeated several time unnecessarily, leading to data redundancy.

The relation exhibits the following transitive dependency: client_name $\rightarrow$ supplier_name, supplier_name $\rightarrow$ location (but not the inverse). The relation CLIENT is clearly in 2NF, but because of the transitive dependency of the nonprime attribute location on the key, it is not in 3NF. This is the cause of the anomalies mentioned above. Eliminating this transitive dependency by splitting the schema into two components will remove these anomalies (see Fig. 88.6). Clearly, the resulting two relations SUP_CLI and SUP_LOC are in 3NF.

Each partial dependency of a nonprime attribute on a key can be expressed as a transitive dependency of a nonprime attribute on a key. Therefore, a schema in 3NF is also in 2NF.

```
SUPPLIES  (client_name,   supplier_name,   location)
           Hosch          Martin           Kenner
           Krad           Martin           Kenner
           Shengru        Evans            Metairie
           Tillis         Rudd             Metairie
           Greene         Evans            Metairie
```

FIGURE 88.5   Instance of SUPPLIES.

```
SUP_CLI (  client_name,  supplier_name)  SUP_LOC(  supplier_name,  location)
           Hosch         Martin                    Martin          Kenner
           Krad          Martin                    Evans           Metairie
           Shengru       Evans                     Rudd            Metairie
           Tillis        Rudd
           Greene        Evans
```

FIGURE 88.6   Decomposition of SUPPLIES into SUP_CLI and SUP_LOC.

BCNF is a stronger version of 3NF. The only difference between BCNF and 3NF is that A_i in the definition of 3NF is no longer required to be nonprime for BCNF. A schema in BCNF is also in 3NF, but the opposite is not necessarily true.

Higher forms, such as 4NF and 5NF, have been suggested. The 4NF and 5NF are based on the concept of *multivalued* dependencies and *join* dependencies, respectively.

Data Definition and Manipulation in Relational Databases

Upon completion of the relational database design, a descriptive language, usually referred to as Data Definition Language (DDL), is used to define the designed schemas and their relationships. The DDL can be used to create new schemas or modify existing ones, but it cannot be used to query the database. Once DDL statements are compiled, they are stored in a special file called the *data dictionary*. A data dictionary is a special file that contains *metadata*, that is, data which describes the different attribute names in each schema, indexes, integrity constraints, and relationships among schemas. During the processing of a query, the DBMS usually checks the data dictionary. The data dictionary can be seen as a relational database of its own. As a result, data manipulation languages that are used to manipulate databases can usually be used to query the data dictionary.

An important function of a DBMS is to provide a Data Manipulation Language (DML) with which a user can retrieve, change, insert, and delete data from the database. DMLs are classified into two types: *procedural* and *nonprocedural*. The main difference between the two types is that in procedural DMLs, a user has to specify the desired data and how to obtain it, while in nonprocedural DMLs, a user has only to describe the desired data. Because they impose less burden on the user, nonprocedural DMLs are normally easier to learn and use.

The component of a DML that deals with data retrieval is referred to as *query language*. A query language can be used interactively in a stand-alone manner, or it can be embedded in a general-purpose programming language such as C and Cobol.

One of the most popular query languages is SQL (Structured Query Language). SQL is a query language based to a large extent on Codd's *relational algebra*. SQL has additional features for data definition and update. Therefore, SQL is a comprehensive relational database language that includes both a DDL and DML.

SQL includes the following commands for data definition: CREATE TABLE, DROP TABLE, and ALTER TABLE. The CREATE TABLE is used to create and describe a new relation. The two relations of Fig. 88.4 can be created in the following manner:

```
CREATE TABLE PRO_INFO ( supplier_name   VARCHAR(12)    NOT NULL,
                        product_name    VARCHAR(8)     NOT NULL,
                        price           DECIMAL(6,2));

CREATE TABLE SUP_LOC ( supplier_name   VARCHAR(12)    NOT NULL,
                       location         VARCHAR(10));
```

The CREATE TABLE command specifies all the attribute names of a relation and their data types (INTEGER, DECIMAL, fixed length character "CHAR", variable length character "VARCHAR", DATE, . . .). The constraint NOT NULL is usually specified for those attributes that cannot have null values. The primary key of each relation in the database is usually required to have a nonnull value.

If a relation is created incorrectly, it can be deleted using the DROP TABLE command. The command is DROP TABLE followed by the name of the relation to be deleted.

The ALTER TABLE is used to add new attribute names to an existing relation, as follows: **ALTER TABLE** SUP_LOC **ADD** zip_code CHAR(5);.

The SUP_LOC relation now contains an extra attribute name, zip_code. In most DBMSs, the zip_code value of existing tuples will automatically be assigned a null value. Other DBMSs allow for the assignment of an initial value to a newly added attribute name.

The DML component of SQL has one basic SELECT query statement that has the following structure:

```
SELECT <attribute_name list>
FROM    <relation_list>
WHERE   <restriction>
```

In the above statement, the SELECT clause specifies the attribute names that are to be retrieved, FROM gives the list of the relations involved, and WHERE is a Boolean predicate that completely specifies the tuples to be retrieved.

Consider the database of Fig. 88.4, and suppose that we want the name of all suppliers that supply either beds or desks. In SQL, this query can be expressed as:

```
SELECT supplier_name
FROM    PRO_INFO
WHERE   product_name = "bed" OR product_name = "sofa"
```

The result of an SQL command may contain duplicate values and is therefore not always a true relation. In fact, the result of the above query, shown below, has duplicate entries.

```
        supplier_name
        Martin
        Martin
        Rudd
```

The entry Martin appears twice in the result, because the supplier Martin supplies both beds and sofas. Removal of duplicates is usually a computationally intensive operation. As a result, duplicate entries are not automatically removed by SQL. To ensure uniqueness, the command DISTINCT should be used. In the above query, if we want the supplier names to be listed only once, the above query should be modified as follows:

```
SELECT DISTINCT supplier_name
FROM            PRO_INFO
WHERE           product_name = "bed" OR product_name = "sofa"
```

In SQL, a query can involve more than one relation. Suppose that we want the list of all suppliers from Metairie who supply beds. Such a query, shown below, involves both PRO_INFO and SUP_LOC.

```
SELECT supplier_name
FROM    PRO_INFO, SUP_LOC
WHERE   PRO_INFO.supplier_name = SUP_LOC.supplier_name
                        AND product_name = "bed"
```

When an SQL expression, such as the one above, involves more than one relation, it is sometimes necessary to qualify attribute names, that is, to precede an attribute name by the relation (a period is placed between the two) it belongs to. Such a qualification removes possible ambiguities.

In SQL, it is possible to have several levels of query nesting; this is done by including a SELECT query statement within the WHERE clause.

The output data can be presented in sorted order by using the SQL ORDER BY clause followed by the attribute name(s) according to which the output is to be sorted.

```
PROFESSOR (  faculty    department           salary )
             Smith      Electrical Eng.      $39,000
             Joe        Mechanical Eng.      $35,000
             Susan      Computer Sc.         $36,000
             Erick      Electrical Eng.      $38,000
             Paul       Electrical Eng.      $37,000
             Johannes   Computer Sc.         $65,000
             Rick       Computer Sc.         $32,000
             Gerard     Computer Sc.         $43,000
             Kenneth    Mechanical Eng.      $40,000
```

FIGURE 88.7   Instance of the relation PROFESSOR.

In database management applications it is often desirable to categorize the tuples of a relation by the values of a set of attributes and extract an aggregated characteristic of each category. Such database management tasks are referred to as *aggregation functions*. For instance, SQL includes the following built-in aggregation functions: SUM, COUNT, AVERAGE, MIN, MAX. The attribute names used for the categorization are referred to as GROUP BY columns. Consider the relation PROFESSOR of Fig. 88.7. Each tuple of the above relation gives the name of a faculty and his department and academic year salary.

Suppose that we want to know the number of faculty in each department and the result to be ordered by department. This query requests for each department a count of the number of faculty. Faculty are therefore categorized according to the attribute name department. As a result, department is referred to as a GROUP BY attribute. In SQL, the above query is formulated as follows:

```
SELECT    department, COUNT (faculty)
FROM      PROFESSOR
GROUP BY  department
ORDER BY  department
```

The result of applying the COUNT aggregation function is a new relation with two attribute names. They are a GROUP BY attribute (department in this case) and a new attribute called COUNT. The tuples are ordered lexicographically in ascending order according to the ORDER BY attribute, which is department in this case:

```
department          COUNT (faculty)
Computer Sc.        4
Electrical Eng.     3
Mechanical Eng.     2
```

Consider now the query: "What is the total salary for all faculty in each department?" The result is to be ordered by department. The SQL expression for this query is

```
SELECT    department, SUM (salary)
FROM      PROFESSOR
GROUP BY  department
ORDER BY  department
```

The result of the SUM aggregation function is

```
department          SUM (salary)
Computer Sc.        $176,000
Electrical Eng.     $114,000
Mechanical Eng.     $75,000
```

The next SQL query involves extracting the average salary in each department and sorting the resulting tuples in ascending order according to the department attribute name.

```
SELECT    department, AVERAGE (salary)
FROM      PROFESSOR
```

```
GROUP BY department
ORDER BY department
```

The above SQL expression yields the following:

```
department            AVERAGE (salary)
Computer Sc.          $44,000
Electrical Eng.       $38,000
Mechanical Eng.       $37,500
```

The remaining two aggregation functions can be used in a similar manner to extract, for example, the maximum salary (for MAX) and minimum salary (for MIN) in each department.

The relations created through the CREATE TABLE command are known as *base relations*. A base relation exists physically and is stored as a file by the DBMS. SQL can be used to create views using the CREATE VIEW command. In contrast to base relations, the creation of a view results in a *virtual relation*, that is, one that does not necessarily correspond to a physical file. Consider the database of Fig. 88.4, and suppose that we want to create a view giving the name of all suppliers located in Metairie, the products each one provides, and the corresponding prices. Such a view, called METAIRIE_SUPPLIER, can be created as follows:

```
CREATE VIEW  METAIRIE_SUPPLIER
AS SELECT    PRO_INFO.supplier_name, product_name, price
FROM         PRO_INFO, SUP_LOC
WHERE        PRO_INFO.supplier_name = SUP_LOC.supplier_name
                          AND location = "Metairie"
```

Because a view is a virtual relation that can be constructed from one or more relations, updating a view may lead to ambiguities. As a result, when a view is generated from more than one relation, there are, in general, restrictions on updating such a view.

Hierarchical Databases

The hierarchical data model [Elmasri and Navathe, 1989] uses a tree data structure to conceptualize associations between different record types. In this model, record types are represented as nodes and associations as links. Each record type, except the root, has only one parent; that is, only parent-child (or one-to-many) relationships are allowed. This restriction gives hierarchical databases their simplicity. Since links are only one way, from a parent to a child, the design of hierarchical database management systems is made simpler, and only a small set of data manipulation commands are needed.

Because only parent-child relationships are allowed, the hierarchical model cannot efficiently represent two main types of relationships: many-to-many relationships and the case where a record type is a child in more than one *hierarchical schema*. These two restrictions can be handled by allowing redundant *record instances*. However, such a duplication requires that all the copies of the. same record should be kept consistent at all times.

The example of Fig. 88.8 shows a hierarchical schema. The schema gives the relationship between a DEPARTMENT, its employees (D_EMPLOYEE), the projects (D_PROJECT) handled by the different departments, and how employees are assigned to these projects. It is assumed that an employee belongs to only one department, a project is handled by only one department, and an employee can be assigned to several projects. Notice that since a project has several employees assigned to it, and an employee can be assigned to more than one project, the relationship between D_PROJECT and D_EMPLOYEE is many-to-many. To model this relationship multiple instances of the same record type D-EMPLOYEE may appear under different projects.

Such redundancies can be reduced to a large extent through the use of *logical links*. A logical link associates a *virtual record* from a hierarchical schema with an actual record from either the same schema or another schema. The redundant copy of the actual record is therefore replaced by a virtual record, which is nothing more than a pointer to the actual one.

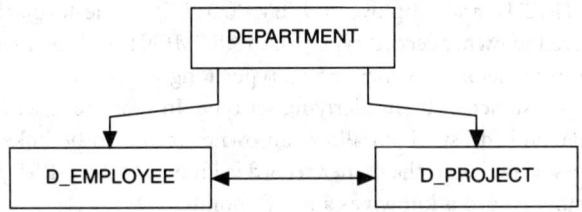

FIGURE 88.8   A hierarchical schema.

Hierarchical DLLs are used by a designer to declare the different hierarchical schemas, record types, and logical links. Furthermore, a root node must be declared for each hierarchical schema, and each record type declaration must also specify the parent record type.

Unlike relational DMLs, hierarchical DMLs such as DL/1 are record at-a-time languages. DL/1 is used by IBM's IMS hierarchical DBMS. In DL/1 a tree traversal is based on a preorder algorithm, and within each tree, the last record accessed through a DL/1 command can be located through a *currency indicator*.

Retrieval commands are of three types:

`GET UNIQUE <record type> WHERE <restrictions>`
Such a command retrieves the leftmost record that meets the imposed restrictions. The search always starts at the root of the tree pointed to by the currency indicator.

`GET NEXT [<record type> WHERE <restrictions>]`
Starting from the current position, this command uses the preodrer algorithm to retrieve the next record that satisfies the restrictions. The clause enclosed between brackets is optional. GET NEXT is used to retrieve the next (preorder) record from the current position.

`GET NEXT WITHIN PARENT [<record type> WHERE <restrictions>]`
It retrieves all records that have the same parent and that satisfy the restrictions. The parent is assumed to have been selected through a previous GET command.

Four commands are used for record updates:

`INSERT`
Stores a new record and links it to a parent. The parent has been already selected through a GET command.

`REPLACE`
The current record (selected through a previous GET command) is modified.

`DELETE`
The current record and all its descendants are deleted.

`GET HOLD`
Locks the current record while it is being modified.

The DL/1 commands are usually embedded in a general-purpose (host) language. In this case, a record accessed through a DL/1 command is assigned to a program variable.

Network Databases

In the network model [Elmasri and Navathe, 1989] associations between record types are less restrictive than with the hierarchy model. Here, associations among record types are represented as graphs.

One-to-one and one-to-many relationships are described using the notion of *set type*. Each set type has an *owner* record type and a *member* record type. In the example of Fig. 88.8, the relationship between DEPARTMENT and employee (D_EMPLOYEE) is one-to-many. This relationship defines a set type where the owner record type is DEPARTMENT and the member record type is D_EMPLOYEE. Each instance of an owner record type along with all the corresponding member records represents a set instance of the underlying set type. In practice, a set is commonly implemented using a circular-linked list which allows an owner record to be linked to all its member records. The pointer associated with the owner record is known as the FIRST pointer, and the one associated with a member record is known as a NEXT pointer

In general, a record type cannot be both the owner and a member of the same set type. Also, a record cannot exist in more than one instance of a specific set type. The latter requirement implies that many-to-many relationships are not directly implemented in the network data model.

The relationship between D_PROJECT and D-EMPLOYEE is many-to-many. In the network model, this relationship is represented by two set types and an intermediate record type. The new record type could be named ASSIGNED_TO (see Fig. 88.9). One set has D_EMPLOYEE as owner and ASSIGNED_TO as member record type, and the other has D_PROJECT as owner and ASSIGNED_TO as member record type.

Standards for the network model's DDL and DML were originally proposed by the CODASYL (Conference on Data Systems Languages) committee in 1971. Several revisions to the original proposal were made later.

In a network DDL, such as that of the IDMS database management system, a set declaration specifies the name of the set, its owner record type, and its member record type. The insertion mode for the set members needs to be specified using combinations of the following four commands:

AUTOMATIC
An inserted record is automatically connected to the appropriate set instance.

MANUAL
In this case, records are inserted into the appropriate set instance by an application program.

OPTIONAL
A member record does not have to be a member of a set instance. The member record can be connected to or disconnected from a set instance using DML commands.

MANDATORY
A member record needs to be connected to a set instance. A member record can be moved to another set instance using the network's DML.

FIXED
A member record needs to be connected to a set instance. A member record cannot be moved to another set instance.

The network's DDL allows member records to be ordered in several ways. Member records can be sorted in ascending or descending order according to one or more fields. Alternatively, a new member record can be inserted next (prior) to the current record (pointed to by the currency indicator) in the set instance. A newly inserted member record can also be placed first (or last) in the set instance. This will lead to a chronological (or reverse chronological) order among member records.

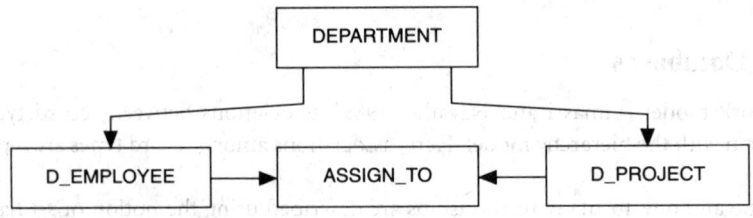

FIGURE 88.9   Representing many-to-many relationships in the network model.

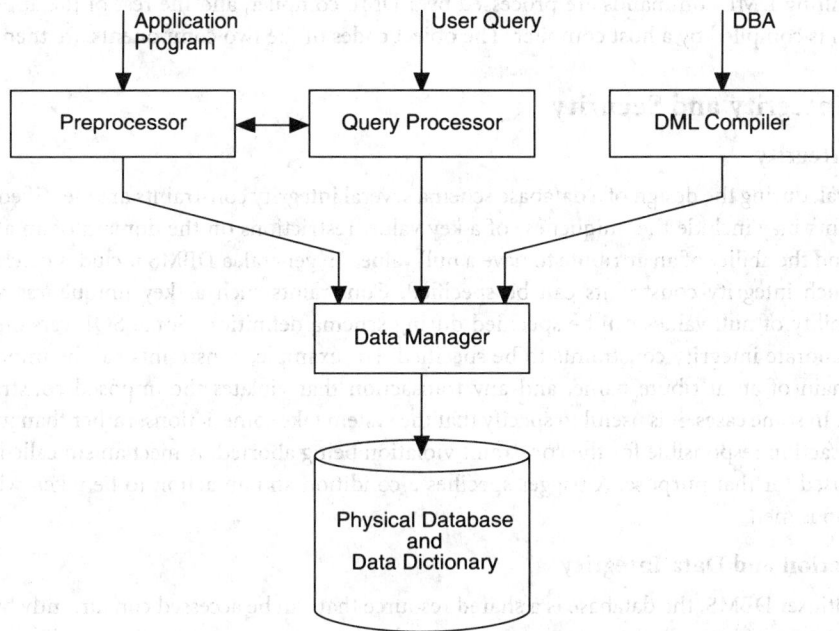

FIGURE 88.10 Simplified architecture of a DBMS.

As with the hierarchy model, network DMLs are record at-a-time languages, and currency indicators are necessary for navigation through the network database. For example, the IDMS main data manipulation commands can be summarized as follows:

CONNECT

Connects a member record to the specified set instance.

DISCONNECT

A member record is disconnected from a set instance (set membership must be manual in this case).

STORE, MODIFY, and DELETE

These commands are used for data storage, modification, and deletion.

FIND

Retrieval command based on set membership.

GET

Retrieval command based on key values.

Architecture of a DBMS

A DBMS is a complicated software structure that includes several components (see Fig. 88.10). The DBMS has to interact with the operating system for secondary storage access. The *data manager* is usually the interface between the DBMS and the operating system. The *DDL compiler* converts schema definitions, expressed using DDL statements, into a collection of metadata tables that are stored in the data dictionary. The design of the schemas is the function of the *database administrator* (DBA). The DBA is also responsible for specifying the data storage structure and access methodology and granting and revoking access authorizations. The *query processor* converts high-level DML statements into low-level instructions that the database manager can interpret. The *DML preprocessor* separates embedded DML statements from the rest of an application program.

The resulting DML commands are processed by a DML compiler, and the rest of the application program is compiled by a host compiler. The object codes of the two components are then linked.

Data Integrity and Security

Data Integrity

In general, during the design of a database schema several integrity constraints are identified. These constraints may include the uniqueness of a key value, restrictions on the domain of an attribute name, and the ability of an attribute to have a null value. In general, a DBMS includes mechanisms with which integrity constraints can be specified. Constraints such as key uniqueness and the admissibility of null values can be specified during schema definition. Some SQL versions allow more elaborate integrity constraints to be specified. For example, constraints can be imposed on the domain of an attribute name, and any transaction that violates the imposed constraints is aborted. In some cases, it is useful to specify that the system take some actions, rather than just have the transaction responsible for the constraint violation being aborted. A mechanism called *trigger* can be used for that purpose. A trigger specifies a condition and an action to be taken when the condition is met.

Transaction and Data Integrity

In a multiuser DBMS, the database is a shared resource that can be accessed concurrently by many users. A *transaction* usually refers to the execution of a retrieval or an update program. Transactions originating from different users may be aimed at the same database records. This situation, if not carefully monitored, may cause the database to become *inconsistent*. Starting from a database in a consistent state, it is obvious that if all transactions are executed one after the other, then the database will remain in a consistent state. In a multiuser DBMS, the serial execution of transactions is wasteful of system resources. In this case, the solution is to interleave the execution of the transactions. However, the interleaving of transactions has to be performed in a way that prevents the database from becoming inconsistent. Suppose that two transactions T_1 and T_2 proceed in the following way:

```
Time            T₁                      T₂
                                        read_account(X)
          read_account(X)              X := X - 20
          X := X - 10                  write_account(X)
          write_account(X)
          read_account(Y)
          Y := Y + 10
          write _account(Y)
```

The first transaction transfers $10 from bank account X to bank account Y. The second transaction withdraws $20 from bank account X. Assume that initially there was $200 in X and $100 in Y. When the two transactions are performed serially, the final amounts in X and Y are $170 and $110, respectively. However, if the two transactions are interleaved as shown, then after the completion of both transactions, there will be $190 in X and $110 in Y. The database is now in an inconsistent state.

It is therefore important to ensure that the interleaving of the execution of transactions leaves the database in a consistent state. One way of preserving data consistency is to ensure that the interleaved execution of transactions is equivalent to their serial execution. This is referred to as *serializable* execution. Therefore, an interleaved execution of transactions is said to be serializable if it is equivalent to a serial execution.

Locking is one of the most popular approachs to achieving serializability. Locking is the process of ensuring that some actions are not performed on a data item. Therefore, a transaction may request a lock on a data item to prevent it from being either accessed or modified by other transactions. There are two basic types of locks. A *shared lock* allows other transactions to read but not write to the data item. An *exclusive lock* allows only a single transaction to read and write a data item. To achieve a high degree of concurrency, the locked data item size must be as small as possible. A data item can range from the whole database to a particular field in a record. Large data items limit concurrency, while small data items result in a large storage overhead and a greater number of lock and unlock operations that the system will have to handle.

Transactions scheduling based on locking achieves serializability in two phases. This is known as *two-phase locking*. During the first phase, the growing phase, a transaction can only lock new data items, but it cannot release any locked ones. During the second phase, the shrinking phase, existing locks can be released, but no new data item can be locked. The two-phase locking scheme guarantees the serializability of a schedule.

Because of its simplicity, the above scheduling scheme is very practical. However, it may lead to *deadlock*. A deadlock occurs when two transactions are waiting for each other to release locks and both cannot proceed. Deadlock prevention (detection) is needed to handle the situation. A deadlock prevention strategy prevents deadlock from occurring. For example, this can be achieved by requiring that a transactions locks all data items it needs for its execution before it can proceed; when the transaction finds that a needed data item is already locked, then it releases all locks. Alternatively, we may execute the transaction without deadlock prevention and check periodically for the existence of a deadlock. In this case, some of the transactions causing the deadlock are *rolled back*, i.e., the effect of these transactions is undone and the data items they locked are unlocked. Deadlock detection can be achieved by the construction of a graph in which each node represents a transaction in the schedule. If a transaction T_i is waiting for a data item that is locked by a transaction T_j, then a directed edge from T_i to T_j is drawn. The existence of a cycle in the resulting graph indicates the existence of a deadlock.

Database Security

A database needs to be protected against unauthorized access. It is the responsibility of the DBA to create account numbers and passwords for legitimate users. The DBA can also specify the type of privileges a particular account has. In relational databases, this includes the privilege to create base relations, create views, alter relations by adding or dropping a column, and delete relations. The DBA can also revoke privileges that were granted previously. In SQL, the command GRANT is used to grant privileges and the REVOKE command to revoke privileges that have been granted.

The concept of views can serve as a convenient security mechanism. Consider a relation EMPLOYEE that gives the name of an employee, date of birth, the department worked for, address, phone number, and salary. A database user who is not allowed to have access to the salary of employees from his own department can have this portion of the database hidden from him. This can be achieved by limiting his access to a view obtained from the relation EMPLOYEE by selecting only those tuples where the department attribute is different from his.

Database security can be enhanced by using *data encryption*. The idea here is to encrypt the data using some coding technique. An unauthorized user will have difficulty deciphering the encrypted data. Only authorized users are provided with keys to decipher the encoded data.

Emerging Trends

Object-Oriented Databases

Object-oriented database systems (OODBMSs) [Brown, 1991] are one of the latest trends in database technology. The emergence of OODBMS is in response to the requirements of nonstandard applications. In general, standard applications, such as commercial and administrative ones, can be effectively modeled using one of the three record-based data models. These applications are characterized by simple data types. Furthermore, for such applications, access and relationships are based on data values. Nonstandard database applications such as those found in engineering CAD/CAM require complex data structures. When these applications are modeled using the relational model, they require an excessive number of relations. In addition, a large number of join operations are usually needed to produce an answer. This has led in most cases to unacceptable performance levels.

The notion of "object" is central to OODBMS. An object can be seen as being an entity consisting of its own *private memory* and *external interface* (or protocol). The private memory is used to store the state of the object, and the external interface consists of a set of operations that can be performed on the object. An object communicates with other objects through messages sent to its external interface. When an object receives a message, it responds by using its own procedures, known as *methods*. The methods are responsible for processing the data in the object's private memory and sending messages to other objects to perform specific tasks and possibly send back appropriate results.

The object-oriented approach provides for a high level of *abstraction*. In addition, this model has constructs that can be used to define new data types and specialized operators that can be applied to them. This feature is known as *encapsulation*.

An object is usually a member of a class. The class specifies the internal structure and the external interface of an object. New object classes can be defined as a *specialization* of existing ones. For example, in a university environment, the object type "faculty" can be seen as a specialization of the object type "employee." Since a faculty is a university employee, it has all the properties of a university employee plus some of its own. For example, some of the general operations that can be performed on an employee could be "raise_salary," "fire_employee," "transfer_employee." For a faculty, specialized operations such as "faculty_tenure" could be defined. Faculty can be viewed as a subclass of employee. As a result, faculty (the subclass) will respond to the same messages as employee (the superclass) in addition to those defined specifically for faculty. This technique is known as *inheritance*. A subclass is said to inherit the behavior of its superclass.

Opponents to the object-oriented paradigm point to the fact that while this model has greater modeling capability, it lacks the simplicity and the strong theoretical foundations of the relational model. Also, the reappearance of the navigational approach is seen by many as a step backward.

Supporters of the object-oriented approach believe that a navigational approach is a necessity in several applications. They point to the rich modeling capability of the model, its high level of abstraction, and its suitability for modular design.

Distributed Databases

A distributed database [Ozsu and Valdurez, 1991] is a collection of interrelated databases spread over the nodes of a computer network. The management of the distributed database is the responsibility of a software system usually known as distributed DBMS (DDBMS). One of the tasks of the DDBMS is to make the distributed nature of the database transparent to the user. A distributed database usually reflects the distributed nature of some applications. For example, a bank may have branches in different cities. A database used by such an organization is usually distributed over all these sites. The different sites are connected by a computer network. A user may access data stored locally or access data stored at other sites through the network.

Distributed databases have several advantages. In distributed databases, the effect of a site failure or data loss at a particular node can be minimized through data replication. However, data replication reduces security and makes the process of keeping the database consistent more complicated.

In distributed databases, data is decomposed into fragments that are allocated to the different sites. A fragment is allocated to a site in a way that maximizes local use. This allocation scheme, which is known as *data localization*, reduces the frequency of remote access. In addition, since each site deals with only a portion of the database, local query processing is expected to exhibit increased performance.

A distributed database is inherently well suited for parallel processing at both interquery and intraquery levels. Parallel processing at the interquery level is the ability to have multiple queries executed concurrently. Parallelism at the intraquery level results from the possibility of a single query being simultaneously handled by many sites, each site acting on a different portion of the database.

The data distribution increases the complexity of DDBMS over a centralized DBMS. In fact, in distributed databases, several research issues in distributed query processing, distributed database design, and distributed transaction processing remain to be solved. It is only then that the potential of distributed databases can be fully appreciated.

Database Machines

The archiving, retrieval, and management of large databases constitute a complex and difficult task. Recent trends in database technology suggest that *database machines* can best undertake the efficient management of large databases [Su, 1988; Abdelguerfi, 1991]. That is, specialized hardware and software configurations aimed primarily at handling large databases and answering complex queries become an attractive solution, called a database machine.

Database machines are classified into several broad classes. Some database machines act as back-end processors. Back-end database processors depend on their host computers. The idea is to off-load DBMS functions from the host to the back-end database machine. This has the advantage of relieving the host from the time-consuming task of data retrieval, update, and enforcement of integrity constraints. RINDA is a back-end relational database processor developed at NTT. RINDA is connected to a general-purpose host (DIPS series mainframe). RINDA accelerates the processing of relational database queries. It has specialized hardware for sorting and searching. It can execute queries 3 to 100 times faster than conventional computers.

Database filters are database machines aimed at solving the so-called I/O bottleneck between the host and the secondary storage. Commands are issued by the host computer to the database filter. The database filter responds by processing records in secondary storage, and only those that meet specific conditions are transferred to the host. Therefore, only relevant records are transferred to the host. CAFS is a database filter connected to ICL mainframes. CAFS performs searching and filtering of records stored in a disk system.

Database engines represent another class of database machines. Database engines are multiprocessor systems that generally support a client/server architecture. Teradata's DBC may include over 1000 processors. Because processors can be added as needed to increase both CPU and input/output capabilities, this database engine can handle very large databases. White Cross, a start-up company, offers a database engine based on the INMOS transputer.

Defining Terms

Database: A shared pool of interrelated data.

Database computer: A special hardware and software configuration aimed primarily at handling large databases and answering complex queries.

Database management system (DBMS): A software system that allows for the definition, construction, and manipulation of a database.

Data model: An integrated set of tools to describe the data and its structure, data relationships, and data constraints.

Distributed database: A collection of multiple, logically interrelated databases distributed over a computer network

References

M. Abdelguerfi and A. K. Sood, Eds., Special Issue on Database Computers, *IEEE Micro,* December 1991.

A. Brown, *Object-Oriented Databases: Applications in Software Engineering,* New York: McGraw-Hill, 1991.

E. F. Codd, "A relational model of data for large shared data banks," *Communications of the ACM,* pp. 377–387, June 1970.

R. Elmasri and S. B. Navathe, *Fundamentals of Database Systems,* Redwood City, Calif.: Benjamin/ Cummings, 1989.

D. Maier, *The Theory of Relational Databases,* New York: Computer Science Press, 1983.

M. T. Ozsu and P. Valdurez, *Principles of Distributed Database Systems,* Englewood Cliffs, N.J.: Prentice-Hall, 1991.

Y. W. S. Su, *Database Computers: Principles, Architecture and Techniques,* New York: McGraw-Hill, 1988.

88.2 Rule-Based Expert Systems

Jay Liebowitz

Expert systems is probably the most practical application of artificial intelligence (AI). Artificial intelligence, as a field, has two major thrusts: (1) to supplement human brain power with intelligent computer power and (2) to better understand how we think, learn, and reason. Expert systems are one application of AI, and they are being developed and used throughout the world [Feigenbaum *et al.,* 1988; Liebowitz, 1990]. Other major applications of AI are robotics, speech understanding, natural-language understanding, computer vision, and neural networks.

Expert systems are computer programs that emulate the behavior of a human expert in a well-bounded domain of knowledge [Liebowitz, 1988]. They have been used in a number of tasks, ranging from sheep reproduction management in Australia, hurricane damage assessment in the Caribbean, boiler plant operation in Japan, computer configuration in the United States, to strategic management consulting in Europe [Liebowitz, 1991b]. Expert systems technology has been around since the late 1950s, but it has been only since 1980–1981 that the commercialization of expert systems has emerged [Turban, 1992].

An expert system typically has three major components: the dialog structure, inference engine, and knowledge base [Liebowitz and DeSalvo, 1989]. The dialog structure is the user interface that allows the user to interact with the expert system. Most expert systems are able to explain their reasoning, in the same manner that one would want human experts to explain their decisions. The inference engine is the control structure within the expert system that houses the search strategies to allow the expert system to arrive at various conclusions. The third component is the **knowledge base,** which is the set of facts and heuristics (rules of thumb) about the specific domain task. The knowledge principle says that the power of the expert system lies in its knowledge base. Expert system shells have been developed and are widely used on various platforms to help one build an expert system and concentrate on the knowledge base construction. Most operational expert systems are integrated with existing databases, spreadsheets, optimization modules, or information systems [Mockler and Dologite, 1992].

The most successful type of expert system is the rule-based, or production, system. This type of expert system is chiefly composed of IF-THEN (condition-action) rules. For example, the infamous MYCIN expert system, developed at Stanford University for diagnosing bacterial infections in the blood (meningitis), is rule-based, consisting of 450–500 rules. XCON, the expert system at Digital Equipment Corporation used for configuring VAX computer systems, is probably the largest rule-based expert system, consisting of over 11,000 rules. There are other types of expert systems that represent knowledge in ways other than rules or in conjunction with rules. Frames, scripts, and semantic networks are popular knowledge representation methods that could be used in expert systems.

The development of rule-based systems is typically called **knowledge engineering.** The knowledge engineer is the individual involved in the development and deployment of the expert system. Knowledge engineering, in rule-based systems, refers primarily to the construction of the knowledge base. As such, there are six major steps in this process, namely (1) problem selection, (2) knowledge acquisition, (3) knowledge representation, (4) knowledge encoding, (5) knowledge testing and evaluation, and (6) implementation and maintenance. The knowledge engineering process typically uses a rapid prototyping approach (build a little, test a little). Each of the six steps in the knowledge engineering process will be briefly discussed in turn.

Problem Selection

In selecting an appropriate application for expert systems technology, there are a few guidelines to follow:

- Pick a problem that is causing a large number of people a fair amount of grief.
- Select a "doable," well-bounded problem (i.e., task takes a few minutes to a few hours to solve)—this is especially important for the first expert system project for winning management's support of the technology.
- Select a task that is performed frequently.
- Choose an application where there is a consensus on the solution of the problem.
- Pick a task that utilizes primarily symbolic knowledge.
- Choose an application where an expert exists and is willing to cooperate in the expert systems development.
- Make sure the expert is articulate and available and a backup expert exists.
- Have the financial and moral support from management.

The problem selection and scoping are critical to the success of the expert systems project. As with any information systems project, the systems analysis stage is an essential and crucial part of the development process. With expert systems technology, if the problem domain is not carefully selected, then difficulties will ensue later in the development process.

Knowledge Acquisition

After the problem is carefully selected and scoped, the next step is knowledge acquisition. Knowledge acquisition involves eliciting knowledge from an expert or multiple experts and also using available documentation, regulations, manuals, and other written reports to facilitate the knowledge acquisition process. The biggest bottleneck in expert systems development has, thus far, been in the ability to acquire knowledge. Various automated knowledge acquisition tools, such as Boeing Computer Services' AQUINAS, have been developed to assist in this process, but there are very few knowledge acquisition tools on the market. The most commonly used approaches for acquiring/eliciting knowledge include: interviewing (structured and unstructured), protocol analysis, questionnaires (structured and open-ended), observation, learning by example/analogy, and other various techniques (Delphi technique, statistical methods).

To aid the knowledge acquisition process, some helpful guidelines are:

- Before interviewing the expert, make sure that you (as the knowledge engineer) are familiar/comfortable with the domain.
- The first session with the expert should be an introductory lecture on the task at hand.
- The knowledge engineer should have a systematic approach to acquiring knowledge.
- Incorporate the input and feedback from the expert (and users) into the system—get the expert and users enthusiastic about the project.
- Pick up manuals and documentation on the subject material.
- Tape the knowledge acquisition sessions, if allowed.

Knowledge Representation

After acquiring the knowledge, the next step is to represent the knowledge. In a rule-based expert system, the IF-THEN (condition-action) rules are used. Rules are typically used to represent knowledge if the preexisting knowledge can best be naturally represented as rules, if the knowledge is procedural, if the knowledge is mostly context-independent, and if the knowledge is mostly categorical ("yes-no" type of answers). Frames, scripts, and semantic networks are used as knowledge representation schemes for more descriptive, declarative knowledge. In selecting an appropriate knowledge representation scheme, try to use the representation method which most closely resembles the way the expert is thinking and expressing his/her knowledge.

Knowledge Encoding

Once the knowledge is represented, the next step is to encode the knowledge. Many knowledge engineers use expert system shells to help develop the expert system prototypes. Other developers may build the expert system from scratch, using such languages as Lisp, Prolog, C, and others. The following general guidelines may be useful in encoding the knowledge:

- Remember that for every shell there is a perfect task, but for every task there is NOT a perfect shell.
- Consider using an expert system shell for prototyping/proof-of-concept purposes—remember to first determine the requirements of the task, instead of force-fitting a shell to a task.
- Try to develop the knowledge base in a modular format for ease of updating.
- Concentrate on the user interface and human factors features, as well as the knowledge base.
- Use an incremental, iterative approach.
- Consider whether uncertainty should play a part in the expert system.
- Consider if the expert reasons in a data-driven manner (forward chaining) or a goal-directed manner (backward chaining), or both.

Knowledge Testing and Evaluation

Once the knowledge is encoded in the system, testing and evaluation need to be conducted. Verification and validation refers to checking for the consistency of the knowledge/logic and checking the quality/accuracy of advice reached by the expert system. Various approaches to testing can be used, such as: performing "backcasting" by running the expert system (using a representative set of test cases) against documented cases and comparing the expert system-generated results with the historical results, using blind verification tests (modified Turing test), having the expert and other experts test the system, using statistical methods for testing, and others. In evaluating the expert system, the users should evaluate the design of the human factors in the system (i.e., instructions, free-text comments, ease of updating, exiting capabilities, response time, display and presentation of conclusions, ability to restart, ability for user to offer degree of certainty, graphics, utility of the system, etc.).

Implementation and Maintenance

Once the system is ready to be deployed within the organization, the knowledge engineer must be cognizant of various institutionalization factors [Liebowitz, 1991a; Turban and Liebowitz, 1992]. Institutionalization refers to implementing and transitioning the expert system into the organization. Frequently, the technology is not the limiting factor—the *management* of the technology is often the culprit. An expert system may be accurate and a technical success, but without careful attention to management and institutionalization considerations, the expert system may be a technology transfer failure. There are several useful guidelines for proper institutionalization of expert systems:

- Know the corporate culture in which the expert system is deployed.
- Planning for the institutionalization process must be thought out well in advance, as early as the requirements analysis stage.
- Through user training, help desks, good documentation, hotlines, etc., the manager can provide mechanisms to reduce "resistance to change."
- Solicit and incorporate users' comments during the analysis, design, development, and implementation stages of the expert system.
- Make sure there is a team/individual empowered to maintain the expert system.
- Be cognizant of possible legal problems resulting from the use and misuse of the expert system.
- During the planning stages, determine how the expert system will be distributed.
- Keep the company's awareness of expert systems at a high level throughout the system's development and implementation, and even after its institutionalization.

Defining Terms

Expert systems: A computer program that emulates a human expert in a well-bounded domain of knowledge.

Knowledge base: The set of facts and rules of thumb (heuristics) on the domain task.

Knowledge engineering: The process of developing an expert system.

References

E.A. Feigenbaum, P. McCorduck, and P. Nii, *The Rise of the Expert Company,* New York: Times Books, 1988.

J. Liebowitz, *Introduction to Expert Systems,* New York: Mitchell/McGraw-Hill Publishing, 1988.

J. Liebowitz, Ed., *Expert Systems for Business and Management,* Englewood Cliffs, N.J.: Prentice-Hall, 1990.

J. Liebowitz, *Institutionalizing Expert Systems: A Handbook for Managers,* Englewood Cliffs, N.J.: Prentice-Hall, 1991a.

J. Liebowitz, Ed., *Operational Expert System Applications in the United States,* New York: Pergamon Press, 1991b.

J. Liebowitz, and D. DeSalvo, Eds., *Structuring Expert Systems: Domain, Design, and Development,* Englewood Cliffs, N.J.: Prentice-Hall, 1989.

R. Mockler and D. Dologite, *An Introduction to Expert Systems,* New York: Macmillan Publishing, 1992.

E. Turban, *Expert Systems and Applied Artificial Intelligence,* New York: Macmillan Publishing, 1992.

E. Turban and J. Liebowitz, Eds., *Managing Expert Systems,* Harrisburg, Pa.: Idea Group Publishing, 1992.

Further Information

There are several journals and magazines specializing in expert systems that should be consulted:

Expert Systems with Applications: An International Journal, New York/Oxford: Pergamon Press.

Expert Systems, Medford, N.J.: Learned Information, Inc.

IEEE Expert, Los Alamitos, Calif.: IEEE Computer Society Press.

AI Expert, San Francisco: Miller Freeman Publications.

Intelligent Systems Report, Atlanta: AI Week, Inc.

Parallel Processors

89.1	Introduction	2052
89.2	Classifications	2052
89.3	Types of Parallel Processors	2053
	Ensemble Processors • Array Processors • Associative Processor	
89.4	System Utilization	2057

Tse-yun Feng
The Pennsylvania State University

89.1 Introduction

A computer usually consists of four major components: the arithmetic-logic unit (ALU), the main memory unit (MU), the input/output unit (I/O), and the control unit (CU). Such a computer is known as a uniprocessor since the processing is achieved by operating on one word or word pair at a time. In order to increase the computer performance, we may improve the device technology to reduce the switching (gate delay) time. Indeed, for the past half century we have seen switching speeds improve from 200 to 300 ms for relays to present-day subnanosecond very large scale integration (VLSI) circuits. As the switching speeds of computer devices approach a limit, however, any further significant improvement in performance is more likely to be in increasing the number of words or word pairs that can be processed simultaneously. For example, we may use one ALU to compute N sets of additions N times in a uniprocessor, or we may design a computer system with N ALUs to add all N sets once. Conceptually, such a computer system may still consist of the four major components mentioned previously except that there are N ALUs. An organization with multiple ALUs under the control of a single CU is called a **parallel processor.** To make a parallel processor more efficient and cost-effective, a fifth major component, called **interconnection networks,** is usually required to facilitate the interprocessor and processor-memory communications. In addition, each ALU requires not only its own registers but also network interfaces; the expanded ALU is then called a **processing element** (PE). Figure 89.1 shows a block diagram of a parallel processor.

89.2 Classifications

Flynn has classified computer systems according to the multiplicity of instruction and data streams, where computers are partitioned into four groups [Flynn, 1966]:

1. Single instruction stream, single data stream (SISD): The conventional, word-sequential architecture including pipelined computers (usually with parallel ALU)
2. Single instruction stream, multiple data stream (SIMD): The multiple ALU-type architectures (e.g., parallel/array processor). The ALU may be either bit-serial or bit-parallel.

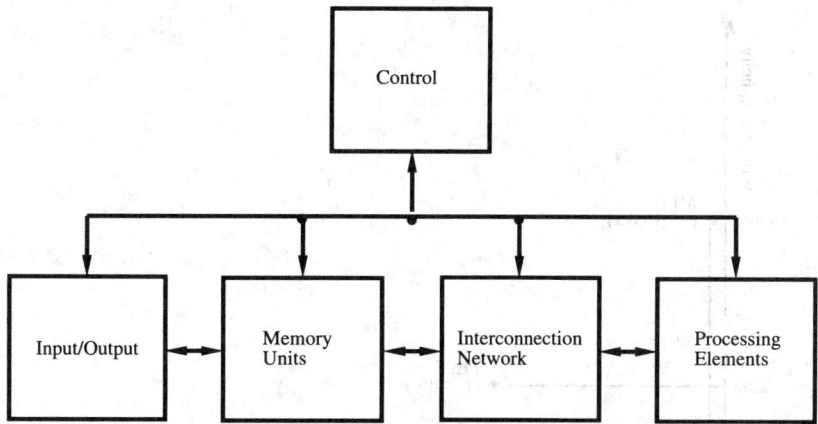

FIGURE 89.1 A basic parallel processor organization.

3. Multiple instruction stream, single data stream (MISD): Not as practical as the other classes.
4. Multiple instruction stream, multiple data stream (MIMD): The multiprocessor system.

As a general rule, one could conclude that SISD and SIMD machines are single CU systems, whereas MIMD machines are multiple CU systems. Flynn's classification does not address the interactions among the processing modules and the methods by which processing modules in concurrent system are controlled. As a result, one can classify both uniprocessors and pipelined computers as SISD machines, because both instructions and data are provided sequentially.

We may also classify computer systems according to the number of bits or bit pairs a computer executes at any instant [Feng, 1972]. For example, a computer may perform operations on one bit or bit pair at a time through the use of a simple serial ALU. For an M-bit word or operand, the operation repeats M times (Point A in Fig. 89.2). To speed up the processing, a parallel ALU is usually used so that all bits of a word can be operated on simultaneously. This is how a conventional word-sequential computer executes on its operands (Point B in Fig. 89.2). In a parallel processor, it may execute either (a) all the ith bits of N operands or operand pairs (i.e., bit slice or bis) or (b) all N M-bit operands or operand pairs simultaneously (Points C and D in Fig. 89.2, respectively). Figure 89.2 also shows some of the systems in this classification. It is seen from this classification that the performance of a computer is proportional to the total number of bits or bit pairs it can execute simultaneously.

Feng's classification [Hwang and Briggs, 1984] was originally intended for parallel processors, and as a result, the number of CUs in a computer system was not specified. Händler extended Feng's classification by adding a third dimension, namely, the number of CUs. Pipelined systems are also included in this classification through additional parameters [Händler, 1977].

89.3 Types of Parallel Processors

Ensemble Processors

An ensemble system is an extension of the conventional uniprocessor systems. It is a collection of N PEs (a PE here consists of an ALU, a set of local registers, and limited local control capability) and N MUs, under the control of a single CU. Thus, the organization of an **ensemble processor** is similar to that shown in Fig. 89.1 except that there are no direct interprocessor and processor-memory communications, i.e., no interconnection networks. When the need for communication arises, it is done through the CU. This slows down the system for applications requiring extensive interproces-

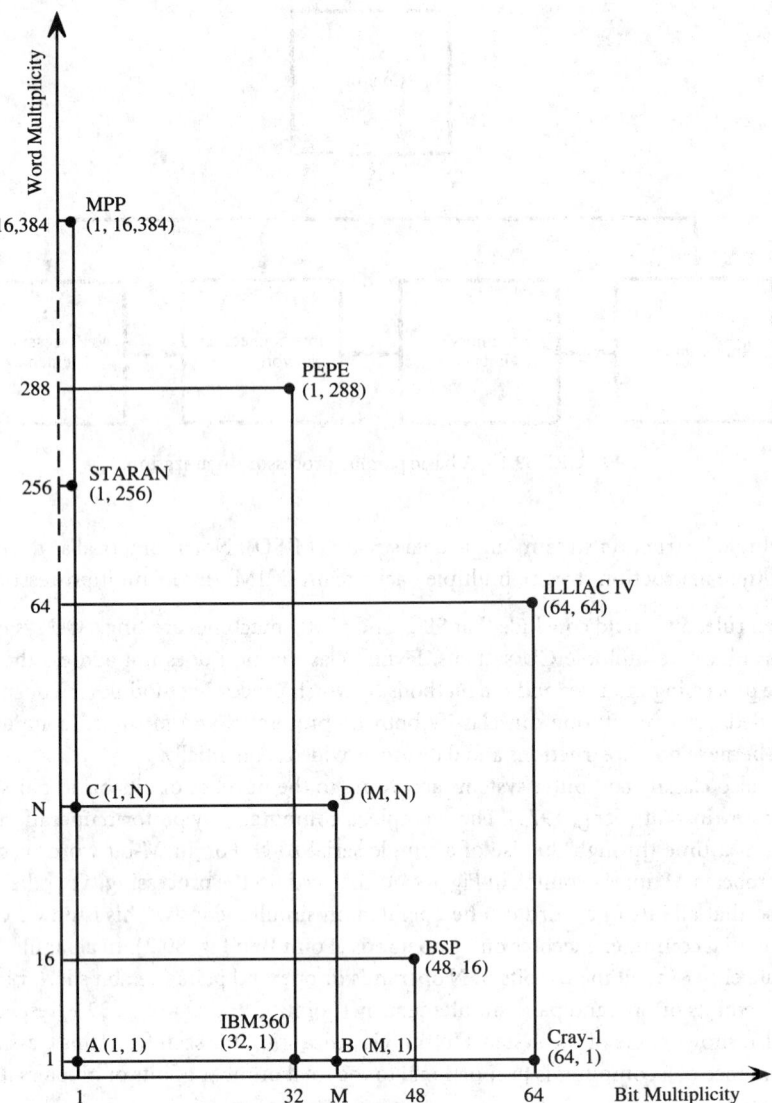

FIGURE 89.2 Feng's classification.

sor and processor-memory communications. For example, the sum of two matrices A and B can be executed in one step, if R^2 PEs are available in an ensemble processor, where R is the rank of the matrices. On the other hand, the product of the same two matrices requires extensive data alignment between the elements of A and B. As a result, it is ineffective for performing matrix multiplications with an ensemble processor. Therefore, while the ensemble processors are capable of executing up to N identical jobs simultaneously, they have very limited applications. Parallel element processing ensemble (PEPE) [Evensen and Troy, 1973] is an example of such parallel processors.

Array Processors

Because of the need for interprocessor and processor-memory communication for most applications, a parallel processor usually has one or more circuits (known as interconnection networks) to support various applications for efficient processing. In general, an **array processor** may consist of

N identical PEs under the control of a single CU and a number of MUs. Within each PE there are circuits for network interface as well as its own local memories. The PEs and MUs communicate with each other through an interconnection network. A typical array processor organization is shown in Fig. 89.3. Depending on the design, each PE may perform serial-by-bit (as in MPP) or parallel-by-bit (as in ILLIAC IV) operations.

As can be seen from Fig. 89.3, the interconnection networks play a very important role in parallel processors. The network usually provides a uniform interconnection among PEs on one hand and PEs and MUs on the other. Different array processor organizations might use different interconnection networks. In general, the interconnection networks can be classified into two categories: static and dynamic, as shown in Fig. 89.4.

ILLIAC IV [Barnes *et al.*, 1968] and MPP [Batcher, 1979] are examples of parallel processors using static interconnections, while STARAN [Batcher, 1973] and BSP [Kuck and Stokes, 1982] are examples using dynamic interconnections.

The CU usually has its own high-speed registers, local memory, and arithmetic unit. Thus, in many cases, it is a conventional computer and the instructions are stored in a main memory, together with data. However, in some machines such as ILLIAC IV, programs are distributed among the local memories of the PEs. Hence, the instructions are fetched from the processors' local memories into an instruction buffer in the CU. Each instruction is either a local type instruction, where it is executed entirely within the CU, or it is a parallel instruction and is executed in the processing array. The primary function of the CU is to examine each instruction as it is to be executed and to determine where the execution should take place.

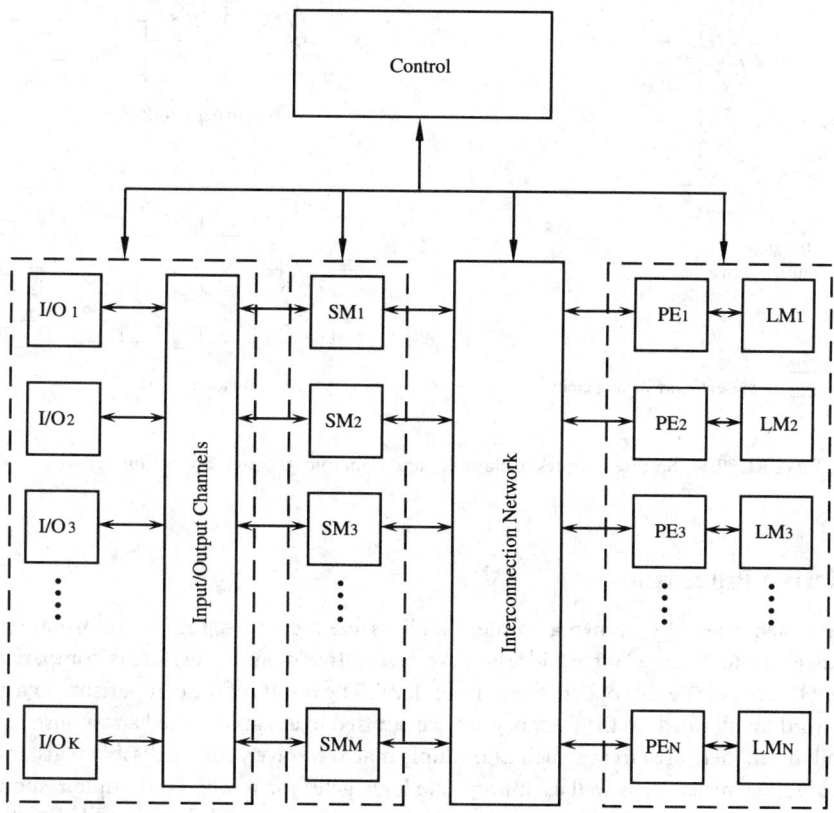

FIGURE 89.3 An array processor organization. I/O, input/output devices; LM, local memory; PE, processing element; SM, shared memory.

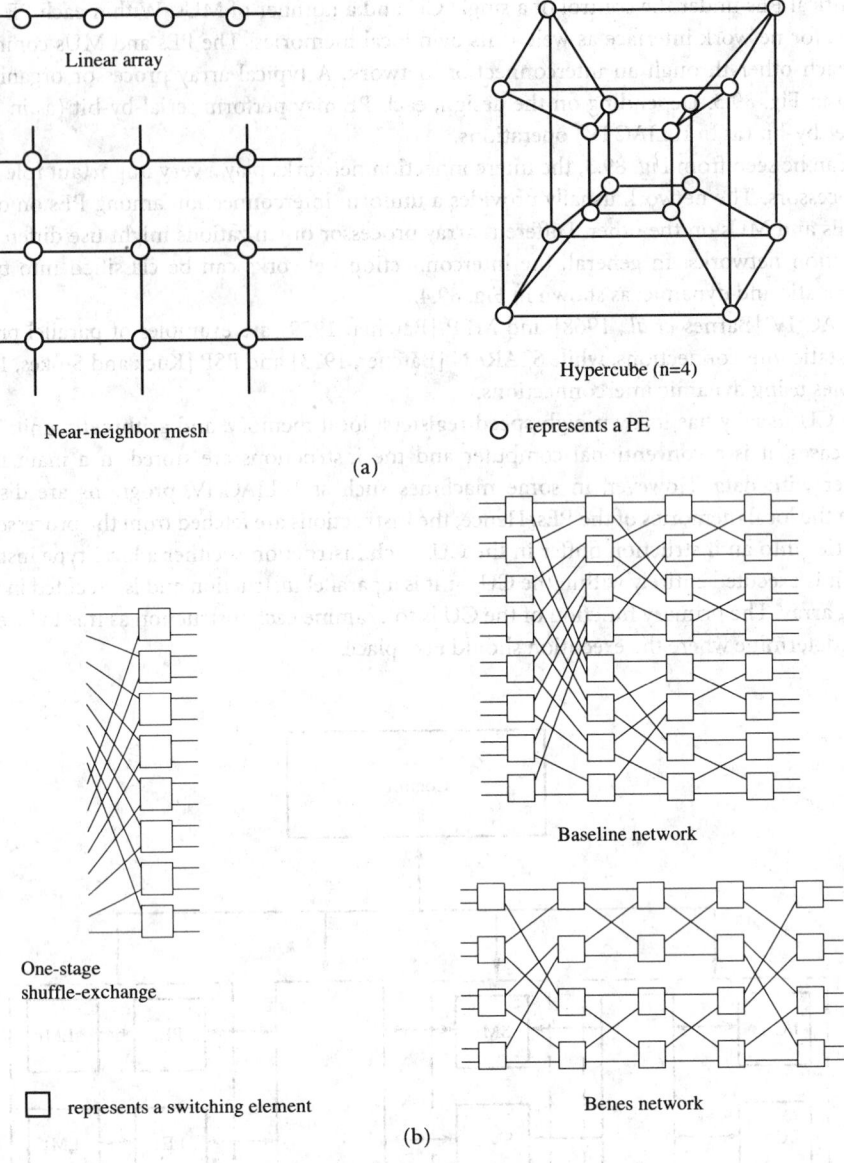

FIGURE 89.4   Some examples of static (a) and dynamic (b) interconnection networks.

Associative Processor

Associative memories, also known as content-addressable memories, retrieve information on the basis of data content rather than addresses. An associative memory performs comparison (i.e., exclusive-OR or equivalence) operations at its bit level. The results of the comparison on a group of bits in a word for all words in the memory are transmitted to a register called a response register or flag. In addition, there are circuits such as multiple match resolver, enable/disable register, a number of temporary registers, as well as appropriate logic gates for resolving multiple responses and information retrieval. For **associative processors,** arithmetic capabilities are added to this unit. The unit can be viewed as consisting of a number of bit-serial PEs. Furthermore, the bit-level logic is moved out of the memory so that the memory part of the processor consists of a number of random-access memories called word modules. A typical associative processor is shown in Fig. 89.5. STARAN and MPP (Fig. 89.2) are representative of this bit-serial, word-parallel SIMD organization. In Fig. 89.5 the common register is where the common operand is stored and the mask register defines the bit positions requiring operation. The enable/disable register provides local control of individual PEs. Because of its simplicity in design the per-PE cost of an associative processor is much lower, but the bit-serial operations slow down the system drastically. To compensate for this, these systems are useful only for applications requiring a large number of PEs.

89.4 System Utilization

As discussed previously, for any computer there is a maximum number of bits or bit pairs that can be processed concurrently, whether it is under single-instruction or multiple-instruction control [Feng, 1972, 1973]. This maximum degree of concurrency, or maximum concurrency (C_m), is an indication of the computer-processing capability. The actual utilization of this capability is indicated by the average concurrency defined to be

$$C_a = \frac{\Sigma c_i \Delta t_i}{\Sigma \Delta t_i}$$

where c_i is the concurrency at Δt_i. If Δt_i is set to one time unit, then the average concurrency over a period of T time units is

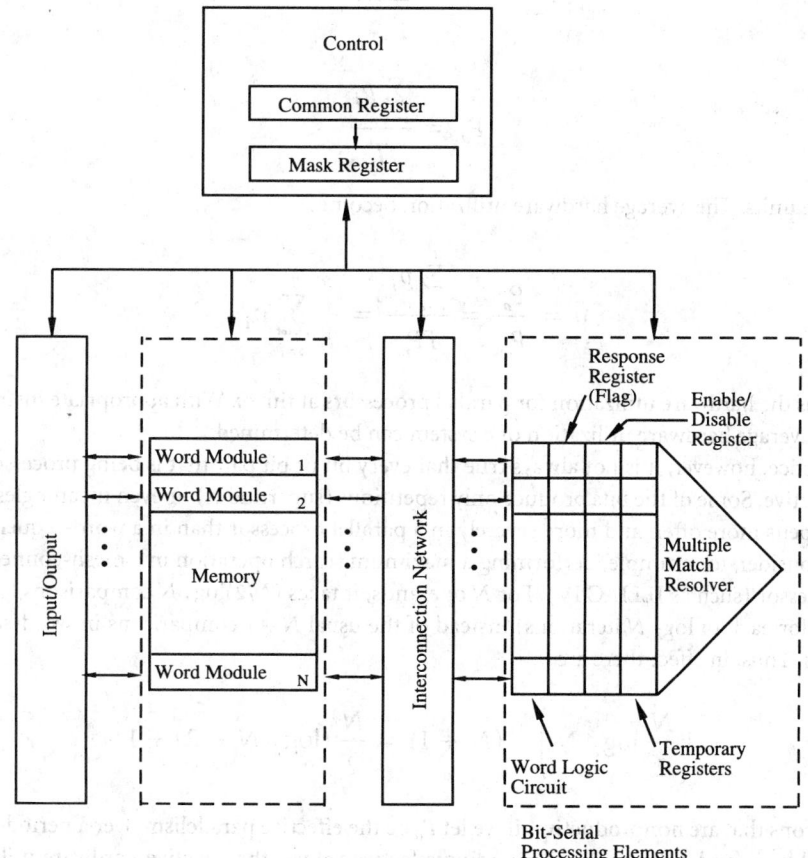

FIGURE 89.5 An associative processor organization.

$$C_a = \frac{\sum\limits_{i=1}^{T} c_i}{T}$$

The average hardware utilization is then

$$\mu = \frac{C_a}{C_m} = \frac{\sum\limits_{i=1}^{T} c_i}{TC_m} = \frac{1}{T}\sum\limits_{i=1}^{T} \sigma_i$$

where σ_i is the hardware utilization at time i. Whereas C_m is determined by the hardware design, C_a or μ is highly dependent on the software and applications. A general-purpose computer should achieve a high μ for as many applications as possible, whereas a special-purpose computer would yield a high μ for at least the intended applications. In either case, maximizing the value of μ for a computer design is important. This equation can also be used to evaluate the relative effectiveness of machine designs.

For a parallel processor, the degree of concurrency is called the degree of parallelism. A similar discussion can be used to define the average hardware utilization of a parallel processor. The maximum parallelism is then P_m, and the average parallelism is

$$P_a = \frac{\sum p_i \Delta t_i}{\sum \Delta t_i}$$

or

$$P_a = \frac{\sum\limits_{i=1}^{T} p_i}{T}$$

for T time units. The average hardware utilization becomes

$$\upsilon = \frac{P_a}{P_m} = \frac{\sum\limits_{i=1}^{T} p_i}{TP_m} = \frac{1}{T}\sum\limits_{i=1}^{T} \rho_i$$

where ρ_i is the hardware utilization for parallel processors at time i. With appropriate instrumentation, the average hardware utilization of a system can be determined.

In practice, however, it is not always true that every bit or bit pair that is being processed would be productive. Some of the bits produce only repetitious (superfluous) or even meaningless results. This happens more often and more severely in a parallel processor than in a word-sequential processor. Consider, for example, performing a maximum search operation in a mesh-connected parallel processor (such as ILLIAC IV). For N operands, it takes $(N/2)\log_2 N$ comparisons ($N/2$ comparisons for each of $\log_2 N$ iterations) instead of the usual $N-1$ comparisons in word-sequential machines. Thus, in effect there are

$$\left(\frac{N}{2}\log_2 N\right) - (N-1) = \frac{N}{2}(\log_2 N - 2) + 1$$

comparisons that are nonproductive. If we let $\hat{P}_a$ be the effective parallelism over a period of T time units and $\hat{\upsilon}$, $\hat{p}_i$, and $\hat{\rho}_i$ be the corresponding effective values, the effective hardware utilization is then

$$\hat{\upsilon} = \frac{\hat{P}_a}{P_m} = \frac{\sum\limits_{i=1}^{T} \hat{p}_i}{T P_m} = \frac{1}{T} \sum\limits_{i=1}^{T} \hat{\rho}_i$$

A successful parallel processor design should yield a high $\hat{\upsilon}$, as well as the required throughput for, at least, the intended applications. This not only involves a proper hardware and software design but also the development of efficient parallel algorithms for these applications.

Suppose T_u is the execution time of an application program using a conventional word-sequential machine, and T_c is the execution time of the same program using a concurrent system; the speed-up ratio is then defined as

$$S = \frac{T_u}{T_c}$$

Naturally, for a specific parallel organization, the speed-up ratio determines how well an application program can utilize the hardware resources. Supporting software has a direct effect on the speed-up ratio.

Defining Terms

Array processor: A parallel processor consisting of a number of processing elements, memory modules, and input/output devices as well as interconnection networks under a single control unit.

Associative processor: A parallel processor consisting of a number of processing elements, memory modules, and input/output devices under a single control unit. The capability of the processing elements is usually limited to the bit-serial operations.

Ensemble processor: A parallel processor consisting of a number of processing elements, memory modules, and input/output devices under a single control unit. It has no interconnection network to provide interprocessor or processor-memory communications.

Interconnection network: A network of interconnections providing interprocessor and processor-memory communications. It may be static or dynamic, distributed, or centralized.

Parallel processor: A computing system consisting of a number of processors, memory modules, input/output devices, and other components under the control of a single control unit. It is known to be a single-instruction-stream, multiple-data-stream (SIMD) machine.

Processing element: A basic processor consisting of an arithmetic-logic unit, a number of registers, network interfaces, and some local control facilities.

References

G. H. Barnes, R. M. Brown, M. Kato, D. J. Kuck, D. L. Slotnick, and R. A. Stokes, "The ILLIAC IV computer," *IEEE Trans. Comput.*, vol. C-7, pp. 746–757, 1968.

K. E. Batcher, "STARAN/RADCAP hardware architecture," *Proc. Sagamore Computer Conf. on Parallel Processing*, pp. 147–152, 1973.

K. E. Batcher, "MPP—A massively parallel processor," *Proc. Int. Conf. on Parallel Processing*, p. 249, 1979.

A. J. Evensen and J. L. Troy, "Introduction to the architecture of a 288-element PEPE," *Proc. Sagamore Computer Conference*, pp. 162–169, 1973.

T. Feng, "An overview of parallel processing systems," *1972 WESCON Tech. Papers*, Session 1— "Parallel Processing Systems," pp. 1–2, 1972.

T. Feng, Parallel Processing Characteristics & Implementation of Data Manipulating Functions, Technical Report RADC-TR-73-189, July 1973.

M. J. Flynn, "Very high speed computing systems," *Proc. IEEE*, vol. 54(12), pp. 1901–1909, 1966.

W. Händler, "The impact of classification schemes on computer architecture," *Proc. Int. Conf. on Parallel Processing*, pp. 7–15, 1977.

K. Hwang and F. A. Briggs, *Computer Architecture and Parallel Processing*, New York: McGraw-Hill, 1984.

D. J. Kuck and R. A. Stokes, "The Borroughs Scientific Processor (BSP)," *IEEE Trans. Comput.*, vol. C-31(5), pp. 363–376, 1982.

Further Information

Proceedings of International Conference on Parallel Processing: An annual conference held since 1972. Recent proceedings published by CRC Press.

IEEE Transactions on Parallel and Distributed Systems: Started in 1990 as a quarterly, now a monthly, published by the IEEE Computer Society.

Journal of Parallel and Distributed Computing: A monthly published by Academic Press.

Computer Architecture and Parallel Processing: A book by K. Hwang and F. A. Briggs published by McGraw-Hill.

90

Operating Systems

90.1 Introduction.. 2061
90.2 Types of Operating Systems 2062
90.3 Distributed Computing Systems................................ 2062
90.4 Fault-Tolerant Systems 2064
90.5 Parallel Processing 2065
90.6 Real-Time Systems... 2066
90.7 Operating System Structure...................................... 2067
90.8 Industry Standards 2069
90.9 Conclusions... 2070

Joseph Boykin
GTE Laboratories

90.1 Introduction

An operating system is just another program running on a computer. It is unlike any other program, however. An operating system's primary function is the management of all hardware and software resources. It manages processors, memory, I/O devices, and networks. It enforces policies such as *protection* of one program from another and *fairness* to ensure that users have equal access to system resources. It is privileged in that it is the only program that can perform specialized hardware operations. The operating system is the primary program upon which all other programs rely.

To understand modern operating systems we must begin with some history [Boykin and LoVerso, 1990]. The modern digital computer is only about 40 years old. The first machines were giant monoliths housed in special rooms, and access to them was carefully controlled. To program one of these systems the user scheduled access time well in advance, for in those days the user had sole access to the machine. The program such a user ran was the *only* program running on the machine.

It did not take long to recognize the need for better control over computer resources. This began in the mid-1950s with the dawn of batch processing and early operating systems that did little more than load programs and manage I/O devices.

In the 1960s we saw more general-purpose systems. New operating systems that provided time-sharing and **real-time computing** were developed. This was the time when the foundation for all modern operating systems was laid.

Today's operating systems are sophisticated pieces of software. They may contain millions of lines of code and provide such services as distributed file access, security, **fault tolerance**, and **real-time** scheduling. In this chapter we examine many of these features of modern operating systems and their use to the practicing engineer.

90.2 Types of Operating Systems

Different operating systems (OS) provide a wide range of functionality. Some are designed as single-user systems and some for multiple users. The operating system, with appropriate hardware support, can protect one executing program from malicious or inadvertent attempts of another to modify or examine its memory. When connected to a storage device such as a disk drive, the OS implements a **file system** to permit storage of files. The file system often includes security features to protect against file access by unauthorized users. The system may be connected to other computers via a *network* and thus provide access to remote system resources.

Operating systems are often categorized by the major functionality they provide. This functionality includes **distributed computing,** *fault tolerance,* **parallel processing,** *real-time,* and *security.* While no operating system incorporates all of these capabilities, many have characteristics from each category.

An operating system does not need to contain every modern feature to be useful. For example, MS-DOS[1] is a single-user system with few of the features now common in other systems. Indeed, this system is little more than a program loader reminiscent of operating systems from the early 1960s. Unlike those vintage systems, there are numerous applications that run under MS-DOS. It is the abundance of programs that solve problems from word processing to spreadsheets to graphics that has made MS-DOS popular. The simplicity of these systems is exactly what makes them popular for the average person.

Systems capable of supporting multiple users are termed *time-sharing* systems; the system is shared among all users, with each user having the view that he or she has all system resources available. Multiuser operating systems provide protection for both the file system and the contents of main memory. The operating system must also mediate access to peripheral devices. For example, only one user may have access to a tape drive at a time.

Fault-tolerant systems rely on both hardware and software to ensure that the failure of any single hardware component, or even multiple components, does not cause the system to cease operation. To build such a system requires that each critical hardware component be replicated at least once. The operating system must be able to dynamically determine which resources are available and, if a resource fails, move a running program to an operational unit.

Security has become more important during recent years. Theft of data and unauthorized access to data are prevented in secure systems. Within the United States, levels of security are defined by a government-produced document known as the *Orange Book.* This document defines seven levels of security, denoted from lowest to highest as *D, C1, C2, B1, B2, B3,* and *A1.* Many operating systems provide no security and are labeled D. Most time-sharing systems are secure enough that they could be classified at the C1 level. The C2 and B1 levels are similar, and this is where most secure operating systems are currently classified. During the 1990s B2 and B3 systems will become readily available from vendors. The A1 level is extremely difficult to achieve, although several such systems are being worked on.

In the next several sections we expand upon the topics of distributed computing, fault-tolerant systems, parallel processing, and real-time systems.

90.3 Distributed Computing Systems

The ability to connect multiple computers through a communications network has existed for many years. Initially, computer-to-computer communication consisted of a small number of systems performing bulk file transfers. The 1980s brought the invention of high-speed *local area networks,* or LANs. A LAN allows hundreds of machines to be connected together. New capabilities

---

[1]MS-DOS is a trademark of Microsoft, Inc.

began to emerge, such as *virtual terminals* that allowed a user to log on to a computer without being physically connected to that system. Networks were used to provide remote access to printers, disks, and other peripherals. The drawback to these systems was the software; it was not sophisticated enough to provide a totally integrated environment. Only small, well-defined interactions among machines were permitted.

Distributed systems provide the view that *all* resources from every computer on the network are available to the user. What's more, access to resources on a remote computer is viewed in the same way as access to resources on the local computer. For example, a file system that implements a directory hierarchy, such as UNIX,[2] may have some directories on a local disk while one or more directories are on a remote system. Figure 90.1 illustrates how much of the directory hierarchy would be on the local system, while user directories (shaded directories) could be on a remote system.

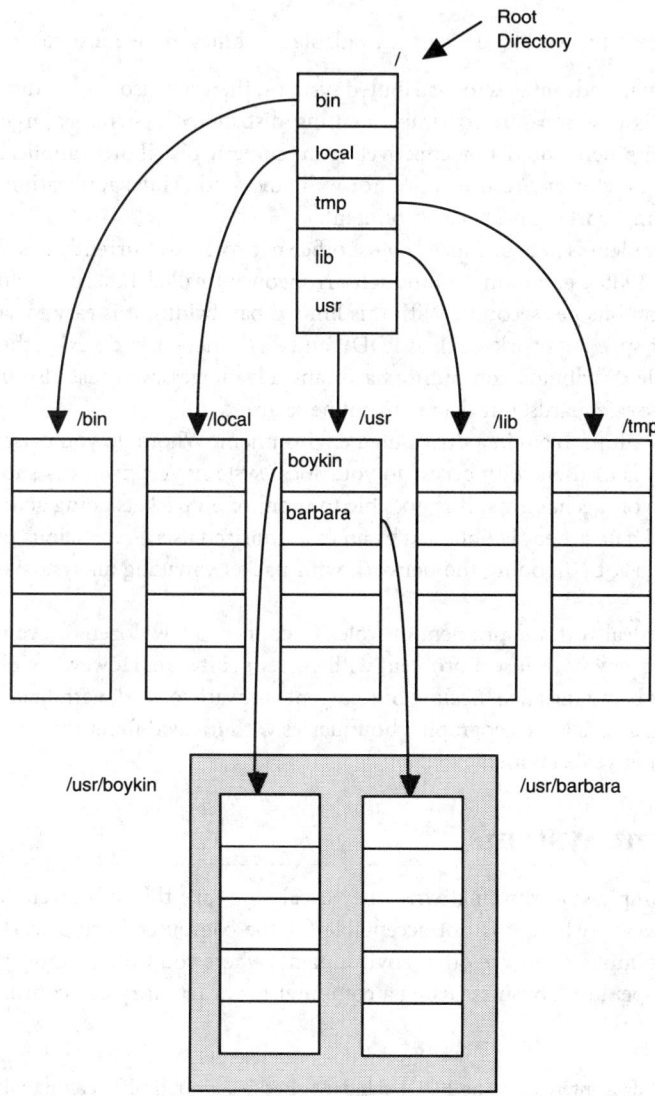

FIGURE 90.1   UNIX file system hierarchy in a distributed environment.

[2]UNIX is a trademark of UNIX Software Laboratories (USL).

There are many advantages of distributed systems. Advantages over centralized systems include [Tanenbaum, 1992]:

- *Economics:* Microprocessors offer a better price/performance than mainframes.
- *Speed:* A distributed system may have more total computing power than a mainframe.
- *Reliability:* If one machine crashes, the system as a whole can still survive.
- *Incremental growth:* Computing power can be added in small increments.

Advantages over nonnetworked personal computers include [Tanenbaum, 1992]:

- *Data sharing:* Allow many users access to a common database.
- *Device sharing:* Allow many users to share expensive peripherals like color printers.
- *Communication:* Make human-to-human communication easier, for example, by electronic mail.
- *Flexibility:* Spread the workload over the available machines in the most cost effective way.

While there are many advantages to distributed systems, there are also several disadvantages. The primary difficulty is that software for implementing distributed systems is large and complex. Small personal computers could not effectively run modern distributed applications. Software development tools for this environment are not well advanced. Thus, application developers are having a difficult time working in this environment.

An additional problem is network speed. Most office networks are currently based on IEEE standard 802.3 [IEEE, 1985], commonly (although erroneously) called Ethernet, which operates at 10 Mb/s (ten million bits per second). With this limited bandwidth, it is easy to saturate the network. While higher-speed networks such as FDDI[3] and ATM[4] networks do exist, they are not yet in common use. While distributed computing has many advantages, we must also understand that without appropriate safeguards, our data may not be secure.

Security is a difficult problem in a distributed environment. Whom do you trust when there are potentially thousands of users with access to your local system? A network is subject to security attack by a number of mechanisms. It is possible to monitor all packets going across the network; hence, unencrypted data are easily obtained by an unauthorized user. A malicious user may cause a *denial-of-service* attack by flooding the network with packets, making all systems inaccessible to legitimate users.

Finally, we must deal with the problem of scale. To connect a few dozen or even a few hundred computers together may not cause a problem with current software. However, global networks of computers are now being installed. Scaling our current software to work with tens of thousands of computers running across large geographic boundaries with many different types of networks is a challenge that has not yet been met.

90.4 Fault-Tolerant Systems

Most computers simply stop running when they break. We take this as a given. There are many environments, however, where it is not acceptable for the computer to stop working. The space shuttle is a good example. There are other environments where you would simply prefer if the system continued to operate. A business using a computer for order entry can continue to operate if

---

[3]Fiber distributed data interface. The FDDI standard specifies an optical fiber ring with a data rate of 100 Mb/s.

[4]Asynchronous transfer mode. A packet-oriented transfer mode moving data in fixed-size packets called *cells.* There is no fixed speed for ATM. Typical speed is currently 155 Mb/s, although there are implementations running at 2 Gb/s.

the computer breaks, but the cost and inconvenience may be high. Fault-tolerant systems are composed of specially designed hardware and software that are capable of continuous operation.

To build a fault-tolerant system requires both hardware and software modifications. Let's take a look at an example of a small problem that illustrates the type of changes that must be made. Remember, the goal of such a system is to achieve continuous operation. That means we can never purposely shut the computer off. How then do we repair the system if we cannot shut it off? First, the hardware must be capable of having circuit boards plugged and unplugged while the system is running; this is not possible on most computers. Second, removing a board must be detected by the hardware and reported to the operating system. The operating system, the manager of resources, must then discontinue use of that resource.

Each component of the computer system, both hardware and software, must be specially built to handle failures. It should also be obvious that a fault-tolerant system must have redundant hardware. If, for example, a disk controller should fail, there must be another controller communicating with the disks that can take over.

One problem with implementing a fault-tolerant system is knowing when something has failed. If a circuit board totally ceases operation, we can determine the failure by its lack of response to commands. Another failure mode exists where the failing component appears to work but is operating incorrectly. A common approach to detect this problem is a *voting* mechanism. By implementing three hardware replicas the system can detect when any one has failed by its producing output inconsistent with the other two. In that case, the output of the two components in agreement is used.

The operating system must be capable of restarting a program from a known point when a component on which the program was running has failed. The system can use *checkpoints* for this purpose. When an application program reaches a known state, such as when it completes a transaction, it stores the current state of the program and all I/O operations; this is known as a checkpoint. Should a component on which this program is running fail, the operating system can restart the program from the most recent checkpoint.

While the advantage of fault-tolerant systems is obvious, they come at a price. Redundant hardware is expensive, and software capable of recovering from faults runs more slowly. As with many other systems, the price may be more than offset by the advantage of continuous computing.

90.5 Parallel Processing

No matter how fast computers become, it seems they are never fast enough. Manufacturers make faster computers by decreasing the amount of time it takes to do each operation. An alternative is to build a computer that performs several operations simultaneously. A parallel computer, also called a multiprocessor, is one that contains more than one CPU.[5]

The advantage of a parallel computer is that it can run more than one program simultaneously. In a general-purpose time-sharing environment, parallel computers can greatly enhance overall system throughput. A program shares a CPU with fewer programs. This approach is similar to having several computers connected on a network but has the advantage that all resources are more easily shared.

To take full advantage of a parallel computer will require changes to the operating system [Boykin and Langerman, 1990] and application programs. Most programs are easily divided into pieces that can each run at the same time. If each of these pieces is a separate thread of control, they could run simultaneously on a parallel computer. By so dividing the application, the program may run in less time than it would on a single-processor (uniprocessor) computer.

Within the application program, each thread runs as if it were the only thread of control. It may call functions, manipulate memory, perform I/O operations, etc. If the threads do not interact with

[5]Central processing unit, the hardware component that does all arithmetic and logical operations.

each other, then, to the application programmer, there is little change other than determining how to subdivide the program. However, it would be unusual for these threads not to interact. It is this interaction that makes parallel programming more complex.

In principle, the solution is rather simple. Whenever a thread will manipulate memory or perform an I/O operation, it must ensure that it is the *only* thread that will modify that memory location or do I/O to that file until it has completed the operation. To do so, the programmer uses a *lock*. A lock is a mechanism that allows only a single thread to execute a given code segment at a time. Consider an application with several threads of control. Each thread performs an action and writes the result to a file—the same file. Within each thread we might have code that looks as follows:

```
thread()
{
        dowork();
        writeresult();
}

writeresult()
{
        lock();
        write(logfid, result, 512);
        unlock();
}
```

In this example the **writeresult** function calls the **lock** function before it writes the result and calls **unlock** afterward. Other threads simultaneously calling **writeresult** will wait at the call to **lock** until the thread that currently holds the lock calls the **unlock** function.

While this approach is simple in principle, in practice it is more difficult. It takes experience to determine how a program may be divided. Even with appropriate experience, it is more difficult to debug a multithreaded application. With several threads of control operating simultaneously, it is not simply a matter of stepping through the program line by line to find a mistake. Most often, it is the interaction between threads that is the problem.

Multithreading a program may not be a trivial matter. As with most types of programming, however, experience makes the process easier. The benefit is significantly enhanced performance.

90.6 Real-Time Systems

Real-time systems are those that guarantee that the system will respond in a predetermined amount of time. We use real-time systems when, for example, computers control an assembly line or run a flight simulator. In such an environment we define an action that must occur and a deadline by which we wish that action to take place. On an assembly line an event may occur, such as a part arriving at a station, and an action, such as painting that part. The deadline we impose will be based on the speed of the assembly line. Obviously, we must paint the part before it passes to the next station. This is called a *hard* real-time system because the system must meet a strict deadline.

Another class of system is termed *soft* real-time. These are environments in which response time is important, but the consequences are not as serious as, for example, on an assembly line. Airline reservation systems are in this category. Rapid response time to an event, such as an agent attempting to book a ticket, is important and must be considered when the system performs other activities.

One way of distinguishing hard and soft real-time systems is by examining the *value* of a response over time. For example, if a computer was controlling a nuclear reactor and the reactor began to overheat, the command to open the cooling valves has extremely high value until a deadline, when the reactor explodes. After that deadline, there is no value in opening the valves (see Fig. 90.2).

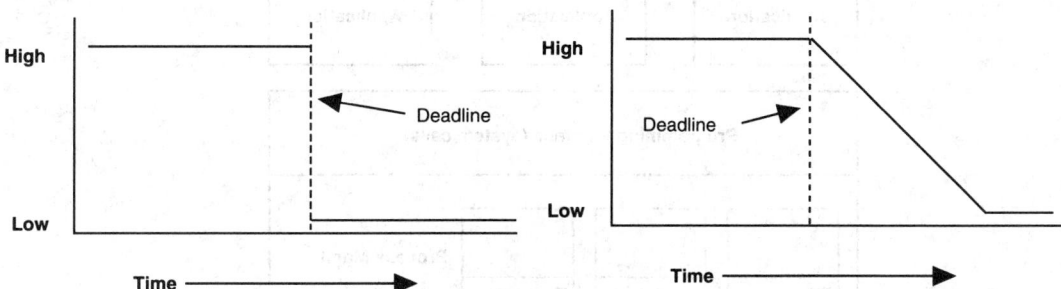

FIGURE 90.2 Relative value of a response over time in a critical situation.

FIGURE 90.3 Relative value of a response over time in a noncritical situation.

Relatively few events require that type of responsiveness. Most events have a deadline, but there continues to be value in responding to that event even past the deadline. In our airline reservation example, the airline may wish to respond to a customer request within, say, 10 seconds. However, if the response comes in 11 seconds, there is still value in the response. The value is lessened because the customer has become upset. As time increases, the customer becomes more and more upset and the value of responding decreases. We illustrate this in Fig. 90.3.

90.7 Operating System Structure

Operating systems are large, complex pieces of software. They must handle asynchronous events such as interrupts from I/O devices, control hardware memory management units (MMUs) to implement virtual memory, support multiple simultaneous users, implement complex network protocols, and much more. As with any software of this magnitude, an operating system is logically divided into smaller pieces. The structure of a typical modern operating system is depicted in Fig. 90.4.

From the user's standpoint, the operating system is a collection of system calls—the *programmers' interface*. Sometimes this is termed an *application program interface,* or API. System calls provide the mechanism for an application program to obtain services from the system. System calls exist to perform file operations such as *create, open, close, read,* and *write.* For terminals, system calls would perform such functions as changing the baud rate and number of parity bits. Network connections may be established or network protocol options, such as the size of network buffers, are also controlled through system calls.

While every operating system provides a system call interface, there is little uniformity to the appearance of that interface. Some systems provide an interface that appears as a simple function call. For example, to open a file under the UNIX operating system, we use the following system call:

```
open("/home/boykin/crc-press/oschapter", O_RDONLY);
```

Other operating systems require a user to fill in complex data structures for various operations. For example, the following code fragment illustrates how to send an IPC message using the Mach operating system's **interprocess communication** (IPC) facility [Boykin *et al.*, 1993]:

```
msg_header_t header;

header.msg_simple = TRUE;
header.msg_size = sizeof(header);
header.msg_type = MSG_TYPE_NORMAL;
```

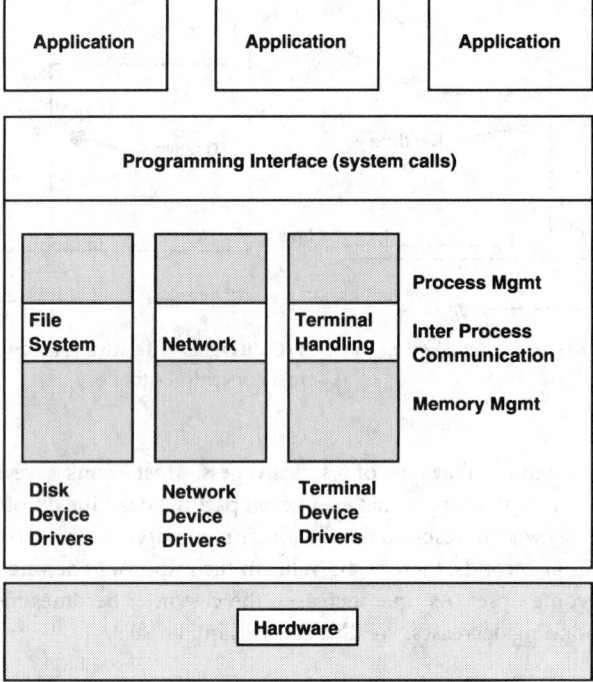

FIGURE 90.4   The structure of a modern operating system.

```
header.msg_local_port = PORT_NULL;
header.msg_remote_port = remote_port;
header.msg_id = 100;
```

Regardless of the interface format, a programmer should become familiar with the parameters, options, and return codes from each system call to use the system proficiently.

Beneath the programming interface lies the heart of the operating system. We can divide the system into two major sections. The first section directly implements the system calls. This includes the file system, terminal handling, etc. The second section provides basic capabilities upon which the rest of the system is built. Interprocess communication, memory management, and **process** management are all examples of these basic capabilities. A brief explanation of each of these sections will be given shortly.

The lowest level of the operating system interfaces directly with the computer hardware. For each physical device, such as a disk, tape, or serial line, a device driver must exist to communicate with the hardware. Device drivers accept requests to read or write data or determine the status of the device. They may do polled I/O or be interrupt driven, although polled I/O is usually only done on small personal computers. Writing a device driver requires a thorough knowledge of the hardware as well as the interface to the operating system.

In addition to I/O devices, the system must also manipulate such hardware as counters, timers and memory management units. Timers are used to satisfy user requests such as terminating an operation after a specified length of time. MMUs provide the ability to protect memory. Each time a program is run, the operating system programs the MMU with the physical memory addresses the program may access. Any attempt to access other memory is not allowed by the MMU.

An MMU is also required to implement *virtual memory.* Virtual memory allows a program to use more memory than is physically present on the machine. The operating system implements virtual memory by using an external device, typically a disk, to store portions of the program that are not currently in use. When a program attempts to access memory temporarily stored on disk, the MMU traps[6] to the operating system, which reads the memory from disk and restarts the program.

In recent years the structure depicted here has been changing. A new concept, called the *micro-kernel,* has begun to emerge. The idea behind a micro-kernel is to dramatically reduce the size of the operating system by placing most OS subsystems in the application layer. A micro-kernel would not be a usable system by itself. A number of programs would be run on top of the micro-kernel to provide such services as a file system and network protocols.

In the micro-kernel architecture shown in Fig. 90.5, notice that subsystems traditionally within the operating system are now at the same level as an application program. An application program wishing to, for example, open a file makes its request to the file system program, rather than the micro-kernel. The file system may call upon other OS subsystems or on the micro-kernel to perform an operation.

From the user standpoint, there is no programming difference between a micro-kernel structure and a traditional structure. There are two advantages of the micro-kernel approach. The first is that programming and debugging at the application layer is inherently simpler than programming at the OS layer. The benefit here is to the OS designer and implementors who can now write and debug OS code faster and easier than before. This benefits the user by having an operating system that is more reliable.

The second advantage stems from the ability to incorporate several different OS environments on top of the same micro-kernel. In this way, the computer acts as though it is running several operating systems. For example, if both MS-DOS and UNIX coexisted on the same micro-kernel, the user could choose to run an MS-DOS spreadsheet or word processor and communicate using UNIX network commands. The user has gained increased flexibility.

90.8 Industry Standards

As computer technologies come into widespread use, users begin to desire standardization. Standardization allows a user to know that a program written to a standard will work without concern

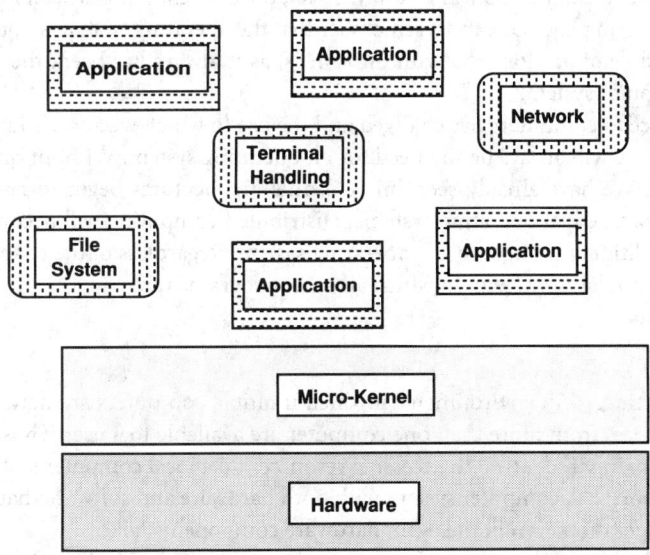

FIGURE 90.5 Micro-kernel structure.

[6]A *trap* is a hardware signal that is received by the operating system. It is very similar to an interrupt from an I/O device.

for which vendor supplies the programming environment. Operating systems are no exception to this general rule, and there are several standards, both industry standards and *de facto* standards, that apply. Porting software from one system to another, often an expensive proposition, becomes a trivial task.

Perhaps the most notable OS standard is POSIX, standard number 1003 [IEEE, 1990], sponsored by the IEEE Computer Society's Technical Committee on Operating Systems. POSIX is a family of standards based on the UNIX operating system that includes the system call interface, user-level commands, real-time extensions, and networking extensions. The POSIX system call interface, 1003.1, was adopted by the U.S. government verbatim as a Federal Information Processing Standard, FIPS 151. Many vendors conform to POSIX; thus, a program that conforms to this standard can be ported to many system platforms without change.

An example of a *de facto* standard is the X/Open Portability Guide (XPG) [X/Open, 1989]. X/Open is not a standards-setting body but is a joint initiative by members of the business community to adopt and adapt existing standards into a consistent environment. The X/Open system interface and headers are based on POSIX 1003.1 but also include extensions to POSIX-defined interfaces as well as additional interfaces.

The importance of such standards is evidenced by the strong support of such organizations as the Open Software Foundation. OSF's OSF/1 operating system conforms to various POSIX standards. Where not superseded by POSIX, it also conforms to XPG and AT&T's System V Interface Definition (SVID) [AT&T, 1985]. Conforming to these standards is considered critical for the success of OSF/1.

Some might consider an operating system such as MS-DOS to be a *de facto* standard. While MS-DOS is in common use, however, it is proprietary software subject to change without notice. Defining a standard implies an open system on which vendors and users agree.

90.9 Conclusions

I have been hearing for the past 15 years about the demise of the operating system. It has been said over and over that the role of the OS will go away. So far, the only change has been to *expand* on the role the operating system plays. One must remember that the operating system is not the user interface it portrays or the applications that run on it. It is, as it always has been, the manager of all resources on a computer system.

While the interface to computers has changed and the use to which we apply computer technology has changed, there will always be the need for an operating system. Without question, the OS will change as well. We have already seen micro-kernel architectures begin to emerge from the research labs into commercial operating systems. Distributed computing will become more widespread and force additional changes to the operating system. Regardless of the changes that come, it will always be the operating system on which all other programs rely.

Defining Terms

Distributed computing: An environment in which multiple computers are networked together and the resources from more than one computer are available to a user. Those resources are accessed in a manner identical to accessing resources on a local computer system.

Fault-tolerant systems: A computer system with both hardware and software that are capable of continuous operation even in the event hardware components fail.

File system: The logical organization of files on a storage device, typically a disk drive. The file system may support a hierarchical structure with directories and subdirectories (sometimes called folders).

Interprocess communication: The transfer of information between two cooperating programs. Communication may take the form of a *signal* (the arrival of an event) or the transfer of data.

Parallel processing: A parallel computer is one that contains more than one CPU. Parallel processing is when a program is divided into multiple threads of control, each of which is capable of running simultaneously. On a parallel computer, multiple threads could be running at the same time, thus resulting in better performance than on a uniprocessor system.

Process: A single executable program. A process is the context in which an operating system places a running program. It contains the program itself as well as allocated memory, open files, network connections, etc.

Real-time computing: Support for environments in which response time to an event must occur within a predetermined amount of time. Real-time systems may be categorized into *hard* and *soft* real-time.

References

AT&T, *System V Interface Definition,* Spring 1985, Issue 1, AT&T Customer Information Center, Indianapolis, Indiana.

J. Boykin, D. Kirschen, A. Langerman, and S. LoVerso, *Programming Under Mach,* Reading, Mass.: Addison-Wesley, 1993.

J. Boykin and A. Langerman, "Mach/4.3BSD: Parallelization without reimplementation," *Computing Systems Journal,* vol. 3, no. 1, 1990.

J. Boykin and S. LoVerso, "Recent developments in operating systems," *Computer,* vol. 23, no. 5, 1990.

H. M. Dietel, *Operating Systems,* 2nd ed., Reading, Mass.: Addison-Wesley, 1990.

IEEE, *Carrier Sense Multiple Access with Collision Detection (CSMA/CD) Access Method and Physical Layer Specifications,* American National Standard ANSI/IEEE Std. 802.3, 1985.

IEEE, *Information Technology—Portable Operating System Interface (POSIX) Part 1: System Application Program Interface (API) [C Language],* New York: IEEE, 1990.

A. Silberschatz, J. L. Peterson, and P. B. Galvin, *Operating Systems Concepts,* 3rd ed., Reading, Mass.: Addison-Wesley, 1991.

A. S. Tanenbaum, *Modern Operating Systems,* Englewood Cliffs, N.J.: Prentice-Hall, 1992.

X/Open Portability Guide, X/Open Company Ltd., Englewood Cliffs, N.J.: Prentice-Hall, 1989.

Further Information

Many textbooks describe operating system concepts. The three cited in the reference section [Dietel, 1990; Silberschatz *et al.,* 1991; and Tanenbaum, 1992] are excellent. The IEEE Computer Society has a number of tutorials on operating system related topics such as fault tolerance, real-time, local area networks and distributed processing. Readers should contact the Computer Society Press office at 10662 Los Vaqueros Circle, Los Alamitos, Calif. 90720. Phone: 714-821-8380.

For those interested in learning more about the implementation of specific operating systems, M. J. Bach, *The Design of the UNIX Operating System,* Prentice-Hall, 1986, describes the implementation of AT&T System V. The 4.3BSD operating system is described in Leffler *et al., The Design and Implementation of the 4.3BSD UNIX Operating System,* Addison-Wesley, 1990.

91

Computer Security

91.1	Introduction	2072
91.2	Physical Security	2074
91.3	Cryptology	2074
91.4	Software Security	2077
91.5	Hardware Security	2080
91.6	Network Security	2081
91.7	Personnel Security	2084

J. Arlin Cooper
Sandia National Laboratories

91.1 Introduction

Computer security is protection of computing assets and computer network communication assets against abuse, unauthorized use, unavailability through intentional or unintentional actions, and protection against undesired information disclosure, alteration, or misinformation. In today's environment, the subject encompasses computers ranging from supercomputers to microprocessor-based controllers and microcomputers, software, peripheral equipment (including terminals, printers), communication media (e.g., cables, antennas, satellites), people who use computers or control computer operations, and networks (some of global extent) that interconnect computers, terminals, and other peripherals.

Widespread publicity about computer crimes (losses estimated at between $300 million and $500 billion per year), **hacker** (cracker) penetrations, and **viruses** has given computer security a high profile in the public eye [1]. The same sorts of technologies that have made computers and computer network communications essential tools for information and control in almost all businesses and organizations have provided new opportunities for adversaries and for accidents or natural occurrences to interfere with crucial functions. Some of the important aspects are industrial/national espionage, loss of functional integrity (e.g., in air traffic control, monetary transfer, and national defense systems), and violation of society's desires (e.g., compromise of privacy).

Fortunately, technological developments also make a variety of controls (proactive and follow-up) available for computer security. These include personal transaction devices (e.g., **smart cards**, and **tokens**), **biometric verifiers**, **port protection devices**, encryption, authentication, and digital signature techniques using symmetrical (single-key) or asymmetrical (**public-key**) approaches, automated auditing, formal evaluation of security features and security products, and decision support through comprehensive system analysis techniques. Although the available technology is sophisticated and effective, no computer security protective measures are perfect, so the goal of prevention (security assurance) is almost always accompanied by detection (early discovery of security penetration) and penalty (denial of goal, e.g., information destruction; or response, e.g., prosecution and punishment) approaches.

The information in this section is intended to survey the major contemporary computer security threats, vulnerabilities, and controls. A general overview of the security environment is shown in Fig. 91.1. The oval in the figure contains an indication of some of the crucial concentrations of resources that exist in many facilities, including digital representations of money; representations of information about operations, designs, software, and people; hardware for carrying out (or peripheral to) computing and communications; people involved in operating the facility; utility connections (e.g., power); and interconnection paths to outside terminals and users, including hard-wired connections, modems for computer (and FAX) communication over telephone lines, and electromagnetic links (e.g., to satellite links, to ground antenna links, and to aircraft, space-craft, and missiles). Each of these points of termination is also likely to incorporate computer (or controller) processing.

Other factors implied include the threats of fire, water damage, loss of climate control, electrical disturbances (e.g., due to lightning or power loss), line taps or **TEMPEST emanations** interception, probes through known or unknown dial-up connections, unauthorized physical entry, unauthorized actions by authorized personnel, and delivery through ordinary channels (e.g., mail) of information (possibly misinformation) and software (possibly containing embedded threat programs). Also indicated is guidance for personnel about acceptable and unacceptable actions through policy and regulations. The subject breadth can be surveyed by categorizing into physical security, cryptology techniques, software security, hardware security, network security, and personnel security (including legal and ethical issues). Because of the wide variety of threats, vulnerabilities, and assets, selections of controls and performance assessment typically are guided by security-specific decision-support analyses, including risk analysis and probabilistic risk assessment (PRA).

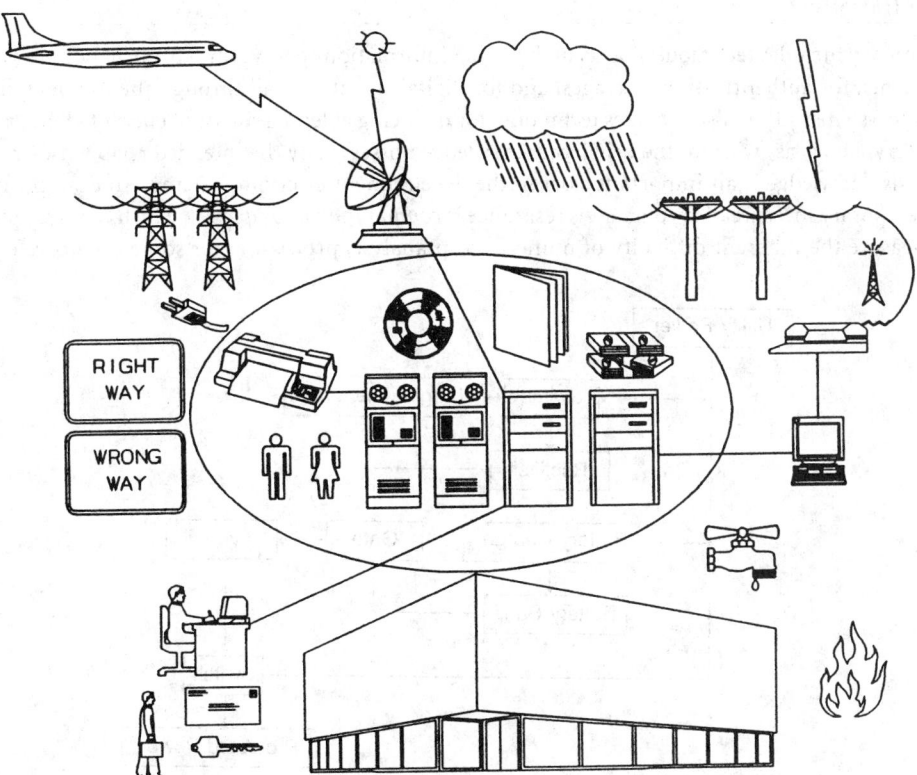

FIGURE 91.1 An overview of the computer and communications security environment.

91.2 Physical Security

Physical access security ranges from facility access control (usually through personal identification or authentication) to access (or antitheft) control for individual items (e.g., diskettes and personal computers). Techniques used generally center around intrusion prevention (or invoking a significant time delay for an adversary) and intrusion detection, which allows a response through security guard, legal or administrative action, or automatic devaluation of the penetration goal (e.g., through information destruction) [2].

Physical environmental security protects against natural threats, such as power anomalies or failures, water damage, fire, earthquake, and lightning damage, among others. An example suited to computer requirements is Halon fire suppression (although Halon use is now being restricted because of environmental concern). Note that some of the natural threats can also be adversary-caused. Since there is potential (in spite of protection) for a loss, contingency planning is essential. This includes provisions for software backup (usually off-site), hardware backup (e.g., using reciprocal agreements, hot sites, or cold sites [2]), and disaster recovery, guided by a structured team that has prepared through tests (most typically simulated).

An example of power protection technology is the widely used uninterruptible power system (UPS). An online UPS implementation is shown in Fig. 91.2. Utility power is shown passed through a switch to a rectifier and gated to an inverter. The inverter is connected to the critical load to be protected. In parallel, continuous charge for a battery bank is provided. Upon loss of utility power, the battery bank continues to run the inverter, thereby furnishing power until graceful shutdown or switching to an auxiliary engine generator can be accomplished. The switch at the lower right protects the UPS by disconnecting it from the load in case of a potentially catastrophic (e.g., short) condition.

91.3 Cryptology

Cryptology includes techniques for securely hiding information (encrypting) from all but intended recipients, for authenticating messages, and for digital signatures, all through the use of ciphers (cryptosystems) [3]. It also includes techniques for deducing at least a subset of encrypted information (cryptanalysis) without the privileged knowledge possessed by the intended recipients. Cryptanalysis knowledge is an important asset in the development of cryptosystems. An example of a contemporary measure of cryptanalysis resistance is *computational complexity,* which can be applied to measure the inherent difficulty of numeric cryptanalysis processing for some cryptosystems.

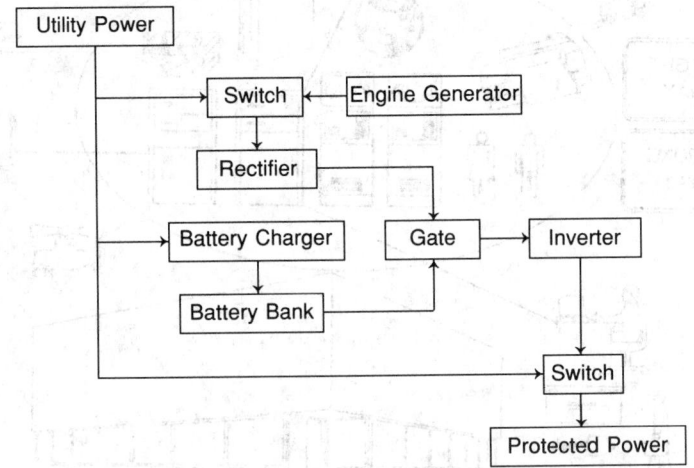

FIGURE 91.2   Uninterruptible power system.

Figure 91.3 shows the main components of cryptology. The information to be protected is called plaintext (cleartext), and protected information is called ciphertext. Adversaries can passively obtain ciphertext, or they might actively interrupt the communication link and attempt to spoof the information recipient.

Some of the objectives of encryption are secrecy, authentication (assurance to recipient of sender identity), and digital signatures (authentication plus assurance to the sender and to any third parties that the recipient could not have created the signature). As in physical security, assurance of integrity means preventing interference in the information-conveying process or, failing that, detecting interference. Here, interference may have the aims of eavesdropping, modifying, introducing misinformation, disavowing messages, and falsely claiming receipt of messages.

Almost all cryptosystems involve transformations (frequently made public and almost always assumed to be known by adversaries) of information based on one or more *keys* (see Fig. 91.3), at least one of which must be kept secret to protect against adversaries. A single-key (symmetric) cryptosystem has only one secret key, which is used to encrypt information by the sender and to decrypt information by the recipient. A prior secure process is necessary so that both sender and recipient know (and no adversary knows) the key.

The most well-known and most widely used single-key cryptosystem in history is the Data Encryption Standard (DES), published by the U.S. National Bureau of Standards [4] (now the National Institute of Standards and Technology, NIST), with National Security Agency (NSA) consultation. DES utilizes a 56-bit key (some *weak* and *semi-weak* keys are excluded) to encipher information in blocks of 64 bits. It involves substitution and permutation, linear and nonlinear transformations, and 16 successive "rounds" of key-dependent processing (general indication of logic shown in Fig. 91.4). The DES cryptosystem is identical for encryption and decryption, except that the order of application of the 16 key extractions is reversed. Like most cryptosystems of this type, DES is usually used with some form of *chaining* (mixing ciphertext or information that produces ciphertext from one block with plaintext or information that produces ciphertext in the subsequent block at the transmitter, and then inverting the process at the receiver). Three chaining techniques specified for DES (and usable in most other cryptosystems) are indicated in Fig. 91.5, along with the basic electronic codebook block form. The *k* bits shown are typically eight bits, and these are shifted into the first *k* positions of a shift-register/buffer after each encryption. Coordinated time stamps or initial values (IVs) are used to prevent identical transformation for each system start.

Although the DES key length was acceptable to most users when the standard was released in 1977, increases in computing power have made exhaustive search less expensive, so the relative security of DES has decreased. NSA now supports some of its own secret algorithms as DES replacements ("COMSEC Commercial Endorsement Program, Type II" devices), although NIST support for DES continues and no algorithmic weaknesses in DES have been publicly revealed.

Public-key cryptosystems [5] use two different keys (asymmetric systems). For example, information can be encrypted with one key and decrypted with a different (but related through a secure process) key. If the aim is secrecy, the decryption key must be secret so only the recipient can

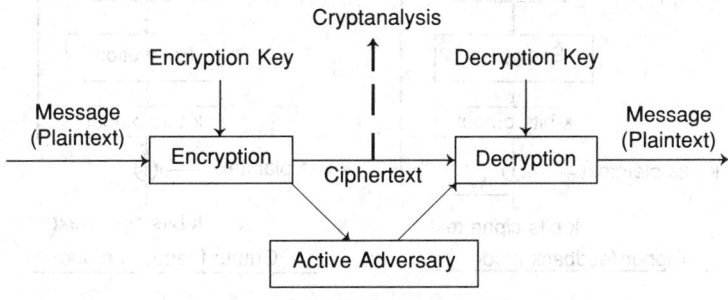

FIGURE 91.3 Basic cryptosystem functions.

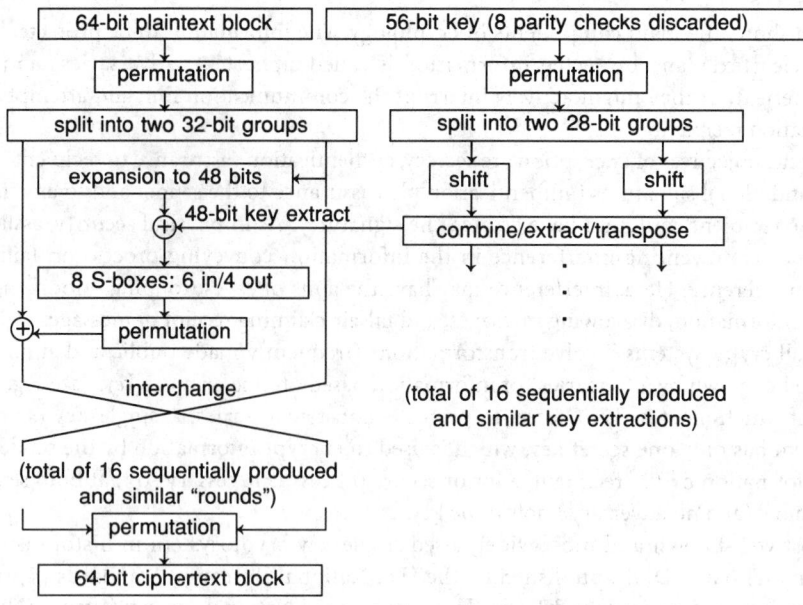

FIGURE 91.4   Basic function of the DES algorithm.

decrypt. In this case, however, the encryption key can be publicly known and known to be associated with a particular potential recipient. Although the sender can be assured of information secrecy in this process, the recipient cannot be assured of sender authenticity. If the secret key of a pair of keys is used by a sender to encrypt, any recipient who knows the sender's public key can be assured of sender authenticity, but there is no assurance of secrecy. If the public-key cryptosystem has commutative transformations (as does the RSA cryptosystem), encryption with the sender's secret key and with the recipient's public key for encipherment, and decryption by the recipient with his or her secret key and with the sender's public key provides both secrecy and authenticity.

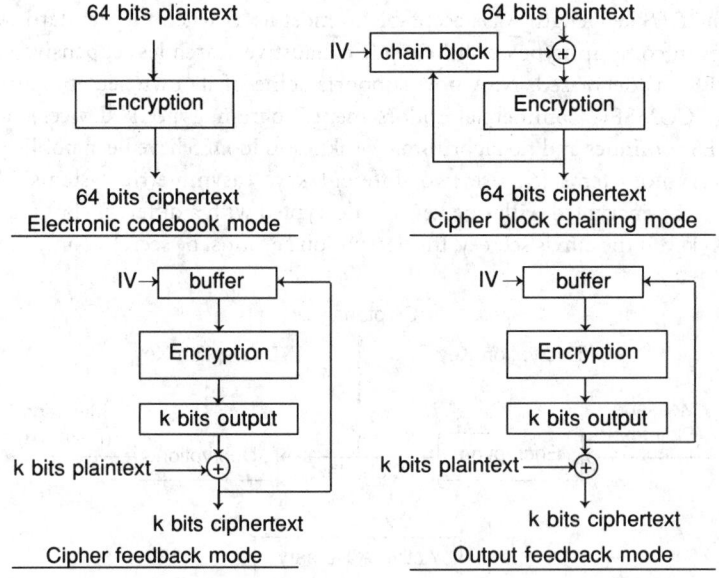

FIGURE 91.5   Modes of use for block cryptosystems.

RSA (named after Rivest, Shamir, and Adleman) is the most well known and most widely used public-key cryptosystem. Unlike DES, the key length of RSA encryption is user-selectable. However, the length chosen must be securely long (long enough that knowledge of the public key is not helpful in determining the secret key). Key selection begins with the choice of two prime numbers, each approximately 100 decimal digits long, giving about a 200-digit number on which the RSA encryption is based [Eq. (91.1)]. The security of the system depends on the difficulty of factoring large numbers that have no relatively small factors. Equation (91.2) shows how a secret modulus is determined, and Eq. (91.3) shows how the modulus is used to relate the secret key and the public key. Equation (91.4) gives the RSA encryption process, and Eq. (91.5) gives the RSA decryption process. An adversary who could factor n could use Eq. (91.2) to determine the modulus, ϕ, and then the secret key, d, from Eq. (91.3), given the public key, e.

$$n = pq \tag{91.1}$$

$$\phi = (p-1)(q-1) \tag{91.2}$$

$$ed = 1 \;(\mathrm{mod}\; \phi) \tag{91.3}$$

$$C = M^e \;(\mathrm{mod}\; n) \tag{91.4}$$

$$M = C^d \;(\mathrm{mod}\; n) \tag{91.5}$$

For equivalent security, the computational burden of RSA and similar public-key cryptosystems is significantly greater than DES and similar single-key cryptosystems. As a result, where large amounts of information must be communicated, public-key systems are frequently used for secure communication of a key intended for a single-key system, which is then in turn used for mainstream encryption.

RSA has well known cryptographic digital signature capabilities (transformed by the sender using the sender's secret key; transformed by the receiver using the sender's public key), which gives assurance that the information was initiated by the signer and that the sender cannot deny creating the information. A relatively new signature technique, Digital Signature Standard (DSS) [6], has been proposed by NIST. The basic differences between DSS and RSA are that DSS is intended only for digital signatures, DSS patents are intended to be government owned, the proposed DSS key lengths will be constrained, and the security of DSS is based on the difficulty of finding logarithms of large numbers.

91.4 Software Security

A number of techniques that are commonly implemented in software can contribute to protection against adversaries. These include password authentication; memory, file, and database access restrictions; restrictions on processing actions; development and maintenance controls; and auditing.

Passwords, which are intended to authenticate a computer user in a cost-effective way, are sometimes user-selected (a technique resulting in a relatively small potential population), sometimes user-selected from a computer-generated collection, sometimes randomly generated, and sometimes randomly generated from a phonetic construction (for pronounceability and memorization ease) [2]. Examples of phonetic passwords are TAMOTUT, OTOOBEC, SKUKOMO, ALTAMAY, and ZOOLTEE. These five were each chosen from a different phonetic construction (five of the approximately 25 commonly used).

Security control can be physical, temporal, logical, or procedural. Two important logical or procedural control principles are part of fundamental multilevel security (multiple levels of sensitivity and multiple user clearance levels on the same system), as described by part of the Bell–La Padula

HISTORY OF FACTORING

Because of the importance of factoring to RSA security, factoring methodology and accomplishments are of considerable interest. Techniques for factoring "hard" numbers were available for only up to about 50 digits in about a day's computing time until 1983, when a match between mathematical development (the quadratic sieve) and computer vector processing capabilities contributed to factoring up to 58-digit numbers in equivalent time. The next year, a 69-digit number was factored in about 32 hours on a Cray 1S. A few months later, a 71-digit number was factored in less than 10 hours on a Cray XMP. By the end of the decade, collections of small computers had been coupled in a worldwide effort to demonstrate that numbers of more than 100 (116 in 1991) digits could be cost-effectively factored. This explosive trend (Fig. 91.6), although not expected to continue because of current mathematical limitations (at present many orders of magnitude more computation time is needed than would threaten 200-digit numbers), demonstrates the importance of factoring prognosis in forecasting the long-term security of RSA.

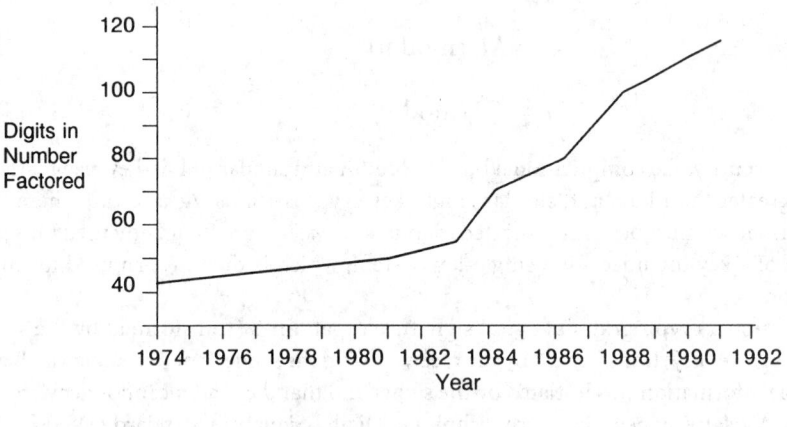

FIGURE 91.6   Factoring history.

model. The simple security principle restricts users of a particular clearance level from reading information that is of a more sensitive (more highly classified) level. The star property prohibits information flow from the level at which its sensitivity has been determined to any lower level (write-down). Analogous integrity protection is provided by the Biba integrity model [7].

Protection rules can be mandatory (used mainly by the government or military) or discretionary (compartmented according to need-to-know regimes of trust typically determined by file owners). The combination of security levels and protection rules at the same level can be associated with a lattice model. In addition to matching the security controls, the lattice model facilitates mathematical verification of security implementations.

A common logical protection rule specification gives the rights of subjects (action initiators) to act on objects (action targets) at any particular time. One way to view these rules (although seldom implemented in this manner) is to consider an access matrix (Table 91.1) containing rows for subject indicators and columns for object indicators. The matrix entries are the *rights* of subjects to objects. Actual implementation may differ, e.g., by using directories, or *capability lists,* or *capability tokens* (row designations for rights of subjects) or *access control lists* (column designation for rights to objects).

These types of rules can be augmented by software (and/or hardware) memory protection through techniques including fences, base/bounds registers, tagged registers, and paging [7].

Database management system (DBMS) security and integrity protections include access controls but generally require finer granularity and greater protection (especially for relational databases) against subtle forms of information deduction such as inference and aggregation. Integrity protection mechanisms include field checks, change logs, two-phase updates, error protection codes, range comparisons, and query controllers [8]. Secrecy depends on access control (e.g., file passwords), query controllers, and encryption.

Processing restrictions can, in addition to those implied by memory, file, and database controls, limit the ability of users to, for example, try multiple passwords or multiple user IDs; make financial transactions; change security parameters; move, rename, or output information; and deliver covert channel information (signaling systematically using authorized actions to codify unauthorized data delivery).

Software development and maintenance controls include standards under which programs (including security features) are designed to meet requirements, coded in structured or modular form, reviewed during development, tested, and maintained. Configuration or change control is also important. Computer auditing is intended to provide computer records about user actions for routine review (a productive application for expert systems) and for detailed investigation of any incidents or suspicious circumstances. It is essential that audit records be tamper-proof.

Software security features (including auditing) can be provided as part of the computer operating system or they can be added to an operating system as an add-on product. A U.S. government multilevel *trusted computing base* development program through NSA's National Computer Security Center (NCSC) resulted in a well known security methodology and assessment scheme for these types of software (and hardware) products [9]. A significant number of operating systems and software security packages have been evaluated and given ratings by NCSC, in addition to hardware–software combinations, encryption devices, and network security systems. The basic evaluation determines the degree of confidence that the system will be resistant to external penetration and internal unauthorized actions. The most secure systems known are classified A1 and utilize a reference monitor (checking every request for access to every resource), a security kernel (concentration of all security-related functions into a module that facilitates protection and validation), and protection against covert channels. Formal analysis is used to assure that the implementation correctly corresponds to the intended security policy. There is an operational efficiency penalty associated with secure multilevel operating systems.

Other classes (in order of progressively fewer security features, which results in decreasing security) are B3, B2, B1, C2, C1, and D (see Table 91.2, where security features generally accumulate, reading up from the table bottom).

In addition to computer activity directly controlled by personnel, a family of software threats can execute without direct human control. These techniques include the **Trojan horse**, the virus, the **worm**, the **logic bomb**, and the **time bomb**. The virus and worm (because they copy themselves and spread) are both capable of global-spanning attacks over relatively short time frames. Protection against these threats includes limiting user threats through background screening, using expert system software scanners that search for adversarial program characteristics, comparators, and authenticators or digital signatures that facilitate detection of software tampering.

Table 91.1 An Access Matrix

Subjects/Objects	O_1	O_2	O_3	O_4	O_5
S_1	Own, write, read	Own, read, execute	Own, read, delete	Read, write, execute	Read
S_2		Read	Execute		Read
S_3	Write		Read		Read

Table 91.2   NCSC Security Evaluation Ratings

Class Name	Summary of Salient Features
Class A1	Formal top-level specification and verification of security features, trusted software distribution, covert channel formal analysis
Class B3	Tamper-proof kernelized security reference monitor (tamper-proof, analyzable, testable), structured implementation
Class B2	Formal security model design, covert channel identification and tracing, mandatory controls for all resources (including communication lines)
Class B1	Explicit security model, mandatory (Bell–La Padula) access control, labels for internal files and exported files, code analysis and testing
Class C2	Single-level protection for important objects, log-in control, auditing features, memory residue erasure
Class C1	Controlled discretionary isolation of users from data, authentication, testing
Class D	No significant security features identified

Other software-intensive threats include trapdoors, superzapping, browsing, asynchronous attacks, and the salami attack [2]. These all usually involve unauthorized actions by authorized people and are most effectively counteracted by insider personnel controls (see Section 91.7, "Personnel Security").

91.5 Hardware Security

In addition to personal authentication through something known (e.g., passwords or PINs), users can be authenticated through something possessed or by something inherent about the user (or by combinations of the three). Hardware devices that contribute to computer security using the approach of something possessed include tokens and smart cards. Biometric verifiers authenticate by measuring human characteristics. Other hardware security devices include encryptor/decryptor units and port protection devices (to make dial-up attacks by hackers more difficult). A generic diagram depicting some of these applied to control of users is shown in Fig. 91.7. The controls can be used individually or in various combinations.

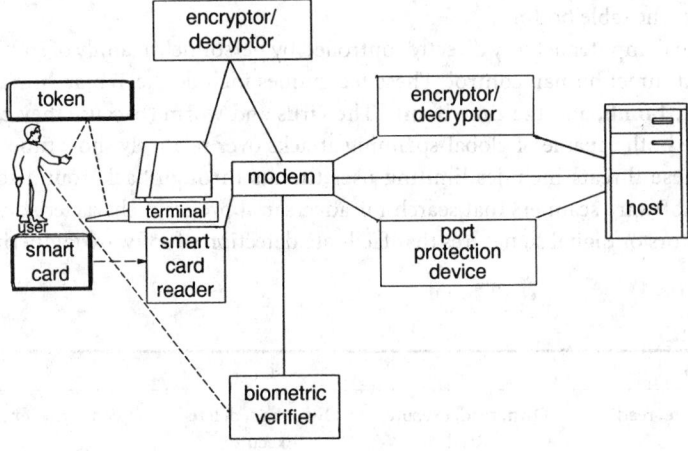

FIGURE 91.7   Depiction of hardware controls.

Tokens are devices that can be hand-carried by authorized computer users and are intended to increase password security by assuring that passwords are used only once, thereby reducing the vulnerability to password compromise. The devices contain an internal algorithm, which either works in synchronization with an identical algorithm in the host computer or transforms an input derived from a computer prompt into a password that matches the computer-transformed result. In order to protect against loss, most also require a user password for token access.

Smart cards are credit-card-sized devices intended to facilitate secure transactions, such as credit card purchases, purchases or cash withdrawals that result in bank account debits, or information interchanges. The most common application uses a card reader/network that exchanges data with the smart card over a serial data bus. User information and security information are stored in encrypted form in the card, and physical access to the internal card circuitry is protected by tamper-proof (self-destructive) sealing. Use of the card is controlled by password access.

Because of the vulnerability of passwords to compromise by disclosure or various forms of information tapping, and because of the vulnerability of loss of carried items (e.g., ROM keys, magnetic stripe cards), biometric devices have been developed to measure human characteristics in ways that are resistant to counterfeiting. These devices include signature verifiers (for examining the velocity, acceleration, and pressure characteristics imparted during signing as a function of time), fingerprint and palmprint readers (for examining print pattern characteristics, for example, with the flesh applied to a glass platen), voice verifiers (which evaluate speech characteristics, usually in response to system prompts), hand geometry (including some three-dimensional aspects), eye retina vessel pattern examination (through infrared reflection), and typing rhythm assessment (for user keyboard inputs).

Systematic cracker attacks on dial-up computer ports frequently include searches for modem tones followed by attempts to guess passwords. In response, port protection devices (PPDs) enhance dial-up security. The basic feature of many PPDs is that no modem tone is provided until an additional security barrier (or barriers) is overcome. Most PPDs require a code before computer port connection. Some also identify the user by the code entered, disconnect the call, and dial the number at which the user is expected to be (typically using a separate line to avoid dial-in intercept of the outgoing call).

Personal computer (PC) security is of contemporary interest because these relatively new tools have contributed to a set of security vulnerabilities that differs substantially from conventional computer security concerns. For example, PC users may be more naive about security in general, PC hardware and software and administrative controls are generally more primitive, the PC physical environment is generally less controlled, and PCs are generally more easily misused (e.g., company PCs used for personal benefit).

An additional hardware security topic is associated with TEMPEST (a program to assess the potential for data processing equipment to inadvertently generate "compromising emanations" that convey information to a surreptitious remote sensor). Although originally of concern because of requirements to protect government and military classified data, industrial espionage is now also a concern. Various forms of protection can be used, such as electromagnetic shielding, physical separation of processing equipment from potential adversary locations, fiber-optic communication, and encrypted data transmission. Some commercial equipment has been certified by NSA to have low emanations.

91.6 Network Security

Many business, informational, and scientific interchanges take place nationally and internationally over networks under computer control. Management of network security is exacerbated by physical dispersal and security philosophy disparity. For example, network adversaries may be harder to identify and locate than local computer adversaries [For an interesting account of overcoming this problem, see Ref. 10]. As another example, a user at a location that would not allow some form of activity (e.g., copying information from one level of security to a lower level) might find a network connection to a facility for which the activity was accepted. The intended local restriction might thereby be circumvented by conducting the restricted activity at the more permissive location. Opportunities for passive interception (tapping or emanations), or for active spoofing (involving misinformation, replay, etc.), or for disruption (including jamming) are also generally greater owing to network utilization.

There are many network topologies, but they can be decomposed into four basic canonical types (Fig. 91.8). The star topology has been traditionally used in centrally controlled networks (e.g., the control portion of the telephone system), and security is typically within the central control. Use of star topology in local-area networks (LANs) is increasing. Mesh topology is not readily amenable to central control but is well tailored to protect wide-area network integrity. Mesh topology accommodates variable-path routing schemes, such as packet transmission. The bus topology is commonly used in complex physically constrained systems, such as computing and communication processor interconnection, and missiles and aircraft. The ring and bus topologies are frequently used in LANs. The shared communication media for LANs can jeopardize secrecy unless communications are encrypted.

Network security considerations include secrecy, integrity, authenticity, and covert channels. Potential controls include cryptosystems (for secrecy, integrity, and authentication); error-protection codes, check sums (and other "signatures"), and routing strategies (for integrity); protocols (for key distribution and authentication); access control (for authentication); and administrative procedures (for integrity and covert channel mitigation). Where encryption is used, network key distribution can be difficult if the physical dimensions of the network are large. Note that several techniques described under hardware security (smart cards, tokens, biometrics, PPDs) are useful for network authentication.

Various network security approaches have been used; they can be basically classified into centralized security and distributed security (although the use of combinations of the two is common). It is difficult to maintain effective centralized administrative control in a network larger than a LAN, because of the logistics of maintaining current security and authentication data. Network efficiency, reliability, and integrity are also limited by the performance of the central security controller. The major weaknesses of distributed control are associated with inconsistent security enforcement and the security-relevant communication burden.

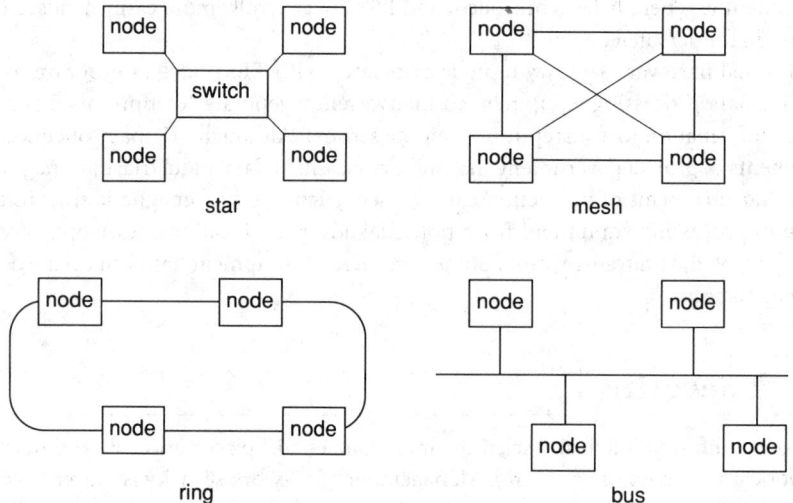

FIGURE 91.8 Basic network topologies.

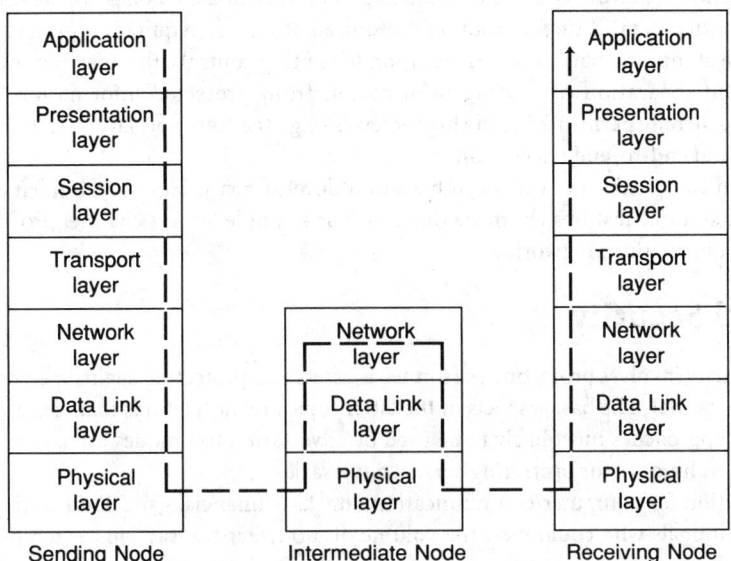

FIGURE 91.9 The ISO network protocol model.

Networks frequently comprise networks of networks (internets), using *bridges* or *routers* for protocol pass-through and filtering and *gateways* for protocol translation and buffering. Bridges, routers, and gateways may also have the role of distributed network security controllers.

Various security protocols (orderly coordinated sequences of steps) are used to authenticate network users to each other. The basic purpose of security protocols is to assure all of the parties involved that the parties with whom they are communicating are behaving as expected. Protocols can be arbitrated, adjudicated, or self-enforcing [8]. Analogously, cryptographic *sealing* and cryptographic *time stamps* (or other sequence identifiers) can prevent message (or message segment) reuse. These approaches can authenticate the message contents as well as the communicating party. Secure key distribution protocol (e.g., by a network key server or key distribution center) is an important application for protocols.

Network communication involves several levels of nonsecurity protocol for the purpose of allowing users to communicate with integrity. For this purpose, a number of network standards have been developed (e.g., TCP/IP, ISDN, GOSIP). An important example is the International Standards Organization Open Systems Interconnection model (OSI). The basic OSI structure is shown in Fig. 91.9.

The physical layer mediates access to the transmission medium. Network systems such as token ring, token bus, and carrier sense multiple access with collision detection (CSMA/CD) work at this level. The data link layer can be used to enhance transmission quality through framing, error correction, and check sums. Link (point-to-point) encryption is typically implemented in hardware at this level. The network layer handles network routing functions. This is the highest layer necessary for an intermediate node, as shown in the figure. Correct routing (and protection for pass-through information from users at the intermediate node) is important to security. The transport layer provides end-to-end (initial source to final destination) interprocess communication facilities. End-to-end encryption can be implemented in software at this level. The session layer manages the overall network connection during a user activity. This connectivity is also important to security. The presentation layer converts between user syntax and network syntax, and the application layer provides user services such as electronic mail and data file transfers. Encryption can be implemented at either of the latter two layers.

Link encryption requires exposure of information at intermediate nodes. While this is essential (at least for routing data) if further routing is required, it may be required to protect information from exposure at intermediate nodes. In addition to routing controls, this may require end-to-end encryption and separation of routing information from protected information. End-to-end encryption is generally performed at the higher levels (e.g., the transport layer). It is not unusual to use both link and end-to-end encryption.

Like isolated computers, networks can have multilevel security. However, implementation and verification of security features are more difficult. For example, covert channel protection is currently quite problematic in networks.

91.7 Personnel Security

Personnel security involves protecting personnel as assets and protecting against personnel because of the threat potential. The basic aspects of the latter topic are motivations that cause various types of threats, the approaches most likely to be used by adversaries, techniques for assessing the threat potential, and techniques for protecting against adversaries.

One motivation for computer/communication attacks is financial gain. This motivation ranges from career criminals who could view the volume of monetary transactions as tempting, and the potential detachment from personnel confrontations as affording low risk of apprehension, to basically honest people who have trouble resisting what they consider to be less serious temptations. Industrial espionage is one of many examples of financial motivation that may result in attempts to evade computer security control.

Another important motivation is information gain or information modification, which could represent no direct financial gain. Some people are curious about information to which they have no right. Some want to modify information (e.g., grades, criminal records, medical records, personnel records) because it reflects an image they want to change.

The motivation of causing personal or public outrage in order to advance a cause is a common motivation. Terrorism is an example, and many acts of terrorism against computers have occurred [2], especially in Europe. Sometimes the cause is related to revenge, which may be manifested through vandalism.

Espionage activities can be motivated by financial gain, national or company loyalty, blackmail, or even love. Hackers are frequently motivated by the challenge of overcoming security barriers. Usually, self-image and image with peers (e.g., through electronic bulletin board proclamations of breakthroughs) is a strong factor.

Personnel adversaries most commonly choose what they perceive to be the easiest and/or the safest avenue to achieve the desired objective. This is analogous to looking for the weakest link to break a chain. Because these avenues may be either inherent or unknown, security barrier uniformity is frequently sought through the application of basic principles (e.g., separation of duties). Some adversaries are motivated enough and skilled enough to use ingenious approaches, and these provide a warning about what unexpected approaches might succeed. One of the most interesting and informative examples was the break by William Friedman of the Vernam cipher used by the Germans in the Second World War [2]. Another was the use of an unintended electronic mail program feature that allowed an adversary to plant a Trojan horse in a privileged area of a computer, which resulted in system privileges when the computer ran routine periodic housekeeping. This was the genesis of the title of the book *The Cuckoo's Egg* [10]. The same adversary was one of several known to have broken one-way transformed password encryption by downloading the transform and the transformed outputs to his own computer and then exhaustively encrypting a dictionary of words and potential passwords, noting where matches to transformed outputs were obtained.

Approaches used to assess the types of unexpected attacks that might be used are mainly to catalog past approaches that have been used and to foresee new approaches through adversarial simulation. An example of this simulation is the "tiger team" approach, where a collection of personnel of various backgrounds synergistically brainstorm approaches, at least some of which may be tested or actually carried out. Tiger teams have a long history of finding unexpected approaches that can be successful, and thereby identifying protection needs.

Protection against personnel attacks generally falls into two categories: protection against insiders (those having some authorization to use resources) and against outsiders (those who gain access to resources in unauthorized ways). Some protective measures are tailored to one or the other of these two groups; some are applicable in either case.

Typical protections against unauthorized insider activities include preemployment screening and background investigation, polygraph examinations (within legal limits), administrative controls (e.g., security plans and access logs), routine examination and monitoring of activities through audit records, ethics and motivational training, and the threat of legal or job-related punishment for improper activities. The ethics of computer use varies from organization to organization because of society's inexperience in weighing the moral aspects of the topic.

Protection against outsiders includes physical and logical access control using the various forms of hardware and software authentication discussed previously and the threat of legal prosecution for transgressions. This threat depends in large measure on the available laws covering computer security violations. Computer security laws and law enforcement have traditionally been weak relative to laws covering other types of activities (largely because of new legal aspects associated with computing), but a large number of legal approaches are now possible because of laws enacted during the past two decades.

Computer laws or laws that apply to computing include the Copyright Act of 1976 (amended in 1980 to allow software copyrights and help protect against **software piracy**), the Patent Act of 1980 (adding firmware and software coverage), "shrinkwrap licenses" (some legal protection, some deterrent), the Computer Crime Statute of 1984 (applicable to U.S. government computers), the Privacy Act of 1974 (for U.S. government applications and similar to privacy laws in a number of other countries), the National Security Decision Directive (NSDD) 145 (for NSA-enhanced protection of "sensitive unclassified" information, largely intended to prevent technology drain to unfriendly countries), the Computer Security Act of 1987 (restoring NIST as the primary agency responsible for sensitive unclassified security), the Right to Financial Privacy Act of 1978, the Freedom of Information Act of 1966, the Electronic Funds Transfer Act, the Fair Credit Reporting Act, the Crime Control Act, the Electronic Communications Privacy Act, the Computer Fraud and Abuse Act, and the Foreign Corrupt Practices Act.

There is now considerable interest in international legal computer communication agreements, especially among countries that interchange significant amounts of computer data. International forums have brought many countries together with the intents of regulatory commonality and transborder data communication control.

Defining Terms

Biometric verifiers: Devices that help authenticate by measuring human characteristics.

Hacker: Person who explores computer and communication systems, usually for intellectual challenge, commonly applied to those who try to circumvent security barriers (crackers).

Logic bomb: Destructive action triggered by some logical outcome.

Port protection device: Device in line with modem that intercepts computer communication attempts and requires further authentication.

Public-key cryptosystem: System that uses a pair of keys, one public and one private, to simplify the key distribution problem.

Smart cards: Credit-card-sized devices containing a microcomputer, used for security-intensive functions such as debit transactions.

Software piracy: Unauthorized copying of software for multiple uses, thereby depriving software vendor of sales.

TEMPEST emanations: Electromagnetic, conductive, etc. leakage of information that can be recovered remotely.

Time bomb: Destructive action triggered by computer calendar/clock reading.

Token: Device that generates or assists in generation of one-time security code/passwords.

Trojan horse: Implanted surreptitious code within an authorized program.

Virus: A self-replicating program that is inserted in an application program or other executable routine.

Worm: Self-replicating code that, once initiated, consumes memory resources.

References

1. K. Hafner and J. Markoff, *Cyberpunk*, New York: Simon and Schuster, 1991.
2. J. A. Cooper, *Computer and Communications Security*, New York: McGraw-Hill, 1989.
3. G. J. Simmons, *Contemporary Cryptology*, Piscataway, N.J.: IEEE Press, 1992.
4. "Data Encryption Standard," National Bureau of Standards FIPSPUB 46, Washington, D.C., January 1977.
5. W. Diffie and M. Hellman, "New directions in cryptography," *IEEE Trans. Inf. Theory*, vol. IT-22, November 1976.
6. "Digital Signature Standard (Draft)," National Institute of Standards and Technology FIPSPUB XX, Washington, D.C., August 1991.
7. M. Gasser, *Building a Secure Computer System*, New York: Van Nostrand Reinhold, 1988.
8. C. P. Pfleeger, *Security in Computing*, Englewood Cliffs, N.J.: Prentice-Hall, 1989.
9. "Department of Defense Trusted Computer System Evaluation Criteria," DOD 5200.28-STD, December 1985.
10. C. Stoll, *The Cuckoo's Egg*, New York: Doubleday, 1989.

Further Information

B. Bloombecker, *Spectacular Computer Crimes*, Homewood, Ill.: Dow-Jones-Irwin, 1990.

P. J. Denning, *Computers Under Attack*, New York: ACM Press, 1990.

D. E. Robling Denning, *Cryptography and Data Security*, Reading, Mass.: Addison-Wesley, 1982.

P. Fites, P. Johnston, and M. Kratz, *Computer Virus Crisis*, New York: Van Nostrand Reinhold, 1989.

J. L. Mayo, *Computer Viruses*, Blue Ridge Summit, Pa.: Windcrest, 1989.

National Research Council, *Computers at Risk*, Washington, D.C.: National Academy Printers, 1991.

E.F. Troy, "Security for Dialup Lines," U.S. Department of Commerce, Washington, D.C., 1986.

"Trusted Network Interpretation," NCSC-TG-005, Ft. George G. Meade, Md.: National Computer Security Center, 1987.

92

Computer Reliability

92.1	Introduction	2087
92.2	Definitions of Failure, Fault, and Error	2087
92.3	Failure Rate and Reliability	2088
92.4	Relationship between Reliability and Failure Rate	2089
92.5	Mean Time to Failure	2090
92.6	Mean Time to Repair	2090
92.7	Mean Time between Failures	2090
92.8	Availability	2091
92.9	Calculation of Computer System Reliability	2091
92.10	Markov Modeling	2092
92.11	Software Reliability	2093
92.12	Reliability Calculations for Real Systems	2094

Chris G. Guy
University of Reading

92.1 Introduction

This chapter outlines the knowledge needed to estimate the **reliability** of any electronic system or subsystem within a computer. The word *estimate* was used in the first sentence to emphasize that the following calculations, even if carried out perfectly correctly, can provide no guarantee that a particular example of a piece of electronic equipment will work for any length of time. However, they can provide a reasonable guide to the probability that something will function as expected over a given time period. The first step in estimating the reliability of a computer system is to determine the likelihood of failure of each of the individual components, such as resistors, capacitors, integrated circuits, and connectors, that make up the system.

92.2 Definitions of Failure, Fault, and Error

A *failure* occurs when a system or component does not perform as expected. Examples of failures at the component level could be a base-emitter short in a transistor somewhere within a large integrated circuit or a solder joint going open circuit because of vibrations. If a component experiences a failure, it may cause a fault, leading to an error, which may lead to a system failure.

A *fault* may be either the outward manifestation of a component failure or a design fault. Component failure may be caused by internal physical phenomena or by external environmental effects such as electromagnetic fields or power supply variations. *Design faults* may be divided into two classes. The first class of design fault is caused by using components outside their rated specification. It should be possible to eliminate this class of faults by careful design checking. The second class, which is characteristic of large digital circuits such as those found in computer systems, is caused by the designer not taking into account every logical condition that could occur during system operation. All computer systems have a software component as an integral part of their operation, and software is especially prone to this kind of design fault.

A fault may be permanent or transitory. Examples of *permanent faults* are short or open circuits within a component caused by physical failures. *Transitory faults* can be subdivided further into two classes. The first, usually called *transient faults*, are caused by such things as alpha-particle radiation or power supply variations. Large random access memory circuits are particularly prone to this kind of fault. By definition, a transient fault is not caused by physical damage to the hardware. The second class is usually called *intermittent faults*. These faults are temporary but reoccur on a regular basis. They are caused by loose physical connections between components or by components used at the limits of their specification. Intermittent faults often become permanent faults after a period of time. A fault may be *active* or *inactive*. For example, if a fault causes the output of a digital component to be stuck at logic 1, and the desired output is logic 1, then this would be classed as an inactive fault. Once the desired output becomes logic 0, then the fault becomes active.

The consequence for the system operation of a fault is an error. As the error may be caused by a permanent or by a transitory fault, it may be classed as a *hard error* or a *soft error*. An error in an individual subsystem may be due to a fault in that subsystem or to the propagation of an error from another part of the overall system.

The terms *fault* and *error* are sometimes interchanged. The term *failure* is often used to mean anything covered by these definitions. The definitions given here are those in most common usage.

Physical faults within a component can be characterized by their external electrical effects. These effects are commonly classified into *fault models*. The intention of any fault model is to take into account every possible failure mechanism, so that the effects on the system can be worked out. The manifestation of faults in a system can be classified according to the likely effects, producing an *error model*. The purpose of error models is to try to establish what kinds of corrective action need be taken in order to effect repairs.

92.3 Failure Rate and Reliability

An individual component may fail after a random time, so it is impossible to predict any pattern of failure from one example. It is possible, however, to quantify the rate at which members of a group of identical components will fail. This rate can be determined by experimental means using accelerated life tests. In a normal operating environment, the time for a statistically significant number of failures to have occurred in a group of modern digital components could be tens or even hundreds of years. Consequently, the manufacturers must make the environment for the tests extremely unfavorable in order to produce failures in a few hours or days and then extrapolate back to produce the likely number of failures in a normal environment. The **failure rate** is then defined as the number of failures per unit time, in a given environment, compared with the number of surviving components. It is usually expressed as a number of failures per million hours.

If $f(t)$ is the number of components that have failed up to time t, and $s(t)$ is the number of components that have survived, then $z(t)$, the *failure rate* or *hazard rate,* is defined as

$$z(t) = \frac{1}{s(t)} \cdot \frac{df(t)}{d(t)} \tag{92.1}$$

Most electronic components will exhibit a variation of failure rate with time. Many studies have shown that this variation follows a common pattern, shown in Fig. 92.1. For obvious reasons this is known as a *bathtub* curve. The first phase, where the failure rate starts high but is decreasing with time, is where the components are suffering infant mortality; in other words, those that had manufacturing defects are failing. This is often called the *burn-in* phase. The second part, where the failure rate is roughly constant, is the useful life period of operation for the component. The final part, where the failure rate is increasing with time, is where the components are starting to wear out.

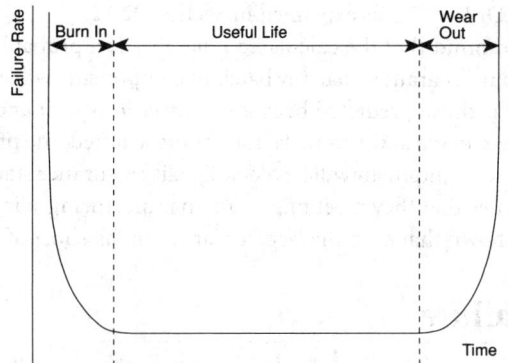

FIGURE 92.1   Variation of failure with time.

Using the same nomenclature as before, if:

$$s(t) + f(t) = N \tag{92.2}$$

i.e., N is the total number of components in the test, then the *reliability* $r(t)$ is defined as

$$r(t) = \frac{s(t)}{N} \tag{92.3}$$

or in words, and using the definition from the *IEEE Standard Dictionary of Electrical and Electronic Terms,* reliability is the probability that a device will function without failure over a specified time period or amount of usage, under stated conditions.

92.4 Relationship between Reliability and Failure Rate

Using Eqs. (92.1), (92.2), and (92.3) then

$$z(t) = -\frac{N}{s(t)} \cdot \frac{dr(t)}{d(t)} \tag{92.4}$$

λ is commonly used as the symbol for the failure rate $z(t)$ in the period where it is a constant, i.e., the useful life of the component. Consequently, we may write Eq. (92.4) as

$$\lambda = -\frac{1}{r(t)} \cdot \frac{dr(t)}{d(t)} \tag{92.5}$$

Rewriting, integrating, and using the limits of integration as $r(t) = 1$ at $t=0$ and $r(t) = 0$ at $t = \infty$ gives the result:

$$r(t) = e^{-\lambda t} \tag{92.6}$$

Remember, this result is true only for the period of operation where the failure rate is a constant.

For most common components, real failure rates can be obtained from such handbooks as the American military MIL-HDBK-217E, as explained in Section 92.12.

It must also be borne in mind that the calculated reliability is a probability function based on lifetime tests. There can be no guarantee that any batch of components will exhibit the same failure rate and hence reliability as those predicted because of variations in manufacturing conditions. Even if the components were made at the same factory as those tested, the process used might have been slightly different and the equipment will be older. Quality assurance standards are imposed on companies to try to guarantee that they meet minimum manufacturing standards, but recent cases in the United States have shown that even the largest plants can fall short of these standards.

92.5 Mean Time to Failure

A figure that is commonly quoted because it gives a readier feel for the system performance is the **mean time to failure** or MTTF. This is defined as

$$\text{MTTF} = \int_0^\infty r(t)\, dt \qquad (92.7)$$

Hence, for the period where the failure rate is constant:

$$\text{MTTF} = \frac{1}{\lambda} \qquad (92.8)$$

92.6 Mean Time to Repair

For many computer systems it is possible to define a **mean time to repair** (MTTR). This will be a function of a number of things, including the time taken to detect the failure, the time taken to isolate and replace the faulty component, and the time taken to verify that the system is operating correctly again. While the MTTF is a function of the system design and the operating environment, the MTTR is often a function of unpredictable human factors and, hence, is difficult to quantify. Figures used for MTTR for a given system in a fixed situation could be predictions based on the experience of the reliability engineers or could be simply the maximum response time given in the maintenance contract for a computer. In either case, they will be subject to wild fluctuations. To take an extreme example, if the service engineer has a flat tire while on the way to effect the repair, then the repair time may be many times the predicted MTTR. For some systems no MTTR can be predicted, as they are in situations that make repair impossible or uneconomic. Computers in satellites are a good example. In these cases and all others where no errors in the output can be allowed, fault tolerant approaches must be used in order to extend the MTTF beyond the desired system operational lifetime.

92.7 Mean Time between Failures

For systems where repair is possible, a figure for the expected time between failures can be defined as

$$\text{MTBF} = \text{MTTF} + \text{MTTR} \qquad (92.9)$$

The definitions given for MTTF and MTBF are the most commonly accepted ones. In some texts, MTBF is wrongly used as mean time before failure, confusing it with MTTF. In many real systems, MTTF is very much greater than MTTR, so the values of MTTF and MTBF will be almost identical, in any case.

92.8 Availability

Availability is defined as the probability that the system will be functioning at a given time during its normal working period.

$$Av = \frac{\text{total working time}}{\text{total time}} \tag{92.10}$$

This can also be written as

$$Av = \frac{\text{MTTF}}{\text{MTTF} + \text{MTTR}} \tag{92.11}$$

Some systems are designed for extremely high availability. For example, the computers used by AT&T to control its telephone exchanges are designed for an availability of 0.9999999, which corresponds to an unplanned downtime of 2 min in 40 years. In order to achieve this level of availability, fault tolerant techniques have to be used from the design stage, accompanied by a high level of monitoring and maintenance.

92.9 Calculation of Computer System Reliability

For systems that have not been designed to be fault tolerant it is common to assume that the failure of any component implies the failure of the system. Thus, the system failure rate can be determined by the so-called parts count method. If the system contains m types of component, each with a failure rate λ_m, then the system failure rate λ_s can be defined as

$$\lambda_s = \sum_1^m N_m \cdot \lambda_m \tag{92.12}$$

where N_m is the number of each type of component.

The system reliability will be

$$r_s(t) = \prod_1^m N_m \cdot r_m \tag{92.13}$$

If the system design is such that the failure of an individual component does not necessarily cause system failure, then the calculations of MTTF and $r_s(t)$ become more complicated.

Consider two situations where a computer system is made up of several subsystems. These may be individual components or groups of components, e.g., circuit boards. The first is where failure of an individual subsystem implies system failure. This is known as the series model and is shown in Fig. 92.2. This is the same case as considered previously, and the parts count method, Eqs. (92.12) and (92.13), can be used. The second case is where failure of an individual subsystem does not imply system failure. This is shown in Fig. 92.3. Only the failure of every subsystem means that the system has failed, and the system reliability can be evaluated by the following method. If $r(t)$ is the reliability (or probability of not failing) of each subsystem, then $q(t) = 1 - r(t)$ is the probability of an individual subsystem failing. Hence, the probability of them all failing is

$$q_s(t) = \left[1 - r(t)\right]^n. \tag{92.14}$$

for n subsystems.

Hence the system reliability will be:

$$r_s(t) = 1 - \left[1 - r(t)\right]^n \tag{92.15}$$

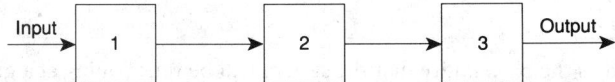

FIGURE 92.2   Series model.

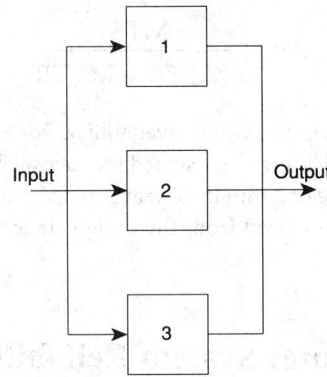

FIGURE 92.3   Parallel model.

In practice, systems will be made up of differing combinations of parallel and series networks; the simplest examples are shown in Figs. 92.4 and 92.5.

Parallel-Series System

Assuming that the reliability of each subsystem is identical, then the overall reliability can be calculated thus. The reliability of one unit is r; hence the reliability of the series path is r^n. The probability of failure of each path is then $q = 1 - r^n$. Hence, the probability of failure of all m paths is $(1 - r^n)^m$, and the reliability of the complete system is

$$r_{ps} = 1 - (1 - r^n)^m \tag{92.16}$$

Series-Parallel System

Making similar assumptions, and using a similar method, the reliability can be written as

$$r_{sp} = \left[1 - (1 - r)^n\right]^m \tag{92.17}$$

It is straightforward to extend these results to systems with subsystems having different reliabilities and in different combinations. It can be seen that these simple models could be used as the basis for a fault tolerant system, i.e., one that is able to carry on performing its designated function even while some of its parts have failed.

92.10 Markov Modeling

Another approach to determining the probability of system failure is to use a Markov model of the system, rather than the combinatorial methods outlined previously. Markov models involve the defining of *system states* and *state transitions.* The mathematics of Markov modeling are well

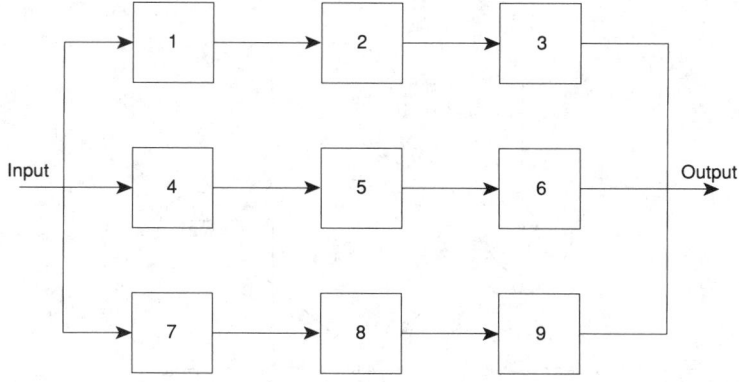

FIGURE 92.4   Parallel series model.

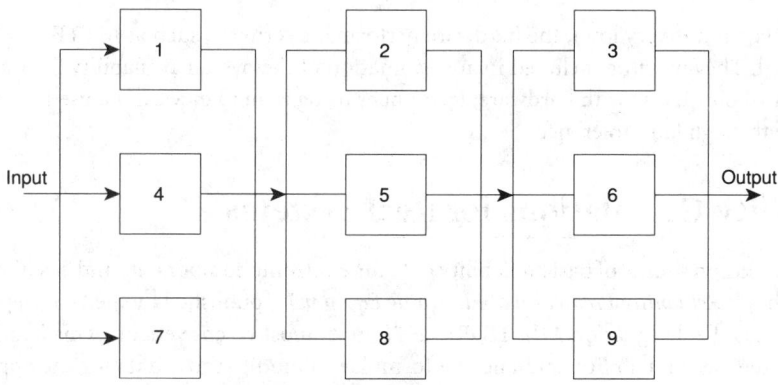

FIGURE 92.5   Series-parallel model.

beyond the scope of this brief introduction, but most engineering mathematics textbooks will cover the proofs.

To model the reliability of any system it is necessary to define the various fault-free and faulty states that could exist. For example, a system consisting of two identical units (A and B), either of which has to work for the system to work, would have four possible states. They would be (1) A and B working; (2) A working, B failed; (3) B working, A failed; and (4) A and B failed. The system designer must assign to each state a series of probabilities that determine whether it will remain in the same state or change to another after a given time period. This is usually shown in a state diagram, as in Fig. 92.6. This model does not allow for the possibility of repair, but this could easily be added.

92.11 Software Reliability

One of the major components in any computer system is its software. Although software is unlikely to wear out in a physical sense, it is still impossible to prove that anything other than the simplest of programs is totally free from bugs. Hence, any piece of software will follow the first and second parts of the normal bathtub curve (Fig. 92.1). The burn-in phase for hardware corresponds to the early release of a complex program, where bugs are commonly found and have to be fixed. The useful life phase for hardware corresponds to the time when the software can be described as stable, even though bugs may still be found. In this phase, where the failure rate can be characterized as

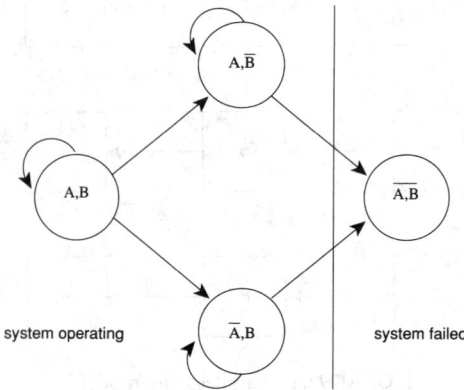

FIGURE 92.6   State diagram for two-unit parallel system.

constant (even if it is very low), the hardware performance criteria, such as MTTF and MTTR can be estimated. They must be included in any estimation of the overall availability for the computer system as a whole. Just as with hardware, techniques using redundancy can be used to improve the availability through fault tolerance.

92.12 Reliability Calculations for Real Systems

The most common source of basic reliability data for electronic components and circuits is the military handbook *Reliability Prediction of Electronic Equipment,* published by the U.S. Department of Defense. It has the designation MIL-HDBK-217E in its most recent version. This handbook provides both the basic reliability data and the formulae to modify those data for the application of interest. For example, the formula for predicting the failure rate, λ_p, of a bipolar or MOS microprocessor is given as

$$\lambda_p = \pi_Q(C_1\pi_T\pi_V + C_2\pi_E)\pi_L \text{ failures per } 10^6 \text{ hours}$$

where π_Q is the part quality factor, with several categories, ranging from a full mil-spec part to a commercial part; π_T is the temperature acceleration factor, related to both the technology in use and the actual operating temperature; π_V is the voltage stress derating factor, which is higher for devices operating at higher voltages; π_E is the application environment factor (the handbook gives figures for many categories of environment, ranging from laboratory conditions up to the conditions found in the nose cone of a missile in flight); π_L is the device learning factor, related to how mature the technology is and how long the production of the part has been going on; C_1 is the circuit complexity factor, dependent on the number of transistors on the chip; and C_2 is the package complexity, related to the number of pins and the type of package.

The following figures are given for a 16-bit microprocessor, operating on the ground in a laboratory environment, with a junction temperature of 51°C. The device is assumed to be packaged in a plastic, 64-pin dual in-line package and to have been manufactured using the same technology for several years:

$$\pi_Q = 20 \qquad \pi_T = 0.89 \qquad \pi_V = 1 \qquad \pi_E = 0.38$$
$$\pi_L = 1 \qquad C_1 = 0.06 \qquad C_2 = 0.033$$

Hence, the failure rate λ_p for this device, operating in the specified environment, is predicted to be 1.32 failures per 10^6 hours. To calculate the predicted failure rate for a system based around this microprocessor would involve similar calculations for all the parts, including the passive components, the PCB, and connectors, and multiplying all the resultant failure rates together. The resulting figure could then be inverted to give a predicted MTTF. This kind of calculation is repetitive, tedious, and therefore prone to errors, so many companies now provide software to perform the calculations. The official Department of Defense program for automating the calculation of reliability figures is called ORACLE. It is regularly updated to include all the changes since MIL-HDBK-217E was released. Versions for VAX/VMS and the IBM PC are available from the Rome Air Defense Center, RBET, Griffiss Air Force Base, NY 13441-5700. Other software to perform the same function is advertised in the publications listed under Further Information.

Defining Terms

Availability: This figure gives a prediction for the proportion of time that a given part or system will be in full working order. It can be calculated from

$$Av = \frac{MTTF}{MTTF + MTTR}$$

Failure rate: The failure rate, λ, is the (predicted or measured) number of failures per unit time for a specified part or system operating in a given environment. It is usually assumed to be constant during the working life of a component or system.

Mean time to failure: This figure is used to give an expected working lifetime for a given part, in a given environment. It is defined by the equation

$$MTTF = \int_{0}^{\infty} r(t)\, dt$$

If the failure rate λ is constant, then

$$MTTF = \frac{1}{\lambda}$$

Mean time to repair: The MTTR figure gives a prediction for the amount of time taken to repair a given part or system.

Reliability: Reliability $r(t)$ is the probability that a component or system will function without failure over a specified time period, under stated conditions.

References

B. W. Johnson, *Design and Analysis of Fault Tolerant Digital Systems,* Reading, Mass.: Addison-Wesley, 1989.

V. P. Nelson and B. D. Carroll, *Tutorial: Fault Tolerant Computing,* Washington, D.C.: IEEE Computer Society Press, 1987.

D. K. Pradhan, *Fault Tolerant Computing, Theory and Techniques,* vols. I and II, Englewood Cliffs, N. J.: Prentice-Hall, 1986.

Further Information

The quarterly magazine *IEEE Transactions on Reliability* contains much of the latest research on reliability estimation techniques.

The monthly magazine *Microelectronics and Reliability* covers the field of reliability estimation and also includes papers on actual measured reliabilities.

Sometimes, manufacturers make available measured failure rates for their devices, although they may only provide them if the answer is good.

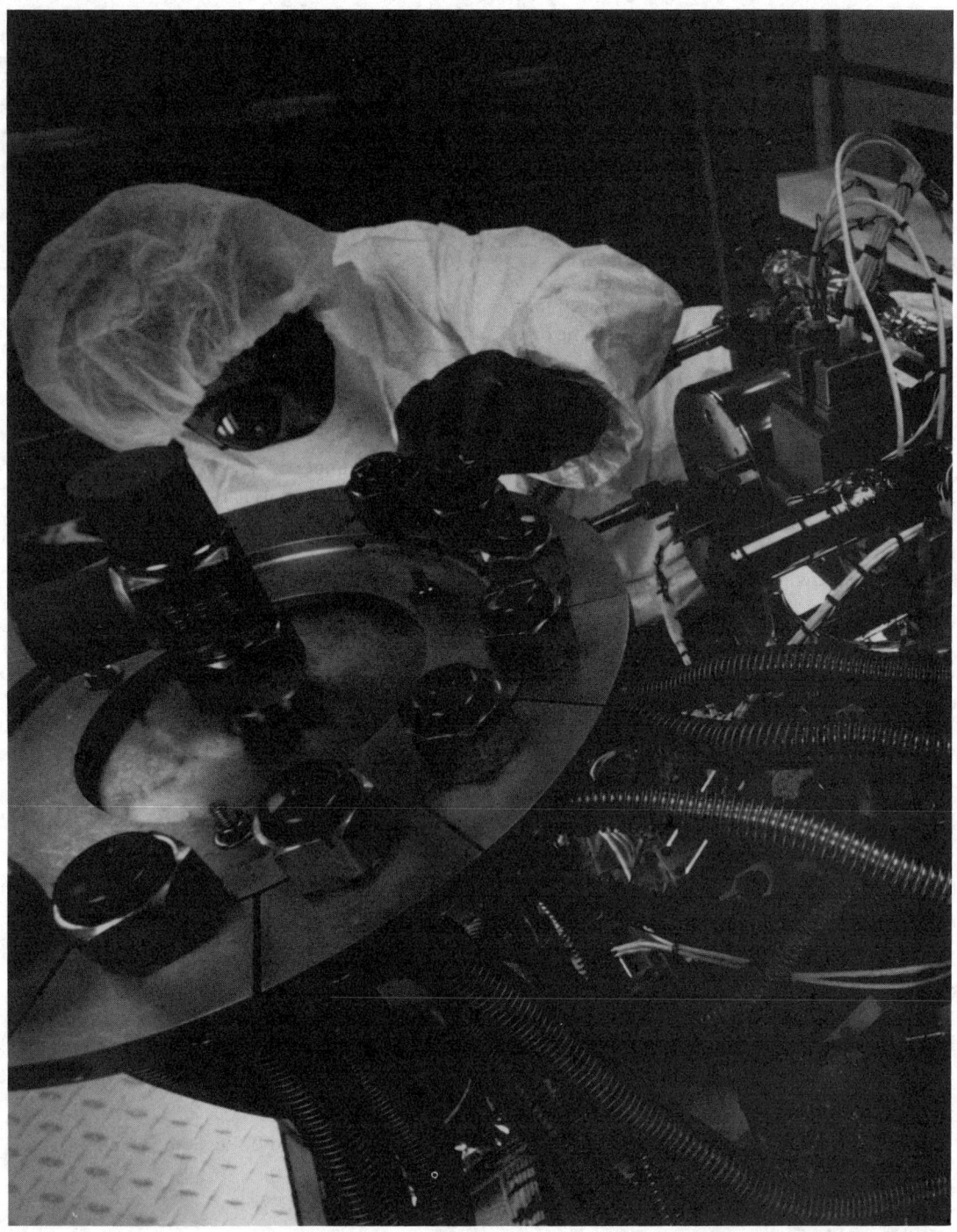

An important surveillance technology at Grumman Corporation is the *infrared mosaic sensor*. Congress and the Defense Department have already recommended the development of a U.S. Air Force Follow-On Early Warning System (FEWS) which could employ mosaic sensors to identify and track intercontinental ballistic missiles by the infrared signals they emit during their boost phase. In the photo, the entire optics package of the infrared mosaic sensor is readied for laboratory testing in an ultra-cold high-vacuum environment. This design is unlike any other proposed system in that it is constantly staring at the entire earth in search of a missile launch. (Photo courtesy of Grumman Corporation.)

Systems

Richard C. Dorf
University of California, Davis

93 Control Systems   *W. Brogan, G. Lee, A. Sage, B. Kuo, C. Phillips, R. Harbor, R. Jacquot, J. McInroy* 2099
Models • Dynamic Response • Frequency Response Methods: Bode Diagram Approach • Root Locus • Compensation • Digital Control Systems

94 Robotics   *T. Lasky, T. Hsia, R. Tummala, N. Odrey* 2154
Robot Configuration • Dynamics and Control • Applications

95 Aerospace Systems   *C. Spitzer, D. Martinec, C. Leondes, A. Rana, W. Check* 2188
Avionics Systems • Communications Satellite Systems: Applications

96 Command, Control, and Communications (C³)   *G. Clapp, D. Sworder* 2211
Background • The Technologies of C³ • The Dynamics of Encounters • The Role of the Human Decisionmaker in C³

97 Industrial Systems   *G. Cook, K. Andersen, R. Barnett, A. Wallace, R. Spée* 2223
Welding and Bonding • Large Drives

98 Man-Machine Systems   *D. McRuer* 2247
Several Natures of Man-Machine Control—A Catalog of Behavioral Complexities • Full-Attention Compensatory Operations—The Crossover Model

99 Vehicular Systems   *R. Dorf* 2255

100 Industrial Illuminating Systems   *K. Chen* 2257
New Concepts in Designing an Industrial Illuminating System • Factors Affecting Industrial Illumination • System Components • Applications • System Energy Efficiency Considerations

101 Instruments   *J. Schmalzel* 2277
Physical Variables • Instrument Elements • Instrumentation System • Modeling Elements of an Instrumentation System • Summary of Noise Reduction Techniques • Personal Computer-Based Instruments • Modeling PC-Based Instruments • The Effects of Sampling • Other Factors

102 Navigation Systems   *M. Kayton* 2285
Coordinate Frames • Categories of Navigation • Dead Reckoning • Radio Navigation • Celestial Navigation • Map-Matching Navigation • Navigation Software • Design Trade-Offs

A SYSTEM, consisting of interconnected components, is built to achieve a desired purpose. In 1990 the National Academy of Engineering identified the ten outstanding engineering achievements of the preceding twenty-five years. These feats included five accomplishments made possible by utilizing modern systems: the Apollo lunar landing, satellites, computer-aided manufacturing, computer axial tomography and the jumbo jet. The present challenge to control engineers is the modeling and control of modern, complex, interrelated systems such as traffic control systems, chemical processes, and robot systems. However, simultaneously, the fortunate engineer has the opportunity to control many very useful and interesting industrial automation systems.

This section is concerned with the analysis and design of ten types of systems. These systems draw their name from the application, for example, vehicular systems. Chapter 93 is concerned with control systems. A control system is an interconnection of components forming a system configuration that will provide a desired system response. The basis for analysis of a system is the foundation provided by a linear system theory, which assumes a cause-effect relationship for the components of a system. This chapter discusses the six key conceptual approaches to the analysis and design of control systems: models, dynamic response, frequency response, root locus, compensation, and digital control.

The next chapter has three articles that describe the key uses for robots. A robot is a system consisting of a manipulator and a computer programmed for use in a variety of tasks. In this chapter the authors describe the configuration, dynamics and control, and applications of a robot.

Chapter 95 describes aerospace systems in avionics and their use in communication satellite systems. Chapter 96 discusses the command and control systems used to monitor and control military aerospace systems. Chapter 97 describes two key industrial systems: welding and bonding, and large drives, while the following chapter is concerned with the man-machine system and models used to describe and analyze this system. The next three chapters look at the key characteristics and electronic controls for vehicular systems, industrial illuminating systems and their use and description, and instruments, which are systems consisting of sensors and electronic circuits, usually for measurement applications. Finally, navigation systems are described, in which navigation on the land, sea, or in the air is considered and modern approaches are described.

Note that related uses such as reliability (Chapter 92), digital devices (Section VIII), and circuits (Section I) are described elsewhere.

Nomenclature

Symbol	Quantity	Unit	Symbol	Quantity	Unit
B	magnetic flux density	Tesla	ω_c	crossover frequency	rad/s
β	phase shift		ω_n	natural frequency	
e_{ss}	steady state error		ω_r	resonance frequency	
$E(s)$	error function		ω_s	angular shaft speed	rad/s
f	supply frequency	Hz	p_d	heat source density	W/m^2
g	gravitational constant		P	number of poles	
$H(s)$	system transfer function		P	potential energy	
I	current	A	Φ	magnetic flux	Wb
J	inertia		σ	conductivity	S/m
K	total kinetic energy		t_m	minimum interaction time	s
K	stiffness		T	shaft torque	
L	Lagrangian		ζ	damping ratio	

93

Control Systems

William L. Brogan
University of Nevada, Las Vegas

Gordon K. F. Lee
North Carolina State University

Andrew P. Sage
George Mason University

Benjamin C. Kuo
*University of Illinois
(Urbana-Champaign)*

Charles L. Phillips
Auburn University

Royce D. Harbor
University of West Florida

Raymond G. Jacquot
University of Wyoming

John E. McInroy
University of Wyoming

93.1 Models 2099
      Classes of Systems to Be Modeled • Two Major Approaches to
      Modeling • Forms of the Model • Nonuniqueness • Approximation
      of Continuous Systems by Discrete Models
93.2 Dynamic Response 2106
      Computing the Dynamic System Response • Measures of the
      Dynamic System Response
93.3 Frequency Response Methods: Bode Diagram Approach ... 2113
      Frequency Response Analysis Using the Bode Diagram • Bode
      Diagram Design-Series Equalizers • Composite Equalizers •
      Minor-Loop Design
93.4 Root Locus 2131
      Root Locus Properties • Root Loci of Digital Control Systems
      • Design with Root Locus
93.5 Compensation 2139
      Control System Specifications • Design • Modern Control Design
      • Other Modern Design Procedures
93.6 Digital Control Systems 2147
      A Simple Example • Single-Loop Linear Control Laws • Propor-
      tional Control • PID Control Algorithm • The Closed-Loop System
      • A Linear Control Example

93.1 Models

William L. Brogan

A naive trial-and-error approach to the design of a control system might consist of constructing a controller, installing it into the system to be controlled, performing tests, and then modifying the controller until satisfactory performance is achieved. This approach could be dangerous and uneconomical, if not impossible. A more rational approach to control system design uses mathematical models. A *model* is a mathematical description of system behavior, as influenced by input variables or initial conditions. The model is a stand-in for the actual system during the control system design stage. It is used to predict performance; to carry out stability, sensitivity, and trade-off studies; and answer various "what-if" questions in a safe and efficient manner. Of course, the validation of the model, and all conclusions derived from it, must ultimately be based upon test results with the physical hardware.

The final form of the mathematical model depends upon the type of physical system, the method used to develop the model, and mathematical manipulations applied to it. These issues are discussed next.

Classes of Systems to Be Modeled

Most control problems are multidisciplinary. The system may consist of electrical, mechanical, thermal, optical, fluidic, or other physical components, as well as economic, biological, or ecological systems. Analogies exist between these various disciplines, based upon the similarity of the equations that describe the phenomena. The discussion of models in this section will be given in mathematical terms and therefore will apply to several disciplines.

Figure 93.1 [Brogan, 1991] shows the classes of systems that might be encountered in control systems modeling. Several branches of this tree diagram are terminated with a dashed line indicating that additional branches have been omitted, similar to those at the same level on other paths.

Distributed parameter systems have variables that are functions of both space and time (such as the voltage along a transmission line or the deflection of a point on an elastic structure). They are described by partial differential equations. These are often approximately modeled as a set of *lumped parameter* systems (described by ordinary differential or difference equations) by using modal expansions, finite element methods, or other approximations [Brogan, 1968]. The lumped parameter continuous-time and discrete-time families are stressed here.

Two Major Approaches to Modeling

In principle, models of a given physical system can be developed by two distinct approaches. Figure 93.2 shows the steps involved in *analytical modeling*. The real-world system is represented by an

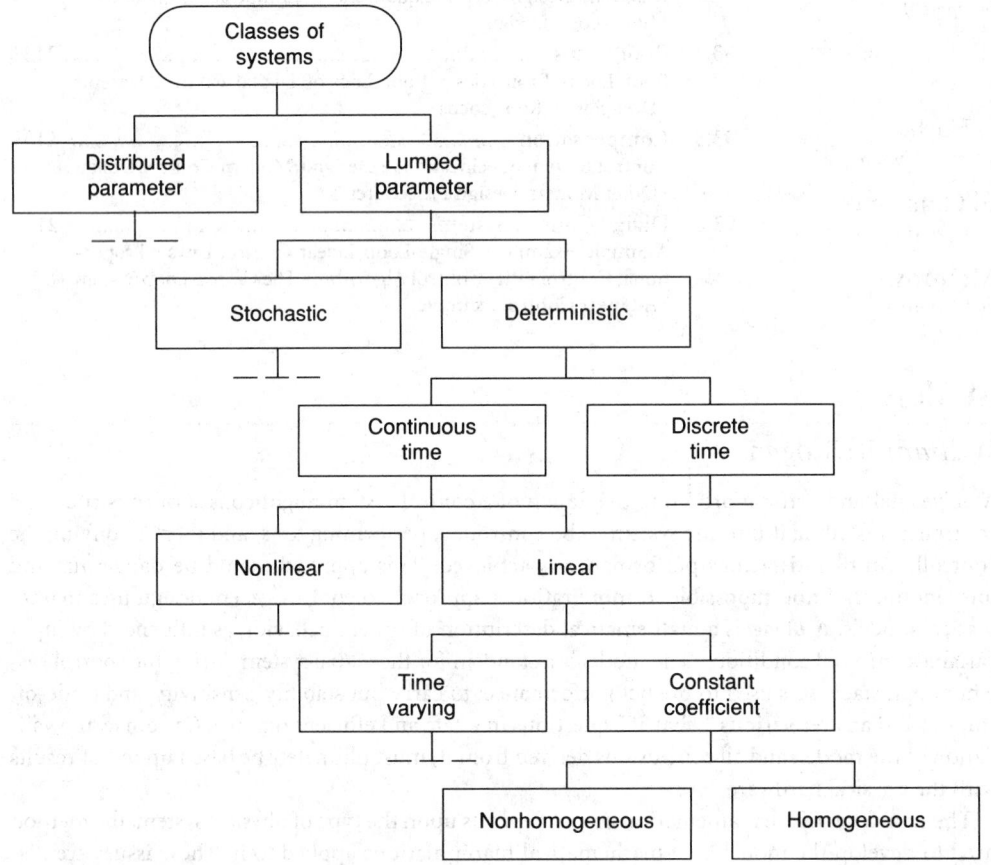

FIGURE 93.1   Major classes of system equations. (*Source:* W.L. Brogan, *Modern Control Theory*, 3rd ed., Englewood Cliffs, N.J.: Prentice-Hall, 1991, p. 13. With permission.)

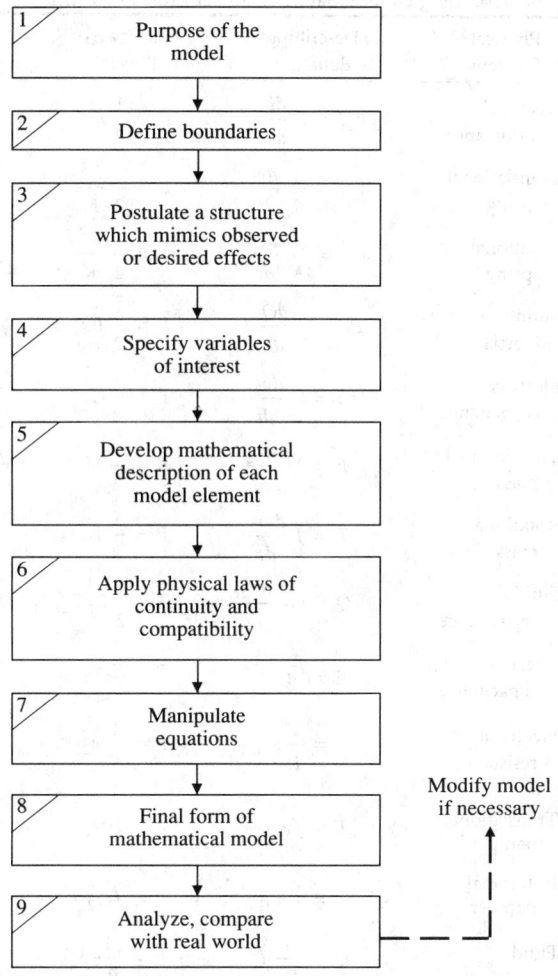

Steps in modeling

FIGURE 93.2 Modeling considerations. (*Source:* W.L. Brogan, *Modern Control Theory,* 3rd ed., Englewood Cliffs, N.J.: Prentice-Hall, 1991, p. 5. With permission.)

interconnection of idealized elements. Table 93.1 [Dorf, 1989] shows model elements from several disciplines and their elemental equations. An electrical circuit diagram is a typical result of this physical modeling step (box 3 of Fig. 93.2). Application of the appropriate physical laws (Kirchhoff, Newton, etc.) to the idealized physical model (consisting of point masses, ideal springs, lumped resistors, etc.) leads to a set of mathematical equations. For a circuit these will be mesh or node equations in terms of elemental currents and voltages. Box 6 of Fig. 93.2 suggests a generalization to other disciplines, in terms of continuity and compatibility laws, using through variables (generalization of current that flows through an element) and across variables (generalization of voltage, which has a differential value across an element) [Shearer *et al.,* 1967; Dorf, 1989].

Experimental or *empirical* modeling typically assumes an *a priori* form for the model equations and then uses available measurements to estimate the coefficient values that cause the assumed form to best fit the data. The assumed form could be based upon physical knowledge or it could be just a credible assumption. Time-series models include autoregressive (AR) models, moving

Table 93.1 Summary of Describing Differential Equations for Ideal Elements

Type of Element	Physical Element	Describing Equation	Energy E or Power $\mathcal{P}$	Symbol
Inductive storage	Electrical inductance	$v_{21} = L\dfrac{di}{dt}$	$E = \dfrac{1}{2}Li^2$	
	Translational spring	$v_{21} = \dfrac{1}{K}\dfrac{dF}{dt}$	$E = \dfrac{1}{2}\dfrac{F^2}{K}$	
	Rotational spring	$\omega_{21} = \dfrac{1}{K}\dfrac{dT}{dt}$	$E = \dfrac{1}{2}\dfrac{T^2}{K}$	
	Fluid inertia	$P_{21} = I\dfrac{dQ}{dt}$	$E = \dfrac{1}{2}IQ^2$	
Capacitive storage	Electrical capacitance	$i = C\dfrac{dv_{21}}{dt}$	$E = \dfrac{1}{2}Cv_{21}^2$	
	Translational mass	$F = M\dfrac{dv_2}{dt}$	$E = \dfrac{1}{2}Mv_2^2$	
	Rotational mass	$T = J\dfrac{d\omega_2}{dt}$	$E = \dfrac{1}{2}J\omega_2^2$	
	Fluid capacitance	$Q = C_f\dfrac{dP_{21}}{dt}$	$E = \dfrac{1}{2}C_f P_{21}^2$	
	Thermal capacitance	$q = C_t\dfrac{d\tau_2}{dt}$	$E = C_t\tau_2$	
Energy dissipators	Electrical resistance	$i = \dfrac{1}{R}v_{21}$	$\mathcal{P} = \dfrac{1}{R}v_{21}^2$	
	Translational damper	$F = fv_{21}$	$\mathcal{P} = fv_{21}^2$	
	Rotational damper	$T = f\omega_{21}$	$\mathcal{P} = f\omega_{21}^2$	
	Fluid resistance	$Q = \dfrac{1}{R_f}P_{21}$	$\mathcal{P} = \dfrac{1}{R_f}P_{21}^2$	
	Thermal resistance	$q = \dfrac{1}{R_t}\tau_{21}$	$\mathcal{P} = \dfrac{1}{R_t}\tau_{21}$	

Source: R.C. Dorf, *Modern Control Systems*, 5th ed., Reading, Mass.: Addison-Wesley, 1989, p. 33. With permission.

average (MA) models, and the combination, called ARMA models. All are difference equations relating the input variables to the output variables at the discrete measurement times, of the form

$$y(k+1) = a_0 y(k) + a_1 y(k-1) + a_2 y(k-2) + \cdots + a_n y(k-n)$$
$$+ b_0 u(k+1) + b_1 u(k) + \cdots + b_p u(k+1-p) + v(k) \qquad (93.1)$$

where $v(k)$ is a random noise term. The z-transform transfer function relating u to y is

$$\frac{y(z)}{u(z)} = \frac{b_0 + b_1 z^{-1} + \cdots + b_p z^{-p}}{1 - (a_0 z^{-1} + \cdots + a_{n-1} z^{-n})} = H(z) \qquad (93.2)$$

In the MA model all $a_i = 0$. This is alternatively called an all-zero model or a finite impulse response (FIR) model. In the AR model all b_j terms are zero except b_0. This is called an all-pole model or an infinite impulse response (IIR) model. The ARMA model has both poles and zeros and also is an IIR model [Makhoul, 1975].

Adaptive and learning control systems have an experimental modeling aspect. The data fitting is carried out on-line, in real time, as part of the system operation. The modeling described above is normally done off-line [Astrom and Wittenmark, 1989].

Forms of the Model

Regardless of whether a model is developed from knowledge of the physics of the process or from empirical data fitting, it can be further manipulated into several different but equivalent forms. This manipulation is box 7 in Fig. 93.2. The class that is most widely used in control studies is the deterministic lumped-parameter continuous-time constant-coefficient system. A simple example has one input u and one output y. This might be a circuit composed of one ideal source and an interconnection of ideal resistors, capacitors, and inductors. The equations for this system might consist of a set of mesh or node equations. These could be reduced to a single nth-order linear ordinary differential equation by eliminating extraneous variables.

$$\frac{d^n y}{dt^n} + a_{n-1}\frac{d^{n-1}y}{dt^{n-1}} + \cdots + a_1\frac{dy}{dt} + a_0 y = b_0 u + b_1\frac{du}{dt} + \cdots + b_m\frac{d^m u}{dt^m} \qquad (93.3)$$

This nth-order equation can be replaced by an input-output transfer function

$$\frac{Y(s)}{U(s)} = H(s) = \frac{b_m s^m + b_{m-1}s^{m-1} + \cdots + b_1 s + b_0}{s^n + a_{n-1}s^{n-1} + \cdots + a_1 s + a_0} \qquad (93.4)$$

The inverse Laplace transform $\mathcal{L}^{-1}\{H(s)\} = h(t)$ is the system impulse response function. Alternatively, by selecting a set of n internal **state variables**, Eq.(93.3) can be written as a coupled set of first-order differential equations plus an algebraic equation relating the states to the original output y. These equations are called state equations, and one possible choice for this example is, assuming $m = n$,

$$\dot{\mathbf{x}}(t) = \begin{bmatrix} -a_{n-1} & 1 & 0 & 0 & \cdots & 0 \\ -a_{n-2} & 0 & 1 & 1 & \cdots & 0 \\ \vdots & \vdots & \vdots & \vdots & \vdots & \vdots \\ -a_1 & 0 & 0 & 0 & \cdots & 1 \\ -a_0 & 0 & 0 & 0 & \cdots & 0 \end{bmatrix} x(t) + \begin{bmatrix} b_{n-1} - a_{n-1}b_n \\ b_{n-2} - a_{n-2}b_n \\ \vdots \\ b_1 - a_1 b_n \\ b_0 - a_0 b_n \end{bmatrix} u(t)$$

and

$$y(t) = [1 \quad 0 \quad 0 \quad \cdots \quad 0]\mathbf{x}(t) + b_n u(t) \qquad (93.5)$$

In matrix notation these are written more succinctly as

$$\dot{\mathbf{x}} = A\mathbf{x} + Bu \quad \text{and} \quad y = C\mathbf{x} + Du \qquad (93.6)$$

Any one of these six possible model forms, or others, might constitute the result of box 8 in Fig. 93.2. Discrete-time system models have similar choices of form, including an nth-order difference equation as given in Eq. (93.1) or a z-transform input-output transfer function as given in Eq. (93.2). A set of n first-order difference equations (state equations) analogous to Eq. (93.5) or (93.6) also can be written.

Extensions to systems with r inputs and m outputs lead to a set of m coupled equations similar to Eq. (93.3), one for each output y_i. These higher-order equations can be reduced to n first-order state differential equations and m algebraic output equations as in Eq. (93.5) or (93.6). The **A** matrix is again of dimension $n \times n$, but **B** is now $n \times r$, **C** is $m \times n$, and **D** is $m \times r$. In all previous discussions, the number of state variables, n, is the order of the model. In transfer function form, an $m \times r$ matrix $H(s)$ of transfer functions will describe the input-output behavior

$$Y(s) = H(s)U(s) \tag{93.7}$$

Other transfer function forms are also applicable, including the left and right forms of the matrix fraction description (MFD) of the transfer functions [Kailath, 1980]

$$H(s) = P(s)^{-1}N(s) \qquad \text{or} \qquad H(s) = N(s)P(s)^{-1} \tag{93.8}$$

Both **P** and **N** are matrices whose elements are polynomials in s. Very similar model forms apply to continuous-time and discrete-time systems, with the major difference being whether Laplace transform or z-transform transfer functions are involved.

When time-variable systems are encountered, the option of using high-order differential or difference equations versus sets of first-order state equations is still open. The system coefficients $a_i(t)$, $b_j(t)$ and/or the matrices $\mathbf{A}(t)$, $\mathbf{B}(t)$, $\mathbf{C}(t)$, and $\mathbf{D}(t)$ will now be time-varying. Transfer function approaches lose most of their utility in time-varying cases and are seldom used. With nonlinear systems all the options relating to the order and number of differential or difference equation still apply.

The form of the nonlinear state equations is

$$\dot{x} = f(\mathbf{x}, \mathbf{u}, \mathbf{t})$$
$$y = h(\mathbf{x}, \mathbf{u}, \mathbf{t}) \tag{93.9}$$

where the nonlinear vector-valued functions $f(\mathbf{x}, \mathbf{u}, \mathbf{t})$ and $h(\mathbf{x}, \mathbf{u}, \mathbf{t})$ replace the right-hand sides of Eq. (93.6). The transfer function forms are of no value in nonlinear cases.

Stochastic systems [Maybeck, 1979] are modeled in similar forms, except the coefficients of the model and/or the inputs are described in probabilistic terms.

Nonuniqueness

There is not a unique correct model of a given system for several reasons. The selection of idealized elements to represent the system requires judgment based upon the intended purpose. For example, a satellite might be modeled as a point mass in a study of its gross motion through space. A detailed flexible structure model might be required if the goal is to control vibration of a crucial onboard sensor. In empirical modeling, the assumed starting form, Eq. (93.1), can vary.

There is a trade-off between the complexity of the model form and the fidelity with which it will match the data set. For example, a pth-degree polynomial can exactly fit to $p + 1$ data points, but a straight line might be a better model of the underlying physics. Deviations from the line might be caused by extraneous measurement noise. Issues such as these are addressed in Astrom [1980].

The preceding paragraph addresses nonuniqueness in determining an input-output system description. In addition, state models developed from input-output descriptions are not unique. Suppose the transfer function of a single-input, single-output linear system is known exactly. The state variable model of this system is not unique for at least two reasons. An arbitrarily high-order state variable model can be found that will have this same transfer function. There is, however, a unique minimal or irreducible order n_{min} from among all state models that have the specified transfer function. A state model of this order will have the desirable properties of **controllability** and **observability**. It is interesting to point out that the minimal order may be less than the actual order of the physical system.

The second aspect of the nonuniqueness issue relates not to order, i.e., the *number* of state variables, but to *choice* of internal variables (state variables). Mathematical and physical methods of

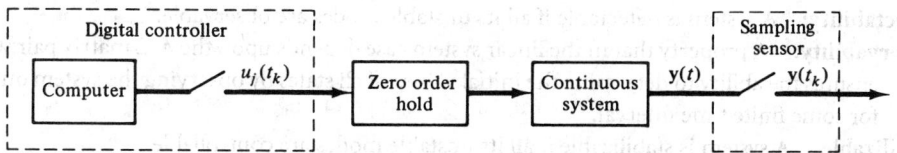

FIGURE 93.3 (*Source:* W.L. Brogan, *Modern Control Theory*, 3rd ed., Englewood Cliffs, N.J.: Prentice-Hall, 1991, p. 319. With permission.)

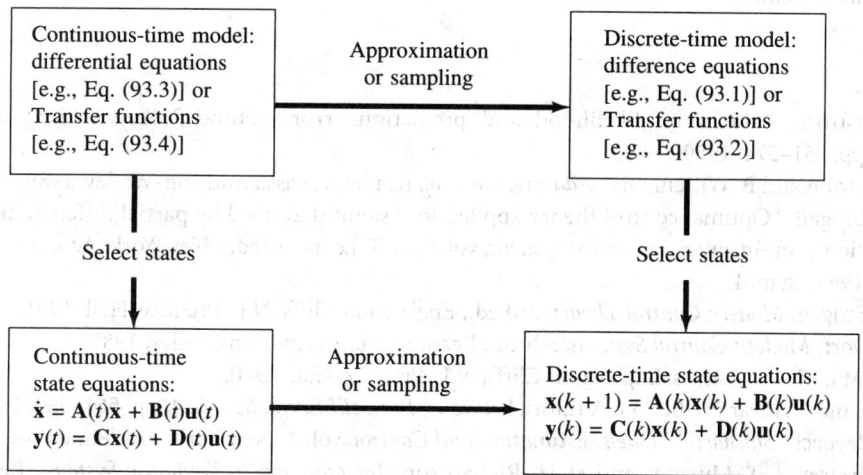

FIGURE 93.4 State variable modeling paradigm. (*Source:* W.L. Brogan, *Modern Control Theory*, 3rd ed., Englewood Cliffs, N.J.: Prentice-Hall, 1991, p. 309. With permission.)

selecting state variables are available [Brogan, 1991]. An infinite number of choices exist, and each leads to a different set {*A*, *B*, *C*, *D*}, called a realization. Some state variable model forms are more convenient for revealing key system properties such as stability, controllability, observability, **stabilizability**, and **detectability**. Common forms include the controllable canonical form, the observable canonical form, the Jordan canonical form, and the Kalman canonical form.

The reverse process is unique in that every valid realization leads to the same model transfer function

$$H(s) = C\{sI - A\}^{-1}B + D \qquad (93.10)$$

Approximation of Continuous Systems by Discrete Models

Modern control systems often are implemented digitally, and many modern sensors provide digital output, as shown in Fig. 93.3. In designing or analyzing such systems discrete-time approximate models of continuous-time systems are frequently needed. There are several general ways of proceeding, as shown in Fig. 93.4. Many choices exist for each path on the figure. Alternative choices of states or of approximation methods, such as forward or backward differences, lead to an infinite number of valid models.

Defining Terms

Controllability: A property that in the linear system case depends upon the **A,B** matrix pair which ensures the existence of some control input that will drive any arbitrary initial state to zero in finite time.

Detectability: A system is detectable if all its unstable modes are observable.

Observability: A property that in the linear system case depends upon the **A,C** matrix pair which ensures the ability to determine the initial values of all states by observing the system outputs for some finite time interval.

Stabilizable: A system is stabilizable if all its unstable modes are controllable.

State variables: A set of variables that completely summarize the system's status in the following sense. If all states x_i are known at time t_0, then the values of all states and outputs can be determined uniquely for any time $t_1 > t_0$, provided the inputs are known from t_0 onward. State variables are components in the state vector. State space is a vector space containing the state vectors.

References

K. J. Astrom, "Maximum likelihood and prediction error methods," *Automatica*, vol. 16, pp. 551–574, 1980.

K. J. Astrom and B. Wittenmark, *Adaptive Control*, Reading, Mass.: Addison-Wesley, 1989.

W. L. Brogan, "Optimal control theory applied to systems described by partial differential equations," in *Advances in Control Systems*, vol. 6, C. T. Leondes (ed.), New York: Academic Press, 1968, chap. 4.

W. L. Brogan, *Modern Control Theory*, 3rd ed., Englewood Cliffs, N.J.: Prentice-Hall, 1991.

R. C. Dorf, *Modern Control Systems*, 5th ed., Reading, Mass.: Addison-Wesley, 1989.

T. Kailath, *Linear Systems*, Englewood Cliffs, N.J.: Prentice-Hall, 1980.

J. Makhoul, "Linear prediction: A tutorial review," *Proc. IEEE*, vol. 63, no. 4, pp. 561–580, 1975.

P. S. Maybeck, *Stochastic Models, Estimation and Control*, vol. 1, New York: Academic Press, 1979.

J. L. Shearer, A. T. Murphy, and H. H. Richardson, *Introduction to Dynamic Systems*, Reading, Mass.: Addison-Wesley, 1967.

Further Information

The monthly *IEEE Control Systems Magazine* frequently contains application articles involving models of interesting physical systems.

The monthly *IEEE Transactions on Automatic Control* is concerned with theoretical aspects of systems. Models as discussed here are often the starting point for these investigations.

Automatica is the source of many related articles. In particular an extended survey on system identification is given by Astrom and Eykhoff in vol. 7, pp. 123–162, 1971.

Early developments of the state variable approach are given by R. E. Kalman in "Mathematical description of linear dynamical systems," *SIAM J. Control Ser.*, vol. A1, no. 2, pp. 152–192, 1963.

93.2 Dynamic Response

Gordon K. F. Lee

Computing the Dynamic System Response

Consider a linear time-invariant dynamic system represented by a differential equation form

$$\frac{d^n y(t)}{dt^n} + a_{n-1} \frac{d^{n-1} y(t)}{dt^{n-1}} + \cdots + a_1 \frac{dy(t)}{dt} + a_0 y(t)$$
$$= b_m \frac{d^m f(t)}{dt^m} + \cdots + b_1 \frac{df(t)}{dt} + b_0 f(t) \tag{93.11}$$

where $y(t)$ and $f(t)$ represent the output and input, respectively, of the system.

Let $p^k(\cdot) \triangleq (d^k/dt^k)(\cdot)$ define the differential operator so that (93.11) becomes

$$(p^n + a_{n-1}p^{n-1} + \cdots + a_1 p + a_0)y(t) = (b_m p^m + \cdots + b_1 p + b_0)f(t) \quad (93.12)$$

The solution to (93.11) is given by

$$y(t) = y_S(t) + y_I(t) \quad (93.13)$$

where $y_S(t)$ is the **zero-input response,** or that part of the response due to the initial conditions (or states) only, and $y_I(t)$ is the **zero-state response,** or that part of the response due to the input $f(t)$ only.

Zero-Input Response: $y_S(t)$

Here $f(t) = 0$, and thus (93.11) becomes

$$(p^n + a_{n-1}p^{n-1} + \cdots + a_1 p + a_0)y(t) = 0 \quad (93.14)$$

That is,

$$D(p)y(t) = 0$$

The roots of $D(p) = 0$ can be categorized as either distinct or multiple. That is, in general,

$$D(p) = \prod_{i=1}^{q} (p - \lambda_i)^{k_i} \prod_{i=1}^{r} (p - \lambda_{q+i})$$

where there are r distinct roots and q sets of multiple roots (each set has multiplicity k_i). Note that $r + \sigma = n$, where $\sigma \triangleq \sum_{i=1}^{q} k_i$. Each distinct root contributes a term to $y_S(t)$ of the form $c_i e^{\lambda_i t}$, where c_i is a constant, while each set of multiple roots contributes a set of terms to $y_S(t)$ of the form $\sum_{j=0}^{k_i-1} c_{i,j} t^j e^{\lambda_i t}$, where $c_{i,j}$ is some constant. Thus, the zero-input response is given by

$$y_S(t) = \sum_{i=1}^{q} \sum_{j=0}^{k_i-1} c_{i,j} t^j e^{\lambda_i t} + \sum_{i=1}^{r} c_{\sigma+i} e^{\lambda_{\sigma+i} t} \quad (93.15)$$

The coefficients $c_{i,j}$ and $c_{\sigma+i}$ are selected to satisfy the initial conditions.

Special Case

If all the roots of $D(p) = 0$ are distinct and the initial conditions for (93.11) are given by

$$\left\{ y(0), \frac{dy(0)}{dt}, \dots, \frac{d^{n-1}y(0)}{dt^{n-1}} \right\}$$

then the coefficients of (93.15) are given by the solution of

$$
\begin{bmatrix}
y(0) \\
\dfrac{dy(0)}{dt} \\
\vdots \\
\dfrac{d^{n-1}y(0)}{dt^{n-1}}
\end{bmatrix}
=
\begin{bmatrix}
1 & 1 & \cdots & 1 \\
\lambda_1 & \lambda_2 & \cdots & \lambda_n \\
\vdots & \vdots & \vdots & \vdots \\
\lambda_1^{(n-1)} & \lambda_2^{(n-1)} & \cdots & \lambda_n^{(n-1)}
\end{bmatrix}
\begin{bmatrix}
c_1 \\
c_2 \\
\vdots \\
c_n
\end{bmatrix}
\quad (93.16)
$$

Zero-State Response: $y_I(t)$

Here the initial conditions are made identically zero. Observing (93.11), let

$$H(p) = \frac{b_m p^m + \cdots + b_1 p + b_0}{p^n + a_{n-1} p^{n-1} + \cdots + a_1 p + a_0}$$

denote a rational function in the p operator. Consider using partial-fraction expansion on $H(p)$ as

$$H(p) = \sum_{i=1}^{q} \sum_{j=1}^{k_i} \frac{g_{i,j}}{(p - \lambda_i)^j} + \sum_{i=1}^{r} \frac{g_{\sigma+i}}{p - \lambda_{q+i}} \tag{93.17}$$

when the first term corresponds to the sets of multiple roots and the second term corresponds to the distinct roots.

Note the constant residuals are computed as

$$g_{\sigma+i} = [(p - \lambda_{q+i}) H(p)]_{p = \lambda_{q+i}}$$

and

$$g_{i,j} = \frac{1}{(k_i - j)!} \frac{d^{(k_i - j)}}{dp^{(k_i - j)}} \left\{ (p - \lambda_i)^{k_i} H(p) \right\} \Big|_{p = \lambda_i}$$

Then

$$h(t) = \sum_{i=1}^{q} \sum_{j=1}^{k_i} \frac{g_{i,j}}{(j - 1)!} t^{j-1} e^{\lambda_i t} + \sum_{i=1}^{r} g_{\sigma+1} e^{\lambda_{\sigma+i} t} \tag{93.18}$$

is the **impulse response** of the system (93.11). Then the zero-state response is given by

$$y_I(t) = \int_0^t f(\tau) h(t - \tau)\, d\tau \tag{93.19}$$

that is, $y_I(t)$ is the time convolution between input $f(t)$ and impulse response $h(t)$.

Measures of the Dynamic System Response

Several measures may be employed to investigate dynamic response performance. These include:

1. Speed of the response—how quickly does the system reach its final value
2. Accuracy—how close is the final response to the desired response
3. Relative stability—how stable is the system or how close is the system to instability
4. Sensitivity—what happens to the system response if the system parameters change

Objectives 3 and 4 may be analyzed by frequency domain methods (Section 93.3). Time-domain measures classically analyze the dynamic response by partitioning the total response into its steady-state (objective 2) and transient (objective 1) components. The **steady-state response** is that part of the response which remains as time approaches infinity; the **transient response** is that part of the response which vanishes as time approaches infinity.

Measures of the Steady-State Response

In the steady state, the accuracy of the time response is an indication of how well the dynamic response follows a desired time trajectory. Usually a test signal (reference signal) is selected to measure accuracy. Consider Fig. 93.5. In this configuration, the objective is to force $y(t)$ to track a ref-

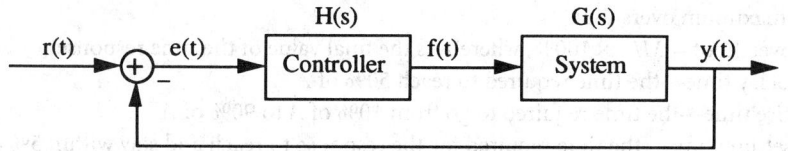

FIGURE 93.5 A tracking controller configuration.

erence signal $r(t)$ as close as possible. The **steady-state error** is a measure of the accuracy of the output $y(t)$ in tracking the reference input $r(t)$. Other configurations with different performance measures would result in other definitions of the steady-state error between two signals.

From Fig. 93.5, the error $e(t)$ is

$$e(t) = |r(t) - y(t)| \tag{93.20}$$

and the steady-state error is

$$e_{SS}(t) = \lim_{t \to \infty} e(t) = \lim_{s \to \infty} sE(s) \tag{93.21}$$

where $E(s)$ is the Laplace transform of $e(t)$, and s is the Laplacian operator. With $G(s)$ the transfer function of the system and $H(s)$ the transfer function of the controller, the transfer function between $y(t)$ and $r(t)$ is found to be

$$T(s) = \frac{G(s)H(s)}{1 + G(s)H(s)} \tag{93.22}$$

with

$$E(s) = \frac{R(s)}{1 + G(s)H(s)} \tag{93.23}$$

Direct application of the steady-state error for various inputs yields Table 93.2. Note $u(t)$ is the unit step function. This table can be extended to an mth-order input in a straightforward manner. Note that for $e_{SS}(t)$ to go to zero with a reference signal $Ct^m u(t)$, the term $G(s)H(s)$ must have at least m poles at the origin (a type m system).

Measures of the Transient Response

In general, analysis of the transient response of a dynamic system to a reference input is difficult. Hence formulating a standard measure of performance becomes complicated. Typically one designs a controller to make the system behave like a second-order system whereby analysis becomes more manageable.

Consider a reference unit step input to a dynamic system (Fig. 93.6). Critical parameters that measure transient response include:

Table 93.2 Steady-State Error Constants

$r(t)$ Test Signal	$e_{SS}(t)$	Error Constant
$Ru(t)$: step function	$\dfrac{R}{1 + K_p}$	$K_p = \lim_{s \to 0} G(s)H(s)$
$Rtu(t)$: ramp function	$\dfrac{R}{K_v}$	$K_v = \lim_{s \to 0} sG(s)H(s)$
$\dfrac{R}{2}t^2 u(t)$: parabolic function	$\dfrac{R}{K_a}$	$K_a = \lim_{s \to 0} s^2 G(s)H(s)$

1. *M:* maximum overshoot
2. % overshoot = $M/A \times 100\%$, where A is the final value of the time response
3. t_d: delay time—the time required to reach 50% of A
4. t_r: rise time—the time required to go from 10% of A to 90% of A
5. t_s: settling time—the time required for the response to reach and stay within 5% of A

To calculate these measures, consider a second-order system

$$T(s) = \frac{\omega_n^2}{s^2 + 2\xi\omega_n s + \omega_n^2} \tag{93.24}$$

where ξ is the damping coefficient and ω_n is the natural frequency of oscillation.

For the range $0 < \xi < 1$, the system response is *underdamped,* resulting in a damped oscillatory output. For a unit step input, the response is given by

$$y(t) = 1 + \frac{e^{-\xi\omega_n t}}{\sqrt{1 - \xi^2}} \sin\left(\omega_n\sqrt{1 - \xi^2}\,t - \tan^{-1}\frac{\sqrt{1 - \xi^2}}{-\xi}\right) \quad (0 < \xi < 1) \tag{93.25}$$

The eigenvalues (poles) of the system [roots of the denominator of $T(s)$] provide some measure of the time constants of the system. For the system under study, the eigenvalues are at

$$-\xi\omega_n \pm j\omega_n\sqrt{1 - \xi^2} \qquad \text{where} \qquad j \triangleq \sqrt{-1}$$

From the expression of $y(t)$, one sees that the term $\xi\omega_n$ affects the rise time and exponential decay time.

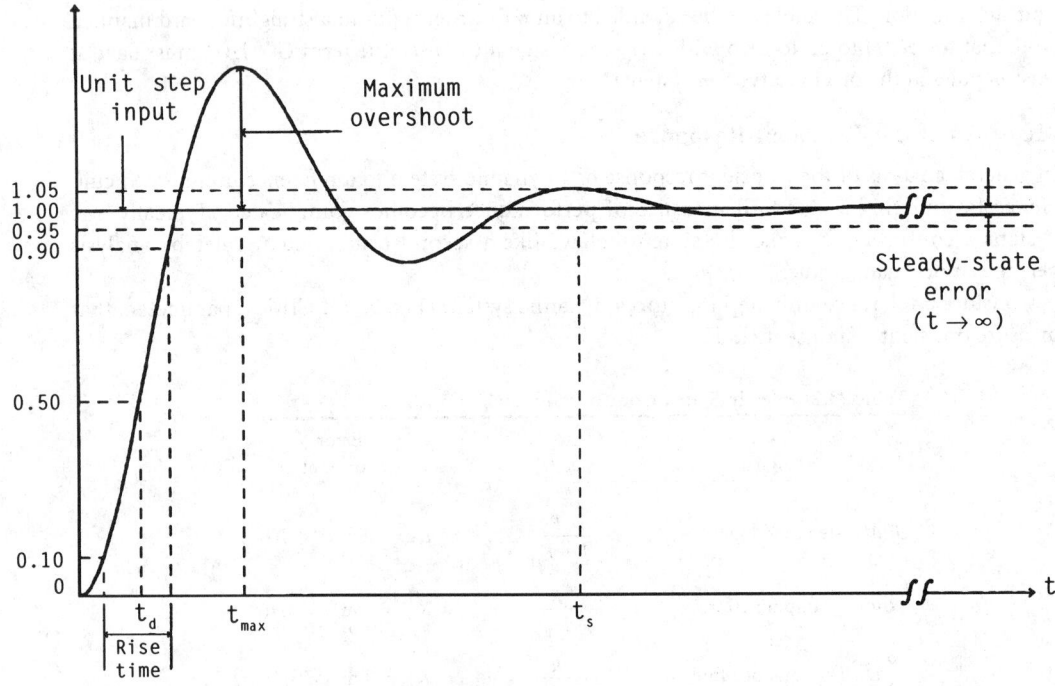

FIGURE 93.6 Step response.

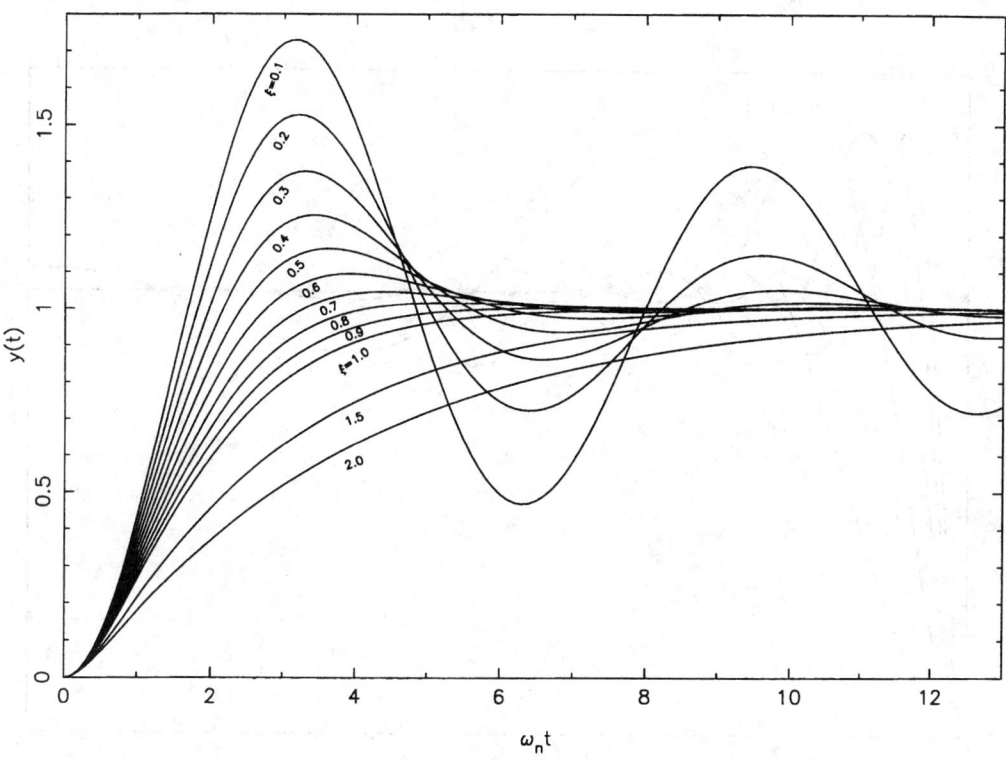

FIGURE 93.7 Effect of the damping coefficient on the dynamic response.

The effects of the damping coefficient on the transient response are seen in Fig. 93.7.

The effects of the natural frequency of oscillation ω_n of the transient response can be seen in Fig. 93.8. As ω_n increases, the frequency of oscillation increases.

For the case when $0 < \xi < 1$, the underdamped case, one can analyze the critical transient response parameters.

To measure the peaks of Fig. 93.6, one finds

$$y_{\text{peak}}(t) = 1 + (-1)^{n-1} \exp \frac{-n\pi\xi}{\sqrt{1-\xi^2}} \qquad n = 0, 1, \ldots \tag{93.26}$$

occurring at

$$t = \frac{n\pi}{\omega_n\sqrt{1-\xi^2}} \qquad \begin{matrix} n\text{: odd (overshoot)} \\ n\text{: even (undershoot)} \end{matrix} \tag{93.27}$$

Hence

$$y_{\text{max}} = 1 + \exp \frac{-\pi\xi}{\sqrt{1-\xi^2}} \tag{93.28}$$

occurring at

$$t_{\text{max}} = \frac{\pi}{\omega_n\sqrt{1-\xi^2}} \tag{93.29}$$

With these parameters, one finds

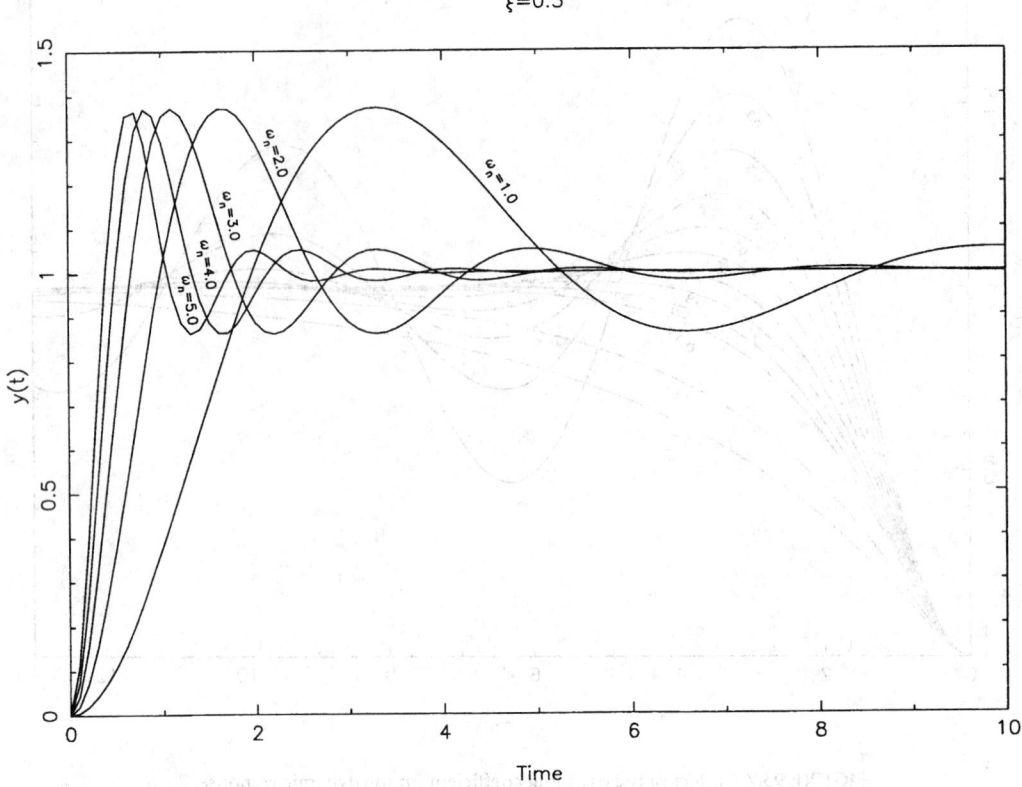

FIGURE 93.8   Effect of the natural frequency of oscillation on the dynamic response.

$$t_d \approx \frac{1 + 0.7\xi}{\omega_n}$$

$$t_r \approx \frac{1 + 1.1\xi + 1.4\xi^2}{\omega_n}$$

and

$$t_s \approx \frac{3}{\xi\omega_n}$$

Note that increasing ξ decreases the % overshoot and decreases the settling time but increases t_d and t_r.

When $\xi = 1$, the system has a double pole at $-\omega_n$, resulting in a *critically damped* response. This is the point when the response just changes from oscillatory to exponential in form. For a unit step input, the response is given by

$$y(t) = 1 - e^{-\omega_n t}(1 + \omega_n t) \qquad (\xi = 1) \qquad (93.30)$$

For the range $\xi > 1$, the system is overdamped due to two real system poles. For a unit step input, the response is given by

$$y(t) = 1 + \frac{1}{c_1 - c_2}\left(\frac{1}{c_1} e^{c_1 \omega_n t} - \frac{1}{c_2} e^{c_2 \omega_n t} \right) \qquad (\xi > 1)$$

$$c_1 = -\xi + \sqrt{\xi^2 - 1} \qquad c_2 = -\xi - \sqrt{\xi^2 - 1} \qquad (93.31)$$

Finally, when $\xi = 0$, the response is purely sinusoidal. For a unit step, the response is given by

$$y(t) = 1 - \cos \omega_n t \qquad (\xi = 0) \qquad (93.32)$$

Defining Terms

Impulse response: The response of a system when the input is an impulse function.

Steady-state error: The difference between the desired reference signal and the actual signal in steady-state, i.e., when time approaches infinity.

Steady-state response: That part of the response which remains as time approaches infinity.

Transient response: That part of the response which vanishes as time approaches infinity.

Zero-input response: That part of the response due to the initial condition only.

Zero-state response: That part of the response due to the input only.

Further Information

J. J. D'Azzo and C. H. Harpis, *Linear Control System Analysis and Design,* New York: McGraw-Hill, 1981.

R. C. Dorf, *Modern Control Systems,* 5th ed., Reading, Mass.: Addison-Wesley, 1989.

M. E. El-Hawary, *Control Systems Engineering,* Reston, Va.: Reston, 1984.

G. H. Hostetter, C. J. Savant, Jr., and R. T. Stefani, *Design of Feedback Control Systems,* Philadelphia: Saunders, 1989.

B. C. Kuo, *Automatic Control Systems,* Englewood Cliffs, N.J.: Prentice-Hall, 1987.

K. Ogata, *Modern Control Engineering,* Englewood Cliffs, N.J.: Prentice-Hall, 1970.

N. K. Sinha, *Control Systems,* New York: Holt, 1986.

93.3 Frequency Response Methods: Bode Diagram Approach

Andrew P. Sage

Our efforts in this section are concerned with analysis and design of linear control systems by frequency response methods. Design generally involves trial-and-error repetition of analysis until a set of design **specifications** has been met. Thus, analysis methods are most useful in the design process, which is one phase of the **systems engineering** life cycle [Sage, 1992]. We will discuss one design method based on **Bode diagrams.** We will discuss the use of both simple **series equalizers** and composite equalizers as well as the use of minor-loop feedback in systems design.

Figure 93.9 presents a flowchart of the frequency response method design process and indicates the key role of analysis in linear systems control design. The flowchart of Fig. 93.9 is applicable to control system design methods in general. There are several iterative loops, generally calling for trial-and-error efforts, that compose the suggested design process. An experienced designer will often be able, primarily due to successful prior experience, to select a system structure and generic components such that the design specifications can be met with no or perhaps a very few iterations through the iterative loop involving adjustment of equalizer or compensation parameters to best meet specifications.

If the parameter optimization, or parameter refinement such as to lead to maximum phase margin, approach shows the specifications cannot be met, we are then assured that no **equalizer** of the specific form selected will meet specifications. The next design step, if needed, would consist of modification of the equalizer form or structure and repetition of the analysis process to determine equalizer parameter values to best meet specifications. If specifications still cannot be met, we will

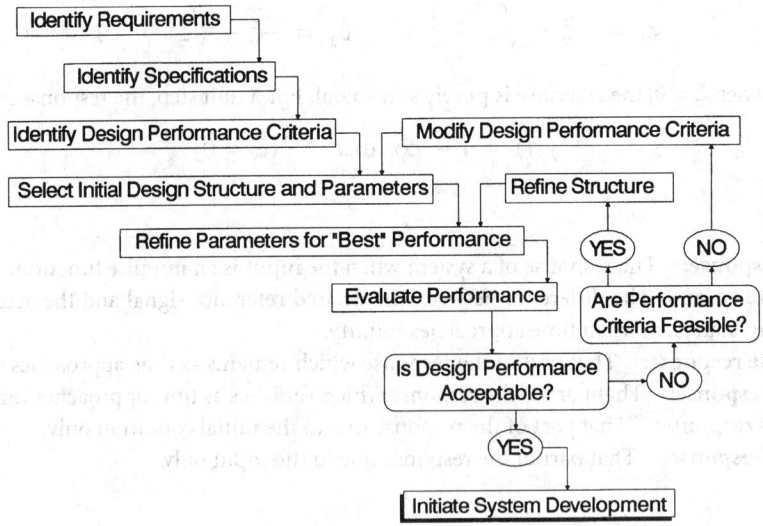

FIGURE 93.9   System design life cycle for frequency-response-based design.

usually next modify generic fixed components used in the system. This iterative design and analysis process is again repeated. If no reasonable fixed components can be obtained to meet specifications, then structural changes in the proposed system are next contemplated. If no structure can be found that allows satisfaction of specifications, either the client must be requested to relax the frequency response specifications or the project may be rejected as infeasible using present technology. As we might suspect, economics will play a dominant role in this design process. Changes made due to iteration in the inner loops of Fig. 93.9 normally involve little additional costs, whereas those made due to iterations in the outer loops will often involve major cost changes.

Frequency Response Analysis Using the Bode Diagram

The steady-state response of a stable linear constant-coefficient system has particular significance as we know from an elementary study of electrical networks and circuits and of dynamics. We consider a stable linear system with input-output transfer function

$$H(s) = \frac{Z(s)}{U(s)}$$

We assume a sinusoidal input $u(t) = \cos \omega t$ so that we have for the Laplace transform of the system output

$$Z(s) = \frac{sH(s)}{s^2 + \omega^2}$$

We expand this ratio of polynomials using the partial-fraction approach and obtain

$$Z(s) = F(s) + \frac{a_1}{s + j\omega} + \frac{a_2}{s - j\omega}$$

In this expression, $F(s)$ contains all the poles of $H(s)$. All of these lie in the left half plane since the system, represented by $H(s)$, is assumed to be stable. The coefficients a_1 and a_2 are easily determined as

$$a_1 = \frac{H(-j\omega)}{2}$$

$$a_2 = \frac{H(j\omega)}{2}$$

We can represent the complex transfer function $H(j\omega)$ in either of two forms,

$$H(j\omega) = B(\omega) + jC(\omega)$$

$$H(-j\omega) = B(\omega) - jC(\omega)$$

The inverse Laplace transform of the system transfer function will result in a transient term due to the inverse transform of $F(s)$, which will decay to zero as time progresses. A steady-state component will remain, and this is, from the inverse transform of the system equation, given by

$$z(t) = a_1 e^{-j\omega t} + a_2{}^{j\omega t}$$

We combine several of these relations and obtain the result

$$z(t) = B(\omega)\left(\frac{e^{j\omega t} + e^{-j\omega t}}{2}\right) - C(\omega)\left(\frac{e^{j\omega t} - e^{-j\omega t}}{2j}\right)$$

This result becomes, using the Euler identity,[1]

$$z(t) = B(\omega) \cos \omega t - C(\omega) \sin \omega t$$
$$= [B^2(\omega) + C^2(\omega)]^{1/2} \cos(\omega + \beta)$$
$$= |H(j\omega)| \cos(\omega t + \beta)$$

where $\tan \beta(\omega) = C(\omega)/B(\omega)$.

As we see from this last result, there is a very direct relationship between the transfer function of a linear constant-coefficient system, the time response of a system to any known input, and the sinusoidal steady-state response of the system. We can always determine any of these if we are given any one of them. This is a very important result. This important conclusion justifies a design procedure for linear systems that is based only on sinusoidal steady-state response, as it is possible to determine transient responses, or responses to any given system input, from a knowledge of steady-state sinusoidal responses, at least in theory. In practice, this might be rather difficult computationally without some form of automated assistance.

Bode Diagram Design-Series Equalizers

In this subsection we consider three types of series equalization:

1. Gain adjustment, normally attenuation by a constant at all frequencies
2. Increasing the phase lead, or reducing the phase lag, at the **crossover frequency** by use of a phase **lead network**
3. Attenuation of the gain at middle and high frequencies such that the crossover frequency will be decreased to a lower value where the phase lag is less, by use of a **lag network**

---

[1]The Euler identity $e^{jx} = \cos x + j \sin x$ is a very useful one and leads immediately to very helpful trigonometric identities such as $\cos(x + y) = \cos x \cos y - \sin x \sin y$ and $\sin(x + y) = \sin x \cos y + \cos x \sin y$.

In the subsection that follows this, we will first consider use of a composite or **lag-lead network** near crossover to attenuate gain only to reduce the crossover frequency to a value where the phase lag is less. Then we will consider more complex composite equalizers and state some general guidelines for Bode diagram design. Here, we will use Bode diagram frequency domain design techniques to develop a design procedure for each of three elementary types of series equalization.

Gain Reduction

Many linear control systems can be made sufficiently stable merely by reduction of the open-loop system gain to a sufficiently low value. This approach ignores all performance specifications, however, except that of phase margin (PM) and is, therefore, usually not a satisfactory approach. It is a very simple one, however, and serves to illustrate the approach to be taken in more complex cases.

The following steps constitute an appropriate Bode diagram design procedure for compensation by gain adjustment:

1. Determine the required PM and the corresponding phase shift $\beta_c = -\pi + \text{PM}$.
2. Determine the frequency ω_c at which the phase shift is such as to yield the phase shift at crossover required to give the desired PM.
3. Adjust the gain such that the actual crossover frequency occurs at the value computed in step 2.

Phase-Lead Compensation

In compensation using a phase-lead network, we increase the phase lead at the crossover frequency such that we meet a performance specification concerning phase shift. A phase-lead-compensating network transfer function is

$$G_c(s) = \left(1 + \frac{s}{\omega_1}\right) \bigg/ \left(1 + \frac{s}{\omega_2}\right) \qquad \omega_1 < \omega_2$$

Figure 93.10 illustrates the gain versus frequency and phase versus frequency curves for a simple lead network with the transfer function of the foregoing equation. The maximum phase lead obtainable from a phase-lead network depends upon the ratio ω_2/ω_1 that is used in designing the network. From the expression for the phase shift of the transfer function for this system, which is given by

$$\beta = \tan^{-1} \frac{\omega}{\omega_1} - \tan^{-1} \frac{\omega}{\omega_2}$$

we see that the maximum amount of phase lead occurs at the point where the first derivative with respect to frequency is zero, or

$$\left. \frac{d\beta}{d\omega} \right|_{\omega = \omega_m} = 0$$

or at the frequency where

$$\omega_m = (\omega_1 \omega_2)^{0.5}$$

This frequency is easily seen to be at the center of the two break frequencies for the lead network on a Bode log asymptotic gain plot. It is interesting to note that this is exactly the same frequency that we would obtain using an arctangent approximation[2] with the assumption that $\omega_1 < \omega < \omega_2$.

---

[2]The arctangent approximation is $\tan^{-1}(\omega/\alpha) = \omega/\alpha$ for $\omega < \alpha$ and $\tan^{-1}(\omega/\alpha) = \pi/2 - \alpha/\omega$ for $\omega > \alpha$. This approximation is rather easily obtained through use of a Taylor series approximation.

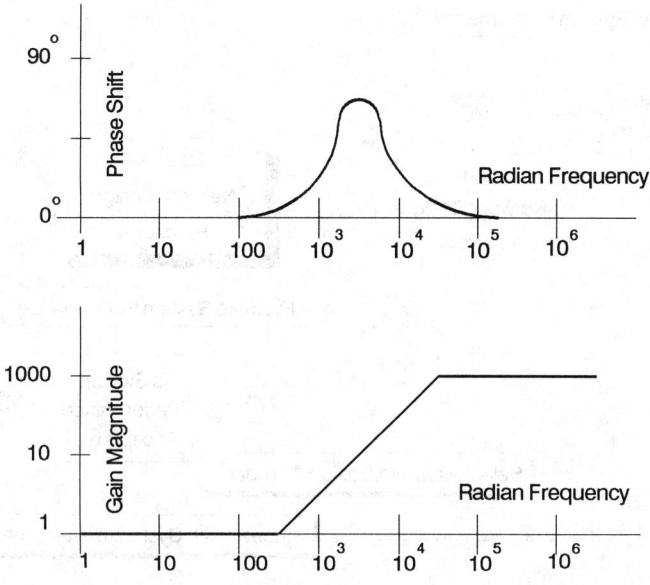

FIGURE 93.10 Phase shift and gain curves for a simple lead network.

There are many ways of realizing a simple phase-lead network. All methods require the use of an active element since the gain of the lead network at high frequencies is greater than 1. A simple electrical network realization is shown in Fig. 93.11.

We now consider a simple design example. Suppose that we have an open-loop system with transfer function

$$G_f(s) = \frac{10^4}{s^2}$$

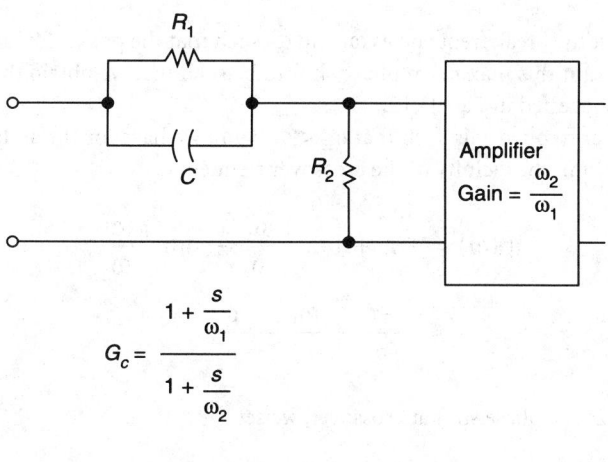

$$G_c = \frac{1 + \dfrac{s}{\omega_1}}{1 + \dfrac{s}{\omega_2}}$$

$$\omega_1 = \frac{1}{R_1 C}, \qquad \omega_2 = \left(1 + \frac{R_1}{R_2}\right)\omega_1$$

FIGURE 93.11 A simple electrical lead network.

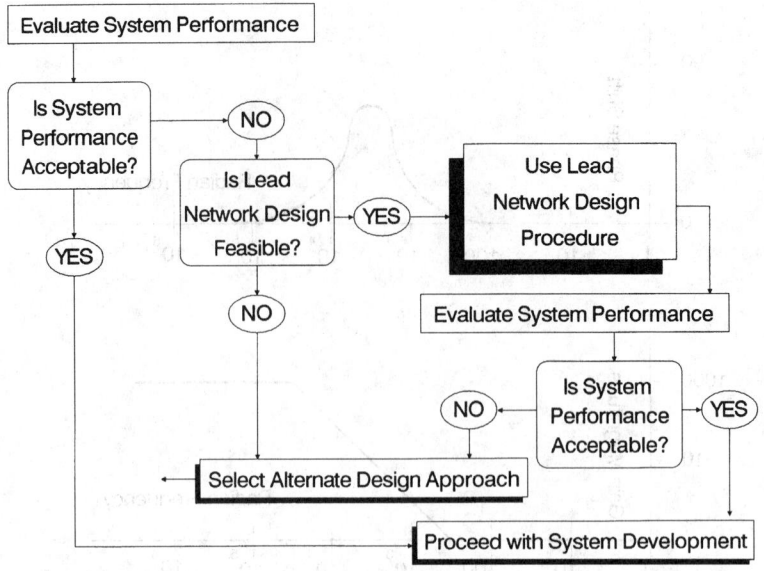

FIGURE 93.12   Life cycle of frequency domain design incorporating lead network compensation.

It turns out that this is often called a type-two system due to the presence of the double integration. This system will have a steady-state error of zero for a constant acceleration input. The crossover frequency, that is to say the frequency where the magnitude of the open-loop gain is 1, is 100 rad/s. The PM for the system without equalization is zero. We will design a simple lead network compensation for this zero PM system. If uncompensated, the closed-loop transfer function will be such that the system is unstable and any disturbance at all will result in a sinusoidally oscillating output.

The asymptotic gain diagram for this example is easily obtained from the open-loop transfer function

$$G_f(s)G_c(s) = \frac{K(1 + s/\omega_1)}{(1 + s/\omega_2)s^2}$$

and we wish to select the break frequencies ω_1 and ω_2 such that the phase shift at crossover is maximum. Further, we want this maximum phase shift to such that we obtain the specified PM. We use the procedure suggested in Fig. 93.12.

Since the crossover frequency is such that $\omega_1 < \omega_c < \omega_2$, we have for the arctangent approximation to the phase shift in the vicinity of the crossover frequency

$$\beta(\omega) = -\pi + \tan^{-1}\frac{\omega}{\omega_1} - \tan^{-1}\frac{\omega}{\omega_2}$$

$$\approx \frac{-\pi}{2} - \frac{\omega_1}{\omega} - \frac{\omega}{\omega_2}$$

In order to maximize the phase shift at crossover, we set

$$\left.\frac{d\beta}{d\omega}\right|_{\omega=\omega_m} = 0$$

and obtain as a result

$$\omega_m = (\omega_1\,\omega_2)^{0.5}$$

We see that the crossover frequency obtained is halfway between the two break frequencies ω_1 and ω_2 on a logarithmic frequency coordinate. The phase shift at this optimum value of crossover frequency becomes

$$\beta_c = \beta(\omega_c) = \frac{-\pi}{2} - 2\left(\frac{\omega_1}{\omega_2}\right)^{0.5}$$

For a PM of $-3\pi/4$, for example, we have $-3\pi/4 = -\pi/2 - 2(\omega_1/\omega_2)^{0.5}$, and we obtain $\omega_1/\omega_2 = 0.1542$ as the ratio of frequencies. We see that we have need for a lead network with a gain of $\omega_1/\omega_2 = 6.485$. The gain at the crossover frequency is 1, and from the asymptotic gain approximation that is valid for $\omega_1 < \omega < \omega_2$, we have the expressions $|G(j\omega)| = K/\omega\omega_1$ and $|G(j\omega_c)| = 1 = K/\omega_c\omega_1$ which for a known K can be solved for ω_c and ω_1.

Now that we have illustrated the design computation with a very simple example, we are in a position to state some general results. In the direct approach to design for a specified PM we assume a single lead network equalizer such that the open-loop system to transfer function results. This approach to design results in the following steps that are applicable for Bode diagram design to achieve maximum PM within an experientially determined control system structure that comprises a fixed plant and a compensation network with adjustable parameters:

1. We find an equation for the gain at the crossover frequency in terms of the compensated open-loop system break frequency.
2. We find an equation of the phase shift at crossover.
3. We find the relationship between equalizer parameters and crossover frequency such that the phase shift at crossover is the maximum possible and a minimum of additional gain is needed.
4. We determine all parameter specifications to meet the PM specifications.
5. We check to see that all design specifications have been met. If they have not, we iterate the design process.

Figure 93.12 illustrates the steps involved in implementing this frequency domain design approach.

Phase-Lag Compensation

In the phase-lag-compensation frequency domain design approach, we reduce the gain at low frequencies such that crossover, the frequency where the gain magnitude is 1, occurs before the phase lag has had a chance to become intolerably large. A simple single-stage phase-lag-compensating network transfer function is

$$G_c(s) = \frac{1 + s/\omega_2}{1 + s/\omega_1} \qquad \omega_1 < \omega_2$$

Figure 93.13 illustrates the gain and phase versus frequency curves for a simple lag network with this transfer function. The maximum phase lag obtainable from a phase-lag network depends upon the ratio ω_2/ω_1 that is used in designing the network. From the expression for the phase shift of this transfer function,

$$\beta = \tan^{-1}\frac{\omega}{\omega_2} - \tan^{-1}\frac{\omega}{\omega_1}$$

we see that maximum phase lag occurs at that frequency ω_c where $d\beta/d\omega = 0$. We obtain for this value

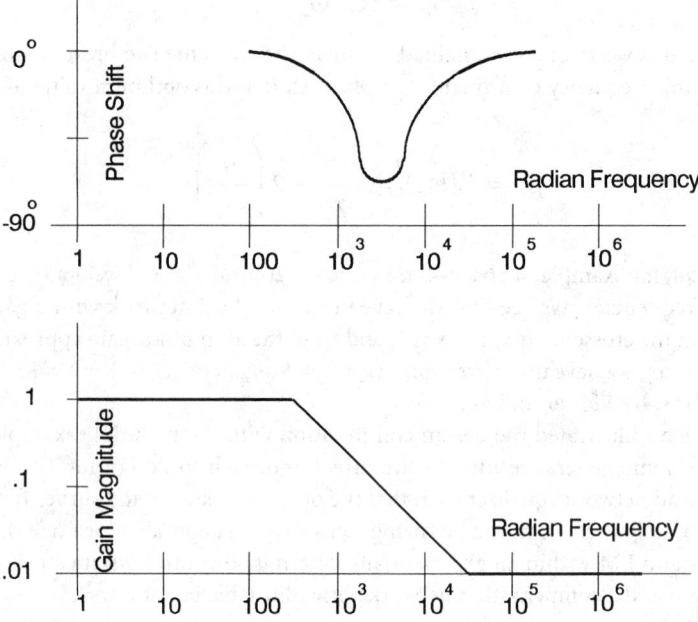

FIGURE 93.13   Phase shift and gain curves for a simple lag network.

$$\omega_m = (\omega_1\omega_2)^{0.5}$$

which is at the center of the two break frequencies for the lag network when the frequency response diagram is illustrated on a Bode diagram log-log asymptotic gain plot.

The maximum value of the phase lag obtained at $\omega = \omega_m$ is

$$\beta_m(\omega_m) = \frac{\pi}{2} - 2\tan^{-1}\left(\frac{\omega_2}{\omega_1}\right)^{0.5}$$

$$= \frac{\pi}{2} - 2\tan^{-1}\left(\frac{\omega_1}{\omega_2}\right)^{0.5}$$

which can be approximated in a more usable form, using the arctangent approximation, as

$$\beta_m(\omega_m) \approx \frac{\pi}{2}\sqrt{\frac{\omega_2}{\omega_1}}$$

The attenuation of the lag network at the frequency of minimum phase shift, or maximum phase lag, is obtained from the asymptotic approximation as

$$\left|G_c(\omega_m)\right| = \left(\frac{\omega_1}{\omega_2}\right)^{0.5}$$

Figure 93.13 presents a curve of attenuation magnitude obtainable at the frequency of maximum phase lag and the amount of the phase lag for various ratios ω_2/ω_1 for this simple lag network.

There are many ways to physically realize a lag network transfer function. Since the network only attenuates at some frequencies, as it never has a gain greater than 1 at any frequency, it can be realized with passive components only. Figure 93.14 presents an electrical realization of the simple lag network. Figure 93.15 presents a flowchart illustrating the design procedure envisioned here for lag network design. This is conceptually very similar to that for a lead network and makes use of the five-step parameter optimization procedure suggested earlier.

The object of lag network design is to reduce the gain at frequencies lower than the original crossover frequency in order to reduce the open-loop gain to unity before the phase shift becomes so excessive that the system PM is too small. A disadvantage of lag network compensation is that the attenuation introduced reduces the crossover frequency and makes the system slower in terms of its transient response. Of course, this would be advantageous if high-frequency noise is present and we wish to reduce its effect. The lag network is an entirely passive device and thus is more economical to instrument than the lead network.

In lead network compensation we actually insert phase lead in the vicinity of the crossover frequency to increase the PM. Thus we realize a specified PM without lowering the medium-frequency system gain. We see that the disadvantages of the lag network are the advantages of the lead network and the advantages of the lag network are the disadvantages of the lead network.

We can attempt to combine the lag network with the lead network into an all-passive structure

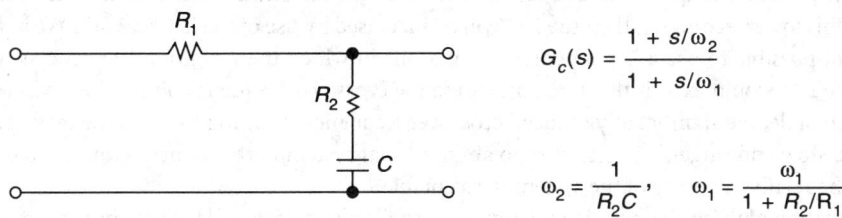

$$G_c(s) = \frac{1 + s/\omega_2}{1 + s/\omega_1}$$

$$\omega_2 = \frac{1}{R_2 C}, \qquad \omega_1 = \frac{\omega_1}{1 + R_2/R_1}$$

FIGURE 93.14   A simple electrical lag network.

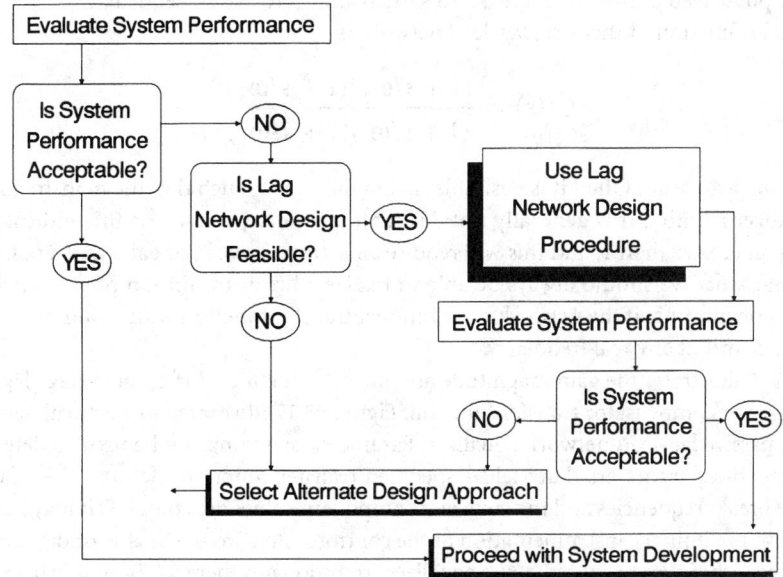

FIGURE 93.15   Life cycle of frequency domain design incorporating lag network compensation.

called a lag-lead network. Generally we obtain better results than we can achieve using either a lead or a lag network. We will consider design using lag-lead networks in our next subsection as well as more complex composite equalization networks.

Composite Equalizers

In the previous subsection we examined the simplest forms of series equalization: gain adjustment, lead network compensation, and lag network compensation. In this subsection we will consider more complex design examples in which composite equalizers will be used for series compensation. The same design principles used earlier in this section will be used here as well.

Lag-Lead Network Design

The prime purpose of a lead network is to add phase lead near the crossover frequency to increase the PM. Accompanied with this phase lead is a gain increase that will increase the crossover frequency. This will sometimes cause difficulties if there is much phase lag in the uncompensated system at high frequencies. There may be situations where use of a phase-lead network to achieve a given PM is not possible due to too many high-frequency poles.

The basic idea behind lag network design is to reduce the gain at "middle" frequencies such as to reduce the crossover frequency to a lower value than for the uncompensated system. If the phase lag is less at this lower frequency, then the PM will be increased by use of the lag network. We have seen that is not possible to use a lag network in situations in which there is not a frequency where an acceptable PM would exist if this frequency were the crossover frequency. Even if use of a lag network is possible, the significantly reduced crossover frequency resulting from its use may make the system so slow and sluggish in response to an input that system performance is unacceptable even though the relative stability of the system is acceptable.

Examination of these characteristics or attributes of lead network and lag network compensation suggests that it might be possible to combine the two approaches to achieve the desirable features of each approach. Thus we will attempt to provide attenuation below the crossover frequency to decrease the phase lag at crossover and phase lead closer to the crossover frequency in order to increase the phase lead of the uncompensated system at the crossover frequency.

The transfer function of the basic lag-lead network is

$$G_c(s) = \frac{(1 + s/\omega_2)(1 + s/\omega_3)}{(1 + s/\omega_1)(1 + s/\omega_4)}$$

where $\omega_4 > \omega_3 > \omega_2 > \omega_1$. Often it is desirable that $\omega_2\omega_3 = \omega_1\omega_4$ such that the high-frequency gain of the equalizer is unity. It is generally not desirable that $\omega_1\omega_4 > \omega_2\omega_3$ as this indicates a high-frequency gain greater than 1, and this will require an active network, or gain, and a passive equalizer. It is a fact that we should always be able to realize a linear minimum phase network using passive components only if the network has a rational transfer function with a gain magnitude that is no greater than 1 at any real frequency.

Figure 93.16 illustrates the gain magnitude and phase shift curves for a single-stage lag-lead network equalizer or compensator transfer function. Figure 93.17 illustrates an electrical network realization of a passive lag-lead network equalizer. Parameter matching can be used to determine the electrical network parameters that yield a specified transfer function. Because the relationships between the break frequencies and the equalizer component values are complex, it may be desirable, particularly in preliminary instrumentation of the control system, to use analog or digital computer programming techniques to construct the equalizer. Traditionally, there has been much analog computer simulation of control systems. The more contemporary approach suggests use of digital computer approaches that require numerical approximation of continuous-time physical systems.

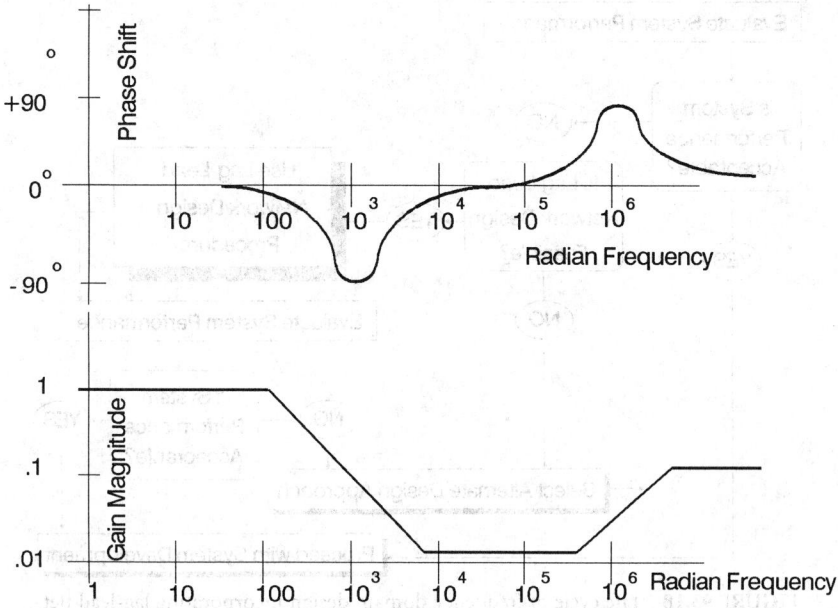

FIGURE 93.16 Phase shift and gain curves for a simple lag-lead network.

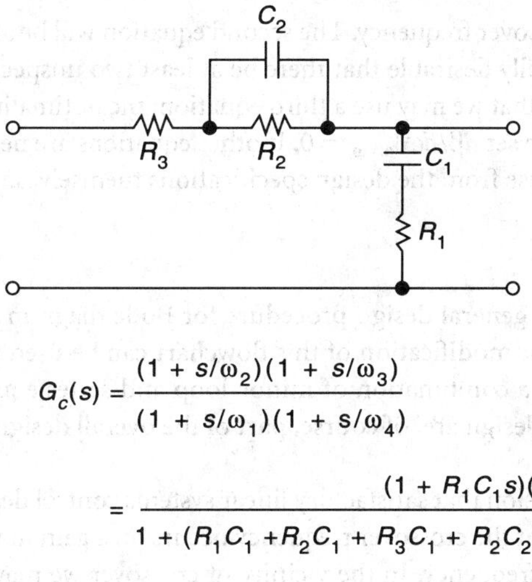

$$G_c(s) = \frac{(1 + s/\omega_2)(1 + s/\omega_3)}{(1 + s/\omega_1)(1 + s/\omega_4)}$$

$$= \frac{(1 + R_1 C_1 s)(1 + R_2 C_2 s)}{1 + (R_1 C_1 + R_2 C_1 + R_3 C_1 + R_2 C_2)s + (R_1 R_2 C_1 C_2 + R_2 R_3 C_1 C_2)s^2}$$

Special case: $R_3 = 0$

$$\omega_2 = \frac{1}{R_1 C_1}, \quad \omega_3 = \frac{1}{R_2 C_2}, \quad \omega_1 \omega_4 = \omega_2 \omega_3, \quad \omega_1 + \omega_4 = \omega_2 + \omega_3 + \frac{1}{R_1 C_2}$$

FIGURE 93.17 Simple electrical lag-lead network.

Figure 93.18 presents a flowchart that we may use for lag-lead network design. We see that this flowchart has much in common with the charts and design procedures for lead network and lag network design and that each of these approaches first involves determining or obtaining a set of desired specifications for the control system. Next, the form of a trial compensating network and the number of break frequencies in the network are selected. We must then obtain a number of equations, equal to the number of network break frequencies plus 1. One of these equations shows

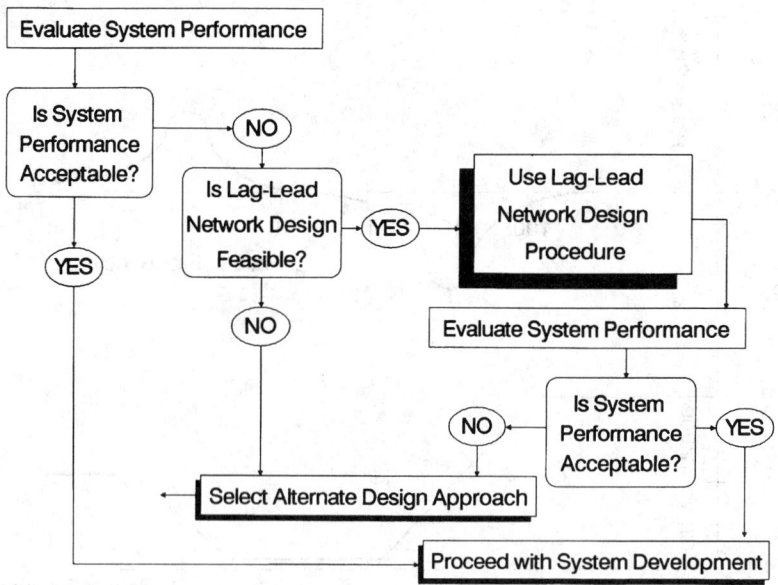

FIGURE 93.18   Life cycle of frequency domain design incorporating lag-lead network compensation.

that the gain magnitude is 1 at the crossover frequency. The second equation will be an equation for the phase shift at crossover. It is generally desirable that there be at least two unspecified compensating network break frequencies such that we may use a third equation, the optimality of the phase shift at crossover equation, in which we set $d\beta/d\omega\big|_{\omega=\omega_c} = 0$. If other equations are needed to represent the design situation, we obtain these from the design specifications themselves.

General Bode Diagram Design

Figure 93.19 presents a flowchart of a general design procedure for Bode diagram design. As we will see in the next subsection, a minor modification of this flowchart can be used to accomplish design using minor-loop feedback or a combination of minor-loop and series equations. These detailed flowcharts for Bode diagram design are, of course, part of the overall design procedure of Fig. 93.9.

Much experience leads to the conclusion that satisfactory linear systems control design using frequency response approaches is such that the crossover frequency occurs on a gain magnitude curve which has a –1 slope at the crossover frequency. In the vicinity of crossover we may approximate any minimum phase transfer function, with crossover on a –1 slope, by

$$G(s) = G_f(s)G_c(s) = \frac{\omega_c \omega_1^{n-1}(1 + s/\omega_1)^{n-1}}{s^n(1 + s/\omega_2)^{m-1}} \quad \text{for } \omega_1 > \omega_c > \omega_2$$

Here ω_1 is the break frequency just prior to crossover and ω_2 is the break frequency just after crossover. It is easy to verify that we have $|G(j\omega_c)| = 1$ if $\omega_1 > \omega_c > \omega_2$. Figure 93.20 illustrates this rather general approximation to a compensated system Bode diagram in the vicinity of the crossover frequency. We will conclude this subsection by determining some general design requirements for a system with this transfer function and the associated Bode asymptotic gain magnitude diagram of Fig. 93.20.

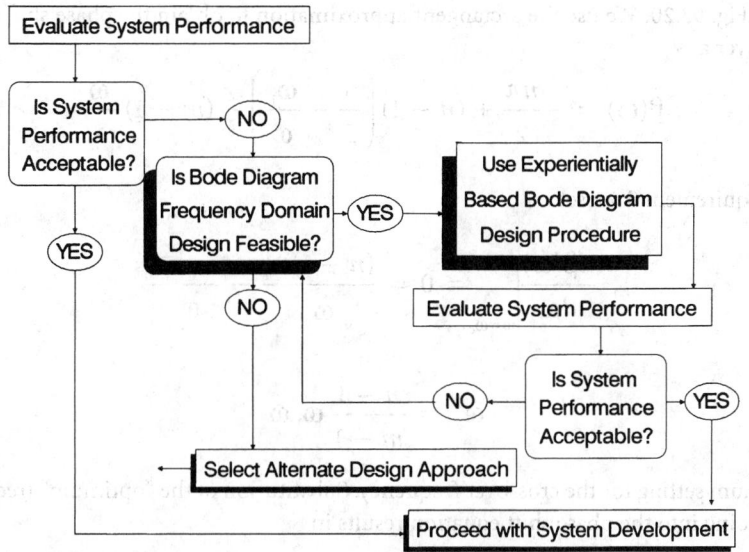

FIGURE 93.19   Life cycle of frequency domain design incorporating general Bode diagram compensation procedure.

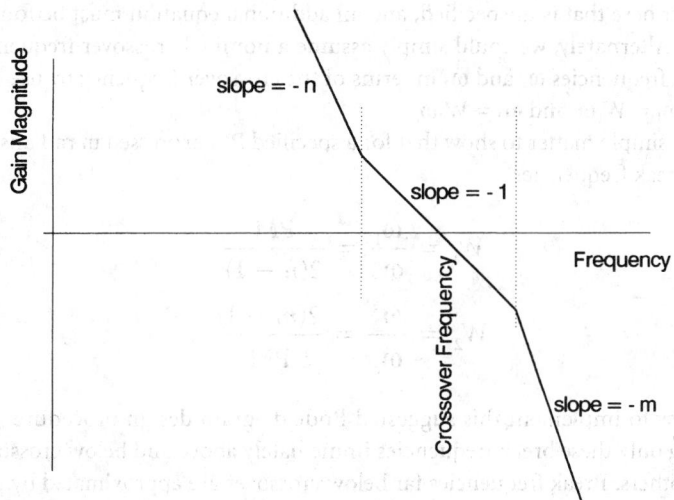

FIGURE 93.20   Illustration of generic gain magnitude in the vicinity of crossover.

There are three unknown frequencies in the foregoing equation. Thus we need three requirements or equations to determine design parameters. We will use the same three requirements used thus far in all our efforts in this section, namely:

1. The gain at crossover is 1.
2. The PM is some specified value.
3. The PM at crossover is the maximum possible for a given ω_2/ω_1 ratio.

We see that the first requirement, that the gain is 1 at the crossover frequency, is satisfied by the foregoing equation if the crossover frequency occurs on the –1 slope portion of the gain curve as assumed in Fig. 93.20. We use the arctangent approximation to obtain the phase shift in the vicinity of crossover as

$$\beta(\omega) = -\frac{n\pi}{2} + (n-1)\left(\frac{\pi}{2} - \frac{\omega_1}{\omega}\right) - (m-1)\frac{\omega}{\omega_2}$$

To satisfy requirement 3 we set

$$\left.\frac{d\beta(\omega)}{d\omega}\right|_{\omega=\omega_c} = 0 = \frac{(n-1)\omega_1}{\omega_c^2} - \frac{m-1}{\omega_2}$$

and obtain

$$\omega_c^2 = \frac{n-1}{m-1}\omega_1\omega_2$$

as the optimum setting for the crossover frequency. Substitution of the "optimum" frequency given by the foregoing into the phase shift equation results in

$$\beta(\omega_c) = \frac{-\pi}{2} - 2\sqrt{(m-1)(n-1)}\sqrt{\frac{\omega_1}{\omega_2}}$$

We desire a specific PM here, and so the equalizer break frequency locations are specified. There is a single parameter here that is unspecified, and an additional equation must be found in any specific application. Alternately, we could simply assume a nominal crossover frequency of unity or simply normalize frequencies ω_1 and ω_2 in terms of the crossover frequency by use of the normalized frequencies $\omega_1 = W_1\omega_c$ and $\omega_2 = W_2\omega_c$.

It is a relatively simple matter to show that for a specified PM expressed in radians, we obtain for the normalized break frequencies

$$W_1 = \frac{\omega_1}{\omega_c} = \frac{PM}{2(n-1)}$$

$$W_2 = \frac{\omega_2}{\omega_c} = \frac{2(m-1)}{PM}$$

It is relatively easy to implement this suggested Bode diagram design procedure which is based upon considering only these break frequencies immediately above and below crossover and which approximate all others. Break frequencies far below crossover are approximated by integrations or differentiations, that is, poles or zeros at $s = 0$, and break frequencies far above the crossover frequency are ignored.

Minor-Loop Design

In our efforts thus far in this section we have assumed that compensating networks would be placed in series with the fixed plant and then a unity feedback ratio loop closed around these elements to yield the closed-loop system. In many applications it may be physically convenient, perhaps due to instrumentation considerations, to use one or more minor loops to obtain a desired compensation of a fixed plant transfer function.

For a single-input–single-output linear system there are no theoretical advantages whatever to any minor-loop compensation to series compensation as the same closed-loop transfer function can be realized by all procedures. However, when there are multiple inputs or outputs, then there may be considerable advantages to minor-loop design as contrasted to series compensation design. Multiple inputs often occur when there is a single-signal input and one or more noise or disturbance inputs present and a task of the system is to pass the signal inputs and reject the noise inputs. Also there may be saturation-type nonlinearities present, and we may be concerned not only with the primary system output but also with keeping the output at the saturation point within bounds such that the system remains linear. Thus there are reasons why minor-loop design may be preferable to series equalization.

We have discussed block diagrams elsewhere in this handbook. It is desirable here to review some concepts that will be of value for our discussion of minor-loop design. Figure 93.21 illustrates a relatively general linear control system with a single minor loop. This block diagram could represent many simple control systems. $G_1(s)$ could represent a discriminator and series compensation and $G_2(s)$ could represent an amplifier and that part of a motor transfer function excluding the final integration to convert velocity to position. $G_3(s)$ might then represent an integrator. $G_4(s)$ would then represent a minor-loop compensation transfer function, such as that of a tachometer.

The closed-loop transfer function for this system is given by

$$\frac{Z(s)}{U(s)} = H(s) = \frac{G_1(s)G_2(s)G_3(s)}{1 + G_2(s)G_4(s) + G_1(s)G_2(s)G_3(s)}$$

It is convenient to define several other transfer functions that are based on the block diagram in Fig. 93.21. First there is the minor-loop gain

$$G_m(s) = G_2(s)G_4(s)$$

which is just the loop gain of the minor loop only. The minor loop has the transfer function

$$\frac{Z_m(s)}{U_m(s)} = H_m(s) = \frac{G_2(s)}{1 + G_2(s)G_4(s)} = \frac{G_2(s)}{1 + G_m(s)}$$

There will usually be a range or ranges of frequency for which the minor-loop gain magnitude is much less than 1, and we then have

$$\frac{Z_m(s)}{U_m(s)} = H_m(s) \approx G_2(s) \qquad \left| G_m(\omega) \right| \ll 1$$

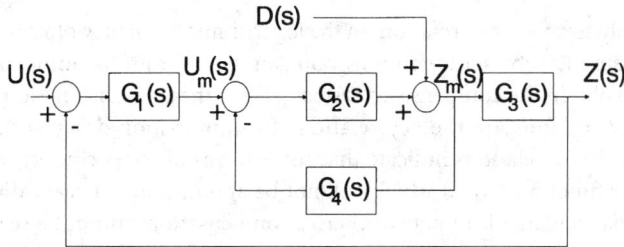

FIGURE 93.21 Feedback control system with a single minor loop and output disturbance.

There will also generally be ranges of frequency for which the minor-loop gain magnitude is much greater than 1, and we then have

$$\frac{Z_m(s)}{U_m(s)} = H_m(s) \approx \frac{1}{G_4(s)} \qquad \left|G_m(\omega)\right| \gg 1$$

We may use these two relations to considerably simplify our approach to the minor-loop design problem. For illustrative purposes, we will use two major-loop gain functions. First we will consider the major-loop gain with the minor-loop-compensating network removed such that $G_4(s) = 0$. This represents the standard situation we have examined in the last subsection. This uncompensated major-loop transfer function is

$$G_{Mu}(s) = G_1(s)G_2(s)G_3(s)$$

With the minor-loop compensation inserted, the major-loop gain, the input-output transfer function with the unity ratio feedback open, is

$$G_{Mc}(s) = \frac{G_1(s)G_2(s)G_3(s)}{1 + G_m(s)}$$

We may express the complete closed-loop transfer function in the form

$$\frac{Z(s)}{U(s)} = H(s) = \frac{G_{Mc}(s)}{1 + G_{Mc}(s)}$$

A particularly useful relationship may be obtained by combining the last three equations into one equation of the form

$$G_{Mc}(s) = \frac{G_{Mu}(s)}{1 + G_m(s)}$$

We may give this latter equation a particularly simple interpretation. For frequencies where the minor-loop gain $G_m(s)$ is low, the minor-loop–closed major-loop transfer function $G_{Mc}(s)$ is approximately that of the minor-loop–open major-loop transfer function G_{Mu} in that

$$G_{Mc}(s) \approx G_{Mu}(s) \qquad \left|G_m(\omega)\right| \ll 1$$

For frequencies where the minor-loop gain $G_m(s)$ is high, the minor-loop–closed major-loop transfer function is just

$$G_{Mc}(s) \approx \frac{G_{Mu}(s)}{G_m(s)} \qquad \left|G_m(\omega)\right| \gg 1$$

This has an especially simple interpretation on the logarithmic frequency plots we use for Bode diagrams for we may simply subtract the minor-loop gain $G_m(s)$ from the minor-loop–open major-loop gain $G_{Mu}(s)$ to obtain the compensated system gain as the transfer function $G_{Mc}(s)$.

The last several equations are the key relations for minor-loop design using this frequency response approach. These relations indicate that some forms of series compensation yield a given major-loop transfer function $G_{Mc}(s)$ which will not be appropriate for realization by minor-loop compensation. In particular, a lead network series compensation cannot be realized by means of equivalent minor-loop compensation. The gain of the fixed plant $G_{Mu}(s)$ is raised at high frequencies due to the use of a lead network compensation. Also, we see that $G_{Mc}(s)$ can only be lowered by use of a minor-loop gain $G_m(s)$.

A lag network used for series compensation will result in a reduction in the fixed plant gain

$|G_{Mu}(\omega)|$ at all high frequencies. This can only be achieved if the minor-loop transfer gain $G_m(s)$ is constant for high frequencies. In some cases this may be achievable but often will not be. It is possible to realize the equivalent of lag network series equalization by means of a minor-loop equalization for systems where the low- and high-frequency behavior of $G_{Mu}(s)$, or $G_f(s)$, and $G_{Mc}(s)$ are the same and where the gain magnitude of the compensated system $|G_{Mc}(s)|$ is at no frequency any greater than is the gain magnitude of the fixed plant $|G_f(s)|$ or the minor-loop–open major-loop transfer function $|G_{Mu}(s)|$. Thus we see that lag-lead network series equalization is an ideal type of equalization to realize by means of equivalent minor-loop equalization. Figures 93.9 and 93.19 represent flowcharts of a suggested general design procedure for minor-loop compensator design as well as for the series equalization approaches we examined previously.

In our work thus far we have assumed that parameters were constant and known. Such is seldom the case, and we must naturally be concerned with the effects of parameter variations, disturbances, and nonlinearities upon system performance. Suppose, for example, that we design a system with a certain gain assumed as K_1. If the system operates open loop and the gain K_1 is in cascade or series with the other input-output components, then the overall transfer function changes by precisely the same factor as K_1 changes. If we have an amplifier with unity ratio feedback around a gain K_1, the situation is much different. The closed-loop gain would nominally be $K_1/(1 + K_1)$, and a change to $2K_1$ would give a closed-loop gain $2K_1/(1 + 2K_1)$. If K_1 is large, say 10_3, then the new gain is 0.99950025, which is a percentage change of less than 0.05% for a change in gain of 100%.

Another advantage of minor-loop feedback occurs when there are output disturbances such as those due to wind gusts on an antenna. We consider the system illustrated in Fig. 93.21. The response due to $D(s)$ alone is

$$\frac{Z(s)}{D(s)} = \frac{1}{1 + G_2(s)\,G_4(s) + G_1(s)\,G_2(s)}$$

When we use the relation for the minor-loop gain

$$G_m(s) = G_2(s)G_4(s)$$

and the major-loop gain

$$G_{Mc}(s) = \frac{G_1(s)G_2(s)}{1 + G_2(s)G_4(s)}$$

we can rewrite the response due to $D(s)$ as

$$\frac{Z(s)}{D(s)} = \frac{1}{[1 + G_m(s)]G_{Mc}(s)}$$

Over the range of frequency where $|G_{Mc}(j\omega)| \gg 1$, such that the corrected loop gain is large, the attenuation of a load disturbance is proportional to the uncorrected loop gain. This is generally larger over a wider frequency range than the corrected loop gain magnitude $|G_{Mc}(j\omega)|$, which is what the attenuation would be if series compensation were used.

Over the range of frequencies where the minor-loop gain is large but where the corrected loop gain is small, that is, where $|G_m(j\omega)| > 1$ and $|G_{Mc}(j\omega)| < 1$, we obtain for the approximate response due to the disturbance

$$\frac{Z(s)}{D(s)} \approx G_m(s)$$

and the output disturbance is therefore seen to be attenuated by the minor-loop gain rather than unattenuated as would be the case if series compensation had been used. This is, of course, highly desirable.

At frequencies where both the minor-loop gain transfer and the major-loop gain are small we have $Z(s)/D(s) \approx 1$, and over this range of frequencies neither minor-loop compensation nor series equalization is useful in reducing the effect of a load disturbance. Thus, we have shown here that there are quite a number of advantages to minor-loop compensation as compared to series equalization. Of course, there are limitations as well.

Summary

In this section, we have examined the subject of linear system compensation by means of the frequency response method of Bode diagrams. Our approach has been entirely in the frequency domain. We have discussed a variety of compensation networks, including:

1. Gain attenuation
2. Lead networks
3. Lag networks
4. Lag-lead networks and composite equalizers
5. Minor-loop feedback

Despite its age, the frequency domain design approach represents a most useful approach for the design of linear control systems. It has been tested and proven in a great many practical design situations.

Defining Terms

Bode diagram: A graph of the gain magnitude and frequency response of a linear circuit or system, generally plotted on log-log coordinates. A major advantage of Bode diagrams is that the gain magnitude plot will look like straight lines or be asymptotic to straight lines. H. W. Bode, a well-known Bell Telephone Laboratories researcher, published *Network Analysis and Feedback Amplifier Design* in 1945. The approach, first described there, has been refined by a number of other workers over the past half-century.

Crossover frequency: The frequency where the magnitude of the open-loop gain is 1.

Equalizer: A network inserted into a system that has a transfer function or frequency response designed to compensate for undesired amplitude, phase, and frequency characteristics of the initial system. *Filter* and *equalizer* are generally synonymous terms.

Lag network: In a simple phase-lag network, the phase angle associated with the input-output transfer function is always negative, or lagging. Figures 93.13 and 93.14 illustrate the essential characteristics of a lag network.

Lag-lead network: The phase shift versus frequency curve in a phase lag-lead network is negative, or lagging, for low frequencies and positive, or leading, for high frequencies. The phase angle associated with the input-output transfer function is always positive, or leading. Figures 93.16 and 93.17 illustrate the essential characteristics of a lag-lead network, or composite equalizer.

Lead network: In a simple phase-lead network, the phase angle associated with the input-output transfer function is always positive, or leading. Figures 93.10 and 93.11 illustrate the essential characteristics of a lead network.

Series equalizer: In a single-loop feedback system, a series equalizer is placed in the single loop, generally at a point along the forward path from input to output where the equalizer itself consumes only a small amount of energy. In Fig. 93.21, $G_1(s)$ could represent a series equalizer. $G_1(s)$ could also be a series equalizer if $G_4(s) = 0$.

Specification: A statement of the design or development requirements to be satisfied by a system or product.

Systems engineering: An approach to the overall life cycle evolution of a product or system. Generally, the systems engineering process comprises a number of phases. There are three essential phases in any systems engineering life cycle: formulation of requirements and specifications, design and development of the system or product, and deployment of the system. Each of these three basic phases may be further expanded into a larger number. For example, deployment generally comprises operational test and evaluation, maintenance over an extended operational life of the system, and modification and retrofit (or replacement) to meet new and evolving user needs.

References

J. L. Bower and P. M. Schultheiss, *Introduction to the Design of Servomechanisms*, New York: Wiley, 1958.

A. P. Sage, *Linear Systems Control*, Champaign, Ill.: Matrix Press, 1978.

A. P. Sage, *Systems Engineering*, New York: Wiley, 1992.

M. G. Singh, Ed., *Systems and Control Encyclopedia*, Oxford: Pergamon, 1987.

Further Information

Many of the practical design situations used to test the frequency domain design approach are described in the excellent classic text by Bower and Schultheiss [1958]. A rather detailed discussion of the approach may also be found in Sage [1978] on which this discussion is, in part, based. A great variety of control systems design approaches, including frequency domain design approaches, are discussed in a recent definitive control systems encyclopedia [Singh, 1987], and there are a plethora of new introductory control systems textbooks that discuss it as well. As noted earlier, frequency domain design, in particular, and control systems design, in general, constitute one facet of systems engineering effort, such as described in Sage [1992].

93.4 Root Locus

Benjamin C. Kuo

Root locus represents a trajectory of the roots of an algebraic equation with constant coefficients when a parameter varies. The technique is used extensively for the analysis and design of linear time-invariant control systems. For linear time-invariant control systems the roots of the characteristic equation determine the stability of the system. For a stable continuous-data system the roots must all lie in the left half of the s plane. For a digital control system to be stable, the roots of the characteristic equation must all lie inside the unit circle $|z| = 1$ in the z plane. Thus, in the s plane, the imaginary axis is the stability boundary, whereas in the z plane the stability boundary is the unit circle. The location of the characteristic equation roots with respect to the stability boundary also determine the relative stability, i.e., the degree of stability, of the system.

For a linear time-invariant system with continuous data, the characteristic equation can be written as

$$F(s) = P(s) + KQ(s) = 0 \tag{93.33}$$

where $P(s)$ is an Nth-order polynomial of s,

$$P(s) = s^N + a_1 s^{N-1} + \cdots + a_{N-1} s + a_N \tag{93.34}$$

and $Q(s)$ is the Mth-order polynomial of s,

$$Q(s) = s^M + b_1 s^{M-1} + \cdots + b_{M-1} s + b_M \tag{93.35}$$

where N and M are positive integers. The real constant K can vary from $-\infty$ to $+\infty$. The coefficients $a_1, a_2, \ldots, a_N, b_1, b_2, \ldots, b_M$ are real. As K is varied from $-\infty$ to $+\infty$, the roots of Eq. (93.33) trace out continuous trajectories in the s plane called the *root loci*.

The above development can be extended to digital control systems by replacing s with z in Eqs. (93.33) through (93.35).

Root Locus Properties

The root locus problem can be formulated from Eq. (93.33) by dividing both sides of the equation by the terms that do not contain the variable parameter K. The result is

$$1 + \frac{KQ(s)}{P(s)} = 0 \tag{93.36}$$

For a closed-loop control system with the loop transfer function $KG(s)H(s)$, where the gain factor K has been factored out, the characteristic equation is known to be the zeros of the rational function

$$1 + KG(s)H(s) = 0 \tag{93.37}$$

Since Eqs. (93.36) and (93.37) have the same form, the general root locus problem can be formulated using Eq. (93.36).

Equation (93.37) is written

$$G(s)H(s) = -\frac{1}{K} \tag{93.38}$$

To satisfy the last equation, the following conditions must be met simultaneously:

$$\text{Condition on magnitude:} \quad \left| G(s)H(s) \right| = \frac{1}{|K|} \tag{93.39}$$

where K varies between $-\infty$ and $+\infty$.

$$\text{Conditions on angles:} \quad \angle G(s)H(s) = (2k+1)\pi \quad K \geq 0$$
$$= \text{odd multiples of } \pi \text{ rad} \tag{93.40}$$
$$\angle G(s)H(s) = 2k\pi \quad K \leq 0$$
$$= \text{even multiples of } \pi \text{ rad} \tag{93.41}$$

where $k = 0, \pm 1, \pm 2, \ldots, \pm$ any integer.

In general, the conditions on angles in Eqs. (93.40) and (93.41) are used for the construction of the root loci, whereas the condition on magnitude in Eq. (93.39) is used to find the value of K on the loci once the loci are drawn. Let $KG(s)H(s)$ be of the form

$$KG(s)H(s) = \frac{K(s + a)}{(s + b)(s + c)} \tag{93.42}$$

Applying Eqs. (93.40) and (93.41) to the last equation, the angles conditions are

$$K \geq 0: \angle G(s)H(s) = \angle(s + a) - \angle(s + b) - \angle(s + c)$$
$$= (2k + 1)\pi \tag{93.43}$$

$$K \leq 0: \angle G(s)H(s) = \angle(s + a) - \angle(s + b) - \angle(s + c)$$
$$= 2k\pi \tag{93.44}$$

where $k = 0, \pm 1, \pm 2, \dots$. The graphical interpretation of the last two equations is shown in Fig. 93.22. For the point s_1 to be a point on the root locus, the angles of the phasors drawn from the poles and zeros of $G(s)H(s)$ to s_1 must satisfy Eq. (93.43) or (93.44) depending on the sign of K. Applying the magnitude condition of Eq. (93.39) to (93.42), the magnitude of K is expressed as

$$|K| = \frac{|s+b||s+c|}{|s+a|} = \frac{B \cdot C}{A} \tag{93.45}$$

where A, B, and C are the lengths of the phasors drawn from the poles and zeros of $G(s)H(s)$ to the point s_1.

The following properties of the root loci are useful for sketching the root loci based on the pole-zero configuration of $G(s)H(s)$. Many computer programs, such as the ROOTLOCI in the ACSP software package [Kuo, 1991b], are available for computing and plotting the root loci. The proofs and derivations of these properties can be carried out from Eqs. (93.39), (93.40), and (93.41) [Kuo, 1991a].

Starting Points (K = 0 Points). The points at which $K = 0$ on the root loci are at the poles of $G(s)H(s)$.

Ending Points (K = ±∞ Points). The points at which $K = \pm\infty$ on the root loci are at the zeros of $G(s)H(s)$. The poles and zeros referred to above include those at $s = \infty$.

Number of Root Loci. The total number of root loci of Eq. (93.37) equals the higher of the number of poles and zeros of $G(s)H(s)$.

Symmetry of Root Loci. The root loci are symmetrical with respect to the axes of symmetry of the pole-zero configuration of $G(s)H(s)$. In general, the root loci are symmetrical at least to the real axis of the complex s plane.

Asymptotes of the Root Loci. Asymptotes of the root loci refer to the behavior of the root loci at $|s| = \infty$ when the number of poles and zeros of $G(s)H(s)$ is not equal. Let N denote the number of finite poles of $G(s)H(s)$ and M be the number of finite zeros of $G(s)H(s)$. In general, $2|N - M|$ of the loci will approach infinity in the s plane. The properties of the root loci at $|s| = \infty$ are described by the angles and the intersects of the asymptotes. When $N \neq M$, the angles of the asymptotes are given by

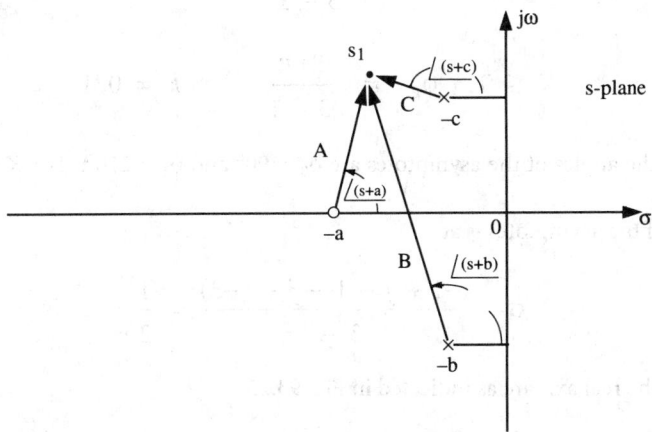

FIGURE 93.22 Graphical interpretation of magnitude and angle conditions of root loci.

$$\phi_k = \begin{cases} \dfrac{(2k + 1)\pi}{|N - M|} & K \geq 0 \qquad\qquad (93.46) \\[4mm] \dfrac{2k\,\pi}{|N - M|} & K \leq 0 \qquad\qquad (93.47) \end{cases}$$

where $k = 0, 1, 2, \ldots, |N - M| - 1$.

The asymptotes intersect on the real axis at

$$\sigma = \frac{\sum \text{finite poles of } G(s)H(s) - \sum \text{finite zeros of } G(s)H(s)}{N - M} \qquad (93.48)$$

Root Loci on the Real Axis. The entire real axis of the s plane is occupied by the root loci. When $K > 0$, root loci are found in sections of the real axis to the right of which the total number of poles and zeros of $G(s)H(s)$ is *odd*. When $K < 0$, root loci are found in sections to the right of which the total number of poles and zeros of $G(s)H(s)$ is *even*.

As a summary of the root locus properties discussed above, the properties of the root loci of the following equation are displayed in Fig. 93.23.

$$s^3 + 2s^2 + 2s + K(s + 3) = 0 \qquad (93.49)$$

Dividing both sides of the last equation by the terms that do not contain K we get

$$1 + KG(s)H(s) = 1 + \frac{K(s + 3)}{s(s^2 + 2s + 2)} \qquad (93.50)$$

Thus, the poles of $G(s)H(s)$ are at $s = 0$, $s = -1 + j$, and $s = -1 - j$. The zero of $G(s)H(s)$ is at $z = -3$.

As shown in Fig. 93.23, the $K = 0$ points on the root loci are at the poles of $G(s)H(s)$, and the $K = \pm\infty$ points are at the zeros of $G(s)H(s)$. Since $G(s)H(s)$ has two zeros at $s = \infty$, two of the three root loci approach infinity in the s plane. The root loci are symmetrical to the real axis of the s plane, since the pole-zero configuration of $G(s)H(s)$ is symmetrical to the axis. The asymptotes of the two root loci that approach infinity are characterized by Eqs. (93.46) through (93.48). The angles of the asymptotes are:

$$K \geq 0: \quad \phi_k = \frac{(2k + 1)\pi}{3 - 1} \qquad k = 0, 1 \qquad (93.51)$$

$$K \leq 0: \quad \phi_k = \frac{2k\pi}{3 - 1} \qquad k = 0, 1 \qquad (93.52)$$

Thus, for $K \geq 0$, the angles of the asymptotes are $\phi_0 = 90°$ and $\phi_1 = 270°$. For $K \leq 0$, $\phi_0 = 0°$ and $\phi_1 = 180°$.

The intersect of the asymptotes is at

$$\sigma = \frac{-1 + j - 1 - j - (-3)}{3 - 1} = \frac{1}{2} \qquad (93.53)$$

The root loci on the real axis are as indicated in Fig. 93.23.

Angles of Departure and Arrival. The slope of the root locus in the vicinity of a pole of $G(s)H(s)$ is measured at the *angle of departure* and that in the vicinity of a zero of $G(s)H(s)$ is measured at the *angle of arrival.*

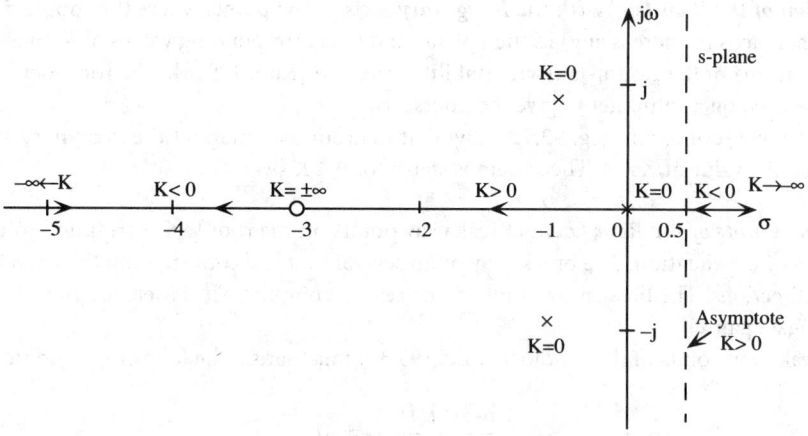

FIGURE 93.23 Some properties of the root loci of $G(s)H(s) = K(s+3)/[s(s^2 + 2s + 2)]$.

The angle of departure (arrival) of a root locus at a pole (zero) of $G(s)H(s)$ is determined by assigning a point s_1 to the root locus that is very close to the pole (zero) and applying the angle conditions of Eqs. (93.40) or (93.41). Figure 93.24 illustrates the calculation of the angles of arrival and departure of the root locus at the pole $s = -1 + j$. We assign a point s_1 that is on the locus for $K > 0$ near the pole and draw phasors from *all* the poles and the zero of $G(s)H(s)$ to this point. The angles made by the phasors with respect to the real axis must satisfy the angle condition in Eq. (93.46). Let the angle of the phasor drawn from $-1 + j$ to s_1 be designated as θ, which is the angle of departure; the angles drawn from the other poles and zero can be approximated by regarding s_1 as being very close to $-1 + j$. Thus, Eq. (93.46) leads to

$$\angle G(s_1)H(s_1) = -\theta - 135° - 90° + 26.6° = -180° \qquad (93.54)$$

or $\theta = -18.4°$. For the angle of arrival of the root locus at the pole $s = -1 + j$, we assign a point s_1 on the root loci for $K < 0$ near the pole. Drawing phasors from all the poles and the zero of $G(s)H(s)$ to s_1 and applying the angle condition in Eq. (93.47), we have

$$\angle G(s_1)H(s_1) = -\theta - 135° - 90° + 26.6° = 0° \qquad (93.55)$$

Thus, the angle of arrival of the root locus for $K < 0$ is $\theta = 198.4°$. Similarly, we can show that the angles of arrival and departure of the root locus at $s = -3$ are $180°$ and $0°$, respectively.

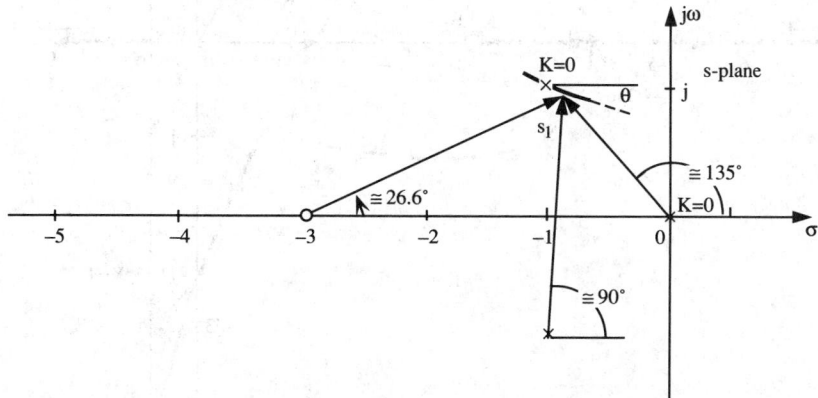

FIGURE 93.24 Angle of arrival and departure calculations.

Intersection of the Root Loci with the Imaginary Axis. The points where the root loci intersect the imaginary axis (if there is any) in the s plane, and the corresponding values of K, may be determined by means of the Routh-Hurwitz stability criterion [Kuo, 1991a]. The root locus program can also be used on a computer to give the intersects.

The complete root loci in Fig. 93.25 show that the root loci intersect the imaginary axis at $s = \pm j2.45$, and the value of K is 4. The system is stable for $0 \le K < 4$.

Breakaway Points of the Root Loci. Breakaway points on the root loci correspond to multiple-order roots of the equation. At a breakaway point several root loci converge and then break away in different directions. The breakaway point can be real or complex. The latter case must be in complex conjugate pairs.

The breakaway points of the root loci of Eq. (93.37) must satisfy the following condition:

$$\frac{dG(s)H(s)}{ds} = 0 \qquad (93.56)$$

On the other hand, not all solutions of Eq. (93.56) are breakaway points. To satisfy as a breakaway point, the point must also lie on the root loci, or satisfy Eq. (93.37). Applying Eq. (93.56) to the function $G(s)H(s)$ given in Eq. (93.50), we have the equation that the breakaway points must satisfy,

$$2s^3 + 11s^2 + 12s + 6 = 0 \qquad (93.57)$$

The roots of the last equation are $s = -4.256$, $-0.622 + j0.564$ and $-0.622 - j0.564$. As shown in Fig. 93.25, only the solution $s = -4.256$ is a breakaway point on the root loci.

Root Loci of Digital Control Systems

The root locus analysis presented in the preceding subsections can be applied to digital control sys-

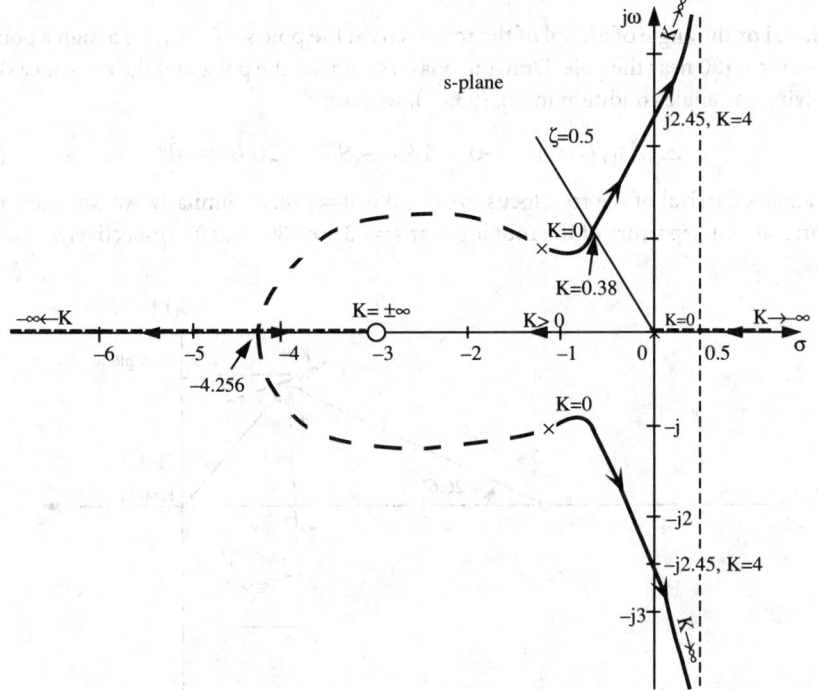

FIGURE 93.25 The complete root loci of $G(s)H(s) = K(s+3)/[s(s^2 + 2s + 2)]$.

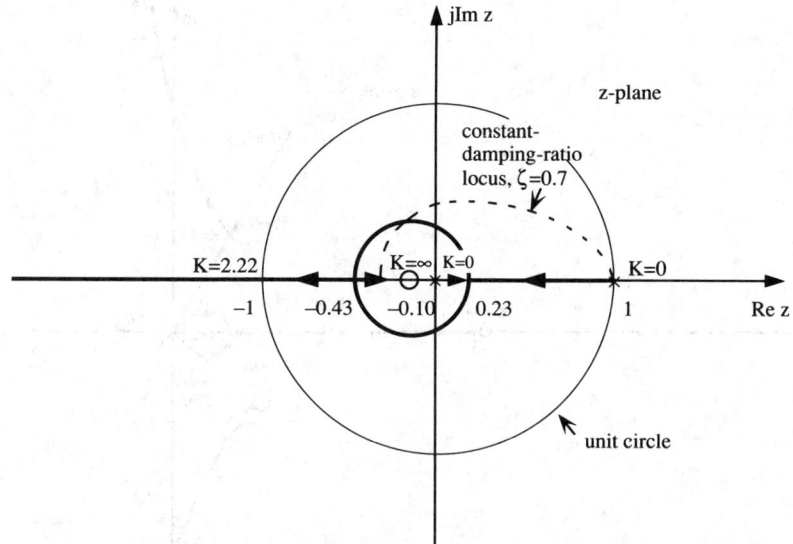

FIGURE 93.26 Root loci in the z plane for a digital control system.

tems without modifying the basic principles. For a linear time-invariant digital control system, the transfer functions are expressed in terms of the z-transform rather than the Laplace transform. The relationship between the z-transform variable z and the Laplace transform variable s is

$$z = e^{Ts} \qquad (93.58)$$

where T is the sampling period in seconds. Typically, the characteristic equation roots are solutions of the equation

$$1 + KGH(z) = 0 \qquad (93.59)$$

where K is the variable gain parameter. The root loci for a digital control system are constructed in the complex z plane. All the properties of the root loci in the s plane apply readily to the z plane, with the exception that the stability boundary is now the unit circle $|z| = 1$. That is, the system is stable if all the characteristic equation roots lie inside the unit circle.

As an illustration, the open-loop transfer function of a digital control system is given as

$$G(z) = \frac{K(z + 0.1)}{z(z - 1)} \qquad (93.60)$$

The characteristic equation of the closed-loop system is

$$z(z - 1) + K(z + 0.1) = 0$$

The root loci of the system are shown in Fig. 93.26. Notice that the system is stable for $0 \le K < 2.22$. When $K = 2.22$, one root is at $z = -1$, which is on the stability boundary.

Design with Root Locus

The root locus diagram of the characteristic equation of a closed-loop control system can be used for design purposes. The roots of the characteristic equation can be positioned in the s plane (or the z plane for digital control systems) to realize a certain desired relative stability or damping of the system. It should be kept in mind that the zeros of the closed-loop transfer function also affect the relative stability of the system, although the absolute stability is strictly governed by the characteristic equation roots.

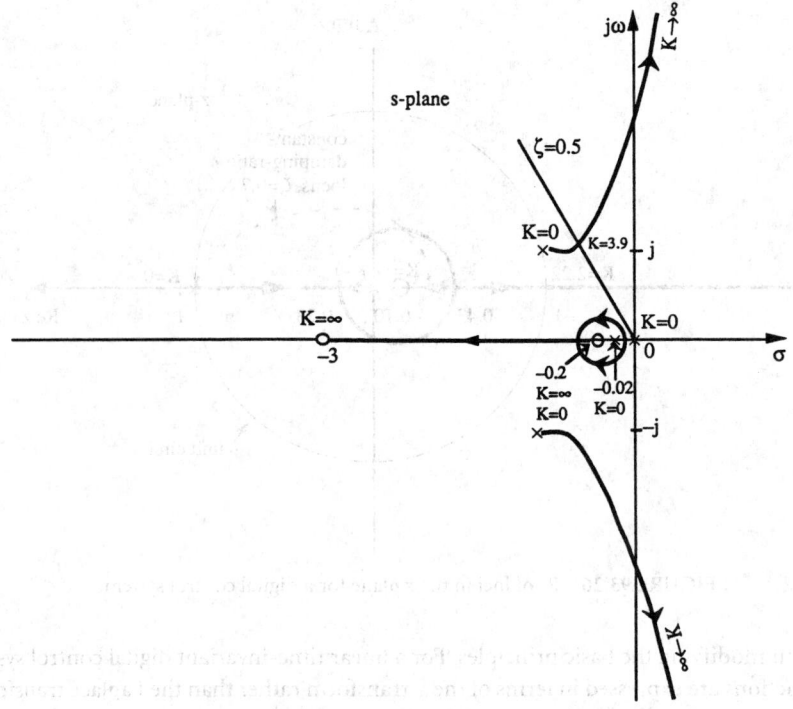

FIGURE 93.27   Root loci of Eq. (93.62).

As an illustrative example, the constant-damping ratio line for $\zeta = 0.5$ is shown in Fig. 93.25. The intersect of the $\zeta = 0.5$ line and the root locus corresponds to $K = 0.38$. Let us assume that we want to keep the relative damping at approximately 0.5 but the gain K should be increased tenfold. The following cascade controller is applied to the system [Evans, 1948]:

$$G_c(s) = \frac{1 + 5s}{1 + 50s} \qquad (93.61)$$

The open-loop transfer function of the compensated system is now

$$G_c(s)G(s)H(s) = \frac{0.1K(s + 3)(s + 0.2)}{s(s + 0.02)(s^2 + 2s + 2)} \qquad (93.62)$$

Figure 93.27 shows the root locus diagram of the compensated system for $K \geq 0$. The shape of the complex root loci is not appreciably affected by the controller, but the value of K that corresponds to a relative damping ratio of 0.5 is now approximately 3.9.

In a similar manner the root loci of digital control systems can be reshaped in the z plane for design. The constant-damping ratio locus in the z plane is shown in Fig. 93.26.

Defining Terms

Angles of departure and arrival: The slope of the root locus in the vicinity of a pole of $G(s)H(s)$ is measured as the angle of departure, and that in the vicinity of a zero of $G(s)H(s)$ is measured as the angle of arrival.

Asymptotes of root loci: The behavior of the root loci at $|s| = \infty$ when the number of poles and zeros of $G(s)H(s)$ is not equal.

Breakaway points of the root loci: Breakaway points on the root loci correspond to multiple-order roots of the equation.

Root locus: The trajectory of the roots of an algebraic equation with constant coefficient when a parameter varies.

References and Further Information

R. C. Dorf, *Modern Control Systems,* 5th ed., Reading, Mass.: Addison-Wesley, 1989.

W. R. Evans, "Graphical analysis of control systems," *Trans. AIEE,* vol. 67, pp. 547–551, 1948.

B. C. Kuo, *Automatic Control Systems,* 6th ed., Englewood Cliffs, N.J.: Prentice-Hall, 1991a.

B. C. Kuo, *ACSP Software and Manual,* Englewood Cliffs, N.J.: Prentice-Hall, 1991b.

B. C. Kuo, *Digital Control Systems,* 2nd ed., New York: Holt, 1992a.

B. C. Kuo, *DCSP Software and Manual,* Champaign, Ill.: SRL, Inc., 1992b.

93.5 Compensation

Charles L. Phillips and Royce D. Harbor

Compensation is the process of modifying a closed-loop control system (usually by adding a *compensator* or *controller*) in such a way that the compensated system satisfies a given set of design specifications. This section presents the fundamentals of compensator design; actual techniques are available in the references.

A single-loop control system is shown in Fig. 93.28. This system has the transfer function from input $R(s)$ to output $C(s)$

$$T(s) = \frac{C(s)}{R(s)} = \frac{G_c(s)G_p(s)}{1 + G_c(s)G_p(s)H(s)} \tag{93.63}$$

and the characteristic equation is

$$1 + G_c(s)G_p(s)H(s) = 0 \tag{93.64}$$

where $G_c(s)$ is the *compensator* transfer function, $G_p(s)$ is the *plant* transfer function, and $H(s)$ is the *sensor* transfer function. The transfer function from the disturbance input $D(s)$ to the output is $G_d(s)/[1 + G_c(s)G_p(s)H(s)]$. The function $G_c(s)G_p(s)H(s)$ is called the *open-loop function.*

Control System Specifications

The compensator transfer function $G_c(s)$ is designed to give the closed-loop system certain specified characteristics, which are realized through achieving one or more of the following:

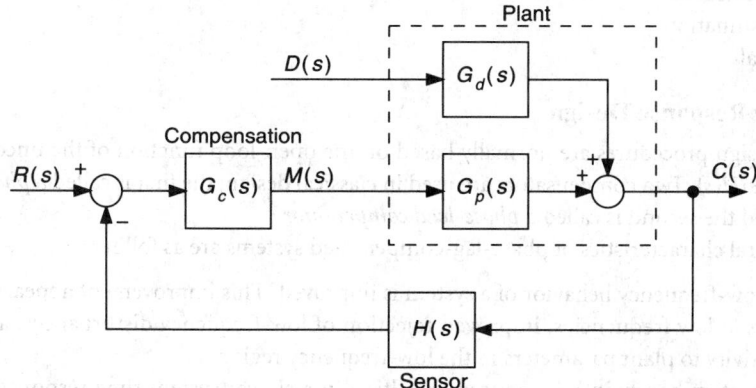

FIGURE 93.28 A closed-loop control system. (*Source:* C.L. Phillips and R.D. Harbor, *Feedback Control Systems,* 2nd ed., Englewood Cliffs, N.J.: Prentice-Hall, 1991, p. 161. With permission.)

1. Improving the transient response. Increasing the speed of response is generally accomplished by increasing the open-loop gain $G_c(j\omega)G_p(j\omega)H(j\omega)$ at higher frequencies such that the system bandwidth is increased. Reducing overshoot (ringing) in the response generally involves increasing the phase margin ϕ_m of the system, which tends to remove any resonances in the system. The phase margin ϕ_m occurs at the frequency ω_1 and is defined by the relationship

$$\left| G_c(j\omega_1)G_p(j\omega_1)H(j\omega_1) \right| = 1$$

with the angle of $G_c(j\omega_1)G_p(j\omega_1)H(j\omega_1)$ equal to $(180° + \phi_m)$.

2. Reducing the steady-state errors. Steady-state errors are decreased by increasing the open-loop gain $G_c(j\omega)G_p(j\omega)H(j\omega)$ in the frequency range of the errors. Low-frequency errors are reduced by increasing the low-frequency open-loop gain and by increasing the type number of the system [the number of poles at the origin in the open-loop function $G_c(s)G_p(s)H(s)$].

3. Reducing the sensitivity to plant parameters. Increasing the open-loop gain $G_c(j\omega)G_p(j\omega)$ $H(j\omega)$ tends to reduce the variations in the system characteristics due to variations in the parameters of the plant.

4. Rejecting disturbances. Increasing the open-loop gain $G_c(j\omega)G_p(j\omega)H(j\omega)$ tends to reduce the effects of disturbances [$D(s)$ in Fig. 93.28] on the system output, provided that the increase in gain does not appear in the direct path from disturbance inputs to the system output.

5. Increasing the relative stability. Increasing the open-loop gain tends to reduce phase and gain margins, which generally increases the overshoot in the system response. Hence, a trade-off exists between the beneficial effects of increasing the open-loop gain and the resulting detrimental effects of reducing the stability margins.

Design

Design procedures for compensators are categorized as either *classical methods* or *modern methods*. Classical methods discussed are:

- Phase-lag frequency response
- Phase-lead frequency response
- Phase-lag root locus
- Phase-lead root locus

Modern methods discussed are:

- Pole placement
- State estimation
- Optimal

Frequency Response Design

Classical design procedures are normally based on the open-loop function of the uncompensated system, $G_p(s)H(s)$. Two compensators are used in classical design; the first is called a *phase-lag compensator*, and the second is called a *phase-lead compensator*.

The general characteristics of phase-lag-compensated systems are as follows:

1. The low-frequency behavior of a system is improved. This improvement appears as reduced errors at low frequencies, improved rejection of low-frequency disturbances, and reduced sensitivity to plant parameters in the low-frequency region.

2. The system bandwidth is reduced, resulting in a slower system time response and better rejection of high-frequency noise in the sensor output signal.

The general characteristics of phase-lead-compensated systems are as follows:

1. The high-frequency behavior of a system is improved. This improvement appears as faster responses to inputs, improved rejection of high-frequency disturbances, and reduced sensitivity to changes in the plant parameters.
2. The system bandwidth is increased, which can increase the response to high-frequency noise in the sensor output signal.

The transfer function of a first-order compensator can be expressed as

$$G_c(s) = \frac{K_c(s/\omega_0 + 1)}{s/\omega_p + 1} \tag{93.65}$$

where $-\omega_0$ is the compensator zero, $-\omega_p$ is its pole, and K_c is its dc gain. If $\omega_p < \omega_0$, the compensator is phase-lag. The Bode diagram of a phase-lag compensator is given in Fig. 93.29 for $K_c = 1$.

It is seen from Fig. 93.29 that the phase-lag compensator reduces the high-frequency gain of the open-loop function relative to the low-frequency gain. This effect allows a higher low-frequency gain, with the advantages listed above. The pole and zero of the compensator must be placed at very low frequencies relative to the compensated-system bandwidth so that the destabilizing effects of the negative phase of the compensator are negligible.

If $\omega_p > \omega_0$ the compensator is phase-lead. The Bode diagram of a phase-lead compensator is given in Fig. 93.30 for $K_c = 1$.

It is seen from Fig. 93.30 that the phase-lead compensator increases the high-frequency gain of the open-loop function relative to its low-frequency gain. Hence, the system has a larger bandwidth, with the advantages listed above. The pole and zero of the compensator are generally difficult to place, since the increased gain of the open-loop function tends to destabilize the system, while the phase lead of the compensator tends to stabilize the system. The pole-zero placement for the phase-lead compensator is much more critical than that of the phase-lag compensator.

A typical Nyquist diagram of an uncompensated system is given in Fig. 93.31. The pole and the zero of a phase-lag compensator are placed in the frequency band labeled *A*. This placement negates the destabilizing effect of the negative phase of the compensator. The pole and zero of a phase-lead compensator are placed in the frequency band labeled *B*. This placement utilizes the stabilizing effect of the positive phase of the compensator.

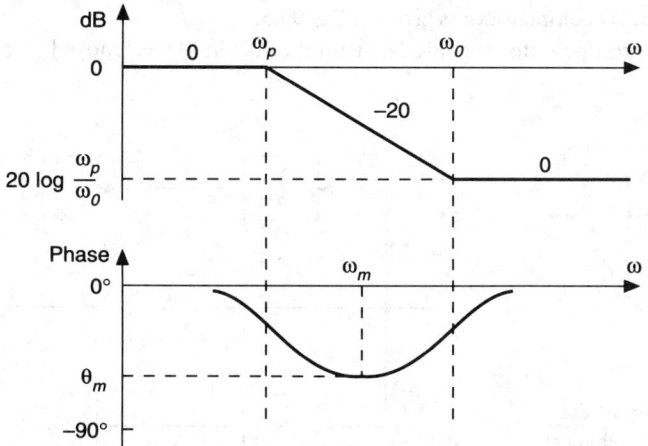

FIGURE 93.29 Bode diagram for a phase-lag compensator. (*Source:* C.L. Phillips and R.D. Harbor, *Feedback Control Systems*, 2nd ed., Englewood Cliffs, N.J.: Prentice-Hall, 1991, p. 358. With permission.)

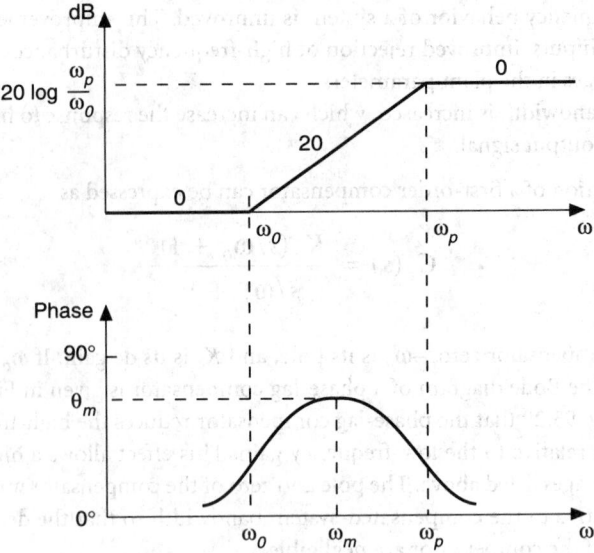

FIGURE 93.30 Bode diagram for a phase-lead compensator. (*Source:* C.L. Phillips and R.D. Harbor, *Feedback Control Systems,* 2nd ed., Englewood Cliffs, N.J.: Prentice-Hall, 1991, p. 363. With permission.)

PID Controllers

Proportional-plus-integral-plus-derivative (PID) compensators are probably the most utilized form for compensators. These compensators are essentially equivalent to a phase-lag compensator cascaded with a phase-lead compensator. The transfer function of this compensator is given by

$$G_c(s) = K_P + \frac{K_I}{s} + K_D s \qquad (93.66)$$

A block diagram portrayal of the compensator is shown in Fig. 93.32. The integrator in this compensator increases the system type by one, resulting in an improved low-frequency response. The Bode diagram of a PID compensator is given in Fig. 93.33.

With $K_D = 0$, the compensator is phase-lag, with the pole in (93.65) moved to $\omega_p = 0$. As a result

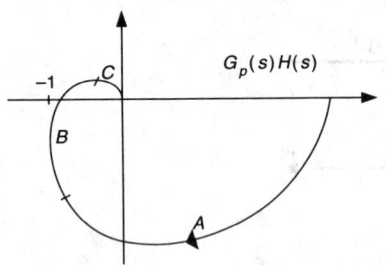

FIGURE 93.31 A typical Nyquist diagram for $G_p(s)H(s)$. (*Source:* C.L. Phillips and R.D. Harbor, *Feedback Control Systems,* 2nd ed., Englewood Cliffs, N.J.: Prentice-Hall, 1991, p. 364. With permission.)

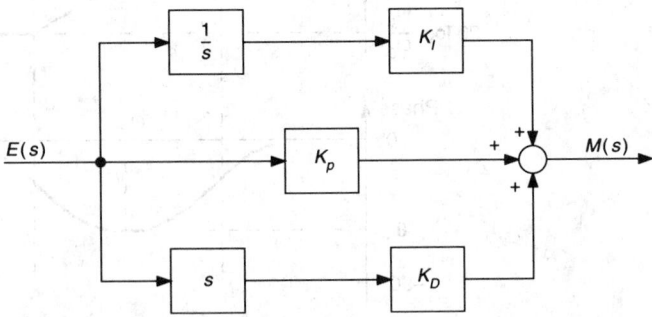

FIGURE 93.32 Block diagram of a PID compensator. (*Source:* C.L. Phillips and R.D. Harbor, *Feedback Control Systems,* 2nd ed., Englewood Cliffs, N.J.: Prentice-Hall, 1991, p. 378. With permission.)

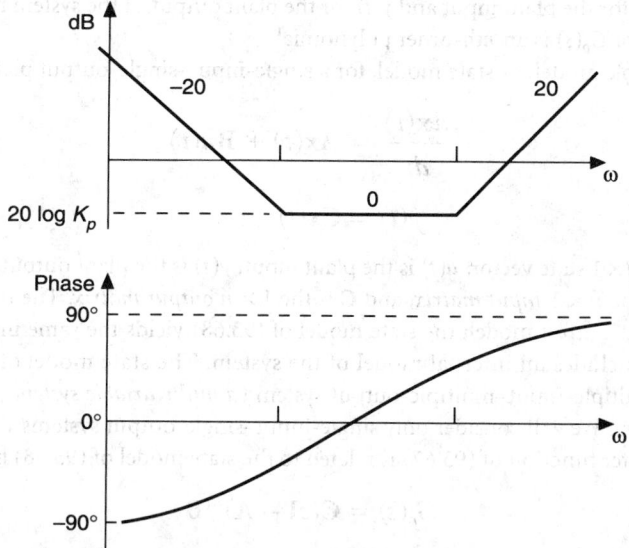

FIGURE 93.33 Bode diagram of a PID compensator. (*Source:* C.L. Phillips and R.D. Harbor, *Feedback Control Systems,* 2nd ed., Englewood Cliffs, N.J.: Prentice-Hall, 1991, p. 382. With permission.)

the compensator is type one. The zero of the compensator is placed in the low-frequency range to correspond to the zero of the phase-lag compensator discussed above.

With $K_I = 0$, the compensator is phase-lead, with a single zero and the pole moved to infinity. Hence, the gain continues to increase with increasing frequency. If high-frequency noise is a problem, it may be necessary to add one or more poles to the PD or PID compensators. These poles must be placed at high frequencies relative to the phase-margin frequency such that the phase margin (stability characteristics) of the system is not degraded. PD compensators realized using rate sensors minimize noise problems [Phillips and Harbor, 1991].

Root Locus Design

Root locus design procedures generally result in the placement of the two dominant poles of the closed-loop system transfer function. A system has two dominant poles if its behavior approximates that of a second-order system.

The differences in root locus designs and frequency response designs appear only in the interpretation of the control-system specifications. A root locus design that improves the low-frequency characteristics of the system will result in a phase-lag controller; a phase-lead compensator results if the design improves the high-frequency response of the system. If a root locus design is performed, the frequency response characteristics of the system should be investigated. Also, if a frequency response design is performed, the poles of the closed-loop transfer function should be calculated.

Modern Control Design

The classical design procedures above are based on a transfer-function model of a system. Modern design procedures are based on a *state-variable model* of the plant. The plant transfer function is given by

$$\frac{Y(s)}{U(s)} = G_p(s) \tag{93.67}$$

where we use $u(t)$ for the plant input and $y(t)$ for the plant output. If the system model is nth order, the denominator of $G_p(s)$ is an nth-order polynomial.

The state-variable model, or state model, for a single-input–single-output plant is given by

$$\frac{d\mathbf{x}(t)}{dt} = \mathbf{A}\mathbf{x}(t) + \mathbf{B}u(t)$$

(93.68)

$$y(t) = \mathbf{C}\mathbf{x}(t)$$

where $\mathbf{x}(t)$ is the $n \times 1$ state vector, $u(t)$ is the plant input, $y(t)$ is the plant output, $\mathbf{A}$ is the $n \times n$ *system matrix*, $\mathbf{B}$ is the $n \times 1$ *input matrix*, and $\mathbf{C}$ is the $1 \times n$ *output matrix*. The transfer function of (93.67) is an input-output model; the state model of (93.68) yields the same input-output model and in addition includes an internal model of the system. The state model of (93.68) is readily adaptable to a multiple-input–multiple-output system (*a multivariable system*); for that case, $u(t)$ and $y(t)$ are vectors. We will consider only single-input–single-output systems.

The plant transfer function of (93.67) is related to the state model of (93.68) by

$$G_p(s) = \mathbf{C}(s\mathbf{I} - \mathbf{A})^{-1}\mathbf{B}$$

(93.69)

The state model is not unique; many combinations of the matrices $\mathbf{A}$, $\mathbf{B}$, and $\mathbf{C}$ can be found to satisfy (93.69) for a given transfer function $G_p(s)$.

Classical compensator design procedures are based on the open-loop function $G_p(s)H(s)$ of Fig. 93.28. It is common to present modern design procedures as being based on only the plant model of (93.68). However, the models of the sensors that measure the signals for feedback must be included in the state model. This problem will become more evident as the modern procedures are presented.

Pole Placement

Probably the simplest modern design procedure is *pole placement*. Recall that root locus design was presented as placing the two dominant poles of the closed-loop transfer function at desired locations. The pole-placement procedure places *all* poles of the closed-loop transfer function, or equivalently, all roots of the closed-loop system characteristic equation, at desirable locations.

The system design specifications are used to generate the desired closed-loop system characteristic equation $\alpha_c(s)$, where

$$\alpha_c(s) = s^n + \alpha_{n-1}s^{n-1} + \cdots + \alpha_1 s + \alpha_0 = 0$$

(93.70)

for an nth-order plant. This characteristic equation is realized by requiring the plant input to be a linear combination of the plant states, that is,

$$u(t) = -K_1 x_1(t) - K_2 x_2(t) - \cdots - K_n x_n(t) = -\mathbf{K}\mathbf{x}(t)$$

(93.71)

where $\mathbf{K}$ is the $1 \times n$ feedback-gain matrix. Hence *all* states must be measured and fed back. This operation is depicted in Fig. 93.34.

The feedback-gain matrix $\mathbf{K}$ is determined from the desired characteristic equation for the closed-loop system of (93.70):

$$\alpha_c(s) = |s\mathbf{I} - \mathbf{A} + \mathbf{B}\mathbf{K}| = 0$$

(93.72)

This equation can be solved for $\mathbf{K}$:

$$\mathbf{K} = [0 \quad 0 \quad \cdots \quad 0 \quad 1][\mathbf{B} \quad \mathbf{A}\mathbf{B} \quad \cdots \quad \mathbf{A}^{n-1}\mathbf{B}]^{-1}\alpha_c(\mathbf{A})$$

(93.73)

where $\alpha_c(\mathbf{A})$ is (93.70) with the scalar s replaced with the matrix $\mathbf{A}$. A plant is said to be *controllable* if the inverse matrix in (93.73) exists. Calculation of $\mathbf{K}$ completes the design process. A simple computer algorithm is available for solving (93.73) for $\mathbf{K}$.

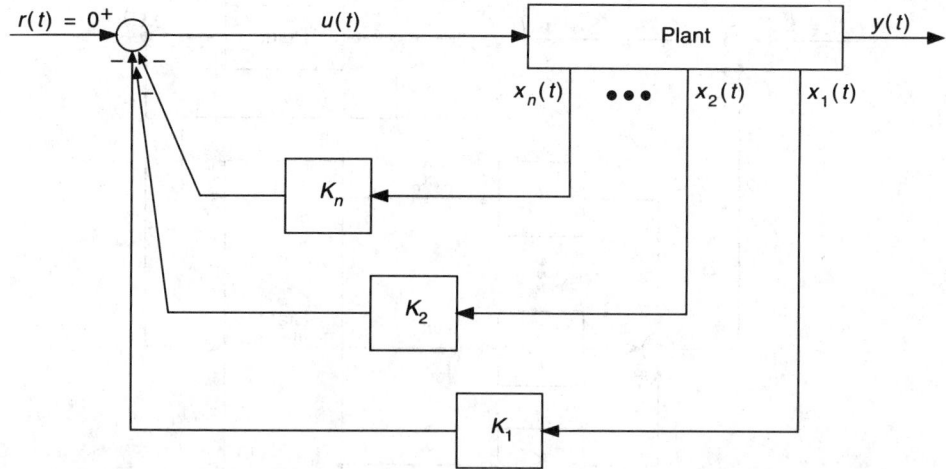

FIGURE 93.34 Implementation of pole-placement design. (*Source:* C.L. Phillips and R.D. Harbor, *Feedback Control Systems,* 2nd ed., Englewood Cliffs, N.J.: Prentice-Hall, 1991, p. 401. With permission.)

State Estimation

In general, modern design procedures require that the state vector $\mathbf{x}(t)$ be fed back, as in (93.71). The measurement of all state variables is difficult to implement for high-order systems. The usual procedure is to estimate the states of the system from the measurement of the output $y(t)$, with the estimated states then fed back.

Let the estimated state vector be $\hat{\mathbf{x}}$. One procedure for estimating the system states is by an *observer,* which is a dynamic system realized by the equations

$$\frac{d\hat{\mathbf{x}}(t)}{dt} = (\mathbf{A} - \mathbf{GC})\hat{\mathbf{x}}(t) + \mathbf{B}u(t) + \mathbf{G}y(t) \tag{93.74}$$

with the feedback equation of (93.71) now realized by

$$u(t) = -\mathbf{K}\hat{\mathbf{x}}(t) \tag{93.75}$$

The matrix $\mathbf{G}$ in (93.74) is calculated by assuming an nth-order characteristic equation for the observer of the form

$$\alpha_e(s) = \left|s\mathbf{I} - \mathbf{A} + \mathbf{GC}\right| = 0 \tag{93.76}$$

This equation can be solved for $\mathbf{G}$:

$$\mathbf{G} = \alpha_e(\mathbf{A})[\mathbf{C}\quad \mathbf{CA}\quad \cdots\quad \mathbf{CA}^{n-1}]^{-T}[0\quad 0\cdots\quad 0\quad 1]^T \tag{93.77}$$

where $[\cdot]^T$ denotes the matrix transpose. A plant is said to be *observable* if the inverse matrix in (93.77) exists. An implementation of the closed-loop system is shown in Fig. 93.35. The observer is usually implemented on a digital computer. The plant and the observer in Fig. 93.35 are both nth-order; hence, the closed-loop system is of order $2n$.

The observer-pole-placement system of Fig. 93.35 is equivalent to the system of Fig. 93.36, which is of the form of closed-loop systems designed by classical procedures. The transfer function of the controller-estimator (equivalent compensator) of Fig. 93.36 is given by

$$\mathbf{G}_{ec}(s) = \mathbf{K}[s\mathbf{I} - \mathbf{A} + \mathbf{GC} + \mathbf{BK}]^{-1}\mathbf{G} \tag{93.78}$$

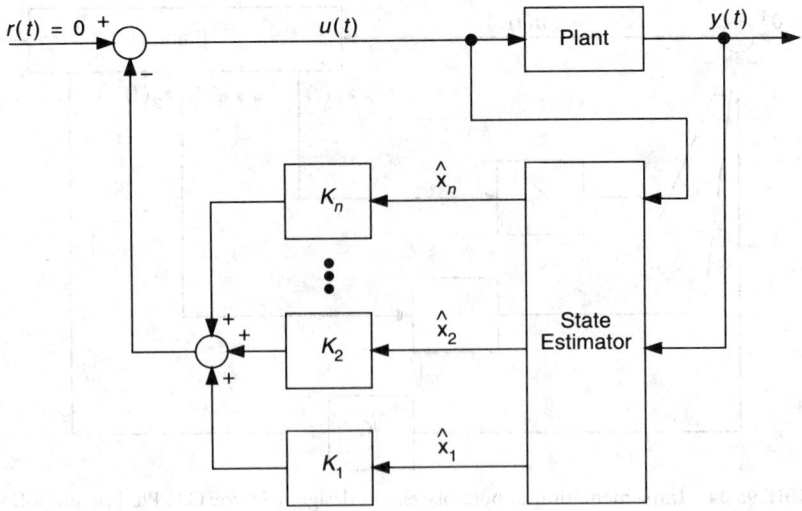

FIGURE 93.35 Implementation of observer-pole-placement design. (*Source:* C.L. Phillips and R.D. Harbor, *Feedback Control Systems,* 2nd ed., Englewood Cliffs, N.J.: Prentice-Hall, 1991, p. 417. With permission.)

This compensator is nth-order for an nth-order plant; hence, the total system is of order $2n$. The characteristic equation for the compensated system is given by

$$|s\mathbf{I} - \mathbf{A} + \mathbf{GC} + \mathbf{BK}| = \alpha_c(s)\alpha_e(s) = 0 \qquad (93.79)$$

The roots of this equation are the roots of the pole-placement design plus those of the observer design. For this reason, the roots of the characteristic equation for the observer are usually chosen to be faster than those of the pole-placement design.

Linear Quadratic Optimal Control

We define an optimal control system as one for which some mathematical function is minimized. The function to be minimized is called the *cost function.* For steady-state linear quadratic optimal control the cost function is given by

$$V_\infty = \int_t^\infty [\mathbf{x}^T(\tau)\mathbf{Q}\mathbf{x}(\tau) + Ru^2(\tau)]\,d\tau \qquad (93.80)$$

where $\mathbf{Q}$ and $\mathbf{R}$ are chosen to satisfy the design criteria. In general, the choices are not straightforward. Minimization of (93.80) requires that the plant input be given by

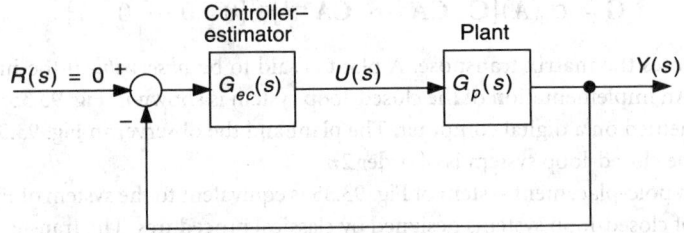

FIGURE 93.36 Equivalent system for pole-placement design. (*Source:* C.L. Phillips and R.D. Harbor, *Feedback Control Systems,* 2nd ed., Englewood Cliffs, N.J.: Prentice-Hall, 1991, p. 414. With permission.)

$$u(t) = -R^{-1}\mathbf{B}^T\mathbf{M}_\infty\mathbf{x}(t) \tag{93.81}$$

where the $n \times n$ matrix $\mathbf{M}_\infty$ is the solution to the *algebraic Riccati equation*

$$\mathbf{M}_\infty\mathbf{A} + \mathbf{A}^T\mathbf{M}_\infty - \mathbf{M}_\infty\mathbf{B}R^{-1}\mathbf{B}^T\mathbf{M}_\infty + \mathbf{Q} = 0 \tag{93.82}$$

The existence of a solution for this equation is involved [Friedland, 1986] and is not presented here. Optimal control systems can be designed for cost functions other than that of (93.80).

Other Modern Design Procedures

Other modern design procedures exist; for example, *self-tuning control systems* continually estimate certain plant parameters and adjust the compensator based on this estimation. These control systems are a type of *adaptive control systems* and usually require that the control algorithms be implemented using a digital computer. These control systems are beyond the scope of this book (see, for example, Astrom and Wittenmark, 1984).

Defining Term

Compensation: The process of physically altering a closed-loop system such that the system has specified characteristics. This alteration is achieved either by changing certain parameters in the system or by adding a physical system to the closed-loop system; in some cases both methods are used.

References and Further Information

K. J. Astrom and B. Wittenmark, *Computer Controlled Systems,* Englewood Cliffs, N.J.: Prentice-Hall, 1984.

W. L. Brogan, *Modern Control Theory,* Englewood Cliffs, N.J.: Prentice-Hall, 1985.

R. C. Dorf, *Modern Control Systems,* 5th ed., Reading, Mass.: Addison-Wesley, 1989.

G. F. Franklin, J. D. Powell, and A. Emami-Naeini, *Feedback Control of Dynamic Systems,* Reading, Mass.: Addison-Wesley, 1986.

B. Friedland, *Control System Design,* New York: McGraw-Hill, 1986.

B. C. Kuo, *Automatic Control Systems,* Englewood Cliffs, N.J.: Prentice-Hall, 1987.

C. L. Phillips and R. D. Harbor, *Feedback Control Systems,* 2nd ed., Englewood Cliffs, N.J.: Prentice-Hall, 1991.

93.6 Digital Control Systems

Raymond G. Jacquot and John E. McInroy

The use of the **digital computer** to control physical processes has been a topic of discussion in the technical literature for over four decades, but the actual use of a digital computer for control of industrial processes was reserved only for massive and slowly varying processes such that the high cost and slow computing speed of available computers could be tolerated. The invention of the integrated circuit microprocessor in the early 1970s radically changed all that; now microprocessors are used in control tasks in automobiles and household appliances, applications where high cost is not justifiable.

When the term *digital control* is used, it usually refers to the process of employing a digital computer to control some process that is characterized by continuous-in-time dynamics. The control can be of the open-loop variety where the control strategy output by the digital computer is dictated without regard to the status of the process variables. An alternative technique is to supply the digital computer with digital data about the process variables to be controlled, and thus the control strategy output by the computer depends on the process variables that are to be controlled. This latter strategy is a **feedback control** strategy wherein the computer, the process, and interface hardware form a closed loop of information flow.

Examples of dynamic systems that are controlled in such a closed-loop digital fashion are flight control of civilian and military aircraft, control of process variables in chemical processing plants, and position and force control in industrial robot manipulators. The simplest form of feedback control strategy provides an on-off control to the controlling variables based on measured values of the process variables. This strategy will be illustrated by a simple example in a following subsection.

In the past decade and a half many excellent textbooks on the subject of digital control systems have been written, and most of them are in their second edition. The texts in the References provide in-depth development of the theory by which such systems are analyzed and designed.

A Simple Example

Such a closed-loop or feedback control situation is illustrated in Fig. 93.37, which illustrates the feedback control of the temperature in a simple environmental chamber that is to be kept at a constant temperature somewhat above room temperature.

Heat is provided by turning on a relay that supplies power to a heater coil. The on-off signal to the relay can be supplied by 1 bit of an output port of the microprocessor (typically the port would be 8 bits wide). A second bit of the port can be used to turn a fan on and off to supply cooling air to the chamber. An analog-to-digital (A/D) converter is employed to convert the amplified thermocouple signal to a digital word that is then supplied to the input port of the microprocessor. The program being executed by the microprocessor reads the temperature data supplied to the input port and compares the binary number representing the temperature to a binary version of the desired temperature and makes a decision whether or not to turn on the heater or the fan or to do nothing. The program being executed runs in a continuous loop, repeating the operations discussed above.

This simple on-off control strategy is often not the best when extremely precise control of the process variables is required. A more precise control may be obtained if the controlling variable levels can be adjusted to be somewhat larger if the deviation of the process variable from the desired value is larger.

Single-Loop Linear Control Laws

Consider the case where a single variable of the process is to be controlled, as illustrated in Fig. 93.38.

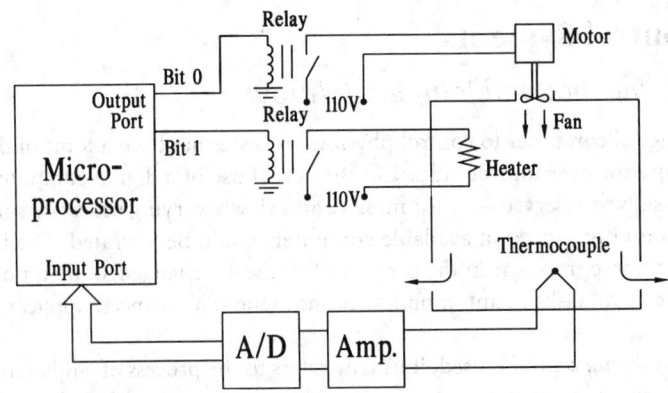

FIGURE 93.37 Microprocessor control of temperature in a simple environmental chamber.

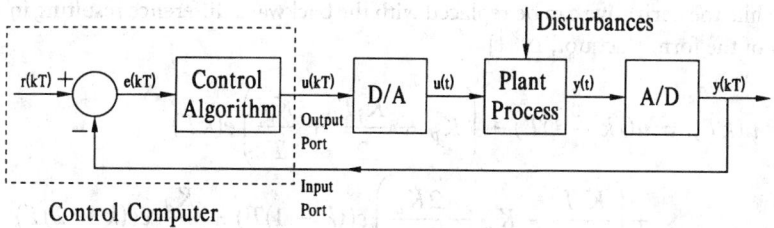

FIGURE 93.38   Closed-loop control of a single process variable.

The output of the plant $y(t)$ is to be sampled every T seconds by an A/D converter, and this sequence of numbers will be denoted as $y(kT)$, $k = 0, 1, 2, \ldots$. The goal is to make the sequence $y(kT)$ follow some desired known sequence [the reference sequence $r(kT)$]. Consequently, the sequence $y(kT)$ is subtracted from $r(kT)$ to obtain the so-called error sequence $e(kT)$. The control computer then acts on the error sequence, using some control algorithms, to produce the control effort sequence $u(kT)$ that is supplied to the digital-to-analog (D/A) converter which then drives the actuating hardware with a signal proportional to $u(kT)$. The output of the D/A converter is then held constant on the current time interval, and the control computer waits for the next sample of the variable to be controlled, the arrival of which repeats the sequence. The most commonly employed control algorithm or control law is a linear difference equation of the form

$$u(kT) = a_n e(kT) + a_{n-1} e((k-1)T) + \cdots + a_0 e((k-n)T)$$
$$+ b_{n-1} u((k-1)T) + \cdots + b_0 u((k-n)T) \qquad (93.83)$$

The question remains as to how to select the coefficients $a_0, \ldots, a_n$ and $b_0, \ldots, b_{n-1}$ in expression (93.83) to give an acceptable degree of control of the plant.

Proportional Control

This is the simplest possible control algorithm for the digital processor wherein the most current control effort is proportional to the current error or using only the first term of relation (93.83)

$$u(kT) = a_n e(kT) \qquad (93.84)$$

This algorithm has the advantage that it is simple to program, while, on the other hand, its disadvantage lies in the fact that it has poor disturbance rejection properties in that if a_n is made large enough for good disturbance rejection, the closed-loop system can be unstable (i.e., have transient responses which increase with time). Since the object is to regulate the system output in a known way, these unbounded responses preclude this regulation.

PID Control Algorithm

A common technique employed for decades in chemical process control loops is that of proportional-plus-integral-plus-derivative (PID) control wherein a continuous-time control law would be given by

$$u(t) = K_p e(t) + K_i \int_0^t e(\tau)d\tau + K_d \frac{de}{dt} \qquad (93.85)$$

This would have to be implemented by an analog filter.

To implement the design in digital form the proportional term can be carried forward as in relation (93.84); however, the integral can be replaced by trapezoidal integration using the error sequence, while the derivative can be replaced with the backward difference resulting in a computer control law of the form [Jacquot, 1981]

$$u(kT) = u((k-1)T) + \left(K_p + \frac{K_i T}{2} + \frac{K_d}{T}\right) e(kT)$$

$$+ \left(\frac{K_i T}{2} - K_p - \frac{2K_d}{T}\right) e((k-1)T) + \frac{K_d}{T} e((k-2)T) \tag{93.86}$$

where T is the duration of the sampling interval. The selection of the coefficients in this algorithm (K_i, K_d, and K_p) is best accomplished by the Ziegler-Nichols tuning process [Franklin *et al.*, 1990].

The Closed-Loop System

When the plant process is linear or may be linearized about an operating point and the control law is linear as in expressions (93.83), (93.84), or (93.86), then an appropriate representation of the complete closed-loop system is by the so-called z-transform. The z-transform plays the role for linear, constant-coefficient difference equations that the Laplace transform plays for linear, constant-coefficient differential equations. This z-domain representation allows the system designer to investigate system time response, frequency response, and stability in a single analytical framework.

If the plant can be represented by an s-domain transfer function $G(s)$, then the discrete-time (z-domain) transfer function of the plant, the analog-to-digital converter, and the driving digital-to-analog converter is

$$G(z) = \left(\frac{z-1}{z}\right) Z\left\{L^{-1}\left[\frac{G(s)}{s}\right]\right\} \tag{93.87}$$

where $Z(\cdot)$ is the z-transform and $L^{-1}(\cdot)$ is the inverse Laplace transform. The transfer function of the control law of (93.83) is

$$D(z) = \frac{U(z)}{E(z)} = \frac{a_n z^n + a_{n-1} z^{n-1} + \cdots + a_0}{z^n - b_{n-1} z^{n-1} - \cdots - b_0} \tag{93.88}$$

For the closed-loop system of Fig. 93.38 the closed-loop z-domain transfer function is

$$M(z) = \frac{Y(z)}{R(z)} = \frac{G(z)D(z)}{1 + G(z)D(z)} \tag{93.89}$$

where $G(z)$ and $D(z)$ are specified above. The characteristic equation of the closed-loop system is

$$1 + G(z)D(z) = 0 \tag{93.90}$$

The dynamics and stability of the system can be assessed by the locations of the zeros of (93.90) (the closed-loop poles) in the complex z plane. For stability the zeros of (93.90) above must be restricted to the unit circle of the complex z plane.

A Linear Control Example

Consider the temperature control of a chemical mixing tank shown in Fig. 93.39. From a transient power balance the differential equation relating the rate of heat added $q(t)$ to the deviation in temperature from the ambient $\theta(t)$ is given as

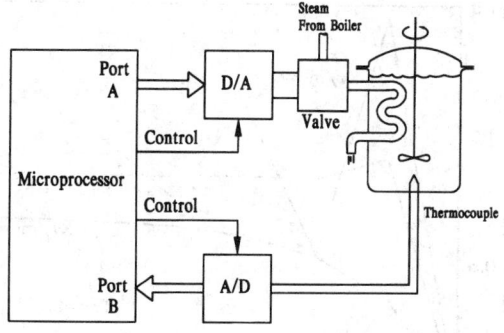

FIGURE 93.39   A computer-controlled thermal mixing tank.

$$\frac{d\theta}{dt} + \frac{1}{\tau}\,\theta = \frac{1}{mc}\,q(t) \tag{93.91}$$

where τ is the time constant of the process and mc is the heat capacity of the tank. The transfer function of the tank is

$$\frac{\Theta(s)}{Q(s)} = G(s) = \frac{1/mc}{s + 1/\tau} \tag{93.92}$$

The heater is driven by a D/A converter, and the temperature measurement is sampled with an A/D converter. The data converters are assumed to operate synchronously, so the discrete-time transfer function of the tank and the two data converters is from expression (93.87):

$$G(z) = \frac{\Theta(z)}{Q(z)} = \frac{\tau}{mc}\,\frac{1 - e^{-T/\tau}}{z - e^{-T/\tau}} \tag{93.93}$$

If a proportional control law is chosen, the transfer function associated with the control law is the gain $a_n = K$ or

$$D(z) = K \tag{93.94}$$

The closed-loop characteristic equation is from (93.90):

$$1 + \frac{K\tau}{mc}\,\frac{1 - e^{-T/\tau}}{z - e^{-T/\tau}} = 0 \tag{93.95}$$

If a common denominator is found, the resulting numerator is

$$z - e^{-T/\tau} + \frac{K\tau}{mc}\,(1 - e^{-T/\tau}) = 0 \tag{93.96}$$

The root of this equation is

$$z = e^{-T/\tau} + \frac{K\tau}{mc}\,(e^{-T/\tau} - 1) \tag{93.97}$$

If this root location is investigated as the gain parameter K is varied upward from zero, it is seen that the root starts at $z = e^{-T/\tau}$ for $K = 0$ and moves to the left along the real axis as K increases. Initially it is seen that the system becomes faster, but at some point the responses become damped and oscillatory, and as K is further increased the oscillatory tendency becomes less damped, and finally a value of K is reached where the oscillations are sustained at constant amplitude. A further

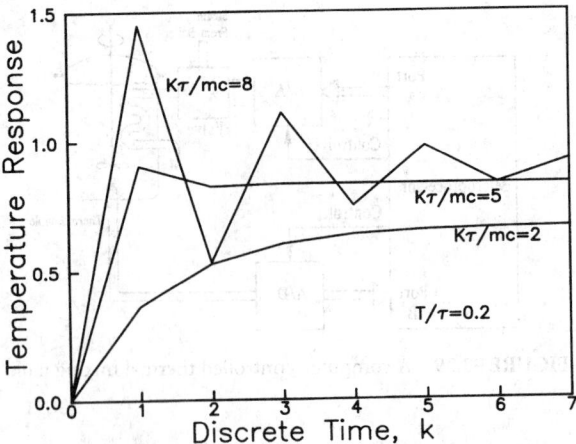

FIGURE 93.40 Step responses of proportionally controlled thermal mixing tank.

increase in K will yield oscillations that increase with time. Typical unit step responses for $r(k) = 1$ and $T/\tau = 0.2$ are shown in Fig. 93.40.

It is easy to observe this tendency toward oscillation as K increases, but a problem that is clear from Fig. 93.40 is that in the steady state there is a persistent error between the response and the reference [$r(k) = 1$]. Increasing the gain K will make this error smaller at the expense of more oscillations. As a remedy for this steady-state error problem and control of the dynamics, a control law transfer function $D(z)$ will be sought that inserts integrator action into the loop while simultaneously canceling the pole of the plant. This dictates that the controller have a transfer function of the form

$$D(z) = \frac{U(z)}{E(z)} = \frac{K(z - e^{-T/\tau})}{z - 1} \tag{93.98}$$

Typical unit step responses are illustrated in Fig. 93.41 for several values of the gain parameter. The control law that must be programmed in the digital processor is

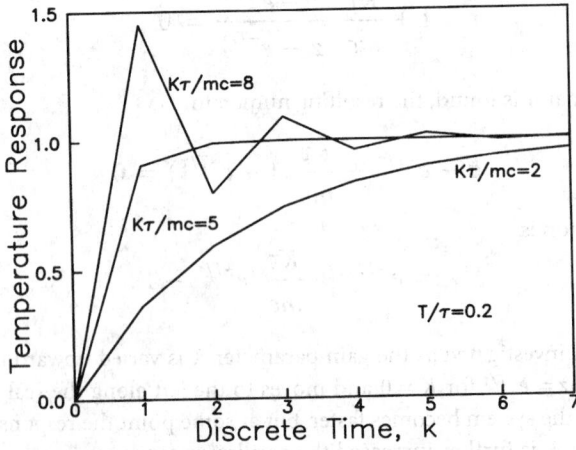

FIGURE 93.41 Step responses of the compensated thermal mixing tank.

$$u(kT) = u((k-1)T) + K[e(kT) - e^{-T/\tau}e((k-1)T)] \qquad (93.99)$$

The additional effort to program this over that required to program the proportional control law of (93.94) is easily justified since K and $e^{-T/\tau}$ are simply constants.

Defining Terms

Digital computer: A collection of digital devices including an arithmetic logic unit (ALU), read-only memory (ROM), random-access memory (RAM), and control and interface hardware.

Feedback control: The regulation of a response variable of a system in a desired manner using measurements of that variable in the generation of the strategy of manipulation of the controlling variables.

References

K. J. Astrom and B. Wittenmark, *Computer Controlled Systems: Theory and Design*, Englewood Cliffs, N.J.: Prentice-Hall, 1984.

G. F. Franklin, J. D. Powell, and M. L. Workman, *Digital Control of Dynamic Systems*, 2nd ed., Reading, Mass.: Addison-Wesley, 1990.

C. H. Houpis and G. B. Lamont, *Digital Control Systems: Theory, Hardware, Software*, 2nd ed., New York: McGraw-Hill, 1992.

R. G. Jacquot, *Modern Digital Control Systems*, New York: Marcel Dekker, 1981.

B. C. Kuo, *Digital Control Systems*, 2nd ed., Orlando, Fla.: Saunders, 1992.

C. L. Phillips and H. T. Nagle, *Digital Control System Analysis and Design*, 2nd ed., Englewood Cliffs, N.J.: Prentice-Hall, 1990.

H. F. Van Landingham, *Introduction to Digital Control Systems*, New York: Macmillan, 1985.

Further Information

The *IEEE Control Systems Magazine* is a useful information source on control systems in general and digital control in particular. Highly technical articles on the state of the art in digital control may be found in the *IEEE Transactions on Automatic Control* and the ASME *Journal of Dynamic Systems, Measurement and Control*.

94

Robotics

Ty A. Lasky
University of California, Davis

Tien C. Hsia
University of California, Davis

R. Lal Tummala
Michigan State University

Nicholas G. Odrey
Lehigh University

94.1 Robot Configuration 2154
Cartesian Configuration • Cylindrical Configuration • Spherical
Configuration • Articulated Configuration • SCARA Configuration
• Gantry Configuration • Additional Information
94.2 Dynamics and Control 2163
Independent Joint Control of the Robot • Dynamic Models • Computed Torque Methods • Adaptive Control • Resolved Motion
Control • Compliant Motion • Flexible Manipulators
94.3 Applications 2175
Justification • Implementation Strategies • Applications in Manufacturing • Emerging Issues

94.1 Robot Configuration

Ty A. Lasky and Tien C. Hsia

Configuration is a fundamental classification for industrial robots. Configuration refers to the geometry of the robot manipulator, i.e., the manner in which the links of the manipulator are connected at each joint. The Robotic Industries Association (RIA) defines a robot as *a manipulator designed to move material, parts, tools, or specialized devices, through variable programmed motions for the performance of a variety of tasks.* With this definition, attention is focused on industrial manipulator arms, typically mounted on a fixed pedestal base. Mobile robots and hard automation [e.g., computer numerical control (CNC) machines] are excluded. The emphasis here is on serial-chain manipulator arms, which consist of a serial chain of linkages, where each link is connected to exactly two other links, with the exception of the first and last links, which are connected to only one other link. Additionally, the first three links, called the major linkages, are focused on, with only a brief mention of the last three links, or wrist joints, also called the minor linkages.

Robot configuration is an important consideration in the selection of a manipulator. Configuration refers to the way the manipulator links are connected at each joint. Each link will be connected to the subsequent link by either a linear (sliding or prismatic) joint, which can be abbreviated with a P, or a revolute (or rotary) joint, abbreviated with an R. Using this notation, a robot with three revolute joints would be abbreviated as RRR, while one with a rotary joint followed by two linear (prismatic) joints would be denoted RPP. Each configuration type is well suited to certain types of tasks and ill suited to others. Some configurations are more versatile than others. In addition to the geometrical considerations, the robot configuration will have an effect on the structural stiffness of the robot, which may be an important consideration. Also, configuration will have an impact on the complexity of the forward and inverse **kinematics**, which are the mappings between the robot actuator (joint) space, and the Cartesian position and orientation of the robot end effector, or tool.

There are six major robot configurations commonly used in industry. Details for each configuration will be presented in subsequent subsections. The simplest configuration is the Cartesian robot, which consists of three orthogonal, linear joints (PPP) so that the robot moves in the x, y, and z directions in the joint space. The cylindrical configuration consists of one revolute and two linear joints (RPP) so that the robot joints correspond to a cylindrical coordinate system. The spherical configuration consists of two revolute joints and one linear joint (RRP) so that the robot moves in a spherical, or polar, coordinate system. The articulated (arm and elbow) configuration consists of three revolute joints (RRR), giving the robot a somewhat human-like range of motion. The SCARA (Selective Compliance Articulated Robot for Assembly) configuration consists of two revolute joints and one linear joint (RRP), arranged in a different fashion than the spherical configuration. It may also be equipped with a revolute joint on the final sliding link. The gantry configuration is essentially a Cartesian configuration, with the robot mounted on an overhead track system. One can also mount other robot configurations on an overhead gantry system to give the robot an extended workspace, as well as free up valuable factory floor space. The percentage usage of the first five configuration types is listed in Table 94.1. This table does not include gantry robots, which are assumed to be included in the Cartesian category. Additionally, this information is from 1988, so it may not accurately represent the current usage.

Table 94.1 Robot Arm Geometry Usage

Arm Geometry	Percent of Use
Cartesian	18
Cylindrical	15
Spherical	10
Articulated	42
SCARA	15

Source: V. D. Hunt, *Robotics Sourcebook*, New York: Elsevier, 1988. With permission.

In general, robots with a rotary base will have a speed advantage. However, they generally have more variation in resolution and dynamics compared to Cartesian robots. This can lead to inferior performance if a fixed controller is used over the robot's entire workspace.

Cartesian Configuration

The Cartesian configuration consists of three orthogonal, linear axes, abbreviated as PPP, as shown in Fig. 94.1. Thus, the joint space of the robot corresponds directly with the standard right-handed Cartesian *xyz*-coordinate system, yielding the simplest possible kinematic equations. The work

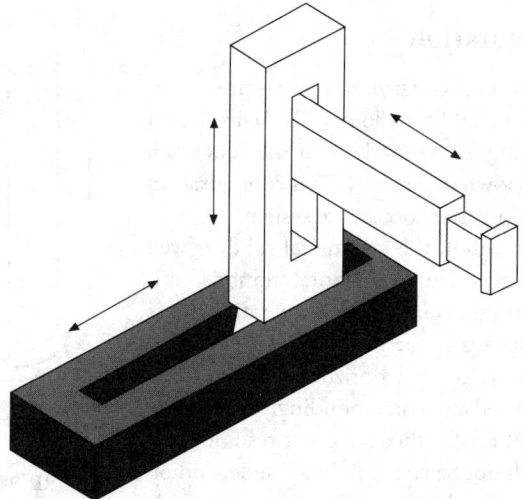

FIGURE 94.1 The Cartesian configuration. (*Source:* T. Owen, *Assembly with Robots*, Englewood Cliffs, N.J.: Prentice-Hall, 1985. With permission.)

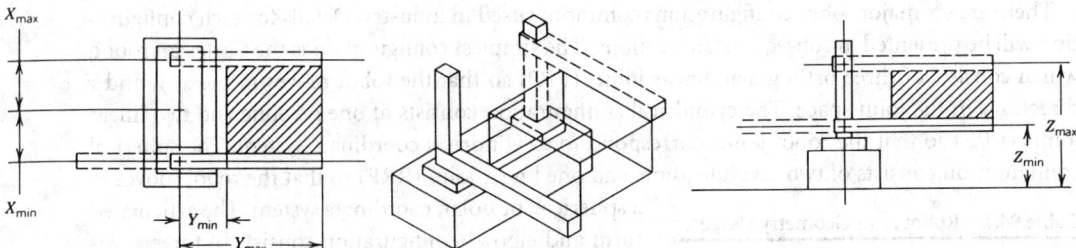

FIGURE 94.2   Cartesian robot work envelope. (*Source:* R. M. Wygant, "Ergonomics, robot selection," in *Concise International Encyclopedia of Robotics*, R. C. Dorf, Ed., New York: Wiley-Interscience, 1990. With permission.)

envelope of the Cartesian robot is shown in Fig. 94.2. The work envelope encloses all the points that can be reached by the robot arm or the mounting point for the end effector or tool. The area reachable by an end effector or tool is not considered part of the work envelope. All interaction with other machines, parts, or processes must take place within this volume of space [Critchlow, 1985]. Here, the workspace of a robot is assumed to be equivalent to the work envelope.

There are several advantages to this configuration. As noted above, the robot is kinematically simple, since motion on each Cartesian axis corresponds to motion of a single actuator. This eases the programming of linear motions. In particular, it is easy to do a straight vertical motion, which is the most common motion in assembly tasks. The Cartesian geometry also yields a constant arm resolution throughout the workspace of the arm; i.e., for any configuration of the arm, the resolution for each axis corresponds directly to the resolution for that joint. The simple geometry of the Cartesian robot leads to correspondingly simple manipulator dynamics. The disadvantages of this configuration are inability to reach objects on the floor or points invisible from the base of the robot and slow speed of operation in the horizontal plane compared to robots with a rotary base. Additionally, the Cartesian configuration requires a large operating volume for a relatively small workspace.

Cartesian robots are used for several applications. As noted above, they are well suited for assembly operations, as they can easily perform vertical straight-line insertions. Because of the ease of straight-line motions, they are also well suited to machine loading and unloading. They are also used in clean room tasks.

Cylindrical Configuration

The cylindrical configuration consists of one vertical revolute joint and two orthogonal linear joints (RPP), as shown in Fig. 94.3. The resulting work envelope of the robot is a cylindrical annulus, as shown in Fig. 94.4. This configuration corresponds with the cylindrical coordinate system.

As with the Cartesian robot, the cylindrical robot is well suited for straight-line vertical and horizontal motions, so it is useful for assembly and machine loading operations. It is also capable of higher speeds in the horizontal plane due to the rotary base. However, general horizontal straight-line motion is more complicated and correspondingly more difficult to coordinate. Additionally, the end-point resolution of the cylindrical robot will not be constant but will depend on the extension of the horizontal linkage. They are not capable of reaching around obstacles. Additionally, if a monomast construction is used on the horizontal linkage, then there can be clearance problems behind the robot.

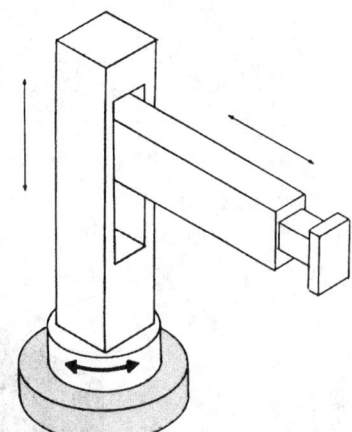

FIGURE 94.3   The cylindrical configuration. (*Source:* T. Owen, *Assembly with Robots*, Englewood Cliffs, N.J.: Prentice-Hall, 1985. With permission.)

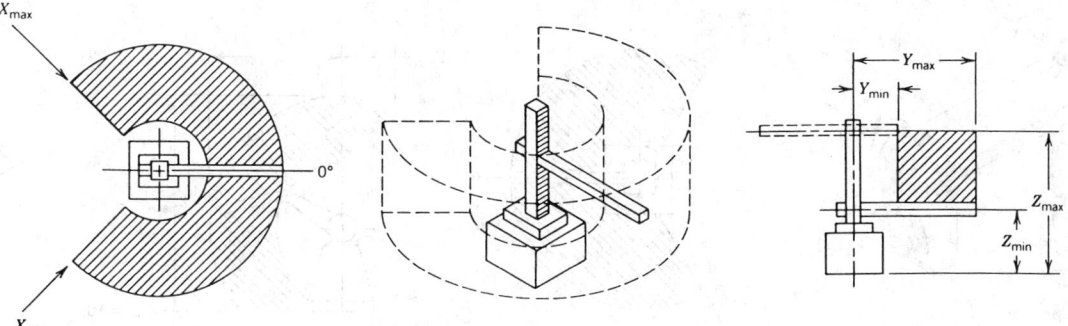

FIGURE 94.4   Cylindrical robot work envelope. (*Source:* R. M. Wygant, "Ergonomics, robot selection," in *Concise International Encyclopedia of Robotics*, R. C. Dorf, Ed., New York: Wiley-Interscience, 1990. With permission.)

Spherical Configuration

The spherical (or polar) configuration consists of two revolute joints and one linear joint (RRP), as shown in Fig. 94.5. This results in a set of joint coordinates that matches with the spherical coordinate system. A typical work envelope for a spherical robot is shown in Fig. 94.6.

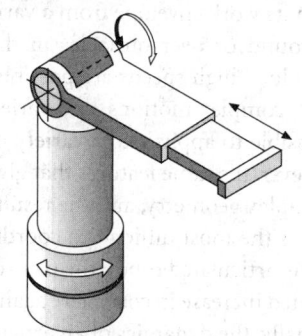

Spherical robots are typically heavy-duty robots. They have the advantages of high speed due to the rotary base and a large work volume, but are more complex kinematically than either the Cartesian or cylindrical robots. Generally, they are used for heavy-duty tasks in, for example, automobile manufacturing. They do not have the flexibility to reach around obstacles in the workspace. Spherical robots also do not have fixed resolution throughout the workspace.

Articulated Configuration

The articulated (or anthropomorphic, jointed, arm-and-elbow) configuration consists of three revolute joints (RRR), as shown in Fig. 94.7. The resulting joint coordinates do not directly match any Cartesian coordinate system. A slice of a typical work envelope for an articulated robot is shown in Fig. 94.8.

FIGURE 94.5   The spherical configuration. (*Source:* T. Owen, *Assembly with Robots*, Englewood Cliffs, N.J.: Prentice-Hall, 1985. With permission.)

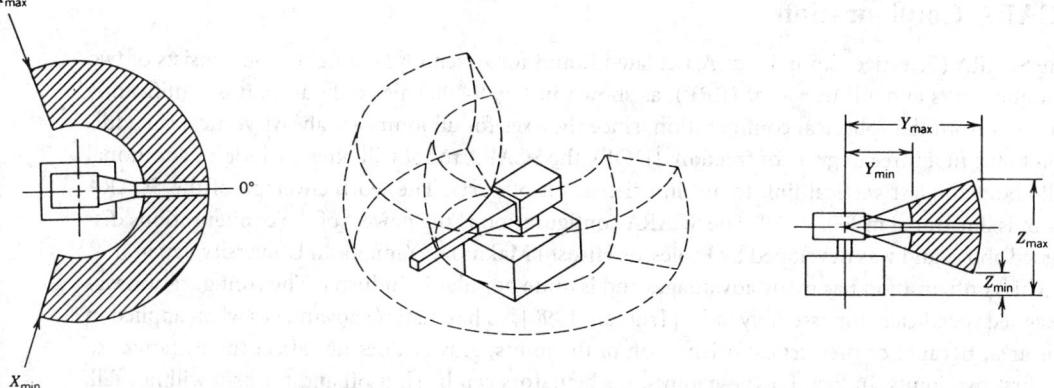

FIGURE 94.6   Spherical robot work envelope. (*Source:* R. M. Wygant, "Ergonomics, robot selection," in *Concise International Encyclopedia of Robotics*, R. C. Dorf, Ed., New York: Wiley-Interscience, 1990. With permission.)

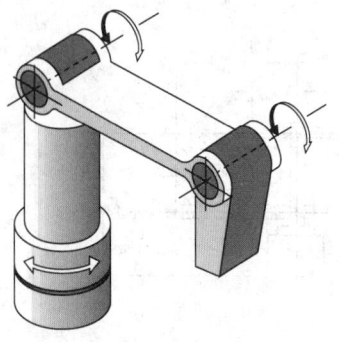

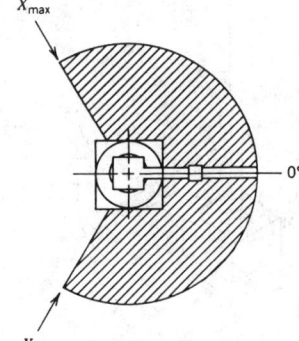

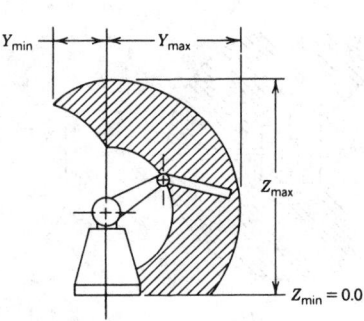

FIGURE 94.7 The articulated configuration. (*Source:* T. Owen, *Assembly with Robots*, Englewood Cliffs, N.J.: Prentice-Hall, 1985. With permission.)

FIGURE 94.8 Articulated robot work envelope. (*Source:* R. M. Wygant, "Ergonomics, robot selection," in *Concise International Encyclopedia of Robotics*, R. C. Dorf, Ed., New York: Wiley-Interscience, 1990. With permission.)

The articulated robot is currently the most commonly used, both in research and industry. It has several advantages over other configurations. It is closest to duplicating a human assembler, so there may be less need to redesign an existing workstation to utilize an articulated robot. It has a very large, dexterous work envelope; i.e., it can reach most points in its work envelope from a variety of orientations. This means that it can more easily reach around or over obstacles in the workspace or into parts or machines. Because all the joints are revolute, high speeds are possible. The articulated arm is good for tasks involving multiple insertions, complex motions, and varied tool orientations. The flexibility of this configuration makes it possible to apply it to a variety of tasks, so the user has fewer limitations on the use of the robot. However, the same features that give this robot its advantages lead to certain disadvantages. This is a complex geometry, and the resulting kinematic equations are quite complicated. This configuration is the most difficult to coordinate the joints to move in a straight line. Due to the complexity of the articulated robot, the control is generally more difficult than for other geometries, with an associated increase in cost. Here again, the arm resolution is not fixed throughout the workspace. Additionally, the dynamics of an articulated arm vary widely throughout the workspace so that performance will vary over the workspace for a fixed controller. In spite of these disadvantages, the articulated arm has been applied to a wide variety of applications, including spray painting, clean room tasks, machine loading, and parts-finishing tasks.

SCARA Configuration

The SCARA (Selective Compliance Articulated Robot for Assembly) configuration consists of two revolute joints and a linear joint (RRP), as shown in Fig. 94.9. This configuration is significantly different from the spherical configuration, since the axes for all joints are always vertical. In addition to the first three **degrees of freedom** (DOF), the SCARA robot will often include an additional roll about the last vertical link to aid in orientation of parts. The work envelope of the SCARA robot is illustrated in Fig. 94.10. The SCARA configuration is the newest of the configurations discussed above and was developed by Professor Hiroshi Makino of Yamanashi University, Japan.

This configuration has many advantages and is quite popular in industry. The configuration was designed specifically for assembly tasks [Truman, 1990], so has distinct advantages when applied in this area. Because of the vertical orientation of the joints, gravity does not affect the dynamics of the first two joints. In fact, for these joints, the actuators can be shut off and the arm will not fall, even without the application of brakes. This allows compliance in the horizontal directions to be selectively varied; therefore, the robot can comply to horizontal forces. Horizontal compliance is

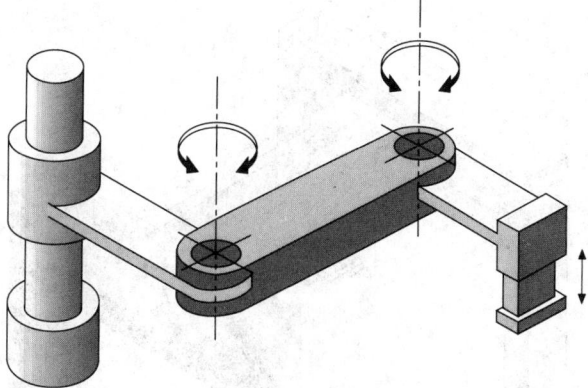

FIGURE 94.9 The SCARA configuration. (*Source:* T. Owen, *Assembly with Robots*, Englewood Cliffs, N.J.: Prentice-Hall, 1985. With permission.)

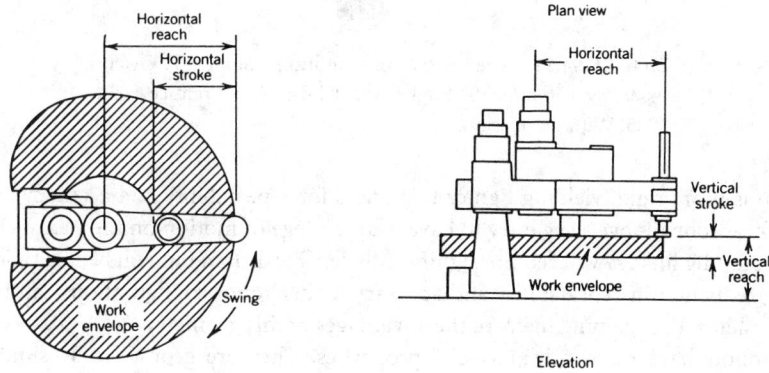

FIGURE 94.10 SCARA robot work envelope. (*Source:* W. R. Tanner, "Classification," in *Concise International Encyclopedia of Robotics*, R. C. Dorf, Ed., New York: Wiley-Interscience, 1990. With permission.)

important for vertical assembly operations. Because of the vertical linear joint, straight-line vertical motions are simple. Also, SCARA robots typically have high positional repeatability. The revolute joints allow high-speed motion. For a typical printed circuit board assembly, the SCARA can insert one component about every 3.5 s, while more complex insertions can take about 5 s. On the negative side, the resolution of the arm is not constant throughout the workspace, and the kinematic equations are relatively complex. In addition, the vertical motion of the SCARA configuration is typically quite limited. While the SCARA robot can reach around objects, it cannot reach over them in the same manner as an articulated arm.

Gantry Configuration

The gantry configuration is essentially the same as the Cartesian configuration geometrically, but it is suspended from an overhead crane and typically can be moved over a large workspace. It consists of three linear joints (PPP), and is illustrated in Fig. 94.11. In terms of work envelope, it will have a rectangular volume that sweeps out most of the inner area of the gantry system, with a height limited by the length of the vertical mast, and the headroom above the gantry system. One consideration in the selection of a gantry robot is the type of vertical linkage employed in the z axis. A mono-

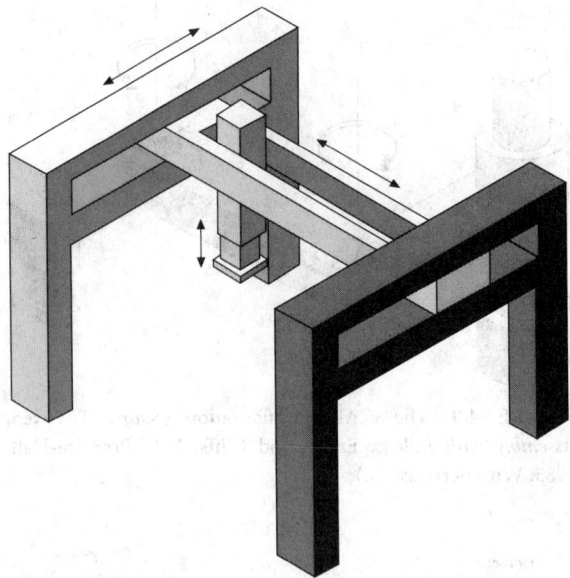

FIGURE 94.11   The gantry configuration. (*Source:* T. Owen, *Assembly with Robots*, Englewood Cliffs, N.J.: Prentice-Hall, 1985. With permission.)

mast design is more rigid, yielding tighter tolerances for repeatability and accuracy, but requires significant headroom above the gantry to have a large range of motion on the z axis. On the other hand, a telescoping linkage will require significantly less headroom but will be less rigid, with corresponding reduction in repeatability and accuracy. Other robot configurations can be mounted on gantry systems, thus gaining many of the advantages of this geometry.

Gantry robots have many advantageous properties. They are geometrically simple, like the Cartesian robot, with the corresponding controller and dynamic simplicity. For the same reasons, the gantry robot has a constant arm resolution throughout the workspace. The gantry robot has better dynamics than the pedestal-mounted Cartesian robot, as its linkages are not cantilevered. One major advantage over revolute-based robots is that its dynamics vary much less over the workspace. This leads to less vibration, less performance degradation, and more even performance than typical pedestal-mounted robots in full extension. Gantry robots are much stiffer than other robot configurations, although they are still much less stiff than numerical control (NC) machines. The gantry robot can straddle a workstation, or several workstations for a large system, so that one gantry robot may be able to perform the work of several pedestal-mounted robots. As with the Cartesian robot, the gantry robot's simple geometry is similar to that of an NC machine, so that technicians will be more familiar with the system and should require less training time. Also, there is no need for special path or trajectory computations. A gantry robot can be programmed directly from a computer-aided design (CAD) system with the appropriate interface, and straight-line motions are particularly simple to program. Large gantry robots have a very high payload capacity. Small, table-top systems can achieve linear speeds of up to 40 in./s (1.025 m/s), with a payload capacity of 5.0 lb (2.26 kg), making them suitable for assembly operations. However, most gantry robot systems are not as precise as some other configurations, such as the SCARA configuration. Additionally, it is sometimes more difficult to apply a gantry robot to an existing workstation, as the workpieces must be brought into the gantry's work envelope, which may be harder to do than for a pedestal-mounted manipulator.

Gantry robots can be applied in many areas. They are used in the nuclear power industry to load and unload reactor fuel rods. Gantry robots are also applied to materials-handling tasks, such as parts transfer, machine loading, palletizing, materials transport, and some assembly applications. In addition, gantry robots are used for process applications such as welding, painting, drilling, routing, cutting, milling, inspection, nondestructive testing, application of sealants and adhesives, and parts marking.

The gantry robot configuration is the fastest-growing segment of the robotics industry. While gantry robots accounted for less than 5% of the units shipped in 1985, they are projected to account for about 30% of the robots by the end of the 1990s. One reason for this is summed up in Long [1990], which contains much more information on gantry robots in general:

> Currently gantry robot cells are being set up which allow manufacturers to place a sheet of material in the gantry's work envelope and begin automatic cutting, trimming, drilling, milling, assembly and finishing operations which completely manufacture a part or sub-assembly using quick-change tools and programmed subroutines.

Additional Information

The above six configurations are the main types currently used in industry. However, there are other configurations used in either research or specialized applications. Some of these configurations have found limited application in industry and may become more prevalent in the future.

All the above configurations are serial chain-link manipulators. An alternative to this standard approach is the parallel configuration, known as the Stewart platform [Waldron, 1990]. This manipulator consists of two platforms connected by three prismatic linkages. This arrangement yields the full six DOF motion (three-position, three-orientation) that can be achieved with a six-axis serial configuration but has a comparably very high stiffness. It is used as a motion simulator for pilot training. The negative aspects of this configuration are its relatively restricted motion capability and its geometric complexity.

The above configurations are restricted to a single manipulator arm. There are tasks that are either difficult or impossible to perform with a single arm. With this realization, there has been significant interest in the use of multiple arms to perform coordinated tasks. Possible applications include carrying loads that exceed the capacity of a single robot and assembling objects without special fixturing. Multiple arms will be particularly useful in zero-gravity environments. While there are significant advantages to the use of multiple robots, the complexity, in terms of both kinematics and dynamics, is quite high. However, the use of multiple robots should open new areas of application for robots.

Typical industrial robots have six or fewer DOF. With six DOF, the robot can, within its work envelope, reach arbitrary positions and orientations. At the edge of the work envelope, a six-DOF robot can attain only one orientation. To increase the geometric flexibility of the manipulator, it is useful to consider the possibility of robots with more than six DOF, i.e., redundant robots. These robots are highly dexterous and can use the presence of extra DOF in many advantageous ways: obstacle avoidance, minimization of joint torques, avoidance of kinematic **singularities** (points where the manipulator cannot move in certain directions), bracing strategies where part of the arm is braced against a structure, which raises the lowest structural resonant frequency of the arm, etc. While the **redundant manipulator** configuration does have many desirable properties, the geometric complexity has limited their application in industry.

For any of the six standard robot configurations, the orientation capability of the major linkages is severely limited. Thus, it is critical to provide additional joints, known as the minor linkages, to provide the capability of varied orientations for a given position. Most robots include a three-DOF revolute joint wrist that is connected to the last link of the major linkages. The three revolute axes will be orthogonal and will usually intersect in a common point, known as the wrist center point. Then, the kinematic equations of the manipulator can be partitioned into locating the Cartesian position of the wrist center point and then determining the orientation of a Cartesian frame fixed to the wrist axes.

Conclusions

Each of the six standard configurations has specific advantages and disadvantages. When choosing a manipulator for a specific task, the properties of the manipulator geometry are one of the most important considerations. If the manipulator will be used for a wide variety of tasks, one may need to trade off performance for any given task for the flexibility that will allow the manipulator to work for the various tasks. In such a case, a more flexible geometry should be considered. The future of robotics will be interesting. With the steady increase in computational capabilities, the more complex geometries, including redundant and multiple robots, should begin to see increased applications in industry.

Defining Terms

Degrees of freedom: The number of degrees of freedom (DOF) of a manipulator is the number of independent position variables that must be specified in order to locate all parts of the manipulator. For a typical industrial manipulator, the number of joints equals the number of DOF.

Kinematics: The kinematics of the manipulator refers to the geometric properties of the manipulator. *Forward kinematics* is the computation of the Cartesian position and orientation of the robot end effector given the set of joint positions. *Inverse kinematics* is the computation of the joint positions given the Cartesian position and orientation of the end effector. The inverse kinematic computation may not be possible in closed form, may have no solution, or may have multiple solutions.

Redundant manipulator: A redundant manipulator contains more than six DOF.

Singularity: A *singularity* is a location in the workspace of the manipulator at which the robot loses one or more DOF in Cartesian space, i.e., there is some direction (or directions) in Cartesian space along which it is impossible to move the robot end effector no matter which robot joints are moved.

References

J. J. Craig, *Introduction to Robotics: Mechanics and Control*, Reading, Mass.: Addison-Wesley, 1986.

A. J. Critchlow, *Introduction to Robotics*, New York: Macmillan, 1985.

E. Long, "Gantry robots," in *Concise International Encyclopedia of Robotics*, R. C. Dorf, Ed., New York: Wiley-Interscience, 1990.

R. Truman, "Component assembly onto printed circuit boards," in *Concise International Encyclopedia of Robotics*, R. C. Dorf, Ed., New York: Wiley-Interscience, 1990.

K. J. Waldron, "Arm design," in *Concise International Encyclopedia of Robotics*, R. C. Dorf, Ed., New York: Wiley-Interscience, 1990.

Further Information

The journal *IEEE Transactions on Robotics and Automation* is a valuable source for a wide variety of robotics research topics, occasionally including new robot configurations. Additionally, IEEE's *Control Systems Magazine* occasionally publishes an issue devoted to robotic systems.

Another journal that often has robotics-related articles is the ASME *Journal of Dynamic Systems, Measurement and Control*.

An additional source of robotics information is *The Proceedings of the IEEE International Conference on Robotics and Automation*. This conference is held annually.

An interesting alternative forum for the exchange of advice and information on robotics is the Usenet news group, "comp.robotics." Consult your system administrator for information on this valuable service.

94.2 Dynamics and Control

R. Lal Tummala

The primary purpose of the robot control system is to issue commands to joint actuators to faithfully execute a planned trajectory in the **tool space**. This may involve position control when the manipulator is following a trajectory through free space or a combination of position and force control if the manipulator is to react continuously to contact forces at the tool or end effector.

Control systems can operate either in open loop or closed loop. In open-loop systems, the output has no effect on the input. On the other hand, closed-loop systems continuously sense the output and make appropriate adjustments to the input in order to keep the output at the desired level.

The majority of the current industrial robots use the **independent joint control** method and close the loop around the joints of the robot. The desired joint positions corresponding to a tool trajectory are either taught by using a *teach box* or generated by solving an inverse kinematics problem. The independent joint control method, however, is effective only at low speeds. As the speeds increase, the coupling effects between the joints increase and warrant the inclusion of these effects in the controller development. Advanced controller development and implementation based on full dynamics is one of the active areas of current research. New advances in sensor technology, faster computers, and industrial competition provide new opportunities and motivation for accelerating the development and implementation of advanced controllers for robots in the near future.

Independent Joint Control of the Robot

The independent joint control method assumes that a single joint of a robot is moving while all the other joints are fixed. A typical joint position control system is shown in Fig. 94.12, where the actuator used is a dc servomotor [Luh, 1983]. In general, any one or a combination of electric motors and hydraulic or pneumatic pistons can be used to move the joint through the desired positions. These motors may be connected directly to the joint or indirectly through gears, chains, cables, or lead screws. The desired joint positions that are inputs to the position loops are obtained from the trajectory planner. The actual position of the joint is obtained by using a position sensor, such as a potentiometer or an optical encoder. An amplifier is used for increasing the system gain, denoted by K_a. The velocity feedback K_v is used to reinforce the effect of back emf for controlling the damp-

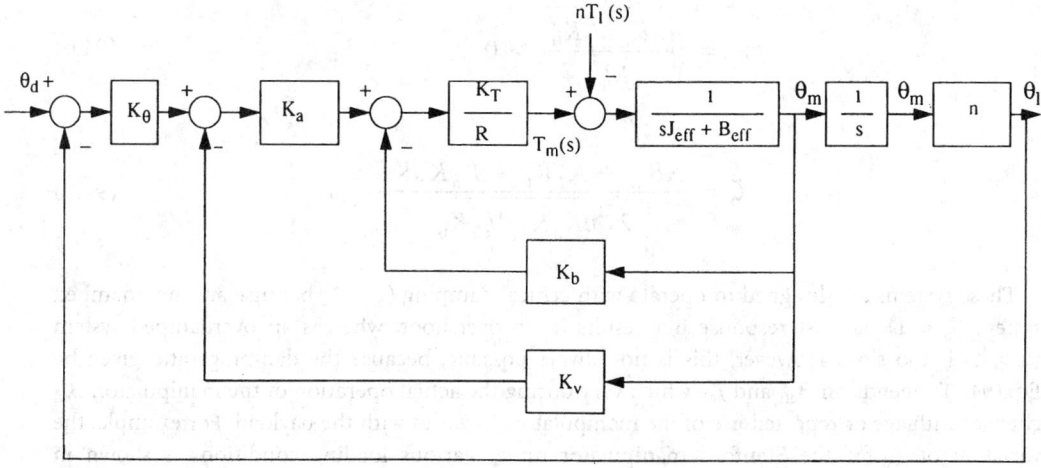

FIGURE 94.12 Closed-loop control of a robot joint. (*Source:* Adapted from J. Y. S. Luh, "Conventional controller design for industrial robots: A tutorial," *IEEE Trans. Systems, Man, Cybernetics,* vol. SMC-13, no. 3, June 1983. © 1983 IEEE.)

ing of the system. This can be done either using a tachometer or computing the difference in angular displacements of the actuator shaft over a fixed time interval.

The design of the control system involves fixing the values of K_a and K_v to achieve the desired response. Consider the closed-loop transfer function of the system shown in Fig. 94.12 (assuming $nT_1 = 0$),

$$\frac{\theta_1(s)}{\theta_d(s)} = \frac{nK_a K_T K_\theta}{s^2 RJ_{\text{eff}} + s(RB_{\text{eff}} + K_T K_b + K_a K_T K_v) + nK_T K_a K_\theta} \tag{94.1}$$

where K_a = gain of the amplifier, K_T = torque constant of the motor, K_b = back emf constant, K_θ = position sensor constant, R = resistance of the motor winding, and n = gear ratio. θ_L = link position (rad) and θ_m = angular displacement at the actuator side (rad).

The effective inertia, J_{eff}, and damping, B_{eff}, are defined as

$$J_{\text{eff}} = J_m + n^2 J_L \tag{94.2}$$

and

$$B_{\text{eff}} = B_m + n^2 B_L \tag{94.3}$$

where J_m = total inertia on the motor side, B_m = damping coefficient at the motor side, J_L = inertia of the robot link, and B_L = damping coefficient at the load side.

This is a second-order system and stable for all values of K_a and K_v. The values of K_a and K_v are selected to achieve a desired transient response by fixing the damping ratio and the natural frequency of the system and are described below.

The characteristic equation for the above system is

$$s^2 + s\frac{RB_{\text{eff}} + K_T K_b + K_a K_T K_v}{RJ_{\text{eff}}} + \frac{nK_T K_a K_\theta}{RJ_{\text{eff}}} = 0 \tag{94.4}$$

This can be conveniently written as

$$s^2 + 2\zeta\omega_n s + \omega_n^2 = 0 \tag{94.5}$$

where the natural frequency ω_n and the damping ratio ζ of the system are given as

$$\omega_n = \sqrt{\frac{nK_a K_T K_\theta}{RJ_{\text{eff}}}} > 0 \tag{94.6}$$

$$\zeta = \frac{RB_{\text{eff}} + K_T K_b + K_a K_T K_v}{2\sqrt{nK_a K_T RJ_{\text{eff}} K_\theta}} \tag{94.7}$$

These systems are designed to operate with critical damping ($\zeta = 1$) because an underdamped system ($\zeta < 1$) has fast response but results in an overshoot, whereas an overdamped system ($\zeta > 1$) is too slow. However, this is not always possible, because the damping ratio given by Eq. (94.7) depends on B_{eff} and J_{eff} which vary during the actual operation of the manipulator. B_{eff} changes with age or repeated use of the manipulator. J_{eff} varies with the payload. For example, the variation of J_{eff} for the Stanford manipulator under various loading conditions is shown in Fig. 94.13. J_{eff} also varies with the configuration of the manipulator during the actual operation. So a compromise solution will be to design the controller such that $\zeta \geq 1$ throughout the intended operation.

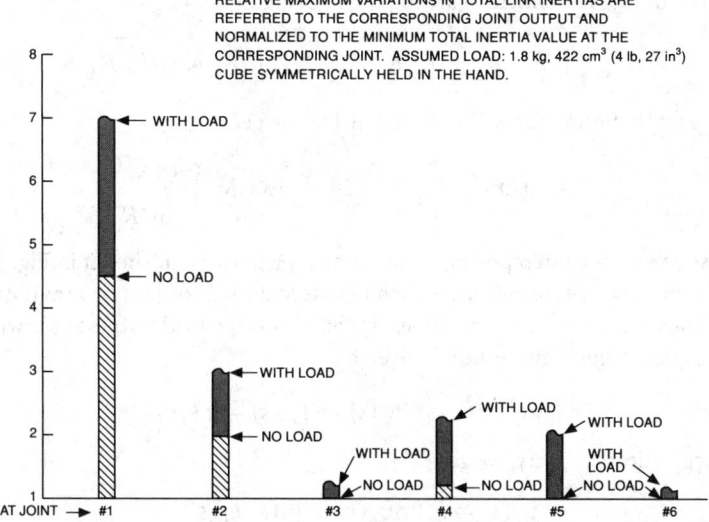

FIGURE 94.13 Variations of link inertias for JPL-Stanford manipulator. (*Source:* A.K. Bejczy, Jet Propulsion Lab, Pasadena, Calif., American Automatic Control Conference Tutorial Workshop, Washington, D.C., June 18, 1982.)

The undamped natural frequency ω_n is selected to be no more than half the resonance frequency of the robot to avoid any structural damage to the robot [Paul, 1981]. These resonances are possible due to the flexibilities associated with the links of the robot and the shafts within the drive system, to name a few. These are called *unmodeled* resonances because they are not explicitly included in the model. In our case, if K_{eff} and J_{eff} are the effective stiffness and the inertias of the joint, respectively, then the resonance frequency ω_r is given by

$$\omega_r = \sqrt{\frac{K_{eff}}{J_{eff}}} \tag{94.8}$$

Since K_{eff} is difficult to estimate but constant for a given joint, we can experimentally determine the resonance frequencies for a known inertia and use this information for fixing the gain. Suppose ω is the resonance frequency for a given value of effective inertia J, then

$$\omega = \sqrt{\frac{K_{eff}}{J}} \tag{94.9}$$

To minimize the effects of unmodeled resonances, we use

$$\omega_n \leq \frac{\omega_r}{2} = \frac{\omega}{2}\sqrt{\frac{J}{J_{eff}}} \tag{94.10}$$

The selection of K_a and K_v depends on selecting ζ and ω_n. Using Eqs. (94.6) and (94.10), we can find an upper bound on K_a given by

$$K_a \leq \frac{J\omega^2 R}{4nK_T K_\theta} \tag{94.11}$$

The upper bound on K_v is obtained by setting $\zeta \geq 1$. Using Eq. (94.7),

$$RB_{\text{eff}} + K_T K_b + K_a K_T K_v \geq 2\sqrt{nK_a K_T RJ_{\text{eff}} K_\theta} \tag{94.12}$$

Substituting the upper bound for K_a from Eq. (94.11), we get

$$K_v \geq \left(\omega R \sqrt{JJ_{\text{eff}}} - RB_{\text{eff}} - K_T K_b \right) \frac{4nK_\theta}{J\omega^2 R} \tag{94.13}$$

The steady-state errors to step position commands for the system shown in Fig. 94.12 are zero. However, in the presence of disturbances such as external load torques or **gravitational torques**, the system will have steady-state errors. For example, if T_L is the load torque as shown in Fig. 94.12, the available torque for the joint motion is given by

$$(J_{\text{eff}} s^2 + B_{\text{eff}} s)\, \theta_m(s) = T_m(s) - nT_L(s) \tag{94.14}$$

Using the superposition property, we get

$$\theta_L(s) = F_1(s)\theta_d(s) + F_2(s)T_L(s) \tag{94.15}$$

where

$$F_1(s) = \frac{nK_a K_T K_\theta}{\Omega(s)}$$

$$F_2(s) = -\frac{n^2 R}{\Omega(s)} \tag{94.16}$$

$$\Omega(s) = RJ_{\text{eff}} s^2 + (RB_{\text{eff}} + K_b K_T + K_v K_a K_T)s + nK_a K_T K_\theta$$

Now if $T_L(s) = C_L/s$ and $\theta_d(s) = C_\theta/s$, then the steady-state error is

$$e_{ss} = \frac{nC_L R}{K_a K_T K_\theta} \tag{94.17}$$

Since the value of K_a has an upper bound, this error cannot be made arbitrarily small. A possible way to reduce this error is to add a feedforward term, as shown in Fig. 94.14 [Luh, 1983]. The feedforward signal $T_d(s)$ is chosen such that the steady-state error is zero. In this case,

$$T_d(s) = \frac{R}{K_T K_a K_R} n\hat{T}_L(s) \tag{94.18}$$

Similar considerations apply for other disturbances such as frictional torques and gravitational

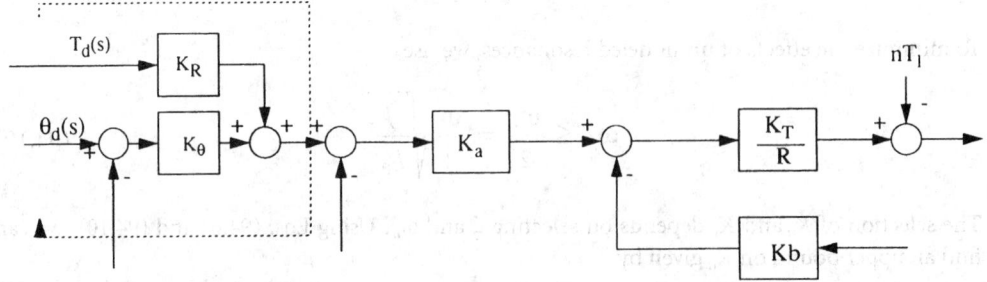

FIGURE 94.14 Feedback compensation method for disturbances. (*Source:* Adapted from J. Y. S. Luh, "Conventional controller design for industrial robots: A tutorial," *IEEE Trans. Systems, Man, Cybernetics,* vol. SMC-13, no. 3, June 1983. ©1983 IEEE.)

torques. Notice from Eq. (94.18) that the feedforward signal is a function of the estimated torque. The burden of determining these torques should not be underestimated. The other factor that was not mentioned earlier is the centrifugal torque, a nonlinear function of velocity. In the case of positioning applications, the velocity tends to zero as $t \rightarrow \infty$. However, if the robot is required to follow a conveyor with constant speed, then the input is a velocity. In this case, the centrifugal contribution will affect the steady-state velocity error. A feedforward compensation can be used in this case as well. Another method of compensating for the steady-state errors caused by gravitational and load torque disturbances is by adding an integral feedback (PID control), which of course increases the order of the system. The system is no longer stable for all values of the gains and thus adds another constraint in the selection of K_a and K_v.

So far we have considered the control of one joint of the robot while the other joints are fixed. Implementation of this control by successively positioning each joint while the other joints are fixed slows the robot operation and can also result in awkward hand motions, which is undesirable especially when the robot is supposed to follow a continuous path. Simultaneous fast motion of the joints, on the other hand, requires the inclusion of dynamic interactions between the joints. The controllers designed without considering these dynamic interactions tend to make the arm move slower and can potentially cause overshoots, oscillations, and path errors. To estimate the dynamic effects, one needs to obtain the equations of motion (dynamic models) of the robot. These equations are, in general, complex and take the form of coupled nonlinear differential equations.

Dynamic Models

Two of the most popular methods used to obtain dynamic models of the robot are the *Newton-Euler method* and the *Lagrangian method*. The equations obtained using Lagrangian formulation are more suitable for the application of modern control theory than the recursive equations obtained using the Newton-Euler method. In the Lagrangian formulation, the dynamic models are obtained using kinetic and potential energies associated with the rigid bodies in motion. The derivation is systematic and simple. This method yields closed-form dynamic equations that explicitly express joint variables in terms of joint torques. To arrive at these equations, one starts with a set of generalized coordinates q_i, $i = 1, 2, 3, \ldots, n$, that completely locate the dynamic system and finds the total kinetic energy K and potential energy P of the system [Paul, 1981]. Then the equations of motion are given by

$$\frac{d}{dt} \frac{\partial L}{\partial \dot{q}_i} - \frac{\partial L}{\partial q_i} = T_i \quad \text{for } i = 1, 2, 3 \ldots n \quad (94.19)$$

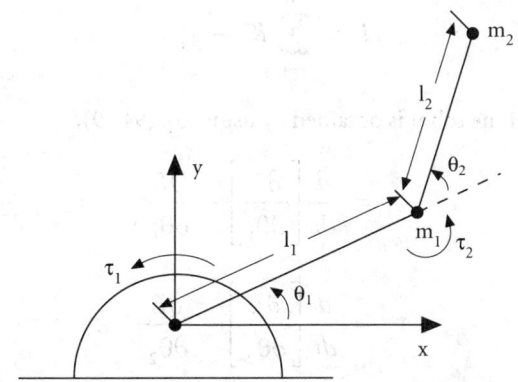

FIGURE 94.15   Two-degree-of-freedom planar manipulator.

where T_i is the generalized force and $L(q,\dot{q}) = K - P$ is the Lagrangian. A simple example is given next to illustrate these ideas.

Example. Consider a planar arm with two degrees of freedom, shown in Fig. 94.15. For simplicity, we assume that the masses m_1 and m_2 of the links are represented by point masses at the end of the links. The link lengths are l_1 and l_2, respectively. The variables θ_1 and θ_2 are the joint angles. We know that the kinetic energy of a mass m moving at a linear velocity v is given by $1/2\ mv^2$ and the potential energy associated with a mass m located at a height h in a gravitational field is given by mgh, where g is the gravitational constant.

The kinetic energy K_1 for mass m_1 is found by observing that

$$
\begin{aligned}
x_1 &= l_1 \cos \theta_1 \\
y_1 &= l_1 \sin \theta_1 \\
v_1^2 &= \dot{x}_1^2 + \dot{y}_1^2 \\
K_1 &= \tfrac{1}{2} m_1 l_1^2 \dot{\theta}_1^2
\end{aligned}
\tag{94.20}
$$

Similarly, the kinetic energy K_2 for mass m_2 is given by

$$
\begin{aligned}
K_2 &= \tfrac{1}{2} m_2 v_2^2 \\
v_2^2 &= \dot{x}_2^2 + \dot{y}_2^2
\end{aligned}
\tag{94.21}
$$

From Fig. 94.15, we have

$$
\begin{aligned}
x_2 &= l_1 \cos \theta_1 + l_2 \cos (\theta_1 + \theta_2) \\
y_2 &= l_1 \sin \theta_1 + l_2 \sin (\theta_1 + \theta_2) \\
v_2^2 &= l_1^2 \dot{\theta}_1^2 + l_2^2 (\dot{\theta}_1 + \dot{\theta}_2)^2 + 2 l_1 l_2 \cos \theta_2 (\dot{\theta}_1^2 + \dot{\theta}_1 \dot{\theta}_2)
\end{aligned}
\tag{94.22}
$$

The potential energies for the masses m_i, $i = 1, 2$, are given by

$$
\begin{aligned}
P_1 &= m_1 g\, l_1 \sin \theta_1 \\
P_2 &= m_2 g [l_1 \sin \theta_1 + l_2 \sin(\theta_1 + \theta_2)]
\end{aligned}
\tag{94.23}
$$

The next step is to form the Lagrangian,

$$
L = \sum_{i=1}^{2} K_i - P_i
$$

The dynamic model of the robot is obtained by using Eq. (94.19),

$$
\tau_1 = \frac{d}{dt}\left[\frac{\partial L}{\partial \dot{\theta}_1}\right] - \frac{\partial L}{\partial \theta_1}
\tag{94.24}
$$

$$
\tau_2 = \frac{d}{dt}\left[\frac{\partial L}{\partial \dot{\theta}_2}\right] - \frac{\partial L}{\partial \theta_2}
\tag{94.25}
$$

where τ_i, $i = 1, 2$, are the joint torques.

The equations for a general n-degrees-of-freedom robot can be derived by following the same procedure and are compactly written in the generalized coordinates q as

$$D(q)\ddot{q} + H(q, \dot{q}) + V\dot{q} + G(q) = \tau \tag{94.26}$$

where $D(q)$ is the $n \times n$ inertia matrix, $H(\cdot)$ is an $n \times 1$ vector describing the centripetal and Coriolis terms, V is the coefficient of friction and $G(q)$ is an $n \times 1$ vector describing gravitational torques. For the above example, $q_1 = \theta_1$ and $q_2 = \theta_2$. Thus,

$$D(\theta) = \begin{bmatrix} l_2^2 m_2 + 2l_1 l_2 m_2 \cos\theta_2 + l_1^2(m_1 + m_2) & l_2^2 m_2 + l_1 l_2 m_2 \cos\theta_2 \\ l_2^2 m_2 + l_1 l_2 m_2 \cos\theta_2 & l_2^2 m_2 \end{bmatrix} \tag{94.27}$$

$$H(\theta, \dot{\theta}) = \begin{bmatrix} -m_2 l_1 l_2 \sin\theta_2 \dot{\theta}_2^2 - 2m_2 l_1 l_2 \sin\theta_2 (\dot{\theta}_1 \dot{\theta}_2) \\ (m_2 l_1 l_2 \sin\theta_2)\dot{\theta}_1^2 \end{bmatrix} \tag{94.28}$$

$$G(\theta) = \begin{bmatrix} m_2 l_2^2 g \cos(\theta_1 + \theta_2) + (m_1 + m_2) l_1 g \cos\theta_1 \\ m_2 l_2 g \cos(\theta_1 + \theta_2) \end{bmatrix} \tag{94.29}$$

where

$$\theta = \begin{bmatrix} \theta_1 \\ \theta_2 \end{bmatrix} \qquad \dot{\theta} = \begin{bmatrix} \dot{\theta}_1 \\ \dot{\theta}_2 \end{bmatrix} \qquad \ddot{\theta} = \begin{bmatrix} \ddot{\theta}_1 \\ \ddot{\theta}_2 \end{bmatrix} \tag{94.30}$$

Notice that the inertia matrix $D(\theta)$ is a function of only the position θ. In general, the inertia matrix is symmetric and positive definite and thus invertible. The diagonal elements of this matrix represent the effective inertias at the respective joints, while the off-diagonal elements represent the coupling inertias. For example, the term $m_2 l_2^2$ represents the effective inertia at the joint 2, and the term $l_2^2 m_2 + l_1 l_2 m_2 \cos\theta_2$ represents the coupling inertia between joints 1 and 2, i.e., the effect of acceleration of joint 1 on joint 2.

The terms in the matrix $H(\cdot)$ contain all the terms associated with the **centripetal** and **Coriolis forces**. The terms that depend upon the square of the joint velocity are *centripetal forces*. The terms that contain the product of joint velocities are *Coriolis forces*. In our example, the term $-(m_2 l_1 l_2 \sin\theta_2)\dot{\theta}_2^2$ represents the centripetal force acting at joint 1 due to the velocity at joint 2. Similarly, the term $(m_2 l_1 l_2 \sin\theta_2)\dot{\theta}_1^2$ represents the centripetal force acting at joint 2 due to the velocity at joint 1. The term $-(2m_2 l_1 l_2 \sin\theta_2)\dot{\theta}_1 \dot{\theta}_2$ is the Coriolis force acting at joint 2 due to the velocities at joints 1 and 2.

The term $G(\theta)$ contains all the terms involving gravitational constant g. Note that these terms depend only on the position of the arm in the gravitational field. If the arm is operating in the gravity-free environment, then these terms become zero. The term $V\dot{\theta}$ reflects the frictional forces present in the robot system. In our example, these terms are assumed to be zero. However, in practical robots a substantial amount of friction can be present that if not considered will overestimate the torque available for accelerating the joints.

The above example illustrates that the existence of significant coupling between the joints, if ignored, can cause positioning and tracking errors when the joints are moving simultaneously. However, all these coupling terms become small at low speeds. In this case, independent joint control with appropriate compensations as discussed earlier may be quite adequate. As the operational speeds increase, one needs to take into consideration the full dynamics in the development of control algorithms.

Computed Torque Methods

Several control algorithms that incorporate dynamics were developed. Many of these are variations of the computed torque method, which is similar to the feedback linearization method used for the control of nonlinear systems [Spong and Vidyasagar, 1989]. In the computed torque method shown in Fig. 94.16, the required input forces or torques are computed as follows:

$$\tau = D(q)\left[\ddot{q}_d + K_v(\dot{q}_d - \dot{q}) + K_p(q_d - q)\right] + H(q, \dot{q}) + V\dot{q} + G(q) \qquad (94.31)$$

where K_v and K_p are diagonal matrices with diagonal elements representing velocity and position gains. If this torque is chosen as the input in Eq. (94.26), we get

$$D(q)\left[\ddot{q}_d - \ddot{q} + K_v(\dot{q}_d - \dot{q}) + K_p(q_d - q)\right] = 0 \qquad (94.32)$$

Since the inertia matrix, $D(q)$ is nonsingular, we get

$$\ddot{E} + K_v\dot{E} + K_p E = 0 \qquad (94.33)$$

which represents a set of decoupled equations where the error $E = q_d - q$. If we select the values of K_v and K_p such that the characteristic roots of Eq. (94.33) have negative real parts, then E approaches zero asymptotically. Effectiveness of this algorithm depends heavily on two factors: (1) the accuracy of the model and (2) the ability to compute the coefficient matrices of the equations of motion in real time.

If the model is not an exact representation of the system, Eq. (94.33) becomes

$$\ddot{E} + K_v\dot{E} + K_p E = R(\ddot{q}, \dot{q}, q) \qquad (94.34)$$

where R is the mismatch between the model and the actual dynamics of the robot. This is given by

$$R(\ddot{q}, \dot{q}, q) = \hat{D}^{-1}(q)\left[(D(q) - \hat{D}(q))\ddot{q} + (H(q, \dot{q}) - \hat{H}(q, \dot{q})) + G(q) - \hat{G}(q)\right] \qquad (94.35)$$

Observe that if the model is an exact match, Eq. (94.34) leads to Eq. (94.33) and the convergence of q to q_d can be guaranteed.

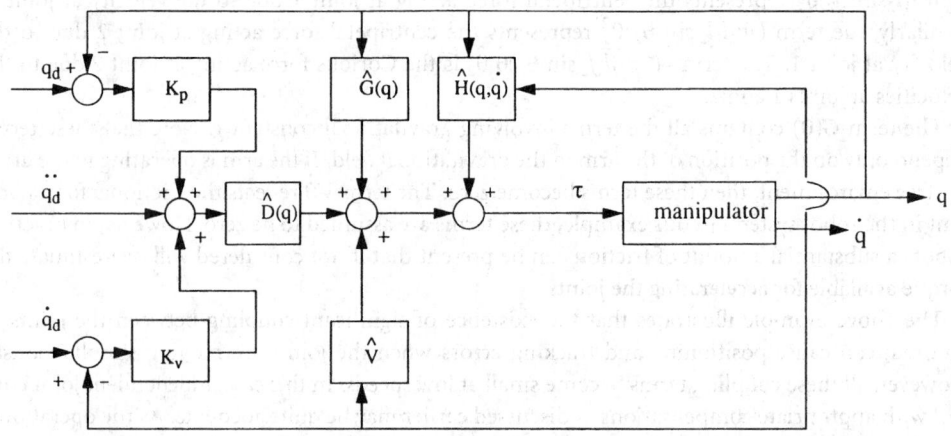

FIGURE 94.16 Computed torque method. (*Source:* M. W. Spong and M. Vidyasagar, *Robot Dynamics and Control*, New York: John Wiley & Sons, 1989. With permission.)

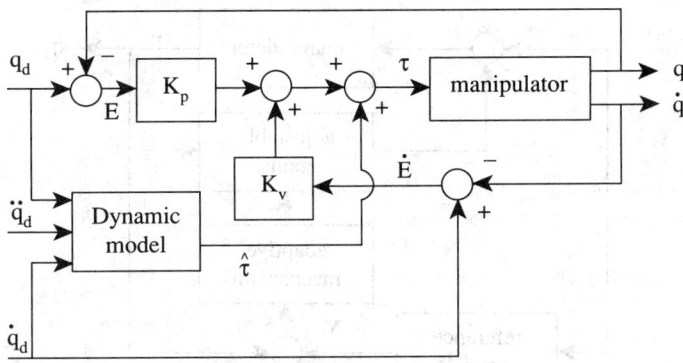

FIGURE 94.17 Dynamic model outside the feedback loop. (*Source:* J. J. Craig, *Introduction to Robotics*, Reading, Mass.: Addison-Wesley, 1989. With permission.)

Even if the model is accurate, the ability to compute the dynamics at sample rate (60 to100 Hz is typical) is still a problem. It is estimated that the Stanford manipulator requires 2000 floating-point additions and 1500 multiplications to compute all joint torques. A way to overcome this problem is to use the control scheme where the model is outside the feedback loop shown in Fig. 94.17 [Craig, 1989]. In this case, the desired torques are calculated *a priori* using the model given in Eq. (94.26) as follows:

$$\hat{\tau} = \hat{D}\ddot{q}_d + \hat{H}(q_d, \dot{q}_d) + \hat{V}\dot{q}_d + \hat{G}(q_d) \tag{94.36}$$

Then from Fig. 94.17, we get

$$D\ddot{q} + H(q, \dot{q}) + V\dot{q} + G(q) = \hat{\tau} + K_v(\dot{q}_d - \dot{q}) + K_p(q_d - q) \tag{94.37}$$

If the mismatch between the model and the robot is small, then we get

$$D(\ddot{q}_d - \ddot{q}) + K_v(\dot{q}_d - \dot{q}) + K_p(q_d - q) = 0 \tag{94.38}$$

Since the inertia matrix is nonsingular, we can rewrite the above equation as

$$\ddot{E} + D^{-1}K_v\dot{E} + D^{-1}K_pE = 0 \tag{94.39}$$

where $E = q_d - q$ and can be made to go to zero asymptotically by selecting the gains K_v and K_p appropriately. This method has a definite advantage over the earlier method because the model need not be evaluated in real time. However, it does not provide complete decoupling because the inertia matrix is not diagonal. Furthermore, since the gains are continuously modified by the inertia matrix, the response is a function of the configuration and the payload. A way to circumvent this problem is to continuously modify the gains K_v and K_p. This obviously suggests an adaptive control approach.

Adaptive Control

In an attempt to reduce the errors caused by the mismatch of the model with the real system, several adaptive control schemes have been investigated. Model reference adaptive control (MRAC) is one such approach. Dubowsky and DesForges [1979] were the first to use this method for manipulator control. This method is illustrated in Fig. 94.18. They have chosen a linear second-order system with desired ζ and ω_n as a reference model for each joint. Their scheme works as long as the manipulator changes configuration slowly relative to the adaptation rate. Since then several researchers have extended the concepts well developed for linear systems to manipulator control. Two aspects that are central to all these methods are identification of the plant or its parameters

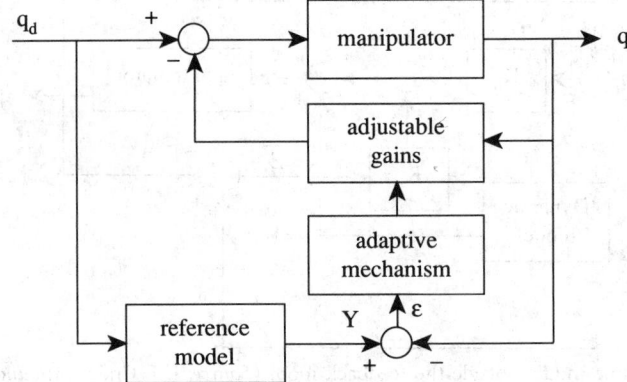

FIGURE 94.18   Model reference adaptive control.

and use of this new information to update the control law. An extensive review of recent work in this area is given by Craig [1988] and Hsia [1986]. In spite of many approaches suggested for this problem, no attempt has been made to implement these methods by the robot industry.

Resolved Motion Control

So far we have discussed the methods to achieve desired joint motion. In practice, the desired motion is specified in terms of hand motions. Resolved motion control methods such as *resolved motion rate control (RMRC)* and *resolved motion acceleration control* have been suggested [papers by Whitney and Luh *et al.* in Brady *et al.*, 1982]. In these methods, the joint motions are coordinated to achieve coordinated hand motion along any world coordinate axis. Given the relationship between the position and orientation of the hand, $x(t)$, and the joint coordinates $q(t)$ as

$$x(t) = f(q(t)) \tag{94.40}$$

then RMRC transforms the linear and angular velocity of the hand (end effector) to joint velocities using the relationship

$$\dot{q} = J^{-1}(q(t))\dot{x} \tag{94.41}$$

where $J(q(t))$ is the **Jacobian** matrix. Using the above equation, the combination of the joint rates for a given hand motion can be obtained. However, special consideration must be given when the inverse of the Jacobian matrix does not exist. This occurs when the dimension of $x(t)$ and $q(t)$ are not the same (robots with redundant degrees of freedom) or when a nonredundant robot loses one or more degrees of freedom in its workspace (singular configurations).

Resolved motion acceleration control extends the concepts of RMRC by including desired acceleration of the hand as well. Differentiating Eq. (94.41) twice with respect to time, we get

$$\ddot{x} = J(q)\ddot{q} + \dot{J}(q, \dot{q})\dot{q}(t) \tag{94.42}$$

where $\ddot{x}$ is the acceleration of the hand and $\ddot{q}$ is the joint acceleration. To reduce the position and orientation errors of the hand to zero,

$$\ddot{x}(t) = \ddot{x}_d(t) + K_v[\dot{x}_d(t) - \dot{x}(t)] + K_p[x_d(t) - x(t)] \tag{94.43}$$

By selecting the gains K_p and K_v we can force the error $e(t) = x_d(t) - x(t)$ to zero as before. The desired joint acceleration can be obtained from Eqs. (94.41) and (94.42), and is given as follows:

$$\ddot{q}(t) = J^{-1}(q)\begin{bmatrix} \ddot{x}_d(t) + K_v\big(\dot{x}_d(t) - \dot{x}(t)\big) \\ + K_p\big(x_d(t) - x(t) - \dot{J}(q, \dot{q})\dot{q}(t)\big) \end{bmatrix} \tag{94.44}$$

Since the inverse of the Jacobian is involved, this method suffers from drawbacks similar to the RMRC method.

Compliant Motion

The position control methods described above are not sufficient when the robot has to react continuously to contact forces at the end effector. Consider, for example, a simple operation of sliding a block of wood on a table along a desired path. Pure position control will not work because any small errors orthogonal to the table may result in the block either losing contact with the surface of the table or forcing the block through the table, which can either damage the table or the end effector. To perform this task, we need to control the position in the plane of the table and control force normal to the table. This is called *compliance motion control* and is required whenever the robot is in contact with its "environment." To perform the above task, for example, a coordinate system called *compliance frame* or *constraint frame* is defined such that at each instant and along each axis the task can be expressed as a pure position control or pure force control. Suppose we associate a coordinate system with the z axis normal to the table surface. Then to perform this task, we need to control the position along the x and y directions and force control in the z direction to maintain continuous contact with the table surface. In this case, the position along the z direction is not controlled because one cannot control both position and force in the same direction, just as we cannot control both voltage and current across a resistor. Hence, this framework will provide a natural separation between the axes that need to be positionally controlled and the axes that need to be force-controlled. This is the idea behind the hybrid position/force control developed by Raibert and Craig [1981].

In general, compliance motion control is very important whenever the robot is required to make contact with its "environment." This is true for many assembly tasks. Apart from the active methods of control discussed above, passive methods can be used to introduce the desired compliance. One such passive scheme is the use of a *remote center compliance* (RCC) device developed at Draper Laboratories. The RCC is a purely mechanical device consisting of a spring with six degrees of freedom that is inserted between the wrist and the end effector. By adjusting the stiffness of the springs, various levels of compliance can be obtained. However, these methods suffer from lack of programmability achieved through active methods. Active method of control, however, requires sensing of contact forces and torques at the end effector. *Joint torque sensors, wrist sensors, fingertip tactile sensors,* and *force pedestals* can be used for this purpose.

Joint torque sensors, as the name implies, are placed at the joints of the manipulator. If **F** represent the vector of forces at the end effector, then the corresponding vector of joint torques is obtained by using $\tau = [J(q)]^T \mathbf{F}$, where $J(q)$ is the Jacobian and q are the generalized joint coordinates. Joint torque sensing has some drawbacks. First, to obtain the endpoint forces **F**, the Jacobian which changes with the configuration has to be inverted in real time. Second, the sensors at the joints not only measure the forces and torques applied at the hand but also those applied at the other points of the manipulator. *Wrist sensors* are better at reducing this uncertainty because they are placed close to the end effector and below the last powered joint of the manipulator. Several wrist sensors are available commercially with necessary electronics to obtain force/torque measurements at high speeds suitable for real-time force control. Another method for providing information about the gripping forces is by mounting *tactile sensors* at the fingertips. However, these may not be suitable in situations where high gripping forces are required. *Force pedestals* are employed when a common platform is used for many tasks. In this case, the platform is instrumented to measure interacting forces and torques.

Flexible Manipulators

The discussion so far assumed that the links of the robot are rigid. These are designed intentionally to minimize the vibrations. Most of the present-day industrial robots fall into this category. These robots, however, cannot handle objects heavier than about 5% their weight. In contrast, lightweight flexible arms consume less energy, achieve faster speeds, and can potentially perform precision assembly tasks. However, it is not possible to move these arms quickly without the onset of structural vibrations due to inadequate structural damping. Efforts have been underway to increase the damping without substantial increase in weight by using composite materials or actively controlling the vibrations or both.

Defining Terms

Centripetal forces: Forces that are present during the robot motion. They depend upon the square of the joint velocities of the robot and tend to reduce the power available from the actuators.

Compliant motion: Motion of the manipulator (robot) when it is in contact with its "environment," such as writing on a chalkboard or assembling parts.

Coriolis forces: Forces/torques that depend upon the product of joint velocities.

Gravitational torques: Torques that depend upon the position of the robot in the gravitational field.

Independent joint control: Control of a single joint of a robot while all the other joints are fixed.

Jacobian of the manipulator: A matrix that maps the joint velocities into end effector velocities.

Tool space: Space of a 6×1 vector representing the positions and orientations of the tool or end effector of the robot.

References

H. Asada and J. J. E. Slotine, *Robot Analysis and Control*, New York: John Wiley & Sons, 1986.

M. Brady, J. M. Hollerbach, T. L. Johnson, T. Lozano-Perez, and M. T. Mason, *Robot Motion: Planning and Control*, Cambridge, Mass.: The MIT Press, 1982.

J. J. Craig, *Adaptive Control of Mechanical Manipulators*, Reading, Mass.: Addison-Wesley, 1988.

J. J. Craig, *Introduction to Robotics*, Reading, Mass.: Addison-Wesley, 1989.

S. Dubowsky and D. T. DesForges, "The application of model-referenced adaptive control of robotic manipulators," *ASME J. Dyn. Syst. Meas. Control*, 1979.

K. Fu, R. Gonzalez, and C.S.G. Lee, *Robotics: Control, Sensing, Vision, and Intelligence*, New York: McGraw-Hill, 1987.

T.C. Hsia, "Adaptive control of robot manipulators—a review," IEEE Conference on Robotics and Automation, San Francisco, 1986.

A. Koivo, *Control of Robotic Manipulators*, New York: John Wiley & Sons, 1989.

J. Y. S. Luh, "Conventional controller design for industrial robots—a tutorial," *IEEE Trans. Syst., Man and Cybern.*, vol. SMC-13, no. 3, June 1983.

R. P. Paul, *Robot Manipulators: Mathematics, Programming and Control*, Cambridge, Mass.: The MIT Press, 1981.

M. Raibert and J. Craig, "Hybrid position/force control of manipulators," *ASME J. Dyn. Syst. Meas. Control*, June 1981.

M. W. Spong and M. Vidyasagar, *Robot Dynamics and Control*, New York: Wiley, 1989.

Further Information

More information about this subject can be obtained by referring to many of the textbooks available on this subject. These are given in the References. Readers who are interested in current research may refer to several journals published by the Institute of Electrical and Electronics Engineers. In particular, *IEEE Transactions on Robotics and Automation*, *IEEE Transactions on Automatic*

Control, and *IEEE Transactions on Systems, Man and Cybernetics* along with the conference proceedings published by the respective societies are useful in this regard.

94.3 Applications

Nicholas G. Odrey

An important utilization of robotics has traditionally been in manufacturing operations. By their very design and reprogrammable features, robots have enhanced the capabilities for flexibility in automation. Robot applications initially focused on replacing repetitive, boring, and hazardous manual tasks. Such initial applications required minimal control, programming, or sensory capability and have evolved to applications that use enhanced controller designs and sophisticated sensory capability. The first recorded commercial application of an industrial robot was at the Ford Motor Company in 1961 that used a Unimate robot to unload a die-casting machine. Since then, robots have been used in various manufacturing processes, fabrication, and assembly operations. Current issues relate to the degree of integration with the total manufacturing system and to the degree of autonomy and/or complexity one wishes to implement for a robotic system. In potential applications, it is necessary to determine the degree of sophistication that one wishes to implement coupled with a detailed economic analysis. The focus in this section is to present a practical implementation strategy for robots within a manufacturing environment, to review particular applications, and to discuss issues relevant to enhancing robot applications on the manufacturing shop floor. Such issues include sensors and their integration within an intelligent control system, the development of grippers for enhanced dexterity, and integration topics within a flexible **cellular manufacturing** system.

Justification

Reprogrammable automated devices such as robots provide the flexible automation capability for modern production systems. To evaluate a potential robotic application within a manufacturing environment, both technical and economic issues must be addressed. Typical technical issues include the choice of the number of **degrees of freedom** to perform a task, the level of controller and programming complexity, end effector and sensor choices, and degree of integration within the overall production system. Economic issues have typically been addressed from a traditional point of view, but it is important to note that other criteria should also be evaluated before a final decision is made to implement a robotic system. Such criteria may be both quantitative and qualitative.

Traditional economic approaches analyze investments and costs to compare alternative projects. Three methods are commonly used: (1) payback period method, (2) equivalent uniform annual costs (EUAC) method, and (3) return on investment (ROI). The payback method balances initial investment cost against net annual cash flow during the life of the project to determine the time required to recoup the investment. Many corporations today require relatively short (1- to 3-year) payback periods to justify an investment. In the current environment with the drive toward shortened product life cycles, it is not unusual to see payback period requirements of no greater than 1 year. The payback technique does not consider the time value of money and should be considered only as a first part attempt at justification.

The EUAC and ROI methods consider the time value of money (continuous or discrete compounding) and convert all investments, cash flows, salvage values, and any other revenues and costs into their equivalent uniform annual cash flow over the anticipated life of the project. In the EUAC method, the interest rate is known and set at a minimal acceptable rate of return, whereas the ROI method has the objective to determine the interest rate earned on the investment. Details to such techniques are presented in various engineering economy texts such as those by White *et al.* [1977] and Thuesen and Fabrycky [1989].

Various more sophisticated approaches have been taken to justify robotic and automated system implementation. Estimates of indirect factors such as taxes, capital gain or losses, variability consideration, and associated expected value analysis along with **decision tree analysis** and Markovian decision analysis [Michel, 1986] are but a few methods to justify such systems. Other factors to be recognized in robotic justification are that robots are reusable from one project to the next and there is a difference in production rates for a robotic implementation over a manual process. A changeover from a manual method to a robotic implementation would have the potential to affect revenues for any project. Many companies have also developed standard investment analysis forms for an economic evaluation of a proposed robot project. These forms are helpful in displaying costs and savings for a project. Groover *et al.* [1986] presents one such proposed form and gives several references to examples of forms specifically designed for projects devoted to robotics and related automation areas.

The aforementioned techniques are important in performing an economic justification for a proposed robotic installation. Still, in general, there are other issues that should be included in the overall analysis. These issues are of particular importance if one is considering installing a more comprehensive system such as a flexible manufacturing system that may include many robots and automated systems. As noted by Proth and Hillion [1990], these issues give rise to criteria that are both quantitative and qualitative. Quantitative criteria include not only reduced throughput time and work-in-process inventory but also criteria related to increased productivity coupled with fewer resources. Another measurable criterion is the reduction in management and monitoring staff as a result of smaller quantities and automatic monitoring by sensors. Quality improvement can also be measured both quantitatively and qualitatively. Qualitative benefits from quality improvement can include increased customer satisfaction, increased competitiveness, simplified production management, and other factors. It should be noted that any benefits and cost reductions for installation of an automated system are difficult to evaluate and reflect a long-term commitment of the corporation.

Strategic factors should be incorporated in the overall economic justification process, but they are difficult to access and incorporate due to their inherent complexity. Verk [1990] proposes a general framework that attempts to integrate both qualitative and quantitative factors in an economic justification process. The approach taken is being tested at Cincinnati Milacron and the Mazak Corporation.

Implementation Strategies

A logical approach is a prerequisite to robotic implementation within a manufacturing firm. The following steps have been proposed by Groover *et al.* [1986] to implement a robotic system:

1. Initial familiarization with the technology.
2. Plant survey to identify potential applications.
3. Selection of an application(s).
4. Selection of a robot(s) for the application(s).
5. Detailed economic analysis and capital authorization.
6. Plan and engineer the installation.
7. Installation.

It should be noted that a particular company may have nuances that could modify the above steps. Also of note is that the underlying issue is systems integration and any robotic application should consider total system impact as well as include the equipment, controllers, sensors, software, and other necessary hardware to have a fully functional and integrated system. Another good source of information on robot implementation is the text by Asfahl [1992].

Critical factors for the introduction of robotics technology within a corporation are management support and production personnel acceptance of the technology. Companies such as General Electric have developed checklists to determine the degree of workforce acceptance. Given that the above two factors are met, a plant survey is conducted to determine suitability for automation or robotic implementation. Two general categories of robot applications may be distinguished: (1) a project for a new plant, or (2) placing a robot project in an existing facility. We focus here on the latter category.

General considerations for a robot installation include hazardous, repetitive, or uncomfortable working conditions, difficult handling jobs, or multishift operations. High- and medium-volume production typically has many examples of repetitive operations. It can prove useful to investigate injury (particularly muscular) reports with medical personnel and ergonomics experts to identify potential manual operations that may be alleviated with the aid of robotics or automation. Multishift operations associated with high demand for a product are likely candidates for robot applications. As compared to manual work that typically has a high variable labor cost, a robot substitution would have a high fixed cost which can be distributed over the number of shifts plus a low variable cost. The overall effect of a robot application would then be to reduce the total operating cost.

Once potential robot applications are identified, one typically must determine which application is the best to pursue. Economic and technical criteria must both be considered. Usually, a simple application that is easy to integrate into the overall system is a good initial choice. A fundamental rule is to implement any straightforward application to minimize the risk of failure. The General Electric Company has been successful in choosing robot applications by considering the following technical criteria:

- Operation is simple and repetitive.
- Cycle time for the operation is greater than five seconds.
- Parts can be delivered with the proper POSE (position and orientation).
- Part weight is suitable (typical upper weight limit is 1100 lb).
- No inspection is required for the operation.
- One to two workers can be replaced in a 24-hour period.
- Setups and changeovers are infrequent.

A choice of a robot for a selected application can be a very difficult decision. Vendor information, expert opinion, and various sources such as the *Robotics Product Database* [Flora, 1989] can aid in the selection. Selection needs to consider the appropriate combination of parameters suitable for the application. These parameters or technical features include the degrees of freedom, the type of drive and control system, sensory capability, programming features, accuracy and precision requirements, and load capacity of the selected robot. Various point or weighing schemes can be applied to rate different robot models.

The planning and engineering of a robot installation must address many issues, including the operational methods to be employed, workcell design and its control, the choice or design of end effectors and other fixturing and tooling requirements, and sensory and programming requirements. In addition, one needs to focus on safety considerations for the workcell as well as overall systems integration. Computer-aided design (CAD) is very helpful to study potential **machine interference** and various layout problems as well as estimating various performance parameters. Various commercial CAD software packages exist to analyze such problems. One such example is McAuto's PLACE System. The study at this stage should consider the basic purpose and function of the planned workcell. Consideration needs to be given to analyzing the cycle time that is basic to determining the production rate. An approach developed by Nof and Lechtman [1982], called Robot Time and Motion (RTM), is useful for analyzing the cycle time of robots.

Applications in Manufacturing

Robots have proven to be beneficial in many industrial and nonindustrial environments. Here, we focus on applications within a traditional manufacturing (shop floor) setting and, in particular, on applications which fall into the following three broad categories:

1. Material handling and machine loading/unloading
2. Processing
3. Assembly and inspection

The discussion that follows is not all-inclusive but rather is intended to present (1) an overview of such applications and (2) a few of the more current topics which are impacting the shop floor, particularly as related to flexible manufacturing systems. In the latter case, such issues include developments in **sensor integration**, mobility, sensory interactive grippers/hands, and issues pertaining to intelligent machines and robots. An important reference for many if not all robotic topics is the *International Encyclopedia of Robotics* edited by Dorf [1988].

Material Handling and Machine Loading/Unloading

Applications in this category pertain to the grasping and movement of a workpart or item from one location to another. General considerations for such applications pertain to the gripper design, distances moved, robot weight capacity, the POSE, and robot-dependent issues pertaining to the configuration, degrees of freedom, accuracy and precision, the controller, and programming features. POSE information is particularly important if there are no sensors (e.g. vision) to provide such information prior to pick-up. Specialized grippers have been designed for various applications in all three of the listed categories [Engelberger, 1980]. Quick-change wrists enabling the robot to change grippers (or tools in processing applications) during the production cycle have also become more common since their introduction [Vranich, 1984], as have multiple grippers mounted turret-like at the end of a robotic arm. Various factors need to be considered in the selection and design of grippers. One such checklist of factors can be found in Groover *et al.* [1986]. It should be noted that certain applications may require a high degree of accuracy and precision whereas others do not. Higher requirements result in more sophisticated drive mechanisms and controllers with associated increased costs.

Material handling applications are typically unsophisticated with minimal control requirements. Two- to four-degrees-of-freedom robots may be sufficient in many tasks. More sophisticated operations such as palletizing may require up to six degrees of freedom with stricter control requirements and more programming features. Various criteria that have proved to contribute to the success of material handling and machine load/unload applications can be found in Groover *et al.* [1986]. In addition, excellent examples and case studies on robotic loading/unloading are given in the text by Asfahl [1992].

Processing

Robotic processing applications are considered here to be those applications in which a robot actually performs work on a part and requires that the end effector is a tool. Examples include spot welding electrodes, arc welding, and spray-painting nozzles. The most common robotic applications in manufacturing processes are listed in Table 94.2 [Odrey, 1992a]. Many more processing applications are possible.

Spot welding and arc welding represent two major applications of industrial robots. It has been noted that industrial robot usage in welding tasks may be as high as 40% [Ross, 1984]. Spot welding robots have found wide use in automotive assembly lines and have been found to improve weld quality and provide more consistent welds and better repeatability of weld locations.

Table 94.2   Most Common Robotic Applications in Manufacturing Processes

Spot welding	Grinding
Continuous arc welding	Deburring
Spray coating	Polishing
Drilling	Wire brushing
Routing	Riveting
Waterjet cutting	Laser machining

Continuous arc welding is a more difficult application than spot welding. Welding of dissimilar materials, variations in weld joints, dimensional variations from part to part, irregular edges, and gap variations are some of the difficulties encountered in the continuous arc welding processes. Typical arc welding processes include gas metal arc welding (GMAW), shielding metal arc welding (SMAW), i.e., the commonly known "stick" welding, and submerged arc welding (SAW). The most heavily employed robotic welding process is GMAW in which a current is passed through a consumable electrode and into a base metal, and a shielding gas (typically CO_2, argon, or helium) minimizes contamination during melting and solidification.

In welding, a worker can compensate automatically by varying welding parameters such as travel speed, deposition rate by current adjustment, weave patterns, and multiple welds where required. Duplicating human welding ability and skill requires that industrial robots have sensor capability and complex programming capability. A wide variety of sensors for robotic arc welding are commercially available and are designed to track the welding seam and provide feedback information for the purpose of guiding the welding path.

Two basic categories of sensors exist to provide feedback information: noncontact sensors and contact sensors. Noncontact sensors include arc-sensing systems and machine vision systems. The former, also referred to as a *through-the-arc* system, uses feedback measurements via the arc itself. Specifically, measurements for feedback may be the current (constant-voltage welding) or the voltage (constant-current welding) obtained by programming the robot to perform a weave pattern. The motion results in measurements that are interpreted as vertical and cross-seam position. Adaptive positioning is possible by regulating the arc length (constant-current systems) as irregularities in gaps or edge variations are encountered.

Vision systems track the weld seam, and any deviations from the programmed seam path are detected and fed back to the controller for automatic tracking. Single-pass systems detect variations and make corrections in one welding pass. Double-pass systems first do a high-speed scan of the joint to record in memory deviations from the programmed seam path, with actual welding corrections occurring on the second "arc-on" pass. Single-pass systems give the advantages of reduced cycle time and of being able to compensate for thermal distortions during the welding operation. One recent example of a microcomputer-based single-pass system using a welding torch and laser-ranging sensor on a six-axis robot is given by Nayak and Ray [1990]. Their system, dubbed ARTIST for adaptive, real-time, intelligent, seam tracker, has a two-level integrated control system in which the high level contains rule-based heuristics and model-based reasoning to arrive at real-time decisions, whereas the low level enables tracking of a three-dimensional welding seam.

It should be noted that arc welding, like many manufacturing processes, is not well enough understood physically that one can formulate an exact mathematical model to describe the process. Attempts to optimize welding schedules for any arc welding process have led to expert systems for such processes [Tonkay and Knott, 1989]. Other examples of such work can be found in publications of the *Welding Journal* [e.g., Lucas, 1987; Fellers, 1987].

A robotic arc welding cell provides several advantages over manual welding operations. These advantages include higher productivity as measured by "arc-on" time, elimination of worker fatigue, decreased idle time, and improved safety. It is also important to correct upstream production operations to reduce variations. This is best accomplished during the design and installation phase of a robotic welding cell. During this phase, issues to consider include delivery of materials to the cell, fixtures and welding positioners, methods required for the processes, and any production and inventory control problems related to the efficient utilization and operation of the cell.

Other processing applications for robot use include spray coating and various machining or cutting operations. Spray coating is a major application in the automotive industry where robots have proven suitable in overcoming various hazards such as fumes, mist, nozzle noise, fire, and possible carcinogenic ingredients. The advantages of robotic spray coating are lower energy consumption, improved consistency of finish, and reduced paint quantities used. To install a robotic painting application, one needs to consider certain manual requirements. These include continuous-path control to emulate the motion of a human operator, a hydraulic drive system to minimize electrical spark hazards, and manual lead-through programming with multiple program storage capability [Groover, *et al.*, 1986]. Newer schemes have considered geometric modeling, painting mechanics, and robot dynamics to output an optimal trajectory based on CAD data describing the objects [Suh *et al.*, 1991]. The objective of such work is to plan an optimal robot trajectory that gives uniform coating thickness and minimizes coating time.

Machining operations utilizing robots typically employ end effectors that are powered spindles attached to the robot wrist. A tool is attached to the spindle to perform the processing operation. Examples of tools would be wire brushes or a grinding wheel. It should be noted that such applications are inherently flexible and have the disadvantage that such operations would be less accurate than a regular machine tool. Finishing operations, such as deburring, have provided excellent opportunities for robotic application. Force control systems have proven particularly useful in regulating the contact force between the tool and the edge of the work to be deburred. One such example for robotic deburring is given by Stepien *et al.* [1987]. In general, force-torque sensors mounted at the robot wrist have proven extremely useful in many applications in processing and assembly operations. The Lord Corporation and JR3 are two manufacturers of such commercial sensors.

Assembly Applications

Automated assembly has become a major application for robotics. Assembly applications consider two basic categories: parts mating and parts joining. Parts mating refers to peg-in-hole or hole-on-peg operations, whereas joining operations are concerned not only with mating but also a fastening procedure for the parts. Typical fastening procedures could include powered screwdrivers with self-tapping screws, glues, or similar adhesives.

In parts-mating applications, remote center compliance (RCC) devices have proven to be an excellent solution. In general, compliance is necessary for avoiding or minimizing impact forces, for correcting positioning error, and for allowing relaxation of part tolerances. In choosing an RCC device, the following parameters need to be determined prior to an application:

- Remote center distance (center of compliance). This is the point about which the active forces are at a minimum. The distance is chosen by considering the length of the part and the gripper.

- Axial force capacity. Maximum designed axial force to function properly.

- Compressive stiffness. Should be high enough to withstand any press fitting requirements.

- Lateral stiffness. Refers to force required to deflect RCC perpendicular to direction of insertion.

- Angular stiffness. Relates to forces that rotate the part about the compliant center (also called the cocking stiffness).

- Torsional stiffness. Relates to moments required to rotate a part about the axis of insertion.

Other parameters also include the maximum allowable lateral and angular errors as determined by the size of the part and by its design. These errors must be large enough to compensate for errors due to parts, robots, and fixturing. Passive and instrumented (IRCC) devices have been developed for assembly applications. One such device that combines a passive compliance with active control is described by Xu and Paul [1992]. In addition, the SCARA (Selective Compliance Articulated Robot for Assembly) class of robots is stiff vertically but relatively compliant laterally.

Many opportunities exist for flexible assembly systems. Many of the issues for such systems have been addressed by Soni [1991]. The reader is also referred to the *Design for Robotic Assembly Handbook* [Boothroyd and Dewhurst, 1985] for quantitative methods to evaluate a product's ease of assembly by robots. Carter [1990] presents a method for determining robot assembly task time as derived from tests and industrial experience. Carter also addresses the relationship between product design and robotic assembly cycle time. Some of the current trends in automated assembly include coordinating multiple robots to increase the flexibility and reliability of an assembly cell [Coupes *et al.*, 1989; Zheng and Sias, 1986], interaction with CAD databases to automatically generate assembly plans [Wolter, 1989; Nnaji, 1989] and the application of sensors to automatic assembly systems [Cook, 1991]. Meijer and Jonker [1991] consider an architecture for an intelligent assembly cell and its subsequent implementation. An article by Jarneteg [1990] considers the strategies necessary for developing adaptive assembly systems.

Inspection

Inspection involves checking of parts, products, and assemblies as a verification of conformation to the specification of the engineering design. With the emphasis on product quality, there is a growing emphasis for 100% inspection. Machine vision systems, robot-manipulated active sensing for inspection, and automatic test equipment are being integrated into total inspection systems. Robot application of vision systems include part location, part identification, and bin picking. Machine vision systems for inspection typically perform tasks which include dimensional accuracy checks, flaw detection, and correctness and completeness of an assembled product. Current vision inspection systems are predominantly two-dimensional systems capable of extracting feature information, analyzing such information, and comparing to known patterns previously trained into the system. As documented by Nurre and Hall [1989], various techniques for three-dimensional measurements have also been developed by many researchers. Primary factors to be considered in the design or application of a vision system include the resolution and field of view of the camera, the type of camera, lighting requirements, and the required throughput of the vision system.

Machine vision application can be considered to have three levels of difficulty, namely, that (1) the object can be controlled in both appearance and position, (2) it can be controlled in either appearance or position, or (3) neither can be controlled. The ability to control both position and appearance requires advanced, potentially three-dimensional vision capabilities. The objective in an industrial setting is to lower the level of difficulty involved. It should be noted that inspection is but one category of robotic applications of machine vision. Two other broad categories are identification and visual servoing and navigation. In the latter case, the purpose of the vision system is to direct the motion of the robot based on visual input. The reader is directed to Groover *et al.* [1986] for further details.

Emerging Issues

Robotics, by definition, is a highly multidisciplinary field. Applications are broad, and even those applications focused on the manufacturing shop floor are too numerous to cover in full here. The reader is referred to the various journals published by the IEEE and other societies and publishers, a few of which have been listed in the references. Still, it is worthwhile to note a few issues relevant to manufacturing shop-floor applications that could have an impact over the next decade. These issues include gripper development, mobility, and intelligent robots. The objective of this work is the overall integration of a flexible manufacturing system.

In a manufacturing process or assembly operation, the actions required of a gripper will vary with the task. Much work has been done in developing multifigured hands such as the Utah-MIT hand, the Salisbury hand, and others with an increasing interest of adding tactile sensory input for dexterous manipulation [Allen *et al.*, 1989]. As noted by Allen and his colleagues, robotic systems need to process multiple source data and be easily programmable for grasping and manipulation tasks. One study focused on capturing a machinist's skill in working with parts and tools and codifying this knowledge in a grip taxonomy has been done by Cutkosky and Wright [1986]. Their study suggests some general principles for the design, construction, and control of hands in a manufacturing (particularly machining) environment. The reader is also referred to the work of Feddema and Ahmad [1986] for the development of an algorithm for a static robot grasp for automated assembly and the work of Cutkosky [1991] on robotic grasping and manipulation. This latter work considers dynamic contact and the application of dynamic tactile sensors in manipulation tasks. An application to identify and locate circuit board fixtures within a robotic workcell that integrates a vision system with a tactile probe is given by DeMeter and Deisenroth [1987].

Automated guided vehicles (AGVs) currently dominate the movement of parts through a flexible manufacturing system (FMS). AGVs typically restrict the path to predetermined routes and subsequently decrease the "flexibility" of the system. Work is being done on mobile robots to address this issue. Research by Arkin and Murphy [1990] focuses on intelligent mobility within a manufacturing environment. The reader is also referred to the research of Wiens and Black [1992] who address a mobile robot system within a manufacturing cell as a means to increase the flexibility, capability, and capacity of a robot-based manufacturing cell.

The issues involved with intelligent robots have been surveyed by Nitzan [1985], where he notes that future proliferation of robotic applications will depend strongly on machine (robotic) intelligence. Such applications will lead to a greater diversity of applications and will not be just manufacturing oriented. The reader is also referred to work on intelligent machines by Weisbin [1986]. It should be noted that particular interest has been directed toward integration of multiple sensors as a means to enhance robot intelligence [Luo and Lin, 1989; Pin *et al.*, 1991]. The text by Klafter *et al.* [1989] categorizes the major sensory needs for robotic tasks and gives valuable insights to current and future robotics applications. **Intelligent control** systems, particularly hierarchical control systems, are being developed by many organizations and research institutes [Odrey, 1992b]. Such systems are expected to have an impact both at the shop-floor level and the management levels of production facilities well into the next century.

Defining Terms

Cellular manufacturing: Grouping of parts by design and/or processing similarities such that the group (family) is manufactured on a subset of machines which constitute a cell necessary for the group's production.

Decision tree analysis: Decomposing a problem into alternatives represented by branches where nodes (branch intersections) represent a decision point or chance event having probabilistic outcome. Analysis consists of calculating expected values associated with the chain of events leading to the various outcomes.

Degrees of freedom: The total number of individual motions typically associated with a machine tool or robot.

Intelligent control: A sensory-interactive control structure incorporating cognitive characteristics that can include artificial intelligence techniques and contain knowledge-based constructs to emulate learning behavior with an overall capacity for performance and/or parameter adaptation.

Machine interference: The idle time experienced by any one machine in a multiple-machine system that is being serviced by an operator (or robot) and is typically measured as a percentage of the total idle time of all the machines in the system to the operator (or robot) cycle time.

Sensor fusion: Combining of multiple sources of sensory information into one representational format.

Sensor integration: The synergistic use of multiple sources of sensory information to assist in the accomplishment of a task.

References

P. K. Allen, P. Michelman, and K. S. Roberts, "An integrated system for dextrous manipulation," IEEE International Conference on Robotics and Automation, 1989, pp. 612–616.

R. C. Arkin and R. R. Murphy, "Autonomous navigation in a manufacturing environment," *IEEE Trans. Robotics Autom.*, vol. 6, no. 4, pp. 445–454, 1990.

C. R. Asfahl, *Robots and Manufacturing Automation*, New York: Wiley, 1992.

G. Boothroyd and P. Dewhurst, "Design for Robotic Assembly," Department of Industrial and Manufacturing Engineering, University of Rhode Island, Kingston, 1985.

P. W. Carter, "Estimating cycle time in design for robotic assembly," *J. Manu. Syst.*, vol. 9, no. 1, pp. 1–12, 1990.

J. W. Cook, "Applying sensors to automatic assembly systems," *IEEE Trans. Ind. Appl.*, vol. 27, no. 2, pp. 282–285, 1991.

D. Coupes, A. Delchambre, and P. Gaspart, "The supervision and management of a two robots flexible assembly cell," Proceedings of IEEE Conference on Robotics and Automation, 1989, pp. 540–550.

M. R. Cutkosky, "Robotic grasping and manipulation," *Proceedings of NSF Design and Manufacturing Systems Conference*, Dearborn, Mich.: Society of Manufacturing Engineers, 1991, pp. 423–430.

M. R. Cutkosky and P. K. Wright, "Modeling manufacturing grips and correlations with the design of robotic hands," IEEE International Conference on Robotics and Automation, San Francisco, Calif., April 7–10, 1986, pp. 1533–1539.

E. C. DeMeter and M. P. Deisenroth, "The integration of visual and tactile sensing for the definition of regions within a robot workcell," Robots 11/17th ISIR, Chicago, Il., April 26–30, 1987, pp. 10-51 to 10-61.

R. C. Dorf, Ed., *International Encyclopedia of Robotics*, vols. 1–3, New York: Wiley, 1988.

J. F. Engelberger, "Robotics in practice," AMA COM: A Division of American Management Associations, 1980.

J. T. Feddema and S. Ahmad, "Determining a static robot grasp for automated assembly," IEEE International Conference on Robotics and Automation, San Francisco, Calif., April 7–10, 1986, pp. 918–924.

K. G. Fellers, "A PC approach to welding variables," *Weld. J.*, vol. 66, pp. 31–40, 1987.

P. C. Flora, Ed., *Robotics Product Database*, 6th ed., Orlando, Fla.: TecSpec, 1989.

M. P. Groover, M. Weiss, R. N. Nagel, and N. G. Odrey, *Industrial Robotics: Technology, Programming, and Applications*, New York: McGraw-Hill, 1986.

B. G. Jarneteg, "FAS control strategies for adaptive assembly systems," 21st CIRP International Seminar on Manufacturing Systems, Stockholm, Sweden, 1990.

R. D. Klafter, T. A. Chmielewski, and M. Negin, *Robotic Engineering: An Integrated Approach*, Englewood Cliffs, N.J.: Prentice-Hall, 1989.

W. Lucas, "Microcomputer systems, software and expert systems for welding engineering," *Weld. J.*, vol. 66, pp. 19–30, 1987.

R. C. Luo and M.-H. Lin, "Intelligent robot multi-sensor data fusion for flexible manufacturing systems," Proceedings of NSF 15th Conference on Production Research and Technology, University of California-Berkeley, Jan. 9–13, 1989, pp. 73–85.

B. R. Meijer and P. P. Jonker, "The architecture and philosophy of the DIAC (Delft Intelligent Assembly Cell)," IEEE Conference on Robotics and Automation, Sacramento, Calif., 1991, pp. 2218–2223.

M. Michel, "Justification models for flexible manufacturing," *Robots' 10 Conference Proceedings*, Dearborn, Mich.: Society of Manufacturing Engineers, 1986, pp. 2-55 to 2-81.

N. Nayak and A. Ray, "An integrated system for intelligent seam tracking in robotic welding: part 1—conceptual and analytical development; part 2—design and implementation," IEEE International Conference on Robotics and Automation, 1990.

D. Nitzan, "Development of intelligent robots: achievements and issues," *IEEE J. Robotics Autom.* vol. RA-1, no. 1, pp. 3–13, 1985.

B. O. Nnaji, "RALPH: An automatic robot assembly language programmer: an overview," Proceedings of Robots 13 Conference, Gaithersburg, Md., May 7–11, 1989, pp. 16-41 to 16-63.

S. Y. Nof and H. Lechtman, "The RTM method of analyzing robot work," *Ind. Eng.*, April 1982, pp. 38–48.

J. H. Nurre and E. L. Hall, "Three dimensional vision for automated inspection," Proceedings of Robots 13 Conference, Gaithersburg, Md., May 7–11, 1989, pp. 16-1 to 16-11.

N. G. Odrey, "Robotics and automation," *Maynard's Industrial Engineering Handbook*, 4th ed., W. K. Hodson, Ed., New York: McGraw-Hill, 1992a.

N. G. Odrey, "Control systems," *1992 McGraw-Hill Yearbook of Science and Technology*, New York: McGraw-Hill, 1992b, pp. 87–90.

F. G. Pin *et al.*, "Robotic learning from distributed sensory sources," *IEEE Trans. Syst. Man and Cybern.*, vol. 21, no. 5, pp. 1216–1223, 1991.

J. M. Proth and H. P. Hillion, *Mathematical Tools in Production Management*, New York: Plenum Press, 1990.

B. Ross, "Machines that can see: here comes a new generation," *Bus. Week.*, January 1984, p. 118.

A. H. Soni, "Flexible assembly systems: Opportunities and challenges," Proceedings of the 1991 NSF Design and Manufacturing Systems Conference, University of Texas at Austin, Jan. 9–11, 1991, pp. 367–373.

T. M. Stepien, L. M. Sweet, M. C. Good, and M. Tomizuka, "Control of tool/workpiece contact force with application of robotic deburring," *IEEE J. Robotics Autom.*, vol. RA-3, no. 1, pp. 7–18, 1987.

S.-H. Suh, I.-K. Woo, and S.-K. Noh, "Automatic trajectory planning system (ATPS) for spray painting robots," *J. Manu. Syst.*, vol. 10, no. 5, pp. 396–406, 1991.

G. J. Thuesen and W. J. Fabrycky, *Engineering Economy*, 7th ed., Englewood Cliffs, N.J.: Prentice-Hall, 1989.

G. L. Tonkay and K. Knott, "Intelligent process specification for robotic arc welding," Proceedings of World Conference on Robotics Research: The Next Five Years and Beyond. Robotics International of the Society of Manufacturing Engineers, Gaithersburg, Md., May 11–17, 1989.

S. Verk, "Strategic optimization cycle as a competitive tool for economic justification of advanced manufacturing systems," *J. Manu. Syst.*, vol. 9, no. 3, pp. 194–205, 1990.

J. M. Vranich, "Quick change system for robots," SME Paper MS84-418, Conference on Robotics Research—The Next Five Years and Beyond, Lehigh University, Bethlehem, 1984.

C. R. Weisbin, "CESAR research in intelligent machines," SME Paper MS586-772, Robotics Research Conference, Scottsdale, Ariz., Aug. 18-21, 1986.

J. A. White, M. H. Agee, and K. E. Case, *Principles of Engineering Economic Analysis*, New York: John Wiley & Sons, 1977.

G. J. Wiens and J. T. Black, "Design for mobility within a manufacturing cell," Proceedings of the NSF Design and Manufacturing Systems Conference, Georgia Institute of Technology, Jan. 8–10, 1992, pp. 1147–1150.

J. D. Wolter, "On the automatic generation of assembly plans," Proceedings of the 1989 IEEE Conference on Robotics and Automation, 1989, pp. 62–68.

Y. Xu and R. P. Paul, "Robotic instrumented compliant wrist," *ASME J. Eng. for Ind.*, vol. 114, pp. 120–123, 1992.

Y. F. Zheng and F. R. Sias, Jr., "Two robot arms in assembly," IEEE Conference on Robotics and Automation, San Francisco, Calif., April 7–10, 1986, pp. 1230–1235.

Further Information

Various journals publish on topics pertaining to robots. Sources include the bimonthly *IEEE Journal of Robotics and Automation,* the quarterly journal *Robotics and Computer-Integrated Manufac-*

turing (published by Pergamon Press), *Robotics* (published by Cambridge University Press since 1983), and the *Journal of Robotic Systems* (published by Wiley).

IEEE has sponsored since 1984 the annual "International Conference on Robotics and Automation." IEEE conference proceedings and journals are available from the IEEE Service Center, Piscataway, N.J.

The Society of Manufacturing Engineers (SME) is another source for robot publications that are concerned with both research issues and applications. Robots 1 through 13 (1989) conference proceedings are available as well as the Robot Research conference proceedings (three to date) of Robotics International (RI) of SME. A directory of robot research laboratories is also available. Contact SME, Dearborn, Mich.

The three-volume *International Encyclopedia of Robotics: Applications and Automation* (R. C. Dorf, ed.), published by Wiley (1988), brings together the various interrelated fields constituting robotics and provides a comprehensive reference.

PARKINSON'S GUN DIRECTOR

A dream come true

The nocturnal insight of a particular engineer led to one of the most effective pieces of air-defense technology in World War II, and to Presidential Medals for Merit and several patents for him, his boss, and other co-workers.

The engineer was David B. Parkinson; the place was Bell Telephone Laboratories in New York City; and the time was spring, 1940. Parkinson was then a 29-year-old member of the technical staff who specialized in electromechanical design. He was intent on improving an instrument called an automatic level recorder, which

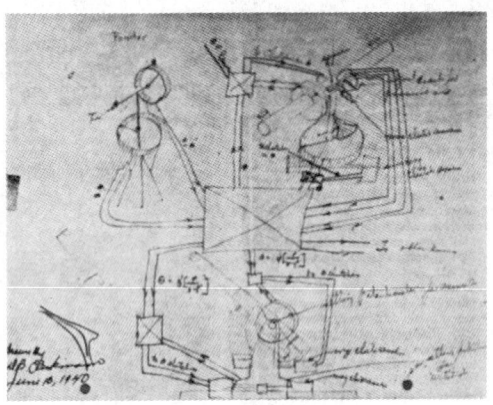

used strip-chart paper to plot the logarithm of an applied, rapidly varying voltage. A critical component was a small potentiometer, which controlled a pair of magnetic clutches that in turn controlled the pen.

Meanwhile, the evacuation of hundreds of thousands of stranded British and French soldiers from the beaches of Dunkirk, France, across the Channel into England was the top story in the U.S. press. Parkinson's dual preoccupations seem to have precipitated a dream, which he later described in an unpublished memoir:

"I found myself in a gun pit or revetment with an anti-aircraft gun crew. . . [A] gun there. . .was firing occasionally, and the impressive thing was that every shot brought down an airplane! After three or four shots one of the men in the crew smiled at me and beckoned me to come closer to the gun. When I drew near he pointed to the exposed end of the left trunnion. Mounted there was the control potentiometer of my level recorder!"

The next morning, he realized the significance of the dream—that "if my potentiometer could control the pen on the recorder, something similar could, with suitable engineering, control an anti-aircraft gun," Parkinson said recently in a telephone interview from his home in Cleveland Heights, Ohio.

At work that morning, Parkinson discussed his idea with his boss, Clarence A. Lovell, one of the early developers of the operational amplifier. They worked for several days and evenings, writing a report explaining how op amps could be used to integrate, differentiate, and manipulate tabular data.

After the report was written, Lovell and Parkinson met with Lovell's boss, Edward C. Wente. Parkinson realized that in meeting with Wente, he would need a diagram to explain his ideas. So just before they met, on June 18, he made a quick sketch on a sheet of plain white typing paper [see illustration].

A proposal for exploratory work on an electromechanical system for directing anti-aircraft guns was submitted to and approved by the Army Signal Corps, and an engineering model was delivered for testing to the Army at Fort Monroe, Md., on Dec. 1, 1941.

Production models began rolling off the assembly lines at Western Electric Co. in Chicago early in 1943.

Inputs to the director were provided at first by an optical rangefinder and later by radar. The director aimed the gun by taking the data for the aircraft's present position and continuously calculating the future position of the target.

Simple arithmetic, differentiation, and integration were performed in the M-9 by 28 of Lovell's operational amplifiers. The output of the op amps went to three 50-centimeter-diameter cast-iron potentiometers, the output voltage of which in turn drove the gun controls.

With the older, mechanical directors, many thousands of shells were fired to bring down a single aircraft; the M-9 brought the number down to about 100 shells per hit on an aircraft, and about 200 for a hit on the smaller V-1 rockets.

After the war, Parkinson enjoyed a brief period of celebrity and today, hanging in a frame over the mantel of his living room at home, is the drawing he did that morning in June 1940.

Source: Adapted from G. Zorpette, *IEEE Spectrum,* p. 43, April 1989. © 1989 IEEE.

95

Cary R. Spitzer
NASA Langley Research Center

Daniel A. Martinec
Aeronautical Radio, Inc.

Cornelius T. Leondes
*University of California,
San Diego*

Abdul Hamid Rana
GTE Spacenet

William Check
GTE Spacenet

Aerospace Systems

95.1 Avionics Systems 2188
A Modern Example System • Data Buses • Displays • Power •
Software in Avionics • Navigation Equipment • Emphasis on
Communications • Avionics in the Cabin • Avionics Standards

95.2 Communications Satellite Systems: Applications 2194
Satellite Launch • Spacecraft and Systems • Earth Stations •
VSAT Communication System • Video • Audio • Next-Gener-
ation Systems

95.1 Avionics Systems

Cary R. Spitzer, Daniel A. Martinec, and Cornelius T. Leondes

Avionics (aviation electronics) systems perform many functions: (1) for both military and civil aircraft, avionics are used for flight controls, guidance, navigation, communications, and surveillance;
and (2) for military aircraft, avionics also may be used for electronic warfare, reconnaissance, fire
control, and weapons guidance and control. These functions are achieved by the application of the
principles presented in other chapters of this handbook, e.g., signal processing, electromagnetic,
communications, etc. The reader is directed to these chapters for additional information on these
topics. This section focuses on the system concepts and issues unique to avionics.

Development of an avionics system follows the traditional systems engineering flow from definition and analysis of the requirements and constraints at increasing level of detail, through detailed
design, construction, **validation,** installation, and maintenance. Like some of the other aerospace
electronic systems, avionics operate in real time and perform mission- and life-critical functions.
These two aspects combine to make avionics system design and **verification** especially challenging.

Although avionics systems perform many functions, there are three elements common to most
systems: data buses, displays, and power. Data buses are the signal interfaces that lead to the high
degree of integration found today in many modern avionics systems. Displays are the primary form
of crew interface with the aircraft and, in an indirect sense, through the display of synoptic information also aid in the integration of systems. Power, of course, is the life blood of all electronics.

The generic processes in a typical avionics system are signal detection and preprocessing, signal
fusion, computation, control/display information generation and transmission, and feedback of
the response to the control/display information. (Of course, not every system will perform all of
these functions.)

A Modern Example System

The B-777 Airplane Information Management System (AIMS) is the first civil transport aircraft
application of the integrated, modular avionics concept, similar to that being used in the U.S. Air

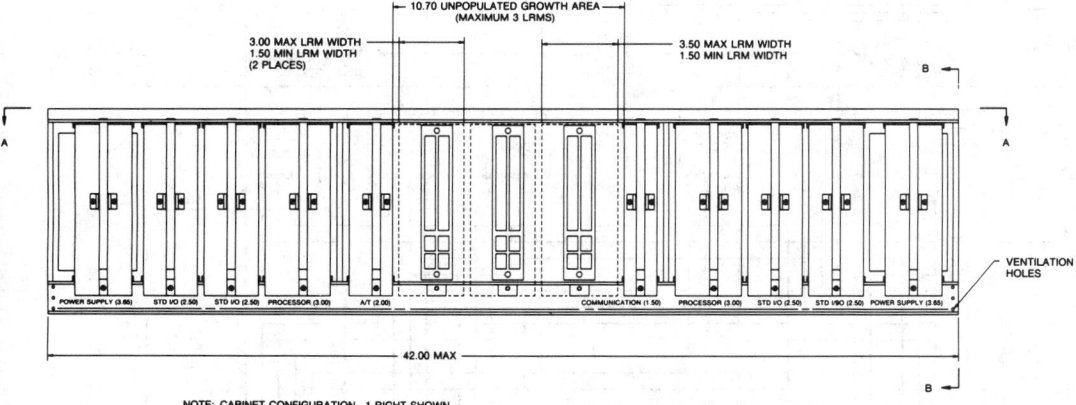

FIGURE 95.1 Cabinet assembly outline and installation (typical installation). (Courtesy of Honeywell, Inc.)

Force F-22. Figure 95.1 shows the AIMS cabinet with ten modules installed and three spaces for additional modules to be added as the AIMS functions are expanded. Figure 95.2 shows the AIMS architecture.

AIMS functions performed in both cabinets include flight management, electronic flight instrument system (EFIS) and engine indicating and crew alerting system (EICAS) displays management, central maintenance, airplane condition monitoring, communications management, data conversion and gateway (ARINC 429 and ARINC 629), and engine data interface. AIMS does not control the engines or operate any internal or external voice or data link communications hardware. Subsequent generations of AIMS may include some of these latter functions.

In each cabinet the line replaceable modules (LRMs) are interconnected by dual ARINC 659 backplane data buses. The cabinets are connected to the triplex redundant ARINC 629 system and fly-by-wire data buses and are also connected via the system buses to the three multifunction control display units (MCDU) used by flight crew and maintenance personnel to interact with AIMS. The cabinets transmit merged and processed data over quadruple redundant ARINC 629 buses to the EFIS and EICAS displays.

In the AIMS the high degree of function integration requires levels of system availability and integrity not found in traditional distributed, federated architectures. These extraordinary levels of availability and integrity are achieved by the extensive use of **fault-tolerant** hardware and software maintenance diagnostics and promise to reduce the chronic problem of unconfirmed removals and low mean time between unconfirmed removals (MTBUR).

Data Buses

As noted earlier, data buses are the key to the emerging integrated avionics architectures. Table 95.1 summarizes the major features of the most commonly used buses. MIL-STD-1553 and ARINC 429 were the first data buses to be used for general aircraft data communications. These are used today widely in military and civil avionics, respectively, and have demonstrated the significant potential of data buses. The others listed in the table build on their success.

Displays

All modern avionics systems use electronic displays, either CRTs or flat-panel LCDs that offer exceptional flexibility in display format and significantly higher reliability than electromechanical displays. Because of the very bright ambient sunlight at flight altitudes the principal challenge for

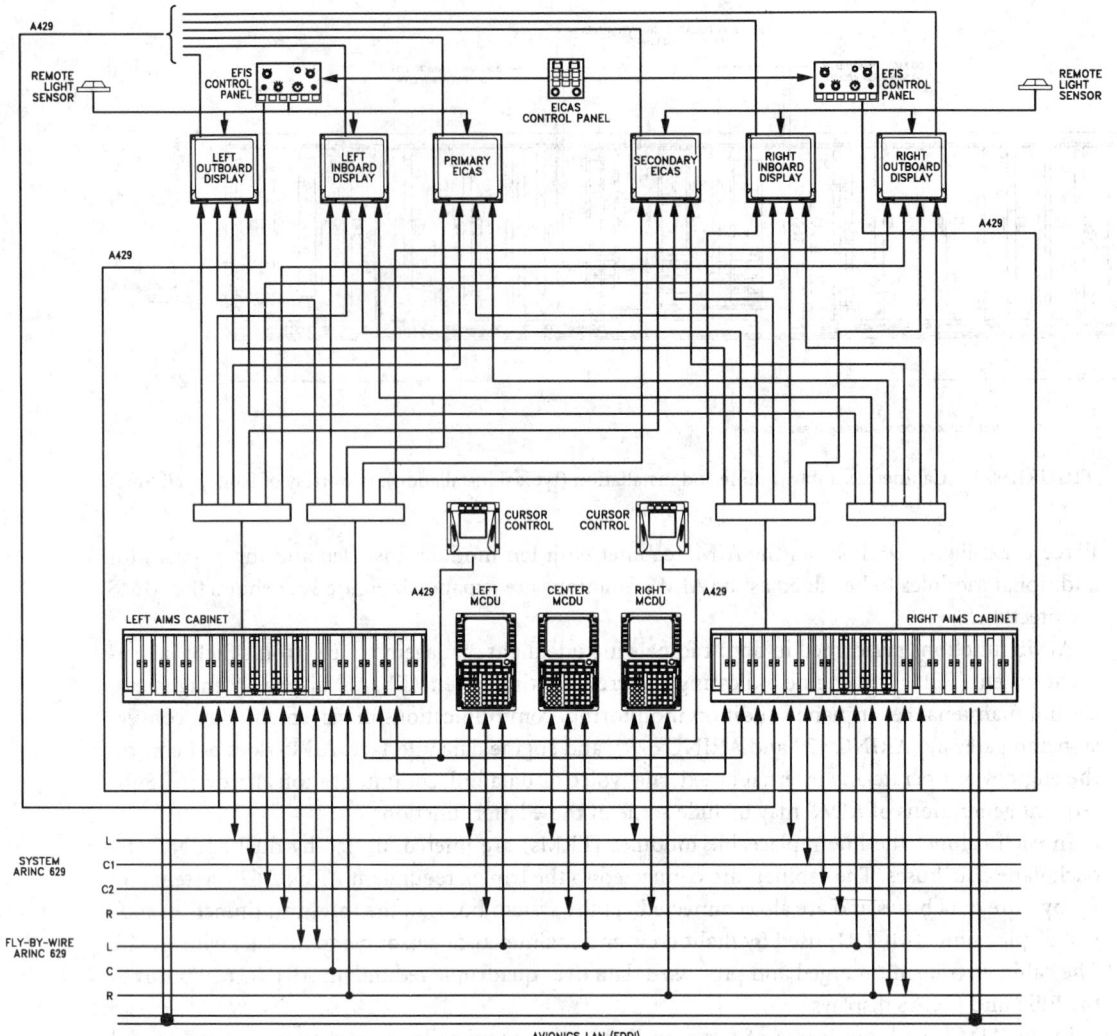

FIGURE 95.2   Architecture for AIMS baseline configuration. (Courtesy of Honeywell, Inc.)

an electronic display is adequate brightness. CRTs achieve the required brightness through the use of a shadow mask design coupled with narrow bandpass optical filters. Flat-panel LCDs also use optical filters and a bright backlight to achieve the necessary brightness.

Because of the intrinsic flexibility of electronic displays, a major issue is the design of display formats. Care must be taken not to place too much information in the display and to ensure that the information is comprehendible in high workload (aircraft emergency or combat) situations.

Table 95.1   Characteristics of Common Avionics Buses

Bus Name	Word Length	Bit Rate	Transmission Media
MIL-STD-1553	20 bits	1 Mb/s	Wire
DOD-STD-1773	20 bits	TBS	Fiber optic
High-speed data bus	32 bits	50 Mb/s	Wire or fiber optic
ARINC 429	32 bits	14.5/100 kb/s	Wire
ARINC 629	20 bits	2 Mb/s	Wire or fiber optic
ARINC 659	32 bits	100 MB/s	Wire

Power

Aircraft power is generally of two types: 28 vdc, and 115 vac, 400 Hz. Some 270 vdc is also used on military aircraft. Aircraft power is, by most standards, of poor quality. Under normal conditions, there can be transients of up to 100% of the supply voltage and power interruptions of up to 1 second. This poor quality places severe design requirements on the avionics power supply, especially where the avionics are performing a full-time, flight-critical function.

Software in Avionics

Most avionics currently being delivered are microprocessor controlled and are software intensive. The "power" achieved from software programs coupled with a sophisticated processor results in very sophisticated avionics with many functions and a wide variety of options. The combination of sophistication and flexibility has resulted in complex procedures for validation and certification. The **brick-walling** of software modules in a system during the initial development process to ensure isolation between critical and noncritical modules has been helpful in easing the certification process.

There are no standard software programs or standard software certification procedures. The Radio Technical Commission for Aeronautics (RTCA) has prepared Document DO-178 to provide guidance regarding preparation and certification of avionics civil software. The standard approaches to developing, categorizing, and documenting avionics civil software in DO-178 are widely used. The document is now undergoing revision.

For military avionics software, the principal document is DOD-STD-2167. This standard establishes the procedures for all phases of software design, development, coding, integration, and testing and presents the outline for all required documentation as a function of program development timeline. Many software languages have been used in the past in avionics applications; however, today there is a strong trend for both military and avionics civil software to use Ada wherever reasonably possible.

Navigation Equipment

A large portion of the avionics on an aircraft are dedicated to the navigation function. The following list provides the types of navigation and related sensing equipment commonly found on aircraft:

- Flight control computer
- Flight management computer (FMC)
- Inertial navigation system (INS)
- Attitude heading and reference system (AHRS)
- Air data computer
- Radio altimeter
- Radar
- **Distance Measuring Equipment** (DME)
- Instrument Landing System (ILS)
- Microwave Landing System (MLS)
- VHF OmniRange (VOR) Receiver
- Transponder
- Global Positioning System (GPS)

Emphasis on Communications

An ever-increasing portion of avionics is being dedicated to communications. Much of the increase comes in the form of digital communications for either digitized voice or data transfer. Military aircraft typically use digital communications for security. Civil aircraft use digital communications to transfer data for improved efficiency of operations and RF spectrum utilization. Both arenas are focusing more on enhanced communications to fulfill the requirements for better operational capability.

Various types of communications equipment are used on aircraft. The following list tabulates typical communications equipment:

- VHF transceiver (118–136 MHz)
- UHF transceiver (225–328 MHz/335–400 MHz for military)
- HF transceiver (2.8–24 MHz)
- Satellite (1530–1559/1626.5–1660.5 MHz)
- Satellite (various frequencies for military)
- **ACARS** (VHF data link)
- **JTIDS** (military spread spectrum)

In the military environment the need for communicating aircraft status and for aircraft reception of crucial information regarding mission objectives are primary drivers behind improved avionics. In the civil environment (particularly commercial transport), the desire for improved passenger services, more efficient aircraft routing and operation, safe operations, and reduced time for aircraft maintenance are the primary drivers for improving the communications capacity of the avionics.

The requirements for digital communications for civil aircraft have grown so significantly that the industry as a whole embarked on a virtually total upgrade of the communications system elements. The goal is to achieve a high level of flexibility in processing varying types of information as well as attaining compatibility between a wide variety of communication devices. The approach bases both ground system and avionics design on the ISO Open System Interconnect (OSI) model. This seven-layer model separates the various factors of communications into clearly definable elements of physical media, protocols, addressing, and information identification.

The implementation of the OSI model requires a much higher level of complexity in the avionics as compared to avionics designed for simple dedicated point-to-point communications. The avionics interface to the physical medium will generally possess a higher bandwidth. The bandwidth is required to accommodate the overhead of the additional information on the communications link for the purpose of system management. The higher bandwidths pose a special problem for aircraft designers due to weight and electromagnetic interference (EMI) considerations. Additional avionics are required to perform the buffering and distribution of the information received by the aircraft. Generally a single unit, commonly identified as the communications management unit (CMU), will perform this function.

The CMU can receive information via RF transceivers operating in conjunction with terrestrial, airborne, or space-based transceivers. The capability also exists for transceiver pairs employing direct wire connections or short-range optical links to the aircraft. The CMU also provides the routing function between the avionics, when applicable. Large on-board databases, such as an electronic library, may be accessed and provide information to other avionics via the CMU.

Avionics in the Cabin

Historically, the majority of avionics have been located in the electronics bay and the cockpit of commercial air transport airplanes. Cabin electronics had generally been limited to the cabin interphone and public address system, the sound and central video system, and the lighting control system. More recently the cabin has been updated with passenger telephones using both terrestrial and satellite systems. The terrestrial telephone system operates in the 900-MHz band in the United States and will operate near 1.6 GHz in Europe. The satellite system, when completely operational, will also operate near 1.6 GHz. Additional services available to the passengers are the ability to send facsimiles (FAXes) and to view virtually real-time in-flight position reporting via connection of the video system with the flight system. Private displays at each seat will allow personal viewing of various forms of entertainment including movies, games, casual reading, news programming, etc.

Avionics Standards

Standards play an important role in avionics. Military avionics are controlled by the various standards (MIL-STDs, DOD-STDs, etc.) for packaging, environmental performance, operating characteristics, electrical and data interfaces, and other design-related parameters. General aviation avionics are governed by fewer and less stringent standards. Technical Standard Orders (TSOs) released by the Federal Aviation Administration (FAA) are used as guidelines to ensure airworthiness of the avionics. TSOs are derived from and, in most cases, reference RTCA documents characterized as Minimum Operational Performance Standards and Minimum Avionics System Performance Standards. EUROCAE is the European counterpart of RTCA.

The commercial air transport industry adheres to multiple standards at various levels. The International Civil Aviation Organization (ICAO) is commissioned by the United Nations to govern aviation systems including but not limited to Instrument Landing Systems, Microwave Landing Systems, VHF OmniRange Systems, and Distance Measuring Equipment. The ICAO Standards and Recommended Practices (SARPS) control system performance, availability requirements, frequency utilization, etc. at the international level. The SARPS in general maintain alignment between the national avionics standards such as those published by EUROCAE and RTCA.

The commercial air transport industry also uses voluntary standards created by the Airlines Electronic Engineering Committee and published by Aeronautical Radio Inc. (ARINC). The ARINC "characteristics" define form, fit, and function of airline avionics.

Defining Terms

ACARS: A digital communications link using the VHF spectrum for two-way transmission of data between an aircraft and ground. It is used primarily in civil aviation applications.

Brickwalling: Generally used in software design in critical applications to ensure that changes in one area of software will not impact other areas of software or alter their desired function.

Distance measuring equipment: The combination of a receiver and a transponder for determining aircraft distance from a remote transmitter. The calculated distance is based on the time required for the return of an interrogating pulse set initiated by the aircraft transponder.

Fault tolerance: The built-in capability of a system to provide continued correct execution in the presence of a limited number of hardware or software faults.

JTIDS: Joint Tactical Information Distribution System using spread spectrum techniques for secure digital communication. It is used for military applications.

Validation: The process of evaluating software at the end of the software development process to ensure compliance with software requirements.

Verification: (1) The process of determining whether the products of a given phase of the software development cycle fulfill the requirements established during the previous phase. (2) Formal proof of program correctness. (3) The act of reviewing, inspecting, testing, checking, auditing, or otherwise establishing and documenting whether items, processes, services, or documents conform to specified requirements (IEEE).

References

Airlines Electronic Engineering Committee Archives, Aeronautical Radio Inc.

Federal Radionavigation Plan, DOT-VNTSC-RSPA-90-3/DOD4650.4, Departments of Transportation and Defense, 1990.

M.J. Morgan, "Integrated modular avionics for next generation commercial airplanes," *IEEE/AES Systems Magazine*, pp. 9–12, August 1991.

C.R. Spitzer, *Digital Avionics Systems*, 2nd ed., New York: McGraw-Hill, 1992.

Further Information

K. Feher, *Digital Communications*, Englewood Cliffs, N.J.: Prentice Hall, 1981.

J.L. Farrell, *Integrated Aircraft Navigation*, New York: Academic Press, 1976.

L.E. Tannas, Jr., *Flat Panel Displays and CRTs*, New York: Van Nostrand Reinhold, 1985.

M. Kayton and W.R. Fried, *Avionics Navigation Systems*, New York: John Wiley and Sons, 1969.

95.2 Communications Satellite Systems: Applications

Abdul Hamid Rana and William Check

The history of satellites began in 1957 when the Soviet Union launched Sputnik I, the world's first satellite. In the 1960s the commercial sector became actively involved in satellite communications with the launch of Telstar I by the Bell System followed by the use of a **geosynchronous orbit**. With this type of an orbit, an object 22,753 miles above the earth will orbit the earth once every 24 hours above the equator, and from the earth's surface appear to be stationary. The first geostationary orbit was achieved by NASA using a SYNCOM in 1963. The Communications Satellite Act was signed by the United States Congress in 1962 and created the Communications Satellite Corporation (COMSAT). This was followed by the formation of INTELSAT, an organization that is composed of over 120 countries and provides global satellite communication services. In the 1970s, multiple companies in the private sector in the United States began to operate their own domestic satellite systems. Today there are five companies providing this service in the United States: GTE Spacenet, GE Americom, Hughes, AT&T, and COMSAT. Other nations such as Canada, Australia, Indonesia, Japan, etc. have their own satellite systems. In addition to INTELSAT, other operators such as Alpha Lyracom have been allowed to offer international satellite services. Several regional satellite systems have also been formed. Examples of these are EUTELSAT, Intersputnik, ARABSAT, AsiaSat, etc. [Pritchard and Sciulli, 1986].

The satellite-based communications systems have significantly evolved over a three-decade period. In the 1960s, satellite communications for commercial use became a viable alternative because of the demand for reliable communications (telephony and voice). In the 1970s, technical innovations made larger, more powerful and more versatile satellites possible. Advanced modulation and multiple-access schemes resulted in smaller, less expensive **earth stations** and better service offerings that were lower cost and higher quality. In the 1980s **very small aperture terminals** (**VSATs**) emerged and the Ku-band frequency spectrum became widely used. Today's satellites support data, voice, and video communications applications. The VSAT industry has given an overall boost to the entire satellite communication industry.

As new satellites are launched, they will have long-term applications which have expanded opportunities. These include private long-haul networks for internal communications, cable TV, pay TV, business voice and data, satellite news gathering, direct broadcast to the home, integrated VSATs, private international satellite service, high-definition TV, mobile service, personal communications, and ISDN. Disaster recovery planning increasingly includes satellites in order to overcome the obvious shortcomings of existing terrestrial networks. With the allocation of frequencies for personal communications, the promise of global communications and the reality of a personal phone is destined to push satellite communications to a new age.

This section describes satellite communications from the application point-of-view. Since VSATs are setting the trends for satellite communication, a significant portion of the section is devoted to this topic. After a review of the satellites' launch and their characteristics, VSAT networks are discussed in detail. Video/audio applications are described next, along with the equipment necessary for these applications. The section is concluded with a summary of next-generation trends.

Satellite Launch

Launching a communications satellite into orbit is a complex and expensive process. This first stage in a satellite's airborne life may cost several million dollars. The cost for launching is primarily a function of the satellite's weight and size. Traditional geosynchronous communications satellites tend to be large and more costly to launch, although the more compact digital communications payloads and longer satellite life will reduce life cycle costs. Future low earth orbit communications satellites will tend to be smaller and more economical to launch.

A shortage of launch vehicles influenced the economics of the launch industry following the 1986 U.S. Space Shuttle *Challenger* disaster. The shortage has now given way to other launch alternatives. The dominant player in the satellite launch business is the French company Arianespace. Major U.S. players in the satellite launch business are Martin Marietta, General Dynamics, and McDonnell Douglas.

The launch of a satellite payload into the geosynchronous orbit involves many complex steps. Using the launch vehicle, the payload is first placed in a parking orbit. This is a nearly circular orbit which places the satellite approximately 300 km above the earth's surface. After reaching this orbit, the next step is to fire a motor known as the payload assist module (PAM) to place the payload in a transfer orbit. The PAM motor is discarded afterwards. The transfer orbit is an elliptical orbit whose perigee matches the parking orbit and whose apogee matches the geostationary orbit.

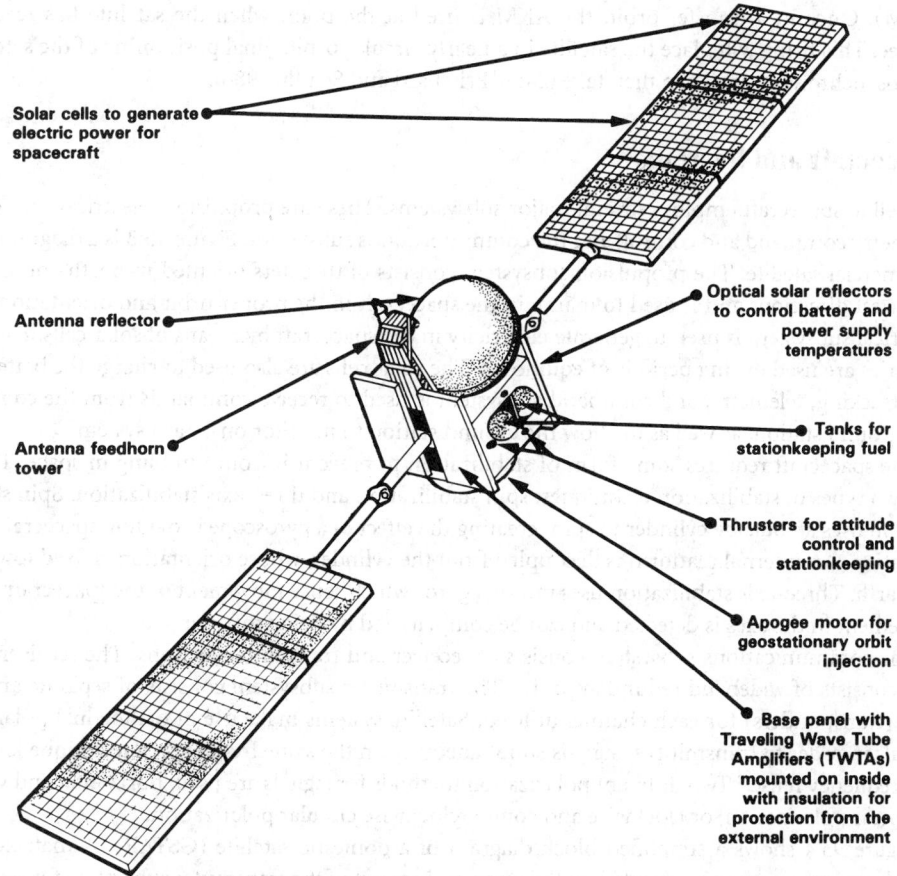

FIGURE 95.3 Simplified block diagram of a GSTAR satellite.

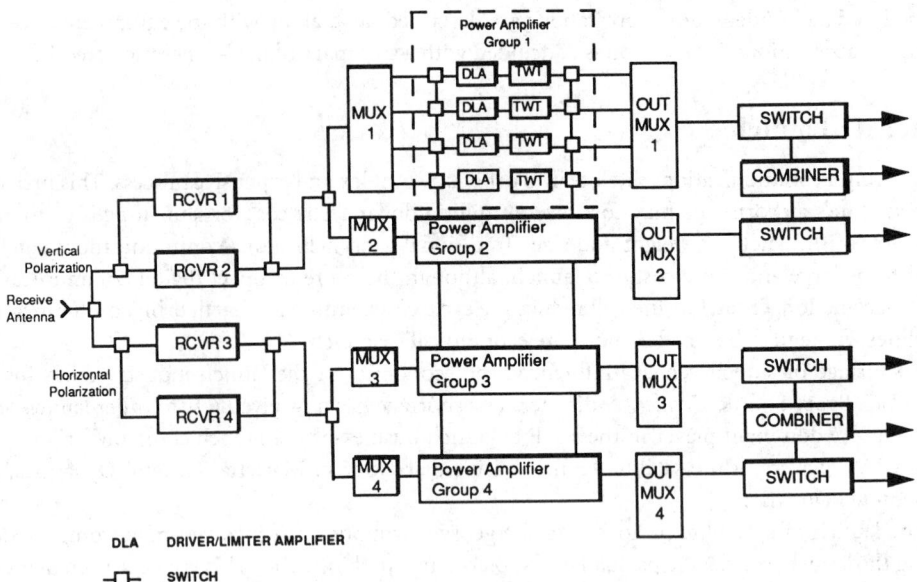

DLA   DRIVER/LIMITER AMPLIFIER

-□-   SWITCH

FIGURE 95.4   Simplified block diagram of GSTAR satellite.

Perigee is defined as the point in the orbit closest to the earth, while apogee is the point in the orbit furthest from the earth. The payload itself consists of the satellite with an apogee kick motor (AKM). Once in a transfer orbit, the AKM is fired at the point when the satellite has reached apogee. This firing will place the satellite in a nearly circular orbit. Final positioning of the satellite in geosynchronous orbit can then take place [Pritchard and Sciulli, 1986].

Spacecraft and Systems

A satellite spacecraft employs several major subsystems. These are propulsion, electrical, tracking, telemetry command and control, and the communications subsystem. Figure 95.3 is a diagram of a commercial satellite. The propulsion subsystem consists of thrusters oriented in north-south and east-west directions and is used to maintain the spacecraft in the proper orbit and orientation. An electrical subsystem is used to generate electricity in the spacecraft by means of solar cells. Backup batteries are used during periods of equinoxes. The solar cells are also used to charge the batteries. The tracking, telemetry, and command subsystem is used to receive commands from the controlling ground station, as well as to allow the ground station to monitor on-board systems.

The spacecraft requires some form of stabilization to prevent it from tumbling in space. There are two types of stabilization techniques: spin stabilization and three-axis stabilization. Spin stabilization uses an outside cylinder to spin, creating the effect of a gyroscope providing spacecraft stabilization. An internal platform is decoupled from the cylinder, whose orientation is fixed towards the earth. Three-axis stabilization uses internal gyros which sense movement of the spacecraft. Any movement in the axes is detected and can be compensated by firing thruster jets.

The communications subsystem consists of receiver and transmitter sections. The receiver system consists of wideband redundant units. The transmitter subsystem consists of separate amplifiers (transponders) for each channel utilized. Satellite systems make use of orthogonal polarized signals in order to transmit two signals simultaneously on the same frequency, a technique known as "frequency reuse." Two different polarization methods for signals are used: horizontal and vertical linear polarization, or clockwise and counterclockwise circular polarization.

Figure 95.4 shows a simplified block diagram of a domestic satellite (GSTAR). A matrix-type switching arrangement is provided on the input and output of the transmitter subsystem for switch-

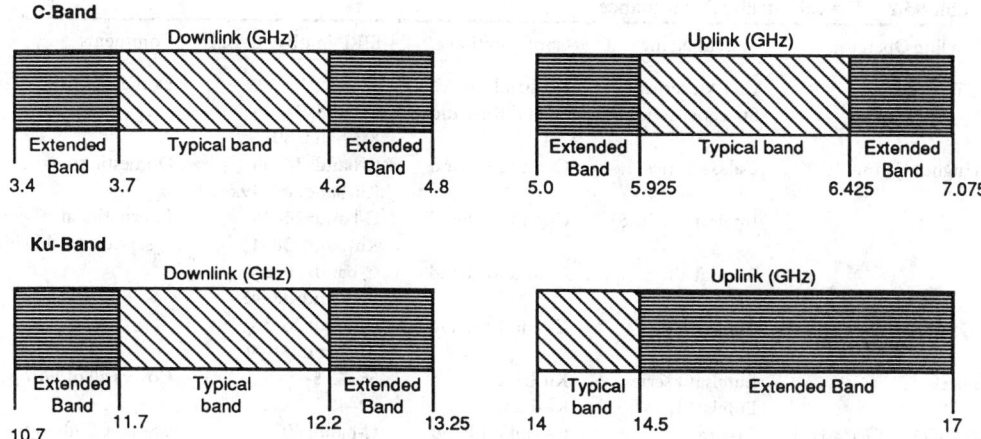

FIGURE 95.5 Ku- and C-band frequency allocation chart.

ing to backup transponders. This satellite is three-axis stabilized and operates at Ku-band. There are 16 operational transponders with a bandwidth of 54 MHz each. The employment of frequency reuse provides nearly 1000 MHz of usable bandwidth. Fourteen of the 16 operational transponders use 20-W traveling wave tube amplifiers (TWTA) to provide ground-commandable east or west regional coverage, for 48-state (CONUS) coverage. The remaining two transponders provide 50-state coverage using 27-W TWTAs. For the 50-state channels, one spare 27-W TWTA provides protection for the two operating TWTAs (3-for-2 redundancy). For the remaining transponder channels, 5 spare 20-W TWTAs provide protection for 14 operating TWTAs (19-for-14 redundancy). Redundant communications receivers are provided on a 4-for-2 basis.

The power radiated from a satellite is described as its effective radiated isotopic power (EIRP) and is the radiated power of the satellite in decibels referenced to one watt of power. The units are in dBW. The strength of the signal received on the ground is a function of the spacecraft location and that of the ground station and will vary depending upon location. A map of the signal strength contours is called the satellite's "footprint."

Geosynchronous Satellites

There are some 253 Ku-band and 256 C-band satellites in geosynchronous orbit. These satellites are typically spaced anywhere between 1 to 3 degrees apart. Older satellites no longer in active service may be spaced less than one degree in an inclined orbit. Many new Ku- and C-band satellite launches are scheduled for the early 1990s.

The frequency plan for C- and Ku-band satellite services is shown in Fig. 95.5 The typical transmit frequency band used for fixed satellite services in the Ku-band is 14.0–14.5 GHz. Receive frequency is 11.7–12.2 GHz. Some satellites also use the extended band. For C-band satellites, the typical operating transmit frequency is 5.925–6.425 GHz and the receive frequency is 3.7–4.2 GHz. The operating band was extended at WARC '79 to 7.075 GHz to be assigned to individual countries for domestic satellite systems. Ka-band satellites have downlinks in the frequency range 17–23 GHz and uplinks in the range 27–31 GHz. Some European and Japanese satellites operate in this range [Long, 1991].

The satellite performance data indicate a wide range of variation in the specifications among the various satellites. Most satellites have a design lifetime of 10 years. The newer satellites planned for the early 1990s tend to have an extended life of 12–15 years. Most domestic U.S. satellites have 24 transponders. The older generation of Asian satellites have very few transponders per satellite. Some planned satellites will have a large number of transponders. Nominal transponder bandwidths include 36, 54, and 72 MHz.

Satellite power is increasing in the newer generation of satellites. Lower-power satellites have an

Table 95.2 Typical Satellite Performance

Satellite Operator	System Name	Configuration	EIRP in dBW at Edge	Comments
GTE Spacenet	GSTAR series	Ku-band	38–48	Domestic coverage
	Spacenet series	C- and Ku-band	C-band: 34–36 Ku-band: 39	
Hughes Comm	Galaxy Series	C- and Ku-band	C-band: 34–38 Ku-band: 45–49.5	Domestic coverage
Intelsat	Intelsat VA (IBS)	C- and Ku-band	C-band: 20–26 Ku-band: 38–41	International service, worldwide
	Intelsat VI	C- and Ku-band	C-band: 20–26 Ku-band: 38–41	
	Intelsat VII	C- and Ku-band	C-band: 26–36 Ku-band: 41–46	
Eutelsat	Eutelsat I series	Ku-band	35–43.5	Covers all of Europe
	Eutelsat II	Ku-band	42–47	
NASDA-NTT (Japan)	Sakura 2	C- and Ka-band	C-band: 30 Ka-band: 37	CS-4a, CS-4b in the Sakura series is scheduled for launch during 1992–94

EIRP in the 20–35 dBW range. There are a significantly large number of medium-power satellites in the 35–45 dBW range. The high-power satellites planned for the early 1990s tend to have power in the 50–60 dBW range. Direct broadcast satellites are planned for transponder power in the 60–120 W range. The power generally varies with polarization, frequency, and beam. Table 95.2 is a profile of typical satellite performance characteristics.

Mobile Satellite Systems

Mobile satellite systems encompass communications on land, in the air, or over the oceans ideally allowing a person to communicate with anyone anywhere [Long, 1991]. The Inmarsat system is a mobile communications system providing global coverage through a variety of communication paths. In the United States, the FCC has authorized a consortium of companies, the American Mobile Satellite Consortium (AMSC), to provide domestic mobile satellite services. AMSC will make use of geostationary satellites to provide a domestic offering similar to the international offering of Inmarsat.

Over the past two decades, there has been active work in the area of low earth orbit (LEO) satellite systems. In general, LEOs are designed to provide a full range of communication services, ranging from two-way voice and data communications to global communications. Proposed systems are designed to complement existing cellular communications technology. Several companies have proposed LEO systems and have made application to the FCC for a "Pioneer's Preference" license. This license allows the use of new and innovative technology. Motorola's Iridium system is potentially the largest, using 77 satellites to provide coverage over the entire globe.

Because of the low altitude of the orbit, LEO systems will use multiple satellites to provide coverage over a regional area or over the entire globe. Satellites operating at a low orbit are less costly due to the reduced launch costs and reduced weight. However, a low orbit requires the use of multiple satellites since the low altitude of the system will result in smaller beam coverage. Since these satellites are not geostationary, ground stations must track an LEO satellite as it passes overhead.

Direct Broadcast Satellites

The direct broadcast satellites (DBS) concept is to transmit programming directly to homes using a small receive-only antenna via high-powered satellites. Through the use of a high-powered satellite, a small receive-only satellite antenna may be used for home reception, with the ultimate goal to offer antennas less than one foot in diameter. High-powered DBS satellites use high-powered transponders, i.e., 60–120 W. To prevent interference into the small receive antennas at these high power levels, the DBS satellites will be spaced further apart in geosynchronous orbit.

The first efforts in DBS began in the early 1980s when COMSAT built several DBS satellites, but did not launch them. Internationally, many countries currently have DBS services. Several European countries have high-powered DBS satellites; many others use medium-powered satellites. The DBS industry in the United States is being revitalized by advances in digital video compression technology and the announcement of new players to offer DBS services. Hughes Communications and United States Satellite Broadcasting (USSB) have begun construction of a high-powered DBS satellite, intended to be in operation by mid-1994. As an alternative to the launch of a high-powered satellite, medium-powered DBS systems make use of existing satellites in orbit. However, larger home antennas are required, approximately 2 feet or greater in diameter. Two medium-powered DBS systems have been announced in the United States. K Prime Partners, a partnership of cable operators and GE Americom, provides a subset of cable offering to homes. SkyPix Television is another system preparing to rollout service to homes, offering programming from the entertainment industry. Digital video compression techniques are used in this system to allow multiple video channels in a transponder.

Earth Stations

Earth stations are the interface point for communications to and from the satellite [Ha, 1986]. An earth station can be divided into two subsystems, the transmit chain and the receive chain. A common element between the transmit and receive chain is the antenna. Because of the large signal attenuation at RF frequencies, the earth station antenna must have high signal gain and be highly directional to focus the power to and from the satellite. A parabolic-shaped reflector antenna is used by earth stations since it can provide these characteristics.

The transmit chain consists of several major components: baseband equipment, modulators, frequency upconverters, high-power amplifiers (HPA), and combiner circuitry used to switch the output of the HPAs to the antenna. The receive chain uses a low-noise amplifier to receive the satellite signals, frequency downconverters, demodulator, and baseband equipment.

In the transit chain the signals are modulated, combined, and frequency-shifted with an upconverter to the desired satellite transmit frequency. After upconversion, the signals are amplified by HPAs. In a large earth station, there may be many HPAs which feed to a single antenna. These signals must be switched and combined appropriately. At microwave frequencies, waveguide combiners are used to route the output of the HPAs to the antenna.

In the receive chain, the counterpart to the HPA is the low-noise amplifier (LNA), which is used to amplify the signals received from the antenna. This amplifier must be designed for maximum gain with a very small noise contribution. The noise generated in this unit contributes significantly to the overall performance of the receive side of the earth station. Gallium arsenide (GaAs) FETs are commonly used in the amplifier section of the LNA because of their low-noise characteristics. The LNA feeds the signal to the frequency downconverter, which converts it to IF frequency suitable for demodulator.

A hub monitoring and control (M&C) system provides the monitoring and control of the RF equipment and baseband equipment. Redundant RF equipment is common at a hub, and the M&C system is used to monitor the components and provide automatic switchover in the event of equipment failure. Switchover between equipment can occur either by operator initiation or automatically by the M&C upon sensing an equipment failure.

Technical characteristics of *large earth stations* have been established for use with the INTELSAT system. INTELSAT categorizes two types of earth stations: multipurpose and special purpose. A multipurpose earth station can be used with any service, while a special-purpose earth station is restricted. Multipurpose standard A, B, and C earth stations have antenna diameters from 11 to 33 meters. Special-purpose standard D, E, and F earth stations have antenna diameters between 3.5 to 11 meters.

In addition to fixed earth stations, "portable" earth stations, called *transportables,* have been manufactured which can be taken to locations originating the programming. These transportables are usually mounted on a truck or trailer and include all the components necessary for an earth station. In the case of the transportable, the antenna size is selected to be as small as 4 meters in diameter. A transportable earth station is designed to be upgraded with "building blocks" to handle heavy, medium, and thin route traffic. Transportable earth stations are designed to meet the requirements for various applications such as temporary business communications, temporary carrier service, backup during the retrofit of an existing earth station, and disaster recovery.

Another type of earth station is the *flyaway.* This is a small remote satellite terminal which can be packed into suitcases for shipment on an airline for delivery anywhere in the world. These systems consist of a small antenna, RF unit, and baseband equipment to provide a complete satellite communications station. An example is an L-band version which provides audio communications via the Inmarsat system. Fitting into a small suitcase, it contains a telephone handset, RF electronics, and antenna that can be assembled to provide audio communications anywhere in the world.

VSAT Communication System

Recent advances in technology have revolutionized the satellite communications industry by deployment of very small aperture terminal (VSAT) networks for data, voice, and video communication. Since the mid-1980s, VSAT networks have become widely used in the oil, lodging, financial, auto, retail, and manufacturing industries. VSATs are making private networks a viable alternative for many companies, for applications such as point-of-sale, reservation systems, remote monitoring and control, branch office administration, financial transactions, etc. A VSAT is a small earth station suitable for installation at a customer's premises. A VSAT typically consists of antenna less than 2.4 m, an outdoor unit to receive and transmit signals, and an indoor unit containing the satellite and terrestrial interface units [Rana *et al.,* 1990].

VSAT networks fall into three general categories: broadcast networks, point-to-point networks, and interactive networks. In a broadcast network, a centralized hub station broadcasts data, audio, and/or video to a group of receive-only VSATs. Low-cost receive-only VSATs can receive news, weather services, and financial information. Music distribution and video broadcast via broadcast networks is widely used. Point-to-point networks provide direct communication between two locations without the requirement of a large hub for data, voice, and image transmission. Variations of these networks include point-to-multipoint dedicated circuits or demand-assigned mesh topologies. Interactive networks are used for two-way communications services between a central hub station and a large number of VSATs in a star topology. Table 95.3 is a summary of the salient features of VSAT networks. VSATs are available for both C- and Ku-band frequency. Most VSAT systems use BPSK modulation with Rate 1/2 FEC. For interactive networks, the inbound channel is shared on contention basis to conserve space segment.

A critical element of VSAT networks is the network availability. The VSAT system availability is affected by three major components: effects of rain attenuation, equipment availability, and software availability. The effects of rain attenuation for Ku-band networks are significant. While link availability is usually specified at 99.5%, link performance can be optimized to nearly any desired value through the use of energy dispersion techniques or large antenna sizes. The network hardware must be highly reliable. Hub hardware should provide for optional redundancy and the ability to achieve better than 99.9% availability. The use of hub diversity and uplink power control can also be used to improve the network availability. The VSAT hardware availability is less catastrophic; the loss of one VSAT does not constitute network failure but may require a service call to rectify the problem. Hence, it is common to use nonredundant but highly reliable VSAT units. Soft-

Table 95.3 Typical VSAT Systems Features

Feature	Interactive	Point-to-point	Broadcast
Topology	Star	Point-to-point, mesh	Point-to-multipoint
Communication	Between hub and VSATs, VSAT to VSAT through hub	VSAT to VSAT	Hub to VSATs
Frequency	Ku-, C-band	Ku-, C-band	Ku-, C-band
Hub antenna	3–11 m	—	3–11m
VSAT antenna	0.9–2.4 m	1.8, 2.4 m	0.5–2.4 m
Hub to remote access	TDM, SCPC, spread spectrum	SCPC	SCPC, spread spectrum, FM[2]
Remote to hub access	ALOHA, reservation stream, CDMA	SCPC	—
Outbound data rate (Kbps)	56–512	9.6–2048	9.6–2048
Inbound data rate (Kbps)	9.6–128	9.6–2048	—
Modulation	BPSK, QPSK, DPSK	BPSK, QPSK	BPSK, QPSK, FM[2]
FEC	Rate 1/2, convolutional or block	Rate 1/2, convolutional	Rate 1/2, convolutional or block
Protocols	SDLC, Bisync, Async, X.25, Burroughs and others	Clear channel	Clear channel, synchronous, HDLC format

ware availability needs to be improved since software failures dominate the overall availability of interactive networks in existing VSAT products.

Interactive networks are becoming by far the most popular for data communication and audio/video overlays. The remaining portion of this section is devoted to these networks. An interactive VSAT system consists of a hub, VSAT, network management system, and associated transmission and processing subsystems. These subsystems along with sophisticated **satellite access protocols** and terrestrial protocol interfaces make interactive networks a flexible and powerful communication medium.

Hub

The hub performs all functions that are necessary to establish and maintain virtual connections between the central location and VSATs. In private dedicated networks, the hub is co-located with the user's data processing facility. In shared hub networks, the hub is connected to the user equipment via terrestrial backhaul circuits. Since the hub is a single point for failure in a star network, it is typically configured with 1:1 or 1:N redundancy. The hub consists of antenna, RF, and baseband equipment (Fig. 95.6). It will handle multiple channels of inbound and outbound data and often one or more channels of audio or video broadcast.

The hub antenna consists of a parabolic reflector and associated electrical and mechanical support equipment. The RF subsystem converts the modulated carrier to RF frequency, provides the necessary signal amplification, and transmits the resulting RF carrier to the antenna subsystem. It also receives RF signals from the antenna subsystem, provides low-noise amplification, RF/IF conversion, and passes the resulting IF carriers to the baseband equipment subsystem. The hub baseband equipment consists of the modem equipment and the processing equipment. The hub modems employ continuous modulators and burst demodulators. The processing equipment interfaces to the modem equipment and provides the satellite access processing and protocol processing for interface to the customer host.

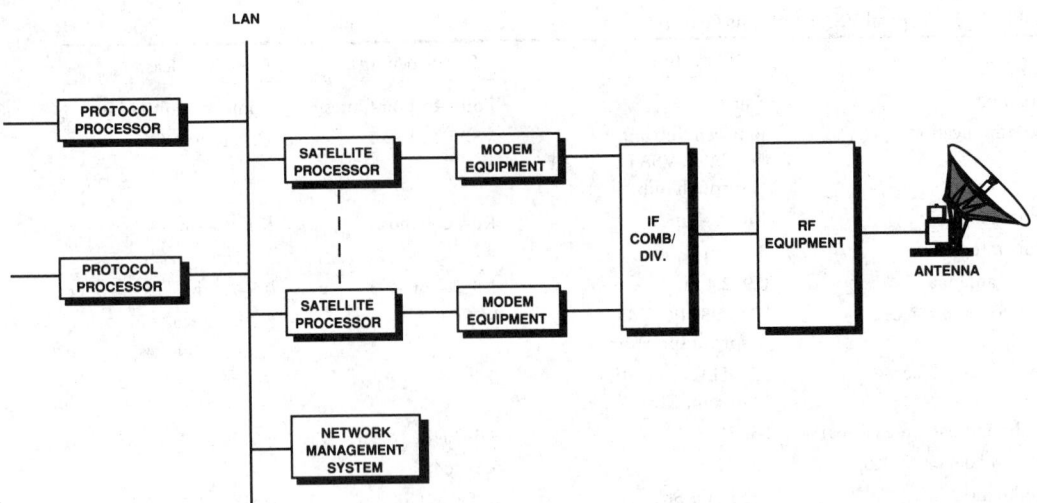

FIGURE 95.6   Block diagram of a hub.

VSAT

The VSAT consists of an antenna, outdoor unit (ODU), interfacility link (IFL), and indoor unit (IDU). The IFL connects the IDU and ODU subsystems, providing the transmit and receive lines, monitor and control signals, and dc power for the ODU electronics. A single-cable IFL, in which all signals are multiplexed on the same cable, is usually used to reduce the cost of IFL. VSATs nominally use a 1.2- or 1.8-m offset feed parabolic antenna. Smaller antenna sizes are preferable to reduce the installation cost. Options for small antennas include the use of either a submeter parabolic reflector or a flat-plate antenna. The choice of antenna is a tradeoff among performance, installation cost, and aesthetic considerations.

The ODU consists of a solid-state power amplifier (SSPA), a low-noise amplifier, upconverter, and a downconverter. VSAT SSPA modules are usually between 1.0 to 3.0 W. The ODU cost can be significantly lowered by utilization of a low-power SSPA (0.1 to 0.5 W) consistent with obtaining the required output power. The VSAT receive side front end can be economically configured using an LNB. Low-cost HEMT LNBs are currently available with 50–60 dB gain and noise figures lower than 1.3 dB.

Direct modulation of the RF carrier may lower the cost of the VSAT IF and RF electronics while consolidating modulation and upconversion functions. Direct modulation allows the design of a VSAT with fewer parts, smaller size, and lower weight than with traditional outdoor units. Figure 95.7 is a block diagram showing a conventional VSAT and a VSAT using direct modulation. An L-band receive interface between the ODU and IDU is preferable in order to receive audio and video overlays.

The IDU is located near the user terminal equipment. Major IDU functions include outbound carrier signal acquisition, tracking, demodulation, bit synchronization, burst modulation, and protocol processing. It also controls the operation of the ODU, monitors VSAT health, and responds to hub commands. The baseband processing system performs satellite channel access and protocol and customer interface processing functions. A video/audio port can be provided with an RF splitter at the IDU to separate the received audio/video signal for the optional video/ audio receiver.

Network Management System

The network management system (NMS) is a critical element of a VSAT network. Through the NMS, the user can have full control of his network, which is usually not possible in the case of terrestrial network facilities. The NMS generally provides a centralized management tool for hub and VSAT equipment configuration control, assignment of inbound and outbound satellite

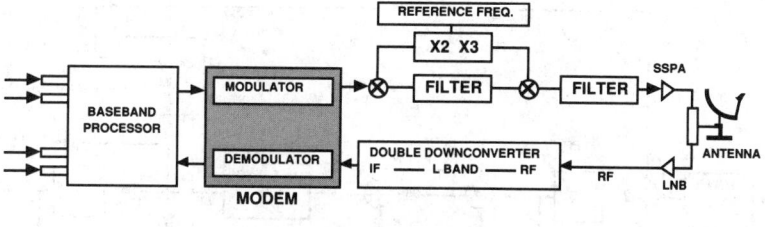

(a) Conventional VSAT

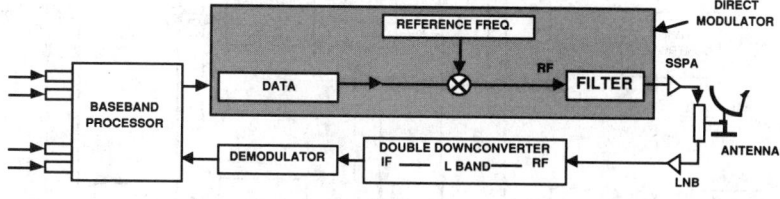

(b) VSAT using direct modulation.

FIGURE 95.7   Simplified VSAT block diagram.

channels, network monitor and control, switchover to back-up equipment, network statistics collection, downline loading of new software, and report generation. In the shared hub environment, the hub operator controls the allocation of resources among various users and controls the RF transmission facility. The user must have the ability to manage his portion of the network transparent to other users. In the case of a dedicated hub, a single management entity can exert full control over the network, including RF transmission facilities.

The network management system standards community has defined five functional areas as requirements for network management systems. These areas are fault management, accounting management, configuration management, performance management, and security management. The VSAT network management system should be capable of interfacing with other network management systems by supporting a standard network management protocol. Protocol standards showing indications of being the most widely accepted are the Common Management Interface Protocol (CMIP) and the Simple Network Management Protocol (SNMP).

Transmission System

Most VSAT systems employ BPSK or QPSK modulation with rate R = 1/2, K = 7 convolutional coding and soft-decision Viterbi decoding on both the inbound and outbound channels. Differential phase shift keying (DPSK) modulation may be used to reduce the demodulator complexity and cost. DPSK is relatively insensitive to phase noise and frequency offset, thus allowing the use of lower-cost LNBs in the VSAT terminals. However, as compared to BPSK, convolutionally encoded DPSK requires about 2 dB greater E_b/N_o at a BER of 10^{-5}. In addition, if operation is required below 10 dB, some form of low-level interleaving may be required.

In lieu of performing the VSAT demodulation function via the traditional analog circuit techniques, an all-digital implementation using digital signal processing (DSP) techniques may be considered. The merits of DSP include the development of a more testable, producible, maintainable, configurable, and cost-effective demodulator. Figure 95.8 presents an illustration of the DSP demodulator functions to be implemented using the DSP processor(s). The functions of the major blocks are as follows: phase locked loop (PLL) for carrier acquisition, narrowband Costas loop for data detection, external automatic gain control (AGC), dynamically advance/retard sampling to achieve optimum data sampling, and A/D converters for signal analog-to-digital conversion.

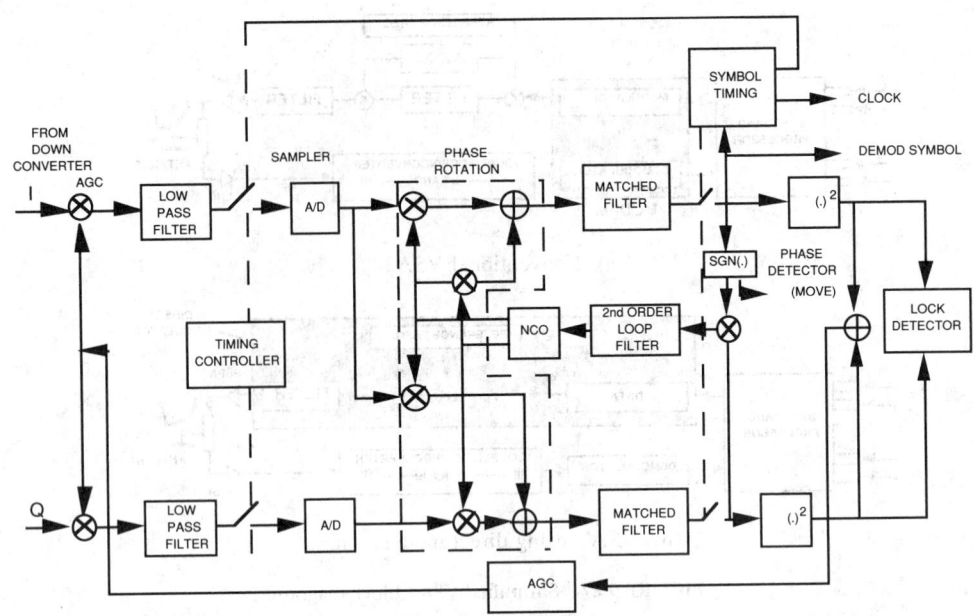

FIGURE 95.8   DSP demodulator functional diagram.

A VSAT system must employ frequency agility in the remote terminal to use an assigned block of frequencies within a transponder. Within the assigned frequency band, one or more outbound carriers and a number of inbound carriers are precisely located. On the VSAT receive or outbound side, the LNB output can be demodulated directly using a synthesizer-controlled local oscillator, or further downconversion can be used under synthesizer control to obtain the demodulator input signal at a standard IF frequency such as 70 or 140 MHz. In the inbound direction, channel selection can be accomplished by two methods. First, the carrier frequency of the modulator can be shifted to select the appropriate channel and a fixed upconverter may be used to obtain the RF signal. Second, the synthesizer output frequency may be multiplied up to RF to obtain the carrier, which may then be modulated directly with the data as described in Cannistraro and McCarter [1990].

Satellite Access Protocols

The multiple satellite access protocol is one of the most critical elements to the performance of a VSAT network. VSAT systems tend to be used in applications where message delay is critical and this protocol is the controlling element to the delay-throughput performance of the system. During the past 15 years, there have been numerous multiple-access protocols developed and simulated in the context of satellite packet communications [Raychaudhuri and Joseph, 1988]. Table 95.4 provides a comparison of throughput vs. delay for various satellite access protocols.

In the outbound or hub-to-VSAT direction, a TDM channel is employed. This channel may be regarded as a point-to-multipoint or broadcast channel with node selectivity being achieved by the use of addressing information embedded in the modulated data stream. The delay performance of this channel is essentially controlled by the queuing behavior of the hub. In the VSAT-to-hub direction, a large number of VSATs share the channel to conserve space segment. Most VSAT networks utilize a combination of slotted ALOHA protocol for the interactive component of the inbound traffic and a reservation TDMA scheme for any bulk data transfers. Most protocols are adaptive in the sense that as the channel traffic increases, they automatically evolve into reservation TDMA systems. Code division multiple access (CDMA) has been used in VSATs operating at C-band. CDMA permits more than one signal to simultaneously utilize the channel bandwidth in a noninterfering manner. This makes it possible to significantly increase the utilization and throughput of the channel.

Table 95.4 Random Multi-Access Protocols Comparison

	Throughput	Comments
Pure ALOHA	0.13–0.18	Low cost, good for variable-length messages
Slotted ALOHA	0.25–0.37	Good for fixed-length messages
Selective reject ALOHA	0.20–0.30	Variation of pure ALOHA with a modified algorithm
Tree CRA	0.40–0.49	Sensing capability for collision resolution, good for fixed-length messages
Announced retransmission random access (ARRA)	0.50–0.60	Uses modified algorithm of slotted ALOHA by announcement of transmission
Random access with notification	0.45–0.55	Uses partition for new and retransmitted message
CDMA	0.10–0.40	Used in spread spectrum systems, low delay

Interface Capabilities

Most VSAT systems support common data communications protocols such as SDLC, X.25, Async, Bisync, Burroughs Poll Select, etc. Coexistence of different protocols is allowed in a network. A VSAT supports multiple ports with common interfaces such as RS232C, RS422, V.35, etc. VSAT networks typically must provide **protocol spoofing** to provide acceptable delay and throughput performance to the end-user application. To minimize the effect of satellite delay, the host computer front-end processor is emulated at the VSAT location, and multiple cluster controllers are emulated at the hub location. The polling associated with the front-end processor to cluster controller communication is not carried on the satellite link, but is instead emulated locally.

Video

Satellites are an excellent medium for video transmission since they can provide a broadcast capability with wide bandwidth. Video on satellites is ideal for applications such as videoconferencing, business TV, distance learning, satellite news gathering, etc.

Video Teleconferencing

Satellite communications provides a cost-effective and flexible means of interactive videoconferencing. Technological improvement in videocompression has resulted in low-cost codecs at data rates less than T1, and good quality videoconferencing is possible at data rates as low as 256 Kbps. Low-cost satellite terminals coupled with low-cost codecs are making videoconferencing via satellite affordable and practical for many organizations. Applications includes all types of business meetings and technical information exchange such as management and staff meetings, new product introductions and updates, sales meetings, training, and market presentations. Videoconferencing allows people at different locations to meet with almost as much ease as being in the same room, providing benefits of increased productivity, reduced travel time and cost, and increased management visibility.

A generic videoconference system is presented in Fig 95.9. The system consists of a specially designed room, video/audio equipment, transmission equipment, monitor and control computer, and space segment. The video and audio feeds from the meeting room pass through the codec and are compressed. From the codec, the signal passes to the satellite modem for modulation. The radio frequency/terminal (RFT) upconverts the modulated carrier and amplifies it for transmission to the satellite. At the other site, the process is reversed.

A videoconferencing network features point-to-point, broadcast, or point-to-multipoint architectures. In a point-to-point system, two sites are configured for interactive conference with duplex audio and full motion video transmission. Videoconferencing broadcast is appropriate for formal presentations where the presenter does not need to see the audience, such as a speech from a senior

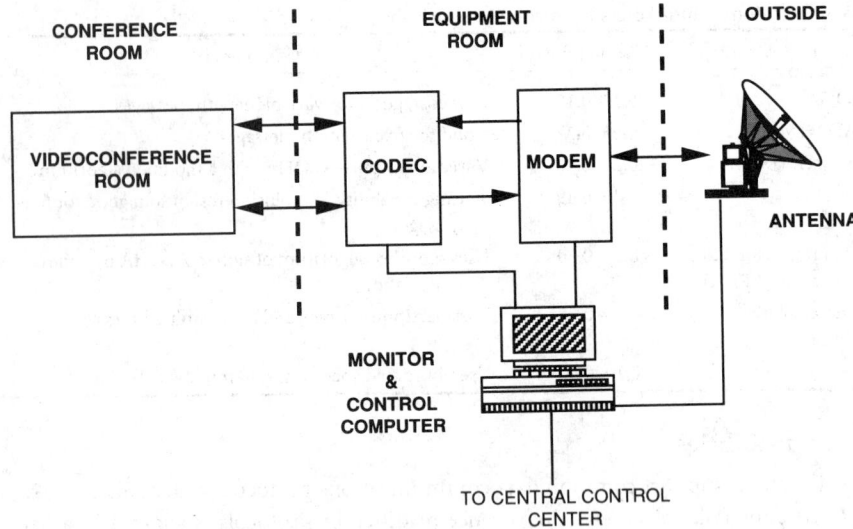

FIGURE 95.9   A generic videoconference system.

corporate executive. In a point-to-multipoint conference, multiple sites can receive a transmitting site. Two of the primary sites are fully interactive with each other. A feature called multipoint switching has been implemented in some commercial systems. This feature allows switching of receive and transmit sites during the conference. The multipoint switching feature can be provided using either a TDMA or SCPC system. A TDMA system allows multiple sites to transmit and receive in a mesh configuration. An economical multipoint switching system is possible with SCPC using only two transmit frequencies. In a "chair" controlled conference, the chair is assigned one of these frequencies for the duration of the conference. Dynamic allocation of the second frequency is controlled by the chair to any of the participating sites at any time during the conference.

Video Broadcast

Video broadcast over satellite is attractive for industry segments such as educational TV, distance learning, business television, and television receive-only (TVRO) applications. Business television allows users to transmit broadcast-quality video programming from a studio to any number of specified locations equipped with TVROs. A video broadcast capability, as an overlay to interactive data networks, is becoming increasingly popular for corporate presentations, education, and training.

A video uplink consists of a video exciter, HPA, antenna, and optionally an encryption system such as B-MAC (multiplexed analog component, version B) encoder, for business video broadcasts. Each remote VSAT must be configured to receive the video transmission. This involves adding a video receiver at each VSAT location that plugs into the VSAT IDU. Audio/video signals from the video receiver can be presented directly or through a B-MAC decoder to a standard TV monitor.

The digital compressed video signal can be used as a replacement for an analog video distribution. Digital coding technology can be used to compress video signals to reduce data rates to 2 Mbps or even lower and reproduce near broadcast-quality video. Distribution of digital video signals at such rates requires less transponder bandwidth and a smaller antenna at remote terminals. Compression techniques used are based on one or a combination of the following: inter/intra-frame prediction, adaptive differential transform, conditional replenishment, discrete cosine transform, adaptive prediction, motion compensation, and vector quantization [Patterson and Delp, 1990].

Satellite News Gathering

Satellite news gathering (SNG) is used for live, on-the-spot coverage and news exchanges with other commercial broadcast stations. This is made possible by the availability of occasional-use

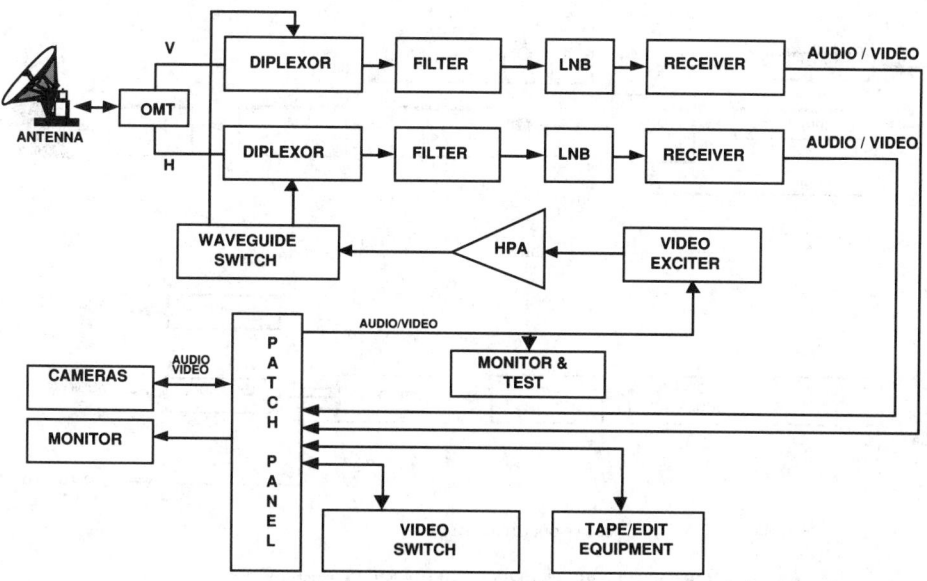

FIGURE 95.10   SNG vehicle video/audio system.

space segment and transportable earth stations on news trucks. An SNG systems consists of a compact earth station and video/audio transmission system on a truck. A duplex voice channel is used to coordinate between the space segment provider, studio, and the SNG truck.

Figure 95.10 presents a block diagram of a typical SNG system. The RF subsystem has a transmit path and two independent receive paths. The transmit path consists of an HPA and a frequency agile video exciter which modulates and upconverts the video signal to the satellite's RF frequency. A waveguide switch is used to select transmit polarization. Camera signals go simultaneously to tape for storage and for transmission over the satellite. A receive path is typically provided for both receive polarizations. Each path consists of a transmit reject filter and an LNB which downconverts to L-band. The received L-band signal passes through a video satellite receiver, from which point it can be routed to various monitor or test points or be routed to a tape device for recording and storage. Although not commonplace yet, the compressed digital video may replace the current SNG analog frequency-modulated video systems in the future.

Audio

The use of commercial broadcast audio transmission via satellite began in the late 1970s with National Public Radio and Mutual Broadcasting using Western Union's WESTAR I satellite. The main application was to send high-quality audio to radio broadcast stations to transmit programming information. This type of system makes use of single channel per carrier (SCPC) satellite transmission, where each satellite channel corresponds to one audio channel. The entire satellite channel is FM modulated. Pre-emphasis is used over the channel to provide additional noise reduction. A variation of this technique, called multiple channel per carrier (MCPC), can be used to transmit multiple channels over a single satellite carrier. Figure 95.11 is a block diagram of the MCPC system.

As the marketplace searched for lower-cost systems, the FM² (or FM/FM) modulation technique evolved, allowing the use of low-cost FM receivers. Through a high-powered FM modulated carrier on the satellite, a low-cost audio and data broadcast receiver can be built. This FM/FM modulation technique is widely used to distribute audio and data on a low-cost basis.

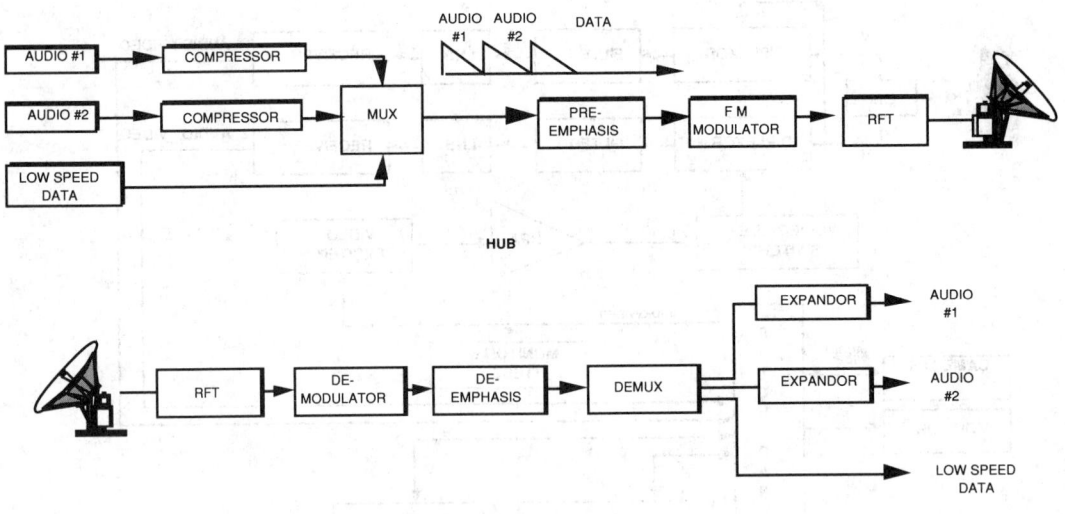

FIGURE 95.11   Block diagram of the MCPC system.

In addition to audio broadcasts, satellite-based voice applications include point-to-point voice, multinode interactive voice, and voice over data VSATs. Point-to-point voice is most prevalently used for high-volume voice trunking for long-distance connectivity or transoceanic connectivity. A multinode, interactive voice architecture is ideal in providing voice connectivity to remote locations that are not serviced by terrestrial voice facilitates. Both mesh and star configurations are used to provide multinode voice connectivity. Automated satellite access control and resource allocation techniques are used to allow for granting requested on-demand availability of voice connectivity. To support voice over data VSATs, an audio encoder is used to accept an analog voice signal, digitize and packetize it, and format it for transmission through the VSAT data network. A voice port may either be implemented as part of a "baseline" data/voice card or as an add-on stand-alone box.

The integrated data/voice system employs a TDM outbound carrier and shared inbound carriers for data and voice transmission. Two types of voice network communications alternatives may be implemented for voice channel communications: a poll/response access scheme and a reservation TDMA access scheme. With the poll/response access scheme, the hub polls the VSAT voice ports on a cyclic basis. The VSATs return their responses in the form of call requests or status updates. The number of sites in the voice network determines the rate at which VSATs are polled. Thus, this scheme is suitable for a small network. Excessive polling delays will be encountered for a network with a relatively large number of remotes.

In reservation TDMA, on the other hand, voice call requests are serviced by the assignment (reservation) of a logical channel for inbound voice traffic. Although various means are implemented to avoid collisions on the satellite link, the time needed to reserve capacity on an inbound carrier may be lengthy, depending on traffic conditions. Therefore, call setup times are not as predictable as they are with the poll/response access scheme.

The VSAT design is ideally suited for digital compressed voice. Coding rates of 32, 16, and 9.6 kbps and lower can presently be achieved, depending on the compression technique employed. There are two classes of digitizing voice signals: waveform coding and vocoding. In waveform coding, the analog voice curve is coded and then reproduced by modeling its physical shape. Data rates are relatively high, i.e., higher than 9.6 kbps. Vocoding attempts to reproduce the analog voice curve by abstractly "identifying" the type and shape of the curve. Only a set of parameters is transmitted, describing the nature of the curve. Achieved data rates can be as low as 1.2 kbps.

Next-Generation Systems

The new wave of satellites planned for the 1990s have much higher power than their predecessors. The Intelsat K satellite, for example, is equipped with 60-W TWTAs and will serve the projected increase in worldwide traffic, video, and VSAT services. Another example is the Telstar 4 satellite which has variable power up to 120 W for Ku-band transmissions and is being promoted to be HDTV compatible in preparation for expected widespread use of HDTV. Other trends in satellite design, i.e., NASA's advanced communication technology satellite (ACTS), include the use of multiple spot beams and onboard IF and/or baseband switching. Onboard switching coupled with electronically hopped spot beams and laser intersatellite links will be tested in the 1990s. Spot beams provide higher satellite EIRP which will permit small, low-cost VSATs to accommodate higher bit rate transmissions. The use of multiple-beam architectures also increases bandwidth availability through frequency reuse. Advances in multibeam satellites with onboard baseband processing allows some of the intelligence in the central hub and VSAT equipment to be moved to the satellite. The result is expected to be improved VSAT-to-VSAT communications and a platform to provide dynamic bandwidth allocation [Naderi and Wu, 1988].

The trend in deploying higher-power satellites has an inverse effect on the size of the earth station antenna. The earth stations are becoming smaller, less complex, and more cost effective. Private hubs are now typically in the range of 3.5 to 7.6 m and are not required to be staffed. Two-way VSATs antennas originally deployed in sizes from 1.2 to 1.8 m are now using elliptical or rectangular-shaped antennas with apertures equivalent to 1.0 m or less. Two-way ultra-small aperture terminals are also emerging. These lower-cost, lower-functionality earth stations are designed for thin route, niche-type applications such as point-of-sale and credit card transaction processing. The advances in DSP technology will continue to enhance the capabilities and performance while at the same time lowering the cost of VSATs. The advances in MMIC technology continue to miniaturize the RF components while increasing reliability.

With advances in digital signal processing and compression techniques, analog video and audio transmission will increasingly be converted to digital transmissions. The advanced compression techniques reduce the bandwidth requirements and allow for smaller and lower-cost VSAT antennas to be used. The continued technological advances in satellite technology and the emerging demand for more flexible communication services will generate new satellite communications applications, such as LAN interconnections and ISDN support [Murthy and Gordon, 1989]. Satellite communications will also play a major role in mobile communications on land, air, and on sea. In addition to telephony services, new services such as global distress and safety applications, global positioning, navigation, voice messaging, and data transmissions will be possible.

Defining Terms

Earth station: The interface point for communications to and from a satellite. An earth station (also known as a hub) consists of an antenna and transmit and receive subsystems.

Geosynchronous orbit: An orbit 22,753 miles above the earth in which an object will orbit the earth once every 24 hours above the equator and will appear to be stationary from the earth's surface.

Protocol spoofing: A technique used by VSAT networks to reduce the network delay. The satellite network emulates the host computer front-end processor at the VSAT location and emulates the multiple cluster controllers at the hub location.

Satellite access protocol: A set of rules by which a number of distributed VSATs communicate reliably over a shared satellite channel.

VSAT: Very small aperture terminal. A small earth station suitable for installation at a customer's premises. A VSAT typically consists of an antenna less than 2.4 m, an outdoor unit to receive and transmit signals, and an indoor unit containing the satellite and terrestrial interface units.

References

J.C.L. Cannistraro and S. McCarter, "Direct modulation lowers VSAT equipment costs," *Microwaves and RF,* pp. 99–102, August 1990.

T.T. Ha, *Digital Satellite Communications,* New York: MacMillan, 1986.

M. Long, *World Satellite Almanac,* 3rd ed., Winter Beach, Fla.: MLE, Inc. 1991.

K.M. Murthy and K.G. Gordon, "VSAT networking concepts and new applications development," *IEEE Communications Magazine,* pp. 43–49, May 1989.

F.M. Naderi and W.W. Wu, "Advanced satellite concepts for future generation VSAT networks," *IEEE Communications Magazine,* vol. 26, pp. 13–22, July 1988.

H.A. Patterson and E.J. Delp, "An overview of digital image bandwidth compression," *Journal of Data and Computer Communications,* pp. 39–49, Winter 1990.

W. Pritchard and J.A. Sciulli, *Satellite Communication Systems Engineering,* Englewood Cliffs, N.J.: Prentice-Hall, 1986.

A.H. Rana, J. McCoskey, and W. Check, "VSAT technology, trends, and applications," *IEEE Proc.,* vol. 78, no. 7, pp. 1087–1095, July 1990.

D. Raychaudhuri and K. Joseph, "Channel access protocols for Ku-band VSAT networks: A comparative evaluation," *IEEE Communications Magazine,* vol. 26, no. 5, pp. 34–44, May 1988.

Further Information

The *World Satellite Almanac* provides a tutorial of the satellite communications industry. It includes the technical characteristics and footprint maps for geosynchronous satellites worldwide. Contact: MLE Inc., P.O. Box 159, Winter Beach, FL 32971.

World Satellite Communications and Earth Station Design is a text which provides an analytical presentation of communication satellites and their applications. Contact: CRC Press, Inc., 2000 Corporate Blvd., N.W., Boca Raton, FL 33431.

The monthly *IEEE Communications Magazine* investigates VSAT communications in a special series spanning several issues between 1988 and 1989. Contact: IEEE Service Center, 445 Hoes Lane, Piscataway, NJ 08854-4150.

96

Command, Control, and Communications (C³)

G. Clapp
Naval Command, Control and Ocean Surveillance Center

D. Sworder
University of California, San Diego

96.1 Scope 2211
96.2 Background 2211
96.3 The Technologies of C³ 2214
96.4 The Dynamics of Encounters 2216
96.5 The Role of the Human Decisionmaker in C³ 2218
96.6 Summary 2221

96.1 Scope

The focus of this chapter is not a detailed profile of a current or planned military C³ **system** but it is rather on the issues and the technologies of the C³ mission. Evolving technology, an evolving world order, and constant programmatic reorderings render such express descriptions to become rapidly outdated. Thus block diagrams of specific military systems (and listings of their acronyms) are de-emphasized. Of paramount interest is not electronics technology in isolation, but rather technology integrated into systems and analysis of these systems operating under complex real world environments that include technologically capable adversaries. The human commander or **decisionmaker,** as the principal action element in a C³ system, is included explicitly in the system analysis.

96.2 Background

Electronics technology is nowhere more intensively and broadly applied than in military systems. Military systems are effective only through their command and control (C²) and this is recognized by the fact that C³ is a critical discipline within the military. Frequently systems will be denoted C²I or C³I rather than command and control. This adds to C² the essential area of **intelligence** and intelligence products derived from surveillance systems. All variants of these acronyms are to be considered equal, whether or not communications, intelligence, or surveillance have been left implicit or made explicit. Likewise the superscript notation is considered optional and interchangeable. The formal discipline of C3 within the military has not been matched by focused technical journals or university curricula due to its highly multidisciplinary nature.

Two definitions from a Joint Chiefs of Staff (JCS) publication [JCS, Pub. 1] capture the breadth of C2. This reference defines command and control as "The exercise of authority and direction by a properly designated commander over assigned forces in the accomplishment of his mission. Command and control functions are performed through an arrangement of personnel, equipment, communications, facilities, and procedures which are employed by a commander in planning, directing, coordinating and controlling forces and operations in the accomplishment of his mission."

C2 systems are defined, with almost equal breadth, as "An integrated system comprised of doctrine, procedures, organizational structure, personnel, equipment, facilities, and communications which provides authorities at all levels with timely and adequate data to plan, direct and control their operations."

Though general, two points emerge from these definitions: (1) C^3 is multidisciplinary and (2) C^3 is a process which, to this point, includes only implicit roles for electronics technology. One military service, however, often refers to C^4 and C^4I or has even used C^4I^2 where the final C and second I refer to computers and interoperability, respectively, as acknowledgment of the increasing reliance on technology.

A C3 system can be visualized as shown in Fig. 96.1. Within the constraints imposed by organization, **doctrine,** and the skills of the personnel of the military unit, the commander plans and controls his forces. At a basic level, command and control is a resource allocation problem, which often must be solved under much tighter time horizons and subject to greater uncertainty levels than exist in civil applications.

The four basic components display overlapped regions to indicate their inseparability. A portion of each category can be designed in isolation; a new antenna or a new radio with decreased size, weight, or power consumption has minimal impact on the other components. However, insertion of a broad new technology (e.g., a radio relay combined with a remotely piloted vehicle (RPV) or the networking of radios) has wide reaching consequences and it may take years to fully integrate into doctrine, training, and organization. The conjunction of the four areas, when specified with some detail, represents or contains an architecture. If the assets, the doctrine, and so on are limited to just one military function, then the aggregation is referred to as a mission architecture. Figure 96.2 depicts two approaches to achieving C3 architectures. The first [Fig. 96.2(a)] is essentially an aggregation and combination of existing assets and is referred to as a "bottom-up" architecture. The "top-down" version of architecture development [Fig. 96.2(b)] begins with earlier and high order perspective (and higher order oversight). Interfaces and interface standards become more important in top-down architectures; instead of numerous custom and unique interfaces, a minimal set of interface standards is desired. When new or updated equipment is designed or acquired it can be integrated without new interface developments, a key property of an **"open system"** **architecture.** A developing architecture of this type is entitled, at the Joint Chiefs of Staff level, "C4I for the Warrior." Service-specific top-down architectures are Copernicus (Navy), AirLand 2000 (Army), MTACCS (Marine Corps Tactical Command and Control System) and a yet unnamed Air Force architecture. Each of these are to be considered as evolving architectures and all reflect the impact and importance of scenarios with highly mobile nodes. The open system or top-down approach promotes interoperability between the developments of each service.

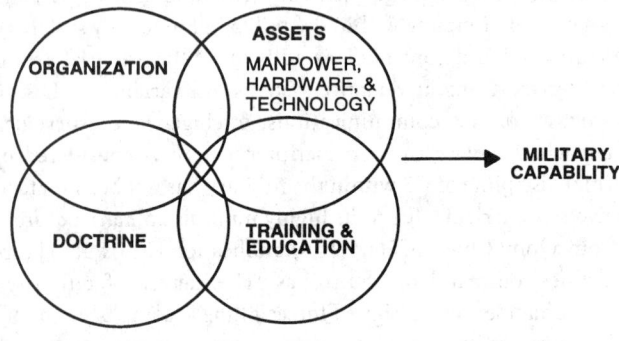

FIGURE 96.1 Components of C^3.

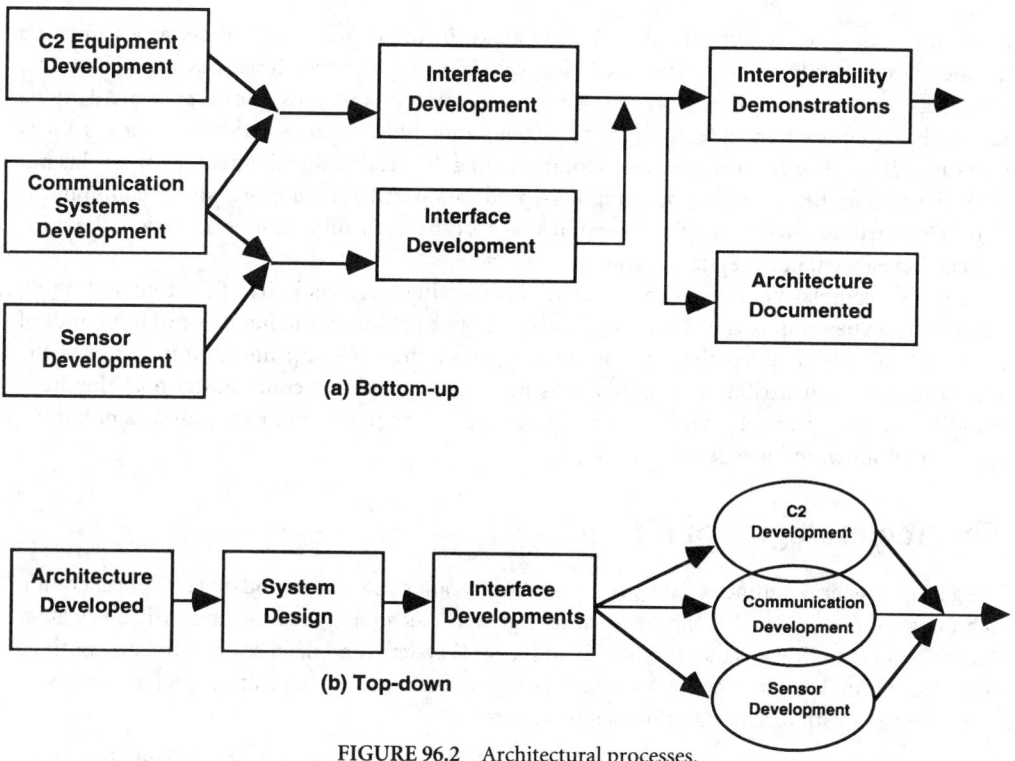

FIGURE 96.2 Architectural processes.

Capital investment constraints limit strict adherence to either architectural approach. MTACCS is a meta-system of seven independently developed systems and is best described as a hybrid architecture. Most communication systems within any of the above architectures existed prior to an architecture and thus have a hybrid nature.

Doctrine is a formalized description of military mission definitions and often includes the procedures to accomplish those missions. Doctrine will also often specify the organizational structure appropriate to the specific missions. Some military establishments adhere to strong doctrinal orientation, even down to strict dictation of technology developments. Other establishments treat doctrine as a loose guideline that can be liberally modified. One foreign military analyst observed that U.S. commanders did not seem to read their own doctrinal publications, and even if they did, would not feel compelled to follow them. A flexible military organization with flexible doctrine, however, can be constrained by inflexible hardware and software. Thus an emerging C3 emphasis is a technical focus on modular equipments, standard interfaces between equipments, "open system" architectures, and (software) programmable equipments.

The best way to understand military C3 is to view it as a set of adaptive control loops. The basic variable is information and most of the effort in C3 synthesis is devoted to information handling and management. The resource allocation problem with feedback found in C3 has obvious similarities to those found in corporate operations and public safety service operations. Each is characterized by multiple priorities, limited resources, timelines, and deadlines for performance. Measures of the consequences of a given action tend to be obscured both by its antecedent actions and by changing external **environments.** The external environment contains both continuous events (i.e., tracking of targets) and discontinuous events (i.e., an equipment failure or the onset of communications jamming).

Command and control systems are examples of perhaps the most complex adaptive systems. In its *static* state, C3 assets are aggregates of sensors, processors, databases, humans (with their attributes and organizations), computer hardware/software, mobile platforms, weapons, and communication equipments distributed over wide areas. In the *dynamic* state these assets must be mapped into capabilities in the presence of uncertain or unexpected threats, evolving missions, changing environments, mixed with unreliable communications and possible deception. All can be expected to occur over extended geographic regions and at high tempos. In short, **C³ maps assets into capabilities**. The control processes require rapid and accurate decisionmaking; from this has come the need for heavy reliance on computer-based data systems and high-reliability communications. Despite the existence of fielded weapon systems capable of autonomous operation, the principal action element in the system is still human.

C3 system complexity arises primarily from the magnitude and mobility of the forces involved; forces that can be composed of up to thousands of mobile platforms and hundreds of thousands of personnel. To this is added the large amount of uncertainty present; uncertainty borne of the adversary, of human attributes, dynamics, hostile environments, and communications. Hundreds of radio frequency channels may be in simultaneous use supporting command, surveillance, intelligence, personnel, and logistics functions.

96.3 The Technologies of C³

The general scenario outlined in the previous sections cannot be accommodated by last generation technology of grease pencils, maps, and visual signaling. Technology covered in nearly every other chapter of this handbook is incorporated in military C3 systems and defense departments continue to support technology developments from sub-micron microprocessing chips to global systems.

Technologies with recent major impact on C3 are

a. Space communications/data links/networking. The newer and critical role of digital (computer-computer) communications has largely been made possible by satellite communication systems. Tactical data links (short-range digital communications) have been enhanced by error control techniques such as coding, automatic repeat requests, and spread spectrum radios. Networking, a well-established commercial technique, is being developed for tactical applications. Networking offers survivability through alternate routing, more efficient (shared) use of channel capacity, and a level of interoperability between interconnected users. Traditional voice communications remain important; recent DoD directives require all voice circuits to be secure or encrypted. Digitized voice techniques, primarily Linear Predictive Coding, offer advantage in digital encryption and compression. Asynchronous transfer mode (ATM) switching will become essential for high-capacity systems.

b. Space surveillance, terrestrial surveillance, data fusion. The quantity and quality of surveillance systems continues rapid growth utilizing sensors from ground-based, airborne, and space-based imagery vantages. A growing base for remote sensors is on Remotely Piloted Vehicles (RPV) which promote the requirement for data in near real time. Unmanned Airborne Vehicles (UAV) and Unmanned Underwater Vehicles (UUV) platform developments continue as a response to a broad range of C3I needs. Two classes of surveillance are active surveillance systems (radar, sonar, and optical) and passive systems (electronic surveillance measuring (ESM), acoustic, infra-red and visual imagery). Passive techniques are preferred as they do not leave a signature that can be exploited by adversaries. A plethora of new sensor systems challenges the currently available communications, processors, and processing systems. Particularly challenging is both the fusion of the outputs of multiple similar sensors and also of dissimilar sensor systems. Fusion protocols and algorithms, software intensive, claim an increasing fraction of available resources. With multiple new sensor systems, a technology challenge is the processing, correlation, and fusing of surveillance data into intelligence products and their distribution in a timely and usable form.

c. Computer-based data and information systems. Rapidly evolving processing technology allows vast amounts of data handling and management with corresponding shortening of control decisionmaking times. The ability to match computer processing capability with high data rate, reliable, and survivable computer-grade communications on a global basis to small mobile platforms is an ongoing challenge. Military information systems, in order to retain trusted functioning, require procedures for input data that may have been delayed, omitted, partial, inaccurate, or irrelevant (DOPII). Expanding amounts of software and staff are needed as a response to increased tempo, data volume, and quality.

d. Architectures and architectural thinking. C3 assets, especially communications, are evolving as assets to be shared, controlled, and rapidly reallocated rather than be dedicated to a specific user. Joint and combined operations, requiring improved interoperability, are becoming common as operations become more regionalized. Two functions of focus, Battle Damage Assessment (BDA) and Indicators and Warnings (I&W), are best implemented when surveillance, communications, and intelligence are architecturally integrated. Integrated systems are also best for timely response to deception and false alarms. A current Navy direction is not to inundate the afloat commander with volumes of unsolicited data but rather have him request what is needed. This style, called information pull, represents a significant change from traditional information push. The impact on supporting communications is to give it a more "bursty" character, driven by external events.

e. Digital signal processing, programmable systems. Single-function C3 hardware is evolving to multifunction capability. Each node or platform will emerge with new capabilities that permit rapid and flexible reallocation. Current generation tactical military aircraft, as delivered, have virtually no additional space or weight allowance for new equipments. A desire is to evolve from costly retrofitting to a state of software insertion and integration. Traditional single-band radio systems will be replaced with **programmable multiband, multiwaveform systems.** Near real-time management and control of highly flexible, programmable systems will become a growing research and development thrust.

f. Interoperability and standards. C3I systems, with many dispersed nodes, rely heavily on computer-computer communications. Standards are being promoted by industry and government to simplify the development, acquisition, and insertion of new technology as well as to promote interoperability between independently developed systems. Two additional motivations for new standards are increased traffic requirements and increased system complexity. C3 applications and users have found significant benefit in increasing communication with programmatically unrelated data sources such as databases and sensors. There is an increase in internal communications as well. Also systems have become more complex, forcing programs to develop modularized architectures. Software is replacing hardware as the most complicated component of communications and C2 systems to design, build, and maintain. Modularized architectures are required to simplify development and enable insertion of new technologies. Finally, as systems become more complex, cost becomes a greater factor; this leads to a desire to acquire available commercial off-the-shelf (COTS) components, if they fulfill a system's requirements.

The primary computer-to-computer communications architecture has been the Open Systems Interconnection (OSI) Reference Model. The OSI Reference Model has been successful as a layered architecture with well-defined interfaces and specified division of functions. The Department of Defense has committed to adopting an enhanced version of the OSI protocols, called the Government OSI Profile (GOSIP). OSI/GOSIP integration into C3 systems is lagging because of delays in accredited vendor implementations and the cost of upgrading the existing communications infrastructure. NATO allies have adopted their independent OSI procurement policies.

OSI brings to C3 a set of application services that had not been previously available. For example, the OSI electronic mail standards (usually called X.400) provide message forwarding, distribution list creation and distribution, and obsolete message extraction among other services to users. In addition to the security protocols contained in the lower layers of the

OSI stack, X.400 has its own security services such as message origin authentication, message flow confidentiality, message content integrity, and nonrepudiation of delivery, services that are highly desirable in C3 environments. OSI also has enhanced file transfer and management capabilities, systems management, directory, and transaction processing, among other application functions, all providing enhanced capability to C3 users.

g. Precision timing and position location (GPS). Navigation/position location historically is important and becomes more so in high dynamic maneuver warfare. With the introduction of the Global Positioning System (GPS), 3-dimensional positioning is available to the smallest of high-mobility nodes. Even with a less than complete satellite constellation, position accuracies can become less than 100 m.

h. Displays and workstations. High-resolution displays combined with programmable workstations and software lead to flexible node functions and consequently to flexible architectures. A C3 workstation could, in principle, support any of a number of C3I functions; a relocation of operators may be the only requirement to physically relocate a command node. Numerous decision aids are now being included within workstations and with their more comprehensive capability are now often described as decision support systems (DSS). Man-machine interface (MMI), as a result, grows in importance.

i. Software techniques. With the growing computational power and memory capability of microprocessor systems, C3 system performance will increasingly be determined by software performance. The cost and complexity of software appears to expand in proportion to host computer capability and is more frequently becoming a system limiting factor. ADA is dictated to be the common programming language of the Defense Department; however, exceptions can be approved. Verification and validation (V&V) of generated software and software maintenance have grown to necessitate organizational changes within the military. Software standards have also increased in importance in new C3 systems. POSIX standards (published as IEEE 1003) govern the software interfaces to operating system services in various computing platforms [NIST, 1990]. As such, they allow application programs written according to the standards to be reused. POSIX standardizes interfaces to security, networking, and diverse system services, including file management, memory and process management, and system administration services. POSIX.5 provides bindings for the ADA programming language.

j. Simulation and modeling. Both techniques are employed with the objective of designing or analyzing the performance of a C3I system. With the advent of faster computation, complex scenarios can be "gamed" in near real time, and modeling will then be within the decision aid realm.

96.4 The Dynamics of Encounters

Within dynamic systems, and the C3 systems that support them, it is important to identify and clarify time scales involved. Military engagements range from sub-second events such as local missile point defense to the long-term development and implementation of global strategy. Each involves basic aspects of decision and control theory: objectives, observations, and feedback and control. In the military environment, the observation aspect is especially complex, requiring the placement, collection, transmission, and aggregation of data from numerous dispersed sources. Control and decision techniques derived for one echelon level may be inappropriate for others, primarily due to the time available for the assessment and feedback process. Often, the impact of a decision will not be measurable before yet another control decision is required. Thus, the relative roles of automation and humans will be different at different levels. The human may have to project a decisionmaking consequence long before the system hardware/software can obtain measures of it.

As an example of encounter space-time domains, surface Navy echelon levels have order-of-magnitude scales as shown:

Organization Level	Time Scale of Interest	Geographic Extent (km)
Platform	seconds-minutes	10's
Battle Group	minutes-hours	100's
Fleet	hours-days	1000's
Theater	days-weeks	1000's +
Service/National	weeks-years	Global

At the platform level, the time scale range reflects engagement times which may include limited or local amounts of tracking. At the Battle Group level, the time scale corresponds to tasks such as maneuver, coordinated engagement, and track management.

Any of the organizational levels may additionally have planning functions that precede the operational time scales by up to months or years. The planning side includes events such as logistics, maintainence, training, and exercises, all of which contribute toward becoming a more capable combatant. Figure 96.3 portrays the planning and the operational or execution phases as well as portraying the adaptive control loop approach to C3. The lighter shaded feedback path is employed when it is required to compare status with the current plan. It is also available for adjustment when plans or objectives are modified. The execution phases are represented by the Stimulus-Hypothesis-Options-Response (SHOR) Paradigm suggested by Wohl [1981]. The control theoretic implications are apparent in the figure; the Stimulus-Hypothesis is a representation of situation assessment with its implicit uncertainty. Quickness and accuracy with which a military command organization can transverse the execution loop is a general measure of performance (MOP). Qualitatively it is generally accepted that the side with the best ability to transverse the SHOR execution loop will have a significant military advantage. In this light, attributes of the execution loop become a measure of effectiveness (MOE) of the C3 system in terms of operational outcomes. Rules of Engagement (ROE) impact tempo by reducing uncertainty or options available to the decisionmaker. Some scenarios develop with such quickness that the C3 system must react nearly reflexively (e.g., without consideration of possible options). One class of rules is made known to all the participants; if a particular manuever is observed, then a specified response will result.

The SHOR paradigm illustrates why counter-communications and counter-command and control are increasingly important operational and technical areas. Counter-C3 need only delay the process rather than disrupt or destroy it in order to be an effective technique. The Navy, for example, is now incorporating **electronic warfare** (EW) as a warfare area on equal status to the traditional anti-submarine (ASW), anti-aircraft warfare (AAW), and anti-surface warfare (ASUW) areas.

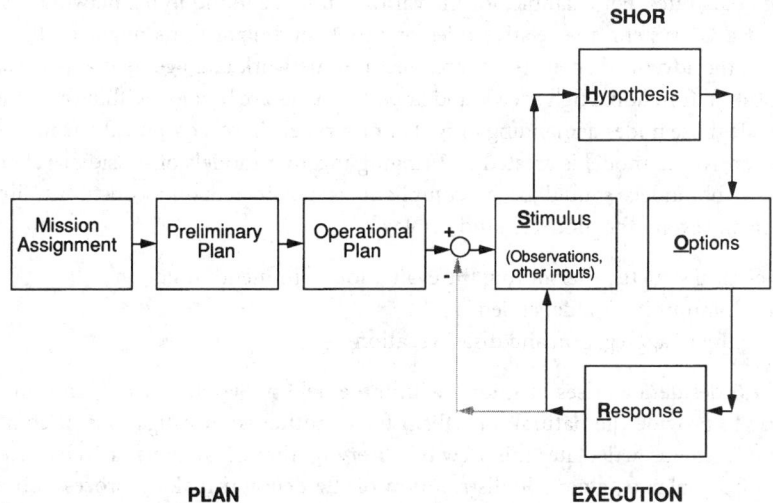

FIGURE 96.3　The planning and execution phases of operations.

Control of the electromagnetic spectrum is becoming as critical as the control of the physical battlefield. Electronic counter-measures (ECM) such as jamming and deception are technical options available to the commander. Either adversary may elect to respond to the ECM threat by a series of electronic counter-counter measure (ECCM) techniques. Anti-jam (AJ) communications can employ a variety of techniques such as spread spectrum, power control, adaptive coding and feedback, multiple routes, and adaptive antenna arrays. A signal may also be protected by making it difficult to intercept; some low probability of intercept (LPI) methods are again spread spectrum, directive antennas, power control, EM propagation strategies, and message brevity.

The SHOR paradigm has important advantages. First, it is generally applicable to all military echelon levels. Second, it represents a control process with its explicit dynamics rather than a relational or physical interconnection of system components. Finally, it puts focus on the roles of controlling and decisionmaking without a pre-bias on whether that function should be performed by humans or computers. The remaining challenge is to be able to describe both human and computer performance with a common type of representational framework.

96.5 The Role of the Human Decisionmaker in C³

Designers of C3 systems often fail to acknowledge the fact that the "central, essential ingredients in any command and control system are not the things which they plan and design; rather they are the commanders and decisionmakers themselves" [Wohl, 1981]. Despite its centrality, designation of human roles is seemingly arbitrary and often controversial. In most system studies, the human decisionmaker is not thought of as an integral part of the system, but is instead given an external position as a "user" of data or an "input" to the rest of the system. Without a means of integrating the behavior of interrelated decisionmakers into a comprehensive description of system response, the proper hominal role is difficult to determine. To justify and support human action, a clear understanding of the benefits and limitations of human intervention is required.

The complexity and unpredictability of a C3 environment prompt the inclusion of hominal blocks. The ability to respond to changing operational conditions requires "intelligence," and in a C3 system this intelligence is distributed between people and algorithms. The human has a marvelous capacity for coping with vague and confusing data, making sense out of information so fragmentary that it would paralyze a computer. A computer information processing algorithm has, in turn, an unexcelled capability to process and display data at a rate that would bewilder a person. Proper marriage of humans and computers yields a robust system, quick to adapt to changes and capable of handling high data rates. For example, for the various subtasks found in the network management component of a C3 system, the relative roles of people and algorithms might be that shown in Fig. 96.4. With the advent of open system architectures, network management appears as a crucial resource allocation function. High speeds and large databases are best left within the domain of the computer, while those nodes demanding insight appropriately have a corporeal flavor.

A comprehensive C2 model is created by bringing together models of subsidiary elements. The form of these submodels should be as compliant as possible within constraints imposed by tractability. In any event, the model should display:

1. An analytical structure permitting the evaluation of influence functions
2. Explicit communication dependence
3. Amenability to aggregation and disaggregation

Each of these desiderata arises in studies within the field of System Science, and this discipline would appear to provide the natural formalism for quantitative investigations of command and control systems. Athans articulates this view by observing that C3 systems "are characterized by a high degree of complexity, a generic distribution of the decision-making process among several decision making 'agents,' the need for reliable operation in the presence of multiple failures, and

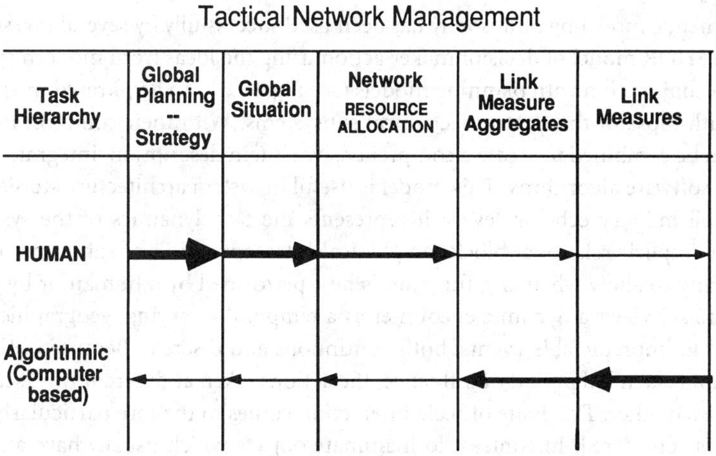

Tactical Network Management

Task Hierarchy	Global Planning -- Strategy	Global Situation	Network RESOURCE ALLOCATION	Link Measure Aggregates	Link Measures
HUMAN					
Algorithmic (Computer based)					

FIGURE 96.4 Control hierarchies.

the inevitable interaction of humans with computer-based decision support systems and decision aids" [Athans, 1987]. It needs to be emphasized, however, that a C2 system differs from those commonly encountered in system theory in at least three primary ways:

1. Because command and control is at its essence a human decisionmaking activity, it is not sufficient to model only the sensors, computers, displays, etc. The hominal dynamics must be integrated with those of the electromechanical elements.
2. Any effective C3 system must have the capacity to evolve over time. Such systems are frequently established with a limited set of elements. Either for a specific operation or during their lifetime a subset of these elements will be modified or replaced, and their roles expanded or constricted as changing demands are placed upon the system. Hence, the system description must be more flexible than those in common use.
3. In contrast to conventional system design problems, there is no single nominal operating condition about which the system is maintained. Indeed, the critical attribute of a C3 system is its ability to respond to major changes in condition or state. In two Middle East naval events (USS *Stark, Vincennes*) the missile defense systems were set for a state that had just immediately changed. Furthermore, the system is often used in environments quite different from those envisioned in its design. Hence, the uncertain circumstances within which the decisionmakers must accomplish their tasks must be properly reflected in any system architecture.

A commander brings special skills to such a system, but some of them are difficult to quantify. For example, people have singular competence in:

1. Decisionmaking in semantically rich problem domains
2. Analogical reasoning and problem structuring
3. Information processing and application of heuristics

To properly identify a specific function for a human decisionmaker, the advantage accruing to his inclusion must be shown. Quantitative models of human responses have been developed in various ways, from ad hoc to purely normative. In the most promising of these, the form of the response dynamics of an individual commander is determined from the solution to an optimization problem. The optimization problem is framed by supposing that the decisionmaker strives to act in the most effective way, but is constrained by both cognitive limitations and temporal pressures. When the decisionmaker's milieu and motivation are expressed in an analytical framework containing both the exogenous influences of the conflict and the endogenous predispositions generated by training and personal inclination, the input-output relation for the commander is, in principle, expressible as a set of differential equations with logical branching.

This fundamental modeling philosophy has been used successfully by several investigators. Wohl developed the SHOR model of decisionmaker action using the ideas from modern systems theory. The SHOR, in conjunction with planning models (see Fig. 96.3) can be phrased in analytical terms compatible with those of the electromechanical subsystems. With their common form, all of the submodels can be combined to create a comprehensive system description, integrating people with hardware and software algorithms. This model is useful in system architecture studies because it is applicable to all military echelon levels; it represents the fast dynamics of the system explicitly rather than by implicit relational blocks or physical interconnection of subsystem elements, and there is flexibility to allow whether a function is best performed by a human or by an algorithm.

A decisionmaker views a dynamic encounter as a temporally varying, geographically dispersed system subject to unpredictable events, both continuous and discrete. Because critical command decisions have an extended period of influence, the actions taken at different time scales cannot be isolated from each other. This issue of scale interaction comes to the fore particularly when hominal modeling is considered. In contrast to inanimate objects which usually have a single, natural time scale, the demands on a commander transcend the time scale divisions. A trained decisionmaker exhibits a wide spectrum of behaviors as both his tasks and operating environments change; the commander is the truly adaptive block in a command and control architecture. Athans referred to C3 systems as "event driven" because major changes in an engagement occur at isolated times and modulate the more frequent local irregularities [Athans, 1987]. He suggested that the proper model would be a hybrid in which "the state variables are both continuous and discrete." In this metapartitioning of the comprehensive state space, the discrete states represent global (or macro) occurrences that modulate the local (or micro) aspects. This decomposition is useful in formulating the human response model because people react differently in different time scales. The reaction to local phenomena has a reflexive quality. It is in this reaction to the infrequent, but pivotal, macroevents that the idiosyncrasies thought to be particularly human are manifest.

To capture hominal behavior analytically, a framework delineating the intrinsic features of a C3 environment is required. At the macrolevel, the important attributes of a command and control environment are tempo, uncertainty, and complexity. The mission directed decisionmaker model (MDDM) described in Clapp and Sworder [1992] decomposes the C2 model in the hybrid form suggested by Athans. One block in the MDDM, the stimulus-hypothesis evaluation model (SHEM), quantifies relevant features of an engagement while representing the observation and situation assessment tendencies of the decisionmaker in terms of a few natural parameters. Because of its simple structure, the SHEM lends itself to the analysis of systems containing human decisionmakers.

To be more specific, the C2 environmental model must be flexible enough to portray the sudden, large-scale variations in circumstances which occur in operations. It is advantageous to phrase the model in such a way as to make explicit its dependence on events of macroscopic scale as well as the decisionmaker's response. The engagement model used in the MDDM has the form:

$$(d/dt)x_p = f(x_p, u_p, r_t) + g(x_p, u_p, r_t)w_t$$

where x_p is the "global" system state vector representing the external environment to which the decisionmaker seeks to respond. The decisionmaker's action variable is u_p. The process $\{w_t\}$ represents only one portion of the primitive randomness in the encounter—that associated with high-frequency uncertainty and various local disturbances. The supplementary process, $\{r_t\}$, indicates the mode of evolution of the encounter. Transitions in $\{r_t\}$ thus signify extensive events. These macro-events tend to have more temporal structure than that displayed by $\{w_t\}$, but the times of occurrence are typically unpredictable. Different values of r_t (sometimes called supervariables) are identified with different hypotheses delineating the macrostatus of the encounter. It is usually assumed that the number of modal hypotheses is finite.

Even with the aggregation implicit in the engagement model, the encounter dynamics are complex and nonlinear. A decisionmaker mentally converts the engagement dynamics into a hybrid equation with separate descriptions of local and global aspects. The input-output dynamics of the commander are expressed as an ordinary differential equation with updates at observation times. In Sworder *et al.* [1992], the ability of the SHEM to predict the response of a trained decisionmaker was investigated. An experiment measuring the proficiency of trained air-defense officers in differentiating hostile from friendly targets confirmed the utility of the SHEM.

96.6 Summary

C3I systems, commanders/decisionmakers, and decision aids all have a common performance objective. They must contribute to accurate and timely *situation assessments and responses* in scenarios that have a wide range of tempos, noise, clutter, uncertainty, and complexity.

The C3 system necessarily has the ability to rapidly acquire, process, and transfer large volumes of data over extended regions. Trained, experienced human decisionmakers excel at assessing complex patterns in highly cluttered environments and determining appropriate responses. Decision aids perform as a "smart" interface between these two dissimilar players. Electronics technology provides the means for designing increasingly capable C3 systems and is at its most effective when the system architecture allows flexible and dynamic interoperation of the various hardware "devices" with their trained and motivated decisionmakers.

Defining Terms

Command, control, communications (C³): The process of mapping assets (resources available to the military commander) into capabilities. This control process is impacted by tempo, noise/clutter, and scenario complexity.

Decisionmaking: A commander's or operator's action that changes the status of his information or other assets under his control.

Doctrine: A formalized description of military mission definitions to include the procedures to accomplish those missions. Doctrine will also often specify the organizational structure appropriate to the specific mission.

Electronic warfare: Contention for the control of the electromagnetic (EM) spectrum, to allow active and passive EM sensing and communications while denying the same ability to adversaries. Includes deceptive EM techniques.

Environment: A set of objects outside the system, a change in whose attributes affects, and is affected by, the behavior of the system.

Intelligence: The aggregated and processed information about the environment, including potential adversaries, available to commanders and their staff.

Open system architecture: A layered architectural design that allows subsystems and/or components to be readily replaced or modified; it is achieved by adherence to standardized interfaces between layers.

Programmable radio system: Radios based on digital waveform synthesis and digital signal processing to allow simultaneous multiband, multiwaveform performance.

System: A set of objects with relations between them and their attributes or properties. It is embedded in an environment containing other interrelated objects.

References

M. Athans, "Command and control (C2) theory: A challenge to control science," *IEEE Trans. on Automatic Control,* vol. AC-32, pp. 286–293, April 1987.

G.A. Clapp and D.D. Sworder, "Command, control and communications: The human role in military C3 systems," in *Control and Dynamic Systems, Advances in Theory and Applications,* vol. 52, New York: Academic Press, 1992, pp. 513–541.

Joint Chiefs of Staff (JCS), Publication 1, "Definitions," undated.

National Institute of Standards and Technology [NIST], FIPS 151-1, POSIX: Portable Operating System Interface for Computer Environments (IEEE 1003.1–1988) March 1990.

D.D. Sworder, G.A. Clapp, and R. Vojak, "A Dynamic Input-Output Model of the Decisionmaking Process," Proceedings of the 1992 Symposium on Command and Control Research, Monterey, Calif., June 1992.

J.W. Wohl, "Force management requirements for air force tactical command and control," *IEEE Trans. on Systems, Man and Cybernetics*, vol. SMC-11, pp. 618–639, Sept. 1981.

Further Information

W. Stallings, *Handbook of Computer-Communications Standards*, vol. 1, The Open Systems Interconnections (OSI) and OSI-Related Standards, New York: Macmillan, 1987.

W. Stallings, *Handbook of Computer-Communications Standards*, vol. 3, Department of Defense (DOD) Protocol Standards, New York: Macmillan, 1988.

Information Technology for Command and Control, S. Andriole and S. Halpern, Eds., IEEE Press, New York, 1991.

SIGNAL, a monthly (trade) magazine published by the Armed Forces Communications-Electronics Association (AFCEA), Annandale, Va. Contains numerous brief articles on current C3I topics of interest.

T.P. Coakley, *Command and Control for War and Peace*, National Defense University, U.S. Government Printing Office, Washington, D.C., 1992.

A.D. Hall, *Metasystems Methodology*, Oxford, England: Pergamon Press, 1989.

97

George E. Cook
Vanderbilt University

Kristinn Andersen
Vanderbilt University

Robert Joel Barnett
Vanderbilt University

Alan K. Wallace
Oregon State University

René Spée
Oregon State University

Industrial Systems

97.1   Welding and Bonding 2223
       Control System Requirements • System Parameters • Welding
       System • Sensing • Modeling • Control • Conclusions
97.2   Large Drives 2237
       Configurations • Selection and Compatibility • Principles and
       Features of Operation • Control Aspects • Future Trends

97.1 Welding and Bonding

George E. Cook, Kristinn Andersen, and Robert Joel Barnett

Most welding processes require the application of heat or pressure, or both, to produce a bond between the parts being joined. The welding control system must include means for controlling the applied heat, pressure, and filler material, if used, to achieve the desired weld microstructure and mechanical properties.

Welding usually involves the application or development of localized heat near the intended joint. Welding processes that use an electric arc are the most widely used in industry. Other externally applied heat sources of importance include electron beams, lasers, and exothermic reactions (oxyfuel gas and thermit). For fusion welding processes, a high energy density heat source is normally applied to the prepared edges or surfaces of the members to be joined and is moved along the path of the intended joint. The power and energy density of the heat source must be sufficient to accomplish local melting.

Control System Requirements

Insight into the control system requirements of the different welding processes can be obtained by consideration of the power density of the heat source, interaction time of the heat source on the material, and effective spot size of the heat source.

A heat source power density of approximately 10^3 W/cm^2 is required to melt most metals [Eagar, 1986]. Below this power density the solid metal can be expected to conduct away the heat as fast as it is being introduced. On the other hand, a heat source power density of 10^6 or 10^7 W/cm^2 will cause vaporization of most metals within a few microseconds, so for higher power densities no fusion welding can occur. Thus, it can be concluded that the heat sources for all fusion welding processes lie between approximately 10^3 and 10^6 W/cm^2 heat intensity. Examples of welding processes that are characteristic of the low end of this range include **oxyacetylene welding, electroslag welding,** and **thermit welding.** The high end of the power density range of welding is occupied by **laser beam welding** and **electron beam welding.** The midrange of heat source power densities is filled in by the various arc welding processes.

For pulsed welding, the interaction time of the heat source on the material is determined by the pulse duration, whereas for continuous welding the interaction time is proportional to the spot diameter divided by the travel speed. The minimum interaction time required to produce melting can be estimated from the relation for a planar heat source given by [Eagar, 1986]

$$t_m = [K/p_d]^2$$

where p_d is the heat source density (watts per square centimeter) and K is a function of the thermal conductivity and thermal diffusivity of the material. For steel, Eagar gives K equal to 5000 W/cm²/s. Using this value for K, one sees that the minimum interaction time to produce melting for the low power density processes, such as oxyacetylene welding with a power density on the order of 10^3 W/cm², is 25 s, while for the high energy density beam processes, such as laser beam welding with a power density on the order of 10^6 W/cm², is 25 μs. Interaction times for arc welding processes lie somewhere between these extremes.

An example of practical process parameters for a continuous **gas tungsten arc weld** (GTAW) are 100 A, 12 V, and travel speed 10 ipm (4.2 mm/s). The peak power density of a 100-A, 12-V gas tungsten arc with argon shielding gas, 2.4-mm diameter electrode, and 50-degree tip angle has been found to be approximately 8×10^3 W/cm². Assuming an estimated spot diameter of 4 mm, the interaction time (taken here as the spot diameter divided by the travel speed) is 0.95 s. At the other extreme, 0.2-mm (0.008-in.) material has been laser welded at 3000 in./min (1270 mm/s) at 6 kW average power. Assuming a spot diameter of 0.5 mm, the interaction time is 3.94×10^{-4} s.

Spot diameters for the high density processes vary typically from 0.2 mm to 1 mm, while the spot diameters for arc welding processes vary from roughly 3 mm to 10 mm or more. Assuming a rule of thumb of 1/10 the spot diameter for positioning accuracy, we conclude that typical positioning accuracy requirements for the high power density processes is on the order of 0.1 mm and for the arc welding processes is on the order of 1 mm. The required control system response time should be on the order of the interaction time and, hence, may vary from seconds to microseconds, depending on the process chosen. With these requirements it can be concluded that the required accuracy and response speed of control systems designed for welding increases as the power density of the process increases. Furthermore, it is clear that the high power density processes *must* be automated because of the human's inability to react quickly and accurately enough.

System Parameters

The variables of the welding process are separated here into **direct weld parameters** (DWP) and **indirect weld parameters** (IWP) [Cook, 1981]. The DWP are those pertaining to the weld reinforcement and fusion zone geometry, mechanical properties of the completed weld, weld microstructure, and discontinuities. The IWP are those input variables that collectively control the DWP. The IWP are the welding equipment setpoint variables, e.g., voltage, current, travel speed, electrode feed rate, travel angle, electrode extension, focused spot size, and beam power.

Welding System

The various DWP, or process variables, that we would like to control and the many possible IWP, or equipment variables, that we may set to achieve the desired output are shown in Fig. 97.1. From the standpoint of feedback control, the welding process depicted in Fig. 97.1 presents two principal problems: (1) in most cases the relationships between the IWP and DWP are nonlinear, and (2) the variables are generally highly coupled.

With most production welding today, the designer of the welded part specifies the desired weld characteristics (the DWP), including acceptable tolerance windows. The job of the welding engi-

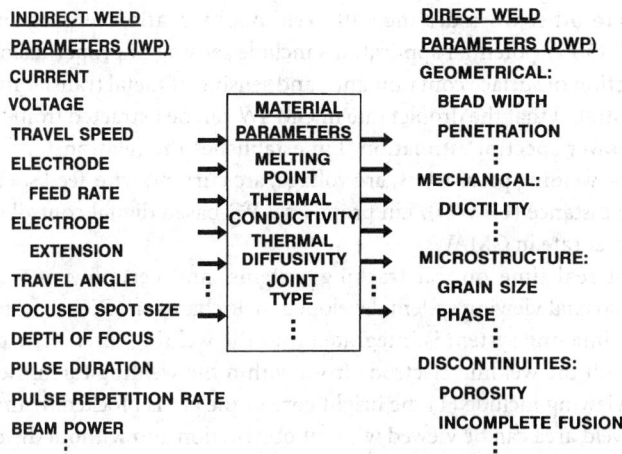

FIGURE 97.1 Input and output variables of welding process.

neer then is to determine a set of IWP that will produce the desired DWP. Most automated welding systems today may be expected to have good control over the IWP, including joint tracking for heat source positioning. Therefore, if production floor conditions do not differ too much from the laboratory conditions under which the weld procedures were developed, then the welding operation can be expected to satisfy quality inspection and control procedures. If not, human operators must be depended upon to provide the necessary feedback to make corrective actions in the welding equipment settings.

The human involvement in this scenario can be reduced or eliminated by sensing selected DWP, comparing the sensed variables with desired values, and implementing a multivariable controller that will reduce automatically the error between the desired and sensed DWP to zero or an acceptably low difference. Dynamic and steady-state process models are required for both design and stable operation of the multivariable feedback control system. However, the models do not need to be as globally accurate as the models required for open-loop control. In exchange for accuracy, the models used for control system purposes must be computable in real time, and generally, it is important that they provide both steady-state and dynamic information of the interrelationships between the coupled variables of the system. It is generally important that these relationships be "tunable" in real time to permit calibrating the multivariable system controller to the actual operating conditions at any given time.

Successful implementation of multivariable weld process control involves (1) sensing, (2) modeling, and (3) control. Issues dealing with each of these will be discussed in the following sections.

Sensing

In recent years, great strides have been made in sensor technology, particularly in the areas of optical sensors, arc sensing, and infrared, acoustic, and ultrasonic sensing.

Optical Sensing

Optical sensing technology has been developed and used for a number of applications, including joint tracking and fill control, sensing of molten pool width, sensing of weld bead profile, arc length sensing and control, sensing and control of electrode extension in **gas metal arc welding** (GMAW), and sensing of weld depth or penetration. Yi [1991] has investigated the ability to estimate GTA weld penetration by means of measuring the weld pool vibration frequency. Yi used the optical signal emitted from the welding arc as a means of sensing the weld pool vibration. Digital signal processing was used to estimate the oscillation frequency of the weld pool from the sensed optical signal. References to other work dealing with weld pool vibration sensing and analysis may be found in Yi [1991]. Other potential applications include sensing of proper fusion characteristics at the sidewalls, detection of surface contaminants, and sensing of metal transfer mode in GMAW. Liu [1991] has demonstrated that the droplet rate in GMAW can be extracted from the arc infrared signal by means of power spectral estimation. Liu establishes the relationships between the metal droplet rate and the welding parameters, arc voltage, arc current, wire-feed speed, and the contact tube-to-workpiece distance (CTWD). Liu proposes a PC-based digital control system for controlling the metal droplet rate in GMAW.

One of the first real-time optical tracking systems, and certainly one of the more novel approaches, was a coaxial viewing system developed by Richardson [Richardson *et al.*, 1984]. With this approach, the imaging system is integrated into the welding torch. The point of welding is viewed coaxially with the welding electrode from within the welding torch. Advantages reported for this system of viewing include (1) the bright core of the arc is blocked by the electrode/contact tip, (2) the entire weld area can be viewed without obstruction and without distortion by the viewing angle, and (3) the system is nonintrusive into the weld area and is nondirectional.

A number of optical tracking systems make use of a projected laser strip or a scanned laser beam to provide structured lighting that permits three-dimensional profiling of the joint, typically in front of the heat source. Several such tracking systems are commercially available and offer robust solutions to the joint tracking problem.

A viewing system that provides remarkably good images of the electrode and molten pool area has been developed from laser and night imaging technology. The system's operation is based on the use of a high-intensity pulsed laser or strobe light synchronized with an image intensifier and camera to suppress the arc light and produce a clear view of the arc area. The excellent image obtained with this system offers a great deal of potential for various types of optical process sensing requirements.

Arc Sensing

Arc sensing (or through-the-arc sensing) has many applications, some, such as automatic voltage control, dating back 30 years or more. The obvious advantage of arc sensing is that use of the arc itself as a sensor means there is not any need for external sensors, with the associated concern for their reliability in the harsh environment of the welding arc.

One of the most widely reported recent applications of arc sensing is for purposes of vertical and lateral tracking and width control [Cook, 1983]. For this application, the sensing method is based on the changes in current and/or voltage when the arc is weaved back and forth across the joint. Inventions have been disclosed for both nonconsumable arc welding processes and consumable arc welding processes (see references in Cook *et al.*, 1990). Applications range from pipe welding to robotic arc welding to turbine blade repair. For submerged arc welding (SAW), for example, current variations of approximately 10% at the sidewalls have been observed while welding in a joint consisting of a 45-degree included angle with a 5-mm root opening. With a nominal current of 580 A at the center of the joint, the current at the sidewalls is approximately 640 A. Variations of this magnitude may be used to implement robust control algorithms for joint tracking and width control. Shepard [1991] presents a thorough treatment of the mechanisms that establish and influence self-regulation in GMAW. Components of a dynamic GMAW process model are identified, including the power source, joule heating in the electrode, electrode burn-off rate, and arc voltage. A numerical simulation of the nonlinear dynamic model for self-regulation is implemented, computing current I and electrode extension in response to CTWD, voltage, and feed rate. The I/CTWD response is shown to be frequency dependent, increasing significantly at higher frequencies. The frequency at which the response increases is shown to be primarily dependent on electrode current density, occurring at lower frequencies for lower current densities. A linearized closed-form model for the I/CTWD frequency response is derived from the simulation equations and is shown to provide accurate results. The closed-form model clearly indicates the relationships between the model parameters that establish the observed characteristics of self-regulation dynamics. Initial implementations of through-the-arc seam tracking methods use simple current levels to identify the lateral limits of the weld joint, adjusting the torch centerline to maintain symmetry. The dynamic model developed by Shepard provides a basis to infer actual joint geometry from position and current information acquired during cross-seam oscillation. The relationships developed by Shepard also refine the basis for selection of welding procedures in GMAW applications, particularly for through-the-arc sensing applications. The models define the relationships to generate surfaces to facilitate selection of electrode diameter, feed rate, voltage, electrode extension, and CTWD to optimize desired characteristics such as low-frequency sensitivity, high-frequency sensitivity, and transition frequency subject to requirements on heat input and deposition rate. These interrelationships may be used as extensions to existing expert systems for selection of welding procedures.

Arc sensing has been proposed as a means of sensing GTA pool motion after excitation from pulsations in the current. The concept of using weld pool motion as a pool geometry sensing method is based on the fluid dynamics of the constrained weld pool, which depend on the properties of the molten pool material, the surface tension, and the shape of the pool.

Another potential application of arc sensing is detection of the metal-transfer mode in GMAW. The droplet transfer mode in the GMAW process has a large effect on weld pool metallurgy, influencing penetration, solidification, heat flow, and mass input. Researchers have attempted to correlate perturbations in the electrical arc signals with droplet transfer. This work has demonstrated the ability to detect the detachment of individual droplets and to distinguish among the three transfer modes: globular, spray, and streaming, as defined by Lancaster [1986].

Measurements of the incremental arc resistance by Shepard [1991] suggest that the metal-transfer mode of the gas metal arc may also be detected by the rapid transitions of the incremental arc resistance at the transition regions of metal transfer (particularly at the spray-to-streaming transition). The incremental resistance was obtained by perturbing the voltage with a 1-V_{p-p}, 15-Hz sinusoidal variation. In the arc resistance measurements, CTWD and electrode extension (and hence arc length) were held constant and data were taken over a wide range of current. A nominal CTWD of 25.4 mm was used, with a 15-mm electrode extension. Feed rate was varied from the globular/spray transition point to the upper ranges imposed by equipment limitations. A small (1 V_{p-p}), "high-frequency" (15 Hz) sinusoidal perturbation was superimposed on the power source voltage to allow measurement of the incremental resistance at each operating point. The frequency was sufficiently high that the electrode extension did not vary significantly. For each data point, an 8-s record was acquired at 1-kHz sample rate. The frequency response function (FRF) was used to compute the incremental resistance by calculating the current produced in response to the sinusoidal voltage perturbation. The FRF gives the magnitude and phase angle of a linear model of the arc *V-I* characteristic about the given operating point, making up the total resistance of arc plus electrode. Results of the incremental arc resistance measurements were plotted as a function of current. The most significant feature of these data was the large peak in incremental resistance in the region of the projected/streaming transition. The height of the peak is roughly twice the nominal resistance at higher currents. The incremental arc resistance increases sharply at the upper end of projected transfer mode, peaking just after the transition to decline to a relatively steady level through the upper end of the streaming transfer range.

For weld procedures that include cross-seam oscillation, or weaving, of the heat source, arc sensing provides a reliable indicator of sidewall/adjacent bead fusion. As the sidewall or adjacent bead is approached in the weave cycle, the electrical signals change in response to the change in CTWD for GMAW or arc gap for GTAW. This change is, of course, the signal used for tracking control in through-the-arc tracking; however, it provides a useful indicator of proper penetration into the sidewall or adjacent bead independently of whether arc sensing is used for tracking purposes.

Andersen *et al.* [1989] have reported the use of arc signal parameters as a potential control means for GMAW, short-circuiting transfer. Digital signal processing was used to extract from the electrical signals various features, including average and peak values of voltage and current, short-circuiting frequency, arc period, shorting period, and the ratio of the arcing to shorting period. Additionally, a joule heating model was derived that accurately predicted the melt-back distance during each short. The ratio of the arc period to short period was found to be a good indicator for monitoring and control of stable arc conditions. Any change in the arcing voltage, for a given power circuit condition, leads to corresponding changes in the arcing/shorting time ratio. Such changes in arcing voltage may occur with change in the shielding gas, in the surface condition (in the form of contaminates) of the electrode wire and work, and in their composition, such as the presence of rare earths, in the wire electrode or work materials that affect the arc characteristics.

Andersen *et al.* [1989] show that if the average arc current may be assumed nominally constant because of constant electrode feed, then the arcing/shorting time ratio serves as a sensitive index of the operation of the GMAW short-circuiting system. The arcing/shorting time ratio can be used to control the short-circuiting gas metal arc in a feedback loop by adjusting the open circuit voltage to compensate for variations in the arcing voltage.

Finally, the electrical arc signals vary as a function of contaminants on the workpiece and/or electrode, and these variations may be sensed and correlated with the changes observed in surface conditions.

Infrared Sensing

Infrared sensing has inherent appeal for weld sensing. Potential applications include cooling rate measurements, discontinuity sensing, penetration estimation, seam tracking, and weld pool geometry measurement.

Acoustical Sensing

The acoustical signals generated by the welding arc are a principal source of feedback for manual welders. Recently, acoustical signals have been studied as a sensing means for automated welding as well.

Sound generated by the electric arc of a gas tungsten arc weld has been used for arc length control. With this system the current is pulsed a small amount at an audible rate to generate an audible tone at the arc. The intensity of the arc-generated tone has been shown to be proportional to the arc length and, hence, can be suitably processed to provide a feedback signal for arc length control.

Acoustical signals generated by gas metal arcs have been correlated with the detachment of individual droplets from the filler wire. Research has demonstrated the ability to detect the detachment of individual droplets and to distinguish among transfer modes: globular, spray, and streaming transfer. This may lead to a means of closed-loop control of the heat and mass input during both pulsed and nonpulsed GMAW.

Acoustical signals have also been reported as a means of plasma monitoring in laser beam welding (LBW). Specifically, experiments have been conducted to characterize the interaction between the incident laser light, the plasma formation, and the target material during pulse welding with an Nd:YAG laser. In the experiments, the acoustical signal, picked up by a microphone, was used to signal plasma initiations and propagation. A correlation was observed between the number of plasmas generated and the weld pool penetration in a target.

Acoustic emission has been used for monitoring LBW in real time. The acoustic sensor has been reported to detect laser misfiring, loss of power, improper focus, and excess root opening.

Ultrasonic Sensing

The use of ultrasonics for weld process sensing has the potential to detect weld pool geometry and discontinuities in real time. However, to be useful in realistic production systems, a means must be developed for injecting the ultrasound and receiving it with noncontacting sensors. Lasers have been proposed as a sound source, and electromagnetic acoustic transducers (EMATs) have been proposed for ultrasound reception. With this proposed approach, the pulsed laser is directed to impinge on the molten pool, setting up stress waves that are transmitted through the workpiece and picked up by the EMAT receiver.

Modeling

Weld process models intended for control purposes are characterized by the need to be computable in real time. This rules out many of the more exact numerical models that have been developed for finite element and finite difference methods. However, these computationally intensive numerical models may be quite useful in developing simpler models that can be used in the control of multivariable weld feedback control systems. Another important aspect of process models used for control purposes is that they generally need to provide both static and dynamic information.

Analytical Models

Since the 1940s considerable research has been focused on developing steady-state models that would predict DWP, given a set of IWP. Easily computed analytical models, based solely on conductive heat transfer, are reasonably accurate but primarily are of value in establishing approximate relationships. Improvements to these early analytical models have been proposed that permit obtaining a better match to actual conditions and that may be calibrated in real time; however, accuracy remains limited in the absence of modeling extensions that require computationally intensive numerical solution.

Empirical and Statistical Models

Other approaches taken to developing steady-state weld process models include: empirically derived relationships between the IWP and DWP, with coefficients chosen to match experimental data and statistically derived relationships. Both of these approaches have proven to possess only a limited range of applicability, and they do not lend themselves to real-time "tuning" in a multivariable control system application.

Artificial Neural Network Models

A promising method based on an artificial neural network (ANN) has been studied and found to be accurate and computationally fast in the application mode. Furthermore, the ANN can be refined at any time with the addition of new training data and thus promises a method of continuously adapting to the actual welding conditions.

Andersen [1992] has reported the application of an ANN to mapping between the IWP's arc current, travel speed, arc length, and plate thickness and the DWP's bead width and penetration for GTAW. A back-propagation network, using 10 nodes in a single hidden layer (Fig. 97.2), was used for the modeling. A variety of different network configurations were initially evaluated for this purpose. Generally, it was found that one hidden layer was sufficient for weld modeling, and the best training rate was obtained with on the order of 5 to 20 nodes in the hidden layer. The same plate material was assumed throughout the experiment, which eliminated the need for specifying any of the material parameters. Otherwise, additional input parameters might have included thermal conductivity, diffusivity, etc.

A total of 72 welds, produced on two material thicknesses of 3.175 and 6.350 mm, were used for the purpose of training and testing the network for modeling purposes. Weld current values of 80, 100, 120, and 140 A, travel speeds of 2.12, 2.75, and 3.39 mm/s, and arc lengths of 1.52, 2.03, and 2.54 mm were used. Eight of the welds, which were randomly selected, were not used in the training phase but were reserved for testing the model. With a learning rate parameter of 0.6 and a momentum term of 0.9, the network was trained for 200,000 iterations.

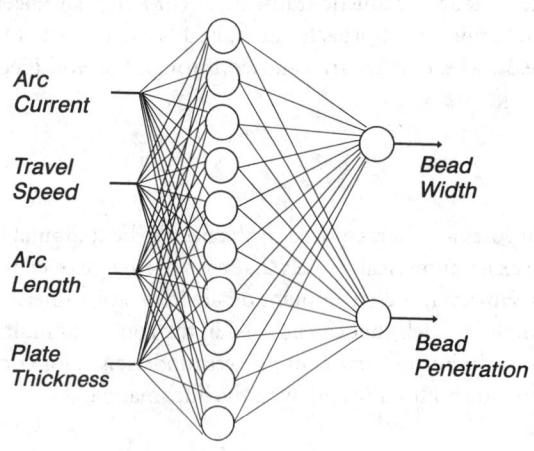

Arc
Current

Travel
Speed

Arc
Length

Plate
Thickness

Bead
Width

Bead
Penetration

FIGURE 97.2   A neural network used for weld modeling.

Once the network had been trained with the 64 training welds, the remaining 8 welds were applied to test the modeling network. The root mean square (RMS) values of the errors were calculated separately for the bead width and penetration, resulting in about 5% and 18% RMS errors, respectively. These results agree with other similar experiments reported by Andersen, in that modeling accuracy is typically on the order of 10-20%.

Weld modeling studies have also been carried out on the **variable polarity plasma arc welding** (VPPAW) process. Modeling of the crown and root width in the keyhole welding mode was of specific interest, and the model inputs were the forward and reverse current values, the torch standoff distance, and the travel speed. The crown and root width errors of the model were generally determined to be on the order of 10–20% or better.

An observation relating to the weld modeling experiments should be noted here. The precision of the bead measurements was 0.1 mm, which corresponds to 2 and 7% precision for the average bead width and penetration, respectively. Furthermore, inaccuracies in measurements of the data, which were used to train the neural network model, tend to degrade the general performance of the model. Width measurements are generally more reliable than penetration measurements, as they are made in several locations along the top of the bead. A penetration measurement is usually made on a single cross section, and it requires chemical etching, which results in a relatively blurred boundary between the bead and the surrounding base metal. This difference is reflected in the consistently lower accuracy of the penetration modeling, compared with the width modeling.

A back-propagation network was also constructed by Andersen [1992] to model the inverse relations, i.e., the DWP-IWP relations, of the weld sample set used in the forward modeling study. A number of neural network configurations were initially used in attempting to train networks to determine the necessary current, travel speed, and arc length for desired bead width and penetration. Preliminary attempts did not result in acceptable training convergence. Closer examination revealed that welds which resulted in full or almost full penetration yielded very irregular bead measurements. It was hypothesized that these irregularities might contribute to the poor training performance. These welds (total of five), which represented the largest pool dimensions on the 3.175-mm test plate, were removed from the training data, and to maintain an equally large data set for the 6.350-mm plate, the five largest welds were ignored there as well. Six welds were randomly selected from the remaining data for each plate thickness for testing only.

Using the revised data set, a network of 50 nodes in a single hidden layer was successfully trained. The learning rate was 0.6, the momentum term was 0.9, and the network was trained for 300,000 iterations. The equipment parameters, or IWP, suggested by the neural network were compared with the actual parameters used to produce the test welds. The RMS deviations between these were current, 9.7%; travel speed, 23.9%; and arc length, 25.5%. Although these deviations between the IWP used to produce the original training set and the IWP suggested by the ANN are rather large, the results are not unexpected because of the nonuniqueness of the inverse problem. The results do not imply that the resulting bead geometries would be accordingly erroneous, because a given width-penetration pair may be attained through multiple nonunique combinations of equipment parameters. For example, an arc current increase may be largely offset by a corresponding increase in travel speed.

To assess the reliability of the ANN for equipment parameter selection, the parameters suggested by the inverse model were used to produce a new set of welds, and bead width and penetration measurements were carried out as before. These widths and penetrations were compared with the original data set. The RMS errors were width, 5.5%, and penetration, 19.9%. These differences between the new geometry parameters and the original ones are approximately the same as the errors observed from the weld model. Again, it is suggested that uncertainty in bead measurements contributes significantly to these errors.

When compared to other control modeling methodologies, neural networks have certain drawbacks as well as advantages. Of the drawbacks, the most notable is the lack of comprehension of the physics of the process. Relating the qualitative effects of the network structure or parameters to the process parameters is usually impossible. On the other hand, other control modeling methods resort to substantial simplifications of either the physical process or more exact numerical models and therefore also trade computability for comprehensibility. The advantages of neural models include relative accuracy and generality. If the training data for a neural network is general enough, spanning the entire ranges of process parameters, the resulting model will capture the complexion of the process, including nonlinearities and parameter cross couplings, over the same ranges. Model development is much simpler than for most other models. Instead of theoretical analysis and development for a new model, the neural network tailors itself to the training data. The network can be refined at any time with addition of new training data. Finally, the neural network can calculate its results relatively quickly, as the input data are only propagated once through the network in the application mode.

The reader is referred to Andersen [1992] for a more thorough discussion of the neural network approach to weld process modeling. Andersen also presents a detailed comparison of neural network modeling to two analytical models and a statistically based multidimensional parameter interpolation approach.

Control

Practical Considerations

The easiest approach to controlling multiple weld process parameters can be realized if input variables can be found that affect only a single output quantity. If the output variable is affected by another input variable as well, then one may be the primary variable while the other may constitute a secondary feedback loop that is capable of controlling the output quantity by a relatively small amount with respect to the basic level set by the primary variable. For example, high-frequency pulsation of the current in GTAW may provide a means of controlling the depth of penetration over a small range without affecting the width of the weld bead. In this case the heat input, as determined by the voltage, current, and travel speed, would be the primary input variable controlling the width and penetration, while the high-frequency pulsation would be the secondary variable capable of producing small corrections to the basic penetration depth.

Even for single-variable weld process control, nonlinearities in the process may call for an adaptive system to automatically adjust the parameters of the controller when the process parameters and disturbances are unknown or change with time. For example, Bjorgvinsson [1992] shows that a simple automatic voltage control (AVC) system may be unstable over a wide range of current settings because of the variation of the arc sensitivity (voltage change per unit change of arc length) with current. A simplified schematic of an AVC system is shown in Fig. 97.3. The arc voltage (proportional to the arc length) is compared with a reference voltage in a simple position servo. If an error exists between the reference voltage and the arc voltage, the servo motor moves the welding torch up or down to reduce the error to zero. If K_a is the gain of the AVC motor drive system and K_s is the arc sensitivity ($K_s = dV_{arc}/dL_{arc}$), then the overall loop gain K is given by $K = K_a K_s$. The closed-loop stability of the position control system is dependent on the loop gain and will obviously vary from its design setting if K_s changes. Bjorgvinsson shows that for helium shielding gas,

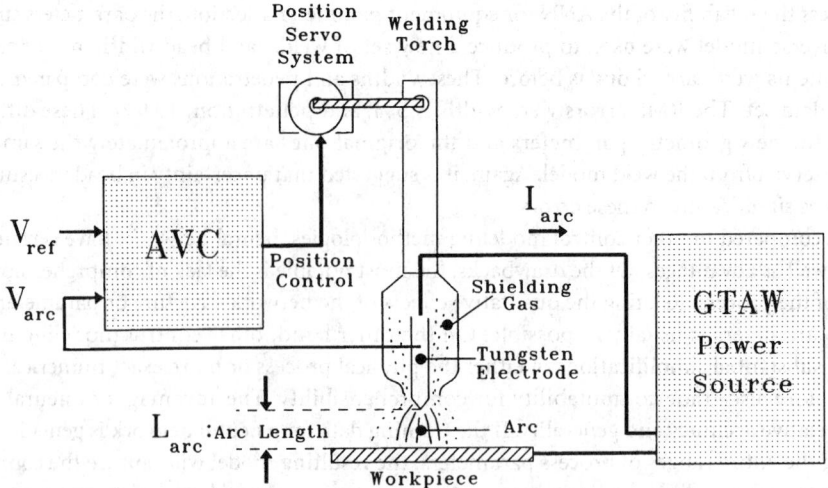

FIGURE 97.3   Simplified gas tungsten arc welding setup.

the arc sensitivity may vary by approximately a 5:1 ratio over a current range of 15 to 150 A. In this case, for a standard proportional controller, the overshoot to a step input at 15 A is approximately 40% if the controller gain K_a is fixed and set for optimum response at 150 A. Bjorgvinsson proposes a gain-setting adaptive controller (see Fig. 97.4) to vary the controller gain in such a manner as to compensate for the changing arc sensitivity for all levels of welding current. Knowing the arc current, the adaptive controller uses information stored in a look-up table or computed from a mathematical model of the arc to adjust K_a in response to changes in K_s such that the product $K_aK_s = K$ is maintained constant independent of the current. The result is uniform closed-loop stability characteristics of the AVC system throughout the complete weld. This includes the up-slope period, when the current is varied from the low arc-initiation value to the nominal welding current, which is maintained until the down-slope period, when the current is brought back to a low value for termination of the arc.

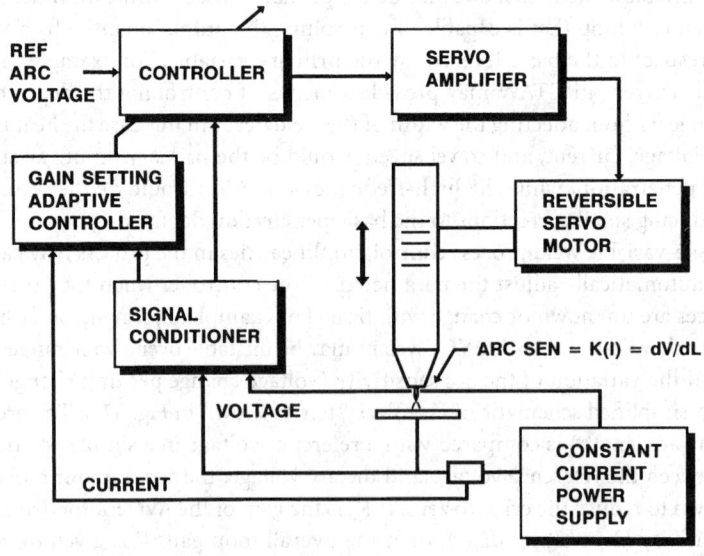

FIGURE 97.4   Gain-setting adaptive automatic voltage control.

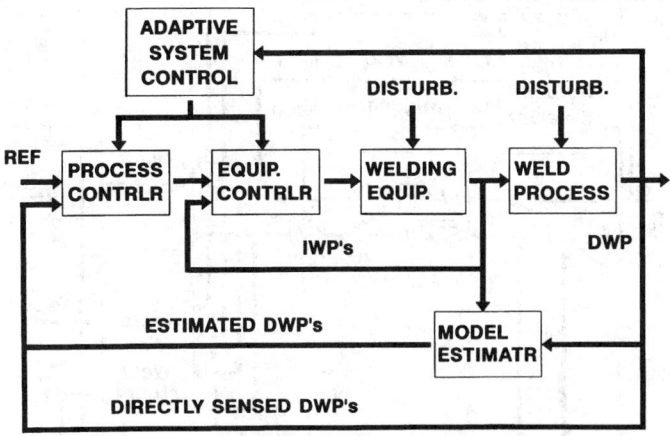

FIGURE 97.5   Multivariable adaptive weld process control system.

General Approaches to Multivariable and Adaptive Weld Process Control

The welding process is generally nonlinear, and the different variables are normally coupled. If we can assume localized linearity, then adaptive control techniques can be used to change the controller characteristics in response to changes in the operating domain. To handle the multivariable control problem, we attempt to decouple the process input–output variables by appropriate controller design in order to reduce the system to a set of essentially noninteracting loops. Controller design can then be carried out using single-loop techniques.

Necessary and sufficient conditions have been derived for decoupling a multivariable system. Unfortunately, the conditions are, in general, unlikely to be satisfied in practice because of model approximations, measurement uncertainties, parameter perturbations, and other causes. Therefore, system decoupleability may be inhibited by constant compensation techniques. In these situations, it is more appropriate to decouple the system in real time using an adaptive controller. It has been shown that such an adaptive controller can be expected to eventually achieve exact decoupling after the system parameters have converged.

A general multivariable adaptive direct weld process control system is shown in Fig. 97.5. It will frequently be the case that not all of the DWP that we wish to control can be directly sensed with available sensors. In this case, we may estimate the DWP(s) that we cannot measure and use the estimated values for feedback information. Control of these parameters will obviously not be any better than the model used to estimate them. However, the model may be continuously tuned, i.e., calibrated, from both the IWP and those DWP that are directly sensed.

Cook *et al.* [1991] have described a multivariable weld process control system that makes use of a model to estimate one of two DWP(s) controlled. The system, shown in Fig. 97.6, was configured to accept weld bead width and weld penetration as its two inputs. The system used width sensing, but penetration was only available as an estimate from the forward process model acting in parallel to the actual process. Conventional time-based up-sloping/down-sloping was used for weld initiation and termination, so an inverse process model was used to provide initial weld IWP(s) (following up-slope) to the weld start sequencer. Referring to Fig. 97.6, the desired bead width and penetration are specified by the user as W_o and P_o, respectively. These parameters, as well as the workpiece thickness H, are routed to a neural network setpoint selector (inverse process model), which produces the nominal travel speed, current, and arc length (v_o, I_o, and L_o, respectively). Arc initiation and stabilization are controlled in an open-loop fashion by the weld start sequencer. Given the desired equipment parameters, the arc is typically initiated and established at a relatively low current, with the other equipment parameters set at some nominal values. Once the arc has

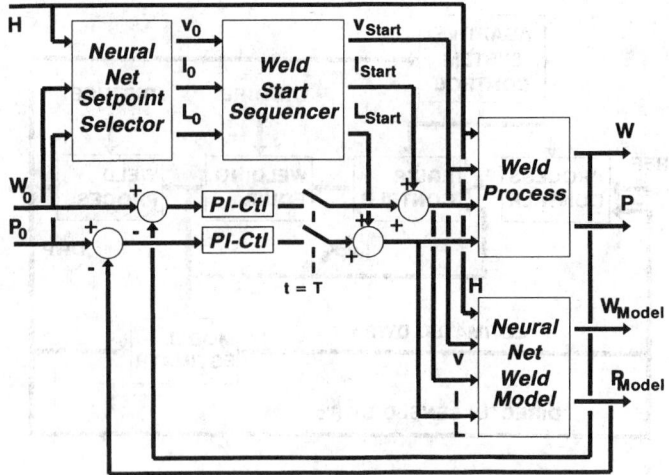

FIGURE 97.6   Closed-loop weld process control system.

been established, the equipment parameters are ramped to the setpoint values specified by the neural network. When the setpoint values have been reached, at time $t = T$, the closed-loop process control is enacted. As stated previously, the bead width from the process was monitored in real time, while a real-time penetration sensor was not used. Therefore, a second neural network (forward process model) is run in parallel with the process to yield estimates of the penetration. The measured bead width and the estimated penetration are subtracted from the respective reference values, processed through proportional-plus-integral controllers, and added to the final values obtained from the setpoint sequencer. When a workpiece thickness variation is encountered in the process, the system adjusts the current and the arc length accordingly to maintain constant bead geometry.

To demonstrate the multivariable weld process control system Cook *et al.* report an experiment using mild steel for the workpiece material. Plates of two thicknesses, 3.175 and 6.35 mm, were joined together, and a bead-on-plate weld using the nominal parameters ($I = 100$ A, $L_{arc} = 2.54$ mm, $v = 2.54$ mm/s) was made across the boundary between the plates, from the thicker section to the thinner one. The bead width and penetration were 3.6 and 0.9 mm, respectively, on the thicker plate. With the controller disabled (equipment parameters maintained constant), the bead width increased to 4.0 mm and the penetration increased to 1.2 mm when the weld pool entered the thinner plate. With the controller enabled, the width and penetration were maintained the same on the thin plate as they were on the thick plate with only a slightly discernible transient.

Intelligent Control

Practical weld process control implementation, particularly with multivariable and adaptive control, involves a substantial body of heuristic knowledge concerning the weld process and the numerous constraints that are involved in its control. The role that intelligent control concepts can play is to provide a systematic approach to dealing with these constraints.

For example, for a given set of material parameters, one may wish to control several geometrical parameters plus cooling rate for the GMAW process, while maintaining operation in the spray transfer mode of the process. Because of the close coupling among the equipment, material, and geometric parameters, and because of the small latitude of permissible variation of one parameter once the others are specified, tight constraints on the control system will be necessary to achieve the desired process quality.

It will be desirable to specify degrees of control permitted over the various parameters in terms of a hierarchy of parameter importance. For example, while the wire feed rate has an influence on bead width in the GTAW process, it would not be desirable to allow the wire feed rate to be varied excessively as a means of controlling bead width. Further, the allowable variation of a given parameter, or parameters, may not be symmetrical about the desired set point. Again, for the GTAW process, an increase in current may be partially offset by an increase in travel speed, whereas a reduction in both parameters would tend to more rapidly force the geometrical parameters outside the desired range.

Consideration of the process dynamics is also necessary, particularly for successful control during the initiation and termination phases of the overall welding operation. In addition to the hierarchical considerations referred to above, the time sequence and rate of change of each parameter should be considered. Intelligent control concepts may be used to handle these practical control issues in a formal and logical manner.

Conclusions

Rapid advances have occurred in the development of sensors and in the development of both steady-state and dynamic models suitable for real-time weld process control applications. In combination with multivariable, adaptive control theory methods, the tools are becoming available for significant progress in multivariable, direct weld process control. Long-range efforts will focus on combining process modeling and microstructural evolution modeling for eventual control of both macro and micro parameters.

Defining Terms

Direct weld parameters (DWP): A collection of parameters that characterize the weld in terms of the weld reinforcement and fusion zone geometry, mechanical properties, weld microstructure, and discontinuities.

Electron beam welding: A welding process that produces coalescence of metals with the heat obtained from a concentrated beam composed primarily of high-velocity electrons impinging on the surfaces to be joined.

Electroslag welding: A welding process that produces coalescence of metals with molten slag that melts the filler metal and the surfaces of the parts to be joined.

Gas metal arc welding (GMAW): A welding process that produces coalescence of metals by heating them with an arc between a consumable filler metal electrode and the parts to be joined. The process is used with shielding gas and without the application of pressure.

Gas tungsten arc welding (GTAW): A welding process that produces coalescence of metals by heating them with an arc between a nonconsumable tungsten electrode and the parts to be joined. The process is used with shielding gas and without the application of pressure. Filler metal may or may not be used.

Indirect weld parameters (IWP): A collection of parameters that establish the welding equipment setpoint values. Examples include voltage, current, travel speed, electrode feed rate, travel angle, electrode geometry, focused spot size, and beam power.

Laser beam welding (LBW): A welding process that produces coalescence of materials with the heat obtained from the application of a concentrated coherent light beam impinging on the surfaces to be joined.

Oxyacetylene welding: An oxyfuel gas welding process that produces coalescence of metals by heating them with a gas flame obtained from the combustion of acetylene with oxygen. The process may be used with or without the application of pressure and with or without the use of filler metal.

Thermit welding: A welding process that produces coalescence of metals by heating them with superheated liquid metal from a chemical reaction between a metal oxide and aluminum, with or without the application of pressure.

Variable polarity plasma arc welding (VPPAW): A welding process that produces coalescence of metals by heating them with a constricted variable polarity arc between an electrode and the parts to be joined (transferred arc) or between the electrode and the constricting nozzle (nontransferred arc). Shielding is obtained from the hot, ionized gas issuing from the torch as well as from a normally employed auxiliary shielding gas source. Pressure is not applied, and filler metal may or may not be added.

References

K. Andersen, *Studies and Implementation of Stationary Models of the Gas Tungsten Arc Welding Process*, M.S. Thesis, Vanderbilt University, 1992.

K. Andersen, G. E. Cook, Y. Liu, D. S. Mathews, and M. D. Randall, "Modeling and control parameters for GMAW, short circuiting transfer," in *Advances in Manufacturing Systems Integration and Processes*, D. A. Dornfeld, Ed., Dearborn, Mich.: Society of Manufacturing Engineers, 1989.

J. B. Bjorgvinsson, *Adaptive Voltage Control in Gas Tungsten Arc Welding*, M.S. Thesis, Vanderbilt University, 1992.

G. E. Cook, "Feedback and adaptive control in automated arc welding systems," *Metal Construction*, vol. 13, no. 9, pp. 551–556, 1981.

G. E. Cook, "Robotic arc welding: Research in sensory feedback control," *IEEE Transactions on Industrial Electronics*, vol. IE-30, no 3, pp. 252–268, 1983.

G. E. Cook, K. Andersen, and R. J. Barnett, "Feedback and adaptive control in welding," in *Recent Trends in Welding Science and Technology*, S. A. David and J. M. Vitek, Eds., Metals Park, Ohio: ASM International, 1990, pp. 891–903.

G. E. Cook, K. Andersen, R. J. Barnett, and J. F. Springfield, "Intelligent gas tungsten arc welding control," in *Automated Welding Systems in Manufacturing*, J. Weston, Ed., Cambridge, England: Abington Publishing, 1991.

T. W. Eagar, "The physics and chemistry of welding processes," in *Advances in Welding Science and Technology*, S. A. David, Ed., Metals Park, Ohio: ASM International, 1986, pp. 291–298.

J. F. Lancaster, *The Physics of Welding*, New York: Pergamon Press, 1986.

Y. Liu, *Metal Droplet Rate Control for Gas Metal Arc Welding*, Ph.D. Dissertation, Vanderbilt University, 1991.

R. W. Richardson, A. Gutow, R. A. Anderson, and D. F. Farson, "Coaxial weld pool viewing for process monitoring and control," *Welding Journal*, vol. 63, no. 3, pp. 43–50, 1984.

M. E. Shepard, *Modeling of Self-Regulation in Gas-Metal Arc Welding*, Ph.D. Dissertation, Vanderbilt University, 1991.

Y. C. Yi, *Weld Pool Vibration Analysis in Gas Tungsten Arc Welding*, M.S. Thesis, Vanderbilt University, 1991.

Further Information

Other recommended reading on welding technology, welding processes, and welding automation and control includes *Welding Handbook, Volume 1—Welding Technology* (American Welding Society, Miami, 1987), *Welding Handbook, Volume 2—Welding Processes* (American Welding Society, Miami, 1991), *Advances in Welding Science and Technology* (edited by S. A. David ASM International, Metals Park, Ohio, 1986), *Recent Trends in Welding Science and Technology* (edited by S. A. David and J. M. Vitek, ASM International, Metals Park, Ohio, 1990), *Developments in Mechanised and Robotic Welding* (edited by G. R. Salter, The Welding Institute, Cambridge, England, 1980),

Modeling and Control of Casting and Welding Processes (edited by S. Kou and R. Mehrabian, The Metallurgical Society, Inc., Warrendale, Penn.), *Developments in Automated and Robotic Welding* (edited by D. N. Waller, The Welding Institute, Cambridge, England, 1987), *Developments and Innovations for Improved Welding Production* (The Welding Institute, Cambridge, England, 1983), *Automated Welding Systems in Manufacturing* (Abington Publishing, Cambridge, England, 1991), and *Robotic Welding* (edited by J. Lane, IFS Publications Ltd., Bedford, England, 1987).

97.2 Large Drives

Alan K. Wallace and René Spée

A drive is a system that converts electrical energy into useful, controlled, mechanical work. As such, it is a vital component in many industrial processes. The adjustable speed and torque of drives, in contrast to the typically uncontrolled values obtainable directly from most electrical motors, have been made possible by the introduction of high-power electronic devices operating in switching modes. Appropriate selection, installation, and operation are essential for the process effectiveness and energy efficiency necessary for industrial competitiveness.

Drives may be considered as consisting of three major subsystems: the motor or machine, which converts electrical energy to the required driving torques over specified speed ranges; the converter, which processes the electrical energy, received from the utility at constant voltage and frequency, into the forms required by the motor; and the controller, which adjusts the operation of the converter based on performance requirements and comparison with measured signals of actual performance. These three subsystems are interlinked by a communications subsystem as shown in Fig. 97.7.

Although the demarcation between large and small drives is somewhat subjective, in general, devices such as positioning actuators and machine tools are examples of small drives, whereas large drives are applied to loads such as pumps, compressors, and bulk material processing. The rating of a large drive is expressed in hundreds or thousands of kilowatts. The supplies for these drives are three-phase power obtained from the utility system at medium or high voltages.

An advanced contemporary industrial drive is the result of an integration of several continually evolving technologies. In machines, improvements in the materials for magnetic circuits and electrical insulation enable higher specific ratings (i.e., better rating per unit mass or volume). In converters, the development of higher power and faster switching semiconductor devices increases ratings and enables more sophisticated operational techniques. In controllers, incorporation of faster, more powerful microprocessors enables the use of adaptive control techniques with such features as self-diagnostics and automatic setup. Many significant developments in these areas are described in compilations of technical papers [Bose, 1981] and appropriate texts [Bose, 1986].

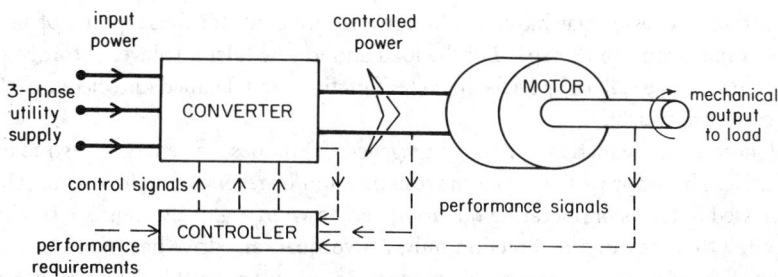

FIGURE 97.7   Typical drive system.

Configurations

In contrast to small drives and servosystems in which many diverse forms of both **direct current (dc) motors** and alternating current (ac) motors are found, large drives are dominated by only four distinct motor types: separate (or shunt) field dc motors (DCM), cage-rotor **induction motors** (CRIM), wound-rotor (or slip-ring) induction motors (WRIM), and **synchronous** (dc field) **motors** (SM).

For adjustable operation the DCM requires a controllable dc source that can be provided by either an ac-to-dc converter, such as a controlled **rectifier** (CR), or a dc-to-dc converter, known as a chopper. The latter is not common in industrial drives, being more appropriate for vehicle traction, and, consequently, will not be considered here. The three ac machines require ac-to-ac converters with frequency adjustability. This is produced by voltage source **inverters** (VSI), current source inverters (CSI), machine commutated inverters (MCI), and **cycloconverters** (CYCLO). Although other combinations may be found in some cases, Table 97.1 summarizes the more commonly used drive configurations. In certain cases the converters do not operate to control the main power supply to the machine but, as described later, perform a slip energy recovery (SER) function. Details of the form and construction of these converters, motors, and drives can be found in appropriate texts [Gyugyi and Pelly, 1976; Sen, 1981; Leonard, 1985].

Table 97.1 Drive Component Combinations

	CR	VSI	CSI	MCI	CYCLO
DCM	X				
CRIM		X	X		
WRIM			X	X	X
SM				X	X

Selection and Compatibility

An appropriately applied drive first must meet the shaft torque range and speed range of its load. From these, the appropriate motor type, number of poles, and (for ac machines) the frequency range can be selected. This selection is based on two basic equations that relate motor armature current (i), supply frequency (f), air gap flux density (B), number of poles (P), angular shaft synchronous speed (ω_s), and shaft torque (T):

$$T \propto PBI \tag{97.1}$$

$$\omega_s \propto \frac{f}{P} \tag{97.2}$$

From the products of torque and speed the motor (output) rating is derived. Large machines have good efficiencies (greater than 95%) and good power factors at rated operating conditions. Consequently, the output ratings of the converters are not substantially higher than those of the motors that they operate. Figure 97.8 shows areas typical of drive system operation; these result from a combination of physical limitations and economic considerations. Figure 97.8 should be interpreted in conjunction with Table 97.1 while noting that SER systems are a special case of WRIM operation. Certain processes may have, in addition, requirements for the response of a drive to follow changes of the torque and/or speed of the load and for the tolerable level of torque pulsations. These requirements may call for special controller functions and detailed knowledge of the interaction of motor and converter.

Electrical motors can be made to operate in *regenerative* modes, i.e., energy is extracted from the load by the drive. This improves the dynamic response and/or reversing performance. This requirement is expressed in terms of operating *quadrants* as shown in Fig. 97.9. Hence, a single-quadrant drive is required to motor in one direction only. A two-quadrant drive has to motor and brake in one direction. A four-quadrant drive has to be regenerative and reversible. The number of required quadrants is reflected in the complexity of the converter.

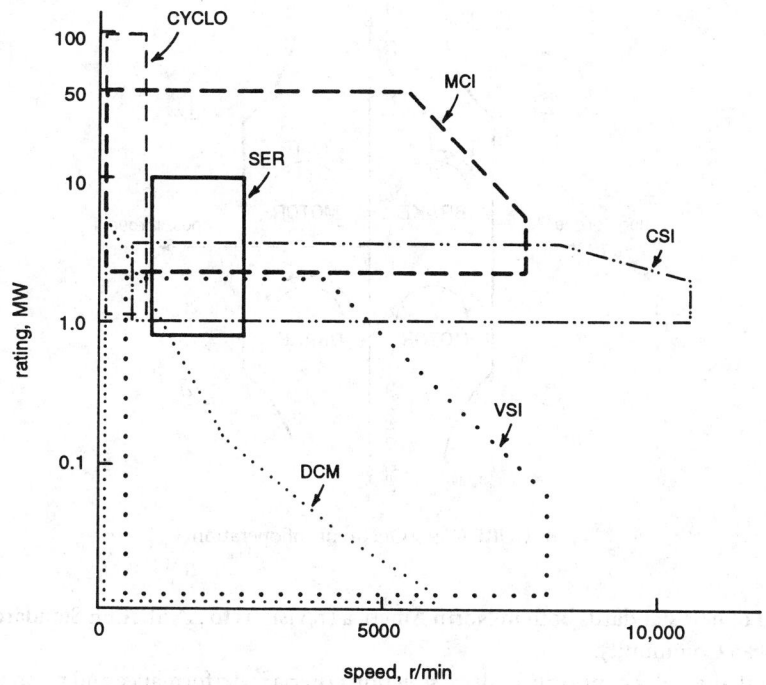

FIGURE 97.8 Classification of drives by rating.

The power and speed envelopes of Fig. 97.8 show considerable areas of overlap. Drive selection in these cases is generally based on required response, the operational environment, and economic considerations. For example, dc motors are larger, more complex and vulnerable, and more costly than their equivalently rated ac counterparts. Depending upon the operational quadrants required, however, a controlled rectifier is substantially cheaper than an inverter. It follows that, in many cases, a dc motor system is more economical than an induction motor equivalent. In damp, dirty, corrosive, or explosive environments, however, the simplicity and robustness of the induction motor makes it preferable for purely practical reasons.

The effects of a drive on its environment are significant in the selection and design process. Converters that are called upon to switch very large currents, hundreds or thousands of amps, at frequencies up to several kilohertz produce serious magnetic fields around the devices themselves and their cables or leads. The *electromagnetic compatibility* (EMC) issue must be addressed to ensure that other equipment, such as controllers and computers, is not adversely affected by the operation of the drive. In addition, power electronic converters present nonresistive, nonlinear loads to the power supply system. Consequently, the currents drawn can be of poor power factor and high *total harmonic distortion* (THD), which is defined in terms of the fundamental and harmonic components of current as

$$\text{THD} = \frac{\sqrt{\sum_{\text{all harmonics}} I_{\text{harmonic}}^2}}{I_{\text{fundamental}}} \cdot 100\% \tag{97.3}$$

Significant THD can result in financial penalties being imposed on the drive/operator by the supplying authority and cause overheating of adjacent equipment. Moreover, *power quality* issues are

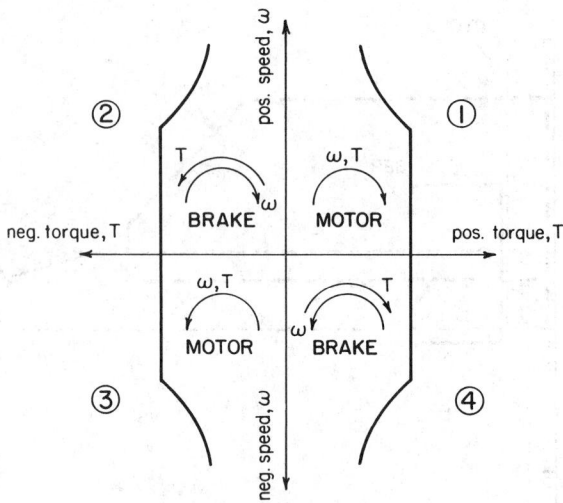

FIGURE 97.9   Quadrants of operation.

the subject of new standards both in North America (revisions to ANSI-IEEE Standard 519) and in the European Community.

In general, the order of priority of drive selection criteria is performance and response; operating environment; power factor and THD; EMC; economics.

Principles and Features of Operation

Before the introduction of power semiconductors, both induction motors and synchronous motors were effectively fixed-speed machines, except where highly expensive rotary frequency conversion sets could be justified. Under these conditions the DCM was traditionally the basis of adjustable speed drives. Figure 97.10 shows schematically the two major components of a dc motor: the armature (rotating) and the field winding (stationary). In a DCM the armature current reacts with the air gap flux produced by the field to develop torque in accordance with Eq. (97.1). For a given constant field winding current, if the armature current is maintained at the rated value, the motor will develop rated torque at all speeds. However, the applied voltage (V) must overcome the internal voltage of the armature (E), which is given by

$$E \propto \omega_r B \qquad\qquad (97.4)$$

where ω_r is the actual speed of the motor. Hence V must be increased to increase motor speed. When the limit of the applied voltage is reached, the motor speed can only be increased further by

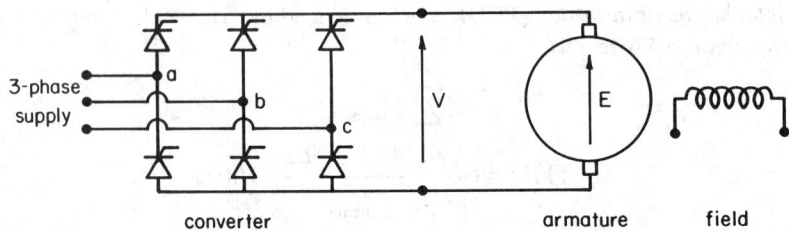

FIGURE 97.10   Schematic of dc motor drive.

reducing the air gap flux to maintain the armature voltage in accordance with Eq. (97.4). This is done by reduction of the field winding current in the field weakening mode of operation. The result is a decreasing torque in accordance with an approximately constant power curve, as shown in the single-quadrant torque-speed characteristic of Fig. 97.11.

The three-phase thyristor bridge converter shown in the schematic of Fig. 97.10 will produce the output voltage waveform shown in Fig. 97.12, which can be shown to produce a mean (dc) voltage of

$$V = \frac{3\sqrt{2}}{\pi} V_L (1 - \sin \delta) \tag{97.5}$$

in which V_L is the rms line voltage of the ac supply and δ is the delay angle. At higher output voltages (i.e., small δ), the ripple content is small and the armature current is constant dc. This causes virtually rectangular current pulses at the three-phase input terminals of the rectifier, a high THD condition. Increasing δ of the rectifier decreases the voltage applied to the armature. In consequence, the conduction periods of the rectifier shift with respect to the ac supply voltages. Thus, at low power levels, in addition to high THD, the displacement power factor is low. When the applied armature voltage (V) is reduced below the internal voltage (E), with the motor in motion, the second quadrant (braking operation) is entered. In order to achieve four-quadrant operation, either a changeover switch (to reverse the polarity of the armature connections to the rectifier output) or a second converter (with thyristors connected in the opposite sense) is required to enable the required current reversal.

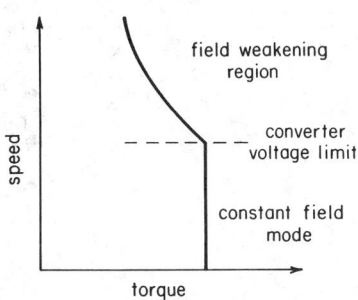

FIGURE 97.11 Controlled operation of dc machine.

The operating speed of a CRIM is best adjusted by control of the terminal supply frequency, in accordance with Eq. (97.2) with a slight adjustment for the operating slip (i.e., the small difference between the synchronous speed, ω_s, and the rotor speed, ω_r)

$$\text{slip} = \frac{\omega_s - \omega_r}{\omega_s} \tag{97.6}$$

A basic induction machine drive system is shown in Fig. 97.13. The operation of the motor at constant slip over a range of controlled frequencies can be represented by considering operation at a number of discrete frequencies (f_1 to f_6) as shown in Fig. 97.14. For each applied frequency the machine assumes operation at the given slip resulting in the operating points (m_1 to m_4) for a constant load torque. Except at low speeds (where the resistance predominates), the impedance of the machine is effectively controlled by the inductive reactance, which is proportional to the applied

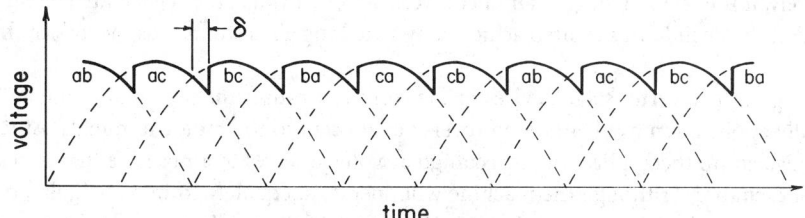

FIGURE 97.12 Phase controlled rectifier voltage.

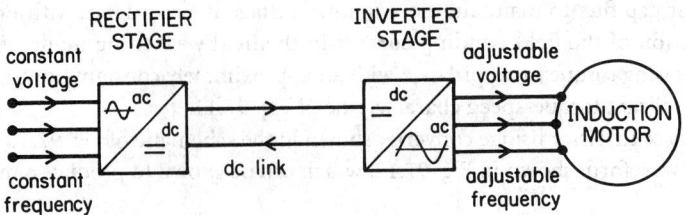

FIGURE 97.13   Induction motor drive.

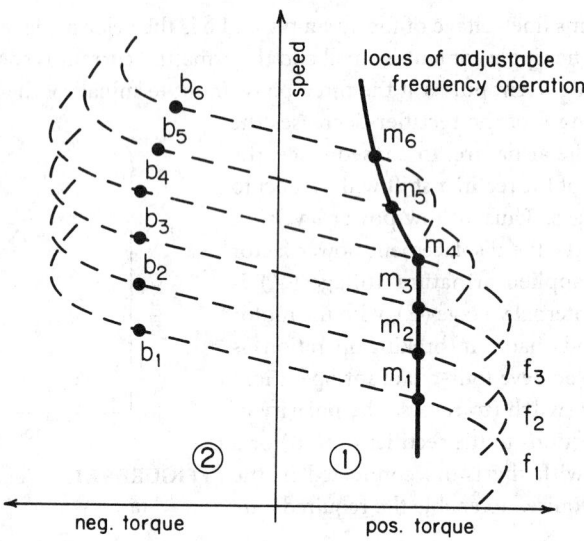

FIGURE 97.14   Development of induction motor drive torques.

frequency. Hence, in order to maintain rated motor current, the voltage must be increased following a constant volts per hertz ratio. However, above a certain frequency, the output voltage of the inverter becomes limited by the dc link voltage developed from the input rectifier. Rated motor current cannot be maintained, and the resultant torque is reduced to typical operating points (m_5 and m_6). The loci of the operating points form a torque-speed characteristic similar to that shown in Fig. 97.11 for the dc drive.

Braking operation of an induction machine drive can be obtained by observing the rotor speed and exciting the machine at a frequency that produces a negative slip, i.e., operating points b_1 to b_6 in Fig. 97.11 result from this strategy. Under these conditions, however, the inverter stage of the converter rectifies the output of the motor. This increases the voltage of the dc link to a level above the normal output of the rectifier stage. If the rectifier has controllable devices, it can be made to invert the energy in the dc link to utility frequency and hence return it to the three-phase supply. Alternatively, if the rectifier stage is an uncontrolled diode bridge, the regenerated energy must be dissipated in the dc link; this is often achieved by switching a resistor across the link in the braking mode.

Switching of the inverter stage devices of the converter causes the potential of the dc link to be sequentially applied, removed, and then reverse connected to the motor terminals. At it simplest, this is equivalent to the application of rectangular voltage waves to a machine that is designed for sinusoidal excitation. Although the machine will operate adequately from rectangular, or overlapping, step-wave excitation, the high harmonic content of the resulting currents cause additional losses in the motor, resulting in a performance derating. In very large drives, where line commutated thyristors are needed to handle the power, or in more moderate-sized drives at high speeds, where the commutation (switching) losses in the semiconductors prevent more sophisticated modes of operation, step-wave excitation may be unavoidable. However, increased ratings of gate-turn-off thyristors (GTO) and the development of MOS-controlled thyristors (MCT) are making voltage modulation techniques available to larger drives. Unlike a regular thyristor, which requires either a natural or forced current zero for turn-off, the more advanced devices can be controlled by relatively small gate (or firing) pulses. This enables numerous commutations during one period of the fundamental frequency. Figure 97.15(a) shows the voltage waveform produced by applying the technique known as *pulse-width modulation* (PWM). Apart from the fundamental, the lowest- order harmonics of this function appear in a sideband around the modulation frequency. The resulting current is much closer to a sinusoid, as shown in Fig. 97.15(b), because high-frequency components are attenuated by the predominantly inductive nature of the motor impedance. Although PWM techniques reduce unnecessary losses in the motors, the higher frequencies may excite mechanical resonances in the audio frequency range. Thus, the motor may become a source of acoustic noise, which, depending on application and existing environment, may be of concern.

For reasons of manufacturability and operational efficiency, the largest induction machines are of the wound-rotor (WRIM) type. These can be, and often are, controlled in the same manner as the CRIM just described. However, access to the rotor circuits via slip-rings enables the alternative form of control known as *slip-energy recovery*, as shown in Fig. 97.16. The advantages of SER are in reduced size and cost of the converter if the required speeds do not extend greatly from the natural synchronous speed of the motor. This is often the case for large drives.

For very large drives, the *cycloconverter* replaces the inverters as the most appropriate converter in either the stator controlled or SER configuration. Cycloconverters develop the adjustable frequencies required by directly forming approximations to ac waveforms from segments of all the

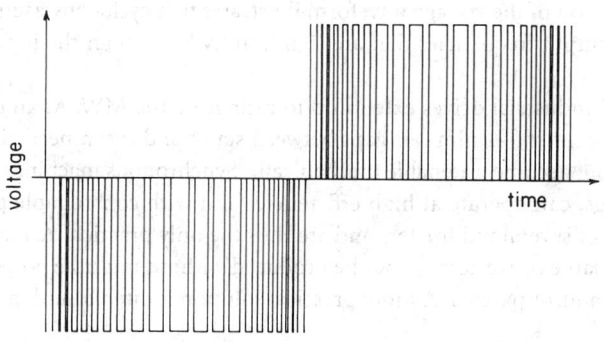

FIGURE 97.15(a)   PWM line-to-line voltage.

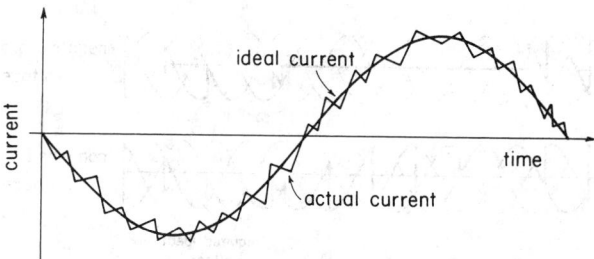

FIGURE 97.15(b)   Motor current due to PWM excitation.

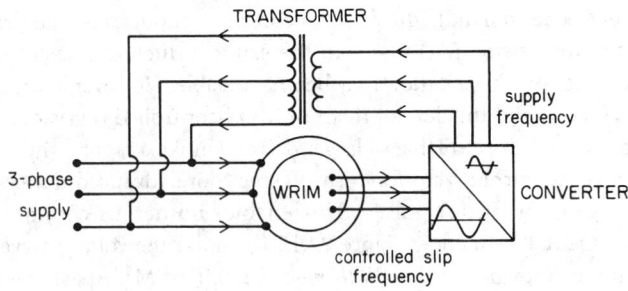

FIGURE 97.16   Slip energy recovery drive.

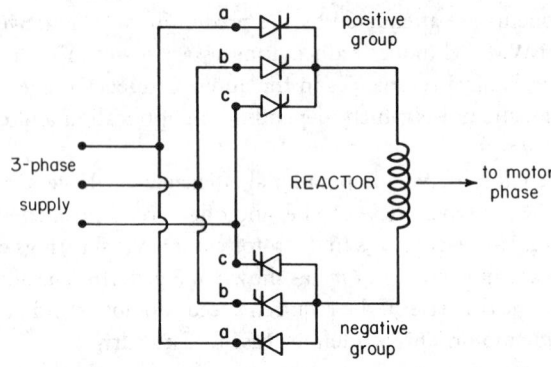

FIGURE 97.17   One phase of cycloconverter.

phases of the supply. Hence, each phase of the input supply needs to be connectable to every phase of the machine with both positive and negative polarities. Figure 97.17 is a schematic of a single phase of a CYCLO power circuit, and a typical voltage waveform development is shown in Fig. 97.18. Examination of the voltage waveform illustrates that cycloconverters are only appropriate for generating output frequencies that are significantly lower than the input supply frequency (typically, $f_{out,max} \approx \frac{1}{3} f_{supply}$).

The largest of all industrial drives extend up to ratings of 100 MW. At an order of magnitude below this rating the short (~1 mm) air gaps between stator and rotor, needed for efficient induction motor operation, become untenable mechanically. Synchronous machines, with dc rotor fields excited via slip rings, can operate at high efficiencies and with controllable power factors while employing air gaps of several millimeters and are thus the only practical ac machine for very large drives. In addition, large converters cannot be produced without multiple power electronic devices connected in series and/or parallel. A more practical solution is often found in the parallel connec-

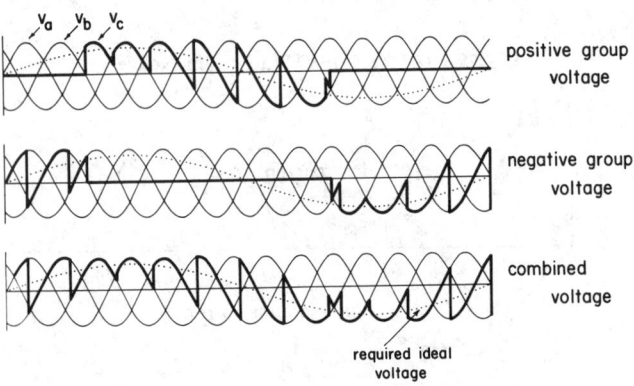

FIGURE 97.18   Cycloconverter output waveforms.

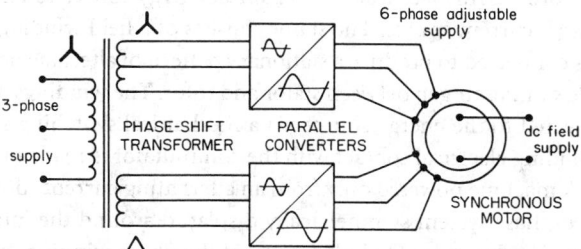

FIGURE 97.19   Very large synchronous motor drive.

tion of whole converters. If parallel converters are justified, parallel motor windings, arranged in a six-phase configuration, can be useful for purposes of isolation and reduced criticality of controls. A typical very large drive is shown schematically in Fig. 97.19. The operation of this and alternative configurations is described in the literature [Stemmler, 1991].

Control Aspects

A comprehensive description of control techniques is significantly beyond the scope of this chapter, but detailed coverage is available in the recommended literature. Almost without exception, large drives are both controlled and protected in response to performance measurements which provide signals for control loops. The type of control strategy, the type of control loop, and the relative importance of the particular loops for a given application depend on the performance requirements. For example, where rapid response to changes in the load torque and/or speed is needed, shaft speed will constitute the major feedback signal and *vector control* can enable an induction motor drive to respond as well as the more traditional dc motor system. For very large drives system inertia is such that dynamic response is not an issue. More likely, the optimization of specific performance parameters, such as efficiency, is of value to the user. For this application the predominant control loops will be based on current and/or power measurements working in self-optimization or other adaptive control strategies.

Future Trends

The most significant future developments in large drives are likely to result from improvements in, and the application of, more advanced power electronic devices. This will enable converter operation at higher ratings and higher frequencies. The most direct initial evidence of this will be the ever-increasing rating at which inverters replace cycloconverters.

The proposed revisions to ANSI/IEEE-519 concerning the tolerable harmonic current pollution levels of the supply will promote control strategies and converter topologies to replace expensive front-end filtering. Advanced control of inverter rectifier stages and new topologies such as *matrix converters* and *resonant converters* will be introduced in lieu of inverters and cycloconverters.

The present cadre of machine types will remain, although the trend away from dc motor drives will continue. For certain highly specialized applications, requiring extremely high efficiency and specific performance regardless of cost, synchronous motors using high-coercivity permanent magnet fields will be used. For work in severe environments and where high specific torque is needed, the switched reluctance motor system shows considerable promise [Greenhough, 1991].

Defining Terms

Cycloconverter: A system of power electronic devices that converts alternating current energy at a constant voltage and constant frequency to an output of adjustable voltage and adjustable frequency. The conversion is done directly without the intermediate direct current stage used in a rectifier/inverter combination.

Direct current (dc) motor: An electrical to mechanical energy conversion machine usually powered from a direct current source. The stator consists of a field winding system of a number of salient poles connected to produce a stationary pattern of alternate north and south polarity magnetic flux in the air gap between stator and rotor. The windings of the rotor (or armature) are connected to the energy source via a mechanical switching system, known as the commutator. Sliding electrical contact with the commutator is made by carbon brushes.

Induction motor: A machine powered only from an alternating current source. The stator windings are a three-phase system symmetrically displaced around the internal periphery. The combination of the physical spatial placement of the phase windings and the time delay or sequence of the currents flowing in them produces a magnetic field pattern of alternate north and south poles that rotates within the air gap. The rotor can take one of two forms: a system of high-current, short-circuited conductors called a squirrel cage or a three-phase winding system with terminals brought out via slip rings and brushes.

Inverter: A system of power electronic devices that converts direct current energy to alternating current energy by controlled sequential switching. Various control techniques have been developed to enable control of both the output frequency and output voltage.

Rectifier: A system of power electronic devices that converts alternating current energy to direct current energy. Two generic forms is common: the uncontrolled rectifier and the controlled rectifier, the output voltage of which can be adjusted. Most rectifiers contain filtering elements, such as series inductors or parallel capacitors, at their outputs to reduce the ripple of the terminal voltage.

Synchronous motor: A machine requiring both direct current and alternating current sources. The stator winding system is three-phase, similar to that of the induction motor. The rotor is a direct current system similar to the stator of the dc motor but with the mechanical freedom to rotate. Access to the rotor field winding is via slip rings and brushes.

References

ANSI/IEEE Standard 519-1992, *Guide for Harmonic Control and Reactive Compensation of Static Power Converters*, December 1992.

B. K. Bose, Ed., *Adjustable Speed AC Drive Systems*, New York: IEEE Press (Wiley), 1981.

B. K. Bose, *Power Electronics and AC Drives*, Englewood Cliffs, N.J.: Prentice-Hall, 1986.

P. Greenhough, *Switched Reluctance Drives for Applications in Hazardous Areas*, 5th International Conference on Electrical Machines and Drives (IEE 341), London: IEE, 1991, pp. 11–16.

L. Gyugyi and B. R. Pelly, *Static Power Frequency Changers*, New York: Wiley, 1976.

W. Leonard, *Control of Electrical Drives*, Berlin: Springer-Verlag, 1985.

P. C. Sen, *Thyristor DC Drives*, New York: Wiley, 1981.

H. Stemmler, *Large Converter Fed Synchronous Machines*, International Conference on the Evolution and Modern Aspects of Synchronous Machines, Swiss Federal Institute of Technology, Zürich, 1991, pp. B1–B7.

Further Information

The Institute of Electrical and Electronics Engineers (IEEE) has three publications reporting on power electronics, electric machines, and drives. The *Transactions on Industry Applications* is published bimonthly, while the *Transactions on Energy Conversion* and the *Transactions on Power Electronics* appear quarterly. The technical IEEE societies associated with these journals also sponsor semiannual or annual conferences. For information, contact IEEE Service Center, 445 Hoes Lane, P.O. Box 1331, Piscataway, NJ 08855-1331.

Other sources of information include the *Proceedings of the Institution of Electrical Engineers (IEE)* and the *European Power Electronics and Drives Journal* in Europe as well as the *Transactions of the Institute of Electrical Engineers of Japan*.

98

Man-Machine Systems

98.1 Introduction 2247
98.2 Several Natures of Man-Machine Control—A Catalog of
      Behavioral Complexities 2247
98.3 Full-Attention Compensatory Operations—The Crossover
      Model 2250
      Crossover Frequency for Full-Attention Operations • Remnant
      • Effects of Changes in the Task Variables • Effects of Divided
      Attention

Duane McRuer
Systems Technology, Inc.

98.1 Introduction

In principle the dynamic behavior of the human element in man-machine systems can be described in terms similar to those used to describe other system elements. There are, however, major complications in quantification because of the enormous versatility of the human engaged, simultaneously, as the on-going *architect* and modifier of the man-machine system itself and as an operating entity within that system. In other words, the adaptive and learning capabilities of the human permit both set-up and modification of the effective system structure and the subsequent self-improvement and tuning of the human dynamic characteristics within that structure.

The situations which are simplest to quantify are those in which the *machine* has time-stationary dynamic properties and the human has, after architectural, learning, and adaptation phases, achieved a similar state. Under these circumstances human dynamic operations can be characterized by quasi-linear describing functions and a remnant [Graham and McRuer, 1961] or operator-induced noise. This is the context here.

98.2 Several Natures of Man-Machine Control—A Catalog of Behavioral Complexities

Figure 98.1 [McRuer and Krendel, 1974] shows a general quasi-linear man-machine system with time-stationary properties. This diagram is suitable for the description of human behavior in an interactive man-machine system wherein the human responds to visually sensed inputs and communicates with the machine via a manipulator of some sort (e.g., control stick, wheel, pedal, etc.). This block diagram indicates the minimum needed number of major functional signal pathways internal to the human operator to characterize different behavioral features. The constituent human sensing, data processing, computing, and actuating elements are connected as internal signal processing pathways which can be "reconfigured" as the situation changes. Such reconfiguration is an aspect of human behavior as a system architect. Functional operations on internal signals within a given pathway may also be modified.

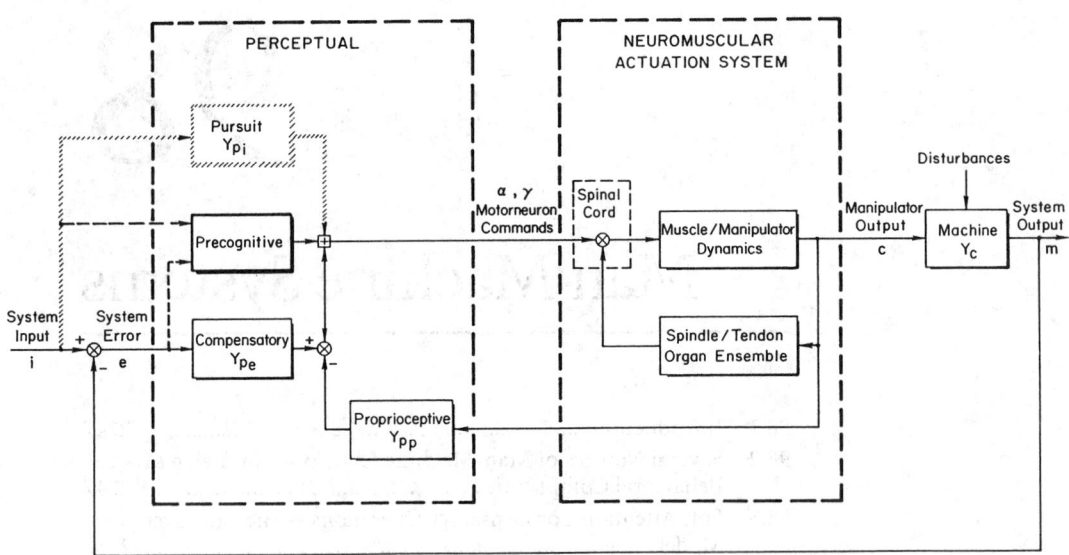

FIGURE 98.1 Major human operator pathways in a man-machine system.

The specific internal signal organizational possibilities depicted in Fig. 98.1 have been discovered by manipulating experimental situations (e.g., by changing system inputs and machine dynamics) to isolate different combinations of the specific blocks shown [McRuer and Jex, 1967; McRuer and Krendel 1974; McRuer 1980].

To describe the parts of the figure start at the far right with the *controlled element*. This is the machine being controlled by the human. To its left is the actual interface between the human and the machine—the neuromuscular actuation system, which is the human's output mechanism. This in itself is a complicated feedback control system capable of operating as an open-loop or combined open-loop/closed-loop system, although that level of complication is not explicit in the simple feedback control system shown here. In the diagram the neuromuscular system comprises limb, muscle, and manipulator dynamics in the forward loop and muscle spindle and tendon organ ensembles as feedback elements. Again, many more biological sensors and other elements are actually involved; this description is intended only to be generally indicative of the minimum level of complexity associated with the *human actuation elements*. All of these elements operate within the human at the level from the spinal cord to the periphery.

There are other sensor systems, such as joint receptors and peripheral vision, which indicate limb output position. These operate through higher centers and are subsumed in the *proprioceptive* feedback loop incorporating a block at the perceptual level further to the left in the diagram. If motion cues are present, these too can be associated in similar proprioceptive blocks with feedbacks from the controlled element output.

The other three pathways shown at the perceptual level correspond to three different types of control operations on the visually presented system inputs. Depending on which pathway is effectively present, the control structure of the man-machine system can appear to be *open-loop*, or *combination open-loop/closed-loop*, or totally *closed-loop* with respect to visual stimuli.

When the *compensatory* block is appropriate at the perceptual level, the human controller acts in response to errors or controlled-element output quantities only. Only the Y_{pe} block "exists", with Y_{pi} and the precognitive block both equal to zero. With the compensatory pathway operational, continuous closed-loop control is exerted on the machine so as to minimize system errors in the presence of commands and disturbances. **Compensatory behavior** will characteristically be present when the commands and disturbances are random-appearing and when the only information displayed to the human controller consists of system errors or machine outputs. In the simple case where the describing function Y_{pe} is defined so as to account for the perceptual and neuromuscular components, the system is single-input/single-output, and the operator-induced noise is neglected, the closed-loop system output/input dynamics will be

$$\frac{m}{i} = \frac{Y_{pe}Y_c}{1 + Y_{pe}Y_c} \tag{98.1}$$

and the error/input

$$\frac{e}{i} = \frac{1}{1 + Y_{pe}Y_c} \tag{98.2}$$

Thus, for compensatory situations, the man-machine system emulates the classic single-input/single-output feedback system. The output can be made to follow the input and the error can be reduced only by making the open-loop describing function large compared to 1 over the operating bandwidth of the system.

When the command inputs can be distinguished from the system outputs by virtue of the display (e.g., i and m are shown or detectable as separate entities relative to a reference) or preview (e.g., as in following a curved course) the *pursuit* block in Fig. 98.1 comes into play and joins the compensatory. The introduction of this new signal pathway provides an open-loop control in conjunction with the compensatory closed-loop error correcting action. The output/input dynamics of the man-machine system will then become

$$\frac{m}{i} = \frac{(Y_{pi} + Y_{pe})Y_c}{1 + Y_{pe}Y_c} \tag{98.3}$$

and the error/input describing function is

$$\frac{e}{i} = \frac{1 - Y_{pi}Y_c}{1 + Y_{pe}Y_c} \tag{98.4}$$

With the pursuit system organization the error can be reduced by the human's operations in two ways: by making the open-loop describing function large compared with 1 and by generating a pursuit path describing function which tends to be the inverse of the controlled element. This can, of course, only be done over a limited range of frequencies. The quality of the overall control in the pursuit case can, in principle, be much superior to that where only compensatory operations are possible.

An even higher level of control is possible. When complete familiarity with the controlled element dynamics and the entire perceptual field is achieved, the highly skilled human operator can, under certain conditions, generate neuromuscular commands which are deft, discrete, properly timed, scaled, and sequenced so as to result in machine outputs which are almost exactly as desired. These neuromuscular commands are selected from a repertoire of previously learned control movements. They are conditioned responses which may be triggered by the situation and the command and control quantities, but they are not continuously dependent on these quantities. This pure open-loop programmed-control-like behavior is called **precognitive**. Like the pursuit pathway, it often appears in company with compensatory follow-up or simultaneous operations. This forms a dual-mode form of control in which the human's manual output is initially dominated by the precognitive action, which does most of the job, and is then completed when needed by compensatory error-reduction actions.

The above description of human action pathways available in man-machine systems has emphasized the visual modality. Similar behavior patterns can be exhibited to some extent in other modalities as well. Thus the human's interactions with machines can be even more extraordinarily varied than described here and can range completely over the spectrum from open-loop to closed-loop in character in one or more modalities.

98.3 Full-Attention Compensatory Operations— The Crossover Model

The compensatory pathways with manual control operations using the visual modality have been extensively studied. Thousands of experiments have been performed, and most of the adaptive features of human behavior associated with these kinds of operations are well understood. There are both classical control [e.g., McRuer and Krendel, 1974; and McRuer *et al.*, 1990] and optimal control [e.g., Baron and Kleinman, 1969; Kleinman *et al.*, 1970; Curry *et al.*, 1976; and Thompson, 1990] theoretical formulations available to predict steady-state and dynamic performance.

By far the simplest human behavioral "law" for compensatory systems is the *crossover model*. This states that, for a particular controlled element transfer function, Y_c, the human operator adopts a describing function, Y_{pe}, such that the open-loop man-machine transfer characteristics appear as

$$Y_p Y_c = \frac{\omega_c e^{-\tau j \omega}}{j\omega} \tag{98.5}$$

The two parameters in the crossover model are the crossover frequency, ω_c, and an effective pure time delay, τ. The model applies only in the immediate region of the crossover frequency. The typical data shown in Fig. 98.2 illustrate how well this relationship is obeyed for a variety of subjects and a particular controlled element. The agreement with the amplitude ratio is excellent over a broad range of frequencies. The phase agreement is good in the region of the crossover frequency, ω_c, but departs somewhat at lower frequencies. Figure 98.2 also shows the *extended crossover model*. Here the effects in the crossover region of a potentially large number of low-frequency lags and leads (in the machine and/or the operator) are represented by a phase contribution given by $\exp(-j\alpha/\omega)$. Here the time constant $1/\alpha$ is a lumped-constant representation of myriad low-frequency phase characteristics. It is an appropriate approximation *only* in the general region of crossover and is not intended to extend to extremely low frequencies.

Fundamentally, the crossover model states that the human's transfer characteristics will be different for each set of machine dynamics, but that the form of the composite total open-loop dynamics will be substantially invariant. The effective time delay in Eq. (98.5) is a low-frequency approximation to the combination of all manner of high-frequency pure delays, lags, and leads, including a component representing the effects of the neuromuscular actuation system reflected to the crossover region. It follows that the effective time delay, τ, is not a constant. Its two major components are (1) the effective composite time delay of the controlled element (including manipulator effects)— the sum of the machine's lags minus leads at frequencies well above crossover and (2) the high-frequency dynamics of the human operator approximated by a pure delay which has an equivalent phase shift at frequencies within

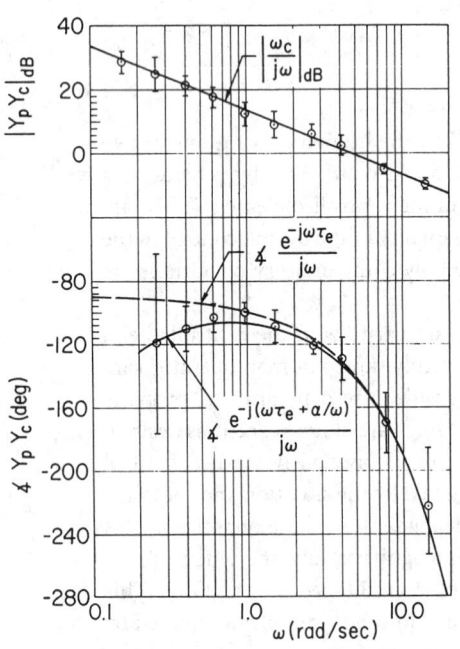

$$\left[\omega_c = 4.75 \text{ rad/sec}, \ \tau_e = 0.18 \text{ sec}, \ \alpha = 0.11 \text{ rad/sec} \right]$$

FIGURE 98.2 Data and crossover models for a simple rate-control-like controlled element.

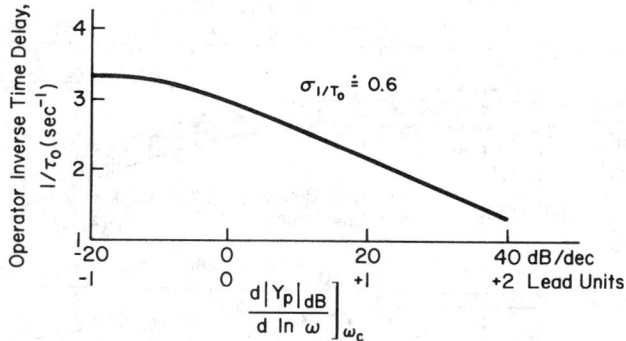

FIGURE 98.3 Variation of crossover model dynamic stimulus-response latency with degree of operator lead equalization.

the crossover region. The latter includes a minimum of 0.1 second for the neuromuscular system and an additional increment which depends on the amount of lead generation required of the human to offset the controlled element deficiencies in order to make good the crossover model form. Figure 98.3 [McRuer and Krendel, 1974] shows this variation for a wide range of controlled elements (the neuromuscular delay component is included). More refined estimates are available [e.g., McRuer *et al.*, 1990], but the above description is suitable for first-order estimates of behavior and dynamic performance.

Crossover Frequency for Full-Attention Operations

The crossover frequency tends to be constant for a given set of task variables (controlled-element form, inputs, disturbances, etc.). For example, as a controlled-element gain is changed, the human will change gain to compensate, resulting in the same crossover frequency. The maximum attainable crossover frequency, ω_u, will be

$$\omega_u = \frac{\pi}{2\tau} \tag{98.6}$$

This corresponds to zero phase margin. The nominal crossover frequency and associated pilot gain can be estimated from the condition to provide minimum mean-squared error in the presence of the appropriate form of continuous attention remnant. "Remnant" is operator-induced noise; as described below it depends on the nature of the operator's equalization and is larger when low-frequency lead is required to make good the crossover model. Thus, the need to generate lead impacts both the effective time delay and the remnant and, accordingly, the crossover frequency for which the minimum mean-squared error is obtained. The nominal crossover frequency for full-attention operations can be estimated [McRuer *et al.*, 1990] using

$$\omega_c/\omega_u$$

	ω_c/ω_u
No Operator Lead	0.78
Low-Frequency Operator Lead	0.66

Remnant

The second component of the operator's response is operator-induced noise or remnant. Remnant can, in principle, result from several sources, but in single-loop systems with ideal linear manipulator characteristics and no significant nonlinearities in the controlled element, the basic cause appears to be random time-varying behavior within the operator, which can be thought of as con-

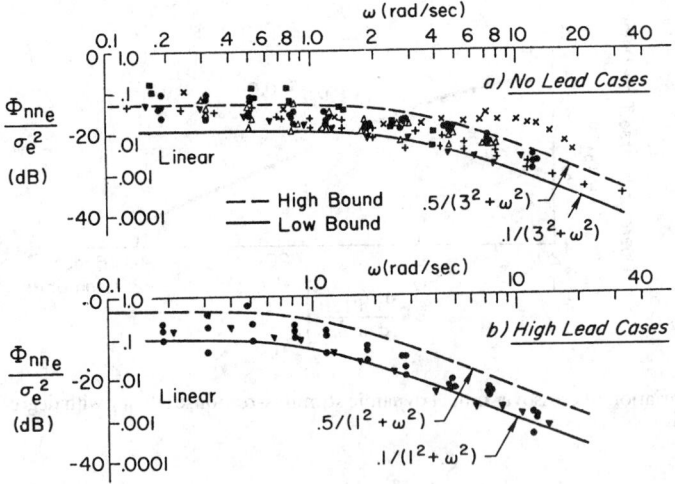

FIGURE 98.4 Normalized remnant spectra.

tinuous random fluctuations in the effective time delay. The remnant can be described as a continuous, relatively broadband, power spectral density. Fig. 98.4 provides a cross-section of remnant data from several sources. It is very important to note that the magnitude of the power spectral density scales approximately with the mean-squared error.

Effects of Changes in the Task Variables

The task variable which has the most important effect on the trained operator's behavior is the controlled element dynamics. Indeed, the natures of human adaptive changes in adjusting to the controlled element is the main thrust of the crossover model and remnant discussion above. More generally, task variables other than the machine dynamics, as well as environmental and operator-centered variables, can change operator gain, and hence crossover frequency, effective time delay, and remnant. Accordingly, ω_c and τ variations become quantification measures of changes or differences in the task, environmental, and operator-centered variables expressed directly in terms of the operator's control actions.

A common example is the reduction of crossover frequency when the amplitude of the command or disturbance signals are very small. This reflects the human's indifference to small errors and constitutes the principal human behavioral nonlinearity in the crossover model context. Another example occurs in measuring the effects of training, where ω_c increases with trials until stable conditions are obtained for that particular subject and set of constant task and environmental conditions. Similarly, operator gain and remnant can be modified as a consequence of changes in operator-centered variables. A notable example is the decrease in gain and increase in remnant which accompanies alcohol ingestion.

Effects of Divided Attention

Human operators in man-machine systems are, in general, involved in two types of operations—control tasks and a diverse combination of monitoring/supervising/communicating/data-gathering/decision making activities referred to as "managerial tasks." While the operator's attention is "divided" between the control and managerial tasks, these are often performed nearly simultaneously as parallel processing operations.

By definition, control workload is highest when the operator's full attention is required for control purposes and when this attention is focused on only the most critical input information needed for closed-loop control. For this reason the full-attention crossover model and remnant for compensatory behavior treated above has received the major attention here. Estimates and considerations based on full-attention compensatory assumptions will generally be conservative. For instance, the dynamic performance of the overall man-machine system will typically be improved when additional cues and information provide the basis for the generation of pursuit behavior.

For a given situation the minimum divided attention level should be established by the demands of the control task. When divided attention conditions are present in compensatory situations the major effects

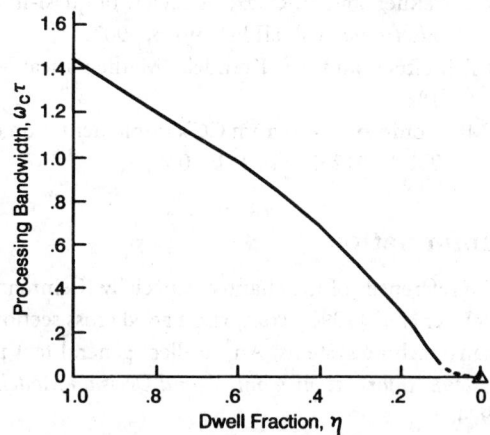

FIGURE 98.5   Effect of divided attention on processing bandwidth.

on the control performance are reduced crossover frequency and increased system error. To a first order the divided attention effects on average crossover frequency are given in Fig. 98.5. Here the "control dwell fraction," is η, the proportion of the total time spent on the control task. There are many other complications and considerations [McRuer *et al.*, 1990], but these require more than handbook treatment.

Defining Terms

Compensatory behavior: Human dynamic behavior in which the operator's actions are conditioned primarily by the closed-loop man-machine system errors.

Compensatory display: For the simplest case, a display which shows only the difference between the desired input command and the system output.

Precognitive behavior: Conditioned responses triggered by the total situation; essentially pure open-loop control.

Pursuit behavior: The human operator's outputs depend on system errors, as in compensatory behavior, but may also be direct functions of system inputs and outputs. The human response pathways make the man-machine system a combined open-loop, closed-loop system.

Pursuit display: In the simplest case, a display which shows input command, system output, and the system error as separable entities.

References

S. Baron and D.L. Kleinman, "The Human As An Optimal Controller and Information Processor," NASA CR-1151, 1969.

R.E. Curry, W.C. Hoffman, and L.R. Young, "Pilot Modeling for Manned Simulation," AFFDL-TR-76-124, 1976.

D. Graham and D. McRuer, *Analysis of Nonlinear Control Systems,* New York: John Wiley & Sons, 1961 (also Dover, 1971).

D.L. Kleinman, S. Baron, and W.H. Levison, "An optimal control model of human response," *Automatica,* vol. 9, no. 3, 1970.

D.T. McRuer, "Human dynamics in man-machine systems," *Automatica,* vol. 16, no. 3, 1980.

D.T. McRuer, W.E. Clement, P.M. Thompson, and R.E. Magdaleno, "Pilot Modeling for Flying Qualities Applications," WRDC-TR-89-3125, vol. II, 1990.

D.T. McRuer and H.R. Jex, "A review of quasi-linear pilot models," *IEEE Trans. Human Factors in Electronics,* vol. HFE-8, no. 3, 1967.

D.T. McRuer and E.S. Krendel, "Mathematical Models of Human Pilot Behavior," AGARD-AG-188, 1974.

P.M. Thompson, "Program CC's Implementation of the Human Optimal Control Model," WRDC-TR-89-3125, vol. III, 1990.

Further Information

The references of the chapter, especially Kleinman *et al.* [1970], McRuer and Krendel [1974], and McRuer *et al.* [1990], comprise a good cross section of detailed information on modeling aspects of man-machine systems. An excellect general text is T.B. Sheridan and W.R. Farrell, *Man-Machine Systems: Information, Control, and Decision Models of Human Performance,* Cambridge: MIT Press, 1974.

Encyclopedic coverage appears in K.R. Boff, L. Kaufman, and J.P. Thomas, *Handbook of Perception and Human Performance,* New York: Wiley, 1986, and K.R. Boff and J.E. Lincoln, "Engineering Data Compendium: Human Perception and Performance," Harry G. Armstrong Aerospace Medical Research Laboratory, Wright-Patterson Air Force Base, Ohio, 1988.

The aperiodic proceedings of the so-called "Annual Manual" contain a great deal of information about man-machine system developments. Since 1965 these have been published by NASA as SP's (NASA Special Publications) under the general heading of *NASA—University Conference on Manual Control.*

A comprehensive summary of models, references, and applications to aerospace vehicle control appears in "Advances in Flying Qualities," *AGARD Lecture Series LS-157,* 1988.

99

Vehicular Systems

Richard C. Dorf
University of California, Davis

Electrical systems are widely used in automobiles and trucks today and electronic systems currently make up about 6% of the value of a car. Vehicle traction was one of the first uses of electric power in the late 1890s.

By 1910 electric automobiles were commonplace. Nevertheless, they were replaced by gasoline-fueled automobiles by 1920 because electric cars operated at lower top speeds and over shorter ranges without recharging than gasoline cars could achieve. However, the availability of electric motive power remained a critical factor in the development of cities. Since the mid-1970s, when the **electric vehicle** reemerged as an appealing transportation option, many have recognized the potential of electric fleet vans. An electric vehicle uses electric energy storage, electric controls, and electric propulsion devices. Because the vans use batteries to drive their electric motors, they are well suited to the short routes and regular schedules followed by vans in a company fleet. One such fleet van, the General Motors Griffon, is produced in England. Because the vans can be recharged regularly at night, they offer electric utilities a new off-peak demand. At the same time, electric vehicles run cleanly and burn no gasoline.

Increasing the distance an electric vehicle can travel on a single charge is the most significant factor in expanding the market for electric vans. The 60-mile (97-km) range of the Griffon makes it a replacement candidate for about 600,000 commercial fleet vehicles now operating in the United States. If advanced batteries doubled the range of a van to 120 miles, the potential market for these vehicles could top 2 million.

Electric vehicles can also benefit from the development of efficient solar cells, which can be used to charge the batteries.

Electronic systems are currently used to control the engine, transmission, steering, braking, suspension, and traction. Many autos incorporate an integrated computer system for controlling the functions mentioned. In addition, electronic systems are used to display information such as speed and engine conditions.

Air bag inflation units use sensors and electronic controls to insure proper inflation within milliseconds after a collision.

By the year 2000, as antilock brakes, **active suspensions,** and other computer-dependent technologies are fully utilized, electronic systems may constitute up to 20% of the value of a car. Much of the added computing power will be used for new technology for smart cars and smart roads, or IVHS (intelligent vehicle/highway systems). The term refers to a varied assortment of electronics that provide real-time information on accidents, congestion, routing, and roadside services to drivers and traffic controllers. IVHS also encompasses devices that would make vehicles more autonomous: collision-avoidance systems and lane-tracking technology that alert drivers to impending disaster or allow a car to drive itself.

Defining Terms

Active suspension: An electronically controlled suspension system for maintaining level suspension of a vehicle.

Electric vehicle: Vehicle using electric energy storage, electric controls, and electric propulsion devices.

Reference

R. K. Jurgen, "Putting electronics to work in the 1991 car models," *IEEE Spectrum,* pp. 75–78, December 1990.

Further Information

Contact the Society of Automotive Engineers, 400 Commonwealth Drive, Warrendale, PA 15096. See the *IEEE Transactions on Vehicle Technology.*

100

Industrial
Illuminating Systems

100.1 New Concepts in Designing an Industrial
Illuminating System... 2257
Determination of Illuminance Levels • Illumination Computa-
tional Methods
100.2 Factors Affecting Industrial Illumination 2264
Basic Definitions • Factors and Remedies • Daylighting
100.3 System Components.. 2268
Light Sources • Ballasts • Luminaires
100.4 Applications 2271
Types of Industrial Illuminating Systems • Selection of the
Equipment
100.5 System Energy Efficiency Considerations 2272
Energy-Saving Lighting Techniques • Lighting Controls •
Lighting and Energy Standards

Kao Chen
Carlsons Consulting Engineers

100.1 New Concepts in Designing an Industrial Illuminating System

Determination of Illuminance Levels

Among the many new concepts for lighting design, the first to be discussed is the new method of determining **illuminance** levels. In the past when illuminating engineers wanted to find the recommended illuminance level for a given task, they would look in the lighting handbook to find a recommended level and then design an illuminating system for the task using the value as a minimum. This procedure provides very little latitude for fine-tuning an illumination design. In the new method, a more comprehensive investigation of required illuminance is performed according to the following steps:

1. Instead of a single recommended illuminance value, a category letter is assigned. Table 100.1 shows different category letters for a selected group of industries (partial only; for complete list see *IES Lighting Handbook* [1987]).
2. The category letters are used to define a range of illuminance. Table 100.2 details illuminance categories and illuminance values for generic types of activities in interiors.
3. From within the recommended range of illuminance, a specific value of illuminance is selected after consideration is given to the average age of workers, the importance of speed and accuracy, and the reflectance of task background.

Table 100.1 Illuminance Categories for Selected Group of Industries

Area/Activity	Illuminance Category	Area/Activity	Illuminance Category
Aircraft maintenance	a	Canning	
Aircraft manufacturing	a	Continuous-belt canning	E
Assembly		Sink canning	E
Simple	D	Hand packing	D
Moderately difficult	E	Olives	E
Difficult	F	Examination of canned samples	F
Very difficult	G	Container handling	
Exacting	H	Inspection	F
Automobile manufacturing		Can unscramblers	E
Bakeries		Labeling and cartoning	D
Mixing room	D	**Casting (see Foundries)**	
Face of shelves	D	**Central stations (see Electric generating stations)**	
Inside of mixing bowl	D	**Chemical plants (see Petroleum and chemical plants)**	
Fermentation room	D	**Clay and concrete products**	
Make-up room		Grinding, filter presses, kiln rooms	C
Bread	D	Molding, pressing, cleaning, trimming	D
Sweet yeast-raised products	D	Enameling	E
Proofing room	D	Color and glazing—rough work	E
Oven room	D	Color and glazing—fine work	F
Fillings and other ingredients	D	**Cleaning and pressing industry**	
Decorating and icing		Checking and sorting	E
Mechanical	D	Dry and wet cleaning and steaming	E
Hand	E	Inspection and spotting	G
Scales and thermometers	D	Pressing	F
Wrapping	D	Repair and alteration	F
Book binding		**Cloth products**	
Folding, assembling, pasting	D	Cloth inspection	I
Cutting, punching, stitching	E	Cutting	G
Embossing and inspection	F	Sewing	G
Breweries		Pressing	F
Brew house	D	**Clothing manufacture (see Sewn Products)**	
Boiling and keg washing	D	Receiving opening, storing, shipping	D
Filling (bottles, cans, kegs)	D	Examining (perching)	I
Candy making		Sponging, decanting, winding,	
Box department	D	measuring	D
Chocolate department		Piling up and marking	E
Husking, winnowing, fat extraction,		Cutting	G
crushing and refining, feeding	D	Pattern making, preparation of trimming,	
Bean cleaning, sorting,		piping, canvas and shoulder pads	E
dipping, packing, wrapping	D	Filling, bundling, shading, stitching	D
Milling	E	Shops	F
Cream making		Inspection	G
Mixing, cooking, molding	D	Pressing	F
Gum drops and jellied forms	D	Sewing	G
Hand decorating	D	**Control rooms (see Electric generating stations—interior)**	
Hard candy		**Corridors (see Service spaces)**	
Mixing, cooking, molding	D	**Cotton gin industry**	
Die cutting and sorting	E	Overhead equipment—separators, driers, grid	
Kiss making and wrapping	E	cleaners, slick machines, conveyers, feed-	
Canning and preserving		ers and catwalks	D
Initial grading raw material samples	D	Gin stand	D
Tomatoes	E	Control console	D
Color grading and cutting rooms	F	Lint cleaner	D
Preparation		Bale press	D
Preliminary sorting		**Dairy farms (see Farms)**	
Apricots and peaches	D	**Dairy products**	
Tomatoes	E	Fluid milk industry	
Olives	F	Boiler room	D
Cutting and pitting	E	Bottle storage	D
Final sorting	E	Bottle sorting	E

a Industry representatives have established a table of single illuminance values which, in their opinion, can be used. Illuminance values for specific operations can also be determined using illuminance categories of similar tasks and activities found in this table and the application of the appropriate weighting factors.

Source: IES Lighting Handbook, Application Volume.

Table 100.2 Illuminance Categories and Illuminance Values for Generic Types of Activities in Interiors

Type of Activity	Illuminance Category	Ranges of Illuminances		Reference Work-Plane
		Lux	Footcandles	
Public spaces with dark surroundings	A	20–30–50	2–3–5	
Simple orientation for short temporary visits	B	50–75–100	5–7.5–10	General lighting throughout spaces
Working spaces where visual tasks are only occasionally performed	C	100–150–200	10–15–20	
Performance of visual tasks of high contrast or large size	D	200–300–500	20–30–50	
Performance of visual tasks of medium contrast or small size	E	500–750–1,000	50–75–100	Illuminance on task
Performance of visual tasks of low contrast or very small size	F	1,000–1,500–2,000	100–150–200	
Performance of visual tasks of low contrast and very small size over a prolonged period	G	2,000–3,000–5,000	200–300–500	Illuminance on task, obtained by a combi-
Performance of very prolonged and exacting visual tasks	H	5,000–7,500–10,000	500–750–1,000	nation of general and local (supplementary
Performance of very special visual tasks of extremely low contrast and small size	I	10,000–15,000–20,000	1,000–1,500–2,000	lighting)

Source: IES Lighting Handbook, Application Volume.

The importance of acknowledging the speed and accuracy with which a task must be performed is readily recognized. Less obvious is the need to consider the age of workers and the reflectance of task background.

To compensate for reduced visual acuity, more illuminance is needed. Using the average age of workers as the age criterion is a compromise between the need of the young and the older workers and, therefore, a valid criterion.

Task background affects the ability to see because it affects **contrast**, an important aspect of visibility. More illuminance is required to enhance the visibility of tasks with poor contrast. Reflectance is calculated by dividing the reflected value by the incident value. The data given in Tables 100.3 and 100.4 are taken from the *IES Lighting Handbook* [1987] and are applied to provide a single value of illuminance from within the range recommended.

Illuminating system design can begin after the desired value of illuminance for a given task has been determined. Based on the *IES Handbook*, the zonal cavity method of determining the number of luminaires and lamps to yield a specified maintained luminance remains unchanged.

Illumination Computational Methods

Zonal Cavity Method. Introduced in 1964, the zonal cavity method of performing lighting computations has gained rapid acceptance as the preferred way to calculate number and placement of luminaires required to satisfy a specified illuminance level requirement. Zonal cavity provides a higher degree of accuracy than does the old lumen method, because it gives individual consideration to factors that are glossed over empirically in the lumen method.

Definition of Cavities. With the zonal cavity method, the room is considered to contain three vertical zones or cavities. Figure 100.1 defines the various cavities used in this method of computation. Height for luminaire to ceiling is designated as the ceiling cavity (h_{cc}). Distance from luminaire to the work plane is the room cavity (h_{rc}), and the floor cavity (h_{fc}) is measured from the work plane to the floor.

To apply the zonal cavity method, it is necessary to determine a parameter known as the "**cavity ratio**" (CR) for each of the three cavities. Following is the formula for determining the cavity ratio:

$$\text{cavity ratio} = \frac{5h\,(\text{room length} + \text{room width})}{(\text{room length} \times \text{room width})} \tag{100.1}$$

where h equals h_{cc} for ceiling cavity ratio (CCR), h_{rc} for room cavity ratio (RCR), h_{fc} for floor cavity ratio (FCR).

Table 100.3 Weighting Factors for Selecting Specific Illuminance Within Ranges A, B, and C

Occupant and Room	Weighting Factor		
Characteristics*	−1	0	+1
Workers' age (average)	Under 40	40 to 55	Over 55
Average room reflectance[1]	>70%	30 to 70%	<30%

Source: IES Lighting Handbook, Application Volume.
Note: This table is used for assessing weighting factors in rooms where a task is not involved.
1. Assign the appropriate weighting factor for each characteristic.
2. Add the two weights; refer to Table 100.2, Categories A through C:
   a. If the algebraic sum is −1 or −2, use the lowest range value.
   b. If the algebraic sum is 0, use the middle range value.
   c. If the algebraic sum is +1 or +2, use the highest range value.
*To obtain average room reflectance: determine the areas of ceiling, walls, and floor; add the three to establish room surface area; determine the proportion of each surface area to the total; multiply each proportion by the pertinent surface reflectance; and add the three numbers obtained.

Table 100.4 Weighting Factors for Selecting Specific Illuminance Within Ranges D through I

Task or Worker	Weighting Factor		
Characteristics	−1	0	+1
Workers' age (average)	Under 40	40 to 55	Over 55
Speed or accuracy*	Not important	Important	Critical
Reflectance of task background, %	>70%	30 to 70%	<30%

Source: IES Lighting Handbook, Application Volume.
Note: Weighting factors are based upon worker and task information.
1. Assign the appropriate weighting factor for each characteristic.
2. Add the two weights; refer to Table 100.2, Categories D through I:
   a. If the algebraic sum is −2 or −3, use the lowest range value.
   b. If the algebraic sum is −1, 0, or +1, use the middle range value.
   c. If the algebraic sum is +2 or +3, use the highest range value.
*Evaluation of speed and accuracy requires that time limitations, the effect of error on safety, quality, and cost, etc. be considered. For example, leisure reading imposes no restrictions on time, and errors are seldom costly or unsafe. Reading engineering drawings or a micrometer requires accuracy and, sometimes, speed. Properly positioning material in a press or mill can impose demands on safety, accuracy, and time.

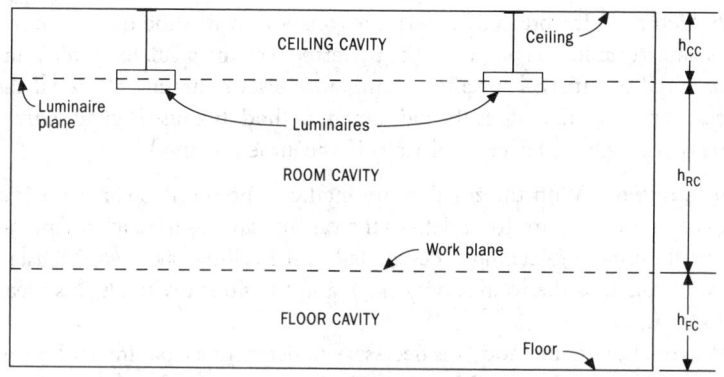

FIGURE 100.1 Basic cavity divisions of space.

Lumen Method Details.   Because of the ease of application of the lumen method which yields the average illumination in a room, it is usually employed for larger areas, where the illumination is substantially uniform. The lumen method is based on the definition of a **footcandle,** which equals one lumen per square foot:

$$\text{footcandle} = \frac{\text{lumen striking an area}}{\text{square feet of area}} \tag{100.2}$$

In order to take into consideration such factors as dirt on the luminaire, general depreciation in lumen output of the lamp, and so on, the above formula is modified as follows:

$$\text{footcandle} = \frac{\text{lamps/luminaire} \times \text{lumens/lp} \times \text{CU} \times \text{LLF}}{\text{area/luminaire}} \tag{100.3}$$

In using the lumen method, the following key steps should be taken:

a. Determine the required level of illuminance.
b. Determine the **coefficient of utilization (CU)** which is the ratio of the lumens reaching the working plane to the total lumens generated by the lamps. This is a factor that takes into account the efficiency and the distribution of the luminaire, its mounting height, the room proportions, and the reflectances of the walls, ceiling, and floor. Rooms are classified according to shape by 10 room cavity numbers. The cavity ratio can be calculated using the formula given in Eq. (100.1). The coefficient of utilization is selected from tables prepared for various luminaires by manufacturers.
c. Determine the **light loss factor (LLF)**. The final light loss factor is the product of all the contributing loss factors. Lamp manufacturers rate filament lamps in accordance with their output when the lamp is new; vapor discharge lamps (fluorescent, mercury, and other types) are rated in accordance with their output after 100 hr of burning.
d. Calculate the number of lamps and luminaires required:

$$\text{no. of lamps} = \frac{\text{footcandles} \times \text{area}}{\text{lumens/lp} \times \text{CU} \times \text{LLF}} \tag{100.4}$$

$$\text{no. of luminaires} = \frac{\text{no. of lamps}}{\text{lamps/luminaire}} \tag{100.5}$$

e. Determine the location of the luminaire—luminaire locations depend on the general architecture, size of bays, type of luminaire, position of previous outlets, and so on.

Point-by-Point Method. Although currently light computations emphasize the zonal cavity method, there is still considerable merit in the point-by-point method. This method lends itself especially well to calculating the illumination level at a particular point where total illumination is the sum of general overhead lighting and supplementary lighting. In this method, information from luminaire **candlepower distribution** curves must be applied to the mathematical relationship. The total contribution from all luminaires to the illumination level on the task plane must be summed.

Direct Illumination Component. The angular coordinate system is most applicable to continuous rows of fluorescent luminaires. Two angles are involved: a longitudinal angle α and a lateral angle β. Angle α is the angle between a vertical line passing through the seeing task (point P) and a line from the seeing task to the end of the rows of luminaires. Angle α is easily determined graphically from a chart showing angles α and β for various combinations of V and H. Angle β is the angle between the vertical plane of the row of luminaires and a tilted plane containing both the seeing task and the luminaire or row of luminaires. Figure 100.2 shows how angles α and β are defined. The direct illumination component for each luminaire or row of luminaires is determined by referring to the table of direct illumination components for the specific luminaire. The direct illumination components are based on the assumption that the luminaire is mounted 6 ft above the seeing task. If this mounting height is other than 6 ft, the direct illumination component shown in

Table 100.5 Direct Illumination Components for Category III Luminaire (Based on F40 Lamps Producing 3100 Lumens)

	Direct Illumination Components															
8	5	15	25	35	45	55	65	75	5	15	25	35	45	55	65	75
α	Vertical Surface Illumination Footcandles at a Point on a Plane Parallel to Luminaires								Vertical Surface Illumination Footcandles at a Point on a Plane Perpendicular to Luminaires							
0–10	.9	2.6	3.6	3.9	3.3	1.9	.7	.1	.9	.8	.7	.5	.3	.1	—	—
0–20	1.8	5.0	7.0	7.7	6.6	3.8	1.5	.2	3.6	3.2	2.7	1.9	1.2	.5	.1	—
0–30	2.6	7.2	10.1	11.3	9.8	5.7	2.3	.3	7.7	7.0	5.8	4.3	2.7	1.1	.3	—
0–40	3.2	9.0	12.8	14.5	12.9	7.7	3.2	.5	12.6	11.6	9.7	7.5	4.9	2.1	.6	—
0–50	3.7	10.3	14.9	17.1	15.7	9.6	4.3	.7	17.8	16.6	14.2	11.2	7.7	3.4	1.1	.1
0–60	4.0	11.2	16.3	18.8	17.6	11.3	5.5	1.0	22.6	21.2	18.4	14.7	10.4	5.1	1.9	.2
0–70	4.1	11.6	17.0	19.8	18.9	12.7	6.8	1.4	26.2	24.7	21.8	17.8	13.1	7.2	3.2	.3
0–80	4.1	11.7	17.3	20.2	19.4	13.3	7.4	1.9	28.2	26.7	23.8	19.7	14.9	8.7	4.3	.8
0–90	4.1	11.7	17.3	20.2	19.4	13.4	7.5	2.0	28.6	27.1	24.2	20.1	15.3	9.1	4.7	1.1

	F.C. at a Point on Work Plane								Category III
0–10	10.6	9.5	7.6	5.5	3.3	1.3	.3	—	
0–20	20.6	18.5	14.9	10.9	6.6	2.6	.7	—	
0–30	29.4	26.5	21.6	16.0	9.8	4.0	1.1	—	
0–40	36.5	33.1	27.4	20.6	12.9	5.4	1.5	—	
0–50	41.8	38.1	31.9	24.3	15.7	6.7	2.0	.1	
0–60	45.2	41.3	34.8	26.8	17.6	7.9	2.6	.2	
0–70	46.9	43.0	36.4	28.3	18.9	8.9	3.2	.3	
0–80	47.4	43.6	36.9	28.8	19.4	9.3	3.5	.4	
0–90	47.5	43.7	37.0	28.8	19.4	9.3	3.5	.4	

2 T-12 Lamps—Any Loading
For T-10 Lamps—CU × 1.02

Luminance Coefficients for 20% Effective Floor Cavity Reflectance

		Reflectances											
Ceiling Cavity		80		50		10		80		50		10	
Walls		50	30	50	30	50	30	50	30	50	30	50	30
WDRC	RCR	Wall Luminance Coefficients						Ceiling Cavity Luminance Coefficients					
.281	1	.246	.140	.220	.126	.190	.109	.230	.209	.135	.124	.025	.023
.266	2	.232	.127	.209	.115	.182	.102	.222	.190	.130	.113	.024	.021
.245	3	.216	.115	.196	.105	.172	.095	.215	.176	.127	.105	.024	.020
.226	4	.202	.102	.183	.097	.161	.088	.209	.164	.124	.099	.023	.019
.212	5	.191	.097	.173	.090	.154	.082	.204	.156	.121	.094	.023	.018
.196	6	.178	.090	.163	.084	.145	.076	.200	.149	.118	.090	.022	.017
.182	7	.168	.083	.153	.078	.136	.071	.194	.144	.115	.087	.022	.017
.170	8	.158	.077	.145	.072	.130	.066	.190	.139	.113	.085	.021	.016
.159	9	.150	.072	.138	.068	.123	.062	.185	.135	.110	.082	.021	.016
.149	10	.141	.068	.130	.064	.116	.059	.180	.131	.107	.080	.020	.016

Table 100.5 must be multiplied by $6/V$, where V is the mounting height above the task. Thus the total direct illumination component would be the product of $6/V$ and the sum of the individual direct illumination components of each row.

Reflected Illumination Components on the Horizontal Surfaces. This is calculated in exactly the same manner as the average illumination using the lumen method, except that the reflected radiation coefficient (RRC) is substituted for the coefficient of utilization.

$$FC_{RH} = \frac{\text{lamps/luminaire} \times \text{lumens/lp} \times \text{RRC} \times \text{LLF}}{\text{area/luminaire}} \qquad (100.6)$$

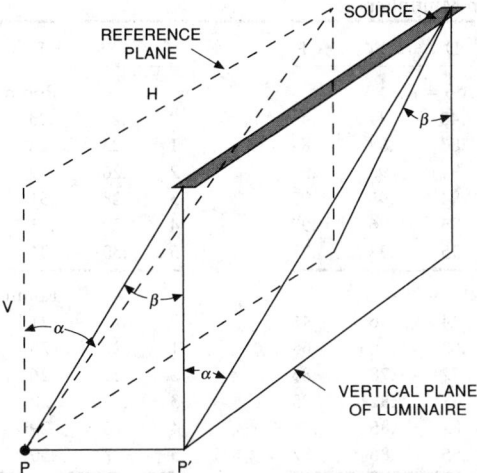

FIGURE 100.2 Definition of angular coordinate systems for direct illumination component.

where RRC = LC_W + RPM (LC_{CC} – LC_W), LC_W = wall luminance coefficient, LC_{CC} = ceiling cavity luminance coefficient, and RPM = room position multiplier.

The wall luminance coefficient and the ceiling cavity luminance coefficient are selected for the appropriate room cavity ratio and proper wall and ceiling cavity reflectances from the table of luminance coefficients in the same manner as the coefficient of utilization. The room position multiplier is a function of the room cavity ratio and of the location in the room of the point where the illumination is desired. Table 100.6 lists the value of the RPM for each possible location of the part in the rooms of all room cavity ratios.

Figure 100.3 shows a grid diagram that illustrates the method of designating the location in the room by a letter and a number.

Reflected Illumination Components on the Vertical Surfaces.   To determine illumination reflected to vertical surfaces, the approximate average value is determined using the same general formula, but substituting WRRC (wall reflected radiation coefficient) for the coefficient of utilization:

$$FC_{RV} = \frac{\text{lamps/luminaire} \times \text{lumens/lp} \times WRRC \times LLF}{\text{area/luminaire (on work plane)}} \qquad (100.7)$$

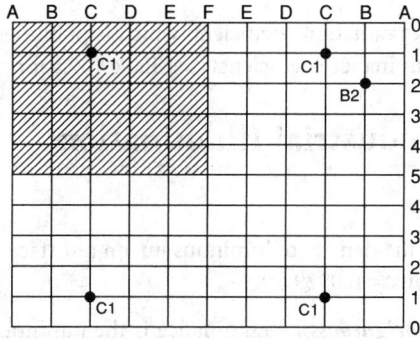

FIGURE 100.3 Grid diagram for locating points on the work plane.

Table 100.6 Room Position Multipliers

	A	B	C	D	E	F		A	B	C	D	E	F
		Room Cavity Ratio = 1							Room Cavity Ratio = 6				
0	.24	.42	.47	.48	.44	.48	0	.20	.23	.26	.28	.29	.30
1	.42	.74	.81	.83	.84	.84	1	.23	.26	.29	.31	.33	.36
2	.47	.81	.90	.92	.93	.93	2	.26	.29	.35	.37	.38	.40
3	.48	.83	.92	.94	.95	.95	3	.28	.31	.37	.39	.41	.43
4	.48	.84	.93	.95	.96	.97	4	.29	.33	.38	.41	.43	.45
5	.48	.84	.93	.95	.97	.97	5	.30	.36	.40	.43	.45	.47
		Room Cavity Ratio = 2							Room Cavity Ratio = 7				
0	.24	.36	.42	.44	.46	.46	0	.18	.21	.23	.25	.26	.27
1	.36	.51	.60	.63	.66	.68	1	.21	.23	.26	.28	.29	.30
2	.42	.60	.68	.72	.78	.83	2	.23	.26	.30	.32	.33	.34
3	.44	.63	.72	.77	.82	.85	3	.25	.28	.32	.34	.35	.36
4	.46	.66	.78	.82	.85	.86	4	.26	.29	.33	.35	.37	.37
5	.46	.68	.83	.85	.86	.87	5	.27	.30	.34	.36	.37	.38
		Room Cavity Ratio = 3							Room Cavity Ratio = 8				
0	.23	.32	.37	.40	.42	.42	0	.17	.18	.21	.22	.22	.23
1	.32	.40	.48	.51	.53	.57	1	.18	.20	.23	.25	.26	.26
2	.37	.48	.58	.61	.64	.67	2	.21	.23	.26	.27	.28	.29
3	.40	.51	.61	.65	.69	.71	3	.22	.25	.27	.29	.30	.30
4	.42	.53	.64	.69	.73	.75	4	.22	.26	.28	.30	.31	.32
5	.42	.57	.67	.71	.75	.77	5	.23	.26	.29	.30	.31	.32
		Room Cavity Ratio = 4							Room Cavity Ratio = 9				
0	.22	.28	.32	.35	.37	.37	0	.15	.17	.18	.19	.20	.20
1	.28	.33	.40	.42	.44	.48	1	.17	.18	.20	.21	.22	.23
2	.32	.40	.48	.50	.52	.57	2	.18	.20	.23	.24	.25	.25
3	.35	.42	.50	.54	.58	.61	3	.19	.21	.24	.25	.26	.26
4	.37	.44	.52	.58	.62	.64	4	.20	.22	.25	.26	.26	.27
5	.37	.48	.57	.61	.64	.66	5	.20	.23	.25	.26	.27	.27
		Room Cavity Ratio = 5							Room Cavity Ratio = 10				
0	.21	.25	.28	.31	.33	.33	0	.14	.16	.16	.17	.18	.18
1	.25	.29	.33	.36	.38	.42	1	.16	.17	.18	.19	.19	.20
2	.28	.33	.40	.42	.44	.48	2	.16	.18	.19	.21	.22	.22
3	.31	.36	.42	.46	.49	.52	3	.17	.19	.21	.22	.23	.23
4	.33	.38	.44	.49	.52	.54	4	.18	.19	.22	.23	.23	.24
5	.33	.42	.48	.52	.54	.56	5	.18	.20	.22	.23	.24	.25

where

$$\text{WRRC} = \frac{\text{wall luminance coefficient}}{\text{average wall reflectance}} - \text{WDRC} \qquad (100.8)$$

where WDRC is the wall direct radiation coefficient, which is published for each room cavity ratio together with a table of wall luminance coefficients (see Table 100.5 for a specific type of luminance).

100.2 Factors Affecting Industrial Illumination

Basic Definitions

Illuminance. Illuminance is the density of luminous lux on a surface expressed in either footcandles (lumens/ft^2) or lux (lx) (lux = 0.0929 fc).

Luminance (or photometric brightness). Luminance is the luminous intensity of a surface in a given direction per unit of projected area of the surfaces, expressed in candelas per unit area or in lumens per unit area.

Reflectance. Reflectance is the ratio of the light reflected from a surface to that incident upon it. Reflection may be of several types, the most common being specular, diffuse, spread, and mixed.

Glare. Glare is any brightness that causes discomfort, interference with vision, or eye fatigue.

Color Rendering Index (CRI). In 1964 the CIE (Commission Internationale de l'Eclairage) officially adopted the IES procedure for rating lighting sources and developed the current standard by which light sources are rated for their color rendering properties. The CRI is a numerical value for the color comparison of one light source to that of a reference light source.

Color Preference Index (CPI). The CPI is determined by a similar procedure to that used for the CRI. The difference is that CPI recognizes the very real human ingredient of preference. This index is based on individual preference for the coloration of certain identifiable objects, such as complexions, meat, vegetables, fruits, and foliage, to be slightly different than the colors of these objects in daylight. CPI indicates how a source will render color with respect to how we best appreciate and remember that color.

Equivalent Sphere Illumination (ESI). ESI is a means of determining how well a lighting system will provide task visibility in a given situation. ESI may be predicted for many points in a lighting system through the use of any of several available computer programs or measured in an installation with any of several different types of meters.

Visual Comfort Probability (VCP). Discomfort glare is most often produced by direct glare from luminances that are excessively bright. Discomfort glare can also be caused by reflected glare, which should not be confused with **veiling reflections**, which cause a reduction in visual performance rather than discomfort. VCP is based in terms of the percentage of people who will be expected to find the given lighting system acceptable when they are seated in the most undesirable location.

Factors and Remedies

Quality of illumination pertains to the distribution of luminaires in the visual environment. The term is used in a positive sense and implies that all luminaires contribute favorably to visual performance. However, glare, diffusion, reflection, uniformity, color, luminance, and **luminance ratio** all have a significant effect on visibility and the ability to see easily, accurately, and quickly. Industrial installations of poor quality are easily recognized as uncomfortable and possibly hazardous. Some of the factors are discussed in more detail below.

Direct Glare. When glare is caused by the source of lighting within the field of view, whether daylight or electric, it is defined as direct glare. To reduce direct glare, the following suggestions may be useful:

a. Decrease the brightness of light sources or lighting equipment, or both.
b. Reduce the area of high luminance causing the glare condition.
c. Increase the angle between the glare source and the line of vision.
d. Increase the luminance of the area surrounding the glare source and against which it is seen.

To reduce direct glare, luminaires should be mounted as far above the normal line of sight as possible and should be designed to limit both the luminance and the quality of light emitted in the 45–85 degree zone because such light may interfere with vision. This precaution includes the use of supplementary lighting equipment. There is such a wide divergence of tasks and environmental conditions that it may not be possible to recommend a degree of quality satisfactory to all needs. In production areas, luminaires within the normal field of view should be shielded to at least 25 degrees from the horizontal, preferably to 45 degrees.

Reflected Glare. Reflected glare is caused by the reflection of high-luminance light sources from shiny surfaces. In the manufacturing area, this may be a particularly serious problem where critical seeing is involved with highly polished sheet metal, vernier scales, and machined metal surfaces. There are several ways to minimize or eliminate reflected glare:

  a. Use a light source of low luminance, consistent with the type of work in process and the surroundings.
  b. If the luminance of the light source cannot be reduced to a desirable level, it may be possible to orient the work so that reflections are not directed in the normal line of vision.
  c. Increasing the level of illumination by increasing the number of sources will reduce the effect of reflected glare by reducing the proportion of illumination provided on the task by sources located in positions causing reflections.
  d. In special cases, it may be practical to reduce the specular reflection by changing the specular character of the offending surface.

Distribution, Reflection, and Shadows. Uniform horizontal illuminance (maximum and minimum not more than one-sixth above or below the average level) is usually desirable for industrial interiors to permit flexible arrangements of operations and equipment and to assure more uniform luminance in the entire area.

Reflections of light sources in the task can be useful provided that the reflection does not create reflected glare. In the machining and inspection of small metal parts, reflections can indicate faults in contours, make scribe marks more visible, and so on.

Shadows from the general illumination systems can be desirable for accenting the depth and forms of various objects, but harsh shadows should be avoided. Shadows are softer and less pronounced when large diffusing luminaires are used or the object is illuminated from many sources. Clearly defined shadows are distinct aids in some specialized operations, such as engraving on polished surfaces, some type of bench layout work, or certain textile inspections. This type of shadow effect can best be obtained by supplementary directional lighting combined with ample diffused general illumination.

Luminance and Luminance Ratios. The ability to see details depends on the contrast between the detail and its background. The greater the contrast difference in luminance, the more readily the seeing task is performed. The eye functions most comfortably and efficiently when the luminance within the remainder of the environment is relatively uniform. In manufacturing, there are many areas where it is not practical to achieve the same luminance relationships as easily as in offices. Table 100.7 is shown as a practical guide to recommended maximum luminance ratios for industrial areas. To achieve the recommended luminance relationships, it is necessary to select the reflectances of all the finishes of the room surfaces and equipment as well as control of the luminance distribution of the lighting equipment. Table 100.8 lists the recommended reflectance values for industrial interiors and equipment. High-reflectance surfaces are desirable to provide the recommended luminance relationships and high utilization of light.

Color Quality of Light. In general, for seeing tasks industrial areas, there appears to be no effect upon visual acuity by variation in color of light. However, where color discrimination or color matching is a part of the work process, such as in the printing and textile industries, the color of light should be carefully selected. Color always has an effect on the appearance of the workplace and on the complexions of people. The illuminating system and the decorative scheme should be properly coordinated.

Veiling Reflections. Figure 100.4 shows that light would reflect into the eyes of the viewer from the "offending zone" and defines the zone of veiling reflection. Veiling reflection would diminish visibility, but the viewer would be unaware of it. The **contrast rendition factor** (**CRF**) can be applied as a measure of the amount of veiling reflection.

Table 100.7   Recommended Maximum Luminance Ratios for Industrial Areas

	Environmental Classification		
	A	B	C
(1) Between tasks and adjacent darker surroundings	3 to 1	3 to 1	5 to 1
(2) Between tasks and adjacent lighter surroundings	1 to 3	1 to 3	1 to 5
(3) Between tasks and more remote darker surfaces	10 to 1	20 to 1	*
(4) Between tasks and more remote lighter surfaces	1 to 10	1 to 20	*
(5) Between luminaires (or windows, skylights, etc.) and surfaces adjacent to them	20 to 1	*	*
(6) Anywhere within normal field of view	40 to 1	*	*

*Luminance ratio control not practical.

A—Interior areas where reflectances of entire space can be controlled in line with recommendations for optimum seeing conditions.

B—Areas where reflectances of immediate work area can be controlled, but control of remote surround is limited.

C—Areas (indoor and outdoor) where it is completely impractical to control reflectances and difficult to alter environmental conditions.

Source: *IES Lighting Handbook, Application Volume.*

Table 100.8   Recommended Reflectance Values for Industrial Interiors and Equipment

Surfaces	Reflectance[1] (%)
Ceiling	80 to 90
Walls	40 to 60
Desk and bench tops, machines and equipment	25 to 45
Floors	not less than 20

[1]Reflectance should be maintained as near as practical to recommended values.

Source: *IES Lighting Handbook, Application Volume.*

Another important factor is the **lighting effectiveness factor** (**LEF**). An overall lighting system efficiency factor considers both the quality of light as reference to equivalent sphere illumination and the effects of veiling reflections. Light patterns such as "batwing" can help solve veiling reflection problems. Figure 100.5 shows the light distribution curve of a typical batwing luminaire.

Daylighting

The daylight contribution should be carefully evaluated and should always be coordinated with a planned electric lighting system.

Fenestration.   **Fenestration** has at least three useful purposes in industrial buildings:

a. For the admission, control, and distribution of daylight.
b. For a distant focus for the eyes, which relaxes the eye muscles.
c. To eliminate the dissatisfaction many people experience in completely closed-in areas.

An adequate electric lighting system should always be provided because of the wide variation in daylight.

Building Orientation.   All fenestration should be equipped with control device appropriate to any luminance problems. Special attention should be given to glare control latitudes where fenestration frequently receives direct sunlight. Diffuse-glaring fixed or adjustable louvers are some of the control means that may be applied.

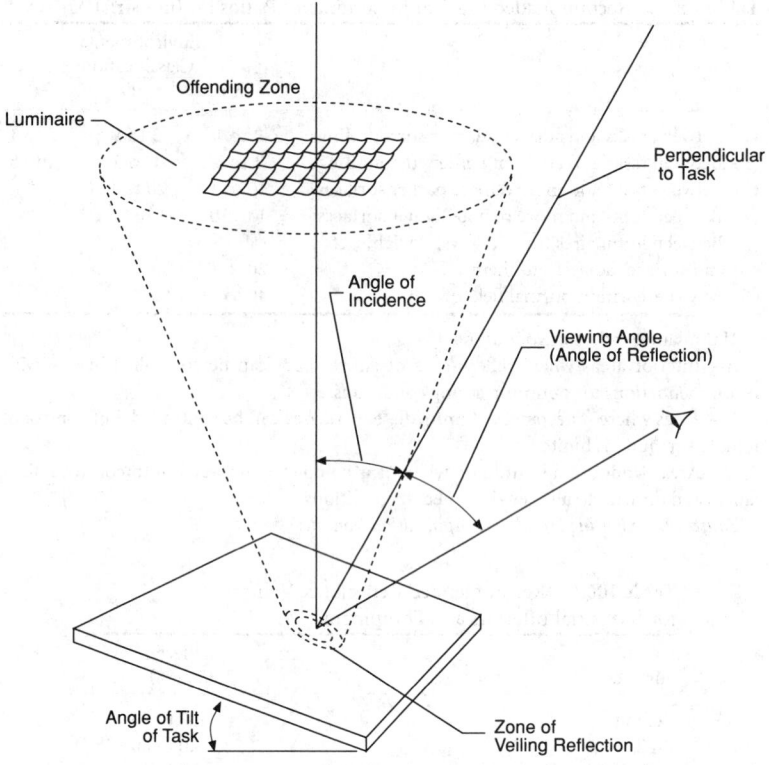

FIGURE 100.4   Diagram showing "offending zone" and zone of veiling reflection.

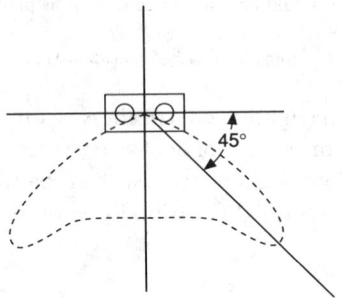

FIGURE 100.5   A typical "batwing" light distribution.

For an industrial building, windows in the sidewalls admit daylight and natural ventilation and afford occupants a view out. However, their uncontrolled luminance may be a problem. There are many control means to make daylight useful to workers' seeing tasks, resulting in energy savings as the ultimate goal.

100.3 System Components

Light Sources

Incandescent

a. Recent technology made possible a line of energy-saving incandescent lamps that use the rare gas krypton as a fill gas.

b. Reflector (R) lamps offer better utilization of the light provided by the lamp compared to a nonreflector type. In this family, there are R lamps, PAR (parabolic aluminized reflector) lamps, and a newer line ER (elliptical reflector) lamps which allow reduction of 50% or more in energy consumption.

c. Infrared Halogen (IR)—PAR lamps combine both the infrared heat-reflection technology and the regenerative halogen cleaning cycle to provide a dramatic increase in lamp efficacy (4% reduction in energy consumption). Available in 30, 60, and 100 W.

Fluorescent

a. Energy-efficient lamps are now available in all popular sizes and colors for most applications. Limitations of energy-saving reduced-wattage lamps are:

- Ambient temperature must be above 60°F.
- Used on high p.f. fluorescent ballasts only.
- Not to be used where drafts of cold air are directed onto the lamp.

b. Typical energy savings are 6 W per lamp for the popular 4-ft 40-W replacement and 15 W per lamp for the 8-ft slimline 75-W replacement.

c. Compact fluorescent lamps are gaining popularity because they are energy efficient, fit into a small enclosed housing, and can be adapted for incandescent socket use.

d. Virtually all compact fluorescent lamps use the "rare earth" phosphors for good color rendition and lumen maintenance characteristics.

e. Utilizing advanced phosphor technology with the optimization of bulb diameter, 40-W lamps are now available that can be retrofitted in a F40 preheat or rapid-start circuit. The new lamp, which could save energy and improve color rendition requirement, has been legislated.

High-Intensity Discharge (HID)

Today HID lamps include mercury vapor, metal halide, high-pressure sodium, and low-pressure sodium lamps. Metal halide lamps offer the best opportunity from a color acceptability point of view. High-pressure sodium lamps offer the highest luminous efficacy in an environment where color distinction is not critical. Since HID lamps have had very few problems in application, they are likely to experience further development in the coming years.

Ballasts

Fluorescent. Electronic ballasts are now available for the F40T12, the slimline, the new T8 lamps, and other energy-saving fluorescent lamps on both 120- and 277-V circuits. Using high-frequency ballasts, the efficacy can be raised by nearly 12%. Although electronic ballasts cost more than the standard core-coil ballasts, operating factors should reflect an appreciable reduction in life-cycle cost for a lighting system. There are two types of dimming ballasts: core and electronic. High-frequency ballasts can readily be used to dim fluorescent lamps over a wide range of light level. All external control wiring is low voltage or fiber-optic wiring.

A recent study indicates that 2 F40T12 lamps operated on an electronic ballast will attain an efficacy of 75–80 LPW versus 62 LPW for the same lamps if operated on a standard core-coil type ballast. With the dimmable electronically ballasted system, energy savings can be as high as 40% with respect to a core ballasted system.

High-Intensity Discharge. The choice of a ballast depends on economic considerations versus performance. A mercury lamp will operate from metal halide ballast, but the converse is not always true. There are several different types of ballasts for high-pressure sodium lamps:

a. Reactor or lag ballast—Inexpensive, low power losses, and small in size.

b. Lead ballast —Fairly good regulation for both line and lamp voltage variation.

   c. Magnetic regulated ballast—Provides best voltage regulation with change of either input voltage or lamp voltage. It is the most costly and has the greatest wattage loss.

   d. Electronic ballast—Maintains a steady constant wattage output with changes in the source impedance as well as excellent regulation. During the life of a high-pressure sodium lamp, it can save 20% more energy by maintaining a constant wattage output in addition to the 15% intrinsic energy savings compared to an equivalent core-coil ballast.

Luminaires

Types of Industrial Luminaires. Selection of a specific type for an installation requires consideration of many factors: candlepower distribution, efficiency, shielding and brightness control, mounting height, lumen maintenance characteristics, mechanical construction, and environmental suitability for use in normal, hazardous, or special areas. In general there are five types in accordance with CIE classifications, namely, direct type, semi-direct type, direct-indirect type, semi-indirect type, and indirect type.

Figure 100.6 shows luminaire types with the percentage of total luminaire output emitted above and below horizontal.

Supplementary Luminaire Types. There are five major types based on the candlepower distribution and luminance:

   Type S-I—directional
   Type S-II—spread, high luminance
   Type S-III—spread, moderate luminance
   Type S- IV—uniform luminance
   Type S-V—uniform luminance with pattern

High-Pressure Sodium. Proper luminaire design is the key to lighting efficiency. Newly developed luminaires use prismatic glass reflectors that are especially made for high-pressure sodium lamps. In addition to achieving maximum light utilization, they redirect the intense light source with excellent light cutoff and high-angle brightness control. Luminaire manufacturers recommend aluminum reflectors for all general-purpose industrial applications and glass-coated reflectors where maintenance practice is compatible with servicing glass.

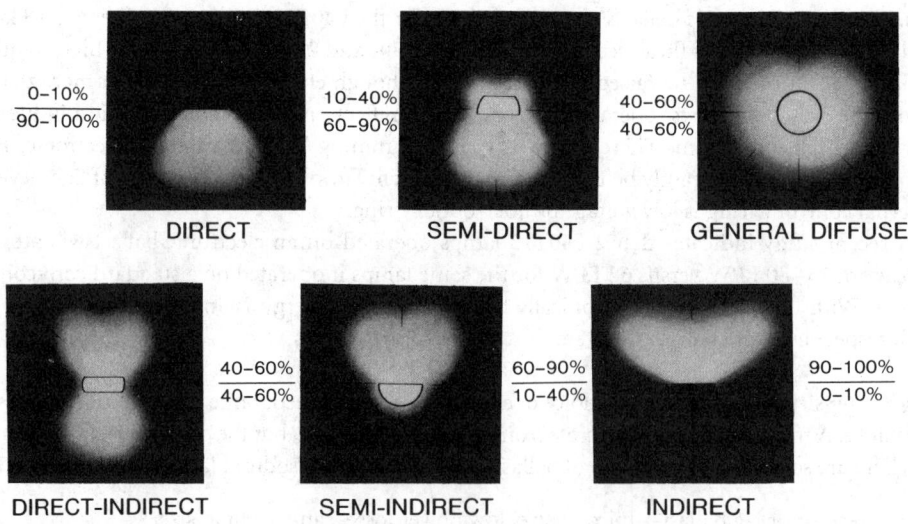

FIGURE 100.6 General lighting luminaire classifications.

Fluorescent. A new trend for lighting new buildings is the increased use of the reflectorized fixtures. This trend may be traced to an increase in the number of state and national lighting efficiency standards in recent years. However, these fixtures can create a "teardrop-like" distribution that may eliminate glare on a computer screen, but also reduces light to other areas.

00.4 Applications

Types of Industrial Illuminating Systems

Factory Illumination for Visual Tasks. The prime requirement for industrial illumination is to facilitate the performance of visual tasks through high-quality illumination. There are three types of lighting used in industrial areas.

- *General Lighting.* It should be designed to provide the desired level of illumination uniformly over the entire area. The variation of light level from point to point within the area should be within 17% of the selected level. A good general lighting system makes it possible to change the location of machinery without rearranging the lighting and also permits full utilization of floor space.

- *Localized General Lighting.* Within a general area there may be a few areas where tasks performed require a greater quantity of light and a different quality of light. When applied, care must be exercised to eliminate direct or reflected glare from the task and from other workers.

- *Supplementary Lighting.* Supplementary lighting is specified for different seeing tasks that require a specific amount or quality of light not readily obtained by standard general lighting methods. Supplementary lighting is a valuable industrial lighting tool. Typical problems arise where work is shielded from the general lighting system by an obstruction or its brightness is otherwise lowered where low contrast, such as scribe marks on steel, may lead to visual errors, and where the product moves too rapidly to be seen clearly by the unaided eye. To attain a good balance, it is important to coordinate the design of supplementary and general lighting with great care.

Security Lighting. Security lighting pertains to the lighting of building exterior and surrounding areas out to and including the boundaries of the property. Security lighting contributes to a sense of personal security and to the protection of property. It may be accomplished through:

- Surveillance lighting to detect and observe intruders.
- Protective lighting to discourage or deter attempts at entrance, vandalism, etc.
- Lighting for safety to permit safe movement of guards and other authorized persons.

Emergency Lighting. Emergency lighting is provided for use when the power supply for the normal lighting fails to ensure that escape routes can be effectively identified and used. Standby lighting is that part of emergency lighting that is sometimes provided to enable normal activities to continue.

The following are recommended minimum illumination requirements for exit signs and egress route:

- *Internally illuminated signs.* An illuminance of 54 lux (5 fc) on the face of the sign is usually specified.
- *Externally illuminated exit sign.* NFPA 101 requires 54 lux (5 fc) on the face of the sign.
- *Egress route.* The horizontal illuminance of any escape route should not be less than 1% of the average provided by the normal lighting, with a minimum average of 5 lux (0.5 fc) at floor level.
- *Location of egress luminaires.* A luminaire should be provided for each exit door and emergency exit door to provide sufficient light to a level of 30 lux (3 fc).

Summaries. In large industrial areas, all these lighting systems may be used. In small areas, localized general lighting may also serve as a substitute for general lighting. In this case, additional supplementary lighting may be required to increase the quantity or improve the quality of the illumination. Many factors must be considered in selecting a lighting system. It is not feasible to recommend one or two systems for all conditions. Because of the relationship of ceiling height to light utilization, most industrial applications call for either direct or semi-direct lighting systems.

Selection of the Equipment

In the selection of equipment, light sources, and luminaires, many variables must be considered. As with any list of variables, it is necessary for purpose of comparison to hold some factors constant. In industrial illumination that factor is usually mounting height and location.

High-Bay Areas. The work generally presents visual tasks that are not difficult because of large machinery and other objects. Illuminance levels for high-bay areas generally range from 50 to 150 fc, although more and more areas are being lighted with 200 and 300 fc. At a high mounting height, it is possible to obtain uniform illumination by using a few high-wattage sources rather than a larger number of low-wattage sources. For luminaires with medium and narrow distribution, greater mounting height or closer spacing is ordinarily required for uniform general illumination.

Regardless of mounting height, wide distribution luminaires are well suited for use in areas that are wide in respect to mounting height. Large machinery and objects tend to cut off light and cast shadows. Since this makes it difficult to see important vertical and angular surfaces, broad light distribution is essential.

High-intensity discharge or fluorescent luminaires for high-bay lighting may be enclosed, ventilated open, or nonventilated open. Enclosed luminaires are usually of a heavy-duty type with a gasketed glass cover to protect the reflector and light source from collection of dirt. The initial luminaire efficiency is lower and the equipment is more costly. Ventilated-open luminaires have largely replaced the nonventilated type.

As far as choices of lamps are concerned, metal halide and HPS are preferred over the mercury type. The use of fluorescent lamps in high-bay areas is limited. Only where the area proportions are such that the room cavity ratios are in the range of 1 to 3 may fluorescent lamps be acceptable. Only high or extra high output fluorescent in 8-ft sizes are recommended.

Medium- and Low-Bay Areas. Seeing tasks in medium- and low-bay areas are usually more difficult than those encountered in the high-bay areas. Increasing the size and reducing the brightness of the luminaires will improve visual comfort and will improve the visibility of specular objects. It may not improve the visibility of diffuse three-dimensional objects.

Luminaires used for general lighting in medium-bay areas are nearly always of the direct or semi-direct type, either fluorescent or wide distribution HID. They may be the ventilated or non-ventilated type and the lamps may be shielded by louvers, baffles, or other devices. For lower mounting, the trend is toward the semi-direct type.

Some of the visual tasks involve specular or semi-specular objects, for which optimum lighting might be an indirect system. The quality of fluorescent sources, with their broad distribution of light, makes them a prime selection for medium- and low-bay lighting. When the proper quality control can be attained, low-wattage HID sources are finding an increasing number of low-bay applications.

100.5 System Energy Efficiency Considerations

Energy-Saving Lighting Techniques

Fluorescent Systems Considerations. Fluorescent lamps are sensitive to ambient temperatures. By using reduced wattage lamps or low-loss ballasts, less heat will be generated and the operating temperature point of the lamp will probably change. The critical area is the coldest spot on the bulb surface. Most fluorescent lamps will peak in light output at around a 100°F cold-spot temperature. For enclosed luminaire types that ordinarily operate the lamp at higher temperature, replacing standard lamps with high-efficacy, reduced wattage lamps may result in a net increase in luminaire output even though the reduced wattage lamps are rated for less output than are standard lamps.

Using Daylight.   Daylight should be dealt with by first analyzing it and then establishing a design technique to integrate it with the electric lighting system. Daylight may be adequate in quantity and quality to reduce the electric lighting load and result in energy conservation. Poor quality of daylight may lead to discomfort and a loss in visibility that may result in a decrease in human performance and productivity.

Daylighting Design from Windows.   The longhand design procedure involves two steps:

- Determine the quantity of illumination coming to the window surface.
- Use that quantity to determine the daylight contribution to the interior part of the space.

Once the contribution of illumination to the window surface has been calculated, two longhand methods are available to determine the illumination contribution to the space. The first method is to follow the point-by-point procedure, which makes two assumptions: (1) interreflected component is ignored and (2) the window is a uniform diffuse emitter. The second method is a lumen method that calculates illumination values at three points defined as the maximum, midway, and minimum. This method includes both the direct and interreflected components of illumination.

Task-Ambient Lighting.   This is a particular form of nonuniform illumination that combines task illuminance and ambient illuminance. One advantage is improved energy efficiency. The task component of task-ambient lighting may take two forms: (1) furniture-mounted lighting built into a workstation or (2) floor-mounted fixtures that can be placed adjacent to a desk. The ambient lighting component may be supplied in two ways: (1) conventional luminaires on the ceiling or (2) indirect fixtures utilizing HID or fluorescent lamps with the output directed to the ceiling and adjacent walls. For ceiling-mounted troffers used for ambient lighting, a plug-in system of wiring should be considered so that luminaires can be relocated as task locations change.

Lighting Controls

In order to save energy, it is essential that minimum acceptable lighting levels be used during off-hours, cleaning periods, and for other nonpeak periods as is practical. The ultimate system of control would be to remotely control every fixture and to program the mode of operation, but this is hardly possible. Solid-state dimmers are available, or ballasts can be circuited in separate groupings. Solid-state controls are available for dimming entire areas of ballasted lights, but special ballasts are required and the controls could be expensive.

Manual control of a lighting system is often the least expensive, but also the least effective alternative. Automatic controls vary from a simple timer to a sophisticated computer system. Figure 100.7 shows a typical programmable lighting control scheme. A price versus benefit cost analysis will be required for each installation. The system should be programmed for normal operation and have a local manual override. A good convenient practice is to have lights switched in distributed groups so that areas can be lighted or darkened as conditions change.

Lighting and Energy Standards

In 1976, the Energy Research and Development Association (ERDA) contracted with the National Conference of States on Building Codes and Standards (NCSBCS) to codify ASHRAE 90-75. The resulting document was called "The Model Code for Energy Conservation in New Buildings." The

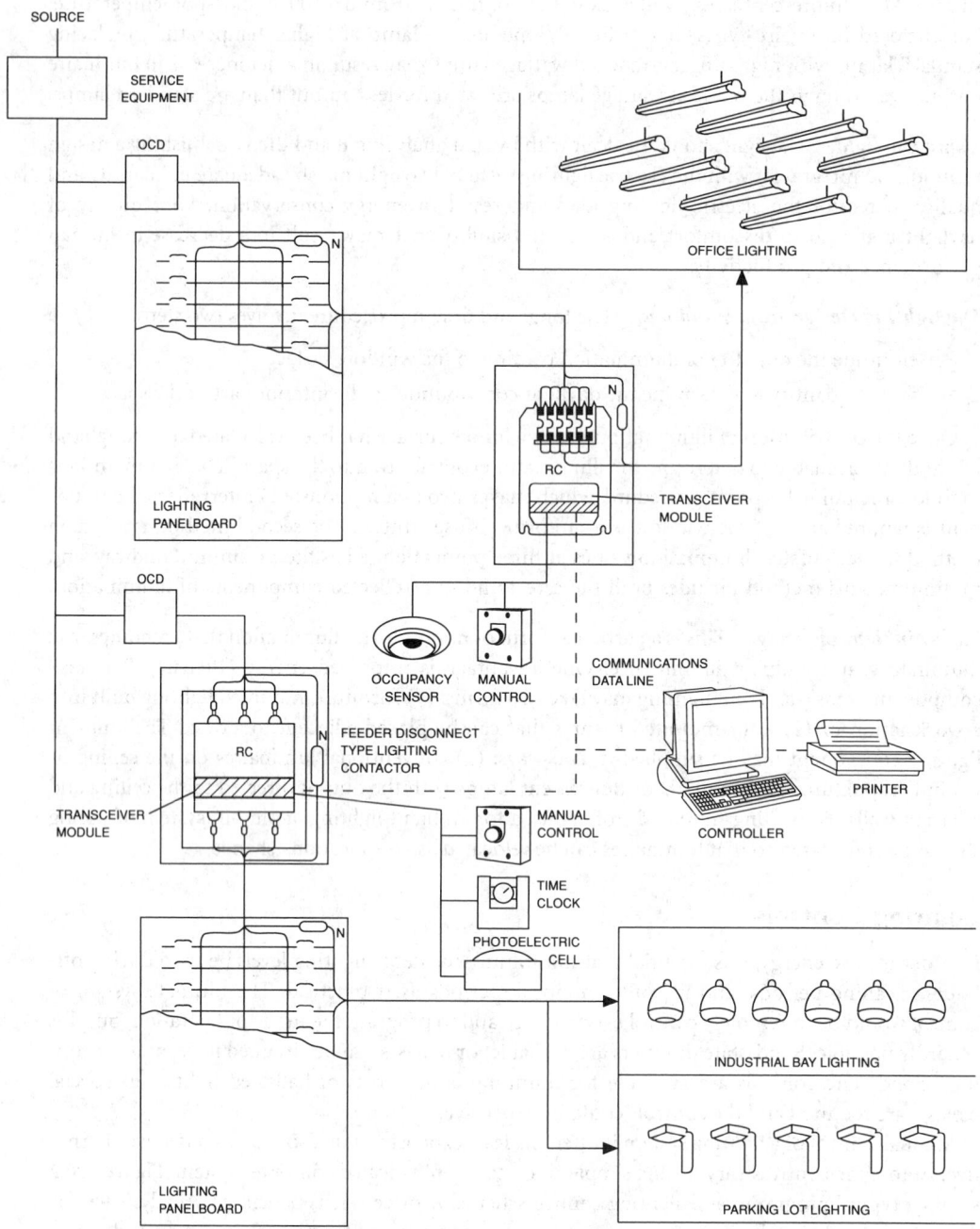

FIGURE 100.7 Programmable lighting control scheme.

model code has been adopted by a number of states to satisfy the requirements of Public Laws 94-163 and 94-385.

There have been several revisions on the ANSI/ASHRAE/IES 90-75 since 1976. All were included in the lighting portion of ANSI/ASHRAE/IES 90A-1980, "Energy Conservation in New Building Design," and in EMS-1981, "IES Recommended Lighting Power Budget Determination Procedure."

ASHRAE/IES 90.1-1989, "Energy Efficient Design of New Buildings Except New Low-Rise

Residential Buildings," is a useful and practical standard for energy-conserving building design and operations. The Department of Energy published in the *Federal Register* of May 6, 1987 a proposed interim rule entitled "Energy Conservation Voluntary Performance Standards for New Commercial and Multi-Family High Rise Residential Buildings." When issued, this rule will be mandatory for all federal buildings and a voluntary recommendation for nonfederal buildings.

Defining Terms

Candlepower distribution: A curve, generally polar, representing the variation of luminous intensity of a lamp or luminaire in a plane through the light center.

Cavity ratio (CR): A number indicating cavity proportions calculated from length, width, and height. It is further defined into ceiling cavity ratio, floor cavity ratio, and room cavity ratio.

Coefficient of utilization (CU): The ratio of the lumens reaching the working plane to the total lumens generated by the lamp. This factor takes into account the efficiency and distribution of the lumanaire, its mounting height, the room proportions, and the reflectances of the walls, ceiling, and floor.

Color preference index (CPI): Measure appraising a light source for appreciative viewing of colored objects or for promoting an optimistic viewpoint by flattery.

Color rendering index (CRI): Measure of the degree of color shift objects undergo when illuminated by the light source as compared with the color of those same objects when illuminated by a reference source of comparable color temperature.

Contrast: The relationship between the luminances of an object and its immediate background. It is equal to $(L_1 - L_2)/L_1$ where L_1 and L_2 are the luminances of the background and object. The ratio L/L_1 is also known as Weber's fraction.

Contrast rendition factor (CRF): The ratio of visual task contrast with a given lighting environment to the contrast with sphere illumination.

Equivalent sphere illumination (ESI): The level of sphere illumination which would produce task visibility equivalent to that produced by a specific lighting environment.

Fenestration: Any opening or arrangement of opening (normally filled with media for control) for the admission of daylight.

Footcandle: The unit of illuminance when the foot is taken as the unit of length. It is the illuminance on a surface one square foot in area on which there is a uniformly distributed flux of one lumen.

Illuminance: The density of luminous flux on a surface expressed in either footcandles (lumens/ft^2) or lux (lx). (lux = 0.0929 fc)

Lighting effectiveness factor (LEF): The ratio of equivalent sphere illumination to ordinary measured or calculated illumination.

Light loss factor (LLF): The ratio of the illumination when it reaches its lowest level at the task just before corrective action is taken, to the initial level if none of the contributing loss factors were considered.

Luminance ratio: The ratio between the luminance of two areas in the visual field.

Veiling reflection: Regular reflections superimposed upon diffuse reflections from an object that partially or total obscure the details to be seen by reducing the contrast.

Visual comfort probability (VCP): This rating is based in terms of the percentage of people who will be expected to find the given lighting system acceptable when they are seated in most undesirable locations.

References

ANSI/IES, "Recommended Practices for Industrial Lighting," Illuminating Engineering Society, New York, 1991.

K. Chen, *Industrial Power Distribution and Illuminating Systems,* New York: Marcel Dekker, 1990.

K. Chen, "New concepts in interior lighting design," *IEEE Trans. Industry Applications,* Sept./ Oct. 1984.

IES Lighting Handbook, Application Volume, Illuminating Engineering Society, New York, 1987.

Lighting Handbook, Westinghouse Electric Corporation, Bloomfield, N.J., 1976.

Further Information

L. Watson, *Lighting Design Handbook,* New York.: McGraw-Hill, 1991. It focuses on the art and process of lighting design and provides invaluable, up-to-date technical details on equipment, color use, scenic projection, lasers, holograms, fiber-optics, computers, and energy conservation.

C.L. Robbins, *Daylighting—Design and Analysis,* New York: Van Nostrand Reinhold, 1986. Organized to correspond to the building design process, the book contains data for calculation of annual cost and energy savings as well as many case studies.

Software—Lighting Calculations by Zonal Cavity Method, Orloff Computer Services, 1820 E. Garry Ave., Santa Ana, CA 92705.

101

Instruments

101.1	Introduction	2277
101.2	Physical Variables	2277
101.3	Instrument Elements	2277
101.4	Instrumentation System	2278
101.5	Modeling Elements of an Instrumentation System	2278
101.6	Summary of Noise Reduction Techniques	2281
101.7	Personal Computer-Based Instruments	2281
101.8	Modeling PC-Based Instruments	2281
101.9	The Effects of Sampling	2282
101.10	Other Factors	2282

John L. Schmalzel
University of Texas, San Antonio

101.1 Introduction

Instruments are the means for monitoring or measuring physical variables. The basic elements of an instrument application are shown in Fig. 101.1. The physical system produces a variable, $s(t)$, shown as time-varying, which is processed by an instrument to yield a desired output information variable, $i(t)$.

101.2 Physical Variables

A variety of physical variables are measured; the type depends on the application. For example, a biomedical experiment usually requires measurement of pressure and temperature. Typical physical variables with corresponding units are summarized in Table 101.1.

101.3 Instrument Elements

Producing meaningful information from physical variables requires conversion and processing. *Electronic instruments* require that physical variables be converted to electrical signals through a process of *transduction*, followed by signal *conditioning* and signal *processing* to obtain useful results. Transducers change one form of signal energy to another: usually the transformation from physical to electrical is useful for instrumentation. Examples of common transducers include thermistors and thermocouples for temperature and strain gages for deflection and pressure measurements. Signal conditioning consists of amplification, filtering, limiting, and other simple operations that prepare the raw transducer output signal for further operations. Signal processing applies some algorithm to the basic signal in order to obtain meaningful information. The use of

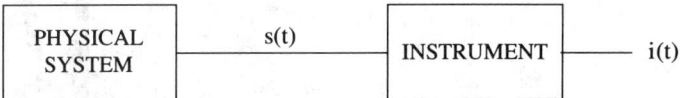

FIGURE 101.1 Generalized block diagram of an instrument applied to physical measurement.

Table 101.1 Representative Physical Variables, Symbols, and Units

Physical Variable	Symbol	SI Units, Abbreviations
Length	L	meter, m
Time	t	second, s
Velocity	V	m/s
Force	F	newton, N
Mass	m	kilogram, kg
Pressure	P	N/m^2
Energy	E	joules, J
Charge	Q	coulomb, C
Capacitance	C	farad, F
Resistance	R	ohm, Ω
Current	I	ampere, A
Voltage	V	volt, V
Frequency	f	hertz, Hz
Conductivity	σ	mho/m

microcomputers within an instrument makes it possible to perform many useful functions including calibration, linearization, conversion, storage, display, and transmission. A block diagram of a representative microcomputer-based instrument in shown in Fig. 101.2.

101.4 Instrumentation System

An instrument is never used in isolation. The instrumentation elements contribute to the overall system response in a number of ways based on the types of **measurement system** elements that are present. Major elements include (1) sources, (2) interconnect, (3) device under test, (4) detectors, and (5) environmental variables. Figure 101.3 shows the elements of a prototype instrumentation system.

101.5 Modeling Elements of an Instrumentation System

Best results are achieved when the instrumentation system is clearly understood, and its effects compensated when practical. Lumped parameter modeling of the elements shown in Fig. 101.3 provides a means for determining the effects of each element on the overall system behavior. Of particular importance are the input and output impedances of each element. In addition, the effects of interconnect and environmental variables can also be modeled to determine their influence on the system. The relative dimensions of the measurement system with respect to the highest frequencies encountered—whether signal or noise—determine whether simplified circuit theory models or generalized solutions to Maxwell's equations must be used. Generally, if measurement system dimensions are on the order of 1/20 of the shortest wavelength, simple circuit theory models can be used. Operation in this regime also allows impedance matching to be largely ignored, e.g., not requiring mandatory use of 50-Ω sources, 50-Ω transmission lines, and 50-Ω terminations, which is commonly encountered in high-frequency communication systems. Table 101.2 summarizes several common instruments and input or output impedance models corresponding to

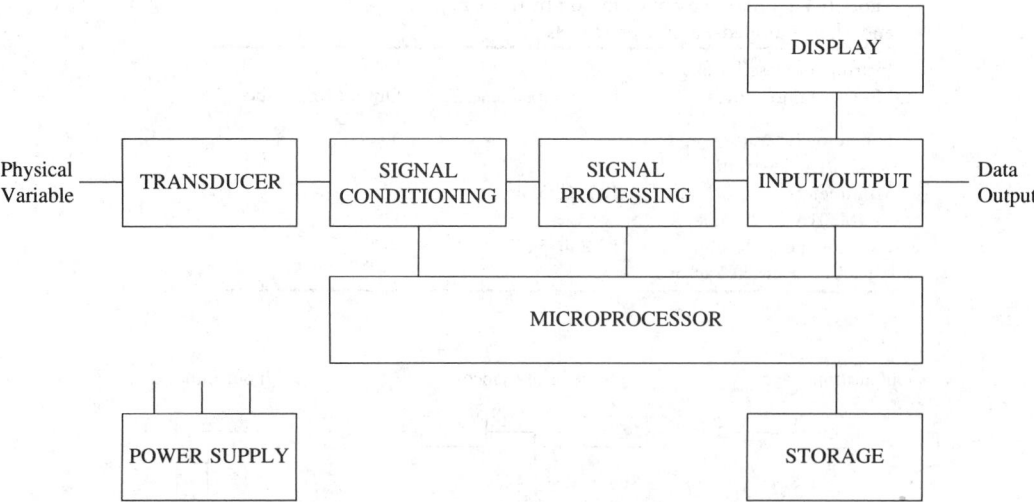

FIGURE 101.2 Block diagram of generalized, microprocessor-based instrument.

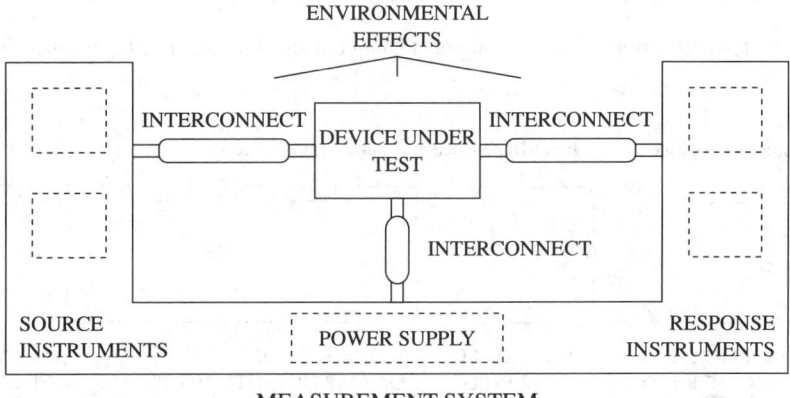

FIGURE 101.3 Fundamental elements of a prototype measurement system.

Fig. 101.4. At low frequencies, interconnect can be modeled by ignoring the very low series resistance and inductance (Z_{s1}, Z_{s2}) terms and considering only the shunt capacitance (Z_p) which varies in the range of 50–150 pf/m for different types of cable. At high frequencies, the characteristic impedance of the interconnect is used, e.g., 50 or 75 Ω for commonly used coaxial cables, 120 Ω for twisted pair.

The response of an entire instrumentation system can be modeled by interconnecting the individual model elements. Figure 101.5 shows an example that was obtained by substituting models for an operational amplifier (op amp) circuit (corresponds to the device under test in Fig. 101.3) that was driven by a function generator for the source and that measured the response with an oscilloscope connected to the output of the op amp using a 10X probe. In this application, the impedance of the interconnect between the source and op amp can be neglected since the frequencies are low and the input impedance of the op amp is much greater than that of the cable. The circuit model of the 10X probe contains a very high series impedance (9 MΩ||13 pf), so it should not be ignored.

Table 101.2 Summary of Common Instruments
and Their Lumped-Parameter Models

Instrument Description, Model, Manufacturer	Input Impedance, Z_i	Output Impedance, Z_o
Function Generator, FG501A, Tektronix		50 Ω
Multimeter, DM501A, Tektronix	10 MΩ (Volts mode)	
Oscilloscope, 54601A, Hewlett Packard	1 MΩ∥13 pf	

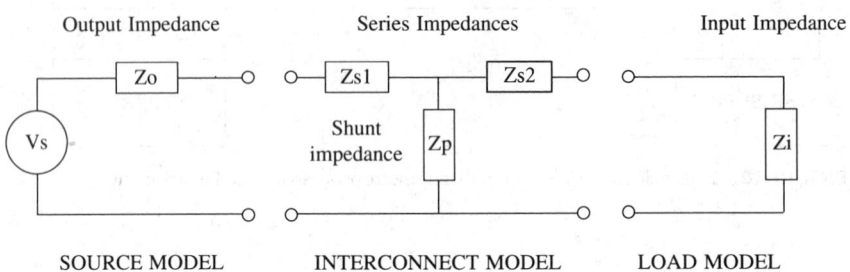

FIGURE 101.4 Simplified output and input models for instrument elements.

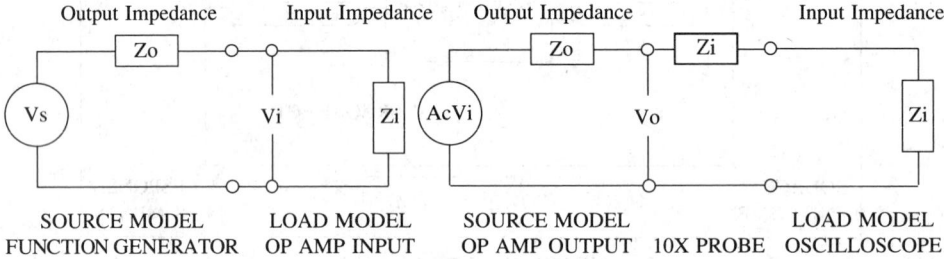

FIGURE 101.5 Model of representative instrumentation system. Each variable would be substituted as required. For example, $V_s = 1.0 \sin 2\pi1000t$ for a 0.707 V_{rms}, 1 kHz sine wave; $Z_o = 50\ \Omega$ for the FG501A; $Z_i = 1$ kΩ for an op amp configured as an inverting amplifier with $R_i = 1$ kΩ; and gain of 10; $A_cV_i = -10.0 \sin 2\pi1000t$; $Z_o = 1\ \Omega$ for low current output; $Z_i = 9$ MΩ∥1.4 pf for a compensated 10X probe; and with $Z_i = 1$ MΩ∥13 pf for the input model of the HP54601A oscilloscope.

The models can be used to determine the *frequency response* of the complete system which describes the magnitude and phase response of the system to sinusoidal, steady-state inputs. This can reveal the contribution of each element to the overall response and helps indicate which elements produce the dominant response. If each of N elements has an individual transfer function, $H_i(j\omega)$, a composite transfer function can be found for the total system, $T(j\omega)$, which is generally not the product of $H_1(j\omega)^*H_2(j\omega)^*\ldots^*H_N(j\omega)$ due to loading effects between elements. The graphical results of the frequency response analysis is termed a *Bode plot*. Analytic techniques can be used; however, use of a circuit simulation program such as *PSpice* (MicroSim Corp.) simplifies the analysis of instrumentation circuit models. A library of subcircuit models can be developed for each instrument and device under test to support routine measurement system loading effects analysis. A PSpice subcircuit definition for the HP54601A oscilloscope input follows.

```
.SUBCKT HP54601A 1 2
Cin 1 2 13p
Rin 1 2 1MEG
.ENDS
```

101.6 Summary of Noise Reduction Techniques

Elimination of undesired measurement errors can benefit from a systematic approach to identifying and solving noise problems. Source, interconnect, and response elements of a measurement system can be treated individually. Some techniques, such as shielding, are applicable to all three. Various combinations of techniques should be tried to find the best results. There are many choices of grounding techniques that vary depending on whether elements are floating or ground-referred and based on bandwidth. In general, multiple ground connections which create *ground loops* should be avoided. Difficult ground loop problems may require isolation or some other technique to interrupt the ground connection between elements. Table 101.3 summarizes a checklist of noise reduction techniques.

101.7 Personal Computer-Based Instruments

Many instrument functions are available for interface to personal computer (PC) systems. These range from plug-in cards that reside on the PC backplane to stand-alone instruments that communicate with the PC over standard interfaces such as RS-232 or IEEE-488. Software to control *data acquisition, analysis,* and *display* completes the computer-based instrument, which is termed a **virtual instrument.** Examples of such software include *Lab View* (National Instruments, Inc.) for Macintosh PCs and *Lab Windows* (National Instruments, Inc.) for IBM-compatible PCs. Figure 101.6 shows a block diagram of an output screen developed using *Lab Windows* for an acoustic measurement application. A menu bar provides pull-down options. Several windows simultaneously display selection options and present results graphically and/or with text.

101.8 Modeling PC-Based Instruments

The approach outlined previously for modeling conventional measurement systems can be extended to PC-based instruments with one major difference: PC-based instruments by their nature are digital machines and perform functions in discrete time. Best performance of PC-based instrument systems must therefore consider *sampled data* effects. Figure 101.7 shows a data acquisition system modeled using an ideal sampler which instantaneously samples a continuous signal,

Table 101.3 Noise Reduction Checklist

Source	Interconnect	Response
Shield enclosures	Shield leads	Shield enclosures
Filter inputs and outputs	Minimize loop areas (twist)	Filter inputs and outputs
Limit bandwidth	Signal leads near ground	Limit bandwidth
Minimize loop areas	Separate low-, high-level signals	Minimize loop areas
	Keep ground leads short	
	Low f: Use single ground	
	High f: Use multiple grounds	
	Keep signal leads short	

Source: H. W. Ott, *Noise Reduction Techniques in Electronic Systems*, 2nd ed., New York: John Wiley & Sons, 1988, Appendix B.

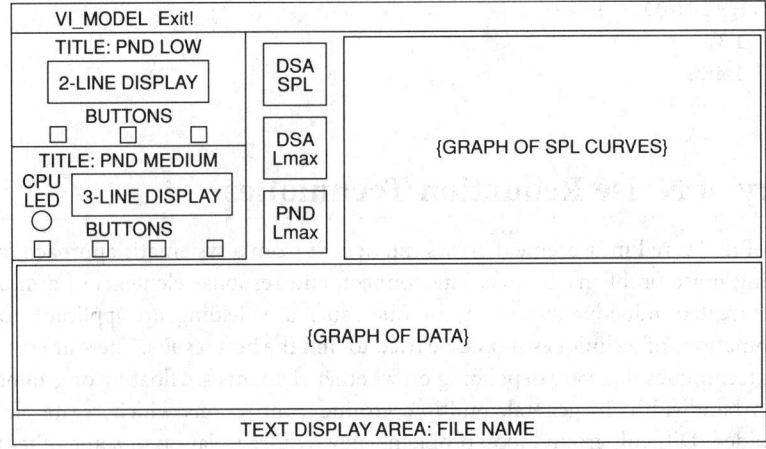

FIGURE 101.6   Example of a *Lab Windows* output screen.

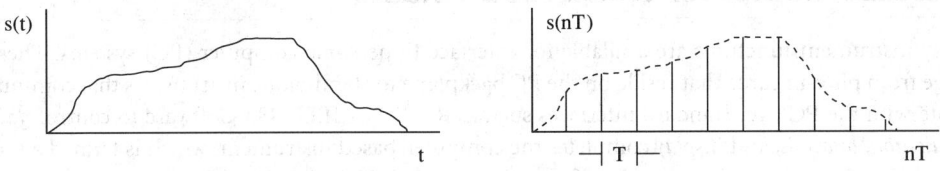

FIGURE 101.7   Sampling a continuous-time signal yields a discrete-time signal.

$s(t)$, every T seconds. This yields a sequence, $s(nT)$, of discrete values that represent the value of the continuous signal at integer multiples of T seconds.

101.9 The Effects of Sampling

The Fourier transform of a sampled signal yields a frequency domain function that is periodic in frequency, with a period that is $1/f_s$. The *sampling theorem* states that in order to unambiguously preserve information, the sampling frequency, $f_s = 1/T$, must be at least twice the highest frequency present in the continuous-time signal. If f_s is less than twice the highest frequency found in the base spectrum, *aliasing* will occur. Aliased frequencies are indistinguishable from one another. A useful method for visualizing this result is through the use of an *aliasing diagram*. An example is shown in Fig. 101.8. The *Nyquist frequency* is defined to be $f_s/2$.

101.10 Other Factors

Other important factors that should be considered when using PC-based instruments over manual counterparts are summarized in Table 101.4. The sampling rate of the data acquisition process must be chosen to meet the requirements of the Nyquist frequency. A low-pass filter that eliminates energy above the Nyquist frequency should be employed and is termed an *anti-aliasing* filter. In order to provide sufficient transition band for the filter, a slightly higher sampling rate should generally be employed. A factor of 1.25^*f_s is a good starting point. Automated equipment may introduce substantial transients into the measurement system. Sufficient time must be provided for the resulting transients to settle to an acceptable error bound, for example, 1%.

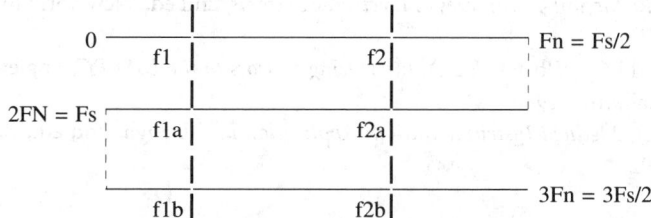

FIGURE 101.8   Aliasing diagram. The two baseband frequencies, f_1 and f_2, have aliases at frequencies that intersect the vertical dashed lines. For example, for a sampling frequency of 10 kHz, and with $f_1 = 1$ kHz, $f_2 = 3.5$ kHz, energy at 9 kHz and 11 kHz would be aliases of 1 kHz, and energy at 6.5 kHz and 13.5 kHz would be aliases of 3.5 kHz.

Table 101.4   Automated Measurement Factors

Factor	Consideration
Leveling	Frequency response measurements require use of a leveled generator. Alternatively, store a calibration curve.
Multiplexing	Measurements from multiple nodes require lead switching to shared instruments; consider these effects.
Sampling frequency	Must exceed the Nyquist frequency.
	Include an anti-aliasing filter. Manual instruments typically use integrating (dual-slope) analog-to-digital converters which give good noise rejection over integer numbers of line cycles. Faster sampling rates for automatic test equipment achieved using successive-approximation or other techniques. User may have to perform averaging as a post-processing step in order to achieve acceptable signal-to-noise ratio.
Settling time	Allow sufficient time for transients to settle for both stimulus/response instruments and device under test.
Storage	Automatic measurements can produce large arrays of data at high speeds. Actual throughput to a hard disk may be much less than the maximum sampling rate of a data acquisition element (plug-in board, external instrument).
Triggering	Choices between free-running, external, internal.

Defining Terms

Instrument: The means for monitoring or measuring physical variables. Usually include transducers, signal conditioning, signal processing, and display.

Measurement system: The sum of all stimulus and response instrumentation, device under test, interconnect, environmental variables, and the interaction among all the elements.

Virtual instrument: An instrument created through computer control of a collection of instrument resources with analysis and display of the data collected.

References

N. Ahmed and T. Natarajan, *Discrete-Time Signals and Systems*, Reston, Va.: Reston Publishing, 1983.

R.S.C. Cobbold, *Transducers for Biomedical Measurements: Principles and Applications*, New York: John Wiley & Sons, 1974.

E.O. Doebelin, *Measurement Systems: Application and Design*, 4th ed., New York: McGraw-Hill, 1990.

J.P. Holman, *Experimental Methods for Engineers*, 5th ed., New York: McGraw-Hill, 1989.

H. W. Ott, *Noise Reduction Techniques in Electronic Systems*, 2nd ed., New York: John Wiley & Sons, 1988.

W.J. Tompkins and J.G. Webster (Eds.), *Interfacing Sensors to the IBM PC*, Englewood Cliffs, N. J.: Prentice-Hall, Inc., 1988.

J.G. Webster (Ed.), *Medical Instrumentation: Application and Design*, 2nd ed., Boston: Houghton Mifflin, 1992.

Further Information

The monthly journal *IEEE Transactions on Biomedical Instrumentation* reports advances in biomedical instrumentation. For subscription information, contact: IEEE Service Center, 445 Hoes Lane, P.O. Box 1331, Piscataway, NJ 08855-1331. Phone (800) 678-IEEE.

Information on automatic test equipment and software for data acquisition, analysis, and display can be obtained from a number of vendors. A summary of a number of software packages can be found in the November, 1991 issue of *IEEE Spectrum* (Schmalzel, "Data handling," vol. 28, no. 11, pp. 38–42, 61).

102

Navigation Systems

102.1 Introduction 2285
102.2 Coordinate Frames 2286
102.3 Categories of Navigation 2286
102.4 Dead Reckoning 2287
102.5 Radio Navigation 2289
102.6 Celestial Navigation 2292
102.7 Map-Matching Navigation 2293
102.8 Navigation Software 2293
102.9 Design Trade-Offs 2294

Myron Kayton
Consulting Engineer

102.1 Introduction

Navigation is the determination of the position and velocity of a moving vehicle on land, at sea, in the air, or in space. The three components of position and the three components of velocity make up a six-component **state vector** that fully describes the translational motion of the vehicle. Surveyors are beginning to use the same sensors as navigators but are achieving higher accuracy as a result of longer periods of observation and more complex, non-real-time data reduction.

In the usual navigation system, the state vector is derived on-board, presented to the crew, recorded on-board, or transmitted to a remote station. Navigation information is usually sent to other on-board subsystems, for example, to the waypoint steering, engine control, communication control, and weapon control computers. Navigation includes the measurement of a target vehicle's state vector by sensors on the ground or in another vehicle. The external sensor usually tracks passive radar returns or a transponder. The technology of externally derived state vectors is being applied to terrestrial vehicles as explained at the end of Section 102.5.

Traditionally, *ship navigation* included the art of pilotage—entering and leaving port, making use of wind and tides, and knowing the coasts and sea conditions. However, in modern usage, navigation is confined to the measurement of the state vector. The handling of the vehicle is called *conning* for ships, *flight control* for aircraft, and *attitude control* for spacecraft.

The term *guidance* has two meanings, both of which are different than navigation:

1. Steering toward a destination of known position from the vehicle's present position. The steering equations on a planet are derived from a plane triangle for nearby destinations and from a spherical triangle for distant destinations.
2. Steering toward a destination without measuring the state vector explicitly. A guided vehicle homes on radio, infrared, or visual emissions. Guidance toward a *moving* target is usually of interest to military tactical missiles in which a steering algorithm assures impact within the

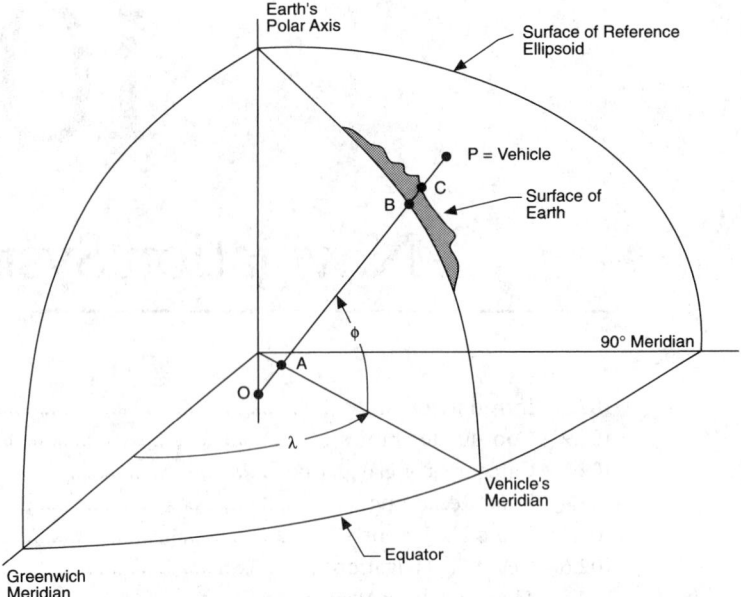

FIGURE 102.1 Latitude-longitude-altitude coordinate frame. ϕ = geodetic latitude; $\overline{OP}$ is normal to the ellipsoid at B; λ = geodetic longitude; h = $\overline{BP}$ = altitude above reference ellipsoid = altitude above mean sea level.

maneuver and fuel constraints of the interceptor. Guidance toward a *fixed* target involves beam riding, as in the Instrument Landing System in Section 102.5.

102.2 Coordinate Frames

Navigation is with respect to a coordinate frame of the designer's choice. Terrestrial vehicles navigate relative to an earth-bound coordinate frame, the most common of which is latitude-longitude-altitude (Fig. 102.1). For local navigation over scores of kilometers, various map grids exist whose coordinates can be calculated from latitude-longitude. Short-range robots navigate with respect to the local terrain or a building's walls. Spacecraft in orbit around the earth navigate with respect to an earth-centered, inertially nonrotating coordinate frame whose z axis coincides with the polar axis of the earth and whose x axis lies along the equator. Interplanetary navigation is with respect to a sun-centered, inertially nonrotating coordinate frame whose z axis is perpendicular to the **ecliptic** and whose x axis points to a convenient star [see Yuen, 1983].

102.3 Categories of Navigation

Navigation systems can be categorized as:

1. *Absolute navigation systems* that measure the state vector without regard to the path traveled by the vehicle in the past. These are of two kinds:
   - Radio systems (Section 102.5). They consist of a network of transmitters (sometimes also receivers) on the ground or in satellites. A vehicle detects the transmissions and computes its position relative to the known positions of the stations in the navigation coordinate frame. The vehicle's velocity is measured from the Doppler shift of the transmissions or from a sequence of position measurements.

- Celestial systems (Section 102.6). They measure the elevation and azimuth of celestial bodies relative to the navigation coordinates. They are used in special-purpose high-altitude aircraft and in spacecraft. Manual celestial navigation was practiced at sea for millenia [see Bowditch].

2. *Dead-reckoning navigation systems* that derive their state vector from a continuous series of measurements relative to an initial position. There are two kinds, those that measure vehicle heading and either speed or acceleration (Section 102.4) and those that measure emissions from continuous-wave radio stations whose signals create ambiguous "**lanes**" (Section 102.5).

   Dead-reckoning systems must be reinitialized as errors accumulate and if power is lost.

3. *Mapping navigation systems* that observe images of the ground, profiles of altitude, sequences of turns, or other external features on-board (Section 102.7).

102.4 Dead Reckoning

The simplest dead-reckoning systems measure vehicle speed and heading, resolve speed into the navigation coordinates, then integrate to obtain position (Fig. 102.3). The oldest heading sensor is the magnetic compass, a magnetized needle or electrically excited toroidal core (called a *flux gate*), as shown in Fig. 102.2. It measures the direction of the earth's magnetic field to an accuracy of 2 degrees at a steady velocity below 60 degrees magnetic latitude. The horizontal component of the magnetic field points toward *magnetic north*. The angle from true to magnetic north is called *magnetic variation* and is stored in the computers of modern vehicles as a function of position over the region of anticipated travel. *Magnetic deviations* caused by iron in the vehicle can exceed 30 degrees and must be compensated in the navigation computer.

Centimeters

FIGURE 102.2 Saturated-core (flux-gate) magnetometer. The drive coil and orthogonal sensing coils wound on the toroidal core measure two components of the magnetic field in the plane of the toroid. (Courtesy of Humphrey, Inc., San Diego, Calif.)

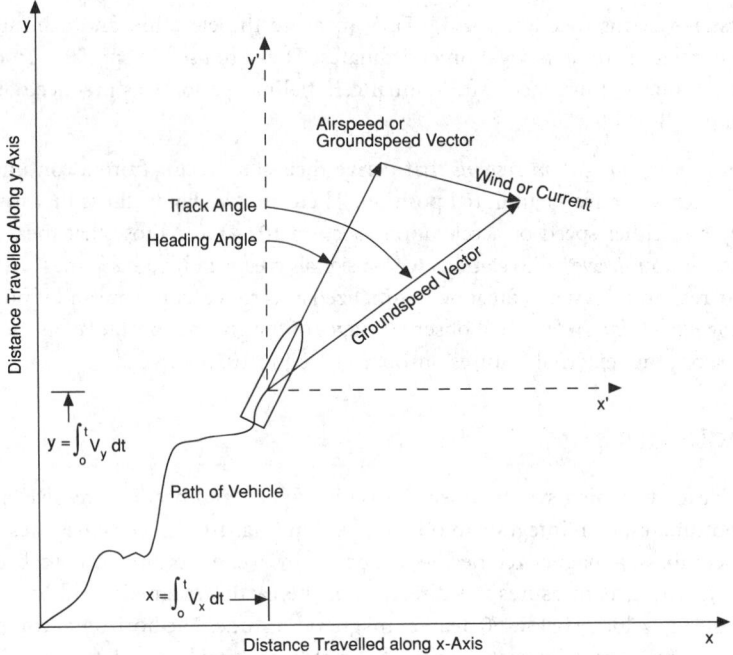

FIGURE 102.3   Geometry of dead reckoning.

A more complex heading sensor is the gyrocompass, consisting of a spinning wheel whose axle is constrained to the horizontal plane (often by a pendulum). The ships' version points north when properly compensated and exhibits errors less than a degree, caused by maneuvering. The aircraft version (also called a directional gyroscope) holds any preset heading relative to earth, drifting at 100 degrees/h or more. Inexpensive gyroscopes are often coupled to magnetic compasses to reduce maneuver-induced errors.

The simplest speed sensor is a wheel odometer that generates electrical pulses. Ships use a dynamic-pressure probe or an electric-field sensor that measures the speed of the hull through the conductive water. Aircraft measure the dynamic pressure of the airstream from which they derive airspeed in an *air-data* computer. Dynamic-pressure sensors are insensitive to the component of airspeed or waterspeed normal to the sensor's axis *(leeway* in a ship, *drift* in an aircraft), thereby introducing an error into the dead-reckoning computation. A Doppler radar measures the frequency shift in radar returns from the ground or water below the aircraft, from which speed is inferred. A Doppler sonar measures a ship's speed relative to the water layer or ocean floor from which the beam reflects. Multibeam Doppler radars or sonars measure all the components of the vehicle's velocity.

The most complex dead-reckoning system is an inertial navigator in which accelerometers measure vehicle acceleration while gyroscopes measure the orientation of the accelerometers. An onboard computer resolves the accelerations into navigation coordinates and integrates them to obtain velocity and position. The gyroscopes and accelerometers are mounted in either of two ways:

1. In servoed gimbals that angularly isolate them from the vehicle.
2. Fastened directly to the vehicle ("strapped-down"), whereupon the sensors are exposed to the maximum angular rates and accelerations of the vehicle (Fig. 102.4).

Most accelerometers consist of a gram-sized proof mass mounted on flexure pivots. Older gyroscopes contained metal wheels rotating in ball bearings or gas bearings. The newest gyroscopes are evacuated cavities in which counterrotating laser beams are compared in phase to measure the sensor's angular velocity relative to **inertial space** about an axis normal to the plane of the beams.

FIGURE 102.4 Inertial reference unit. Two laser gyroscopes (flat discs), electrical connectors, and shock mounts are visible; the accelerometers are not. This unit is used in the Airbus A-310 and many military aircraft such as the F-14, F-15, and F-18. (Courtesy of Litton Guidance and Control Systems.)

Vibrating hemispheres and rotating vibrating bars are also sold as navigation-quality gyroscopes (drift rates less than 0.1 degree/h). The gyroscopes measure vehicle orientation within 0.1 degree for steering and pointing.

Fault-tolerant configurations of cleverly oriented supplementary gyroscopes and accelerometers (typically four to six) detect and correct sensor failures. Inertial navigators are used aboard naval ships, in large jet airliners, in most military aircraft, in space boosters and entry vehicles, in manned spacecraft, in tanks, and on large mobile artillery pieces.

102.5 Radio Navigation

Scores of radio navigation aids have been invented and many of them have been widely deployed, as summarized in Table 102.1.

The most precise is the global positioning system (GPS), a network of 24 satellites and a dozen ground stations for monitoring and control. A vehicle derives its three-dimensional position and velocity from ranging signals at 1.575 GHz received from four or more satellites (military users also receive 1.227 GHz; see Table 102.1). The former Soviet Union began to deploy a similar system, called GLONASS. GPS offers 100-m ranging errors to civil users and 15-m ranging errors to military users. Receivers are available for less than $2000 in 1992. They are being used on highways, in low-rise cities, on boats, in aircraft, and in low-flying spacecraft. GPS will be fully operational in 1995, though in 1993 it already provided continuous worldwide navigation for the first time in history. GPS will undoubtedly diminish the importance of dead reckoning, thereby reducing the cost of many navigation systems.

The most widely used marine radio aid in 1992 was Loran-C (see Table 102.1). The 100-kHz signals are usable within 1000 **nautical miles (nm)** of a "chain" consisting of three to four stations (Fig. 102.5). Chains cover the United States, part of western Europe, and a few other areas. The former Soviet Union has a compatible system called

Table 102.1 Worldwide Radio Navigation Aids

System	Frequency		Number of Stations	Number of Users in 1991		
	Hz	Band		Air	Marine	Space
Omega	10–13 kHz	VLF	8	15,000	10,000	0
Loran-C/Chaika	100 kHz	LF	50	80,000	450,000	0
Decca	70–130 kHz	LF	150	2,000	20,000	0
Beacons*	200–1600 kHz	MF	4000	200,000	500,000	0
Instrument Landing System (ILS)*	108–112 MHz	VHF	1500	130,000	0	0
	329–335 MHz	UHF				
VOR*	108–118 MHz	VHF	1500	200,000	0	0
SARSAT/COSPAS	121.5 MHz	VHF	5 satellites			
	243, 406 MHz	UHF				
Transit	150, 400 MHz	VHF	7 satellites	0	95,000	0
PLRS	420–450 MHz	UHF	In development			
JTIDS	960–1213 MHz	L	In development			
DME*	962–1213 MHz	L	1500	90,000	0	4
Tacan*	962–1213 MHz	L	850	15,000	0	4
Secondary Surveillance Radar (SSR)*	1030, 1090 MHz	L	800	250,000	0	0
Identification Friend or Foe (IFF)						
GPS-GLONASS	1227, 1575 MHz	L	24 satellites	1,000	5,000	6
Satellite Control Network (SCN)	1760–1850 MHz	S	10	0	0	200
	2200–2300 MHz	S				
Spaceflight Tracking and Data Network (STDN)	2025–2150 MHz	S	2 satellites	0	0	50
	2200–2300 MHz	S				
Radio Altimeter	4200 MHz	C	None	20,000	0	0
MLS*	5031–5091 MHz	C	2	10	0	0
FPQ–6, FPQ-16 radar	5.4–5.9 GHz	C	10	0	0	0
Weather/map radar	10 GHz	X	None	3,000	0	0
Shuttle rendezvous radar	13.9 GHz	Ku	None	0	0	4
Airborne Doppler radar	13–16 GHz	Ku	None	5,000	0	0
SPN-41/46 landing aid	15 GHz	Ku	50	500	0	0
SPN-42 carrier-landing radar	33 GHz	Ka	50	500	0	0

*Standardized by International Civil Aviation Organization.

Chaika. The vehicle-borne receiver measures the difference in time of arrival of pulses emitted by two stations, thus locating the vehicle on one branch of a hyperbola. Two or more station pairs give a two-dimensional position fix whose typical accuracy is 0.25 nm, limited by propagation uncertainties over the terrain between the transmitting station and the user. Loran is also used by general aviation aircraft for en-route navigation and for nonprecision approaches to airports (in which the cloud bottoms are more than 500 ft above the runway).

The most widely used aircraft radio aid is VORTAC, whose stations offer three services:

1. Analog bearing measurements at 108 to 118 MHz (called VOR) (see Table 102.1). The vehicle compares the phases of a rotating cardioid pattern and an omnidirectional sinusoid emitted by the station.
2. Pulse distance measurements (DME) at 1 GHz by measuring the time delay for an aircraft to interrogate a VORTAC station and receive a reply.
3. Tacan bearing information conveyed in the amplitude modulation of the DME replies from the VORTAC stations.

Throughout the western world, civil aircraft use VOR/DME and military aircraft use Tacan/DME. The successor states to the Soviet Union may adopt VORTAC as well.

Omega is a worldwide radio aid consisting of eight radio stations that emit continuous sine

FIGURE 102.5   Loran-C Station at Kure, Hawaii. (Official photograph of U.S. Coast Guard.)

waves at 10 to 13 kHz. Vehicles with precise clocks measure their range to a station by observing the absolute time of reception. Other vehicles measure the range differences between two stations in the form of phase differences between the received sinusoids. Omega creates hyperbolic "lanes" that are 10 to 150 nm wide. The lanes are indistinguishable from each other by measuring phase; hence the vehicle must count lanes from a point of known position. Errors are about 2 nm due to radio propagation irregularities. Omega is used by submarines, over-ocean general-aviation aircraft, and a few international air carriers.

Landing guidance throughout the western world is with the Instrument Landing System (ILS). It creates a horizontal guidance signal near 110 MHz and a vertical guidance signal near 330 MHz. Their intersection is a line in space that leads an aircraft from a distance of about 10 nm to within 50 ft above touchdown (ILS gives no information about where the aircraft is located along the beam). U.S. Navy aircraft use a microwave scanning system at 15.6 GHz to land on aircraft carriers; NASA's space shuttle uses the Navy system to land at its spaceports. Another microwave landing system (MLS) at 5 GHz is being deployed and may gradually replace the ILS in civil operations. Within 50 ft above the runway, a radio altimeter measures altitude and guides the flare maneuver.

All the space-faring nations operate worldwide radio networks that track spacecraft, compute their state vectors, and predict future state vectors using complex models of gravity, atmospheric drag, and lunisolar perturbations. NASA operates two tracking and data relay satellites (TDRS) that track spacecraft in low earth orbit. Accuracies of 10 to 50 m and 0.3 m/s are being achieved. Specialized ground-based tracking stations monitor and reposition the world's many communication satellites. Other specialized stations track and communicate with deep space probes. They achieve accuracies of 30 m and a few centimeters per second, even at enormous interplanetary distances, due to long periods of observation and precise orbit equations. Kayton [1990] discusses these networks.

Position-tracking and position-reporting systems are navigation-communication systems that monitor the state vectors of many vehicles and usually display the data in a control room. They have long been in use for airplane traffic control and space vehicles. Table 102.1 lists the *Secondary Surveillance Radars* that receive coded replies from aircraft so they can be identified by human controllers and by collision-avoidance algorithms. The table also lists the U.S. NASA and military spacecraft-tracking networks (STDN and SCN). Position tracking and reporting systems are slowly coming into use for terrestrial vehicles, e.g., fire trucks, police cars, ambulances, and truck fleets. In 1993, communications links in several cities transmitted vehicle-derived position (radio or dead-reckoned) to a control center. Several navigation-communication satellites are being developed to offer similar services worldwide. The aeronautical bureaucracy calls them Automatic Dependent Surveillance (ADS) systems. They may replace the ground-based air-traffic-control systems of the developed world and will provide instant air traffic control systems over oceans and undeveloped land areas. Military systems that measure the position of vehicles on battlefields and report to headquarters have been in development for decades. Examples are the American Joint Tactical Information Distribution Systems (JTIDS) and the Position Location Reporting System (PLRS).

A worldwide network of SARSAT-COSPAS stations monitors signals from satellite-based transponders listening on 121.5, 243, and 406 MHz, the three international distress frequencies. Software at the stations allows the position of emitters on earth to be calculated within 20 kilometers from the Doppler shift history so that rescue vehicles can be dispatched.

102.6 Celestial Navigation

Human navigators use sextants to measure the elevation angle of celestial bodies above the visible horizon. The peak elevation angle occurs at local noon or midnight:

$$\text{elev angle (degrees)} = 90 - \text{latitude} + \text{declination}$$

Thus at local noon or midnight, latitude can be calculated by simple arithmetic. Declination, the angle of the sun or star above the earth's equatorial plane, was part of the ancient navigator's proprietary lore. Tables of the declination of the sun were first tabulated in the fifteenth century in Spain. When time became measurable at sea, with a chronometer in the nineteenth century and by radio in the twentieth century, off-meridian observations of the elevation of two or more celestial bodies were possible at any known time of night (cloud cover permitting). These fixes were hand-calculated using logarithms. In the 1930s, hand-held sextants were built that measured the elevation of celestial bodies from an aircraft using a bubble-level reference instead of the horizon. The accuracy of celestial fixes was 3–10 miles at sea and 5–20 miles in the air, limited by the uncertainty in the horizon and the inability to make precise angular measurements on a pitching, rolling vehicle. Kayton [1990] reviews the history of celestial navigation at sea and in the air.

The first automatic star trackers were built in the late 1950s. They measured the azimuth and elevation of stars relative to a gyroscopically stabilized platform. Approximate position measurements by dead reckoning allowed the telescope to point within a fraction of a degree of the desired star. Thus, a narrow field-of-view was possible, permitting the telescope and photodetector to track stars in the daytime. An on-board computer stored the right ascension and declination of 20–100 selected stars and computed the vehicle's position. Automatic star trackers are used in long-range military aircraft and on space shuttles.

Spacecraft use the line-of-sight to the sun and stars to measure orientation (for *attitude control*). Earth-pointing spacecraft usually carry horizon scanners that locate the center of the earth's carbon-dioxide disc. All spacecraft navigate by radio tracking from external stations, usually on earth. When interplanetary spacecraft approach the target planet, the navigation computers (on earth) transform from sun-centered to planet-centered coordinates by observing star occultations and transmitting the images to earth for human interpretation. Even near the earth, celestial navigation has been the subject of many experiments. During the Apollo translunar missions, crews measured the angle between celestial bodies and the earth or moon with a specially designed manual sextant coupled to a digital computer which calculated the state vector. Other experiments have been made in which American and Soviet crews used manual sextants to observe the angle between celestial bodies and landmarks on earth, from which state vectors were calculated. Autonomous vehicles on other planets and certain military spacecraft may also need celestial navigation.

102.7 Map-Matching Navigation

Mapping radars present a visual image of the terrain to a vehicle's crew, who can "pickle" prominent landmarks with a cursor. Automatic map-matchers have been built that correlate the observed radar image to stored images and thereby update the dead-reckoned state vector. More commonly, aircraft and cruise missiles measure the vertical profile of the terrain below the vehicle and match it to a stored profile. Periodic terrain matching reduces the long-term drift of their inertial navigators. The profile of the terrain is measured by subtracting the readings of a baro-inertial altimeter (calibrated for altitude above sea level) and a radio altimeter (measuring terrain clearance). An onboard computer calculates the autocorrelation function between the measured profile and each of many stored profiles on possible parallel paths of the vehicle. The highest autocorrelation identifies the actual path and the longitudinal position of the vehicle along that path. The on-board inertial navigator usually contains a digital filter that slowly corrects the drift of the azimuth gyroscope as a sequence of fixes is obtained. Hence the direction of the route through the stored map is usually known, saving the considerable computation time that would be needed to correlate for an unknown azimuth of the flight path. Marine versions profile the seafloor with a sonar and compare the measured profile to stored bottom maps.

Experimental city-street map matchers are being developed as periodic updates for the dead reckoning of automobiles and trucks. Dead reckoning of a land vehicle is usually by means of distance and heading (as inferred from odometers on opposite wheels). As in any dead-reckoning system, position errors will accumulate rapidly. Thus, a periodic correction is made when the vehicle executes a large turn into a street whose position is unambiguously identifiable on the map outside the ellipse of probable dead-reckoning error. An accuracy of several meters is possible in such a system if all streets are included on the stored map (e.g., alleys, driveways, and parking garages). This kind of map matcher must be initialized by the driver of the car or truck.

The most complex mapping systems observe their surroundings, usually by digitized video, and create a map of the surrounding terrain. Guidance software then directs the vehicle over the terrain on a trajectory that optimizes a selected goal. Such systems are in the early stages of development, for example, for hazardous sites such as nuclear plants, waste-disposal facilities, and battlefields, and for unmanned planetary exploration.

102.8 Navigation Software

Navigation software is sometimes embedded in a central processor with other avionic-system software, sometimes confined to one or more navigation computers. The navigation software processes the measurements of each sensor (e.g., inertial or air data). It contains sensor-unique algorithms and data, calibration constants, initialization sequences, self-test algorithms, reasonability tests, and alternative algorithms for periods when sensors have failed or are not receiving information. The state vector can be calculated independently from each sensor; more often, the navigation software contains multisensor algorithms that calculate the best estimate of position and velocity from several sensors. Prior to 1970, the best estimate was calculated from a least squares algorithm with constant weighting functions or from a linear filter with constant coefficients. Now, a *Kalman filter* often calculates the time-varying weighting functions from a mathematical model of the expected dynamic behavior of each sensor.

   Digital maps, often stored on compact disc, are carried on some aircraft and land vehicles so that a position can be visually displayed to the crew. Section 102.7 discusses the use of digital maps to improve position estimates by matching observations to the stored terrain data. Algorithms for waypoint steering and for control of the vehicle's attitude are contained in the software of other avionic subsystems; they are not included in navigation.

   Specially equipped aircraft (and sometimes ships) are often used for the routine calibration of radio navigation aids, dynamic-pressure speed sensors, heading sensors, and new algorithms.

102.9 Design Trade-Offs

The designers of a navigation system conduct trade-offs for each vehicle to determine which navigation systems to use. Tradeoffs consider the following attributes:

- *Cost,* including the construction and maintenance of transmitter stations and the purchase of on-board electronics and software. Users are concerned only with the costs of on-board hardware and software and with access charges (if any) to external transmitters/receivers.
- *Accuracy* of position and velocity, which is specified as a circular error probable (CEP, in meters or nautical miles). The maximum allowable CEP is often based on the calculated risk of collision on a typical mission.
- *Autonomy,* the extent to which the vehicle determines its own position and velocity without external aids. Autonomy is important to certain military vehicles and to civil vehicles operating in areas of inadequate radio-navigation coverage.
- *Time delay* in calculating position and velocity, caused by computational and sensor delays.
- *Geographic coverage.* Radio systems operating below 100 kHz can be received beyond line of sight on earth; those operating above 100 MHz are confined to line of sight. On other planets, new navigation aids—perhaps navigation satellites or ground stations—will be installed.
- *Automation.* The operator (on-board crew or ground controller) receives a direct reading of position, velocity, and equipment status, usually without human intervention. The navigator's crew station disappeared in aircraft in the 1970s, and human navigators are scarce in the 1990s, even on ships, because electronic equipment automatically selects stations, measures position and velocity, calculates waypoint steering, and accommodates failures.

Defining Terms

Ecliptic: Plane of earth's orbit around the sun.

Inertial space: Any coordinate frame whose origin is on a freely falling (orbiting) body and whose axes are nonrotating relative to the fixed stars. It is definable within 10^{-7} degree/h.

Lanes: Hyperbolic bands on the earth's surface in which continuous-wave radio signals repeat in phase.

Nautical mile (nm):   1852 m, exactly. Approximately 1 minute of arc on the earth's surface.

State vector: Six-component vector, three of whose elements are position and three of whose elements are velocity.

Update: The intermittent resetting of the dead-reckoned state vector based on absolute navigation measurements (see Section 102.3).

References

R. H. Battin, *An Introduction to the Mathematics and Methods of Astrodynamics,* Washington: AIAA Press, 1987, 796 pp.

P. Berlin, *The Geostationary Applications Satellite,* Cambridge: Cambridge University Press, 1988, 214 pp.

N. Bowditch, *The American Practical Navigator*, Washington, D.C.: U.S. Government Printing Office, latest edition, 1524 pp.

M. Kayton, *Navigation: Land, Sea, Air, and Space*, New York: IEEE Press, 1990, 461 pp.

M. Kayton and W.R. Fried, *Avionic Navigation Systems*, New York: Wiley, 1969, 666 pp.

R. A. Minzner, *The U.S. Standard Atmosphere 1976*, NOAA Report 76-1562, NASA SP-390, 1976 or latest edition, 227 pp.

NASA, *Space Network Users Guide*, Greenbelt, Md.: Goddard Space Flight Center, 1988 or latest edition, 500 pp.

U.S. Air Force, *NAVSTAR-GPS Interface Control Document*, Annapolis, Md.: ARINC Research, 1991, 115 pp.

U.S. Government, *Federal Radio Navigation Plan*, Department of Transportation, 1990, 169 pp., issued biennially.

J. Yuen, *Deep Space Telecommunication Systems Engineering*, New York: Plenum Press, 1983, 603 pp.

Further Information

IEEE Transactions on Aerospace and Electronic Systems, bimonthly through 1991, now quarterly.

Proceedings of the IEEE Position Location and Navigation Symposium (PLANS), biennially.

Navigation, journal of the U.S. Institute of Navigation, quarterly.

Journal of Navigation, Royal Institute of Navigation (UK), quarterly.

AIAA Journal of Guidance and Control, bimonthly.

Proceedings of the IEEE Vehicle Navigation and Information Systems Conference, annually.

Commercial aeronautical standards produced by International Civil Aviation Organization (ICAO, Montreal) and by ARINC, Annapolis, Md.

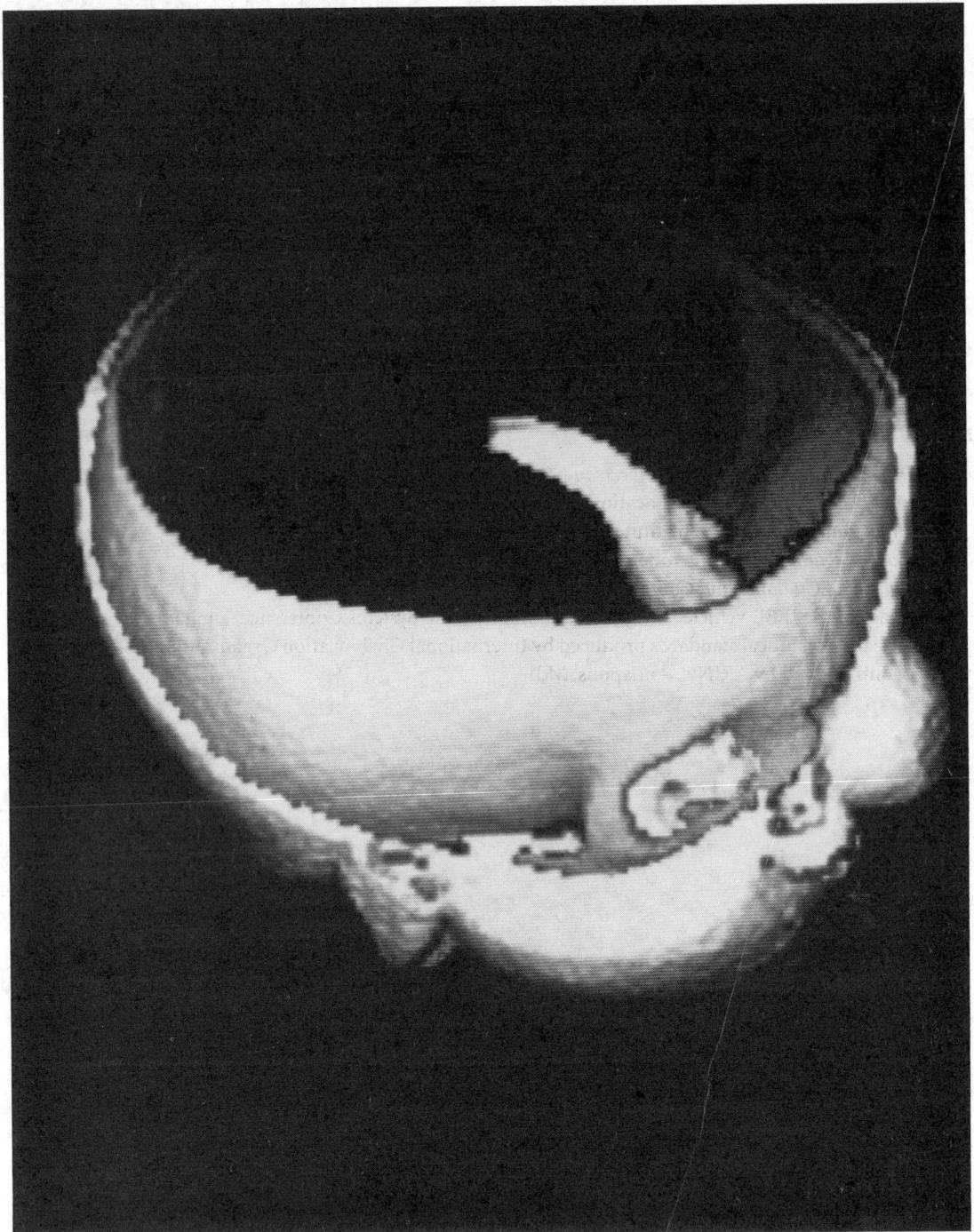

A computer with graphics software assembled 58 axial scans from a CT to produce this three-dimensional view of the brain. At the University of Kansas at Kansas City, Kansas this view was used to diagnose a massive tumor causing partial blindness in the 58-year-old male. The long, tubular tumor extends from the vicinity of the optic nerve at the front of the head almost to the mid-brain. This new application of CT, known as three-dimensional reconstruction imaging, was used to diagnose the tumor and treat the patient with the aid of radiation therapy. No invasive surgical procedures were needed to restore this man's vision. (Photo from Howard Sochurek.)

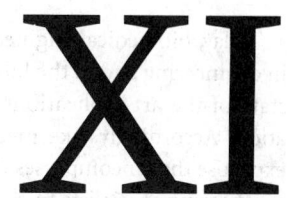

Biomedical Systems

Joseph D. Bronzino
Trinity College

103　Bioelectricity　*J. Reilly, L. Geddes, C. Polk* 2301
　　Neuroelectric Principles • Bioelectric Events • Application of Electric and Magnetic Fields in
　　Bone and Soft Tissue Repair

104　Biomedical Sensors　*M. Neuman* 2342
　　Physical Sensors • Chemical Sensors • Bioanalytical Sensors • Applications

105　Bioelectronics and Instruments　*J. Bronzino, E. Berbari* 2351
　　Quantitative Analysis of the Electroencephalograms • The Electrocardiograph

106　Medical Imaging　*M. Fox, L. Frizzell* 2374
　　Tomography • Ultrasound

107　Rehabilitation Engineering　*C. Robinson* 2387
　　Rehabilitation Concepts • Engineering Concepts in Sensory Rehabilitation • Engineering
　　Concepts in Motor Rehabilitation • Engineering Concepts in Communications Disorders •
　　Appropriate Technology • An Example of Rehabilitation Engineering • The Future of Electrical
　　Engineering in Rehabilitation

108　Biocomputing　*L. Kun, M. Baretich* 2397
　　Clinical Information Systems • Hospital Information Systems

109　Safety and Risk-Control Issues　*Y. David* 2408
　　The Biomedical Equipment Risk Causes • Risk-Control Programs

T ECHNOLOGICAL INNOVATION in the twentieth century has progressed at such an accelerated pace that it has permeated almost every facet of our lives. This is especially true in the field of
medicine and the delivery of health care services. Although the art of medicine has a long history,
the evolution of a health care system capable of providing a wide range of positive therapeutic
treatments in the prevention and cure of illnesses is a decidedly new phenomenon. Of particular
importance in this evolutionary process has been the establishment of the modern hospital as the
center of a technologically sophisticated health care system. In the process, the discipline of
biomedical engineering has emerged as an integrating medium for two dynamic professions,
medicine and engineering, assisting in the struggle against illness and diseases by providing materials, tools, and techniques (such as signal and image processing and artificial intelligence) that can
be utilized for research, diagnosis, and treatment by health care professionals.

Today, biomedical engineering is an interdisciplinary branch of engineering heavily based both in engineering and in the life sciences. It ranges from theoretical, nonexperimental undertakings to state-of-the-art applications. It can encompass research, development, implementation, and operation. Accordingly, like medical practice itself, it is unlikely that any single person can acquire expertise that encompasses the entire field. As a result, there are now a great number of biomedical engineering specialists to cover this broad spectrum of activity. Yet because of the interdisciplinary nature of this activity, there is considerable interplay and overlapping of interest and effort between them. For example, biomedical engineers engaged in the development of biosensors may interact with those interested in prosthetic devices to develop a means to detect and use the same bioelectric signal to power a prosthetic device. Those engaged in automating the clinical chemistry laboratory may collaborate with those developing expert systems to assist clinicians in making clinical decisions based upon specific laboratory data. The possibilities are endless.

There are seven major career areas in biomedical engineering: (1) application of engineering system analysis and modeling (computer simulation) to biological problems; (2) measurement or monitoring of physiological signals; (3) diagnostic interpretation via signal processing techniques of bioelectric data; (4) therapeutic and rehabilitation procedures and devices; (5) prosthetic devices for replacement or augmentation of bodily functions; (6) computer analysis of patient-related data; and (7) medical imaging, i.e., the graphic display of anatomical detail or physiological function. Biomedical engineers, therefore, engage in the following pursuits:

- Design of instrumentation for human physiology research
- Monitoring astronauts and maintenance of life in space
- Research in new materials for implanted artificial organs
- Development of new diagnostic instruments for blood analysis
- Computer modeling of the function of the human heart
- Writing software for analysis of medical research data
- Analysis of medical device hazards for the U.S. government
- Monitoring the physiological functions of animals
- Development of new diagnostic imaging systems
- Design of telemetry systems for patient monitoring
- Design of biomedical sensors for measurement of human physiological systems variables
- Research on artificial intelligence (AI) and development of expert systems for diagnosis of diseases
- Design of closed-loop control systems for drug administration
- Modeling of the physiological systems of the human body
- Design of instrumentation for sports medicine
- Development of new dental materials
- Design of computers and communication aids for the handicapped
- Research in pulmonary fluid dynamics (biorheology)
- Study of the biomechanics of the human body

This list is not intended to be all-inclusive, for there are many other applications that utilize the talents and skills of the biomedical engineer. In fact, the list of activities of biomedical engineers depends upon the medical environment in which they work. This is especially true for the "clinical engineers," i.e., biomedical engineers employed in hospitals or clinical settings.

The utilization of biomedical engineers offers great potential benefit in the identification of problems and needs of our present health care delivery system that can be solved using existing engineering technology and systems methodology. Consequently, the field of biomedical engineering offers hope in the continuing battle to provide high-quality health care at reasonable cost. The purpose of this section, therefore, is to provide a broad overview of biomedical engineering topics of interest to electrical engineers.

Nomenclature

Symbol	Quantity	Unit	Symbol	Quantity	Unit
C	resistivity	Ω	n	valence of ion	
C	proton density		p	pressure	N/m
C_m	membrane capacitance	F	Q_T	threshold charge	C
d	diameter	m	R	universal gas constant	
D	lateral beamwidth		R	transmembrane resistance	Ω
f_d	doppler shift	Hz	σ	conductivity	S/m
f_L	Larmour frequency	Hz	t_d	decay time	s
F	Faraday constant		t_r	rise time	s
Γ	pressure reflection coefficient		T	absolute temperature	K
I_{mt}	minimum threshold current	A	V	membrane potential	V
I_t	threshold current	A	W	nodal gap width	
k	propagation constant		Z	valence of substance	
L	internodal distance	m	Z	acoustic impedance	Ω
m	mass	g			

103

Bioelectricity

103.1 Neuroelectric Principles 2301

103.2 Bioelectric Events 2311
Origin of Bioelectricity • Law of Stimulation • Recording Action
Potentials • The Electrocardiogram (ECG) • Electromyography
(EMG) • Electroencephalography (EEG)

103.3 Application of Electric and Magnetic Fields in Bone and Soft
Tissue Repair 2329
History • Devices for Bone and Cartilage Repair • Soft Tissue
Repair and Nerve Regeneration • Mechanisms and Dosimetry

J. Patrick Reilly
The Johns Hopkins University

L. A. Geddes
Purdue University

C. Polk
University of Rhode Island

103.1 Neuroelectric Principles

J. Patrick Reilly

Natural bioelectric processes are responsible for nerve and muscle function. These processes can be affected by externally applied electric currents that are intentionally introduced through medical devices or unintentionally introduced through accidental exposure (electric shock). A thorough treatment of this topic is given in Reilly [1992].

Externally applied electric currents can excite nerve and muscle cells. Muscle can be stimulated directly or indirectly through the nerves that enervate the muscle. Thresholds of stimulation of nerve are generally well below thresholds for direct stimulation of muscle. An understanding of neuroelectric principles is a valuable foundation for investigation into both sensory and muscular responses to electrical stimulation.

Figure 103.1 illustrates functional components of sensory and motor (muscle) **neurons**. The illustrated nerve fibers are **myelinated**, i.e., covered with a fatty layer of insulation called *myelin* and having *nodes of Ranvier* where the myelin is absent. The conducting portion of the nerve fiber is a long, hollow structure known as an **axon**. The axon plus myelin sheath is frequently referred to as a nerve **fiber**, or neuron. Bundles of neurons are called *nerves*.

The body is equipped with a vast array of sensors (receptors) for monitoring its internal and external environment. Electrical stimulation generally involves the *somatosensory* system, i.e., the system of receptors found in the skin and internal organs. Other specialized receptors include those in the visual and auditory systems and chemical receptors by which neurons communicate with one another.

The somatosensory receptors can be classified as mechanoreceptors, thermoreceptors, chemoreceptors, and nociceptors. Numerous specializations of mechanoreceptors respond to specific attributes of mechanical stimulation. Thermoreceptors are specialized to respond to either heat or cold stimuli. Nociceptors are unresponsive until the stimulus reaches the point where tissue damage is imminent and are usually associated with pain. Many nociceptors are responsive to a broad

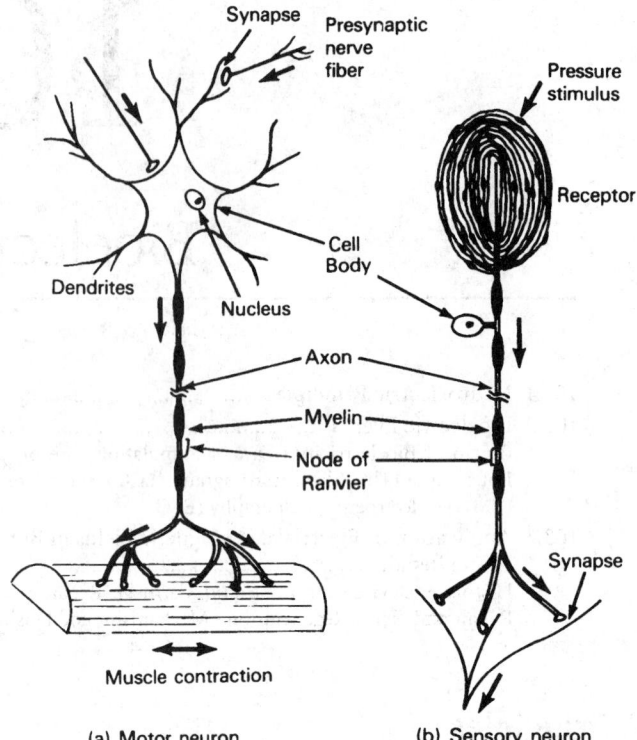

FIGURE 103.1 Functional components of (a) motor and (b) sensory neurons. Arrows indicate the direction of information flow. Signals are propagated across synapses via chemical neurotransmitters and elsewhere by membrane depolarization. Synapses are inside the spinal column. The sizes of the components are drawn on a distorted scale to emphasize various features.

spectrum of noxious levels of mechanical, heat, and chemical stimuli. The muscles are equipped with specialized receptors to monitor and control muscle movement and posture. Figure 103.1 illustrates a *pacinian corpuscle*, which responds to the onset or termination of a pressure stimulus applied to the skin.

When a sensory receptor is stimulated, it produces a voltage change called a *generator potential*. The generator potential is graded: if you squeeze a pacinian corpuscle, for example, it produces a voltage; if you squeeze it harder, it produces a greater voltage. The generator potential initiates a sequence of events that leads to a propagating **action potential** (a "nerve impulse" in common parlance).

The functional boundary of the biological cell is a thin (about 10 nm) bimolecular lipid and protein structure called a **membrane**. Electrochemical forces across the membrane regulate chemical exchange across the cell. The medium within the cell (the *plasm*) and outside the cell (the *interstitial fluid*) is composed largely of water containing various ions. The difference in the concentration of ions inside and outside the cell causes an electrochemical force across the cell membrane. The membrane is a semipermeable dielectric that allows some ionic interchange. Under conditions of electrochemical equilibrium (no net force in either direction), the membrane will attain a potential described by the *Nernst* equation

$$V_m = \frac{RT}{FZ} \ln \frac{[S]_o}{[S]_i} \tag{103.1}$$

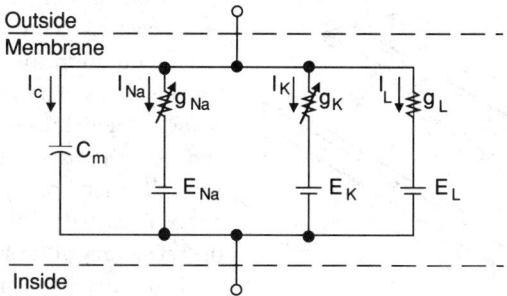

FIGURE 103.2 Hodgkin-Huxley membrane model.

where $[S]_i$ and $[S]_o$ represent the concentrations of ionic substance S inside and outside the cell, R is the universal gas constant, T is absolute temperature, F is the Faraday constant (number of coulombs per mole of charge), and Z is the valence of substance S. Using the values $R = 8.31$ J/mol K, $T = 310$ K, $F = 96{,}500$ C/mol, and $Z = +1$ (for a monovalent cation), converting to the base 10 logarithm, and expressing V_m in millivolts, we obtain

$$V_m = 61 \log \frac{[S]_o}{[S]_i} \tag{103.2}$$

In a quiescent state, nerve and muscle cells maintain a membrane potential typically around -60 to -90 mV, with the inside of the cell negative relative to the outside. Two ions that are involved in the electrical response of nerve and muscle are Na^+ and K^+. The concentration of these ions inside and outside the cell dictates the Nernst potential according to Eq. (103.2). Example concentrations in $\mu M/cm^3$ for a nerve axon would be $[Na^+]_i = 50$, $[Na^+]_o = 460$, $[K^+]_i = 400$, and $[K^+]_o = 10$. The Na^+ potential is found to be around $+60$ mV; the K^+ potential is found to be somewhat more negative than the resting potential. Obviously, the cell maintains in a state of electrochemical disequilibrium. The energy that maintains this force is derived from the metabolism of the cell—a dead cell will eventually revert to a state of equilibrium. Considering the transmembrane potential (≈ 100 mV), and its small thickness (≈ 10 nm), the electric field across the membrane is enormous (≈ 10 MV/m).

The membrane is semipermeable; that is, it is a lossy dielectric which allows the passage of certain ions. The ionic permeability varies substantially from one ionic species to another. The ionic channels in the excitable membrane will vary their permeability in response to the transmembrane potential; this property distinguishes the excitable membrane from the ordinary cellular membrane, and it supports propagation of nerve impulses.

The electrodynamics of the excitable membrane of unmyelinated nerves were first described in detail in the Nobel prize work of Hodgkin and Huxley [1952]. This work was later extended to the myelinated nerve membrane by Frankenhaeuser and Huxley [1964]. Figure 103.2 illustrates an electrical model of the Hodgkin-Huxley membrane, which consists of nonlinear conductances for Na^+ and K^+ and a linear leakage element. The potential sources shown in the diagram are the Nernst potentials for the particular ions as given by Eq. (103.2). The capacitance term C_m is formed by the dielectric membrane separating the conductive media on either side. The conductances g_{Na} and g_K apply to Na^+ and K^+ channels; the conductance g_L is a general "leakage" channel that is not specific to any particular ion. The g_{Na} and g_K conductivities are highly dependent on the voltage applied across the membrane as described by a set of nonlinear differential equations. When the membrane is in the resting state, $g_{Na} \ll g_K$, and the membrane potential moves toward the Nernst potential for Na^+. In this depolarized state, the membrane is said to be excited. The transition between the resting and excited condition of the membrane occurs rather abruptly when the membrane potential has been depolarized by roughly 15 mV. After excitation, the ionic channel conduc-

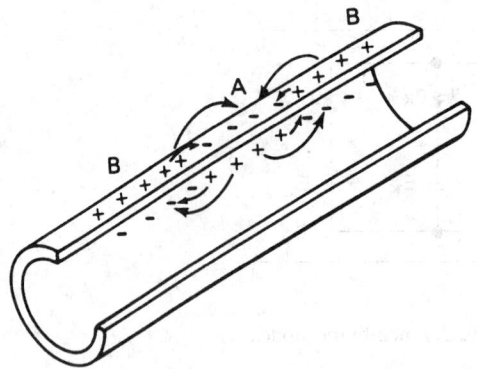

FIGURE 103.3 Spread of the depolarization wave front along an axon. Depolarization occurring in region A results in charge transfer from the adjacent regions.

tances vary again, causing the membrane to revert back to its resting potential.

The duration of the excited state lasts roughly 1 ms. The progression of the membrane voltage during the period of excitation and recovery is termed an *action potential.* After the membrane has been excited, it cannot be reexcited until a recovery period, called the **refractory period**, has passed.

Figure 103.3 illustrates the processes that support the propagation of an action potential. Consider that point A on the axon is depolarized. The local depolarization causes ionic transfer between adjacent points on the axon, thus propagating the region of depolarization. If depolarization were initiated from an external electrical source on a resting membrane at point A, an action potential would propagate in both directions away from the site of stimulation. The body's natural condition, however, is to initiate an action potential at the terminus of the axon, which then propagates in only one direction.

Electrical Model for Nerve Excitation

Myelinated fibers have much lower thresholds of excitation than unmyelinated fibers. Accordingly, the myelinated fiber is an appropriate choice for electrical stimulation studies.

Figure 103.4 illustrates an electrical model for myelinated nerve as originally formulated by McNeal [1976]. The myelin internodes are treated as perfect insulators and the nodes as individual circuits consisting of capacitance C_m and an ionic conductance term. The nodes are interconnected through the internal axon medium by conductances G_a. The current flowing in the biological medium creates voltage disturbances $V_{e,n}$ at the exterior of the nodes.

The current emanating from the nth node is the sum of capacitive and ionic currents described by

$$C_m \frac{dV_n}{dt} + I_{i,n} = G_a(V_{i,n-1} - 2V_{i,n} + V_{i,n+1}) \qquad (103.3)$$

where C_m is the membrane capacitance at the node, V_n is the transmembrane potential, $I_{i,n}$ is the total ionic current, and $V_{i,n}$ is the internal voltage. In this expression, V_n is taken relative to the resting potential, such that $V_n = 0$ applies to the membrane resting potential. The ionic current flux is the sum of individual ionic terms (similar to the representation in Fig. 103.4),

$$I_{i,n} = \pi dW(J_{Na} + J_K + J_L + J_P) \qquad (103.4)$$

where the J terms are ionic current densities as described by a set of nonlinear differential equations developed by Frankenhaeuser and Huxley [1964] for a myelinated nerve membrane. Other relationships are

$$G_a = \frac{\pi d^2}{4\rho_i L} \qquad (103.5)$$

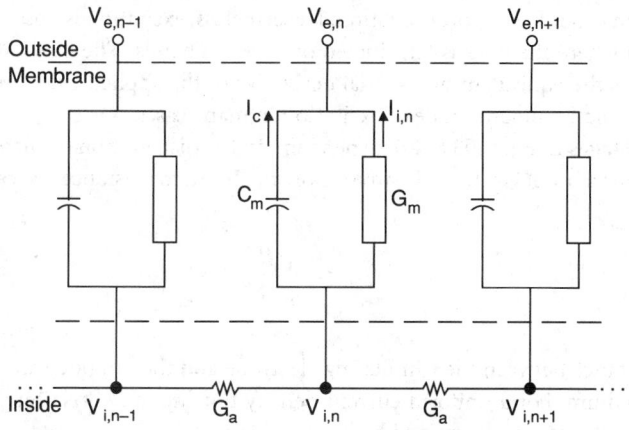

FIGURE 103.4 Equivalent circuit model for electrical excitation of myelinated nerve fiber. The membrane conductance G_m is described by nonlinear ionic conductances, similar to the representation in Fig. 103.2.

$$C_m = c_m \pi d W \qquad (103.6)$$

where d is the axon diameter at the node, ρ_i is the resistivity of the internal axon medium, L is the internodal distance, W is the nodal gap width, and c_m is the membrane capacitance per unit area. The relationship between the axon diameter d and the fiber diameter D (including myelin) is $d \approx 0.7D$. The voltage V_n across the membrane is

$$V_n = V_{i,n} - V_{e,n} \qquad (103.7)$$

where $V_{i,n}$ and $V_{e,n}$ are the internal and external nodal voltages with reference to a distant point in the conducting medium outside the axon. Substituting Eq. (103.7) into (103.3) results in

$$\frac{dV_n}{dt} = \frac{1}{C_m}[G_a(V_{n-1} - 2V_n + V_{n+1} + V_{e,n} - 2V_{e,n} + V_{e,n+1}) - I_{i,n}] \qquad (103.8)$$

Equation (103.8) may be analogously expressed in continuous form as

$$\tau_m \frac{\partial V}{\partial t} - \lambda^2 \frac{\partial^2 V}{\partial x^2} + V = \lambda^2 \frac{\partial^2 V_e}{\partial x^2} \qquad (103.9)$$

where V and V_e are membrane voltage and external voltage, respectively, at longitudal position x, τ_m is the membrane time constant given by C_m/G_m, λ is the membrane space constant given by $\sqrt{r_m/r_i}$, and G_m is membrane conductance. Equation (103.9) connects continuous and discrete spatial derivatives by $\partial^2 V/\partial x^2 \approx V_{n-1} - 2V_n + V_{n+1}$ and $\partial^2 V_e/\partial x^2 \approx V_{e,n-1} - 2V_{e,n} + V_{e,n}$. An additional relationship is $I_{i,n} = V/G_m$. The left-hand side of Eq. (103.9) is the so-called cable equation that was developed by Oliver Heaviside over 100 years ago in connection with the analysis of the first transatlantic telegraphy cable. The right-hand side is a driving function due to the external field in the biological medium. For additional information on cable theory as applied to the excitable membrane, the reader is directed to Jack *et al.* [1983].

One conclusion that can be drawn from Eqs. (103.8) and (103.9) is that a second spatial derivative of voltage (or equivalently a first derivative of the electric field) must exist along the long axis of an excitable fiber in order to support excitation. Nevertheless, excitation is possible in a locally constant electric field where the fiber is terminated or where it bends. The orientation change or the termination creates the equivalent of a spatial derivative of the applied field. Stimulation at "ends and bends" can be the dominant mode of excitation in many cases.

The external voltages in Eq. (103.8) are dependent on the distribution of current within the biological medium. For a point electrode in an isotropic medium, for instance, we can determine these voltages by

$$V_{e,n} = \frac{\rho_e I}{4\pi r_n} \tag{103.10}$$

where r_n is the distance between the stimulating electrode and the nth node and ρ_e is the resistivity of the external medium. For a uniform current density flowing in a direction parallel to the fiber axis, the external voltages are determined by

$$V_{e,n} = V_{e,1} + ELn \tag{103.11}$$

where $V_{e,1}$ is a reference voltage at the terminal node, L is the internodal distance, n is the node number, and E is the electric field in the medium. The electric field is related to current density by $J = E\sigma$, where $\sigma = 1/\rho_e$ is the conductivity of the medium and J is the current density. Since the response of the electrical model is independent of $V_{e,1}$, we may assume $V_{e,1} = 0$ for convenience in Eq. (103.11).

The internodal distance L is proportional to fiber diameter D through the relationship $L/D \approx 100$. Other fiber diameter relationships are expressed in Eqs. (103.5) and (103.6). Because of these relationships, thresholds of electrical stimulation will vary inversely with fiber diameter. The distribution of myelinated nerve diameters found in human peripheral nerve or skeletal muscle typically ranges from 5 to 20 µm.

Figure 103.5 illustrates the response of the myelinated nerve model of Fig. 103.4 to a rectangular current stimulus [Reilly et al., 1985]. The example is for a small cathodal electrode that is 2 mm radially distant from a 20-µm fiber and directly above a central node. The transmembrane voltage ΔV is scaled relative to the resting potential. The solid curves show the response at the node nearest the stimulating electrode. Response a is for a pulse that is 80% of the threshold current, b is at threshold, and c is 20% above threshold. The threshold stimulus pulse in this example has an amplitude I_T of 0.68 mA. Response a is similar to that of a linear network with a parallel resistor and capacitor and charged by a brief current pulse. Responses b and c demonstrate the highly nonlinear response of the excitable membrane. The dashed curves in Fig. 103.5 show the membrane response to a threshold stimulus at the three nodes adjacent to the one nearest the stimulating electrode. The time delay implies a propagation velocity of 43 m/s, which is typical of a 20-µm fiber. The membrane response seen in curves b through f illustrates the action potential described earlier. The action potential is typically described as an "all-or-nothing" response; that is, its amplitude is not normally graded—either the axon is excited, or it is not.

The threshold current needed for excitation is highly dependent on its duration and waveshape. A common format for representing the response of a nerve is through **strength-duration curves**, i.e., the plot of the threshold of excitation versus the duration of the stimulating current. We can determine the threshold of excitation by "titrating" the stimulus current between a threshold and no-threshold condition.

Figure 103.6 illustrates strength-duration curves derived from the myelinated nerve model described previously under the same conditions applying to Fig. 103.5. Three types of stimulus current apply to Fig. 103.6: a monophasic constant current pulse, a symmetric biphasic rectangular current, and a single cycle of a sine wave. The phase duration indicated on the horizontal axis applies to the initial cathodal half cycle for the two biphasic waves. Stimulus magnitude is given in

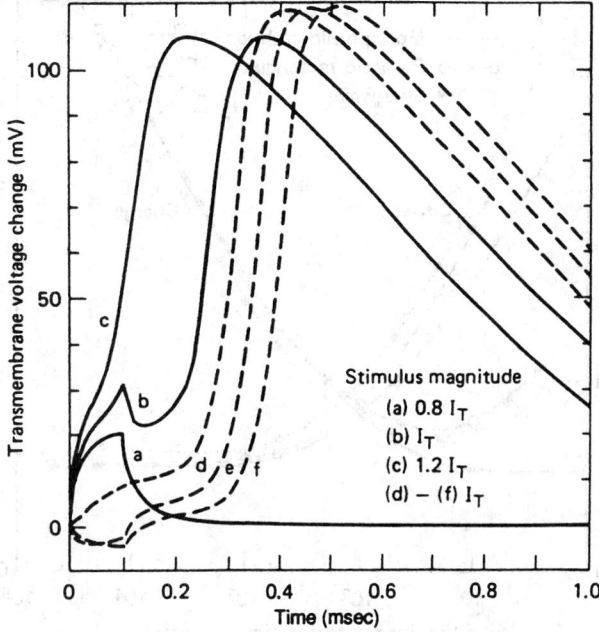

FIGURE 103.5 Response of myelinated nerve model to rectangular monophasic current of 100 ms duration, 20-μm diameter fiber, point electrode 2 mm from central node. Solid lines show response at node nearest electrode for three levels of current. I_T denotes threshold current. Dashed lines show propagated response at next three adjacent nodes for a stimulus at threshold. (*Source:* J. P. Reilly, V. T. Freeman, and W. D. Larkin, "Sensory effects of transient electrical stimulation—Evaluation with a neuroelectric model," *IEEE Trans. Biomed. Eng.*, vol. BME-32, no. 12, pp. 1001–1011, © 1985 IEEE.)

terms of peak current on the right vertical axis and in terms of the charge in a single monophasic phase of the stimulus on the left vertical axis. The charge is computed by $Q = It_p$ for the rectangular waveforms and $Q = (2/\pi)It_p$ for the sinusoidal waveforms (I is threshold current and t_p is phase duration).

The solid curve labeled "current" is of the type that is most often represented as a strength-duration curve. For this curve, the minimum threshold current occurs for long-stimulus durations and is called the *rheobasic current*, or simply **rheobase**. The duration consistent with twice the rheobase is called the **chronaxie**. The solid curve in Fig. 103.6 labeled "charge" gives the area under the rectangular current pulse. The threshold charge is a minimum for short-duration stimuli.

Mathematical curve fits to the strength-duration curves for monophasic rectangular stimuli are

$$\frac{I_T}{I_o} = \frac{1}{1 - e^{-t/\tau e}} \tag{103.12}$$

and

$$\frac{Q_T}{Q_o} = \frac{t/\tau_e}{1 - e^{-t/\tau e}} \tag{103.13}$$

where I_T is threshold current, Q_T is threshold charge, I_o is the minimum threshold current for long-duration stimuli, Q_o is the minimum threshold charge for short-duration stimuli, and τ_e is an

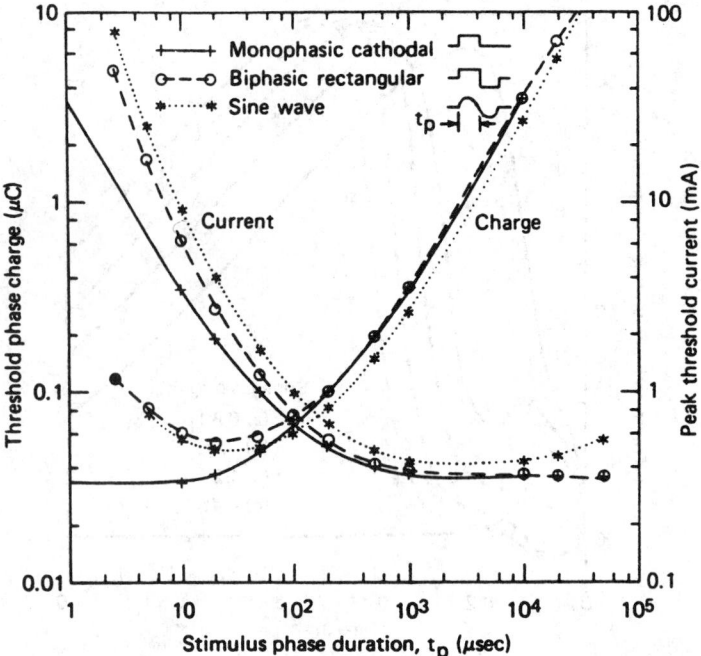

FIGURE 103.6 Strength/duration relationships derived from the myelinated nerve model: current thresholds and charge thresholds for single-pulse monophasic and for single-cycle biphasic stimuli with initial cathodal phase, point electrode 2 mm distant from 20 μm fiber. Threshold current refers to the peak of the stimulus waveform. Charge refers to a single phase for biphasic stimuli. (*Source:* J. P. Reilly, V. T. Freeman, and W. D. Larkin, "Sensory effects of transient electrical stimulation—Evaluation with a neuroelectric model," *IEEE Trans. Biomed. Eng.*, vol. BME-32, no. 12, pp. 1001–1011, © 1985 IEEE.)

experimentally determined strength-duration time constant. It is readily shown that chronaxie = τ_e ln 2 = $0.693\tau_e$ in this formulation.

Values of I_o and Q_o vary considerably with experimental parameters such as electrode size and location and the size of the neuron. Values of τ_e also vary considerably with experimental conditions: a value around 250 μs is typical for both sensory and motor nerve excitation via cutaneous electrodes, and values around 125 μs are observed for stimulation of axons by small electrodes. Much longer time constants are associated with direct stimulation of muscle cells.

The current reversal of a biphasic stimulus can reverse a developing action potential that was elicited by the initial phase. As a result, a biphasic pulse may have a higher threshold than a monophasic pulse as suggested by the biphasic thresholds in Fig. 103.6. The degree of biphasic threshold elevation is magnified as the stimulus duration is reduced.

A sinusoidal current is a special case of a biphasic stimulus. Sinusoidal threshold response can be represented by strength-frequency curves, as shown by the solid curves in Fig. 103.7 for the myelinated nerve model. Several experimental curves have been included in the figure; these have been shifted vertically to facilitate comparisons. Notice that the myelinated nerve model predicts a lower threshold for stimulation by a continuous sine wave as compared with a single cycle.

The strength-frequency curve follows a U-shaped function, with a minimum at mid frequencies and an upturn at both low and high frequencies. At low frequencies the slow rate of change of the sinusoid prevents the membrane capacitance from building up a depolarizing voltage because membrane capacitance is counteracted by membrane leakage. This process describes the neural property known as *accommodation*, i.e., the adaptation of a nerve to a slowly varying or constant

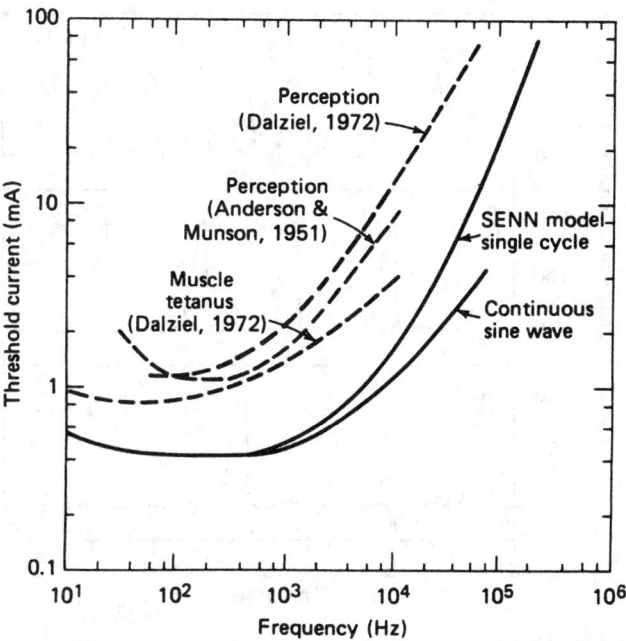

FIGURE 103.7 Strength-frequency curves for sinusoidal current stimuli. Dashed curves are from experimental data. Solid curves apply to myelinated nerve model. Experimental curves have been shifted vertically to facilitate comparisons.

stimulus. The high-frequency upturn occurs because of the canceling effects of a current reversal on the membrane voltage change. An empirical fit to strength-frequency curves is

$$I_t = I_o K_H K_L \tag{103.14}$$

where I_t is the threshold current, I_o is the minimum threshold current, and K_H and K_L are high- and low-frequency terms, defined, respectively, as

$$K_H = \left[1 - \exp\left(-\frac{f_e}{f} \right) \right]^{-a} \tag{103.15}$$

and

$$K_L = \left[1 - \exp\left(-\frac{f}{f_o} \right) \right]^{-b} \tag{103.16}$$

where f_e and f_o are constants that determine the points of upturn in the strength-frequency curve at high and low frequencies, respectively. An upper limit of $K_L \leq 4.6$ is assumed for Eq. (103.16) to account for the fact that excitation may be obtained with finite dc currents. An empirical fit of Eqs. (103.15) and (103.16) to the mylinated nerve model thresholds indicates that $a = 1.45$ for a single-cycle stimulus and $a = 0.9$ for a continuous stimulus; $b = 0.8$ regardless of stimulus duration. The value of I_o will depend on various conditions of stimulation, including the size of the electrode, its location on the body, and the location of the stimulated nerve.

With continuous sinusoidal stimulation, it is possible to produce a series of action potentials that are phase-locked to the individual sinusoidal cycles, as noted in Fig. 103.8. This makes the sinusoidal stimulus much more potent than a single pulse of the same phase duration. This potency is

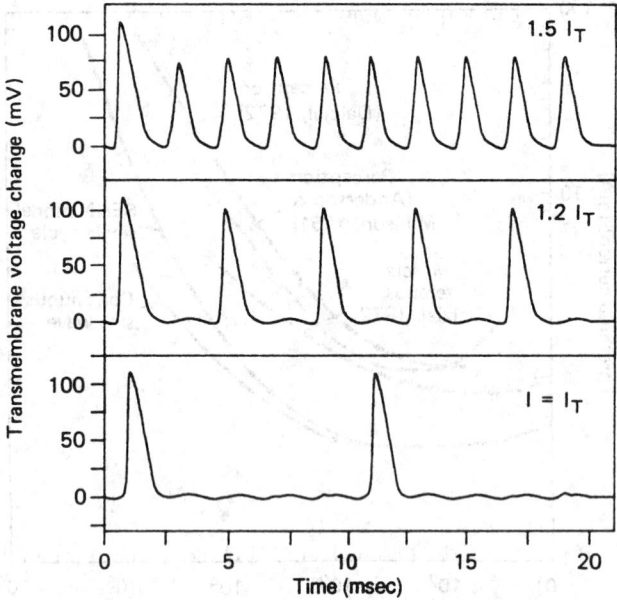

FIGURE 103.8   Model response to continuous sinusoidal stimulation at 500 Hz. The lower panel depicts the response to a stimulus current set at threshold level (I_T) for a single-cycle stimulus. Upper panels show responses for stimulation 20 and 50% above the single-cycle threshold. (*Source:* J. P. Reilly, V. T. Freeman, and W. D. Larkin, "Sensory effects of transient electrical stimulation—Evaluation with a neuroelectric model," *IEEE Trans. Biomed. Eng.*, vol. BME-32, no. 12, pp. 1001–1011, © 1985 IEEE.)

a consequence of the fact that perceived magnitude for neurosensory stimulation and muscle tension for neuromuscular stimulation both increase with the rate of action potential production.

Defining Terms

Action potential: A propagating change in the conductivity and potential across a nerve cell's membrane; a nerve impulse in common parlance.

Axon: The conducting portion of a nerve fiber—a roughly tubular structure whose wall is composed of the cellular membrane and which is filled with an ionic medium.

Chronaxie: The minimum duration of a unidirectional square-wave current needed to excite a nerve when the current magnitude is twice rheobase.

Fiber, nerve: A single nerve cell; a neuron—classified on the presence or absence of myelin. Myelinated nerve cells have diameters typically in the range 2 to 20 μm and conduction velocities of 5 to 120 m/s; unmyelinated nerves have diameters from 0.3 to 1.3 μm and conduction velocities of 0.6 to 2.3 m/s. Fiber lengths may be up to 1 m. The term nerve usually refers to a bundle of nerve fibers.

Membrane: The functional boundary of a cell. Nerve cells possess membranes that are excitable by virtue of their nonlinear electrical conductance properties (see Action potential).

Myelinated nerve: A nerve fiber insulated with a fatty substance called myelin and having periodically exposed *nodes of Ranvier.*

Neuron: A nerve cell. Sensory neurons carry information from sensory receptors in the peripheral nervous system to the brain; motor neurons carry information from the brain to the muscles.

Refractory period: A period of time after the initiation of an action potential during which further excitation is impossible (absolute refractory period) or requires a greater stimulus (relative refractory period).

Rheobase: The minimum current necessary to cause nerve excitation—applicable to a long-duration current (e.g., several milliseconds).

Strength-duration curve: A curve expressing the functional relationship between the threshold of excitation of a nerve fiber and the duration of a unidirectional square-wave electrical stimulus.

References

B. Frankenhaeuser and A. F. Huxley, "The action potential in the myelinated nerve fiber of *Xenopus laevis* as computed on the basis of voltage clamp data," *J. Physiol.*, vol. 171, pp. 302–315, 1964.

A. L. Hodgkin and A. F. Huxley, "A quantitative description of membrane current and its application to conduction and excitation in nerve," *J. Physiol.*, vol. 117, pp. 500–544, 1952.

J. J. B. Jack, D. Noble, and R. W. Tsien, *Electric Current Flow in Excitable Cells*, Oxford: Clarendon Press, 1983.

D. R. McNeal, "Analysis of a model for excitation of myelinated nerve," *IEEE Trans. Biomed. Eng.*, vol. BME-22, pp. 329–337, 1976.

J. P. Reilly, V. T. Freeman, and W. D. Larkin, "Sensory effects of transient electrical stimulation—Evaluation with a neuroelectric model," *IEEE Trans. Biomed. Eng.*, vol. BME-32, no. 12, pp. 1001–1011, 1985.

J. P. Reilly, *Electrical Stimulation and Electropathology*, New York: Cambridge University Press, 1992.

Further Information

For further information, the reader is directed to the references listed at the end of this chapter. Additional references are:

R. Plonsey and R. C. Barr, *Biolectricity—A Quantitive Approach*, New York: Plenum, 1988.

W. Agnew and D. McCreery, *Neural Prostheses*, Englewood Cliffs, N.J.: Prentice-Hall, 1990.

E. R. Kandel, J. H. Schwartz, and T. M. Jessell (Eds.), *Principles of Neural Science*, 3rd ed., New York: Elsevier, 1991.

Several journals treat engineering applications of neuroelectric principles, such as *IEEE Transactions on Biomedical Engineering*, *Medical and Biological Engineering and Computing*, and *Annals of Biomedical Engineering*.

Of the many conferences treating bioelectric responses, one having a broad range of applications is the IEEE Annual Conference on Engineering in Medicine and Biology.

103.2 Bioelectric Events

L. A. Geddes

Bioelectric signals are exploited for the diagnostic information that they contain. Such signals are often used to monitor and guide therapy. Although all living cells exhibit bioelectric phenomena, a small variety produce potential changes that reveal their physiological function. The most familiar bioelectric recordings are the electrocardiogram, ECG (which reflects the excitation and recovery of the whole heart), the electromyogram, EMG (which reflects the activity of skeletal muscle), and the electroencephalogram, EEG (which reflects the activity of the outer layers of the brain, the cortex). The following paragraphs will describe (1) the origin of all bioelectric phenomena; (2) the nature of the electrical activity of the heart, skeletal muscle, and the brain; and (3) the characteristics of instrumentation used to display these events.

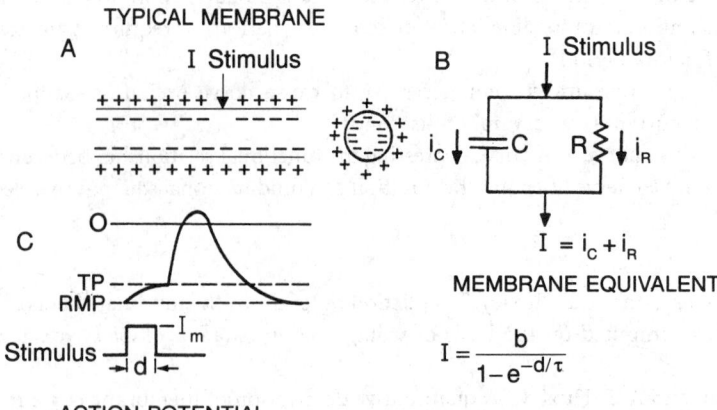

FIGURE 103.9 (A) Typical charged membrane, (B) its equivalent circuit, and (C) action potential resulting from a stimulus I of duration d.

Origin of Bioelectricity

Cell membranes resemble charged capacitors operating near the dielectric breakdown voltage. Assuming a typical value of 90 mV for the transmembrane potential and a membrane thickness of 100 Å, the voltage gradient across the membrane is 0.9×10^5 V/cm. A typical value for the capacitance is about 1 μF/cm^2.

The transmembrane charge is the result of a **metabolic process** that creates ionic gradients with a high concentration of potassium ions (K$^+$) inside and a high concentration of sodium ions (Na$^+$) outside. There are concentration gradients for other ions, the cell wall being a semipermeable membrane that obeys the Nernst equation (60 mV/decade concentration gradient for univalent ions). The result of the ionic gradient is the transmembrane potential that, in the cells referred to earlier, is about 90 mV, the interior being negative with respect to the exterior. Figure 103.9 illustrates this concept for a cylindrical cell.

The transmembrane potential is stable in inexcitable cells, such as the red blood cell. However, in excitable cells, a reduction in transmembrane potential (either physiological or induced electrically) results in excitation, characterized by a transmembrane ion flux, resulting from a membrane permeability change. When the transmembrane potential is reduced by about one-third, Na$^+$ ions rush in; K$^+$ ions exit slightly later while the cell depolarizes, reverse polarizes, then repolarizes. The resulting excursion in transmembrane potential is a propagated action potential that is characteristic for each type of cell. In Fig. 103.10 are shown the action potentials of (A) a single cardiac ventricular muscle cell, (C) a skeletal muscle cell, and (E) a nerve cell. In (B) and (D), the ensuing muscular contractions are shown. An important property of the action potential is that it is propagated without decrement over the entire surface of the cell, the depolarized region being the stimulus for adjacent polarized regions. In contractile cells it is the action potential that triggers release of mechanical energy as shown in Figs. 103.10(B) and (D).

Law of Stimulation

Although action potentials are generated physiologically, it should be obvious that excitable cells can be made to respond by the application of a negative pulse of sufficient current density (I) and duration (d) to reduce the transmembrane potential to a critical value by removing charge, thereby reducing the membrane potential to the threshold potential (TP), as shown in Fig. 103.9. The law of stimulation is $I = b/(1 - e^{-d/\tau})$, where b is the threshold current density for an infinitely long-

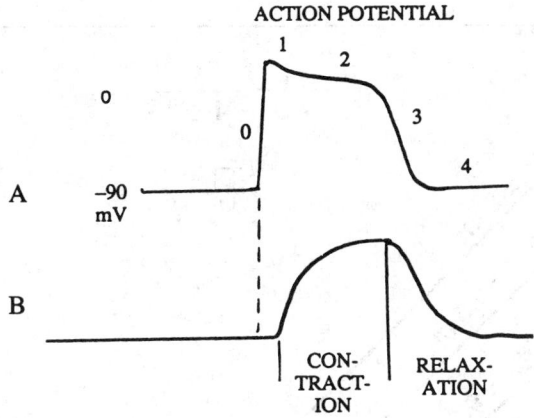

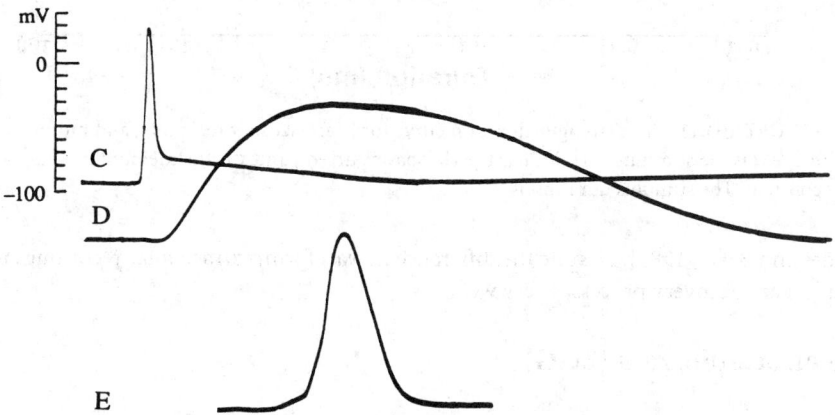

FIGURE 103.10 The action of (A) cardiac muscle and (B) its contraction, (C) skeletal muscle and (D) and its contraction. The action potential of nerve is shown in (E).

duration pulse and τ is the cell membrane time constant, being different for each type of excitable tissue. Figure 103.11 is a plot of the threshold current (I) versus duration (d) for mammalian cardiac muscle, sensory receptors, and motor nerve. This relationship is known as the *strength-duration curve.*

Recording Action Potentials

Action potentials of single excitable cells are recorded with transmembrane electrodes (micron diameter) only in research studies. When action potentials are used for diagnostic purposes, extra-cellular electrodes are used that are both large and distant from the population of cells which become active and recover. The depolarization and repolarization processes send small currents through the conducting environmental tissues and fluids, resulting in a time-varying potential field. Appropriately placed electrodes allow recording the electrical activity of the bioelectric generators. However, the waveforms of such recordings are vastly different from those of the transmembrane action potentials shown in Fig. 103.10. By using cable theory, it is possible to show that such extracellular recordings resemble the second derivative of the excursion in transmembrane poten-

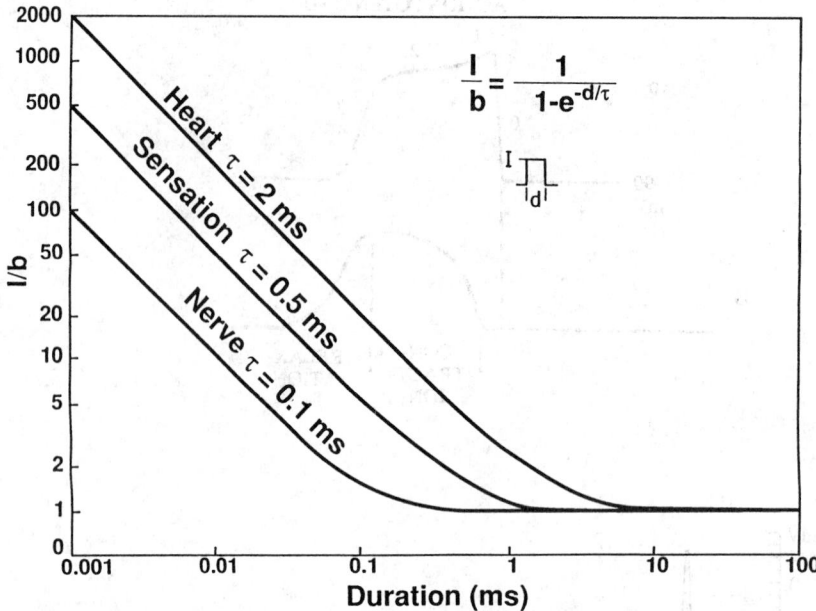

FIGURE 103.11   The strength-duration curve for heart, sensory receptors, and motor nerve. *I* is the stimulus current, *b* is the rheobasic current, and τ is the membrane time constant. The stimulus duration is *d*.

tial [Geddes and Baker, 1989]. Despite the difference in waveform, extracellular recordings identify the excitation and recovery processes very well.

The Electrocardiogram (ECG)

Origin

The heart is two double-muscular pumps. The atria pump blood into the ventricles, then the two ventricles contract. The right ventricle pumps venous blood into the lungs, and the left ventricle pumps oxygen-rich blood into the aorta. Figure 103.12 is a sketch of the heart and great vessels, along with genesis of the ECG.

The ECG consists of two parts: the electrical activity of the atria and that of the ventricles. Both components have an excitation wave and a recovery wave. Within the right atrium is a specialized node of modified cardiac muscle, the sinoatrial (SA) node, that has a spontaneously decreasing transmembrane potential which reaches the threshold potential (TP), resulting in self-excitation (Fig. 103.12, upper left). Therefore the SA node is the cardiac pacemaker, establishing the heart rate. The SA node action potential stimulates the adjacent atrial muscle, completely exciting it and giving rise to the first event in the cardiac cycle, the P wave, the trigger for atrial contraction. Atrial excitation is propagated to another specialized node of tissue in the base of the ventricles, the atrioventricular (AV) node, the bundle of His and the Purkinje fibers. Propagation of excitation over the ventricles gives rise to the QRS, or simply the R wave, which triggers ventricular contraction. Meanwhile during the QRS wave, the atria recover, giving rise to the T_p wave, following which the atria relax. The T_p wave is not ordinarily seen in the ECG because it is obscured by the ventricular QRS wave. During the QRS wave the ventricles contract, then relax following their recovery potential, the T wave; Fig. 103.12 summarizes this sequence.

Ordinarily the T_p wave is not visible. However, if the propagation of excitation from the atria to the ventricles is blocked, the T_p wave can be seen. Figure 103.13 is a record of the ECG from a limb

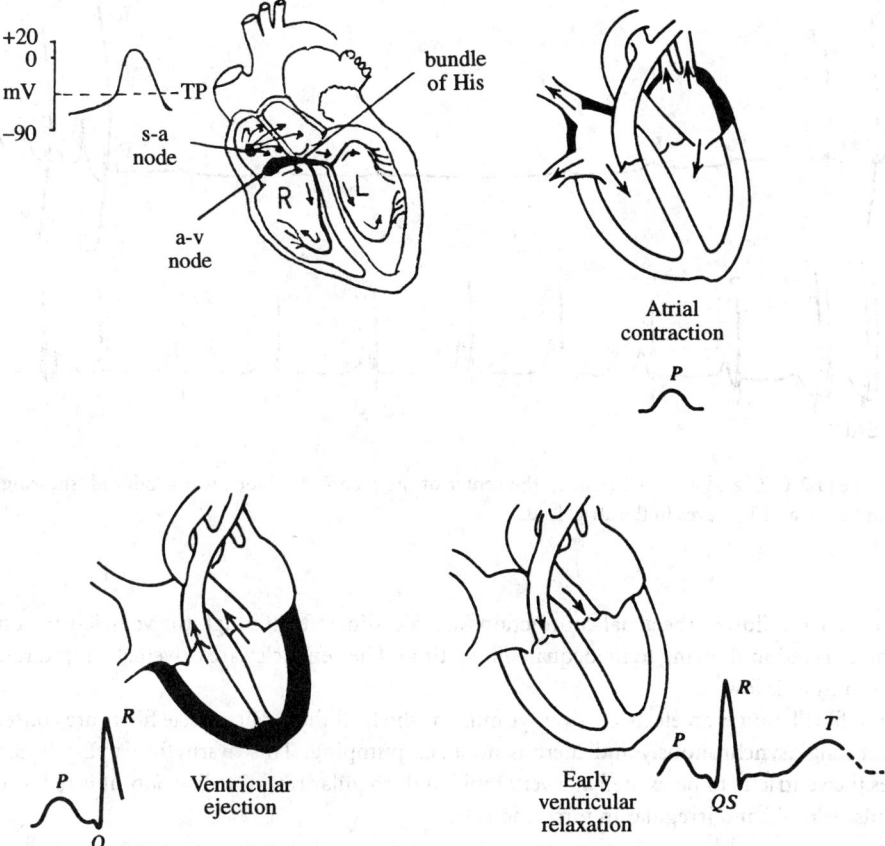

FIGURE 103.12 Genesis of the ECG. The SA node is the pacemaker, setting the rate. Excitation is propagated from the atria to the AV node, then to the bundle of His, and to the ventricular muscle via the Purkinje fibers. The SA node has a decreasing membrane potential that reaches the threshold potential (TP), resulting in spontaneous excitation (inset).

lead and a recording from a lead within the right atrium in a subject with transient AV block. Note that the sharp P wave in the atrial lead coincides with the P wave in the limb recording and that the atrial lead shows both P and T_p waves, easily identified during AV block.

Clinical Significance

From the foregoing it can be seen that the ECG is only a timing signal; there is no dynamic information in its amplitude. Nonetheless, by observing the orderly P-QRS-T sequence it is possible to determine if the excitatory and recovery processes in the heart are functioning normally.

Disturbances in the orderly timing of the cardiac cycle are elegantly displayed by the ECG. For example, each atrial excitation may not be delivered to the AV node. AV block exists when there is less than a 1/1 correspondence between the P and QRS complexes (Fig. 103.13).

Figure 103.14(1) shows a normal ECG and Fig. 103.14(2) illustrates a 2/1 AV block with two P waves for each QRS-T complex. Complete AV block exists when none of the atrial excitations reach the AV node, as shown in Fig. 103.14(3). In this case the ventricles developed their own rhythm, which was slow; cardiac output is low, and in such a situation an artificial pacemaker must be implanted.

For many reasons, the atria develop a rapid rate called *atrial tachycardia* or *supraventricular tachycardia*. A very rapid atrial rate is designated *atrial flutter* [Fig. 103.14(4)]. With both atrial

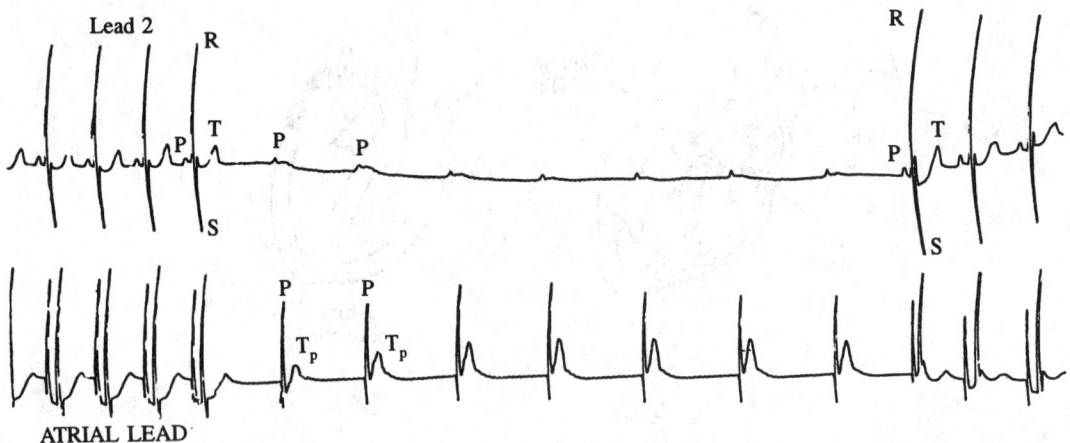

FIGURE 103.13 Lead 2 ECG and an atrial lead. In the center of the record AV block was produced, showing the P waves in lead 2 and the P and T_p waves in the atrial lead.

tachycardia and flutter, the atrial contractions are coordinated, although the ventricular pumping capability is reduced owing to inadequate filling time. The ventricles are driven at a rapid rate, and cardiac output is low.

Atrial fibrillation is an electrical dysrhythmia in which all the atrial muscle fibers are contracting and relaxing asynchronously and there is no atrial pumping. This dysrhythmia [Fig. 103.14(5)] causes the ventricles to be excited at a very rapid and irregular rate. Cardiac output is reduced, and the pulse is rapid and irregular in force and rate.

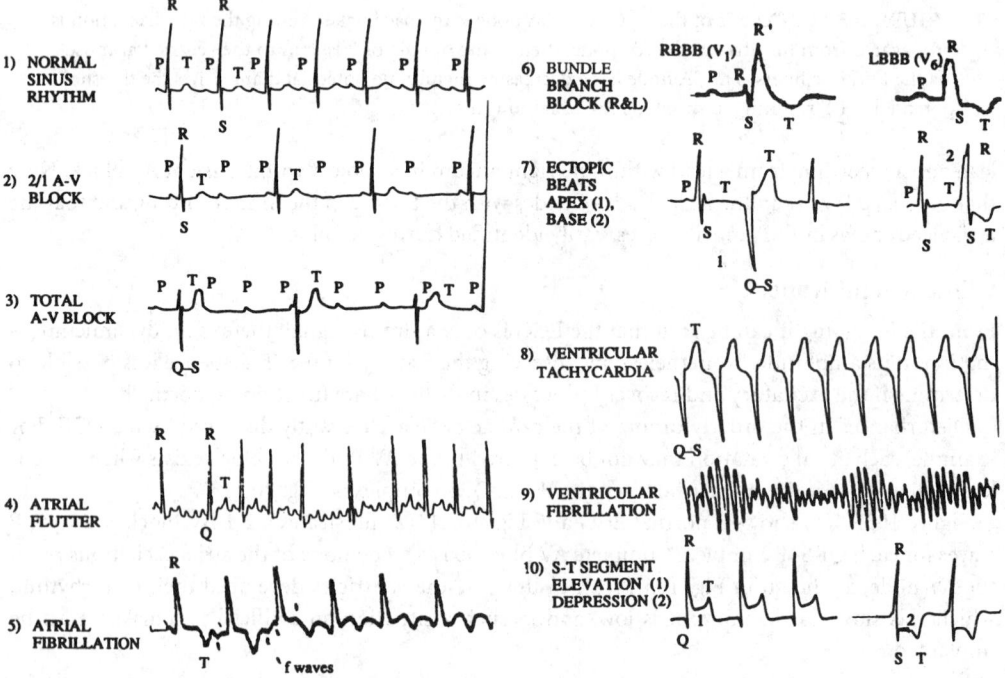

FIGURE 103.14 ECG waveforms.

If the propagation of excitation in the ventricles is impaired by damage to the bundle of His, the coordination of excitation and contraction is impaired and reveals itself by a widening of the QRS wave, and often a notch is present; Fig. 103.14(6) illustrates right (RBBB) and left (LBBB) bundle-branch block. These waveforms are best identified in the chest (V) leads.

All parts of the heart are capable of exhibiting rhythmic activity, there being a rhythmicity hierarchy from the SA node to the ventricular muscle. In abnormal circumstances the atria and ventricles can generate spontaneous beats. Such ectopic excitations do not ordinarily propagate normally, and therefore the ECG waveforms are different. Figure 103.14(7) illustrates ventricular **ectopic beats** in which the first (1) Q-S,T wave arose at the apex and the second (2) R-S,T wave arose at the base of the ventricles. The coupled (bigeminus) beat usually produces no arterial pulse because of inadequate time for filling of the ventricles and poor coordination of the contraction.

The ventricles may become so excitable that they develop a rapid rhythm called *ventricular tachycardia,* as shown in Fig 103.14(8). In this situation the pumping capability is diminished owing to the high rate that impairs filling and to impaired coordination of contraction. Ventricular fibrillation is a condition in which all of the ventricular muscle fibers contract and relax independently and asynchronously. Pumping ceases and cardiac output falls to zero. The ECG [Fig. 103.14(9)] exhibits rapid oscillations of waxing and waning amplitude at a rate of 800 to 1500 per minute. Ventricular fibrillation is lethal unless the circulation is restored within a few minutes, first by cardiopulmonary resuscitation (CPR) and then by electrical defibrillation. The latter technique employs the delivery of a substantial pulse of current through the heart applied directly or with transchest electrodes. Figure 103.15 illustrates ventricular tachycardia (left), ventricular fibrillation (center), and defibrillation (right), with the restoration of pumping.

When a region of the ventricles is deprived of its coronary artery blood supply, the cells in this region lose their ability to generate action potentials and to contract. These cells remain depolarized while they are dying and do not contribute to genesis of the QRS-T complex. Instead, there appears a shift in the portion of the ECG between the S and T waves, i.e., there is an S-T segment shift. This is the cardinal sign of a **myocardial infarction** (heart attack) and is almost always accompanied by chest pain (angina pectoris). Figure 103.14(10) illustrates the ECG in myocardial infarction. Whether the S-T segment displacement is up (1) or down (2) depends on the region of the ventricles injured, as well as the lead used to record the ECG.

ECG Leads

The spread of excitation and recovery over the atria and ventricles varies in direction with time. Therefore, excitation and recovery are vectors, and the location of body-surface electrodes is

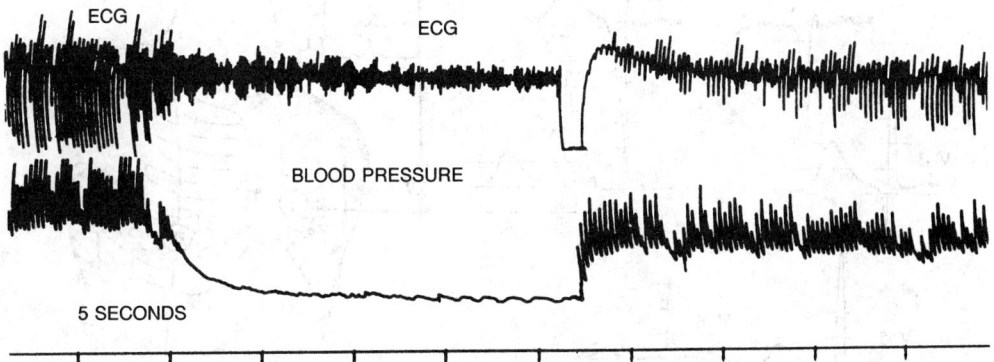

FIGURE 103.15 The electrocardiogram (ECG) and blood pressure during ventricular tachycardia (left), which progressed to ventricular fibrillation (center). A strong transchest shock was applied to defibrillate the ventricles that resumed pumping with the tachycardia returning (right).

important. For this reason, standard electrode sites have been adopted, as shown in Fig. 103.16. There are three standard limb leads, three augmented (a) limb leads, and six chest (V) leads, the latter being monopolar. The reference for the monopolar chest leads is the centerpoint of three resistors (r), each joined to one of the limb electrodes. The right leg is used to ground the subject. Each lead "sees" a different region of the heart. The use of so many leads allows quick and easy identification of the direction of propagation of excitation (and recovery) by merely inspecting the amplitudes of the waveforms in the various leads. If excitation (or recovery) travels orthogonal to a lead axis, the net amplitude will be zero or very small. If excitation (or recovery) travels parallel to a lead axis, the amplitude of the wave will be maximum. Figure 103.16 illustrates the amplitudes of the P, QRS, and T waves for the 12 ECG leads. Note that leads 1, 2, 3, aVR, aVL, and aVF identify the vector projections in the frontal plane. Leads V_{1-6} identify the vector components in a quasi-horizontal plane. There are normal values for the amplitudes and durations for the P, QRS, and T waves as

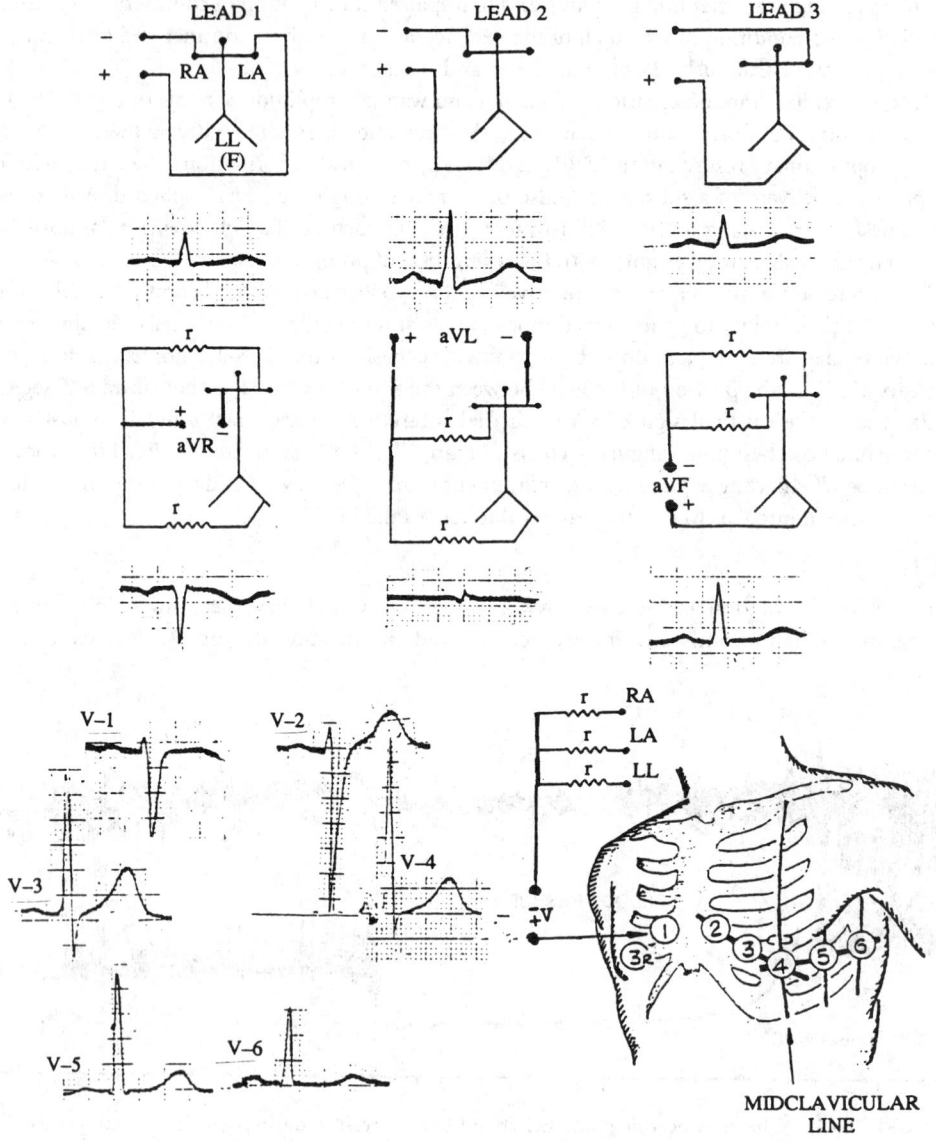

FIGURE 103.16   The limb and chest (V) leads.

well as their vectors. The interested reader can find more on ECG in the many handbooks on this subject. Two good, recently published texts are by Chou [1991] and Phillips and Feeney [1990].

Instrumentation

Standards of performance evolved from recommendations by the American Medical Association in 1950 and the American Heart Association in 1954 and 1967. These recommendations have been collected and expanded into an American National Standard, published by the Association for the Advancement of Medical Instrumentation (AAMI) [1991]. The title of the document is "Diagnostic Electrocardiographic Devices." This document not only lists all the performance and labeling requirements, but also provides useful information on testing ECGs and should be consulted by those contemplating construction of an ECG. Only some of the highlights of the standard will be presented here.

The ECG is displayed by a direct-writing pen that employs a heated stylus writing on thermosensitive paper. Two chart speeds are used, 25 and 50 mm/s. The rulings on the paper represent 40 ms when the standard speed (25 mm/s) is used. The amplitude sensitivity is 10 mm for a 1-mV input signal. The sinusoidal frequency response extends from 0.05 to 100 Hz for the 30% attenuation points. The input stage is a differential amplifier with an input impedance in excess of 2.4 MΩ The common-mode rejection ratio (CMRR) is measured with a 20-V (rms) 60-Hz generator with an output impedance of 51,000 Ω connected in series with a 10-pF capacitor. The 60-Hz CMRR should be in excess of 5000. The maximum dc leakage current through any patient electrode is 0.2 μA.

Electromyography (EMG)

The electrical activity of skeletal muscle is monitored to assess the integrity of the motor nerve that supplies it and to evaluate recovery of the motor nerve following injury to it. The EMG is also characteristically altered in many degenerative muscle diseases. Although muscle action potentials can be detected with skin-surface electrodes, a monopolar or bipolar needle electrode is used in clinical EMG. The electrical activity is displayed on an oscilloscope screen and monitored aurally with a loudspeaker.

Contraction of Skeletal Muscle

The functional unit of the muscular system is the motor unit, consisting of a nerve cell located within the spinal cord, its axon (nerve fiber), and the group of muscle fibers that it innervates, as shown in Fig. 103.17. Between the nerve fiber and the muscle fibers is the myoneural junction, the site where acetylcholine is liberated and transmits excitation to the muscle fibers. The number of muscle fibers per nerve fiber is called the innervation ratio, which ranges from 1:1 to about 1000:1;

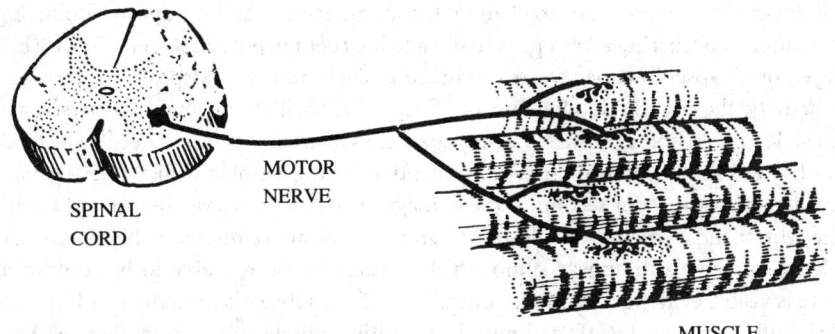

FIGURE 103.17 The functional unit of the muscular system, the motor unit, consisting of a nerve cell located within the spinal cord, its axon, and the muscle fibers that it innervates.

the former ratio is characteristic of the extraocular muscles, and the latter is typical for the postural muscles.

A single stimulus received by the nerve fiber physiologically, or a single stimulus delivered to it electrically, will cause all the innervated muscle fibers to contract and relax; this response is called a *twitch*. Figure 103.10(C) and (D) illustrates the relationship between the muscle action potential and twitch. Note that the action potential is almost over before contraction begins and the contraction far outlasts the duration of the action potential.

If multiple stimuli are delivered to a single motor-nerve fiber with an increasing frequency, the twitches fuse into a sustained (tetanic) contraction whose force is much more than that of a twitch. This occurs because each action potential liberates contractile energy. The critical fusion frequency depends on the type of muscle, but in general it is about 25 to 40 per second.

The force developed by a whole muscle consisting of thousands of motor units is graded in two ways: (1) by the frequency of nerve impulses in each nerve fiber and (2) by the number of motor units that are activated.

Clinical EMG

When the electrical activity of skeletal muscle is examined for diagnostic purposes, an insulated needle electrode, bare only at the tip [Fig. 103.18(A)], is inserted into the muscle and paired with a skin-surface electrode. Another skin-surface electrode is used to ground the subject. Occasionally a coaxial needle electrode [Fig. 103.18(B)] or a bipolar hypodermic needle electrode [Fig. 103.18(C)] is used. In the latter case the outer sleeve is used as the ground. A high-gain, differential amplifier, oscilloscope, and loudspeaker are used, as shown in Figure 103.18(D).

In a normal subject at rest, the electrical activity monitored during insertion of the needle electrode consists of a short burst of muscle action potentials displayed on the oscilloscope and heard in the loudspeaker. These action potentials are called *insertion* potentials and subside quickly in the normal muscle. When the muscle is at rest, there is no electrical activity (electrical silence). If the muscle is contracted voluntarily, the frequency of action potentials increases with the force developed by the muscle. However, there is no linear or constant relationship between these two events. Each action potential, called a *normal motor-unit* potential, lasts a few milliseconds to the first zero crossing, as shown in Fig. 103.18(E).

There is considerable art associated with placing the exploring electrode. If the electrode tip is not adjacent to contracting muscle fibers, the sound of the action potential is muffled and electrode adjustment is required. The same is true for detecting fibrillation potentials (see below).

If the nerve cell in the spinal cord or the nerve fiber supplying a muscle is damaged, the muscle cannot be contracted voluntarily or reflexly (by tapping the tendon) and is therefore paralyzed. In the absence of therapeutic intervention and with the passage of time, the nerve beyond the damaged site dies and the muscle fibers start to degenerate. In about 2 1/2 to 3 weeks in humans, the individual muscle fibers start to contract and relax spontaneously and randomly, producing short-duration, randomly occurring action potentials called *fibrillation* potentials [Fig. 103.18(F)], which are displayed on the oscilloscope screen and heard as clicks in the loudspeaker. Although there is electrical activity, the muscle develops no net force. The fibrillation potentials persist as long as there are viable muscle fibers. In such a denervated muscle, insertion of the needle electrode elicits a vigorous train of short-duration insertion potentials that resemble fibrillation potentials with a frequency of about 1 to 10 per second. If the damaged ends of the nerve are brought together surgically, the central end of the nerve will start to grow slowly and reinnervate the muscle. Gradually the fibrillation potentials disappear, although the muscle is still not able to be contracted. Long before there is visible evidence of muscle contraction, if the subject is presented with the EMG display and asked to contract the affected muscle, primitive muscle action potentials can be elicited. With the passage of time, the fibrillation potentials disappear and there is electrical silence at rest and primitive (nascent) motor-unit activity occurs with voluntary contraction. Later when reinnervation is complete, only normal motor-unit potentials are present with voluntary contraction and electrical silence at rest.

The EMG is also used to diagnose some degenerative muscle and related nerve disorders. *Myotonia* is a degenerative disease of muscle fibers in which the muscle relaxes poorly. Insertion of the needle electrode elicits an intense burst of insertion potentials that sound like a thunderstorm in the loudspeaker. A similar response is obtained by tapping the muscle. When relaxation does occur, there is electrical silence. Voluntary contraction produces normal action potentials along with shorter-duration action potentials from the diseased muscle fibers.

Myasthenia gravis is a disease in which there is impairment of transmission of acetylcholine across the myoneural junctions to the muscle fibers. As a result, muscle contraction cannot be sustained. Because the muscle fibers are normally innervated, there are no fibrillation potentials. With voluntary contraction, normal action potentials occur, and if the disease is severe, the action potentials decrease in frequency as the force of contraction decreases and soon sustained muscle contraction cannot be maintained.

Muscular dystrophy is a degenerative disease of muscle fibers in which there is **atrophy** of some fibers, swelling in others, and an increase in sarcolemmal and connective tissue with the deposition of fat. Insertion of the needle electrode elicits a vigorous burst of short-duration, high-frequency action potentials. Typically at rest there are no fibrillation potentials. With voluntary contraction, the action potentials are short in duration, high in frequency, and produce a whirring sound in the loudspeaker. As fatigue supervenes, the frequency and amplitude decrease.

The reader who is interested in obtaining more information on EMG will find it in books by Cohen and Brumlik [1969] and Marinacci [1955]. Both contain a wealth of clinical information.

Instrumentation

As yet there is no American National Standard for EMG, although steps are being taken in this direction. As shown in Fig. 103.18, the EMG is displayed in two ways: (1) visually with an oscillo-

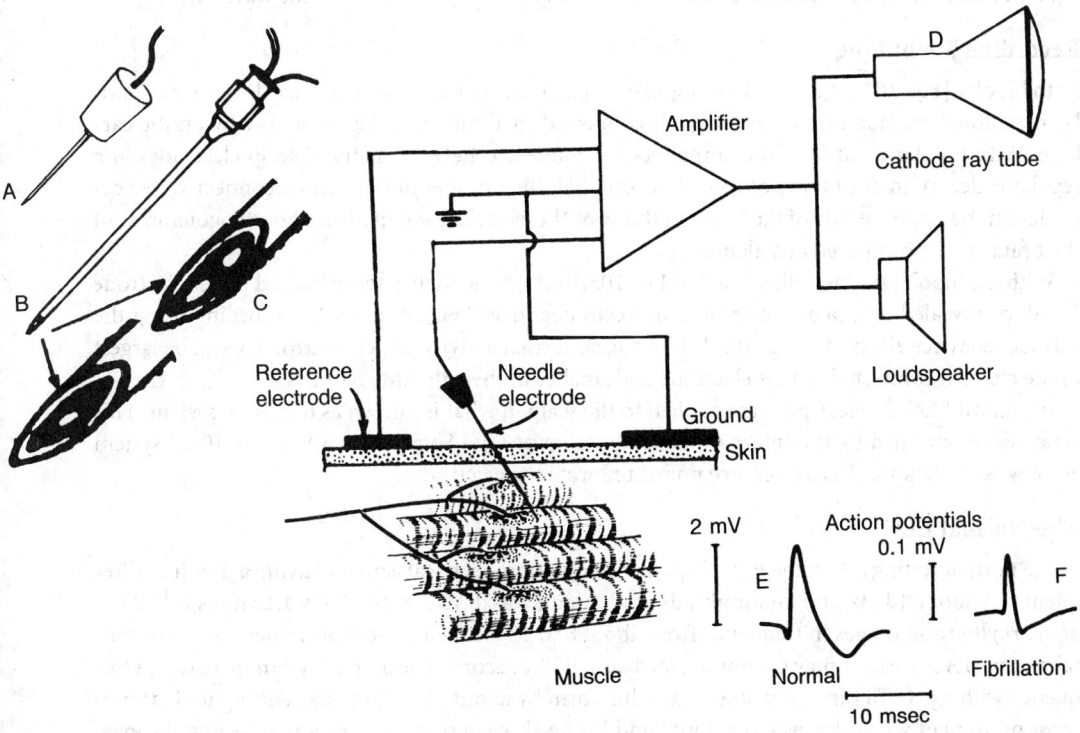

FIGURE 103.18 Equipment used for electromyography. (A) Needle electrode, (B) hypodermic monopolar and (C) bipolar electrodes, (D) the recording apparatus, (E) skeletal muscle action potential, and (F) fibrillation potential.

scope and (2) aurally with a loudspeaker. Both are needed to enable acquisition and analysis of the EMG.

Buchtal *et al.* [1954] stated that the principal frequency components for the human EMG require a bandwidth of 1 Hz to 10 kHz. It has been found that a time constant of about 50 ms is satisfactory, which corresponds to a low-frequency −3-db point of 3 Hz. For needle electrodes with a tip diameter of 0.1 mm or larger, the input impedance (one side to ground) should not be lower than that of a 500-kΩ resistor in parallel with less than 25-pF capacitance.

Smaller-area electrodes require a higher input impedance [Geddes *et al.*, 1967]. The cable used to connect the needle electrode to the amplifier should not add more than 250 pF to the input capacitance. The common-mode rejection ratio (CMRR) should be in excess of 5000.

Electroencephalography (EEG)

The electrical activity of the brain can be recorded with electrodes on the scalp, on the exposed brain, or inserted into the brain. The latter method is used only in research studies. When recordings are made with brain-surface (cortex) electrodes, the recording is called an electrocorticogram (ECoG). With scalp electrodes, the recording is designated an electroencephalogram (EEG) that is displayed by direct-inking pens using a chart speed of 3 cm/s. Typically 8 to 12 channels are recorded simultaneously.

Although the brain consists of about 10^{14} neurons, the EEG reflects the electrical activity of the outer layer, the cortex, which is the seat of consciousness. The type of electrical activity depends on the location of the electrodes and the level of alertness. The frequency and amplitude are profoundly affected by alertness, drowsiness, sleep, hyperventilation, anesthesia, the presence of a tumor, head injury, and epilepsy. The clinical correlation between cerebral disorders and the voltage and frequency spectra is well ahead of the physiological explanations for the waveforms.

Recording Technique

Both bipolar [Fig. 103.19(A)] and monopolar [Fig. 103.19(B)] techniques are used. With monopolar recording, one side of each amplifier is connected to a reference electrode, usually on the earlobe. With bipolar recording, the amplifiers are connected between pairs of scalp electrodes in a regular order. With both types of recording, one-half the number of channels is connected to electrodes on the opposite side of the head. In this way, the electrical activity from homologous areas of the brain can be compared at a glance.

With the bipolar method illustrated in Fig. 103.19(A), abnormal activity located under electrode X will be revealed as a phase reversal in adjacent channels. With monopolar recording using the earlobe reference electrode [Fig. 103.19(B)] the abnormal activity under electrode X will be largest in the channel connected to that electrode and smaller in the adjacent channels.

In clinical EEG, 21 electrodes are applied to the scalp in what is known as the 10-20 system. This array was established by the International Federation of EEG Societies in 1958. The 10-20 system employs skull landmarks as reference points to locate the electrodes.

The Normal EEG

In the normal resting adult, the EEG displays a fluctuating electrical activity having a dominant frequency of about 10 Hz and an amplitude in the range of 20 to 200 μV. This activity is called the *alpha rhythm* and ranges in frequency from about 8 to 12 Hz, being most prominent in the **occipital** and **parietal** areas. It may occupy as much as half the record. The alpha rhythm increases in frequency with age from birth and attains its adult form by about 15 to 20 years. The alpha rhythm is most prominent when the eyes are closed and in the absence of concentration. Opening the eyes, engaging in patterned vision, or performing such cerebral activity as mental arithmetic diminishes or abolishes the alpha rhythm. Figure 103.20 presents a good example of this phenomenon.

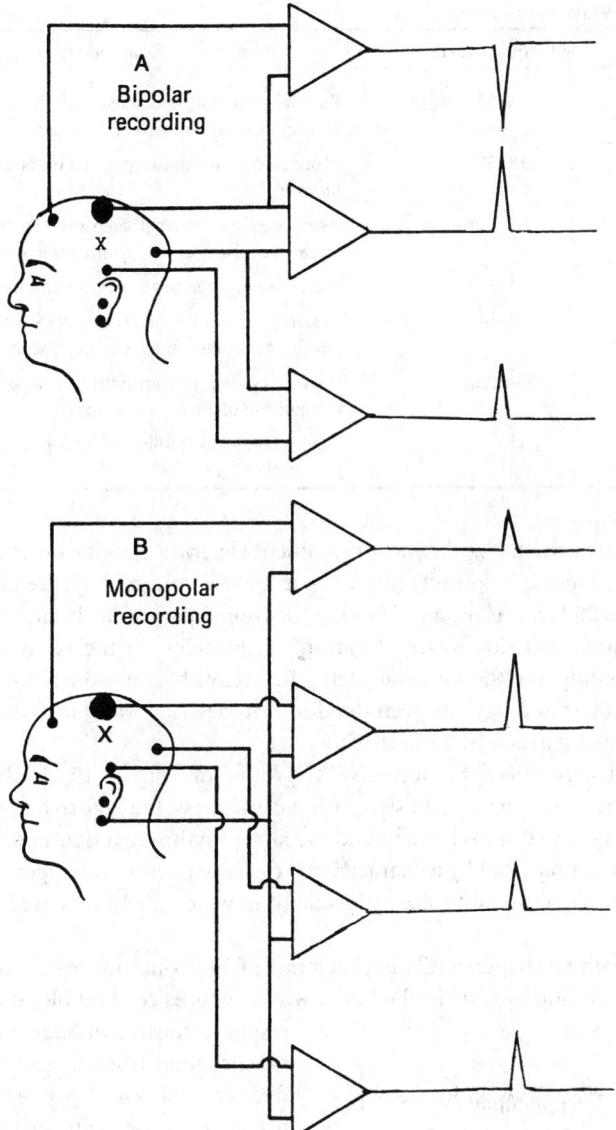

FIGURE 103.19   Methods of recording the EEG. (A) The bipolar and
(B) the monopolar method. Note how abnormal activity under electrode X is revealed by the two techniques.

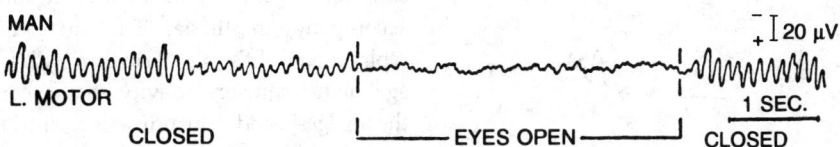

FIGURE 103.20   The EEG of a relaxed human subject with eyes closed and open. Note that the record is
dominated with alpha rhythm (8–12 Hz) when the eyes are closed. (*Source:* Derived in part from M.A.B.
Brazier, *The Electrical Activity of the Nervous System,* London: Sir Isaac Pitman & Sons, Ltd., 1951. With
permission.)

Table 103.1   ECG Waveform Terminology

Waveform	Frequency, Hz	Conditions
Alpha	8–12	Parietal-occipital, associated with the awake and relaxed subject, prominent with eyes closed.
Beta	18–30	More evident in frontal-parietal leads, seen best when alpha is blocked.
Delta	1–3.5	Associated with normal sleep and present in children less than 1 year old, also seen in organic brain disease.
Theta	4–7	Parietal-temporal, prominent in children 2 to 5 years old.
Sleep spindle (sigma)	12–14	Waxing and waning of a sinusoidal-like wave having the envelope that resembles a spindle, seen during sleep.
Lambda	Transient	Visually evoked, low-amplitude, occipital wave, resulting from recognition of a new visual image.
Spike and wave	ca. 3	Sharp wave (spike) followed by rounded wave associated with petit mal epilepsy.

Although the alpha rhythm is the most prominent electrical activity, other frequencies are present. For example, there is a considerable amount of low-voltage, high-frequency (beta) activity ranging from 18 to 30 Hz. It is usually found in the frontal part of the brain. However, the normal electroencephalogram contains waves of various frequencies (in the range of 1 to 60 Hz) and amplitudes, depending on the cerebral state. To establish communication among electroencephalographers, a terminology has been developed to describe waveforms and their frequencies; Table 103.1 presents a glossary of these terms.

Drowsiness and sleep affect the normal EEG profoundly. Figure 103.21 illustrates the typical changes that occur as a subject goes to sleep. With drowsiness, the higher-frequency activity which is associated with alertness or excitement and the alpha rhythm that dominates the waking record in the relaxed state are replaced by a characteristic cyclic sequence of changes which constitute the focus of a new specialty devoted to sleep physiology, in which the EEG is used to identify different stages of sleep.

Rapid, deep breathing (hyperventilation) at a rate of 30 per minute for about 3 min reduces the partial pressure of carbon dioxide in the blood which reduces cerebral blood flow. A typical EEG response consists of large-amplitude, bilaterally synchronous, frontally prominent waves with a frequency of 4 to 7 per second. The frequency usually decreases with increasing hyperventilation. The lack of bilateral symmetry is an indication of abnormality.

Anesthesia dramatically alters the EEG in a manner that depends on the type and amount of anesthetic given. Despite differences among anesthetic agents, some important similarities accompany anesthesia. The first change is replacement of the alpha rhythm with low-voltage high-frequency activity that accompanies the analgesia and delirium stages. Thus the EEG resembles that of an alert or excited subject, although the subject is not appropriately responsive to stimuli; usually the response is excessive and/or inappropriate. From this point on, the type of EEG obtained with deepening

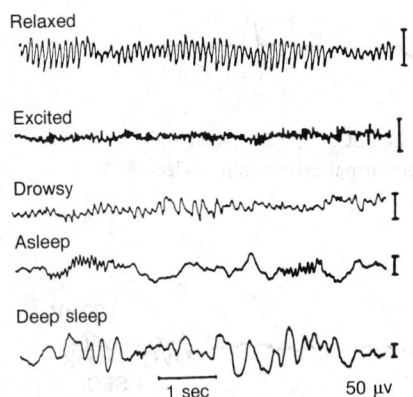

FIGURE 103.21   The EEG of a subject going to sleep. (*Source:* H. Jasper, in *Epilepsy and Cerebral Localization*, W.G. Penfield and T.C. Erickson, Eds., Springfield, Ill.: Charles C Thomas, 1941. With permission.)

anesthesia depends on the type of anesthetic. However, when a deeper level of anesthesia is reached, the EEG waveform becomes less dependent on the type of anesthetic. Large-amplitude low-frequency waves begin to dominate the record, and with deepening anesthesia their frequency is reduced and they begin to occur intermittently. With very (dangerously) deep anesthesia, the record is flat (i.e., isoelectric). Complicating interpretation of the EEG in anesthesia are the effects of **hypoxia, hypercapnia**, and hypoglycemia, all of which mimic deep anesthesia.

Clinical EEG

The EEG plays a valuable role in identifying intracranial pathology. The clinical utility relies on recognition of patterns of frequency, voltage, and waveform. Localization of abnormal areas is provided by the multiple scalp electrodes and recording channels.

The EEG has its greatest value as an aid in the diagnosis and differentiation of the many types of epilepsy, a condition in which groups of neurons in the brain become hyperexcitable and, depending on their location, produce sensory, motor, and/or **autonomic** manifestations. The epilepsies associated with cortical lesions are often detected by the scalp EEG. The EEG in epileptics is usually abnormal between, as well as during, attacks. The EEG provides information on the location of the area (or areas) of abnormal neuronal activity.

Petit mal epilepsy is characterized by a transient loss (few to 20 s) of conscious thought, although motor activity may continue. Often there are eye movements and blinking. The EEG shows a characteristic 3 per second spike-and-wave pattern [Fig. 103.22(A)]. Psychomotor epilepsy is characterized by sensory hallucinations and abnormal thoughts, often with stereotyped behavior. During the attack, the subject is stuporous and the EEG [Fig. 103.22(B)] has a characteristic pattern. Jacksonian, or motor, epilepsy starts in a specific area of the motor cortex and is preceded by an aura, a characteristic sensation perceived by the subject. The convulsion starts with localized muscle twitching that often starts in the face, hand, arm, then spreads over the entire body as a generalized convulsion; Fig. 103.22(C) shows the onset of a convulsion. Consciousness is lost during and for a short time after the fit. The EEG provides information on the origin of the abnormal discharge in the motor cortex. Grand mal epilepsy is characterized by a contraction all the muscles (tonic phase), then jerking (clonic phase). Consciousness is lost, and the subject is in a coma for some time following the attack. The EEG [Fig. 103.22(D)] shows high-voltage, high-frequency waves that progress over the entire cortex.

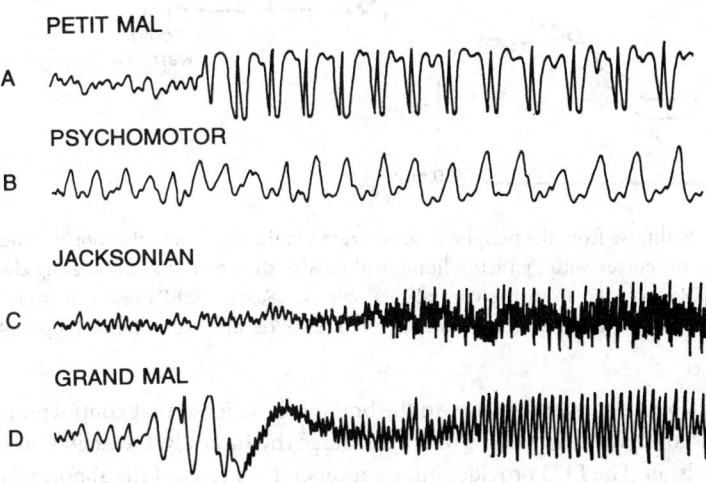

FIGURE 103.22 EEG waveforms in epilepsy: (A) petit mal, (B) psychomotor, (C) Jacksonian, and (D) grand mal. (*Source:* Derived from F. A. Gibbs and E. L. Gibbs, *Atlas of Encephalography*, London: Addison-Wesley, 1952. With permission.)

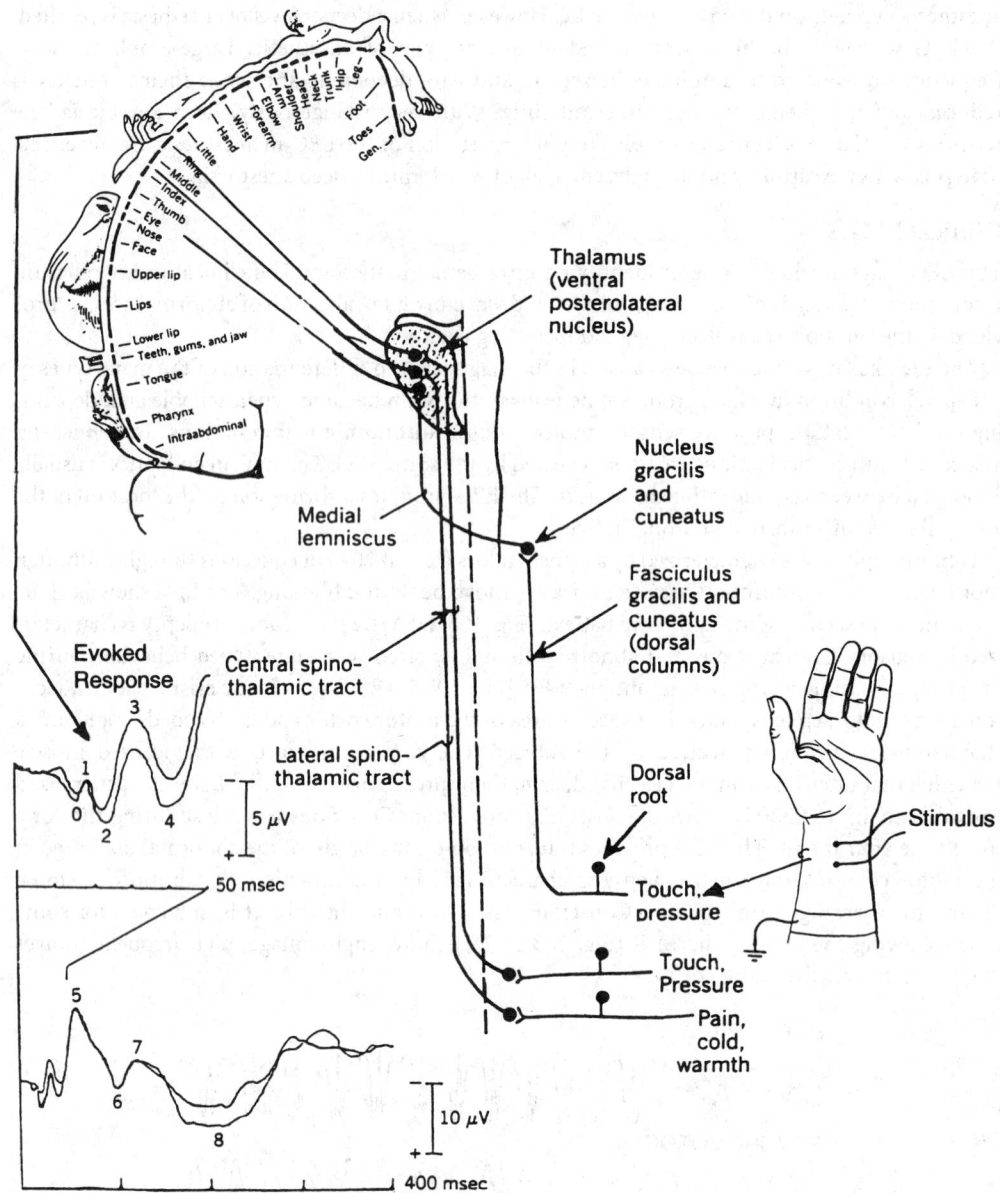

FIGURE 103.23 Pathways from the peripheral sense organs to the cortex and the topographical distribution of sensation along the cortex with Penfield's homunculus. Also shown are the stimulating electrodes on the wrist and the SSEPs recorded from the contralateral cortex. (*Source:* SSEPs redrawn from T. W. Picton, "Evoked cortical potentials, how? what? and why?," *Am. J. EEG Technol.,* vol. 14, no. 4, pp. 9–44, 1974. With permission.)

Traumatic epilepsy results from injury to the brain. It is believed that contraction of scar tissue acts as a stimulus to adjacent nerve cells which discharge rhythmically, the excitation spreading to a grand mal convulsion. The EEG provides information on the origin of the abnormal discharge.

Tumors are associated with low-frequency (delta) waves. However, other intracranial lesions also produce slow waves. Although the EEG can identify the location of tumors, usually it cannot differentiate between brain injury, infection, and vascular accident, all of which produce low-frequency waves. Interpretation of the EEG always includes other clinical information.

For those wishing to delve deeper into EEG, additional information can be found in most text-books of medical physiology. The three-volume *Atlas of EEG*, authored by Gibbs and Gibbs [1952], contains a wealth of information on EEG in epilepsy and includes a vast array of eight-channel EEGs.

Instrumentation

The American EEG Society [1986] published guidelines for the performance of EEG machines. The guidelines recommended a minimum of eight channels. Chlorided silver disks or gold electrodes, adhered to the scalp with collodion, are recommended; needle electrodes are not. A chart speed of 3 cm/s is standard, and a recording sensitivity of 5 to 10 μV/mm is recommended. The frequency response extends from 1 to 70 Hz for the −3-dB points.

Evoked Potentials

With the availability of signal averaging using a digital computer, it is possible to investigate the integrity of the neural pathways from peripheral sense organs to the cortex by using appropriate stimuli (e.g., clicks, light flashes, or current pulses). Usually the stimulus consists of a few hundred to about 1000 pulses, averaged to produce the somatosensory-evoked potential (SSEP). Likewise, it is possible to evaluate the integrity of the neural pathways from the motor cortex to peripheral muscles by applying multiple short-duration current pulses to scalp electrodes and recording nerve and/or muscle action potentials with skin-surface electrodes. Such recordings are called motor-evoked potentials (MEPs). With both SSEPs and MEPs, the largest responses appear on the opposite side of the body from the stimulus.

Because the responses are in digital form, they can be written out in hard-copy format. With SSEPs, the response consists of many waves occurring at various times after the stimulus. To permit close examination of the various waveforms, several displays are presented, each with a different time axis. Figure 103.23 presents a sketch of the neural pathways from the periphery to the cortex, showing the topographic distribution of sensation along the cortex using the homunculus created by Penfield, described in detail in 1968. Also shown in Fig. 103.23 is a typical SSEP obtained by stimulating the median nerve with skin-surface electrodes connected to an isolated (i.e., not grounded) stimulator output circuit. Note the remarkably low amplitude of the responses that were obtained by averaging the response to 240 stimuli. Note also that the first display showed the responses from 0 to 50 ms and the second display presented the responses in the 0- to 400-ms interval.

Figure 103.24 shows the motor pathways from the motor cortex to a typical muscle. The cortical motor areas are represented by Penfield's homunculus. A train of 250 stimuli were applied between electrodes on the scalp and in the mouth. The motor-evoked potentials were recorded with skin-surface electrodes over the spinal column. The MEP of a patient in whom the muscles were paralyzed is also shown in Fig. 103.24. Because the muscles were paralyzed, the MEPs shown in the figure represent action potentials in the spinal cord. Note the prominent peaks at 7 and 14 ms. These peaks provide information on the path taken by the nerve impulses initiated in the motor cortex.

Although there is no ANSI standard for evoked-potential recording, the American EEG Society [1986] published guidelines for equipment performance and recording techniques. This information should be consulted by those contemplating entry into this field.

Conclusion

This section has focused only on the three most prominent bioelectric events, those of the heart, skeletal muscle, and brain. The eye, ear, sweat glands, and many types of smooth muscle produce action potentials that are used for their diagnostic value, as well as being the subject of on-going research. The reader interested in delving deeper into this field can find such information in a book by Geddes and Baker [1989].

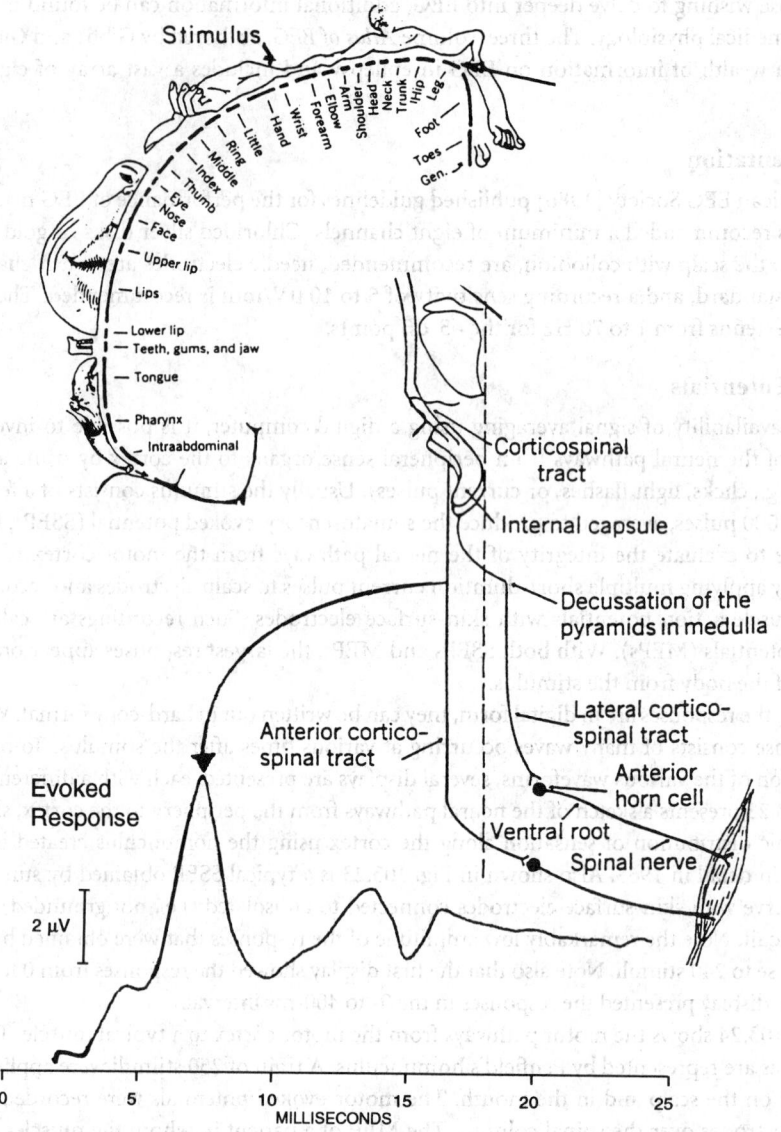

FIGURE 103.24 Neural motor pathways from the cortex to a typical muscle. The motor areas are represented by Penfield's homunculus. A train of 240 stimuli delivered to the motor cortex provided the average MEP detected with electrodes placed over the spinal column in this patient to whom a muscle paralytic drug was given; therefore the MEPs are from the spinal cord. (*Source:* Redrawn from Levy *et al.*, 1984, and Penfield and Rasmussen, 1968.)

Defining Terms

Atrophy: Wasting of cells deprived of nourishment.

Autonomic: That part of the nervous system which controls the internal organs.

Ectopic beat: A heart beat that originates from other than the normal site.

Hypocapnia: A condition of reduced carbon dioxide in the blood.

Hypoxia: A reduced amount of oxygen.

Metabolic process: The method by which cells use oxygen and produce carbon dioxide and heat.

Myocardial infarction: A heart attack in which a region of the heart muscle is deprived of blood and soon dies.

Occipital: The back of the brain.

Parietal: The side of the brain.

References

American EEG Society, "Guidelines in EEG and evoked potentials," *Amer. J. Clin. Neurophysiol.,* vol. 3 (Suppl.), 1986.

Association for the Advancement of Medical Instrumentation (AAMI), Diagnostic ECG Devices, ANSI-AAMI Standard EC-101-1991.

F. Buchtal, C. Guld, and P. Rosenflack, "Action potential parameters of normal human muscle and their dependence on physical variables," *Acta Physiol. Scand,.* vol. 32, pp. 200–220, 1954.

T-C. Chou, *Electrocardiography in Clinical Practice,* 3d ed. Philadelphia: W. B. Saunders, 1991.

H.L. Cohen and F. Brumlik, *Manual of Electromyography,* New York: Hoeber Medical Division, Harper & Row, 1969.

L.A. Geddes, L.E. Baker, and M. McGoodwin, "The relationship between electrode area and amplifier input impedance in recording muscle action potentials," *Med. Biol. Eng. Comput.,* vol. 5, pp. 561–568, 1967.

L.A. Geddes and L.E. Baker, *Principles of Applied Biomedical Instrumentations,* 3rd ed., New York: Wiley, 1989.

F.A. Gibbs and E.L. Gibbs, *Atlas of Electroencephalography,* London: Addison-Welsey, 1952.

International Federation of EEG Societies, J. Knott, Chairman, *EEG Clin. Neurophysiol.,* vol. 10, pp. 378–380, 1958.

International Federation for Electroencephalography and Clinical Neurophysiology, *EEG. Clin. Neurophysiol.,* vol. 10, pp. 371–375, 1958.

W.J. Levy, D.H. York, M. McCaffery, and F. Tanzer, "Evoked potentials from transcranial stimulation of the motor cortex in humans," *Neurosurgery,* vol. 15, no. 3, pp. 287–302, 1983.

A.A. Marinacci, *Clinical Electromyography,* Los Angeles: San Lucas Press, 1955.

W. Penfield and T. Rasmussen, *The Cerebral Cortex of Man,* New York: Hafner, 1968.

R.E. Phillips and M.K. Feeney, *The Cardiac Rhythms: A Systematic Approach to Interpretation,* 3rd ed., Philadelphia: W. B. Saunders, 1990.

T.W. Picton, "Evoked cortical potentials, how? what? and why?," *Am. J. EEG Technol.,* vol. 14, no. 4, pp. 9–44, 1974.

103.3 Application of Electric and Magnetic Fields in Bone and Soft Tissue Repair

C. Polk

History

As early as 1962 in the United States [2], and even earlier—1957—in Japan [1] it was shown that electric potential differences appear across both living and dead bone subjected to mechanical stress. C. A. L. Bassett and R. O. Becker observed that these stress-generated electrical signals decayed very slowly in comparison with similarly initiated signals in piezoelectric crystals and concluded [2] that piezoelectric phenomena "while probably present, were not the sole cause of these potentials." Later analysis and experiments established that the observed signals were primarily due to ion displacement within the porous regions and multiple fluid-filled channels present in all bone. The early observations already suggested that direct application of an externally generated voltage might have an effect on bone development. This was shown to be the case by Bassett *et al.* [3] who found that a dc current of the order of 1 μA (corresponding to a current density of approximately 0.01 A/m²) produced massive osteogenesis near the cathode when electrodes were implanted into the femur of living dogs.

Having shown that application of a dc electric field to nonexcitable, connective tissue cells can produce effects similar to those elicited by mechanical stress, Bassett and his co-workers realized that clinical exploitation of these phenomena would require surgical implantation of electrodes with attendant danger of infection. They proceeded therefore to explore whether noninvasive, inductive coupling that gave waveforms similar to those endogenously produced by mechanical stress could lead to beneficial bone development. In 1974 they reported favorable results obtained with pulsed electromagnetic fields on dogs [4]. Signals of this type have generally been identified as **PEMF** (pulsed electromagnetic field) in the orthopedics/electrical stimulation community for the last 20 years and have been applied successfully in a large number of cases for the repair of **nonunions** [5]. In Germany relatively large-amplitude, low-frequency (<20 Hz) sinusoidal magnetic fields have been used for both bone repair and wound healing [6].

Although the noninvasive PEMF treatment for nonunions (fractures that fail to heal) became—at least in the United States—the most widely used clinical application of subradio frequency fields, several investigators pursued the application of dc electric fields through implanted electrodes and the application of higher-frequency currents through electrode contacts placed on the skin surface to enhance bone repair [7]. At the same time mostly laboratory investigations, *in vitro* and on animals, explored the application of all three modalities—PEMF, implanted dc electrodes, and higher-frequency coupling through skin electrodes—to produce blood vessel regeneration (**angiogenesis**), soft tissue healing, nerve repair or regeneration, and regression of tumors. Motivated by the suggested clinical applications, a large number of basic science investigations have been initiated, and are continuing today, with the object of understanding the mechanisms through which low-intensity electric and magnetic fields affect cells and living tissue. Some of these are reviewed briefly in the subsection Mechanisms and Dosimetry.

Devices for Bone and Cartilage Repair

In the United States medical devices are approved for clinical use only after it has been shown to the satisfaction of the U.S. Food and Drug Administration (FDA) that they are not only safe, but also effective. This is a much more stringent requirement than the mere demonstration of safety demanded presently in most European countries (although some EEC countries are considering moving toward approval criteria that are similar to those used in the United States). The United States laws governing the sale of medical devices for clinical use are also much more restrictive than the controls over implementation of new surgical procedures that involve only informal medical peer review. Even organized multi-patient clinical trials require not more than approval by a local hospital-based institutional review board. The FDA Office of Device Evaluation does approve some new devices, after careful review, for clearly limited clinical trials. However, information on such limited, temporary approval is not made available. Table 103.2 therefore shows only those devices which are presently (as of April 13, 1992) approved and does not include experimental systems which may be undergoing currently limited clinical trials. Some of the latter are discussed further on, based only on information furnished by manufacturers or available from the published medical literature.

The devices listed in Table 103.2 are approved for one of three applications: the treatment of *nonunions* (fractures that have failed to heal after standard treatment involving setting and stabilization with casts), congenital **pseudoarthroses** [8], and promotion of spinal fusion. Although many animal experiments (and possibly a few human trials, especially in Europe) have evaluated the application of electric or magnetic fields for acceleration of fresh fracture healing and for reversal of osteoporosis, no devices are currently approved in the United States for these purposes.

Table 103.2 Electrical Bone Growth Stimulators Approved by the U.S. FDA as of April 13, 1992

Manufacturer	Device	Indications	Technology	Approved	Text References
Electro-Biology, Inc.	EBI Bone Healing System*	Nonunion; congential pseudoarthroses; failed fusions	Noninvasive pulsed electro-magnetic field (PEMF)	November 1979	A B
American Medical Electronics, Inc.	Physio-Stim	Nonunions (excluding vertebrae and flat bones)	Noninvasive PEMF	February 1986	C
American Medical Electronics, Inc	Spinal-Stim	To promote spinal fusion as an adjunct to surgery or as nonop-erative treatment when 9 months have elapsed since the last surgery	Noninvasive PEMF	February 1990	D
Biolectron, Inc.	Orthopak BGS System	Nonunions (excluding vertebrae and flat bones)	Noninvasive/cap-acitively coupled	February 1986	E
Electro-Biology, Inc.	Orthogen/Osteogen	Nonunion of long bones	Implantable dc	January 1980	F
Electro-Biology, Inc.	Sp F-4 (2) Implantable BGS†	Spinal fusion adjunct	Implantable dc	April 1987	G
Biolectron, Inc.	Zimmer Direct Current Bone Growth Stimulator (DCGBS)	Nonunion fractures	Semi-invasive dc with percutaneous pins	November 1979	H

*Also known as Bi-Osteogen Systems 204.
†Also known as BGS-Osteostim HS 11.

Classified by electrical and mechanical characteristics, the devices in Table 103.2 are either:

1. Noninvasive:
   a. Generating time-varying magnetic fields applied by coils to the affected body part (A, B, C, D)
   b. Generating time-varying electric fields applied through skin-surface electrodes (**capacitively coupled**) (E)
2. Invasive or semi-invasive: dc applied from an implanted battery (F, G) or (semi-invasive) dc applied with percutaneous pins (H).

A signal typical for some PEMF (A, C, D) devices is illustrated in Fig. 103.25. Part A shows the magnetic field versus time and Part B the corresponding electric field induced into a linear, isotropic medium. The waveform shown on Part B can be measured by a probe coil having a sufficiently large number of turns. The frequency spectrum of the electric field is shown in Fig. 103.26. Signals used by the different manufacturers are protected by patents, and FDA approval is for particular sig-

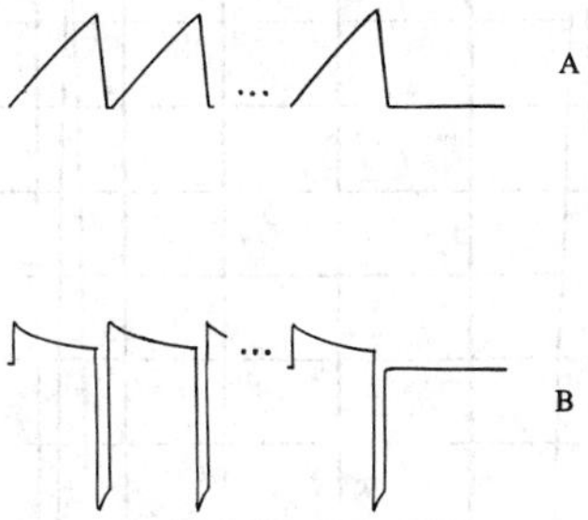

FIGURE 103.25 A typical PEMF signal (signal A of Table 103.2). (A) Magnetic field versus time. (B) Electric field ($\propto \partial B/\partial t$) versus time. Signal consists of 15 pulse bursts per second. Each burst is 4.5 ms long and contains 20 pulses. In each pulse the magnetic field increases from 0 to approx. 2 mT during 200 μs, decreases to 0 again during 23 μs, and is equal to 0 for 2 μs before the next 225-μs sequence begins.

nal parameters within specified tolerances on time and amplitude. The pseudoarthrosis signal used by Electrobiology, Inc. (EBI) (B in Table 103.2) consists of single pulses repeated at a rate of 72 pps rather than the pulse bursts illustrated in Figs. 103.25 and 103.26. Each magnetic field pulse increases from zero to 3.5 mT in 380 μs and then decreases slowly to zero in approximately 4.5 ms. The signals that are now in use have evolved considerably from those employed in the initial studies, and some have little resemblance to the endogenous electrical signals elicited by mechanical stress.

The PEMF signals employed by the various manufacturers in the United States and Europe can have several different pulse shapes, rise and decay times, pulse widths, pulse repetition rates, and amplitudes. Since it has been shown (see Mechanisms and Dosimetry) that all these variables can have a profound effect on the biological action of a particular signal, it is essential that reports on effectiveness or lack of effectiveness of PEMF give an exact description of the signal which was used.

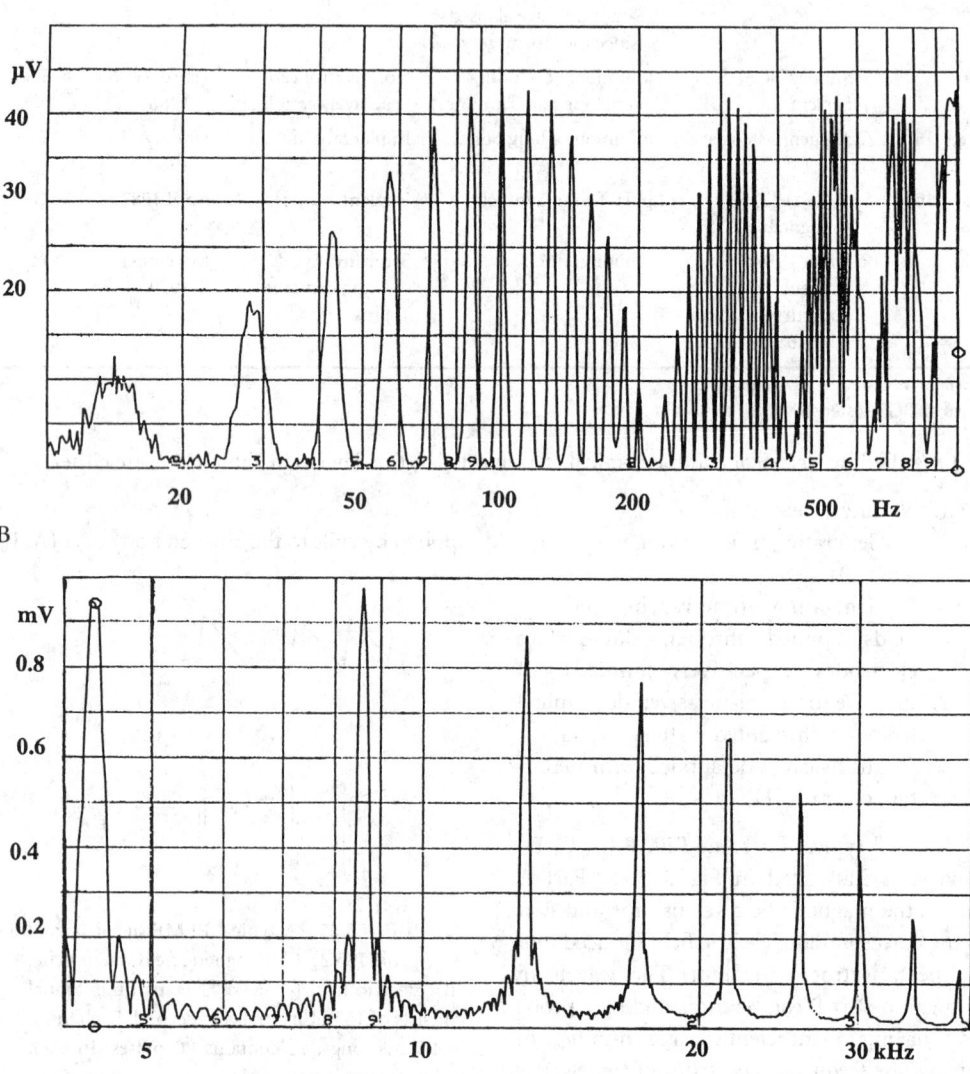

FIGURE 103.26　Electric field spectrum $|E(\omega)|$ of signal in Fig. 103.25 as measured by the output from an air-core coil (0.6 cm mean diameter, 65 turns). (A) 10 Hz to 1 kHz (50 μV full scale); (B) 4 to 40 kHz (1 mV full scale).

Unfortunately the medical literature is replete with examples where this information is either incomplete or completely absent. Details of shape, orientation, and location of the application coil or coils are also important, since these parameters, together with pulse amplitude and shape, determine the nature of the magnetic and electric field at the location of the injured tissue. If the amplitude of the axially directed magnetic flux density B is constant over some region of radius R within the cross section of a circular cylinder, the induced electric field is

$$\bar{E} = -\left(\frac{\partial B}{\partial t}\right) \frac{r}{2} \hat{\phi} \tag{103.17}$$

provided the material of the cylinder is electrically homogeneous and isotropic. In Eq. (103.17) $r < R$ is the distance from the center of the cylinder and $\hat{\phi}$ is a unit vector in the circumferential direction. For magnetic fields varying sinusoidally as $B_0 \cos \omega t$, $\bar{E} = \omega B_0 (r/2) \sin \omega t \, \hat{\phi}$. Since most biological objects are neither homogeneous nor isotropic, the actual induced electric fields at various points in the tissue or cells may deviate substantially from the values given by Eq. (103.17) [9, 10]. Equation (103.17) is useful only for *estimating* the spatial *average* value of the induced electric field, which depends in the bone environment on the point-to-point variation of the electrical properties of muscle, fat, cartilage, periosteum (outer bone membrane), and bone marrow.

Current pulses of the PEMF devices are usually produced by the discharge of capacitor banks controlled by a timing network. The applicator coil cannot be interchanged among different devices because its inductance and resistance are a part of the discharge network. While earlier bone growth stimulators employed Helmholtz coil-pair arrangements, most present devices have single coils which can be custom-shaped for particular limbs. Figure 103.27 shows a typical system sold by EBI. This unit is driven by rechargeable batteries, and the control unit (shown in Fig. 103.27) includes an elapsed-time clock to measure the total time of stimulation of the fracture being treated. A typical treatment time can be between 2 and 10 h/day over a period of 6 months.

The so-called capacitively coupled device (E in Table 103.2) generates a continuous sine wave at a frequency of 60 kHz. The total current through the skin contains a not negligible conduction component since conductive contact is made between the applicator electrodes and the skin that represents a "leaky" capacitor. Electric fields produced at the tissue level by this device are between 1 and 50 V/m [11]. These levels are very much higher than the average amplitude of the electric

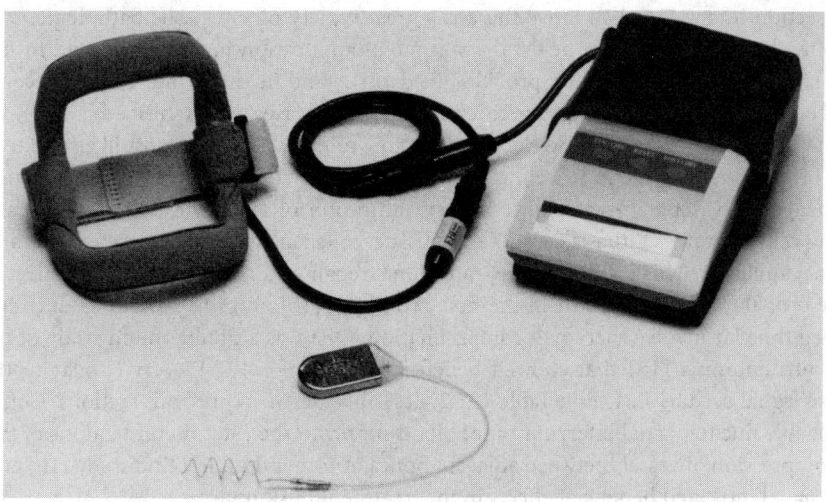

FIGURE 103.27 PEMF applicator and control unit and implantable dc stimulating device (battery with wire electrode) manufactured by Electrobiology, Inc. (systems A and G of Table 103.2). (Photograph courtesy of Electrobiology, Inc.)

fields produced in tissue by PEMF devices and also higher than the instantaneous peak values produced by some of the PEMF systems. It is interesting to note here that *in vitro* experiments with truly capacitively coupled fields showed enhancement of bone cell proliferation at very much lower amplitude (10^{-5} V/m) when the frequency of the continuous sine wave was 10 Hz rather than 60 kHz [12].

An invasive (implantable) dc device (F, G in Table 103.2) is also shown on Fig. 103.27. The small (approximately $4 \times 2 \times 0.5$ cm) titanium case contains a long-life battery that is connected to it. The case acts as the anode, and two (shown) or four titanium wires act as the cathode. The amplitude of the continuous current is between 5 μA (for some spinal fusion applications) and 20 μA (for nonunion of long bones). The cathodes are placed at the location where bone growth is to be enhanced, for example, at the vertebrae that are surgically fused, while the case is placed in a convenient location at some distance from the bone. Treatment details and success rates, in comparison with surgical procedures without use of dc stimulation, are discussed in the medical literature [13].

Although not used or approved for use in the United States, the German "Magnetodyn" system [6] is interesting not only because it employs sinusoidal magnetic fields between 2 and 20 Hz but also because it relies on metallic implants that are used for fixation of the bone to act as the "secondary" of a "transformer" whose primary is the external applicator. Sometimes an implanted pickup coil ("secondary") is connected to fixation bars or to screws (electrically insulated from the bars) on each side of a pseudoarthrotic gap. With this system, peak electric fields of 40 V/m and current densities of 5 A/m^2 have been produced in the gap.

A PEMF signal (similar to B in Table 103.2) has also been used experimentally to arrest **osteonecrosis** (bone death possibly due to vascular impairment or toxic agents) of the femoral head. Application for 8 h/day over 12 months gave substantially better results than the standard surgical (decompression) treatment [14].

Soft Tissue Repair and Nerve Regeneration

No electric or magnetic system to aid nerve regeneration or soft tissue repair is approved by the FDA at the present time for nonexperimental therapy. However, considerable animal and *in vitro* experimentation in this country and abroad suggests the clinical usefulness of electric currents for soft tissue repair [15] and possibly also to enhance repair of nerve fibers that have sustained crush or transsection injury [16, 17]. Since there is a great variety of soft tissue pathologies that could respond to electric or magnetic fields, the volume of application in this area could in the future become larger than in orthopedics, provided field-tissue and field-cell interactions become better understood and clinical benefits for specific injuries and diseases are established.

Beneficial effects of time-varying electric fields in wound healing are most likely related to promotion of *angiogenesis*. Wound healing consists of several stages, the first being inflammation, when changes in vascular permeability occur; infiltration of leucocytes and macrophages takes place; and cells migrate, synthesize granulation tissue, collagen, and proteoglycans, and initiate formation of capillaries. This is followed by transitional repair and remodeling phases. Electrical currents are probably only important in the first two stages [15] and the experimental clinical trials performed thus far involve therapy for inflammation. One was a double-blind study of persistent rotator cuff tendinitis [18] that showed beneficial effects of a PEMF very similar to the pseudarthrosis signal used by EBI (B in Table 103.2, described above in previous section). Cuff tendinitis is probably due to partial interruption of blood supply to the rotator cuff tendons of the shoulder by compression of vessels between adjacent bones. Another double-blind study [19] employing a PEMF signal indicated beneficial effects in the treatment of skin ulcers.

Clinical trials [20] and animal experiments also involved irradiation of wounds with pulsed high and very high frequencies between 3 and 44 MHz, at power levels from 73 μW to 15 W, employing pulse widths between 65 and 100 μs and pulse repetition frequencies between 200 Hz and 1 kHz. A commonly used animal model for wound healing is the McFarlane skin flap involving partial excision of a rectangular skin section on the back of the rat. Survival of the flap depends mainly on blood supply, with vascularization of the skin flap being an indirect measure of treatment success. The EBI signal shown on Fig. 103.26 (A in Table 103.2) is reported to have decreased skin flap necrosis when exposure was for 6 h/day for 3 days, while exposure for 18 h on the first day after injury had no observable effect [21]. Exposure with a triangular, symmetric and almost continuous magnetic field (18-ms triangular pulse, followed by 2-ms pause) at a frequency of 50 Hz (8-mT peak to peak value) produced a significant increase in wound contraction in rats (in comparison with controls) who were exposed to the field for 30 min immediately after surgery and for the same period thereafter every 12 h. In an effort to determine what type of signals are most beneficial for the acceleration of wound healing, skin flap necrosis was observed under exposure to sinusoidal magnetic fields at constant $\partial B/\partial t$ of 0.5 T/s using frequencies of 20, 72, and 500-Hz [22]. While signals at 20 and 72 Hz significantly decreased necrosis after 7 days, the 500-Hz signal was ineffective.

Very soon after some types of bone injury and pathology were first treated with PEMF, the effects of PEMF on peripheral nerve regeneration became subjects of investigation. Improved neural function that appeared as an unintended "side effect" in the clinical treatment of nonunions led Kort *et al.* to a systematic investigation of PEMF effects on neural regeneration in rats [23]. Other investigators employed other animal models and also compared PEMF with direct current as agents for neural regeneration. More recent work [17] employed a PEMF signal consisting of 20-ms pulses at a repetition rate of 2 pulses per second with **exponential rise** and **decay times** of, respectively, 0.5 and 1 ms. Amplitudes were 0.3 mT for experiments with crush lesions of the sciatic nerve in rats and 0.05 mT for stimulation *in vitro* of neurite outgrowth in dorsal root ganglia. Estimated values for the mean induced electric field pulses were 5 mV/m in the animal experiments and 0.7 mV/m in the 60-mm-diameter culture dishes of the *in vitro* work. Stimulation for 2 h per day of the *in vitro* cultures produced approximately 50% enhancement of neurite outgrowth in comparison with controls after 2 days.

The *in vivo* experiments using the 0.3-mT pulse produced "a 22% increase in the rate of regeneration relative to controls" as measured by a standardized test of reflex response. A very interesting observation was that animals exposed to the 0.3-mT field for 4 h/day for 7 days *prior* to the crush injury, who received *no* treatment after injury, responded similarly to those treated postinjury. Analysis of extracts from sciatic nerve segments after sacrifice of the animals showed that treatment with PEMF changed the molecular weight distribution of synthesized polypeptides.

A report on a very limited (13 subjects) clinical trial [24] of spinal nerve stimulation in para- and quadriplegics employing pulsed electric current introduced by needle electrodes produced "encouraging results" in terms of increased sensory perception and motor function. The signal obtained from a Chinese multipurpose therapy apparatus was described as follows: "The pulsed-wave generator produced a biphasic wave form of 2 ms duration with an initial slow positive deflection followed by an 80-μs rise time to its maximal negative deflection and subsequent asymptotic decay. This wave shape was pulsed in ramped bursts of 200 pulses per second for 1.5 s with 0.75 s rest time before the next ramped burst. Peak-to-peak voltages of approximately 30 V were common, and current flow was on the order of 1 mA."

Mechanisms and Dosimetry

Although clinical effects of electric currents, PEMF, and sinusoidal electric and magnetic fields are well documented, and although some specific biochemical results have been obtained *in vitro* or in animal experiments that suggest explanation of bone and soft tissue effects, the mechanism of field-to-cell or field-to-protein transduction is presently not understood. As a consequence, optimum "dose" (what field magnitude for how long) and optimum waveshapes or frequencies for particular clinical applications are unknown, and dosimetry relies largely on trial and error methods.

It is known that electrokinetic or **streaming potentials** rather than piezoelectricity make the principal contribution to electric potentials generated by mechanically stressed bone [25, 26]. Thus potential differences appear when mechanical loading displaces fluid that contains "counterions" which normally reside opposite ions fixed to cell or intercellular matrix surfaces. These potentials are likely to play a role in intercellular signaling and in bone as well as cartilage and soft tissue development. While the original intent of electrical bone therapy was to simply mimic endogenously generated fields, a much wider range of signals was found to be clinically useful. Furthermore it was found later that some weak ionic currents ($\approx 5 \times 10^{-2}$ A/m^2) [26] appear endogenously without mechanical stress and that extremely weak sinusoidal electric fields can produce profound effects on cells *in vitro*. For example, "transcription" (information transfer from DNA to RNA) and "translation" (formation of new proteins initiated by RNA) have been modified by electric fields estimated at 10^{-5} V/m that were generated inductively by a sinusoidal 60-Hz magnetic field of 5.7 µT [27]. Similarly, calcium metabolism was affected significantly in mitogen-activated lymphocytes by a 4-µT, 16-Hz magnetic field (in the presence of a 23.4-µT dc magnetic field) that induced an average electric field of about (2) 10^{-5} V/m [28]. Sinusoidal 15-Hz magnetic fields at the 0.5-mT level, giving an estimated mean electric field in the affected tissue of less than 10^{-3} V/m, significantly affected cartilage development in immature rats [29].

It is likely that the mechanism involved when direct currents are directly applied to injured bone (or other tissue) differs from the transduction sequence which must be acting when low-intensity alternating fields are employed. Even the 5-µA continuous current of the implantable devices (F, G in Table 103.2) when distributed over an (estimated) 5-cm^2 area corresponds to a steady electric field of 1 V/m in bone tissue of 0.01-S/m conductivity. This value is large compared with the average (but not the peak) fields induced by some PEMF devices and very large compared with the average fields induced in tissue by other PEMF devices [30] or the mV/m ELF sinusoidal fields that affect cartilage and bone development [29, 31]. For example, if one assumes a mean radius of 4 cm for a particular human bone fracture, one obtains from Eq. (103.17) the electric field values between 0.18 and 1.74 V/m shown on Table 103.3. When electrodes are implanted, as for the dc signals, chemical reactions at the electrodes may play a role in bone and cartilage formation. For example, the reaction at a stainless steel cathode involves consumption of dissolved oxygen and increase in local pH [26].

If the mechanism of interaction were to involve simple charge transfer by the applied electric field, it would be useful to compare the magnitude of the charge transferred by a single pulse, within a specified volume, with the random charge fluctuation due to thermal excitation during the pulse. An equation due to Einstein [32] gives the mean square value of the charge fluctuation δq in terms of Boltzmann's constant k, the absolute temperature T, the conductance G of the current-carrying region, and the observation time t:

$$<\delta q^2> = 2GkTt \tag{103.18}$$

It is then easy to find the electric field required to transfer during time t a charge at least equal to δq over the length of a conductance of volume v (v = length × cross-sectional area) and uniform conductivity σ. One obtains

$$E \geq \left(\frac{2kT}{v\sigma t} \right)^{1/2} \tag{103.19}$$

Assuming bone tissue with conductivity 10^{-2} S/m, a physiological temperature of 37°C, an interaction volume of 10^{-14} m^3 (about equal to the volume of a cell with 10-µm radius), and an observation time of 200 µs equal to the duration of the positive pulse of device A, one obtains from Eq. (103.19) a minimum value of 6.6 V/m. Unless v and σ can be assumed to be much larger, this would indicate that the values given in Table 103.3 should be below thermal noise. If one considers, instead of charge transfer over some as yet unknown path, the voltage induced by the applied field across the membrane of the idealized spherical cell with 10-µm radius, and compares the energy of

Table 103.3 Electric Field (V/m) Induced by PEMF Signals at Radius of 4 cm into Electrically Uniform Medium; B Perpendicular to Plane in which Radius is Defined

	Positive Peak	Negative Peak	Average of Rectified Signal
PEMF device A (Table 103.2, Figs. 103.25, 103.26)	0.2	1.74	0.024
PEMF device B (Table 103.2)	0.18	0.015	$(1.4) 10^{-4}$

the repetitive pulse below 100 Hz—where biological action apparently occurs [31]—with the thermal noise voltage given by

$$<V_n^2> = 4kTR\,(\Delta f) \qquad (103.20)$$

where (Δf) is the bandwidth and R the transmembrane resistance, one finds again that the electric field due to the PEMF devices would be at best only marginally above thermal noise. The induced electric fields of the *in vitro* experiments mentioned above are also clearly below thermal noise.

It is possible to obtain somewhat better signal-to-noise ratios if one considers either larger interaction volumes (assuming electrical phenomena involving the intercell volume), elongated cells, or cells connected by gap junctions [33]. It is also important to note that in the extremely inhomogeneous biological system, the actual electric field at a particular point can be considerably larger or smaller than its spatial average. Nevertheless very substantial improvement of signal-to-noise ratios would require signal averaging and limitation of bandwidth by resonance phenomena [34, 35]. Weak steady and time-varying magnetic fields could also be detected above thermal equilibrium noise by ferrimagnetic single-domain particles that have recently been detected in the human brain [36]. In addition, the applicability of Eqs. (103.18) and (103.20) is questionable, because living systems are often far from thermal equilibrium. For example, only mitogen-stimulated—and not quiescent—lymphocytes are affected by weak electric and magnetic fields [28]. Also, some molecules inside cells may at times be involved in systematic and guided rather than random thermal motion [37].

Several attempts have been made to construct theoretical models that would explain narrowband resonances in biological systems. Experiments to confirm or reject these hypotheses have thus far given ambiguous results. One theory assumes that ion transfer through cell membranes is affected by cyclotron resonance [38]. It is based on the fact that the cyclotron resonance frequency $\omega_c = 2\pi f_c$ of several physiologically important ions of charge Q and mass m, in the steady magnetic field B_0 of the earth, falls into the ELF range:

$$\omega_c = \frac{QB_0}{m} \qquad (103.21)$$

For example, the Ca^{2+} ion has a resonance frequency of 16 Hz in a dc field of 23.4 µT. However, it has been pointed out that the collision frequency in the physiological environment would be very much larger than the cyclotron frequency and would therefore wipe out any resonance motion, that the usually hydrated ions would have a total mass larger than m, and that the energy gain caused by an alternating field of frequency ω_c (as in a cyclotron) would require an orbital radius larger by many orders of magnitude than a typical cell radius [39]. Another theory postulates that the binding of Ca^{2+} to the protein calmodulin (ubiquitous in all vertebrates) should be affected by magnetic fields at frequencies ω_c and ω_c/n (where n is an integer) [40, 41]. This mechanism involves Zeeman splitting at ELF, due to B_0, of infrared vibrational modes that are chemically or thermally excited.

Some experiments showing a resonant cell response at the frequency given by (103.21) could not be replicated, while others were performed only over a very narrow frequency range. Nevertheless, apparently successful attempts have been made to stimulate bone cell proliferation at these field combinations [42]. However, in the reported experiments the alternating frequency was neither moved away from the resonance value nor changed by varying the dc field. Thus these attempts may just confirm other work showing that magnetic fields at frequencies below about 50 Hz affect bone development [29, 31].

To clarify the sequence of biological events that occurs when PEMF signals are applied to developing bone, the following *in vivo* experiment was performed [43]. Twenty-five milligrams of demineralized rat bone matrix in powdered form was implanted along the thoracic musculature of immature rats. This powder recruits cells from the surrounding tissue leading to formation of cartilage within 6 to 10 days; thereafter progressive calcification occurs, leading to formation of fibrous particles by days 12 to 14 and formation of a small bone ("ossicle"). These developments were compared in a large number of paired rats, with equal numbers unexposed and exposed (8 h/day) to the PEMF signal illustrated on Figs. 103.25 and 103.26 (A in Table 103.2). Estimates of the mean electric fields in the exposed tissues give values equal to about one-fourth of those listed on line 1 in Table 103.3. Chemical and histological analysis of ossicles harvested from animals, sacrificed on every second day, showed that exposure to this PEMF signal at the applied level significantly increased both rate and quantity of cartilage formation and enhanced maturation of the subsequent bone. The experimenters concluded that field exposure either enhanced recruitment or proliferation of cartilage precursor cells, increased differentiation of precursor to cartilage cells, or accelerated maturation of cartilage cells.

Getting even closer to fundamental events at the cellular level, both signals A and B (Table 103.2) were used to expose cultured mouse bone cells and mouse skin fibroblasts, as well as explanted mouse pineal cells in organ culture [44]. In all three cases various chemical procedures were employed to examine "beta-adrenergic **receptors**." These are cell surface protein strands that span the cell membrane and emerge from it; they mediate cell response to agents such as epinephrine (= adrenaline) or norepinephrine through so-called **G proteins** which act essentially as molecular amplifiers at the interior surface of the cell. Other G proteins are involved in the response to growth factors. The arrangement of the exposure coils and culture plates was such as to give mean electric fields equal to about one-tenth of the values shown in Table 103.3. Exposures were of 4 h duration. Specific types of G proteins were stimulated by the A signal, others by the B signal. The total number of binding sites on the cells was not affected, but the affinity of the receptors for specific hormones was changed, suggesting a change in receptor conformation. It is interesting to compare this work, which employed peak values of the order of 10^{-2} V/m and time average values not greater than $2 (10^{-3})$ V/m, with other *in vitro* experiments showing effects on enzyme activity at the cell surface by ELF sinusoidal fields between $5 (10^{-4})$ V/m and 30 V/m [45]. Related experiments at higher field intensities and the theory of field effects on catalysis [46] also show that electric field action at the exterior of the cell surface can be translated via enzyme-catalyzed chemical reactions to the cell interior.

Defining Terms

Angiogenesis: Formation of blood vessels.

Capacitively coupled fields or currents: Fields applied to the affected limb by electrodes touching the skin (the current from the electrodes has both displacement and conduction components).

Chondrogenesis: Formation of cartilage.

Exponential decay time t_d: Defined by $B = B_0 \exp(-t/t_d)$

Exponential rise time t_r: Defined by $B = B_0[1 - \exp(-t/t_r)]$

G protein: See Receptor.

Nonunion: Bone fracture that fails to heal within normally expected period with conventional management.

Osteonecrosis: Death of bone within a living vertebrate.

PEMF: Pulsed electromagnetic field.

Pseudoarthrosis: Formation of a pseudojoint in a broken or not completely formed bone (usually of congenital origin).

Receptor: Large protein molecule that protrudes from, and is embedded in, the membrane of a eucaryotic (nucleus-containing) cell. The part of the receptor outside the cell binds only to selected molecules that then cause chemical activity of G proteins bound to the end at the cell interior. Activity continues as long as a single molecule (for example, of a hormone, the first "messenger") is bound to the exterior part, and many molecules of a "second messenger" are released on the cell interior.

Streaming potential: Potential difference produced when liquid pressure displaces "counterions" that are normally held by electrostatic forces near ions of the opposite sign embedded in the surface of a stationary material.

References

1. E. Fukada and I. Yasuda, "On the piezoelectric effect in bone," *J. Phys. Soc. Japan,* vol. 12, p. 1158, 1957.

2. C.A.L. Bassett and R.O. Becker, "Generation of electric potentials by bone in response to mechanical stress," *Science,* vol. 13, p. 1063, 1963.

3. C.A.L. Bassett, R.J. Pawluk, and R.O. Becker, "Effects of electric currents on bone *in vivo*," *Nature,* vol. 204, p. 652, 1964.

4. C.A.L. Bassett, R.J. Pawluck, and A.A. Pilla, "Augmentation of bone repair by inductively coupled electromagnetic fields," *Science,* vol. 184, pp. 575–577, 1974.

5. H.R. Gossling, R.A. Bernstein, and J. Abbott, "Treatment of ununited tibial fractures, a comparison of surgery and pulsed electromagnetic fields (PEMF)," *Orthopaedics,* vol. 15, no. 6, pp. 711–719, 1992.

6. W. Kraus, "Magnetfeld Therapie und magnetisch induzierte Elektrostimulation in der Orthopadie," *Orthopade,* vol. 13, pp. 78–92, 1984.

7. C.T. Brighton, Z.B. Friedenberg, and J. Black, "Evaluation of the use of constant direct current in the treatment of nonunion," in *Electrical Properties of Bone and Cartilage,* C.T. Brighton, J. Black, and S.R. Pollack (eds.), New York: Grune and Stratton, 1979, pp. 519–545.

8. C.A.L. Bassett, "Biology of fracture repair, nonunion and pseudoarthrosis," in *Compilations of Fracture Management,* H.R. Gossling and S.L. Pillsbury (eds.), New York: J. B. Lippincott, 1984, pp. 1–8.

9. C. Polk and J.H. Song, "Electric fields induced by low frequency magnetic fields in inhomogeneous biological structures that are surrounded by an electric insulator," *Bioelectromagnetics,* vol. 11, pp. 235–249, 1990.

10. A.M.J. Van Amelsfort, *An Analytical Algorithm for Solving Inhomogeneous Electromagnetic Boundary-Value Problems for a Set of Coaxial Circular Cylinders,* Eindhoven, The Netherlands: James Clerk Maxwell Foundation, 1990.

11. S.R. Pollack and C.T. Brighton, "Dosimetry in electrical stimulation," *Trans. Bioelectric Repair and Growth Society,* vol. IX, p. 40, 1989.

12. R.J. Fitzsimmons, J. Farley, W.R. Adey, and D.J. Baylink, "Embrionic bone matrix formation is increased after exposure to a low amplitude capacitively coupled electric field *in vitro*," *Biochimica et Biophysica Acta,* vol. 882, pp. 51–56, 1986.

13. J. Nerubay, B. Marganit, J.J. Bubis, A. Tadmar, and A. Katznelson, "Stimulation of bone formation by electrical current on spinal fusion," *Spine,* vol. 11, p. 167, 1986.

14. R.K. Aaron and E. Steinberg, "Electrical stimulation of the femoral head," *Seminars in Arthroplasty,* vol. 2, no. 3, pp. 214–224, 1991.

15. D.J. Canaday and R.C. Lee, "Scientific basis for clinical applications of electric fields in soft tissue repair," in *Electromagnetics in Medicine and Biology,* C.T. Brighton and S.R. Pollack (eds.), San Francisco: San Francisco Press, 1991, pp. 275–280.

16. H. Ito and C.A.L. Bassett, "Effect of weak, pulsing electromagnetic fields on neural regeneration in the rat," *Clinical Orthopaedics,* vol. 181, pp. 283–290, 1983.

17. B.F. Siskin, M. Kanje, G. Lundborg, and W. Kurtz, "Pulsed electromagnetic fields stimulate nerve regeneration *in vitro* and *in vivo,*" *Restorative Neurology and Neuroscience,* vol. 1, pp. 303–309, 1990.

18. A. Binder, G. Parr, B. Hazelman, and S. Fitton-Jackson, "Pulsed electromagnetic field therapy of persistent rotator cuff tendinitis. A double-blind controlled clinical assessment," *Lancet,* pp. 695–698, March 31, 1984.

19. M. Ieran, S. Zaffuto, M. Bagnacani, M. Annovi, A. Moratti, and R. Cadossi, "Effect of low frequency pulsing electromagnetic fields on skin ulcers of venous origin in humans: A double-blind study," *J. Orthop. Research,* vol. 8, no. 2, pp. 276–282, 1990.

20. R.H.C. Bentall, "Low level pulsed radiofrequency fields and the treatment of soft-tissue injuries," *Bioelectrochemistry and Bioenergetics,* vol. 16, pp. 531–548, 1986.

21. E.A. Luce and G.C. Bryant, "Dose-response of electromagnetic field current in rat skin flap survival," *Trans. Bioelectrical Repair and Growth Society,* vol. 6, p. 72, 1986.

22. B.F. Sisken and E. Herbst, "Wound healing: electrical and electromagnetic fields," *Proc. 12th Ann. Intern. Conf. IEEE Engineering in Medicine and Biology Society,* vol. 4, no. 5, p. 1533, 1990.

23. J. Kort, H. Ito, and C.A.L. Bassett, "Effects of pulsing electromagnetic fields on peripheral nerve regeneration," *J. Bone Joint. Surg. Orthop. Trans.,* vol. 4, p. 238, 1980.

24. W. Ellis, "Pulsed subcutaneous electrical stimulation in spinal cord injury," *Bioelectromagnetics,* vol. 8, pp. 159–164, 1987.

25. D. Gross and W.S. Williams, "Streaming potential and the electromechanical response of physiologically moist bone," *J. Biomechanics,* vol. 15, pp. 227–295, 1982.

26. L.S. Lavine and A.J. Grodzinsky, "Electrical stimulation of repair of bone," *J. Bone Joint Surgery,* vol. 69, no. 4, pp. 626–630, 1987.

27. R. Goodman and S.A. Henderson, "Transcription and translation in cells exposed to extremely low frequency electromagnetic fields," *Bioelectrochemistry and Bioenergetics,* vol. 25, pp. 335–355, 1991.

28. M.G. Yost and R.P. Liburdy, "Time-varying and static magnetic fields act in combination to alter calcium signal transduction in the lymphocyte," *FEBS,* vol. 296, no. 2, pp. 117–122, 1992.

29. D.M. Ciombor, R. Aaron, H. Fisher, C. Polk, D. Gautreau, and D. Cherlin, "Effect of 15 Hz Sinusoidal Magnetic Field on Cartilage Development *In Vivo* Depends Non-Linearly on Duration of Daily Stimulation," *Project Resumes, The Annual Review of Research on Biological Effects of 50 and 60 Hz Electric and Magnetic Field,* U.S. Dept. of Energy, Office of Energy Management, 1991, p. P-8.

30. C.T. Rubin, K.J. McLeod, and L.E. Lanyon, "Prevention of osteoporosis by pulsed electromagnetic fields," *J. Bone and Joint Surgery,* vol. 71A, no. 3, pp. 411–418, 1989.

31. K.J. McLeod and C.T. Rubin, "Frequency specific modulation of bone adaptation by induced electric fields," *J. Theor. Biol.,* vol. 145, pp. 385–396, 1990.

32. A. Einstein, in *Investigation on the Theory of Brownian Movement,* M.R. Furth and A.D. Cowper, (eds.), New York: Dover Publications, 1956 (originally published 1905 in German), p. 33.

33. C. Polk, "Dosimetric extrapolations across biological systems: dosimetry of ELF magnetic fields," *Bioelectromagnetics,* vol. 13 (S1), 1992.

34. J.C. Weaver and R.D. Astumian, "The response of living cells to very weak electric fields: the thermal noise limit," *Science,* vol. 247, pp. 459–562, January 26, 1990.

35. K.R. Adair, "Constraints on biological effects of weak extremely low frequency electromagnetic fields," *Physical Review A,* pp. 1039–1048, 1991.

36. J.L. Kirschvink, A. Kobayashi-Kirschvink, and B.J. Woodford, "Magnetic biomineralization in the human brain," *Proc. Natl. Acad. Sci. USA,* vol. 89, 1992.

37. M. Hoffman, "Motor molecules on the move," *Science*, vol. 256, pp. 1758–1760, June 26, 1992.
38. A.R. Liboff and B.R. McLeod, "Kinetics of channelized membrane ions in magnetic fields," *Bioelectromagnetics*, vol. 9, pp. 39–51, 1988.
39. C. Polk, "Physical mechanisms by which low-frequency magnetic fields can affect the distribution of counterions on cylindrical biological cell surfaces," *J. Biol. Phys*, vol. 14, pp. 3–8, 1986.
40. V.V. Lednev, "Possible mechanism for the influence of weak magnetic fields on biological systems," *Bioelectromagnetics*, vol. 12, pp. 71–75, 1991.
41. R.K. Adair, "Criticism of Lednev's mechanism for the influence of weak magnetic fields on biological systems," *Bioelectromagnetics*, vol. 13, pp. 231–235, 1992.
42. R. Fitzsimmons, D. Baylink, F.P. Magee, and A.M. Weinstein, "Electromagnetic field stimulated bone cell proliferation" (Abstract), *J. Bone and Mineral Research*, vol. 6 (Supplement 1), August 1991.
43. R.K. Aaron, D.M. Ciombor, and J. Grant, "Stimulation of experimental endochondral ossification by low-energy pulsing electromagnetic fields," *J. Bone and Mineral Research*, vol. 4, no. 2, pp. 227–233, 1989.
44. R.A. Luben, "Effects of low energy electromagnetic fields on signal transduction by G protein linked receptors," in *Proc. First World Congress for Electricity and Magnetism in Biology and Medicine*, San Francisco: San Francisco Press, 1992 (in press); abstract on p. 4 of Abstract Book, World Congress EMBM, 1992.
45. M. Blank, "Na, K-ATPase function in alternating electric fields," *FASEB Journal*, vol. 6, pp. 2434–2438, April 1992.
46. B. Robertson and R.D. Astumian, "Frequency dependence of catalyzed reactions in a weak oscillating field," *J. Chem. Phys.*, vol. 94, pp. 7414–7419, 1991.

Further Information

W.R. Adey, "Electromagnetic fields, cell membrane amplification, and cancer promotion," in *Extremely Low Frequency Electromagnetic Fields: The Question of Cancer*, B.W. Wilson, R.G. Stevens, and L. E. Anderson (eds.), Columbus, Ohio: Battelle Press, 1990, pp. 211–249.

C.A.L. Bassett, "Bioelectromagnetics in service of medicine," *Bioelectromagnetics*, 13:7–17, 1992.

Bioelectromagnetics, the bi-monthly (formerly quarterly) journal of the Bioelectromagnetics Society. Published by Wiley-Liss, New York, since 1980.

J. Black, *Electrical Stimulation*, New York: Praeger, 1987.

M. Blank (ed.), *Electricity and Magnetism in Biology and Medicine*, San Francisco: San Francisco Press, 1993.

C. Branden and J. Tooze, *Introduction to Protein Structure*, New York: Garland Publishing, 1991.

C.T. Brighton, J. Black, and S. Pollack (eds.), *Electrical Properties of Bone and Cartilage*, New York: Grune and Stratton, 1979.

C.T. Brighton and S.R. Pollack (eds.), *Electromagnetics in Medicine and Biology*, San Francisco: San Francisco Press, 1991.

C. Polk and E. Postow, *CRC Handbook of Biological Effects of Electromagnetic Fields*, Boca Raton, Fla.: CRC Press, 1986. (A second edition should become available late in 1993 or in 1994.)

Transactions of the Bioelectric Repair and Growth Society. Vols. I through XI. Published annually since 1980 by the Bioelectric Repair and Growth Society, Dresher, Pennsylvania.

104

Biomedical Sensors

104.1 Introduction 2342
104.2 Physical Sensors 2343
104.3 Chemical Sensors 2345
104.4 Bioanalytical Sensors 2347
104.5 Applications 2348
104.6 Summary 2349

Michael R. Neuman
Case Western Reserve
University

104.1 Introduction

Any instrumentation system can be described as having three fundamental components: a sensor, a signal processor, and a display and/or storage device. Although all these components of the instrumentation system are important, the sensor serves a special function in that it interfaces the instrument with the system being measured. In the case of biomedical instrumentation a **biomedical sensor** (which in some cases may be referred to as a biosensor) is the interface between the electronic instrument and the biologic system. There are some general concerns that are very important for any sensor in an instrumentation system regarding its ability to effectively carry out the interface function. These concerns are especially important for biomedical sensors, since the sensor can affect the system being measured as well as the system can affect the sensor. Sensors must be designed so that they minimize their interaction with the biologic host. It is important that the presence of the sensor does not affect the variable being measured in the vicinity of the sensor as a result of the interaction between the sensor and the biologic system. If the sensor is placed in a living organism, that organism will probably recognize the sensor as a foreign body and react to it. This may in fact change the quantity being sensed in the vicinity of the sensor so that the measurement reflects the foreign body reaction rather than a central characteristic of the host.

Similarly, the biological system can affect the performance of the sensor. The foreign body reaction might cause the host to attempt to break down the materials of the sensor as a way to remove it. This may, in fact, degrade the sensor package so that the sensor can no longer perform in an adequate manner. Even if the foreign body reaction is not strong enough to affect the measurement, just the fact that the sensor is placed in a warm, aqueous environment may cause water to eventually invade the package and degrade the function of the sensor.

Finally, as will be described below, sensors that are implanted in the body are not accessible for calibration. Thus, such sensors must be extremely stable so that frequent calibrations are not necessary.

Biomedical sensors can be classified according to how they are used with respect to the biologic system. Table 104.1 shows that sensors can range from noninvasive to invasive as far as the biologic host is concerned. The most noninvasive of biomedical sensors do not even contact the biological system

Table 104.1 Classification of
Biomedical Sensors According to
Their Interface with the
Biologic Host

Noninvasive	Noncontacting
	Body surface
Invasive	Indwelling
	Implanted

Table 104.1 Classification of Biomedical Sensors According to Their Interface with the Biologic Host

being measured. Sensors of radiant heat or sound energy coming from an organism are examples of noncontacting sensors. **Noninvasive sensors** can also be placed on the body surface. Skin surface thermometers, biopotential electrodes, and strain gauges placed on the skin are examples of noninvasive sensors. Indwelling sensors are those which can be placed into a natural body cavity that communicates with the outside. These are sometimes referred to as minimally invasive sensors and include such familiar sensors as oral-rectal thermometers, intrauterine pressure transducers, and stomach pH sensors. The most invasive sensors are those that need to be surgically placed and that require some tissue damage associated with their installation. For example, a needle electrode for picking up electromyographic signals directly from muscles; a blood pressure sensor placed in an artery, vein, or the heart itself; or a blood flow transducer positioned on a major artery are all examples of invasive sensors.

We can also classify sensors in terms of the quantities that they measure. **Physical sensors** are used in measuring physical quantities such as displacement, pressure, and flow, while **chemical sensors** are used to determine the concentration of chemical substances within the host. A subgroup of the chemical sensors that are concerned with sensing the presence and the concentration of biochemical materials in the host are known as **bioanalytical sensors**, or sometimes they are referred to as biosensors.

In the following paragraphs we will look at each type of sensor and present some examples as well as describe some of the important issues surrounding these types of sensors.

104.2 Physical Sensors

Physical variables associated with biomedical systems are measured by a group of sensors known as physical sensors. A list of typical variables that are frequently measured by these devices is given in Table 104.2. These quantities are similar to physical quantities measured by sensors for non-biomedical applications, and the devices used for biomedical and nonbiomedical sensing are, therefore, quite similar. There are, however, two principal exceptions: pressure and flow sensors.

The measurement of blood pressure and blood flow in humans and other animals remains a difficult problem in biomedical sensing. Direct blood pressure measurement refers to evaluation of

Table 104.2 Physical Variables Sensed by Biomedical Sensors

Displacement (linear and angular)
Temperature
Force (weight and mass)
Pressure
Flow
Radiant energy (optical)

the blood pressure using a sensor that is in contact with the blood being measured or contacts it through an intermediate fluid such as a physiologic saline solution. Direct blood pressure sensors are invasive. Indirect blood pressure measurement involves a sensor that does not actually contact the blood. The most familiar indirect blood pressure measurement is the sphygmomanometer cuff that is usually used in most medical examinations. It is a noninvasive instrument. Until recently, the primary sensor used for direct blood pressure measurement was the unbonded strain gauge pressure transducer shown in Fig. 104.1. The basic principle of this device is that a differential pressure seen across a diaphragm will cause that diaphragm to deflect. This deflection is then measured by a displacement transducer. In the unbonded strain gauge sensor a closed chamber is covered by a flexible diaphragm. This diaphragm is attached to a structure that has four fine gauge wires drawn between it and the chamber walls. A dome with the appropriate hardware for coupling to a pressure source covers the diaphragm on the side opposite the chamber such that when the pressure in the dome exceeds the pressure in the chamber, the diaphragm is deflected into the chamber. This causes two of the fine wires to stretch by a small amount while the other two wires contract by the same amount. The electrical resistance of the wires that are stretched increases while that of the wires that contract decreases. By connecting these wires, or

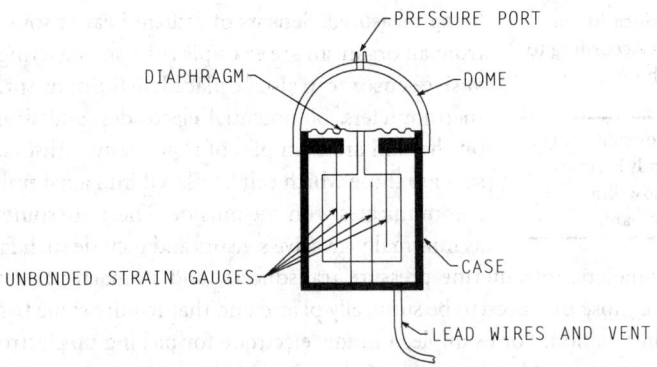

FIGURE 104.1   An unbonded strain gauge pressure transducer.

more correctly these unbonded strain gauges, into a Wheatstone bridge circuit, a voltage proportional to the deflection of the diaphragm can be obtained.

In recent years semiconductor technology has been applied to the design of pressure transducers. Silicon strain gauges that are much more sensitive than their wire counterparts are formed on a silicon chip, and micromachining technology is used to form this portion of the chip into a diaphragm with the strain gauges integrated into its surface. This structure is then incorporated into a plastic housing and dome assembly. The entire sensor can be fabricated and sold inexpensively so that disposable, single-use devices can be made. These have the advantage that they are only used on one patient and they do not have to be cleaned and sterilized between patients. By using them on only one patient, the risk of transmitting blood-borne infections is eliminated.

In biomedical applications pressure is generally referenced to atmospheric pressure. Therefore, the pressure in the chamber of the pressure transducer must be maintained at atmospheric pressure. This is done by means of a vent in the chamber wall or a fine bore, flexible capillary tube that couples the chamber to the atmosphere. This tube is usually included in the electrical cable connecting the pressure transducer to the external instrumentation such that the tube is open to the atmosphere at the cable connecter.

In using this sensor to measure blood pressure the dome is coupled to a flexible plastic tube, and the dome and tube are filled with a physiological saline solution.[1] As described by Pascal's Law, the pressure in the dome, and hence against the diaphragm, will be the same as that at the tip of the tube provided the tip of the tube is at the same horizontal level as the dome. Thus by threading the tube into a blood vessel, an invasive procedure, the blood pressure in that vessel can be transmitted to the dome and hence the diaphragm of the pressure transducer. The pressure transducer will, therefore, sense the pressure in the vessel. This technique is known as external direct blood pressure measurement, and the flexible plastic tube that enters the blood vessel is known as a catheter. It is important to remember that the horizontal level of the blood pressure transducer dome must be the same as that of the tip of the catheter in the blood vessel to accurately measure the pressure in that vessel without adding an error due to the hydrostatic pressure in the catheter.

In addition to problems due to hydrostatic pressure differences between the chamber and the dome, catheters introduce pressure errors as a result of the dynamic properties of the catheter, fluid, dome, and diaphragm. These properties as well as air bubbles in the catheter, or obstructions due to clotted blood or other materials, introduce resonances and damping. These problems can be minimized by utilizing miniature pressure transducers fabricated using microelectronic semiconductor technology that are located at the tip of a catheter rather than at the end that is external to

[1] It must be pointed out that the use of such a sensor is not limited to blood pressure measurement. The strain gauge pressure sensor can be used to measure the pressure of any fluid to which it is appropriately coupled.

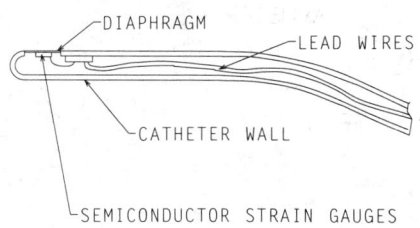

FIGURE 104.2 A catheter tip pressure transducer.

the body. A general arrangement for such a pressure transducer is shown in Fig. 104.2. As with the disposable sensors, strain gauges are integrated into the diaphragm of the transducer such that they detect very small deflections of this diaphragm. Because of the small size, small diaphragm displacement, and lack of a catheter with a fluid column, these sensors have a much broader frequency response, give a clearer signal, and do not have any hydrostatic pressure error.

Although the indwelling catheter tip pressure transducer appears to solve many of the problems associated with the external pressure transducer, there are still important problems in pressure transducer design that need to be addressed. Long-term stability of pressure transducers is not very good. This is especially problematic for venous pressure measurements which are carried out at relatively low pressure. Long-term changes in baseline pressure require pressure transducers to be frequently adjusted to be certain of zero pressure. While this can be done relatively easily for external and indwelling pressure transducers, there is no way to carry out this procedure for implanted transducers, since there is not a way to establish zero pressure at the sensor. Thus devices that have very low long-term baseline drift are essential for implantable applications.

The packaging of the pressure transducer also represents a problem that needs to be addressed. Packaging must both protect the transducer and be biocompatible. It also must allow the appropriate pressure to be transmitted from the biologic fluid to the diaphragm. The amount of packaging material required should be kept at a minimum so as not to substantially increase the size of implantable or indwelling sensors. Furthermore, the material must be mechanically stable so that it does not swell or contract, since this will most likely change the baseline pressure seen by the sensor. These problems need to be overcome before miniature pressure transducers can be used reliably in implantable applications.

104.3 Chemical Sensors

There are many biomedical problems where it is necessary to know the concentration of a particular substance in a biological sample. Chemical sensors provide the interface between an instrument and the specimen to allow one to determine this concentration. These sensors can be used on a biological specimen taken from the host and tested in a laboratory, or they can be used for *in vivo* measurements either as noninvasive or invasive sensors, the latter being the most frequently used.

Table 104.3 Classifications of Chemical Biomedical Sensors

1. Electrochemical
   a. Amperometric
   b. Potentiometric
   c. Coulometric
2. Optical
   a. Colorimetric
   b. Emission and absorption spectroscopy
   c. Fluorescence
   d. Chemiluminescence
3. Thermal methods
   a. Calorimetry
   b. Thermoconductivity
4. Nuclear magnetic resonance

There are many types of chemical sensors used in biomedical instrumentation. Table 104.3 lists some general categories of sensors. Electrochemical and optical sensors are most frequently used for biomedical measurements both *in vivo* and *in vitro*. An example of an electrochemical sensor is the Clark electrode illustrated in Fig. 104.3. This consists of an electrochemical cell separated from the specimen being measured by an oxygen-permeable membrane. The cell is driven at a fixed potential of 600 mV, and under these conditions the following reaction occurs at the noble metal cathode:

$$O_2 + 4e^- + H_2O \rightarrow 4OH^-$$

This reaction involves the reduction of molecular oxygen that diffuses into the cell through the oxygen-permeable membrane. Since the other components of the reaction are in abundance, the rate

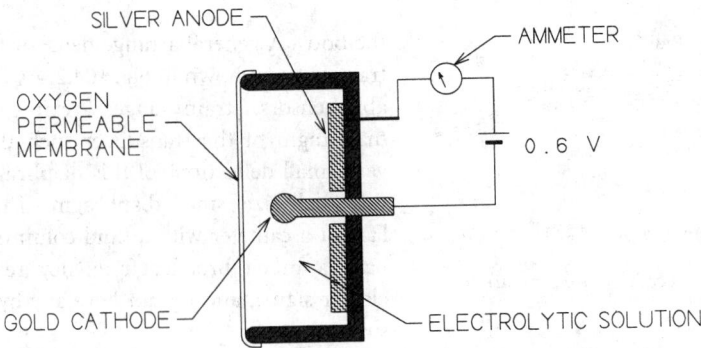

FIGURE 104.3 The Clark electrode, an amperometric electrochemical sensor of oxygen.

of the reaction is limited by the amount of oxygen available. Thus, the rate of electrons used at the cathode is directly related to the available oxygen. In other words, the cathode current is proportional to the partial pressure of oxygen in the specimen being measured.

The electrochemical cell is completed by the silver anode. The reaction at the anode involves forming the low-solubility salt, silver-chloride, from the anode material itself and the chloride ion contained in the electrolyte. The cell is designed so that these materials are also in abundance so that their concentration does not affect the sensor performance. This type of sensor is an example of an **amperometric** electrochemical sensor.

Another type of electrochemical sensor that is frequently used in biomedical laboratories is the glass pH electrode illustrated in Fig. 104.4. The acidity or alkalinity of a solution is characterized by its pH. This quantity is defined as

$$pH = -\log_{10}[H^+]$$

where $[H^+]$ is the activity of the hydrogen ions in solution, a quantity that is related to the concentration of the hydrogen ions. This sensor only works in an aqueous environment. It consists of an inner chamber containing an electrolytic solution of known pH and an outer solution with an unknown pH that is to be measured. The membrane consists of a specially formulated glass that will in essence allow hydrogen ions to pass in either direction but will not pass other chemical species. If the concentration of hydrogen ions in the external solution is greater than that in the internal solution, there will be a gradient forcing hydrogen ions to diffuse through the membrane

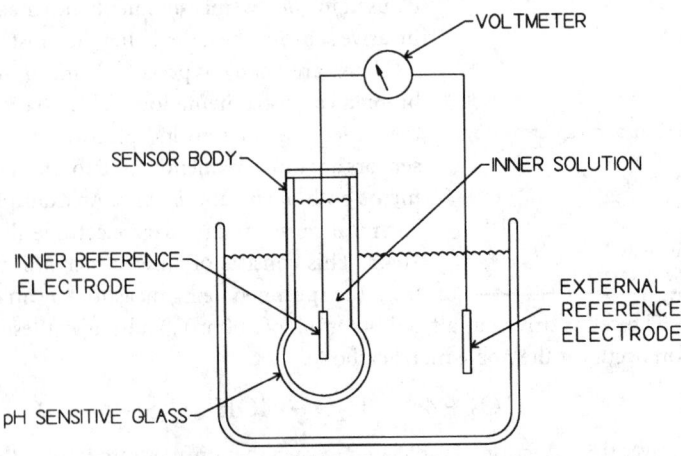

FIGURE 104.4   A glass electrode pH sensor.

into the internal solution. This will cause the internal solution to have a greater positive charge than the external solution so that an electrical potential and, hence, an electric field will exist across the membrane. This field will counteract the diffusion of hydrogen ions due to the concentration difference and so an equilibrium will be eventually established. The potential across the membrane at this equilibrium condition will be related to the hydrogen ion concentration difference (or more accurately the activity difference) between the inner and outer solutions. This potential is given by the Nernst equation

$$E = -\frac{RT}{nF} \ln\left(\frac{a_1}{a_2}\right)$$

where E is the potential measured, R is the universal gas constant, T is the absolute temperature, n is the valence of the ion, and a_1 and a_2 are the activities of the ions on each side of the membrane. Thus the potential measured across the glass membrane will be proportional to the pH of the solution being studied. At room temperature the sensitivity of the electrode is approximately 60 mV/pH. It is not practical to measure the potential across the membrane directly and so reference electrodes, sensors that can be used to measure electrical potential of an electrolytic solution, are used to contact the solution on either side of the membrane to measure the potential difference across it. The reference electrodes and the glass membrane are incorporated into the structure shown in Fig. 104.4 known as a glass pH electrode. This is an example of a **potentiometric** measurement made using an ion-selective membrane.

There are other types of ion-selective membrane potentiometric chemical sensors that are used for biomedical applications. The membranes of these sensors determine the ion being sensed. The membrane can be based upon glass or a polymeric material such as polyvinyl chloride, but the key component is the substance that is added to the membrane that allows it to selectively pass a single ion.

Important problems in the development of chemical biomedical sensors are similar to those discussed above for the pressure sensor. Issues of long-term stability and packaging are critical to the success of a chemical sensor. The package is even more critical in chemical sensors than it was in pressure sensors in that the package must protect portions of the sensor that require isolation from the solutions being measured while it provides direct contact of the chemically sensitive portions of the sensor to the solution. The maintenance of a window through the package for this contact represents a critical aspect of sensor development. Frequent calibration is also necessary for chemical sensors. Just about every type of chemical sensor requires some sort of calibration using a standard solution with known concentration of the **analyte** being sensed. The best calibration procedure is a two-point procedure where two standards are used to establish the slope and the intercept of the calibration line. Some chemical sensors have stable slopes but need to be calibrated in terms of the baseline or intercept. In this case a single-point calibration can be used.

104.4 Bioanalytical Sensors

A special class of sensors of biological molecules has evolved in recent years. These bioanalytical sensors take advantage of one of the following biochemical reactions: (1) enzyme-substrate, (2) antigen-antibody, or (3) ligand-receptor. The advantage of using these reactions in a sensor is that they are highly specific for a particular biological molecule, and sensors with high sensitivity can be developed based upon these reactions. The basic structure of a bioanalytical sensor is shown in Fig. 104.5. There are two principal portions of the sensor. The first contains one component of the biological sensing reaction such as the enzyme or the antibody, and the second component involves a means of detecting whether the biological reaction has taken place. This second portion of a bioanalytical sensor is made up of either a physical or chemical sensor that serves as the detector of the biological reaction. As illustrated in Fig. 104.5, this detector can consist of an electrical

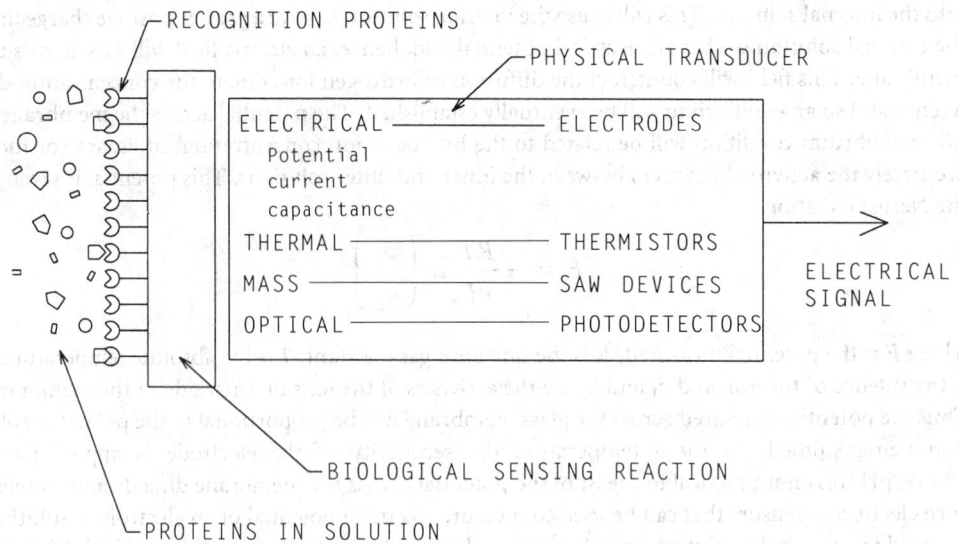

FIGURE 104.5 A generalized bioanalytical sensor.

sensor such as used in electrochemical sensors, a thermal sensor, a sensor of changes in capacitance, a sensor of changes in mass, or a sensor of optical properties.

An example of a bioanalytical sensor is a glucose sensor. The first portion of the sensor contains the enzyme glucose oxidase. This enzyme promotes the oxidation of glucose to glucuronic acid and consumes oxygen in the process. Thus, by placing an oxygen sensor along with the glucose oxidase in the bioanalytical sensor, one can determine the amount of glucose oxidized by measuring the amount of oxygen consumed. An even better approach is to have two identical sensor structures in the same package. The only difference is that only one of the sensors contains the enzyme. When there is no glucose present, both sensors will measure the same oxygen partial pressure. The presence of glucose, however, will cause the sensor with the glucose oxidase to have a reduced partial pressure of oxygen due to the oxygen consumption of the reaction. By making a differential measurement of oxygen partial pressure with both sensors, other factors that can cause an apparent change in oxygen partial pressure such as temperature will have a much lower effect than if a single sensor was used.

Stability problems are important for bioanalytical sensors, especially those that are used for long-term measurements. Not only are the stability issues the same as for the physical and chemical sensors, but they are also related to preservation of the biological molecules used in the first stage of the sensor. These molecules can often be degraded or destroyed by heat or exposure to light. Even aging can degrade some of these molecules. Thus, an important issue in dealing with bioanalytical sensors is the preservation of the biochemical components of the sensor.

104.5 Applications

Biomedical sensors and instrumentation are used in biomedical research and patient care applications. In terms of patient care, sensors are used as a part of instruments that carry out patient screening by making measurements such as blood pressure using automated apparatus. Specimen analysis is another important application of biomedical sensors in patient care. This can include analyses that can be carried out by the patients themselves in their homes such as is done with home blood glucose analyzers. Instrumentation based upon biomedical sensors can be used in the physician's office for carrying out some chemical analyses of patient specimens such as urinalysis or elementary blood chemistries such as serum glucose and electrolytes. Sensors also are a part of large multicomponent automatic blood analyzers used in the central clinical laboratory of major medical centers.

Another application for biomedical sensors is in patient monitoring. Sensors represent the front end of critical care monitors used in the intensive care unit and in the operating and recovery rooms. Measurements cover a wide range of biomedical variables such as continuous recordings of blood pressure and transcutaneous measurement of the partial pressure of carbon dioxide in the blood. The performance of these instruments is strongly dependent on biomedical sensors. Patient monitoring can also be carried out in the various clinical units of the hospital. Devices such as ambulatory cardiac monitors that allow patients to be observed while they are free to move around if they desire are becoming important in clinical care in "step-down" units for patients who have completed their stay in the intensive care unit. Patient monitoring has even made its way into the home. Home cardiorespiratory monitors are thought to have some potential value in identifying infants at risk of sudden infant death.

104.6 Summary

Sensors serve an important function in biomedical instrumentation systems in that they provide the interface between the electronic instrument and the biologic system being measured. Very often the quality of the instrument is based upon the quality of the sensor at the instrument's front end. Although electronic signal processing has been developed to a high level, the signals are no better than the quality of the sensors that provide them. Although there have been many advances in biomedical sensor technology, many problems remain. Biomedical sensors will continue to be an important area for research and development in biomedical engineering.

Defining Terms

Amperometric sensor: An electrochemical sensor that determines the amount of a substance by means of an oxidation-reduction reaction involving that substance. Electrons are transferred as a part of the reaction, so that the electrical current through the sensor is related to the amount of the substance seen by the sensor.

Analyte: The substance being measured by a chemical or bioanalytical sensor and instrumentation system.

Bioanalytical sensor: A special case of a chemical sensor for determining the amount of a biochemical substance. This type of sensor usually makes use of one of the following types of biochemical reactions: enzyme-substrate, antigen-antibody, or ligand-receptor.

Biomedical sensor: A device for interfacing an instrumentation system with a biological system such as a biological specimen or an entire organism. The device serves the function of detecting and measuring in a quantitative fashion a physiological property of the biologic system.

Chemical sensor: The interface device for an instrumentation system that determines the concentration of a chemical substance.

Noninvasive sensor: The interface device of an instrumentation system that measures a physiologic variable from an organism without interrupting the integrity of that organism. This device can be in direct contact with the surface of the organism or it can measure the physiologic quantity while remaining remote from the organism.

Physical sensor: An interface device at the input of an instrumentation system that quantitatively measures a physical quantity such as pressure or temperature.

Potentiometric sensor: A chemical sensor that measures the concentration of a substance by determining the electrical potential between a specially prepared surface and a solution containing the substance being measured.

References

R. S. C. Cobbold, *Transducers for Biomedical Measurements: Principles and Applications*, New York: John Wiley, 1974.

D. G. Fleming, W. H. Ko, and M. R. Neuman, Eds., *Indwelling and Implantable Pressure Transducers*, Cleveland: CRC Press, 1977.

L. A. Geddes, *The Direct and Indirect Measurement of Blood Pressure*, Chicago: Year Book Medical Publishers, 1970.

L. A. Geddes, *Electrodes and the Measurement of Bioelectric Events*, New York: John Wiley, 1972.

W. Göpel, J. Hesse and J. N. Zemel, *Sensors; A Comprehensive Survey*, Weinheim, Germany: VCH Verlagsgesellschaft, 1989.

J. Janata, *Principles of Chemical Sensors*, New York: Plenum Press, 1989.

R. Pallas-Areny and J. G. Webster, *Sensors and Signal Conditioning*, New York: John Wiley, 1991.

J. I. Peterson and G. G. Vurek, "Fiber-optic sensors for biomedical applications," *Science*, vol. 224, pp. 123–127, 1984.

P. Rolfe, "Review of chemical sensors for physiological measurement," *J. Biomed. Eng.*, vol. 10, pp. 138–145, 1988.

J. G. Webster, Ed., *Encyclopedia of Medical Devices and Instrumentation*, New York: John Wiley, 1988.

Further Information

Research reports on biomedical sensors appear in many different journals ranging from those that are concerned with clinical medicine through those that are engineering and chemistry oriented. Three journals, however, represent major sources of biomedical sensor papers. These are listed as follows:

The *IEEE Transactions on Biomedical Engineering* is a monthly journal devoted to research papers on biomedical engineering. Papers on biomedical sensors frequently appear, and the February 1986 issue was devoted entirely to the topic of biomedical sensors. For more information or subscriptions, contact IEEE Service Center, 445 Hoes Lane, P.O. Box 1331, Piscataway, NJ 08855-1331.

The international journal *Medical and Biological Engineering and Computing* is published bi-monthly by the International Federation for Medical and Biological Engineering. This journal also contains frequent reports on biomedical sensors and related topics. Subscription information can be obtained from Peter Peregrinus Ltd., P.O. Box 96, Stevenage, Herts SG12SD, United Kingdom.

The journal *Biomedical Instrumentation and Technology* is published by the Association for the Advancement of Medical Instrumentation. This bimonthly journal has reports on biomedical instrumentation for clinical applications, and these include papers on biomedical sensors. Subscription information can be obtained from Hanley and Belfus, 210 S. 13th Street, Philadelphia, PA 19107.

There are also several scientific meetings that include biomedical sensors. The major meeting in the area is the international conference of the IEEE Engineering in Medicine and Biology Society. An extensive book of extended abstracts for this meeting is published each year by the IEEE. Further information can be obtained by contacting the IEEE at the address listed above.

105

Bioelectronics and Instruments

Joseph D. Bronzino
Trinity College

Edward J. Berbari
University of Oklahoma

105.1 Quantitative Analysis of the Electroencephalograms 2351
The Language of the Brain • Historical Perspective • Frequency
Analysis • EEG Recording Techniques • Topographic Mapping
105.2 The Electrocardiograph 2362
Physiology • Instrumentation • Conclusions

105.1 Quantitative Analysis of the Electroencephalograms

Joseph D. Bronzino

Electroencephalograms (EEGs) are recordings of the minute (generally less than 300 μV) electrical potentials produced by the brain. Since 1924, when Hans Berger reported the measurements of rhythmic electrical activity on the human scalp, it has been suggested that these patterns of bioelectrical origin may provide clues regarding the neuronal bases for specific behaviors and has offered great promise to reveal correlations between pathological processes and the electrical activity of specific regions of the brain.

Over the years, EEG analyses have been conducted primarily in clinical settings, to detect gross organic pathologies and the epilepsies, and in research facilities to quantify the central effect of new pharmacological agents. As a result of these efforts, cortical EEG patterns have been shown to be modified by a variety of conditions which may or may not be pathologic. For example, biochemical, metabolic, circulatory, hormonal, neuroelectric, and behavioral factors are all variables that affect EEG activity. Alteration of these variables produces a large variety of electrical patterns with different frequency and voltage characteristics. In the past, this large variation in brain electrical activity has limited the interpretation of the EEG to visual inspection by a trained electroencephalographer capable of distinguishing normal activity from localized or generalized abnormalities of particular types from relatively long EEG records. This approach has left clinicians and researchers alike lost in a sea of EEG paper records. The application of computer technology to analyze EEG data, however, has permitted significant changes in the EEG caused by various insults to the brain such as alcohol, malnutrition, and lead.

With this in mind, this section provides an introduction to some of the basic concepts underlying the generation of the EEG, a review of the basic approaches used in quantifying alterations in the EEG, and some insights regarding quantitative electrophysiology techniques.

The Language of the Brain

The mass of brain tissue is composed of bundles of nerve cells (neurons) which constitute the fundamental building blocks of the nervous system. Figure 105.1 is a schematic drawing of just such a cell.

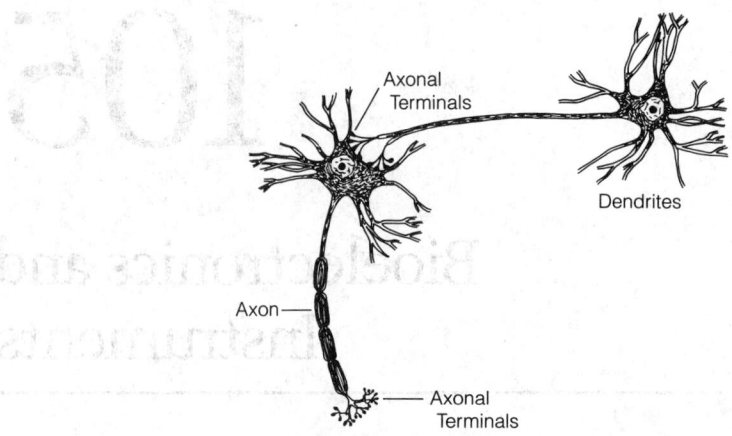

FIGURE 105.1 Basic structure of the neuron.

It consists of three major components: the cell body (or soma), the receptor zone (or dendrites), and the axon, which carries electrical signals from the soma to target sites such as muscles, glands, or other neurons. Numbering approximately 20 billion in each human being, these tiny cells come in a variety of sizes and shapes. Although neurons are anatomically distinct units having no physical continuity between their processes, the axon ends on the soma and the dendrites of other cells in what is called a synapse. Under the microscope this often stands out as a spherical enlargement at the end of the axon to which various names have been given, for example, boutons, end-plate, or synaptic terminals. This ending does not actually make physical contact with the soma or dendrite but is separated by a narrow cleft (gap) of approximately 100 to 200 Å (10^{-9} m) wide. This is known as the synaptic cleft. Each of these synaptic endings contains a large number of submicroscopic spherical structures (synaptic vesicles) that can be detected only under an electron microscope. These synaptic vesicles, in turn, are essentially "chemical carriers" containing transmitter substance that is released into the synaptic cleft on excitation.

When an individual neuron is excited, an electrical signal is transmitted along its axon to many tiny branching, diverging fibers near its far end. These axonal terminals end as synapse on a large number of other neurons. When an electrical pulse arrives at the synapse, it triggers the release of a tiny amount of transmitter substance which crosses the synaptic cleft thereby altering the membrane potential of the receiving neuron. If the change is above a certain threshold value, the neuron is activated and generates an action potential of its own which is propagated along its axon, and the process is repeated.

Neurons are involved in every conceivable action taken by the body, whether it is to control its own internal environment or to respond to changes in the external world. As a result, they are responsible for such essential functions as:

- Accepting and converting sensory information into a form that can be processed within the nervous system by other neurons.
- Processing and analyzing this information so that an "integrated portrait" of the incoming data can be obtained.
- Translating the final outcome or "decision" of this analysis process into appropriate electrical or chemical form needed to stimulate glands or activate muscles.

Evolution has played a role in the development of these unique neurons and in the arrangement and development of interconnections between nerve cells in the various parts of the brain. Since the brain is a most complex organ, it contains numerous regions designed for specific tasks. One might, in fact, consider it to be a collection of organs arranged together to act in the harmony of activity we recognize as the individual's state of consciousness or as life itself. Over the years, anatomists and physiologists have identified and named most pathways (tracts), most groups of neurons (nuclei), and most of the major parts of the human brain. Such attention to detail is certainly not necessary here. It will serve our purpose to simply provide a broad overview of the organization of the brain and speak of three general regions: the brainstem, cerebellum, and the cerebral cortex.

The brainstem, or old brain, is really an extension and elaboration of the spinal chord. This section of the brain evolved first and is the location of all the centers that control the regulatory systems, such as respiration, necessary for physical survival of the organism. In addition, all sensory pathways find their way into the brainstem, thereby permitting the integration of complex input patterns to take place within its domain.

Above the brainstem is a spherical mass of neuronal tissue called the cerebellum. This remarkable structure is a complex monitor and modifier of body movements. The cerebellum does not initiate movements, but only modifies motor control activated in other areas. Cerebellar operation is not only dependent on evolutionary development, but relies heavily on actual use and patterns of learned motor behavior acquired throughout life. It is for this reason that the movements of a gymnast are smooth and seemingly effortless.

The most conspicuous part of all in the human brain is the cerebral cortex. Compared to most mammals, it is so large in man that it becomes a covering that surrounds and hides most of the other regions of the brain. Wrinkled and folded, the cerebral tissue is literally pressed into the limited space allocated to it. Although it has been possible to ascertain that certain cortical areas such as visual cortex, the sensory projection area, and the motor strip are associated with specific functions, the overall operation of this complex structure is still not completely understood. However, for the sake of convenience, it has been arbitrarily divided (based primarily on anatomical considerations) into the following areas: frontal lobe, parietal lobe, temporal lobe, and occipital lobe (Fig. 105.2). Each of these segments of the cortex, which is the source of intellectual and imaginative capacities, includes millions of neurons and a host of interconnections.

It is generally agreed that brain function is based on the organization of the activity of large numbers of neurons into coherent patterns. Since the primary mode of activity of these nerve cells is electrical in nature, it is not surprising that a composite of this activity can be detected in the form of electrical signals. Of extreme interest, then, are the actual oscillations, rhythms, and patterns seen in the cryptic flow of electrical energy coming from the brain itself, i.e., in the EEG.

Historical Perspective

In 1875, Caton published the initial account of the recording of the spontaneous electrical activity of the brain from the cerebral cortex of an experimental animal. The amplitude of these electrical

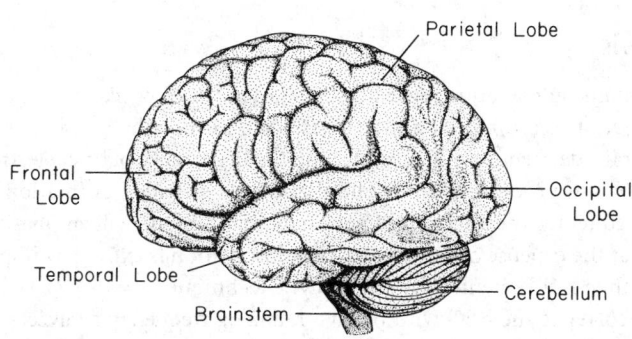

FIGURE 105.2 Major divisions of the cerebral cortex.

oscillations was so low, that is, on the order of microvolts, that Caton's discovery is all the more amazing because it was made 50 years before suitable electronic amplifiers became available. In 1924, Hans Berger, of the University of Jena in Austria, carried out the first human EEG recordings using electrical metal strips pasted to the scalps of his subjects as electrodes and a sensitive galvanometer as the recording instrument. Berger was able to measure the irregular, relatively small electrical potentials (i.e., 50 to 100 μV) coming from the brain. By studying the successive positions of the moving elements of the galvanometer recorded on a continuous roll of paper, he was able to observe the resultant patterns in these brain waves as they varied with time. From 1924 to 1938, Berger laid the foundation for many of the present applications of electroencephalography. He was the first to use the word electroencephalogram in describing these brain potentials in man. Berger noted that these brain waves were not entirely random, but instead displayed certain periodicities and regularities. For example, he observed that although these brain waves were slow (i.e., exhibited a synchronized pattern of high amplitude and low frequency, <3 Hz) in sleep and states of depressed function, they were faster (i.e., exhibited a desynchronized pattern of low amplitude and high frequency, 15–25 Hz) during waking behavior. He suggested, quite correctly, that the brain's activity changed in a consistent and recognizable fashion when the general status of the subject changed, as from relaxation to alertness. Berger also concluded that these brain waves could be greatly affected by certain pathological conditions after noting the marked increase in the amplitude of these brain waves brought about by convulsive seizures. However, in spite of the insights provided by these studies, Berger's original paper published in 1929 did not excite much attention. In essence, the efforts of this most remarkable pioneer were largely ignored until similar investigations were carried out and verified by British investigators.

It was not until 1934 when Adrian and Matthews published their classic paper verifying Berger's findings that the reality of human brain waves was accepted and EEG studies were put on a firmly established basis. One of their primary contributions was the identification of certain rhythms in the EEG, regular oscillations at approximately 10–12 Hz in the occipital lobes of the cerebral cortex. They found that this alpha rhythm in the EEG would disappear when the brain displayed any type of attention or alertness or focused on objects in the visual field. The physiological basis for these results, the "arousing influence" of external stimuli on the cortex, was not formulated until 1949 when Moruzzi and Magoun demonstrated the existence of widely spread pathways through the central reticular core of the brainstem capable of exerting a diffuse activating influence on the cerebral cortex. This reticular activating system has been called the brain's response selector because it alerts the cortex to focus on certain incoming information while ignoring other. It is for this reason that a sleeping mother will immediately be awakened by her crying baby or the smell of smoke, and yet ignore the traffic outside her window or the television still playing in the next room. An in-depth discussion of these early studies is beyond the scope of this presentation; however, for the interested reader an excellent historical review of this early era in brain research has been recorded in a fascinating text by Brazier [1968].

Frequency Analysis

In general, the EEG contains information regarding changes in the electrical potential of the brain obtained from a given set of recording electrodes. These data include the characteristic waveform with its variation in amplitude, frequency, phase, etc. and the occurrence of brief electrical patterns, such as spindles. Any analysis procedure cannot simultaneously provide information regarding all of these variables. Consequently, the selection of any analytic technique will emphasize changes in one particular variable at the expense of the others. This observation is extremely important if one is to properly interpret the results obtained by any analytic technique.

In early attempts to correlate the EEG with behavior, analog frequency analyzers were used to examine single channels of EEG data. Although disappointing, these initial efforts did introduce the utilization of frequency analysis to study gross brain wave activity. Although **power spectral**

analysis, i.e., the magnitude square of Fourier transform, provides a quantitative measure of the frequency distribution of the EEG, it does so as mentioned above, at the expense of other details in the EEG such as the amplitude distribution, as well as the presence of specific patterns in the EEG.

The first systematic application of power spectral analysis by general-purpose computers was reported in 1963 by Walter; however, it was not until the introduction of the **fast Fourier transform (FFT)** by Cooley and Tukey in the early 1970s that machine computation of the EEG became commonplace. Although an individual FFT is ordinarily calculated for a short section of EEG data (e.g., from 1 to 8 s epoch), such segmentation of a signal with subsequent averaging over individual modified periodograms has been shown to provide a consistent estimator of the power spectrum, and an extension of this technique, the compressed spectral array, has been particularly useful for computing EEG spectra over long periods of time. A detailed review of the development and use of various methods to analyze the EEG is provided by Givens and Redmond [1987].

Figure 105.3 provides an overview of the computational processes involved in performing spectral analysis of the EEG, i.e., including computation of auto and **cross spectra** [Bronzino, 1984]. It is to be noted that the power spectrum is the autocorrellogram, i.e., the correlation of the signal with itself. As a result, the power spectrum provides only magnitude information in the frequency domain; it does not provide any data regarding phase. The power spectrum is computed by:

$$P(f) = \text{Re}^2[X(f)] + \text{Im}^2[X(f)] \qquad (105.1)$$

where $X(f)$ is the Fourier transform of the EEG.

Power spectral analysis not only provides a summary of the EEG in a convenient graphic form, but also facilitates statistical analysis of EEG changes which may not be evident on simple inspection of the records. In addition to absolute power derived directly from the power spectrum, other measures calculated from absolute power have been demonstrated to be of value in quantifying various aspects of the EEG. Relative power expresses the percent contribution of each frequency band to the total power and is calculated by dividing the power within a band by the total power across all bands. Relative power has the benefit of reducing the intersubject variance associated with absolute power that arises from intersubject differences in skull and scalp conductance. The disadvantage of relative power is that an increase in one frequency band will be reflected in the calculation by a decrease in other bands; for example, it has been reported that directional shifts between high and low frequencies are associated with changes in cerebral blood flow and metabolism. Power ratios between low (0–7 Hz) and high (10–20 Hz) frequency bands have been demonstrated to be an accurate estimator of changes in cerebral activity during these metabolic changes.

Although the power spectrum quantifies activity at each electrode, other variables derivable from FFT offer a measure of the relationship between activity recorded at distinct electrode sites. Coherence (which is a complex number), calculated from the cross-spectrum analysis of two signals, is similar to cross-correlation in the time domain. The **magnitude squared coherence (MSC)** values range from 1 to 0, indicating maximum or no synchrony, respectively, and are independent of power. The temporal relationship between two signals is expressed by phase, which is a measure of the lag between two signals for common frequency components or bands. Phase is expressed in units of degrees, 0° indicating no time lag between signals or 180° if the signals are of opposite

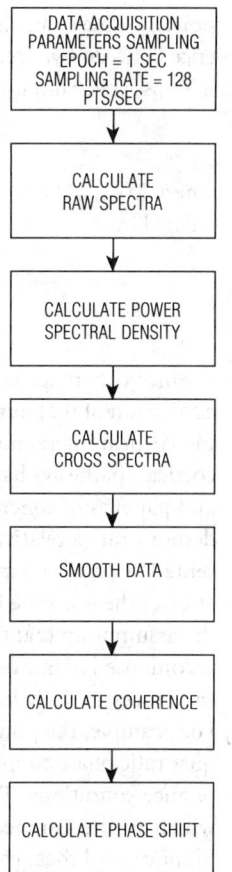

FIGURE 105.3 Block diagram of measures determined from spectral analysis.

polarity. Phase can also be transformed into the time domain, giving a measure of the time difference between two frequencies.

Cross spectrum is computed by:

$$\text{Cross spectrum} = X(f)\,Y^*(f) \tag{105.2}$$

where $X(f)$, $Y(f)$ are Fourier transforms and $*$ indicates complex conjugates and coherence is calculated by

$$\text{Coherence} = \frac{\text{Cross spectrum}}{\sqrt{PX(f) - PY(f)}} \tag{105.3}$$

Since coherence is a complex number, the phase is simply the angle associated with the polar expression of that number. MSC and phase represent measures that can be employed to investigate the cortical interactions of cerebral activity. For example, short (intracortical) and long (cortico-cortical) pathways have been proposed as the anatomic substrate underlying the spatial frequency and patterns of coherence. Therefore, discrete cortical regions linked by such fiber systems should demonstrate a relatively high degree of synchrony, whereas the time lag between signals, as represented by phase, quantifies the extent to which one signal leads another.

Over the years the EEG has been extensively studied using spectral analysis approaches based on the assumption that the amplitude distribution of the EEG is Gaussian (normal) and that the EEG is composed of mutually uncorrelated frequency components. However, for non-Gaussian/nonlinear processes, only limited information can be recovered by the use of conventional approaches. For example, the power spectrum cannot distinguish nonlinearly coupled waves (the existence of quadratic phase coupling) from spontaneously excited independent waves possessing the same resonance conditions. This is an especially important consideration when examining EEG patterns which are the summed result of more than one EEG generator, as is the case in the generation of the hippocampal theta rhythm.

Evidence has emerged indicating that the EEG recorded during different behavioral states exhibits varying degrees of deviation from a Gaussian distribution as well as nonlinear interactions between different frequency bands. Such findings indicate the importance of quantifying the non-Gaussian behavior of the EEG and identifying the nonlinear interactions that may play a role in the generation of specific EEG patterns.

In recent years, recognition of the unique capabilities of **bispectra** and the improvement in computing capabilities have increased interest in bispectral analysis and its use to analyze the EEG. For example, using bispectral analysis on the EEG obtained from the hippocampal formation in the adult rat, the existence of significant **quadratic phase coupling** in the 6- to 8-Hz (theta) frequency range between hippocampal regions CA1 and the dentate gyrus during REM sleep in the rat has been demonstrated [Ning and Bronzino, 1989]. The result of this nonlinear coupling is the appearance, in the frequency spectrum, of a small peak centered at approximately 13–14 Hz (beta range) which reflects the summation of the two theta frequency waves (Fig. 105.4). Conventional power spectral approaches are incapable of distinguishing the fact that this peak results from the interaction of these two generators, i.e., it is not an intrinsic wave.

Recently, there has been considerable interest in using knowledge-based technology to automate the scanning of the EEG during different vigilance states or other behaviors. An excellent review of the failure of present methods to achieve a high degree of success in this area, and how smart system and chaos approaches may yield fruitful results in the future, is provided by Jansen [1991].

EEG Recording Techniques

Scalp recordings of spontaneous neuronal activity in the brain, identified as the EEG, allow measurement of potential changes over time between a signal electrode and a reference electrode

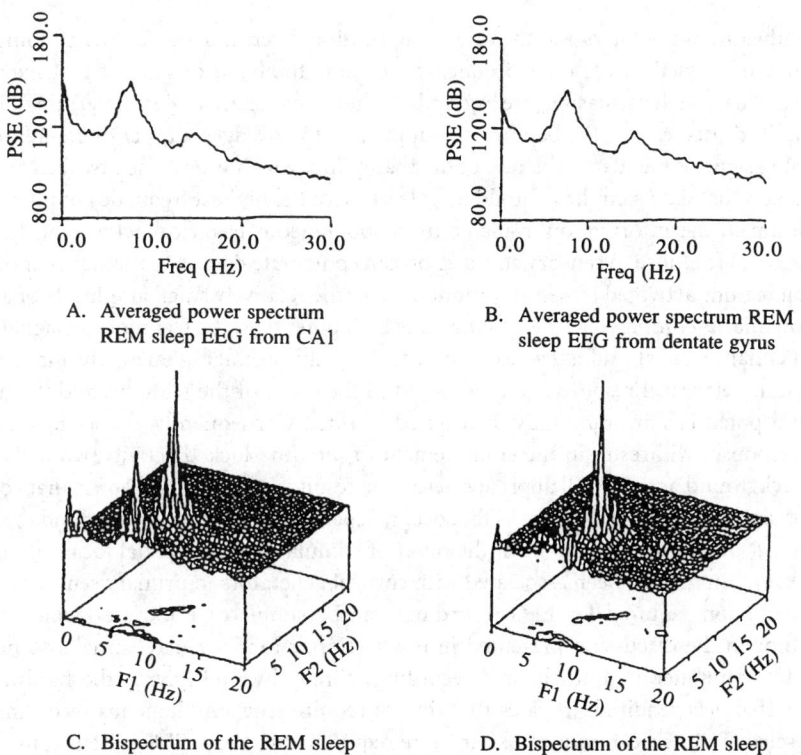

A. Averaged power spectrum
REM sleep EEG from CA1

B. Averaged power spectrum REM
sleep EEG from dentate gyrus

C. Bispectrum of the REM sleep
EEG from CA1

D. Bispectrum of the REM sleep
EEG from dentate gyrus

FIGURE 105.4   A and B represent the averaged power spectra of 80 4-s epochs of REM sleep (sampling rate = 128 Hz) obtained from hippocampal CA1 and the dentate gyrus, respectively. Note that both spectra exhibit clear power peaks at 7-Hz (theta) and 14-Hz (beta) frequencies. C and D represent the bispectrum of these same epochs from CA1 and the dentate gyrus, respectively. Computation of the bicoherence index at 7 Hz shows significant quadratic phase coupling at this frequency, indicating that the 14-Hz peak is not spontaneously generated, but results from quadratic phase coupling.

[Kondraski, 1986]. Compared to other biopotentials, such as the electrocardiogram, the EEG is extremely difficult for an untrained observer to interpret. As might be expected, partially as a result of the spatial mapping of functions onto different regions of the brain, correspondingly different waveforms are visible, depending on electrode placement. Recognizing that some standardization was necessary for comparison of research as well as clinical EEG records, the International Federation in Electroencephalography and Clinical Neurophysiology adopted the 10–20 electrode placement system [Jasper, 1958]. Additional electrodes to monitor extracerebral contaminants of the EEG such as eye movement, EKG, and muscle activity are essential. The acquisition of EEG for quantitative analysis should also require the ability to view the EEG during collection on a polygraph or high-resolution video display.

Since amplification, filtering, and digitization determine the frequency characteristics of the EEG and the source of potential artifacts, the acquisition parameters must be chosen with an understanding of their effects on signal acquisition and subsequent analysis. Amplification, for example, increases the amplitude range (volts) of the analog-to-digital (A/D) converter. The resolution of the A/D converter is determined by the smallest amplitude of steps that can be sampled. This is calculated by dividing the voltage range of the A/D converter by 2 to the power of the number of bits of the A/D converter. For example, an A/D converter with a range of ±5 V with 12-bit resolution can resolve samples as small as ±2.4 mV. Appropriate matching of amplification and A/D converter sensitivity permits resolution of the smallest signal while preventing clipping of the largest signal amplitudes.

The bandwidth of the filters and the rate of digitization determine the frequency components of interest that are passed, while other frequencies outside the band of interest that may represent potential artifacts, such as aliasing, are rejected. A filter's characteristics are determined by the rate of the amplitude decrease at the bandwidth's upper and lower edges. Proper digital representation of the analog signal depends on the rate of data sampling, which is governed by the Nyquist theorem that states that data sampling should be at least twice the highest frequency of interest.

In addition to the information available from spontaneous electrical activity of the EEG, the brain's electrical response to sensory stimulation can contribute data as to the status of cortical and subcortical regions activated by sensory input. Due to the relatively small amplitude of a stimulus-evoked potential as compared to the spontaneous EEG potentials, the technique of signal averaging is used to enhance the stimulus-evoked response. Stimulus averaging takes advantage of the fact that the brain's electrical response is time-locked to the onset of the stimulus and the nonevoked background potentials are randomly distributed in time. Consequently, the average of multiple stimulus responses will result in the enhancement of the time-locked activity, while the averaged random background activity will approach zero. The result is an evoked response that consists of a number of discrete and replicable peaks that occur, depending upon the stimulus and the recording parameters, at predicted latencies from the onset of stimulation. The spatial localization of maximum peak amplitudes has been associated with cortical generators in primary sensory cortex.

Instrumentation required for EEG recordings can be simple or elaborate [Kondraski, 1986]. (Note: Although the discussion presented in this section is for a single-channel system it can be extended to simultaneous multichannel recordings simply by multiplying the hardware by the number of channels required. In cases that do not require true simultaneous recordings, special electrode selector panels can minimize hardware requirements.) Any EEG system consists of electrodes, amplifiers (with appropriate filters) and a recording device.

Commonly used scalp electrodes consist of Ag-AgCl disks, 1 to 3 mm in diameter, with a very flexible long lead that can be plugged into an amplifier. Although it is desirable to obtain a low-impedance contact at the electrode ski interface (less than 10 kΩ), this objective is confounded by hair and the difficulty of mechanically stabilizing the electrodes. Conductive electrode paste helps obtain low impedance and keep the electrodes in place. A type of cement (collodion) is used to fix small patches of gauze over electrodes for mechanical stability, and leads are usually taped to the subject to provide some strain relief. Slight abrasion of the skin is sometimes used to obtain better electrode impedances, but this can cause irritation and sometimes infection (as well as pain in sensitive subjects).

For long-term recordings, as in seizure monitoring, electrodes present major problems. Needle electrodes, which must be inserted into the tissue between the surface of the scalp and skull, are sometimes useful. However, the danger of infection increases significantly. Electrodes with self-contained miniature amplifiers are somewhat more tolerant because they provide a low-impedance source to interconnecting leads, but they are expensive. Despite numerous attempts to simplify the electrode application process and to guarantee long-term stability, none has been widely accepted.

Instruments are available for measuring impedance between electrode pairs. The procedure is recommended strongly as good practice, since high impedance leads to distortions that may be difficult to separate from actual EEG signals. In fact, electrode impedance monitors are built into some commercial devices for recording EEGs. Standard dc ohmmeters should not be used, since they apply a polarizing current that causes build-up of noisy electrode potential at the skin-electrode interface. Commercial devices apply a known-amplitude sinusoidal voltage (typically 1 kHz) to an electrode pair circuit and measure root mean square (rms) current, which is directly related to the magnitude of the impedance.

From carefully applied electrodes, signal amplitudes of 1 to 10 μV can be obtained. Considerable amplification (gain = 10^6) is required to bring these levels up to an acceptable level for input to recording devices. Because of long electrode leads and the common electrically noisy environment where recordings take place, differential amplifiers with inherently high input impedance and high common mode rejection ratios are essential for high-quality EEG recordings.

In some facilities, special electrically shielded rooms minimize environmental electrical noise, particularly 60-Hz alternating current (ac) line noise. Since much of the information of interest in the EEG lies in the frequency bands less than 40 Hz, low-pass filters in the amplifier can be switched in to attenuate 60-Hz noise sharply.

For attenuating ac noise when the low-pass cutoff is greater than 60 Hz, many EEG amplifiers have notch filters that attenuate only frequencies in a narrow band centered around 60 Hz. Since important signal information may also be attenuated, notch filtering should be used as a last resort; one should try to identify and eliminate the source of interference instead.

In trying to identify 60-Hz sources to eliminate or minimize their effect, it is sometimes useful to use a dummy source, such as a fixed 100-kΩ resistor attached to the electrodes. An amplifier output represents only contributions from interfering sources. If noise can be reduced to an acceptable level (at least by a factor of 10 less than EEG signals) under this condition, one is likely to obtain uncontaminated EEG records.

Different types of recording instruments obtain a temporary or permanent record of the EEG. The most common recording device is a pen or chart recorder (usually multichannel) that is an integral part of most commercially available EEG instruments. The bandwidth of clinical EEGs is relatively low (less than 40 Hz) and therefore within the frequency response capabilities of these devices. Recordings are on a long sheet of continuous paper (from a folded stack), fed past the moving pen at one of several selectable constant speeds. The paper speed translates into distance per unit time or cycles per unit time, to allow EEG interpreters to identify different frequency components or patterns within the EEG. Paper speed is selected according to the monitoring situation at hand: slow speeds (10 mm/s) for observing the spiking characteristically associated with seizures and faster speeds (up to 120 mm/s) for the presence of individual frequency bands in the EEG.

In addition to (or instead of) a pen recorder, the EEG may be recorded on a multichannel frequency modulated (FM) analog tape recorder. During such recordings, a visual output device such as an oscilloscope or video display is necessary to allow visual monitoring of signals, so that corrective action (reapplying the electrodes and so on) can take place immediately if necessary.

Sophisticated FM cassette recording and playback systems allow clinicians to review long EEG recordings over a greatly reduced time, compared to that required to flip through stacks of paper or observe recordings as they occur in real time. Such systems take advantage of time compensation schemes, whereby a signal recorded at one speed (speed of the tape moving past the recording head of the cassette drive) is played back at a different, faster speed. The ratio of playback to recording speed is known, so the appropriate correction factor can be applied to played-back data to generate a properly scaled video display. A standard ratio of 60:1 is often used. Thus, a trained clinician can review each minute of real-time EEG in 1 s. The display appears to be scrolled at a high rate horizontally across the display screen. Features of these instruments allow the clinician to freeze a segment of EEG on the display and to slow down or accelerate tape speed from the standard playback as needed. A time mark channel is usually displayed as one of the traces as a convenient reference (vertical "tick" mark displayed at periodic intervals across the screen).

Computers can also be recording devices, digitizing (converting to digital form) one or several amplified EEG channels at a fixed rate. In such sampled data systems, each channel is repeatedly sampled at a fixed time interval (sample interval) and this sample is converted into a binary number representation by an A/D converter. The A/D converter is interfaced to a computer system so that each sample can be saved in the computer's memory. A set of such samples, acquired at a sufficient sampling rate (at least two times the highest frequency component in the sampled signal), is sufficient to represent all the information in the waveform. To ensure that the signal is band-limited, a low-pass filter with a cutoff frequency equal to the highest frequency of interest is used. Since physically realizable filters do not have the ideal characteristics, the sampling rate is usually greater than two times the filter's cutoff frequency. Furthermore, once converted to a digital format, digital filtering techniques can be used.

On-line computer recordings are only practical for short-term recordings or for situations in which the EEG is immediately processed. This limitation is primarily due to storage requirements. For example, a typical sampling rate of 128 Hz yields 128 new samples per second that require storage. For an 8-channel recording, 1,024 samples are acquired per second. A 10-minute recording period yields 614,400 data points. Assuming 8-bit resolution per sample, over 0.5 megabyte (MB) of storage is required to save the 10-minute recording.

Processing can consist of compression for more efficient storage (with associated loss of total information content), as in data record or epoch averaging associated with evoked responses, or feature extraction and subsequent pattern recognition, as in automated spike detection in seizure monitoring.

Topographic Mapping

Computerized tomography (CT) and magnetic resonance imaging (MRI) have demonstrated the impact of spatial displays on data interpretation and analysis. Similarly, mapping techniques have been applied to electrophysiologic data to depict the spatial information available from multielectrode recordings. This effort has been assisted by the development and implementation of low-cost, high-resolution graphic displays on microcomputer systems. The data are frequently presented as two-dimensional topographic color maps [Zappulla, 1991]. In the time domain, color values depict the changes in potential across the scalp at each time point. This is exemplified by mapping peaks of an evoked potential or the spatial distribution of an epileptic spike. Temporal changes in the spatial distribution of voltage can be presented graphically as a series of maps constructed at adjacent time points or by cartooning the topographic maps over the time interval of interest. In the frequency domain, color coding can be used to spatially map power, covariance, and phase values. These maps may be constructed for the broadband activity or for selective frequency components.

Unlike CT and MRI displays where each picture element or pixel value represents real data, most of the pixels comprising an EEG and ER topographic map consist of interpolated values. This is because the activity from a finite number of electrodes represents a sampling of the spatial activity over the scalp. Consequently, the remaining values of the map located outside the electrode positions must be estimated from this sampled activity. One technique for deriving these values is linear interpolation. In the case of a four-point interpolation, the map is divided into boxes whose corners are defined by real data. The interpolated points within the boxes are calculated by the weighted sum of the four real data points, based on their distance from the interpolated point. Although linear interpolation is the most popular technique, polynomial regression and surface spline interpolation have been employed as alternative procedures. These methods reduce the discontinuities inherent in linear interpolation and offer better estimates of extreme values. Polynomial regression has the additional advantage of permitting quantitative comparisons between maps by taking into account the topographic information represented in the map.

Maps can be presented in any of several projections to assist in interpretation [Zappulla, 1991]. The most common projection is the top view which presents the spatial distribution of variables from all leads simultaneously. Lateral, posterior, and anterior projections highlight focal areas of interest. Although mapping presents a method by which spatial information can be efficiently communicated, it is important to be alert to the artifacts that can arise from map construction and manipulation. Topographic spatial artifacts that can lead to misinterpretation include ring enhancement around a spike using source-derivation references, spatial aliasing arising from linear interpolation which causes maximal activity to be mapped at electrode sites, the enhancement of activity away from the midline, and the attenuation of midline activity on amplitude asymmetry maps (centrifugal effect).

The quality of the spatial information derivable from EEG recordings depends upon the number of recording electrodes, the choice of the reference electrode, and the conductive properties of intracranial and extracranial structures. The localization of cortical activity from scalp recordings assumes that the potentials recorded from the scalp reflect cortical activity generated in proximity to the recording electrode. Therefore, the greater the density of recording electrodes, the more accurate the estimate of the spatial distribution of scalp potentials and the localization of cortical generators. However, since the distance between the cortical source and recording electrode, as well as the low conductivity of the skull, results in a selective attenuation of small dipole fields, most available EEG information can be obtained with an average scalp-electrode spacing of 2 cm.

Topographic maps are constructed from monopolar electrodes referenced to a common cephalic (linked ears or mandible, chin and nose) or noncephalic (linked clavicles or a balanced sternum-vertebra) electrode. Although the reference electrode should be free of any EEG activity, in practice most cephalic electrodes contain some EEG activity, while noncephalic electrodes are a potential source of EKG or muscle activity. Differential amplification of an EEG-contaminated reference electrode can decrease or cancel similar activity in neighboring electrodes, while at electrodes distant from the reference, the injected activity will be present as a potential of opposite polarity. Similarly, noncerebral potentials can be injected into scalp electrodes and misinterpreted as cerebral activity. Therefore, a nonneutral reference electrode can result in misleading map configurations. Several techniques have been applied to circumvent this problem. The construction of multiple maps using several different references can sometimes assist in differentiating active and reference electrode activity. This can be accomplished by acquiring serial EEG records using different references. Alternatively, various references can be acquired simultaneously during acquisition, and various montages can be digitally reconstructed, post hoc.

A more computationally intensive method for localizing a source at an electrode involves calculating the local source activity at any one electrode based on the average activity of its neighbors, weighted by their distance from the source. The technique has the advantage of suppressing potentials that originate outside the measurement area and weighing factors for implementing source deviation techniques for each of the electrodes in the 10–20 system are available.

Another reference technique, the average head reference, uses the average activity of all active electrodes as the common reference. In this approach, the activity at any one electrode will vary depending upon the activity at the site of the reference electrode, which can be anywhere on the recording montage. Therefore, for N number of recording electrodes, each being a potential reference, there are $N-1$ possible voltage measurements at each instant of time for each electrode. Maps constructed using the average head reference represent a unique solution to the problem of active reference electrodes in that the average reference produces an amplitude-weighted reference-free map of maximal and minimal field potentials. Power maps constructed from the average reference best depict the spatial orientation of the generating field, and the areas with extreme values are closest to the generating processes [Zappulla, 1991].

Topographical maps represent an efficient format for displaying the extensive amount of data generated by quantitative analysis. However, for reasons discussed above, the researcher and clinician must be cautious in deriving spatial and functional conclusions from mapped data. Although the replicability of map configurations across subjects or experimental conditions may represent a useful basis for experimental and diagnostic classification, judgments concerning the localization of cortical generators or functional localization of cerebral activity are less certain and more controversial. Research continues on defining models and validating assumptions that relate scalp potentials to cortical generators in an attempt to arrive at accurate mathematical solutions that can be applied to mapping functions.

Defining Terms

Bispectra: Computation of the frequency distribution of the EEG exhibiting nonlinear behavior.

Cross spectra: Computation of the energy in the frequency distribution of two different electrical signals.

Electroencephalogram (EEG): Recordings of the electrical potentials produced by the brain.

Fast Fourier transform (FFT): Algorithms that permit rapid computation of the Fourier transform of an electrical signal, thereby representing it in the frequency domain.

Magnitude squared coherence (MSC): A measure of the degree of synchrony between two electrical signals at specific frequencies.

Power spectral analysis: Computation of the energy in the frequency distribution of an electrical signal.

Quadratic phase coupling: A measure of the degree to which specific frequencies interact to produce a third frequency.

References

M. Brazier, *Electrical Activity of the Nervous System,* 3rd ed., Baltimore: Williams and Wilkins, 1968.

J.D. Bronzino, "Quantitative analysis of the EEG: General concepts and animal studies," *IEEE Transactions on BME,* vol. 31, no. 12, pp. 850–856, 1984.

A.S. Givens and A. Remond, Eds., *Methods of Analysis of Brain Electrical and Magnetic Signals, EEG Handbook,* Revised Series, vol. 1, Amsterdam: Elsevier, 1987.

B.H. Jansen, "Quantitative analysis of EEGs: Is there chaos in the future," *International J. Biomed. Comp.,* vol. 27, pp. 95–123, 1991.

G.V. Kondraski, "Neurophysiological measurements," in *Biomedical Engineering and Instrumentation,* J.D. Bronzino, Ed., Boston: PWS Publishing, 1986, pp. 138–179.

T. Ning and J.D. Bronzino, "Bispectral analysis of the rat EEG during different vigilance states," *IEEE Trans. in BME,* vol. 34, no. 4, pp. 497–499, 1989.

J.R. Smith, "Automated analysis of sleep EEG data," in *Clinical Applications of Computer Analysis of EEG and Other Neurophysiological Signals, EEG Handbook,* revised series, vol. 2, Amsterdam: Elsevier, 1986, pp. 93–130.

R.A. Zappulla, "Fundamentals and applications of quantitative electrophysiology," in *Windows on the Brain,* R.A. Zappulla, F. F. LeFever, J. Jaeger, and R. Bilder, Eds., *Annals N.Y. Acad. Sci.,* vol. 620, pp. 1–21, 1991.

Further Information

Windows on the Brain (R.A. Zappulla, F.F. LeFever, J. Jaeger, and R. Bilder, Eds.), *Annals N.Y. Acad. Sci.,* vol. 620, 1991.

Journals: *IEEE Transactions in Biomedical Engineering* and *Electroencephalography and Clinical Neurophysiology.*

105.2 The Electrocardiograph

Edward J. Berbari

The electrocardiogram (**ECG**) is the recording on the body surface of the electrical activity generated by the heart. It was originally observed by Waller in 1889 using his pet bulldog as the signal source and the capillary electrometer as the recording device. In 1903 Einthoven enhanced the technology by using the string galvanometer as the recording device and using human subjects with a variety of cardiac abnormalities. Einthoven is chiefly responsible for introducing some concepts still in use today including the labeling of the various waves, defining some of the standard recording sites using the arms and legs, and developing the first theoretical construct whereby the heart is modeled as a single time varying dipole. We also owe the "EKG" acronym to Einthoven's native Dutch language where the root word "cardio" is spelled with a "k".

In order to record an ECG waveform, a differential recording between two points on the body is made. Traditionally each differential recording is referred to as a lead. Einthoven defined three leads numbered with the Roman numerals I, II, and III. They are defined as:

$$I = V_{RA} - V_{LA} \tag{105.4}$$

$$II = V_{RA} - V_{LL} \tag{105.5}$$

$$III = V_{LA} - V_{LL} \tag{105.6}$$

where RA = right arm, LA = left arm, and LL = left leg. Because the body is assumed to be purely resistive, at ECG frequencies, the four limbs can be thought of as wires attached to the torso. Hence lead I could be recorded from the respective shoulders without a loss of cardiac information. Note that these are not independent and the following relationship holds: $II = I + III$.

For 30 years the evolution of the ECG proceeded when F. N. Wilson [1934] added concepts of a "unipolar" recording. He created a reference point by tying the three limbs together and averaging their potentials so that individual recording sites on the limbs or chest surface would be differentially recorded with the same reference point. Wilson extended the biophysical models to include the concept of the cardiac source enclosed within the volume conductor of the body. He erroneously thought that the central terminal was a true zero potential. However, from the mid-1930s until today the 12 leads composed of the three limb leads, three leads in which the limb potentials are referenced to a modified **Wilson terminal** (the augmented leads [Goldberger, 1942]), and six leads placed across the front of the chest and referenced to the Wilson terminal form the basis of the standard **12-lead ECG**. Figure 105.5 summarizes the 12-lead set. These sites are historically based, have a built in redundancy, and are not optimal for all cardiac events. The voltage difference from any two sites will record an ECG, but it is these standardized sites with the massive 90-year collection of empirical observations that has firmly established their role as the standard. Figure 105.6 is a typical or stylized ECG recording from lead II. Einthoven chose the letters of the alphabet from P to U to label the waves and to avoid conflict with other physiologic waves being studied at the turn of the century. The ECG signals are typically in the range of ±2 mV and require a recording bandwidth of 0.05–150 Hz. Full technical specification for ECG equipment has been proposed by both the American Heart Association [1984] and the Association for the Advancement of Medical Instrumentation [Bailey *et al.*, 1990].

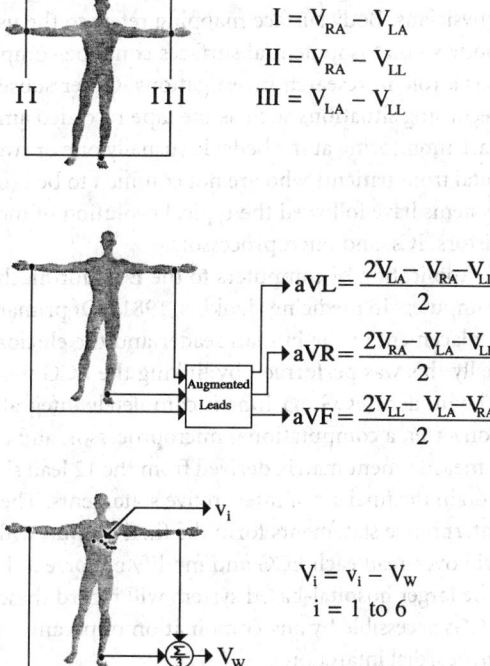

FIGURE 105.5 The 12-lead ECG is formed by the three bipolar surface leads: I, II, and III; the augmented Wilson terminal referenced limb leads: *aVR, aVL, aVF*; and the Wilson terminal referenced chest leads: V_1, V_2, V_3, V_4, V_5, and V_6.

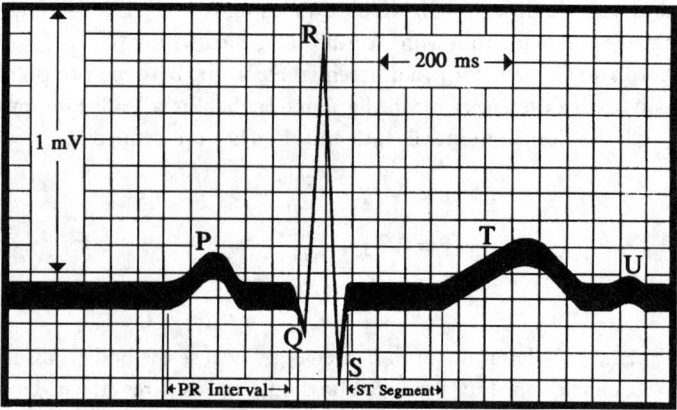

FIGURE 105.6 Stylized version of a normal lead II recording showing the
P wave, QRS complex, and the T and U waves. The PR interval and the ST
segment are significant time windows. The peak amplitude of the QRS is
about 1 mV. The vertical scale is usually 1 mV/cm. The time scale is usually
based on mm/s scales with 25 mm/s being the standard form. The small
boxes of the ECG are 1 × 1 mm.

There have been several attempts to change the approach for recording the ECG. The vectorcar-
diogram used a weighted set of recording sites to form an orthogonal *XYZ* lead set. The advantage
here was minimum lead set but in practice it gained only a moderate degree of enthusiasm among
physicians. Body surface mapping refers to the use of many recording sites (>64) arranged on the
body so that isopotential surfaces could be computed and analyzed over time. This approach still
has a role in research investigations. Other subsets of the 12-lead ECG are used in limited mode
recording situations such as the tape recorded ambulatory ECG (usually two leads) or in intensive
care monitoring at the bedside (usually one or two leads) or telemetered within regions of the hospital from patients who are not confined to bed (one lead). The recording electronics of these ECG
systems have followed the typical evolution of modern instrumentation, e.g., vacuum tubes, transistors, ICs, and microprocessors.

Application of computers to the ECG for machine interpretation was one of the earliest uses of
computers in medicine [Jenkins, 1981]. Of primary interest in the computer-based systems was the
replacement of the human reader and the elucidation of the standard waves and intervals. Originally this was performed by linking the ECG machine to a centralized computer via phone lines.
The modern ECG machine is completely integrated with an analog front end, a 12- to 16-bit A/D
converter, a computational microprocessor, and dedicated I/O processors. These systems compute
a measurement matrix derived from the 12 lead signals and analyze this matrix with a set of rules to
obtain the final set of interpretive statements. The depiction of the 12 analog signals and this set of
interpretive statements form the final output, with an example shown in Fig. 105.7. The physician
will over-read each ECG and modify or correct those statements which are deemed inappropriate.
The larger hospital-based system will record these corrections and maintain a large database of all
ECGs accessible by any combination of parameters, e.g., all males, older than 50, with an inferior
myocardial infarction.

More recently the high-resolution ECG (**HRECG**) as been developed whereby the digitized ECG
is signal averaged to reduce random noise (Berbari *et al.*, 1973, 1977). This approach, coupled with
post averaging high-pass filtering, is used to detect and quantify low-level signals (≈1.0 μV) not
detectable with standard approaches. This computer-based approach has enabled the recording of
events which are predictive of future life-threatening cardiac events [Berbari *et al.*, 1978; Simson,
1981].

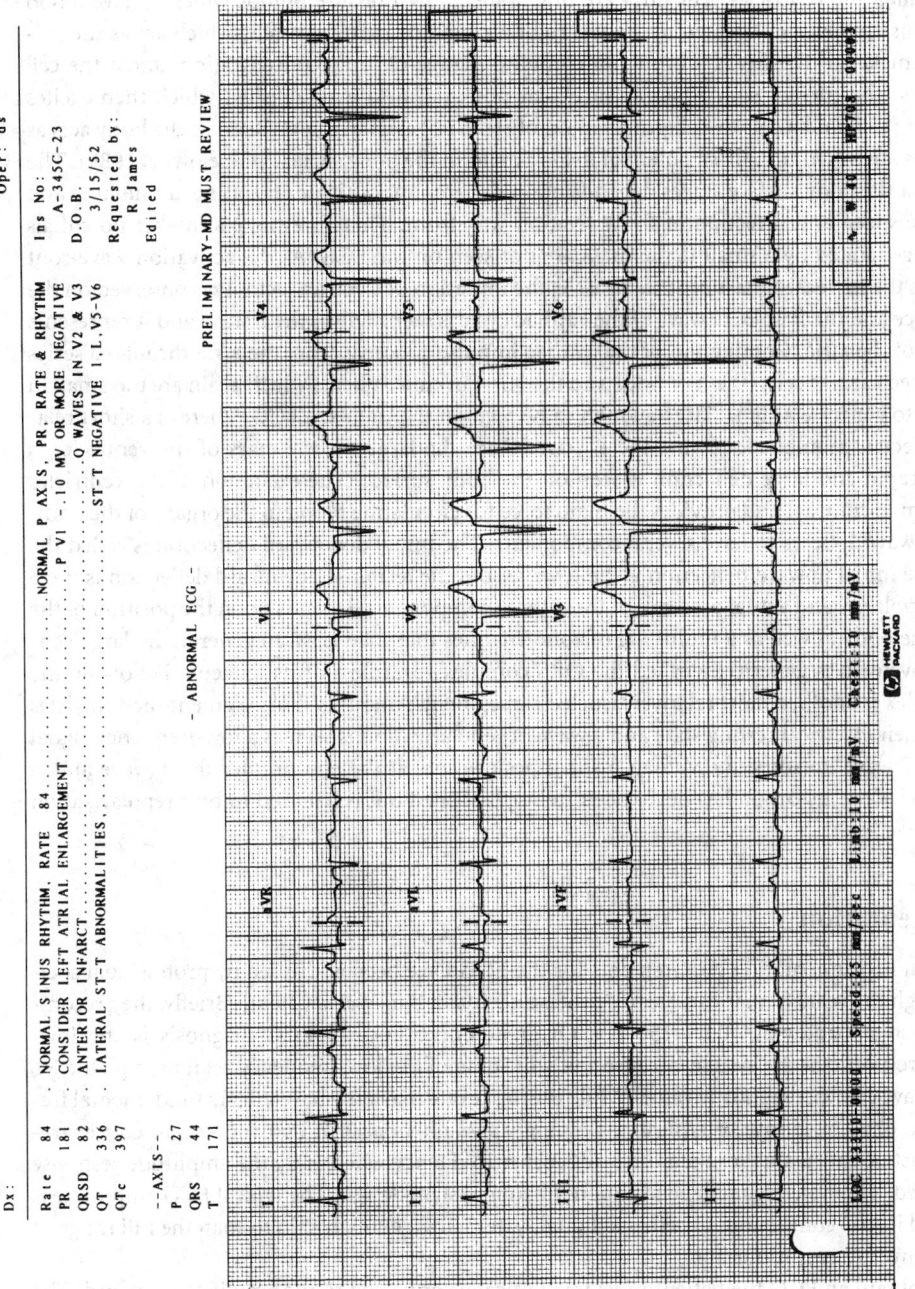

FIGURE 105.7 Example of an interpreted 12-lead ECG. A 2½-second recording is shown for each of the 12 leads. The bottom trace is a continuous 10-second rhythm strip of lead II. Patient information is given in the top area, below which is printed the computerized interpretive statements. (Tracing is courtesy of the Hewlett-Packard Co., Palo Alto, CA.)

Physiology

The heart has four chambers; the upper two chambers are called the atria and the lower two chambers are called the ventricles. The atria are thin-walled, low-pressure pumps which receive blood from venous circulation. Located in the top right atrium are a group of cells which act as the primary pacemaker of the heart. Through a complex change of ionic concentration across the cell membranes (the current source) an extracellular potential field is established which then excites neighboring cells and a cell-to-cell propagation of electrical events occurs. Because the body acts as a purely resistive medium, these potential fields extend to the body surface [Geselowitz, 1989]. The character of the body surface waves depends upon the amount of tissue activating at one time and the relative speed and direction of the activation wavefront. Therefore the pacemaker potentials which are generated by a small tissue mass are not seen on the ECG. As the activation wavefront encounters the increased mass of atrial muscle, the initiation of electrical activity is observed on the body surface and the first ECG wave of the cardiac cycle is seen. This is the P wave and it represents activation of the atria. Conduction of the cardiac impulse proceeds from the atria through a series of specialized cardiac cells (the A-V node and the His-Purkinje system) which again are too small in total mass to generate a signal large enough to be seen on the standard ECG. There is a short relatively isoelectric segment following the P wave. Once the large muscle mass of the ventricles is excited, a rapid and large deflection is seen on the body surface. The excitation of the ventricles causes them to contract and provides the main force for circulating blood to the organs of the body. This large wave appears to have several components. The initial downward deflection is called the Q wave, the initial upward deflection is the R wave, and the terminal downward deflection is the S wave. The polarity and actual presence of these three components depends upon the position of the leads on the body as well as a multitude of abnormalities that may exist. In general, the large ventricular waveform is generically called the QRS complex regardless of its makeup. Following the QRS complex is another short relatively isoelectric segment. After this short segment the ventricles return to their electrical resting state and a wave of repolarization is seen as a low-frequency signal called the T wave. In some individuals a small peak occurs at the end or after the T wave and is called the U wave. Its origin has never been fully established but is believed to be a repolarization potential.

Instrumentation

The general instrumentation requirements for the ECG have been addressed by professional societies through the years [American Heart Association, 1984; Bailey *et al.*, 1990]. Briefly, they recommend a system bandwidth 0.05–150 Hz. Of great importance in ECG diagnosis is the low-frequency response of the system because shifts in some of the low-frequency regions, e.g., the ST segment, have critical diagnostic value. While the heart rate may only have a 1-Hz fundamental frequency, the phase response of typical analog high-pass filters is such that the system corner frequency must be much smaller than the 3-dB corner frequency where only the amplitude response is considered. The system gain depends upon the total system design. The typical ECG amplitude is $\pm 2\,mV$ and if A/D conversion is used in a digital system, then enough gain to span the full range of the A/D converter is appropriate.

To first obtain an ECG the patient must be physically connected to the amplifier front end. The patient/amplifier interface is formed by a special bioelectrode which converts the ionic current flow of the body to the electron flow of the metallic wire. These electrodes typically rely on a chemical paste or gel with a high ionic concentration. This acts as the transducer at the tissue-electrode interface. For short-term applications silver-coated suction electrodes or "sticky" metallic foil electrodes are used. Long-term recordings, such as the case for the monitored patient, require a stable electrode/tissue interface and special adhesive tape material surrounds the gel and a Ag^+/Ag^+Cl electrode.

At any given time, the patient may be connected to a variety of devices, e.g., respirator, blood pressure monitor, temporary pacemaker, etc., some of which will invade the body and provide a low resistance pathway to the heart. It is essential that the device not act as a current source and inject the patient with enough current to stimulate the heart and cause it to fibrillate. Some bias currents are unavoidable for the system input stage and recommendations are that these leakage currents be less than 10 μA per device. This not only applies to the normal setting but if a fault condition arises whereby the patient comes in contact with the high voltage side of the ac power lines, then the isolation must be adequate to prevent 10 μA of fault current as well. This mandates that the ECG reference ground not be connected physically to the low side of the ac power line or its third wire ground. For ECG machines the solution has typically been to AM modulate a medium-frequency carrier signal (≈400 kHz) and use an isolation transformer with subsequent demodulation. Other methods of signal isolation can be used but the primary reason for the isolation is to keep the patient from being part of the ac circuit in the case of a patient to power line fault. In addition, with many devices connected in a patient monitoring situation it is possible that ground loop currents will be generated. To obviate this potential hazard a low-impedance ground buss is often installed in these rooms and each device chassis will have an external ground wire connected to the buss. Another unique feature of these amplifiers is that they must be able to withstand the high-energy discharge of a cardiac defibrillator.

Figure 105.8 shows a three-channel ECG amplifier schematic used in a high-resolution ECG system. The patient is dc coupled to the front end differential, instrumentation amplifier. The first stage of gain is relatively low (≈100) because there can be a significant signal drift due to a high static charge on the body or low-frequency offset potentials generated by the electrolyte in the tissue/electrode interface. In this particular amplifier the signal is bandpass filtered prior to the isolation stage. To further limit the high floating potential of the patient and to improve the system common mode rejection a driven ground is usually used. This ground is simply an average of the limb potentials inverted by a single amplifier and connected to the right leg.

Older style ECG machines recorded one lead at a time, then evolved to three simultaneous leads. This necessitated the use of switching circuits as well as analog weighting circuits to generate the various 12 leads. This is usually eliminated in modern digital systems by using an individual single-ended amplifier for each electrode on the body. Each potential signal is then digitally converted and all of the ECG leads can be formed mathematically in software. This would necessitate a nine-amplifier system. By performing some of the lead calculations with the analog differential amplifiers this can be reduced to an eight-channel system. Thus only the individual chest leads V_1 through V_6 and any 2 of the limb leads, e.g., I and III, are needed to calculate the full 12-lead ECG. Figure 105.9 is a block diagram of a modern digital-based ECG system. This system uses up to 13 single-ended amplifiers and a 16-bit A/D converter, all within a small lead wire manifold or amplifier lead stage. The digital signals are optically isolated and sent via a high-speed serial link to the main ECG instrument. Here the 32-bit CPU and DSP chip perform all of the calculations and a hard copy report is generated (Fig. 105.7). Notice that each functional block has its own controller and the system requires a real-time, multitasking operating system to coordinate all system functions. Concomitant with the data acquisition is the automatic interpretation of the ECG. These programs are quite sophisticated and are continually evolving. It is still a medical/legal requirement that these ECGs be over-read by the physician.

High-resolution capability is now a standard feature on most digitally based ECG systems or as a stand-alone microprocessor-based unit [Berbari, 1988]. The most common application of the HRECG is to record very low-level (≈1.0 μV) signals which occur after the QRS complex but are not evident on the standard ECG. These "late potentials" are generated from abnormal regions of the ventricles and have been strongly associated with the substrate responsible for a life-threatening rapid heart rate (ventricular tachycardia). The typical HRECG is derived from three bipolar leads configured in an anatomic *XYZ* coordinate system. These three ECG signals are then digitized at a rate of 1000–2000 Hz/channel, time aligned via a real-time QRS correlator, and summed in the

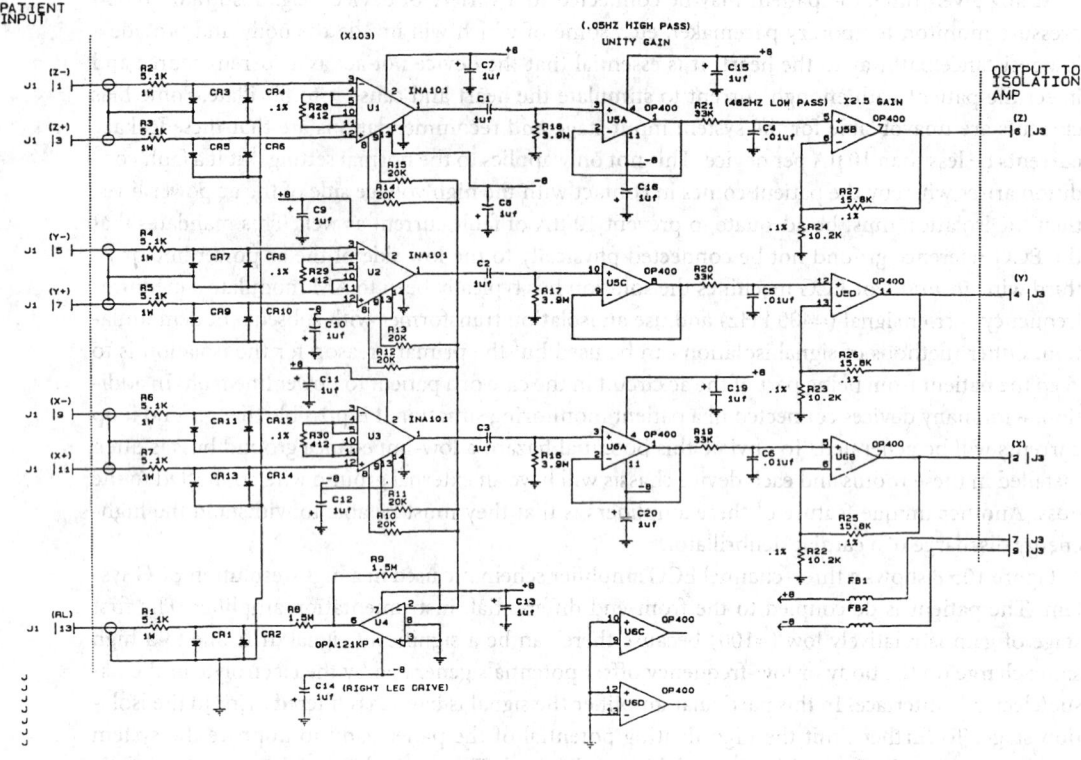

FIGURE 105.8 This schematic represents a typical three-lead *XYZ* amplifier set used in a high-resolution ECG. The instrumentation amplifier (INA101) and bandpass filter (OP400) for each channel are on the isolated side of the power supply. The diode pairs and 4.1 kΩ resistors on each lead wire provide high-voltage defibrillation protection. The outputs of each differential amplifier are averaged through an amplifier (OPA121) and provide the right leg drive. (Schematic is courtesy of Corazonix Corp., Oklahoma City.)

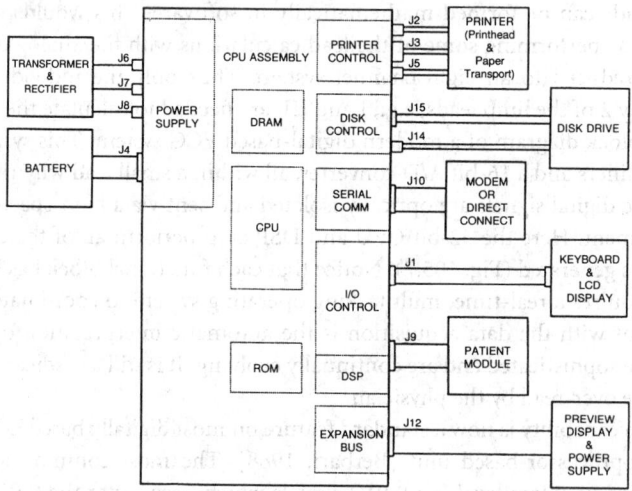

FIGURE 105.9 Block diagram of a microprocessor-based ECG system. It includes all of the elements of a personal computer class system, e.g., 80386 processor, 2 Mbytes of RAM, disk drive, 640 × 480 pixel LCD display, and is battery operable. In addition, it includes a DSP56001 chip and multiple controllers which are managed with a real-time, multitasking operating system. (Diagram is courtesy of the Hewlett-Packard Co., Palo Alto, Calif.)

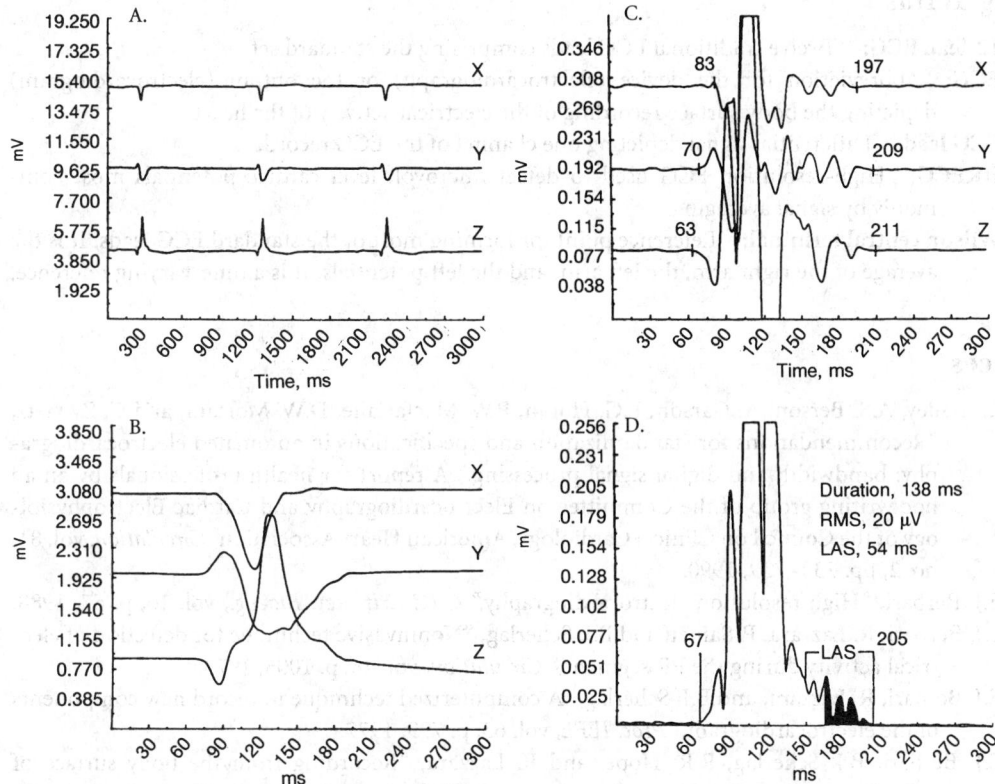

FIGURE 105.10 The signal processing steps typically performed to obtain a high-resolution ECG are shown in panels A–D. See text for a full description.

form of a signal average. Signal averaging will theoretically improve the signal-to-noise ratio by the square root of the number of beats averaged. The underlying assumptions are that the signals of interest do not vary, on a beat-to-beat basis, and that the noise is random. Figure 105.10 has four panels depicting the most common sequence for processing the HRECG to measure the late potentials. Panel A depicts a three-second recording of the *XYZ* leads close to normal resolution. Panel B was obtained after averaging 200 beats and with a sampling frequency of 10 times that shown in panel A. The gain is also five times greater. Panel C is the high-pass filtered signal using a partially time reversed digital filter having a second-order Butterworth response and a 3-dB corner frequency of 40 Hz [Simson, 1981]. Note the appearance of the signals at the terminal portion of the QRS complex. A common method of analysis, but necessarily optimal, is to combine the filtered *XYZ* leads into a vector magnitude $(X^2 + Y^2 + Z^2)^{1/2}$. This waveform is shown in panel D. From this waveform several parameters have been derived such as total QRS duration, including late potentials, the rms voltage value of the terminal 40 ms, and the low-amplitude signal (LAS) duration from the 40-μV level to the end of the late potentials. Abnormal values for these parameters are used to identify patients at high risk of ventricular tachycardia following a heart attack.

Conclusions

The ECG is one of the oldest instrument-bound measurements in medicine. It has faithfully followed the progression of instrumentation technology. Its most recent evolutionary step, to the microprocessor-based system, has allowed for an enhanced, high-resolution ECG which has opened new vistas of ECG analysis and interpretation.

Defining Terms

12-lead ECG: Twelve traditional ECG leads comprising the standard set.

ECG: Abbreviation for the device (electrocardiograph) or the output (electrocardiogram) depicting the body surface recording of the electrical activity of the heart.

ECG lead: Differential signal depicting one channel of the ECG record.

HRECG: High-resolution ECG used to detect microvolt-level cardiac potentials most commonly by signal averaging.

Wilson central terminal: Reference point for forming most of the standard ECG leads. It is the average of the right arm, the left arm, and the left potentials. It is a time-varying reference.

References

J.J. Bailey, A.S. Berson, A. Garson, L.G. Horan, P.W. Macfarlane, D.W. Mortara, and C. Zywietz, "Recommendations for standardization and specifications in automated electrocardiography: bandwidth and digital signal processing," A report for health professionals by an ad hoc writing group of the Committee on Electrocardiography and Cardiac Electrophysiology of the Council on Clinical Cardiology, American Heart Association, *Circulation*, vol. 81, no. 2, pp. 730–739, 1990.

E.J. Berbari, "High resolution electrocardiography," *CRC Crit. Rev. Bioeng.*, vol. 16, p. 67, 1988.

E.J. Berbari, R. Lazzara, P. Samet, and B.J. Scherlag, "Noninvasive technique for detection of electrical activity during the PR segment," *Circulation*, vol. 48, p. 1006, 1973.

E.J. Berbari, R. Lazzara, and B.J. Scherlag, "A computerized technique to record new components of the electrocardiogram," *Proc. IEEE*, vol. 65, p. 799, 1977.

E.J. Berbari, B.J. Scherlag, R.R. Hope, and R. Lazzara, "Recording from the body surface of arrhythmogenic ventricular activity during the ST segment," *Am. J. Cardiol.*, vol. 41, p. 697, 1978.

W. Einthoven, "Die galvanometrische Registrirung des menschlichen Elektrokardiogramms, zugleich eine Beurtheilung der Anwendung des Capillar-Elecktrometers in der Physiologie," *Pflugers Arch. Ges. Physiol.*, vol. 99, p. 472, 1903.

D.B. Geselowitz, "On the theory of the electrocardiogram," *Proc. IEEE*, vol. 77, p. 857, 1989.

E. Goldberger, "A simple, indifferent, electrocardiographic electrode of zero potential and a technique of obtaining augmented, unipolar, extremity leads," *Amer. Heart J.*, vol. 23, p. 483, 1942.

J.M. Jenkins, "Computerized electrocardiography," *CRC Crit. Rev. Bioeng.*, vol. 6, p. 307, 1981.

M.B. Simson, "Use of signals in the terminal QRS complex to identify patients with ventricular tachycardia after myocardial infarction," *Circulation*, vol. 64, p. 235, 1981.

"Voluntary standard for diagnostic electrocardiographic devices," ANSI/AAMI EC11a, Arlington, Va.: Association for the Advancement of Medical Instrumentation, 1984.

A.D. Waller, "On the electromotive changes connected with the beat of the mammalian heart, and the human heart in particular," *Phil. Trans. B.*, vol. 180, p. 169, 1889.

F.N. Wilson, F.S. Johnston, and I.G.W. Hill, "The interpretation of the galvanometric curves obtained when one electrode is distant from the heart and the other near or in contact with the ventricular surface," *Amer. Heart J.*, vol. 10, p. 176, 1934.

Further Information

Comprehensive Electrocardiology: Theory and Practice in Health and Disease, Volumes 1–3, P. W. Macfarlane and T.D. Veitch Lawrie, Eds., England: Pergamon Press, 1989.

High-Resolution Electrocardiography, M.D. Nabil El-Sherif and M.D. Gioia Turitto, Eds., Mount Kisco, N.Y.: Futura Publishing Company, 1992.

Medical Instrumentation: Application and Design, 2nd ed., J. G. Webster, Ed., Boston: Houghton Mifflin, 1992.

MEDICAL ELECTRONICS

• *Monitoring blood gas in real time*
• *Fast temperature readings by ear*
• *Interest revives in artificial heart research*

One noteworthy sign of progress in medical electronics is in blood gas monitoring used to assess the cardiovascular, respiratory, and metabolic performance of patients during surgery and in emergency situations. It is now possible to do such monitoring in real time with the PB3300 intra-arterial blood gas monitoring system developed by Puritan-Bennett Corp., Overland Park, Kan., which will offer the system later this year. Among other companies expected to market such units are Optical Sensors for Medicine, Abbott Laboratories, C.R. Bard, and Pfizer.

In the PB3300 system, special sensing fibers are part of a microprocessor-based unit. The sensor [see illustration] contains three small optical fibers, each about the diameter of a human hair, and a thermocouple in a biocompatible package. The entire sensor is small enough to be inserted into a radial artery through an arterial catheter less than 1 mm in diameter.

Each optical fiber has at its tip a special fluorescent dye. The dye on one fiber is sensitive to partial pressures of oxygen, to partial pressures of carbon dioxide on another, and to hydrogen ion concentration on the third. (Engineers at the National Institutes of Health developed the first probes of this type.) The high-energy photons in light sent through the fibers by the system are absorbed by the dyes. They then re-emit the light at a different wavelength and intensity. Depending on the particular dye, the intensity of the re-emitted light is changed by nearby concentrations of oxygen, carbon dioxide, or hydrogen ions. This change is converted by the instrument to readings of partial pressures of O_2, CO_2, and pH.

The thermocouple measures blood temperature in the radial artery which is then displayed along with the blood gas values.

Even the mundane task of taking a patient's temperature benefits from modern technology. Introduced last year was a new tympanic or ear thermometer that measures infrared heat generated by the eardrum and surrounding tissue and displays the temperature in about 2 seconds.

Manufacturers include Diatek, IVAC, Thermoscan, Exeregen, and Intelligent Medical Systems. The thermometers sell for US $350–$600 and require a single-use disposable probe cover costing a few cents.

Support for the Artificial Heart
In July 1992, the Institute of Medicine, Washington, D.C. released a report, "The Artificial Heart:

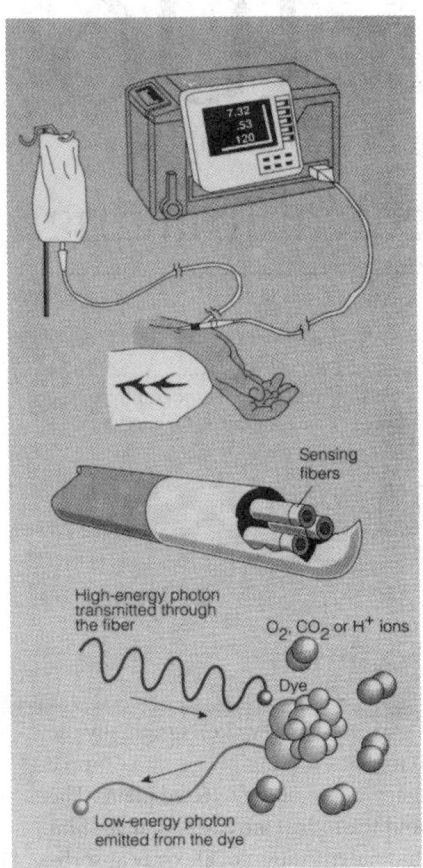

Sensing
fibers

High-energy photon
transmitted through
the fiber

O_2, CO_2 or H^+ ions

Dye

Low-energy photon
emitted from the dye

The intra-arterial blood gas monitoring system from Puritan-Bennett Corp. uses dye-tipped optical sensors. Light intensities, re-emitted from the dyes, are proportional to the concentrations of nearby oxygen, carbon dioxide, and hydrogen ion concentrations.

Prototypes, Policies, and Patients," which calls for the National Heart, Lung, and Blood Institute (NHLBI) to continue research contracts in artificial heart technology beyond their expiration dates in 1993. The continuance, the report said, should be for an interim period to support work on ventricular assist devices and the total artificial heart. NHLBI is the only U.S. government source of funds for artificial heart research.

In 1988 research contracts on the total artificial heart were suspended after NHLBI decided the money would be better spent for research on ventricular assist devices, which were further along in development. That decision met strong resistance from congressional leaders and other officials and was later rescinded.

The first model of the total artificial heart is not likely to be approved by the U.S. Food and Drug Administration before the year 2005. Ventricular assist devices should be available by the late 1990s. Clinical trials on a long-term, fully implantable unit were scheduled to begin in 1992. Units are typically implanted in the abdomen, connected to tubes through which the blood is drawn from one of the heart's ventricles, and pumped into the circulatory system.

What complicates the funding issue is the uncertainty as to whether the cost of a total artificial heart will ever fall into line with the benefit to the patient.

Source: Adapted from R. K. Jurgen, *IEEE Spectrum,* p. 61, January 1992. © 1992 IEEE.

106

Medical Imaging

M. D. Fox
University of Connecticut

Leon A. Frizzell
University of Illinois

106.1 Tomography.. 2374
Computerized Tomography • Positron Emission Tomography • Single Photon Emission Computed Tomography • Magnetic Resonance Imaging • Imaging
106.2 Ultrasound... 2380
Fundamentals of Acoustics • Principles of Pulse-Echo Ultrasound • Future Developments

106.1 Tomography

M. D. Fox

The term **tomography** derives from the Greek *tomos* (cutting) and *grapho* (to write). Originally the term was applied to sectional radiography achieved by a synchronous motion of the x-ray source and detector in order to blur undesired data while creating a sharp image of the selected plane. The term *tomography* was used to distinguish between such slices and the more conventional plain film radiograph, which represents a two-dimensional shadowgraphic superposition of all x-ray absorbing structures within a volumetric body.

Computerized tomography, also known as **computerized axial tomography,** was introduced by EMI, Ltd. in 1973 and transformed medical imaging by obviating the superposition of intervening structures present in conventional radiographic images. Initially, the clinical application was for imaging the head, but soon the technique found wide application in body imaging.

As medical imaging has evolved into a multimodality field, the meaning of tomography has broadened to include any images of thin cross-sectional slices, regardless of the modality utilized to produce them. Thus, tomographic images can be generated by **magnetic resonance imaging** (MRI), ultrasound (US), computerized tomography (CT), or such nuclear medicine techniques as **positron emission tomography** (PET) or **single photon emission computerized tomography** (SPECT). For the purposes of this discussion we will cover all of the foregoing modalities with the exception of ultrasound, which will be treated separately.

Since the power of such computerized techniques was recognized, the practice of radiology has been revolutionized by making possible much more precise diagnosis of a wide range of conditions. In this necessarily brief discussion we will describe the basic physical principles of the major tomographic modalities as well as their key clinical applications.

Computerized Tomography

The basic concept of computerized tomography can be described by consideration of Fig. 106.1. An x-ray source is passed through an aperture to produce a fan-shaped beam that passes through the body of interest with absorption along approximately parallel lines. The natural logarithm of the

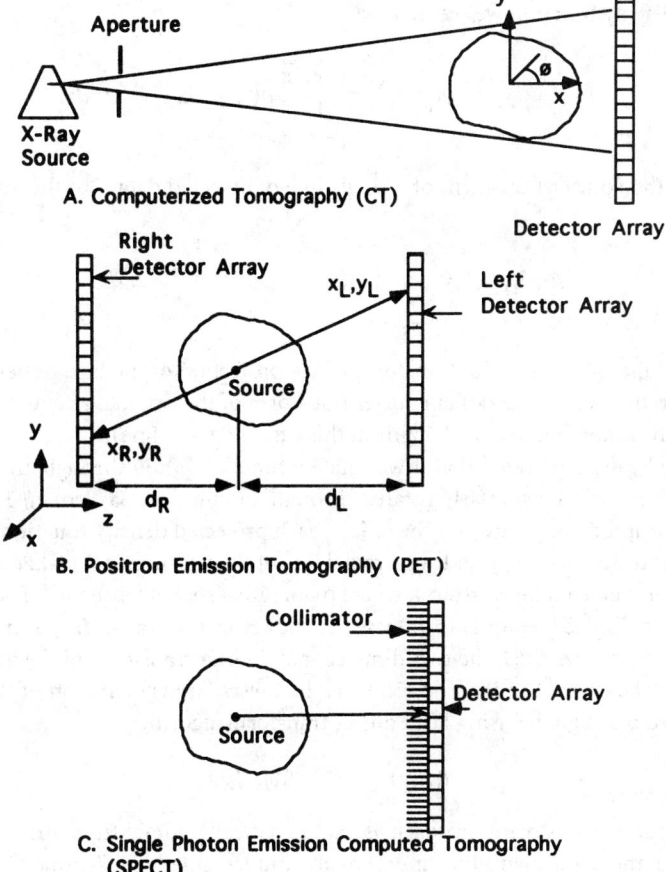

A. Computerized Tomography (CT)

B. Positron Emission Tomography (PET)

C. Single Photon Emission Computed Tomography (SPECT)

FIGURE 106.1 Comparison of three photon-based tomographic imaging modalities.

detected intensity will be the integral of the linear attenuation coefficient of the object along the ray directed from the source to the detector element. If the source and the detector array are synchronously rotated about a point within the object, a number of lines of data can be collected, each representing the projected density of the object as a function of lateral position and angle.

A number of mathematical techniques can and have been used to recover the two-dimensional distribution of the linear attenuation coefficient from this array of measurements. These include iterative solution of a set of simultaneous linear equations, Fourier transform approaches, and techniques utilizing back-projection followed by deconvolution [Macovski, 1983]. Conceptually, the Fourier transform approach is perhaps the most straightforward, so we will describe it in some detail.

Using the coordinate system of Fig. 106.1(A) and assuming parallel rays, the intensity picked up by the detector array can be expressed as

$$I_d(y) = I_0 \exp[-\int a(x,y)dx]$$

where $a(x,y)$ represents the linear attenuation coefficient to x-ray photons within the body as a function of x,y position, and I_0 is the source intensity. Rearranging, we see that

$$a_p(y) = \int_{-\infty}^{\infty} a(x, y)dx = \ln[I_d(y)/I_0]$$

where $a_p(y)$ is the projected attenuation function. Taking a one-dimensional Fourier transform of this projected density function we see that

$$F[a_p(y)] = A_p(f_y) = \int\limits_{-\infty}^{\infty} \int\limits_{-\infty}^{\infty} a(x,y)dx\,e^{-j2\pi f_y y}dy$$

where $A_p(f_y)$ is the Fourier transform of a single line of detected data. But this can also be written

$$A_p(0, f_y) = \int\limits_{-\infty}^{\infty} \int\limits_{-\infty}^{\infty} a(x, y)dx\,e^{-j2\pi(0x+f_y y)}dy$$

Thus, the one-dimensional Fourier transform of the projection of the linear attenuation function, $a_p(y)$, is equal to the two-dimensional Fourier transform of the original attenuation function evaluated along a line in the frequency domain (in this case the $f_x = 0$ line).

It can readily be demonstrated that if we rotate a function $a(x,y)$ through an angle ϕ in the x,y plane, its transform will be similarly rotated through an angle ϕ [Castleman, 1979]. Thus as we rotate the source and detector around the object, each projected density function detected $a_p(\rho,\phi_i)$ can be Fourier transformed to provide one radial line of the two-dimensional Fourier transform of the desired reconstructed image, $A(\rho,\phi_i)$, where ρ is a radial spatial frequency. The set of all $A(\rho,\phi_i)$ for small angular displacements ϕ_i form a set of spokes in the transform domain which can be interpolated to estimate $A(f_x,f_y)$, the two-dimensional Fourier transform of the image in rectangular coordinates. The image can then be recovered by inverse transformation of $A(f_x,f_y)$, which can readily be carried out digitally using fast Fourier transform algorithms, i.e,

$$a(x,y) = F^{-1}[A(f_x,f_y)]$$

While the Fourier transform approach is mathematically straightforward, many commercial scanners utilize the equivalent but more easily implemented back-projection/deconvolution approach, where each ray is traced back along its propagation axis. When all rays have been back-projected and the result summed, one obtains an approximate (blurred) image of that plane. This image can then be sharpened (deblurred) through the use of an appropriate filter, which is usually implemented by convolving with an appropriate two-dimensional deblurring function. Refer to Macovski [1983] for the details of this process.

Clinically, the impact of computerized tomography was dramatic due to the vastly increased density resolution, coupled with the elimination of the superposition of overlying structures, allowing enhanced differentiation of tissues with similar x-ray transmittance, such as blood, muscle, and organ parenchyma. CT scans of the head are useful for evaluation of head injury and detection of tumor, stroke, or infection. In the body, CT is also excellent in detecting and characterizing focal lesions, such as tumors and abscesses, and for the evaluation of the skeletal system. [Axel *et al.*, 1983]. In recent years the advent of magnetic resonance systems has provided even greater soft tissue contrast, and thus the role of CT has been constrained by this at times competing modality.

Positron Emission Tomography

Unlike computerized tomography, which relies on photons produced by an external source, in the modalities of positron emission tomography (PET) and single photon emission computed tomography (SPECT), the source of radiation is a radioisotope that is distributed within the body, and thus these modalities are sometimes referred to as forms of emission computed tomography (ECT). While conventional CT can produce images based upon anatomy of organs, emission CT techniques can quantitate the distribution of tracer materials that can potentially elucidate physiologic function.

The positron or positive electron is a positively charged particle that can be emitted from the nucleus of a radionuclide. The positron travels at most a few millimeters before being annihilated by interaction with a negative electron from the surrounding tissue. The product of this event is the emission of 511-keV gamma ray photons which travel in almost exactly opposite directions. The detectors themselves can be either discrete detectors or a modified Anger camera like those used in conventional nuclear imaging. A coincidence detector is employed to limit recorded outputs to cases in which events are detected simultaneously in both detector arrays, thus reducing the pickup of noise or scattering.

A possible detection scheme is illustrated in Fig. 106.1(B). The detector arrays shown can be made energy selective to eliminate lower energy scattered gamma rays. While the distribution of radioactivity can be reconstructed using the reconstruction from projection techniques described in the section on CT [Hurculak, 1987], the x,y source position of an event can be determined directly from the detection geometry as follows [Macovski, 1983]:

$$x \approx x_L \, d_R/(d_R + d_L) + x_R \, d_L/(d_R + d_L)$$

$$y \approx y_L \, d_R/(d_R + d_L) + y_R \, d_L/(d_R + d_L)$$

Typically a single plane is studied, and no collimators are required. A drawback of PET has been that because of the short half-lives of positron-producing radioisotopes, the use of this modality has required the presence of an expensive cyclotron facility located near the hospital.

One important radionuclide commonly used in PET is oxygen 15 with a half-life of 2.07 minutes, which can be bonded to water for measurement of cerebral blood flow or to O_2/CO_2 to assess cerebral oxygen utilization. Another is carbon 11 with a half-life of 20.4 minutes, which can be bonded to glucose to trace glucose utilization. F-18 fluorodeoxyglucose (FDG) has been used to demonstrate the degree of malignancy of primary brain tumors, to distinguish necrosis from tumor, and to predict outcome [Coleman, 1991]. Perhaps the most unusual feature of this modality is the ability to quantitate the regional metabolism of the human heart [Schelbert, 1990].

Single Photon Emission Computed Tomography

In contrast to PET, SPECT can be utilized with any radioisotope that emits gamma rays, including such common radioisotopes as Tc-99m, I-125, and I-131 which have been utilized in conventional nuclear imaging for the last 30–35 years and which due to their relatively long half-lives are available at reasonable cost at nearly every modern hospital. Due to the need for direction sensitivity of the detector, a collimator must be used to eliminate gamma rays from other than the prescribed direction, thus resulting in a 1–2 order of magnitude decrease in quantum efficiency as compared with PET scanning [Knoll, 1983].

The basic concept of SPECT is illustrated in Fig. 106.1(C). A gamma ray photon from a radionuclide with energy above 100 keV will typically escape from the body without further interaction, and thus the body can be regarded as a transparent object with luminosity proportional to the concentration of the radionuclide at each point. The reconstruction mathematics are similar to those derived for absorption CT, with the exception that the variable reconstructed is a source distribution rather than an attenuation coefficient. Some errors can be introduced in the reconstruction because of the inevitable interaction of gamma rays with overlying tissue, even at energies above 100 keV, although this can be compensated for to some extent. Detection of scattered radiation can be reduced through the use of an energy acceptance window in the detector.

Technetium 99m can be used to tag red blood cells for blood pool measurements, human serum albumin for blood pool and protein distribution, or monoclonal antibodies for potential detection of individual tumors or blood cells. Emission computed tomography techniques such as PET and SPECT follow the recent trend toward imaging techniques that image physiologic processes as opposed to anatomic imaging of organ systems. The relatively low cost of SPECT systems has led to a recent resurgence of interest in this modality.

Magnetic Resonance Imaging

The basic magnetic resonance concept has been used as a tool in chemistry and physics since its discovery by Bloch in 1946, but its use expanded tremendously in the 1980s with the development of means to represent magnetic resonance signals in the form of tomographic images. Magnetic resonance imaging is based on the magnetic properties of atomic nuclei with odd numbers of protons or neutrons, which exhibit magnetic properties because of their spin. The predominant source of magnetic resonance signals in the human body is hydrogen nuclei or protons. In the presence of an external magnetic field, these hydrogen nuclei align along the axis of the field and can precess or wobble around that field direction at a definite frequency known as the Larmour frequency. This can be expressed:

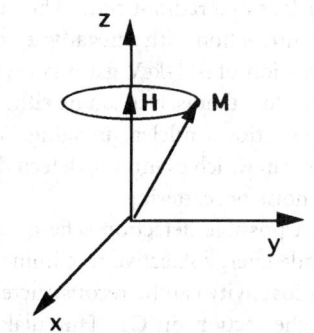

FIGURE 106.2 Geometry of precessing proton in a static magnetic field oriented in the *z* direction.

$$f_0 = \gamma H$$

where f_0 is the Larmour frequency, γ is the gyromagnetic ratio which is a property of the atomic element, and H is the magnitude of the external magnetic field. For example, given a gyromagnetic ratio of 42.7 MHz/tesla for hydrogen and a field strength of 1 tesla (10 kilogauss), the Larmour frequency would be 42.7 MHz, which falls into the radio frequency range.

The magnetic resonance effect occurs when nuclei in a static magnetic field H are excited by a rotating magnetic field H_1 in the x,y plane, resulting in a total vector field **M** given by

$$\mathbf{M} = H\mathbf{z} + H_1(\mathbf{x} \cos \omega_0 t + \mathbf{y} \sin \omega_0 t)$$

Upon cessation of excitation, the magnetic field decays back to its original alignment with the static field H, emitting electromagnetic radiation at the Larmour frequency, which can be detected by the same coil that produced the excitation [Macovski, 1983].

Imaging

As shown in Fig. 106.3, one method for imaging utilizes a transmit/receive coil to emit a magnetic field at frequency f_0 which is the Larmour frequency of plane P. Subsequently, magnetic gradients are applied in the y and x directions. The detected signal during the data collection window can be expressed as

$$S(t_x, t_{yi}) = \int_{-\infty}^{\infty} \int_{-\infty}^{\infty} s(x, y) \exp[-i\gamma(G_x x t_x + G_y y t_{yi})] \, dx \, dy$$

where $s(x,y)$ represents the magnetic resonance signal at position (x,y) (G_x, G_y) are the x and y gradients, t_x is time within the data collection window, t_{yi} is the y direction gradient application times, and γ is the gyromagnetic ratio. The two-dimensional spatial integration is obtained by appropriate geometry of the detection coil. Collecting a number of such signals for a range of t_{yi}, we can obtain the two-dimensional function $S(t_x, t_y)$. Comparing this to the two-dimensional Fourier transform relation

$$F(u,v) = \int_{-\infty}^{\infty} \int_{-\infty}^{\infty} f(x,y) \exp[-i2\pi(ux + vy)] \, dx \, dy$$

we see that the detected signal $S(t_x, t_y)$ is the two-dimensional Fourier transform of the magnetic resonance signal $s(x,y)$ with $u = \gamma G_x t_x / 2\pi$, $v = \gamma G_y t_y / 2\pi$. The magnetic resonance signal $s(x,y)$

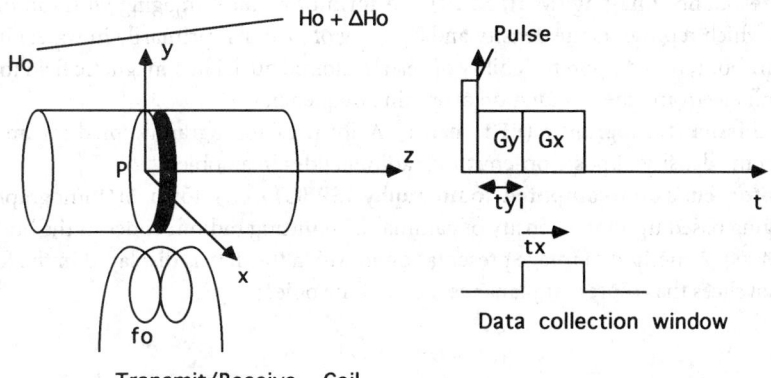

FIGURE 106.3   Concept of magnetic resonance imaging. The static magnetic field H_0 has a gradient such that excitation at frequency f_0 excites only the plane P. Gradient G_y in the y direction is applied for time t_{yi}, causing a phase shift along the y direction. Gradient G_x in the x direction is applied for time t_x, causing a frequency shift along the x direction. Repetition of this process for different t_{yi} allows the receive coil to pick up a signal which is the two-dimensional Fourier transform of the magnetic resonance effect within the slice.

depends on the precise sequence of pulses of magnetic energy used to perturb the nuclei. For a typical sequence known as spin-echo consisting of a 90-degree pulse followed by a 180-degree pulse spaced at time τ with the data collection at $t_e = 2\tau$, and t_r being the repetition time between 90-degree pulses, the detected magnetic resonance signal can be expressed

$$s(x,y) = \rho(1 - e^{-tr/T_1})(e^{-te/T_2})$$

where ρ is the proton density, and T_1 (the spin-lattice decay time) and T_2 (the spin-spin decay time) are constants of the material related to the bonding of water in cells [Wolf and Popp, 1984]. Typically T_1 ranges from 0.2 to 1.2 seconds, while T_2 ranges from 0.05 to 0.15 seconds.

By modification of the repetition and orientation of excitation pulses, an image can be made T_1, T_2, or proton density dominated. A proton density image shows static blood and fat as white and bone as black, while a T_1 weighted image shows fat as white, blood as gray, and cerebrospinal fluid as black. T_2 weighted images tend to highlight pathology since pathologic tissue tends to have longer T_2 than normal.

In general, magnetic resonance imaging has greater intrinsic ability to distinguish between soft tissues than computerized tomography. It also has some ability to visualize moving blood. As the preceding discussion indicates, magnetic resonance is a richer and more complex modality than CT. Typically MRI has been more expensive than CT. Both MRI and CT have been used primarily for anatomic imaging, but MRI has the potential through spectroscopy (visualization of other nuclei than hydrogen) to become a factor in physiologic imaging. Thus, it can be anticipated that magnetic resonance imaging will continue to increase and become an even more important modality in the next decade.

Defining Terms

Computerized axial tomography (CAT scan, CT):   A form of medical imaging based upon the linear attenuation coefficient of x-rays in which a tomographic image is reconstructed from computer-based analysis of a multiplicity of x-ray projections taken at different angles around the body.

Magnetic resonance imaging (MRI, NMR):   A form of medical imaging with tomographic display which represents the density and bonding of protons (primarily in water) in the tissues of the body, based upon the ability of certain atomic nuclei in a magnetic field to absorb and reemit electromagnetic radiation at specific frequencies.

Positron emission tomography (PET scan):   A form of tomographic medical imaging based upon the density of positron-emitting radionuclides in an object.

Single photon emission computed tomography (SPECT):   A form of tomographic medical imaging based upon the density of gamma ray-emitting radionuclides in the body.

Tomography: A method of image presentation in which the data is displayed in the form of individual slices that represent planar sections of the object.

References

L. Axel, P.H. Arger, and R. Zimmerman, "Applications of computerized tomography to diagnostic radiology," *Proceedings of the IEEE*, vol. 71, no. 3, p. 293, March 1983.

K.R. Castleman, *Digital Image Processing*, Englewood Cliffs, N.J.: Prentice-Hall, 1979.

R.E. Coleman, "Single photon emission computed tomography and positron emission tomography," *Cancer*, vol. 67 (4 Suppl.), pp. 1261–1270, Feb. 1991.

P.M. Hurculak, "Positron emission tomography," *Canadian Journal of Medical Radiation Technology*, vol. 18, no. 1, March 1987.

G.F. Knoll, "Single-photon emission computed tomography," *Proceedings of the IEEE*, vol. 71, no. 3, p. 320, March 1983.

A. Macovski, *Medical Imaging Systems*, Englewood Cliffs, N.J.: Prentice-Hall, 1983.

H.R. Schelbert, "Future perspectives: Diagnostic possibilities with positron emission tomography," *Roentgen Blaetter*, vol. 43, no. 9, pp. 384–390, Sept. 1990.

G.L. Wolf and C. Popp, *NMR, A Primer for Medical Imaging*, Thorofare, N.J.: Slack, Inc., 1984.

Further Information

The journal *IEEE Transactions on Medical Imaging* describes advances in imaging techniques and image processing. *Investigative Radiology,* published by the Association of University Radiologists, emphasizes research carried out by hospital-based physicists and engineers. *Radiology,* published by the North American Society of Radiologists, contains articles which emphasize clinical applications of imaging technology. *Diagnostic Imaging,* publishing by Miller Freeman, Inc., is a good source of review articles and information on the imaging marketplace.

106.2 Ultrasound

Leon A. Frizzell

Ultrasound, acoustic waves at frequencies higher than those audible by humans, has developed over the past 35 years into an indispensable clinical diagnostic tool. Currently, ultrasound is used to image most parts of the body. More than half of all pregnant women in the United States are examined with ultrasound. This widespread utilization has resulted from ultrasound's proven clinical utility for imaging soft tissues compared to more expensive imaging techniques. The development of ultrasound, particularly for fetal examinations, has also been fostered by its safety record; no case of an adverse biological effect induced by diagnostic ultrasound has ever been reported in humans [AIUM, 1988].

Diagnostic ultrasound systems are used primarily for soft tissue imaging, motion detection, and flow measurement. Except for some Doppler instruments, these systems operate in a **pulse-echo** mode. A brief summary of some of the fundamentals of acoustic wave propagation and the principles of ultrasound imaging follows.

Table 106.1 Approximate Ultrasonic Attenuation Coefficient,
Speed, and Characteristic Impedance for Water and Selected Tissues at 3.5 MHz

Tissue	Attenuation Coefficient (m^{-1})	Speed (m/s)	Characteristic Impedance (10^6 Pa s/m)
Water	0.2	1520	1.50
Amniotic fluid	0.7	1510	1.51
Blood	7	1550	1.60
Liver	35	1580	1.74
Muscle	50	1560	1.72
Bone	800	3360	5.70
Lung	1000	340	0.25

Fundamentals of Acoustics

Unlike electromagnetic waves, acoustic waves require a medium for propagation. The acoustic wave phenomenon causes displacement of particles (consisting of many molecules), which results in pressure and density changes within the medium. For a traveling sinusoidal wave, the variation in acoustic pressure (the difference between the total and ambient pressure), excess density, particle displacement, particle velocity, and particle acceleration can all be represented by the form

$$p = P e^{-\alpha x} \cos(\omega t - kx) \tag{106.1}$$

for a wave propagating in the positive x direction, where p is the pressure (or one of the other parameters listed above), P is its amplitude, ω is the angular frequency, and $\omega = 2\pi f$ where f is the frequency in hertz, k is the propagation constant and $k = \omega/c$ where c is the propagation speed, α is the attenuation coefficient, and t is the time. The wave can experience significant attenuation, as represented by the exponential decay of amplitude with distance, during propagation in tissues. The attenuation coefficient varies greatly among tissues [Goss *et al.*, 1978, 1980; Haney and O'Brien, 1986] but is low for most body fluids, much higher for solid tissues, and very high for bone and lung (see Table 106.1). The skin depth is the distance that the wave can propagate before being attenuated to e^{-1} of its original amplitude and is thus simply the inverse of the attenuation coefficient.

Ultrasound is typically used to image soft body tissues such as liver, but the sound beam often travels through fluids, for example, through amniotic fluid when imaging the fetus. Generally, bone and lung are not imaged with ultrasound. The attenuation processes include absorption, which is the conversion of acoustic energy to heat, and scattering, which will be addressed later. The attenuation increases roughly linearly with frequency in the 2- to 10-MHz range typically used for medical imaging. This range represents a compromise between increased penetration at lower frequencies (because of decreased attenuation) and improved resolution associated with higher frequencies as discussed below. Thus, the lower frequencies are used when greater penetration is required, such as for fetal imaging in the obese patient, and higher frequencies for lesser penetration, such as the examination of peripheral vascular flow.

When an acoustic wave impinges on an interface between two media of different specific acoustic impedance, a portion of the incident energy is reflected. For normal incidence on an infinite plane interface, the pressure reflection coefficient is given by [Kinsler *et al.*, 1982]

$$R = \frac{z_2 - z_1}{z_2 + z_1} \tag{106.2}$$

where z_1 and z_2 are the specific acoustic impedance of the incident and transmitting media, respectively. For a plane wave the specific acoustic impedance is equal to the characteristic impedance which is the product of the density and acoustic speed in the medium (see Table 106.1). The speed

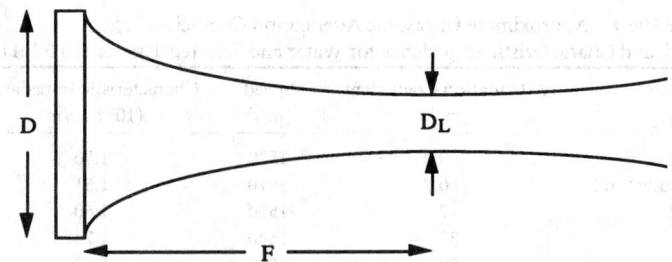

FIGURE 106.4 Cross section of a typical focused circular ultrasound source of aperture diameter D and focal length F, showing the focused beam of lateral beam width at the focus D_L.

is dependent upon the density and the elastic properties of the medium. Thus, at an interface between media exhibiting different densities or elastic properties, i.e., compressibility, some acoustic energy will be reflected. Although the reflection coefficient at an interface between muscle and bone is large (approximately 0.54) the reflection coefficient between two soft tissues such as liver and muscle is quite small (approximately 0.006). Reflection at oblique incidence obeys Snell's law in the same way it applies to electromagnetic waves.

In addition to the specular reflection that occurs at an interface between two media of different specific acoustic impedance as described above (where any curvature along the interface is negligible over distances comparable to a wavelength), energy may also be scattered in all directions by inhomogeneities in the medium. An acoustic image is formed by using this scattered energy as well as specular reflections. The fraction of the incident energy reflected or scattered is very small for soft tissues.

Although it is convenient to consider plane waves of infinite lateral extent, as was done above, real sources generate finite beams of ultrasound. These sources may be unfocused, but for the typical diagnostic system they are focused. Figure 106.4 shows the acoustic field from a typical focused source. The source consists of a piezoelectric transducer which converts electrical to acoustic energy and vice versa. Most transducers for medical applications are made from ceramic materials such as a lead zirconate titanate (PZT) mixture. For a circular aperture these may be circular disks with a plano-concave lens mounted in front to produce spherical focusing. Alternatively, the transducer itself may be a spherical segment that produces a focused field without a lens. Some probes utilize electronic focusing methods. Such a phased array probe consists of many individual elements which can be excited with signals having a controlled delay with respect to one another such that the signals constructively interfere at the desired focal region. At a receiver the signals are combined with delays associated with various elements to provide reinforcement of the signals from a receiving focal region.

The lateral beam width D_L is directly dependent upon the wavelength λ and focal length F and inversely related to the aperture diameter (diameter of the transducer) D [Hueter and Bolt, 1955]:

$$D_L = 2.44 \frac{F\lambda}{D} \tag{106.3}$$

Because $f\lambda = c$, the higher the frequency the smaller is λ and D_L. The smaller D_L, the better the lateral resolution near the focus, but the beam spread is greater with distance from the focus. Thus, the strength of the focusing varies among transducers so that the user may choose very good resolution over a short region or somewhat poorer resolution that is maintained over a greater depth. With phased array transducers, the focal region can be varied dynamically to optimize lateral resolution at all distances.

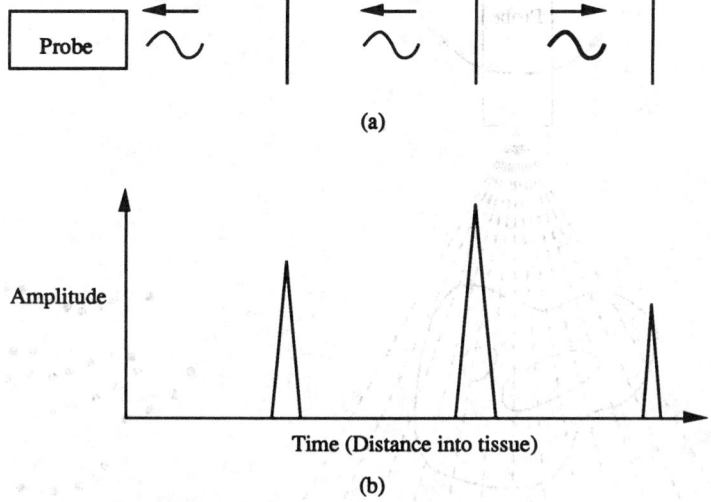

FIGURE 106.5 (a) The transmitted pulse (heavy wave) and echoes from reflecting structures; (b) the resulting A-mode display.

Principles of Pulse-Echo Ultrasound

Ultrasound imaging usually employs frequencies in the 2- to 10-MHz range, though some of the new intravascular probes use higher frequencies. Images are formed by using a transducer within a probe to generate a short pulse (typically on the order of 1 μs in duration) of ultrasound which is propagated through the tissue. A portion of the energy in this pulse is reflected back toward the transducer from specular reflectors and from scatterers in the tissue. These acoustic echoes, with amplitudes much lower than the transmitted pulse, are converted by the transducer to electrical signals which are converted to a (rectified) video signal, amplified by a time gain controlled amplifier, and displayed. The **A-mode display** is rarely used but simply involves display of the received echoes as amplitude versus time of arrival. The time of arrival is related by the wave speed to the tissue depth from which the echo returns, i.e., $d = ct/2$. Figure 106.5 provides a very simple representation of this process where the A-mode display associated with specular reflection from three different interfaces is illustrated. For clinical imaging the interfaces would not necessarily be perpendicular to the axis of the sound beam, and there would be a continuum of echoes, a continuous received signal, due to energy backscattered from within the tissues. Since the ultrasound pulse is attenuated as it propagates, all ultrasonic imaging systems use a logarithmic variation of amplifier gain with time to compensate the exponential attenuation of the tissue. Thus, echoes from structures reflecting or backscattering the same fraction of the incident signal will have the same amplitude after passing through the time gain controlled amplifier.

A **B-mode display** is typically used for ultrasound imaging. It involves display of the echoes at various brightness or gray levels corresponding to their amplitude. A **two-dimensional B-mode display** involves movement of the transducer (manually or automatically), movement of a mirror to change the direction of the field (automatically), or movement of the ultrasound beam directly (electrically) such that it scans a plane through the body. Figure 106.6 provides a simplified representation (again, echoes are shown as arising from interfaces only) of the formation of a B-mode image. The direction of the beam is monitored so that the received signals along each path are placed in their correct location on the display. Typically, the orientation information and echoes are processed by a digital scan converter for appropriate display of the two-dimensional image on a cathode ray tube in the standard format used for television picture display. Most B-mode systems in use today create an image in 0.1 s or less, so that the image is displayed in real-time for viewing

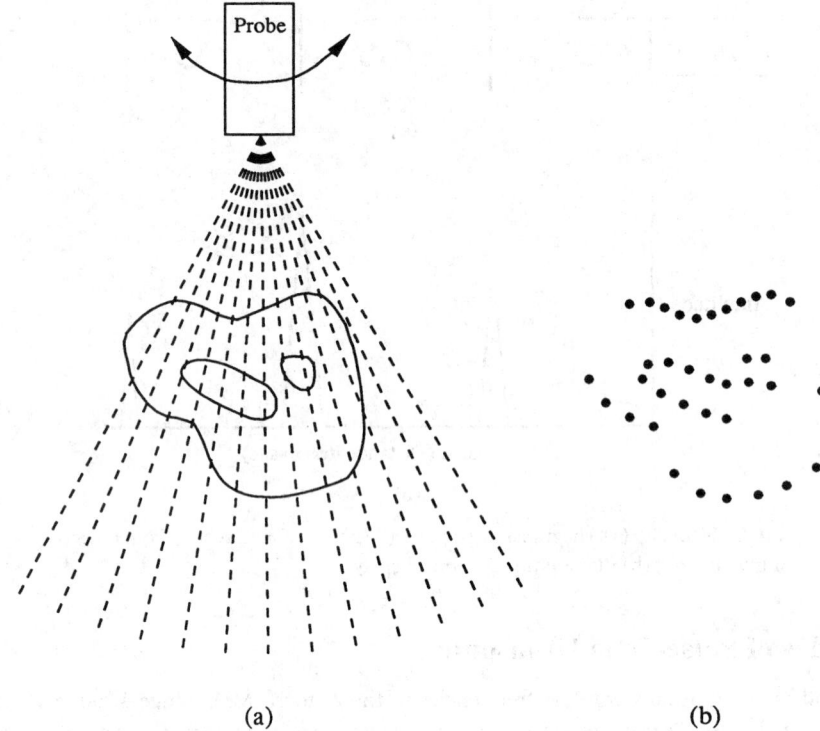

(a)   (b)

FIGURE 106.6   (a) The transmitted pulse paths for a rotating transducer probe; (b) the resulting two-dimensional B-mode display of echoes from the interfaces only.

of moving structures, such as structures in the heart or the fetus moving within the womb. This is not possible with the typical magnetic resonance or computed tomography system.

Many systems now use digital processing to enhance portions of the image. For example, it is possible to emphasize the large amplitude, small amplitude, or midrange signals. It is also possible to perform a more sophisticated analysis to enhance edges.

Many specialty probes have been designed for intracavitary examination. Examples include examination of the fetus with a vaginal probe, the prostate with a rectal probe, and blood vessel walls with intravascular probes. The intracavitary probe offers the advantage of decreasing the distance from the transducer to the tissue of interest and thus decreasing attenuation such that higher frequencies can be used for greater resolution. The lateral resolution of a focused probe is improved with frequency as discussed in the preceding subsection, but the axial (along the ultrasound beam) resolution also improves with frequency. The shorter the transmitted pulse, the better the axial resolution. Shorter pulses are generated by sources with a larger bandwidth, which corresponds to a higher center frequency when the sources have bandwidths which are approximately the same fraction of the center frequency.

The use of ultrasound for motion detection and measurement has increased tremendously in recent years. Most of these systems use the Doppler principle, but some use time domain detection. In Doppler detection, if the ultrasound is reflected from a target moving at some speed v_t toward (away from) the source at an angle θ with respect to the beam axis, the frequency of the transmitted signal f is shifted up (down) by an amount f_D, the Doppler shift, according to the following relation:

$$f_D = \frac{2 f v_t \cos \theta}{c} \tag{106.4}$$

In principle a measurement of f_D, when f, c, and θ are known, will yield the speed of the target v_t. However, it is often difficult to determine θ because the angle the transducer axis makes with a blood vessel, for example, is often unknown. Even when that angle is known, the flow is not necessarily along the direction of the vessel at every location and for all times. Time domain detection of motion, by measuring the movement of specific echoes from one pulse to another, is a recently developed alternative to Doppler detection that is not currently widely used.

For many years **duplex** systems, which provide both a two-dimensional image and a Doppler signal, showing the change of target speed with time, from a particular selected target area, have been in wide use. More recently, **color flow imaging** has been employed, which provides a two-dimensional color (typically red or blue) image of flow toward and away from the transducer superimposed on the gray scale image of stationary tissue structures. For these systems the speed, whether from Doppler or time domain detection schemes, is indicated by color saturation, hue, or luminance. These systems have proven very valuable for detecting the existence of flow in a region, detecting obstructions to flow and the turbulence associated with this, detecting reduced flow, and so on. Some other systems add to the color flow image a display of speed versus time for a region that is defined by a user-movable box.

Future Developments

It seems clear that the continuing development of intracavitary transducers, particularly for intravascular imaging, and the use of ultrasound intraoperatively will lead to more high-frequency commercially available probes that will produce better resolution images for these applications. The ever-increasing computer power/cost ratio means that the development of useful three-dimensional image display for ultrasound systems should not be far in the future.

Defining Terms

A-mode display: Returned ultrasound echoes displayed as amplitude versus depth into the body.

B-mode display: Returned ultrasound echoes displayed as brightness or gray scale levels corresponding to the amplitude versus depth into the body.

Color flow imaging or color Doppler: Two-dimensional image showing color-coded flow toward and away from the transducer displayed with the two-dimensional gray scale image of stationary targets.

Duplex ultrasound: Simultaneous display of speed versus time for a chosen region and the two-dimensional B-mode image.

Pulse-echo ultrasound: Using a probe containing a transducer to generate a short ultrasound pulse and receive echoes of that pulse, associated with specular reflection from interfaces between tissues or scattering from inhomogeneities within the tissue, to form a display of the tissue backscatter properties.

Two-dimensional B-mode display: Echoes from a transducer, or beam, scanned in one plane displayed as brightness (or gray scale) versus location for the returned echo to produce a two-dimensional image.

References

American Institute of Ultrasound in Medicine, "Bioeffects considerations for the safety of diagnostic ultrasound," *J. Ultrasound Med.*, vol. 7, no. 9 (supplement), 1988.

S.A. Goss, R.L. Johnston, and F. Dunn, "Comprehensive compilation of empirical ultrasonic properties of mammalian tissues," *J. Acoust. Soc. Am.*, vol. 64, pp. 423–457, 1978.

S.A. Goss, R.L. Johnston, and F. Dunn, "Compilation of empirical ultrasonic properties of mammalian tissues. II," *J. Acoust. Soc. Am.*, vol. 68, pp. 93–108, 1980.

M.J. Haney, and W.D. O'Brien, Jr., "Temperature dependency of ultrasonic propagation properties in biological materials," in *Tissue Characterization with Ultrasound*, vol. 1, J. Greenleaf, Ed., Boca Raton, Fla.: CRC Press, 1986, pp. 15–55.

T.F. Hueter and R.H. Bolt, *Sonics*, New York: John Wiley, 1955, p. 267.

L.E. Kinsler, A.R. Frey, A.B. Coppens, and J.V. Sanders, *Fundamentals of Acoustics*, 3rd ed., New York: John Wiley, 1982.

Further Information

The American Institute of Ultrasound in Medicine publishes monthly the *Journal of Ultrasound in Medicine*, which contains largely clinically oriented articles, and many clinically oriented ultrasound texts are available covering almost any medical discipline. However, there are only a few books that provide more than a cursory treatment of the basic physics and instrumentation of ultrasound imaging.

One text that has been regularly updated since the first volume appeared in 1980 is *Diagnostic Sonography: Principles and Instruments*, 4th edition, by F.W. Kremkau (W.B. Saunders, Philadelphia, 1993). Though this text is designed primarily to train sonographers who do not have a technical background, it provides the fundamentals of ultrasound imaging in a format that is very easy to read and understand.

Other texts that provide a more technical background (though some are a bit dated) include:

Biomedical Ultrasonics, by P. N. T. Wells (Academic Press, New York, 1977).

New Techniques and Instrumentation in Ultrasonography, edited by P.N.T. Wells and M.C. Ziskin (Churchill Livingstone, New York, 1980).

Medical Physics of CT and Ultrasound, edited by G.D. Fullerton and J.A. Zagzebski (American Institute of Physics, New York, 1980).

Physical Principles of Medical Ultrasonics, edited by C. R. Hill (John Wiley, New York, 1986).

Ultrasonic Bioinstrumentation, by D.A. Christensen (John Wiley, New York, 1988).

The Physics of Medical Imaging, edited by S. Webb (IOP Publishing, New York, 1988).

Principles of Medical Imaging, by B. Tsui, M. Smith, and K.K. Shung (Academic Press, New York, 1992).

107

Rehabilitation Engineering

107.1 Introduction 2387
107.2 Rehabilitation Concepts 2388
107.3 Engineering Concepts in Sensory Rehabilitation 2388
107.4 Engineering Concepts in Motor Rehabilitation 2390
107.5 Engineering Concepts in Communications Disorders 2391
107.6 Appropriate Technology 2392
107.7 An Example of Rehabilitation Engineering 2392
107.8 The Future of Electrical Engineering in Rehabilitation 2395

Charles J. Robinson
University of Pittsburgh

107.1 Introduction

Rehabilitation engineering is the application of science and technology to ameliorate the handicaps of individuals with disabilities [Reswick, 1982]. Note that these individuals have a **disability,** but the source of the **handicap** might well reside elsewhere. For instance, steps are a handicapping barrier to a person in a wheelchair, while an appropriately designed ramp removes this handicap.

The human sensory and motor (movement) systems are marvelously engineered, both within a given system and integrated across systems. Thus the rehabilitation engineer faces a daunting task in trying to design augmentative or replacement systems. Engineering advances have resulted in enormous strides in the field of rehabilitation. The blind can be given "sight," the deaf provided with a sense of their surroundings, the mute aided to communicate again, and those without full control of a limb (or with the limb missing) can regain the lost function by artificial means. But the present level of available functional restoration still pales in comparison to the capabilities of nondisabled individuals.

There is a core body of knowledge that defines an electrical engineer. Biomedical engineering is less precisely defined (even if it involves electrical engineering), but in general a biomedical engineer must be proficient not only in a traditional engineering discipline but must also have a working knowledge of things biological or medical. The rehabilitation engineer must not only be technically proficient as an engineer and know biology and medicine, but must also integrate artistic, social, financial, psychological, and physiological considerations to develop a device, technique, or concept that meets the needs of the disabled population which the engineer is serving.

Imagine being the design engineer on a project that has an unknown, highly nonlinear plant, with coefficients whose variations in time appear to follow no known or solvable model, where time (yours and your client's) and funding are severely limited, and where no known solution has been developed (or if it has, will need modification for nearly every client so no economy of scale exists). Further, there will be severe impedance mismatches between available appliances and your client's needs. Or, the low residual channel capacity of one of your client's senses will require enormous signal compression to get a signal with any appreciable information content through it. Welcome to the world of the rehabilitation engineer!

107.2 Rehabilitation Concepts

Rehabilitation engineers generally work in a team setting in collaboration with physical and occupational therapists, orthopedic surgeons, physical medicine specialists and/or neurologists. Some rehabilitation engineers are interested in certain activities that we do in the course of a normal day that could be summarized as **activities of daily living (ADL)**. These include eating, toileting, combing hair, brushing teeth, and reading. Other engineers focus on *mobility* and the limitations to mobility. Mobility can be personal (e.g., within a home or office) or public (automobile, public transportation, accessibility questions in buildings). Mobility also includes the ability to move functionally through the environment. Thus, the question of mobility is not limited to that of getting from place to place, but also includes such questions as whether one can reach an object in a particular setting or whether a paralyzed urinary bladder can be made functional again. Barriers that limit mobility are also studied. For instance, an ill-fitted wheelchair cushion or support system will most assuredly limit mobility by reducing the time that an individual can spend in a wheelchair before he or she must vacate it to avoid serious and difficult-to-heal pressure sores. Other groups of rehabilitation engineers deal with *sensory disabilities*, such as sight or hearing, or with *communications disorders*, both in the production side (e.g., the nonvocal) or in the comprehension side. For any given client, a rehabilitation engineer might have all these concerns to consider (i.e., ADLs, mobility, sensory and communication dysfunctions).

A key concept in physical or sensory rehabilitation is that of **residual function or residual capacity.** Such a concept implies that the function or sense can be quantified, that the performance range of that function or sense is known in a nonimpaired population, and that the use of residual capacity by a disabled individual should be encouraged. These measures of human performance can be made subjectively by clinicians or objectively by some rather clever computerized test devices.

A rehabilitation engineer asks three key questions: Can a diminished function or sense be successfully augmented? Is there a substitute way to return the function or to restore a sense? Is the solution appropriate and cost-effective? These questions give rise to two important rehabilitation concepts: orthotics and prosthetics. An **orthosis** is an appliance that aids an existing function. A **prosthesis** provides a substitute.

An artificial limb is a prosthesis, as is a wheelchair. An ankle brace is an orthosis. So are eyeglasses. In fact, eyeglasses might well be the ultimate rehabilitation device. They are inexpensive, have little social stigma, and are almost completely unobtrusive to the user. They have let many millions of individuals with correctable vision problems lead productive lives. In essence, however, a pair of eyeglasses is an optical device, governed by traditional equations of physical optics. Eyeglasses can be made out of simple glass (from a raw material as abundant as the sands of the earth) or complex plastics such as those that are ultraviolet-sensitive. They can be ground by hand or by sophisticated computer-controlled optical grinders. Thus, crude technology can restore functional vision. Increasing the technical content of the eyeglasses (either by material or manufacturing method) in most cases will not increase the amount of function restored, but it might make the glasses cheaper, lighter, and more prone to be used.

107.3 Engineering Concepts in Sensory Rehabilitation

Of the five traditional senses, vision and hearing most define the interactions that permit us to be human. These two senses are the main input channel through which data with high information content can flow. We read; we listen to speech or music; we view art. A loss of one or the other of these senses (or both) can have a devastating impact on the individual affected. Rehabilitation engineers attempt to restore the functions of these senses either through augmentation or via sensory substitution systems. Eyeglasses and hearing aids are examples of augmentative devices that can be used if some residual capacity remains. A major area of rehabilitation engineering research deals with *sensory substitution systems* [Kaczmarek *et al.*, 1991].

The visual system has the capability to detect a single photon of light yet also has a dynamic range that can respond to intensities many orders of magnitude greater. It can work with high-contrast items and with those of almost no contrast, and across the visible spectrum of colors. Millions of parallel data channels form the optic nerve that comes from an eye; each channel transmits an asynchronous and quasi-random (in time) stream of binary pulses. While the temporal coding on any one of these channels is not fast (on the order of 200 bits/s or less), the capacity of the human brain to parallel process the entire image is faster than any supercomputer yet built.

If sight is lost, how can it be replaced? A simple pair of eyeglasses will not work, since either the sensor (the retina), the communication channel (the optic nerve and all its relays to the brain), or one or more essential central processors (the occipital part of the cerebral cortex for initial processing; the parietal and other cortical areas for information extraction) has been damaged. For replacement within the system, one must determine where the visual system has failed and whether a stage of the system can be artificially bypassed. If one uses another sensory modality (e.g., touch or hearing) as an alternate input channel, one must determine whether there is sufficient bandwidth in that channel and whether the higher-order processing hierarchy is plastic enough to process information coming via a different route.

While the above discussion might seem just philosophical, it is more than that. We normally read printed text with our eyes. We recognize words from their (visual) letter combinations. We comprehend what we read via a mysterious processing in the parietal and temporal parts of the cerebral cortex. Could we perhaps read and comprehend this text or other forms of writing through our fingertips with an appropriate interface? The answer surprisingly is yes! And, the adaptation actually goes back to one of the earliest applications of coding theory—that of the development of Braille. Braille condenses all text characters to a raised matrix of 2 by 3 dots (2^6 combinations), with certain combinations reserved as indicators for the next character (such as a number indicator) or for special contractions. Trained readers of Braille can read over 250 words per minute of grade 2 Braille (as fast as most sighted readers can read printed text!). Thus, the Braille code is, in essence, a rehabilitation engineering concept where an alternate sensory channel is used as a substitute and where a recoding scheme has been employed.

Rehabilitation engineers and their colleagues have designed other ways to read text. To replace the retina as a sensor element, a modern high-resolution, high-sensitivity, fast-imaging sensor (CCD, etc.) is employed to capture a visual image of the text. One method, used by the Kurzweil reading machine, converts the scanned image to text by using optical character recognition schemes and then outputs the text as speech via text-to-speech algorithms. This machine essentially recites the text, much as a sighted helper might do when reading aloud to the blind individual. The user of the Kurzweil device is thus freed of the absolute need for a helper. Such *independence* is often the goal of rehabilitation.

Another device (the OPTICON) presents the pattern of text scanned by a small hand-held scanner to a matrix of vibrating pins on which the index finger rests [Bliss *et al.*, 1970]. This technique works at the input stage because the fingertip has a high density of vibration detectors and there is a good match of signal content to the spatial and temporal characteristics of the receiver. While this input mode is completely different from vision or hearing, some users can achieve good reading speeds. This implies that the central processor for text pattern content (i.e., comprehension) does not rely solely on visual neural pathways. A goal of rehabilitation is to find alternate means to perform a task.

Perhaps the most interesting method presents an image of the scanned data directly to the visual cortex via an array of implantable electrodes that are used to electrically activate nearby cortical structures. The visual cortex is laid out in a topographic fashion such that there is an orderly mapping of the signal from different parts of the retina to corresponding parts of the occipital cortex. The goal of stimulation is to mimic the neural activity that would have been evoked had the signal come through normal channels. And, such stimulation does produce the sensation of light. Since the "image" stays within the visual system, the rehabilitation solution is said to be **modality-specific**. However, substantial problems remain in the design of the electrode arrays that serve as the interface between the electronics and neurological tissue in this application, and the surgery required is very invasive.

Deafness is another manifestation of a loss of a communication channel, this time for the sense of hearing. Hearing aids are now commercially available that can adaptively filter out background noise (a predictable signal) while amplifying speech (unpredictable) using autoregressive, moving average (ARMA) signal processing. Totally deaf individuals use vision as a substitute input channel when communicating via sign language (also a substitute code) and can sign at information rates that match or exceed those of verbal communication.

Deafness is often brought on (or occurs congenitally) by damage to the cochlea. The cochlea normally transduces variations in sound pressure intensity at a given frequency into patterns of neural discharge. This neural code is then carried by the auditory (eighth cranial) nerve to the brainstem where it is preprocessed and relayed to the auditory cortex for initial processing and on to the parietal and other cortical areas for information extraction. Similar to the case for the visual system, the cochlea, auditory nerve, auditory cortex, and all relays in-between maintain a topological map, this time based on tone frequency (tonotopic).

If deafness is solely due to cochlear damage (as is often the case) and if the auditory nerve is still intact, a newly commercialized system called a cochlear implant can substitute for the regular transducer array (the cochlea) while still sending the signal through the normal auditory channel (to maintain modality specificity). A few individuals with cochlear implants have been able to comprehend speech sent over a telephone, but most users simply gain a crude, but needed, sense of their (auditory) environment. This application is discussed in greater detail in Section 107.7.

An exciting new development is occurring outside the field of rehabilitation that could have a profound impact on the ability of the deaf to comprehend speech. Electronics companies are beginning to market universal translation aids for travelers, where a phrase spoken in one language is captured, parsed, translated, and restated (either spoken or displayed) in another language. The deaf would simply require that the display be in the language that they use for writing.

107.4 Engineering Concepts in Motor Rehabilitation

Limitations in mobility can severely restrict the quality of life of an individual so affected. A wheelchair is a prime example of a prosthesis that can restore personal mobility to those who cannot walk. Given the proper environment (fairly level floors, roads, etc.), modern wheelchairs can be highly efficient. In fact, the fastest times in one of man's greatest tests of endurance, the Boston marathon, are achieved by the wheelchair racers. Although they do gain the advantage of being able to roll, they still must climb the same hills, and do so with only one-fifth the muscle power available to an able-bodied marathoner.

While a wheelchair user could certainly go down a set of steps (not recommended), climbing steps in a normal manual or electric wheelchair is a virtual impossibility. Ramps or lifts are engineered to provide accessibility in these cases, or special climbing wheelchairs can be purchased. Wheelchairs also do not work well on surfaces with high rolling resistance or viscous coefficients (e.g., mud, rough terrain), so alternate mobility aids must be found if access to these areas is to be provided to the physically disabled. Hand-controlled cars, vans, tractors, and even airplanes are now driven by wheelchair users. The design of appropriate control modifications falls to the rehabilitation engineer.

Loss of a limb can greatly impair functional activity. The engineering aspects of artificial limb design increase in complexity as the amount of residual limb decreases, especially if one or more joints are lost. As an example, a person with a mid-calf amputation could use a simple wooden stump to extend the leg and could ambulate reasonably well, but such a leg is not cosmetically appealing and completely ignores any substitution for ankle function.

Immediately following World War II, the U.S. government began the first concerted effort to foster better engineering design for artificial limbs. Dynamically lockable knee joints were designed for artificial limbs for above-knee amputees. In the ensuing years, energy-storing artificial ankles have been designed, some with prosthetic feet so realistic that beach thongs could be worn with them! Artificial hands, wrists, and elbows were designed for upper-limb amputees. Careful design of the actuating cable system also provided for a sense of hand grip force so that the user had some feedback and did not need to rely on vision alone for guidance.

Perhaps the most transparent (to the user) artificial arms are the ones that use electrical activity generated by the muscles remaining in the stump to control the actions of the elbow, wrist, and hand [Stein *et al.*, 1988]. This electrical activity is known as myoelectricity and is produced as the muscle contraction spreads through the muscle. Note that these muscles, if intact, would have controlled at least one of these joints (e.g., the biceps and triceps for the elbow). Thus, a high level of modality specificity is maintained since the functional element is substituted only at the last stage. All the batteries, sensor electrodes, amplifiers, motor actuators, and controllers (generally analog) reside entirely within these myoelectric arms. An individual trained in the use of a myoelectric arm can perform some impressive tasks with this arm. Current engineering research efforts involve the control of simultaneous multijoint movements (rather than the single-joint movement now available) and the provision for sensory feedback from the end effector of the artificial arm to the skin of the stump via electrical means.

107.5 Engineering Concepts in Communications Disorders

Speech is a uniquely human means of interpersonal communication. Problems that affect speech can occur at the initial transducer (the larynx) or at other areas of the vocal tract. They can be of neurological (due to cortical, brainstem, or peripheral nerve damage), structural, and/or cognitive origin. A person might only be able to make a halting attempt at talking or might not have sufficient control of other motor skills to type or write.

If only the larynx is involved, an externally applied artificial larynx can be used to generate a resonant column of air that can be modulated by other elements in the vocal tract. If other motor skills are intact, typing can be used to generate text, which in turn can be spoken via text-to-speech devices described in Section 107.3. The rate of typing (either whole words or via coding) might be fast enough so that reasonable speech rates could be achieved.

The rehabilitation engineer often becomes involved in the design or specification of *augmentative communication aids* for individuals who do not have good muscle control, either for speech or for limb movement. A whole industry has developed around the design of symbol or letter boards, where the user can point out (often painstakingly) letters, words, or concepts. Some of these boards now have speech output. Linguistics and information theory have been combined in the invention of predictive spellers, where a letter or two selected gives rise to hints of the word desired. Such hints are based on sentence structure and increasingly on content.

Some individuals can produce speech, but it is dysarthric and very hard to understand. Yet the utterance does contain information. Can this limited information be used to figure out what the individual wanted to say and then voice it by artificial means? Research labs are now employing neural network theory to determine which pauses in an utterance are due to content (i.e., between a word or sentence) and which are due to unwanted halts in speech production.

107.6 Appropriate Technology

Rehabilitation engineering lies at the interface of a wide variety of technical, biological, and other concerns. A user might (and often does) put aside a technically sophisticated rehabilitation device in favor of a simpler device that is cheaper and easier to use and maintain. The cosmetic appearance of the device (or cosmesis) sometimes becomes the overriding factor in acceptance or rejection of a device. A key design factor often lies in the use of the **appropriate technology** to accomplish the task adequately given the extent of the resources available to solve the problem and the residual capacity of the client. Adequacy can be verified by determining that increasing the technical content of the solution results in disproportionately diminishing gains or escalating costs. Thus, a rehabilitation engineer must be able to distinguish applications where high technology is required from those where such technology results in an incremental gain in cost, durability, acceptance, and other factors. Further, appropriateness very much depends on location. What is appropriate to a client near a major medical center in a highly developed country might not be appropriate to one in a rural setting or in a developing country.

This is not to say that rehabilitation engineers should shun advances in technology. In fact, a fair proportion of rehabilitation engineers work in a research setting where state-of-the-art technology is being applied to the needs of the disabled. However, it is often difficult to transfer complex technology from a laboratory to disabled consumers not directly associated with that laboratory. Such devices are often designed for use only in a structured environment, are difficult to repair properly in the field, and often require a high level of user interaction or sophistication.

Technology transfer in the rehabilitation arena is difficult due to the limited and fragmented market. Advances in rehabilitation engineering are often piggybacked onto advances in commercial electronics. For instance, the exciting developments in text-to-speech and speech-to-text devices mentioned earlier are being driven by the commercial marketplace, not by the rehabilitation arena. But such developments will be welcomed by rehabilitation engineers no less.

107.7 An Example of Rehabilitation Engineering

In this short article, it is difficult to convey just how much engineering actually can go into the design of a rehabilitation device. Clearly, mechanical and materials engineering must play an important role in the design of modern orthopedic aids (seating surfaces, artificial limbs or joints, wheelchairs, braces, etc.), but the influence of electrical and electronics engineering (as opposed to technology) is often more subtle. For instance, there are many rehabilitation applications that use computer technology, but very few of these actually involve applying engineering principles to the rehabilitation problem itself. There are, however, classes of rehabilitation problems where engineering techniques are essential to the development of an appropriate solution. One of these classes involves the use of residual neuronal pathways to restore a sense lost by injury, disease, or genetic deficit.

A good example of this latter class is the cochlear implant, which was briefly described in a preceding section. At first glance, the design of a cochlear prosthesis to restore hearing appears daunting. The hearing range of a healthy young individual is 20 to 16,000 Hz. The transducing structure, the cochlea, has 3500 inner and 12,000 outer hair cells, each best activated by a specific frequency that causes a localized mechanical resonance in the basilar membrane of the cochlea [see Fig. 107.1(A)]. Deflection of a hair cell causes the cell to fire an all-or-none (i.e., pulsatile) neuronal discharge whose rate of repetition depends to a first approximation on the amplitude of the stimulus. The outputs of these hair cells have an orderly convergence on the 30,000 to 40,000 fibers that make up the auditory portion of the eighth cranial nerve. These afferent fibers in turn go to brainstem neurons that process and relay the signals on to higher brain centers [Klinke, 1983]. For many causes of deafness, the hair cells are destroyed, but the eighth nerve remains intact. Thus, if one

A. Normal Transduction

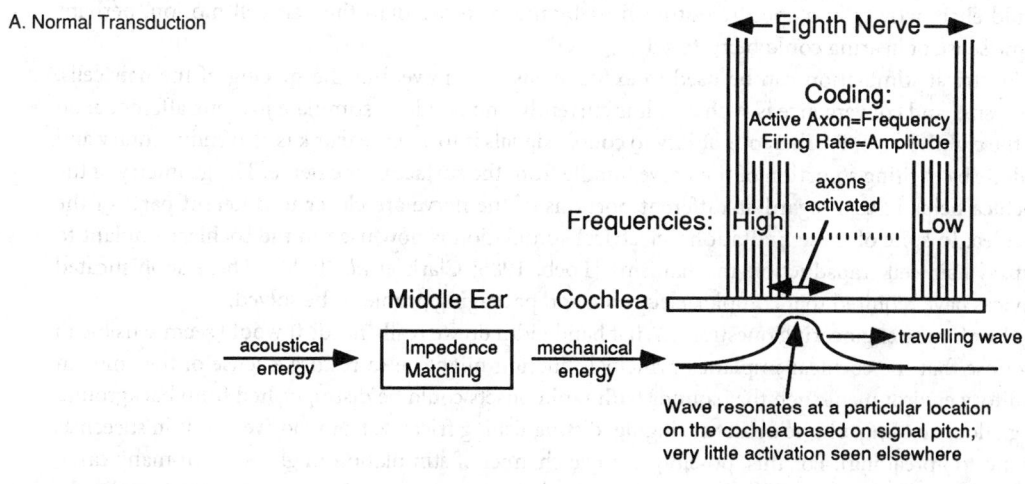

B. Single Monopolar
   Implanted Electrode

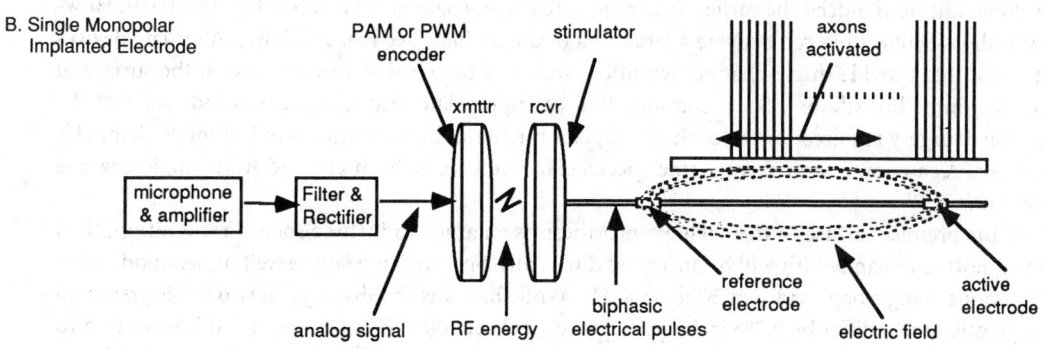

C. Multichannel Implant
   with Bipolar Electrodes

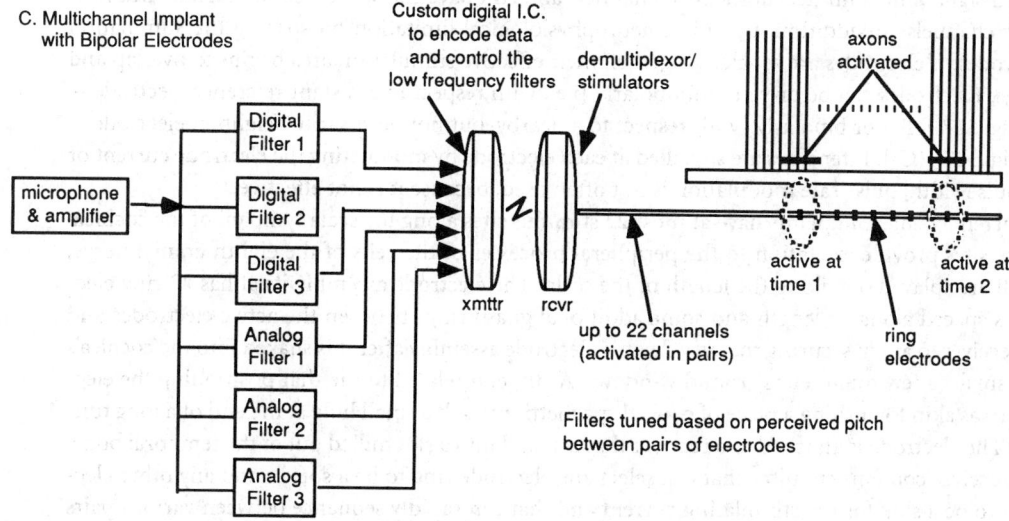

FIGURE 107.1 (A) Mechanisms of normal hearing, and electrical stimulation via (B) a single- or (C) multichannel cochlear implant.

could elicit activity in a specific output fiber by means other than the hair cell motion, perhaps some sense of hearing could be restored.

Electrical stimulation can be used to excite axons in a nerve, but the spacing of the hair cells, their size, and convergence is such that it is currently impossible to stimulate just one afferent axon in the eighth nerve. Further, our ability to couple signals into a nerve trunk is also rudimentary and is done by exciting a portion of the nerve bundle from the surface of the nerve. The geometry of the cochlea helps in this regard as different portions of the nerve are closer to different parts of the cochlea. In spite of these limitations, electrical stimulation is now used in the cochlear implant to bypass hair cell transduction mechanisms [Loeb, 1985; Clark *et al.*, 1990]. These sophisticated devices have required that complex electronic and packaging problems be solved.

Now for the engineering questions. What bandwidth do we really need? It would seem sensible to propose that the cochlear implant's principal function might be to restore a sense of the ambient auditory environment such that sounds with rapid onsets could be distinguished from background (e.g., doors closing, phone or alarm ringing, distinguishing fricative from plosive words in speech as an aid to lipreading). For this, possibly a single channel of stimulation might work in many cases, especially if the amplitude of the sound is encoded via some stimulation parameter (amplitude, pulsewidth, etc.). Indeed the earliest implants were single-channel devices [see Fig. 107.1(B)]. However, the stimulus used can activate a large area of the cochlea [see Fig. 107.1(B)]. Alternate devices do exist that provide single-channel sensation, such as a buzzer that vibrates against the surface of the sternum (breastbone). So, a consideration of appropriate technology could indicate that the invasive surgery required for the cochlear implant might not be warranted if only a single channel is desired. As a counter argument, some spectacular results have been claimed from single-channel devices.

If the premise for use of the cochlear prosthesis is enlarged to include speech recognition, then the question of bandwidth will be answered differently. Speech can be quite well understood over a telephone using a bandwidth of 30 to 3500 Hz. While humans can distinguish two tones presented sequentially that differ by 0.3% in frequency (e.g., 3 Hz at 1000 Hz), they require at least one-third of an octave difference (i.e., a critical bandwidth) to distinguish two pure tones presented simultaneously [Klinke, 1983]. This appears to occur because of the way energy travels down the cochlea [see Fig. 107.1(A)]. The human auditory range has about 24 such bands. Speech itself has energy focused in three principal frequency bands (formants). Thus, it might be possible to compress the speech signal into a limited number of channels and still have the speech be intelligible. But how many channels? In addition, there is a neurophysiological limitation on spacing the stimulation electrodes. If electrodes are set close together, their effective stimulation area begins to overlap and merge. Electrodes can be driven monopolarly [i.e., with respect to a distant reference electrode—see Fig. 107.1(B)] or bipolarly [with respect to a nearby, but not necessarily neighbor, electrode—see Fig. 107.1(C)]. Intensities are signalled at each electrode by modulating the electrode current or the pulsewidth; pulse rate modulation is not often used, because it is not effective.

Current cochlear implants have at most 22 stimulus sites along the scala tympani of the cochlea. Those sites provide excitation to the peripheral processes of the cells of the eighth cranial nerve, which are splayed out along the length of the scala. The electrode assembly itself has 22 ring electrodes spaced along its length and some additional guard rings between the active electrodes and the receiver to aid in securing the very flexible electrode assembly after it is snaked into the cochlea's very small (a few millimeters) round window. (A surgeon related to me that positioning the electrode was akin to pushing a piece of cooked spaghetti through a small hole at the end of a long tunnel.) The electrode is attached to a receiver that is inlaid into a slot milled out of the temporal bone. The receiver contains circuitry that can select any electrode ring to be a source and any other electrode to be a sink for the stimulating current and that can rapidly sequence between various pairs of electrodes [see Fig. 107.1(C)]. The receiver is powered and controlled by a radio-frequency link with an external transmitter whose alignment is maintained by means of a permanent magnet embedded in the receiver.

A digital signal processor stores information about a specific user and his or her optimal electrode locations for specific frequency bands. The signal frequencies are decomposed into bands by
means of analog and digital filters—the voicing frequency and the first and second formant frequencies are filtered digitally, while the frequencies above 2 kHz are filtered with three analog filters. The transmitter receives input from these tunable filters [see Fig. 107.1(C)]. But these filters
are assigned to particular electrodes after a very exhaustive testing program that attempts to determine the relative pitch perception produced by each electrode ring or pair. The object is to determine what pair of electrodes best produces the subjective perception of a certain pitch *in the
implanted individual himself or herself* and then to associate a particular filter with that pair via the
controller. Within the limits of threshold and saturation, the amplitude of the stimulus pulse delivered to the pair is determined by the energy in the filter band. The amplitudes are translated into
biphasic pulse amplitudes and widths that can be changed dynamically over a wide range as determined by the characteristics of the particular user. The pulse amplitudes can be varied from 15 μA
to 1.5 mA, the widths from 200 to 400 μs. Two schemes are used to determine pulse repetition frequency (PRF): the PRF can be set equal to the voicing frequency or it can be fixed at about 250 Hz.

An enormous amount of compression occurs in taking the frequency range necessary for speech
comprehension and reducing it to a few discrete channels. At present, the optimum compression
algorithm is unknown. What is amazing is that some totally deaf individuals (albeit a small minority of from 5 to 10% of users) can relearn to comprehend speech exceptionally well without
lipreading through the use of these implants. Other individuals find that the implant aids in
lipreading. For some, only an awareness of environmental sounds is apparent; for another group,
the implant appears to have had little effect. If you could (as I have been able to) finally converse in
unaided speech with an individual who had been rendered totally blind and deaf by a traumatic
brain injury, you begin to appreciate the power of rehabilitation engineering.

107.8 The Future of Electrical Engineering in Rehabilitation

Electrical engineering permeates many aspects of rehabilitation. Signal processing, control and
information theory, materials design, and computers are all in widespread use. Neural networks,
microfabrication, fuzzy logic, virtual reality, image processing, and other emerging electrical engineering tools are increasingly being applied. The challenge to rehabilitation engineers is to find
advances in *any* field, engineering or otherwise, that will aid their disabled clients. With our society's
increasing push into the information age, my bet is that electrical engineering advances should have
the greatest impact in the future on enhancing the quality of life of these disabled individuals.

Defining Terms

Activities of daily living (ADL): Personal activities that are done by almost everyone in the course
of a normal day, including eating, toileting, combing hair, brushing teeth, and reading. ADLs
are distinguished from hobbies and from work-related activities (e.g., typing).

Appropriate technology: The technology that will accomplish a task adequately given the
resources available. Adequacy can be verified by determining that increasing the technological content of the solution results in diminishing gains or increasing costs.

Disability: Lack of a competent ability to perform a particular function or functions in the manner in which these functions are normally performed, due to a physical, medical, or mental
incapacity or loss. Compare with handicap.

Handicap: A barrier to normal function, a disadvantage to the performance of a task. A
person might or might not be handicapped by a disability, or even need a disability to be
handicapped. Additionally, the handicap might be imposed from outside the disability.
Indeed, everyone is occasionally handicapped by a step that is too high, but people in
wheelchairs are handicapped by steps in general if they need to climb them.

Modality-specific: A task that is specific to a single sense or movement pattern.

Orthosis: A modality-specific appliance that aids the performance of a function or movement by augmenting or assisting the residual capabilities of that function or movement. An orthopedic brace is an orthosis.

Prosthesis: An appliance that substitutes for the loss of a particular function, generally by involving a different modality as an input and/or output channel. An artificial limb, a sensory substitution system, or an augmentative communication aid are prosthetic devices.

Rehabilitation engineering: The application of science and technology to ameliorate the handicaps of individuals with disabilities [Reswick, 1982].

Residual function or residual capacity: Residual function is a measure of the ability to carry out one or more general tasks using the methods normally used. Residual capacity is a measure of the ability to carry out these tasks using any means of performance. These residual measures are generally more subjective than other more quantifiable measures such as residual strength.

References

J.C. Bliss, M.H. Katcher, C.H. Rogers, and R.P. Shepard, "Optical-to-tactile image conversion for the blind." *IEEE Trans. Man-Machine Systems,* vol. MMS-11, pp. 58–65, 1970.

G.M. Clark, Y.C. Tong, and J.F. Patrick, *Cochlear Prostheses,* Edinburgh: Churchill Livingstone, 1990.

K.A. Kaczmarek, J.G. Webster, P. Bach-y-Rita, and W.J. Tompkins, "Electrotactile and vibrotactile displays for sensory substitution," *IEEE Trans. Biomed. Eng.,* vol. 38, pp. 1–16, 1991.

R. Klinke, "Physiology of the sense of equilibrium, hearing and speech," in *Human Physiology,* R.F. Schmidt and G. Thews, Eds., Berlin: Springer-Verlag, 1983, Chap. 12.

G.E. Loeb, "The functional replacement of the ear," *Scientific American,* vol. 252, pp. 104–111, 1985.

J. Reswick, "What is a rehabilitation engineer?" *Annual Review of Rehabilitation,* vol. 2, 1982.

R.B. Stein, D. Charles, and K.B. James, "Providing motor control for the handicapped: a fusion of modern neuroscience, bioengineering, and rehabilitation," in *Advances in Neurology,* vol. 47, *Functional Recovery in Neurological Disease,* S.G. Waxman, Ed., New York: Raven Press, 1988.

Further Information

Readers interested in rehabilitation engineering can contact RESNA, an interdisciplinary association for the advancement of rehabilitation and assistive technologies, 1101 Connecticut Ave., N.W., Suite 700, Washington, DC 20036.

The U.S. Department of Veterans Affairs puts out a quarterly, *Journal of Rehabilitation R&D.* The January issue of each year contains an overview of most of the rehabilitation engineering efforts occurring in the United States and Canada, with over 500 listings.

The IEEE Engineering in Medicine and Biology Society has begun publishing the *IEEE Transactions on Rehabilitation Engineering,* a quarterly journal. The reader should contact the IEEE for further details.

108

Biocomputing

Luis Kun
Cedars-Sinai Medical Center

Matthew F. Baretich
University of Colorado

108.1 Clinical Information Systems 2397
Computer-Based Record • Clinical Information Standards •
Bedside Terminals/Point-of-Care Systems • Imaging and the
CIS • Systems Integration • Smart/Optical Cards
108.2 Hospital Information Systems 2405
The Clinical Environment • Healthcare Codes and Standards

108.1 Clinical Information Systems

Luis Kun

The main objective of this section is to provide the reader with a summary of areas that relate to clinical information systems. Since this field is so wide, the following topics will be covered mainly because of their importance within the field of medical informatics and the impact that these areas will have in healthcare delivery in the near future. At the end of this section there is a list of definitions that should help the reader not used to related acronyms and a list of suggested bibliographic references which should allow those interested to further increase their knowledge.

Computer-Based Record

Besides improvements in patient care, enhancing the productivity of physicians, nurses, and all healthcare-related personnel is very high on the agenda of all hospitals. Hospitals, clinics, HMOs, doctors' offices, emergency care centers, group practices, laboratories, radiology clinics, and nursing homes among others have a need to share patients' records. Aside from the direction that all of these medical-related centers will have with a required connection to the insurance companies/agencies to speed up payments and their accuracy, the growing need is to have the ability to transfer patients' medical files electronically anywhere in the world. As medical centers become more competitive, they will become worldwide centers of excellence for their given specialties. In turn then, their services will be marketed to the entire world population, becoming true global resources.

The trend of converting hospitals into "paperless hospitals" is becoming one of the most important topics of the 1990s. In 1970, chartered by the National Academy of Sciences, the Institute of Medicine working under the Policy Matters for Public Health has actively pursued the creation of a computer-based record (CBR). In July of 1991 a book was published by the Institute of Medicine in regards to the CBR. The requirements to compile an all-digital medical record (ADMR) will require ways to combine data, graphics, voice, signals, and images, both clinical and document.

The architecture that will accommodate all these forms of information for capturing, storing, communicating, and displaying is extremely complex. Some of the technologies involved include optical fibers, LANs, compact/optical disks, bedside terminals, medical image display stations, image diagnostic workstations, and picture archival and communications systems to name a few.

The High Performance Computing and Communications Initiative (HPCCI) was signed into law in December of 1991. Although most of the emphasis for this initiative was from a research and academic sense, some of the true practical values of these highways of information will occur at the clinical level. While advances are taking place in different parts of the world in fighting diseases such as cancer, AIDS, heart disease, cystic fibrosis, Alzheimer's, Parkinson's, Gaucher's, and malignant hyperthermia, not sharing the knowledge learned by all the groups would be a terrible underutilization of extremely costly resources, causing duplication of effort and enormous waste of time and resources.

The four technologies that have been considered critical by the National Institutes of Health for the coming years are molecular medicine, vaccine development, structural biology, and biotechnology. The four will greatly be affected by the HPCCI. Finally, the integration of all medical-related information will be the most complex task that the healthcare arena will face this decade.

Clinical Information Standards

One of the most demanding and key areas for successfully integrating the hospital information system (HIS) with the clinical information systems (CIS) from multiple clinical departments and/or clinical areas deals with clinical information standards. Two of the driving forces behind the automation of the patient record deal with national concerns related to healthcare costs and quality of healthcare. These concerns have generated demand for managed care. The automated patient record could then be one of the vehicles to achieve managed care.

Clinical information standards are constantly evolving. They were developed (some are still in the process of development; e.g., IEEE/MIB P1073) by very different sets of requirements. What follows is a brief description and structure of most of these standards.

Communications/Storage (e.g., HL/7, IEEE/MEDIX P1157, ANSI ASC X12, ACR/NEMA, IEEE/MIB P1073)

The HL/7 standards group aimed to define vendor-independent communications standards among components of hospital information systems. The IEEE, ANSI, ACR/NEMA, and ASTM have been very active in creating standards through subcommittees from organizations within. As an example, ASTM has the following Healthcare Automation Committees (E31.XX):

- E31.10: Computer automation in the Hospital Pharmacy
- E31.11: Data exchange standards for Clinical Laboratory results
- E31.12: Medical Informatics
- E31.13: Clinical Laboratory Systems
- E31.14: Clinical Laboratory Instrument Interface
- E31.15: Health Knowledge Representation

The MEDIX mission was to establish a robust and flexible communications standard for the exchange of data between heterogeneous healthcare information systems. The MIB was created mainly to allow the exchange of data from medical instrumentation, e.g., monitoring devices and hospital information systems. Many of the manufacturers of these devices have proprietary hardware, e.g., buses and/or software, which complicates this exchange. Bedside terminals in the intensive care environment will benefit immensely from such a standard, since most hospitals' ICUs and CCUs have many vendors' equipment in their units. To effectively integrate and manage the data are major goals of the MIB.

Classification/Reimbursement (e.g., ICD, DRG, SNOMED, CPT, DSM, RCS, UMLS)

ICDs were originally used for public health morbidity statistics; now in the United States they are primarily used for reimbursement. Its structure is numbered classification of diseases grouped by anatomical areas. The DRGs facilitate the definition of case-mix for hospital reimbursement. Its structure is multi-axial: severity of illness, prognosis, treatment difficulty, need for intervention, and resource intensity. SNOMED provides description of pathological tests related to patient identification. It has four axes: function (primary symptoms), etiology (cause of disease), morphology (description of disease form), and topology (area of body). CPT is primarily used for reimbursement and utilization review. It derives codes from specialty nomenclatures divided into chapters: systemic (medicine, anesthesia, etc.), topological (cardiovascular, lymphatic, etc.), and technological (radiology, laboratory, etc.). DSM provides consistent abbreviations for prescription and administrative use. It facilitates psychiatric education and research. Its structure is multi-axial: clinical syndromes, developmental and personality disorders, physical disorder, severity psychological stresses, and global assessment functioning. RCS is a comprehensive nomenclature and classification of medical terms for computerized records. UMLS facilitates the unification of clinical data classification systems into a single unified medical language system. It will also facilitate the creation of data into compatible automated patient record systems. Its structure reconciles clinical terminology, semantics, and formats of the major clinical coding and reference systems.

Knowledge (e.g., ARDEN SYNTAX)

The ARDEN SYNTAX is a standard for sharing medical knowledge bases in the form of medical logic modules (MLM). Its structure is derived from the HELP (LDS Hospital) and the CARE (Regenstrief MC) systems. The MLMs accommodate alerts, management critiques, therapy suggestions, diagnosis scoring, etc. Each MLM is limited to the knowledge to make a single decision.

HCFA (e.g., UCDS, WARP, UHDDS)

UCDS provides an electronic clinical data set that Medicare can use to perform clinical quality reviews. The quality evaluation is done by using algorithms related to surgical procedures, disease specific, organ specific, discharge status and disposition, etc. The UCDS permits the hospital to enter the data into a personal computer; then this information can be sent electronically to the HCFA. WARP provides an epidemiologic approach to quality assurance. It hopes to overcome about 50% of ICD miscoding and its initial focus is on ambulatory chart review rather than real-time patient care. It is not a diagnostic or procedural classification system. It basically provides a model for encoding clinical information. It is an object-oriented case tool. UHDDS was created for studies on quality of care and fraud. It is also used for auditing Medicare and Medicaid subsystems.

Bedside Terminals/Point-of-Care Systems

Patient information is generated on an ongoing basis, wherever the patient may be. Almost two decades ago with the creation of the first programmable calculators, a trend started in terms of calculating hemodynamic variables in the OR, etc. This approach was improved with the creation of personal computers, ending with the development of what are now called bedside terminals. Companies such as Clinicom, Emtek, Hewlett-Packard, Hospitronics, and Spacelabs offer systems that can go from doing simply patient monitoring, to a complete data acquisition, data management, and data analysis system that incorporates in some cases diagnosis and treatment therapy.

From the patients' point of view, it is critical to integrate their demographic information with their clinical data. Usually the HIS contains all the ADT, orders, laboratory, pharmacy, etc. while the CIS may be more of a departmental system such as ICU/CCU, which contains hemodynamic variables, i.e., blood pressure, stroke volume, heart rate, etc. Both systems need to coexist. Point-of-care systems, many times known as bedside terminals, include both general med/surgery and the

ICU/CCU type. The general type include functions such as patient assessment, nursing diagnosis, patient care plans, kardex, discharge planning, discharge summary, medication administration record, I/O, vital signs, activities of daily living, patient classification/acuity, etc. The ICU/CCU systems in addition contain information regarding drug administration, fluid analysis, hemodynamic analysis (i.e., blood gas report, ECG, blood pressures, pulse oximeters, cardiac output), respiratory analysis (i.e., ventilator data, O_2/CO_2 analyzer), and real-time monitoring. Today's trends are incorporating imaging devices in both at the regular nursing stations, at the operating rooms, and at the recovery room/ICU/CCU. The motivation is to incorporate all patients' information and have it available wherever they may be. As a patient moves from a regular bed to the OR, back to an ICU, and later to a regular nursing station, the electronic record follows the patient. The one big difference with paper charts is that the electronic record can be shared simultaneously within and outside the institution.

Having the ability to look at electronic images in all of these locations not only opens the doors for consultation within the institution but also with outside institutions and/or expert individuals.

Imaging and the CIS

Imaging plays two very important roles within the context of a computer-based record (CBR). Document imaging allows for all those records that exist today in storage for the medical records departments to be scanned and incorporated electronically with the rest of the patient's current records existing in the HIS and CIS. The second role is from the perspective of clinical images. Most imaging experts will call this PACS, which stands for picture archival and communications system and is mostly associated with the Radiology Department of the hospital. We can view clinical images as a form of data which can be generated in any department.

Some of these typical clinical departments utilizing clinical images are radiology, cardiology (e.g., echocardiography, fluoroscopic techniques, cine cameras, 3D modeling, gamma cameras), orthopedic surgery, plastic surgery, obstetrics/gynecology, laboratories (e.g., genetics, chromosome analysis, cytology, hematology, clinical chemistry, pathology, histology, electron microscope), maxillofacial clinics, sports medicine, and oncology (e.g., radiation therapy, chemotherapy), emergency rooms, intensive care units, etc.

There are five imaging modalities: x-ray, magnetic resonance imaging (MRI), computer tomography (CT), nuclear medicine (NM), and ultrasound (US). These modalities create images which are very different not only in medical terms but in their size and content. As a result, there are three main areas under PACS which are critical in succeeding with such systems: communications (i.e., network, transmission protocol, and image format), archiving (i.e., database and storage media), and image processing (i.e., display, user interface, and IP algorithms).

Systems Integration

As an example of systems integration in the emergency care environment (see Fig. 108.1), from an information-flow point of view we see the following:

1. Information coming and going to the HIS, e.g., laboratory, pharmacy, orders, etc.
2. Information going to outpatient clinics for referring services, admissions to the hospital, or even to the patient's physician at home.
3. In the emergency room, the utilization of an intensive care type of bedside terminal allowing data collection, analysis and management, and also the ability to view clinical images in the ER.
4. From a consulting point of view, the whole electronic patient record, under an integrated diagnostic system, allows for any (department) consulting physician within or outside the hospital to review the case.

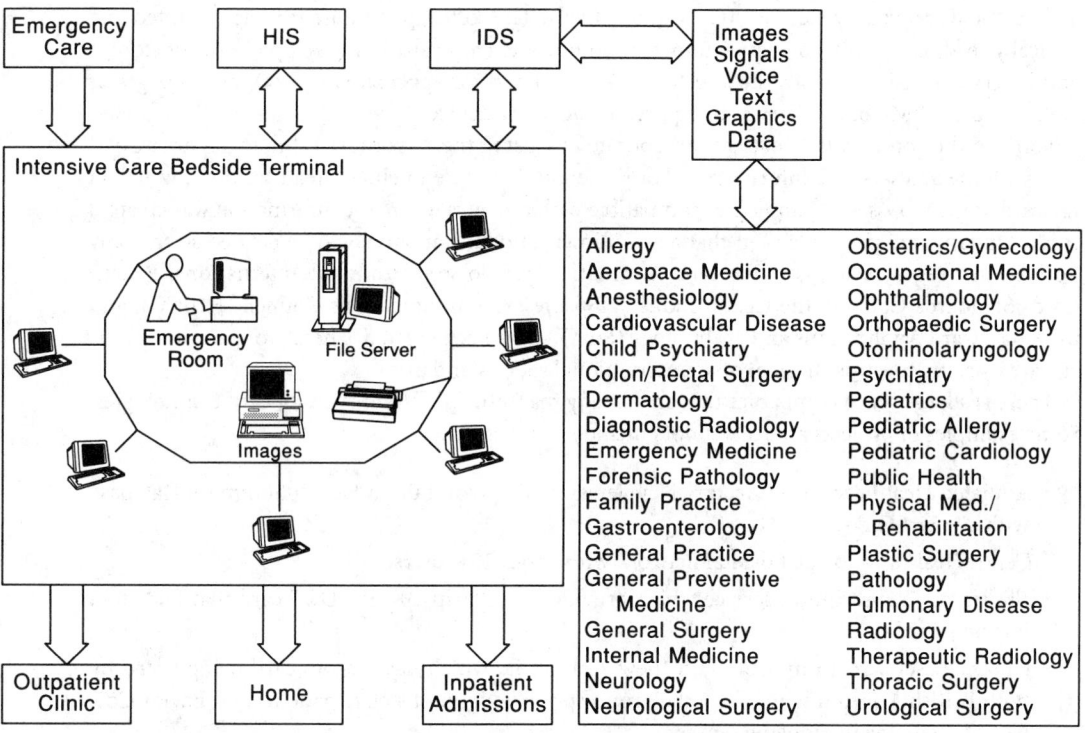

Allergy
Aerospace Medicine
Anesthesiology
Cardiovascular Disease
Child Psychiatry
Colon/Rectal Surgery
Dermatology
Diagnostic Radiology
Emergency Medicine
Forensic Pathology
Family Practice
Gastroenterology
General Practice
General Preventive
    Medicine
General Surgery
Internal Medicine
Neurology
Neurological Surgery

Obstetrics/Gynecology
Occupational Medicine
Ophthalmology
Orthopaedic Surgery
Otorhinolaryngology
Psychiatry
Pediatrics
Pediatric Allergy
Pediatric Cardiology
Public Health
Physical Med./
    Rehabilitation
Plastic Surgery
Pathology
Pulmonary Disease
Radiology
Therapeutic Radiology
Thoracic Surgery
Urological Surgery

FIGURE 108.1 An example of medical systems integration.

Smart/Optical Cards

Smart/optical cards provide a wide range of applications in the medical field. The patient, the provider (e.g., physician, dentist, etc.), the hospital, and the insurer can all benefit from such a card. The card will eventually contain all data forms—voice, text, graphics, clinical images, document images, signals, and data values collected from medical instrumentation. Besides patient identification/demographics, medical history, medications, allergies, and insurance verification, the system could contain the patient's picture, fingerprint, digital signature, voice signature, and even genetic/blood information for security reasons.

The patient is admitted and treatment is provided more quickly, historical information is more accurate, and personal physicians and specialists can be consulted more quickly. Less testing may be a direct result, and faster diagnosis is accomplished. Since information needs to be entered only once, patients do not need to rely on their memory, particularly in emergency situations.

The hospital identifies the patient and accesses all the medical records information from multiple departments more quickly. It needs fewer staff to find records from the hospital/clinics (even from other institutions), and this could reduce the length of stay.

The provider is better informed for a quicker diagnosis by getting all the available history at admission and can consult with the patient's personal physician and specialist by having their respective phone numbers. All prior records from the same or a different set of institutions coexist in the card. It also can reduce exposure to malpractice.

The insurer reduces fraudulent claims, reduces costs for data entry, and has more complete and accurate claims data. Also, by eliminating redundant tests costs are reduced.

Most of the cards can be classified into five groups by the type of technologies used: microfilm, magnetic strip, softstrip, chip, and laser/optical. Microfilm is hard to change and can be damaged by both temperature and humidity. Magnetic strip contains little information, approximately 2K, and can be destroyed by electric and magnetic fields. The softstrip, because it is laser printed and optically read, is difficult to change information on. The chip card has only up to 10K of storage and is very expensive. Finally, the laser/optical card allows for approximately 1000 typed pages or approximately 4 Mb of memory and requires a read/write device.

Some of the complexities that are incorporated by using these types of technologies are associated with the access to the information. For someone to be able to either "read" and/or "write" in the card, it must possess technologies compatible with the ones where the information was created. It is a fundamental principle then that a set of international standards will be created so that any hospital that requires access to the card information can do so. Already the International Patient Cards Standards Council, the Health Industry Business Communications Council (HIBCC), and the Smart Card Applications and Technology (SCAT) have been created. These groups, among others, are working towards the goal of an international set of standards.

There is a large set of companies that are already marketing different types of card technologies. Some examples of projects and/or vendors include:

- Affiliated Healthcare in Princeton, New Jersey, which maintains a Health Summary Database with a Smart Card.
- CentraHealth, a Florida hospital network with about 12K users.
- Clinicard, a subscription service which provides a softstrip, 3K, PC DOS card that folds to a business card size.
- Drexler LaserCard from Mountain View, California, which has 4.1-Mb card being tested by both British Telecom with a hospital group specializing in obstetric patients and Baylor College of Medicine in Houston, Texas.
- Eltrax in St. Paul, Minnesota, which is associated with several HIS manufacturers (Spectrum, McDonald Douglas, SMS, and Meditech) and provides a magnetic strip card with about 900-character capacity.
- IMSG/INFODYNE from Englewood, Colorado, which has a medical information card on magnetic strip carrying up to 600 characters.
- IntelliScan from American Medical Data Corp. in Atlanta, Georgia, which has information stored as 350 characters of readable text printed on the top of the card and up to 850 characters of detailed medical information optically encoded at the bottom. It is being used by hospitals in both Texas and Mississippi. The cards are customized to each hospital's database.
- Lifecard, early pilot (1985) card for electronic claims provided by Blue Cross/Shield of Maryland.
- Medfirst credit card, combining both medical information and financial credit, in test by Humana and Discovery Card.
- Medi-Card, a chip card from MediData Systems in Allston, Massachusetts.
- MedKey, from Biloxi Regional Medical Center, Biloxi, Mississippi.
- Medical Information Systems, in St. Louis, Missouri, which has a microfilm card that can contain up to 18 pages of information, including signals, text, images, data, and color photos.
- Ulticard, a 64K RAM memory chip in a credit card sized pack being tested in Houston at Baylor and Methodist hospitals.

Some of the cards are being tested in different countries. Sweden, the leader for about 20 years, has been using a patient card which is issued at birth by the government together with an ID number. Sweden has a socialized medicine program and it has been in their best interest to develop uniform standards so that the information can be accessed by every institution in the country. Belgium, Canada, France, Great Britain, Spain, and Switzerland all have several systems on trial.

Acronyms

ACR/NEMA: American College of Radiology/National Equipment Manufacturers Association
ANSI ASC X12: American National Standards Institute Accredited Standards Committee
ARDEN SYNTAX: Syntax for Medical Logic Modules
ASTM: American Society for Testing and Materials
CIS: Clinical Information System
CPT: Current Procedural Terminology
DRG: Diagnostic Related Group
DSM: Diagnostic and Statistical Manual of Mental Disorders
EDI: Electronic Data Interchange
HCFA: Healthcare Financing Administration
HIS: Hospital Information System
HL/7: Health Level/7
ICD: International Classification of Diseases
IDS: Integrated Diagnostic System
IEEE: Institute of Electrical and Electronics Engineers
IEEE/MEDIX P1157: Medical Data Interchange
IEEE/MIB P1073: Medical Information Bus
OSI: Open Systems Interconnection
RCS: Read Classification System
SNOMED: Systemized Nomenclature of Medicine
UCDS: Uniform Clinical Data Set
UHDDS: Uniform Hospital Discharge Data Set
UMLS: Unified Medical Language System
WARP: Wisconsin Ambulatory Review Project

References

Bedside Terminals/Point-of-Care

W. Donovan and S. Corrales, *The Book on Bedside Computing*, Long Beach, Calif.: Inside Healthcare Computing, 1991.

L. Kun, "The use of a personal computer for patient-condition-treatment in a CUU/ICU environment," *IEEE Transactions on Biomedical Engineering*, vol. BME-30, no. 8, August 1983.

L. Kun, "Rapid assessment of hemodynamic cardiorespiratory function for the critically ill with a personal computer," *IEEE Transactions on Biomedical Engineering*, vol. BME-30, no. 8, August 1983.

D. O'Boyle, G. Feiherr, and R. Gough, *The Buyer's Guide to Bedside Computer Systems*, Rockville, Md.: National Report of Computers & Health, 1991.

M.M. Shabot *et al.*, "Rapid bedside computation of cardiorespiratory variables with a programmable calculator," *Critical Care Med.*, vol. 5, p. 105, 1977.

Classification Systems Standards

C. Chute, "Tutorial 19: Clinical data representation," in *Proceedings of SCAMC 91*, November 1991.

B. Humphreys, *Building the Unified Medical Language System*, Bethseda, Md.: National Library of Medicine, 1989.

Communications Standards

J. Harrington, IEEE/EMBS P1158, "Medical Data Interchange (MEDIX) overview and status report," in *Proceedings of SCAMC 90*, November 1990.

National Electrical Manufacturers Association, "Digital Imaging and Communications," ACR-NEMA Standards Publication No. 300-1988, 1988.

R.E. Norden-Paul, IEEE Proposed Standard 1073, "Medical Information Bus: An Introduction and Progress Report," in *Proceedings of the 9th Annual Conference of the IEEE-EMBS*, vol. 2, MIB Symposium, Boston, pp. 1209–1211, 1987.

Knowledge Base Standards

"The ARDEN SYNTAX for medical logic modules," in *Proceedings of SCAMC 90*, November 1990.

"Emerging standards for medical logic," in *Proceedings of SCAMC 90*, November 1990.

Clinical Imaging/PACS

Y. Kim and F.A. Spelman, Eds., "Images of the twenty-first century," in *Proceedings of the Annual International IEEE-EMBS*, vol. 11, part 2, track 2, Imaging, pp. 345–630; Track 23, Picture Archiving and Communications Systems, pp. 775–793, 1989.

L. Kun, "Imaging and the clinical information system," in *Proceedings of '91 International Workshop on Medical Imaging*, Korea Institute of Science and Technology, Seoul, Korea.

Computerized Medical Record

M. Ball and M. Collin, Eds., *Aspects of the Computer-Based Patient Record*, New York: Springer-Verlag, 1992.

J. Blair, "Overview of clinical information representation and standard organization," in *Proceedings of the Fall 92 ECHO Meeting*, Palm Beach, Calif., 1992.

Institute of Medicine, *Computer-Based Patient Record*, Washington, D.C.: National Academy Press, 1991.

C.J. McDonald *et al.*, "The benefits of automated medical record systems for ambulatory care," in *Proceedings of the Computer Applications in Medical Care Conference*, New York: IEEE Computer Society, pp. 157–171, October 1986.

W.W. Stead *et al.*, "Practicing nephrology with a computerized medical record," *Kidney Int.*, vol. 24, pp. 446–454, 1983.

Q.E. Whiting-O'Keefe *et al.*, "A computerized summary medical record system can produce more information than the standard medical record," in *Proceedings of MedInfo '86*, Washington, D.C., 1986.

High-Performance Computing and Communications (HPCC)

D.A. Bromley, "The Federal High-Performance Computing Program," Washington, D.C.: Executive Office of the President, Office of Science and Technology Policy, 1989.

"National High-Performance Computer Technology Act," Congressional Record, U.S. Senate 101st Congress, First Session 5/18/89, Washington, D.C.

Smart/Optical Cards

Handbook of Optical Memory Systems. Bi-monthly updating service. Boston: Medical Records Institute.

Proceedings of the 13th Annual International Conference IEEE/EMBS, Track 21: Session 5, Medical Informatics V: Optical and Smart Cards, Orlando, Fla., pp. 1387–1392, October 1991.

1989 Smart Card Industry Directory, Palo Alto, Calif.: Palo Alto Management Inc., 1989.

108.2 Hospital Information Systems

Matthew F. Baretich

What does an electrical engineer need to know to be part of a team designing and implementing a hospital information system? For the most part, the necessary skills are those required to design and implement any comprehensive information system in a complex organization. Hospitals do, however, have unique characteristics that must be taken into account. These characteristics are described in the following pages.

The Clinical Environment

Hospitals are, indeed, complex organizations. They perform a vital function (patient care) but are subject to strict regulation and operate under severe financial constraints. Quality of patient care is the highest value, but a competitive marketplace demands efficient operation. Hospital information systems range from nonexistent to antique to state-of-the-art.

Hospitals are highly professionalized. Each professional group has a particular area of expertise and a unique perspective regarding the healthcare delivery system. Hospital administrators are much like administrators of other organizations. Recent graduates essentially have standard MBA (Master of Business Administration) degrees with some extent of healthcare specialization. However, many administrators in positions of authority received MHA (Master of Hospital Administration) degrees from programs more closely affiliated with medical schools than with business schools.

Hospitals also have large clinical staffs which include nurses and technologists (who are hospital employees) and medical doctors (who are usually not hospital employees). Clinicians are educated in the biological and medical sciences, and their preparation generally includes a large component of practical experience in the hospital as well as theoretical study in the classroom. As hospital employees, nurses and technologists (respiratory, laboratory, etc.) are part of the administrative structure of the hospital. Medical doctors (physicians and surgeons), on the other hand, are part of a separate medical staff structure that is largely independent of the hospital's administrative structure. However, medical doctors control the admission and discharge of the hospital's patients, and many hospital activities are the result of medical orders for patient services.

The number of hospital employees with an engineering background is limited. For the electrical engineer who is involved in the implementation of a hospital information system, hospital-based technical support may include an information systems department and a clinical engineering (or biomedical engineering) department.

The following aspects of the healthcare delivery system are worthy of study by an electrical engineer working in the clinical environment:

- The healthcare delivery system in the United States [Williams and Torrens, 1984]
- The organizational structure of hospitals [Goldberg and Buttaro, 1990]
- The characteristics of hospital information systems [Austin, 1988; Minard, 1991]

With this background information the electrical engineer will be better prepared to translate the concerns of hospital administrators and clinicians into the technical specifications of the hospital information system.

Healthcare Codes and Standards

The healthcare delivery system is a highly regulated industry. Numerous governmental and non-governmental organizations have established codes and standards intended to promote safe and effective patient care. Although there can be significant differences in the regulatory environment from one hospital to another, the major codes and standards are relatively uniform.

The **National Electrical Code** (NFPA 70), promulgated by the National Fire Protection Association (NFPA: Quincy, Massachusetts) applies to hospitals. Specifically, Article 517 deals with "Health Care Facilities." A more focused document, however, is the *Standard for Health Care Facilities* (NFPA 99). The most accessible format for this information is the NFPA's *Health Care Facilities Handbook* [Klein, 1990] which includes the full text of NFPA 99 as well as interpretive and explanatory material.

Many of the healthcare-related provisions of the electrical code are based on two concerns. First, many patients in surgery and intensive care depend on electrical equipment for life support. Such equipment ranges from heart-lung bypass devices to mechanical ventilators. Therefore, much attention is devoted to ensuring the availability of electrical power in the event that the primary power distribution system fails. A hospital information system that provides life-support functions may be subject to these provisions.

Second, because of the use of invasive medical procedures, many patients are considered to be "electrically susceptible." Under certain conditions, electrical currents on the order of microamperes can cause ventricular fibrillation, a potentially fatal disruption of normal cardiac function. Therefore, the NFPA and other organizations have established strict standards for grounding, "leakage" current, and other electrical parameters. These standards apply to devices and cabling in patient-care locations of the hospital.

The **Joint Commission on Accreditation of Healthcare Organizations** (Chicago, Illinois) is another major source of standards affecting hospitals. The JCAHO's *Accreditation Manual for Hospitals* [JCAHO, 1993] covers the entire spectrum of hospital activities. Pursuit of JCAHO accreditation is voluntary but, in practice, essentially all hospitals seek accreditation to ensure eligibility for reimbursement under certain governmental programs. At present, JCAHO standards include little reference to information systems. However, this is expected to change and, therefore, familiarity with the latest edition of the *Accreditation Manual for Hospitals* is advisable.

Another standard unique to the healthcare system is **Health Level 7** (HL7) which is a data communications protocol intended to facilitate the interfacing of various components in a hospital information system [Walker, 1989]. These components range from accounting systems (financial data) to clinical laboratory information systems (laboratory test results) to medical records systems (documentation of patient care services) to patient data management systems (physiological data). In the recent past, each such component was independent and generally incompatible with other components. However, to achieve high quality in patient care at the lowest cost, both administrators and clinicians need integrated, comprehensive access to a wide variety of information.

HL7 is an attempt to specify the types of data (and their formats) to be shared within a hospital information system. For example, if all components of the system use a common format for a patient's name, then it is possible for a single database query to gather all data regarding that patient. This also allows automation of certain activities such as billing (through the accounting system) for laboratory tests ordered by clinicians (through the clinical laboratory system). Unfortunately, HL7 has not achieved its promise but it does represent a significant step away from the chaotic past [Bond *et al.*, 1990].

Summary

The electrical engineer will be only one of many professionals involved in the implementation of a hospital information system. Successful participation in this team will depend on more than the electrical engineering skills that are applicable to any information system project. The critical success factor is an understanding of the hospital—the people (clinicians and administrators), their objectives (low cost and high quality), and the environment within which they work.

Defining Terms

HL7: A data communications protocol for interfacing components of a hospital information system.

JCAHO: The Joint Commission on Accreditation of Healthcare Organizations, an organization that promulgates standards affecting hospital operations.

NEC: The National Electrical Code, an NFPA standard that is commonly adopted by governmental units and, therefore, having the force of law.

NFPA: The National Fire Protection Association, an organization that promulgates standards affecting electrical systems in hospitals.

References

C.J. Austin, *Information Systems for Health Services Administration*, 3rd ed., Ann Arbor, Mich.: Health Administration Press, 1988.

V. Bond, J. Lenahan, and W. Wagner, "HL7: A practical perspective," *Healthcare Informatics*, vol. 7, no. 10, p. 46, 1990.

A.J. Goldberg and R.A. Buttaro, Eds., *Hospital Departmental Profiles*, 3rd ed., Chicago: American Hospital Publishing, 1990.

JCAHO, *Accreditation Manual for Hospitals*, 1993 ed., Chicago: Joint Commission on Accreditation of Healthcare Organizations, 1993.

B.R. Klein, Ed., *Health Care Facilities Handbook*, 3rd ed., Quincy, Mass.: National Fire Protection Association, 1990.

B. Minard, *Health Care Computer Systems for the 1990s*, Ann Arbor, Mich.: Health Administration Press, 1991.

J.M. Walker, "Integrating information systems with HL7," *Hospitals*, vol. 63, no. 13, p. FB60, 1989.

S.J. Williams and P.R. Torrens, *Introduction to Health Services*, 2nd ed., New York: John Wiley & Sons, 1984.

Further Information

Many of the major professional societies dealing with computer science and engineering have healthcare–related divisions. Further information can be obtained from each professional society.

The Healthcare Information and Management Systems Society is a division of the American Hospital Association that deals with information systems, telecommunications, and management engineering. For further information contact the American Hospital Association, Chicago, Illinois.

Major periodicals that focus on hospital information systems include *National Report on Computers and Health* and *Healthcare Informatics*. These publications, and other healthcare-related literature, can be found in the libraries of academic medical centers.

109

Safety and Risk-Control Issues

109.1 Introduction 2408
109.2 The Biomedical Equipment Risk Causes 2409
    Equipment Function • Equipment Malfunction • User Error
    • Utility Factor • Risk-Control Issues
109.3 Risk-Control Programs 2411
109.4 Conclusion 2411

Yadin David
Texas Children's Hospital

109.1 Introduction

The risk of injury, damage, or loss is present in all human activities. Consequently, the public is reminded about all sorts of risks, and the perception of these risks plays an important role in governmental legislation and in establishing organizations that control technical standards to reduce that risk.

To reduce the probability of the risk and to achieve a rational approach to risk control, a formal process is required that incorporates the analysis of the desired safety level, the designed method of protection, the acceptable probability threshold, any previous victims' experience, and the ramifications of the activity involved. This analysis is usually supported by a consensus of experts that define acceptable methods and quantifiable levels of risk control. There is a tendency to assume that technology itself is an overall threat to our health and safety. However, comparison of life expectancies in technologically developed countries to those in underdeveloped ones does not support this theory.

While biological effects associated with human-machine interface, like those associated with the exposure to radiation from utility power lines (high voltage and very low frequency) or with long exposure duration to video display terminals, were studied for some time, within the health-care delivery system that interface is very complex [1] and rapidly changing. In this environment, the margin of safety is the product of an interaction between smart facilities, equipment, materials, and a full range of human interventions. It is in this clinical environment that patients in various conditions, a busy staff, a temporary labor force, and the wide variant of technology converge. This interaction can lead to a degree of risk that reaches an unacceptable level when means for monitoring, controlling, and educating all of the entities involved are not supported and committed to by qualified professionals.

The purpose of this chapter is to familiarize readers with the issues involved with the safe use of medical equipment. Distinctions are made between various risks and the means to control them while equipment is deployed.

109.2 The Biomedical Equipment Risk Causes

The health-care delivery system, in general, and the practice of modern interventional medicine, in particular, are impacted by a continuous process of technological evolution. This process incorporates a wide variety of human-machine interfaces. From home-care equipment on one end of the spectrum to the transplantation of artificial organs on the other, changes in the state of the art are leading to an increased dependence on and the utilization of instrumentation in the daily delivery of patient care. Biomedical equipment is being utilized in many different ways. For example, the automatic implanted defibrillator is an invasive device having direct contact with the cardiovascular system; on the other hand, the infusion pump controls the rate of drugs infused into the body while being external to the body, without direct electrical contact with the patient; and still, the computed axial tomography scanner, while considered a noninvasive system, radiates ionized energy into the body during its search for clinical information.

Today's hospitals are feeling the pressure not only to provide a safe environment to their patients but also to protect their staff from various occupation-related hazards. Comprehensive risk-control programs are implemented that include **equipment management programs**.

The quest for safety and for sound scientific information about medical equipment-related risks and the various aspects of managing and controlling them has placed members of the industrial, equipment users/maintainers, regulatory, and academic communities at times on opposite sides of safety, especially those risks for which the hazard's magnitude cannot be measured directly [2]. In these cases risk assessors use assumptions to bridge gaps in knowledge.

Risk, the probability or likelihood that an adverse effect will occur [3], has one or more of the following causes.

Equipment Function

Harm can be caused by operating equipment that was developed without scientific knowledge available within the health-care or the scientific community. The health effect of "normal function" computer use was an issue of public debate last year. The potential hazards to which the operators of video display terminals (VDTs) were exposed was studied by the National Institute for Occupational Safety and Health (NIOSH); however, no significant difference of risk was found. The widespread use of gastric hypothermia in the early 1960s to cure ulcers medically and avoid surgical intervention is another example. Since clinical experience later demonstrated a failure to meet the therapeutic goal, the technology was abandoned [4].

Another potential source for unsafe conditions can be the result of a suboptimal equipment design. Design errors or suboptimal design are often close to each other. However, a failure to provide a panel guard on an instrument control panel, or a lock on an otherwise easily bumped switch, or poor labeling of a position of a critical switch can increase the risk level associated with that equipment function.

Equipment Malfunction

The litigation morass surrounding injuries and death due to faulty medical equipment obscures valuable safety-related data [5]. Equipment failure can be attributed to a single component, material, or a complete subsystem failure. Malfunction can be the result of manufacturing error, lack of or poor maintenance procedures, random unpredictable failure, packaging effect, inadequate environmental conditions such as temperature and humidity, or electromagnetic (EMI) or radiofrequency (RFI) interferences. The close proximity of many microprocessor-based **medical devices** at the patient bedside can contribute to the failure of life-support devices such as a ventilator due to an EMI radiated from another device such as a hospital's central clock controller. Similarly, an infant warmer unit can malfunction due to a RFI emitted from an electrocautery unit used in the same operating room.

Equipment can also malfunction due to tampering by either a trained or untrained operator.

Table 109.1   Experts' Ranking Factors of Device Failures

Contributing Factor	Percent Ranking It as #1 Cause
Improper use	64
Inadequate maintenance and repair	13
Faulty design	8
Improper labeling and instruction	5
Defective components	2
Other	8

Source: Adapted from "Federal Regulation of Medical Devices—Problems Still to Be Overcome," Report to the Congress of the United States, September 30, 1983.

User Error

Many medical devices present risks if they are not designed, set up, tested, applied, cleaned, or serviced properly. User error is the single most common cause of medical equipment malfunctions. It accounts for over 50% of medical equipment-related injuries. These errors result from inadequate training, lack of experience and supervision, and inadequate or unavailability of instruction manuals. They are reinforced by time and psychological pressures on staff and by the mismatch between equipment sophistication and users' skills.

Utility Factor

Power sources, such as electricity, compressed air, and medical gas supply, support most of the medical equipment in the hospital. To reduce the risk associated with a utility failure, hospitals are required to provide a local emergency source of power to avoid unsafe conditions. After power source switching or as a result of power transient, some medical devices will change to an unsafe condition because they will not automatically resume their last selected mode of operation. In such cases, users need to be instructed to manually reselect the equipment mode after such a switch.

In a report to the Congress of the United States by the Comptroller General [6], more than half (64%) of the medical device experts believed that improper use and inadequate maintenance and repair are the leading causes of device failures and patient injuries. A 1973 ECRI study found that 66% of the respondents believed that operator error was responsible for device failure [7]. Table 109.1 shows the experts' ranking of the most predominant factors contributing to medical device failures.

Risk-Control Issues

When health-care professionals such as physicians, nurses, engineers, and technicians apply medical technology in support of diagnosing or treating patients, the impact of an accident may have serious consequences, including injury, death, or property loss. The mechanism of injury or death associated with medical equipment accidents falls into one or more of the following categories:

- Overdose
- Embolism
- Electrocution (burn)
- Performance failure
- Suffocation
- Skin lesion
- Fire
- Mechanical impact

Research conducted at the Massachusetts General Hospital in Boston showed that in certain fields, such as anesthesiology, user errors account for over 70% of the failures analyzed [8]. User errors range from a mistake in the operation or application of equipment to serious errors in judgment affecting the care outcome of a particular patient. Therefore, if design engineers incorporate effective means for safe user interface and environmental compatibility into their equipment design, we all will benefit as hospitals will operate equipment that is safer.

109.3 Risk-Control Programs

Clinical risk management is an organized and systematic activity aimed at controlling and diminishing the probability of a harmful incident which otherwise may lead to the deterioration of care outcome, decline of staff performance, and worsening of the hospital environment [9]. In a successful implementation of a hospital-wide safety agenda, the equipment management program plays a critical role. An equipment management program continuously improves equipment performance, particularly as it relates to safety, quality, and cost of ownership. To reduce the probability of an incident several means of protection can be put in place. Throughout the medical equipment life cycle [10], from the early stage of equipment design, through manufacturing, hospital installation, users' credentialing, application, maintenance, and upgrades or relocation, provisions for safe equipment use should be emphasized. This is particularly important because medical equipment is interfacing with a variety of patients, each having changing biological conditions. This interface is unique to the hospital environment. It consists of the patient-machine, the user-machine, the accessories/transducer-machine, and the utilities-machine interfaces. Each one of these interfaces needs to be safe and controlled.

The regulatory community established mandatory as well as voluntary standards and recommended guidelines aimed at risk reduction. The Food and Drug Administration (FDA) with its Good Manufacturing Practice (GMP) and more recently with the **Safe Medical Devices Act (SMDA)** [11], the National Fire Protection Association (NFPA) with its Health Care Facility code [12], and the Joint Commission on Accreditation of Health Care Organizations (JCAHO) through its accreditation process [13] contributed to the availability and operation of safer medical equipment.

In order to control risk associated with the use of medical devices, an equipment management program should monitor equipment performance through the selection of appropriate indicators. Once the acceptable safety threshold for the indicators is determined, their variances need to be trended. The indicators themselves should reflect the risk intensity, and their variance should be measurable. A significant variance requires a corrective action. Once applied to a variance, the impact of the corrective action should be an improvement in the equipment performance and a good demonstration for the effectiveness of the equipment management program. Some indicators frequently used are the number of devices that failed to pass maintenance inspection, the number and intensity of equipment-related patient injury or death, the frequency and intensity of equipment that suffered physical abuse, mean time between failures, and the number of devices for which failure was reported when, however, no problem was found. The latter one may be used also as an indicator for users' training need, as the problem may be with untrained users. The equipment management program uses these indicators to adjust its planned maintenance schedule, to initiate users' training, to control cost of ownership, and to select [14] the next equipment to be acquired by the hospital.

109.4 Conclusion

The hospital presents a complex environment in which patients, various health-care professionals, equipment, and facility are interacting. In such an environment, proper attention must be devoted to the equipment's technical design, adequate instruction and users' training, reliable utility and stable environment, and maintenance and repair services. From design to the clinical application phase, the safe performance of medical equipment should be planned for and monitored. A comprehensive equipment management program [15] provides a systematic approach to risk control and is an important asset for providing safe clinical environments.

Defining Terms

Equipment hazard: A possible source of peril, danger, risk, or difficulty.

Equipment management program: An accountable and systematic approach to assuring that cost-effective and safe equipment is available to meet the demands of patient-care services.

Medical device: According to federal statute a medical device is "an instrument, apparatus, implement, machine, contrivance, implant, *in vitro* reagent, or other similar or related article, including any component, part, or accessory, which is (1) recognized in the official National Formulary, or in the United States Pharmacopoeia, or any supplement to them, (2) intended for use in the diagnosis of disease or other conditions, or in the cure, mitigation, treatment, or prevention of disease in man or other animals, or (3) intended to affect the structure or any function of the body of man or other animals and which is not dependent upon being metabolized for the achievement of any of its principal intended purposes."

Safe Medical Devices Act (SMDA):   A public law which imposes reporting requirements on "device-user facilities" including hospitals, ambulatory surgical facilities, nursing homes, and outpatient clinics. They are required to report information that "reasonably suggests" the probability that a medical device has caused or contributed to the death, serious injury, or serious illness of a patient at that facility.

References

1. E. P. Stanley, *Handbook of Hospital Safety*, CRC Series in Engineering in Medicine and Biology, Boca Raton, Fla.: CRC Press, 1981.
2. *Hazards: Technology and Fairness*, National Academy of Engineering, Series on Technology and Social Priorities, Washington, D.C.: National Academy Press, 1986.
3. *Compilation of ASTM Standard Definitions*, 7th ed., 19103, Philadelphia: American Society for Testing and Materials, 1990.
4. *Medical Device Reporting under the Safe Medical Devices Act: A Guide for Healthcare Facilities*, ECRI Report, 19462, Plymouth Meeting, Penn., 1991.
5. H. Newcomb Morse, "Is manufacturer liable for possible failure of equipment?" *Medical Electronics Products*, October 1990.
6. Federal Regulation of Medical Devices—Problems Still to be Overcome, General Accounting Office, Report to the Congress of the United States, GAO/HRD-83-53, September 1983.
7. ECRI, Medical Device Experience Monitoring System, Contract Report by Emergency Care Research Institute, Plymouth Meeting, Penn., June 28, 1973.
8. R. C. Schwing and W. A. Albers, *Societal Risk Assessment. How Safe Is Safe Enough?* New York: Plenum Press, 1980.
9. J. B. Cooper, R. S. Newbower, and C. D. Long, "Learning from anesthesia mishaps," *QRB/Quality Review Bulletin*, pp. 10–16, March 1981.
10. Y. David, "The medical instrument life cycle strategy for clinical engineers," *IEEE/EMBS Magazine*, vol. 4, no. 2, pp. 25–27, June 1985.
11. Code Federal Register, "Proposed Rules," Department of Health and Human Services, Part VI, Food and Drug Administration, 21 CFR Parts 803 and 807, vol. 56, no. 228, Tuesday, November 26, 1991.
12. B. R. Klein, Ed., *Healthcare Facilities Handbook*, Quincy, Mass.: National Fire Protection Association, May 1990.

13. *Implementing the 1989 PTSM Standards: Case Studies*, Plant, Technology & Management Series, No. 2, The Joint Commission on Accreditation of Healthcare Organization, Chicago, 1989.

14. J. Bronzino, *Management of Medical Technology: A Primer for Clinical Engineers*, Stoneham, Mass.: Butterworth Heinemann, 1992.

15. Y. David, "Safety and risk control in the clinical environment," in *A New Horizon on Medical Physics and Biomedical Engineering*, Proceedings of the Tutorial Session in the World Congress on Medical Physics and Biomedical Engineering, Amsterdam: Elsevier, 1991.

Further Information

Hospital safety program, in *Encyclopedia of Medical Devices and Instrumentation*, J. G. Webster, Ed., New York: John Wiley & Sons, 1988, pp. 2575–2585.

Health Technology Management, Emergency Care Research Institute, Plymouth Meeting, Penn., 1991.

Safety Management for Health Care Facilities, Management and Compliance Series, American Society of Hospital Engineering of the American Hospital Association, Chicago, 1989.

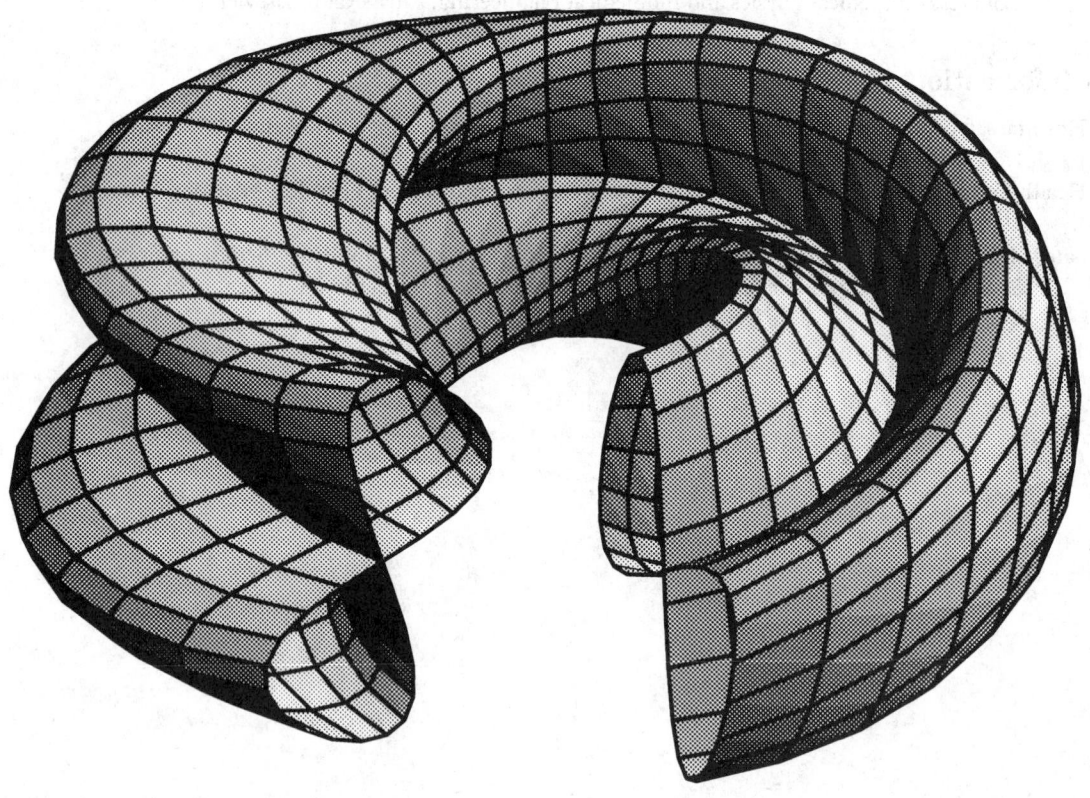

The Klein bottle.

The surface of the Klein bottle results from rotating a figure eight about an axis and putting a twist in it; its surface is compact and nonorientable. Mathematically, the parameters of the Klein bottle can be explained in a complex equation. By using the symbolic manipulation computer program *Mathematica*, the curves of the Klein bottle can be illustrated. Computations that are very complicated to do by hand can frequently be performed with ease in *Mathematica*.

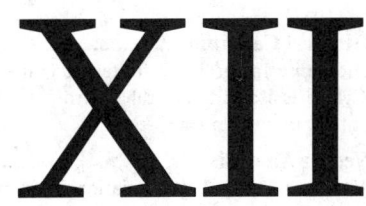

XII

Mathematics, Symbols, and Physical Constants

Ronald J. Tallarida
Temple University

Greek Alphabet 2419

International System of Units (SI) 2419
Definitions of SI Base Units • Names and Symbols for the SI Base Units • SI Derived Units with Special Names and Symbols • Units in Use Together with the SI

Conversion Constants and Multipliers 2422
Recommended Decimal Multiples and Submultiples • Conversion Factors—Metric to English • Conversion Factors—English to Metric • Conversion Factors—General • Temperature Factors • Conversion of Temperatures

Physical Constants 2424
General • π Constants • Constants Involving e • Numerical Constants

Symbols and Terminology for Physical and Chemical Quantities 2425
Classical Mechanics • Electricity and Magnetism • Electromagnetic Radiation • Solid State

Elementary Algebra and Geometry 2430
Fundamental Properties (Real Numbers) • Exponents • Fractional Exponents • Irrational Exponents • Logarithms • Factorials • Binomial Theorem • Factors and Expansion • Progression • Complex Numbers • Polar Form • Permutations • Combinations • Algebraic Equations • Geometry

Determinants, Matrices, and Linear Systems of Equations 2436
Determinants • Evaluation by Cofactors • Properties of Determinants • Matrices • Operations • Properties • Transpose • Identity Matrix • Adjoint • Inverse Matrix • Systems of Linear Equations • Matrix Solution

Trigonometry 2442
Triangles • Trigonometric Functions of an Angle • Trigonometric Identities • Inverse Trigonometric Functions

Analytic Geometry 2446
Rectangular Coordinates • Distance between Two Points; Slope • Equations of Straight Lines • Distance from a Point to a Line • Circle • Parabola • Ellipse • Hyperbola ($e > 1$) • Change of Axes

Series 2454
Bernoulli and Euler Numbers • Series of Functions • Error Function • Series Expansion

Differential Calculus 2461
Notation • Slope of a Curve • Angle of Intersection of Two Curves • Radius of Curvature • Relative Maxima and Minima • Points of Inflection of a Curve • Taylor's Formula • Indeterminant Forms • Numerical Methods • Functions of Two Variables • Partial Derivatives

Integral Calculus 2466
Indefinite Integral • Definite Integral • Properties • Common Applications of the Definite Integral •
Cylindrical and Spherical Coordinates • Double Integration • Surface Area and Volume by Double
Integration • Centroid

Vector Analysis 2472
Vectors • Vector Differentiation • Divergence Theorem (Gauss) • Stokes' Theorem • Planar Motion in
Polar Coordinates

Special Functions 2475
Hyperbolic Functions • Laplace Transforms • z-Transform • Fourier Series • Functions with Period Other
than 2π • Bessel Functions • Legendre Polynomials • Laguerre Polynomials • Hermite Polynomials •
Orthogonality

Statistics 2486
Arithmetic Mean • Median • Mode • Geometric Mean • Harmonic Mean • Variance • Standard Deviation
• Coefficient of Variation • Probability • Binomial Distribution • Mean of Binomially Distributed Variable
• Normal Distribution • Poisson Distribution

Tables of Probability and Statistics 2490
Areas Under the Standard Normal Curve • Poisson Distribution • t-Distribution • χ^2 Distribution •
Variance Ratio

Table of Derivatives 2496

Integrals 2497
Elementary Forms • Forms Containing $(a + bx)$

The Fourier Transforms 2500
Fourier Transforms • Finite Sine Transforms • Finite Cosine Transforms • Fourier Sine Transforms •
Fourier Cosine Transforms • Fourier Transforms

Numerical Methods 2507
Solution of Equations by Iteration • Finite Differences • Interpolation

Probability 2517
Definitions • Definition of Probability • Marginal and Conditional Probability • Probability Theorems •
Random Variable • Probability Function (Discrete Case) • Cumulative Distribution Function (Discrete
Case) • Probability Density (Continuous Case) • Cumulative Distribution Function (Continuous Case) •
Mathematical Expectation

Positional Notation 2520
Change of Base • Examples • $10^{\pm n}$ in Octal Scale • 2^n in Decimal Scale • $n \log_{10} 2$, $n \log_2 10$ in Decimal
Scale • Addition and Multiplication Tables • Mathematical Constants in Octal Scale

Fundamental Physical Constants 2524

Periodic Table of the Elements 2525

Classification of Electromagnetic Radiation 2526
Letter Designations of Microwave Bands

Electrical Resistivity 2527
Electrical Resistivity of Pure Metals • Electrical Resistivity of Selected Alloys • Electrical Resistivity of
Selected Alloy Cast Irons • Resistivity of Selected Ceramics

Dielectric Constants 2532
Dielectric Constants of Solids • Dielectric Constants of Ceramics • Dielectric Constants of Glasses

Properties of Semiconductors 2533
Semiconducting Properties of Selected Materials • Band Properties of Semiconductors • Resistivity of
Semiconducting Minerals

Properties of Magnetic Alloys 2536
High-Permeability Magnetic Alloys • Cast Permanent Magnetic Alloys • Properties of Antiferromagnetic
Compounds • Saturation Constants and Curie Points of Ferromagnetic Elements • Magnetic Properties
of Transformer Steels • High Silicon Transformer Steel • Saturation Constants for Magnetic Substances •
Initial Permeability of High Purity Iron for Various Temperatures • Magnetic Materials

Resistance of Wires 2543

Credits 2545

THE GREAT ACHIEVEMENTS in engineering deeply affect the lives of all of us and also serve to remind us of the importance of mathematics. Interest in mathematics has grown steadily with these engineering achievements and with concomitant advances in pure physical science. Whereas scholars in nonscientific fields, and even in such fields as botany, medicine, geology, etc., can communicate most of the problems and results in nonmathematical language, this is virtually impossible in present-day engineering and physics. Yet it is interesting to note that until the beginning of the twentieth century engineers regarded calculus as something of a mystery. Modern students of engineering now study calculus, as well as differential equations, complex variables, vector analysis, orthogonal functions, and a variety of other topics in applied analysis. The study of systems has ushered in matrix algebra and, indeed, most engineering students now take linear algebra as a core topic early in their mathematical education.

This section contains concise summaries of relevant topics in applied engineering mathematics and certain key formulas, that is, those formulas that are most often needed in the formulation and solution of engineering problems. Whereas even inexpensive electronic calculators contain tabular material (e.g., tables of trigonometric and logarithmic functions) that used to be needed in this kind of handbook, most calculators do not give symbolic results. Hence, we have included formulas along with brief summaries that guide their use. In many cases we have added numerical examples, as in the discussions of matrices, their inverses, and their use in the solutions of linear systems. A table of derivatives is included, as well as key applications of the derivative in the solution of problems in maxima and minima, related rates, analysis of curvature, and finding approximate roots by numerical methods. A list of infinite series, along with the interval of convergence of each, is also included.

Of the two branches of calculus, integral calculus is richer in its applications, as well as in its theoretical content. Though the theory is not emphasized here, important applications such as finding areas, lengths, volumes, centroids, and the work done by a nonconstant force are included. Both cylindrical and spherical polar coordinates are discussed, and a table of integrals is included. Vector analysis is summarized in a separate section and includes a summary of the algebraic formulas involving dot and cross multiplication, frequently needed in the study of fields, as well as the important theorems of Stokes and Gauss. The part on special functions includes the gamma function, hyperbolic functions, Fourier series, orthogonal functions, and both Laplace and z-transforms. The Laplace transform provides a basis for the solution of differential equations and is fundamental to all concepts and definitions underlying analytical tools for describing feedback control systems. The z-transform, not discussed in most applied mathematics books, is most useful in the analysis of discrete signals as, for example, when a computer receives data sampled at some prespecified time interval. The Bessel functions, also called cylindrical functions, arise in many physical applications, such as the heat transfer in a "long" cylinder, whereas the other orthogonal functions discussed—Legendre, Hermite, and Laguerre polynomials—are needed in quantum mechanics and many other subjects (e.g., solid-state electronics) that use concepts of modern physics.

The world of mathematics, even applied mathematics, is vast. Even the best mathematicians cannot keep up with more than a small piece of this world. The topics included in this section, however, have withstood the test of time and, thus, are truly *core* for the modern engineer.

This section also incorporates tables of physical constants and symbols widely used by engineers. While not exhaustive, the constants, conversion factors, and symbols provided will enable the reader to accommodate a majority of the needs that arise in design, test, and manufacturing functions.

Mathematics, Symbols, and Physical Constants

Greek Alphabet

Greek letter			Greek name	English equivalent	Greek letter			Greek name	English equivalent
A	α		Alpha	a	N	ν		Nu	n
B	β		Beta	b	Ξ	ξ		Xi	x
Γ	γ		Gamma	g	O	o		Omicron	ŏ
Δ	δ		Delta	d	Π	π		Pi	p
E	ε		Epsilon	ĕ	P	ρ		Rho	r
Z	ζ		Zeta	z	Σ	σ	ς	Sigma	s
H	η		Eta	ē	T	τ		Tau	t
Θ	θ	ϑ	Theta	th	Y	υ		Upsilon	u
I	ι		Iota	i	Φ	φ	φ	Phi	ph
K	κ		Kappa	k	X	χ		Chi	ch
Λ	λ		Lambda	l	Ψ	ψ		Psi	ps
M	μ		Mu	m	Ω	ω		Omega	ō

International System of Units (SI)

The International System of units (SI) was adopted by the 11th General Conference on Weights and Measures (CGPM) in 1960. It is a coherent system of units built from seven *SI base units,* one for each of the seven dimensionally independent base quantities: they are the meter, kilogram, second, ampere, kelvin, mole, and candela, for the dimensions length, mass, time, electric current, thermodynamic temperature, amount of substance, and luminous intensity, respectively. The definitions of the SI base units are given below. The *SI derived units* are expressed as products of powers of the base units, analogous to the corresponding relations between physical quantities but with numerical factors equal to unity.

In the International System there is only one SI unit for each physical quantity. This is either the appropriate SI base unit itself or the appropriate SI derived unit. However, any of the approved decimal prefixes, called *SI prefixes,* may be used to construct decimal multiples or submultiples of SI units.

It is recommended that only SI units be used in science and technology (with SI prefixes where appropriate). Where there are special reasons for making an exception to this rule, it is recommended always to define the units used in terms of SI units. This section is based on information supplied by IUPAC.

Definitions of SI Base Units

Meter—The meter is the length of path traveled by light in vacuum during a time interval of 1/299 792 458 of a second (17th CGPM, 1983).

Kilogram—The kilogram is the unit of mass; it is equal to the mass of the international prototype of the kilogram (3rd CGPM, 1901).

Second—The second is the duration of 9 192 631 770 periods of the radiation corresponding to the transition between the two hyperfine levels of the ground state of the cesium-133 atom (13th CGPM, 1967).

Ampere—The ampere is that constant current which, if maintained in two straight parallel conductors of infinite length, of negligible circular cross-section, and placed 1 meter apart in vacuum, would produce between these conductors a force equal to 2×10^{-7} newton per meter of length (9th CGPM, 1948).

Kelvin—The kelvin, unit of thermodynamic temperature, is the fraction 1/273.16 of the thermodynamic temperature of the triple point of water (13th CGPM, 1967).

Mole—The mole is the amount of substance of a system which contains as many elementary entities as there are atoms in 0.012 kilogram of carbon-12. When the mole is used, the elementary entities must be specified and may be atoms, molecules, ions, electrons, or other particles, or specified groups of such particles (14th CGPM, 1971).

Examples of the use of the mole:

1 mol of H_2 contains aboaut 6.022×10^{23} H_2 molecules, or 12.044×10^{23} H atoms
1 mol of HgCl has a mass of 236.04 g
1 mol of Hg_2Cl_2 has a mass of 472.08 g
1 mol of Hg_2^{2+} has a mass of 401.18 g and a charge of 192.97 kC
1 mol of $Fe_{0.91}S$ has a mass of 82.88 g
1 mol of e^- has a mass of 548.60 µg and a charge of −96.49 kC
1 mol of photons whose frequency is 10^{14} Hz has energy of about 39.90 kJ

Candela—The candela is the luminous intensity, in a given direction, of a source that emits monochromatic radiation of frequency 540×10^{12} hertz and that has a radiant intensity in that direction of (1/683) watt per steradian (16th CGPM, 1979).

Names and Symbols for the SI Base Units

Physical quantity	Name of SI unit	Symbol for SI unit
length	meter	m
mass	kilogram	kg
time	second	s
electric current	ampere	A
thermodynamic temperature	kelvin	K
amount of substance	mole	mol
luminous intensity	candela	cd

SI Derived Units with Special Names and Symbols

Physical quantity	Name of SI unit	Symbol for SI unit	Expression in terms of SI base units	
frequency[1]	hertz	Hz	s^{-1}	
force	newton	N	$m\ kg\ s^{-2}$	
pressure, stress	pascal	Pa	$N\ m^{-2}$	$= m^{-1}\ kg\ s^{-2}$
energy, work, heat	joule	J	$N\ m$	$= m^2\ kg\ s^{-2}$
power, radiant flux	watt	W	$J\ s^{-1}$	$= m^2\ kg\ s^{-3}$
electric charge	coulomb	C	$A\ s$	

Physical quantity	Name of SI unit	Symbol for SI unit		Expression in terms of SI base units
electric potential, electromotive force	volt	V	J C^{-1}	= m^2 kg s^{-3} A^{-1}
electric resistance	ohm	Ω	V A^{-1}	= m^2 kg s^{-3} A^{-2}
electric conductance	siemens	S	Ω^{-1}	= m^{-2} kg^{-1} s^3 A^2
electric capacitance	farad	F	C V^{-1}	= m^{-2} kg^{-1} s^4 A^2
magnetic flux density	tesla	T	V s m^{-2}	= kg s^{-2} A^{-1}
magnetic flux	weber	Wb	V s	= m^2 kg s^{-2} A^{-1}
inductance	henry	H	V A^{-1} s	= m^2 kg s^{-2} A^{-2}
Celsius temperature[2]	degree Celsius	°C	K	
luminous flux	lumen	lm	cd sr	
illuminance	lux	lx	cd sr m^{-2}	
activity (radioactive)	becquerel	Bq	s^{-1}	
absorbed dose (of radiation)	gray	Gy	J kg^{-1}	= m^2 s^{-2}
dose equivalent (dose equivalent index)	sievert	Sv	J kg^{-1}	= m^2 s^{-2}
plane angle	radian	rad	1	= m m^{-1}
solid angle	steradian	sr	1	= m^2 m^{-2}

[1] For radial (circular) frequency and for angular velocity the unit rad s^{-1}, or simply s^{-1}, should be used, and this may not be simplified to Hz. The unit Hz should be used only for frequency in the sense of cycles per second.

[2] The Celsius temperature θ is defined by the equation:

$$\theta/°C = T/K - 273.15$$

The SI unit of Celsius temperature interval is the degree Celsius, °C, which is equal to the kelvin, K. °C should be treated as a single symbol, with no space between the ° sign and the letter C. (The symbol °K, and the symbol °, should no longer be used.)

Units in Use Together with the SI

These units are not part of the SI, but it is recognized that they will continue to be used in appropriate contexts. SI prefixes may be attached to some of these units, such as milliliter, ml; millibar, mbar; megaelectronvolt, MeV; kilotonne, ktonne.

Physical quantity	Name of unit	Symbol for unit	Value in SI units
time	minute	min	60 s
time	hour	h	3600 s
time	day	d	86 400 s
plane angle	degree	°	(π/180) rad
plane angle	minute	'	(π/10 800) rad
plane angle	second	"	(π/648 000) rad
length	ångstrom[1]	Å	10^{-10} m
area	barn	b	10^{-28} m^2
volume	litre	l, L	dm^3 = 10^{-3} m^3
mass	tonne	t	Mg = 10^3 kg
pressure	bar[1]	bar	10^5 Pa = 10^5 N m^{-2}
energy	electronvolt[2]	eV (= e × V)	≈1.60218 × 10^{-19} J
mass	unified atomic mass unit[2,3]	u (= m_a(^{12}C)/12)	≈1.66054 × 10^{-27} kg

[1] The ångstrom and the bar are approved by CIPM for "temporary use with SI units," until CIPM makes a further recommendation. However, they should not be introduced where they are not used at present.

[2] The values of these units in terms of the corresponding SI units are not exact, since they depend on the values of the physical constants e (for the electronvolt) and N_A (for the unified atomic mass unit), which are determined by experiment.

[3] The unified atomic mass unit is also sometimes called the dalton, with symbol Da, although the name and symbol have not been approved by CGPM.

Conversion Constants and Multipliers

Recommended Decimal Multiples and Submultiples

Multiples and submultiples	Prefixes	Symbols	Multiples and submultiples	Prefixes	Symbols
10^{18}	exa	E	10^{-1}	deci	d
10^{15}	peta	P	10^{-2}	centi	c
10^{12}	tera	T	10^{-3}	milli	m
10^{9}	giga	G	10^{-6}	micro	μ (Greek mu)
10^{6}	mega	M	10^{-9}	nano	n
10^{3}	kilo	k	10^{-12}	pico	p
10^{2}	hecto	h	10^{-15}	femto	f
10	deca	da	10^{-18}	atto	a

Conversion Factors—Metric to English

To obtain	Multiply	By
Inches	Centimeters	0.3937007874
Feet	Meters	3.280839895
Yards	Meters	1.093613298
Miles	Kilometers	0.6213711922
Ounces	Grams	$3.527396195 \times 10^{-2}$
Pounds	Kilograms	2.204622622
Gallons (U.S. Liquid)	Liters	0.2641720524
Fluid ounces	Milliliters (cc)	$3.381402270 \times 10^{-2}$
Square inches	Square centimeters	0.1550003100
Square feet	Square meters	10.76391042
Square yards	Square meters	1.195990046
Cubic inches	Milliliters (cc)	$6.102374409 \times 10^{-2}$
Cubic feet	Cubic meters	35.31466672
Cubic yards	Cubic meters	1.307950619

Conversion Factors—English to Metric*

To obtain	Multiply	By
Microns	Mils	**25.4**
Centimeters	Inches	**2.54**
Meters	Feet	**0.3048**
Meters	Yards	**0.9144**
Kilometers	Miles	**1.609344**
Grams	Ounces	28.34952313
Kilograms	Pounds	**0.45359237**
Liters	Gallons (U.S. Liquid)	**3.785411784**
Millimeters (cc)	Fluid ounces	29.57352956
Square centimeters	Square inches	**6.4516**
Square meters	Square feet	**0.09290304**
Square meters	Square yards	**0.83612736**
Milliliters (cc)	Cubic inches	**16.387064**
Cubic meters	Cubic feet	$2.831684659 \times 10^{-2}$
Cubic meters	Cubic yards	0.764554858

* Boldface numbers are exact; others are given to ten significant figures where so indicated by the multiplier factor.

Conversion Factors—General*

To obtain	Multiply	By
Atmospheres	Feet of water @ 4°C	2.950×10^{-2}
Atmospheres	Inches of mercury @ 0°C	3.342×10^{-2}
Atmospheres	Pounds per square inch	6.804×10^{-2}
BTU	Foot-pounds	1.285×10^{-3}
BTU	Joules	9.480×10^{-4}
Cubic feet	Cords	**128**
Degree (angle)	Radians	57.2958
Ergs	Foot-pounds	1.356×10^{7}
Feet	Miles	**5280**
Feet of water @ 4°C	Atmospheres	33.90
Foot-pounds	Horsepower-hours	1.98×10^{6}
Foot-pounds	Kilowatt-hours	2.655×10^{6}
Foot-pounds per min	Horsepower	3.3×10^{4}
Horsepower	Foot-pounds per sec	1.818×10^{-3}
Inches of mercury @ 0°C	Pounds per square inch	2.036
Joules	BTU	1054.8
Joules	Foot-pounds	1.35582
Kilowatts	BTU per min	1.758×10^{-2}
Kilowatts	Foot-pounds per min	2.26×10^{-5}
Kilowatts	Horsepower	0.745712
Knots	Miles per hour	0.86897624
Miles	Feet	1.894×10^{-4}
Nautical miles	Miles	0.86897624
Radians	Degrees	1.745×10^{-2}
Square feet	Acres	**43560**
Watts	BTU per min	17.5796

Temperature Factors

$$°F = 9/5 \, (°C) + 32$$

Fahrenheit temperature = 1.8 (temperature in kelvins) – 459.67

$$°C = 5/9 \, [(°F) - 32)]$$

Celsius temperature = temperature in kelvins – 273.15

Fahrenheit temperature = 1.8 (Celsius temperature) + 32

Conversion of Temperatures

From	To	
°Celsius	°Fahrenheit	$t_F = (t_C \times 1.8) + 32$
	Kelvin	$T_K = t_C + 273.15$
	°Rankine	$T_R = (t_C + 273.15) \times 18$
°Fahrenheit	°Celsius	$t_C = \dfrac{t_F - 32}{1.8}$
	Kelvin	$T_k = \dfrac{t_F - 32}{1.8} + 273.15$
	°Rankine	$T_R = t_F + 459.67$
Kelvin	°Celsius	$t_C = T_K - 273.15$
	°Rankine	$T_R = T_K \times 1.8$
°Rankine	°Fahrenheit	$t_F = T_R - 459.67$
	Kelvin	$T_K = \dfrac{T_R}{1.8}$

* Boldface numbers are exact; others are given to ten significant figures where so indicated by the multiplier factor.

Physical Constants

General

Equatorial radius of the earth = 6378.388 km = 3963.34 miles (statute).
Polar radius of the earth, 6356.912 km = 3949.99 miles (statute).
1 degree of latitude at 40° = 69 miles.
1 international nautical mile = 1.15078 miles (statute) = 1852 m = 6076.115 ft.
Mean density of the earth = 5.522 g/cm^3 = 344.7 lb/ft^3
Constant of gravitation (6.673 ± 0.003) × 10^{-8} cm^3 gm^{-1} s^{-2}.
Acceleration due to gravity at sea level, latitude 45° = 980.6194 cm/s^2 = 32.1726 ft/s^2.
Length of seconds pendulum at sea level, latitude 45° = 99.3575 cm = 39.1171 in.
1 knot (international) = 101.269 ft/min = 1.6878 ft/s = 1.1508 miles (statute)/h.
1 micron = 10^{-4} cm.
1 ångstrom = 10^{-8} cm.
Mass of hydrogen atom = (1.67339 ± 0.0031) × 10^{-24} g.
Density of mercury at 0°C = 13.5955 g/ml.
Density of water at 3.98°C = 1.000000 g/ml.
Density, maximum, of water, at 3.98°C = 0.999973 g/cm^3.
Density of dry air at 0°C, 760 mm = 1.2929 g/l.
Velocity of sound in dry air at 0°C = 331.36 m/s = 1087.1 ft/s.
Velocity of light in vacuum = (2.997925 ± 0.000002) × 10^{10} cm/s.
Heat of fusion of water 0°C = 79.71 cal/g.
Heat of vaporization of water 100°C = 539.55 cal/g.
Electrochemical equivalent of silver 0.001118 g/s international amp.
Absolute wavelength of red cadmium light in air at 15°C, 760 mm pressure = 6438.4696 Å.
Wavelength of orange-red line of krypton 86 = 6057.802 Å.

π Constants

$$\pi = 3.14159\ 26535\ 89793\ 23846\ 26433\ 83279\ 50288\ 41971\ 69399\ 37511$$
$$1/\pi = 0.31830\ 98861\ 83790\ 67153\ 77675\ 26745\ 02872\ 40689\ 19291\ 48091$$
$$\pi^2 = 9.8690\ 44010\ 89358\ 61883\ 44909\ 99876\ 15113\ 53136\ 99407\ 24079$$
$$\log_e\pi = 1.14472\ 98858\ 49400\ 17414\ 34273\ 51353\ 05871\ 16472\ 94812\ 91531$$
$$\log_{10}\pi = 0.49714\ 98726\ 94133\ 85435\ 12682\ 88290\ 89887\ 36516\ 78324\ 38044$$
$$\log_{10}\sqrt{2\pi} = 0.39908\ 99341\ 79057\ 52478\ 25035\ 91507\ 69595\ 02099\ 34102\ 92128$$

Constants Involving *e*

$$e = 2.71828\ 18284\ 59045\ 23536\ 02874\ 71352\ 66249\ 77572\ 47093\ 69996$$
$$1/e = 0.36787\ 94411\ 71442\ 32159\ 55237\ 70161\ 46086\ 74458\ 11131\ 03177$$
$$e^2 = 7.38905\ 60989\ 30650\ 22723\ 04274\ 60575\ 00781\ 31803\ 15570\ 55185$$
$$M = \log_{10}e = 0.43429\ 44819\ 03251\ 82765\ 11289\ 18916\ 60508\ 22943\ 97005\ 80367$$
$$1/M = \log_e 10 = 2.30258\ 50929\ 94045\ 68401\ 79914\ 54684\ 36420\ 76011\ 01488\ 62877$$
$$\log_{10}M = 9.63778\ 43113\ 00536\ 78912\ 29674\ 98645\ -10$$

Numerical Constants

$$\sqrt{2} = 1.41421\ 35623\ 73095\ 04880\ 16887\ 24209\ 69807\ 85696\ 71875\ 37695$$
$$\sqrt[3]{2} = 1.25992\ 10498\ 94873\ 16476\ 72106\ 07278\ 22835\ 05702\ 51464\ 70151$$
$$\log_e 2 = 0.69314\ 71805\ 59945\ 30941\ 72321\ 21458\ 17656\ 80755\ 00134\ 36026$$
$$\log_{10}2 = 0.30102\ 99956\ 63981\ 19521\ 37388\ 94724\ 49302\ 67881\ 89881\ 46211$$
$$\sqrt{3} = 1.73205\ 08075\ 68877\ 29352\ 74463\ 41505\ 87236\ 69428\ 05253\ 81039$$
$$\sqrt[3]{3} = 1.44224\ 95703\ 07408\ 38232\ 16383\ 10780\ 10958\ 83918\ 69253\ 49935$$
$$\log_e 3 = 1.09861\ 22886\ 68109\ 69139\ 52452\ 36922\ 52570\ 46474\ 90557\ 82275$$
$$\log_{10}3 = 0.47712\ 12547\ 19662\ 43729\ 50279\ 03255\ 11530\ 92001\ 28864\ 19070$$

Symbols and Terminology for Physical and Chemical Quantities

Name	Symbol	Definition	SI unit
Classical Mechanics			
mass	m		kg
reduced mass	μ	$\mu = m_1 m_2/(m_1 + m_2)$	kg
density, mass density	ρ	$\rho = m/V$	kg m^{-3}
relative density	d	$d = \rho/\rho^\theta$	1
surface density	ρ_A, ρ_S	$\rho_A = m/A$	kg m^{-2}
specific volume	v	$v = V/m = 1/\rho$	m^3 kg^{-1}
momentum	$\boldsymbol{p}$	$\boldsymbol{p} = m\boldsymbol{v}$	kg m s^{-1}
angular momentum, action	$\boldsymbol{L}$	$\boldsymbol{L} = \boldsymbol{r} \times \boldsymbol{p}$	J s
moment of inertia	I, J	$I = \Sigma m_i r_i^2$	kg m^2
force	$\boldsymbol{F}$	$\boldsymbol{F} = \mathrm{d}\boldsymbol{p}/\mathrm{d}t = m\boldsymbol{a}$	N
torque, moment of a force	$\boldsymbol{T}, (\boldsymbol{M})$	$\boldsymbol{T} = \boldsymbol{r} \times \boldsymbol{F}$	N m
energy	E		J
potential energy	E_p, V, Φ	$E_p = -\int \boldsymbol{F} \cdot \mathrm{d}\boldsymbol{s}$	J
kinetic energy	E_k, T, K	$E_k = (1/2)mv^2$	J
work	W, w	$W = \int \boldsymbol{F} \cdot \mathrm{d}\boldsymbol{s}$	J
Hamilton function	H	$H(q, p)$ $= T(q, p) + V(q)$	J
Lagrange function	L	$L(q, \dot{q})$ $= T(q, \dot{q}) - V(q)$	J
pressure	p, P	$p = F/A$	Pa, N m^{-2}
surface tension	γ, σ	$\gamma = \mathrm{d}W/\mathrm{d}A$	N m^{-1}, J m^{-2}
weight	$G, (W, P)$	$G = mg$	N
gravitational constant	G	$F = Gm_1 m_2/r^2$	N m^2 kg^{-2}
normal stress	σ	$\sigma = F/A$	Pa
shear stress	τ	$\tau = F/A$	Pa
linear strain, relative elongation	ε, e	$\varepsilon = \Delta l/l$	1
modulus of elasticity, Young's modulus	E	$E = \sigma/\varepsilon$	Pa
shear strain	γ	$\gamma = \Delta x/d$	1
shear modulus	G	$G = \tau/\gamma$	Pa
volume strain, bulk strain	θ	$\theta = \Delta V/V_0$	1
bulk modulus, compression modulus	K	$K = -V_0(\mathrm{d}p/\mathrm{d}V)$	Pa
viscosity, dynamic viscosity	η, μ	$\tau_{x,z} = \eta(\mathrm{d}v_x/\mathrm{d}z)$	Pa s
fluidity	ϕ	$\phi = 1/\eta$	m kg^{-1} s
kinematic viscosity	ν	$\nu = \eta/\rho$	m^2 s^{-1}
friction coefficient	$\mu, (f)$	$F_{frict} = \mu F_{norm}$	1
power	P	$P = \mathrm{d}W/\mathrm{d}t$	W
sound energy flux	P, P_a	$P = \mathrm{d}E/\mathrm{d}t$	W
acoustic factors			
reflection factor	ρ	$\rho = P_r/P_0$	1
acoustic absorption factor	$\alpha_a, (\alpha)$	$\alpha_a = 1 - \rho$	1
transmission factor	τ	$\tau = P_{tr}/P_0$	1
dissipation factor	δ	$\delta = \alpha_a - \tau$	1

Symbols and Terminology for Physical and Chemical Quantities (continued)

Name	Symbol	Definition	SI unit

Electricity and Magnetism

Name	Symbol	Definition	SI unit
quantity of electricity, electric charge	Q		C
charge density	ρ	$\rho = Q/V$	C m^{-3}
surface charge density	σ	$\sigma = Q/A$	C m^{-2}
electric potential	V, ϕ	$V = dW/dQ$	V, J C^{-1}
electric potential difference	$U, \Delta V, \Delta\phi$	$U = V_2 - V_1$	V
electromotive force	E	$E = \int(F/Q) \cdot ds$	V
electric field strength	E	$E = F/Q = -\text{grad } V$	V m^{-1}
electric flux	Ψ	$\Psi = \int D \cdot dA$	C
electric displacement	D	$D = \varepsilon E$	C m^{-2}
capacitance	C	$C = Q/U$	F, C V^{-1}
permittivity	ε	$D = \varepsilon E$	F m^{-1}
permittivity of vacuum	ε_0	$\varepsilon_0 = \mu_0^{-1} c_0^{-2}$	F m^{-1}
relative permittivity	ε_r	$\varepsilon_r = \varepsilon/\varepsilon_0$	1
dielectric polarization (dipole moment per volume)	P	$P = D - \varepsilon_0 E$	C m^{-2}
electric susceptibility	χ_e	$\chi_e = \varepsilon_r - 1$	1
electric dipole moment	p, μ	$p = Qr$	C m
electric current	I	$I = dQ/dt$	A
electric current density	j, J	$I = \int j \cdot dA$	A m^{-2}
magnetic flux density, magnetic induction	B	$F = Qv \times B$	T
magnetic flux	Φ	$\Phi = \int B \cdot dA$	Wb
magnetic field strength	H	$B = \mu H$	A M^{-1}
permeability	μ	$B = \mu H$	N A^{-2}, H m^{-1}
permeability of vacuum	μ_0		H m^{-1}
relative permeability	μ_r	$\mu_r = \mu/\mu_0$	1
magnetization (magnetic dipole moment per volume)	M	$M = B/\mu_0 - H$	A m^{-1}
magnetic susceptibility	$\chi, \kappa, (\chi_m)$	$\chi = \mu_r - 1$	1
molar magnetic susceptibility	χ_m	$\chi_m = V_m\chi$	m^3 mol^{-1}
magnetic dipole moment	m, μ	$E_p = -m \cdot B$	A m^2, J T^{-1}
electrical resistance	R	$R = U/I$	Ω
conductance	G	$G = 1/R$	S
loss angle	δ	$\delta = (\pi/2) + \phi_I - \phi_U$	1, rad
reactance	X	$X = (U/I)\sin\delta$	Ω
impedance (complex impedance)	Z	$Z = R + iX$	Ω
admittance (complex admittance)	Y	$Y = 1/Z$	S

Symbols and Terminology for Physical and Chemical Quantities (continued)

Name	Symbol	Definition	SI unit

Electricity and Magnetism (continued)

Name	Symbol	Definition	SI unit
susceptance	B	$Y = G + iB$	S
resistivity	ρ	$\rho = E/j$	Ω m
conductivity	κ, γ, σ	$\kappa = 1/\rho$	S m^{-1}
self-inductance	L	$E = -L(dI/dt)$	H
mutual inductance	M, L_{12}	$E_1 = L_{12}(dI_2/dt)$	H
magnetic vector potential	A	$B = \nabla \times A$	Wb m^{-1}
Poynting vector	S	$S = E \times H$	W m^{-2}

Electromagnetic Radiation

Name	Symbol	Definition	SI unit
wavelength	λ		m
speed of light			
in vacuum	c_0		m s^{-1}
in a medium	c	$c = c_0/n$	m s^{-1}
wavenumber in vacuum	$\tilde{\nu}$	$\tilde{\nu} = \nu/c_0 = 1/n\lambda$	m^{-1}
wavenumber (in a medium)	σ	$\sigma = 1/\lambda$	m^{-1}
frequency	ν	$\nu = c/\lambda$	Hz
circular frequency, pulsatance	ω	$\omega = 2\pi\nu$	s^{-1}, rad s^{-1}
refractive index	n	$n = c_0/c$	1
Planck constant	h		J s
Planck constant/2π	$\hbar$	$\hbar = h/2\pi$	J s
radiant energy	Q, W		J
radiant energy density	ρ, w	$\rho = Q/V$	J m^{-3}
spectral radiant energy density			
in terms of frequency	ρ_ν, w_ν	$\rho_\nu = d\rho/d\nu$	J m^{-3} Hz^{-1}
in terms of wavenumber	$\rho_{\tilde{\nu}}, w_{\tilde{\nu}}$	$\rho_{\tilde{\nu}} = d\rho/d\tilde{\nu}$	J m^{-2}
in terms of wavelength	ρ_λ, w_λ	$\rho_\lambda = d\rho/d\lambda$	J m^{-4}
Einstein transition probabilities			
spontaneous emission	A_{nm}	$dN_n/dt = -A_{nm}N_n$	s^{-1}
stimulated emission	B_{nm}	$dN_n/dt = -\rho_{\tilde{\nu}}(\tilde{\nu}_{nm}) \times B_{nm}N_n$	s kg^{-1}
stimulated absorption	B_{mn}	$dN_n/dt = \rho_{\tilde{\nu}}(\tilde{\nu}_{nm})B_{mn}N_m$	s kg^{-1}
radiant power, radiant energy per time	Φ, P	$\Phi = dQ/dt$	W
radiant intensity	I	$I = d\Phi/d\Omega$	W sr^{-1}
radiant exitance (emitted radiant flux)	M	$M = d\Phi/dA_{\text{source}}$	W m^{-2}
irradiance (radiant flux received)	$E, (I)$	$E = d\Phi/dA$	W m^{-2}
emittance	ε	$\varepsilon = M/M_{\text{bb}}$	1
Stefan–Boltzmann constant	σ	$M_{\text{bb}} = \sigma T^4$	W m^{-2} K^{-4}

Symbols and Terminology for Physical and Chemical Quantities (continued)

Name	Symbol	Definition	SI unit

Electromagnetic Radiation (continued)

Name	Symbol	Definition	SI unit
first radiation constant	c_1	$c_1 = 2\pi hc_0^2$	W m^2
second radiation constant	c_2	$c_2 = hc_0/k$	K m
transmittance, transmission factor	τ, T	$\tau = \Phi_{tr}/\Phi_0$	1
absorptance, absorption factor	α	$\alpha = \Phi_{abs}/\Phi_0$	1
reflectance, reflection factor	ρ	$\rho = \Phi_{refl}/\Phi_0$	1
(decadic) absorbance	A	$A = \lg(1 - \alpha_i)$	1
napierian absorbance	B	$B = \ln(1 - \alpha_i)$	1
absorption coefficient			
(linear) decadic	a, K	$a = A/l$	m^{-1}
(linear) napierian	α	$\alpha = B/l$	m^{-1}
molar (decadic)	ε	$\varepsilon = a/c = A/cl$	m^2 mol^{-1}
molar napierian	κ	$\kappa = \alpha/c = B/cl$	m^2 mol^{-1}
absorption index	k	$k = \alpha/4\pi\tilde{v}$	1
complex refractive index	$\hat{n}$	$\hat{n} = n + ik$	1
molar refraction	R, R_m	$R = \dfrac{(n^2 - 1)}{(n^2 + 2)} V_m$	m^3 mol^{-1}
angle of optical rotation	α		1, rad

Solid State

Name	Symbol	Definition	SI unit
lattice vector	R, R_0		m
fundamental translation vectors for the crystal lattice	$a_1; a_2; a_3,$ $a; b; c$	$R = n_1 a_1 + n_2 a_2 + n_3 a_3$	m
(circular) reciprocal lattice vector	G	$G \cdot R = 2\pi m$	m^{-1}
(circular) fundamental translation vectors for the reciprocal lattice	$b_1; b_2; b_3,$ $a^*; b^*; c^*$	$a_i \cdot b_k = 2\pi\delta_{ik}$	m^{-1}
lattice plane spacing	d		m
Bragg angle	θ	$n\lambda = 2d \sin \theta$	1, rad
order of reflection	n		1
order parameters			
short range	σ		1
long range	s		1
Burgers vector	b		m
particle position vector	r, R_j		m
equilibrium position vector of an ion	R_0		m

Symbols and Terminology for Physical and Chemical Quantities (continued)

Name	Symbol	Definition	SI unit
		Solid State (continued)	
displacement vector of an ion	u	$u = R - R_0$	m
Debye–Waller factor	B, D		1
Debye circular wavenumber	q_D		m^{-1}
Debye circular frequency	ω_D		s^{-1}
Grüneisen parameter	γ, Γ	$\gamma = \alpha V/\kappa C_V$	1
Madelung constant	$\alpha, \mathcal{M}$	$E_{coul} = \dfrac{\alpha N_A z_+ z_- e^2}{4\pi\varepsilon_0 R_0}$	1
density of states	N_E	$N_E = dN(E)/dE$	$J^{-1}\ m^{-3}$
(spectral) density of vibrational modes	N_ω, g	$N_\omega = dN(\omega)/d\omega$	$s\ m^{-3}$
resistivity tensor	ρ_{ik}	$E = \rho \cdot j$	$\Omega\ m$
conductivity tensor	σ_{ik}	$\sigma = \rho^{-1}$	$S\ m^{-1}$
thermal conductivity tensor	λ_{ik}	$J_q = -\lambda \cdot \mathrm{grad}\ T$	$W\ m^{-1}\ K^{-1}$
residual resistivity	ρ_R		$\Omega\ m$
relaxation time	τ	$\tau = l/v_F$	s
Lorenz coefficient	L	$L = \lambda/\sigma T$	$V^2\ K^{-2}$
Hall coefficient	A_H, R_H	$E = \rho \cdot j + R_H(B \times j)$	$m^3\ C^{-1}$
thermoelectric force	E		V
Peltier coefficient	Π		V
Thomson coefficient	$\mu, (\tau)$		$V\ K^{-1}$
work function	Φ	$\Phi = E_\infty - E_F$	J
number density, number concentration	$n, (p)$		m^{-3}
gap energy	E_g		J
donor ionization energy	E_d		J
acceptor ionization energy	E_a		J
Fermi energy	E_F, ε_F		J
circular wave vector, propagation vector	k, q	$k = 2\pi/\lambda$	m^{-1}
Bloch function	$u_k(r)$	$\psi(r) = u_k(r)\exp(ik \cdot r)$	$m^{-3/2}$
charge density of electrons	ρ	$\rho(r) = -e\psi^*(r)\psi(r)$	$C\ m^{-3}$
effective mass	m^*		kg
mobility	μ	$\mu = v_{drift}/E$	$m^2\ V^{-1}\ s^{-1}$
mobility ratio	b	$b = \mu_n/\mu_p$	1
diffusion coefficient	D	$dN/dt = -DA(dn/dx)$	$m^2\ s^{-1}$
diffusion length	L	$L = \sqrt{D\tau}$	m
characteristic (Weiss) temperature	ϕ, ϕ_W		K
Curie temperature	T_C		K
Néel temperature	T_N		K

Elementary Algebra and Geometry

Fundamental Properties (Real Numbers)

$a+b=b+a$	Commutative Law for Addition
$(a+b)+c=a+(b+c)$	Associative Law for Addition
$a+0=0+a$	Identity Law for Addition
$a+(-a)=(-a)+a=0$	Inverse Law for Addition
$a(bc)=(ab)c$	Associative Law for Multiplication
$a\left(\dfrac{1}{a}\right)=\left(\dfrac{1}{a}\right)a=1,\ a\neq0$	Inverse Law for Multiplication
$(a)(1)=(1)(a)=a$	Identity Law for Multiplication
$ab=ba$	Commutative Law for Multiplication
$a(b+c)=ab+ac$	Distributive Law

DIVISION BY ZERO IS NOT DEFINED

Exponents

For integers m and n

$$a^n a^m = a^{n+m}$$

$$a^n/a^m = a^{n-m}$$

$$\left(a^n\right)^m = a^{nm}$$

$$\left(ab\right)^m = a^m b^m$$

$$\left(a/b\right)^m = a^m/b^m$$

Fractional Exponents

$$a^{p/q} = \left(a^{1/q}\right)^p$$

where $a^{1/q}$ is the positive qth root of a if $a>0$ and the negative qth root of a if a is negative and q is odd. Accordingly, the five rules of exponents given above (for integers) are also valid if m and n are fractions, provided a and b are positive.

Irrational Exponents

If an exponent is irrational, e.g., $\sqrt{2}$, the quantity, such as $a^{\sqrt{2}}$ is the limit of the sequence, $a^{1.4}, a^{1.41}, a^{1.414}, \ldots$.

Operations with Zero

$$0^m = 0;\ a^0 = 1$$

Logarithms

If x, y, and b are positive and $b \neq 1$

$$\log_b(xy) = \log_b x + \log_b y$$

$$\log_b(x/y) = \log_b x - \log_b y$$

$$\log_b x^p = p \log_b x$$

$$\log_b(1/x) = -\log_b x$$

$$\log_b b = 1$$

$$\log_b 1 = 0 \qquad Note: b^{\log_b x} = x.$$

Change of Base ($a \neq 1$)

$$\log_b x = \log_a x \log_b a$$

Factorials

The factorial of a positive integer n is the product of all the positive integers less than or equal to the integer n and is denoted $n!$. Thus,

$$n! = 1 \cdot 2 \cdot 3 \cdot \ldots \cdot n.$$

Factorial 0 is defined: $0! = 1$.

Stirling's Approximation

$$\lim_{n \to \infty} (n/e)^n \sqrt{2\pi n} = n!$$

Binomial Theorem

For positive integer n

$$(x+y)^n = x^n + nx^{n-1}y + \frac{n(n-1)}{2!}x^{n-2}y^2 + \frac{n(n-1)(n-2)}{3!}x^{n-3}y^3 + \cdots + nxy^{n-1} + y^n.$$

Factors and Expansion

$$(a+b)^2 = a^2 + 2ab + b^2$$

$$(a-b)^2 = a^2 - 2ab + b^2$$

$$(a+b)^3 = a^3 + 3a^2b + 3ab^2 + b^3$$

$$(a-b)^3 = a^3 - 3a^2b + 3ab^2 - b^3$$

$$(a^2 - b^2) = (a-b)(a+b)$$

$$(a^3 - b^3) = (a-b)(a^2 + ab + b^2)$$

$$(a^3 + b^3) = (a+b)(a^2 - ab + b^2)$$

Progression

An *arithmetic progression* is a sequence in which the difference between any term and the preceding term is a constant (d):

$$a, a+d, a+2d, \ldots, a+(n-1)d.$$

If the last term is denoted l $[=a+(n-1)d]$, then the sum is

$$s = \frac{n}{2}(a+l).$$

A *geometric progression* is a sequence in which the ratio of any term to the preceding term is a constant r. Thus, for n terms

$$a, ar, ar^2, \ldots, ar^{n-1}$$

The sum is

$$S = \frac{a - ar^n}{1 - r}$$

Complex Numbers

A complex number is an ordered pair of real numbers (a, b).

Equality: $(a, b) = (c, d)$ if and only if $a = c$ and $b = d$

Addition: $(a, b) + (c, d) = (a+c, b+d)$

Multiplication: $(a, b)(c, d) = (ac - bd, ad + bc)$

The first element (a, b) is called the *real* part; the second the *imaginary* part. An alternate notation for (a, b) is $a + bi$, where $i^2 = (-1, 0)$, and $i = (0, 1)$ or $0 + 1i$ is written for this complex number as a convenience. With this understanding, i behaves as a number, i.e., $(2 - 3i)(4 + i) = 8 - 12i + 2i - 3i^2 = 11 - 10i$. The conjugate of $a + bi$ is $a - bi$ and the product of a complex number and its conjugate is $a^2 + b^2$. Thus, *quotients* are computed by multiplying numerator and denominator by the conjugate of the denominator, as illustrated below:

$$\frac{2+3i}{4+2i} = \frac{(4-2i)(2+3i)}{(4-2i)(4+2i)} = \frac{14+8i}{20} = \frac{7+4i}{10}$$

Polar Form

The complex number $x + iy$ may be represented by a plane vector with components x and y

$$x + iy = r(\cos\theta + i\sin\theta)$$

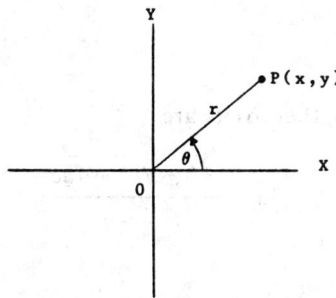

FIGURE 1 Polar form of complex number.

(see Figure 1). Then, given two complex numbers $z_1 = r_1(\cos \theta_1 + i \sin \theta_1)$ and $z_2 = r_2(\cos \theta_2 + i \sin \theta_2)$, the product and quotient are

product: $\qquad z_1 z_2 = r_1 r_2 [\cos(\theta_1 + \theta_2) + i \sin(\theta_1 + \theta_2)]$

quotient: $\qquad z_1/z_2 = (r_1/r_2)[\cos(\theta_1 - \theta_2) + i \sin(\theta_1 - \theta_2)]$

powers: $\qquad z^n = [r(\cos \theta + i \sin \theta)]^n = r^n[\cos n\theta + i \sin n\theta]$

roots: $\qquad z^{1/n} = [r(\cos \theta + i \sin \theta)]^{1/n}$

$$= r^{1/n} \left[\cos \frac{\theta + k.360}{n} + i \sin \frac{\theta + k.360}{n} \right], \qquad k = 0, 1, 2, \ldots, n-1$$

Permutations

A permutation is an ordered arrangement (sequence) of all or part of a set of objects. The number of permutations of n objects taken r at a time is

$$p(n,r) = n(n-1)(n-2) \ldots (n-r+1)$$

$$= \frac{n!}{(n-r)!}$$

A permutation of positive integers is "even" or "odd" if the total number of inversions is an even integer or an odd integer, respectively. Inversions are counted relative to each integer j in the permutation by counting the number of integers that follow j and are less than j. These are summed to give the total number of inversions. For example, the permutation 4132 has four inversions: three relative to 4 and one relative to 3. This permutation is therefore even.

Combinations

A combination is a selection of one or more objects from among a set of objects regardless of order. The number of combinations of n different objects taken r at a time is

$$C(n,r) = \frac{P(n,r)}{r!} = \frac{n!}{r!(n-r)!}$$

Algebraic Equations

Quadratic

If $ax^2 + bx + c = 0$, and $a \neq 0$, then roots are

$$x = \frac{-b \pm \sqrt{b^2 - 4ac}}{2a}$$

Cubic

To solve $x^3 + bx^2 + cx + d = 0$, let $x = y - b/3$. Then the *reduced cubic* is obtained:

$$y^3 + py + q = 0$$

where $p = c - (1/3)b^2$ and $q = d - (1/3)bc + (2/27)b^3$. Solutions of the original cubic are then in terms of the reduced cubic roots y_1, y_2, y_3:

$$x_1 = y_1 - (1/3)b \qquad x_2 = y_2 - (1/3)b \qquad x_3 = y_3 - (1/3)b$$

The three roots of the reduced cubic are

$$y_1 = (A)^{1/3} + (B)^{1/3}$$

$$y_2 = W(A)^{1/3} + W^2(B)^{1/3}$$

$$y_3 = W^2(A)^{1/3} + W(B)^{1/3}$$

where

$$A = -\frac{1}{2}q + \sqrt{(1/27)p^3 + \frac{1}{4}q^2},$$

$$B = -\frac{1}{2}q - \sqrt{(1/27)p^3 + \frac{1}{4}q^2},$$

$$W = \frac{-1 + i\sqrt{3}}{2}, \qquad W^2 = \frac{-1 - i\sqrt{3}}{2}.$$

When $(1/27)p^3 + (1/4)q^2$ is negative, A is complex; in this case A should be expressed in trigonometric form: $A = r(\cos\theta + i\sin\theta)$ where θ is a first or second quadrant angle, as q is negative or positive. The three roots of the reduced cubic are

$$y_1 = 2(r)^{1/3}\cos(\theta/3)$$

$$y_2 = 2(r)^{1/3}\cos\left(\frac{\theta}{3} + 120°\right)$$

$$y_3 = 2(r)^{1/3}\cos\left(\frac{\theta}{3} + 240°\right)$$

Geometry

Figures 2 to 12 are a collection of common geometric figures. Area (A), volume (V), and other measurable features are indicated.

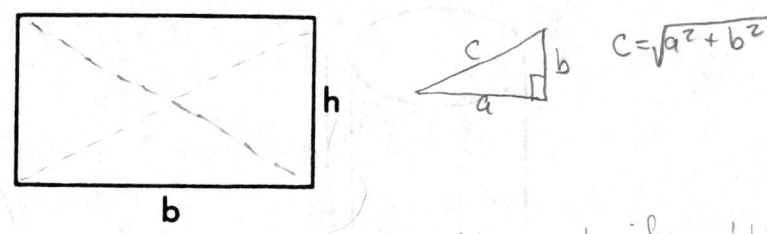

FIGURE 2   Rectangle. $A = bh.$ = 2 triangles ∴ triangle = $\frac{1}{2}bh$

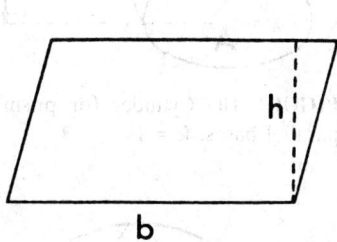

FIGURE 3   Parallelogram. $A = bh.$

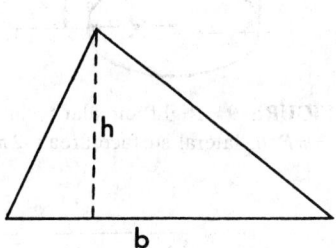

FIGURE 4   Triangle. $A = \dfrac{1}{2}bh.$

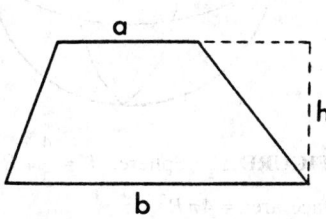

FIGURE 5   Trapezoid. $A = \dfrac{1}{2}(a+b)h.$

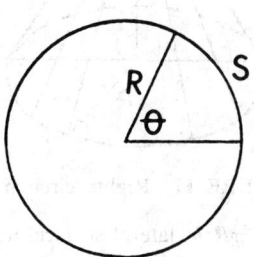

FIGURE 6   Circle. $A = \pi R^2$; circumference $= 2\pi R$; arc length $S = R\theta$ (θ in radians).

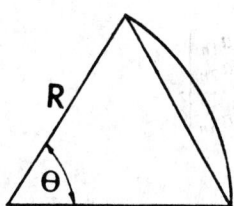

FIGURE 7   Sector of circle. $A_{\text{sector}} = \dfrac{1}{2}R^2\theta$; $A_{\text{segment}} = \dfrac{1}{2}R^2(\theta - \sin\theta).$

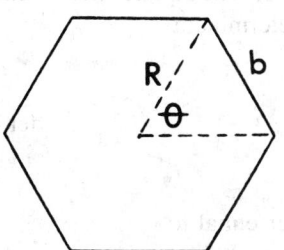

FIGURE 8   Regular polygon of n sides. $A = \dfrac{n}{4}b^2 \operatorname{ctn}\dfrac{\pi}{n}$; $R = \dfrac{b}{2}\csc\dfrac{\pi}{n}.$

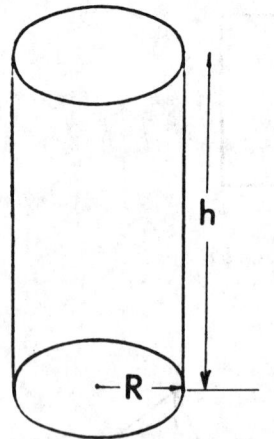

FIGURE 9 Right circular cylinder. $V = \pi R^2 h$; lateral surface area $= 2\pi Rh$.

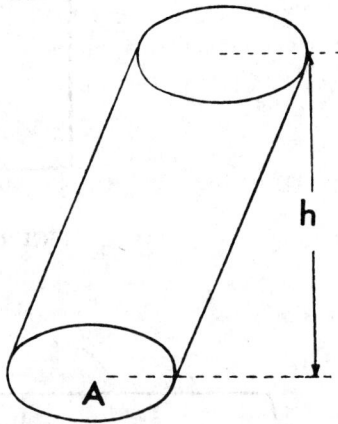

FIGURE 10 Cylinder (or prism) with parallel bases. $V = Ah$.

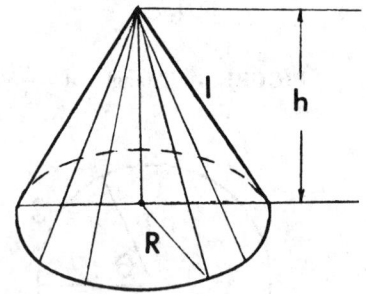

FIGURE 11 Right circular cone. $V = \frac{1}{3}\pi R^2 h$; lateral surface area $= \pi Rl = \pi R\sqrt{R^2 + h^2}$.

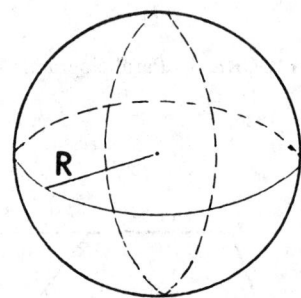

FIGURE 12 Sphere. $V = \frac{4}{3}\pi R^3$; surface area $= 4\pi R^2$.

Determinants, Matrices, and Linear Systems of Equations

Determinants

Definition. The square array (matrix) A, with n rows and n columns, has associated with it the determinant

$$\det A = \begin{vmatrix} a_{11} & a_{12} & \cdots & a_{1n} \\ a_{21} & a_{22} & \cdots & a_{2n} \\ \cdots & \cdots & \cdots & \cdots \\ a_{n1} & a_{n2} & \cdots & a_{nn} \end{vmatrix},$$

a number equal to

$$\sum (\pm) a_{1i} a_{2j} a_{3k} \cdots a_{nl}$$

where $i, j, k, \ldots, l$ is a permutation of the n integers $1, 2, 3, \ldots, n$ in some order. The sign is plus if the permutation is *even* and is minus if the permutation is *odd*. The 2×2 determinant

$$\begin{vmatrix} a_{11} & a_{12} \\ a_{21} & a_{22} \end{vmatrix}$$

has the value $a_{11}a_{22} - a_{12}a_{21}$ since the permutation $(1,2)$ is even and $(2,1)$ is odd. For 3×3 determinants, permutations are as follows:

1,	2,	3	even
1,	3,	2	odd
2,	1,	3	odd
2,	3,	1	even
3,	1,	2	even
3,	2,	1	odd

Thus,

$$\begin{vmatrix} a_{11} & a_{12} & a_{13} \\ a_{21} & a_{22} & a_{23} \\ a_{31} & a_{32} & a_{33} \end{vmatrix} = \begin{cases} +a_{11} & \cdot & a_{22} & \cdot & a_{33} \\ -a_{11} & \cdot & a_{23} & \cdot & a_{32} \\ -a_{12} & \cdot & a_{21} & \cdot & a_{33} \\ +a_{12} & \cdot & a_{23} & \cdot & a_{31} \\ +a_{13} & \cdot & a_{21} & \cdot & a_{32} \\ -a_{13} & \cdot & a_{22} & \cdot & a_{31} \end{cases}$$

A determinant of order n is seen to be the sum of $n!$ signed products.

Evaluation by Cofactors

Each element a_{ij} has a determinant of order $(n-1)$ called a *minor* (M_{ij}) obtained by suppressing all elements in row i and column j. For example, the minor of element a_{22} in the 3×3 determinant above is

$$\begin{vmatrix} a_{11} & a_{13} \\ a_{31} & a_{33} \end{vmatrix}$$

The cofactor of element a_{ij}, denoted A_{ij}, is defined as $\pm M_{ij}$, where the sign is determined from i and j:

$$A_{ij} = (-1)^{i+j} M_{ij}.$$

The value of the $n \times n$ determinant equals the sum of products of elements of any row (or column) and their respective cofactors. Thus, for the 3×3 determinant

$$\det A = a_{11}A_{11} + a_{12}A_{12} + a_{13}A_{13} \text{ (first row)}$$

or

$$= a_{11}A_{11} + a_{21}A_{21} + a_{31}A_{31} \text{ (first column)}$$

etc.

Properties of Determinants

a. If the corresponding columns and rows of A are interchanged, det A is unchanged.

b. If any two rows (or columns) are interchanged, the sign of det A changes.

c. If any two rows (or columns) are identical, det $A = 0$.

d. If A is triangular (all elements above the main diagonal equal to zero), $A = a_{11} \cdot a_{22} \cdot \ldots \cdot a_{nn}$:

$$\begin{vmatrix} a_{11} & 0 & 0 & \cdots & 0 \\ a_{21} & a_{22} & 0 & \cdots & 0 \\ \cdots & \cdots & \cdots & \cdots & \cdots \\ a_{n1} & a_{n2} & a_{n3} & \cdots & a_{nn} \end{vmatrix}$$

e. If to each element of a row or column there is added C times the corresponding element in another row (or column), the value of the determinant is unchanged.

Matrices

Definition. A matrix is a rectangular array of numbers and is represented by a symbol A or $[a_{ij}]$:

$$A = \begin{bmatrix} a_{11} & a_{12} & \cdots & a_{1n} \\ a_{21} & a_{22} & \cdots & a_{2n} \\ \cdots & \cdots & \cdots & \cdots \\ a_{m1} & a_{m2} & \cdots & a_{mn} \end{bmatrix} = [a_{ij}]$$

The numbers a_{ij} are termed *elements* of the matrix; subscripts i and j identify the element as the number in row i and column j. The order of the matrix is $m \times n$ ("m by n"). When $m = n$, the matrix is square and is said to be of order n. For a square matrix of order n the elements $a_{11}, a_{22}, \ldots, a_{nn}$ constitute the main diagonal.

Operations

Addition. Matrices A and B of the same order may be added by adding corresponding elements, i.e., $A + B = [(a_{ij} + b_{ij})]$.

Scalar multiplication. If $A = [a_{ij}]$ and c is a constant (scalar), then $cA = [ca_{ij}]$, that is, every element of A is multiplied by c. In particular, $(-1)A = -A = [-a_{ij}]$ and $A + (-A) = 0$, a matrix with all elements equal to zero.

Multiplication of matrices. Matrices A and B may be multiplied only when they are conformable, which means that the number of columns of A equals the number of rows of B. Thus, if A is $m \times k$ and B is $k \times n$, then the product $C = AB$ exists as an

$m \times n$ matrix with elements c_{ij} equal to the sum of products of elements in row i of A and corresponding elements of column j of B:

$$c_{ij} = \sum_{l=1}^{k} a_{il}b_{lj}$$

For example, if

$$\begin{bmatrix} a_{11} & a_{12} & \cdots & a_{1k} \\ a_{21} & a_{22} & \cdots & a_{2k} \\ \cdots & \cdots & \cdots & \cdots \\ a_{m1} & \cdots & \cdots & a_{mk} \end{bmatrix} \cdot \begin{bmatrix} b_{11} & b_{12} & \cdots & b_{1n} \\ b_{21} & b_{22} & \cdots & b_{2n} \\ \cdots & \cdots & \cdots & \cdots \\ b_{k1} & b_{k2} & \cdots & b_{kn} \end{bmatrix} = \begin{bmatrix} c_{11} & c_{12} & \cdots & c_{1n} \\ c_{21} & c_{22} & \cdots & c_{2n} \\ \cdots & \cdots & \cdots & \cdots \\ c_{m1} & c_{m2} & \cdots & c_{mn} \end{bmatrix}$$

then element c_{21} is the sum of products $a_{21}b_{11} + a_{22}b_{21} + \ldots + a_{2k}b_{k1}$.

Properties

$$A + B = B + A$$

$$A + (B + C) = (A + B) + C$$

$$(c_1 + c_2)A = c_1 A + c_2 A$$

$$c(A + B) = cA + cB$$

$$c_1(c_2 A) = (c_1 c_2) A$$

$$(AB)(C) = A(BC)$$

$$(A + B)(C) = AC + BC$$

$$AB \neq BA \text{ (in general)}$$

Transpose

If A is an $n \times m$ matrix, the matrix of order $m \times n$ obtained by interchanging the rows and columns of A is called the *transpose* and is denoted A^T. The following are properties of A, B, and their respective transposes:

$$\left(A^T\right)^T = A$$

$$(A + B)^T = A^T + B^T$$

$$(cA)^T = cA^T$$

$$(AB)^T = B^T A^T$$

A *symmetric* matrix is a square matrix A with the property $A = A^T$.

Identity Matrix

A square matrix in which each element of the main diagonal is the same constant a and all other elements zero is called a *scalar* matrix.

$$\begin{bmatrix} a & 0 & 0 & \cdots & 0 \\ 0 & a & 0 & \cdots & 0 \\ 0 & 0 & a & \cdots & 0 \\ \cdots & \cdots & \cdots & \cdots & \\ 0 & 0 & 0 & \cdots & a \end{bmatrix}$$

When a scalar matrix multiplies a conformable second matrix A, the product is aA; that is, the same as multiplying A by a scalar a. A scalar matrix with diagonal elements 1 is called the *identity*, or *unit* matrix and is denoted I. Thus, for any nth order matrix A, the identity matrix of order n has the property

$$AI = IA = A$$

Adjoint

If A is an n-order square matrix and A_{ij} the cofactor of element a_{ij}, the transpose of $[A_{ij}]$ is called the *adjoint* of A:

$$adj A = [A_{ij}]^T$$

Inverse Matrix

Given a square matrix A of order n, if there exists a matrix B such that $AB = BA = I$, then B is called the *inverse* of A. The inverse is denoted A^{-1}. A necessary and sufficient condition that the square matrix A have an inverse is det $A \neq 0$. Such a matrix is called *nonsingular*; its inverse is unique and it is given by

$$A^{-1} = \frac{adj A}{\det A}$$

Thus, to form the inverse of the nonsingular matrix A, form the adjoint of A and divide each element of the adjoint by det A. For example,

$$\begin{bmatrix} 1 & 0 & 2 \\ 3 & -1 & 1 \\ 4 & 5 & 6 \end{bmatrix} \text{ has matrix of cofactors } \begin{bmatrix} -11 & -14 & 19 \\ 10 & -2 & -5 \\ 2 & 5 & -1 \end{bmatrix},$$

$$\text{adjoint} = \begin{bmatrix} -11 & 10 & 2 \\ -14 & -2 & 5 \\ 19 & -5 & -1 \end{bmatrix} \text{ and determinant 27.}$$

Therefore,

$$A^{-1} = \begin{bmatrix} \dfrac{-11}{27} & \dfrac{10}{27} & \dfrac{2}{27} \\ \dfrac{-14}{27} & \dfrac{-2}{27} & \dfrac{5}{27} \\ \dfrac{19}{27} & \dfrac{-5}{27} & \dfrac{-1}{27} \end{bmatrix}.$$

Systems of Linear Equations

Given the system

$$
\begin{aligned}
a_{11}x_1 + a_{12}x_2 + \cdots + a_{1n}x_n &= b_1 \\
a_{21}x_1 + a_{22}x_2 + \cdots + a_{2n}x_n &= b_2 \\
\vdots \qquad\qquad \vdots \qquad\qquad \vdots \qquad\quad \vdots \\
a_{n1}x_1 + a_{n2}x_2 + \cdots + a_{nn}x_n &= b_n
\end{aligned}
$$

a unique solution exists if $\det A \neq 0$, where A is the $n \times n$ matrix of coefficients $[a_{ij}]$.

Solution by Determinants (Cramer's Rule)

$$
x_1 =
\begin{vmatrix}
b_1 & a_{12} & \cdots & a_{1n} \\
b_2 & a_{22} & & \\
\vdots & \vdots & & \vdots \\
b_n & a_{n2} & & a_{nn}
\end{vmatrix}
\div \det A
$$

$$
x_2 =
\begin{vmatrix}
a_{11} & b_1 & a_{13} & \cdots & a_{1n} \\
a_{21} & b_2 & \cdots & & \cdots \\
\vdots & \vdots & & & \\
a_{n1} & b_n & a_{n3} & & a_{nn}
\end{vmatrix}
\div \det A
$$

$$
\vdots
$$

$$
x_k = \frac{\det A_k}{\det A},
$$

where A_k is the matrix obtained from A by replacing the kth column of A by the column of b's.

Matrix Solution

The linear system may be written in matrix form $AX = B$ where A is the matrix of coefficients $[a_{ij}]$ and X and B are

$$
X =
\begin{bmatrix}
x_1 \\
x_2 \\
\vdots \\
x_n
\end{bmatrix}
\qquad
B =
\begin{bmatrix}
b_1 \\
b_2 \\
\vdots \\
b_n
\end{bmatrix}
$$

If a unique solution exists, $\det A \neq 0$; hence A^{-1} exists and

$$
X = A^{-1}B.
$$

Trigonometry

Triangles

In any triangle (in a plane) with sides a, b, and c and corresponding opposite angles A, B, C,

$$\frac{a}{\sin A} = \frac{b}{\sin B} = \frac{c}{\sin C}.$$
(Law of Sines)

$$a^2 = b^2 + c^2 - 2cb \cos A.$$
(Law of Cosines)

$$\frac{a+b}{a-b} = \frac{\tan\frac{1}{2}(A+B)}{\tan\frac{1}{2}(A-B)}.$$
(Law of Tangents)

$$\sin\frac{1}{2}A = \sqrt{\frac{(s-b)(s-c)}{bc}},$$
where $s = \frac{1}{2}(a+b+c)$.

$$\cos\frac{1}{2}A = \sqrt{\frac{s(s-a)}{bc}}.$$

$$\tan\frac{1}{2}A = \sqrt{\frac{(s-b)(s-c)}{s(s-a)}}.$$

$$\text{Area} = \frac{1}{2}bc \sin A$$
$$= \sqrt{s(s-a)(s-b)(s-c)}.$$

If the vertices have coordinates $(x_1, y_1), (x_2, y_2), (x_3, y_3)$, the area is the *absolute value* of the expression

$$\frac{1}{2}\begin{vmatrix} x_1 & y_1 & 1 \\ x_2 & y_2 & 1 \\ x_3 & y_3 & 1 \end{vmatrix}$$

Trigonometric Functions of an Angle

With reference to Figure 13, $P(x, y)$ is a point in either one of the four quadrants and A is an angle whose initial side is coincident with the positive x-axis and whose terminal side contains the point $P(x, y)$. The distance from the origin $P(x, y)$ is denoted by r and is positive. The trigonometric functions of the angle A are defined as:

$$
\begin{aligned}
\sin A &= \text{sine } A &&= y/r \\
\cos A &= \text{cosine } A &&= x/r \\
\tan A &= \text{tangent } A &&= y/x \\
\text{ctn } A &= \text{cotangent } A &&= x/y \\
\sec A &= \text{secant } A &&= r/x \\
\csc A &= \text{cosecant } A &&= r/y
\end{aligned}
$$

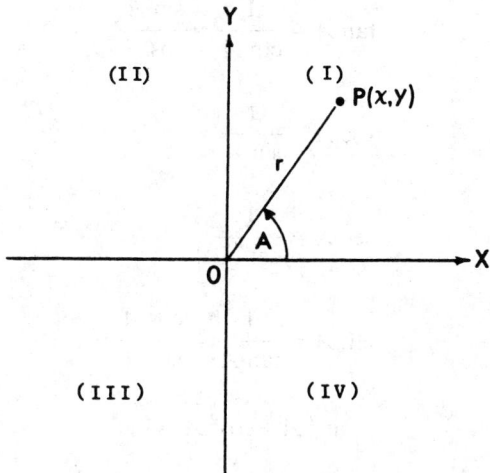

FIGURE 13 The trigonometric point. Angle A is taken to be positive when the rotation is counterclockwise and negative when the rotation is clockwise. The plane is divided into quadrants as shown.

Angles are measured in degrees or radians; $180° = \pi$ radians; 1 radian $= 180°/\pi$ degrees.

The trigonometric functions of 0°, 30°, 45°, and integer multiples of these are directly computed.

	0°	30°	45°	60°	90°	120°	135°	150°	180°
sin	0	$\dfrac{1}{2}$	$\dfrac{\sqrt{2}}{2}$	$\dfrac{\sqrt{3}}{2}$	1	$\dfrac{\sqrt{3}}{2}$	$\dfrac{\sqrt{2}}{2}$	$\dfrac{1}{2}$	0
cos	1	$\dfrac{\sqrt{3}}{2}$	$\dfrac{\sqrt{2}}{2}$	$\dfrac{1}{2}$	0	$-\dfrac{1}{2}$	$-\dfrac{\sqrt{2}}{2}$	$-\dfrac{\sqrt{3}}{2}$	-1
tan	0	$\dfrac{\sqrt{3}}{3}$	1	$\sqrt{3}$	∞	$-\sqrt{3}$	-1	$-\dfrac{\sqrt{3}}{3}$	0
ctn	∞	$\sqrt{3}$	1	$\dfrac{\sqrt{3}}{3}$	0	$-\dfrac{\sqrt{3}}{3}$	-1	$-\sqrt{3}$	∞
sec	1	$\dfrac{2\sqrt{3}}{3}$	$\sqrt{2}$	2	∞	-2	$-\sqrt{2}$	$-\dfrac{2\sqrt{3}}{3}$	-1
csc	∞	2	$\sqrt{2}$	$\dfrac{2\sqrt{3}}{3}$	1	$\dfrac{2\sqrt{3}}{3}$	$\sqrt{2}$	2	∞

Trigonometric Identities

$$\sin A = \frac{1}{\csc A}$$

$$\cos A = \frac{1}{\sec A}$$

$$\tan A = \frac{1}{\operatorname{ctn} A} = \frac{\sin A}{\cos A}$$

$$\csc A = \frac{1}{\sin A}$$

$$\sec A = \frac{1}{\cos A}$$

$$\operatorname{ctn} A = \frac{1}{\tan A} = \frac{\cos A}{\sin A}$$

$$\sin^2 A + \cos^2 A = 1$$

$$1 + \tan^2 A = \sec^2 A$$

$$1 + \operatorname{ctn}^2 A = \csc^2 A$$

$$\sin(A \pm B) = \sin A \cos B \pm \cos A \sin B$$

$$\cos(A \pm B) = \cos A \cos B \mp \sin A \sin B$$

$$\tan(A \pm B) = \frac{\tan A \pm \tan B}{1 \mp \tan A \tan B}$$

$$\sin 2A = 2 \sin A \cos A$$

$$\sin 3A = 3 \sin A - 4 \sin^3 A$$

$$\sin nA = 2 \sin(n-1)A \cos A - \sin(n-2)A$$

$$\cos 2A = 2 \cos^2 A - 1 = 1 - 2 \sin^2 A$$

$$\cos 3A = 4 \cos^3 A - 3 \cos A$$

$$\cos nA = 2 \cos(n-1)A \cos A - \cos(n-2)A$$

$$\sin A + \sin B = 2 \sin \frac{1}{2}(A+B) \cos \frac{1}{2}(A-B)$$

$$\sin A - \sin B = 2 \cos \frac{1}{2}(A+B) \sin \frac{1}{2}(A-B)$$

$$\cos A + \cos B = 2 \cos \frac{1}{2}(A+B) \cos \frac{1}{2}(A-B)$$

$$\cos A - \cos B = -2 \sin \frac{1}{2}(A+B) \sin \frac{1}{2}(A-B)$$

$$\tan A \pm \tan B = \frac{\sin(A \pm B)}{\cos A \cos B}$$

$$\operatorname{ctn} A \pm \operatorname{ctn} B = \pm \frac{\sin(A \pm B)}{\sin A \sin B}$$

$$\sin A \sin B = \frac{1}{2}\cos(A-B) - \frac{1}{2}\cos(A+B)$$

$$\cos A \cos B = \frac{1}{2}\cos(A-B) + \frac{1}{2}\cos(A+B)$$

$$\sin A \cos B = \frac{1}{2}\sin(A+B) + \frac{1}{2}\sin(A-B)$$

$$\sin \frac{A}{2} = \pm \sqrt{\frac{1-\cos A}{2}}$$

$$\cos \frac{A}{2} = \pm \sqrt{\frac{1+\cos A}{2}}$$

$$\tan \frac{A}{2} = \frac{1-\cos A}{\sin A} = \frac{\sin A}{1+\cos A} = \pm \sqrt{\frac{1-\cos A}{1+\cos A}}$$

$$\sin^2 A = \frac{1}{2}(1 - \cos 2A)$$

$$\cos^2 A = \frac{1}{2}(1 + \cos 2A)$$

$$\sin^3 A = \frac{1}{4}(3\sin A - \sin 3A)$$

$$\cos^3 A = \frac{1}{4}(\cos 3A + 3\cos A)$$

$$\sin ix = \frac{1}{2}i(e^x - e^{-x}) = i \sinh x$$

$$\cos ix = \frac{1}{2}(e^x + e^{-x}) = \cosh x$$

$$\tan ix = \frac{i(e^x - e^{-x})}{e^x + e^{-x}} = i \tanh x$$

$$e^{x+iy} = e^x(\cos y + i \sin y)$$

$$(\cos x \pm i \sin x)^n = \cos nx \pm i \sin nx$$

Inverse Trigonometric Functions

The inverse trigonometric functions are multiple valued, and this should be taken into account in the use of the following formulas.

$$\sin^{-1} x = \cos^{-1}\sqrt{1-x^2}$$

$$= \tan^{-1}\frac{x}{\sqrt{1-x^2}} = \text{ctn}^{-1}\frac{\sqrt{1-x^2}}{x}$$

$$= \sec^{-1}\frac{1}{\sqrt{1-x^2}} = \csc^{-1}\frac{1}{x}$$

$$= -\sin^{-1}(-x)$$

$$\cos^{-1} x = \sin^{-1}\sqrt{1-x^2}$$

$$= \tan^{-1}\frac{\sqrt{1-x^2}}{x} = \text{ctn}^{-1}\frac{x}{\sqrt{1-x^2}}$$

$$= \sec^{-1}\frac{1}{x} = \csc^{-1}\frac{1}{\sqrt{1-x^2}}$$

$$= \pi - \cos^{-1}(-x)$$

$$\tan^{-1} x = \text{ctn}^{-1}\frac{1}{x}$$

$$= \sin^{-1}\frac{x}{\sqrt{1+x^2}} = \cos^{-1}\frac{1}{\sqrt{1+x^2}}$$

$$= \sec^{-1}\sqrt{1+x^2} = \csc^{-1}\frac{\sqrt{1+x^2}}{x}$$

$$= -\tan^{-1}(-x)$$

Analytic Geometry

Rectangular Coordinates

The points in a plane may be placed in one-to-one correspondence with pairs of real numbers. A common method is to use perpendicular lines that are horizontal and vertical and intersect at a point called the *origin*. These two lines constitute the coordinate axes; the horizontal line is the x-axis and the vertical line is the y-axis. The positive direction of the x-axis is to the right whereas the positive direction of the y-axis is up. If P is a point in the plane one may draw lines through it that are perpendicular to the x- and y-axes (such as the broken lines of Figure 14). The lines intersect the x-axis at a point with coordinate x_1 and the y-axis at a point with coordinate y_1. We call x_1 the x-coordinate or *abscissa* and y_1 is termed the y-coordinate or *ordinate* of the point P. Thus, point P is associated with the pair of real numbers (x_1, y_1) and is denoted $P(x_1, y_1)$. The coordinate axes divide the plane into quadrants I, II, III, and IV.

Distance between Two Points; Slope

The distance d between the two points $P_1(x_1, y_1)$ and $P_2(x_2, y_2)$ is

$$d = \sqrt{(x_2 - x_1)^2 + (y_2 - y_1)^2}$$

In the special case when P_1 and P_2 are both on one of the coordinate axes, for instance, the x-axis,

$$d = \sqrt{(x_2 - x_1)^2} = |x_2 - x_1|,$$

or on the y-axis,

$$d = \sqrt{(y_2 - y_1)^2} = |y_2 - y_1|.$$

The midpoint of the line segment P_1P_2 is

$$\left(\frac{x_1 + x_2}{2}, \frac{y_1 + y_2}{2} \right).$$

The slope of the line segment P_1P_2, provided it is not vertical, is denoted by m and is given by

$$m = \frac{y_2 - y_1}{x_2 - x_1}.$$

The slope is related to the angle of inclination α (Figure 15) by

$$m = \tan \alpha$$

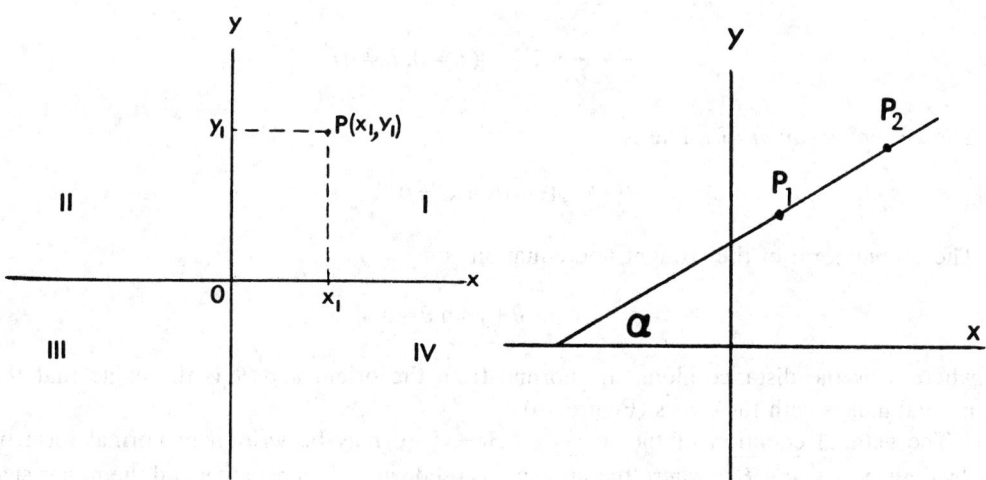

FIGURE 14 Rectangular coordinates.

FIGURE 15 The angle of inclination is the smallest angle measured counterclockwise from the positive x-axis to the line that contains P_1P_2.

Two lines (or line segments) with slopes m_1 and m_2 are perpendicular if

$$m_1 = -1/m_2$$

and are parallel if $m_1 = m_2$.

Equations of Straight Lines

A *vertical* line has an equation of the form

$$x = c$$

where $(c, 0)$ is its intersection with the x-axis. A line of slope m through point (x_1, y_1) is given by

$$y - y_1 = m(x - x_1)$$

Thus, a *horizontal line* (slope = 0) through point (x_1, y_1) is given by

$$y = y_1.$$

A nonvertical line through the two points $P_1(x_1, y_1)$ and $P_2(x_2, y_2)$ is given by either

$$y - y_1 = \left(\frac{y_2 - y_1}{x_2 - x_1}\right)(x - x_1)$$

or

$$y - y_2 = \left(\frac{y_2 - y_1}{x_2 - x_1}\right)(x - x_2).$$

A line with x-intercept a and y-intercept b is given by

$$\frac{x}{a} + \frac{y}{b} = 1 \qquad (a \neq 0, b \neq 0).$$

The *general equation* of a line is

$$Ax + By + C = 0$$

The *normal form* of the straight line equation is

$$x \cos \theta + y \sin \theta = p$$

where p is the distance along the normal from the origin and θ is the angle that the normal makes with the x-axis (Figure 16).

The general equation of the line $Ax + By + C = 0$ may be written in normal form by dividing by $\pm\sqrt{A^2 + B^2}$, where the plus sign is used when C is negative and the minus sign is used when C is positive:

$$\frac{Ax + By + C}{\pm\sqrt{A^2 + B^2}} = 0,$$

so that

$$\cos \theta = \frac{A}{\pm \sqrt{A^2 + B^2}}, \qquad \sin \theta = \frac{B}{\pm \sqrt{A^2 + B^2}}$$

and

$$p = \frac{|C|}{\sqrt{A^2 + B^2}}.$$

Distance from a Point to a Line

The perpendicular distance from a point $P(x_1, y_1)$ to the line $Ax + By + C = 0$ is given by d

$$d = \frac{Ax_1 + By_1 + C}{\pm \sqrt{A^2 + B^2}}.$$

Circle

The general equation of a circle of radius r and center at $P(x_1, y_1)$ is

$$(x - x_1)^2 + (y - y_1)^2 = r^2.$$

Parabola

A parabola is the set of all points (x, y) in the plane that are equidistant from a given line called the *directrix* and a given point called the *focus*. The parabola is symmetric about a line that contains the focus and is perpendicular to the directrix. The line of symmetry intersects the parabola at its *vertex* (Figure 17). The eccentricity $e = 1$.

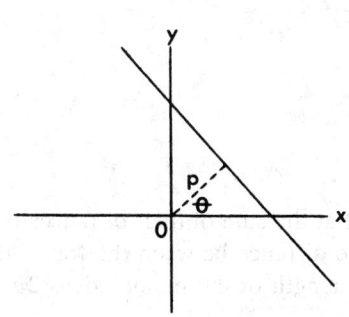

FIGURE 16 Construction for normal form of straight line equation.

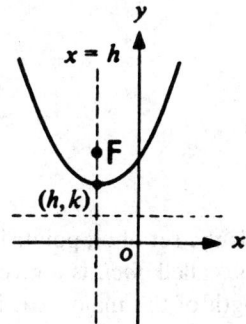

FIGURE 17 Parabola with vertex at (h, k). F identifies the focus.

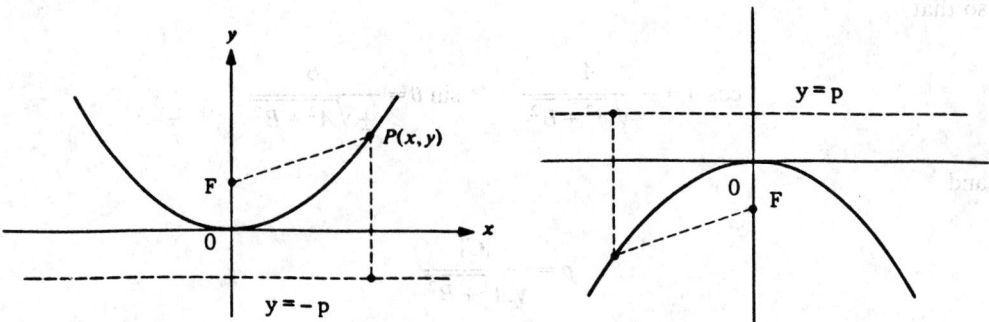

FIGURE 18 Parabolas with y-axis as the axis of symmetry and vertex at the origin.
(Left) $y = \dfrac{x^2}{4p}$; (right) $y = -\dfrac{x^2}{4p}$.

The distance between the focus and the vertex, or vertex and directrix, is denoted by $p(>0)$ and leads to one of the following equations of a parabola with vertex at the origin (Figures 18 and 19):

$$y = \frac{x^2}{4p} \qquad \text{(opens upward)}$$

$$y = -\frac{x^2}{4p} \qquad \text{(opens downward)}$$

$$x = \frac{y^2}{4p} \qquad \text{(opens to right)}$$

$$x = -\frac{y^2}{4p} \qquad \text{(opens to left)}$$

For each of the four orientations shown in Figures 18 and 19, the coresponding parabola with vertex (h, k) is obtained by replacing x by $x - h$ and y by $y - k$. Thus, the parabola in Figure 20 has the equation

$$x - h = -\frac{(y - k)^2}{4p}.$$

Ellipse

An ellipse is the set of all points in the plane such that the sum of their distances from two fixed points, called *foci*, is a given constant $2a$. The distance between the foci is denoted $2c$; the length of the major axis is $2a$, whereas the length of the minor axis is $2b$ (Figure 21) and

$$a = \sqrt{b^2 + c^2}.$$

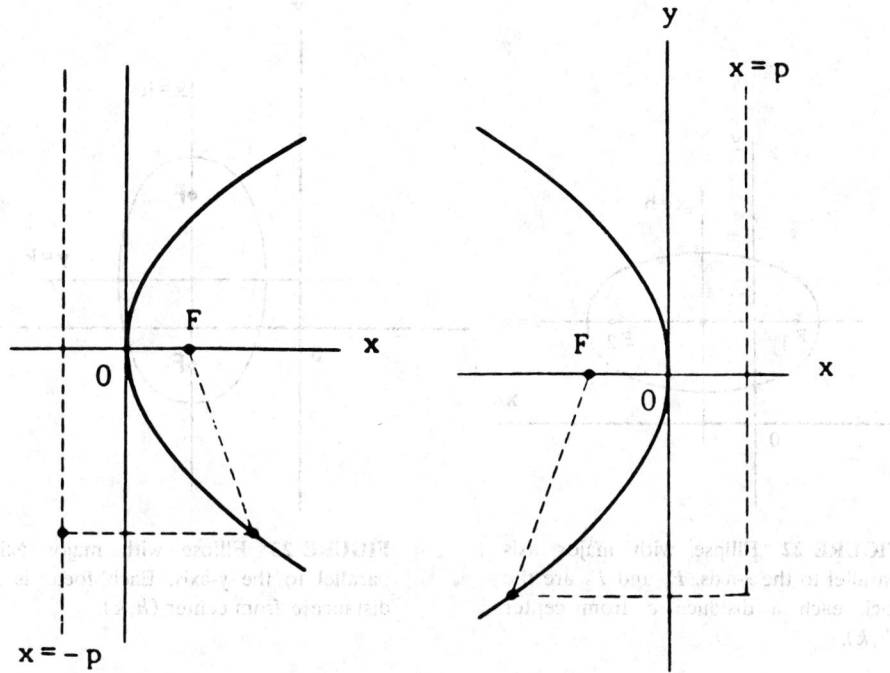

FIGURE 19 Parabolas with x-axis as the axis of symmetry and vertex at the origin. (Left) $x = \dfrac{y^2}{4p}$; (right) $x = -\dfrac{y^2}{4p}$.

The eccentricity of an ellipse, e, is < 1. An ellipse with center at point (h, k) and major axis *parallel to the x-axis* (Figure 22) is given by the equation

$$\frac{(x-h)^2}{a^2} + \frac{(y-k)^2}{b^2} = 1.$$

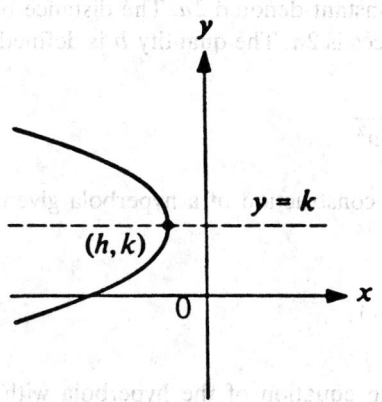

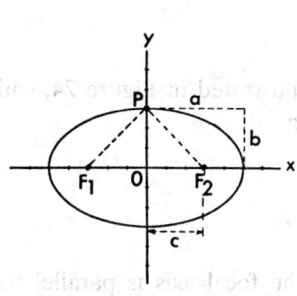

FIGURE 20 Parabola with vertex at (h, k) and axis parallel to the x-axis.

FIGURE 21 Ellipse; since point P is equidistant from foci F_1 and F_2 the segments $F_1 P$ and $F_2 P = a$; hence $a = \sqrt{b^2 + c^2}$.

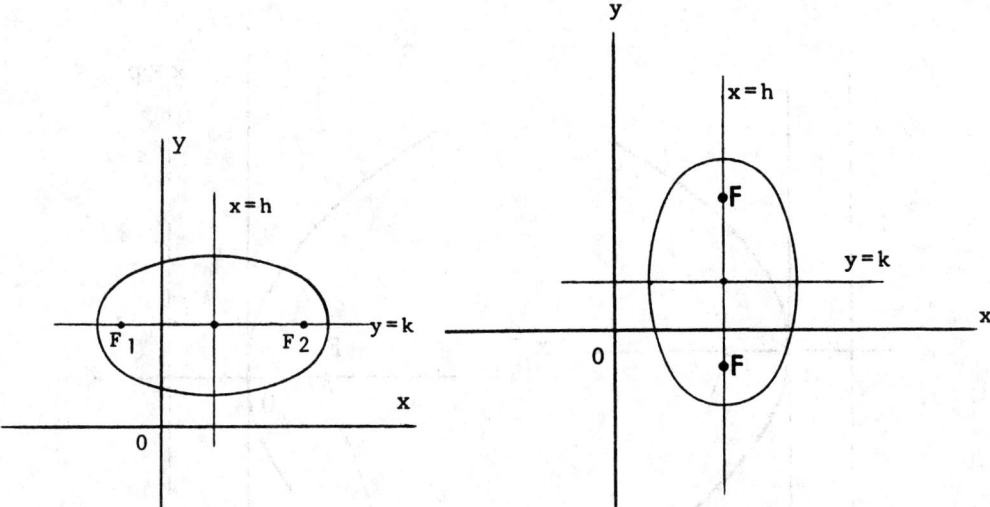

FIGURE 22 Ellipse with major axis parallel to the x-axis. F_1 and F_2 are the foci, each a distance c from center (h,k).

FIGURE 23 Ellipse with major axis parallel to the y-axis. Each focus is a distance c from center (h,k).

An ellipse with center at (h,k) and major axis parallel to the y-axis is given by the equation (Figure 23)

$$\frac{(y-k)^2}{a^2} + \frac{(x-h)^2}{b^2} = 1.$$

Hyperbola ($e > 1$)

A hyperbola is the set of all points in the plane such that the difference of its distances from two fixed points (foci) is a given positive constant denoted $2a$. The distance between the two foci is $2c$ and that between the two vertices is $2a$. The quantity b is defined by the equation

$$b = \sqrt{c^2 - a^2}$$

and is illustrated in Figure 24, which shows the construction of a hyperbola given by the equation

$$\frac{x^2}{a^2} - \frac{y^2}{b^2} = 1.$$

When the focal axis is parallel to the y-axis the equation of the hyperbola with center (h,k) (Figures 25 and 26) is

$$\frac{(y-k)^2}{a^2} - \frac{(x-h)^2}{b^2} = 1.$$

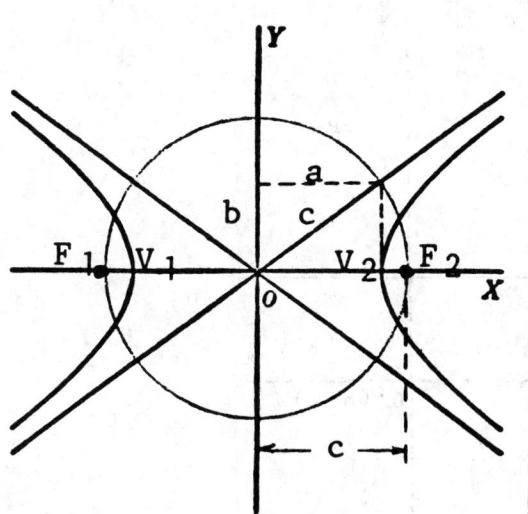

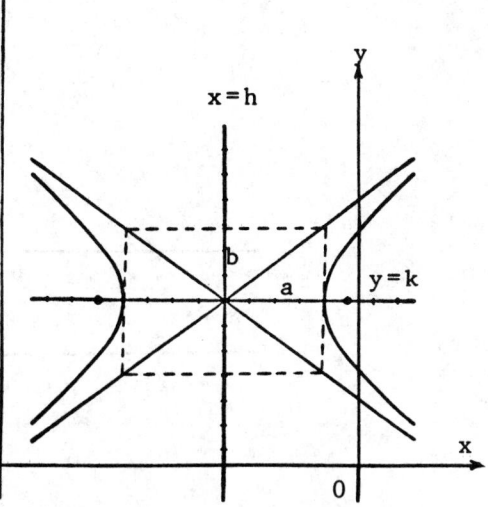

FIGURE 24 Hyperbola; V_1, V_2 = vertices; F_1, F_2 = foci. A circle at center 0 with radius c contains the vertices and illustrates the relation among a, b, and c. Asymptotes have slopes b/a and $-b/a$ for the orientation shown.

FIGURE 25 Hyperbola with center at (h, k):
$$\frac{(x-h)^2}{a^2} - \frac{(y-k)^2}{b^2} = 1; \quad \text{slopes of asymptotes } \pm b/a.$$

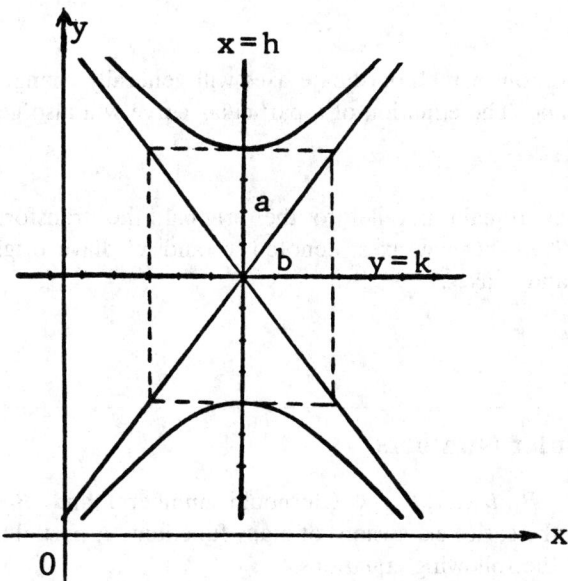

FIGURE 26 Hyperbola with center at (h, k): $\dfrac{(y-k)^2}{a^2} - \dfrac{(x-h)^2}{b^2} = 1$; slopes of asymptotes $\pm a/b$.

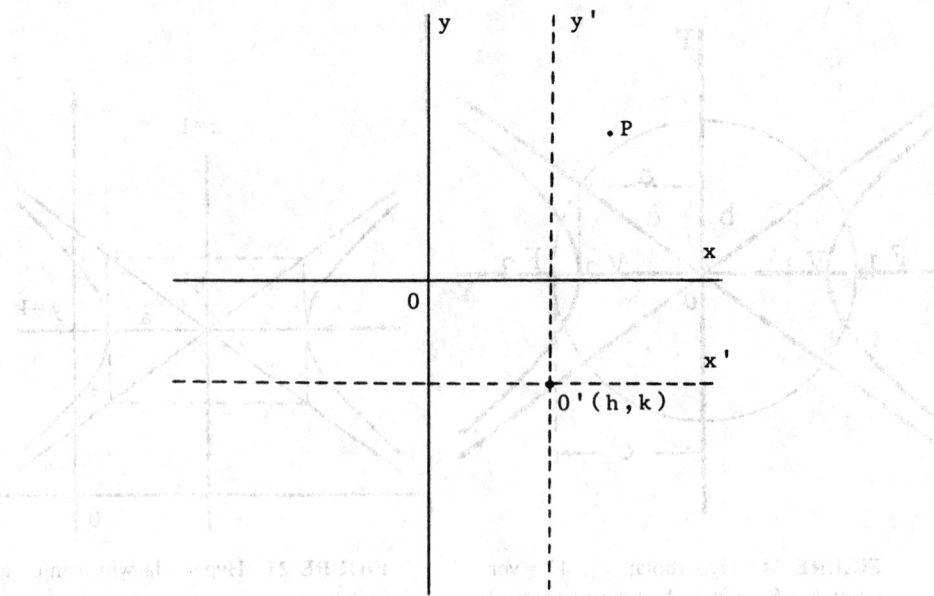

FIGURE 27 Translation of axes.

If the focal axis is parallel to the x-axis and center (h, k), then

$$\frac{(x-h)^2}{a^2} - \frac{(y-k)^2}{b^2} = 1$$

Change of Axes

A change in the position of the coordinate axes will generally change the coordinates of the points in the plane. The equation of a particular curve will also generally change.

Translation

When the new axes remain parallel to the original, the transformation is called a *translation* (Figure 27). The new axes, denoted x' and y', have origin $0'$ at (h, k) with reference to the x and y axes.

Series

Bernoulli and Euler Numbers

A set of numbers, $B_1, B_3, \ldots, B_{2n-1}$ (Bernoulli numbers) and $B_2, B_4, \ldots, B_{2n}$ (Euler numbers) appear in the series expansions of many functions. A partial listing follows; these are computed from the following equations:

$$B_{2n} - \frac{2n(2n-1)}{2!} B_{2n-2} + \frac{2n(2n-1)(2n-2)(2n-3)}{4!} B_{2n-4} - \cdots + (-1)^n = 0,$$

and

$$\frac{2^{2n}(2^{2n}-1)}{2n}B_{2n-1}=(2n-1)B_{2n-2}-\frac{(2n-1)(2n-2)(2n-3)}{3!}B_{2n-4}+\cdots+(-1)^{n-1}.$$

$$
\begin{array}{ll}
B_1 = 1/6 & B_2 = 1 \\
B_3 = 1/30 & B_4 = 5 \\
B_5 = 1/42 & B_6 = 61 \\
B_7 = 1/30 & B_8 = 1385 \\
B_9 = 5/66 & B_{10} = 50521
\end{array}
$$

$$
\begin{array}{ll}
B_{11} = 691/2730 & B_{12} = 2702765 \\
B_{13} = 7/6 & B_{14} = 199360981
\end{array}
$$
$$\vdots \qquad\qquad \vdots$$

Series of Functions

In the following, the interval of convergence is indicated, otherwise it is all x. Logarithms are to the base e. Bernoulli and Euler numbers (B_{2n-1} and B_{2n}) appear in certain expressions.

$$(a+x)^n = a^n + na^{n-1}x + \frac{n(n-1)}{2!}a^{n-2}x^2 + \frac{n(n-1)(n-2)}{3!}a^{n-3}x^3 + \cdots$$

$$+ \frac{n!}{(n-j)!j!}a^{n-j}x^j + \cdots \qquad\qquad [x^2 < a^2]$$

$$(a-bx)^{-1} = \frac{1}{a}\left[1 + \frac{bx}{a} + \frac{b^2x^2}{a^2} + \frac{b^3x^3}{a^3} + \cdots\right] \qquad [b^2x^2 < a^2]$$

$$(1\pm x)^n = 1 \pm nx + \frac{n(n-1)}{2!}x^2 \pm \frac{n(n-1)(n-2)x^3}{3!} + \cdots \qquad [x^2 < 1]$$

$$(1\pm x)^{-n} = 1 \mp nx + \frac{n(n+1)}{2!}x^2 \mp \frac{n(n+1)(n+2)}{3!}x^3 + \cdots \qquad [x^2 < 1]$$

$$(1\pm x)^{\frac{1}{2}} = 1 \pm \frac{1}{2}x - \frac{1}{2\cdot4}x^2 \pm \frac{1\cdot3}{2\cdot4\cdot6}x^3 - \frac{1\cdot3\cdot5}{2\cdot4\cdot6\cdot8}x^4 \pm \cdots \qquad [x^2 < 1]$$

$$(1\pm x)^{-\frac{1}{2}} = 1 \mp \frac{1}{2}x + \frac{1\cdot3}{2\cdot4}x^2 \mp \frac{1\cdot3\cdot5}{2\cdot4\cdot6}x^3 + \frac{1\cdot3\cdot5\cdot7}{2\cdot4\cdot6\cdot8}x^4 \mp \cdots \qquad [x^2 < 1]$$

$$(1\pm x^2)^{\frac{1}{2}} = 1 \pm \frac{1}{2}x^2 - \frac{x^4}{2\cdot4} \pm \frac{1\cdot3}{2\cdot4\cdot6}x^6 - \frac{1\cdot3\cdot5}{2\cdot4\cdot6\cdot8}x^8 \pm \cdots \qquad [x^2 < 1]$$

$$(1\pm x)^{-1} = 1 \mp x + x^2 \mp x^3 + x^4 \mp x^5 + \cdots \qquad [x^2 < 1]$$

$$(1\pm x)^{-2} = 1 \mp 2x + 3x^2 \mp 4x^3 + 5x^4 \mp \cdots \qquad [x^2 < 1]$$

$$e^x = 1 + x + \frac{x^2}{2!} + \frac{x^3}{3!} + \frac{x^4}{4!} + \cdots$$

$$e^{-x^2} = 1 - x^2 + \frac{x^4}{2!} - \frac{x^6}{3!} + \frac{x^8}{4!} - \cdots$$

$$a^x = 1 + x \log a + \frac{(x \log a)^2}{2!} + \frac{(x \log a)^3}{3!} + \cdots$$

$$\log x = (x-1) - \frac{1}{2}(x-1)^2 + \frac{1}{3}(x-1)^3 - \cdots \qquad [0 < x < 2]$$

$$\log x = \frac{x-1}{x} + \frac{1}{2}\left(\frac{x-1}{x}\right)^2 + \frac{1}{3}\left(\frac{x-1}{x}\right)^3 + \cdots \qquad \left[x > \frac{1}{2}\right]$$

$$\log x = 2\left[\left(\frac{x-1}{x+1}\right) + \frac{1}{3}\left(\frac{x-1}{x+1}\right)^3 + \frac{1}{5}\left(\frac{x-1}{x+1}\right)^5 + \cdots\right] \qquad [x > 0]$$

$$\log(1+x) = x - \frac{1}{2}x^2 + \frac{1}{3}x^3 - \frac{1}{4}x^4 + \cdots \qquad [x^2 < 1]$$

$$\log\left(\frac{1+x}{1-x}\right) = 2\left[x + \frac{1}{3}x^3 + \frac{1}{5}x^5 + \frac{1}{7}x^7 + \cdots\right] \qquad [x^2 < 1]$$

$$\log\left(\frac{x+1}{x-1}\right) = 2\left[\frac{1}{x} + \frac{1}{3}\left(\frac{1}{x}\right)^3 + \frac{1}{5}\left(\frac{1}{x}\right)^5 + \cdots\right] \qquad [x^2 > 1]$$

$$\sin x = x - \frac{x^3}{3!} + \frac{x^5}{5!} - \frac{x^7}{7!} + \cdots$$

$$\cos x = 1 - \frac{x^2}{2!} + \frac{x^4}{4!} - \frac{x^6}{6!} + \cdots$$

$$\tan x = x + \frac{x^3}{3} + \frac{2x^5}{15} + \frac{17x^7}{315} + \cdots + \frac{2^{2n}(2^{2n}-1)B_{2n-1}x^{2n-1}}{(2n)!} \qquad \left[x^2 < \frac{\pi^2}{4}\right]$$

$$\operatorname{ctn} x = \frac{1}{x} - \frac{x}{3} - \frac{x^3}{45} - \frac{2x^5}{945} - \cdots - \frac{B_{2n-1}(2x)^{2n}}{(2n)!x} - \cdots \qquad [x^2 < \pi^2]$$

$$\sec x = 1 + \frac{x^2}{2!} + \frac{5x^4}{4!} + \frac{61x^6}{6!} + \cdots + \frac{B_{2n}x^{2n}}{(2n)!} + \cdots \qquad \left[x^2 < \frac{\pi^2}{4}\right]$$

$$\csc x = \frac{1}{x} + \frac{x}{3!} + \frac{7x^3}{3\cdot5!} + \frac{31x^5}{3\cdot7!} + \cdots + \frac{2(2^{2n+1}-1)}{(2n+2)!}B_{2n+1}x^{2n+1} + \cdots \qquad [x^2 < \pi^2]$$

$$\sin^{-1} x = x + \frac{x^3}{6} + \frac{(1\cdot3)x^5}{(2\cdot4)5} + \frac{(1\cdot3\cdot5)x^7}{(2\cdot4\cdot6)7} + \cdots \qquad [x^2 < 1]$$

$$\tan^{-1} x = x - \frac{1}{3}x^3 + \frac{1}{5}x^5 - \frac{1}{7}x^7 + \cdots \qquad [x^2 < 1]$$

$$\sec^{-1} x = \frac{\pi}{2} - \frac{1}{x} - \frac{1}{6x^3} - \frac{1\cdot3}{(2\cdot4)5x^5} - \frac{1\cdot3\cdot5}{(2\cdot4\cdot6)7x^7} - \cdots \qquad [x^2 > 1]$$

$$\sinh x = x + \frac{x^3}{3!} + \frac{x^5}{5!} + \frac{x^7}{7!} + \cdots$$

$$\cosh x = 1 + \frac{x^2}{2!} + \frac{x^4}{4!} + \frac{x^6}{6!} + \frac{x^8}{8!} + \cdots$$

$$\tanh x = (2^2 - 1)2^2 B_1 \frac{x}{2!} - (2^4 - 1)2^4 B_3 \frac{x^3}{4!} + (2^6 - 1)2^6 B_5 \frac{x^5}{6!} - \cdots \quad \left[x^2 < \frac{\pi^2}{4} \right]$$

$$\text{ctnh } x = \frac{1}{x}\left(1 + \frac{2^2 B_1 x^2}{2!} - \frac{2^4 B_3 x^4}{4!} + \frac{2^6 B_5 x^6}{6!} - \cdots \right) \quad [x^2 < \pi^2]$$

$$\text{sech } x = 1 - \frac{B_2 x^2}{2!} + \frac{B_4 x^4}{4!} - \frac{B_6 x^6}{6!} + \cdots \quad \left[x^2 < \frac{\pi^2}{4} \right]$$

$$\text{csch } x = \frac{1}{x} - (2 - 1)2 B_1 \frac{x}{2!} + (2^3 - 1)2 B_3 \frac{x^3}{4!} - \cdots \quad [x^2 < \pi^2]$$

$$\sinh^{-1} x = x - \frac{1}{2}\frac{x^3}{3} + \frac{1 \cdot 3}{2 \cdot 4}\frac{x^5}{5} - \frac{1 \cdot 3 \cdot 5}{2 \cdot 4 \cdot 6}\frac{x^7}{7} + \cdots \quad [x^2 < 1]$$

$$\tanh^{-1} x = x + \frac{x^3}{3} + \frac{x^5}{5} + \frac{x^7}{7} + \cdots \quad [x^2 < 1]$$

$$\text{ctnh}^{-1} x = \frac{1}{x} + \frac{1}{3x^3} + \frac{1}{5x^5} + \cdots \quad [x^2 > 1]$$

$$\text{csch}^{-1} x = \frac{1}{x} - \frac{1}{2 \cdot 3x^3} + \frac{1 \cdot 3}{2 \cdot 4 \cdot 5x^5} - \frac{1 \cdot 3 \cdot 5}{2 \cdot 4 \cdot 6 \cdot 7x^7} + \cdots \quad [x^2 > 1]$$

$$\int_0^x e^{-t^2}\, dt = x - \frac{1}{3}x^3 + \frac{x^5}{5 \cdot 2!} - \frac{x^7}{7 \cdot 3!} + \cdots$$

Error Function

The following function, known as the error function, erf x, arises frequently in applications:

$$\text{erf } x = \frac{2}{\sqrt{\pi}} \int_0^x e^{-t^2}\, dt$$

The integral cannot be represented in terms of a finite number of elementary functions, therefore values of erf x have been compiled in tables. The following is the series for erf x:

$$\text{erf } x = \frac{2}{\sqrt{\pi}} \left[x - \frac{x^3}{3} + \frac{x^5}{5 \cdot 2!} - \frac{x^7}{7 \cdot 3!} + \cdots \right]$$

There is a close relation between this function and the area under the standard normal curve (Table 1 in the Tables of Probability and Statistics). For evaluation it is convenient to use z instead of x; then erf z may be evaluated from the area $F(z)$ given in Table 1 by use of the relation

$$\text{erf } z = 2F(\sqrt{2}\,z)$$

Example

$$\text{erf}(0.5) = 2F[(1.414)(0.5)] = 2F(0.707)$$

By interpolation from Table 1, $F(0.707) = 0.260$; thus, $\text{erf}(0.5) = 0.520$.

Series Expansion

The expression in parentheses following certain of the series indicates the region of convergence. If not otherwise indicated it is to be understood that the series converges for all finite values of x.

Binomial

$$(x+y)^n = x^n + nx^{n-1}y + \frac{n(n-1)}{2!}x^{n-2}y^2 + \frac{n(n-1)(n-2)}{3!}x^{n-3}y^3 + \cdots \quad (y^2 < x^2)$$

$$(1\pm x)^n = 1 \pm nx + \frac{n(n-1)x^2}{2!} \pm \frac{n(n-1)(n-2)x^3}{3!} + \cdots \text{ etc.} \quad (x^2 < 1)$$

$$(1\pm x)^{-n} = 1 \mp nx + \frac{n(n+1)x^2}{2!} \mp \frac{n(n+1)(n+2)x^3}{3!} + \cdots \text{ etc.} \quad (x^2 < 1)$$

$$(1\pm x)^{-1} = 1 \mp x + x^2 \mp x^3 + x^4 \mp x^5 + \cdots \quad (x^2 < 1)$$

$$(1\pm x)^{-2} = 1 \mp 2x + 3x^2 \mp 4x^3 + 5x^4 \mp 6x^5 + \cdots \quad (x^2 < 1)$$

Reversion of Series

Let a series be represented by

$$y = a_1 x + a_2 x^2 + a_3 x^3 + a_4 x^4 + a_5 x^5 + a_6 x^6 + \cdots \quad (a_1 \neq 0)$$

to find the coefficients of the series

$$x = A_1 y + A_2 y^2 + A_3 y^3 + A_4 y^4 + \cdots$$

$$A_1 = \frac{1}{a_1} \qquad A_2 = -\frac{a_2}{a_1^3} \qquad A_3 = \frac{1}{a_1^5}(2a_2^2 - a_1 a_3)$$

$$A_4 = \frac{1}{a_1^7}(5a_1 a_2 a_3 - a_1^2 a_4 - 5a_2^3)$$

$$A_5 = \frac{1}{a_1^9}(6a_1^2 a_2 a_4 + 3a_1^2 a_3^2 + 14a_2^4 - a_1^3 a_5 - 21a_1 a_2^2 a_3)$$

$$A_6 = \frac{1}{a_1^{11}}(7a_1^3 a_2 a_5 + 7a_1^3 a_3 a_4 + 84a_1 a_2^3 a_3 - a_1^4 a_6 - 28a_1^2 a_2^2 a_4 - 28a_1^2 a_2 a_3^2 - 42a_2^5)$$

$$A_7 = \frac{1}{a_1^{13}} (8a_1^4 a_2 a_6 + 8a_1^4 a_3 a_5 + 4a_1^4 a_4^2 + 120a_1^2 a_2^3 a_4 + 180a_1^2 a_2^2 a_3^2 + 132a_2^6 - a_1^5 a_7$$

$$- 36a_1^3 a_2^2 a_5 - 72a_1^3 a_2 a_3 a_4 - 12a_1^3 a_3^3 - 330a_1 a_2^4 a_3)$$

Taylor

1.
$$f(x) = f(a) + (x-a)f'(a) + \frac{(x-a)^2}{2!} f''(a) + \frac{(x-a)^3}{3!} f'''(a)$$

$$+ \cdots + \frac{(x-a)^n}{n!} f^{(n)}(a) + \cdots \text{ (Taylor's Series)}$$

(Increment form)

2.
$$f(x+h) = f(x) + hf'(x) + \frac{h^2}{2!} f''(x) + \frac{h^3}{3!} f'''(x) + \cdots$$

$$= f(h) + xf'(h) + \frac{x^2}{2!} f''(h) + \frac{x^3}{3!} f'''(h) + \cdots$$

3. If $f(x)$ is a function possessing derivatives of all orders throughout the interval $a \leq x \leq b$, then there is a value X, with $a < X < b$, such that

$$f(b) = f(a) + (b-a)f'(a) + \frac{(b-a)^2}{2!} f''(a) + \cdots$$

$$+ \frac{(b-a)^{n-1}}{(n-1)!} f^{(n-1)}(a) + \frac{(b-a)^n}{n!} f^{(n)}(X)$$

$$f(a+h) = f(a) + hf'(a) + \frac{h^2}{2!} f''(a) + \cdots + \frac{h^{n-1}}{(n-1)!} f^{(n-1)}(a)$$

$$+ \frac{h^n}{n!} f^{(n)}(a + \theta h), \qquad b = a + h, 0 < \theta < 1.$$

or

$$f(x) = f(a) + (x-a)f'(a) + \frac{(x-a)^2}{2!} f''(a) + \cdots + (x-a)^{n-1} \frac{f^{(n-1)}(a)}{(n-1)!} + R_n,$$

where

$$R_n = \frac{f^{(n)}[a + \theta \cdot (x-a)]}{n!} (x-a)^n, \qquad 0 < \theta < 1.$$

The above forms are known as Taylor's series with the remainder term.

4. Taylor's series for a function of two variables

If $\left(h\dfrac{\partial}{\partial x} + k\dfrac{\partial}{\partial y} \right) f(x,y) = h\dfrac{\partial f(x,y)}{\partial x} + k\dfrac{\partial f(x,y)}{\partial y}$;

$$\left(h\frac{\partial}{\partial x} + k\frac{\partial}{\partial y} \right)^2 f(x,y) = h^2\frac{\partial^2 f(x,y)}{\partial x^2} + 2hk\frac{\partial^2 f(x,y)}{\partial x \partial y} + k^2\frac{\partial^2 f(x,y)}{\partial y^2}$$

etc., and if $\left(h\dfrac{\partial}{\partial x} + k\dfrac{\partial}{\partial y} \right)^n f(x,y)\Big|_{\substack{x=a \\ y=b}}$ with the bar and subscripts means that after differentiation we are to replace x by a and y by b,

$$f(a+h, b+k) = f(a,b) + \left(h\frac{\partial}{\partial x} + k\frac{\partial}{\partial y} \right) f(x,y)\Big|_{\substack{x=a \\ y=b}} + \cdots$$

$$+ \frac{1}{n!}\left(h\frac{\partial}{\partial x} + k\frac{\partial}{\partial y} \right)^n f(x,y)\Big|_{\substack{x=a \\ y=b}} + \cdots$$

MacLaurin

$$f(x) = f(0) + xf'(0) + \frac{x^2}{2!}f''(0) + \frac{x^3}{3!}f'''(0) + \cdots + x^{n-1}\frac{f^{(n-1)}(0)}{(n-1)!} + R_n,$$

where

$$R_n = \frac{x^n f^{(n)}(\theta x)}{n!}, \qquad 0 < \theta < 1.$$

Exponential

$$e = 1 + \frac{1}{1!} + \frac{1}{2!} + \frac{1}{3!} + \frac{1}{4!} + \cdots$$

$$e^x = 1 + x + \frac{x^2}{2!} + \frac{x^3}{3!} + \frac{x^4}{4!} + \cdots \qquad \text{(all real values of } x)$$

$$a^x = 1 + x\log_e a + \frac{(x\log_e a)^2}{2!} + \frac{(x\log_e a)^3}{3!} + \cdots$$

$$e^x = e^a\left[1 + (x-a) + \frac{(x-a)^2}{2!} + \frac{(x-a)^3}{3!} + \cdots \right]$$

Logarithmic

$$\log_e x = \frac{x-1}{x} + \frac{1}{2}\left(\frac{x-1}{x} \right)^2 + \frac{1}{3}\left(\frac{x-1}{x} \right)^3 + \cdots \qquad (x > \tfrac{1}{2})$$

$$\log_e x = (x-1) - \tfrac{1}{2}(x-1)^2 + \tfrac{1}{3}(x-1)^3 - \cdots \qquad (2 \geq x > 0)$$

$$\log_e x = 2\left[\frac{x-1}{x+1} + \frac{1}{3}\left(\frac{x-1}{x+1}\right)^3 \frac{1}{5}\left(\frac{x-1}{x+1}\right)^5 + \cdots\right] \qquad (x > 0)$$

$$\log_e(1+x) = x - \tfrac{1}{2}x^2 + \tfrac{1}{3}x^3 - \tfrac{1}{4}x^4 + \cdots \qquad (-1 < x \le 1)$$

$$\log_e(n+1) - \log_e(n-1) = 2\left[\frac{1}{n} + \frac{1}{3n^3} + \frac{1}{5n^5} + \cdots\right]$$

$$\log_e(a+x) = \log_e a + 2\left[\frac{x}{2a+x} + \frac{1}{3}\left(\frac{x}{2a+x}\right)^3 + \frac{1}{5}\left(\frac{x}{2a+x}\right)^5 + \cdots\right]$$
$$(a > 0, \, -a < x < +\infty)$$

$$\log_e\frac{1+x}{1-x} = 2\left[x + \frac{x^3}{3} + \frac{x^5}{5} + \cdots + \frac{x^{2n-1}}{2n-1} + \cdots\right], \qquad -1 < x < 1$$

$$\log_e x = \log_e a + \frac{(x-a)}{a} - \frac{(x-a)^2}{2a^2} + \frac{(x-a)^3}{3a^3} - + \cdots, \qquad 0 < x \le 2a$$

Trigonometric

$$\sin x = x - \frac{x^3}{3!} + \frac{x^5}{5!} - \frac{x^7}{7!} + \cdots \qquad \text{(all real values of } x\text{)}$$

$$\cos x = 1 - \frac{x^2}{2!} + \frac{x^4}{4!} - \frac{x^6}{6!} + \cdots \qquad \text{(all real values of } x\text{)}$$

$$\tan x = x + \frac{x^3}{3} + \frac{2x^5}{15} + \frac{17x^7}{315} + \frac{62x^9}{2835} + \cdots$$

$$+ \frac{(-1)^{n-1}2^{2n}(2^{2n}-1)B_{2n}}{(2n)!}x^{2n-1} + \cdots, \qquad \left[x^2 < \frac{\pi^2}{4}, \text{ and } B_n \text{ represents the } n\text{th Bernoulli number.}\right]$$

$$\cot x = \frac{1}{x} - \frac{x}{3} - \frac{x^2}{45} - \frac{2x^5}{945} - \frac{x^7}{4725} - \cdots$$

$$- \frac{(-1)^{n+1}2^{2n}}{(2n)!}B_{2n}x^{2n-1} - \cdots, \qquad \left[x^2 < \pi^2, \text{ and } B_n \text{ represents the } n\text{th Bernoulli number.}\right]$$

Differential Calculus

Notation

For the following equations, the symbols $f(x)$, $g(x)$, etc., represent functions of x. The value of a function $f(x)$ at $x = a$ is denoted $f(a)$. For the function $y = f(x)$ the derivative of y with respect to x is denoted by one of the following:

$$\frac{dy}{dx}, \quad f'(x), \quad D_x y, \quad y'.$$

Higher derivatives are as follows:

$$\frac{d^2 y}{dx^2} = \frac{d}{dx}\left(\frac{dy}{dx}\right) = \frac{d}{dx} f'(x) = f''(x)$$

$$\frac{d^3 y}{dx^3} = \frac{d}{dx}\left(\frac{d^2 y}{dx^2}\right) = \frac{d}{dx} f''(x) = f'''(x), \text{ etc.}$$

and values of these at $x = a$ are denoted $f''(a)$, $f'''(a)$, etc. (see Table of Derivatives).

Slope of a Curve

The tangent line at a point $P(x, y)$ of the curve $y = f(x)$ has a slope $f'(x)$ provided that $f'(x)$ exists at P. The slope at P is defined to be that of the tangent line at P. The tangent line at $P(x_1, y_1)$ is given by

$$y - y_1 = f'(x_1)(x - x_1).$$

The *normal line* to the curve at $P(x_1, y_1)$ has slope $-1/f'(x_1)$ and thus obeys the equation

$$y - y_1 = [-1/f'(x_1)](x - x_1)$$

(The slope of a vertical line is not defined.)

Angle of Intersection of Two Curves

Two curves, $y = f_1(x)$ and $y = f_2(x)$, that intersect at a point $P(X, Y)$ where derivatives $f_1'(X)$, $f_2'(X)$ exist, have an angle (α) of intersection given by

$$\tan \alpha = \frac{f_2'(X) - f_1'(X)}{1 + f_2'(X) \cdot f_1'(X)}.$$

If $\tan \alpha > 0$, then α is the acute angle; if $\tan \alpha < 0$, then α is the obtuse angle.

Radius of Curvature

The radius of curvature R of the curve $y = f(x)$ at point $P(x, y)$ is

$$R = \frac{\{1 + [f'(x)]^2\}^{3/2}}{f''(x)}$$

In polar coordinates (θ, r) the corresponding formula is

$$R = \frac{\left[r^2 + \left(\dfrac{dr}{d\theta}\right)^2\right]^{3/2}}{r^2 + 2\left(\dfrac{dr}{d\theta}\right)^2 - r\dfrac{d^2 r}{d\theta^2}}$$

The *curvature K* is $1/R$.

Relative Maxima and Minima

The function f has a relative maximum at $x=a$ if $f(a) \geq f(a+c)$ for all values of c (positive or negative) that are sufficiently near zero. The function f has a relative minimum at $x=b$ if $f(b) \leq f(b+c)$ for all values of c that are sufficiently close to zero. If the function f is defined on the closed interval $x_1 \leq x \leq x_2$, and has a relative maximum or minimum at $x=a$, where $x_1 < a < x_2$, and if the derivative $f'(x)$ exists at $x=a$, then $f'(a)=0$. It is noteworthy that a relative maximum or minimum may occur at a point where the derivative does not exist. Further, the derivative may vanish at a point that is neither a maximum or a minimum for the function. Values of x for which $f'(x)=0$ are called "critical values." To determine whether a critical value of x, say x_c, is a relative maximum or minimum for the function at x_c, one may use the second derivative test

1. If $f''(x_c)$ is positive, $f(x_c)$ is a minimum

2. If $f''(x_c)$ is negative, $f(x_c)$ is a maximum

3. If $f''(x_c)$ is zero, no conclusion may be made

The sign of the derivative as x advances through x_c may also be used as a test. If $f'(x)$ changes from positive to zero to negative, then a maximum occurs at x_c, whereas a change in $f'(x)$ from negative to zero to positive indicates a minimum. If $f'(x)$ does not change sign as x advances through x_c, then the point is neither a maximum nor a minimum.

Points of Inflection of a Curve

The sign of the second derivative of f indicates whether the graph of $y=f(x)$ is concave upward or concave downward:

$$f''(x) > 0: \text{concave upward}$$

$$f''(x) < 0: \text{concave downward}$$

A point of the curve at which the direction of concavity changes is called a point of inflection (Figure 28). Such a point may occur where $f''(x)=0$ or where $f''(x)$ becomes infinite. More precisely, if the function $y=f(x)$ and its first derivative $y'=f'(x)$ are continuous in the interval $a \leq x \leq b$, and if $y''=f''(x)$ exists in $a < x < b$, then the graph of $y=f(x)$ for $a < x < b$ is concave upward if $f''(x)$ is positive and concave downward if $f''(x)$ is negative.

Taylor's Formula

If f is a function that is continuous on an interval that contains a and x, and if its first $(n+1)$ derivatives are continuous on this interval, then

$$f(x) = f(a) + f'(a)(x-a) + \frac{f''(a)}{2!}(x-a)^2 + \frac{f'''(a)}{3!}(x-a)^3 + \cdots + \frac{f^{(n)}(a)}{n!}(x-a)^n + R,$$

where R is called the *remainder*. There are various common forms of the remainder:

Lagrange's Form

$$R = f^{(n+1)}(\beta) \cdot \frac{(x-a)^{n+1}}{(n+1)!}; \ \beta \text{ between } a \text{ and } x.$$

Cauchy's Form

$$R = f^{(n+1)}(\beta) \cdot \frac{(x-\beta)^n (x-a)}{n!}; \ \beta \text{ between } a \text{ and } x.$$

Integral Form

$$R = \int_a^x \frac{(x-t)^n}{n!} f^{(n+1)}(t) \, dt.$$

Indeterminant Forms

If $f(x)$ and $g(x)$ are continuous in an interval that includes $x=a$ and if $f(a)=0$ and $g(a)=0$, the limit $\lim_{x \to a}(f(x)/g(x))$ takes the form "0/0", called an *indeterminant form*. *L'Hôpital's rule* is

$$\lim_{x \to a} \frac{f(x)}{g(x)} = \lim_{x \to a} \frac{f'(x)}{g'(x)}.$$

Similarly, it may be shown that if $f(x) \to \infty$ and $g(x) \to \infty$ as $x \to a$, then

$$\lim_{x \to a} \frac{f(x)}{g(x)} = \lim_{x \to a} \frac{f'(x)}{g'(x)}.$$

(The above holds for $x \to \infty$.)

Examples

$$\lim_{x \to 0} \frac{\sin x}{x} = \lim_{x \to 0} \frac{\cos x}{1} = 1$$

$$\lim_{x \to \infty} \frac{x^2}{e^x} = \lim_{x \to \infty} \frac{2x}{e^x} = \lim_{x \to \infty} \frac{2}{e^x} = 0$$

Numerical Methods

a. *Newton's method* for approximating roots of the equation $f(x)=0$: A first estimate x_1 of the root is made; then provided that $f'(x_1) \neq 0$, a better approximation is x_2

$$x_2 = x_1 - \frac{f(x_1)}{f'(x_1)}.$$

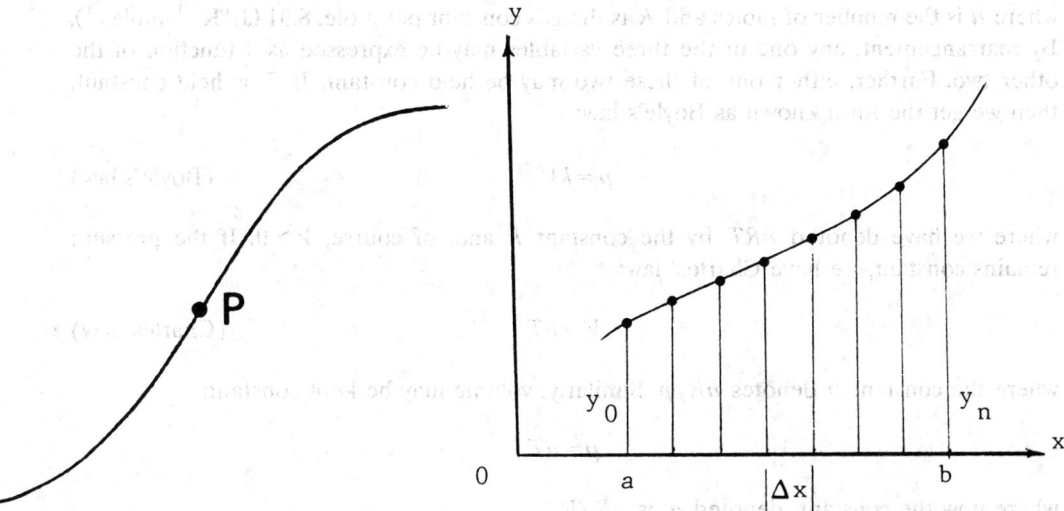

FIGURE 28 Point of inflection.   **FIGURE 29** Trapezoidal rule for area.

The process may be repeated to yield a third approximation x_3 to the root:

$$x_3 = x_2 - \frac{f(x_2)}{f'(x_2)}.$$

provided $f'(x_2)$ exists. The process may be repeated. (In certain rare cases the process will not converge.)

b. *Trapezoidal rule for areas* (Figure 29): For the function $y = f(x)$ defined on the interval (a, b) and positive there, take n equal subintervals of width $\Delta x = (b-a)/n$. The area bounded by the curve between $x = a$ and $x = b$ (or definite integral of $f(x)$) is approximately the sum of trapezoidal areas, or

$$A \sim \left(\frac{1}{2} y_0 + y_1 + y_2 + \cdots + y_{n-1} + \frac{1}{2} y_n \right) (\Delta x)$$

Estimation of the error (E) is possible if the second derivative can be obtained:

$$E = \frac{b-a}{12} f''(c)(\Delta x)^2,$$

where c is some number between a and b.

Functions of Two Variables

For the function of two variables, denoted $z = f(x, y)$, if y is held constant, say at $y = y_1$, then the resulting function is a function of x only. Similarly, x may be held constant at x_1, to give the resulting function of y.

The Gas Laws

A familiar example is afforded by the ideal gas law that relates the pressure p, the volume V and the absolute temperature T of an ideal gas:

$$pV = nRT$$

where n is the number of moles and R is the gas constant per mole, 8.31 $(\mathrm{J \cdot {}^\circ K^{-1} \cdot mole^{-1}})$. By rearrangement, any one of the three variables may be expressed as a function of the other two. Further, either one of these two may be held constant. If T is held constant, then we get the form known as Boyle's law:

$$p = kV^{-1} \qquad \text{(Boyle's law)}$$

where we have denoted nRT by the constant k and, of course, $V > 0$. If the pressure remains constant, we have Charles' law:

$$V = bT \qquad \text{(Charles' law)}$$

where the constant b denotes nR/p. Similarly, volume may be kept constant:

$$p = aT$$

where now the constant, denoted a, is nR/V.

Partial Derivatives

The physical example afforded by the ideal gas law permits clear interpretations of processes in which one of the variables is held constant. More generally, we may consider a function $z = f(x, y)$ defined over some region of the x-y-plane in which we hold one of the two coordinates, say y, constant. If the resulting function of x is differentiable at a point (x, y) we denote this derivative by one of the notations

$$f_x, \qquad \delta f/dx, \qquad \delta z/dx$$

called the *partial derivative with respect to x*. Similarly, if x is held constant and the resulting function of y is differentiable, we get the *partial derivative with respect to y*, denoted by one of the following:

$$f_y \qquad \delta f/dy \qquad \delta z/dy$$

Example

Given $z = x^4 y^3 - y \sin x + 4y$, then

$$\delta z/dx = 4(xy)^3 - y \cos x;$$

$$\delta z/dy = 3x^4 y^2 - \sin x + 4.$$

Integral Calculus

Indefinite Integral

If $F(x)$ is differentiable for all values of x in the interval (a, b) and satisfies the equation $dy/dx = f(x)$, then $F(x)$ is an integral of $f(x)$ with respect to x. The notation is $F(x) = \int f(x)\, dx$ or, in differential form, $dF(x) = f(x)\, dx$.

For any function $F(x)$ that is an integral of $f(x)$ it follows that $F(x) + C$ is also an integral. We thus write

$$\int f(x)\, dx = F(x) + C.$$

Definite Integral

Let $f(x)$ be defined on the interval $[a, b]$ which is partitioned by points $x_1, x_2, \ldots, x_j, \ldots, x_{n-1}$ between $a = x_0$ and $b = x_n$. The jth interval has length $\Delta x_j = x_j - x_{j-1}$, which may vary with j. The sum $\sum_{j=1}^{n} f(v_j)\Delta x_j$, where v_j is arbitrarily chosen in the jth subinterval, depends on the numbers $x_0, \ldots, x_n$ and the choice of the v as well as f; but if such sums approach a common value as all Δx approach zero, then this value is the definite integral of f over the interval (a, b) and is denoted $\int_a^b f(x)\, dx$. The *fundamental theorem of integral calculus* states that

$$\int_a^b f(x)\, dx = F(b) - F(a),$$

where F is any continuous indefinite integral of f in the interval (a, b).

Properties

$$\int_a^b [f_1(x) + f_2(x) + \cdots + f_j(x)]\, dx = \int_a^b f_1(x)\, dx + \int_a^b f_2(x)\, dx + \cdots + \int_a^b f_j(x)\, dx.$$

$$\int_a^b cf(x)\, dx = c \int_a^b f(x)\, dx, \text{ if } c \text{ is a constant.}$$

$$\int_a^b f(x)\, dx = -\int_b^a f(x)\, dx.$$

$$\int_a^b f(x)\, dx = \int_a^c f(x)\, dx + \int_c^b f(x)\, dx.$$

Common Applications of the Definite Integral

Area (Rectangular Coordinates)

Given the function $y = f(x)$ such that $y > 0$ for all x between a and b, the area bounded by the curve $y = f(x)$, the x-axis, and the vertical lines $x = a$ and $x = b$ is

$$A = \int_a^b f(x)\, dx.$$

Length of Arc (Rectangular Coordinates)

Given the smooth curve $f(x, y) = 0$ from point (x_1, y_1) to point (x_2, y_2), the length between these points is

$$L = \int_{x_1}^{x_2} \sqrt{1 + (dy/dx)^2}\, dx,$$

$$L = \int_{y_1}^{y_2} \sqrt{1 + (dx/dy)^2}\, dy.$$

Mean Value of a Function

The mean value of a function $f(x)$ continuous on $[a, b]$ is

$$\frac{1}{(b-a)} \int_a^b f(x)\,dx.$$

Area (Polar Coordinates)

Given the curve $r=f(\theta)$, continuous and non-negative for $\theta_1 \le \theta \le \theta_2$, the area enclosed by this curve and the radial lines $\theta=\theta_1$ and $\theta=\theta_2$ is given by

$$A = \int_{\theta_1}^{\theta_2} \frac{1}{2} [f(\theta)]^2\,d\theta.$$

Length of Arc (Polar Coordinates)

Given the curve $r=f(\theta)$ with continuous derivative $f'(\theta)$ on $\theta_1 \le \theta \le \theta_2$, the length of arc from $\theta=\theta_1$ to $\theta=\theta_2$ is

$$L = \int_{\theta_1}^{\theta_2} \sqrt{[f(\theta)]^2 + [f'(\theta)]^2}\,d\theta.$$

Volume of Revolution

Given a function $y=f(x)$ continuous and non-negative on the interval (a,b), when the region bounded by $f(x)$ between a and b is revolved about the x-axis the volume of revolution is

$$V = \pi \int_a^b [f(x)]^2\,dx.$$

Surface Area of Revolution
(Revolution about the x-axis, between a and b)

If the portion of the curve $y=f(x)$ between $x=a$ and $x=b$ is revolved about the x-axis, the area A of the surface generated is given by the following:

$$A = \int_a^b 2\pi f(x)\{1 + [f'(x)]^2\}^{1/2}\,dx$$

Work

If a variable force $f(x)$ is applied to an object in the direction of motion along the x-axis between $x=a$ and $x=b$, the work done is

$$W = \int_a^b f(x)\,dx.$$

Cylindrical and Spherical Coordinates

a. Cylindrical coordinates (Figure 30)

$$x = r\cos\theta$$

$$y = r\sin\theta$$

element of volume $dV = r\,dr\,d\theta\,dz$.

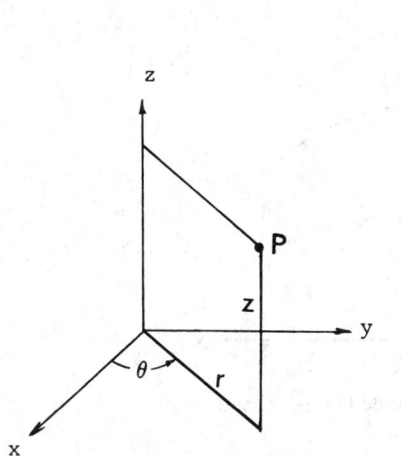

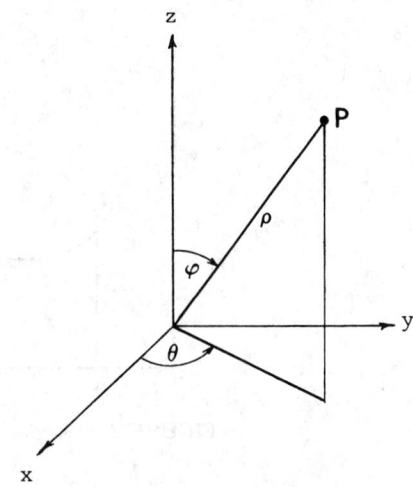

FIGURE 30 Cylindrical coordinates.   **FIGURE 31** Spherical coordinates.

b.   Spherical coordinates (Figure 31)

$$x = \rho \sin \phi \cos \theta$$

$$y = \rho \sin \phi \sin \theta$$

$$z = \rho \cos \phi$$

element of volume $dV = \rho^2 \sin \phi \, d\rho, d\phi \, d\theta$.

Double Integration

The evaluation of a double integral of $f(x, y)$ over a plane region R

$$\iint_R f(x, y) \, dA$$

is practically accomplished by iterated (repeated) integration. For example, suppose that a vertical straight line meets the boundary of R in at most two points so that there is an upper boundary, $y = y_2(x)$, and a lower boundary, $y = y_1(x)$. Also, it is assumed that these functions are continuous from a to b (see Figure 32). Then

$$\iint_R f(x, y) \, dA = \int_a^b \left(\int_{y_1(x)}^{y_2(x)} f(x, y) \, dy \right) dx$$

If R has left-hand boundary, $x = x_1(y)$, and a right-hand boundary, $x = x_2(y)$, which are continuous from c to d (the extreme values of y in R) then

$$\iint_R f(x, y) \, dA = \int_c^d \left(\int_{x_1(y)}^{x_2(y)} f(x, y) \, dx \right) dy$$

Such integrations are sometimes more convenient in polar coordinates, $x = r \cos \theta$, $y = r \sin \theta$; $dA = r \, dr \, d\theta$.

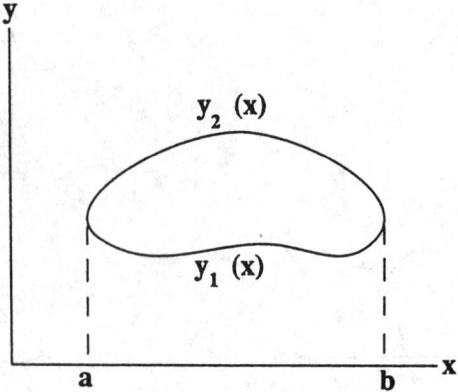

FIGURE 32   Region R bounded by $y_2(x)$ and $y_1(x)$.

Surface Area and Volume by Double Integration

For the surface given by $z = f(x, y)$, which projects onto the closed region R of the x–y-plane, one may calculate the volume V bounded above by the surface and below by R, and the surface area S by the following:

$$V = \iint_R z\, dA = \iint_R f(x, y)\, dx\, dy$$

$$S = \iint_R [1 + (\delta z/\delta x)^2 + (\delta z/\delta y)^2]^{1/2}\, dx\, dy$$

[In polar coordinates, (r, θ), we replace dA by $r\, dr\, d\theta$].

Centroid

The centroid of a region R of the x–y-plane is a point (x', y') where

$$x' = \frac{1}{A} \iint_R x\, dA; \qquad y' = \frac{1}{A} \iint_R y\, dA$$

and A is the area of the region.

Example.   For the circular sector of angle 2α and radius R, the area A is αR^2; the integral needed for x', expressed in polar coordinates is

$$\iint x\, dA = \int_{-\alpha}^{\alpha} \int_0^R (r \cos \theta) r\, dr\, d\theta$$

$$= \left[\frac{R^3}{3} \sin \theta \right]_{-\alpha}^{+\alpha} = \frac{2}{3} R^3 \sin \alpha$$

Centroids

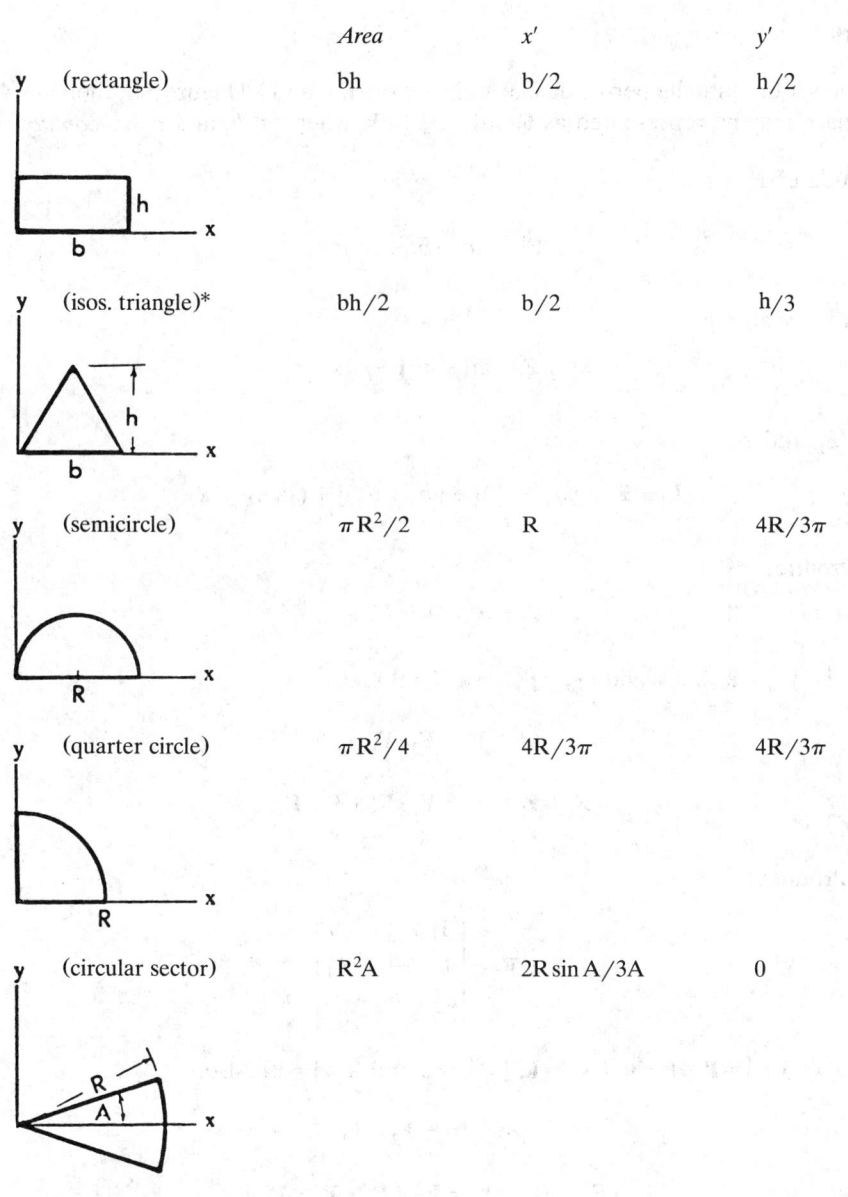

		Area	x'	y'
y	(rectangle)	bh	b/2	h/2
y	(isos. triangle)*	bh/2	b/2	h/3
y	(semicircle)	$\pi R^2/2$	R	$4R/3\pi$
y	(quarter circle)	$\pi R^2/4$	$4R/3\pi$	$4R/3\pi$
y	(circular sector)	R^2A	$2R\sin A/3A$	0

*$y' = h/3$ for any triangle of altitude h.

FIGURE 33

and thus,

$$x' = \frac{\frac{2}{3}R^3 \sin \alpha}{\alpha R^2} = \frac{2}{3}R\frac{\sin \alpha}{\alpha}.$$

Centroids of some common regions are shown in Figure 33.

Vector Analysis

Vectors

Given the set of mutually perpendicular unit vectors **i**, **j**, and **k** (Figure 34), then any vector in the space may be represented as $\mathbf{F} = a\mathbf{i} + b\mathbf{j} + c\mathbf{k}$, where a, b, and c are *components*.

Magnitude of *F*

$$|\mathbf{F}| = (a^2 + b^2 + c^2)^{\frac{1}{2}}$$

Product by Scalar *p*

$$p\mathbf{F} = pa\mathbf{i} + pb\mathbf{j} + pc\mathbf{k}.$$

Sum of F_1 and F_2

$$\mathbf{F}_1 + \mathbf{F}_2 = (a_1 + a_2)\mathbf{i} + (b_1 + b_2)\mathbf{j} + (c_1 + c_2)\mathbf{k}$$

Scalar Product

$$\mathbf{F}_1 \cdot \mathbf{F}_2 = a_1 a_2 + b_1 b_2 + c_1 c_2$$

(Thus, $\mathbf{i} \cdot \mathbf{i} = \mathbf{j} \cdot \mathbf{j} = \mathbf{k} \cdot \mathbf{k} = 1$ and $\mathbf{i} \cdot \mathbf{j} = \mathbf{j} \cdot \mathbf{k} = \mathbf{k} \cdot \mathbf{i} = 0$.) Also

$$\mathbf{F}_1 \cdot \mathbf{F}_2 = \mathbf{F}_2 \cdot \mathbf{F}_1$$

$$(\mathbf{F}_1 + \mathbf{F}_2) \cdot \mathbf{F}_3 = \mathbf{F}_1 \cdot \mathbf{F}_3 + \mathbf{F}_2 \cdot \mathbf{F}_3$$

Vector Product

$$\mathbf{F}_1 \times \mathbf{F}_2 = \begin{vmatrix} \mathbf{i} & \mathbf{j} & \mathbf{k} \\ a_1 & b_1 & c_1 \\ a_2 & b_2 & c_2 \end{vmatrix}$$

(Thus, $\mathbf{i} \times \mathbf{i} = \mathbf{j} \times \mathbf{j} = \mathbf{k} \times \mathbf{k} = 0$, $\mathbf{i} \times \mathbf{j} = \mathbf{k}$, $\mathbf{j} \times \mathbf{k} = \mathbf{i}$, and $\mathbf{k} \times \mathbf{i} = \mathbf{j}$.) Also,

$$\mathbf{F}_1 \times \mathbf{F}_2 = -\mathbf{F}_2 \times \mathbf{F}_1$$

$$(\mathbf{F}_1 + \mathbf{F}_2) \times \mathbf{F}_3 = \mathbf{F}_1 \times \mathbf{F}_3 + \mathbf{F}_2 \times \mathbf{F}_3$$

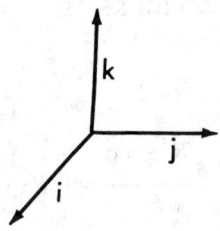

FIGURE 34   The unit vectors **i**, **j**, and **k**.

$$\mathbf{F}_1 \times (\mathbf{F}_2 + \mathbf{F}_3) = \mathbf{F}_1 \times \mathbf{F}_2 + \mathbf{F}_1 \times \mathbf{F}_3$$

$$\mathbf{F}_1 \times (\mathbf{F}_2 \times \mathbf{F}_3) = (\mathbf{F}_1 \cdot \mathbf{F}_3)\mathbf{F}_2 - (\mathbf{F}_1 \cdot \mathbf{F}_2)\mathbf{F}_3$$

$$\mathbf{F}_1 \cdot (\mathbf{F}_2 \times \mathbf{F}_3) = (\mathbf{F}_1 \times \mathbf{F}_2) \cdot \mathbf{F}_3$$

Vector Differentiation

If $\mathbf{V}$ is a vector function of a scalar variable t, then

$$\mathbf{V} = a(t)\mathbf{i} + b(t)\mathbf{j} + c(t)\mathbf{k}$$

and

$$\frac{d\mathbf{V}}{dt} = \frac{da}{dt}\mathbf{i} + \frac{db}{dt}\mathbf{j} + \frac{dc}{dt}\mathbf{k}.$$

For several vector functions $\mathbf{V}_1, \mathbf{V}_2, \ldots, \mathbf{V}_n$

$$\frac{d}{dt}(\mathbf{V}_1 + \mathbf{V}_2 + \cdots + \mathbf{V}_n) = \frac{d\mathbf{V}_1}{dt} + \frac{d\mathbf{V}_2}{dt} + \cdots + \frac{d\mathbf{V}_n}{dt},$$

$$\frac{d}{dt}(\mathbf{V}_1 \cdot \mathbf{V}_2) = \frac{d\mathbf{V}_1}{dt} \cdot \mathbf{V}_2 + \mathbf{V}_1 \cdot \frac{d\mathbf{V}_2}{dt},$$

$$\frac{d}{dt}(\mathbf{V}_1 \times \mathbf{V}_2) = \frac{d\mathbf{V}_1}{dt} \times \mathbf{V}_2 + \mathbf{V}_1 \times \frac{d\mathbf{V}_2}{dt}.$$

For a scalar valued function $g(x, y, z)$

(**gradient**) $\quad \text{grad } g = \nabla g = \frac{\delta g}{\delta x}\mathbf{i} + \frac{\delta g}{\delta y}\mathbf{j} + \frac{\delta g}{\delta z}\mathbf{k}.$

For a vector valued function $\mathbf{V}(a, b, c)$, where a, b, c are each a function of x, y, and z,

(**divergence**) $\quad \text{div } \mathbf{V} = \nabla \cdot \mathbf{V} = \frac{\delta a}{\delta x} + \frac{\delta b}{\delta y} + \frac{\delta c}{\delta z}$

(**curl**) $\quad \text{curl } \mathbf{V} = \nabla \times \mathbf{V} = \begin{vmatrix} \mathbf{i} & \mathbf{j} & \mathbf{k} \\ \dfrac{\delta}{\delta x} & \dfrac{\delta}{\delta y} & \dfrac{\delta}{\delta z} \\ a & b & c \end{vmatrix}$

Also,

$$\operatorname{div}\operatorname{grad} g = \nabla^2 g = \frac{\delta^2 g}{\delta x^2} + \frac{\delta^2 g}{\delta y^2} + \frac{\delta^2 g}{\delta z^2}.$$

and

$$\operatorname{curl}\operatorname{grad} g = \mathbf{0}; \qquad \operatorname{div}\operatorname{curl}\mathbf{V} = 0;$$

$$\operatorname{curl}\operatorname{curl}\mathbf{V} = \operatorname{grad}\operatorname{div}\mathbf{V} - (\mathbf{i}\nabla^2 a + \mathbf{j}\nabla^2 b + \mathbf{k}\nabla^2 c).$$

Divergence Theorem (Gauss)

Given a vector function F with continuous partial derivatives in a region R bounded by a closed surface S, then

$$\iiint_R \operatorname{div}\mathbf{F}\, dV = \iint_S \mathbf{n}\cdot\mathbf{F}\, dS,$$

where $\mathbf{n}$ is the (sectionally continuous) unit normal to S.

Stokes' Theorem

Given a vector function with continuous gradient over a surface S that consists of portions that are piecewise smooth and bounded by regular closed curves such as C, then

$$\iint_S \mathbf{n}\cdot\operatorname{curl}\mathbf{F}\, dS = \oint_C \mathbf{F}\cdot d\mathbf{r}$$

Planar Motion in Polar Coordinates

Motion in a plane may be expressed with regard to polar coordinates (r, θ). Denoting the position vector by $\mathbf{r}$ and its magnitude by r, we have $\mathbf{r} = r\mathbf{R}(\theta)$, where $\mathbf{R}$ is the unit vector. Also, $d\mathbf{R}/d\theta = \mathbf{P}$, a unit vector perpendicular to $\mathbf{R}$. The velocity and acceleration are then

$$\mathbf{v} = \frac{dr}{dt}\mathbf{R} + r\frac{d\theta}{dt}\mathbf{P};$$

$$\mathbf{a} = \left[\frac{d^2 r}{dt^2} - r\left(\frac{d\theta}{dt}\right)^2\right]\mathbf{R} + \left[r\frac{d^2\theta}{dt^2} + 2\frac{dr}{dt}\frac{d\theta}{dt}\right]\mathbf{P}.$$

Note that the component of acceleration in the $\mathbf{P}$ direction (transverse component) may also be written

$$\frac{1}{r}\frac{d}{dt}\left(r^2\frac{d\theta}{dt}\right)$$

so that in purely radial motion it is zero and

$$r^2 \frac{d\theta}{dt} = C \text{ (constant)}$$

which means that the position vector sweeps out area at a constant rate [see Area (Polar Coordinates) in the section entitled Integral Calculus].

Special Functions

Hyperbolic Functions

$$\sinh x = \frac{e^x - e^{-x}}{2} \qquad\qquad \operatorname{csch} x = \frac{1}{\sinh x}$$

$$\cosh x = \frac{e^x + e^{-x}}{2} \qquad\qquad \operatorname{sech} x = \frac{1}{\cosh x}$$

$$\tanh x = \frac{e^x - e^{-x}}{e^x + e^{-x}} \qquad\qquad \operatorname{ctnh} x = \frac{1}{\tanh x}$$

$$\sinh(-x) = -\sinh x \qquad\qquad \operatorname{ctnh}(-x) = -\operatorname{ctnh} x$$

$$\cosh(-x) = \cosh x \qquad\qquad \operatorname{sech}(-x) = \operatorname{sech} x$$

$$\tanh(-x) = -\tanh x \qquad\qquad \operatorname{csch}(-x) = -\operatorname{csch} x$$

$$\tanh x = \frac{\sinh x}{\cosh x} \qquad\qquad \operatorname{ctnh} x = \frac{\cosh x}{\sinh x}$$

$$\cosh^2 x - \sinh^2 x = 1 \qquad\qquad \cosh^2 x = \frac{1}{2}(\cosh 2x + 1)$$

$$\sinh^2 x = \frac{1}{2}(\cosh 2x - 1) \qquad\qquad \operatorname{ctnh}^2 x - \operatorname{csch}^2 x = 1$$

$$\operatorname{csch}^2 x - \operatorname{sech}^2 x = \operatorname{csch}^2 x \operatorname{sech}^2 x \qquad \tanh^2 x + \operatorname{sech}^2 x = 1$$

$$\sinh(x + y) = \sinh x \cosh y + \cosh x \sinh y$$

$$\cosh(x + y) = \cosh x \cosh y + \sinh x \sinh y$$

$$\sinh(x - y) = \sinh x \cosh y - \cosh x \sinh y$$

$$\cosh(x - y) = \cosh x \cosh y - \sinh x \sinh y$$

$$\tanh(x + y) = \frac{\tanh x + \tanh y}{1 + \tanh x \tanh y}$$

$$\tanh(x - y) = \frac{\tanh x - \tanh y}{1 - \tanh x \tanh y}$$

Laplace Transforms

The Laplace transform of the function $f(t)$, denoted by $F(s)$ or $L\{f(t)\}$, is defined

$$F(s) = \int_0^\infty f(t)e^{-st}\,dt$$

provided that the integration may be validly performed. A sufficient condition for the existence of $F(s)$ is that $f(t)$ be of exponential order as $t \to \infty$ and that it is sectionally continuous over every finite interval in the range $t \geq 0$. The Laplace transform of $g(t)$ is denoted by $L\{g(t)\}$ or $G(s)$.

Operations

$f(t)$ $\qquad\qquad\qquad$ $F(s) = \int_0^\infty f(t)e^{-st}\,dt$

$af(t) + bg(t)$ $\qquad\qquad$ $aF(s) + bG(s)$

$f'(t)$ $\qquad\qquad\qquad$ $sF(s) - f(0)$

$f''(t)$ $\qquad\qquad\qquad$ $s^2F(s) - sf(0) - f'(0)$

$f^{(n)}(t)$ $\qquad\qquad\quad$ $s^nF(s) - s^{n-1}f(0) - s^{n-2}f'(0) - \cdots - f^{(n-1)}(0)$

$tf(t)$ $\qquad\qquad\qquad$ $-F'(s)$

$t^nf(t)$ $\qquad\qquad\quad$ $(-1)^nF^{(n)}(s)$

$e^{at}f(t)$ $\qquad\qquad\quad$ $F(s-a)$

$\int_0^t f(t-\beta)\cdot g(\beta)\,d\beta$ $\qquad$ $F(s)\cdot G(s)$

$f(t-a)$ $\qquad\qquad\quad$ $e^{-as}F(s)$

$f\left(\dfrac{t}{a}\right)$ $\qquad\qquad\quad$ $aF(as)$

$\int_0^t g(\beta)\,d\beta$ $\qquad\qquad$ $\dfrac{1}{s}G(s)$

$f(t-c)\delta(t-c)$ $\qquad\quad$ $e^{-cs}F(s), c > 0$

where

$\delta(t-c) = 0$ if $0 \leq t < c$

$\qquad\quad\; = 1$ if $t \geq c$

$f(t) = f(t+\omega)$ $\qquad\qquad$ $\dfrac{\displaystyle\int_0^\omega e^{-s\tau}f(\tau)\,d\tau}{1 - e^{-s\omega}}$

(periodic)

Table of Laplace Transforms

$f(t)$	$F(s)$		$f(t)$	$F(s)$	
1	$1/s$		$\sinh at$	$\dfrac{a}{s^2-a^2}$	
t	$1/s^2$		$\cosh at$	$\dfrac{s}{s^2-a^2}$	
$\dfrac{t^{n-1}}{(n-1)!}$	$1/s^n$	$(n=1,2,3,\ldots)$	$e^{at}-e^{bt}$	$\dfrac{a-b}{(s-a)(s-b)},$	$(a\neq b)$
$\sqrt{t}$	$\dfrac{1}{2s}\sqrt{\dfrac{\pi}{s}}$		$ae^{at}-be^{bt}$	$\dfrac{s(a-b)}{(s-a)(s-b)},$	$(a\neq b)$
$\dfrac{1}{\sqrt{t}}$	$\sqrt{\dfrac{\pi}{s}}$		$t\sin at$	$\dfrac{2as}{(s^2+a^2)^2}$	
e^{at}	$\dfrac{1}{s-a}$		$t\cos at$	$\dfrac{s^2-a^2}{(s^2+a^2)^2}$	
te^{at}	$\dfrac{1}{(s-a)^2}$		$e^{at}\sin bt$	$\dfrac{b}{(s-a)^2+b^2}$	
$\dfrac{t^{n-1}e^{at}}{(n-1)!}$	$\dfrac{1}{(s-a)^n}$	$(n=1,2,3,\ldots)$	$e^{at}\cos bt$	$\dfrac{s-a}{(s-a)^2+b^2}$	
$\dfrac{t^x}{\Gamma(x+1)}$	$\dfrac{1}{s^{x+1}},$	$x>-1$	$\dfrac{\sin at}{t}$	$Arc\tan\dfrac{a}{s}$	
$\sin at$	$\dfrac{a}{s^2+a^2}$		$\dfrac{\sinh at}{t}$	$\dfrac{1}{2}\log_e\left(\dfrac{s+a}{s-a}\right)$	
$\cos at$	$\dfrac{s}{s^2+a^2}$				

z-Transform

For the real-valued sequence $\{f(k)\}$ and complex variable z, the z-transform, $F(z)=Z\{f(k)\}$ is defined by

$$Z\{f(k)\}=F(z)=\sum_{k=0}^{\infty}f(k)z^{-k}$$

For example, the sequence $f(k)=1$, $k=0,1,2,\ldots$, has the z-transform

$$F(z)=1+z^{-1}+z^{-2}+z^{-3}\cdots+z^{-k}+\cdots.$$

z-Transform and the Laplace Transform

When $F(t)$, a continuous function of time, is sampled at regular intervals of period T the usual Laplace transform techniques are modified. The diagramatic form of a simple sampler together with its associated input-output waveforms is shown in Figure 35.

Defining the set of impulse functions $\delta_\tau(t)$ by

$$\delta_\tau(t) \equiv \sum_{n=0}^{\infty} \delta(t - nT)$$

the input-output relationship of the sampler becomes

$$F^*(t) = F(t) \cdot \delta_\tau(t)$$

$$= \sum_{n=0}^{\infty} F(nT) \cdot \delta(t - nT).$$

While for a given $F(t)$ and T the $F^*(t)$ is unique, the converse is not true.

For function $U(t)$ the output of the ideal sampler $U^*(t)$ is a set of values $U(kT)$, $k = 0, 1, 2, \ldots$, that is,

$$U^*(t) = \sum_{k=0}^{\infty} U(t)\,\delta(t - kT)$$

The Laplace transform of the output is

$$\mathscr{L}\{U^*(t)\} = \int_0^\infty e^{-st} U^*(t)\,dt = \int_0^\infty e^{-st} \sum_{k=0}^{\infty} U(t)\,\delta(t - kT)\,dt$$

$$= \sum_{k=0}^{\infty} e^{-skT} U(kT)$$

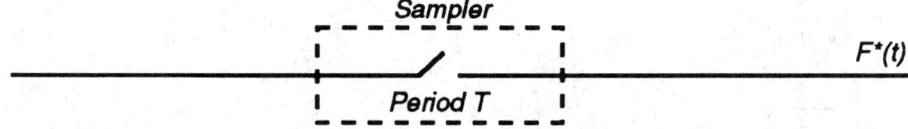

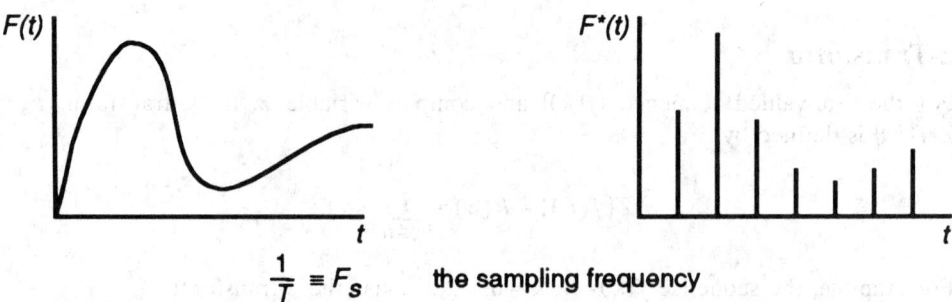

$$\frac{1}{T} \equiv F_s \qquad \text{the sampling frequency}$$

FIGURE 35

Defining $z = e^{sT}$ gives

$$\mathscr{L}\{U^*(t)\} = \sum_{k=0}^{\infty} U(kT)z^{-k}$$

which is the z-transform of the sampled signal $U(kT)$.

Properties

Linearity: $Z\{af_1(k) + bf_2(k)\} = aZ\{f_1(k)\} + bZ\{f_2(k)\} = aF_1(z) + bF_2(z)$

Right-shifting property: $Z\{f(k-n)\} = z^{-n}F(z)$

Left-shifting property: $Z\{f(k+n)\} = z^n F(z) - \sum_{k=0}^{n-1} f(k)z^{n-k}$

Time scaling: $Z\{a^k f(k)\} = F(z/a)$

Multiplication by k: $Z\{kf(k)\} = -z\,dF(z)/dz$

Initial value: $f(0) = \lim_{z \to \infty} (1 - z^{-1})F(z) = F(\infty)$

Final value: $\lim_{k \to \infty} f(k) = \lim_{z \to 1} (1 - z^{-1})F(z)$

Convolution: $Z\{f_1(k)^*f_2(k)\} = F_1(z)F_2(z)$

z-Transforms of Sampled Functions

$f(k)$	$Z\{f(kT)\} = F(z)$
1 at k; else 0	z^{-k}
1	$\dfrac{z}{z-1}$
kT	$\dfrac{Tz}{(z-1)^2}$
$(kT)^2$	$\dfrac{T^2 z(z+1)}{(z-1)^3}$
$\sin \omega kT$	$\dfrac{z \sin \omega T}{z^2 - 2z \cos \omega T + 1}$
$\cos \omega T$	$\dfrac{z(z - \cos \omega T)}{z^2 - 2z \cos \omega T + 1}$
e^{-akT}	$\dfrac{z}{z - e^{-aT}}$
kTe^{-akT}	$\dfrac{zTe^{-aT}}{(z - e^{-aT})^2}$
$(kT)^2 e^{-akT}$	$\dfrac{T^2 e^{-aT} z(z + e^{-aT})}{(z - e^{-aT})^3}$

$e^{-akT}\sin \omega kT$ $\dfrac{ze^{-aT}\sin \omega T}{z^2-2ze^{-aT}\cos \omega T+e^{-2aT}}$

$e^{-akT}\cos \omega kT$ $\dfrac{z(z-e^{-aT}\cos \omega T)}{z^2-2ze^{-aT}\cos \omega T+e^{-2aT}}$

$a^k \sin \omega kT$ $\dfrac{az\sin \omega T}{z^2-2az\cos \omega T+a^2}$

$a^k \cos \omega kT$ $\dfrac{z(z-a\cos \omega T)}{z^2-2az\cos \omega T+a^2}$

Fourier Series

The periodic function $f(t)$ with period 2π may be represented by the trigonometric series

$$a_0 + \sum_1^\infty (a_n \cos nt + b_n \sin nt)$$

where the coefficients are determined from

$$a_0 = \frac{1}{2\pi} \int_{-\pi}^\pi f(t)\, dt$$

$$a_n = \frac{1}{\pi} \int_{-\pi}^\pi f(t)\cos nt\, dt$$

$$b_n = \frac{1}{\pi} \int_{-\pi}^\pi f(t)\sin nt\, dt \qquad (n=1,2,3,\ldots)$$

Such a trigonometric series is called the Fourier series corresponding to $f(t)$ and the coefficients are termed Fourier coefficients of $f(t)$. If the function is piecewise continuous in the interval $-\pi \le t \le \pi$, and has left- and right-hand derivatives at each point in that interval, then the series is convergent with sum $f(t)$ except at points t_i at which $f(t)$ is discontinuous. At such points of discontinuity, the sum of the series is the arithmetic mean of the right- and left-hand limits of $f(t)$ at t_i. The integrals in the formulas for the Fourier coefficients can have limits of integration that span a length of 2π, for example, 0 to 2π (because of the periodicity of the integrands).

Functions with Period Other Than 2π

If $f(t)$ has period P the Fourier series is

$$f(t) \sim a_0 + \sum_1^\infty \left(a_n \cos \frac{2\pi n}{P}t + b_n \sin \frac{2\pi n}{P}t \right),$$

where

$$a_0 = \frac{1}{P} \int_{-P/2}^{P/2} f(t)\, dt$$

$$a_n = \frac{2}{P} \int_{-P/2}^{P/2} f(t) \cos \frac{2\pi n}{P}t\, dt$$

$$b_n = \frac{2}{P} \int_{-P/2}^{P/2} f(t)\sin \frac{2\pi n}{P}t\, dt.$$

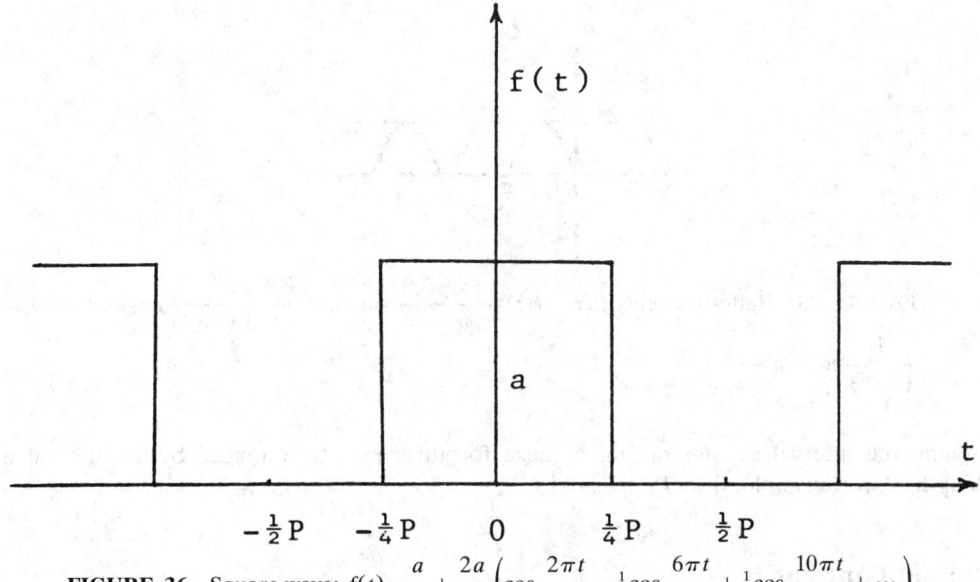

FIGURE 36 Square wave: $f(t) \sim \dfrac{a}{2} + \dfrac{2a}{\pi}\left(\cos\dfrac{2\pi t}{P} - \tfrac{1}{3}\cos\dfrac{6\pi t}{P} + \tfrac{1}{5}\cos\dfrac{10\pi t}{P} + \cdots\right)$.

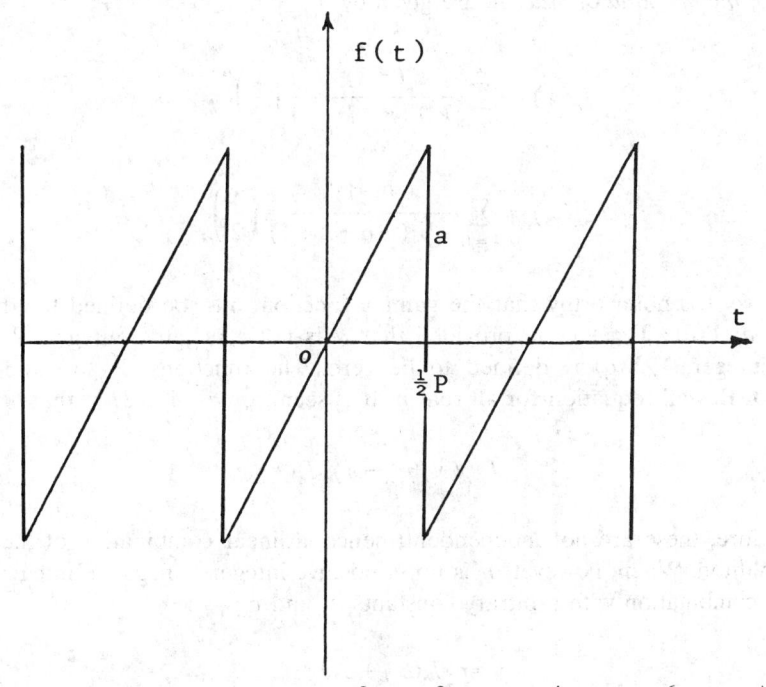

FIGURE 37 Sawtooth wave: $f(t) \sim \dfrac{2a}{\pi}\left(\sin\dfrac{2\pi t}{P} - \tfrac{1}{2}\sin\dfrac{4\pi t}{P} + \tfrac{1}{3}\sin\dfrac{6\pi t}{P} - \cdots\right)$.

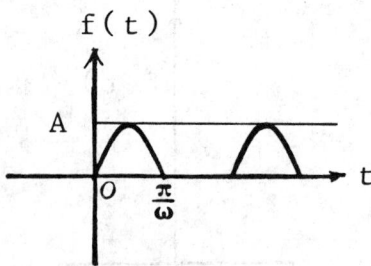

FIGURE 38   Half-wave rectifier: $f(t) \sim \dfrac{A}{\pi} + \dfrac{A}{2} \sin \omega t - \dfrac{2A}{\pi} \left(\dfrac{1}{(1)(3)} \cos 2\omega t + \dfrac{1}{(3)(5)} \cos 4\omega t + \cdots \right).$

Again, the interval of integration in these formulas may be replaced by an interval of length P, for example, 0 to P.

Bessel Functions

Bessel functions, also called cylindrical functions, arise in many physical problems as solutions of the differential equation

$$x^2 y'' + xy' + (x^2 - n^2)y = 0$$

which is known as Bessel's equation. Certain solutions of the above, known as *Bessel functions of the first kind of order n*, are given by

$$J_n(x) = \sum_{k=0}^{\infty} \frac{(-1)^k}{k!\,\Gamma(n+k+1)} \left(\frac{x}{2} \right)^{n+2k}$$

$$J_{-n}(x) = \sum_{k=0}^{\infty} \frac{(-1)^k}{k!\,\Gamma(-n+k+1)} \left(\frac{x}{2} \right)^{-n+2k}$$

In the above it is noteworthy that the gamma function must be defined for the negative argument q: $\Gamma(q) = \Gamma(q+1)/q$, provided that q is not a negative integer. When q is a negative integer, $1/\Gamma(q)$ is defined to be zero. The functions $J_{-n}(x)$ and $J_n(x)$ are solutions of Bessel's equation for all real n. It is seen, for $n = 1, 2, 3, \ldots$ that

$$J_{-n}(x) = (-1)^n J_n(x)$$

and, therefore, these are not independent; hence, a linear combination of these is not a general solution. When, however, n is not a positive integer, a negative integer, nor zero, the linear combination with arbitrary constants c_1 and c_2

$$y = c_1 J_n(x) + c_2 J_{-n}(x)$$

is the general solution of the Bessel differential equation.

The zero order function is especially important as it arises in the solution of the heat equation (for a "long" cylinder):

$$J_0(x) = 1 - \frac{x^2}{2^2} + \frac{x^4}{2^2 4^2} - \frac{x^6}{2^2 4^2 6^2} + \cdots$$

while the following relations show a connection to the trigonometric functions:

$$J_{\frac{1}{2}}(x) = \left[\frac{2}{\pi x}\right]^{1/2} \sin x$$

$$J_{-\frac{1}{2}}(x) = \left[\frac{2}{\pi x}\right]^{1/2} \cos x$$

The following recursion formula gives $J_{n+1}(x)$ for any order in terms of lower order functions:

$$\frac{2n}{x} J_n(x) = J_{n-1}(x) + J_{n+1}(x)$$

Legendre Polynomials

If Laplace's equation, $\nabla^2 V = 0$, is expressed in spherical coordinates, it is

$$r^2 \sin\theta \frac{\delta^2 V}{\delta r^2} + 2r \sin\theta \frac{\delta V}{\delta r} + \sin\theta \frac{\delta^2 V}{\delta\theta^2} + \cos\theta \frac{\delta V}{\delta\theta} + \frac{1}{\sin\theta} \frac{\delta^2 V}{\delta\phi^2} = 0$$

and any of its solutions, $V(r, \theta, \phi)$, are known as *spherical harmonics*. The solution as a product

$$V(r, \theta, \phi) = R(r)\Theta(\theta)$$

which is independent of ϕ, leads to

$$\sin^2\theta\Theta'' + \sin\theta\cos\theta\Theta' + [n(n+1)\sin^2\theta]\Theta = 0$$

Rearrangement and substitution of $x = \cos\theta$ leads to

$$(1 - x^2)\frac{d^2\Theta}{dx^2} - 2x\frac{d\Theta}{dx} + n(n+1)\Theta = 0$$

known as *Legendre's equation*. Important special cases are those in which n is zero or a positive integer, and, for such cases, Legendre's equation is satisfied by polynomials called Legendre polynomials, $P_n(x)$. A short list of Legendre polynomials, expressed in terms of x and $\cos\theta$, is given below. These are given by the following general formula:

$$P_n(x) = \sum_{j=0}^{L} \frac{(-1)^j (2n - 2j)!}{2^n j! (n-j)! (n-2j)!} x^{n-2j}$$

where $L = n/2$ if n is even and $L = (n-1)/2$ if n is odd. Some are given below:

$P_0(x) = 1$

$P_1(x) = x$

$P_2(x) = \dfrac{1}{2}(3x^2 - 1)$

$P_3(x) = \dfrac{1}{2}(5x^3 - 3x)$

$P_4(x) = \dfrac{1}{8}(35x^4 - 30x^2 + 3)$

$P_5(x) = \dfrac{1}{8}(63x^5 - 70x^3 + 15x)$

$P_0(\cos\theta) = 1$

$P_1(\cos\theta) = \cos\theta$

$P_2(\cos\theta) = \dfrac{1}{4}(3\cos 2\theta + 1)$

$P_3(\cos\theta) = \dfrac{1}{8}(5\cos 3\theta + 3\cos\theta)$

$P_4(\cos\theta) = \dfrac{1}{64}(35\cos 4\theta + 20\cos 2\theta + 9)$

Additional Legendre polynomials may be determined from the *recursion formula*

$$(n+1)P_{n+1}(x) - (2n+1)xP_n(x) + nP_{n-1}(x) = 0 \qquad (n = 1, 2, \dots)$$

or the *Rodrigues formula*

$$P_n(x) = \frac{1}{2^n n!}\frac{d^n}{dx^n}(x^2 - 1)^n$$

Laguerre Polynomials

Laguerre polynomials, denoted $L_n(x)$, are solutions of the differential equation

$$xy'' + (1-x)y' + ny = 0$$

and are given by

$$L_n(x) = \sum_{j=0}^{n}\frac{(-1)^j}{j!}C_{(n,j)}x^j \qquad (n = 0, 1, 2, \dots)$$

Thus,

$$L_0(x) = 1$$

$$L_1(x) = 1 - x$$

$$L_2(x) = 1 - 2x + \frac{1}{2}x^2$$

$$L_3(x) = 1 - 3x + \frac{3}{2}x^2 - \frac{1}{6}x^3$$

Additional Laguerre polynomials may be obtained from the recursion formula

$$(n+1)L_{n+1}(x) - (2n+1-x)L_n(x) + nL_{n-1}(x) = 0$$

Hermite Polynomials

The Hermite polynomials, denoted $H_n(x)$, are given by

$$H_0 = 1, \quad H_n(x) = (-1)^n e^{x^2} \frac{d^n e^{-x^2}}{dx^n}, \qquad (n = 1, 2, \ldots)$$

and are solutions of the differential equation

$$y'' - 2xy' + 2ny = 0 \qquad (n = 0, 1, 2, \ldots)$$

The first few Hermite polynomials are

$$H_0 = 1 \qquad\qquad H_1(x) = 2x$$
$$H_2(x) = 4x^2 - 2 \qquad\qquad H_3(x) = 8x^3 - 12x$$
$$H_4(x) = 16x^4 - 48x^2 + 12$$

Additional Hermite polynomials may be obtained from the relation

$$H_{n+1}(x) = 2xH_n(x) - H_n'(x),$$

where prime denotes differentiation with respect to x.

Orthogonality

A set of functions $\{f_n(x)\}$ $(n = 1, 2, \ldots)$ is orthogonal in an interval (a, b) with respect to a given weight function $w(x)$ if

$$\int_a^b w(x) f_m(x) f_n(x)\, dx = 0 \qquad \text{when } m \neq n$$

The following polynomials are orthogonal on the given interval for the given $w(x)$:

Legendre polynomials:	$P_n(x)$	$w(x) = 1$
		$a = -1, b = 1$
Laguerre polynomials:	$L_n(x)$	$w(x) = \exp(-x)$
		$a = 0, b = \infty$
Hermite polynomials:	$H_n(x)$	$w(x) = \exp(-x^2)$
		$a = -\infty, b = \infty$

The Bessel functions *of order n*, $J_n(\lambda_1 x)$, $J_n(\lambda_2 x),\ldots,$ are orthogonal with respect to $w(x)=x$ over the interval $(0,c)$ provided that the λ_i are the positive roots of $J_n(\lambda c)=0$:

$$\int_0^c x J_n(\lambda_j x) J_n(\lambda_k x)\, dx = 0 \qquad (j \neq k)$$

where n is fixed and $n \geq 0$.

Statistics

Arithmetic Mean

$$\mu = \frac{\Sigma X_i}{N},$$

where X_i is a measurement in the population and N is the total number of X_i in the population. For a *sample* of size n the sample mean, denoted $\bar{X}$, is

$$\bar{X} = \frac{\Sigma X_i}{n}.$$

Median

The median is the middle measurement when an odd number (n) measurements is arranged in order; if n is even, it is the midpoint between the two middle measurements.

Mode

It is the most frequently occurring measurement in a set.

Geometric Mean

$$\text{geometric mean} = \sqrt[n]{X_1 X_2 \ldots X_n}$$

Harmonic Mean

The Harmonic mean H of n numbers $X_1, X_2,\ldots, X_n$, is

$$H = \frac{n}{\Sigma(1/Xi)}$$

Variance

The mean of the sum of squares of deviations from the mean (μ) is the population variance, denoted σ^2

$$\sigma^2 = \Sigma(X_i - \mu)^2/N.$$

The sample variance, s^2, for sample size n is

$$s^2 = \Sigma(X_i - \overline{X})^2/(n-1).$$

A simpler computational form is

$$s^2 = \frac{\Sigma X_i^2 - \dfrac{(\Sigma X_i)^2}{n}}{n-1}$$

Standard Deviation

The positive square root of the population variance is the standard deviation. For a population

$$\sigma = \left[\frac{\Sigma X_i^2 - \dfrac{(\Sigma X_i)^2}{N}}{N} \right]^{1/2} ;$$

for a sample

$$s = \left[\frac{\Sigma X_i^2 - \dfrac{(\Sigma X_i)^2}{n}}{n-1} \right]^{1/2} .$$

Coefficient of Variation

$$V = s/\overline{X}.$$

Probability

For the sample space U, with subsets A of U (called "events"), we consider the probability measure of an event A to be a real-valued function p defined over all subsets of U such that:

$0 \le p(A) \le 1$

$p(U) = 1$ and $p(\Phi) = 0$

If A_1 and A_2 are subsets of U

$p(A_1 \cup A_2) = p(A_1) + p(A_2) - p(A_1 \cap A_2)$

Two events A_1 and A_2 are called mutually exclusive if and only if $A_1 \cap A_2 = \phi$ (null set). These events are said to be independent if and only if $p(A_1 \cap A_2) = p(A_1)p(A_2)$.

Conditional Probability and Bayes' Rule

The probability of an event A, given that an event B has occurred, is called the conditional probability and is denoted $p(A/B)$. Further

$$p(A/B) = \frac{p(A \cap B)}{p(B)}$$

Bayes' rule permits a calculation of *a posteriori* probability from given *a priori* probabilities and is stated below:

If $A_1, A_2, \ldots, A_n$ are n mutually exclusive events, and $p(A_1) + p(A_2) + \cdots + p(A_n) = 1$, and B is any event such that $p(B)$ is not 0, then the conditional probability $p(A_i/B)$ for any one of the events A_i, *given that B has occurred* is

$$p(A_i/B) = \frac{p(A_i)p(B/A_i)}{p(A_1)p(B/A_1) + p(A_2)p(B/A_2) + \cdots + p(A_n)p(B/A_n)}$$

Example. Among 5 different laboratory tests for detecting a certain disease, one is effective with probability 0.75, whereas each of the others is effective with probability 0.40. A medical student, unfamiliar with the advantage of the best test, selects one of them and is successful in detecting the disease in a patient. What is the probability that the most effective test was used?

Let B denote (the event) of detecting the disease, A_1 the selection of the best test, and A_2 the selection of one of the other 4 tests; thus, $p(A_1) = 1/5$, $p(A_2) = 4/5$, $p(B/A_1) = 0.75$ and $p(B/A_2) = 0.40$. Therefore

$$p(A_1/B) = \frac{\dfrac{1}{5}(0.75)}{\dfrac{1}{5}(0.75) + \dfrac{4}{5}(0.40)} = 0.319$$

Note, the *a priori* probability is 0.20; the outcome raises this probability to 0.319.

Binomial Distribution

In an experiment consisting of n independent trials in which an event has probability p in a single trial, the probability P_X of obtaining X successes is given by

$$P_X = C_{(n,X)} p^X q^{(n-X)}$$

where

$$q = (1-p) \text{ and } C_{(n,X)} = \frac{n!}{X!(n-X)!}.$$

The probability of between a and b successes (both a and b included) is $P_a + P_{a+1} + \cdots + P_b$, so if $a = 0$ and $b = n$, this sum is

$$\sum_{X=0}^{n} C_{(n,X)} p^X q^{(n-X)} = q^n + C_{(n,1)} q^{n-1} p + C_{(n,2)} q^{n-2} p^2 + \cdots + p^n = (q+p)^n = 1.$$

Mean of Binomially Distributed Variable

The mean number of successes in n independent trials is $m = np$ with standard deviation $\sigma = \sqrt{npq}$.

Normal Distribution

In the binomial distribution, as n increases the histogram of heights is approximated by the bell-shaped curve (normal curve)

$$Y = \frac{1}{\sigma\sqrt{2\pi}} e^{-(x-m)^2/2\sigma^2}$$

where $m =$ the mean of the binomial distribution $= np$, and $\sigma = \sqrt{npq}$ is the standard deviation. For any normally distributed random variable X with mean m and standard deviation σ the probability function (density) is given by the above.

The *standard* normal probability curve is given by

$$y = \frac{1}{\sqrt{2\pi}} e^{-Z^2/2}$$

and has mean $= 0$ and standard deviation $= 1$. The total area under the standard normal curve is 1. Any normal variable X can be put into standard form by defining $Z = (X - m)/\sigma$; thus the probability of X between a given X_1 and X_2 is the area under the standard normal curve between the corresponding Z_1 and Z_2 (Table 1 in the Tables of Probability and Statistics). The standard normal curve is often used instead of the binomial distribution in experiments with discrete outcomes. For example, to determine the probability of obtaining 60 to 70 heads in a toss of 100 coins, we take $X = 59.5$ to $X = 70.5$ and compute corresponding values of Z from mean $np = 100 \frac{1}{2} = 50$, and the standard deviation $\sigma = \sqrt{(100)(1/2)(1/2)} = 5$. Thus, $Z = (59.5 - 50)/5 = 1.9$ and $Z = (70.5 - 50)/5 = 4.1$. From Table 1, area between $Z = 0$ and $Z = 4.1$ is 0.5000 and between $Z = 0$ and $Z = 1.9$ is 0.4713; hence, the desired probability is 0.0287. The binomial distribution requires a more lengthy computation

$$C_{(100,60)}(1/2)^{60}(1/2)^{40} + C_{(100,61)}(1/2)^{61}(1/2)^{39} + \cdots + C_{(100,70)}(1/2)^{70}(1/2)^{30}.$$

Note that the normal curve is symmetric, whereas the histogram of the binomial distribution is symmetric only if $p = q = 1/2$. Accordingly, when p (hence q) differ appreciably from $1/2$, the difference between probabilities computed by each increases. It is usually recommended that the normal approximation not be used if p (or q) is so small that np (or nq) is less than 5.

Poisson Distribution

$$P = \frac{e^{-m} m^r}{r!}$$

is an approximation to the binomial probability for r successes in n trials when $m = np$ is small (<5) and the normal curve is not recommended to approximate binomial probabilities (Table 2 in the Tables of Probability and Statistics). The variance σ^2 in the Poisson distribution is np, the same value as the mean. **Example**: A school's expulsion rate is 5 students per 1000. If class size is 400, what is the probability that 3 or more will be expelled? Since $p = 0.005$ and $n = 400$, $m = np = 2$, and $r = 3$. From Table 2 we obtain for $m = 2$ and $r(=x) = 3$ the probability $p = 0.323$.

Tables of Probability and Statistics

TABLE 1: Areas Under the Standard Normal Curve

z	0.00	0.01	0.02	0.03	0.04	0.05	0.06	0.07	0.08	0.09
0.0	0.0000	0.0040	0.0080	0.0120	0.0160	0.0199	0.0239	0.0279	0.0319	0.0359
0.1	0.0398	0.0438	0.0478	0.0517	0.0557	0.0596	0.0636	0.0675	0.0714	0.0753
0.2	0.0793	0.0832	0.0871	0.0910	0.0948	0.0987	0.1026	0.1064	0.1103	0.1141
0.3	0.1179	0.1217	0.1255	0.1293	0.1331	0.1368	0.1406	0.1443	0.1480	0.1517
0.4	0.1554	0.1591	0.1628	0.1664	0.1700	0.1736	0.1772	0.1808	0.1844	0.1879
0.5	0.1915	0.1950	0.1985	0.2019	0.2054	0.2088	0.2123	0.2157	0.2190	0.2224
0.6	0.2257	0.2291	0.2324	0.2357	0.2389	0.2422	0.2454	0.2486	0.2517	0.2549
0.7	0.2580	0.2611	0.2642	0.2673	0.2704	0.2734	0.2764	0.2794	0.2823	0.2852
0.8	0.2881	0.2910	0.2939	0.2967	0.2995	0.3023	0.3051	0.3078	0.3106	0.3133
0.9	0.3159	0.3186	0.3212	0.3238	0.3264	0.3289	0.3315	0.3340	0.3365	0.3389
1.0	0.3413	0.3438	0.3461	0.3485	0.3508	0.3531	0.3554	0.3577	0.3599	0.3621
1.1	0.3643	0.3665	0.3686	0.3708	0.3729	0.3749	0.3770	0.3790	0.3810	0.3830
1.2	0.3849	0.3869	0.3888	0.3907	0.3925	0.3944	0.3962	0.3980	0.3997	0.4015
1.3	0.4032	0.4049	0.4066	0.4082	0.4099	0.4115	0.4131	0.4147	0.4162	0.4177
1.4	0.4192	0.4207	0.4222	0.4236	0.4251	0.4265	0.4279	0.4292	0.4306	0.4319
1.5	0.4332	0.4345	0.4357	0.4370	0.4382	0.4394	0.4406	0.4418	0.4429	0.4441
1.6	0.4452	0.4463	0.4474	0.4484	0.4495	0.4505	0.4515	0.4525	0.4535	0.4545
1.7	0.4554	0.4564	0.4573	0.4582	0.4591	0.4599	0.4608	0.4616	0.4625	0.4633
1.8	0.4641	0.4649	0.4656	0.4664	0.4671	0.4678	0.4686	0.4693	0.4699	0.4706
1.9	0.4713	0.4719	0.4726	0.4732	0.4738	0.4744	0.4750	0.4756	0.4761	0.4767
2.0	0.4772	0.4778	0.4783	0.4788	0.4793	0.4798	0.4803	0.4808	0.4812	0.4817
2.1	0.4821	0.4826	0.4830	0.4834	0.4838	0.4842	0.4846	0.4850	0.4854	0.4857
2.2	0.4861	0.4864	0.4868	0.4871	0.4875	0.4878	0.4881	0.4884	0.4887	0.4890
2.3	0.4893	0.4896	0.4898	0.4901	0.4904	0.4906	0.4909	0.4911	0.4913	0.4916
2.4	0.4918	0.4920	0.4922	0.4925	0.4927	0.4929	0.4931	0.4932	0.4934	0.4936
2.5	0.4938	0.4940	0.4941	0.4943	0.4945	0.4946	0.4948	0.4949	0.4951	0.4952
2.6	0.4953	0.4955	0.4956	0.4957	0.4959	0.4960	0.4961	0.4962	0.4963	0.4964
2.7	0.4965	0.4966	0.4967	0.4968	0.4969	0.4970	0.4971	0.4972	0.4973	0.4974
2.8	0.4974	0.4975	0.4976	0.4977	0.4977	0.4978	0.4979	0.4979	0.4980	0.4981
2.9	0.4981	0.4982	0.4982	0.4983	0.4984	0.4984	0.4985	0.4985	0.4986	0.4986
3.0	0.4987	0.4987	0.4987	0.4988	0.4988	0.4989	0.4989	0.4989	0.4990	0.4990

Source: R. J. Tallarida and R. B. Murray, *Manual of Pharmacologic Calculations with Computer Programs*, 2nd ed., New York: Springer-Verlag, 1987. With permission.

TABLE 2: Poisson Distribution

Each number in this table represents the probability of obtaining at least X successes, or the area under the histogram to the right of and including the rectangle whose center is at X.

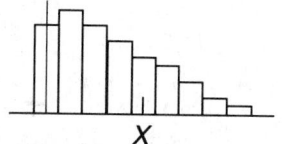

m	$X=0$	$X=1$	$X=2$	$X=3$	$X=4$	$X=5$	$X=6$	$X=7$	$X=8$	$X=9$	$X=10$	$X=11$	$X=12$	$X=13$	$X=14$
.10	1.000	.095	.005												
.20	1.000	.181	.018	.001											
.30	1.000	.259	.037	.004											
.40	1.000	.330	.062	.008	.001										
.50	1.000	.393	.090	.014	.002										
.60	1.000	.451	.122	.023	.003										
.70	1.000	.503	.156	.034	.006	.001									
.80	1.000	.551	.191	.047	.009	.001									
.90	1.000	.593	.228	.063	.013	.002									
1.00	1.000	.632	.264	.080	.019	.004	.001								
1.1	1.000	.667	.301	.100	.026	.005	.001								
1.2	1.000	.699	.337	.120	.034	.008	.002								
1.3	1.000	.727	.373	.143	.043	.011	.002								
1.4	1.000	.753	.408	.167	.054	.014	.003	.001							
1.5	1.000	.777	.442	.191	.066	.019	.004	.001							
1.6	1.000	.798	.475	.217	.079	.024	.006	.001							
1.7	1.000	.817	.507	.243	.093	.030	.008	.002							
1.8	1.000	.835	.537	.269	.109	.036	.010	.003	.001						
1.9	1.000	.850	.566	.296	.125	.044	.013	.003	.001						
2.0	1.000	.865	.594	.323	.143	.053	.017	.005	.001						
2.2	1.000	.889	.645	.377	.181	.072	.025	.007	.002						
2.4	1.000	.909	.692	.430	.221	.096	.036	.012	.003	.001					
2.6	1.000	.926	.733	.482	.264	.123	.049	.017	.005	.001					
2.8	1.000	.939	.769	.531	.308	.152	.065	.024	.008	.002	.001				
3.0	1.000	.950	.801	.577	.353	.185	.084	.034	.012	.004	.001				
3.2	1.000	.959	.829	.620	.397	.219	.105	.045	.017	.006	.002				
3.4	1.000	.967	.853	.660	.442	.256	.129	.058	.023	.008	.003	.001			
3.6	1.000	.973	.874	.697	.485	.294	.156	.073	.031	.012	.004	.001			
3.8	1.000	.978	.893	.731	.527	.332	.184	.091	.040	.016	.006	.002			
4.0	1.000	.982	.908	.762	.567	.371	.215	.111	.051	.021	.008	.003	.001		
4.2	1.000	.985	.922	.790	.605	.410	.247	.133	.064	.028	.011	.004	.001		
4.4	1.000	.988	.934	.815	.641	.449	.280	.156	.079	.036	.015	.006	.002	.001	
4.6	1.000	.990	.944	.837	.674	.487	.314	.182	.095	.045	.020	.008	.003	.001	
4.8	1.000	.992	.952	.857	.706	.524	.349	.209	.113	.056	.025	.010	.004	.001	
5.0	1.000	.993	.960	.875	.735	.560	.384	.238	.133	.068	.032	.014	.005	.002	.001

Source: H. L. Adler and E. B. Roessler, *Introduction to Probability and Statistics*, 6th ed., New York: W. H. Freeman, 1977. With permission.

TABLE 3: *t*-Distribution

deg. freedom, f	90% ($P = 0.1$)	95% ($P = 0.05$)	99% ($P = 0.01$)
1	6.314	12.706	63.657
2	2.920	4.303	9.925
3	2.353	3.182	5.841
4	2.132	2.776	4.604
5	2.015	2.571	4.032
6	1.943	2.447	3.707
7	1.895	2.365	3.499
8	1.860	2.306	3.355
9	1.833	2.262	3.250
10	1.812	2.228	3.169
11	1.796	2.201	3.106
12	1.782	2.179	3.055
13	1.771	2.160	3.012
14	1.761	2.145	2.977
15	1.753	2.131	2.947
16	1.746	2.120	2.921
17	1.740	2.110	2.898
18	1.734	2.101	2.878
19	1.729	2.093	2.861
20	1.725	2.086	2.845
21	1.721	2.080	2.831
22	1.717	2.074	2.819
23	1.714	2.069	2.807
24	1.711	2.064	2.797
25	1.708	2.060	2.787
26	1.706	2.056	2.779
27	1.703	2.052	2.771
28	1.701	2.048	2.763
29	1.699	2.045	2.756
inf.	1.645	1.960	2.576

Source: R. J. Tallarida and R. B. Murray, *Manual of Pharmacologic Calculations with Computer Programs*, 2nd ed., New York: Springer-Verlag, 1987. With permission.

TABLE 4: χ^2–Distribution

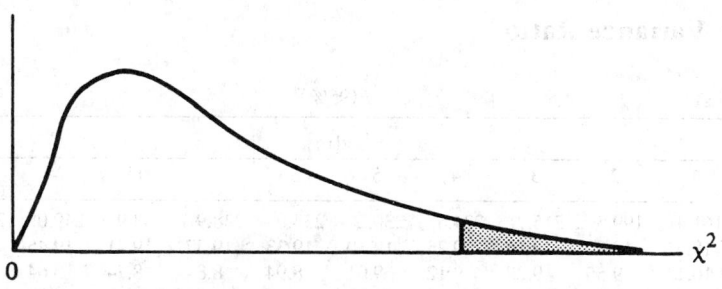

v	0.05	0.025	0.01	0.005
1	3.841	5.024	6.635	7.879
2	5.991	7.378	9.210	10.597
3	7.815	9.348	11.345	12.838
4	9.488	11.143	13.277	14.860
5	11.070	12.832	15.086	16.750
6	12.592	14.449	16.812	18.548
7	14.067	16.013	18.475	20.278
8	15.507	17.535	20.090	21.955
9	16.919	19.023	21.666	23.589
10	18.307	20.483	23.209	25.188
11	19.675	21.920	24.725	26.757
12	21.026	23.337	26.217	28.300
13	22.362	24.736	27.688	29.819
14	23.685	26.119	29.141	31.319
15	24.996	27.488	30.578	32.801
16	26.296	28.845	32.000	34.267
17	27.587	30.191	33.409	35.718
18	28.869	31.526	34.805	37.156
19	30.144	32.852	36.191	38.582
20	31.410	34.170	37.566	39.997
21	32.671	35.479	38.932	41.401
22	33.924	36.781	40.289	42.796
23	35.172	38.076	41.638	44.181
24	36.415	39.364	42.980	45.558
25	37.652	40.646	44.314	46.928
26	38.885	41.923	45.642	48.290
27	40.113	43.194	46.963	49.645
28	41.337	44.461	48.278	50.993
29	42.557	45.722	49.588	52.336
30	43.773	46.979	50.892	53.672

Source: J. E. Freund and F. J. Williams, *Elementary Business Statistics: The Modern Approach*, 2nd ed., Englewood Cliffs, N.J.: Prentice-Hall, 1972. With permission.

TABLE 5: Variance Ratio

$F(95\%)$

n_2	n_1									
	1	2	3	4	5	6	8	12	24	∞
1	161.4	199.5	215.7	224.6	230.2	234.0	238.9	243.9	249.0	254.3
2	18.51	19.00	19.16	19.25	19.30	19.33	19.37	19.41	19.45	19.50
3	10.13	9.55	9.28	9.12	9.01	8.94	8.84	8.74	8.64	8.53
4	7.71	6.94	6.59	6.39	6.26	6.16	6.04	5.91	5.77	5.63
5	6.61	5.79	5.41	5.19	5.05	4.95	4.82	4.68	4.53	4.36
6	5.99	5.14	4.76	4.53	4.39	4.28	4.15	4.00	3.84	3.67
7	5.59	4.74	4.35	4.12	3.97	3.87	3.73	3.57	3.41	3.23
8	5.32	4.46	4.07	3.84	3.69	3.58	3.44	3.28	3.12	2.93
9	5.12	4.26	3.86	3.63	3.48	3.37	3.23	3.07	2.90	2.71
10	4.96	4.10	3.71	3.48	3.33	3.22	3.07	2.91	2.74	2.54
11	4.84	3.98	3.59	3.36	3.20	3.09	2.95	2.79	2.61	2.40
12	4.75	3.88	3.49	3.26	3.11	3.00	2.85	2.69	2.50	2.30
13	4.67	3.80	3.41	3.18	3.02	2.92	2.77	2.60	2.42	2.21
14	4.60	3.74	3.34	3.11	2.96	2.85	2.70	2.53	2.35	2.13
15	4.54	3.68	3.29	3.06	2.90	2.79	2.64	2.48	2.29	2.07
16	4.49	3.63	3.24	3.01	2.85	2.74	2.59	2.42	2.24	2.01
17	4.45	3.59	3.20	2.96	2.81	2.70	2.55	2.38	2.19	1.96
18	4.41	3.55	3.16	2.93	2.77	2.66	2.51	2.34	2.15	1.92
19	4.38	3.52	3.13	2.90	2.74	2.63	2.48	2.31	2.11	1.88
20	4.35	3.49	3.10	2.87	2.71	2.60	2.45	2.28	2.08	1.84
21	4.32	3.47	3.07	2.84	2.68	2.57	2.42	2.25	2.05	1.81
22	4.30	3.44	3.05	2.82	2.66	2.55	2.40	2.23	2.03	1.78
23	4.28	3.42	3.03	2.80	2.64	2.53	2.38	2.20	2.00	1.76
24	4.26	3.40	3.01	2.78	2.62	2.51	2.36	2.18	1.98	1.73
25	4.24	3.38	2.99	2.76	2.60	2.49	2.34	2.16	1.96	1.71
26	4.22	3.37	2.98	2.74	2.59	2.47	2.32	2.15	1.95	1.69
27	4.21	3.35	2.96	2.73	2.57	2.46	2.30	2.13	1.93	1.67
28	4.20	3.34	2.95	2.71	2.56	2.44	2.29	2.12	1.91	1.65
29	4.18	3.33	2.93	2.70	2.54	2.43	2.28	2.10	1.90	1.64
30	4.17	3.32	2.92	2.69	2.53	2.42	2.27	2.09	1.89	1.62
40	4.08	3.23	2.84	2.61	2.45	2.34	2.18	2.00	1.79	1.51
60	4.00	3.15	2.76	2.52	2.37	2.25	2.10	1.92	1.70	1.39
120	3.92	3.07	2.68	2.45	2.29	2.17	2.02	1.83	1.61	1.25
∞	3.84	2.99	2.60	2.37	2.21	2.10	1.94	1.75	1.52	1.00

TABLE 5: Variance Ratio (continued)

$F(99\%)$

n_2	n_1									
	1	2	3	4	5	6	8	12	24	∞
1	4,052	4,999	5,403	5,625	5,764	5,859	5,982	6,106	6,234	6,366
2	98.50	99.00	99.17	99.25	99.30	99.33	99.37	99.42	99.46	99.50
3	34.12	30.82	29.46	28.71	28.24	27.91	27.49	27.05	26.60	26.12
4	21.20	18.00	16.69	15.98	15.52	15.21	14.80	14.37	13.93	13.46
5	16.26	13.27	12.06	11.39	10.97	10.67	10.29	9.89	9.47	9.02
6	13.74	10.92	9.78	9.15	8.75	8.47	8.10	7.72	7.31	6.88
7	12.25	9.55	8.45	7.85	7.46	7.19	6.84	6.47	6.07	5.65
8	11.26	8.65	7.59	7.01	6.63	6.37	6.03	5.67	5.28	4.86
9	10.56	8.02	6.99	6.42	6.06	5.80	5.47	5.11	4.73	4.31
10	10.04	7.56	6.55	5.99	5.64	5.39	5.06	4.71	4.33	3.91
11	9.65	7.20	6.22	5.67	5.32	5.07	4.74	4.40	4.02	3.60
12	9.33	6.93	5.95	5.41	5.06	4.82	4.50	4.16	3.78	3.36
13	9.07	6.70	5.74	5.20	4.86	4.62	4.30	3.96	3.59	3.16
14	8.86	6.51	5.56	5.03	4.69	4.46	4.14	3.80	3.43	3.00
15	8.68	6.36	5.42	4.89	4.56	4.32	4.00	3.67	3.29	2.87
16	8.53	6.23	5.29	4.77	4.44	4.20	3.89	3.55	3.18	2.75
17	8.40	6.11	5.18	4.67	4.34	4.10	3.79	3.45	3.08	2.65
18	8.28	6.01	5.09	4.58	4.25	4.01	3.71	3.37	3.00	2.57
19	8.18	5.93	5.01	4.50	4.17	3.94	3.63	3.30	2.92	2.49
20	8.10	5.85	4.94	4.43	4.10	3.87	3.56	3.23	2.86	2.42
21	8.02	5.78	4.87	4.37	4.04	3.81	3.51	3.17	2.80	2.36
22	7.94	5.72	4.82	4.31	3.99	3.76	3.45	3.12	2.75	2.31
23	7.88	5.66	4.76	4.26	3.94	3.71	3.41	3.07	2.70	2.26
24	7.82	5.61	4.72	4.22	3.90	3.67	3.36	3.03	2.66	2.21
25	7.77	5.57	4.68	4.18	3.86	3.63	3.32	2.99	2.62	2.17
26	7.72	5.53	4.64	4.14	3.82	3.59	3.29	2.96	2.58	2.13
27	7.68	5.49	4.60	4.11	3.78	3.56	3.26	2.93	2.55	2.10
28	7.64	5.45	4.57	4.07	3.75	3.53	3.23	2.90	2.52	2.06
29	7.60	5.42	4.54	4.04	3.73	3.50	3.20	2.87	2.49	2.03
30	7.56	5.39	4.51	4.02	3.70	3.47	3.17	2.84	2.47	2.01
40	7.31	5.18	4.31	3.83	3.51	3.29	2.99	2.66	2.29	1.80
60	7.08	4.98	4.13	3.65	3.34	3.12	2.82	2.50	2.12	1.60
120	6.85	4.79	3.95	3.48	3.17	2.96	2.66	2.34	1.95	1.38
∞	6.64	4.60	3.78	3.32	3.02	2.80	2.51	2.18	1.79	1.00

Source: R. A. Fisher and F. Yates, *Statistical Tables for Biological, Agricultural and Medical Research*, London: The Lingman Group, Ltd. With permission.

Table of Derivatives

In the following table, a and n are constants, e is the base of the natural logarithms, and u and v denote functions of x.

1. $\dfrac{d}{dx}(a) = 0$

2. $\dfrac{d}{dx}(x) = 1$

3. $\dfrac{d}{dx}(au) = a\dfrac{du}{dx}$

4. $\dfrac{d}{dx}(u+v) = \dfrac{du}{dx} + \dfrac{dv}{dx}$

5. $\dfrac{d}{dx}(uv) = u\dfrac{dv}{dx} + v\dfrac{du}{dx}$

6. $\dfrac{d}{dx}(u/v) = \dfrac{v\dfrac{du}{dx} - u\dfrac{dv}{dx}}{v^2}$

7. $\dfrac{d}{dx}(u^n) = nu^{n-1}\dfrac{du}{dx}$

8. $\dfrac{d}{dx}e^u = e^u\dfrac{du}{dx}$

9. $\dfrac{d}{dx}a^u = (\log_e a)a^u\dfrac{du}{dx}$

10. $\dfrac{d}{dx}\log_e u = (1/u)\dfrac{du}{dx}$

11. $\dfrac{d}{dx}\log_a u = (\log_a e)(1/u)\dfrac{du}{dx}$

12. $\dfrac{d}{dx}u^v = vu^{v-1}\dfrac{du}{dx} + u^v(\log_e u)\dfrac{dv}{dx}$

13. $\dfrac{d}{dx}\sin u = \cos u\dfrac{du}{dx}$

14. $\dfrac{d}{dx}\cos u = -\sin u\dfrac{du}{dx}$

15. $\dfrac{d}{dx}\tan u = \sec^2 u\dfrac{du}{dx}$

16. $\dfrac{d}{dx}\operatorname{ctn} u = -\csc^2 u\dfrac{du}{dx}$

17. $\dfrac{d}{dx}\sec u = \sec u \tan u\dfrac{du}{dx}$

18. $\dfrac{d}{dx}\csc u = -\csc u \operatorname{ctn} u\dfrac{du}{dx}$

19. $\dfrac{d}{dx}\sin^{-1} u = \dfrac{1}{\sqrt{1-u^2}}\dfrac{du}{dx}, \quad (-\tfrac{1}{2}\pi \le \sin^{-1} u \le \tfrac{1}{2}\pi)$

20. $\dfrac{d}{dx}\cos^{-1} u = \dfrac{-1}{\sqrt{1-u^2}}\dfrac{du}{dx}, \quad (0 \le \cos^{-1} u \le \pi)$

21. $\dfrac{d}{dx}\tan^{-1} u = \dfrac{1}{1+u^2}\dfrac{du}{dx}$

22. $\dfrac{d}{dx}\operatorname{ctn}^{-1} u = \dfrac{-1}{1+u^2}\dfrac{du}{dx}$

23. $\dfrac{d}{dx}\sec^{-1} u = \dfrac{1}{u\sqrt{u^2-1}}\dfrac{du}{dx},$

$\qquad (-\pi \le \sec^{-1} u < -\tfrac{1}{2}\pi; 0 \le \sec^{-1} u < \tfrac{1}{2}\pi)$

24. $\dfrac{d}{dx}\csc^{-1} u = \dfrac{-1}{u\sqrt{u^2-1}}\dfrac{du}{dx},$

$\qquad (-\pi < \csc^{-1} u \le -\tfrac{1}{2}\pi; 0 < \csc^{-1} u \le \tfrac{1}{2}\pi)$

25. $\dfrac{d}{dx}\sinh u = \cosh u\dfrac{du}{dx}$

26. $\dfrac{d}{dx}\cosh u = \sinh u\dfrac{du}{dx}$

27. $\dfrac{d}{dx}\tanh u = \operatorname{sech}^2 u\dfrac{du}{dx}$

28. $\dfrac{d}{dx}\operatorname{ctnh} u = -\operatorname{csch}^2 u\dfrac{du}{dx}$

29. $\dfrac{d}{dx}\operatorname{sech} u = -\operatorname{sech} u \tanh u\dfrac{du}{dx}$

30. $\dfrac{d}{dx}\operatorname{csch} u = -\operatorname{csch} u \operatorname{ctnh} u\dfrac{du}{dx}$

31. $\dfrac{d}{dx}\sinh^{-1} u = \dfrac{1}{\sqrt{u^2+1}}\dfrac{du}{dx}$

32. $\dfrac{d}{dx}\cosh^{-1} u = \dfrac{1}{\sqrt{u^2-1}}\dfrac{du}{dx}$

33. $\dfrac{d}{dx}\tanh^{-1} u = \dfrac{1}{1-u^2}\dfrac{du}{dx}$

34. $\dfrac{d}{dx}\operatorname{ctnh}^{-1} u = \dfrac{-1}{u^2-1}\dfrac{du}{dx}$

35. $\dfrac{d}{dx}\operatorname{sech}^{-1} u = \dfrac{-1}{u\sqrt{1-u^2}}\dfrac{du}{dx}$

36. $\dfrac{d}{dx}\operatorname{csch}^{-1} u = \dfrac{-1}{u\sqrt{u^2+1}}\dfrac{du}{dx}$

Additional Relations with Derivatives

$$\frac{d}{dt}\int_a^t f(x)\,dx = f(t) \qquad \frac{d}{dt}\int_t^a f(x)\,dx = -f(t)$$

If $x = f(y)$, then $\dfrac{dy}{dx} = \dfrac{1}{\dfrac{dx}{dy}}$

If $y = f(u)$ and $u = g(x)$, then $\dfrac{dy}{dx} = \dfrac{dy}{du}\cdot\dfrac{du}{dx}$ \qquad (chain rule)

If $x = f(t)$ and $y = g(t)$, then $\dfrac{dy}{dx} = \dfrac{g'(t)}{f'(t)}$, and $\dfrac{d^2y}{dx^2} = \dfrac{f'(t)g''(t) - g'(t)f''(t)}{[f'(t)]^3}$

(*Note*: exponent in denominator is 3.)

Integrals

Elementary Forms

1. $\displaystyle\int a\,dx = ax$

2. $\displaystyle\int a\cdot f(x)\,dx = a\int f(x)\,dx$

3. $\displaystyle\int \phi(y)\,dx = \int \frac{\phi(y)}{y'}\,dy, \qquad$ where $y' = \dfrac{dy}{dx}$

4. $\displaystyle\int (u+v)\,dx = \int u\,dx + \int v\,dx, \qquad$ where u and v are any functions of x

5. $\displaystyle\int u\,dv = u\int dv - \int v\,du = uv - \int v\,du$

6. $\displaystyle\int u\frac{dv}{dx}\,dx = uv - \int v\frac{du}{dx}\,dx$

7. $\displaystyle\int x^n\,dx = \frac{x^{n+1}}{n+1}, \qquad$ except $n = -1$

8. $\displaystyle\int \frac{f'(x)\,dx}{f(x)} = \log f(x), \qquad (df(x) = f'(x)\,dx)$

9. $\displaystyle\int \frac{dx}{x} = \log x$

10. $\displaystyle\int \frac{f'(x)\,dx}{2\sqrt{f(x)}} = \sqrt{f(x)}, \qquad (df(x) = f'(x)\,dx)$

11. $\displaystyle\int e^x\,dx = e^x$

12. $\int e^{ax} dx = e^{ax}/a$

13. $\int b^{ax} dx = \dfrac{b^{ax}}{a \log b}$, $(b > 0)$

14. $\int \log x \, dx = x \log x - x$

15. $\int a^x \log a \, dx = a^x$, $(a > 0)$

16. $\int \dfrac{dx}{a^2 + x^2} = \dfrac{1}{a} \tan^{-1} \dfrac{x}{a}$

17. $\int \dfrac{dx}{a^2 - x^2} = \begin{cases} \dfrac{1}{a} \tanh^{-1} \dfrac{x}{a} \\ \text{or} \\ \dfrac{1}{2a} \log \dfrac{a+x}{a-x}, & (a^2 > x^2) \end{cases}$

18. $\int \dfrac{dx}{x^2 - a^2} = \begin{cases} -\dfrac{1}{a} \coth^{-1} \dfrac{x}{a} \\ \text{or} \\ \dfrac{1}{2a} \log \dfrac{x-a}{x+a}, & (x^2 > a^2) \end{cases}$

19. $\int \dfrac{dx}{\sqrt{a^2 - x^2}} = \begin{cases} \sin^{-1} \dfrac{x}{|a|} \\ \text{or} \\ -\cos^{-1} \dfrac{x}{|a|}, & (a^2 > x^2) \end{cases}$

20. $\int \dfrac{dx}{\sqrt{x^2 \pm a^2}} = \log(x + \sqrt{x^2 \pm a^2})$

21. $\int \dfrac{dx}{x\sqrt{x^2 - a^2}} = \dfrac{1}{|a|} \sec^{-1} \dfrac{x}{a}$

22. $\int \dfrac{dx}{x\sqrt{a^2 \pm x^2}} = -\dfrac{1}{a} \log \left(\dfrac{a + \sqrt{a^2 \pm x^2}}{x} \right)$

Forms Containing ($a + bx$)

For forms containing $a + bx$, but not listed in the table, the substitution $u = \dfrac{a+bx}{x}$ may prove helpful.

23. $\int (a + bx)^n \, dx = \dfrac{(a+bx)^{n+1}}{(n+1)b}$, $(n \neq -1)$

24. $\int x(a + bx)^n \, dx = \dfrac{1}{b^2(n+2)}(a+bx)^{n+2} - \dfrac{a}{b^2(n+1)}(a+bx)^{n+1}$, $(n \neq -1, -2)$

25. $\int x^2(a+bx)^n\,dx = \dfrac{1}{b^3}\left[\dfrac{(a+bx)^{n+3}}{n+3} - 2a\dfrac{(a+bx)^{n+2}}{n+2} + a^2\dfrac{(a+bx)^{n+1}}{n+1}\right]$

26. $\int x^m(a+bx)^n\,dx = \begin{cases} \dfrac{x^{m+1}(a+bx)^n}{m+n+1} + \dfrac{an}{m+n+1}\int x^m(a+bx)^{n-1}\,dx \\[4pt] \text{or} \\[4pt] \dfrac{1}{a(n+1)}\left[-x^{m+1}(a+bx)^{n+1}\right. \\[10pt] \qquad\qquad \left. +(m+n+2)\int x^m(a+bx)^{n+1}\,dx\right] \\[4pt] \text{or} \\[4pt] \dfrac{1}{b(m+n+1)}\left[x^m(a+bx)^{n+1} - ma\int x^{m-1}(a+bx)^n\,dx\right] \end{cases}$

27. $\displaystyle\int \dfrac{dx}{a+bx} = \dfrac{1}{b}\log(a+bx)$

28. $\displaystyle\int \dfrac{dx}{(a+bx)^2} = -\dfrac{1}{b(a+bx)}$

29. $\displaystyle\int \dfrac{dx}{(a+bx)^3} = -\dfrac{1}{2b(a+bx)^2}$

30. $\displaystyle\int \dfrac{x\,dx}{a+bx} = \begin{cases} \dfrac{1}{b^2}[a+bx-a\log(a+bx)] \\[4pt] \text{or} \\[4pt] \dfrac{x}{b} - \dfrac{a}{b^2}\log(a+bx) \end{cases}$

31. $\displaystyle\int \dfrac{x\,dx}{(a+bx)^2} = \dfrac{1}{b^2}\left[\log(a+bx) + \dfrac{a}{a+bx}\right]$

32. $\displaystyle\int \dfrac{x\,dx}{(a+bx)^n} = \dfrac{1}{b^2}\left[\dfrac{-1}{(n-2)(a+bx)^{n-2}} + \dfrac{a}{(n-1)(a+bx)^{n-1}}\right], \qquad n \neq 1,2$

33. $\displaystyle\int \dfrac{x^2\,dx}{a+bx} = \dfrac{1}{b^3}\left[\dfrac{1}{2}(a+bx)^2 - 2a(a+bx) + a^2\log(a+bx)\right]$

34. $\displaystyle\int \dfrac{x^2\,dx}{(a+bx)^2} = \dfrac{1}{b^3}\left[a+bx - 2a\log(a+bx) - \dfrac{a^2}{a+bx}\right]$

35. $\displaystyle\int \dfrac{x^2\,dx}{(a+bx)^3} = \dfrac{1}{b^3}\left[\log(a+bx) + \dfrac{2a}{a+bx} - \dfrac{a^2}{2(a+bx)^2}\right]$

36. $\displaystyle\int \dfrac{x^2\,dx}{(a+bx)^n} = \dfrac{1}{b^3}\left[\dfrac{-1}{(n-3)(a+bx)^{n-3}}\right.$

$\left. + \dfrac{2a}{(n-2)(a+bx)^{n-2}} - \dfrac{a^2}{(n-1)(a+bx)^{n-1}}\right], \qquad n \neq 1,2,3$

37. $\int \dfrac{dx}{x(a+bx)} = -\dfrac{1}{a}\log\dfrac{a+bx}{x}$

38. $\int \dfrac{dx}{x(a+bx)^2} = \dfrac{1}{a(a+bx)} - \dfrac{1}{a^2}\log\dfrac{a+bx}{x}$

39. $\int \dfrac{dx}{x(a+bx)^3} = \dfrac{1}{a^3}\left[\dfrac{1}{2}\left(\dfrac{2a+bx}{a+bx}\right)^2 + \log\dfrac{x}{a+bx}\right]$

40. $\int \dfrac{dx}{x^2(a+bx)} = -\dfrac{1}{ax} + \dfrac{b}{a^2}\log\dfrac{a+bx}{x}$

41. $\int \dfrac{dx}{x^3(a+bx)} = \dfrac{2bx-a}{2a^2x^2} + \dfrac{b^2}{a^3}\log\dfrac{x}{a+bx}$

42. $\int \dfrac{dx}{x^2(a+bx)^2} = -\dfrac{a+2bx}{a^2x(a+bx)} + \dfrac{2b}{a^3}\log\dfrac{a+bx}{x}$

The Fourier Transforms

For a piecewise continuous function $F(x)$ over a finite interval $0 \leq x \leq \pi$, the *finite Fourier cosine transform* of $F(x)$ is

$$f_c(n) = \int_0^\pi F(x)\cos nx\,dx \qquad (n=0,1,2,\dots). \tag{1}$$

If x ranges over the interval $0 \leq x \leq L$, the substitution $x' = \pi x/L$ allows the use of this definition, also. The inverse transform is written

$$\bar{F}(x) = \dfrac{1}{\pi}f_c(0) + \dfrac{2}{\pi}\sum_{n=1}^{\infty} f_c(n)\cos nx \qquad (0 < x < \pi) \tag{2}$$

where $\bar{F}(x) = \dfrac{[F(x+0)+F(x-0)]}{2}$. **We** observe that $\bar{F}(x) = F(x)$ at points of continuity. The formula

$$f_c^{(2)}(n) = \int_0^\pi F''(x)\cos nx\,dx$$

$$= -n^2 f_c(n) - F'(0) + (-1)^n F'(\pi) \tag{3}$$

makes the finite Fourier cosine transform useful in certain boundary value problems. Analogously, the *finite Fourier sine transform* of $F(x)$ is

$$f_s(n) = \int_0^\pi F(x)\sin nx\,dx \qquad (n=1,2,3,\dots) \tag{4}$$

and

$$\overline{F}(x) = \frac{2}{\pi} \sum_{n=1}^{\infty} f_s(n) \sin nx \qquad (0 < x < \pi) \tag{5}$$

Corresponding to (6) we have

$$f_s^{(2)}(n) = \int_0^{\pi} F''(x) \sin nx \, dx$$

$$= -n^2 f_s(n) - nF(0) - n(-1)^n F(\pi). \tag{6}$$

Fourier Transforms

If $F(x)$ is defined for $x \geq 0$ and is piecewise continuous over any finite interval, and if

$$\int_0^{\infty} F(x) \, dx$$

is absolutely convergent, then

$$f_c(\alpha) = \sqrt{\frac{2}{\pi}} \int_0^{\infty} F(x) \cos(\alpha x) \, dx \tag{7}$$

is the *Fourier cosine transform* of $F(x)$. Furthermore,

$$\overline{F}(x) = \sqrt{\frac{2}{\pi}} \int_0^{\infty} f_c(\alpha) \cos(\alpha x) \, d\alpha. \tag{8}$$

If $\lim_{x \to \infty} \dfrac{d^n F}{dx^n} = 0$, an important property of the Fourier cosine transform

$$f_c^{(2r)}(\alpha) = \sqrt{\frac{2}{\pi}} \int_0^{\infty} \left(\frac{d^{2r} F}{dx^{2r}} \right) \cos(\alpha x) \, dx$$

$$= -\sqrt{\frac{2}{\pi}} \sum_{n=0}^{r-1} (-1)^n a_{2r-2n-1} a^{2n} + (-1)^r \alpha^{2r} f_c(\alpha) \tag{9}$$

where $\lim_{x \to 0} \dfrac{d^r F}{dx^r} = a_r$, makes it useful in the solution of many problems.

Under the same conditions,

$$f_s(\alpha) = \sqrt{\frac{2}{\pi}} \int_0^{\infty} F(x) \sin(\alpha x) \, dx \tag{10}$$

defines the *Fourier sine transform* of $F(x)$, and

$$\overline{F}(x) = \sqrt{\frac{2}{\pi}} \int_0^\infty f_s(\alpha) \sin(\alpha x) \, d\alpha. \tag{11}$$

Corresponding to (12) we have

$$f_s^{(2r)}(\alpha) = \sqrt{\frac{2}{\pi}} \int_0^\infty \frac{d^{2r}F}{dx^{2r}} \sin(ax) \, dx$$

$$= -\sqrt{\frac{2}{\pi}} \sum_{n=1}^r (-1)^n \alpha^{2n-1} a_{2r-2n} + (-1)^{r-1} \alpha^{2r} f_s(\alpha). \tag{12}$$

Similarly, if $F(x)$ is defined for $-\infty < x < \infty$, and if $\int_{-\infty}^\infty F(x) \, dx$ is absolutely convergent, then

$$f(\alpha) = \frac{1}{\sqrt{2\pi}} \int_{-\infty}^\infty F(x) e^{i\alpha x} \, dx \tag{13}$$

is the *Fourier transform* of $F(x)$, and

$$\overline{F}(x) = \frac{1}{\sqrt{2\pi}} \int_{-\infty}^\infty f(\alpha) e^{-i\alpha x} \, d\alpha. \tag{14}$$

Also, if

$$\lim_{|x| \to \infty} \left| \frac{d^n F}{dx^n} \right| = 0 \quad (n = 1, 2, \ldots, r-1),$$

then

$$f^{(r)}(\alpha) = \frac{1}{\sqrt{2\pi}} \int_{-\infty}^\infty F^{(r)}(x) e^{i\alpha x} \, dx = (-i\alpha)^r f(\alpha). \tag{15}$$

Finite Sine Transforms

$f_s(n)$	$F(x)$
1 $f_s(n) = \int_0^\pi F(x) \sin nx \, dx \quad (n = 1, 2, \ldots)$	$F(x)$
2 $(-1)^{n+1} f_s(n)$	$F(\pi - x)$
3 $\dfrac{1}{n}$	$\dfrac{\pi - x}{\pi}$

Finite Sine Transforms (continued)

	$f_s(n)$	$F(x)$
4	$\dfrac{(-1)^{n+1}}{n}$	$\dfrac{x}{\pi}$
5	$\dfrac{1-(-1)^n}{n}$	1
6	$\dfrac{2}{n^2}\sin\dfrac{n\pi}{2}$	$\begin{cases} x & \text{when } 0 < x < \pi/2 \\ \pi-x & \text{when } \pi/2 < x < \pi \end{cases}$
7	$\dfrac{(-1)^{n+1}}{n^3}$	$\dfrac{x(\pi^2-x^2)}{6\pi}$
8	$\dfrac{1-(-1)^n}{n^3}$	$\dfrac{x(\pi-x)}{2}$
9	$\dfrac{\pi^2(-1)^{n-1}}{n}-\dfrac{2[1-(-1)^n]}{n^3}$	x^2
10	$\pi(-1)^n\left(\dfrac{6}{n^3}-\dfrac{\pi^2}{n}\right)$	x^3
11	$\dfrac{n}{n^2+c^2}[1-(-1)^n e^{c\pi}]$	e^{cx}
12	$\dfrac{n}{n^2+c^2}$	$\dfrac{\sinh c(\pi-x)}{\sinh c\pi}$
13	$\dfrac{n}{n^2-k^2}(k\neq 0,1,2,\dots)$	$\dfrac{\sinh k(\pi-x)}{\sin k\pi}$
14	$\begin{cases} \dfrac{\pi}{2} & \text{when } n=m \\ & \quad (m=1,2,\dots) \\ 0 & \text{when } n\neq m \end{cases}$	$\sin mx$
15	$\dfrac{n}{n^2-k^2}[1-(-1)^n\cos k\pi]$ $(k\neq 1,2,\dots)$	$\cos kx$
16	$\begin{cases} \dfrac{n}{n^2-m^2}[1-(-1)^{n+m}] \\ \quad\text{when } n\neq m=1,2,\dots \\ 0 \quad\quad \text{when } n=m \end{cases}$	$\cos mx$
17	$\dfrac{n}{(n^2-k^2)^2}(k\neq 0,1,2,\dots)$	$\dfrac{\pi\sin kx}{2k\sin^2 k\pi}-\dfrac{x\cos k(\pi-x)}{2k\sin k\pi}$
18	$\dfrac{b^n}{n}(\lvert b\rvert\leq 1)$	$\dfrac{2}{\pi}\arctan\dfrac{b\sin x}{1-b\cos x}$
19	$\dfrac{1-(-1)^n}{n}b^n \quad (\lvert b\rvert\leq 1)$	$\dfrac{2}{\pi}\arctan\dfrac{2b\sin x}{1-b^2}$

Finite Cosine Transforms

$f_c(n)$	$F(x)$
1 $f_c(n) = \displaystyle\int_0^\pi F(x)\cos nx\,dx \quad (n = 0,1,2,\ldots)$	$F(x)$
2 $(-1)^n f_c(n)$	$F(\pi - x)$
3 0 when $n = 1,2,\ldots;\ f_c(0) = \pi$	1
4 $\dfrac{2}{n}\sin\dfrac{n\pi}{2};\ f_c(0) = 0$	$\begin{cases} 1 \text{ when } 0 < x < \pi/2 \\ -1 \text{ when } \pi/2 < x < \pi \end{cases}$
5 $-\dfrac{1-(-1)^n}{n^2};\ f_c(0) = \dfrac{\pi^2}{2}$	x
6 $\dfrac{(-1)^n}{n^2};\ f_c(0) = \dfrac{\pi^2}{6}$	$\dfrac{x^2}{2\pi}$
7 $\dfrac{1}{n^2};\ f_c(0) = 0$	$\dfrac{(\pi - x)^2}{2\pi} - \dfrac{\pi}{6}$
8 $3\pi^2\dfrac{(-1)^n}{n^2} - 6\dfrac{1-(-1)^n}{n^4};\ f_c(0) = \dfrac{\pi^4}{4}$	x^3
9 $\dfrac{(-1)^n e^c\pi - 1}{n^2 + c^2}$	$\dfrac{1}{c}e^{cx}$
10 $\dfrac{1}{n^2 + c^2}$	$\dfrac{\cosh c(\pi - x)}{c\sinh c\pi}$
11 $\dfrac{k}{n^2 - k^2}[(-1)^n\cos\pi k - 1]$ $(k \neq 0,1,2,\ldots)$	$\sin kx$
12 $\dfrac{(-1)^{n+m} - 1}{n^2 - m^2};\ f_c(m) = 0 \quad (m = 1,2,\ldots)$	$\dfrac{1}{m}\sin mx$
13 $\dfrac{1}{n^2 - k^2} \quad (k \neq 0,1,2,\ldots)$	$-\dfrac{\cos k(\pi - x)}{k\sin k\pi}$
14 0 when $n = 1,2,\ldots;$ $f_c(m) = \dfrac{\pi}{2} \quad (m = 1,2,\ldots)$	$\cos mx$

Fourier Sine Transforms

$F(x)$	$f_s(\alpha)$
1 $\begin{cases} 1 & (0<x<a) \\ 0 & (x>a) \end{cases}$	$\sqrt{\dfrac{2}{\pi}}\left[\dfrac{1-\cos\alpha}{\alpha}\right]$
2 $x^{p-1}(0<p<1)$	$\sqrt{\dfrac{2}{\pi}}\dfrac{\Gamma(p)}{\alpha^p}\sin\dfrac{p\pi}{2}$
3 $\begin{cases} \sin x & (0<x<a) \\ 0 & (x>a) \end{cases}$	$\dfrac{1}{\sqrt{2\pi}}\left[\dfrac{\sin[a(1-\alpha)]}{1-\alpha}-\dfrac{\sin[a(1+\alpha)]}{1+\alpha}\right]$
4 e^{-x}	$\sqrt{\dfrac{2}{\pi}}\left[\dfrac{\alpha}{1+\alpha^2}\right]$
5 $xe^{-x^2/2}$	$\alpha e^{-\alpha^2/2}$
6 $\cos\dfrac{x^2}{2}$	$\sqrt{2}\left[\sin\dfrac{\alpha^2}{2}C\left(\dfrac{\alpha^2}{2}\right)-\cos\dfrac{\alpha^2}{2}S\left(\dfrac{\alpha^2}{2}\right)\right]^{*}$
7 $\sin\dfrac{x^2}{2}$	$\sqrt{2}\left[\cos\dfrac{\alpha^2}{2}C\left(\dfrac{\alpha^2}{2}\right)+\sin\dfrac{\alpha^2}{2}S\left(\dfrac{\alpha^2}{2}\right)\right]^{*}$

*$C(y)$ and $S(y)$ are the Fresnel integrals

$$C(y)=\frac{1}{\sqrt{2\pi}}\int_0^y\frac{1}{\sqrt{t}}\cos t\,dt,$$

$$S(y)=\frac{1}{\sqrt{2\pi}}\int_0^y\frac{1}{\sqrt{t}}\sin t\,dt.$$

Fourier Cosine Transforms

$F(x)$	$f_c(\alpha)$
1 $\begin{cases} 1 & (0<x<a) \\ 0 & (x>a) \end{cases}$	$\sqrt{\dfrac{2}{\pi}}\dfrac{\sin a\alpha}{\alpha}$
2 $x^{p-1}\quad(0<p<1)$	$\sqrt{\dfrac{2}{\pi}}\dfrac{\Gamma(p)}{\alpha^p}\cos\dfrac{p\pi}{2}$
3 $\begin{cases} \cos x & (0<x<a) \\ 0 & (x>a) \end{cases}$	$\dfrac{1}{\sqrt{2\pi}}\left[\dfrac{\sin[a(1-\alpha)]}{1-\alpha}+\dfrac{\sin[a(1+\alpha)]}{1+\alpha}\right]$
4 e^{-x}	$\sqrt{\dfrac{2}{\pi}}\left(\dfrac{1}{1+\alpha^2}\right)$
5 $e^{-x^2/2}$	$e^{-\alpha^2/2}$
6 $\cos\dfrac{x^2}{2}$	$\cos\left(\dfrac{\alpha^2}{2}-\dfrac{\pi}{4}\right)$
7 $\sin\dfrac{x^2}{2}$	$\cos\left(\dfrac{\alpha^2}{2}+\dfrac{\pi}{4}\right)$

Fourier Transforms

	$F(x)$	$f(\alpha)$
1	$\dfrac{\sin ax}{x}$	$\begin{cases} \sqrt{\dfrac{\pi}{2}} & \lvert\alpha\rvert < a \\ 0 & \lvert\alpha\rvert > a \end{cases}$
2	$\begin{cases} e^{iwx} & (p < x < q) \\ 0 & (x < p, x > q) \end{cases}$	$\dfrac{i}{\sqrt{2\pi}} \dfrac{e^{ip(w+\alpha)} - e^{iq(w+\alpha)}}{(w+\alpha)}$
3	$\begin{cases} e^{-cx+iwx} & (x > 0) \\ 0 & (x < 0) \end{cases} \quad (c > 0)$	$\dfrac{i}{\sqrt{2\pi}\,(w + \alpha + ic)}$
4	$e^{-px^2} \quad R(p) > 0$	$\dfrac{1}{\sqrt{2p}} e^{-\alpha^2/4p}$
5	$\cos px^2$	$\dfrac{1}{\sqrt{2p}} \cos\left[\dfrac{\alpha^2}{4p} - \dfrac{\pi}{4}\right]$
6	$\sin px^2$	$\dfrac{1}{\sqrt{2p}} \cos\left[\dfrac{\alpha^2}{4p} + \dfrac{\pi}{4}\right]$
7	$\lvert x \rvert^{-p} \quad (0 < p < 1)$	$\sqrt{\dfrac{2}{\pi}} \dfrac{\Gamma(1-p)\sin\dfrac{p\pi}{2}}{\lvert\alpha\rvert^{(1-p)}}$
8	$\dfrac{e^{-a\lvert x\rvert}}{\sqrt{\lvert x\rvert}}$	$\dfrac{\sqrt{\sqrt{(a^2+\alpha^2)} + a}}{\sqrt{a^2+\alpha^2}}$
9	$\dfrac{\cosh ax}{\cosh \pi x} \quad (-\pi < a < \pi)$	$\sqrt{\dfrac{2}{\pi}} \dfrac{\cos\dfrac{a}{2}\cosh\dfrac{\alpha}{2}}{\cosh\alpha + \cos a}$
10	$\dfrac{\sinh ax}{\sinh \pi x} \quad (-\pi < a < \pi)$	$\dfrac{1}{\sqrt{2\pi}} \dfrac{\sin a}{\cosh\alpha + \cos a}$
11	$\begin{cases} \dfrac{1}{\sqrt{a^2-x^2}} & (\lvert x\rvert < a) \\ 0 & (\lvert x\rvert > a) \end{cases}$	$\sqrt{\dfrac{\pi}{2}}\, J_0(a\alpha)$
12	$\dfrac{\sin\left[b\sqrt{a^2+x^2}\right]}{\sqrt{a^2+x^2}}$	$\begin{cases} 0 & (\lvert\alpha\rvert > b) \\ \sqrt{\dfrac{\pi}{2}}\, J_0(a\sqrt{b^2-\alpha^2}) & (\lvert\alpha\rvert < b) \end{cases}$
13	$\begin{cases} P_n(x) & (\lvert x\rvert < 1) \\ 0 & (\lvert x\rvert > 1) \end{cases}$	$\dfrac{i^n}{\sqrt{\alpha}} J_{n+\frac{1}{2}}(\alpha)$

Fourier Transforms (continued)

	$F(x)$	$f(\alpha)$
14	$\begin{cases} \dfrac{\cos\left[b\sqrt{a^2-x^2}\right]}{\sqrt{a^2-x^2}} & (\lvert x\rvert < a) \\ 0 & (\lvert x\rvert > a) \end{cases}$	$\sqrt{\dfrac{\pi}{2}}\, J_0(a\sqrt{a^2+b^2}\,)$
15	$\begin{cases} \dfrac{\cosh\left[b\sqrt{a^2-x^2}\right]}{\sqrt{a^2-x^2}} & (\lvert x\rvert < a) \\ 0 & (\lvert x\rvert > a) \end{cases}$	$\sqrt{\dfrac{\pi}{2}}\, J_0(a\sqrt{\alpha^2-b^2}\,)$

The following functions appear among the entries of the tables on transforms.

Function	Definition	Name
$Ei(x)$	$\displaystyle\int_{-\infty}^{x} \frac{e^v}{v}\,dv$; or sometimes defined as $$-Ei(-x) = \int_x^\infty \frac{e^{-v}}{v}\,dv$$	
$Si(x)$	$\displaystyle\int_0^x \frac{\sin v}{v}\,dv$	
$Ci(x)$	$\displaystyle\int_\infty^x \frac{\cos v}{v}\,dv$; or sometimes defined as negative of this integral	
$erf(x)$	$\dfrac{2}{\sqrt{\pi}}\displaystyle\int_0^x e^{-v^2}\,dv$	Error function
$erfc(x)$	$1 - erf(x) = \dfrac{2}{\sqrt{\pi}}\displaystyle\int_x^\infty e^{-v^2}\,dv$	Complementary function to error function
$L_n(x)$	$\dfrac{e^x}{n!}\dfrac{d^n}{dx^n}(x^n e^{-x}), \quad n = 0,1,\dots$	Laguerre polynomial of degree n

Numerical Methods

Solution of Equations by Iteration

Fixed-Point Iteration for Solving $f(x) = 0$

Transform $f(x) = 0$ into the form $x = g(x)$. Choose an x_0 and compute $x_1 = g(x_0)$, $x_2 = g(x_1)$, and in general

$$x_{n+1} = g(x_n), \qquad n = 0,1,2,\dots$$

Newton-Raphson Method for Solving $f(x) = 0$

f is assumed to have a continuous derivative f'. Use an approximate value x_0 obtained from the graph of f. Then compute

$$x_1 = x_0 - \frac{f(x_0)}{f'(x_0)}, \qquad x_2 = x_1 - \frac{f(x_1)}{f'(x_1)}$$

and in general

$$x_{n+1} = x_n - \frac{f(x_0)}{f'(x_n)}$$

Secant Method for Solving $f(x) = 0$

The secant method is obtained from Newton's method by replacing the derivative $f'(x)$ by the difference quotient

$$f'(x_n) = \frac{f(x_n) - f(x_{n-1})}{x_n - x_{n-1}}$$

Thus

$$x_{n+1} = x_n - f(x_n) \frac{x_n - x_{n-1}}{f(x_n) - f(x_{n-1})}$$

The secant method needs two starting values x_0 and x_1.

Method of Regula Falsi for Solving $f(x) = 0$

Select two starting values x_0 and x_1. Then compute

$$x_2 = \frac{x_0 f(x_1) - x_1 f(x_0)}{f(x_1) - f(x_0)}$$

If $f(x_0) \cdot f(x_2) < 0$, replace x_1 by x_2 in formula for x_2, leaving x_0 unchanged, and then compute the next approximation x_3; otherwise, replace x_0 by x_2, leaving x_1 unchanged, and compute the next approximation x_3. Continue in a similar manner.

Finite Differences

Uniform Interval h

If a function $f(x)$ is tabulated at a uniform interval h, that is, for arguments given by $x_n = x_0 + nh$, where n is an integer, then the function $f(x)$ may be denoted by f_n.

This can be generalized so that for all values of p, and in particular for $0 \le p \le 1$,

$$f(x_0 + ph) = f(x_p) = f_p,$$

where the argument designated x_0 can be chosen quite arbitrarily.

The following table lists and defines the standard operators used in numerical analysis.

Symbol	Function	Definition
E	Displacement	$Ef_p = f_{p+1}$
Δ	Forward difference	$\Delta f_p = f_{p+1} - f_p$
∇	Backward difference	$\nabla f_p = f_p - f_{p-1}$
$\wedge$	Divided difference	
δ	Central difference	$\delta f_p = f_{p+\frac{1}{2}} - f_{p-\frac{1}{2}}$
μ	Average	$\mu f_p = \frac{1}{2}(f_{p+\frac{1}{2}} + f_{p-\frac{1}{2}})$
Δ^{-1}	Backward sum	$\Delta^{-1} f_p = \Delta^{-1} f_{p-1} + f_{p-1}$
∇^{-1}	Forward sum	$\nabla^{-1} f_p = \nabla^{-1} f_{p-1} + f_p$
δ^{-1}	Central sum	$\delta^{-1} f_p = \delta^{-1} f_{p-1} + f_{p-\frac{1}{2}}$
D	Differentiation	$Df_p = \dfrac{d}{dx} f(x) = \dfrac{1}{h} \cdot \dfrac{d}{dp} f_p$
$I(=D^{-1})$	Integration	$If_p = \displaystyle\int^{x_p} f(x)\, dx = h \int^{p} f_p\, dp$
$J(=\Delta D^{-1})$	Definite integration	$Jf_p = h \displaystyle\int_{p}^{p+1} f_p\, dp$

I, Δ^{-1}, ∇^{-1} and δ^{-1} all imply the existence of an arbitrary constant which is determined by the initial conditions of the problem.

Where no confusion can arise the f can be omitted as, for example, in writing Δ_p for Δf_p.

Higher differences are formed by successive operations, e.g.,

$$\Delta^2 f_p = \Delta_p^2$$

$$= \Delta \cdot \Delta_p$$

$$= \Delta(f_{p+1} - f_p)$$

$$= \Delta_{p+1} - \Delta_p$$

$$= f_{p+2} - f_{p+1} - f_{p+1} + f_p$$

$$= f_{p+2} - 2f_{p+1} + f_p$$

Note that $f_p \equiv \Delta_p^0 \equiv \nabla_p^0 \equiv \delta_p^0$.

The disposition of the differences and sums relative to the function values is as shown (the arguments are omitted in these cases in the interest of clarity).

Calculus of Finite Differences

Forward difference scheme

$$
\begin{array}{cccc}
\Delta_{-1}^{-2} & f_{-2} & \Delta_{-2}^{2} & \\
& \Delta_{-1}^{-1} \quad \Delta_{-2} \quad \Delta_{-3}^{3} & & \\
\Delta_{0}^{-2} & f_{-1} & \Delta_{-2}^{2} & \\
& \Delta_{0}^{-1} \quad \Delta_{-1} \quad \Delta_{-2}^{3} & & \\
\Delta_{1}^{-2} & f_{0} & \Delta_{-1}^{2} & \\
& \Delta_{1}^{-1} \quad \Delta_{0} \quad \Delta_{-1}^{3} & & \\
\Delta_{2}^{-2} & f_{1} & \Delta_{0}^{2} & \\
& \Delta_{2}^{-1} \quad \Delta_{1} \quad \Delta_{0}^{3} & & \\
\Delta_{3}^{-2} & f_{2} & \Delta_{1}^{2} & \\
\end{array}
$$

Backward difference scheme

$$
\begin{array}{cccc}
\nabla_{-3}^{-2} & f_{-2} & \nabla_{-1}^{2} & \\
& \nabla_{-2}^{-1} \quad \nabla_{-1} \quad \nabla_{0}^{2} & & \\
\nabla_{-2}^{-2} & f_{-1} & \nabla_{0}^{2} & \\
& \nabla_{-1}^{-1} \quad \nabla_{0} \quad \nabla_{1}^{3} & & \\
\nabla_{-1}^{-2} & f_{0} & \nabla_{1}^{2} & \\
& \nabla_{0}^{-1} \quad \nabla_{1} \quad \nabla_{2}^{3} & & \\
\nabla_{0}^{-2} & f_{1} & \nabla_{2}^{2} & \\
& \nabla^{-1} \quad \nabla_{2} \quad \nabla_{3}^{3} & & \\
\nabla_{1}^{-2} & f_{2} & \nabla_{3}^{2} & \\
\end{array}
$$

Central difference scheme

$$
\begin{array}{cccc}
\delta_{-2}^{-2} & f_{-2} & \delta_{-2}^{2} & \delta_{-2}^{4} \\
& \delta_{-1\frac{1}{2}}^{-1} \quad \delta_{-1\frac{1}{2}} \quad \delta_{-1\frac{1}{2}}^{3} & & \\
\delta_{-1}^{-2} & f_{-1} & \delta_{-1}^{2} & \delta_{-1}^{4} \\
& \delta_{-\frac{1}{2}}^{-1} \quad \delta_{-\frac{1}{2}} \quad \delta_{-\frac{1}{2}}^{3} & & \\
\delta_{0}^{-2} & f_{0} & \delta_{0}^{2} & \delta_{0}^{4} \\
& \delta_{\frac{1}{2}}^{-1} \quad \delta_{\frac{1}{2}} \quad \delta_{\frac{1}{2}}^{3} & & \\
\delta_{1}^{-2} & f_{1} & \delta_{1}^{2} & \delta_{1}^{4} \\
& \delta_{1\frac{1}{2}}^{-1} \quad \delta_{1\frac{1}{2}} \quad \delta_{1\frac{1}{2}}^{3} & & \\
\delta_{2}^{-2} & f_{2} & \delta_{2}^{2} & \delta_{2}^{4} \\
\end{array}
$$

In the forward difference scheme the subscripts are seen to move forward into the difference table and no fractional subscripts occur. In the backward difference scheme the subscripts lie on diagonals slanting backwards into the table while in the central difference scheme the subscripts maintain their position and the odd order subscripts are fractional.

All three however are merely alternative ways of labeling the same numerical quantities as any difference is the result of subtracting the number diagonally above it in the preceding column from that diagonally below it in the preceding column or, alternatively, it is the sum of the number diagonally above it in the subsequent column with that immediately above it in its own column.

In general $\Delta_{p-\frac{1}{2}n}^{n} \equiv \delta_{p}^{n} \equiv \nabla_{p+\frac{1}{2}n}^{n}$.

If a polynomial of degree r is tabulated exactly i.e., without any round-off errors, then the rth differences are constant.

The following table enables the simpler operators to be expressed in terms of the others:

	E	Δ	δ, μ	∇
E	—	$1 + \Delta$	$1 + \mu\delta + \frac{1}{2}\delta^2$	$(1 - \nabla)^{-1}$
Δ	$E - 1$	—	$\mu\delta + \frac{1}{2}\delta^2$	$\nabla(1 - \nabla)^{-1}$
δ	$E^{\frac{1}{2}} - E^{-\frac{1}{2}}$	$\Delta(1 + \Delta)^{-\frac{1}{2}}$	$2(\mu^2 - 1)^{\frac{1}{2}}$	$\nabla(1 - \nabla)^{-\frac{1}{2}}$
∇	$-E^{-1}$	$\Delta(1 + \Delta)^{-1}$	$\mu\delta - \frac{1}{2}\delta^2$	—
μ	$\frac{1}{2}(E^{\frac{1}{2}} + E^{-\frac{1}{2}})$	$\frac{1}{2}(2 + \Delta)(1 + \Delta)^{-\frac{1}{2}}$	$(1 + \frac{1}{4}\delta^2)^{\frac{1}{2}}$	$\frac{1}{2}(2 - \nabla)(1 - \nabla)^{-\frac{1}{2}}$

In addition to the above there are other identities by means of which the above table can be extended, viz.,

$$E = e^{hD} = \Delta\nabla^{-1}$$

$$\mu = E^{-\frac{1}{2}} + \frac{1}{2}\delta = E^{\frac{1}{2}} - \frac{1}{2}\delta = \cosh(\tfrac{1}{2}hD)$$

$$\delta = E^{-\frac{1}{2}}\Delta = E^{\frac{1}{2}}\nabla = (\Delta\nabla)^{\frac{1}{2}} = 2\sinh(\tfrac{1}{2}hD).$$

Note the emergence of Taylor's series from

$$f_p = E^p f_0$$

$$= e^{phD} f_0$$

$$= f_0 + phDf_0 + \frac{1}{2!}p^2 h^2 D^2 f_0 + \cdots.$$

Interpolation

Finite difference interpolation entails taking a given set of points and fitting a function to them. This function is usually a polynomial. If the graph of $f(x)$ is approximated over one tabular interval by a chord of the form $y = a + bx$ chosen to pass through the two points

$$(x_0, f(x_0)), \qquad (x_0 + h, f(x_0 + h))$$

the formula for the interpolated value is found to be

$$f(x_0 + ph) = f(x_0) + p[f(x_0 + h) - f(x_0)]$$

$$= f(x_0) + p\Delta f_0$$

If the graph of $f(x)$ is approximated over two successive tabular intervals by a parabola of the form $y = a + bx + cx^2$ chosen to pass through the three points

$$(x_0, f(x_0)), \qquad (x_0 + h, f(x_0 + h)), \qquad (x_0 + 2h, f(x_0 + 2h))$$

the formula for the interpolated value is found to be

$$f(x_0+ph)=f(x_0)+p[f(x_0+h)-f(x_0)]$$

$$+\frac{p(p-1)}{2!}[f(x_0+2h)-2f(x_0+h)+f(x_0)]$$

$$=f_0+p\Delta f_0+\frac{p(p-1)}{2!}\Delta^2 f_0$$

Using polynomial curves of higher order to approximate the graph of $f(x)$, a succession of interpolation formulas involving higher differences of the tabulated function can be derived. These formulas provide, in general, higher accuracy in the interpolated values.

Newton's Forward Formula

$$f_p=f_0+p\Delta_0+\frac{1}{2!}p(p-1)\Delta_0^2+\frac{1}{3!}p(p-1)(p-2)\Delta_0^3\cdots. \qquad 0\leq p\leq 1$$

Newton's Backward Formula

$$f_p=f_0+p\nabla_0+\frac{1}{2!}p(p+1)\nabla_0^2+\frac{1}{3!}p(p+1)(p+2)\nabla_0^3\cdots. \qquad 0\leq p\leq 1$$

Gauss' Forward Formula

$$f_p=f_0+p\delta_{\frac{1}{2}}+G_2\delta_0^2+G_3\delta_{\frac{1}{2}}^3+G_4\delta_0^4+G_5\delta_{\frac{1}{2}}^5\cdots. \qquad 0\leq p\leq 1$$

Gauss' Backward Formula

$$f_p=f_0+p\delta_{-\frac{1}{2}}+G_2^*\delta_0^2+G_3\delta_{-\frac{1}{2}}^3+G_4^*\delta_0^4+G_5\delta_{-\frac{1}{2}}^5\cdots \qquad 0\leq p\leq 1$$

$$\text{In the above } G_{2n}=\begin{pmatrix}p+n-1\\2n\end{pmatrix}$$

$$G_{2n}^*=\begin{pmatrix}p+n\\2n\end{pmatrix}$$

$$G_{2n+1}=\begin{pmatrix}p+n\\2n+1\end{pmatrix}$$

Stirling's Formula

$$f_p=f_0+\tfrac{1}{2}p(\delta_{\frac{1}{2}}+\delta_{-\frac{1}{2}})+\tfrac{1}{2}p^2\delta_0^2+S_3\left(\delta_{\frac{1}{2}}^3+\delta_{-\frac{1}{2}}^3\right)+S_4\delta_0^4+\cdots. \qquad -\tfrac{1}{2}\leq p\leq\tfrac{1}{2}$$

Steffenson's Formula

$$f_p = f_0 + \tfrac{1}{2}p(p+1)\delta_{\frac{1}{2}} - \tfrac{1}{2}(p-1)p\delta_{-\frac{1}{2}} + (S_3 + S_4)\delta_{\frac{1}{2}}^3 + (S_3 - S_4)\delta_{-\frac{1}{2}}^3 \cdots \quad -\tfrac{1}{2} \leq p \leq \tfrac{1}{2}.$$

In the above $S_{2n+1} = \dfrac{1}{2}\begin{pmatrix} p+n \\ 2n+1 \end{pmatrix}$

$$S_{2n+2} = \frac{p}{2n+2}\begin{pmatrix} p+n \\ 2n+1 \end{pmatrix}$$

$$S_{2n+1} + S_{2n+2} = \begin{pmatrix} p+n+1 \\ 2n+2 \end{pmatrix}$$

$$S_{2n+1} - S_{2n+2} = -\begin{pmatrix} p+n \\ 2n+2 \end{pmatrix}$$

Bessel's Formula

$$f_p = f_0 + p\delta_{\frac{1}{2}} + B_2\left(\delta_0^2 + \delta_1^2\right) + B_3\delta_{\frac{1}{2}}^3 + B_4(\delta_0^4 + \delta_1^4) + B_5\delta_{\frac{1}{2}}^5 + \cdots \qquad 0 \leq p \leq 1$$

Everett's Formula

$$f_p = (1-p)f_0 + pf_1 + E_2\delta_0^2 + F_2\delta_1^2 + E_4\delta_0^4 + F_4\delta_1^4 + E_6\delta_0^6 + F_6\delta_1^6 + \cdots \quad 0 \leq p \leq 1$$

The coefficients in the above two formulae are related to each other and to the coefficients in the Gaussian formulae by the identities

$$B_{2n} \equiv \tfrac{1}{2}G_{2n} \equiv \tfrac{1}{2}(E_{2n} + F_{2n})$$

$$B_{2n+1} \equiv G_{2n+1} - \tfrac{1}{2}G_{2n} \equiv \tfrac{1}{2}(F_{2n} - E_{2n})$$

$$E_{2n} \equiv G_{2n} - G_{2n+1} \equiv B_{2n} - B_{2n+1}$$

$$F_{2n} \equiv G_{2n+1} \equiv B_{2n} + B_{2n+1}$$

Also for $q \equiv 1 - p$ the following symmetrical relationships hold:

$$B_{2n}(p) \equiv B_{2n}(q)$$

$$B_{2n+1}(p) \equiv -B_{2n+1}(q)$$

$$E_{2n}(p) \equiv F_{2n}(q)$$

$$F_{2n}(p) \equiv E_{2n}(q)$$

as can be seen from the tables of these coefficients.

Bessel's Formula (Unmodified)

$$f_p = f_0 + p\delta_{\frac{1}{2}} + B_2\left(\delta_0^2 + \delta_1^2\right) + B_3\delta_{\frac{1}{2}}^3 + B_4(\delta_0^4 + \delta_1^4) + B_5\delta_{\frac{1}{2}}^5 + B_6\left(\delta_0^6 + \delta_1^6\right) + B_7\delta_{\frac{1}{2}}^7 + \cdots$$

Lagrange's Interpolation Formula

$$f(x) = \frac{(x-x_1)(x-x_2)\ldots(x-x_n)}{(x_0-x_1)(x_0-x_2)\ldots(x_0-x_n)}f(x_0)$$

$$+ \frac{(x-x_0)(x-x_2)\ldots(x-x_n)}{(x_1-x_0)(x_1-x_2)\ldots(x_1-x_n)}f(x_1)$$

$$+ \cdots + \frac{(x-x_0)(x-x_1)\ldots(x-x_{n-1})}{(x_n-x_0)(x_n-x_1)\ldots(x_n-x_{n-1})}f(x_n)$$

Newton's Divided Difference Formula

$$f(x) = f_0 + (x+x_0)f[x_0,x_1] + (x-x_0)(x-x_1)f[x_0,x_1,x_2]$$

$$+ \cdots + (x-x_0)(x-x_1)\ldots(x-x_{n-1})f[x_0,x_1,\ldots,x_n]$$

where

$$f[x_0,x_1] = \frac{f_1-f_0}{x_1-x_0},$$

$$f[x_0,x_1,x_2] = \frac{f[x_1,x_2]-f[x_0,x_1]}{x_2-x_0},$$

$$\ldots$$

$$f[x_0,x_1,\ldots,x_k] = \frac{f[x_1,x_2,\ldots,x_k]-f[x_0,x_1,\ldots,x_{k-1}]}{x_k-x_0}$$

The layout of a divided difference table is similar to that of an ordinary finite difference table.

$$
\begin{array}{ccccccc}
x_{-1} & f_{-1} & & \Delta_{-1}^{2} & & \Delta_{-1}^{4} & \\
& & \Delta_{-\frac{1}{2}} & & \Delta_{-\frac{1}{2}} & & \\
x_0 & f_0 & & \Delta_{0}^{2} & & \Delta_{0}^{4} & \\
& & \Delta_{\frac{1}{2}} & & \Delta_{\frac{1}{2}}^{3} & & \\
x_1 & f_1 & & \Delta_{1}^{2} & & \Delta_{1}^{4} & \\
\end{array}
$$

where the Δ's are defined as follows:

$$\Delta_r^0 \equiv f_r, \qquad \Delta_{r+\frac{1}{2}} \equiv (f_{r+1}-f_r)/(x_{r+1}-x_r),$$

and in general

$$\Delta_r^{2n} \equiv \left(\Delta_{r+\frac{1}{2}}^{2n-1} - \Delta_{r-\frac{1}{2}}^{2n-1} \right)/(x_{r+n}-x_{r-n})$$

and

$$\Delta_{r+\frac{1}{2}}^{2n+1} \equiv \left(\Delta_{r+1}^{2n} - \Delta_{r}^{2n} \right)/(x_{r+1+n}-x_{r-n}).$$

Iterative Linear Interpolation

Neville's modification of Aiken's method of iterative linear interpolation is one of the most powerful methods of interpolation when the arguments are unevenly spaced as no prior knowledge of the order of the approximating polynomial is necessary nor is a difference table required.

The values obtained are successive approximations to the required result and the process terminates when there is no appreciable change. These values are of course useless if a new interpolation is required when the procedure must be started afresh.

Defining
$$f_{r,s} \equiv \frac{(x_s - x)f_r - (x_r - x)f_s}{(x_s - x_r)}$$

$$f_{r,s,t} \equiv \frac{(x_t - x)f_{r,s} - (x_r - x)f_{s,t}}{(x_t - x_r)}$$

$$f_{r,s,t,u} \equiv \frac{(x_u - x)f_{r,s,t} - (x_r - x)f_{s,t,u}}{(x_u - x_r)},$$

the computation is laid out as follows:

$$
\begin{array}{llllll}
x_{-1} & (x_{-1} - x) & f_{-1} & & & \\
 & & & f_{-1,0} & & \\
x_0 & (x_0 - x) & f_0 & & f_{-1\,0,1} & \\
 & & & f_{0,1} & & f_{-1\,0\,1,2} \\
x_1 & (x_1 - x) & f_1 & & f_{0,1,2} & \\
 & & & f_{1,2} & & \\
x_2 & (x_2 - x) & f_2 & & & \\
\end{array}
$$

As the iterates tend to their limit the common leading figures can be omitted.

Gauss' Trigonometric Interpolation Formula

This is of greatest value when the function is periodic, i.e., a Fourier series expansion is possible.

$$f(x) = \sum_{r=0}^{n} C_r f_r,$$

where $C_r = N_r(x)/N_r(x_r)$ and

$$N_r(x) = \left[\sin\frac{(x - x_0)}{2}\right]\left[\sin\frac{(x - x_1)}{2}\right]\cdots\left[\sin\frac{(x - x_{r-1})}{2}\right]\left[\sin\frac{(x - x_{r+1})}{2}\right]\cdots\left[\sin\frac{(x - x_n)}{2}\right].$$

This is similar to the Lagrangian formula.

Reciprocal Differences

These are used when the quotient of two polynomials will give a better representation of the interpolating function than a simple polynomial expression.

A convenient layout is as shown below:

$$
\begin{array}{llll}
x_{-1} & f_{-1} \\
& & \rho_{-\frac{1}{2}} \\
x_0 & f_0 & & \rho_0^2 \\
& & \rho_{\frac{1}{2}} & & \rho_{\frac{1}{2}}^3 \\
x_1 & f_1 & & \rho_1^2 & & \rho_1^4 \\
& & \rho_{1\frac{1}{2}} & & \rho_{1\frac{1}{2}}^3 \\
x_2 & f_2 & & \rho_2^2 \\
& & \rho_{2\frac{1}{2}} \\
x_3 & f_3
\end{array}
$$

where

$$
\rho_{r+\frac{1}{2}} \equiv \frac{x_{r+1}-x_r}{f_{r+1}-f_r}
$$

and

$$
\rho_r^2 \equiv \frac{x_{r+1}-x_{r-1}}{f_{r+\frac{1}{2}}-f_{r-\frac{1}{2}}}+f_r
$$

In general

$$
\rho_{r+\frac{1}{2}}^{2n+1} \equiv \frac{x_{r+n+1}-x_{r-n}}{\rho_{r+1}^{2n}-\rho_r^{2n}}+\rho_{r+\frac{1}{2}}^{2n-1}
$$

$$
\rho_r^{2n} \equiv \frac{x_{r+n}-x_{r-n}}{\rho_{r+\frac{1}{2}}^{2n-1}-\rho_{r-\frac{1}{2}}^{2n-1}}+\rho_r^{2n-2}.
$$

The interpolation formula is expressed in the form of a continued fraction expansion.

The expansion corresponding to Newton's forward difference interpolation formula, in the sense of the differences involved, is

$$
f(x)=f_0+\cfrac{(x-x_0)}{\rho_{\frac{1}{2}}+\cfrac{(x_2-x_1)}{\rho_1-f_0+\cfrac{(x-x_2)}{\rho_{1\frac{1}{2}}^3-\rho_{\frac{1}{2}}+\cfrac{(x_4-x_3)}{\rho_2^4-\rho_1^2+(x-x_4)}}}}
$$
$$
\text{etc.}
$$

while that corresponding to Gauss' forward formula is

$$
f(x)=f_0+\cfrac{(x-x_0)}{\rho_{\frac{1}{2}}+\cfrac{(x_2-x_1)}{\rho_0^2-f_0+\cfrac{(x_3-x_{-1})}{\rho_{\frac{1}{2}}^3-\rho_{\frac{1}{2}}+\cfrac{(x_4-x_2)}{\rho_0^4-\rho_0^2+(x-x_{-2})}}}}
$$
$$
\text{etc.}
$$

Probability

Definitions

A sample space S associated with an experiment is a set S of elements such that any outcome of the experiment corresponds to one and only one element of the set. An event E is a subset of a sample space S. An element in a sample space is called a sample point or a simple event (unit subset of S).

Definition of Probability

If an experiment can occur in n mutually exclusive and equally likely ways, and if exactly m of these ways correspond to an event E, then the probability of E is given by

$$P(E) = \frac{m}{n}.$$

If E is a subset of S, and if to each unit subset of S a non-negative number, called its probability, is assigned, and if E is the union of two or more different simple events, then the probability of E, denoted by $P(E)$, is the sum of the probabilities of those simple events whose union is E.

Marginal and Conditional Probability

Suppose a sample space S is partitioned into rs disjoint subsets where the general subset is denoted by $E_i \cap F_j$. Then the marginal probability of E_i is defined as

$$P(E_i) = \sum_{j=1}^{s} P(E_i \cap F_j)$$

and the marginal probability of F_j is defined as

$$P(F_j) = \sum_{i=1}^{r} P(E_i \cap F_j)$$

The conditional probability of E_i, given that F_j has occurred, is defined as

$$P(E_i/F_j) = \frac{P(E_i \cap F_j)}{P(F_j)}, \qquad P(F_j) \neq 0$$

and that of F_j, given that E_i has occurred, is defined as

$$P(F_j/E_i) = \frac{P(E_i \cap F_j)}{P(E_i)}, \qquad P(E_i) \neq 0.$$

Probability Theorems

1. If ϕ is the null set, $P(\phi) = 0$.

2. If S is the sample space, $P(S) = 1$.

3. If E and F are two events

$$P(E \cup F) = P(E) + P(F) - P(E \cap F).$$

4. If E and F are mutually exclusive events,

$$P(E \cup F) = P(E) + P(F).$$

5. If E and E' are complementary events,

$$P(E) = 1 - P(E').$$

6. The conditional probability of an event E, given an event F, is denoted by $P(E/F)$ and is defined as

$$P(E/F) = \frac{P(E \cap F)}{P(F)},$$

where $P(F) \neq 0$.

7. Two events E and F are said to be independent if and only if

$$P(E \cap F) = P(E) \cdot P(F).$$

E is said to be statistically independent of F if $P(E/F) = P(E)$ and $P(F/E) = P(F)$.

8. The events $E_1, E_2, \ldots, E_n$ are called mutually independent for all combinations if and only if every combination of these events taken any number at a time is independent.

9. *Bayes Theorem*.

If $E_1, E_2, \ldots, E_n$ are n mutually exclusive events whose union is the sample space S, and E is any arbitrary event of S such that $P(E) \neq 0$, then

$$P(E_k/E) = \frac{P(E_k) \cdot P(E/E_k)}{\sum\limits_{j=1}^{n} \left[P(E_j) \cdot P(E/E_j) \right]}$$

Random Variable

A function whose domain is a sample space S and whose range is some set of real numbers is called a random variable, denoted by $\mathbf{X}$. The function $\mathbf{X}$ transforms sample points of S into points on the x-axis. $\mathbf{X}$ will be called a discrete random variable if it is a random variable that assumes only a finite or denumerable number of values on the x-axis. $\mathbf{X}$ will be called a continuous random variable if it assumes a continuum of values on the x-axis.

Probability Function (Discrete Case)

The random variable $\mathbf{X}$ will be called a discrete random variable if there exists a function f such that $f(x_i) \geq 0$ and $\sum\limits_{i} f(x_i) = 1$ for $i = 1, 2, 3, \ldots$ and such that for any event E,

$$P(E) = P[\mathbf{X} \text{ is in } E] = \sum\limits_{E} f(x)$$

where $\sum_E$ means sum $f(x)$ over those values x_i that are in E and where $f(x) = P[\mathbf{X} = x]$. The probability that the value of $\mathbf{X}$ is some real number x, is given by $f(x) = P[\mathbf{X} = x]$, where f is called the probability function of the random variable $\mathbf{X}$.

Cumulative Distribution Function (Discrete Case)

The probability that the value of a random variable $\mathbf{X}$ is less than or equal to some real number x is defined as

$$F(x) = P(\mathbf{X} \le x)$$

$$= \Sigma f(x_i), \qquad -\infty < x < \infty,$$

where the summation extends over those values of i such that $x_i \le x$.

Probability Density (Continuous Case)

The random variable $\mathbf{X}$ will be called a continuous random variable if there exists a function f such that $f(x) \ge 0$ and $\int_{-\infty}^{\infty} f(x)\,dx = 1$ for all x in interval $-\infty < x < \infty$ and such that for any event E

$$P(E) = P(\mathbf{X} \text{ is in } E) = \int_E f(x)\,dx.$$

$f(x)$ is called the probability density of the random variable $\mathbf{X}$. The probability that $\mathbf{X}$ assumes any given value of x is equal to zero and the probability that it assumes a value on the interval from a to b, including or excluding either end point, is equal to

$$\int_a^b f(x)\,dx.$$

Cumulative Distribution Function (Continuous Case)

The probability that the value of a random variable $\mathbf{X}$ is less than or equal to some real number x is defined as

$$F(x) = P(\mathbf{X} \le x), \qquad -\infty < x < \infty$$

$$= \int_{-\infty}^x f(x)\,dx.$$

From the cumulative distribution, the density, if it exists, can be found from

$$f(x) = \frac{dF(x)}{dx}.$$

From the cumulative distribution

$$P(a \le \mathbf{X} \le b) = P(\mathbf{X} \le b) - P(\mathbf{X} \le a)$$

$$= F(b) - F(a)$$

Mathematical Expectation

Expected Value

Let **X** be a random variable with density $f(x)$. Then the expected value of **X**, $E(X)$, is defined to be

$$E(\mathbf{X}) = \sum_x xf(x)$$

if **X** is discrete and

$$E(\mathbf{X}) = \int_{-\infty}^{\infty} xf(x)\, dx$$

if **X** is continuous. The expected value of a function g of a random variable **X** is defined as

$$E[g(\mathbf{X})] = \sum_x g(x) \cdot f(x)$$

if **X** is discrete and

$$E[g(\mathbf{X})] = \int_{-\infty}^{\infty} g(x) \cdot f(x)\, dx$$

if **X** is continuous.

Positional Notation

In our ordinary system of writing numbers, the value of any digit depends on its position in the number. The value of a digit in any position is ten times the value of the same digit one position to the right, or one-tenth the value of the same digit one position to the left. Thus, for example,

$$173.246 = 1 \times 10^2 + 7 \times 10^1 + 3 + 2 \times \frac{1}{10} + 4 \times \frac{1}{10^2} + 6 \times \frac{1}{10^3}.$$

There is no reason that a number other than 10 cannot be used as the *base*, or *radix*, of the number system. In fact, bases of 2, 8, and 16 are commonly used in working with digital computers. When the base used is not clear from the context, it is usually indicated as a parenthesized subscript or merely as a subscript. Thus

$$743_{(8)} = 7 \times 8^2 + 4 \times 8 + 3 = 7 \times 64 + 4 \times 8 + 3 = 448 + 32 + 3 = 483_{(10)}$$

$$1011.101_{(2)} = 1 \times 2^3 + 0 \times 2^2 + 1 \times 2 + 1 + 1 \times \tfrac{1}{2} + 0 \times \tfrac{1}{4} + 1 \times \tfrac{1}{8} = 11.625_{(10)}$$

Change of Base

In this section, it is assumed that all calculations will be performed in base 10, since this is the only base in which most people can easily compute. However, there is no logical reason that some other base could not be used for the computations.

To convert a number from another base into base 10:

Simply write down the digits of the number, with each one multiplied by its appropriate positional value. Then perform the indicated computations in base 10, and write down the answer.

For examples, see the two examples in the previous section.

To convert a number from base 10 into another base:

The part of the number to the left of the point and the part to the right must be operated on separately. For the integer part (the part to the left of the point):

a. Divide the number by the new base, getting an integer quotient and remainder.

b. Write down the remainder as the last digit of the number in the new base.

c. Using the quotient from the last division in place of the original number, repeat the above two steps until the quotient becomes zero.

For the fractional part (the part to the right of the point):

a. Multiply the number by the new base.

b. Write down the integral part of the product as the first digit of the fractional part in the new base.

c. Using the fractional part of the last product in place of the original number, repeat the above two steps until the product becomes an integer, or until the desired number of places have been computed.

Examples

These examples show a convenient method of arranging the computations.

1. Convert $103.118_{(10)}$ to base 8.

$$
\begin{array}{rl}
8 & \lfloor 103 \rfloor \quad 7 \\
8 & \lfloor 12 \rfloor \quad 4 \\
& \quad 1 \qquad\qquad 147.074324\ldots
\end{array}
$$

$$
\begin{array}{r}
.118 \\
\underline{8} \\
.944 \\
\underline{8} \\
7.552 \\
\underline{8} \\
4.416 \\
\underline{8} \\
3.328 \\
\underline{8} \\
2.624 \\
\underline{8} \\
4.992
\end{array}
$$

The calculation of the fractional part could be carried out as far as desired. It is a non-terminating fraction which will eventually repeat itself.

$$103.118_{(10)} = 147.074324\ldots_{(8)}$$

The calculations may be further shortened by not writing down the multiplier and divisor at each step of the algorithm, as shown in the next example.

2. Convert $275.824_{(10)}$ to base 5.

5	$\lfloor 275 \rfloor$	0		.824
	$\lfloor 55 \rfloor$	0		4.120
	$\lfloor 11 \rfloor$	1		0.600
	2			3.000

$$275.824_{(10)} = 2100.403_{(5)}$$

To convert from one base to another (neither of which is 10):

The easiest procedure is usually to convert first to base 10, and then to the desired base. However, there are two exceptions to this:

1. If computational facility is possessed in either of the bases, it may be used instead of base 10, and the appropriate one of the above methods applied.

2. If the two bases are different powers of the same number, the conversion may be done digit-by-digit to the base which is the common root of both bases, and then digit-by-digit back to the other base.

Example: Convert $127.653_{(8)}$ to base 16. (For base 16, the letters A–F are used for the digits $10_{(10)}-15_{(10)}$.)

The first step is to convert the number to base 2, simply by converting each digit to its binary equivalent:

$$127.653_{(8)} = 001 \quad 010 \quad 111 \quad \cdot \quad 110 \quad 101 \quad 011_{(2)}$$

Now by simply regrouping the binary number into groups of four binary digits, starting at the point, we convert to base 16:

$$127.653_{(8)} = 101 \quad 0111 \quad \cdot \quad 1101 \quad 0101 \quad 1_{(2)} = 57.D58_{(16)}$$

$10^{\pm n}$ in Octal Scale

10^n	n	10^{-n}	10^n	n	10^{-n}
1	0	1.000 000 000 000 000	112 402 762 000	10	0.000 000 000 006 676
12	1	0.063 146 314 631 463	1 351 035 564 000	11	0.000 000 000 000 500
144	2	0.005 075 341 217 270	16 432 451 210 000	12	0.000 000 000 000 043
1 750	3	0.000 406 111 564 571	221 441 634 520 000	13	0.000 000 000 000 003
23 420	4	0.000 032 155 613 531	2 657 142 036 440 000	14	0.000 000 000 000 000
303 240	5	0.000 002 476 132 611	34 327 724 461 500 000	15	0.000 000 000 000 000
3 641 100	6	0.000 000 206 157 364	434 157 115 760 200 000	16	0.000 000 000 000 000
46 113 200	7	0.000 000 015 327 745	5 432 127 413 542 400 000	17	0.000 000 000 000 000
575 360 400	8	0.000 000 001 257 144	67 405 553 164 731 000 000	18	0.000 000 000 000 000
7 346 545 000	9	0.000 000 000 104 560			

2^n in Decimal Scale

n	2^n	n	2^n	n	2^n
0.001	1.00069 33874 62581	0.01	1.00695 55500 56719	0.1	1.07177 34625 36293
0.002	1.00138 72557 11335	0.02	1.01395 94797 90029	0.2	1.14869 83549 97035
0.003	1.00208 16050 79633	0.03	1.02101 21257 07193	0.3	1.23114 44133 44916
0.004	1.00277 63359 01078	0.04	1.02811 38266 56067	0.4	1.31950 79107 72894
0.005	1.00347 17845 09503	0.05	1.03526 49238 41377	0.5	1.4121 35623 73095
0.006	1.00416 75432 38973	0.06	1.04246 57608 41121	0.6	1.51571 65665 10398
0.007	1.00486 38204 23785	0.07	1.04971 66836 23067	0.7	1.62450 47927 12471
0.008	1.00556 05803 98468	0.08	1.05701 80405 61380	0.8	1.74110 11265 92248
0.009	1.00625 78234 97782	0.09	1.06437 01824 53360	0.9	1.86606 59830 73615

$n \log_{10} 2$, $n \log_2 10$ in Decimal Scale

n	$n \log_{10} 2$	$n \log_2 10$	n	$n \log_{10} 2$	$n \log_2 10$
1	0.30102 99957	3.32192 80949	6	1.80617 99740	19.93156 85693
2	0.60205 99913	6.64385 61898	7	2.10720 99696	23.25349 66642
3	0.90308 99870	9.96578 42847	8	2.40823 99653	26.57542 47591
4	1.20411 99827	13.28771 23795	9	2.70926 99610	29.89735 28540
5	1.50514 99783	16.60964 04744	10	3.01029 99566	33.21928 09480

Addition and Multiplication Tables

Binary Scale		Octal Scale	
Addition	Multiplication	Addition	Multiplication
		0 01 02 03 04 05 06 07	1 02 03 04 05 06 07
		1 02 03 04 05 06 07 10	2 04 06 10 12 14 16
$0 + 0 = 0$	$0 \times 0 = 0$	2 03 04 05 06 07 10 11	3 06 11 14 17 22 25
$0 + 1 = 1 + 0 = 1$	$0 \times 1 = 1 \times 0 = 0$	3 04 05 06 07 10 11 12	4 10 14 20 24 30 34
$1 + 1 = 10$	$1 \times 1 = 1$	4 05 06 07 10 11 12 13	5 12 17 24 31 36 43
		5 06 07 10 11 12 13 14	6 14 22 30 36 44 52
		6 07 10 11 12 13 14 15	7 16 25 34 43 52 61
		7 10 11 12 13 14 15 16	

Mathematical Constants in Octal Scale

$\pi = (3.11037\ 552421)_{(8)}$ $e = (2.55760\ 521305)_{(8)}$ $\gamma = (0.44742\ 147707)_{(8)}$

$\pi^{-1} = (0.24276\ 301556)_{(8)}$ $e^{-1} = (0.27426\ 530661)_{(8)}$ $\log_6 \gamma = -(0.43127\ 233602)_{(8)}$

$\sqrt{\pi} = (1.61337\ 611067)_{(8)}$ $\sqrt{e} = (1.51411\ 230704)_{(8)}$ $\log_2 \gamma = -(0.62573\ 030645)_{(8)}$

$\log_6 \pi = (1.11206\ 404435)_{(8)}$ $\log_{10} e = (0.33626\ 754251)_{(8)}$ $\sqrt{2} = (1.32404\ 746320)_{(8)}$

$\log_2 \pi = (1.51544\ 163223)_{(8)}$ $\log_2 e = (1.34252\ 166245)_{(8)}$ $\log_6 2 = (0.54271\ 027760)_{(8)}$

$\sqrt{10} = (3.12305\ 407267)_{(8)}$ $\log_2 10 = (3.24464\ 741136)_{(8)}$ $\log_6 10 = (2.23273\ 067355)_{(8)}$

Fundamental Physical Constants

Summary of the 1986 Recommended Values of the Fundamental Physical Constants

Quantity	Symbol	Value	Units	Relative uncertainty (ppm)
Speed of light in vacuum	c	299 792 458	ms^{-1}	(exact)
Permeability of vacuum	μ_o	$4\pi \times 10^{-7}$	N A^{-2}	
		$= 12.566\ 370614...$	10^{-7} N A^{-2}	(exact)
Permittivity of vacuum	ϵ_o	$1/\mu_o c^2$		
		$= 8.854\ 187\ 817...$	10^{-12} F m^{-1}	(exact)
Newtonian constant of gravitation	G	6.672 59(85)	10^{-11} m^3 kg^{-1} s^{-2}	128
Planck constant	h	6.626 0755(40)	10^{-34} J s	0.60
$h/2\pi$	$\hbar$	1.054 572 66(63)	10^{-34} J s	0.60
Elementary charge	e	1.602 177 33(49)	10^{-19} C	0.30
Magnetic flux quantum, $h/2e$	Φ_o	2.067 834 61(61)	10^{-15} Wb	0.30
Electron mass	m_e	9.109 3897(54)	10^{-31} kg	0.59
Proton mass	m_p	1.672 6231(10)	10^{-27} kg	0.59
Proton-electron mass ratio	m_p/m_e	1836.152701(37)		0.020
Fine-structure constant, $\mu_o ce^2/2h$	α	7.297 353 08(33)	10^{-3}	0.045
Inverse fine-structure constant	α^{-1}	137 035 9895(61)		0.045
Rydberg constant, $m_e c\alpha^2/2h$	R_∞	10 973 731.534(13)	m^{-1}	0.0012
Avogadro constant	N_A,L	6.022 1367(36)	10^{23} mol^{-1}	0.59
Faraday constant, $N_A e$	F	96 485.309(29)	C mol^{-1}	0.30
Molar gas constant	R	8.314 510(70)	J mol^{-1} K^{-1}	8.4
Boltzmann constant, R/N_A	k	1.380 658(12)	10^{-23} J K^{-1}	8.5
Stefan-Boltzmann constant, $(\pi^2/60)k^4/\hbar^3 c^2$	σ	5.670 51(19)	10^{-8} W m^{-2} K^{-4}	34
Non-SI units used with SI				
Electronvolt, $(e/C)J = \{e\}J$	eV	1.602 17733(40)	10^{-19} J	0.30
(Unified) atomic mass unit, $1\ u = m_u = 1/12m(^{12}C)$	u	1.660 5402(10)	10^{-27} kg	0.59

Note: An abbreviated list of the fundamental constants of physics and chemistry based on a least-squares adjustment with 17 degrees of freedom. The digits in parentheses are the one-standard-deviation uncertainty in the last digits of the given value. Since the uncertainties of many entries are correlated, the full covariance matrix must be used in evaluating the uncertainties of quantities computed from them.

Periodic Table of the Elements

1 **Group** **IA**	**2**	International notation										**13**	**14**	**15**	**16**	**17**	**18** **VIIIA**
1 H 1.00797	**IIA**	Previous international notation Common U.S. notation										**IIIB** **IIIA**	**IVB** **IVA**	**VB** **VA**	**VIB** **VIA**	**VIIB** **VIIA**	**2** He 4.0026
3 Li 6.941	**4** Be 9.01218											**5** B 10.81	**6** C 12.011	**7** N 14.0067	**8** O 15.9994	**9** F 18.9984	**10** Ne 20.179
11 Na 22.9898	**12** Mg 24.305	**3** **IIIA** **IIIB**	**4** **IVA** **IVB**	**5** **VA** **VB**	**6** **VIA** **VIB**	**7** **VIIA** **VIIB**	**8**	**9** **VIIIA** **VIII**	**10**	**11** **IB**	**12** **IIB**	**13** Al 26.9815	**14** Si 28.086	**15** P 30.9738	**16** S 32.064	**17** Cl 35.453	**18** Ar 39.948
19 K 39.0983	**20** Ca 40.08	**21** Sc 44.956	**22** Ti 47.90	**23** V 50.942	**24** Cr 51.996	**25** Mn 54.9380	**26** Fe 55.847	**27** Co 58.9332	**28** Ni 58.69	**29** Cu 63.546	**30** Zn 65.38	**31** Ga 69.72	**32** Ge 72.59	**33** As 74.9216	**34** Se 78.96	**35** Br 79.904	**36** Kr 83.80
37 Rb 85.47	**38** Sr 87.62	**39** Y 88.905	**40** Zr 91.22	**41** Nb 92.906	**42** Mo 95.94	**43** Tc 98.906	**44** Ru 101.07	**45** Rh 102.905	**46** Pd 106.4	**47** Ag 107.868	**48** Cd 112.40	**49** In 114.82	**50** Sn 118.69	**51** Sb 121.75	**52** Te 127.60	**53** I 126.9044	**54** Xe 131.30
55 Cs 132.905	**56** Ba 137.33	**71** Lu 174.97	**72** Hf 178.49	**73** Ta 180.948	**74** W 183.85	**75** Re 186.2	**76** Os 190.2	**77** Ir 192.2	**78** Pt 195.09	**79** Au 196.967	**80** Hg 200.59	**81** Tl 204.37	**82** Pb 207.19	**83** Bi 208.980	**84** Po 210.05	**85** At **210**	**86** Rn 222.00
87 Fr **223**	**88** Ra 226.02	**103** Lr **262**	**104** Rf **261**	**105** Ha **262**	**106** **263**	**107** **262**											

Inner Transition Elements

Lanthanide series

57 La 138.91	**58** Ce 140.12	**59** Pr 140.907	**60** Nd 144.24	**61** Pm **145**	**62** Sm 150.4	**63** Eu 151.96	**64** Gd 157.25	**65** Tb 158.925	**66** Dy 162.50	**67** Ho 164.930	**68** Er 167.26	**69** Tm 168.934	**70** Yb 173.04

Actinide series

89 Ac 227.0278	**90** Th 232.038	**91** Pa 231.036	**92** U 238.03	**93** Np 237.048	**94** Pu 239.13	**95** Am 243.13	**96** Cm **247**	**97** Bk **247**	**98** Cf **251**	**99** Es **252**	**100** Fm **257**	**101** Md **258**	**102** No **259**

Atomic weight is shown below symbol. Boldface indicates isotope with the longest known half-life.

Classification of Electromagnetic Radiation

Basic Conversions: $c = \lambda\nu = \nu/k;\ \nu = c/\lambda = ck;\ \lambda = c/\nu = 1/k;\ k = \nu/c = 1/\lambda$

$c = $ speed of light $= 2.99792458 \times 10^8$ m/s

Frequency (ν)	Wavelength (λ)	Wave number (k)	Names of bands	Approximate photon energies
$3 \times 10^0 - 3 \times 10^1$ Hz 3 — 30 Hz	$10^8 - 10^7$ m 100 — 10 Mm	$10^{-8} - 10^{-7}$ m^{-1} 10 — 100 Gm^{-1}	ELF-(ELF 1), ITU band no. 1	
$3 \times 10^1 - 3 \times 10^2$ Hz 30 — 300 Hz	$10^7 - 10^6$ m 10 — 1 Mm	$10^{-7} - 10^{-6}$ m^{-1} 100 Gm^{-1} — 1 Mm^{-1}	SLF-(ELF 2), ITU band no. 2, mega-meter waves	
$3 \times 10^2 - 3 \times 10^3$ Hz 300 Hz — 3 kHz	$10^6 - 10^5$ m 1 Mm — 100 km	$10^{-6} - 10^{-5}$ m^{-1} 1 — 10 Mm^{-1}	ULF-(ELF 3), ITU band no. 3	
$3 \times 10^3 - 3 \times 10^4$ Hz 3 — 30 kHz	$10^5 - 10^4$ m 100 — 10 km	$10^{-5} - 10^{-4}$ m^{-1} 10 — 100 Mm^{-1}	VLF, ITU band no. 4, myriameter waves	
$3 \times 10^4 - 3 \times 10^5$ Hz 30 — 300 kHz	$10^4 - 10^3$ m 10 — 1 km	$10^{-4} - 10^{-3}$ m^{-1} 100 Mm^{-1} — 1 km^{-1}	LF, ITU band no. 5, kilometer waves	
$3 \times 10^5 - 3 \times 10^6$ Hz 300 kHz — 3 MHz	$10^3 - 10^2$ m 1 km — 100 m	$10^{-3} - 10^{-2}$ m^{-1} 1 — 10 km^{-1}	MF, ITU band no. 6, hectometer waves	
$3 \times 10^6 - 3 \times 10^7$ Hz 3 — 30 MHz	$10^2 - 10^1$ m 100 — 10 m	$10^{-2} - 10^{-1}$ m^{-1} 10 — 100 km^{-1}	HF, ITU band no. 7, decameter waves	
$3 \times 10^7 - 3 \times 10^8$ Hz 30 — 300 MHz	$10^1 - 10^0$ m 10 — 1 m	$10^{-1} - 10^0$ m^{-1} 100 km^{-1} — 1 m^{-1}	VHF, ITU band no. 8, meter waves	
$3 \times 10^8 - 3 \times 10^9$ Hz 300 MHz — 3 GHz	$10^0 - 10^{-1}$ m 1 m — 100 mm	$10^0 - 10^1$ m^{-1} 1 — 10 m^{-1}	UHF, ITU band no. 9, decimeter waves[a]	
$3 \times 10^9 - 3 \times 10^{10}$ Hz 3 — 30 GHz	$10^{-1} - 10^{-2}$ m 100 — 10 mm	$10^1 - 10^2$ m^{-1} 10 — 100 m^{-1}	SHF, ITU band no. 10, centimeter waves[a]	
$3 \times 10^{10} - 3 \times 10^{11}$ Hz 30 — 300 GHz	$10^{-2} - 10^{-3}$ m 10 — 1 mm	$10^2 - 10^3$ m^{-1} 100 m^{-1} — 1 mm^{-1} (1 — 10 cm^{-1})	EHF, ITU band no. 11, millimeter waves	
$3 \times 10^{11} - 3 \times 10^{12}$ Hz 300 GHz — 3 THz	$10^{-3} - 10^{-4}$ m 1 mm — 100 μm	$10^3 - 10^4$ m^{-1} 1 — 10 mm^{-1} (10 — 100 cm^{-1})	Part of micrometer waves, includes part of far or thermal infrared; ITU band no. 12	
$3 \times 10^{12} - 3 \times 10^{13}$ Hz 3 — 30 THz	$10^{-4} - 10^{-5}$ m 100 — 10 μm	$10^4 - 10^5$ m^{-1} 10 — 100 mm^{-1} (100 — 1000 cm^{-1})	Part of micrometer waves includes part of far (thermal) infrared	
$3 \times 10^{13} - 3 \times 10^{14}$ Hz 30 — 300 THz	$10^{-5} - 10^{-6}$ m 10 — 1 μm (100,000 — 10,000 Å)	$10^5 - 10^6$ m^{-1} 100 mm^{-1} — 1 μm^{-1}	Part of μm waves, part of infrared	$(1.6 - 16) \times 10^{-20}$ joule {0.1 — 1 eV}
$3 \times 10^{14} - 3 \times 10^{15}$ Hz **300 THz — 3 PHz**	$10^{-6} - 10^{-7}$ m 1 μm — 100 nm (10,000 — 1000 Å)	$10^6 - 10^7$ m^{-1} 1 — 10 μm^{-1}	Near infrared, visible, near ultraviolet	$(1.6 - 16) \times 10^{-19}$ joule {1 — 10 eV}
$3 \times 10^{15} - 3 \times 10^{16}$ Hz 3 — 30 PHz	$10^{-7} - 10^{-8}$ m 100 — 10 nm (1000 — 100 Å)	$10^7 - 10^8$ m^{-1} 10 — 100 μm^{-1}	Part of "vacuum" - ultraviolet	$(1.6 - 16) \times 10^{-18}$ joule {10 — 100 eV}
$3 \times 10^{16} - 3 \times 10^{17}$ Hz 30 — 300 PHz	$10^{-8} - 10^{-9}$ m 10 — 1 nm (100 — 10 Å)	$10^8 - 10^9$ m^{-1} 100 μm^{-1} — 1 nm^{-1}	Part of soft X-rays	$(1.6 - 16) \times 10^{-17}$ joule {100 — 1000 eV)}
$3 \times 10^{17} - 3 \times 10^{18}$ Hz **300 PHz — 3 EHz**	$10^{-9} - 10^{-10}$ m 1 nm — 100 pm (10 — 1 Å)	$10^9 - 10^{10}$ m^{-1} 1 — 10 nm^{-1}	Part of soft X-rays	$(1.6 - 16) \times 10^{-16}$ joule {1 — 10 keV}
$3 \times 10^{18} - 3 \times 10^{19}$ Hz 3 — 30 EHz	$10^{-10} - 10^{-11}$ m 100 — 10 pm (1 — 0.1 Å)	$10^{10} - 10^{11}$ m^{-1} 10 — 100 nm^{-1}	Hard X-rays and part of soft γ-rays	$(1.6 - 16) \times 10^{-15}$ joule {10 — 100 keV}
$3 \times 10^{19} - 3 \times 10^{20}$ Hz 30 — 300 EHz	$10^{-11} - 10^{-12}$ m 10 — 1 pm (0.1 — 0.01 Å)	$10^{11} - 10^{12}$ m^{-1} 100 nm^{-1} — 1 pm^{-1}	Part of soft and part of hard γ-rays (limit at 510 keV)	$(1.6 - 16) \times 10^{-14}$ joule {100 keV — 1 MeV}
$3 \times 10^{20} - 3 \times 10^{21}$ Hz **300 — 3000 EHz**	$10^{-12} - 10^{-13}$ m 1 pm — 100 fm (0.01 — 0.001 Å)	$10^{12} - 10^{13}$ m^{-1} 1 — 10 pm^{-1}	Part of hard γ-rays and part of "cosmic" γ-rays	$(1.6 - 16) \times 10^{-13}$ joule {1 — 10 MeV}
$3 \times 10^{21} - 3 \times 10^{22}$ Hz **3000 — 30,000 EHz**	$10^{-13} - 10^{-14}$ m 100 — 10 fm (0.001 — 0.0001 Å)	$10^{13} - 10^{14}$ m^{-1} 10 — 100 pm^{-1}	γ-rays produced by cosmic rays	$(1.6 - 16) \times 10^{-12}$ joule {10 — 100 MeV}

Classification of Electromagnetic Radiation (continued)

Note: Abbreviations used in this table: Å—ångstrom (1Å = 10^{-10} m); EHF—extremely high frequency; ELF—extremely low frequency; eV—electronvolt (1 eV = 1.60219×10^{-19} joule); fm —femtometer (10^{-15} m); GHz—gigahertz (10^9 hertz); Gm—gigameter (10^9 m); HF—high frequency; Hz—hertz (s^{-1}); ITU—International Telecommunications Union; keV—kiloelectronvolt (10^3 eV); km—kilometer (10^3 m); LF—low frequency; m—meter; MeV—megaelectronvolt (10^6 eV); MF—medium frequency; MHz—megahertz (10^6 hertz); Mm—megameter (10^6 meter); mm—millimeter (10^{-3} meter); μm—micrometer (10^{-6} meter); nm—nanometer (10^{-9} meter); pm—picometer (10^{-12} meter); SHF super high frequency; SLF—super low frequency; THz—terahertz; UHF—ultra high frequency; ULF—ultra low frequency; VHF—very high frequency; VLF—very low frequency.

[a]   Also called "microwaves;" not to be confused with "micrometer waves."

Compiled by Hans Dolezalek

Letter Designations of Microwave Bands

Frequency (GHz)	Wavelength (cm)	Wavenumber (cm^{-1})	Band
1–2	30–15	0.033–0.067	L-Band
2–4	15–7.5	0.067–0.133	S-Band
4–8	7.5–3.7	0.133–0.267	C-Band
8–12	3.7–2.5	0.267–0.4	X-Band
12–18	2.5–1.7	0.4–0.6	Ku-Band
18–27	1.7–1.1	0.6–0.9	K-Band
27–40	1.1–0.75	0.9–1.33	Ka-Band

Electrical Resistivity

Electrical Resistivity of Pure Metals

The first part of this table gives the electrical resistivity, in units of 10^{-8} Ω m, for 28 common metallic elements as a function of temperature. The data refer to polycrystalline samples. The number of significant figures indicates the accuracy of the values. However, at low temperatures (especially below 50 K) the electrical resistivity is extremely sensitive to sample purity. Thus the low-temperature values refer to samples of specified purity and treatment.

The second part of the table gives resistivity values in the neighborhood of room temperature for other metallic elements that have not been studied over an extended temperature range.

Electrical Resistivity in 10^{-8} Ω m

T/K	Aluminum	Barium	Beryllium	Calcium	Cesium	Chromium	Copper
1	0.000100	0.081	0.0332	0.045	0.0026		0.00200
10	0.000193	0.189	0.0332	0.047	0.243		0.00202
20	0.000755	0.94	0.0336	0.060	0.86		0.00280
40	0.0181	2.91	0.0367	0.175	1.99		0.0239
60	0.0959	4.86	0.067	0.40	3.07		0.0971
80	0.245	6.83	0.075	0.65	4.16		0.215
100	0.442	8.85	0.133	0.91	5.28	1.6	0.348
150	1.006	14.3	0.510	1.56	8.43	4.5	0.699
200	1.587	20.2	1.29	2.19	12.2	7.7	1.046
273	2.417	30.2	3.02	3.11	18.7	11.8	1.543
293	2.650	33.2	3.56	3.36	20.5	12.5	1.678
298	2.709	34.0	3.70	3.42	20.8	12.6	1.712
300	2.733	34.3	3.76	3.45	21.0	12.7	1.725
400	3.87	51.4	6.76	4.7		15.8	2.402
500	4.99	72.4	9.9	6.0		20.1	3.090
600	6.13	98.2	13.2	7.3		24.7	3.792
700	7.35	130	16.5	8.7		29.5	4.514
800	8.70	168	20.0	10.0		34.6	5.262
900	10.18	216	23.7	11.4		39.9	6.041

Electrical Resistivity in 10^{-8} Ω m (continued)

T/K	Gold	Hafnium	Iron	Lead	Lithium	Magnesium	Manganese
1	0.0220	1.00	0.0225		0.007	0.0062	7.02
10	0.0226	1.00	0.0238		0.008	0.0069	18.9
20	0.035	1.11	0.0287		0.012	0.0123	54
40	0.141	2.52	0.0758		0.074	0.074	116
60	0.308	4.53	0.271		0.345	0.261	131
80	0.481	6.75	0.693	4.9	1.00	0.557	132
100	0.650	9.12	1.28	6.4	1.73	0.91	132
150	1.061	15.0	3.15	9.9	3.72	1.84	136
200	1.462	21.0	5.20	13.6	5.71	2.75	139
273	2.051	30.4	8.57	19.2	8.53	4.05	143
293	2.214	33.1	9.61	20.8	9.28	4.39	144
298	2.255	33.7	9.87	21.1	9.47	4.48	144
300	2.271	34.0	9.98	21.3	9.55	4.51	144
400	3.107	48.1	16.1	29.6	13.4	6.19	147
500	3.97	63.1	23.7	38.3		7.86	149
600	4.87	78.5	32.9			9.52	151
700	5.82		44.0			11.2	152
800	6.81		57.1			12.8	
900	7.86					14.4	

T/K	Molybdenum	Nickel	Palladium	Platinum	Potassium	Rubidium	Silver
1	0.00070	0.0032	0.0200	0.002	0.0008	0.0131	0.00100
10	0.00089	0.0057	0.0242	0.0154	0.0160	0.109	0.00115
20	0.00261	0.0140	0.0563	0.0484	0.117	0.444	0.0042
40	0.0457	0.068	0.334	0.409	0.480	1.21	0.0539
60	0.206	0.242	0.938	1.107	0.90	1.94	0.162
80	0.482	0.545	1.75	1.922	1.34	2.65	0.289
100	0.858	0.96	2.62	2.755	1.79	3.36	0.418
150	1.99	2.21	4.80	4.76	2.99	5.27	0.726
200	3.13	3.67	6.88	6.77	4.26	7.49	1.029
273	4.85	6.16	9.78	9.6	6.49	11.5	1.467
293	5.34	6.93	10.54	10.5	7.20	12.8	1.587
298	5.47	7.12	10.73	10.7	7.39	13.1	1.617
300	5.52	7.20	10.80	10.8	7.47	13.3	1.629
400	8.02	11.8	14.48	14.6			2.241
500	10.6	17.7	17.94	18.3			2.87
600	13.1	25.5	21.2	21.9			3.53
700	15.8	32.1	24.2	25.4			4.21
800	18.4	35.5	27.1	28.7			4.91
900	21.2	38.6	29.4	32.0			5.64

T/K	Sodium	Strontium	Tantalum	Tungsten	Vanadium	Zinc	Zirconium
1	0.0009	0.80	0.10	0.000016		0.0100	0.250
10	0.0015	0.80	0.102	0.000137	0.0145	0.0112	0.253
20	0.016	0.92	0.146	0.00196	0.039	0.0387	0.357
40	0.172	1.70	0.751	0.0544	0.304	0.306	1.44
60	0.447	2.68	1.65	0.266	1.11	0.715	3.75
80	0.80	3.64	2.62	0.606	2.41	1.15	6.64
100	1.16	4.58	3.64	1.02	4.01	1.60	9.79
150	2.03	6.84	6.19	2.09	8.2	2.71	17.8
200	2.89	9.04	8.66	3.18	12.4	3.83	26.3
273	4.33	12.3	12.2	4.82	18.1	5.46	38.8
293	4.77	13.2	13.1	5.28	19.7	5.90	42.1
298	4.88	13.4	13.4	5.39	20.1	6.01	42.9
300	4.93	13.5	13.5	5.44	20.2	6.06	43.3
400		17.8	18.2	7.83	28.0	8.37	60.3

Electrical Resistivity in 10^{-8} Ω m (continued)

T/K	Sodium	Strontium	Tantalum	Tungsten	Vanadium	Zinc	Zirconium
500		22.2	22.9	10.3	34.8	10.82	76.5
600		26.7	27.4	13.0	41.1	13.49	91.5
700		31.2	31.8	15.7	47.2		104.2
800		35.6	35.9	18.6	53.1		114.9
900			40.1	21.5	58.7		123.1

Element	T/K	Electrical resistivity 10^{-8} Ω m
Antimony	273	39
Bismuth	273	107
Cadmium	273	6.8
Cerium	290–300	82.8
Cobalt	273	5.6
Dysprosium	290–300	92.6
Erbium	290–300	86.0
Europium	290–300	90.0
Gadolinium	290–300	131
Gallium	273	13.6
Holmium	290–300	81.4
Indium	273	8.0
Iridium	273	4.7
Lanthanum	290–300	61.5
Lutetium	290–300	58.2
Mercury	273	94.1
Neodymium	290–300	64.3
Niobium	273	15.2
Osmium	273	8.1
Polonium	273	40
Praseodymium	290–300	70.0
Promethium	290–300	75
Protactinium	273	17.7
Rhenium	273	17.2
Rhodium	273	4.3
Ruthenium	273	7.1
Samarium	290–300	94.0
Scandium	290–300	56.2
Terbium	290–300	115
Thallium	273	15
Thorium	273	14.7
Thulium	290–300	67.6
Tin	273	11.5
Titanium	273	39
Uranium	273	28
Ytterbium	290–300	25.0
Yttrium	290–300	59.6

Electrical Resistivity of Selected Alloys

Values of the resistivity are given in units of 10^{-8} Ω m. General comments in the preceding table for pure metals also apply here.

Alloy Aluminum-Copper

Wt % Al	273 K	293 K	300 K	350 K	400 K
99[a]	2.51	2.74	2.82	3.38	3.95
95[a]	2.88	3.10	3.18	3.75	4.33
90[b]	3.36	3.59	3.67	4.25	4.86
85[b]	3.87	4.10	4.19	4.79	5.42
80[b]	4.33	4.58	4.67	5.31	5.99
70[b]	5.03	5.31	5.41	6.16	6.94
60[b]	5.56	5.88	5.99	6.77	7.63
50[b]	6.22	6.55	6.67	7.55	8.52
40[c]	7.57	7.96	8.10	9.12	10.2
30[c]	11.2	11.8	12.0	13.5	15.2
25[f]	16.3[aa]	17.2	17.6	19.8	22.2
15[h]	—	12.3	—	—	—
19[g]	10.8[aa]	11.0	11.1	11.7	12.3
5[e]	9.43	9.61	9.68	10.2	10.7
1[b]	4.46	4.60	4.65	5.00	5.37

Alloy—Aluminum-Magnesium

Wt % Al	273 K	293 K	300 K	350 K	400 K
99[c]	2.96	3.18	3.26	3.82	4.39
95[c]	5.05	5.28	5.36	5.93	6.51
90[c]	7.52	7.76	7.85	8.43	9.02
85	—	—	—	—	—
80	—	—	—	—	—
70	—	—	—	—	—
60	—	—	—	—	—
50	—	—	—	—	—
40	—	—	—	—	—
30	—	—	—	—	—
25	—	—	—	—	—
15	—	—	—	—	—
10[b]	17.1	17.4	17.6	18.4	19.2
5[b]	13.1	13.4	13.5	14.3	15.2
1[a]	5.92	6.25	6.37	7.20	8.03

Alloy—Copper-Gold

Wt % Cu	273 K	293 K	300 K	350 K	400 K
99[c]	1.73	1.86[aa]	1.91[aa]	2.24[aa]	2.58[aa]
95[c]	2.41	2.54[aa]	2.59[aa]	2.92[aa]	3.26[aa]
90[c]	3.29	4.42[aa]	3.46[aa]	3.79[aa]	4.12[aa]
85[c]	4.20	4.33	4.38[aa]	4.71[aa]	5.05[aa]
80[c]	5.15	5.28	5.32	5.65	5.99
70[c]	7.12	7.25	7.30	7.64	7.99
60[c]	9.18	9.13	9.36	9.70	10.05
50[c]	11.07	11.20	11.25	11.60	11.94
40[c]	12.70	12.85	12.90[aa]	13.27[aa]	13.65[aa]
30[c]	13.77	13.93	13.99[aa]	14.38[aa]	14.78[aa]
25[c]	13.93	14.09	14.14	14.54	14.94
15[c]	12.75	12.91	12.96[aa]	13.36[aa]	13.77
10[c]	10.70	10.86	10.91	11.31	11.72
5[c]	7.25	7.41[aa]	7.46	7.87	8.28
1[c]	3.40	3.57	3.62	4.03	4.45

Alloy—Copper-Nickel

Wt % Cu	273 K	293 K	300 K	350 K	400 K
99[c]	2.71	2.85	2.91	3.27	3.62
95[c]	7.60	7.71	7.82	8.22	8.62
90[c]	13.69	13.89	13.96	14.40	14.81
85[c]	19.63	19.83	19.90	2032	20.70
80[c]	25.46	25.66	25.72	26.12[aa]	26.44[aa]
70[i]	36.67	36.72	36.76	36.85	36.89
60[i]	45.43	45.38	45.35	45.20	45.01
50[i]	50.19	50.05	50.01	49.73	49.50
40[c]	47.42	47.73	47.82	48.28	48.49
30[i]	40.19	41.79	42.34	44.51	45.40
25[c]	33.46	35.11	35.69	39.67[aa]	42.81[aa]
15[c]	22.00	23.35	23.85	27.60	31.38
10[c]	16.65	17.82	18.26	21.51	25.19
5[c]	11.49	12.50	12.90	15.69	18.78
1[c]	7.23	8.08	8.37	10.63[aa]	13.18[aa]

Alloy—Copper-Palladium

Wt % Cu	273 K	293 K	300 K	350 K	400 K
99[c]	2.10	2.23	2.27	2.59	2.92
95[c]	4.21	4.35	4.40	4.74	5.08
90[c]	6.89	7.03	7.08	7.41	7.74
85[c]	9.48	9.61	9.66	10.01	10.36
80[c]	11.99	12.12	12.16	12.51[aa]	12.87
70[c]	16.87	17.01	17.06	17.41	17.78
60[c]	21.73	21.87	21.92	22.30	22.69
50[c]	27.62	27.79	27.86	28.25	28.64
40[c]	35.31	35.51	35.57	36.03	36.47
30[c]	46.50	46.66	46.71	47.11	47.47
25[c]	46.25	46.45	46.52	46.99[aa]	47.43[aa]
15[c]	36.52	36.99	37.16	38.28	39.35
10[c]	28.90	29.51	29.73	31.19[aa]	32.56[aa]
5[c]	20.00	20.75	21.02	22.84[aa]	24.54[aa]
1[c]	11.90	12.67	12.93[aa]	14.82[aa]	16.68[aa]

Alloy—Copper-Zinc

Wt % Cu	273 K	293 K	300 K	350 K	400 K
99[b]	1.84	1.97	2.02	2.36	2.71
95[b]	2.78	2.92	2.97	3.33	3.69
90[b]	3.66	3.81	3.86	4.25	4.63
85[b]	4.37	4.54	4.60	5.02	5.44
80[b]	5.01	5.19	5.26	5.71	6.17
70[b]	5.87	6.08	6.15	6.67	7.19
60	—	—	—	—	—
50	—	—	—	—	—
40	—	—	—	—	—
30	—	—	—	—	—
25	—	—	—	—	—
15	—	—	—	—	—
10	—	—	—	—	—
5	—	—	—	—	—
1	—	—	—	—	—

Alloy—Gold-Palladium

Wt % Au	273 K	293 K	300 K	350 K	400 K
99[c]	2.69	2.86	2.91	3.32	3.73
95[c]	5.21	5.35	5.41	5.79	6.17
90[i]	8.01	8.17	8.22	8.56	8.93
85[b]	10.50[aa]	10.66	10.72[aa]	11.100[aa]	11.48[aa]
80[b]	12.75	12.93	12.99	13.45	13.93
70[b]	18.23	18.46	18.54	19.10	19.67
60[a]	26.70	26.94	27.02	27.63[aa]	28.23[aa]
50[a]	27.23	27.63	27.76	28.64[aa]	29.42[aa]
40[a]	24.65	25.23	25.42	26.74	27.95
30[b]	20.82	21.49	21.72	23.35	24.92
25[b]	18.86	19.53	19.77	21.51	23.19
15[a]	15.08	15.77	16.01	17.80	19.61
10[a]	13.25	13.95	14.20[aa]	16.00[aa]	17.81[aa]
5[a]	11.49[aa]	12.21	12.46[aa]	14.26[aa]	16.07[aa]
1[a]	10.07	10.85[aa]	11.12[aa]	12.99[aa]	14.80[aa]

Electrical Resistivity of Selected Alloys (continued)

	273 K	293 K	300 K	350 K	400 K
Alloy—Gold-Silver					
Wt % Au					
99[b]	2.58	2.75	2.80[aa]	3.22[aa]	3.63[aa]
95[a]	4.58	4.74	4.79	5.19	5.59
90[j]	6.57	6.73	6.78	7.19	7.58
85[j]	8.14	8.30	8.36[aa]	8.75	9.15
80[j]	9.34	9.50	9.55	9.94	10.33
70[j]	10.70	10.86	10.91	11.29	11.68[aa]
60[j]	10.92	11.07	11.12	11.50	11.87
50[j]	10.23	10.37	10.42	10.78	11.14
40[j]	8.92	9.06	9.11	9.46[aa]	9.81
30[a]	7.34	7.47	7.52	7.85	8.19
25[a]	6.46	6.59	6.63	6.96	7.30[aa]
15[a]	4.55	4.67	4.72	5.03	5.34
10[a]	3.54	3.66	3.71	4.00	4.31
5[i]	2.52	2.64[aa]	2.68[aa]	2.96[aa]	3.25[aa]
1[b]	1.69	1.80	1.84[aa]	2.12[aa]	2.42[aa]
Alloy—Iron-Nickel					
Wt % Fe					
99[a]	10.9	12.0	12.4	—	18.7
95[c]	18.7	19.9	20.2	—	26.8
90[c]	24.2	25.5	25.9	—	33.2
85[c]	27.8	29.2	29.7	—	37.3
80[c]	30.1	31.6	32.2	—	40.0
70[b]	32.3	33.9	34.4	—	42.4

	273 K	293 K	300 K	350 K	400 K
60[c]	53.8	57.1	58.2	—	73.9
50[d]	28.4	30.6	31.4	—	43.7
40[d]	19.6	21.6	22.5	—	34.0
30[c]	15.3	17.1	17.7	—	27.4
25[b]	14.3	15.9	16.4	—	25.1
15[c]	12.6	13.8	14.2	—	21.1
10[c]	11.4	12.5	12.9	—	18.9
5[c]	9.66	10.6	10.9	—	16.1[aa]
1[b]	7.17	7.94	8.12	—	12.8
Alloy—Silver-Palladium					
Wt % Ag					
99[b]	1.891	2.007	2.049	2.35	2.66
95[b]	3.58	3.70	3.74	4.04	4.34
90[b]	5.82	5.94	5.98	6.28	6.59
85[k]	7.92[aa]	8.04[aa]	8.08	8.38[aa]	8.68[aa]
80[k]	10.01	10.13	10.17	10.47	10.78
70[k]	14.53	14.65	14.69	14.99	15.30
60[i]	20.9	21.1	21.2	21.6	22.0
50[k]	31.2	31.4	31.5	32.0	32.4
40[m]	42.2	42.2	42.2	42.3	42.3
30[b]	40.4	40.6	40.7	41.3	41.7
25[k]	36.67[aa]	37.06	37.19	38.1[aa]	38.8[aa]
15[i]	27.08[aa]	26.68[aa]	27.89[aa]	29.3[aa]	30.6[aa]
10[i]	21.69	22.39	22.63	24.3	25.9
5[b]	15.98	16.72	16.98	18.8[aa]	20.5[aa]
1[a]	11.06	11.82	12.08[aa]	13.92[aa]	15.70[aa]

[a] Uncertainty in resistivity is ± 2%.
[b] Uncertainty in resistivity is ± 3%.
[c] Uncertainty in resistivity is ± 5%.
[d] Uncertainty in resistivity is ± 7% below 300 K and ± 5% at 300 and 400 K.
[e] Uncertainty in resistivity is ± 7%.
[f] Uncertainty in resistivity is ± 8%.
[g] Uncertainty in resistivity is ± 10%.
[h] Uncertainty in resistivity is ± 12%.
[i] Uncertainty in resistivity is ± 4%.
[j] Uncertainty in resistivity is ± 1%.
[k] Uncertainty in resistivity is ± 3% up to 300 K and ± 4% above 300 K.
[m] Uncertainty in resistivity is ± 2% up to 300 K and ± 4% above 300 K.
[a] Crystal usually a mixture of α-hcp and fcc lattice.
[aa] In temperature range where no experimental data are available.

Electrical Resistivity of Selected Alloy Cast Irons

Description	Electrical resistivity ($\mu\Omega \cdot m$)
Abrasion-resistant white irons	
Low-C white iron	0.53
Martensitic nickel–chromium iron	0.80
Corrosion-resistant irons	
High-silicon iron	0.50
High-chromium iron	
High-nickel gray iron	1.0[a]
High-nickel ductile iron	1.0[a]
Heat-resistant gray irons	
Medium-silicon iron	
High-chromium iron	
High-nickel iron	1.4–1.7
Nickel–chromium–silicon iron	1.5–1.7
High-aluminum iron	2.4
Heat-resistant ductile irons	
Medium-silicon ductile iron	0.58–0.87
High-nickel ductile (20 Ni)	1.02
High-nickel ductile (23 Ni)	1.0[a]

[a] Estimated.

Source: Data from *ASM Metals Reference Book,* 2nd ed., American Society for Metals, Metals Park, Ohio, 1984.

Resistivity of Selected Ceramics (Listed by Ceramic)

Ceramic	Resistivity (Ω-cm)
Borides	
Chromium diboride (CrB_2)	21×10^{-6}
Hafnium diboride (HfB_2)	$10–12 \times 10^{-6}$ at room temp.
Tantalum diboride (TaB_2)	68×10^{-6}
Titanium diboride (TiB_2) (polycrystalline)	
85% dense	$26.5–28.4 \times 10^{-6}$ at room temp.
85% dense	9.0×10^{-6} at room temp.
100% dense, extrapolated values	$8.7–14.1 \times 10^{-6}$ at room temp.
	3.7×10^{-6} at liquid air temp.
Titanium diboride (TiB_2) (monocrystalline)	
Crystal length 5 cm, 39 deg. and 59 deg. orientation with respect to growth axis	$6.6 \pm 0.2 \times 10^{-6}$ at room temp.
Crystal length 1.5 cm, 16.5 deg. and 90 deg. orientation with respect to growth axis	$6.7 \pm 0.2 \times 10^{-6}$ at room temp.
Zirconium diboride (ZrB_2)	9.2×10^{-6} at 20°C
	1.8×10^{-6} at liquid air temp.
Carbides: boron carbide (B_4C)	$0.3–0.8$

Dielectric Constants

Dielectric Constants of Solids

These data refer to temperatures in the range 17–22°C.

Material	Freq. (Hz)	Dielectric constant	Material	Freq. (Hz)	Dielectric constant
Acetamide	4×10^8	4.0	Phenanthrene	4×10^8	2.80
Acetanilide	—	2.9	Phenol (10°C)	4×10^8	4.3
Acetic acid (2°C)	4×10^8	4.1	Phosphorus, red	10^8	4.1
Aluminum oleate	4×10^8	2.40	Phosphorus, yellow	10^8	3.6
Ammonium bromide	10^8	7.1	Potassium aluminum		
Ammonium chloride	10^8	7.0	sulfate	10^6	3.8
Antimony trichloride	10^8	5.34	Potassium carbonate		
Apatite $\perp$ optic axis	3×10^8	9.50	(15°C)	10^8	5.6
Apatite $\parallel$ optic axis	3×10^8	7.41	Potassium chlorate	6×10^7	5.1
Asphalt	$<3 \times 10^6$	2.68	Potassium chloride	10^4	5.03
Barium chloride (anhyd.)	6×10^7	11.4	Potassium chromate	6×10^7	7.3
Barium chloride ($2H_2O$)	6×10^7	9.4	Potassium iodide	6×10^7	5.6
Barium nitrate	6×10^7	5.9	Potassium nitrate	6×10^7	5.0
Barium sulfate (15°C)	10^8	11.4	Potassium sulfate	6×10^7	5.9
Beryl $\perp$ optic axis	10^4	7.02	Quartz $\perp$ optic axis	3×10^7	4.34
Beryl $\parallel$ optic axis	10^4	6.08	Quartz $\parallel$ optic axis	3×10^7	4.27
Calcite $\perp$ optic axis	10^4	8.5	Resorcinol	4×10^8	3.2
Calcite $\parallel$ optic axis	10^4	8.0	Ruby $\perp$ optic axis	10^4	13.27
Calcium carbonate	10^6	6.14	Ruby $\parallel$ optic axis	10^4	11.28
Calcium fluoride	10^4	7.36	Rutile $\perp$ optic axis	10^8	86
Calcium sulfate ($2H_2O$)	10^4	5.66	Rutile $\parallel$ optic axis	10^8	170
Cassiterite $\perp$ optic axis	10^{12}	23.4	Selenium	10^8	6.6
Cassiterite $\parallel$ optic axis	10^{12}	24	Silver bromide	10^6	12.2
d-Cocaine	5×10^8	3.10	Silver chloride	10^6	11.2
Cupric oleate	4×10^8	2.80	Silver cyanide	10^6	5.6
Cupric oxide (15°C)	10^8	18.1	Smithsonite $\perp$ optic	10^{12}	9.3
Cupric sulfate (anhyd.)	6×10^7	10.3	axis		
Cupric sulfate ($5H_2O$)	6×10^7	7.8	Smithsonite $\parallel$ optic	10^{10}	9.4
Diamond	10^8	5.5	axis		
Diphenylmethane	4×10^8	2.7	Sodium carbonate (an-	6×10^7	8.4
Dolomite $\perp$ optic axis	10^8	8.0	hyd.)		
Dolomite $\parallel$	10^8	6.8	Sodium carbonate	6×10^7	5.3
Ferrous oxide (15°C)	10^8	14.2	($10H_2O$)		
Iodine	10^8	4	Sodium chloride	10^4	6.12
Lead acetate	10^6	2.6	Sodium nitrate	—	5.2
Lead carbonate (15°C)	10^8	18.6	Sodium oleate	4×10^8	2.75
Lead chloride	10^6	4.2	Sodium perchlorate	6×10^7	5.4
Lead monoxide (15°C)	10^8	25.9	Sucrose (mean)	3×10^8	3.32
Lead nitrate	6×10^7	37.7	Sulfur (mean)	—	4.0
Lead oleate	4×10^8	3.27	Thallium chloride	10^6	46.9
Lead sulfate	10^6	14.3	p-Toluidine	4×10^8	3.0
Lead sulfide (15°)	16^6	17.9	Tourmaline $\perp$ optic	10^4	7.10
Malachite (mean)	10^{12}	7.2	axis		
Mercuric chloride	10^6	3.2	Tourmaline $\parallel$ optic	10^4	6.3
Mercurous chloride	10^6	9.4	axis		
Naphthalene	4×10^8	2.52	Urea	4×10^8	3.5
			Zircon $\perp, \parallel$	10^8	12

Dielectric Constants of Ceramics

Material	Dielectric constant 10^6 Hz	Dielectric strength Volts/mil	Volume resistivity Ohm-cm (23°C)	Loss factor*
Alumina	4.5—8.4	40—160	10^{11}—10^{14}	0.0002—0.01
Corderite	4.5—5.4	40—250	10^{12}—10^{14}	0.004—0.012
Forsterite	6.2	240	10^{14}	0.0004
Porcelain (dry process)	6.0—8.0	40—240	10^{12}—10^{14}	0.0003—0.02
Porcelain (wet process)	6.0—7.0	90—400	10^{12}—10^{14}	0.006—0.01
Porcelain, zircon	7.1—10.5	250—400	10^{13}—10^{15}	0.0002—0.008
Steatite	5.5—7.5	200—400	10^{13}—10^{15}	0.0002—0.004
Titanates (Ba, Sr, Ca, Mg, and Pb)	15—12.000	50—300	10^8—10^{15}	0.0001—0.02
Titanium dioxide	14—110	100—210	10^{13}—10^{18}	0.0002—0.005

Dielectric Constants of Glasses

Type	Dielectric constant at 100 MHz (20°C)	Volume resistivity (350°C megohm-cm)	Loss factor[a]
Corning 0010	6.32	10	0.015
Corning 0080	6.75	0.13	0.058
Corning 0120	6.65	100	0.012
Pyrex 1710	6.00	2,500	0.025
Pyrex 3320	4.71	—	0.019
Pyrex 7040	4.65	80	0.013
Pyrex 7050	4.77	16	0.017
Pyrex 7052	5.07	25	0.019
Pyrex 7060	4.70	13	0.018
Pyrex 7070	4.00	1,300	0.0048
Vycor 7230	3.83	—	0.0061
Pyrex 7720	4.50	16	0.014
Pyrex 7740	5.00	4	0.040
Pyrex 7750	4.28	50	0.011
Pyrex 7760	4.50	50	0.0081
Vycor 7900	3.9	130	0.0023
Vycor 7910	3.8	1,600	0.00091
Vycor 7911	3.8	4,000	0.00072
Corning 8870	9.5	5,000	0.0085
G. E. Clear (silica glass)	3.81	4,000—30,000	0.00038
Quartz (fused)	3.75 4.1 (1 MHz)	—	0.0002 (1 MHz)

[a] Power factor × dielectric constant equals loss factor.

Properties of Semiconductors

Semiconducting Properties of Selected Materials

Substance	Minimum energy gap (eV) R.T.	Minimum energy gap (eV) 0 K	$\dfrac{dE_g}{dT} \times 10^4$ eV/°C	$\dfrac{dE_g}{dP} \times 10^6$ eV·cm²/kg	Density of states electron effective mass m_{d_n} (m_o)	Electron mobility and temperature dependence μ_n (cm²/V·s)	$-x$	Density of states hole effective mass m_{d_p} (m_o)	Hole mobility and temperature dependence μ_p (cm²/V·s)	$-x$
Si	1.107	1.153	−2.3	−2.0	1.1	1,900	2.6	0.56	500	2.3
Ge	0.67	0.744	−3.7	±7.3	0.55	3,800	1.66	0.3	1,820	2.33
αSn	0.08	0.094	−0.5		0.02	2,500	1.65	0.3	2,400	2.0
Te	0.33				0.68	1,100		0.19	560	

Semiconducting Properties of Selected Materials (continued)

Substance	Minimum energy gap (eV)		$\dfrac{dE_g}{dT}$ $\times 10^4$ eV/°C	$\dfrac{dE_g}{dP}$ $\times 10^6$ eV·cm²/kg	Density of states electron effective mass m_{d_n} (m_o)	Electron mobility and temperature dependence μ_n (cm²/V·s)	$-x$	Density of states hole effective mass m_{d_p} (m_o)	Hole mobility and temperature dependence μ_p (cm²/V·s)	$-x$
	R.T.	0 K								
III–V Compounds										
AlAs	2.2	2.3				1,200			420	
AlSb	1.6	1.7	−3.5	−1.6	0.09	200	1.5	0.4	500	1.8
GaP	2.24	2.40	−5.4	−1.7	0.35	300	1.5	0.5	150	1.5
GaAs	1.35	1.53	−5.0	+9.4	0.068	9,000	1.0	0.5	500	2.1
GaSb	0.67	0.78	−3.5	+12	0.050	5,000	2.0	0.23	1,400	0.9
InP	1.27	1.41	−4.6	+4.6	0.067	5,000	2.0		200	2.4
InAs	0.36	0.43	−2.8	+8	0.022	33,000	1.2	0.41	460	2.3
InSb	0.165	0.23	−2.8	+15	0.014	78,000	1.6	0.4	750	2.1
II–VI Compounds										
ZnO	3.2		−9.5	+0.6	0.38	180	1.5			
ZnS	3.54		−5.3	+5.7		180			5 (400°C)	
ZnSe	2.58	2.80	−7.2	+6		540			28	
ZnTe	2.26			+6		340			100	
CdO	2.5 ± 0.1		−6		0.1	120				
CdS	2.42		−5	+3.3	0.165	400		0.8		
CdSe	1.74	1.85	−4.6		0.13	650	1.0	0.6		
CdTe	1.44	1.56	−4.1	+8	0.14	1,200		0.35	50	
HgSe	0.30				0.030	20,000	2.0			
HgTe	0.15		−1		0.017	25,000		0.5	350	
Halite Structure Compounds										
PbS	0.37	0.28	+4		0.16	800		0.1	1,000	2.2
PbSe	0.26	0.16	+4		0.3	1,500		0.34	1,500	2.2
PbTe	0.25	0.19	+4	−7	0.21	1,600		0.14	750	2.2
Others										
ZnSb	0.50	0.56			0.15	10				1.5
CdSb	0.45	0.57	−5.4		0.15	300			2,000	1.5
Bi_2S_3	1.3					200			1,100	
Bi_2Se_3	0.27					600			675	
Bi_2Te_3	0.13		−0.95		0.58	1,200	1.68	1.07	510	1.95
Mg_2Si		0.77	−6.4		0.46	400	2.5		70	
Mg_2Ge		0.74	−9			280	2		110	
Mg_2Sn	0.21	0.33	−3.5		0.37	320			260	
Mg_3Sb_2		0.32				20			82	
Zn_3As_2	0.93					10	1.1		10	
Cd_3As_2	0.55				0.046	100,000	0.88			
GaSe	2.05			3.8					20	
GaTe	1.66	1.80	−3.6			14	−5			
InSe	1.8					9000				
TlSe	0.57		−3.9		0.3	30		0.6	20	1.5
$CdSnAs_2$	0.23				0.05	25,000	1.7			
Ga_2Te_3	1.1	1.55	−4.8							
$\alpha\text{-}In_2Te_3$	1.1	1.2			0.7				50	1.1
$\beta\text{-}In_2Te_3$	1.0								5	
$Hg_5In_2Te_8$	0.5								11,000	
SnO_2									78	

Band Properties of Semiconductors

Part A. Data on Valence Bands of Semiconductors (Room Temperature)

	Band curvature effective mass (expressed as fraction of free electron mass)				Measured (light)
Substance	Heavy holes	Light holes	"Split-off" band holes	Energy separation of "split-off" band (eV)	hole mobility $(cm^2/V \cdot s)$
Semiconductors with Valence Band Maximum at the Center of the Brillouin Zone ("F")					
Si	0.52	0.16	0.25	0.044	500
Ge	0.34	0.043	0.08	0.3	1,820
Sn	0.3				2,400
AlAs					
AlSb	0.4			0.7	550
GaP				0.13	100
GaAs	0.8	0.12	0.20	0.34	400
GaSb	0.23	0.06		0.7	1,400
InP				0.21	150
InAs	0.41	0.025	0.083	0.43	460
InSb	0.4	0.015		0.85	750
CdTe	0.35				50
HgTe	0.5				350

Semiconductors with Multiple Valence Band Maxima

		Band curvature effective masses			Measured (light)
Substance	Number of equivalent valleys and direction	Longitudinal m_L	Transverse m_T	Anisotropy $K = m_L/m_T$	hole mobility $cm^2/V \cdot s$
PbSe	4 "L" [111]	0.095	0.047	2.0	1,500
PbTe	4 "L" [111]	0.27	0.02	10	750
Bi_2Te_3	6	0.207	~0.045	4.5	515

Part B. Data on Conduction Bands of Semiconductors (Room Temperature Data)

Single Valley Semiconductors

Substance	Energy gap (eV)	Effective mass (m_o)	Mobility $(cm^2/V \cdot s)$	Comments
GaAs	1.35	0.067	8,500	3 (or 6?) equivalent [100] valleys 0.36 eV above this maximum with a mobility of ~50
InP	1.27	0.067	5,000	3 (or 6?) equivalent [100] valleys 0.4 eV above this minimum
InAs	0.36	0.022	33,000	Equivalent valleys ~1.0 eV above this minimum
InSb	0.165	0.014	78,000	
CdTe	1.44	0.11	1,000	4 (or 8?) equivalent [111] valleys 0.51 eV above this minimum

Multivalley Semiconductors

			Band curvature effective mass			
Substance	Energy Gap	Number of equivalent valleys and direction	Longitudinal m_L	Transverse m_T	Anisotropy $K = m_L/m_T$	Comments
Si	1.107	6 in [100] "Δ"	0.90	0.192	4.7	
Ge	0.67	4 in [111] at "L"	1.588	0.0815	19.5	
GaSb	0.67	as Ge (?)	~1.0	~0.2	~5	
PbSe	0.26	4 in [111] at "L"	0.085	0.05	1.7	
PbTe	0.25	4 in [111] at "L"	0.21	0.029	5.5	
Bi_2Te_3	0.13	6			~0.05	

Resistivity of Semiconducting Minerals

Mineral	ρ (ohm · m)	Mineral	ρ (ohm · m)
Diamond (C)	2.7	Gersdorffite, NiAsS	1 to 160 × 10^{-6}
Sulfides		Glaucodote, (Co, Fe)AsS	5 to 100 × 10^{-6}
Argentite, Ag$_2$S	1.5 to 2.0 × 10^{-3}	Antimonide	
Bismuthinite, Bi$_2$S$_3$	3 to 570	Dyscrasite, Ag$_3$Sb	0.12 to 1.2 × 10^{-6}
Bornite, Fe$_2$S$_3$ · nCu$_2$S	1.6 to 6000 × 10^{-6}	Arsenides	
Chalcocite, Cu$_2$S	80 to 100 × 10^{-6}	Allemonite, SbAs$_2$	70 to 60,000
Chalcopyrite, Fe$_2$S$_3$ · Cu$_2$S	150 to 9000 × 10^{-6}	Lollingite, FeAs$_2$	2 to 270 × 10^{-6}
Covellite, CuS	0.30 to 83 × 10^{-6}	Nicollite, NiAs	0.1 to 2 × 10^{-6}
Galena, PbS	6.8 × 10^{-6} to 9.0 × 10^{-2}	Skutterudite, CoAs$_3$	1 to 400 × 10^{-6}
Haverite, MnS$_2$	10 to 20	Smaltite, CoAs$_2$	1 to 12 × 10^{-6}
Marcasite, FeS$_2$	1 to 150 × 10^{-3}	Tellurides	
Metacinnabarite, 4HgS	2 × 10^{-6} to 1 × 10^{-3}	Altaite, PbTe	20 to 200 × 10^{-6}
Millerite, NiS	2 to 4 × 10^{-7}	Calavarite, AuTe$_2$	6 to 12 × 10^{-6}
Molybdenite, MoS$_2$	0.12 to 7.5	Coloradoite, HgTe	4 to 100 × 10^{-6}
Pentlandite, (Fe, Ni)$_9$S$_8$	1 to 11 × 10^{-6}	Hessite, Ag$_2$Te	4 to 100 × 10^{-6}
Pyrrhotite, Fe$_7$S$_8$	2 to 160 × 10^{-6}	Nagyagite, Pb$_6$Au(S, Te)$_{14}$	20 to 80 × 10^{-6}
Pyrite, FeS$_2$	1.2 to 600 × 10^{-3}	Sylvanite, AgAuTe$_4$	4 to 20 × 10^{-6}
Sphalerite, ZnS	2.7 × 10^{-3} to 1.2 × 10^4	Oxides	
Antimony-sulfur compounds		Braunite, Mn$_2$O$_3$	0.16 to 1.0
Berthierite, FeSb$_2$S$_4$	0.0083 to 2.0	Cassiterite, SnO$_2$	4.5 × 10^{-4} to 10,000
Boulangerite, Pb$_5$Sb$_4$S$_{11}$	2 × 10^3 to 4 × 10^4	Cuprite, Cu$_2$O	10 to 50
Cylindrite, Pb$_3$Sn$_4$Sb$_2$S$_{14}$	2.5 to 60	Hollandite, (Ba, Na, K)Mn$_8$O$_{16}$	2 to 100 × 10^{-3}
Franckeite, Pb$_5$Sn$_3$Sb$_2$S$_{14}$	1.2 to 4	Ilmenite, FeTiO$_3$	0.001 to 4
Hauchecornite, Ni$_9$(Bi, Sb)$_2$S$_8$	1 to 83 × 10^{-6}	Magnetite, Fe$_3$O$_4$	52 × 10^{-6}
Jamesonite, Pb$_4$FeSb$_6$S$_{14}$	0.020 to 0.15	Manganite, MnO · OH	0.018 to 0.5
Tetrahedrite, Cu$_3$SbS$_3$	0.30 to 30,000	Melaconite, CuO	6000
Arsenic-sulfur compounds		Psilomelane, KMnO · MnO$_2$ · nH$_2$O	0.04 to 6000
Arsenopyrite, FeAsS	20 to 300 × 10^{-6}	Pyrolusite, MnO$_2$	0.007 to 30
Cobaltite, CoAsS	6.5 to 130 × 10^{-3}	Rutile, TiO$_2$	29 to 910
Enargite, Cu$_3$AsS$_4$	0.2 to 40 × 10^{-3}	Uraninite, UO	1.5 to 200

Source: Carmichael, R. S., Ed., *Handbook of Physical Properties of Rocks,* Vol. I, Boca Raton, FL: CRC Press, 1982.

Properties of Magnetic Alloys

Name	Composition,* weight percent					Remanence, B_r (Gauss)	Coercive force, H_e (Oersteds)	Maximum energy product, $(BH)_{max}$ (Gauss-Oersteds × 10^{-6})
	Al	Ni	Co	Cu	Other			
U.S.A.								
Alnico I	12	20–22	5			7,100	440	1.4
Alnico II	10	17	12.5	6		7,200	540	1.6
Alnico III	12	24–26		3		6,900	470	1.35
Alnico IV	12	27–28	5			5,500	700	1.3
Alnico V**	8	14	24	3		12,500	600	5.0
Alnico V DG**	8	14	24	3		13,100	640	6.0
Alnico VI**	8	15	24	3	1.25 Ti	10,500	750	3.75
Alnico VII**	8.5	18	24	3	5 Ti	7,200	1,050	2.75
Alnico XII	6	18	35		8 Ti	5,800	950	1.6
					1 Mn	10,000	50	0.2
					0.9 C			
Chromium steel					3.5 Cr	9,700	65	0.3
					0.9 C			
					0.3 Mn			

Properties of Magnetic Alloys (continued)

Name	Composition,* weight percent					Remanence, B_r (Gauss)	Coercive force, H_e (Oersteds)	Maximum energy product, $(BH)_{max}$ (Gauss-Oersteds × 10^{-6})
	Al	Ni	Co	Cu	Other			
Cobalt steel			17		2.5 Cr 8 W 0.75 C	9,500	150	0.65
Cunico		21	29	50		3,400	660	0.80
Cunife		20		60		5,400	550	1.5
Ferroxdur 1			$BaFe_{12}O_{19}$			2,200	1,800	1.0
Ferroxdur 2			$BaF_{12}O_{19}$ (oriented)			3,840	2,000	3.5
Platinum–cobalt			23		77 Pt	6,000	4,300	7.5
Remalloy			12		17 Mo	10,500	250	1.1
Silmanol	4.4				86.6 Ag 8.8 Mn	550	6,000	0.075
Tungsten steel					5 W 0.3 Mn 0.7 C	10,000	70	0.32
Vicalloy I			52		10 V	8,800	300	1.0
Vicalloy II (wire)			52		14 V	10,000	510	3.5
Germany								
Alni 90	12	21				8,000	350	1.2
Alni 120	13	27				6,000	570	1.2
Alnico 130	12	23	5			6,300	620	1.4
Alnico 160	11	24	12	4		6,200	700	1.6
Alnico 190	12	21	15	4		7,000	700	1.8
Alnico 250	8	19	23	4	6 Ti	6,500	1,000	2.2
Alnico 400**	9	15	23	4		12,000	650	4.8
Alnico 580** (semicolumnar)	9	15	23	4		13,000	700	6.0
Oerstit 800	9	18	19	4	4 Ti	6,600	750	1.95
Great Britain								
Alcomax I	7.5	11	25	3	1.5 Ti	12,000	475	3.5
Alcomax II	8	11.5	24	4.5		12,400	575	4.7
Alcomax IISC (semicolumnar)	8	11	22	4.5		12,800	600	5.15
Alcomax III	8	13.5	24	3	0.8 Nb	12,500	670	5.10
Alcomax IIISC (semicolumnar)	8	13.5	24	3	0.8 Nb	13,000	700	5.80
Alcomax IV	8	13.5	24	3	2.5 Nb	11,200	750	4.30
Alcomax IVSC (semicolumnar)	8	13.5	24	3	2.5 Nb	11,700	780	5.10
Alni, high B_r	13	24		3.5		6,200	480	1.25
Alni, normal						5,600	580	1.25
Alni, high H_e	12	32			0–0.5 Ti	5,000	680	1.25
Alnico, high B_r	10	17	12	6		8,000	500	1.70
Alnico, normal						7,250	560	1.70
Alnico, high H_e	10	20	13.5	6	0.25 Ti	6,600	620	1.70
Columax (columnar)	similar to Alcomax III or IV					13,000–14,000	700–800	7.0–8.5
Hycomax	9	21	20	1.6		9,500	830	3.3

* Remainder of unlisted composition is either iron or iron plus trace impurities.
** Cast anisotropic. Unmarked ones are cast isotropic.

Properties of Magnetic Alloys (continued)

High-Permeability Magnetic Alloys

Name	Composition,* weight percent	Sp. gr., g/cm^3	Tensile strength kg/mm^2**	Form	Remark	Use
Silicon iron AISI M 15	Si 4	7.68–7.64	44.3	—	Annealed 4 h 802–1093°C	Low core losses
Silicon iron AISI M 8	Si 3	7.68–7.64	44.2	Grain oriented	Annealed 4 h 802–1204°C	
45 Permalloy	Ni 45; Mn 0.3	8.17	—	—	—	Audio transformer, coils, relays
Monimax	Ni 47; Mo 3	8.27	—	—	—	High-frequency coils
4-79 Permalloy	Ni 79; Mo 4; Mn 0.3	8.74	55.4	—	H$_2$ annealed 1121°C	Audio coils, transformers, magnetic shields
Sinimax	Ni 43; Si 3	7.70	—	—	—	High-frequency coils
Nu-metal	Ni 75; Cr 2; Cu 5	8.58	44.8	—	H$_2$ annealed 1221°C	Audio coils, magnetic shields, transformers
Supermalloy	Ni 79; Mo 5; Mn 0.3	8.77	—	—	—	Pulse transformers, magnetic amplifiers, coils
2-V Permendur	Co 40; V 2	8.15	46.3	—	—	DC electromagnets, pole tips

* Iron is additional alloying metal.
** kg/mm^2 × 1422.33 = lbs/in.^{2}

Cast Permanent Magnetic Alloys

Alloy name, country of manufacture*	Composition,** weight percent	Sp. gr., g/cm^3	Thermal expansion Cm × 10^{-6} cm × °C	Between °C	Tensile strength ***kg/ mm^2	Form	Remark†	Use
Alnico I (USA)	Al 12; Ni 20–22; Co 5	6.9	12.6	20–300	2.9	Cast	i.	Permanent magnets
Alnico II (USA)	Al 10; Ni 17; Cu 6; Co 12.5	7.1	12.4	20–300	2.1 45.7	Cast Sintered	i.	Temperature controls, magnetic toys and novelties
Alnico III (USA)	Al 12; Ni 24–26; Cu 3	6.9	12	20–300	8.5	Cast	i.	Tractor magnetos
Alnico IV (USA)	Al 12; Ni 27–28; Co 5	7.0	13.1	20–300	6.3 42.1	Cast Sintered	i.	Application requiring high coercive force
Alnico V (USA)	Al 8; Ni 14; Co 24; Cu 3	7.3	11.3	—	3.8 35	Cast Sintered	a.	Application requiring high energy
Alnico V DG (USA)	Al 8; Ni 14; Co 24; Cu 3	7.3	11.3	—	—	—	a., c.	
Alnico VI (USA)	Al 8; Ni 15; Co 24; Cu 3; Ti 1.25	7.3	11.4	—	16.1	Cast	a.	Application requiring high energy
Alnico VII (USA)	Al 8.5; Ni 18; Cu 3; Co 24; Ti 5	7.17	11.4	—	—	—	a.	

* USA—United States; GB—Great Britain; Ger—Germany.
** Iron is the additional alloying metal for each of the magnets listed.
*** kg/mm^2 × 1422.33 = lb/in.^{2}
† i. = isotropic; a. = anisotropic; c. = columnar; sc. = semicolumnar.

Properties of Magnetic Alloys (continued)

Cast Permanent Magnetic Alloys (continued)

Alloy name, country of manufacture*	Composition,** weight percent	Sp. gr., g/cm³	Thermal expansion		Tensile strength		Form	Remark†	Use
			$Cm \times 10^{-6}$ cm × °C	Between °C	***kg/mm²				
Alnico XII (USA)	Al 6; Ni 18; Co 35; Ti 8	7.2	11	20–300	—	—	—	Permanent magnets
Comol (USA)	Co 12; Mo 17	8.16	9.3	20–300	88.6	—	—	Permanent magnets
Cunife (USA)	Cu 60; Ni 20	8.52	—	—	70.3	—	—	Permanent magnets
Cunico (USA)	Cu 50; Ni 21	8.31	—	—	70.3	—	—	Permanent magnets
Barium ferrite Feroxdur (USA)	Ba $Fe_{12}O_{19}$	4.7	10	—	70.3	—	—	Ceramics
Alcomax I (GB)	Al 7.5; Ni 11; Co 25; Cu 3; Ti 1.5	—	—	—	—	—	a.	Permanent magnets
Alcomax II (GB)	Al 8; Ni 11.5; Co 24; Cu 4.5	—	—	—	—	—	a.	Permanent magnets
Alcomax II SC (GB)	Al 8; Ni 11; Co 22; Cu 4.5	7.3	—	—	—	—	a., sc.	
Alcomax III (GB)	Al 8; Ni 13.5; Co 24; Nb 0.8	7.3	—	—	—	—	a.	Magnets for motors, loudspeakers
Alcomax IV (GB)	Al 8; Ni 13.5; Cu 3; Co 24; Nb 2.5	—	—	—	—	—	—	Magnets for cycle-dynamos
Columax (GB)	Similar to Alcomax III or IV	—	—	—	—	—	a., sc.	Permanent magnets, heat treatable
Hycomax (GB)	Al 9; Ni 21; Co 20; Cu 1.6	—	—	—	—	—	a.	Permanent magnets
Alnico (high H_c) (GB)	Al 10; Ni 20; Co 13.5; Cu 6; Ti 0.25	7.3	—	—	—	—	i.	
Alnico (high B_r) (GB)	Al 10; Ni 17; Co 12; Cu 6	7.3	—	—	—	—	i.	
Alni (high H_c) (GB)	Al 12; Ni 32; Ti 0–0.5	6.9	—	—	—	—	i.	
Alni (high B_r) (GB)	Al 13; Ni 24; Cu 3.5	—	—	—	—	—	i.	
Alnico 580 (Ger)	Al 9; Ni 15; Co 23; Cu 4	—	—	—	—	—	i.	
Alnico 400 (Ger)	Al 9; Ni 15; Co 23; Cu 4	—	—	—	—	—	a.	
Oerstit 800 (Ger)	Al 9; Ni 18; Co 19; Cu 4; Ti 4	—	—	—	—	—	i.	Permanent magnets
Alnico 250 (Ger)	Al 8; Ni 19; Co 23; Cu 4; Ti 6	—	—	—	—	—	i.	
Alnico 190 (Ger)	Al 12; Ni 21; Cu 4; Co 15	—	—	—	—	—	i.	

* USA—United States; GB—Great Britain; Ger—Germany.
** Iron is the additional alloying metal for each of the magnets listed.
*** kg/mm² × 1422.33 = lb/in.²
† i. = isotropic; a. = anisotropic; c. = columnar; sc. = semicolumnar.

Properties of Magnetic Alloys (continued)

Cast Permanent Magnetic Alloys (continued)

Alloy name, country of manufacture*	Composition,** weight percent	Sp. gr., g/cm³	Thermal expansion		Tensile strength	Form	Remark†	Use
			Cm × 10⁻⁶ cm × °C	Between °C	***kg/ mm²			
Alnico 160 (Austria)	Al 11; Ni 24; Co 12; Cu 4	—	—	—	—		i.	Permanent magnets, sintered
Alnico 130 (Ger)	Al 12; Ni 23; Co 5	—	—	—	—		i.	
Alni 120 (Ger)	Al 13; Ni 27	—	—	—	—		i.	
Alni 90 (Ger)	Al 12; Ni 21	—	—	—	—		i.	

* USA—United States; GB—Great Britain; Ger—Germany.
** Iron is the additional alloying metal for each of the magnets listed.
*** $kg/mm^2 \times 1422.33 = lb/in.^2$
† i. = isotropic; a. = anisotropic; c. = columnar; sc. = semicolumnar.

Properties of Antiferromagnetic Compounds

Compound	Crystal symmetry	θ_N (K)	θ_P (K)	$(P_A)_{eff}$ μ_B	P_A μ_B
$CoCl_2$	Rhombohedral	25	−38.1	5.18	3.1 ± 0.6
CoF_2	Tetragonal	38	50	5.15	3.0
CoO	Tetragonal	291	330	5.1	3.8
Cr	Cubic	475			
Cr_2O_3	Rhombohedral	307	485	3.73	3.0
$CrSb$	Hexagonal	723	550	4.92	2.7
$CuBr_2$	Monoclinic	189	246	1.9	
$CuCl_2 \cdot 2H_2O$	Orthorhombic	4.3	4–5	1.9	
$CuCl_2$	Monoclinic	~70	109	2.08	
$FeCl_2$	Hexagonal	24	−48	5.38	4.4 ± 0.7
FeF_2	Tetragonal	79–90	117	5.56	4.64
FeO	Rhombohedral	198	507	7.06	3.32
$\alpha\text{-}Fe_2O_3$	Rhombohedral	953	2940	6.4	5.0
$\alpha\text{-}Mn$	Cubic	95			
$MnBr_2 \cdot 4H_2O$	Monoclinic	2.1	$\left\{ \begin{matrix} 2.5 \\ 1.3 \end{matrix} \right\}$	5.93	
$MnCl_2 \cdot 4H_2O$	Monoclinic	1.66	1.8	5.94	
MnF_2	Tetragonal	72–75	113.2	5.71	5
MnO	Rhombohedral	122	610	5.95	5.0
$\beta\text{-}MnS$	Cubic	160	982	5.82	5.0
$MnSe$	Cubic	~173	361	5.67	
$MnTe$	Hexagonal	310–323	690	6.07	5.0
$NiCl_2$	Hexagonal	50	−68	3.32	
NiF_2	Tetragonal	78.5–83	115.6	3.5	2.0
NiO	Rhombohedral	533–650	~2000	4.6	2.0
$TiCl_3$		100			
V_2O_3		170			

1. θ_N = Néel temperature, determined from susceptibility maxima or from the disappearance of magnetic scattering.
2. θ_P = a constant in the Curie–Weiss law written in the form $\chi_A = C_A/(T + \theta_P)$, which is valid for antiferromagnetic material for $T > \theta_N$.
3. $(P_A)_{eff}$ = effective moment per atom, derived from the atomic Curie constant $C_A = (P_A)^2_{eff}(N^2/3R)$ and expressed in units of the Bohr magneton, $\mu_B = 0.9273 \times 10^{-20}$ erg gauss⁻¹.
4. P_A = magnetic moment per atom, obtained from neutron diffraction measurements in the ordered state.

Properties of Magnetic Alloys (continued)

Saturation Constants and Curie Points of Ferromagnetic Elements

Element	σ_s (20°C)	M_s (20°C)	σ_s (0 K)	n_B	Curie point (°C)
Fe	218.0	1,714	221.9	2.219	770
Co	161	1,422	162.5	1.715	1,131
Ni	54.39	484.1	57.50	0.604	358
Gd	0	0	253.5	7.12	16

σ_s = saturation magnetic moment/gram; M_s = saturation magnetic moment/cm^3, in cgs units; n_B = magnetic moment per atom in Bohr magnetons.

Source: *American Institute of Physics Handbook*, McGraw-Hill, 1963.

Magnetic Properties of Transformer Steels

Ordinary Transformer Steel

B (Gauss)	H (Oersted)	Permeability = B/H
2,000	0.60	3,340
4,000	0.87	4,600
6,000	1.10	5,450
8,000	1.48	5,400
10,000	2.28	4,380
12,000	3.85	3,120
14,000	10.9	1,280
16,000	43.0	372
18,000	149	121

High Silicon Transformer Steels

B	H	Permeability
2,000	0.50	4,000
4,000	0.70	5,720
6,000	0.90	6,670
8,000	1.28	6,250
10,000	1.99	5,020
12,000	3.60	3,340
14,000	9.80	1,430
16,000	47.4	338
18,000	165	109

Saturation Constants for Magnetic Substances

Substance	Field intensity	Induced magnetization	Substance	Field intensity	Induced magnetization
	(For saturation)			(For saturation)	
Cobalt	9000	1300	Nickel, hard	8000	400
Iron, wrought	2000	1700	annealed	7000	515
cast	4000	1200	Vicker's steel	15000	1600
Manganese steel	7000	200			

Initial Permeability of High Purity Iron for Various Temperatures

L. Alberts and B. J. Shepstone

Temperature °C	Permeability (gauss/oersted)
0	920
200	1040
400	1440
600	2550
700	3900
770	12580

Magnetic Materials

High-Permeability Materials

Material	Form	Fe	Ni	Co	Mo	Other	Typical heat treatment °C	Permeability at $B = 20$ gausses	Maximum permeability	Saturation flux density B gausses	Hysteresis ‡ loss, W_s ergs/cm³	Coercive ‡ force H_c oersteds	Resistivity microhm cm	Density, g/cm³
Cold rolled steel	Sheet	98.5	—	—	—	—	950 Anneal	180	2,000	21,000	—	1.8	10	7.88
Iron	Sheet	99.91	—	—	—	—	950 Anneal	200	5,000	21,500	5,000	1.0	10	7.88
Purified iron	Sheet	99.95	—	—	—	—	1480 H₂ + 880	5,000	180,000	21,500	300	.05	10	7.88
4% Silicon-iron	Sheet	96	—	—	—	4 Si	800 Anneal	500	7,000	19,700	3,500	.5	60	7.65
Grain oriented*	Sheet	97	—	—	—	3 Si	800 Anneal	1,500	30,000	20,000	—	.15	47	7.67
45 Permalloy	Sheet	54.7	45	—	—	.3 Mn	1050 Anneal	2,500	25,000	16,000	1,200	.3	45	8.17
45 Permalloy †	Sheet	54.7	45	—	—	.3 Mn	1200 H₂ Anneal	4,000	50,000	16,000	—	.07	45	8.17
Hipernik	Sheet	50	50	—	—	—	1200 H₂ Anneal	4,500	70,000	16,000	220	.05	50	8.25
Monimax	Sheet	—	—	—	—	—	1125 H₂ Anneal	2,000	35,000	15,000	—	.1	80	8.27
Sinimax	Sheet	—	—	—	—	—	1125 H₂ Anneal	3,000	35,000	11,000	—	—	90	—
78 Permalloy	Sheet	21.2	78.5	—	—	.3 Mn	1050 + 600 Q§	8,000	100,000	10,700	200	.05	16	8.60
4-79 Permalloy	Sheet	16.7	79	—	4	.3 Mn	1100 + Q	20,000	100,000	8,700	200	.05	55	8.72
Mu metal	Sheet	18	75	—	—	2 Cr, 5 Cu	1175 H₂	20,000	100,000	6,500	—	.05	62	8.58
Supermalloy	Sheet	15.7	79	—	5	.3 Mn	1300 H₂ + Q	100,000	800,000	8,000	—	.002	60	8.77
Permendur	Sheet	49.7	—	50	—	.3 Mn	800 Anneal	800	5,000	24,500	12,000	2.0	7	8.3
2V Permendur	Sheet	49	—	49	—	2 V	800 Anneal	800	4,500	24,000	6,000	2.0	26	8.2
Hiperco	Sheet	64	—	34	—	Cr	850 Anneal	650	10,000	24,200	—	1.0	25	8.0
2-81 Permalloy	Insulated powder	17	81	—	2	—	650 Anneal	125	130	8,000	—	<1.0	10⁶	7.8
Carbonyl iron	Insulated powder	99.9	—	—	—	—	—	55	132	—	—	—	—	7.86
Ferroxcube III	Sintered powder	MnFe₂O₄ + ZnFe₂O₄					—	1,000	1,500	2,500	—	.1	10⁸	5.0

*Properties in direction of rolling.
† Similar properties for Nicaloi, 4750 alloy, Carpenter 49, Armco 48.
‡ At saturation.
§ Q, quench or controlled cooling.

Permanent Magnet Alloys

Material	Percent composition (remainder Fe)	Heat treatment* (temperature, °C)	Magnetizing force H_{max} oersteds	Coercive force H_c oersteds	Residual induction B_r gausses	Energy product BH_{max} ×10⁻⁶	Method of fabrication†	Mechanical properties‡	Weight lb/in³
Carbon steel	1 Mn, 0.9 C	Q 800	300	50	10,000	.20	HR, M, P	H, S	.280
Tungsten steel	5 W, 0.3 Mn, 0.7 C	Q 850	300	70	10,300	.32	HR, M, P	H, S	.292
Chromium steel	3.5 Cr, 0.9 C, 0.3 Mn	Q 830	300	65	9,700	.30	HR, M, P	H, S	.280
17% Cobalt steel	17 Co, 0.75 C, 2.5 Cr, 8 W	—	1,000	150	9,500	.65	HR, M, P	H, S	—
36% Cobalt steel	36 Co, 0.7 C, 4 Cr, 5 W	Q 950	1,000	240	9,500	.97	HR, M, P	H, S	.296
Remalloy or Comol	17 Mo, 12 Co	Q 1200, B 700	1,000	250	10,500	1.1	HR, M, P	H	.295
Alnico I	12 Al, 20 Ni, 5 Co	A 1200, B 700	2,000	440	7,200	1.4	C, G	H, B	.249
Alnico II	10 Al, 17 Ni, 2.5 Co, 6 Cu	A 1200, B 600	2,000	550	7,200	1.6	C, G	H, B	.256
Alnico II (sintered)	10 Al, 17 Ni, 2.5 Co, 6 Cu	A 1300	2,000	520	6,900	1.4	Sn, G	H	.249
Alnico IV	12 Al, 28 Ni, 5 Co	Q 1200, B 650	3,000	700	5,500	1.3	Sn, C, G	H	.253
Alnico V	8 Al, 14 Ni, 24 Co, 3 Cu	AF 1300, B 600	2,000	550	12,500	4.5	C, G	H, B	.264
Alnico VI	8 Al, 15 Ni, 24 Co, 3 Cu, 1 Ti	—	3,000	750	10,000	3.5	C, G	H, B	.268
Alnico XII	6 Al, 18 Ni, 35 Co, 8 Ti	—	3,000	950	5,800	1.5	C, G	H, B	.26
Vicalloy I	52 Co, 10 V	B 600	1,000	300	8,800	1.0	C, CR, M, P	D	.295
Vicalloy II (wire)	52 Co, 14 V	CW + B 600	2,000	510	10,000	3.5	C, CR, M, P	D	.292
Cunife (wire)	60 Cu, 20 Ni	CW + B 600	2,400	550	5,400	1.5	C, CR, M, P	D, M	.311
Cunico	50 Cu, 21 Ni, 29 Co	—	3,200	660	3,400	.80	C, CR, M, P	D, M	.300
Vectolite	30 Fe₂O₃, 44 Fe₃O₄, 26 Co₂O₃	—	3,000	1,000	1,600	.60	Sn, G	W	.113
Silmanal	86.8 Ag, 8.8 Mn, 4.4 Al	—	20,000	6,000ᵃ	550	.075	C, CR, M, P	D, M	.325
Platinum-cobalt	77 Pt, 23 Co	Q 1200, B 650	15,000	3,600	5,900	6.5	C, CR, M	D	—
Hyflux	Fine powder	—	2,000	390	6,600	.97	—	—	.176

ᵃ Value given is intrinsic H_c.
* Q—Quenched in oil or water. A—Air cooled. B—Baked. F—Cooled in magnetic field. CW—Cold worked.
† HR—Hot rolled or forged. CR—Cold rolled or drawn. M—Machined. G—Must be ground. P—Punched. C—Cast. Sn—Sintered.
‡ H—Hard. B—Brittle. S—Strong. D—Ductile. M—Malleable. W—Weak.

Resistance of Wires

The following table gives the approximate resistance of various metallic conductors. The values have been computed from the resistivities at 20°C, except as otherwise stated, and for the dimensions of wire indicated. Owing to differences in purity in the case of elements and of composition in alloys, the values can be considered only as approximations.

The following dimensions have been adopted in the computations.

B. & S. gauge	Diameter mm	Diameter mils 1 mil = .001 in	B. & S. gauge	Diameter mm	Diameter mils 1 mil = .001 in
10	2.588	101.9	26	0.4049	15.94
12	2.053	80.81	27	0.3606	14.20
14	1.628	64.08	28	0.3211	12.64
16	1.291	50.82	30	0.2546	10.03
18	1.024	40.30	32	0.2019	7.950
20	0.8118	31.96	34	0.1601	6.305
22	0.6438	25.35	36	0.1270	5.000
24	0.5106	20.10	40	0.07987	3.145

*Advance (0°C) ρ = 48. × 10⁻⁶ ohm cm

B. & S. No.	Ohms per cm	Ohms per ft
10	.000912	.0278
12	.00145	.0442
14	.00231	.0703
16	.00367	.112
18	.00583	.178
20	.00927	.283
22	.0147	.449
24	.0234	.715
26	.0373	1.14
27	.0470	1.43
28	.0593	1.81
30	.0942	2.87
32	.150	4.57
34	.238	7.26
36	.379	11.5
40	.958	29.2

Aluminum ρ = 2.828 × 10⁻⁶ ohm cm

B. & S. No.	Ohms per cm	Ohms per ft
10	.0000538	.00164
12	.0000855	.00260
14	.000136	.00414
16	.000216	.00658
18	.000344	.0105
20	.000546	.0167
22	.000869	.0265
24	.00138	.0421
26	.00220	.0669
27	.00277	.0844
28	.00349	.106
30	.00555	.169
32	.00883	.269
34	.0140	.428
36	.0223	.680
40	.0564	1.72

Eureka (0°C) ρ = 47. × 10⁻⁶ ohm cm

B. & S. No.	Ohms per cm	Ohms per ft
10	.000893	.0272
12	.00142	.0433
14	.00226	.0688
16	.00359	.109
18	.00571	.174
20	.00908	.277
22	.0144	.440
24	.0230	.700
26	.0365	1.11
27	.0460	1.40
28	.0580	1.77
30	.0923	2.81
32	.147	4.47
34	.233	7.11
36	.371	11.3
40	.938	28.6

Excello ρ = 92. × 10⁻⁶ ohm cm

B. & S. No.	Ohms per cm	Ohms per ft
10	.00175	.0533
12	.00278	.0847
14	.00442	.135
16	.00703	.214
18	.0112	.341
20	.0178	.542
22	.0283	.861
24	.0449	1.37
26	.0714	2.18
27	.0901	2.75
28	.114	3.46
30	.181	5.51
32	.287	8.75
34	.457	13.9
36	.726	22.1
40	1.84	56.0

Brass ρ = 7.00 × 10⁻⁶ ohm cm

B. & S. No.	Ohms per cm	Ohms per ft
10	.000133	.00406
12	.000212	.00645
14	.000336	.0103
16	.000535	.0163
18	.000850	.0259
20	.00135	.0412
22	.00215	.0655
24	.00342	.104
26	.00543	.166
27	.00686	.209
28	.00864	.263
30	.0137	.419
32	.0219	.666
34	.0348	1.06
36	.0552	1.68
40	.140	4.26

Climax ρ = 87. × 10⁻⁶ ohm cm

B. & S. No.	Ohms per cm	Ohms per ft
10	.00165	.0504
12	.00263	.0801
14	.00418	.127
16	.00665	.203
18	.0106	.322
20	.0168	.512
22	.0267	.815
24	.0425	1.30
26	.0675	2.06
27	.0852	2.60
28	.107	3.27
30	.171	5.21
32	.272	8.28
34	.432	13.2
36	.687	20.9
40	1.74	52.9

German silver ρ = 33. × 10⁻⁶ ohm cm

B. & S. No.	Ohms per cm	Ohms per ft
10	.000627	.0191
12	.000997	.0304
14	.00159	.0483
16	.00252	.0768
18	.00401	.122
20	.00638	.194
22	.0101	.309
24	.0161	.491
26	.0256	.781
27	.0323	.985
28	.0408	1.24
30	.0648	1.97
32	.103	3.14
34	.164	4.99
36	.260	.794
40	.659	20.1

Gold ρ = 2.44 × 10⁻⁶ ohm cm

B. & S. No.	Ohms per cm	Ohms per ft
10	.0000464	.00141
12	.0000737	.00225
14	.000117	.00357
16	.000186	.00568
18	.000296	.00904
20	.000471	.0144
22	.000750	.0228
24	.00119	.0363
26	.00189	.0577
27	.00239	.0728
28	.00301	.0918
30	.00479	.146
32	.00762	.232
34	.0121	.369
36	.0193	.587
40	.0487	1.48

Constantan (0°C) ρ = 44.1 × 10⁻⁶ ohm cm

B. & S. No.	Ohms per cm	Ohms per ft
10	.000838	.0255
12	.00133	.0406
14	.00212	.0646
16	.00337	.103
18	.00536	.163
20	.00852	.260
22	.0135	.413
24	.0215	.657
26	.0342	1.04
27	.0432	1.32
28	.0545	1.66
30	.0866	2.64
32	.138	4.20
34	.219	6.67
36	.348	10.6
40	.880	26.8

Copper, annealed ρ = 1.724 × 10⁻⁶ ohm cm

B. & S. No.	Ohms per cm	Ohms per ft
10	.0000328	.000999
12	.0000521	.00159
14	.0000828	.00253
16	.000132	.00401
18	.000209	.00638
20	.000333	.0102
22	.000530	.0161
24	.000842	.0257
26	.00134	.0408
27	.00169	.0515
28	.00213	.0649
30	.00339	.103
32	.00538	.164
34	.00856	.261
36	.0136	.415
40	.0344	1.05

Iron ρ = 10. × 10⁻⁶ ohm cm

B. & S. No.	Ohms per cm	Ohms per ft
10	.000190	.00579
12	.000302	.00921
14	.000481	.0146
16	.000764	.0233
18	.00121	.0370
20	.00193	.0589
22	.00307	.0936
24	.00489	.149
26	.00776	.237
27	.00979	.299
28	.0123	.376
30	.0196	.598
32	.0312	.952
34	.0497	1.51
36	0.789	2.41
40	.200	6.08

Lead ρ = 22. × 10⁻⁶ ohm cm

B. & S. No.	Ohms per cm	Ohms per ft
10	.000418	.0127
12	.000665	.0203
14	.00106	.0322
16	.00168	.0512
18	.00267	.0815
20	.00425	.130
22	.00676	.206
24	.0107	.328
26	.0171	.521
27	.0215	.657
28	.0272	.828
30	.0432	1.32
32	.0687	2.09
34	.109	3.33
36	.174	5.29
40	.439	13.4

Trade mark.

Resistance of Wires (continued)

Magnesium $\rho = 4.6 \times 10^{-6}$ ohm cm

B. & S. No.	Ohms per cm	Ohms per ft
10	.0000874	.00267
12	.000139	.00424
14	.000221	.00674
16	.000351	.0107
18	.000559	.0170
20	.000889	.0271
22	.00141	.0431
24	.00225	.0685
26	.00357	.109
27	.00451	.137
28	.00568	.173
30	.00903	.275
32	.0144	.438
34	.0228	.696
36	.0363	1.11
40	.0918	2.80

Manganin $\rho = 44. \times 10^{-6}$ ohm cm

B. & S. No.	Ohms per cm	Ohms per ft
10	.000836	.0255
12	.00133	.0405
14	.00211	.0644
16	.00336	.102
18	.00535	.163
20	.00850	.259
22	.0135	.412
24	.0215	.655
26	.0342	1.04
27	.0431	1.31
28	.0543	1.66
30	.0864	2.63
32	.137	4.19
34	.218	6.66
36	.347	10.6
40	.878	26.8

Platinum $\rho = 10. \times 10^{-6}$ ohm cm

B. & S. No.	Ohms per cm	Ohms per ft
10	.000190	.00579
12	.000302	.00921
14	.000481	.0146
16	.000764	.0233
18	.00121	.0370
20	.00193	.0589
22	.00307	.0936
24	.00489	.149
26	.00776	.237
27	.00979	.299
28	.0123	.376
30	.0196	.598
32	.0312	.952
34	.0497	1.51
36	.0789	2.41
40	.200	6.08

Silver (18°C) $\rho = 1.629 \times 10^{-6}$ ohm cm

B. & S. No.	Ohms per cm	Ohms per ft
10	.0000310	.000944
12	.0000492	.00150
14	.0000783	.00239
16	.000124	.00379
18	.000198	.00603
20	.000315	.00959
22	.000500	.0153
24	.000796	.0243
26	.00126	.0386
27	.00160	.0486
28	.00201	.0613
30	.00320	.0975
32	.00509	.155
34	.00809	.247
36	.0129	.392
40	.0325	.991

Molybdenum $\rho = 5.7 \times 10^{-6}$ ohm cm

B. & S. No.	Ohms per cm	Ohms per ft
10	.000108	.00330
12	.000172	.00525
14	.000274	.00835
16	.000435	.0133
18	.000693	.0211
20	.00110	.0336
22	.00175	.0534
24	.00278	.0849
26	.00443	.135
27	.00558	.170
28	.00704	.215
30	.0112	.341
32	.0178	.542
34	.0283	.863
36	.0450	1.37
40	.114	3.47

Monel Metal $\rho = 42. \times 10^{-6}$ ohm cm

B. & S. No.	Ohms per cm	Ohms per ft
10	.000798	.0243
12	.00127	.0387
14	.00202	.0615
16	.00321	.0978
18	.00510	.156
20	.00811	.247
22	.0129	.393
24	.0205	.625
26	.0326	.994
27	.0411	1.25
28	.0519	1.58
30	.0825	2.51
32	.131	4.00
34	.209	6.36
36	.331	10.1
40	.838	25.6

Steel, piano wire (0°C) $\rho = 11.8 \times 10^{-6}$ ohm cm

B. & S. No.	Ohms per cm	Ohms per ft
10	.000224	.00684
12	.000357	.0109
14	.000567	.0173
16	.000901	.0275
18	.00143	.0437
20	.00228	.0695
22	.00363	.110
24	.00576	.176
26	.00916	.279
27	.0116	.352
28	.0146	.444
30	.0232	.706
32	.0368	1.12
34	.0586	1.79
36	.0931	2.84
40	.236	7.18

Steel, invar (35% Ni) $\rho = 81. \times 10^{-6}$ ohm cm

B. & S. No.	Ohms per cm	Ohms per ft
10	.00154	.0469
12	.00245	.0746
14	.00389	.119
16	.00619	.189
18	.00984	.300
20	.0156	.477
22	.0249	.758
24	.0396	1.21
26	.0629	1.92
27	.0793	2.42
28	.100	3.05
30	.159	4.85
32	.253	7.71
34	.402	12.3
36	.639	19.5
40	1.62	49.3

*Nichrome $\rho = 150. \times 10^{-6}$ ohm cm

B. & S. No.	Ohms per cm	Ohms per ft
10	.0021281	.06488
12	.0033751	.1029
14	.0054054	.1648
16	.0085116	.2595
18	.0138383	.4219
20	.0216218	.6592
22	.0346040	1.055
24	.0548088	1.671
26	.0875760	2.670
28	.1394328	4.251
30	.2214000	6.750
32	.346040	10.55
34	.557600	17.00
36	.885600	27.00
38	1.383832	42.19
40	2.303872	70.24

Nickel $\rho = 7.8 \times 10^{-6}$ ohm cm

B. & S. No.	Ohms per cm	Ohms per ft
10	.000148	.00452
12	.000236	.00718
14	.000375	.0114
16	.000596	.0182
18	.000948	.0289
20	.00151	.0459
22	.00240	.0730
24	.00381	.116
26	.00606	.185
27	.00764	.233
28	.00963	.294
30	.0153	.467
32	.0244	.742
34	.0387	1.18
36	.0616	1.88
40	.156	4.75

Tantalum $\rho = 15.5 \times 10^{-6}$ ohm cm

B. & S. No.	Ohms per cm	Ohms per ft
10	.000295	.00898
12	.000468	.0143
14	.000745	.0227
16	.00118	.0361
18	.00188	.0574
20	.00299	.0913
22	.00476	.145
24	.00757	.231
26	.0120	.367
27	.0152	.463
28	.0191	.583
30	.0304	.928
32	.0484	1.47
34	.0770	2.35
36	.122	3.73
40	.309	9.43

Tin $\rho = 11.5 \times 10^{-6}$ ohm cm

B. & S. No.	Ohms per cm	Ohms per ft
10	.000219	.00666
12	.000348	.0106
14	.000553	.0168
16	.000879	.0268
18	.00140	.0426
20	.00222	.0677
22	.00353	.108
24	.00562	.171
26	.00893	.272
27	.0113	.343
28	.0142	.433
30	.0226	.688
32	.0359	1.09
34	.0571	1.74
36	.0908	2.77
40	.230	7.00

Tungsten $\rho = 5.51 \times 10^{-6}$ ohm cm

B. & S. No.	Ohms per cm	Ohms per ft
10	.000105	.00319
12	.000167	.00508
14	.000265	.00807
16	.000421	.0128
18	.000669	.0204
20	.00106	.0324
22	.00169	.0516
24	.00269	.0820
26	.00428	.130
27	.00540	.164
28	.00680	.207
30	.0108	.330
32	.0172	.524
34	.0274	.834
36	.0435	1.33
40	.110	3.35

Zinc (0°C) $\rho = 5.75 \times 10^{-6}$ ohm cm

B. & S. No.	Ohms per cm	Ohms per ft
10	.000109	.00333
12	.000174	.00530
14	.000276	.00842
16	.000439	.0134
18	.000699	.0213
20	.00111	.0339
22	.00177	.0538
24	.00281	.0856
26	.00446	.136
27	.00563	.172
28	.00710	.216
30	.0113	.344
32	.0180	.547
34	.0286	.870
36	.0454	1.38
40	.115	3.50

Credits

Material in Section XII was reprinted from the following sources:

D. R. Lide, Ed., *CRC Handbook of Chemistry and Physics,* 73rd ed., Boca Raton, Fla.: CRC Press, 1992: International System of Units (SI), conversion constants and multipliers (conversion of temperatures), symbols and terminology for physical and chemical quantities, fundamental physical constants, classification of electromagnetic radiation, electrical resistivity (pure metals, selected alloys) , dielectric constants, properties of semiconductors, properties of magnetic alloys, resistance of wires.

W. H. Beyer, Ed., *CRC Standard Mathematical Tables and Formulae,* 29th ed., Boca Raton, Fla.: CRC Press, 1991: Greek alphabet, conversion constants and multipliers (recommended decimal multiples and submultiples, metric to English, English to metric, general, temperature factors), physical constants, series expansion, integrals, the Fourier transforms, numerical methods, probability, positional notation.

R. J. Tallarida, *Pocket Book of Integrals and Mathematical Formulas,* 2nd ed., Boca Raton, Fla.: CRC Press, 1992: Elementary algebra and geometry; determinants, matrices, and linear systems of equations; trigonometry; analytic geometry; series; differential calculus; integral calculus; vector analysis; special functions; statistics; tables of probability and statistics; table of derivatives.

J. F. Pankow, *Aquatic Chemistry Concepts,* Chelsea, Mich.: Lewis Publishers, 1991: Periodic table of the elements.

J. Shackelford and W. Alexander, Eds., *CRC Materials Science and Engineering Handbook,* Boca Raton, Fla.: CRC Press, 1992: Electrical resistivity of selected alloy cast irons, resistivity of selected ceramics.

Credits

Material in Section XII was compiled from the following sources:

D.R. Lide, Ed., *CRC Handbook of Chemistry and Physics*, 74th ed. Boca Raton, Fla.: CRC Press, 1993. International System of Units (SI), conversion constants and multipliers (conversion of temperatures), symbols and terminology for physical and chemical quantities, fundamental physical constants, classification of electromagnetic radiation, electrical resistivity, galvanomagnetic effects, dielectric constants, properties of semiconductors, electronic properties, resistance of wires.

W.H. Beyer, Ed., *CRC Standard Mathematical Tables and Formulae*, 29th ed. Boca Raton, Fla.: CRC Press, 1991. Greek alphabet, conversion factors and multipliers, recommended decimal multiples and submultiples, metric to English, English to metric, general, temperature factors, conversion of temperatures, physical constants, series expansion, the Fourier transforms, numerical mathematical probability, positional angular.

R.J. Tallarida, *Pocket Book of Integrals and Mathematical Formulas*, 2nd ed. Boca Raton, Fla.: CRC Press, 1992. Elementary algebra and geometry, determinants, matrices, and linear systems of equations, trigonometry, analytic geometry, series, differential calculus, integral calculus, vector analysis, special functions, statistics, tables of probability and statistics, table of derivatives.

J. R. Zastrow, *A journal of Chemistry*. Chelsea, Mich.: Lewis Publishers, 1990. Properties of elements.

I. Shmulenson and W. Abramowitz, Eds., *CRC Handbook, Science, and Mathematics Handbook*, Boca Raton, Fla.: CRC Press, 1991. Historical resistivity of selected alloys, thermal resistivity of selected materials.

Associations and Societies

American Society of Mechanical Engineers (ASME)
345 East 47th St.
New York, NY 10017-2392

Tel. # (212) 705-7722 · FAX # (212) 705-7674

Founded in 1880, the ASME currently has over 118,000 members. The ASME is concerned with all aspects of mechanical engineering and related fields. The society is divided geographically into 12 regions and 200 local sections in the U.S., Canada, and Mexico. There are over 21,000 students in 300 student sections at colleges and universities.

The ASME carries out its goals through the following five councils:

Council on Member Affairs: This council is concerned with all regions, sections, subsections, local groups, student sections, and mechanical engineering clubs. It deals with members' interests and standards, grades, recruitment, and retention.

Council on Education: Responsibilities include the development and accreditation of engineering curricula, the development of professional understanding among engineering students, and continuing education programs for engineers.

Council on Codes and Standards: This council supervises the technical standards-setting program and is also responsible for a system of accreditation for manufacturers of equipment. The ASME currently has nearly 600 codes and standards in print. The internationally accepted standards conform to the procedures set by the American National Standards Institute (ANSI).

Council on Public Affairs: This council directs the programs relating to government and international activities and the dissemination of information to the public.

Council on Engineering: This council directs the technical interests of ASME, overseeing 35 technical divisions, three institutes, a research center, and several interdisciplinary programs which keep members up to date on the latest engineering advances. The technical divisions are set up to speed the transfer of technological developments to industrial applications. Activities of the technical divisions include technical conferences, workshops, exhibits, publications, and research. The following is a list of the ASME technical divisions:

- *Basic Engineering:* Applied Mechanics, Bioengineering, Fluids Engineering, Heat Transfer, Tribology.

- *Energy Conversion:* Internal Combustion Engines, Fuels and Combustion Technologies, Nuclear Engineering, Power.

- *Energy Resources:* Advanced Energy Systems, Ocean Engineering, Petroleum, Solar Energy.

- *Environment and Transportation:* Aerospace, Environmental Control, Noise Control and Acoustics, Rail Transportation, Solid Waste Processing.

- *General Engineering:* Management, Safety, Technology and Society.

- *Materials and Structures:* Materials, Pressure Vessels and Piping, NDE Engineering, Offshore Mechanics and Arctic Engineering.

- *Manufacturing:* Materials Handling Equipment, Plant Engineering and Maintenance, Process Industries, Production Engineering, Textile Engineering.

- *Systems and Design:* Computers in Engineering, Design Engineering, Dynamic Systems and Control, Electrical and Electronic Packaging, Fluid Power Systems and Technology.

Association for Computing Machines (ACM)

1515 Broadway
New York, NY 10036

Tel. # (212) 869-7440 • FAX # (212) 869-1228
E-Mail: ACMHELP@ACNVM.BITNET

Founded in 1947, the ACM is the oldest and largest educational and scientific computer organization in the world. The ACM has 80,000 members and more than 700 chapters, student chapters, and local special interest groups (SIGs). The ACM also publishes 12 major computing journals, numerous conference proceedings, and special publications.

Membership is open to computing professionals and students involved in programming, management, systems analysis/design/engineering, research and development, education consulting, operations, sales/marketing, personal computing, and CAD/CAM/CAE. There are four types of membership:

- *Voting Membership:* To qualify you must have earned at least a Bachelor's Degree or academic equivalent, or have four years of full-time experience in information processing, and must subscribe to the purposes of ACM.

- *Associate Membership:* To qualify you must subscribe to the purposes of the ACM. You do not need to meet the other qualifications for Voting Members. Associate Members may not vote or hold office in the ACM.

- *Student Membership:* To qualify a student must ascribe to the purposes of the ACM and be registered at an accredited educational institution on a full-time basis. An application form must be accompanied by a certificate of attendance signed by an instructor. Student members have the same privileges as Associate Members.

- *Institutional Membership:* Established as a service to companies, governmental agencies, colleges, and universities.

ACM maintains 32 SIGs specializing in specific computing disciplines. ACM members and nonmembers may join SIGs. The SIGs publish regular newsletters, sponsor or co-sponsor more than 75 conferences annually, and host local meetings and workshops.

The following is a list of ACM's SIGs:

SIGACT
Automata & Computability Theory

SIGADA
Ada Programming Language

SIGAPL
APL Programming Language

SIGARCH
Architecture of Computer Systems

SIGART
Artificial Intelligence

SIGBDP
Business Data Processing and Management

SIGBIO
Biomedical Computing

SIGCAPH
Computers and the Physically Handicapped

SIGCAS
Computers and Society

SIGCHI
Computers and Human Interaction

SIGCOMM
Data Communications

SIGCPR
Computer Personal Research

SIGCSE
Computer Science Education

SIGCUE
Computer Uses in Education

SIGDA
Design Automation

SIGDOC
Systems Documentation

SIGFORTH
Forth Programming Language

SIGGRAPH
Computer Graphics

SIGIR
Information Retrieval

SIGMETRICS
Measurement and Evaluation

SIGMICRO
Microprogramming

SIGMOD
Management of Data

SIGNUM
Numerical Mathematics

SIGOIS
Office Information System

SIGOPS
Operating Systems

SIGPLANS
Programming Languages

SIGSAC
Security, Audit and Control

SIGSAM
Symbolic and Algebraic Manipulation

SIGSIM
Simulation

SIGSMALL/PC
Small and Personal Computing Systems and
Applications

SIGSOFT
Software Engineering

SIGUCCS
University and College Computing Services

Association of Energy Engineers (AEE)
4025 Pleasantdale Rd., Suite 420
Atlanta, GA 30340

Tel. # (404) 447-5083

The AEE, with 6200 members, is an organization of professionals interested in energy engineering, energy management, and cogeneration. The AEE maintains a division, The Cogeneration Institute, with 1400 members. The AEE has committees involved with educational activities, energy consulting, government engineers, plant and building energy management, solar power, and wind engineering. The AEE also supports a council on National Energy Policy.

Audio Engineering Society (AES)
60 East 42nd St., Room 2520
New York, NY 10065

The AES, with more than 10,000 members, is the largest organization in the world that deals with the science and engineering of recording and reproducing sound. Established in 1948, all members of the sound industry and related fields are eligible for membership consideration.

Computer Measurement Group (CMG)
6397 Little River Turnpike
Alexandria, VA 22312

Established in 1968, the CMG has 2000 members. It is involved with activities involving computer performance evaluation.

Electronic Connector Study Group (ECSG)
104 Wilmot Rd., Suite 201
Deerfield, IL 60015

Founded in 1958, this group has over 1500 members. It serves engineers, designers, technicians, manufacturers, and users of electronic connections.

Eta Kappa Nu
Box HKN
University of Missouri–Rolla
Rolla, MO 65401

Eta Kappa Nu is an honor fraternity for electrical engineers. Founded in 1904, Eta Kappa Nu has more than 150,000 members.

Institute of Electrical and Electronics Engineers (IEEE)

Headquarters Office
345 East 47th St.
New York, NY 10017-2394

Tel. # (212) 705-7900 • FAX # (212) 752-4929 • Telex: 236-411

Service Center (includes Membership Services)
445 Hoes Lane, P.O. Box 1331
Piscataway, NJ 08855-1331

Tel. # (908) 981-0060 • Tel. # (800) 678-IEEE (toll free)
FAX # (908) 981-0027 • Telex: 833-233

United States Services (IEEE-USA)
1828 L St., N.W., Suite 1202
Washington, DC 20036-5104

Tel. # (202) 785-0017 • FAX # (202) 785-0835

The IEEE is the world's largest technical professional society. Founded in 1884, the IEEE has membership today of more than 320,000 who conduct and participate in its activities in 147 countries. The IEEE focuses on advancing the theory and practice of electrical, electronics, and computer engineering and computer science. The IEEE annually sponsors approximately 300 conferences; conducts diverse educational programs (including individual learning programs and satellite courses); conducts over 4500 local meetings, symposia, and special events; publishes more than 75 transactions, magazines, and journals; and has developed and revises more than 600 standards.

Interest in electrical/electronics engineering is the basic requirement for membership in the IEEE. The grade of membership on admission depends on the extent of involvement and contribution to IEEE-designated fields. The IEEE has five membership grades:

- *Student Membership:* Available to students who must carry at least 50% of a normal full-time academic program as a registered undergraduate or graduate student in a regular course of study in an IEEE-designated field. Student members, upon graduation with at least a baccalaureate degree or its equivalent from a recognized educational program, shall be transferred to Member grade. Student members graduating with at least a two-year degree shall be transferred to Associate grade.

- *Associate Membership:* Available for technical and nontechnical applicants who do not presently meet the qualifications for Member grade and for those who are progressing through continuing education and work experience, towards the qualifications for Member grade.

- *Membership:* A professional grade limited to those who have demonstrated professional competence in IEEE-designated fields. For admission or elevation to the grade of Member, a candidate shall be either: (1) an individual in IEEE-designated fields who shall have received a baccalaureate degree or its equivalent in those fields from a recognized educational program or shall have had at least three years of experience in a position normally requiring the qualifications of a baccalaureate engineer; (2) a designated teacher of a subject in an IEEE-designated field holding a baccalaureate degree or its equivalent in that field, or one who has had at least three years of professional teaching experience and has participated in planning and conducting courses; (3) a person regularly employed in IEEE-designated fields for at least six years who, by experience, has demonstrated competence in work of a professional character; or (4) an executive who, for at least six years, has under his/her direction important technical, engineering, or research work in IEEE-designated fields.

- *Senior Membership:* The candidate shall have been in active professional practice for at least ten years and shall have shown significant performance over a period of five of those years.
- *Fellowship:* The grade of Fellow recognizes unusual distinction in the profession and shall be conferred only by invitation of the Board of Directors.

IEEE members have the option of joining one or more of 35 specialized technical societies. Each society focuses in depth on specific areas of technical interest. Generally, each society publishes one or more technical publications and society newsletters. The following is a list of the IEEE societies:

Aerospace and Electronic Systems	Industry Applications
Antennas and Propagation	Information Theory
Broadcast Technology	Instrumentation and Measurement
Circuits and Systems	Laser and Electro-Optics
Communications	Magnetics
Components, Hybrids and Manufacturing Technology	Microwave Theory and Techniques
	Nuclear and Plasma Sciences
Computer (Separate entry for the IEEE Computer Society below)	Oceanic Engineering
	Power Electronics
Consumer Electronics	Power Engineering
Control Systems	Professional Communication
Dielectrics and Electrical Insulation	Reliability
Education	Robotics and Automation
Electromagnetic Compatibility	Signal Processing
Electron Devices	Social Implications of Technology
Engineering Management	Systems, Man and Cybernetics
Engineering in Medicine and Biology	Ultrasonics, Ferroelectrics and Frequency Control
Geoscience and Remote Sensing	Vehicular Technology
Industrial Electronics	

IEEE Computer Society

10662 Los Vaqueros Circle
P.O. Box 3014
Los Alamitos, CA 90720-1264

Tel. # (714) 821-8380

The IEEE Computer Society, founded in 1951, is the largest society within the IEEE, with 95,000 members. It is the world's largest organization of computer professionals. It promotes the development of computer and information sciences. The society publishes 14 special interest publications. It also conducts and supports symposia, workshops, technical committees, and standards groups and maintains active local chapters.

The Institution of Electrical Engineers (IEE)
Savoy Place
London WC2R 0BL
United Kingdom

Founded over 100 years ago, the IEE has a worldwide membership of close to 100,000. The IEE sets professional qualification standards and accredits university and college courses within the U.K., issues regulations governing the safe installation of electrical and electronic equipment, advises on the formulation of British national and international standards, sets standards for professional conduct of its members, organizes conferences and meetings, provides educational services and information to the young, and is the voice of the profession to government and other agencies of concern.

The IEE also maintains *INSPEC*, which is the largest English-language database of information on physics, electrotechnology, computer science, and engineering. The *INSPEC* database currently contains records for almost four million scientific and technical papers, and is increasing by over 250,000 records per year. The contents of over 4000 journals and 1000 conference proceedings, as well as numerous books, reports, and dissertations, are regularly examined by *INSPEC* for relevant articles to abstract and index for the database. To carry out its work *INSPEC* has a staff of 140. *INSPEC* is available in a variety of media, including online, CD-ROM, magnetic tape, and a variety of printed products.

International Electronics Packaging Society (IEPS)
114 N. Hale St.
Wheaton, IL 60187

Tel. # (312) 260-1044

The IEPS, founded in 1977, has over 1500 members. The society is involved with the technology and materials utilized in all phases of electronic packaging. The society serves, but is not limited to, electronic system design engineers.

International Society for Hybrid Microelectronics (ISHM)
P.O. Box 2698
1861 Wiehle Ave., Suite 340
Reston, VA 22090

Founded in 1967, the ISHM has 7000 members. The society includes materials scientists, electrical and electronics engineers, and physical scientists involved with all aspects of the technologies of ceramics, thick and thin films, semiconductor packaging, semiconductor devices, and monolithic circuits. The ISHM supports educational activities for institutions and businesses involved with microelectronics. The society also sponsors numerous workshops, seminars, and conferences.

The International Society for Optical Engineering (SPIE)
SPIE
P.O. Box 10
Bellingham, WA 98227-0010

Tel. # (206) 676-3290 • FAX # (206) 647-1445

Founded in 1955, the SPIE has over 10,000 members. The SPIE is a nonprofit society dedicated to advancing knowledge in optical and optoelectronic applied science and engineering. The society has an extensive series of conferences and workshops and maintains a very active publishing program. The SPIE maintains and fosters particularly close relationships between researchers and end-users of optical devices and instrumentation.

Materials Research Society (MRS)
9800 McKnight Rd., Suite 327
Pittsburgh, PA 15237

Tel. # (412) 367-3003

Founded in 1973, the MRS with over 6000 active members is involved in all aspects of materials science research and materials engineering. The MRS holds two annual meetings and sponsors numerous short courses and seminars. Areas of research interest include all classes of inorganic and organic materials.

National Society of Professional Engineers (NSPE)
1420 King St.
Alexandria, VA 22314

Founded in 1934, the NSPE has over 75,000 members. The society consists of professional engineers and engineers in training in all fields of engineering who are registered in accordance with state and territorial regulations in the U.S. and Canada. The society is also open to qualified graduate engineers, students, and registered land surveyors. It is involved in all aspects of engineering as a profession. The NSPE is actively involved in following all governmental and regulatory action as it pertains to professional engineering.

The Optical Society of America (OSA)
2010 Massachusetts Ave., NW
Washington, DC 20036

Tel. # (202) 223-8130 • FAX # (202) 223-1096
Membership Tel. # (800) 762-6960

The OSA, established in 1916, is a not-for-profit international society for engineers and scientists working in optics, photonics, or related fields. Its more than 11,000 members are engaged in the supply and applications of optics and photonics, including electro-optics, lasers, and fiber optics. The OSA conducts conferences, courses, exhibitions, specialized topical meetings, and educational programs for teachers. The society publishes eight journals and numerous digests and proceedings. The OSA has six technical groups. Each member and student member can participate in up to three technical groups. These currently include groups in divisions such as Information Processing, Optical Science, Optical Technology, Photonics, Quantum Electronics, and Vision and Medical Optics. All members of OSA also become members of the American Institute of Physics (AIP).

The Society of Manufacturing Engineers (SME)
One SME Drive
P.O. Box 930
Dearborn, MI 48121

Tel. # (313) 271-1500 • FAX # (313) 271-2861

The SME supports over 75,000 manufacturing professionals in 70 countries with more than 300 active senior chapters and 200 student chapters. In addition, the SME hosts 16 expositions each year with 2500 exhibitors and 175,000 attendees. The society also serves all manufacturing interest areas.

The SME has created specific associations and groups to help members keep abreast of their areas of specialty. Of particular interest to electrical engineers and computer scientists are the following: Computer and Automated Systems Association of SME (CASA) (12,800 members), which focuses on the development of totally integrated manufacturing; Association for Electronics Manufacturing of SME (EM) (3600 members); Machine Vision Association of SME (MVA) (3300 members), which promotes the effective utilization of machine vision technology for quality and productivity improvement in manufacturing; Robotics International of SME (RI) (7600 members), dedicated to the advancement of robot technology; Networking and Communications in Manufacturing Group of SME, which provides professionals with the latest information on applications of networking and communications in the areas of manufacturing management and the creation and operation of a network.

The Society of Women Engineers (SWE)
United Engineering Center, Room 305
345 East 47th St.
New York, NY 10017

Tel. # (212) 705-7855 • FAX # (212) 319-0947

The SWE, founded in 1949, has an international membership of 14,000 women and men. SWE student sections have been chartered at more than 240 colleges, universities, and engineering institutes. The SWE is a nonprofit educational service organization of graduate engineers and men and women with equivalent engineering experience. The specific objectives of the society are to inform young women, their parents, counselors, and the general public of the qualifications and achievements of women engineers and the opportunities open to them; to assist women engineers in preparing themselves for a return to active work after temporary retirement; to serve as a center for information on women in engineering; and to encourage women engineers to attain high levels of education and professional achievement.

Tau Beta Pi Association

P.O. Box 8840
University Station
Knoxville, TN 37996

Tel. # (615) 546-4578

Founded in 1885, Tau Beta Pi is the engineering honor society. Current membership is 330,000, with 15 groups and 196 active collegiate chapters and 55 alumni chapters.

Indexes

Author Index 2559

Index of Key Tables 2562

Index of Key Figures 2564

Index of Key Equations 2570

Subject Index 2577

Indexes

Author Index

Abdelguerfi, M., 2032
Agbo, Samuel O., 975
Allebach, Jan, 329
Amin, Ahmed, 1099
Andersen, Kristinn, 2223
Angelopoulos, Nick, 39
Argila, Carl A., 1976

Bahill, A. Terry, 207
Bahl, I. J., 1004
Balabanian, Norman, 79, 88
Ballou, Glen, 15
Bannister, B. R., 1721
Bannister, Joseph, 1460
Bar-Cohen, Avram, 784
Baretich, Matthew F., 2405
Barnett, Robert Joel, 2223
Bartnikas, R., 1132
Batalama, Stella N., 321
Bate, Geoffrey, 811
Bavarian, Behnam, 420
Becker, R. A., 759
Belcher, Melvin L., 949
Berbari, Edward J., 2362
Bhat, Ashoka K. S., 711
Bhutta, Imran A., 460
Bickart, Theodore A., 79
Bitler, Bill, 385
Bogart, Theodore F., Jr., 132
Bolton, Martin, 1735
Bomar, Bruce W., 238
Bose, Anjan, 1344
Bose, Bimal K., 729
Bose, N. K., 359

Bouman, Charles A., 329
Boykin, Joseph, 2061
Brews, John R., 567
Brogan, William L., 2099
Bronzino, Joseph D., 2297, 2351
Bush, Marcia A., 306

Cadzow, James A., 251
Carpenter, Gordon L., 634
Carroll, Bill D., 1741
Carter, G. Clifford, 406
Chan, Shu-Park, 1, 92
Chassaing, Rulph, 385
Check, William, 2194
Chen, Kao, 2257
Chen, Mo-Shing, 1217, 1223
Chen, Sue-Ling, 1748
Chen, Wai-Kai, 169
Cherin, Allen H., 987
Choma, John, Jr., 639
Ciletti, Michael D., 58
Clapp, G., 2211
Clegg, Almon H., 1397
Cogdell, J. R., 53
Compton, R. C., 861
Cook, George E., 2223
Cooper, J. Arlin, 2072
Cover, Thomas M., 1517
Czeck, Edward W., 1878

Daigle, J. N., 1447
Darcie, T. E., 1417
David, Yadin, 2408

Delin, Kevin A., 1114
Delp, Edward J., 329
Demarest, Kenneth, 849
Dervisoglu, Bulent I., 1816
DiFonzo, Daniel F., 1532
Doelitzsch, Dennis F., 1367
Dorf, Richard C., 47, 149, 178,
    184, 189, 1359, 1405, 1410,
    2097, 2255
Durbeck, Robert C., 1958

Ehrlich, Alexander C., 1106
El-Hawary, Mohamed E., 1242
Elshabini-Riad, Aicha, 460
Ephraim, Yariv, 287
Etter, Delores M., 225
Etzold, K. F., 1087

Farnell, Gerald W., 1077
Feaster, William M., 1300
Feisel, Lyle D., 1051
Feldman, James M., 1878
Feng, Tse-yun, 2052
Fitch, J. Patrick, 869
Fox, M. D., 2374
Frenzel, James F., 1753
Frizzell, Leon A., 2380
Fussell, Jesse W., 298

Galler, Donald, 1333
Garrod, Susan A. R., 771
Geddes, L. A., 2311

Gelmont, Boris, 435
Gerla, Mario, 1460
Gibson, Jerry D., 279
Gildenblat, Gennady Sh., 435
Ginsberg, Gerald L., 603
Giri, Jay C., 1344
Glover, J. Duncan, 1287
Goodman, James R., 1927
Graham, Peter, 1622
Gross, Charles A., 1252, 1296, 1302, 1303, 1307
Gungor, R. B., 1279
Guy, Chris G., 2087

Hamacher, V. Carl, 1865
Harbor, Royce D., 2139
Hecht, Jeff, 738
Hemming, Leland H., 903
Hinton, H. S., 1641
Hoeppner, Conrad H., 1578
Hoole, S. Ratnajeevan H., 1799
Hsia, Tien C., 94.1
Huber, Manfred N., 1441
Hudgins, Jerry L., 126

Irwin, J. David, 63

Jacquot, Raymond G., 2147
Johnson, Barry W., 2020
Johnson, David E., 158
Jones, Capers, 1985

Kaminow, Ivan P., 1434
Karady, George G., 1193, 1310
Katz, Randy H., 1658
Kayton, Myron, 2285
Kazakos, Dimitri, 321
Kennedy, E. J., 616
Kersting, William H., 1191
Kerwin, William J., 111
Kolias, N. J., 861
Kong, Jin Au, 803
Kosbar, Kurt L., 1593
Kraus, Allan D., 69
Kryder, Mark H., 826
Kun, Luis, 2397
Kuo, Benjamin C., 2131
Kurumbalapitiya, Dhammika, 1799

Lai, K. C., 1223
Lall, Pradeep, 5
Lasky, Ty A., 2154
Lee, Gordon K. F., 2106
Lee, Peter A., 1670
Lee, William C. Y., 1546
Leondes, Cornelius T., 2188
Lewis, Ted G., 81.25
Liebowitz, Jay, 2048
Lightner, Michael, 653
Lindsey, Jefferson F. III, 1367
Liu, Chen-Ching, 1321
Looney, Carl G., 1488, 1499

Maddy, Steven L., 1567
Malocha, Donald C., 1062
Mansuripur, M., 1675
Marks, R. J. II, 1510
Martin, André, 1778
Martin, Johannes J., 1915, 1996
Martinec, Daniel A., 2188
Massara, Robert E., 674
Mayaram, Kartikeya, 139
McInroy, John E., 2147
McRuer, Duane, 3247
Mehta, Sanjay K., 406
Milkovic, Miran, 454
Miller, E. K., 1028
Morris, James E., 1763, 1772
Moss, Gregory L., 1613

Needham, Wayne, 490
Neelakanta, P. S., 1173
Nessmith, Josh T., 949
Neudorfer, Paul, 198
Neuman, Michael R., 2342

Odrey, Nicholas G., 94.22
Oklobdzija, Vojin G., 1858
Oldfield, John V., 1839
Orlando, Terry P., 1114
Ovan, Mil, 1557

Palais, Joseph C., 1427
Parhi, Keshab K., 370
Parks, Harold G., 475
Paul, Clayton R., 50
Pecht, Michael, 5
Phadke, Arun G., 1269

Phillips, Charles L., 2139
Pillai, S. Unnikrishna, 315
Polk, C., 2329
Poor, H. Vincent, 1478
Poularikas, Alexander D., 229
Preparata, Franco P., 1695
Pricer, W. David, 1651
Pu, Yuan, 1162

Rajala, Sarah A., 345
Rajaram, S., 499
Rajashekara, Kaushik, 694, 702
Ramakumar, R., 1207
Rana, Abdul Hamid, 2194
Rawat, Banmali S., 799
Raymond, Jacques, 1870
Reilly, J. Patrick, 2301
Robertazzi, Thomas G., 2015
Robinson, Charles J., 2387
Robrock, Richard B. II, 1468
Roden, Martin S., 1394
Rogers, Peter H., 1055
Roman, John M., 919
Rozanski, Evelyn P., 2004
Rubinstein, Marcos, 935

Sadiku, Matthew N. O., 837
Sage, Andrew P., 2113
Salek, Stanley, 1397
Sandige, Richard S., 1161, 1635, 1711
Sankaran, C., 33
Schmalzel, John L., 2277
Schroeter, Juergen, 395
Seely, J. Leland, 591
Serra, Micaela, 1808
Shaw, Leonard, 1355
Sherr, Solomon, 1938
Shim, Theodore I., 315
Sibul, L. H., 359
Smith, L. Montgomery, 238
Smith, Rosemary L., 1152
Soclof, Sidney, 530
Sohi, Gurindar S., 1927
Spée, René, 2237
Spitzer, Cary R., 2188
Stanton, K. Neil, 1344
Staudhammer, John, 1748
Steadman, J. W., 431, 683

Steer, Michael B., 882
Stephenson, F. W., 460
Sworder, D., 2211
Szidarovszky, Ferenc, 207

Tallarida, Ronald J., 2415
Tariyal, Basant K., 987
Thallam, Rao S., 1227
Thomas, Joy A., 1517
Tinder, Richard F., 1843
Tragoudas, Spyros, 581
Tranter, William H., 1593
Trew, Robert J., 891
Trowbridge, C. W., 1018
Tummala, R. Lal, 2163

Uman, Martin A., 935
Ungvichian, Vichate, 919

Vranesic, Zvonko G., 1865
Vu, Khoi Tien, 1321

Wait, John V., 625
Wallace, Alan K., 2237
Wan, Zhen, 47, 149, 178, 184,
189, 1359, 1405, 1410
Wang, Chih-Lin, 1162
Watkins, Laurence S., 742
Watson, J., 545
Weber, Larry F., 1786

Whatmore, Roger W., 1126
Whitaker, Jerry, 1379
Whitehead, D. G., 1721
Wilamowski, B. M., 683
Wilcox, Lynn D., 306
Wiltse, James C., 964
Windley, Phillip J., 1753

Young, David, 1162
Yu, Yixin, 1321

Zaky, Safwat G., 1865
Zargham, Mehdi R., 581
Ziemer, Rodger E., 1554

Index of Key Tables

1.1   Color code table for resistors, 11
4.1   Butterworth denominator polynomials, 114
4.2   Thomson denominator polynomials, 116
4.3   Chebyshev denominator polynomials, 116
6.1   Laplace transform pairs, 151
6.2   Laplace transform properties, 154
6.3   Inverse Laplace transform determination, 155
6.4   One-sided Laplace transform properties, 160
8.2   Key z-transform pairs, 182
13.3   Structured matrices, 265
14.2   Speech coding standards, 284
14.3   Speech coder performance comparison, 285
18.1   Expressions for sound speed, 409
18.2   Underwater acoustics applications, 417
21.7   Interconnection limits for telecommunications systems, 502
23.2   Commercially available design of integrated circuit programs, 600
24.1   Integrated circuit packaging technology comparison, 605
24.3   Characteristics of common integrated circuit packages, 608
27.1   Solving circuit equations, 654
27.2   Gaussian elimination, 658
27.3   LU decomposition, 659
27.4   Newton–Raphson, 662
30.1   Important commercial lasers, 739
31.1   D/A and A/D integrated circuits, 772
32.1   Thermal conductivities of typical packaging materials, 790
34.1   Units in magnetism, 813
34.4   "Hard" and "soft" magnetic materials, 821
35.5   Dimensions of standard rectangular waveguides, 857
38.2   Selection of filters used in transmitters, 916
39.1   Radar bands, 950
39.2   Radar transmitter technology, 952

43.1 Model types in computational electromagnetics, 1030
43.2 Developing computer models, 1030
43.3 Desirable attributes in computer models, 1031
43.4 Approximations in computer model development, 1032
43.6 Applicability of integral equation and differential equation based computer models, 1035
45.2 Common SAW material properties, 1064
46.1 Typical acoustic properties, 1079
47.1 Ferroelectric, piezoelectric, and electrostrictive materials, 1088
50.2 Material parameters for type I superconductors, 1121
50.3 Material parameters for conventional type II superconductors, 1122
50.4 Material parameters for high-temperature type II superconductors, 1122
51.1 Pyroelectric properties of selected materials, 1130
52.1 Electrical properties of insulating liquids, 1132
53.1 Physical and chemical transduction principles, 1153
56.1 Power plant technical data, 1195
58.1 Electrical properties of metals used in transmission lines, 1219
58.2 Standard system voltage, kV, 1221
58.4 High-voltage direct-current projects data, 1229
58.6 Typical transmission equipment loading criteria, 1292
58.7 Typical minimum transmission voltage criteria, 1292
60.1 Typical primary feeder voltages, 1311
60.2 Secondary voltages and connections, 1314
61.1 Typical synchronous generator parameters, 1328
66.1 Characteristics of standard local-area networks, 1467
67.1 Continuous/discrete classification of stochastic processes, 1501
68.2 List of satellite frequency allocations, 1543
73.2 Logic families and subfamilies, 1614
73.6 Comparison of TTL 2-input NAND gates, 1629
73.8 Comparison of three types of CMOS, 1633
74.1 Memory hierarchy, 1670
75.1 Switching algebra summary, 1703
75.8 Typical integrated circuit arithmetic and logic devices, 1746
77.1 Plasma display attributes, 1788
80.1 Binary-to-decimal conversion, 1845
80.3 BCH and BCO number systems, 1848
80.4 Recommended methods for integer conversion by noncomputer means, 1850
80.5 Recommended methods for fraction conversion by noncomputer means, 1851
83.1 List of input devices, 1939
83.6 Input devices—advantages and disadvantages, 1956
84.1 Origins and causes of software defects, 1986
93.1 Summary of describing differential equations for ideal elements, 2102
94.2 Robotic applications in manufacturing processes, 2178
95.1 Characteristics of common avionics buses, 2190
100.1 Illuminance categories for selected group of industries, 2258
102.1 Worldwide radio navigation aids, 2290
103.1 Electroencephalography waveform terminology, 2324
104.3 Classification of chemical biomedical sensors, 2345
109.1 Experts' ranking factors of biomedical device failures, 2410

Index of Key Figures

1.3   Resistor with Ohm's law relation, 6
1.9   Power rating curve for a resistor, 10
1.21   Reactance chart, 31
3.1   Graph representation of a linear circuit, 59
4.3   Singly terminated Butterworth filter, 117
4.4   Doubly terminated Butterworth filter, 118
4.5   Singly terminated Thomson filter, 118
4.6   Doubly terminated Thomson filter, 119
4.7   Singly terminated Chebyshev filter, 119
4.8   Doubly terminated Chebyshev filter, 120
4.9   Two-zero, three-pole inverse Chebyshev filter, 120
4.10   Two-zero, four-pole inverse Chebyshev filter, 121
5.4   Single-phase full-wave rectifier circuit, 129
5.5   Single-phase full-wave rectifier circuit with output filter, 129
5.6   Single- and three-phase bridge rectifier circuits, 130
5.7   Three-phase rectifier output compared to input signals, 130
5.8   Cockroft–Walton circuit, 131
5.9   Limiting circuits, 132
5.10   Clipping, another form, 133
5.12   Parallel clipping circuits, 134
5.13   Operational amplifier limiting circuit, 135
5.14   Operational amplifier limiting circuits using Zener diodes, 136
5.15   Double-ended clipping, 137
5.17   Precision rectifying circuit, 138
5.18   Precision rectifying circuit, 138
5.21   Simple diode circuit, 146
6.1   Time-domain differentiation, 152
6.2   Time shift, 152
6.3   Time-convolution property, 153
6.7   Transformed circuit elements, 163
6.10   Thévenin's and Norton's theorems, 164

2564

6.11   Circuit for obtaining open-circuit voltage and short-circuit current, 165
6.12   Thévenin equivalent circuit terminated in a resistor, 165
6.14   Unstable circuit, 168
9.1   T and Π two-port networks, 185
9.2   Balanced three-phase voltage source, 186
9.3   Wye (Y)-connected loads, 186
9.4   Delta-connected loads, 186
9.5   General wye- and delta-connected loads, 187
10.1   Magnitude functions of ideal filters: low-pass; high-pass; bandpass; bandstop, 190
10.2   Magnitude functions of an ideal comb filter, 190
10.4   Phase function of ideal linear-phase bandpass filter, 192
10.5   Causal filter magnitude functions, 192
11.1   Single-input/single-output linear system, 199
11.2   Bode diagrams, 200
12.3   Simple electrical system, 219
12.5   Simple transistor circuit, 220
14.1   Differential pulse code modulation, 281
14.4   Subband coders, 283
14.5   Gaussian autoregressive speech model, 289
14.6   Composite source model for speech signals, 289
14.9   Speech production model, 298
14.14 Linear predictive analysis, 302
14.15 Linear predictive analysis, 303
14.16 Homomorphic (cepstral) analysis, 304
14.17 Homomorphic (cepstral) analysis, 304
16.1   Digital image formation, 331
16.21 Illustration of block matching, 355
17.9   Multirate filter, 391
17.11 Video signal display, 392
17.12 Video line signal, 392
18.8   Typical underwater acoustical signal processing scenario, 407
18.10 Typical sound paths between source and receiver, 411
19.1   Processing element and sigmoid function, 421
19.2   Three-layer perceptron, 422
19.3   Hopfield crossbar neural network, 423
19.4   Architecture of Hopfield neural circuit, 424
21.1   Oxidation process, 476
21.16 History of increasing scales of integration in IC technology, 499
21.18 Rent's rule for LSI and VLSI chips, 501
21.20 Representation of a transmission line, 504
21.21 Classical microstrip design of a PCB structure, 506
21.22 Classical stripline design of a PCB structure, 506
21.41 Model for noise evaluation in connectors and wirebonds, 523
22.25 Bipolar transistor, 546
22.27 Transistor biasing circuit, 547
22.29 Complete common-emitter stage, 550
22.30 Signal or ac load line, 550

22.31   Base spreading resistance, 552

22.37   Complete frequency response, 560

22.39   Difference amplifier, 563

22.43   High-performance *n*-channel MOSFET, 567

23.2   Layout and fabrication of MOS transistors, 582

23.3   Fabrication steps for MOS transistors, 583

23.5   Placement and routing techniques, 584

23.11   Routing techniques, initial configuration, 589

23.12   Routing techniques, first step, 589

23.13   Routing techniques, final step, 590

23.17   Schematic entry design system, 597

23.18   Hardware description language design system, 598

24.2   Integrated circuit packages, 609

24.4   Discrete electronic component packages, 611

25.1   Ideal operational amplifier, 617

25.2   Inverting and noninverting amplifiers, 618

25.7   Typical operational amplifier environment, 625

25.8   Conventional operational amplifier symbol, 625

26.1   Amplifier circuits, 635

27.2   Tableau formulation, 656

27.3   Modified nodal equations, 657

27.5   Newton–Raphson example, 662

27.7   Diode linearized model, 664

27.8   Linearized network, 664

27.11   Integration methods, 667

28.13   First-order high-pass filter, 686

28.18   Sallen–Key bandpass filter, 688

29.1   Thyristor symbol, 695

29.2   Triac symbol, 696

29.3   Gate turn-off thyristor symbol, 696

29.4   Basic Darlington configuration, 698

29.5   Power MOSFET circuit symbol, 698

29.16   dc-dc converter configurations, 709

32.5   External thermal resistances, 794

33.1   Electromagnetic wave spectrum, 809

34.10   Fundamental magnetic recording configuration, 827

35.2   Modes of wave propagation, 841

35.5   Geometry of spherical earth reflection, 845

35.7   Uniform waveguide, 849

35.8   Rectangular waveguide, 852

35.11   Circular waveguide, 854

36.1   Antenna patterns for a short dipole, 862

36.6   Log-periodic dipole array, 869

36.7   Several types of antennas, 870

37.3   Waveguide and microstrip discontinuities, 885

37.4   Terminations and attenuators, 886

37.5   Microwave resonators, 887

37.6   Tuning elements, 888

37.12   IMPATT diode, 895

38.16   Frequency spectrum chart, 920

39.1   Pulse radar, 950

40.13   Schematic diagram of a lightwave communications system, 989

40.14   Geometry of single-mode and multimode fibers, 989

40.15   Characteristics of fiber types, 990

40.25   Fiber drawing process, 997

40.26   Fiber cable designs, 1000

41.1   Amplifier circuits configurations, 1005

41.10   Microwave control circuits, 1014

42.2   Switched reluctance motor, 1022

45.7   Multiphase SAW unidirectional transducers, 1071

45.8   Unit cell of a 3PUDT and the basic equivalent circuit, 1072

45.9   3PUDT requiring analysis of acoustic transducer responses and electrical phasing and matching networks, 1072

45.10   Single-phase unidirectional transducers, 1073

54.1   Kerr magnetooptic effect, 1166

55.1   Smart/intelligent structures, 1174

55.2   Smart material applications, 1177

55.3   Smart/intelligent systems, 1184

56.1   Components of fossil power plants, 1194

56.2   Flow diagram of a drum-type boiler, 1196

56.4   Typical turbine arrangement, 1197

56.9   Boiling water reactor arrangement, 1201

56.11   Hydroelectric power plant arrangement, 1203

58.3   Generalized conductor model, 1221

58.4   Pipe-type cable system, 1223

58.6   Back-to-back dc system, 1228

58.26   Basic static VAR compensator configurations, 1247

58.34   Fault types in power systems, 1256

58.39   Elements of a protection system for a power system, 1268

58.51   Typical power angle-time relations, 1280

59.8   Transformer circuit data from short-circuit tests, 1306

59.9   Autotransformer connection, 1308

60.1   Electric energy system, 1311

60.2   Radial primary distribution system, 1313

61.10   Classification of ac and dc motors for industrial applications, 1333

62.4   Operator training simulator block diagram, 1352

64.6   Linear transversal equalizer, 1410

67.3   Noise process, 1489

67.5   Thermal noise in a resistor, 1491

67.12   Example of the Huffman algorithm, 1519

68.1   Several types of satellite links, 1533

69.3   Evolution of wireless communications, 1557

73.6   Circuit interfacing requirements, 1621

73.7   Definitions of switching times, 1623

73.15   Graphic classification of bistable devices, 1635

73.16   Basic S-R NOR latch implementation, 1636

73.22   Classification of optical logic devices, 1641

73.30   Fundamental limitations of optical logic devices, 1648

74.2   Cross section, trench capacitors and "stacked" capacitors, 1653

74.10   Maximal areal density law, 1658

74.11   Disk terminology, 1659

74.21   General features of an optical disk, 1677

74.22   Micrographs of several types of optical storage media, 1680

74.30   Thermomagnetic recording process, 1688

74.32   Magneto-optical Kerr effect, 1690

75.1   Binary adder, 1696

75.15   Graphic classification of logic circuits, 1712

75.28   Control timing parameters, 1723

76.9   Digital signal processor architecture, 1761

77.19   Typical electron gun unipotential lens structure, 1779

77.20   Typical electron gun bipotential lens structure, 1780

77.21   Principle of electrostatic deflection, 1781

77.22   Principle of electromagnetic deflection, 1781

77.23   Family tree of plasma display products, 1787

77.24   Gas discharge reactions, 1790

78.1   Data acquisition system, 1800

79.1   Taxonomy of IC testing methods, 1809

80.1   IEEE standard bit format for normalized floating-point representation, 1857

80.3   Two-bus structure, 1868

80.5   Levels of programming in a computer system, 1871

80.6   Computer system levels, 1872

82.2   Simple interleaved memory system, 1929

82.3   Complex interleaved memory system, 1930

83.8   Optical interrupter, 1946

83.21   Six basic electrophotographic printer process steps, 1959

84.4   Software engineering methods overview, 1979

86.1   Bus-type local area network, 2016

86.2   Token ring local area network, 2017

87.2   General concept of standby sparing, 2021

87.5   Time redundancy concept, 2025

89.1   Basic parallel processor organization, 2053

91.1   Computer and communications security environment, 2073

93.1   Major classes of system equations, 2100

93.11   Simple electrical lead network, 2117

93.14   Simple electrical lag network, 2121

93.17   Simple electrical lag-lead network, 2123

93.34   Implementation of pole-placement design, 2145

94.1   Cartesian configuration, 2155

94.5   Spherical configuration, 2157

94.7   Articulated configuration, 2158

94.9   SCARA configuration, 2159

94.11   Gantry configuration, 2160

95.9   Generic videoconference system, 2206

96.2   Two approaches to achieving C3 architectures, 2213

97.1 Input and output variables of welding process, 2225
100.1 Basic cavity divisions of space, 2260
100.6 General lighting luminaire classifications, 2270
101.3 Fundamental elements of a prototype measurement system, 2279
102.3 Geometry of dead reckoning, 2288
103.1 Functional components of (a) motor and (b) sensory neurons, 2302
103.2 Hodgkin–Huxley membrane model, 2303
103.12 Genesis of the electrocardiogram, 2315
103.18 Equipment used for electromyography, 2321
104.5 Generalized bioanalytical sensor, 2348
105.1 Basic structure of the neuron, 2352
105.2 Major divisions of the cerebral cortex, 2353
106.3 Concept of magnetic resonance imaging, 2379

Index of Key Equations

A

Absorption of light, 744
Acoustic fundamentals, 2381
ac quantities, 1234
ac steady-state power, 81
Active filters, 674
Algebra, 1700, 1703
Ampère's law, 1297
Amplifiers, see also specific types
  applications of, 625
  common-base, 648
  common-collector, 650
  common-emitter, 644
  difference, 562
  ideal, 617
  operational, 617, 619, 625
  practical, 619
Analog signal conditioning, 1803
Analysis of arrays
  equally spaced linear arrays, 866
  identical elements, 866
  planar (2-D) arrays, 867
Antenna directivity, 955
Approximate expressions for sound speed, 409
Arithmetic, 1859
Arithmetic logic units, 1744
Arrays, see also specific types
  analysis of, 866, 867
  discrete, 364
  equally spaced, 866
  linear, 866
  planar (2-D), 867
  spatial, 360
Attenuation constant, 841
Availability, 2091

B

Backward Euler (BE), 667
Bandpass filters, 191
Bandwidth of memory systems, 1927
Basic filter specifications, 1064
Basic per-unit scaling equations, 1297
Bayesian estimation scheme, 323
Binary number systems, 1853
Biot Savart law, 1025
Bipolar transistors, 545, 899
Bode diagrams, 2114, 2124
Boltzmann relation, 447
Boltzmann transport equation, 440
Brightness contrast of color CRTs, 1783
Butterworth denominator polynomials, 114
Butterworth filters, 192

C

Canonical equations, 1058
Capacitance matrix, 1220
Capacitors, 16
Cathode ray tube (CRT) brightness contrast, 1783
Causal filters, 191
Celestial navigation, 2292
Characteristic impedance of interconnection
  medium, 504
Characterization of passive elements, 882
Chebyshev denominator polynomials, 116
Chebyshev filters, 193
Chebyshev functions, 115
Chips per wafer yield, 594
Christoffel equations, 1078
Circuits
  applications of, 161

common-collector, 555
  equations for, 655
  simulators and, 655
  three-phase, 88
  transformed, 162
Circular waveguides, 854
Clausius–Mossotti equation, 1136
Closed-loop systems, 2150, 2249
Coherence processing, 416
Collin's notation, 861
Color CRT brightness contrast, 1783
Common-base amplifiers, 648
Common-collector (CC) amplifiers, 650
Common-collector (CC) circuits, 555
Common-emitter (CE) amplifiers, 644
Common-mode rejection ratio, 1803
Compensation, 2139
Complex power, 82
Computer arithmetic, 1859
Computer control law, 2150
Computerized tomography, 2375
Computer system reliability, 2091
Constitutive relations, 804, 838
Contrast ratio of color CRTs, 1783
Control design, 2143
Conversion between number systems, 1848
Convolution equations, 386
Core loss, 1302
Cross modulation, 144
Crossover frequency of man-machine systems, 2251
Crosstalk noise, 514, 516
Current laws, 60, 61
Current mirror, 564

D

Data compression, 1518, 1519
Data windowing, 233
dc motors, 1334
dc quantities, 1234
dc signals, 49
Device efficiency, 1766
Dielectric breakdown, 1137
Dielectric losses, 1133
Dielectrics, 1132
Diendorfer and Uman model, 943
Difference amplifiers, 562
Differential equations, 666, 1042, 2149
  ideal elements, 2102
Differential modeling, 1042
Differentiation theorems, 158
Diffraction theory, 1683
Digital filters, 238, 246
Dimensionless instability factor, 1965
Diodes, see also specific types
  equations for, 665
  forward-biased, 450, 451
  light-emitting, 750, 1766
  pn-junction, 447

Directivity, 955, 1060
Discrete arrays, 364
Discrete Fourier transform, 229
  amplitude spectrum of, 232
  phase spectrum of, 233
  power spectrum of, 232
  properties of, 230
  relation with other Fourier transforms, 232
Distance protection, 1272
Distortion, see also specific types
  cross modulation, 144
  Hamming, 1522
  harmonic, 139
  intermodulation, 143
  rate, 1522
  triple-beat, 144
Doppler principle, 2384
Dynamic models in robotics, 2167
Dynamic system response, 2106, 2107, 2108

E

Effective resistance, 1218
Electrical loss, 1302
Electroacoustic coupling, 1058
Electromagnetic equations, 35, 803, 1325
Electromechanical equations on operating
      frequency, 1325
Empirical formulas, 470
Encoders, 294, see also specific types
Energy distribution load characteristics, 1316
Entropy, 1518, 1519
Enzyme sensors, 1159
Equally spaced linear arrays, 866
Equations of state, 1099
Equivalent circuit models, 461
Estimation, see also specific types
  Bayesian, 323
  maximum likelihood, 326
  mean-square, 323
  minimax, 326
  parameter, 321, 323, 326
External stability, 216

F

Failure rate, 2026, 2089, 2090, 2091
Faraday's law, 1296
Fast Fourier transform, 235
Fault types in power systems, 1256
Ferroelectric materials, 1088, 1089
Fibers, 986, 1432
Fick's Laws, 479
Field equations, 1018
Figure of merit for sonar, 408
Filters, see also specific types
  active, 674
  bandpass, 191
  basic, 1064

basic specifications for, 1064
Butterworth, 192
causal, 191
Chebyshev, 193
design of, 122, 238
digital, 238, 246
ideal, 189
implementation of, 246
linear-phase, 190, 191
low-pass, 113, 190
passive, 674
specifications for, 1064
transformation rules design of, 122
"First Law in Disk Density," 1658
First London equation, 1116
Floating-point number systems, 1856
Forms of the model, 2103
Forward-biased diodes, 450, 451
Forward Euler method (FE), 667
Fourier transforms, 300
    discrete, 229, 230, 232, 233
    fast, 235
Frequency
    crossover, 2251
    Larmour, 2378
    operating, 1325
Frequency domain, 1041, 1042
Frequency response, 541
    analysis of, 2114
    of JFETs, 541
    plotting of, 199
Frequency-shift keying, 1364
Friis transmission equation, 842

G

Gaussian elimination equations, 658
Gaussian white noise, 1491
General electromagnetic properties, 1115
Graphical approach to large signal analysis, 636
Graph theory
    flowgraph approach to, 91, 98
    k-tree approach to, 94, 102

H

Hall effect, 440, 1106
Hamming distortion, 1522
Harmonic distortion factors, 139, 140
Heat transfer fundamentals, 787
Heisenberg Uncertainty Principle, 1766
Helmholtz equations, 838, 1116
Hidden Markov models, 309
High-frequency response, 558
Hopfield network, 423

I

Ideal elements, 2102
Ideal filters, 189
Ideal operational amplifiers, 617

Identical elements, 866
Illumination computational methods,
    2259
Impedance, see also specific types
    of capacitors, 16
    of inductors, 24
    internal, 1219
    intrinsic, 841
    media, 841
    radiation, 1059
    series, 1218, 1219
Impulse responses, 166
Independent joint control in robotics, 2164
Induction motors, 1337
Inductors, 24
Instability factor, 1965
Integral equation modeling, 1041, 1042
Integral modeling, 1041
Integrated circuit cost equations, 594
Interconnection elements, 524
Intermodulation distortion, 143
Internal impedance, 1219
Intrinsic impedance, 841
Inverse Laplace transform, 154

J

Josephson equations, 1118
Junction field-effect transistors
    biasing, 533
    frequency and time-domain response of,
        541
    output resistance, 538
    source-follower voltage gain, 539
    transfer characteristics of, 536
    voltage-variable resistor of, 543

K

Kirchhoff's current law, 60, 61
Kirchhoff's voltage law, 62
Known signals, 1480

L

Lag-lead network design, 2122
Laplace transform, 111, 149, 154, 160
    inverse, 154
    properties of, 151
Large signal analysis, 634, 636
Larmour frequency, 2378
Law of stimulation, 2312
Learning rule, 420
Least significant bit, 1803
Light absorption, 744
Light-emitting diodes (LEDs), 750, 1766
Light phase velocity, 743
Likelihood ratio, 1479
Linear arrays, 866
Linear control example, 2151
Linear equations, 657, 2149

Linear-phase filters, 190, 191
Load characteristics, 1316
London equations, 1116
Look-ahead technique, 377, 383
Loop filter equations, 1569
Loop gain, 1567
Low-frequency performance, 553
Low-pass filters, 113, 190
LU decomposition, 659
Lyapunov theory of stability, 209

M

Magnetic loss, 1302
Magnetooptic effects, 1162, 1163
Man-machine systems, 2251
M-ary phase-shift keying, 1365
Matrices, see also specific types
    algebraic properties of, 262
    capacitance, 1219, 1220
    positive-semidefinite, 263
    structural properties of, 265
Maximum likelihood estimation, 326
Maximum transfer power, 85
Maxwell's equations, 803, 815, 838, 1025
Mean-square estimation, 323
Mean time between failures, 2090
Mean time to failure, 2090
Measurement
    on CRTs, 1783
    of thermal noise, 1493
Measurement noise, 1497
Memory system bandwidth, 1927
MESFET noise figures, 901
Mesh analysis, 66
Metal-oxide semiconductor field-effect transistors
    (MOSFET), 570, 577
Metric space formulation, 252
Microwave resonators, 886
Miniaturization of MOSFETs, 577
Minimax estimation scheme, 326
Minor-loop design, 2127
Modeling, see also specific types
    Diendorfer and Uman, 943
    differential equation, 1042
    dynamic, 2167
    equivalent circuit, 461
    forms in, 2103
    hidden Markov, 309
    integral equation, 1041, 1042
    in robotics, 2167
    SAW transducer, 1066
    transducer, 1066
    transmission line, 942
Modern control design, 2143
Modified nodal approach, 656
Modulation, 144, 1359
Motors
    dc, 1334
    induction, 1337
    synchronous, 1335

N

Navigation, 2292
Nernst equation, 2302, 2347
Networks, see also specific types
    equations for, 174, 184, 188, 1218
    functions of, 165
    Hopfield, 423
    lag-lead, 2122
    neural, 420
    reciprocity theorem for, 77
    resistor, 6
    superposition theorem for, 69
    T, 184, 188
    Tellegen's theorem for, 74
    theorems for, 69, 74, 77
Neural networks, 420
Neuroelectric equations, 2302
Newton–Raphson equation, 662, 663
Nodal approach, 656
Node analysis, 64
Noise
    crosstalk, 514, 516
    Gaussian, 1491
    measurement, 1497
    quantization, 1497
    in semiconductor devices, 758
    thermal, 1493
    white, 1491
Noise margins of digital circuits, 1618
Noise power, 1489
Nonlinear equations, 661, 1325
Normal form of state equations, 170
Norton's theorem, 164
Number systems
    binary, 1853
    conversion between, 1848
    floating-point, 1856
    polynomial representations and, 1844
    positional representations and, 1844
    signed binary, 1853
Numerical calculations of swing equations, 1284,
    1286
Numerical methods, 1020
Numerical power ratios, 1540

O

Ohm's law, 6, 64
Operating freuqency, 1325
Operational amplifiers, 617, 619, 625
Optimal encoders, 294
Output/input dynamics, 2249

P

Parabolic graded-index fiber, 986
Parallel processors, 2057
Parameter estimation, 321
    Bayesian, 323
    maximum likelihood, 326

mean-square, 323
minimax, 326
Parameterized signals, 1480
Passive element characterization, 882
Passive filters, 674
Peak elevation angle for celestial navigation, 2292
Perceptrons, 421
Per-unit scaling equations, 1297
Phase constants, 841
Phase-lag compensation, 2119
Phase-lead compensation, 2116
Phase voltages, 185
Photovoltaics, 1208
Physical system stability, 218
Piezoelectric excitation, 1080
Piezoelectric materials, 1088, 1089
Π network equations, 184, 188
Planar (2-D) arrays, 867
Planck's law, 748, 1492
Planck's theory, 751
pn-junction, 448
pn-junction diode, 447
Poisson's equation, 469
Polarization coupling factor, 1538
Polynomials, see also specific types
    Butterworth denominator, 114
    Chebyshev denominator, 116
    representations of, 1844
    Thomson denominator, 116
Positional representation, 1844
Power, 79
    ac steady-state, 81
    complex, 82
    maximum transfer, 85
    reactive, 82
    Tellegen's theorem for, 80
Power factor, 83
Power transformers, 1296, 1297
Practical operational amplifiers, 619
Prandtl number, 792
Processing elements, 420
Propagation, 1077
Protection, 1273
Pyroelectric effect, 1127
Pyroelectric materials, 1129

Q

Quantization noise, 1497
Quantization of signals, 347

R

Radar range equations, 954, 958
Radiation
    impedance and, 1059
    Planck's law of, 1492
Radon transform, 341
Ramp function, 48

Random signals, 1482
Rate distortion theory, 1522
Reactive power, 82
Reciprocity theorem, 77
Recording process, 828
Rectangular waveguides, 849
Reliability
    computer system, 2091
    failure rate and, 2089, 2090, 2091
    of real systems, 2094
    of series systems, 2029
Resistance
    effective, 1218
    of resistors, 5
Resistors
    characteristics of, 5, 28
    impedance of, 28
    networks of, 6
    Ohm's law and, 6
    resistance of, 5
Resonators, 886
Reynolds number, 792
Rice–Holmberg thunderstorm ratio, 848
Risetime degradation of interconnection elements,
    524
Robotics, 2164, 2167
Root locus, 2131, 2132, 2138

S

Sampling functions, 1035
Sampling theorems, 346, 1510
    generalizations of, 1515
    proof of, 1511
    sources of error, 1513
Satellite orbits and pointing angles, 1535
SAW transducer modeling, 1066
Scaling equations, 1297
Scharfetter–Gummel empirical formula, 470
Second London equation, 1116
Semiconductor devices, 758, see also specific types
Semiconductor materials, 1766
Semiconductors
    Boltzmann transport equation, 440
    electrons and holes, 438
    energy bands, 435
    Hall effect, 440
    transport properties, 438
Sensors, 1159
Series impedance, 1218, 1219
Shockley diode equation, 127
Shunt admittance of transmission line parameters,
    1219, 1220
Signals, see also specific types
    dc, 49
    detection of, 1479, 1480, 1482, see Signal
        detection
    known, 1480
    parameterized, 1480

quantization of, 347
  random, 1482
  speech, 290
Signal-to-noise ratio, 1432, 1803
Signed binary numbers, 1853
Simulators, 655
Slab waveguide, 986
Small-signal analysis, 641
  common base amplifier, 648
  common collector amplifier, 650
  common emitter amplifier, 644
  design considerations for, 646
Small-signal operation, 551
Snell's law, 976
Sonar, 408
Sound speed, 409
Spatial arrays, 360
Spectral analysis, 315, 316
Speech signal estimation, 290
Speed of sound, 409
Speed-power product, 1620
Stability, 208
  external, 216
  Lyapunov theory of, 209
  of physical systems, 218
  of time-invariant linear systems, 210
Stable operation of power systems, 1279
State equations, 1099
  for networks, 174
  in normal form, 170
  writing of, 171
Static magnetic fields, 811
Step function, 47
Step-index fiber, 986
Step responses, 166
Stimulation law, 2312
Stochastic processes, 1499
  classifications of, 1501
  examples of, 1502
Superposition theorem, 69
Swing equations, 1286, 1325
Switching algebra, 1700, 1703
Synchronous motors, 1335
System utilization of parallel processors, 2057

T

Tableau approach, 655
Tellegen's theorem, 74, 80
Thermal noise, 1493
Thévenin's theorem, 164
Thomson denominator polynomials, 116
Three-phase circuits, 88
Thunderstorm ratio, 848
Time domain, 1042
Time-domain response, 541
Time-invariant linear systems, 210
T network equations, 184, 188
Total harmonic distortion, 140

Total pulse spread, 986
Transducers, 1066
Transformations, see also specific types
  of fourth rank polar tensor, 1100
  from low-pass to other filter types, 684
  wye-delta, 186
Transformed circuits, 162
Transformers
  electrical loss and, 1302
  magnetic (core) loss and, 1302
  performance of, 1302, 1303
  power, 1296, 1297
  voltage regulation and, 1303
Transforms, see also specific types
  Fourier, 230, 232, 233, 235
  inverse Laplace, 154
  Laplace, 111, 149, 151, 154, 160
  z-, 178, 179, 180, 181
Transistors, see also specific types
  bipolar, 545, 899, 901
  junction field-effect, 533, 536, 538, 539, 543
  metal-oxide semiconductor field-effect,
    570, 577
Transmission compensation, 1242
Transmission lines
  equations for, 943
  modeling of, 942
  series impedance and, 1218, 1219
Transport equations, 440
Trapezoidal (TR), 667
Triple-beat distortion, 144

U

Underground power cable parameters, 1223
Uniform waveguides, 849
Unilateral z-transform, 180, 181
Unit impulse function, 47

V

Vector Helmholtz equation, 1116
Velocity, 743
Velocity filtering, 365
Voltage
  Kirchhoff's law of, 62
  phase, 185
  regulation of, 1303
Voltage-variable resistor of JFET, 543

W

Wave equations, 807, 977, 1116
Waveguides, see also specific types
  circular, 854
  rectangular, 849
  slab, 986
  uniform, 849
White noise, 1491

Wiener–Khinchin theorem, 317
Wind-electric conversion, 1210
Wye-delta transformations, 186

Z

Zero-input response, 2107
Zero-state response, 2108

z-transform
    convolution, 180
    inversion of, 181
    linearity, 178
    multiplication by a^n, 180
    time reversal, 180
    translation, 179
    unilateral, 180, 181

Subject Index

A

AAL, see Asynchronous transfer
mode adaptation layer
ABS, see Alternate billing services;
Anti-lock braking systems
Absolute convergence, **157**, 150–151
Absolutely stable circuits, **168**, 167
Absolute magnitude difference
function (AMDF), 300, 301
Absolute navigation systems,
2286–2287
Absolute power, 2355
Absorption coefficients, 744, 1767
Absorption of light, 744
Abstract classes, **1913**, 1911
Abstract data types (ADTs), **1914,
1925**, 1908, 1909, 1915,
1916, 1918–1919
Abstractions, **1913**, 1902, 1905
algorithm, 1906
data, 1998–1999, see also
Abstract data types (ADTs)
in database management systems,
2033
object-oriented programming
and, 1912
ac-dc plasma displays, 1795–1796
ac energy measurement, 85–87
ac generators, 1321–1329
ac machines, 731–736, see also
specific types
ac motors, **1341**, 1333, 1334,
1340–1341, 2238
ac plasma displays, **1798**, 1794–1795
ac power measurement, 85–87
ac power supplies, 727–728
ac schematic diagrams, **651**, 644,
647

ac steady-state power, **87**, 81–85
ac-to-dc converters, 702–705
ac transmission
asynchronous, **1241**, 1228
dc transmission vs., 1233
economics of, 1233
harmonics of, 1235–1236
overhead, 1217–1222
synchronous, 1228
underground, 1223–1227
ACARS, **2193**
Acceleration sensors, 1155
Accelerometers, 1155, 1161, 2288
Acceptance angles, 976, 977
Acceptors, **446, 458**, 438, 447
Access lines, **1458**, 1449
Access time, **1675**, 1670, 1681–1682
Accuracy
in radar, 956–957
in robotics, 2160
ACE, see Annular control electrode;
Area control errors
ACM, see ac motors; Association
for Computing Machines
Acoustic advantage, 408
Acoustic data fusion, 417
Acoustic engineering, 1187
Acoustic phonetic symbols, 304
Acoustics, 2381–2382
nonlinear, 1057
signal processing and, see
Acoustic signal processing
ultrasound and, 1079
Acoustic sensors, 2228
Acoustic signal processing, 395–418
active noise control and, 403–404
active sound control and,
403–404
advanced, 416–417

applications of, 417
audio coding and, 400–401
data fusion in, 417
echo cancellation and, 401–403
hearing aids and, 397–398
normalization in, 416
performance limitations of,
411–412
spatial processing and, 398–400
steerable microphone arrays and,
395–397
underwater, see Underwater
acoustic signal processing
Acoustic smart materials, 1175
Acoustic waves, 1083, 1085, 2380,
2381–2382, see also
Acoustics
Acousto-optic devices, 1083, see
also specific types
Acousto-optic interactions, 1085
Action potentials, **2310**, 2302, 2304,
2312
developing, 2308
fibrillation, 2320
insertion, 2320
muscle, 2319
propagation of, 2306
recording of, 2313–2314
Activation potential of neurons, 423
Active devices, **901**
Active filters, **682, 690**, 674–691,
771, see also specific types
bandpass, 683, 684, 688–689
bandstop, 683, 684–685, 689–690
cascade, 685
cascaded second-order sections
and, 676–680
Chebyshev, 678
classification of, 675–676

high-pass, 683, 684, 686, 687–688
for integrated circuits, 681–682
loop, 1569
low-pass, 674–682, 685, 686
operational amplifiers and, **632**, 628
passive ladder simulation and, 680–681
RC, **631**, 628, 674, 675, 676, 678
realization of, 683–690
sensitivity of, **682**, **690**, 675–676, 683
switched-capacitor, 682
transformation of, 683–685
Active load circuits, 539
Active microwave devices, **901**, 891–902, see also specific types
semiconductor material properties and, 892–893
three-terminal, 896–901
two-terminal, 894–896
Active networks, 102, 1437, see also specific types
Active noise control (ANC), 403–404
Active region, **544**, 532, 1083
Active sonar, 408
Active sound control (ASC), 403–404
Active suspension in vehicles, **2256**, 2255
Activities of daily living (ADL), **2395**, 2388
ACTS, see Advanced communication technology satellite
Acyclic precedence graphs, 372, 373
AD558 digital-to-analog converter, 774–775
AD7524 digital-to-analog converter, 775, 776
Adaption sublayers, **1458**, 1454
Adaptive beamforming, 416
Adaptive clustering, 425
Adaptive controllers, 2233, 2237
Adaptive control systems, 218, 2147, 2171–2172, 2213, 2217, see also specific types
welding and, 2233–2234
Adaptive differential pulse-code modulation (ADPCM), 284, 779, 1401, 1525
Adaptive equalization, 1412
Adaptive filters, 379, 1415, see also specific types
Adaptive relaying, 1277–1278
Adaptive resonance theory (ART), 426–428
Adaptive structures, 1174

Adaptive systems, 1175, see also specific types
Adaptive weld process control, 2233–2234
ADC, see Analog-to-digital converters
ADCS, see Attitude determination and control system
Adder overflow limit cycle, 249
Adders, 1695, 1696, 1736, 1742–1745, see also specific types
Addition, 1862, 2432, 2438, 2523
Additive polarity, 1299
Addressability, 2005
Addressable latches, 1723
Address errors, **1901**, 1883
Addressing, 1882, 1886–1888
Address space, 1927
Adequacy of transmission, **1295**, 1287
Adjoint, 2440
ADL, see Activities of daily living
Admittance, 111
ADMRs, see All-digital medical records
ADPCM, see Adaptive differential pulse-code modulation
ADP crystals, 1055
ADS, see Automatic Dependent Surveillance
ADTs, see Abstract data types
Advanced communication technology satellite (ACTS), 2209
Advanced intelligent networks (AINs), 1473–1475
A&E, see Audio and electroacoustics
AEE, see Association of Energy Engineers
Aerospace systems, 2188–2210
avionics in, 2188–2193
earth stations in, 2199–2200
network management systems in, 2202–2203
next-generation, 2209
satellite, see Satellites
spacecraft and, 2196–2199
transmission in, 2203–2204
video transmission in, 2205–2207
VSATs in, **2209**, 1533, 2194, 2200–2205, 2209
Aeroturbines, 1210
AES, see Audio Engineering Society
A/F, see Air-fuel ratio
AGC, see Automatic gain control; Automatic generation control
AGVs, see Automated guided vehicles

AI, see Artificial intelligence
AIB, see Analog interface board
AIC, see Analog interface chip
AIEE standards for insulators, 1221
Aiken's method, 2514
AIMS, see Airplane Information Management System
AINs, see Advanced intelligent networks
Air bags, 2255
Airborne vehicles, 2214, see also specific types
Air capacitors, **30**
Air conditioning, 2263
Aircraft control systems, 2020, 2285
Airplane Information Management System (AIMS), 2188–2189, 2190
AKMs, see Apogee kick motors
Algebra, 2430–2434
Boolean, **1711**
switching, 1699–1703
Algebraic equations, 2434
Algebraic properties of matrices, 262–265
Algebraic Ricati equation, 2147
Algorithms, see also specific types
analysis of, 1997
Baum iterative, 292
booth recoding, 1863
in computer arithmetic, **1864**, 1858, 1861–1864
D-, 1812
FAN, 1812
Huffman, **1526**, 339, 340, 350, 1519, 1520
learning, 423
least-mean square, 402, 403, 404
least-squares, 403
Lempel–Ziv, **1526**, 1520–1521
linear congruential, 1600, 1601
Lloyd, 1524
noise averaging, 393
in pattern recognition, 343
PODEM, 1812
recursive least-squares, 403
recursive signal processing, 377
in signal restoration, 261–262
speech detection, 395
SRT, 1864
transformations of, 371, see also specific types
universal coding, 1520
Viterbi, **313**, 312, 1403
Aliasing, 358, **1516**, **1815**, 346, 1813, 1936
diagrams of, 2282
errors in, 332
sampling theorems and, **1516**, 1510, 1515

Alkyl benzenes, 1144
All-digital medical records
    (ADMRs), 2397
Alleviate acoustic feedback
    (howling), 401
All-optical devices, 1641–1643, see
    also specific types
All-optical networks, 1440
Alloys, 1178, 1180, 1181, see also
    specific types
    magnetic, 2536–2541, 2542
    properties of, 2536–2541
    resistivity of, 2530–2531
All-pole models, 302, 303, 2102
Alpha particles, 1652
Alternate billing services (ABS),
    1472–1473
Alternating current, see ac
Altimeters, 971
ALU, see Arithmetic-logic unit
Alumina, 1145
Aluminum electrolytic capacitors,
    21, 22
AM, see Amplitude modulation
Ambient temperature, **30**, 21
AMDF, see Absolute magnitude
    difference function
American Society of Mechanical
    Engineers, 2547
A-mode displays, **2385**, 2383
Amorphous magnetic materials,
    825–826
Ampère, 1003
Ampere squared seconds, **45**, 40
Ampere-turns, **30**, 26
Ampère's law, 804, 812
Amperometric sensors, **2349**, 2346
Amplification, 752, see also
    Amplifiers
Amplifiers, 634–652, see also
    specific types
    in aerospace systems, 2199
    bipolar-junction transistor, see
        Bipolar-junction transistor
        (BJT) amplifiers
    broadband, 1005, 1007
    capacitors coupled with, 634
    circuits for, 635
    common-base, 634, 648–650
    common-collector, 650–651
    common-emitter, 634, 644–647
    common-source, 536, 538–539,
        541–542
    complementary symmetry diode
        compensated, 638, 639
    detector element, 1128
    difference, **566**, 561, 562–564, see
        also Long-tailed pair
    differential, see Difference
        amplifiers

emitter-follower, 634, 637
erbium-doped fiber, **1425**, 1430
gain characteristics of, 553
graphical approach to, 636–637
high-input impedance, 1127
high-power, 2199, 2206, 2207
IMPATT-diode, 1005
instrumentation, 771, 1803
intermediate power, 1388
inverter, 618, 625, 626
isolation, 771
large-signal analysis and,
    634–639
low-noise, 2199
negative resistance, 1005
noise figure of, 1422
noise performance of, 1006
noninverting, 618, 627–631
operational, see Operational
    amplifiers
operational transconductance,
    674
optical, 1430
power, 637–639, 1005, 1006,
    1008, 1388
sample-and-hold, **1807**, 771,
    1804, 1805
in satellites, 1541, 2199, 2206,
    2207
small-signal, 541–542, 899
small-signal analysis and, see
    Small-signal analysis
solid-state, 2202
solid-state circuits and,
    1005–1008
transistors in, 1005
ultrabroadband, 1005
valve, 675
Amplitude modulation (AM), 144,
    2263
    phase, 1363, 1410
    quadrature-, 1365, 1366, 1419,
        1534, 1542, 1555
    telemetry and, 1580
Amplitude modulation (AM) radio,
    1367–1372, 1374
Amplitude modulation (AM)-VSB
    video, 1418–1419, 1422,
    1423, 1424, 1425
Amplitude spectrum, 232–233
Analog comparators, 778
Analog-to-digital converters (ADC),
    631, **1761**, **1807**, 386, 388,
    389, 424, 771–783, see also
    specific types
    AD571, 780
    ADC-208, 779–780
    bypassing on, 780–781
    conversion processes in, 775–778
    coupling and, 1805

data acquisition systems and,
    1807, 1799, 1802, 1803,
    1804, 1805, 1806
digital audio broadcasting and,
    1399
digital image processing and, 331,
    332
electroencephalography and,
    2357
flash, **781**, 775, 777, 778, 779
grounding on, 780–781
integrated circuits in, 771,
    778–781
microprocessors and, **1761**, 1756
Nyquist, **782**, 772
oversampling, **782**, 772, 776
performance criteria for,
    771–773
sampling, 778
Analog evaluation fixtures, 388
Analog filters, 2395
Analog interface board (AIB), 386
Analog interface chip (AIC), 386
Analog interfaces, 771
Analog lightwave systems, 1419
Analog signal conditioning,
    1802–1803
Analog signal interface, 1801–1802
Analog transmission, 768
Analysis-by-synthesis, **286**, 280, 282,
    283
Analyte, **2349**, 2347
Analytic geometry, 2446–2454
Analytical modeling, 2100, 2229, see
    also specific types
ANC, see Active noise control
Anderson–Mott transition, 445
Anesthesia, 2324
Angiogenesis, **2338**, 2330, 2334
Animation, 2005
Anisotropic etching, 488
Anisotropic materials, 805–806,
    1078, see also specific types
Anisotropy, 822, 826, 1018
    crystalline, 822, 823, 831
    magnetocrystalline, 819, 823
    magnetoelastic, 822, 823–825
    shape, 822, 823, 831
ANN, see Artificial neural network
Annealing, 484, 587–588
Annular control electrode (ACE)
    pulsing, 1388
Anodes, **30**, 20, 1770
Antenna array, 1029
Antenna noise, 1494
Antennas, **878**, 861–878, see also
    specific types
    in aerospace systems, 2199, 2200,
        2206
    in AM radio, 1372, 1374

Page on which term is defined is indicated in bold.

aperture, see Aperture antennas
aperture area for, 955
array, see Array antennas
axial slot on cylinder, 870
Cassegrain, 875, 876
classes of, 870
conical horn, 870
continuous current distributions
    and, 875–876
directivity of, **878**, 862–863,
    876–877, 955
earth stations and, 2199
efficiency of, 863
electronic scanning, 951
end loading of, 865
in FM radio, 1375
Fourier transforms and, 875–876,
    951
gain of, **869**, **878**, 862–863,
    876–877
geometric designs of, 874–875
horn, 870
input impedance of, 863
isolation of, 964
linear, 1185
magnetic dipole, 863
Maxwell's equations and, 875,
    876
oscillators and, 870–873
parabolic reflector, 870, 875
parameters for, 876–878
patterns in, 861
purpose of, 870
pyramidal horn, 870
in radar, 951–952, 955, 956,
    964
in radio, 1372, 1374, 1375
receiving, 842
reflector, 956
resonant half-wavelength, 864
in satellites, 1542, 2206
scanning, 951
short dipole, 861–862
smart, 1185
in telemetry, 1586–1587,
    1588–1589
in television, 1390
temperature and, 1494
transmitting, 842
wire, see Wire antennas
Anthropomorphic (articulated)
    configuration in robotics,
    2157–2158
Anti-aircraft warfare (AAW), 2217
Antibodies, 1158, 1159
Antiferromagnetic compounds,
    2540
Antifuses, **1657**, 1654
Antigens, 1159
Anti-jam communications, 2218

Anti-lock braking systems (ABS),
    2255
Anti-submarine warfare (ASW),
    2217
Anti-surface warfare (ASUW), 2217
APD, see Avalanche photodiodes
Aperture antennas, **878**, 870–878,
    1185
APL, see Application program
    interface
Apodization, 1069–1070
Apogee kick motors (AKMs), 2196
Apparent power, 82
Application program interface
    (APL), 2067
Application-specific ICs (ASICs),
    601, **1657**, 591–602, 1735,
    1829
  categories of, 591
  chip size in, 596
  commercially available programs,
    600
  computer graphics and, 2008
  custom, 591, 592–593
  design of, 591, 597–599
  development cost of, 592
  development time for, 596–597
  disadvantages of, 591
  hardware description language
    and, 598–599
  manufacturing cost of, 593–596
  in memory devices, **1657**, 1654
  prototypes in, 596
  reasons for using, 591–597
  registered, 1733
  schematic entry and, 597–598
  standard, 591
  surface mount technology and,
    607
  types of, 591–597
  unit cost of, 593–596
Appropriate technology, **2395**, 2392
Approximation, see also Estimation
  arctangent, 2116
  Butterworth, 679
  Chebyshev, 679
  in computational
    electromagnetics, 1031,
    1032, 1035
  of continuous systems, 2105
  far-field, 875, 876
  Fresnel, 875
  near-field, 875
  nonnegative sequence, 268–270
  quasi-static, 1080
  reduced-rank, 262–263
  Stirling's, 2431, 2506
  straight-line, 201
  successive, **782**, 775, 776, 777,
    780

Toeplitz–Hermitian matrix,
    269–270
AR, see Autoregressive
Arcback, 1239
Architecture, **1869**, 1865–1870, see
    also specific types
  of arithmetic processors, 381–382
  bus structure and, 1868
  of C3, 2212, 2213, 2215
  of computer communication
    networks, **1458**, 1447–1448,
    1451–1455
  of computer relays, 1277
  of database management systems,
    2043–2044
  of databases, 2043–2044
  in digital signal processing, 385
  digital storage, 1662
  disk system, see Disk system
    architecture
  distributed computing and, 1869
  functional units in, 1865–1869
  Harvard, 385, 1761
  hypercube, 1760
  of intelligent networks,
    1469–1470
  of magnetic disks, 1659–1664
  of microprocessors, 1760–1761
  open system, **2221**, 2212
  operational concepts and,
    1866–1867
  parallel computing and, 1869
  of radar, 951
  virtual, 1870
Arc lamps, 749
Arc sensors, 2226–2228
Arctangent approximation, 2116
Arc voltage, 2231
Area control errors (ACE), 1347
Areas under the standard normal
    curve, 2490
Arithmetic, 382, 1760, see also
    specific operations
  computer, see Computer
    arithmetic
Arithmetic coding, 1520, 1526, 2023
Arithmetic-logic unit (ALU), **1746**,
    1869, 1741–1747, 1761,
    1866, 2052
  multifunction, 1745
  standard integrated circuit, 1746
Arithmetic mean, 2486
Arithmetic processor architectures,
    381–382
Arithmetic progression, 2432
ARMA, see Autoregressive moving
    average
Armature circuits, **1331**, 1321
Armature reaction, **1331**, 1322,
    1330

Page on which term is defined is indicated in bold.

Armortisseur windings, 1326
Armstrong, Edwin Howard, 1576–1577
ARPANET, 2018
Array antennas, **869**, **878**, 865, 866–869, 870, 872
Array factor, 866
Array gain, 413
Array polynomials, 866
Array processors, 2059, 2054–2055
Arrays, **366**, 1029, see also specific types
  analysis of, 865
  broadside, 866–867
  character, 1881
  charge-coupled device, 1953
  data types and, 1917–1918
  detector, 2375, 2377
  discrete, 361–365
  disk, 1668
  equally spaced linear, 866–867
  field-programmable gate, **1740**, 1733, 1735, 1736, 1739
  of identical elements, 866
  log-periodic dipole, 868–869
  n-element, 872
  one-dimensional, 360
  page-filling, 1878
  phased, **869**, 865, 867, 1762, 2382
  photosensor, 1554
  planar, 867–868
  programmable, 1735–1741
  programmable gate, see Programmable gate arrays (PGAs)
  programmable logic, **1741**, 1716, 1735, 1736–1739, 1740
  pyroelectric, 1129
  semiconductor detector, 755–757
  spatial, 360
  two-dimensional, 360, 867–868
  of wire antennas, 865, 866–869
  Yagi–Uda, 868
Arrhenius relationship, 785, 786
ART, see Adaptive resonance theory
Articulated configuration in robotics, 2157–2158
Artificial heart research, 2373
Artificial intelligence, 343, 2048
Artificial limbs, 2391, 2392
Artificial neural network (ANN) models, 2229–2231
ASC, see Active sound control
ASCRs, see Asymmetrical silicon-controlled rectifiers
ASICs, see Application-specific ICs
A-site, **1096**
ASME, see American Society of Mechanical Engineers
Aspect ratio, **1397**

Assembler language, 1905, 1906
Assemblers, **1901**, **1914**, 1879
Assembly, **601**, 595
Assembly language, 1878–1902
  addressing and, 1886–1888
  calling conventions and, 1889–1892
  compiler optimization and, 1894–1900
  high-level languages and, 1898–1900
  memory and, 1882–1886
  number count and, 1878–1881
  registers and, 1882–1886
  transactional paradigms and, 1892–1894
Associated reference directions, **671**
Association for Computing Machines, 2548
Association of Energy Engineers, 2550
Associative (content-addressable) memory, 2056–2057
Associative processors, **2059**, 2056–2057, see also specific types
Associativity transformation, 380–381
A-stable methods, 669
ASW, see Anti-submarine warfare
Asymmetric silicon-controlled rectifiers (ASCRs), 697
Asymmetric thyristors, 721
Asymptotes of root loci, **2138**, 2133
Asymptotic stability, **222**, 208, 209, 211, 212, 215, 216
Asynchronous ac transmission, **1241**, 1228
Asynchronous circuits, **1720**, 1711
Asynchronous flip-flops, 1721
Asynchronous operations, 376
Asynchronous sequential logic circuits, 1719, 1720
Asynchronous transfer mode (ATM) adaptation layer (AAL), **1446**, 1444–1445, 1457, 1458
Asynchronous transfer mode (ATM) networks, **1446**, **1458**, 1442–1443, 1444–1445, 2064
  computer, 1448, 1456, 1457
ATE, see Automatic test equipment
ATM, see Asynchronous transfer mode
Atmospheric hydrometeors, 847–848
Atomic displacement, 1178
ATPG, see Automatic test pattern generation

Atrophy, **2328**, 2321
Attachment process, **945**, 936
Attention, 2252–2253
Attenuation, **1001**, 837
  frequency vs., 990
  lightwaves and, **1001**, 984–986
  of microwaves, 847
  rain, 2200
  stopband, 239
  ultrasound and, 1080
Attenuation coefficients, 984, 2375, 2381
Attenuation constants, 838, 841, 858, 1080
Attenuation index, 744
Attenuators, 885
Attitude, 1544, 2285
Attitude determination and control system (ADCS), 1543, 2285
Attraction, 16
Attributes, 251, see also specific types
Attribute sets, **276**, 251, 252–253, 255–260
Audio broadcasting, 2208, see also Audio transmission; Digital audio broadcasting (DAB); Radio
Audio and electroacoustics (A&E), 395–406
  active noise control and, 403–404
  active sound control and, 403–404
  audio coding and, 400–401
  echo cancellation and, 401–403
  hearing aids and, 397–398
  spatial processing and, 398–400
  steerable microphone arrays and, 395–397
Audio Engineering Society, 2550
Audio oscillator, 419
Audio recording, 776, 826, 827, 835
Audio transmission, 405, see also Audio broadcasting
  coding in, 400–401
  compression of, 1401–1403
  satellites and, 2207–2208
Augmentative communication aids, 2391
Aural output power, **1392**, 1382
Autocorrelation, **320**, **1498**, **1509**, 153, 299, 300, 316, 317, 319
  noise and, 1488, 1489–1490, 1493
  stochastic processes and, 1501, 1504, 1505, 1515
Autocovariance function, 1501
Automated guided vehicles (AGVs), 2182
Automated measurement, 2283

Page on which term is defined is indicated in bold.

Automatic Dependent Surveillance
(ADS) systems, 2292
Automatic focusing, **1693**,
1684–1685
Automatic gain control (AGC),
2203
Automatic generation control
(AGC), 1346–1348
Automatic test equipment (ATE),
595
Automatic test pattern generation
(ATPG), 1812, 1819, 1820,
1822
Automatic tracking, **1693**,
1685–1687
Automatic voltage control (AVC),
2231, 2232
Autonomic, **2328**, 2325
Autonomous operation, **1734**, 1727
Autoregressive (AR) models, **297**,
288, 289, 292, 294, 295,
2101
linear predictive analysis and, 302
Autoregressive moving average
(ARMA) models, 302, 2102,
2103
Autoregressive moving average
(ARMA) signal processing,
2390
Autoregressive processes, 316
Autotransformers, **1308**, 1307–1308
Auxiliary memory, **1675**, 1671
Availability, **2030**
in computer systems, **2095**, 2091
fault tolerance and, **2030**, 2020,
2029
of local-area networks, 1465
steady-state, 2030
of VSAT systems, 2200
Avalanche, 127
Avalanche breakdown, **458**, 456
Avalanche breakdown voltage, 128
Avalanche diodes, 457–458,
1430–1431
Avalanche gain, 753
Avalanche injection, **1657**, 1655
Avalanche photodiodes, 753,
1430–1431, 1432, 1435
AVC, see Automatic voltage control
Avionics, 2188–2193
Axes, 2454
Axial slot on cylinder antennas, 870
Axons, **2310**, 2301, 2352

B

Backbone, **1566**, 1562
Backlighting, 1774
Backoff condition, **1544**, 1541

Back-projection, 2376
Back-propagation networks, 2229
Back-scattered energy, 2383
Back-to-back dc systems, 1228
Backward Euler, 667
Backward substitution, 659
Baker, Walter R. G., 1477
Balanced three-phase fault, 1253,
1256, 1259–1260
Balanced voltages, **188**
Ballasts, 2269–2270
Bandgap, 1763, 1764
Bandgap energy, **458**, 450
Bandlimited signals, 1510, 1511,
1513, 1514
Bandpass filters, **124**, 117, 123, 239,
see also specific types
active, 683, 684, 688–689
ideal, 189, 190, 191, 192, 1506
operational amplifiers and, 629
Band-reject filters, 239
Bandstop filters, 189, 190, 683,
684–685, 689–690
Bandwidth, **528**, **1001**, **1516**, **1668**,
see also Frequency; specific
types
disk system architectures and,
1664
memory and, **1936**, 1927
memory devices and, **1668**, 1664
sampling theorems and, **1516**,
1511
3-dB, **195**, 652, 192, 195
transition, 239
in wireless local-area networks,
1565
Bank busy time, 1928
Bardeen, Cooper, and Schrieffer
(BCS) theory, 1114
Bardeen, John, 652
Bar graphs, 1768–1769
Barium strontium titanate, 1131
Barium titanate, 1055, 1056, 1127,
1145, 1181–1182
Barkhousen noise, 1497
Barrier layer, **1001**, 994
Barrier voltage, **458**, 447
Bartlet window, 233
Base, 545
Baseband modulation, 1462
Base change, 2520–2521
Base-collector region, 899
Base-emitter pn junction, 899
Base-index addressing, 1886, 1887
Base spreading resistance, 552
Basic charge distribution function
(BCDF), 1068
Basis function, 1037, 1038–1039
Batch processing, 2061
Batteries, 51, 1586, 1587, 2255

Baum iterative algorithm, 292
Baum–Welch algorithm, **313**, 312
Bayesian detectors, **1486**, 1479
Bayesian estimation, **328**, 292, 323,
324
Bayes' Rule, 2487–2488
Bayes theorem, 2518
BCD, see Binary coded decimal
BCDF, see Basic charge distribution
function
BCH, see Binary-coded hexadecimal
BCL, see Block carry lookahead
BCLAs, see Block carry lookahead
adders
BCM, see Body computer module
BCNF, see Boyce–Codd normal
form
BCO, see Binary-coded octal
BCS, see Bardeen, Cooper, and
Schrieffer
BE, see Backward Euler
Beam energy, 482
Beamformers, **366**, 359, 360–364,
413, 416
Beam index cathode ray tubes,
1782–1783
Beam pulsing, **1392**, 1388
Beam shaping, 1682–1683
Beam splitters, 1691
Beat frequency, 143
Bedside terminals, 2399–2400
Behavior
compensatory, **2253**, 2248,
2250–2254
human, 2247, 2248, 2249,
2250–2254
modeling of, 654
precognitive, **2253**, 2249
pursuit, **2253**
BEM, see Boundary element
method
Benzenes, 1144
Benzoic acid, 767
BER, see Bit error rate
Bernoulli number, 2454–2455
Bernoulli source, 1523
Bernoulli wave equation, 315
Bessel equation, 2482
Bessel formula, 2513
Bessel functions, 1219, 1225,
2482–2483, 2486
Beta-adrenergic receptors, 2338
Beta cutoff frequency, 642
BG, see Bruce–Golde
Bianisotropic media, 805–806
Biased diodes, **138**, 133
Biasing, **328**, 322
amplifiers and, 636
of bipolar transistors, 546–550,
564

Page on which term is defined is indicated in bold.

circuit, **566**
current mirror, 564
JFET, 533–536
of *pn* junction, 750
self-, 534
stability and, 534–536, 548
of transistor at cutoff, 638
voltage source, 533–534
BIBO, see Bounded input–bounded
output
BICFETs, see Bipolar inversion
channel heterojunction
field-effect transistors
Bidirectional filter response, 1070–
1071
Bidirectional transducers, **1075**
BIG, see Bismuth-substituted
yttrium-iron-garnet
Biisotropic media, 806–807
BILBO, see Built-in logic block
observation
Bilinear transform, 243–244
Binary code, 2022
Binary-coded decimal (BCD)
number systems, 1846–1848
Binary-coded hexadecimal (BCH)
number systems, 1846–1848
Binary-coded octal (BCO) number
systems, 1848
Binary digits, 340
Binary functions, 1695–1697
Binary number systems, **1858**, 1843,
1845–1846, see also specific
types
decimal, 1846–1848
hexadecimal, 1846–1848
octal, 1848
signed, 1853–1856
unsigned, 1845–1848
Binary-phase-shift-keyed (BPSK)
signaling, 1407, 1408, 2200,
2203
Binary searchtrees, **1926**, 1923, see
also Binary trees
Binary sequence, 1727
Binary shift registers (BSR), 1600
Binary-to-decimal conversion, 1845
Binary trees, **1926**, 340, 1923
Binary variables, 1695–1697
Binaural attributes, **405**, 398, 399
Binomial distribution, 2488
Binomially distributed variables,
2488
Binomial theorem, 2431
Bioanalytical sensors, **2349**,
2347–2348
Biocomputing, 2397–2407
hospital information systems and,
2400, 2405–2407
medical records and, 2397–2403

patient monitoring and,
2397–2403
smart cards and, 2401–2402
systems integration and, 2400
Bioelectricity, 2301–2341
action potential recording and,
2313–2314
in bone repair, 2329–2334,
2335–2338
electrocardiography and, see
Electrocardiography (ECG)
electroencephalography and, see
Electroencephalography
(EEG)
electromyography and, 2311,
2319–2322
nerve excitation and, 2304–2311
neuroelectricity and, see
Neuroelectricity
origin of, 2312
in soft tissue repair, 2329–2339
devices for, 2330–2334
dosimetry in, 2335–2338
history of, 2329–2330
mechanisms in, 2335–2338
stimulation law and, 2312–2313
Bioelectronics, see Electrocardiography (ECG);
Electroencephalography
(EEG)
Biogas, 1213
Biomass energy, **1215**, 1207,
1213–1214
Biomedical engineering, 1187
Biomedical equipment, 2408–2413,
see also Biomedical sensors;
Medical; specific types
Biomedical sensors, **2349**,
1158–1159, 1161,
2342–2350, see also specific
types
applications of, 2348–2349
bioanalytical, **2349**, 2347–2348
chemical, **2349**, 2345–2347
indwelling, 2343
invasive, 2343
noncontacting, 2343
noninvasive, **2349**, 2343
optical, 2345
physical, **2349**, 2343–2345
potentiometric, **2349**, 2347
Biometric verifiers, **2085**, 2072
Biosensors, see Biomedical sensors
Biot Savart law, **1025**, 1019
Bipolar devices, **131**, see also
specific types
Bipolar diodes, 126, 127, 128
Bipolar inversion channel
heterojunction field-effect
transistors (BICFETs), 1646

Bipolar-junction transistor (BJT)
amplifiers, 538, 639, 640,
641, 644
beta cutoff frequency of, 642
common-base, 648–650
common-collector, 650–651
internal emitter resistance of, 646
monolithic, 643
Bipolar-junction transistors (BJTs),
see also Bipolar transistors
classification of, 1624
CMOS and, 1616
current sources and, 53, 54, 56
dc equivalent circuit models of,
465–467
equivalent circuit models and,
460
large-signal equivalent circuit
models of, 461–462, 463
logic gates and, 1624–1626
microwave devices and, 892, 893,
896, 898, 1004
small-signal equivalent circuit
models of, 464, 465
TTL logic and, 1615
voltage sources and, 53, 54, 56
Bipolar transistors, 545–566, see
also Bipolar-junction
transistors (BJTs); specific
types
biasing of, 546–550, 564
common-collector (CC) circuit
and, 555–557
complete response of, 560
current sources and, 56
design of, 560
emitter-follower and, 555–557
frequency performance of,
553–555
frequency response of, 558–559,
560
heterojunction, 892, 901
high-frequency response of,
558–559
insulated-gate, 737, 699–700, 706,
734
integrated circuits and, 561
logic elements and, 1613
low-frequency performance of,
553–555
microwave devices and,
898–901
noise and, 901
small-signal equivalent circuits
and, 552–553
small-signal operation of,
550–552
voltage sources and, 56
Bipoles, **1241**
Biquads, **682**, 248, 676

B-ISDN, see Broadband-integrated services digital network
Bismuth-substituted yttrium-iron-garnet (BIG), 1167, 1168
Bispectra, **2361**, 2356
Bispectral analysis, 2356
BIST, see Built-in self-test
Bistable devices, **1640**, 1635–1641, 1718, see also Flip-flops; Latches; specific types
BIT, see Built-in tests
Bit error rate (BER), 1407, 1435, 1594, 1602, 1603, 1604
  in mobile radio communications, 1547
  satellites and, 1540
Bit line capacitance, 1652
Bit-parallel processing, 371
Bit-serial processing, **383**, 371, 374, 376, 380, 381, 382
Bit/word slice chips, 1749
BJT, see Bipolar-junction transistors
Blackbodies, 747, 748, 749
Black-box testing, 1990
Blackman–Harris window, 234
Blackman window, 234
Blanking, **1392**, 1380, 1384
Blind via, **614**
Block carry lookahead adders (BCLAs), **1747**, 1745
Block carry lookahead (BCL) logic, 1745
Block codes, **1409**, 1405, 1406
Block matching, 353–355
Block truncation coding, 338
Blood-gas monitoring, 2372
Blood pressure sensors, 2343
B-mode displays, **2385**, 2383
Bode diagrams, **206**, **2130**
  design of, 2124–2126
  frequency response and, 199, 200–205, 2113, 2114–2115
  composite equalizers and, 2122–2126
  design and, 2124–2126
  design-series equations and, 2115–2122
  phase-lag compensators and, 2141
  phase-lead compensators and, 2116–2119, 2142
  PID compensators and, 2143
  in stability analysis, 214
Boella effect, 7
Boilers, **1205**, 1194–1197
Boiling-water reactors (BWRs), 1201
Boltzmann constant
  bioelectricity and, 2336
  bipolar transistors and, 546

dielectrics and, 1135
distributed power generation and, 1209
operational amplifiers and, 621
optoelectronics and, 758
photovoltaics and, 1209
pyroelectrics and, 1128
satellites and, 1537
semiconductors and, 436, 440, 470
superconductivity and, 1114
Boltzmann relation, **458**, 447
Boltzmann transport equation (BTE), 440, 471–472, 1107
Boltzmann voltage, 640
Bonding, 2223–2237
  control systems for, 2223–2224, 2231–2235
  parameters for, 2224
  sensors in, 2225–2229
Bone repair, 2329–2334, 2335–2338
Boning–Garton effect, 1137
Boolean algebra, **1711**
Boolean connectives, 1707–1710
Boolean expressions, **1711**, 1699, 1703–1707, 1892, see also specific types
Boolean functions, **1711**
Boolean variables, 1938
Boost converters, 709
Booth recoding algorithm, 1863
Bottlenecks, 1451
Boundary conditions, 470–471
Boundary element method (BEM), 1023
Boundary scan, **1836**, 1826–1828
Bounded function, 320
Bounded input–bounded output (BIBO) stability, **168**, **222**, 167, 216–218
Boyce–Codd normal form (BCNF), 2035, 2037
Boyle model, **623**, 622
BPSK, see Binary-phase-shift-keyed
Bragg cells, **1171**, 1169–1170
Bragg diffraction, 1169
Brain language, 2351–2353
Braking, **737**, 730, 732, 2255
Branch currents, **63**, 65
Branch relationships, **671**, 655
Branch voltage, **63**
Brattain, Walter H., 652
Breakaway points of root loci, **2138**, 2136
Breakdown strength, **38**, 34
Breakpoints, **206**, 201
Brickwalling of software, **2193**, 2191
Bridge rectifiers, 129, 130
Bridging faults, 1810

Brightness, 746–747
  photometric, see Luminance
Brillouin scattering, 1425
Brillouin zone boundary, 1110
Broadband amplifiers, 1005, 1007
Broadband communications, **1446**, 1441
Broadband emissions, **934**
Broadband-integrated services digital network (B-ISDN), **1458**, 1441–1447, 1448, 1457
Broadband signal processing, 1083
Broadcasting, 1359–1404, see also specific types
  audio, see Audio broadcasting
  channel classifications in, 1368–1369
  demodulation in, 1359–1365, 1386
  digital audio, see Digital audio broadcasting (DAB)
  field strength in, 1369–1370
  frequency allocations in, 1368, 1373, 1383
  frequency-shift keying and, 1364
  interference in, 1398
  M-ary phase shift keying and, 1365
  modulation in, **1367**, 1359–1365
  phase amplitude modulation and, 1363
  pulse-code modulation and, 1363–1364
  quadrature amplitude modulation and, 1365, 1366
  radio, see Radio
  station classifications in, 1368–1369, 1373–1375
  superheterodyne technique in, 1360–1362
  television, see Television
  video, see Video broadcasting
Broadside arrays, 866–867
Brown, Charles E. L., 1309
Brown–Boveri, beginnings of, 1309
Bruce–Golde (BG) model, 941
Bruton transformation, 681
BS, see Beam splitters
B-site, **1096**, 1092
BSR, see Binary shift registers
BTE, see Boltzmann transport equation
Bubble-jet printers, 1968
Buck-boost converters, 710, 713
Buck converters, 708–709
Buffer stage, 556
Building orientation, 2267
Built-in logic block observation (BILBO), 1832, 1833

Page on which term is defined is indicated in bold.

Built-in self-test (BIST), **498**, **1815**, **1836**, 490–491, 497, 1808, 1830–1832
  output response analysis and, 1812, 1813
  test pattern generation and, 1811
Built-in-tests (BIT), 953
Bulk acoustic waves (BAW), 1070
Bulk modes, 1083
Buried-channel MOSFET, 567
Buried via, **614**
Burst noise, 1496–1497
Bus, **1544**, **1869**, 1868
  carrier-sense, 2015–2016
  fault level in MVA at, 1267
  in local-area networks, 1461
  structure of, 1868–1869
  time-slotted, 1464
  token, **1459**, 1455, 1463, 2016–2017
Bus load forecasting, 1351
Butterworth approximation, 679
Butterworth denominator polynomials, 114
Butterworth filters, 117, 122, 123, 124, see also specific types
  low-pass, 114, 115
  transfer function of, 192–193
Butterworth function, 113, 114, 115, 117, 121
BWRs, see Boiling-water reactors
Bypass capacitors, 553, 558
Bypassing on converters, 780–781
Bypass switches, 1463
Bytes, 1846, 1882, 1885

C

C3 system, see Command, control, and communications
Cable, 503
  coaxial, see Coaxial cable
  fiber-optic, see Fiber-optic cable
  grounding of, 905, 906
  oil-filled, 1223
  parameters for, 1223–1226
  pipe-type, 1223, 1224
  self-contained oil-filled, 1223
  shielding of, 913
  single-core, 1224
  standards for, 1227
  underground, 1223–1227
  in wireless local-area networks, 1559–1561
Cable shield grounding, 905
Cable television, 1418
Cable theory, 2313
Cache memory, **1936**, 1929, 1932, 1933

Cache references, 1894
Caches, 1668
CACS, see Computer-aided circuit simulation
CAD, see Charge area development; Computer-aided design
Cadmium telluride, 1129
CAE, see Computer-aided engineering
CAFS database, 2047
Cage-rotor induction motors (CRIM), 2238, 2241, 2243
Calculus
  differential, 2461–2466
  of finite differences, 2510
  integral, 2466–2472
Calibration, 771, 1058, 1587
Call blocks, 1889
Calling Card Service, 1472
Calling conventions, 1889–1892
Caltech Intermediate Form (CIF), 581
Camcorders, 757–759
Candlepower distribution, **2275**, 2261
Canonical equations, 1058–1059
Canonical form, 2105
Capacitance, **30**, 16, 17, 18
  bit line, 1652
  depletion, 455–456
  diffusion, **458**, 454–455
  gate, 574
  inductor, 28, 29
  input, 1128
  interconnection, 575
  junction, **458**, 454
  mutual, 513
  net, 641
  parasitic, 1492
  pyroelectrics and, 1128
  word line, 1652
Capacitance matrix, 1219–1220
Capacitive coupling, 1945
Capacitive crosstalk, 518
Capacitively coupled fields, **2338**, 2331
Capacitive noise, 522
Capacitive overlay technology, 1948
Capacitor banks, **1250**, **1319**, 1312
Capacitors, **30**, 15–18, 19–24, see also specific types
  air, **30**
  aluminum electrolytic, 21, 22
  amplifiers coupled with, 634
  bypass, 553, 558
  ceramic, 19
  common-emitter bypass, 558
  coupling, 550

dielectric absorption and, 18
dielectric constants for, 17, 19
dielectrics of, **30**, 1144
disk, **30**
electrolytic, 31, 20–24
in energy distribution, 1318–1319
film, 19
foil, 20, 23
full-wave, 23
impedance of, 20
mica, 19
paper-foil-filled, 20
parallel, 16, 17
passive signal processing and, 111
polarized, **32**
power factor and, 18
quality factor in, 18
serial, 17, 1243–1244
shunt, 1244–1247
solid-electrolyte sintered-anode tantalum, 23
tantalum, 21, 23, 24
thin-film, 1146
thyristor-switched shunt, 1247
types of, 19–24
wet-electrolyte sintered-anode (wet-slug) tantalum, 23, 24
Capture range, **1574**
Capture registers, **1762**, 1755
Carbon microphones, 1057
Cardinal series, **1516**, 1510–1511, 1512, 1514, 1515
Cardioid receivers, 1060
Cardiopulmonary resuscitation (CPR), 2317
Carlson, Chester, 1309
Carriers, see also specific types
  chip, **614**, 607–608
  drift of, 447
  interference ratio to, 1424
  leaded chip, **614**, 607
  leadless chip, **614**, 607
  lifetime of, **458**, 456
  majority, **458**, 447
  minority, 456, 1763, 1767
  multiple channel per, 2207
  noise ratio to, see Carrier-to-noise ratio (CNR)
  sine wave, 1504
  single channel per, 2206, 2207
  temperature of, 469
Carrier-sense buses, 2015–2016
Carrier-sense multiple-access (CSMA), **1458**, 1455, 1463
Carrier sense multiple-access (CSMA) with collision detection (CSMA/CD), 2015
Carrier-to-interference ratio (CIR), 1424

Page on which term is defined is indicated in bold.

Carrier-to-noise ratio (CNR), 1418, 1419, 1420, 1424
  digital audio broadcasting and, 1400
  magnetooptics and, 1171
  satellites and, 1537, 1540–1541
Carry-completion adders, 1743
Carry-free operation, 382
Carry lookahead adders (CLAs), 1743, 1744
Carry lookahead logic, 1744
Carry ripple operation, 382
Carry-save operations, 382
Carry-select adders, 1743
Cartesian configuration in robotics, 2155–2156
Cartilage formation and repair, 2330–2334, 2338
Cascaded second-order sections, 676–680
Cascade filters, 685
Cascade realizations, 248–249
CASE, see Computer-aided software engineering
Cassegrain antennas, 875, 876
Castellations, **614**
Castor oil, 1145
Catalysis, 2338
Catastrophic thermal failure, **796**, 784
Catheters, 2344, 2345
Cathode ray tubes (CRTs), **1785**, 1778–1786, 1866, 1941, 1942, see also Video display terminals (VDTs)
  beam index, 1782–1783
  brightness of, 1783, 1784
  color, 1782–1784
  computer graphics and, 2004, 2006–2007
  contrast in, 1783
  digital image processing and, 330, 332, 334
  dynamic RAM and, 1652
  electromagnetic deflection and, 1780–1782
  electron gun in, 1779
  electrostatic deflection and, 1780, 1781
  measurements on, 1783–1784
  monochrome, 1778–1782
  penetration, 1783
  projection screen, 1784–1785
  screens for, 1782
  standard, 2006–2007
  in television, 1380, 1390, 1391
Cathodes, 30, 754, 1770, 1779–1780, see also Cathode ray tubes (CRTs)

CAT scans, see Computerized tomography (CT)
CATV, see Community antenna television
Cauchy–Schwarz inequality, 1 500
Cauchy's form, 2464
Cauer function (elliptic function) filters, 679
Causal filters, **195**, 191–192, 239
Cavity-based devices, 1641, 1642–1643, see also specific types
Cavity ratio (CR), **2275**, 2259
CB, see Circuit breakers; Common-base
CBR, see Computer-based records
CC, see Common-collector
CCCS, see Current-controlled current sources
CCD(s), see Charge-coupled devices
CCIS, see Common-channel interoffice signaling
CCM, see Continuous current mode
C compiler, **393**, 390
CCR, see Ceiling cavity ratio
CCS, see Common-channel signaling
CCS7 Network, 1470–1471, 1473
CCVS, see Current-controlled voltage sources
CD, see Circular dichroism; Common-drain; Compact disks
CDMA, see Code division multiple access
CDR, see Critical design review
CDV, see Compressed digital video
CE, see Common-emitter
Ceiling cavity luminance coefficient, 2263
Ceiling cavity ratio (CCR), 2259
Celestial navigation systems, 2292–2293
CELL, see Surface-emitting laser logic
Cells, **1446**, see also specific types
  Bragg, **1171**, 1169–1170
  contractile, 2312
  fuel, **1215**, 1212
  libraries of, **601**
  magnetooptic, **1171**, 1169–1170
  in networks, 1442
  semiconductor storage, 1866
  solar, 445, 447, 458, 753, 1156, 1208, 2255, see also Photovoltaics
  splitting of, **1553**, 1549

standard, **602**, 591, 593
  structured, **602**, 599
Cellular communications, **1539**, **1566**, 279, 1062, 1535, 1546–1553, 1557, see also Mobile radio; Wireless communications; specific types
Cellular manufacturing, **2182**, 2175
Cellular membranes, 2302, 2303
Cellular radios, see Mobile radio
Cellulose, 1149, 1150
CELP, see Code-excited linear prediction
CEM, see Computational electromagnetics
Central limit theorem, 1490
Central processing unit (CPU), **1869**, 1865, 1875, 1879, 1902
Centripetal forces, **2174**, 2169
Centroids, 2470–2472
Cepstral (homomorphic) analysis, 303, 307–308
Cepstrum, **305**, 303, 304, 308
Ceramic capacitors, 19
Ceramics, 1055, 1127, 1131, 1180, 2532, 2533, see also specific types
CET, see Constant elapsed time
CFR, see Confirmation to receiver
CG, see Common-gate
Channel capacity theorem, 1523
Channel decoders, 1401
Channel encoders, **1403**, 1400
Channel length modulation, 538
Channels, **579**, 567, 568, 1601–1602
Character arrays, 1881
Characteristic impedance, **528**, **891**, **1085**, 504–513, 1081, 2381
  manufacturing tolerances and, 510–513
  printed circuit board structure and, 505–510
Charge amplification, 752
Charge area development (CAD), 1960
Charge carriers, **902**
Charge-coupled device (CCD) arrays, 1953
Charge-coupled devices (CCDs), **759**, 330, 392, 756, 1726, see also specific types
Charges, 813–814
Charge-transport layer (CTL), 1960
Chebyshev approximation, 679
Chebyshev denominator polynomials, 116
Chebyshev filters, 118, 119, 120, 121, 122

Page on which term is defined is indicated in bold.

active, 678
  frequency curves for, 194
  transfer function of, 193
Chebyshev functions, 115–117
Chebyshev transform, 1599
Chemical engineering, 1187
Chemical quantities, 2425–2429
Chemical sensors, **2349**, 1157–1158,
  2345–2347
Chemical transduction, 1153
Chemical vapor deposition (CVD),
  488, 1001, 486–487, 993,
  994–995
Chemoreceptors, 2301
Chip carriers, **614,** 607–608
Chip resistors, 610
Chips, **601,** see also Integrated
  circuits
  analog interface, 386
  for ASICs, 596
  bit/word slice, 1749
  in digital signal processing, 1760,
    1761
  for microprocessors, 1751–1752
  semiconductor, 1866
  temperature of, 789
  in thermal management, 786
  thermal resistance in, 789–796
  for VLSI, 385–393
  word-slice, 1749
Chiral media, 806
Chirp, **1425,** 1420
Chlorine, 478
Cholesteric crystals, **1777,** 1775
Chondrogenesis, **2338**
CHP, see Combined heat and
  power
Christoffel equations, 1078
Chromatography, 1158
Chrominance, 1384
Chronaxie, **2310,** 2307, 2308
Chronometers, 2292
CIF, see Caltech Intermediate Form
CIR, see Carrier-to-interference
  ratio
Circles, 219
Circuit breakers, 1240–1241, 1269,
  1270, 1312, 1314, see also
  Protection
Circuits, **1113,** see also Filters;
  specific types
  absolutely stable, 167
  in ac generators, 1323–1326
  active load, 539
  Ampère's law of, 804, 812
  amplifier, 635
  analysis of, 56–57
  armature, **1331,** 1321
  asynchronous, **1720,** 1711
  biasing, **566**

branches of, **63,** 58, 59
clipping, see Limiters
Cockroft–Walton, 130, 131
combinational, 1613, 1818, 1825
combinational logic, 1711–1718
combinatorial, see Combinational
  circuits
common-collector, **566,** 555–557,
  899
common-emitter, 548
control, 1011–1013
converter, 1233–1234
coupled tuned, 1084
data acquisition systems and,
  1802
dedicated, 371
diode-resistor, 452
distribution, 1344
double-rank, 1720
equivalent, see Equivalent circuits
field, **1331, 1321**
in generators, 1323–1326,
  1330–1331
graph theory and, **1113**
grounding of, 906
high-Q resonant, 1084
hybrid, 887–888
hybrid-pi equivalent, **651,**
  640–642, 643
infinite gain high-pass, 687
integrated, see Integrated circuits
isolated, **728**
Laplace transform and,
  161–162
limiting, 132–135, 136, 137
logic, see Logic circuits
modeling of, 51, 54, 60,
  1254–1255
negative feedback, 647
noninverting, 627–631
nonlinear, see Nonlinear circuits
operational amplifiers and, 622
in optoelectronics, see Optoelectronics
output, 1626
parallel clipping, 134
parallel resonant, 1084
passive, 122–124
peripheral, 1748
precision rectifying, 136–138
pulse-mode, 1720
realizations of, 685–690
reflection-type, 1005
sequential, 1613, 1711
series resonant, 1084
short, see Short circuits
simulation of, see Computer-
  aided circuit simulation
  (CACS)
small-signal equivalent, 552–553

solid state, see Solid state circuits
stability of, 167–168
subtransmission, 1344
in surface mount technology,
  606–611
synchronous, 1711
theory of, 53
three-phase, 88–92
topology of, 57
totem-pole output, 1626
transformed, **168,** 162–164
unbalanced, 90
unconditionally stable, 167
unstable, 167, 168
Circuit-set (loop-set), **109,** 93, see
  also Edges
Circuit switching, **1459,** 1437, 1438,
  1449
Circuit under test (CUT), 1817
Circular convolution, 247
Circular dichroism (CD), 1163
Circular waveguides, 854–855,
  858
Circulators, **1171,** 889–890,
  1166–1169
CIRF, see Cochannel interference
  reduction factor
CIS, see Clinical information
  systems
CISCs, see Complex instruction set
  computers
CLA, see Carry lookahead adders
Cladding, **1001,** 988, 994
Clamping, 133
Clark electrodes, 2345
Classes, **1913, 1914,** 1909, 1911
Classical mechanics, 2425
Classifiers, **297**
Clausius–Mossotti equation, 1136
Clinical information systems (CIS),
  2398–2399, 2400
Clipping, 134, 137, 1424
Clipping circuits, see Limiters
Clock input, 1640
Clocks, 1062, 1574, 1827
Closed convex sets, **276,** 251,
  258–260
Closed-loop control systems, 215,
  2137, 2150, 2163
Closed-loop PLLs, 1570, 1571
Closed-loop transfer function, 1567,
  1571, 2118, 2126–2127,
  2137
Closed projection operators,
  260–262
Closed subspaces, 255–258
Clustering, 307, 338, 425
Clutter, 949, 959
CLV, see Constant linear velocity
CM, see Cross modulation

CMG, see Computer Measurement
  Group
CMIP, see Common Management
  Interface Protocol
CML, see Current-mode logic
CMOS, see Complementary MOS
CMRR, see Common-mode
  rejection ratio
CMT, see Cadmium telluride
CMUs, see Communication
  management units
C/N, see Carrier-to-noise ratio
CNR, see Carrier-to-noise ratio
Coarticulation, 306
Coates formula, 101
Coates graphs, **109**, 100, 101, 102,
  103
Coaxial cable, **1566**, **2018**, 2015
  in B-ISDN signal transmission,
    1443
  grounding of, 906
  in local-area networks, 1461,
    1869
  magnetism and, 812
  in wireless local-area networks,
    1559
Cobalt, 819, 822, 823
Cochannel interference reduction
  factor (CIRF), **1553**, 1548
Cochlear implants, 2392, 2393
Cockroft–Walton circuit, 130, 131
Codd's relational algebra, 2037
Code, **1526**, 1518
Code division multiple access
  (CDMA), **1553**, 1462, 1542,
  2204
  in mobile radio, 1548, 1549,
    1552–1553
Coded surface acoustic wave filters,
  1073–1074
Code-excited linear prediction
  (CELP), 285, 286
Coding theory, 2389
Coefficient of thermal expansion
  (CTE) mismatch, **614**
Coefficient of utilization (CU),
  2275, 2261, 2263
Coefficient of variation, 2487
Coefficient quantization error, 249
Coercive fields, **1096**, **1131**, 1093,
  1127
Coercivity, **826**, **835**, 822, 823, 831,
  832
  magnetic, 1687
  in magneto-optical disk data
    storage, 1687
  maximum, 824
  nucleation-controlled, 822
  remanent, **835**, 824
Cofactors, 2437–2438

Coherence, **2362**, 416, 1507–1508,
  2355, 2356
Coherence lengths, 1114, 1120
Coherent detection, **1434**, 1480
Coherent light, **742**, 738, 740, 741,
  745
Coherent processing interval (CPI),
  953, 955, 957, 959, 963
Coherent pulse radar, **963**, 955
Coherent rotation, 824
Coil inductance, 26
Coils, **30**
Coincidence detectors, 2377
Collectors, 545, 640
Collision integrals, 440
Color cathode ray tubes, 1782–1784
Color coding of resistors, 10–11
Color computer graphics, 2006
Color flow imaging, **2385**
Color-notch filters, 1388
Color plasma displays, 1797–1798
Color preference index (CPI), **2275**,
  2265
Color rendering index (CRI), **2265**
Color signal decoding, 1386
Color signal encoding, 1384–1386
Color switches, 1775
Color video displays, 1390,
  1782–1784
Combinational circuits, 1613, 1818,
  1825
Combinational lock, **1711**
Combinational logic circuits, **1720**,
  1711–1718, 1741
Combinational networks,
  1651–1667
  binary variables and, 1695–1697
  Boolean connectives and,
    1707–1710
  Boolean expressions and, **1711**,
    1699, 1703–1707
Combinations, 2433
Combinatorial circuits, see
  Combinational circuits
Combined heat and power (CHP)
  units, 1212
Command, control, and communications (C3), **2221**,
  2211–2222
  architecture of, 2212, 2213, 2215
  background on, 2211–2214
  decisionmakers and, 2211,
    2218–2221
  dynamics of encounters and,
    2216–2218
  interoperability of, 2215
  standards for, 2215
  technologies of, 2214–2216
Common-base (CB) amplifiers, 634,
  648–650

Common-base (CB) configuration,
  899
Common-base (CB) current gain,
  900
Common-channel interoffice
  signaling (CCIS), **1459**,
  1450, 1470
Common-channel signaling (CCS),
  1475, 1470
Common-collector (CC) amplifiers,
  650–651
Common-collector (CC) circuits,
  566, 555–557, 899
Common-drain (CD) JFET, 536
Common-emitter (CE) amplifiers,
  634, 644–647
Common-emitter (CE) bypass
  capacitors, 558
Common-emitter (CE) current
  gain, 642
Common-emitters (CE), **566**, 548
  current gain in, 546
  current sources and, 54
  degenerate, **566**, 561–562
  microwave devices and, 899
  voltage sources and, 54
Common-gate (CG) JFET, 536
Common Management Interface
  Protocol (CMIP), 2203
Common-mode input impedance,
  620
Common-mode input signals, 562
Common-mode rejection, 564
Common-mode rejection ratio
  (CMRR), **1807**, 617, 619,
  1803, 2322
Common-source (CS) amplifiers,
  536, 538–539, 541–542
Communication aids, 2391, see also
  specific types
Communication link, **1607**, 1594
Communication management units
  (CMUs), 2192
Communications disorders, 2388,
  2391
Communications systems, 769, see
  also specific types
  in avionics, 2188, 2191–2192
  broadband, **1446**, 1441
  cellular, see Cellular communica-
    tions
  command, control and, see
    Command, control, and
    communications (C3)
  computer-aided design of, **1607**,
    1593–1609
  channels and, 1601–1602
  limitations of, 1595–1596
  low-pass models and,
    1598–1599

Page on which term is defined is indicated in bold.

pseudorandom generators and, 1599–1601
receivers and, 1601–1602
role of, 1594–1595
structure of, 1596–1597
transmitters and, 1601–1602
computer network, see Computer communication networks
connection-mode, 1453
data acquisition systems and, 1804–1806
digital, see Digital communication
fiber-optic cable in, see Fiber-optic cable
instrument modeling and, 2278
interprocess, 2070, 2067
lightwaves in, 989
long distance, 1427–1434
networks of, 1594
optical, see Optical communication
optical fibers in, see Optical fibers
personal, see Personal communications systems (PCS)
protocols in, 1981
satellite, see Satellites
space, 2214
telephone, see Telecommunications
wireless, see Wireless communications
Community antenna television (CATV), 1462
Commutation, 710, 1241, 1331, 702, 1330
forced-, 737, 733
inverters and, 733
self-, 737, 734
Commutation (overlap) angle, 1241
Compact disks (CDs), 1693, 400, 1676, 1679
Companding, 778–779
Compandors, 348
Companion matrix, 176, 175
Compatibility
electromagnetic, see Electromagnetic compatibility (EMC)
electromagnetic spectrum and, 919–921
of large drives, 2238–2240
laws of, 2101
in local-area networks, 1563
measurement of, 932–933
specifications for, 921–931
in European Community, 928–931
in military, 925–927
in U.S., 921–927

in wireless local-area networks, 1563
Compensation, 2147, 1242–1252, 2139–2147
design and, 2140–2143
feedforward, 2167
modern control design and, 2143–2147
phase-lag, 2119–2122, 2140, 2141
phase-lead, 2116–2119, 2140, 2142
proportional-plus-integral-plus-derivative, 2142
shunt capacitors and, 1244–1246
shunt reactors and, 1246
static VAR, 1246–1249
synchronous, 1244
temperature in, 826
theorems on, 78
Compensation transfer function, 2127
Compensators, see Compensation
Compensatory behavior, 2253, 2248, 2250–2254
Compensatory display, 2253
Compilers, 1901, 1914, 1878, 1904, see also specific types
C, 393, 390
high-level language, 1880, 1894
operations of, 1894–1898
optimization of, 1894–1900
silicon, 602, 598
in SPARC computers, 413, 1894, 1895
variable allocation of, 1894–1896
Complemental variables, 1703–1704
Complementary error function, 846
Complementary MOS (CMOS) technology, 601, 568, 692–693, 1613, 1614, 1616–1617
advanced, 1634
ASICs and, 591, 593
design of, 1633–1634
displays and, 1772
inverter design in, 583
logic circuits and, 1712
logic gates and, 1622, 1631–1634
memory devices and, 1651, 1653, 1654
microprocessors and, 1748
in networks, 1443
testing of, 1809, 1810, 1811
thermal management and, 786
Complementary symmetry diode compensated (CSDC) amplifiers, 638, 639
Complement notation, 1847
Complement representation, 1858, 1853–1856, 1860–1861

Complete set of state variables, 176, 171
Complex-conjugate transfer function, 674
Complex depth of penetration, 1219, 1225
Complex domain, 61–62, 63
Complex instruction set computers (CISCs), 1901, 1749, 1880, 1882, 1889, 1899
call conventions in, 1891
transactional paradigms in, 1894
Complex numbers, 2432
Complex permittivity, 1133, 1136
Complex power, 82, 84–85
Complex relative permittivity, 840, 844
Complex variables, 181
Compliant motion control in robotics, 2174, 2173
Component mounting site, 614
Composite equalizers, 2122–2126
Composite second-order (CSO) distortion, 1425, 1422
Composite source model (CSM), 289, 292, 293
Composite triple beat (CTB), 1425, 1422
Composite video, 1392, 1382, 1383–1384
Composition resistors, 12
Compressed digital video (CDV), 1419
Compression, 343, 358, 141
audio, 1401–1403
data, see Data compression
image, 1526
pulse, 963, 953, 958–959
speech, 1525
video, 358, 345, 349, 1526, 2205
Compression ratios, 1401
Compressive strain, 1105
Computational complexity, 2074
Computational electromagnetics (CEM), 1047, 1028–1049
analytical issues in, 1031–1034
approximation in, 1031, 1032, 1035
background on, 1029–1031
basis function and, 1037, 1038–1039
classification of model types in, 1030
computer time in, 1044–1046
differential equations in, 1031, 1033, 1035, 1042–1043
error analysis in, 1046–1047
error checking in, 1046–1047
geometrical optics model and, 1034

Page on which term is defined is indicated in bold.

goal of, 1029
integral equations in, see Integral
    equations
method of moments in, **1048**,
    1031, 1036–1040
modal-expansion model in,
    1033–1034
numerical issues in,
    1035–1040
practical considerations in,
    1041–1044
sampling in, **1048**, 1035–1036,
    1043–1044
validation in, 1046–1047
Computational energy, 423
Computed tomography, see
    Computerized tomography
    (CT)
Computed torque, 2170
Computer-aided circuit simulation
    (CACS), **671**, 653–672
circuit equations and, 654
differential equation solving in,
    666–670
equation formulation and,
    655–657
hierarchy of, 654
Kirchhoff's laws and, 63
for large circuits, 670–671
linear equation solving in,
    657–661
modified nodal approach to, **672**,
    656–657
MOS circuits and, 671
Newton–Raphson procedure in,
    661, 662, 663–666
nonlinear equation solving in,
    661–663
Tableau approach to,
    655–656
Computer-aided design (CAD),
    671, **1607**, 371, 382, 581,
    653, see also Computer-
    aided circuit simulation;
    Computer graphics
architecture and, 1869
of communications systems,
    1593–1609, see also
    Communications systems
computer graphics and, 2004,
    2005, 2007, see also
    Computer graphics
Kirchhoff's laws and, 63
magnetic tape and, 1671
memory devices and, 1671
microwave, 1004
programmable arrays and, 1740
in robotics, 2160, 2177
Computer-aided engineering (CAE),
    601, 598

Computer-aided software engineer-
    ing (CASE), **1984**, 1977,
    1983–1984, 1991
Computer architecture, see
    Architecture
Computer arithmetic, 1858–1865
addition in, 1862
algorithms in, **1864**, 1858,
    1861–1864
division in, 1863–1864, 1866
multiplication in, 1862–1863,
    1866
number representation in,
    1859–1861
subtraction in, 1862
Computer-based instruments, 2281,
    see also specific types
Computer-based records (CBR),
    2397–2398, 2400
Computer clocks, 1062, 1574
Computer communication
    networks, **1459**, 1447–1460,
    see also Computer
    networks; specific types
architecture of, **1458**, 1447–1448,
    1451–1455
design of, 1451
general concepts of, 1448–1451
intelligent, see Intelligent
    networks
local-area, see Local-area
    networks (LANs)
recent developments in,
    1456–1458
Computer compilers, see Compilers
Computer disks, see Disk drives;
    Disk system architectures
Computer errors, 2088
Computer facsimile, 1556
Computer faults, 2088
Computer graphics, **2013**, 1869,
    1943–1945, 2004–2014, see
    also Computer-aided design
    (CAD)
cathode ray tubes and, 2004,
    2006–2007
color, 2006
displays and, 2004, 2006–2007
hardware for, 2005–2007
interaction and, 2012
resolution in, 2005
software for, 2007–2012
Computer hackers, **2085**, 2072
Computer hardware, see Hardware
Computer input-output devices, see
    Input-output devices
Computer instruction register,
    1867
Computer instructions, 1865,
    1866–1867

Computerized axial tomography,
    see Computerized
    tomography (CT)
Computerized tomography (CT),
    2356, 2360, 2374–2376,
    2400
emission, 2376
single photon emission, **2380**,
    2374, 2376, 2377
Computer keyboards, **1957**, 1865,
    1938–1941, 2012
Computer languages, see also
    Computer programming;
    specific types
assembler, 1905, 1906
assembly, see Assembly
    language
data definition, 2032
data manipulation, see Data
    manipulation languages
    (DMLs)
dedicated simulation,
    1608, 1597
design of, 2005
fourth-generation, 2007
hardware description, 598–599,
    2008
higher-order, see High-level
    languages (HLLs)
high-level, see High-level
    languages (HLLs)
human-crafted, 1902
interpreted, 1905
machine, 1878
macro assembler, 1905
page description, 2005, 2012
prototyping, 1968
register transfer, 1727–1728,
    1729
simulation, **1608**, 1597
structured query, 2037, 2038,
    2039, 2044
Computer magnetic tape, see
    Magnetic tape
Computer Measurement Group,
    2550
Computer memory, see Memory
Computer microprocessors, see
    Microprocessors
Computer networks, **2018**, 1434,
    1435–1437, 2015–2018, see
    also specific types
communications, see Computer
    communication networks
security of, 2072, 2081–2084
Computer operating systems, see
    Operating systems
Computer output, see Input-output
    devices
Computer printers, see Printers

Computer programming,
    1878–1926, see also
    Computer languages;
    Software; specific types
  abstract data types in, 1916
  addressing and, 1886–1888
  application, 2067
  assembly language for, see
    Assembly language
  calling conventions in,
    1889–1892
  compiler optimization and,
    1894–1900
  complexity of, 1915
  data type constructors in,
    1917–1919
  dynamic data types in,
    1919–1924
  fundamental data types in,
    1916
  high-level languages in, see
    High-level languages
    (HLLs)
  languages in, see Computer
    languages
  methodology in, 1996–2003
  micro-, see Microprogramming
  N self-checking, 2025
  object, **1915**
  object-oriented, see Object-
    oriented programming
    (OOP)
  paradigms in, 1906–1912
  self-checking, 2025
  source, **1915**, 1904
  structured, 1909
  transactional paradigms and,
    1892–1894
  translating, 1904
Computer reliability, **2095**,
    2087–2095
  calculation of, 2091–2092
  errors and, 2087–2088
  failure and, 2087–2090
  fault and, 2087–2088
  Markov modeling and,
    2092–2093
  for real systems, 2094–2095
  software and, 2093–2094
Computers, **1869**, **2153**, 1759–1760,
    1865, see also Biocom-
    puting; Microprocessors;
    specific aspects of comput-
    ers; specific parts; specific
    types
  complex instruction set, see
    Complex instruction set
    computers (CISCs)
  failure in, **2095**, 2087–2090
  first, 2019

  hardware for, see Hardware
  input-output devices for, see
    Input-output devices
  interconnections in, 1460
  languages of, see Computer
    languages
  operational concepts of,
    1866–1867
  programming for, see Computer
    programming
  radar and, 953
  reduced instruction set, **1901**,
    1882, 1888, 1892, 1938,
    1939
  relaying with, **1278**, 1276–1278
  reliability of, see Computer
    reliability
  security for, see Computer
    security
  software for, see Software
  SPARC, see SPARC computers
  super-, 1866, 1931
  universal, 1525
  in vehicles, 2255
Computer scanners, **1957**, 1953,
    1954
Computer security, 2072–2086
  C3 system and, 2215
  cryptology in, 2074–2077
  hardware, 2080–2081
  network, 2081–2084
  personnel, 2084–2085
  physical, 2074
  software, 2077–2080
Computer simulation, see
    Simulation
Computer software, see Software
Computer tape drives, 826–827, 830
Computer viruses, **2086**, 2072, 2079
Computer vision, 343
Computer words, 1859, 1866, 1882
Computer worms, **2086**, 2079
Concentration gradients, **458**, 448
Concurrency control mechanisms,
    2032
Concurrent checking (on-line
    testing), **1816**, 1810
Condensation (vapor-phase) energy,
    613
Condenser microphones, 1056
Condensers, 1199, see also specific
    types
Conditional expected penalty, 323
Conditional factorization, 1502
Conditional probability, 2487–2488,
    2517
Conditional-sum adders, 1743
Condition codes, **1901**, 1888
Conductance, 1067
  drain, 576

  drain-to-source, 543
  dynamic drain-to-source, 538
  frequency-dependent, 1068
  matrix of, 655
  nonlinear, 55
  transfer, 536–537
Conducted emissions, **934**, 933
Conducting channel, 530
Conduction, 787, 921, 1179
Conduction bands, **446**, **1110**, 436,
    1106, 1763, 2535
Conductive heat transfer, **796**
Conductive resistance, 787, 789
Conductivity, **1150**, 1018
  of crystals, 892
  dielectrics and, **1150**, 1132, 1133,
    1135
  infinite, 1296
  perfect, 1115, 1122, 1123
  of superconductors, 1118
  thermal, 790, 1130
Conductivity modulation, 640
Conductors, **30**, 16, 751–752,
    1218–1219, 1959–1962
Confidence interval, 1602
Configurational freezing, 825
Confirmation to receiver (CFR)
    signals, 1555
Conical horn antennas, 870
Connectedness of graphs, **109**
Connectionless service, **1459**, 1450,
    1463
Connection-mode communications,
    1453
Connection-oriented service, **1459**,
    1450, 1463
Connection weight, **428**, 420, 421
Connectors, 503, 514, 522–524,
    1430, see also specific types
Conservation of energy, 81, 442
Conservation of momentum, 442,
    1764
Consistent estimators, **328**, 322
Constant-current source, 539
Constant elapsed time (CET), 1755
Constant linear velocity (CLV),
    1659
Constant-luminance principle, 1386
Constant voltage, 133
Constitutive parameters, 807
Constitutive relations, 56
Constraining core, **614**
Constructive methods in placement
    problem solving, 584, 585
Contact noise, 1496–1497
Contact resistances, 790
Contact tube-to-workpiece distance
    (CTWD), 2226, 2227
Content-addressable (associative)
    memory, 2056–2057

Page on which term is defined is indicated in bold.

Contingency analysis, 1350
Continuity, 2101
Continuous current distributions, 875–876
Continuous current mode (CCM), 712, 721
Continuous ink-jet printers, 1965–1966
Continuous sampling, 1515–1516
Continuous speech recognition, 313, 306, 310, 312
Continuous systems, 2105, see also specific types
Continuous time (CT) signals, 182, 1598
Continuous-time measurements, 1486
Continuous wave (CW), 1006
Continuous wave (CW) lasers, **742**, 739
Continuous wave (CW) radar, 964–974
    applications of, 970–973
    Doppler, 965–967
    frequency modulation, 967–969
    interrupted, 955, 969
Continuum-mechanical variables, 807
Contractile cells, 2312
Contrast, **1785**, **2275**, 1783, 2259
Contrast enhancement, 334
Contrast ratio, 1783
Contrast rendition factor (CRF), **2275**, 2266
Contrast sensitivity, 357–358
Contrast stretching, 334, 335
Control circuits, 1011–1013
Controllability, **2105**, 2104
Controlled current sources, **57**, 53–57, 73
Controlled impedance networks, 773
Controlled rectifiers, 2238
Controlled voltage sources, **57**, 53–57, 73
Controllers, 1876, 2099, 2139, see also Control systems; specific types
    adaptive, 2233, 2237
    direct-memory access, 1868
    fault tolerance in, 2020
    gain-setting adaptive, 2232
    memory, 1930
    micro-, **1752**, **1762**, 1748, 1749–1750, 1751, 1755, 1756
    polarization, **770**, 765
    proportional-integral-derivative, 1757, 2142–2143, 2149–2150, 2167

Control memory, **1877**, 1872
Control systems, 2099–2153, see also Command, control, and communications (C3); Controllers; specific types
    adaptive, see Adaptive control systems
    aircraft, 2020, 2285
    attitude determination and, 1543
    for bonding, 2223–2224, 2231–2235
    closed-loop, 215, 2137, 2150, 2163
    compensation and, see Compensation
    composite equalizers and, 2122–2126
    design of, 2099, 2113, 2143–2147
    design-series equalizers and, 2115–2122
    digital, see Digital control systems
    dynamic response in, 2106–2113
    excitation, 1326
    feedback, see Feedback control systems
    flight, 2020, 2285
    frequency response in, see Frequency response
    in illuminating systems, 2273
    impulse response in, **2113**, 2103, 2108
    intelligent, **2182**, 2234–2235
    learning, 2103
    linear, 2113, 2150–2153
    microprocessors in, 1756–1759
    modeling of, 2099–2106
    modern design of, 2143–2147
    nonuniqueness and, 2104–2105
    phase-lead compensation and, 2116–2119
    in robotics, 2163–2167, 2171–2172
    root locus and, 2131–2139
    self-tuning, 2147
    single-loop linear control laws and, 2148–2149
    specifications for, 2139–2140
    steady-state response in, **2113**, 2108–2109
    time step, 666, 669–670
    transient response in, **2113**, 2108, 2109–2113, 2140
    trial-and-error approach to, 2099, 2113
    for welding, 2223–2224, 2231–2235
Control units, 1866
Convection, 787, 793
Convective heat transfer, **796**

Convective resistance, 787, 789, 791
Convergence, 663, 1513
Convergence zone (CZ), 410
Converse piezoelectric tensor, **1104**
Conversion constants, 2422–2423
Converter circuits, 1233–1234
Converters, **728**, 1237–1239, 2237, see also Inverters; Power supplies; specific types
    ac-to-dc, 702–705
    analog-to-digital, see Analog-to-digital converters (ADC)
    boost, 709
    buck, 708–709
    buck-boost, 710, 713
    cyclo-, see Cycloconverters
    dc-to-ac, 705–708
    dc-to-dc, 708–710, 771
    digital-to-analog, see Digital-to-analog converters (DAC)
    digital-to-RF, 1419
    double-ended PWM, 714–717
    electrical-to-optical, 1437
    full-bridge, 717
    generalized impedance, 674
    half-bridge, 715–717
    in machine control, 709–716
    monotonic, 773
    negative impedance, 680
    nonisolated single-ended PWM, 712–714
    optical-to-electrical, 1437
    parallel resonant, 719
    phase-controlled, 702, 729, 730, see also ac-to-dc converters
    positive impedance, 680
    principles of operation of, 1233–1237
    pulse number of, **1241**
    pulse-width modulation, **728**, 712–717, 730–731
    push-pull, 714–715, 716
    quasi-resonant, 717–718
    series-parallel resonant, 719, 720, 723–725
    series resonant, 719
    voltage-to-frequency, 771
Convolution, 180, 231, 245, 247, 336, 953
Convolutional coding, **1409**, 1403, 1405, 1407
Coolidge, William L., 1151
Cooling systems, 1200
Cooper pairs, 1114
Coordinate frames in navigation systems, 2286
Coordination principles, 1271
Copper twisted pair fiber, **1566**, 1559
CORE package, 2007

Page on which term is defined is indicated in bold.

Cores, **614**, **1001**, 35–36,
  1224–1225, 1300–1301, see
  also specific types
Coriolis forces, **2174**, 2169
Correlation, 1500
Correlation coefficient, 1500
Correlation detectors, **1487**, 1480
Cost function, **328**, 323, 2146
Cotree, **176**
Cotree-link voltages, 172
Cotton–Mouton effect, **1171**, 1163,
  1164–1165
Coulomb, Charles, 1002
Counters, 387, 1727, 1731–1733,
  1867, 1890, see also specific
  types
Coupled inductance, 25
Coupled tuned circuits, 1084
Couplers, see Coupling
Coupling, **1914**, 888, see also
    specific types
  for analog-to-digital converters,
    1805
  capacitive, **2338**, 1945, 2331
  in data acquisition systems, 1805
  directional, 887–888
  electroacoustic, 1058–1059
  exchange, 832
  far-field, 906
  high-level languages and, 1907
  inductive, 513, 2330
  mutual capacitance, 513
  near-field, 906
  quadratic phase, **2362**, 2356
  for SHA systems, 1805
  smart materials and, 1177
  source, 1430
  spin-orbit, 819, 823
  ultrasound and, 1083
Coupling capacitors, 550
Coupling coefficients, 1064, 1089
Coupling constants, 1080, 1081,
  1083–1084
Covariance, 1500
Covariance matrix, 1502
CPI, see Coherent processing
  interval; Color preference
  index
CPR, see Cardiopulmonary
  resuscitation
CPU, see Central processing unit
CR, see Cavity ratio; Controlled
  rectifiers
Cramer's Rule, 99, 2441
Crash recovery strategy, 2032
Creeping code, 1727
CRF, see Contrast rendition factor
CRI, see Color rendering index
CRIM, see Cage-rotor induction
  motors

Critical angle, 746
Critical band filters, 400
Critical bands, **405**
Critical clearing angle, **1284**, 1283
Critical clearing time, **1284**
Critical current, 1114, 1118, 1123
Critical design review (CDR), 1989
Critical field, 1114, 1327
Critical magnetic field, 1120, 1121
Critical path, 372, 373
Critical race, **1640**, 1636
Critical temperature, 1114
CrossCheck technique, 1828–1830
Cross-correlation, 1506–1507, 2355
Cross-linked polyethylene, 1146
Cross modulation, **147**, 144–145
Crossover distortion, 145
Crossover frequency, **2130**, 2115,
  2250, 2251
Crossover models, 2250–2254
Cross-polarization distortion, 847
Cross spectra, **2362**, 2355
Cross-spectral analysis, 2355, 2356
Cross-spectral density function,
  1507
Crosstalk, **528**, 513–524
  capacitive, 518
  forward (far-end), 518
  impedance and, 520–521
  inductive, 513
  maximum, 516, 517, 521
  reduction in, 522
Crosstalk noise, 513, 514, 515,
  522–524
CRTs, see Cathode ray tubes
Cryptology, 2074–2077
Crystalline anisotropy, 822, 823,
  831
Crystalline phases, **1096**, 1092
Crystal oscillators, 1089, 1572
Crystal point group, 1100–1101
Crystals, 805, see also specific types
  ADP, 1055
  cholesteric, **1777**, 1775
  conductivity of, 892
  dielectrics and, 1138
  liquid, 1127, 1775
  nematic, **1777**, 1772
  nickel, 819
  smectic, 1775
  symmetry properties of, 1087
  transduction and, 1055
  twisted nematic, **1777**, 1772, 1774
  ultrasound and, 1078
CS, see Common-source
CSDC, see Complementary
  symmetry diode compen-
  sated
CSI(s), see Current source inverters
CSM, see Composite source model

CSMA, see Carrier-sense multiple
  access
CSMA/CD, see Carrier sense
  multiple-access with
  collision detection
CSO, see Composite second-order
CT, see Computerized tomography;
  Continuous time
CTB, see Composite triple beat
CTE, see Coefficient of thermal
  expansion
CTL, see Charge-transport layer
CTWD, see Contact tube-to-
  workpiece distance
CU, see Coefficient of utilization
Cumulative distribution, 2519
Curie temperature, **826**, **1096**, **1131**,
  1092
Current, see also specific types
  alternating, see ac
  branch, **63**, 65
  common-emitter gain in, 546
  commutation of, **710**, **1331**, 702,
    1330
  continuous, 712, 721
  critical, 1114, 1118, 1123
  dark, 754
  dc bias, 620
  dc offset, **1267**, 1265–1266
  delta transformation and, 187
  density of, 1099
  diffusion, 449, 452, 469
  direct, see dc
  discontinuous, 712, 721
  displacement, 815, 1019
  drain saturation, **545**
  eddy, 36, 1019, 1024
  equivalent input, 622
  forward knee, 641
  generation-recombination, **459**
  high forward, 452–453
  input, 620, 622
  input bias, 620
  instantaneous, 81
  Kirchhoff's law of, see
    Kirchhoff's laws, for current
  leakage, **32**, 20–21
  magnetic field production of,
    814–815
  in MOSFET, 568–570
  Norton, 57
  offset, 620
  phase, 1262–1265
  phasor, 82
  quiescent collector, 640
  rate of change in, 630
  resistors and, 5–6
  reverse, 450
  reverse generation-recombina-
    tion, **459**

Page on which term is defined is indicated in bold.

reverse leakage, **32**
reverse saturation, **459**, 449
ripple, **32**, 21
saturation, **459**, 449
sequence, 1261–1264
sources of, see Current sources
superconductivity and, 1118
terminal, 1156
voltage vs., 1156
wye transformation and, 187
zero, 707
Current-controlled current sources (CCCS), 54
Current-controlled voltage sources (CCVS), 54
Current-derived negative feedback, 547
Current-fed inverters, 733–734
Current-fed pulse-width modulation inverters, 734
Current-feedback operational amplifiers, 622
Current mirror, **566**, 561, 564, 565–566
Current-mode logic (CML), 1618
Current noise, 9
Current phasors, 89
Current-ratio transfer function, 95
Current regulator diodes, 539
Current source inverters (CSIs), 705, 706–707, 2238
Current sources, 47–57
  controlled, 53–57, 73
  current-controlled, 54
  dc signals and, 49
  ideal, 50–51
  impulse signals and, 47–48
  independent, **63**
  network theorems and, 73
  practical, 51–53
  ramp signals and, 48–49
  single, 95
  sinusoidal signals and, 49
  step signals and, 47
  traveling, 941
  voltage-controlled, 54
Curvature, 2462
Custom ASICs, **601**, 591, 592–593
Custom Calling telephone services, 1468
CUT, see Circuit under test
Cut-off conditions, 548
Cut-off frequency, **859**, 113, 239, 851, 853, 858
Cut-off wavelength, **1001**
Cutset, **176**, 169
CVD, see Chemical vapor deposition
CW, see Continuous wave
Cyclic codes, **1409**, 1406–1407

Cycloconverters, **728**, **2246**, 705, 983, 2238, 2243, 2244
  induction motor control by, 734, 735
Cylinders, **1668**, 1660
Cylindrical configuration in robotics, 2156
Cylindrical coordinates, 2468–2469
Cylindrical fibers, 980–981
Cylindrical functions, see Bessel functions
CZ, see Convergence zone

D

DA, see Dielectric absorption
DAB, see Digital audio broadcasting
DAC, see Digital-to-analog converters
DAD, see Discharge area development
D-algorithm, 1812
DAM, see Diagnostic acceptability measure
DAMA, see Demand assigned multiple access
Damper windings, 1326
Damping coefficient, 2110, 2111
Damping factor, **1574**, 1569
Damping ratio, **206**, 202
Dark current, 754
Darlington configuration, 694, 697–698
Dart leader, **945**, 936
DASDs, see Direct-access storage devices
Data, see also Information; specific types
  abstractions in, 1998–1999, see also Abstract data types (ADTs)
  acquisition of, see Data acquisition systems
  C3 system and, 2214–2215
  collection of, 1753–1755
  compaction of, 1812, 1815
  compression of, see Data compression
  definition, 2037–2040
  excitation-response, 274
  fusion of, 417, 2214
  independence of, 2032
  integrity of, 2044–2045
  localization of, 2046
  mobile systems for, 1552
  modeling of, **2047**, 2033
  recording of, 1806
  recursive modeling of, 274–276
  sampling of, **183**, 182–183, 2281

security of, 2044–2045
signal-enhanced modeling of, 275–276
structure of, see Data structures
test, 1827, 1828
transfer time for, 1664
types of, see Data types
Data acquisition systems, 771, 777, 778, 1799–1807
  analog signal conditioning and, 1802–1803
  analog signal interface in, 1801–1802
  block diagram of, 1800
  communication interface of, 1804–1806
  data recording in, 1806
  digital signal interface in, 1801–1802
  in power generation, 1345–1346
  sample-and-hold techniques in, 1804
  signal-to-noise ratio in, **1807**, 1803
  software for, 1806
  supervisory control and, 1345, 1348, 1349, 1351
Database administration systems (DBAS), **1475**, 1473, 2043, see also Database management systems (DBMS); Databases
Database computers, **2047**, see also Databases
Database filters, 2047
Database management systems (DBMS), **2047**, 2032, see also Database administration systems (DBAS); Databases
  abstraction in, 2033
  architecture of, 2043–2044
  database machines and, 2047
  data definition and manipulation in, 2037
  data integrity in, 2044
  data models and, 2033
  distributed, 2046–2047
  multiuser, 2044
  offloading functions of, 2047
  security of, 2079
Databases, **2047**, 2032–2047, see also Database administration systems (DBAS); Database management systems (DBMS)
  abstractions in, 2032–2033
  architecture of, 2043–2044
  in avionics, 2188, 2189
  biocomputing and, 2406

computer communication
networks and, 1449
data integrity in, 2044–2045
data models and, 2033
design of, 2005, 2034–2037
distributed, **2047**, 2046–2047
800 service, 1471, 1472
hierarchical, 2040–2041
high-level languages and, 1902
in hospitals, 2406
in intelligent networks, 1468,
1469, 1470
INWATS, 1470
line information, **1475**, 1472
network, 2041–2043
object-oriented, 2045–2046
relational, 2033–2040
schemes in, 2032
security in, 2044–2045
trends in, 2045–2047
Data compression, 1517–1527
arithmetic coding and, 1520,
1526
entropy and, **1526**, 1517–1520
Huffman algorithm and, **1526**,
1519, 1520
Kolmogorov complexity and,
1526, 1524–1525
Lempel–Ziv algorithm and, **1526**,
1520–1521
in practice, 1525–1526
quantization and, **1526**,
1523–1524
rate distortion theory and,
1521–1523
vector quantization and,
1523–1524
Data definition languages (DDLs),
2032, 2037, 2043
Data direction registers, 1729
Data gloves, 2005
Datagrams, 1450, 1451
Data link control (DLC), 1456
Data link entities, **1459**, 1452,
1454
Data link layers, 1463–1464
Data manipulation, 2037–2040
Data manipulation languages
(DMLs), 2032, 2037, 2038,
2041, 2043, 2044
Data noise, 1513–1514
Data processing, see Computers
Data storage devices, see Memory
devices
Data structures, 1915–1926, 1999,
see also specific types
abstraction of, 1905
data constructors and,
1917–1919
dynamic, 1919–1924

object-oriented programming
and, 1924–1925
Data tablets, **1956**, 1943–1945,
1948
Data types, 1915–1926, see also
specific types
abstract, see Abstract data types
(ADTs)
arrays and, 1917–1918
constructors of, 1917–1919
dynamic, 1919–1924
enumerated, 1917
functions and, 1923–1924
fundamental, 1916
graphs and, 1923–1924
object-oriented programming
and, 1924–1925
records and, 1917
relations and, 1923–1924
sets and, 1923–1924
variant records and, 1918
Data windowing, 233–234
Datum alignment, 1882
Daylighting, 2267–2268, 2273
DBAS, see Database administration
systems
DBMS, see Database management
systems
DBSs, see Direct broadcast satellites
dc bias currents, 620
dc circuit breakers, 1240–1241
dc equivalent circuit models,
465–467
dc gain, 1505
dc generators, 1329–1331
dc machines, 729–731, see also
specific types
dc motors, **1341**, **2246**, 1333–1334,
1340–1341, 2238
analysis of, 1334–1335
large drives and, **2246**
schematic of, 2240
dc offset, **1267**, 1265–1266
dc operating point, 634–636
dc parallel plate reactors, 485
dc plasma displays, **1798**, 1788,
1791–1794
dc power supplies, 711–727
design of, 725–727
fixed-frequency operation of,
725
pulsewidth-modulated, 712–717
resonant
double-ended, 719–727
single-ended, 717–719
variable-frequency operation of,
720–722
dc signals, 49, 452
dc-to-ac converters, 705–708
dc-to-dc converters, 708–710, 771

dc transfer characteristics, 140
dc transmission
ac transmission vs., 1233
back-to-back, 1228
configuration of, 1228–1232
high-voltage, see High-voltage dc
(HVDC) transmission
multiterminal, 1231–1232
point-to-point, 1228–1231
two-terminal, 1228–1231
DCM, see dc motors; Discontinu-
ous current mode
DCT, see Discrete cosine transforms
DCUs, see Disk control units
DD, see Divided differences
DDLs, see Data definition languages
DE, see Differential equations
Dead-reckoning navigation systems,
2287–2289
Decade, **206**, 201
Decibels, **206**, **418**, 201, 408
Decimal scale, 2523
Decimation operation, 390–391
Decision-feedback equalizers
(DFEs), 1412–1413,
1414–1415
Decisionmakers, **2221**, 2211,
2218–2221
Decision support systems (DSS),
424, 2216
Decision threshold, 1479
Decision tree analysis, **2182**
Declarative paradigms, **1914**, 1908
Declarative reasoning, 1908
Decoding, 338, 1386, 1401,
1555–1556, 2203, see also
specific types
Decoupling, 781, 1628–1629, 2233
Dedicated circuits, 371
Dedicated simulation languages,
1608, 1597
Defect density, 594
Definite integrals, 2467–2468
Deflection, 1780–1782, 1941, 2006
Deformation potential, 1102
Degenerate common emitters, **566**,
561–562
Degrees of freedom (DOF), **2182**,
960
of electrons, 816
in robotics, **2162**, **2182**, 2158,
2161, 2172, 2175
Delay (firing) angle, **1241**
Delayed binding, **1914**, 1904
Delayed branching, 1897
Delay faults, 1810
Delay lines, 1084
Delay relaxation, 379
Delta circuits, **92**
Delta-delta circuits, 91

Delta function, 1039, see Unit
    impulse
Delta modulation, **781**, 775
Delta quantities, 1272
Delta transformation, 186–188
Delta voltage, 1272
Delta-wye circuits, 91
Demagnetizing field, **835**, 829
Demand assigned multiple access
    (DAMA), 1542
Demand factor (DF), 1316
Demodulation
    in broadcasting, 1359–1365, 1386
    in facsimile, 1555–1556
    in phase-locked loops, 1574
    in satellites, 2203
    synchronous, 1386
De Morgan's laws, 1704
Dendrites, 2352
Dependability, **2030**, 1270
Dependability evaluation,
    2025–2030
Depletion capacitance, 455–456
Depletion layer, 571, 577
Depletion mode, **1657**, 1655
Depletion-mode MOSFET, 567
Depolarization of radio waves, 847
Derating curves, 10
Derivatives, 2496–2497
Design for test (DFT), 1809,
    1816–1837
    ad-hoc techniques in, 1819–1823
    future of, 1835
    path-delay testing and,
        1832–1835
    structured techniques in,
        1823–1832
    synchronous vs. asynchronous,
        1820–1821
Design-series equalizers, 2115–2122
Desired impulse response (DIR),
    1415
Desk checking, 1988
Desktop publishing, 337, 2004, 2012
Detectability, **2106**, 2105
Detect/emit devices, 1646–1647, see
    also specific types
Detection threshold estimation, 953
Detectivity, 758–759, 1129
Detect/modulate devices, 1643–
    1646, see also specific types
Detector arrays, 2375, 2377
Detector element amplifiers, 1128
Detectors, 751–759, see also specific
    types
    Bayesian, **1486**, 1479
    coincidence, 2377
    correlation, **1487**, 1480
    discrete, 2377
    distributed, 1485–1486

envelope, **1487**, 1481
generalized likelihood ratio, 1483
imaging, 755–759
infrared, 1126, 1129
light, 1951
likelihood, 413, 1481, 1483
motion, 1395, 2380, 2384, 2385
Neyman–Pearson, **1487**, 1479
nonparametric, 1485
phase, 1567, 1572, 1573
photo-, 1417, 1420, 1430–1431,
    1435, 1946
photoemissive, 754–755
quadratic, **1487**, 1482
robust, 1485
sample-and-hold, 1573
semiconductor, 751–759
sequence, 1484
sequential, 1485–1486
Determinants, 2436–2438, 2441
Deuterated isomorph (DTGS), 1130
Device under test (DUT), **498**, 490,
    1817
DF, see Demand factor; Dissipation
    factor; Drive factor
DFEs, see Decision-feedback
    equalizers
DFT, see Design for test; Discrete
    Fourier transforms
DG, see Distributed power
    generation
Diagnostic acceptability measure
    (DAM), 281, 285
Diagnostic rhyme test (DRT), 281,
    285
Diagnostic systems, 1876, 2380, see
    also specific types
Diagonal microinstructions, 1872
Diamagnetics, 816
Dichroism, 1163
di/dt effects, **701**, 695
Dielectric absorption, 18
Dielectric breakdown voltage, 2312
Dielectric constants, **30**, **1150**, 1132,
    1133, 1144, 1146,
    2532–2533
    capacitor, 17, 19
    of ceramics, 2533
    effective, 506
    of glasses, 2533
    of insulation, 16
    interconnections and, 504, 506
    magnetism and, 813
    of solids, 2532
Dielectric losses, **1150**, 1128,
    1133–1137, 1143, 1150
Dielectrics, **38**, **1150**, 1132–1151
    breakdown and, 1137–1140
    of capacitors, **30**, 1144
    insulation aging and, 1140–1142

liquid, 1139, 1143, 1144–1145
materials in, 1143–1150
polar, 1126–1127
semiconductor manufacturing
    and, 484
of smart materials, **1188**, 1179
solid, 1137, 1139, 1143
transformers and, 34
wave equation and, 977
Dielectric strengths, **1150**, 1138
Dielectric waveguides, 975, 984
Diendorfer–Uman (DU) model,
    941, 942–943
Dies, **1752**, see also Chips
Difference amplifiers, **566**, 561,
    562–564, see also Long-
    tailed pair
Difference-in-to-difference-out
    voltage gain, 563
Differential amplifiers, see
    Difference amplifiers
Differential calculus, 2461–2466
Differential detection, 1691
Differential equation propagator,
    1031
Differential equations, see also
    specific types
    in computational electro-
        magnetics, 1031, 1033,
        1035, 1042–1043
    semiconductor manufacturing
        and, 480
    in stability analysis, 218–219
Differential errors, 141–142
Differential input resistance, 620
Differential linearity, 773
Differential phase shift keying
    (DPSK) modulation,
    2203
Differential pulse code modulation
    (DPCM), 281, 282, 283,
    338, 350, 1407
Differential resistance, 454
Differential signal, 617
Differentiation theorems, 158–160
Diffraction, 1034, 1070, 1169,
    1691
Diffraction constants, 1059
Diffraction grating, 1083, 1085
Diffraction-limited focusing, 1682
Diffusion, **458**, **489**
    diodes and, 447
    of electrons, 447, 449
    Fick's laws of, 479
    Gaussian, 480
    of holes, 447
    interstitial, 479
    limited source, 480
    mechanism of, 479–480
    practical, 480

Page on which term is defined is indicated in bold.

in semiconductor manufacturing,
489, 479–481
substitutional, 479
Diffusion capacitance, **458**, 454–455
Diffusion coefficients, 479, 480, 481
Diffusion constants, **458**, 449, 479
Diffusion current, 449, 452, 469
Diffusion gradients, 449
Digital audio broadcasting (DAB),
400, 1397–1404
Digital audio recordings, 776
Digital communication, 1405–1416,
1865
coding in, 1405–1409
equalization in, **1416**,
1410–1416
Digital computers, see Computers
Digital control systems, 2136–2137,
2147–2153
closed-loop, 2150
example of, 2148
linear, 2150–2153
single-loop linear control laws
and, 2148–2149
Digital displays, 1763, see also
Light-emitting diodes
(LEDs); specific types
Digital electronics, 1118
Digital filters, 2360, see also specific
types
causal, 239
digital signal processing and, see
Digital signal processing
(DSP)
finite impulse response type, see
Finite impulse response
(FIR) filters
frequency-selective, 239
infinite impulse response type,
see Infinite impulse
response (IIR) filters
in rehabilitation engineering,
2395
transfer functions and, 238
Digital hearing aids, 397–398
Digital image compression, 336–341
Digital image processing, **343**,
329–345
computer vision and, 343
edge detection in, 342–343
image analysis and, 343
image capture and, 330–334
image compression and, 336–341
image enhancement and,
335–336
image reconstruction and, **343**,
341–342
point operations and, 334–335
Digital photography, 335
Digital priority encoders, 778

Digital signal interface, 1801–1802
Digital signal processing (DSP),
393, 229–278, 370, 382,
385–386, 395–406
active noise control and, 403–404
in aerospace systems, 2203, 2209
analog interfaces to, 771
analog-to-digital converters and,
779
architecture in, 385
audio coding and, 400–401
in audio and electroacoustics,
395–406
audio coding and, 400–401
microphone arrays and,
395–397
sound control and, 403–404
spatial processing and,
398–400
C3 system and, 2215
chips in, 1760, 1761
computer graphics and, 2008
data windowing and, 233–234
digital filters and, 238–250
finite impulse response type,
239–242, 244–247
infinite impulse response type,
242–244, 247–250
echo cancellation and, 401–403
fast Fourier transform and,
234–236
first-generation, 370
fixed-point, **393**, 385, 386–387
floating-point, **393**, 387
future of, 382, 393
Harvard architecture in, 385
hearing aids and, 397–398
microphone arrays and, 395–397
microprocessors in, 1760–1761
programmable, 371, 372
radar and, 953
satellites and, 2203, 2209
second-generation, 370
signal restoration and, see Signal
restoration
simulation and, 1597
spatial processing and, 398–400
special-purpose, **393**
steerable microphone arrays and,
395–397
transforms and, 229–238
amplitude spectrum and,
232–233
data windowing and, 233–234
phase spectrum and, 232–233
power spectrum and, 232–233
properties of, 230–231
Digital storage architecture (DSA),
1662
Digital tachometers, 1755

Digital-to-analog converters (DAC),
631, **1762**, 386, 390,
771–783, see also specific
types
AD558, 774–775
AD7524, 775, 776
bypassing on, 780–781
control systems and, 2149
conversion process in, 773–774
in digital audio broadcasting,
1401
digital control systems and, 2149
digital image processing and, 332
fixed reference, **781**, 773, 774
grounding on, 780–781
integrated circuits in, 771, 772,
774–775, 780–781
integrating, **781**
microprocessors and, **1762**, 1755,
1756
multiplying, **781**, 773, 774
performance criteria for, 771–773
varying reference, 775
Digital-to-RF conversion, 1419
Digit complement (DC) system,
1861
Digitizers, **1956**, 329, 1943–1945,
2012, see also specific types
Digit-serial processing, **383**, 371,
374
Diminished radix complement
representation, 1855–1856
Diode-resistor circuits, 452
Diodes, **131**, 126–128, 447–459, see
also specific types
applications of, 447, see also
specific types
avalanche, 457–458, 1430–1431
biased, **138**, 133
bipolar, 126, 127, 128
complementary symmetry, 638
current regulator, 539
emitter-base junction, 640, 641
forward-biased, 133, 450–451
freewheeling, 702, 703
Gunn, 892, 894, 1004
high forward current and,
452–453
high-voltage, 128
IMPATT, 892, 894, 895–896,
1004, 1005, 1010
large-signal model of, 452, 453
laser, 1429, 1435, 1436, 1682,
1683, 1684
light-emitting, see Light-emitting
diodes (LEDs)
metal-semiconductor, see
Schottky diodes
in microwave devices, 892,
894–896, 1004

Page on which term is defined is indicated in bold.

photo-, see Photodiodes
PIN, 752–753, 890, 1013, 1430,
　　1435
planar, 448
pn-junction, see *pn*-junction
　　diodes
rectifier, 447
reverse breakdown of, **458**,
　　456–457
Schottky, see Schottky diodes
semiconductor, 738, 890
silicon, 134
small-signal incremental model
　　of, 454–456
tunnel, 447, 458, 892, 894, 1004
unipolar, 126
varactor, 447, 458
voltage and, 127
Zener, 135, 137, 447, 457–458
Diode-transistor logic (DTL), 1624
Diode voltage, 127
DIP, see Dual inline packaging
Dipole orientation losses, 1134
Dipoles, 1178
DIR, see Desired impulse response
Dirac distribution (unit impulse),
　　50, 47–48
Dirac function, 232
Direct-access storage devices
　　(DASDs), 1681, see also
　　specific types
Direct access testing, 492–493
Direct broadcast satellites (DBSs),
　　1534, 2198–2199
Direct current, see dc
Direct-form realizations, 247
Direct glare, 2265
Directional comparison blocking
　　scheme, 1276–1277
Directional couplers, 887–888
Directional overcurrent relays,
　　1271–1272
Direction cosines, 1078
Directivity
　　of antennas, **878**, 862–863,
　　　876–877, 955
　　of electroacoustic devices,
　　　1060
　　of microwave devices, 888
　　of transducers, 1060
Direct matrix, 1039–1040
Direct memory access (DMA),
　　1807, 1804
Direct memory access (DMA)
　　controller, 1868
Direct modulation, **1425**,
　　1417–1418
Direct overwrite, 1688, 1689, 1690
Direct view storage tubes (DVSTs),
　　2007

Direct weld parameters (DWP),
　　2235, 2224, 2225, 2229,
　　2230, 2233
Direct *z*-transform, 178
Dirichlet partition, 1524
Disabilities, **2395**, 2387, 2388, see
　　also specific types
Disassemblers, **1901**, 1879
Discharge area development (DAD),
　　1960
Discontinuous current mode
　　(DCM), 712, 721
Discrete arrays, 361–365
Discrete cosine transforms (DCTs),
　　338, 352, 1419, 1526
Discrete detectors, 2377
Discrete Fourier transforms (DFT),
　　229, 301, 302, 307
　　FIR filter implementation and,
　　　246–247
　　Fourier transforms and, 232
　　inverse, 236–238
　　properties of, 230–231
　　sensor array processing and, 361
Discrete models, 2105, see also
　　specific types
Discrete probability, 2518–2519
Discrete radiators, 870–873
Discrete sequence, **250**, 238
Discrete spectrum, 319
Discrete time, 1349
Discrete-time signals, 178, 182
Discrete-time white Gaussian noise,
　　1487, 1480
Discrete transform, 283
Discriminators, 1586
Disk arm, **1668**, 1659
Disk arrays, 1668
Disk caches, 1668
Disk capacitors, **30**
Disk control units (DCUs), 1662
Disk density, 1666–1667
Disk drives, **1668**, 1659, see also
　　Disk system architectures;
　　specific types
　　fixed-head, 1666
　　magnetic, 1659–1664
　　optical, 1676
　　parallel transfer, 1666
Disk rotation speed, 1681
Disks, see also specific types
　　compact, **1693**, 400, 1676, 1679
　　laser, 1682
　　magneto-optical, see Magneto-
　　　optical disk data storage
　　optical, 1676, 1677, 1679
Disk scheduling, 1668
Disk system architectures,
　　1658–1670, see also Disk
　　drives

extensions to, 1666–1668
input/output workloads and,
　　1664–1665
for magnetic disks, 1659–1664
Disk track times, 1658
Disordered semiconductors, 445
Dispatch, **1353**, 1346, 1347–1348,
　　1350
Dispersion, **1001**, **1425**, 985, 991, 992
Dispersion-shifted fiber, 1429
Dispersive filters, 1073
Dispersive media, 1547, see also
　　specific types
Displacement, 1080, 1088, 1089,
　　1178
Displacement current, 815, 1019
Displacement sensors, 1155, 1156
Displays, 1763–1798, see also
　　specific types
　　A-mode, **2385**, 2383
　　B-mode, **2385**, 2383
　　cathode ray tube, see Cathode ray
　　　tubes (CRTs)
　　color, 1390, 1782–1784
　　compensatory, **2253**
　　computer graphics and, 2004,
　　　2006–2007
　　digital, 1763
　　electroluminescent, 2007
　　electromagnetic deflection system
　　　and, 1780–1782
　　electronic, 2188, 2189–2190
　　electrostatic deflection and, 1780,
　　　1781
　　gas discharge, 1789–1791
　　gray scale and, 1796–1797
　　head-mounted, 2005
　　interfaces and, 1768–1770,
　　　1775–1777
　　LEDs in, see Light-emitting
　　　diodes (LEDs)
　　liquid-crystal, see Liquid-crystal
　　　displays (LCDs)
　　plasma, see Plasma displays
　　principles of, 1772–1775
　　projection screen, 1784–1785
　　pursuit, **2253**
　　vacuum fluorescent, 1942
　　in vehicles, 2255
　　video, see Video display terminals
　　　(VDTs)
Dissipation factor (DF), **31**, **1150**,
　　18, 1133, 1137, 1140
Dissociation, 1136
Distance, 1520, 2023, 2290,
　　2447–2448, 2449
Distance measuring equipment
　　(DME), **2193**, 2191
Distance protection, **1278**,
　　1272–1274

Distance relays, 1272–1275
Distortion, 139–147, 1523, see also
    specific types
  clipping and, 1424
  composite second-order, **1425**,
    1422
  cross modulation and, **147**,
    144–145
  crossover, 145
  cross-polarization, 847
  differential-error method and,
    141–142
  failure-to-follow, **147**, 145
  five-point method and, 142–143
  frequency, **147**
  Hamming, 1522, 1523
  harmonic, **147**, 139–140, 141,
    142, 143
  intermodulation, **147**, 143–144
  Itakura–Saito measures of, 290
  measurement of, 1522
  phase, **147**, 145
  power-series method and,
    140–141
  pre-, 1425
  rate, **1526**, 1521–1523
  resonance-enhanced, 1423
  second-order, 1422
  signal ratio to, 1411
  simulation of, 145–147
  squared error, 1522
  surface acoustic wave filters and,
    1070
  third-order, 1422
  three-point method and, 142
  total harmonic, **147**, 2239, 2240,
    2241
  triple-beat, 144
Distributed arithmetic, 382
Distributed computing, **1869**, **2070**,
    1869, 2062–2064
Distributed databases, **2047**,
    2046–2047
Distributed detection, 1485–1486
Distributed file access, 2061
Distributed parameter systems, 2100
Distributed power generation, **1215**,
    1207–1215
  biomass energy in, 1207,
    1213–1214
  fuel cells in, 1212
  geothermal power plants in,
    1215, 1207, 1211–1212
  hydroelectric power plants in,
    1215
  hydropower in, 1211
  integrated renewable energy
    systems in, 1207, 1215
  photovoltaics in, **1215**,
    1208–1210

  potential for, 1208
  solar-thermal-electric conversion
    in, 1213
  thermionics in, **1215**, 1214–1215
  thermoelectrics in, 1214
  tidal energy in, **1215**, 1212
  wind-electric conversion in, **1215**,
    1210–1211
Distributed processing, 1761
Distributed queue dual bus
    (DQDB) interface, 1456,
    1464, 2017, 2018
Distribution circuits, 1344
Distribution laws, 1704
Distributivity transformation, 381
Disturbance, **1285**, 2140, see also
    Fault
Divergence, 740, 843, 844, 2473,
    2474
Divided attention, 2252–2253
Divided differences (DD), 669
Division, 1863–1864, 1866
DLC, see Data link control
DMA, see Direct memory access
DME, see Distance measuring
    equipment
DMLs, see Data manipulation
    languages
Doctrine, **2221**
DOES, see Double heterostructure
    optoelectronic switch
DOF, see Degrees of freedom
Domains, **1096**, 1087, 1092, 1171,
    1696, see also specific types
  complex, 61–62, 63
  frequency, see Frequency domain
  solution, **1048**
  time, see Time domain
Donors, **446**, **458**, 437, 438, 447
Dopant-ion control, 579
Dopants, 478, 479, 481
Doping, **15**, 13
Doppler compensators, 414
Doppler detection, 2384, 2385
Doppler filters, 953, 955, 958, 959
Doppler frequency, **973**, 959, 964,
    968, 969
Doppler instruments, 2380
Doppler modulation, 959
Doppler navigators, 971–972
Doppler principle, 407, 416, 2384
Doppler radar, **973**, 955, 964,
    965–967, 2288
Doppler shift, **418**, 1085, 2384
Doppler sidebands, 956
Doppler sonar, 2288
Doppler term, 874
Dosimetry, 2335–2338
Dot matrix printers, 1972–1973
Double-bit errors, 2023

Double-ended clipping, 137
Double-ended limiting circuits, 135,
    137
Double-ended PWM converters,
    714–717
Double heterostructure optoelec-
    tronic switch (DOES), 1646
Double integration, 2469, 2470
Double phase-to-ground fault,
    1257–1258, 1261, 1272
Double-rank circuits, 1720
Double-sided assembly, **614**
Doubly terminated filters, 113, 114,
    115, 117, 118, 119, 120
DPCM, see Differential pulse code
    modulation
DPSK, see Differential phase shift
    keying
DQDB, see Distributed queue dual
    bus
Drains, **579**, 530
  barrier lowering induced by, 577
  bipolar transistors and, 548
  characteristics of, 533
  common, 536
  conductance and, 576
  lightly doped, 578
  MOSFET and, 22, 46, 567, 576
  response time and, 575
Drain saturation current, **545**
Drain-to-source conductance, 538,
    543
Drain-to-source resistance, 538, 539
DRAM, see Dynamic RAM
Drift, **458**, 447, 469
Drift-diffusion equation, 440
Drive factor (DF), 1624
Driving ability, 574–576, 577
Driving capability, 1623
Driving point impedance, **651**
Drop in voltage, **15**, 6, 59, 62, 1316
DRT, see Diagnostic rhyme test
Dry etching, **489**, 488
Dry oxides, 476, 477
DSA, see Digital storage architecture
DSP, see Digital signal processing
DSS, see Decision support systems
DT, see Discrete time
DTGS, see Deuterated isomorph
DTL, see Diode-transistor logic
DTW, see Dynamic time warping
DU, see Diendorfer–Uman
Dual inline packaging (DIP), **614**,
    501, 607
Duality, 1702
Duplexers, 889
Duplex ultrasound, **2385**
DUT, see Device under test
Duty cycle, **710**
dv/dt effects, **701**, 695

DVSTs, see Direct view storage tubes
DWP, see Direct weld parameters
Dynamic allocation, 1884
Dynamic brakes, **737**, 732
Dynamic data types, 1919–1924
Dynamic drain-to-source conductance, 538
Dynamic drain-to-source resistance, 538, 539
Dynamic forward transfer conductance, 536
Dynamic models of robotics, 2167–2173
Dynamic-pressure sensors, 2288
Dynamic RAM (DRAM), 1652, 1655
Dynamic response in control systems, 2106–2113
Dynamic simulation, 1352
Dynamic stacks, 1887
Dynamic storage, 1884
Dynamic systems, 420, 2106–2113, see also specific types
Dynamic time warping (DTW), **313**, 307, 308–309
Dynodes, 755

E

Early voltage, 538, 641
Earphones, 1055
Earth electrode system, **918**
Earth's magnetic field, 2287
Earth stations, **2209**, 2199–2200
Earth and wave propagation and, 843–847
EBC, see Extended boundary condition
EBCDIC, see Extended Binary Coded Decimal Interchange Code
Eber and Moll equations, 55
Ebullient heat transfer rate, **796**, 788
ECC, see Error correction coding
ECCM, see Electronic counter-counter measures
ECG, see Electrocardiography
Echo cancellation, 401–403
Eckart filters, 414
ECL, see Emitter-coupled logic
Ecliptic, **2294**, 2286
ECMs, see Electronic counter-measures
ECoG, see Electrocorticography
Economic dispatch (ED), **1353**, 1346, 1347–1348
Economizers, **1205**, 1194

ECSG, see Electronic Connector Study Group
ECT, see Emission computed tomography
Ectopic beat, **2328**
Eddy current, 36, 1019, 1024
EDFA, see Erbium-doped fiber amplifiers
Edge elements, 1022–1023
Edge emitting LEDs, 750
Edges, **343**, 58, 59, 62, 93, 335, see also Circuit-set
    detection of, 342–343
    in digital image processing, **343**
    enhancement of, 392, 393
    linking of, 343
    in moving objects, 343
Edge-triggered flip-flops, **1640**, 1638–1639, 1718, 1719, 1723
Edison, Thomas A., 223, 1206
Edison effect, 1214
EEG, see Electroencephalography
EEPROMs, 1655, 1656
EF, see Emitter-follower
Effective isotropic radiated power (EIRP), 863, 2197, 2209
Effective mass, **1110**, 1109
Effective noise, 1494
Effective radiated power (ERP), **1379**, **1392**, 863, 1370, 1382, 2197, 2209
Efficient parameter estimate, 322
EFIS, see Electronic flight instrument system
EICAS, see Engine indicating and crew alerting system
Eigenvalues, 207, 211, 214, 217, 219, 220
800 service database, 1471, 1472
EIRP, see Effective isotropic radiated power
Elastic buffers, 1727
Elasticity, 1077, 1099
Elastoresistance tensor, 1101–1102
Electret materials, **1096**, 1087, 1179, see also specific types
Electrical defibrillation, 2317
Electrical fuses, see Fuses
Electrical machines, see specific types
Electrical properties of metals, 1219
Electrical resistance, see Resistance
Electrical systems in vehicles, 2255
Electrical telemetry, 1579
Electrical-to-optical converters, 1437
Electrical trees, 1141
Electric displacement, 1080
Electric energy, 1115

distribution of, see Energy distribution
generation of, see Power generation
storage of in vehicles, 2255
Electric fields, **810**, 807
    atomic displacements vs., 1178
    in bone repair, 2329–2334
    computational electromagnetics and, 1032
    electron velocity vs., 893
    equation of state and, 1099
    Gauss' laws of, 804
    Hall, 1107
    intensity of, **1379**, 1018, 1369
    from lightning, 935, 937–940, 941
    magnetic field production by, 815
    multiplication of, 1036
    piezoelectric excitation and, 1080
    piezoresistivity and, 1099
    in radio, 1369
    reverse bias, 899
    in soft tissue repair, 2329–2339
    strength of, 838
    superconductivity and, 1115
    time-dependent, 812
    ultrasound and, 1080, 1083
Electric scalar potential, 1018
Electric vehicles, **2256**, 2255
Electrified electrorheological (ER) fluid, 1178
Electroacoustic coupling, 1058–1059
Electroacoustic devices, 1055–1060, see also Electroacoustics; specific types
Electroacoustics, **405**, **1060**, **1188**, 1055, 1178–1179, see also Audio and electroacoustics; Electroacoustic devices
Electroacoustic smart materials, **1188**, 1178–1179, 1181
Electroacoustic smart sensors, 1183
Electrocardiography (ECG), **2370**, 2311, 2314–2319, 2362–2371
    clinical significance of, 2315–2317
    high-resolution, **2370**, 2364, 2367
    instrumentation for, 2319, 2366–2369
    origin of, 2314–2315
    physiology and, 2366
    12-lead, **2370**, 2363
Electrochemical degradation, 1775
Electrochemical sensors, 2345, 2346, 2348
Electrocorticography (ECoG), 2322
Electrodes, see also specific types
    annular control, 1388
    in bioelectricity, 2336

Clark, 2345
earth, **918**
in EEG, 2327
in electroencephalography, 2358
extracellular, 2313
fabrication of, 767
glass, 2346
ion-selective, 1157–1158, 1159
pH, 2346, 2347
push-pull, 763
in rehabilitation engineering, 2389, 2394
resistance of, 1070
transmembrane, 2313
transparent, 1775
in welding, 2224
Electrodynamic transducers, 1056
Electrodynamometers, 85
Electroencephalography (EEG), **2362**, 2311, 2322–2327, 2351–2362
brain language and, 2351–2353
clinical, 2325–2327
evoked potentials in, 2327
frequency analysis in, 2354–2356
history of, 2353–2354
instrumentation for, 2327
normal, 2322–2325
recording techniques in, 2322, 2323, 2356–2360
topographic mapping and, 2360–2361
waveform terminology in, 2324
Electroluminescence, **1771**, 1764, 1789, 1942, 2007
Electrolytes, **31**, 1179, 1182, 1185
Electrolytic capacitors, **31**, 20–24
Electromagnetic acoustic transducers (EMATs), 2229
Electromagnetic active surfaces, 1176, 1182, 1185
Electromagnetic aperture, 1185–1186
Electromagnetic compatibility (EMC), **918**, 912, 919–921, 2239, 2240
Electromagnetic deflection, 1780
Electromagnetic design, 2008
Electromagnetic engineering, 1187
Electromagnetic equations, 34–35, 1325–1326
Electromagnetic fields, 803–810
anisotropic media and, 805–806
bianisotropic media and, 805–806
biisotropic media and, 806–807
constitutive matrices and, 807
constitutive relations and, 804–807

Maxwell's equations and, 803–804, 807, 815, 1018, 1019
pulsed, see Pulsed electromagnetic fields (PEMF)
strengths of, 807
three-dimensional analysis of, see Three-dimensional analysis
wave equations and, 807–810
wavenumbers and, 810
wave solutions and, 807–810
wave vectors and, 808–810
Electromagnetic induction, 1098, 1325
Electromagnetic interference (EMI), **934**, 707, 712, 780–781
causes of, 904
compatibility and, 903, 904, 913
optical fibers and, 988
reduction of, 913
suppression of, 903, 913
Electromagnetic propagation, 1029
Electromagnetic pulse (EMP), **918**
Electromagnetic radiation, 742, 743, 2427–2428, see also Light
classification of, 2526–2527
sources of, 1494
Electromagnetics, 1029
computational, see Computational electromagnetics
essence of, 1029
propagation and, 505
smart material applications to, 1185–1186
of smart materials, 1179–1180
superconductivity and, 1115–1118
theory of, 880–881
Electromagnetic sensors, 1182, 1183
Electromagnetic shielding, 1182, 1185
Electromagnetic simulation, 2008
Electromagnetic smart materials, **1188**, 1175–1176, 1179–1180, 1181–1182
Electromagnetic spectrum, 919–921, 1564
Electromagnetic wave propagation, see Wave propagation
Electromechanical coupling constants, 1081, 1083–1084
Electromechanical energy conversion, 1321
Electromechanical equations, 1325–1326
Electromechanical relays, **1278**
Electromotive force, 27
Electromyography (EMG), 2311, 2319–2322
Electron avalanche, 1139

Electron bands, **1110**
Electron beam welding, **2235**, 2223
Electron guns, 1779, 2006
Electron multiplication, **759**, 755, 757
Electrons, 437–438, 445, see also Holes
control grid for, 2006
degrees of freedom of, 816
diffusion of, 447, 449
electric fields and, 893
magnetism and, 816
recombination of holes and, 1763
temporarily depleted, 1488
velocity of, 893
Electronic Connector Study Group, 2550
Electronic counter-counter measures (ECCMs), 953, 2218
Electronic counter-measures (ECMs), 2218
Electronic displays, 2188, 2189–2190, see also specific types
Electronic filters, see Filters
Electronic flight instrument system (EFIS), 2189
Electronic scanning antennas, 951
Electronic structure, 1108–1109, 1110
Electronic surveillance measuring (ESM), 2214
Electronic switching, **631**
Electronic systems in vehicles, 2255
Electronic warfare, **2221**, 1188, 2217
Electro-optic integrated optic phase modulators, 763
Electro-optic properties, **1188**, 761, 1178
Electro-optic smart materials, **1188**, 1178, 1181
Electro-optic switching, 1439
Electrophotography, 1958, 1959–1964
Electrophotonic devices, 1646, see also specific types
Electroplastic effects (EPE), **1188**, 1177, 1180
Electroplastic smart materials, **1188**, 1177, 1180
Electrorheological properties, **1188**, 1181
Electrorheological smart fluids, **1188**
Electrorheological smart materials, **1188**, 1178, 1181
Electroslag welding, **2235**, 2223
Electrostatic deflection, 1780, 1781

Page on which term is defined is indicated in bold.

Electrostatic dissipative/conductive
  surfaces, 1185
Electrostatic plotters, 2005
Electrostatic sources, 1056
Electrostriction, 1090
Electrostrictive materials, **1097,**
  1088, 1090, see also specific
  types
Ellipses, 2450–2452
Elliptic function (Cauer function)
  filters, 679
EM, see Electromagnetic; Expecta-
  tion-maximization
EMATs, see Electromagnetic
  acoustic transducers
Embedded memory block,
  1822–1823
Embedded speech coders, 284
EMC, see Electromagnetic
  compatibility
Emergency lighting, 2271
EMG, see Electromyography
EMI, see Electromagnetic interfer-
  ence
Emission computed tomography
  (ECT), 2376
Emissions, see also specific types
  broadband, **934**
  conducted, **934,** 933
  light, **759,** 750
  radiated, **934**
  spontaneous, 1421
Emission tomography, 342, 2376
Emissivity, 748, 1128
Emitter-base junction diffusion
  resistance, 640
Emitter-base junction diodes, 640,
  641
Emitter-base junction injection
  coefficient, 640
Emitter-coupled logic (ECL), 1613,
  1614, 1618, 1622,
  1629–1631, 1712
  networks and, 1443
Emitter degeneration resistance, 646
Emitter-follower, **566,** 555–557
Emitter-follower amplifiers, 634,
  637
Emitter resistance, 640
Emitters, 545
EMP, see Electromagnetic pulse
Empirical modeling, 2101, 2229, see
  also specific types
EMS, see Energy management
  systems
E&M transmission lines, 1062, 1063
Emulation, 1874–1875
Emulators, **1877,** 1874, see also
  specific types
Enclosure theory, 906–908

Encoding, **297, 1762,** see also
  specific types
  channel, **1403,** 1400
  color signal, 1384–1386
  digital priority, 778
  image, 338
  knowledge, 2050
  Manchester baseband, 1462
  multiple sub-Nyquist, **1397,** 1394
  optical, 1947
  phase, 1674
  predictive, 338, 350–352
  run-length, **1556,** 1555
  shaft, 1755
  source, **1404,** 1400, 1401–1403
  voice, 776
Encounter dynamics, 2216–2218
Encryption, 337
End loading, 865
Energy, **87,** 79–87, see also specific
  types
  ac, 85–87
  available, 822
  average stored, 84
  back-scattered, 2383
  bandgap, **458,** 450
  beam, 482
  biomass, **1215,** 1207, 1213–1214
  computational, 423
  condensation (vapor-phase), 613
  conservation of, 81, 442
  distribution of, see Energy
    distribution
  electric, see Electric energy
  electromechanical, 1321
  Fermi, **473,** 436, 438, 444
  flow of, 1078, 1080, 1081
  geothermal, **1215,** 1202–1203,
    1207, 1211–1212
  heat, 1197
  impact ionization threshold, 443
  implant, 482
  incident, 906
  infrared, 613
  ionization, 445
  kinetic, 1114, 1115, 1494, 2168
  magnetic, 24, 504, 1115
  management of, see Energy
    management systems (EMS)
  mechanical, 1197
  potential, 1494, 2168
  quadratic, 424
  quasi-Fermi, **473,** 469, 471
  radio frequency, 921, 923, 949,
    1564
  reaction, 487
  resonant, 444
  of room noise, 396
  scattered, 1082, 2382, 2383
  solar, **1215,** 1213

sources for, 1207
stored, 16, 822, 886, 1344
  average, 84
  effects of, 140, 141
  magnetic, 504
  switching, 1648
  thermal, 1213
  tidal, **1215,** 1212
  transfer of, 505
  transistor switching, 786
  transmission of, 82
  vapor-phase, 613
  wind, **1215,** 1210–1211
Energy bands, **446,** 435–436
Energy control systems (ECS), 1351
Energy distribution, 1310–1320
  capacitors in, 1318–1319
  load characteristics in, 1316–1317
  primary, 1312–1313
  radial, 1314–1315
  secondary, 1314
  secondary networks in,
    1315–1316
  voltage regulators in, 1318–1319
Energy distribution constant, 1138
Energy efficiency in illuminating
  systems, 2272–2275
Energy gaps, **446,** 436
Energy management systems
  (EMS), **1353,** 1344–1353
  automatic generation control
    and, 1346–1348
  energy control systems and, 1351
  load management in, 1344,
    1348–1349
  operator training simulators in,
    1351–1352
  security in, 1344, 1350–1351
Energy products, 822
Energy recovery, 2238, 2243
Energy-saving lighting, 2272–2273
Engine indicating and crew alerting
  system (EICAS), 2189
Engines, 2255
Enhanced SMR, 1547
Enhancement mode, **1657,** 568,
  1655
ENI, see Equivalent noise current
ENIAC, 2019
Ensemble average, 316
Ensemble processors, **2059,**
  2053–2054
Entities, **1459,** 1452, 1454
Entropy, **343, 1526,** 339, 1517–1518,
  1519–1520
Entropy coding, **1397,** 339, 1396
ENV, see Equivalent noise voltage
Envelope detectors, **1487,** 1481
Environments, **2221,** 2213
Enzymes, 2338

Page on which term is defined is indicated in bold.

Enzyme sensors, 1158, 1159
EPE, see Electroplastic effects
Epilepsy, 2325, 2326
Epitaxial growth techniques, 444
Epitaxial layer, **15**, 13
Epoxy resins, 1149
EPR, see Ethylene-propylene rubber
EPROMs, 1655
Equal-area criterion, 1282
Equality, 2432
Equalization, 1410–1416
Equalizer coefficients, 1410, 1411
Equalizers, **1416**, **2130**, 2113, see
    also specific types
  composite, 2122–2126
  decision-feedback, 1412–1413,
    1414–1415
  design-series, 2115–2122
  fractionally spaced, 1413
  least-mean-squared, 1411
  linear transversal, 1410–1412
  nonlinear, 1412–1413
  series, **2130**
  zero-forcing, 1411
Equalizing pulses, **1392**, 1381
Equally spaced linear arrays,
    866–867
Equal pulse spacing control, 1239
Equal ripple, **124**, 115, 117
Equation of state, 1099–1100
Equilibrium
  diodes and, 447
  dopant segregation and, 478
  in stability analysis, 208, 209,
    210, 221
  thermal, 2337
Equipment hazard, **2412**
Equipment management programs,
    2412, 2409
Equivalent black blackbodies, 749
Equivalent circuits, 538–539,
    552–553, 635, 639, see also
    specific types
  of capacitors, 20
  current sources and, 55
  in dc motors, 1334, 1335
  hybrid-pi, **651**, 640–642, 643
  in induction motors, 1336, 1338
  large-signal models of, 461–463
  lightning and, 944
  modeling of, 460–468, 944
    dc, 465–467
    large-signal, 461–463
    small-signal, 463–465, 466
    software for, 467–468
  Norton, 57
  per-phase, 1305–1307
  for resistors, 7
  sequence, 1253–1255
  small-signal, 56, 463–465, 466

Thévenin, 57
  three-winding transformer,
    1296–1300
  voltage sources and, 55
Equivalent impedance, 72–74
Equivalent input noise, 1495
Equivalent noise current (ENI), **623**,
    620, 622
Equivalent noise voltage (ENV),
    623, 620, 622
Equivalent series inductance (ESI),
    18
Equivalent series resistance (ESR),
    32, 18, 20
Equivalent sphere illumination
    (ESI), **2275**, 2265
Equivalent uniform annual costs
    (EUAC) method, 2175
ER, see Electrified electrorheological
Erasure, 1688
Erbium-doped fiber amplifiers
    (EDFA), **1425**, 1430
Erbium-doped fiber lasers, 1429
Ergodicity, 1508–1509
ERP, see Effective radiated power
Error control coding, 337
Error correction coding (ECC),
    1693, 1678, 2022
Error detecting codes, 2022
Error-reduction actions, 2249
Errors, **2030**, 2457–2458, see also
    Fault; specific types
  address, **1901**, 1883
  aliasing, 332
  analysis in computational
    electromagnetics, 1046–1047
  area control, 1347
  biomedical equipment user, 2410
  bit, see Bit error rate (BER)
  checking of in computational
    electromagnetics, 1046–1047
  coefficient quantization, 249
  complementary, 846
  in computational
    electromagnetics, 1031, 1046
  in computer systems, 2088
  differential, 141–142
  double-bit, 2023
  estimation, 1509
  fault tolerance and, **2030**, 2020
  focus, 1685
  global truncation, 667, 670
  in local-area networks, 1465
  local truncation, 667–668, 669,
    670
  in long distance communications,
    1432
  loop, 1568, 1571
  mean-square, 290, 323, 1411,
    1412, 1509

  measurement, 961–962, 1497
  minimum mean-square, 290,
    291, 292, 293–295, 296
  in modeling, **1048**, 1046
  modeling of, 2088
  numerical modeling, 1046
  physical modeling, 1046
  probability of, 1483
  processing, 1498
  quantization, 331, 348
  root mean square, 1500
  rounding off, 1852
  single-bit, 2023
  soft, 1652
  sources of, 1513–1514
  squared, 1522
  steady-state, **2113**, 2109, 2118,
    2140
  track, 1679
  truncation, 672, **1516**, 667–668,
    669, 670, 1514
  unit control, 1347
Error signal, 617
ES, see Expanded storage
ESI, see Equivalent series induc-
    tance; Equivalent sphere
    illumination
ESM, see Electronic surveillance
    measuring
ESR, see Equivalent series resistance
Essential modeling, **1984**, 1977,
    1980
Estimation, **328**, 321, see also
    Approximation; specific
    types
  in acoustic signal processing,
    414–415
  Bayesian, **328**, 292, 323, 324
  consistent, **328**, 322
  detection threshold, 953
  efficient, 322
  error in, 1509
  homogeneous, **328**, 325
  linear, **328**, 325
  maximum *a posteriori*, 323
  maximum likelihood, **328**, 292,
    326–327, 1415
  mean-square, **328**, 323–325
  minimax, **328**, 326
  minimum mean-square, 323
  motion, 358, 345, 354–356
  nonhomogeneous linear, **328**,
    325
  nonlinear mean-square, **328**, 324
  nonparametric, **328**, 322
  parameter, **328**, 321–328
  problems in, 1479
  in radar, 961–962
  robust, **328**, 322
  signal, 290–294, 1478

Page on which term is defined is indicated in bold.

spectral, see Spectral estimation
and modeling
state, 1350, 2145–2146
strongly consistent, 322
symbol error rate, **1608**,
1602–1604
unbiased, **328**, 322
in underwater acoustic signal
processing, 414–415
Eta Kappa Nu, 2550
Etching, **489**, 488
Ethernet, 2064
Ethylene-propylene rubber (EPR),
1146
EUAC, see Equivalent uniform
annual costs
Euclidean-induced metrics, 252
Euler identity, 61, 2115
Euler number, 2454–2455
Euler's constant, 1225
Eureka 95 and EU95, **1397**, 1394,
1395
Eureka-147/DAB, 1399, 1403
Evaluation module (EVM), 388
Evanescent fields, 978, 980
Even parity, 2023
Everett's formula, 2513
EVM, see Evaluation module
EW, see Electronic warfare
Excess (offset) representations, 1856
Exchange coupling, 832
Excitation control system, 1326
Excitation matrix, 274
Excitation-response data, 274
Exciters for radar, 952–953
Expanded storage (ES), 1667
Expectation-maximization (EM),
292, 296
Expected value, **321**, 1488, 1499,
2520
Experimental modeling, 2101
Expert systems, **2051**, 1908, 1909,
2048–2051, see also specific
types
Explicit methods, 1042, see also
specific types
Explicit values, **1864**, 1860
Exponential decay time, **2338**, 2335
Exponential failure law, 2027
Exponential rise time, **2338**, 2335
Exponential series, 2460
Exponential signals, 270–273
Exponents, 1856, 2430
Extended Binary Coded Decimal
Interchange Code (EB-
CDIC), 1672
Extended boundary condition
(EBC), 1033, 1034
Extended source, **759**
External cavity klystron, **1392**, 1388

External load torque, 2166
External modulation, **1425**,
1424–1425
External quantum efficiency, **1771**,
1767
External stability, **222**, 216
External thermal resistance,
791–793, 794
Extinction angle, **1241**
Extracellular electrodes, 2313
Extrapolation, 1516
Extra-signal fluctuations, 1488
Extrinsic absorption, 985
Extrinsic magnetic properties, 818,
819–825
Eye diagrams, 1605
Eyeglasses, 2388
Eyephones, 2005

F

Fabry-Perot structures, **1649**, 1641,
1643
Facsimile, **1556**, 1554–1557
Factorials, 2431
Factoring, 2078
Factorization, 155–156, 1502
Factors, 2431
Failure
computer, **2095**, 2087–2090
mean time between, 785, 2028,
2090
mean time to, see Mean time to
failure (MTTF)
rate of, 2025, 2026
Failure density function, 2026
Failure-to-follow distortion, **147**,
145
Failure unit (FIT), 502, 503
False alarms, **1487**, 959, 1479
False contouring, 334
FAN algorithm, 1812
Fan-out, **1621**, **1720**, 1620, 1624,
1699
Farad, **32**, 19
Faraday, Michael, 1098
Faraday rotation, **1171**,
1163–1164
Faraday's laws, 804, 1018, 1296
Far-end crosstalk, 518
Far-end noise, 514, 517
Far-field coupling, 906
Far-field region, **878**, **934**, 862, 875,
876
Fast convolution, 953
Fast Fourier transform (FFT), **238**,
2362, 385, 778, 1761, 2355
digital signal processing and, **238**,
229, 234–236

electroencephalography and,
2355
inverse, 236
microprocessors and, 1761
radar and, 953
in sensor array processing, 361
Fast ion (superionic electric)
conduction, 1179
Fast packet adaption (FFA)
sublayer, 1456
Fast packet networks (FPNs), **1459**,
1456–1457
Fast packet relay (FPR) sublayer,
1456
Fast packets, 1456
Fault, **1267**, **1285**, 1252, 2020, see
also Errors
analysis of, see Fault analysis
avoidance of, **2030**
balanced three-phase, 1253,
1259–1260
computer, 2088
detection of, 2021
double phase-to-ground,
1257–1258, 1261, 1272
grading in semiconductor
manufacturing, 496–497
location of, 2021
masking of, 2021
modeling of, 1810–1811
multiphase, 1272
phase-to-ground, 1256,
1260–1261, 1272
phase-to-phase, 1256–1257, 1261,
1272
recovery from, 2021
sequential, **1816**, 1810
simulation of, **1816**, 1812,
1819
single phase-to-ground, 1256,
1260–1261, 1272
three-phase, 1253, 1256,
1259–1260
tolerance of, see Fault tolerance
types of, 1253–1258, 1272
Fault analysis, 1252–1268
example of, 1258–1265
fault types in, 1253–1258
system model simplifications in,
1253
Fault coverage, **498**, **1815**, 1810,
2028
Fault MVA, **1267**
Fault tolerance, **2030**, **2070**,
2020–2031, 2061, 2062,
2064–2065
active (dynamic) approaches to,
2020, 2021
applications of, 2020
in avionics, **2193**, 2189

Page on which term is defined is indicated in bold.

dependability evaluation and,
2025–2030
dynamic (active) approaches to,
2020, 2021
hardware redundancy and,
2020–2021
hybrid approach to, 2021, 2022
information redundancy and,
2021–2023
in navigation, 2289
passive approaches to, 2020, 2021
software redundancy and,
2024–2025
time redundancy and, 2024
FAX, see Facsimile
FBSOA, see Forward bias safe
operating area
FCR, see Floor cavity ratio
FC-TCR, see Fixed-capacitor,
thyristor-controlled reactors
FDDI, see Fiber-distributed data
interface
FDM, see Finite difference methods;
Frequency-division
multiplexing
FDMA, see Frequency-division
multiple access
FDNR, see Frequency-dependent
negative resistance
FDTD, see Finite-difference time
domain
Feature extraction, 338
Feature overload, 1759
Feature size, **1752**
Feedback, see also specific types
allevitae acoustic, 401
cancellation of, 404
control of, see Feedback control
systems
negative, 215, 547, 556
neural networks and, 426
single-input/single-output, 2126,
2249
in stability analysis, 215, 216
Feedback coefficients, 1412
Feedback control systems, **2153**,
2127, 2148, see also specific
types
closed-loop, 215
integral, 2167
man-machine systems and, 2248
smart materials and, 1173
Feedback division ratio, 1567
Feedback loops, 99, 372, 378, 1186
Feedback shift registers (FSRs),
1727
Feeders, **1319**, 1310, 1311, 1312,
1314, 1316, 1318
Feeder switching, 1349
Feedforward, 1424

Feedforward compensation, 2167
Feedforward cutsets, 371
Feedforward neural network, **428**,
421
Feedwater systems in power
generation, 1200
Fejer window, 233
FEM, see Finite element methods
Fenestration, 2275, 2267
Fermi energy, **473**, 436, 438, 444
Fermi surface, 1110
Ferrimagnetism, 816–818
Ferrites, 1180
Ferroelectric ceramics, 1180
Ferroelectric materials, **1097**,
1087–1097, see also specific
types
applications of, 1089–1091
description of, 1092–1097
electrical characteristics of,
1093–1096
mechanical characteristics of,
1088–1092
structure of, 1091–1092
Ferroelectrics, **1131**, 1126, 1127,
1179, 1185
Ferromagnetic materials, 1179,
1497, 1675, see also specific
types
Ferromagnetism, 816–818
FETs, see Field-effect transistors
FFA, see Fast packet adaption
FFOL, see Fiber-distributed data
interface (FDDI) follow on
LANs
FFT, see Fast Fourier transform
Fiber-distributed data interface
(FDDI), 1435–1437, 1456,
2017, 2018
in local-area networks, 1464
operating systems and, 2064
Fiber-distributed data interface
(FDDI) follow on LANs
(FFOL), 1435–1437
Fiber-optic cable, **2018**, 987–1001,
2016, 2018, see also Optical
fibers
applications of, 1001
coating of, 998
design of, 999–1001
magnetooptics and, 1166
manufacturing of, 993–999
packaging of, 999–1001
proof testing of, 998–999
Fiber-optic gyros, 769
Fiber-optic sensors, 1057, 1182
Fiber-optic wiring, 2269
Fibers, see also specific types
applications of, 990
coating of, 998

copper twisted pair, **1566**, 1559
cylindrical, 980–981
design of, 999–1001
dispersion of, **1425**
drawing of, 996–999
graded-index, **1001**, 975, 976, 989
in long distance, 1428–1429
multimode, 989, 992
nonlinearity of, **1425**
optical, see Optical fibers
packaging of, 999–1001
proof testing of, 998–999
single-mode, **1434**, 989, 992,
1428, 1435, 1436
step-index, 975, 976, 981, 989
Fibrillation potentials, 2320
Fick's laws, 479
Field circuits, **1331**, 1321
Field-effect transistors (FETs), 617,
620, 892
in aerospace systems, 2199
bipolar inversion channel
heterojunction, 1646
CrossCheck technique and, 1828
dc equivalent circuit models of,
467
dielectrics and, 1146
equivalent circuit models and,
460
large-signal equivalent circuit
models of, 462–463
in memory devices, 1651, 1653,
1655, 1656
in microwave devices, 896–898
optical communication and, 1421
photo-, 753
pyroelectrics and, 1127, 1129
in satellites, 2199
small-signal equivalent circuit
models of, 464, 465, 466
video transmission and, 1421
Field equations in three-dimensional analysis, 1018–1020
Field-programmable gate arrays
(FPGAs), 601, **1740**, 591,
592, 597, 1735, 1736
ASICs and, 1733
CAD and, 1740
description of, 1739
Field propagators, **1047**, 1029,
1031–1034
Fields, 32, **1392**
Field scan, 1380
Field strength, **918**, **934**, 1369–1370,
1375
Field weakening, 1335
Field windings, 1331
FIFO, see First-in, first-out
Figure of merit (FOM), **418**, 408,
409

File systems, **2070**, 2032, 2062
Fill-in, **671**, 660
Film capacitors, 19
Film resistors, 8
Filters, **682**, **690**, 903–904, 913–917,
    see also Circuits; specific
    types
  active, see Active filters
  active RC, **631**, 628, 674
  adaptive, 379, 1415
  analog, 2395
  bandpass, see Bandpass filters
  band-reject, 239
  bandstop, 189, 190, 683,
    684–685, 689–690
  Butterworth, see Butterworth
    filters
  cascade, 685
  causal, **195**, 191–192, 239
  Chebyshev, see Chebyshev filters
  coded SAW, 1073–1074
  color-notch, 1388
  critical band, 400
  design of, **250**, 121–124, 913–914,
    see also specific filters
  digital, see Digital filters
  in digital communication, 1414,
    1415
  dispersive, 1073
  Doppler, 953, 955, 958, 959
  doubly terminated, 113, 114, 115,
    117, 118, 119, 120
  Eckart, 414
  in electroencephalography, 2358,
    2360
  elliptic function, 679
  finite impulse response, see Finite
    impulse response (FIR)
    filters
  frequency-selective, 239
  guidance in selection of, 913
  high-pass, see High-pass filters
  history of theory of, 674
  ideal, **195**, 189–191
  implementation of, **250**, see also
    specific filters
  infinite impulse response, see
    Infinite impulse response
    (IIR) filters
  interpolation, 391
  inverse Chebyshev, 118–121
  Kalman, 963, 2293
  lattice digital, 375, 380
  least mean square adaptive,
    379
  linear, 1504–1506
  loop, **1574**, 1568–1570, 1573
  low-pass, see Low-pass filters
  matched, 1414, 1482
  maximally flat delay, 125

  maximally flat magnitude, **125**,
    113
  in microwave devices, 887, 888
  multipole, 1084
  narrowband, 1082, 1084
  in navigation systems, 2293
  nonlinear interference, 1643
  nonrecursive digital, 372
  octave band, 391
  one-third-octave, 391
  output, 129
  passive, **690**, 674–675, 683,
    1569
  Rayleigh wave, see Surface
    acoustic wave (SAW)
    filters
  RC, **631**, 628, 674
  in receivers, 917
  rectifiers and, 129
  recursive digital, 378
  in rehabilitation engineering,
    2395
  Sallen–Key, 677–678, 687, 688
  singly terminated, 114, 116, 117,
    118, 119
  special-purpose, 915
  speech signal processing and, 293
  surface acoustic wave, see Surface
    acoustic wave (SAW) filters
  synthesis of, 674
  testing of, 915
  Thomson, 118, 119
  tracking, 963
  transfer functions of, see Transfer
    functions
  in transmitters, 916
  transversal, 1063
  types of, 913, see also specific
    types
  ultrasound and, 1082, 1084
  velocity, **366**, 360, 365–366
  waveguide, 887
  Wiener, 1508
Final test, **601**, 595
Fin efficiency, **796**, 788
Fingertip tactile sensors, 2173
Finite cosine Fourier transforms,
    2504
Finite difference methods (FDM),
    471, 1037, 2508–2511
  interpolation in, 2511–2516
Finite-difference time domain
    (FDTD), 1042
Finite element methods (FEM), 471,
    1020–1021, 1022, 1023,
    1024
Finite impulse response (FIR)
    filters, **250**, 371, 373
  acoustic signal processing and,
    396, 399, 402

  design of, 239–242
  frequency response of, 390
  implementation of, 244–247,
    386–387, 388–390
  optimal, 241
  in sensor array processing,
    360–361, 362
  surface acoustic wave filters and,
    1063
Finite impulse response (FIR)
    models, 2102
Finite sine Fourier transforms,
    2502–2503
Finite-state machine (FSM), 1738,
    1827
Finite wordlength effects, 250, 242,
    249–250
Finned surfaces, 788
FIR, see Finite impulse response
Firing (delay) angle, **1241**
Firmware, **1877**, 1870, 1874
First-in, first-out (FIFO) memory,
    1727
First Law in Disk Density, 1658
First-order logic, 1903
FIT, see Failure unit
Five-point method, 142–143
Fixed-capacitor, thyristor-controlled
    reactor (FC-TCR),
    1247–1248
Fixed-head disks, 1666
Fixed-point arithmetic, 1760
Fixed-point DSP, **393**, 385, 386–387
Fixed-point iteration, 2507
Fixed-point number systems, 1844,
    1856
Fixed reference digital-to-analog
    converters, **781**, 773, 774
Fixed resistors, 11–12
Fixed satellite service (FSS), 1533
Fixed-to-fixed radio communica-
    tion, 1546
Flag registers, 1722
Flags, 1721, 1894, see also Condi-
    tion codes
Flash analog-to-digital converters,
    781, 775, 777, 778, 779
Flash EEPROMs, 1655
Flatband voltage, 573
Flat packs of integrated circuits,
    614, 607, 608–610
Flat tension mask (FTM), 1782
Fletcher–Munson threshold-of-
    hearing chart, 1402
Flexible manufacturing systems
    (FMS), 2182
Flicker, 1380
Flicker noise, 1496–1497
Flight control systems, 2020, 2285
Flight instruction systems, 2189

Flip-flops, **1640**, 1613, 1635, 1637–1640, 1718, 1721, see also Bistable devices; specific types
asynchronous, 1721
clocked, 1738
collection of, 1722
edge-triggered, **1640**, 1638–1639, 1718, 1719, 1723
implementation of, 1718
inputs of, 1738
master-slave, 1732
set-reset, 1721
synchronous, 1722
types of, 1723, 1733
use of, 1639–1640
Floating gates, 1655
Floating inductors, 680
Floating-point arithmetic, 1760
Floating-point DSP, **393**, 387
Floating-point number systems, **1858**, 1854, 1856–1858, 1864
Floor cavity ratio (FCR), 2259
Floppy disk drives, 826, 830
Flow
laminar, 792
measurement of, 2380
of power, 89, 1288–1289, 1350, 1351
resistance of, 793
turbulent, 793
Flowgraph G₁ (modified Coates graph), **109**
Flowgraphs, 98–101
k trees vs., 102–106
signal-, **109**, 99
Fluids, **1188**, 1178, 1181, see also Liquids; specific types
Fluorescent lights, 745, 749–750, 2269, 2271, 2272–2273
plasma displays and, 1797
Fluorocarbons, 1144
Flux, **1785**, 1785
Flux density, 747, 1018
FM, see Frequency modulation
FMS, see Flexible manufacturing systems
Focus error signals, 1685
Foil tantalum capacitors, 23
Folded bit line, 1652
Folding transformation, **383**, 374–376
FOM, see Figure of merit
Footcandles, **2275**
Forced-commutation, **737**, 733
Force of attraction, 16
Force pedestals, 2173
Force sensors, 1155, 1156
Force-torque sensors, 2180

Formal parameters, **1459**, 1453
Formants, 302
Form of equations, **1085**, 1077
Forward-backward procedure, **313**, 311
Forward-biased base-emitter junction, 538
Forward-biased diodes, 133, 450–451
Forward bias safe operating area (FBSOA), 698
Forward (far-end) crosstalk, 518
Forward knee current, 641
Forward paths, 99
Forward substitution, 659
Forward transconductance, 641
Forward voltage, **701**
Fossil power plants, 1194–1200
Fourier analysis, 1607, see also specific types
of ac waveforms, 1235–1236
of distortion, 145, 146, 147
speech signal processing and, 300–302
Fourier coefficients, 316
Fourier cosine series, 142
Fourier equations, 787, 790
Fourier series, 1512, 1515, 2480
Fourier slice theorem, 342
Fourier transforms, **1516**, 239, 2500–2507, see also specific types
antennas and, 875–876, 951
aperture antennas and, 875–876
cosine, 2500, 2501, 2504, 2505
digital image processing and, 342
digital signal processing and, 385
discrete, see Discrete Fourier transforms (DFT)
fast, see Fast Fourier transform (FTT)
finite cosine, 2504
finite sine, 2502–2503
inverse, 290, 303, 1512
in medical imaging, 2376
noise and, 1489, 1490
nonnegative-definite, 268
one-dimensional, 2376
radar and, 951, 953
sampling theorems and, **1516**, 1511, 1512, 1513
in sensor array processing, 360–361
sine, 2502–2503, 2505
speech analysis and synthesis and, 300, 301, 303
speech signal processing and, 303
stochastic processes and, 1505
surface acoustic wave filters and, 1065

two-dimensional, 2376
Four-mesh networks, 67
Four-node networks, 64, 65
Four-quadrant operation, 737, 731
Fourth harmonic distortion, 142
FP, see Floating-point
FPGAs, see Field-programmable gate arrays
FPNs, see Fast packet networks
FPR, see Fast packet relay
FR, see Frequency response
Fractals, **2013**, 2004
Fractional exponents, 2430
Fractionally spaced equalizers (FSE), 1413
Fraction conversion, 1850–1853
Frame relay, 1448
Frames, **1392**, 1379, 1454
Fraunhofer propagation region, **878**, 875, 876
Free electron model, 1108
Free-field voltage sensitivity, 1057
Free-run frequency, **1574**
Free-space loss, 843
Freewheeling diodes, 702, 703
Frequency, see also Bandwidth; Frequency response
allocation of, see Frequency allocations
attenuation vs., 990
beat, 143
beta cutoff, 642
bipolar transistors and, 553–555
of Chebyshev filters, 194
conductance dependent on, 1068
crossover, **2130**, 2115, 2250, 2251
cut-off, 113, 239, 851, 853, 858
diversity of, 1400
Doppler, **973**, 959, 964, 968, 969
electroencephalography and analysis of, 2354–2356
free-run, **1574**
glottal excitation, 300
harmonic, **38**, 36
intermediate, 1360
load, 1347, 1348
natural, **1575**, 201, 1569, 2110, 2112
Nyquist, 2282
of oscillation, 2110, 2112
output, 143, 144
phase shift dependent on, 884
propagation constant and, 851
pulse repetition, 955, 959, 2395
radial, 1133
radio, see Radio frequency (RF)
Raman, 1179
resonant, **32**, 30, 1084
sideband, 1179
sinusoidal, 198, 679

Page on which term is defined is indicated in bold.

sonar system dependency on, 408
of speech (pitch), **305**, 298
synthesis of, 1574
unity-gain, 576, 629
voltage ratio to, 732, 733
Frequency allocations
  for broadcasting, 1368, 1373,
    1383
  for radio, 1368, 1373
  for satellites, 1542, 1543
  for telemetry, 1587–1588
  for television, 1383
  for wireless local-area networks,
    1564–1565
Frequency-dependent negative
  resistance (FDNR), 674, 681
Frequency distortion, **147**
Frequency-division multiple access
  (FDMA), **1539**, 1534, 1541,
  1542
Frequency-division multiplexing
  (FDM), 1418, 1439, 1534,
  1542
Frequency domain, 2108
  in computational
    electromagnetics, 1041, 1042
  design of, 2119, 2124, 2125
  instruments and, 2282
Frequency-domain differentiation
  formulas, 159
Frequency domain speech coders,
  282–283, 299
Frequency modulation (FM), 400
  in electroencephalography, 2359
  phase-locked loops and, 1574
  satellites and, 2207
  spectrum of, 1378–1379
  telemetry and, 1580
  in video, 1418
Frequency modulation (FM)/
  continuous wave (CW)
  radar, 967–969
Frequency modulation (FM)/
  frequency modulation (FM)
  telemetry, 1582, 1585, 1589
Frequency modulation (FM) radio,
  1372–1379, 1576
  transmitters for, 1375, 1583
Frequency multipliers, 1010
Frequency response, **206**, 198–206,
  2113–2131
  of bipolar transistors, 558–559,
    560
  Bode diagrams and, see Bode
    diagrams
  compensation and, 2140–2141
  of FIR filters, 390
  instrument modeling and, 2280
  of JFET, 540–543
  linear curves of, 200

linear plotting of, 199–200, 205
  in linear systems, 198
  methods of plotting of, 206
  of one-third-octave filters, 391
  phase-lead compensation and,
    2116–2119
  sinusoidal, 679
  source-follower, 540–541
  ultrasound and, 1083, 1084
  welding and, 2227
Frequency reuse, **1544**
Frequency scaling, 122
Frequency-selective filters, 239
Frequency-shift keying (FSK), 1364,
  1410
Frequency spectrum, 920
Frequency synthesizers, 1572
Frequency-to-voltage converters,
  771
Fresnel approximation, 875
Fresnel reflection coefficients, 843,
  844, 1165
Frictional torque, 2166
Friis transmission equation, 842
Frobenius norm distance metrics,
  252
FSE, see Fractionally spaced
  equalizers
FSK, see Frequency-shift keying
FSM, see Finite-state machine
FSRs, see Feedback shift registers
FSS, see Fixed satellite service
FTM, see Flat tension mask
Fuel, **1205**, 1196, see also specific
  types
Fuel cells, **1215**, 1212
Fuel scheduling, 1350
Full adders, **1747**, 1742
Full-attention operations, 2251
Full-bridge converters, 717
Full-wave capacitors, 23
Full-wave control, **710**
Full-wave rectifiers, 128
Functional block diagrams, 390, 391
Functional dependencies, 2034
Functionality, **1926**
Functional paradigms, **1914**, 1907
Fundamental mode, **1640**, 1639
Fundamental products, 1704
Fundamental theorem of expecta-
  tion, 1500
Fuse map generation, 1718
Fuses, 39–45, see also specific types
  coordination of, 40
  high rupturing capacity, 43–44
  interrupting rating of, **45**, 39
  performance of, 40
  ratings of, 39–40
  renewable, 43
  selective coordination of, 40

standards for, 41–43
  trends in, 44
Fusion center, 1485
Fusion splices, 1430
Fusion welding, 2223
Fusion zone geometry, 2224

G

Gain, **902**, see also specific types
  adjustment of, 2115
  as amplifier characteristic, 553
  antenna, **869**, **878**, 862–863,
    876–877
  array, 413
  automatic control of, 2203
  avalanche, 753
  common-base (CB) current, 900
  common-emitter current, 546,
    642
  control of, 2203
  difference-in-to-difference-out
    voltage, 563
  infinite, 687
  loop, **1574**, 620, 1567
  open-loop, 617, 620
  phase detector, **1575**
  reduction in, 2116
  small-signal ac voltage, 537–538
  source-follower voltage, 539–540
  spectral, 291
  static common emitter current,
    640
  terminal voltage, 551
  voltage-controlled oscillator,
    1575
Gain margin, 2140
Gain-setting adaptive controllers,
  2232
Galerkin's method, 1039
Gallium-arsenide, 126, 2199
Gamma radiation, 2377
Gamma ray photons, 2377
Gantry configuration in robotics,
  2159–2161
Gap filler transmitters, **1403**
Gap junctions, 2337
Gas chromatography, 1158
Gas discharge displays, 1789–1791
Gas discharge reactions, 1789–1791
Gases, 1132, 1139, 1143, 1196, 1213,
  see also specific types
Gas Laws, 2465–2466
Gas metal arc welding (GMAW),
  2235, 2225, 2226, 2227,
  2228, 2234
Gas tungsten arc welding (GTAW),
  2235, 2224, 2227, 2229,
  2231, 2232

Gate arrays, 591, 592, see also
   specific types
Gate capacitance, 574
Gate delay/stage, 575
Gated latches, 1637
Gated registers, 1722–1724
Gate insulators, 568
Gate response time of MOSFET,
   574
Gates, **579**, 530, 1698, see also
   specific types
   capacitance of, 574
   common, 536
   effective loading of, 1623
   floating, 1655
   insulation of, 568
   logic, see Logic gates
   MOSFET and, **579**, 567
   specification parameters for,
      1622–1624
   unused, 1629
Gate-turn-off (GTO) thyristors,
   696–697, 734, 2243
Gauges, 1018, 1019, see also specific
   types
Gauss' formulas, 804, 2512, 2515
Gaussian autoregressive model, see
   Autoregressive (AR) model
Gaussian channel noise, 287
Gaussian diffusion, 480
Gaussian distribution, 482, 845,
   1488
Gaussian elimination, **671**, 657, 658,
   660
Gaussian mapping, 1601
Gaussian noise, 287, 288, 1485,
   1490, 1601
   white, **1487**, 1480, 1484, 1490,
      1491–1492, 1506
Gaussian PDF, 961
Gaussian probability density, 1497
Gaussian processes, 1483,
   1501–1502, see also specific
   types
Gaussian profiles, 481, 483
Gaussian quadrature, 1036
Gaussian random variables, 1504,
   1523
Gaussian stochastic process, 1601
Gaussian subsources, 293
Gaussian system, 1133
Gaussian window, 234
Gauss–Seidel/Jacobi methods, 471
GDS2, **601**
Generalized finite element methods,
   1025
Generalized impedance converters
   (GIC), 674
Generalized likelihood ratios, 1481,
   1483

Generalized transfer function, 1029
General register machines, 1883
General trees, **1926**, 1923
Generation-recombination current,
   459
Generators, 1197–1199, 1321–1331,
   see also specific types
   ac, see ac generators
   alternating, see ac generators
   circuit models of, 1323–1326,
      1330–1331
   dc, 1329–1331
   fault analysis and, 1258
   induction, 1321, 1329
   mathematical models of,
      1323–1326, 1330–1331
   modeling of, 1323–1326,
      1330–1331
   noise, 391, 1599–1601
   principles of operation of, 1322,
      1329–1330
   pseudorandom, **1608**, 391,
      1599–1601
   sequence, 1601
   signal, 1599–1601
   superconducting, 1327–1328
   synchronous, see Synchronous
      generators
   test pattern, see Test pattern
      generation
   types of, 1321
   Van de Graaff, 1342–1343
Generic flow control (GFC), 1443
Geometric corrections, 1101–1102
Geometric mean, 2486
Geometric modeling, 2005, 2011
Geometric optics, 745–746, 1034
Geometric progression, 2432
Geometric theory of diffraction
   (GTD), 1034
Geometry, 2434–2436
   analytic, see Analytic geometry
Geopressured source, 1203
Geosynchronous satellites, **2209**,
   2197–2198
Geothermal power plants, **1215**,
   1202–1203, 1207,
   1211–1212
GFC, see Generic flow control
GIC, see Generalized impedance
   converters
Glare, 2265, 2266
Glass electrodes, 2346
Glasses, 1145, 2533
Glitch, **1720**
Global asymptotic stability, **222**,
   208, 209
Global convergence, 663
Global positioning system (GPS),
   2289

Global registers, 1883
Global truncation errors (GTE),
   667, 670
Glottal excitation frequency, 300
Glucose, 1158
Glucose sensors, 2348
GMAW, see Gas metal arc welding
GMP, see Good Manufacturing
   Practice
Good Manufacturing Practice
   (GMP), 2411
Governors, 1326
G proteins, **2338**
GPS, see Global positioning system
Graded-index (GRIN) fibers, **1001**,
   975, 976, 982–984, 989
Gradient descent, 422
Graphical approach to amplifiers,
   636–637
Graphical user interfaces, 1912
Graphics, see Computer graphics
Graphics tablets, 2012
Graphic symbols, 1733–1734
Graphs, **109**, 92–93, see also Graph
   theory
   Coates, **109**, 100, 101, 102, 103
   connectedness of, **109**
   data types and, 1923–1924
   flow-, see Flowgraphs
   linear, **109**
   Mason, 102
   modified Coates, **109**, 100
   sub-, **109**, 110, 93
Graph theory, 64, 92–110, see also
      Graphs; Topology
   flowgraphs and, 98–101
   k trees vs., 102–106
   k trees and, **109**, 93, 94–98
   flowgraphs vs., 102–106
   network, 92
Gravitational constant, 2168
Gravitational torque, **2174**, 2166
Gray scale, 1796–1797, 2385
Grazing angle, 844, 845
Greek alphabet, 2419
Green's function, 1030, 1032, 1033
Green's theorem, 1032
Grids, **601**, **614**
GRIN, see Graded-index
Grooved media, **1693**, 1679, 1686
Grounding, 903–906
   on analog-to-digital converters,
      780–781
   cable, 906
   cable shield, 905
   circuit, 906
   design of, 905–906
   on digital-to-analog converters,
      780–781
   logic gates and, 1628–1629

Page on which term is defined is indicated in bold.

multipoint, 906
principles of, 904–905
safety and, 905
single-point, 906
Groundwave propagation, 1370,
1371
Group velocity, 744
Growth factors, 2338
G/T, 1538–1540
GTAW, see Gas tungsten arc
welding
GTD, see Geometric theory of
diffraction
GTE, see Global truncation errors
GTO, see Gate-turn-off
Guide wavelength, **859**, 851
Gummel–Poon representation, 462
Gunn diodes, 892, 894, 1004
Gyrators, 680, 681, 806
Gyrocompasses, 2288
Gyroscopes, 2288, 2289

H

Hafnia, 1146
Half-bridge converters, 715–717
Half-wave dipoles, 864
Halfwords, 1882
Hall angle, 1106
Hall coefficient, 440, 1106, 1110
Hall effect, 1106–1111
electronic structure and,
1108–1109, 1110
in magnetic field measurement,
440–441
magnetism and, 814
semiconductors and, 440–442,
1110
theory of, 1107–1108
Hall effect sensors, 1941
Hall electric field, 1107
Hall resistance, 441, 442
Halogen lamps, 2269
Halstead's metric, 1906
Hamming distance, 2022
Hamming distortion, 1522, 1523
Hamming window, 233, 301, 302
Handicaps, **2395**, 2387
Handoff, **1539**, 1547, 1548
Hard copy technologies, 2005–2006
Hardware, see also Computers;
specific types
for computer graphics,
2005–2007
construction of, 1902
fault tolerance and, 2020–2021
limitations of, 1981
redundancy of, 2020–2021
security of, 2080–2081

Hardware description languages
(HDLs), 598–599, 2008
Harmonic balance, 147
Harmonic distortion, **147**, 139–140,
141, 142, 143
total, see Total harmonic
distortion (THD)
Harmonic frequency, **38**, 36
Harmonic mean, 2486
Harmonic oscillators, 218
Harmonic resonance, 1246
Harmonics, 139, 147, 1235–1237
Hartley, Ralph V. L., 46
Harvard architecture, 385, 1761
Hash coding, 1924
Hash function, 1924
Hashing, 1924
Hazard, **1720**
HBA, see Host bus adapters
HBTs, see Heterojunction bipolar
transistors
HDA, see Head disk assembly
HDLs, see Hardware description
languages
HDTV, see High-definition
television
Head disk assembly (HDA), **1669**,
1660
Head-mounted displays (HMDs),
2005
Heads, **1669**, 832–834, 1658, 1659,
1666
Healthcare codes and standards,
2405–2406
Heap, 1884–1886
Hearing aids, 397–398
Hearing protective devices (HPDs),
403
Heat energy, 1197
Heat fluxes, 786
Heat sinks, 793
Heat source power density, 2223,
2224
Heat sources, 786, 791, 997,
2223
Heat spreaders, 791
Heat transfer
conductive, **796**
convective, **796**
ebullient, **796**, 788
fundamentals of, 787–788
mechanisms of, 793, 794
radiative, **796**, 787
Heat transfer coefficients, **796**, 787,
788, 791, 792, 793, 794, 795
Height-of-burst (HOB) sensors,
970, 971
Helium-neon lasers, 738
Helmholtz coil, 2333
Helmholtz's wave equations, 838

HEMTs, see High electron mobility
transistors
Hermite polynomials, 2485
Hermiticity and Onsager relation,
1162
Hertz, Heinrich, 1026–1027
Heterodyne, 1010
Heterojunction bipolar transistors
(HBTs), 892, 901, 1004
Heterojunction LEDs, 1768
Heuristics, 584
Hexadecimal number systems, **1720**,
1858, 1846–1848, 1888
Hexamethyldisilazane (HMDS), 487
HICs, see Hybrid integrated circuits
HID, see High-intensity discharge
Hidden Markov models (HMMs),
297, **313**, 290, 291, 293, 295,
296
speech recognition and, 307,
309–313
topology of, 310
Hierarchical databases, 2040–2041
High-capacity switching, 1443
High-definition television (HDTV),
1394–1397, 1419, 2209
satellites and, 1534
video signal processing and, 345,
353
High electron mobility transistors
(HEMTs), 892, 898, 1004,
1005
Higher-order languages (HOLs),
see High-level languages
(HLLs)
Higher-order spectra, 1483
High-fidelity sound reproduction,
1058
High forward current, 452–453
High-frequency smart shielding
materials, 1175
High-input impedance amplifiers,
1127
High-intensity discharge (HID)
lamps, 2269
High-level languages (HLLs), **1877**,
1901, **1914**, 1874, 1876,
1878, 1902–1915
applications of, 1879
assembly language and,
1898–1900
compilers and, 1880, 1894
description of, 1903–1905
optimizers of, 1882
paradigms and, 1906–1912
High-pass filters, **124**, 117, 239, see
also specific types
active, 683, 684, 686, 687–688
ideal, 189, 190
transformation to, 684

High Performance Computing and Communications Initiative (HPCCI), 2398
High-power amplifiers (HPAs), 2199, 2206, 2207
High-Q resonant circuits, 1084
High-resolution electrocardiography (HRECG), **2370**, 2364, 2367
High rupturing capacity (HRC) fuses, **44**, 43–44
High-voltage dc (HVDC) transmission, 1227–1242
  configuration of, 1228–1232
  economics of, 1233
High-voltage diodes, 128
High-voltage networks, 1310
High-voltage substations, 1310
High-voltage transmission networks, 1344
Hilbert space setting, 255, 256
Hilbert transform susceptance, 1067, 1068
Histograms, 335
HL7, **2406**, 2398
HLLs, see High-level languages
HMDS, see Hexamethyldisilazane
HMDs, see Head-mounted displays
HMICs, see Hybrid microwave integrated circuits
HMMs, see Hidden Markov models
HOB, see Height-of-burst
Hodgkin–Huxley membrane model, 2303
HOL (higher-order language), see High-level languages (HLLs)
Holes, **446**, **1110**, 437–438, 445, 1109, see also Electrons
  diffusion of, 447
  diffusion constant for, 449
  recombination of electrons and, 1763
  temporarily depleted, 1488
Homogeneity conditions, 69
Homogeneous estimates, 325
Homogeneous linear estimators, **328**
Homomorphic (cepstral) analysis, 303, 307–308
Hooke's law, 1077
Hopfield crossbar neural network, 423–424
Hop-off resistance, 12
Horizontal, **1392**
Horizontal microinstructions, **1877**, 1871–1872
Hormones, 1158
Horn antennas, 870
Horns, 1060
Hospital information systems (HIS), 2400, 2405–2407

Host bus adapters (HBA), 1664
Hot-electron effects, 578
Howling (alleviate acoustic feedback), 401
HPAs, see High-power amplifiers
HPCCI, see High Performance Computing and Communications Initiative
HPDs, see Hearing protective devices
H-plane, 861
HRC, see High rupturing capacity
HRECG, see High-resolution electrocardiography
HSPICE, 621
Hue, **1392**, 1382
Huffman algorithm, **1526**, 339, 340, 350, 1519, 1520
Human actuation elements, 2248
Human behavior, 2247, 2248, 2249, 2250–2254
Human-machine interface, see Man-machine interface (MMI)
Human visual system (HVS), 357, 358
Hum bars, **1392**, 1381
HVDC, see High-voltage dc
HVS, see Human visual system
Hybrid integrated circuits (HICs), 502, 887–888, 892, 1004
Hybrid microwave integrated circuits (HMICs), **1014**, 1004, 1011
Hybrid number systems, 382, 394
Hybrid-pi equivalent circuits, **651**, 640–642, 643
Hybrid-pi transistors, 552, 553, 554, 559
Hydraulic telemetry, 1579
Hydraulic transducers, 1057
Hydrodynamic modeling, 440
Hydroelectric power plants, **1215**, 1203–1204, 1211
Hydrofluoric acid, 1146
Hydrogenerators, 1205
Hydrometeors, 847–848
Hydrophones, **418**, 1055, 1059
Hydrophone sensors, 412
Hydrostatic pressure, 1102, 1104
Hydrothermal scheduling, 1350
Hydrothermal source, 1202
Hyperbola, 2452–2454
Hyperbolic functions, 2475
Hypercapnia, 2325
Hypercube architecture, 1760
Hypocapnia, **2328**
Hypoxia, **2328**, 2325
Hysteresis, **1097**, 1018, 1087, 1090, 1091, 1094

Hysteresis-band method, 737, 736
Hysteresis loop, 822
Hysteresis loss, 35

I

ICs, see Integrated circuits
ICW, see Interrupted continuous wave
Ideal current sources, 50–51
Ideal filters, **195**, 189–191, 192, 1506, see also specific types
Ideality factor, **458**, 450
Ideal operational amplifiers, **623**, 616, 617–619
Ideal transformers, 806, 1296–1298
Ideal voltage sources, 50–51
Identity matrix, 2439–2440
IE, see Integral equations
IEE, see Institution of Electrical Engineers
IEEE, see Institute of Electrical and Electronics Engineers
IEEE Computer Society Press, 2552
IEEE standards, **2018**, 2015, 2016
  in biocomputing, 2398
  for large drives, 2240, 2245
  for local-area networks, 1466, 1467
  for number systems, 1857
  for operating systems, 2064
  for testing, 1827
IEPS, see International Electronics Packaging Society
IF, see Intermediate frequency
IGBTs, see Insulated-gate bipolar transistors
IIR, see Infinite impulse response
ILE, see Internal latch enable
Illuminance, **2275**, 747, 2257–2259, 2264, see also Illuminating systems
Illuminating systems, 2257–2276, see also Light
  applications of, 2271–2272
  components of, 2268–2271
  control systems in, 2273
  daylighting and, 2267–2268, 2273
  design of, 2257–2264
  energy efficiency in, 2272–2275
  energy standards for, 2273–2275
  equipment selection for, 2272
  factors affecting, 2264–2268
  illuminance level determination and, 2257–2259
  illumination computational methods for, 2259–2264
  light sources and, 2268–2269

Page on which term is defined is indicated in bold.

luminaires in, 2270–2271
types of, 2271–2272
Illumination, **1785**, 1784
Illumination computational
    methods, 2259–2264
Image analysis, 343
Image capture, 330–334
Image coding, **343**, 329, 337
Image compression, **343**, 426, 1526
Image digitizers, 329
Image dissector tubes, 758
Image encoding, 338
Image enhancement, 335–336
Image features, **343**
Image-intensified tubes, 756–758
Image orthicon tubes, 757
Image quality, 356–357
Image reconstruction, **343**, 341–342
Imaginary power, 82
Imaging, 2378–2379, see also
    specific types
    clinical information systems and,
        2400
    color flow, 2385
    magnetic resonance, **2380**, 2360,
        2374, 2378, 2379, 2400
    medical, see Medical imaging
    soft tissue, 2380
Imaging detectors, 755–759, see also
    specific types
Immunosensors, 1158–1159, see
    also specific types
Impact avalanche and transit time
    (IMPATT) diodes, 892, 894,
    895–896, 1004, 1005, 1010
Impact ionization, 127, 443
Impact printers, 1970–1975
IMPATT, see Impact avalanche and
    transit time
Impedance, 32
    of capacitors, 20
    characteristic, see Characteristic
        impedance
    common-mode input, 620
    crosstalk and, 520–521
    driving point, **651**
    equivalent, 72–74
    of inductors, 24–25, 28–29
    input, 620, 863, 1084
    intrinsic, 839
    line, **934**, 933
    linear circuit analysis and, 67
    lumped, 1082
    matching of, 2278
    mismatches in, 915
    mutual, 1218, 1224
    output, 57
    radiation, 1059–1060
    reference, **891**, 883
    self-, 1218, 1224

series, 1218–1219
short-circuit, 1306
steady-state, 111
surge, **1222**
terminating, 1084
Thévenin, 73, 74, 85, 164, 165
transformation of, 542–543
transformer, 38, 806, 1236, 1306
ultrasound and, 1081, 1082
wave, **859**, 851
Impedance converter/inverter
    networks, 680, 682
Impedance matrix, 1036
Impedance networks, 773
Impedance scaling, **125**, 122
Implant energy, 482
Implementation modeling, 1977,
    1981–1983
Implementation part, **1914**, 1910
Implicit methods, 1042, see also
    specific types
Implicit values, **1864**, 1860
Impregnated-paper insulation,
    1149
Impulse function, 232
Impulse response, **168**, 166–167,
    238, 1066, 1508
    in control systems, **2113**, 2103,
        2108
    finite, see Finite impulse response
        (FIR)
    infinite, see Infinite impulse
        response (IIR)
Impulse signals, 47–48
Incandescent lights, 748, 2268–2269
Incident energy, 906
In-circuit emulators, 1751
Incoherent light, 746–750
Incremental available power, 1493,
    1494
Incremental input resistance, 551
Incremental modeling, **458**,
    454–456
Incremental resistance, **458**, 454
Indefinite integrals, 2466–2467
Independent joint control in
    robotics, **2174**, 2163–2167
Independent nodes, 57, 94
Independent variables, 1500
Independent verification and
    validation (IV&V), 1989
Indeterminant forms, 2464
Indirect weld parameters (IWP),
    2235, 2224, 2225, 2229,
    2230, 2233
Indium-tin oxide (ITO), **1777**, 1773
Inductance, **32**, 29
    coil, 26
    coupled, 25
    equivalent series, 18

mutual, **32**, 25, 513, 522
simulation of, 680, 681
Induction, electromagnetic, 1098
Induction generators, 1321, 1329
Induction law of Faraday, 804
Induction motors, **1341**, **2246**, 1334,
    1336, 2238, 2241, 2242,
    2243
    analysis of, 1337–1340
    current-fed inverter, 733–734
    current-fed PWM inverter, 734
    slip power recovery control of,
        735–736
    voltage-fed inverter, 731–732
    wound-rotor, **1341**, 735, 1334,
        2243
Inductive coupling, 513, 2330
Inductive crosstalk, 513
Inductive noise, 515, 523
Inductive reactance, **32**, 24
Inductors, **32**, 15–18, 24–30, see
    also specific types
    impedance of, 24–25, 28–29
    quality factor for, 27
    resonant frequency of, 30
    serial, 24
    time constant for, 27–28
Industrial illuminating systems, see
    Illuminating systems
Industrial systems, see specific
    types
Indwelling sensors, 2343
Inequality, 1500, 1518
Inertial reference units, 2289
Inertial space, **2294**, 2288
Inertia matrix, 2169
Infinite conductivity, 1296
Infinite gain high-pass circuits, 687
Infinite impulse response (IIR)
    filters, **250**, 390
    design of, 242–244
    implementation of, 247–250
Infinite impulse response (IIR)
    models, 2103
Infinite memory, 1503
Inflection points of curves, 2463,
    2465
Information modeling, **1984**, 1980,
    see also Information theory
Information-preserving coders,
    349–350
Information redundancy,
    2021–2023
Information sources, 1517
Information systems, see also Data;
    Databases; Information
    theory; specific types
    C3 system and, 2214–2215
    clinical, 2398–2399, 2400
    hospital, 2400, 2405–2407

Page on which term is defined is indicated in bold.

input-output devices for, see
  Input-output devices
traveler, 2255
vehicle driver, 2255
Information technology equipment
  (ITE), 929
Information theory, 1478–1527, see
  also Information modeling;
  Information systems
  birth of, 1528–1531
  coherence and, 1507–1508
  data compression and, see Data
    compression
  entropy and, 1517–1518
  ergodicity and, 1508–1509
  linear transformations and,
    1489–1490
  noise and, see Noise
  parametrized signals and,
    1480–1482
  random variables and, 1499–1500
  sampling theorems and,
    1510–1517
  signal detection, see Signal
    detection
  stochastic processes and, see
    Stochastic processes
Infrared detectors, 1126, 1129
Infrared energy, 613
Infrared halogen lamps, 2269
Infrared radiation, 1126
Infrared sensors, 752, 2228
Infrared systems, 1564
Inheritance, 1910
Injection electroluminescence, **1771**,
  1764
Injection molding, 1146
Ink-jet plotters, 2005
Ink-jet printers, 1965–1968, 2006
Inner product space, 255
Inorganic solids, 1145–1146
In-phase components, 1481
Input backoff, 1541
Input bias current, 620
Input capacitance, 1128
Input current, 620, 622
Input impedance, 620, 863, 1084
Input offset voltage source, 620
Input-output devices, **1869**, 1866,
  1875, 1975, see also specific
  types
  advantages of, 1956
  computer graphics and, 2004,
    2012
  data tablets, **1956**, 1943–1945,
    1948
  disadvantages of, 1956
  functional evaluation of, 1955
  graphics tablets, 2012
  joysticks, **1956**, 1947–1948, 2012

keyboards, **1957**, 1865,
    1938–1941, 2012
  light pens, **1957**, 1941–1943, 2012
  mouse, **1957**, 1945–1947, 2012
  performance parameters for,
    1955
  printers, see Printers
  scanners, **1957**, 1953, 1954
  touch, **1957**, 1948–1953, 2012
  trackballs, **1957**, 1945, 1947,
    2012
  voice, **1957**, 1954, 2012
Input-output ports, **1734**,
  1728–1731
Input-output transfer function,
  2103, 2114, 2128
Input-output units, see also Output
  devices
Input-output workloads, 1664–1665
Input power, 1541
Input resistance, 551, 620
Input signals, 562
Input switching, 773
Input transducers, 1580
Input vector, **176**, 170
Input voltage, 620, 622, 725, 1068
Insertion loss, **891**, 883, 884
Insertion potentials, 2320
Insolation, **1215**, 1207, 1208, 1209,
  see also Photovoltaics
Instability, **222**, 207, 211
Instance variables, **1914**, 1910
Instantaneous overcurrent relays,
  1271
Institute of Electrical and Electron-
  ics Engineers,
  2551–2552
Institution of Electrical Engineers,
  2553
Instrumentation amplifiers, 771,
  1803
Instruments, **2283**, 2277–2284, see
  also specific types
  computer-based, 2281, see also
    specific types
  Doppler, 2380
  in electrocardiography, 2319,
    2366–2369
  in electroencephalography, 2327
  electronic flight, 2189
  elements of, 2277–2281
  medical, 778, see also specific
    types
  modeling of elements of,
    2277–2281
  noise reduction techniques and,
    2281
  physical variables and, 2277
  sampling and, 2282
  virtual, **2283**

Insulated-gate bipolar transistors
    (IGBTs), **737**, 699–700, 706,
    734
Insulation, 1221–1222
  aging and, 1140–1142
  dielectric constants of, 16
  gate, 568
  liquid materials for, 1144–1145
  resistance of, 18
  solid-liquid materials for,
    1149–1150
  solid materials for, 1145–1149
  synthetic, 1221
  transmission lines, 1221
Integers, 1848–1850, 1860–1861
Integral calculus, 2466–2472
Integral cavity klystron, **1392**, 1388
Integral equation propagator, **1048**,
    1032–1033, 1036, 1037,
    1040, 1041–1042, 1043
Integral equations, **1048**,
    1032–1033, 1036, 1037,
    1040, 1041–1042, 1043, see
    also specific types
  frequency domain and, 1041
  time domain and, 1042
Integral feedback, 2167
Integral form, 2464
Integral methods, 1023
Integrals, 2466–2468, 2497–2500
Integrated circuits, **614**, 581–602,
    632–633, see also Chips
  active filters and, 674, 675,
    681–682
  in analog-to-digital converters,
    771, 778–781
  application-specific, see
    Application-specific ICs
    (ASICs)
  arithmetic-logic units and, 1741
  assemby of, 605
  bipolar transistors and, 561
  bistable devices and, 1635
  design of, 591
  in digital-to-analog converters,
    771, 772, 774–775, 780–781
  fabrication of, 475, 484
  flat packs of, **614**, 607, 608–610
  history of, 616
  hybrid, 502, 892, 1004, 1011
  large-scale, 603, 1572, 1622,
    1631
  layout of, **590**, **601**, 581–584
  logic elements of, see Logic
    elements
  medium-scale, 1614, 1741
  in memory devices, 1651–1657
  microwave, see Microwave
    integrated circuits (MICs)
  monolithic, 892, 896

monolithic microwave, **1015**, 1004, 1006, 1007, 1008, 1009, 2209
in MOSFET, 568
noise in leads of, 522–524
in optoelectronics, 760, 765
packaging of, 604, 607, 608–610
phase-locked loops and, 1572
placement routines and, **590**, 583, 584–589
quality of, 1808
reliability of, 1808
in resistors, 13
routing routines and, **590**, 583, 589–590
in semiconductor manufacturing, 475, 484, 499
in sensors, 1160
serial interface, 771
small-outline, 610, 611
small-scale integration, 1614, 1622
switching logic of, 1613
testing of, 1808–1816
fault models and, 1810–1811
taxonomy of, 1808–1810
Integrated injection logic, 1712
Integrated optics, **770**, 759, 760–765
Integrated renewable energy systems (IRES), **1215**, 1207
Integrated services digital network (ISDN), **1446**, 1441, 1450, 1451, 2209
broadband, see Broadband-integrated services
Integrating digital-to-analog converters, **781**
Integrodifferential equations, 160–161
Intelligence, **2221**, 2211
Intelligent control systems, **2182**, 2234–2235
Intelligent materials, see Smart materials
Intelligent networks, 1468–1476
advanced, 1473–1475
alternate billing services and, 1472–1473
architecture of, 1469–1470
CCS7, 1470–1471
history of, 1468–1469
systems of, 1470
Intelligent structures, see Smart structures
Intelligent vehicle/highway systems (IVHS), 2255
Intelligibility, **297**, 280–281, 282, 287
maximization of, **405**, 398

Intensity modulation, 763, 769, 1419–1420
Intentional radiators, 923–925
Interactive-real-time systems, 2004
Interaural attributes, **405**
Intercellular signaling, 2336
Interchanges in energy management systems, **1353**, 1346, 1348
Interconnection capacitance, 575
Interconnection networks, **2059**, 2052
Interconnections, 499–529
characteristic impedance and, **528**, 504–513
manufacturing tolerances and, 510–513
printed circuit board structure and, 505–510
of computers, 1460
costs of, 501, 502
crosstalk and, see Crosstalk
delay in, 575
distribution of, 502
limits of, 502
of local-area networks, 1442
logic gates and, 1629
metrics of, 501–503
patterns of, **601**, 591
propagation delay and, **528**, 503–504
reliability of, 503
risetime degradation and, **528**, 524–527
Interconnections and packaging (I&P), **528**, **615**, 500, 603
Interdigital transducers, **1075**
Interfaces, **631**, **1025**, 1081, 1082, see also specific types
analog signal, 1801–1802
application program, 2067
common management, 2203
digital signal, 1801–1802
distributed queue dual bus, 1456, 1464, 2017, 2018
fiber-distributed data, see Fiber-distributed data interface (FDDI)
graphical user, 1912
light-emitting diodes and, 1768–1770
liquid-crystal displays and, 1775–1777
between log families, 1621
man-machine, 2216, 2408, 2409
network-node, 1443, 1445
numerical control, 2004
specifications for, **1914**, 1910
user, 1981, 2005, 2007

user-network, 1443, 1445
in VSAT systems, 2205
Interference, 1593, see also specific types
in broadcasting, 1398
carrier ratio to, 1424
cochannel, **1553**, 1548
electromagnetic, see Electromagnetic interference (EMI)
intersymbol, **1416**, 1410, 1411, 1415, 1484
in mobile radio, **1553**, 1548
multipath, 1421
radio frequency, **918**
in robotics, **2182**, 2177
Interferometers, 763
Interframe coding, 1419
Interlaced scanning, **1392**, **1397**, 1380–1381
Interleaved memory, **1936**, 1928–1931
Intermediate frequency (IF), 1360
Intermediate power amplifiers (IPA), 1388
Intermittent faults, 2088
Intermodulation distortion, **147**, 143–144
Internal base resistance, 640
Internal emitter resistance, 646
Internal latch enable (ILE) input, 1733
Internal quantum efficiency, **1771**, 1767
Internal reflection, 975, 1767
Internal resistance, 52
Internal stability, **222**, 208
Internal state variables, 1818
Internal thermal resistance, 790–791, 793
International Electronics Packaging Society, 2553
International Society for Hybrid Microelectronics, 2553
International Society for Optical Engineering, 2554
International Standards Organization reference model (ISORM), **1458**, 1447, 1448, 1451, 1452
International System of Units, 2419–2421
Internets, **1459**, 1455–1456
Interoffice signaling, **1459**, 1450, 1470
Interpolation, **1516**, 1515, 2511–2516
Interpolation filters, 391
Interpolation operation, 391
Interpreters, **1914**, 1904, 1905

Page on which term is defined is indicated in bold.

Interprocess communication (IPC), **2070**, 2067
Interrupted continuous wave (ICW) radar, 955, 969
Interrupting rating of fuses, **45**, 39
Interrupts, **1762**, 1748, 1875
Interrupt signals, 1867
Interstitial diffusion, 479
Intersymbol interference (ISI), **1416**, 1410, 1411, 1415, 1484
Intersystems, 903, see also specific types
Intervalley scattering, 1104
Intrasystems, 903
Intrinsic absorption, 985
Intrinsic emitter-base junction, 641
Intrinsic impedance, 839, 841
Intrinsic magnetic properties, 818–819
Invasive sensors, 2343
Inverse Chebyshev filters, 118–121
Inverse Chebyshev function, 117
Inverse discrete transform, 283
Inverse Fourier transforms, 290, 303, 1512
Inverse Laplace transform, 154–157, 2103
Inverse matrix, 2440
Inverse power law, 1142
Inverse-time characteristics, 1270
Inverse trigonometric functions, 2446
Inverse *z*-transform, 178, 181–182
Inversion, 1040
Inverter amplifiers, 618, 625, 626
Inverter operational amplifiers, 625, 626
Inverters, **728**, **2246**, 584, 983, 2238, see also Converters; dc- to-ac converters; specific types
   CMOS and, 583
   current-fed, 733–734
   current source, 705, 706–707, 2238
   design of, 583
   machine commutated, 2238
   pulse-width modulation, 705, 707, 708, 734
   resonant-link, 707–708
   symbols for, 1698
   voltage-fed, 731–732
   voltage source, 705–706, 2238
Inverting amplifiers, 618
INWATS database, 1470
Ion etching, 488
Ionic relaxation loss, 1134
Ion implantation, **489**, 479, 481–484, 571, 572
Ionization, 443, 1789

Ionization enery, 445
Ion jump probability, 1135
Ionographic printers, 1964
Ion-selective electrodes (ISE), 1157–1158, 1159
Ion-selective membranes, 2347
I&P, see Interconnections and packaging
IPA, see Intermediate power amplifiers
IPC, see Interprocess communication
IR, see Interrupting rating
IRES, see Integrated renewable energy systems
IRE units, **1392**, 1384
Irrational exponents, 2430
Irrelevancy, **405**, 337, 400
Irvin's curves plot surface dopant, 481
ISDN, see Integrated services digital network
ISE, see Ion-selective electrodes
ISHM, see International Society for Hybrid Microelectronics
ISI, see Intersymbol interference
Isolated circuits, **728**
Isolated single-ended topology, 713–714
Isolated word recognition, **313**, 306, 309, 310, 311
Isolation amplifiers, 771
Isolators, 889–890, 1430, see also specific types
   optical, **1172**, 1166–1169
   waveguide, 1168
ISORM, see International Standards Organization reference model
Isotropic media, 1078, 1082
IS-SPICE, 621
Itakura–Saito distortion measures, 290
ITE, see Information technology equipment
Iteration, 1040, 2507–2508
Iteration period, 373, 374, 375
Iterative linear interpolation, 2514–2515
Iterative methods, 584, 587–589, 657, see also specific types
IT network, **188**, 184–188
ITO, Indium-tin oxide
IVHS, see Intelligent vehicle/ highway systems
IV&V, see Independent verification and validation
IWP, see Indirect weld parameters

J

Jacobian, **2174**, 212, 2173
Jacobian matrix, 2172
Jacquard, Joseph, 1650
JAD, see Joint application design
JFET, see Junction field-effect transistors
Jitter, **1516**, 1513, 1514
JND, see Just-noticeable difference
Johnson noise, see Thermal noise
Joint Tactical Information Distribution System (JTIDS), **2193**, 2292
Joint test action group (JTAG), 493–494, 497
Joint torque sensors, 2173
Josephson equations, 1118
Josephson junction, 1120
Joysticks, **1956**, 1947–1948, 2012
JSR, see Jump subroutines
JTAG, see Joint test action group
JTIDS, see Joint Tactical Information Distribution System
Jump subroutines (JSR), 1889, 1890
Junction capacitance, **458**, 454
Junction field-effect transistors (JFET), 530–545
   biasing of, 533–536
   common-drain, 536
   common-gate, 536
   frequency response of, 540–543
   gate insulators and, 568
   n-channel, 530, 531, 532
   output resistance of, 538–539
   p-channel, 530–531
   small-signal ac voltage gain and, 537–538
   source follower and, 539–541, 542–543
   time-domain response of, 541–543
   transfer characteristics of, 536–538
   voltage-variable resistance and, 533, 543–544
Junction operating temperature, 640
Just-noticeable difference (JND), 357

K

Kaiser–Bessel window, 234
Kalman canonical form, 2105
Kalman filters, 963, 2293
Karnaugh maps, 1716
KCL, see Kirchhoff's laws, for current

KDP, see Potassium dihydrogen phosphate
Kerr effects, **1171**, 1165–1166, 1178
  all-optical devices and, 1641
  magneto-optical, **1693**, 1690
  polar, 1690
Keyboards, **1957**, 1865, 1938–1941, 2012
Kilby, Jack, 632–633
Kinematics, **2162**, 2154, 2161
Kinetic energy, 1114, 1115, 1494, 2168
Kirchhoff integrals, 1032
Kirchhoff's laws
  for current, **68, 672**
    CACS and, 655
    in complex domain, 61–62
    importance of, 63
    in linear circuit analysis, 58–62
    mesh analysis and, 63–69
    networks and, 63–69
    node analysis and, 63–69
    nonlinear circuits and, 133
    passive signal processing and, 112
    state variables and, 169, 171, 172
    Tellegen's theorem and, 80
    three-phase circuits and, 89
  current sources and, 56
  for radiation, 748
  for voltage, **68, 672**
    CACS and, 655
    in complex domain, 63
    importance of, 63
    Laplace transform and, 162, 163
    in linear circuit analysis, 58–59, 62–63
    mesh analysis and, 63–69
    networks and, 63–69
    node analysis and, 63–69
    Tellegen's theorem and, 80
  voltage sources and, 56
Klystrode, **1392**, 1388
Klystron, **1392**, 1387, 1388
Knowledge acquisition, 2049
Knowledge base, **2051**, 2048
Knowledge encoding, 2050
Knowledge engineering, **2051**, 2032–2051
  databases and, see Databases
  rule-based expert systems and, 2048–2051
Knowledge representation, 2050
Knowledge testing, 2050
Kolmogorov complexity, **1526**, 1524–1525
Kraft inequality, 1518
Kramer's generalization, **1516**, 1515

k trees, **109**, 93, 94–98
  admittance product of, **109**, 93
  flowgraphs vs., 102–106
  Maxwell, 102, 105
Kullback Leibler distance, 1520
KVL, see Kirchhoff's laws, for voltage

L

Ladder networks, 69
Ladder transformation, 680–681
Lagging power factor, 83
Lag-lead networks, **2130**, 2116, 2122–2124
Lag networks, **2130**, 2115, 2119
Lagrange's formula, 2464, 2514
Lagrangian interpolation, **1516**, 1515
Lagrangian method, 2167, 2168
Laguerre polynomials, 2484–2485
Lambda, 582
Lambert, 747
Lame constants, 1078
Laminar flow, 792
LAN, see Local-area networks
Landau levels, 441
Land pattern, **614**
Langmuir, Irving, 1151
Language
  computer, see Computer languages
  modeling of, 306, 307, 310, 311
Laplace equation, 1080
Laplace operator, 111
Laplace transform, **157**, 111, 149–168, 183, 368–369, 2476–2477
  absolute convergence and, **157**, 150–151
  anticausal components of, 157
  applications of, 158–168
    differentiation theorems and, 158–160
    to electric circuits, 161–162
    to integrodifferential equations, 160–161
    Norton's theorems and, 164–165
    Thévenin's theorems and, 164–165
    to transformed circuits, 162–164
  autocorrelation function of, 153
  causal components of, 157
  of continuous-time signal, 182
  control systems and, 2103, 2137
  derivatives of, 158, 159

  of first-order causal exponential signal, 150
  impulse responses and, **168**, 166–167
  inverse, 154–157, 2103
  linearity of, 151
  network functions and, **168**, 165–166
  Norton's theorems and, 164–165
  one-sided, 158, 160
  pair tables for, 150–151
  properties of, 151–154
  region of absolute convergence and, **157**, 150–151
  semiconductor manufacturing and, 480
  stability of, 167–168
  state variables and, 169
  step responses and, 166–167
  table of, 160, 2477
  Thévenin's theorems and, 164–165
  time-convolution property of, 152–154
  time-correlation property of, 153
  time-domain differentiation and, 152
  time shift and, 152
  transfer function and, **168**, 165–166
  z-transform and, 2478–2479
Laplace transform integral, 149–150
Large disturbance, **1285**
Large drives, 2237–2246, see also specific types
Large-scale integration (LSI), 603, 1572, 1622, 1631
Large-signal analysis, 634–639
Large-signal equivalent circuit models, 461–463
Large-signal models, 452, 453, 461–463, see also specific types
Large-signal switching, 456
Laser beam welding (LBW), **2235**, 2223, 2228
Laser diodes, 1429, 1435, 1436, 1682, 1683, 1684
Laser disks, 1682
Laser medium, **742**, 738
Laser power modulation (LPM), 1688–1689
Laser printers, 2005, 2006
Lasers, 738–742, 1641, 1646, see also specific types
  commercial, 739, 741–742
  continuous wave, **742**, 739
  erbium-doped fiber, 1429
  fiber-optics and, 997
  focusing of beam of, 745

Page on which term is defined is indicated in bold.

helium-neon, 738
invention of, 1124–1125
linearity in, 1423–1424
liquid dye, 738
long distance and, 1427, 1429
in magneto-optical disk data
    storage, 1682
optical communication and, 1417
pulsed, 738
ruby, 738
semiconductor, 738, 740, 741
single-mode, 745
solid-state, **742**, 740
in surface mount technology, 613
in welding, 2223, 2228
yttrium, aluminum, garnet, 760,
    1424
Last-in–first-out (LIFO) structure,
    1884
Latches, **1640**, 1613, 1635–1637,
    1721, see also Bistable
    devices; specific types
addressable, 1723
gated, 1637
internal enable of, 1733
transparent, 1723
use of, 1639–1640
Latency of memory, **1669**, **1936**,
    1667, 1681, 1927, 1931
Lattice damage, 484
Lattice digital filters, 375, 380
Lattice realization, 249
Lattice temperature, 1138
Lauffen–Frankfurt project, 1309
Layers, 1451, 1452
data link, 1463–1464
in local-area networks,
    1461–1464
management, 1464
physical media-dependent, 1457
Layout editors, 581
Layout of integrated circuits, **590**,
    601, 581–584
Layout-versus-schematic (LVS)
    programs, 598
LBW, see Laser beam welding
LCA, see Linear congruential
    algorithms
LCARs, see Linear cellular automata
    registers
LCDs, see Liquid-crystal displays
LCLVs, see Liquid-crystal light
    valves
LC oscillators, 1572
LCP, see Left circular polarization
LCR ladder networks, 674, 675, 676,
    681
Leaded chip carriers, **614**, 607
Leading power factor, 83
Lead iron niobate, 1130

Leadless chip carriers, **614**, 607
Lead magnesium niobate (PMN),
    1097, 1090
Lead metaniobate, 1055
Lead scandium tantalate, 1131
Lead titanate, 1130
Lead zirconate, 1130
Lead zirconate titanate (PZT), **1097**,
    1055, 1056, 1057, 1087, 1127
applications of, 1089, 1090, 1091
electrical characteristics of, 1096
mechanical characteristics of,
    1088, 1089
in medical imaging, 2382
smart materials and, 1180, 1181
structure of, 1091, 1092
in ultrasound imaging, 2382
Leakage current, **32**, 20–21
Learning algorithms, 423
Learning control systems, 2103
Learning problems, 426
Learning rule, **428**, 421
Least-mean square (LMS) adaptive
    filters, 379
Least-mean square (LMS) algo-
    rithm, 402, 403, 404
Least-mean square (LMS) equaliz-
    ers, 1411
Least-recently-used (LRU) memory,
    1932
Least significant bit (LSB), **1807**,
    1845
Least significant digit (LSD), 1844
Least-squares algorithms, 403
LEDs, see Light-emitting diodes
LEF, see Lighting effectiveness
    factor
Left circular polarization (LCP), 1164
Legendre equation, 2483
Legendre polynomials, 1033,
    2483–2484, 2485
Lempel–Ziv algorithm, **1526**,
    1520–1521
Level-sensitive scan design (LSSD),
    1823–1825, 1834
Level shifters, 773
Levinson polynomials, 320
Lexicons, 307, 310
LF, see Load factor
LFC, see Load frequency control
LFSRs, see Linear feedback shift
    registers
LGF, see Loudness-growth function
LIDB, see Line information
    databases
Life cycle modeling, 1978
LIFO, see Last-in–first-out
Light, see also Illuminating systems;
    Lightwaves; Optoelectronics
absorption of, 744

attenuation of, 744
brightness of, 746–747
coherent, **742**, 738, 740, 741, 745
detection of, **759**
emergency, 2271
emission of, **759**, 750
fluorescent, see Fluorescent lights
geometric optics and, 745–746
incandescent, 748, 2268–2269
incoherent, 746–750
intensity of, 753, 763
modulation of, 761
monochromatic, **742**, 738, 740,
    741
phase velocity and, 743
polarization of, 744, 765
propagation of, 743, 975, 977
properties of, 742–744, 755
properties of rays of, 745–746
refraction of, 743, 745–746
security, 2271
sources of, 742–750
illuminating systems and,
    2268–2269
long distance and, 1429
statistical properties of, 755
thermal sources of, 747–748
velocity of, 743, 744
Light adaptation, 357
Light-amplifying optical switch
    (LAOS), **1649**
Light beam interruption,
    1950–1951
Light detectors, 1951
Light-emitting diodes (LEDs),
    750–751, 1429, 1435, 1436,
    1646, 1647, 1763–1772
efficiency of, 1766–1768
input-output devices and, 1946,
    1951
interfaces and, 1768–1770
logic gates and, 1627
optical devices and, 1641
principles of, 1763–1764
semiconductor materials and,
    1764–1766
seven-segment common cathode,
    1713
Lighting effectiveness factor (LEF),
    2275, 2267
Light loss factor (LLF), **2275**, 2261
Lightly doped drains, 578
Lightning, 935–948
cloud-to-ground, 935
electric fields from, 935, 937–940,
    941
flashes of, **946**, 935, 936
magnetic fields from, 935,
    937–940, 941
occurrence statistics for, 936–937

overhead wire interactions with, 943–944
physics of, 935–936
protection from, 905
return stroke of, **946**, 936, 941–943
terminology of, 935–936
voltage induced by, 944, 946, 947
Lightning rods, 905
Light pens, **1957**, 1941–1943, 2012
Light sensors, 447
Lightwaves, **1426**, 975–1001, 1417–1418, see also Light
attenuation and, **1001**, 984–986
in communications systems, 989
extrinsic absorption and, 985
fiber-optic cable and, see Fiber-optic cable
intrinsic absorption and, 985
in local-area networks, 1462
Mie scattering and, 985
optical fibers and, see Optical fibers
propagation of, 977
Rayleigh scattering and, 985
ray theory and, 976–977
wave equations and, 977
waveguides of, see Lightwave waveguides
Lightwave waveguides, 975–987
attenuation and, 984–986
cylindrical fibers and, 980–981
dispersion of, 985–986
graded-index fibers and, 975, 976, 982–984
pulse spreading and, 985, 986–987
ray theory and, 976–977
slab, 978–980
step-index fibers and, 975, 976, 981
wave equations and, 977
Likelihood detectors, 413, 1481, 1483
Likelihood ratios, **1487**, 1479, 1481, 1483
Limit cycles, 249
Limiters, **138**, 132–138
Limiting circuits, 132–135, 136, 137
Limiting methods, 665, see also specific types
Linear cellular automata registers (LCARs), 1813, 1815
Linear circuit analysis, 58–110
ac energy measurement and, 85–87
ac power measurement and, 85–87
ac steady-state power and, 81–85

graph theory and, see Graph theory
Kirchhoff's current law in, 58–62
Kirchhoff's voltage law in, 58–59, 62–63
maximum power transfer and, 85
mesh analysis and, **68**, 66–67
network theorems and, see Network theorems
node analysis and, **69**, 64–66
Tellegen's theorem and, 80–81
three-phase circuits and, 88–92
Linear congruential algorithms (LCAs), 1600, 1601
Linear dependence, 1831
Linear equations, 2441
Linear feedback shift registers (LFSRs), **1734**, **1816**, 1727, 1813–1815, 1830, 1831, 1832
Linear filtering, 1504–1506
Linear graphs, see Graphs
Linearity, **1426**
analog-to-digital converters and, 771, 773
differential, 773
digital-to-analog converters and, 771, 773
of Laplace transform, 151
in lasers, 1423–1424
network theorems and, 69–71
in optical communication, 1422–1423
of photomultipliers, 755
of transfer functions, 112–113
in video transmission, 1422–1423
of z-transform, 178
Linearization, 57
Linear networks, **78**
Linear position sensors, 753
Linear predictive analysis, 302–303
Linear predictive coding (LPC), 285, 286, 307, 308, 1401, 1525
Linear quadratic optimal control, 2146–2147
Linear rate constant, 477
Linear receivers, 1414
Linear regression analysis, 316
Linear transformations, 1489–1490
Linear transversal equalizers, 1410–1412
Linear variety property sets, 258
Line impedance stabilization networks (LISNs), **934**, 933
Line information databases (LIDBs), **1475**, 1472
Line printers, **1975**, 1970–1973
Liners, 1878
Linewidth, **987**

Link, **176**, 171
Linkers, **1901**
Lin–Uman–Standler (LUS) model, 941
Liouville–Von Neumann equation, 472
Liquid-crystal displays (LCDs), 445
color, 1942
computer graphics and, 2007
interfaces and, 1775–1777
principles of, 1772–1775
Liquid-crystal light valves (LCLVs), **1649**
Liquid crystals, 1127, 1775
Liquid dye lasers, 738
Liquids, 1139, 1143, 1144–1145, see also Fluids; specific types
LISNs, see Line impedance stabilization networks
Lithography, **489**, 487–488
LLC, see Logical link control
LLF, see Light loss factor
Lloyd algorithm, 1524
Lloyd–Max scalar quantizers, 339
LMS, see Least-mean square
LNAs, see Low-noise amplifiers
Load
current mirror, 565–566
in energy distribution, 1316–1317
forecasting of, 1349, 1351
management of, 1344, 1348–1349
three-phase circuits and, 90
Loadability of relays, 1274
Load factor (LF), 1624
Load-flow programs, 1288–1289
Load frequency control (LFC), 1347, 1348
Load lines, 548, 550, 636
Local-area networks (LANs), **1459**, 1434, 1435–1437, 1448, 1455–1456, 1457, 1460–1468, 2015–2017
aerospace systems and, 2209
architecture and, 1869
availability of, 1465
cost of, 1560, 1563
end user reaction to, 1563
errors in, 1465
failures in, 1465
growth of, 1559
interconnection of, 1442
management layer of, 1464
market factors in, 1558–1559
multichannel, 1462
operating systems and, 2062
optimized service area for, 1562
physical layer of, 1462–1463
reliability of, 1465, 1563
satellite communications systems and, 2209

Page on which term is defined is indicated in bold.

security of, 1563, 2082
service model for, 1461–1466
specialized services in, 1464–1466
standards for, 1466, 1467
topology of, **1467**, 1461
transmission in, 1455, 1461
user reaction to, 1563
wireless, see Wireless local-area
    networks
workgroup sizes for, 1562
Local convergence, 663
Localized heat sources, 786, 791
Local truncation errors (LTE),
    667–668, 669, 670
Local variables, 1881, 1884, 1890,
    1938
Lock range, **1574**
Logarithmic number systems, 382
Logarithmic series, 2460–2461
Logarithms, 2431
Logic, see also Logic circuits; Logic
    elements
  block carry lookahead, 1745
  carry lookahead, 1744
  choice of family of, 1634
  CMOS, see Complementary MOS
    technology
  current-mode, 1618
  diode-transistor, 1624
  emitter-coupled, 1613, 1614,
    1618, 1622, 1629–1631,
    1712
  families and subfamilies of,
    1614–1615, 1621, 1634
  integrated injection, 1712
  interfaces between families of,
    1621
  levels of, **1621**, 1616, 1618, see
    also specific types
  nonsaturated, 1624, 1627
  parameters of, 1618–1620
  programmable array, **1740**, 1716,
    1718, 1737, 1738, 1740
  random, 1713
  resistor-transistor, 1624
  saturated, 1624
  surface-emitting laser, **1649**
  switching, 1613
  transistor-transistor, see
    Transistor-transistor logic
    (TTL)
Logical devices, see Arithmetic logic
    units (ALUs); Combina-
    tional networks; Logic
    circuits; Programmable
    arrays; Registers; specific
    types
Logical link control (LLC), 1456,
    1463
Logic bombs, **2085**, 2079

Logic circuits, see also Logic
  asynchronous sequential, 1719,
    1720
  combinational, **1720**, 1711–1718,
    1741
  graphic classification of, 1711,
    1712
  sequential, **1720**, 1711,
    1718–1719, 1720
  synchronous sequential,
    1718–1719
  two-state, 1711
Logic elements, 1613–1650, see also
    Logic; Logic gates; specific
    types
  bistable devices and, 1635–1641,
    see also specific types
  classification of, 1613
  combinatorial, 1613
  optical devices and, see Optical
    devices
  sequential, 1613
Logic gates, **1634**, 1613, 1622–1635,
    see also specific types
  all-optical, 1641
  bipolar junction transistors and,
    1624–1626
  effective loading of, 1623
  optical, 1643
  Sagnac, **1649**
  specification parameters for,
    1622–1624
  truth tables for, 1614
  unused, 1629
Logic partitioning, 1822
Log-periodic dipole arrays, 868–869
London equations, 1116
Long-channel devices, 569, 574, see
    also specific types
Long distance, 1427–1434
Longitudinal Kerr effect, 1166
Longitudinal waves, 1078
Long-tailed pair, **566**, 561, see also
    Difference amplifiers
Long-term memory (LTM), 421
Longwords, 1882
Look-ahead technique, **383**,
    377–380
Loop, see Circuits
Loop (mesh) analysis, **68**, 57, 66–67
Loop error function, 1568, 1571
Loop filters, **1574**, 1567, 1568–1570,
    1573
Loop gain, **1574**, 620, 1567
Loop-set (circuit-set), **109**, 93, see
    also Edges
Loop systems, **176**, 169
Loop transmission, see Loop gain
Lorentz gauge, 1018
Lossless compression, **343**, 337

Loss tangent, 840
Lossy compression, **343**, 337
Loudness-growth function (LGF),
    398
Loudness restoration, **405**, 398
Loudspeaker reproduction, 399
Loudspeakers, 403, 1055, 1057,
    1059
Low-frequency fields, 1019
Low-noise amplifiers (LNAs),
    2199
Low-pass-equivalent (LPE)
    waveforms, **1608**, 1598,
    1599, 1601
Low-pass filters, **125**, **1574**,
    113–121, 239, 674–682, see
    also specific types
  active, 674–682, 685, 686
  cascaded second-order sections
    and, 676–680
  classification of, 675–676
  first-order, 685
  ideal, 189, 190–191
  magnitude, 193
  operational amplifiers and, 628,
    629, 630
  passive, 674–675
  phase-locked loops and, **1574**,
    1573
  sensitivity of, 675–676
  transformation from, 683–685
  VLSI and, 392
Low-pass to bandpass transforma-
    tion, 123–124
Low-pass to high-pass transforma-
    tion, 122–123
Low-power television (LPTV), **1393**,
    1383
Low probability of intercept (LPI),
    2218
LPC, see Linear predictive coding
LPE, see Low-pass-equivalent
LPI, see Low probability of
    intercept
LPM, see Laser power modulation
LPTV, see Low-power television
LRU, see Least-recently-used
LSB, see Least significant bit
LSD, see Least significant digit
LSI, see Large-scale integration
LSSD, see Level-sensitive scan
    design
LTE, see Local truncation errors
LTM, see Long-term memory
LU decomposition, 659
LU factorization, 658
Lumens, 747, 2260
Luminaires, 2261, 2265, 2270–2271,
    2272
Luminance, 747, 2264, 2266

of cathode ray tubes, 1783, 1784
of plasma displays, 1789, 1797
television and, 1384, 1386
Luminance ratios, **2275**, 2265, 2266,
2267
Luminescence, **1771**, 1764, 1942,
2007
Luminous efficiency, **1798**, 1789,
1797
Lumped impedance, 1082
Lumped parameter systems, 2100
Lumped-π equivalent networks,
1220
LUS, see Lin–Uman–Standler
LVS, see Layout-versus-schematic
Lyapunov stability theory, 209–210

M

MA, see Moving average
MAC, see Media-access control
Machine commutated inverters
(MCI), 2238
Machine interference, **2182**, 2177
Machine language, 1878, see also
Computer languages
Machine-level instructions, 1902
Mach operating system, 2067
Mach–Zehnder interferometers, 763
Mach–Zehnder modulation, 769,
1424
MacLaurin series, 2460
Macro assembler language, 1905
Macros, **1901**, 1879, 1899
MAD, see Maximal areal density
Magnesium oxide, 1145
Magnetic alloys, 2536–2541, 2542,
see also specific types
Magnetic circular birefringence
(MCB), 1163–1164
Magnetic circular dichroism
(MCD), 1163
Magnetic coercivity, 1687
Magnetic dipole, 863
Magnetic disks, 1659–1664, 1866
Magnetic energy, 24, 504, 1115
Magnetic field modulation (MFM),
1689–1690
Magnetic fields, **38**, **810**, 807
in bone repair, 2329–2334
computational electromagnetics
and, 1032
from constant currents, 811–812
critical, 1120, 1121
current produced by, 814–815
earth's, 2287
electric fields produced by, 815
Gauss' laws of, 804
generators and, 1321

Hall effect and, 440–441
intensity of, 1018
as intermediate medium, 1321
large drives and, 2239
from lightning, 935, 937–940, 941
measurements of, 440–441
navigation systems and, 2287
quasi-static, 1183
in soft-tissue repair, 2329–2330
static, 811–812
strength of, 838
superconductivity and, 1115
time-dependent, 812
time-varying, 814–815
ultraquantum, 442
uniform, 440
Magnetic flux, **38**, **810**
Magnetic flux density, **810**, 813,
1018, 1115
Magnetic linear birefringence
(MLB), 1164–1165
Magnetic linear dichroism (MLD),
1163
Magnetic materials, 820, 821, 822,
825–826, 2542, see also
specific types
Magnetic properties, see also
specific types
extrinsic, 818, 819–825
intrinsic, 818–819
of smart materials, **1188**,
1179–1180
Magnetic recording, 826–835, see
also specific types
fundamentals of, 827–828
media for, 830–832
noise in, 831
readback process in, 830
record process in, 828–830
Magnetic recording heads, 832–834
Magnetic resonance imaging (MRI),
2380, 2360, 2374, 2378,
2379, 2400
Magnetic shielding, 1175, 1185
Magnetic stress constant, 1177
Magnetic tape, **1675**, 1670–1675,
1866
format of, 1672–1673
Magnetic transducers, 1056
Magnetism, 811–826
amorphous magnetic materials
and, 825–826
chemical quantities in,
2426–2427
diamagnetics and, 816
electricity and, 1002–1003,
1026–1027
extrinsic properties and, 818,
819–825
ferri-, 816–818

ferro-, 816–818
forces on moving charges and,
813–814
intrinsic properties and, 818–819
magnetic flux density and, 813
Maxwell's equations, 815–816
para-, **836**, 816, 831, 835
physical quantities in, 2426–2427
relative permeabilities and, 813,
814
static magnetic fields and,
811–812
time-dependent electric fields
and, 812
time-dependent magnetic fields
and, 812
time-varying magnetic fields and,
814–815
units in, 813
Magnetization, **835**, **836**, 829, 831,
1021, 1162
Magnetocrystalline anisotropy, 819,
823
Magnetoelastic anisotropy, 822,
823–825
Magnetographic printers, 1964
Magnetomotive force, **32**, 26
Magneto-optical disk data storage,
1675–1694
access time in, 1681–1682
automatic focusing in, 1684–1685
automatic tracking in, 1685–1687
description of, 1677–1682
disk rotation speed in, 1681
materials of, 1691–1693
optical path in, 1682–1684
readout in, 1690–1691
thermomagnetic recording and,
1687–1690
tracks in, 1679–1680
Magneto-optical Kerr effect, **1693**,
1690
Magnetooptic Bragg cells, **1171**,
1169–1170
Magnetooptics, 1162–1172
applications of, 1166–1172
classification of, 1163–1166
recording in, **1171**, 1171
Magnetoresistance, **835**
Magnetostatic wave (MSW)-based
guided-wave magnetooptic
Bragg cells, 1169–1170
Magnetostriction, **1188**, 819, 826,
1056, 1177, 1180, 1943
Magnetostriction coefficient, 823
Magnetostrictive sensors, 1182–1183
Magnetostrictive smart materials,
1188, 1177, 1180
Magnetostrictive transducers, 1056,
1180

Page on which term is defined is indicated in bold.

Magnetron systems, 485
Magnitude squared coherence
　　(MSC), **2362**, 2355, 2356
Main memory, 1931
Maintainability, **2030**
Main thermocline, 409
Majority carriers, 458, 447, see also
　　specific types
MAN, see Metropolitan area
　　networks
Manchester baseband encoding,
　　1462
Manipulators in robotics, **2162**,
　　2161, 2173–2174
Man-machine interface (MMI),
　　2216, 2408, 2409
Man-machine systems, 2247–2254,
　　see also specific types
Mantissa, **1858**, 1856
MAP, see Maximum *a posteriori*
Map-matching navigation systems,
　　2293
MAR, see Memory address
　　registers
Marconi, Guglielmo, 1592
Marginal probability, 2517
Markov modeling, 2092–2093
　　hidden, see Hidden Markov
　　models (HMMs)
Markov processes, 1501–1502, 1506,
　　see also specific types
Markowitz count, 661
Markowitz criteria, 661
M-ary data transmission, 1484
M-ary phase shift keying (MPSK),
　　1365, 1400, 1542
Masking, 358, 400, 1728
Masking threshold, 400, 401
Masks, **601**, 591
Mason graphs, 102
Mass, **1110**, 1060, 1109
Mass flow rate sensors, 1155,
　　1161
Master-slave flip-flops, 1637–1638,
　　1732
Matched filters, 1414, 1482
Material dispersion, **1434**
Material handling, 2178
Materials Research Society, 2554
Mathematical constants, 2523
Mathematical expectation, 2520
Mathematical models, see also
　　specific types
　　of ac generators, 1323–1326
　　of control systems, 2099
　　of generators, 1323–1326,
　　　1330–1331
Mathematical operations,
　　2438–2439, see also specific
　　types

Matrices, 2438, 2441, see also
　　specific types
　　algebraic properties of, 262–265
　　capacitance, 1219–1220
　　covariance, 1502
　　excitation, 274
　　exponential signals and, 270–273
　　identity, 2439–2440
　　inverse, 2440
　　multiplication of, 2438–2439
　　positive-semidefinite, 263–265
　　potential coefficient, 1220
　　rank-reduced, 272–273
　　structural properties of, 265–268
　　Toeplitz–Hankel, 267, 268
　　Toeplitz–Hermitian, 269–270
Matrix addressing, 1776
Matrix fraction description (MFD),
　　2104
Maximal areal density (MAD),
　　1669, 1658
Maximally flat delay (MFD) filters,
　　125
Maximally flat magnitude (MFM)
　　filters, 125, 113, see also
　　Butterworth function
Maximum *a posteriori* (MAP)
　　decision rule, 295, 296
Maximum *a posteriori* (MAP)
　　estimate, 323
Maximum-likelihood (ML) array
　　processing, 416
Maximum-likelihood (ML)
　　estimation (MLE), **328**, 292,
　　326–327, 1415
Maximum power transfer, **78**,
　　76–77, 85, 1242
Maxwell, James Clerk, 880–881, 974
Maxwell *k* trees, 102, 105
Maxwell's equations, **1025**
　　antennas and, 875, 876
　　in computational
　　　electromagnetics, 1029
　　electromagnetic fields and, **1025**,
　　　803–804, 807, 815, 1018,
　　　1019
　　graded-index fibers and, 982
　　in instrument modeling, 2278
　　lightwaves and, 982
　　magnetism and, 815–816
　　piezoelectric excitation and, 1080
　　semiconductors and, 469
　　in three-dimensional analysis,
　　　1025, 1018, 1019
　　ultrasound and, 1080
Maxwell's rule, 28
Maze routers, 589
M&C, see Monitoring and control
MCB, see Magnetic circular
　　birefringence

MCD, see Magnetic circular
　　dichroism
MCI, see Machine commutated
　　inverters
McLaurin's series, 1099
MCPC, see Multiple channel per
　　carrier
MCT, see Monte Carlo simulation
MCVD, see Modified chemical
　　vapor deposition
MDCT, see Modified discrete cosine
　　transform
MDDM, see Mission directed
　　decisionmaker model
Mealy-type outputs, 1738
Mean, 1508, 2486, 2488
Mean free path, **732**, 469
Mean opinion score (MOS), 280,
　　285, 286
Mean-square ergodic property, 1509
Mean-square errors (MSE), 290,
　　323, 1411, 1412, 1509
　　minimum, 290, 291, 292,
　　　293–295, 296
Mean-square estimation, **328**,
　　323–325
Mean-square value, 1500
Mean time between failures
　　(MTBF), 785, 2028, 2090
Mean time between unconfirmed
　　removals (MTBUR), 2189
Mean time to failure (MTTF), **2095**,
　　2090, 2091, 2094
　　fault tolerance and, 2027, 2028,
　　　2030
Mean time to repair (MTTR), **2095**,
　　2028, 2090, 2091, 2094
Measure of effectiveness (MOE),
　　2217
Measurement, **2283**, see also specific
　　types
　　of ac power, 85–87
　　automated, 2283
　　on cathode ray tubes, 1783–1784
　　of compatibility, 932–933
　　continuous-time, 1486
　　of distortion, 1522
　　errors in, 961–962, 1497
　　in-phase components of, 1481
　　instrument modeling and, 2279
　　quadrature components of, 1481
　　synchronized phasor, 1278
　　telemetry and, 1582–1583, 1589
　　of thermal noise, 1493–1494
　　time domain, 2108
Measure of performance (MOP),
　　2217
Mechanical energy, 1197
Mechanical engineering, 1187
Mechanical resonance, 1084

Page on which term is defined is indicated in bold.

Mechanical telemetry, 1579
Mechanics, 2425
Mechanoreceptors, 2301
Media-access control (MAC), **1459**, **1467**, 1455, 1463
Median, 2486
Medical devices, 2412, see also Biomedical equipment; specific types
Medical imaging, 777, 1760, 2374–2386, see also specific types
  soft tissue, 2380
  tomographic, 2374–2380, see also Tomography
  ultrasound, 2380–2385
Medical instrumentation, 778
Medical logic modules (MLM), 2399
Medical records, 2397–2403
Medium-scale integration (MSI), **1720**, 1614, 1622, 1714, 1741
Meissner effect, 1115, 1116, 1121
Member fields, **1914**, 1910
Member functions, **1914**, 1910
Membranes, **2310**, 2302, 2303, 2312, 2338, 2347
Membrane time constant, 2313
Membrane voltage, 2309
Memory, **1798**, 1695, 1865, 1927–1937, see also Memory devices; Memory element; specific types
  assembly language and, 1882–1886
  associative (content-addressable), 2056–2057
  auxiliary, **1675**, 1671
  bandwidth and, 1927
  cache, **1936**, 1929, 1932, 1933
  control, **1877**, 1872
  devices for, see Memory devices
  direct access to, **1807**, 1804
  embedded, 1822–1823
  enhanced, 607
  first-in, first-out, 1727
  future access to, 1932
  hierarchies of, 1931–1934
  interleaved, **1936**, 1928–1931
  latency of, **1669**, **1936**, 1667, 1681, 1927, 1931
  least-recently-used, 1932
  main, 1931
  nonvolatile, **1675**, 1654–1655, 1670
  on-chip, **393**
  parallel, 1928–1931
  in plasma displays, **1798**, 1788, 1794–1795

  primary, 1866
  programmable, 1654–1655, 1715
  random-access, see Random-access memory (RAM)
  read-only, 1651, 1656, 1715, 1822–1823, 1830, 1874
  secondary, 1866
  tag, 1932
  virtual, **1936**, 1876, 1934–1936, 2068
Memory address, 1927
Memory address registers (MARs), 1867
Memory aliasing, 1936
Memory bank, 1928
Memory bank busy time, 1928
Memory controllers, 1930
Memory cycle, 1866
Memory data registers (MDR), 1867
Memory devices, 1651–1694, 1718, see also Memory; Memory element; specific types
  access time for, 1670
  disk system architectures and, see Disk system architectures
  history of, 1671
  input/output workloads and, 1664–1665
  integrated circuits in, 1651–1657
  magnetic tape, **1675**, 1670–1675
  magneto-optical, see Magneto-optical disk data storage
  nonvolatile, **1675**, 1654–1655, 1670
  programmable, 1654–1655
  thermomagnetic recording and, 1687–1690
Memory element, **1640**, 1635, see also Memory; Memory devices
Memory management units (MMUs), 1671
Memory module, 1928
Memory units (MUs), **1869**, 1866, 2052
Memory word, 1927
MEPs, see Motor-evoked potentials
Mercury cadmium telluride, 1129
Meridional rays, **987**
MESFETs, see Metal-semiconductor field-effect transistors
Mesh (loop) analysis, **68**, 57, 66–67
Message switching, **1459**, 1449–1450
Metabolic process, **2328**, 2312
Metafiles, 2005
Metalanguages, **1914**, 1903
Metal-film resistors, 12
Metal-in-gap (MiG) recording, 832, 833
Metallic glasses, 1056

Metal-oxide semiconductor field-effect transistors (MOSFET), 567–579
  control of, 577–578
  current in, 568–570
  driving ability and, 574–576, 577
  hot-electron effects and, 578
  integrated circuits in, 568
  limitations of, 577–579
  logic elements and, 1616, 1617
  miniaturization and, 577–579
  power, 698–699, 736
  programmable arrays and, 1735
  pyroelectrics and, 1127
  rugged, 699
  strong-inversion and, 569–570
  subthreshold and, 570, 577–578
  threshold voltage in, 570–573
  transconductance and, 576
  voltage in, 568–573
Metal-oxide semiconductor (MOS) transistors, 1613
Metal-oxide semiconductors (MOS), 1145, 1729
  CACS and, 671
  complementary, see Complementary MOS (CMOS)
  fabrication of, 582, 583
  layout of, 582
  n-type, 1748
  specialized techniques for circuits in, 671
  thyristors controlled by, 700–701, 2243
Metal-oxide-silicon devices, see Metal-oxide semiconductors (MOS)
Metals, 1179, 1219, 1691, 2527–2529, see also specific types
Metal-semiconductor diodes, see Schottky diodes
Metal-semiconductor field-effect transistors (MESFETs), 892, 896, 897, 901, 1004, 1005
Meta-signaling, 1445
Method of moments (MoM), **1048**, 440, 1031, 1036–1040
Method of residues, 181
Method of successive projections, **276**, 251
Methods, **1914**, 1910
Metrics, **1914**, 1906
  Euclidean-induced, 252
  Frobenius norm distance, 252
  Halstead's, 1906
  of interconnections, 501–503
Metric space formulation, 252

Page on which term is defined is indicated in bold.

Metropolitan-area networks
  (MANs), **1459**, 1456, 1457,
  1461, 2017–2018
MFD, see Matrix fraction descrip-
  tion; Maximally flat delay
MFM, see Maximally flat magnitude
MIC, see Microwave integrated
  circuits; Monolithic
  integrated circuits
Mica, 1145
Mica capacitors, 19
Microassemblers, 1873–1874
Microbending, **1001**, 991
MICRO-CAP, 630
Microcells, **1566**
Microcellular networks, 1564–1565
Microcommands, **1877**, 1871
Microcomputers, 1759–1760, see
  also Computers
Microcontrollers, **1752**, **1762**, 1748,
  1749–1750, 1751, 1755,
  1756
Microinstructions, **1877**, 1871–1872
Microkernels, 2069
Micromachining, **1161**, 1159
Microphones, **1060**, 207, 1055,
  1056, 1057, 1059
Microprocessors, **1752**, 1748–1762,
  see also Computers; specific
  types
  architecture of, 1760–1761
  chips for, 1751–1752
  components of, 1754
  in computing, 1759–1761
  in consumer electronics, 1757
  control applications of, 1756–1759
  cost of, 1750–1751
  in data collection, 1753–1755
  development support for, 1751
  in digital signal processing,
    1760–1761
  in digital tachometers, 1755
  in manufacturing, 1757–1758
  packaging of, 1750–1751
  in point-of-sale terminals,
    1754–1755
  practical, 1748–1752
  programming of, 1751
  social issues in use of, 1759
  software and, 1750
  in transportation, 1758
  trends in, 1752
  types of, 1749–1750
  in vehicles, 2255
Microprogramming, 1870–1877, see
  also Computer program-
  ming
  applications of, 1875–1876
  development in, 1873–1874
  emulation in, 1874–1875

by hand, 1873
high-level languages for, **1877**,
  1874, 1876
levels of, 1870–1871
microinstructions and,
  1871–1872
optimization in, 1872
Microsensors, **1161**, 1159–1161
Microstrip discontinuities, 884
Microwave CAD, 1004
Microwave devices, 882–902, see
  also Microwaves; specific
  types
  active, see Active microwave
    devices
  attenuators and, 885
  characterization of elements of,
    882–884
  circulators in, 889–890
  components of, 882, 891
  diodes in, 892, 894–896, 1004
  directional couplers and, 887–888
  directivity of, 888
  discontinuities and, 884
  filters in, 887, 888
  hybrid circuits and, 887–888
  impedance transformers and, 885
  isolation and, 888
  isolators in, 889–890
  passive, see Passive microwave
    devices
  reference impedance and, 883
  resonators and, 886–887, 890
  semiconductor, 890
  semiconductor materials and,
    892–893
  terminations and, 885
  three-terminal, 896–901
  transferred electron devices and,
    892, 894–895, 896
  transistors in, 892, 896–901, 1004
  transmission lines and, 883, 884
  tuning elements in, 887
  two-terminal, 894–896
  waveguides in, 882, 884, 889
Microwave integrated circuits
  (MICs), **1014**, 892, 1004,
  1011
  monolithic, **1015**, 1004, 1006,
    1007, 1008, 1009, 2209
Microwave reflection, 1182
Microwave resonators, 886–887
Microwaves, 847, 1185, 1417, 2527,
  see also Microwave devices
Mie scattering, 985
MiG, see Metal-in-gap
Miller effect, **651**, 558, 647
Millimeter-wave seeker, 972–973
MIMD, see Multiple instruction
  stream, multiple data stream

Mineral oil, 1144, 1178
Miniaturization, 577–579
Minimax estimation, **328**, 326
Minimum mean-square errors
  (MMSE), 290, 291, 292,
  293–295, 296
Minimum mean-square estimate,
  323
Minimum phase factor, 318, 320
Minimum phase transfer functions,
  320
Minority carriers, 456, 1763, 1767
Minor-loop design, 2126–2130
MISD, see Multiple instruction
  stream, single data stream
Mismatch, 1082
MISR, see Multi-input signature
  register; Multiple-input shift
  registers
Miss, 1479
Missile-launched space probes,
  1578
Missile telemetry, 1581
Missile terminal guidance seeker,
  973, 969, 972–973
Mission directed decisionmaker
  model (MDDM), 2220
Miss probability, **1487**, 1479
Mixed technology, **614**, 603, 611,
  see also specific types
Mixers, 1010
ML, see Maximum likelihood
MLB, see Magnetic linear birefrin-
  gence
MLD, see Magnetic linear dichro-
  ism
MLE, see Maximum-likelihood
  estimation
MLM, see Medical logic modules
MMI, see Man-machine interface
MMICs, see Monolithic microwave
  integrated circuits
MMSE, see Minimum mean-square
  error
MMUs, see Memory management
  units
MNRU, see Modulated noise
  reference unit
Mobile charge, **473**
Mobile data systems, 1552
Mobile radio, **1539**, 279, 287, 1062,
  1535, 1546–1553, see also
  Cellular communications;
  Wireless communications
  description of, 1547–1549
  fixed-to-fixed radio communica-
    tion vs., 1546
  problems in, 1546–1547
  specialized, 1547
  spectrum allocation for, 1549

Page on which term is defined is indicated in bold.

Mobile satellite services (MMS), 1534
Modal-expansion model, 1033–1034
Modality-specific rehabilitation, **2396**, 2390
Modal system, 169
Mod-anode pulsing, 1388
Mode, **1001**, 2486, see also specific types
Mode charts, **987**, 979
Modeling, see also specific types
  of ac generators, 1323–1326
  all-pole, 2102
  analytical, 2100, 2229
  artificial neural network, 2229–2231
  autoregressive, see Autoregressive (AR) model
  autoregressive moving average, 302, 2102, 2103
  behavioral, 654
  Boyle, **623**, 622
  Bruce–Golde (BG), 941
  C3 system and, 2216
  of channels, 1601–1602
  choice of model type in, 1517
  circuit, 51, 54, 60, 1254
  complexity in, 2104
  composite source, 289, 292, 293
  of control systems, 2099–2106
  crossover, 2250–2254
  data, **2047**, 2033
  Diendorfer–Uman, 941, 942–943
  discrete, 2105
  dynamic, 2167–2173
  empirical, 2101, 2229
  of equivalent circuit, see Equivalent circuits
  errors in, **1048**, 1046, 2088
  essential, **1984**, 1977, 1980
  experimental, 2101
  fault, 1810–1811
  finite impulse response, 2102
  of generators, 1323–1326, 1330–1331
  geometric, 1034, 2005, 2011
  hidden Markov, see Hidden Markov models (HMMs)
  Hodgkin–Huxley membrane, 2303
  hydrodynamic, 440
  implementation, 1977, 1981–1983
  impulse response, see Impulse response
  incremental, **458**, 454–456
  infinite impulse response (IIR), 2103
  information, **1984**, 1980, see also Information theory

of instrument elements, 2277–2281
  language, 306, 307, 310, 311
  large-signal, see Large-signal models
  life cycle, 1978
  Lin–Uman–Standler, 941
  Markov, 2092–2093, see also Hidden Markov models (HMMs)
  mathematical, see Mathematical models
  mission directed decisionmaker, 2220
  modal-expansion, 1033–1034
  moving average, 2101–2102
  of noise reduction, 288–290
  object collaboration, **1985**, 1980, 1981
  of oxidation, 476
  physical, 2101
  piecewise linear, 453
  of receivers, 1601–1602
  recursive, 274–276
  of return stroke of lightning, 941–943
  in robotics, 2167–2173
  signal-enhanced, 275–276
  signal-present, 1479
  small-signal, 454–456, 463–465, 466
  solid, **2013**, 2004, 2011
  spectral, see Spectral estimation and modeling
  of speech enhancement, 288–290, 292–294
  state-variable, 2143
  statistical, 2229
  stimulus-hypothesis evaluation, 2220, 2221
  Swerling, 960
  time-invariant, 1507
  time-series, 2101
  transmission line, 941, 942, 1021
  transmitter, 1601–1602
  traveling current source, 941
  of welding processes, 2229–2231
Model reference adaptive control (MRAC), 2171
Modems, 2201
  of waveguides, **859**, 849–852
Modified chemical vapor deposition (MCVD), 993, 994–995
Modified Coates graph (flowgraph G₁), **109**, 100
Modified discrete cosine transform (MDCT), 401
Modified nodal approach to CACS, 672, 656–657

Modulated noise reference unit (MNRU), 280
Modulation, **1367**, see also specific types
  adaptive differential pulse code, 284, 779, 1525
  amplitude, see Amplitude modulation (AM)
  baseband, 1462
  in broadcasting, **1367**, 1359–1365
  channel length, 538
  complex envelope function for types of, 1361
  conductivity, 640
  cross, **147**, 144–145
  delta, **781**, 775
  differential phase shift keying, 2203
  differential pulse-code, 281, 282, 283, 338, 350
  direct, **1425**, 1417–1418
  Doppler, 959
  effects of, 1178–1179
  external, **1425**, 1424–1425
  in facsimile, 1555
  frequency, see Frequency modulation (FM)
  index of, 144
  intensity, 763, 769, 1419–1420
  laser power, 1688–1689
  light, 761
  in long distance, 1429
  Mach–Zehnder, 769, 1424
  magnetic field, 1689–1690
  offset quadrature phase-shift keyed, 1604–1605
  phase, 770, 762, 763, 1555, 1574, 1580
  phase amplitude, 1363, 1446
  in phase-locked loops, 1574
  pulse, 1410
  pulse-code, see Pulse-code modulation (PCM)
  pulse-width, 705, 706, 1689, 2243
  quadrature-amplitude, 1365, 1366, 1419, 1534, 1542, 1555
  satellites and, 1541–1542
  telemetry and, 1580, 1583, 1589–1590
  trellis-coded, 1408–1409
Modulation codes, 1679
Modulation transfer function (MTF), 1783–1784
Module structure charts, **1984**, 1983
MOE, see Measure of effectiveness
M-of-N systems, 2029
Moment method, 1046
Momentum, 442, 1764
Momentum relaxation time, 438
Monaural attributes, **405**, 398

Page on which term is defined is indicated in bold.

Monitoring and control (M&C) systems, 2199

Monitoring of patients, 2397–2403

Monochromatic light, **742**, 738, 740, 741

Monochrome cathode ray tubes, 1778–1782

Monolithic integrated circuits (MICs), 892, 896
history of, 616
microwave, **1015**, 1004, 1006, 1007, 1008, 1009, 2209

Monolithic microwave integrated circuits (MMICs), **1015**, 1004, 1006, 1007, 1008, 1009, 2209

Monolithic power amplifiers, 1006

Monophase microinstructions, 1872

Monotonic converters, 773

Monte Carlo (MC) simulation, **1608**, 440, 472, 1602, 1603, 1604

Moore-type outputs, 1738

MOP, see Measure of performance

Morphing, 2005

Morphological operators, 336

Morphotropic phase boundary (MPB), **1097**, 1092

MORS, see Multiple overlapping register set

Morse code, 1517

MOS, see Mean opinion score; Metal-oxide semiconductor

MOSFET, see Metal-oxide semiconductor field-effect transistors

Most significant bit (MSB), 1845

Most significant digit (MSD), 1844

Motional-feedback loudspeakers, 403

Motion analysis, 415–416

Motion-compensated predictive coding, 351–352

Motion detectors, 1395, 2380, 2384, 2385

Motion estimation, **358**, 345, 354–356

Motor-evoked potentials (MEPs), 2327, 2328

Motor neurons, 2301

Motor rehabilitation, 2390–2391

Motors, 1333–1341, 2237, see also specific types
ac, **1341**, 1333, 1334, 1340–1341, 2238
analysis of, 1334–1340
apogee kick, 2196
applications of, 1333–1334
cage-rotor induction, 2238, 2241, 2243

dc, see dc motors
induction, see Induction motors
permanent magnet, **1341**, 736, 1333
separately excited, **1341**, 1334
squirrel cage induction, **1341**, 1334
synchronous, see Synchronous motors
wound-rotor, **1341**, 735, 1334, 2243

Mouse, **1957**, 1945–1947, 2012

Moving average (MA) models, 2101–2102

Moving charges, 813–814

Moving coil loudspeaker, 1059

Moving coil microphones, 1059

Moving coil transducers, 1058

MPB, see Morphotropic phase boundary

MPSK, see *M*-ary phase shift keying

MQW, see Multiple quantum well

MRAC, see Model reference adaptive control

MRI, see Magnetic resonance imaging

MRS, see Materials Research Society

MS, see Mean-square

MSB, see Most significant bit

MSC, see Magnitude squared coherence

MSD, see Most significant digit

MSDC, see Multistage depressed collectors

MS-DOS, 2069, 2070

MSE, see Mean-square error

MSI, see Medium-scale integration

MSS, see Mobile satellite services

MSW, see Magnetostatic wave

MTBF, see Mean time between failures

MTBUR, see Mean time between unconfirmed removals

MTDC, see Multiterminal dc

MTTF, see Mean time to failure

MTTR, see Mean time to repair

MU, see Memory units

Multicasting, 1461

Multichip modules, 786, 789, 795

Multidimensional sampling theorems, 1515

Multidimensional signal processing, 329–367, see also specific types
applications of, 359–360
beamformers in, **366**, 359, 360–364
computer vision and, 343
digital image compression and, 336–341

digital image processing and, see Digital image processing
edge detection in, 342–343
high-definition television and, 353
image analysis and, 343
image capture and, 330–334
image enhancement and, 335–336
image quality and, 356–357
image reconstruction and, **343**, 341–342
information-preserving coders in, 349–350
motion estimation in, 345, 354–356
point operations and, 334–335
predictive coding in, 350–352
quantization in, 345, 347–349
sensor array, see Sensor array processing
subband coding in, 352
token matching in, 356
transform coding in, 352
velocity filtering in, **366**, 360, 365–366
video, see Video signal processing
visual perception and, 356–359

Multifunction arithmetic-logic units, 1745

Multi-input signature register (MISR), 1814, 1828

Multimedia, 1398, 1441, 1442

Multimode fibers, 989, 992

Multimode pulse spreading, 986

Multipath effects, **848**, 838

Multipath geometry, 843

Multipath interference, 1421

Multiphase faults, 1272

Multiphase unidirectional transducers, 1071–1073

Multiple access
carrier-sense, **1458**, 1455, 1463, 2015
code division, see Code division multiple access (CDMA)
demand assigned, 1542
frequency-division, **1539**, 1534, 1541, 1542
satellites and, 1534, 1541, 1542
time-division, see Time-division multiple access (TDMA)

Multiple channel per carrier (MCPC), 2207

Multiple fan-out, 1699

Multiple-input shift registers (MISR), 1814, 1828

Multiple instruction stream, multiple data stream (MIMD) systems, 2053

Multiple instruction stream, single data stream (MISD) systems, 2053

Multiple overlapping register set (MORS), 1883, 1884

Multiple-pass devices, 1641, see also specific types

Multiple quantum well (MQW), **1649**, 1643, 1644, 1647

Multiple signals, 1483–1484

Multiple sub-Nyquist encoding (MUSE), **1397**, 1394

Multiplexing, **1807**, 776, 778, see also specific types

in B-ISDN signal transmission, 1443

data acquisition systems and, 1799, 1801

digital image processing and, 337

displays and, 1775

frequency-division, 1418, 1439, 1534, 1542

liquid-crystal displays and, 1775

in networks, 1443

optical frequency-division, 1403, 1431, 1439

optical time-division, 1439

programmable arrays and, 1737

space-division, 1439

statistical, 1451

in telemetry, 1580, 1585, 1589–1590

time-, 1585

time-division, see Time-division multiplexing

wavelength-division, 1431, 1439

Multiplication, 2432

in computer arithmetic, 1862–1863, 1866

of electric fields, 1036

electron, **759**, 755, 757

of matrices, 2438–2439

pattern, 866

scalar, 2438

tables for, 2523

z-transform and, 180

Multiplier round-off limit cycle, 249

Multipliers, 1010, 1011, 2422–2423, see also specific types

Multiplying digital-to-analog converters, **781**, 773, 774

Multipoint grounding, 906

Multipoint switching, 2206

Multipole filters, 1084

Multiprocessing, 1760

Multirate simulation, 1599

Multistage depressed collectors (MSDC), **1393**, 1388

Multistep methods, **672**, 666, 667–668, 669–670

Multiterminal dc transmission, 1231–1232

Multiuser transmission, 1484

Multivalley semiconductors, 1102–1104

Multivariable systems, 2144

Multivariable weld process control, 2233–2234

Mu-metal, 1179

Muscle action potentials, 2319

Muscular dystrophy, 2321

MUSE, see Multiple sub-Nyquist encoding

Musical noise, 291

Mutual capacitance, 513

Mutual capacitance coupling, 513

Mutual coupling coefficients, 513

Mutual impedance, 1218, 1224

Mutual inductance, **32**, 25, 513, 522

Mutual information, 1522

Mutual transfer conductance, 536

MUX, see Multiplexing

Myasthenia gravis, 2321

MYCIN expert system, 2048

Myelinated nerve fibers, **2310**, 2301, 2304, 2305, 2306, 2307

Myocardial infarction, **2329**, 2317

Myoelectricity, 2391

Myotonia, 2321

N

Nagoka's equation, 26

Nanostructure engineering, 444–445

Narrowband filters, 1082, 1084

n-ary trees, **1926**, 1923

National Electrical Code (NEC), **2407**, 2406

National Fire Protection Association (NFPA), **2407**, 2406, 2411

National Society of Professional Engineers, 2554

Natural binary coded decimal (NBCD) number systems, 1846–1847

Natural frequency, **1575**, 201, 1569, 2110, 2112

Nautical miles, **2294**

Navigation systems, 1062, 2188, 2191, 2285–2295, see also specific types

absolute, 2286–2287

aircraft, 2020, 2285

automation in, 2294

celestial, 2292–2293

coordinate frames in, 2286

dead-reckoning, 2287–2289

design of, 2294

map-matching, 2293

radios in, 2286, 2289–2292

ship, 2285

software for, 2293–2294

spacecraft, 2285

state vector, **2294**, 2285

types of, 2286–2287

update, 2294

NBCD, see Natural binary coded decimal

n-channel JFET, 530, 531, 532

n-channel metal-oxide semiconductors (NMOS), 1631

NDCs, see Normalized device coordinate systems

NDE, see Nondestructive evaluation

Near-end noise, 514

Nearest-neighbor classification, 343

Near-field approximation, 875

Near-field coupling, 906

NEC, see National Electrical Code

Negative feedback, 215, 547, 556

Negative feedback circuits, 647

Negative impedance converters (NIC), 680

Negative impedance inverters (NII), 680

Negative-positive-zero (NPO), **32**, 19

Negative resistance amplifiers, 1005

Negative temperature coefficient (NTC) resistors, 14

Neighborhood function, 425

n-element arrays, 872

Nematic crystals, **1777**, 1772

NEP, see Noise equivalent power

Nernst equation, 2302, 2312, 2347

Nernst potential, 1157, 2303

Nerve fibers, **2310**, 2301

Nerves, 2301, 2351

excitation of, 2304–2311

regeneration of, 2334–2335

repair of, 2330

sciatic, 2335

spinal, 2335

stimulated, 2309

Nested inverse radix, 1851

Nested radix, 1848

Net capacitance, 641

Network databases, 2041–2043

Network functions, 168, 165–166

Network graph theory, 92

Network management systems (NMSs), 2202–2203

Network-node interface (NNI), 1443, 1445

Networks, 674, 1441–1476, see also specific types

active, 102, 1437

advanced intelligent, 1473–1475

Page on which term is defined is indicated in bold.

all-optical, 1440
artificial neural, 2229–2231
asynchronous transfer mode, see
    Asynchronous transfer
    mode (ATM) networks
back-propagation, 2229
broadband-integrated services
    digital, see Broadband-
    integrated services digital
    network
C3 system and, 2214, 2218, 2219
carrier-sense multiple access,
    1455
CCS7, 1470–1471, 1473
combinational, see Combina-
    tional networks
communication, 1594
computer, see Computer
    networks
computer communication, see
    Computer communication
    networks
controlled impedance, 773
design of, 107
electromagnetic fields and, 806
fast packet, **1459**, 1456–1457
four-mesh, 67
four-node, 64, 65
future of, 2018
general concepts of, 1448–1451
high-voltage, 1310
high-voltage transmission, 1344
impedance, 773
impedance converter/inverter,
    680, 682
with independent nodes, 94
integrated services digital, see
    Integrated services digital
    network (ISDN)
interconnection, **2059**, 2052
Kirchhoff's laws and, 63–69
ladder, 69
lag, **2130**, 2115, 2119
lag-lead, **2130**, 2116, 2122–2124
LCR ladder, 674, 675, 676, 681
linear, **78**
line impedance stabilization, **934**,
    933
local-area, see Local-area
    networks (LANs)
lumped-π equivalent, 1220
management of, 2218, 2219
metropolitan-area, **1459**, 1456,
    1457, 1461, 2017–2018
microcellular, 1564–1565
microwave devices and, 883
neural, see Neural networks
nondegenerate, **176**, 171
nonlinear, 175–176
one-port, **902**, 894

passive, 93, 95, 98
personal communication, see
    Personal communications
    systems (PCS)
phase-lead, **2130**, 2115, 2116
photonic, 1434–1440
Π, **188**, 184–188
precision resistor ladder, 778
private, 1443, 1473
public-domain, 1461
reactive, 1280–1281
reduced reactive, 1280–1281
resistor, 6–7
secondary energy distribution,
    1315–1316
security of, 2072, 2081–2084
signaling, 1443, 1469
with single current source, 95
state equations for, 174–175
state-variable, 679
switching, 1443
synchronous optical, 1437
theorems for, see Network
    theorems
three-node, 64
time-varying, 175–176, 218
token ring, 1463, 2016
topology and, 92, 107
transfer functions in, 95
two-mesh, 66
two-port, **902**
underground cable, 1344
virtual circuit, 1451
virtual private, 1443, 1473
wide-area, **1460**, 1447, 1457,
    1461, 2018
wireless local-area, see Wireless
    local-area networks
Network theorems, 69–79, see also
    specific types
    compensation, 78
    linearity and, 69–71
    maximum power transfer, **78**,
        76–77
    of Norton, **79**, 71–74, 164–165
    reciprocity, **79**, 77–78
    substitution, 78
    superposition, **79**, 69–71
    of Tellegen, **87**, 74–75, 80–81,
        84–85
    of Thévenin, **79**, 71–74, 164–165
Neural accommodation, 2308
Neural networks, 420–429
    adaptive resonance theory and,
        426–428
    feedforward, **428**, 421
    Hopfield crossbar, 423–424
    perceptrons and, 421–423
    recurrent, **428**
    self-organizing, 425, 426

topology of, **428**, 421, 425
topology-preserving, 424–426
Neuroelectricity, 2301–2311
Neuromuscular actuation system,
    2248
Neurons, **2310**, 420, 423, 2301, 2351
Newton–Euler method, 2167
Newton–Raphson (NR) procedure,
    672, 661, 662, 663–666,
    2508
Newton's formulas, 471, 2464–2465,
    2512, 2514
Neyman–Pearson detectors, **1487**,
    1479
NFPA, see National Fire Protection
    Association
NIC, see Negative impedance
    converters
Nichols charts, **206**, 199
Nickel crystals, 819
Nickel-titanium alloys (Nitinol),
    1178, 1180, 1181
Night vision, 756, 1126
NII, see Negative impedance
    inverters
Nitinol (nickel-titanium alloys),
    1178, 1180, 1181
Nitrogen, 1143
NLFP, see Nonlinear Fabry–Perot
NLIFs, see Nonlinear interference
    filters
NMOS, see *n*-channel metal-oxide
    semiconductors
NNI, see Network-node interface
Nociceptors, 2301
Nodal systems, 176, see also Nodes
Node analysis, **69**, 57, 64–66
Nodes, **63**, 58, 59, see also specific
    types
    independent, 94
    nonreference, 65
    output, 99
    processing, 1451
    reference, **69**, 64, 94, 98
    super-, **69**, 66
    switching, **1459**, 1449
Node voltage, 63
Noise, **297**, **1498**, 287, 1488–1499,
    see also specific types
    active control of, 403–404
    amplifier, 1006
    antenna, 1494
    autocorrelation and, 1488,
        1489–1490, 1493
    Barkhousen, 1497
    bipolar transistors and, 901
    burst, 1496–1497
    capacitive, 522
    carrier ratio to, see Carrier-to-
        noise ratio (CNR)

in connectors, 522–524
contact, 1496–1497
coping with, 1498
crosstalk, 513, 514, 515, 522–524
current, 9
data, 1513–1514
discrete-time white Gaussian,
    1487, 1480
effective, 1494
energy of typical room, 396
equivalent input, 1495
far-end, 514, 517
flicker, 1496–1497
Gaussian, see Gaussian noise
generation of, 391, 1599–1601
inductive, 515, 523
in integrated circuit leads,
    522–524
Johnson, see Thermal noise
linear transformations and,
    1489–1490
in magnetic recording, 831
in medical imaging, 2377
MESFETs and, 901
in mobile radio communications,
    1547
musical, 291
near-end, 514
non-Gaussian, 414, 1484, 1593
Nyquist, see Thermal noise
in optical communication,
    1420–1422
in optoelectronics, 758
partition, 1496
phase, 1010, 1570
in phase-locked loops,
    1570–1572
pink, 1490
popcorn (burst), 1496–1497
power spectral density and,
    1489–1490
pseudorandom generation of,
    391, 1599–1601
quantization, 331, 1497
random, 1488
in receivers, 1422
reduction of, 287–297, 2281
    model-based approach to,
        292–294
    models of, 288–290
    performance measures of,
        288–290
    signal classification in,
        295–296
    signal estimation and, 290–294
    source coding and, 294–295
relative intensity, **1426**, 1417,
    1421–1422
room, 396
roundoff, 249

shot, **1426**, 758, 1417, 1420–1421,
    1432, 1496
signal ratio to, see Signal-to-noise
    ratio (SNR)
sources of, 1488, 1498
statistics of, 1488
thermal, see Thermal noise
in video transmission, 1420–1421
white, see White noise
in wirebonds, 522–524
Noise averaging algorithms, 393
Noise equivalent power (NEP), 759,
    1128
Noise factor (NF), 1494–1495
Noise figure, **902**, **1426**, 1005, 1422
Noise immunity, **1621**, 1618
Noise margins, 1618, 1624
Noise power, 1488–1489, 1495
Noise ratio, 1494–1495
Noncharacteristic harmonics, 1237
Noncoherent pulse radar, **963**, 955
Noncontacting sensors, 2343
Nondecaying memory, 1504
Nondegenerate network, **176**, 171
Nondestructive evaluation (NDE),
    1082
Non-Gaussian noise, 414, 1484,
    1593
Nonhomogeneous linear estimates,
    328, 325
Nonideal operational amplifiers,
    619–621
Nonimpact printers, 1965–1970
Noninvasive sensors, **2349**, 2343
Noninverting circuits, 627–631
Noninverting operational amplifi-
    ers, 618, 627–631
Nonisolated single-ended PWM
    converters, 712–714
Nonlinear circuits, 126–147, see also
    specific types
    diodes and, 126–128
    distortion and, see Distortion
    limiters and, 132–138
    output waveform from, 142
    rectifiers and, 128–131
    superconductivity and, 1118
Nonlinear conductance, 55
Nonlinear dielectric properties,
    1188, 1179
Nonlinear dynamic systems, 420
Nonlinear electroacoustic proper-
    ties, **1188**, 1178–1179
Nonlinear electromagnetic
    properties, 1179–1180
Nonlinear electro-optic properties,
    1188, 1178
Nonlinear equalizers, 1412–1413
Nonlinear Fabry–Perot (NLFP)
    structures, **1649**, 1641

Nonlinear interference filters
    (NLIFs), 1643
Nonlinear magnetic properties,
    1188, 1179–1180
Nonlinear mean-square estimate,
    328, 324
Nonlinear networks, 175–176, see
    also specific types
Nonlinear receivers, 1414–1415
Nonnegative-definite Toeplitz–
    Hermitian matrix approxi-
    mation, 269–270
Nonnegative sequence approxima-
    tion, 268–270
Nonparametric detection, 1485
Nonparametric estimation, **328**, 322
Nonrational systems, 320
Nonrecurring engineering (NRE),
    591, 592–593
Nonrecursive digital filters, 372
Nonredundant number systems,
    1864, 1859, 1860
Nonreference nodes, 65
Non-return-to-zero (NRZ) mode,
    1674
Non-return-to-zero (NRZ) pulse
    codes, 1433
Nonsaturated logic, 1624, 1627
Nonsaturated region, **545**, 532
Nonseparable code, 2023
Nonstate variables, **176**, 172
Nonstationary processes, 1502,
    1504, see also specific types
Nonuniform sampling, 1510, 1514,
    1515
Nonunions, **2338**, 2330
Nonuniqueness, 2104–2105
Nonvolatile memory, **1675**,
    1654–1655, 1670
Nonweighted number systems, 1860
Normal distribution, 2489
Normalization, 113, 2034
Normalized device coordinate
    systems (NDCs), 2011
Normalized values, 113, 115
Normally-on MOSFET, 567–568
Normal trees, **176**, 171
Normed vector space formulation,
    253–254
Norton current, 57
Norton equivalent, 74
Norton equivalent circuits, 57
Norton's theorems, **79**, 71–74,
    164–165
NPN transistors, 638
NPO, see Negative-positive-zero
NR, see Newton–Raphson
NRE, see Nonrecurring engineering
NRZ, see Non-return-to-zero
N self-checking programming, 2025

Page on which term is defined is indicated in bold.

NSPE, see National Society of Professional Engineers
NTC, see Negative temperature coefficient
*n*th harmonic factor, 139
N-type metal-oxide semiconductor, 1748
Nuclear auxiliary power, 1214
Nuclear medicine, 2374, 2400, see also Medical imaging; specific types
Nuclear power plants, 1200–1201
Nucleation-controlled coercivity, 822
Number systems, 382, 1843–1877, see also specific types
   binary, see Binary number systems
   binary-coded decimal, 1846–1848
   binary-coded hexadecimal, 1846–1848
   binary-coded octal, 1848
   complex, 2432
   for computer arithmetic, 1859–1861
   conversion between, 1848–1853
   desirable charcteristics in, 1843
   fixed-point, 1844, 1856
   floating-point, **1858**, 1854, 1856–1858, 1864
   fraction conversion in, 1850–1853
   hcxadecimal, **1720, 1858,** 1846–1848, 1888
   integer conversion in, 1848–1850
   natural binary coded decimal, 1846–1847
   nonredundant, **1864,** 1859, 1860
   nonweighted, 1860
   octal, **1858,** 1846–1848
   polynomial representation in, 1844–1845
   positional representation in, 1844–1845
   redundant, **1864,** 1859, 1860
   residue, 1860
   signed binary, 1853–1856
   unsigned binary, 1845–1848
Numerical aperture, **1001,** 976, 977, 991, 1683, 1684
Numerical constants, 2424
Numerical control interfaces, 2004
Numerical integration, 1040
Numerical methods, 2464–2465, 2507–2518, see also specific types
   finite differences and, 2508–2511
   iteration and, 2507–2508
Numerical problems, 1902
NVRAM, 1655

Nyquist analog-to-digital converters, **782,** 772
Nyquist density, 1515
Nyquist diagrams, 214
Nyquist frequency, 2282
Nyquist intervals, 1515
Nyquist noise, see Thermal noise
Nyquist plots, **206,** 199, 205
Nyquist rates, **1516,** 1510, 1515, 1524
Nyquist sampling, 307, 332, 772, 780, 1598
Nyquist stability criterion, 216

O

Object code, **1901,** 1878
Object collaboration models, **1985,** 1980, 1981
Object interface specification, **1985,** 1980
Objective lens, **1693,** 1683–1684, 1691
Object-oriented design (OOD), 1910
Object-oriented programming (OOP), **2013,** 2001–2002, 2005, 2011, 2036
   abstraction and, 1912
   for databases, 2045–2046
   data types and, 1924–1925
   hybrid, 1912
   paradigm of, **1914,** 1903, 1909
Object programs, **1915**
Objects, **1914, 1984,** 1909, 1913, 1980
Observability, **2106,** 2104, 2105
Occam's razor, 1525
Occasional designer, **601,** 599
Occipital lobe, **2329,** 2322
OCR, see Optical character recognition
Octal number systems, **1858,** 1846–1848
Octal scale, 2523
Octave, **125**
Octave band filters, 391
Odd parity, 2023
Odometers, 2288, 2293
Oersted, Hans Christian, 1003
OFDM, see Optical frequency-division multiplexing
Off-line testing, **1816,** 1810
Offset current, 620
Offset (excess) representations, 1856
Offset quadrature phase-shift keyed (OQPSK) modulation, 1604–1605

Offset voltage source, 620
O(f(n)) performance, **1926**
Ohmic region, 545, 532
Ohm's law, **69,** 64, 65, 112, 438, 533
Oil-filled cable, 1223
OLE, see Optical logic etalon
On-chip memory, **393**
One-dimensional propagation, 1081–1082
One-port networks, **902,** 894
One-third-octave filters, 391
On-line testing, **1816,** 1810
OOD, see Object-oriented design
OOP, see Object-oriented programming
Op amp, see Operational amplifiers
Open collector outputs, 1627
Open-loop gain, 617, 620
Open-loop systems, 2139, 2163
Open-loop transfer function, 214, 1568
Open system architecture, 2221, 2212
Open System Interconnect (OSI) systems, 2192, 2215
Operating conditions, 546
Operating systems, 1876, 2061–2071, see also specific types
   distributed computing and, 2062–2064
   fault-tolerance systems and, 2061, 2062, 2064–2065
   high-level languages and, 1902
   industry standards for, 2069–2070
   MS-DOS, 2069, 2070
   parallel processing and, 2062, 2065–2066
   real-time computing and, 2061, 2062, 2066–2067
   standards for, 2069–2070
   structure of, 2067–2069
   types of, 2062
   UNIX, 2063, 2069
Operational amplifiers, **623, 631,** **690,** 134, 135, 136, 137, 616–631, 771
   in active filter realization, 683
   active filters and, 674
   applications of, 625–631
   circuits-oriented approach to, 622
   history of, 616
   ideal, **623,** 616, 617–619
   inverter, 625, 626
   inverting, 618
   nonideal, 619–621
   noninverting, 618, 627–631

Page on which term is defined is indicated in bold.

practical, 619–621
SPICE and, **623**, 616, 621–623, 624, 630
Operational transconductance amplifiers (OTA), 674
Operator training simulators (OTS), 1351–1352
Opinion equivalent, 280
Optical amplifiers, 1430
Optical character recognition (OCR), 329, 1953
Optical circulators, **1171**, 1166–1169
Optical communication, 1417–1440
clipping in, 1424
external modulation and, 1424–1425
intensity modulation and, 1419–1420
lightwave technology and, 1417–1418
linearity in, 1422–1423
long distance and, 1427–1434
new approaches to, 1438–1440
noise in, 1420–1422
photonic networks in, 1434–1440
video transmission and, see Video transmission
Optical devices, 1641–1650, see also specific types
all-optical type, 1641–1643
limitations of, 1647–1648
optoelectronic type, 1643–1647
Optical disks, 1676, 1677, 1679
Optical encoders, 1947
Optical fiber cable, see Fiber-optic cable
Optical fibers, **1566**, 975, 976, 987–1001, see also Fiber-optic
applications of, 988, 990, 1001
in B-ISDN signal transmission, 1443
classification of, 988–991
coating of, 998
design of, 999–1001
drawing of, 996–999
features of, 988–991
in local-area networks, 1461, 1869
packaging of, 999–1001
proof testing of, 998–999
in telemetry, 1579
transmission characteristics of, 990, 991–993, 1417
in wireless local-area networks, 1559
Optical flow methods, 356
Optical frequency-division multiplexing (OFDM), 1403, 1431, 1439

Optical frequency routing, 1440
Optical guided-wave devices, **770**, 759, 760, see also specific types
Optical isolators, **1172**, 1166–1169
Optical logic etalon (OLE), **1649**
Optical logic gates, 1643
Optical path, **1693**, 1682–1684
Optical permittivity tensors, 1162
Optical polarizers, 761, 762
Optical properties of semiconductors, 443–444
Optical radiation sensors, 1155, 1156
Optical repeaters, **1001**, 988
Optical sensors, 2225–2226, 2345, 2348
Optical Society of America, 2554
Optical surface materials, 1176
Optical switching, 763, 765, 768
Optical telemetry, 1579
Optical time-division multiplexing (OTDM), 1439
Optical time-domain reflectometers (OTDR), 769
Optical-to-electrical converters, 1437
Optical waves, 1085, see also Light
OPTICON, 2389
Optimization, 424
Optoelectronic devices, 1641, 1643–1647, see also specific types
Optoelectronics, 738–770, see also Light
applications of, 768–770
circuits in, 759–770
applications of, 768–770
device fabrication and, 765–767
integrated optics and, 759, 760–765
packaging and, 767–768
detectors in, see Detectors
device fabrication in, 765–767
geometric optics and, 745–746
integrated circuits in, 760, 765
lasers in, 738–742
light absorption and, 744
noise in, 758–759
phase velocity and, 743
OQPSK, see Offset quadrature phase-shift keyed
Organic esters, 1145
Organic solids, 1146–1149
Ornstein–Uhlenbeck process, 1504, 1506
Orthogonality, **328**, 324, 2485–2486
Orthopedic aids, 2392
Orthosis, **2396**, 2388

OS, see Operating systems
OSA, see Optical Society of America
Oscillation, 215, 249, 2110, 2112, see also Oscillators
Oscillators, **878**, 870–873, see also Oscillation; specific types
audio, 419
crystal, 1089, 1572
harmonic, 218
LC, 1572
phase-shift, 632–633
ring, 575, 577
solid state circuits and, 1009–1010
in telemetry, 1586
tunable, 1009
voltage-controlled, 1567, 1568, 1570, 1573, 1574, 1586
OSI, see Open System Interconnect
Osteogenesis, 2330
Osteonecrosis, **2338**, 2334
Osteoporosis, 2330
OTA, see Operational transconductance amplifiers
OTDM, see Optical time-division multiplexing
OTDR, see Optical time-domain reflectometers
OTH, see Over-the-horizon
OTS, see Operator training simulators
Outliers, 327
Output characteristics, see Drains
Output circuits, 1626
Output equations, **176**
Output filters, 129
Output frequency, 143, 144
Output impedance, 57
Output latch, 778
Output nodes, 99
Output rate of change, 630
Output resistance, 538–539, 576
Output response analysis, 1812–1815
Output vector, **176**, 171
Output voltage, 146, 526, 539, 630, 638
Outside vapor deposition (OVD), 993, 996
OVD, see Outside vapor deposition
Overcurrent protection, 1270–1272
Overcurrent relays, 1270–1272
Overflow oscillation, 249
Overhead ac transmission, 1217–1222
Overhead wire interactions with lightning, 943–944
Overlap-add method, 247
Overlap (commutation) angle, **1241**
Overlap-save method, 246

Page on which term is defined is indicated in bold.

Overlay technology, 1948, 1949–1950
Oversampling analog-to-digital converters, **782**, 772, 776
Over-the-horizon (OTH) radar, 953
Oxidation, **489**, 476–478
Oxide cathodes, 1779
Oxides, 476, 477–478
Oxyacetylene welding, **2235**, 2223

P

PA, see Public address
Packaged-component subassemblies, 603–604
Packaging and interconnecting (P&I), **528**, **615**, 500, 603
Packets, 1454
Packet switching, **1459**, 1437, 1448, 1449, 1450, 2018
PACS, 2400
PAE, see Power-added efficiency
Page description languages (PDLs), 2005, 2012
Page-filling arrays, 1878
Page printers, **1975**, 1958–1965
Pagers, 1062
Paging, 1876
PAL, see Programmable array logic
Paley–Wiener (physical realizability) criterion, 318
PAM, see Payload assist module; Phase amplitude modulation
Paper-foil-filled capacitors, 20
Papoulis' generalization, **1516**, 1515
Parabolas, 2449–2450
Parabolic rate constant, 477
Parabolic reflector antennas, 870, 875
Paradigms, **1915**, see also specific types
  in computer programming, 1906–1912
  declarative, **1914**, 1908
  functional, **1914**, 1907
  high-level languages and, 1906–1912
  of object-oriented programming, **1914**, 1903, 1909
  procedural, **1915**, 1906
  shifts in, 1902, 1909
  Stimulus-Hypothesis-Options-Response, 2217, 2218, 2220
  of switching, 1449
  transactional, 1892–1894
Paraelectrics, **1131**
Parallel capacitors, 16, 17
Parallel clippers, 134

Parallel clipping circuits, 134
Parallel computing, 1869, see also Parallel processors
Parallel configuration in robotics, 2161
Parallel conversion, 775
Parallel element processing ensemble (PEPE), 2054
Parallel memory, 1928–1931
Parallel operation, **1735**
Parallel processors, **383**, **1869**, **2059**, **2071**, 374, 2052–2060, see also specific types
  array, **2059**, 2054–2055
  associative, **2059**, 2056–2057
  ensemble, **2059**, 2053–2054
  operating systems and, 2062, 2065–2066
  VLSI and, 371, 372–373, 2052
Parallel projection, 2011
Parallel realizations, 248–249
Parallel resonant circuits, 1084
Parallel resonant converters (PRCs), 719
Parallel-series systems, 2092, 2093
Parallel systems, 2028, 2029, see also specific types
Parallel transfer disks, 1666
Paramagnetic state, 1163
Paramagnetism, **836**, 816, 831, 835
Parameter estimation, **328**, 321–328
Parametric sonars, 1057
Parametrized signals, 1480–1482
Parasitic capacitance, 1492
Parietal lobe, **2329**, 2322
Parity code, 2023
Parkinson, David B., 2186–2187
Parks–McClellan program, 242
Parseval's theorem, 231, 232, 1514
Partial derivatives, 2466
Partial fraction approach, 2114
Partial fraction expansion, 155–157
Partitioning methods in placement problem solving, 584, 585–587
Partition noise, 1496
Paschen curves, 1139
Passband, **125**, 113
Passband ripple, 239, 243, 244
Pass-by-pointer, 1889
Pass-by-reference, 1889, 1895
Pass-by-value, 1889, 1895
Passive circuits, 122–124
Passive components, see Capacitors; Electrical fuses; Inductors; Resistors; Transformers; specific types
Passive filters, **690**, 674–675, 683, 1569, see also specific types

Passive ladder simulation, 680–681
Passive microwave devices, 882–891, see also specific types
  characterization of elements of, 882–884
  components of, 882
  directional couplers and, 887–888
  discontinuities and, 884
  filters in, 888
  hybrid circuits and, 887–888
  resonators and, 886–887, 890
  semiconductor, 890
  superconductivity and, 1119
  transmission lines and, 883, 884
Passive networks, 93, 95, 98
Passive signal processing, 111–125
  filter design and, 121–124
  low-pass filters and, 113–121
Passive sonar, 407, 408
Passive telemetry, 1590
Path-delay testing, **1836**, 1832–1835
Paths, **109**, 93, 99, see also specific types
  critical, 372, 373
  forward, 99
  mean free, **732**, 469
  optical, **1693**, 1682–1684
Path-set, **109**
Patient monitoring, 2397–2403
Pattern generation, **1816**, 494, 495, 1812
  test, see Test patterns, generation of
Pattern mapping, 425
Pattern multiplication, 866
Pattern recognition, 343, 422
Pattern transfer, 487–488
Payback period method, 2175
Payload assist module (PAM), 2195
PBS, see Polarizing beam splitters
PBXs, see Private branch exchanges
PC, see Program counters
PCBs, see Polychlorinated biphenyls; Printed circuit boards
p-channel JFET, 530–531
PCM, see Pulse-code modulation
PCS, see Personal communications systems
PCVD, see Plasma chemical vapor deposition
PDF, see Probability density function
PDLs, see Page description languages
PDR, see Preliminary design review
PDUs, see Protocol data units
PE, see Phase encoding; Processing elements
Pellat–Debye equations, 1134

Pel recursive methods, 354
Pels, **1556**, 1547
PEMF, see Pulsed electromagnetic fields
Penalty, **328**, 323
Pen-based computing, 1951–1952
Penetration cathode ray tubes, 1783
Penetration depth, 839, 1116, 1120
Penning ionization, 1789
Penstock, **1205**, 1203
PEPE, see Parallel element processing ensemble
Perceptrons, 421–423
Perfect conductivity, 1115, 1122, 1123
Perfect induction, 1701
Performability, **2030**
Periodic signal, 315
Periodic table of elements, 2525
Periodograms, 316
Peripheral circuits, 1748
Peripherals, **1752**, see also Input-output devices; specific types
Permalloy, 819
Permanent faults, 2088
Permanent magnet (PM) motors, **1341**, 736, 1333
Permeabilities, 813, 814, 1018, 1115, 2303
Permeability tensors, 805
Permittivity, 1018
  benzene and, 1144
  complex, 840, 844, 1133, 1136
  dielectrics and, 1132, 1133, 1135, 1136, 1144
  superconductivity and, 1115
Permittivity tensors, 805, 1080, 1162
Permutations, 2433
Perovskites, 1104, 1131
Personal communications systems (PCS), 1456, 1462, 1535, 1546–1566, see also specific types
  cellular, see Cellular communications
  facsimile in, 1554–1557
  mobile radio in, see Mobile radio
  problems in, 1546–1547
  service systems in, 1552–1553
  wireless LANs in, see Wireless local-area networks
Personnel security, 2084–2085
PET, see Positron emission tomography
Petrothermal source, 1202
PF, see Power factor
PGAs, see Programmable gate arrays
pH electrodes, 2346, 2347

pH sensors, 2346
Phase, **32**, 2355, 2356
Phase amplitude modulation (PAM), 1363, 1446
Phase angles, 61
Phase coefficients, 61
Phase constant, 838, 841
Phase-controlled converters, **711**, 702, 729, 730, see also ac-to-dc converters
Phase-controlled thyristors, 705, 729
Phase currents, 1262–1265
Phased-array radar, **963**, 951, 956, 1762
Phased arrays, **869**, 865, 867, 1762, 2382
Phase detector gain, **1575**
Phase detectors, 1567, 1572, 1573
Phase difference, 82
Phase distortion, **147**, 145
Phase encoding (PE) mode, 1674
Phase-lag compensation, 2119–2122, 2140, 2141
Phase-lead compensation, 2116–2119, 2140, 2142
Phase-lead networks, **2130**, 2115, 2116
Phase loading, 1312
Phase-locked loops (PLLs), 1567–1575, 2203
  applications of, 1567, 1574
  components of, 1572–1574
  design of, 1572
  noise in, 1570–1572
Phase margin (PM), 2116, 2118, 2119, 2121, 2122, 2126, 2140
Phase modulation (PM), **770**, 762, 763, 1555, 1574, 1580
Phase noise, 1010, 1570
Phase shifters, 951
Phase-shift keying (PSK), 1400, 1410, 1534
Phase shifts, 215, 884, 1305
Phase spectrum, **238**, 232–233
Phase-to-ground fault, 1256, 1260–1261, 1272
Phase-to-phase fault, 1256–1257, 1261, 1272
Phase transfer functions, 320
Phase velocity, **1085**, 743, 1078, 1081
Phase voltages, 705, 1261–1265
Phasor current, 82
Phasor diagrams, 82–83
Phasors, **92**, 61, 67, 89, 90, 1081, 1278, see also specific types
Phasor voltage, 82
Phonetic coding, 1401

Phonons, **446**, 438, 1764
Phosphor, 750, 756, 1942, 2006
Photocathodes, 754
Photoconductors, 751–752, 1959–1962
Photodetectors, 1417, 1420, 1430–1431, 1435, 1946, see also specific types
Photodiodes, 447, 458, 752–753, 1156, see also specific types
  avalanche, 1430–1431, 1432, 1435
  vacuum, 754, 755
Photoelastic coefficients, 1085
Photoelectrons, 755
Photoemissive detectors, 754–755
PhotoFETs, 753
Photolithography, 1159
Photometric brightness, see Luminance
Photomultiplier tubes, 755
Photonic networks, 1434–1440
Photons, **446**, 443, 1763, 1767
  counting of, 755
  gamma ray, 2377
  momentum of, 1764
Photosensor arrays, 1554
Phototransistors, 753
Photovoltaics, **1215**, 1208–1210, see also Solar cells
Physical constants, 2424, 2524, see also specific types
Physical media-dependent (PMD) layer, **1467**, 1457, 1462
Physical models, 2101, see also specific types
Physical placement of logic (PPL), **602**, 591, 593, 599–601
Physical quantities, 2425–2429
Physical realizability (Paley–Wiener) criterion, 318
Physical sensors, **2349**, 1155–1156, 2343–2345, see also specific types
Physical systems, 218–222, 2099, see also specific types
Physical transduction, 1153
Physical vapor deposition (PVD), **489**, 484–485
Physical variables, 2277
P&I, see Packaging and interconnecting
PIC, see Positive impedance converters
Pickup setting of relays, 1270
π constants, 2424
PID, see Proportional-integral-derivative
Piecewise linear models, 453
Piezoelectric coefficient, 1176–1177
Piezoelectric coupling, 1083

Page on which term is defined is indicated in bold.

Piezoelectric coupling constants, 1080

Piezoelectric excitation, 1080–1081

Piezoelectric hydrophones, 1059

Piezoelectric ink-jet printers, 1967

Piezoelectric materials, **1097**, 1087–1098, see also specific types
  applications of, 1089–1091
  mechanical characteristics of, 1088–1092
  structure of, 1091–1092

Piezoelectric polymers, 1056

Piezoelectrics, **1131**, 1126
  bone repair and, 2329
  electroacoustic devices and, 1055–1056
  input-output devices and, 1948, 1950
  of smart materials, **1189**, 1176–1177, 1179, 1180
  soft tissue repair and, 2329

Piezoelectric sensors, 1182, 1948

Piezoelectric strain coefficient, 1059

Piezoelectric tensor, **1104**

Piezoelectric transducers, **1085**

Piezoresistivity, 1099–1105

Piezoresistivity tensor, **1104**, 1100

Piezoresistors, 1154

Pigtailing, 768

PII, see Positive impedance inverters

Pilot protection, 1274–1276

Pinch-off voltage, **545**, 530

PIN diodes, 752–753, 890, 1013, 1430, 1435

Pink noise, 1490

Pipelining transformation, **383**, **393**, 371–372, 377

Pipe-type cables, 1223, 1224

Pitch, **305**, 298

Pivot element, 658, 659

Pivoting, **672**, 659–660

Pivot selection, 659

Pixels, 344, **1393**, **1556**, **1975**, 1379, 1554, 1775
  in cathode ray tubes, 1778
  in digital image processing, **344**, 331, 332, 335
  in plasma displays, 1795
  printers and, **1975**, 1962

PLA, see Programmable logic arrays

Placement routines, **590**, 583, 584–589

Planar arrays, 867–868

Planar diodes, 448

Planar motion, 2474–2475

Planck's constant, 472, 1114, 1135, 1763

Planck's law, 748, 1492

Planck's theory, 751

Planetary exploration, 2293

Plane waves, 1077, 1080, 1081, 1083

Plasma, **1798**

Plasma chemical vapor deposition (PCVD), 993, 995

Plasma displays, 1786–1798
  ac, **1798**, 1794–1795
  ac-dc, 1795–1796
  attributes of, 1787–1789
  color, 1797–1798
  computer graphics and, 2007
  current limiting for, 1791
  dc, **1798**, 1788, 1791–1794
  gas discharge characteristics of, 1789–1791
  gray scale and, 1796–1797
  lifetime of, 1788
  memory in, **1798**, 1788, 1794–1795
  types of, 1791–1796

Plasma etching, 488

Plasma modified chemical vapor deposition (PMCVD), 993, 995–996

Platters, **1669**, 1659

PLDs, see Programmable logic devices

PLLs, see Phase-locked loops

Plotting, 2005, 2012

PLRS, see Position Location Reporting System

PLZT/Si, **1649**

PM, see Permanent magnet; Phase margin; Phase modulation

PMCVD, see Plasma modified chemical vapor deposition

PMD, see Physical media-dependent

PMN, see Lead magnesium niobate

pn junction, 899

pn-junction diodes, **131**, 126, 128, 447–448, 750, 753
  with applied voltage, 448–450
  large-signal switching behavior of, 456
  sensors and, 1156, 1160

pnpn devices, 1647, see also specific types

PNP transistors, 638

Pockel's effect, 1178

PODEM algorithm, 1812

Point-by-point method, 2261

Point groups, 1101

Point-of-care systems, 2399–2400

Point-of-sale terminals, 1754–1755

Point operations, **344**, 334–335

Point source, 759, 746

Point-to-point dc transmission, 1228–1231

Poisson counting, 1486

Poisson distribution, 2489, 2491

Poisson's equation, 469, 470

Polar (spherical) configuration in robotics, 2157

Polar coordinates, 2468, 2474–2475

Polar dielectrics, 1126–1127

Polar form, 2432–2433

Polarity of transformers, **1308**, 1299

Polarization, **1097**, 744, 1166
  dielectrics and, 1134, 1136, 1137
  internal, 1091
  left circular, 1164
  liquid-crystal displays and, 1772
  losses of, 1134, 1137
  magneto-optical disk data storage and, 1690
  negative, 1087
  positive, 1087
  remanent, **1097**, **1131**
  right circular, 1164
  saturation, **1131**
  spontaneous, **1131**
  telemetry and, 1588
  wave, 844

Polarization controllers, 770, 765

Polarization isolation, **1544**

Polarization vector, 1078

Polarized capacitors, **32**

Polarizers, 761, 762

Polarizing beam splitters (PBS), 1691

Polar Kerr effect, 1166, 1690

Pole placement design, 2144

Poles, 112, 200, 201, see also specific types
  active filters and, 675
  Laplace transform and, 167
  operational amplifiers and, 620
  speech signal processing and, 302

Police radar, 971

Poling, 1127

Polybutenes, 1144

Polycarbonates, 1149

Polychlorinated biphenyls (PCBs), 1144

Polydimethylsiloxane fluids, 1144

Polyesters, 1146

Polyethylene, 1138, 1146

Polyimides, 1146

Polymers, 487, 1056, 1127, 1138, 1180, 1181, see also specific types

Polynomials, 112, 193, see also specific types
  array, 866
  Butterworth denominator, 114
  Chebyshev denominator, 116
  factorization of, 155–156
  frequency response and, 198

Page on which term is defined is indicated in bold.

Hermite, 2485
Laguerre, 2484–2485
Legendre, 1033, 2483–2484, 2485
Levinson, 320
number systems and, 1844–1845
sensor array processing and, 364–365
Thomson denominator, 115, 116
trigonometric, 1515
Polyphase microinstructions, 1872
Polypropylene, 1146, 1150
Polysilicon, **1657**, 1654
Polytetrafluoroethylene (PTFE), 1146
Polyvinylidene fluoride (PVDF), 1056, 1127, 1180
Popcorn (burst) noise, 1496–1497
Porcelain, 1145
Portable telephones, see Cellular communications; Mobile radio
Port protection devices, **2085**, 2072
POS, see Product-of-sum
Positional notation, 2520–2523
Position Location Reporting System (PLRS), 2292
Positive impedance converters (PIC), 680
Positive impedance inverters (PII), 680
Positive-semidefinite matrices, 263–265
Positive temperature coefficient (PTC) of resistivity (PTCR), 1099, 1104
Positive temperature coefficient (PTC) resistors, 14
Positron emission tomography (PET), **2380**, 2374, 2376–2377
Positrons, 2377
POSIX, 2070
Postprocessing, 1596
PostScript, 2005, 2007, 2008
Potassium dihydrogen phosphate (KDP), 1127, 1181
Potential, 1081
action, see Action potentials
activation, 423
deformation, 1102
electric scalar, 1018
fibrillation, 2320
motor-evoked, 2327, 2328
Nernst, 1157
scalar, 1018, 1080
somatosensory-evoked, 2327
streaming, **2339**
surface, 569, 571
threshold, 2312
transmembrane, 2312

Potential coefficient matrix, 1220
Potential energy, 1494, 2168
Potentiometers, 12
Potentiometric sensors, **2349**, 2347
Power, **87**, 79–87
absolute, 2355
ac, **87**, 81–87
apparent, 82
aural output, **1392**
average, 82
backoff of, 1541
complex, 82, 84–85
consumption of, 1623
conversion of, see Power conversion
diagrams of, 82–83
dissipation of, 1620
effective radiated, **1379**, **1392**, 863, 1370, 1382, 2197, 2209
fault analysis and, see Fault analysis
flow of, 89, 1288–1289, 1350, 1351
generation of, see Power generation
imaginary, 82
incremental available, 1493, 1494
input, 1541
instantaneous, 81
loss of, 637
maximum transfer of, 76–77, 85, 1242
measurement of, 85–87
noise, 1488–1489, 1495
noise equivalent, 759, 1128
optimal flow of, 1351
phase difference and, 82
pulsed, 1006
reactive, **87**, 82, 1242
real, **87**, 82
relative, 2355
in satellites, 1543
sources of, 1156
splitting of, 761
steady-state, **87**, 81–85
for telemetry, 1584, 1585
transfer of, 76–77, 85, 1082, 1242
transmission of, see Transmission
Power-added efficiency (PAE), 1007
Power amplifiers, 637–639, 1005, 1006, 1008, 1388
Power angle, **1285**, 1279–1280, 1284, 1285
Power conversion, 702–710
Power density spectrum, **305**
Power dissipation, 786
Power factor (PF), 32, **87**, 18, 83–84, 711
Power factor (PF) angle, 83

Power generation, 1193–1205, see also specific types
data acquisition and control in, 1345–1346
distributed, see Distributed power generation
distribution and, see Energy distribution
dynamic simulation of, 1352
fossil power plants in, 1194–1200
geothermal power plants in, **1215**, 1202–1203, 1207, 1211–1212
hydroelectric power plants in, **1215**, 1203–1204, 1211
nuclear power plants in, 1200–1201
Power MOSFETs, 698–699, 736
Power rating of resistors, 10
Power semiconductor devices, 694–701, see also specific types
Power-series method, 140–141
Power sources for telemetry, 1584
Power spectral analysis, **2362**, 2354–2355
Power spectral density, **1498**, 317, 318, 1489–1490, 1505
Power spectrum, **238**, **321**, 232–233, 1489
Power supplies, 711–728, see also Converters; specific types
ac, 727–728
dc, see dc power supplies
decoupling of, 781
design of, 725–727
filter, 21
fixed-frequency operation of, 725
for lasers, 738
logic gates and, 1628–1629
pulsewidth-modulated, 712–717
regulation of, 711
resonant
double-ended, 719–727
single-ended, 717–719
special, 728
uninterruptible, 728, 727, 2074
variable-frequency operation of, 710–722
voltages for, 625
Power-supply rejection ratio (PSRR), 617, 619
Power supply units (PSUs), **1807**, 1800
Power transfer function, 1506
Power transformers, see Transformers
Power transistors, 638, 697–698
Poynting vectors, 839, 1078

Page on which term is defined is indicated in bold.

PPL, see Physical placement of logic
PRA, see Probabilistic risk assessment
Practical diffusion, 480
Practical operational amplifiers, 619–621
Prandtl number, **796**, 792
Precision rectifying circuits, 136–138
Precision resistor ladder networks, 778
Precognitive behavior, **2253**, 2249
Prediction, 1516
Predictive coding, 338, 350–352
Predictive speech coders, **286**, 281–282, 283
Predistortion, 1425
Preemphasis, 1375
Prefix condition, 340
Preformatting, **1693**, 1678, 1679
Pregnancy test sensors, 1158
Pregrooves, 1679, 1686
Preliminary breakdown, **946**, 935
Preliminary design review (PDR), 1989
Preprocessors, 1596
Pressure, 1102, 1104, 1183
Pressure electricity, 1088
Pressure reflection coefficients, 2381
Pressure sensors, 1155, 1161
Pressure transducers, 2343, 2344
Pressurized water reactors (PWRs), 1200
Prewarping, 243
Prewhitening, 1484
PRF, see Pulse repetition frequency
Primary, **1308**, 1299
Primary energy distribution, 1312–1313
Primary memory, 1866
Primary service area for radio, **1379**, 1367
Primary side, **615**
Primary system of equations, **176**, 169
Principal system, 805
Printed circuit boards (PCBs), 500, 501, 502, 503, 1828
  boundary scan and, 1826
  crosstalk and, 514
  fine-line, 502
  high-speed, 528
  layout of, 501
  multilayer, 528
  risetime degradation and, 524
  signal degradation and, 505
  structure of, 505–510
Printers, 1958–1975, 2005, see also specific types
  bubble-jet, 1968

classification of, 1958
  dot matrix, 1972–1973
  electrophotographic, 1112–1113
  impact, 1970–1975
  ink-jet, 1965–1968, 2006
  ionographic, 1964
  laser, 2005, 2006
  line, **1975**, 1970–1973
  magnetographic, 1964
  nonimpact, 1965–1970
  page, **1975**, 1958–1965
  resistive ribbon, 1969–1970
  serial, see Serial printers
  thermal, 1964–1965, 1968–1970
  typeball, 1974
Private branch exchanges (PBXs), 2017
Private networks, 1443, 1473
Private virtual networks (PVNs), 1443, 1473
Probabilistic risk assessment (PRA), 2073, 2077
Probability, 2487–2488, 2517–2520, see also Statistics
  conditional, 2487–2488, 2517
  of detection, 413
  discrete, 2518–2519
  of error, 1483
  false-alarm, **1487**, 413, 1479
  Gaussian, 1497
  marginal, 2517
  mathematical expectation in, 2520
  miss, **1487**, 1479
  tables of, 2490–2495
  theorems for, 2517–2518
  theory of, 845
Probability density function (PDF), 321, 316, 407, 961, 1499, 2519
Problem selection, 2049
Procedural paradigms, **1915**, 1906
Process, **2071**
Processing elements (PEs), **428**, **2059**, 420, 421, 2052, see also specific types
Processing errors, 1498
Processing nodes, 1451
Processor, see Central processing unit (CPU)
Product-of-sum (POS) expressions, 1706, 1709
Product relaxation, 379
Product terms, 1736
Program counters (PCs), 387, 1867, 1890
Programmable array logic (PAL), **1740**, 1716, 1718, 1737, 1738, 1740
Programmable arrays, 1735–1741

Programmable counters, 1732
Programmable gate arrays (PGAs), 1718, 1739
  field-, see Field-programmable gate arrays (FPGAs)
Programmable logic arrays (PLAs), **1741**, 1716, 1735, 1736–1739, 1740, see also specific types
Programmable logic devices (PLDs), **1741**, 1716, 1718, 1735, 1740, see also specific types
Programmable memory, 1654–1655, 1715
Programmable radio systems, **2221**
Programmable read only memory (PROM), 1715
Programming, see Computer programming; Software
Programming languages, see Computer languages
Program status word registers, 1722
Progression, 2432
Projection, **344**
Projection mapping, **276**, 261
Projection operators, 256, 260–262
Projection screen cathode ray tubes, 1784–1785
Projectors, 1055
PROM, see Programmable read only memory
Proof testing of fiber-optic cable, 998–999
Propagation
  of acoustic waves, 2381
  of action potential, 2306
  delay in, see Propagation delay
  electromagnetic, 1029
  of electromagnetic waves, 505
  of groundwaves, 1370, 1371
  of light, 743, 975, 977
  of lightwaves, 975, 977
  navigation systems and, 2290
  one-dimensional, 1081–1082
  radio, 1370
  in solids, 1077–1080
  of sound under water, 409–412
  space, see Space propagation
  of ultrasound, 1077–1080, 1081–1082
  uncertainties about, 2290
  wave, see Wave propagation
Propagation constant, **848**, 838, 851
Propagation delay, 528, 503–504, 1742
Propagation delay time, **1621**, 1620, 1623
Propagation factors, **848**, 842

Propagators, **1047**, 1029,
     1031–1034, see also specific
     types
Proper subgraphs, **109**, 93
Proportional-integral-derivative
     (PID) controllers, 1757,
     2142–2143, 2149–2150,
     2167
Prosthesis, **2396**, 2388, 2390, 2392,
     see also specific types
Protection, 1268–1278, see also
     Security
  computer relaying and,
     1276–1278
  directional comparison blocking
     scheme in, 1275–1276
  distance, **1278**, 1272–1274
  overcurrent, 1270–1272
  pilot, 1274–1276
  principles of, 1268–1270
  reliability of, **1278**, 1270
  speed of, 1269–1270
  step-distance, 1273
  zones of, 1269
Protocol data units (PDUs), **1459**,
     1453
Protocol spoofing, **2209**, 2205
Proton exchange method, 766
Prototyping, **1915**, 382, 1968
Pseudoarthrosis, **2339**, 2330,
     2332
Pseudo-coloring, 334
Pseudo-exhaustive testing, 1822
Pseudo-inputs, 1818
Pseudo-outputs, 1818
Pseudo-random generators, **1608**,
     391, 1599–1601
Pseudo-random pattern generation,
     1816, 1812
Pseudo-random testing, **1836**, 1830
Pseudo-registers, 1894
PSK, see Phase-shift keying
PSPICE, 621, 630
PSRR, see Power-supply rejection
     ratio
PSUs, see Power supply units
PTC, see Positive temperature
     coefficient
PTCR, see Positive temperature
     coefficient of resistivity
PTFE, see Polytetrafluoroethylene
Public address systems, 207
Public-domain networks, 1461
Public-key cryptosystems, **2085**,
     2072
Pulse-code modulation (PCM), **782**,
     283, 775, 776, 778–779,
     1363–1364
  adaptive differential, 284, 779,
     1401

in digital audio broadcasting,
     1401
Pulse coding, 1580
Pulse compression, **963**, 953,
     958–959
Pulsed electromagnetic fields
     (PEMF), 2330, 2334, 2335,
     2336, 2337, 2338
  effectiveness of, 2332
  typical signal from, 2331
Pulsed lasers, 738
Pulse-Doppler radar, **973**, 964
Pulsed power, 1006
Pulsed welding, 2224
Pulse-echo ultrasound, **2385**,
     2383–2385
Pulse-mode circuits, 1720
Pulse-modulation systems, 1410, see
     also Pulse-code modulation
     (PCM)
Pulse number of converters, **1241**
Pulse radar
  accuracy in, 956–957
  antennas in, 951–952, 956
  applications of, 949–951
  architecture of, 951
  basic concept of, 949
  coherent, **963**, 955
  data processing and, 953
  design of, 951–953
  error sources in, 961–962
  estimation in, 961–962
  exciters for, 952–953
  noncoherent, **963**, 955
  overview of, 949–951
  performance prediction for,
     953–958
  range of, 953–955
  receivers for, 952–953
  resolution in, 956–957
  signal processing and, 953
  subsystems for, 951–953
  tracking with, 961–962
  transmitters for, 952
  waveforms in, 958–959
Pulse repetition frequency (PRF),
     955, 959, 2395
Pulse spreading, 985, 986–987
Pulse-triggered flip-flops,
     1637–1638
Pulse width, 1720
Pulse-width modulation (PWM),
     706, 1689, 2243
Pulse-width modulation (PWM)
     converters, **728**, 712–717,
     730–731
Pulse-width modulation (PWM)
     inverters, 705, 707, 708,
     734
Pulsing, **1392**, 1388

Pure longitudinal and shear waves,
     1085
Pure mode, 1080
Pursuit behavior, **2253**
Pursuit display, **2253**
Pushdown stacks, **1915**, 1904
Push-pull converters, 714–715, 716
Push-pull electrodes, 763
Push-pull method, 1679, 1686
PVD, see Physical vapor deposition
PVDF, see Polyvinylidene fluoride
PVNs, see Private virtual networks
PWM, see Pulse-width modulation
PWRs, see Pressurized water
     reactors
Pyramidal horn antennas, 870
Pyroelectric array, 1129
Pyroelectric coefficient, 1127, 1130
Pyroelectrics, **1131**, 1087,
     1126–1131
Pyrosensitive smart materials, **1189**,
     1176, 1179, 1182
PZT, see Lead zirconate titanate

Q

QAM, see Quadrature-amplitude
     modulation
QFD, see Quality function
     deployment
QRCs, see Quasi-resonant
     converters
Quadratic detectors, **1487**, 1482
Quadratic energy, 424
Quadratic phase coupling, **2362**,
     2356
Quadrature-amplitude modulation
     (QAM), 1365, 1366, 1419,
     1534, 1542, 1555
Quadrature components, 1481
Quadriplegics, 2335
Quad trees, 1923
Qualitative robustness, **328**
Quality assurance software, 1989
Quality factor, **32**, 18, 27, 886
Quality function deployment
     (QFD), 1986
Quality management, 1986
Quality ratings, 357
Quantization, **344**, **1546**, 331,
     1523–1524
  errors in, 331, 348
  sampling and, **344**
  vector, **344**, **1546**, 338, 339,
     348–349, 1523–1524, see
     also Vector quantization
     (VQ)
  in video signal processing, 345,
     347–349

Page on which term is defined is indicated in bold.

Quantization noise, 331, 1497
Quantum efficiency, **1434**, **1771**, 754, 1767
Quantum mechanics, 472–473, 818, 1486
Quantum transport theory, 440
Quarter-wave thickness, 1082
Quartz, 1055, 1063
Quasi-analytic simulation, 1602
Quasi-bidirectional ports, 1728, 1731
Quasi-Fermi energy, **473**, 469, 471
Quasi-linear man-machine systems, 2247
Quasi-longitudinal waves, 1078
Quasi-resonant converters (QRCs), 717–718
Quasi-shear waves, 1078
Quasi-static approximation, 1080
Quasi-static magnetic fields, 1183
Quasi steady-state, 469
Quaternary (quad) trees, 1923
Queues, 1451, 1919, 1922
Quiescent collector current, 640
Quiescent conditions, 546
Quiescent voltage, 641

R

Race-free state assignments, **1720**, 1719
Radar, 408, 873, 874, 949–984, see also specific types
  accuracy in, 956–957
  altimeters and, 971
  antennas for, 951–952, 955, 956, 964
  applications of, 949–951, 970–973
  architecture of, 951
  coherent, **963**, 955
  continuous wave, see Continuous wave (CW) radar
  data processing and, 953
  design of, 951–953
  detection with, 959–961
  Doppler, **973**, 955, 964, 965–967, 2288
  error sources in, 961–962
  estimation in, 961–962
  exciters for, 952–953
  false alarms in, 959
  FM/CW, 967–969
  interrupted continuous wave, 955
  line-of-sight, 953
  loss estimation in, 955–956
  microprocessors in, 1760, 1762
  noncoherent, **963**, 955
  over-the-horizon, 953

  performance prediction for, 953–958
  phased-array, **963**, 951, 956, 1762
  police, 971
  pulse, see Pulse radar
  pulse-Doppler, **973**, 964
  range of, 953–955, 958
  receivers for, 952–953
  resolution in, 956–957
  search, 949, 958, 959–961
  signal processing and, 953
  signal-to-noise ratio in, 953, 969
  subsystems for, 951–953
  synthetic aperture, 950, 957
  temperature in, 955–956
  tracking, 950, 958, 961–962
  transmitters for, 952
  ultrasound and, 1082, 1085
  waveforms in, 958–959
Radar-absorbing materials (RAMs), 1176, 1182, 1185, see also specific types
Radar cross section (RCS), **963**, 954, 955, 956, 958, 960
Radar detection systems, 777
Radar digitization, 778
Radar proximity fuzes, 970–971
Radial energy distribution, 1314–1315
Radial frequency, 1133
Radiated emissions, **934**
Radiation, see also specific types
  blackbody, 749
  electromagnetic, see Electromagnetic radiation
  gamma, 2377
  infared, 1126
  Kirchhoff's law of, 748
  optical, 1155, 1156
  patterns of, **869**, 861, 862
  Planck's law of, 1492
  pyroelectrics and, 1128
  radio frequency energy and, 921
  reflected, 2262
  sensors and, 1155
  shielding and, 906
  sources of, 1128
  thermal, 787, 788
  X-ray, 488
Radiational thermal resistance, 788
Radiation choppers, 1128
Radiation efficiency factor, 877
Radiation impedance, 1059–1060
Radiation mass, 1060
Radiation reactance, 1059
Radiation resistance, 863, 864
Radiative efficiency, 1767
Radiative heat transfer, **796**, 787
Radiators, see also specific types
  discrete, 870–873

  intentional, 923–925
  unintentional, 921, 922, 923
Radio, 1367–1379
  AM, 1367–1372, 1374
  antennas for, 1372, 1374, 1375
  channel classifications in, 1368–1369
  field strength of, 1369–1370, 1375
  FM, see Frequency modulation (FM) radio
  frequency allocations for, 1368, 1373
  mobile, see Mobile radio
  in navigation systems, 2286, 2289–2292
  primary service area for, **1379**, 1367
  propagation of, 1370
  secondary service area for, **1379**, 1367–1368
  station classifications in, 1368–1369, 1373–1375
  transmitters for, 1370–1372, 1375, 1574
Radio Button System (RBS), 1977, 1980, 1981, 1982, 1983
Radio frequency (RF), 1417, 1419, 1420
Radio frequency (RF) energy, 921, 923, 949, 1564
Radio frequency (RF) equipment, 2199, 2200
Radio frequency (RF) interference (RFI), **918**
Radio frequency/terminals (RFT), 2205
Radiometers, 1483
Radionuclides, 2377
Radio wave depolarization, 847
Radix, **1858**, 1844, 1848, 1851, 1859
Radix complement representation, 1854–1856, 1861
Radix divide method, 1849
Radix multiply method, 1851
Rad Lab, 1016–1017
Radon transform, 341, 342
Raised-drain designs, 575
RAM, see Radar-absorbing materials; Random-access memory
Raman active media, 1181
Raman frequencies, 1179
Ramp signals, **50**, 48–49
Ramsey's reaction concept, 1032
Random-access memory (RAM), 1651, 1686
  block in, 1822–1823
  dynamic, 1652, 1655
  programmable arrays and, 1735
  static, 1653–1654, 1655

Page on which term is defined is indicated in bold.

Random-access scan, 1825–1826
Random coding, 1523
Random deflection, 1941
Random initial condition, 1502
Random logic, 1713
Random multi-access protocols, 2205
Random noise, 1488
Random processes, see Stochastic processes
Random signals, **1499**, 1482–1483
Random testing, **1816**
Random variables, **321**, 316, 1499–1500, 2518
  entropy of, 1517
  Gaussian, 1504, 1523
Range complement (RC) system, 1861
Rank-reduced matrices, 272–273
Rao–Cramèr bound, 322, 327
Rapid prototyping, 382
Rapid thermal annealing (RTA), 484
Rapid thermal processing, 579
Raster, **1393**, **1785**, 1380, 1941
Rate distortion function, **1526**, 1522, 1523
Rate distortion theory, 1521–1523
Rate-length product, 1427
Rational systems, 320
Rayleigh backscatter, **1426**, 1421
Rayleigh criterion, 846
Rayleigh resolution, 873, 876
Rayleigh–Ritz variational method, 1032
Rayleigh scattering, 985
Rayleigh wave filters, see Surface acoustic wave (SAW) filters
Ray theory, 976–977
Ray tracing, 409, 2005
RBS, see Radio Button System
RBSOA, see Reverse-bias safe operating area
RC active filters, **631**, 628, 674, 675, 676, 678
RCAs, see Ripple-carry adders
RCC, see Remote center compliance
RCP, see Right circular polarization
RCR, see Room cavity ratio
RCS, see Radar cross section
RCTs, see Reverse-conducting thyristors
Reactance, **30**, **32**, 24, 31, 1279
Reaction energy, 487
Reactive ion beam etching (RIBE), 488
Reactive ion etching (RIE), 488
Reactive near field, **878**
Reactive networks, 1280–1281
Reactive power, **87**, 82, 1242

Reactors, **1205**, **1250**, 485, 1200, 1201, see also specific types
  fixed-capacitor, thyristor-controlled, 1247–1248
  shunt, **1250**, 1246
  thyristor-controlled, 1247–1250
  thyristor-switched capacitor, thyristor-controlled, 1249–1250
  thyristor-switched shunt, 1247
Readback process, 830
Read-only memory (ROM), 1651, 1656, 1715, 1822–1823, 1830, 1874
Read/write heads, 1658, 1659
Realization, **1499**, 1501, 2105
  of active filters, 683–690
  cascade, 248–249
  of circuits, 685–690
  of noise process, 1488
Real power, **87**, 82
Real systems, 2094–2095, see also specific types
Real-time computing, **2071**, 2061, 2062, 2066–2067
Real-time scheduling, 2061
Receiver operating characteristics (ROC), **418**, 413, 414
Receivers, see also specific types
  cardioid, 1060
  digital audio broadcasting, 1398, 1400
  electromagnetic acoustic transducer, 2229
  filters in, 917
  linear, 1414
  in long distance, 1427, 1431
  modeling of, 1601–1602
  noise in, 1422
  nonlinear, 1414–1415
  for radar, 952–953
  superheterodyne, **1367**, 1360
  in telemetry, 1580, 1586, 1590–1591
Receiving antennas, 842
Receptors, **2339**, 2338, see also specific types
Reciprocal differences, 2515–2516
Reciprocity, 1058
Reciprocity theorem, **79**, 77–78
Reclosers, **1319**
Recombination, 443–444, 1763, 1766
Record copy in facsimile, **1556**
Recording, see also specific types
  of action potentials, 2313–2314
  audio, 776, 826, 827, 835
  digital audio, 776
  in electroencephalography, 2322, 2323, 2356–2360

  facsimile, 1556
  by laser power modulation, 1688–1689
  magnetic, see Magnetic recording
  by magnetic field modulation, 1689–1690
  magnetic tape, 1674
  magnetooptic, **1171**
  metal-in-gap, 832, 833
  thermomagnetic, 1687–1690
  video, 835
  video cassette, 826–827, 1090, 1676
Rectangular cavity resonators, 887
Rectangular coordinates, 2467
Rectangular waveguides, 852–853, 857, 859
Rectifier diodes, 447
Rectifiers, **2246**, 128–131, 136, 732, 2238, see also specific types
  asymmetrical silicon-controlled, 697
  bridge, 129, 130
  filters and, 129
  full-wave, 128
  silicon-controlled, see Thyristors
  uncontrolled, **131**, 128
Recurrent neural network, **428**
Recursion, **1915**
Recursion formula, 2484
Recursive digital filters, 378
Recursive equations, 2167
Recursive filters, 379
Recursive least-squares (RLS) algorithm, 403
Recursive modeling, 274–276
Recursive signal processing algorithms, 377
Reduced characteristic tables, **1640**, 1636
Reduced instruction set computers (RISC), **1901**, 1882, 1888, 1892, 1938, 1939
Reduced-rank approximation, 262–263
Reduced reactive networks, 1280–1281
Redundancy, 337
  hardware, 2020–2021
  information, 2021–2023
  in logic, 1821
  signal, **405**, 400
  software, 2024–2025
  time, 2024
  triple modular, 2021, 2029
Redundant manipulators, **2162**, 2161
Redundant number systems, **1864**, 382, 1859, 1860
Redundant robots, 2161

Page on which term is defined is indicated in bold.

Reed–Solomon code, 1607
Reference impedance, **891**, 883
Reference nodes, 69, 64, 94, 98
Reflectance, 1165, 2259, 2265
Reflected glare, 2266
Reflected illumination, 2263–2264
Reflected radiation coefficient
    (RRC), 2262
Reflected waves, 1082
Reflection, 846
    Fresnel coefficient of, 844
    geometry of, 845
    internal, 975, 1767
    microwave, 1182
    microwave devices and, 882, 883
    minimum, 526
    pressure, 2381
    Snell's law of, 843
    ultrasound and, 1081, 1082
    veiling, **2275**, 2265, 2266
    voltage, 1082
Reflection coefficients, 320, 882, 883
Reflection-type circuits, 1005
Reflectivity, 1643
Reflectivity tensor, 1165
Reflectometers, 518, 769
Reflector antennas, 956
Reflector lamps, 2269
Refraction, 745–746
Refractive index, **987**, 743, 975,
    1085, 1643
    cladding and, **1001**, 988, 994
    dielectrics and, 1136
    irradiance-dependent, 1641
    single-mode, 993
    wave propagation and, 840
Refractory period, **2311**, 2304
Regenerative braking, **737**, 730
Regenerative feedback circuits, 1576
Region of absolute convergence,
    157, 150–151
Registered ASICs, 1733
Registers, **1735**, 1695, 1721–1735,
    1866, 1867, see also specific
    types
    assembly language and,
        1882–1886
    counters and, 1731–1733
    data direction, 1729
    feedback shift, 1727
    flag, 1722
    flip-flops in, see Flip-flops
    gated, 1722–1724
    global, 1883
    graphic symbols and, 1733–1734
    input-output ports and,
        1728–1731
    linear feedback shift, **1734**, 1727
    memory address, 1867
    memory data, 1867

program status word, 1722
    scratch, 1889
    shift, 1725–1727, 1731
Register transfer language (RTL),
    1727–1728, 1729
Regression analysis, 316
Regula falsi method, 2508
Regulation, **38**
Rehabilitation engineering, **2396**,
    2387–2396
    appropriate technology in, **2395**,
        2392
    in communications disorders,
        2388, 2391
    in motor rehabilitation,
        2390–2391
    in sensory rehabilitation,
        2388–2390
Relational databases, 2033–2040
Relations and data types, 1923–1924
Relative entropy, 1520
Relative intensity noise (RIN), **1426**,
    1417, 1421–1422
Relative maxima and minima, 2463
Relative permeabilities, 813, 814
Relative power, 2355
Relative refractive index difference,
    987, 975
Relative stability, 2131, 2137, 2140
Relaxation processes, 1135
Relaxation times, **1110**, 438, 1109,
    1135, 1136
Relaxed look-ahead, 378, 380
Relays, 1269–1271, see also
    Protection; specific types
    adaptive, 1277–1278
    computer, **1278**, 1276–1278
    directional overcurrent,
        1271–1272
    distance, 1272–1275
    electromechanical, **1278**
    instantaneous overcurrent, 1271
    inverse-time characteristics of,
        1270
    loadability of, 1274
    overcurrent, 1270–1272
    pickup setting of, 1270
    reliability of, **1278**, 1270
    solid-state, **1278**
Reliability, **2030**
    computer, see Computer
        reliability
    fault tolerance and, **2030**, 2026,
        2027, 2028, 2029
    improvement in, 503
    of integrated circuits, 1808
    interconnection costs and, 501
    of interconnections, 503
    of local-area networks, 1465,
        1563

    of protection, **1278**, 1270
    of relays, **1278**, 1270
    of thermal management, 785
    of transmission, **1295**, 1287
    of wireless local-area networks,
        1563
Remanence, 826
Remanent coercivity, 835, 824
Remanent magnetization, 835, 829,
    831
Remanent polarization, **1097**, **1131**
Remez exchange algorithm, 242
Remnant, 2251–2252
Remote center compliance (RCC),
    2173, 2180
Remotely piloted vehicles (RPVs),
    2212, 2214
Remote sensors, 337
Remote terminal units (RTUs),
    1353, 1345, 1348
Rendering, **2013**, 2011
Renewable fuses, 43
Repeatability, **1161**, 1154, 2160
Repeaters, **1001**, 988, 1430, 1543,
    see also specific types
Reserve monitoring, 1348
Residual function or capacity, **2396**,
    2388
Residue number systems (RNSs),
    382, 1860
Residues, method of, 181
Resins, 1149
Resistance
    area-specific, 795
    base spreading, 552
    collector, 640
    conduction thermal, 787
    conductive, 787, 789
    contact, 790
    convective, 787, 789, 791
    differential, 454, 620
    dynamic drain-to-source, 538,
        539
    electrode, 1070
    emitter, 640
    emitter-base junction diffusion,
        640
    emitter degeneration, 646
    equivalent series, **32**, 18
    external thermal, 791–793, 794
    frequency-dependent negative,
        674, 681
    Hall, 441, 442
    hop-off, 12
    incremental, **458**, 454, 551
    of inductors, 28, 29
    input, 551, 620
    insulation, 18
    internal, 52
    internal base, 640

internal emitter, 646
internal thermal, 790–791, 793
magneto-, **835**
output, 538–539, 576
radiation, 863, 864
radiational thermal, 788
relative changes of, 1105
sheet, 481
spreading, 791
temperature coefficients of, **15**, 7, 1218
thermal, see Thermal resistance
Thévenin, 85, 647
Thévenin equivalent, 636
total thermal, 793–794
transmembrane, 2337
voltage coefficients of, **15**, 8
voltage-variable, 533, 543–544
of wires, 2543–2544
Resistance reflection factor, 648
Resistance thermometers, 1155
Resistive overlay technology, 1949–1950
Resistive ribbon printers, 1969–1970
Resistivity, **15**, 5, 6, 1099, 1104, 2527–2532
of alloys, 2530–2531
of ceramics, 2532
of metals, 2527–2529
of semiconductor materials, 2536
Resistivity tensor, **1104**
Resistors, 5–15, see also specific types
characteristics of, 5–11
chip, 610
color coding of, 10–11
composition, 12
current characteristics of, 5–6
equivalent circuits for, 7
film, 8
fixed, 11–12
high-frequency effects of, 7–8
integrated circuits in, 13
metal-film, 12
negative temperature coefficient, 14
networks of, 6–7
piezo-, 1154
polysilicon load, 1654
positive temperature coefficient, 14
power rating of, 10
rectangular, 6
special-purpose, 13–14
stability of, 12
thick film, 1104
types of, 11–14
variable, 8, 12–13
voltage characteristics of, 5–6

voltage-dependent, 890
voltage rating of, **15**, 10
wire-wound, 11–12
Resistor-transistor logic (RTL), 1624
Resolution, 956–957, 2005
Resolved motion control, 2172–2173
Resolved motion rate control (RMRC), 2172, 2173
Resonance, 206, **1250**, 202, 1084, 1243, 2165, 2337
harmonic, 1246
subsynchronous, **1250**, 1243
Resonance-enhanced distortion (RD), 1423
Resonant energy, 444
Resonant frequency, **32**, 30, 1084
Resonant half-wavelength antennas, 864
Resonant-link dc-to-dc converters, 710
Resonant-link inverters, 707–708
Resonant power supplies, 717–727
double-ended, 719–727
single-ended, 717–719
Resonators, **742**, 738, 886–887, 890, 1074–1075, see also specific types
Response diagrams, 216
Retiming transformation, **383**, 373–374, 376
Return address, 1889, 1890
Return loss, **891**, 883, 884
Return stroke of lightning, **946**, 936, 941–943
Return-to-zero (RZ) codes, 1433
Reuse, **1915**, 1910
Reverberation/clutter, **418**
Reverse-bias electric fields, 899
Reverse-bias gate-to-channel, 538
Reverse-bias safe operating area (RBSOA), 698
Reverse breakdown of diodes, **458**, 456–457
Reverse-conducting thyristors (RCTs), 697
Reverse current, 450
Reverse generation-recombination current, **459**
Reverse leakage current, **32**
Reverse saturation current, **459**, 449
Reverse voltage, **701**, 21, 694
Revolving fields, **1331**, 1322
Reynolds number, **796**
RF, see Radio frequency
RFT, see Radio frequency/terminals
Rheobase, **2311**, 2307
Rheostats, 13
Ribbon microphones, 1060

RIBE, see Reactive ion beam etching
Ridge waveguides, 856
RIE, see Reactive ion etching
Right circular polarization (RCP), 1164
Right-hand rule, 27
RIN, see Relative intensity noise
RINDA database, 2047
Ring oscillators, 575, 577
Ripple, **131**, 129
Ripple-carry adders (RCAs), **1747**, 1742
Ripple current, **32**, 21
RISC, see Reduced instruction set computers
Risetime degradation, **528**, 524–527
Risk analysis, 2073
Risk assessment, 2073, 2077
Risk control, 2410–2411
RLS, see Recursive least-squares
RMRC, see Resolved motion rate control
RNSs, see Residue number systems
Robotics, 329, 426, 1757, 2154–2185
adaptive control in, 2171–2172
anthropomorphic configuration in, see Articulated configuration
applications of, 2175–2185
implementation of, 2175, 2176–2177
justification for, 2175–2176
in manufacturing, 2178–2181
processing, 2178–2180
articulated configuration in, 2157–2158
Cartesian configuration in, 2155–2156
cellular manufacturing in, **2182**, 2175
compliant motion control in, **2174**, 2173
control systems in, 2163–2167, 2171–2172
cylindrical configuration in, 2156
dynamic models of, 2167–2173
flexible manipulators in, 2173–2174
gantry configuration in, 2159–2161
independent joint control in, **2174**, 2163–2167
issues in, 2181–2182
kinematics and, **2162**, 2154, 2161
manipulators in, **2162**, 2161, 2173–2174
manufacturing applications of, 2178–2181

Page on which term is defined is indicated in bold.

in material handling, 2178
modeling in, 2167–2173
parallel configuration in, 2161
polar configuration in, see
  Spherical configuration
processing applications of,
  2178–2180
resolved motion control in,
  2172–2173
SCARA configuration in,
  2158–2159
sensors in, **2182**, 2173, 2178,
  2180
spherical configuration in, 2157
workforce acceptance of, 2177
Robot Time and Motion (RTM),
  2177
Robust detection, 1485
Robust estimation, **328**, 322, 328
Robust threshold selection, 343
ROC, see Receiver operating
  characteristics
Rochelle salt, 1055
Rodrigues formula, 2484
ROE, see Rules of Engagement
ROM, see Read-only memory
Room cavity ratio (RCR), 2259
Room position multipliers, 2263,
  2264
Root locus, **2139**, 214, 2131–2139
  design with, 2137–2138, 2143
  of digital control systems,
    2136–2137
Root mean square (rms) error, 1500
Rotational latency, **1669**, 1664
Rotational position sensing, **1669**,
  1664
Rotation sensors, 769
Rotors, **1341**
Roughness coefficient, 843
Rounding off errors, 1852
Roundoff noise, 249
Routh and Hurwitz methods, 213
Routing routines, **590**, 583, 589–590
RPVs, see Remotely piloted vehicles
RRC, see Reflected radiation
  coefficient
RTA, see Rapid thermal annealing
RTL, see Register transfer language;
  Resistor-transistor logic
RTM, see Robot Time and Motion
RTUs, see Remote terminal units
Rubber, 1146, 1149
Ruby lasers, 738
Rugged MOSFETs, 699
Rule-based expert systems,
  2048–2051
Rules of Engagement (ROE), 2217
Run-length encoding, **1556**, 1555
RZ, see Return-to-zero

S

Safe Medical Devices Act (SMDA),
  2412, 2411
Safety, **2030**
  of biomedical equipment,
    2408–2413
  fault tolerance and, **2030**
  grounding and, 905
Sagnac logic gates, **1649**
Sallen–Key filters, 677–678, 687, 688
Sample-and-hold amplifiers (SHA),
  1807, 771, 1804, 1805
Sample-and-hold phase detectors,
  1573
Sample-and-hold techniques, 1804
Sampled-servo concept, 1680, 1686
Sampled tracking, 1686–1687
Sample functions, **1509**, 1501, 1504
Sampling, **358**
  in computational electromag-
    netics, **1048**, 1035–1036,
    1043–1044
  continuous, 1515–1516
  data, **183**, 182–183, 2281
  in digital image processing, 330,
    331
  instruments and, 2282
  nonuniform, 1510, 1514, 1515
  Nyquist, 332, 772, 780, 1598
  quantization and, **344**
  rate of, **1516**
  theorems on, see Sampling
    theorems
  theory of, 1510, see also
    Sampling theorems
  in video signal processing, **358**,
    345, 346–347
Sampling analog-to-digital
  converters, 778
Sampling density, 1036
Sampling period, **183**, 182
Sampling theorems, **1516**, 346,
  1510–1517
  cardinal series and, 1510–1511,
    1512, 1514, 1515
  generalizations of, 1515–1516
  multidimensional, 1515
  nonuniform, 1510, 1514, 1515
  proof of, 1511–1513
  time-bandwidth product and,
    1513
SAR, see Successive approximation
  register; Synthetic aperture
  radar
Satellite news gathering (SNG),
  2206–2207
Satellites, 1532–1545, 2194–2209
  access to, 1541–1542
  access protocols for, **2209**, 2204

antennas in, 1542, 2206
applications of, 1532–1533
audio and, 1398, 2207–2208
carrier-to-noise ratio and, 1537,
  1540–1541
communications link and,
  1537–1538
digital audio broadcasting and,
  1398
digital links and, 1540
direct broadcast, 1534,
  2198–2199
earth stations and, 2199–2200
frequency allocations for, 1542,
  1543
functions of, 1533–1535
geosynchronous, **2209**,
  2197–2198
G/T and, 1538–1540
launchings of, 2195–2196
mobile communications and,
  2198
modulation and, 1541–1542
navigation systems and, 2291
network management systems
  and, 2202–2203
next-generation, 2209
orbits of, 1533, 1535–1537
pointing angles of, 1535–1537
protocol spoofing and, **2209**,
  2205
Seasat, 873
spacecraft and, 2196–2199
subsystems in, 1542–1544
system noise temperature and,
  1538–1540
television and, 1532
transmission and, 2203–2204
trends in, 1544
video and, 1533, 2205–2207
VSATs and, **2209**, 2194,
  2200–2205, 2209
Saturated logic, 1624
Saturated region, 532
Saturation, **1393**, 569, 1382
Saturation current, **459**, 449
Saturation magnetization, **836**
Saturation polarization, **1131**
Saturation temperature, 788
Saturation velocity, 574
Saturation voltage, 548
SAW, see Submerged arc welding;
  Surface acoustic wave
SBS, see Stimulated Brillouin
  scattering
SC, see Switched-capacitor
SCA, see Subsidiary communication
  authorization
SCADA, see Supervisory control
  and data acquisition

Scalar differential state equations, 174–175
Scalar multiplication, 2438
Scalar potential, 1018, 1080
Scales of integration, 501
Scaling, **125**, 117, 121–122
Scan design, **1836**
Scanning, 358, **1393**, **1556**, **1957**, 1554, 1953, 1954
Scanning antennas, 951
Scan/set logic, 1825
Scan testing in semiconductor manufacturing, **498**, 491–492
SCARA, see Selective Compliance Articulated Robot for Assembly
Scattered energy, 1082, 2382, 2383
Scattered look-ahead, **383**, 379
Scattered waves, 1085
Scharfetter–Gummel empirical formula, 470
Schawlow, Arthur, 1124–1125
Scheduling, 424
Schematic capture, **602**
Schematic entry, 597–598
Schottky contacts, **474**, 470
Schottky diodes, **131**, 126, 127, 128, 447, 458, 1004, 1010
  logic gates and, 1604, 1627
Schottky junction, 895
Schottky transistor symbol, 1628
Schottky TTL, 1624
Schrodinger wave equation (SWE), 472–473
Sciatic nerve, 2335
Scientific visualization, **2013**, 2004, 2005, 2008
SCP, see Service control points
SCPC, see Single channel per carrier
SCR (silicon-controlled rectifiers), see Thyristors
Scratch registers, 1889
Screen coordinate systems (SCSs), 2011
SCS, see Screen coordinate systems
SCSI, 1660, 1664
S/D, see Signal-to-distortion ratio
SDH, see Synchronous digital hierarchy
SDM, see Space-division multiplexing
SDR, see System design review
SDS, see Structured distribution systems
Search radar, 949, 958, 959–961
Seasat satellite, 873
Seasonal thermocline, 409
Secant method, 2508
Secondary, **1308**, 1299

Secondary energy distribution, 1314, 1315–1316
Secondary marketing research, **1566**, 1561
Secondary memory, 1866
Secondary service area for radio, **1379**, 1367–1368
Secondary side, **615**
Secondary systems of equations, **176**, 169
Second harmonic factor, 139, 141, 142, 143
Second-order intermodulation distortion, 143
Sectors, **1669**, **1693**, 1659, 1679, 1681
Security, 2061, 2062, see also Protection
  C3 system and, 2215
  computer, see Computer security
  of computer networks, 2081–2084
  data, 2044–2045
  of databases, 2044–2045, 2079
  of hardware, 2080–2081
  of local-area networks, 1563
  of networks, 2072, 2081–2084
  personnel, 2084–2085
  physical, 2074
  of protection, 1270
  of software, 2077–2080
  of transmission, **1295**, 1287
  of wireless local-area networks, 1563
Security constrained dispatch, 1350
Security control, **1353**, 1344, 1350–1351
Security lighting, 2271
Seebeck voltage, 1214
SEEDs, see Self-electro-optic effect devices
Seek time, **1669**, 1664, 1681
Seismology, 359
Selective Compliance Articulated Robot for Assembly (SCARA), 2158–2159
Self-aligned contact, 577
Self-biasing, 534
Self-checking programming, 2025
Self-commutation, **737**, 734
Self-contained oil-filled cables, 1223
Self-electro-optic effect devices (SEEDs), **1649**, 753–754, 1643–1646, see also specific types
Self-impedance, 1218, 1224
Self term, 1036
Self-tuning control systems, 2147
Semantics, **1915**, 1903

Semi-analytic (SA) simulation, 1602, 1603, 1607
Semiconducting perovskites, 1104
Semiconductor body, **579**, 567
Semiconductor chips, 1866, see also Chips; Integrated circuits
Semiconductor detector arrays, 755–757
Semiconductor detectors, 751–759
Semiconductor devices, see also Semiconductors; specific types
  efficiency of, 1766–1768
  power, 694–701
  principles of, 1763–1764
Semiconductor diodes, 890
Semiconductor lasers, 738, 740, 741
Semiconductor manufacturing, 475–529, see also Semiconductors
  annealing and, 484
  built-in self-test in, **498**, 490–491, 497
  crosstalk and, see Crosstalk
  deposition and, 484–487
  diffusion in, **489**, 479–481
  direct access testing in, 492–493
  dopant segregation and, 478
  fault coverage in, **498**
  fault grading in, 496–497
  interconnections and, see Interconnections
  ion implantation and, **489**, 479, 481–484
  joint test action group and, 493–494, 497
  junction formation and, 483–484
  lithography in, **489**, 487–488
  masking and, 483–484
  pattern transfer in, 487–488
  scan testing in, **498**, 491–492
  temperature in, 495–496
  testing in, **498**, 490–497
  test program flow in, 497
  thermal oxidation in, **489**, 476–478
  tolerances and, 510–513
  voltage in, 495–496
Semiconductor materials, 126, 892–893, 1764–1766, see also Semiconductors; specific types
  properties of, 2533–2534
  resistivity of, 2536
Semiconductors, 435–474, see also Semiconductor devices; Semiconductor materials; specific types
  band properties of, 2535

---

boundary conditions and, 470–471
conduction bands of, 2535
disordered, 445
electrons and, 437–438
energy bands and, 435–436
equivalent circuit models and, see Equivalent circuit models
Hall effect and, 440–442, 1110
holes and, 437–438
manufacturing of, see Semiconductor manufacturing
multivalley, 1102–1104
nanostructure engineering and, 444–445
optical properties of, 443–444
properties of, 2533–2536
recombination processes and, 443–444
in simulation, 468–473
transport properties of, 438–440
Semiconductor storage cells, 1866
Sensitivity, **682**, **690**, 675–676, 683
  of electroacoustic devices, 1057–1058
  of sensors, **1161**, 1154
Sensor array processing, 359–367
  beamformers in, **366**, 359, 360–364
  discrete arrays in, 361–365
  finite impulse response filters in, 360–361, 362
  polynomials and, 364–365
  velocity filtering in, **366**, 360, 365–366
Sensors, **1161**, 1152–1161, see also specific types
  acceleration, 1155
  acoustic, 2228
  amperometric, **2349**, 2346
  arc, 2226–2228
  bio-, see Biomedical sensors
  bioanalytical, **2349**, 2347–2348
  biomedical, see Biomedical sensors
  block diagram of, 1154
  in bonding, 2225–2229
  chemical, **2349**, 1157–1158, 2345–2347
  displacement, 1155, 1156
  dynamic-pressure, 2288
  electroacoustic smart, 1183
  electrochemical, 2345, 2346, 2348
  electromagnetic, 1182, 1183
  enzyme, 1158, 1159
  fiber-optic, 1057, 1182
  fingertip tactile, 2173
  force, 1155, 1156
  force-torque, 2180
  fusion of, **2182**

glucose, 2348
Hall effect, 1941
height-of-burst, 970, 971
hydrophone, 412
image capture and, 330
immuno-, 1158–1159
indwelling, 2343
infrared, 752, 2228
integrated circuits in, 1160
integration of, **2182**, 2178
invasive, 2343
joint torque, 2173
light, 447
linear position, 753
magnetostrictive, 1182–1183
mass flow rate, 1155, 1161
micro-, **1161**, 1159–1161
noncontacting, 2343
noninvasive, **2349**, 2343
optical, 1155, 1156, 2225–2226, 2345, 2348
pH, 2346
physical, **2349**, 1155–1156, 2343–2345
piezoelectric, 1182, 1948
potentiometric, **2349**, 2347
pressure, 1155, 1161
in protection, 1270
radiation, 1155, 1156
remote, 337
repeatability in, **1161**, 1154
in robotics, **2182**, 2173, 2178, 2180
rotation, 769
sensitivity of, **1161**, 1154
shape-memory effects, 1183
smart, **1189**, 1181, 1182–1183
smart materials and, 1173, 1174, 1175
smart-skin, 1182, 1187
stability of, **1161**, 1154
temperature, 1154, 1155
thermal, 2348
through-the-arc, 2226
ultrasonic, 2228–2229
in vehicles, 2255
velocity, 1155
voltage, 769
in welding systems, 2225–2229
wrist, 2106
Sensory disabilities, 2388
Sensory neurons, 2301
Sensory rehabilitation, 2388–2390
Sensory signals, 424
Sensory structures, 1174
Sensory systems, 1175
Separable code, 2023
Separately excited dc motors, **1341**, 1334
Sequence currents, 1261–1264

Sequence detection, 1484
Sequence equivalent circuits, 1253–1255
Sequence generators, 1601
Sequence quantities, **1267**
Sequencers, **1741**, 1737, 1738
Sequence voltages, 1261–1264
Sequential circuits, 1613
Sequential detection, 1485–1486
Sequential faults, **1816**, 1810
Sequential logic circuits, **1720**, 1711, 1718–1719, 1720
Sequential probability ratio test, 1486
SER, see Slip energy recovery
Serial capacitors, 17, 1243–1244
Serial inductors, 24
Serial interface ICs, 771
Serial operation, **1735**
Serial printers, **1975**, see also specific types
  impact, 1973
  nonimpact, 1965–1970
  thermal, 1968–1970
  wire matrix, 1974–1975
Serial processing, **383**, 371, 374, 376, 380, 381, 382
Serial wire matrix printers, 1974–1975
Series, 2454–2461, see also specific types
  cardinal, **1516**, 1510–1511, 1512, 1514
  expansion of, 2458–2461
  Fourier, 1512, 1515, 2480
  of functions, 2455–2457
  logarithmic, 2460–2461
  reversion of, 2458–2459
  Volterra, 147, 1599
Series capacitors, 17, 1243–1244
Series equalizers, **2130**
Series impedance, 1218–1219
Series-parallel resonant converters (SPRCs), 719, 720, 723–725
Series-parallel systems, 2092, 2093
Series resonant circuits, 1084
Series resonant converters (SRCs), 719
Series systems, 2028, 2029, see also specific types
Servers, 1913
Service control points (SCPs), **1475**, 1470, 1471, 1473, 1475
Service-creation capabilities, 1473
Service management systems (SMSs), **1475**, 1471
Service primitives, **1459**, 1452, 1453
Service switching points (SSPs), 1471
Set-reset flip-flops, 1721

Page on which term is defined is indicated in bold.

Sets, 1923–1924
Settling, 1178
Setup and hold time requirement,
1640, 1718, 1723
Setup in television, **1393**, 1384
SFD, see Switching field distribution
SHA, see Sample-and-hold
amplifiers
Shadowing function, 843, 846
Shadows, 2266
Shaft encoders, 1755
Shannon, Claude E., 1528–1531
Shannon code, 1519
Shannon theorem, 339
Shape anisotropy, 822, 823, 831
Shape functions, 1020
Shape-memory alloys (SMA), 1180,
1181
Shape-memory effects (SME), **1189**,
1178, 1180–1181
Shape-memory effects (SME)
sensors, 1183
Shape-memory hybrid composites
(SMHC), 1180, 1181
Shape-memory polymers (SMP),
1180, 1181
Shape-memory smart materials,
1189, 1180–1181
SHB, see Spatial-hole burning
Shear force, 1178
Shear waves, 1078
Sheath self-impedance, 1224
Sheet resistance, 481
Sheet resistivity, 6
SHEM, see Stimulus-hypothesis
evaluation model
Shielding, 903–904, 906–913
cable, 913
effectiveness of, **918**, 906, 907,
908, 911
electromagnetic, 1182, 1185
enclosure theory and,
906–908
good practices in, 911–913
high-frequency, 1175
magnetic, 1175, 1185
penetrations and, **918**, 908–911
testing of, 911
Shift-register latch (SRL), 1823,
1824, 1825
Shift registers, 1725–1727, 1731
Ship navigation systems, 2285
Shockley, William, 652
Shockley–Read–Hall relationship,
444, 470
SHOR, see Stimulus-Hypothesis-
Options-Response
Short-channel devices, 569, 574, see
also specific types
Short-channel effects, 575

Short circuit gain-bandwidth
products, **651**, 641
Short circuits, 642, 1289
in energy distribution, 1314
in energy management systems,
1351
fault and, 1252
impedance of, 1306
ultrasound and, 1084
welding and, 2227
Short dipole, 861–862
Short-term memory (STM), 421
Shot noise, **1426**, 758, 1417,
1420–1421, 1432, 1496
Shunt admittance, 1217, 1219–1220
Shunt capacitors, 1244–1247
Shunt reactors, **1250**, 1246
Shunts, **1250**, 1242
Sideband frequencies, 1179
Sigma-delta conversion, **782**, 775,
776
Sigmoid function, 421, 422
Signal detection, 1478–1487
for known signals, 1480
multiple signals and, 1483–1484
for parametrized signals,
1480–1482
for random signals, 1482–1483
Signal-enhanced modeling,
275–276
Signal estimation, 1478
Signal-flowgraphs, **109**, 99
Signal generators, 1599–1601
Signaling, **1446**, **1459**, 1445, 1450,
1470, see also Signals
Signaling networks, 1443, 1469
Signaling system no. 7 (SS7), **1475**,
1445, 1451, 1470
Signal-present model, 1479
Signal processing, see also specific
types
acoustic, see Acoustic signal
processing
autoregressive, moving average,
2390
bit-serial, see Bit-serial processing
broadband, 1083
digital, see Digital signal
processing (DSP)
multidimensional, see Multidimensional signal processing
parallel, see Parallel processing
passive, see Passive signal
processing
radar and, 953
recursive algorithms for, 377
serial, see Serial processing
speech, see Speech signal
processing
ultrasound and, 1082

very large scale integration for,
see Very large scale
integration (VLSI)
video, 777
Signal recovery, see Signal restoration
Signal reference subsystems, **918**,
905
Signal restoration, **276**, 251–276
algebraic properties of matrices
and, 262–265
algorithms in, 261–262
attribute sets and, **276**, 251,
252–253, 255–260
closed convex sets and, **276**,
258–260
closed subspaces and, 255–258
exponential signals and, 270–273
Hilbert space setting and, 255,
256
linear variety property sets and,
258
metric space formulation and,
252
nonnegative sequence approximation and, 268–270
normed vector space formulation
and, 253–254
recursive modeling of data and,
274–276
structural properties of matrices
and, 265–268
subsequence, 273
vector space formulation and,
253–254
Signals, **297**, see also Signaling;
specific types
averaging of, 2337, 2358
bandlimited, 1510, 1511, 1513,
1514
B-ISDN, 1443–1444
classification of, 295–296
conditioning of, 1160
confirmation to receiver, 1555
constellation of, 1605, 1607
continuous time, 182, 1598
dc, 49, 452
degradation of, 505, 525
detection of, see Signal detection
differential, 617
error, 617, 1685
estimation of, 290–294
excess of, 408
exponential, 270–273
focus error, 1685
impulse, 47–48
intercellular, 2336
interrupt, 1867
irrelevance in, **405**, 400
known, 1480

multiple, 1483–1484
parametrized, 1480–1482
passive sonar, 407
periodic, 315
pre-processing of, **313**, 306, 307–308
processing of, see Signal processing
ramp, **50**
random, **1499**, 1482–1483
redundancy in, **405**, 400
restoration of, see Signal restoration
sensory, 424
sinusoidal, **50**, 49, 271–272
step, 47
synchronizing video, 1381–1382
track error, 1686
unit-step, **8**, 6
video, 1381–1382
Signal-to-distortion ratio (S/D), 1411
Signal-to-noise power ratio, 1498
Signal-to-noise ratio (SNR), **297**, **418**, **1807**, 287, 293, 296
acoustic signal processing and, 413, 414, 415
in bioelectricity, 2337
in data acquisition systems, **1807**, 1803
in digital communication, 1415
in long distance communications, 1432
in optical communication, 1418
in radar, 953, 969
signal detection and, 1483
simulation and, 1594
Signal-to-noise voltage ratio, 1497
Signal transfer points (STPs), **1475**, 1470, 1471
Signature, 1812
Signature analysis, **1816**, 1814–1815
Signed binary number systems, 1853–1856
Signed-digit number systems, 382
Signed integers, 1860–1861
Signed-magnitude (SM) representation, 1853–1854, 1860, 1862
Sign-select redundant-to-two's complement conversion method, 382
SIL, see Surge impedance loading
Silica glass, 1428
Silicon, 1775
Silicon-carbide, 126
Silicon compilers, **602**, 598
Silicon-controlled rectifiers (SCRs), see Thyristors
Silicon diodes, 134
Silicon dioxide, 1145

Silicon liquids, 1144
Silicon nitride, 1146
Silicon V-grooves, 767
Silicone rubber, 1149
SIMD, see Single instruction stream, multiple data stream
Simple Mail Transfer Protocol (SMTP), 1450
Simple Network Management Protocol (SNMP), 2203
Simulated annealing, 587–588
Simulation, **1607**, 1593, 1718, 1869, see also Simulators; specific types
C3 system and, 2216
circuit modeling and, 60
computer-aided circuit, see Computer-aided circuit simulation (CACS)
computer graphics and, 2008
dedicated languages for, **1608**, 1597
design of, 1597
of distortion, 145–147
dynamic, 1352
electromagnetic, 2008
fault, **1816**, 1812, 1819
of inductance, 680, 681
interdisciplinary nature of, 1597
languages for, **1608**, 1597
limitations of, 1595
low-pass type, 1598–1599
Monte Carlo, **1608**, 440, 472, 1602, 1603, 1604
motivation for use of, 1595
multirate, 1599
passive ladder, 680–681
of power generation, 1352
products of, 1604–1607
quasi-analytic, 1602
role of, 1594–1595
semi-analytic, 1602, 1603, 1607
structure of, 1596–1597
validation of results of, **1608**, 1604
Simulation exercisors, 1596
Simulators, **602**, 468–473, see also Simulation; specific types
Sine wave carriers, 1504
Single-bit errors, 2023
Single-bus systems, 1868
Single channel per carrier (SCPC) systems, 2206, 2207
Single-core cable, 1224
Single inline packaging (SIP), **615**
Single-input/single-output systems, 2126, 2249
Single instruction stream, multiple data stream (SIMD) systems, 2052, 2053

Single instruction stream, single data stream (SISD) systems, 2052, 2053
Single-mode fiber (SMF), **1434**, 989, 992, 1428, 1435, 1436
Single-pass devices, 1641, 1642, see also specific types
Single phase-to-ground fault, 1256, 1260–1261, 1272
Single-phase unidirectional transducers (SPUDT), 1073
Single photon emission computerized tomography (SPECT), **2380**, 2374, 2376, 2377
Single-point grounding, 906
Single-sided assembly, **615**
Singly terminated filters, 114, 116, 117, 118, 119
Singularities, **2162**, 2161
Singular value decomposition (SVD), **277**, 262
Sinusoidal frequency, 198, 679
Sinusoidal input signals, 139
Sinusoidal outputs, 198
Sinusoidal signals, **50**, 49, 271–272
Sinusoidal steady-state, **125**, 61, 1081, 2115
Sinusoidal stimulation, 2309, 2310
SIP, see Single inline packaging
SISD, see Single instruction stream, single data stream
Skeletal muscle contraction, 2319–2320
Skin depth, 839, 1116, 1117
Skin effect, 1219
Slab waveguides, 978–980
Slip energy recovery (SER), 2238, 2243
Slip power recovery control of induction motors, 735–736
Slope of curves, 2462
Slowness surface, **1085**, 1078
SM, see Signed-magnitude; Synchronous motors
SMA, see Shape-memory alloys
Small disturbance, **1285**
Small-outline integrated circuits (SOICs), 610, 611
Small-scale integration (SSI), 1614, 1622
Small-signal ac voltage gain, 537–538, 540
Small-signal amplifiers, 899
Small-signal analysis, 639–652, see also specific types
common-base (CB) amplifiers and, 648–650
common-collector (CC) amplifiers and, 650–651

common-emitter amplifiers and, 644–647
current sources and, 55
hybrid-pi equivalent circuits and, 640–642, 643
voltage sources and, 55
Small-signal common-source amplifiers, 541–542
Small-signal equivalent circuits, 56, 552–553
Small-signal models, 454–456, 463–465, 466, see also specific types
Small-signal operation, 550–552
Small-signal transition, 576
Small-signal variables, 57, 54, 55
Smart cards, **2085**, 2072, 2081, 2401–2402
Smart cars, 2255
Smart materials, **1189**, 1173–1190, see also specific types
acoustic, 1175
applications of, 1176–1180, 1183–1186
high-tech, 1186–1188
classification of, 1175–1176
dielectrics of, **1188**, 1179
electroacoustic, **1188**, 1178–1179, 1181
electromagnetic, **1188**, 1175–1176, 1179–1180, 1181–1182
electro-optic, **1188**, 1178, 1181
electroplastic, **1188**, 1177, 1180
electrorheological, **1188**, 1178, 1181
ferroelectrics of, 1179
future of, 1186–1188
high-tech applications of, 1186–1188
magnetic properties of, **1188**, 1179–1180
magnetostrictive, **1188**, 1177, 1180
material properties for, 1176–1180
nonlinear dielectric properties of, **1188**, 1179
nonlinear electroacoustic properties of, **1188**, 1178–1179
nonlinear electromagnetic properties, 1179–1180
nonlinear electro-optic properties of, **1188**, 1178
nonlinear magnetic properties of, **1188**, 1179–1180
optical surface, 1176
piezoelectrics of, **1189**, 1176–1177, 1179, 1180

pyrosensitive, **1189**, 1176, 1179, 1182
radar-absorbing, 1176
shape-memory, **1189**, 1178, 1180–1181
smart structures and, 1174–1175
state-of-the-art, 1180–1182
structural, 1175
structural engineering applications of, 1184–1185
thermal, **1189**, 1175
Smart roads, 2255
Smart sensors, **1189**, 1181, 1182–1183
Smart skins, 1182, 1187
Smart structures, **1188**, **1189**, 1174–1175, see also Smart materials
Smart systems, 1183–1186, see also specific types
SMDA, see Safe Medical Devices Act
SMDS, see Switched multimegabit data service
SME, see Shape-memory effects; Society of Manufacturing Engineers
Smectic crystals, 1775
SMF, see Single-mode fiber
SMHC, see Shape-memory hybrid composites
Smoothing operation, 336
SMR, see Specialized mobile radio
SMSs, see Service management systems
SMT, see Surface mount technology
SMTP, see Simple Mail Transfer Protocol
SN, see Switching networks
SNAP, see Systems for nuclear auxiliary power
Snell's law, 745–746, 843, 1165
SNG, see Satellite news gathering
SNMP, see Simple Network Management Protocol
SNR, see Signal-to-noise ratio
Society of Manufacturing Engineers, 2555
Society of Women Engineers, 2555
Soft errors, 1652
Soft tissue imaging, 2380
Soft tissue repair, 2329–2339
devices for, 2330–2334
dosimetry in, 2335–2338
history of, 2329–2330
mechanisms of, 2335–2338
Software, 1865, 1867, 1976–2003, see also Computer programming; specific types
abstraction of, 1998–1999

algorithm analysis for, 1997
in avionics, 2191
brickwalling of, **2193**, 2191
bugs in, **1995**, 1985
C3 system and, 2216
causes of defects in, 1986–1987
computer-aided, **1984**, 1977, 1983–1984, 1991
for computer graphics, 2007–2012
"crisis" in, 1976
in data acquisition systems, 1806
debugging of, **1995**, 1987
defect prevention in, **1995**, 1985, 1992–1993
defect removal in, **1995**, 1985, 1987–1988
optimal series of, 1992–1993
post-release, 1993–1994
pre-test, 1988–1990, 1993
efficiency of, 1996
engineering, 2008–2009
for equivalent circuit models, 467–468
essential modeling in, **1984**, 1977, 1980
flow of control in, 1997–1998
hierarchical structuring of, 1999–2001
implementation modeling in, 1977, 1981–1983
information modeling in, **1984**, 1980
for intelligent networks, 1468
library, 2010–2011
maintenance of, **2003**, 1996
methods in, 1977–1980
microprocessors and, 1750
modularity of, 1999
for navigation systems, 2293–2295
object collaboration models in, **1985**, 1980, 1981
object interface specification in, 1980
object-oriented, **2013**, 2011
object-oriented programming of, 2001–2002
optical character recognition, 329, 1953
origins of defects in, 1986–1987
plotting, 2012
portability of, **2003**, 1996
post-release defect removal in, 1993–1994
pre-test defect removal in, 1988–1990, 1993
program correctness in, **2003**, 1996
redundancy in, 2024–2025

Page on which term is defined is indicated in bold.

reliability of, 2093–2094
security of, 2077–2080
self-checking, 2025
state transition diagrams in, 1980
structure of, 1999–2001
for telecommunications, 1468
testing of, 1987, 1990–1992, 2002
verification of, 1987, 1989
waterfall methods in, 1976
Software piracy, **2086**
Software quality assurance (SQA),
    1989
SOICs, see Small-outline integrated
    circuits
Solar cells, 445, 447, 458, 753, 1156,
    1208, 2255, see also
    Photovoltaics
Solar energy, **1215**, 1213
Soldering, 613
Solid dielectrics, 1137, 1139
Solid electrolytes, 1179, 1182, 1185
Solid-electrolyte sintered-anode
    tantalum capacitors, 23
Solid insulating materials,
    1145–1149
Solid-liquid insulating systems,
    1149–1150
Solid modeling, **2013**, 2004, 2011,
    see also specific types
Solid oxide fuel cells, 1212
Solids, see also specific types
    dielectric constants of, 2532
    dielectrics and, 1137, 1139, 1143
    inorganic, 1145–1146
    organic, 1146–1149
Solid-state circuits, 1004–1015, see
    also specific types
    amplifiers and, 1005–1008
    control, 1011–1013
    mixers and, 1010
    multipliers and, 1010, 1011
    oscillators and, 1009–1010
Solid-state disks (SSDs), 1667–1668
Solid-state lasers, **742**, 740
Solid-state power amplifiers (SSPA),
    2202
Solid-state relays, **1278**
Solid states, 2428–2429
Solitons, **1649**
Solution domain, **1048**
Solution matrix, 1040
Somatosensory-evoked potential
    (SSEP), 2327
Somatosensory system, 2301
Sonar, **418**, 408–409, 1056, 1057,
    2288
SONET, see Synchronous optical
    network
SOP, see Sum-of-products
Sound, 409–410, 1058

Sound speed profile (SSP), 409
Sound velocity profile (SVP), **418**,
    409
Sound waves, 1057
Source, **579**, 530, 1517, 1881
    common, 536
    constant-current, 539
    MOSFET and, 567
Source coding, 294–295, 339
Source couplers, 1430
Source decoders, 1401
Source encoders, **1404**, 1400,
    1401–1403
Source entropy, 339
Source follower, 539–541, 542–543
Source-follower voltage gain,
    539–540
Source programs, **1915**, 1904
Source stepping methods, 666
Source voltages, 1029
Space charge losses, 1134
Space communications, 2214
Spacecrafts, 2196–2199, 2285
Space-division multiplexing (SDM),
    1439
Space probes, 1578
Space propagation, 837–848
    in atmosphere, 840–843
    earth and, 843–847
    in simple media, 838–840
Space surveillance, 2214
Space-time coordinates, 807
SPARC computers, 1879, 1880,
    1882, 1883, 1885
    addressing in, 1887, 1888
    assembly language and, 1900
    call conventions in, 1889
    compilers in, 413, 1894, 1895
    return addresses in, 1890
Sparse equations, **672**
Sparse matrices, 660–661
Spatial arrays, 360
Spatial-hole burning (SHB), 1423,
    1424
Spatial processing, 398–400
Spatial sampling, 331, 332
Specialized mobile radio (SMR),
    1547
Special-purpose DSP, **393**
Special-purpose resistors, 13–14
Specifications, **2130**, 2113
    for compatibility, see Compati-
      bility
    for control systems, 2139–2140
    for data structures, 1915
    for data types, 1915
    interface, **1914**, 1910
    logic gate parameter, 1622–1624
SPECT, see Single photon emission
    computerized tomography

Spectral analysis, 315–321
Spectral correlation coefficient, 1507
Spectral correlation density
    function, 1507
Spectral estimation and modeling,
    315–328, 1607
    parameter estimation in,
      321–328
    spectral analysis in, 315–321
Spectral gain functions, 291
Spectral lines, 1506
Spectral subtraction method,
    290–291
Spectral width, **1434**
SPECTRE, 147
Spectrograms, 307, 308
Spectrum, **305**, 315, see also specific
    types
Spectrum reuse, **1566**, 1564
Speech, 2391, see also Speech signal
    processing
    compression of, 1525
    enhancement of, see Speech
      enhancement
    frequency of (pitch), **305**, 298
    intelligibility of, **297**, 280–281,
      282, 287
    processing of, 1760
    production of, 298
    quality of, **297**, 280–281, 287
    recognition of, see Speech
      recognition
    spectrograms of, 307, 308
    storage of, 279–286
    synthesis of, **305**, 304–305
    transmission of, 279–286
Speech analysis, **305**, 298–305
Speech coding, 279–286, see also
    specific types of coders
Speech detection algorithms, 395
Speech enhancement, **297**, 287–297
    modeling of, 288–290, 292–294
    performance measures of, 288–
      290
    signal classification in, 295–296
    signal estimation and, 290–294
    source coding and, 294–295
Speech recognition, 306–313, 1954
    continuous, **313**
Speech signal processing, 279–314
    analysis of, 298–305
    coding and, 279–286
    enhancement of, see Speech
      enhancement
    homomorphic (cepstral) analysis
      of, 303
    intelligibility in, 280–281, 282,
      287
    linear predictive analysis of,
      302–303

Page on which term is defined is indicated in bold.

noise reduction in, see Noise, reduction of
performance comparisons in, 285–286
quality of, 280–281, 287
signal classification in, 295–296
signal estimation and, 290–294
source coding and, 294–295
speech recognition and, 306–313
speech synthesis and, 304–305
storage and, 279–286
synthesis and, 304–305
transmission and, 279–286
Speed, **1331**, 1321, 1324, 1755, 2287
Speed-power product, **1621**, 1620
Spherical configuration in robotics, 2157
Spherical coordinates, 2468–2469
SPICE, **623**, **672**, 143, 146, 147, 148
  CACS and, **672**, 653, 654, 655, 669
  equivalent circuit models and, 467–468
  operational amplifiers and, 616, 621–623, 624, 630
SPIE, see International Society for Optical Engineering
Spin, 816
Spinal fusion, 2330
Spinal nerves, 2335
Spindle, **1669**
Spin-orbit coupling, 819, 823
Splices, 1430
Spoken language processing, 306, see also Speech recognition
Spontaneous emissions, 1421
Spontaneous polarization, **1131**
SPRCs, see Series-parallel resonant converters
Spreading resistance, 791
Spread-spectrum, 1400
SPUDT, see Single-phase unidirectional transducers
Sputtering, 484, 1692, 1693
SQA, see Software quality assurance
SQL, see Structured query language
Squared error distortion, 1522
SQUIDs, see Superconducting quantum interference devices
Squirrel cage induction motors, 1341, 1334
SRAM, see Static RAM
SRL, see Shift-register latch
SRR, see System requirements review
SRT algorithm, 1864
SS7, see Signaling system no. 7
SSDs, see Solid-state disks

SSEP, see Somatosensory-evoked potential
SSI, see Small-scale integration
SSP, see Service switching points; Sound speed profile
SSPA, see Solid-state power amplifiers
SSR, see Subsynchronous resonance
Stability, **222**, **672**, **1285**, 207, 1292–1293, 2105
  absolute, 2137
  of ac generators, 1324
  analysis of, see Stability analysis
  asymptotic, see Asymptotic stability
  bounded input-bounded output (BIBO), **168**, **222**, 167, 216–218
  critical clearing angle for, 1283
  degree of, 2131
  external, **222**, 216
  global asymptotic, **222**, 208, 209
  internal, **222**, 208
  Laplace transform and, 167–168
  Lyapunov theory of, 209–210
  relative, 2131, 2137, 2140
  of resistors, 12
  of sensors, **1161**, 1154
  state of the system and, 208
  steady-state, **1286**, 1279
  time, 15, 12
  of time-invariant linear systems, 210–216
  transient, **1286**, 1279–1280, 1289
  transient operation and, 1280
Stability analysis, 207–223
  bounded-input, bounded-output (BIBO) stability in, 216–218
  Lyapunov theory in, 209–210
  of physical systems, 218–222
  state of the system and, 208
  with state-space notation, 210–213
  of time-invariant linear systems, 210–216
  transfer functions in, 213–216
Stabilizability, **2106**, 2105
Stable operations of transmission, 1279
Stack crash, 1884
Stack pointers, 1884
Stacks, **1915**, 1884, 1887, 1891, 1896, 1904, 1922
Staebler–Wromski effect, 445
Standard broadcasting, see Amplitude modulation
Standard cells, **602**, 591, 593
Standard deviation, 1489, 1500, 2487

Standard graphic symbols, 1733–1734
Standard integrated circuit arithmetic-logic units, 1746
Standard voltages, 1220, 1227
Standby sparing, 2021
Star trackers, 2292
State, **176**, 208, see also specific types
  change of, 1721
  concept of, 171
  estimation of, 1350, 2145–2146
  steady, see Steady state
State equations, **176**
  control systems and, 2103
  for networks, 174–175
  nonlinear, 2104
  in normal form, **176**, 170–171, 173, 174
  scalar differential, 174–175
  state variables and, 170–171
  systematic procedure in writing, 171–174
State-space notation, 210–213
State-space realizations, 249
State transition diagrams, **1985**, 1980
State variables, **176**, **2106**, 169–177, 2103, 2143
  biquads and, 679–680
  complete set of, **176**, 171
  concept of, 171
  internal, 1818
  networks of, 679
  nonlinear networks and, 175–176
  state equations and, 170–175
  time-varying networks and, 175–176
State vector, **176**, 170
State vector navigation systems, **2294**, 2285
Static common emitter current gain, 640
Staticizers, 1723
Static magnetic fields, 811–812
Static RAM (SRAM), 1653–1654, 1655
Statics limit, 1019–1020
Static VAR compensators (SVC), 1246–1249
Stationarity, 1501
Stationary processes, 1519
Statistical models, 2229, see also Statistics; specific types
Statistical multiplexing, 1451
Statistics, 2486–2497, see also Probability
  tables of, 2490–2495
Stators, **1341**

Page on which term is defined is indicated in bold.

Steady state
  in control systems, **2113**,
    2108–2109
  protection and, 1268
  quasi, 469
  sinusoidal, **125**, 61, 1081, 2115
  three-phase transmission lines
    and, 1220
Steady-state availability, 2030
Steady-state errors, **2113**, 2109,
    2118, 2140
Steady-state impedance, 111
Steady-state power, **87**, 81–85
Steady-state sinusoidal outputs, 198
Steady-state stability, **1286**, 1279
Steady-state transistors, 641
Steerable microphone arrays,
    395–397
Steffenson's formula, 2513
Step-distance protection, 1273
Step-index fibers, 975, 976, 981, 989
Stepped leader, **946**, 935
Step responses, **168**, 166–167
Step signals, 47
Step sizes, **672**, 666
Stewart–McCumber parameter,
    1120
Stewart platform, 2161
Stilb, 747
Stick balancing problem, 219–220
Stiffened elastic constants, 1081
Stiffness tensor, 1077
Stiff systems, **672**, 668–669
Stimulated Brillouin scattering
    (SBS), 1425
Stimulation, 2309, 2310, 2312–2313
Stimulus-hypothesis evaluation
    model (SHEM), 2220, 2221
Stimulus-Hypothesis-Options-
    Response (SHOR) Para-
    digm, 2217, 2218, 2220
Stirling's approximation, 2431, 2512
STM, see Short-term memory;
    Synchronous transfer mode
Stochastic processes, **1509**,
    1499–1510, 1514, 1515, see
    also specific types
  classifications of, 1501
  coherence and, 1507–1508
  cross-correlation of, 1506–1507
  ergodicity and, 1508–1509
  examples of, 1502–1504
  Gaussian, 1501–1502
  Markov, 1501–1502
  random variables and, 1499–1500
  stationarity of, 1501
Stochastic sequence, 1502
Stochastic systems, **321**, 2104, see
    also specific types
Stokes' theorem, 812, 2474

Stopband, **125**, 113, 192
Stopband attenuation, 239
Stored-program control switches,
    1468, 1469
STPs, see Signal transfer points
Straight-line approximation, 201
Straight-line equations, 2448–2449
Strain, 1088, 1089, 1105, 1155
Strain tensor, 1077
Streaming potential, **2339**
Strength-duration curves, **2311**,
    2306, 2308, 2313
Strength-frequency curves, 2308,
    2309
Stress, 1080, see also specific types
  applied, 823
  ferroelectric materials and, 1088,
    1089
  internal, 823
  magnetic, 1177
  piezoelectric materials and, 1088,
    1089
  strain and, 1088, 1089
  tensile, 1102
  uniaxial tensile, 1102
Stress tensor, 1077
Strictness, **1926**
Strobing, 1769
Strong inversion, **579**, 569–570
Strongly consistent estimators, 322
Strongly stationary processes, 1502
Structural engineering, 1184–1185,
    1187
Structural smart materials, 1175
Structured cells, **602**, 599
Structured distribution systems
    (SDS), **1566**, 1561
Structured matrix sets, 277
Structured programming, 1909
Structured query language (SQL),
    2037, 2038, 2039, 2044
Stuck-at fault, **1816**, 1810
Subband coding, 352, 1401
Subband speech coders, 283, 284
Subgap processes, 443
Subgraphs, **109**, 110, 93
Subject copy in facsimile, **1556**
Submerged arc welding (SAW),
    2226
Subsequence signal restoration, 273
Subsidiary communication
    authorization (SCA), **1379**,
    1373
Substations, 1319, 1310
Substitutional diffusion, 479
Substitution theorem, 78
Substrates, **579**, 485, 611–612, see
    also specific types
Subsynchronous resonance (SSR),
    1250, 1243

Subthreshold, **580**, 570, 577–578
Subtracters, 1742–1743
Subtraction, 1862
Subtractive polarity, 1299
Subtransient period, 1279
Subtransient reactance, 1279
Subtransmission circuits, 1344
Successive approximation, **782**, 775,
    776, 777, 780
Successive approximation register
    (SAR), 777
Successive projections, **276**, 251
Sulfur hexafluoride, 1143
Sum-of-products (SOP) form, **1721**,
    1704, 1706, 1708, 1714,
    1715, 1736
Sum relaxation, 379, 380
Supercomputers, 1866, 1931
Superconducting generators,
    1327–1328
Superconducting quantum
    interference devices
    (SQUIDs), 1120
Superconducting wire, 1123
Superconductivity, **1123**,
    1114–1123, see also
    Superconductors
  electromagnetism and,
    1115–1118
  electronics and, 1118–1120
  material properties for, 1121,
    1122
  superconductor types and,
    1120–1123
Superconductors, see also Super-
    conductivity
  conductivity of, 1118
  inductance of, 1116, 1117, 1118
  material properties for, 1121,
    1122
  types of, 1120–1123
Superheaters, **1205**, 1194, 1196
Superheterodyne receivers, **1367**,
    1360, 1576
Superheterodyne technique,
    1360–1362
Superionic electric (fast ion)
    conduction, 1179
Supermalloy, 1179
Supernodes, **69**, 66
Superparamagnetism, **836**, 831, 835
Superposition, **79**, 69–71
Supervised learning, **428**
Supervisory control and data
    acquisition (SCADA)
    systems, 1345, 1348, 1349,
    1351
Supporting plane, **615**
Surface acoustic wave (SAW) filters,
    1075, 761, 1009, 1062–1075

Page on which term is defined is indicated in bold.

bidirectional response of, 1070–1071

coded, 1073–1074

dispersive, 1073

distortion in, 1070

in input-output devices, 1952–1953

material properties for, 1063–1064

resonators and, 1074–1075

schematic diagram of, 1064

second-order effects and, 1070

sensors and, 1159

specifications of, 1064–1066

transducers and, 1066–1073

ultrasound and, 1083

Surface-emitting laser logic (CELL), **1649**

Surface layer, 409

Surface mount technology (SMT), **615**, 603–615

assembly processing in, 611–614

chip carriers in, 607–608

circuits in, 606–611

cleaning in, 613

packaged-component subassemblies and, 603–604

repairing in, 614

reworking in, 614

soldering in, 613

substrates in, 611–612

technology of, 604–606

Surface potential, 569, 571

Surge impedance loading (SIL), **1222**

Surge tanks, **1205**

Surge voltage, 21

Surveillance, 2188, 2211, 2214, 2292

Suspension in vehicles, 2255

SVC, see Static VAR compensators

SVD, see Singular value decomposition

SVP, see Sound velocity profile

SWE, see Schrodinger wave equation; Society of Women Engineers

Swerling models, 960

Swing equations, **1331**, 1282, 1286, 1325

Switched-capacitor (SC) active filters, 682

Switched multimegabit data service (SMDS), 1456

Switching, 130, see also specific types

bypass, 1463

circuit, **1459**, 1437, 1438, 1449

color, 1775

in computer communication networks, 1447

double heterostructure optoelectronic, 1646

electrooptic, 1439

energy of, 1648

feeder, 1349

high-capacity, 1443

input, 773

large-signal, 456

light-amplifying optical, **1649**

logic of, 1613

message, **1459**, 1449–1450

multipoint, 2206

one-variable, 1697–1698

optical, 763, 765, 768

packet, **1459**, 1437, 1448, 1449, 1450, 2018

paradigms of, 1449

speed of, 1620, 1647

stored-program control, 1468, 1469

in telecommunications, 1468

theory of, **1711**

two-variable, 1697–1698

virtual circuit, 1437, 1438, 1450

zero-current, 717

zero-voltage, 717, 718–719

Switching algebra, 1699–1703

Switching field distribution (SFD), 824, 825

Switching networks (SNs), 1443

Switching nodes, **1459**, 1443, 1449

Symbol error rate estimation, **1608**, 1602–1604

Symbolic problems, 1902

Symmetrical components method, 1218, 1252

Synapses, 2352

Synaptic efficacies, 421

Synchronized phasor measurements, 1278

Synchronizing video signals, 1381–1382

Synchronous ac transmission, 1228

Synchronous circuits, **1721**, 1711

Synchronous compensators, 1244

Synchronous demodulation, 1386

Synchronous detection, **1393**

Synchronous digital hierarchy (SDH), **1446**, 1437, 1444

Synchronous flip-flops, 1722

Synchronous generators, 1279, 1280, 1283, 1322–1327, 1328

mathematical/circuit models of, 1323–1326

parameters for, 1326–1327

principle of operation of, 1322

Synchronous motors, **1341**, **2246**, 1334, 1341, 2238, 2245

analysis of, 1335–1337

performance of, 1338

permanent magnet, 736

wound field, 736

Synchronous optical network (SONET), 1437

Synchronous reactance, 1279, 1324, 1328

Synchronous sequential logic circuits, 1718–1719

Synchronous speed, **1331**, 1321, 1324

Synchronous transfer mode (STM), 1442

Syntax, **1915**, 1903

Synthesizers, **602**, 598, 1062, 1572, 2204

Synthetic aperture radar (SAR), 950, 957

System design life cycle, 2114

System design review (SDR), 1989

System load forecast, 1349

System noise temperature, 1538–1540

System requirements review (SRR), 1989

Systems, **2221**, see also specific types

adaptive, 1175

continuous, 2105

distributed parameter, 2100

dynamic, 420, 2106–2108, 2113

inter-, 903

intra-, 903

loop, **176**, 169

lumped parameter, 2100

multivariable, 2144

nodal, **176**

nonlinear dynamic, 420

nonrational, 320

open-loop, 2139, 2163

oscillating, 215

parallel, 2028, 2029

parallel-series, 2092, 2093

physical, 218–222, 2099

rational, 320

real, 2094–2095

rule-based expert, 2048–2051

series, 2028, 2029

series-parallel, 2092, 2093

simple, 69

single-input/single-output, 2126, 2249

smart, 1183–1186

state of, 208, see also State

stochastic, **321**, 2104

time-variable, 2104

two-bus, 1868

type-two, 2118

unstable, 207

Systems engineering, **2131**, 2113

Page on which term is defined is indicated in bold.

Systems for nuclear auxiliary power (SNAP), 1214
Systems integration, 2400
Systolic array design, **383**, 375, 376

T

TAB, see Tape-automated bonding
Tableau formulation, **672**, 655–656
Taboo channels, **1397**, 1395
Tachometers, 1755
Tag memory, 1932
Tangential continuity condition, 1019
Tantalum capacitors, 21, 23, 24
Tantalum pentoxide, 1146
TAP, see Test access port
Tap changers, **1319**, 1312
Tape-automated bonding (TAB), 609–610
Target motion analysis (TMA), 415–416
Task-ambient lighting, 2273
Task variables, 2252
Tau Beta Pi Association, 2556
Taylor series, 662, 1035, 2459–2460
Taylor's formula, 2463–2464
TB, see Triple-beat
TBP, see Time bandwidth product
TC, see Temperature coefficients; True and complement
TCA, see Trichloroethane
TCD, see Temperature coefficient of delay
TCE, see Trichloroethylene
TCM, see Trellis-coded modulation
TCR, see Thyristor-controlled reactors
TCS, see Traveling current source
TDI, see Test data input
t-distribution, 2492
TDM, see Time-division multiplexing
TDMA, see Time-division multiple access
TDO, see Test data output
TDRs, see Test data registers; Time domain reflectometers
TE, see Transverse-electric
TEDs, see Transferred electron devices
Telecommunications, 776, 779, 1760, see also Teleconferencing
    long distance, 1427–1434
    mobile, see Mobile radio
    networks for, 401, 1468
    software for, 1468
    switching in, 1468

Teleconferencing, 395, 2205–2206, see also Telecommunications
Telemetry, 1578–1591
    antennas in, 1586–1587, 1588–1589
    applications of, 1583–1584
    batteries in, 1586
    calibration in, 1587
    description of, 1578–1582
    discriminators in, 1586
    electrical, 1579
    FM/FM, 1582, 1585, 1589
    frequency allocations for, 1587–1588
    hydraulic, 1579
    limitations of, 1585–1586
    measurement with, 1582–1583, 1589
    mechanical, 1579
    missile, 1581
    modulation and, 1580, 1583, 1589–1590
    multiplexing in, 1580, 1585, 1589–1590
    optical, 1579
    oscillators in, 1586
    passive, 1590
    power for, 1584, 1585
    receivers in, 1580, 1586, 1590–1591
    time-division, 1581
    transducers in, 1580, 1583
    transmission and, 1582–1583, 1589
    transmitters in, 1580, 1586, 1587
Telemetry tracking and command (TT&C) system, 1543
Telephones, see Telecommunications
Television, **759**, 1379–1394, see also Video broadcasting; Video transmission
    advent of, 1477
    antennas in, 1390
    cable, 1418
    color signal decoding in, 1386
    color signal encoding in, 1384–1386
    community antenna, 1462
    frequency allocation for, 1383
    high-definition, see High-definition television (HDTV)
    industry standards for, 1382–1386
    interlaced scanning fields in, 1380–1381
    low-power, **1393**, 1383
    reception in, 1390–1391

    satellites and, 1532
    scanning lines and fields in, 1380
    synchronizing video signals and, 1381–1382
    telemetry and, 1580, 1583
    transmission in, 345
    transmitters for, 1382, 1386–1390
Television cameras, 757
Television receive-only (TVRO) systems, 2206
Tellegen's theorem, **87**, 74–75, 80–81, 84–85
TEM, see Transverse-electromagnetic
Temperature
    absolute junction operating, 640
    ambient, 30, 21
    antenna, 1494
    avalanche breakdown voltage and, 128
    batteries and, 1587
    in bipolar diodes, 128
    carrier, 469
    chip, 789
    compensation, **826**
    conversion of, 2423
    critical, 1114
    Curie, **826**, 1096, **1131**, 1092
    dielectrics and, 1138
    equivalent input noise, 1495
    forward-biased diodes and, 450–451
    forward voltage drop and, 128
    integrated circuits and, 589
    junction operating, 640
    lattice, 1138
    measurement of, 1582
    in radar, 955–956
    saturation, 788
    in semiconductor manufacturing, 495–496
    superconductivity and, 1114
    system noise, 1538–1540
    telemetry and, 1582, 1587
    thermal management and, 786, 787, 788
    thermal noise and, 1493
    transmitters and, 1587
Temperature coefficients, **15**, **32**, **459**, 7, 14, 451, 1218
Temperature coefficients of delay (TCD), 1064, 1074
Temperature sensors, 1154, 1155, see also specific types
TEMPEST, **918**, **2086**, 2073
Template matching, see Dynamic time warping (DTW)
Temporal effect, 358
Tendinitis, 2334
Tensile strain, 1105

Tensile stress, 1102
Tensors, see also specific types
  converse piezoelectric, **1104**
  elastoresistance, 1101–1102
  optical permittivity, 1162
  permeability, 805
  permittivity, 805, 1080, 1162
  piezoelectric, **1104**
  piezoresistivity, **1104**, 1100
  reflectivity, 1165
  resistivity, **1104**
  stiffness, 1077
  strain, 1077
  stress, 1077
Terminal current, 1156
Terminal guidance missiles, **973**,
    969, 972–973
Terminal voltage, 551, 1325, 1330,
    1334
Terminating codes, 1555
Terminating impedance, 1084
TES, see Track error signals
Testability problem, 1817–1819
Test access port (TAP), 1827, 1829
Test clocks, 1827
Test data input (TDI), 1827
Test data output (TDO), 1827, 1828
Test data registers (TDRs), 1827
Testing, 1808–1837, see also specific
    types
  ad-hoc techniques in, 1819–1823
  automatic equipment for, 595
  black-box, 1990
  built-in-, 953
  built-in self-, see Built-in self-test
    (BIST)
  of CMOS, 1809, 1810, 1811
  conducted emission, 933
  design for, see Design for test
    (DFT)
  diagnostic rhyme, 281, 285
  digital IC, 1808–1816
  direct access, 492–493
  of embedded memory blocks,
    1822–1823
  fault models and, 1810–1811
  of fiber-optic cable, 998–999
  final, **601**, 595
  of integrated circuits, 1808–1816
  knowledge, 2050
  linear cellular automata registers
    and, 1813, 1815
  linear feedback shift registers
    and, 1813–1815
  off-line, **1816**, 1810
  on-line, **1816**, 1810
  output response analysis and,
    1812–1815
  path-delay, **1836**, 1832–1835
  pregnancy, 1158

  proof, of fiber-optic cable,
    998–999
  pseudo-exhaustive, 1822
  pseudo-random, **1836**, 1830
  random, **1816**
  scan, 498, 491–492
  in semiconductor manufacturing,
    498, 490–497
  sequential probability ratio, 1486
  of shielding, 911
  signature analysis and, **1816**,
    1814–1815
  of software, 1987, 1990–1992,
    2002
  structured techniques in,
    1823–1832
  taxonomy of, 1808–1810
  testability problem in,
    1817–1819
  verification and, 1808
Test mode select (TMS), 1827
Test patterns, **602**, **1816**
  generation of, 1809, 1811–1812,
    1819, 1830
  automatic, 1812, 1819, 1820,
    1822
Test scripts, 1991
Test vectors, see Test patterns
Text compression, 1525
TF, see Transfer functions
TFEL, see Thin-film electrolumines-
    cent
TFT, see Thin-film transistors
TGS, see Triglycine sulphate
THD, see Total harmonic distortion
Thermal breakdown, 1140
Thermal conduction, 787
Thermal conductivity, 790, 1130
Thermal degradation, 1142
Thermal design, 785, 788
Thermal energy, 1213
Thermal engineering, 1187
Thermal equilibrium, 2337
Thermal expansion, 1152
Thermal expansion mismatch, **615**
Thermal failure, 784
Thermal fins, **796**, 788
Thermal imagers, 1126
Thermal management, **796**,
    784–798
  chips in, 786, 789–796
  heat transfer fundamentals and,
    787–788
  motivation for, 784–786
  multichip modules in, 786, 789,
    795
  nomenclature in, 796
  reliability of, 785
  requirements for, 786–787
  thermal resistance and, 789–796

Thermal materials, **1189**, 1175, see
    also specific types
Thermal noise, 9, 621, 758, 1432
  as Gaussian white noise,
    1491–1492
  measurement of, 1493–1494
  pyroelectrics and, 1128
  in radar, 956
  resistors and, 8
  sources of, 1495
Thermal oxidation, **489**, 476–478
Thermal plotters, 2005
Thermal printers, 1964–1965,
    1968–1970
Thermal processing, 579
Thermal radiation, 787, 788
Thermal resistance, **796**, 788
  area-specific, 795
  chip module, 789–796
  conductive, 789
  convective, 787, 789, 791
  external, 791–793, 794
  flow, 793
  internal, 790–791, 793
  total, 793–794
Thermal runaway, 784
Thermal sensors, 2348
Thermal smart materials, 1175
Thermal sources of light, 747–748
Thermal stress, 786
Thermal transport, 787, 790, see
    also Heat transfer
Thermionics, **1215**, 1214–1215
Thermistors, 14, 1155
Thermit welding, **2236**, 2223
Thermocline, 409
Thermocouples, 1155
Thermodynamical variables, 807
Thermoelectrics, 1214
Thermomagnetic process, **1693**
Thermomagnetic recording,
    1687–1690
Thermometers, 1152, 1155
Thermoreceptors, 2301
Thévenin equivalents, 57, 71, 73,
    636
Thévenin impedance, 73, 74, 85,
    164, 165
Thévenin resistance, 85, 647
Thévenin's theorems, **79**, 71–74,
    164–165, 196–197
Thévenin voltage, 57
Thick-film resistors, 1104
Thin-film capacitors, 1146
Thin-film electroluminescent
    (TFEL) units, 1942
Thin-film ferroelectric materials,
    1086
Thin-film transistors (TFT), 445
Thin oxides, 578

Page on which term is defined is indicated in bold.

Third harmonic distortion, 142
Third harmonic factor, 139, 141, 143
Third-order intermodulation distortion, 144
Thomson, Joseph J., 1151
Thomson denominator polynomials, 115, 116
Thomson filters, 118, 119
Thomson functions, 115
3-dB bandwidth, **195**, **652**, 192, 195
Three-dimensional analysis of electromagnetic fields, 1018–1025
 field equations for, 1018–1020
 modern design and, 1023–1025
 numerical methods for, 1020–1023
Three-node networks, 64
Three-phase circuits, 88–92
Three-phase connections, **188**, 185–186, 1303–1307
Three-phase fault, 1253–1256, 1259–1260
Three-phase power systems, 1272
Three-phase thyristors, 704, 730
Three-phase transformers, 1305
Three-phase transmission lines, 1218, 1220
Three-point method, 142
Three-terminal devices, **902**, 896–901, 1004, see also Transistors
Three-winding transformer equivalent circuits, 1296–1300
Threshold, **580**, 568
Threshold function, 423
Threshold-of-hearing chart, 1402
Threshold potential (TP), 2312
Threshold voltage, 570–573
Throughput, **1669**, 1665
Through-the-arc (arc) sensors, 2226–2228
Through via, **615**
Thyristor-controlled reactors (TCR), 1247–1250
Thyristors, **1250**, 694–695, 702, 706, 1239–1240, 2241, 2243, see also specific types
 asymmetric, 721
 gate-turn-off, 696–697, 734
 MOS-controlled, 700–701
 phase-controlled, 705, 729
 reverse-conducting, 697
 three-phase, 704, 730
 types of, 694–695
Thyristor-switched capacitor, thyristor-controlled reactors (TSC-TCR), 1249–1250

Thyristor-switched shunt capacitors (TSC), 1247
Thyristor-switched shunt reactors, 1247
Thyristor valves, **1241**, 1239
Tidal energy, **1215**, 1212
Ti indiffusion method, 765, 766
Time averaging, **1509**, 1508
Time-bandwidth product (TBP), **1516**, 958–959, 1513
Time bombs, **2086**, 2079
Time constants, **32**, 17, 27–28
Time-convolution property of Laplace transform, 152–154
Time-correlation property of Laplace transform, 153
Time delay, **418**, 115, 414–415
Time-dependent electric fields, 812
Time-dependent magnetic fields, 812
Time-division multiple access (TDMA), **1553**, 1541, 2204, 2206, 2208
 in mobile radio, **1553**, 1548, 1549
Time-division multiplexing (TDM), 1419, 1431, 1436, 1439, 1440, 1464
 optical, 1439
 satellites and, 1534, 2204, 2208
Time-division telemetry, 1581
Time domain
 in computational electromagnetics, 1042–1043
 detection of, 2384, 2385
 differentiation of, 152
 finite-difference, 1042
 JFET and, 541–543
 measurement of, 2108
Time domain reflectometers (TDRs), 518, 769
Time-invariant linear systems, 210–216
Time-invariant models, 1507
Time-multiplexing, 1585
Time redundancy, 2024
Time reversal, 180
Time-series models, 2101
Time-sharing systems, 2062
Time shift, 152
Time-slotted bus, 1464
Time stabilities, **15**, 12
Time step control, 666, 669–670
Time to market (TTM), 1817
Time-variable systems, 2104
Time-varying magnetic fields, 814–815
Time-varying networks, 175–176, 218
Time-varying potential field, 2313
Time warping, **313**, 307, 308–309

TLB, see Translation lookaside buffer
TLM, see Transmission line modeling
TM, see Transverse-magnetic
TMA, see Target motion analysis
T-matrix approach, 1034
TMR, see Triple modular redundancy
TMS, see Test mode select
TMS32010 DSP system, 385
TMS32020 DSP system, 385
TMS320C25 DSP system, 385, 386–387
TMS320C30 DSP system, 385, 387, 388–390
T network, **188**, 184–188
Toeplitz–Hankel matrices, 267, 268
Toeplitz–Hermitian matrix approximation, 269–270
Toeplitz matrices, 318
Toggle, **1640**, 1639
Token bus, **1459**, 1455, 1463, 2016–2017
Token matching, 356
Token ring, **1459**, 1435–1437
Token ring networks, 1455, 1463, 2016
Tokens, **2086**, 2072, 2081
Tolerance, 549
Tomographic-type images, 1082
Tomography, **344**, **2380**, 2374–2380, see also specific types
 computerized, **2356**, 2360, 2374–2376, 2377, 2400
 digital image processing and, **344**, 341, 342
 emission, 342, 2376
 positron emission, **2380**, 2374, 2376–2377
 single photon emission computerized, **2380**, 2374, 2376, 2377
Tool space, **2174**, 2163
Topographic mapping, 2360–2361
Topology, 92, 1596, see also Graph theory
 of circuits, 57
 isolated single-ended, 713–714
 of local-area networks, **1467**, 1461
 in network analysis and design, 107
Topology-preserving neural networks, 424–426
Topology processors, 1350
Torque, 2170–2171
 computed, 2170
 control of, 730
 external load, 2166

frictional, 2166
gravitational, **2174**, 2166
Torque converters, 2258
Total harmonic distortion (THD),
    147, 140, 2239, 2240, 2241
Total quality management (TQM),
    1986
Total thermal resistance, 793–794
Totem-pole output circuits, 1626
Touch input devices, **1957**,
    1948–1953, 2012
Touchscreens, 1948
Toughness coefficient, 845
Tourmaline, 1055, 1126
Townsend breakdown, 1139
TP, see Threshold potential
TPG, see Test pattern generation
TPO, see Transmitter power output
TQM, see Total quality manage-
    ment
Trackballs, **1957**, 1945, 1947, 2012
Track buffers, **1669**, 1665
Track-error detection, 1679
Track error signals (TES), 1686
Track-following procedures, 1679,
    1680, 1689
Tracking, **1693**, 1658, 1685–1687
Tracking filters, 963
Tracking radar, 950, 958, 961–962
Track pitch, 1679
Tracks, **1669**, **1693**, 1658,
    1679–1680
Traction control in vehicles, 2255
Transactional paradigms,
    1892–1894
Transaction integrity, 2044–2045
Transactions in energy management
    systems, 1353, 1348
Transceivers, 2192
Transconductance, **902**, 546, 551,
    576, 641
Transconductance curve, 547
Transducers, **1278**, 1152, see also
    specific types
  bidirectional, **1075**
  directivity of, 1060
  electrodynamic, 1056
  electromagnetic acoustic, 2229
  hydraulic, 1057
  image capture and, 330
  input, 1580
  interdigital, **1075**
  magnetic, 1056
  magnetostrictive, 1056, 1180
  in medical imaging, 2382
  moving coil, 1058
  multiphase unidirectional,
    1071–1073
  piezoelectric, **1085**
  pressure, 2343, 2344

protection and, 1268
  single-phase unidirectional, 1073
  surface acoustic wave filters and,
    1066–1073
  in telemetry, 1580, 1583
  ultrasound and, 1082–1085
  unidirectional, **1075**, 1071–1073
Transduction, 1055–1057, 1152,
    1153, 1156
Transfer characteristics of JFET,
    536–538
Transfer equations, 536
Transfer functions, **125**, **168**,
    111–113, 165–166, 2119, see
    also specific types
  acoustic signal processing and,
    404
  active filters and, 674, 675, 676,
    678, 685
  biquad, 676
  closed-loop, 1567, 1571, 2118,
    2126–2127, 2137
  compensation, 2127
  complex-conjugate, 674
  in computational electro-
    magnetics, 1029
  control systems and, 2104
  current-ratio, 95
  digital filters and, 238
  of external ear, 399
  of filters, 189–195
    digital, 238
    finite impulse response, 239
    loop, 1568
  FIR filter design and, 239
  frequency response and, 198
  generalized, 1029
  input-output, 2103, 2114, 2128
  light and, 764
  linearity of, 112–113
  of loop filters, 1568
  magnitude of, 112
  modulation, 1783–1784
  in networks, 95
  open-loop, 214, 1568
  optoelectronics and, 764
  power, 1506
  in stability analysis, 213–216
  stochastic processes and, 1506
  VLSI and, 379
  voltage, 674, 675, 677, 678
  voltage-ratio, 95
  z-transform, 2102
Transfer operator, 1727
Transferred electron devices
    (TEDs), 892, 894–895, 896
Transfer time, **1669**, 1664
Transformations, **125**, see also
    specific types
  algorithm, 371

associativity, 380–381
  Bruton, 681
  delta, 186–188
  diagrams of, **1985**
  distributivity, 381
  folding, **383**, 374–376
  gauge, 1018
  impedance, 542–543
  ladder, 680–681
  linear, 1489–1490
  of low-pass filters, 683–685
  low-pass to bandpass, 123–124
  low-pass to high-pass, 122–123
  pipelining, **383**, **393**, 371–372,
    377
  principle of, 34
  retiming, **383**, 373–374, 376
  rules of, 122–124
  unfolding, **383**, 374
  voltage, 1344
  wye, 186–188
Transform coding, 352
Transform coefficients, 283, 352
Transformed circuits, **168**,
    162–164
Transformer ratio, 1083
Transformers, **1308**, 33–39,
    1296–1309, see also specific
    types
  auto-, **1308**, 1307–1308
  connections in, 36–37
  construction of, 1300–1302
  core of, 35–36, 1300–1301
  delta connection in, 37
  electromagnetic equation and,
    34–35
  in energy distribution, 1314
  fault analysis and, 1258
  fundamentals of, 1296–1300
  harmonic frequency of, 36
  ideal, 806, 1296–1298
  impedance of, 38, 885, 1236,
    1306
  losses in, 36
  magnetic properties of steels in,
    2541
  open-delta connection in, 37
  performance of, 1302–1303
  phase shifts and, 1305
  polarity of, **1308**, 1299
  steels in, 2541
  T connection in, 37
  three-phase, 1305
  in three-phase connections,
    1303–1307
  three-winding, 1296–1300
  transformation principle and, 34
  two-winding, 1300
  types of, 33–34
  windings in, 1301–1302

Page on which term is defined is indicated in bold.

Y connection in, 36–37
zigzag connection in, 37
Transforms, see also specific types
bilinear, 243–244
Chebyshev, 1599
digital signal processing and, 229–238
discrete, 283
discrete-cosine, 338, 352, 1419, 1526
discrete Fourier, see Discrete Fourier transforms (DFT)
fast Fourier, see Fast Fourier transform (FTT)
Fourier, see Fourier transforms
Laplace, see Laplace transform
modified discrete cosine, 401
Radon, 341, 342
z, see z-transform
Transform speech coders, 282
Transient faults, 2088
Transient operation, **1286**, 1279–1287
Transient reactance, 1279
Transient response in control systems, **2113**, 2108, 2109–2113, 2140
Transient stability, **1286**, 1279–1280, 1289
Transistors, 783, see also specific types
in amplifiers, 1005
bias of at cutoff, 638
bipolar, see Bipolar transistors
bipolar-junction, see Bipolar-junction transistors (BJT)
characteristic curves for, 635
Darlington, 694, 697–698
density of, **602**
enhancement mode, **1657**, 1655
field-effect, see Field-effect transistors (FETs)
heterojunction, 892, 901, 1004
high electron mobility, 1004, 1005
hybrid-π, 552, 553, 554, 559
insulated-gate bipolar, **737**, 699–700, 706, 734
junction field-effect, see Junction field-effect transistors (JFET)
metal-oxide semiconductor, 1613
metal-oxide semiconductor field-effect, see Metal-oxide semiconductor field-effect transistors
metal-semiconductor field-effect, 892, 896, 897, 901, 1004, 1005

in microwave devices, 892, 896–901, 1004
MOSFET, see Metal-oxide semiconductor field-effect transistors
NPN, 638
photo-, 753
PNP, 638
power, 638, 697–698
steady-state, 641
thin-film, 445
Transistor switching energy, 786
Transistor-transistor logic (TTL), 1613, 1614, 1615–1616, 1622, 1729
bistable devices and, 1635
design of, 1627–1629
logic circuits and, 1712
logic gates and, 1624–1629
nonsaturated, 1627
Schottky, 1624
subfamilies of, 1629
Transition bandwidth, 239
Transition metals, 1691, see also specific types
Transition region, **195**, 192
Transition time, 1623
Transitory faults, 2088
Translating programs, 1904
Translation, **1393**, 179, 1383
Translation lookaside buffer (TLB), 1936
Transmembrane electrodes, 2313
Transmembrane potential, 2312
Transmembrane resistance, 2337
Transmission, 1217–1295, see also Power; Transmission lines; Transmitters
ac, see ac transmission
adequacy of, **1295**, 1287
in aerospace systems, 2203–2204
analog, 768
audio, see Audio transmission
of B-ISDN signals, 1443–1444
capacity for, 1242
compensation and, see Compensation
computer relaying and, 1276–1278
dc, see dc transmission
of energy, 82
in facsimile, 1555
Friis equation for, 842
functions of, 1287
high-voltage dc, see High-voltage dc (HVDC) transmission
high-voltage networks for, 1344
loading criteria for equipment in, 1291

in local-area networks, 1455, 1461
loop, see Loop gain
loss of, 1350, 1351
microwave devices and, 882, 883
multiuser, 1484
optical fiber, 990, 991–993, 1417
overcurrent protection and, 1270–1272
pilot protection and, 1274–1276
planning for, 1287–1295
protection and, see Protection
reliability of, **1295**, 1287
satellites and, 2203–2204
security of, **1295**, 1287
shunt reactors and, 1246
speech, 279–286
telemetry and, 1582–1583, 1589
in television, 345
token ring, 1455
transient operation and, 1279–1287
ultrasound and, 1081
in vehicles, 2257–2258
video, see Video transmission
voice, 2208
voltage criteria for, 1292
Transmission coefficients, 882, 883, 1082
Transmission line model of lightning, 941, 942
Transmission lines, 883, 884, see also Transmission
ac overhead, 1217–1222
circular, 505
coverage of 100%, 1275
E&M, 1062, 1063
in energy distribution, 1310
energy management systems and, 1344
equations for, 943–944
fault analysis and, 1258
grounding and, 906
instrument modeling and, 2278
interconnections and, 504, 505
lightning and, 941, 942, 943–944
low-level, 906
metals used in, 1219
modeling of, 1021
one-dimensional, 1081, 1083
for overhead ac, 1217–1222
rectangular, 505
in telemetry, 1583
three-phase, 1218, 1220
ultrasound and, 1081, 1082, 1083
Transmission zeros, 679
Transmitted waves, 1082
Transmitter power output (TPO), 1382

Transmitters, 1427, 1435, see also Transmission; specific types
AM radio, 1370–1372
design of, 1387–1388
digital audio broadcasting, 1399, 1403
filters used in, 916
FM radio, 1375, 1583
gap filler, **1403**
modeling of, 1601–1602
radar, 952
radio, 1370–1372, 1375, 1574
in telemetry, 1580, 1586, 1587
television, 1382, 1386–1390
temperature and, 1587
Transmitting antennas, 842
Transmitting current response, 1057
Transparent electrodes, 1775
Transparent latches, 1723
Transparent mode, **1640**, 1637
Transponders, 2285
Transport properties of semiconductors, 438–440
Transposition, 1502, 2439
Transversal filters, 1063
Transverse-electric (TE) mode, 850
Transverse-electromagnetic (TEM) mode, 850
Transverse-electromagnetic (TEM) waves, 743, 745
Transverse Kerr effect, 1166
Transverse-magnetic (TM) mode, 850
Transverse mode patterns, 979, 980, 982
Trapezoidal rule for areas, 2465
Trapezoids, 667
Traveler information systems, 2255
Traveling current source (TCS) model, 941
Tree-branch voltages, 172
Tree codes, **1409**, 1405
Tree-finding programs, 107
Trees, **110**, **176**, 171, see also specific types
binary, **1926**, 340, 1923
branches of, **109**, 93
decision, **2182**
dielectrics and, 1141
electrical, 1141
general, **1926**, 1923
graph theory and, 93
k, see k trees
n-ary, **1926**, 1923
normal, **176**, 171
quaternary (quad), 1923
water, 1141
Trellis-coded modulation (TCM), 1408–1409
Triacs, 694–695

Trichloroethane (TCA), 478
Trichloroethylene (TCE), 478
Triglycine sulphate (TGS), 1127, 1130
Trigonometric interpolation, 2515–2516
Trigonometric moment problems, 319
Trigonometric polynomials, 1515
Trigonometric series, 2461
Trigonometry, 2442–2446
Triode region, **545**, 532
Triple-beat distortion, 144
Triple modular redundancy (TMR), 2021, 2029
Triple transit echo (TTE), **1075**, 1070
Trojan horse, **2086**, 2079
True and complement (TC) representation, 1860–1861, 1862
Truncation, 1514
Truncation errors, 672, **1516**, 667–668, 669, 670, 1514
Trunked mobile systems, 1547–1548
Trunks, **1460**, 1449
Truth tables, **1621**, 1613, 1614, 1697, 1714, 1742
TSC, see Thyristor-switched shunt capacitors
TSC-TCR, see Thyristor-switched capacitor, thyristor-controlled reactors
TT&C, see Telemetry tracking and command
TTE, see Triple transit echo
TTL, see Transistor-transistor logic
TTM, see Time to market
Tumors, 2326, 2330
Tuners, 887
Tungsten filament lamps, 748
Tungsten-impregnated cathodes, 1780
Tuning elements, 887
Tunnel diodes, 447, 458, 892, 894, 1004
Tunneling, **1657**, 1118, 1655
Tunnel junction, 1118
Turbines, 1197, 1198, 1210
Turbulent flow, 793
Turing machines, 1525
TV, see Television
TVRO, see Television receive-only
Twisted nematic devices, **1777**, 1772, 1774, see also specific types
Twisted-ring counters, 1727
Two-dimensional analysis, 1020, 1023, see also specific types
Two-dimensional arrays, 867–868

Two-mesh networks, 66
Two-port networks, **902**
Two-quadrant operation, **737**, 730
Two's complement, 1741, 1743, 1854–1855, 1861
Two-terminal dc transmission, 1228–1231
Two-terminal devices, **902**, 894–896, 1004, see also Diodes
Two-winding transformers, 1300
Typeball printers, 1974
Type-two systems, 2118

U

UART, see Universal asynchronous receiver/transmitter
UAVs, see Unmanned airborne vehicles
UCE, see Unit control errors
UDT, see Unidirectional transducers
ULSI, see Ultra large scale integration
Ultrabroadband amplifiers, 1005
Ultra large scale integration (ULSI), 475, 488
Ultraquantum magnetic fields, 442
Ultra-small aperture terminals, 2209
Ultrasonic sensors, 2228–2229
Ultrasound, 1077–1086, 2374, 2400
acoustic properties and, 1079
acoustic waves and, 1083, 2380, 2381–2382
diagnostic, 2380
duplex, **2385**
filters and, 1082, 1084
in medicine, 2380–2385
narrowband filters and, 1082
one-dimensional propagation of, 1077–1080
piezoelectric excitation and, 1080–1081
propagation of, 1077–1080, 1081–1082
pulse-echo, **2385**, 2383–2385
signal processing and, 1082
transducers and, 1082–1085
Unbalanced circuits, 90
Unbiased estimators, **328**, 322
Unbounded problems, 1908
Uncomplemental variables, 1703–1704
Unconditionally stable circuits, 167
Uncontrolled rectifiers, **131**, 128
Underground ac transmission, 1223–1227

Page on which term is defined is indicated in bold.

Underground cable, 1223–1227,
    1344
Underwater acoustic signal
    processing, 406–419
  advanced, 416–417
  applications of, 406, 417
  estimation in, 414–415
  localization in, 414–415
  performance limitations of,
    411–412
  processing functions in, 413–416
  propagation and, 409–412
  sonar and, 408–409
  technical overview of, 407
Underwater vehicles, 2214, see also
    specific types
Unfolding transformation, **383**, 374
UNI, see User-network interface
Uniaxial tensile stress, 1102
Unidirectional transducers (UDT),
    1075, 1071–1073
Uniform convergence, 1513
Uniform intervals, 2508–2511
Unilateral z-transform, 180–181
Unintentional radiator exempted
    devices, 921–922
Unintentional radiators, 921, 922,
    923
Uninterruptible power supply
    (UPS), **728**, 727, 2074
Unipolar diodes, 126
Uniprocessors, 2052
Unit control errors (UCE), 1347
Unit delay, 239, 494
Unit impulse, **50**, 47–48
Unit-step processes, 47, 475, see
    also specific types
Unit-step signals, **8**, 6
Unity-gain frequency, 576, 629
Universal asynchronous receiver/
    transmitter (UART), **1762**,
    1754
Universal coding algorithm, 1520
Universal computers, 1525
Universal function approximation,
    422
Universal product codes (UPCs),
    1754–1755
UNIX, 2063, 2069
Unmanned airborne vehicles
    (UAVs), 2214
Unmanned underwater vehicles
    (UUVs), 2214
Unmodeled resonances, 2165
Unsigned binary number systems,
    1845–1848
Unstable circuits, 167, 168
Unstable systems, 207
Unsupervised learning, **428**
Unvoiced excitation, 298

UPCs, see Universal product codes
Update navigation systems, 2294
UPS, see Uninterruptible power
    supply
US, see Ultrasound
User interfaces, 1981, 2005, 2007
User-network interface (UNI),
    1443, 1445
UUVs, see Unmanned underwater
    vehicles

V

Vacuum evaporation, 484
Vacuum fluorescent displays
    (VFDs), 1942
Vacuum photodiodes, 754, 755
VAD, see Vapor axial deposition
Valence bands, **446**, 436, 1763
Validation, **2193**, 1046–1047, 1986,
    2188
Valve amplifiers, 675
Van de Graaff generator, 1342–1343
Vandermonde matrix, 319
Vapor axial deposition (VAD), 993,
    996
Vapor deposition, **488, 489**,
    484–485, 486–487
  chemical, see Chemical vapor
    deposition (CVD)
  outside, 993, 996
Vapor-phase energy, 613
Varactor diodes, 447, 458
Varactor multipliers, 1010
Varactors, 890
Variable-length coding, 339
Variable phase shifters, 887
Variable polarity plasma arc welding
    (VPPAW), **2236**, 2230
Variable resistors, 8, 12–13
Variables, 2101, see also specific
    types
  binary, 1695–1697
  binomially distributed, 2488
  Boolean, 1938
  complemental, 1703–1704
  complex, 181
  continuum-mechanical, 807
  functions of two, 2465–2466
  independent, 1500
  instance, **1914**, 1910
  internal state, 1818
  local, 1881, 1884, 1890, 1938
  non-state, **176**, 172
  physical, 2277
  random, see Random variables
  small-signal, **57**, 54, 55
  state, see State variables
  task, 2252

theory of functions of complex,
    181
thermodynamical, 807
uncomplemental, 1703–1704
Variable-speed constant-frequency
    (VSCF) generation, 1211
Variance, 1499, 2486–2487
Variance ratios, 2494–2495
Variant records, 1918
Varistors, 14
Varying reference digital-to-analog
    converters, 775
VAX computers, 1880, 1882, 1885
  addressing in, 1886, 1888
  compilers in, 1894, 1895
  program counters in, 1890
  return addresses in, 1890
  transactional paradigms in, 1892,
    1894
  vocabulary in, 1899
VCCS, see Voltage-controlled
    current sources
VCI, see Virtual channel identifier
VCOs, see Voltage-controlled
    oscillators
VCP, see Visual comfort probability
VCRs, see Video cassette recorders
VCS, see Virtual circuit switch
VCVS, see Voltage-controlled
    voltage sources
VDTs, see Video display terminals
Vector analysis, 2472–2475
Vector basis function, 1022
Vector differentiation, 2473
Vector processors, 1930
Vector products, 2472–2473
Vector quantization (VQ), **344**,
    1546, 307, 308, 338, 339,
    348–349, 1419
  image compression and, 425–426
  information theory and,
    1523–1524
Vectors, **498**, 2472–2474, see also
    specific types
  flux density, 1018
  input, **176**, 170
  normed, 253–254
  output, **176**, 171
  polarization, 1078
  Poynting, 839, 1078
  quantization of, see Vector
    quantization (VQ)
  semiconductor manufacturing
    and, 490
  state, **176**, 170
  test, see Test patterns
  voltage, 1036
  wave, 808–810, 1081
Vectorscope, **1393**, 1386
Vector space formulation, 253–254

Vector sum excited linear prediction (VSELP), 285, 286
Vector supercomputers, 1931
Vehicle driver information systems, 2255
Vehicles, 2255–2256, see also specific types
  automated guided, 2182
  electric, **2256**, 2255
  remotely piloted, 2212, 2214
  unmanned, 2214
Veiling reflection, 2275, 2265, 2266
Velocity
  constant linear, 1659
  group, 744
  of light, 743, 744
  phase, **1085**, 743, 1078, 1081
  sensors of, 1155
Velocity filtering, **366**, 360, 365–366
Velocity saturation, 569, 574
Verification, **2193**, 1808, 2188
Verilog language, 598
Vertical microinstructions, **1877**, 1872
Vertical-to-surface transmission electrophotonic devices (VSTEPs), 1646
Vertices, 92
Very large scale integration (VLSI), 370–393, 475, 499, 503
  applications of, 385–393
  arithmetic processor architectures and, 381–382
  assembly language and, 1899
  associativity and, 380–381
  CAD and, 371, 382
  chips for, 385–393
  computer graphics and, 2004, 2005
  computer networks and, 2018
  digital image processing and, 336
  distributivity and, 381
  folding and, **383**, 374–376
  future of, 382, 392
  integrated circuits and, 581
  logic elements and, 1617, 1622
  look-ahead technique and, **383**, 377–380
  parallel processors and, 371, 372–373, 2052
  pipelining and, **383**, **393**, 371–372, 377
  real-time applications of, 390–392
  retiming and, **383**, 373–374, 376
  surface mount technology and, 603
  thermal management and, 786
  unfolding and, **383**, 374

Very small aperture terminals (VSATs), **2209**, 1533, 2194, 2200–2205, 2209
VFDs, see Vacuum fluorescent displays
V-grooves, 767
VHDL language, 598–599
Via, **615**
Video broadcasting, 2206, see also Television
Video cameras, 757–759
Video cassette recorders (VCRs), 826–827, 1090, 1676
Video display terminals (VDTs), 1390, 2409, see also Cathode ray tubes (CRTs)
Video recording, 835
Video signal processing, 345–359, 777
  compression in, 349
  high-definition television and, 345, 353
  image quality and, 356–357
  information-preserving coders in, 349–350
  motion estimation in, 345, 354–356
  predictive coding in, 350–352
  quantization in, 345, 347–349
  sampling in, 358, 345, 346–347
  subband coding in, 352
  synchronizing of signals in, 1381–1382
  token matching in, 356
  transform coding in, 352
  vector quantization in, 348–349
  visual perception and, 356–359
Video teleconferencing, 2205–2206
Video transmission, 756, 1405–1415, 2205–2207, see also Television
  in aerospace systems, 2205–2207
  AM, 1406–1407
  AM-VSB, 1418–1419, 1422, 1423, 1424, 1425
  applications of, 1418–1419
  composite, **1392**, 1382, 1383–1384
  compressed digital, 1419
  compression and, **358**, 345, 349, 2205
  compression of, 1526
  external modulation and, 1424–1425
  field-effect transistors and, 1421
  formats for, 1418–1419
  frequency modulation, 1418
  intensity modulation and, 1419–1420
  linearity in, 1422–1423

  noise in, 1420–1421
  rate analysis of, 392
  satellites and, 2205–2207
  signals from, 392, 393
Vidicon camera tubes, 757–759
Viewports, 2011
View reference point, 2011
Virtual architecture, 1870
Virtual channel identifier (VCI), 1443
Virtual circuit networks, 1451
Virtual circuit switching (VCS), 1437, 1438, 1450
Virtual functions, 1911
Virtual instruments, **2283**
Virtual memory, **1936**, 1876, 1934–1936, 2068
Virtual path identifier (VPI), 1443
Virtual private networks, 1443, 1473
Virtual reality systems, **2013**, 2004, 2005
Virtual records, 2040
Viscosity, 1080, 1178
Visibility, 2259
Visual comfort probability (VCP), **2275**, 2265
Visual perception, 356–358, 356–359
Visual thresholding, 357–358
Viterbi algorithm, **313**, 312, 1403
Viterbi decoding, 2203
VLSI, see Very large scale integration
VMOS, 634
Voiced excitation, 298, 299–300
Voice encoding, 776
Voice input to computers, **1957**, 1954, 2012
Voice transmission, 2208
Voicing, 305, 298
Voigt notation, 1162
Volatile devices, **1640**, 1635, see also specific types
Volta, Alessandro, 1002
Voltage
  arc, 2231
  automatic control of, 2231, 2232
  avalanche breakdown, 128
  balanced, **188**
  barrier, **458**, 447
  Boltzmann, 640
  branch, **63**
  clamping to, 133
  constant, 133
  control of, 1245
  cotree-link, 172
  criteria for, 1292
  current vs., 1156
  delta, 1272
  delta transformation and, 187
  dielectric breakdown, 1138, 2312

difference-in–to–difference-out gain in, 563
diode, 127
drop in, **15**, 6, 59, 62, 1316
early, 538, 641
equivalent input, 620, 622
feeder, 1311
flatband, 573
forward, **701**
forward drop in, 128
frequency ratio to, 732, 733
input, 620, 622, 725, 1068
instantaneous, 81
Kirchhoff's law of, see Kirchhoff's laws, for voltage
lightning-induced, 944, 946, 947
maximum working, 18
membrane, 2309
in MOSFET, 568–573
multiplication of, 131
node, **63**
offset, 620
output, 146, 526, 539, 630, 638
phase, 705, 1261–1264
phasor, 82
pinch-off, **545**, 530
pn-junction diodes with applied, 448–450
power supply, 625
quiescent, 641
regulation of, 1303, 1317–1319
resistors and, 5–6
reverse, **701**, 21, 694
rise in, 59
saturation, 548
Seebeck, 1214
in semiconductor manufacturing, 495–496
sequence, 1261–1264
small-signal ac gain in, 540
source, 1029
source-follower gain in, 539–540
sources of, see Voltage sources
standard, 1220, 1227
superconductivity and, 1118
surge, 21
terminal, 551, 1325, 1330, 1334
terminal current vs., 1156
Thévenin, 57
Thévenin equivalent, 636
threshold, 570–573
transformation of, 1344
tree-branch, 172
wye transformation and, 187
zero, 707
Voltage coefficients of resistance, **15**, 8
Voltage-controlled current sources (VCCS), 54

Voltage-controlled oscillators (VCOs), **1575**, 1567, 1568, 1570, 1573, 1574, 1586
Voltage-controlled voltage sources (VCVS), 54
Voltage-dependent resistors, 890
Voltage-derived negative feedback, 556
Voltage-fed inverters, 731–732
Voltage gradient, 2312
Voltage phasors, 89
Voltage rating of resistors, **15**, 10
Voltage-ratio transfer function, 95
Voltage references, 771, 773
Voltage reflection, 1082
Voltage regulators, 1303, 1317–1319
Voltage responsivity, 1128
Voltage sensors, 769
Voltage source inverters (VSIs), 705–706, 2238
Voltage sources, 47–57
    biasing of, 533–534
    controlled, 53–57, 73
    current-controlled, 54
    dc signals and, 49
    ideal, 50–51
    impulse signals and, 47–48
    independent, **63**
    network theorems and, 73
    practical, 51–53
    ramp signals and, 48–49
    sinusoidal signals and, 49
    step signals and, 47
    voltage-controlled, 54
Voltage standing wave ratio (VSWR), **891**, 883
Voltage-to-frequency converters, 771
Voltage transfer function (VTF), 674, 675, 677, 678, 681
Voltage-variable resistance (VVR), 533, 543–544
Voltage vector, 1036
Volterra series, 147, 1599
Volume, 2468, 2470
Von Neumann, John, 1937
Voronoi partition, 1524
Vortex, 1121
VPI, see Virtual path identifier
VPPAW, see Variable polarity plasma arc welding
VQ, see Vector quantization
VR, see Virtual reality
VSATs, see Very small aperture terminals
VSCF, see Variable-speed constant-frequency
VSELP, see Vector sum excited linear prediction
VSIs, see Voltage source inverters

VSTEPs, see Vertical-to-surface transmission electrophotonic devices
VSWR, see Voltage standing wave ratio
VTF, see Voltage transfer function
VVR, see Voltage-variable resistance

W

Wafer, 602
Wafer patterning, 487–488
Wafer sort, **602**, 594
Wagner absorption factor, 1136
Wall direct radiation coefficient, 2264
Wall luminance coefficient, 2263, 2264
Wall reflected radiation coefficient (WRRC), 2263
WAN, see Wide-area networks
Wanlass, Frank, 692–693
Warfare systems, **2221**, 1188, 2217
Waterfall methods, 1976
Water trees, 1141
Wattmeters, 85, 86
Wave equations, 807–810
Wavefront processors, 376
Waveguide filters, 887
Waveguide isolators, 1168
Waveguides, **859**, 849–859, see also specific types
    circular, 854–855, 858
    commercially available, 856
    dielectric, 975, 984
    discontinuities in, 884
    dispersion of, 985–986, 992
    double-edged, 856
    EH mode of, 850
    HE mode of, 850
    hybrid, 850, 884, 889
    lightwave, see Lightwave waveguides
    losses of, 856–858
    in microwave devices, 882, 884, 889
    mode launching of, 858–859
    modes of, **859**, 849–852
    rectangular, 852–853, 857, 859
    ridge, 856
    single-edged, 856
    slab, 978–980
    superconductivity and, 1119, 1164
    surface acoustic wave filters and, 1062
    transverse-electric mode of, 850

transverse-electromagnetic mode
    of, 850
transverse-magnetic mode of, 850
Wave impedance, 859, 851
Wavelength, 859, 1001, 839, 851
Wavelength-division multiplexing
    (WDM), 1431, 1439
Wavenumbers, 366, 810
Wave polarization, 844
Wave propagation, 848, 837–859,
    1026, 1027
    in atmosphere, 840–843
    earth and, 843–847
    modes of, 841
    in simple media, 838–840
    in space, see Space propagation
    waveguides and, see Waveguides
    waveguides mode launching and,
        858–859
Waves, see also specific types
    acoustic, 1083, 1085, 2380,
        2381–2382
    light-, see Lightwaves
    longitudinal, 1078
    micro-, see Microwave
    optical, 1085
    plane, 1077, 1080, 1081, 1083
    propagation of, see Wave
        propagation
    pure longitudinal and shear,
        1085
    quasi-longitudinal, 1078
    quasi-shear, 1078
    reflected, 1082
    scattered, 1085
    shear, 1078
    sound, 1057
    transmitted, 1082
Wave solutions, 807–810
Wavetilt formula, 940
Wave vectors, 808–810, 1081
Wave velocity, 839
WCS, see World coordinate system
WDM, see Wavelength-division
    multiplexing
Weak inversion, 570, 572
Weakly stationary (WS) processes,
    1499, 1509, 1488, 1489,
    1501, 1502, 1504, see also
    specific types
    cross-correlation and, 1507
    linear filtering of, 1504–1506
Weight functions, 1038–1039
Welding, 2223–2237
    control systems for, 2223–2224,
        2231–2235
    modeling of, 2229–2231
    parameters for, 2224
    robotics in, 2178–2179
    sensors in, 2225–2229

Wenzel, Kramers and Brillouin
    (WKB) method, 982–983
Wet-electrolyte sintered-anode
    (wet-slug) tantalum
    capacitors, 23, 24
Wet etching, 489, 488
Wet oxidation, 478
Wet oxides, 476
Wheelchairs, 2388, 2390, 2392
Wheeler's equation, 26
White noise, 318, 1490, 1506
    Gaussian, 1487, 1480, 1484, 1490,
        1491–1492, 1506
Wide-area networks (WANs), 1460,
    1447, 1457, 1461, 2018
Wide-sense, see Wealky stationary
Wide sense stationary processes, 317
Wiener factor, 319
Wiener filters, 1508
Wiener–Khinchin relations, 1489,
    1506
Wiener–Khinchin theorem, 317
Wiener process, 1504
Wilson central terminal, 2370, 2363
Winchester disks, 1669, 1659
Wind-electric conversion, 1215,
    1210–1211
Windings, 32
Window-blinding, 1883, 1889, 1896
Window operations, 344, 335, 336
Windows, 238, 233–234, 240–241,
    see also specific types
Window size selection, 343
Wire, 1123, 2543–2544
Wire antennas, 861–869
Wireless communications, 1557,
    1558, 1559–1561, see also
    Cellular communications;
    Wireless local-area
    networks; specific types
Wireless local-area networks, 1566,
    1557–1566
    cable in, 1559–1561
    cost of, 1560, 1563
    end user reaction to, 1563
    frequency allocation for,
        1564–1565
    growth of, 1559
    market research on, 1566,
        1558–1559, 1561
    optimized service area for, 1562
    performance of, 1563
    reliability of, 1563
    security of, 1563
    user reaction to, 1563
    user requirements for, 1561–1562
    workgroup sizes for, 1562
Wire-wound resistors, 11–12
WKB, see Wenzel, Kramers and
    Brillouin

Word line capacitance, 1652
Word-parallel processing, 383, 371,
    374
Word recognition, 313, 306, 309,
    310, 311, 313
Word-serial processing, 371, 374
Word-slice chips, 1749
Wordspotting, 313, 306, 310
Work, 2468
Working conditions, 546
World coordinate system (WCS),
    2011
WORM, see Write-once read-many
Wound-rotor induction motors
    (WRIMs), 1341, 735, 1334,
    2243
WRIMs, see Wound-rotor
    induction motors
Wrist sensors, 2106
Write-once read-many (WORM)
    technology, 1676
WRRC, see Wall reflected radiation
    coefficient
WS, see Weakly stationary
Wye circuits, 92, 90, 92
Wye-delta circuits, 91
Wye transformation, 186–188
Wye-wye circuits, 91

X

Xerography, 445, 1112–1113
XLPE, see Cross-linked polyethylene
X-ray photons, 2375
X-ray radiation, 488

Y

YAG, see Yttrium, aluminum,
    garnet
Yagi–Uda arrays, 868
Yield, 498
Yield enhancement, 490
YIG, see Yttrium-iron-garnet
Yttrium, aluminum, garnet (YAG)
    lasers, 760, 1424
Yttrium-iron-garnet (YIG), 890,
    1009, 1162, 1167, 1168

Z

ZCS, see Zero-current switching
Zener breakdown, 459, 127–128
Zener diodes, 135, 137, 447,
    457–458
Zero crossing rate, 299
Zero current, 707

Page on which term is defined is indicated in bold.

Zero-current switching (ZCS), 717
Zero dispersion wavelength, 1429
Zero-forcing (ZF) equalizers, 1411
Zero-input response, **2113**, 2107
Zero/nonzero structure, 660
Zeros, 112, 200, 201, 302, 679
Zero-state response, **2113**, 2107, 2108
Zero voltage, 707
Zero-voltage switching (ZVS), 717, 718–719
ZF, see Zero-forcing

Zonal cavity method, 2259
ZSPICE, 621
z-transform, **183**, 178–183, 239, 2477–2480
   causal and anticausal pairs of, 179
   control systems and, 2102
   convolution property of, 180
   direct, 178
   inverse, 178, 181–182
   Laplace transform and, 2478–2479

   linearity of, 178
   multiplication property of, 180
   properties of, 178–180, 2479
   for sampled data, 182–183
   of sampled functions, 2479–2480
   sampling period and, 182
   of sequences shifted in time, 179
   time reversal and, 180
   translation property of, 179
   unilateral, 180–181
ZVS, see Zero-voltage switching

Page on which term is defined is indicated in bold.